Dictionary of Science & Technology

The Wordsworth

Dictionary of Science & Technology

Wordsworth Reference

First published as *Chambers Science and Technology Dictionary*
by W&R Cambers Ltd, Edinburgh, and The Press Syndicate
of the University of Cambridge, 1988.

This edition published 1995 by Wordsworth Editions Ltd,
Cumberland House, Crib Street, Ware, Hertfordshire SG12 9ET.

ISBN 1-85326-351-6

Printed and bound in Denmark by Nørhaven.

The paper in this book is produced from pure wood
pulp, without the use of chlorine or any other substance
harmful to the environment. The energy used in its
production consists almost entirely of hydroelectricity
and heat generated from waste materials, thereby
conserving fossil fuels and contributing little to the
greenhouse effect.

Contents

Preface

Arrangement

The entries in this dictionary are strictly alphabetical with single letter entries occurring at the beginning of each letter. Greek letters will be found under the nearest anglicized equivalent, 'omega' under 'o' and 'psi' under 'p'.

Numbers in chemical nomenclature: the convention is for any leading numerals to be ignored, so that '2,4,5,-T' will be found at the position determined by 'T'; in other areas the entry occurs at the position determined by the number spelt out, thus '32mo' will be found at the position of 'thirtytwomo'.

Italic and Bold

Italic is used for:

(1) alternative forms of, or alternative names for, the headword. Entries for conditions in Medicine and Veterinary Medicine frequently list synonyms at the beginning of the entry. In other entries synonyms may be added in italics at the end of an entry after 'also' or 'also called';

(2) terms derived from the headword, often after 'adj.' or 'pl.';

(3) variables in mathematical formulae;

(4) generic and specific names in binomial classification in the Biological Sciences;

(5) for emphasis.

Bold is used for:

(1) cross-references, either after 'see', 'cf.' etc. or in the body of the entry. Such a cross-reference indicates that there is a headword elsewhere which amplifies the original entry;

(2) vector notation in Physics etc.

Appendices

These contain tables of chemical formulae, chemical elements, SI conversion factors, physical constants, plant and animal kingdom classifications, SI units, geological eras and paper sizes.

Wherever the symbol ⇒ occurs at the end of an entry a diagram representing its chemical formula will be found in an appendix which contains the formulae of most of the main ring compounds together with representative sugars, amino acids and nucleic acid bases.

Trade Names

These are shown by initial capitals, the prefix TN or some other statement. If we have failed to acknowledge trade names we will be happy to make amends in future editions.

Subject categories

Abbreviations where appropriate are shown in brackets.

Acoustics (Acous.)
Aeronautics (Aero.)
Architecture (Arch.)
Astronomy (Astron.)
Automobiles (Autos.)
  including
  Internal Combustion Engines
Behaviour (Behav.)
  including
  Animal Behaviour
  Ethology
  Psychiatry
  Psychology
Biology (Biol.)
  including
  Bacteriology
  Biochemistry
  Cell Biology
  Cytology
  Genetics
  Histology
  Microscopy
  Molecular Biology
Botany (Bot.)
Building (Build.)
  including
  Carpentry
  Joinery
  Painting
  Plumbing
  Sanitary Engineering
Chemical Engineering (Chem. Eng.)
Chemistry (Chem.)
Civil Engineering (Civ. Eng.)
  including
  Railways
Computers (Comp.)
Crystallography (Crystal.)
Ecology (Ecol.)
Electrical Engineering (Elec. Eng.)
Electronics
Engineering (Eng.)
  including
  Heating
  Instruments
  Metallurgy
  Powder Metallurgy
  Tools
Forestry (For.)
Foundry Practice (Foundry)
General (Genrl.)
Geology (Geol.)
  including
  Geophysics
  Oceanography
Glass
Hydraulic Engineering (Hyd. Eng.)
Image Technology (Image Tech.)
  including
  Cinematography
  Photography
  Television
Immunology (Immun.)
Mathematics (Maths.)
Medicine (Med.)
  including
  Anatomy
  Nutrition
  Pharmacology
  Physiology
  Surgery
Meteorology (Meteor.)
Mineralogy (Min.)
Mineral Extraction (Min. Ext.)
  including
  Mineral Processing
  Mining
  Oils
Nuclear Engineering (Nuc. Eng.)
Paper
Physics (Phys.)
  including
  Heat
  Hydraulics
  Light
  Magnetism
  Mechanics
  Nucleonics
  Optics
Plastics
Powder Technology (Powder Tech.)
Printing (Print.)
  including
  Bookbinding
  Typography
Radar
Radiology (Radiol.)
Ships
  including
  Navigation

Space
Statistics (Stats.)
Surveying (Surv.)
Telecommunications (Telecomm.)
  including
  Cables
  Radio
  Telegraphy
  Telephony

Textiles
  including
  Spinning
  Weaving
Veterinary Science (Vet.)
Zoology (Zool.)

Contributors

General Editor: Professor Peter M. B. Walker, former Professor of Natural History, University of Edinburgh; former Director of Medical Research Council Mammalian Genome Unit

Contributors to the Science and Technology Dictionary include:

Acoustics
Dr Maria Heckl
Department of Engineering
University of Cambridge

Aeronautics
Prof J. E. Allen
School of Aeronautics
Cranfield Institute of Technology

Architecture
Ms I. McGeoch
Doune, Perthshire.
Practising architect

Astronomy
Dr S. Mitton
Fellow of St Edmund's College
University of Cambridge,
and Editorial Director,
Cambridge University Press

Automobiles
Mr A. Nahum
Department of Aeronautics,
Science Museum, London

Behaviour
Dr J. Herrmann
Lecturer in Psychology,
Napier College of Commerce
and Technology, Edinburgh

Biology
Dr A. P. Bird
MRC Clinical and Population
Cytogenetics Unit,
Western General Hospital, Edinburgh

Dr D. J. Bower
MRC Clinical and Population
Cytogenetics Unit,
Western General Hospital, Edinburgh

Prof D. S. Falconer, FRS
former Head of Department of
Genetics,
University of Edinburgh

Dr A Maddy
Senior Lecturer,
Department of Zoology,
University of Edinburgh

Botany
Mr A. J. Tulett
Lecturer, Department of Botany,
University of Edinburgh

Building
Mr John S. Young
Head of Building and
Head of Department of Wood
and Finishing Trades,
Telford College of Further Education,
Edinburgh

Mr Jeff Brown
Senior Lecturer, Painting and
Decorating,
Telford College

Mr W. A. Crighton
Head of Department of
Civil Engineering,
Telford College

Mr John Macdonald
Senior Lecturer,
Painting and Decorating,
Telford College

Mr J. M. McGowan
Senior Lecturer,
Civil Engineering and
Construction Technology,
Telford College

Chemistry
Dr R. O. Gould
Senior Lecturer,
Department of Chemistry,
University of Edinburgh

Computers
Ms J. van Rijsbergen
Glasgow

Ecology/Forestry
Dr J. Grace, Reader
Department of Forestry and
Natural Resources,
University of Edinburgh

Electrical Engineering
Dr H. W. Whittington
Senior Lecturer,
Department of Electrical Engineering,
University of Edinburgh

Electronics
Mr E. C. Davies
Department of Engineering,
Science Museum, London

Engineering
Mr W. Thomas
Head of Department of Engineering,
Telford College, Edinburgh

Mr G. Morrison
Head of Technology Division,
West Lothian College, Bathgate

Geology
Prof G. Y. Craig
former Head of Department,
Grant Institute of Geology,
University of Edinburgh

Image Technology
Mr B. Happé
consultant in
motion picture technology

Immunology
The late Prof J. H. Humphrey, FRS
Royal Postgraduate Medical School,
University of London,
former Deputy Director of
National Institute for
Medical Research

Mathematics
Mr P. Jackson
Mathematics sub-editor,
Cambridge University Press

Medicine/Radiology
Dr A. L. Muir
Reader, Department of Medicine,
University of Edinburgh

Meteorology
Mr R. P. Waldo Lewis
former Head of Library and
Publications Branch at the
Meteorological Office

Mineral Extraction
Dr A. Park
Department of Geology,
University of Glasgow

Mineralogy
Dr P. A. Sabine
former Chief Scientific Officer
British Geological Survey,
Chairman International Union of
Geological Sciences Commission on
Systematics in Petrology

Nuclear Engineering
Prof H. W. Wilson
former Director,
Scottish Universities Research and
Reactor Centre

Paper
Mr R. Bray
former Director of the
National Association of Paper
Merchants, London

Physics
Dr D. F. Grant
former Senior Lecturer in Physics,
University of St Andrews

Printing
Mr I. M. Campbell
Head of Department of Print Media,
Publishing and Communication,
Napier College, Edinburgh

Radar
Mr E. C. Davies
Department of Engineering,
Science Museum, London

Space
Dr D. Shapland
Head of ESA Astronaut Office,
European Space Agency, Paris

Statistics
Mr J. C. Duffy
Lecturer, Department of Statistics,
University of Edinburgh

Telecommunications
Mr E. C. Davies
Department of Engineering,
Science Museum, London

Textiles
Dr J. Honeyman
former Research Director of
the Shirley Institute

Veterinary Science
Mr J. Chapman
Kippen, Stirlingshire.
Practising veterinary surgeon

Zoology
Dr A. W. Ewing
Lecturer, Department of Zoology,
University of Edinburgh

Abbreviations used in the dictionary

The following are the more common abbreviations used in this dictionary. Many others (including contractions, prefixes and symbols) occur in their alphabetical position in the text, especially at the beginning of each letter of the alphabet. For SI and other symbols, see the Tables on pages 1002–1004 and 1008.

abbrev(s).	abbreviation(s)
adj(s).	adjective(s)
approx.	approximately
asym.	asymmetrical
at. no.	atomic number
bp	boiling point
c.	century
°C	degree(s) Celsius (centigrade)
ca	circa
Can.	Canada
cf.	compare
C.G.	centre of gravity
CGS	centimetre-gram(me)-second
circum.	circumference
CNS	central nervous system
colloq.	colloquially
conc.	concentrated
contr.	contraction
CRO	cathode-ray oscilloscope
CRT	cathode-ray tube
deg.	degree(s)
dim.	diminutive
E.	East
elec.	electrical
esp.	especially
°F	degree(s) Fahrenheit
ft	foot, feet
g	acceleration due to gravity; gram(me)(s)
G	constant of gravitation
Gk.	Greek
h., hr	hour(s)
in	inch(es)
kg	kilogram(me)(s)
km	kilometre(s)
K	kelvin(s)
L.	Latin
lb	pound(s)
m	metre(s)
min.	minute(s)
MKS(A)	metre-kilogram(me)-second (-ampere)
mp	melting point
n.	noun
N.	North
pfx.	prefix
pl(s).	plural(s)
°R	degree(s) Rankine
r.a.m.	relative atomic mass
rel.d.	relative density
r.m.s.	root-mean-square
S.	South
S., sec.	second(s)
SI	Système International (d'Unités)
sing.	singular
sp	species (*sing.*)
spec.	specific
spp	species (*pl.*)
sq.	square
s.t.p.	standard temperature and pressure
sym.	symmetrical
syn.	synonym
temp.	temperature
TN	trade (proprietary) name
TR	transmit-receive (tube)
UK	United Kingdom
US	United States (of America)
W.	West
y., yr	year(s)
yd	yard(s)

Greek alphabet

The letters of the Greek alphabet, frequently used in technical terms, are given here for purposes of convenient reference. The roman letters refer to the dictionary letter at which any headwords beginning with Greek letters will be found.

Α	α	alpha	a	Ν	ν	nu	n
Β	β	bēta	b	Ξ	ξ	xi	x
Γ	γ	gamma	g	Ο	ο	omīcron	o
Δ	δ	delta	d	Π	π	pi	p
Ε	ϵ	epsīlon	e	Ρ	ϱ	rho	r
Ζ	ζ	zēta	z	Σ	σ ς	sigma	s
Η	η	ēta	e	Τ	τ	tau	t
Θ	θ ϑ	thēta	t	Υ	υ	upsīlon	u
Ι	ι	iōta	i	Φ	φ	phi	p
Κ	ϰ	kappa	k	Χ	χ	chi	k
Λ	λ	lambda	l	Ψ	ψ	psi	p
Μ	μ	mu	m	Ω	ω	ōmega	o

Acknowledgements

I would like to express on behalf of the publishers and myself grateful thanks to all who have revised the entries in their subjects. In many cases this was clearly an immense task, and it says much for the hard and skilful work of the contributors that the whole revision was completed in under two years.

I would also like to thank the Computing Laboratory of the University of St Andrews for organizing the original data entry and for their technical support and backup. My particular thanks are due to Dr Maurice Shepherd, Director of the Industrial Liaison Office of the University, for arranging the link between the publishers and the laboratory, writing the software for the addition of the printer codes and for guiding my efforts in writing the utility programs needed for this work.

Peter M. B. Walker

A

a- *(Genrl.)*. Prefix signifying *on*. Also shortened form of *ab-, ad-, an-, ap-*.

a *(Genrl.)*. Symbol for: acceleration; relative activity; linear absorption coefficient; amplitude.

***a*-** *(Chem.)*. An abbrev. for: (1) asymmetrically substituted; (2) *ana-*, i.e. containing a condensed double aromatic nucleus substituted in the 1.5 positions.

α *(Genrl.)*. See under **alpha**. Symbol for: absorption coefficient; attenuation coefficient; acceleration, angular acceleration; fine structure constant; helium nucleus.

α- *(Chem.)*. Symbol for: (1) substitution on the carbon atom of a chain next to the functional group; (2) substitution on a carbon atom next to one common to two condensed aromatic nuclei; (3) substitution on the carbon atom next to the hetero-atom in a hetero-cyclic compound; (4) a stereo-isomer of a sugar.

A *(Eng.)*. Same as **Ae**. The equilibrium temperature of a *phase transformation*.

Å *(Phys.)*. Symbol for Ångström.

A *(Genrl.)*. Symbol for: area; ampere; absolute temperature; relative atomic mass (atomic weight); magnetic vector potential; Helmholtz function.

[A] *(Phys.)*. A strong absorption band in deep red of the solar spectrum (wavelength 762.128 nm) caused by oxygen in the earth's atmosphere. The first of the Fraunhofer lines.

$[\alpha]_D^t$ *(Chem.)*. Symbol for the specific optical rotation of a substance at t°C, measured for the D line of the sodium spectrum.

A & AEE *(Aero.)*. Abbrev. for *Aeroplane & Armament Experimental Establishment*, at Boscombe Down, UK.

A-amplifier *(Acous.)*. One associated with, or immediately following, a high-quality microphone, as in broadcasting studios. N.B. Not the same as **class-A amplifier**.

A and R display *(Radar)*. See **r-display**.

ab *(Build.)*. Abbrev. for *As Before* in bills of quantities, etc.

ab- *(Elec.Eng.)*. Prefix to name of unit, indicating derivation in the CGS system.

ABA *(Bot.)*. Abbrev. for *ABscisic Acid.*

abactinal *(Zool.)*. See **abambulacral**.

abacus *(Arch.)*. The uppermost part of a column capital or pilaster, on which the architrave rests.

abacus *(Maths.)*. A bead frame. Used as an arithmetical calculating aid.

abambulacral *(Zool.)*. Pertaining to that part of the surface of an Echinoderm lacking tube feet.

abamurus *(Arch.)*. A supporting wall or buttress, built to add strength to another wall.

abandonment *(Min.Ext.)*. Voluntary surrender of legal rights or title to a mining claim.

abapical *(Zool.)*. Pertaining to, or situated at, the lower pole: remote from the apex.

abatjour *(Arch.)*. An opening to admit light, and generally to deflect it downwards; a skylight.

abaxial *(Bot.)*. The surface of a leaf, petal etc., that during early development faced away from the axis. Usually it is therefore the under-surface of an expanded leaf.

abaxial *(Image Tech.,Phys.)*. Said of rays of light which do not coincide with the optical axis of a lens system.

abaxial *(Zool.)*. Remote from the axis.

Abbe refractometer *(Chem.)*. An instrument for measuring directly the refractive index of liquids.

abdominal air sac *(Zool.)*. Posterior part of the lung in Birds.

abdominal cavity *(Med.)*. See **peritoneal cavity**.

abdominal gills *(Zool.)*. In the aquatic larvae of many Insects, paired segmental leaflike or filamentous expansions of the abdominal cuticle for respiration.

abdominal limbs *(Zool.)*. Segmented abdominal appendages in most Crustacea which are used for swimming, setting up currents of water for feeding and/or respiration or carrying eggs and young. In Diplopoda, segmented ambulatory appendages on the abdomen.

abdominal pores *(Zool.)*. Apertures leading from the coelom to the exterior in certain Fish and in Cyclostomata.

abdominal reflex *(Zool.)*. Contraction of the abdominal wall muscles when the skin over the side of the abdomen is stimulated.

abdominal regions *(Med.)*. Nine regions into which the human abdomen is divided by two horizontal and two vertical imaginary planes, i.e. right and left hypochondriac, right and left lumbar, right and left iliac, epigastric, umbilical and hypogastric.

abducens *(Med.)*. In Vertebrates, the 6th cranial nerve, purely motor in function, supplying the rectus externus muscle of the eye.

abduction *(Med.)*. The action of pulling a limb or part away from the median axis.

abductor *(Zool.)*. Any muscle that draws a limb or part away from the median axis by contraction; e.g. the abductor pollicis, which moves the thumb outward.

Abegg's rule *(Chem.)*. Empirical rule that solubility of salts of alkali metals with strong acids decreases from lithium to caesium, i.e. with increase of rel. at. mass, and those with weak acids follow the opposite order. Sodium chloride is an exception to this rule, being less soluble than potassium chloride.

Abegg's rule of eight *(Chem.)*. The sum of the maximum positive and negative valencies of an element is eight, e.g. S in SF_6 and H_2S.

Abel flash-point apparatus *(Min.Ext.)*. A petroleum-testing apparatus for determining the flash-point.

Abelian group *(Maths.)*. A group in which the group operation is commutative. It is important in the study of rings and vector spaces.

abelite *(Chem.)*. Explosive, composed mainly of ammonium nitrate and trinitrotoluene.

aberrant *(Bot.,Zool.)*. Having characteristics not strictly in accordance with the type.

aberration *(Astron.)*. An apparent change of position of a heavenly body, due to the velocity of light having a finite ratio to the relative velocity of the source and the observer.

aberration *(Phys.)*. In an image-forming system e.g., an optical or electronic lens, failure to produce a true image i.e. a point object as a point image etc. Five geometrical aberrations are recognized by von Seidel, viz., spherical aberration, coma, astigmatism, curvature of the field and distortion. See also **chromatic aberration**.

abiogenesis *(Bot.,Zool.)*. *Spontaneous generation.* The development of living organisms from non-living matter; either the spontaneous generation of yeasts, bacteria etc. believed in before Pasteur, or the gradual process postulated for the early Precambrian in modern theories of the origin of life.

abiotic *(Bot.)*. Pertaining to non-living things.

ablation *(Med.)*. Removal of body-tissue by surgery.

Abney law *(Phys.)*. If a spectral colour is de-saturated by the addition of white light, and if its wavelength is less than 570 nm, its hue then moves towards the red end of the spectrum, while if the wavelength is more than 570 nm its hue moves towards the blue.

Abney level *(Surv.)*. Hand-held instrument in which angles of steep sights are measured while simultaneously viewing a spirit-level bubble.

Abney mounting *(Phys.)*. A form of mounting for a concave diffraction grating, in which the eye-piece (or photographic plate holder) is fixed at the centre of curvature of the grating and the slit can move around the circumference of the Rowland circle, to bring different orders of spectrum into view.

abnormal glow discharge *(Electronics)*. One carrying current in excess of that which is required to cover the cathode completely with visible radiation.

abnormal reflections *(Telecomm.)*. Those from the ionosphere for frequencies in excess of *critical frequency*.

ABO blood group substances *(Immun.)*. Large glycopeptides with oligosaccharide side chains bearing ABO antigenic determinants identical to those of the erythrocytes of the same individual, present in mucous secretions of persons who possess the secretor gene.

ABO blood group system *(Immun.)*. The most important of the antigens of human red blood cells for blood transfusion serology. Humans belong to one of four groups; A, B, AB, and O. The red cells of each group carry respectively the A antigen, the B antigen, both A and B antigens, or neither. Natural antibodies (resulting from immunization by bacteria in the gut) are present in the blood against the blood group antigen which is absent from the red cells. Thus persons of group A have anti-B, of group B have anti-A, of group O have anti-A and anti-B, and group AB have neither. Before blood transfusion the blood must be cross-matched to ensure that red cells of one group are not given to a person possessing antibodies against them.

abomasitis *(Vet.)*. Inflammation of the **abomasum**.

abomasum *(Zool.)*. In ruminant Mammals, the fourth or true stomach. Also called *reed, rennet*.

A-bomb *(Phys.)*. See **atomic bomb**.

aboral *(Zool.)*. Opposite to, leading away from, or distant from, the mouth. See **abambulacral**.

abort *(Space)*. The termination of a vehicle's flight either by failure or deliberate action to prevent dangerous consequences; if manned, a predetermined sequence of events is followed to ensure the safety of the crew.

abortifacient *(Med.)*. Anything which causes artificial abortion; a drug which does this.

abortion *(Med.,Zool.)*. (1) Expulsion of the foetus from the uterus during the first 3 months of pregnancy. Abortion may be spontaneous or induced. (2) Termination of the development of an organ.

abradant *(Eng.)*. A substance, usually in powdered form, used for grinding. See **abrasive**.

abrade *(Eng.)*. Cut or tear, at two surfaces in contact and relative motion.

Abram's 'law' *(Civ.Eng.)*. Ratio of water to cement for chemical action to impart strength to concrete is 0.35:1.

abranchiate *(Zool.)*. Lacking gills.

abrasion *(Med.)*. A rubbed-away area of the surface-covering of the body; i.e. of skin or of mucous membrane.

abrasion hardness *(Min.Ext.)*. Resistance to abrasive wear, under specified conditions, of metal or mineral.

abrasive *(Chem.)*. A substance used for the removal of matter by scratching and grinding (abrasion); e.g. silicon carbide (carborundum).

abrasive blast cleaning *(Build.)*. A method for preparing steel for painting whereby abrasive particles, e.g. copper slag, are projected under air pressure through a nozzle. Very effective in removing rust and mill scale leaving an *anchor pattern* on the substrate affording good paint adhesion.

abrasive papers *(Build.)*. Special papers coated in grit used for flatting down. Supplied in a range of grits from very fine to coarse in two main types. (1) Dry abrasive papers. (2) Waterproof abrasive papers.

abreaction *(Behav.)*. In psychoanalytic theory, an intense emotional outburst to a previously repressed experience, the therapeutic effect is called *catharsis*.

A & B roll printing *(Image Tech.)*. Method of film printing with alternate scenes assembled in two rolls, each having black spacing equivalent in length to the omitted scene; double printing from the two allows the inclusion of *fade* and *dissolve* effects and avoids visible splice marks between scenes in 16 mm printing.

ABS *(Plastics)*. *Acrilonitrile-Butadiene-Styrene*. Range of copolymers based on cyanoethene/but 1,2:3,4-diene/phenylethene.

ABS brake *(Autos.)*. Commonly used for **anti-lock brake**. Germ. *anti blockier system*.

abscess *(Med.)*. A localised collection of pus in infected tissue, usually confined within a capsule.

abscisic acid *(Bot.)*. *Dormin*. A sesquiterpenoid plant growth substance, ($C_{15}H_{20}O_4$) with a variety of reported effects, e.g. inhibiting growth, causing stomatal closure, and promoting senescence, abscission and dormancy. Abbrev. *ABA*.

abscissa *(Maths.)*. For rectilineal axes of co-ordinates, the distance of a point from the axis of ordinates measured in a direction parallel to the axis of abscissae, which is usually horizontal. The sign convention is that measurements to the right from the axis of ordinates are positive, measurements to the left negative. pl. *abscissae*. Cf **Cartesian co-ordinates**.

abscission *(Bot.)*. The organised shedding of parts of a plant by means of an abscission layer.

abscission layer, separation layer *(Bot.)*. In the **abscission zone**, a layer of cells the disjunction or breakdown of which causes abscission.

abscission zone *(Bot.)*. Zone at the base of leaf, petal, fruit etc. that contains the **abscission layer** and **protective layer**.

absolute *(Maths.)*. In general there are two points at infinity on every line. The assemblage of these points at infinity is a conic (a quadric in three dimensions) called the absolute. Its form determines the metrical properties of the geometrical system being operated. Thus in Euclidean geometry, the absolute is the degenerate conic comprising the line at infinity taken twice, while in non-Euclidian geometry, the absolute is either a real conic (hyperbolic geometry) or an imaginary conic (elliptic geometry).

absolute address *(Comp.)*. Code designation of a specific store location as determined by the **hardware**. Cf. **relative address**.

absolute age *(Geol.)*. The geological age of a fossil, mineral, rock or event, generally given in years. Preferred synonym *radiometric age*. See **radiometric dating**.

absolute alcohol *(Chem.)*. Water-free ethanol; rel.d. 0.793 (15.5°C); bp 78.4°C; obtained from rectified spirit by adding benzene and refractionating. Very hygroscopic.

absolute ampere *(Phys.)*. The standard MKS unit of electric current; replaced the international ampere in 1948. See **ampere**.

absolute block system *(Civ.Eng.)*. See **block system**.

absolute ceiling *(Aero.)*. The height at which the rate of climb of an aircraft, in standard atmosphere, would be zero; the maximum height attainable under standard conditions.

absolute coefficient *(Maths.)*. A coefficient with an absolute value, that is a multiplier which is numerical rather than symbolic.

absolute configuration *(Chem.)*. The arrangement of groups about an asymmetric atom, especially a tetrahedrally bonded atom with four different substituents. See **chirality** and **Cahn-Ingold-Prelog system**.

absolute convergence *(Maths.)*. A series Σa_r is absolutely convergent if the series $\Sigma|a_r|$ is convergent.

absolute electrometer *(Phys.)*. High-grade attracted-disk electrometer in which an absolute measurement of potential can be made by 'weighing' the attraction between two charged disks against gravity.

absolute humidity *(Meteor.)*. Alternative to **vapour concentration**.

absolute instrument *(Phys.)*. An instrument which measures a quantity directly in absolute units, without the necessity for previous calibration.

absolute magnitude *(Astron.)*. See **magnitudes**.

absolute permeability *(Elec.Eng.)*. See **permeability**.

absolute potential *(Chem.)*. The theoretical true potential difference between an electrode and a solution of its ions, measured against a hypothetical reference electrode, having an absolute potential of zero, with reference to the same solution.

absolute pressure *(Phys.)*. Pressure measured with respect to zero pressure, in units of force per unit of area.

absolute reaction rates *(Chem.)*. Theory that if the rate of a chemical reaction is governed by the rate of crossing an energy barrier or of forming an **activated complex**, then it can be calculated from statistical thermodynamics. See **Arrhenius theory of dissociation**.

absolute-rest precipitation tanks *(Build.)*. Those for batch treatment of sewage, as opposed to tanks taking a continuous flow. After 2- or 3-hr settlement, the top water is drawn off from above and the precipitated sludge from below.

absolute temperature *(Phys.)*. A temperature measured with respect to **absolute zero**, i.e. the zero of the **Kelvin thermodynamic scale**, a scale which cannot take negative values. See **kelvin, Rankine scale**.

absolute threshold *(Behav.)*. The minimal intensity of a physical stimulus required to produce a response.

absolute units *(Phys.)*. Those derived directly from the fundamental units of a system and not based on arbitrary numerical definitions. The differences between absolute and international units were small; both are now superseded by the definitions of the SI.

absolute value *(Maths)*. See **modulus**.

absolute viscosity *(Phys.,Eng.)*. See **coefficient of viscosity**.

absolute wavemeter *(Telecomm.)*. One in which the frequency of the injected RF signal is by calculation of physical properties (circuit elements or dimensions) of a resonant circuit line or cavity.

absolute weight *(Phys.)*. The weight of a body in a vacuum.

absolute zero *(Phys.)*. The least possible temperature for all substances. At this temperature the molecules of any substance possess no heat energy. A figure of -273.15°C is generally accepted as the value of absolute zero.

absorbance *(Chem.)*. The logarithm of the ratio of the intensity of light incident on a sample to that transmitted by it. It is usually directly proportional to the concentration of the absorbing substance in a solution. See **Beer's Law**.

absorbed dose *(Radiol.)*. The energy absorbed by the patient from the decay of a radionuclide given for diagnostic or therapeutic purposes. The unit is a gray (*GY*). 1 GY = 1 Joule/kg.

absorbency tests *(Paper)*. Any test methods intended to measure the capacity of a paper to absorb liquids or fluids. Results are usually expressed as the gain in weight of the test piece, the capillary rise in a test strip in given time, or the time required to reach a predetermined capillary rise.

absorber *(Phys.)*. Any material which converts energy of radiation or particles into another form, generally heat. Energy transmitted is not absorbed. Scattered energy is often classed with absorbed energy. See **total absorption coefficient, true absorption coefficient**.

absorber rod *(Nuc.Eng.)*. Alternative name for *control rod*.

absorbing material *(Phys.)*. Any medium used for absorbing energy from radiation of any type.

absorbing well *(Civ.Eng.)*. A shaft sunk through an impermeable stratum to allow water to drain through to a permeable one.

absorptance *(Phys.)*. A measure of the ability of a body to absorb radiation; the ratio of the radiant flux absorbed by the body to that incident on the body. Formerly *absorptivity*.

absorptiometer *(Chem.)*. An apparatus for determining the solubilities of gases in liquids or the absorption of light.

absorption *(Immun.)*. Used in immunology to describe the use of reagents to remove unwanted antibodies or antigens from a mixture.

absorption band *(Phys.)*. A dark gap in the continuous spectrum of white light transmitted by a substance which exhibits selective absorption.

absorption capacitor *(Elec.Eng.)*. One connected across spark gap to damp the discharge.

absorption coefficient *(Chem.)*. The volume of gas, measured at s.t.p., dissolved by unit volume of a liquid under normal pressure (i.e. 1 atmosphere).

absorption coefficient *(Phys.)*. (1) At a discontinuity (*surface absorption coefficient*), (*a*) the fraction of the energy which is absorbed, or (*b*) the reduction of amplitude, for a beam of radiation or other wave system incident on a discontinuity in the medium through which it is propagated, or in the path along which it is transmitted. (2) In a medium (*linear absorption coefficient*), the natural logarithm of the ratio of incident and emergent energy or amplitude for a beam of radiation passing through unit thickness of a medium. (The *mass absorption coefficient* is defined in the same way but for a thickness of the medium corresponding to unit mass per unit area.) N.B. *True absorption coefficients* exclude scattering losses, *total absorption coefficients* include them. See **atomic absorption coefficient**.

absorption discontinuity *(Phys.)*. See **absorption edge**.

absorption dynamometer *(Eng.)*. A dynamometer which absorbs and dissipates the power which it measures; e.g. the ordinary rope brake and the Froude hydraulic brake. Cf. **transmission dynamometer**.

absorption edge *(Phys.)*. The wavelength at which there is an abrupt discontinuity in the intensity of an absorption spectrum for electromagnetic waves, giving the appearance of a sharp edge in its photograph. This transition is due to one particular energy-dissipating process becoming possible or impossible at the limiting wavelength. In X-ray spectra of the chemical elements the K-absorption edge for each element occurs at a wavelength slightly less than that for the K-emission spectrum. Also *absorption discontinuity*.

absorption hygrometer *(Meteor.)*. An instrument by which the quantity of water vapour in air may be measured. A known volume of air is drawn through tubes containing a drying agent such as phosphorus pentoxide; the increase in weight of the tubes gives the weight of water vapour in the known volume of air.

absorption inductor *(Elec.Eng.)*. See **interphase transformer**.

absorption lines *(Phys.)*. Dark lines in a continuous spectrum caused by absorption by a gaseous element. The positions (i.e., the wavelengths) of the dark absorption lines are identical with those of the bright lines given by the same element in emission.

absorption plant *(Min.Ext.)*. Plant where oils are removed from natural gas by absorption in suitable oil.

absorption refrigerator *(Eng.)*. A plant in which ammonia is continuously evaporated from an aqueous solution under pressure, condensed, allowed to evaporate (so absorbing heat), and then reabsorbed.

absorption spectrum *(Phys.)*. The system of absorption bands or lines seen when a selectively absorbing substance is placed between a source of white light and a spectroscope. See **Kirchhoff's law**.

absorption tubes *(Chem.)*. Tubes filled with solid absorbent for the absorption of moisture (e.g. silica gel) and gases (e.g. charcoal).

absorption wavemeter *(Elec.Eng.)*. One which depends on a resonance absorption in a tuned circuit, constructed with very stable inductance and capacitance.

absorptivity, absorptive power *(Phys.)*. See **absorptance**.

abundance *(Ecol.)*. See **relative abundance, frequency**.

abundance, abundance ratio *(Phys.)*. For a specified element, the proportion or percentage of one isotope to the total, as occurring in nature.

abutment *(Civ.Eng.)*. A point or surface provided to withstand thrust; e.g. end supports of an arch or bridge. See **knapsack-**.

abutment load *(Min.Ext.)*. In stoping or other deep-level excavation, weight transferred to the adjacent solid rock by unsupported roof.

abutting joint *(Build.)*. A joint whose plane is at right angles to the fibres, the fibres of both joining pieces being in the same straight line.

abyssal *(Geol.)*. Refers to the ocean floor environment between ca 4000-6000 m. See **littoral, bathyal**.

abyssal deposits *(Geol.)*. Pelagic marine sediments,

accumulating in depths of more than 2000 m including, with increasing depth, calcareous oozes, siliceous oozes and red clay (> 500 m).

abyssopelagic *(Ecol.)*. Refers to the region of deep water which excludes the ocean floor: floating in the ocean depths.

ac *(Chem.)*. Symbol for Actinium.

a.c. *(Elec.Eng)*. Abbrev. for *alternating current*.

Ac *(Eng.)*. The *transformation temperature* on heating of the phase changes of iron or steel, subscripts indicating the designated change.

ac- *(Chem)*. Abbrev. indicating substitution in the alicyclic ring.

acacia *(For.)*. A member of the Leguminosae giving a coarse-textured hardwood, reddish-brown in colour. Used for tool handles, vehicle parts, walking sticks and turned articles.

acacia gum *(Chem.)*. See **gum arabic**.

acanthite *(Min.)*. An ore of silver, Ag_2S, crystallizing in the monoclinic system. Cf. **argentite**.

acantho- *(Bot.)*. Prefix from Gk. *akantha*, spine, thorn.

Acanthocephala *(Zool.)*. A phylum of elongate worms with rounded body and a protrusible proboscis, furnished with recurved hooks; there is no mouth or alimentary canal; the young stages are parasitic in various Crustaceans, the adults in Fish and aquatic Birds and Mammals. Thorny-headed worms.

acanthoma *(Med.)*. A tumour of epidermal cells.

acanthosis nigricans *(Med.)*. A rare disease characterized by pigmentation and warty growths on the skin, often associated with cancer of the stomach or uterus.

acanthozooid *(Zool.)*. In Cestoda, the proscolex, or head-portion, of a bladder-worm. Cf. **cystozooid**.

acariasis *(Vet.)*. Contagious skin disease caused by mites (*acari*).

Acarina *(Zool.)*. An order of small *Arachnida*, with globular, undivided body. The immature stages (hexapod larvae) have 6 legs. A large worldwide group, occupying all types of habitat, and of great economic importance. Many are ectoparasitic. Mites and ticks.

acarophily, acarophytism *(Bot.)*. A symbiotic association between plants and mites.

acaulescent *(Bot.)*. Having a short stem.

acauline, acaulose *(Bot.)*. Stemless or nearly so.

a.c. balancer *(Elec.Eng.)*. An arrangement of transformers or reactors used to equalize the voltages between the wires of a multiple-wire system. Also called a *static balancer*.

a.c. bias *(Electronics)*. A high-frequency signal applied to a magnetic tape recording head along with the signal to be recorded. This stabilizes magnetic saturation and improves frequency response, at the same time reducing noise and distortion. The bias signal frequency has to be many times the highest recording frequency.

AC-boundary layer *(Acous.,Aero.)*. See **Stokes layer**.

accelerated ageing test *(Elec.Eng.)*. A stability test for cables using twice normal working voltage. It is claimed this gives quick results that correlate with service records.

accelerated fatigue test *(Eng.)*. The application, mechanically, of simulated operating loads to a machine or component to find the limit of safe fatigue life before it is reached in service.

accelerate-stop distance *(Aero.)*. The total distance, under specified conditions, in which an aircraft can be brought to rest after accelerating to **critical speed** for an engine failure at takeoff.

accelerating chain *(Electronics)*. That section of an electron beam tube or system, e.g. CRT or electron microscope, in which electrons are accelerated by voltages on accelerating electrodes. Also used in *particle accelerators*.

accelerating contactor *(Elec.Eng.)*. One of the contactors of an electric-motor control panel which cuts out starting resistance, thereby causing the motor to accelerate.

accelerating electrode *(Electronics)*. One in a thermionic valve or CRT maintained at a high positive potential with respect to the electron source. It accelerates electrons in their flight to the anode but does not collect a high proportion of them.

accelerating machine, accelerator *(Electronics)*. Machine used to accelerate charged particles to very high energies. See also **betatron, cyclotron, linear accelerator, synchrocyclotron**, and **synchrotron**.

accelerating potential *(Electronics)*. That applied to an electrode to accelerate electrons from a cathode.

acceleration *(Phys.)*. The rate of change of velocity, expressed in metres (or feet) per second squared. It is a vector quantity and has both magnitude and direction.

acceleration due to gravity *(Phys.)*. Acceleration with which a body would fall freely under the action of gravity in a vacuum. This varies according to the distance from the earth's centre, but the internationally adopted value is $9.806\,65\ m/s^2$ or $32.1740\ ft/s^2$. See **Helmert's formula**.

acceleration error *(Aero.)*. The error in an airborne magnetic compass due to manoeuvring; caused by the vertical component of the earth's magnetic field when the centre of gravity of the magnetic element is displaced from normal.

acceleration stress *(Space)*. The influence of acceleration (or deceleration) on certain physiological parameters of the human body. Man can withstand transverse accelerations better than longitudinal ones which have a profound effect on the cardiovascular system. The degree of tolerance also depends on the magnitude and duration of the acceleration.

acceleration tolerance *(Space)*. The maximum of *g* forces an astronaut can withstand before 'blacking out' or otherwise losing control.

accelerator *(Aero.)*. A device, similar to a **catapult**, but generally mounted below deck level, for assisting the acceleration of aircraft flying off aircraft carriers. Land versions have been tried experimentally.

accelerator *(Autos.)*. A pedal connected to the carburettor throttle valve of a motor vehicle, or to the fuel injection control.

accelerator *(Build.)*. A hardener or catalyst mixed with synthetic resins in **two pack materials** to speed up the hardening rate.

accelerator *(Chem.)*. (1) A substance which increases the speed of a chemical reaction. See **catalysis**. (2) A substance which increases the efficient action of an enzyme. (3) Any substance effecting acceleration of the vulcanization process of rubber. The principal types are aldehyde derivatives of Schiff's bases, butyraldehyde-butylidene-aniline, di-orthotolyl-guanidine, diphenyl-guanidine, benzthiazyl disulphide, tetramethylthiuran disulphide, zinc dimethyl-dithiocarbamate.

accelerator *(Civ.Eng.)*. Any substance mixed with cement concrete for the purpose of hastening hardening.

accelerator *(Electronics)*. See **accelerating machine**.

accelerator *(Image Tech.)*. A chemical used to increase the rate of development; e.g. sodium carbonate or borax.

accelerator *(Plastics)*. Substance which increases catalytically the hardening rate of a synthetic resin.

accelerator *(Zool.)*. Any muscle or nerve which increases rate of action.

accelerator pump *(Autos.)*. A small cylinder and piston fitted to some types of carburettor, and connected to the throttle so as to provide a momentarily enriched mixture when the engine is accelerated.

accelerometer *(Acous.,Electronics)*. Transducer used to provide a signal proportional to the rate of acceleration of a vibrating or other body, usually employing the piezoelectric principle. See also **pick-up head**.

accelerometer *(Aero.)*. An instrument, carried in aircraft, guided missiles and spacecraft for measuring acceleration in a specific direction. Main types are *indicating*, *maximum reading*, *recording* (graphical), and *counting* (digital, totalling all accelerations above a set value). See **impact accelerometer, vertical-gust recorder**.

accentuation *(Telecomm.)*. See **pre-emphasis**.

acceptance angle *(Electronics)*. The solid angle within which all incident light reaches the photocathode of a phototube.

acceptor *(Chem.)*. (1) The reactant in an induced reaction whose rate of reaction with a third substance is increased by the presence of the inductor. (2) The atom which accepts electrons in a co-ordinate bond.

acceptor *(Electronics)*. Impurity atoms introduced in small quantities into a crystalline semiconductor and having a lower valency than the semiconductor, from which they attract electrons. In this way *holes* are produced, which effectively become positive charge carriers; the phenomenon is known as *p-type conductivity*. See also **donor, impurity**.

acceptor circuit *(Telecomm.)*. Tuned circuit responding to a signal of one specific frequency.

acceptor level *(Electronics)*. See **energy levels**.

access eye *(Build.)*. A screwed plug provided in soil, waste, and drain pipes at bends and junctions, to clear a stoppage.

accessorius *(Zool.)*. A muscle which supplements the action of another muscle: in Vertebrates, the eleventh cranial nerve or spinal accessory.

accessory *(Zool.)*. See **accessorius**.

accessory bud *(Bot.)*. A bud additional to a normal axillary bud.

accessory cell *(Bot.)*. A **subsidiary cell** in a stomatal complex.

accessory cell *(Immun.)*. Cell other than a lymphocyte which takes part in an immune reaction, e.g. in *antigen presentation* and/or by modulating the function of the lymphocyte. Usually **macrophages** or **dendritic cells**.

accessory chromosome *(Biol.)*. See **sex chromosome**.

accessory gearbox *(Aero.)*. A gearbox, driven remotely from an aero-engine, on which aircraft accessories, e.g. hydraulic pump, electrical generator, are mounted.

accessory glands *(Zool.)*. Glands of varied structure and function in connection with genitalia, especially of Arthropoda.

accessory hearts *(Zool.)*. See **accessory pulsatory organs**.

accessory minerals *(Geol.)*. Minerals which occur in small, often minute, amounts in igneous rocks; their presence or absence makes no difference to classification and nomenclature.

accessory pigments *(Bot.)*. Pigments found in chloroplasts and blue-green algae which transfer their absorbed energy to *chlorophyll a* during photosynthesis. They include *cholorophylls b, c* and *d*, the *carotenoids* and the *phycobilins*.

accessory plates *(Min.)*. Quartz-wedge, gypsum plate, mica plate. Used with petrological microscope to help determine the optical character of a mineral as an aid in its examination.

accessory pulsatory organs *(Zool.)*. In some Insects and Molluscs, saclike contractile organs, pulsating independent hearts; variously situated on the course of the circulatory system. Also *accessory hearts*.

accessory shoe *(Image Tech.)*. Mounting forming part of the camera body on to which separate units such as flash-guns, range-finders etc. may be fitted.

access selector *(Telecomm.)*. A selector used to connect common equipment (e.g. in routine tests of artificial traffic equipment) sequentially to corresponding points in each of a series of identical circuits.

access time *(Comp.)*. Time interval between the instant at which data is called from store and the instant at which the data can be used. It can vary from microseconds with **fast store** to minutes with **magnetic tape**.

access to store *(Comp.)*. Entry or extraction of data from a memory location. The method and speed of access depends on the type of memory. See **fast store, random access memory, serial access memory, backing store**.

Accipitriformes *(Zool.)*. Order comprising the birds of prey. Hooked bill with fleshy cere, long talons with opposable hind toe, predaceous. Eagles, harriers, vultures, buzzards.

a.c. circuit *(Elec.Eng.)*. One which passes a.c., e.g. it may have a capacitor in series, which blocks direct current.

acclimation *(Biol.)*. See **acclimatization**.

acclimatization, acclimation *(Behav.)*. Reversible alteration of an individual's tolerance for environmental conditions, e.g. for temperature conditions.

acclimatization, acclimation *(Biol.)*. Becoming adapted to environmental stress; physiological adjustment in an organism when moved to a new environment, usually taking days or weeks. See **hardening**.

accommodation *(Behav.)*. For Piaget, one of the two main biological forces responsible for *cognitive* development; it refers to the process of modifying pre-existing cognitive organisations in order to absorb information that no longer fits into existing *schemata*. See also **assimilation**.

accommodation *(Med.)*. Ability of the eye to change its effective focal length to see objects distinctly at varying distances. The range of vision for a human eye is from about 250 mm to infinity. Power of accommodation usually diminishes with advancing age.

accommodation rig *(Min.Ext.)*. Offshore rig with sleeping, supply and recreational facilities.

a.c. commutator motor *(Elec.Eng.)*. An a.c. motor which embodies a commutator as an essential part of its construction. See **a.c. series motor, compensated induction motor, repulsion motor, Schrage motor**.

accordion fold *(Print.)*. Method of folding a leaflet or insert so that it opens out and closes in a zig-zag fashion. Also called *concertina fold*.

accoucheur *(Med.)*. A physician who practises midwifery, an *obstetrician*.

accretion *(Astron.)*. Process in which a celestial body, particularly an evolved dwarf star or a planet, is enlarged by the accumulation of extraneous matter falling in under gravity.

accretion *(Geol.)*. The process of enlargement of a continent by the tectonic coalescence of exotic crustal fragments.

accretion *(Zool.)*. External addition of new matter: growth by such addition.

accumulated temperature *(Meteor.)*. The integrated product of the excess of air temperature above a threshold value and the period in days during which such excess is maintained.

accumulation point *(Maths.)*. Of a set of points: one such that every neighbourhood of it includes at least one point of the set. Also called *limit point*.

accumulator *(Comp.)*. Special storage register associated with the **arithmetic logic unit**, used for holding the results of a computation or data transfer.

accumulator *(Elec.Eng.)*. Voltaic cell which can be charged and discharged. On charge, when an electric current is passed through it into the positive and out of the negative terminals (according to the conventional direction of flow of current), electrical energy is converted into chemical energy. The process is reversed on discharge, the chemical energy, less losses both in potential and current, being converted into useful electrical energy. Accumulators therefore form a useful portable supply of electric power, but have the disadvantages of being heavy and of being at best 70% efficient. More often known as *battery*, also called *reversible cell, secondary cell, storage battery*.

accumulator box *(Elec.Eng.)*. A vessel, usually made of glass, lead-lined wood, or celluloid, for containing the plates and electrolyte of an accumulator.

accumulator grid *(Elec.Eng.)*. The lead grid which forms one of the plates of a lead-acid accumulator having pasted plates.

accumulator switchboard *(Elec.Eng.)*. A switchboard upon which are mounted all necessary switches and instruments for controlling charging and discharging of a battery of accumulators.

accumulator traction *(Elec.Eng.)*. See **battery traction**.

accumulator vehicle *(Elec.Eng.)*. See **battery vehicle**.

acellular *(Bot.)*. Not partitioned into cells. Sometimes used for *unicellular* but also for *multinucleate* or **coenocytic**.

acenaphthenequinone *(Chem.)*. Chemical, crystallizing in yellow needles, sparingly soluble in water. Forms the basis of scarlet and red vat dyes of the 'Ciba' type.

acentric *(Biol.)*. Having no centromere, applied to chromosomes and chromosome segments.

acentrous *(Zool.)*. Having a persistent notochord with no vertebral centra, as in the Cyclostomata.

acephalous *(Zool.)*. Showing no appreciable degree of cephalization: lacking a head-region, as *Pelecypoda*.

acervulus *(Bot.)*. A dense cushionlike mass of conidiophores and conidia formed by some fungi. adj. *acervulate*.

acetabular bone *(Zool.)*. In Crocodiles, the separate lower end of the ilium: in some Mammals, as *Galeopithecus*, an additional bone lying between the ilium and pubis and frequently fused with one or the other.

acetabulum *(Zool.)*. In Platyhelminthes, Hirudinea, and Cephalopoda, a circular muscular sucker: in Insects, a thoracic aperture for insertion of a leg: in Vertebrates, a facet or socket of the pelvic girdle with which articulates the pelvic fin or head of the femur: in ruminant Mammals, one of the cotyledons of the placenta.

acetal *(Chem.)*. $CH_3CH(OC_2H_5)_2$, bp 104°C, *1,1-diethoxy ethane*. The term *acetal* is applied to any compound of the type $R.CH(OR')_2$, where R and R′ are organic radicals and R may be hydrogen.

acetaldehyde *(Chem.)*. *Ethanal*. CH_3CHO, colourless, ethereal, pungent liquid, bp 21°C, mp −121°C, rel.d. 0.8, oxidation product of ethanol; intermediate for production of ethanoic acid, important raw material for the synthesis of organic compounds.

acetal resins *(Plastics)*. Term applied to stable high polymers of formaldehyde widely used as engineering plastics. See also **Delrin**.

acetals *(Chem.)*. The dehydration products of aldehydes, with an excess of alcohols present. Acetals may be termed 1,1-dialkoxyalkanes.

acetamide *(Chem.)*. *Ethanamide*. $CH_3.CO.NH_2$, needles, soluble in water and alcohol, mp 82°C, bp 222°C, a primary amide of ethanoic acid.

acetate fibres *(Textiles)*. Continuous filaments and staple fibres manufactured from cellulose acetate produced from cotton linters or wood pulp. Between 74% and 92% of the hydroxyl groups in the original cellulose are acetylated. For *triacetate fibres* the cellulose is more highly acetylated.

acetate film *(Image Tech.)*. Film with its photographic emulsion coated on a **base** of cellulose triacetate, of low inflammability. Also *non-flam film, safety film*.

acetates *(Chem.)*. *Ethanoates*. Salts of acetic (ethanoic) acid; e.g. sodium acetate. Also esters of acetic acid.

acetic acid *(Chem.)*. *Ethanoic acid*. $CH_3.COOH$, synthesized from acetylene (ethyne), also obtained by the destructive distillation of wood; and by the oxidation of ethanol. Acetic acid has a mp 16.6°C, bp 118°C, rel.d. (20°C) 1.0497.

acetic anhydride *(Chem.)*. *Ethanoic anhydride*. The anhydride of acetic acid, $(CH_3CO)_2O$. Colourless liquid bp 137°C. Used industrially for preparation of cellulose acetate and acetylsalicylic acid. Valuable laboratory acetylating agent.

acetic fermentation *(Chem.)*. The fermentation of dilute ethanol solutions by oxidation in presence of bacteria, especially *Bacterium aceti*. Acetic acid is formed.

acetin *(Chem.)*. *Monoacetin, glyceryl monoacetate*, $CH_3COO.C_3H_5(OH)_2$, bp 130°C, rel.d. 1.22, a colourless hygroscopic liquid, used as intermediate for explosives, solvent for basic dyestuffs, tanning agent.

acetone *(Chem.)*. *Propanone*. CH_3COCH_3, bp 56°C, of ethereal odour, a very important solvent, basis for organic synthesis. Acetone is the simplest saturated ketone. Useful solvent for acetylene.

acetone *(Med.)*. Found in the blood and urine of patients with uncontrolled diabetes mellitus or in starvation.

acetone cyanhydrin *(Chem.)*. *2-hydroxy 2-methyl propanonitrile*. $(CH_3)_2C(OH)CN$. Addition product of acetone and hydrogen cyanide.

acetone resin *(Chem.)*. A synthetic resin formed by the reaction of acetone with another compound, such as phenol or formaldehyde.

acetonitrile *(Chem.)*. CH_3CN, a polar organic solvent.

acetonuria *(Med.)*. See **Ketonuria**.

acetophenone *(Chem.)*. $C_6H_5COCH_3$, *phenyl methyl ketone*, mp 20°C, bp 202°C., rel.d. 1.03, large colourless crystals or liquid, soluble in most organic solvents, insoluble in water; used for organic synthesis and in perfumery.

acetoxyl group *(Chem.)*. The group $CH_3.CO.O$-.

acetylation *(Chem.)*. *Ethanoylation*. Reaction which has the effect of introducing an acetyl radical (CH_3CO) into an organic molecule.

acetylcelluloses *(Chem.)*. See **cellulose acetates**.

acetyl chloride *(Chem.)*. *Ethanoyl chloride*. CH_3COCl, mp −112°C, bp 51°C, rel.d. 1.105. Colourless liquid, of pungent odour, used for synthesis, in particular for introducing the acetyl group into other compounds.

acetyl choline *(Med.)*. One of the substances which transmits impulses from one nerve to another, particularly in the parasympathetic nervous system but also with an important rôle in the brain. A deficiency has been demonstrated in **Alzheimer's disease**.

acetylene *(Chem.)*. *Ethyne*. HC|CH, a colourless, poisonous gas, owing its disagreeable odour to impurities, soluble in ethanol, in acetone (25 times its volume at s.t.p.), and in water. Bp −84°C, rel.d. 0.91. Prepared by the action of water on calcium carbide and catalytically from naphtha. Used for welding, illuminating, acetic acid synthesis and for manufacturing derivatives particularly by Reppe synthesis.

acetyl group *(Chem.)*. *Ethanoyl group*. The radical of acetic acid, viz., CH_3CO—.

acetylide *(Chem.)*. *Ethynide*. Carbide formed by bubbling acetylene through a solution of a metallic salt, e.g. cuprous acetylide, Cu_2C_2. Violently explosive compounds.

acetylsalicylic acid *(Chem.)*. $C_6H_4(O.COCH_3)$ COOH, mp approximately 128°C. Used in medical and veterinary practice as an analgesic, antipyretic and antirheumatic. The acid or its salts are the active components of aspirin.

aceval *(Aero.)*. Abbrev. for *Air Combat EVALuation*.

ACF diagram *(Geol.)*. A triangular diagram used to represent the chemical composition of metamorphic rocks. The three corners of the diagram are Al_2O_3, CaO, and FeO + MgO.

a.c. generator *(Elec.Eng.)*. An electromagnetic generator for producing alternating e.m.f. and delivering a.c. to an outside circuit. See **alternator, induction generator**.

achalasia *(Med.)*. Failure to relax.

achalasia of the cardia *(Med.)*. Failure to relax on the part of the sphincter round the opening of the oesophagus into the stomach.

achene *(Bot.)*. A dry indehiscent, 1-seeded fruit, formed from a single carpel, and with the seed distinct from the fruit wall. Also called *achaenocarp, akene*.

Achilles tendon *(Zool.)*. In Mammals, the united tendon of the soleus and gastrocnemius muscles; the hamstring.

achlorhydria *(Med.)*. Absence of hydrochloric acid from gastric juice.

acholuric jaundice *(Med.)*. See **spherocytosis**.

achondrite *(Geol.)*. A type of stony meteorite which compares closely with some basic igneous rocks such as eucrite.

achondroplasia *(Med.)*. Dwarfism characterized by shortness of arms and legs, with a normal body and head. adj. *achondroplastic*.

achroglobin *(Zool.)*. A colourless respiratory pigment occurring in some Mollusca and some Urochorda.

achroite *(Min.)*. See **tourmaline**.

achromatic *(Build.)*. Schemes of decoration using white and natural greys i.e. free from colour.

achromatic lens *(Phys.)*. A lens designed to minimize chromatic aberration. The simplest form consists of two component lenses, one convergent, the other divergent, made of glasses having different dispersive powers, the ratio of their focal lengths being equal to the ratio of the dispersive powers. Also *antispectroscopic lens*.

achromatic prism *(Phys.)*. An optical prism with a minimum of dispersion but a maximum of deviation.

achromatic sensation *(Phys.)*. A visual perception of grey. Represented by the equal energy point on a chromaticity diagram.

achromatic stimulus *(Phys.)*. One which produces an **achromatic sensation**.

achylia gastrica *(Med.)*. Complete absence of pepsin and hydrochloric acid from the gastric secretion.

acicle *(Bot.)*. A stiff bristle, or slender prickle; sometimes with a glandular tip.

acicular *(Bot.,Min.)*. (1) Needle-shaped. (2) The needle-like habit of crystals.

aciculum *(Zool.)*. A stout internal chaeta in the parapodium of Polychaeta, acting as a muscle attachment.

acid *(Chem.)*. Normally, a substance which (a) dissolves in water with the formation of hydrogen ions, (b) dissolves metals with the liberation of hydrogen gas, or (c) reacts with a **base** to form a **salt**. More generally, a substance which tends to lose a proton (**Brönsted-Lowry theory**) or to accept an electron pair (**Lewis theory**).

acid amides *(Chem.)*. A group of compounds derived from an acid by the introduction of the amino group in place of the hydroxyl radical of the carboxyl group.

acid anhydride *(Chem.)*. Compound generating a hydroxylic acid on addition of (or derived from the acid by removal of) one or more molecules of water, e.g. sulphur (VI) oxide $SO_3 + H_2O = H_2SO_4$.

acid azides *(Chem.)*. The acyl derivatives of hydrazoic acid, obtainable from **acid hydrazides** by treatment with nitrous acid. They are very unstable.

acid brittleness *(Eng.)*. That developed in steel in pickling bath, through evolution of hydrogen.

acid chlorides *(Chem.)*. Compounds derived from acids by the replacement of the hydroxyl group by chlorine.

acid cure *(Min.Ext.)*. In extraction of uranium from its ores, lowering of gangue carbonates by puddling with sulphuric acid before leach treatment.

acid drift *(Min.Ext.)*. Tendency of ores, pulps, products, to become acidic through pick-up of atmospheric oxygen through standing.

acid dyes *(Image Tech.)*. Dyes which have their colour associated with the negative ion or radical.

acid egg *(Chem.Eng.)*. A pump for sulphuric acid, of simple and durable construction, with few moving parts. The acid is run into a pressure vessel, usually egg-shaped, from which it can be forcibly expelled by compressed air.

acid esters *(Chem.)*. Compounds derived from acids in which part of the replaceable hydrogen has been exchanged for an alkyl radical.

acid etching *(Build.)*. A specialised branch of signwriting in which glass is etched by hydrofluoric acid.

acid fixer *(Image Tech.)*. Fixing solution (hypo) with the addition of an acid (sodium bisulphite or potassium metabisulphite) to prevent staining.

acid growth hypothesis *(Bot.)*. The hypothesis that auxin-stimulated cell elongation results from increased proton extrusion (see **proton-translocating ATPase**) with an increased wall extensibility as a result of its lower pH.

acid hydrazides *(Chem.)*. Hydrazine derivatives into which an acyl group has been introduced.

acidimetry *(Chem.)*. The determination of acids by titration with a standard solution of alkali. See **titration, volumetric analysis**.

acidity *(Chem.)*. Loosely, the extent to which a solution is acid. See **pH-value**. More strictly, the concentration of any species titratable by strong base in a solution.

acidizing an oil well *(Min.Ext.)*. To improve the flow of oil from a limestone formation by pumping acid into it.

acid mine water *(Min.Ext.)*. Water containing sulphuric acid as a result of the breakdown of the sulphide minerals in rocks. Acid mine water causes corrosion of mining equipment, and may contaminate water supplies into which it drains.

acidosis *(Med.)*. A condition where the hydrogen ion concentration of blood and body tissues is increased, (normal range 36–43 nmol.l). *Respiratory acidosis* is caused by retention of carbon dioxide by the lungs. *Metabolic acidosis* by retention of non-volatile acids (renal failure, diabetic ketosis) or loss of base (severe diarrhoea).

acid process *(Eng.)*. A steel-making process, either Bessemer, open-hearth, or electric, in which the furnace is lined with a siliceous refractory, and for which pig iron low in phosphorus is required, as this element is not removed. See **basic process**.

acid process *(Paper)*. Any pulp digestion process utilizing an acid reagent e.g. a bisulphite liquor with some free sulphur dioxide.

acid radical *(Chem.)*. A molecule of an acid minus replaceable hydrogen.

acid rain *(Ecol.)*. Rain that is unnaturally acid (pH 3 to 5.5) as a result of pollution of the atmosphere with oxides of nitrogen and sulphur from the burning of coal and oil.

acid refractory *(Eng.)*. See **silica**.

acid resist foils *(Eng.)*. Blocking foils for use in etching metal. The foil is stamped on to paper and the excess foil blocked on to the metal rule or other object which is then exposed to an acidic etching fluid such as ferric chloride.

acid resisting paints *(Build.)*. A range of surface coating specially formulated to resist acid attack. For use in building and surfaces where acid spillage or exposure to an acid atmosphere is likely.

acid rock *(Geol.)*. An igneous rock with $<10\%$ quartz.

acid salts *(Chem.)*. Salts formed by replacement of part of the replaceable hydrogen of the acid.

acid slag *(Eng.)*. Furnace slag in which silica and alumina exceed lime and magnesia.

acid sludge *(Chem.Eng.)*. Mixture of 'green acid' and 'mahogany soap' salts; reaction product in sulphuric acid treatment of oil refinery heavy fractions, used in froth flotation process for recovering iron minerals.

acid soil complex *(Bot.)*. Combination of aluminium and/or manganese toxicity with calcium deficiency which affects a relatively calcicole (or noncalcifuge) plant growing on an acid soil; preventable horticulturally by liming. Cf. **lime-induced chlorosis**.

acid solution *(Chem.)*. An aqueous solution containing more hydroxonium ions than hydroxyl ions, i.e. has a **pH-value** <7; one which turns blue litmus red.

acid steel *(Eng.)*. Steel made by an acid process.

acid stop *(Image Tech.)*. Weak acid processing solution used immediately after the **developer** to halt its chemical activity and neutralize it before fixing. Also *stop bath*.

acid value *(Chem.)*. The measure of the free acid content of vegetable oils, resins etc., indicated by the no. of mg of potassium hydroxide (KOH) required to neutralize 1 g of the substance.

aciniform *(Zool.)*. Berry-shaped; e.g. in Spiders, the *aciniform* glands producing silk and leading to the median and posterior spinnerets.

actinostele *(Bot.)*. A protostele in which the xylem is star-shaped in cross section with phloem between the arms as in the roots of most seed plants and the stems of *Lycopodium*.

Ackermann steering *(Autos.)*. Arrangement whereby a line extended from the track-arms, when the wheels are set straight ahead, should meet on the chassis centre-line at 2/3 of the wheelbase from the front, allowing the inner stub-axle to move through a greater angle than the outer.

Acker process *(Chem.)*. Process for production of sodium hydroxide. Molten sodium chloride is electrolysed, using a molten lead cathode, and the resulting lead-sodium alloy is decomposed by water, yielding pure lead and pure sodium hydroxide. Obsolete.

acknowledgement signal *(Telecomm.)*. A signal transmitted along a circuit from B to A when triggered by a signal from A to B.

A-class insulation *(Elec.Eng.)*. A class of insulating material to which is assigned a temperature of 105°C. See **class-A, -B, -C**, etc., **insulating materials**.

ACM *(Comp.)*. Association for Computing Machinery. US professional association.

a.c. magnet *(Elec.Eng.)*. Electromagnet excited by a.c. having normally a laminated magnetic circuit. See **shaded pole**.

acme screw-thread *(Eng.)*. A thread having a profile angle of 29 degrees and a flat crest, used for lathe lead screw, etc., for easy engagement by a split nut.

acmite *(Min.)*. A variety of aegirine; also used for the $NaFe^{+3}Si_2O_6$ end member.

a.c. motor *(Elec.Eng.)*. An electric motor which operates from a single or polyphase a.c. supply. See **capacitor motor, induction motor, synchronous motor**.

acne *(Med.)*. Inflammation of a sebaceous gland. Pimples in adolescents are commonly due to infection with the acne bacillus.

acnode *(Maths.)*. See **double point**.

acoelomate, acoelomatous *(Zool.)*. Without a true coelom.

acoelomate triploblastica *(Zool.)*. Animals with three embryonic cell layers but no coelom. They consist of the Platyhelminthes, Nematoda, and some minor phyla, i.e. all the helminth phyla.

acoelous *(Zool.)*. Lacking a gut-cavity.

acontia *(Zool.)*. In Anthozoa, free threads, loaded with nematocysts, arising from the mesenterics or the mesenteric filaments, and capable of being discharged via the mouth or via special pores.

acotyledonous *(Bot.)*. The embryo of a vascular plant, having no cotyledons.

acoustextile *(Build.)*. A unit of material specially designed for increasing acoustic absorption of walls.

acoustic absorption *(Acous.)*. Transfer of energy into thermal energy when sound is incident at an interface.

acoustic absorption factor, coefficient *(Acous.)*. The measure of the ratio of the acoustic energy absorbed by a surface to that which is incident on the surface. For an open window this can be 1.00, for painted plaster 0.02. The value varies with the frequency of the incident sounds; e.g., for 2 cm glass fibre it is 0.04 at 125 Hz, 0.80 at 4000 Hz.

acoustical mass, inertia *(Acous.)*. Given by M, where ωM is that part of the acoustical reactance which corresponds to the inductance of an electrical reactance: ω is the pulsatance $= 2\pi \times$ frequency.

acoustical reproduction *(Acous.)*. See **electrical reproduction**.

acoustical stiffness *(Acous.)*. For an enclosure of volume V, given by $S = \rho c^2/V$, where c is velocity of propagation of sound and ρ is density. It is assumed that the dimensions of the enclosure are small compared with the sound wavelength and that the walls around the volume do not deflect.

acoustic amplifier *(Acous.)*. One amplifying mechanical vibrations.

acoustic branch *(Phys.)*. The lattice dynamics of a crystal containing n atoms per unit cell show that the dispersion curve (frequency ω against wavenumber q) has $3n$ branches of which 3 are acoustic branches. The branches are characterized by different patterns of movement of the atoms. For the acoustic branches, ω is proportional to q for small q. See **optic branch**.

acoustic centre *(Acous.)*. The effective 'source' point of the spherically divergent wave system observed at distant points in the radiation field of an acoustic transducer.

acoustic compliance *(Acous.)*. Reciprocal of acoustical stiffness.

acoustic construction *(Build.)*. Also called *discontinued construction*. Building construction which aims at the control of transmission of sound, or of mechanical vibration giving rise to sound, particularly unwanted noises. The parts of the structure are separated by air-spaces or acoustic absorbing material and can be decoupled by the interposing of springs.

acoustic coupler *(Comp.)*. A device which enables a digital signal to be transmitted over the telephone network using an ordinary telephone handset.

acoustic delay line *(Telecomm.)*. A device, magnetostrictive or piezoelectric, e.g. a quartz bar or plate of suitable geometry, which reflects an injected sound pulse many times within the body.

acoustic distortion *(Acous.)*. Distortion in sound-reproducing systems.

acoustic feedback *(Telecomm.)*. Instability or oscillation in a second reproduction system caused by the microphone or pick-up receiving vibrations from the loudspeaker.

acoustic filter *(Acous.)*. One which uses tubes and resonating boxes in shunt and series as reactance elements, providing frequency cutoffs in acoustic wave transmission, as in an electric wave filter.

acoustic grating *(Acous.)*. A diffraction grating for production of directive sound. Spacings are much larger than in optical gratings due to the longer wavelength of sound waves. Both transmission and reflection gratings are used.

acoustic impedance, reactance and resistance *(Acous.)*. Impedance is given by complex ratio of sound pressure on surface to sound flux through surface, and has imaginary (reactance) and real (resistance) components, respectively. Unit is the **acoustic ohm**.

acoustic interferometer *(Acous.)*. Instrument in which measurements are made by study of interference pattern set up by two sound or ultrasonic waves generated at the same source.

acoustic lens *(Acous.)*. A system of slats or disks to spread or converge sound waves.

acoustic models *(Acous.)*. Scale models of rooms (e.g. concert halls) or structures which are used to measure qualities important for architectural acoustics and noise control, e.g. sound distribution. The scale is typically between 1:10 and 1:20. In order to adjust the wavelength, the frequency has to be increased by a factor 10 to 20.

acoustic ohm *(Acous.)*. Unit of acoustic resistance, reactance, and impedance. 10^5 Pa s/m^3.

acousticolateral system *(Zool.)*. In Vertebrates, afferent nerve-fibres related to the neuromast organs and to the ear, receptors in aquatic forms of relatively slow vibrations.

acoustic perspective *(Acous.)*. The quality of depth and localization inherent in a pair of ears, which is destroyed in a single channel for sound reproduction. It is transferable with 2 microphones and 2 telephone ear-receivers with matched channels, and more adequately realized with 3 microphones and 3 radiating receivers with 3 matched channels.

acoustic plaster *(Build.)*. Rough or flocculent plaster which has good acoustic absorbing properties and which can be used for covering walls. Added to the mix is fine aluminium, which evolves gas on contact with water and so aerates the mass. These tiny holes lower the acoustic impedance and so reduce the reflection of incident sound waves.

acoustic pressure *(Acous.)*. See **sound pressure**.

acoustic radiator *(Acous.)*. Device to generate and radiate sound. The most common radiators are: (1) Vibrating elastic systems (membrane, string, vocal cord) which cause a fluctuating pressure in the surrounding medium. (2) Electrically driven membranes and plates (loud-speaker, sonar transducer). (3) Vortices in turbulent fluid flow.

acoustic ratio *(Acous.)*. The ratio between the directly radiated sound intensity from a source, at the ear of a listener (or a microphone), and the intensity of the reverberant sound in the enclosure. The ratio depends on the distance from the source, the polar distribution of the radiated sound power, and the period of reverberation of the enclosure.

acoustic reactance *(Acous.)*. See under **acoustic impedance**.

acoustic resonance *(Acous.)*. Enhancement of response to an acoustic pressure of a frequency equal or close to the **eigenfrequency** of the responding system. When a system is at resonance, the imaginary part of its impedance is zero. Prominent in Helmholtz resonators, organ and other pipes, and vibrating strings.

acoustics *(Genrl.)*. The science of mechanical waves including production and propagation properties.

acoustic saturation *(Acous.)*. The aural effectiveness of a source of sound amid other sounds; it is low for a violin, but high for a triangle. The relative saturation of instruments indicates the number required in an auditorium of given acoustic properties.

acoustic scattering *(Acous.)*. Irregular and multi-directional reflection and diffraction of sound waves produced by multiple reflecting surfaces the dimensions of which are small, cf. a **wavelength**; or by certain discontinuities in the medium through which the wave is propagated.

acoustic spectrometer *(Acous.)*. An instrument designed to analyse a complex sound signal into its wavelength components and measure their frequencies and relative intensities. See **real-time analyser**.

acoustic spectrum *(Phys.)*. Graph showing frequency distribution of sound energy emitted by source.

acoustic streaming *(Acous.)*. Generation of constant flows by a strong sound wave. Acoustic streaming is a non-linear effect. It is responsible for the motion of the light particles (Lycopodium spores) in a **Kundt's tube**. Also see **quartz wind**.

acoustic suspension *(Acous.)*. Sealed-cabinet system of loudspeakers in which the main restoring force of the diaphragm is provided by the acoustic stiffness of the enclosed air.

acoustic telescope *(Acous.)*. Array of microphones. The signals of the microphones are added with certain phase-delays so as to generate desired directivities. See also **directional microphone**.

acoustic tile *(Build.)*. One made of a soft, sound-absorbing substance.

acquired behaviour *(Behav.)*. Refers to a variety of processes which underlie attempts to adapt to environmental change; definitions are various and reflect the interest of different disciplines (e.g. psychology, physiology); most definitions assume long term changes in the central nervous system and exclude short term behaviour changes due to maturation, fatigue, sensory adaptation or habituation.

acquired character *(Zool.)*. A modification of an organ during the lifetime of an individual due to use or disuse, and not inherited.

acquired characters, inheritance of *(Biol.)*. See **Lamarckism**.

acquired immunity *(Immun.)*. Immunity resulting from exposure to foreign substances or microbes. Cf. **natural immunity**.

acquired immunodeficiency syndrome *(Immun.)*. *AIDS*. Severe immunodeficiency due to infection by a retrovirus (HIV I) which infects predominantly **helper T-lymphocytes** and leads to failure of **cell-mediated immunity**. This results in susceptibility to many bacterial, protozoal, fungal and viral infections to which normal persons are resistant or from which they recover completely. The virus can also infect some other cells such as macrophages and brain cells. It is transmitted by sexual intercourse or by inoculation of blood, e.g. via infected needles or by blood transfusion. High risk groups are homosexuals and drug addicts but the infection now appears widespread among heterosexuals in central Africa. At present there is no cure and little hope of recovery once the full syndrome has developed.

acquired variation *(Biol.)*. Any departure from normal structure or behaviour, in response to environmental conditions, which becomes evident as an individual develops.

ACR *(Aero.)*. Abbrev. for *Approach Control Radar*.

Acrasiomycetes *(Bot.)*. *Dictyosteliomycetes*. Cellular slime moulds. Class of slime moulds (Myxomycota) feeding phagotrophically as *myxamoebae* which aggregate to form a migratory **pseudoplasmodium** which eventually develops into a fruiting body liberating most of the cells as spores. E.g. *Dictyostelium*.

acre *(Surv.)*. A unit of area, equal to 10 sq. chains (1 chain = 66 ft) or 4840 sq.yd. = 0.4047 hectare. The following are now obsolete: Cheshire acre, 10 240 sq.yd; Cunningham acre, 6250 sq.yd; Irish acre, 7840 sq.yd; Scottish acre, 6150.4 sq.yd.

acridine *(Chem.)*. $C_{13}H_9N$, a basic constituent of the crude anthracene fraction of coal-tar. It crystallizes in colourless needles and has a very irritating action upon the epidermis. Chemically it may be considered an analogous compound to anthracene, in which one of the CH groups of the middle ring is replaced by N. Certain amino-acridines have valuable bactericidal powers. Used as an intermediate in the preparation of dyestuffs.

acriflavine *(Med.)*. A deep orange, crystalline substance possessing antiseptic (bacteriostatic and bactericidal) properties; used in wound dressings.

Acrilan *(Chem.)*. TN for a synthetic polyacrylonitrile fibre obtained by copolymerizing acrylonitrile (85%) with vinyl acetate (15%).

acritarch *(Geol.)*. A unicellular microfossil of unknown biological affinity, abundant in Precambrian and Palaeozoic strata.

acro- *(Genrl.)*. Prefix from Gk. *Akros*, topmost, farthest, terminal.

acrocarp *(Bot.)*. A moss in which the main axis is terminated by the development of reproductive organs. Any subsequent growth must be sympodial. Most are erect in habit. Cf. **pleurocarp**.

acrocentric *(Biol.)*. Having centromere at the end, applied to chromosomes; a rod-shaped chromosome.

acrocyanosis *(Med.)*. A vascular disorder (usually of young women) in which there is persistent blueness of the extremities.

acrodont *(Zool.)*. Said of teeth which are fixed by their bases to the summit of the ridge of the jaw.

acrolein *(Chem.)*. $CH_2{=}CH.CHO$, *acrylaldehyde, propenal*, a colourless liquid, bp 52.5°C, of pungent odour, obtained by dehydrating glycerine in the presence of a catalyst.

acromegaly *(Med.)*. A disease in which occurs enlargement of hands and feet, and thickening of nose, jaw, ears, and brows, due to over-production of growth hormone after the **epiphyses** of the long bones have fused. Usually due to an adenoma (tumour) of the pituitary gland.

acromion *(Zool.)*. In higher Vertebrates, a ventral process of the spine of the scapula. adj. *acromial*.

acron *(Zool.)*. In Insects, the embryonic, presegmental region of the head.

acroparaethesia, acroparesthesia *(Med.)*. Numbness and tingling of the fingers, tending to persist, in middle-aged women.

acropetal *(Bot.)*. Transport or differentiation towards the apex, away from the base.

acropodium *(Zool.)*. That part of the pentadactyl limb of land Vertebrates which comprises the digits and includes the phalanges.

acrosome *(Zool.)*. Structure forming the tip of a mature spermatozoon. adj. *acrosomal*.

acroterium *(Arch.)*. A base or mounting, on the apex and/or extremities of a pediment, for the support of an ornamental figure or statuary. pl. *acroteria*.

acrotrophic *(Zool.)*. In Insects, said of ovarioles in which nutritive cells occur at the apex.

acrylaldehyde *(Chem.)*. See **acrolein**.

acrylamide gel electrophoresis *(Biol.)*. A gel for the electrophoretic separation of proteins and RNA according to their molecular weight. The monomer can be cast in the form of sheets or cylinders by polymerization *in situ* to give a clear gel. See **sodium dodecyl sulphate**.

acrylic acid *(Chem.)*. *Prop-2-enoic acid*. $CH_2{=}CH.COOH$, mp 7°C, bp 141°C, of similar odour to acetic acid; a very reactive substance, the acid belongs to the series of alkene-monocarboxylic, or oleic, acids.

acrylic ester *(Chem.)*. An ester of acrylic acid or of a structural derivative of acrylic acid, e.g. methacrylic acid or its chemical derivatives.

acrylic fibres *(Textiles)*. Continuous filaments or, more usually, staple fibres made from linear polymers which are synthesized from several monomers containing at least 85% by weight of acrylonitrile.

acrylic resin paints *(Build.)*. A range of water thinned paints including emulsions, wood primers, undercoats and microporous coatings. These paints have the advantage of good adhesion, excellent durability, speed of drying and good colour retention.

acrylic resins *(Plastics)*. Resins formed by the polymerization of the monomeric derivatives, generally esters or amides, of acrylic acid or *a*-methylacrylic acid. They are transparent, water-white and thermoplastic; resistant to age, light, weak acids, alkalis, alcohols, alkanes and fatty oils; but attacked by oxidizing acids, aromatic hydrocarbons, chlorinated hydrocarbons, ketones and esters. They are mainly used for optical purposes, as lenses and instrument covers. Polymethyl methacrylate is widely known under the TNs *Perspex, Lucite* etc.

acrylonitrile *(Chem.)*. *Vinyl cyanide* (1-cyano-ethene), used as raw material for synthetic acrylic fibres, e.g. Acrilan, Orlon, Courtelle.

ACS *(Aero.)*. Abbrev. for (1) *Active Control System*. (2) *Attitude Control System*. (3) *Air Conditioning System*.

a.c. series motor *(Elec.Eng.)*. A series motor designed for operation from an a.c. supply; it is characterized by a laminated field structure and usually a compensating winding.

ACT *(Aero.)*. Abbrev. for *Active Control Technology*.

ACTH *(Med.)*. *Adrenocorticotrophic hormone, corticotrophin*. A protein hormone of the anterior pituitary gland controlling many secretory processes of the adrenal cortex. Used medically to stimulate cortisol production as an anti-inflammatory measure.

actin *(Biol.)*. A globular protein (G actin) which can polymerize into long fibres (F actin). Originally discovered in muscle in association with myosin it is now known to be widely distributed at sites of cellular movement.

actinal *(Zool.)*. In Echinodermata, see **ambulacral**: in Anthozoa, pertaining to the crown, including the mouth and tentacles; star-shaped.

actinic radiation *(Radiol.)*. Ultraviolet waves, which have enhanced biological effect by inducing chemical change; basis of the science of photochemistry.

actinic rays *(Image Tech.)*. Electromagnetic waves of wavelengths that can cause a latent image, potentially developable, in a photographic emulsion. They include an extension at each end of the visible spectrum and X-rays.

actinides *(Chem.)*. A group name for the series of elements of atomic numbers 89–104 (inclusive), which indicates that they have similar properties to actinium.

actinin *(Biol.)*. A protein frequently associated with actin, serving as a terminus for actin filaments. It was first found in the Z bands of striated muscle.

actinium *(Chem.)*. A radioactive element in the third group of the periodic system. Symbol Ac, at. no. 89, r.a.m. 227; half-life 21.7 years. Produced from natural radioactive decay of the ^{235}U isotope or by neutron bombardment of ^{226}Ra. Gives its name to the actinium $(4n+3)$ series of radioelements.

actino- *(Genrl.)*. Prefix from Gk. *aktis*, ray.

actinobacillosis *(Vet.)*. A chronic granulomatous disease of Cattle caused by the bacterium *Actinobacillus lignieresii* and characterized by infection of the tongue ('wooden tongue') and occasionally of the stomach, lungs, and lymph glands. In sheep the bacterium causes abscesses involving the head and neck or internal organs.

actinobiology *(Radiol.)*. The study of the effects of radiation upon living organisms.

actinodermatitis *(Med.)*. Inflammation of the skin arising from the action of radiation, usually applied to over-exposure to ultraviolet light (sunburn).

actinodromous *(Bot.)*. Leaf venation having 3 or more primary veins originating at the base of the lamina and running out towards the margin. (Formerly, *palmate* or *digitate venation*).

actinoid *(Zool.)*. Star-shaped.

actinolite *(Min.)*. A monoclinic calcium magnesium member of the amphibole group, green in colour and usually showing an elongated or needlelike habit; occurs in metamorphic and altered basic igneous rocks.

actinomorphic *(Bot.,Zool.)*. *Star-shaped*. Organisms which are radially symmetrical; divisible into two similar parts by any one of several longitudinal planes passing though the centre. Includes starfish, sea urchins and plants in which the stamens are helically arranged rather than whorled.

Actinomycetales *(Biol.)*. An order of Bacteria producing a fine mycelium and sometimes arthrospores and conidia. Some members are pathogenic in animals and plants. Some produce antibiotics, e.g. **streptomycin**.

actinomycosis *(Med.)*. A chronic granulomatous infection most commonly affecting the head and neck caused by infection by various species of filamentous bacteria of the genus *Actinomyces*. Actinomycosis also occurs in cattle and pigs but the infection is not transmitted to man.

actinomycosis *(Vet.)*. A chronic granulomatous disease of cattle and swine, caused by infection by the actinomycete *Actinomyces bovis*. In cattle the lower jaw is commonly affected ('lumpy jaw'); in swine the mammary gland is the common site of infecion.

Actinopterygii *(Zool.)*. Subclass of Gnathostomata in which the basal elements of the paired fins do not project outside the body wall, the fin webs being supported by rays alone. Ganoid scales are diagnostic of the group and the skeleton is fully ossified in most members. The most widespread modern group of fishes, including cod, herring, etc.

actinotherapy *(Med.)*. Treatment by means of ultraviolet, infrared and luminous radiations.

Actinozoa *(Zool.)*. See **Anthozoa**.

action *(Horol.)*. The functioning of the escapement of a watch, clock etc.

action *(Image Tech.)*. (1) 'Action!', the director's command to start the performance of a scene. (2) The performance itself to be recorded by the camera. (3) The film record of this performance as picture only, separate from the sound record.

action *(Phys.)*. Time integral of kinetic energy (E) of a conservative dynamic system undergoing a change, given by

$$2\int_{t_1}^{t_2} E\,dt.$$

action potential *(Med.)*. That produced in a nerve by a stimulus. It is a voltage pulse arising from sodium ions entering the axon and changing its potential from -70 mV to $+40$ mV. With a continuing stimulus the pulses are repeated at up to several hundred times a second, leading in motor nerves to continuous muscular response (tetanus).

action spectrum *(Bot.)*. The relationship, usually plotted as a graph, between the rate of a light-dependent physiological process (e.g. photosynthesis or photomorphogenesis) and the wavelength of light. Comparison with the spectral absorption of known pigments may suggest which is involved.

activated *(Immun.)*. Term applied to lymphocytes or macrophages which have undergone differentiation from a resting state, and have acquired new capacities such as the ability to secrete **lymphokines** or in the case of macrophages increased ability to kill and digest microbes.

activated carbon *(Chem.)*. Carbon obtained from vegetable matter by carbonization in the absence of air, preferably in a vacuum. Activated carbon has the property of adsorbing large quantities of gases. Important for gas masks, adsorption of solvent vapours, clarifying of liquids and in medicine.

activated cathode *(Electronics)*. Emitter in thermionic devices comprising a filament of basic tungsten metal, alloyed with thorium, which is brought to the surface by process of activation, such as heating without electric field.

activated charcoal *(Chem.)*. *Charcoal* treated with acid etc. to increase its adsorptive power.

activated complex *(Chem.)*. In the Eyring theory of **absolute reaction rates** activated molecules of the reactants collide to form a high energy complex which can lose energy either by decomposing to reform the reactants or by rearranging to form the products.

activated sintering *(Eng.)*. Sintering of a **compact** in the presence of a gaseous chemical which will react with part of the metal content, and thus ease and speed the migration necessary for joining the particles into a body.

Because the metal salt is unstable, purging of the heated reaction vessel results in reduction to metal. Halides, which also deoxidize the metal particle surfaces, are commonly used for this purpose.

activated sludge *(Build.)*. Sludge through which compressed air has been blown, or which has been aerated by mechanical agitation as part of the sewage treatment process.

activating agent *(Min.Ext.)*. See **activator**.

activation *(Build.)*. The process of sewage purification by mixing with air and activated sludge.

activation *(Chem.)*. (1) The heating process by which the capacity of carbon to adsorb vapours is increased. (2) An increase in the energy of an atom or molecule, rendering it more reactive.

activation *(Eng.)*. Altering the surface of a metal to a chemically active state. Cf. **passivation**.

activation *(Nuc.Eng.)*. Induction of radioactivity in otherwise nonradioactive atoms, e.g. in a cyclotron or reactor.

activation *(Zool.)*. A step in the fertilization process triggered by the incorporation of the spermatozoon into the egg cytoplasm, by which the secondary oocyte is stimulated to complete its division and becomes haploid.

activation cross-section *(Phys.)*. Effective cross-sectional area of a target nucleus undergoing bombardment by neutrons, etc. for radioactivation analysis. Measured in *barns*.

activation energy *(Chem.)*. The excess energy over that of the ground state which an atomic system must acquire to permit a particular process, such as emission or reaction, to occur.

activator *(Biol.)*. (1) Of an enzyme, a small molecule which binds to it and increases its activity. (2) Of DNA transcription, a protein which, by binding to a specific sequence, increases the production of a gene product. (3) Any agency bringing about *activation*.

activator *(Min.Ext.)*. Surface-active chemical used in flotation process to increase the attraction to a specific mineral in an aqueous pulp of collector ions from the ambient liquid and increase its aerophilic quality. Also *activating agent*.

activator *(Phys.)*. An impurity, or displaced atom, which augments luminescence in a material, i.e. a sensitizer such as copper in zinc sulphide.

active array *(Radar)*. Antenna array in which the individual elements are separately excited by integrated circuit or transistor amplifiers.

active centres *(Chem.)*. Centres of higher catalytic activity formed by peak or loosely bound atoms on the surface of an adsorbent.

active chromatin *(Biol.)*. See **transcriptionally active chromatin**.

active component *(Phys.)*. The accepted term for denoting the component of the vector representing an alternating quantity which is in phase with some reference vector; e.g. the active component of the current, commonly called the active current. See **active current**, **active voltage**, **active volt-amperes**.

active control *(Acous.)*. Modern technique of noise or vibration control employing one or more sources that generate signals with the aim of making the resulting total signal smaller. Used e.g. for the control of low-frequency air-borne noise and vibration of machinery. See **anti-sound**.

active control system *(Aero.)*. An advanced automatic flight control system designed to provide several special features, e.g. activation of flight control surfaces to minimize gust loads and bending stresses in the wing by detection and response to normal accelerations, provision of stability to a naturally unstable aircraft and implementation of pilot manoeuvre demands. All these characteristics improve aircraft behaviour and performance but the ACS demands extensive integration between aerodynamics, structure and electronic system design to achieve these advantages with reliability and safety.

active current *(Phys.)*. That component of a vector representing the a.c. in a circuit which is in phase with the voltage of the circuit. The product of this and the voltage gives power.

active device *(Electronics)*. A component capable of controlling voltages or currents, to produce gain or switching action in a circuit; valves, diodes and transistors, and integrated circuits are all classed as active devices or components.

active electrode *(Elec.Eng.)*. The electrode of an electrical precipitator which is kept at a high potential. Also *discharge electrode*.

active filter *(Electronics)*. One which combines amplification with conventional passive filter components (capacitance, inductance, resistance) to enhance fixed or tunable passband or rejection characteristics.

active homing *(Radar)*. A guidance system where the missile contains the transmitter for illuminating the target and the receiver for the reflected energy.

active hydrogen *(Chem.)*. Atomic form of hydrogen obtained when molecular hydrogen is dissociated by heating, or by electrical discharge at low pressure.

active lattice *(Nuc.Eng.)*. Regular pattern of arrangement of fissionable and nonfissionable materials in the core of a lattice reactor.

active lines *(Image Tech.)*. Those which are effective in establishing a picture.

active margin *(Geol.)*. A continental margin characterized by earthquakes, igneous activity and mountain building as a result of convergent- or transform-plate movements. See **passive margin**.

active mass *(Chem.)*. Molecular concentration generally expressed as moles/dm^3; in the case of gases, active masses are measured by partial pressures. See **activity (2)**.

active materials *(Phys.)*. (1) General term for essential materials required for the functioning of a device, e.g. iron or copper in a relay or machine, electrode materials in a primary or secondary cell, emitting surface material in a valve, or photocell, phosphorescent and fluorescent material forming a phosphor in a CRT, or that on the signal plate of a TV camera. (2) Term applied to all types of radioactive isotopes.

active power *(Elec.Eng.)*. (1) See **active volt-amperes**. (2) The time average over one cycle of the instantaneous input powers at the points of entry of a polyphase circuit.

active resistance *(Bot.)*. Host resistance to a pathogen built up as result of the previous presence of the pathogen or its metabolites.

active satellite *(Space)*. Satellite equipped for sending out probing signals and receiving returned information. A *passive satellite* only receives information on the state of the target.

active space *(Zool.)*. The area surrounding an animal within which it can communicate.

active transducer *(Telecomm.)*. Any transducer in which the applied power controls or modulates locally supplied power, which becomes the transmitted signal, as in a modulator, a radiotransmitter, or a carbon microphone.

active transport *(Biol.)*. The transport of a solute across a cell membrane against a concentration gradient (uncharged molecules) or electrochemical potential gradient (ions) and, therefore, requiring the input of energy. It is *primary* if the energy comes from a chemical reaction (see **proton-translocating ATPase**), *secondary* if the return across the membrane of previously transported ions is coupled to and drives the active transport of a second ion or molecule. See also **pump**. Cf. **facilitated diffusion**.

active voltage *(Elec.Eng.)*. That component of a vector representing the voltage which is in phase with the current in a circuit.

active volt-amperes *(Elec.Eng.)*. Product of the active voltage and the amperes in a circuit, or of the active current (amperes) and the voltage of the circuit; equal to the power in watts. Also termed *active power*.

activity *(Chem.)*. (1) See **optical activity**. (2) The ideal or thermodynamic concentration of a substance the substitution of which for the true concentration permits the application of the law of mass action.

activity *(Elec.Eng.)*. The magnitude of the oscillations of a piezoelectric crystal relative to the exciting voltage.

activity *(Radiol.)*. Attribute of an amount of radionuclide. Describes the rate at which transformations occur. The unit is a **becquerel** (*Bq*).

activity coefficient *(Chem.)*. The ratio of the **activity** to the true concentration of a substance.

activity constant *(Chem.)*. The **equilibrium constant** written in terms of activities instead of molar concentrations.

a.c. transformer *(Elec.Eng.)*. An electromagnetic device which alters the voltage and current of an a.c. supply in inverse ratio to one another. It has no moving parts and is very efficient.

Aculeata *(Zool.)*. Stinging hymenoptera, e.g. bees, ants and some wasps.

aculeate *(Bot.)*. Bearing prickles, or covered with needle-like outgrowths.

acuminate *(Bot.)*. Having a long point bounded by hollow curves; usually descriptive of a leaf-apex. dim. *acuminulate*.

acupuncture *(Med.)*. The practice of puncturing the skin with needles to produce analgesia, anaesthesia or for wider therapeutic purposes. The practice originated in China and the mechanism of action is not clear but it may stimulate the body to produce its own analgesic substances called **endorphins**.

acutance *(Image Tech.)*. Objective formulation of the sharpness of a photographic image, expressed as:

$$\bar{G}_x/(D_B - D_A)$$

where

$$\bar{G}_x^2 = \frac{\sum(\Delta D/\Delta x)^2}{N}$$

N = no. of increments between A and B, $D_B - D_A$ = average gradient of density curve, $\Delta D/\Delta x$ = maximum gradient curve.

acute *(Bot.)*. Bearing a sharp and rather abrupt point: said usually of leaf-tip.

acute *(Med.)*. Said of a disease which rapidly develops to a crisis. Cf. **chronic**.

acute angle *(Maths.)*. An angle of less than 90°. Cf. **obtuse angle**.

acute phase substances *(Immun.)*. Proteins which appear in the blood in increased amounts shortly after the onset of infections or tissue damage. They are made in the liver and include **C-Reactive protein**, fibrinogen, proteolytic enzyme inhibitors, transferrin. The stimulus is interleukin-1 (IL-1) released by macrophages. These proteins probably serve to counteract some of the effects of tissue damage.

ACV *(Aero.)*. Abbrev. for *Air Cushion Vehicle* (hovercraft).

acyclic compound *(Chem.)*. See **aliphatic compound**.

acylation *(Chem.)*. Introduction of an acyl group into a compound, by treatment with a carboxylic acid, its anhydride or its chloride.

acyl-CoA *(Biol.)*. Coenzyme A conjugated by a thioester bond to an acyl group, e.g. acetyl-CoA, succinyl-CoA. These compounds are intermediates in the transfer of the acyl groups, e.g. the formation of citric acid by the interaction of acetyl-AcA with oxaloacetic acid.

acyl group *(Chem.)*. Carboxylic radical RCO (R being aliphatic), e.g. CH_3CO.

acylic *(Bot.)*. Having the parts of the flower arranged in spirals, not in whorls.

A/D *(Comp.)*. See **analogue-to-digital converter**.

ad- *(Genrl.)*. Prefix signifying to, at.

A-D *(Image Tech.)*. Analogue-to-Digital, referring to the conversion of signals.

ADA *(Comp.)*. Programming language designed for complex **on-line**, **real time** monitoring (e.g. in military applications). Named in honour of Ada Lovelace.

adamantine *(Min.)*. See **lustre**.

adamantine compound *(Chem.)*. Compound with the same tetrahedral covalent crystal structure as the diamond, e.g. zinc sulphide (sphalerite).

adambulacral *(Zool.)*. In Echinodermata, adjacent to the ambulacral areas.

adamellite *(Geol.)*. A type of granite with approximately equal amounts of alkali-feldspar and plagioclase.

Adam's apple *(Zool.)*. In Primates, a ridge on the anterior or ventral surface of the neck, caused by the protuberance of the thyroid cartilage of the larynx.

Adams' catalyst *(Chem.)*. A hydrogenation catalyst based on platinum oxide.

Adams sewage lift *(Build.)*. An apparatus employed to force sewage from a low-level sewer into a nearby high-level sewer by using the sewage in the latter from a point that will give the air-pressure necessary to secure the lift of sewage.

adaptation *(Behav.)*. Various meanings: *evolutionary adaptation*, adjustment to environmental demands through the long term process of natural selection acting on the genotype; *sensory adaptation*, a short term change in the response of a sensory system as a consequence of repeated or protracted stimulation; *adaptation* (child psychology), a term used by Jean Piaget to describe the developmental process underlying the child's growing awareness and interactions with the physical and social world. The process of **assimilation**, **accommodation**, and **equilibration** are fundamental to this concept of psychological adaptation.

adaptation *(Bot.,Zool.)*. Any morphological, physiological or behavioural characteristic which fits an organism to the conditions under which it lives; the genetic or developmental processes by which such characteristics arise.

adaptation of the eye *(Biol.)*. The sensitivity adjustment effected after considerable exposure to light (*light adapted*), or darkness (*dark adapted*).

adapter *(Elec.Eng.)*. Accessory used in electrical installations for connecting a piece of apparatus fitted with one size or type of terminals to a supply point fitted with another size or type.

adapter *(Image Tech.)*. (1) An arrangement for using types of photographic material in a camera different from that for which it was designed; e.g. filmpack in a plate camera, or a smaller plate than normal. (2) A device for the interchange of lenses between different types of camera.

adaptive array *(Radar)*. A radar antenna (either a *phased array* or an *active array*) whose gain, directivity and side lobes can be adjusted automatically to optimise the radar's performance under specific operating conditions.

adaptive radiation *(Ecol.)*. Evolutionary diversification of species from a common ancestral stock, filling available ecological niches. Also *divergent adaptation*.

adaptor hypothesis *(Biol.)*. The prediction that some molecule would be needed to adapt the 4 base genetic code to the 20 amino-acid product. **tRNA** fulfills the prediction.

adaxial *(Bot.)*. That surface of a leaf, petal etc. that during early development faced towards the axis (and usually, therefore, the upper surface of an expanded leaf). Cf. **abaxial**.

Adcock antenna *(Telecomm.)*. Directional antenna consisting of pairs of vertical wires, spaced by one half wavelength or less, and fed in phase opposition; a figure-of-eight radiation pattern results, and arrays of Adcock antennae can be used for direction-finding.

ADD *(Aero.)*. Abbrev. for *Airstream Direction Detector* (for stall protection).

addend *(Maths.)*. See **addition**.

addendum *(Eng.)*. Radial distance between the major and pitch cylinders of an external thread; the radial distance between the minor and pitch cylinders of an internal thread; also the height from the pitch circle to the tip of the tooth on a gear wheel.

adder *(Comp.)*. Device which adds digital signals. It can also be applied to an amplifier in analogue computing. See **full-adder**, **half-**.

addict *(Med.)*. Someone physically dependent on a drug and who will experience withdrawal effects if the drug is discontinued.

Addison's disease *(Med.)*. A disease on which there is progressive destruction of the supra-renal cortex; characterized by extreme weakness, wasting, low blood-pressure, and pigmentation of the skin. Not to be confused with *Addison's anaemia*.

addition *(Maths.)*. The process of finding the sum of two quantities, which are called the *addend* and the *augend*. Denoted by the plus sign +. Numbers are added in accordance with the usual rules of arithmetic, but addition of other mathematical entities has to be defined specifically for the entities concerned, e.g. the parallelogram rule for the addition of two vectors.

addition agent *(Elec.Eng.)*. A substance added to the electrolyte in an electrodeposition process in order to improve the character of the deposit formed. The agent does not take part in the main electrochemical reaction.

additive constant *(Surv.)*. A term used in the computation of distance by tacheometric methods. It is that length (usually constant and small) which must be added to the product of staff intercept and multiplying constant to give the true distance of the object. See also **anallatic lens**.

additive function *(Maths.)*. A function $f(x)$ such that $f(x+y)=f(x)+f(y)$. If $f(x+y)<f(x)+f(y)$, the function is said to be *sub-additive*. If $f(x+y)>f(x)+f(y)$, the function is said to be *super-additive*.

additive genetic variance *(Biol.)*. That part of the genetic variance of a *quantitative character* that is transmitted and so causes resemblance between relatives.

additive printer *(Image Tech.)*. Photographic or motion picture printer or enlarger in which the intensity and colour of the exposing light is controlled by the separate variation of its red, green and blue components.

additive process *(Image Tech.)*. Colour reproduction system in which the picture is presented by the combination (addition) of red, green and blue light representing these three components in the original subject; effectively obsolete for general photography and cinematography but is the basis for colour TV display.

additive property *(Chem.)*. One whose value for a given molecule is equal to the sum of the values for the constituent atoms and linkages.

address *(Comp.)*. Code identifying a **store location**. See **absolute-, relative-, symbolic-**.

addressable cursor *(Comp.)*. One for which a program can specify the position usually by giving the coordinates.

address calculation *(Comp.)*. Process of determining an **absolute address** from the contents of the address field in a **machine code instruction**. See **indexed address, address modification**.

address field *(Comp.)*. Part of a **machine-code instruction** which contains addresses. Also *operand field* or *operand*. See also **address calculation**.

address modification *(Comp.)*. Process of changing the address in a machine instruction, so that each time the instruction is executed, it can refer to a different storage location. See **indexed address**.

address register *(Comp.)*. Register which holds the address part of the instruction being executed.

address space *(Comp.)*. The number of locations that can be addressed directly, as determined by the design of the **instruction set**.

adduct *(Chem.)*. The addition product of a reaction between molecules.

adductor *(Zool.)*. A muscle that draws a limb or part inwards, or towards another part; e.g. *adductor mandibulae* in *Amphibia* is a muscle which assists in closing the jaws.

adelphous *(Bot.)*. Said of an androecium in which the stamens are partly or wholly united by their filaments.

adendritic *(Zool.)*. Without dendrites.

adenine *(Chem.)*. *6-aminopurine*, one of the five bases in nucleic acids in which it pairs with thymine in DNA and uracil in RNA. See **genetic code**. ⇨

adenitis *(Med.)*. Inflammation of a gland.

adeno- *(Bot.)*. Prefix from Gk. *aden*, gland.

adenohypophysis *(Med.)*. The glandular lobe of the pituitary gland, derived from buccal ectoderm (Rathke's pouch).

adenoid *(Med.)*. Lymphoid tissue in nasopharynx of children which may become enlarged as a result of repeated upper respiratory tract infection.

adenoid *(Zool.)*. Glandlike.

adenoma *(Med.)*. A benign tumour with a glandlike structure or developed from glandular epithelium.

adenomyoma *(Med.)*. See **endometrioma**.

adenopathy *(Med.)*. Disease or disorder of glandular tissue. The term is usually used in reference to lymphatic gland enlargement.

adenosine triphosphate, ATP *(Biol.)*. The triphosphate of the nucleotide adenosine (adenine + ribose). As the predominant high-energy phosphate compound of all living organisms it has a pivotal role in cell energetics, mediating, by interconversion with the diphosphate (ADP), the energy transfer between *exergonic* and *endergonic* metabolic reactions.

adenyl cyclase *(Chem.)*. An enzyme which catalyses the formation of cyclic adenylic acid from ATP.

ADF *(Aero.)*. Abbrev. for *Automatic Direction Finding*.

adhesion *(Build.)*. (1) The securing of a bond between plaster and backing, by physical means as opposed to mechanical keys. (2) The inherent ability of a surface coating of paint to adhere to the underlying surface. Lack of adhesion is the cause of such defects as flaking, blistering and cissing. Good surface preparation and selection of suitable primers are important factors affecting adhesion.

adhesion *(Elec.Eng.)*. Mutual forces between two magnetic bodies linked by magnetic flux, or between two charged nonconducting bodies which keeps them in contact.

adhesion *(Eng.)*. The intimate sticking together of surfaces under compressive stresses by metallic bonds which form as a function of stress, time and temperature. The speed of formation is related to slip dislocations, and may occur instantaneously under high shear stresses. See **cold welding**.

adhesion *(Med.)*. Abnormal union of two parts which have been inflamed: a band of fibrous tissue which joins such parts.

adhesion *(Phys.)*. Intermolecular forces which hold matter together, particularly closely contiguous surfaces of neighbouring media, e.g., liquid in contact with a solid. US *bond strength*.

adhesion plaque *(Biol.)*. A specialized region of the plasma membrane where stress fibres terminate.

adhesive-bonded non-woven fabric *(Textiles)*. A fabric made from **batt** or **web** of fibres stuck together with an adhesive.

adhesive cells *(Zool.)*. Glandular cells producing a viscous adhesive secretion for attachment use, as on the pedal disk of *Hydra*, the tentacles of Ctenophora, and in the epidermis of Turbellaria.

adhesive wear *(Eng.)*. The welding together and subsequent shearing of small parts of two sliding surfaces.

adiabatic *(Phys.)*. Without loss or gain of heat.

adiabatic change *(Phys.)*. A change in the volume and pressure of the contents of an enclosure without exchange of heat between the enclosure and its surroundings.

adiabatic curve *(Phys.)*. The curve obtained by plotting P against V in the adiabatic equation.

adiabatic demagnetization *(Phys.)*. A method of obtaining very low temperatures. A paramagnetic salt is cooled to 1 K by liquid helium. The salt is magnetized under isothermal conditions and then magnetized under adiabatic conditions. As a result the temperature falls. Temperatures below 10^{-2}K can be obtained in this way.

adiabatic efficiency *(Eng.)*. (1) Of a steam-engine or turbine, the ratio of the work done per unit mass of steam to the available energy represented by adiabatic heat drop. (2) Of a compressor, the ratio of that work required to compress a gas adiabatically to the work actually done by the compressor piston or impeller.

adiabatic equation *(Phys.)*. PV_γ = constant, an equation expressing the law of variation of pressure (P) with the volume (V) of a gas during an adiabatic change, γ being the ratio of the specific heat of the gas at constant pressure to that at constant volume. The value of γ is approximately 1.4 for air at S.T.P.

adiabatic expansion *(Phys.)*. See **adiabatic change**.

adiabatic lapse rate *(Meteor.)*. The rate of decrease of temperature which occurs when a parcel of air rises adiabatically through the atmosphere.

adiabatic process *(Phys.)*. Process which occurs without interchange of heat with surroundings.

adiactinic *(Phys.)*. Said of a substance which does not transmit photochemically-active radiation, e.g., safelights for dark-room lamps.

A-digit hunter *(Telecomm.)*. In a director system, a hunting selector which, before dialling tone is returned to a calling subscriber, selects and connects a free A-digit selector.

A-digit selector *(Telecomm.)*. In a director system, a 2-motion selector which returns dialling tone to a calling subscriber, steps vertically to the level indicated by the first digit dialled, then rotates automatically into that level to select and connect the BC-digit selector of a free director.

ad infinitum *(Maths.)*. Latin phrase meaning continuing (in a similar manner) indefinitely.

adipamide *(Chem.)*. *1,4-Butanedicarboxamide*, NH_2-$CO(CH_2)_4CONH_2$. Used in synthetic fibre manufacture.

adipic acid *(Chem.)*. *Butanedicarboxylic acid*, HOOC. $(CH_2)_4$.COOH. Colourless needles; mp 149°C, bp 265°C; formed by the oxidation of *cyclo*-hexanone, or by the treatment of oleic acid with nitric acid. Used in the preparation of 6.6 **nylon**.

adipo- *(Genrl.)*. Prefix from L. *adeps*, fat.

adipocere, mortuary fat *(Med.)*. White or yellowish waxy substance formed by the post-mortem conversion of body fats to higher fatty acids.

adipose tissue *(Zool.)*. A form of connective tissue consisting of vesicular cells filled with fat and collected into lobules.

adiposis dolorosa *(Med.)*. A condition characterized by the development of painful masses of fat under the skin and by extreme weakness.

Adiprene *(Plastics)*. TN for a US polyurethane elastomer which combines high abrasion resistance with hardness, resilience and a good load-bearing capacity.

A-display *(Radar)*. Coordinate display on a CRT in which a level time base represents distance and vertical deflexions of beam indicate echoes.

adit *(Civ.Eng.)*. An access tunnel (usually nearly horizontal) leading to a main tunnel, and frequently used (in lieu of a shaft) in the excavation of the latter, or for exploration or drainage.

adjacent channel *(Telecomm.)*. One whose frequency is immediately above or below that of the required signal.

adjective dyes *(Chem.)*. Dyes which have no direct affinity for the particular textile fibre but can be affixed to it by a **mordant**.

adjoint *(Maths.)*. Of a square matrix or determinant: the **transpose of the matrix** or determinant obtained by replacing each element by its cofactor. Some writers use an untransposed adjoint. Also called *adjugate*.

adjugate *(Maths.)*. See **adjoint**.

adjustable-pitch propeller *(Aero.)*. See **propeller**.

adjustable-port proportioning valve *(Eng.)*. Air and fuel valves for oil or gas burners, motor operated in unison by automatic temperature-control equipment.

adjusting rod *(Horol.)*. An instrument for testing the pull of the mainspring. It is a rod having at one end an adjustable clamp for attaching to a fusee or barrel arbor, and provided with sliding weights for balancing the pull exerted by the mainspring. Its use is now confined to the adjustment of chronometers and English full-plate watches.

adjuvants *(Immun.,Med.)*. In general remedies which assist others, more particularly substances which increase the immunogenicity of antigens when administered with them. One action is to provide a depot from which the antigen is released slowly, and another is to activate macrophages in the neighbourhood so as to ensure more effective antigen presentation. Some absorb the antigen onto mineral particles, such as aluminium hydroxide, others use water-in-oil emulsions with or without macrophage stimulatory substances such as muramyl dipeptide.

adlacrimal *(Zool.)*. The lacrimal bone of Reptiles, so called to indicate that it is not homologous with the lacrimal bone of Mammals.

Admiralty brass *(Eng.)*. See **Tobin bronze**.

admission *(Eng.)*. The point in the working cycles of a steam or I.C. engine at which the inlet valve allows entry of the working fluid into the cylinder.

admittance *(Phys.)*. Property which permits the flow of current under the action of a potential difference. The reciprocal of impedance.

adnate *(Bot.)*. Joined to another organ of a differant kind, as when stamens are fused to the petals. Cf. **connate**.

adnexa *(Med.)*. Appendages; usually refers to ovaries and Fallopian tubes.

adobe (clay) *(Build.)*. A name for any kind of mud which when mixed with straw can be sun-dried into bricks.

adoral *(Zool.)*. Adjacent to the mouth.

ADP *(Comp.)*. Automatic data processing. See **data processing**.

Adrastea *(Astron.)*. A tiny natural satellite of **Jupiter**, discovered in 1979 by the Voyager 2 mission.

adrectal *(Zool.)*. Adjacent to the rectum.

adrenal *(Zool.)*. Adjacent to the kidney: pertaining to the adrenal gland.

adrenal cortex *(Zool.)*. Outer portion of the adrenal gland which produces glucocorticoids.

adrenal gland *(Zool.)*. See **suprarenal body**.

adrenaline *(Biol.)*. See **catecholamine**. syn. *epinephrine*.

adrenal medulla *(Zool.)*. The inner region of the adrenal gland which produces catecholamines.

adrenergic *(Med.)*. Pertaining to or causing stimulation of the sympathetic nervous system; applied to sympathetic nerves which act by releasing an adrenaline-like substance from their nerve endings.

adrenocorticotrophic hormone *(Biochem.)*. See **ACTH**.

adsorbate *(Chem.)*. A substance, usually in gaseous or liquid solution, which is to be removed by adsorption.

adsorbate *(Phys.)*. Substance adsorbed at a phase boundary.

adsorbent *(Chem.)*. The substance, either solid or liquid, on whose surface **adsorption** of another substance takes place.

adsorption *(Chem.)*. The taking up of one substance at the surface of another.

adsorption catalysis *(Chem.)*. The catalytic influence following adsorption of the reactants, exercised upon many reactions, often upon a specific adsorbent, attributed to the free residual valencies and hence higher reactivity of molecules at an interface (e.g. the ammonia oxidation process).

adsorption chromatography *(Chem.)*. See **chromatography**.

adsorption isotherm *(Chem.)*. The relation between the amount of a substance adsorbed and its pressure or concentration, at constant temperature.

adsorption potential *(Chem.)*. Change of potential in an ion in passing from a gas or solution phase on to the surface of an adsorbent.

adsorption surface area *(Powder Tech.)*. The surface area of a particle of powder calculated from data got by a stated adsorption method.

adularescence *(Min.)*. A milky or bluish sheen shown by moonstone.

adularia *(Min.)*. A transparent or milky-white variety of potassium feldspar, distinguished by its morphology.

advance *(Civ.Eng.)*. The length of railway track beyond a signal which is covered by that signal.

advanced gas-cooled reactor *(Nuc.Eng.)*. Gas-cooled (carbon dioxide), graphite moderated reactor using slightly enriched uranium oxide fuel clad in stainless steel, in use in the UK.

advance metal *(Eng.)*. Copper-base alloy with 45% nickel.

advance workings *(Min.Ext.)*. In flattish seams, mining in which the whole face is carried forward, no support pillars being left.

advancing colours *(Build.)*. Colours, generally in the red to yellow range which have the visual effect of making surfaces appear closer or more prominent.

advantage ratio *(Nuc.Eng.)*. Ratio between the relative radiation dosage received at any point in a nuclear reactor and that of a reference position for the same time of exposure.

advection *(Meteor.)*. The transference of any quantity by horizontal motion of the air.

advection fog *(Meteor.)*. Fog produced by the **advection** of warm moist air across cold ground.

adventitia *(Zool.)*. Accidental or inessential structures: the superficial layers of the wall of a blood-vessel. adj. *adventitious*.

adventitious *(Bot.)*. Applied to a plant-part developed out of the usual order or in an unusual position. An *adventitious bud* is any bud except an **axillary** bud; it gives rise to an *adventitious branch*. An *adventitious root* develops from some part of a plant other than a pre-existing root.

adventive *(Bot.)*. A plant not permanently established in a given habitat or area.

advertisement *(Behav.)*. A conspicuous display that can involve coloration, posture or sound, and that serves to convey some information about the sender, e.g. age, sex, status, motivation.

adze *(Build.)*. A cutting tool with an arched blade at right angles to the handle, used like a double handed axe for shaping timber. Mainly in boat construction.

adze-eye hammer *(Build)*. Type of claw hammer with pronounced curve on the claw.

Ae *(Eng.)*. The *transformation temperature* at equilibrium of the phase changes in iron and steel, subscripts indicating the designated change.

aecidiospore, aeciospore *(Bot.)*. The dikaryotic spore formed in an **aecidium**.

aecidium, aecium *(Bot.)*. A cup-shaped, spore-forming structure (sorus) characteristic of some rust fungi (Uredinales).

aedeagus *(Zool.)*. In male Insects the intromittent organ.

aedicule *(Arch.)*. A shrine, set into a wall and framed by two columns, **entablature** and **pediment**; the framing of a window or door in this manner.

aegirine *(Min.)*. Green sodium-iron member of the pyroxene group of minerals, essentially $NaFe^{+3}Si_2O_6$. Characteristic of the alkaline igneous rocks. *Acmite* is a brown variety and also used for the pure NaFe end-member.

aegirine-augite *(Min.)*. Minerals intermediate between **aegirine** and **augite**.

aegithognathous *(Zool.)*. Of Birds, having a type of palate in which the maxillopalatines do not meet the vomer or each other, and in which the vomer is broad and truncate anteriorly; the palatines and pterygoids articulate with the basisphenoid rostrum.

aegophony, egophony *(Med.)*. The bleating quality of voice heard through the stethoscope when fluid and air are present in the pleural cavity.

aenigmatite *(Min.)*. A complex silicate of sodium, iron, and titanium; occurs as reddish black triclinic crystals in alkaline igneous rocks. Also known as *cossyrite*.

aeolian deposits *(Geol.)*. Sediments deposited by wind and consisting of sand or dust (**loess**).

aeolian tone *(Acous.)*. A musical note set up by vortex action on a stretched string when it is placed in a stream of air. See **Strouhal number**.

aeolotropic *(Phys.)*. Having physical properties which vary according to the direction or position in which they are measured. Synonymous with *anisotropic* in this sense.

aerated concrete *(Build.)*. Concrete made by adding constituents to the mix which, by chemical reaction, liberate gases which are entrapped in the concrete and thereby reduce its density and increase its heat insulation value.

aerating root *(Bot.)*. See **pneumatophore**.

aerating tissue *(Bot.)*. See **aerenchyma**.

aeration test burner *(Eng.)*. For measuring the combustion characteristics of commercial gases. Abbrev. *ATB*.

aerenchyma *(Bot.)*. Tissue with particularly well-developed, air-filled, intercellular spaces. Characteristic of the cortex of roots and stems of hydrophytes, where it probably facilitates gas exchange between the roots and the leaves. Cf. **pneumatophore**.

aerial *(Telecomm.)*. Original UK term for *antenna* but most technical publications refer to antenna. Reference is still made to *aerial* in domestic use, e.g. *television aerial, car radio aerial*.

aerial fog *(Image Tech.)*. Fog caused by exposure of portions of the film to air in a processing machine.

aerial root *(Bot.)*. Adventitious roots rising above ground, especially the long roots hanging from some tropical epiphytes, and the short roots acting as attaching organs for many climbers (e.g. ivy) and epiphytes. See also **prop root**.

aerial ropeway *(Civ.Eng.)*. An apparatus for the overhead transport of materials in carriers running along an overhead cable or cables supported on towers.

aerial surveying *(Surv.)*. A process of surveying by photographs taken from the air, the photographs being of two types: (a) those giving a vertical or plan view, (b) those giving an oblique or bird's-eye view. See **vertical aerial photograph, oblique aerial photograph**.

AERO *(Aero.)*. Abbrev. for *Air Education and Recreation Organisation* (UK).

aeroacoustics *(Acous.)*. Branch of acoustics that treats sound generation and transmission by fluid flow.

aerobe *(Biol.)*. An organism which can live and grow only in the presence of free oxygen: an organism which uses aerobic respiration.

aerobic *(Bot.)*. Characterized by, or occurring in, the presence of free oxygen.

aerobic *(Med.)*. Requiring oxygen for respiration.

aerobic respiration *(Biol.)*. See **respiration**.

aerodynamic balance *(Aero.)*. (1) A balance, usually but not necessarily in a wind tunnel, designed for measuring aerodynamic forces or moments. (2) Means for balancing air loads on flying control surfaces, so that the pilot need not exert excessive force, particularly as speed increases. The principle is to use aerodynamic forces, either directly on a portion of the control surface ahead of the hinge line, or indirectly through a small auxiliary surface with a powerful moment arm, to counterbalance the main airloads. Examples of the first are **horn balance**, *inset hinge* and *Frise balances*, and of the second the **balance tab**.

aerodynamic braking *(Space)*. Use of a planet's atmosphere to reduce the speed of space vehicles.

aerodynamic centre *(Aero.)*. The point about which the pitching moment coefficient is constant for a range of aerofoil incidence.

aerodynamic co-efficient *(Aero.)*. A non-dimensional measure of aerodynamic force, pressure or moment that expresses the characteristics of a particular shape at a given incidence to the airflow. Typically lift co-efficient $C_L = \text{Lift}/\frac{1}{2}\rho V^2 S$, where ρ is air density, *V* is air speed and *S* is a typical area of the body (e.g. wing area). Similarly for drag co-efficient.

aerodynamic damping *(Aero.)*. The suppression of oscillations by the inherent stability of an aircraft or of its control surfaces.

aerodynamic heating *(Space)*. The heating of a vehicle passing through the atmosphere, caused by friction and compression of air (or other gas).

aerodynamics *(Aero.)*. That part of the mechanics of fluids that deals with the dynamics of gases. Particularly, the study of forces acting upon bodies in motion in air.

aerodynamic sound *(Acous.)*. See **flow noise**.

aerodyne *(Aero.)*. Any form of aircraft deriving lift in flight principally from aerodynamic forces. Commonly called *heavier-than-air aircraft*; e.g. **aircraft, glider, kite, helicopter**.

aeroelastic divergence *(Aero.)*. Aeroelastic instability which occurs when aerodynamic forces, or moments, increase more quickly than the elastic restoring forces or couples in the structure. Generally applied to wing weakness where the incidence at the tips increases under load, so tending to twist the wings off.

aeroelasticity *(Aero.)*. The interaction of aerodynamic forces and the elastic reactions of the structure of an aircraft. Phenomena are most prevalent when manoeuvring at very high speed.

aeroembolism *(Aero.)*. Release of nitrogen bubbles into the blood stream resulting from too rapid a reduction in ambient air pressure. Cf. **caisson disease**, the *bends*, encountered by undersea divers.

aero-engine *(Aero.)*. The power unit of an aircraft. Originally a lightweight reciprocating internal-combustion engine, usually Otto cycle, as a general rule either air-cooled radial, in-line, vee or liquid-cooled vee; gas turbines gradually superseded reciprocating engines from 1945 for large civil and military aircraft but reciprocating engines are still widely used in small aircraft. See **ramjet, gas turbine, turbofan, turbojet, turboprop, variable cycle engine, turboramjet, turborocket**.

Aerofall mill *(Min.Ext.)*. Dry grinding mill with large diameter and short cylindrical length in which ore is mainly or completely ground by large pieces of rock; mill is swept by air-currents which remove finished particles.

aerofoil *(Aero.)*. A body shaped so as to produce an aerodynamic reaction (lift) normal to its direction of motion, for a small resistance (drag) in that plane. A wing, plane, aileron, tail plane, rudder, elevator, etc.

aerofoil section *(Aero.)*. The cross-sectional shape or profile of an aerofoil.

aero-isoclinic wing *(Aero.)*. A sweptback wing which has its torsional and flexural stiffness so adjusted that the angle of attack remains constant as the wing bends under flight loads, instead of decreasing with deflexion towards the tip, which is the normal geometric effect.

aerolites *(Geol.)*. A general name for stony as distinct from iron meteorites.

aerological diagram *(Meteor.)*. A thermodynamic diagram used for plotting the results of upper-air soundings usually containing, as reference lines, isobars, isotherms, **dry adiabatics, saturated adiabatics** and lines of constant **saturation humidity mixing ratio**.

aerology *(Aero.,Meteor.)*. The study of the **free atmosphere**.

aeronautical engineering *(Aero.)*. That branch of engineering concerned with the design, production and maintenance of aircraft structures, systems, and power units.

aeronautical fixed services *(Aero.)*. A telecommunication service between fixed stations for the transmission of aeronautical information, particularly navigational safety, and flight planning messages. Abbrev. *AFS*.

aeronautics *(Aero.)*. All activities concerned with aerial locomotion.

aerophagy *(Med.)*. The swallowing of air, with consequent inflation of the stomach.

aerophare *(Telecomm.)*. See **radio beacon**.

aerophone *(Acous.)*. Group of musical instruments in which the air in a tube-shaped resonator is excited to vibrate.

aeroplane *(Aero.)*. See **aircraft**.

aerosol *(Chem.)*. (1) A colloidal system, such as a mist or a fog, in which the dispersion medium is a gas. (2) Pressurized container with built-in spray mechanism used for packaging insecticides, deodorants, paints etc.

aerospaceplane *(Aero.)*. Aircraft-like vehicle which can take-off from and land on runways, manoeuvre in the atmosphere, operate in space and re-enter the atmosphere.

aerostat *(Aero.)*. Any form of aircraft deriving support in the air principally from its buoyancy, e.g. a balloon or airship.

aerothermodynamics *(Space)*. Particular branch of thermodynamics relating to the heating effects associated with the dynamics of a gas; in particular the physical effects produced in the air flowing over a vehicle during launch and re-entry.

aestival *(Bot.,Zool.)*. Occurring in summer, or characteristic of summer.

aestivation *(Bot.)*. (1) Also *preloration*. Arrangement of unexpanded sepals and petals in the flower bud. (2) The spatial organization of unexpanded leaves, sepals or petals. Types include *imbricate*, *valvate*, *quincuncial*, *contorted*, *intricate*. Cf. **vernation**.

aestivation *(Zool.)*. Prolonged summer torpor, as in some Insects. Cf. **hibernation**.

aether *(Genrl.)*. See **ether**.

aetiology, etiology *(Med.)*. The medical study of the causation of disease.

AF *(Telecomm.)*. See **audio frequency**.

AFC *(Telecomm.)*. Abbrev. for *Automatic Frequency Control*.

AFCS *(Aero.)*. *Automatic Flight Control System*. A category of **automatic pilot** for the control of an aircraft while *en-route*. It can be monitored by speed and altitude data signals, signals from *ILS* and *VOR*, has automatic approach capability and is disengaged before landing. Cf. **autoflare, autoland** and **autothrottle**.

afebrile *(Med.)*. Without fever.

affective behaviour *(Behav.)*. Refers to a wide range of behaviour in which the emotional aspects of social interactions are salient and often fundamental, e.g. mother-infant interactions.

affective disorders *(Behav.)*. A group of disorders whose primary characteristic is a disturbance of mood; feelings of elation or sadness become intense and unrealistic.

afferent *(Zool.)*. Carrying towards, as blood vessels carrying nervous impulses to the central nervous system. Cf. **efferent**.

afferent arc *(Zool.)*. The sensory or receptive part of a reflex arc, including the adjustor neurone(s).

affine *(Phys.)*. Said of characteristic curves of apparatus when these curves differ only in the scales of one or both coordinates.

affine transformation *(Maths.)*. Geometrically, a transformation which preserves parallel lines. Algebraically, an invertible linear transformation. The two are equivalent.

affinity *(Chem.)*. The extent to which a compound or a *functional group* is reactive with a given reagent. See also **electron affinity**.

affinity *(Immun.)*. Measure of the strength of interaction or binding between antigen and antibody or between a receptor and its ligand.

affinity *(Textiles)*. The quantitative expression of *substantivity* which is the attraction between fibres and dye or other substance. It results in fabric attracting colour when immersed in a solution of dye.

aflagellar *(Zool.)*. Lacking flagella.

aflatoxins *(Bot.)*. Group of secondary metabolites produced by *Aspergillus flatus* and *A. parasiticus* which commonly grow on stored food especially peanuts, rice and cotton seed. Some are highly toxic to cattle and are suspected of causing liver cancer in Africa.

A-frame *(Eng.)*. See **sheers**.

African glanders *(Vet.)*. See **epizootic lymphangitis**.

African horse-sickness *(Vet.)*. A viral disease of Horses and other equidae, occurring in Africa and the Middle East, and transmitted by midges (*Culicoides*). Occurs in an acute, pulmonary form called locally *dunkop* meaning 'thin head') and due to pulmonary oedema, or in a subacute cardiac form called *dikkop* (meaning 'thick head') which is characterized by oedematous swelling of head, neck, and, sometimes, body.

African mahogany *(For.)*. West African hardwood from the genus *Khaya* which seasons without much difficulty. The wood is liable to infestation by powder-post beetles. It works easily and finishes to a good surface.

African swine fever *(Vet.)*. An acute, contagious, viral disease of Pigs in Africa; characterized by fever, diarrhoea, and multiple haemorrhages. Wart hogs are symptomless carriers of the virus.

African whitewood *(For.)*. See **obeche**.

afrormosia *(For.)*. A very durable wood from West Africa, used as a substitute for teak.
aft cg limit *(Aero.)*. See **cg limits**.
afterbirth *(Med.,Zool.)*. The placenta and membranes expelled from the uterus after delivery of the foetus. See **decidua**.
afterbody *(Aero.)*. Rear portion of a flying-boat hull, aft of the main step.
afterburner *(Aero.)*. See **reheat**.
afterburning *(Autos.)*. In an internal-combustion engine, persistence of the combustion process beyond the period proper to the working cycle, i.e. into the expansion period.
afterburst *(Min.Ext.)*. Delayed further collapse of underground workings after a rockburst.
aftercooler *(Min.Ext.)*. Chamber in which heat generated during compression of air is removed, allowing cool air to be piped underground.
afterdamp *(Min.Ext.)*. The non-inflammable heavy gas, carbon dioxide, left after an explosion in a coal mine. The chief gaseous product produced by the combustion of coal-gas. See **black damp, choke damp, fire damp, white damp**.
afterglow *(Electronics)*. See **persistence**.
afterheat *(Nuc.Eng.)*. That which comes from fission products in a reactor after it has been shut down.
after-image *(Phys.)*. Formation of image on retina of eye after removal of visual stimulus, in colour complementary to this stimulus. See **complementary after-image**.
after-pains *(Med.)*. Pains occurring after child-birth due to contraction of the uterus.
afterpeak *(Ships)*. Space abaft the aftermost bulkhead. Lower part frequently used as fresh-water tank, upper part may be used as storeroom.
after-ripening *(Bot.)*. The poorly-understood chemical and/or physical changes which must occur inside the dry seeds of some plants after shedding or harvesting, if germination is to take place after the seeds are moistened.
after tack *(Build.)*. A defect in a surface coating whereby the coating appears to dry normally up to a point but thereafter retains a slight degree of tackiness.
afwillite *(Min.)*. Hydrated calcium silicate occurring in natural rocks and set cements.
Ag *(Chem.)*. The symbol for silver (*argentum*).
agalactia *(Med.)*. Failure of the breast to secrete milk. Also *agalacia*.
agalmatolite *(Min.)*. See **pagodite**.
agamic *(Zool.)*. See **parthenogenetic**.
agamogenesis *(Bot.,Zool.)*. Asexual reproduction.
agamogony *(Zool.)*. See **schizogony**.
agamont *(Zool.)*. See **schizont**.
agamospermy *(Bot.)*. Reproduction by seed formed without sexual fusion. See **apomixis**.
agarics *(Bot.)*. The *Agaricales*, an order of the Hymenomycetes containing the mushrooms and toadstools; ca 3000 species. The spores are borne on the surface of gills or in the lining of pores.
agarose gel electrophoresis *(Biol.)*. Standard method for fractionating DNA fragments produced by restriction endonuclease digestion. Fragments migrate through the gel matrix under the influence of an electric field.
agate *(Min.)*. A cryptocrystalline variety of silica, characterized by parallel, and often curved, bands of colour.
AGC *(Telecomm.)*. Abbrev. for *Automatic Gain Control*.
age distribution *(Ecol.)*. The relative frequency, in an animal or plant population, of individuals of different ages. Generally expressed as a polygon or age pyramid, the number or percentage of individuals in successive age classes being shown by the relative width of horizontal bars.
age equation *(Nuc.Eng.)*. See **age theory**.
age hardening *(Eng.)*. The production of structural change spontaneously after some time; normally it is useful in improving mechanical properties in some respect, particularly hardness.
ageing *(Eng.)*. Gradual rise in strength due to physical change in metals and alloys, in which there is breakdown from supersaturated solid solution and lattice precipitation over a period of days at atmospheric temperature.
ageing *(Phys.)*. Change in the properties of a substance with time. A change in the magnetic properties of iron, e.g. increase of hysteresis loss of sheet-steel laminations; also the process whereby the subpermanent magnetism can be removed in the manufacture of permanent magnets.
ageing *(Textiles)*. The exposure of freshly printed fabrics to steam to produce fully developed colours.
agenesia, agenesis. *(Med.)*. Imperfect development (or failure to develop) of any part of the body.
ageotropic *(Bot.)*. Not reacting to gravity. See **tropism**.
age theory *(Nuc.Eng.)*. In nuclear reactor theory, the slowing-down of neutrons by elastic collisions. The *age equation* relates the spatial distribution of neutrons to their energy. The equation is given by

$$\nabla^2 q - \frac{\partial q}{\partial \tau} = 0$$

where q is the slowing-down density and τ is the **Fermi age**. It was first formulated by Fermi who assumed that the slowing-down process was continuous and so is least applicable to media containing light elements.
agglomerate *(Geol.)*. An indurated rock built of angular rock-fragments embedded in an ashy matrix, and resulting from explosive volcanic activity. Occurs typically in volcanic vents.
agglomerate *(Powder Tech.)*. Assemblage of particles rigidly joined together, as by partial fusion (sintering) or by growing together.
agglomerating value *(Min.Ext.)*. Index of the binding (sintering) qualities of coal which has been subjected to a prescribed heat treatment.
agglutination *(Chem.)*. The coalescing of small suspended particles to form larger masses which are usually precipitated.
agglutination *(Immun.)*. Clumping of particles such as bacteria or red cells when linked by antibodies binding antigenic determinants present on the particles.
agglutination *(Zool.)*. The formation of clumps by some *Protozoa* and spermatozoa.
agglutinin *(Med.)*. Specifically a constituent of the blood plasma of 1 individual which causes agglutination by reacting with a specific receptor in the red corpuscles in the blood of another individual.
aggregate *(Civ.Eng.)*. The sand and broken stone and brick or gravel which together form one of the constituents of concrete, the others being cement and water. See **coarse-aggregate, fine-aggregate**.
aggregate *(Geol.,Min.)*. A mass consisting of rock or mineral fragments.
aggregate *(Powder Tech.)*. Assemblage of particles which are loosely coherent.
aggregate fruit *(Bot.)*. (1) The fruit-like structure formed by a single flower with several free carpels. (2) Sometimes a *multiple fruit*.
aggregate ray *(Bot.)*. A group of small rays closely spaced so as to appear to be one large ray.
aggregate species *(Bot.)*. Abbrev. *agg*. A group of 2 or more closely similar species denoted, for convenience, by a single shared name, e.g. the blackberry, *Rubus fruticosus* agg.
aggregation *(Ecol.)*. A type of animal and plant dispersion in which individuals are closer to each other than they would be if they were randomly dispersed. See **contagious distribution**.
aggressive behaviour *(Behav.)*. Various and often non-overlapping meanings; most often refers to intentionally delivered noxious stimuli to a **conspecific**, but it can also refer to any behaviour that intimidates or damages the conspecific, e.g. scent marking. See also **agonistic behaviour**.
aggressive mimicry *(Zool.)*. Resemblance to a harmless species in order to facilitate attack.
agitation *(Image Tech.)*. Vigorous movement of film and solutions during processing to ensure that fresh chemicals are brought in contact with the emulsion.

agitator *(Min.Ext.)*. Tank, usually cylindrical, near bottom of which is a mixing device such as a propellor or airlift pump. Finely ground mineral slurries (the aqueous component perhaps being a leaching solution) are exposed to appropriate chemicals for purpose of extraction of gold, uranium, or other valuable constituents. Types are **Pachuca** or *Brown*, *Dorr*, and *Devereux*.

A-glass *(Glass.)*. A glass containing 10–15% of alkali (calculated as Na_2O) and used for the manufacture of glass fibre.

aglomerular *(Zool.)*. Devoid of glomeruli; as the kidney in certain fishes.

aglossate, aglossal *(Zool.)*. Lacking a tongue. n. *aglossia*, congenital absence of the tongue.

agmatite *(Geol.)*. A **migmatite** with a brecciated appearance.

Agnatha *(Zool.)*. A superclass of eel-shaped chordates without jaws or pelvic fins. Lampreys and Hagfishes.

agnathous, agnathostomatous *(Zool.)*. Having a mouth without jaws, as in the Lampreys.

Agnesi *(Maths.)*. See **witch of Agnesi**.

agnosia *(Med.)*. Loss of ability to recognise the nature of an object through the senses of the body.

agonic line *(Geol.)*. The line joining all places with no magnetic declination, i.e. those where true north and magnetic north coincide on a compass.

agonistic behaviour *(Behav.)*. A broad class of behaviour patterns, including all types of attack, threat, appeasement and flight, between members of the same species in response to a conflict between aggression and fear. Behaviour often alternates between attack and escape, e.g. across a territory boundary.

agoraphobia *(Behav.)*. A phobia characterised by a fear of open spaces.

AGR *(Nuc.Eng.)*. Abbrev. for *Advanced Gas-cooled Reactor*.

agranulocytosis *(Med.)*. A pathological state in which there is a marked decrease in the number of granulocytes in the blood.

agraphia *(Med.)*. Loss of power to express thought in writing, as a result of a lesion in the brain.

agrestal *(Bot.)*. Growing in cultivated ground, but not itself cultivated; e.g., a weed.

agroforestry *(Ecol.)*. Form of land use in which herbaceous crops and tree crops co-exist in an integrated scheme of farming. See **taungya**.

AGS *(Aero.)*. Abbrev. for *Aircraft General Standard*.

a.h.m. *(Elec.Eng.)*. See **ampere hour meter**.

AI *(Comp.)*. See **artificial intelligence**.

AIAA *(Aero.)*. Abbrev. for *American Institute of Aeronautics and Astronautics*.

AIDS *(Immun.,Med.)*. Abbrev. for *Acquired Immuno Deficiency Syndrome*.

aiguille *(Build)*. A stone-boring tool.

aileron droop *(Aero.)*. The rigging of ailerons so that under static conditions their trailing edges are below the wing trailing-edge line, pressure and suction causing them to rise in flight to the aerodynamically correct position.

ailerons *(Aero.)*. Surfaces at the trailing edge of the wing, controlled by the pilot, which move differentially to give a rolling motion to the aircraft about its longitudinal axis.

air absorption *(Acous.)*. Absorption of sound waves propagating in air. It is caused by molecular relaxation processes and viscosity.

air bells *(Image Tech.)*. Minute bubbles which have adhered to the emulsion during processing, leaving small circular spots where it has been protected from chemical action.

air bladder *(Zool.)*. In Fish, an air-containing sac developed as a diverticulum of the gut, with which it may retain connection by the pneumatic duct in later life; usually it has a hydrostatic function, but in some cases it may be respiratory or auditory, or assist in phonation.

air-blast circuit-breaker (switch) *(Elec.Eng.)*. A form of circuit-breaker (switch) in which an arc is deliberately drawn between two contacts. The arc is cooled by a blast of high pressure air which removes ions, thereby extinguishing the arc and breaking the circuit.

air brake *(Aero.)*. An extendable device, most commonly a hinged flap on wing or fuselage, controlled by the pilot, to increase the drag of an aircraft. Originally a means of slowing bombers to enable them to dive more steeply, it is an essential flight control on clean jet aircraft and sailplanes.

air brake *(Eng.)*. (1) A mechanical brake operated by air-pressure acting on a piston. (2) An absorption dynamometer in which the power is dissipated through the rotation of a fan or airscrew.

air break *(Elec.Eng.)*. Term applied to a switch or circuit-breaker which has the contacts in air.

air brick *(Build.)*. A perforated cast-iron, concrete, or earthenware brick built into a wall admitting air under the floors or into rooms.

air brush *(Build.)*. A small spray gun used in decorative artwork and signwork.

air cap *(Build.)*. The frontmost part of a spray gun which directs compressed air into the fluid stream to 'atomize' it and project it onto the surface in the form of a spray pattern.

air capacitor *(Elec.Eng.)*. One in which the dielectric is nearly all air, for tuning electrical circuits with minimum dielectric loss.

air cell *(Eng.)*. A small auxiliary combustion chamber used in certain types of compression-ignition engines, for promoting turbulence and improving combustion.

air chamber *(Bot.)*. An air-filled cavity e.g. towards the upper surface of the thallus of some liverworts, opening externally by a pore and containing photosynthetic cells, or in some hydrophytes.

air classifier *(Min.Ext.)*. Appliance in which vertical, horizontal, or cyclonic currents of air sort falling ground particles into equal-settling fractions or separate relatively coarse falling material from finer dust which is carried out. Also *air elutriator*.

air cleaner *(Autos.)*. A filter placed at the intake of an internal-combustion engine to remove dust from the air entering the cylinders.

air compressor *(Eng.)*. A machine which draws in air at atmospheric pressure, compresses it, and delivers it at higher pressure.

air conduction *(Acous.)*. The passing of noise energy along an air path, as contrasted with structure-borne conduction of vibrational energy.

air-cooled engine *(Autos.)*. An internal-combustion engine in which the cylinders, finned to increase surface area, are cooled by an airstream. See **cowling**.

air-cooled machine, transformer *(Elec.Eng.)*. A machine, transformer or other piece of apparatus, in which the heat occasioned by the losses is carried away solely by means of air. The flow of air over the heated surfaces may be due to natural convection or may be produced by a fan.

air cooler *(Phys.)*. The cold 'accumulator' used in the Linde process of air liquefaction for the preliminary cooling of the air.

air cooling *(Eng.)*. The cooling of hot bodies by a stream of cold air; distinct from water cooling.

aircraft *(Aero.)*. Any mechanically driven heavier-than-air flying machine with wings of fixed or variable sweep angle. Sub-divisions: landplane, seaplane (float seaplane and flying-boat), amphibian.

aircraft engine *(Aero.)*. See **aero-engine**.

aircraft flutter *(Telecomm.)*. Term used for the rapid fluctuations in VHF reception, affecting sound and vision; due to a secondary transmission path, or rapidly shifting phase, set up by reflection from an aircraft.

Aircraft General Standard *(Aero.)*. Term referring to small parts or items such as bolts, nuts, rivets, fork joints etc. which are common to all types of aircraft. Abbrev. *AGS*.

aircraft noise *(Aero.)*. Noise from propeller, engine, exhaust, and that generated aerodynamically over the surfaces; characterized by unstable low frequencies. See **jet noise**.

air data system *(Aero.)*. A centralized unit into which are fed the essential physical measurements for flight, e.g. air speed, **Mach number**, pitot and static pressure, barometric altitude, stagnation air temperature. From this central source data are transmitted to the cockpit dials, to flight and navigational instruments and to computers, etc. Abbrev. *ADS*.

air door *(Min.Ext.)*. In mine ventilating system, door which admits air or varies its direction.

air dose *(Radiol.)*. Radiation dose in röntgens delivered at a point in free air.

airdox *(Min.Ext.)*. American system for breaking coal in fiery mine by use of injected high-pressure air.

air drag *(Space)*. Resistance to the motion of a body passing through the Earth's atmosphere, most serious in the lower regions, producing changes in the geometry of the orbit, even causing the body to re-enter. More generally the term *atmospheric drag* is used in reference to other planets.

air dry *(Min.Ext.)*. Said of minerals in which moisture content is in equilibrium with that of atmosphere.

air dry *(Paper)*. Pulp or paper, the moisture content of which is in equilibrium with the surrounding atmosphere. The basis of sale for wood pulp, i.e. pulp with a conventionally accepted theoretical moisture content is usually 10% on total mass.

air drying *(Build.)*. The process by which a dry film, able to be handled, is formed from oil or paints under normal atmospheric conditions only. This may occur by solvent evaporation and/or by the oxidation and polymerization of the film constituents on exposure to the oxygen in the air. See also **stoving**.

air ducts *(Eng.)*. Pipes or channels through which air is distributed throughout buildings or machinery heating and ventilation.

aired up *(Min.Ext.)*. Said of an oil plunger pump which no longer sucks because gas or air has filled the suction chamber.

air ejector *(Eng.)*. A type of air pump used for maintaining a partial vacuum in a vessel through the agency of a high-velocity steam jet which entrains the air and exhausts it against atmospheric pressure.

air elutriator *(Min.Ext.)*. See **air classifier**.

air engine *(Eng.)*. (1) An engine in which air is used as the working substance. Rapid heating from an external source expands the air in the cylinder with consequent motion being imparted to a piston. After transfer to a compression cylinder, for rapid cooling, the air is returned to the working cylinder for the next cycle. Also *hot air engine*. See **Stirling engine**. (2) A small reciprocating engine driven by compressed air.

air-entraining agent *(Eng.)*. Resin added to either cement or concrete in order to trap small air bubbles.

air equivalent *(Phys.)*. The thickness of an air column at 15°C and 1 atmosphere pressure which has the same absorption of a beam of radiation as a given thickness of a particular substance.

air escape *(Build.)*. For releasing excess air from a water pipe. A valve is opened by a float when sufficient air has accumulated and closed in time to prevent loss of water.

air exhauster *(Eng.)*. (1) A suction fan. (2) A vacuum pump.

air filter *(Autos.)*. Attachment to the air intake of a carburettor for cleaning air drawn in to the engine.

air-float table *(Min.Ext.)*. Shaking table in which concentration of heavy fraction in sand-sized feed is promoted by air blown up through the deck, which is porous. Used in desert work.

airflow meter *(Aero.)*. An instrument, mainly experimental, for measuring the airflow in ducts.

air flue *(Build.)*. A flue which is built into a chimney-stack so as to withdraw vitiated air from a room.

airframe *(Aero.)*. The complete aircraft structure without power plant, systems, equipment, furnishings and other readily removable items.

air frost *(Meteor.)*. A screen temperature below 0°C. See **wind frost**.

air-fuel ratio *(Eng.)*. The proportion of air to fuel in the working charge of an internal-combustion engine, or in other combustible mixtures, expressed by weight for liquid fuels and by volume for gaseous fuels.

air-gap *(Elec.Eng.)*. Gap with points or knobs, adjusted to breakdown at a specified voltage and hence limit voltages to this value.

air-gap *(Phys.)*. Section of air, usually short, in a magnetic circuit, especially in a motor or generator, a relay, or a choke. The main flux passes through the gap, with leakage outside depending on dimensions and permeability.

air-gap torsion meter *(Eng.)*. A device for measuring the twist in a shaft by causing the relative rotation of two sections to alter the air-gap between a pair of electromagnets, the resulting change in the current flowing being indicated by an ammeter.

air, gas drilling *(Min.Ext.)*. The use of air instead of mud as the cooling and debris removal medium. Faster and easier than mud drilling, it cannot prevent water ingress and emergency mud equipment will then be necessary.

air gate *(Foundry)*. Passage from interior of a mould to allow the escape of air and other gases as the metal is poured in. See **riser**.

airglow *(Astron.)*. The faint permanent glow of the night sky, due to light-emission from atoms and molecules of sodium, oxygen and nitrogen, activated by sunlight during the day.

air-hardening steel *(Eng.)*. Steel with sufficient carbon and other alloying elements to allow sections over 5 cm (2 in) to harden fully when cooled in air or other gas from above its transformation temperature. Also *self-hardening steel*.

air heater *(Eng.)*. (1) Direct-fired heaters, in which the products of combustion are combined with the air. (2) Indirect-fired heaters, in which the combustion products are excluded from the air flow. Both can be operated in the recirculation system, by which a proportion of the heated air is returned to and passed through the heating chamber. See **air preheater**.

air hoist *(Min.Ext.)*. Air winch or other mechanical hoist actuated by compressed air.

airing *(Eng.)*. Removal of sulphur from molten copper in wirebar furnace, together with slag-forming impurities.

air insulation *(Elec.Eng.)*. Term applied when the insulation for part of an electrical circuit is provided by atmospheric air, e.g., a high voltage transmission line, which is suspended between transmission towers (pylons), is insulated for the section between the towers by atmospheric air.

air intake *(Aero.)*. Any opening introducing air into an aircraft, but that for the main engine air is usually implied if unqualified.

air intake *(Eng.)*. Vent in a carburettor through which air is sucked to mix with the petrol vapour from the jet.

air-intake guide vanes *(Aero.)*. Radial, toroidal or volute vanes which guide the air into the compressor of a gas turbine, or the super-charger of a reciprocating engine.

air jet spinning *(Textiles)*. A specialized system in which staple fibres are converted into yarn by being spun together by the action of air jets which strike the fibres tangentially and cause them to rotate.

air jet texturing *(Textiles)*. See **textured yarn**.

air jig *(Min.Ext.)*. In waterless countries, use of pulses of air to stratify crushed ore into heavy and light layers.

airlance *(Min.Ext.)*. Length of piping used to work compressed air into settled sand or to free choked sections of process plant, restoring aqueous flow.

air layering *(Bot.)*. Horticultural method of vegetative propagation of especially shrubby house plants, in which an aerial shoot is induced to form roots, by wounding and packing the wound with e.g. sphagnum moss, while still attached to the plant. It is severed and potted up after rooting.

air laying *(Textiles)*. Method for forming a **batt** or **web** by collecting fibres from an air stream on a lattice ready for manufacturing a **non-woven fabric**.

air leg *(Min.Ext.)*. Telescopic cylindrical prop expanded by compressed air, used to support rock drill.

airless injection *(Eng.)*. Injection of liquid fuel into the cylinder of an oil engine by a high-pressure fuel pump, so dispensing with the compressed air necessary in the early diesel engines. Also called *solid injection*.

airless spraying *(Build.)*. A system of spray application in which paint at extremely high fluid pressure is forced though a precision orifice in the spray gun when it 'atomizes' in a cloud of fine particles.

airlift *(Nuc.Eng.)*. General use of air or neutral gas to blow material (particularly radioactive) as solid or liquid in processing, to avoid pumps.

airlift pump *(Min.Ext.)*. An air-operated displacement pump for elevating or circulating pulp in cyanide plants.

airline *(Phys.)*. Straight line drawn on the magnetization curve of a motor, or other electrical apparatus, expressing the magnetizing force necessary to maintain the magnetic flux across an air gap in the magnetic circuit.

air liquefier *(Phys.)*. A type of gas refrigerating machine based on the 'Stirling' or hot-air engine cycle.

air lock *(Civ.Eng.)*. Device by which access is obtained to the working chamber (filled with compressed air to prevent entry of water) at base of a hollow caisson. The workman at surface enters and is shut in an air-tight chamber filled with air at atmospheric pressure. Pressure within this air-lock is gradually raised to that used in the working chamber, so that the workman can pass out through another door and communicate with the working chamber. See **caisson**.

air lock *(Eng.)*. An air pocket or bubble in a pipeline which obstructs the flow of liquid. See **vapour lock**.

air log *(Aero.)*. An instrument for registering the distance travelled by an aircraft relative to the air, not to the ground.

air manometer *(Phys.)*. A pressure gauge in which the changes in volume of a small quantity of air enclosed by mercury in a glass tube indicate changes in the pressure to which it is subjected.

airmanship *(Aero.)*. Skill in piloting an aircraft.

air mass *(Meteor.)*. A part of the atmosphere where the horizontal temperature gradient at all levels within it is very small, perhaps of the order of 1°C/100 km. See **frontal zone**.

air mass flow *(Aero.)*. In a gas turbine power plant, the quantity of air which is ingested by the compressor, normally expressed in pounds or kilograms per second.

air meter *(Eng.)*. An apparatus used to measure the rate of flow of air or gas.

air-mileage unit *(Aero.)*. An automatic instrument which derives the air distance flown and feeds it into other automatic navigational instruments.

air miles per gallon *(Aero.)*. The number of miles flown through the air for each gallon of fuel burnt by the propulsion unit(s).

air monitor *(Radiol.)*. Radiation (e.g. γ-ray) measuring instrument used for monitoring contamination or dose rate in air.

air plant *(Bot.)*. See **epiphyte**.

air pocket *(Aero.)*. A colloquialism for a localized region of rising or descending air current. Causes an abrupt vertical acceleration as an aircraft passes through it, severity increasing with speed and also with low wing loading. Also *bump*. See **vertical gust**.

airpore *(Bot.)*. See **stoma, airchamber, lenticel**.

airport markers *(Aero.)*. Particoloured boards defining areas on an airfield, e.g. *boundary markers* which indicate the limits of the landing area, *taxi-channel markers* for taxi tracks, *obstruction markers* for ground hazards, and *runway visual markers*, situated at equal distances, by which visibility is gauged in bad weather.

airport meteorological minima *(Aero.)*. The minimum cloud base (vertical) and visibility (horizontal) in which landing or takeoff is permitted at a particular aerodrome. *ICAO* standards: Cat. 1, 200 ft (60 m) height, 2600 ft (800 m) RVR; Cat. 2, 100 ft (30 m) height, 1300 ft (400 m) RVR; Cat. 3, zero height, *(a)* 700 ft (210 m), *(b)* 150 ft (45 m), *(c)* zero RVR. See **runway visual range** (RVR).

air position *(Aero.)*. The geographical position which an aircraft would reach in a given time if flying in still air.

air-position indicator *(Aero.)*. An automatic instrument which continually indicates air position, incorporating alterations of course and speed.

air preheater *(Eng.)*. System of tubes or passages, heated by flue gas, through which combustion air is passed for preheating before admission to the combustion chamber, thus appreciably raising flame temperatures and returning to the combustion chamber some utilizable heat that would otherwise be lost. See also **recuperative air-heater, regenerative air-heater**.

air pressure *(Build.)*. An important feature in spray painting being one of the factors affecting the paint application.

air pump *(Eng.)*. A reciprocating or centrifugal pump used to remove air, and sometimes the condensate, from the condenser of a steam plant. See **air ejector**. Any device used for transferring air from one place to another. A **compressor** is a pump used for increasing the pressure on the high-pressure side; a **vacuum pump** is one in which the object of pumping is to reduce the pressure in the low-pressure side. A **blower** is a pump used for obtaining a rapidly moving air blast.

air receiver *(Build.,Eng.)*. A pressure vessel in which compressed air is stored facilitating moisture removal and the equalizing of pressure fluctuations before air is conveyed to the spray gun or other equipment, e.g. drills, hammers etc.

Air Registration Board *(Aero.)*. The airworthiness authority of the U.K. until its functions were taken over in 1972 by the **Civil Aviation Authority**. Abbrev. *ARB*.

air route *(Aero.)*. In organized flying, a defined route between two aerodromes; usually provided with direction-finding facilities, lighting, emergency-landing grounds, etc. See also **airway**

air-sacs *(Zool.)*. In Insects, thin-walled distensible dilations of the tracheae, occurring especially in rapid fliers, which increase the oxygen capacity of the respiratory system and otherwise assist the act of flight: in Birds, expansions of the blind ends of certain bronchial tubes, which project into the general body cavity and assist in respiration, as well as lightening the body. Also *aerostats*.

airscrew *(Aero.)*. Defined in 1951 to be 'any type of screw designed to rotate in air'. Term now obsolete and replaced by **propeller**, a device for propelling aircraft and **fan**, a rotating bladed device for moving air in ducts or e.g. wind tunnels. See **rotor**.

air seal *(Eng.)*. Curtain of air maintained in front of kiln or furnace door to aid retention of heat.

air shaft *(Civ.Eng.)*. An air passage, usually vertical or nearly vertical, which provides for the ventilation of a tunnel or mine.

airship *(Aero.)*. Any power-driven **aerostat**. Types: *non-rigid airship*, one with the envelope so designed that the internal pressure maintains its correct form without the aid of a built-in structure; small, and used for naval patrol work; *rigid airship*, one having a rigid structure to maintain the designed shape of the hull, and to carry the loads; usually a number of ballonets or gas bags inside the frame; large, used for military purposes in World War I, and having limited commercial use until 1938; *semi-rigid airship*, one having a partial structure, usually a keel only, to distribute the load to, and maintain the designed shape of, the envelope or ballonets; intermediate size.

air shooting *(Min.Ext.)*. (1) Charging of shot-hole so as to leave pockets of air, thus reducing shatter-effect of blast. (2) In seismic prospecting, explosion in air, above rock formation under examination, as method of propagating seismic wave.

air sinuses *(Zool.)*. In Mammals, cavities connected with the nasal chambers and extending into the bones of the skull, especially the maxillae and frontals.

air space *(Aero.)*. That part of the atmosphere which lies above a nation and which is therefore under the jurisdiction of that nation.

air space *(Bot.)*. Air-filled *intercellular spaces*, esp. large ones.

air-spaced cable *(Telecomm.)*. See **dry-core cable**.

air-spaced coil *(Elec.Eng.)*. Inductance coil in which the adjacent turns are spaced (instead of being wound close together) to reduce self-capacitance and dielectric loss.

air speed *(Aero.)*. Speed measured relative to the air in which the aircraft or missile is moving, as distinct from ground speed. See **equivalent air speed, indicated air speed, true air speed.**

air standard cycle *(Autos.)*. A standard cycle of reference by which the performance of different internal-combustion engines may be compared, and their relative efficiencies calculated.

air standard efficiency *(Autos.)*. The thermal efficiency of an internal-combustion engine working on the appropriate air standard cycle.

airstrip *(Aero.)*. A unidirectional landing area, usually of grass or of a makeshift nature.

air superiority fighter *(Aero.)*. Combat aircraft intended to remove hostile aircraft from a volume of airspace and so establish control of the air.

air surveying *(Surv.)*. See **aerial surveying.**

air-swept mill *(Min.Ext.)*. In dry grinding of rock in ball mill, use of modulated current of air to remove sufficiently pulverized material from the charge in the mill.

air table *(Min.Ext.)*. See **air-float table.**

air-traffic control *(Aero.)*. The organised control, by visual and radio means, of the traffic on air routes and into and out of aerodromes. ATC is divided into general *area control*, including defined *airways*; *control zones*, of specified area and altitude, round busy aerodromes; *approach control* for regulating aircraft landing and departing; and *aerodrome control* for directing aircraft movements on the ground and giving permission for take-off. Air-traffic control operates under two systems: *visual flight rules* and, more severely, *instrument flight rules*. Since World War II great advances in radar technology have enabled *air traffic controllers* to be given very complete 'pictures' of the position of aircraft, not only in flight, but also when manoeuvring on the ground. Abbrev. *ATC*.

air-traffic control centre *(Aero.)*. An organization providing (1) air-traffic control in a control area and (2) **flight information** in a region.

air-traffic controller *(Aero.)*. One who is licensed to give instructions to aircraft in a control zone.

air transformer *(Build.)*. A device situated between the air receiver and spray guns incorporating an air pressure regulator, a filter system and attachment for air hose.

Air Transport Association *(Aero.)*. A US organization noted particularly for its specification which sets a standard to which manufacturers of aircraft and associated equipment are required to produce technical manuals for the aircraft operator's use. The specification is accepted by *IATA* as the basis for international standardization. Abbrev. *ATA*.

air trap *(Build.)*. A trap which, by a water-seal, prevents foul air from rising from sinks, wash basins, drains, sewers etc. Also *drain trap*, *stench trap*, *U-bend*.

air valve *(Build.)*. A valve in the spray gun which controls the flow of air by the operation of the trigger.

air volume spraying *(Build.)*. A method of spray application which involves higher volume and lower pressure of air than high pressure air spraying.

air wall *(Nuc.Eng.)*. Wall of ionization chamber designed to give same ionization intensity inside chamber as in open space. This means the wall is made of elements with atomic numbers similar to those for air constituents.

airway *(Aero.)*. A specified 3-dimensional corridor (the lower as well as the upper boundary being defined) between **control zones** which may only be entered by aircraft in radio contact with **air-traffic control.**

airway *(Min.Ext.)*. Underground passage used mainly for ventilation.

airworthy *(Aero.)*. (1) Fit for flight aircraft, aero-engine, instrument or equipment. (2) Complying with the regulations laid down for ensuring the fitness of an aircraft for flight. (3) Possessing a **certificate of airworthiness.**

Airy disk *(Image Tech.)*. Circular image of a point source of light formed by a lens. After Sir George Airy 1801-92.

Airy points *(Phys.)*. The best points for supporting a bar horizontally so that bending shall be a minimum. The distance apart of the points is equal to $\frac{l}{\sqrt{n^2-l}}$, where l is the length of the bar and n the number of supports.

Airy's differential equation *(Maths.)*. One of the form

$$\frac{d^2y}{dx^2}=xy.$$

Airy's integral *(Phys.)*. The factor 1.22, by which the dimensions of the diffraction pattern produced by a slit must be multiplied to obtain the dimensions of the pattern due to a circular aperture.

Airy spirals *(Phys.)*. The spiral interference patterns produced when quartz, cut perpendicularly to the axis, is examined in convergent light circularly polarized.

aisle *(Arch.)*. A side division of the *nave* or other part of a church or similar building, generally separated off by pillars; (loosely) any division of a church, or a small building attached: (loosely) a passage between rows of seats.

akaryote *(Zool.)*. A cell lacking a nucleus, or one in which the nucleoplasm is not aggregated to form a nucleus.

akinesia *(Med.)*. Absence of, or diminished spontaneous, movement characteristic of diseases such as **Parkinsonism.**

akinete *(Bot.)*. A non-motile thick-walled resting spore containing food reserves, formed without division by the direct modification of a vegetative cell in some Cyanophyceae.

Akulon *(Plastics)*. TN for Dutch nylon-6 polymer used for mouldings and fibres.

Al *(Chem.)*. The symbol for aluminium.

Ala *(Chem.)*. Symbol for **alanine.**

ala *(Zool.)*. Any flat, wing-like process or projection, especially of bone. adj. *alar, alary*.

alabaster *(Min.)*. A massive form of gypsum, often pleasingly blotched and stained. Because of its softness it is easily carved and polished, and is widely used for ornamental purposes. Chemically it is a $CaSO_4.2H_2O$. *Oriental alabaster*, onyx marble is a beautifully banded form of stalagmitic *calcite*.

alalia *(Med.)*. See **aphonia.**

alanine *(Chem.)*. *2-aminopropanoic acid*, $CH_3C(NH_2)$-COOH. The L- or S-isomer is a common constituent of proteins. Symbol Ala, short form A. ⇨

ALARA *(Nuc.Eng.)*. *As low as reasonably achievable* said, e.g. of radiation levels, decontamination, etc.

alarm flag *(Elec.Eng.)*. See **flag indicator.**

alary muscles *(Zool.)*. In Insects, pairs of striated muscles arising from the terga and spread out fanwise over the surface of the dorsal diaphragm. Also *aliform muscles*.

ala spuria *(Zool.)*. See **bastard wing.**

alastrim *(Med.)*. *Variola minor*; a mild form of smallpox differing from it in certain features, mainly nonfatal.

alate *(Bot.)*. Winged; applied to stems when decurrent leaves are present.

alate *(Zool.)*. Having a broad lip (especially of shells): in Porifera, a type of triradiate spicule with unequal angles.

Albada viewfinder *(Image Tech.)*. One with a lightly silvered plano-concave objective which reflects frame marks placed on the eyepiece and at the focus of the mirror.

albedo *(Astron.)*. A measure of the reflecting power of a non-luminous surface, expressed as the ratio of energy reflected in all directions to total incident energy.

albedo *(Phys.)*. (1) Photometeric term for the fraction of incident light *diffusely* reflected from a surface. (2) Ratio of the neutron flow density out of a medium free from sources, to the neutron flow density into it, i.e. reflection factor of a surface for neutrons.

Albers-Schönberg disease *(Med.)*. See **osteo-petrosis.**

albert *(Paper)*. A former standard size of note-paper, 192 × 102 mm (6 × 4 in).

albertite *(Min.)*. A pitch-black solid bitumen of the asphalite group.

Albian *(Geol.)*. A stage of the Cretaceous System,

comprising the rocks between the Aptian stage below and the Cenomanian stage above.

albinism *(Zool.)*. The state of being an **albino**.

albino *(Bot.,Zool.)*. An abnormal plant lacking chlorophyll or normal plant pigments. An animal deficient in pigment in hair, skin, eyes etc. adj. *albinotic*.

albite *(Min.)*. The end-member of the plagioclase group of minerals. Ideally a silicate of sodium and aluminium; but commonly contains small quantities of potassium and calcium in addition and crystallizes in the triclinic system.

albitization *(Geol.)*. In igneous rocks, the process by which a soda-lime feldspar (plagioclase) is replaced by albite (soda-feldspar).

albumen *(Zool.)*. White of egg containing a number of soluble proteins, mainly ovalbumin. adj. *albuminous*.

albumen process *(Print.)*. One in which dichromated albumen is used as a light-sensitive coating when preparing **surface plates** for lithography and line blocks for relief printing.

albumin *(Biol.)*. A general term for proteins soluble in water as distinct from saline. Specific albumins are designated by their sources, e.g. egg albumin from egg white, serum albumin from blood serum.

albuminous *(Bot.)*. **Endospermic**.

albuminous cell *(Bot.)*. Specialized parenchyma cell in gymnosperm phloem, associated with a sieve cell but not originating from the same precursor cell. Cf. **companion cell**.

albuminuria *(Med.)*. Albumin in the urine.

alburnum *(Bot.)*. Sapwood.

Alchlor process *(Chem.)*. Used in refining lubricants, by removal of impurities with aluminium chloride.

Alclad *(Eng.)*. Composite sheets consisting of an alloy of the Duralumin type (to give strength) coated with pure aluminium (to give corrosion resistance).

alcohol *(Chem.)*. A general term for compounds formed from hydroxyl groups attached to carbon atoms in place of hydrogen atoms. The general formula is *R*.OH, wherein *R* signifies an aliphatic radical. In particular, *ethanol*. Hydroxyl groups attached to aromatic rings give *phenols*. See **methanol-**, **absolute alcohol**.

alcohol fuel *(Autos.)*. Volatile liquid fuel consisting wholly or partly of alcohol, able to withstand high-compression ratios without detonation.

alcoholic fermentation *(Bot.)*. A form of anaerobic **respiration** in which a sugar is converted to alcohol and carbon dioxide. Alcoholic fermentation by yeasts is important in baking, brewing and wine-making.

alcoholism *(Behav.)*. Disease produced by addiction to alcohol, manifesting itself in a variety of psychotic disorders, e.g. hallucinosis, delirium tremens.

alcoholometry *(Chem.)*. The quantitative determination of alcohol in aqueous solutions.

Alcomax *(Eng.)*. British equivalent of *Alnico*.

aldehyde acids *(Chem.)*. Products of the partial oxidation of dihydric alcohols, containing both an aldehyde group and a carboxyl group.

aldehyde resins *(Plastics)*. Highly polymerized resinous condensation products of aldehydes obtained by treatment of aldehydes with strong caustic soda.

aldehydes *(Chem.)*. Alkanals. A group of compounds containing the CO- radical attached to both a hydrogen atom and a hydrocarbon radical, viz. *R*.CHO

alder *(For.)*. A tree (*Alnus*) producing a straight-grained fine-textured hardwood noted for its durability under water. It is used for cabinet making, plywood, shoe heels, clogs, bobbins, wooden cogs and small turned items.

aldimines *(Chem.)*. Condensation products of phenols with hydrocyanic acid, formed in the presence of gaseous hydrogen chloride.

aldohexoses *(Chem.)*. The most important group of **monosaccharides**, including **glucose** and **galactose**. All have a formula which can be expressed: $OH.CH_2.(CH.OH)_4.CHO$. ⇨

aldol *(Chem.)*. *2-Hydroxybutanal*, a condensation product of acetaldehyde, viz. $H_3C.CH(OH).CH_2.CHO$.

aldolase *(Biol.)*. The enzyme which catalyses the cleavage of fructose-1-6-diphosphate into glyceraldehyde-3-phosphate and dihydroxyacetone.

aldol condensation *(Chem.)*. The condensation of 2 aldehyde molecules in such a manner that the oxygen of the 1 molecule reacts with the hydrogen of the other molecule, forming a hydroxyl group, with the simultaneous formation of a new link between the 2 carbon atoms. Water is eliminated.

aldoses *(Chem.)*. A group of monosaccharides with an aldehydic constitution, e.g. glucose. See **aldohexose**.

aldosterone *(Chem.,Med.)*. A potent mineralocorticoid secreted by *zona glomerulosa* of the adrenal cortex which promotes the retention of sodium ions and water.

aldoximes *(Chem.)*. *Hydroxyimino alkanes*. A group of compounds in which the oxygen of the aldehyde group is substituted by the radical =N.OH, derived from hydroxylamine $H_2N.OH$ and an aldehyde by dehydration. The general formula is R.CH=N.OH.

aldrin *(Chem.)*. A chloro-derivative of naphthalene used as a contact insecticide, incorporated in plastics to make cables resistant to termites; persistent toxicity; formerly used in agriculture, especially against wireworm.

alecithal *(Zool.)*. Of ova, having little or no yolk.

aleph-0 *(Maths.)*. Cardinal number of the set of natural numbers; first infinite cardinal; written $\aleph$.

aleurone *(Bot.)*. Reserve protein occurring in seed granules, usually in the outermost layer, *aleurone layer*, of the endosperm, e.g. in cereals and other grasses.

Aleutian disease *(Vet.)*. A chronic, fatal disease of mink, with characteristic changes in the liver, kidneys, and other organs; it occurs especially in mink homozygous for the aleutian gene controlling fur colour.

alexandrite *(Min.)*. A variety of chrysoberyl, the colour varying, with the conditions of lighting, between emerald green and red.

alexia *(Med.)*. Word blindness; loss of the ability to interpret written language due to a lesion in the brain.

alexin *(Immun.)*. Obsolete term for complement used by Bordet who first described complement.

Alford antenna *(Telecomm.)*. One comprising a vertical cylindrical tube with longitudinal slots, often used to transmit VHF or UHF.

algae *(Bot.)*. Prokaryotic and eukaryotic photosynthetic organisms with *chlorophyll a* and other photosynthetic pigments, releasing O_2. Plant body unicellular, colonial, filamentous, siphoneous or parenchymatous, never with roots, stems or leaves. Sex organs unicellular or multicellular with all cells fertile (except Charales). The zygote does not develop into a multicellular embryo within the female sex organ. Includes Cyanophyceae and Prochlorophyceae (both prokaryotic), and Rhodophyceae, Cryptophyceae, Dinophyceae, Heterokontophyta, Haptophyceae, Euglenophyceae and Chlorophyta (all eukaryotic). Not a natural group, but the word is useful in many contexts.

algae poisoning *(Vet.)*. A form of poisoning affecting farm livestock in the US due to the ingestion of toxins in decomposing algae; characterized by nervous symptoms and death.

algal corrosion *(Aero.)*. Impairment of structure and systems by algae and other micro-organisms.

algal layer, algal zone *(Bot.)*. A layer of algal cells lying inside the thallus of a heteromerous lichen. Also called *gonidial layer*, *gonimic layer*.

algebra *(Maths.)*. Originally the abstract investigation of the properties of numbers by means of symbols (x,y, etc.). In this context typical algebraic problems are the solving of equations, the summation of series, permutations and combinations, and matrices. More recently extended to include the study of sets in general with operations. In this context there is the study of mappings, groups, rings, integral domains, fields and vector spaces. See also **mathematics**.

algebraic function *(Maths.)*. One which can be defined by a finite number of algebraic operations, including root extraction, e.g. the quotient of 2 polynomials is an algebraic function, but $\sin x$ is not.

algebraic number *(Maths.)*. A root of a polynomial equation with rational coefficients.

Algerian onyx *(Min.)*. Another name for *oriental alabaster*. See under **alabaster**.

algesis *(Med.)*. The sense of pain.

alginic acid *(Chem.)*. *Norgine*, $C_3H_8O_6$, occurs both in the free state and the calcium salt in the larger brown algae (*Phaeophyceae*). The sodium salt gives a very viscous solution in water even at a concentration of only 2%, and is used in the dyeing, textile, plastics and explosives industries, in making waterproofing and insulating materials, foodstuffs, adhesives, cosmetics, and in medicine, for its sodium absorption ability by a cation exchange reaction.

Algol *(Astron.)*. A star, β, in the constellation of Perseus which is the prototype of the eclipsing binary, where one component passes in front of the other at each revolution, causing an eclipse and a systematic fluctuation of magnitude.

ALGOL-68 *(Comp.)*. *ALGOrithmic Language, 1968*. Very powerful language with structured programming features. Like the earlier ALGOL-60 designed to aid the programming of algorithms.

algology *(Bot.)*. The study of algae.

algorithm *(Comp.)*. A set of rules which specify a sequence of actions to be taken to solve a problem. Each rule is precisely and unambiguously defined so that in principle it can be carried out by machine. See **formal language theory**.

aliasing *(Acous.)*. Error in making real time spectra of short signals or of directivity in sound fields. Caused by insufficient number of data points.

aliasing *(Image Tech.)*. Image imperfections resulting from limited detail in a **raster** display, for example, resulting in diagonal lines appearing stepped.

A-licence *(Aero.)*. Basic private pilot's licence.

alicyclic *(Chem.)*. Abbrev. for *aliphatic-cyclic*. A ring compound not containing **aromatic** groups.

alidade *(Surv.)*. An accessory instrument used in plane-table surveying, consisting of a rule fitted with sights at both ends, which gives the direction of objects from the plane-table station. Also called *sight rule*.

alien *(Bot.)*. Believed on good evidence to have been introduced by man and now more of less naturalized.

alien tones *(Acous.)*. Frequencies, harmonic and sum-and-difference products, introduced on sound reproduction because of nonlinearity in some part of the transmission path.

aliform muscles *(Zool.)*. See **alary muscles**.

alignment *(Civ.Eng.)*. (1) A setting in line (usually straight) of, e.g. successive lengths of a railway which is to be constructed. (2) The plan of a road or earthwork.

alignment *(Eng.)*. The setting in a true line of a number of points, e.g. the centres of the bearings supporting an engine crankshaft.

alignment *(Phys.)*. Process of orientating e.g. spin axes of atoms, during magnetization and similar operations.

alignment *(Telecomm.)*. Adjustment of preset tuned circuits to give optimum performance.

alignment chart *(Maths.)*. See **nomogram**.

alimentary *(Med.)*. Pertaining to the nutritive functions or organs.

alimentary canal *(Med.)*. The passage from the mouth to the anus which receives, digests and assimilates nutrients from foodstuffs: the digestive tract: the gut. Also *alimentary tract*.

alimentary system *(Med.)*. All the organs connected with digestion, absorption and nutrition, comprising the digestive tract and associated glands and masticatory mechanisms.

aliphatic compounds *(Chem.)* Methane derivatives of fatty compounds; open-chain or ring carbon compounds not having *aromatic* properties.

aliquot *(Phys.)*. A small sample of a material assayed to determine the properties of the whole. Term often applied to radioactive material.

aliquot part *(Maths.)*. A number or quantity which exactly divides a given number or quantity.

aliquot part *(Min.Ext.)*. In sampling for process control, representative fraction, from quantitative analysis of which information as to the assay grade is given.

aliquot scaling, -tuning *(Acous.)*. In a piano, the provision of extra wires above the normal wires. These are not struck, but are tuned very slightly above the octave of the struck strings below, so that by sympathetic vibration the musical quality of the note is enhanced.

Alismatidae *(Bot.)*. Subclass or superorder of monocotyledons. Aquatic and semi-aquatic herbs, typically with perianth segments free, sometimes differentiated into sepals and petals, mostly apocarpous, pollen trinucleate, mostly lacking endosperm in the mature seeds. Contains ca 500 spp. in 14 families.

alisphenoid *(Zool.)*. A winglike cartilage bone of the Vertebrate skull, forming part of the lateral wall of the cranial cavity, just in front of the foramen lacerum: one of a pair of dorsal bars of cartilage in the developing Vertebrate skull, lying in front of the basal plate, parallel to the trabeculae: one of the sphenolateral cartilages.

alizarin *(Chem.)*. $C_{14}H_6O_2(OH)_2$, *1,2-dihydroxyanthraquinone*, one of the most important natural and synthetic dyes; red prisms, or needles, mp 289°C, soluble in alcohol and ether, very slightly soluble in water, soluble in caustic soda; insoluble stains are formed with the oxides of aluminium, tin, chromium, and iron. Alizarin can be nitrated and forms the basis of a series of other dyestuffs.

alkali *(Chem.)*. A hydroxide which dissolves in water to form an *alkaline* or *basic solution* which has pH > 7 and contains hydroxyl ions, OH^-.

alkali disease *(Vet.)*. (1) *Western duck sickness*. A form of botulism causing death of wild duck and other water fowl in America, due to the ingestion of vegetation contaminated with toxin produced by *Clostridium botulinum*, type C. (2) A chronic disease of domestic animals, characterized by emaciation, stiffness, and anemia, due to an excess of selenium in the diet. An acute form of the disease is called *blind staggers*.

alkali granite *(Geol.)*. An acid, coarse-grained (plutonic) rock carrying free quartz and characterized by a large excess of alkali-feldspar over plagioclase. Cf. **adamellite-**, **granodiorite**. In general, the prefix used with a rock name implies a preponderance of soda- or potash- feldspar or feldspathoid over plagioclase, e.g. alkali dolerite.

alkali metals *(Chem.)*. The elements lithium, sodium, potassium, rubidium, caesium and francium, all metals in the first group of the periodic system. In most compounds they occur as univalent ions.

alkalimetry *(Chem.)*. The determination of alkali by titration with a standard solution of acid as in volumetric analysis. See **titration** and **volumetric analysis**.

alkaline earth metals *(Chem.)*. The elements calcium, strontium, barium and radium, all divalent metals in the second group of the periodic system.

alkaline phosphatase *(Biol.)*. Enzyme commonly conjugated with antibodies for use in **indirect immunoassay**. It catalyses a reaction which deposits dye at the site of the bound antibody. Abbrev. *AP*.

alkalinity *(Bot.)*. Of e.g. a lake, $[HCO_3^-] + 2[CO_3^{2-}] + [OH^-] - [H^+]$. (Square brackets indicate molar concentrations.)

alkalinity *(Chem.)*. The extent to which a solution is alkaline. See **pH value**, **acidity**.

alkali resisting paints *(Build.)*. Surface coatings formulated to withstand exposure to alkaline materials, substrates and atmospheres. Primary paints for use on alkaline substrates where an oil paint system is specified are normally based on tung oil with coumarone or phenolic resin.

alkaloids *(Chem.)*. Natural organic bases found in plants; characterized by their specific physiological action and toxicity; used by many plants as a defence against herbivores, particularly insects. Alkaloids may be related to various organic bases, the most important ones being pyridine, quinoline, *iso*quinoline, pyrrole and other more complicated derivatives. Most alkaloids are crystalline solids, others are volatile liquids, and some are gums.

They contain nitrogen as part of a ring, and have the general properties of amines.

alkalosis *(Med.)*. Where there is a decrease in hydrogen ion concentration in blood and tissue. May be respiratory due to excessive loss of CO^2 in the lungs (hyperventilation) or metabolic due to loss of non-volatile acids from the body, i.e. vomiting.

alkane *(Chem.)*. General name of hydrocarbons of the methane series, of general formula C_nH_{2n+2}.

Alkathene *(Chem.)*. Tradename for *polyethylene*.

alkene *(Chem.)*. *Olefin*. General name for unsaturated hydrocarbons of the ethene series, of general formula C_nH_{2n}.

alkyd resins *(Plastics)*. Formerly known as glyptal resins, i.e. condensation products derived from glycerol and phthalic anhydride. Now the term also covers diallyl esters and various polyesters used as resin binders in alkyd moulding materials.

alkyl *(Chem.)*. A general term for monovalent aliphatic hydrocarbon radicals.

alkylating drug *(Med.)*. Cytotoxic drugs which act by damaging DNA and interfering with cell replication. *Cyclophosphamide, chlorambucil, busulphan* and *mustine* are common examples.

alkylene *(Chem.)*. A general term for divalent hydrocarbon radicals.

alkyne *(Chem.)*. An aliphatic hydrocarbon with a triple bond. The simplest is ethyne or acetylene, $HC{\equiv}CH$.

allanite *(Min.)*. A cerium-bearing epidote occurring as an accessory mineral in igneous and other rocks.

allantoic *(Zool.)*. See **allantois**.

allantoin *(Med.)*. Diureide of glyoxylic acid, product occurring in allantoic fluid and excreted in urine of certain Mammals and, in Insects, by the blowfly larvae.

allantois *(Zool.)*. In the embryos of higher Vertebrates, a saclike diverticulum of the posterior part of the alimentary canal, having respiratory, nutritive, or excretory functions. It develops to form one of the embryonic membranes. adj. *allantoic*.

Allan valve *(Eng.)*. Once popular slide-valve design with an internal passage designed to reduce valve travel and wear.

all-burnt *(Aero.)*. The moment at which the fuel of a missile or spacecraft is completely consumed.

alleghanyite *(Min.)*. A hydrated manganese silicate, crystallizing in the monoclinic system.

allele *(Biol.)*. Abbrev. for *allelomorph*. Any one of the alternative forms of a specified gene. Different alleles usually have different effects on the phenotype. Any gene may have several different alleles, called *multiple alleles*. Genes are *allelic* if they occupy the same *locus*. adj. *allelic*.

all-electric signalling *(Elec.Eng.)*. A railway signalling system in which the signals and points are operated electrically by solenoids or motors, and are also controlled electrically. See **electropneumatic signalling**.

allelopathy *(Biol.)*. The condition when one strain is harmful to another of the same species.

allelopathy *(Ecol.)*. Adverse influence exerted by one individual plant over another by the production of a chemical inhibitor, often a terpenoid or phenolic.

allemontite *(Min.)*. A solid solution of antimony and arsenic.

Allen cone *(Min.Ext.)*. Conical tank used for continuous sedimentation of liquids at constant level, the solids being removed from the base of the cone and the clear liquid drawn off from the top.

allene *(chem.)*. *Propadiene, dimethylene methane*, $CH_2{=}C{=}CH_2$ obtained by the electrolysis of itaconic acid.

Allen equation *(Min.Ext.)*. One applied to sedimentation of finely ground particles intermediate between streamline and turbulent in settling mode.

$$\rho = Kr^n p\mu^{2-n} v^n$$

where ρ is fluid resistance; *K*, a constant for shape and velocity of fall; *r*, radius of an equivalent sphere; *n*, a coefficient of velocity, *v*; *p*, density of fluid; and μ the kinematic viscosity *b*/*p*, *b* being absolute viscosity.

allenes *(Chem.)*. Generic term for a series of non-conjugated and di-olefinic hydrocarbons, of which *allene* is the first, and which have the general formula $C_{2n}H_{2nb2}$. They consist mostly of colourless liquids with strong garlic odour.

Allen's law *(Zool.)*. An evolutionary generalization stating that feet, ears, and tails of mammals tend to be shorter in colder climates, when closely allied forms are compared.

Allen's loop test *(Elec.Eng.)*. A modification of the Varley loop test for localizing a fault in an electric cable; it is particularly suitable for high-resistance faults in short lengths of cable.

allergen *(Immun.)*. Antigenic substances which provoke an allergic response (see **allergy**). Commonly used to describe those which cause immediate type **hypersensitivity** such as pollens or insect venoms.

allergic, allergy *(Immun.,Med.)*. (1) Showing altered responsiveness to an antigen as the result of previous contact with that antigen. Responsiveness is usually increased, but can be decreased. (2) The reaction of the body to a substance to which it has become sensitive, characterized by oedema, inflammation and destruction of tissue.

alliaceous *(Bot.)*. Looking, or smelling like an onion.

alligator *(Min.Ext.)*. See **jaw breaker**.

alligatoring *(Build.)*. See **crocodiling**.

all insulated switch *(Elec.Eng.)*. See **shockproof switch**.

allithium *(Aero.)*. Aluminium-lithium alloys.

all-moving tail *(Aero.)*. A one-piece **tail plane**, also controlled by the pilot as is the **elevator**. Also *flying tail, stabilator*, and see **T-tail**.

allo- *(Chem.)*. A prefix used to show that a compound is a **stereoisomer** of a more common compound.

allo- *(Genrl.)*. Prefix from Gk. *allos*, other.

allo- *(Immun.)*. Term used to describe a gene product, tissue etc. from a different individual of the same species, e.g. *allograft*.

allobar *(Phys.)*. A mixture of isotopes of an element differing in proportion from that naturally occurring.

allochromatic *(Electronics)*. Having photoelectric properties which arise from micro-impurities, or from previous specific irradiation.

allochromy *(Phys.)*. Fluorescent reradiation of light of different wavelength from that incident on a surface. See **Stokes' law**.

allochthonous *(Geol.)*. Refers to a block of rock that is exotic to its environment, e.g. a block of limestone that has slid down a submarine slope into a muddy environment or a tectonically moved block.

allogamy *(Bot.)*. Fertilization involving pollen and ovules from: (1) different flowers (whether on the same plant or not), including *geitonogamy* and *xenogamy*; (2) genetically distinct individuals of the same species (i.e. from another **genet**.). See **cross-fertilization, cross-pollination**. Cf. **autogamy**.

allograft *(Immun.)*. See **homograft**.

allomeric *(Biol.)*. Having the same crystalline form but a different chemical composition.

allometry *(Biol.)*. The relationship between the growth rates of different parts of an organism.

allomone *(Zool.)*. A chemical signal produced by one species of animal which influences the behaviour of members of another to the advantage of the signaller. Cf. **pheromone**.

allopatric *(Bot.)*. Two species or populations not growing in the same geographical area; unable to interbreed by reason of distance or geographical barrier. Cf. **sympatric**.

allopatric speciation *(Bot.)*. The accumulation of genetic differences in an isolated population leading to the evolution of a new species.

allophane *(Min.)*. Hydrous aluminium silicate, apparently amorphous.

allopolyploid *(Bot.)*. A polyploid of hybrid origin containing sets of chromosomes from two or more different species, often self-fertile but not interbreeding with the parental species. Allopolyploidy is important in speciation and in the evolution of some crop plants e.g. wheat, *brassica*. See **amphidiploid**. Cf. **autopolyploid**.

allopurinol *(Med.)*. Drug which is a xanthine oxidase inhibitor and blocks the breakdown of xanthine to uric acid. It is therefore of use in the prophylaxis of **gout**.

all-or-nothing piece *(Horol.)*. A piece of the mechanism of a repeating watch which either allows the striking of the hours and quarters or entirely prevents it. Also called *stop slide*.

all-or-nothing response *(Zool.)*. In many irritable protoplasmic systems, response to stimuli is either with full intensity or not at all; e.g. in lower animals, nematocysts; in higher animals, nerve fibres, cardiac and voluntary muscle fibres.

allose *(Chem.)*. An aldohexose, an optical stereoisomer of glucose. ⇨

allosteric protein *(Biol.)*. Protein which alters its 3-dimensional conformation as a result of the binding of a smaller molecule, often leading to altered activity, e.g. of an enzyme.

allosteric site *(Biol.)*. An enzymic site, distinct from the **active site**, which by binding molecules other than the substrate induces a *conformational change* which effects the enzyme's activity.

alloter *(Telecomm.)*. A uniselector used to improve the efficiency of distribution of line-finders, by automatically pre-selecting and pre-connecting the first available line finder in the group to which it has access.

allotetraploid *(Bot.)*. See **amphidiploid**.

allotriomorphic *(Geol.)*. A textural term used for igneous rocks describing crystals which show a form related to surrounding previously crystallized minerals rather than to their own internal structure.

allotropous flower *(Bot.)*. A flower in which the nectar is accessible to all kinds of insect visitors.

allotropy *(Chem.)*. The existence of an element in two or more solid, liquid, or gaseous forms, in one phase of matter, called *allotropes*.

allotype *(Immun.)*. Used to describe identifiable differences between immunoglobulin molecules that are inherited as alleles of a single genetic locus. They are due to single amino acid substitutions in light or heavy chains, and are useful in population studies.

allotype *(Zool.)*. (1) An additional type-specimen of the opposite sex to the original type specimen. (2) An animal or plant fossil selected as a species or subspecies, illustrating morphological details not shown in the holotype.

allowable deficiencies *(Aero.)*. Aircraft systems or certain items of their equipment, tabulated in the flight or operating manual, which even if unserviceable will not prevent an aircraft from being flown or create a hazard in flight.

allowances *(Aero.)*. In airline terminology, fuel reserves are frequently referred to as allowances, and are usually specified as time factors under certain conditions, as distance plus descent, or as a percentage (by weight or volume) of the cruising fuel for a given stage.

allowed band *(Phys.)*. Range of energy levels permitted to electrons in a molecule or crystal. These may or may not be occupied.

allowed transition *(Phys.)*. Electronic transition between energy levels which is not prohibited by any quantum selection rule.

alloy *(Chem.)*. A mixture of metals, or of a metal with a non-metal in which the metal is the major component. Alloys may be compounds, eutectic mixtures or solid solutions.

alloy *(Eng.)*. Metal prepared by adding other metals or nonmetals to a basic metal to secure desirable properties. Solid solution in which maximum difference in radius of component atoms is below 15%. Lower mp, decreased electrical conductivity and increased hardness are characteristic.

alloy cast-iron *(Eng.)*. Cast-iron containing alloying elements. Usually some combination of nickel, chromium, copper and molybdenum. These elements may be added to increase the strength of ordinary irons, to facilitate heat treatment, or to obtain martensitic, austenitic or ferritic irons.

alloy junction *(Electronics)*. One formed by alloying one or more impurity metals with a semiconductor. Small buttons of impurity metal are placed at desired locations on a semiconductor wafer; heating to melting point and rapidly cooling again produces regions of *p-type* or *n-type* conductivity, according to choice of impurity. Also *fused junction*.

alloy reaction limit *(Eng.)*. Concentration in alloy of a specific component, below which corrosion occurs in a given environment.

alloy steel *(Eng.)*. A steel to which elements not present in carbon steel have been added, or in which the content of manganese or silicon is increased above that in carbon steel. See **high speed steel, nickel steel, stainless steel**.

allozymes *(Biol.)*. Different forms of an enzyme specified by *allelic genes*.

all-pass network *(Telecomm.)*. One which introduces a specified phase-shift response without appreciable attenuation for any frequency.

Allström relay *(Elec.Eng.)*. See **relay types**.

alluvial mining *(Min.Ext.)*. Exploitation of alluvial or placer deposits. Minerals thus extracted include tin, gold, gemstones, rare earths, platinum. Term embraces beach deposits, eluvials, riverine, and offshore workings.

alluvial values *(Min.Ext.)*. Those shown by panning or assay to be recoverable from an alluvial deposit.

alluvium *(Geol.)*. Sand, silt and mud deposited by a river or floods; geologically recent in age.

allyl alcohol *(Chem.)*. *l-hydroxy prop-2-ene*, $H_2C=CH.CH_2OH$, an unsaturated primary alcohol, present in wood spirit, made from glycerine and oxalic acid. Mp −129°C, bp 96°C, rel.d. 0.85, of very pungent odour; an intermediate for organic synthesis.

allyl chloride *(Chem.)*. $Cl.CH_2CH\text{-}CH_2$. Important intermediate in the manufacture of synthetic glycerol. Formed by the high temperature (400–600°C) chlorination of propene.

allylene *(Chem.)*. *Propyne*, $CH_3.C{\equiv}CH$.

allyl group *(Chem.)*. The unsaturated monovalent aliphatic group $H_2C=CH.CH_2$—.

allyl resin *(Chem.)*. One formed by the polymerization of chemical compounds of the **allyl group**.

allyl sulphide *(Chem.)*. Oil of garlic; bp 139°C; $(CH_2=CH.CH_2)_2S$. Colourless liquid, found in garlic and largely responsible for its odour. It possesses antiseptic qualities.

Almagest *(Astron.)*. The Arabic form of the title of Claudius Ptolemy's great astronomical treatise, 'The Mathematical Syntaxis', written in Greek about AD 140.

almandine *(Min.)*. Iron-aluminium garnet, occurring in mica-schists and other metamorphic rocks. Commonly forms well-developed crystals, often with 12 or 24 faces.

almandine spinel *(Min.)*. See **ruby spinel**.

almond oil *(Chem.)*. Used for fruit essences, in perfumery and soap making; two grades are known: *bitter almond* and *sweet almond oil*.

almucantar *(Astron.)*. A small circle of the celestial sphere parallel to the horizontal plane. The term is also applied to an instrument for measuring altitudes and azimuths.

Alnico *(Eng.)*. US TN for a high-energy permanent magnet material, an alloy of aluminium, nickel, cobalt, iron and copper.

alopecia *(Med.)*. Baldness.

alopecia areata *(Med.)*. A condition in which the hair falls out in patches, leaving smooth, shiny, bald areas.

Aloxite *(Eng.)*. TN designating a proprietary fused alumina and associated abrasive products.

alpaca *(Textiles)*. The fine, strong hair of the alpaca (*Lama pacos*) of South America; the fabric made from such hair. This animal belongs to the camel family and is a close relative of the Llama (*L. glama*) and the Vicuna (*L. vicugna*).

alpha (α) activation *(Chem.)*. The influence of organic radicals and groups in directing the course of chemical reactions, e.g. the carbonyl function in ketones which leads to 2-halogenation predominating.

alpha-beta brass *(Eng.)*. Copper-zinc alloy containing 38–46% (usually 40%) of zinc. It consists of a mixture of the α-constituent (see **alpha brass**) and the β-constituent (see **beta brass**).

alpha-brass *(Eng.)*. A copper-zinc alloy containing up to 38% of zinc. Consists constitutionally of a solid solution of zinc in copper. Commercial alpha brasses of several compositions are made. The most widely used contain 30–37%; others contain 5–20% of zinc. All are used mainly for cold-working.

alpha-bronze *(Eng.)*. A copper-tin alloy consisting of the alpha solid solution of tin in copper. Commercial forms contain 4 or 5% of tin. This alloy, which differs from gun metal and phosphor bronze in that it can be worked, is used for coinage, springs, turbine blades etc.

Alpha Centauri *(Astron.)*. The brightest star in the constellation Centaurus, actually three stars, the faintest of which, **Proxima Centauri**, is the nearest star to the sun.

alpha-chain *(Immun.)*. Heavy chain of **IgA**. Alpha-chain disease is a rare disease in which the intestine is infiltrated by lymphoma which makes alpha chains but no light chains, owing to a deletion involving the site required to link the two.

alpha chamber *(Phys.)*. Ionization chamber for measurements of α-radiation intensity.

alpha counter *(Phys.)*. Tube for counting α-particles, with pulse selector to reject those arising from β- and γ- rays.

alpha counter tube *(Phys.)*. See **alpha chamber**.

alpha cut-off *(Electronics)*. Frequency at which the current amplification of a transistor has fallen by more than 3 dB (0.7) of its low-frequency value.

alpha decay *(Phys.)*. Radioactive disintegration resulting in emission of α-particle. Also *alpha disintegration*.

alpha decay energy *(Phys.)*. The sum of the kinetic energies of the α-particle emitted and the recoil of the product atom in a radioactive decay.

alpha diversity *(Ecol.)*. See **diversity**.

alpha emitter *(Phys.)*. Natural or artificial radio-active isotope which disintegrates through emission of α-rays.

alphafetoprotein *(Immun.)*. A plasma protein made by the foetus, but not by adults unless they have primary liver cancer or some other tumours in which foetal genes are expressed. Alphafetoprotein escapes into the maternal blood during pregnancy, and an abnormally high concentration at certain stages has been found to indicate that normal closure of the spinal canal of the foetus is incomplete (spina bifida).

alpha helix *(Biol.)*. Important element of protein structure formed when a polypeptide chain turns regularly about itself to form a rigid cylinder stabilized by hydrogen bonding.

alpha iron *(Eng.)*. The polymorphic form of iron, stable below 906°C. Has a body-centred cubic lattice, and is magnetic up to 768°C.

alphameric *(Comp.)*. See **alphanumeric**.

alphanumeric *(Comp.)*. From the set of characters consisting of the alphabet and the numerals 0–9. Most other typewriter characters are usually excluded and are reserved for **programming**, as **control characters** or as **prompts**. Also *alphameric*.

alphanumeric printers *(Print.)*. Printers using characters of the alphabet and the numbers from 0 to 9.

alpha particle *(Phys.)*. Nucleus of helium atom of mass number four, consisting of two neutrons and two protons and so doubly positively charged. Emitted from natural or radioactive isotopes. Often written α-particle.

alpha pulp *(Paper)*. Wood pulp processed so that only a very small percentage of hemicellulose remains. Also called *dissolving pulp*.

alpha radiation *(Phys.)*. **Alpha particles** emitted from radioactive isotopes.

alpha rays *(Phys.)*. Stream of **alpha particles**.

alpha-ray spectrometer *(Nuc.Eng.)*. Instrument for measuring energy distribution of α-particles emitted by a radioactive source.

alpha rhythm *(Med.)*. The regular electro-encephalographic pattern of ca. 10 Hz obtained from waking, inactive individual. The pattern is disrupted by such changes as falling asleep, concentrating etc.

alpha wave *(Med.)*. The principal slow wave (frequency ca. 10 Hz) produced by the human brain and recorded as the electro-encephalogram (EEG).

alpha-wrap *(Image Tech.)*. Tape path on the drum of a **helical-scan** VTR giving full 360° contact.

alpine *(Bot.)*. Vegetation or plants of high mountains.

Alpine orogeny *(Geol.)*. The fold movements during the Tertiary period which lead to the development of the Alps and associated mountain chains.

alstonite *(Min.)*. A double carbonate of calcium and barium.

altar tomb *(Arch.)*. A raised tomb or monument usually standing detached or in a position against a wall, and sometimes supporting an effigy. In appearance it resembles a solid altar, but it is never used as one.

altazimuth *(Astron.)*. A type of telescope in which the principal axis can be moved independently in altitude (swinging on a horizontal axis) and azimuth (swinging on a vertical axis). Used in very large *optical telescopes* and *radio telescopes*.

altazimuth *(Surv.)*. An instrument similar to the **theodolite** but generally larger and capable of more precise work.

alternant *(Maths.)*. A determinant whose elements are functions of $x_1,...x_n$ such that interchanging the variables x_i and x_k interchanges the ith row and the kth rows and thus changes the sign of the determinant.

alternate *(Bot.)*. Leaves, branches etc. placed singly on the parent axis, i.e. not in pairs (opposite), not whorled. See also **alternate host**.

alternate airfield *(Aero.)*. One designated in a *flight plan* at which a pilot will land if prevented from alighting at his destination.

alternate angles *(Maths.)*. If a transversal cuts two straight lines, the alternate angles lie on either side of it, each angle having one of the lines as an arm, the other arm being in both cases the transversal. If the angles are between the two lines, they are interior alternate angles. If neither of the angles is interior, they are said to be exterior alternate angles.

alternate host *(Bot.)*. One of the two (rarely more) hosts of a parasite which has the different stages of its life cycle in unrelated hosts. Cf. **alternative host**.

alternating cleavage *(Zool.)*. See **spiral cleavage**.

alternating current *(Elec.Eng.)*. Generally abbreviated to **a.c.** Electric current whose flow alternates in direction; the time of flow in one direction is a half-period, and the length of all half-periods is the same. The normal waveform of a.c. is sinusoidal, which allows simple vector or algebraic treatment. Provided by **alternators** or electronic **oscillators**.

alternating function *(Maths.)*. A function of two or more variables such that interchanging any two changes the sign but not the absolute value of the function e.g., $f(x,y,z) = -f(y,x,z)$. Also called *antisymmetric function*.

alternating-gradient focusing *(Electronics)*. The principle follows the optical analogy whereby a series of alternate converging and diverging lenses may lead, under suitable conditions, to a net focusing effect since the rays will strike the diverging lenses nearer to the axis. Using magnetic or electrostatic lenses, the idea has been used for the design of electron synchrotrons and ion linear accelerators.

alternating gradient synchrotron *(Phys.)*. A *synchrotron* modified by having magnetic-field gradients around the orbit alternating towards and away from the centre of the orbit. This produces a focusing effect which reduces beam divergence caused by the mutual repulsion of the particles in the beam. Proton energies of up to 500 GeV and electron energies of about 10 GeV can be achieved.

alternating light *(Ships.)*. A navigation mark identified during darkness by a light showing alternating colours. See **flashing light**, **occulting light**.

alternating series *(Maths.)*. A series whose terms are alternately positive and negative. If the terms of an alternating series decrease monotonically then a necessary and sufficient condition for convergence is that they tend to zero.

alternating stress *(Phys.)*. The stress induced in a

material by a force which acts alternately in opposite directions.

alternation of generations *(Biol.)*. The regular alternation of two (rarely, three) types of individual in the life history of an animal or plant; typically in plants a diploid **sporophyte** and a haploid **gametophyte** or in animals a sexually- and an asexually-produced form. They may be morphologically similar (isomorphic) or different (heteromorphic).

alternative host *(Bot.)*. One of two or more possible hosts for a given stage in the life cycle of a parasite particularly when it is not the commonest or the most important economically. Cf. **alternate host**.

alternative medicine *(Med.)*. Systems of medicine like acupuncture, chiropractice, homeopathy and osteopathy, which are able to alleviate symptoms for reasons which are poorly understood. The methods have not usually been subjected to test by **clinical trial**.

alternative pathway of complement activation *(Immun.)*. See **complement**.

alternative routing *(Telecomm.)*. The manual or automatic diversion, to a prearranged secondary route, of traffic which originates at an instant when the primary route is not available.

alternator *(Elec.Eng.)*. A type of a.c. generator, driven at a constant speed corresponding to the particular frequency of the electrical supply required from it. Also *synchronous generator*.

altimeter *(Aero.,Phys.)*. An aneroid barometer used for measuring altitude by the decrease in atmospheric pressure with height. The dial of the instrument is graduated to read the altitude directly in feet or metres, the zero being set to ground or aerodrome level. See **encoding-, radio-, recording-**.

altitude *(Aero.,Surv.)*. The height in feet or metres above sea level. For precision in determining the performance of an aircraft, this must be corrected for the deviation of the meteorological conditions from that of the standardized atmosphere (*International Standard Atmosphere*). See **cabin altitude, pressure altitude**.

altitude *(Astron.)*. The angular distance of a heavenly body measured on that great circle which passes, perpendicular to the plane of the horizon, through the body and through the zenith. It is measured positively from the horizon to the zenith, from 0° to 90°.

altitude *(Maths.)*. (1) The line through the vertex of a geometrical figure or solid perpendicular to its base. (2) The length of this line.

altitude level *(Surv.)*. Sensitive spirit level which ensures that theodolite is truly horizontal with respect to the telescope when vertical angles are measured.

altitudes by barometer *(Phys.)*. See **Babinet's formula for altitude**.

altitude switch *(Aero.)*. A switching device generally comprising electrical contacts, actuated by an aneroid capsule which in turn is deflected by change in atmospheric pressure. The contacts are adjusted to make or break a warning circuit at the pressure corresponding to a predetermined altitude.

altitude valve *(Aero.)*. A manually- or automatically-operated valve fitted to the carburettor of an aero-engine for correcting the mixture-strength as air density falls with altitude.

altocumulus *(Meteor.)*. White and/or grey patch, *sheet* or *layer* of *cloud*, generally with shading, composed of laminae, rounded masses, rolls, etc., which are partly fibrous or diffuse and which may or may not be merged; most of the irregularly arranged small elements usually have an apparent width of between 1° and 5°. Occurs between 3000–7500 m. Abbrev. *Ac*.

altometer *(Surv.)*. See **theodolite**.

altostratus *(Meteor.)*. Greyish or bluish *cloud sheet* or *layer* of striated, fibrous or uniform appearance, totally or partly covering the sky, and having parts thin enough to reveal the Sun at least vaguely, as through ground glass. Altostratus does not show halo phenomenon and occurs at 3000–7500 m. Abbrev. *As*.

altrices *(Zool.)*. Birds whose young are hatched in a very immature condition, generally blind, naked, or with down feathers only, unable to leave the nest, fed by the parents; e.g. the Perching Birds, Passeriformes.

altrose *(Chem.)*. An aldohexose, an optical stereoisomer of glucose. ⇨

altruism *(Ecol.)*. Broad class of animal behaviour in which an individual benefits another at the risk of its own life or expense.

ALU *(Comp.)*. See **arithmetic logic unit**.

alula *(Zool.)*. See **alia spuria**.

alum *(Min.)*. Hydrated aluminium potassium sulphate and related compounds.

alumina *(Min.)*. Aluminium oxide. See **corundum**.

aluminate *(Chem.)*. Salt of aluminic acid, H_3AlO_3, a tautomeric form of aluminium hydroxide, which acts as a weak acid. Ortho-aluminates have the general formula M_3AlO_3 or $M_3Al(OH)_6$, and meta-aluminates, $MAlO_2$ or $MA(OH)_4$, where M is a monovalent metal. *Sodium aluminate*, Na_3AlO_3, is used as a coagulent in water purification and softening.

aluminium *(Chem.)*. Silver-white metallic element, forming a protective film of oxide. Symbol Al, at no. 13, r.a.m. 26.9815, mp 659.7°C, bp 1800°C, rel.d. 2.58. Obtained from bauxite, it has numerous uses and is the basis of light alloys for use in, e.g. structural work; alloyed with silicon for transformer laminations, and iron and cobalt in many types of permanent magnet. Polished aluminium reflects well beyond the visible spectrum in both directions, and does not corrode in sea water. Foil aluminium is much used for capacitors. The metal can be used as a window in X-ray tubes and as sheathing for reacting fuel rods. Aluminium is produced on a large scale where electric power is cheap when bauxite is electrolysed in fused cryolite. ^{28}Al and ^{29}Al are strong γ-ray emitters of very short half-life. As as electrode in gas-discharge tube, it does not sputter like other metals. In US *aluminum*.

aluminium alloys *(Eng.)*. Those in which aluminium is the basis (i.e. predominant) metal, e.g. aluminium-copper and aluminium-silicon alloys etc. Also called *light alloys*.

aluminium anode cell *(Elec.Eng.)*. One with an aluminium anode immersed in an electrolyte which does not attack aluminium. The cathode may also be of aluminium or some other metal, e.g. lead. Such cells can be used as rectifiers or as high-capacitance capacitors. See **electrolytic capacitor**.

aluminium antimonide *(Electronics)*. A semi-conducting material used for transistors up to a temperature of 500°C.

aluminium-brass *(Eng.)*. Brass to which aluminium has been added to increase its resistance to corrosion. Used for condenser tubes. Contains 1–6% Al, 24–42% Zn, 55–71% Cu.

aluminium bronze *(Eng.)*. Copper-aluminium alloys which contain 4-11% aluminium, and may also contain up to 5% each of iron and nickel or ½% tin. These alloys have high tensile strength, are capable of being cast or cold-worked, and are resistant to corrosion.

aluminium foil *(Build.)*. Used in conjunction with plaster board for insulation purposes in walls or roofs.

aluminium leaf *(Build.)*. Thin foil similar to but thicker than gold leaf, used for decorative work. Normally sold in books of 25 leaves with silver leaf sizes from 82 mm to 152 mm^2.

aluminium paints *(Build.)*. Comprise paste or powder in a suitable medium, their main uses being 1. as decorative finishing materials where their high opacity and reflectivity are important; 2. as primary paints where their qualities of mechanical strength, moisture resistance and sealing properties are invaluable.

aluminium powder *(Build.)*. Classified as flake or granulated. Flake powders give a **leafing** effect and are used when sealing properties are required.

aluminium-steel cable *(Elec.Eng.)*. See **steel-cored aluminium**.

aluminon *(Chem.)*. Ammonium aurine-tricarboxylate. Reagent for the colorimetric detetection and estimation of aluminium, with which it forms a characteristic red colour.

alumino-silicates *(Chem.,Min.)*. Compounds of alumina, silica and bases, with water of hydration in some cases. They include clays, mica, zeolites, constituents of glass and porcelain.

aluminothermic process *(Chem.)*. The reduction of metallic oxides by the use of finely divided aluminium powder. An intimate mixture of the oxide to be reduced and aluminium powder is placed in a refractory crucible; a mixture of aluminium powder and sodium peroxide is placed over this and the mass fired by means of a fuse or magnesium ribbon. The aluminium is almost instantaneously oxidized, at the same time reducing the metallic oxide to metal. This process, also known as the *thermite process*, is used especially for the oxides of metals which are reduced with difficulty (e.g. titanium, molybdenum). On ignition, the mass may reach a temperature of 3500°C. Magnesium incendiary bombs have thermite as the igniting agent.

aluminous cement *(Civ.Eng.)*. A cement containing 30–50% of lime, 30–50% of alumina, and not more than 30–50% of silica, iron oxide etc. The aluminous cements are less susceptible than ordinary Portland cements to low temperature during setting, and to the action of seawater and acids; and they harden rapidly, due to heat generated in setting.

aluminum *(Eng.)*. See **aluminium**

alums *(Chem.)*. A large number of isomorphous compounds whose general formula is: $R'R'''(SO_4)_2.12H_2O$; or $R'_2SO_4.R'''(SO_4)_3.24H_2O$ where R′ represents an atom of a univalent metal or radical—potassium, sodium, ammonium, rubidium, caesium, silver, thallium; and R‴ represents an atom of a tervalent metal—aluminium, iron, chromium, manganese, thallium. See also **pseudo-alums**.

alumstone *(Min.)*. See **alunite**.

alunite *(Min.)*. Hydrated sulphate of aluminium and potassium, resulting from the alteration of acid igneous rocks by solfataric action; used in the manufacture of alum. Also *alumstone*.

alunogen *(Min.)*. Hydrated aluminium sulphate, occurring as a white incrustation or efflorescence formed in two different ways: either by volcanic action, or by the decomposition of pyrite in carbonaceous or alum shales.

alveolar, alveolate *(Bot.,Zool.)*. Having pits over the surface, and resembling honeycomb.

alveolitis *(Med.)*. Inflammation of the pulmonary alveoli, usually by external source, e.g. inhaled mouldy hay.

alveolus *(Zool.)*. A small pit or depression on the surface of an organ: the cavity of a gland: a small cavity of the lungs: in higher Vertebrates, the tooth socket in the jaw bone: in Echinodermata, part of Aristotle's lantern, one of five pairs of grooved ossicles which grasp the teeth: in Gastropoda, the glandular end-portion of the tubules of the digestive gland, secreting enzymes.

ALVEY *(Comp.)*. British programme, began 1982, for research and development in computing and information technology.

Alzheimer's disease *(Med.)*. A degenerative brain disease, manifesting itself in premature ageing, with speech disorder.

Am *(Chem.)*. Symbol for americium.

amagat *(Phys.)*. The unit of density of a gas at 0°C and 1 atmos. pressure; usually 1 amagat = 1 mole/22.4 dm^3.

Amagat's law of combining volumes *(Chem.)*. The volume of a mixture of gases is equal to the sum of the volumes of the different gases, as existing each by itself at the same temperature and pressure.

amalgam *(Chem.)*. The alloy of a metal with mercury.

amalgam *(Min.Ext.)*. The pasty amalgam of gold and mercury, about 1/3 gold by weight obtained from the plates in a mill treating gold ores.

amalgamating table *(Min.Ext.)*. Flat sheet of metal to which mercury has adhered to form a thin soft film, used to catch metallic gold as mineral sands are washed gently over it. Also *amalgamated plate*.

amalgamation pan *(Min.Ext.)*. Circular cast-iron pan in which finely-crushed gold-bearing ore or concentrate is ground with mercury, the valuable metal thus being amalgamated before separate retrieval.

amalgam barrel *(Min.Ext.)*. Small ball mill used to regrind gold-bearing concentrates, and then give them prolonged rubbing contact with mercury.

amalgam retort (still) *(Eng.)*. Iron vessel in which the mercury is distilled off from gold or silver amalgam obtained in amalgamation.

Amalthea *(Astron.)*. The fifth natural satellite of **Jupiter**, discovered in 1892.

amaurosis *(Med.)*. Blindness due to a lesion of the retina, optic nerve, or optic tracts.

amazonstone, amazonite *(Min.)*. A green variety of microcline, sometimes cut and polished as a gemstone, and falsely called 'Amazon Jade'.

amber *(Min.)*. A fossil resin containing succinic acid in addition to resin acid and volatile oils. See **succinite**.

ambergris *(Zool.)*. A greyish white fatty substance with a strong but agreeable odour, obtained from the intestines of diseased sperm whales; sometimes found floating on the surface of the sea. It is used in perfumery as a fixative; on suitable treatment it yields ambreic acid.

amber mutation *(Biol.)*. A base change in a **coding sequence** of DNA, which gives the **stop codon** UAG, resulting in a shortened gene product.

amberoid *(Min.)*. See **ambroid**.

ambi- *(Genrl.)*. Latin form of **amphi-**.

ambient illumination *(Image Tech.)*. Background uncontrollable light level at a location.

ambient noise *(Acous.)*. The noise existing in a room or any other environment, e.g. the ocean.

ambient noise level *(Acous.)*. Random uncontrollable and irreducible noise level at a location or in a valve or circuit.

ambient temperature *(Elec.Eng.)*. Term used to denote temperature of surrounding air.

ambiophony *(Acous.)*. Technique of sound reproduction which creates an illusion to the listener of being in a very large room.

ambipolar *(Electronics)*. Said of any condition or property which applies equally to positive and negative charge carriers (e.g. positive or negative ions, holes, electrons) in a plasma or semiconductor.

ambisexual, ambosexual *(Biol.)*. Pertaining to both sexes; activated by both male and female hormones.

amblygonite *(Min.)*. Fluophosphate of aluminium and lithium, a rare white or greenish mineral, crystallizing in the triclinic system and found in pegmatites.

amblyopia *(Med.)*. Dimness of vision, from the action of noxious agents on the optic nerve or retina.

ambroid *(Min.)*. A synthetic amber formed by heating and compressing pieces of natural amber too small to be of value in themselves. Also called *amberoid, pressed amber*.

ambrosia *(Zool.)*. Certain Fungi which are cultivated for food by some Beetles (See **ambrosia beetle**): the pollen of flowers collected by social Bees and used in the feeding of the larvae.

ambrosia beetle *(Zool.)*. Beetles of the family Scotylidae which cultivate the fungus *Monilia candida* in galleries in wood to feed their larvae and themselves.

ambulacra *(Zool.)*. In Echinodermata, the radial bands of locomotor tube feet. adj. *ambulacral*

ambulacral grooves *(Zool.)*. Radially-arranged grooves containing the tube feet in Asteroidea.

ambulatory *(Arch.)*. Covered promenade area, particularly the curved aisle behind the altar in an apse.

ambulatory *(Zool.)*. Having the power of walking; used for walking.

ameiosis *(Biol.)*. Nonpairing of the chromosomes in synapsis (meiosis).

amelia *(Med.)*. A congenital abnormality in which one or more limbs are completely absent.

amelification *(Zool.)*. The formation of enamel.

ameloblast *(Zool.)*. A columnar cell forming part of a layer immediately covering the surface of the dentine, and secreting the enamel prisms in the teeth of higher Vertebrates.

amenorrhoea, amenorrhea *(Med.)*. Absence or suppression of menstruation.

ament *(Med.)*. One suffering from **amentia**; a mentally deficient person.

amentia *(Med.)*. Mental deficiency: failure of the mind to develop normally, whether due to inborn defect, or to injury or disease.

Amentiferae *(Bot.)*. See **Hamamelidae**.

amentum *(Bot.)*. A **catkin**. adj. *amentiform*.

American bond *(Build.)*. A form of bond in which every 5th or 6th course consists of headers, the other courses being stretchers. Very much used because it can be quickly laid.

american caisson *(Civ.Eng.)*. See **stranded caisson**.

american filter *(Min.Ext.)*. See **disk filter**.

American red gum *(For.)*. *Liquidambar styraciflua*, a tree of silky surface and irregular grain, Not very durable; used mainly for furniture, fittings and panelling. See **satin walnut**.

American Standard Wire Gauge *(Eng.)*. See **Brown and Sharpe Wire Gauge**.

American water turbine *(Eng.)*. See **mixed-flow water turbine**.

americium *(Chem.)*. Transuranic element, symbol Am, at. no. 95, half life 475 years. Of great value as a long life α-particle emitter, free of criticality hazards and γ-radiation, e.g. in laboratory neutron sources.

amesite *(Min.)*. A variety of septechlorite rich in magnesium and aluminium.

ametabolic *(Zool.)*. Having no obvious metamorphosis.

amethyst *(Min.)*. A purple form of quartz, used as a semiprecious gemstone.

amianthus *(Min.)*. A fine silky asbestos.

amicable numbers *(Maths.)*. See **friendly numbers**.

amicrons *(Chem.)*. Particles, of the order of nanometers, invisible in the ultramicroscope; they act as a nuclei for larger submicron particles.

amides *(Chem.)*. *Alkanamides*. A group of compounds in which the hydroxyl of the carboxyl group of acids has been replaced by the amide group $—NH_2$. They may be regarded as ammonia derivatives in which the hydrogen has been replaced by an acyl group.

amidines *(Chem.)*. Compounds derived from amides $R.CO.NH_2$, R.CO.NHR′, and $R.CO.NR'_2$ in which the oxygen has been replaced by the divalent imido residue NH or NR. Amidines are crystalline bases forming stable salts but are themselves readily hydrolysed.

amido group *(Chem.)*. $CO.NH_2$ group in amides, when replacing the hydroxyl in a carboxyl group.

amines *(Chem.)*. *Aminoalkanes*. Organic derivatives of ammonia NH_3 in which one or more hydrogen atoms are replaced by organic radicals. See **amino group**.

aminoacetic acid *(Chem.)*. *Aminoethanoic acid*. NH_2CH_2COOH (glycocoll, glycine), mp 230°C, colourless crystals, soluble in water, slightly in alcohol but not in ether. The simplest of the **amino acids**.

amino acids *(Chem.)*. *Aminoalkanoic acids*, generally with the amino group on the carbon adjacent to the carboxyl group. Some 24 of these are the building blocks of proteins: see appendix for formulae. All except glycine are chiral, and usually have the L- or S-configuration. Some amino acids can be synthesized in an animal body; these vary somewhat from species to species. Those which cannot be synthesized are called *essential amino acids*. For man, these are arginine, histidine, isoleucine, leucine, lysine, methionine, phenylalanine, threonine, tryptophan and valine.

aminoaldehydic resins *(Plastics)*. See **urea resins**

aminoglycosides *(Med.)*. Group of antibiotics particularly effective against **gram-negative bacteria**. Common members of this group are *gentamycin, neomycin, streptomycin* and *tobramycin*.

amino group *(Chem.)*. Essential component of the **amino acids**. See the appendix for the formulae of all amino acids.

aminoplastic resin *(Plastics)*. One derived from the reaction of urea, thiourea, melamine, or allied compounds (e.g. cyanamide polymers and diaminotriazines) with aldehydes, particularly formaldehyde (methanal).

amiodarone *(Med.)*. Iodine containing drug used in the treatment of complex arrhythmias, in particular those associated with **Wolf-Parkinson-White syndrome**.

amitosis, amitotic division *(Biol.)*. Direct division of the nucleus by constriction, without the formation of a spindle and chromosomes; direct nuclear division, occurring in the meganuclei of the *Ciliophora*. Cf. **mitosis, meiosis**.

amitryptyline *(Med.)*. **Tricyclic anti-depressant** drug used in the treatment of moderate to severe depression. Also tends to sedate.

ammeter *(Elec.Eng.)*. An indicating instrument for measuring the current in an electric circuit.

ammines *(Chem.)*. Complex inorganic compounds which result from the addition of one or more ammonia molecules to a molecule of a salt or similar compound.

ammonia *(Chem.)*. NH_3. A colourless, pungent gas, bp −33.5°C, extremely soluble in water and very soluble in alcohol. Formed by bacterial decomposition of protein, purines, and urea. Obtained mainly by the **Haber process**. Forms salts with most acids, and nitrides with metals. The liquefied gas is used as a refrigerant.

ammonia clock *(Phys.)*. An accurate clock controlled by the periodic inversion of the ammonia molecule with a frequency of 2.3786×10^{10} Hz. See **atomic clock**.

ammonia oxidation process *(Chem.)*. An important process for producing nitric acid by the catalytic oxidation of ammonia gas with air on the surface of platinum-rhodium gauze on 900°C.

ammonia-soda process *(Chem.)*. See **Solvay's-**.

ammonification, ammonization *(Bot.)*. The release of ammonia from amino acids (and ultimately protein) in decaying organic matter by soil bacteria. A step in the mineralization of nitrogen. Commonly followed, except in waterlogged and/or acid soils, by **nitrification**.

ammonite *(Geol.)*. An extinct fossil cephalopod abundant in Mesozoic rocks. The shell is chambered and most frequently coiled in a plane.

ammonium *(Chem.)*. The ion NH_4^+, which behaves in many respects like an alkali metal ion.

ammonium alum *(Chem.)*. See **alums**.

ammonium chloride *(Chem.)*. NH_4Cl. A white salt formed by the reaction of ammonia with hydrochloric acid. See **sal-ammoniac**.

ammonium dihydroxide phosphate *(Chem.)*. Piezo-electric crystal used in microphones and other transducers; it can withstand a temperature higher than can Rochelle salt.

ammonium hydroxide *(Chem.)*. NH_4OH. A solution of ammonia in water. Also *ammonia hydrate*.

ammonolysis *(Chem.)*. Solvolysis in liquid ammonia solution.

amnesia *(Med.)*. *Loss of memory*. Common in dissociation states of hysteria. In a concussed patient *retrograde amnesia* is loss of memory of events immediately preceding the concussion.

amniocentesis *(Biol.)*. Method for diagnosing foetal abnormalities in which foetal cells removed from amniotic fluid at about the 16th week of gestation are cultured and used in diagnostic assays, including **probing** with DNA sequences to detect disease-associated alleles.

amnion *(Zool.)*. In Insects, the inner cell envelope covering and arising from the edge of the germ band: in higher Vertebrates, one of the embryonic membranes, the inner fold of blastoderm covering the embryo, formed of ectoderm internally and somatic mesoderm externally.

Amniota *(Zool.)*. Those higher Vertebrates which possess an amnion during development, i.e. Reptiles, Birds, and Mammals. adj. *amniote*.

amniotic cavity *(Zool.)*. In *Amniota*, the space between the embryo and the amnion.

amniotic fluid *(Zool.)*. Liquid filling the amniotic cavity.

amniotic folds *(Zool.)*. Protrusions round the periphery of the blastoderm which give rise to the amnion and the chorion.

amoebiasis *(Med.)*. Disease caused by a Rhizopod parasite (*Entamoeba histolytica*), producing dysentery.

Amoebida *(Zool.)*. An order of Sarcodina the members of which extrude lobose pseudopodia and generally lack a skeleton, or have only a simple shell; their ectoplasm is never vacuolated.

amoebocyte *(Zool.)*. A metazoan cell having some of the characteristics of an amoeboid cell, especially as regards form and locomotion; in Porifera, a wandering cell of varied function; in Echinodermata, a wandering coelomic cell of excretory function; a leucocyte.

amoeboid *(Bot.,Zool.)*. Of a cell, having no fixed form, creeping, and putting out pseudopodia.

amoeboid movement *(Zool)*. Locomotion of an individual cell by means of pseudopodia.

amorphous *(Crystal.)*. Noncrystalline.

amorphous metal *(Phys.)*. A material with good conductivity, electrical and thermal, and with other metallic-like properties but with atomic arrangements that are not periodically ordered as in crystalline metal solids, e.g. metallic glass.

amorphous semiconductor *(Phys.)*. Semiconductors prepared in the amorphous state. They tend to have much lower electrical conductivities than those of their crystalline counterparts.

amorphous sulphur *(Chem.)*. Formed when sulphur vapour is cooled quickly. If the product so formed is treated with carbon disulphide and filtered, the amorphous sulphur is left on the filter as a white substance. It consists of long chains of sulphur atoms and reverts to *rhombic sulphur* at room temperature.

amosite *(Min.)*. A monoclinic amphibole form of **asbestos**, the name embodying the initials of the company exploiting this material in the Transvaal, viz. the 'Asbestos Mines of South Africa'.

amp *(Phys.)*. Deprecated abbrev. for *ampere*.

amperage *(Phys.)*. Current in amperes, more especially the rated current of an electrical apparatus, e.g. fuse or motor.

ampere *(Phys.)*. SI unit of electric current. Defined as that current which, if maintained in two parallel conductors of infinite length, of negligible cross-section, and placed 1 metre apart in vacuum, would produce between the conductors a force equal to 2×10^{-7} newtons per metre of length. One of the SI fundamental units.

ampere hour *(Phys.)*. Unit of charge, equal to 3600 coulombs, or 1 ampere flowing for 1 hour.

ampere-hour capacity *(Phys.)*. Capacity of an accumulator battery measured in ampere-hours, usually specified at a certain definite rate of discharge. Also applicable to primary cells.

ampere-hour efficiency *(Phys.)*. In an accumulator, the ratio of the ampere hour output during discharge to the ampere hours input during charge.

ampere-hour meter *(Elec.Eng.)*. A meter designed to record the product of current and time (ampere hours) for a given circuit or passing at a given point. If the voltage is constant, the meter can be calibrated as an energy (kilowatt hour) meter.

Ampère's law *(Phys.)*. That which states the magnetic field (H) in the neighbourhood of a conductor, length l, carrying current I; $\int H dl = I$.

Ampère's rule *(Phys.)*. Rule for the direction of the magnetic field associated with a current. The direction of the field is that of an advancing righthand screw when turning with the current. Alternatively, if the conductor is grasped with the right hand, the thumb pointing in the direction of the current, the fingers will curl around the conductor in the direction of the field.

Ampère's theory of magnetization *(Phys.)*. A theory based on the assumption that the magnetic property of a magnet is due to currents circulating in the molecules of the magnet.

ampere-turn *(Phys.)*. SI unit of magneto-motive force, which drives flux through magnetic circuits, arising from 1 ampere flowing round one turn of a conductor. Abbrev. *At*.

ampere-turn amplification, gain *(Phys.)*. Ratio of the load ampere-turns to the control ampere-turns in a magnetic amplifier.

ampere-turns per metre *(Phys.)*. SI unit of magnetizing force, magnetic field intensity.

amphetamine *(Med.)*. The sulphate is used as a drug for its vasomotor, respiratory, and stimulant effects. Popularly 'purple hearts'.

amphi- *(Chem.)*. Containing a condensed double aromatic nucleus substituted in the 2,6-positions. Gk. prefix, on both sides, around.

amphiaster *(Biol.)*. During cell-division by meiosis or mitosis, the two asters and the spindle connecting them.

Amphibia *(Zool.)*. Class of semi-aquatic chordates with larvae possessing gills and anamniotic eggs. Frogs, Toads, Salamanders.

amphibian *(Aero.)*. Aircraft capable of taking off and alighting on land or water; e.g. seaplane or flying boat with retractable landing gear, or landplane with **hydroskis**.

amphibious *(Bot.,Zool.)*. Adapted for both terrestrial and aquatic life.

amphiblastic *(Zool.)*. Of ova, showing complete but unequal segmentation.

amphiboles *(Min.)*. An important group of dark-coloured rock-forming silicates, including hornblende, the commonest.

amphibolic *(Zool.)*. Capable of being turned backwards or forwards, as the 4th toe of Owls.

amphibolite *(Geol.)*. A crystalline, coarse grained rock, containing amphibole as an essential constituent, together with feldspar and frequently garnet; like hornblende-schist, formed by regional metamorphism of basic igneous rocks, but not foliated.

amphicoelous *(Zool.)*. Having both ends concave, as the centra of the vertebrae of Selachii and a few Reptiles.

amphicondylous, amphicondylar *(Zool.)*. Having two occipital condyles.

amphicribal bundle *(Bot.)*. A vascular bundle in which a central strand of xylem is surrounded by phloem.

amphidentate *(Chem.)*. Of a ligand, capable of coordinating through one of two different atoms; e.g. $S{=}C{=}N^-$.

amphidiploid *(Bot.)*. An **allopolyploid** containing the diploid set of chromosomes from each of two species.

amphimixis *(Bot.)*. True sexual reproduction, with the fusion of two gametes to form a zygote. Cf. **apomixis**.

Amphineura *(Zool.)*. A class of bilaterally symmetrical Mollusca in which the foot, if present, is broad and flat, the mantle is undivided, and the shell is absent or composed of eight valves. Coat-of-Mail Shells etc.

amphiont *(Zool.)*. See **zygote**.

amphipathic *(Chem.,Phys.)*. Descriptive of unsymmetrical molecular group, one end being hydrophilic and the other hydrophobic (wetting and nonwetting).

amphiphloic *(Bot.)*. A stele, having phloem on both sides of the xylem, e.g. solenostele.

amphiplatyan *(Zool.)*. Having both ends flat, as in certain types of vertebral centrum.

amphipneustic *(Zool.)*. Possessing both gills and lungs: in dipterous larvae, having the prothoracic and posterior abdominal spiracles only functional.

Amphipoda *(Zool.)*. An order of Malacostrica in which the carapace is absent, the eyes are sessile, and the uropods styliform; the body is laterally compressed. They show great variety of habitat, being found on the shore, in the surface waters of the sea, in fresh water, and in the soil of tropical forests. Some are parasitic. Whale Lice, Sandhoppers, Skeleton Shrimps etc.

amphipodous *(Zool.)*. Having both ambulatory and natatory appendages.

amphiprotic *(Chem.)*. Having both protophilic (i.e. basic) and protogenic (i.e. acidic) properties.

amphiprotic solvent *(Chem.)*. Solvent capable of showing either protophilic (i.e. acid generating) or protogenic (base generating) properties to different solutes, e.g. water with HCl and NH_3, respectively.

amphirhinal *(Zool.)*. Having two external nares.

amphistomatal, amphistomatic *(Bot.)*. Leaf etc., having stomata on both surfaces. Cf. **epistomatal**.

amphistomous *(Zool.)*. Having a sucker at each end of the body; as leeches.

amphitheatre *(Arch.)*. An oval or circular building in which the spectators' seats surround the arena or open space in which the spectacle is presented, the seats rising away from the arena.

amphithecium *(Bot.)*. Outer layer(s) of a developing sporophyte of a Bryophyte giving rise to capsule wall. Cf. **endothecium**.

amphitrichous *(Bot.,Zool.)*. Having a flagellum at each end of the cell.

amphitropous *(Bot.)*. An ovule bent like a 'V' and attached, to its stalk, near the middle of its concave side.

amphivasal bundle *(Bot.)*. A vascular bundle in which a central strand of phloem is surrounded by xylem.

ampholines *(Biol.)*. Mixtures of aliphatic amino acids with a range of iso-electric points which are used to establish the pH gradients used in **iso-electric focusing**.

amphoric *(Acous.)*. Like the sound made by blowing across a narrow-necked vase.

amphoteric *(Chem.)*. Having both acidic and basic properties, e.g. aluminium oxide, zinc oxide, which form salts with acids and with alkalis.

ampicillin *(Med.)*. A semisynthetic penicillin with a broad range of activity against those bacteria causing bronchitis, pneumonia, gonorrhoea, certain forms of meningitis, enteritis, biliary and urinary tract infections.

amplexicaul *(Bot.)*. Said of a sessile leaf with its base clasping the stem horizontally.

amplexiform *(Zool.)*. A type of wing-coupling formed in some Lepidoptera, whereby the wings are coupled simply by overlapping basally.

amplexus *(Zool.)*. See **copulation**.

amplidyne *(Elec.Eng.)*. A rotary magnetic amplifier of high gain which is used in servo systems.

amplification *(Biol.)*. The process by which multiple copies of genes or DNA sequences are formed.

amplifier *(Elec.Eng.,Telecomm.)*. A circuit or assembly which uses a valve, transistor or solid-state device, magnetic contrivance or any active device to increase the strength of a signal without appreciably altering its characteristics. An amplifier transfers power from an external source to the signal, unlike a transformer; therefore, an amplifier with equal input and output impedances might exhibit current or voltage gain, or it may have unity current or voltage gain with an increase or decrease in impedance from input to output. Analogous devices employing pneumatic and hydraulic systems are sometimes used.

amplitude *(Maths.)*. A confusing term occasionally used for the argument of a complex number. It would more naturally be taken to mean the modulus, and is best avoided.

amplitude *(Phys.)*. The maximum value of a periodically varying quantity during a cycle; e.g., the maximum displacement from its position of a vibrating particle, the maximum value of an alternating current (see **peak value**), or the maximum displacement of a sine wave. See also **double amplitude**.

amplitude discriminator *(Telecomm.)*. See **pulse-height discriminator**.

amplitude distortion *(Telecomm.)*. Distortion of waveform arising from the nonlinear static or dynamic response of a part of a communication system, the output amplitude of the signal at any instant not having a constant proportionality with the corresponding input signal.

amplitude limiter, discriminator *(Image Tech.)*. One which separates synchronizing signals in a TV signal from the video (picture) signal.

amplitude peak *(Telecomm.)*. Maximum positive or negative excursion from zero of any periodic disturbance.

ampoule *(Med.)*. A small, sealed glass capsule for holding measured quantities of vaccines, drugs, serums etc. ready for use.

ampoule tubing *(Glass)*. Tubing of special composition suited to the manufacture of ampoules. It must work well in the blowpipe flame, and must resist the action of the materials stored in the ampoule.

ampulla *(Med.)*. Dilated end of a canal or duct.

ampulla *(Zool.)*. Any small membranous vesicle: in Vertebrates, the dilation housing the sensory epithelium at one end of a semicircular canal of the ear; in Mammals, part of a dilated tubule in the mammary gland; in Fish, the terminal vesicle of a neuromast organ; in Echinodermata, the internal expansion of the axial sinus below the madreporite; in Ctenophora, one of a pair of small sacs forming part of the aboral sense organ. adj. *ampullary*.

amu *(Phys.)*. Abbrev. for *atomic mass unit*.

amydricaine hydrochloride *(Med.)*. Used as a local anaesthetic, particularly in ophthalmic practice.

amyelinate *(Zool.)*. Of nerve-fibres, nonmedullated, lacking a myelin sheath.

amygdala *(Zool.)*. A lobe of the cerebellum: one of the palatal tonsils.

amygdale, amygdule *(Geol.)*. An almond-shaped infilling (by secondary minerals such as agate, zeolites, calcite, etc.) of elongated steam cavities in igneous rocks.

amygdalin *(Chem.)*. $C_{20}H_{27}O_{11}N$, colourless prisms, mp 200°C, a glucoside found in bitter almonds, in peach and cherry kernels.

amyl acetate *(Chem.)*. *Pentyl ethanoate*. $CH_3.CO.O.C_5.H_{11}$, colourless liquid, of ethereal pear-like odour, bp 138°C *iso*amyl acetate is also known under the name of *pear oil*. It is used for fruit essences and is an important solvent for nitrocellulose.

amyl alcohol *(Chem.)*. *Pentanol*. $C_5H_{11}OH$, the fraction of fusel oil that distils about 131°C. There are 8 isomers possible and known, viz. 4 primary, 3 secondary, 1 tertiary amyl alcohol. The important isomers are: *iso*amyl alcohol, (−)amyl alcohol, tertiary amyl alcohol (amylene hydrate).

amyl-, amylo- *(Bot.)*. Prefix used to mean of starch.

amylase *(Biol.)*. Enzyme which hydrolyses the internal 1,4-glycosidic bonds of starch. The amylase found in human saliva is known as *ptyalin*.

amyl group *(Chem.)*. The pentyl radical C_5H_{11}.

amyl nitrite *(Chem.)*. $C_5H_{11}.O.NO$, the nitrous acid ester of *iso*-amyl alcohol, a yellowish liquid, bp 98°C, of pleasant odour. Intermediate for the preparation of nitroso- and of diazo-compounds. Used in medicine as a vasodilator, esp. for acute angina.

amylobarbitone *(Med.)*. The sodium compound was formerly used as an intermediate-acting hypnotic and sedative.

amyloid *(Chem.)*. A starchlike cellulose compound, produced by treatment of cellulose with concentrated sulphuric acid for a short period.

amyloid *(Immun.)*. An insoluble fibrillary material deposited around blood vessels. This can arise in different ways, but one form which occurs with myelomas making light chains or in very prolonged infections with raised immunoglobulin levels, results from aggregates of free light chains.

amyloidosis *(Med.)*. The deposition of amyloid in the organs and tissues of the body, often as a result of chronic infection.

amylolytic *(Zool.)*. Starch-digesting.

amylopectin *(Bot.)*. A polymer, α-1→4 linked with α1→6 branches, of glucose. A constituent of starch.

amylose *(Bot.,Chem.)*. The sol constituent of starch paste; linear polymer of glucose units having α-1→4 glucosidic bonds. Cf. **amylopectin**.

amylum *(Bot.)*. Starch.

amyotrophic lateral sclerosis *(Med.)*. A nervous disease in which atrophy of the muscle follows degenerative changes in the motor fibres of the spinal cord and brain.

amyotrophy *(Med.)*. Wasting or atrophy of muscle.

an *(Genrl.)*. Prefix from Gk. *an*, not. See also **ap-**.

ana *(Genrl.)*. Prefix from Gk. *ana*, up, anew.

anabatic wind *(Meteor.)*. A local wind blowing up a slope heated by sunshine, and caused by the difference in density between the warm air in contact with the ground and the cooler air at corresponding heights in the **free atmosphere**.

anabiosis *(Biol.)*. A temporary state of reduced metabolism in which metabolic activity is absent or undetectable. See also **cryptobiosis**.

anabolic *(Biol.)*. Metabolic events which lead to the synthesis of body constituents.

anabolism *(Biol.)*. The chemical changes proceeding in living organisms with the formation of complex substances from simpler ones. adj. *anabolic*.

anabolite *(Biol.)*. A substance participating in anabolism.

anadromous *(Zool.)*. Having the habit of migrating from more dense to less dense water to breed, generally from oceanic to coastal waters, or from salt water to fresh water; as the Salmon.

anaemia *(Vet.)*. Lit. *no blood. Primary anaemia*, failure to produce red blood cells or haemoglobin. *Secondary anaemia*, blood loss. See **feline infectious anaemia, infectious anaemia of horses**.

anaemia, anemia *(Med.)*. Diminution of the amount of total circulating haemoglobin in the blood.

anaerobe *(Biol.)*. An organism which can grow in the absence or near absence of oxygen. Facultative anaerobes can utilize free oxygen; obligate anaerobes are poisoned by it. adj. *anaerobic*.

anaerobic *(Med.)*. Living in the absence of oxygen. Anaerobic respiration is the liberation of energy which does not require the presence of oxygen.

anaerobic respiration *(Biol.)*. See **respiration**.

anaerobiosis *(Biol.)*. Existence in the absence of oxygen. adj. *anaerobic*.

anaesthesia, anesthesia *(Med.)*. Correctly, loss of feeling but often applied to the technique of pain relief for surgical procedures.

anaesthetic, anesthetic *(Med.)*. Insensible to touch (loosely, also to pain and temperature): a drug which produces insensibility to touch, pain, and temperature, with or without loss of consciousness.

anaesthetist *(Med.)*. One skilled in the administration of anaesthetic drugs. US, *anesthesiologist*.

anafront *(Meteor.)*. A situation at a front, warm or cold, where the warm air is rising relative to the **frontal zone**.

anaglyph *(Image Tech.)*. Pair of stereoscopic images reproduced in two colours, generally red and blue-green, for viewing with corresponding colour filters, one for each eye, to give a three-dimensional sensation.

anal *(Zool.)*. See **anus**.

anal cerci *(Zool.)*. In Insects, sensory appendages of one of the posterior abdominal somites, generally the 11th, retained throughout life.

anal character *(Behav.)*. In psychoanalytic theory, an adult personality derived from unresolved conflicts (**fixation**) during the anal stage of psychosexual development. It is characterised by a **reaction-formation** against impulsiveness, resulting in personality traits of an extremely controlled and compulsive type (e.g. extreme tidiness, parsimony). See **anal phase**.

analcime, analcite *(Min.)*. Hydrated aluminium silicate, a member of the zeolite group, occurring in some igneous and sedimentary rocks, but particularly in cavities in lavas.

analeptic *(Med.)*. Having restorative or strengthening properties.

analgesia *(Med.)*. Loss of sensibility to pain.

analgesic *(Med.)*. A drug which relieves pain.

anal-gland disease *(Vet.)*. Inflammation of the perianal glands, and inflammation or impaction of the canine anal sacs.

anallatic lens *(Surv.)*. Special lens which, when correctly placed between the object glass and the eyepiece lens of a tacheometric telescope, optically reduces the additive constant for the tacheometer to zero.

anallatic telescope *(Surv.)*. One which, when used in tacheometry, has a zero additive constant.

anallatism *(Surv.)*. See **centre of anallatism**.

analogous colour scheme *(Build.)*. A decorative colour scheme comprising colours which lie close to one another in the chromatic circle.

analogous organs *(Bot.)*. Organs which are similar in appearance and/or function but which are neither equivalent morphologically nor of common evolutionary origin e.g. foliage leaves and cladodes; the wings of birds and insects. Cf. **homologous organs**.

analogue *(Chem.)*. A compound which may be considered to be derived from another by the substitution of saturated aliphatic groups for hydrogen, e.g. ethanol is an analogue of methanol.

analog(ue) *(Meteor.)*. A previous *weather map* similar to the current map. The developments following the analogue aid forecasting.

analogue *(Telecomm.)*. Any form of transmission of information where the transmitted signal's information-bearing characteristic (usually amplitude or frequency) is varied in direct proportion to the intensity of the sound, or brightness of pictures etc., which it is desired to communicate. Cf. **digital**.

analogue computer *(Comp.)*. Computer which uses continuous physical variables such as voltage or pressure to represent and manipulate the measurements it handles. Now usually a special purpose computer.

analogue filter *(Telecomm.)*. One suitable for filtering analogue signals, i.e. those which are continuous with time. Cf. **digital filter**.

analogue to digital converter *(Comp.,Elec.Eng.)*. Device for converting analogue signals into digital ones for subsequent computer processing or for transmission over data links. Also *digitizer*, *A to D converter*.

analogue watch, clock *(Horol.)*. The traditional watch or clock with rotating hands. Cf. **digital watch, clock**.

analogy *(Bot.,Zool.)*. Likeness in function but not in evolutionary origin, e.g. tendrils, which may be modified leaves, branches, inflorescences; the wings of Birds and of Insects. adj. *analogous*. Cf. **homology**.

analogy *(Electronics)*. Correspondence of pattern or form between mechanical and electrical quantities, or vice versa; e.g. a network of resistance, capacitance and inductance can be made to represent a complex mechanical system, or a stretched rubber membrane for the potential distributions between electrodes in electronic tubes.

anal phase (stage) *(Behav.)*. In psychoanalytic theory, the second phase of *psychosexual development*, during which the focus of pleasure is on activities related to retaining and expelling the faeces; occurs in the second year of life. See **anal character**.

analplerotic *(Biol.)*. Reactions which replenish deficiencies of metabolic intermediates, e.g. the formation of oxaloacetate by the carboxylation of pyruvate.

anal suture *(Zool.)*. In the posterior wings of some Insects, a line of folding, separating the anal area of the wing from the main area.

analyser *(Chem.)*. The second Nicol prism in a polarimeter; when rotated 90° relative to polarizer it will not allow polarized light to pass.

analysis *(Maths.)*. The theory of limits, including sequences, series, differentiation and integration. The practical methods of calculus are applied analysis.

analysis meter *(Telecomm.)*. A registering meter used to determine the loading of groups of circuits with calls, particularly for determining the correctness or otherwise of grading.

analysis of variance *(Stats.)*. The partition of the total variation in a set of observations into components corresponding to differences between and within sub-classifications of the data, used as a method of comparing sub-classification means.

analyst *(Chem.)*. A person who carries out any process of analysis.

analytical engine *(Comp.)*. Name for the first general purpose **digital computer** designed about 1835 by Charles Babbage but only partly built.

analytical geometry *(Maths.)*. The application of the methods of algebra to geometry by utilizing the concept of co-ordinates.

analytical reagent *(Chem.)*. Also *AR*. An indication of a definite standard of purity.

analytic continuation *(Maths.)*. A process for extending the range of definition of a function of a complex variable. Starting with a first Taylor series about a point A representing the function, a second Taylor series about a point B within the first circle of convergence can be obtained and, providing the line AB does not intersect the first circle of convergence at a singularity, the circle of convergence of the second Taylor series will extend beyond the first circle. This process can be repeated indefinitely and it can be shown that, if a particular region can be reached by two different routes, the two Taylor series obtained will be identical, providing no singularities lie between the two routes. Some functions have their own *natural boundaries* beyond which analytic continuation is impossible. Such a function is

$$g(z) = 1 + \sum_{r=1}^{\infty} z^{2^r}$$

which has an infinite number of singularities on every arc of its circle of convergence $|z| = 1$.

analytic function *(Maths.)*. A function of a complex variable is analytic in a region if it is single-valued and differentiable at all points of the region. The terms *holomorphic, monogenic* and *regular* are sometimes used. Sometimes a function is said to be analytic in a region if it is single termed and differentiable at all but a finite number of points (*singularities*) of the region. In this case the term *regular* is frequently contrasted with *analytic* to mean that there are no singularities. When the singularities are all poles the term *meromorphic* is sometimes used. A function that is analytic throughout the whole complex plane, except possibly at a finite number of singularities, is called an *entire* or an *integral* function.

anamnesis *(Med.)*. The recollection of past things: the patient's recollections of symptoms and past illnesses.

anamnestic *(Immun.)*. Antibody response following antigenic stimulation in which antibodies occur reactive with an antigen previously encountered but different from that which elicited the response.

Anamniota *(Zool.)*. Vertebrates without an amnion during development. i.e. Amphibia and Fish.

anamniotic *(Zool.)*. Lacking an amnion during development. Also *anamniote*.

anamorph *(Bot.)*. An asexual or imperfect stage of a fungus (especially Deuteromycotina). Cf. **teliomorph**.

anamorphic lens *(Image Tech.)*. Lens with cylindrical elements giving different magnification in horizontal and vertical directions. In **wide-screen** cinematography the image is compressed laterally in the camera and expanded to compensate in projection. The equivalent term *anamorphotic* is rare.

Ananke *(Astron.)*. The twelfth natural satellite of **Jupiter**.

anaphase *(Biol.)*. The stage in mitotic or meiotic nuclear division when the chromosomes or *half-chromosomes* move away from the equatorial plate to the poles of the spindle; more rarely, all stages of mitosis leading up to the formation of the chromosomes.

anaphoresis *(Chem.)*. The migration of suspended particles towards the anode under the influence of an electric field.

anaphylatoxin *(Immun.)*. Peptides released from **complement** components C3 and C5 during complement activation which act on mast cells to release histamine etc. as in anaphylaxis.

anaphylaxis, anaphylactic shock *(Immun.)*. An acute immediate hypersensitivity reaction following administration of an antigen to a subject resulting from combination of the antigen with IgE on mast cells or **basophils** which causes these cells to release histamine and other vasoactive agents. An acute fall in blood pressure may be so severe as to be fatal. Other symptoms include bronchospasm, laryngeal oedema and urticaria.

anaplasia *(Med.)*. Loss of the differentiation of a cell associated with proliferative activity; a characteristic of a malignant tumour.

anaplasmosis *(Vet.)*. *Gall-sickness*. A disease of cattle caused by infection by protozoa of the genus *Anaplasma*, and characterized by fever, anaemia, and jaundice. The protozoa are found in the red blood corpuscles and are transmitted by ticks, biting flies, and mosquitoes. Infections of sheep and swine also occur.

anapophysis *(Zool.)*. In higher Vertebrates, a small process just below the postzygapophysis which strengthens the articulation of the lumber vertebrae.

anapsid *(Zool.)*. Having the skull completely roofed over, i.e. having no dorsal foramina other than the nares, the orbits, and the parietal foramen.

anarthrous *(Zool.)*. Without distinct joints.

anasarca *(Med.)*. Excessive accumulation of fluid (dropsy) in the skin and subcutaneous tissues.

Anaspida *(Zool.)*. A subclass of the reptiles containing the oldest known forms, characterized by the temporal region of the skull lacking the fenestrae. Turtles.

anastigmat lens *(Phys.)*. A photographic objective designed to be free from astigmatism on at least one extra-axial zone of the image plane.

anastomosis *(Bot.)*. A cross-connection.

anastomosis *(Med.)*. A communication between two blood vessels: an artificial commmunication, made by operation, between any two parts of the alimentary canal. pl. *anastomoses*.

anastomosis *(Zool.)*. The formation of a meshwork of blood vessels or nerves: the union of blood vessels or nerves formed by the splitting of a common trunk.

anatase *(Min.)*. One of the three naturally occurring forms of crystalline titanium dioxide, of tabular or bipyramidal habit. See also **octahedrite**.

anatomy *(Bot.,Zool.)*. (1) The study of the form and structure of animals and plants; it includes the study of minute structures, and thus includes **histology**. (2) Dissection of an organized body in order to display its physical structure.

anatropous *(Bot.)*. An inverted ovule, so that the micropyle is next to the stalk.

anaxial *(Zool.)*. Asymmetrical.

anchor bolt *(Build.)*. A bolt used to secure frameworks, stanchion bases, etc., to piers or foundations, and having usually a large plate washer built into the latter as anchorage.

anchor clamp *(Elec.Eng.)*. A fitting attached to the overhead contact wire of a tramway or railway to support the wire, and also to take the longitudinal tension and prevent movement of the wire in a direction parallel to the track.

anchor escapement *(Horol.)*. See **recoil escapement**.

anchor-gate *(Civ.Eng.)*. A heavy gate, such as a canal lock gate, which is supported at its upper bearing by an anchorage in the masonry such as an *anchor and collar*.

anchoring *(Print.)*. A method of fastening plates to mounts when the usual flange is not available; a thin bolt is passed through the mount and secured with a nut. Superseded by sweating to metal mount or by using adhesive film on all kinds of mount.

anchor string *(Min.Ext.)*. Length of **casing** run into the top of wells and often cemented in to prevent a **blowout** outside the casing. It provides fixings for the well-head equipment.

anconeal *(Zool.)*. Pertaining to, or situated near, the elbow.

anconeus *(Zool.)*. An extensor muscle of the arm attached in the region of the elbow.

AND *(Comp.)*. A logical operator such that (p AND q) written $\overline{p \cdot q}$ takes the value TRUE if p is TRUE and q is TRUE otherwise (p AND q) takes the value FALSE. See **logical operation**.

andalusite *(Min.)*. One of several crystalline forms of aluminium silicate; a characteristic product of the contact metamorphism of argillaceous rocks. Orthorhombic. See also **chiastolite**.

andesine *(Min.)*. A member of the plagioclase group of minerals, with a small excess of sodium over calcium: typical of the intermediate igneous rocks.

andesite *(Geol.)*. A fine-grained igneous rock (usually a lava), of intermediate composition, having plagioclase as the dominant feldspar.

AND gate *(Comp.)*. Gate producing an output signal only if all inputs are energized simultaneously, i.e. output signal is 1 when all input signals are 1. See **logical operations**. Also *AND element*.

andiron *(Build.)*. A metal support for wood in an open fire. Also called a *firedog*.

andradite *(Min.)*. Common calcium-iron garnet. Mainly dark brown to yellow or green. See also **demantoid, melanite, topazolite**.

andro- *(Genrl.)*. Prefix from Gk. *aner, andros*, man, male.

androconia *(Zool.)*. In certain male Lepidoptera, scent scales serving to disseminate the **pheromones** which serve the purpose of sexual attraction.

androcyte *(Bot.)*. Cells in an antheridium which will metamorphose to form antherozoids.

androdioecious *(Bot.)*. A species, having some individuals with male flowers only and others hermaphrodite flowers only. Cf. **dioecious**.

androecium *(Bot.)*. The male part of a flower, consisting of one or more stamens. Cf. **gynoecium**.

androgen *(Biol.)*. General term for a group of male sex hormones, which e.g. stimulate the growth of male secondary sex characteristics. Cf. **oestrogen**.

androgenesis *(Bot.,Zool.)*. (1) Development from a male cell. (2) Development of an egg after entry of male germ cell without the participation of the egg nucleus.

androgenic *(Med.)*. Having the effects of a male sex hormone.

androgynophore *(Bot.)*. Same as androphore.

androgynous *(Bot.)*. Bearing staminate and pistillate flowers on distinct parts of the same inflorescence: having the male and female organs on or in the same branch of the thallus.

Andromeda nebula *(Astron.)*. The spiral galaxy M31 in Andromeda, visible to the naked eye. Of comparable size with our own Galaxy, it is about 2 million light-years distant.

andromonoecious *(Bot.)*. A species in which all of the plants bear both male and hermaphrodite flowers. Cf. **monoecious**.

androphore *(Bot.)*. An elongation of the receptacle of the flower between the corolla and the stamens.

androsporangium *(Bot.)*. Sporangium in which androspores are produced.

androspore *(Bot.)*. In heterosporous plants, same as **microspore**. Cf. **gynospore**.

anecdysis *(Zool.)*. The intermoult period in Arthropoda.

anechoic room *(Acous.)*. One in which internal sound reflections are reduced to an ineffective value by extremely high sound absorption (e.g. glass-fibre wedges).

anelasticity *(Phys.)*. Any deviation from an ideal internal structure of a body which would dampen or attenuate an elastic wave therein.

anelectric *(Phys.)*. Term once used for a body which does not become electrified by friction.

anemia, anesthesia *(Med.)*. See **anaemia, anaesthesia**.

anemochorous *(Bot.)*. Seeds or other *propagules* dispersed by wind.

anemograph *(Meteor.)*. See **anemometer**.

anemometer *(Eng.)*. An instrument for measuring the rate of flow of a gas, either by mechanical or electrical methods.

anemometer *(Meteor.)*. An instrument for measuring the speed of the wind. A common type consists of four hemispherical cups carried at the ends of four radial arms pivoted so as to be capable of rotation in a horizontal plane, the speed of rotation being indicated on a dial calibrated to read wind speed directly. An *anemograph* records the speed and sometimes the direction.

anemophily *(Bot.)*. Pollination by means of wind. Dispersal of spores by wind. adj. *anemophilous*.

anemotaxis *(Zool.)*. Orientation to an odour source based upon wind direction.

anencephaly *(Med.)*. Neural tube defect in which skull and cerebral hemispheres fail to develop. adj. *anencephalic*.

anergy *(Immun.)*. Absence of ability to give the expected allergic responses, especially delayed-type hypersensitivity. Occurs when the lymphocytes or monocytes needed are absent or suppressed.

aneroid barometer *(Meteor.,Surv.)*. One having a vacuum chamber or syphon bellows of thin corrugated metal, one end diaphragm of which is fixed, the other being connected by a train of levers to a scale pointer which records the movements of the diaphragm under changing atmospheric pressure.

anethole *(Chem.)*. *p-Propenyl anisole*; $CHMe{:}CH.C_6H_4.OMe$. An ether forming the chief constituent of oil of aniseed, an essential oil.

aneuploid *(Biol.)*. A cell or individual with missing or extra chromosomes or parts of chromosomes, then called a *segmental aneuploid*.

aneurysm *(Med.)*. Pathological dilatation, fusi-form or saccular, of an artery.

angel beam *(Arch.)*. A horizontal member of a mediaeval roof truss, usually decorated with angels carved on the member.

angels *(Radar)*. Radar echoes from an invisible and sometimes undefined origin. High-flying birds, insect swarms and certain atmospheric conditions can be responsible.

angina pectoris *(Med.)*. A condition characterized by the sudden onset of pain or crushing sensation in the chest which may radiate to the throat and arms. Frequently provoked by exercise and due to the narrowing of the coronary arteries.

angio- *(Genrl.)*. Prefix from Gk. *angion*, denoting a case or vessel.

angioblast *(Zool.)*. An embryonic mesodermal cell from which the vessels and early blood cells are derived.

angiocardiography *(Radiol.)*. The radiological examination of the heart and great vessels after injection of a **contrast medium**.

angiography *(Radiol.)*. The study of the cardio-vascular system by means of radio-opaque media.

angiology *(Biol.)*. The study or scientific account of the anatomy of blood and lymph vascular systems.

angioma *(Med.)*. See **haemangioma**.

angioneurotic oedema *(Med.)*. An immunologically mediated disease which produces dramatic and sometimes life-threatening swelling of the eyelids, lips, mucous membranes of the mouth and respiratory tract.

angioplasty *(Radiol.)*. A procedure where a balloon **catheter** is inserted into a blood vessel and the balloon inflated to widen a narrowed segment.

angiosperms, flowering plants *(Bot.)*. Group containing those *seed plants* in which the ovules are enclosed by carpels, the pollen germinating on a stigma and the pollen tube growing to the ovule; there is characteristically double fertilisation; the xylem usually has vessels. Contains ca 220 000 spp. in two classes, *Dicotyledons* and *Monocotyledons*. Alternatively classified as a class, Angiospermae, or as a division Anthophyta or Magnoliaphyta.

angiotensin-converting enzyme inhibitors (ACE inhibitors) *(Med.)*. Drugs that inhibit the enzyme that converts the inactive form of **angiotensin** (I) to the active form (angiotensin II) and are used in the treatment of hypertension and heart failure. Captopril and Enalapril are common examples.

angle *(Eng.)*. See **angle iron**.

angle *(Maths.)*. The inclination of one line to another, measured in degrees (of which there are 360 to one complete revolution), or in radians (of which there are 2π to one complete revolution).

angle bars *(Print.)*. On rotary presses, bars at an angle to transfer one or more webs of paper over each other, or the web to the other side of the press, or at right angles to its previous direction. Sometimes called *turner bars*.

angle bead *(Build.)*. A small rounded moulding placed at an angle formed by plastered surfaces to protect from damage.

angle bearing *(Eng.)*. A shaft-bearing in which the joint between base and cap is not perpendicular to the direction of the load, but is set at an angle.

angle block *(Build.)*. A small wooden block used in woodwork to make joints, especially right-angle joints, more rigid.

angle board *(Build.)*. One used as a gauge by which to plane boards to a required angle between two faces.
angle brace *(Build.)*. (1) Any bar fixed across the inside of an angle in a framework to render the latter more rigid. Also *angle tie, dragon tie.* (2) A special tool for drilling in corners where there is not room to use the cranked handle of the ordinary brace.
angle bracket *(Build.)*. A bracket projecting from the corner of a building beneath the eaves, and not at right-angles to the face of the wall.
angle bracket *(Eng.)*. A bracket consisting of two sides set at right angles, often stiffened by a gusset.
angle cleat *(Build.)*. A small bracket formed of angle iron, used to support or locate a member in a structural framework.
angle closer *(Build.)*. Loose term for **closer** cut at an angle.
angle cutter *(Paper)*. A machine in which the cross cut knife is not at a right angle to the edge of the reel, for cutting sheets of paper from the reel. The parallelogram-shaped sheets were originally intended for conversion into banker envelopes.
angled deck *(Aero.)*. The flight deck of an aircraft carrier prolonged diagonally from one side of the ship, so that aircraft may fly-off and land-on without interference to or from aircraft parked at the bows. U.S. *canted deck.*
angledozer *(Civ.Eng.)*. See **bulldozer**.
angle drilling *(Min.Ext.)*. Technique for drilling at an angle to an existing bore, achieved by special **downhole** equipment, either to straighten a bore, gather oil from a wide area to a production platform or to reach otherwise inaccessible formations, e.g. under a city. Also *deviated drilling*. See **slant rig**.
angle elevation *(Surv.)*. The vertical angle measured above the horizontal, from the surveyor's instrument to the point observed.
angle float *(Build.)*. A plasterer's trowel, specially shaped to fit into the angle between adjacent walls of a room.
angle gauge *(Build.)*. A tool which is used to set off and test angles in carpenter's, bricklayer's and mason's work.
angle iron *(Eng.)*. Mild steel bar rolled to the cross-section of the letter L, much used for light structural work. Also called *angle, angle bar, angle steel.*
angle modulation *(Telecomm.)*. Any system in which the transmitted signal varies the phase-angle of an otherwise steady carrier frequency, i.e. phase and frequency modulation.
angle of acceptance *(Phys.)*. The horizontal angle within which light rays should reach a window to ensure adequate penetration.
angle of advance *(Eng.)*. (1) The angle in excess of 90° by which the eccentric throw of a steam-engine valve gear is in advance of the crank. (2) The angle between the position of ignition and outer dead centre in a spark-ignition engine; optimizes combustion of the fuel.
angle of approach light *(Aero.)*. A light indicating an approach path in a vertical plane to a definite position in the landing area.
angle of arrival *(Telecomm.)*. Angle of elevation of a downcoming wave.
angle of attack *(Aero.)*. The angle between the *chord line* of an aerofoil and the relative airflow, normally the immediate flight path of the aircraft. Also erroneously called *angle of incidence.*
angle-of-attack indicator *(Aero.)*. An instrument which senses the true angle of incidence to the relative airflow and presents it to the pilot on a graduated dial or by means of an indicating light.
angle of bank *(Aero.)*. See **angle of roll**.
angle of bite *(Eng.)*. Maximum angle obtainable between the roll radius where it first contacts the metal and the line joining the centres of the two opposing rolls, when rolling metal. Also *angle of nip.*
angle of contact *(Eng.)*. The angle subtended at the centre of a pulley by that part of the rim in contact with the driving belt.
angle of contact *(Phys.)*. The angle made by the surface separating two fluids (one of them generally air) with the wall of the containing vessel, or with any other solid surface cutting the fluid surface. For liquid-air surfaces, the angle of contact is measured in the liquid.
angle of cut-off *(Phys.)*. The largest angle below the horizontal at which a reflector allows the light-source to be visible when viewed from a point outside the reflector.
angle of deflection *(Electronics)*. That of the electron beam in a CRT.
angle of departure *(Telecomm.)*. Angle of elevation of maximum emission of electromagnetic energy from an antenna.
angle of depression *(Surv.)*. The vertical angle measured below the horizontal, from the surveyor's instrument to the point observed; also *plunge angle.*
angle of deviation *(Phys.)*. The angle which the incident ray makes with the emergent ray when light passes through a prism or any other optical device.
angle of dip *(Geol.)*. See **dip**.
angle of flow *(Elec.Eng.)*. Angle, or fraction of alternating cycle, during which current flows, e.g. in a thyristor. Also *conduction angle.*
angle of friction *(Eng.)*. The angle between the normal to the contact surfaces of two bodies, and the direction of the resultant reaction between them, when a force is just tending to cause relative sliding.
angle of heel *(Ships)*. The angle through which a floating vessel or pontoon tilts owing to eccentric placing of loads etc.; the angle of inclination of a ship due to 'rolling' or to a 'list'. It is the angle formed between the transverse centre line of the ship when on 'even keel' and when inclined.
angle of incidence *(Aero.)*. Angular setting of any aerofoil to a reference axis. See **angle of attack**.
angle of incidence *(Phys.)*. The angle which a ray makes with the normal to a surface on which it is incident.
angle of lag, angle of lead *(Elec.Eng.)*. In a.c. circuit theory the phase angle by which the current lags behind, or leads ahead of, the voltage. See also **phase angle**.
angle of minimum deviation *(Phys.)*. The minimum value of the angle of deviation for a ray of light passing through a prism. By measuring this angle (θ) and also the angle of the prism (α), the refractive index of the prism may be calculated by means of the expression:

$$n = \frac{\sin \frac{1}{2}(\alpha + \theta)}{\sin \frac{1}{2}\alpha}$$

angle of nip *(Min.Ext.)*. The maximum included angle between two approaching faces in a crushing appliance, such as a set of rolls, at which a piece of rock can be seized and entrained.
angle of obliquity *(Eng.)*. The deviation of the direction of the force between two gear teeth in contact, from that of their common tangent.
angle of pressure *(Eng.)*. The angle between a gear tooth profile and a radial line at its pitch point.
angle of reflection *(Phys.)*. The angle which a ray, reflected from a surface, makes with the normal to the surface. The angle of reflection is equal to the *angle of incidence.*
angle of refraction *(Phys.)*. The angle which is made by a ray refracted at a surface separating two media with the normal to the surface. See **refractive index, Snell's law**.
angle of relief *(Eng.)*. The angle between the back face of a cutting tool and the surface of the material being cut.
angle of repose *(Civ.Eng.,Powder Tech.)*. The greatest angle to the horizontal which is made naturally by the inclined surface of a heap of loose material or embankment.
angle of roll *(Aero.)*. The angle through which an aircraft must be turned about its longitudinal axis to bring the lateral axis horizontal. Also horizontal *angle of bank.*
angle of slide *(Min.Ext.)*. Slope at which heaped rock commences to break away.
angle of stall *(Aero.)*. The angle of attack which corresponds with the maximum lift coefficient.
angle of twist *(Eng.)*. The angle through which one section of a shaft is twisted relative to another section when a torque is applied.

angle of view *(Image Tech.)*. The angle subtended at the centre of the lens by the limits of the image recorded; in still photography this is taken as the diagonal of the negative area but in motion picture and TV work it is the width of the frame.

angle plate *(Eng.)*. Cast iron plate with the faces machined truly square and having slots on each face for clamping bolts. Used to hold work when marking off on a **surface plate** or when machining on a lathe face plate or machine tool table.

angle rafter *(Build.)*. See **hip rafter**. The rafter at the hip of a roof to receive the **jack-rafters**. Also *angle ridge*.

angle shaft *(Build.)*. An angle bead which is enriched with, e.g. a capital base.

anglesite *(Min.)*. Orthorhombic sulphate of lead, a common lead ore; named after the original locality, Anglesey.

angle staff *(Build.)*. A strip of wood placed at an angle formed by plastered surfaces to protect from damage. A rounded staff is called an *angle bead*.

angle steel *(Eng.)*. See **angle iron**.

angle stone *(Build.)*. A **quoin**.

angle support *(Elec.Eng.)*. A transmission line tower or pole placed at a point where the line changes its direction. Such a tower or pole differs from a normal tower or pole in that it has to withstand a force tending to overturn it (due to the resultant pull of the conductors).

angle tie *(Build.)*. See **angle brace**.

angora *(Textiles)*. The hair of the angora rabbit or the soft yarn and fabric made from it.

Ångström *(Phys.)*. Unit of wavelength for electromagnetic radiation covering visible light and X-rays. Equal to 10^{-10} m. The unit is also used for interatomic spacings. Symbol Å. Superseded by nanometre ($=10^{-9}$ m). Named after the Swedish physicist A.J.Ångström (1814-74).

Anguilliformes *(Zool.)*. Order of Osteichthyes with pelvic fins and girdle absent or reduced; elongate forms. *Eels*.

angular acceleration *(Phys.)*. The rate of change of angular velocity; usually expressed in radians per second squared.

angular acceleration *(Space)*. The acceleration of a spacecraft around an axis, resulting in pitch, roll or yaw.

angular contact bearing *(Eng.)*. A ball-bearing for radial and thrust loads in which a high shoulder on one side of the outer race takes the thrust.

angular diameter *(Astron.)*. Observed diameter of any celestial object expressed as the angle subtended by its diameter as perceived by the observer.

angular displacement *(Phys.)*. The angle turned through by a body about a given axis, or the angle turned through by a line joining a moving point to a given fixed point.

angular distance of stars *(Astron.)*. Observed angular separation of two stars as perceived by the observer.

angular distribution *(Phys.)*. The distribution relative to the incident beam of scattered particles or the products of nuclear reactions.

angular divergence *(Bot.)*. The angle subtended at the mid-line of an apical meristem of a shoot by the mid-point of two successive leaf primordia. This varies between species but, where the phyllotaxis is spiral it is commonly the Fibonacci angle 137.5°.

angular frequency *(Phys.)*. Frequency of a steady recurring phenomenon, expressed in radians per second, i.e., frequency in Hz multiplied by 2π. Symbol p or ω; also called *pulsatance, radian frequency*.

angular magnification *(Phys.)*. Defined as the ratio of the angle subtended at the eye by an image formed by an optical instrument to the angle subtended by the object at the unaided eye.

angular momentum *(Maths.)*. See **momentum**.

angular momentum *(Phys.)*. The moment of the linear momentum of a particle about an axis. Any rotating body has an angular momentum about its centre of mass, its *spin angular momentum*. The angular momentum of the centre of mass of a body relative to an external axis is its *orbital angular momentum*. In atomic physics, the orbital angular momentum of an electron is *quantized* and can only have values which are exact multiples of the **Dirac constant**. In particle physics, the angular momentum of particles which appear to have spin energy is quantized to values that are multiples of half the Dirac constant.

angular thread *(Eng.)*. See **vee thread**.

angular velocity *(Phys.)*. The rate of change of angular displacement, usually expressed in radians per second.

Angus-Smith process *(Build.)*. An anticorrosion process applied to sanitary ironwork; this is heated to about 316°C immediately after casting, and then plunged into a solution of 4 parts coal-tar or pitch, 3 parts prepared oil, and 1 part paranaphthalene heated to about 149°C. See **Bower-Barff process**.

anharmonic *(Electronics)*. Said of any oscillation system in which the restoring force is nonlinear with displacement, so that the motion is not simple harmonic.

anharmonic ratio *(Maths.)*. See **cross-ratio**.

anhedral *(Aero.)*. See **dihedral angle**.

anhedral *(Geol.)*. A term used in petrography to denote a crystal which does not show any crystal faces, i.e. one which is irregular in shape.

anhidrosis *(Med.)*. Absence of secretion of sweat.

anhidrosis *(Vet.)*. *Nonsweating, drysweating*. An affection of horses in humid tropical countries characterized by inability to sweat after exercise; believed to be due to prolonged overstimulation of the sweat glands by adrenaline.

anhydrides *(Chem.)*. Substances, including organic compounds and inorganic oxides, which either combine with water to form acids, or which may be obtained from the latter by the elimination of water.

anhydrite *(Build.,Min.)*. Naturally occurring anhydrous calcium sulphate which readily forms gypsum and from which anhydrite plaster is made by grinding to powder with a suitable accelerator.

anhydrite process *(Chem.Eng.)*. A process for the manufacture of sulphuric acid from anhydrite $CaSO_4$. The mineral is roasted with a reducing agent and certain other minerals in large kilns, so that SO_2 gas in relatively low concentration is recovered and after cleaning is passed to a specially designed **contact process**. The solid residue is, under normal conditions, readily converted into cement and this forms an economic factor in the process. In Britain the process has a special significance as it provides a large potential of sulphuric acid from an indigenous source of sulphur.

anhydrous *(Chem.)*. A term applied to oxides, salts etc. to emphasize that they do not contain water of crystallization or water of combination.

anhydrous lime *(Build.)*. See **lime**.

anilides *(Chem.)*. *N-phenyl amides*. A group of compounds in which the hydrogen of the amino group in aniline is substituted by organic acid radicals. The most important compound of this class is *acetanilide*.

aniline *(Chem.)*. $C_6H_5NH_2$, *phenylamine, aminobenzene*, a colourless oily liquid, mp −8°C, bp 189°C, rel.d. 1.024, slightly soluble in water; manufactured by reducing nitrobenzene with iron shavings and hydrochloric acid at 100°C. Basis for the manufacture of dyestuffs, pharmaceuticals, plastic (with methanal) and many other products.

aniline black *(Chem.)*. An azine dye, produced by the oxidation of aniline on the fabric.

aniline dyes *(Chem.)*. A general term for all synthetic dyes having aniline as their base.

aniline foils *(Print.)*. Blocking foils which contain dyestuff; used chiefly for leather.

aniline formaldehyde (methanal) *(Chem.)*. Synthetic resin formed by the polycondensation of aniline with formaldehyde.

aniline oil *(Chem.)*. A coal-tar fraction consisting chiefly of crude aniline.

aniline printing *(Print.)*. See **flexographic printing**.

anilinium chloride *(Chem.)*. *Phenylammonium chloride*, $C_6H_5NH_2.HCl$, mp 198°C, bp 245°C, rel.d. 1.22, white crystals, soluble in most organic solvents and water.

anima, animus *(Behav.)*. Term used in Jungian psychology to denote the unconscious feminine component in men and the unconscious masculine component in women.

animal charcoal *(Chem.)*. The carbon residue obtained from carbonization of organic matter such as blood, flesh etc.

animal electricity *(Zool.)*. A term used to denote the ability possessed by certain animals (e.g. electric eel) of giving powerful electric shocks.

animal field *(Zool.)*. In developing blastulae, a region distinguished by the character of the contained yolk granules, and representing the first rudiment of the germ band.

animal pole *(Zool.)*. In the developing ovum, the apex of the upper hemisphere, which contains little or no yolk; in the blastula, the corresponding region, wherein the micromeres lie.

animal-sized *(Paper)*. Paper which has been sized by passing the sheet or web through a bath containing a solution essentially of gelatine and then drying. See **tub sizing**.

animation *(Image Tech.)*. Apparent movement produced by recording step-by-step a series of still drawings, three-dimensional objects or computer-generated images.

animism *(Behav.)*. Attributing feelings and intentions to non-living things. In Piagetian theory children's thinking is characterised by animism in the years two to six.

anion *(Phys.)*. Negative ion, i.e. atom or molecule which has gained one or more electrons in an electrolyte, and is therefore attracted to an anode, the positive electrode. Anions include all nonmetallic ions, acid radicals, and the hydroxyl ion. In a primary cell, the deposition of anions on an electrode makes it the negative pole. Anions also exist in gaseous discharge.

anisaldehyde *(Chem.)*. *4-methoxybenzaldehyde*. Colourless liquid; bp 248°C, occurring in aniseed, and used in perfumery.

anisidines *(Chem.)*. *Amino-anisoles, methoxyanilines*, $CH_3O.C_6H_4.NH_2$, bases similar to aniline. Intermediates for dyestuffs.

aniso- *(Genrl.)*. Prefix from Gk. *an*, not; *isos*, equal.

anisocercal *(Zool.)*. Having the lobes of the tail-fin unequal.

anisodactylous *(Zool.)*. Of Birds, having 3 toes turned forward and 1 turned backward when perching, as in the Passeriformes.

anisodesmic structure *(Crystal.)*. One giving a crystal marked difference between its bond strengths in the intersecting axial planes.

anisogamete *(Biol.)*. A gamete differing from the other conjugant in form or size. adj. *anisogamous*.

anisogamy *(Biol.)*. Sexual fusion of gametes that differ in size but not necessarily in form. See **heterogamy, isogamy, oögamy**.

anisokont *(Bot.)*. Having 2 flagella unequal in length but otherwise more or less similar. Cf. **isokont, heterokont**.

anisole *(Chem.)*. *Phenyl methyl ether*, $C_6H_5.O.CH_3$, a colourless liquid, bp 155°C.

anisomeric *(Chem.)*. Not isomeric.

anisopleural *(Zool.)*. Bilaterally asymmetrical.

anisotonic *(Chem.)*. Not isotonic.

anisotropic *(Min.,Phys.)*. Said of crystalline material for which physical properties depend upon direction relative to crystal axes. These properties normally include elasticity, conductivity, permittivity, permeability etc.

anisotropic *(Zool.)*. Of ova, having a definite polarity, in relation to the primary axis passing from the animal pole to the vegetable pole. n. *anisotropy*.

anisotropic conductivity *(Phys.)*. Body which has a different conductivity for different directions of current flow, electric or thermal.

anisotropic dielectric *(Phys.)*. One in which electric effects depend on the direction of the applied field, as in many crystals.

anisotropic liquids *(Chem.)*. See **liquid crystals**.

anisotropy *(Chem.)*. Describes a property of a substance when that property depends on direction as revealed by measurement, e.g. **crystals** and **liquid crystals** in which the refractive index is different in different directions.

ankerite *(Min.)*. A carbonate of calcium, magnesium, and iron.

ankylosing spondylitis *(Med.)*. Rheumatoid arthritis of the spine, which may progress to cause complete spinal and thoracic rigidity.

ankylosis *(Med.)*. Fixation of a joint by fibrous bands within it, or by pathological union of the bones forming the joint.

ankylosis, anchylosis *(Zool.)*. The fusion of two or more skeletal parts, especially bones.

ankylostomiasis *(Med.)*. *Hookworm disease*; infection by two parasitic nematode worms in the small intestine (*Ankylostoma duodenale* and *A. americanum*), which produces iron deficiency for those on an inadequate diet.

anlage *(Zool.)*. See **primordium**.

annabergite *(Min.)*. Hydrated nickel arsenate, apple-green monoclinic crystals, rare, usually massive. Associated with other ores of nickel. Also called *nickel bloom*.

anneal *(Biol.)*. To reform the duplex structure of a nucleic acid.

annealing *(Phys.)*. Process of maintaining a material at a known elevated temperature to reduce dislocations, vacancies and other metastable conditions, e.g. steel or glass. In ferrous alloys the metal is held at a temperature above the upper critical temperature for a variable time and then cooled at a predetermined rate, depending on the alloy and the particular properties of hardness, machineability etc. which are needed. The term is usually qualified, e.g. *quench annealing, isothermal annealing, graphitizing*.

annealing furnace *(Eng.)*. Batch-worked or continuous oven or furnace with controllable atmosphere in which metal, alloy or glass is annealed.

Annelida *(Zool.)*. A phylum of metameric Metazoa, in which the perivisceral cavity is coelomic, and there is only one somite in front of the mouth; typically there is a definite cuticle and chitinous setae arising from pits of the skin; the central nervous system consists of a pair of preoral ganglia connected by commissures to a postoral ventral ganglionated chain; if a larva occurs it is a trochophore. Earthworms, Ragworms, Leeches.

annihilation *(Phys.)*. Spontaneous conversion of a particle and its antiparticle into radiation, e.g. positron and electron yielding two γ-ray photons each of energy 0.511 MeV.

annihilation radiation *(Phys.)*. The radiation produced by the annihilation of an elementary particle with its corresponding antiparticle.

annihilator *(Maths.)*. *An* annihilator of x is y such that $yx = 0$. Here x and y may be elements of rings, functions etc. and need not be the same type of object so long as xy is defined. An annihilator of a set X is y which is an annihilator of every element of X. *The* annihilator is the set of all such individual annihilators.

annite *(Min.)*. The ferrous iron end-member of the biotite series of micas.

annoyance *(Acous.)*. The psychological effect arising from excessive noise. There is no absolute measure, but the annoyance caused by specified classes of noise can be correlated.

annual *(Bot.)*. A plant that flowers and dies within a period of one year from germination. Cf. **ephemeral, biennial, perennial**.

annual equation *(Astron.)*. One of four terms describing the orbit of the Moon, which arises from the eccentricity of the Earth's orbit round the Sun. Its period is one year.

annual load factor *(Elec.Eng.)*. The load factor of a generating station, supply-undertaking, or consumer, taken over a whole year.

annual parallax *(Astron.)*. The motion of the Earth round the Sun causes minute changes in the apparent positions of the stars. The regular annual displacement is the *annual parallax*. It is largest, at 0.71 seconds of arc, for the star **Proxima Centauri**.

annual ring *(Bot.)*. A **growth ring** formed over a year.

annular bit *(Build.)*. A bit which cuts an annular (ring-shaped) channel and leaves intact a central cylindrical plug.

annular borer *(Civ.Eng.)*. A rock-boring tool which does the work of an **annular bit**, and provides a means of obtaining a core showing a section of the strata.

annular combustion chamber *(Aero.)*. A gas turbine combustion chamber in which the perforated **flame tube** forms a continuous annulus within a cylindrical outer casing.

annular eclipse *(Astron.)*. See **eclipse**.

annular gear *(Eng.)*. A ring in the shape of an annulus with gear teeth cut on the periphery for engagement with a pinion. Usually shrunk fit onto a mating diameter, e.g. starter ring on automobile flywheel.

annular space *(Min.Ext.)*. The space between the **casing** and the producing or drilling bore.

annular thickening *(Bot.)*. The secondary wall deposited in the form of discrete transverse rings or hoops, in tracheids and vessel elements of xylem, especially **protoxylem**.

annular vault *(Build.)*. See **barrel vault**.

annulated column *(Arch.)*. A column formed of slender shafts clustered together, or sometimes around a central column, and secured by stone or metal bands.

annulus *(Bot.)*. (1) A membranous frill present on the stipe of some agarics. (2) A patch or a crest of cells with thickened walls occurring in the wall of the sporangium of ferns, and bringing about dehiscence by setting up a strain as they dry. (3) A zone of cells beneath the operculum of the sporangium of a moss, which break down and assist in the liberation of the operculum.

annulus *(Maths.)*. A plane surface bounded by two concentric circles, i.e. like a washer.

annulus *(Zool.)*. Any ring-shaped structure: the fourth digit of a pentadactyl forelimb: in Arthropoda, subdivision of a joint forming jointlets; in Hirudinea, a transverse ring subdividing a somite externally. adj. *annular, annulate*.

annunciator *(Civ.Eng.)*. Any device for indicating audibly the passage of a train past a point.

annunciator *(Elec.Eng.)*. Arrangement of indicators which display details on operational condition and functioning of complex plant.

anode *(Electronics)*. *Plate* (US). In a valve or tube, the electrode held at a positive potential with respect to a cathode, and through which positive current generally enters the vacuum or plasma, through collection of electrons. Also positive electrode of battery or cell. See **cathode, ultor**.

anode breakdown voltage *(Electronics)*. That required to trigger a discharge in a cold-cathode glow tube when the starter gap (if any) is not conducting. It is measured with any grids or other electrodes earthed to cathode.

anode brightening *(Eng.)*. See **electrolytic polishing**.

anode characteristic *(Electronics)*. Graph relating anode current and anode voltage for an electron tube.

anode dark space *(Electronics)*. Dark zone near the anode in a glow-discharge tube.

anode dissipation *(Electronics)*. Generally, the energy produced at the anode of a thermionic tube and wasted as heat owing to the bombardment by electrons; specifically, the maximum permissible power which may be dissipated at the anode.

anode drop (fall) *(Electronics)*. The voltage between the positive column and the anode of a gas discharge tube. It may be positive, zero, or negative, depending on the gas pressure, but not the discharge current.

anode efficiency *(Electronics)*. Ratio of a.c. power in the load circuit to the d.c. power supplied to the anode of a valve amplifier or oscillator.

anode feed *(Electronics)*. Supply of direct current to anode of valve, generally decoupled, so that the supply circuit does not affect the condition of operation of the valve.

anode glow *(Electronics)*. Luminous zone on anode side of positive column in a gas-discharge tube.

anode modulation *(Electronics)*. Insertion of the modulating signal into the anode circuit of a valve, which is oscillating or is rectifying the carrier. Also called *plate modulation*.

anode mud *(Eng.)*. See **anode slime**.

anode polishing *(Eng.)*. Same as **electrolytic polishing**.

anode saturation *(Electronics)*. Limitation of current through the anode of a valve, arising from current, voltage, temperature, or space charge.

anode shield *(Electronics)*. Electrode used in high-power gas tubes to shield the anode from damage by ion bombardment.

anode slime *(Eng.)*. Residual slime left when anode has been electrolytically dissolved. It may contain valuable byproduct metals. Also called *anode mud*.

anode strap *(Telecomm.)*. Connecting strip between alternate anode segments of a multi-cavity magnetron. Used for mode selection and control.

anode tap *(Electronics)*. Tapping point on the inductance coil of a tuned-anode circuit, to which the anode is connected. The position of the tap is adjusted so that the tube operates into the optimum impedance.

anodic etching *(Elec.Eng.)*. A method of preparing metals for electrodeposition by making them the anode in a suitable electrolyte and at a suitable current density.

anodic oxidation *(Chem.)*. Oxidation, i.e. removal of electrons from a substance by placing it in the anodic region of an electrolytic cell. The substance to be oxidized may be either a part of the electrolyte or the anode itself. See **anodizing**.

anodic protection *(Eng.)*. System for passivating steel by making it the anode in a protective circuit. Cf. **cathodic protection**.

anodic treatment *(Chem.)*. See **anodizing**.

anodized *(Eng.)*. Said of metal surface protected by chemical or electrolytic action.

anodizing *(Chem.)*. A process by which a hard, noncorroding oxide film is deposited on aluminium or light alloys. The aluminium is made the anode in an electrolytic cell containing chromic (VI) or sulphuric acid. Also called *anodic treatment*.

anodontia *(Zool.)*. Absence of teeth.

anoestrus *(Zool.)*. In Mammals, a resting stage of the oestrus cycle occurring between successive heat periods.

anomalistic month *(Astron.)*. The interval (amounting to 27.554 55 days) between two successive passages of the moon in its orbit through perigee.

anomalistic year *(Astron.)*. The interval (equal to 365.259 64 mean solar days) between two successive passages of the sun, in its apparent motion, through perigee.

anomaloscope *(Phys.)*. An instrument for detection and classification of defective colour vision. Two colours are mixed, and the result matched with a third.

anomalous dispersion *(Phys.)*. The type of dispersion given by a medium having a strong absorption band, the value of the refractive index being abnormally high on the longer wave side of the band, and abnormally low on the other side. In the spectrum produced by a prism made of such a substance the colours are, therefore, not in their normal order.

anomalous magnetization *(Elec.Eng.)*. Irregular distribution of magnetization, e.g. when consequent poles exist as well as main poles on a magnetic circuit.

anomalous scattering *(Phys.)*. See **scattering**.

anomalous secondary thickening *(Bot.)*. The production of new vascular tissue by a secondarily formed cambium.

anomalous viscosity *(Phys.)*. A term used to describe liquids which show a decrease in viscosity as their rate of flow or velocity gradient increases. Such liquids are also known as *non-Newtonian liquids*.

anomaly *(Astron.)*. The angle between the radius vector of an orbiting body and the major axis of the orbit, measured from perihelion (periastron, etc.) in the direction of motion.

anomaly *(Genrl.)*. Any departure from the strict characteristics of the type.

anomer *(Chem.)*. In carbohydrate chemistry, one of two isomers differing in conformation at the aldehydic carbon in a ring form.

anomeristic *(Zool.)*. Of metameric animals, having an indefinite number of somites.

anorectic *(Med.)*. Substance that suppresses feeling of hunger.

anorexia *(Med.)*. Loss of appetite.

anorexia nervosa *(Behav.,Med.)*. Chronic failure to eat for fear of gaining weight or emotional disturbance; results in malnutrition, semistarvation and sometimes death.

anorthic system *(Crystal.)*. See **triclinic system**.

anorthite *(Min)*. The calcium end-member of the plagioclase group feldspars; silicate of calcium and aluminium, occurring in some basic igneous and other rocks.

anorthoclase *(Min.)*. A triclinic sodium-rich high-temperature sodium-potassium feldspar; occurs typically in volcanic rocks, and is also known from a syenite, larvikite, from S. Norway, which is widely used for facing buildings.

anorthosite *(Geol.)*. A coarse-grained plutonic igneous rock, consisting almost entirely of plagioclase, near labradorite in composition.

anosmatic *(Zool.)*. Lacking the sense of smell.

anosmia *(Med.)*. Loss, partially or completely, of the sense of smell.

anoxia *(Med.)*. Lack of oxygen.

anoxia, anoxaemia, anoxemia *(Med.,Zool.)*. Deficiency of oxygen in the blood: any condition of insufficient oxygen supply to the tissues: any condition which retards oxidation processes in the tissues and cells.

anoxybiosis *(Zool.)*. Life in absence of oxygen.

Anseriformes *(Zool.)*. An order of Birds with webbed feet. The members of this order are unusual in the possession of an evaginable penis; they are all aquatic forms, living on the animals found living in the mud at the bottom of shallow waters and in marshes; some are powerful fliers. Geese, Ducks, Screamers, Swans.

answer-back code *(Telecomm.)*. The identifying code combination(s) which a teleprinter automatically transmits in response to an incoming 'who-are-you' signal.

answer-back unit *(Telecomm.)*. An individually-modified teleprinter subassembly, which transmits identifying code combinations, whenever it is triggered by an incoming 'who-are-you' signal.

answer print *(Image Tech.)*. First print from the edited negative of a film shown to the producer for approval before release.

antacids *(Med.)*. Group of compounds given in the treatment of dyspepsia. *Magnesium trisilicate* and *aluminium hydroxide* are common examples.

antagonism *(Bot.)*. A relationship between different organisms in which one partly or completely inhibits the growth of, or kills, a second, especially when due to a toxic metabolite. See **antibiotic, allelopathy**.

antagonist *(Med.)*. A word with many meanings where one factor or structure opposes another, e.g. a drug which works in opposition to a hormone or a muscle which opposes the action of another.

antagonizing screws *(Surv.)*. See **clip screws**.

antapex *(Astron.)*. See **solar-antapex**.

ante- *(Genrl.)*. Prefix from L. *ante*, before.

antebrachium *(Zool.)*. The region between the brachium and the carpus in land Vertebrates; the fore-arm.

antecedent *(Maths.)*. (1) In a ratio a:b, a is the antecedent, b is the consequent. (2) In material implication (if p, then q), p is the antecedent, q is the consequent.

antecedent drainage *(Geol.)*. A river system that has maintained its original course despite subsequent folding or uplift.

antechamber *(Eng.)*. A small auxiliary combustion-chamber, used in some compression-ignition engines, in which partial combustion of the fuel is used to force the burning mixture into the cylinder, so promoting more perfect combustion. See **precombustion chamber**.

antecubital *(Zool.)*. In front of the elbow.

antefixae *(Arch.)*. Ornaments placed at the eaves and cornices of ancient buildings to hide the ends of the roof tiles; sometimes perforated to convey water away from the roof.

antenna *(Zool.)*. In Arthropoda, one of a pair of anterior appendages, normally many-jointed and of sensory function: in Angler Fish the elongate first dorsal fin-ray, which bears terminally a skinny flap, used by the fish to attract prey. pl. *antennae*. adjs. *antennary, antennal*.

antenna changeover switch *(Telecomm.)*. Switch used for transferring an antenna from the transmitting to the receiving equipment, and vice versa, protecting the receiver.

antenna downlead *(Telecomm.)*. Wire running from the elevated part or conductor of an antenna down to the transmitting or receiving equipment.

antenna effect *(Telecomm.)*. (1) Errors arising when a directional antenna, used in an electronic navigation system, picks up radiation from a non-intended direction, as a result of imperfections in the radiation pattern. (2) Spurious effects in radio-direction finding systems caused by stray capacitance between a loop antenna and earth.

antenna efficiency *(Telecomm.)*. Same as *radiation efficiency*.

antenna feeder *(Telecomm.)*. The transmission line or cable by which energy is fed from the transmitter to the antenna.

antenna field *(Telecomm.)*. Map showing electromagnetic field strength produced by antenna in the form of contour lines joining points of equal field intensity. See **radiation pattern**.

antenna gain *(Telecomm.)*. Ratio of maximum energy flux from antenna, to that which would have been received from a non-directional aerial radiating the same power. See **directivity**.

antenna impedance *(Telecomm.)*. Complex ratio of voltage to current at the point where the feeder is connected.

antennal glands *(Zool.)*. The principal organs of Crustacea. They open at the bases of the osmoregulatory appendages from which they take their name. Also called *maxillary glands*.

antenna load *(Elec.Eng.)*. Same as **dummy load**.

antenna noise temperature *(Telecomm.)*. The temperature of a **black body** which, when placed around an antenna similar to the real one, but loss-free and perfectly matched to the receiver, produces the same noise power, within a specified frequency band, as the real antenna in its operating environment.

antenna resistance *(Telecomm.)*. Total power supplied to an antenna system divided by the square of a specified current, e.g. in the feeder, or at the earth connection of an open-wire antenna.

antenna-shortening capacitor *(Telecomm.)*. One connected in series with an antenna to allow operation at a frequency other than its natural resonant one. See **loaded antenna**.

antennule *(Zool.)*. A small antenna: in some Arthropoda (as the Crustacea) which possess two pairs of antennae, one of the first pair.

antepetalous, antipetalous *(Bot.)*. Inserted opposite to the petals.

anteposition *(Bot.)*. Situation opposite, and not alternate to, another plant member.

anterior *(Bot.)*. (1) The side of a flower next to the bract, or facing the bract. (2) That end of a motile organism which goes first during locomotion.

anterior *(Zool.)*. In animals in which cephalization has occurred, nearer the front or cephalad end of the longitudinal axis: in human anatomy, ventral.

antero- *(Genrl.)*. Prefix from *anterior*, former.

anterograde amnesia *(Behav.)*. Loss of memory for events after injury to the brain or mental trauma; with little effect on information acquired previously.

antesepalous, antisepalous *(Bot.)*. Inserted opposite to the sepals.

anthelion *(Meteor.)*. A mock sun appearing at a point in the sky opposite to and at the same altitude as the sun. The phenomenon is caused by the refraction of sunlight by ice crystals.

anthelminthic *(Med.)*. Drugs used against parasitic worms.

anther *(Bot.)*. Fertile part of a stamen, usually containing 4 sporangia, and producing pollen.

anther culture *(Bot.)*. The aseptic culture on suitable medium of anthers, with the possibility of the production of haploid callus, embryoid and plantlets. If treated with e.g. colchicine these plantlets may give rise to diploid plants, autodiploids, that are completely homozygous.

antheridiophore, antheridial receptacle *(Bot.)*. A specialized branch of a thallus bearing antheridia.

antherozoid *(Bot.)*. A motile male gamete, spermatozoid, or sperm.

anthesis *(Bot.)*. The opening of a flower bud: by extension, the duration of life of any one flower, from the opening of the bud to the setting of fruit.

Anthocerotopsida, Anthocerotae *(Bot.)*. Class of the Bryophyta containing the hornworts, which differ from the liverworts (Hepaticae) in that the gametophyte is always thalloid and that the sporophyte show a relatively long-continued growth, and spore production, from an intercalary meristem near its base.

anthocyanins *(Bot.)*. A large group of water-soluble, flavonoid, glycoside pigments in cell vacuoles responsible for the red, purple and blue colours of flowers, fruit and leaves in most flowering plants. Cf. **betamains**.

anthogenesis *(Zool.)*. A form of parthenogenesis in which both males and females are produced by asexual forms, as in some Aphidae.

anthophilous *(Bot.)*. Flower-loving; feeding on flowers.

anthophore *(Bot.)*. An elongation of the floral receptacle between the calyx and corolla.

anthophyllite *(Min.)*. An orthorhombic amphibole, usually massive, and normally occurring in metamorphic rocks; magnesium iron silicate, of low aluminium content.

Anthophyta *(Bot.)*. (1) Usually Angiospermae, see **angiosperms**. (2) Rarely Spermatophyta.

Anthozoa *(Zool.)*. A class of Cnidaria in which alternation of generations does not occur, the medusoid phase being entirely suppressed; the polyps may be solitary or colonial; the gonads are of endodermal origin. Also *Actinozoa*.

anthracene *(Chem.)*. $C_{14}H_{10}$, colourless, blue fluorescent crystals, mp 218°C, bp 340°C, a valuable raw material for dyestuffs obtained from the fraction of coal-tar boiling above 270°C. Anthracene represents a group of polycyclic compounds with a series of 3 benzene rings condensed together. Carcinogenic. Useful as a scintillator in photoelectric detection of β-particles. ⇨

anthracene oil *(Chem.)*. A coal-tar fraction boiling above 270°C, consisting of anthracene, phenanthrene, chrysene, carbazole and other aromatic hydrocarbon oils.

anthracite *(Geol.)*. The highest metamorphic rank of coal. See **rank of coal**.

anthracnose *(Bot.)*. One of a number of plant diseases characterized by black, usually sunken, lesions; mostly caused by one of the fungi of the Melanconiales.

anthracosis *(Med.)*. 'Coal-miner's lung', produced by inhalation of coal dust.

anthraflavine *(Chem.)*. An anthraquinone vat dye-stuff, which dyes cotton greenish yellow, obtained by heating 2-methylanthraquinone with alcoholic potassium hydroxide at 150°C.

anthranil *(Chem.)*. The intramolecular anhydride of anthranilic acid (2-aminobenzoic acid), intermediate in the synthesis of indigo.

anthranilic acid *(Chem.)*. $C_6H_4(COOH).NH_2$, *2-aminobenzoic acid*, obtained from phthalimide by the Hofmann reaction, an oxidation product of indigo.

anthraquinone *(Chem.)*. *Diphenylene diketone*. $C_6H_4(CO)_2C_6H_4$, yellow needles or prisms, which sublime easily, mp 285°C, bp 382°C. More closely related to diketones than to quinones. Obtained by the oxidation of anthracene with sulphuric acid and chromic (VI) acid. Parent substance of an important group of dyes, including alizarin.

anthrax *(Med.,Vet.)*. An acute infective disease caused by the anthrax bacillus, communicable from animals to man in whom it causes cutaneous malignant pustules and lung, intestinal and nervous system infection. Also *woolsorter's disease*. Notifiable disease in animals for which vaccines are available.

anthraxolite *(Min.)*. A member of the asphaltite group.

anthraxylon *(Min.)*. One of the constituents of coal, derived from the lignin of the plants forming the seam.

anthropogenic *(Bot.)*. Vegetational change etc., resulting from or influenced by man's activities.

anthropoid *(Zool.)*. Resembliing Man: pertaining to, or having the characteristics of the Anthropoidea.

anthropomorph *(Zool.)*. A conventional design of the human figure; resembling a human in form or in attributes.

anthropophyte *(Bot.)*. A plant introduced incidentally in the course of cultivation.

anti- *(Genrl.)*. Prefix from Gk. *anti*, against.

anti-aldoximes *(Chem.)*. The stereoisomeric form of aldoximes in which the H and the OH groups are far removed from each other. See **synaldoximes**.

anti-aliasing *(Image Tech.)*. Treatment of video picture signal elements to reduce the effects of **aliasing**.

anti-auxin *(Bot.)*. Compound that in low concentrations will directly interfere with **auxin** action, e.g. tri-chlorobenzoic acid.

antibaryon *(Phys.)*. Antiparticle of a baryon, i.e. a hadron with a baryon number of −1. The term **baryon** is often used generically to include both.

antibiosis *(Biol.)*. A state of mutual antagonism. Cf. **symbiosis**.

antibiotic resistance *(Biol.)*. The property of micro-organisms or cells, which can survive high concentrations of a normally lethal agent. Normally acquired by the selection of a rare resistant mutant in the presence of low concentrations of the agent, but can be added by **genetic manipulation**.

antibody *(Immun.)*. Immunoglobulin with combining site able to combine specifically with antigenic determinants on an antigen. See **immunoglobulin**.

antibonding orbital *(Phys.)*. Orbital electron of 2 atoms, which increases in energy when the atoms are brought together, and so acts against the closer bonding of a molecule.

antical *(Bot.)*. The upper surface of a thallus, stem, or leaf.

anticapacitance switch *(Elec.Eng.)*. One designed to have very little capacitance between the terminals when in the open condition.

anti-capillary groove *(Build.)*. See **check throat**.

anticathode *(Med.,Phys.)*. The anode target of an X-ray tube on which the cathode rays are focused, and from which the X-rays are emitted.

anti-climb scheme *(Build.)*. Paints which are specially formulated to permanently remain wet after application to deter vandals.

anticlinal *(Bot.)*. Perpendicular to the nearest surface. If a cell divides anticlinally the daughter cells will be separated by an anticlinal wall. Cf. **periclinal**.

anticline *(Geol.)*. A type of fold, comparable with an arch, the strata dipping outwards, away from the fold-axis.

anti-clutter *(Radar)*. Refers to a circuit or part of a radar system designed to eliminate unwanted echoes, *clutter*, and permit the display of signals which might otherwise be obscured. Often takes the form of a gain control which automatically reduces gain immediately after the transmitted pulse and gradually restores it during the interval leading up to the anticipated return echo.

anticoagulant *(Med.)*. Any chemical substance which hinders normal clotting of blood, e.g. heparin, warfarin sodium, phenindione.

anticodon *(Biol.)*. The sequence of three bases on **tRNA** which binds to the codon of **mRNA**. The complement of the coding triplet.

anticoincidence circuit *(Electronics)*. One which delivers a pulse if one of two pulses is independently applied, but not when both are applied together or non-simultaneously within an assigned time interval.

anticoincidence counter *(Electronics)*. System of counters and circuits which record only if an ionizing particle passes through particular counters but not through the others.

anticollision beacon *(Aero.)*. A flashing red or blue light which is mounted above and below an aircraft to make it conspicuous when flying in **control zones** or other busy areas.

anticondensation paints *(Build.)*. A range of coatings specially formulated to form an isolating barrier between cold substrates and moisture laden air.

anticyclone *(Meteor.)*. A distribution of atmospheric pressure in which the pressure increases towards the centre. Winds in such a system circulate in a clockwise direction in the northern hemisphere and in a counter-clockwise direction in the southern hemisphere. Anti-cyclones give rise to fine, calm weather conditions, although in winter fog is likely to develop.

anticyclotron tube *(Electronics)*. A type of travelling-wave tube.

antidazzle mirror *(Autos.)*. One having a two-position setting, providing a dim partial reflection of headlamps behind for night driving.

antidiazo compounds *(Chem.)*. The stereoisomeric form of diazo compounds in which the groups attached to the nitrogen atoms are far removed from each other.

antidromic *(Biol.)*. Contrary to normal direction, e.g. applied to nerve cells, when the impulse is conducted along the axon towards the cell body.

antifading antenna *(Telecomm.)*. One which confines radiation mainly to small angles of elevation, to minimize radiation of *sky waves* which are prone to fading. For medium-wave transmitters, the antenna is usually a vertical mast about 0.6 of a wavelength high. **Adaptive arrays** are also used to combat fading in higher-frequency applications.

antiferromagnetism *(Phys.)*. Phenomenon in some magnetically ordered materials in which there is an antiparallel alignment of spins in two interpenetrating structures so that there is no overall bulk spontaneous magnetization. Antiferromagnetics have a positive susceptibility. The antiparallel alignment is disturbed as the temperature increases until at the **Néel temperature** the material becomes paramagnetic.

antiflood and tidal valve *(Build.)*. A valve consisting of a cast-iron box containing a floating ball, fitted near a drain outlet to prevent back flow.

antifouling composition *(Civ.Eng.)*. A substance applied in paint form to ships' bottoms and structures subject to the action of sea water, to discourage marine growths.

antifouling paints *(Build.)*. Highly poisonous paints applied to the hulls of ships to minimize the accumulation of barnacles.

antifreeze *(Chem.,Eng.)*. Solutes which lower the freezing point of water, usually in automobile engine cooling. Most commonly ethylene or propylene glycol, or methanol with a few per cent. of corrosion inhibitor such as phosphates. A 35% solution of ethene glycol or 30% of methanol and water will not freeze at temperatures above $-5°F(-20.6°C)$.

antifriction bearing *(Eng.)*. Used to describe a wide range of bearings such as *ball-*, *roller-*, *special metallic alloy-*, and *plastic based bearings*. All designed to reduce friction between moving parts. The choice depending on the *duty*.

antifriction metal *(Eng.)*. See **white metal**.

antigen *(Immun.)*. A substance which has determinant groups which can interact with specific receptors on lymphocytes or on antibodies released from them. The term is often used to include substances which can stimulate an immune response, although these are more correctly termed *immunogens*.

antigenic determinant *(Immun.)*. A small part of the antigen which has a structure complementary to the recognition site on a T-cell receptor or an antibody. Most antigens are large molecules with several different antigenic determinants, which interact with lymphocytes with differing specific recognition sites.

antigenic variation *(Immun.)*. Many viruses, bacteria and protozoa can develop new antigenic determinants as a result of genetic mutation and selection during multiplication in their hosts. If the variation involves the antigenic component which stimulates protective immunity, the variant can cause infection in subjects who would otherwise be immune to that microbe.

antiglobulin *(Immun.)*. Term used to describe antibodies against immunoglobulins which can be used to detect the presence of immunoglobulins bound to the surface of cells or microbes, and so to detect the binding of antibodies to them.

antigorite *(Min.)*. One of three minerals which are collectively known as *serpentine*, hydrated magnesium silicate. It is abundant in the rock type, serpentinite.

anti-*g* suit *(Aero.)*. A close fitting garment covering the legs and abdomen, which is inflated, either automatically or at will by the wearer, so that counter-pressure is applied when blood is displaced away from the head and heart during high-speed manoeuvres. Colloquially *g-suit*.

anti-*g* valve *(Aero.)*. (1) A spring-loaded mass type of air valve which automatically regulates the inflation of an *anti-g suit* according to the acceleration (*g*) loads being imposed. (2) A valve incorporated in some aircraft fuel systems to prevent engines being starved of fuel under specific *g* loads.

antihalation *(Image Tech.)*. The use of backing to reduce halation in plates or films.

antihistamine *(Med.)*. A substance or drug which inhibits the actions of histamine by blocking its site of action.

antihunting *(Telecomm.)*. In a d.c. closed loop feedback system, the use of a transformer in series with the load current, so that a rate-of-change signal can be injected into the system to ensure stability and avoid self-oscillation.

anti-icing *(Aero.)*. Protection of aircraft against icing by preventing ice formation on e.g. wind-shield panels, leading edges of wings, tail units and turbine engine air intakes. The most common methods are to apply continuous heating by hot air tapped from an engine, by electrical heating elements or periodically inflating rubber bags. Cf. **de-icing**.

anti-idiotype *(Immun.)*. Antibody which recognizes the combining site of an antibody against an antigenic determinant on an antigen. The combining site of the anti-idiotype may thus be expected to resemble the shape of the determinant on the original antigen. If anti-idiotypes are used in turn to elicit antibodies against them, it is possible to obtain anti-anti-idiotypes which recognize the determinant on the original antigen, even though this was never administered. In principle such a method could be used to immunize against antigens which in themselves are dangerous or impractical.

anti-incrustator *(Eng.)*. A substance used to prevent the formation of scale on the internal surfaces of steam boilers.

anti-induction network *(Telecomm.)*. A network connected between circuits to minimize crosstalk.

antiknock substances *(Autos.)*. Substances added to petrol to lessen its tendency to detonate or 'knock' in an engine; e.g. lead (IV) ethyl.

antiknock value *(Autos.)*. The relative immunity of a volatile liquid fuel from detonation, or 'knocking', in a petrol engine, as compared with some standard fuel. See **knock rating**, **octane number**.

antilepton *(Phys.)*. An antiparticle of a **lepton**. Positron, positive muon, antineutrinos and the tau-plus particle are antileptons.

antilock brake *(Autos.)*. System which prevents the locking of road wheels under braking, giving improved control and stopping on poor road surfaces. A sensor detects over-rapid deceleration of the wheels and signals for a reduction in braking effort. Also *ABS brake*.

antilogarithm *(Maths.)*. A number whose logarithm is the given number.

anti-lymphocytic serum *(Immun.)*. Serum containing antibodies reactive with surface antigens on lymphocytes and capable of killing or otherwise suppressing their capabilities. Used, usually as Ig concentrate, as an immunosuppressive agent to prevent tissue graft rejection or in severe autoimmune disease.

antimetabolite *(Med.)*. Drugs used in treatment of cancer which are incorporated into new nuclear material and prevent normal cell division. Common examples are methotrexate, cytoarabinose and fluorouracil.

antimonial lead *(Chem.Eng.)*. A lead-antimony alloy of controlled analysis with much improved chemical strength and retaining good chemical corrosion resistance. For sheet and pipe antimony may be as high as 12%, for machined castings up to 20–25%.

antimonial lead ore *(Min.)*. See **bournonite**.

antimoniates *(Chem.)*. The antimonic acids give antimoniates with aqueous solutions of potassium hydroxide.

antimonite *(Min.)*. See **stibnite**.

antimony *(Chem.)*. Metallic element, symbol Sb, at.no. 51, r.a.m. 121.75, mp 630°C, rel.d. 6.6. Used in alloys for cable covers, batteries etc.; also as a donor impurity in germanium. Has several radioactive isotopes which emit very penetrating γ-radiation. These are used in laboratory neutron sources.

antimony alloys *(Eng.)*. Antimony is not used as the basis of important alloys, but it is an essential constituent in type metals, bearing metals (which contain 3–20%), in lead for shrapnel (10%), storage battery plates (4–12%), roofing, gutters and tank linings (6–12%).

antimony black *(Eng.)*. Finely powdered antimony, gives plaster casts a metallic look.

antimony glance *(Min.)*. Obs. name for *stibnite*.

antimony halides *(Chem.)*. Antimony (III) fluoride SbF_3, and (V) fluoride SbF_5. (III) chloride $SbCl_3$, and (V) chloride $SbCl_5$. (III) bromide and (III) iodide.

antimony hydrides *(Chem.)*. Two hydrides, stibine [(III) hydride] SbH_3, and the solid dihydride Sb_2H_2. See also *stibine*.

antimonyl *(Chem.)*. The monovalent radical SbO—. (Sb (III) present).

antimuon *(Phys.)*. Antiparticle of a **muon**.

antimutagen *(Biol.)*. A compound which inhibits the action of a mutagen.

antineutrino *(Phys.)*. Antiparticle to the **neutrino**. As there are four types of neutrino there are also four types of antineutrino.

antineutron *(Phys.)*. Antiparticle with spin and magnetic moment oppositely orientated to those of neutron.

antinode *(Phys.)*. At certain positions in a standing wave system of acoustic or electric waves or vibrations, the location of maxima of some wave characteristic, e.g. amplitude, displacement, velocity, current, pressure, voltage. At the **nodes** these would have minimum values.

anti-nuclear factor *(Immun.)*. Auto-antibody reactive with nucleic acids (DNA or RNA) present in the blood of subjects with **systemic lupus erythematosus** (SLE) and some other auto-immune conditions. Used diagnostically.

antioxidants *(Chem.)*. These are substances which delay the oxidation of paints, plastics, rubbers etc. Raw vegetable oils contain natural antioxidants which reduce the speed of drying of paints. Deliberately added antioxidants, generally phenol derivatives, delay the skinning of paints in the can at the cost of slightly slower drying. Similar substances added to plastics, rubbers, foods and drugs delay degradation by oxidation.

antiparallax mirror *(Phys.)*. Mirror positioned on an arc adjacent to the scale of an indicating instrument, so that the parallax error in reading the indication of the pointer is avoided by aligning the eye with the pointer and its image.

antiparallel *(Maths.)*. Said of vectors which are parallel, but operate in opposite directions.

antiparticle *(Phys.)*. A particle that has the same mass as another particle but has opposite values for its other properties such as charge, baryon number, strangeness etc. The antiparticle to a fundamental particle is also fundamental. For example, the electron and positron are particle and antiparticle. Interaction between such a pair means simultaneous **annihilation**, with the production of energy in the form of radiation.

antiperistaltic *(Zool.)*. Said of waves of contraction passing from anus to mouth, along the alimentary canal; Cf. **peristaltic**. n. *antiperistalsis*.

antiperthite *(Min.)*. An intergrowth of plagioclase and potassium feldspars with plagioclase as the dominant phase. See **perthite**.

antipetalous *(Bot.)*. See **antepetalous**.

antipodal cells *(Bot.)*. Wall-less cells, usually three, typically haploid, derived by mitotic division of the magaspore, lying in the embryo sac at the end remote from the micropyle.

antipodal points *(Maths.)*. The points at either end of a diameter of a sphere.

antipodes *(Geol.)*. On a sphere, e.g. the earth, points on the surface at either extremity of a diameter.

antipolarizing winding *(Elec.Eng.)*. One on a transformer or choke which carries a d.c. to neutralize the magnetizing effect of another d.c.

antipriming pipe *(Eng.)*. A pipe placed on the steam space of a boiler, so as to collect the steam while excluding entrained water. See **priming** (1).

antiproton *(Phys.)*. Short-lived particle, half-life 0.05 μsec, identical with proton, but with negative charge; annihilating with normal proton, it yields mesons.

antipyretic *(Med.)*. Counteracting fever; a remedy for fever.

antipyrine *(Med.)*. Formerly used as analgesic and antipyretic.

antiqua *(Print.)*. The German name for **roman type**.

Antiquarian *(Arch.)*. The general term to describe the final phase of the **Renaissance** style when architects reverted to ancient models as a source of inspiration; Greek, Roman, Gothic and to a certain extent Egyptian architecture were studied and it became fashionable to travel and to compile portfolios of drawings of ancient ruins. It is regarded as a reaction to the flamboyant **Baroque** phase which preceded.

antiquark *(Phys.)*. The antiparticle of a **quark**.

antique *(Paper)*. The surface finish originally applied to machine-made papers made in imitation of handmade printings. The term is now used to describe any rough-surfaced paper which bulks well, e.g. book or cover paper.

antique *(Print.)*. A bold type face known as *Antique Roman*. The lines of the letters are almost uniform in thickness.

antiresonance frequency *(Electronics)*. Frequency at which the parallel impedance of a tuned circuit rises to a maximum.

antiroll bar *(Autos.)*. Torsion bar mounted transversely in the chassis in such a way as to counteract the effect of opposite spring deflections.

antisag bar *(Build.)*. A vertical rod connecting the main tie of a roof truss to the ridge to support it against sagging under its own weight.

antisepsis *(Med.)*. The inhibition of growth, or the destruction of bacteria in the field of operation by chemical agent: the principle of antiseptic treatment.

antiseptic *(Med.)*. Counteracting sepsis or contamination with bacteria: an agent which destroys bacteria or prevents their growth.

anti-set-off spray *(Print.)*. To prevent **set-off** the surface of each freshly-printed sheet receives a layer of fine particles, which prevent contact with the succeeding sheet.

anti-set-off tympan cover *(Print.)*. A top cover for the second cylinder of any perfecting press, flat-bed or rotary, consisting of a material coated with very small glass beads.

antisolar glass *(Glass)*. Glass which absorbs heat from sunshine and reduces glare, but transmits most of the light.

antisound *(Acous.)*. Sound signal with same amplitude but opposite phase of some unwanted sound signal so that both signals cancel each other when superimposed. Used in **active control**.

antispin parachute *(Aero.)*. A small parachute, normally in a canister, which may be fixed to the tail (occasionally to the wing tips) of an aircraft or glider for release in emergency to lower the nose into a dive and so assist recovery from a spin. It is jettisoned after use. Colloq. *spin chute*.

antispray film *(Elec.Eng.)*. An oil film placed on the surface of accumulator cells to prevent the formation of acid spray due to the bursting of gas bubbles during the charging process.

antistatic agent *(Textiles)*. A substance applied to a textile to render it less prone to becoming charged with static electricity by friction during processing or in wear.

antistatic fluid *(Acous.)*. That applied to a direct-recording disk before cutting a record, to obviate swarf adhering to the surface, because of generation of electric charge. Also applied to finished gramophone disks to avoid the attraction of dust, and to sound film, to avoid static discharges and consequent scratches on the sound track.

anti-Stokes lines *(Phys.)*. Those in scattered or fluorescent light with frequencies greater than that in the incident radiation, because of departure of atoms or molecules from their normal states.

antisurge valve *(Aero.)*. A valve for bleeding off surplus compressor air to suppress the unstable airflow due to **surge** in a gas turbine engine.

antisymmetric *(Phys.)*. Pattern or waveform in which symmetry is complete except for one particular, e.g., sign of electric charge, direction of current, or of components in waveform. A system containing several electrons must be described quantum mechanically by an *antisymmetric eigenfunction*.

antisymmetric dyadic *(Maths.)*. See **conjugate dyadics**.

antisymmetric function *(Maths.)*. See **alternating function**.

antithetic alternation of generations *(Bot.)*. (1) See **antithetic theory of alternation**. (2) Sometimes the same as *heteromorphic alternation of generations*.

antithetic theory of alternation *(Bot.)*. The hypothesis that the sporophyte is a novel phase in the life cycle resulting from the postponement of meiosis. Cf. **homologous theory of alternation**.

antitoxin *(Immun.)*. Antibody capable of neutralizing toxins which are made usually by microbes, e.g. tetanus antitoxin, diphtheria antitoxin. Used in treatment when the main damaging agent is the toxin. Produced by immunizing horses or other animals, but liable to cause **serum sickness**. Nowadays antibodies from pre-immunized humans are used when possible, and in future human monoclonal antibodies may become available.

antitrades *(Meteor.)*. Winds, at a height of 900 m or more, which sometimes occur in regions where tradewinds are prevalent, their direction being opposite to that of the tradewinds.

anti-transmit-receive tube *(Radar)*. Gas discharge tube which isolates a pulsed radar transmitter from the antenna so that echoes can be received. Abbrev. *ATR tube*. Cf. **transmit-receive tube**.

antitranspirant *(Bot.)*. Substance which reduces transpiration (and, usually, also photosynthesis) e.g. by causing stomatal closure or by forming a more or less impermeable surface film; of some use horticulturally when transplanting.

antivivisectionists *(Med.)*. Those who oppose experiments on live animals.

Antonoff's rule *(Chem.)*. The interfacial tension between two liquid phases in equilibrium is equal to the difference of the surface tensions of the two phases.

antorbital *(Zool.)*. In front of the orbit. In Vertebrates, a small bone in the nasal region.

antrorse *(Zool.)*. Directed or bent forward.

antrum *(Zool.)*. A sinus, as the maxillary sinus in Vertebrates: a cavity, as the **antrum of Highmore**. pl. *antra*.

antrum of Highmore *(Med.)*. An air-containing cavity in the maxilla which communicates with the nasal cavity.

Antrycide *(Vet.)*. TN for drug which shows low toxicity and effective trypanocidal powers in cases of *T. congolense*, *T. vivax*, *T. evansi*, *T. brucei*, *T. simiae*, in cattle and various domestic animals.

Anura *(Zool.)*. See **Salientia**.

anural, anurous *(Zool.)*. Without a tail: pertaining to *Anura*.

anuria *(Med.)*. Complete failure to secrete urine.

anus *(Zool.)*. The opening of the alimentary canal by which indigestible residues are voided, generally posterior. adj. *anal*.

anvil *(Eng.)*. A block of iron, sometimes steel-faced, on which work is supported during forging.

anvil *(Med.)*. One of the three small bones (ossicles) which transmit mechanical vibrations between the outer ear drum and the inner ear. See also **incus**.

anvil cloud *(Meteor.)*. A common feature of a thundercloud, consisting of a wedge-shaped projection of cloud suggesting the point of an anvil.

anvil cutter, anvil chisel *(Eng.)*. A chisel with a square shank for insertion in the hardy hole of a smith's anvil, the cutting edge being uppermost.

anxiety *(Behav.)*. An unpleasant psychological state most commonly identified with intense fear; the situational determinants are not always apparent.

anxiolytic *(Med.)*. Drug used for relieving anxiety states.

aorta *(Zool.)*. In Arthropoda, Mollusca, and most Vertebrates, the principal arterial vessel(s) by which the oxygenated blood leaves the heart and passes to the body: in Amphibians, the principal artery by which blood passes to the posterior part of the body, formed by the union of the systemic arteries: in Fish (*ventral aorta*), the vessel by which the blood passes from the heart to the gills, and also (*dorsal aorta*) the vessel by which the blood passes from the gills to the body. adj. *aortic*.

aortic arches *(Zool.)*. In Vertebrates, a series of pairs of vessels arising from the *ventral aorta*.

aortic incompetence *(Med.)*. A defect in aortic valve function which allows blood to regurgitate from the aorta to the left ventricle during diastole when the valve should be tight shut. Occurs in **rheumatic heart disease** and in **infective endocarditis** but may also complicate **syphilis**, **ankylosing spondylitis** and **Reiter's syndrome**.

aortitis *(Med.)*. Inflammation (usually syphilitic) of the aorta.

AP *(Biol.)*. See **alkaline phosphatase**.

ap- *(Genrl.)*. Another form of *an-*.

AP *(Surv.)*. Abbrev. for *Amsterdamsch Peil* (Amsterdam level), i.e. the datum, or mean level, used as a basis for levels in Holland, Belgium and N. Germany.

apatetic coloration *(Zool.)*. See **cryptic coloration**.

apatite *(Min.)*. Phosphate of calcium, also containing fluoride, chloride, hydroxyl or carbonate ions, according to the variety. It is a major constituent of sedimentary phosphate rocks, of the bones and teeth of vertebrate animals (including Man), and it is usually present as an accessory mineral in igneous rocks.

aperient *(Med.)*. A drug having a laxative or purgative effect.

aperiodic *(Phys.)*. Said of any potentially vibrating system, electrical, mechanical, or acoustic, which, because of sufficient damping, does not vibrate when impulsed. Used particularly of the pointers of indicating instruments, which having no natural period of oscillation, do not oscillate before coming to rest in the final position, and so give their ultimate reading as fast as possible.

aperiodic *(Telecomm.)*. Said of any device or circuit (e.g. antenna, amplifier) which does not exhibit any variation in characteristics with varying frequency of applied signals.

aperiodic antenna *(Telecomm.)*. One with useful efficiency over a range of radio frequencies, terminated to minimize resonance by reflection, e.g. **Beverage antenna**, **rhombic antenna**. Also called *non-resonant antenna*.

aperturate *(Bot.)*. Pollen grains having one or more apertures, that is, areas of the wall where the exine is thinner or absent and through which the pollen tube may emerge. Cf. **colpus**, **pore**.

aperture *(Image Tech.,Phys.)*. (1) The opening, usually circular, through which light enters an optical system, such as a camera lens; its area may be varied by an **iris** diaphragm to control the amount of light passing. See **aperture number**, **f-number**, **numerical aperture**, **stop**. (2) The rectangular opening at which motion picture film is exposed in a camera or projector.

aperture *(Radar,Telecomm.)*. The effective area over which an aerial extracts power from an incident plane wave. The aperture (A) and gain (G) are related by the equation $G=4\pi A/\lambda^2$.

aperture correction *(Image Tech.)*. One form of enhancement of signal differences at image boundaries to increase apparent sharpness.

aperture distortion *(Image Tech.)*. That arising from the scanning spot having finite, instead of infinitely small, dimensions.

aperture efficiency *(Telecomm.)*. The ratio of an antenna's actual directivity to the theoretical figure which would be obtained with ideal aperture illumination, i.e. with uniform electromagnetic field strength over its aperture.

aperture number *(Image Tech.)*. Same as **f-number**.

aperture plate *(Image Tech.)*. Plate carrying the opening at which film is exposed or projected.

aperture synthesis *(Astron.)*. Two or more radio telescope antennas are connected as pairs of **interferometers** in this technique. The amplitude and phase of the interference pattern is continuously recorded. The interferometer baseline is normally variable, and the rotation of the Earth changes the position angle with respect to a distant radio source. The **Fourier transform** of the amplitude and phase patterns are then used to compute a map of the radio source. By means of long baselines, achieved by linking telescopes on different continents, it is possible to achieve a **resolving power** of 0.001 arc seconds or so.

apetaly *(Bot.)*. Absence of petals. adj. *apetalous*.

apex *(Bot.)*. The end of an organ or plant part remote from its point of attachment or origin, e.g. root tip or shoot tip.

apex *(Genrl.)*. The top or pointed end of anything. adj. *apical*.

apex *(Med.)*. Said of the root of a tooth, of the top of the upper lobe of a lung, or of the rounded end of the left ventricle of the heart.

apex *(Min.Ext.)*. The 'outcrop' (exposure) or upper edge of a vein reef or lode.

apex beat *(Med.)*. The point that is furthest out and furthest down the chest where the heart beat is visible or palpable.

apex law *(Min.Ext.)*. The law entitling the discoverer of an outcrop or exposure of ore to exploit it in depth beyond its lateral boundaries.

apex stone *(Build.)*. Triangular stone at the summit of a gable, often decorated with a carved trefoil. Also termed *saddle stone*.

apgar score *(Med.)*. A scoring system for assessing a baby's condition at birth of which a value of 0, 1 or 2 is given to each of five signs; colour, heart-rate, muscle tone, breathing effort and response to stimulation. A score of 10 indicates the baby is in excellent condition.

aphagia *(Med.,Zool.)*. Inability to swallow or feed.

aphakia *(Med.)*. Absence of the lens of the eye.

aphasia *(Behav.,Med.)*. Loss of, or defect in, language function due to a lesion in certain association areas of the brain. May be unable to comprehend speech or written word (receptive) or be defective in writing or speech (expressive).

aphelion *(Astron.,Space)*. The farthest point from the Sun on a planet's, comet's or spacecraft's orbit. pl. *aphelia*.

apheliotropic *(Bot.)*. Turning away from the sun.

aphids *(Zool.)*. Insects of the family Aphididae (order Hemiptera). Reproduction is either sexual or parthenogenetic, oviparous of viviparous, giving rise to a complex life cycle. Greenfly.

aphonia *(Med.)*. Loss of voice.

aphotic zone *(Ecol.)*. The zone of the sea below about 1500 metres which is essentially dark. Cf. **photic zone**.

aphototropic *(Bot.)*. (1) Usually *not phototropic*. (2) Less commonly and confusingly, *negatively phototropic*.

aphthous fever *(Vet.)*. See **foot-and-mouth disease**.

aphthous ulcer *(Med.)*. A small grey ulcer in the mouth.

Apiaceae *(Bot.)*. Same as **Umbelliferae**.

apical body *(Zool.)*. See **acrosome**.

apical cell *(Bot.)*. A single cell at the apex of a filament, multicellular thallus or organ from which all the cells of the filament or organ are descended.

apical cells *(Zool.)*. In some Invertebrates, e.g. the Limpet (*Patella*), during cleavage of the ovum, a quartette of small cells at the apex of the egg.

apical dome *(Bot.)*. The usually dome-shaped part of an apical meristem distal to the most recently formed leaf primordium.

apical dominance *(Bot.)*. The influence of a terminal bud in inhibiting or controlling the growth of buds or lateral branches on the shoot below it, ceasing if the terminal bud is destroyed.

apical growth *(Bot.)*. (1) The elongation of tubular cell or hypha by continued growth at the apex only (the normal pattern for root hairs, pollen tubes and fungal hyphae). (2) The condition in which the only transverse divisions in a filament of cells take place in the apical cell.

apical meristem *(Bot.)*. A group of meristematic cells at the tip of a thallus, stem or root, which divide to produce the precursor of the cells of the thallus, or of the primary tissues of root or shoot. There may or may not be a distinct apical cell.

apical placentation *(Bot.)*. The condition in which the ovule or ovules is/are inserted at the top of the ovary.

apical plate *(Zool.)*. In various pelagic larval forms, such as trochophores, tornariae, echinoplutei, and larvae of some Crinoidea, an aggregation of columnar ectoderm cells at the apical pole, usually bearing cilia.

apical sense organ *(Zool.)*. In Ctenophora, an elaborate sensory structure formed of small otoliths united into a morula, supported on 4 pillars of fused cilia and covered by a roof of fused cilia.

apicolysis *(Med.)*. An operation for compressing or collapsing the apex of the lung; as in the treatment of pulmonary tuberculosis.

apiculate *(Bot.)*. Ending in a short, sharp point.

API scale *(Phys.)*. American Petroleum Institute scale. Scale of relative density, similar to Baumé scale. Degrees $API=(141.5/s)-131s$, where s is the rel.d. of the oil against water at 15°C.

apituitarism *(Med.)*. Absence or deficiency of pituitary gland secretion; also *hypohypophysism*. See also **hypopituitarism**.

Apjohn's formula *(Phys.)*. A formula which may be used for determining the pressure of water vapour in the air from readings of the wet and dry bulb hygrometer. The formula is: $p_t=p_w-0.000\,75H(t-t_w)\;[1-0.008(t-t_w)]$, where p_w is the saturated vapour pressure at the temperature (t_w) of the wet bulb, H is the barometric height, and t is the temperature of the dry bulb.

APL *(Comp.)*. Scientific programming language using a special character set, and syntax designed to aid the programming of mathematics. *A Programming Language*.

aplacental *(Zool.)*. Without a placenta.

aplanatic *(Phys.)*. Said of an optical system which produces an image free from spherical aberration.

aplanatic refraction *(Phys.)*. Refraction at a surface under conditions in which there is no spherical aberration and in which the sine condition is satisfied.

aplanetic *(Bot.)*. Organisms which are non-motile or lack a motile stage.

aplanogamete *(Bot.,Zool.)*. A nonmotile gamete.

aplanospore *(Bot.)*. A non-motile spore e.g. autospore, hypnospore. Cf. **zoospore**.

aplasia *(Med.)*. Defective structural development.

aplite *(Geol.)*. A fine textured, light coloured, igneous rock. It may range in composition from granitic to gabbroic, but more commonly the former.

apneusis *(Med.)*. State of maintained inspiration.

apneustic *(Zool.)*. Possessing no organs specialized for respiration; in some aquatic insect larvae, having no functional spiracles, respiration taking place through the general body surface or by means of gills.

apneustic centre *(Zool.)*. In the higher Vertebrates, that part of the brain which controls the inflation of the lungs.

apnoea *(Med.)*. Cessation of breathing. Recognized to

occur in sleep in obese and other persons. *Sleep apnoea syndrome*.

apnoea *(Zool.)*. In forms with pulmonary respiration, cessation of respiratory movements, due to diminution of carbon dioxide tension in the alveolar air.

apo- *(Genrl.)*. Prefix from Gk. *apo*, away.

apocarpous *(Bot.)*. A gynoecium consisting of two or more free (i.e. not fused) carpels. Cf. **syncarpous**.

apochromatic lens *(Phys.)*. A lens so designed that it is corrected for chromatic aberration for three wavelengths thus reducing the secondary spectrum.

apochromatic objective *(Phys.)*. Microscope objective in which spherical and chromatic aberrations have been corrected as completely as possible.

Apoda *(Zool.)*. (1) An order of Amphibians having a cylindrical snakelike body without limbs, reduced eyes, and an anterior sensory tentacle; burrowing forms, living near water and feeding chiefly upon earthworms. Caecilians.

apodal, apodous *(Zool.)*. Without feet: without locomotor appendages.

apodeme *(Zool.)*. In Arthropoda, an ingrowth of the cuticle forming an internal skeleton and serving for the insertion of muscles: in Insects, more particularly, an internal lateral chitinous process of the thorax.

apodous larva *(Zool.)*. A type of insect larva in which the trunk appendages are completely suppressed; formed in some Coleoptera, Diptera, Hymenoptera and Lepidoptera.

apogamy *(Bot.)*. The development of a sporophyte directly from a cell of the gametophyte without fusion of gametes so that the resulting sporophyte has the same chromosome number as the parent gametophyte. Cf. **apospory, apomixis**. adj. *apogamous*.

apogee *(Astron.,Space)*. The point in the orbit of the Moon or an artificial satellite which is furthest from the Earth. Also the highest altitude attained by a missile.

apogee motor *(Space)*. Engine fired at the apogee of an elliptical orbit to establish a circular orbit whose altitude is that of the apogee of the original orbit. Similarly a *perigee motor* for transforming a circular orbit into an eccentric one.

A-point *(Eng.)*. *Ae point*. Temperature above which steel can be hardened. The equilibrium point of the *transformation temperature*.

Apollo *(Space)*. The NASA program for putting a man on the Moon and returning him to Earth. The Apollo spacecraft consisted of a *Command Module* (CM), a *Service Module* (SM) and a *Lunar Excursion Module* (LEM). The CM-SM combination remained in lunar orbit (with one astronaut) while the lunar landing (with two astronauts) was effected by the LEM. Earth re-entry being effected by the CM only, the crew and CM being recovered from the sea. The ensemble was launched by a Saturn V rocket. Using this system, Neil Armstrong and Buzz Aldrin became the first men to step on the Moon. (July 1969).

Apollo asteroid *(Astron.)*. An **asteroid** whose orbit brings it within 1 AU of the Sun.

Apollonius' circle *(Maths.)*. If A and B are two fixed points, then the locus of the point P, which moves so that the ratio $PA{:}PB$ is a constant, is the *circle of Apollonius*.

Apollonius' theorem *(Maths.)*. If the base BC of a triangle ABC is divided by a point P, such that

$$\frac{BP}{PC}=\frac{r}{s}, \text{ then}$$

$$s.\,AB^2+r.\,AC^2=(s+r).\,AP^2+s.\,PB^2+r.\,PC^2.$$

Apollo program *(Astron.)*. Manned spaceflight program of the USA 1968-72, leading to the first lunar landing on 20 July 1969.

apomecometer *(Surv.)*. Instrument, based on optical square, for measuring heights and distances.

apomixis *(Bot.)*. (1) *Agamospermy*, reproduction by seeds formed without sexual fusion. (2) Any form of asexual reproduction, including vegetative propagation.

apomorphine *(Med.)*. An alkaloid of the morphine series, obtained from morphine by dehydration. It is not a narcotic, but is an expectorant and emetic.

apophyllite *(Min.)*. A secondary mineral occurring with zeolites in amygdales in basalts and other igneous rocks. Composition: hydrated fluorosilicate of potassium and calcium.

apophysis *(Geol.)*. A veinlike offshoot from an igneous intrusion.

apophysis *(Zool.)*. In Vertebrates, a process from a bone, usually for muscle attachment: in Insects, a ventral chitinous ingrowth of the thorax for muscle insertion.

apoplast *(Bot.)*. That part of the plant body which is external to the living protoplasts i.e. the cell walls, the intercellular spaces and the lumina of dead cells such as xylem vessels and tracheids or in some contexts, the water filled parts of this space. Cf. **symplast**.

apoplexy *(Med.)*. Sudden loss of consciousness and paralysis as a result of haemorrhage into the brain or of thrombosis of a cerebral artery.

apoprotein *(Biol.)*. The protein component of a conjugated protein. The *globin* of haemoglobin.

aporogamy *(Bot.)*. The entrance of the pollen tube into the ovule by a path other than through the micropyle.

aposematic coloration *(Zool.)*. Warning coloration, often yellow and black as in some stinging insects.

apospory *(Bot.)*. The development of a gametophyte directly from a sporophyte cell without meiosis and the formation of spores. The resulting gametophyte has the same chromosome number as the parent sporophyte. Cf. **apogamy, apomixis**.

apostilb *(Phys.)*. A unit of surface luminance used in the case of diffusing surfaces, numerically equal to 1/10 000 **lambert** ($1/\pi\,\mathrm{cd\,m^{-2}}$).

apostrophe *(Bot.)*. The position assumed by chloroplasts in bright light, when they lie against the radial walls of the cells of the palisade layer of the mesophyll.

apothecaries' weight *(Chem.)*. Obsolete system based on the *grain* as unit: 1 grain = 15.4 g.

apothecium *(Bot.)*. An open ascocarp, often cup- or saucer-shaped.

Appalachian orogeny *(Geol.)*. A fold period in eastern N. America that extended from the Devonian to the Permian. It was caused by *subduction* which led to the closure of the Atlantic ocean. See **subduction zone**.

apparent cohesion *(Civ.Eng.)*. Cohesion of silts and sands due to surface tension in the enclosed films of water, which films tend to pull the silt grains together.

apparent expansion, coefficient of *(Phys.)*. See **coefficient of apparent expansion**.

apparent horizon *(Surv.)*. See **visible horizon**.

apparent magnitude *(Astron.)*. See **magnitudes**.

apparent particle density *(Powder Tech.)*. The mass of a particle of powder divided by the volume of the particle, excluding open pores but including closed pores.

apparent powder density *(Powder Tech.)*. The mass of the powder divided by the volume occupied by it under specified conditions of packing.

apparent power *(Phys.)*. The volt-amperes, i.e., the product of volts and amperes in an a.c. circuit or system.

apparent solar day *(Astron.)*. The interval, not constant owing to the earth's elliptical orbit, between two successive transits of the true sun over the meridian.

apparent solar time *(Astron.)*. Time as measured by the apparent position of the Sun in the sky, e.g. by a sundial.

appearing *(Print.)*. Term referring to the depth of the actual printed matter on a page, exclusive of traditional white line at foot.

appeasement behaviour *(Behav.)*. Submissive behaviour which inhibits attack by a conspecific, often by minimizing threat signals or by mimicking sexual or infantile behaviours, e.g. crouching or sexual invitation.

appendage *(Bot.)*. General term for any external outgrowth which does not appear essential to growth or reproduction of the plant.

appendage *(Zool.)*. A projection of the trunk, as the parapodia and tentacles of Polychaeta, sensory tentacle of Apoda, fins of Fish, and limbs of land Vertebrates; in

Arthropoda, almost exclusively one of the paired, metamerically arranged, jointed structures with sensory, masticatory or locomotor function, but also used for the wings of Insecta.

appendicectomy, appendectomy *(Med.)*. The surgical removal of the **appendix vermiformis**.

appendicitis *(Med.)*. Inflammation of the appendix vermiformis.

appendicular *(Zool.)*. Pertaining to, or situated on, an appendage.

appendix *(Zool.)*. An outgrowth.

appendix vermiformis *(Med.,Zool.)*. In some Mammals, the distal rudiment of the caecum of the intestine, which in Man is a narrow, blind tube of gut, from 25–250 mm in length. See **appendicitis**.

appetitive behaviour *(Behav.)*. A term used to refer to the active exploratory phase that precedes the presumed goal of a behaviour sequence. Traditionally the appetitive phase is said to lead to **consummatory behaviour**.

Appleby-Frodingham process *(Chem.Eng.)*. A fluidized bed process using specially prepared iron oxide for removal of sulphur from coke oven, and coal gas and vapourized oils. Removes H_2S, CS_2, mercaptans and thiophenes, converting sulphur direct to sulphur dioxide for sulphuric acid.

Applegate diagram *(Electronics)*. Presentation of the **bunching** and **debunching** of an electron beam in a velocity-modulation tube, e.g. a klystron.

Appleton layer *(Phys.)*. Same as **F-layer**.

applications programmer *(Comp.)*. Programmer who writes for specific user applications.

applications software *(Comp.)*. Specialized **software** designed to help carry out a real life task such as *stock control*.

applicator *(Elec.Eng.)*. Electrodes used in industrial high-frequency heating or medical diathermy; often specially shaped to fit the sample or body. See also **heating inductor**.

applied geology *(Geol.)*. Geology studied in relation to human activity.

applied mathematics *(Genrl.)*. Originally the application of mathematics to physical problems, differing from physics and engineering in being concerned more with mathematical rigour and less with practical utility. More recently, also includes numerical analysis, statistics and probability, and applications of mathematics to biology, economics, insurance etc.

applied potential tomography *(Phys.)*. A system of medical imaging based on the measurement of the electrical impedance, at about 50 kHz frequency, between many electrodes placed around the body.

applied power *(Phys.)*. That *applied* to an electrical transducer is not equal to the actual power received, because of the reflection arising from nonequality of impedance matching. The *applied power* is the power which would be received if the load matched the source in impedance.

applied psychology *(Genrl.)*. That part of psychology which puts its knowledge to work in practical situations, e.g. in vocational guidance and assessment, education and industry.

applied stress *(Civ.Eng.)*. The stress induced in a member under load.

appliqué *(Textiles)*. Ornament, frequently of fabric or plastic, attached to the surface of a fabric to give a three-dimensional effect.

apposition *(Bot.)*. The addition of new material to a cell wall at the surface next to the plasmalemma. Cf. **intussusception**.

appraisal well *(Min.Ext.)*. Oil well drilled near a discovery well to find out the size of the field.

appressed *(Bot.)*. Flattened, and pressed close to, but not united with, another organ.

appressorium *(Bot.)*. A flattened outgrowth which attaches a parasite to its host; especially a modified hypha, closely applied to the host epidermis. A narrow infection hypha or penetration tube is pushed into the cell or space below the attachment.

approach control radar *(Aero.)*. A surveillance radar which shows on a CRT display the positions of aircraft in an aerodrome's traffic control area. Abbrev. *ACR*.

approach lights *(Aero.)*. Lights indicating the desired approach to a runway, usually of sodium or high-intensity type and laid in a precise pattern of a lead-in line with cross-bars at set distances from the **runway threshold**.

approach speed *(Aero.)*. The indicated air speed at which an aircraft approaches for landing.

approximate integration *(Maths.)*. Method of approximating to the area under a curve by considering a small number of its ordinates. See **Simpson's rule**, **three-eighths rule** and **Weddle's rule**.

appulse *(Astron.)*. Seemingly close approach of two celestial objects as perceived by an observer, particularly the close approach of a planet or asteroid to a star without the occurrence of an eclipse.

apron *(Aero.)*. A firm surface of concrete or 'tarmac' laid down adjacent to aerodrome buildings to facilitate the movement, loading and unloading of aircraft.

apron *(Build.)*. The lead sheeting or other weather resistant material raggled into a wall to divert water into a gutter or drain.

apron *(Eng.)*. In a lathe, that part of the saddle enclosing the gear operated by the lead screw.

apron *(Hyd.Eng.)*. (1) The protecting slope on the downstream side of the sluices of a lock gate or dam provided to withstand the force of the falling water. (2) The bags of concrete, blocks of masonry, etc., deposited around the toe of a sea wall to protect its base from scour caused by the returning wave.

apron *(Image Tech.)*. Flexible strip used as film support in some types of processing tank.

apron *(Paper)*. A strip of rubber, metal or other material at the outlet from the flow box to seal the gap between it and the machine wire.

apron conveyor *(Eng.)*. A conveyor for transporting packages or bulk materials, consisting of a series of metal or wood slats (also rubber, cotton, felt, wire etc.) attached to an endless chain. Also called *slat conveyor*.

apron feeder *(Min.Ext.)*. Short endless conveyor belt, sturdily built of articulated plates, used to draw ore at regulated rate from bottom of stockpile or ore bin.

apron lining *(Build.)*. A lining of wrought boarding covering the apron piece at a staircase landing.

apron piece *(Build.)*. The horizontal timber carrying the upper ends of the carriage pieces or rough-strings of a wooden staircase. Also *pitching-piece*.

aprotic *(Chem.)*. Term normally restricted to solvents such as acetonitrile, which have high **relative permittivities** and hence aid the separation of electric charges but do not provide protons.

apse *(Arch.)*. The semicircular or polygonal recess, either arched or dome-roofed, terminating the choir or chancel of a church. Also called *apsis*.

apse line *(Astron.)*. The diameter of an elliptic orbit which passes through both foci and joins the points of greatest and least distance of the revolving body from the centre of attraction. Also called *line of apsides*.

apterism *(Zool.)*. The condition of winglessness, either primitive or secondary, found in many Insects.

apterous *(Zool.)*. Without wings.

apterygial *(Zool.)*. Without wings: without fins.

Apterygota *(Zool.)*. A subclass of small, primitively wingless Insects showing litle metamorphosis. Bristletails and Springtails.

Aptian *(Geol.)*. A stage of the Cretaceous System lying between the Barremian below and the Albian above.

aptitude *(Behav.)*. A specific ability or capacity to learn. *General aptitude* refers to the capacity for acquiring knowledge in a wide range of areas, as opposed to a *specific aptitude*, such as the ability to acquire musical skills. Aptitude testing is an attempt to measure individual differences in potential for learning, as opposed to achievement testing, which measures present levels of competence in a given area.

APU *(Aero.)*. See **auxiliary power unit**.

apyrexia *(Med.)*. Absence of fever.

aq *(Chem.)*. A symbol representing a large volume of water.

aqua fortis *(Chem.)*. Ancient name for concentrated nitric acid.

aquamarine *(Min.)*. A variety of beryl, of attractive blue-green colour, used as a gemstone.

aqua regia *(Chem.)*. A mixture consisting of 1 volume of concentrated nitric acid to 3 volumes of concentrated hydrochloric acid.

aquatic *(Bot.)*. See **Hydrophyte**.

aquatint *(Print.)*. An intaglio printing process using a copper plate with a ground of resin particles. Half-tone effects are produced by etching and progressive stopping-out.

aqueduct *(Civ.Eng.)*. An artificial conduit, generally elevated, used to convey water, generally for long distances, for supply purposes.

aqueduct *(Zool.)*. A channel or passage filled with or conveying fluid: in higher Vertebrates, the reduced primitive ventricle of the midbrain.

aqueductus Sylvii *(Zool.)*. In Vertebrates, the ventricle of the midbrain; the iter.

aqueductus vestibuli *(Zool.)*. In Craniata, a narrow tube arising from the auditory sac and opening on the dorsal surface of the head, as in some fishes, or ending blindly. The endolymphatic duct.

aqueous *(Chem.)*. Consisting largely of water; dissolved in water.

aqueous humour *(Zool.)*. In Vertebrates, the watery fluid filling the space between the lens and the cornea of the eye.

aqueous tissue *(Bot.)*. Water-storage tissue, made up of large, thin-walled, hyaline cells.

aquiculture *(Zool.)*. Augmentation of aquatic animals of economic importance by direct methods: cultivation of the resources of sea and inland waters as distinct from exploitation.

aquifer *(Geol.)*. Rock formation containing water in recoverable quantities.

Ar *(Chem.)*. (1) Symbol for *argon*. (2) A general symbol for an aryl, or aromatic, radical.

Ar *(Eng.)*. The *transformation temperature* on cooling of the phase changes in iron and steel, subscripts indicating the appropriate change.

AR *(Image Tech.)*. See **aspect ratio**.

AR *(Chem.)*. Abbrev. for *Analytical Reagent*, indicating a definite standard of purity of a chemical.

arabesque *(Arch.)*. An ornamental work used in decorative design for flat surfaces; consists usually of interlocked curves which may be painted, inlaid, or carved in low relief.

arabic numbers *(Maths.)*. The numbers 1, 2, 3 etc., as opposed to the roman numbers I, II, III etc. Actually derived from India.

arabinose *(Chem.)*. *L-arabinose*, $C_5H_{10}O_5$, is produced by boiling gum-arabic, cherry gum, or beetroot chips with dilute sulphuric acid; prisms soluble in water forming a dextrorotatory solution. Arabinose is a monosaccharide belonging to the pentose group. Used as a culture medium for certain bacteria. See **pentoses**. ⇨

arabitol *(Chem.)*. A pentahydric alcohol, $OH.CH_2.(CH.OH)_3.CH_2.OH$. This is the alcohol corresponding to **arabinose**, and obtained from it by reduction with sodium tetrahydroborate.

arachis oil *(Chem.)*. Peanut oil.

Arachnida *(Zool.)*. A class of mainly terrestrial Arthropoda in which the head and thorax are continuous (**prosoma**). The head bears pedipalps and chelicerae but no antennae. There are four pairs of ambulatory legs. Spiders, harvest-men, mites, ticks, scorpions, etc.

arachnidium *(Zool.)*. In Spiders, the spinnerets and silk glands.

arachnodactyly *(Med.)*. Extreme length of fingers and toes seen in **Marfan's syndrome**.

arachnoid *(Bot.,Zool.)*. Cobweblike. Formed of entangled hairs or fibres: pertaining to or resembling the Arachnida: 1 of the 3 membranes which envelop the brain and spinal cord of Vertebrates, lying between the dura mater and the pia mater.

araeostyle *(Arch.)*. A colonnade in which the space between the columns is equal to or greater than 4 times the lower diameter of the columns.

aragonite *(Min.)*. The relatively unstable, orthorhombic form of crystalline calcium carbonate, deposited from warm water, but prone to inversion into calcite; also stable at high pressures. See also **flos ferri**.

Arago point *(Phys.)*. The bright spot found along the axis in the shadow of a disk illuminated normally.

Arago's rotation *(Elec.Eng.)*. Experiments (conducted by Arago before the discovery of electromagnetic induction by Faraday) in which a rotating copper disk was made to cause rotation of a pivoted magnet.

Araldite *(Plastics)*. TN for range of epoxy resins used for adhesives, encapsulation of electrical components etc., because of its chemical resistance.

aramid fibres *(Plastics)*. Fibres made from linear polymers containing recurring amide groups (-CO-NH-) joined directly to two aromatic rings (aramid is derived from *aromatic amide*). The fibres have exceptionally high strength and are frequently used in composite materials.

Araneae *(Zool.)*. An order of Arachnida in which the prosoma is joined to the apparently unsegmented opisthosoma by a waist. Spinnerets and several kinds of spinning glands occur. The pedipalps are modified in the male for the transmission of sperm. Spiders.

araneous *(Zool.)*. Cobweblike.

arbitration bar *(Eng.)*. Test bar, cast with a given heat of metal, to determine whether the main casting is to specification.

arbor *(Eng.)*. (1) Cylindrical or conical shaft on which a cutting tool or part to be machined is mounted. (2) See **mandrel**.

arbor *(Horol.)*. The axis or shaft upon which a rotatable part is mounted: the shaft upon which a wheel or pinion is mounted. See **barrel arbor, fusee**.

arboretum *(Bot.)*. An area devoted to the cultivation of trees and other woody plants.

arbuscule *(Bot.)*. (1) A dwarf tree or shrub of treelike habit. (2) A much-branched haustorium formed within the host cells by some endophytic fungi. See **vesicular arbuscular mycorrhiza**.

ARC *(Aero.)*. Abbrev. for (1) *Aeronautical Research Council*, UK. (2) *Ames Research Centre*, US.

arc *(Elec.Eng.)*. Ionic gaseous discharge maintained between electrodes, characterized by low voltage and high current. See **mercury-arc rectifier**.

arc *(Maths.)*. A portion of a curve.

arc *(Phys.)*. See **arc lamp**. Also *iron arc, mirror arc, open arc, rotary arc, sun arc*.

arc absorber *(Elec.Eng.)*. Same as **spark absorber**, but referring to a discharge likely to be destructive if not extinguished.

arcade *(Arch.)*. (1) A series of arches, usually in the same plane, supported on columns, e.g. the nave arcades in churches. When filled in with masonry, it becomes a 'blind arcade'. (2) An arched passage, especially one having shops on one or both sides.

arc-back *(Electronics)*. Flow of electrons, opposite to that intended, in a mercury-arc rectifier. Caused by a heated spot on the anode acting as a cathode, leading to possible damage.

arc baffle *(Electronics)*. Means of preventing liquid mercury contacting an anode in a mercury-arc rectifier. Also called *splash baffle*.

arc balance *(Horol.)*. See **balance arc**.

arc-control device *(Elec.Eng.)*. A device fitted to the contacts of a circuit breaker to facilitate the extinction of the arc.

arc cosh x *(Maths.)*. Also written $cosh^{-1}x$. See **inverse hyperbolic functions**.

arc cos x *(Maths.)*. Also written $cos^{-1}x$. See **inverse trigonometrical functions**.

arc crater *(Elec.Eng.)*. Depression formed in electrodes between which an electric arc has been maintained. In arc welding, the depression which occurs in the weld metal.

arc deflector *(Elec.Eng.)*. Magnetic arrangement for controlling the position of the arc in an arc lamp. Also *arc shield*.

arc duration *(Elec.Eng.)*. Time during which an arc exists between the contacts of an opening switch or circuit breaker. In a.c. circuits usually measured in cycles, varying between half a cycle and perhaps 20 cycles.

arc furnace *(Elec.Eng.)*. An electric furnace in which the heat is produced by an electric arc between carbon electrodes, or between a carbon electrode and the furnace charge.

arch *(Civ.Eng.)*. A form of structure having a curved shape, used to support loads or to resist pressures.

arch *(Zool.)*. A curved or arch-shaped skeletal structure supporting, covering, or enclosing an organ or organs, as *haemal arch, neural arch, zygomatic arch*.

archaeo- *(Genrl.)*. Prefix from Gk. *archaios*, ancient.

archaeostomatous *(Zool.)*. Having a persistent blastopore, which gives rise to the mouth.

arch brick *(Build.)*. A brick having a wedge shape, especially one with a curved face suitable for wells and other circular work.

arch bridge *(Civ.Eng.)*. A bridge that depends on the principle of the arch for its stability. See **rigid arch, three-hinged arch, two-hinged arch**.

arch dam *(Civ.Eng.)*. In which the abutments are solid in rock at sides of impounding area.

arche- *(Genrl.)*. Prefix from Gk. *archē*, beginning.

Archean *(Geol.)*. A term applied to the oldest Pre-Cambrian rocks in a region, identified as those which have suffered the greatest degree of metamorphism.

archecentra *(Zool.)*. In Vertebrates, centra formed by the enlargement of the bases of the arched elements which grow around the notochord outside its primary sheath; cf. **chordacentra**. adj. *archecentrous, arcicentrous, arcocentrous*.

archegonial chamber *(Bot.)*. A small cavity at the micropylar end of the female gametophyte of some gymnosperms, e.g. cycads, into which the spermatozoids are liberated to swim to the archegonia.

Archegoniatae *(Bot.)*. In some classifications, one of the main groups within the plant kingdom, including the Bryophyta and Pteridophyta. Characterized by the presence of the archegonium as the female organ, and by the regular alternation of gametophyte and sporophyte in the life cycle.

archegoniophore, archegonial receptacle *(Bot.)*. A specialized branch of a thallus bearing archegonia.

archegonium *(Bot.)*. A sessile or stalked organ, bounded by a multicellular wall, and flask-shaped in general outline. It consists of a chimney-like neck containing an axial series of neck-canal cells, and a swollen venter below, containing a single egg and a ventral-canal cell. The archegonium is the female organ of Bryophyta and Pteridophyta, and, in a slightly simplified form, of most Gymnospermae.

archencéphalon *(Zool.)*. In Vertebrates, the primitive forebrain; the cerebrum.

archenteron *(Zool.)*. Cavity in the gastrula, enclosed by endoderm. It opens to the exterior at the blastopore.

archesporium *(Bot.)*. The tissue in a sporangium that gives rise to the spore mother cells, including the region of the nucellus giving rise to the megaspore mother cells.

archetype *(Behav.)*. A notion associated with the psychology of Carl Jung; it refers to an emotionally laden image assumed to be present in the unconscious mind of all human beings throughout history; an aspect of the *collective unconscious*.

archi- *(Genrl.)*. Prefix from Gk. *archi-*, first, chief.

Archiannelida *(Zool.)*. A class of Annelida, of small size and marine habit, which usually lack setae and parapodia and have part of the epidermis ciliated; the nervous system retains a close connection with the epidermis; they resemble the Polychaeta in many of their characteristics.

archiblastic *(Zool.)*. Exhibiting total and equal segmentation: pertaining to the protoplasm of the egg: pertaining to an **archiblastula**.

archiblastula *(Zool.)*. A regular spherical blastula, having cells of approximately equal size.

archicoel *(Zool.)*. See **blastocoel**.

Archimedean drill *(Eng.)*. A drill in which to-and-fro axial movement of a nut on a helix causes an alternating rotary motion of the bit.

Archimedean screw *(Hyd.Eng.)*. An ancient water-lifting contrivance: a hollow inclined screw (or a pipe wound in helix fashion around an inclined axis) which had its lower end in water so that, on rotation of the 'screw', water rose to a higher level.

Archimedes' principle *(Phys.)*. When a body is wholly or partly immersed in a fluid it experiences an upthrust equal to the weight of fluid it displaces; the upthrust acts vertically through the centre of gravity of the displaced fluid.

Archimedes' spiral *(Maths.)*. A spiral with polar equation $r = a\theta$.

archinephric *(Zool.)*. In Vertebrates, pertaining to the archinephros (see **pronephros**): in Invertebrates, pertaining to the larval kidney or **archinephridium**.

archinephridium *(Zool.)*. In Invertebrates, the larval excretory organ, usually a solenocyte.

archinephros *(Zool.)*. See **pronephros**.

archipallium *(Zool.)*. In Vertebrates, that part of the cerebral hemispheres not included in the olfactory lobes and corpora striata, and comprising the hippocampus and the olfactory tracts and associated olfactory matter: that part of the pallium excluding the neopallium.

archipelago *(Geol.)*. A sea area thickly interspersed with islands, originally applied to the part of the Mediterranean which separates Greece from Asia.

architectural acoustics *(Acous.)*. The study of propagation of sound waves in buildings, the results being applied to the design of studios and auditoria for optimum audition and to the noise isolation of buildings.

architrave *(Arch.)*. The lowest part of an entablature in immediate contact with the abacus on the capital of a column. Also called *epistyle*.

architrave *(Build.)*. The mouldings surrounding a door or window opening, including the lintel.

architrave block *(Build.)*. The block, placed at the foot of the side moulding around a door opening, into which the skirting fits.

architrave jambs *(Build.)*. The mouldings at the sides of a door or window opening.

architype *(Zool.)*. A primitive type from which others may be derived.

archive file *(Comp.)*. Copy of file held for safety on a secure and stable storage medium.

archivolt *(Arch.)*. An ornamental moulding carried around the face of an arch.

Archosauria *(Zool.)*. Sub-class of diaspid reptiles which were the dominant forms in the Mesozoic. Only surviving group are the Crocodiles and Alligators.

arch piece *(Ships)*. See **stern frame**.

arch stone *(Civ.Eng.)*. A stone shaped like a wedge, and used as a constituent part on the centre line of an arch. Also called *voussoir*.

arcicentrous *(Zool.)*. See **archecentra**.

arcing contact *(Elec.Eng.)*. An auxiliary contact fitted to a switch or circuit breaker, arranged so that it opens after and closes before the main contact, thereby bearing the brunt of any burning due to the arc which occurs when a circuit is interrupted. Designed for easy replacement. Also called *arcing tips*.

arcing-ground suppressor *(Elec.Eng.)*. See **arc suppressor**.

arcing ring *(Elec.Eng.)*. Circular or oval ring conductor, placed concentrically with a pin insulator or a string of insulators, for deflecting an arc from the insulator surface which could be damaged.

arcing shield *(Elec.Eng.)*. See **grading shield**.

arcing tips *(Elec.Eng.)*. See **arcing contact**.

arcing voltage *(Elec.Eng.)*. That below which a current cannot be maintained between 2 electrodes.

arc lamp *(Phys.)*. A form of electric lamp which makes use of an electric arc between 2 carbon electrodes as the source of light. It has an extremely high intrinsic brilliance, and is therefore used for searchlights, spotlights, etc. See **carbon arc-lamp**.

arc-lamp carbon *(Elec.Eng.)*. A cylindrical stick of carbon used as the electrode of a carbon arc lamp. The diameter is usually between about ¼ in. and 1 in. (6 mm and 25 mm).

arcocentrous *(Zool.)*. See **archecentra**.

arc of approach *(Eng.)*. The arc on the pitch circle of a gear wheel over which 2 teeth are in contact and approaching the pitch point.

arc of contact *(Eng.)*. The arc on the pitch circle of a gear wheel over which 2 teeth are in contact.

arc of recess *(Eng.)*. The arc on the pitch circle of a gear wheel over which 2 teeth are in contact while receding from the pitch point.

arc shield *(Elec.Eng.)*. See **arc deflector**.

arc sinh x *(Maths.)*. Also written $sinh^{-1}x$. See **inverse hyperbolic functions**.

arc sin x *(Maths.)*. Also written $sin^{-1}x$. See **inverse trigonometrical functions**.

arc spectrum *(Phys.)*. A spectrum originating in the nonionized atoms of an element; usually capable of being excited by the application of a comparatively low stimulus, such as the electric arc. See **spark spectra**.

arc spraying *(Eng.)*. Method of fusing refractory ceramic or metal powders by blowing them through an electric arc or plasma.

arc-stream voltage *(Elec.Eng.)*. Voltage drop along the arc stream of an electric arc, excluding the voltage drops at the anode and cathode.

arc suppression coil *(Elec.Eng.)*. See **Petersen coil**.

arc suppressor *(Elec.Eng.)*. A device for automatically earthing the neutral point of an insulated-neutral transmission or distribution line in the event of an arcing ground being set up. Also called *arcing-ground suppressor*.

arc tanh x *(Maths.)*. Also written $tanh^{-1}x$. See **inverse hyperbolic functions**.

arc tan x *(Maths.)*. Also written $tan^{-1}x$. See **inverse trigonometrical functions**.

arc therapy *(Radiol.)*. That in which the angle of rotation of therapeutic radiation is limited, to avoid certain important regions, e.g. lungs.

arc-through *(Electronics)*. Overflow of electron stream into an intended nonconducting period.

arcuate *(Bot.,Zool.)*. Bent like a bow.

arcus senilis *(Med.)*. Degeneration of the periphery of the cornea in old people. May occur earlier in **hypercholesterolaemia**.

arc voltage *(Elec.Eng.)*. The total voltage across an electric arc, i.e. the sum of the arc stream voltage, the voltage drop at the anode and the voltage drop at the cathode. The term is frequently used in connection with arc welding, and with the arc in a switch or circuit breaker.

arc welding *(Elec.Eng.)*. Process for joining of metal parts by fusion in which the heat necessary for fusion is produced by an electric arc struck between 2 electrodes or between an electrode and the metal.

are *(Surv.)*. A metric unit of area used for land measurement. 1 are = 100 m^2, = 119.6 yd^2. See also **hectare**.

area *(Build.)*. The sunken space around the basement of a building, providing access and natural lighting and ventilation.

area *(Surv.)*. In plane surveying, the superficial content of a ground surface of definite extent, as projected on to a horizontal plane.

areal velocity *(Astron.)*. The rate, constant in elliptic motion, at which the radius vector sweeps out unit area.

area-moment method *(Civ.Eng.)*. A method of structural analysis based on the slope and displacement of any part of the structure.

area monitoring *(Radiol.)*. The survey and measurement of types of ionizing radiation and dose levels in an area in which radiation hazards are present or suspected.

area opaca *(Zool.)*. In developing tetrapods, a whitish peripheral zone of blastoderm, in contact with the yolk.

area pellucida *(Zool.)*. In developing tetrapods, a central clear zone of blastoderm, not in direct contact with the yolk.

area rule *(Aero.)*. An aerodynamic method of reducing drag at transonic speeds by maintaining a smooth cross-sectional variation throughout the length of an aircraft. Because of the effect of the wing, this often results in a 'wasp-waist' on the fuselage or the addition of bulges to the wing or fuselage.

area vasculosa *(Zool.)*. In developing tetrapods, part of the extraembryonic blastoderm, in which the blood vessels develop.

Arecaceae *(Bot.)*. Same as **Palmae**.

Arecidae *(Bot.)*. Subclass or superorder of monocotyledons. Trees, shrubs, terrestrial herbs and a few free-floating aquatics, mostly with broad, petiolate leaves often net-veined, inflorescence of usually numerous small flowers generally subtended by a spathe and often aggregated into a spadix. Contains ca. 6400 spp. in 5 families including Palmae, Araceae, Pandanaceae and Lemnaceae.

arenaceous, arenicolous *(Bot.,Zool.)*. (1) Plants growing best in sandy soil. (2) Animals occurring in sand. (3) Composed of sand or similar particles, as the shells of some kinds of Radiolaria.

arenaceous rocks *(Geol.)*. Sedimentary rocks in which the principal constituents are sand grains, including the various sorts of sands and sandstones.

Arenig *(Geol.)*. The lowest (oldest) stage of rocks in the Ordovician System, taking their name from Arenig mountain in N. Wales, where they were originally described by Adam Sedgwick.

arenite *(Geol.)*. The general term for any sedimentary rock with sand-sized grains.

areola *(Zool.)*. (1) One of the spaces between the cells and fibres in certain kinds of connective tissue. (2) In the Vertebrate eye, that part of the iris bordering the pupil. (3) In Mammals, the dark-coloured area surrounding the nipple. pl. *areolae*.

areolar, areolate *(Bot.,Zool.)*. (1) Divided into small areas or patches. (2) Pitted. (3) Pertaining to an areola.

areolar tissue *(Zool.)*. A type of connective tissue consisting of cells separated by a mucin matrix in which are embedded bundles of white and yellow fibres.

areole *(Bot.)*. (1) A small area delimited in some way, especially (a) an island into which a reticulated and veined leaf is divided by the veins, and (b) an area demarcated by the network of cracks in a lichen thallus. (2) A cushion, representing a condensed lateral shoot from which spines, branches and flowers arise in cacti.

arfvedsonite *(Min.)*. A monoclinic iron-rich alkali-amphibole.

Arg *(Chem.)*. Symbol for **arginine**.

Argand burner *(Eng.)*. A form of gas- or oil-burner in which air is admitted to the inside of a cylindrical wick, ensuring a large area of contact between the flame and the fuel.

Argand diagram *(Elec.Eng.)*. The vector diagram for showing the magnitude and phase angle of a vector quantity with reference to some other vector quantity.

Argand diagram *(Maths.)*. A plane diagram, named after its description by the Swiss J.R.Argand in 1806 but actually first described by the Norwegian Caspar Wessel in 1797, in which every complex number can be represented uniquely by a single point. The diagram represents the complex number $z = x + iy$ by the point (x,y) referred to rectangular cartesian axes Oxy, the axis Ox being called the *real axis* and the axis Oy the *imaginary axis*.

argentate *(Bot.)*. Of silvery appearance.

argentic (silver (II)) oxide *(Chem.)*. AgO, an oxide of silver.

argentiferous *(Min.)*. Containing silver.

argentite *(Min.)*. An important ore of silver, having the composition Ag_2S (silver sulphide); crystallizes in the cubic system. Cf. **acanthite**. Also called *silver glance*.

argentous (silver (I)) oxide *(Chem.)*. Formula, Ag_2O. A lower oxide of silver.

argillaceous rocks *(Geol.)*. Sediments of silt or clay-particle size. Common clay-minerals are kaolinite and montmorillonite.

argillicolous *(Bot.)*. Living on a clayey soil.

argillite *(Geol.)*. A slightly metamorphosed siltstone or mudstone which lacks fissility.

arginine *(Chem.)*. *2-amino-5-guanidopentanoic acid*, $H_2N.C(NH).NH.(CH_2)_3.CH(NH_2).COOH$. The L- or S-isomer is an essential amino acid. Symbol Arg, short form R. ⇨

argon *(Chem.)*. An element which forms no known compound, one of the rare gases. Symbol Ar, at.no. 18, r.a.m. 39.948. A colourless, odourless, monatomic gas; mp −189.2°C; bp −185.7°C; density 1.7837 g/dm^3 at s.t.p. Argon constitutes about 1% by volume of the atmosphere, from which it is obtained by the fractionation of liquid air. It is used in gas-filled electric lamps, radiation counters, fluorescent tubes etc.

argon laser *(Phys.)*. One using singly-ionized argon. It gives strong emission at 488.0, 514.5 and 496.5 nm.

argument *(Comp.)*. Input parameter to a program.

argument *(Maths.)*. Of a function $f(x)$: the variable x. Of a complex number $x+iy$: the angle $\tan^{-1} y/x$ which is the angle in the Argand diagram between the line from the origin to the point representing the complex number and the positive direction of the real axis.

Argyll-Robertson pupil *(Med.)*. An irregular eccentric pupil which reacts to accommodation, but not light.

argyrodite *(Min.)*. A double sulphide of germanium and silver, the mineral in which the element germanium was first discovered.

arid zone *(Ecol.)*. A zone of latitude 15°–30°N and S in which the rainfall is so low that only desert and semi-desert vegetation occurs, and irrigation is necessary if crops are to be grown.

aril *(Bot.)*. An outgrowth on a seed, formed from the stalk or from near the micropyle. It may be spongy or fleshy, or may be a tuft of hairs.

ARINC *(Aero.)*. *Aeronautical Radio INCorporated*, an organization (US) whose membership includes airlines, aircraft constructors and *avionics* component manufacturers. It publishes technical papers and agreed standards, and finances research.

Aristotle's lantern *(Zool.)*. In Echinoidea, the framework of muscles and ossicles supporting the teeth, and enclosing the lower part of the oesophagus.

arithmetic *(Genrl.)*. The science of numbers, including such processes as addition, subtraction, multiplication, division and the extraction of roots.

arithmetical operation *(Comp.)*. An **operation** performed using the laws of arithmetic. Cf. **logical operation**.

arithmetical progression *(Maths.)*. A sequence of numbers, each term being obtained from the preceding one by the addition of the common difference, e.g. $a,(a+d),(a+2d)\ldots(a+(n-1)d),\ldots$ where d is the common difference. The sum of the series of the first n terms

$$\sum_{r=0}^{n-1}(a+rd)=\frac{n}{2}\{2a+(n-1)d\}.$$

arithmetic continuum *(Maths.)*. The aggregate of all real numbers, rational and irrational.

arithmetic logic unit *(Comp.)*. Circuits within the central processing unit where arithmetic/logic operations are performed. Also *arithmetic unit*. Abbrev. *ALU*.

arithmetic mean *(Maths.)*. Of n numbers a_r: their sum divided by n, i.e.

$$\frac{1}{n}\sum_{1}^{n}a_r.$$

arithmetic operator *(Comp.)*. Symbol used to indicate the arithmetic operation to be performed (e.g. '+' in 5+7). See **operator**.

arithmetic register *(Comp.)*. Special store location usually part of the **arithmetic logic unit**, used to hold operands and results temporarily during processing.

arithmetic shift *(Comp.)*. One where, if bits are shifted from the right of the location, they are lost and copies of the sign bit are shifted in at the opposite end. If the data represents a number, this operation preserves the positive/negative sign.

arithmetic unit *(Comp.)*. See **arithmetic logic unit**.

arkose *(Geol.)*. A feldspar-rich, coarse-grained sandstone, derived from the erosion of granites and gneisses.

ARM *(Aero.)*. Abbrev. for *Anti-Radiation Missile*.

arm *(Telecomm.)*. See **branch**.

arm *(Zool.)*. In Echinodermata, a prolongation of the body in the direction of a radius: in Cephalopoda, one of the tentacles surrounding the mouth: in bipedal Mammals, one of the upper limbs.

armature *(Elec.Eng.)*. (1) Moving part which closes a magnetic circuit and which indicates the presence of electric current as the agent of actuation, as in all relays, electric bells, sounders, telephone receivers. (2) Piece of low-reluctance ferromagnetic material (*keeper*) for temporarily bridging the poles of a permanent magnet, to reduce the leakage field and preserve magnetization. (3) The rotating part (**rotor**) of a d.c. motor or generator.

armature *(Telecomm.)*. The magnetic part of a *rocking-armature* telephone earpiece; the free end is attached to the diaphragm.

armature bars *(Elec.Eng.)*. Rectangular copper bars forming the conductors on the armature in large electric machines having only a few conductors per slot.

armature coil *(Elec.Eng.)*. An assembly of conductors ready for placing in the slots of the armature of an electric machine.

armature conductor *(Elec.Eng.)*. One of the wires or bars on the armature of an electric machine.

armature core *(Elec.Eng.)*. The assembly of laminations forming the magnetic circuit of the armature of an electric machine. The thickness of each lamination is usually of the order of 0.5 mm.

armature end connections *(Elec.Eng.)*. The portion of the armature conductors which project beyond the end of the armature core, and which are used for making the connections among the various conductors. Also called *overhang*.

armature end plate *(Elec.Eng.)*. The end plate of a laminated armature core. It is of sufficient mechanical strength to enable the laminations to be clamped together tightly to prevent vibration. Sometimes also called *armature head*.

armature ratio *(Elec.Eng.)*. Ratio of distance moved by the spring buffer of an electromagnetic relay, to that moved by the armature.

armature reactance *(Elec.Eng.)*. A reactance associated with the armature winding of a machine, caused by armature leakage flux, i.e. flux which does not follow the main magnetic circuit of the machine.

armature reaction *(Elec.Eng.)*. The magnetic field in an electrical machine produced by the armature current.

armature relay *(Elec.Eng.)*. A relay operated electromagnetically, thus causing the armature to be magnetically attracted.

armature winding *(Elec.Eng.)*. The complete assembly of conductors carried on the armature and connected to the commutator or to the terminals of the machine.

Armco *(Eng.)*. TN for a soft iron with less than 1% impurities. Resistivity 6.2 by volume and 5.4 by mass, compared with copper.

armed *(Bot.)*. Protected by prickles, thorns, spines, barbs etc.

armillary sphere *(Astron.)*. Celestial globe, first used by the Greek astronomers, in which the sky is represented by a skeleton framework of intersecting circles, the Earth being at the centre. In antiquity, of major importance for measuring star positions.

arming press *(Print.)*. A form of blocking press used for stamping designs on book covers.

armour-clad switchgear *(Elec.Eng.)*. See **metal-clad switchgear**.

armour clamp *(Elec.Eng.)*. A fitting designed to grip the armouring of a cable where it enters a box. Also called *armour gland, armour grip*.

armour plate *(Eng.).* Specially heavy alloy steel plate forged in hydraulic presses, hardened on the surface; used for the protection of warships. Approximate composition: C 0.2–0.4%, Cr 1.0–3.5%, Ni 1.5–3.5% and Mo 0–0.5%.

Armstrong oscillator *(Telecomm.).* The original oscillator, in which tuned circuits in the anode and grid circuits of a valve are coupled.

Arndt-Eistert reaction *(Chem.).* Used for converting a carboxylic acid to a higher homologue. The acid chloride is added to an excess of diazomethane to form a diazoketone. The ketone undergoes catalytic rearrangement to the higher homologue or a derivative.

aromatic compounds *(Chem.).* Compounds related to benzene. Ring compounds containing **conjugated double bonds**.

aromatic hydrogenation *(Chem.).* Hydrogenation in the naphthalene series, of such nature that hydrogenation takes place only in the unsubstituted benzene ring.

aromatic properties *(Chem.).* The characteristic properties of aromatic compounds, e.g. reaction with concentrated nitric acid, forming nitro derivatives, reaction with concentrated sulphuric acid, forming sulphonated derivatives. The homologues of benzene differ from alkanes with regard to oxidation by readily forming benzene carboxylic acids. There are many other distinguishing characteristics between aromatic hydrocarbons and alkanes.

arousal *(Behav.).* A general psychophysiological concept referring to the effect of various non-specific stimulation or motivational factors on a number of physiological variables, e.g. heart rate, skin resistance. It is used to describe differences in responsiveness to general stimulation, usually along a continuum from drowsiness to alertness, for example.

ARPA *(Comp.). Advanced Research Projects Agency.* Supported by US government grant money and now renamed DARPA, *Defence Advanced Research Projects Agency.*

ARPA Internet *(Comp.).* A linkage of several US networks including **ARPANET** and **MILNET**. It exists to facilitate sharing resources and collaboration by participating research organisations as well as to provide a testbed for new developments in networking.

ARPANET *(Comp.).* A long-distance **packet switching** US network used by research interests funded by **ARPA**.

array *(Comp.).* Set of storage locations referenced by a single identifier. Individual elements of the array are referenced by combining one or more *subscripts* with the identifier, e.g. NICK(20) is an element in the array NICK, and JOS(3,5) is an element in the two-dimensional array JOS.

array *(Stats.).* A set of values for a particular variate.

array *(Telecomm.).* Used to describe an assembly of two or more individual radiating elements, appropriately spaced and energized to achieve desired directional properties. See **beam antenna**.

array bounds *(Comp.).* Limits on the number of items in an array.

array dimension *(Comp.).* Number of subscripts necessary to identify an item in an array (e.g. CLAR(26,3) has dimension 2).

array processor *(Comp.).* One designed to allow any machine instruction to operate on a number of data locations simultaneously.

arrectores pilorum *(Zool.).* In Mammals, unstriated muscles attached to the hair follicles, which cause the hair to stand on end by their contraction.

arrested crushing *(Min.Ext.).* Crushing so conducted that the rock falling through the machine is free to drop clear of the zone of comminution when broken smaller than the exit orifice or set.

arrested failure *(Elec.Eng.).* The taking of a cable off voltage, and examination before failure is complete. This is very instructive in determining the mechanism of breakdown.

arrester *(Elec.Eng.).* See **lightning arrester**.

arrester gear *(Aero.).* (1) A device on aircraft carriers and some military aerodromes, usually consisting of a number of individual transverse cables held by hydraulic shock-absorbers, which stop an aircraft when its **arrester hook** catches a cable, (2) A barrier net, usually of nylon or webbing attached to heavy drag weights, which stops fast aircraft from over-running the end of the runway in an emergency.

arrester hook *(Aero.).* A hook extended from an aircraft to engage the cable of an arrester gear, mainly on aircraft carriers.

arrest points *(Eng.).* Discontinuities on heating and cooling curves, due to absorption of heat during heating or evolution of heat during cooling, and indicating structural (phase) changes occurring in a metal or alloy.

Arrhenius theory of dissociation *(Chem.).* The description of aqueous solutions in terms of acids, which dissociate to give hydrogen ions, and bases, which dissociate to give hydroxyl ions. The product of the reaction of an acid and a base is a salt and water. The dissociation of these species gives their solutions the property of conducting electricity.

arrhenotoky *(Zool.).* Parthenogenetic production of males.

arrhythmia *(Med.).* Abnormal rhythm of the heart beat.

arris *(Build.).* The (generally) sharp exterior edge formed at the intersection of two surfaces not in the same plane (e.g. the meeting of two sides of a stone block). See also **external angle**.

arris edge *(Glass).* Small bevel, of width not exceeding 1/16 in (1.5 mm), at an angle of approximately 45° to the surface of the glass.

arris fillet *(Build.).* A small strip of wood of triangular cross-section packed beneath the lower courses of slates or tiles on a roof to throw off the water which might otherwise get under the flashing.

arris gutter *(Build.).* A V-shaped gutter, usually made of wood.

arris rail *(Build.).* A rail, with triangular cross-section, secured to posts for fences in such a manner as to show the arris in front.

arris tile *(Build.).* Purpose-made angular tile used to cover the intersections at hips and ridges in slated and tiled roofs. See also **bonnet tile**.

arris-wise *(Build.).* A term used to describe the sawing of square timber diagonally.

arrow *(Surv.).* Light steel wire pin, bent into ring at one end and perhaps flagged with piece of bright cloth, used to mark measured lengths in chain traversing.

arsenic *(Chem.).* Symbol As, at. no. 33, r.a.m. 74.9216, oxidation states 3, 5. An element which occurs free and combined in many minerals. An impurity of several commercial metals. Called grey or γ-arsenic to distinguish it from the other allotropic modifications. Mp 814°C (36 atm.), bp 615°C (sublimes), rel.d. 5.73 at 15°C. Used in alloys and in the manufacture of lead shot. It is important as donor impurity in germanium semiconductor devices. The arsenic of commerce, As_2O_3, *Arsenious oxide, arsenic oxide.* Obtained from the roasting of arsenical ores. It is highly poisonous, and its presence in foods and drinks is subject to severe restriction. Medical uses, once important, have much declined, but still used as a herbicide and rodenticide.

arsenic acid *(Chem.).* H_3AsO_4. Formed by the action of hot dilute nitric acid upon arsenic, or by digesting arsenic (III) oxide with nitric acid. Arsenic acid is also formed when arsenic (V) oxide is dissolved in water.

arsenical copper *(Eng.).* Copper containing up to about 0.6% arsenic. This element slightly increases the hardness and strength and raises the recrystallization temperature.

arsenical pyrites *(Min.).* See **arsenopyrite**.

arsenic halides *(Chem.).* Arsenic (V) fluoride, AsF_5; arsenic (III) fluoride, AsF_3; arsenic (III) chloride, $AsCl_3$; arsenic (III) bromide, $AsBr_3$; arsenic (III) iodide, AsI_3.

arsenide *(Chem.).* Arsenic unites with most metals to form *arsenides*; e.g. iron – $FeAs_2$. Arsenides are decomposed by water or dilute acids with the formation of the hydride **arsine**.

arsenious acid *(Chem.).* Solution of arsenious oxide. See **white arsenic**.

arsenites *(Chem.).* Arsenates (III). Salts of arsenious acid.

arseniuretted hydrogen *(Chem.).* See **arsine**.

arsenolite *(Min.).* Arsenic oxide, a decomposition product of arsenical ores; occurring commonly as a white incrustation, rarely as octahedral crystals.

arsenopyrite *(Min.).* Sulphide of iron and arsenic; the chief ore of arsenic. Also known as *mispickel*.

arsine *(Chem.).* AsH_3. Arsenic (III) hydride. Produced by the action of nascent hydrogen upon solutions of the element, or by the action of dilute sulphuric acid upon sodium or zinc arsenide. Very poisonous. Also called *arseniuretted hydrogen*. Arsines are organic derivatives of AsH_3 in which one or more hydrogen atoms is replaced by an alkyl radical; other hydrogen atoms may also be replaced by halogen, etc.

Art Deco *(Arch.).* A style of design which took its name from the Exposition des Arts Decoratif, an international trade fair held in Paris in 1925. In the 1920s the style was characterized by curvilinear shapes and by stylized human and animal forms, but by the 1930s a commitment to industrial technology dictated shapes which tended to lend themselves to methods of mass production; hence the sleek, angular, linear forms in stainless steel, aluminium, chromed metals and colourful glazed tiles and plastics. Architects saw Art Deco as an alternative to the unadorned glass, concrete and steel structures of contemporary architectural design, and motifs, often reminiscent of Egyptian art forms, were employed structurally and as surface decoration.

artefact *(Zool.).* Any apparent structure which does not represent part of the actual specimen, but is due to faulty preparation.

artefact, artifact *(Genrl.).* A simple man-made stone, wood or metal implement.

arterial drainage *(Build.).* A system of drainage in which the flow from a number of branch drains is led into one main channel.

arterial system *(Zool.).* That part of the vascular system which carries the blood from the heart to the body.

arteriography *(Radiol.).* The radiological examination of arteries following direct injection of a **contrast medium**, e.g. coronary arteriography, renal arteriography, carotid arteriography.

arteriole *(Zool.).* A small artery.

arteriosclerosis *(Med.).* Hardening or stiffening of the arteries due to thickening and loss of elasticity of arterial walls. Commonly but incorrectly used to imply **atherosclerosis**.

arteritis *(Med.).* Inflammation of an artery.

artery *(Zool.).* One of the vessels of the vascular system, that conveys the blood from the heart to the body. adj. *arterial*.

Artesian well *(Civ.Eng.).* A well sunk into a permeable stratum which has impervious strata above and below it, and which outcrops at places higher than the place where the well is sunk, so that the hydrostatic pressure of the water in the permeable stratum is alone sufficient to force the water up out of the well. Named from Artois (France).

arthralgia *(Med.).* Pain in a joint.

arthrectomy *(Med.).* Excision of a joint.

arthritic *(Zool.).* Pertaining to the joints: situated near a joint.

arthritis *(Med.).* Inflammation of a joint.

arthrodesis *(Med.).* The surgical immobilization of a joint by fusion of the joint surfaces.

arthrodia *(Zool.).* A joint.

arthrodial membranes *(Zool.).* In Arthropoda, flexible membranes connecting adjacent body sclerities and adjacent limb joints, and occurring also at the articulation of the appendages.

arthrography *(Radiol.).* The radiological examination of a joint cavity after direct injection of air or other **contrast media**.

Arthrophyta *(Bot.).* Division of the plant kingdom, the horsetails and allies, here treated as the class Sphenopsida.

Arthropoda *(Zool.).* A phylum of metameric animals having jointed appendages (some of which are specialized for mastication) and a well-developed head; there is usually a hard chitinous exoskeleton; the coelom is restricted, the perivisceral cavity being haemocoelic. Centipedes, Millipedes, Insects, Crabs, Lobsters, Shrimps, Spiders, Scorpions, Mites, Ticks, etc.

arthrospore *(Bot.).* Spore resulting from hyphal fragmentation.

arthrotomy *(Med.).* Surgical incision into a joint.

Arthus reaction *(Immun.).* A type III **allergic reaction** named after the person who first described it.

articular(e) *(Zool.).* Pertaining to, or situated at, or near, a joint. In Vertebrates, a small cartilage at the angle of the mandible, derived from the Meckelian, and articulating with the quadrate forming the lower half of the jaw hinge. pl. *articularia*.

articulated *(Bot.).* Jointed or segmented; divided into portions that may easily be separated.

articulated blade *(Aero.).* A rotorcraft blade which is mounted on one or more hinges to permit flapping and movement about the **drag axis**.

articulation *(Arch.).* The means by which an architect gives definition to the individual elements of a building.

articulation *(Eng.).* The connection of 2 parts in such a way (usually by a pin joint) as to permit relative movement.

articulation *(Telecomm.).* Percentage of specified speech components (usually **logatoms**) received over a communication system; may be (a)*word*: percentage of words correctly received; (b)*syllable*: percentage number of meaningless syllables correctly recognized; (c)*sound*: percentage number of fundamental speech-sounds (consonant, vowel, initial or final consonant) correctly recognized.

articulation *(Zool.).* The movable or immovable connection between 2 or more bones.

artifact *(Biol.,Radiol.).* An error in an image which has no counterpart in reality.

artificial ageing *(Eng.).* See **temper-hardening**.

artificial antenna *(Telecomm.).* Combination of resistances, capacitances, and inductances with the same characteristics as an antenna except that it does not radiate energy. It is used in place of the normal antenna for purposes such as repair and checking of a transmitter, or for re-tuning of the transmitter onto a different frequency. Also *dummy antenna, phantom antenna*.

artificial classification *(Bot.,Zool.).* A classification based on one or a few arbitrarily chosen characters, and giving no attention to the natural relationships of the organism; the old grouping of plants into trees, shrubs, and herbs was an *artificial classification*.

artificial community *(Bot.).* A plant community kept in existence by artificial means; e.g., a garden habitat or a cloche.

artificial daylight *(Phys.).* Artificial light having approximately the same spectral distribution curve as daylight, i.e., having a colour temperature of about 4000 K.

artificial disintegration *(Phys.).* The transmutation of nonradioactive substances brought about by the bombardment of the nuclei of their atoms by high-velocity particles, such as α-particles, protons, or neutrons.

artificial ear *(Acous.).* Device for testing earphones which presents an acoustic impedance similar to the human ear and includes facilities for measuring the sound pressure produced at the ear.

artificial earth *(Telecomm.).* See **counterpoise**.

artificial feel *(Aero.).* In an aircraft flying control system, especially with *automatic control of flying surfaces*, in which the pilot's control actions are modified to provide forces moving the flying controls, a natural feel, opposing the pilot's actions, is fed back from the controls. Since these forces vary mostly with dynamic air pressure ($q = \frac{1}{2}\varrho v^2$) artificial feel is sometimes known as '*q*'-feel.

artificial flags *(Build.).* See **concrete paving slabs**.

artificial horizon *(Aero.).* See **gyro horizon**.

artificial horizon *(Surv.)*. An apparatus, for example a shallow trough filled with mercury, used in order to observe altitudes of celestial bodies with a sextant on land, i.e. where there is no visible horizon. The reflection of the object in the artificial horizon is viewed directly and the object itself indirectly by reflection from the index glass of the sextant.

artificial insemination *(Vet.)*. See **insemination**.

artificial intelligence *(Comp.)*. (1) The concept that computers can be programmed to assume capabilities thought to be like human intelligence such as learning, reasoning, adaptation and self-correction. (2) An extensive branch of **computer science** embracing **pattern recognition, knowledge based systems, computer vision, robotics**, *scene analysis*, **natural language processing, mechanical theorem proving** with more areas being added all the time. Abbrev. *AI*.

artificial kidney *(Med.)*. Machine which is used to replace the function of the body's own organ, when the latter is faulty or has ceased to function. The patient's blood is circulated through sterile semipermeable tubing lying in a suitable solution and is purified by dialysis.

artificial larynx *(Med.)*. A reed actuated by the air passing through an opening in front of throat to assist articulation of person who has undergone tracheotomy operation.

artificial line *(Telecomm.)*. Repeated network units which have collectively some or all of the transmission properties of a line; also *simulated line*.

artificial planet *(Space)*. A space probe which has sufficient velocity to escape from the Earth's gravitational field, but not from that of the Sun, so that it revolves about that body as a member of the solar system.

artificial pneumothorax *(Med.)*. See **pneumothorax** (2).

artificial radioactivity *(Phys.)*. Radiation from isotopes after high energy bombardment in an accelerator by α-particles, protons and other light nuclei, or by neutrons in a nuclear reactor. Discovered by I. Curie-Joliot and F. Joliot in 1933. See **activation**.

artificial rubber *(Plastics)*. See **synthetic rubber**.

artificial satellite *(Space)*. Man-made space vehicle whose velocity is sufficient to maintain it in orbit about another body. Earth satellites are used for the purposes of observation of the surface, the atmosphere, the Sun and deep space, as communication links, in weather forecasting and for the performance of microgravity and other technological experiments.

artificial stone *(Build.)*. A precast imitation of natural stone made in block moulds.The interior of the block is of concrete, the required exterior face of cement mixed with dust or chippings of the natural stone to be imitated.

artificial traffic *(Telecomm.)*. Automatically generated calls which are deliberately mixed with subscriber-originated traffic to sample the over-all service provided by the switching equipment of an automatic exchange, by recording or holding faults recognized by test equipment.

artificial voice *(Acous.)*. Loudspeaker and baffle for simulating speech in testing of microphones.

Artinskian *(Geol.)*. A stratigraphical stage in the Lower Permian rocks of Russia and eastern Europe.

artiodactyl *(Zool.)*. Possessing an even number of digits.

Artiodactyla *(Zool.)*. The one order of the Mammalia containing the 'even-toed' hooved 'Ungulates', i.e. those with a **paraxonic foot**. Includes pigs, peccaries, hippopotami, camels, llamas, giraffes, sheep, buffaloes, oxen, deer, gazelles, and antelopes.

Art Nouveau *(Arch.)*. A decorative movement in European art which occurred during the late 19th and early 20th centuries. It was basically the stylized adaption of plant forms and was used by artists and architects mainly for surface decoration, although the characteristic curvilinear forms were incorporated into the structure, notably in France and Spain.

art paper *(Paper)*. Paper coated on one or both sides with one or more applications of an aqueous suspension of adhesive and mineral matter, such as china clay, to provide a surface(s) suitable for high-class colour print reproduction.

arundinaceous *(Bot.)*. Reedlike and thin.

aryl *(Chem.)*. A term for aromatic monovalent hydrocarbon radicals; e.g. C_6H_5Cl is an aryl halide.

aryl amines *(Chem.)*. Amino derivatives of the aromatic series, e.g. $C_6H_5NH_2$ (aniline).

arytaenoid *(Zool.)*. (1) Vertebrates, one of a pair of anterior lateral cartilages, forming part of the framework of the larynx. (2) In general, pitcher-shaped.

ASA speed *(Image Tech.)*. *American Standards Association* photographic speed rating, expressed on an arithmetic scale.

asbestos *(Min.)*. The fibrous form of minerals which are resistant to heat and mineral attack and may be used for making fireproof fabrics, brake linings, electrical and heating insulation, etc. Toxic as inhaled dust. Minerals used for asbestos are principally chrysotile (serpentine), and the amphiboles actinolite, amosite, anthophyllite and crocidolite (blue asbestos).

asbestos cement *(Build.)*. An inexpensive, but brittle, fire-resisting and weather-proof, non-structural building material, made from Portland cement and asbestos; it is rolled into various forms such as plain sheets, corrugated sheets, roofing slates, rainwater goods etc.

asbestosis *(Med.)*. Disease of the lungs due to inhalation of asbestos particles, causing severe pulmonary fibrosis and may be associated with malignant growth of the pleura (mesothelioma).

asbestos shingles *(Build.)*. A fire-resisting roof-covering, consisting of asbestos cement, made into the form of shingles.

asbolane, asbolite *(Min.)*. A form of **wad** – soft earthy manganese dioxide, containing cobalt.

asbolite *(Min.)*. Synonym of **asbolane**.

ascending letters *(Print.)*. Letters the top portions of which rise above the general level of the line; e.g., *b, d, f, h*.

ascending node *(Space)*. For Earth, the point at which a satellite crosses the equatorial plane travelling from south to north.

ascertainment *(Biol.)*. In human genetics, the way by which families come to the notice of the investigator. The method of ascertainment may lead to biassed data.

Aschelminthes *(Zool.)*. Phylum of invertebrate animals which have in common the possession of a pseudocoelom and an unsegmented elongate body with terminal anus and a non-muscular gut.

Aschoff's nodes, Aschoff's bodies *(Med.)*. Inflammatory nodules found in rheumatic inflammation of the heart.

ascidium *(Bot.)*. A pitcher-shaped leaf or part of a leaf.

ASCII *(Comp.)*. See **character code**.

ASCII keyboard *(Comp.)*. Provides the full range of ASCII characters including **control characters**.

ascites *(Med.)*. Accumulation of fluid in the peritoneal cavity.

ascocarp, ascoma *(Bot.)*. The fruiting body of the Ascomycotina consisting of a sterile wall more or less enclosing the asci. See also **apothecium, perithecium**.

ascolichen *(Bot.)*. One of the majority of *lichens* in which the fungal constituent is an ascomycete.

ascoma *(Bot.)*. See **ascocarp**.

ascomycete *(Bot.)*. A fungus of the Ascomycotina.

Ascomycotina, Ascomycetes *(Bot.)*. Subdivision or class of those Eumycota or true fungi in which the sexual spores are formed in **asci** usually within ascocarps. No motile stages. Usually mycelial with hyphae with simple septa; some are yeasts. Asexual reproduction by conidia. Includes the Hemiascomycetes, Plectomycetes, Pyrenomycetes and Discomycetes.

ascorbic acid *(Biol.)*. See **vitamin C**.

ascospore *(Bot.)*. Spore, typically, uninucleate and haploid, formed within an ascus.

ascus *(Bot.)*. Specialised, usually more or less cylindrical cell within which (usually 8) ascospores are formed following fusion of 2 heterokaryotic nuclei in the ascomycete reproduction.

asdic *(Acous.)*. Underwater acoustic detecting system which transmits a pulse and receives a reflection from underwater objects, particularly submarines, at a distance. Also used by trawlers to detect shoals of fish. Equivalent to US *sonar*. (*Allied Submarine Detection Investigation Committee*).

asepalous *(Bot.)*. Devoid of sepals.

asepsis *(Med.)*. Freedom from infection.

aseptate *(Bot.)*. Not divided into segments or cells by septa.

asexual *(Biol.)*. Without sex; lacking functional sexual organs.

asexual reproduction *(Biol.)*. Any form of reproduction not depending on a sexual process or on a modified sexual process.

ash *(Chem.)*. Nonvolatile inorganic residue remaining after the ignition of an organic material.

ash *(For.)*. A tree (*Fraxinus*) yielding a tough and elastic timber. Typical uses are for ladders, hammer and tool handles, spokes, oars, poles, camp furniture, coffins and artificial limbs.

ash *(Geol.)*. See **volcanic ash**.

ash curve *(Min.Ext.)*. Graph which shows result of sink-and-float laboratory test in form of relationship between specific gravity of crushed small particles and the ash content at that gravity.

ashen light *(Astron.)*. A faint glow sometimes seen in that part of the disc of Venus that is not directly illuminated by the Sun.

Ashgill *(Geol.)*. The youngest epoch of the Ordovician.

ashlar *(Build.)*. (1) Masonry work in which the stones are accurately squared and dressed to given dimensions so as to make very good joints over the whole of the touching surfaces. (2) A thin facing of squared stones or thin slabs laid in courses, with close-fitting joints, to cover brick, concrete, or rubble walling.

Asn *(Chem.)*. Symbol for **asparagine**.

Asp *(Chem.)*. Symbol for **aspartic acid**.

asparagine *(Chem.)*. $NH_2.CO.CH_2CH(NH_2).COOH$. The monoamide of aspartic acid. Symbol Asn, short form N. ⇨

asparagus stone *(Min.)*. Apatite of a yellowish-green colour, thus resembling asparagus.

aspartic acid *(Chem.)*. *2-aminobutanedioic acid*, HOOC. $CH_2.CH(NH_2).COOH$. The L- or S- isomer is a constituent of proteins. Mp 271°. An **amino acid**, formed by the hydrolysis of **asparagine**. Symbol Asp, short form D. ⇨

aspect *(Aero.)*. See under **attitude**.

aspect *(Bot.)*. (1) Degree of exposure to sun, wind etc. of a plant habitat. (2) Effect of seasonal changes on the appearance of vegetation.

aspect *(Civ.Eng.)*. On railways the indication given by a coloured light signal, as contrasted with that of a semaphore arm signal. A multiple-aspect signal (MAS) conveys more information.

aspect ratio *(Aero.)*. The ratio of span/mean chord line of an aerofoil (usually in wing); defined as $(span)^2/area$. Important for **induced drag** and range/speed characteristics. Normal figure between 6 and 9, lesser values than 6 being *low aspect ratios*, greater than 9 *high aspect ratios*.

aspect ratio *(Image Tech.)*. Ratio of the width to the height of the reproduced picture, e.g. 4×3, often expressed with the height as unity, 1.33:1. **Wide-screen** systems have aspect ratios between 1.65:1 and 2.35:1. Abbrev. *AR*.

aspect ratio *(Nuc.Eng.)*. In fusion machines, the ratio of the major to minor radii of a torus.

asperate, asperous *(Bot.)*. Having a rough surface due to short, upstanding stiff hairs.

aspergillosis *(Med.)*. A disease of the lungs caused by the fungus *Aspergillus fumigatus*. May cause allergic reaction in the bronchioles to give asthma or may infect old cavities to give a ball-like growth or *aspergilloma*. In severely immune compromised patients may spread beyond the lung.

aspergillosis *(Vet.)*. *Pneumomycosis, brooder pneumonia*. An infection of the respiratory organs of birds by fungi of the genus *Aspergillus*. In cattle the fungus is a cause of abortion.

Aspergillus *(Bot.)*. A form-genus of Deuteromycotina. Includes parasites (causing e.g. *aspergillosis*), saprophytes, food-spoilage organisms (see **aflatoxins**). Used to prepare soy sauce and industrial enzymes.

asperity *(Phys.)*. Actual region of contact between two surfaces, elastically and plastically flattened to take the load (normal force).

aspermia *(Med.)*. Complete absence of spermatozoa.

asphalt *(Geol.)*. A bituminous deposit formed in oil-bearing strata by the removal, usually through evaporation, of the volatiles. Occurs in the 'tar pools' of California and elsewhere and in the 'pitch lake' in Trinidad, whence enormous quantities are exported.

asphalt *(Min.)*. The name given to various bituminous substances which may be (1) of natural occurrence (see above), (2) a residue in petroleum distillation, (3) a mixture of asphaltic bitumen and granite chippings, sand or powdered limestone. Asphalt is used extensively for paving, road-making, damp-proof courses, in the manufacture of roofing felt and paints and as the raw material for certain moulded plastics. See **bitumen, mastic asphalt**.

asphaltenes *(Chem.)*. Such constituents of asphaltic bitumens as are soluble in carbon disulphide but not in petroleum spirit. See **carbenes, malthenes**.

asphaltite *(Min.)*. A group name for the organic compounds albertite, anthraxolite, grahamite, impsonite, libollite, nigrite, and uintaite.

aspheric surface *(Phys.)*. A lens surface which departs to a greater or lesser degree from a sphere, e.g. one having a parabolic or elliptical section.

asphyxia *(Med.)*. Suffocation due to lack of inspired oxygen.

aspirated psychrometer *(Meteor.)*. A **psychrometer** which uses a forced draught of at least 8 mi/h (12 km/h) over the wet bulb.

aspiration *(Med.)*. The removal of fluids or gases from the body by suction.

aspiration pneumonia *(Med.)*. Pneumonia due to inhalation of food, drink and gastric contents.

aspirator *(Chem.)*. A device for drawing a stream of air or oxygen or liquid through an apparatus by suction.

aspirin *(Chem.)*. The common name for acetylsalicylic acid.

asplanchnic *(Zool.)*. Having no gut.

assay *(Chem.)*. The quantitative analysis of a substance to determine the proportion of some valuable or potent constituent, e.g. the active compound in a pharmaceutical or metals in an ore. See **dry assay, wet assay**.

assay balance *(Chem.)*. A balance specially made for weighing the small amounts of matter met with in assaying. See also **chemical balance**.

assayer *(Chem.)*. A person who carries out the process of assay. See also **dry assay, wet assay**.

assay ton *(Eng.)*. Used in assaying precious metals. It is equivalent to 29.16 g and 32.67 g for the short and long ton respectively. The number of mg of precious metal in an assay ton of ore indicates the **assay value**, since 1 mg of precious metal per assay ton = 1 troy oz of precious metal per avoirdupois ton of ore.

assay value *(Eng.)*. Troy ounces of precious metal per avoirdupois ton of ore.

assemble edit *(Image Tech.)*. Videotape editing in which a new scene is added to follow directly on existing material.

assembler *(Comp.)*. Program, usually provided by the computer manufacturer, to translate a program written in **assembly language** into **machine code**. In general, each assembly language instruction is changed into one machine-code instruction. Also *assembly program*. Cf. **compiler, disassembler**.

assembly *(Comp.)*. Process of converting a program written in **assembly language** into **machine code**.

assembly language *(Comp.)*. **Low-level** programming language, generally using **symbolic addresses**, which is translated into machine code by an **assembler**.

assign *(Comp.)*. Place a value in the memory location corresponding to a given variable.

assigned frequency *(Telecomm.)*. That assigned as centre frequency of a class of transmission, with tolerance, by authority.

assigning authority *(Ships)*. A national body authorized to assign **load lines** to ships.

assimilation *(Behav.)*. In Piagetian theory, one of the two main biological forces responsible for cognitive development; it refers to the ability to absorb new information into pre-existing cognitive organizations (*schemata*); see also **accommodation**.

assimilation *(Bot.)*. The metabolic processes, mostly anabolic, by which the mostly inorganic substances, taken up by plants, are converted into the constituents of the plant body. Includes photosynthesis.

assimilation *(Geol.)*. The incorporation of extraneous material in *igneous magma*.

assimilation *(Zool.)*. (1) Conversion of food material into protoplasm, after it has been ingested, digested, and absorbed. (2) Resemblance of an animal to its surroundings, not only by coloration but also by configuration.

assimilatory quotient *(Bot.)*. Same as **photosynthetic quotient**.

assisted take-off *(Aero.)*. Supplementing the full power of the normal engines by auxiliary means, which may or may not be jettisonable. Small turbojet or rocket motor units, powder, or liquid rockets may be used. See **JATO, RATOG**.

assize *(Build.)*. A cylindrical block of stone forming part of a column, or of a layer of stone in a building.

associate Bertrand curves *(Maths.)*. See **conjugate Bertrand curves**.

associated emission *(Electronics)*. That which brings about equilibrium between incident photons and secondary electrons in ionization.

associated liquid *(Chem.,Phys.)*. One in which molecules of same kind form complex structure, e.g. water. See **hydrogen bond**.

association *(Bot.)*. A plant community usually occupying a wide area, consisting of a definite population of species, having a characteristic appearance and habitat, and stable in its duration.

association *(Print.)*. In rotary printing, the bringing together of separate webs, after printing, to pass through the folder as a complete product.

association *(Zool.)*. In certain Sporozoa, adherence of individuals without fusion of nuclei: a characteristic set of animals, belonging to a particular habitat.

associative *(Maths.)*. A binary operation * is associative if $(a*b)*c = a*(b*c)$ for all a,b,c in the set concerned. Thus in ordinary arithmetic + and × are associative, − and ÷ are not.

associative learning *(Behav.)*. Refers to learning through the formation of associations between ideas or events based on their co-occurrence in past experience. The term originated in philosophy, but it is now most often used as a synonym for learning through both *classical* and *operant conditioning* procedures.

associative memory *(Comp.)*. A type of memory in which the location of each item of data is derived from its content, thus avoiding the use of an address. See **key**.

associative storage *(Comp.)*. Storage which is identified by means of content rather than by an address. Also *content-addressable storage*.

assortative mating *(Bot.)*. Non-random mating caused by e.g. pollinating insects, which may cause preferential inbreeding or outbreeding.

astable circuit *(Electronics)*. An active circuit, having two quasi-stable states, which alternates automatically and continuously between them, e.g. certain *multivibrators* and *blocking oscillators*.

A-stage *(Plastics)*. Stage at which a synthetic resin of the phenol formaldehyde type is fusible and wholly soluble in alcohols and acetone.

astatic galvanometer *(Phys.)*. Moving-magnet galvanometer in which adjustable magnets form an astatic system.

astatic system *(Phys.)*. Ideally an arrangement of two or more magnetic needles on a single suspension so that in a uniform magnetic field, such as the earth's field, there is no resultant torque on the suspension.

astatine *(Chem.)*. Radioactive element, the heaviest halogen. Symbol At; at. no. 85, mass nos. 202–212, 214–219, half-lives 2×10^{-6} s to 8 hr. Isotopes occur naturally as members of the actinium, uranium or neptunium series, or may be produced by the α-bombardment of bismuth.

astelic *(Bot.)*. Not having a stele.

aster *(Biol.)*. A group of radiating fibrils formed of cytoplasmic granules surrounding the centrosome, seen immediately prior to and during cell division.

Asteraceae *(Bot.)*. Same as **Compositae**.

astereognosis *(Med.)*. Loss of ability to recognize, by the sense of touch, the 3-dimensional nature of an object.

Asteridae *(Bot.)*. Subclass or superorder of dicotyledons. Some trees and shrubs but mostly herbs, mostly sympetalous, stamens as many as corolla lobes or fewer, mostly with 2 fused carpels or with a pseudomonomerous ovary. Contains ca. 56 000 spp. in 43 families including Solanaceae, Scrophulariaceae, Labiatae, Verbenaceae, Rubiaceae and Compositae.

asterism *(Astron.)*. A conspicuous or memorable group of stars, smaller in area than a constellation, such as the *Plough*.

asterism *(Min.)*. The star effect, with four-, six-, or twelve-rayed stars, seen by reflected light in gemstones e.g. ruby and sapphire cut *en cabochon*, and produced by rod-like inclusions.

asteroid *(Astron.)*. One of thousands of rocky objects normally found between the orbits of Mars and Jupiter, ranging in size from 1 to 1000 km. A few, e.g. Eros, have passed close to the earth.

Asteroidea *(Zool.)*. A class of Echinodermata, having a dorsoventrally flattened body of pentagonal or stellate form; the arms merge into the disk; the tube feet possess ampullae and lie in grooves on the lower surface of the arms; the anus and madreporite are aboral; a well-developed skeleton; free-living carnivorous forms. Starfish.

asthenia *(Med.)*. Loss of muscular strength.

asthenosphere *(Geol.)*. The shell of earth below the lithosphere. It is identifiable by low seismic wave velocities and high seismic attenuation. It is soft, probably partly molten and a zone of magma generation. It is equivalent to the *upper mantle*.

asthma *(Immun.,Med.)*. A chronic disease characterized by difficulty in breathing, accompanied by wheezing and difficulty in expelling air from the lungs. This is due to constriction of the bronchi and their blocking by viscid mucous secretions. In some cases — so-called extrinsic asthma — the condition is due to inhaled or ingested allergens and the main symptoms are caused by histamine and other mediators released from mast cells in a type I allergic reaction.

astigmatism *(Med.)*. Unequal curvature of the refracting surfaces of the eye, which prevents the focusing of light rays to a common point on the retina.

astigmatism *(Phys.)*. A defect in an optical system on account of which, instead of a point image being formed of a point object, 2 short line images (focal lines) are produced at slightly different distances from the system and at right angles to each other. Astigmatism is always present when light is incident obliquely on a simple lens or spherical mirror.

astomatous *(Bot.,Zool.)*. (1) Lacking stomata. (2) Without a mouth.

Aston dark space *(Electronics)*. That in the immediate vicinity of a cathode, in which the emitted electrons have velocities insufficient to ionize the gas.

Aston whole-number rule *(Phys.)*. Empirical observation that relative atomic masses of isotopes are approximately whole numbers. See **mass-spectrograph**.

Astrafoil *(Print.)*. A thin, dimensionally stable transparent plastic sheet used for mounting lithographic negatives or positives.

astragal *(Build.)*. A small convex moulding having a semicircular cross-section sometimes plain and sometimes curved. The smaller cross-section bars in windows separating glass panes.

astragal plane *(Build.)*. A plane adapted for cutting astragal mouldings.

astragal tool *(Build.)*. A special tool, with a semicircular cutting edge, used in wood-turning for turning beads and astragals.

astragalus *(Zool.)*. In tetrapod Vertebrates, one of the ankle bones, corresponding to the lunar in the wrist.

astrakhan *(Textiles)*. A curled-pile woven, warp-knitted, or weft-knitted fabric designed to resemble the fleece of a still-born or very young astrakhan lamb.

astringent *(Med.)*. Having the power to constrict or contract organic tissues: that which does this.

astrobleme *(Geol.)*. A circular impact structure on the Earth's crust, caused by a meteorite.

astrocompass *(Aero.)*. A nonmagnetic instrument that indicates true north relative to a celestial body.

astrocyte *(Zool.)*. A much branched, star-shaped neuroglia cell.

astrodome *(Aero.)*. A transparent dome, fitted to some aircraft usually on the top of the fuselage, with calibrated optical characteristics, for astronomical observations.

astroid *(Maths.)*. Four-cusped starlike curve with cartesian equation $x^{2/3}+y^{2/3}=a^{2/3}$. Envelope of a straight line whose ends move along the co-ordinate axes. A hypocycloid in which the radius of the rolling circle is $\frac{1}{4}$ or $\frac{3}{4}$ times the radius of the fixed circle. Also *asteroid*. Cf. **roulette** and **glissette**.

astrolabe *(Astron.)*. Ancient instrument (ca 200 BC) for showing the positions of the Sun and bright stars at any time and date. If fitted with sights, also used for measuring the altitude above the horizon of celestial objects, and in this mode a 15th C. fore-runner of the **sextant**.

astrometry *(Astron.)*. The precise measurement of position in astronomy, generally deduced from the coordinates of images on photographic plates.

astronaut *(Space)*. A man or woman who flies in space; early astronauts had to be very fit, whereas nowadays more benign conditions lead to requirements of normal fitness and high technical qualifications. Also *cosmonaut, spationaut*.

astronautics *(Space)*. The science of space flight.

Astronomer Royal *(Astron.)*. Formerly the title of the Director of the Royal Greenwich Observatory. Since 1972, a purely honorary title awarded to a distinguished British astronomer.

astronomical clock *(Astron.)*. An elaborate clock showing astronomical phenomena such as the phases of the Moon and principally found in medieval cathedrals e.g. Wells, Somerset and Strasbourg. In modern observatories the term is applied to any clock displaying **sidereal time**.

Astronomical Ephemeris *(Astron.)*. Annual handbook (*ephemeris*) published a few years in advance by Her Majesty's Nautical Almanac Office, essentially identical to *The American Ephemeris* and issued in an abridged form as the *Nautical Almanac*.

astronomical telescope *(Astron.)*. Any telescope specifically designed to collect, detect and record electromagnetic radiation from any cosmic source, including radio waves, infrared radiation, light, X-rays and gamma rays. See **radio-**, **Newtonian-**, **Cassegrain-**, **coudé-**, **X-ray-**, **Maksutov-**, **refracting-**, **reflecting-**, **neutrino-**, **altazimuth** and **equatorial** mounts. See also **Schmidt optical system**.

astronomical triangle *(Astron.)*. Triangle on the celestial sphere formed by a heavenly body S, the zenith Z, and the pole P. The 3 angles are the hour-angle at P, the azimuth at Z, and the parallactic angle at S.

astronomical twilight *(Astron.)*. The interval of time during which the sun is between 12° and 18° below the horizon, morning and evening. See **civil twilight, nautical twilight**.

astronomical unit *(Astron.)*. Mean distance of the earth from the sun, 1.496×10^8 km or about 93 million miles. Abbreviated AU and commonly used as a unit of distance within the solar system. There are 63 240 AU in one light year.

astronomy *(Genrl.)*. The science of the heavens in all its branches.

astrophyllite *(Min.)*. A complex hydrated silicate of potassium, iron, manganese, titanium, and zirconium; occurs in brown laminae in alkaline, igneous rocks.

astrophysics *(Astron.)*. That branch of astronomy which applies the laws of physics to the study of interstellar matter and the stars, their constitution, evolution, luminosity, etc.

astrosclereide *(Bot.)*. A sclereide with radiating branches ending in points.

asulam *(Ecol.)*. Translocated herbicide used widely to control bracken in grasslands, upland pastures and forest plantations.

asymmeter *(Elec.Eng.)*. An instrument having 3 movements so arranged that any lack of symmetry when these are connected to a 3-phase system can be observed by a single reading.

asymmetric *(Bot.)*. Irregular in form: not divisible into halves about any longitudinal plane.

asymmetrical conductivity *(Elec.Eng.)*. Phenomenon whereby a substance, or a combination of substances as in a rectifier, conducts electric current differently in opposite directions.

asymmetrical, dissymmetrical, nonsymmetrical *(Elec. Eng.)*. Said of circuits, networks, or transducers when the impedance (image impedance, or iterative impedance) differs in the two directions.

asymmetric atom *(Chem.)*. An atom bonded to three or more other atoms in such a way the arrangement cannot be superimposed on its mirror image. In particular, a carbon atom attached to four different groups. Most *chiral* molecules can be described in terms of specific asymmetric atoms, e.g. the alpha-carbon atoms in amino acids. See **chirality**.

asymmetric conductor *(Elec.Eng.)*. Conductor which has a different conductivity for currents flowing in different directions through it, e.g. a diode.

asymmetric flight *(Aero.)*. The condition of flying with asymmetrically balanced thrust, weight, drag or lift forces, as could occur, for example, with one external weapon mounted under one wing or in a twin-engined aircraft with one engine inoperative.

asymmetric reflector *(Phys.)*. A reflector in which the beam of light produced is not symmetrical about a central axis.

asymmetric refractor *(Phys.)*. A refractor in which the light is redirected, unsymmetrically, about a central axis.

asymmetric synthesis *(Chem.)*. The synthesis of optically active compounds from racemic mixtures. This can be carried out in some cases by chemical methods in which one component is more reactive than the other one. In other cases asymmetric synthesis occurs in the presence of enzymes.

asymmetric system *(Crystal.)*. See **triclinic system**.

asymmetric top *(Chem.)*. A model of a molecule having no 3- or higher-fold axis of symmetry.

asymmetry *(Genrl.)*. The condition of being *asymmetrical*.

asymmetry *(Zool.)*. The condition of the animal body in which no plane can be found which will divide the body into two similar halves; as in Snails.

asymmetry potential *(Elec.Eng.)*. The potential difference between the inside and outside surface of a hollow electrode.

asymptote *(Maths.)*. Of a given curve: usually a line, but sometimes a curve, frequently called a *curvilinear asymptote*; such that the distance of a point on the given curve from it tends to zero as the point tends to infinity in some direction along the curve. It can be considered as a tangent at infinity.

asymptotic breakdown voltage *(Elec.Eng.)*. See **serving**.

asymptotic curve *(Maths.)*. One on a surface such that at every point on it, its tangent lies in an asymptotic direction at that point.

asymptotic directions *(Maths.)*. See **hyperbolic point on a surface** and **conjugate directions**.

asynapsis *(Biol.)*. Absence of pairing of chromosomes at meiosis.

asynchronous data transmission *(Comp.)*. The transmission of data in which the end of the transmission of one character initiates the transmission of the next.

asynchronous motor *(Elec.Eng.)*. See **non-synchronous motor**.

asynchronous (nonsynchronous) computer *(Comp.)*. One in which operations are not all timed by a master clock. The signal to start an operation is provided by the completion of the previous operation.

asynergia, asynergy *(Med.)*. Lack of co-ordinated movement between muscles with opposing actions, due to a lesion in the nervous system.

asystole *(Med.)*. Arrest of heart contraction.

At *(Chem.)*. The symbol for astatine.

AT *(Elec.Eng.)*. Abbrev. for *Ampere-Turn*.

ATA *(Aero.)*. Abbrev. for **Air Transport Association**.

atacamite *(Min.)*. A hydrated chloride of copper, widely distributed in S. America, Australia, India, etc. in the oxidation zone of copper deposits; occurring also at St. Just, Cornwall, UK.

atactic *(Plastics)*. Term used to denote certain linear hydrocarbon polymers in which substituent groups are arranged at random around the main carbon chain. See also **isotactic** and **syndiotactic**.

atactosol *(Chem.)*. A colloidal sol not containing **tactoids**.

atactostele *(Bot.)*. Stele characteristic of the stems of monocotyledons, consisting of many vascular bundles apparently scattered throughout the ground tissue.

atavism *(Biol.)*. The appearance in an individual of characteristics believed to be those of its distant ancestors.

ataxia, ataxy *(Med.)*. Inco-ordination of muscles, leading to irregular and uncontrolled movements; due to lesions in the nervous system.

ataxia telangiectasia *(Biol.)*. Human clinical syndrome in which spontaneous chromosome rearrangements occur at a high rate, preferentially involving non-homologous chromosomes.

ATB *(Eng.)*. Abbrev. for *Aeration Test Burner*.

ATC *(Aero.)*. See **air traffic control**.

ATCRBS *(Radar)*. Abbrev. for *Air Traffic Control Radar Beacon System*. A direct development of the WWII **IFF** system. Operating at about 1 GHz, it gives air traffic controllers three-dimensional positional information and full identification of aircraft.

atelectasis *(Med.)*. Failure to expand and collapse of part or all of the lung.

atenolol *(Med.)*. Beta receptor antagonist or blocker which acts more on the β_2-receptors. Thus its action is more towards the heart with fewer effects on airway tone.

athermal solutions *(Chem.)*. Solutions formed without production or absorption of heat on mixing the components.

athermal transformation *(Eng.)*. A reaction in e.g. the phase transformation of steel in which thermal activation is not required and it is driven by increasing pressure as the phase changes.

atheroma *(Med.)*. Same as **atherosclerosis**.

atherosclerosis *(Med.)*. Thickening of and rigidity of the intima of the arteries, caused by deposition of *atheroma*. Commonest form of arterial disease in Western societies. Also *atheroma*.

athetosis *(Med.)*. Slow, involuntary, spontaneous, repeated writhing movements of the fingers and of the toes, due to a brain lesion.

Atkinson cycle *(Autos.)*. A working cycle for internal-combustion engines, in which the expansion ratio exceeds the compression ratio; more efficient than the Otto cycle, but mechanically impracticable.

Atlas *(Astron.)*. The fifteenth natural satellite of **Saturn**, discovered in 1980.

Atlas *(Comp.)*. See **second generation computer**.

atlas *(Zool.)*. The first cervical vertebra.

atm *(Phys.)*. Abbrev. for **standard atmosphere**, a practical unit of pressure defined as 101.325 kN/m^2 or 1.013 25 bar.

atmolysis *(Chem.,Phys.)*. The method of separation of the components of a mixture of two gases, which depends on their different rates of diffusion through a porous partition.

atmometer *(Bot.)*. Apparatus, like a potometer, but designed to measure water loss from a wet, non-living surface e.g. a porous pot.

atmospheric absorption *(Acous.)*. Diminution of intensity of a sound wave in passing through the air, apart from normal inverse square relation, and arising from transfer of sound energy into heat.

atmospheric absorption *(Astron.)*. The absorption of the light of the stars by the earth's atmosphere; it is practically negligible above 45° altitude, but the extinction amounts to about half a magnitude at 20°, 1 magnitude at 10°, and 2 magnitudes at 4° altitude.

atmospheric acoustics *(Acous.)*. Concerned with the propagation of sound in the atmosphere, of importance in sound ranging and aircraft noise.

atmospheric electricity *(Meteor.)*. That causing increasing potential with height, about 100 V/m, in calm conditions, altered considerably by thunder-clouds. See **lightning**.

atmospheric engine *(Eng.)*. Earliest form of practicable steam engine, in which a partial vacuum created by steam condensation allowed atmospheric pressure to drive down the piston.

atmospheric gas-burner system *(Eng.)*. A natural-draught burner injector, in which the momentum of the gas passing into the injector throat inspirates part of the air required for combustion.

atmospheric line *(Eng.)*. A datum line drawn on an indicator diagram by allowing atmospheric pressure to act on the indicator piston or diaphragm.

atmospheric pressure *(Phys.)*. The pressure exerted by the atmosphere at the surface of the earth is due to the weight of the air. Its standard value is $1.013\,25 \times 10^5$ N/m^2, or 14.7 lbf/in^2. Variations in the atmospheric pressure are measured by means of the barometer. See **barometric pressure, standard atmosphere**.

atmospheric radio wave *(Telecomm.)*. Any radio wave which reaches its destination after reflection from the upper ionized layers of the atmosphere.

atmospherics *(Telecomm.)*. Interfering or disturbing signals of natural origin. Also called *spherics, strays*. See also **static**.

atmospheric tides *(Meteor.)*. The changes of atmospheric pressure arising directly from changes in temperature due to the Earth's rotation. See **diurnal range**.

atmospheric waveguide duct *(Telecomm.)*. Atmospheric layer which acts as a waveguide for high-frequency (<20 MHz) radio waves under certain conditions of temperature and humidity, giving reception far outside the normal service area.

atokous *(Zool.)*. Having no offspring: sterile.

atom *(Chem.)*. The smallest particle of an element which can take part in a chemical reaction. See **atomic structure, Dalton's atomic theory**.

atom *(Comp.)*. A primitive syntactic unit, an indivisible unit of data.

atomic absorption coefficient *(Phys.)*. For an element, the fractional decrease in intensity of radiation per number of atoms per unit area. Symbol μ_a. Related to the linear absorption coefficient μ by

$$\mu = \frac{1}{V}\sum_i n_i(\mu_a)_i,$$

where the material contains n_i atoms of element i in a volume V.

atomic bomb *(Phys.)*. Bomb in which the explosive power, measured in terms of equivalent TNT, is provided by nuclear fissionable material such as ^{235}U or ^{239}Pu. The bombs dropped on Hiroshima and Nagasaki (1945) were of this type. Also called *A-bomb, atom bomb* and *fission bomb*. See also **hydrogen bomb**.

atomic bond *(Chem.)*. See **covalent bond**.

atomic clock *(Phys.)*. A clock whose frequency of

operation is controlled by the frequency of an atomic or molecular process. The inversion of the ammonia molecule with a frequency of 23 870 Hz provides the basic oscillations of the *ammonia clock*. The difference in energy between two states of a caesium atom in a magnetic field giving a frequency of 9 192 631 770 Hz is the basis of the *caesium clock* which has an accuracy of better than 1 in 10^{13}.

atomic disintegration *(Phys.)*. Natural decay of radioactive atoms, with radiation, into chemically different atomic products.

atomic energy *(Chem.)*. Strictly the energy (chemical) obtained from changing the combination of atoms originally in fuels. Now normally applied to energy obtained from breakdown of fissile atoms in nuclear reactors. See **nuclear energy**.

atomic frequency *(Phys.)*. A natural vibration frequency in an atom used in the atomic clock.

atomic heat *(Chem.)*. Product of specific heat capacity and r.a.m. in grams; approx. the same for most solid elements at high temperatures.

atomic hydrogen *(Chem.)*. See **active hydrogen**.

atomicity *(Chem.)*. The number of atoms contained in a molecule of an element.

atomic mass unit *(Chem.)*. Abbrev. *u*. Exactly one twelfth the mass of a neutral atom of the most abundant isotope of carbon, ^{12}C. $u = 1.660 \times 10^{-27}$ kg. Before 1960 *u* was defined in terms of the mass of the ^{16}O isotope and *u* was 1.6599×10^{-27} kg. See **atomic weight**.

atomic number *(Chem.)*. The order of an element in the periodic (Mendeleev) chemical classification, and identified with the number of unit positive charges in the nucleus (independent of the associated neutrons). Equal to the number of external electrons in the neutral state of the atom, and determines its chemistry. Symbol *Z*.

atomic orbital *(Chem.)*. A wave-function defining the energy of an electron in an atom.

atomic plane *(Phys.)*. A solid is crystalline because its atoms are ordered in intersecting planes (*atomic planes*) corresponding to the planes of the crystal. See **X-ray crystallography**.

atomic radii *(Chem.)*. Half of the internuclear distance between the nuclei of two identical non-bonded atoms at such a separation that they neither attract nor repel one another.

atomic refraction *(Chem.)*. The contribution made by a mole of an element to the molecular refraction of a compound.

atomic scattering *(Phys.)*. That of radiation, usually electrons or X-rays, by the individual atoms in the medium through which it passes. The scattering is by the electronic structure of the atom in contrast to *nuclear scattering* which is by the nucleus.

atomic scattering factor *(Phys.)*. The ratio of the amplitude of coherent scattered X-radiation from an atom to that of a single electron placed at the atomic centre. The atomic scattering factor depends on the electron-density distribution in the atom and is a function of the scattering angle.

atomic spectrum *(Phys.)*. Electronic transitions between the discrete energy states of an atom involve either the emission or absorption of *photons*. Such emission or absorption spectra are *line spectra*. The spectrum is characteristic of the atom involved.

atomic structure *(Phys.)*. The chemical behaviour of the various elements arises from the differences in the electron configuration of the atoms in their normal electrically neutral state. Each atom consists of a heavy nucleus with a positive charge produced by a number of **protons** equal to its atomic number. There are an equal number of electrons outside the nucleus to balance this charge. The nucleus also contains electrically neutral **neutrons**; protons and neutrons are collectively referred to as **nucleons**. The Sommerfeld model, modified by the wave mechanical concept of orbitals, describes the electron configuration of the atom. Electrons are fermions which must conform to the Pauli exclusion principle and no two electrons in the same atom can be in the same quantum state, i.e. have the same set of four quantum numbers. The principle quantum number indicates the shell to which the orbital belongs and varies from 1 (*K*-shell) closest to the nucleus to 7 (*Q*-shell), the most remote. In general, the closer an electron is to the nucleus the greater the coulomb attraction and so the greater the binding energy retaining the electron in the atom. Nuclear binding forces tend to give greatest stability when the neutron number and the proton number are approximately equal. Due to electrostatic repulsion between protons, the heavier nuclei are most stable when more than half their nucleons are neutrons; elements with more than 83 protons are unstable and undergo radioactive disintegration. Those with more than 92 protons are not found naturally on earth, but can be synthesized in high-energy laboratories. These are the trans-uranic elements which have short half-lives. Most elements exist with several stable **isotopes** and the chemical atomic weight gives the average of a normal mixture of these isotopes.

atomic transmutation *(Phys.)*. The change of one type of atom to another as a result of a nuclear reaction. The transmutation can be produced by high-energy radiation or particles and is most easily produced by neutron irradiation. The change in atomic number means the chemical nature of the atom has been changed, e.g. gold can be transmuted into mercury; the converse of the ancient alchemist's goal.

atomic volume *(Chem.)*. Ratio for an element of the rel. at. mass to the density; this shows a remarkable periodicity with respect to atomic number.

atomic weight *(Chem.)*. *Relative at. mass*. Mass of atoms of an element in *atomic mass units* on the **unified scale** where $u = 1.660 \times 10^{-27}$ kg. For natural elements with more than one isotope, it is the average for the mixture of isotopes.

atomized powder *(Powder Tech.)*. One produced by the dispersion of molten metal or other material by spraying under conditions such that the material breaks down into powder.

atomizer *(Eng.)*. A nozzle through which oil fuel is sprayed into the combustion chamber of an oil engine or boiler furnace. Its function is to break up the fuel into a fine mist so as to ensure good dispersion and combustion.

atony *(Med.)*. Loss of muscular tone.

atopy *(Immun.)*. A constitutional or hereditary tendency to develop high levels of IgE and immediate hypersensitivity to allergens, especially those which are absorbed across the respiratory mucosa.

ATP *(Chem.)*. Abbrev. for *adenosine triphosphate*.

ATPase *(Biol.)*. Enzyme which converts ATP to ADP. In this process the free energy change of the exergonic hydrolysis is used to drive an endergonic reaction, e.g. muscle myosin possesses ATPase activity and the ATP breakdown is coupled to the movement of the *myosin* fibres relative to the *actin*.

atrate, atratous *(Bot.)*. Blackened; blackening.

atresia *(Med.)*. Pathological narrowing of any channel of the body.

atresia *(Zool.)*. Disappearance by degeneration; as the follicles in the Mammalian ovary. adjs. *atresic, atretic*.

atrial *(Zool.)*. Pertaining to the **atrium**.

atrial fibrillation *(Med.)*. Atrial arrhythmia marked by rapid randomized contractions of the atrial myocardium, causing a totally irregular, and often rapid, ventricular rate.

atriopore *(Zool.)*. The opening by which the atrial cavity communicates with the exterior.

atrioventricular *(Zool.)*. See **auriculoventricular**.

atrium *(Arch.)*. The entrance court in a Roman house, open to the sky in the centre, but with a roofed perimeter walkway. The concept is used in contemporary building where several storeys of offices or shops frequently enclose a void or *atrium*, glazed at roof level, thus allowing daylight to reach the surrounding accommodation.

atrium *(Zool.)*. In Platyhelminthes, a space into which open the ducts from the male and female genital organs:

in pulmonate Mollusca, a cavity into which the vagina and the penis open and which itself opens to the exterior: in Protochordata, the cavity surrounding the respiratory part of the pharynx: in Vertebrates, the anterior part of the nasal tract: in Reptiles and Birds, the cavity connecting the bronchus with the lung chambers: in the developing Vertebrate heart, the division between the sinus venosus and the ventricle, which will later give rise to the auricles.

atrophic rhinitis *(Vet.)*. Infectious disease of pigs, characterized by chronic rhinitis and deformity of the snout. Various infectious agents, plus other factors, are involved. *Pasteurella multocida* and *Bordatella bronchiseptiсaca* are commonly isolated. Incidence can be limited by vaccination.

atrophy *(Med.)*. Wasting of a cell or of an organ of the body.

atrophy *(Zool.)*. Degeneration, i.e. diminution in size, complexity, or function, through disuse.

atropine *(Med.)*. An alkaloid of the tropane group, obtained from solanaceous plants in which it occurs in traces. It causes dilatation of the pupil of the eye. Atropine constitutes the active principle of belladonna (obtained from *Atropa belladonna*, 'deadly nightshade'). It has sedative effects on the central nervous system, and blocks activity of the vagus nerve causing decreased respiratory tract secretions and increased heart rate; hence its use before general anaesthesia.

atropus *(Bot.)*. See **orthotropous**.

ATR tube *(Radar)*. See **anti-transmit-receive tube**.

attached column *(Arch.)*. A column partially built into a wall, instead of standing detached.

attachment theory *(Behav.)*. A wide variety of research efforts which centres around the proposal, put forward by Bowlby, that the emotional bond between human infants and their caretakers has an evolutionary history and that various attachment behaviours, e.g. smiling, crying, have physical and psychological functions of enhancing survival. The theory includes perspectives from ethology, psychoanalysis and developmental psychology.

attapulgite *(Min.)*. See **palygorskite**.

attar of roses (otto de rose) *(Chem.)*. Oil distilled from fresh roses for perfumery purposes.

attention *(Behav.)*. *Selective attention.* That aspect of perception which implies a readiness to respond to a particular stimulus or aspects of it.

attenuated vaccine *(Immun.)*. Live bacterial or virus vaccine in which the microbes have been selected or otherwise treated in such a way as to greatly diminish their capacity to cause disease but still to retain their ability to evoke protective immunity. Examples are *poliomyelitis, measles*, and *yellow fever* vaccines.

attenuation *(Bot.)*. Lessening of the capacity of a pathogen to cause disease.

attenuation *(Phys.,Telecomm.)*. General term for reduction in magnitude, amplitude, or intensity of a physical quantity, arising from absorption, scattering, or geometrical dispersion. The latter, arising from diminution by the inverse-square law, is not generally considered as attenuation proper. See also **absorption**.

attenuation coefficient *(Phys.)*. See **total absorption coefficient**.

attenuation compensation *(Telecomm.)*. The use of networks to correct for frequency dependent attenuation, e.g. in transmission lines. See **pre-emphasis**.

attenuation constant *(Telecomm.)*. The real part of α in the relationship $\rho = \rho e^{-\alpha x}$, where ρ is a physical quantity, such as the amplitude of a wave propagating along a transmission path, x is the distance along the path. The imaginary part of α is known as the **phase constant**. More simply, but less commonly defined by $\mu = \alpha\lambda$ where μ is the attenuation and λ is wavelength, i.e. α is the attenuation per wavelength distance of propagation. See **decibel, neper, propagation constant**.

attenuation distortion *(Telecomm.)*. Distortion of a complex waveform resulting from the differing attenuation of each separate frequency component in the signal. This form of distortion is difficult to avoid, e.g. in transmission lines.

attenuation of X-rays *(Radiol.)*. The term covering both absorption and scattering of X-rays as they pass through an object.

attenuator *(Telecomm.)*. An arrangement of fixed or variable resistive elements designed to reduce the strength of any signal (audio- or radio-frequency) without reducing appreciable distortion. Attenuators also incorporate impedance matching to the transmission lines or circuits to which they are connected, regardless of the attenuation they introduce. For lower-frequency applications they may be simply variable or fixed resistances; for high-frequencies they may be pieces of resistive material, introduced into transmission lines, stripline or waveguide. Fixed attenuators are sometimes referred to as *pad*.

attitude *(Aero.)*. The *attitude* or *aspect* of an aircraft in flight, or on the ground, is defined by the angles made by its axes with the relative airflow, or with the ground, respectively.

attitude *(Behav.)*. An inferred disposition to feel, think and act in certain ways which is used to explain the variation between individuals in their response to similar situations; attitudes are assumed to represent the effects of past experience on behaviour through their effects on the cognitive and emotional structuring of perception.

attitude angle *(Autos.)*. See **slip angle**.

attitude control *(Space)*. The provision of a desired orientation to satisfy mission requirements; it is usually effected by a low thrust system in conjunction with a measuring instrument, such as a star sensor, and maintained by a stabilizing device, such as a gyroscope. Attitude control can also be maintained by spinning the spacecraft about one of its axes.

attitude indicator *(Aero.)*. A *gyro horizon* which indicates the true attitude of the aircraft in pitch and roll throughout 360° about these axes. See **heading indicator**.

attitude scale *(Behav.)*. Standard procedure for measuring attitudes.

attracted-disk electrometer *(Elec.Eng.)*. Fundamental instrument in which potential is measured by the attraction between two oppositely charged disks.

attribution theories *(Behav.)*. Theories concerned with how individuals explain everyday events by attributing the cause of a person's behaviour to either situational or personal factors, or some combination of each.

attrition test *(Civ.Eng.)*. A test for the determination of the wear-resisting properties of stone, particularly stone for roadmaking. Pieces of the stone are placed in a closed cylinder, which is then rotated for a given time, after which the loss of weight due to wear is found.

Attwood's formula *(Ships)*. A formula for determining the moment of static stability at large angles of heel of a ship. Taking angle of heel θ, and the weight of the ship W, moment

$$= W\left(\frac{v \times hh_1}{V} \pm BG \sin\theta\right) \text{ foot tons,}$$

where v = volume of emerged wedge; hh_1 = distance between C.G.'s of emerged and immersed wedges; V = volume of displacement; B = centre of transverse buoyancy; G = centre of gravity.

at.wt. *(Genrl.)*. *Atomic weight*, now **relative atomic mass**.

AU *(Astron.)*. Abbrev. for *astronomical unit*.

Au *(Chem.)*. The symbol for gold (*aurum*).

audibility *(Acous.)*. Ability to be heard; said of faint sounds in the presence of noise. The extreme range of audibility is 20 to 20 000 Hz in frequency, depending on the applied intensity; and from 2×10^{-5} N/m² (rms) at 1000 Hz (the zero of the phon scale, selected as the average for good ears) to 120 dB.

audible ringing tone *(Telecomm.)*. An audible tone fed back to a caller as an indication that ringing current has been remotely extended to the called subscriber's telephone. On circuits in UK it is heard as a double beat recurring at 2 second intervals. Also called *audible signal*.

AU diode *(Electronics)*. See **backward diode**.

audiofrequency *(Acous.,Telecomm.)*. Frequency which, in an acoustic wave, makes it audible. In general, any wave motion including frequencies in the range of, e.g. 30 Hz to 20 kHz.

audiofrequency amplifier *(Telecomm.)*. Amplifier for frequencies within the audible range.

audiofrequency choke *(Elec.Eng.)*. Inductor with appreciable reactance at audiofrequencies.

audiofrequency shift modulation *(Telecomm.)*. Method of facsimile transmission in which tone values from black to white are represented by a graded system of audiofrequencies.

audiofrequency transformer *(Telecomm.)*. Transformer for use in a communication channel or amplifier, designed with a specified, normally uniform, response for frequencies used in sound reproduction.

audiogram *(Acous.)*. Standard graph or chart which indicates the hearing loss (in **bels**) of an individual ear in terms of frequency. See **noise-**, **sound-level meter**.

audiometer *(Acous.)*. Instrument for measurement of acuity of hearing. Specifically to measure the minimum intensities of sounds perceivable by an ear for specified frequencies. See also **gramophone-audiometer**, **noise audiometer**.

auditory, aural *(Zool.)*. Pertaining to the sense of hearing or to the apparatus which subserves that sense: the 8th cranial nerve of Vertebrates, supplying the ear.

auditory canal *(Acous.,Med.)*. Duct connecting the ear drum with the external ear (**pinna**), by which sound waves are transmitted from outer to inner ear.

auditory ossicles *(Zool.)*. Three small bones, the *incus*, *malleus*, and *stapes*, bridging the tympanic cavity of the middle ear in mammals.

auditory perspective *(Acous.)*. See **stereophony**.

audit trail *(Comp.)*. Record of the file updating which takes place during a specific transaction. It enables a trace to be kept of all operations on files.

augend *(Maths.)*. See **addition**.

augen-gneiss *(Geol.)*. A coarsely crystalline rock of granitic composition, containing lenticular, eye-shaped masses of feldspar or quartz embedded in a finer matrix. A product of regional metamorphism.

auger *(Build)*. A tool for boring holes, especially in wood or in the earth.

Auger effect *(Phys.)*. An atom ionized by the ejection of an inner electron can lose energy either by the emission of an X-ray photon as an outer electron makes a transition to the vacancy in the inner shell *or* by the ejection of an outer electron, the *Auger effect*. The energies of the Auger electrons emitted are characteristic of the atomic energy levels.

Auger yield *(Phys.)*. For a given excited state of an atom of a given element, the probability of de-excitation by Auger process intead of by X-ray emission.

augite *(Min.)*. A pyroxene, a complex aluminous silicate of calcium, iron, and magnesium, crystallizing in the monoclinic system, and occurring in many igneous rocks, particularly those of basic composition; it is an essential constituent of basalt, dolerite, and gabbro.

augmentor *(Aero.)*. Means of increasing forces: (a) afterburning in a gas turbine; (b) by induced airflow in a rocket; (c) in a wing of STOL aircraft by ducting compressed airflow from a gas-turbine into circulation-increasing slots and flaps to create high lift co-efficients and thereby giving slow landing speeds.

Aujesky's disease *(Vet.)*. *Pseudorabies, mad itch, infectious bulbar paralysis.* An encephalomyelitis affecting cattle, sheep, pigs, dogs, cats, and rats, and caused by a thermostable herpes virus. Very widespread on farms but can be eradicated. Notifiable disease, controlled by vaccination or a slaughter policy.

aura *(Med.)*. A diffuse fear, movement, sensation or mental disturbance, which precedes an epileptic convulsion or an attack of migraine.

aural *(Zool.)*. See **auditory**.

aural masking *(Acous.)*. See **masking**.

auramines *(Chem.)*. Dyestuffs of the diphenylmethane series.

aureole *(Elec.Eng.)*. Luminous glow from outer portion of electric arc. This has different spectral distribution from that from the highly-ionized core.

aureole *(Geol.)*. Area surrounding an igneous intrusion affected by metamorphic changes.

aureole *(Meteor.)*. (1) The reddish ring round the sun or moon, forming the inner part of a corona. (2) The bright indefinite ring round the sun in the absence of clouds.

auric acid, auric oxide *(Chem.)*. Gold (III) oxide. Formula Au_2O_3. An amphoteric oxide of gold.

auricle *(Bot.)*. A small ear-shaped lobe at the base of a leaf or other organ.

auricle *(Zool.)*. A chamber of the heart connecting the afferent blood vessels with the ventricle: the external ear of Vertebrates: any lobed appendage resembling the external ear. Also called *auricula*.

auricled, auriculate *(Zool.)*. Having auricles.

auricularia *(Zool.)*. In Holothuria and Asteroidea, a pelagic ciliated larva, having the cilia arranged in a single band, produced into a number of short processes.

auriculoventricular *(Zool.)*. Pertaining to, or connecting, the auricle and ventricle of the heart; e.g. the *auriculo-ventricular connection*, a bundle of muscle fibres which transmits the wave of contraction from the auricle to the ventricle, in higher Vertebrates.

auriferous deposit *(Geol.)*. A natural repository of gold, in the general sense, including gold-bearing lodes and sediments such as sands and gravels, or their indurated equivalents, which contain gold in detrital grains or nuggets. See **blanket**, **lode**, **placers**.

auriferous pyrite *(Min.)*. Iron sulphide in the form of pyrite, carrying gold, probably in solid solution.

aurine *(Chem.)*. *Pararosolic acid*, $(HO.C_6H_4)_2{=}C{=}C_6H_4{=}O$, made from phenol, oxalic acid, and sulphuric acid. It is similar to rosolic acid in properties. Basis of aurine dyes.

aurora *(Astron.)*. Luminous curtains or streamers of light seen in the night sky at high latitudes, caused when electrically charged particles from the Sun are guided by the Earth's magnetic field to the polar regions, there colliding with atoms in the upper atmosphere. In the northern hemisphere known as *aurora borealis* and in the southern *aurora australis*.

auroral zone *(Telecomm.)*. Zone where radio transmission is affected by aurora.

aurous *(Chem.)*. Containing gold (I)

aurum *(Chem.)*. See **gold**.

ausculation *(Med.)*. The act of listening to the sounds produced in the body.

austempering *(Eng.)*. Heating a steel to transform it to austenite, followed by quenching to a temperature above the martensitic change point, but below the critical range, so that pearlite and bainite are directly formed. It is then cooled to atmospheric temperature.

austenite *(Eng.)*. Originally, a solid solution of carbon in γ-iron; now includes all solid solutions based on γ-iron.

austenitic steels *(Eng.)*. Steels containing sufficient amounts of nickel, nickel and chromium, or manganese to retain austenite at atmospheric temperature; e.g. *austenitic stainless steel* and *Hadfield's manganese steel*.

Austin Moore prosthesis *(Med.)*. A prosthesis which is inserted into the upper end of the femur to reconstruct the hip joint when the original head has been removed.

Australasian region *(Zool.)*. One of the primary faunal regions into which the land surface of the globe is divided; includes Australia, New Guinea, Tasmania, New Zealand, and the islands south and east of Wallace's line.

australites *(Min.)*. See **tektites**.

Austrian cinnabar *(Chem.)*. See **basic lead chromate**.

aut- *(Genrl.)*. Prefix. See **auto-**.

autacoid, autocoid *(Med.)*. General name for an endocrine secretion; a specific organic substance formed by the cells of one organ, and passed by them into the circulating fluid, to produce effects upon other organs. See **hormones**.

autecology *(Ecol.)*. The study of the ecology of any individual species. Cf. **synecology**.

authigenic *(Geol.)*. Pertaining to minerals which have crystallized in a sediment during or after its deposition.

authoritarian personality *(Behav.)*. A term introduced into social psychology in a famous post-war study of anti-semitism; it refers to a cluster of social attitudes and personality attributes that are hostile and rigid. See **F-scale**.

autism, infantile *(Behav.,Med.)*. A *childhood psychosis* originating in infancy (before thirty months of age), characterized by a lack of responsiveness in social relationships, language abnormality, and a need for constant environmental input, or *sameness*; stereotypic motor habits, overactivity, and epilepsy are often associated with it.

autoallogamy *(Bot.)*. The condition of a species in which some individual plants are capable of self-pollination and others of cross-pollination.

auto-antibody *(Immun.)*. Antibody which reacts specifically with an antigen present on normal constituents of the body of the individual in whom the antibody was made. B-lymphocytes able to make auto-antibodies are present in healthy individuals, but are suppressed or not normally stimulated. In auto-immune disease the regulation mechanisms break down and auto-antibodies are present in the blood, which may or may not be responsible for the disease process.

auto-assemble *(Image Tech.)*. System of videotape editing in which selected scenes are transferred in their required sequence according to a pre-selected programme of time-code information.

auto-, aut- *(Genrl.)*. Prefix from Gk. *autos*, self.

autocapacitance coupling *(Elec.Eng.,Telecomm.)*. Coupling of two circuits by a capacitor included in series with a common branch.

autocatalysis *(Chem.)*. The catalysis of a reaction by the product of that reaction, e.g. MnO_4^- and $C_2O_4^{2-}$ by Mn^{2+} ions.

autocatalysis *(Zool.)*. Reaction or disintegration of a cell or tissue, due to the influence of one of its own products.

autochthonous *(Zool.)*. In an aquatic community, said of food material produced within the community: more generally, indigenous, inherited, hereditary (e.g. *autochthonous* species, *autochthonous* characteristics). Cf. **allochthonous behaviour**.

autocidal control *(Zool.)*. Method of insect pest control by release of sterile or genetically altered individuals into the wild population.

autoclave *(Chem.)*. A vessel, constructed of thick-walled steel (usually alloy steel or frequently nickel alloys), for carrying out chemical reactions under pressure and at high temperatures. Pressure gauge, safety valve and thermometer pocket are provided for control.

autoclave *(Med.)*. An apparatus for sterilization by steam at high pressure

auto coarse pitch *(Aero.)*. The setting of the blades of a propeller to the minimum drag position if there is a loss of engine power during takeoff.

autocollimator *(Phys.)*. (1) An instrument for accurately measuring small changes in the inclination of reflecting surfaces. Principally used for engineering metrology measurements. (2) A convex mirror used to produce a parallel beam of light from a reflecting telescope. It is placed at the focus of the main mirror.

autoconverter *(Elec.Eng.)*. A special form of converter used with certain types of electric-battery vehicle; it is arranged to operate from a constant-voltage battery supply and give an output voltage inversely proportional to the current.

autocorrelation *(Telecomm.)*. Technique for detecting weak signals against a strong background level. Signal is subjected to controlled delay, the original delay signals then being fed to the autocorrelation unit which responds strongly only if delay is an exact multiple of signal period.

autodiploid *(Bot.)*. See **anther culture**.

autodyne *(Telecomm.)*. Of an electrical circuit in which the same elements and valves are used both as oscillator and detector. Also called *endodyne*, *self-heterodyne*.

autodyne receiver *(Telecomm.)*. One utilizing the principle of beat reception and including an autodyne oscillator.

autoecious, autoxenous *(Bot.)*. A parasite or pest living only in a single host species. Cf. **heteroecious**.

autoerotism *(Behav.)*. A condition where sensual pleasure is sought and gratified in one's own person, without the aid of an external love object; e.g. masturbation, thumb sucking. See also **narcissism**.

autoflare *(Aero.)*. An automatic landing system which operates on the **flare out** part of the landing, using an accurate radio-altimeter.

autogamy *(Bot.)*. Fertilization involving pollen and ovules from (1) the same flower or sometimes (2) (more widely) the same plant or genetically identical individuals (same **genet** or **clone**). See **self-pollination, self-fertilisation**. Cf. **allogamy**.

autogamy *(Zool.)*. (1) Self-fertilization. (2) The fusion of sister-cells, or of 2 sister-nuclei.

autogenic *(Bot.)*. (1) A change or development resulting from the internal processes of a system; cf. **allogenic**. (2) See **autonomic**.

autograft *(Biol.,Med.)*. A mass of tissue, or an organ, moved from one region to another within the same organism.

auto-ignition *(Autos.)*. The self-ignition or spontaneous combustion of a fuel when introduced into the heated air charge in the cylinder of a compression-ignition engine. See **spontaneous ignition temperature**.

auto-immune diseases *(Med.)*. A group of diseases caused by antigen/antibody reactions to the host's own tissues. See **auto-immunity**.

auto-immunity *(Immun.)*. A condition in which T- or B-lymphocytes capable of recognising 'self' constituents are present and activated so as to cause damage to cells by cell-mediated immunity or to release auto-antibodies, and so to cause *auto-immune diseases*. There are many such conditions, the clinical manifestations of which depend upon the nature of the cells damaged and whether circulating **immune complexes** are formed.

auto-inductive coupling *(Elec.Eng.)*. Coupling of two circuits by an inductance included in series with a common branch.

auto-infection *(Zool.)*. Re-infection of a host by its own parasites.

auto-intoxication *(Med.)*. Poisoning of the body by toxins produced within it.

autoland *(Aero.)*. A landing in which the descent, forward speed, **flare-out**, alignment with the runway and touch-down are all automatically controlled. See **autoflare, autothrottle**.

autolithography *(Print.)*. The drawing by the artist of his design direct on the stone or plate.

autolysis *(Biol.)*. The breakdown of living matter caused by the action of enzymes produced in the cells concerned; self-digestion. adj. *autolytic*. See **lysosomes**.

automatic arc lamp *(Phys.)*. An arc lamp in which the feeding of the carbons into the arc and the striking of the arc are done automatically, by electromagnetic or other means.

automatic arc welding *(Elec.Eng.)*. Arc welding carried out in a machine which automatically moves the arc along the joint to be welded, feeds the electrode into the arc, and controls the length of the arc.

automatic beam control *(Image Tech.)*. System in a TV camera which momentarily alters the beam current in the camera tube to reduce the tailing effects on moving highlights. Abbrev. *ABC*. Also known as the *automatic beam optimizer*, *ABO*.

automatic blankets *(Print.)*. On letterpress rotary machines, a covering for the impression cylinder with a felt base and a surface of rubber on plastic.

automatic brightness control *(Image Tech.)*. Circuit used in some television receivers to keep average brightness level of screen constant.

automatic camera *(Image Tech.)*. Camera in which focus, lens aperture and shutter speed are selected automatically; film advance by motor drive may also be included. Priority selection may be available, for example, exposure based on either general or spot areas and with aperture or shutter speed limitations.

automatic circuit-breaker *(Elec.Eng.)*. A circuit-breaker which automatically opens the circuit as soon as certain predetermined conditions (e.g. an overload) occur.

automatic computer *(Comp.)*. Obsolete term for one type of **first generation computer**.

automatic contrast control *(Image Tech.)*. Form of automatic gain control used in video signal channel of a television receiver

automatic control *(Eng.)*. (1) Switching system which operates control switches in correct sequence and at correct intervals automatically. (2) Control system incorporating servomechanism or similar device, so that feedback signal from output of system is used to adjust the controls and maintain optimum operating conditions.

automatic cut-out *(Elec.Eng.)*. A term frequently applied to a small automatic circuit-breaker suitable for dealing with currents of a few amperes.

automatic data processing *(Comp.)*. See **data processing**.

automatic direction finding *(Aero.)*. Airborne *navaid* tuned to radio source of known position. Using rotatable loop aerial mounted above an aircraft to detect the direction of the radio source by rotating until the signal is zero. Abbrev. *ADF*.

automatic exposure *(Image Tech.)*. Control system which by means of a photo-electric device in the camera automatically sets the lens aperture/shutter speed combination. Selection of a particular image area, optimum depth of field, shutter speed range etc. may be provided. Abbrev. *AE*.

automatic flushing cistern *(Build.)*. Uses the filling of a siphon tube to cause periodic discharge from a cistern. Used for flushing urinals; also for drains having insufficient fall to ensure self-cleansing.

automatic focusing *(Image Tech.)*. Control system for automatically setting the lens focus to the subject distance; in a simple form, this may be by means of a coupled range-finder but advanced types employ completely automatic examination of the image. Abbreviated *AF*. In an enlarger or rostrum camera, lens focus is mechanically set by the distance from the base.

automatic frequency control *(Telecomm.)*. Electronic or mechanical means for automatically compensating, in a receiver, frequency drifts in transmission carrier or local oscillator. Usually abbrev. to *AFC*.

automatic gain control *(Telecomm.)*. System in amplifiers which compensates for a wide range of input signals to give a more uniform level of output and thus accommodate for a wide range of conditions including fading, masking of antenna and ambient light.

automatic gate, door *(Build.)*. A door-system (sliding, hinged, folding) electrically, pneumatically or hydraulically powered for self-opening and closing on impulse triggered by stepping-pad, photoelectric beam or similar device, as in a lift door.

automatic gate lock *(Elec.Eng.)*. A lock on the gate (or door) of a lift-car or landing, which is arranged so that it can only be released, and the gate or door opened, when the car is in a position of safety at the landing concerned.

automatic generating plant *(Elec.Eng.)*. A small generating station, being a petrol (or diesel) engine-driven generator along with a battery; the engine is automatically started up when the battery voltage falls below a certain value and stopped when it is fully charged. The term is also applied to the plant in small unattended hydroelectric generating stations.

automatic mixture control *(Aero.)*. A device for adjusting the fuel delivery to a reciprocating engine in proportion to air density.

automatic observer *(Aero.)*. An apparatus for recording, photographically or electronically, the indications of a large number of measuring instruments on experimental research aircraft.

automatic parachute *(Aero.)*. A parachute for personnel which is extracted from its pack by a static line attached to the aircraft.

automatic phase control *(Image Tech.)*. In reproducing colour TV images, the circuit which interprets the phase of the chrominace signal as a signal to be sent to a matrix.

automatic pilot *(Aero.)*. A device for guiding and controlling an aircraft on a given path. It may be set by the pilot or externally by radio control. Colloquially *George*.

automatic pipette *(Chem.)*. See **pipette**.

automatic quadder *(Print.)*. Linotype and Intertype line-casting machines can automatically centre each line or **quad** it to left or right; on the Monotype one depression of the quadder key can result in either 5 or 10 quads.

automatic quiet gain control *(Telecomm.)*. Joint use of automatic gain control and muting.

automatic reel change *(Print.)*. On rotary machines, equipment to attach a new reel to an old web, without stopping the machine and severing the butt end of the old web. Also called *autopaster, flying paster*.

automatic screw machine *(Eng.)*. Fully-automatic single-spindle or multiple-spindle bar stock turret lathe.

automatic shutter *(Image Tech.)*. In a film projector, a shutter which cuts off the light when the mechanism stops, to protect the film from heat.

automatic signalling *(Civ.Eng.)*. A system of railway signalling, usually with electric control, in which the signals behind a train are automatically put to 'danger' as soon as the train has passed, and held in that position until the train has attained the next section of line.

automatic stabilizer *(Aero.)*. A form of automatic pilot, operating about one or more axes, adjusted to counteract dynamic instability. Colloquially *autostabilizer*. See also **damper**.

automatic starter *(Elec.Eng.)*. A starter for an electric motor which automatically performs the various starting operations (e.g. cutting out steps of starting resistance) in the correct sequence, after being given an initial impulse by means of a push-button or other similar device.

automatic stoker *(Eng.)*. See **mechanical stoker**.

automatic substation *(Elec.Eng.)*. A substation containing rotating machinery (and, therefore, normally requiring the presence of an attendant), which, as occasion demands, is started and stopped automatically; e.g. by a voltage relay which operates when the voltage falls below or rises above a certain predetermined value.

automatic synchronizer *(Elec.Eng.)*. A device which performs the process of synchronization in an a.c. circuit automatically.

automatic tap-changing equipment *(Elec.Eng.)*. A voltage-regulating device which automatically changes the tapping on the winding of a transformer to regulate the voltage in a desired manner.

automatic tracking *(Radar)*. Servo control of radar system operated by a received signal, to keep antenna aligned on target.

automatic train stop *(Elec.Eng.)*. A catch, used in conjunction with an automatic signalling system, which engages a trip-cock on the train if the train passes a signal at danger.

automatic transmission *(Eng.)*. A power transmission system for road vehicles, in which the approximately optimum engine speed is maintained through mechanical or hydraulic speed changing devices which are automatically selected and operated by reference to the road speed of the vehicle. Additional features meet special requirements for braking, reversing, parking, and unusual driving conditions.

automatic trolley reverser *(Elec.Eng.)*. An arrangement of the overhead contact line of a tramway, located at terminal points, which ensures that the trolley collector is reversed when the direction of motion of the car is reversed.

automatic tuning *(Telecomm.)*. (1) System of tuning in which any of a number of predetermined transmissions may be selected by means of push-buttons or similar devices. (2) Fine tuning of receiver circuits by electronic means, following rough tuning by hand.

automatic voltage regulator *(Elec.Eng.)*. A voltage regulator which automatically holds the voltage of a distribution circuit or an alternator constant within certain limits, or causes it to vary in a predetermined manner. See **automatic tap-changing equipment, moving-coil regulator**.

automatic volume compression *(Telecomm.)*. Reduction of signal voltage range from sounds which vary widely in volume, e.g. orchestral music. This is necessary before they can be recorded or broadcast but ideally requires corresponding expansion in the reproducing system to compensate.

automatic volume control *(Telecomm.)*. Alteration of the contrast (dynamics) of sound during reproduction by any means. By compression (*compounder*) a higher level of average signal is obtained for modulation of a carrier, the expansion (*expander*) performing the reverse function at the receiver. In high-fidelity reproduction, arbitrary expansion can be disturbing because of variation in background noise, if present. Abbrev. *AVC*.

automatic volume expansion *(Telecomm.)*. Expansion of dynamic range, e.g. by keeping peak level constant and automatically reducing the lower levels. Used to counteract loss of dynamic range through studio or recording equipment, or during transmission.

automatic weather station *(Meteor.)*. Transistorized and packaged apparatus which measures and transmits weather data for electronic computation.

automation *(Genrl.)*. Industrial closed-loop control system in which manual operation of controls is replaced by servo operation.

automatism *(Behav.)*. An automatic act done without the full co-operation of the personality, which may even be totally unaware of its existence. Commonly seen in hysterical states, such as fugues and somnambulism, but may also be a local condition as in automatic writing.

automaton *(Comp.)*. A device which can take a finite number of states.

automixte system *(Elec.Eng.)*. A system of operation of petrol-electric vehicles in which a battery, connected in parallel with the generator, supplies current during starting and heavy-load periods and is charged by the generator during light-load periods. Also *Pieper sytem*.

automorphism *(Maths.)*. A one-to-one homomorphic mapping from a set on to itself, i.e. an **isomorphism** from a set on to itself. See **homomorphism**.

autonomic, autonomous *(Bot.,Zool.)*. Independent: self-regulating: spontaneous.

autonomic movement *(Bot.)*. Movement in parts of organisms maintained by an internal stimulus, e.g. beating of flagella, cyclosis, chromosome movements, circumnutation. Cf. **paratonic movement**.

autonomic nervous system *(Zool.)*. In Vertebrates, a system of motor nerve fibres supplying the smooth muscles and glands of the body. See **parasympathetic nervous system, sympathetic nervous system**.

autonomics *(Electronics)*. Study of self-regulating systems for process control, optimizing performance.

autonomous vehicle *(Aero.)*. Generally unmanned aircraft operating without external assistance.

autopaster *(Print.)*. See **automatic reel change**.

autopilot *(Aero)*. See **automatic pilot**.

autoplasma *(Zool.)*. In tissue culture, a medium prepared with plasma from the same animal from which the tissue was taken; cf. **heteroplasma, homoplasma**. adj. *autoplastic*.

autoplastic transplantation *(Zool.)*. Reinsertion of a transplant or graft from a particular individual in the same individual. Cf. **heteroplastic, homoplastic, zenoplastic**.

autoplate *(Print.)*. A machine which can deliver a curved stereoplate for rotary printing every 15 sec; built to suit the requirements of each particular rotary machine.

autopodium *(Zool.)*. In Vertebrates, the hand or foot.

autopolyploid *(Biol.)*. A **polyploid** containing 3 or more **basic chromosome sets** all from the same species.

autopsy *(Med.)*. See **necropsy**.

autoradiograph *(Image Tech.)*. Photographic record, usually of a biological specimen, produced by exposure to radiation from self-contained radioactive material which has been injected or absorbed.

autoradiography *(Biol.)*. Originally used to show the distribution of radioactive molecules in cells and tissues after injecting the organism with, or growing the cells in a medium containing a radioactive precursor. It is now widely used to show the distribution of radiolabelled molecules separated on the basis of size, charge etc. Photographic film or emulsion is exposed after applying it to the section, fixed cell or separating medium and the distribution of developed grains viewed directly or under the microscope. Similar procedures exist for fluorescent and other labels.

autoradiography *(Min.)*. Examination of radioactive minerals in thin and polished sections by allowing them to rest on photographic film to record their radioactive emissions.

auto-reclose circuit-breaker *(Elec.Eng.)*. A circuit-breaker which, after tripping due to a fault, automatically recloses after a time interval which may be adjusted to have any value between a fraction of a second and 1 or 2 min.

autorotation *(Aero.)*. (1) The spin; continuous rotation of a symmetrical body in a uniform air-stream due entirely to aerodynamic moments. (2) Unpowered rotorcraft flight, i.e. a helicopter with engine stopped, in which the symmetrical aerofoil rotates at high incidence parallel with the airflow.

autoscoper *(Min.Ext.)*. Pneumatic rock drill mounted on long cylinder with extendable ram, so as to be held firmly across opening (*stope*) when packed out by compressed air.

autoset level *(Surv.)*. A form of dumpy level for rapid operation, in which the essential features are a quick-levelling head, and an optical device which neutralizes errors of levelling so that the bubbles need not be central while an observation is being made.

autoshaping *(Behav.)*. The classical conditioning of an *operant response* that is not reinforced by instrumental conditioning.

autoshaver *(Print.)*. A machine which trims the curved stereoplate to the required thickness.

autosome *(Biol.)*. A chromosome that is not one of the *sex-determining* chromosomes.

autospasy *(Zool.)*. The casting of a limb or part of the body when it is pulled by some outside agent, as when the Slow-worm casts its tail.

autospore *(Bot.)*. A non-motile spore, one of many formed within the parent algal cell and having all the characteristics of the parent in miniature before it is set free. Characteristic of some Chlorococcales.

autostabilizer *(Aero.)*. See **automatic stabilizer**.

autostyly *(Zool.)*. In Craniata, in Dipnoi and all tetrapods, a type of jaw suspension in which the hyoid arch is broken up and the hyomandibular attached to the skull. adj. *autostylic*.

autosynchronous motor *(Elec.Eng.)*. A term frequently used to denote a **synchronous induction motor**.

autotetraploid *(Biol.)*. A polyploid containing 4 similar sets of chromosomes all from the same species.

autothrottle *(Aero.)*. A device for controlling the power of an aëro-engine to keep the approach path angle and speed constant during an automatic blind landing.

autotomy *(Zool.)*. Voluntary separation of a part of the body (e.g. limb, tail), as in certain Worms, Arthropods, and Lizards.

autotransductor *(Elec.Eng.)*. One in which the same winding is used for power transfer and control.

autotransformer *(Elec.Eng.)*. Single winding on a laminated core, the coil being tapped to give desired voltages.

autotransformer starter *(Elec.Eng.)*. A starter for squirrel-cage induction motors, in which the voltage, applied to the motor at starting is reduced by means of an autotransformer.

autotransplantation *(Zool.)*. See **autoplastic transplantation**.

autotrophic *(Bot.)*. Able to elaborate all its chemical constituents from simple, inorganic compounds (especially, all its carbon compounds from CO_2). Cf. **heterotrophic**.

autotrophic bacteria *(Biol.)*. Bacteria which obtain their energy from light and inorganic compounds, and which are able to utilize carbon dioxide in assimilation.

autoxidation *(Chem.)*. (1) The slow oxidation of certain substances on exposure to air. (2) Oxidation which is induced by the presence of a second substance, which is itself undergoing oxidation.

autoxidator *(Chem.)*. An alkene-oxygen compound acting as a carrier or intermediate agent during oxidation, in particular during **autoxidation**.

autumnal equinox *(Astron.)*. See **equinox**.

autumn wood *(Bot.)*. See **late wood**.

autunite *(Min.)*. Hydrated calcium uranium phosphate, yellow in colour.

auxanometer *(Bot.)*. A device for recording the elongation of a plant stem, leaf etc., traditionally by means of a lever and smoked drum.

auxiliary air intake *(Aero.)*. (1) An air intake for accessories, cooling, cockpit air, etc. (2) Additional intake for turbojet engines when running at full power on the ground, usually spring-loaded so that it will open only at a predetermined suction value.

auxiliary attachment *(Horol.)*. A special attachment to a compensation balance, for the purpose of reducing the middle-temperature error.

auxiliary circle *(Maths.)*. Of an ellipse, the circle whose diameter is the major axis of the ellipse. Of a hyperbola, the circle whose diameter is the transverse axis.

auxiliary contact *(Elec.Eng.)*. See **auxiliary switch**.

auxiliary equation *(Maths.)*. Of an ordinary linear differential equation $F(D)y = f(x)$, where D is the differential operator d/dx: the algebraic equation $F(m) = 0$.

auxiliary lift-motor *(Elec.Eng.)*. A small motor forming part of the driving equipment of an electric lift; used for operating the lift at reduced speeds.

auxiliary plant *(Elec.Eng.)*. A term used in generating-station practice to cover the condenser pumps, mechanical stokers, feed-water pumps, and other equipment used with the main boiler, turbine and generator plant.

auxiliary pole *(Elec.Eng.)*. See **compole**.

auxiliary power unit *(Aero.)*. An independent airborne engine to provide power for ancillary equipment, electrical services, starting, etc. May be a small reciprocating or turbine. Abbrev. *APU*.

auxiliary rotor *(Aero.)*. A small rotor mounted at the tail of a helicopter, usually in a perpendicular plane, which counteracts the torque of the main rotor; used to give directional and rotary control to the aircraft.

auxiliary spark gap *(Elec.Eng.)*. A small spark gap for automobile ignition, placed in series with the main gap of a sparking-plug; it improves the quality and certainty of the spark.

auxiliary store *(Comp.)*. See **backing store**.

auxiliary switch *(Elec.Eng.)*. A small switch operated mechanically from a main switch or circuit-breaker; used for operating such auxiliary devices as alarm bells, indicators, etc.

auxiliary tanks *(Aero.)*. See under **fuel tanks**.

auxiliary winding *(Elec.Eng.)*. A special winding on a machine or transformer, additional to the main winding.

auxin *(Bot.)*. A plant growth substance, *indole-3-acetic acid*, (IAA), or any of a number of natural or artificial substances with similar effects. Auxins promote root initiation, cell elongation, xylem differentiation and may be involved in apical dominance and tropism. At high concentrations some synthetic auxins are used as herbicides. See also **2-4-dichlorophenoxyacetic acid**, **indole-3-butyric acid**, **naphthalene acetic acid**, **2-4-5-trichlorophenoxyacetic acid**.

auxochromes *(Chem.)*. A chromophore (or group of atoms) introduced into dyestuffs to give full effectiveness to the colouring properties. The pricipal auxochromes are Cl, Br, SO_3H, NO_2, NH_2, OH. Auxochromes can also permit, by being present, the formation of salts and the creation of a dyestuff. They have a selective absorption frequency for radiation, and function as colour carriers for a frequency determined mainly by the compound of which they are part.

auxocyte *(Biol.)*. Any cell in which meiosis has begun: an androcyte, sporocyte, spermatocyte, or oöcyte, during the period of growth.

auxometer *(Phys.)*. An apparatus for measuring the magnifying power of an optical system.

auxotonic *(Zool.)*. Of muscle contraction, of or against increasing force.

auxotroph *(Biol.)*. A variant organism requiring the addition of special nutrients or growth factors before they will grow, e.g. amino-acid requiring bacteria. Cf. **prototroph**.

available *(Bot.)*. That part of e.g. water or mineral nutrient in the soil or fertilizer, which can be drawn upon by a plant. Cf. **unavailable**.

available light photography *(Image Tech.)*. Photography carried out without the use of flash or artificial lighting.

available line *(Image Tech.)*. Percentage of total length of scanning line on a CRT screen on which information can be displayed.

available potential energy *(Meteor.)*. That part of the total potential energy of the atmosphere available for conversion into kinetic energy by adiabatic redistribution of its mass so that the density stratification becomes horizontal everywhere.

available power efficiency *(Elec.Eng.)*. The ratio of electrical power available at the terminals of an electro-acoustic transducer to the acoustical power output of the transducer. The latter should conform with the reciprocity principle so that the efficiency in sound reception is equal to that in transmission.

available power gain *(Telecomm.)*. The ratio of the available power output of an amplifier to the input power; equal to *power gain* only when the output of the device or circuit is correctly matched to the load.

available power response *(Elec.Eng.)*. For an electro-acoustic transducer, the ratio of mean square sound pressure at a distance of 1 m, in a defined direction from the 'acoustic centre' of the transducer, to the available electrical power input. The response will be expressed in dB above the reference response of $1\,\mu bar^2$/watt of available electrical power.

avalanche *(Phys.)*. Self-augmentation of ionization. See **Townsend avalanche**, **Zener effect**.

avalanche diode *(Electronics)*. A semiconductor breakdown diode, usually silicon, in which avalanche breakdown occurs across the entire p-n junction, giving a voltage drop which is constant and independent of current. Avalanche diodes break down much more sharply than **Zener diodes**. Used in high-speed switching circuits and microwave oscillators.

avalanche effect *(Electronics)*. Cumulative multiplication of carriers in a semiconductor because of avalanche breakdown. This occurs when the electric field across the barrier region is strong enough to allow production and cumulative multiplication of carriers by ionization.

avalanche transistor *(Electronics)*. One depending on avalanche breakdown to produce hole-electron pairs. It can give very high gain in the common-emitter mode or very rapid switching.

avascular *(Med.)*. Not having blood vessels.

AVC *(Telecomm.)*. Abbrev. for *Automatic Volume Control*.

aventurine feldspar *(Min.)*. A variety of plagioclase, near albite-oligoclase in composition, characterized by minute disseminated particles of red iron oxide which cause firelike flashes of colour. Also called *sunstone*.

aventurine quartz *(Min.)*. A form of quartz spangled, sometimes densely, with minute inclusions of either mica or iron oxide. Used in jewellery. Sometimes falsely called 'Indian Jade'.

average *(Ships)*. Loss or damage of marine property, less than total: compensation payment in proportion to amount insured.

average *(Stats.)*. General term for *mean*, *median*, and *mode*.

average current *(Elec.Eng.)*. That current obtained by adding together the products of currents flowing in a circuit and the times for which they flow and dividing by the total time considered. i.e.

$$\frac{1}{n}\sum_{j=1}^{n}|x_j-\bar{x}|,$$

where the mean

$$\bar{x}=\frac{1}{n}\sum_{j=1}^{n}x_j.$$

For direct current the average value is constant: for true alternating current the average value is zero.

average curvature *(Maths.)*. See **curvature** (3).

average haul distance *(Civ.Eng.)*. The distance between the centre of gravity of a cutting and that of the embankment formed from material excavated from the cutting.

average power output *(Telecomm.)*. In an amplitude-modulated transmission, the radio-frequency power delivered by a transmitter, averaged over one cycle or other specified interval of the modulating signal.

aversive, aversion therapy *(Behav.)*. A form of *behaviour therapy* in which the undesirable response is paired with an aversive stimulus.

aversive stimulus *(Behav.)*. A stimulus that an animal will attempt to avoid or escape from; it can be used experimentally to *punish* or *negatively reinforce* a response.

Aves *(Zool.)*. A class of Chordata adapted for aerial life. The forelimbs are modified as wings, the sternum and pectoral girdle are modified to serve as origins for the wing-muscles, and the pelvic girdle and hind limbs to support the entire weight of the body on the ground. The body is covered with feathers and there are no teeth. Respiratory and vascular systems are modified for homoiothermy. Birds.

avgas *(Aero.)*. *Aviation Gasoline*. See **aviation spirit**.

avian big liver disease *(Vet.)*. See **avian leucosis**.

avian diphtheria *(Vet.)*. See **fowl pox**.

avian erythroblastosis *(Vet.)*. See **avian leucosis**.

avian erythroid leucosis *(Vet.)*. See **avian leucosis**.

avian favus *(Vet.)*. *White comb*. A fungal disease of fowls affecting the skin of the head and comb, due to *Trichophyton gallinae*.

avian gout *(Vet.)*. A sympton of nephritis in the domestic chicken, in which urates become deposited on the surface of internal organs (*visceral gout*) and in the joints.

avian granuloblastosis *(Vet.)*. See **avian leucosis**.

avian leucosis *(Vet.)*. *Lymphoid leukosis*; occurs naturally only in the chicken, caused by members of the leukosis/sarcoma group of avian retroviruses. The virus is transmitted efficiently through the embryo, and infected birds become depressed prior to death. There are few typical clinical symptoms, but diffuse or nodular tumours are found in most organs. Involvement of the *bursa* is considered **pathognomonic** and the disease is controlled by eradication. The virus is important in cancer research.

avian monocytosis *(Vet.)*. *Blue comb, pullet disease*. An acute or subacute disease of chickens characterized by depression, loss of appetite, and diarrhoea. The main pathological changes are focal necrosis of liver cells, enteritis, nephritis, and an increased number of monocytes in the blood. The cause is unknown, but a similar disease in turkeys is caused by a coronavirus.

avian myeloblastosis *(Vet.)*. See **avian leucosis**.

avian paratyphoid *(Vet.)*. A contagious disease of birds due to infection by *B. aertrycke*.

avian spirochaetosis *(Vet.)*. *Avian sleeping sickness*. An acute and highly fatal septicaemia of domestic, and certain wild birds, due to infection by the spirochaete *Borrelia anserina (B. gallinarum)*; transmitted by ticks of the genera *Argas* and *Ornithodorus*.

avian typhoid *(Vet.)*. A contagious disease of birds due to infection by *Bacterium gallinarum*.

avian visceral lymphomatosis *(Vet.)*. See **avian leucosis**.

aviation kerosene *(Aero.)*. Finely filtered paraffin for turbine engines, abbrev. *avtur*. See **aviation spirit** and **wide-cut fuel**.

aviation spirit *(Aero.)*. A motor fuel with a low initial boiling point and complying with a certain specification, for use in aircraft. Ranges from 73 to 120/130 octane rating; abbrev. *avgas*. See **aviation kerosene, wide-cut fuel**.

aviatrix *(Aero.)*. Female aviator.

avidin *(Immun.)*. A protein which binds very strongly to biotin. It can be labelled by fluorescence or by attachment of enzymes, and is used to reveal antibodies to which biotin has been conjugated.

avidity *(Immun.)*. A measure of the strength of binding between an antigen and an antibody. Since antigens are liable to have several combining sites the avidity is an average and less precise than affinity.

avionics *(Aero.,Space)*. The collective word for a space- or aircraft's subsystem elements which involve electronic principles. A contraction of *AVIation and electrONICS*.

avitaminosis *(Med.)*. The condition of being deprived of vitamins: any deficiency disease caused by lack of vitamins.

Avogadro number, constant *(Chem.)*. The number of atoms in 12 g of the pure isotope ^{12}C; i.e. the reciprocal of the *atomic mass unit* in grams. It is also by definition the number of molecules (or atoms, ions, electrons) in a **mole** of any substance and has the value $6.022\,52\times10^{23}\,mol^{-1}$. Symbol N_A or L.

Avogadro's law *(Chem.)*. Equal volumes of different gases at the same temperature and pressure contain the same number of molecules.

avpin *(Aero.)*. *Aviation isopropyl nitrate*.

avpol *(Aero.)*. *Aviation Petrol, Oil and Lubricant*.

avtag *(Aero.)*. *Aviation widecut turbine fuel*. See **wide-cut fuel**.

avtur *(Aero.)*. *Aviation Turbine fuel*. See **aviation kerosene**.

avulsion *(Med.)*. The tearing away of a part.

AWACS *(Aero.)*. Abbrev. for *Airborne Warning And Control System*.

awl *(Build.)*. A small pointed or edged tool for making holes which are to receive nails or screws.

awn *(Bot.)*. (1) A long bristle borne on the glumes and/or lemmas of some grasses e.g. barley. (2) A similar structure on another organ.

awning deck *(Ships)*. A superstructure deck, as the name implies. In its simplest form, it is the top deck of a 2-deck ship, and places the ship in a certain category for scantling amd freeboard.

axe *(Build)*. A pointed hammer used for dressing stone.

axed arch *(Build.)*. An arch built from bricks cut to a wedge shape.

axed work *(Build.)*. Hard building stone dressed with an axe to leave a ribbed face.

axenic culture *(Bot.)*. A culture of a single species in the absence of all others; pure culture.

axes *(Crystal.)*. See **crystallographic axes**.

axes *(Genrl.)*. pl. of *axis*.

axes *(Maths.)*. Of a conic: that pair of conjugate diameters which are mutually perpendicular. For an ellipse they are referred to as *major* and *minor* in accordance with their length. For a hyperbola the one which does not cut the curve is called the *conjugate* axis, and the other the *transverse* axis. Of co-ordinates: the fixed reference lines used in a system of co-ordinates.

axial *(Arch.)*. A term which implies that a building is symmetrical about a central axis.

axial *(Bot.)*. Relating to the axis of a plant or organ; longitudinal.

axial compressor *(Eng.)*. A multistage, high-efficiency compressor comprising alternate rows of moving and fixed blades attached to a rotor and its casing respectively.

axial engine *(Aero.)*. Turbine engine with an axial-flow compressor.

axial-flow compressor *(Aero.)*. A compressor in which alternate rows of radially-mounted rotating and fixed aerofoil blades pass the air through an annular passage of decreasing area in an axial direction.

axial-flow turbine *(Aero.)*. Characteristic aero-engine turbine, usually of 1 to 3 rotating stages, in which the gas flow is substantially axial.

axial pitch *(Eng.)*. The distance from any point on one thread or helix to the correspoding point on the next thread or helix measured along the axis of the screw or helix.

axial-plane cleavage *(Geol.)*. **Cleavage** parallel to the axial plane of a fold.

axial ratio *(Phys.)*. Ratio of major to minor axis of polarization ellipse for wave propagated in waveguide, polarized light, etc.

axial response *(Acous.)*. The response of a microphone or loudspeaker, measured with the sound-measuring device on the axis of the apparatus being tested.

axial runout *(Eng.)*. Variation from the plane normal to its axis of a rotating part. Its *wobble* rather than its *eccentricity*. Cf. **radial runout**.

axial skeleton *(Zool.)*. The skeleton of the head and trunk: in Vertebrates, the cranium and vertebral column, as opposed to the appendicular skeleton.

axiate pattern *(Zool.)*. The morphological differentiation of the parts of an organism, with reference to a given axis.

axil *(Bot.)*. The upper hollow where the adaxial surface of leaf or bract attaches to a stem.

axile *(Bot.)*. Coinciding with the longitudinal axis.

axilemma *(Zool.)*. In medullated nerve fibres, the whole of the medullary sheath.

axile placentation *(Bot.)*. That in which the placentas are in the angles formed where the septa along the central axis of an ovary meet the 2 or more locules, e.g. tomato.

axilla *(Med.)*. *The arm-pit*: the angle between the fore-limb and the body. adj. *axillary*.

axillary *(Bot.)*. Situated in or arising from an axil; especially of buds, shoots, flowers, inflorescences etc.

axillary air sac *(Zool.)*. In Birds, one of the paired air sacs, lying in the axillary position. It communicates with the median interclavicular air sac.

axinite *(Min.)*. A complex borosilicate of calcium and aluminium, with small quantities of iron and manganese, produced by pneumatolysis and occurring as brown wedge-shaped triclinic crystals.

axiom *(Maths.)*. An assumption; usually one of the basic assumptions underlying a particular branch of mathematics. Cf. **postulate**.

axiotron *(Electronics)*. Valve in which the electron stream to the anode is controlled by the magnetic field of the heating current.

axis *(Aero.)*. The 3 axes of an aircraft are the straight lines through the centre of gravity about which change of attitude occurs: *longitudinal* or *drag* axis in the plane of symmetry (roll); *normal* or *lift* axis vertically in the plane of symmetry (yaw) and the *lateral* or *pitch* axis transversely (pitch). See **wind axes**.

axis *(Biol.)*. (1) A central line of symmetry of an organ or organism. (2) A stem or root. (3) A rachis. (4) In higher vertebrates, the second cervical vertebra.

axis *(Image Tech.)*. Of a lens, the line of symmetry of the optical system; the line along which there is no refraction.

axis *(Maths.)*. A line, usually notional, which has a peculiar importance in relation to a particular problem or set of circumstances. Thus the *axis of symmetry* of a figure is a line which divides it symmetrically, and the *axis of rotation* is the line about which a body rotates. pl. *axes*.

axis of symmetry *(Maths.)*. See **symmetry**.

axle *(Eng.)*. The cross-shaft or beam which carries the wheels of a vehicle; they may be either attached to and driven by it, or freely mounted thereon.

axle-box *(Eng.)*. Box-shaped housing containing the axle bearings and lubricant. Constrained laterally on guides and supports the weight of vehicle through springs.

axle pulley *(Build.)*. A pulley set in a sash-frame so that the sash-cord connecting the window and its balancing weight may run over it.

axle weight *(Eng.)*. That part of the all-up weight of a vehicle which is borne by the wheels on one particular axle.

axon *(Zool.)*. The process of a typical nerve cell or neuron which transmits an impulse or action potential away from the cell body.

axoneme *(Biol.)*. The central core of a **cilium** consisting of **microtubules** and associated proteins.

axoneme *(Zool.)*. In certain Ciliophora, the central strand of the stalk: the axial filament of a flagellum.

axonometric projection *(Arch.)*. A three dimensional representation of an architectural drawing in which all parallel lines remain parallel and to scale, unlike a perspective, the lines of which converge. Axonometric drawings fall into various categories depending on the angle of projection. See **isometric projection**.

axonometry *(Crystal.)*. Measurement of the axes of crystals.

azeleic acid *(Chem.)*. Heptan 1,7-dicarboxylic acid $COOH.(CH_2)_7.COOH$, mp 106°C. Found in rancid fat. Prepared by the oxidation of oleic acid.

azeotropic distillation *(Chem.Eng.)*. A process for separating, by distillation, products not easily separable otherwise. The essential is the introduction of another substance, called the **entrainer**, which then forms an **azeotropic mixture**, with one or sometimes both initial substances but using only small quantities, then is distilled off leaving the pure product.

azeotropic mixtures *(Chem.)*. Liquid compounds whose boiling point, and hence composition, does not change as vapour is generated and removed on boiling. The boiling point of the azeotropic mixture may be lower (e.g. water-ethanol) or higher (e.g. water-hydrochloric acid) than those of its components. Also *constant-boiling mixtures*.

azides *(Chem.)*. (1) See **acid azides**. (2) Salts of hydrazoic acid. The heavy metal azides are explosive.

azimino compounds *(Chem.)*. Heterocyclic compounds containing 3 adjacent nitrogen atoms in 1 ring, very stable. They are prepared by the action of nitrous acid on 1,2-diamines or 1,8-diaminonaphthalenes.

azimuth *(Astron.,Maths.,Surv.)*. The azimuth of a line or celestial body is the angle between the vertical plane containing the line or celestial body and the plane of the meridian, conventionally measured from north through east in astronomical computations, and from south through west in triangulation and precise traverse work. See **azimuth angle**.

azimuth *(Image Tech.)*. The angle, normally 90°, between the direction of motion of the film or tape and the slit or gap in the optical or magnetic head.

azimuth *(Telecomm.)*. See **bearing**.

azimuthal power instability *(Nuc.Eng.)*. Abnormal neutron behaviour which results in uneven nuclear conditions in the reactor.

azimuth angle *(Surv.)*. Horizontal angle of observed line with reference to true North.

azimuth marker *(Radar)*. Line on radar display made to pass through target so that the bearing may be determined.

azimuth stabilized PPI *(Radar)*. Form of plan position indicator display which is stabilized by a gyrocompass, so that the top of the screen always corresponds to North.

azines *(Chem.)*. Organic bases containing a heterocyclic aromatic ring of 4 carbon and 2 nitrogen atoms, the nitrogen atoms being in the *para*-position with respect to one another.

azobenzene *(Chem.)*. $C_6H_5.N{=}N.C_6H_5$. Orange-red crystals, mp 68°C. Prepared by the partial reduction of nitrobenzene.

azo dyes *(Chem.)*. Derivatives of azobenzene, obtained as the reaction products of diazonium salts with tertiary amines or phenols (hydroxy-benzenes). Usually coloured yellow, red, or brown, they have acidic or basic properties.

azo group *(Chem.)*. The group —N=N—, generally combined with 2 aromatic radicals. The azo group is a chromophore, and a whole class of dyestuffs is characterized by the prescence of this group.

Azoic *(Geol.)*. That part of the Precambrian without life.

azomethane *(Chem.)*. $CH_3.N{:}N.CH_3$, bp 1.5°C, a yellow liquid, obtained by the oxidation of *sym* dimethyl hydrazine with chromic (VI) acid.

azomide *(Chem.)*. See **hydrazoic acid**.

azonal soil *(Bot.)*. Immature soil. Cf. **zonal soil**.

azonium bases *(Chem.)*. A group of bases including **azines** and **quinoxalines**.

azoospermia *(Med.)*. Complete absence of spermatozoa in the semen.

azotaemia *(Med.)*. An excess of nitrogenous breakdown products in the blood (*uraemia*).

azote *(Chem.)*. The French name for nitrogen.

azothioprine *(Med.)*. Drug used to suppress the immune response. Used to suppress rejection after transplantation and in a variety of connective tissue disorders.

Azotobacter *(Biol.)*. Free-living genus of bacteria in soil and water which are able to fix free nitrogen in the presence of carbohydrates.

azoturia *(Vet.)*. *Paralytic equine myoglobinuria.* An acute degeneration of muscles in horses of unknown cause, occurring particularly during exercise after a few days of idleness. Characterized by stiffness, lameness and paralysis, hardness of affected muscles, and urine coloured red to dark brown due to myoglobin.

azoxy compounds *(Chem.)*. Mostly yellow or red crystalline substances obtained by the action of alcoholic potassium hydroxide upon the nitro compounds, or by the oxidation of azo compounds.

azurite *(Min.)*. A deep-blue hydrated basic carbonate of copper, occurring either as monoclinic crystals or as kidney-like masses built of closely packed radiating fibres.

azusa *(Telecomm.)*. US radio tracking system for missile guidance.

azygomatous *(Zool.)*. Lacking a zygomatic arch.

azygos *(Zool.)*. An unpaired structure. adj. *azygous.*

azygospore *(Bot.)*. A structure resembling a zygospore in morphology, but not resulting from a previous sexual union of gametes or of gametangia.

B

b- *(Chem.)*. A symbol for: (1) substitution on the carbon atom of a chain next but one to the functional group; (2) substitution on a carbon atom next but one to an atom common to two condensed aromatic nuclei; (3) substitution on the carbon atom next but one to the hetero-atom in a heterocyclic compound; (4) a stereo-isomer of a sugar.

β *(Genrl.)*. For β-brass, function, particles, waves, etc., see under **beta**. Symbol for: phase constant; ratio of velocity to velocity of light.

β- *(Phys.)*. The intermediate refractive index in a biaxial crystal.

B *(Genrl.)*. Symbol for: boron; susceptance in an a.c. circuit (unit = siemens; measured by the negative of the reactive component of the admittance); magnetic flux density in a magnetic circuit (unit = tesla = $Wb/m^2 = Vs/m^2$).

[B] *(Phys.)*. A Fraunhofer line in the red of the solar spectrum, due to absorption by the earth's atmosphere. [B] is actually a close group of lines having a head at a wavelength 686.7457 nm.

Ba *(Chem.)*. The symbol for barium.

BA *(Genrl.)*. Abbrev. for *British Association screw-thread.*

Babbitt's metal *(Eng.)*. A bearing alloy originally patented by Isaac Babbitt, composed of 50 parts tin, 5 antimony, and 1 copper. Modern addition of lead greatly extends range of service. Composition varies widely, with tin 5% to 90%; copper 1.5% to 6%; antimony 7% to 10%; lead 5% to 48.5%.

Babcock and Wilcox boiler *(Eng.)*. A water-tube boiler consisting in its simplest form of a horizontal drum from which is suspended a pair of headers carrying between them an inclined bank of straight tubes.

Babesia *(Vet.)*. A genus of protozoal parasites which occur in the erythrocytes of mammals.

babesiosis *(Vet.)*. *Piroplasmosis.* Disease caused by protozoa of the genus *Babesia.*

Babinet's compensator *(Phys.)*. A device used, in conjunction with a Nicol prism, for the analysis of elliptically polarized light. It consists of two quartz wedges having their edges parallel and their optic axes at right angles to each other.

Babinet's formula for altitude *(Phys.)*.

$$\text{Altitude} = \frac{32(500 + t_1 + t_2)(B_1 - B_2)}{B_1 + B_2}\ \text{metres},$$

t_1 and t_2 being the respective temperatures in degrees Celsius, and B_1 and B_2 the barometric heights at sea level and at the station whose altitude is required.

Babinet's principle *(Phys.)*. The radiation field beyond a screen which has apertures, added to that produced by a complementary screen (in which metal replaces the holes, and spaces the metal) is identical to the field which would be produced by the unobstructed beam of radiation, i.e. the two diffraction patterns will also be complementary.

Babinski's sign *(Med.)*. Extension of big toe and fanning of other toes on stimulation of sole of foot; a sign of organic disease of the nervous system. It is normal in infants. Now more usually termed *extensor plantar response.*

Babo's law *(Phys.)*. The vapour pressure of a liquid is lowered when a nonvolatile substance is dissolved in it, by an amount proportional to the concentration of the solution.

baby *(Image Tech.)*. A small incandescent spotlight used in film and television production.

baccate *(Bot.)*. Resembling a berry.

Bacillaceae *(Biol.)*. A family of bacteria included in the order *Eubacteriales.* Many are able to produce highly resistant endospores; large Gram-positive rods; aerobic or anaerobic; includes many pathological species, e.g. *Bacillus anthracis* (anthrax). See **clostridium**.

bacillaemia, bacillemia *(Med.)*. Presence of bacilli in the blood.

Bacillariophyceae *(Bot.)*. *Diatomophyceae.* The diatoms, a class of eukaryotic algae in the division Heterokontophyta. The cell wall or *frustule* contains silica. Mostly unicellular, some colonies and chains of cells. Mostly phototrophic (sometimes auxotrophic); some are heterotrophs. Ubiquitous; fresh water and marine, planktonic, benthic and epiphytic, in soils. Fossil deposits of frustules constitute diatomaceous earth or kieselguhr. Two orders; Centrales and Pennales.

bacillary necrosis *(Vet.)*. See **necrobacillosis**.

bacillary white diarrhoea *(Vet.)*. See **pullorum disease**.

bacilluria *(Med.)*. Presence of bacilli in the urine.

bacillus *(Biol.)*. (1) A rod-shaped member of the Bacteria. (2) Genus in the family *Bacillaceae.* pl. *bacilli.*

back *(Build.)*. The upper part of a hand rail, roof rafter, or dome rib. The back of a window is the part between the sill of the sash frame and the floor.

back *(Genrl.)*. A large vat used in various industries, such as dyeing, soap making, brewing. Also *beck.*

back ampere-turns *(Elec.Eng.)*. That part of the armature ampere-turns which produces a direct demagnetizing effect on the main poles. Also called *demagnetizing ampere-turns.*

back band *(Build.)*. The outside member of a door or window casing.

back boiler *(Build.)*. A domestic hot-water boiler fitted in the back of an open fire or range; usually made of cast iron or copper.

back cock *(Horol.)*. In a clock, the bracket (on the back plate) from which the pendulum is suspended.

back coupling *(Telecomm.)*. Any form of coupling which permits the transfer of energy from the output circuit of an amplifier to its input circuit. See **feedback, regeneration**.

backcross *(Biol.)*. The mating of an individual to one of its parents or parental strains. In *Mendelian genetics* a mating of a *heterozygote* to the *recessive homozygote,* producing a 1:1 ratio in the progeny.

back edging *(Build.)*. A method of cutting a tile or brick by chipping away the biscuit below the glazed face, the front itself being scribed.

back e.m.f *(Elec.Eng.)*. That which arises in an inductance (because of rate of change of current), in an electric motor (because of flux cutting) or in a primary cell (because of polarization), or in a secondary cell (when being charged). Also *counter e.m.f.*

back-e.m.f. cells *(Elec.Eng.)*. Cells connected into an electric circuit in such a way that their e.m.f. opposes the flow of current in the circuit.

back emission *(Electronics)*. That of electrons from the anode.

backer *(Build.)*. A narrow slate laid on the back of a broad, square headed slate, at the place where a course of slates begins to narrow.

back-fire *(Autos.)*. (1) Premature ignition during the starting of an internal combustion engine, resulting in an explosion before the end of the compression stroke, and consequent reversal of the direction of rotation. (2) an explosion of live gases accumulated in the exhaust system due to incomplete combustion in the cylinder.

backfitting *(Nuc.Eng.)*. Making changes to nuclear (and other) plants already designed or built, e.g. to cater for changes in safety criteria.

back-flap *(Build.)*. The part of a shutter which folds up behind; also *back-fold, back-shutter.*

back-flap hinge *(Build.)*. A hinge in two square leaves, screwed to the face of a door which is too thin to permit the use of a butt hinge.

back flow *(Build.,Hyd.Eng.)*. Water or sewage flow in a direction contrary to normal.

back focus *(Image Tech.)*. Distance between the rear surface of a lens and the image of an object at infinity.

back-fold *(Build.)*. See **back-flap**.

back gauge *(Build.)*. The distance from the centre of a rivet- or bolt-hole to the back edge of an angle cleat or channel.

back gear *(Eng.)*. A speed-reducing gear fitted to the headstock of a belt-driven metal-turning lathe. It consists of a simple layshaft, which may be brought into gear with the coned pulley and mandrel when required.

background *(Phys.)*. A general problem in physical measurements, which limits the ability to detect or accurately measure any given phenomenon. Background consists of extraneous signals arising from any cause which might be confused with the required measurements, e.g., in electrical measurements of nuclear phenomena and of radioactivity, it would include counts emanating from amplifier noise, cosmic rays, insulator leakage etc. Cf. **signal/noise ratio**.

background (ground) noise *(Acous.)*. Extraneous noise contaminating sound measurements and which cannot be separated from wanted signals. Residual output from microphones, pickups, lines etc., giving a **signal/noise ratio**.

background job *(Comp.)*. Job having a low priority within a multiprogramming system. See **foreground/background**, **job queue**, **time sharing**.

background radiation *(Radiol.)*. Radiation coming from sources other than that being observed.

backhand welding *(Eng.)*. That in which the torch or electrode hand faces the direction of travel, thus *post-heating* the existing weld. Cf. **forehand welding**.

back hearth *(Build.)*. That part of a hearth under the grate.

backheating *(Electronics)*. Excess heating of a cathode due to bombardment by high-energy electrons returning to the cathode. In magnetrons, it may be sufficient to keep the cathode at operating temperature without external heating.

backing *(Build.)*. The operation of packing up a joist so that its upper surface shall be in line with those of deeper joists employed under the same floor.

backing *(Image Tech.)*. Light-absorbent layer on the rear surface of photographic film or plate to reduce **halation**.

backing *(Meteor.)*. The changing of a wind in a counterclockwise direction. Cf. **veering**.

backing *(Print.)*. The process by which one half of the sections of a volume are bent over to the right and the other half to the left at the back. The projections formed are **joints**, to which the covers are hinged, by hand or machine. Some machines also perform **rounding**.

backing boards *(Print.)*. Wedge-shaped wooden boards between which an unbound book is held in the lyingpress, while the joints are being formed for attaching the cover.

backing coil *(Elec.Eng.)*. See **bucking coil**.

backings *(Build.)*. Also *strappings*. Wooden battens, secured to rough walls, for the fixing of wood linings etc.

backing store *(Comp.)*. Means of storing large amounts of data outside the immediate access store. Also *secondary storage*. Will be a combination of **magnetic disk**, **drum**, **tape**, **optical storage**, containing **archive files**, **back-up files**.

backing-up *(Build.)*. The use of inferior bricks for the inner face of a wall.

backing-up *(Print.)*. (1) Printing on the second side of a sheet. (2) Backing a printing plate to required height.

back inlet gulley *(Build.)*. A trapped gulley in which the inlets discharge under the grating and above the level of the water in the trap, so that splashing and blocking of the grating are avoided.

back iron *(Build.)*. The stiffening plate screwed to the cutting iron of a plane. Also *cap iron*, *cover iron*.

back joint *(Build.)*. The part of the back of a stone step which is dressed to fit into the rebate of the upper step.

back-kick *(Eng.)*. Term applied to the violent reversal of an internal combustion engine during starting, due to a **back-fire**.

backlash *(Eng.)*. The lost motion between two elements of a mechanism, i.e., the amount the first has to move, owing to imperfect connection, before communicating its motion to the second.

backlash *(Telecomm.)*. (1) Mechanical deficiency in a tuning control, with a difference in dial reading between clockwise and anti-clockwise rotation. (2) Property of most regenerative and oscillator circuits, by which oscillation is maintained with a smaller positive feedback than is required for inception.

back lighting *(Image Tech.)*. Lighting illuminating the subject from behind, opposite the camera, often to provide rim light or halo effects.

back lining *(Build.)*. (1) The part of a cased sash-frame next to the wall, opposite the pulley-stile. (2) The piece of framing which forms the back of a recess for boxing shutters.

back lining *(Print.)*. See **hollows**.

back lobe *(Telecomm.)*. Lobe of polar diagram for antenna, microphone, etc., which points in the reverse direction to that required.

backlocking *(Civ.Eng.)*. Holding a signal lever partially restored until completion of a predetermined sequence of operations.

back-mixing *(Chem.Eng.)*. In a chemical reactor, jet engine or other apparatus through which material flows and thereby undergoes a change in some property, back mixing is said to occur when some of the material is turned back so that it mixes with that which has entered the reactor after it.

back-mutation *(Biol.)*. See **reversion**.

back observation *(Surv.)*. One made with instrument on station just left. Also *back sight*.

back-off *(Min.Ext.)*. (1) Raising the drilling bit or *down-hole assembly* for a short distance from the bottom of the hole. (2) To unscrew drilling components.

back-planing *(Print.)*. Reducing duplicate plates to the necessary thickness, by shaving metal from the back of steros and electros, and by grinding, in the case of rubber or plastic plates.

backplate lampholder *(Elec.Eng.)*. A lampholder fitted with a support which enables it to be screwed on to a flat surface. Also called a *batten-lampholder*.

back porch *(Image Tech.)*. A short period of **black level** signal transmitted at the end of the horizontal **sync pulse** before the picture information.

back-porch effect *(Telecomm.)*. The prolonging of the collector current in a transistor for a brief time after the input signal (particularly if large) has decreased to zero.

back pressure *(Build.)*. Air pressure in drainage pipes exceeding atmospheric pressure.

back pressure *(Eng.)*. The pressure opposing the motion of the piston of an engine on its exhaust stroke: the exhaust pressure of a turbine. Increased by clogged or defective exhaust system.

back pressure *(Med.)*. Proximal pressure produced by an obstruction to fluid flowing through the cardio-vascular or urinary systems.

back pressure *(Min.Ext.)*. The hydrostatic pressure exerted by the fluids in the bore of a well which acts against that of the oil and gas in the strata. Control of this pressure by valves and **chokes** maintains an even and productive flow of oil.

back-pressure turbine *(Eng.)*. A steam turbine from which the whole of the exhaust steam, at a suitable pressure, is taken for heating purposes.

back priming *(Build.)*. The technique of coating surfaces before erection, which, after erection will be inaccessible but may be subject to deterioration if left unprotected.

back projection *(Image Tech.)*. (1) Projection of a picture, from film, transparency or video, on to a translucent screen to be viewed from the opposite side. (2) A form of motion picture composite photography in which the projected picture forms the background to action taking place in front of it, both being photographed together.

back rake *(Eng.)*. In a lathe tool, the inclination of the top surface or face to a plane parallel to the base of the tool.

back rest *(Eng.)*. See **steady**.

back saw *(Build.)*. A saw stiffened by a thickened back; e.g. a tenon saw.

back scatter *(Phys.)*. The deflection of radiation or particles by scattering through angles greater than 90° with reference to the original direction of travel.

backsetting *(Build.)*. A stone with a boasted or broached face, but with smooth border all round.

back shift *(Min.Ext.)*. In colliery, the afternoon shift.

back shore *(Build.)*. One of the outer members of an arrangement of raking shores or props, for supporting temporarily the side of a building. The *back shore* supports the raking shore, or raker, which takes the thrust from the highest part of the building.

back-shutter *(Build.)*. See **back-flap**.

back sight *(Surv.)*. See **back observation**.

back-stay *(Eng.)*. See **steady**.

back-step *(Print.)*. See **black-step marks**.

backstop *(Elec.Eng.)*. The structure of a relay which limits the travel of the armature away from the pole-piece or core.

back stopes *(Min.Ext.)*. **Overhand (or overhead) stopes**, worked up-slope. Ant. *underhand stopes*.

back titration *(Chem.)*. In volumetric analysis, the technique of adding a reagent in excess, to exceed the end-point, then determining the end-point by titrating the excess.

back-to-back *(Electronics)*. Parallel connection of valves, with the anode of one connected to the cathode of the other, or transistors in parallel in opposite directions, to allow control of a.c. current without rectification. Frequently used with **thyratrons** and **ignitrons**.

back-to-back test *(Elec.Eng.)*. The arrangement, originated by Hopkinson, for testing two substantially similar electrical machines on full load, by coupling them mechanically and loading them by regulating the electrical circuit, the total power supply accounting for total losses only; it has been extended to transformers and mechanical gearing.

back-up file *(Comp.)*. Copy of a file, held inside a computer system, to be used in the event of the current file being corrupted.

Backus-Naur form *(Comp.)*. See **BNF**.

backward busying *(Telecomm.)*. Applying busy condition at the incoming end of a trunk or junction (usually during testing or fault-clearance) to indicate at outgoing end that circuit must not be used.

backward diode *(Electronics)*. One with characteristic of reverse shape to normal. Also *back diode*, *AU diode*.

backward hold *(Telecomm.)*. A method of interlocking the links of a switching chain by originating a locking condition in the final link and extending it successively backwards to each of the preceding links.

backward shift, backward lead *(Elec.Eng.)*. Movement of the brushes of a commutating machine around the commutator, from the neutral position, and in a direction opposite to that of the rotation of the commutator, so that the brushes short-circuit zero e.m.f. conductors when the load current, through armature reaction, results in a rotation of the neutral axis of the air-gap flux. Shifting the brushes in this way reduces sparking on the commutator.

backward signalling *(Telecomm.)*. Signalling from the called to the calling end of a circuit.

backward wave *(Electronics)*. In a travelling-wave tube, a wave with group-velocity in the opposite direction to the electron stream. Cf. **forward wave**.

backward-wave tube *(Telecomm.)*. General term for a family of microwave **travelling wave tubes** in which energy on a slow-wave circuit or structure, linked closely to the electron beam, flows in the opposite direction to the electrons. They can be used as stable, low-noise amplifiers or as oscillators; as the latter, they can be easily tuned over a wide frequency range by altering the beam voltage.

back-water *(Hyd.Eng.)*. Water dammed back in a stream or reservoir by some obstruction.

backwater *(Paper)*. Water, containing fine fibres, loading and other additives, removed in the forming section of a paper or board-making machine. It is generally re-used within the system or clarified in a saveall to recover suspended matter.

back-water curve *(Hyd.Eng.)*. The longitudinal profile of the water surface in the case of non-uniform flow in an open channel, when the water surface is not parallel to the invert, owing to the depth of water having been incrased by the placing of a weir across the flow.

back wave *(Telecomm.)*. See **spacing wave**.

bacteraemia, bacteremia *(Med.)*. Presence of bacteria in the blood.

bacteria *(Biol.)*. A large group of unicellular or multicellular organisms, lacking chlorophyll, with a simple nucleus, multiplying rapidly by simple fission, some species developing a highly resistant resting ('spore') phase; some species also reproduce sexually, and some are motile. In shape spherical, rodlike, spiral, or filamentous, they occur in air, water, soil, rotting organic material, animals and plants. Saprophytic forms are more numerous than parasites, but the latter include both animal and plant pathogens. A few species are autotrophic.

bacteria beds *(Build.)*. Layers of a filtering medium such as broken stone or clinker, used in the final or oxidizing stage in sewage treatment. See **contact bed, percolating filter**.

bacterial leaching *(Min.Ext.)*. See **microbiological mining**.

bacterial recovery *(Min.Ext.)*. See **microbiological mining**.

bactericide *(Biol.)*. A substance which destroys bacteria. adj. *bactericidal*. Cf. **bacteriostat**.

bacteriocin *(Biol.)*. A toxin produced by one class of bacteria which kills another, usually related class.

bacteriology *(Genrl.)*. The scientific study of bacteria.

bacteriophage *(Biol.)*. A **virus** which infects bacteria. syn. *phage*.

bacteriostat *(Biol.)*. A substance which inhibits growth but does not kill bacteria.

bacteriotoxin *(Biol.)*. A toxin destructive to bacteria.

bacteriotropin *(Biol.)*. A substance, usually of blood serum, which renders bacteria more subject to the action of antitoxin or more readily phagocytable; e.g. opsonin.

bacteroid *(Bot.)*. The enlarged, X- or Y-shaped, nitrogen-fixing form of a *Rhizobium* bacterium within a legume root nodule.

Bacteroidaceae *(Biol.)*. A family of pleomorphic bacteria included in the order *Eubacteriales*. Usually Gram-negative rods; anaerobic; inhabit the intestine and mucous membrane of higher Vertebrates.

bacteruria *(Med.)*. The presence of bacteria in the urine.

baddeleyite *(Min.)*. Zirconium dioxide, found in Brazil, where it is exploited as a source of zirconium.

badger *(Build.)*. An implement used to clear mortar from a drain after it has been laid.

badger plane *(Build)*. Large wooden plane with a rebated offside edge, used for wide rebates.

badger softener *(Build.)*. A brush with badger hair filling, used in water colour graining.

Badischer process *(Chem.)*. See **contact process**.

Baeyer's tension (strain) theory *(Chem.)*. Since the four valencies of the tetravalent carbon atom are symmetrically distributed in space, one may predict the strain involved in the formation of a ring compound from a chain of carbon atoms, and also estimate the stability of a ring compound from the number of carbon atoms in the ring.

baffle *(Acous.)*. Extended surface surrounding a diaphragm of a sound source (loudspeaker) so that an acoustic short-circuit is prevented.

baffle *(Aero.)*. Any device to impede or divide a fluid flow, (a) in a tank, to reduce sloshing of liquid propellants, (b) plates fitted between cylinders of air-cooled engines to assist cooling.

baffle *(Electronics)*. Internal structure or electrode, with no external connection, used in gas-filled tubes to control the discharge or its decay.

baffle *(Min.Ext.)*. In hydraulic or rake classifier, a plate set across and dipping into the pulp pool; in mechanically agitated flotation cell, one so set as to reduce centrifugal movement in the upper part of the cell.

baffle loudspeaker *(Acous.)*. An open-diaphragm loudspeaker, in which the radiation of sound power is enhanced by surrounding it with a large plane baffle, generally of wood.

baffle plate *(Eng.)*. A plate used to prevent the movement of a fluid in the direction which it would normally follow, and to direct it into the desired path.

baffle plate *(Telecomm.)*. Plate inserted into waveguide to produce change in mode of transmission.

bagassosis *(Immun.,Med.)*. Respiratory disease similar to **farmer's lung** occurring in persons who inhale dust from mouldy sugar cane. Due to Type III hypersensitivity reaction induced by thermophilic mould spores.

bag plug *(Build.)*. A drain plug consisting of a cylindrical canvas bag which is placed in the drain pipe and inflated.

bag pump *(Hyd.Eng.)*. A form of bellows pump, in which the valved disk taking the place of the bucket is connected to the base of the barrel by an elastic bag, distended at intervals by rings.

bailer *(Min.Ext.)*. Length of piping closed at bottom by clack valve, used to remove drilling sands and muds from borehole.

bailey *(Arch.)*. The open area within the fortified walls of a mediaeval castle and usually surmounting a mound or **motte**.

Bailey bridge *(Civ.Eng.)*. A temporary bridge made by assembling portable prefabricated panels. A 'nose' is projected over rollers across the stream, being followed by the bridge proper, with roadway. Used also over pontoons.

Bailey test for sulphur *(Chem.)*. After fusion with sodium carbonate, the aqueous solution gives a blood-red colour with sodium nitroprusside.

bailiff *(Min.Ext.)*. See **overman** (1).

Baily furnace *(Elec.Eng.)*. An electric-resistance furnace in which the resistance material is crushed coke placed between carbon electrodes; used for heating ingots and bars in rolling mills, for annealing, etc.

Baily's beads *(Astron.)*. A phenomenon, first observed by Baily in 1836, in which, during the last seconds before a solar eclipse becomes total, the advancing dark limb of the moon appears to break up into a series of bright points.

bainite *(Eng.)*. Bainite forms when austenite is transformed at temperatures intermediate between the pearlite and martensite ranges. It is an acidular mixture of cementite and ferrite, varying in its proportions with the formation temperature.

baize *(Textiles)*. A light-weight woollen felt used to cover tables (e.g. for billiards) and notice-boards.

baked (caked) *(Print.)*. Said of type which, having become stuck together, is difficult to distribute.

baked core *(Foundry)*. A dry sand core baked in the oven to render it hard and to fix its shape. See **core sand**.

baked images *(Print.)*. The technique of heating a printing plate (mainly lithographic) to harden the printing image and thus increase the image's resistance to wear, hence lengthening the run expectancy on the press.

Bakelite *(Plastics)*. TN for synthetic resin (named after L.H. Baekeland), the product of the condensation of cresol or phenol with formaldehyde. Now largely superseded.

bake-out *(Elec.Eng.)*. Preliminary heating of electrodes and container of an electronic valve to ensure freedom from the later release of gases.

Baker's cyst *(Med.)*. A cyst found behind the knee, which may or may not communicate with the synovial joint.

baking soda *(Chem.)*. *Sodium hydrogen carbonate, bicarbonate of soda.*

BAL *(Med.)*. *British anti-lewisite*, dithioglycerol. Antidote for poisoning by **lewisite** and other poisons, particularly arsenic and mercury.

balance *(Acous.)*. Adjustment of sources of sound in studios so that the final transmission adheres to an artistic standard.

balance *(Biol.)*. Equilibrium of the body; governed from the cerebellum, in response to stimuli from the eyes and extremities and especially from the **semicircular canals** of the ears.

balance *(Chem.)*. See **chemical balance**.

balance *(Elec.Eng.)*. A balance in bridge measurements is said to be obtained when the various impedances forming the arms of the bridge have been adjusted, so that no current flows through the detector. See also **current weigher**.

balance *(Horol.)*. The vibrating member of a watch, chronometer or clock (with platform escapement). In conjunction with the balance-spring it forms the time-controlling element.

balance arc *(Horol.)*. The portion of the vibration during which the balance is not detached from the escapement.

balance arm *(Horol.)*. The portion of the balance connecting the rim to the staff.

balance bar *(Hyd.Eng.)*. The heavy beam by which a canal-lock gate may be swung on its pintle, and which partially balances the outer end of the gate.

balance box *(Eng.)*. A box, filled with heavy material, used to counterbalance the weight of the jib and load of a crane of the cantilever type.

balance-bridge *(Civ.Eng.)*. See **bascule bridge**.

balance cock *(Horol.)*. The detachable bracket which carries the upper pivot of the balance staff.

balance-crane *(Eng.)*. A crane with two arms, one having counterpoise arrangements to balance the load taken by the other.

balanced amplifier *(Telecomm.)*. One in which there are two identical signal-handling branches operating in phase opposition, with input and output connections balanced to earth. A **push-pull amplifier** is an example.

balanced-armature pick-up *(Acous.)*. A pick-up in which the reproducing needle is held by a screw in a magnetic arm, which is pivoted so that its motion diverts magnetic flux from one arm of a magnetic circuit to another, thereby inducing e.m.f.s in coils on these arms.

balanced-beam relay *(Elec.Eng.)*. One having two coils arranged to exert their forces on plungers at each end of a beam pivoted about its central point.

balanced circuit *(Elec.Eng.)*. For a.c. and d.c., one which is balanced to earth potential, i.e. the two conductors are at equal and opposite potentials with reference to earth at every instant. See **unbalanced circuit**.

balanced draught *(Eng.)*. A system of air-supply to a boiler furnace, in which one fan forces air through the grate, while a second, situated in the uptake, exhausts the flue gases. The pressure in the furnace is thus kept atmospheric, i.e. is *balanced*.

balanced equation *(Chem.)*. The equation for a chemical reaction in which the correct relative numbers of moles of each reactant and product are shown.

balanced line *(Telecomm.)*. One in which the impedances to earth of the two conductors are, or are made to be, equal. Also *balanced system*.

balanced load *(Elec.Eng.)*. A load connected to a polyphase system, or to a single-phase or d.c. 3-wire system, in such a way that the currents taken from each phase, or from each side of the system, are equal and at equal power factors.

balanced mixer *(Telecomm.)*. A mixer, which may be made of discrete components or formed in stripline or waveguide, in which the local oscillator breakthrough in the output is minimized and certain harmonics suppressed. The contribution of local oscillator noise to the receiver's overall performance is also reduced by such a mixer.

balanced modulator *(Telecomm.)*. One in which the carrier and modulating signal are combined in such a way that the output contains the two sidebands but not the carrier. Used in colour television to modulate subcarriers, and in suppressed-carrier communication systems.

balanced network *(Telecomm.)*. One arranged for insertion into a **balanced circuit** and therefore symmetrical electrically about the mid-points of its input and output pairs of terminals.

balanced-pair cable *(Telecomm.)*. One with two conductors forming a loop circuit, the wires being electrically balanced to each other and earth (shield), e.g. an open-wire antenna feeder. Cf. **coaxial cable**.

balanced pedal *(Acous.)*. In an organ console, the foot-operated plate, pivoted so that it stays in any position, for remote control of the shutter of the chambers in which ranks of organ pipes are situated; it also serves for bringing in all the stops in a graded series. See **swell pedal**.

balanced protective system *(Elec.Eng.)*. A form of protective system for electric transmission lines and other apparatus, in which the current entering the line or apparatus is balanced against that leaving it; so long as there is no fault on the line or apparatus this balance will be maintained, but an upsetting of the balance owing to a fault causes relays to trip the faulty circuit. Also called *differential protective system* or (coll.) *earth leak relay* or *earth trip*.

balanced sash *(Build.)*. See **sliding sash**.

balanced solution *(Chem.)*. A solution of two or more salts, in such proportions that the toxic effects of the individual salts are mutually eliminated; sea water is a *balanced solution*.

balanced step *(Build.)*. See **dancing step**.

balanced system *(Telecomm.)*. See **balanced line**.

balanced termination *(Telecomm.)*. A two-terminal load in which both terminals present the same impedance to ground.

balanced voltage (current) *(Elec.Eng.)*. A term used, in connection with polyphase circuits, to denote voltages or currents which are equal to all the phases. Also applied to d.c. 3-wire systems.

balanced weave *(Textiles)*. A weave in which the length of free yarn between the intersections is the same in the warp and weft directions and on both sides of the fabric.

balance equation *(Meteor.)*. An equation expressing the balance between the nondivergent part of the horizontal windfield and the corresponding field of *geopotential* on a constant pressure surface. If Ψ is the stream function, Φ the geopotential, and f the **Coriolis parameter**, the balance equation is

$$f\nabla^2\psi + \nabla\psi\cdot\nabla f + 2\left\{\frac{\partial^2\psi}{\partial x^2}\frac{\partial^2\psi}{\partial y^2} - \left(\frac{\partial^2\psi}{\partial x\partial y}\right)^2\right\} = \nabla^2\phi$$

Winds derived from the balance equation, which requires numerical solution on a computer, are closer to their actual values than those derived from the **geostrophic approximation**, especially in regions where the isobars are markedly curved.

balance gate *(Hyd.Eng.)*. A flood gate which revolves about a central vertical shaft, and which may be made self-opening or self-closing as the current sets in or out of a channel by giving a preponderating area to the inner leaves of the gate.

balance pipe *(Eng.)*. A connecting pipe between two points at which pressure is to be equalized.

balance piston *(Eng.)*. See **dummy piston**.

balance point *(Civ.Eng.)*. Any point where a *mass-haul curve* cuts the datum line, showing that up to this point all excavated material has been used up in embankment.

balancer *(Elec.Eng.)*. A device used on polyphase or 3-wire systems to equalize the voltages between the phases or the sides of the system, when unbalanced loads are being delivered.

balance rim *(Horol.)*. The circular rim of the balance. It may be monometallic or bimetallic. See **compensation balance**.

balancer transformer *(Elec.Eng.)*. An auto-transformer connected across the outer conductors of an a.c. 3-wire system, the neutral wire being connected to an intermediate tapping.

balance spring *(Horol.)*. A very fine metal ribbon, forming a flat spiral, cylindrical or helical spring round the balance staff. It regulates the movement of the balance.

balance staff *(Horol.)*. The staff which carries the balance, and the collet to which the balance spring is attached.

balance tab *(Aero.)*. A *tab* whose movement depends upon that of the main control surface. It helps to balance the aerodynamic loads and reduces the *stick forces*. Cf. **servo tab, spring tab** and **trimming tab**.

balance theories *(Behav.)*. Theories concerned with how an individual's attitudes and perceptions of other people, events or objects, relate to each other.

balance weights *(Elec.Eng.)*. Small weights threaded on radial arms on the movement of an indicating instrument, so adjusted that the pointer gives the same indication whatever the orientation of the instrument.

balance weights *(Eng.)*. A weight used to counterbalance some part of a machine; e.g. the weights applied to a crankshaft to minimize or neutralize the inertia forces due to reciprocating and rotating masses of the engine.

balance wheel *(Horol.)*. A **balance** in the form of a wheel, normally combined with a spiral hairspring.

balancing *(Image Tech.)*. In colour reproduction, control of the levels of the three colour components to achieve a satisfactory picture without obvious colour bias, especially in the representation of neutral grey tones.

balancing *(Surv.)*. The process of adjusting a traverse, i.e. applying corrections to the different survey lines and bearings so as to eliminate the closing error.

balancing *(Telecomm.)*. See **neutralization**.

balancing antenna *(Telecomm.)*. Auxiliary reception antenna which responds to interfering but not to the wanted signals. The interfering signals thus picked up are balanced against those picked up by the main antenna, leaving signals more free from interference.

balancing capacitance *(Telecomm.)*. See **neutralizing capacitance**.

balancing machine *(Eng.)*. A machine for testing the extent to which a revolving part is out of balance, and to determine the weight and position of the masses to be added, or removed, to obtain balance.

balancing ring *(Elec.Eng.)*. See **equalizer ring**.

balancing speed *(Elec.Eng.)*. See **free-running speed**.

balanitis *(Med.)*. Inflammation of the glans penis.

balanoposthitis *(Med.)*. Inflammation of the glans penis and prepuce.

balanorrhagia *(Med.)*. Gonorrhoeal inflammation of the glans penis, with discharge of pus.

balantidiosis *(Vet.)*. An enteric infection, especially of swine, by ciliate protozoa of the genus *Balantidium*.

balanus *(Med.)*. The terminal bulbous portion of the penis; the glans penis.

balas ruby *(Min.)*. A misnomer for the rose-red variety of the mineral spinel. See **false ruby**.

balata *(Chem.)*. The coagulated latex of the bullet tree of S. America. Similar in properties to gutta-percha, but softer and more ductile. Used extensively in the manufacture of golf balls and for impregnating cotton duck belting.

Balbach process *(Chem.)*. Electrolytic separation of gold and silver from a metal, by making it the anode in a bath of silver nitrate, the silver being discharged at the cathode.

Balbiani rings *(Biol.)*. Very large *puffs* at specific sites on Dipteran (fruit-fly) salivary gland chromosomes. Occur when RNA, coding for secretory proteins, is being transcribed at these sites.

balbuties *(Med.)*. Stammering, stuttering.

balconet *(Arch.)*. A low ornamental railing to a door or window, projecting very little beyond the threshold or sill; mainly used in the Swiss style of architecture.

balcony *(Arch.)*. A stage or platform projecting from the wall of a building within or without, supported by pillars or consoles, and surrounded with a balustrade or railing.

baldacchino *(Arch.)*. A canopy suspended above altar, tomb etc., or supported in such position by columns. Also *baldachin, baldaquin*.

bale breaker *(Textiles)*. A machine that opens up the highly compressed cotton from a bale, producing a large number of tufts ready for subsequent blending, cleaning and further opening.

baleen *(Zool.)*. In certain Whales, horny plates arising from the mucous membrane of the palate and acting as a food strainer.

balk *(Civ.Eng.)*. The material between two excavations. Also *baulk*.

Balkan frame *(Med.)*. A frame, with pulleys attached, for supporting the leg in the treatment of fractures.

balking *(Elec.Eng.)*. See **crawling**.

ball-and-socket head *(Image Tech.)*. Camera mounting allowing universal movement in rotation and tilt before fixing by clamping; usually fitted to the top of a tripod.

ball-and-socket joint *(Eng.)*. A joint between two rods, permitting considerable relative angular movement in any plane. A ball formed on the end of one rod is embraced by a spherical cup on the other. Used in light control systems (e.g. in connecting a pair of bell-cranks which operate in planes at right angles) and in the steering mechanism of motor vehicles, in which both ball and cups are of case-hardened metals.

ball-and-socket joint *(Med.)*. *Enarthrosis*. A joint in which the hemispherical end of one bone is received into the socket of another.

ballast *(Civ.Eng.)*. (1) A layer of broken stone, gravel, or other material deposited above the formation level of road or railway; it serves as foundation for road-metal or permanent-way respectively. (2) Sandy gravel used as a coarse aggregate in making concrete.

ballast *(Ships)*. Gravel, stone, or other material placed in the hold of a ship to increase her stability when floating without cargo or with insufficient cargo.

ballast lamp *(Elec.Eng.)*. Normal incandescent lamp used as a ballast resistor, current limiter, alarm, or to stabilize a discharge lamp.

ballast resistance *(Elec.Eng.)*. A term used in electric-railway signalling to denote the resistance between the two track rails across the **ballast** on which the track is laid. If allowed to fall too low, it will have the effect of shunting the signal from a train's wheels.

ballast resistor *(Elec.Eng.)*. One inserted into a circuit to swamp or compensate changes, e.g. those arising through temperature fluctuations. One similarly used to swamp the negative resistance of an arc or gas discharge. See also **barretter**.

ballast tube *(Elec.Eng.)*. Same as *barretter* or *ballast resistor*.

ball-bearing *(Eng.)*. A shaft bearing consisting of a number of hardened steel balls which roll in spherical grooves (ball tracks) formed in an inner race fitted to the shaft and in an outer race carried in a housing. Balls are spaced and held by a light metal or plastic cage. Usually first choice for general engineering application. See also **linear ball bearing, recirculating ball thread**.

ball catch *(Build.)*. A door-fastening in which a spring-controlled ball, projecting through a smaller hole, engages with a striking plate.

ball clay *(Geol.)*. A fine-textured and highly plastic clay — a reworked china clay. So named because it used to be rolled into balls. Also known as *potters clay*.

ball-cock, ball-valve *(Build.)*. A self-regulating cistern-tap which, through a linkage system, is turned off and on by the rise and fall of a floating ball.

ball-ended magnet *(Elec.Eng.)*. A permanent magnet, consisting of a steel wire with a steel ball attached to each end; this gives a close approximation to a unit pole.

ball flower *(Arch.)*. An ornament, like a ball enclosed within three or four petals of a flower, set at regular intervals in a hollow moulding.

balling *(Eng.)*. (1) A process that occurs in the cementite constituent of steels on prolonged annealing at 650°–700°C. (2) The operation of forming balls in a puddling furnace.

ballistic circuit-breaker *(Elec.Eng.)*. A very high-speed circuit breaker, in which the pressure produced by the fusing of an enclosed wire causes interruption of the circuit.

ballistic method *(Elec.Eng.)*. A method of high-grade testing used in electrical engineering, a ballistic galvanometer being used.

ballistic missile *(Aero.,Space)*. See **missile**.

ballistic pendulum *(Phys.)*. A heavy block suspended by strings so that its swings are restricted to one plane. If a bullet is fired into the block, the velocity of the bullet may be calculated from a measurement of the angle of swing of the pendulum.

ballistics *(Phys.)*. The study of the dynamics of the path taken by an object moving under the influence of a gravitational field.

ballistospore *(Bot.)*. A spore that is violently projected e.g. the basidiospore of the basiomycete fungi.

ball joint *(Eng.)*. Same as *ball-and-socket joint*.

ball lightning *(Meteor.)*. A slowly-moving luminous ball, which is occasionally seen at ground level during a thunderstorn. It appears to measure about 0.5 m in diameter.

ball mill *(Chem.Eng.)*. Mill consisting of a horizontal cylindrical vessel in which a given material is ground by rotation with steel or ceramic balls.

ballonnet *(Aero.)*. An air compartment in the envelope of an aerostat, used to adjust changes of volume in the filler gas.

balloon *(Aero.)*. A general term for aircraft supported by buoyancy and not driven mechanically.

balloon *(Arch.)*. A spherical ball or globe crowning a pillar, pier etc.

balloon barrage *(Aero.)*. An anti-aircraft device consisting of suitably disposed tethered *balloons*.

balloon former *(Print.)*. On rotary presses, an additional former mounted above the others, from which folded webs are gathered to make up the sections of multi-sectioned newspapers or magazines. See **length fold collection**.

ballooning of yarn *(Textiles)*. The shape taken up by yarns on the spinning or doubling machines.

ballotini *(Civ.Eng.)*. Small glass beads. Used to increase reflectivity in paints.

ballottement *(Med.)*. Method of diagnosing pregnancy by manual displacement of the foetus in the fluid which surrounds it in the uterus.

ball-pane hammer *(Eng.)*. A fitter's hammer, the head of which has a flat face at one end, and a smaller hemi-spherical face or pane at the other; used chiefly in riveting.

ball race *(Eng.)*. (1) The inner or outer steel ring forming one of the ball-tracks of a ball-bearing. (2) Commonly, the complete *ball bearing*.

ball sizing *(Eng.)*. Forcing a suitable ball through a hole to finish size it, usually part of a **broach** with a series of spherical lands of increasing size arranged along it.

ballstone *(Geol.)*. The name applied to masses of fine unstratified limestone, occurring chiefly in the Wenlock Limestone of Shropshire, and representing colonies of corals in position of growth.

ball-track *(Eng.)*. See **ball-bearing**.

ball valve *(Eng.)*. A single nonreturn valve consisting of a ball resting on a cylindrical seating; used in small water and air pumps.

Balmer series *(Phys.)*. A group of lines in the hydrogen spectrum named after the discoverer and given by the formula:

$$\nu = R_H\left(\frac{1}{2^2} - \frac{1}{n^2}\right),$$

where n has various integral values, ν is the wave number and R_H is the hydrogen Rydberg number ($= 1.0967758 \times 10^7\,m^{-1}$).

balneology *(Med.)*. The scientific study of baths and bathing, and of their application to disease.

BALPA *(Aero.)*. Abbrev. for *British Airline Pilots Association*.

balsam of fir *(Chem.)*. See **Canada balsam**.

balsam of Peru *(Chem.)*. An oleoresin containing esters of benzoic and cinnamic acids, obtained from a South American papilionaceous tree. Used in perfumery and chocolate manufacture.

balsam of Tolu *(Chem.)*. An oleoresin containing esters of benzoic and cinnamic acids, obtained from a South American evergreen tree (*Myroxylon toluiferum*). Used in perfumery, chocolate manufacture, and medically as a mild expectorant.

balsa wood *(For.)*. The wood of *Ochroma lagopus* ('West Indian corkwood'); it is a highly porous wood, and is valued for its lightness; used for vibration isolation.

Baltic redwood *(For.)*. See **Scots pine**.

BALUN *(Telecomm.)*. Abbrev. for *BALance to UNbalance transformer*, usually a resonating section of transmission line, for coupling balanced to unbalanced lines. Also *bazooka.*

baluster *(Arch.)*. A small pillar supporting the coping of a bridge parapet, or the handrail of a staircase.

balustrade *(Arch.)*. A coping or handrail with its supporting balusters.

banak *(For.)*. A tree (*Virola*) having a rather featureless wood, pinkish-brown in colour. It grows in Central America and is useful for turnery purposes. The wood has a slight tendency to split. It is also used for packing-cases and roofing shingles.

banana plug *(Elec.Eng.)*. A single conductor plug which has a spring metal tip, in the shape of a banana. The corresponding socket or jack is termed a *banana jack.*

band *(Biol.)*. **Metaphase** chromosomes, when stained with a variety of **banding techniques** exhibit a pattern of transverse bands of varying width characteristic of specific chromosomes. The bands appear to be made by distinctive interactions between specific types of DNA sequence and proteins. Factors such as base and sequence composition, e.g. blocks of short repeats, are involved. **Polytene** chromosomes also exhibit characteristic bands, with or without staining. They are much more numerous than those of metaphase chromosomes and the more stretched state of the polytene chromosomes would be expected to show a greater number of bands. The two kinds of banding patterns may therefore be related. The combination of size, shape and banding pattern gives each chromosome a unique appearance, which has made **eukaryotic** gene mapping possible.

band *(Build.)*. A flat horizontal member, occasionally ornamented, separating a series of mouldings or dividing a wall surface.

band *(Telecomm.)*. See **frequency band**.

band-and-hook, crook hook *(Build.)*. See **strap hinge**.

B and BB *(Eng.)*. Brand-marks signifying *Best* and *Best Best*, placed on wrought iron to indicate the maker's opinion of its quality.

band (belt) conveyor *(Eng.)*. An endless band passing over, and driven by, horizontal pulleys, thus forming a moving track which is used to convey loose material or small articles.

band brake *(Eng.)*. A flexible band wrapped partially round the periphery of a wheel or drum. One end is anchored, and the braking force is applied to the other.

band chain *(Surv.)*. *Steel tape*. More accurate than ordinary chain.

band clutch *(Eng.)*. A friction clutch in which a fabric-lined steel band is contracted on to the periphery of the driving member by engaging gear. See **friction clutch**.

band cramp *(Build)*. A flexible steel band held in position by clamping screws. Used for shaped work, e.g. chair-backs.

band-edge energy *(Electronics)*. The energy of the edge of the conduction or valence band in a solid. It is the lowest energy required by an electron to remain free, or the maximum permissible energy of a valence electron.

banded precipitate *(Chem.)*. See **Liesegang rings**.

banded structure *(Geol.)*. A structure developed in igneous and metamorphic rocks due to the alternation of layers of different texture and/or composition.

band ignitor tube *(Electronics)*. A valve of mercury pool type in which the control electrode is a metal band outside the glass envelope. Also termed *capacitron.*

banding *(Image Tech.)*. Defect in videotape recording heads causing visible horizontal bands in the picture.

banding plane *(Build.)*. A plane used for cutting out grooves and inlaying strings and bands in straight and circular work.

banding techniques *(Biol.)*. Methods of treating chromosomes to produce patterns of **bands** characteristic of an individual chromosome, as an aid to recognition. All involve staining fixed **metaphase** chromosomes and fall into 5 main groups. (1) Q-banding, staining with quinacrine mustard or 33258 Hoechst; (2) G-banding, removal of some protein followed by staining with Giemsa; (3) R-banding, heat or alkali treatment followed by Giemsa or acridine orange staining (gives reverse pattern of G banding); (4) T-banding, variant of R-banding which mainly stains the **telomeric** regions; (5) C-banding, treatment which mainly stains **constitutive heterochromatin**. In mammals, G- and Q-banding give generally similar patterns, but in plants Q- and C-bands appear to be more closely related.

band merit *(Telecomm.)*. See **gain-bandwidth product**.

band-pass filter *(Telecomm.)*. One which freely passes currents having frequencies within specified nominal limits, and highly attenuates currents with frequencies outside these limits.

band-pass tuning *(Telecomm.)*. See **double-hump effect**.

bands *(Print.)*. The tapes, or cords, placed across the back of a book, to which the sections are attached by sewing. The ends of the bands are subsequently secured to the boards of the cover. See **flexible bands, raised bands**.

bandsaw *(Eng.)*. A narrow endless strip of saw-blading running over and driven by pulleys, as a belt; the strip passes a work table placed normal to the straight part of the blade. The work piece, in wood or metal, is forced against the blade and intricate shapes can be cut.

band spectrum *(Phys.)*. Molecular optical spectrum consisting of numerous very closely spaced lines which are spread through a limited band of frequencies.

band-spreading *(Telecomm.)*. (1) Use of a relatively small tuning capacitor in parallel with the main tuning capacitor of a radio receiver, so that fine tuning control can be done with the smaller; useful when the frequency band is crowded. (2) Mechanical means, like reduction gearing to achieve the same result.

band-stop filter *(Telecomm.)*. One which attenuates signals having frequencies within a certain range or *band*, while freely passing those outside this range. Also *band-rejection filter.*

B and S wire gauge *(Eng.)*. Abbrev. for *Brown and Sharpe wire gauge.*

band theory of solids *(Phys.)*. For atoms brought together to form a crystalline solid, their outermost electrons are influenced by a *periodic* potential function, so that their possible energies form *bands* of allowed values separated by bands of forbidden values (in contrast to the discrete energy states of an isolated atom). These electrons are not localized or associated with any particular atom in the solid. This band structure is of fundamental importance in explaining the properties of metals, semiconductors and insulators. See **energy band, conduction band, valence band**.

bandwidth *(Telecomm.)*. (1) The width, or spread, of the range of frequencies used for a given purpose, e.g. the width of individual channels allotted to speech or to television transmissions. (2) The space occupied in the frequency-domain by signals of a specified nature, e.g. telephone-quality speech, broadcast quality stereophonic music, television, radar transmission etc. (3) The range of frequencies within which the characteristics of a device (filter, amplifier etc.) are within specified limits, often the points at which the performance has changed by 3 dB from a mean level, or the **half-power points**.

Bang's bacillus *(Biol.)*. *Brucella abortus*; the cause of contagious abortion in animals and of undulant fever in man.

Bang's disease *(Vet.)*. See **bovine contagious abortion**.

banister *(Arch.)*. Alternative term for *baluster.*

banjo axle *(Autos.)*. The commonest form of rear-axle casing, in which the provision of the differential casing in the centre produces a resemblance to a banjo with two necks.

bank *(Eng.)*. A number of similar pieces of equipment grouped in line and connected; e.g. a *bank* of engine cylinders, a *bank* of coke-ovens, a *bank* of transformers.

bank *(Print.)*. A bench on which sheets are placed as printed, or on which standing type-matter rests.

bank *(Telecomm.).* An assemblage of fixed electrical contacts arranged (generally in arc formation) to facilitate the selective connection of a traversing wiper (or wipers).

Banka drill *(Min.Ext.).* One widely used in shallow testing of alluvial deposits. Portable, hand-operated, it consists essentially of an assembly of 5-ft pipes, 4 in. in diameter, which are worked into the ground, the material traversed being recovered by a sand pump or **bailer**.

banked fire, boiler *(Eng.).* A boiler furnace in which the rate of combustion is purposely reduced to a very low rate for a period during which the demand for steam has ceased. See **banking loss, banking-up, dead bank.**

banker *(Build.).* A bench upon which bricklayers and stonemasons shape their materials.

banket *(Geol.).* The term originally applied by the Dutch settlers to the gold-bearing conglomerates of the Witwatersrand. It is now used for any metamorphosed conglomerate, containing barren quartz pebbles cemented with siliceous matrix bearing gold.

banking *(Aero.).* Angular displacement of the wings of an aircraft about the longitudinal axis, to assist turning.

banking *(Eng.).* Process of suspending operation in a smelter, by feeding fuel only into the furnace until as much metal and slag as possible have been removed, after which all air inlets are closed.

banking *(Min.Ext.).* The operations involved in removing full trucks, tubs, or wagons and replacing them by empty ones at the top of a shaft.

banking pins *(Horol.).* Vertical pins in the bottom plate of a watch which limit the motion of the lever. In the cylinder escapement, the action of the balance is limited by a single banking pin on the balance. In some watches part of the plate itself forms a solid banking.

banking-up *(Eng.).* Reducing the rate of combustion in a boiler furnace by covering the fire with slack or fine coal.

bank multiple *(Telecomm.).* The multiple connecting the bank of contacts in selectors or uniselector switches in *Strowger* telephone exchanges. The switches are usually detachable from the contacts and multiples for repair, without disturbing the large number of connections in the multiple.

bank paper *(Paper).* A thin writing paper of less than 50 g/m^2, intended for typewriting or correspondence purposes.

bank protector *(Hyd.Eng.).* Any device for minimizing erosion of river banks by water; e.g. groins, pitching etc.

bannisterite *(Min.).* A hydrated silicate of manganese, crystallizing in the monoclinic system, and occurring in manganese deposits in North Wales and New Jersey.

banquette *(Arch.).* A narrow window-seat in masonry, brickwork or wood.

banquette *(Civ.Eng.).* (1) A raised footway inside a bridge parapet. (2) A ledge on the face of a cutting. See **berm.**

BAP *(Bot.).* See **6-Benzylaminopurine.**

bar *(Civ.Eng.).* A pivoted bar, parallel to a running rail, which, being depressed by the wheels of a train, is capable of holding points or giving information about a train's position.

bar *(Eng.).* Material of uniform cross-section, which may be cast, rolled or extruded.

bar *(Horol.).* A narrow detachable plate.

bar *(Meteor.,Phys.).* Unit of pressure or stress, 1 bar = 10^5 N/m^2 or pascals = 750.07 mm of mercury at 0°C and lat. 45°. The *millibar* (1 mbar = 100 N/m^2 or 10^3 dyn/cm^2) is used for barometric purposes. (N.B. Std. atmos. pressure = 1.013 25 bar.) The *hectobar* (1 hbar = 10^7 N/m^2, approx. 0.6475 tonf/in^2) is used for some engineering purposes.

baragnosis *(Med.).* Loss of the ability to judge differences between the weights of objects.

bar-and-yoke *(Elec.Eng.).* Method of magnetic testing in which the sample is in the form of a bar, clamped into a yoke of relatively large cross-section, which forms a low-reluctance return path for the flux.

Bárány's tests *(Med.).* Tests for assessing the functions of the semicircular canals in the inner ear.

barathea *(Textiles).* Woven fabric used for coats and suits and made from silk, worsted, or man-made fibres. Interesting surface appearance arising from the twill or broken-rib weave used in its manufacture.

barb *(Bot.).* A hooked or doubly-hooked hair.

barb *(Zool.).* Any hooked, bristlelike structure: in Birds, one of the lateral processes of the rachis of the feather which form the vane.

Barbados Earth *(Geol.).* A siliceous accumulation of remains of Radiolaria, formed originally in deep water and later raised above sea level.

Barba's law *(Eng.).* Concerned with the plastic deformation of metal test pieces when strained to fracture in a tensile test, it states that test pieces of identical size deform in a similar manner.

barbate *(Bot.).* Bearded; bearing tufts of long hairs.

barbel *(Zool.).* In some Fish, a finger-shaped tactile or chemosensitive appendage arising from one of the jaws.

barber's rash *(Med.).* Infection of the beard region of the face with a bacterium or a fungus.

barbitone *(Med.). Diethyl-malonyl-urea, 5,5-diethylbarbituric acid.* White crystalline solid; mp 191°C. Once widely used as a sedative and hypnotic, usually in the form of the sodium salt. Also called *barbital, veronal.*

barbituric acid *(Chem.). Malonyl urea,* $CO(NH.CO)_2CH_2$, crystallizing in large colourless crystals. The hydrogen atoms of the methylene group are reactive and can be replaced by halogen. Basis of important derivatives with therapeutic action.

barbule *(Zool.).* In Birds, one of the processes borne on the barbs of a feather, by which the barbs are bound together.

barchan *(Geol.).* An isolated crescentic sand-dune.

bar code *(Comp.).* Code consisting of parallel thick and thin lines. Used on a label attached to an item to give machine readable information. See **Universal Product Code.**

bar code reader *(Comp.).* A scanning device which reads a **bar code.** Also *wand.*

bar cramp *(Build).* Metal or wooden bar with a fixed and a moving jaw which can clamp two pieces of wood together for e.g. gluing planks edgewise.

bare *(Eng.).* A term signifying slightly smaller than the specified dimension. Cf. **full.**

bare conductor *(Elec.Eng.).* A conductor not continuously covered with insulation, but supported intermittently by insulators, e.g. bus-bars and overhead lines.

bare electrodes *(Elec.Eng.).* Electrodes used in welding that are not coated with a basic slag-forming substance.

bareface tenon *(Build.).* A tenon which has a shoulder on one face only; used when jointing a rail which is thinner than the stile.

barge *(Print.).* A small case with boxes for sorts such as figures and spaces for use when correcting.

barge board *(Build.).* A more or less ornamental board fixed at the gable end of a roof. It hides the ends of the horizontal timbers, and protects from the weather the underside of the **barge-course.**

barge couple *(Build.).* The outer couples, *pair of rafters,* of a roof which project over the gable of a roof.

barge course *(Build.).* (1) That part of the roof of a house which projects slightly over the gable end, and is made up underneath with mortar to keep out rain, etc. (2) A coping course of bricks laid edge-wise and transversely on a wall.

bar generator *(Image Tech.).* Source of pulse signals, giving a bar pattern for testing TV cathode-ray tubes.

barite *(Min.Ext.). Barium sulphate, barytes.* Used to increase the density of **drilling mud** and thus the **back pressure** during drilling. See **barytes.**

barium *(Chem.).* A heavy element in the second group of the periodic system, an alkaline earth metal. Symbol Ba, at. no. 56, r.a.m. 137.34, mp 850°C. In most of its compounds it occurs as Ba^{2+}.

barium concrete *(Build.).* Concrete containing high proportion of barium compounds, which has a high absorption for radiation and thus used as a shield.

barium enema *(Radiol.).* A radiological examination of

the lower gastro-intestinal tract using barium sulphate as a **contrast medium**.

barium feldspar *(Min.)*. A collective term for barium-bearing feldspars, including celsian and hyalophane.

barium hydroxide *(Chem.)*. $Ba(OH)_2$. See **barium oxide**.

barium oxide *(Chem.)*. BaO. When freshly obtained from the calcined carbonate it is even more reactive with water than calcium oxide and forms barium hydroxide (alkaline). Also called *baryta*.

barium plaster *(Build.)*. A cement-sand plaster containing barium salts, used for lining hospital and experimental X-ray rooms to absorb radiation and minimize back-scattering.

barium sulphate *(Chem.)*. $BaSO_4$. Formed as a heavy white precipitate when sulphuric acid is added to a solution of a barium salt. Very nearly insoluble in water. Although of little pigmentary value, it is much used in paint manufacture and in the preparation of lake pigments. Used in *barium meals*. See **barytes**.

barium titanate *(Chem.)*. $BaTiO_3$; a crystalline ceramic with outstanding dielectric, piezo-electric and ferro-electric properties. Used in capacitors and as a piezo-electric transducer. Has a higher **Curie point** than **Rochelle Salt**.

bark *(Bot.)*. A non-technical term applied to all the tissues outside the cambium, i.e. the corky and other material which can be peeled from a woody stem.

bar keel *(Ships)*. See **keelson**.

Barker index *(Crystal.)*. A method of identification of crystalline substances from measurements of interfacial angles.

barkevikite *(Min.)*. A member of the amphibole group, resembling basaltic hornblende but having a higher total iron content and a low ferric-ferrous iron ratio. Occurs in alkaline plutonic rocks, as at Barkevik, Norway.

Barkhausen effect *(Phys.)*. The tendency for magnetization to occur in discrete steps rather than by continuous change.

Barkhausen-Kurz oscillator *(Telecomm.)*. One with a triode valve having its grid more positive than the anode. Electrons oscillate about the grid before reaching the anode. Output frequency depends on the transit time of electrons though the tube.

bar lathe *(Eng.)*. A small lathe of which the bed consists of a single bar of circular, triangular, or rectangular section.

Barlow lens *(Phys.)*. A plano-convex lens between the objective and eye-piece of a telescope to increase the magnification by increasing the effective focal length.

Barlow's wheel *(Elec.Eng.)*. A primitive electric motor formed from a pivoted starwheel, intermittently dipping into mercury, torque being obtained from interaction of the radial current and a perpendicular magnetic field supplied by a permanent magnet.

bar magnet *(Elec.Eng.)*. A straight bar-shaped permanent magnet, with a **pole** at each end.

bar mill *(Eng.)*. A rolling mill with grooved rolls, for producing round, square or other forms of bar iron of small section.

bar mining *(Min.Ext.)*. Alluvial mining of sand-banks, river bars, or submerged deposits.

bar movement *(Horol.)*. A watch movement in which the upper pivots are carried in bars.

barn *(Phys.)*. Unit of effective cross-sectional area of nucleus equal to $10^{-28}\,m^2$. So called, because it was pointed out that although one barn is a very small unit of area, to an elementary particle the size of an atom which could capture it was 'as big as a barn door'.

Barnard's star *(Astron.)*. A red dwarf star in Ophiuchus, found in 1916 to have the largest proper motion yet measured, amounting to 10 seconds of arc per annum.

barn door *(Image Tech.)*. Pair of adjustable flaps on a studio lamp for controlling the light.

Barnett effect *(Phys.)*. Magnetization of a ferromagnetic material by rapid rotation of the specimen. Used to measure magnetic susceptibility. See **Einstein-de Haas effect**.

barney *(Image Tech.)*. A soft cover to reduce noise from a film camera.

baroclinic atmosphere *(Meteor.)*. An atmosphere which is not **barotropic**.

barograph *(Meteor.)*. A recording barometer, usually of the aneroid type, in which variations of atmospheric pressure cause movement of a pen which traces a line on a clockwork-driven revolving drum.

barometer *(Meteor.)*. An instrument used for the measurement of atmospheric pressure. The **mercury barometer** is preferable if the highest accuracy of readings is important, but where compactness has to be considered, the **aneroid barometer** is often used. For *altitudes by barometer*, see **Babinet's formula**.

barometric corrections *(Meteor.)*. Necessary corrections to the readings of a mercury barometer for index error, temperature, latitude and height.

barometric error *(Horol.)*. The error in the time of swing of a pendulum due to change of air pressure. Though small, it is avoided in precision clocks by causing the pendulum to swing in an atmosphere of constant (low) pressure.

barometric pressure *(Meteor.)*. The pressure of the atmosphere as read by a barometer. Expressed in *millibars* (see **bar**), the height of a column of mercury, or (SI) in hectopascals.

barometric tendency *(Meteor.)*. The rate of change of atmospheric pressure with time. The change of pressure during the previous three hours.

barophil *(Biol.)*. Used of organisms that grow and metabolize as well (or better) at increased pressures as at atmospheric pressure.

barophoresis *(Chem.)*. Diffusion of suspended particles at a speed dependent on external forces.

Baroque *(Arch.)*. One of the later phases of the Renaissance style of architecture prominent in the 17th century and based on classical features but employing lighting, sculpture and painting to produce a theatrical, monumental effect. Typically, forms were manipulated to create focal points rather than being dispersed to achieve the rhythmic consistency which characterized the High Renaissance style. It is regarded as the ultimate fulfilment of the **Renaissance** movement as, by this stage, a distinctive style had emerged.

baroque organ *(Acous.)*. A type of pipe organ in which low fundamentals are obtained by the subjective difference tones arising from pipes operating at the musical interval of the fifth.

baroreceptor *(Med.)*. A sensory receptor sensitive to stimulus of pressure, e.g. receptors located in the walls of arteries and veins responding to changes of intraluminal pressure, such as the carotid and aortic arterial, or right arterial atrial, baroreceptors.

barostat *(Aero.)*. A device which maintains constant atmospheric pressure in a closed volume, e.g. the input and output pressure of fuel metering device of a gas turbine to compensate for atmospheric pressure variation with altitude.

barotropic atmosphere *(Meteor.)*. An atmosphere with zero horizontal temperature gradient at all levels so that the **isopleths** of density and pressure coincide and the **thickness chart** has no pattern.

barrage *(Hyd.Eng.)*. An artificial obstruction placed in a water-course in order to secure increased depth for irrigation, navigation, or some other purpose.

barrage balloon *(Aero.)*. A small captive kite balloon the cable of which is intended to destroy low-flying aircraft.

barrage-fixe *(Hyd.Eng.)*. A dam provided with sluices to control the flow of water.

Barr body *(Biol.)*. The densely-staining **heterochromatin** of the inactive X chromosome. See **X-inactivation**.

barré *(Textiles)*. Undesirable stripes in fabrics. In weft-knitted fabrics sometimes caused by irregularities in the texturing of the yarn resulting in variation in dye affinity.

barred code *(Telecomm.)*. Any dialled code that automatic exchange apparatus is primed to reject by connecting the caller no further than **NU tone**.

barrel *(Eng.)*. A hollow, usually cylindrical, machine part; often revolving, sometimes with wall apertures.

barrel *(Horol.)*. The cylindrical container for housing a mainspring. See **fusee, going barrel, pin barrel**.

barrel *(Min.Ext.)*. US barrel of 42 US gallons (=35 Imperial gallons), frequently employed as a unit of capacity, especially in the oil industry.

barrel amalgamation *(Eng.)*. Recovery of gold from rich concentrates by prolonged gentle tumbling with mercury in a steel barrel.

barrel arbor *(Horol.)*. The arbor carrying the barrel; upon its centre portion is coiled the mainspring during winding. In a 'detachable barrel' the arbor and barrel can be withdrawn from between the plates without dismantling the plates.

barrel bolt *(Build.)*. Hand-operated door fastening comprising a metal rod sliding in cylindrical guides.

barrel cam *(Eng.)*. A cylindrical cam with circumferential or end track.

barrel distortion *(Image Tech.)*. **Curvilinear distortion** of an optical or electronic image in which horizontal and vertical straight lines appear convex, bowed outwards.

barrel distortion *(Phys.)*. A type of curvilinear distortion produced by a lens or lens system whereby a square array is imaged as barrel-shaped i.e. the lines are concave. Also called *positive distortion*. See also **pincushion distortion**.

barrel drain *(Build.)*. A cylindrical drain.

barrel hopper *(Eng.)*. A machine for unscrambling, orientating, and feeding small components during a manufacturing process, in which a revolving barrel tumbles the components on to a sloping, vibrating feeding-blade.

barrel nipple *(Build.)*. See **shoulder nipple**.

barrel-plating *(Elec.Eng.)*. A process of electroplating in which the articles to be plated are placed in a rotating container provided with suitable negative contacts.

barrel printer *(Comp.)*. **Line printer** where the complete character set is provided at each printing position, embossed on the surface of a horizontal barrel or cylinder. See **bi-directional-, chain-**.

barrel-type crankcase *(Autos.)*. A petrol-engine crankcase so constructed that the crankshaft must be removed from one end; in more normal construction the crankcase is split. See **split crankcase**.

barrel vault *(Build.)*. A vault of approximately semicircular cross section, whose length exceeds its diameter. Also called *annular vault, tunnel vault, wagon vault*.

barrel-vault roof *(Civ.Eng.)*. A roof formed of reinforced concrete in the shape of an open cylindrical shell, generally with lateral stiffening diaphragms and edge beams. The roof itself is frequently very thin in section.

barrel winding *(Elec.Eng.)*. See **drum winding**.

barren *(Geol.)*. Without fossils.

barren solution *(Min.Ext.)*. In chemical extraction of metals from their ores, solution left after these have been removed. Cf. **pregnant solution**.

barretter *(Elec.Eng.)*. Iron-wire resistor mounted in a glass bulb containing hydrogen, and having a temperature variation so arranged that the change of resistance ensures that the current in the circuit in which it is connected remains substantially constant over a wide range of voltage.

barrier *(Elec.Eng.)*. (1) In transformers, the solid insulating material which provides the main insulation, apart from the oil. (2) The refractory material intended to localize or direct any arc which may arise on the operation of a circuit-breaker.

barrier coat *(Build.)*. A coating applied to a substrate to protect subsequent coating from active constituents in the substrate.

barrier gear *(Elec.Eng.)*. An arrangement for moving heavy electrical plant, using man power. Rotating machines and transformers are equipped with wheels and movement is possible by inserting crowbars at suitable points and levering the equipment.

barrier layer *(Electronics,Telecomm.)*. (1) In semiconductor junctions, see **depletion layer**. (2) In an optical-fibre cable, an intermediate layer of glass between the low refractive index core and the high refractive index cladding.

barrier-layer capacitance *(Electronics)*. Same as *depletion-layer capacitance*. See **depletion layer**.

barrier penetration *(Acous.)*. The passage of a sound wave, at an angle for which **Snell's law** predicts zero transmission, through a very thin layer.

barrier pillar *(Min.Ext.)*. A pillar of solid coal left in position to protect a main road from subsidence, or as a division, or to protect workings from flooding.

barrier reef *(Geol.)*. A coral-reef developed parallel with the shore-line and enclosing a lagoon between itself and the land. It marks a stage between a fringing-reef and an atoll.

barring motor *(Elec.Eng.)*. A small motor which can be temporarily connected, by a gear or clutch, to a large machine to turn it slowly for adjustment or inspection.

Barrovian metamorphism, Barrovian zones *(Geol.)*. Regional metamorphism of the type first described in the Scottish Highlands by G. Barrow in 1893. Zones of increasing metamorphism are characterized by the presence of a series of index minerals: chlorite, biotite, garnet, staurolite, kyanite, sillimanite. This type of metamorphism has since been recognized in many other parts of the world.

bars of foot *(Vet.)*. Part of the sole of the horse's foot formed by reflexion of the wall on each side of the **frog**.

bar suspension *(Elec.Eng.)*. A method of mounting the motor on an electrically propelled vehicle. One side of the motor is supported on the driving axle and the other side by a spring-suspended bar lying transversely across the truck. Also called *yoke suspension*.

Bartholin's duct *(Zool.)*. An excretory duct of the sublingual gland.

Bartholin's glands *(Zool.)*. In some female Mammals, glands (corresponding with Cowper's glands in the male) lying on either side of the upper end of the vagina.

bartonellosis *(Vet.)*. Infection by organisms of the genus *Bartonella*, affecting man, cattle, dogs and rodents.

bar tracery *(Arch.)*. Window-tracery characteristic of *Gothic work*, resembling more a bar of iron twisted into various forms, than stone.

bar-type current transformer *(Elec.Eng.)*. A **current transformer** in which the primary consists of a single conductor that passes centrally through the iron core upon which the secondary is wound.

bar winding *(Elec.Eng.)*. An armature winding for an electric machine whose conductors are formed of copper bars.

bar-wound armature *(Elec.Eng.)*. An armature with large sectioned conductors which are insulated and fixed in position and connected, in contrast with former-wound conductors which are sufficiently thin to be inserted, after shaping in a suitable jig.

barye *(Phys.)*. See **microbar**.

baryon *(Phys.)*. A *hadron* with a baryon number of +1. Baryons are involved in strong interactions. Baryons include neutrons, protons and hyperons.

baryon number *(Phys.)*. An intrinsic property of an elementary particle. The baryon number of a baryon is +1, of an antibaryon −1. The baryon number of mesons, leptons, and gauge bosons is zero. Baryon number is conserved in all types of interaction between particles. **Quarks** have a baryon number of +1/3 and antiquarks of −1/3.

baryta *(Chem.)*. See **barium oxide**.

baryta paper *(Paper)*. Paper coated on one side with an emulsion of barium sulphate and gelatine. Used in moving pointer recording apparatus and for photographic printing papers.

baryta water *(Chem.)*. A suspension of barium hydroxide in distilled water; it is a fairly strong alkali.

barytes, barite *(Min.)*. Barium sulphate, typically showing tabular orthorhombic crystals. It is a common mineral in association with lead ores, and occurs also as nodules in limestone and locally as a cement of sandstones. Also called *heavy spar*.

barytes concrete *(Nuc.Eng.)*. See **loaded concrete**.

barytocalcite *(Min.)*. A double carbonate of calcium and barium, $BaCa(CO_3)_2$, crystallizing on the monoclinic system, and occurring typically in lead veins.

basal area *(Ecol.)*. A measure of the extent of trees in an area, being the total cross-sectional area of the trunks.

basal body *(Biol.,Bot.)*. (1) A cylindrical structure found at the base of cilia composed of 9 sets of triplet **microtubules** which serves as a centre for the growth of microtubules in culture. In flagellate or ciliate *Protozoa*, zoospores or spermatozoids, it occurs as a small, deeply-staining granule at the base of the locomotory *organelle*. Also *basal granule*, *blepharoplast*. (2) Part of a **thallus** fixed to the substrate by rhizoids.

basal conglomerate *(Geol.)*. A first stage of sedimentation resting on a plane of erosion.

basal corpuscle *(Bot.,Zool.)*. See **basal body**.

basal ganglia *(Med.)*. A localized concentration of grey matter deep in the cerebral hemispheres and the mid brain. They are concerned with the regulation of movement and are often referred to as the extra-pyramidal system. Disease of the basal ganglia gives **Parkinsonism** and **chorea**.

basal ganglia *(Zool.)*. In Vertebrates, ganglia connecting the cerebrum with other nerve-centres.

basal lamina *(Biol.)*. A thin sheet of extracellular matrix underlying epithelia. It contains, in addition to collagen and other proteins, the distinctive glycoprotein laminin.

basal metabolic rate *(Med.)*. The minimal quantity of heat produced by an individual at complete physical and mental rest, but not asleep, 12–18 hr after eating, expressed in milliwatts per square metre of body surface.

basal placentation *(Bot.)*. Placentation in which the ovules are attached to the bottom of the locule in an ovary.

basal planes *(Crystal.)*. The name applied to the faces representing the terminating *pinacoid* in all the crystal systems exclusive of the cubic system.

basal plates *(Zool.)*. In the developing Vertebrate skull, a plate of cartilage formed by the fusion of the parachordals and the trabeculae; in Crinoidea, certain plates situated at or near the top of the stalk; in Echinoidea, certain plates forming part of the apical disk.

basalt *(Geol.)*. A fine-grained igneous rock, dark colour, composed essentially of basic plagioclase feldspar and pyroxene, with or without olivine. In the field, the term is generally restricted to lavas, but many minor intrusions of basic composition show identical characters, and therefore cannot be distinguished in the laboratory. The extrusive equivalent of **gabbro**.

basalt glass *(Geol.)*. See **tachylite**.

basaltic hornblende *(Min.)*. A variety of hornblende with a high ferric-ferrous iron ratio and a low hydroxyl content, occurring chiefly in volcanic rocks. Also called *oxyhornblende*.

basanite *(Geol.)*. A basaltic rock containing plagioclase, augite, olivine, and a feldspathoid (nepheline, leucite, or analcite).

bascule bridge *(Civ.Eng.)*. A counterpoise bridge which can be rotated in a vertical plane about axes at one or both ends. The roadway over the river rises while the counterpoise section descends into a pit. Also called *balance-bridge*.

base *(Bot.)*. The end of an organ or plant part nearest to its point of attachment or origin.

base *(Chem.)*. Generally, a substance which tends to donate an electron pair or co-ordinate an electron. In particular, a substance which dissolves in water with the formation of hydroxyl ions and reacts with acids to form salts.

base *(Electronics)*. (1) The region between the emitter and collector of a transistor, into which minority carriers are injected. It is essentially the *control* electrode of the transistor, in the same way as the grid in a valve. (2) The part of an electron tube which has pins, leads, or terminals through which connections are made to the internal electrodes.

base *(Image Tech.)*. The thin flexible support on which a photographic emulsion or magnetic coating is carried.

base *(Maths.)*. See **cone**, **cylinder**, **logarithm**, **radix**, **triangle**.

baseband *(Telecomm.)*. The frequency band occupied by the signal in modulation.

baseboard *(Build.)*. See **skirting board**.

base bullion *(Eng.)*. Ingot base metal containing sufficient silver or gold to repay recovery, e.g. argentiferous lead.

base circle *(Eng.)*. The circle used in setting out the profiles of gear-wheel teeth of involute form.

base course *(Build.)*. The lowest course of masonry in a building.

Basedow's disease *(Med.)*. See **Graves disease**. Thyrotoxicosis.

base exchange *(Chem.)*. Chemical method used in soil mechanics to strengthen clays by replacing their H-ions with Na-ions.

base level *(Geol.)*. The lowest level towards which erosion progresses.

baseline *(Print.)*. The bottom alignment of type; below it is the *beard* which accommodates the descenders of *f, g* etc.

baseline *(Surv.)*. A survey line the length of which is very accurately measured by precise methods; used as a basis for subsequent triangulation.

base load *(Elec.Eng.)*. That part of the total load on an electrical power system which is applied, where possible, by the most efficient connected generating stations, the remaining **peak load** being supplied intermittently by the more expensive stations.

basement *(Geol.)*. (1) a complex of igneous and metamorphic rocks covered by sediments. (2) the crust of the Earth extending downwards to the Mohorovicic discontinuity.

basement membrane *(Biol.)*. A membrane lying between an epithelium and the underlying connective tissue.

base metal *(Chem.)*. A metal with a relatively negative electrode potential (on IUPAC system). Cf. **noble metal**.

base metal *(Eng.)*. In electrometallurgy, the metals at the end of the electrochemical series remote from the **noble metals**.

base resistance *(Electronics)*. Total resistance to base current, including *spreading effect*.

base unit *(Genrl.)*. The International System of Units (SI) is a coherent system based on seven base units. All derived units are obtained from the base units by multiplication without introducing numerical factors, and approved prefixes are used in the construction of sub-multiples and multiples. There is only one base or derived unit for each physical quantity. The base units are **metre**, **kilogram**, **second**, **ampere**, **kelvin**, **candela**, and **mole**.

base vector *(Maths.)*. In a co-ordinate system, unit vectors taken in the positive direction of the co-ordinate axes are called base vectors, and any vector may be expressed as a linear combination of base vectors.

basher *(Image Tech.)*. A small studio lamp placed close to or on the camera mounting.

Bashkirian *(Geol.)*. The oldest epoch of the Pennsylvanian period.

basi- *(Genrl.)*. Prefix from Gk. *basis*, base.

BASIC *(Comp.)*. A simple programming language (*Beginners' All-purpose Symbolic Instruction Code*). Usually available on microcomputers with an **interpreter**.

basic chromosome number *(Bot.)*. See **basic chromosome set**.

basic chromosome set *(Bot.)*. The haploid set of chromosomes as found in the gametes. Because species may have evolved by polyploidy, aneuploidy or chromosome rearrangement, it may also be possible to infer a basic chromosome set for the ancestor. See also **chromosome complement**.

basicity *(Chem.)*. The number of hydrogen ions of an acid which can be neutralized by a base.

basic lead carbonate *(Chem.)*. Approximate composition $2PbCO_3.Pb(OH)_2$. See **white lead**.

basic lead chromate *(Chem.)*. $PbCrO_4.Pb(OH)_2$. Also called *Austrian cinnabar*. Used as a pigment. Produced when lead chromate is boiled with aqueous ammonia or potassium hydroxide.

basic lead sulphate *(Chem.)*. $2PbSO_4.PbO$. A fine powder, obtained by roasting galena (PbS).

basic loading *(Elec.Eng.)*. The limiting mechanical load, per unit length, on an overhead line conductor.

basic number *(Bot.)*. See **chromosome complement**.

basiconic *(Zool.)*. In Insects, said of certain sub-conical and immobile sensilla arising from the general surface of the cuticle.

basic process *(Eng.)*. A steel-making process, either Bessemer, open-hearth, or electric, in which the furnace is lined with a basic refractory, a slag rich in lime being formed and phosphorus removed.

basicranial *(Med.)*. Pertaining to, or situated at, the base of the skull.

basic rocks *(Geol.)*. Igneous rocks with a low silica content. The limits are usually placed at 45% silica, below which rocks are described as ultrabasic, and 52% silica, above which they are described as intermediate. Basic igneous rocks include basalt, the commonest type of lava, and gabbro, its plutonic equivalent.

basic six *(Aero.)*. The group of instruments essential for the flight handling of an aircraft and consisting of the *airspeed indicator*, *vertical speed indicator*, *altimeter*, *heading indicator*, *gyro horizon* and *turn and bank indicator*.

basic size *(Print.)*. The sheet size used in the US to determine a particular paper's **basis weight**. Some of the more common basic sizes are: *bond* (17″ × 22″); *coated, text, book* and *offset* (25″ × 38″); *cover* (20″ × 26″).

basic slag *(Eng.)*. Furnace slag rich in phosphorus (as calcium phosphate) which, with silicate and lime, is produced in steel making, ground and sold for agricultural purposes, e.g. grassland improvement.

basic solvent *(Chem.)*. A protophilic solvent, hence one which enhances the acidic (i.e. proton donating) properties of the solute.

basic steel *(Eng.)*. Steel which has reacted with a basic lining or additive to produce a phosphorus-rich slag and a low-phosphorus steel.

basic T *(Aero.)*. A layout of flight instruments standardized for aircraft instrument panels in which four of the essential instruments are arranged in the form of a T. The pitch and roll attitude display is located at the junction of the T flanked by airspeed on the left and attitude on the right. The vertical bar portion of the T is taken up by directional information.

basidiocarp *(Bot.)*. The fruiting body of the Basidiomycotina.

basidioma *(Bot.)*. Same as **basidiocarp**.

Basidiomycotina, Basidiomycetes *(Bot.)*. Subdivision or class of those Eumycota or true fungi in which the sexual spores are formed on a basidium. No motile stages. Usually mycelial with septate hyphae. Sexual reproduction typically involves the fusion of a fruiting body in which the basidia develop. Includes the Tiliomycetes comprising the rusts (Uredinales) and smuts (Ustilaginales), the Hymenomycetes and the Gasteromycetes.

basidiospore *(Bot.)*. Spore, typically uninucleate and haploid, formed at the end of a sterigma on a basidium.

basidium *(Bot.)*. Specialized, usually more or less club-shaped, cell on which (typically 4) basidiospores are formed following the fusion of 2 heterokaryotic nuclei and meiosis, in the reproduction of the Basidiomycetes.

basifixed *(Bot.)*. Said of an anther which is attached by its base to the filament.

basifugal *(Bot.)*. Transport, differentiation etc., in the direction away from the base, towards the apex.

basilar *(Bot.,Zool.)*. Situated near, pertaining to, or growing from the base.

basilar membrane *(Acous.,Zool.)*. In Mammals, a flat membrane, part of the partition of the cochlea, containing the collection of auditory nerves in the inner ear which translate mechanical vibrations of differing frequencies into nerve impulses, which are passed to the brain.

basin *(Geol.)*. A large depression in which sediments may be deposited. Alternatively, a gently folded structure in which beds dip inwards from the margin towards the centre.

basin-and-range *(Geol.)*. A structural area of fault-block mountains separated by alluvium-filled basins.

basion *(Med.)*. The midpoint of the anterior margin of the foramen magnum.

basipetal *(Bot.)*. Transport, differentiation etc., in the direction towards the base, away from the apex.

basiphil *(Zool.)*. Having a marked affinity for basic dyes. Also *basophil(e)*.

basiphil, basophil cells *(Med.)*. White blood cells, forming 0–1% of the granulocytes.

basiphilia *(Med.)*. See **basophilia**.

basipodium *(Med.)*. The wrist or ankle.

basis *(Maths.)*. (1) A complete collection of base vectors. (2) A collection of open sets such that every open set is a union of open sets in the basis.

basis cranii *(Zool.)*. In Craniata, the floor of the cranium, formed from the basal plate of the embryo.

basis vector *(Maths.)*. Alternative name for *base vector*.

basis weight *(Paper)*. US method for identifying various papers. The basis weight is the weight in pounds of a ream (500 sheets) of a particular paper in the **basic size** for the grade. The metric system, expressed as g/m^2 is now the preferred system.

basket coil *(Elec.Eng.)*. One with criss-cross layers, so designed to minimize self-capacitance.

basophil *(Med.)*. White blood cell with an affinity for basic stains whose granules contain vasoactive amines and has an important role in infection.

basophilia *(Med.)*. An increase of basophil cells in the blood.

basophil leucocyte *(Immun.)*. Cell present in the blood with properties similar to **mast cells** which binds IgE and can release histamine and other mediators on contact with specific antigen. So called because its granules bind basic dyes.

bas relief *(Arch.)*. Sculpture or carved work in which the figures project less than their true proportions from the surface on which they are carved.

bass boost *(Acous.)*. Amplifier circuit adjustment which regulates the attenuation of the lowest frequencies in the audio scale, usually to offset the progressive loss towards low frequencies.

bass compensation *(Acous.)*. Differential attenuation introduced into a sound-reproducing system when the loudness of the reproduction is reduced below normal, to compensate for the diminishing sensitivity of the ear towards the lowest frequencies reproduced.

bass frequency *(Acous.)*. One towards the lower limit of frequency in an audiofrequency signal or a channel for such, e.g. below 250 Hz.

basswood *(For.)*. A North American tree (*Tilia*) that may grow to over 30 m giving a hardwood with straight-grained fine and uniform texture, creamy white to lightish brown in colour. During seasoning process the wood shrinks considerably but finally stabilizes and is suitable for such purposes as pattern making.

bast *(Bot.)*. Phloem.

bastard *(Genrl.)*. A general term for anything abnormal in shape, size, appearance, etc.

bastard ashlar *(Build.)*. (1) Stones, intended for ashlar work, which are merely rough-scabbled to the required size at the quarry. (2) The face-stones of a rubble wall selected, squared, and dressed to resemble ashlar.

bastard-cut *(Build)*. Describes file teeth of a medium degree of coarseness.

bastard fount *(Print.)*. See **long-bodied type**.

bastard size *(Paper)*. Paper or board not of a standard size.

bastard thread *(Eng.)*. A screw-thread which does not conform to any recognized standard dimensions.

bastard title *(Print.)*. See **half-title**.

bastard tuck pointing *(Build.)*. Pointing in which a slight projection is given to the stopping on each joint.

bastard wing *(Zool.)*. In Birds, quill feathers, usually three in number, borne on the thumb or first digit of the wing. Also *ala spuria, alula*.

bast fibre *(Textiles)*. Obtained from the stems of various plants often by a rotting (*retting*) stage followed by beating. Examples are flax, hemp and jute.

bastite *(Min.)*. A variety of serpentine, essentially

hydrated silicate of magnesium, resulting from the alteration of orthorhombic pyroxenes. Also known as *schillerspar*.

bastnaesite *(Min.)*. Fluocarbonate of lanthanum and cerium. An ore mineral of rare earth elements.

bat *(Build.)*. A portion of a brick, large enough to be used in constructing a wall. See **closer**.

batch *(Glass)*. The mixture of raw materials from which glass is produced in the furnace. A proportion of cullet is either added to the mixture, or placed in the furnace previous to the charge. Also called *charge*.

batch box *(Civ.Eng.)*. See **gauge box**.

batch culture *(Biol.)*. A culture initiated by the inoculation of cells into a finite volume of fresh medium and terminated at a single harvest after the cells have grown. Cf. **continuous culture**.

batch distillation *(Chem.Eng.)*. See under **continuous distillation**.

batch furnace *(Eng.)*. A furnace in which the charge is placed and heated to the requisite temperature. The furnace may be maintained at the operating temperature, or heated and cooled with the charge. Distinguished from **continuous furnace**.

batching *(Min.Ext.)*. See **blocking**.

batching sphere *(Min.Ext.)*. When miscible oil fractions are being sent down a pipe line, an inflatable hard rubber sphere, fitting the pipe, is used to separate the fractions.

batch mill *(Eng.)*. Cylindrical grinding mill into which a quantity of material for precise grinding treatment is charged and worked till finished.

batch process *(Eng.)*. Any process or manufacture in which operations are completely carried out on specific quantities or a limited number of articles, as contrasted to continuous or mass-production.

batch processing *(Comp.)*. Computing jobs are run to completion in sequence. Has the disadvantage that turn-around time is long compared with actual processing time. Cf. **time sharing**, **interactive computing**.

batement light *(Arch.)*. A window, or one division of a window, having vertical sides, but with the sill not horizontal, as where it follows the rake of a staircase.

Batesian mimicry *(Zool.)*. Convergent resemblance between two animals, advantageous in some way to one of them.

bath lubrication *(Eng.)*. A method of lubrication in which the part to be lubricated, such as a chain or gear-wheel, dips into an oil-bath.

batho-, bathy- *(Genrl.)*. Prefix from Gk. *bathys*, deep, used esp. with relation to sea-depths.

bathochrome *(Chem.)*. A radical which shifts the absorption spectrum of a compound towards the red end of the spectrum. Cf. **hypsochrome**.

bathoflare *(Chem.)*. A radical which shifts the fluorescence of a compound towards the red end of the spectrum. Cf. **hypsoflare**.

batholith, bathylith *(Geol.)*. A large body of intrusive igneous rock, frequently granite, with no visible floor.

bathophilous *(Zool.)*. Adapted to an aquatic life at great depths.

bathotonic *(Chem.)*. Tending to diminish surface tension. Cf. **hypotonic**.

B.A. thread *(Eng.)*. See **British Association screw-thread**.

bathyal *(Geol.)*. Refers to the ocean-floor environment between ca 200 and 4000 m. The three zones of increasing depth are **littoral**, **abyssal** and **bathyal**. There are numerous definitions of the depth range of these zones.

bathybic *(Biol.)*. Relating to, or existing in, the deep sea, e.g. plankton floating well below the surface.

bathylimnetic *(Zool.)*. Living in the depths of lakes and marshes.

bathymetric *(Zool.)*. See **abyssopelagic**.

bathysmal *(Zool.)*. See **abyssal zone**.

batik dyeing, printing *(Textiles)*. The fabric is treated with wax to form a pattern that is left unaffected by a dye. The wax may then be removed and a different dye applied to give interesting colour effects.

batiste *(Textiles)*. A soft, fine plain-woven fabric often of flax or cotton.

batrachian *(Zool.)*. Relating to the Salientia (i.e. Frogs and Toads).

batt *(Textiles)*. Loosely coherent sheet of fibres used for the manufacture of non-woven fabrics (Same as a *web*).

batten *(Build.)*. (1) A piece of square-sawn converted timber, 2–4 in (50–100 mm) thickness and 5–8 in (125–200 mm) width, used for flooring or as a support for laths. See also **slating and tiling battens**. (2) A bar fastened across a door, or anything composed of parallel boards, to secure them and to add strength and/or reduce warping.

batten *(Elec.Eng.)*. A fixed or hanging row of lamps used in stage lighting.

battenboard *(Build.)*. See **coreboard**.

batten door *(Build.)*. A door formed of battens placed side by side and secured by others fastened across them.

battened wall *(Build.)*. See **strapping**.

batten-lampholder *(Elec.Eng.)*. See **backplate lampholder**.

batter *(Build.)*. Slope (e.g. of the face of a structure) upwards and backwards.

batter *(Print.)*. Broken or damaged type.

battered baby syndrome *(Med.)*. Describes the state of a baby or child subject to repeated assault from its parent(s) or guardian(s). The child may have multiple bruises of varying age and may have evidence of new and old fractures. Also called *non-accidental injury of childhood*.

batter level *(Surv)*. A form of clinometer for finding the slope of cuttings and embankments.

batter pile *(Civ.Eng.)*. A pile which is driven in at an angle to the vertical.

batter post *(Build.)*. One of the inclined side-timbers supporting the roof of a tunnel.

battery *(Elec.Eng.)*. General term for a number of objects co-operating together, e.g. a number of accumulator cells, dry cells, capacitors, radars, boilers, etc.

battery booster *(Elec.Eng.)*. A motor-generator set used for giving an extra voltage, to enable a battery to be charged from a circuit of a voltage equal to the normal voltage of the battery.

battery coil ignition *(Autos.)*. High-tension supply for sparking plugs in automobiles, in which the interruption of a primary current from a battery induces a high secondary e.m.f. in another winding on the same magnetic circuit, the high tension being distributed in synchronism with the contact-breaker in the primary circuit.

battery cut-out *(Elec.Eng.)*. An automatic switch for disconnecting a battery during its charge, if the voltage of the charging circuit falls below that of the battery.

battery regulating switch *(Elec.Eng.)*. A switch to regulate the number of cells connected in a series in a battery.

battery spear *(Elec.Eng.)*. A special form of spike used to connect a voltmeter to the plates of the accumulator cells for battery-testing under load. The voltmeter incorporates a low resistance in shunt which simulates a heavy load on the battery, thus testing its work capability. The heavy current passed for this purpose necessitates special heavy duty battery connectors.

battery traction *(Elec.Eng.)*. An electric-traction system in which the current is obtained from batteries on the vehicles (*battery vehicles*).

battery vehicle *(Elec.Eng.)*. An electrically-propelled vehicle which derives its energy from a battery carried on the vehicle.

baud *(Comp.,Telecomm.)*. A measure of the signalling speed in a digital communication system; the speed in bauds is the number of discrete conditions or signal events per second, e.g. one baud equals one bit/second in a train of binary signals. Since many digital systems transmit additional information for control and signalling, the baud rate is not necessarily the same as the **data signalling rate**.

Baudot code *(Telecomm.)*. One in which five equal-length bits represent one character; sometimes used for teleprinters where one *start* and one *stop* element are added to each group of five bits.

Baudouin reaction *(Chem.)*. A test for certain vegetable oils which give with alcoholic furfural and concentrated HCl, or with $SnCl_2$ and HCl, a characteristic red colour.

Bauhaus *(Arch.)*. The name of a school founded in 1919 by the architect Walter Gropius, in Weimar, Germany. It aimed to achieve modern design in all aspects of art, in a manner which acknowledged the advancement of industrial technology. The school is especially noted for pioneering a style of architecture which defined clearly the various functional elements of the accommodation within a building.

baulk *(Civ.Eng.)*. See **balk**.

baulk *(For.)*. A piece of timber square-sawn from the log to a size greater than 6 by 6 in (150 by 150 mm).

Baumé hydrometer scale *(Phys.)*. The continental Baumé hydrometer has the rational scale proposed by Lunge, in which 0° is the point to which it sinks in water and 10° the point to which it sinks in a 10% solution of sodium chloride, both liquids being at 12.5°C.

Baum jig *(Min.Ext.)*. Pneumatically pulsed **jig** used in coal-washing plants to lift and remove a lighter and low-ash fraction from a denser one containing shale and high-ash material (dirt), which is stratified downward by the effect of pulsed water, and separately withdrawn.

bauxite *(Geol.)*. A residual rock composed almost entirely of aluminium hydroxides formed by weathering in tropical regions. The most important ore of *Al*. See also **laterite**.

bay *(Arch.)*. Any division or compartment of an arcade, roof, building etc. or space from column to column in a building.

bay *(Telecomm.)*. (1) Unit of racks designed to accommodate numbers of standard-sized panels, e.g. repeaters or logical units. (2) Unit of horizontally extended antenna, e.g. between masts.

Bayard and Alpert gauge *(Phys.)*. One for measuring very low gas pressure by collecting ions on a fine wire inside a helical grid.

Bayer process *(Chem.)*. A process for the purification of bauxite, as the first stage in the production of aluminium. Bauxite is digested with a sodium hydroxide solution which dissolves the alumina and precipitates oxides of iron, silicon, titanium etc. The solution is filtered and the aluminium precipitated as the hydroxide.

Bayesian *(Stats.)*. Proceeding from a definition of probability as a measurement of belief; considering statistical inference as a process of re-evaluating such probabilities on the basis of empirical observation.

bayonet cap *(Elec.Eng.)*. (Common abbrev. BC). A type of cap fitted to an electric lamp, consisting of a cylindrical outer wall fitted with two or three pins for engaging in slots in a lampholder (*bayonet holder*). Within the wall are two contacts connected to the filament, which make contact with two pins in the lampholder. See **centre-contact cap, small bayonet cap**.

bayonet fitting *(Eng.)*. A cylindrical fastening in which a plug is pushed and twisted into a socket, against spring pressure. Two or more pins in the plug engage in corresponding L-shaped slots in the socket.

bayonet holder *(Elec.Eng.)*. See **bayonet cap**.

bay-stall *(Build.)*. See **carol**.

baywood *(For.)*. See **Honduras mahogany**.

bazooka *(Telecomm.)*. See **BALUN**.

B battery *(Elec.Eng.)*. US term for **high-tension battery**.

B.C. *(Elec.Eng.)*. See **bayonet cap**.

BCC *(Chem.)*. Abbrev. for *Body Centred Cubic*.

BCD *(Comp.)*. See **binary coded decimal**.

B-cell *(Immun.)*. See **B-lymphocyte**.

BCF *(Chem.)*. See **bromochloro difluoromethane**.

BCG *(Immun.)*. *Bacille Calmette Guerin*. A living attenuated strain of *Mycobacterium tuberculosis* used as a vaccine to protect against tuberculosis. Induces sensitivity to *purified protein derivative* (See **tuberculin**). Developed by Calmette and Guérin in France in 1909 and first used there in 1921.

B-chromosomes *(Biol.)*. Accessory non-essential chromosomes present in variable numbers in addition to the normal *A-chromosomes*. They are usually small and heterochromatic.

B-class insulation *(Elec.Eng.)*. A class of insulating material to which is assigned a temperature of 130°C. See **class-A, -B, -C, etc., insulating materials**.

BCS *(Comp.)*. British Computer Society. Professional association.

BCS theory *(Phys.)*. Bardeen, Cooper and Schrieffer theory of superconductivity. See **Cooper pairs**.

B-display *(Radar)*. Rectangular radar display with target bearing indicated by horizontal coordinate and target distance by the vertical coordinate, the targets appearing as bright spots.

BDV *(Elec.Eng.)*. Abbrev. for *BreakDown Voltage*.

Be *(Chem.)*. The symbol for beryllium.

beaching gear *(Aero.)*. Floatable, detachable, temporary trolleys which enable a seaplane to be run on and off the shore or slipway.

beacon *(Aero.,Ships)*. (1) System of visual lights indicating fixed features, e.g. masts, reefs. (2) *Radio Beacons*, which can be of any frequency but are usually VHF, and can be omni-directional or of directional beam type. *Vertical fan marker* radio beam type beacons are used to identify particular spots in control zones and on approach patterns. A *non-directional beacon* (abbrev. *NDB*) is a transmitter, the bearing of which can only be determined by an aircraft equipped for direction finding. See **instrument landing system**.

bead *(Build.)*. A small convex moulding formed on wood or other material.

bead-and-quirk *(Build.)*. A bead formed with a narrow groove separating it from the surface which it is decorating. Also a *quirk-bead*.

bead-jointed *(Build.)*. Said of that form of jointing in which one of the butting edges has a bead.

bead-tool *(Build.)*. A specially shaped cutting-tool used in wood-turning for forming convex mouldings.

beak *(Build.)*. The crooked end of a bench hold-fast.

beak *(Zool.)*. See **rostrum**

beak iron, beck iron, bick iron, bickern *(Eng.)*. (1) The pointed, or horn-shaped, end of a blacksmith's anvil; used in forging rings, bends etc. (2) A T-shaped stake, similarly shaped, fitting in the hardy-hole of the anvil.

beam *(Eng.)*. (1) A bar which is loaded transversely. (2) Rolled or extruded sections of certain profiles, e.g. I-beam.

beam *(Phys.)*. A collimated, or approximately unidirectional, flow of electromagnetic radiation (radio, light, X-rays), or of particles (atoms, electrons, molecules). The angular beam width is defined by the half-intensity points.

beam *(Textiles)*. A wooden or metal cylinder having large flanges at each end. Warp yarns are wound on the beam from cones or cheeses correctly arranged for inserting into the loom or warp-knitting machine. Beams are also used to furnish thread during lace making.

beam antenna *(Telecomm.)*. Generally, any antenna which has **directivity**. Most commonly used to describe *short-wave* or VHF antenna, rather than microwave antenna which are almost invariably directional.

beam balance *(Chem.)*. Balance in which the weight of the sample contributes to the balance of moments of a beam about a central fulcrum.

beam compasses *(Eng.)*. An instrument for describing large arcs. It consists of a beam of wood or metal carrying two beam heads, adjustable for position along the beam, and serving as the marking points of the compasses. Also called *trammels*.

beam-coupling coefficient *(Elec. Eng.)*. The ratio of the a.c. signal current produced to the d.c. beam current in beam coupling.

beam current *(Electronics)*. That portion of the gun current in CRT which passes through the aperture in the anode and impinges on the fluorescent screen.

beam-engine *(Eng.)*. A form of construction used in early steam-engines, now obsolete. The vertical steam-cylinder acted at one end of a pivoted beam, the work load being connected to the other.

beam-filling *(Build.)*. Brick, masonry or concrete work used between joists carried upon a wall.

beam-forming electrode *(Electronics)*. Electrode to which a potential is applied to concentrate the electron stream into one or more beams. Used in beam tetrodes and cathode-ray tubes.

beam hole *(Nuc.Eng.)*. Hole in shield of reactor, or that around a cyclotron, for extracting a beam of neutrons or γ-rays or to insert equipment or samples for irradiation.

beam relay *(Elec.Eng.)*. An electromagnetic relay in which the contacts are mounted on a balanced beam with energizing coils acting on each end and tending to tilt it one way or the other.

beam rider (riding) *(Telecomm.)*. System in which a guided missile maintains and returns to a course of maximum signal on a radio beam. See **guided missile**.

beam splitter *(Image Tech.)*. Optical device for dividing a light beam into two or more paths. In particular, a prism system in a camera to produce three colour-separation images from a single objective lens.

beam system *(Telecomm.)*. A point-to-point radio system in which highly directive transmitting and receiving antennae are used.

beam tetrode *(Electronics)*. Tetrode having an additional pair of plates, normally connected internally to the cathode, so designed as to concentrate the electron beam between the screen grid and anode, and thus reduce secondary emission effects.

bearded *(Bot.)*. Having an awn; bearing long hairs like a beard.

bearded needle *(Textiles)*. See **spring needle**.

bearding *(Image Tech.)*. Picture defect in which dark image areas spread into adjacent light areas.

Beard protective system *(Elec.Eng.)*. A form of balanced protective system in which the current entering the winding of an alternator is balanced against that leaving it by passing the conductor at the two ends round the core of a single current transformer, in opposite directions, so that there is normally no flux in the transformer core.

bearer cable *(Elec.Eng.)*. See **messenger wire**.

bearer ring *(Print.)*. See **cylinder bearers**.

bearing *(Build.)*. The part of a beam or girder which actually rests on the supports.

bearing *(Surv.)*. The horizontal angle between any survey line and a given reference direction.

bearing *(Telecomm.)*. Angle of direction in horizontal plane in degrees from true north, e.g. of an arriving radio wave as determined by a direction-finding system.

bearing current *(Elec.Eng.)*. A stray current, induced by magnetic flux linking the shaft of an electrical machine, that flows between the shaft and bearings and may injure the bearing surfaces.

bearing distance *(Build.)*. The unsupported length of a beam between its bearings. Also *clear span*.

bearing metals *(Eng.)*. Metals (alloys) used for that part of a bearing which is in contact with the **journal**; e.g. bronze or white metal, used on account of their low coefficient of friction when used with a steel shaft.

bearing pile *(Civ.Eng.)*. A column which is sunk or driven into the ground to support a vertical load by transmitting it to a firm foundation lower down, or by consolidating the soil so that its bearing power is increased. Generally made of reinforced timber or prestressed concrete.

bearings *(Eng.)*. Supports provided to locate a revolving or reciprocating shaft.

bearing surface *(Eng.)*. That portion of a bearing in direct contact with the journal; the surface of the journal. See **brasses**.

bearing wall *(Civ.Eng.)*. The supporting or abutment wall of a bridge or arch.

beat *(Horol.)*. The blow given by a tooth of an escape wheel as it strikes the pallets. An escapement is said to be *in beat* when this blow is uniform on both pallets.

beat *(Telecomm.)*. Periodic variation in the amplitude of a summation wave containing 2 sinusoidal components of nearly equal frequencies.

beater *(Paper.)*. A vat containing a heavy cylindrical roll (*beater roll*), fitted with bars, parallel to the journal, which rotates against a fixed set of bars (**bedplate**). The paper fibres in suspension in water pass between these bars in preparation for sheet making.

beater *(Textiles)*. High-speed revolving shaft having arms equipped with blades or pins. These beat out the heavy impurities in matted raw fibres in opening and scutching processes.

beater mill *(Min.Ext.)*. *Hammer mill, disintegrating mill, impactor*. In rockbreaking, mill with swinging hammers, disks, or heavy plates, which revolve fast and hit a falling stream of ore with breaking force.

beat frequency *(Telecomm.)*. Generally, the difference frequency produced by the intermodulation of two frequencies. Specifically, the intermediate frequency in a superhet receiver.

beat-frequency oscillator *(Telecomm.)*. Same as *heterodyne oscillator*. Abbrev. *BFO*.

beat-frequency wavemeter *(Telecomm.)*. Same as **heterodyne wavemeter**.

beating *(Textiles)*. The spare threads available during the weaving of wool to replace missing warp threads in the mending process.

beating-up *(Textiles)*. The process in weaving by which the newly inserted weft thread is pushed against the edge of the woven fabric.

beat pins *(Horol.)*. The pins projecting from the ends of the gravity arms of the gravity escapement. These pins, one on either side of the pendulum rod, give impulse to the pendulum and enable the pendulum to raise the gravity arms for unlocking.

beats *(Acous.)*. The subjective difference tone when two sound waves of nearly equal frequencies are simultaneously applied to one ear. It appears as a regular increase and decrease of the combined intensity.

beat screws *(Horol.)*. Screws which provide for the adjustment of the relative position of the crutch and pendulum, so that the escapement may be brought in **beat**.

Beattie-Bridgeman equation of state *(Phys.)*. A semi-empirical equation of state for the compressibility of gases. It is most conveniently stated in the virial form:

$$\frac{pV}{n} = RT + \frac{n\beta}{V} + \frac{n^2\gamma}{V^2} + \frac{n^3\delta}{V^3},$$

where β, γ, and δ are virial coefficients.

Beaufort notation *(Meteor.)*. A code of letters used for indicating the state of the weather; for example, *b* stands for *blue sky*, *o* for *overcast*, *r* for *rain*.

Beaufort scale *(Meteor.)*. A numerical scale of wind force, ranging from 0 for winds less than 1 knot to 12 for winds within the limits 110 to 118 knots. Where V is the mean wind speed in miles/hour, and B is the Beaufort wind force, then $V = 1.87\sqrt{AB^3} = (1.52B)^{3/2}$.

beauty *(Phys.)*. See **bottomness**.

beaver board *(Build.)*. A soft building board made of wood-fibre material.

beavertail antenna *(Telecomm.)*. One producing a broad, flat, radar beam.

Bechgaard salt *(Phys.)*. $(TMTSF)_2X$ where X is an inorganic anion $(PF_6)^-$, $(AlO_4)^-$, $(ReO_4)^-$, etc. and TMTSF is the tetramethyl selenium derivative of TTF (tetrathiofulvalene). These salts are *organic electrical conductors*.

beck *(Genrl.)*. See **back**.

Becke line *(Min.)*. A narrow line of light seen under the microscope at the junction of two minerals (or a mineral and the mount) in contact in a microscope section.

Beck hydrometer *(Phys.)*. Hydrometer for measuring the relative density of liquids less dense than water. Graduated in degrees Beck, where °Beck = 200 (1 − rel.d.).

Beckmann apparatus *(Eng.)*. Apparatus used for measuring the freezing and boiling points of solutions (e.g. in the **cryoscopic method**).

Beckmann molecular transformation *(Chem.)*. The transformation and rearrangement of ketoxime molecules into acid amides or anilides under the influence of reagents, such as acetyl chloride. An important reaction for determining the configuration of stero-isomeric **ketoximes**.

Beckmann thermometer *(Eng.)*. A limited range mercury thermometer with a large bulb. It is used to measure small changes of temperature with great precision. Its mean range can be altered by moving mercury from a reservoir in or out of the bulb.

becquerel *(Phys.)*. SI unit of radioactivity, 1 becquerel is the activity of a quantity of radioactive material in which 1 nucleus decays per second. Abbrev. *Bq*. Replaces the **curie**. $1\ \mathrm{Bq} = 2.7 \times 10^{-11}$ Ci.

Becquerel cell *(Genrl.)*. See **photochemical cell**.

becquerelite *(Min.)*. Hydrated oxide of uranium, an alteration product of uraninite.

bed *(Build.)*. The upper or lower surface of a building-stone or ashlar when it is built into a wall; the horizontal surface upon which a course of bricks is laid in mortar.

bed *(Chem.)*. A packed, porous mass of solid reagent, adsorbent or catalyst through which a fluid is passed for the purpose of chemical reaction.

bed *(Geol.)*. A small rock-unit.

bed *(Print.)*. The heavy steel table of a machine or press on which the forme of type or plates is placed for printing. See also **flat bed**.

bedding *(Build.)*. Material on which an underground pipe is laid, providing support for the pipe. Can be concrete, granular material or the prepared trench bottom.

bedding *(Geol.)*. A term commonly used for **stratification**.

bedding, bedding plane *(Geol.)*. Surface that separates layers of rock. It is caused by changes in mineralogy, grain-size, colour etc.

bedding course *(Civ.Eng)*. See **subcrust**.

bedding-in *(Eng.)*. The process of accurately fitting a bearing to its shaft by scraping the former until contact occurs uniformly over the surface.

bedding-stone *(Build.)*. The flat marble slab used by the bricklayer to test the face of a rubbed brick for flatness.

bed dowel *(Build.)*. A dowel placed in the centre of a stone bed.

bed joints *(Build.)*. The horizontal joints in brickwork or masonry: the radiating joints of an arch.

bed-moulding *(Build.)*. Any moulding used to fill up the bare space beneath a projecting cornice.

bedplate *(Eng.)*. A cast-iron or fabricated steel base, to which the frame of an engine or other machine is attached.

bedplate *(Paper)*. A plate into which metal bars are inserted; situated beneath the roll in a beater.

bedrock *(Min.Ext.)*. Barren formation (seat earth, clay, 'farewell rock') underlying the exploitable part of a mining deposit.

beech *(For.)*. A tree (*Fagus*) yielding a hardwood with straight grain and uniform texture. Its colour ranges from whitish to a light reddish-brown. It bends well and is an excellent wood for turnery purposes. Used for pulley-blocks, tool handles, athletic goods, gymnasium equipment and cabinet making.

beef *(Geol.)*. Fibrous calcite occurring in veins in sedimentary rocks. Rarely, other minerals with the same structure and occurrence.

beekite *(Min.)*. A variety of chalcedony, commonly occurring as an incrustation on pebbles.

Beer's law *(Chem.)*. The degree of absorption of light varies exponentially with the thickness of the layer of absorbing medium and the molar concentration in the latter.

beeswax *(Chem.)*. A white or yellowish plastic substance obtained from honeycomb of the bee, mp 63°–65°C. It consists of the myricyl (melissyl) ester of palmitic acid $C_{15}H_{31}.COOC_{30}H_{61}$, free cerotic acid $C_{25}H_{51}COOH$, and other homologues. Used in polishes, modelling, ointments etc.

beetle *(Build)*. A heavy mallet, or wooden hammer, used for driving wedges, consolidating earth etc. Also called a *mall* or *maul*.

beetle *(Textiles)*. A machine consisting of a row of wooden or metal hammers, which fall on a roll of damp cloth as it revolves. The operation closes the spaces between the warp and the weft yarns, and imparts a soft glossy finish to cotton and linen.

beetle *(Zool.)*. A member of the insectan order of Coleoptera.

beetle-stones *(Geol.)*. Coprolitic nodules akin to septaria which, when broken open, give a fancied resemblance to a fossil beetle.

beet sugar *(Chem.)*. Sucrose derived from sugar beet. Identical with sucrose derived from sugar cane.

Beggiatoales *(Biol.)*. Chemosynthetic sulphur-oxidizing bacteria, some of which resemble filamentous algae, and which may in fact be more closely related to them than to the true bacteria. Sulphur granules occur intracellularly.

behaviourism *(Behav.)*. An approach to psychology that considers only observable behaviour as appropriate subject matter for study, and which views with distrust explanations which refer to non-observable mental events, not directly available for objective verification (e.g. consciousness, imagery).

behaviour therapy (modification) *(Behav.)*. A general approach to psychological treatment which holds that behaviour disorders are the result of maladaptive learning and are best remedied by re-education based on the principles of learning theory. The focus is on behaviour, rather than a hypothetical unconscious process.

behind the pipe *(Min.Ext.)*. Refers to a gas or oil reservoir outside the *casing string*. It may have been deliberately drilled through in the search for more productive regions lower down.

beidellite *(Min.)*. A variety of the montmorillonite group (smectites) of clay minerals.

Beilby layer *(Min.Ext.)*. Flow layer produced by polishing a metal or mineral surface, in which the true lattice structure is modified or destroyed by incipient fusion.

Beilstein test *(Chem.)*. Negative test for the presence of a halogen in an organic compound. The latter is heated in an oxidizing flame on a copper wire; if no halogen is present there is no green colour to the flame. Volatile N compounds also give a green colour.

bel *(Maths.)*. A nondimensional unit used to express the common logarithm of a number, so that the multiplication of the number by a factor can be accomplished by the addition of the logarithm of the factor to the measure in bels. If N is the measure of a in bels, $N = \log_{10}a$, $raN = \log_{10}r + \log_{10}a$, the measure of $r^2N = 2\log_{10}r + \log_{10}a$, etc. Useful wherever constant coefficients are applied to variable quantities, e.g. amplifiers or attenuators in electrical circuits. See **decibel, neper**.

belemnite *(Geol.)*. An extinct Cephalopod, similar to an Octopus in appearance. The portion commonly found as a fossil is the 'guard', the shape and often the size of a rifle bullet.

belemnoid *(Zool.)*. Dart-shaped.

Belfast truss *(Build.)*. A timber bowstring truss having a double bow rafter and a double tie connected by a lattice of cross-members, with a bituminous felt or corrugated iron roof-covering.

belfry *(Arch.)*. A tower, either detached or forming part of a building, containing suspended bells.

Belgian truss *(Civ.Eng.)*. See **French truss**.

bell *(Acous.)*. Hollow metallic vessel with a flared mouth which, when struck, vibrates with a fundamental frequency determined by its mass, dimensions etc.

bell-and-spigot joint *(Civ.Eng.)*. US for *spigot-and-socket joint*.

bell centre punch *(Eng.)*. A centre punch whose point is automatically located centrally on the end of circular work by a sliding hollow conical guide.

bell chuck *(Eng.)*. See **cup chuck**.

bell-crank lever *(Eng.)*. A lever consisting of two arms, generally at right angles, with a common fulcrum at their junction.

bell gable *(Arch.)*. A gable built above the roof in a church having no belfry, and pierced to accommodate a bell.

Bellini's ducts *(Zool.)*. In the kidney of Vertebrates, ducts formed by the union of the primary collecting tubules and opening into the base of the ureter at the pelvis of the kidney.

Bellini-Tosi antenna *(Telecomm.)*. Directional antenna comprising two crossed loops. The direction of maximum reception is controlled by a **radio-goniometer** which varies the relative couplings of the two loops to the receiver.

bell metal *(Eng.)*. High tin bronze, containing up to 30% tin and some zinc and lead. Used in casting bells.

bell-metal ore *(Min.)*. See **stannite**.

bell-mouthed *(Eng.)*. Said of a hole or bore when its diameter gradually increases towards one or both open ends, the bore profile in section being curved. Usually a manufacturing fault.

bellows *(Eng.)*. A flexible, corrugated tubular machine element used for pumping, for transmitting motion, as an expansion joint, etc.

bellows *(Image Tech.)*. The flexible connection between parts of a camera or enlarger, necessarily light-tight, to permit delicate adjustments, usually of focusing.

Bell's palsy *(Med.)*. Sudden paralysis of the muscles of one side of the face, due to impaired conduction in the lower part of the facial nerve. The cause is unknown, but the majority of cases recover.

bell transformer *(Elec.Eng.)*. A small transformer used for obtaining, from the public mains, low-voltage ringing current for house trembler-bells.

bell-type furnace *(Eng.)*. A portable inverted furnace or heated cover operated in conjunction with a series of bases upon which the work to be heated can be loaded and then left to cool after heat treatment. Used chiefly for bright-annealing of nonferrous metals and bright-hardening of steels.

belly *(Print.)*. The side of a type-letter which bears the nick; placed uppermost in the setting stick by the compositor.

belly tank *(Aero.)*. See **ventral tank**.

belt *(Build.)*. A projecting course of stones or bricks. Also *belt course* or *string course*.

belt *(Eng.)*. A strip of leather, cotton, plastic, reinforced rubber etc., generally of rectangular cross section, used for lifting slings and strengthening bands. In endless form used as driving-, conveyor-, abrasive- and other belts.

Belt, Beltian Series *(Geol.)*. A great thickness (perhaps 12 000 m) of younger Pre-Cambrian rocks occurring in the Little Belt Mts., Montana, Idaho, and British Columbia. Argillaceous strata predominate, accompanied by algal limestones. Correlated with the Grand Canyon Series in Colorado and the Uinta Quartzite Series in the Uinta Mts.

belt conveyor *(Eng.)*. See **band conveyor**.

belt drive *(Eng.)*. The transmission of power from one shaft to another by means of an *endless belt* running over pulleys having correspondingly shaped rims.

belt fork, belt striker *(Eng.)*. Two parallel prongs attached at right angles to a sliding rod, used to slide a flat belt from a fast to a loose pulley, and vice versa.

belting *(Eng.)*. A general term descriptive of materials from which driving belts are made, e.g. leather, cotton, balata, woven hair, plastics etc.

belt slip *(Eng.)*. The slipping of a driving belt on the face of a pulley, due to insufficient frictional grip to overcome the resistance to motion.

belt transect *(Bot.)*. A strip of ground marked between two parallel lines so that its vegetation may be recorded and studied. See also **transect, quadrat**.

belvedere *(Arch.)*. A room from which to view scenery; it is built on the top of a house, the sides being either open or glazed.

Bence-Jones protein *(Immun.)*. A protein present in the urine of some persons with **myelomas**, which precipitates on heating to 60°C but redissolves at 80°C. It consists of light chain dimers.

bench *(Telecomm.)*. Fixed rails with adjustable and slidable supports for a waveguide system.

benched foundation *(Build.,Civ.Eng.)*. A foundation which is stepped at the base to safeguard aginst sliding on sloping sites.

bench hook *(Build.)*. A flat piece of wood having a wooden block at the back edge of the top and a similar block fixed on the underside along the front edge, used to steady the work and prevent injury to the bench top.

benching iron *(Surv.)*. A small steel plate sometimes used to provide a solid support for the staff at a change point. It is formed usually of a triangular plate, with the corners turned down so that they may be driven into the ground surface to fix the plate in position, while the staff rests upon a raised central portion.

benchmark *(Comp.)*. Standard program, used to compare the performance of different makes of computer.

bench mark *(Surv.)*. A fixed point of reference for use in levelling, the reduced level of the point with respect to some assumed datum being known. See **Ordnance Bench Mark**.

bench plane *(Build.)*. A plane for use on flat surfaces. See **jack plane, smoothing plane, trying plane**.

bench screw *(Build.)*. The vice fixed at one end of a bench.

bench stop *(Build.)*. A metal or wooden stop, adjustable for height, set in the top of a bench, at one end; used to hold work while it is being planed.

bench test *(Eng.)*. A complete functional test of a piece of apparatus, when new or after repair, carried out in a workshop or laboratory. It is undertaken to ensure correct and satisfactory operation prior to installation in a situation where repair may be difficult.

bench work *(Build.,Eng.)*. Work executed at the bench with hand tools or small machines, as distinct from that done at the machines.

bench work *(Foundry)*. Small moulds made on a bench in the foundry.

bend *(Eng.)*. A curved length of tubing or conduit used to connect the ends of two adjacent straight lengths which are at an angle to one another.

bend *(Telecomm.)*. Alteration of direction of a rigid or flexible waveguide. It is *E* or *minor* when electric vector is in plane of arc of bending and *H* or *major* when at right angles to this. Also *corner*.

Ben Day tints *(Print.)*. Celluloid sheets with a patterned surface, and inked impression from which is used to lay a **mechanical stipple**.

bending iron (pin) *(Build.)*. A tool for straightening or expanding lead pipe.

bending moment *(Eng.)*. The *bending moment* at any imaginary transverse section of a beam is equal to the algebraic sum of the moments of all the forces to either side of the section.

bending moment diagram *(Eng.)*. One representing the variation of bending moment along a beam. It is a graph of bending moment (*y-axis*) against distance along the beam axis (*x-axis*).

bending of strata *(Geol.)*. See **folding**.

bending rollers *(Print.)*. Rollers at the nose of rotary presses. Also called *forming rollers*.

bending rolls *(Eng.)*. Usually three rolls with axes arranged in a triangle so that adjusting one relative to the others forms a curve on a strip or sheet of metal passed between them.

bending strength *(Eng.)*. The ability of a beam, or other structural member, to resist a **bending moment**.

bending test *(Eng.)*. (1) A test made on a beam to determine its deflection under load. (2) A forge test in which flat bars etc. are bent through 180° as a test of ductility.

bending wave *(Acous.)*. Wave observed on thin plates and bars. The motion is perpendicular to the direction of propagation. Important for sound radiation from walls and enclosures.

bendrofluazide *(Med.)*. Thiazide diuretic used in the treatment of oedema and hypertension.

bends *(Med.)*. See **caisson disease**.

Benedict's test *(Chem.,Med.)*. Test for glucose (and for reducing disaccharides), involving the oxidation of the sugar by an alkaline copper sulphate solution, in the presence of sodium citrate, to give a red copper (I) oxide. Used in testing of urine in treatment of diabetes.

beneficiation *(Min.Ext.)*. See **mineral processing**.

Benioff zone *(Geol.)*. A plane beneath the trenches of the Pacific dipping under the continents; the site of earthquake activity.

benitoite *(Min.)*. A strongly dichroic mineral, varying in tint from sapphire blue to colourless, discovered in San Benito Co., California. Silicate of barium and titanium.

bent chisel *(Build)*. Carving tool for recessing backgrounds, made in three types: right-angled, and right and left corner bent.

bent gouge *(Build.)*. A curved gouge for hollowing out concave work.

benthic *(Zool.)*. Living at the soil-water interface at the bottom of a sea or lake.

benthon, benthos *(Ecol.)*. Collectively, the sedentary animal and plant life living on the sea bottom; cf. **nekton**, **plankton**. adj. *benthic*.

bent knees *(Vet.)*. Flexion of the carpus of horses or dogs due to permanent contraction of the flexor tendons or to chronic arthritis.

bentonite *(Geol.,Min.Ext.)*. A valuable clay, similar in its properties to fuller's earth, formed by the decomposition of volcanic glass, under water. Consists largely of montmorillonite. Used as a bond for sand, asbestos etc.; also in the paper, soap and pharmaceutical industries. Thixotropic properties exploited for altering the viscosity of oil **drilling muds**.

bent-tail carrier *(Eng.)*. A **lathe carrier** having a bent shank projection into, and engaged by, a slot in the driving plate or chuck.

benzal chloride *(Chem.)*. *Dichloromethylbenzene*. $C_6H_5.CHCl_2$. bp 207°C. A chlorination product of toluene, intermediate for the production of benzaldehyde. Also *benzylidene chloride*.

benzaldehyde *(Chem.)*. *Benzene carbaldehyde, oil of bitter almonds*. $C_6H_5.CHO$, mp 13°C, bp 179°C, rel.d. 1.05, a colourless liquid, with aromatic odour, soluble in alcohol, ether, slightly in water. Flavouring agent.

benzaldoximes *(Chem.)*. $C_6H_5.CH = N.OH$, formed from benzaldehyde and hydroxylamine; there are 2 stereoisomeric forms. The *alpha* or *antiform*, mp 35°C, can be transformed by means of acids into the *beta* or *syn*-form, mp 125°C.

benzamide *(Chem.)*. *Benzene carboxamide*. $C_6H_5.CO.NH_2$, the amide of benzoic acid, obtainable from benzoyl chloride and ammonia or ammonium carbonate; lustrous plates, mp 130°C.

benzanilide *(Chem.)*. *N-phenylbenzamide*. $C_6H_5.CO.NHC_6H_5$, colourless plates, mp 158°C; the anilide of benzoic acid or benzoyl chloride.

benzanthrone *(Chem.)*. Yellow, crystalline powder. An intermediate widely used in manufacture of vat dyestuffs.

benzene *(Chem.)*. C_6H_6, mp 5°C, bp 80°C, rel.d. 0.879; a colourless liquid, soluble in alcohol, ether, acetone, insoluble in water. Produced from coal-tar and coke-oven gas; can also be synthesized from open-chain hydrocarbons. Basis for benzene derivatives. A solvent for fats, resins etc.; very inflammable. Benzene is the simplest member of the aromatic series of hydrocarbons. Carcinogenic. Its structure was established by Kekulé in 1858. See **benzol**. ⇨

benzene carboxylic acids *(Chem.)*. Aromatic acids originating from benzene.

benzene formula *(Chem.)*. The generally recognized formula for benzene was established by Kekulé. It represents a closed chain of 6 carbon atoms, to each of which a hydrogen atom is attached, the carbon atoms being linked alternately by single and double bonds. It is a stable structure, showing resonance energy of ca 160 kJ/mol.

benzene hexachloride *(Chem.)*. Abbrev. *BHC*. 1,2,3,4,5,6-hexachlorocyclohexane. See **Gammexane**.

benzene hydrocarbons *(Chem.)*. Homologues of benzene of the general formula C_nH_{2nb6}.

benzene nucleus *(Chem.)*. The group of 6 carbon atoms which, with the hydrogen atoms, form the benzene ring. See also **side-chains**.

benzene ring *(Chem.)*. See **benzene formula** and **benzene nucleus**.

benzene-sulphonic acids *(Chem.)*. Aromatic acids formed from compounds of the benzene series by sulphonation. The acid characteristics are given by the group $-SO_2OH$. Important intermediates for dyestuffs.

benzhydrol *(Chem.)*. $(C_6H_5)_2CH.OH$. Reduction product of benzophenone prepared by treatment with zinc and aqueous alcoholic alkali.

benzidine *(Chem.)*. *4-4′-diamino-biphenyl*, $NH_2.C_6H_4.C_6H_4.NH_2$, mp 127°C. White to pinkish crystals, soluble in alcohol, ether, insoluble in water. It is an important intermediate for azodyestuffs. It is a known carcinogen.

benzidine transformation *(Chem.)*. The transformation of benzene-hydrazo-compounds into benzidine derivatives by strong acids.

benzil *(Chem.)*. $C_6H_5.CO.CO.C_6H_2$, mp 95°C, large 6-sided prisms, a diketone of the diphenyl group. Syns. *bibenzoyl, diphenyl-glyoxal*.

benzocaine *(Chem.)*. *Ethyl para-aminobenzoate*. White crystalline powder, insoluble in water; used as a local anaesthetic and for internal treatment of gastritis.

benzodiazepines *(Med.)*. Class of drug used as an **anxiolytic** or **hypnotic**. *Diazepam* (Valium) is commonly used for relieving anxiety and *nitrazepam* (Mogadon) for inducing hypnosis, although the hazards of addiction with these drugs is being increasingly recognized.

benzoic acid *(Chem.)*. *Benzenecarboxylic acid*. $C_6H_5.COOH$, mp 121°C, bp 250°C, colourless glistening plates or needles, sublimes readily, volatile in steam. Used as a fruit preservative.

benzoin *(Chem.)*. $C_6H_5.CHOH.CO.C_6H_5$, mp 137°C, colourless prisms, a condensation product of benzaldehyde. It is both a secondary alcohol and a ketone and can react accordingly. Occurs as a natural resin obtained from a Javanese tree. Chief constituent of Friar's Balsam.

benzol *(Autos.)*. Crude benzene; has been used as a motor spirit and valued for its anti-knock properties.

benzol scrubber *(Chem.)*. A device for washing gases and absorbing the benzol contained therein by means of a high-boiling mineral oil.

benzonitrile *(Chem.)*. *Cyanobenzene*. $C_6H_5.CN$, bp 191°C, the *nitrile* of benzoic acid.

benzophenone *(Chem.)*. *Diphenyl ketone*, $C_6H_5.CO.C_6H_5$, mp 49°C, bp 307°C, colourless prisms, soluble in alcohol and ether. It is dimorphous, mp of the unstable modification 26°C.

benzoquinones *(Chem.)*. $C_6H_4O_2$ (2 isomers, see **quinones**).

benzoyl chloride *(Chem.)*. *Benzene carboxyl chloride*, C_6H_5COCl, a colourless liquid, of pungent odour, bp 198°C, obtained by the action of PCl_5 on benzoic acid, commercially prepared by chlorinating benzaldehyde.

benzoyl peroxide *(Chem.)*. $C_6H_5.CO.O.O.CO.C_6H_5$. Bleaching agent and catalyst for free radical reactions. Mp 108°C. Prepared by the action of sodium peroxide on benzoyl chloride.

benzpinacol *(Chem.)*. $(C_6H_5)_2C(OH).C(OH)(C_6H_5)_2$. Reduction product of benzophenone.

1,2-benzpyrene *(Chem.)*. A polycyclic hydrocarbon isolated from coal tar as pale yellow crystals, mp 177°C. It has strong carcinogenic properties.

benzyl alcohol *(Chem.)*. *Phenylmethanol*. $C_6H_5.CH_2.OH$, a colourless liquid, bp 204°C, the simplest homologue of the aromatic alcohols.

benzylamine *(Chem.)*. *Phenylmethylamine*. $C_6H_5.CH_2.NH_2$, colourless liquid, bp 183°C, a primary amine of the aromatic series.

6-benzylaminopurine *(Bot.)*. *6-benzyladenine, PAB*. A synthetic **cytokinin**.

benzyl chloride *(Chem.)*. *Chloromethyl benzene*. $C_6H_5.CH_2Cl$, colourless liquid, bp 178°C, obtained by the action of chlorine on boiling toluene. Intermediate for benzyl derivatives.

benzylidene chloride *(Chem.)*. *Dichloromethyl benzene, Benzal chloride*.

benzyl penicillin *(Med.)*. The first of the *penicillins*; it has the disadvantage of being inactivated by bacterial penicillinases. It remains, however, the drug of choice for streptococcal, gonococcal and meningococcal infections. Also used to treat syphilis, yaws, tetanus, anthrax, actinomycosis and diphtheria.

benzyne *(Chem.)*. C_6H_4. An unstable intermediate formed by the removal of two ortho-hydrogens from benzene.

beraunite *(Min.)*. Hydrated phosphate of iron, red, found in iron ore deposits.

berber *(Textiles)*. A carpet square hand-woven by north Africans from hand-spun yarns from the natural coloured wool of local sheep. Commonly misused to describe machine-made carpets considered to have a similar appearance.

berberine *(Chem.)*. $C_{20}H_{19}O_5N_1H_2O$, chief alkaloid present in *Hydrastis*. Has been used as an amoebicide and in the treatment of cholera. Also known as *jamaicin, xanthopicrite*.

bergamot oil *(Chem.)*. Yellow-green volatile essential oil from the rind of *Citrus bergamia* (*Rutaceae*). Used in perfumery.

Bergeron-Findeisen theory *(Meteor.)*. That the initiation of precipitation in a cloud consisting mainly of super-cooled water droplets is due to the presence of ice crystals which grow at the expense of the droplets because the saturation vapour pressure with respect to ice is lower than that with respect to liquid water at the same temperature.

Bergius process *(Chem.)*. See **hydrogenation** of coal.

Bergmann's law *(Zool.)*. In warm-blooded animals and within a species southern forms are smaller than northern forms.

Bergstrom's method *(Aero.)*. A method of assessing the stresses in concrete pavements with particular reference to aerodrome runways and taxiing tracks.

beri-beri *(Med.)*. A disease causing peripheral nerve lesions and/or heart failure due to a deficiency of the vitamin B_1 (thiamine).

Berkefeld filter *(Build.)*. Domestic filter using diatomite for removing bacteria. Microporous cellulose or other material now generally used.

berkelium *(Chem.)*. Element, symbol Bk, at. no. 97, synthesized by helium ion bombardment on americium isotope 241.

berm *(Civ.Eng.)*. A horizontal ledge on the side of an embankment or cutting, to intercept earth rolling down the slopes, or to add strength to the construction. Also called a *bench*.

berm ditch *(Civ.Eng.)*. A channel cut along a berm to drain off excess water.

Bernoulli disk *(Comp.)*. See **exchangeable disk pack**.

Bernoulli equation *(Maths.)*. A differential equation of the form $\frac{dy}{dx} + py = qy^n$, where p and q are functions of x alone.

Bernoulli's law *(Phys.)*. For a non-viscous, incompressible fluid in steady flow the sum of the pressure, potential and kinetic energies per unit volume is constant at any point. It is a fundamental law of fluid mechanics.

Bernoulli's numbers *(Maths.)*. The number B_1, B_2, B_3 etc., defined by the expansion

$$\frac{x}{1-e^{-x}} = 1 + \tfrac{1}{2}x + \frac{B_1}{2!}x^2 - \frac{B_2}{4!}x^4 + \frac{B_3}{6!}x^6 - \ldots$$

The first few values are $B_1 = \frac{1}{6}$, $B_2 = \frac{1}{30}$, $B_3 = \frac{1}{42}$, $B_4 = \frac{1}{30}$, $B_5 = \frac{5}{66}$, after which they continue to increase to infinity. Other definitions vary in initial terms, numbering and signs.

Bernoulli's polynomials *(Maths.)*. The coefficients $\varphi_n(z)$ of $\frac{t^n}{n!}$ in the expansion

$$t\frac{e^{zt}-1}{e^t-1} = \sum_{n=1}^{\infty} \varphi_n(z)\frac{t^n}{n!}.$$

Bernoulli's theorems *(Maths.)*.

$$(1)\ \sum_{s=1}^{n} s^r = \frac{n^{r+1}}{r+1} + \frac{1}{2}n^r + \frac{r}{2!}B_1 n^{r-1} - \frac{r(r-1)(r-2)}{4!}B_2 n^{r-3} + \frac{r(r-1)(r-2)(r-3)(r-4)}{6!}B_3 n^{r-5}$$

there being $\frac{1}{2}(r + 3)$ terms if r is odd, and $\frac{1}{2}(r + 4)$ if r is even.

$$(2)\ \sum_{s=1}^{\infty} \frac{1}{s^{2m}} = \frac{(2\pi)^{2m}}{2.(2m)!}B_m,\ m \geqq 1.$$

where B_1, B_2, B_3 etc. are Bernoulli's numbers.

berry *(Bot.)*. A fleshy fruit, without a stony layer, usually containing many seeds.

berry *(Zool.)*. (1) The eggs of Lobster, Crayfish, and other macruran Crustacea. (2) Part of the bill in Swans.

Bertrand curves *(Maths.)*. See **conjugate Bertrand curves**.

bertrandite *(Min.)*. A hydrated beryllium silicate, occurring in pegmatites.

beryl *(Min.)*. A beryllium aluminium silicate, occurring in pegmatites as hexagonal colourless, green, blue, yellow, or pink crystals. Important ore of beryllium, also used as a gemstone. See **aquamarine, emerald**.

berylliosis *(Med.)*. Chronic beryllium poisoning, the main symptom of which is serious, and usually permanent lung damage.

beryllides *(Eng.)*. Compounds of other metals with beryllium.

beryllium *(Chem.)*. Steely uncorrodible white metallic element (Wöhler, 1828). Gk. and Lat. *beryl* = the old mineral name. Symbol Be, at. no. 4, r.a.m. 9.0122, mp 1281°C, bp 2450°C, rel.d. 1.93. Main use is for windows in X-ray tubes and as an alloy for hardening copper. Used as a powder for fluorescent tubes until found poisonous. The metal can be evaporated on to glass, forming a mirror for ultraviolet light. As a slight alloy with nickel, it has the highest coefficient for secondary electron emission, 12.3. Alpha-particles projected into beryllium make it a useful source of neutrons, from which they were discovered by Chadwick in 1932. The oxide (*beryllia*) is a good reflector of neutrons and is also used in thermoluminescent dating; highly toxic. Once called *glucinum*,

beryllium bronze *(Eng.)*. A copper-base alloy containing 2.25% of beryllium. Develops great hardness (i.e. 300–400 Brinell) after quenching from 800°C followed by heating to 300°C; see **temper-hardening**.

beryllonite *(Min.)*. A rare mineral, found as colourless to yellow crystals. Phosphate of beryllium and sodium.

Berzelius theory of valency *(Chem.)*. Chemical affinity is electrical in character, and depends on the mutual attraction of positive (metallic) and negative (non-metallic) elements. A forerunner of the modern **electronic theory of valency**.

Bessel functions *(Maths.)*. Of the first kind of order n:

$$J_n(x) = \sum_{r=0}^{\infty} \frac{(-1)^r}{r!\,\Gamma(n+r+1)}\left(\frac{x}{2}\right)^{n+2r},$$

where Γ indicates the **gamma function**. $J_n(x)$ is a solution of *Bessel's differential equation*. (2) Of the second kind of order n:

$$Y_n(x) = \frac{J_n(x)\cos n\pi - J_{-n}(x)}{\sin n\pi}.$$

There is also Hankel's type of Bessel function of the second kind, namely

$$Y_n(x) = \frac{\pi e^{n\pi i}}{\cos n\pi} Y_n(x).$$

This is a second independent solution of Bessel's differential equation. (3) Of the third kind of order *n*:

$$H_n^{(1)}(x) = J_n(x) + iY_n(x)$$

and

$$H_n^{(2)}(x) = J_n(x) + iY_n(x).$$

These are also called *Hankel functions* of the first and second kind respectively. (4) Modified Bessel functions: (*a*) Of the first kind of order *n*:

$$I_n(x) = \sum_{r=0}^{\infty} \frac{1}{r!\Gamma(n+r+1)} \left(\frac{x}{2}\right)^{n+2r}.$$

(*b*) Of the second kind of order *n*:

$$K_n(x) = \frac{\pi}{2} \frac{I_{-n}(x) - I_n(x)}{\sin n\pi}.$$

$$x^2 \frac{d^2y}{dx^2} + x\frac{dy}{dx} + (x^2 - n^2)y = 0.$$

For all these, notations vary, however. Bessel functions have application to vibration and heat flow.

Bessel's differential equation *(Maths.)*. The equation

$$x^2 \frac{d^2y}{dx^2} + x\frac{dy}{dx} + (x^2 - n^2)y = 0.$$

Satisfied by Bessel functions $J_n(x)$ and $Y_n(x)$.

Bessemer converter *(Eng.)*. Large barrel-shaped tilting furnace, charged while fairly vertical with molten metal, and 'blown' by air introduced below through tuyères. Discharged by tilting.

Bessemer pig iron *(Eng.)*. Pig iron which has been dephosphorized in Bessemer converter lined with basic refractory material.

Bessemer process *(Eng.)*. Removal of impurities from molten metal or matte by blowing air through molten charge in Bessemer converter. Used to remove carbon and phosphorus from steel, sulphur and iron from copper matte.

Best and Best Best *(Eng.)*. See **B and BB**.

best selected copper *(Eng.)*. Metal of a lower purity than high-conductivity copper. Generally contains over 99.75% of copper.

beta-adrenoceptor blocking drugs *(Med.)*. Group of drugs which block β-adrenoreceptors of the heart, peripheral vasculature, bronchi, pancreas and liver. Used in the treatment of angina, hypertension and also migraine, thyrotoxicosis and anxiety states. Commonly subgrouped into those which are unselective *(propranolol)* and those which largely act on the β_1-receptors *(atenolol)*.

beta backscattering sedimentometer *(Powder Tech.)*. One in which the mass of sediment on the bottom of the sedimentation chamber is measured by the amount of radiation it scatters back from a 1 mCi ^{90}Sr radioactive source.

beta brass *(Eng.)*. Copper-zinc alloys, containing 46–49% of zinc, which consist (at room temperature) of the intermediate constituent (or intermetallic compound) known as β.

betacaine *(Chem.)*. See **benzamine**.

Betacam *(Image Tech.)*. TN for a broadcast-standard VTR system using 1/2 in tape for TV recording in a combined camera/recorder.

betacyanins *(Bot.)*. Red pigments of the betalain type, e.g. the red pigment of the beetroot.

beta decay *(Phys.)*. Radioactive disintegration with the emission of an electron or positron accompanied by an uncharged antineutrino or neutrino. The mass number of the nucleus remains unchanged but the atomic number is increased by one or decreased by one depending on whether an electron or positron is emitted. See **electron capture**.

beta detector *(Nuc.Eng.)*. A radiation detector specially designed to measure β-radiation.

beta disintegration *(Phys.)*. See **beta decay**.

beta disintegration energy *(Phys.)*. For electron, β^-, emission it is the sum of the energies of the particles, the neutrino and the recoil atom. For positron, β^+, emission there is in addition the energy of the rest masses of two electrons.

beta diversity *(Ecol.)*. See **diversity**.

BET adsorption theory *(Chem.)*. Brunauer, Emmett and Teller postulated the building up of multimolecular adsorption layers on the catalyst surface, and extended Langmuir's derivation for single molecular layers to obtain an isothermal equation for multimolecular adsorption.

betafite *(Min.)*. A hydrated, niobate, tantalate, and titanate of uranium.

beta function *(Maths.)*. The function *B(p,q)* defined by

$$\int_0^1 x^{p-1}(1-x)^{q-1}dx.$$

Its main property is that

$$B(p, q) = \frac{\Gamma(p).\Gamma(q)}{\Gamma(p+q)}.$$

betaine *(Chem.)*. $(CH_3)_3N^+.CH_2.COO$- Betaine crystallizes with 1 molecule of H_2O; mp of anhydrous betaine 293°C with decomposition. Occurs naturally in plants and animals.

betaines *(Chem.)*. A class name for the **zwitterions** exemplified by betaine.

beta-iron *(Eng.)*. Iron in the temperature range 750°–860°C, in which a change from the magnetic (*alpha*) state to the paramagnetic occurs at about 760°C. With carbon in solution the transition is lowered toward 720°C, and when cooling recalescence is more marked.

betalains *(Bot.)*. A group of nitrogen-containing pigments functionally replacing other pigments including anthocyanins.

Betamax *(Image Tech.)*. TN for a domestic VTR system using 1/2 in tape in a cassette.

beta-microglobulin *(Immun.)*. A protein which forms part of the structure of class I major **histocompatibility antigens**, but is present in small amounts in blood, urine and seminal plasma.

beta-oxidation *(Biol.)*. The oxidative degradation of the fatty acid chains of lipids into 2-carbon fragments by cleavage of the penultimate C-C bonds.

beta particle *(Phys.)*. An electron or positron emitted in beta decay from a radioactive isotope.

beta-pleated sheet *(Biol.)*. Important element of protein structure resulting from hydrogen bonding between parallel polypeptide chains.

beta-ray gauge *(Paper)*. Equipment installed on a paper or converting machine to obtain an indication of grammage by measuring the absorption of β-rays passing through the web.

beta rays *(Phys.)*. Streams of *β-particles*.

beta-ray spectrometer *(Nuc.Eng.)*. One which determines the spectral distribution of energies of β-particles from radioactive substances or secondary electrons.

beta thickness gauge *(Nuc.Eng.)*. Instrument measuring thickness, based on absorption and backscattering (reflection) by material or sample being measured of β-particles from a radioactive source.

betatopic *(Phys.)*. Said of atoms differing in atomic number by one unit. One atom can be considered as ejecting an electron (β-particle) to produce the other one.

betatron *(Phys.)*. Machine used to accelerate electrons to energies of up to 300 MeV in pulsed output. The electrons move in an orbit or constant radius between the poles of an electromagnet, and a rapidly alternating magnetic field provides the means of acceleration. See **cyclotron**.

beta value *(Nuc.Eng.)*. In fusion, the ratio of the outward pressure exerted by the plasma to the inward pressure which the magnetic field is capable of exerting. Also *plasma beta*.

beta waves *(Med.)*. Higher frequency waves (15–60 Hz) produced in human brain.

betaxanthins *(Bot.)*. Yellow pigments of the **betalain** type.

Bethe cycle *(Astron.)*. See **carbon cycle**. (H.A. Bethe, American astrophysicist.)

Bethell's process *(Build.)*. For preserving timber, which is first dried, then subjected to a partial vacuum within a special cylinder, and finally impregnated with creosote under pressure.

béton *(Civ.Eng.)*. Fr. originally for *lime concrete*, now for any kind of concrete.

béton armé *(Civ.Eng.)*. Fr. for *reinforced concrete*.

BET surface area *(Powder Tech.)*. Surface area of a powder calculated from gas adsorption data, by the method devised by *Brunauer, Emmett and Teller*.

Betts process *(Eng.)*. An electrolytic process for refining lead after drossing. The electrolyte is a solution of lead silica fluoride and hydrofluorsilicic acid, and both contain some gelatine. Impurities are all more noble than lead and remain on the anode. Gold and silver are recovered from anode sponge.

between-lens shutter *(Image Tech.)*. One located between the components of a camera lens, actuated by a spring-loaded mechanism which opens a series of metal blades pivoted around the periphery of the aperture, the length of exposure being regulated by an air brake or a clockwork escapement which controls the closing action.

between perpendiculars *(Ships)*. The length between the **forward perpendicular** and after perpendicular (after side of sternpost). Abbrev. *BP*.

Beutler method *(Image Tech.)*. Use of slow film and fast developer to produce the same results as a fast film and slow fine-grain developer, given the same exposure.

BeV *(Nuc)*. See **GeV**.

Bevatron *(Phys.)*. A synchroton at Berkeley, USA which gives a beam of 6.4 GeV protons.

bevel *(Build)*. A light hardwood stock, slotted at one end to take the blade, which is fastened by a clamping screw passing through the stock and the slot in the blade, enabling the latter to be set at any desired angle to the former.

bevel *(Print.)*. The slope on the type from the **face** to the **shoulder**.

bevel gear *(Eng.)*. A system of toothed wheels connecting shafts whose axes are at an angle to one another but in the same plane.

bevelled boards *(Print.)*. Boards intended for covers, with the edges at the head, foot and fore-edge cut at an angle.

bevelled-edge chisel *(Build)*. One with the blade bevelled all the way up to the handle. Used for light work.

bevelled halving *(Build.)*. A halving joint in which the meeting surfaces are not cut parallel to the plane of the timbers but at an angle, so that when they are forced together, the timbers may not be pulled apart by a force in their own plane.

Beverage antenna *(Telecomm.)*. See **wave antenna**.

bezel *(Autos.)*. (1) A retaining rim, e.g. speedometer outer rim, panel light-retaining rim. (2) A small indicator light (e.g. for direction flashers) on instrument panel or dashboard.

bezel *(Build)*. The sloped cutting-edge of a chisel or other cutting tool.

bezel *(Horol.)*. The grooved ring holding the glass of a watch or an instrument dial.

BFO *(Telecomm.)*. Abbrev. for *Beat Frequency Oscillator*.

BGA *(Aero.)*. Abbrev. for *British Gliding Association*.

b-group *(Phys.)*. A close group of Fraunhofer lines in the green of the solar spectrum, due to magnesium.

BHA *(Min.Ext.)*. See **bottom-hole assembly**.

bhang *(Bot.)*. See **cannabis**.

B/H curve *(Elec.Eng.)*. See **magnetization curve**.

B/H loop *(Elec.Eng.)*. See **hysteresis loop**.

BHN *(Eng.)*. Abbrev. for *Brinell Hardness Number*, obtained in the **Brinell hardness test**.

BHP *(Eng.)*. Abbrev. for *Brake HorsePower*.

Bi *(Chem.)*. The symbol for bismuth.

bialternant *(Maths.)*. The quotient of two alternants. A homogeneous symmetric polynomial.

bias *(Electronics)*. Adjustment of a relay so that it operates for currents greater than a given current (against which it is *biased*), or for a current of one polarity.

bias binding *(Textiles)*. A non-fraying narrow fabric made by cutting full-width woven cloth into strips at 45° to the selvedge. The material is used for binding curved seams and garment edges.

bias current *(Electronics)*. Nonsignal current supplied to electrode of semiconductor device, magnetic amplifier, tape recorder, etc., to control operation at optimal working point.

biased protective system *(Elec.Eng.)*. A modification of a balanced protective system, in which the amount of out-of-balance necessary to produce relay operation is increased as the current in the circuit being protected is increased.

biased result *(Surv.)*. In observations, sampling etc., introduction of a systematic error through some malfunction of instrument or weakness in method used, so that error accumulates in a series of measurements.

biasing *(Electronics)*. Polarization of a recording head in magnetic tape recording, to improve linearity of amplitude response, using d.c. or using a.c. much higher than the maximum audio-frequency to be reproduced.

biasing transformer *(Elec.Eng.)*. A special form of transformer used in one form of biased protective system.

biaxial *(Crystal.,Min.)*. Said of a crystal having two optical axes. Minerals crystallizing in the orthorhombic, monoclinic and triclinic systems are biaxial. Cf. **uniaxial**.

bib-cock *(Eng.)*. A draw-off tap for water-supply, consisting of a plug-cock having a downward curved extension for discharge.

bibenzoyl *(Chem.)*. See **benzil**.

Bible paper *(Paper)*. Heavily loaded, strong, thin, printing paper, generally of 20–40 g/m^2.

biblio *(Print.)*. Details of a book's 'history', i.e. original publication date and dates of subsequent reprints and revised editions; usually placed on the verso of the title page.

bib-valve *(Eng.)*. A draw-off tap of the kind used for domestic water-supply; closed by screwing down a rubber-washered disk on to a seating in the valve body.

bicarbonates *(Chem.)*. Hydrogen carbonates. The acid salts of carbonic acid; their aqueous solutions contain the ion $(HCO_3)^-$.

bicarpellary *(Bot.)*. Said of an ovary consisting of two carpels.

biceps *(Zool.)*. A muscle with two insertions. adj. *bicipital*.

bicipital *(Zool.)*. See **biceps**.

bicipital groove *(Zool.)*. A groove between the greater and lesser tuberosities of the humerus in Mammals.

bicollateral bundle *(Bot.)*. A vascular bundle with two strands of phloem adjacent to the single strand of xylem; one placed centrifugally, the other centripetally. Cf. **collateral bundle**.

bicomponent fibre *(Textiles)*. A synthetic fibre made from two fibre-forming polymers which may be arranged to lie side-by-side or as a sheath surrounding a core. By suitable heat treatment a crimped fibre is produced because of the different shrinkage properties of the two polymers. Advantage may also be taken of the sheath having a lower softening temperature than the core to produce a non-woven fabric.

biconcave *(Image Tech.,Phys.)*. Said of a lens having both surfaces concave.

biconical horn *(Telecomm.)*. Two flat cones apex to apex, for radiating uniformly in horizontal directions when driven from a coaxial line. See **discone antenna**.

biconvex *(Image Tech.,Phys.)*. Said of a lens which is convex on both surfaces.

bicuspid, bicuspidate *(Zool.)*. Having two cusps, as the premolar teeth of some Mammals.

bicuspid valve *(Zool.)*. The valve in the left auriculo-ventricular aperture in the Mammalian heart. Also *mitral valve*.

bidentate *(Chem.)*. Complexing molecule or ion with two donating atoms.

bidirectional microphone *(Acous.)*. One which is most sensitive in both directions along one axis.

bi-directional printer *(Comp.)*. **Line printer** in which the right-to-left return movement of the print head is used to print a second line, thus increasing print speed.

bidirectional waveform *(Telecomm.)*. One which shows reversal of polarity, e.g. a bidirectional pulse generator produces both positive and negative pulses.

bieberite *(Min.)*. Hydrated cobalt sulphate, also known as *cobalt vitriol*. Found as stalactites and encrustations in old mines containing other cobalt minerals.

biennial *(Bot.)*. A plant that flowers and dies between its first and second years from germination and which does not flower in its first year. Cf. **annual, perennial**.

bifacial leaf *(Bot.)*. A dorsiventral leaf; typically with palisade mesophyll towards the upper surface and spongy mesophyll below. The commonest sort. Ant. *unifacial leaf*.

bifid *(Bot.,Zool.)*. Divided halfway down into two lobes; forked.

bifilar micrometer *(Astron.)*. An instrument attached to the eyepiece end of a telescope to enable the angular separation and orientation of a visual double star to be measured.

bifilar pendulum *(Phys.)*. See **bifilar suspension**.

bifilar resistor *(Elec.Eng.)*. One formed by winding a resistor with a hairpin-shaped length of resistance wire, thus reducing the total inductance.

bifilar suspension *(Phys.)*. The suspension of a body by two parallel vertical wires or threads which give a considerable controlling torque. If the threads are of length l and are distance d apart, the period of torsional vibration of a suspended body of moment of inertia I and mass m is

$$T = 4\pi\sqrt{\frac{Il}{mgd^2}}.$$

bifurcate *(Bot.,Zool.)*. Twice-forked: forked. v. *bifurcate*. n. *bifurcation*.

bifurcated rivet *(Eng.)*. A rivet with a split shank, used for holding together sheets of light material; it is closed by opening and tapping down the two halves of the shank.

Big Bang *(Astron.)*. Hypothetical model of the universe which postulates that all matter and energy were once concentrated into an unimaginably dense state, or *primeval atom* from which it has been expanding from a *creation event* some 13–20 × 10^9 years ago. Main evidence favouring this model is the **microwave background** (*cosmic background radiation*) and the **redshift** of galaxies.

Big Dipper *(Astron.)*. The asterism (chiefly US) in the constellation Ursa Major, known by a great variety of popular names.

bigeminal pulse *(Med.)*. A pulse in which the beats occur in pairs, each pair being separated from the other by an interval; due to a disturbed action of the heart.

big-end *(Autos.)*. That part of the connecting rod which is attached to the crankshaft.

big-end bolts *(Eng.)*. See **connecting-rod bolts**.

bigeneric *(Zool.)*. Of hybrids, produced by crossing two distinct genera.

bigeneric hybrid *(Bot.)*. A hybrid resulting from a cross between individuals from two different genera, e.g. *Triticale*, a hybrid between wheat (*Triticum*) and rye (*Secale*).

big head disease of horses *(Vet.)*. See **osteodystrophia fibrosa**.

big head disease of sheep *(Vet.)*. *Swelled head.* (1) Infection of the head and neck by *Clostridium septicum*. (2) A form of light sensitization occurring in sheep. Controlled by vaccination.

big head disease of turkeys *(Vet.)*. See **infectious sinusitis of turkeys**.

biguanides *(Med.)*. Drugs used in the treament of maturity onset diabetes. They seem to act by increasing peripheral utilization of glucose and are of particular value in obese diabetics. *Metformin*.

biharmonic equation *(Maths.)*. The equation

$$\frac{\partial^4 z}{\partial x^4} + \frac{2\partial^4 z}{\partial x^2 \partial y^2} + \frac{\partial^4 z}{\partial y^4} = 0.$$

The solutions of this equation are called *biharmonic functions*.

bilabiate *(Bot.)*. With two lips.

bilateral *(Med.)*. Having, or pertaining to, two sides.

bilateral cleavage *(Zool.)*. The type of cleavage of the zygote formed in *Chordata*.

bilateral impedance *(Elec.Eng.)*. Any electrical or electromechanical device through which power can be transmitted in either direction.

bilateral slit *(Phys.)*. A slit used in a spectrometer and consisting of two metal strips whose separation can be accurately adjusted.

bilateral symmetry *(Biol.)*. The condition when an organism is divisible into similar halves by one plane only. Cf. **radial symmetry**.

bilateral tolerance *(Eng.)*. A tolerance with dimensional limits above and below the basic size.

bile *(Med.)*. A viscous liquid produced by the liver. Human bile has an alkaline reaction and possesses a green or golden-yellow colour and a bitter taste. It consists of water, bile salts, mucin and pigments, cholesterol, fats and fatty acids, soaps, lecithin and inorganic compounds.

bile duct *(Zool.)*. The duct formed by the junction of the hepatic duct and the cystic duct, leading into the intestine.

bile pigments *(Med.)*. The chief are bilirubin (reddish-yellow) and its oxidation product biliverdin (green). Produced by the breakdown of haemoglobin, they consist of an open chain of four substituted pyrrole nuclei joined by two methene (= =CH–) bridges and one mythylene (–CH_2–) bridge.

bile salts *(Biol.)*. Breakdown products of cholesterol which accumulate in bile and are responsible for the detergent activity of bile.

bile salts *(Med.)*. Sodium salts of the bile acids, a group of hydroxyl steroid acids, some unsaturated, condensed with taurine or glycine; commonest are the salts of taurocholic and glycocholic acids; secreted in the bile; important in aiding absorption of fats from intestine, due to their surface tension, reducing and emulsification properties.

bilge *(Ships)*. (1) The curved part of the shell joining the bottom to the sides. (2) The space inside this, at the sides of the cellular double bottom, into which unwanted water drains.

bilge keel *(Ships)*. A projecting fin attached to the shell plating for about half the length of the ship, amidships, at the turn of the bilge, to reduce rolling.

bilharziasis *(Med.,Vet.)*. Parasitic disease of man and domestic animals caused by **blood flukes**. Endemic in Africa and Far East. See **schistosomiasis**.

biliary fever *(Vet.)*. A form of **piroplasmosis** affecting the horse and dog, caused by *Babesia equi* or *Babesia canis* respectively. It is characterized by fever, jaundice, and haemoglobinuria.

bilicyanin *(Med.)*. An oxidation product of bilirubin, blue in colour.

bilinear transformation *(Maths.)*. A one-to-one transformation between the w and z planes, given by

$$w = \frac{az+b}{cz+d}.$$

A special form of linear transformation. Also called *Möbius transformation*.

bilirubin *(Med.)*. A reddish pigment occurring in bile; believed to be formed as a breakdown product of haemoglobin. See **bile pigments**.

billitonites *(Geol.)*. See **tektites**.

biliverdin *(Med.)*. A green pigment occurring in bile. It is an oxidation product of **bilirubin**. See **bile pigments**.

bill *(Print.)*. A detailed inventory of a fount of type showing the number of each character.

billet *(Build.)*. A piece of timber which has three sides sawn and fourth left round.

billet *(Eng.)*. Semifinished solid product which has been hot-worked by extrusion, forging and rolling. Smaller than a **bloom**.

billet mills *(Eng.)*. The rolling-mills used in reducing steel ingots to billets. Also called *billet rolls*.

Billet split lens *(Phys.)*. A device used to produce interference fringes. The two halves of the lens are separated so that two images of a slit source provide the coherent sources.

billiard cloth *(Textiles)*. Woollen cloth manufactured from finest quality wool, with a closely cropped dress-face finish to render it perfectly smooth, and damp resistant.

billion *(Genrl.)*. In the US and now more generally, a thousand million, or 10^9. Previously elsewhere, a million million or 10^{12}.

billion-electron-volt *(Phys.)*. See **GeV**.

bill of quantities *(Build.,Civ.Eng.)*. A list of items giving the quantities of material and brief descriptions of work comprised in an engineering or building works contract; it forms the basis for a comparison of tenders.

bilocular *(Bot.)*. Consisting of two loculi or chambers.

bimanous *(Zool.)*. Having the distal part of the two forelimbs modified as hands, as in some Primates.

bimanual *(Med.)*. Performed with both hands, e.g. *bimanual* examination of the female genital organs.

bimastic *(Zool.)*. Having two nipples.

bimetal-fuse *(Elec.Eng.)*. A fuse element composed of two different metals e.g. a copper wire coated with tin or lead.

bimetallic balance *(Horol.)*. A **compensation-balance** in which temperature compensation is obtained by means of a bimetallic strip.

bimetallic plates *(Print.)*. Lithographic plates of great durability, in which the ink-accepting and ink-rejecting parts are of two different metals, e.g. copper and chromium; called *polymetallic* when a supporting metal, e.g. steel, is used.

bimetallic strip *(Eng.,Elec.Eng.)*. Relies on differing temperature coefficients of expansion; the strip deflects when one side of the strip expands more than the other, used in thermal switches, thermostats, thermometers, etc.

bimirror *(Phys.)*. A pair of plane mirrors slightly inclined to one another. Used for the production of two coherent images in interference experiments.

bimorph *(Elec.Eng.)*. Unit in microphones and vibration detectors in which two piezoelectric plates are cemented together in such a way that application of p.d. causes one to contract and the other to expand, so the combination bends as in a bimetallic strip.

binary *(Astron.)*. See **binary star**.

binary *(Chem.)*. Consisting of two components, etc.

binary *(Maths.)*. See **binary scale**.

binary arithmetic *(Maths.)*. Arithmetic using numbers expressed in the binary scale.

binary code *(Comp.)*. A representation of information using a sequence of 0s and 1s. See **character code**.

binary counter *(Telecomm.)*. Flip-flop or toggle circuit which gives one output pulse for two input pulses, thus dividing by two.

binary fission *(Biol.)*. Division of the nucleus into two daughter nuclei, followed by similar division of the cell-body.

binary notation *(Comp.)*. System of representing numbers on the **binary scale**.

binary point *(Comp.,Maths.)*. The **radix point** in calculations carried out on the **binary scale**.

binary scale *(Maths.)*. Scale of counting with radix 2; used because the figures 1 and 0 can be represented by *on* and *off*, *pulse* or *no-pulse*, in electronic circuits. A specific form of binary code.

binary search *(Comp.)*. Strategy for locating a record in an ordered list by repeatedly comparing the key with the mid-item in the list and discarding that half of the list which cannot contain the required record.

binary star *(Astron.)*. A double star in which the two components revolve about their common centre of mass under the influence of their gravitational attraction. See **eclipsing binary**, **spectroscopic binary**, **visual binary**.

binary system and diagram *(Eng.)*. The alloys formed by two metals constitute a *binary alloy system*, which is represented by the *binary constitutional diagram* for the system.

binary tree *(Comp.)*. A **tree** where any node may have no more than two branches.

binary vapour-engine *(Eng.)*. The name given to a heat-engine using two separate working fluids, generally mercury vapour and steam, for the high and low temperature portions of the cycle respectively, thus enabling a large temperature range to be used, with improved thermal efficiency.

binaural *(Acous.)*. Listening with two ears, the result of which is a sense of directivity of the arrival of a sound wave. Said of a stereophonic system with two channels (matched) applying sound to a pair of ears separately, e.g. by earphones. The effect arises from relative phase delay between wavefronts at each ear.

binder *(Build.)*. The medium or vehicle in a surface coating which retains the solids in suspension during storage and on application dries to form the film.

binder *(Powder Tech.)*. Carbon products, organic brake linings, sintered metals, tar macadam etc., employ binder components in the mix to impart cohesion to the body to be formed. The binder may have cold setting properties, or subsequently be heat-treated to give it permanent properties as part of the body or to remove it by volatilization.

binder coat *(Build.)*. A surface coating applied to a friable or unstable surface to provide a stable base for subsequent coatings.

binding-beam *(Build.)*. A timber tie serving to bind together portions of a frame.

binding energy *(Electronics)*. (1) That required to remove a particle from a system, e.g. electron, when it is the **ionization potential**. (2) Energy required to overcome forces of cohesion and disperse a solid into constituent atoms.

binding energy of a nucleus *(Phys.)*. All nuclei have *rest* masses less than the total rest mass of their constituent protons and neutrons. The mass difference m is the *mass decrement* or *mass defect*. This arises because all nucleons bound to the nuclei must have negative energy (potential well). So if a system of free nucleons are combined to form a nucleus the total energy of the system must decrease by an amount B, the binding energy of the nucleus. The decrease B is accompanied by a decrease in the mass, the mass decrement, $M = B/C^2$ (**mass-energy equation**). The binding energy per nucleon is B/A where A is the atomic mass number. See **packing fraction**, **fusion**, **fission**.

binding joist *(Build.)*. See **binder**.

binding wire *(Elec.Eng.)*. See **tie wire**.

Binet-Cauchy theorem *(Maths.)*. If A is a matrix of order $m \times n$, and B is a matrix of order $n \times p$, then any determinant of order r of the product matrix AB is equal to a sum of terms, each of which is the product of a determinant of order r of A, and a determinant of order r of B (in that order).

Binet's formulae for log $\Gamma(z)$ *(Maths.)*. The first formula is

$$\log \Gamma(z) = (z - \tfrac{1}{2}) \log z - z + \tfrac{1}{2} \log(2\pi) + \int_0^\infty \left(\frac{1}{e^t - 1} - \frac{1}{t} + \frac{1}{2} \right) \frac{e^{-tz}}{t} \, dt.$$

The second formula is

$$\log \Gamma(z) = (z - \tfrac{1}{2})\log z - z + \tfrac{1}{2}\log(2\pi)$$

$$+2\int_0^{\infty} \frac{\arctan\left(\frac{t}{z}\right)}{e^{2\pi t} - 1}\,dt.$$

$$(\Gamma(z) > 0).$$

Bingham flow *(Phys.)*. Certain substances when sheared show little or no tendency to flow until a certain shear stress is reached when the flow rate increases sharply, e.g. modelling clay.

binocular camera *(Image Tech.)*. See **stereocamera**.

binoculars *(Phys.)*. A pair of telescopes for use with both eyes simultaneously. Essential components are an objective, an eyepiece and some system of prisms to invert and reverse the image.

binomial *(Maths.)*. An expression containing the sum (or difference) of two terms, e.g. $x+y$.

binomial array *(Radar,Telecomm.)*. A linear array in which the current amplitudes are proportional to the coefficients of a binomial expansion. Such an array has no side lobes.

binomial (binominal) nomenclature *(Biol.)*. The system (introduced by Linnaeus) of denoting an organism by 2 Latin words, the first name of the genus, the second the specific epithet. The 2 words constitute the name of the species; e.g. *Homo sapiens; Bellis perennis*. See **species**.

binomial coefficients *(Maths.)*. See **binomial theorem**.

binomial distribution *(Stats.)*. The probability distribution of the total number of outcomes of a particular kind in a predetermined number of trials, the probability of the outcome being constant at each trial, and the different trials being statistically independent.

binomial theorem *(Maths.)*. The expansion

$$(1+x)^n = 1 + \frac{n}{1}x + \frac{n(n-1)}{1.2}x^2 + \frac{n(n-1)(n-2)}{1.2.3}x^3$$

$$+ \frac{n(n-1)(n-2)(n-3)}{1.2.3.4}x^4 + \ldots$$

which is valid either if n is a positive integer (in which case the series terminates) or if $|x| < 1$. The coefficients of the powers of x are called *binomial coefficients*, that of x^r being written $\binom{n}{r}$. The series can thus be written

$$(1+x)^n = \sum_{r=0}^{n} \binom{n}{r} x^r.$$

Binomial coefficients are used in the construction of probability models for discrete distributions. See **binomial distribution**.

binormal *(Maths.)*. See **moving trihedral**.

binovular twins *(Med.)*. Twins resulting from the fertilization of 2 separate ova.

binucleate phase *(Bot.)*. Same as **dikaryophase**.

bio-aeration *(Build.)*. A system of sewage purification by oxidation; aeration of the crude sewage is effected by passing it through specially designed centrifugal pumps. See also **activated sludge**, **activation**.

bioassay *(Biol.,Ecol.)*. Quantitative determination of a substance by measuring its biological effect on e.g. growth, that is the use of an organism to test the environment.

bio-assay *(Med.)*. Determination of the power of a drug or of a biological product by testing its effect on an animal of standard size.

bio-availability *(Med.)*. The extent and rate at which the active substance in a drug is taken up by the body. Differences in tablet formulation and the rate of absorption by the gut will alter bio-availability.

biochemistry *(Genrl.)*. The chemistry of living things; physiological chemistry.

bioclastic limestone *(Geol.)*. A carbonate made up of broken fragments of organic material, especially shells.

bioclimatology *(Biol.)*. The study of the effects of climate on living organisms.

biocoenosis *(Ecol.)*. The association of animals and plants together, especially in relation to any given feeding area; adj. *coenotic*.

biodegradation *(Biol.)*. The breaking-down of substances by bacteria.

bio-electricity *(Biol.)*. Electricity of organic origin.

bio-engineering *(Chem.Eng.)*. Engineering methods of achieving biosynthesis of animal and plant products, e.g. for fermentation processes. Also *biological engineering*.

bio-engineering *(Med.)*. Provision of artificial means (electronic, electrical etc.) to assist defective body functions, e.g. hearing aids, limbs for thalidomide victims etc. Cf. **genetic engineering**.

biofeedback *(Behav.)*. Refers to procedures whereby subjects are given information about physiological functions that are not normally available to conscious experience (e.g. heart rate, blood pressure etc.) with the object of gaining some conscious control over them.

biogas *(Bot.)*. Gas, mostly methane and carbon dioxide, produced in suitable equipment by bacterial fermentation of organic matter (see biomass) and used as fuel.

biogenesis *(Bot.)*. The formation of living organisms from their ancestors and of organelles from their precursors.

biogeographic regions *(Ecol.)*. Regions of the world containing recognizably distinct and characteristic endemic fauna or flora.

bioherm *(Geol.)*. A reef or mound made up of the remains of organisms growing *in situ*.

biological clock *(Bot.)*. An internal, physiological time-keeping system underlying e.g. circadian rhythms and photoperiodism.

biological constraint *(Behav.)*. A general term in learning theory that refers to the fact that certain behaviours are more easily learned by some organisms than by others, and conversely, that some behaviours are not easily learned by some organisms.

biological containment *(Biol.)*. Alteration of the genetic constitution of an organism so as to minimize its ability to grow in a non-laboratory environment.

biological control *(Ecol.)*. Use of an organism to control a disease, pest or weed.

biological form *(Bot.)*. Same as *physiological race*.

biological half-life *(Phys.)*. Time interval required for half of a quantity of radioactive material absorbed by a living organism to be eliminated naturally.

biological hole *(Nuc.Eng.)*. A cavity within a nuclear reactor in which biological specimens are placed for irradiation experiments.

biological magnification *(Ecol.)*. Process whereby the concentration of a pollutant, within living tissues, increases at each link in a food chain.

biological oxygen demand *(Chem.)*. An indication of the amount of oxygen needed to oxidize fully, by biological means, reducing material in a water sample. It is expressed as the concentration of oxygen gas in parts per million chemically equivalent to the reducing agents in the water. Abbrev. *BOD*.

biological race *(Bot.,Zool.)*. *Physiological race*. A race occurring within a taxonomic species; distinguished from the rest of the species by slight or no morphological differences, but by evident differences of habitat, food-preference or occupation which inhibit interbreeding.

biological shield *(Biol.)*. Screen made of material highly absorbent to radiation used as a protection.

biological warfare *(Genrl.)*. The use of bacteriological (biological) agents and toxins as weapons. Cf. **chemical warfare**.

bioluminescence *(Biol.)*. The production of light by living organisms, as glow-worms, some deep-sea fish, some bacteria, some fungi.

biomass *(Ecol.)*. (1) The total dry mass of an animal or plant population. (2) Organic matter (mostly from plants) harvested as a source of energy (by burning or **biogas** production) and/or as a chemical feedstock.

biomass, pyramid of *(Ecol.)*. A diagrammatic representation of the biomass of a series of organisms, organized according to *trophic layers* to form a pyramid.

biome *(Ecol.)*. The largest land community region recognized by ecologists, e.g., tundra, savanna, grassland, desert, temperate and tropical forest.

biometeorology *(Biol.)*. The study of the effects of atmospheric conditions on living things.

biometrical genetics *(Biol.)*. Same as **quantitative genetics**.

biometry *(Biol.)*. Statistical methods applied to biological problems.

biomining *(Min.Ext.)*. See **microbiological mining**.

bionics *(Genrl.)*. The various phenomena and functions which characterize biological systems with particular reference to electronic systems.

biophysics *(Biol.)*. The physics of vital processes; study of biological phenomena in terms of physical principles.

biopsy *(Med.)*. Diagnostic examination of tissue (e.g. tumour) removed from the living body.

biosphere *(Ecol.)*. That part of the earth (upwards at least to a height of 10 000 m, and downwards to the depths of the ocean, and a few hundred metres below the land surface) and the atmosphere surrounding it, which is able to support life. A term that may be extended theoretically to other planets. Cf. **hydrosphere**, **lithosphere**.

biosynthesis *(Biol.)*. The synthesis of complex molecules using enzymes and biological structures like ribosomes and chromosomes either within or without the cell.

biosystematics *(Biol.)*. The study of relationships with reference to the laws of classification of organisms; *taxonomy*.

biota *(Biol.)*. Fauna and flora of a given region.

biotechnology *(Bot.)*. The use of organisms or their components in industrial or commercial processes, which can be aided by the techniques of **genetic manipulation** in developing e.g. novel plants for agriculture or industry.

Biot-Fourier equation *(Eng.)*. The equation representing the nonsteady conduction of heat through a solid. In the one dimensional case,

$$\frac{\partial\varphi}{\partial t} = a\frac{\partial^2\varphi}{\partial x^2},$$

where temperature of a section at right angles to the flow $=\varphi$; distance in flow direction $=x$; time $=t$; and thermal diffusivity ($k/\rho s$, where k is thermal conductivity, ρ density and s is specific heat capacity) $=a$.

biotic *(Biol.)*. Relating to life.

biotic barrier *(Ecol.)*. Biotic limitations affecting dispersal and/or survival of animals and plants.

biotic climax *(Bot.)*. A community that is maintained in a stable condition because of some biotic factor e.g. grazing. See also **climax**.

biotic factor *(Bot.)*. The activities of any organisms that determine which plants grow where.

biotin *(Biol.)*. See **vitamin B**.

biotinylation *(Biol.)*. Labelling a probe with conjugated biotin, whose high affinity for avidin or anti-biotin antibodies is exploited to mark the spot to which the probe binds by indirect immunoassay.

biotite *(Min.)*. Black mica widely distributed in igneous rocks (particularly in granites) as lustrous black crystals, with a perfect cleavage. Also occurs in mica-schists and related metamorphic rocks. In composition, it is a complex silicate, chiefly of aluminium, iron and magnesium, together with potassium and hydroxyl.

Biot laws *(Phys.)*. The rotation produced by optically active media is proportional to the length of path, to the concentration (for solutions) and to the inverse square of the wavelength of the light.

Biot modulus *(Eng.)*. The heat transfer to a wall by a flowing medium, giving the ratio of heat transfer by convection to that by conduction. Defined as $\alpha\theta/\lambda$ where heat transfer coefficient $=\alpha$, thermal conductivity of medium $=\lambda$, characteristic length of apparatus $=\theta$.

biotope *(Ecol.)*. A small habitat in a large community, e.g., a cattle dropping on a grass prairie, whose several short seral stages comprise a microsere.

biotroph *(Bot.)*. A parasite which feeds off the living cells of its host and therefore needs their continued functioning, e.g. rust and smut fungi. Cf. **necrotroph**. See also **obligate parasite**.

Biot-Savart's law *(Phys.)*. Defines the intensity of magnetic flux density produced at a point a distance from a current-carrying conductor.

biotype *(Biol.)*. A group of individuals within a species with identical, or almost identical, genetic constitution.

bipack *(Image Tech.)*. Two films run through a camera in contact, emulsion-to-emulsion.

biparous *(Zool.)*. Having given birth to 2 young.

bipedal *(Zool.)*. Using only 2 limbs for walking.

bi-phase *(Elec.Eng.)*. See **two-phase**.

bipinnate *(Bot.)*. Said of a compound pinnate leaf with its main segments pinnately divided.

bipolar *(Zool.)*. Having two poles: having an axon at each end, as some nerve cells.

bipolar coordinates *(Maths.)*. A system of co-ordinates in which the position of a point in a plane is specified by its two distances r and r' from two fixed reference points.

bipolar disorder *(Behav.)*. Affective disorder characterized by shifts from one emotional extreme (euphoria, intense activity) to another (depressive episodes). Formerly *manic-depressive psychosis*.

bipolar electrode *(Elec.Eng.)*. An electrode in an electroplating bath not connected to either the anode or cathode; sometimes **secondary electrode**.

bipolar germination *(Bot.)*. Germination of a spore by the formation of 2 germ tubes, one from each end.

bipolar transistor *(Electronics)*. A transistor that uses both positive and negative charge carriers. Both *p-n-p* and *n-p-n* types of bipolar transistor can be manufactured, as discrete devices, or for incorporation into integrated circuits.

biprism *(Image Tech.)*. A prism with a very obtuse angle, used for beam-splitting.

bipropellant *(Space)*. Rocket propellant made up of two liquids, one being the fuel and the other the oxidizer, which are kept separate prior to combustion.

bipyramid *(Crystal)*. A crystal form consisting of two pyramids on a common base, the one being the mirror-image of the other. Each pyramid is built of triangular faces, 3, 4, 6, 8, or 12 in number. See **pyramid**.

biquartz *(Phys.)*. A quartz cut perpendicular to the axis, one half of the disk being right-handed and the other left-handed quartz. The thickness is such that each half rotates the plane of vibration of yellow light through 90° but in opposite directions. The device is used as a sensitive analyser for **saccharimeters**.

biradial symmetry *(Zool.)*. The condition in which part of the body shows radial, part bilateral symmetry; as in some Ctenophora.

biramous *(Zool.)*. Having two branches; forked, as some Crustacean limbs. Cf. **uniramous**.

birch *(For.)*. Common European tree, *Betula*, yielding a lightish to very light brown wood which polishes satisfactorily; due to its lack of colour and figuring is useful for imitating superior wood. It also turns well.

bird's mouth *(Build.)*. A re-entrant angle cut into the end of a timber, so as to allow it to rest over the arris of a cross timber.

birefrigence *(Textiles)*. The difference between the refractive index of a fibre measured parallel to the fibre axis and perpendicular to it. This gives an expression for the orientation of the molecules in the fibre.

birefringence *(Min.)*. Double refraction of light in a crystal: the splitting of incident light into two rays vibrating at right angles to each other, and causing two images to appear, e.g. in calcite. The difference between the greatest and least refractive index is a measure of the birefingence.

birefringent filter *(Phys.)*. One based on the polarization of light which enables a narrow spectral band of <0.1 nm to be isolated, i.e. effectively a **monochromator**; for photographing solar flares etc.

Birmingham Gauge, Birmingham Wire Gauge. *(Eng.)*. Systems of designating the diameters of rods and wires by numbers. Obsolescent, being replaced by preferred metric sizes. Abbrev. *B.G., B.W.G.*

birnessite *(Min.)*. Hydrated manganese oxide with sodium and calcium, and enriched in other elements. It is one of the dominant minerals of the deep-sea **manganese nodules**.

birth-mark *(Med.)*. See **naevus**.

bis-azo dyes *(Chem.)*. See **disazo dyes**.

bischofite *(Min.)*. Hydrated magnesium chloride. A constituent of salt deposits, such as those of Stassfurt in Germany; decomposes on exposure to the atmosphere.

bisector *(Maths.)*. The straight line which divides another line, angle or figure into two equal parts.

biseriate *(Bot.)*. In 2 whorls, cycles, rows or series.

biserrate *(Bot.)*. Of a leaf margin, having a series of sawlike teeth, which are themselves serrated.

bisexual *(Bot.Zool.)*. Possessing both male and female sexual organs. See **hermaphrodite**.

bisexuality *(Behav.)*. (1) Having the physical or psychological attributes of both sexes. (2) Being sexually attracted to both sexes.

Bismarck brown *(Chem.)*. A brown dyestuff obtained by the action of nitrous acid on *m*-phenylenediamine. It contains triamino-azo-benzene, $H_2N.C_6H_4.N{=}N.C_6H_3(NH_2)_2$, and a more complex disazo compound, $C_6H_4{=}[N{=}N.C_6H_3(NH_2)]_2$.

bismite *(Min.)*. The monoclinic phase of bismuth trioxide. Cf. **sillénite**, the cubic phase.

bismuth *(Chem.)*. Element, symbol Bi, at. no. 83, r.a.m. 208.98, mp 268°C. Used as a component of fusible alloys with lead. ^{209}Bi is the heaviest stable nuclide. It is strongly diamagnetic and has a low capture cross-section for neutrons, hence its possible use as a liquid metal coolant for nuclear reactors. A high absorption for γ-rays makes it a useful filter or window for these, while transmitting neutrons.

bismuth glance *(Min.)*. See **bismuthinite**.

bismuth (III) chloride *(Chem.)*. $BiCl_3$. Formed by the direct combination of chlorine and bismuth. Treated with an excess of water, it forms bismuth oxychloride, BiOCl, once used as a pigment called **pearl white**.

bismuth (III) hydride *(Chem.)*. BiH_3. Volatile, unstable compound. Also called *bismuthine*.

bismuth (III) oxide *(Chem.)*. Bi_2O_3. Formed when bismuth is heated in air or when the hydroxide, carbonate, or nitrate is calcined.

bismuthine *(Chem.)*. See **bismuth (III) hydride**.

bismuthinite *(Min.)*. Sulphide of bismuth, rarely forming crystals. Also called *bismuth glance*.

bismuth ochre *(Min.)*. A group name for undetermined oxides and carbonates of bismuth occurring as shapeless masses or earthy deposits.

bismuth spiral *(Elec.Eng.)*. Flat coil of bismuth wire used in magnetic flux measurements; the change of flux is measured by observing the change in resistance of the bismuth wire, which increases with increasing fields.

bismutite *(Min.)*. Bismuth carbonate $(BiO)_2CO_3$, a tetragonal mineral.

bisphenoid *(Crystal.,Min.)*. A crystal form consisting of 4 faces of triangular shape, 2 meeting at the top and 2 at the base in chisellike edges, at right-angles to one another; hence the name, meaning 'double edged'.

bisphenol A *(Plastics)*. *2,2-Bis(4-hydroxyphenyl)-propane*. Intermediate in the manufacture of **epoxy resins**.

bisporangiate *(Bot.)*. Said of a strobilus which consists of megasporophylls and microsporophylls, with megasporangia and microsporangia.

bistable *(Comp.)*. Device which has two stable states, which can be used to represent 0 and 1. See **flip flop**.

bistable circuit *(Telecomm.)*. Valve or transistor circuit which has two stable states which can be decided by input signals; much used in counters and scalers.

bistoury *(Med.)*. A long, narrow surgical knife for cutting abscesses etc.

bisulphites *(Chem.)*. *Hydrogen sulphites, hydrogen sulphates (IV)*. Acid salts of sulphurous acid. Useful as preservatives and as a source of sulphur dioxide. See also **sulphurous acid**.

bisynchronous motor *(Elec.Eng.)*. A motor similar to an ordinary synchronous motor but capable of being made to run at twice synchronous speed.

bit *(Build)*. (1) A boring tool which fits into the socket of a brace, by which it is rotated. (2) The cutting-iron of the plane.

bit *(Comp.)*. A digit in binary notation, i.e. 0 or 1. It is the smallest unit of storage. *Binary digIT.*

bit *(Min.Ext.)*. Cutting end of length of drill steel used in boring holes in rock. See **drill bit, diamond bit, roller bit**.

bit *(Telecomm.)*. (1) Abbrev. for *BInary digiT*; a single occurrence in a code, or language, employing only two kinds of characters (e.g. 0 and 1). (2) A unit of **information content** equal to that of a message, the *a priori* probability of which is one-half.

bitch *(Build.)*. A kind of **dog** in which the ends are bent so as to point in opposite directions.

bit gauge *(Build)*. An attachment to a bit which limits drilling or boring to a given depth. Also *bit stop*.

bit-mapped display *(Comp.)*. A display for which an exact image is kept in **main memory** and is changed by changing the memory and then mapping it to the screen.

BITNET *(Comp.)*. *Because It's Time NETwork*. A cooperative network originating in the US, serving over 1300 hosts located at several hundred sites in many countries, mostly in universities. See **NETNORTH, EARN**.

bitter almond oil *(Chem.)*. **Benzaldehyde**; also occurs naturally in *almond oil*.

bittern *(Chem.)*. The residual liquor remaining from the evaporation of sea water, after the removal of the salt crystals.

Bitter pattern *(Phys.)*. Pattern showing boundaries of magnetic domains in ferromagnetic materials, formed by a magnetic powder concentrated along the boundaries.

bittiness *(Build.)*. A defect where particles of grit, dust and other foreign matter have a detrimental effect on the appearance of a finished paint film. This can only be rectified by rubbing down when dry and re-coating with clean material and tools.

bitumen *(Chem.)*. The nonmineralized substances of coal, lignite etc., and their distillation residues.

bitumen varnishes *(Build.)*. These contain asphalts, driers and aliphatic hydrocarbons.

bituminous felt *(Build.)*. A manufactured material incorporating asbestos flax or other fibres, and bitumen, generally about $\frac{1}{8}$ in. (3 mm) thick. It is produced in rolls, is impervious to water, and is largely used for roof coverings and damp proof courses.

bituminous paints *(Build.)*. A range of surface coatings from cheap tar blacks to good quality **black japan**. Made from pitch, asphaltum or coumarone resin these materials possess good anticorrosive properties, moisture and chemical resistance together with good flexibility and durability. The colour range is restricted to black, brown, dark reds and drab greens.

bituminous plastics *(Plastics)*. Compositions made from natural bitumens (e.g. Trinidad and gilsonite), petroleum pitches, or certain types of coal-tar pitch, along with a suitable filler; formerly used for accumulator cases, where nonabsorbent acid-resisting material is required.

bituminous shale *(Geol.,Min.Ext)*. Shale rich in hydrocarbons, which are recoverable as gas or oil by distillation.

bi-uniform correspondence, mapping, transformation *(Maths.)*. See **one-to-one correspondence**.

biuret *(Chem.)*. $NH_2.CO.NH.CO.NH_2$, colourless needles, crystallizes with 1 molecule of H_2O; mp of the anhydrous compound 190°C. It is formed from urea at 150°–170°C with liberation of NH_3. Has a peptide bond (CONH).

biuret reaction *(Chem.)*. The alkaline solution of biuret gives a reddish-violet coloration on the addition of copper (II) sulphate. As biuret is readily formed from urea, this reaction serves to identify the latter.

bivalent *(Biol.)*. One of the pairs of homologous chromosomes present during meiosis.

bivalent *(Chem.)*. See **divalent**.

bivalve *(Zool.)*. Having the shell in the form of two plates, as in Bivalvia.

Bivalvia *(Zool.)*. Class of Mollusca with the body usually enclosed by paired shell valves joined by a hinge and closed by adductor muscles. Ctenidia or gills are used for filter feeding and there are inhalent and exhalent siphons. Mostly sessile aquatic forms. Mussels, Clams, Oysters, Scallops. Also called *Pelecypoda*.

bivariant *(Chem.)*. Having 2 degrees of freedom.

bivoltine *(Zool.)*. having two broods in a year. Cf. **univoltine**.

Bjerknes circulation theorem *(Meteor.)*. For relative motion with respect to the Earth, the rate of change $\frac{dC}{dt}$ of the circulation C along a closed curve always consisting of the same fluid particles is equal to the number N of isobaric – isosteric (i.e. pressure – density) solenoids enclosed by the curve, minus the product of twice the angular velocity of the Earth and the rate of change $\frac{dA}{dt}$ of the area A defined by the projection of the curve on the equatorial plane. If **V** is the velocity of the fluid, **dl** an element of the curve, p the pressure, and ρ the density, then C and N are defined by

$$C = \oint \mathbf{V} \cdot \mathbf{dl}$$

and

$$N = -\oint \frac{dp}{\rho}.$$

Bk *(Chem.)*. The symbol for berkelium.

black *(Eng.)*. Of parts of castings and forgings not finished by machining; it refers to the dark coating of iron-oxide retained by the surface.

black *(Print.)*. A blemish on a printed sheet caused by a space or lead rising to the height of type.

black ash *(Min.Ext.)*. *Soda ash.* Impure sodium carbonate, containing some calcium sulphide and carbon. Important pH regulator in flotation process.

black-band iron-ore *(Min.Ext.)*. A carbonaceous variety of clay-ironstone, the iron being present as carbonate (siderite); occurs in the English Coal Measures.

black bean *(For.)*. A tree of the genus *Castanospermum*, having extremely hard wood, liable to warping and collapse, but naturally resistant to the attack of wood-rotting fungi and to termites. Used for internal fittings and high-class furniture.

black body *(Phys)*. A body which completely absorbs any heat or light radiation falling upon it. A *black body* maintained at a steady temperature is a full radiator at that temperature, since any black body remains in equilibrium with the radiation reaching and leaving it.

black-body radiation *(Phys.)*. Radiation that would be radiated from an ideal *black body*. The energy distribution is dependent only on the temperature and is described by **Planck's radiation law**. See **Stefan-Boltzmann law, Wien's laws.**

black-body temperature *(Phys.)*. The temperature at which a **black body** would emit the same radiation as is emitted by a given radiator at a given temperature. The *black-body temperature* of carbon-arc crater is about 3500°C, whereas its true temperature is about 4000°C.

black box *(Aero.)*. See **flight recorder**.

black box *(Genrl.)*. A generalized colloquial term for a self-contained unit of electronic circuitry; not necessarily black.

black burst *(Image Tech.)*. A signal without picture information but with **sync pulses, colour burst** and black level.

blackbutt *(For.)*. Tree of the genus *Eucalyptus*, giving a hardwood, of interlocked grain and coarse but uniform texture. Colour light brown, with pinkish markings. The wood both looks and feels greasy and yields an abundant supply of oil. Used for heavy-duty flooring, wood blocks, cabinet work and panelling.

black copper *(Eng.)*. Impure metal, carrying some iron, lead and sulphur. Produced from copper ores by blast furnace reduction.

black crush *(Image Tech.)*. Tonal distortion in the picture whereby varying dark tones are all reproduced as black.

black damp *(Min.Ext.)*. Mine air which has lost part of its oxygen as result of fire, and is dangerously high in carbon dioxide. See **fire damp**.

black death *(Med.)*. Popular name for the bubonic/pneumonic plague that swept Europe in the middle of the 14th Century.

black diamond *(Min.)*. A variety of crypto-crystalline massive diamond, but showing no crystal form. Highly prized, for its hardness as an abrasive. Occurs only in Brazil. Also called *carbonado*.

black disease *(Vet.)*. A toxaemic disease of sheep caused by *Clostridium novyi* (Cl. *oedematiens*). Infection is associated with the presence of liver damage caused by immature liver flukes. Controlled by vaccination.

blackening *(Paper)*. A mottled appearance which may occur on the surface when paper of high moisture content is calendered.

black fever *(Med.)*. See **kala-azar, Rocky Mountain fever.**

black frost *(Meteor.)*. An air frost with no deposit of **hoar frost**.

blackhead *(Vet.)*. *Infectious entero-hepatitis, histomoniasis.* An infectious disease affecting turkeys and occasionally domestic and wild fowl, caused by infection of the liver and caecae by the protozoon, *Histomonas meleagridis*.

blackheart *(For.)*. An abnormal black or dark brown coloration that may occur in the heart-wood of certain timbers.

black hole *(Astron.)*. A region of **spacetime** from which matter and energy cannot escape. A black hole could be a star or galactic nucleus which has collapsed in on itself to the point where its **escape velocity** exceeds the speed of light. Some **binary stars** which strongly emit X-rays may have black hole companions. Black holes are also invoked as the energy sources of **quasars**.

blacking *(Foundry)*. Carbonaceous material applied as a powder or wash to the internal surface of a mould to protect the sand and improve the finish of the casting. Prepared in a *blacking mill*.

black iron oxide *(Build.)*. The only inorganic black pigment used on any scale, mainly in anticorrosive and antifouling paints. It consists mainly of natural or artificial iron (II,III) (ferro-ferric) oxide Fe_3O_4.

black jack *(Min.)*. A popular name for the mineral sphalerite or zinc blende.

black japan *(Build.)*. A semitransparent, quick-drying black varnish, based on asphaltum and drying oil.

black lava glass *(Geol.)*. Massive natural glass of volcanic origin, jet black and vitreous.

black lead *(Min.)*. A commercial form of **graphite**.

blackleg *(Vet.)*. *Blackquarter, quarter ill.* An acute infection of cattle and sheep due to *Clostridium chauvoei*; characterized by fever and usually crepitant swelling of the infected muscles and is controlled by vaccination.

black letter *(Print.)*. A style of type including Old English, Ancient Black, Tudor Black and many others.

black level *(Image Tech.)*. Signal level corresponding to zero illumination in the display.

black liquor *(Paper)*. The reagent used for digesting a fibrous raw material at the end of the digestion process.

Blackman theory of specific heat of solids *(Phys.)*. A theory based on the dynamics of a crystal lattice of particles developed by Born and von Kārmān. It is more exact, but much more complicated than the *Debye theory* which treats the crystal as a continuous isotropic medium.

black mortar *(Build.)*. A low strength mortar containing a proportion of ashes. Sometimes used for pointing, where a dark colour is required.

black opal *(Min.)*. Includes all opals of dark tint, although the colour is rarely black; the fine Australian blue opal, with flame-coloured flashes, is typical.

blackout *(Med.)*. Temporary loss of vision, perhaps with loss of consciousness, due to sudden reduction of blood supply to the brain.

black powder *(Min.Ext.)*. Gun powder used in quarry work. Standard contains 75% potassium nitrate, slow 59% and blasting 40%, the balance being charcoal and sulphur.

blackquarter *(Vet.)*. See **blackleg**.

black red heat *(Eng.)*. Temperature at which hot metal is just seen to glow in subdued daylight – about 540°C.

black sand *(Foundry)*. (1) A mixture of sand and powdered coal forming the floor of an iron foundry.

black-step marks *(Print.)*. Collating marks printed on the fold between the first and last pages of each section, as an alternative or addition to the **signature**. The recommended size is 12-points deep by 5-points wide and the position is stepped down for each successive section, giving an immediate visual check on the **gathering**. Sometimes called *back-step*.

black-tongue *(Vet.)*. A disease of dogs, due to a deficiency of nicotinic acid in the diet.

blackwater *(Vet.)*. See **redwater**.

blackwater fever *(Med.)*. *Haemoglobinuric (haematuric) fever*. An acute disease prevalent in tropical regions, especially Africa, with feverishness, bilious vomiting, and passages of red or dark-brown urine. Thought to be related to malignant tertian malaria infection (*Plasmodium falciparum*).

blackwood *(For.)*. Tree of the genus *Acacia*, a native of Australia and Tasmania. It is a hardwood, the grain being either straight, interlocked, or wavy, but the texture is fine and even. Typical uses include furniture-making, interior fittings, high-class joinery, gun stocks, and cooperage.

bladder *(Bot.)*. A small hollow spherical trap for catching and digesting small animals on the bladder-wort, *Utricularia*.

bladder *(Zool.)*. Any membranous sac containing gas or fluid; especially the urinary sac of Mammals.

bladderworm *(Zool.)*. See **cysticercus**.

blade *(Bot.)*. The flattened part of a leaf (the lamina), sepal, petal, or thallus.

blade *(Elec.Eng.)*. The moving part of a knife-switch which carries the current and makes contact with the fixed jaws.

blade activity factor *(Aero.)*. The capacity of an propellor blade for absorbing power, expressed as a nondimensional function of the surface and expressed by the formula

$$AF = \left[\frac{5}{R}\right]^5 \int_{0 \cdot 2R}^{R} cr^3 dr,$$

where R equals diameter, and c equals blade chord at any radius r.

blade angle *(Aero.)*. The angle between blade chord and plane of rotation at any radius. It is not constant because of the higher air speed toward the tip, the incidence being progressively reduced to maintain optimum thrust. Change of blade angle from root to tip is called *blade twist*.

blade loading *(Aero.)*. The thrust of a helicopter rotor divided by the total area of the blades.

blade twist *(Aero.)*. See **blade angle**.

Blagden's law *(Chem.)*. The lowering of the freezing point of a given solvent is proportional to the molar concentration of the solute.

Blaine fineness tester *(Powder Tech.)*. **Permeameter** in which the powder bed is connected to the low pressure side of a U-tube manometer. Air is allowed to flow through the powder bed to equalize the pressure in the two arms of the manometer. The time required for a specified movement of the manometer is used to calculate the surface area.

Blake crusher *(Min.Ext.)*. See **jaw breaker**.

blank *(Acous.)*. The lacquer-coated disc ready for placing on a recording machine for making records with a stylus.

blank *(Eng.)*. A piece of metal, shaped roughly to the required size, on which finishing processes are carried out.

blanket *(Nuc.Eng.)*. (1) Region of **fertile** material surrounding reactor core in which neutrons leaking from the core breed more fissile fuel, e.g. ^{233}U from thorium. (2) Structure surrounding a fusion reactor core within which fusion neutrons are slowed down, heat is transferred to a primary coolant and tritium is bred from lithium.

blanket *(Print.)*. (1) The rubber-covered clothing of the blanket cylinder on an offset press. (2) The rubber, textile, cork, plastic or composition covering of the impression cylinder on a rotary press.

blanket *(Textiles)*. A thick woven, knitted, or non-woven fabric giving good thermal insulation. Traditionally blankets were made from wool fabrics that were milled and raised but cotton materials of an open construction are also in common use.

blanket bog *(Ecol.)*. Bog vegetation forming an extensive and continuous layer of peat over flat and undulating landscape, deriving its mineral nutrients largely from rainfall, and found in cold wet parts of the world.

blanket clamp *(Print.)*. On impression or offset cylinders, the means of locating the fixed end of the blanket.

blanket cylinder *(Print.)*. The rubber-covered cylinder between the plate cylinder and the impression cylinder on an offset press.

blanket pins *(Print.)*. Pins use to attach the blanket to another sheet on the **blanket wind**.

blanket strake *(Min.Ext.)*. Table or sluice with gentle down-slope, with bottom lining of material on which heavy mineral (e.g. metallic gold) is caught. Originally of rough blanket, now usually corduroy with ribs across flow, or chequered rubber.

blanket-to-blanket press *(Print.)*. A **perfector** machine commonly use in **web offset**, whereby two blanket cylinders oppose each other and use each other to obtain printing impression. A sheet of web passing between the two blanket cylinders is thus printed on both sides.

blanket wind *(Print.)*. The means of attaching the loose end of a blanket round the cylinder, usually a rotating **blanket bar**.

blank flange *(Eng.)*. A disk, or solid flange, used to blank off the end of a pipe.

blank groove *(Acous.)*. Unmodulated groove on disk recording.

blanking *(Electronics,Image Tech.)*. (1) Blocking or disabling a circuit for a required interval of time. (2) Suppression of the picture information while the scanning spot of a CRT returns after each line, *horizontal blanking*, or after each field, *vertical blanking*, taking place during the *blanking interval*.

blank wall *(Build.)*. A wall having no opening in it.

blast *(Eng.)*. Air under pressure, blown into a furnace.

-blast *(Genrl.)*. Suffix from Gk. *blastos*, bud.

blastema *(Zool.)*. Anlage; a mass of undifferentiated tissue; the protoplasmic part of an egg as distinguished from the yolk.

blast-furnace *(Eng.)*. Vertical shaft furnace into top of which mineral fuel, and slag-forming rock is charged. Air, sometimes oxygen-enriched and pre-heated, is blown through from below and products are separately tapped (slag higher and metal lower). Used to smelt iron ore, copper and other minerals.

blast-furnace Portland cement *(Build.)*. Cement made by grinding ordinary Portland cement clinker with granulated blast-furnace slag. It has lower setting properties than ordinary Portland cement.

blasting *(Acous.)*. A marked increase in amplitude distortion due to overloading the capacity of some part of a sound-reproducing system; e.g., attempt to exceed 100% depth of modulation in a radio transmitter, or break of continuity in carbon granules in a carbon transmitter.

blasting *(Civ.Eng.)*. The operation of disintegrating rock etc. by boring a hole in it, filling with gunpowder or other explosive charge, and firing it.

blasting fuse *(Civ.Eng.)*. Compound designed to burn at a regulated speed when closed in a tube, used to ignite detonator or explode blasting charge. Types include 'safety' (slow or instantaneous) and detonating.

blast joint *(Min.Ext.)*. The specially made bottom joint of a producing well's **tubing string**, able to withstand the abrasive action of the oil entering from a high-pressure formation.

blast main *(Eng.)*. The main blast air-pipes supplying air to a furnace.

blasto- *(Genrl.)*. Prefix from Gk. *blastos*, bud.

blastochyle *(Zool.)*. Fluid in the blastocoel.

blastocoel *(Zool.)*. The cavity formed within a segmenting ovum: cavity within a blastula: primary body cavity: segmentaion cavity. Also called *archicoel.*

blastocyst *(Zool.)*. *Germinal vesicle.* In Mammalian development, a structure resulting from the cleavage of the ovum; it consists of an outer hollow sphere and an inner solid mass of cells.

blastoderm *(Zool.)*. In eggs with much yolk, the disk of cells formed on top of the yolk by cleavage.

blastodisc *(Zool.)*. In a developing ovum, the germinal area.

blastomere *(Zool.)*. One of the cells formed during the early stages of cleavage of the ovum.

blastopore *(Zool.)*. The aperture by which the cavity of the gastrula retains communication with the exterior.

blastospore *(Bot.)*. A spore produced by budding.

blast pipe *(Eng.)*. Device located in the smoke box of a steam locomotive used to improve the draft through the fire-tubes. Exhaust steam passing from the nozzle of the blast pipe reduces smoke-box pressure and induces the draft.

blastula *(Zool.)*. A hollow sphere, the wall of which is composed of a single layer of cells, produced as a result of the cleavage of an ovum. Also *blastodermic vesicle, blastosphere.*

blastulation *(Zool.)*. A form of cleavage resulting in the production of a blastula.

blast wave *(Phys.)*. See **shock wave**.

Blatthaller loudspeaker *(Acous.)*. First electrodynamically driven loudspeaker used for high-quality sound reproduction. It has a flat surface which is large compared to the wavelength of the radiated sound and it generates sound of high intensity.

Blavier's test *(Elec.Eng.)*. A method of locating a fault on an electric cable; resistance measurements are taken with the far end of the cable free, and again with it earthed.

blazar *(Astron.)*. A type of extremely luminous extragalactic object, similar to a **quasar** except that the optical spectrum is almost featureless.

blaze *(Surv.)*. Temporary survey mark, such as slash on tree trunk, to guide prospector or explorer.

blazer cloth *(Textiles)*. An all-wool milled and raised fabric used in the manufacture of jackets.

bleach *(Image Tech.)*. Conversion of a developed silver image into a white or colourless compound, often a silver halide; also the chemical solution used for this process.

bleach-fix *(Image Tech.)*. Processing solution which both bleaches a silver image to silver halide and dissolves it away, as in **fixing**.

bleaching *(Build.)*. (1) A paint film defect resulting in a whitening of the film due to exposure to light or chemical agents. (2) A preparatory treatment for wood to remove exposure stains or to balance variations in natural colour prior to staining or varnishing.

bleaching *(Image Tech.)*. The removal of reduced silver after development, so that the remaining silver halide, which has not been developed because of its insufficient exposure to light, can be further developed. The resulting image is a positive.

bleaching *(Paper)*. That part of the pulp purification process intended to bring the raw material to the desired whiteness. Bleaching may be achieved by oxidation methods (e.g. using free and/or combined chlorine) or by reduction.

bleaching *(Textiles)*. A series of wet processes for removing residual impurities, colour and fatty or waxy substances from fibres, yarns, or fabrics. This improves the whiteness and promotes brighter colours after dyeing or printing. Hydrogen peroxide is often used for this purpose.

bleaching powder *(Chem.)*. Commercial bleaching powder is obtained from calcium hydroxide and chlorine. The commercial value depends on the amount of available chlorine.

bleach-out process *(Image Tech.)*. A system of colour printing involving the decolorizing of dyes by exposing them through transparencies.

bleb *(Med.)*. A small vesicle containing clear fluid.

bleed *(Print.)*. (1) Illustration whose edges are cut away after printing are said to *bleed off.* (2) The accidental mutilation of type matter by trimming the paper too closely. (3) The running of printing ink on posters and or packages as a result of weather or chemical action.

bleeder resistor *(Elec.Eng.)*. Resistor placed across secondary of transformer to regulate its response curve, especially when the transformer is not loaded with a proper terminating resistance. One placed in a power supply or rectifier circuit to control its regulation.

bleeder type spray gun *(Build.)*. A spray gun without an air valve. Operation of the trigger controls the flow of paint into the air stream which flows or 'bleeds' constantly through the gun.

bleeding *(Aero.)*. The tapping of air from a gas turbine compressor (1) to prevent **surging** or (2) to feed some other equipment, e.g. cabin pressurization or a de-icing system.

bleeding *(Bot.)*. The exudation of xylem sap, phloem sap or latex from wounds. See also **root pressure**.

bleeding *(Build.)*. A paint defect which occurs when some constituent in the underlying surfaces discolours subsequent surface coatings. It is necessary to apply a buffer or barrier coat to the offending surface to isolate the bleeding constituent before subsequent coatings are applied.

bleeding *(Eng.)*. (1) A method of improving the thermal efficiency of a steam plant by withdrawing a small part of the steam from the higher-pressure stages of a turbine to heat the feed-water. (2) Removing undesirable entrapped air from a hydraulic (e.g. braking) system.

bleeding *(Image Tech.)*. Diffusion of dye from an image.

bleeding *(Textiles)*. In fibres, yarns or fabrics of two or more colours, the running of the darker colours, and consequent staining of the lighter colours, during finishing, washing or solvent cleaning.

bleed line *(Min.Ext.)*. Used in the **blow-out preventer stack** to remove excess gas pressure.

blend *(Textiles)*. An intimate mixture of different qualities or kinds of natural or man-made staple fibres.

blender *(Build.)*. A signwriting brush with a short square sable or ox filling used for blending colours in shading effects.

blende (zinc blende) *(Min.)*. See **sphalerite**.

blending *(Build.)*. A term used in graduation from one colour to another. This effect may be created using a blender, a hair stippler or a spray gun.

blennorrhagia *(Med.)*. Discharge of mucus, usually from the genital organs, due to gonorrhoea.

blennorrhoea *(Med.)*. See **blennorrhagia**. *Acute blennorrhoea,* purulent conjunctivitis, due usually to infection with the gonococcus.

blephar-, blepharo- *(Genrl.)*. Prefix from Gk. *blepharon,* eyelid.

blepharism *(Med.)*. Spasm of the eyelids.

blepharitis *(Med.)*. Chronic inflammation of the eyelids.

blepharoplast *(Biol.)*. See **basal body** (1).

blepharoplegia *(Med.)*. Paralysis of the muscles of the eyelid resulting in drooping of the upper eyelid *blepharoptosis.*

blepharospasm *(Med.)*. Spasm of the orbicular muscle of the eyelid.

BLEU *(Aero)*. The *Blind Landing Experimental Unit* operated by the Royal Aircraft Establishment which developed a fully automatic blind landing system. See **autothrottle** and **flight director**.

B-licence *(Aero.)*. Commercial pilot's licence.

blight *(Bot.)*. The common name for a number of plant diseases characterized by the rapid infection and death of the leaves or the whole plant, e.g. potato late blight (caused by the fungus *Phytophthora infestans*).

blimp *(Aero.)*. Colloquial for *nonrigid airship*.
blimp *(Image Tech.)*. A noise-reducing enclosure for a film camera.
blind apex *(Min.Ext.)*. Outcrop of mineral vein or lode where ore deposit dies out before reaching the surface (*suboutcrop*), leaving an apparently barren deposit.
blind arcade *(Build.)*. See **arcade**.
blind arch *(Arch.,Civ.Eng.)*. A closed arch which does not penetrate the structure; used for ornamentation, to make one face of a building harmonize with another in which there are actual arched openings.
blind area *(Build.)*. A sunken space round the basement of a building, broken up into lengths by small cross-walls, which support the earth-retaining wall but restrict ventilation.
blind blocking *(Print.)*. Applying lettering or design to a book cover without gold leaf or alternative. See **blocking**.
blind flying, blind landing *(Aero.)*. The flying and landing of an aircraft by a pilot who, because of darkness or poor visibility, must rely on the indication of instruments. See **instrument landing system, ground-controlled approach**.
blind flying instruments *(Aero.)*. A group of instruments, often on an individual central panel, essential for blind flying. Commonly airspeed indicator, altimeter, vertical speed, turn-and-slip, artificial horizon, directional gyro. See **basic six, basic T**.
blind image *(Print.)*. Term applied in lithographic printing to a plate image which loses its affinity for ink.
blinding *(Civ.Eng.)*. The process of sprinkling small chippings of stone over a tar dressed road surface.
blinding *(Print.)*. See **blind tooling**.
blind lode *(Min.Ext.)*. *Blind vein*. A lode which does not outcrop to the surface.
blind mortise *(Build.)*. A mortise which does not pass right through the piece in which it is cut.
blind page *(Print.)*. A page which has no printed folio, but is included in the pagination; usually found in the preliminary matter.
blind rivet *(Eng.)*. A type of rivet which can be clinched as well as placed by access to one side only of a structure. Usually based on a tubular or semitubular rivet design, e.g. **Chobert rivet, explosive rivet**.
blind spot *(Telecomm.)*. Point within normal range of a transmitter at which field-strength is abnormally small. Usually results from interference pattern produced by surrounding objects, or geographical features, e.g. valleys.
blind spot *(Zool.)*. In Vertebrates, an area of the retina where there are no visual cells (due to the exit of the optic nerve) and over which no external image is perceived.
blind staggers *(Vet.)*. See **alkali disease**.
blind tooling *(Print.)*. Applying lettering or design to the cover of a book using a press with a heated relief die. Gold leaf is used for special work only; **blocking foil** for **edition binding**. The platen may be horizontal and the press hand-operated or partly mechanized, but for long runs a press with a vertical platen and automatic feeding of case and foil(s) is used.
blind vein *(Min.Ext.)*. See **blind lode**.
blink comparator *(Astron.)*. An instrument in which two photographic plates of the same region are viewed simultaneously, one with each eye, any difference being detected by a device which alternately conceals each plate in rapid succession.
blinking *(Radar)*. Modification of a loran transmission, so that a fluctuation in display indicates incorrect operation.
blip *(Radar)*. Spot on CRT screen indicating radar function.
blister *(Acous.)*. A defect in a gramophone record consequent on the release of gases (e.g. water vapour) during pressing.
blister *(Build.)*. See under **blistering**.
blister *(Eng.)*. A raised area on the surface of solid metal produced by the emanation of gas from within the metal while it is hot and plastic.
blister *(Med.)*. A thin-walled circumscribed swelling in the skin containing clear or blood-stained serum; caused by irritation.
blister *(Vet.)*. An irritant drug applied to the skin to cause inflammation and assist in the healing of deep-seated diseased tissues by reflex nervous action.
blister bar *(Eng.)*. Wrought-iron bars, impregnated with carbon by heating in charcoal. Used in making crucible steel.
blister copper *(Eng.)*. An intermediate product in the manufacture of copper. It is produced in a converter, contains 98.5–99.5% of copper, and is subsequently refined to give commercial varieties, e.g. *tough pitch, deoxidized copper*.
blistering *(Build.)*. A defect in a finished paint film where areas of film rise or blister away from the underlying surface. The causes are lack of adhesion, trapped moisture or resin and heat.
blister steel *(Eng.)*. Wrought-iron bars impregnated with carbon by heating in charcoal. Before 1740 this was the only steel available. Since then, most blister steel has been melted to give crucible steel, most of which is now made, however, from other materials.
blivet *(Aero.)*. Flexible bag for transporting fuel, often slung beneath a helicopter.
bloat *(Vet.)*. *Dew-blown, hoven*. An acute digestive disorder of ruminants in which excessive amounts of gas accumulate in the rumen, associated usually with the ingestion of lush herbage.
Bloch band *(Phys.)*. See **band theory of solids**.
Bloch function *(Phys.)*. The electrons in a crystalline solid move under the influence of a periodic potential function. The solutions of the *Schrödinger equation* must therefore include a factor which has the same periodicity as the potential; these are called *Bloch functions*. See **band theory of solids**.
block *(Comp.)*. Group of records treated as a complete unit during transfer to or from backing.
block *(Eng.)*. The housing holding the pulley or pulleys over which the rope or chain passes in a lifting tackle.
block *(Min.Ext.)*. Rectangular panel of ore defined by drives, raises, and winzes, giving all-round physical access for sampling, testing, and mining purposes.
block *(Print.)*. A term applied to any letterpress printing plate, duplicate or original, brought to type height by mounting.
blockboard *(Build.)*. Board composed of softwood strips bonded together and sandwiched between two outer layers of veneer whose grain runs across them. Also *battenboard, coreboard, laminboard*.
block brake *(Eng.)*. A vehicle brake in which a block of cast-iron is forced against the rim of the revolving wheel, either by hand-power, electromagnetic mechanism, or fluid-pressure acting on a piston. See **air brake, electromagnetic brake**.
block caving *(Min.Ext.)*. Method of mining in which block of ore is undercut, so that it caves in and the fragments gravitate to withdrawal points.
block clutch *(Eng.)*. A friction clutch in which friction blocks or shoes are forced inwards into the grooved rim of the driving member, or expanded into contact with the internal surface of a drum. See **friction clutch**.
block coefficient *(Ships)*. The ratio of the underwater volume of a ship to the volume of an enclosing rectangular block.
block diagram *(Comp.)*. Diagram made up of squares and rectangles representing different hardware and software components with lines to show their interconnections.
block diagram *(Eng.)*. An illustration in which parts of a machine or a process are represented notionally by blocks or similar symbols.
blocked impedance *(Phys.)*. That of the input of a transducer when the output load is infinite, e.g., when the mechanical system, as in a loudspeaker, is prevented from moving.
blocked-out ore *(Min.Ext.)*. See **block**.
block gauge *(Eng.)*. A block of hardened steel having its opposite faces accurately ground and lapped flat and parallel. The faces are separated by a specified distance, the *gauge distance*, used for checking the accuracy of other gauges etc.

block-in-course *(Build.)*. A type of masonry, used for heavy engineering construction, in which the stones are carefully squared and finished to make close joints, and the faces are hammer-dressed.

blocking *(Build.)*. The operation of securing together two pieces of board by gluing blocks of wood in the interior angle.

blocking *(Electronics)*. Cutoff of anode current in a valve because of the application of a high negative voltage to the grid; used in **gating** or **blanking**.

blocking *(Print.)*. Applying lettering or design to the cover of a book using a press with a heated relief die. See **blocking foil**.

blocking action *(Meteor.)*. The effect of a well established, extensive high-pressure area in blocking the passage of a **depression**.

blocking antibody *(Immun.)*. Antibody which combines preferentially with an antigen so as to prevent it from combining with IgE on mast cells, and thereby prevents type I allergic reactions. Blocking antibodies are usually IgG, and the aim of hyposensitization treatment of acute allergies is to stimulate blocking antibody production.

blocking capacitor *(Elec.Eng.)*. One in signal path to prevent d.c. continuity. Also called **buffer capacitor**.

blocking course *(Build.)*. A course of stones laid on the top of a cornice.

blocking factor *(Comp.)*. Number of records in a block.

blocking foil *(Print.)*. A film base with a layer of gold, other metal, or coloured material, used as a substitute for gold leaf. See **blocking**.

blocking-out *(Image Tech.)*. The use of Indian ink or other opaque pigment for covering parts of negatives so that they print white.

blocking-out *(Min.Ext.)*. Method of estimating tonnage of mineral reserves in a volume established by drilling, with reference to the grade of the representative sample.

blocking press *(Print.)*. See **blocking**.

block lava *(Geol.)*. See under **ropy lava**.

block plan *(Build.)*. A plan of a building site, showing the outlines of existing and proposed buildings.

block plane *(Build.)*. A small plane about 6 in (150 mm) long which has no cap iron and has the cutting-bevel reversed; used for planing end grain, and fine work.

block prism *(Image Tech.)*. A cube of glass, slit along its diagonal and half-silvered, for splitting the beam in a 3-colour beam-splitting camera.

block section *(Civ.Eng.)*. The length of track in a railway system that is limited by stop signals.

block system *(Civ.Eng.)*. The system of controlling the movements of trains by signals and by independent communication between block posts, where are situated the instruments indicating the position of trains, condition of the block sections, and controlling levers for signals, points etc. It is *absolute* if one train alone is permitted within a block section, and *permissive* if trains are allowed to follow into a block section already occupied by a train.

block time *(Aero.)*. The time elapsed from the moment an aircraft starts to leave its loading point to the moment when it comes to rest; also *chock-to-chock*, *buoy-to-buoy* (seaplanes) and *flight time*. It is an important factor in airline organization and scheduling.

block tin *(Build.)*. Pure tin.

blockwork *(Build.)*. Walling constructed of pressed or cast blocks with a basic constituent of cement, the other constituents being generally to improve the insulating qualities of the wall, e.g. clinker ash, framed blast furnace slag or an air entraining agent.

blonde *(Image Tech.)*. TN for a 2 kW variable-beam floodlight for studio use.

Blondel-Rey law *(Phys.)*. Used to assess the apparent point brilliance B of a flashing light. $B = B_0\ [f/(a + f)]$, where B_0 is the point brilliance during the flash, f is the duration of the flash in seconds and a is constant of value about 0.2 secs, and B is near the threshold value for white light. The law holds for flashing frequencies of less than 5 Hz.

blondin *(Civ.Eng.)*. See **cable-way**.

blood *(Med.)*. A fluid circulating through the tissues of the body, performing the functions of transporting oxygen, nutrients, and hormones, and carrying waste products to the organs of excretion. It plays an important rôle in maintaining a uniform temperature in the body in warm-blooded organisms. Its relative density in Man is about 1.054–1.060, and it has an alkaline reaction. Its chief constituents are: red cells, white cells, platelets, water (77.5–79%), solids including proteins, lipids, enzymes, hormones and immune bodies, blood sugar, vitamins and inorganic substances. See **ABO blood group system, rhesus blood group system**.

blood albumin *(Med.)*. See **serum albumin**.

blood cell *(Med.)*. See **haematoblast**.

blood-clotting factors *(Med.)*. An internationally agreed scale of discernible factors concerned in blood-clotting; indicated by roman numerals, e.g. I, fibrinogen factor; II, prothrombin factor; III, thromboplastin factor; VIII, antihaemophilic factor A; IX, antihaemophilic factor B (Christmas factor), etc.

blood corpuscle *(Med.)*. A cell normally contained in suspension in the blood. See **erythrocyte, leucocyte**.

blood count *(Med.)*. The number of red or white corpuscles in the blood.

blood donors *(Med.)*. Volunteers who donate their blood for administration to others.

blood flukes *(Zool.)*. Trematodes of the genus *Schistosoma*, parasitic on man and domestic animals *via* various species of water snail as intermediate host. They can attack the liver, spleen, intestines, and urinary system, and occasionally the brain. See **bilharziasis**.

blood groups *(Med.)*. See **ABO blood group system, rhesus blood group system**.

blood islands *(Zool.)*. In developing Vertebrates, isolated syncytial accumulations of reddish mesoderm cells containing primitive erythroblasts, which give rise respectively to the walls of the blood vessels and to the red corpuscles.

blood plasma *(Med.)*. See **plasma**.

blood pressure *(Med.)*. The pressure of blood in the arteries, usually measured by sphygmomanometry. The maximum occurs in *systole* and the minimum in *diastole*. Normal young adults will have a blood pressure of approximately 120/80 mmHg.

blood red heat *(Eng.)*. Dark red glow from heated metal, in temperature range 550°–630°C.

blood serum *(Med.)*. See **serum**.

bloodstone *(Min.)*. Cryptocrystalline silica, a variety of chalcedony, coloured deep-green, with flecks of red jasper. Also *heliotrope*.

blood substitutes *(Med.)*. Plasma, albumin and dextran can be used to substitute volume for loss of blood. Newer substances are being investigated which may be able to transport oxygen.

blood sugar *(Med.)*. The level of glucose in the blood, normally between 3.2–5.2 mmol/l in the fasting state.

blood transfusion *(Med.)*. See **transfusion**.

blood vessel *(Med.)*. An enclosed space, with well-defined walls, through which blood passes. See **artery, capillary, vein**.

bloom *(Bot.)*. (1) A covering of waxy material occurring on the surface of some leaves and fruits, resulting in a whitish cast. (2) A visible, often seasonal occurrence of very large numbers of algae in the plankton of fresh water or sea. See **red tide**.

bloom *(Build.)*. (1) Efflorescence on a brick wall. (2) For paint, see **blooming**.

bloom *(Chem.)*. The colour of the fluorescent light reflected from some oils when illuminated. The colour is usually different from that shown by the oil with transmitted light.

bloom *(Eng.)*. Semi-finished metal, rectangular in cross section and for steel not more than twice as long as it is thick. Cf. **billet**.

bloom *(Glass)*. (1) Obsolete term for the coating of thin dielectric layers on a lens to alter its reflectance properties. (See **coated lens**). (2) A surface film caused by weathering. (3) A surface film of sulphites and sulphates formed during the annealing process.

bloom *(Min.Ext.)*. Efflorescence of altered metallic salt at surface of ore exposure, e.g. cobalt bloom.

blooming *(Build.)*. A paint defect affecting gloss finish materials such as varnish enamel and gloss paint. This defect is normally attributable to moisture, often in vapour or fine droplet form, affecting the film during the drying process.

blooming *(Electronics)*. (1) Spread of spot on CRT phosphor due to excessive beam current. (2) Coating of dielectric surfaces to reduce reflection of EM waves.

blooming *(Image Tech.)*. Treatment of the glass-air surfaces of a lens with a deposit of magnesium fluoride or other substance, which reduces internal reflection and increases light transmission.

blooming mills *(Eng.)*. The rolling mills used in reducing steel ingots to blooms. Called *cogging mills* in UK, and not always distinguished from **billet (slab) mills**.

Bloom's syndrome *(Biol.)*. Human clinical syndrome in which frequent chromosomal rearrangements involving homologous chromosomes occur.

bloop *(Image Tech.)*. Dull thud in sound-film reproduction caused by a join in the sound-track.

blooping patch *(Image Tech.)*. Triangular black patch applied or painted over a splice in the sound track of a film print to prevent it causing a noise on projection.

blotting, blot *(Biol.)*. The method by which biological molecules are transferred from, usually, a gel to a membrane filter. In the former they can be separated physically by, for example, size and in the latter they can be tested for the presence of specific sequences by radioactive or similar **probes**. See **Southern blot**, **immunoblot**.

blotting paper *(Paper)*. Weak, free beaten, unsized paper intended for the absorption of aqueous inks from the surface of documents.

blow *(Eng.)*. In Bessemer converter, passage of air through molten charge.

blow *(Min.Ext.)*. (1) Ejection of part of the explosive charge (unfired) from hole. (2) Sudden rush of gas from coal seam or ore body.

blow-and-blow machines *(Glass)*. Machines in which the glass is shaped in two stages, but each time by blowing, as opposed, for example, to pressing or sucking.

blow back *(Autos.)*. The return, at low speeds, of some of the induced mixture through the carburettor of a petrol-engine; due to late closing of the inlet valve during compression, or by worn or sticking valves.

blowdown stack *(Min.Ext.)*. Container into which refinery vessels can be emptied during emergencies and into which steam or water can be injected to prevent ignition of volatile components.

blower *(Eng.)*. (1) A rotary air-compressor for supplying a relatively large volume of air at low pressure. See **air compressor**, **supercharger**. (2) A ring-shaped perforated pipe, encircling the top of the **blast pipe** in the smokebox, to which steam is supplied while the engine is standing, the jets providing sufficient draught to keep the fire going.

blower *(Min.Ext.)*. (1) A fissure or thin seam which discharges a quantity of coal gas. (2) An auxiliary ventilating appliance, e.g. a fan or venturi tube, for supplying air to subsidiary working places or to dead-ends.

blowfly myiasis *(Vet.)*. *Strike*. Infestation of the skin of sheep, especially in the breech region, by blowfly larvae.

blow-hole *(Geol.)*. An aperture near a cliff-top through which air, compressed in a sea-cave by breaking waves, is forcibly expelled.

blowholes *(Build.)*. Pock marks existing on the surface of *in situ* or precast concrete when the shuttering is removed. Due to air entrapped during the process of placing the concrete and not released normally due to inadequate vibration.

blowholes *(Eng.)*. Gas-filled cavities in solid metals. They are usually formed by the trapping of bubbles of gas evolved during solidification (see **gas evolution**), but may also be caused by steam generated at the mould surface, air entrapped by the incoming metal, or gas given off by inflammable mould dressings.

blowing *(Build.)*. A plastering defect in which a conical piece may be blown out of a finished plastered surface owing to moisture getting to an imperfectly slaked particle of quicklime in the work. Also called *pitting*.

blowing a well *(Min.Ext.)*. Temporary removal of pressure at the well-head to allow tubing and casings to be blown free of debris, water etc.

blowing current *(Elec.Eng.)*. A term used in connection with fuse links to denote the current (dc or rms) which will cause the link to melt.

blowing engine *(Eng.)*. The combined steam- or gas-engine and large reciprocating air-blower for supplying air to a blast-furnace.

blowing-iron *(Glass)*. See **blowpipe**.

blowing-out *(Eng.)*. The operation of stopping down a blast-furnace.

blowing road *(Min.Ext.)*. In colliery, the main ventilation ingress.

blowing room *(Textiles)*. In a cotton-spinning mill, the room containing the bale breakers, openers, and scutchers, in which revolving beaters and exhaust fans remove motes and dust from the fibres.

blow moulding *(Plastics)*. Process for 1-piece articles, e.g. bottles, which combines extrusion and blowing into a split mould.

blown *(Autos.)*. (1) Supercharged. See **boost**. (2) Of cylinder head or manifold gaskets, having sprung a leak under pressure.

blown casting *(Foundry)*. Casting spoilt by porosity.

blown flap *(Aero.)*. A **flap**, the efficiency of which is improved by blowing air or other gas over its upper surface to maintain attached airflow even at high angles of deflection.

blown oil *(Chem.)*. Linseed oil treated by heating and aeration. Used in the manufacture of linoleum.

blown sand *(Geol.)*. Sand which has suffered transportation by wind, the grains in transit developing a perfectly spherical form (*millet-seed sand*); grain-size is dependent upon the wind velocity. See also **sand dunes**.

blowout *(Min.Ext.)*. The sudden eruption of gas and oil from a well, which then has to be controlled before drilling can recommence. Also *wild well*.

blowout coil *(Elec.Eng.)*. See **magnetic blowout**.

blowout magnet *(Elec.Eng.)*. A permanent or an electromagnet used to extinguish more rapidly the arc (in a switch, etc.) caused by breaking an electric circuit.

blowout preventer *(Min.Ext.)*. The stack of heavy-duty valves attached to the casing at a well head, designed to control the pressure in the bore. Includes the **kelly**. Also *Christmas tree*.

blowpipe *(Glass.)*. A metal tube, some 2 m in length, with a bore of 2–4 mm and a thickened nose which is dipped into molten glass and withdrawn from the furnace. The glass is subsequently manipulated on the end of the blowpipe and blown out to shape. Also *blowing-iron*.

blow-up *(Image Tech.)*. Enlarging the image photographically, especially from one gauge of film to a larger one.

blow-up *(Print.)*. Enlargement of copy, artwork, photographic and text.

blub *(Build.)*. A swelling on the surface of newly plastered work.

blubber *(Zool.)*. In marine Mammals, a thick fatty layer of the dermis.

blue asbestos *(Min.)*. Crocodilite, a fibrous variety of riebeckite. The best known occurrences are in South Africa.

blue-backing shot *(Image Tech.)*. Scene in which the foreground action is shot against a uniform blue background for combination with another scene, either by *travelling matte* (Cine) or *colour separation overlay* (TV).

blue billy *(Eng.)*. The residue left after burning off the sulphur from iron sulphide ores.

blue bricks *(Build.)*. Bricks of high strength and durability, highest quality engineering bricks.

blue brittleness *(Eng.)*. Temperature effect due to increase in pearlite in steel between 205° and 315°C. Both brittleness and tensile strength rise.

blue comb *(Vet.)*. See **avian monocytosis**.

blue-glass lamp *(Image Tech.)*. An incandescent lamp with a coloured bulb giving an approximation to daylight illumination.

blue-green algae, blue-green bacteria *(Bot.)*. See **Cyanophyceae**.

blue ground *(Geol.,Min.Ext.)*. *Peridotite, Kimberlite*. Decomposed agglomerate, usually found as a breccia and occurring in volcanic pipes in S. Africa and Brazil; it contains a remarkable assemblage of ultramafic plutonic rock-fragments (many of large size) and diamonds.

blue gum *(For.)*. A strong, brownish-coloured wood from Australia, used for piles, heavy framing etc.

blueing *(Eng.)*. The production of a blue oxide film on polished steel by heating in contact with saltpetre or wood ash; either to form a protective coating, or incidental to annealing.

blueing *(Horol.)*. Hands, screws etc. need only have the appearance of blueing, but for springs, where it is necessary to obtain the desired elastic properties, it must be produced by thermal treatment.

blueing *(Textiles)*. The process of neutralizing a yellowish tint in fabric by adding a blue colour, in order to obtain a better white appearance.

blueing salts *(Eng.)*. Caustic solution of sodium nitrate, used hot to produce a blue oxide film on surface of steel.

blue john *(Min.)*. A massive, blue and white variety of the mineral fluorite, occurring near Castleton, in Derbyshire, and worked for ornamental purposes.

blue key *(Print.)*. Blue images produced photographically on film from an assembly of the key negatives or positives. Used as a guide in the film assembly of separation negatives or positives required to process colour printing.

blue lead *(Chem.)*. A name used in the industry for metallic lead, to distinguish it from other lead products such as white lead, orange lead, red lead etc.

blue-line key *(Print.)*. Used when making sets of lithographic plates for colour printing. A key image of each page in position is prepared on Astrafoil using a blue dye, and the appropriate positives or negatives for each page are patched up in register with the key. When printed down to make a plate, the blue does not affect the result.

blue metal *(Eng.)*. Condensed metallic fume resulting from distillation of zinc from its ore concentrates. Blue tint is due to slight surface-oxidation of the fine particles.

blue nose disease *(Vet.)*. A form of photosensitization occurring in the horse, characterized by oedema and purplish discoloration around the nostrils, nervous symptoms, and sometimes sloughing of unpigmented skin.

blue of the sky *(Meteor.)*. Sunlight is 'scattered' by molecules of the gases in the atmosphere and by dust particles. Since this scattering is greater for short waves than for long waves, there is a predominance of the shorter waves of visible light (i.e., blue and violet) in the scattered light which we see as the blue of the sky.

blueprint *(Image Tech.)*. Process for reproducing plans and engineering drawings, based on the principle that ferric salts are reduced to ferrous when exposed to light; Also *cyanotype*.

blueprint paper *(Paper)*. A paper coated with a solution of potassium ferricyanide and ammonium ferric citrate bound together with gelatine or gum arabic.

blue schist *(Geol.)*. A metamorphic rock formed under conditions of high pressure and relatively low temperature. Characteristic minerals are glaucophane and kyanite.

blue stain *(For.)*. A form of sapstain producing a bluish discoloration; caused by the growth of fungi which, however, do not greatly affect the strength of the wood.

bluestone *(Chem.)*. See **copper (II) sulphate**.

bluetongue *(Vet.)*. *Malarial catarrhal fever of sheep*. A febrile disease of sheep and cattle, occurring in parts of Africa and the Near East, caused by a virus and transmitted by mosquitoes; characterized by haemorrhagic inflammation of the buccal mucosa and cyanosis and swelling of the tongue.

blue vitriol *(Min.)*. The mineral chalcanthite, hydrated copper suphate.

blunt-ended DNA *(Biol.)*. DNA cleaved straight across the double stranded molecule, without forming any single stranded ends. An effect of some **restriction enzymes**.

blunt start *(Eng.)*. End of a threaded screw which is rounded or coned to facilitate insertion.

blurb *(Print.)*. A short note by the publisher recommending a book or its author. Usually printed on the dust-jacket or at the beginning of the preliminary matter.

blushing *(Build.)*. A condition in which a cloudy film appears on a newly lacquered surface; due usually to too rapid drying or to a damp atmosphere.

B-lymphocyte *(Immun.)*. Lymphocytes derived from precursors in the bone marrow (or in birds the *Bursa of Fabricius*, a tissue budding off from the hind gut) which do not undergo differentiation in the thymus. They make immunoglobulins, which are present at the cell surface and act as specific receptors for antigens and, when stimulated, B-lymphocytes manufacture and secrete large amounts of their characteristic immunoglobulin into the circulation. This constitutes the antibody response. Stimulation by **Thymus-dependent antigens** requires cooperation of helper T-lymphocytes, which release B-cell growth and differentiation factors, whereas stimulation by thymus-independent antigens does not need such cooperation. Some **mitogens** such as *pokeweed mitogen* can stimulate B-cells irrespective of the antigen specificity of their Ig receptors. B-lymphocytes are produced continuously throughout life. Most die soon unless stimulated by antigen. Following such stimulation they either differentiate towards antibody secretion or become **B-memory cells**.

BM *(Surv.)*. Abbrev. for *Bench Mark*.

B-memory cell *(Immun.)*. A resting B-cell which is derived from a B-cell which has been stimulated by a specific antigen in a **germinal centre** so as to multiply without going on to secrete antibody. B-memory cells live and recirculate through the blood between **lymphoid tissues** for many weeks. When they encounter the appropriate antigen again, with T-cell help, they rapidly differentiate into antibody secreting cells (or into more B-memory cells). This is the basis of a secondary or booster antibody response. One of the aims of prophylactic immunization is to elicit B-memory cells.

BMEP *(Eng.)*. Abbrev. for *Brake Mean Effective Pressure*.

BMEWS *(Radar)*. Abbrev. for *Ballistic Missile Early Warning System*. An *over the horizon* radar system for the detection of inter-continental ballistic missiles, with linked sites in the UK, Alaska and Greenland.

BNA *(Med.)*. *Basle Nomina Anatomica*, an international anatomical terminology accepted at Basle in 1895 by the Anatomical Society to standardize terms used in describing parts of the anatomy. Since 1955, *Nomina Anatomica*.

BNF *(Comp.)*. *Backus-Naur Form*. Notation for defining the syntax of a programming language.

BNFL *(Nuc.Eng.)*. British Nuclear Fuels plc. Organisation involved in uranium enrichment, fabrication of fuel elements, reprocessing of irradiated nuclear fuel and production of plutonium. It also operates reactors at Chapelcross and at Calder Hall.

board *(For.)*. Timber cut to a thickness of less than 2 in (50 mm), and to any width from 4 in (100 mm) upwards.

board-and-brace work *(Build.)*. Work consisting of boards grooved along both edges, alternating with thinner boards fitting into the grooves.

boarding *(Textiles)*. Heat-treatment of knitted garments, especially nylon stockings, on a former in order to give the desired shape and size.

Board of Trade Unit *(Elec.Eng.)*. The commercial unit of electrical energy, equal to 1 kilowatt-hour.

boards *(Paper)*. Stiff, thick paper, generally of 220 or 250 g/m^2 or more.

boards *(Print.)*. A general term for millboards, strawboards, etc., used for book-covers.

boart *(Min.)*. See **bort**.

boasted ashlar *(Build.)*. See **chiselled ashlar**.

boasted joint surface *(Build.)*. The surface of a stone which has been worked over with a boasting chisel until it is covered with a series of small parallel grooves, thus forming a key for the mortar at the joint.

boasted work *(Build.)*. See **drove work**.

boaster *(Build.)*. See **boasting chisel**.

boasting *(Build.)*. The operation of dressing stone with a broad chisel and mallet.

boasting chisel *(Build.)*. A steel chisel having a fine broad cutting edge, 2 in (50 mm) wide; used for preparing a stone surface prior to finish-dressing with a broad tool. Also *boaster*.

boat *(Chem.)*. The conformation of a six-membered ring in which atoms 1, 2, 4 and 5 are essentially co-planar, while atoms 3 and 6 extend on the same side of the plane. It is less stable than the **chair** conformation.

boat scaffold *(Build.)*. See **cradle scaffold**.

bobbin *(Elec.Eng.)*. A flanged structure intended for the winding of a coil. Also called a *spool*.

bobbin *(Textiles)*. A light spool on which **slubbings**, **roving** or yarn is wound ready for the next process. A weft bobbin or *pirn* is loaded with yarn suitable for use as the weft of woven fabrics. A brass bobbin is formed from two brass disks rivetted together to carry the threads used in the manufacture of lace e.g. on **Leavers machines**. Bobbin lace is a hand-made lace produced from threads fed from small bobbins.

bobbin winding *(Elec.Eng.)*. A term used to denote a transformer winding in which all the turns are arranged on a bobbin, as opposed to a winding in which the turns are in the form of a disk. Generally used for the high-voltage windings of small transformers.

bob-weight *(Eng.)*. A weight used to counterbalance some moving part of a machine. See **balance weights**.

BOD *(Chem.)*. Abbrev. for *Biological Oxygen Demand*.

Bode's law *(Astron.)*. A numerical relationship linking the distances of planets from the Sun, discovered by J. Titius (1766) and published by J. Bode (1772). The basis of this relationship is the series 0,3,6,12,...,384, in which successive numbers are obtained by doubling the previous one. If 4 is added to create the new series 4,7,10,16,...388, the resulting numbers correspond reasonably with the planetary distances on a scale with the Earth's distance equal to 10 units. It is believed to be a chance coincidence.

Bodoni rules *(Print.)*. See **swelled rules**.

body *(Build.)*. (1) The degree of opacity possessed by a pigment. (2) The apparent viscosity of a paint or varnish. (3) The ability of a paint to give a good, uniform film over an irregular or porous surface.

body *(Print.)*. (1) The measurement from top to bottom of a type, rule, etc. The unit is the **point**, 72 points amounting to (approx.) 1 in. (2) The solid part of a piece of type below the printing surface or *face*. Also called *shank, stem*. (3) Body of a work, the text of a volume, distinguished from the preliminary matter, such as title and contents, and the end matter, such as appendices and index.

body cavity *(Zool.)*. The perivisceral space, or cavity, in which the viscera lie; a vague term, sometimes used incorrectly to mean **coelom**.

body cell *(Bot.)*. The cell that divides to give the two sperm cells in the gymnosperm pollen tube.

body cell *(Zool.)*. Somatic cell. Cf. **germ cells**.

body-centred cubic *(Chem.)*. A crystal lattice with a cubic unit cell, the centre of which is identical in environment and orientation to its vertices. Specifically a common structure of metals, in which the unit cell contains 2 atoms, based on this lattice. Abbrev. *BCC*.

body louse *(Med.)*. A surface parasite (*Pediculosis humanus humanus*) which infests man in overcrowded insanitary conditions. It is a vector for **typhus**.

body-section radiography *(Radiol.)*. See **tomography**.

body wall *(Zool.)*. The wall of the perivisceral cavity, comprising the skin and muscle layers.

boehmite *(Min.)*. Orthorhombic form of aluminium monohydrate, AlO.OH. An important constituent of some bauxites.

BOE sill *(Build.)*. Abbrev. for *Brick-On-Edge sill*.

bog *(Ecol.)*. Wetland vegetation forming an acid peat which is mainly composed of dead individuals of the moss, *Sphagnum*. See also **blanket bog**, **raised bog**, **valley bog**.

boghead coal *(Min.)*. Coal which is non-banded and translucent, with high yield of tar and oil on distillation. Consists largely of resins, waxes, wind-borne spores, and pollen cases. Originated in deeper, more open parts of the coal swamps than ordinary household coals. Essentially a spore-coal. Also called *parrot coal*. See also **torbanite**.

bogie, bogie truck *(Eng.)*. (1) A small truck of short wheel-base running on rails. Commonly used for the conveyance of coal, gold or other ores, concrete, etc. (2) A 4- or 6-wheel undercarriage of short wheelbase, which forms a pivoted support at one or both ends of a long rigid vehicle such as a locomotive or coach. US *truck*.

bogie landing gear *(Aero.)*. A main landing gear carrying a pair or pairs of wheels in tandem and pivoted at the end of the shock strut or *oleo*. This arrangement helps to spread the weight of an aircraft over a larger area and also allows the wheel size to be minimized for easier stowage after retraction.

bog iron ore *(Min.)*. Porous form of **limonite** often mixed with vegetable matter. Found in marshes.

boglame *(Vet.)*. See **osteomalacia**.

bog spavin *(Vet.)*. Dilation of the capsule of the tibio-tarsal joint of the horse.

Bohemian garnet *(Min.)*. Reddish crystals of the garnet *pyrope*, occurring in large numbers in the Mittelgebirge in Bohemia.

Bohemian gem-stone *(Min.)*. These consist of the garnet *pyrope*, the false ruby **rose quartz**, and the false topaz **citrine**.

Bohr atom *(Phys.)*. Concept of the atom, with electrons moving in a limited number of circular orbits about the nucleus. These are *stationary states*. Emission or absorption of electromagnetic radiation results only in a **transition** from one orbit (state) to another.

Bohr magneton *(Phys.)*. Unit of magnetic moment, for electron, defined by

$$\mu_B = eh/4\pi m_e c,$$

where e = charge,
h = Planck's constant,
m_e = rest mass,
c = velocity of light,

so that $\mu_B = 9{\cdot}27 \times 10^{-24}\,\mathrm{JT}^{-1}$

The *nuclear Bohr magneton* is defined by

$$\mu_N = eh/4\pi Mc = \frac{\mu_B}{1836} = 5{\cdot}05 \times 10^{-27}\,\mathrm{JT}^{-1},$$

M being rest mass of the proton.

Bohr radius *(Phys.)*. According to the Bohr model of the hydrogen atom, the electron when in its lowest energy state, moves round the nucleus in a circular orbit of radius

$$a_0 = \frac{4\pi\varepsilon_0 \hbar^2}{m_e e^2} = 0{\cdot}5292 \times 10^{-10}\,\mathrm{m},$$

where $\hbar$ is Planck's constant divided by 2π, m_e is the mass of the electron, e is the electronic charge and ε_0 the permittivity of free space. The Bohr radius is a fundamental distance in atomic phenomena.

Bohr-Sommerfeld atom *(Phys.)*. Atom obeying modifications of Bohr's laws suggested by Sommerfeld and allowing for possibility of elliptic electron orbits.

Bohr theory *(Phys.)*. A combination of the Rutherford model of the atom with the quantum theory. The Bohr theory is based on four postulates: (1) An electron in an atom moves in a circular orbit about the nucleus under the influence of the coulombic attraction between the electron and the nucleus. (2) It is only possible for an electron to move in an orbit for which its orbital angular momentum is an integral multiple of $h/2\pi$, where h is Planck's constant. (3) An electron moving in such an orbit does not radiate electromagnetic energy and so its total energy E remains constant. (4) Electromagnetic radiation is emitted if an electron makes a transition from an orbit of energy E_i to one of *lower* energy E_f, and the frequency of the emitted radiation is $v = (E_i - E_f)/h$.

boil *(Med.)*. Infection (with *Staphylococcus aureus*) of a hair follicle, resulting in a painful red swelling, which eventually suppurates.

boiled oil *(Build.)*. Linseed-oil raised to a temperature of from 400–600°F (200–300°C) and admixed with driers.

boiler *(Eng.)*. Describes a wide range of pressure vessels in which water or other fluid is heated and then discharged, e.g. either as hot water for heating or as high pressure steam for power generation.

boiler capacity *(Eng.)*. The weight of steam, usually expressed in kg or lb/hr, which a boiler can evaporate when steaming at full load output.

boiler compositions *(Eng.)*. Chemicals introduced into boiler feed-water to inhibit scale-formation and corrosion, or to prevent priming or foaming. Examples are sodium compounds (such as soda ash), organic matter, and barium compounds.

boiler covering *(Eng.)*. See **lagging**.

boiler cradles *(Ships)*. See **keelson** (2).

boiler crown *(Eng.)*. The upper rounded plates of a boiler of the shell type.

boiler efficiency *(Eng.)*. The ratio of the heat supplied by a boiler in heating and evaporating the feed water to the heat supplied to the boiler in the fuel. It may vary from 60% to 90%.

boiler feed-water *(Eng.)*. The water pumped into a boiler for conversion into steam, usually consisting of condensed exhaust steam and 'makeup' fresh water treated to remove air and impurities.

boiler fittings and mountings *(Eng.)*. See **feed check-valve, pressure gauge, safety valve, stop valve, water gauge**.

boilermaker's hammer *(Eng.)*. One with ball or straight and cross panes; used for caulking, fullering, and scaling boilers.

boiler plate *(Eng.)*. Mild steel plate, generally produced by the open-hearth process; used mainly for the shells and drums of steam-boilers.

boiler pressure *(Eng.)*. The pressure at which steam is generated in a boiler. It may vary from little over atmospheric pressure for heating purposes, to 1500 lb/in.^{2} (10 000 kN/m^2) and over for high-pressure turbines.

boiler scale *(Eng.)*. A hard coating, chiefly calcium sulphate, deposited on the surfaces of plates and tubes in contact with the water in a steam-boiler. If excessive, it leads to overheating of the metal and ultimate failure.

boiler setting *(Eng.)*. The supporting structure on which a boiler rests; usually of brick for land boilers and of steel for marine boilers.

boiler stays *(Eng.)*. Screwed rods or tubes provided to support the flat surfaces of a boiler against the bursting effect of internal pressure.

boiler test *(Eng.)*. (1) A hydraulic-pressure test applied to check watertightness under pressure greater than the working pressure. (2) An efficiency test carried out to determine evaporative capacity and the magnitude of losses.

boiler trial *(Eng.)*. An efficiency test of a steam-boiler, in which the weight of feed-water and of fuel burnt are measured, and various sources of loss assessed.

boiler tubes *(Eng.)*. Steel tubes forming part of the heating surface in a boiler. In water-tube boilers the hot gases surround the tube; in locomotive and some marine boilers (fire-tube boilers), the gases pass through the tube.

boiling *(Phys.)*. The very rapid conversion of a liquid into vapour by the violent evolution of bubbles. It occurs when the temperature reaches such a value that the saturated vapour pressure of the liquid equals the pressure of the atmosphere.

boiling bed *(Chem.)*. In gas fluidized beds, two separate phases are formed at high gas velocities: gas, containing a relatively small proportion of suspended solids, bubbles through a higher density fluidized phase with the result that the system closely resembles in appearance a boiling liquid, hence *boiling bed*.

boiling-point *(Phys.)*. The temperature at which a liquid boils when exposed to the atmosphere. Since, at the boiling-point, the saturated vapour pressure of a liquid equals the pressure of the atmosphere, the boiling-point varies with pressure; it is usual, therefore, to state its value at the standard pressure of 101.325 kN/m^2. Abbrev. *b.p.*

boiling table *(Elec.Eng.)*. A table incorporating in its construction two or more boiling plates.

boiling-water reactor *(Nuc.Eng.)*. Light water reactor in which the water is allowed to boil into steam which drives the turbines directly. Abbrev. *BWR*.

Bok globule *(Astron.)*. Small dark nebula (10^3–10^5 AU in diameter) in the Milky Way, thought to be regions of star formation.

bold face *(Print.)*. A type face heavier than the normal with which it can be used as a companion, e.g. **bold face**.

bole *(Bot.)*. The trunk of a tree.

bole *(Print.)*. A compact clay, a reddish variety of which is used in powdered form (with water and a small quantity of gilding size) as a foundation for gilt edges.

bolection (balection) moulding *(Build.)*. A moulding fixed round the edge of a panel and projecting beyond the surface of the framing in which the panel is held.

bolide *(Astron.)*. A brilliant **meteor**, generally one that explodes: a fire-ball.

boll *(Bot.)*. A capsule, especially of cotton.

boll *(Textiles)*. The seed case of the cotton plant that opens as ripening proceeds. It contains the cotton seeds and the attached fine fibres or **lint**.

bollard *(Ships)*. On a quay or vessel, a short upright post round which ropes are secured for purposes of mooring. Generally, any similar post used, e.g. to prevent vehicular access.

Bollinger bodies *(Vet.)*. Intracytoplasmic inclusion bodies occurring in epithelial cells infected with fowl pox virus; they are made up of aggregates of smaller bodies (*Borrel bodies*), which are the actual masses of the virus collected in the cells.

bolometer *(Elec.Eng.,Telecomm.)*. A device for measuring microwave or infra-red energy, consisting of a temperature dependent resistance used in a bridge circuit which gives an indication when power heats the resistor. Used for power measurement, standing wave detectors and infra-red search and guidance systems.

bolster *(Build)*. A form of cold chisel with a broad splayed-out blade, used in cutting stone slabs etc.

bolster *(Civ.Eng.)*. The actual support for a truss-bridge, at its abutment.

bolster *(Eng.)*. (1) A steel block which supports the lower part of the die in a pressing or punching machine. (2) The rocking steel frame by which the bogie (US *truck*) supports the weight of a locomotive or other rolling stock.

bolt *(Build.)*. (1) A bar used to fasten a door. (2) The tongue of a lock.

bolt *(Eng.)*. A cylindrical, partly screwed bar provided with a head. With a nut, a common means of fastening two parts together.

bolt *(Textiles)*. A specified length of fabric as agreed between seller and buyer.

bolting *(Bot.)*. Premature flowering and seed production especially of a biennial crop plant during its first year; 'running to seed'.

bolt-making machine *(Eng.)*. A machine which forges bolts by forming a head on a round bar.

bolts *(Print.)*. The folded edges at the head, fore-edge, and tail of a sheet before cutting.

Boltzmann equation *(Phys.)*. A fundamental diffusion equation based on particle conservation. The rate of losses, including leakage out of the region of interest and the rate of disappearance by reactions of all kinds, is equal to the rate of production from sources within the region and the rate of scattering into the region.

Boltzmann principle *(Phys.)*. Statistical distribution of large numbers of small particles when subjected to thermal agitation and acted upon by electric, magnetic or gravitational fields. In statistical equilibrium, number of particles n per unit volume in any region is given by $n = n_0 \exp(-E/kT)$, where k = Boltzmann's constant, T = absolute temperature, E = potential energy of particle in given region, n_0 = number per unit volume when $E = 0$.

Boltzmann's constant *(Phys.)*. Given by $k = R/N = 1.380\,5 \times 10^{-23}\,\text{JK}^{-1}$, where R = ideal gas constant, N = Avogadro constant.

bombardment *(Phys.)*. Process of directing a beam of neutrons or high-energy charged particles on to a target material in order to produce nuclear reactions.

bomb calorimeter *(Eng.)*. An apparatus used for determining the calorific values of solid or liquid fuels. The *bomb* consists of a thick-walled, highly polished, steel vessel in which a weighed quantity of the fuel is electrically ignited in an atmosphere of compressed oxygen. The bomb is immersed in a known volume of water to which the combustion heat is transferred, and from the rise of temperature of which the calorific value is calculated.

bomb sampler *(Powder Tech.)*. Device for obtaining samples of dispersed particles at predetermined depths within a suspension, consisting of a closed cylindrical vessel with an automatic valve which opens when an extension tube hits the bottom of the suspension container. The sampler fills with suspension and then closes when the vessel is lifted.

bomb, volcanic *(Geol.)*. See **volcanic bomb**.

bonanza *(Min.Ext.)*. Rich body of ore.

bond *(Build.)*. The system under which bricks or stones are laid in overlapping courses in a wall in such a way that the vertical joints in any one course are not immediately above the vertical joints of an adjacent course.

bond *(Civ.Eng.)*. The adhesion between concrete and its reinforcing steel, due partly to the shrinkage of the concrete in setting and partly to the natural adhesion between the surface particles of steel and concrete. See **mechanical bond**.

bond *(Phys.)*. Link between atoms, considered to be electrical and arising from electrons as distributed around the nucleus of atoms so bonded. See **chemical bond**.

bond angle *(Chem.)*. The angle between the lines connecting the nucleus of one atom to the nuclei of two other atoms bonded to it, e.g. in water the H-O-H angle is about 105°.

bond distance *(Chem.)*. See **bond length**.

bonded fabrics *(Paper)*. A material made by fabricating fibres into sheet form with the aid of a binder. Used for polishing cloths, curtains, filter cloths etc.

bonded-fibre fabric *(Textiles)*. A non-woven fabric made from a mass of fibres held together by adhesive or by processes such as needling or stitching.

bonded wire *(Elec.Eng.)*. Enamelled insulated wire also coated with a thin plastic; after forming a coil, it is heated by a current or in an oven or both for the plastic to set and the coil to attain a solid permanent form.

bond energy *(Chem.)*. The energy in joules released on the formation of a chemical bond between atoms, and absorbed on its breaking.

bonder *(Build.)*. See **bondstone**.

bonding *(Aero.)*. (1) The electrical interconnection of metallic parts of an aircraft normally at earth potential for the safe distribution of electrical charges and currents. Protects against charges due to precipitation, static, and electrostatic induction due to lightning strikes. Reduces interference and provides a low resistance electrical return path for current in earth-return systems. (2) Joining structural parts by adhesive. May be performed at high temperature and pressure.

bonding *(Elec.Eng.)*. An electrical connection between adjacent lengths of armouring or lead sheath, or across a joint. See also **cross bonding**.

bonding clip *(Elec.Eng.)*. A clip used in wiring systems to make connection between the earthed metal sheath of different parts of the wiring, in order to ensure continuity of the sheath.

bond length *(Civ.Eng.)*. The minimum length of reinforcing bar required to be embedded in concrete to ensure that the bond is sufficient for anchorage purposes.

bond length (distance) *(Chem.)*. Distance between bonded atoms in a molecule. Specifically used for distances between atoms in a covalent compound. Typical lengths are 0-H 96 pm, C-H 107 pm, C-C 154 pm, C=C 133 pm, C≡C 120 pm.

bond paper *(Paper)*. A paper similar to bank paper but of $50\,\text{g/m}^2$ or more.

bondstone *(Build.)*. A long stone laid as a header through a wall. Also called a *bonder*.

bond strength *(Phys.)*. See **adhesion**.

bone *(Min.Ext.)*. Coal containing ash (bone) in very fine layers along the cleavage planes. Also called *bony coal*.

bone *(Zool.)*. A variety of connective tissue in which the matrix is impregnated with salts of lime, chiefly phosphate and carbonate.

bone bed *(Geol.)*. A sediment characterized by abundant fossil bones or bone fragments, scales, teeth, coprolites etc.

bone conduction *(Med.)*. The conduction of sound-waves from the bones of the skull to the inner ear, rather than through the ossicles from the outer ear.

bone-dry paper *(Paper)*. Paper dried completely to contain no moisture. Also *oven-dry paper*.

bone marrow grafting *(Med.)*. Grafting or transplantation of bone marrow to patients with bone marrow failure. Used in aplastic anaemia or after therapeutic destruction of the marrow in **leukaemia**.

bone seeker *(Chem.)*. Radioelement similar to calcium, e.g. Sr, Ra, Pu, which can pass into bone where it continues to radiate.

bone-setter *(Med.)*. A medically unqualified person who treats disorders of joints by manipulation. See **osteopathy**.

bone tolerance dose *(Radiol.)*. The dose of ionizing radiation which can safely be given in treatment without bone damage.

bone turquoise *(Min.)*. Fossil bone or tooth, coloured blue with phosphate of iron; widely used as a gemstone. It is not true turquoise, and loses its colour in the course of time. Also called *odontolite*.

boning-in *(Surv.)*. The process of locating and driving in pegs so that they are in line and have their tops also in line; carried out by sighting between a near and a far peg previously set in the gradient desired.

boning-rods *(Surv.)*. T-shaped rods used, in sets of three, to facilitate the process of boning-in; two of the rods are held on the near and far pegs to establish a line of sight between them in the desired gradient, while the third is used to fix intermediate pegs in line.

Bonne's projection *(Geol.)*. A derivative **conical projection** in which the parallels are spaced at true distances along the meridians, which are plotted as curves.

bonnet *(Build.)*. A wire-netting cowl covering the top of a ventilating pipe or a chimney.

bonnet *(Eng.)*. A movable protecting cover, e.g. (1) the cap of the valve-box of a pump; (2) the cover plate of a valve chamber; (3) the hood of a forge; (4) the cover over the engine of a motor vehicle.

bonnet tile *(Build.)*. Special rounded tile used to cover the external angles at hips and ridges on tiled roofs. See **arris tile**.

bookbinding *(Print.)*. Arranging in proper order the sections, etc. of a book and securing them in a cover. The bookbinder distinguishes bound book from cased book, and both from brochure and paperback.

book chase *(Print.)*. A chase with two crossbars, dividing it into quarters; used for bookwork.

book cloth *(Print.)*. The usual covering for edition binding. There are two main kinds: thin, hard-glazed cloth containing starch and other fillings, and matt-surfaced cloth with less starch.

book gill *(Zool.)*. See **gill book**.

book lung *(Zool.)*. See **lung book**.

book plate *(Print.)*. A label pasted on the inside front cover of a book bearing the owner's name, crest, coat-of-arms or peculiar device.

Boolean *(Comp.)*. A **data type** which may take only two values, *true* or *false*; see **truth value**.

Boolean algebra *(Comp.,Maths.)*. An algebra, named after George Boole (1815–64), in which the elements are of two kinds only, *true* or *false*, and in which the basic operations are the logical AND and OR operations, which are normally symbolized as multiplication and addition respectively. Its main applications are to the design of switching networks and to mathematical logic and, in computing, it has formed the basis for the development of **logical operations** and the use of **gates**. See also **stitching network, truth value**.

boom *(Acous.)*. Enhanced reverberation or resonance in an enclosed space at low frequencies, due to reduced acoustic absorption of the surfaces for low frequencies.

boom *(Eng.,Ships)*. Any long beam; more especially (1) the flange of a built-up girder, also *chord*; (2) the main spar of a lifting-tackle; (3) the spar holding the lower part of a fore-and-aft sail; (4) a spar attached to a yard to lengthen it; (5) a barrier of logs to prevent the passage of a vessel; (6) a line of floating timbers used to form a floating harbour; (7) a pole marking a channel.

boom *(Image Tech.)*. Long moveable arm to carry a microphone or lamp above the action in film or television shooting.

boost *(Autos.)*. The amount by which the induction pressure of a supercharged internal-combustion engine exceeds atmospheric pressure expressed in kN/m^2 or lbf/in^2.

boost control *(Aero.)*. A device regulating reciprocating-engine manifold pressure so that supercharged engines are not over-stressed at low altitude.

boost control over-ride *(Aero.)*. In a supercharged piston aero-engine fitted with **boost control**, a device (sometimes lightly wire-locked so that its emergency use can be detected), which allows the normal maximum manifold pressure to be exceeded: also *boost control cut-out*.

booster *(Space)*. A rocket engine, or cluster of engines, part of a launch system, either the first stage or auxiliary stage, used to provide an initial thrust greater than the total lift-off weight.

booster amplifier *(Acous.)*. One used specially to compensate loss in mixers and volume-controls, so as to obviate reduction in signal/noise ratio.

booster coil *(Aero.)*. A battery-energized induction coil which provides a starting spark for aero-engines.

booster fan *(Eng.)*. A fan for increasing the pressure of air or gas; used for restoring the pressure drop in transmission pipes, and for supplying air to furnaces.

booster pump *(Aero.,Eng.)*. A pump which maintains positive pressure between the fuel tank and the engine, thus intensifying the flow. Any pump to increase the pressure of the liquid in some part of a pipe circuit.

booster response *(Immun.)*. The heightened response to readministration of an antigen, or to a microbial infection, which occurs when prior contact with the antigen has elicited **B-memory cells** and **helper T-lymphocyte** cells.

booster rocket *(Aero.)*. See **take-off rocket**.

booster station *(Min.Ext.)*. In long-distance transport by pipeline of oils, other liquids, mineral slurries or water-carried coal, an intermediate pumping station where lost pressure energy is restored.

booster station *(Telecomm.)*. One which rebroadcasts a received transmission directly on the same wavelength.

booster transformer *(Elec.Eng.)*. A transformer connected in series with a circuit to raise or lower the voltage of that circuit.

boost gauge *(Aero.)*. An instrument for measuring the manifold pressure of a supercharged aero-engine in relation to ambient atmosphere or in absolute terms. Also used for racing and some sports in cars.

boost transformer *(Elec.Eng.)*. Same as **buck transformer**.

boot *(Glass)*. See **potette**.

booted *(Zool.)*. Having the feet protected by horny scales.

bootstrap *(Comp.)*. A short section of a program which can be used to get the rest of the program running in a machine. After a computer has been switched off, the **operating system** has to be put back into the **main memory**. To reload the **loader** a bootstrap is used, a small number of instructions often loaded and executed by pushing a button.

bootstrap *(Eng.)*. A self-sustaining system in liquid rocket engines by which the main propellants are transferred by a turbo-pump which is driven by hot gases. In turn the gas generator is fed by propellants from the pump.

bootstrap circuit *(Electronics)*. A feedback circuit in which part of the output is fed back across the input, giving effectively infinite input impedance and unity gain. Often used to improve the linearity of a voltage sweep generator.

bootstrap cold-air unit *(Aero.)*. A unit of the compressor-turbine type in which the air charging an aircraft cabin passes through the compressor and, via an **intercooler**, the turbine.

bootstrapping *(Electronics)*. The technique of using bootstrap feedback. See **bootstrap circuit**.

boot tapping *(Ships)*. The demarcation line between the two main colours of a ship's paintwork, at or near the waterline.

BOP *(Min.Ext.)*. Abbrev. for *Blow Out Preventer*.

boracic acid *(Chem.)*. See **boric acid**.

boracite *(Min.)*. The orthorhombic, pseudocubic form of magnesium borate and chloride, found in beds of gypsum and anhydrite.

borane *(Chem.)*. General name for a hydride of boron, the simplest being diborane, B_2H_6. They are high energy compounds, yielding water and boric acid on oxidation. There are three principal series of boranes: (a) *closo*-, with formulae B_nH_{n+2}, (these mainly occur as anions with fomulae $B_nH_n^{2-}$), (b) *nido*-, with formulae B_nH_{n+4}, and (c) *arachno*-, with formulae B_nH_{n+6}. See **multi-centred bonding**.

borates *(Chem.)*. See **boric oxide**.

borax *(Min.)*. Hydrated sodium borate, deposited by the evaporation of alkaline lakes.

borax bead *(Chem.)*. Borax, when heated, fuses to a clear glass. Fused borax dissolves some metal oxides giving glasses with a characteristic colour. The use of borax bead in chemical analysis is based on this fact.

borazole *(Chem.)*. $B_3H_6N_3$ a compound isoelectronic with **benzene**.

bord-and-pillar *(Min.Ext.)*. A method of mining coal by excavating a series of chambers, rooms, or stalls, leaving pillars of coal in between to support the roof. Also *room-and-pillar*.

Bordeaux B *(Chem.)*. An azo-dyestuff derived from 1-naphthylamine coupled with **R-acid**.

bordered pit *(Bot.)*. A **pit** in which the secondary wall overarches the pit membrane, markedly narrowing the cavity towards the cell lining. Characteristic of tracheids and vessel elements. Cf. **simple pit**.

border effect *(Image Tech.)*. A faint dark line on the denser side of a boundary between a lightly exposed and a heavily exposed region on a developed emulsion. See **Mackie lines**.

border-pile *(Civ.Eng.)*. A pile driven to support the sides of a coffer-dam.

bore *(Eng.)*. The circular hole along the axis of a pipe or machine part; the internal wall of a cylinder; the diameter of such a hole.

bore *(Hyd.Eng.)*. A great tide-wave, with crested front, travelling rapidly up a river; it occurs on certain rivers having obstructed channels.

boreal *(Ecol.)*. Of the North. The *boreal zone* is the geographical region where short summers and long cold winters occur. The *boreal period* in northern Europe extended from 7500 to 5500 BC and had warm summers and cold winters. *Boreal forests* occur in the boreal zone and boreal period.

borehole *(Civ.Eng.)*. A sinking made in the ground by the process of **boring**.

borehole *(Min.Ext.)*. The hole made by a drill for a well; its whole length.

borehole survey *(Min.Ext.)*. Check for deviation from required line of deep borehole. Methods include camera records of compass, plumbline or gyroscope.

boric acid *(Chem.)*. H_3BO_3. A tribasic acid. On heating it loses water and forms metaboric acid, $H_2B_4O_4$, and on further heating it forms tetraboric acid, or the so-called pyroboric acid, $H_2B_4O_7$. On heating at a still higher temperature it forms anhydrous boron (III) oxide, or boric oxide. It occurs as tabular triclinic crystals deposited in the neighbourhood of fumaroles, and known also in solution in the hot lagoons of Tuscany and elsewhere. Also called *boracic acid, sassolite*.

boric oxide *(Chem.)*. *Boron (III) oxide*. B_2O_3. An 'intermediate oxide' like aluminium oxide, having feeble acidic and basic properties. As a weak acid it forms a series of borates. See **boric acid**.

boride *(Chem.)*. Any of a class of substances, some of which are extremely hard and heat-resistant, made by combining boron chemically with a metal.

borine radical *(Chem.)*. The radical BH_3, capable of forming compounds through coordination of a lone pair of electrons to the electron deficient boron, e.g. borine carbonyl, BH_3CO.

boring *(Eng.)*. The process of machining a cylindrical hole, performed in a lathe, boring machine or boring mill; for large holes, or when great accuracy is required, it is preferable to using a **drill**.

boring *(Min.Ext.)*. The drilling of deep holes for the exploitation or exploration of oil fields. The term *drilling* is used similarly in connection with metalliferous deposits.

boring bar *(Eng.)*. A bar clamped to the saddle of a lathe or driven by the spindle of a boring machine, and carrying the boring tool.

boring machine *(Eng.)*. A machine on which boring operations are performed, comprising a head, carrying a driving-spindle, and a table to support the work.

boring mill *(Eng.)*. A vertical boring machine in which the boring bar is fixed, the work being carried by the rotating table.

boring tool *(Eng.)*. The cutting tool used in boring operations. It is held in a boring bar.

Borna disease *(Vet.)*. An infectious encephalomyelitis of horses, caused by a virus, and characterized by mild fever, a variety of nervous symptoms and a high death rate. Sheep may occasionally be affected.

Borneo camphor *(Chem.)*. **Borneol**.

borneol *(Chem.)*. $C_{10}H_{17}.OH$, mp 208°C, bp 212°C, it is a secondary alcohol, and yields camphor on oxidation.

bornite *(Min.)*. Sulphide of copper and iron occurring in Cornwall and many other localities. Develops a brilliant iridescent red and blue tarnish, hence, also called *erubescite, peacock ore, variegated copper ore*.

Born-Oppenheimer approximation *(Phys.)*. Used in considering the electronic behaviour of molecules. The problems of the electronic and nuclear motion are treated separately.

bornyl chloride *(Chem.)*. $C_{10}H_{17}Cl$, mp 148°C, white crystals, identical with pinene hydrochloride, obtained from **borneol** by treatment with PCl_5.

borofluorides *(Chem.)*. See **fluoroboric acid**.

borolanite *(Geol.)*. A basic igneous rock occurring near Loch Borolan, Assynt, in N.W. of Scotland; it consists essentially of feldspar, green mica, garnet, together with conspicuous rounded white aggregates of 'pseudo-leucite', consisting of orthoclase feldspar and altered nepheline.

boron *(Chem.)*. Amorphous yellowish-brown element discovered by Davy, 1808, also Gay-Lussac and Thénard. Symbol B, at no. 5, r.a.m. 10.811, mp 2300°C, rel.d. 2.5. Can be formed into a conducting metal. Most important in reactors, because of great cross-section (absorption) for neutrons; thus, boron steel is used for control rods. The isotope ^{10}B on absorbing neutrons breaks into two charged particles ^7Li and ^4He which are easily detected, and is therefore most useful for detecting and measuring neutrons.

boron carbide *(Chem.)*. B_4C. Obtained from B_2O_3 and coke at about 2500°C. Very hard material, and for this reason used as an abrasive in cutting tools where extreme hardness is required. Extremely resistant to chemical reagents at ordinary temperatures.

boron chamber *(Nuc.Eng.)*. Counter tube containing boron fluoride, or boron-covered electrodes, for the detection and counting of low velocity neutrons, which eject α-particles from the isotope ^{10}B.

boron nitride *(Chem.)*. BN, a compound isoelectronic with elemental carbon, and having two polymorphs, one similar to graphite and the other (called borazon) similar to diamond.

boron trihalides *(Chem.)*. All the four halogens unite with boron to form (III) halides as follows: BF_3,BCl_3, BBr_3,BI_3.

borosilicate glasses *(Glass)*. Heat and chemical-resisting glasses based on the fusion of silica-borax mixtures to produce complex polymers of sodium borosilicate.

Borrel bodies *(Vet.)*. See **Bollinger bodies**.

borrow pit *(Civ.Eng.)*. Excavation which provides extra material to serve as fill when required.

borsic *(Aero.)*. Boron fibre coated with silicon carbide.

bort (boart) *(Min.)*. A finely crystalline form of diamond in which the crystals are arranged without definite orientation. Possessing the hardness of diamond, bort is exceedingly tough, and is used as the cutting agent in rock drills.

Bosch process *(Chem.)*. Production of hydrogen by the catalytic reduction of steam with carbon monoxide at 500°C: $CO+H_2O \rightarrow H_2+CO_2$.

Bose-Einstein distribution law *(Phys.)*. A **distribution law** which is applicable to a system of particles with symmetric wave functions, this being the characteristic of most neutral gas molecules. It can be stated as

$$\bar{n}_i = \frac{g_i}{\frac{1}{A}\exp E_i/kT - 1}$$

where $\bar{n}_i$ = average number of molecules with energy E_i, g_i = degeneracy factor, A = constant.

Bose-Einstein statistics *(Phys.)*. Statistical mechanics law obeyed by a system of particles whose wave function is unchanged when two particles are interchanged.

bosh *(Eng.)*. The tapering portion of a blast-furnace, between the largest diameter (at the bottom of the stack) and the smaller diameter (at the top of the hearth).

bosh *(Glass)*. A water tank for cooling glassmaking tools or quenching glass.

boson *(Phys.)*. A particle which obeys Bose-Einstein statistics but not the Pauli exclusion principle. Bosons have a total spin angular momentum of $n\hbar$ where n is an integer and $\hbar$ is the Dirac constant. Photons, α-particles and all nuclei having an even mass number are bosons.

boss *(Eng.)*. A projection, usually cylindrical, on a machine part in which a shaft or pin is to be supported; e.g. the thickened part at the end of a lever, provided to give a longer bearing to the pin.

boss *(Geol.)*. An igneous intrusion of cylindrical form, less than 100 km^2 in area; otherwise like a bathylith.

bossage *(Build.)*. Roughly dressed stones, such as quoins and corbels, which are built in so as to project, and are finish-dressed in position.

bosset *(Zool.)*. In Deer, the rudiment of the antlers in the first year.

bossing *(Build.)*. The operation of shaping malleable metal, particularly sheet-lead, to make it conform to surface irregularities, by tapping with special mallets, known as *bossing mallets* or *dressers*.

Bostock sedimentation balance *(Powder Tech.)*. A **sedimentation technique** in which a torsion balance is used directly to weigh the accumulation of particles. The high sensitivity enables low concentrations to be used, giving increased accuracy.

bostonite *(Geol.)*. A fine-grained intrusive igneous rock allied in composition to syenite; essentially feldspathic, and deficient in coloured silicates; type-locality, Boston, Mass.

bot *(Vet.)*. The larva of flies of the genus *Gastrophilus*; bots parasitize the membrane of the stomach of horses, rarely of other animals.

botry-, botryo- *(Bot.)*. Prefix meaning like a bunch of grapes.

botryoidal *(Min.)*. Of minerals, formation resembling grapes.

botryoidal *(Zool.)*. Shaped like a bunch of grapes.

botryomycosis *(Vet.)*. A suppurative, granulomatous infection of horses due to *Staphylococcus aureus*. When the spermatic cord becomes infected, following castration, the term *scirrhous cord* is applied.

botryose, botryoid, botrytic *(Bot.,Zool.)*. Branched; like a bunch of grapes. See **racemose**. See also **botryoidal**.

bottle battery *(Elec.Eng.)*. A term used to denote a bichromate cell when electrodes and electrolyte are placed in a glass bottle-shaped container.

bottle glass *(Glass)*. Glass used for the manufacture of common bottles, made from a **batch** comprising essentially sand, limestone and alkali. A typical percentage glass composition may be taken as SiO_2 74.0, Al_2O_3 0.6, CaO 9.0, Na_2O 16.4.

bottle jack *(Eng.)*. A screw-jack in which the lower part is shaped like a bottle.

bottle-making machines *(Glass)*. These may operate in various ways, the bottle being formed in 2 stages, i.e. the parison and the finished bottle. Wide-mouth ware may be formed by pressing the parison and then blowing, narrow-mouth by blowing and blowing or sucking and blowing. In the last method, the glass is gathered by suction into the parison mould, in the other two it is dropped by hand or more probably by a mechanical-feeding device, hence the terms suction-fed and feeder-fed machines.

bottle-nose drip *(Build.)*. The shaped edge formed in sheet-lead work at a step on a roof, when jointing the lead across the direction of fall.

bottle-nosed step *(Build.)*. A step which has the edge and ends rounded.

bottom dead-centre *(Eng.)*. See **outer dead-centre**.

bottom gate *(Foundry)*. An **in-gate** leading from the runner into the bottom of a mould.

bottom-hole assembly *(Min.Ext.)*. The drilling string attached to the bottom of the **drilling pipe**. It will comprise the drill bit and collars used to maintain directionality and may contain several stabilizers and reamers in more difficult conditions, when it is called a *packed-hole assembly*. Abbrev. *BHA*.

bottom-hole pump *(Min.Ext.)*. Electric or hydraulic pump placed at the bottom of a well.

bottoming *(Civ.Eng.)*. The lowest layer of foundation material for a road or other engineering works including structures.

bottoming tap *(Eng.)*. See **plug tap**.

bottomness *(Phys.)*. A property that characterizes *quarks* and so hadrons. The bottomness of leptons and gauge bosons is zero. Bottomness is conserved in strong and electromagnetic interactions between particles but not in weak interactions. Also *beauty*.

bottom plate *(Horol.)*. In a watch, the plate to which the pillars are fixed, generally referred to as the **dial plate**.

bottoms *(Eng.)*. A term used in connection with the Orford process for separating nickel and copper as sulphides. When the mixed sulphides are fused with sodium sulphide, the nickel sulphide separates to the bottom. Hence *bottoms* as distinct from *tops*. In reverberatory furnace, the heaviest molten material at bottom of pool.

bottomset bed *(Geol.)*. Fine-grained sediment laid down at the front of a growing delta. Cf. **foreset** and **topset**.

bottom shore *(Build.)*. One of the members of an arrangement of raking shores to support temporarily the side of a building; it is the one nearest the wall face.

bottom structure *(Geol.)*. See **sole mark**.

bottom yeast *(Bot.)*. Brewer's yeast, *Saccharmyces cerevisiae*, which accumulate at the bottom of the medium during fermentation and used in brewing lagers. Cf. **top yeast**.

botulism *(Med.,Vet.)*. Severe and often fatal poisoning due to eating bottled and canned food contaminated by the anaerobic bacterium (*Clostridium botulinum*), which secretes a potent neurotoxin, *botulin*. Causes many cases of food poisoning in man but all types of animals can be effected by eating contaminated food. Fish farms have outbreaks of *C. botulinum* Type E, which can grow in temperatures as low as 5°C.

Boucherot circuit *(Elec.Eng.)*. An arrangement of inductances and capacitances, whereby a constant-current supply is obtained from a constant-voltage circuit.

bouchon *(Horol.)*. A hollow plug, or bush, inserted in watch or clock plates to form the pivot holes. To repair a worn hole, the hole is enlarged and a bouchon pressed in. In certain cases the jewels are held in bouchons which are a press fit in the plates.

bouchon wire *(Horol.)*. Hollow wire, generally of hard brass, which is cut off to the required lengths to form bouchons.

bouclé *(Textiles)*. Fabric made from fancy yarns and having a rough, textured surface, mainly used for women's garments. Yarn made from a core thread with an outer yarn wrapped round it to give a knobbly appearance.

boudinage *(Geol.)*. Structure found in sedimentary series subjected to folding. It consists of strike-elongated 'sausages' of more rigid rock, enclosed between relatively plastic rocks.

bougie *(Med.)*. A tube or a rod for dilating narrowed passages in the body.

Bouguer anomaly *(Geol.)*. A gravity anomaly which has been corrected for the station height and for the gravitational effect of the slab of material between the station height and datum. Cf. **free-air anomaly**.

Bouguer law of absorption *(Phys.)*. The intensity p of a parallel beam of monochromatic radiation entering an absorbing medium is decreased at a constant rate by each infinitesimally thin layer db,

$$\frac{-dp}{p} = k\,db,$$

where k is a constant that depends on the nature of the medium and on the wavelength.

boulder *(Geol.)*. The unit of largest size occurring in sediments and sedimentary rocks, the limit between pebble and boulder being placed at 256 mm, although some authorities recognize **cobbles** between pebbles and boulders. Boulders may consist of any kind of rock, may be subangular or well rounded, may have originated in place or have been transported by running water or ice. Accumulations of boulders are *boulder beds*.

boulder clay *(Geol.)*. See **till**.

boulder paving *(Build.)*. Paving constructed with rounded boulders laid on a gravel foundation.

boulder wall *(Build.)*. A wall built of boulders or flints set in mortar.

boule *(Min.)*. The pear-shaped or cylindrical drop of synthetic mineral, commonly ruby, sapphire or spinel, produced by the Verneuil flame-fusion process.

Bouma cycle *(Geol.)*. A sedimentary succession of five intervals that makes up a complete turbidite deposit. Typically incomplete.

bounce mark *(Geol.)*. A short depression caused by an object (e.g. shell, pebble) bouncing over the surface of a sediment, especially a turbidite. Also called a prod mark.

bouncing-pin detonation meter *(Eng.)*. An apparatus for determining quantatively the degree of detonation occurring in the cylinder of a petrol-engine; used for fuel testing.

boundary films *(Eng.)*. Films of one constituent of an alloy surrounding the crystals of another.

boundary layer *(Aero.)*. The thin layer of fluid (air) adjacent to the surface in which viscous forces exert a noticeable influence on the motion of the fluid and in which the transition between still air and the body's velocity occurs.

boundary layer *(Chem.Eng.)*. When a fluid flows past a solid surface, the layer of fluid next to the solid surface is brought to rest, setting up a viscous motion in adjacent layers. The boundary layer is the total thickness of fluid over which the surface exerts a differential effect. It governs the rate of heat, mass, or momentum between the solid surface and the homogeneous bulk of the fluid. The velocity of the layer differs significantly from that of the main fluid stream, and is therefore of considerable importance in heat transfer problems, as in nuclear reactors.

boundary layer control *(Aero.)*. Modification of the airflow in the boundary layer to increase lift and/or decrease drag. Means include: removal of the boundary layer by sucking through slots or porous surfaces; use of vortex generators to re-energize sluggish surface flow; ejection of high speed air through slits; blowing, by propulsion efflux, over wing surfaces.

boundary layer, unstirred layer *(Bot.)*. Surface layer of gas or liquid across which molecular movement is diffusion limited. This has a significant effect on the uptake of CO_2 by leaves or of some solutes by cells.

boundary lights *(Aero.)*. Lights defining the boundary of the landing area.

boundary lubrication *(Eng.,Phys.)*. A state of partial lubrication which may exist between two surfaces in the absence of a fluid oil film, due to the existence of adsorbed mono-molecular layers of lubricant on the surfaces.

boundary markers *(Aero.)*. See **airport markers**.

bound book *(Print.)*. Must be hand-sewn on cords or tapes which are firmly attached to the boards (by lacing-in or using split boards) before applying the covering, normally of leather. Cf. **cased book**. See **full-bound, half-bound, quarter-bound, three-quarter-bound**.

bound charge *(Elec.Eng.)*. That induced static charge which is 'bound' by the presence of the charge of opposite polarity which induces it. Also, in a dielectric, the charge arising from polarization. Also *surface charge*. See also **free charge**.

bounded function *(Maths)*. A function is said to be bounded above if it has an upper bound, and bounded below if it has a lower bound. If bounded both above and below, the function is said to be bounded. Cf. **bounds of function**.

bounded set (of numbers) *(Maths.)*. A set of numbers lying between two particular numbers.

bounded set (of points) *(Maths.)*. A set of points for which the set of distances between pairs of points is a bounded set.

bounds of a function *(Maths.)*. An upper bound M of a set S of numbers is a number such that $x \leqslant M$ for all x in S. The *supremum* of S is its least upper bound. Lower bound and *infimum* are defined similarly.

bound state *(Phys.)*. Quantum mechanical state of a system in which the energy is *discrete* and the wave function is localized, e.g. that of an electron in an atom, where transitions between the bound states give rise to atomic spectrum lines.

bound vector *(Maths.)*. A vector with a particular point of application. Also *localized vector*.

bound water *(Bot.)*. Water held by matric forces. Cf. **matric potential**.

bouquet stage *(Biol.)*. See **pachytene**.

Bourdon gauge *(Eng.)*. See **pressure gauge**.

bourgeois *(Print.)*. An old type size, approximately 9-point.

bourne *(Geol.)*. An intermittent or seasonal stream.

bournonite *(Min.)*. Lead copper antimony sulphide; commonly occurs as wheel-shaped twins, hence known as *cog-wheel ore*, or *wheel-ore*. Also *antimonal lead ore*.

Boussinesq approximation *(Meteor.)*. An approximation to the equations of motion in which variations of density from the mean state are ignored except when they are multiplied by the acceleration of gravity.

bouyant density *(Biol.)*. The density of molecules, particles or viruses as determined by flotation in a suitable liquid. A gradient of CsCl will separate DNA according to its base composition, different DNA molecules banding at discrete positions.

bouyoucous hydrometer *(Powder Tech.)*. A special hydrometer used in particle size analysis of soil grains. It gives readings direct in concentration of soil colloids per litre of suspension in water.

bovine acetonaemia *(Vet.)*. *Ketosis, Slow Fever*. Negative energy balance associated with ketonaemia and hypoglycaemia. Affects mainly lactating cows, particularly over the winter period. Ketones are detected in milk, urine and breath. Characterized by reduction in appetite and milk production; sometimes with nervous signs.

bovine brucellosis *(Vet.)*. See **bovine contagious abortion**.

bovine contagious abortion *(Vet.)*. *Bang's disease, bovine brucellosis*. A contagious bacterial infection with *Brucella abortus*. Infects horses, sheep, dogs and man (*undulant fever*), as well as cattle. Main symptom in cattle is abortion. Eradication programmes are possible and the disease is now rare in some countries.

bovine cutaneous streptothricosis *(Vet.)*. A chronic, exudative dermatitis affecting cattle in tropical Africa and elsewhere, caused by a fungus *Dermatophilus congolensis*. A similar disease occurs in goats and horses.

bovine cystic haematuria *(Vet.)*. *Enzootic haematuria, urinary bladder neoplasia*. Symptoms are of intermittent and then continuous haematuria. Various neoplasias can occur but oropharangeal papillomas are also present. Occurs in adult cattle grazed on bracken areas.

bovine farcy *(Vet.)*. *Nocardia farcinia* is involved. Characterized by generalized purulent and granulomatous nodular lesions. Nocardial mastitis has been the predominant infection reported in cattle by this soil-borne organism, usually a chronic problem.

bovine hyperkeratosis *(Vet.)*. *X-disease*. A disease of cattle, characterized by emaciation, loss of hair, and thickening of the skin, due to poisoning by chlorinated naphthalene compounds.

bovine hypomagnesaemia *(Vet.)*. *Lactational tetany, grass staggers, Hereford disease*. A reduction in blood magnesium levels. Clinical disease occurs in spring (and autumn) after turnout onto lush grass or young cereal crops. Climate associated. Symptoms include *hyperaesthesia*, tetany and sudden death. Controlled by mineral supplementation.

bovine infectious petechial fever *(Vet.)*. *Ondiri disease*. A disease of cattle in Kenya characterized by fever, petechial haemorrhages of mucous membranes, diarrhoea, and often death; believed to be caused by a rickettsia-like organism.

bovine ketosis *(Vet.)*. See **bovine acetonaemia**.

bovine lipomatosis *(Vet.)*. A diffuse growth of lipomata in cattle, usually involving the abdominal mesenteries and viscera.

bovine pasteurellosis *(Vet.)*. See **haemorrhagic septicaemia**.

bovine pyelonephritis *(Vet.)*. A specific infection of the kidneys of cattle by the bacterium *Corynebacterium renale*.

bow *(Elec.Eng.)*. A sliding type of current collector, used on electric vehicles to collect the current from an overhead contact-wire. It consists of a bow-shaped contact strip, mounted on a hinged framework.

bow *(Horol.)*. (1) The ring of a pocket-watch case, to which the watch or fob chain is attached. (2) A flexible strip of whalebone or cane, the ends of which are drawn together to give tension to a thread or line which is given a single turn round a pulley of a pair of turns, drill, or mandrel. It is used as a sensitive drive for these tools, and by many it is considered to be the best way to produce very fine accurate pivots.

Bowden gauge *(Elec.Eng.)*. Form of pressure-sensitive transducer.

Bowden-Thomson protective system *(Elec.Eng.)*. A form of protective system for feeders, in which special cables, with the cores surrounded by metallic sheaths, are employed; a fault causes current to flow in the sheath and operate a relay to trip the circuit.

Bowditch's rule *(Surv.)*. A rule for the adjustment of closed compass traverses, in which it may reasonably be assumed that angles and sides are equally liable to error in measurement. According to this rule, the correction in latitude (or departure) of any line is:

$$\frac{\textit{Length of that line}}{\textit{Perimeter of traverse}} \times \begin{array}{l}\textit{Total error in}\\ \textit{latitude (or departure)}\end{array}.$$

bowel oedema disease *(Vet.)*. *Oedema disease, gut oedema.* A disease of pigs characterized by nervous symptoms and oedema in many tissues; believed to be due either to immunological hypersensitivity to *Escherichia coli* or to the production of toxins by this bacterium. Can be limited by anti-sera.

bowenite *(Min.)*. A compact, finely granular, massive form of serpentine, formerly thought to be nephrite, and used for the same purposes.

Bowen ratio *(Meteor.)*. The ratio of the amount of sensible heat (enthalpy) to latent heat lost by a surface to the atmosphere by conduction and turbulence.

Bower-Barff process *(Build.)*. An anti-corrosion process applied to sanitary ironwork; this, when red-hot, has superheated steam passed over it in a closed space, so that a protective layer of black magnetic oxide is formed on the iron-work. See **Angus-Smith process**.

bowk *(Min.Ext.)*. A large iron barrel used when sinking a shaft. Also called *kibble*.

bowl *(Print.)*. The enclosed part of letters such as B and R, and the upper portion of g, distinguishing it from the loop or tail. See **counter**.

bowlingite *(Min.)*. See **saponite**.

Bowman's capsule *(Zool.)*. In the Vertebrate kidney, the dilated commencement of an uriniferous tubule.

Bowman's glands *(Zool.)*. In some Vertebrates, serous glands of the mucous membranes of the olfactory organs.

bow nut *(Eng.)*. See **cap nut**.

bow propeller *(Ships)*. Propeller whose thrust can be directed at right angles to the ship's axis, used in docking and manoeuvring in a confined space. Angular thrust can also be provided to the stern.

bow-saw *(Build)*. A thin-bladed saw which is kept taut by a bow or special frame.

bows, bow compasses *(Eng.)*. See under **spring bows**.

bow sheaves *(Elec.Eng.)*. The sheaves at the bow of cable-laying ship over which the cable passes when it is being laid in the sea or raised for repair.

Bow's notation *(Eng.)*. A method of notation for forces acting at a point, the spaces between the forces being lettered in order, so that any force is described in terms of the letters referring to the two adjacent spaces. By this device the force polygon can be lettered correspondingly. Also called *Henrici's notation*.

bowstring bridge *(Civ.Eng.)*. An arched bridge in which the horizontal thrust on the arch is taken by a horizontal tie joining the two ends of the arch.

bowstring suspension *(Elec.Eng.)*. A form of suspension for the overhead contact-wire of an electric-tramway system, in which the contact-wire is suspended from a short cross-wire attached to the bracket-arm of the pole.

bow strip *(Elec.Eng.)*. See **contact strip**.

box annealing *(Eng.)*. See **close annealing**.

box baffle *(Acous.)*. Box, with or without apertures and damping, one side fitted with an open diaphragm loudspeaker unit, generally coil-driven.

box chronometer *(Ships)*. The marine chronometer. The chronometer is normally supported on gimbals, inside a wooden box with a hinged lid.

box cloth *(Textiles)*. A woven woollen fabric, milled and finished with a smooth surface like felt. **Billiard cloth** is an example.

box culvert *(Civ.Eng.)*. A culvert having a rectangular opening.

box dam *(Civ.Eng.)*. A coffer-dam built to surround an area in which works are to proceed.

box drain *(Build.)*. A small rectangular section drain, usually built in brickwork or concrete.

boxed frame *(Build.)*. A **cased frame**.

boxed mullion *(Build.)*. A hollow mullion in a sash window-frame, arranged to accommodate the counter-weights connected to the vertically moving sashes.

box-frame motor *(Elec.Eng.)*. A traction motor in which the frame is cast in one piece instead of being split.

box girder *(Build.)*. A cast-iron, or mild steel, girder of hollow rectangular section. See **box plate girder**.

box gutter *(Build.)*. A wooden gutter, lined with sheet-lead, zinc, or asphalt, and having upright sides; used along roof valleys or parapets.

box-in *(Print.)*. To surround text with rule, the printed matter appearing in a frame.

boxing *(Build.)*. The part of a window-frame which receives the folded shutter.

boxing shutters *(Build.)*. Shutters at the interior side of a window, hung so as to fold back into a recess in the jambs. Also *folding shutters*.

box nut *(Eng.)*. See **cap nut**.

box plate girder *(Build.)*. A built-up steel girder, similar to the plate girder, but having two web plates at a distance apart, so that flanges and webs enclose a rectangular space.

boxplot *(Stats.)*. The graphical representation of the frequency distribution of a set of values by means of a strip showing the relative positions of the smallest and largest values, the median and the quartiles of a set of values.

box spanner *(Build)*. One composed of a cylindrical piece of steel shaped at one or both ends to a hexagon in order to fit over the appropriate nut. The spanner is turned by a steel rod (*tommy bar*) inserted through a diametrically drilled hole at the opposite end to the hexagon in use. Used for nuts inaccessible to an ordinary spanner.

box-staple *(Build.)*. The part on a door-post into which the bolt of a lock engages.

box stones *(Geol.)*. Hollow concretions. Nodules of sandstone containing molluscan casts found in the Pleistocene deposits of East Anglia.

box tool *(Eng.)*. A single-point cutting tool, set radically or tangentially, used in automatic screw machines and in capstan and turret lathes.

box-type brush-holder *(Elec.Eng.)*. See **brush-box**.

box-type (cage-type) negative plate *(Elec.Eng.)*. A form of negative plate for an accumulator which is made up by riveting two lead grids together and placing the active material in the spaces between them.

boxwood *(For.)*. The pale-yellow, close-grained, hard and tough wood of the box tree, used for drawing scales, tool handles, blocks for wood-cuts etc.; it requires several years of seasoning.

Boyle's law *(Phys.)*. The volume of a given mass of gas is inversely proportional to its pressure at constant temperature. There are deviations from this law at low and high pressures and according to the nature of the gas.

Boyle temperature *(Phys.)*. The temperature at which the second virial coefficient of a gas changes sign. Close to this temperature, Boyle's law provides a good approximation to the equation of state of the gas.

Boy's camera *(Image Tech.)*. A camera for photographing lightning flashes, gyrating lenses separating the strokes.

bp *(Chem.)*. Abbrev. for *boiling-point*.

BP *(Ships)*. Abbrev. for *Between Perpendiculars*.

BPA *(Aero.)*. Abbrev. for *British Parachute Association*.

Br *(Chem.)*. The symbol for bromine.

braccate *(Zool.)*. Of Birds, having feathered legs or feet.

brace *(Build.)*. A tool used to hold a bit and give it rotary motion. The bit is secured axially in a socket at one end, the other end (to which pressure is applied) being in line with it, while the middle part of the brace is cranked out so that the whole may be rotated. Also called a *bit-stock*.

brace *(Eng.)*. A tie rod or bar connecting two parts of a structure for stiffening purposes; it is always subjected to a tensile force.

brace *(Print.)*. ⏞ Usually cast to a definite em measurement; sectional braces are built up to the required length; should point towards the number of lines.

braced girder *(Build.)*. A girder formed of two flanges connected by a web consisting of a number of bars dividing the girder into triangles or trapeziums and transmitting the vertical loads from one flange to another.

brace jaws *(Build).* The parts of the socket of a brace which clamp upon the shank of the brace bit to secure it while drilling.

brachial *(Zool.).* See **brachium**.

brachiate, brachiferous *(Bot.,Zool.).* Branched: having widely spreading branches: bearing arms.

brachi-, brachio- *(Genrl.).* Prefix from L. *brachium*, arm.

brachiocephalic *(Med.).* Pertaining to the arm and head.

Brachiopoda *(Zool.).* A phylum of solitary non-metameric Metazoa, with a well-developed coelom; sessile marine forms, with a lophophore in the form of a double vertical spiral, and usually with a bivalve shell. Brachiopods range from early geological periods up to the present time; they occur in all seas, often at great depths.

brachium *(Zool.).* The proximal region of the fore-limb in land Vertebrates: a tract of nerve-fibres in the brain: more generally, any armlike structure, as the rays of Starfishes. adj. *brachial*.

brachy- *(Genrl.).* Prefix from Gk. *brachus*, short.

brachycephalic *(Med.).* Short-headed; said of skulls whose breadth is at least four-fifths of the length.

brachycerous *(Zool.).* Having short antennae, as some Diptera.

brachydactyly, brachydactylia *(Med.).* Abnormal shortness of fingers or toes.

brachyodont *(Zool.).* Said of Mammals having low-crowned grinding teeth in which the bases of the infoldings of the enamel are exposed: used also of the teeth. Also *brachydons*. Cf. **hypsodont**.

brachypterism *(Zool.).* In Insects, the condition of having wings reduced in length. adj. *brachypterous*.

brachysclereid *(Bot.).* *Stone cell.* A more or less isodiametric cell with a thick lignified wall, e.g. in the flesh of the pear fruit. See **sclereid**.

brachyurous, brachyural *(Zool.).* Said of decapodan Crustacea, in which the abdomen is reduced and bent forward underneath the laterally expanded cephalothorax by which it is completely hidden.

bracing *(Civ.Eng.).* The staying or supporting rods or ties which are used in the construction or strengthening of a structure.

bracing wires *(Aero.).* The wires used to brace the wings of biplanes and the earlier monoplanes. See also **drag wires**, and **landing wires**.

bracken poisoning *(Vet.).* A disease occurring in cattle and horses due to the ingestion of bracken. In cattle, the main symptoms are multiple haemorrhages and high fever, associated with bone marrow damage and later tumours of the gut. In horses, nervous symptoms are shown and the disease is essentially an induced thiamin deficiency.

bracket *(Build.).* A projecting support for a shelf or other part.

bracket arms *(Elec.Eng.).* The transverse projecting arms on the poles, for supporting the overhead contact wire equipment for a tramway or railway system.

bracket baluster *(Build.).* An iron baluster, bent at its foot and fixed into the side of the step, usually when the latter is made of stone or of concrete.

bracketed step *(Build.).* A step supported by a **cut string** which is shaped on its lower edge to form an ornamental bracket.

bracket fungus *(Bot.).* Basidiomycotina which have the fruiting body projecting as a rounded bracket from the side of a tree-trunk or stump.

bracketing *(Build.).* The shaped timber supports forming a basis for plasterwork and moulding of ceilings and parts near ceilings.

Brackett series *(Phys.).* A group of spectral lines of atomic hydrogen in the infrared given by the formula

$$\nu = R_H\left(\frac{1}{n_1^2} - \frac{1}{n_2^2}\right)$$

in which ν = the wave number; $R_H = 1.0967758 \times 10^7\ \text{m}^{-1}$; $n_1 = 4$; n_2 = various integral values.

brackish *(Ecol.).* Salty, but not as salty as sea water. Brackish water occurs in estuaries, creeks and deep wells.

bract *(Bot.).* A leaf, often modified or reduced, which subtends a flower or an inflorescence.

bracteate *(Bot.).* Having bracts.

bracteole *(Bot.).* A small leaf-like organ occurring along the length of a flower stalk between the true subtending bract and the calyx.

bract scale *(Bot.).* The structure in conifers that subtends the *ovuliferous scale* and may be more or less fused to it.

brad *(Build.).* A nail with a small head projecting on one side, or with the head flush with the sides. See **sprig**.

bradawl *(Build.).* A small chisel-edged tool, used to make holes for the insertion of nails and screws.

bradsot *(Vet.).* See **braxy**.

brady- *(Genrl.).* Prefix from Gk. *bradys*, slow.

bradyarthria *(Med.).* Abnormally slow delivery of speech.

bradycardia *(Med.).* Slowness of the beating of the heart.

bradykinesia *(Med.).* Abnormal slowness of the movements of the body.

Bragg angle *(Phys.).* The angle the incident and diffracted X-rays make with a crystal plane when the Bragg equation is satisfied for maximum diffracted intensity.

Bragg curve *(Phys.).* Graph giving average number of ions per unit distance along beam of initially monoenergetic α-particles (or other ionizing particles) passing through a gas.

Bragg equation *(Phys.).* If X-rays of wavelength λ are incident on a crystal, diffracted beams of maximum intensity occur in only those directions in which constructive interference takes place between the X-rays scattered by successive layers of atomic planes. If d is the interplanar spacing, the Bragg equation, $n\lambda = 2d \sin\theta$, gives the condition for these diffracted beams; θ is the angle between the incident and diffracted beams and the planes, and n is an integer. Also applied to electron, neutron and proton diffraction.

Bragg rule *(Phys.).* An empirical relationship according to which the mass stopping power of an element for α-particles (also applicable to other charged particles) is proportional to (relative atomic mass)$^{-0.5}$.

braid *(Textiles).* A wide range of narrow fabric woven on smallware looms and used as a trimming for the dress material, upholstery or coach and car interiors. The product obtained by **braiding**.

braided stream *(Geol.).* A stream which consists of several channels which separate and join in numerous places. Braided streams occur where the gradient is steep and where seasonal floods are liable to occur. They generally have wide beds filled with loose detritus.

braiding *(Textiles).* The process of plaiting in which three or more threads are interlaced to give a flat or tubular fabric.

brain *(Zool.).* A term used loosely to describe the principal ganglionic mass of the central nervous system: in Invertebrates, the pre-oral ganglia: in Vertebrates, the expanded and specialized region at the anterior end of the spinal cord, developed from the three primary cerebral vesicles of the embryo.

brain stem *(Zool.).* In Vertebrates, regions of the brain conforming to the organization of the spinal cord, as distinct from such suprasegmental structures as the cerebral cortex and the cerebellum.

brain stimulation *(Behav.).* Technique for studying the neurophysiological basis of some behaviour patterns thought to be under central nervous system (CNS) control; involves stimulation of the CNS by electrical or chemical means, sometimes producing behaviour that appears *motivated*.

brain voltage *(Med.).* Electric signal waves generated in human brain. Usually classed as alpha, beta, and delta waves according to frequency.

brake *(Eng.).* A device for applying resistance to the motion of a body, either (1) to retard it, as with a vehicle brake, or (2) to absorb and measure the power developed by an engine or motor.

brake *(Print.).* The manual or automatic mechanism to

control the tension of the reel when a rotary press is running.

brake bands *(Print.)*. Strips of leather, fabric, or metal acting on a pulley on the reel shaft of a rotary press.

brake drum *(Eng.)*. A steel or cast-iron drum attached to a wheel or shaft so that its motion may be retarded by the application of an external band or internal brake shoes. See **band-brake, expanding brake**.

brake drum *(Print.)*. A flat or V-shaped wheel at one end of the reel spindle by which the web tension is controlled.

brake efficiency *(Autos.)*. The retarding force expressed as a percentage of the total vehicle weight.

brake-fade *(Autos.)*. Condition caused by over-heating due to excessive use, resulting in decreased efficiency and sometimes in complete absence of stopping-power.

brake horsepower *(Eng.)*. The effective or useful horse-power developed by a prime-mover or electric motor, as measured by a brake or dynamometer and given in kilowatts.

brake lining *(Eng.)*. Strips of asbestos-base friction fabric riveted to the shoes of internal expanding brakes in order to increase the friction between them and the drum and provide a renewable surface. Modern practice is to bond lining to shoe. See **brake shoe**.

brake magnet *(Elec.Eng.)*. A permanent magnet or electromagnet which produces a braking effect, either by inducing eddy currents in a moving conductor or by operating a mechanical brake by means of a solenoid.

brake mean effective pressure *(Eng.)*. That part of the **indicated mean effective pressure** developed in an engine cylinder output equal to the brake horsepower of the engine; the product of *I.M.E.P.* and mechanical efficiency. Abbrev. *B.M.E.P.*

brake pads *(Autos.)*. The friction material in a disk brake, corresponding to the brake shoes in ordinary drum-type brakes.

brake parachute *(Aero.)*. One attached to the tail of some high-performance aircraft and streamed as a brake for landing. Sometimes a ribbon canopy is used for greater strength and on large aircraft a cluster of two or three is required to give sufficient area with convenient stowage. Also called *landing parachute*.

brake shoe *(Eng.)*. (1) Unit which carries the renewable rubbing surface of a block brake. (2) The segmental member which is pressed against the inner surface of a brake drum.

brake thermal efficiency *(Eng.)*. The efficiency of an engine reckoned in terms of the brake horsepower; given by the ratio of the heat equivalent of the brake output to the heat supplied to the engine in the fuel or steam.

braking notches *(Elec.Eng.)*. Positions of the handle of a drum-type controller which apply some form of electric braking.

brammallite *(Min.)*. A variety of illite with sodium as the inter-layer cation.

bran *(Med.)*. Unprocessed wheat bran is used to produce bulky faeces and relieve constipation.

branch *(Comp.)*. See **jump**.

branch *(Phys.)*. Alternative modes of radioactive decay.

branch *(Telecomm.)*. Electric components comprising a minimum path between junction points of common connection in a network. Also *arm*.

branch-circuit *(Elec.Eng.)*. A circuit branched off a main circuit.

branch exchange *(Telecomm.)*. See **private branch exchange**.

branch gap *(Bot.)*. A region of parenchyma in the vascular cylinder of the stem, located above the level where the *branch traces* bend out towards the branch. Cf. **leaf gap**.

branchia *(Zool.)*. In aquatic animals, a respiratory organ consisting of a series of lamellar or filamentous outgrowths; a gill. adj. *branchial*.

branchial arch *(Zool.)*. In Vertebrates, one of a series of bony or cartilaginous structures lying in the pharyngeal wall posterior to the hyoid arch; it prevents the gill-slits from collapsing.

branchial basket *(Zool.)*. (1) In Cyclostomata and cartilaginous Fish, the skeletal framework supporting the gills. (2) In the larvae of certain Dragonflies (Anisoptera), an elaborate modification of the rectum associated with respiration.

branchial chamber *(Zool.)*. In Urochordata, cavity of the pharynx.

branchial clefts *(Zool.)*. See **gill-slits**.

branchial heart *(Zool.)*. In Vertebrates, a heart such as that of the Cyclostomata, in which all the blood entering the heart is deoxygenated and passes thence directly to the respiratory organs: in Cephalopoda, special muscular dilations which pump blood through the capillaries of the ctenidia.

branchial rays *(Zool.)*. Branches of the hyoid and branchial arches which support the gills and gill-septa.

branching *(Phys.)*. The existence of two or more modes by which a radionuclide can undergo radioactive decay, e.g., ^{64}Cu can undergo β^-, β^+ and electron-capture decay.

branching ratio *(Chem.)*. Where a radioactive element can disintegrate in more than one way, the ratio of the quantities of the element undergoing each type of disintegration is called the *branching ratio*.

Branchiopoda *(Zool.)*. Subclass of Crustacea, the members of which are distinguished by the possession of numerous pairs of flattened, leaf-like, lobed swimming feet which also serve as respiratory organs; mainly fresh-water forms including the Fairy Shrimps, Brine Shrimps, Tadpole Shrimps, Clan Shrimps, Water Fleas.

branchiostegal *(Zool.)*. Pertaining to the gill-covers.

branchiostegal membrane *(Zool.)*. In Fish, the lower part of the opercular fold below the operculum.

branchiostege *(Zool.)*. See **branchiostegal membrane**.

branch jack *(Telecomm.)*. See **jack**.

branch of a curve *(Maths.)*. Any section or portion of a curve which is separated by a discontinuity (sometimes any singularity) from another section, e.g. a hyperbola has two branches.

branch of a function *(Maths.)*. See **Riemann surface**.

branch point *(Maths.)*. See **Riemann surface**.

branch switch *(Elec.Eng.)*. A term used in connection with electrical installation work to denote a switch of any type for controlling the current in a branch circuit.

brandering *(Build.)*. The process of nailing small fillets of wood in a counter direction and on the underside of floor joists. Plasterboards or metal lath are fixed to the branders to take plaster. Also *counter lathing*.

brand fungi *(Bot.)*. See **Ustilaginales**.

bran disease *(Vet.)*. See **osteodystrophia fibrosa**.

brand spore *(Bot.)*. The thick-walled resting spore of the brand fungi; it is black or brown, and forms sooty masses.

brass *(Eng.)*. Primarily, name applied to an alloy of copper and zinc, but other elements such as aluminium, iron, manganese, nickel, tin and lead are frequently added. There are numerous varieties.

brasses *(Eng.)*. Those parts of a bearing which provide a renewable wearing surface; they consist of a sleeve or bored block of brass split diametrally, the two halves being clamped into the bearing block by a cap.

Brassica *(Bot.)*. A genus of the Cruciferae which includes cabbage, broccoli, kale, rape, turnip, swede, mustard.

Brassicaceae *(Bot.)*. Same as **Cruciferae**.

brass rule *(Print.)*. Type-high brass strip, preferred to type-metal rule for heavy work.

brattice, brattice cloth *(Min.Ext.)*. A partition for diverting air, for the purpose of ventilation, into a particular working place or section of a mine.

bratticing, brattishing *(Build.)*. See **cresting**.

Braun Blanquet system *(Ecol.)*. Method of classifying vegetation in the European school of phytosociology, first enunciated by Braun Blanquet in 1921. With the advent of computers, its use has declined.

braunite *(Min.)*. A massive, or occasionally well-crystallized, ore of manganese, Composition $3Mn_2O_3.MnSiO_3$.

Braun tube *(Electronics)*. Original name for **cathode-ray tube**, after Braun (1850–1918) the inventor.

Bravais lattices *(Crystal.)*. The 14 distinct lattices which can be formed by the array of representative points in the study of crystal structure.

braxy *(Vet.)*. *Bradsot.* An acute and fatal toxaemia of sheep due to infection of the abomasum by *Clostridium septicum*. Controlled by vaccination.

Brayton cycle *(Eng.)*. A constant-pressure cycle of operations used in gas turbines.

brazier *(Build.)*. A portable iron container for a lighted fire, used to dry off building work in a room, or as a source of warmth for outside night-watchmen.

Brazilian emerald *(Min.)*. A misnomer for pure-green, deeply coloured variety of tourmaline, occurring in Brazil; used as a gemstone.

Brazilian mahogany *(For.)*. Hardwood from the genus *Cariniana*, not a true mahogany, though it may be used for the same purposes. It remains very stable after seasoning.

Brazilian pebble *(Min.)*. The name applied to Brazilian quartz or rock-crystal.

Brazilian peridot *(Min.)*. A misnomer for green crystals of tourmaline or chrysoberyl from Brazil having the typical colour of peridot (olivine).

Brazilian ruby *(Min.)*. A misnomer for pink topaz, or topaz which has become red after heating, or red tourmaline.

Brazilian sapphire *(Min.)*. TN for the beautiful clear blue variety of tourmaline mined in Brazil; used as gemstone, but not a true sapphire.

Brazilian topaz *(Min.)*. True topaz varying in colour from pure white to blue and yellow; mined chiefly in the state of Minas Geraes, Brazil.

brazing *(Eng.)*. The process of joining two pieces of metal by fusing a layer of brass or spelter between the adjoining surfaces.

brazing solders *(Eng.)*. Alloys used for brazing. They include copper-zinc (50–55% copper), copper-zinc-silver (16–52% copper, 4–38% zinc and 10–80% siver), also nickel-silver alloys.

BRC fabric *(Build.,Civ.Eng.)*. A very open, electrically welded, wire mesh with apertures about 3 × 12 in. (75 × 300 mm), used as a reinforcing medium for concrete roads, floor slabs etc. (*British Reinforced Concrete.*).

bread-crust bomb *(Geol.)*. A type of **volcanic bomb** having a compact outer crust and a spongy vesicular interior.

breadth coefficient *(Elec.Eng.)*. See **distribution factor**.

breadth factor *(Elec.Eng.)*. See **distribution factor**.

break *(Build.)*. (1) Any projection from, or recess into, the surface of a wall. (2) To nail the laths so that the joints are staggered, i.e. not in the same vertical line. See also **break-joint** and **breaking joint** below.

break *(Elec.Eng.)*. The shortest distance between the contacts of a switch, circuit-breaker, or similar apparatus, when the contacts are in the fully open position.

break *(Min.Ext.)*. (1) A jointing plane in a coal-seam. (2) Optimum range of size to which ore should be ground before further processing.

breakaway *(Image Tech.)*. A film studio set designed to come to pieces easily.

break-before-make *(Telecomm.)*. Classification of switch and relay wipers where existing contacts are opened before new ones close.

breakdown *(Elec.Eng.)*. The sudden passage of current through an insulating material at **breakdown voltage**.

breakdown crane *(Eng.)*. A portable jib crane carried on a railway truck or motor lorry, for rapid transit to the scene of an accident.

breakdown diode *(Electronics)*. See **Zener diode**.

breakdown voltage *(Elec.Eng.)*. Voltage at which a marked increase in the current through an insulator or semiconductor occurs. Abbrev. *BDV*. See also **disruptive voltage**.

breaker *(Elec.Eng.)*. See **circuit-breaker**.

breaker fabric *(Textiles)*. In a conveyor belt a layer of fabric placed between the main fabric of the belt and the outer rubber or plastic surface. Such fabrics are also part of cross-ply tyres.

breakeven *(Nuc.Eng.)*. In fusion, when the power produced exceeds the power input for heating and confinement.

break impulse *(Telecomm.)*. An impulse formed by interrupting a current in a circuit.

breaking *(Bot.)*. The development of streaks and stripes in flowers owing to virus infection e.g. in Rembrandt tulips.

breaking capacity *(Elec.Eng.)*. The capacity of a switch, circuit-breaker, or other similar device to break an electric circuit under certain specified conditions.

breaking current *(Elec.Eng.)*. The maximum current which a switch, circuit-breaker, or other similar device will interrupt without damage to itself.

breaking down, out *(Min.Ext.)*. Unscrewing the drillpipe preparatory to storage.

breaking elongation, extension *(Textiles)*. The elongation of fibre, yarn, or fabric that has occurred immediately before a breakage.

breaking joint *(Build.)*. The principle of laying bricks of building stones in such a manner that vertical joints are not continuous.

breaking length *(Paper)*. The length beyond which a strip of paper of uniform width would break under its own weight if suspended from one end. Usually expressed in metres.

breaking of the meres *(Bot.)*. The sudden development of large masses of blue-green algae (Cyanophyceae) in small bodies of fresh water.

breaking piece *(Eng.)*. An easily replaceable member of a machine subject to sudden overloads; made weaker than the remainder, so that in breaking it protects the machine from extensive damage.

breaking stress *(Eng.)*. The stress necessary to break a material, either in tension or compression. See **ultimate tensile stress**.

break iron *(Build.)*. See **back iron**.

break jack *(Telecomm.)*. See **jack**.

break joint *(Build.)*. See **breaking point**.

break line *(Print.)*. See **club line**.

break-out *(Foundry)*. Failure of mould of furnace wall to retain molten metal which escapes from the containing system.

breakpoint *(Comp.)*. Temporary halt inserted in a program, in order for the programmer to inspect the contents of registers, storage locations etc. to aid debugging.

break rolls *(Eng.)*. Grooved chilled-steel rollers, set in pairs, for shearing open wheat before separation of various parts of the grain.

break spinning *(Textiles)*. The term open-end spinning is now more usual.

breakthrough *(Min.Ext.)*. In industrial ion-exchange recovery of metal (e.g. uranium) from solution, the point at which traces of metal begin to arrive in the last of a series of resin-filled stripping columns.

breakwater *(Civ.Eng.)*. A natural or artificial coastal barrier serving to break the force of the wave so as to provide safe harbourage behind; it differs from the bulwark in that it has the sea on both sides of it.

breakwater-glacis *(Civ.Eng.)*. An inclined stone paving of piers and breakwaters, designed to take the force of impact of the waves.

breast *(Build.)*. The wall between a window and the floor. See also **chimney-breast**.

breast *(Med.)*. An accessory gland of the generative system, rudimentary in the male and secreting milk in the female. Extending from the third to the sixth rib in the front of the chest, it consists of fatty, fibrous and glandular tissue, the ducts of which end in the nipple.

breast *(Min.Ext.)*. (1) The working coal-face in a colliery. (2) Underground working face. In flat lodes, breast stopes are those from which detached ore will not gravitate without help.

breast bone *(Zool.)*. In higher Vertebrates, the sternum.

breast lining *(Build.)*. Panelling between window board and skirting.

breast mouldings *(Build.)*. Moulding on the part of the wall between a window and the floor.

breast roll *(Paper)*. Roll for carrying the wire cloth at the breast box end of the paper machine.
breastsummer *(Build.)*. See **bressummer**.
Breathalyzer *(Chem.)*. TN for apparatus designed to measure alcohol content of the blood by a chemical analysis of alveolar air (e.g. by reduction of potassium dichromate in sulphuric acid solution).
breather pipe *(Eng.)*. A vent pipe from the crank-case of an internal-combustion engine, to release pressure resulting from blow-by.
breathing *(Image Tech.)*. Variation of sharpness of the image caused by movement of the film in and out of the correct plane of focus in a camera, printer or projector.
breathing *(Zool.)*. An activity of many animals, resulting in the rapid movement of the environment (water or air) over a respiratory surface. Now usually referred to the more general concept of **respiration**.
breathing apparatus *(Min.Ext.)*. Mine rescue equipment in which oxygen is fed to a face mask carried by the wearer, via a demand valve. See **Weg rescue apparatus**.
breathing root *(Bot.)*. See **pneumatophore**.
breccia *(Geol.)*. A coarse-grained clastic rock consisting largely of angular fragments of pre-existing rocks. According to its mode of origin, a breccia may be a *fault-breccia*, a *crush-breccia*, an *intrusion-breccia*, or a *flow-breccia*.
Bredig's arc process *(Eng.)*. Process for making colloidal suspensions of metals in a liquid by striking an arc in the liquid between two electrodes of the metal.
breech block *(Eng.)*. A movable block used for closing and opening an aperture, originally in guns but now also in machines.
breeder *(Nuc.Eng.)*. Fusion machine in which further fuel (tritium) is bred from lithium. See **blanket**.
breeder reactor *(Nuc.Eng.)*. One which produces more fissile material than is consumed in operation. **Fast reactors** can be so designed.
breeding ratio *(Nuc.Eng.)*. Symbol b_r. The number of fissionable atoms produced in fertile material per fissionable atom destroyed in a nuclear reactor. $(b_r - 1)$ is known as the *breeding gain*.
breeze *(Build.)*. A general term for furnace ashes, or for *coke breeze*, *pan breeze*, and *furnace clinker*.
breeze concrete *(Build.)*. A concrete made of 3 parts coke breeze, 1 of sand, and 1 of Portland cement. It is cheap, and nails can be driven into it, but it has poor fire-resisting qualities.
breeze fixing brick *(Build.)*. A brick made from cement and breeze, built into the surface of a wall to take nails.
bregma *(Med.)*. The point of junction of the coronal and sagittal sutures of the skull.
Bréguet spring *(Horol.)*. A special form of balance spring, in which the outer coil of the spring is raised above the plane of the spiral, the end of the spring being bent to a special form before it enters the stud.
breithauptite *(Min.)*. Nickel antimonide, occurring as bright coppery-red hexagonal crystals, widely distributed in sulphide ore deposits in small amounts.
Breit-Wigner formula *(Phys.)*. A theoretical expression for the dependence of the cross-section σ of a particular nuclear reaction on the energy E of the bombarding particle and the width Γ of the resonant energy E_0. σ is proportional to

$$\sigma(A,B) = (2l+1)\frac{\lambda^2}{4\pi}\cdot\frac{W_A W_B}{(E-E_r)^2+(\frac{1}{2}W)^2}$$

The formula has been used with considerable success for many nuclear reactions particularly those involving neutron bombardment.
bremsstrahlung *(Phys.)*. Electromagnetic radiation emitted when a charged particle changes its velocity. Thus when electrons collide with a target, and suffer large decelerations, the X-radiation emitted constitutes the continuous *X-ray spectrum*.
bressumer *(Build.)*. A beam or lintel spanning a wide opening in a wall with whose surface it is flush. Also called a *breastsummer*.
breunnerite *(Chem.,Min.)*. Variety of **magnesite** containing some iron. Used in manufacture of magnesite bricks.
brevi- *(Genrl.)*. Prefix from L. *brevis*, short.
brevier *(Print.)*. An old type size, approximately 8-point.
Brewster angle *(Phys.,Telecomm.)*. A plane wave polarized in the plane of incidence is totally transmitted when incident on a plane dielectric boundary at the Brewster angle θ (measured from the normal to the boundary) is given by $\tan\theta = \sqrt{\varepsilon_2/\varepsilon_1}$ where ε_1, ε_2 are the permittivities of the two media. For optical wavelengths, $\tan\theta = n$, the refractive index between the media. Also *polarizing angle*. See **Brewster windows**.
brewsterite *(Min.)*. A rare strontium-barium zeolite.
Brewster law *(Phys.)*. That relating the Brewster angle (θ) to the refractive index (n) of the medium for a particular wavelength, viz., $\tan\theta = n$. For sodium light incident on particular glass, $n = 1.66$, θ is 51°.
Brewster's bands *(Phys.)*. Interference fringes which are visible when white light is viewed through two parallel and parallel-sided plates, whose thicknesses are in a simple ratio (1:1, 2:1, 1:3, etc.).
Brewster windows *(Phys.)*. Windows attached in certain designs of gas laser to reduce the reflection losses which would arise from the use of external mirrors. Their operation depends on the setting of the windows at the Brewster angle to the incident light.
Brianchon's theorem *(Maths.)*. The lines joining pairs of opposite vertices of a hexagon circumscribed about a conic are concurrent. The dual of **Pascal's theorem**.
brick *(Build.)*. A shaped and burnt block of special clay, used for building purposes.
brick-and-stud work *(Build.)*. See **bricknogging**.
brick-axe *(Build.)*. The two-bladed axe used by bricklayers in dressing bricks to special shapes.
brick clay *(Geol.)*. An impure clay, containing iron and other ingredients. In industry, the term is applied to any clay, loam, or earth suitable for the manufacture of bricks or coarse pottery. See **brick earths**.
brick-core *(Build.)*. Rough brickwork filling between a timber lintel and the soffit of a relieving arch.
brick earths *(Build.)*. Earths used for the manufacture of ordinary bricks; they consist generally of clayey silt interstratified with the fluvioglacial gravels of southern England, frequently exploited in brick manufacture.
bricking *(Build.)*. Work on plastered or stuccoed surfaces, in imitation of brickwork.
bricklayer's hammer *(Build.)*. A hammer having both a hammer-head and a sharpened peen; used for dressing bricks to special shapes.
bricklayer's scaffold *(Build.)*. A scaffold used in the erection of brick buildings, a characteristic being that one end of the **putlogs** is supported in holes left in the wall.
bricknogging *(Build.)*. The type of work used for walls or partitions which are built up of brickwork laid in spaces between timber. Also called *brick-and-stud work*.
brick-on-edge coping *(Build.)*. A coping finish to the exposed top of a wall; formed of bricks built on the edge in cement in courses 4½ instead of 3 in. high, so that the frogs are concealed and only a few joints are exposed to the weather.
brick-on-edge sill *(Build.)*. An external sill to window or door, formed in the manner of the **brick-on-edge coping**.
brick trowel *(Build.)*. A flat triangular-shaped tool used by bricklayers for picking up and spreading mortar.
bridge *(Elec.Eng.)*. Circuit often used for measurement of the ohmic value of passive components, for both a.c. and d.c. Four *arms* of the bridge are arranged in a diamond-shaped configuration, three comprising accurately known impedances and the fourth, the unknown. A voltage supply is connected to two opposite corners of the diamond and a detector between the other two. By adjusting the known component values, *balance* can be achieved in the bridge, where there is null signal indicated by the detector. For this condition, equations are available for the *unknowns* in terms of the other three arms of the bridge.
bridge *(Horol.)*. A raised platform or support, generally with two feet.

bridge board *(Build.)*. See **notch board**.

bridged-T filter *(Telecomm.)*. One consisting of a T-network, with a further arm bridging the two series arms; used for phase compensation.

bridge fuse *(Elec.Eng.)*. A fuse in which the fusible wire is carried in a holder, supported by spring contacts at its two ends; it is thus easily removable for renewing the fuse wire.

bridge gauge *(Eng.)*. A measuring device for detecting the relative movement of two parts of a machine due to wear at bearings, etc.

bridge hanger *(Elec.Eng.)*. A form of hanger of small vertical dimensions, for supporting the overhead contact-wire of a traction system under bridges or tunnels.

bridge-megger *(Elec.Eng.)*. A portable instrument for measuring large resistances on the Wheatstone-bridge principle. A megger contains a source of e.m.f. and the instrument dial on which the balance is indicated.

bridge network *(Telecomm.)*. Same as *lattice network*.

bridge neutralizing *(Elec.Eng.)*. Method for overcoming the adverse effects of interelectrode capacitances in thermionic valve amplifiers. Two valves are connected in push-pull with the anodes and grids cross-connected through balancing capacitors, the whole forming a balanced bridge.

bridge oscillator *(Elec.Eng.)*. One in which positive feedback and limitation of amplitude is determined by a bridge, which contains a quartz crystal for determining the frequency of oscillation. Devised by Meachan for high stability of operation in crystal clocks, etc.

bridge rectifier *(Elec.Eng.)*. Type of full-wave rectifier employing four rectifiers in the form of a bridge. The alternating supply is connected across one diagonal, and the direct-current output is taken from the other.

bridge stone *(Build.)*. A flat stone spanning a narrow area or gutter.

bridge transformer *(Telecomm.)*. The same as **hybrid coil**.

bridge transition *(Elec.Eng.)*. A method, employed in connection with the series-parallel control of traction motors, in which the change from series to parallel is effected without interrupting the main circuit, and without any change in the current flowing in each of the motors.

bridging *(Build.)*. The principle of diminishing lateral distortion of adjacent floor-joists by connecting them together with short cross-pieces.

bridging *(Min.Ext.)*. Arching of jammed rock so as to obstruct flow of ore; clogging of filtering septum by tiny particles which are individually small enough to pass, but which form such arches.

bridging floor *(Build.)*. A floor supported by bridging joists, without girders.

bridging joist *(Build.)*. A timber beam immediately supporting the floor-boards in a floor. Also *common joist*.

bridging ligand *(Chem.)*. A **ligand** that is bonded to more than one metal at a time. In Au_2Cl_6, each gold atom is coordinated by four chlorines, two of which bridge the two gold atoms.

bridle *(Build.)*. Whipcord or wire fastened half-way up the bristle of a new brush, when it is too long or flexible.

bridle *(Elec.Eng.)*. A portion of an overhead contact-wire system. It extends longitudinally between supporting structures and is attached at intervals to the contact-wire, in order to retain the latter in its proper lateral position.

bridle joint *(Build.)*. The converse of the **mortise-and-tenon joint**. The central part on the first member is cut away to leave two side tongues projecting, and the second member is cut away at the sides to receive these tongues.

Briggs logarithms *(Maths.)*. See **logarithms**.

bright annealing *(Eng.)*. The heating and slow cooling of steel or other alloys in a carefully controlled atmosphere, so that oxidation of the surface is reduced to a minimum and the metal surface retains its bright appearance. See **box annealing**.

bright emitter *(Electronics)*. A thermionic valve with a pure tungsten cathode, heated by a d.c. current to 2600 K in order to emit electrons. Originally used on all thermionic valves; now superseded by treated cathodes which emit at much lower temperatures.

brightener *(Elec.Eng.)*. An addition agent added to an electroplating solution to produce bright deposits.

brightening agent *(Textiles)*. A compound that on addition to a white or coloured textile material increases its brightness by converting some of the ultra-violet radiation into visible light. Also *fluorescent brightener* and *optical brightener*.

bright-field illumination *(Bot.)*. The common method of illumination in microscopy in which the specimen appears more or less dark on a bright background. Cf. **phase-contrast microscopy, interference microscopy, dark-ground illumination**.

bright-line viewfinder *(Image Tech.)*. Type of **Albada viewfinder** which reflects a luminous frame outlining the field of view.

brightness *(Phys.)*. See **luminance**. As a quantitative term, *brightness* is deprecated.

brightness control *(Image Tech.)*. That which alters the brightness on a cathode-ray tube screen.

Brighton system *(Elec.Eng.)*. A name sometimes given to the maximum-demand method of charging for an electricity supply; derived from the name of the town in which it was first used.

bright plating *(Elec.Eng.)*. The production of a fairly bright deposit from an electroplating plant. Such surfaces require little finishing.

Bright's disease *(Med.)*. A term formerly used for acute and chronic nephritis.

brilliance *(Acous.)*. The presence of considerable numbers of high harmonics in musical tone, or the enhancement of these in sound reproduction.

brilliant *(Print.)*. Old type size about 4-point.

brilliant green *(Chem.)*. The sulphate of *tetra-ethyl-diamino-triphenyl-methanol anhydride*; a green dye used as a disinfectant.

brilliant viewfinder *(Image Tech.)*. One comprising a reflector between two small lenses. An inclined mirror gives waist-height viewing; a prism gives eye-level viewing. The image is laterally inverted unless an extra lens called an *erector* is added.

Brillouin formula *(Phys.)*. A quantum mechanical analogue in paramagnetism of the Langevin equation in classical theory of magnetism.

Brillouin scattering *(Phys.)*. The scattering of light by the acoustic modes of vibration in a crystal, i.e. *photon-phonon* scattering.

Brillouin zone *(Phys.)*. Polyhedron in k-space, k being position wave vector of the groups or bands of electron energy-states in the band theory of solids. Often constructed by consideration of crystal lattices and their symmetrics.

brindled bricks *(Build.)*. Bricks which, owing to their chemical composition, show a striped surface; when they are otherwise satisfactory they are frequently used in cases where appearance is not an important consideration.

Brinell hardness test *(Eng.)*. A method of measuring the hardness of a material by measuring the area of indentation produced by a hard steel ball under standard conditions of loading. Expressed as *Brinell Hardness Number* which is the quotient of the load on the ball in kgf divided by the area of indentation in mm^2.

brine pump *(Eng.)*. The pump used to circulate brine through the evaporator of a refrigerator, the working parts being of corrosion-resisting alloy.

brise soleil *(Arch.)*. Screening in front windows to interrupt sun glare. It is used in the tropics and in Mediterranean countries and can take the form of precast concrete slats or of a permanent trellis supporting climbing plants.

brisket *(Vet.)*. The breast or anterior sternal region of an animal.

bristle *(Build.)*. The hair of the hog or boar used as a filling material in good quality paint brushes.

Bristol board *(Paper)*. A fine-quality cardboard made by pasting several sheets together, the middle sheets usually being of an inferior grade.

Bristol diamonds *(Min.)*. Small lustrous crystals of quartz i.e. rock crystal, occurring in the Bristol district.

Britannia metal *(Eng.)*. Alloy series of tin (80–90%) with antimony, copper, lead or zinc, or a mixture of these.

British Association (BA) screw thread *(Eng.)*. A system of symmetrical vee threads of 47½° included angle with rounded roots and crests. It is designated by numbers from 0 to 25, ranging from 6.0 mm to 0.25 mm in diameter and from 1 mm to 0.07 mm pitch. Used in instrument work, but now being superseded by standard metric sizes. Even numbers are preferred sizes.

British Columbian pine *(For.)*. See **Douglas Fir**.

British Standard brass (BSB) thread *(Eng.)*. A screw thread of Whitworth profile used for thin-walled tubing; it has 26 threads per inch for a given diameter. See **British Standard Whitworth thread**.

British Standard fine (BSF) thread *(Eng.)*. A screw-thread of Whitworth profile, but of finer pitch for a given diameter; largely used in automobile work.

British Standard Institution *(Genrl.)*. A national organization for the preparation and issue of standard specifications.

British Standard pipe (BSP) thread *(Eng.)*. *British Standard gas thread.* A screw-thread of Whitworth profile, but designated by the bore of the pipe on which it is cut (e.g. ⅜ in Gas) and not by the full diameter, which is a decimal one, slightly smaller than that of the pipe. See **British Standard Whitworth thread**.

British Standard specification *(Genrl.)*. A specification of efficiency, grade, size etc. drawn up by the British Standards Institution, referenced so that the material required can be briefly described in a bill or schedule of quantities. The definitions are legally acceptable.

British Standard Whitworth (BSW) thread *(Eng.)*. The pre-metric British screw thread, still widely used in the US, having a profile angle of 55 degrees and a radius at root and crest of 0.1373 × pitch; 1/6th of the thread cut off. The pitch is standardized with respect to the diameter of the bar on which it is cut.

British Standard Wire Gauge *(Eng.)*. See **Standard Wire Gauge**.

British Thermal Unit *(Phys.)*. The amount of heat required to raise the temperature of 1 lb of water by 1 Fahrenheit degree (usually taken as 60°–61°F). Abbrev. *BTU*. Equivalent to 252 calories, 778.2 ft lbf, 1055 J. 10^5 Btu = 1 therm.

brittle fracture *(Eng.)*. Stress failure occurring suddenly in mild-steel vessels, thought to arise from coalescence of multiple dislocations at the boundaries of the component grains.

brittle micas *(Min.)*. A group of minerals (the clintonite and margarite group) resembling the true micas in crystallographic characters, but having the cleavage flakes less elastic. Chemically, they are distinguished by containing calcium as an essential constituent.

brittleness *(Eng.)*. The tendency to fracture without appreciable deformation and under low stress. It is indicated in tensile test by low ultimate tensile stress and very low elongation and reduction in area. The notched-bar test may, however, reveal brittleness in metals that give a high ultimate tensile stress. See **toughness**.

brittle silver ore *(Min.)*. A popular name for **stephanite**.

Brix *(Chem.Eng.)*. Scale of densities used in the sugar industry. Hydrometers are market in 'degrees Brix', representing the density of a corresponding pure sugar solution in units equivalent to the percentage of sugar in the solution, either by volume ('volume Brix') of by mass ('mass Brix').

BRM *(Comp.)*. See **binary-rate-multiplier**.

broach *(Arch.)*. The sloping timber or masonry pyramid at the projecting corner of the square tower from which springs a *broach spire*.

broach *(Build.)*. The locating pin, within a lock, about which the barrel of the key passes.

broach *(Eng.)*. A metal-cutting tool for machining holes, often non-circular; it consists of a tapered shaft carrying transverse cutting edges, which is driven or pulled through the roughly finished hole.

broache work *(Build.)*. The finish given to a building-stone by dressing it with a punch so that broad diagonal grooves are left.

broach spire *(Arch.)*. An octagonal spire springing from a square tower without a parapet, and having the triangular corners of the tower covered over by short sloping pyramids blending into the spire.

broad *(Build.)*. A wood-turning tool, often consisting of a flat disk with sharpened edges fixed at right angles to a stem; used for shaping the insides and bottoms of cylinders.

broad *(Image Tech.)*. A studio light-source giving a wide angle of illumination.

broadband *(Telecomm.)*. (1) Said of a device (amplifier, mixer, transistor etc.) which is capable of operating with consistent efficiency over a wide range of frequencies. See **wideband amplifier**. (2) Used as a verb to imply the process of making a circuit or device operate over a wide range of frequencies. (3) Description of signals, noise, interference etc., which spreads over a wide range of frequencies.

broad-base tower *(Elec.Eng.)*. A transmission-line tower with each leg separately anchored.

broad beam *(Radiol.)*. Said of a gamma- or X-ray beam when scattered radiation makes a significant contribution to the radiation intensity or dose rate at a point in the medium traversed by the beam.

broadcast channel *(Telecomm.)*. Any specified frequency band used for broadcasting; chosen with regard to freedom from mutual and other forms of interference, consistency of propagation and reception, intended range of broadcasting (i.e. local, international, satellite) and bandwidth of programme material (i.e. sound or vision).

broadcasting *(Telecomm.)*. The transmission of a programme of sound, vision, or fascimile for general reception.

broadcast standard *(Image Tech.)*. The highest quality of video recording and reproduction, suitable for international broadcast transmission, in contrast to the lower quality acceptable for domestic application.

broadcast transmitter *(Telecomm.)*. Radio transmitter designed with broadcasting as one of the primary design criteria.

broadcloth *(Textiles)*. A woollen cloth, woven from fine yarns in a twill weave, heavily milled, and finished with a dress face; originally made in dark colours for suitings but now available in pastel shades.

broadcloth, cotton *(Textiles)*. In US a light-weight poplin shirting fabric.

broad gauge *(Civ.Eng.)*. A railway gauge in excess of the standard 4 ft 8½ in. (1.435 m). In particular, the gauge of 7 ft (2.134 m) laid down by Brunel.

broad irrigation *(Build.)*. A process of sewage purification in which the effluent is distributed over a large area of carefully levelled land, and allowed to soak through it and drain away as ordinary subsoil water down the natural water-courses. Cf. **intermittent filtration**.

broadsheet *(Print.)*. (1) The sheet before it is folded. (2) In rotary printing, the size of newspapers printed **columns around** the cylinder.

broadside *(Print.)*. A large sheet printed on one side, such as a poster.

broadside antenna *(Telecomm.)*. Array in which the main direction of the reception or radiation of electromagnetic energy is normal to the line of radiating elements.

broad-spectrum *(Med.)*. See **wide spectrum**. Said of drugs.

broadstone *(Build.)*. An **ashlar**.

broad tool *(Build.)*. A steel chisel having a cutting edge 3½ in (90 mm) in width, used for finish-dressing stone.

brocade *(Textiles)*. Jacquard designed dress or furnishing fabrics. The design is developed by floating the warp and/or weft threads in irregular order on a simple ground fabric.

Broca's area *(Med.)*. The left inferior convolution of the frontal lobe of the brain; the 'speech centre'.

brochanite *(Min.)*. A basic sulphate of copper occurring in green fibrous masses, or as incrustations; occurs in the oxidization zone of copper deposits.

brochure *(Print.)*. A booklet, pamphlet, or other short work with its pages stitched but not bound.

Brockenspectre *(Meteor)*. See **spectre of the Brocken**.

brockram *(Geol.)*. A sedimentary rock occurring in the Permian strata in N.W. England; consists of angular blocks which probably accumulated as scree material.

Brocot suspension *(Horol.)*. A form of pendulum suspension in which adjustment to the length of the pendulum can be made from the front of the dial.

brog *(Build.)*. An awl.

Broglie wavelength *(Phys.)*. See **de Broglie wavelength**

broke *(Paper)*. Wet or dry paper removed during the paper making or finishing processes and re-used within the mill.

broken colour *(Build.)*. A range of multi-colour decorative effects produced using colour glazes and washes and manipulating them using stipplers, combs, rags, sponges, etc.

broken ends *(Textiles)*. Warp threads which have broken during weaving.

broken-over *(Print.)*. The term used to indicate that plates or other separate sheets to be inserted in a book have been given a narrow fold on the inner edge, so that they will lie flat and turn easily when fixed.

broken picks *(Textiles)*. Defects in weaving due to breaking of the weft.

broken-space saw *(Build.)*. A handsaw having usually 6 teeth to the inch with spaces between each group of teeth.

broken twills *(Textiles)*. Fabrics in which the diagonal line forming the *twill* is broken, or broken and reversed in direction, at intervals.

broken wind *(Vet.)*. A chronic emphysema of the lungs of horses; sometimes associated with *chronic obstructive pulmonary disease*.

brom-cresol green *(Chem.)*. Indicator used in determination of pH values, suitable for ranges 3.6–5.2.

brom-cresol purple *(Chem.)*. Indicator used in determination of pH values within the range 5.2–6.8.

Bromeliaceae *(Bot.)*. Family of ca 2500 spp. of monocotyledonous flowering plants (superorder Commelinidae). Terrestrial and, especially, epiphytic herbs (including tank epiphytes and atmospheric plants) from tropical and subtropical America. Many are **CAM plants**. The flowers often have showy bracts and are bird- or insect-pollinated. Includes the pineapple (the only major CAM crop-plant) and some plants grown for fibre.

bromic acid *(Chem.)*. $HBrO_3$; with bases it forms *bromates* (V). A powerful oxidizing agent.

bromides *(Chem.)*. Salts of hydrobromic acid. Silver bromide is extensively used in photography.

bromination *(Chem.)*. The substitution by bromine in or addition of bromine to organic compounds.

bromine *(Chem.)*. A nonmetallic element in the seventh group of the periodic system, one of the halogens. Symbol Br, at no. 35, r.a.m. 79.909, oxidation states 1, 3, 5, 7, mp −7.3°C, bp 58.8°C, rel.d. 3.19. A dark red liquid, giving off a poisonous vapour, Br_2, with an irritating smell. In combination with various metals it is widely but sparingly distributed. The chief commercial source is sea water from which bromine is manufactured by treating the 'bittern' with chlorine. Bromine is used extensively in synthetic organic chemistry, as an anti-knock additive to motor fuel, in medicine and in halogen-quenched Geiger tubes.

bromochlorodifluoromethane *(Chem.)*. BCF. $CHBrClF_2$. Bp −4°C. Organic substance used as a fire extinguishing fluid, particularly for fires in confined spaces. Low toxicity vapour, 5.7 times as dense as air.

bromoform *(Chem.)*. $CHBr_3$, *tribromomethane*, mp 5°C, bp 151°C, rel.d. 2.9; a colourless liquid, of narcotic odour. Much used in laboratory separation of minerals into *floats*, rel.d. less than 2.9, and *sinks*, greater than 2.9.

bromoil process *(Image Tech.)*. Printing process in which a bleached and tanned bromide print is brushed with oil pigment which adheres to the shadow portion and is repelled by the highlights.

bromoil transfer *(Image Tech.)*. Print made by transferring a bromoil print to another sheet of paper by passing the two in contact through a press.

bromothymol blue *(Chem.)*. Indicator used in acid-alkali titrations, having a pH range of 6.0 to 7.6, changing from yellow to blue.

bronch-, broncho- *(Genrl.)*. Prefix from Gk. *bronchos*, windpipe.

bronchi *(Zool.)*. See **bronchus**.

bronchia *(Zool.)*. The branches of the bronchi. adj. *bronchial*.

bronchiectasis *(Med.)*. A chronic bronchopulmonary infection associated with an abnormal dilatation of the bronchial tree.

bronchiole *(Zool.)*. One of the terminal subdivisions of the bronchia.

bronchitis *(Med.)*. Inflammation of the bronchi. May be acute or chronic, the latter being a major cause of morbidity and mortality in the community.

bronchitis *(Vet.)*. See **husk**.

bronchogenic carcinoma *(Med.)*. Lung cancer arising from the epithelium of the bronchial tract.

bronchography *(Med.)*. The radiological examination of the trachea, bronchi, or the bronchial tree after the introduction of a **contrast medium**.

bronchography *(Radiol.)*. The instillation of **contrast medium** into the bronchi or air passages of the lung to improve X-ray visualization.

bronchoscopy *(Med.)*. Endoscopic examination of the tracheobronchial tree.

bronchus *(Zool.)*. One of two branches into which the trachea divides in higher Vertebrates and which lead to the lungs. pl. *bronchi*. adj. *bronchial*.

Brönsted-Lowry theory *(Chem.)*. Defines as an acid every molecule or ion able to produce a proton and as a base every molecule or ion able to take up a proton. Thus acid $\rightleftharpoons$base + proton. The acid and the corresponding base are called *conjugates*.

Brönsted's relation *(Chem.)*. Expression for the catalytic activity *k* of acids and bases in terms of their dissociation constants viz.,

$$k_{\text{acid}} = G_a K_a^{\alpha}$$

$$k_{\text{base}} = G_b K_b^{\beta}$$

where G_a or G_b is constant for a series of analogous catalysts of a given reaction in a given solvent and at a given temperature.

bronze *(Eng.)*. Primarily an alloy of copper and tin, but the name is now applied to other alloys not containing tin; e.g. aluminium bronze, manganese bronze and beryllium bronze. For varieties and uses of tin bronze, see **alpha bronze, bell metal, gunmetal, leaded** bronze, **phosphor-bronze**.

bronzed diabetes *(Med.)*. See **haemochromatosis**.

bronze powders *(Build.)*. Metallic powders made from alloys of copper and zinc or aluminium. Normally mixed with goldsize or bronze medium immediately prior to use in decorative work.

bronzing *(Build.)*. (1) Certain blue pigments, particularly of the *Prussian blue* and *monastral blue* types, which exhibit a metallic lustre when ground at fairly high concentrations into paint media. (2) Application of imitation gold or other metals by mixing powders with gold size or bronzing medium.

bronzing *(Print.)*. Dusting freshly-printed sheets, by hand or machine, with any suitable metallic powder, bronze-coloured or otherwise.

bronzite *(Min.)*. A form of orthopyroxene, more iron-rich than enstatite and more magnesian than hypersthene; often has metallic sheen, due to the reflection of light from planes of minute metallic inclusions in the surface layers.

bronzitite *(Geol.)*. A rock composed of bronzite with smaller amounts of angite and calcic plagioclase. A common constituent of layered basic igneous intrusions, such as those of the Bushveld, South Africa and Stillwater, Montana.

brood *(Zool.)*. A set of offspring produced at the same birth or from the same batch of eggs.

brookite *(Min.)*. One of the three naturally occurring forms of crystalline titanium dioxide.

Broughton countersink *(Build)*. Countersink which can be fixed to a shell bit, thus enabling the screw-hole and countersink to be bored in one operation.

Brouncker's series for $\log_e 2$ *(Maths.)*.

$$\log_e 2 = \frac{1}{1\cdot 2} + \frac{1}{3\cdot 4} + \frac{1}{5\cdot 6} + \frac{1}{7\cdot 8} + \cdots$$

brow *(Min.Ext.)*. The top of the shaft or 'pit'; hence also *pit-brow*.

Brown agitator *(Min.Ext.)*. See **agitator**.

brown algae *(Bot.)*. The *Phaeophyceae*.

Brown and Sharpe wire gauge *(Eng.)*. A system of designating the diameter of wire by numbers; it ranges from 4/0 (0.46 in) to 48 (0.001 24 in). Also *American Standard Wire Gauge*.

brown coal (lignite) *(Min.Ext.)*. Intermediate between peat and true coals, with high moisture content, the calorific value ranging from about 4000 to 8300 Btu/lb (9.5 to 20 MJ/kg).

brown earths, brown forest soils *(Bot.)*. A range of brown soils, with weakly developed horizons, **mull** humus and pH of say 5–7, formed under deciduous forests at humid, temperate latitudes. They make good agricultural soils. Cf. **podsol**.

brown forest soil *(Ecol.)*. A dark brown and friable soil with no visible layering, which is well aerated with only a thin litter layer. Also called *brown earth*. In Britain, a common soil type with good potential for agriculture.

brown hematite *(Min.)*. A misnomer, the material bearing this name being **limonite**, a hydrous iron oxide, whereas true hematite is anhydrous.

Brownian movement *(Phys.)*. Small movements of light suspended bodies such as galvanometer coils, or a colloid in a solution, due to statistical fluctuations in the bombardment by surrounding molecules of the dispersion medium. See **colloidal state**.

brown nose disease *(Vet.)*. *Copper nose*. A form of photosensitization occurring in cattle, characterized by brown discolouration and irritation of the muzzle and teats.

brown podzolic soil *(Ecol.)*. An acid soil, usually formed from a brown forest soil in areas of high rainfall, with a pale layer from which elements and particles have been leached to a deeper zone where iron is often precipitated to form an impenetrable layer.

brown rot *(Bot.)*. (1) Fungal diseases of plum and other fruit trees, infecting shoots and fruit. (2) The fungal decay of timber in which celluloses are preferentially attacked. Cf. **white rot**.

Brucellaceae *(Biol.)*. A family of obligate parasites belonging to the order *Eubacteriales*. Gram-negative cocci or rods: aerobic or facultatively anaerobic: many pathogenic species, e.g. *Pasturella pestis* (plague), *Pasturella multocida* (fowl cholera, swine plague, haemorrhagic septicaemia), *Brucella abortus* (undulant fever in man, contagious abortion in cattle, goats and pigs).

brucellosis *(Med,.Vet.)*. The diseases caused by infection with organisms of the genus *Brucella*. See **undulant fever**.

brucine *(Chem.)*. $C_{23}H_{26}O_4N_2$, a strychnine base alkaloid, mp of the anhydrous compound 178°C; it contains two methoxyl groups, and is a monoacidic tertiary base. Less physiologically active than *strychnine*.

brucite *(Min.)*. Magnesium hydroxide, occurring as fibrous masses in serpentinite and metamorphosed dolomite.

bruise *(Med.)*. Rupture of blood vessels in a tissue, with extravasation of blood, as a result of a blow which does not lacerate the tissue.

bruit *(Med.)*. A sound or murmur due to vascular blood flow, heard over heart, blood vessels and vascularized organs.

Brunt-Väisälä frequency *(Meteor.)*. The frequency $N/2\pi$ of small vertical oscillations of a parcel of air about its equilibrium position in a stable atmosphere. N is given by

$$N^2 = \frac{g}{\theta}\frac{\partial\theta}{\partial z}$$

where g is the acceleration of gravity and $\frac{\partial\theta}{\partial z}$ is the vertical gradient of **potential temperature**.

brush *(Elec.Eng.)*. A rubbing contact on a commutator, switch or relay. Also *wiper*.

brush border *(Biol.)*. See **microvillus**.

brush-box *(Elec.Eng.)*. That portion of the brush-holder of an electrical machine in which the brush slides or in which it is clamped.

brush coating *(Paper)*. A paper coating method in which the freshly applied wet coating is regulated and smoothed by means of brushes, some stationary and some oscillating, before drying.

brush contact *(Elec.Eng.)*. See **laminated contact**.

brush curve *(Elec.Eng.)*. The voltage drop between the brush arm and the segment beneath the brush at points along the brush width plotted against brush width as an indication of the correctness of compole flux density in a d.c. machine.

brushed fabric *(Textiles)*. A fabric, usually woven, that has been brushed or plucked so that some of the fibres stand out from the constituent yarns. The process is also known as raising and is carried out by machines with rollers covered with wire hooks, emery paper, or teazles.

brush gear *(Elec.Eng.)*. A general term used to denote all the equipment associated with brushes of a commutating or slip-ring machine.

brush-holder *(Elec.Eng.)*. The portion of an electrical machine or other piece of apparatus which holds a brush. See **box-type brush-holder**.

brush-holder arm *(Elec.Eng.)*. The rod or arm supporting one or more brush-holders. Also called **brush spindle**, brush stud.

brushing *(Vet.)*. *Cutting*. An injury to the inside of a horse's leg caused by the shoe of the opposite foot.

brush(ing) discharge *(Elec.Eng.)*. Discharge from a conductor when the p.d. between it and its surroundings exceeds a certain value but is not enough to cause a spark or an arc. It is usually accompanied by hissing high noise. Also *corona*.

brush(ing) discharge *(Min.Ext.)*. Electrical discharge from points along a bar charged to between 18 000 and 80 000 volts to create electrostatic field through which mineral particles fall and acquire polarity, in the high-intensity separation process.

brush lead *(Elec.Eng.)*. See **brush shift**.

brushmarks *(Build.)*. A paintwork defect characterized by visible depressed lines in the direction in which the paint has been brushed on. They are due to insufficient flow or levelling of the liquid paint. Also *tramlines, ropiness*.

brush plating *(Eng.)*. Method in which the anode carries a pad or brush containing concentrated electrolyte or gel which is worked over the surface to be plated. Similar methods are used for *brush polishing*.

brush-rocker *(Elec.Eng.)*. A support for the brushes of an electrical machine which enables them to be moved bodily round the commutator. Also called a *brush-rocker ring*.

brush shift *(Elec.Eng.)*. The amount by which the brushes of a commutating machine are moved from the centre of the neutral zone. Also called *brush lead*. See **backward shift, forward shift**.

brush spindle *(Elec.Eng.)*. See **brush-holder arm**.

brush spring *(Elec.Eng.)*. A spring in a brush-holder which presses the brush against the commutator or slip-ring surface.

brush stud *(Elec.Eng.)*. See **brush-holder arm**.

Brutalism *(Arch.)*. A term developed in UK in the 1950s to describe a style of architecture, popular at that time, characterized by expanses of unrelieved concrete and by the juxtapositioning of massive forms.

brute *(Image Tech.)*. A very large high-intensity spotlight.

bruzze *(Build).* A vee-shaped edge tool used by wood turners as a parting tool and by wheelwrights for chopping mortices in wheels.

Bryophyta *(Bot.).* (1) Division of the plant kingdom containing ca 25 000 spp. of small, rootless, thalloid or leafy, non-vascular plants; includes the liverworts (*Hepaticopsida*), the hornworts (*Anthoceropsida*) and the mosses (*Bryopsida*), having alternation of generations in which the gametophyte is the dominant generation, the sex organs are archegonia and antheridia and the sporophyte is more or less parasitic on the gametophyte. (2) In some, confusing, usages, the mosses alone. See **Bryopsida**.

bryophyte *(Bot.).* See **Bryophyta**.

Bryopsida *(Bot.). Musci.* The mosses. Class of those Bryophyta, ca 15 000 spp., which have a leafy (not thalloid) gametophyte with the leaves not strictly in 2 or 3 ranks, multicellular rhizoids and, in most, a capsule (sporophyte) with a columella and a lid (operculum). Includes *Sphagnum*. See also **acrocarp** and **pleurocarp**.

Bryozoa *(Zool.).* See **Ectoprocta**.

BSB thread *(Eng.).* Abbrev. for *British Standard Brass thread.*

BSF thread *(Eng.).* Abbrev. for *British Standard Fine thread.*

BSI *(Genrl.).* Abbrev. for *British Standards Institution.*

BSP thread *(Eng.).* Abbrev. for *British Standard Pipe thread.*

B-stage *(Plastics).* Transition stage through which a thermosetting synthetic resin of the phenol formaldehyde type passes during the curing process, characterized by softening to rubber-like consistency when heated, and insolubility in ethanol or acetone (propanone).

BSW thread *(Eng.).* Abbrev. for *British Standard Whitworth thread.*

BTU *(Phys.). Board of Trade Unit* = 1 kWh.

Bu *(Chem.).* A symbol for the butyl radical, C_4H_9—.

bubble *(Surv.).* The bubble of air and spirit vapour within a **level tube**: loosely, the level tube itself.

bubble chamber *(Phys.).* A device for making visible the paths of charged particles moving through a liquid. The liquid, often liquid hydrogen, heated above its normal boiling point, becomes superheated if the applied pressure is suddenly released. Bubbles of vapour are formed on the ions produced by the passage of charged particles through the chamber. The tracks of the particles are revealed by the bubbles, which if suitably illuminated, can be photographed. See **cloud chamber**.

bubble point *(Chem.).* The temperature at which the first bubble appears on heating a mixture of liquids. See **dew point**.

bubbles, pressure in *(Phys.).* See **pressure in bubbles**.

bubble store *(Comp.).* See **magnetic bubble memory**.

bubble trier, tube *(Surv.).* See **level trier, tube**.

bubo *(Med.).* An inflamed and swollen lymphatic gland, especially in the groin.

bubonic plague *(Med.).* A form of plague in which there is great swelling of a lymphatic gland, especially those in the groin. See **plague**.

buccal *(Med.).* Belonging to the mouth.

buccal *(Zool.).* Pertaining to, or situated in or on, the cheek or the mouth.

buccal cavity *(Zool.).* The cavity within the mouth opening but prior to the commencement of the pharynx.

buccal glands *(Zool.).* Glands opening into the buccal cavity in terrestrial Craniata; the most important are the salivary glands.

buccal respiration *(Zool.).* See **buccopharyngeal respiration**.

buccinator *(Med.).* A broad, thin muscle at the side of the face, between the upper and lower jaw.

bucco *(Genrl.).* Prefix from L. *bucca*, cheek.

buccopharyngeal respiration *(Zool.).* Breathing by means of the moist vascular lining of the mouth cavity or diverticula thereof, as in some Amphibians and certain Fish which have become adapted to existence on land.

Buchholz relay *(Elec.Eng.).* A protective relay for use with transformers or other oil-immersed apparatus; it embodies a float which becomes displaced and operates the relay contacts if gas bubbles are generated by a fault within the equipment being protected.

buchite *(Geol.).* A glassy rock which represents the result of partial fusion and recrystallization at very high temperatures of clay and shale material. It often occurs as xenoliths within igneous rocks.

Buchmann-Meyer effect *(Acous.).* The special type of reflection of light from the sound-track on a disk record whereby the lateral velocity of the track can be determined.

Buchner funnel *(Chem.).* A stout porcelain funnel having at its base a fixed horizontal perforated plate to act as a support over which a piece of filter paper is placed, thus ensuring a large area of filtration.

bucket *(Hyd.Eng.).* A dredging scoop, usually capable of being opened and shut for convenience in depositing and taking up a load.

bucket conveyor *(Eng.).* A conveyor or elevator consisting of a pair of endless chains running over toothed wheels, and carrying a series of buckets which, on turning over, discharge their contents at the delivery end.

bucket-dredge *(Min.Ext.).* System with 2 pontoons between which a chain of buckets digs through alluvium to mineral-bearing sands and delivers these to concentrating appliances on the pontoon decks. Dredge excavates and floats in a pond in which it traverses the deposit.

bucket-ladder dredger *(Civ.Eng.).* A **bucket dredge** reaching down into the material to be dredged, and lifting it for discharge into the vessel itself or into an attendant vessel.

bucket-ladder excavator *(Civ.Eng.).* A mechanical excavator working on the same principle as a **bucket-ladder dredger** but adapted for use on land. Also *dredger excavator.*

bucket valve *(Eng.).* A nonreturn (delivery) valve fitted in the bucket or piston of some types of reciprocating pump.

bucking coil *(Elec.Eng.).* A winding on an electromagnet to oppose the magnetic field of the main winding. Such a device, known as the *hum-bucking coil*, is sometimes used in electromagnetic loudspeakers to smooth out voltage pulsations in the power supply.

buckle *(Eng.).* (1) To twist or bend out of shape; said usually of plates or of the deformation of a structural member under compressive load. (2) A metal strap. (3) A swelling on the surface of a mould due to steam generated below the surface.

buckle-fold *(Print.).* A fold made in the paper parallel to its leading edge or to the fold previously made in it; the paper is brought to a sudden stop causing it to buckle into, and be folded by, the rollers. Cf. **knife-fold**.

Buckley gauge *(Electronics).* A sensitive ionization gauge for measuring very low gas pressures.

buckling *(Elec.Eng.).* A distortion of accumulator plates caused by uneven expansion, usually as a result of heavy discharges or other maltreatment.

buckling *(Image Tech.).* Pile-up of film in a camera or projector mechanism as the result of a break or incorrect threading.

buckling *(Nuc.Eng.).* A term in reactor diffusion theory giving a measure of curvature of the neutron density distribution. The *geometric buckling* depends only on the shape and dimensions of the assembly while the *material buckling* provides a measure of the multiplying properties of an assembly as a function of the materials and their disposition.

buckram *(Textiles).* A strong linen or cotton fabric stiffened by starch, gum, or latex and used for linings, hats, and in bookbinding.

buck saw *(Build).* A large frame-saw having one bar of the frame extended to form a handle.

buck transformer *(Elec.Eng.).* One with secondary in mains circuit to regulate voltage according to a controlling circuit feeding the primary. Also called *boost transformer*.

buckwheat rash *(Vet.).* See **fagopyrism**.

bud *(Bot.)*. (1) An unexpanded shoot consisting of a short rudimentary stem bearing immature and primordial leaves and/or flowers. At least in extant spermatophytes, buds are expected at shoot tips (terminal or apical buds) and in leaf axils (axillary buds); other buds are accessory or adventitious. (2) An outgrowth of a parent organism that becomes detached and develops into a new individual, especially from some yeasts and other fungi.

budding *(Bot.)*. (1) Asexual reproduction especially by some yeasts. (2) Bud grafting used especially for the propagation of fruit trees and some woody ornamentals including roses, in which a bud, together with more or less of the underlying stem, from the scion variety is grafted into a suitable root-stock.

budding *(Zool.)*. A primitive method of asexual reproduction by growth and specialization and separation by constriction of a part of the parent.

buddle *(Min.Ext.)*. A shallow annular pit for concentrating finely crushed, slimed, base-metal ores.

buddle-work *(Min.Ext.)*. Treatment of finely ground tin-bearing sands by gentle sluicing, in which a heavier fraction of the fed pulp is built up (*buddled*) while the lighter fraction flows to discard. This is continued until a satisfactory concentrate is produced.

bud scale *(Bot.)*. A simplified leaf or stipule on the outside of a bud, forming part of a covering which protects the contents of the bud.

bud sport *(Bot.)*. A shoot, branch, inflorescence or flower differing markedly from the rest of the plant with the differences persisting in vegetatively propagated off-spring; due to nuclear or cytoplasmic mutation when the sport will often be chimeric, or in horticulture to a change in the structure of a pre-existing chimera.

buff *(Eng.)*. A revolving disk composed of layers of cloth charged with abrasive powder; used for polishing metals.

buffer *(Comp.)*. A store area used temporarily to hold data being transmitted between a peripheral device and the central processor, to allow for differences in their working speeds. Buffering can also be used between two peripheral devices. See **double buffering**.

buffer *(Elec.Eng.)*. An electronic amplifier, often with unity gain, which is designed to decouple input from output. Normally designed to have high input impedance so that it does not load the driving stage and low output impedance such that it can provide current drive.

buffer *(Eng.)*. A spring-loaded pad attached to the framework of railway rolling-stock to minimize the shock of collision; any resilient pad used for a resilient purpose.

buffer action *(Chem.)*. The action of certain solutions in opposing a change of composition, especially of hydrogen ion concentration. See **buffer solution**.

buffer battery *(Elec.Eng.)*. A battery of accumulators arranged in parallel with a d.c. generator to equalize the load on the generator by supplying current at heavy-load periods and taking a charge during light-load periods.

buffer capacitor *(Elec.Eng.)*. See **blocking capacitor**.

buffer capacity of developer *(Image Tech.)*. The capacity of an alkali (e.g. sodium metaborate) in a developing solution to maintain a slow rate of decrease in pH value.

buffer circuit *(Acous.)*. The resistance-capacitor unit which determines the rate of rise or fall of the envelope of the waveform of emitted sounds which has been generated in electrostatic circuits in electronic organs.

buffer coat *(Build.)*. A coating applied to a surface to isolate the surface or constituents therein from subsequent coating.

buffer reagent *(Elec.Eng.)*. A substance added to an electrolyte solution which prevents rapid changes in the concentration of a given ion. Also called *buffer*.

buffer resistance *(Elec.Eng.)*. See **discharge resistance**.

buffer solution *(Chem.)*. A solution whose pH-value is not appreciably changed by additions of acid or alkali. Normally it is the solution of a weak acid or base with one of its salts.

buffer spring *(Eng.)*. The part lending resiliency to railway *buffing gear*.

buffer stage *(Telecomm.)*. An amplifying stage coming between the master oscillator and the modulating stage of a radio transmitter, to prevent the changing load of the modulated output from affecting the frequency of the master drive.

buffet boundary *(Aero.)*. The limiting values of **Mach number** and altitude at which an aircraft can be flown without experiencing buffet in unaccelerated flight.

buffeting *(Aero.)*. An irregular oscillation of any part of an aircraft, caused and maintained by an eddying wake from some other part; commonly, tail buffeting in the downwash of the main planes, which gives warning of the approach of the **stall**.

buffing *(Build.)*. The grinding down of a surface to remove extrusions or to expose the underlying material.

bug *(Comp.)*. Defect in a program, or a fault in equipment. See **error, diagnostic**.

bug-eye lens *(Image Tech.)*. See **fish-eye lens**.

bug key *(Telecomm.)*. A Morse key that permits higher transmission speeds than a normal key. A lever, moved horizontally by the hand, sends dashes in one direction when held over. Dots are sent by a spring contact attached to the lever, when the lever is released from sending a dash.

buhl saw *(Build)*. A kind of frame-saw in which the back of the frame is so spaced from the saw itself as to allow the latter to cut well into the work.

buhr (burr) mill *(Min.Ext.)*. One in which material is ground by passage between a fixed and a rubbing surface. Types include old-fashioned flour mill, with a circular grindstone rotating above a fixed lower one, radially grooved to facilitate passage of grist from centre to peripheral discharge. Also, rotating cone in fixed casing, material gravitating through the intervening space. Used for softish material, e.g. grain, food processing, and such minerals as soft limestone.

builders' level *(Build.)*. (1) A spirit-level tube set in a long straightedge; for testing and adjusting levels. (2) A simple form of dumpy or tilting level, used on building works or for running the levels of drains.

builders' staging *(Build.)*. A robust type of scaffold, formed of square timbers strongly braced together, capable of being used for the handling of heavy materials.

building board *(Build.)*. Board manufactured from various materials and supplied with various finishes; used for lining walls and ceilings. See **fibre-board**.

building certificates *(Build.)*. Certificates made out by the architect during the progress, or after completion, of the works on a building contract, to enable the contractors to obtain payments on account or in settlement from the employer.

building line *(Build.)*. The line beyond which a building may not be erected on any given plot.

building paper *(Build.)*. Sandwich of fibre and bitumen between two sheets of heavy paper, used in damp-proofing and for insulation between the soil and road surfacing.

build-up *(Radiol.)*. Increased radiation intensity in an absorber over what would be expected on a simple exponential absorption model. It results from scattering in the surface layers and increases with increasing width of the radiation beam.

build-up sequence *(Eng.)*. The order in which the welding beads are applied along a chamfered seam joining thick metals in order to give maximum strength.

build-up time *(Telecomm.)*. See **rise-time**.

bulb *(Bot.)*. An organ of storage and perennation, usually underground, consisting of a short stem bearing a number of overlapping swollen fleshy leaf bases and/or scale leaves, with or without a *tunic*, the whole enclosing next year's bud e.g. onion. Cf. **corm, rhizome**.

bulb *(Elec.Eng.)*. The gas-tight envelope, usually glass, which encloses the electrodes of a thermionic valve or of an electric discharge lamp, or the filament of an electric filament lamp.

bulb *(Zool.)*. Any bulb-shaped structure. adj. *bulbar*.

bulb bar *(Eng.)*. A rolled, or extruded, bar of strip form in which the section is thickened along one edge.

bulbiferous *(Bot.)*. Having, on the stem, bulbs or bulbils in place of ordinary buds.

bulbil *(Bot.)*. (1) Small bulb or tuber developing above ground from an axillary bud or in an inflorescence and functioning in asexual reproduction.

bulbil *(Zool.)*. (1) A contractile dilatation of an artery. (2) Any small bulblike structure.

bulbus *(Zool.)*. See below and **bulb**.

bulbus arteriosus *(Zool.)*. In many Vertebrates, a strongly muscular region following the conus arteriosus.

bulbus oculi *(Zool.)*. The eyeball of Vertebrates.

bulimia *(Med.)*. An abnormal increase in the appetite, often part of the symptoms of **anorexia nervosa**.

bulk *(Paper)*. A measure of the reciprocal of the density of paper, being the ratio of thickness to substance. A loose synonym for thickness.

bulk concrete *(Civ.Eng.)*. See **mass concrete**.

bulk density *(Powder Tech.)*. The value of the apparent powder density when measured under stated freely poured conditions.

bulked yarn *(Textiles)*. Yarns that have been treated chemically or physically to increase their loftiness. Although the term includes yarns made from staple fibres it is more generally applied to continuous filament yarns as *bulked continuous filaments, BCF*. The term *textured yarn* is now preferred.

bulk factor *(Plastics)*. Ratio of the density of sound moulding to the powder density of the moulding material from which it was made.

bulk flotation *(Min.Ext.)*. Froth flotation process so applied as to concentrate more than one valuable mineral in one operation.

bulkhead *(Aero.)*. In fuselages, a major structural transverse dividing wall providing access between several internal sections, or a strengthened and sealed wall at the front and rear designed to withstand the differential pressure required for pressurization. In power plant nacelles, a structure serving as a firewall.

bulkhead *(Autos.)*. On a public service vehicle, partition at the front between driver and passenger accommodation. The partition betweeen engine and passenger compartments in any vehicle.

bulkhead *(Civ.Eng.)*. A masonry or timber partition to retain earth , as in a tunnel or along a waterfront.

bulkhead *(Ships)*. A partition within a ship's hull or superstructure. It may be transverse or longitudinal, watertight, oiltight, gastight or partially open. It may form part of the ship's subdivision for seaworthiness or otherwise.

bulkhead deck *(Ships)*. The uppermost deck up to which watertight transverse bulkheads are carried.

bulkiness *(Powder Tech.)*. A term used to describe the properties of a bed of powder. It is defined as the reciprocal of the apparent density of the powder under the stated conditions.

bulking *(Build.)*. See also **moisture expansion**: Expression used to describe the increase in volume of damp sand compared to dry or saturated sand.

bulk modulus *(Eng.)*. Relationship of applied stress and volumetric strain which occurs on the application of a uniform stress to all parts of a body.

bulk sample *(Min.Ext.)*. One composed of several portions taken from different locations within a bulk quantity of the material under test.

bulk test *(Nuc.Eng.)*. The large sample usually required for a shield radiation test of a material having a high attenuation.

bulla *(Med.)*. A blister or bleb. A circumscribed elevation above the skin, containing clear fluid; larger than a **vesicle**.

bulla *(Zool.)*. In Vertebrates, with a flask-shaped tympanum, the spherical part of that bone which usually forms a protrusion from the surface of the skull.

bullate *(Bot.)*. (1) Having a blistered or puckered surface. (2) Bubblelike. (3) Bearing one or more small hemispherical outgrowths.

bull-chain *(For.)*. The endless chain used in a log haul for conveying logs.

bulldog calf *(Vet.)*. A lethal form of achondroplasia occurring mainly when Dexter cattle are mated together; due to the inheritance of a pair of semi-dominant genes responsible for the achondroplasia trait.

bulldozer *(Civ.Eng.)*. A power-operated machine, provided with a blade for spreading and levelling material. Also called *angledozer*.

bullet catch *(Build.)*. See **ball catch**.

bulletwood *(For.)*. Timber from a tree of the genus *Mimusops* which also yields *balata*, a soft rubber-like material used in golf balls and for impregnating belts. Renowned for its strength and durability, it is used for structural work, boat building, furniture and cabinet making, tool handles, wheel spokes and railway sleepers.

bull-headed rail *(Civ.Eng.)*. A rail section once used widely in the UK, having the shape roughly of a short dumb-bell in outline, but with unequal heads, the larger being the upper part in use. See **flanged rail**.

bull header *(Build.)*. A brick with one corner rounded, laid with the short face exposed, as a quoin or for sills, etc.

bullhead tee *(Build.)*. A tee having a branch which is longer than the run.

bull-holder *(Vet.)*. Forceps for grasping the nasal septum of cattle as a means of restraint.

bulliform cell *(Bot.)*. An enlarged epidermal cell present, with other similar cells, in longitudinal rows in the leaves of some grasses and alleged to be *motor* cells causing the rolling and unrolling of leaves in response to changes of water status.

bulling *(Civ.Eng.,Mining)*. The operation of detaching a piece of loosened rock by exploding blasting charges inserted in the surrounding fissures.

bulling bar *(Civ.Eng.,Mining)*. An iron bar used to force clay into the crevices in the sides of a bore hole.

bullion *(Eng.)*. (1) Gold or silver in bulk, i.e. as produced at the refineries, not in the form of coin. (2) The gold-silver alloy produced before the metals are separated.

bullion content *(Eng.)*. In parcel of metal or minerals being sold, where the main value is that of the base metal which forms the bulk of the parcel, the contained gold or other precious metal of minor value included in the sale.

bullion point *(Glass)*. The centre piece of a sheet of glass made by the old method of spinning a hot glass vessel in a furnace until it opened out under centrifugal action to a circular sheet. The centre piece bears the mark of attachment to the rod used to spin the sheet. The method is obsolete now, but is revived for 'antique' effects.

bull-nose *(Build.)*. A small metal rebating plane having the mouth for the cutting iron near the front.

bull-nosed step *(Build.)*. A step which, in plan, is half-round or quarter-round at the end.

bull-ring *(Elec.Eng.)*. a metal ring used in the construction of overhead contact wire systems for electric schemes; it forms the junction of three or more straining wires.

bull's eye lens *(Image Tech.)*. A small thick lens, used for condensing light from a source.

bull stretcher *(Build.)*. A brick with one corner rounded, laid, with the long face exposed, as a quoin.

bull wheel *(Min.Ext.)*. The driving pulley for the camshaft of a stamp battery; one on which bull rope of drilling rig is wound.

bulwark *(Civ.Eng.)*. A sea-wall built to withstand the force of the waves; in some cases the reinforcement of the natural **breakwater**.

bumblefoot *(Vet.)*. A cellulitis of the foot of birds due to infection by *Staphylococcus aureus*.

bumetanide *(Med.)*. A **'loop' diuretic** used in the treatment of oedema.

bump *(Aero.)*. See **air pocket**.

bumping-up *(Print.)*. Interlaying the half-tone areas of rotary printing plates to provide heavier impression, there being no provision for **make-ready** on the rubber-covered impression cylinder.

Buna *(Plastics)*. Synthetic rubber manufactured (at first in Germany) by polymerization of butadiene with sodium (hence the name *Bu* + *Na*). Buna-N (NBR, GR-N), made from inter-polymerization of butadiene with acrylonitrile, has good ageing and oil-resisting properties; Buna-S (SBR, GR-S), made from butadiene and styrene, has good mechanical, electrical and ageing properties; especially used for tyres.

buncher *(Electronics).* Arrangement which velocity-modulates and thereby forms bunches of electrons in the electron beam current passing through it. Bunching would be *ideal* if the bunches contained electrons all having the same beam velocity. Also *input gap, buncher gap.* See **catcher, debunching, rhumbatron.**

bunching *(Electronics).* The process of forming a steady electron beam into a succession of electron groups, or *bunches.* The result of interaction between an alternating electric field at the mouth of a cavity (see **rhumbatron**) and an electron beam passing close by. See also **velocity modulation.**

bunching angle *(Electronics).* Transit delay or phase angle between modulation and extraction of energy in a bunched beam of electrons.

bundle *(Bot.).* See **vascular bundle.**

bundle *(Med.).* Fibres collected into a band in the nervous system or in the heart.

bundle cap *(Bot.).* Strand of sclerenchyma or parenchyma adjacent to the xylem and/or phloem sides of a vascular bundle.

bundle conductor *(Elec.Eng.).* Two or more overhead line conductors, suitably spaced to avoid *corona* loss, forming a phase; replaces a single large conductor.

bundle divertor *(Nuc.Eng.).* See **divertor.**

bundle end *(Bot.).* The much simplified termination of a small vascular bundle in the mesophyll of a leaf.

bundle of His *(Med.).* In the mammalian heart, a bundle of small specialized conducting muscle fibres extending from the wall of the right atrium to the septum between the ventricles; responsible for transmitting electrical impulses from atrium to ventricle.

bundle sheath *(Bot.).* A sheath of one or more layers of parenchymatous or sclerenchymatous cells, surrounding a vascular bundle in a leaf.

bungalow *(Arch.).* A single-storey house.

bunion *(Med.).* An enlarged deformed joint of the big toe where it joins the foot, as a result of pressure of tight-fitting shoes or boots, with overlying **bursitis.**

bunker *(Eng.).* A storage space for coal or oil fuel.

bunker capacity *(Ships).* The capacity of a space in a ship used for carrying fuel. It is calculated at a fixed rate of stowage per unit volume, according to fuel, and allowances for obstructions are made in percentage.

bunodont *(Zool.).* Said of mammalian teeth in which the cusps remain separate and rounded. Also *bunoid.* Cf. **lophodont, selenodont.**

Bunsen burner *(Chem.).* A gas burner consisting of a tube with a small gas jet at the lower end, and an adjustable air inlet by means of which the heat of the flame can be controlled; used as a source of heat for laboratory work and formerly, in conjunction with an incandescent mantle, as the usual form of gas burner for illuminating purposes. Invented by Robert Bunsen.

Bunsen flame *(Chem.).* The flame produced when a mixture of a hydrocarbon gas and air is ignited in air, as in a Bunsen burner. It consists of an inner cone, in which carbon monoxide is formed, and an outer one, in which it is burnt.

Bunsen photometer *(Phys.).* See **grease-spot photometer.**

bunt *(Aero.).* A manoeuvre in which an aircraft performs half an inverted loop, i.e. the pilot is on the outside where he experiences **negative g.**

bunt *(Bot.).* Disease of cereals caused by a *smut* fungus in which the grain of infected plants is transformed into a mass of spores.

buoy *(Hyd.Eng.).* A floating vessel, capable of being illuminated at night, moored in estuaries and ship-canals to mark the position of minor shoals, and to show the navigable channel.

buoyancy *(Aero.).* The vertical thrust on an aircraft due to its immersion, either wholly or partly, in a fluid. Equal to the weight of air displaced by the gas-bags in the case of airship: equal to the weight of water displaced by the immersed portions of the floats of a seaplane, or the body of a flying-boat. See also **reserve-.**

buoyancy *(Phys.).* The apparent loss in weight of a body when wholly or partly immersed in a fluid; due to the upthrust exerted by the fluid. See **Archimedes' principle, correction for buoyancy.**

bupivacaine *(Med.).* A powerful local anaesthetic used for regional nerve block anaesthaesia, particularly epidural.

burden *(Elec.Eng.).* A term used to signify the load on an instrument transformer. It is usually expressed as the normal rated load in volt-amperes, or as the impedance of the circuit fed by the secondary winding.

burden *(Eng.).* See **oncosts.**

burden *(Min.Ext.).* (1) Amount of rock to be shattered in blasting between drill-hole and nearest free face. (2) See **overburden.**

Burdizzo pincers *(Vet.).* A castrating instrument which crushes the spermatic cord.

burette *(Chem.).* A vertical glass tube with a fine tap at the bottom and open at the top, usually holding up to 50 cm^3 of reagent solution; used in **volumetric analysis.** The tube is usually graduated in tenths of a cm^3, so that the amount of liquid allowed to run out through the tap can be estimated to one-twentieth of a cm^3.

Burger's vector *(Crystal.).* Translation vector of crystal lattice, representing a displacement which creates a lattice dislocation.

burial site *(Nuc.Eng.).* Place for the deposition, usually in suitable containers of radioisotopes after use, contaminated material or radioactive products of the operation of nuclear reactors. Also *graveyard.*

Burkitt lymphoma *(Immun.,Med.).* A malignant tumour of B-cells, especially affecting the jaw and the gut, common in children in hot humid regions of Africa but not confined to these regions. **Epstein-Barr virus** is present and may be responsible for malignant transformation occurring in a B-cell population subject to constant antigenic stimulation. Associated with a specific chromosomal rearrangement affecting chromosome 8q24.

burlap *(Textiles).* A coarse jute, hemp, or flax fabric.

Burma lancewood *(For.).* A durable wood from the genus *Homalium,* used in India for the making of agricultural implements as well as being a structural timber.

burmite *(Min.).* An amber-like mineral occurring in the upper Hukong Valley, Burma, differing from ordinary amber by containing no succinic acid. A variety of retinite.

burn *(Electronics).* See **ion burn.**

burn *(Space).* Controlled firing of rocket engine for adjusting course, re-entry initiation etc.

burnable poison *(Nuc.Eng.).* Neutron absorber introduced into reactor system to reduce initial reactivity but becoming progressively less effective as burn-up proceeds. This helps to counteract the fall in reactivity as the fuel is used up.

burner firing block *(Eng.).* Unit made from refractory material that fits into a furnace wall at the burner position, having a nozzle-protecting recess at back and a tunnel on the firing side. It is called *quarl* in oil-firing practice.

burner loading *(Eng.).* Potential heat that can be liberated efficiently from a burner. Expressed in kilowatts or Btu/hour.

burner turndown factor *(Eng.).* Minimum gas rate at which a burner is capable of stable flame propagation without the flame flashing back to the air-gas mixing point or blowing off from the burner nozzle or head.

burning *(Eng.).* The heating of an alloy to too high a temperature, causing local fusion or excessive penetration of oxide, and rendering the alloy weak and brittle.

burning *(Min.Ext.).* Changing the colour of certain precious stones by exposing them to heat.

burning-in kiln *(Glass).* A kiln in which stain or enamel colour painted on glass-ware or sheet-glass is fired to cause it to adhere more or less permanently; usually of muffle type.

burning-on *(Foundry).* The process of adding a piece to an existing casting by making a mould round the point of juncture and pouring metal into it.

burnishing *(Print.)*. The operation of applying a brilliant finish to gilt or coloured edges by means of a burnishing tool, which is applied under great pressure.

burnout *(Electronics)*. Sudden failure of any device, caused by excessive current, leading in turn to overheating; may also be due to failure of artificial cooling in any electronic assembly or sub-assembly.

burn-out mask *(Print.)*. See **print-out mask**.

burn-out velocity *(Space)*. The maximum velocity achieved by a rocket when all the propellant has been consumed.

burnt coal *(Min.)*. Sooty product of weathering of a coal outcrop.

burnt deposit *(Elec.Eng.)*. A loose powdery deposit obtained in electroplating, if the rate of deposition is allowed to be too great.

burnt lime *(Build.,Chem.)*. See **lime**.

burnt metal *(Eng.)*. Metal which has become oxidized by overheating, and so is rendered useless for engineering purposes.

burnt sienna *(Build.)*. Orange-brown pigments used in **graining** and **marbling**.

burnt umber *(Build)*. Rich brown pigment used in graining and marbling.

burnup *(Nuc.Eng.)*. (1) In nuclear fuel, amount of fissile material burned up as a percentage of total fissile material originally present. (2) Of fuel element performance, the amount of heat released from a given amount of fuel, expressed as megawatt- (or gigawatt-) days per tonne.

burr *(Bot.)*. A fruit covered with hooks to aid in dispersal by animals.

burr *(Eng.)*. (1) The edge of the material which is turned over during various operations like punching and cutting. (2) A rotary tool with cutting teeth like a file.

burr *(For.)*. An excrescence in tree growth which, when sliced, produces strong contrasts in the form and colour of markings.

burrs *(Build.)*. Lumps of brick, often mis-shapen, which in the burning have fused together, and which are used for rough-walling, artificial rock-work, etc.

bursa *(Med.)*. A synovial sac located at points of friction in the body.

bursa *(Zool.)*. Any saclike cavity: more particularly, in Vertebrates, a sac of connective tissue containing a viscid, lubricating fluid, and interposed at points of friction between skin and bone and between muscle, ligament, and bone.

bursa copulatrix *(Zool.)*. A special genital pouch of various animals acting generally as a female copulatory organ.

bursa inguinalis *(Zool.)*. The cavity of the scrotal sac in Mammals.

bursa of fabricius *(Immun.)*. A sac-like structure arising as a diverticulum from the cloaca of young birds, composed of primary follicles containing B-lymphocyte precursors. The bursa is the only source of these cells in birds and removal of the bursa at hatching (or by certain viral infections) results in a severe B-cell deficiency.

bursa omentalis *(Zool.)*. In Mammals, a sac formed by the epiploon or great omentum.

bursattee, bursati *(Vet.)*. *Cutaneous habronemiasis*. A disease of the skin of horses caused by nematode larvae of the genus *Habronema*; characterized by granulomatous nodules in the skin.

bursicon *(Zool.)*. In Insects, a hormone produced by neurosecretory cells of the brain and released by neurochaemal organs in the thoracic and abdominal ganglia. It effects many post-ecdysal processes such as cuticular tanning.

bursiform *(Bot.)*. Resembling a bag or pouch.

bursitis *(Med.)*. An inflammation of a bursa.

burst *(Image Tech.)*. See **colour burst**.

burst *(Phys.)*. Unusually large pulse arising in an ionization chamber caused by a cosmic-ray shower.

burst *(Telecomm.)*. Sudden increase in strength in received radio signals caused by sudden changes in ionosphere.

burst-can (-cartridge) detector *(Nuc.Eng.)*. An instrument for the early detection of ruptures of the sheaths of fuel elements inside a reactor. Also *leak detector*.

burst cartridge, slug *(Nuc.Eng.)*. Fuel element with a small leak, emitting fission products.

bursting *(Comp.)*. Separating continuous stationery.

bursting disk *(Chem.Eng.)*. A protective device for process vessels in which hazardous operations are performed, consisting of a thin disk of noble or corrosion resisting metal, carefully controlled as to thickness, and designed to burst in event of excess internal pressure, giving a large opening for rapid release of the pressure.

burst test *(Paper)*. A physical test method to determine the limiting pressure (applied normally to the paper surface by means of a rubber diaphragm) that a test piece will withstand when fixed horizontally between two clamps, under the prescribed conditions of test. The Mullen burst tester is frequently used for this purpose.

bus *(Comp.)*. High speed pathway shared by signals from several components of the computer (e.g. all input/output devices would be connected to the I/O bus). Also *highway*, *trunk*. Several linked *highways* are called a *channel*.

bus *(Space)*. Physical means (usually a good electrical conductor) of distributing and/or collecting electrical energy or signals. Sometimes, a universal platform for diverse space experiments and applications.

bus-bar *(Elec.Eng.)*. (1) Length of constant-voltage conductor in a power circuit. Normally of rigid copper construction and located in a power station or substation. (2) Supply rail maintained in a constant potential (including zero or earth) in electronic equipment.

bus-coupler switch *(Elec.Eng.)*. A switch or circuit-breaker serving to connect two sets of duplicate bus-bars.

bush *(Build.)*. A reducing adapter or screwed piece for connecting together in the same line two pipes of different sizes.

bush *(Eng.)*. (1) A cylindrical sleeve, usually inserted in a machine part to form a bearing surface for a pin or shaft. (2) A hardened cylindrical insert in a drilling jig to position a drill or reamer accurately.

bush-hammering *(Civ.Eng.)*. The operation of dressing the surface of stone with a special hammer having rows of projecting points on its striking face.

bushing *(Elec.Eng.)*. An insulator which enables a live conductor to pass through an earthed wall or tank (e.g., the wall of a switch house or the tank of a transformer).

bushing *(Glass)*. A small electric melting unit, usually made of platinum, with numerous holes in the base, used for the manufacture of glass fibres.

bushing current transformer *(Elec.Eng.)*. A current transformer built into a bushing.

bush sickness *(Vet.)*. See **pine**.

bus-line *(Elec.Eng.)*. A cable, extending the whole length of an electric train, which connects all the collector shoes of like polarity. Sometimes called a *power line*.

bus-line couplers *(Elec.Eng.)*. Plug-and-socket connectors to join the bus-line of one coach of an electric train to that of the next.

bustamite *(Min.)*. A triclinic silicate of manganese, calcium, and iron.

bustle pipe *(Eng.)*. Main air pipe surrounding blast furnace, which delivers low-pressure compressed air to tuyères.

bus-wire coupler *(Elec.Eng.)*. A flexible connection between the coaches of an electric train for maintaining the continuity of any bus-wires which have to run throughout the train-length.

busy *(Telecomm.)*. A term applied to any line or equipment signifying *engaged*. See **busy tone**.

busy tone *(Telecomm.)*. An audible signal which is fed back to a caller to indicate that the switching equipment, line, or required subscriber's instrument is already engaged.

butadiene *(Chem.)*. *Butan 1,2 : 3,4 diene*. Divinyl, $CH_2=CH.CH=CH_2$, a di-alkene with conjugate linking. An **isoprene** homologue, an important compound in the manufacture of synthetic rubbers. See **Buna, chloroprene, GR-S**.

butanal *(Chem.)*. *Butyraldehyde*. $CH_3.CH_2.CH_2.CHO$. Bp 76°C. Made by the catalytic dehydrogenation of butan-1-ol.

butane *(Chem.)*. C_4H_{10}, an alkane hydrocarbon, bp 1°C, rel.d. at 0°C 0.600, contained in natural petroleum, obtained from casing head gases in petroleum distillation. Used commercially in compressed form, and supplied in steel cylinders for domestic and industrial purposes, e.g. Calor gas.

butanoic acids *(Chem.)*. See **butyric acids**.

butanol *(Chem.)*. See **butyl alcohol**.

butanoyl *(Chem.)*. The monovalent acyl radical $C_3H_7.CO$—.

butenes *(Chem.)*. C_4H_8, alkene hydrocarbons, the next higher homologues to propylene. Three isomers are possible and known, normally gaseous, b.p. between −6°C and +3°C.

Butex *(Nuc.Eng.)*. TN for diethylene glycol dibutyl ether, used for separating U and Pu from fission products.

buthocrome *(Phys.)*. Particular groups of atoms in organic compounds which have the effect of lowering frequency of the radiation absorbed by these compounds.

butobarbitone *(Med.)*. A barbiturate hypnotic and sedative; 5-butyl-5-ethylbarbituric acid. TNs *Butomet, Soneryl* etc.

butte *(Geol.)*. A small flat-topped steep-sided hill. See **mesa**.

butted *(Print.)*. Said of (1) rules joined at a corner by bringing the end of one to the inside edge of another, not mitring the joint; (2) two or more slugs joined end to end to form a wide line of type.

butter *(Genrl.)*. A fat emulsion containing in solution sugar, albumen, salts, whereas fats and casein are present in colloidal dispersion.

butterfly *(Image Tech.)*. A diffuser used to soften the light from the sun or from lamps; made from silk stretched on a frame.

butterfly circuit *(Elec.Eng.)*. One in which both the inductance and capacitance between opposite re-entrant sections of a punched disk are varied by rotating an insulated coaxial vane of the same shape as the cavity. See **differential capacitor**.

butterfly diagram *(Astron.)*. A graphical presentation of the occurrence of **sunspots** in the 11-year sunspot cycle.

butterfly flower *(Bot.)*. A flower pollinated by butterflies.

butterfly nut *(Eng.)*. See **wing nut**.

butterfly tail *(Aero.)*. See **vee-tail**.

butterfly valve *(Eng.)*. (1) A disk turning on a diametral axis inside a pipe; used as a throttle valve in petrol and gas engines. (2) A valve consisting of a pair of semicircular plates hinged to a common diametral spindle in a pipe; by hingeing axially, the plates permit flow in one direction only.

buttering *(Build.)*. The operation of spreading mortar on the edges of a brick before laying it.

buttering trowel *(Build.)*. A flat tool similar to, but smaller than, the **brick trowel**; used for spreading mortar on a brick before placing it in position.

buttermilk *(Genrl.)*. The aqueous liquid occluded in butter which is removed during the manufacturing process.

butternut *(For.)*. A North American timber of the Walnut family, of economic importance for its fruits as well as its wood. It is brownish in colour with normally straight grain but a somewhat coarse texture. It is mostly used for cabinet making and panelling.

Butterworth filter *(Telecomm.)*. Constant-*k* filter designed to give response of maximum flatness through pass band. Cf. **Chebyshev filter**.

butt gauge *(Build.)*. One used in fitting the butt hinges on doors, having three markers, one for the thickness of the butt, one for the depth of the door, and one the depth of the jamb.

butt hinge *(Build.)*. A hinge formed by 2 leaves, which are secured to the door and door-frame in such a manner that when the door is shut the 2 leaves are folded into contact.

butt joint *(Build.)*. A joint formed between the squared ends of the 2 jointing pieces, which come together but do not overlap.

butt joint *(Eng.)*. A joint between 2 plates whose edges abut, and are covered by a narrow strip or 'strap' riveted or welded to them.

butt-jointed *(Elec.Eng.)*. See **joint**.

buttock *(Min.Ext.)*. Coal from which an undercut has been removed, in readiness for bringing it down.

buttock planes *(Ships)*. Longitudinal sectional planes drawn through a ship's form; used for laying-off in the moulding loft, and for calculation of volumes etc.

button *(Horol.)*. The serrated knob by means of which a keyless watch or similar movement is wound.

button-headed screws *(Eng.)*. Screws having hemispherical heads, slotted for a screwdriver; known also as *half-round screws*.

button microphone *(Acous.)*. Small microphone which can be fitted in the buttonhole.

buttonwood *(For.)*. The N. American equivalent of the sycamore or plane of Europe. It is somewhat difficult to work either by hand or by machine, but it finishes with a good surface and responds to decorative treatment.

buttress *(Civ.Eng.)*. A supporting pier built on the exterior of a wall to enable it to resist outward thrust.

buttress root *(Bot.)*. Form of *prop root* which thickens unevenly to produce a flat, apparently supporting, structure something like a buttress.

buttress screw-thread *(Eng.)*. A screw-thread designed to withstand heavy axial thrust in one direction. The back of the thread slopes at 45°, while the front or thrust face is perpendicular to the axis.

butt-welded tube *(Eng.)*. Tube made by drawing mild steel strip through a bell, so that the strip is coiled into a tube, the edges being then pressed together and welded.

butt-welding *(Eng.)*. The joining of two plates or surfaces by placing them together, edge to edge, and welding along the seam thus formed. See **welding**.

butyl acetate (ethanoate) *(Chem.)*. $CH_3COOC_4H_9$ commercial product has a boiling range of 124°–128°C, rel.d. 0.885, a colourless liquid, of fruity odour, soluble in ethanol, ether, acetone, benzene, turpentine, slightly in water. A very important lacquer solvent.

butyl alcohols *(Chem.)*. $C_4H_9.OH$. There are 4 isomers possible and known, viz.: butan-1-ol $CH_3(CH_2)_3OH$, bp 117°C; butan-2-ol $CH_3CH_2CH(OH)CH_3$, bp 110°C; 2-methyl propan-1-ol, bp 107°C; 2,2-dimethyl-ethanol, mp 25°C, bp 83°C.

butylenes *(Chem.)*. See **butenes**.

butyl group *(Chem.)*. The aliphatic group C_4H_9—, with four isomeric forms: *primary, secondary, iso-* and *tertiary*.

butyl rubber *(Chem.)*. Form of synthetic rubber obtained by copolymerizing isobutylene and isoprene. Used chiefly for tubeless tyres and inner tubes, because of very low air permeability.

butyraldehyde *(Chem.)*. See **butanal**.

butyric acids *(Chem.)*. Butanoic acids. $C_3H_7.COOH$, two isomers, viz., *normal* and *iso*-butyric acid. Only *n*-butyric (butanoic) acid is of importance, m.p. −8°C, b.p.162°C; it is a thick liquid of rancid odour and occurs in rancid butter.

Buxton certification *(Elec.Eng.,Min.Ext.)*. The certification of the suitability of electrical equipment for use in atmosphere in which fire or explosion hazards are present.

Buys Ballot's law *(Meteor.)*. If an observer stands with his back to the wind, the lower atmospheric pressure is on his left in the northern hemisphere, on his right in the southern hemisphere.

buzz *(Aero.)*. (1) Severe vibration of a control surface in transonic or supersonic flight caused by separation of the airflow due to compressibility effects. (2) To interfere with an aircraft in flight by flying very close to it .

buzz track *(Image Tech.)*. (1) A test film used to set the correct position of the slit in an optical sound reproducer. (2) A sound recording of low background level used to fill silent gaps in commentary or dialogue.

BVU *(Image Tech.)*. TN for a high-band VTR system using 3/4 inch tape in a cassette.

BWD *(Vet.)*. See **pullorum disease**.

BWG *(Eng.)*. Abbrev. for *Birmingham Wire Gauge*.

BWR *(Nuc.Eng.)*. See **boiling water reactor**.

bye-channel *(Civ.Eng.)*. Waterway dug round the side of a reservoir or dam to carry off surplus water from the streams entering it.

by-pass *(Build.)*. Any device for directing flow around a fixture, connection or pipe, instead of through it.

by-pass capacitor *(Elec.Eng.,Electronics)*. A capacitor having a low reactance for frequencies of interest connected in shunt with other components so as to short-circuit them for signal frequency currents.

by-pass ratio *(Aero.)*. The ratio of the by-passed airflow to the combustion airflow in a *dual-flow turbojet* having a single air intake.

by-pass turbojet *(Aero.)*. A turbojet in which part of the compressor delivery is by-passed round the combustion zone and turbine to provide a cool, slow propulsive jet when mixed with the residual efflux from the turbine. See **turbofan**.

by-pass valve *(Elec.Eng.)*. A switching device (**SCR** or, in the past, mercury arc valve), connected across the converter switching devices of a high-voltage d.c. transmission system, normally not conducting but able to maintain flow of current whenever the main conducting devices have to be interrupted.

by-pass valve *(Eng.)*. A valve by which the flow of fluid in a system may be directed past some part of the system through which it normally flows, e.g. an oil-filter in a lubrication system.

B-Y signal *(Image Tech.)*. Component of colour TV chrominance signal. Combined with luminance (Y) signal, it gives primary blue component.

bysmalith *(Geol.)*. A form of igneous intrusion bounded by a circular fault and having a domeshaped top; described by Iddings from the Yellowstone Park. Cf. **plug**.

byssinosis *(Immun.)*. Respiratory disease among workers in the vegetable fibre industry, characterized by chest tightness on returning to work after a period of absence. Due to sensitization by substances present in the fibre dust.

byssus *(Zool.)*. In Bivalvia, a tuft of strong filaments secreted by a gland in a pit (**byssus pit**) in the foot and used for attachment. adj. *byssogenous, byssal*.

byte *(Comp.)*. Fixed number of **bits**, often corresponding to a single **character** and operated on as a unit.

bytownite *(Min.)*. A variety of plagioclase feldspar, containing 70 to 90% of the anorthite molecule; occurs in basic igneous rocks.

Bz *(Chem.)*. A symbol for the benzoyl (benzenecarboxyl) radical $C_6H_5.CO$.

C

c *(Chem.)*. A symbol for *concentration.*

c *(Phys.)*. The symbol used for the velocity of electromagnetic radiation *in vacuo*. Its value, according to the most accurate recent measurements, is $2.997\,924\,56 \times 10^8\,\mathrm{m\,s^{-1}}$.

c- *(Chem.)*. Abbrev. for (1) *cyclo-*, i.e. containing an alicyclic ring, (2) *cis-*, i.e. containing the two groups on the same side of the plane of the double bond or ring.

C *(Chem.)*. The symbol for carbon.

C *(Comp.)*. A **high level language** similar to **BCPL**. The UNIX operating system is written in it.

C *(Immun.)*. Symbol for **complement**. The component proteins for complement are designated C1–C9' and in activated form as $C1^-$–$C5^-$.

C *(Phys.)*. (1) Symbol for coulomb. (2) When used after a number of degrees thus: 45°C, the symbol indicates a temperature on the Celsius or Centigrade scale.

C *(Chem.)*. A symbol for: (1) *concentration*; (2) (with subscript) *molar heat capacity*: C_p, at constant pressure: C_v, at constant volume.

C- *(Chem.)*. Containing the radical attached to a carbon atom.

[C] *(Phys.)*. One of the Fraunhofer lines in the red of the solar spectrum. Its wavelength is 656.3045 nm; it is due to hydrogen.

$C1^-$-inhibitor *(Immun.)*. An inhibitor of activated esterase formed from complement, C1. It also inhibits some other esterases activated during blood clotting. Normally present in blood but congenitally absent or inactive in some persons, who are liable to attacks of angioneurotic oedema due to vasoactive peptides released by the esterase.

C3a, C5a *(Immun.)*. Peptide fragments split off from **complement** proteins, C3 and C5, respectively during conversion to their enzymically active forms. The peptides are chemotactic for leucocytes and cause local increase in vascular permeability. They act as *anaphylatoxins* causing histamine release from local mast cells.

C3b receptors *(Immun.)*. Receptors present on cell membranes which can bind C3b, the activated form of complement C3, or its breakdown products, C3bi or C3d. The receptors are designated CR1, CR2 or CR3. The most important, CR1 is present on mononuclear phagocytes, granulocytes, B-lymphocytes and some other cells. The receptor enables immune complexes or microbes which have bound complement to become attached to the cells and increases their ingestion. It is also probably involved in modulating the activation of B-lymphocytes by immune complexes.

Ca *(Chem.)*. The symbol for calcium.

CAA *(Aero.)*. Abbrev. for *Civil Aviation Authority*, UK.

CAB *(Aero.)*. Abbrev. for *Civil Aeronautics Board*, US.

CAB *(Plastics)*. Abbrev. for *Cellulose Acetate Butyrate.*

cabin *(Min.Ext.)*. A fireman's station underground in a coal-mine.

cabin altitude *(Aero.)*. The nominal pressure altitude maintained in the cabin of a pressurized aircraft.

cabin blower *(Aero.)*. An engine-driven pump, usually of displacement type, for maintaining an aircraft cockpit or cabin above atmospheric pressure. Also *cabin supercharger.*

cabin differential pressure *(Aero.)*. The pressure in excess of that of the surrounding atmosphere which is needed to maintain comfortable conditions at high altitude. For an aircraft flying at 9000 m this differential would be about $60\,\mathrm{kN/m^2}$.

cabinet-file *(Build)*. A single-cut smooth file used by joiners and cabinet makers.

cabinet screwdriver *(Build)*. One with a round shank flattened at the end.

cabinet-work *(Build.)*. Fine joinery used in the construction of furniture and fixtures.

cabin hook *(Build.)*. A hooked bar and eye, serving as a fastener for doors and casements.

cabin supercharger *(Aero.)*. See **cabin blower**.

cable *(Eng.)*. A general term for rope or chain used for engineering purposes. Specifically, a ship's anchor cable.

cable-angle indicator *(Aero.)*. An indicator showing the vertical angle between the longitudinal axis of a glider and its towing cable, also its yaw and roll attitude relative to the towing aircraft.

cable buoy *(Ships)*. A buoy attached to an anchor and serving to mark its position.

cablecar *(Civ.Eng.)*. Tram deriving motive power from an underground cable, in the same manner as the **cable railway**.

cable ducts *(Elec.Eng.)*. Earthenware, steel, plastic or concrete pipes through which cables are drawn, and in which they lie.

cable form *(Elec.Eng.)*. The normal scheme of cabling between units of apparatus. The bulk of the cable is made up on a board, using nails at the appropriate corners, each wire of the specified colour identification being stretched over its individual route with adequate **skinner**. When the cable is bound with twine and waxed, it is fitted to the apparatus on the racks and the skinners connected, by soldering, to the **tag blocks**.

cable grip *(Elec.Eng.)*. A flexible cone of wire which is put on the end of a cable. When the cone is pulled, it tightens and bites into the lead sheath of the cable, and can be used to pull the cable into a duct.

cable-laid rope *(Eng.)*. A rope formed of several strands laid together so that the twist of the rope is in the direction to the twist of the strands.

cable-length *(Genrl.)*. One tenth of a nautical mile. (608 ft, 185 m).

cable railway *(Civ.Eng.)*. Means of transport whereby carriages are pulled up an incline by an endless overhead or underground cable.

cable release *(Image Tech.)*. Device for releasing a camera shutter in which the trigger is actuated by a length of stiff wire cable in a flexible tube.

cable stitch *(Textiles)*. In knitting, the rope-like appearance obtained by passing groups of adjacent wales under and over one another.

cable tool drilling *(Min.Ext.)*. Method of drilling in which a heavy sharpened tool bit (*churn-, percussion-drill*) is reciprocated in a borehole by a steel cable attached to a **walking beam**.

cable tools *(Min.Ext.)*. The variously shaped drilling tools used in **cable tool drilling**.

cable wax *(Chem.)*. A solid wax formed by the ionic bombardment of the oil in a cable. It is a good insulator, and cables operate very successfully even when much wax is present. It is produced by a condensation process such as $C_6H_{14}+C_5H_{12}\rightarrow C_{10}H_{22}+CH_4$.

cable-way *(Civ.Eng.)*. A construction consisting of cables slung over and between two towers, so that a skip suspended from the cables may be raised and lowered and moved to any position along the cables. It is used for transport of soil and materials. Also *blondin.*

cabling *(Arch.)*. A round moulding used to decorate the lower parts of the flutes of columns.

cabling *(Telecomm.)*. The collection of cables required for distributing the power supplies in a telephone exchange. See **trunking**.

cabling *(Textiles)*. Twisting together two or more doubled or *folded* yarns. The result in most cases is a balanced cord of four, six or more yarns. *Tyre cord* is one example.

cache memory *(Comp.)*. Extremely fast part of main store.

cachexia *(Med.)*. A marked wasting and emaciation of the body, with an 'earthy' complexion, may occur in some patients with cancer.

cacodyl *(Chem.)*. $As_2(CH_3)_4$, a colourless liquid; bp 170°C; of horribly nauseous odour. It combines directly with oxygen, sulphur, chlorine etc. Cacodyl and cacodyl oxide form the basis for other secondary arsines. Cacodyl derivatives are important as rubber accelerators.

Cactaceae *(Bot.)*. The Cactus family, ca 2000 spp. of dicotyledonous flowering plants (superorder Caryophillidae). Most are leafless, spiny, stem-succulent, CAM plants of the semi-deserts of the New World. Of little economic importance other then as ornamentals, and some edible fruit.

CAD *(Comp.)*. See **computer aided design**.

cadastral survey *(Surv.)*. Land survey, boundary delineation.

cadaver *(Med.)*. A dead body.

cadaverine *(Chem.)*. *1,5-diaminopentane*. $NH_2.(CH_2)_5.NH_2$, a colourless syrupy liquid; bp 178°C. Formed by the bacterial decomposition of diamino acids. Ring formation occurs by elimination of NH_2 with the formation of the heterocyclic base **piperidine**. See also **putrescine**.

cadmium *(Chem.)*. White metallic element, symbol Cd, at. no. 48, r.a.m. 112.40, mp 320.9°C, bp 767°C, rel.d. 8.648. Cadmium plating is widely used as a corrosion protective for steel and its alloys. It is a powerful absorber of neutrons and is used in control elements in nuclear reactors. Films of cadmium are photosensitive in the ultraviolet between 250 and 295 nm, with a peak at 260 nm.

cadmium cell *(Electronics)*. A reference voltage standard, giving 1.0186 V at 20°C. Also known as *Weston standard cadmium cell*.

cadmium copper *(Eng.)*. A variety of copper containing 0.7–1.0% of cadmium. Used for trolley, telephone and telegraph wires because it gives high strength in cold-drawn condition combined with good conductivity.

cadmium photocell *(Electronics)*. A photoconductive cell using cadmium disulphide or cadmium selenide as the photosensitive semiconductor. Sensitive to longer wavelengths and infra-red. It has a rapid response to changes in light intensity.

cadmium red line *(Phys.)*. Spectrum line formerly chosen as a reproducible standard of length. Wavelength = 643.8496 nm.

cadmium-sulphide detector *(Radiol.)*. Radiation detector equivalent to a solid-state ionization chamber, but with amplifier effect (due to hole trapping).

caducibranchiate *(Zool.)*. Possessing gills at one period of the life-cycle only, as in some Caudata.

caducous *(Bot.)*. Falling off at an early stage. Cf. **deciduous**.

caducous *(Zool.)*. See **deciduate**

CAE *(Eng.)*. See **computer aided engineering**.

caecostomy, cecostomy *(Med.)*. Formation of an artificial opening into the caecum.

caecum *(Zool.)*. Any blind diverticulum or pouch, especially one arising from the alimentary canal.

caenogenesis *(Zool.)*. A phenomenon whereby features which are adaptations to the needs of the young stages develop early and disappear in the adult stage. adj. *caenogenetic*.

Caenorhabditis elegans *(Biol.)*. A small nematode worm which is ideally suited for genetic, molecular and cellular studies of eukaryotic development.

Caerfai *(Geol.)*. The oldest epoch of the Cambrian period.

Caesarean, Cesarean section *(Med.)*. Delivery of a foetus through the incised abdomen and uterus.

caesious, caesius *(Bot.)*. Bearing a bluish-grey waxy covering (bloom).

caesium *(Chem.)*. Metallic element, symbol Cs, at. no. 55, r.a.m. 132.905, mp 28.6°C, bp 713°C, rel.d. 1.88. As a photosensitor it has a peak response at 800 nm in the infrared, both thermal- and photo-emission being high. Caesium, when alloyed with antimony, gallium, indium, and thorium, is generally photosensitive.

caesium cell *(Elec.Eng.)*. One having a cathode consisting of a thin layer of caesium deposited on minute globules of silver; particularly sensitive to infrared radiation, but generally approximating to that of the eye.

caesium clock *(Elec.Eng.)*. Frequency-determining apparatus used on caesium-ion resonance of 9 192 631 770 Hz.

caesium-oxygen cell *(Elec.Eng.)*. One in which the vacuum is replaced by an atmosphere of oxygen at very low pressure. It is more sensitive to red light than the caesium cell.

caesium unit *(Radiol.)*. Source of radioactive caesium (*half-life* 33 yr) mounted in a protective capsule.

caespitose, cespitose *(Bot.)*. Growing from the root in tufts, as many grasses. dim. *caespitulose*.

caffeine *(Med.)*. A weak central nervous stimulant found in coffee and tea.

cage *(Civ.Eng.)*. The platform on which goods are hoisted up or lowered down a vertical shaft or guides.

cage *(Eng.)*. The part of a ball or roller bearing which separates the balls or rollers and keeps them correctly spaced along the periphery of the bearing.

cage *(Min.Ext.)*. Steel box used in vertical mine shaft to raise and lower men, materials or trucks (trams, tubs). May have two or three decks.

cage antenna *(Telecomm.)*. One comprising a number of wires connected in parallel, and arranged in the form of a cage to reduce the copper losses and increase the effective capacitance.

cage rotor *(Elec.Eng.)*. A form of rotor, used for induction motors, having on it a cage winding. Also called *squirrel-cage rotor*.

cage-type negative plate *(Elec.Eng.)*. See **box-type negative plate**.

cage winding *(Elec.Eng.)*. A type of winding used for rotors of some types of induction motors, and for the starting or damping windings of synchronous machines. It consists of a number of bars of copper or other conducting material, passing along slots in the core and welded to rings at each end. Also called a *squirrel-cage winding*.

Cahn-Ingold-Prelog system *(Chem.)*. A system for describing stereoisomerism unambiguously. It is based on the identification of a *priority* among substituents, the atom of higher atomic number having higher priority. When the two atoms are of the same element, then *their* substituents are examined etc. For geometrical isomerism, the substituents of highest priority on each side of the bond in question are examined. If they are on the same side of the bond, the prefix Z- is given (Ger. *zusammen*), otherwise E- (Ger. *entgegen*). Chiral centres are viewed with the substituent of lowest priority away from the viewer. The configuration is said to be R- (Lat. *rectus*) if the other substituents in order of decreasing priority are arranged clockwise, otherwise S- (Lat. *sinister*).

CAI *(Comp.)*. See **computer assisted instruction**.

Cailletet process *(Phys.)*. A method for the liquefaction of gases based on the free expansion of a gas from a higher to a lower pressure.

Cailletet's and Mathias' law *(Chem.)*. The arithmetical mean of the densities, d, of a pure unassociated liquid, l, and its saturated vapour, v, is a linear function of the temperature, t. Mathematically expressed, $0.5(d_l + d_v) = A + Bt$.

cairngorm *(Min.)*. Smoky-yellow or brown varieties of quartz; named from Cairngorm in the Scottish Grampians, the more attractively coloured varieties being used as semi-precious gemstones, See *smoky quartz*.

caisson *(Build.)*. A deeply recessed sunk panel in a soffit or ceiling.

caisson disease *(Med.)*. *Decompression sickness*, also called *the bends, diver's palsy, diver's paralysis*. Pains in the joints and paralysis, occurring in workers in compressed air who are suddenly subjected to atmospheric pressure after compression; it is due to accumulation of bubbles of nitrogen in the nervous system.

cake *(Eng.)*. The rectangular casting of copper or its alloys before rolling into sheet or strip.

caking coal *(Min.Ext.)*. Coal which cakes or forms coke when heated in the absence of air.

CAL *(Comp.)*. See **computer assisted learning**.

calamine *(Min.)*. A name formerly used in Britain for **smithsonite** and in America for **hemimorphite**.

Calamitales *(Bot.)*. Order of extinct, mainly Carboniferous Sphenopsida. Sporophytes were mostly large trees with substantial hollow trunks having secondary xylem, and whorled or opposite branches. Co-dominant with the Lepidodendrales in the Carboniferous swamps.

calamus *(Zool.)*. The proximal hollow part of the scapus of a feather; quill. pl. *calami*.

calandria *(Chem.Eng.,Nuc.Eng.)*. Closed vessel penetrated by pipes so that liquids in each do not mix. In evaporating plant the tubes carry the heating fluid and in certain types of nuclear reactor the sealed vessel is called a calandria.

calcaneum *(Zool.)*. In some Vertebrates, the fibulare, or large tarsal bone forming the heel; more generally, the heel itself; in Birds, a process of the metatarsus.

calcar *(Zool.)*. In Insects, a tibial spine; in Amphibians, the prehallux; in Birds, a spur of the leg, or more occasionally, of the wing; in Bats, a bony or cartilaginous process of the calcaneum supporting the interfemoral part of the patagium.

calcarate *(Bot.)*. Bearing one or more spurs.

calcarenite *(Geol.)*. A limestone consisting of detrital lime particles (> 50%) of sand size. See **calcilutite**.

calcareous *(Bot.)*. Containing, or coated with, calcium carbonate (lime).

calcareous *(Chem.)*. Containing compounds of calcium, particularly minerals.

calcareous clay *(Build.)*. See **marl**.

calcareous rock *(Geol.)*. Sediment containing a large amount of calcium carbonate (e.g. limestone, chalk, shelly sandstone, calc-tufa).

calc-flinta *(Geol.)*. A hard fine-grained rock composed of calcium silicate minerals and produced by the contact metamorphism of an impure limestone.

calcicole *(Bot.)*. Plants found on or confined to soils containing free calcium carbonate. Cf. **calcifuge**.

calciferol *(Biol.)*. See **vitamin D**.

calciferous, calcigerous *(Zool.)*. Producing or containing calcium salts.

calcification *(Bot.)*. The accumulation of calcium carbonate on or in cell walls.

calcification *(Zool.)*. The deposition of lime salts, e.g. in diseased or dead tissue such as the walls of arteries.

calcifuge *(Bot.)*. Plants not normally found on, or intolerant of soils containing free calcium carbonate. Cf. **calcicole**.

calcigerous *(Zool.)*. See **calciferous**.

calcigerous glands *(Zool.)*. In some Oligochaeta, a pair of oesophageal glands producing a limy secretion to control the acid/base balance of the body: in some Amphibians, the glands of Swammerdam, calcareous concretions lying on either side of the vertebrae, close to the points of exit of the spinal nerves.

calcilutite *(Geol.)*. A limestone consisting of (> 50%) clay or silt particles of lime. See **calcarenite**.

calcination *(Chem.)*. The subjection of a material to prolonged heating at fairly high temperatures.

calcination *(Min.Ext.)*. The operation of heating ores to drive off water and carbon dioxide, frequently not distinguished from **roasting**.

calcine *(Min.Ext.)*. Ore, carbonate, mineral or concentrate which has been roasted, perhaps in an oxidizing atmosphere, to remove sulphur as SO_2 (*sweet roasting*) or carbon dioxide (*dead roasting*).

calcined powder *(Powder Tech.)*. One produced or modified by heating to a high temperature.

calcinosis *(Med.)*. Deposits of calcium salts in various tissues of the bodies.

calciphile *(Bot.)*. Same as *calcicole*.

calciphobe *(Bot.)*. Same as *calcifuge*.

calcite, calcspar *(Min.)*. The commonest crystalline form of calcium carbonate, showing trigonal symmetry and a great variety of crystal habits. It is the principal constituent of limestone and many marbles, and occurs extensively in other rocks.

calcitonin *(Med.)*. Hormone secreted by parafollicular cells of the thyroid gland in response to high blood calcium, opposing the action of a parathyroid hormone. It is of value in the treatment of hypercalcaemia and **Paget's disease of the bone**.

calcium *(Chem.)*. Metallic element, symbol Ca, at. no. 20, r.a.m. 40.08, mp 850°C, bp 1440°C, rel.d. 1.58. Occurs in nature in the form of several compounds, the form of carbonate predominating. Produced by electrolysis of fused calcium chloride. Used as a reducing metal and as a getter in low-noise valves.

calcium carbide *(Chem.)*. *Ethynide dicarbide*. CaC_2. A compound of calcium and carbon usually prepared by fusing lime and hard coal in an electric furnace. See also **acetylene**.

calcium carbonate *(Chem.)*. $CaCO_3$. Very abundant in nature as chalk, limestone and calcite. Almost insoluble in water, unless the water contains dissolved carbon dioxide, when solution results in the form of calcium hydrogen carbonate, causing the temporary hardness of water.

calcium channel blocking *(Med.)*. Name given to a group of drugs which interfere with intracellular calcium flux. Most are potent vasodilators and some have an anti-arrhythmic effect. Common examples are Nifedipine, Verapamil and Diltiazem.

calcium chloride *(Chem.)*. $CaCl_2$. Formed by the action of hydrochloric acid on the metal and its common compounds. It absorbs moisture from the atmosphere, and is extensively used for drying gases. Tubes filled with granular calcium chloride are used in industry and laboratories for producing dry air.

calcium fluoride *(Chem.)*. CaF_2. In the form of *fluorspar* it is used for the manufacture of hydrofluoric acid. It is also an important constituent of opal glass. It has high thermoluminescent sensitivity.

calcium phosphate precipitation *(Biol.)*. Technique for introducing foreign DNA or chromosomes into cells; co-precipitation with calcium phosphate facilitates the uptake of DNA or chromosomes.

calcium plumbate *(Build.)*. A lead based priming paint with exceptional weather resistance suitable for use on ferrous metals, galvanized or zinc sheeting and composite metal/timber substrates.

calcium tungstate screen *(Electronics)*. A fluorescent screen used in CRT; it gives a blue and ultra-violet luminescence.

calcspar *(Min.)*. See **calcite**.

calculus *(Maths.)*. See **differential-, integral-**.

calculus *(Med.)*. A concretion of mineral or of organic matter in certain organs of the body, e.g. the kidney, the gallbladder.

calculus of variations *(Maths.)*. The determination of one or more functions of one or more independent variables, in order that a definite integral of a known function of these functions and their derivatives shall be stationary.

caldera *(Geol.)*. A large volcanic crater produced by the collapse of underground lava reservoirs or by ring fracture, possibly as a surface expression of **cauldron subsidence**.

Caledonian *(Geol.)*. Appertaining to the great mountain-building episode of late Silurian to early Devonian time.

Caledonoid direction *(Geol.)*. The direction assumed by the Caledonian (Siluro-Devonian) mountain folds and associated structures in Britain and Scandinavia. Commonly N.E.–S.W., but subject to considerable variations.

calendar month, calendar year *(Genrl.)*. Popularly, a month or year as defined in a calendar, particularly the **Gregorian calendar**. A calendar month differs from the synodic (or lunar) month. The terms are also used for periods equivalent to a month or year, e.g. from July 9 to August 9, or July 9 of one year to July 9 of the next.

calender *(Paper,Textiles)*. A vertical arrangement of horizontal rolls of metal or fibrous composition, carried in side frames and providing nips through which the fabric or the paper web may be passed to impart the required degree of finish or to control its thickness and compression. Calenders may form part of a complete processing machine or exist separately.

calendered paper *(Paper)*. Paper that has been calendered. If the calenders were part of the paper making machine the resultant effect is known as machine finished (m.f.). A higher degree of gloss can be obtained by means of a supercalender, separate from the paper machine and containing some rolls of fibrous composition. This is known as supercalendered finish. Abbrev. *sc*.

calf *(Print.)*. Superior quality calfskin used for covering books, finished either smooth or rough.

calf diphtheria *(Vet.)*. *Necrotic or malignant stomatitis*. Ulceration and necrosis of the mouth and pharynx of calves due to infection by the bacterium *Sphaerophorus necrophorus (fusiformis necrophorus)*.

calf scour *(Vet.)*. See **white scour**.

calf tetany *(Vet.)*. *Milk tetany*. A form of **bovine hypomagnesaemia**, occurring in calves, caused by a deficiency of magnesium.

Calgon *(Chem.)*. TN for sodium hexametaphosphate, used in water-softening because of its marked property of forming soluble complexes with calcium. Also used in textile and laundry work.

caliber *(Eng.)*. See **calibre**.

calibrated airspeed *(Aero.)*. Indicated airspeed corrected for *position error* and instrument error only. Not to be confused with **equivalent airspeed** or **true airspeed**. Abbrev. *CAS*. Also *rectified airspeed*.

calibration *(Phys.)*. The process of determining the absolute values corresponding to the graduations on an arbitrary or inaccurate scale on an instrument.

calibre, caliber *(Eng.)*. The internal diameter or bore of a pipe, esp. the barrel of a fire-arm.

calibre, caliber *(Horol.)*. The arrangement of the various components of a watch or clock.

caliche *(Geol.)*. Concretions of calcium carbonate, sodium nitrate, and other minerals occurring in soil in arid regions; due to surface evaporation of subsurface waters.

calico *(Textiles)*. A plain cotton cloth heavier than muslin.

Californian jade *(Min.)*. A compact form of green vesuvianite (idocrase) obtained from California, and used as an ornamental stone and in jewellery. Also known as *californite*.

Californian stamp *(Min.Ext.)*. See **gravity stamp**.

californite *(Min.)*. See **Californian jade**.

californium *(Chem.)*. Manmade element, symbol Cf, at. no. 98, produced in a cyclotron. Its longest lived isotope is ^{251}Cf with 800 years half-life.

caliper *(Horol.)*. The size of a watch movement.

calipers *(Eng.)*. See **callipers**.

calked ends *(Build.)*. The ends of built-in iron ties, split and splayed to provide more secure anchorage. Also *fish tailed*.

calking *(Civ.Eng.)*. See **caulking**.

call *(Comp.)*. In a program, a statement that transfers **control** to a **subroutine** or **co-routine**.

Callier coefficient *(Image Tech.)*. The ratio of **specular density** to **diffuse density** in a photographic image, the difference resulting from the grain structure. Also termed *Callier quotient* or *Callier Q factor* and denoted by *Q*.

Callier effect *(Image Tech.)*. Because of the scatter of light by the granular structure of a photographic image, its transmission is greater for a directional (specular) beam than for a diffuse one. This results in greater image contrast from optical printing or enlarging with a condenser light source than from contact printing or use of a diffused source.

call-indicator *(Elec.Eng.)*. An electric indicating device used in conjunction with a system of electric bells, or in a lift, to indicate the point from which a call has been made.

callipers *(Eng.)*. An instrument, consisting of a pair of hinged legs, used to measure external and internal dimensions.

calliper splint *(Med.)*. A splint fitted so that the patient may walk without any pressure on the foot, the weight of the body being taken by the hip bone.

Callisto *(Astron.)*. The fourth natural satellite of **Jupiter**, 4800 km in diameter, discovered by Galileo.

callose *(Bot.)*. A polymer of glucose, $\beta1\rightarrow3$ linked, occurring e.g. as a cell wall constituent, especially in the sieve areas of sieve elements.

callosity *(Med.)*. A thickening of the skin as a result of irritation or friction.

callous, callose *(Bot.,Med.)*. Hardened, usually thickened and often like horn in appearance.

call sign *(Telecomm.)*. Letter and/or numeral for a transmitting station or one of its authorized channels. Used for calling or identification.

callus *(Bot.)*. (1) A tissue consisting of large, thin-walled parenchymatous cells developing as a result of injury, as in wound healing and grafting, or in tissue culture. (2) An accumulation of *callose*. (3) Hard basal projection at the base of the floret or spikelet of some grasses.

callus *(Med.)*. (1) Callosity. (2) Newly formed bony tissue between the broken ends of a fractured bone.

calmodulin *(Biol.)*. A calcium binding protein which is virtually ubiquitous in eukaryotic cells. It plays a central role in controlling the calcium levels in cytoplasm.

calomel *(Chem.)*. *Mercury (I) chloride*, Hg_2Cl_2; found naturally in whitish or greyish masses, associated with cinnabar. Used in physical chemistry as a reference electrode (i.e. a half-cell comprising a mercury electrode in a solution of potassium chloride saturated with calomel).

calorescence *(Phys.)*. The absorption of radiation of a certain wavelength by a body, and its re-emission as radiation of shorter wavelength. The effect is familiar in the emission of visible rays by a body which has been heated to redness by focusing infrared heat rays on to it.

calorie *(Phys.)*. The unit of quantity of heat in the CGS system. The 15°C calorie is the quantity of heat required to raise the temperature of 1 g of pure water by 1°C at 15°C; this equals 4.1855 J. By agreement, the International Table calorie (cal_{IT}) equals 4.186 J exactly, the thermochemical calorie equals 4.184 J exactly. There are other designations, e.g. gramme calorie, mean calorie, and large or kilocalorie (= 1000 cal, used particularly in nutritional work). The calorie has now been largely replaced by the SI unit of the joule.

calorific value *(Eng.)*. The number of heat units obtained by the combustion of unit mass of a fuel. The numerical value obtained for the calorific value depends on the units used; e.g. lb-calories/lb, British thermal units (Btu)lb or MJ/kg for solid and liquid fuels, and Btu/ft^3 or MJ/m^3 for gaseous fuels. In fuels containing hydrogen, which burns to water vapour, there are two heating values, *gross* and *net*. Also *higher (HCF)* and *lower (LCF) calorific values*. The *gross value* of a fuel is the total heat developed after the products are cooled to the starting point, and the water vapour condensed. The *net value* is the heat produced on combustion of the fuel at any given temperature (for gas, 60°F), with the flue products cooled to the initial temperature, the water vapour remaining uncondensed.

calorifier *(Eng.)*. An apparatus for heating water in a tank, the source of heat being a separate coil of heated pipes immersed in the water in the tank.

calorimeter *(Phys.)*. The vessel containing the liquid used in calorimetry. The name is also applied to the complete apparatus used in measuring thermal quantities.

calorimetry *(Phys.)*. The measurement of thermal constants, such as specific heat, latent heat, or calorific value. Such measurements usually necessitate the determination of a quantity of heat, by observing the rise of temperature it produces in a known quantity of water or other liquid.

calorizing *(Eng.)*. A process of rendering the surface of steel or iron resistant to oxidation by spraying the surface with aluminium and heating to a temperature of 800 to 1000°C.

calotte *(Build.)*. A small dome in the ceiling of a room, used to increase head room.

calotype *(Image Tech.)*. An early wet-plate process using silver iodide, patented in 1841 by Fox Talbot.
calvarium *(Med.)*. The dome of the cranium, above the ears, eyes, and occipital protuberance.
Calvé's disease *(Med.)*. Aseptic necrosis of a vertebral body, usually in children, producing a vertebral collapse.
Calvin cycle *(Bot.)*. Cyclical sequence of reactions in which carbon dioxide is fixed by **ribulose bisphosphate carboxylase** and reduced to produce e.g. sugars. Occurs in all photosynthetic plants and algae and most other autotrophic organisms. See **photosynthesis**. Cf. **Hatch-Slack pathway**.
calving fever *(Vet.)*. See **milk fever**.
calx *(Chem.)*. Burnt or quick-lime.
calx *(Med.)*. Calcaneum; os calcis; the heel.
calycle *(Zool.)*. A synonym for *calyx* (senses 1–3).
calyon *(Build.)*. Flint or pebble stone used in wall construction.
Calypso *(Astron.)*. The fourteenth natural satellite of **Saturn**, discovered in 1980 and associated with **Telesto** and **Tethys**.
calypter *(Zool.)*. One or two small lobes at the posterior margin of the base of the wing in some Diptera.
calyptra *(Bot.)*. The layer of cells, developed from part of the archegonium wall that protects the developing sporophyte in mosses and liverworts.
calyptrate *(Zool.)*. Of Diptera, possessing **calypters**.
calyptrogen *(Bot.)*. The layer of meristematic cells that gives rise to the root cap.
calyptron *(Zool.)*. In calyptrate Diptera, the enlarged squama which covers the haltere. Also called *calypter*.
calyx *(Bot.)*. Outer whorl of the perianth, often green and protective, composed of free or fused sepals.
calyx *(Zool.)*. (1) A pouch of an oviduct, in which eggs may be stored. (2) In some Hydrozoa, the cuplike exoskeletal structure surrounding a hydroid. (3) In Crinoidea, the body as distinct from the stalk and arms. (4) In some mammals, part of the pelvis of the kidney.
calyx tube *(Bot.)*. The tube formed by fused sepals.
cam *(Eng.)*. Linear or rotary device, machined to a predetermined profile, whose movement imparts a linear motion to another component, the *cam follower*. The profile can be complex giving e.g. a variable slow forward and rapid reverse movement to a cutting tool on an automatic lathe or, with several on a shaft, opening and closing the valves of an internal combustion engine in the desired sequence.
CAM *(Eng.)*. See **computer aided machining**.
camber *(Aero.)*. The curvature of an aerofoil, relative to the chord line. Colloq. the curved surface of an aerofoil.
camber *(Autos.)*. Inclination of the wheels, where there is an angle between the plane of the wheel and the vertical.
camber *(Civ.Eng.)*. An upward curvature to e.g. allow for settlement or to facilitate run-off of water.
camber *(Eng.)*. A convexity applied for some specific purpose, e.g. to girders to allow for deflection under load, or to road surfaces for drainage.
camber *(Hyd.Eng.)*. The recess in the side of the entrance to a basin, lock, or graving dock, accommodating the **sliding caisson**.
camber *(Ships)*. The convexity of a deck line in a transverse section, normally 2 cm to each metre of breadth. Its purpose is to assist drainage and provide strength. Also called *round of beam*.
camber arch *(Build.)*. An arch having a flat horizontal extrados and a cambered intrados, with a rise of about 1 cm per metre of span.
camber-beam *(Build.)*. A beam having an arched upper surface, or one sloping down towards each end, so as to form a support for roof covering on a flat roof.
camber flap *(Aero.)*. See **plain flap**.
camber-slip *(Build.)*. A strip of wood having a slightly cambered upper surface, upon which the brickwork of a flat arch is laid, so that after settlement the **soffit** shall be straight.
cambial initial *(Bot.)*. One of the permanently meristematic cells of a cambium.
cambium *(Bot.)*. A layer of meristematic cells, lying parallel to the surface of a stem or root, which undergo **periclinal** divisions to give secondary tissues, See **vascular cambium**, **cork cambium**.
Cambrian *(Geol.)*. The lowest division of the Palaeozoic era, covering an approximate time span of from 590–500 million years. Named after Cambria the roman name for Wales. The corresponding system of rocks.
cambric *(Elec.Eng.)*. Varnished cotton-cambric is much used for insulation purposes in an enormous range of electrical equipment, cables, etc.
cambric *(Textiles)*. A closely-woven fine linen cloth, used chiefly for handkerchiefs; the name is also applied to a plain weave fine quality cotton cloth.
Cambridge plate *(Print.)*. A plastic **duplicate plate** with a hard-wearing face and a resilient backing layer, which was developed at Cambridge University Press.
Cambridge ring *(Comp.)*. A **local area network** designed so that packets of data are entered and removed from frames which move continuously around the ring.
camcorder *(Image Tech.)*. A compact video camera with integral VTR.
came *(Build.)*. A bar of lead suitably grooved to hold and connect adjacent panes of glass in a window.
camel hair *(Build.)*. An extremely soft hair obtained from squirrel tail. Used for gilders mops and size brushes used in glass gilding.
camel hair *(Textiles)*. A silky fibre from the haunch and underpart of the camel or dromedary; used for dress fabrics, warm coverings etc.
cameo *(Arch.)*. (1) Carving or modelling in relief. (2) A striated shell or precious stone carved in relief to show different colours in the layers.
camera *(Image Tech.)*. Apparatus for forming an image of an external scene or subject on a light-sensitive surface, such as a photographic emulsion or the target in a TV **camera-tube**.
camera channel *(Image Tech.)*. In a television studio, the camera, with all its supplies, monitor, control position and communication to the operator, which forms a unit, with others, for supplying video signal to the control room.
camera lucida *(Phys.)*. A device for facilitating the drawing of an image seen in a microscope or other optical instrument. In its simplest form, it consists of a thin plate of unsilvered glass, placed above the eyepiece at an angle of 45° with the axis of the instrument, so as to reflect into the eye of the observer an image of the drawing surface, which is seen simultaneously with the microscope image.
camera marker *(Image Tech.)*. System in a cinematograph camera for simultaneously producing a fogged area on the picture negative and a short tone on the sound record at the start of each take for subsequent synchronization.
camera obscura *(Phys.)*. A darkened room in which an image of surrounding objects is cast on a screen by a long-focus convex lens.
camera tube *(Image Tech.)*. One which converts an image of an external scene into a video signal. Essential component in a TV camera channel. In U.S., a *pick-up tube*.
camouflage *(Geol.)*. Describes the relationship between a trace element and a major element whose ionic charge and ionic radius are similar, as a result of which the trace element always occurs in the minerals of the major element but does not form separate minerals of its own. Thus gallium can be considered to be camouflaged by aluminium.
camouflage *(Radar)*. Treatment of objects so that there is ineffective reflection of radar waves.
cAMP *(Biol.)*. See **cyclic adenosine monophosphate**.
campaniform *(Zool.)*. Dome-shaped; as *campaniform sensilla* of certain Insects, which are mechanoreceptors, occurring widely on the body.
campanile *(Arch.)*. A bell-tower, often detached.
campanulate *(Bot.)*. Bell-shaped.
Campbell bridge *(Elec.Eng.)*. An electrical network, designed by Campbell, for comparing mutual inductances.

Campbell gauge *(Elec.Eng.)*. Electrical bridge, one arm being the filament of a lamp located in the low gas pressure to be measured.

Campbell's formula *(Elec.Eng.)*. That which gives the effective attenuation of a coil-loaded transmission line in terms of the constants of the line and the magnitude of the loading.

Campbell-Stokes recorder *(Meteor.)*. See **sunshine recorder**.

camp ceiling *(Build.)*. A ceiling having two opposite parts sloping in line with the rafters, the middle part being horizontal.

camphane *(Chem.)*. $C_{10}H_{18}$, white crystals; mp 154°C. It is a saturated terpene hydrocarbon, the parent substance of the camphor group; see **camphor**.

camphene *(Chem.)*. $C_{10}H_{16}$, a white solid; mp 50°C. An unsaturated terpene hydrocarbon, occurring in various essential oils; it can be prepared from pinene hydrochloride.

camphor *(Chem.)*. *Common (Japan) camphor*, $C_{10}H_{16}O$, colourless transparent prisms of chacteristic odour; mp 175°C, bp 204°C, rel.d. 0.985. Can be sublimed readily and is volatile in steam. Natural camphor is obtained from the camphor tree (*Cinnamomum camphora*). Widely cultivated in Taiwan, China and Japan. Synthetic camphor (optically inactive) is derived from α-pinene. Camphor is an ingredient of various lotions and liniments; it is used in industry as a plasticizer. See **borneol**.

CAM plant *(Bot.)*. A plant which photosynthesizes by means of **crassulacean acid metabolism**.

cam profile *(Eng.)*. The shape of the cam as determined by the form of the flanks and tip; in general, the cam outline.

camp sheathing *(Civ.Eng.)*. An earth-retaining wall formed of timber piles placed 2 to 3 m apart and connected by stout timber **walings**; often used to support river banks.

camp sheeting *(Build.)*. Sheet piling used in foundation work to retain sandy earth.

camptonite *(Geol.)*. An igneous rock occurring in minor intrusions, and belonging to the family of lamprophyres. It consists essentially of plagioclase feldspar and brown hornblende.

campylotropous *(Bot.)*. An ovule curved so that the micropyle and the stalk are approximately at right angles and the stalk appears to be attached to the side.

camshaft *(Autos.,Eng.)*. (1) Shaft with lobed cams to operate the inlet and exhaust valves of a 4-stroke engine. It is driven at half crankshaft speed by various means. See **timing gear**. (2) Any shaft to which cams are keyed or formed integrally.

camshaft controller *(Elec.Eng.)*. A form of control equipment for electric motors (usually in locomotives), in which the contactors are operated mechanically by cams on a rotating shaft.

cam-type steering-gear *(Autos.)*. Steering gear in which the steering column carries a pair of opposed volute cams, which engage with a peg or roller carried by a short arm attached to the drop-arm spindle.

can *(Nuc.Eng.)*. Cover for reactor fuel rods, usually metallic (aluminium, magnox, stainless steel). See **cladding**.

Canada balsam *(Chem.)*. Balsam of fir, or Canada turpentine. A yellowish liquid, of pine-like odour, soluble in ethoxyethane, trichloromethane, benzene; obtained from *Abies balsamica*. Used for lacquers and varnishes, and as an adhesive for lenses, instruments etc., its refractive index being approximately the same as that of most optical glasses.

Canadian asbestos *(Min.)*. See **chrysotile**.

Canadian shield *(Geol.)*. The name applied to the vast area of Precambrian rocks which cover 5 million square kilometres in E.Canada.

Canadian spruce *(For.)*. Name for the wood of several trees of the *Picea* genus, the most important commercial timber in Canada. Typical uses include paper, sounding boards for musical instruments, food containers etc.

Canadian Standard Freeness *(Paper)*. *CSF*. A laboratory control test of beating in which diluted pulp is allowed to drain through a sieve. A wet beaten pulp gives a low CSF value.

canal *(Bot.)*. An elongated intercellular space containing air, water or a secretory product such as resin or oil.

canal *(Nuc.Eng.)*. Water-filled trench into which the highly-active elements from a reactor core can be discharged. The water acts as a shield against radiation but allows objects to be easily inspected.

canal cell *(Bot.)*. One of the short-lived cells present in the central cavity of the neck of the archegonium.

canalette blind *(Build.)*. See **Italian blind**.

canaliculate *(Bot.)*. Marked longitudinally by a channel or groove.

canaliculus *(Zool.)*. Any small channel; in the liver, an intercellular bile-channel; in bone, one of the ramified passages uniting the lacunae; in nerve-cells, a fine channel penetrating the cytoplasm of the cell-body. Adj. *canalicular*.

canalization *(Med.)*. The formation of a new channel in a clot blocking the lumen of a blood vessel.

canard *(Aero.)*. See **tail-first aircraft**.

canaries *(Image Tech.)*. Extraneous high-frequency noises reproduced from a recording channel.

cancellous, cancellated *(Zool.)*. Having a spongy structure, with obvious interstices.

cancels *(Print.)*. Pages printed to substitute for book pages containing errors.

cancer *(Med.)*. Any *malignant neoplasm*. An uncontrolled growth of cells which exhibits invasiveness and remote growth.

cancerophobia *(Med.)*. Morbid fear of contracting cancer.

cancerous *(Med.)*. Pertaining to cancer or carcinoma.

can coiler *(Textiles)*. A device for feeding a sliver into a revolving can in spiral form so allowing easy withdrawal for subsequent processing.

cancrinite *(Min.)*. A hydrated silicate of aluminium, sodium and calcium, also containing the carbonate ion. Found in alkaline plutonic igneous rocks. Cf. **vishnevite**.

cancrum oris *(Med.)*. *Noma*. A destructive ulceration of the cheek in debilitated children, usually during convalescence from an infectious disease.

candela *(Phys.)*. Fundamental SI unit of luminous intensity. If, in a given direction, a source emits monochromatic radiation of frequency 540×10^{12} Hz, and the radiant intensity in that direction is 1/683 watt per steradian, then the luminous intensity of the source is 1 candela. Abbrev. *cd*.

candidiasis *(Vet.)*. See **moniliasis**.

candle *(Phys.)*. Unit of luminous intensity. See **candela**.

candle-fitting *(Elec.Eng.)*. An electric-light fitting consisting of a opal glass tube with a lamp at the top imitating an ordinary wax candle.

candle-lamp *(Elec.Eng.)*. An electric filament lamp shaped to imitate the flame of a wax candle.

candle power *(Phys.)*. A former unit of luminous intensity, now replaced by the **candela**.

candlewick *(Textiles)*. A coarse folded yarn made from cotton. As well as being used in candles the yarn is used for the manufacture of fabrics suitable for bedspreads. See also **wick**.

CANDU *(Nuc.Eng.)*. Type of thermal nuclear power reactor developed by and widely used in Canada. It uses natural (unenriched) uranium oxide fuel canned in Zircaloy and heavy water as moderator and coolant.

cane-sugar *(Chem.)*. See **sucrose**.

canicola fever *(Med.,Vet.)*. The disease of man caused by infection by *Leptospira canicola*, the natural host of which is the dog.

canine *(Zool.)*. Pertaining to, or resembling, a dog; in Mammals, a pointed tooth with single cusp, adapted for tearing, and occurring between the incisors and premolars; pertaining to a canine tooth; pertaining to a ridge or groove on the surface of the maxillary.

canine distemper *(Vet.)*. *Hard pad*. Paramyxovirus infection with secondary bacterial complications. Symptoms include pyrexia, malaise, conjunctivitis, respiratory, enteric and nervous signs. Hyperkeratosis occasionally also present. Ferrets, foxes and mink are also susceptible.

canine leptospiral jaundice *(Vet.)*. See **canine leptospirosis**.

canine leptospirosis *(Vet.)*. *Canine leptospiral jaundice, yellows*. Infection of dogs by *Leptospira icterohaemorrhagiae* causes an acute septicaemia and hepatitis characterized by fever, widespread haemorrhages and jaundice. *L. canicola* causes an acute septicaemia and nephritis. Controlled by vaccination.

canine parvovirus infection *(Vet.)*. Common cause of death in young pups. Symptoms include depression, loss of apetite, diarrhoea, vomiting, dehydration leading to death. Vaccines available.

canine typhus *(Vet.)*. See **Stuttgart disease**.

canine venereal granulomata *(Vet.)*. Infectious venereal sarcoma of dogs. A contagious, viral infection of dogs characterized by tumour-like formations on the genital mucosae. Spread by both venereal and non-venereal contact.

canker *(Bot.)*. A plant disease characterized by well defined necrotic lesions of a main root, stem or branch in which the tissues outside the xylem disintegrate.

canker *(Vet.)*. (1) A chronic inflammation of the keratogenous membrane of the frog and sole of a horse's foot. (2) Chronic eczema of the ear of dogs. (3) Abscess or ulcer in the mouth, eyelids, ear or cloaca of birds.

cannabis *(Bot.)*. *Indian hemp*. The plant *Cannabis sativa* or *Cannabis indica*. The dried flowers, exuded resin and leaves are used to produce the drug hashish, marihuana or bhang. Cannabis is widely recognized as a drug of addiction.

cannibalism *(Vet.)*. A habit, mainly of poultry raised in captivity and characterized by varying degrees of tissue loss by picking of the vent, toes, feathers and around the head. Contributory factors thought to be overcrowding, excessive light and temperature, nutritional imbalance and variations in feeding regime. Reduced by proper husbandry.

Cannizzaro reaction *(Chem.)*. Used for the *disproportionation* of aromatic aldehydes to alcohols and acids, heating with concentrated alkali, e.g. benzaldehyde→ benzyl alcohol + a benzoate.

cannon bone *(Zool.)*. In the more advanced Artiodactyla, the characteristic bone formed by the fusion of the two metapodials in the limb, associated with the reduction of the number of toes to two.

cannon pinion *(Horol.)*. A wheel in the motion work; the wheel or pinion with an extended pipe to which the minute hand is attached.

cannula *(Med.)*. A tube, usually fitted with a **trocar**, for insertion into the body for the injection or removal of fluids or gases.

cannular combustion chamber *(Aero.)*. A gas turbine combustion system with individual flame tubes inside an annular casing.

canonical assembly *(Phys.)*. Term used in statistical thermodynamics to designate a single assembly of a large number of systems which are such that the number of systems with energies lying between E and $E+dE$ is proportional to $e^{-E}\theta$, where θ is a parameter characteristic of the assembly.

canopy *(Aero.)*. (1) The transparent cover of a cockpit. (2) The fabric (nylon, silk or cotton) body of a parachute, which provides high air drag. Usually hemispherical, but may be lobed or rectangular in shape. See also **ribbon parachute**.

canopy *(Bot.)*. The leaves, stems and branches of a plant or an area of vegetation, considered as a whole.

canopy *(Build.)*. A roof or balcony projecting from a wall or supported on pillars.

canopy cover *(Ecol.)*. Percentage of the ground occupied by the vertical projection of all the individuals of one plant species. The sum of such percentages for all plant species gives the total canopy cover.

cant *(Build.)*. A moulding having plane surfaces and angles instead of curves.

cant *(Surv.)*. Slope of rail or road curve whereby outer radius is superelevated, to counteract centrifugal thrust of traffic.

cant bay *(Arch.)*. A bay window having three sides, the outer two being splayed from the wall sides.

cant board *(Build.)*. A board laid on each side of a valley gutter to support the sheet-lead.

cant brick *(Build.)*. A **splay brick**.

canted column *(Arch.)*. A column having faceted sides instead of curved flutes.

canted deck *(Aero.)*. See **angled deck**.

canted wall *(Build.)*. A wall built at an angle to the surfaces of another wall.

Canterbury hammer *(Build.)*. Type of hammer with thick and shallow-curved claw.

cantharadin *(Med.)*. A pharmaceutical product obtained from the dried elytra of the Spanish Fly (species of *Lytta* and *Mylabris*). Formerly used as a counterirritant.

cantilever *(Civ.Eng.)*. A beam or girder fixed at one end and free at the other.

cantilever bridge *(Civ.Eng.)*. A bridge formed of self-supporting projecting arms built inwards from the piers and meeting in the middle of the span, where they are connected together. See also **suspended span**.

cantilever deck *(Civ.Eng.)*. A type of **cantilever bridge** in which the loads are carried by the upper chord.

cantilevered steps *(Build.)*. See **hanging steps**.

cantilever-through *(Civ.Eng.)*. A type of **cantilever bridge** in which the load is carried by the lower chord.

canting *(Eng.)*. Tilting over from the proper position; as in the canting of a piston in its cylinder under the oblique thrust of the connecting-rod.

canting strip *(Build.)*. A projecting sloping member of wood or masonry fitted around a building to deflect water from the wall. Also *water table*.

cantling *(Build.)*. The lower of two courses of burnt brick enclosing a clamp for firing bricks.

canton *(Build.)*. A pilaster, or quoin forming a salient corner, which projects from the wall face.

canvas *(Textiles)*. A heavy closely-woven fabric made from cotton, flax, hemp or jute for uses where strength and firmness are required e.g. interlinings, sails, tents.

canyon *(Geol.)*. A deep narrow, steep-sided valley.

canyon *(Nuc.Eng.)*. US for long narrow space often partly underground with heavy shielding for essential processing of wastes from reactors.

caoutchouc *(Genrl.)*. Raw *rubber*.

cap *(Biol.)*. A modified base added to the 5′ ends of eukaryotic messenger RNA molecules.

cap *(Build.)*. (1) The upper member of a column. (2) A wall coping. (3) The head added to top of a pile. (4) A planted piece on top of a post for weathering or ornamentation. (5) A hand rail supported on balusters.

cap *(Meteor.)*. (1) The covering of cloud which congregates at the top of a mountain. (2) The transient top of detached clouds above an increasing cumulus. Also *pileus*.

capacitance *(Phys.)*. The capacitance of an isolated conductor is defined as the ratio of the total charge on it to its potential. $C=Q/V$. See **farad, stray capacitance**.

capacitance bridge *(Elec.Eng.)*. An a.c. bridge network for the measurement of capacitance. See **Schering bridge, Wien bridge**.

capacitance coefficients *(Phys.)*. Charges ($q_{1 \dots qn}$) of system of conductors can be expressed in terms of coefficients of electric induction (C_{ij}) by the following equations:

$$q_1 = C_{1\infty}V_1 + C_{12}(V_1 - V_2) + \dots + C_{1n}(V_1 - V_n)$$

$$q_2 = C_{21}(V_2 - V_1) + C_{2\infty}V_2 + \dots + C_{2n}(V_2 - V_n)$$

$$q_n = C_{n1}(V_n - V_1) + C_{n2}(V_n - V_2) + \dots + C_{n\infty}V_n$$

where

$$C_{km} = -C_{mk}\ (m \neq k),$$

and

$$C_{m\infty} = C_{m1} + C_{m2} + \ldots + C_{m(n-1)} + C_{mn}.$$

Fundamental relation for partial capacitances of a number of conductors, e.g., electrodes in valves, conductors in cables, variable air-capacitors.

capacitance coupling *(Elec.Eng.,Electronics)*. Inter-stage coupling through a series capacitance or by a capacitor in a common branch of a circuit.

capacitance grading *(Elec.Eng.)*. Grading of the properties of a dielectric, so that the variation of stress from conductor to sheath is reduced. The inner dielectric has the higher permittivity. Ideally, the grading is continuous and the permittivity varies as the reciprocal of the distance from the centre. See **condenser bushing**.

capacitance integrator *(Elec.Eng.)*. Resistance capacitance circuit whose output voltage is approximately equal to the time integral of the input voltage.

capacitative load *(Elec.Eng.)*. Terminating impedance which is markedly capacitive, taking an a.c. leading in phase on the source e.m.f., e.g. electrostatic loudspeaker.

capacitive reactance *(Elec.Eng.)*. Impedance associated with a capacitor. Has a magnitude in ohms equal to the reciprocal of the product of the capacitance (in farad) and the angular frequency of the supply (in rads per second). Also introduces a 90° phase angle such that the current through the device leads the applied voltage.

capacitor *(Elec.Eng.)*. Electric component having **capacitance**; formed by conductors (usually thin and extended) separated by a dielectric, which may be vacuum, paper (waxed or oiled), mica, glass, plastic foil, fused ceramic, or air, etc. Maximum p.d. which can be applied depends on the electrical breakdown of dielectric used. Modern construction uses sheets of metal foil and insulating material wound into a compact assembly. Air capacitors, of adjustable parallel vanes, are used for tuning high-frequency oscillators. Formerly *condenser*.

capacitor bushing *(Elec.Eng.)*. See **condenser bushing**.

capacitor loudspeaker *(Acous.)*. See **electrostatic loudspeaker**.

capacitor microphone *(Acous.)*. See **electrostatic microphone**.

capacitor modulator *(Acous.)*. Capacitor microphone, or similar transducer, which, by variation in capacitance, modulates an oscillation either in amplitude or frequency.

capacitor motor *(Elec.Eng.)*. A single-phase induction motor arranged to start as a 2-phase motor by connecting a capacitor in series with an auxiliary starting winding. The capacitor may be automatically disconnected when the motor is up to speed (*capacitor-start motor*) or it may be left permanently in circuit for power-factor improvement (*capacitor start-and-run motor*).

capacitor start *(Elec.Eng.)*. Starting unit for electric motor using series capacitance to advance phase of current.

capacitor terminal *(Elec.Eng.)*. See **condenser bushing**.

capacity *(Elec.Eng.)*. The output of an electrical apparatus, e.g. that of a motor or generator in kW. In an accumulator (secondary battery), *capacity* is measured by the ampere-hours of charge it can deliver. Capacity of a switch is the current it can break under specified circuit conditions. Also used for **capacitance**.

cap-and-pin type insulator *(Elec.Eng.)*. A special form of the **suspension insulator**.

Cape asbestos *(Min.)*. Blue asbestos from South Africa. See **crocidolite**.

Cape diamond *(Min.)*. A name used in grading diamonds to designate an off-colour stone of a yellowish tint.

Cape ruby *(Min.)*. A misnomer for the red garnet **pyrope**, obtained in the diamond mines of South Africa. See also **false ruby**.

Cape walnut *(For.)*. See **stinkwood**.

capillariasis *(Vet.)*. Inflammation of the alimentary tract of animals or birds due to infection by nematode worms of the genus *Capillaria*.

capillarity *(Phys.)*. A phenomenon associated with surface tension, which occurs in fine bore tubes or channels. Examples are the elevation (or depression) of liquids in capillary tubes and the action of blotting paper and wicks. The elevation of liquid in a capillary tube above the general level is given by the formula

$$h = \frac{2T\cos\theta}{\rho g r},$$

where T is the surface tension, θ is the angle of contact of the liquid with the capillary, ρ is the liquid density, g is the acceleration due to gravity, r is the capillary radius.

capillary *(Biol.,Zool.)*. (1) Of very small diameter; slender, hairlike. (2) Any thin-walled vessel of small diameter, forming part of a network, which aids rapid exchange of substances between the contained fluid and the surrounding tissues; as *bile capillaries, blood capillaries, lymph capillaries*.

capillary action *(Chem.)*. See **capillarity**.

capillary condensation *(Chem.)*. Hypothesis that adsorbed vapours can condense under the incidence of capillary forces to form liquid inside the pores of the adsorbate.

capillary electrometer *(Chem.)*. An instrument in which small electric currents are detected by movement of a mercury meniscus in a capillary tube.

capillary fitting *(Build.)*. A bend, tee or other fitting whose internal bores are a close fit over the tube. Solder is drawn in to the joint by capillary action or the bores may have been previously *tinned* with solder. Common joint in copper work. Also *compression fitting*.

capillary pressure *(Chem.)*. The pressure developed by **capillarity**. Mathematically expressed as $p = 2T\cos\theta/r$, where T is the surface tension, θ the angle of contact, and r the radius of the capillary.

capillary pyrite *(Min.)*. See **millerite**.

capillary soil water *(Bot.)*. Water held between the particles of the soil by capillarity.

cap iron *(Build.)*. See **back iron**.

capital *(Build.)*. The upper member of a column, pier, or pilaster.

capitate *(Bot.)*. Head-like.

capitate *(Zool.)*. Having an enlarged tip, as *capitate antennae*.

capitellum *(Zool.)*. An enlargement or boss at the end of a bone, for articulation with another bone; more particularly, the smaller of the two articular surfaces on the distal end of the mammalian humerus, for articulation with the radius; the distal knoblike extremity of the haltere in Diptera.

capitulum *(Bot.)*. An inflorescence on which the sessile flowers or florets are crowded on the surface of the enlarged apex of the **peduncle**, the whole group being surrounded, and covered in the bud by an envelope of bracts forming an **involucre**; the whole inflorescence superficially appearing to be one flower, as in the daisy, *Bellis*. See **Compositae**.

capitulum *(Zool.)*. A terminal expansion, as that of some shaft bones, some tentacles, and some hairs.

cap nut *(Eng.)*. A nut whose outer end is closed, so protecting the end of the screw, and giving a neat appearance. Also **dome nut, box nut**.

caponizing *(Vet.)*. Castration of a cock bird.

capped elbow *(Vet.)*. A swelling of the olecranon bursa of animals.

capped hock *(Vet.)*. A swelling over the point of the hock of animals.

capping *(Build.)*. A copper strip rolled over the ridge or roll and welted to the underlying sheets.

capping *(Immun.)*. The phenomenon whereby proteins at cell membranes are caused to accumulate in clusters and then to move to one end of the cell when crosslinked, e.g. by antibody against them. Thus antibodies can cause proteins to disappear from the cell surface (though they may be reformed if antibody is removed).

capping *(Min.Ext.)*. The fixing of a shackle or a swivel to the end of a hoisting rope. Overburden lying above valuable seam or bed of mineral.

capping-brick *(Build.)*. A **coping brick**.

capping-plane *(Build.)*. A plane for giving a slight rounding to the upper surface of a wooden hand-rail.

capric acid *(Chem.)*. *Decanoic acid.* $CH_3(CH_2)_8COOH$.

caprification *(Bot.,Zool.)*. The fertilization of the flowers of fig-trees by the agency of Fig Insects, a family of Chalcids (Agaonidae); the process of hanging caprifigs in the female trees.

cap rock *(Geol.,Min.Ext.)*. Impervious stratum overlying natural gas or oil deposits.

caproic acid *(Chem.)*. *Hexanoic acid.* $CH_3.(CH_2)_4.COOH$. Oily liquid; solidification point about 2°C; bp 205°C. Occurs as glycerides in milk, palm oil etc.

capryl alcohol *(Chem.)*. *Octan-2-ol.* $CH_3.(CH_2)_5.CH(OH).CH_3$. It is obtained by distilling castor oil with strong alkali. Liquid, colourless, strong smell. bp 179°C. Used as a foam reducing agent.

caprylic acid *(Chem.)*. *Octanoic acid,* $C_7H_{15}.COOH$; bp 237°C.

cap screw *(Eng.)*. See **socket head screw**.

capsomere *(Biol.)*. Proteins which form regular structures on the surface of a virus.

caps & small caps *(Print.)*. Small capitals with the first letter in capitals. The first word of a chapter is often set in CAPS & SMALL CAPS.

capstan *(Eng.)*. A vertical drum or spindle on which rope is wound (e.g. for warping a ship alongside a wharf); it is rotated by manpower or by hydraulic or electric motor.

capstan *(Image Tech.)*. Roller providing the constant speed drive in a magnetic tape recorder.

capstan-head screw *(Eng.)*. A screw having a cylindrical head provided with radial holes in its circumference. It is tightened by a tommy bar inserted in these holes.

capstan lathe *(Eng.)*. A lathe in which the tools required for successive operations are mounted radially in a tool-holder resembling a capstan; by revolving this, each tool in turn may be brought into position in exact location.

capstan nut *(Eng.)*. A nut which is tightened in the same way as a **capstan-head screw**.

capsular polysaccharides *(Immun.)*. Those present as constituents of bacterial capsules which often resist engulfment by granulocytes. Antibodies against these polysaccharides can coat the bacteria so that they become susceptible to phagocytosis, and are usually protective.

capsule *(Bot.)*. (1) That part of the sporophyte of a bryophyte which contains the spores. (2) A fruit, dry when mature, composed of more than one carpel, which splits at maturity to release the seeds. (3) A coating of mucilaginous material outside the wall of a bacterial cell.

capsule *(Med.)*. (1) A fibrous or membranous covering enclosing an abdominal structure. (2) A soluble case of gelatine or similar substance in which a medicine may be enclosed.

capsule *(Zool.)*. Any fibrous or membranous covering of a viscus, e.g. the kidney. The name is applied also to certain areas in the brain which are formed by nerve fibres.

captan *(Chem.)*. *N-(trichloromethylthio)cyclohex-4-ene-1,2-dicarboxyimide*; a fungicide.

caption *(Print.)*. The descriptive wording under an illustration. Often called *legend.*

captive balloon *(Aero.)*. A balloon anchored or towed by a line. Usually the term refers to spherical balloons only. Special shapes for stability, etc. are called *kite balloons.*

captive nut *(Eng.)*. A nut (loosely) fastened to an adjacent machine member so as to retain it in position when the corresponding screw element is absent.

captive screw *(Eng.)*. A screw (loosely) fastened in position by its head or shank so as to be retained when unscrewed from the matching machine member.

captive tape *(Print.)*. Tapes which operate only at the slow speeds when the web is being threaded on rotary presses.

capture *(Phys.)*. Any process in which an atomic or nuclear system acquires an additional particle. In a nuclear radiative capture process there is an emission of electromagnetic radiation only, e.g. the emission of γ-rays subsequent to the capture of a neutron by a nucleus.

capture *(Telecomm.)*. In frequency and phase modulation, the diminution to zero of a weak signal (noise) by a stronger signal. See **k-capture**.

caput *(Zool.)*. An abrupt swelling at the distal end of a structure. pl. *capita.* adj. *capitate.*

caput medusae *(Med.)*. Dilated subcutaneous veins around the umbilicus in cirrhosis of the liver.

car *(Aero.)*. In an airship, the part intended for the carrying of the load (crew, passengers, goods, engines etc.). It may be suspended below, or may be inside the hull or envelope.

caracole *(Build.)*. A helical staircase.

Caradoc *(Geol.)*. An epoch of the Ordovician period.

carapace *(Zool.)*. An exoskeleton shield covering part or all of the dorsal surface of an animal; as the bony dorsal shield of a tortoise, the chitinous dorsal shield of some Crustacea.

carat *(Min.)*. Also karat. (1) A standard weight for precious stones. The *metric carat*, standardized in 1932, equals 200 mg (3.086 grain). (2) The standard of fineness for gold. The standard for pure gold is 24 carats; 22 carat gold has two parts of alloy; 18 carat gold 6 parts of alloy.

carbamazepine *(Med.)*. Anti-epileptic drug of particular value in children and in the control of complex partial seizures (temporal lobe epilepsy).

carbamic acid *(Chem.)*. $NH_2.CO.OH$; is not known to occur free, being known only in the form of derivatives, e.g. the ammonium salt, NH_2COONH_4. The esters are known as *urethanes.*

carbamide *(Chem.)*. **Urea**.

carbamyl chloride *(Chem.)*. $NH_2.CO.Cl$; colourless needles of pungent odour; mp 50°C, bp 61°C; formed by the action of hydrochloric acid (gaseous) on cyanic acid; it serves for the synthesis of organic acids.

carbamyl phosphate *(Biol.)*. The phosphate ester of carbamic acid which is an intermediate in the biosynthesis of urea and pyrimidines.

carbanilide *(Chem.)*. *Diphenylurea,* $CO(NHC_6H_5)_2$.

carbanion *(Chem.)*. A short-lived, negatively-charged intermediate formed by the removal of a proton from a $\rightarrow$C—H bond, e.g. in Grignard reactions.

carbazole *(Chem.)*. $(C_6H_4)_2NH$; colourless plates: mp 238°C; sublimes readily; contained in coal-tar and crude anthracene oil. It is the imine (intramolecular) of diphenyl and is formed from diphenylamine by passing the vapour through red-hot tubes, or by distilling *o*-aminodiphenyl over lime at about 600°C.

carbenes *(Chem.)*. (1) Reactive uncharged intermediates of formula: CXY, where X and Y are organic radicals or halogen atoms. (2) Such constituents of asphaltic material as are soluble in carbon disulphide but not in carbon tetrachloride. See **asphaltenes, malthenes**.

carbenium ion *(Chem.)*. A (usually) short-lived intermediate in a reaction with a positive charge on a carbon atom. Also *carbonium ion* and *carbocation.*

carbenoxolone *(Med.)*. Synthetic derivative of glycyrrhizic acid which is useful in healing gastric ulcers.

carbethoxy *(Chem.)*. The group $-COOC_2H_5$ in organic compounds.

carbides *(Chem.)*. Binary compounds of metals with carbon. Carbides of group IV to VI metals (e.g. silicon, iron, tungsten) are exceptionally hard and refractory. In groups I and II, calcium carbide (ethynide) is the most useful. See **cemented carbides, cementite**.

carbide tools *(Eng.)*. Cutting and forming tools used for hard materials or at high temperatures. They are made of carbides of tungsten, tantalum and other metals held in a matrix of cobalt, nickel etc, and are very hard with good compressive strength.

carbimazole *(Med.)*. Drug which inhibits thyroxine synthesis and is used in the treatment of **thyrotoxicosis**.

carbinol *(Chem.)*. *Methanol.* The nomenclature of alcohols is often based on their homologous relation to methanol, e.g. tertiary butyl alcohol, $(CH_3)_3C.OH$, is termed *trimethyl carbinol.*

carbocyclic compounds *(Chem.)*. Also *isocyclic.* These are closed-chain or ring compounds in which the closed chain consists entirely of carbon atoms.

carbohydrates *(Chem.)*. A group of compounds represented by the general formula $C_x(H_2O)_y$. Substances found in plants and animals; e.g. sugars, starch, cellulose. The carbohydrates also comprise other compounds of a different general formula but closely related to the above substances, e.g. rhamnose, $C_6H_{12}O_5$. Carbohydrates are divided into **monosaccharides, oligosaccharides, polysaccharides.** The carbohydrate element in diet supplies energy, provided by the oxidation of the constituent elements.

carbolic acid *(Chem.)*. **Phenol.**

carbolic oils *(Chem.)*. See **middle oils.**

carbomethoxy *(Chem.)*. The group $-CH_3O.CO$ in organic compounds.

carbon *(Chem.)*. Amorphous or crystalline (graphite and diamond) element, symbol C, at. no. 6, r.a.m. 12.011, mp above 3500°C, bp 4200°C. Its allotropic modifications are diamond and graphite. The assumption that its atom is tetravalent, the bonds being directed towards the vertices of a regular tetrahedron, is the basis for all theoretical organic chemistry (see **carbon compounds**). Widely used in brushes for electric generators and motors, and alloyed with iron for steel. Colloidal carbon or graphite is used to coat cathode-ray tubes and electrodes in valves, to inhibit photoelectrons and secondary electrons. High-purity carbon, crystallized to graphite in a coke furnace for many days, is used in many types of nuclear reactors, particularly for moderation of neutrons.

carbonaceous *(Chem.)*. Said of material containing carbon as such or as organic (plant or animal) matter.

carbonaceous rocks *(Geol.)*. Sedimentary deposits of which the chief constituent is carbon, derived from plant residues. Under this heading are included peat, lignite, or brown coal, and the several varieties of true coal (bituminous coal, anthracite, etc.).

carbonado *(Min.)*. See **black diamond.**

carbon arc *(Elec.Eng.)*. An arc between carbon electrodes; usually limited to pure carbon rather than flame carbon electrodes.

carbon-arc lamp *(Elec.Eng.)*. A lamp employing as the light source an arc between carbon electrodes.

carbon-arc welding *(Elec.Eng.)*. Arc welding carried out by means of an arc between a carbon electrode and the material to be welded.

carbonate *(Chem.)*. A compound containing the acid radical of carbonic acid (CO_3 group). Bases react with carbonic acid to form carbonates, e.g. $CaCO_3$, calcium carbonate.

carbonate-apatite *(Min.)*. A variety of apatite containing appreciable CO_2.

carbonatite *(Geol.)*. An igneous rock composed largely of carbonate minerals (calcite and dolomite). Carbonatites are invariably associated with alkaline igneous rocks such as nepheline-syenite, and are rich in a number of unusual minerals, especially those of the rare earth elements.

carbon black *(Chem.)*. Finely divided carbon produced by burning hydrocarbons (e.g. methane) in conditions in which combustion is incomplete. Widely used in the rubber, paint, plastic, ink and other industries. It forms a very fine pigment containing up to 95% carbon, giving a very intense black; prepared by burning natural gas and letting the flame impinge on a cool surface. Also *channel black, gas black.*

carbon brush *(Elec.Eng.)*. A small block of carbon used in electrical equipment to make contact with a moving surface.

carbon compounds *(Chem.)*. Compounds containing one or more carbon atoms in the molecule. They comprise all organic compounds and include also compounds, e.g. carbides, carbonates, carbon dioxide etc., which are usually dealt with in inorganic chemistry. Carbon compounds are the basis of all living matter.

carbon contact *(Elec.Eng.)*. In a switch, an auxiliary contact designed to break contact after and to make contact before the main contact to prevent burning of the latter; it is of carbon and designed to be easily removable.

carbon cycle *(Astron.)*. A chain of nuclear fusion reactions, believed to take place in stars more massive than the Sun, the net effect of which is to transmute protons in helium nuclei, the carbon atoms effectively acting as a catalyst. The reaction is strongly dependent on temperature and for this reason is believed to be the main source of energy in hot massive stars.

carbon cycle *(Biol.)*. The biological circulation of carbon from the atmosphere into living organisms and, after their death, back again. See **carbon dioxide, photosynthesis.**

carbon dating *(Phys.)*. *Radiocarbon dating.* Atmospheric carbon dioxide contains a constant proportion of radioactive ^{14}C, formed by cosmic radiation. Living organisms absorb this isotope in the same proportion. After death it decays with a half-life 5.57×10^3 years. The proportion of ^{12}C to the residual ^{14}C indicates the period elapsed since death.

carbon dioxide *(Chem.)*. CO_2. A colourless gas; density at s.t.p. 1.976 kg/m^3, about 1.5 times that of air. Produced by the complete combustion of carbon, by the action of acids on carbonates (e.g. Kipp's apparatus), by the thermal decomposition of carbonates (e.g. lime burning) and during fermentation. It plays an essential part in metabolism, being exhaled by animals and absorbed by plants (see **photosynthesis**). May be liquefied at 20°C at a pressure of 5.7 MN/m^2, but at atmospheric pressure it sublimes at −78.5°C. Liquid and solid CO_2 are much used as refrigerants, notably for ice-cream, and in fire extinguishers. CO_2 dissolves in water to form unstable carbonic acids; the pressurized solution produces the effervescent 'sparkle' in carbonated beverages. When solid it has a convenient temperature for testing electronic components. High-pressure carbon dioxide has found a considerable use as a coolant in carbon-moderated nuclear reactors.

carbon-dioxide laser *(Phys.)*. One in which the active gaseous medium is a mixture of carbon dioxide and other gases. It is excited by glow-discharge and operates at a wavelength of 10.6 μm. Carbon-dioxide lasers are capable of pulsed output with peak power up to 100 MW or continuous output up to 60 kW.

carbon-dioxide welding *(Eng.)*. *Metal arc welding* using CO_2 as the shielding gas.

carbon disulphide *(Chem.)*. CS_2. Sulphur vapour passed over heated charcoal combines with the carbon to form *carbon disulphide.* Used as a solvent for sulphur and rubber. The disagreeable smell associated with commercial carbon disulphide is due to impurities.

carbon fibres *(Chem.,Textiles)*. A high strength, highly oriented fibre of about 8 μm in diameter, consisting almost exclusively of carbon atoms. It is made as continuous filament by the pyrolysis in an inert atmosphere of organic fibres such as cellulose but usually of polyacrylonitrile. It is used as a reinforcing material with epoxy or polyester resins to form composites which have a higher strength/weight ratio than metals. Boron and glass are alternative fibre materials.

carbon film technique *(Biol.)*. One of the methods used in electron microscopy to provide a supporting film for the specimen. The film is prepared by subliming carbon in a vacuum and is itself supported on a metal grid.

carbon fixation *(Bot.)*. The synthesis of organic compounds from carbon dioxide, most notably in photosynthesis.

carbon gland *(Eng.)*. A type of gland used to prevent leakage along a shaft. It consists of carbon rings cut into segments and pressed into contact with the shaft by an encircling helical spring. See **garter spring.**

carbonic acid *(Chem.)*. H_2CO_3. A weak acid formed when carbon dioxide is dissolved in water.

carbonic acid derivatives *(Chem.)*. Carbonic acid forms both normal and acid salts. The esters, chlorides and amides form two series, viz. normal compounds, in which both hydroxyl groups are substituted; and acid compounds, in which only one hydroxyl group is substituted. The acid compounds are unstable in the free state, but form stable salts.

carbonic acid gas *(Chem.)*. Carbon dioxide effervescing from liquids which have been saturated with carbon dioxide under pressure. The gas escapes when the pressure is withdrawn.

carbonic anhydrase *(Biol.)*. An enzyme in blood corpuscles catalysing the decomposition of carbonic acid. It is essential for the effective transport of carbon dioxide from the tissues to the lungs.

carbonic anhydride *(Chem.)*. A synonym for *carbon dioxide*. See **carbonic acid**.

Carboniferous System *(Geol.)*. A geological period extending from app. 360–280 million years. Divided in the US into the Mississipian and Pennsylvanian periods. In Britain comprises the Carboniferous Limestone, Millstone Grit and Coal Measures. The corresponding system of rocks.

carbon-in-pulp *(Min.Ext.)*. The use of carbon as the adsorbent for the gold leached from ore in the **cyaniding** process. This has largely replaced the zinc-dust method and allows commercial recovery down to 2 ppm of gold.

carbonium ion *(Chem.)*. See **carbenium ion**.

carbonization *(Chem.)*. The destructive distillation of substances out of contact with air, accompanied by the formation of carbon, in addition to liquid and gaseous products. Coal yields coke, while wood, sugar etc. yield charcoal.

carbonization *(Eng.)*. See **cementation**.

carbonization *(Geol.)*. The conversion of fossil organic material to a residue of carbon. Plant material is often preserved in this way.

carbonization *(Textiles)*. The steeping of wool in a dilute solution of sulphuric acid, or its treatment by hydrochloric acid gas (dry process). This converts any cellulosic impurities into carbon dust and thereby facilitates their removal.

carbonized filament *(Electronics)*. Thoriated tungsten filament coated with tungsten carbide to reduce loss of thorium from the surface.

carbon microphone *(Acous.)*. A microphone in which a normally d.c. energizing current is modulated by changes in the resistance of a cavity filled by granulated carbon which is compressed by the movement of the diaphragm. The diameter of the cavity is frequently very much less than that of the diaphragm, and it is then known as a **carbon button**.

carbon monoxide *(Chem.)*. CO. Formed when carbon is heated in a limited supply of air, when carbon dioxide is heated in an excess of carbon, when carbon dioxide is passed over some hot metals, or by dehydration of methanoic acid. It is a product of incomplete combustion. Poisonous. Its properties as a reducing agent render it valuable in industrial processes. See also **carbonyl**.

carbon monoxide-haemoglobinaemia *(Med.)*. Haemoglobin combines with carbon monoxide instantaneously and is thus deprived of its oxygen-exchanging properties. This leads to poisoning of the body and death by asphyxia without the more normal cyanosis.

carbon-nitrogen cycle *(Astron.)*. See **carbon cycle**.

carbon paper *(Paper)*. Paper coated with waxes containing dyes or carbon black, used for making duplicate copies in typewriting, etc.

carbon pile voltage transformer *(Elec.Eng.)*. Variable electrical resistor made from disks or plates of carbon arranged to form a pile.

carbon process *(Image Tech.)*. Printing process using a relief made by the solvent action of warm water on an emulsion of bichromated pigmented gelatine, the image being ultimately transferred to a suitable paper base.

carbon replica technique *(Biol.)*. A method used in electron microscopy for making a surface replica of a specimen. It is coated with a structureless carbon film, and the film and specimen are subsequently removed by dissolving in an appropriate solvent.

carbon resistor *(Elec.Eng.)*. Negative temperature coefficient, non-inductive resistor formed of powdered carbon with ceramic binding material. Used for low-temperature measurements because of the large increase in resistance as temperature decreases.

carbon star *(Astron.)*. Rare giant star showing molecular bands of carbon compounds in its spectrum.

carbon steel *(Eng.)*. A steel whose properties are determined primarily by the percentage of carbon present. Besides iron and carbon, carbon steel contains manganese (up to 1.6%), silicon (up to 0.6%), sulphur and phosphorus (up to 0.1%), but no chromium, nickel, molybdenum etc.

carbon suboxide *(Chem.)*. *Malonic anhydride*, C_3O_2, $O{=}C{=}C{=}C{=}O$; a colourless liquid or gas; mp −107°C, bp +7°C; formed by heating malonic acid to 140°–150°C. Example of a **ketene**.

carbon tetrachloride *(Chem.)*. *Tetrachloromethane*, CCl_4; a colourless liquid, bp 76°C; prepared by the exhaustive chlorination of methane or carbon disulphide. Solvent for fats and oils. Not hydrolysed by water. Because of toxicity, its use in fire-extinguishers and dry-cleaning has now much declined.

carbon tissue *(Image Tech.)*. Paper coated with a mixture of gelatine and a pigment (sometimes carbon powder). Used in the **carbon process**.

carbon tissue *(Print.)*. A photosensitive gelatine layer on a paper backing which is printed down with positives and the photogravure screen, and, after developing, is stripped down on the photogravure cylinder to act as an etchant resist.

carbon value *(Chem.)*. The figure obtained empirically as a measure of the tendency of a lubricant to form carbon when in use.

carbonyl *(Chem.)*. When carbon monoxide acts as a radical, as it appears to do in many reactions, it is called the *carbonyl* group. Carbon monoxide combines with certain metals to form carbonyls, e.g. $Co(CO)_3$, $Ni(CO)_4$, $Fe(CO)_5$, $Mo(CO)_6$. *Aldehydes* and *ketones* are organic carbonyl compounds.

carbonyl chloride *(Chem.)*. **Phosgene**.

carbonyl powders *(Eng.)*. Metal powders produced by reacting carbon monoxide with the metal to form the gaseous carbonyl. This is then decomposed by heat to yield powder of high purity.

Carborundum *(Eng.)*. TN for **silicon carbide** products.

Carborundum wheel *(Eng.)*. See **grinding wheel**.

carboxydismutase *(Bot.)*. Same as *ribulose bisphosphate carboxylase*.

carboxy-haemoglobinaemia *(Med.)*. See **carbon monoxide-haemoglobinaemia**.

carboxylase *(Biol.)*. An enzyme which by the mediation of biotin, its **prosthetic group**, carboxylates its substrate, e.g. pyruvate carboxylate converts the monocarboxylic acid, pyruvate into *dicarboxylic oxaloacetate*.

carboxylase *(Bot.)*. An enzyme that catalyses the incorporation of carbon dioxide into its substrate, e.g. ribulose bisphosphate carboxylase, PEP carboxylase.

carboxyl group *(Chem.)*. The acid group –COOH.

carboxylic acid *(Chem.)*. $R\text{–}(COOH)_n$. An organic compound having one or more carboxyl radicals. A carboxyl radical is usually designated: $-C{\begin{smallmatrix}\nearrow O\\ \searrow OH\end{smallmatrix}}$, but shows resonance.

carboy *(Glass)*. Large, narrow-necked container, usually of balloon shape, having a capacity of 20 l or more.

Carbro process *(Image Tech.)*. Colour print process in which relief images are forced in pigmented tissues by contact with colour-separation bromide prints, a set of three colour positives resulting for mounting in superimposition on a paper backing.

carbuncle *(Med.)*. Circumscribed staphylococcal infection of the subcutaneous tissues.

carbuncle *(Min.)*. A gem variety of garnet cut *en cabochon*. It has a deep red colour. See **almandine**.

carburation *(Eng.)*. The mixing of air with volatile fuel to form a combustible mixture for using in internal-combustion (petrol) engines.

carburettor *(Eng.)*. A device for mixing air and a volatile fuel in correct proportions, in order to form a combustible mixture. It consists essentially of a jet, or jets, discharging the fuel into the air stream under the pressure difference created by the velocity of the air as it flows through a nozzle shaped constriction. *Syn.* carburetter, carburetor.

carburizing *(Eng.)*. A method of **case hardening** of steel in which the metal is heated above its ferrite-austenite transition in contact with a suitable carbon material. It gives a gradient of carbon extending from the surface which can be hardened by quenching either directly or after re-heating.

carbylamines *(Chem.)*. See **isocyanides**.

carcass *(Build.)*. The shell of a house in construction, consisting of walls and roof only, without floors, plastering or joiner's work.

carcassing timber *(Build.)*. Timber for the framing of a building or other structure.

carcinogenesis *(Med.)*. The production and development of cancer.

carcinoma *(Med.)*. A disorderly growth of epithelial cells which invade adjacent tissue and spread via lymphatics and blood vessels to other parts of the body. See also **sarcoma**.

carcinoma en cuirasse *(Med.)*. Extensive carcinomatous infiltration of the skin, characterized by hardness and gross thickening.

carcinomatosis *(Med.)*. Also *carcinosis*. Cancer widely spread throughout the body.

carcinomatous *(Med.)*. Of the nature of cancer.

carcinosis *(Med.)*. See **carcinomatosis**.

carcinotron *(Telecomm.)*. Same as **backward-wave tube**.

card *(Comp.)*. See **punched card**.

card *(Electronics)*. A board containing a circuit with a plug or connector.

card *(Surv.)*. The graduated dial or face of a magnetic compass to which the card and needle are firmly connected.

card *(Textiles)*. Machine for carrying out the **carding** process.

Cardan mount *(Eng.)*. Type of gimbal mount used for compasses and gyroscopes.

card chase *(Print.)*. A small chase for imposing small jobs such as cards.

card clothing *(Textiles)*. A strong material (e.g. a fabric-rubber laminate) fitted with masses of strong, flexible wire teeth, pins or spikes. Often in the form of flats or rotating cylinders.

card deck *(Comp.)*. A set of **punched cards** comprising a program or a set of data.

carded yarns *(Textiles)*. Yarns spun from slivers taken straight from the card.

cardiac *(Med.)*. Anything pertaining to the heart and upper part of the stomach.

cardiac aneurysm *(Med.)*. A fibrous dilatation of one or other ventricle due to destruction of cardiac muscle.

cardia, cardiac sphincter *(Med.)*. The physiological sphincter surrounding the opening of the oesophagus into the stomach.

cardiac arrest *(Med.)*. The sudden cessation of the heart's action as a pump. May be due to ventricular fibrillation or asystole.

cardiac asthma *(Med.)*. A sensation of breathlessness and wheeze due to pulmonary congestion brought on by failure of the left side of the heart.

cardiac massage *(Med.)*. The manual procedure whereby the heart action is maintained after cardiac arrest, externally by massage on the sternum, or by massaging the heart directly with a hand around it.

cardiac muscle *(Biol.)*. The contractile tissue forming the wall of the heart of Vertebrates.

cardiac pacemaker *(Med.)*. Electronic device implanted into chest wall, usually by means of a transvenously introduced catheter in the right atrium or ventricle. It senses and regulates abnormally slow heart rhythms.

cardiac tamponade *(Med.)*. When fluid or blood fills the pericardial sac and interferes with the pumping action of the heart.

cardiac valve *(Zool.)*. A valve at the point of junction of the fore- and mid-intestine in many insects. Also *oesophageal valve*.

cardinal *(Zool.)*. In Bivalvia and Brachiopoda, pertaining to the hinge; more generally, primary, principal, as the *cardinal sinuses* or *veins*, being the principal channel for the return of blood in the lower vertebrates.

cardinal number *(Maths.)*. A property associated with a set such that two sets X and Y have the same cardinal number if and only if X and Y can be arranged in one-to-one correspondence. Cardinal number of a set differs from ordinal number in that the former relates to quantity and the latter to order or arrangement. This distinction, however, becomes of real interest only for transfinite numbers, as is suggested by the fact that for finite sets the same symbol is used for both the cardinal and the ordinal number. For a finite set, the cardinal number is just the number of elements in the set. In the content of infinite sets, a set may have the same cardinal number as some of its proper subsets; but this is not the case for finite sets.

cardinal planes *(Phys.)*. In a lens, planes perpendicular to the principal axis, and passing through the cardinal points of the lens.

cardinal points *(Astron.)*. Name given to the four principal points of the horizon, north, south, east and west, corresponding to **azimuths** 0°, 180°, 90° and 270°. See **quadrantal points**.

cardinal points *(Phys.)*. For a lens or lens system, the two *principal foci*, the two **nodal points** and the two **principal points** are the *cardinal points*. For a lens used in air, the principal points coincide with the corresponding nodal points. For a lens of negligible thickness the principal points and the nodal points all coalesce at a single point at the optical centre of the lens.

carding *(Textiles)*. The process of passing fibres through a *card* so that they are separated and cleaned to emerge in the form of a light fluffy *web* or *sliver*.

cardioblast *(Zool.)*. A mesodermal cell in an embryo, destined to take part in the formation of the heart.

cardiocentesis *(Med.)*. Puncture of the heart with a needle.

cardiogram *(Med.)*. Trace produced by electrocardiograph (*ECG*) showing voltage waveform generated during heart beats.

cardiograph *(Med.)*. See **electrocardiograph**.

cardioid *(Maths.)*. A heart shaped curve with polar equation $r = 2a(1+\cos\theta)$. An epicycloid in which the rolling circle equals the fixed circle. Cf. **limaçon**.

cardioid directivity *(Acous.)*. Special shape of a directivity. It is produced by superimposing the fields of a monopole and a dipole, and has the shape of a cardioid.

cardioid reception *(Telecomm.)*. Describing a receiving system with a directional antenna, the **radiation pattern** of which is heart-shaped.

cardiolipin *(Immun.)*. Phospholipid hapten purified from beef heart which is the active antigen in the Wassermann reaction and other serological tests for syphilis.

cardiology *(Med.)*. That part of medical science concerned with the function and diseases of the heart. n. *cardiologist*.

cardiolysis *(Med.)*. Operative freeing of the heart from the chest wall when it is adherent to it in chronic adhesive pericarditis.

cardiomalacia *(Med.)*. Pathological weakening of the heart muscle.

cardiomyopathy *(Med.)*. Any disease affecting the heart muscle.

cardiopulmonary bypass *(Med.)*. Apparatus including a pump and an oxygenator for artificial maintenance of circulation during operations on the heart.

cardiospasm *(Med.)*. Spasm of the cardia or cardiac sphincter of the stomach.

cardiotachometer *(Med.)*. An electronic amplifying instrument for recording and timing the heart rate.

cardiovascular *(Med.)*. Pertaining to the heart and blood vessels.

carditis *(Med.)*. Inflammation of the muscle and coverings of the heart.

cardo *(Zool.)*. The hinge of the bivalve shell. pl. *cardines*.

card reader *(Comp.)*. Component part of a computing system which scans **punched cards** and delivers signals corresponding to the information recorded by the holes.

caret *(Print.)*. A symbol (⁁) used in proof correcting to indicate that something is to be inserted at that point.

car-floor contact *(Elec.Eng.)*. A contact attached to the false floor of an electrically controlled lift; it is usually arranged to prevent operation of the lift by anyone outside the car while a passenger is in the lift.

caries *(Med.)*. Decay of bone or teeth. adj. *carious*.

carina *(Zool.)*. A median dorsal plate of the exoskeleton of some Cirripedia; a ridge of bone resembling the keel of a boat, as that of the sternum of flying Birds.

carinate *(Bot.,Zool.)*. Shaped like a keel; having a projection like a keel. Also *tropeic*.

carious *(Bot.,Med.)*. Also *cariose*. Appearing as if decayed.

Carme *(Astron.)*. The eleventh natural satellite of **Jupiter**.

carminative *(Med.)*. Relieving gastric flatulence; medicine which does this.

carnallite *(Min.)*. The hydrated chloride of potassium and magnesium, occurring in bedded masses with other saline deposits, as at Stassfurt. Such deposits arise from the desiccation of salt lakes. It is used as a fertilizer.

carnassial *(Zool.)*. In terrestial Carnivora, large flesh-cutting teeth from the first lower molar and the last upper premolars.

carnelian *(Min.)*. A translucent variety of chalcedony (silica) of red or reddish-brown colour.

Carnivora *(Zool.)*. An order of primarily carnivorous Mammals, terrestial or aquatic; usually with three pairs of incisors in each jaw and large prominent canines; the last upper premolar and the first lower molar frequently modified as carnassial teeth; collar-bone reduced or absent; four or five unguiculate digits on each limb. Cats, Lions, Tigers, Panthers, Dogs, Wolves, Jackals, Bears, Raccoons, Skunks, Seals, Sea-Lions and Walruses.

carnivorous *(Zool.)*. Flesh eating.

carnivorous plant *(Bot.)*. *Insectivorous plant*. One of ca. 400 species belonging to several unrelated families, mostly growing on substrates poor in mineral nutrients, which trap and digest insects and other small animals.

Carnot cycle *(Eng.)*. An ideal heat engine cycle of maximum thermal efficiency. It consists of isothermal expansion, adiabatic expansion, isothermal compression, and adiabatic compression to the initial state.

carnotite *(Min.)*. A hydrated vanadate of uranium and potassium, found as a yellow impregnation in sandstones. It is an important source of uranium.

Carnot's theorem *(Phys.)*. No heat engine can be more efficient than a reversible engine working between the same temperatures. It follows that the efficiency of a reversible engine is independent of the working substance and depends only on the temperatures between which it is working.

carol *(Build.)*. *Caroll*. A seat built into the opening of a bay window. Also *bay-stall*.

Caro's acid *(Chem.)*. **Permonosulphuric acid**.

carotenes *(Bot.)*. Red to yellow **carotenoids**, unsaturated tetraterpene hydrocarbons ($C_{40}H_{56}$). Accessory and photoprotective pigments in chloroplasts and in chromoplasts in some fruits and in carrot roots.

carotenoids *(Bot.)*. Red, orange and yellow *terpenoids*, based on tetraterpenes (C_{40}), including the *carotenes* (simple hydrocarbons) and the *xanthophylls* (oxygenated hydrocarbons) in chloroplasts and chromoplasts in all the plant and algal groups as **accessory pigments** or as pigments protecting the major photosynthetic pigments against photo-oxidation. See also **sporopollenin**.

carotid arteries *(Zool.)*. In Vertebrates, the principal arteries carrying blood forward to the head region.

carotid bodies *(Med.)*. Two small oval masses situated close to the carotid sinus, richly vascularized and containing chemoreceptor sensory receptors composed of an epithelioid cell and associated endings of sensory fibres of the glossopharangeal nerve. The chemoreceptors are sensitive to the tension of oxygen and carbon dioxide and to the pH of the arterial blood flowing through them, bringing about important compensatory respiratory and cardiovascular reflexes.

carotid sinus *(Med.)*. The dilated portion of the common carotid artery at its division into internal and external branches. The wall of the sinus is richly endowed with sensory endings of the glossopharangeal nerve which act as baroreceptors, the nerve impulses generated in these sensory receptors producing important compensatory cardiovascular and respiratory reflexes.

carpal *(Zool.)*. Also carpale. One of the bones composing the **carpus** in Vertebrates. pl. *carpals, carpalia*.

carp-, carpo-, -carp, -carpous *(Genrl.)*. Prefix and suffix from Gk. *karpos*, fruit.

carpel *(Bot.)*. A female organ in a flower, bearing and enclosing one or more ovules, and forming singly or with others the **gynaecium**. Typically like a leaf, folded longitudinally so the edges come together, and bearing one or more ovules on a placenta along the line of the junction. Comprises ovary, style (usually) and stigma.

carpellate *(Bot.)*. Of a flower, female.

carpet strip *(Build.)*. A strip of wood secured to the floor below a door.

carpus *(Zool.)*. In land Vertebrates, the basal podial region of the fore-limb; the wrist.

carr *(Ecol.)*. Fen vegetation with a conspicuous component of tree species. See also **fen**.

carriage *(Build.)*. A timber joist giving intermediate support, between the wall string and the outer string, to the treads of wide wooden staircases. Also *carriage-piece* or *rough-string*.

carriage *(Print.)*. The reciprocating assembly on a cylinder printing machine made up of a bed on which the forme lies, bearers at each side, and an ink table.

carriage maker's plane *(Build)*. Special rebate plane for wide rebates, fitted with a back iron to minimize tearing of the grain.

carriage-piece *(Build.)*. See **carriage**.

carriage spring *(Eng.)*. See **laminated spring**.

carriage-type switchgear *(Elec.Eng.)*. See **truck-type switchgear**.

carrier *(Biol.)*. (1) In human genetics particularly, a *heterozygote* for a recessive disorder. (2) A non-radioactive compound added to a *tracer* quantity of the same compound, which is radio-labelled, to assist in its recovery after some chemical or biological process or precipitation. (3) A molecule or molecular system which brings about the transport of a solute across a cell membrane either by **active transport**, or **facilitated diffusion**. See also **pump**. (4) An organism harbouring a parasite but showing no symptoms of disease, especially if it acts as a source of infection. (5) A more or less inert material used either as a diluent or vehicle for the active ingredient of e.g. a fungicide or as a support e.g. for cells in a bioreactor.

carrier *(Electronics)*. See **carriers**.

carrier *(Eng.)*. A device for conveying the drive of a face-plate of a lathe to a piece of work which is being turned between centres. It is clamped to the work and driven by a pin projecting from the face-plate.

carrier *(Image Tech.)*. A frame for holding a negative in an enlarger or slides in a projector.

carrier *(Phys.)*. Nonactive material mixed with, and chemically identical to, a radioactive compound. Carrier is sometimes added to carrier-free material.

carrier *(Telecomm.)*. (1) A vehicle for communicating information, when the chosen medium itself cannot convey the information but can convey a carrier, onto which the information is impressed by **modulation**. (2) In radio transmission, the output of the transmitter before it is modulated. See **frequency modulation**. (3) The frequencies chosen for sending many signals simultaneously along a single communication channel by **frequency-division multiplex**.

carrier *(Textiles)*. Compound added to dye bath to assist the dyeing of hydrophobic fibres particularly with disperse dyes.

carrier beat *(Telecomm.)*. Audible note produced by heterodyne process between two carriers or between a carrier and a reference oscillator.

carrier-controlled approach *(Radar)*. System used for landings on aircraft carriers.

carrier filter *(Telecomm.)*. Filter suitable for discriminating between currents used in carrier telephony according to their frequency, as a means of channel separation.

carrier mobility *(Phys.)*. The mean drift velocity of the charge carriers per unit electric field.

carrier noise *(Acous.)*. Noise which has been introduced into the carrier of a transmitter before modulation.

carrier power *(Phys.)*. Power radiated by a transmitter in absence of modulation.

carriers *(Electronics)*. In a crystal of semi-conductor material thermal agitation will cause a number of electrons to dissociate from their parent atoms; in moving about the crystal they act as *carriers* of negative charge. Other electrons will move from neighbouring atoms to fill the space left behind, thus causing the holes where no electrons exist in the lattice to be transferred from one atom to another. As these holes move around they can be considered as *carriers* of positive charge. See **impurity**.

carrier suppression *(Telecomm.)*. The process of eliminating the **carrier** in amplitude-modulated signals in order to produce *suppressed-carrier modulation*. Achieved through filtering and/or the use of balanced modulators; used in some forms of radio communication and **frequency-division multiplex** telephony giving a better signal-to-noise ratio. See **suppressed-carrier system**.

carrier system *(Telecomm.)*. See **frequency-division multiplex**.

carrier telegraphy *(Telecomm.)*. Use of modulated frequencies, usually in the 5-unit code originated for teleprinters, and transmitted, with others, as a voice-frequency signal in telephone circuits. Also called *voice-frequency telegraphy*.

carrier-to-noise ratio *(Telecomm.)*. Ratio of the received carrier signal to noise voltage immediately before demodulation or limiting stage.

carry *(Comp.,Maths.)*. In computing, as in arithmetic, to put a digit in a position adjacent to the working space to signify a sum equal to the number base.

carry flag *(Comp.)*. Single bit which is set to 1 when a carry occurs during a binary addition.

carrying capacity *(Ecol.)*. Maximum number of individuals of a species which can live on an area of land, usually calculated from food requirements.

carrying current *(Elec.Eng.)*. See **instantaneous carrying-current**.

carstone *(Geol.)*. Brown sandstone in which the grains are cemented by limonite. Iron-pan.

cartesian co-ordinates *(Maths.)*. A system of co-ordinates in which the position of a point in a plane or in space is specified by its distances from two lines or three planes respectively. The two lines, and the three lines in which the three planes intersect are called *co-ordinate axes* and each distance is measured parallel to an axis. If the axes are mutually perpendicular the system is said to be *rectangular*. Devised by Descartes and published in 1637.

Cartesian hydrometer *(Powder Tech.)*. An improved form of hydrometer used in the particle size analysis of soils, having a rubber membrane as part of the wall. Adjustment of pressure above the suspension enables the specific gravity of the hydrometer to be adjusted.

cartesian ovals *(Maths.)*. Curves, named after Descartes, defined by the bipolar equation $mr \pm nr' = \text{const.}$

cartilage *(Biol.)*. A form of connective tissue in which the cells are embedded in a stiff matrix of **chondrin**.

cartography *(Surv.)*. The preparation and drawing of maps which show, generally, a considerable extent of the earth's surface.

carton board *(Paper)*. Any board intended for conversion into cartons.

cartouch, cartouche *(Arch.)*. (1) An ornamental block supporting the eave of a house. (2) An ornamental scroll to receive an inscription.

cartridge *(Comp.)*. Removable module containing software, often permanently stored as an integrated circuit.

cartridge *(Image Tech.)*. (1) Sealed electromechanical transducer unit as in a *pick-up*. (2) Light-tight container for loading unexposed photographic film in a camera. (3) Container for a single spool of motion picture film or magnetic tape feeding into a reproducer and taken up on a separate spool. (4) A container for a continuous loop of motion picture film or magnetic tape.

cartridge *(Nuc.Eng.)*. See **can**.

cartridge brass *(Eng.)*. Copper-zinc alloy containing approximately 30% zinc. Possesses high ductility; capable of being severely cold-worked without becoming brittle. Used for cartridges, tubes etc.

cartridge-operated hammer *(Build)*. One using the force of a small explosive cartridge to drive nails and bolts into concrete, brickwork etc.

cartridge paper *(Paper)*. Originally a tough paper intended for winding the tubes of shotgun cartridges (ammunition cartridge). Now also a paper made for drawing purposes (drawing cartridge) or for lithographic printing (offset cartridge or matt-coated cartridge).

cartridge starter *(Aero.)*. A device for starting aero-engines in which a slow-burning cartridge is used to operate a piston or turbine unit which is geared to the engine shaft.

caruncle *(Bot.)*. An outgrowth from the neighbourhood of the micropyle of a seed. The seed is said to be *carunculate*.

caruncle *(Med.)*. Any small fleshy excrescense; a small growth at the external orifice of the female urethra. (*pl.*) epithelial nodules found at the end of pregnancy on the placenta and amnion.

caruncle *(Zool.)*. Any fleshy outgrowth; in some *Polychaeta*, a fleshy dorsal sense-organ; in some *Acarina*, a tarsal sucker; in embryo chicks, a horny knob at the tip of the beak.

carvacrol *(Chem.)*. $C_{10}H_{14}O$, *1-methyl-4-isopropyl-2-hydroxy-benzene*, an isomer of **thymol**; mp 0°C, bp 236°C; obtained from camphor by heating with iodine; present in *Origanum hirtum*.

carvone *(Chem.)*. Δ-6.8-Terpadiene-2-one, an unsaturated ketone of the terpene series, the principal constituent of caraway seed oil; liquid of bp 228°C; readily forms **carvacrol**.

caryatid *(Arch.)*. A sculptured female form used as a column.

cary-, caryo- *(Genrl.)*. See **kary-, karyo-**.

Caryophyllaceae *(Bot.)*. Family of ca 2000 spp. of dicotyledonous flowering plants (superorder Caryophyllidae). Mostly herbs, mostly temperate. Typically with opposite leaves, flowers with five free petals, and a superior ovary with free-central placentation. Of little economic importance other than as ornamentals, e.g. pinks and carnations (Dianthus).

caryophyllenes *(Chem.)*. Mixture of isomeric **sesquiterpenes** forming the chief constituents of clove oil.

Caryophyllidae *(Bot.)*. Subclass or superorder of dicotyledons. Mostly herbs, most have trinucleate pollen (binucleate is commoner in flowering plants), most have betalains rather than anthocyanins (except Caryophyllaceae) and/or free-central or basal placentation. Contains ca 11 000 spp. in 14 families including Caryophyllaceae, Chenopodiaceae and Cactaceae. (Corresponds approximately to older group Centrospermae.)

caryopsis *(Bot.)*. A dry, indehiscent, one-seeded fruit, characteristic of the grasses, with the ovary wall (pericarp) and seed-coat (testa) united, e.g. a grain of wheat.

CAS *(Aero.)*. Abbrev. for *Collision Avoidance System*.

cascade *(Build.)*. An outflow falling in a series of steps.

cascade *(Telecomm.)*. Number of devices connected in such a way that each operates the next one in turn, as transistors or valves in an amplifier.

cascade generator *(Electronics)*. High voltage generator using a series of voltage multiplying stages, especially when designed for X-ray tubes or low energy accelerators.

cascade impactor *(Powder Tech.)*. Device for sampling aerosols or dusty air, which automatically fractionates the particles or spray droplets by drawing the gas stream through a series of jet impactors of decreasing nozzle size.

cascade particle *(Phys.)*. See **Xi particle**.

cascades *(Aero.)*. Fixed aerofoil blades which turn the airflow round a bend in a duct, e.g. in wind tunnels or engine intakes.

cascade shower *(Phys.)*. Manifestations of cosmic rays in which high-energy mesons, protons and electrons create high-energy photons, which produce further electrons and positrons, thus increasing the number of particles until the energy is dissipated.

cascading of insulators *(Elec.Eng.)*. Flashover of a string of suspension insulators; initiated by the voltage across 1 unit exceeding its safe value and flashing over, thereby imposing additional stress across the other units, and resulting in a complete flashover of the string.

cascara sagrada *(Med.)*. Dried bark of *Rhamnus purshiana*, a shrub of N. California, formerly used (as extract) as a cathartic and laxative drug.

case *(Build.)*. The external facings of a building when these are of better material than the backing.

case *(Eng.)*. That part near the surface of a ferrous alloy which has been so altered as to allow *case hardening*, typically by **carburizing**.

case *(Print.)*. A wooden tray divided into many compartments to accommodate individual letters. Originally in pairs, the **upper case** and **lower case**, they are now largely replaced by the **double case**.

caseation *(Med.)*. The process of becoming cheese-like, e.g. in tissue infected with tubercle bacillus the cells break down into an amorphous cheese-like mass.

case bay *(Build.)*. The space between two binders under a floor.

cased book *(Print.)*. The cover or case is made separately and attached to the book proper as a separate operation. The usual method for **publisher's binding, edition binding**.

cased frame *(Build.)*. The wooden box-frame containing the sash weights of a window.

case-hardening *(Eng.)*. The production of a hard surface layer in steel by heating in a carbonaceous medium to increase the carbon content, then quenching. See **cyanide hardening, pack hardening**.

case-hardening *(For.)*. In seasoning, a condition in which the surface of timber becomes set in an expanded condition and remains under compression whilst the interior is in tension.

case-hardening *(Min.Ext.)*. Cement-like surface on porous rock caused by evaporation.

casein *(Biol.)*. The principal albuminous constituent of milk, of which it is present as a calcium salt. Transformed into insoluble paracasein (cheese) by enzymes. Casein is a raw material for thermoplastic materials used for insulators, handles, buttons, artificial fibres and bristles etc. Also used in adhesives, nerve tonics, and for priming artists' canvases.

Casella automatic microscope *(Powder Tech.)*. A microscope whose stage moves automatically to enable the images of particles to be measured electronically.

casement *(Build.)*. Window sash hinged on one vertical edge to open out or in.

casement *(Textiles)*. A plain-woven cotton (or man-made) fabric used for curtains.

caseous *(Med.)*. Cheeselike; having undergone **caseation**.

caseous lymphadenitis *(Vet.)*. Disease of sheep and goats due to infection with *Corynebacterium ovis* and characterized by nodule formation in the lymph nodes, lungs, skin and other organs. A chronic respiratory problem can ensue.

cashmere *(Textiles)*. Fine down-like fibres obtained from the undercoat of the Asiatic goat, or similar material obtained from goats in Australasia and Scotland. Frequently blended with wool and used for the manufacture of cardigans and sweaters.

casing *(Min.Ext.)*. (1) Piping used to line a drill hole. (2) The steel lining of a circular shaft.

casinghead *(Min.Ext.)*. The part of the well which is above the surface and to which the flow lines and control valves are attached.

casinghead gas *(Min.Ext.)*. Gas produced as a by-product from an oil well.

cask, casket *(Nuc.Eng.)*. See **flask**.

Casogrande hydrometer *(Powder Tech.)*. An improved form of hydrometer used in the particle size analysis of soils. Its main feature is its smooth symmetrical outline as compared with the **bouyoucous hydrometer**.

casparian strip, casparian band *(Bot.)*. A band, running round the cell in which apparently the whole thickness of the primary wall is impregnated with suberin and/or lignin making it impermeable to water and solutes. Typical of the root *endodermis* where it occurs in all radial and transverse walls, preventing the movement in the **apoplast** of water and solutes between the cortex and the stele.

Cassegrain antenna *(Telecomm.)*. One in which radiation from a focus is collimated at one surface, e.g. a parabola, and reflected at another surface, e.g. a plane, or in reverse.

Cassegrain telescope *(Astron.)*. A form of reflecting telescope in which the rays after reflection at the main mirror fall on a small convex mirror placed inside the prime focus. The rays are thus reflected back through a hole in the main mirror, forming an image beyond it. It is similar to the **Gregorian telescope**.

Cassel's yellow *(Chem.)*. Commercial name for lead oxychloride made by heating lead (II) oxide and ammonium chloride.

cassette *(Image Tech.)*. (1) Holder for magnetic tape or film which contains two reels, the tape or film moving from one to another as it passes the reading head or film gate. Reversing or rewinding the cassette allows repeated use and the holder protects tape or film outside the apparatus. (2) A holder for X-ray film, often incorporating an image-intensifying screen, preventing exposure to light but allowing exposure to X-rays.

cassia oil *(Chem.)*. An oil obtained from the bark of *Cinnamomum cassia*, a yellow or brown liquid of cinnamonlike odour; bp 240°–260°C, rel.d. 1.045–1.063.

Cassini's division *(Astron.)*. A dark ring concentric with the ring of Saturn and dividing it into two parts; first observed by G.D.Cassini in 1675.

Cassini's ovals *(Maths.)*. See **ovals of Cassini**.

Cassiopeia A *(Astron.)*. Strongest radio source in the sky, after the Sun, located in the constellation Cassiopeia. About 3 **kiloparsec** away, it is the remnant of a **supernova** seen to explode about 1667.

cassiterite *(Min.)*. Also *tin-stone*. Oxide of tin, crystallizing in the tetragonal system; it constitutes the most important ore of this metal. It occurs in veins and impregnations associated with granitic rocks; also as 'stream-tin' in alluvial gravels.

Cassius *(Chem.)*. See **purple of Cassius**.

cast *(Geol.)*. The impression which fills a natural *mould*. Most frequently a fossil shell, where the infilling is both a cast of the original animal and an internal mould of the shell. Term also used for impressions of sedimentary structures.

cast *(Med.)*. A mould of cellular or organic matter shed from tubular structures of the body, e.g. the bronchi or the tubules of the kidney,

caste *(Zool.)*. In some social Insects, one of the types of polymorphic individuals composing the community.

caster action *(Autos.)*. The use of inclined swivel axes or king-pins by which the steerable front wheels are given fore-and-aft stability and a self-centring tendency after angular deflection by road shocks, on the principle of the domestic caster. See **trail**.

cast holes *(Foundry)*. Holes made in cast objects by the use of cores, in order to reduce the time necessary for machining, and to avoid metal wastage.

casting *(Foundry)*. (1) The operation of pouring molten metals into sand or metal moulds in which they solidify. (2) A metallic article cast in the shape required, as distinct from one shaped by working.

casting *(Print.)*. (1) Operating the Monotype casting machine, which is controlled by perforated spool (see **keying**), and produces, from molten metal, single characters assembled into words and lines. (2) Making a stereotype plate from molten metal.

casting *(Vet.)*. (1) The process of throwing and securing an animal from the upright to the prone position. (2) The pellet of undigested feathers, fur, or bones disgorged by a raptorial bird.

casting box *(Print.)*. A machine for casting stereotype plates, with built-in metal pot or hand-poured. Curved plates require a specific machine. For flat plates the box consists essentially of two heavy iron plates clamped together with interchangeable thickness gauges to enclose the matrix or mould.

casting copper *(Eng.)*. Metal of lower purity than **best selected copper**. Generally contains about 99.4% of copper.

casting ladle *(Foundry)*. A steel ladle, lined with refractory material, in which molten metal is carried from the furnace to the mould in which the casting is to be made.

casting-on *(Foundry)*. See **burning-on**.

casting resins *(Plastics)*. Term applied to liquid resins which can be introduced into a moulded shape and polymerized *in situ*, with or without heating. Originally the term applied to high methanal ratio phenolic resins but is now applied widely to certain acrylic, polyester, epoxy and polyurethane resins.

casting wheel *(Eng.)*. Large wheel on which ingot moulds are arranged peripherally and filled from stream of molten metal issuing from furnace or pouring ladle.

cast-in-situ concrete piles *(Civ.Eng.)*. A type of pile formed by driving a steel pipe into the ground and filling it with concrete, using the pipe as a mould. The mould is withdrawn as the concrete is consolidated by heavy tamping. Also known as *moulded-in-place concrete piles.*

cast iron *(Eng.)*. An iron-carbon alloy in which the carbon content exceeds the solubility of carbon in austenite at the eutectic temperature. Carbon content is usually above 2%. Some kinds are difficult to machine but see **grey iron**, **ductile cast iron**.

castle nut *(Eng.)*. A 6-sided nut in the top of which six radial slots are cut. Two of these line up with a hole drilled in the bolt or screw, a split pin being inserted to prevent turning.

Castner-Kellner process *(Chem.)*. The electrolysis of brine to produce sodium hydroxide, chlorine and hydrogen. The cathode is of mercury, in which the sodium dissolves, the amalgam being removed continuously and treated with water to liberate mercury and sodium hydroxide.

Castner's process *(Chem.)*. The production of sodium cyanide from sodium. As the temperature is gradually raised, dry ammonia is blown into metallic sodium melted with charcoal, the resulting sodamide reacting quickly with the charcoal to form cyanamide. This is heated to ca 850°C, and combines further with carbon to form sodium cyanide.

cast-off *(Print.)*. Calculating the amount of space that a manuscript will occupy when set in a given typeface and measure.

castor oil *(Chem.)*. Oil obtained from the seeds of *Ricinus communis*, a yellow or brown, syrupy, nondrying liquid; mp − 10°C, rel.d. 0.960–0.970, saponification value 178, iodine value 85, acid value 19.21. Main constituent, ricinoleic acid. Uses: *pure*, in medicine and hydraulic fluids; *dehydrated*, as a drying oil in paints; *hydrogenated*, as a fat in cosmetics; *sulphonated*, see **Turkey-red oil**.

castration *(Med.,Vet.)*. Removal or surgical destruction of the testicles.

castration anxiety *(Behav.)*. In psychoanalytic theory, the male child's fear that his penis will be cut off as punishment for his sexual desire for the mother.

cast steel *(Eng.)*. Steel as cast, i.e. not shaped by mechanical working. Originally applied to steel made by the crucible process as distinguished from that made by cementation of wrought-iron.

cast stone *(Build.)*. Precast artificially manufactured building components, e.g. blocks, lintels, sills, copes. Basically precast concrete with a facing of fine material and cement intended to look like natural stone.

cast-up *(Print.)*. The overall *en* content of a job after making necessary additions to the **cast-off** for headlines, folios, etc.

cast welded rail joint *(Civ.Eng.)*. A joint between the ends of two adjacent rails in position using the **thermite** process.

casual species *(Bot.)*. An introduced plant which occurs but is not established in places where it is not cultivated.

cata- *(Genrl.)*. See **kata-**.

catabolism *(Biol.,Med.)*. A metabolic process of breaking down complex molecules into simpler ones and releasing energy.

catacaustic *(Maths.)*. See **caustic curve**.

cataclasis *(Geol.)*. The process of rock deformation involving fracture and rotation of mineral grains without chemical reconstitution.

catadioptric *(Phys.)*. An optical system using a combination of refracting and reflecting surfaces designed to reduce aberrations in a telescope.

catadromous *(Bot,.Zool.)*. See **katadromous**.

catalan process *(Eng.)*. Reduction of haematite to wrought iron by smelting with charcoal.

catalase *(Biol.)*. An enzyme which catalyses the oxidation of many substrates by hydrogen peroxide. In the absence of a suitable substrate it destroys hydrogen peroxide converting it to water.

catalepsy *(Med.)*. The condition in which any posture of a limb can be maintained without movement for a period of time longer than normal; occurring in disease of the cerebellum and in hysteria, also in deep hypnotic states and in certain types of schizophrenia. In animals is used to indicate the action known as 'feigning death' which can be induced in some animals by any sudden disturbance. adj. *cataleptic.*

catalysis *(Chem.)*. The acceleration or retardation of a chemical reaction by a substance which itself undergoes no permanent chemical change, or which can be recovered when the chemical reaction is completed. It lowers the energy of activation.

catalyst *(Build.)*. Substances like driers used in oleo/resinous surface coatings and catalyst additives in two pack coating.

catalyst *(Chem.)*. A substance which catalyses a reaction. See **catalysis**.

catalytic cracking *(Min.Ext.)*. A process of breaking down the heavy hydrocarbons of crude petroleum, using silica or aluminium gel as a catalyst. See **cracking**.

catalytic poison *(Chem.)*. A substance which inhibits the activity of a catalyst. Also anticatalyst.

cataphyll *(Bot.)*. A non-foliage leaf inserted low on a shoot, e.g. a scale on a rhizome or a bud scale. Cf. **hypsophyll** .

cataplexy *(Med.)*. Sudden attack of weakness, following some expression of emotion; the patient falls to the ground, immobile, speechless, but conscious.

catapult *(Aero.)*. An accelerating device for launching an aircraft in a short distance. It may be fixed or rotatable to face the wind. It is usually used on ships which have no landing deck, having been superseded on aircraft carriers by the **accelerator**. During World War II, fighters were carried on *CAMS* (Catapult Armed Merchant Ships) for defence against long-range bombers. Land catapults have been tried but have been superseded by **RATOG** and **STOL** aircraft.

cataract *(Med.)*. Opacity of the lens of the eye as a result of degenerative changes in it.

catarrh *(Med.)*. Inflammation of a mucous membrane, with discharge of mucus, commonly associated with coryza.

catastrophe theory *(Maths.)*. Theory of sudden, as opposed to continuous, changes.

catastrophism *(Geol.)*. The theory that the Earth's geological history has been fashioned by infrequent violent events. See **uniformitarianism**.

catch basin *(Civ.Eng.)*. See **catch pit**.

catch-bolt *(Build.)*. A door lock having a spring-loaded bolt which is normally in the locking position (i.e. extended), but which is automatically and momentarily retracted in the process of shutting the door.

catcher *(Electronics)*. The element in a velocity-modulated UHF or microwave beam tube which abstracts, or *catches*, the energy in a bunched electron stream as it passes through it. See **Buncher**.

catcher foil *(Nuc.Eng.)*. Aluminium sheet used for

measuring power levels in nuclear reactor by absorption of fission fragments.

catching diode *(Electronics)*. One used to *clamp* a voltage or current at a predetermined value. When it becomes *forward-biased* it prevents the applied potential from increasing any further.

catch-line *(Print.)*. A temporary head-line inserted on slip proofs, etc.

catchment area (basin) *(Civ.Eng.)*. The area from which water runs off to any given river valley or collecting reservoir.

catch mounts *(Print.)*. Individual mounts for bookwork printing to hold duplicate plates, and from which they can be removed and replaced without unlocking the forme.

catch muscle *(Zool.)*. A set of smooth muscle fibres which form part of the adductor muscle in bivalve Molluscs, and are capable of keeping the valves closed by means of a sustained tonus: any set of smooth muscle fibres associated with striated muscle fibres for a similar purpose. Also called *arrest muscle*.

catch net *(Elec.Eng.)*. A wire netting placed under high-voltage transmission lines where they pass over public roadways, railways, etc., to prevent danger due to a broken live conductor.

catch pit *(Civ.Eng.)*. A small pit constructed at the entrance to a length of sewer or drain pipe to catch and retain matter which would not easily pass through the pipes. Also *catch basin*. See also **sump**.

catch plate *(Eng.)*. A disk on the spindle nose of a lathe, driving a carrier locked to the work.

catch points *(Civ.Eng.)*. Trailing points placed on an up-gradient for the purpose of derailing rolling stock accidently descending the gradient. See **spring points**.

catch props *(Min.Ext.)*. In a coal mine, props put in advance of the main timbering for safety, i.e. *watch props* or *safety props*.

catch-up *(Print.)*. Term used in lithographic printing when the non-image areas of the printing plate begin to accept ink due to insufficient fount solution being fed to the plate.

catch-water drain *(Civ.Eng.)*. A drain to catch water on a hillside, with open joints or multiple perforations to take in water in as many places as possible.

catch word *(Print.)*. (1) At the foot of the page, the first word of the next, as in manuscripts and books printed prior to the 19th century. (2) At the head of the page a guide word in such books as dictionaries. (3) A guide word at the head of a galley of type.

catch work *(Hyd.Eng.)*. A system of water channels which may be used for flooding land.

cat cracker *(Min.Ext.)*. Refinery vessel in which hydrocarbon fractions are processed in the presence of a catalyst.

cat E *(Aero.)*. Category E damage to an aircraft; equivalent to a total loss or 'write-off'.

catechol *(Chem.)*. *2-hydroxyphenol, 1,2-hydroxybenzene*, $C_6H_4(OH)_2$ (1,2); colourless crystals; mp 104°C, bp 240°C. Occurs in fresh and fossil plant matter and in coal-tar. Important derivatives are **guaiacol** and **adrenaline**.

catecholamines *(Biol.)*. A series of compounds derived from dihydroxyphenylalanine (DOPA), dopamine, noradrenaline (norepinephrine) and adrenaline (epinephrine). They function as *neurotransmitters*, adrenaline also acting as a hormone.

category *(Maths.)*. A collection of objects and arrows such that (1) each arrow leads from an object of the collection to another (or the same one), (2) a composite arrow in the collection is defined in all cases where 2 arrows 'join together' — e.g. f goes from a to b, g goes from b to c, composite arrow goes from a to c, written $f.g$ or $g.f$ according to convention, (3) arrow composition is *associative*, (4) for each object there is defined an identity arrow from that object to itself such that any composite of it with another arrow is equal to that other arrow. For example, in a category of groups the objects will be groups and the arrows will be *group homomorphisms*.

catenary *(Maths.)*. The curve assumed by a perfectly flexible, uniform, inexstensible string when suspended from its ends. Its intrinsic equation is $spi = c \tan\psi$ and its cartesian equation $y = c \cosh x/c$. A string carrying a continuous or discrete load (e.g., as in a suspension bridge) hangs in a parabola if the load is uniformly distributed horizontally.

catenary construction *(Elec.Eng.)*. A method of construction used for overhead contact wires of traction systems. A wire is suspended, in the form of a catenary, between two supports, and the contact wire is supported from this by droppers of different lengths, arranged so that the contact wire is horizontal. See **compound catenary construction**.

catenation *(Biol.)*. The arrangement of chromosomes in chains or in rings.

catenoid *(Maths.)*. The surface of revolution generated by the rotation of a catenary about its internal axis.

Caterpillar *(Eng.)*. TN for vehicles with endless *tracks*.

caterpillar *(Zool.)*. A type of cruciform larva, found in Lepidoptera, which typically possesses abdominal locomotor appendages (prolegs).

cat 'flu *(Vet.)*. See **feline pneumonitis**.

catgut *(Med.)*. Sterilized strands of sheep's or other animal's intestines used as ligatures.

cathartic *(Med.)*. Purgative. A drug which promotes evacuation of the bowel.

cathead *(Eng.)*. *Spider*. A lathe accessory consisting of a turned sleeve having four or more radial screws at each end; used for clamping on to rough work of small diameter and running in the **steady** while centring.

cathead *(Min.Ext.)*. Rotating drum around which a rope can be coiled to provide pull for various operations on a drilling rig.

cathelectrotonus *(Med.)*. Physiological excitability produced in muscle tissue by passage of electric current.

Catherine wheel *(Arch.)*. See **rose window**.

catheter *(Med.)*. A rigid or flexible tube for introduction into vessels or organs of the body. Commonly used to drain the urinary bladder, to measure intra-cardiac pressures or to inject radio-opaque material into the blood vessels (*angiography*).

cathetometer *(Phys.)*. An optical instrument for measuring vertical distances not exceeding a few decimetres. A small telescope, held horizontally can move up and down a vertical pillar. The difference in position of the telescope when the images of the two points whose separation is being measured are lined up with the cross-wires of the telescope, is obtained from the difference in vernier readings on a scale marked on the pillar. Also called *reading telescope* or *reading microscope*.

cathexis *(Behav.)*. A charge of mental energy attached to any particular idea or object.

cathode *(Electronics)*. (1) In an electronic tube or valve, an electrode through which a primary stream of electrons enters the inter-electrode space. During conduction, the cathode is negative with respect to the anode. Such a cathode may be *cold*, electron emission being due to electric fields, photo-emission, or impact by other particles, or *thermionic*, where the cathode is heated by some means. (2) In a semiconductor diode, the electrode to which the forward current flows. (3) In a thyristor, the electrode by which current leaves the thyristor when it is in the ON state. (4) In a light-emitting diode, the electrode to which forward current flows within the device. (5) In electrolytic applications, the electrode through which current leaves any conductor of the non-metallic variety; an electrode at which positive ions are discharged, or negative ions formed.

cathode coating *(Electronics)*. A low work-function surface layer applied to a thermionic or photo-cathode in order to enhance electron emission or to control spectral characteristics. The cathode coating impedance is between the base metal and this layer.

cathode copper *(Eng.)*. The product of electrolytic refining, after which the cathodes are melted, oxidized, poled, and cast into wire-bars, cakes, billets etc.

cathode efficiency *(Electronics)*. Ratio of emission current to energy supplied to cathode. Also emission efficiency.

cathode follower *(Electronics)*. A valve circuit in which the input is connected between the grid and earth, and the output is taken from between the cathode and earth, the anode being earthed to signal frequencies. It has a high input impedance, low output impedance, and unity voltage gain. See **emitter follower**.

cathode glow *(Electronics)*. Glow near the surface of a cathode, its colour depending on the gas or vapour in the tube. If an arc takes place in a partial vacuum, it may fill the greater part of the discharge tube.

cathode luminous sensitivity *(Electronics)*. Ratio of cathode current of photoelectric cell to luminous intensity.

cathode modulation *(Electronics)*. Modulation produced by signal applied to cathode of valve through which carrier wave passes.

cathode poisoning *(Electronics)*. Reduction of thermionic emission from a cathode as a result of minute traces of adsorbed impurities.

cathode ray *(Electronics)*. A stream of negatively charged particles (electrons) emitted normally from the surface of a cathode in vacuum or low-pressure gas. The velocity of the electrons is proportional to the square root of the accelerating potential, being 595 kM/sec for 1 volt. They can be deflected and formed into beams by the application of electric or magnetic fields, or a combination of both, and are widely used in oscilloscopes and TV (in cathode-ray tubes), electron microscopes and electron-beam welding, and electron-beam tubes for high frequency amplifiers and oscillators.

cathode-ray oscillograph *(Electronics)*. An oscillograph in which a permanent (photographic or other) record of a transient or time-varying phenomenon is produced by means of an electron beam in a cathode-ray tube. Deprecated term for **oscilloscope**.

cathode-ray oscilloscope *(Electronics)*. See **oscilloscope**.

cathode-ray tube *(Electronics)*. An electronic tube in which a well-defined and controllable beam of electrons is produced and directed on to a surface to give a visible or otherwise detectable display or effect. Abbrev. **CRT**.

cathode spot *(Electronics)*. Area on a cathode where electrons are emitted into an arc, the current density being much higher than with simple thermionic emission.

cathodic chalk *(Electronics,Ships)*. A coating of magnesium and calcium compounds formed on a steel surface during **cathodic protection** in sea water.

cathodic chalk *(Ships)*. A coating of magnesium and calcium compounds formed on a steel surface during **cathodic protection** in sea water.

cathodic etching *(Electronics)*. Erosion of a cathode by a glow discharge through positive-ion bombardment, in order to show microstructure.

cathodic protection *(Electronics, Ships)*. Protection of a metal structure against electrolytic corrosion by making it the cathode (electron receiver) in an electrolytic cell, either by means of an impressed e.m.f. or by coupling it with a more electronegative metal. In ships and offshore structures, corrosion can be prevented by passing sufficient direct current through the sea water to make the metal hull a cathode. See **sacrificial anode**.

cathodoluminescence *(Phys.)*. The emission of light, with a possible afterglow, from a material when irradiated by an electron beam.

cathodophone *(Acous.)*. Microphone utilizing the silent discharge between a heated oxide-coated filament in air and another electrode. This discharge is modulated directly by the motion of the air particles in a passing sound wave. Also *ionophone*.

catholyte *(Phys.)*. See **catolyte**.

cation *(Phys.)*. Ion in an electrolyte which carries a positive charge and which migrates towards the cathode under the influence of a potential gradient in electrolysis. It is the deposition of the cation in a primary cell which determines the *positive terminal*.

cationic detergents *(Chem.)*. Types of ionic synthetic detergents in which the surface active part of the molecule is the cation, unlike soap and most of the widely used synthetic detergents. Sometimes called *invert soaps*. Typified by the quaternary ammonium salts such as **cetrimide**, benzalkonium chloride, domiphen bromide; cetylpyridinium is similar. All have powerful bactericidal activity, and in addition to skin and utensil cleaning are used in lozenges, creams, mouth-washes etc.

catkin *(Bot.)*. An inflorescence with the flowers sessile on a common axis and typically pendulous, unisexual and wind-pollinated. A common inflorescence of deciduous, north temperate trees.

catolyte, catholyte *(Electronics)*. That portion of the electrolyte of an electrolytic cell which is in the immediate neighbourhood of the cathode.

catophorite *(Min.)*. See **katophorite**.

catoptric element *(Phys.)*. A component of an optical system that uses reflection, not refraction, in the formation of an image.

catoptric lens *(Image Tech.)*. See **mirror lens**.

CAT scanner *(Radiol.)*. See **computed tomography**.

cat's eye *(Glass)*. A crescent-shaped blister.

cat's eye *(Min.)*. A variety of fibrous quartz which shows chatoyancy when suitably cut, as an ornamental stone. The term is also applied to crocidolite when infiltrated with silica (see **tiger's eye, hawk's eye**). A more valuable form is *chrysoberyl cat's eye*; see **cymophane**.

cattle plague *(Vet.)*. See **rinderpest**.

CATV *(Telecomm.)*. Abbrev. for *Community Antenna Television*.

catworks *(Min.Ext.)*. Assemblage of motors and **catheads** which provides power for the many secondary activities on drilling platforms like pipe-hoisting.

Cauchy-Riemann equations *(Maths.)*. The equations

$$\frac{\partial u}{\partial x}=\frac{\partial v}{\partial y} \text{ and } \frac{\partial u}{\partial y}=-\frac{\partial v}{\partial x}$$

which must be satisfied for a function of a complex variable $f(z)=u(x,y)+iv(x,y)$, u and v being real, to be differentiable. Functions u and v are said to be *conjugate*.

Cauchy's convergence tests *(Maths.)*. (1) If $\sqrt[n]{u_s}\to l$ as $n\to\infty$, then the series of positive terms Σu_n converges if $l<1$ and diverges if $l>1$. (2) If $\log(1/u_n)/\log n\to l$ as $n\to\infty$, then the series of positive terms Σu_n converges if $l>1$ and diverges if $l<1$.

Cauchy's dispersion formula *(Phys.)*.

$$\mu=A+\frac{B}{\lambda^2}+\frac{C}{\lambda^{21}}+\ldots$$

An empirical expression for the relation between the refractive index n of a medium and the wavelength λ of light; A, B and C are the constants for a given medium.

Cauchy's distribution *(Maths.)*. A probability distribution of the form

$$p(x)=\frac{1}{\pi[1+(x-a)^2]},$$

where the distribution is symmetric about $x=a$.

Cauchy sequence *(Maths.)*. See **complete metric space**.

Cauchy's inequality *(Maths.)*.

$$\left(\sum_1^n a_r b_r\right)^2 \leqq \left(\sum_1^n a_r^2\right)\cdot\left(\sum_1^n b_r^2\right).$$

The equality occurs when $a_r=kb_r$ for all r.

Cauchy's integral formula *(Maths.)*. If the function $f(z)$ is analytic within and on a closed contour C, and if a is any point within C, then

$$f(a)=\frac{1}{2\pi i}\oint_C\frac{f(z)dz}{z-a}.$$

Cauchy's theorem *(Maths.)*. If a function $f(z)$ is analytic inside and on a contour C, then the integral of $f(z)$ taken round the contour is equal to zero.

cauda *(Zool.)*. The tail, or region behind the anus; any tail-like appendage; the posterior part of an organ, as the *cauda equina*, a bundle of parallel nerves at the posterior end of the spinal cord in Vertebrates. adj. *caudal, caudate*.

caudad *(Zool.)*. Situated near, facing towards, or passing to, the tail region.

Caudata *(Zool.)*. See **Urodela**.

caudate *(Bot.,Zool.)*. Bearing a tail-like appendage.

caudex *(Bot.)*. A trunk or stock.

caul- *(Bot.)*. Prefix or suffix meaning stem.

caul *(Zool.)*. In the higher Vertebrates, the amnion; more generally any enclosing membrane.

cauldron subsidence *(Geol.)*. The subsidence of a cylindrical mass of the earth's crust, bounded by a circular fault up which the lava has commonly risen to fill the cauldron. Good examples have been described from Scotland (Ben Nevis and the Western Isles).

caulescent *(Bot.)*. Having a stalk or a stem.

cauliflory *(Bot.)*. Production of flowers on trunks, branches and old stems of woody plants rather than near the ends of smaller twigs. Occurs in some trees of tropical forest, e.g. cocoa.

cauline *(Bot.)*. (1) Pertaining to a stem. (2) Leaves borne on an obvious stem, not at the extreme base, and well above soil level, i.e. not *radical*.

caulking *(Eng.)*. The process of closing the spaces between overlapping riveted plates or other joints by hammering the exposed edge of one plate into intimate contact with the other. A filler material is also used.

caulking tool *(Eng.)*. A tool, similar in form to a cold chisel but having a blunt edge, for deforming the metal rather than cutting it.

causalgia *(Med.)*. Intense burning pain in the skin after injury to the nerve supplying it.

caustic *(Med.)*. Destructive or corrosive to living tissue; an agent which burns or destroys living tissue.

caustic curve *(Maths.)*. The caustic, or catacaustic, of a given curve *C*, with a given point *P* as radiant point, is the envelope of the rays from *P* after reflection by *C*. If the rays are refracted, the envelope is called the *diacaustic*.

caustic curve *(Phys.)*. A curve to which rays of light are tangential after reflection or refraction at another curve.

caustic embrittlement *(Eng.)*. The intergranular corrosion of steel in hot alkaline solutions, e.g. in boilers.

caustic lime *(Chem.)*. The residue of calcium oxide, obtained from freshly calcined calcium carbonate; it reacts with water, evolving much heat, and producing slaked lime (calcium hydroxide, hydrate of lime, or hydrated lime). Also *quicklime*. See also **lime**.

caustic paint removers *(Build.)*. Combinations of strong caustic solutions and a thickening agent such as flour or starch. Surfaces must be thoroughly washed down and neutralized with a mild acid solution before subsequent decoration. Caustic paint removers have an adverse effect on aluminium and zinc substrates.

caustic pickle *(Build.)*. A tank containing strong caustic soda solution in which used paint pots are immersed for cleaning.

caustic potash *(Chem.)*. *Potassium hydroxide*. the name *potash* is derived from 'ash' (meaning the ash from wood) and 'pot' from the pots in which the aqueous extract of the ash was formerly evaporated.

caustic soda *(Chem.)*. See **sodium hydroxide**.

caustic surface *(Phys.)*. A surface to which rays of light are tangential after reflection or refraction at another surface.

Cavendish experiment *(Phys.)*. An experiment, carried out by Henry Cavendish in 1798, to determine the constant of *gravitation*. A form of torsion balance was used to measure the very small forces of attraction between lead spheres.

cavern *(Geol.)*. A chamber in a rock. Caverns are of varying size, and are due to several causes, the chief being the solution of calcareous rocks by underground waters, and marine action.

cavernosus, cavernous *(Zool.)*. Honeycombed; hollow; containing cavities, e.g. *corpora cavernosa*.

cavetto *(Arch.)*. A hollow moulding, quarter round.

cavil *(Build.)*. A small stone axe resembling a **jedding axe**.

cavilling *(Min.Ext.)*. The drawing of lots for working places (usually for 3 months) in the coal-mine.

caving *(Min.Ext.)*. Controlled collapse of roof in deep mines to relieve pressure. Undercut stoping.

cavitation *(Acous.,Eng.)*. Generation of cavities (e.g. bubbles) in liquids by rapid pressure changes as those induced by ultrasound. When cavity bubbles implode, they produce shock waves in the liquid. Components can be damaged by cavitation if it is induced by turbulent flow.

cavitation *(Bot.)*. The sudden development of a gas bubble in a previously sap-filled xylem *conduit* as a result of excessive tension. See also **embolism**.

cavitation *(Med.)*. The formation of cavities in any structure of the body, especially the lungs.

cavity effect *(Acous.)*. The enhancement in a microphone due to acoustic resonance in a shallow cavity in front of the diaphragm.

cavity-frequency meter *(Electronics)*. One for the use in coaxial or waveguide systems. The frequency is related to the wavelength for resonance in the cavity.

cavity magnetron *(Telecomm.)*. General term (now deprecated) for **magnetrons** having cavities cut into the inner faces of a solid cylindrical anode.

cavity modes *(Phys.)*. Stable electromagnetic or acoustic fields in a cavity which can exhibit resonance. They are **degenerate** if, having similar frequencies, they have fields which differ in pattern from that of main resonance.

cavity radiation *(Phys.)*. The radiation emerging from a small hole leading to a constant temperature enclosure. Such radiation is identical to **black-body** radiation at the same temperature, no matter what the nature of the inner surface of the enclosure.

cavity resonance *(Phys.)*. The enhancement of air flow for certain frequencies, due to neutralization of the mass (or inertia) reactance with the stiffness reactance of air in a partially enclosed space.

cavity resonator *(Phys.)*. Any nearly closed section of waveguide or coaxial line in which a pattern of electric and magnetic fields can be established. Also applies to sound fields.

cavity walls *(Build.)*. Hollow walls, normally built with two $4\frac{1}{2}$-in stretcher bond walls with a 2-in gap between, tied together with wall ties. Cavity walls increase the thermal resistance and prevent rain from reaching the inner face.

cavum *(Zool.)*. A hollow or cavity: it is a division of the concha.

cavus *(Med.)*. See **pes cavus**.

CB *(Elec.Eng.)*. Abbrev. for *Citizens' Band radio*, used by amateur radio operators.

C-banding *(Biol.)*. See **banding techniques**.

CBT *(Comp.)*. Computer based training. See **computer aided training**.

cc *(Genrl.)*. An abbrev. for *cubic centimetre*, the unit of volume in the CGS metric system. Also cm^3.

CCD *(Comp.,Electronics)*. See **charge-coupled device**.

CCD array *(Electronics)*. An array of many thousands of photodiodes, whose response to an image focused on the surface of the array can be converted into a video signal by employing CCD electronic circuits. An alternative to vacuum tubes in television cameras.

CC filter *(Image Tech.)*. *Colour Correction* or *Colour Conversion* filter: a gelatine filter used to change the colour temperature of a light source by a specific amount, e.g. from incandescent 3200 K to daylight 5400 K.

CCIR *(Telecomm.)*. Abbrev. for *Comité Consultatif International des Radiocommunications*.

CCITT *(Telecomm.)*. Abbrev. for *Comite Consultatif International Telegraphique et Telephonique*. United Nations established committee in Geneva for tariffs, technical standards and conformity in order to facilitate international telecommunications.

C-class insulation *(Elec.Eng.)*. A class of insulating material to which is assigned a temperature of over 180°C. See class -A, -B, -C, etc., insulating materials.

CCP *(Chem.)*. Abbrev. for *Cubic Close Packing*.

CCR *(Image Tech.)*. *Cassette Camera Recorder*, a video camera with an integral videotape recorder using a cassette.

CCTV *(Image Tech.)*. *Closed Circuit Television*, a non-broadcast TV system with the cameras and display receivers directly linked.

CCV *(Aero.)*. Abbrev. for *Control-Configured Vehicle*.

Cd *(Chem.)*. The symbol for cadmium.

cd *(Genrl.)*. Symbol for candela.

CD *(Immun.)*. *Cluster of Differentiation*. Term approved by international agreement to designate molecules on leucocytes which are recognized by various different specific antibodies (usually monoclonal). In some cases the molecules are known and identify subsets of lymphocytes or stages of differentiation, e.g. CD1 identifies T-cells in the thymus cortex, CD4 and CD8 identify helper and cytotoxic subsets of T-cells.

C-display *(Radar)*. Display in which bright spot represents the target, with horizontal and vertical displacements representing the bearing and elevation, respectively.

cDNA cloning *(Biol.)*. Procedure by which DNA complementary to mRNA is inserted into a **vector** and propagated. Because mRNA has no **introns**, such cDNA clones can be made to produce a normal polypeptide product.

cDNA, complementary DNA *(Biol.)*. A DNA sequence complementary to any RNA. Formed naturally in the life-cycle of RNA viruses by **reverse transcriptase**, which are widely used in the laboratory to make DNA complements of mRNA.

CdS meter *(Image Tech.)*. An exposure meter making use of a cadmium sulphide photo-resistor powered by a separate battery.

Ce *(Chem.)*. The symbol for **cerium**.

cedar *(For.)*. Genus of tree, *Cedrus*, which is a softwood, with straight grain medium fine and uniform texture. It is used for pattern-making, boat building and cheap furniture. It is of European and Asian origin.

cedar-tree laccolith *(Geol.)*. A multiple laccolith, i.e. a series of laccoliths, one above the other, forming part of a single mass of igneous rock.

Ceefax *(Comp.)*. See **teletext**.

ceiling *(Ships)*. Heavy planking laid on top of the inner double bottom under **hatchways** to protect the tank top.

ceiling joist *(Build.)*. A joist to which a ceiling is fixed.

ceiling plate *(Elec.Eng.)*. Metal plate, fixed to a ceiling, from which light fitting may be mechanically supported. Arrangements are also incorporated to allow electrical connection between the light fitting and the mains supply.

ceiling rose *(Elec.Eng.)*. An enclosure of insulating material (usually plastic) attached to a ceiling to facilitate the installation of an electric light fitting. Terminals are provided to allow connection of the light to the mains supply.

ceiling switch *(Elec.Eng.)*. A switch located in the ceiling but operated by a pull on a cord. Also called a *pull switch*.

ceiling voltage *(Elec.Eng.)*. A term used to denote the maximum voltage which a machine is capable of giving.

celdecor process *(Paper)*. A pulping method for cereals and similar agricultural residues consisting of a mild digestion in caustic soda followed by treatment with gaseous chlorine, caustic soda extraction and bleaching.

-cele *(Med.)*. A suffix derived from the Greek *kēlē*, tumour, hernia.

celestial equator *(Astron.)*. The great circle in which the plane of the earth's equator cuts the celestial sphere; the primary circle to which the coordinates, right ascension and declination are referred.

celestial mechanics *(Astron.)*. The study of the motions of celestial objects in gravitational fields. This subject, founded by Newton, classically deals with satellite and planetary motion within the solar system, using Newtonian gravitational theory, which is much simpler than the **general theory of relativity**.

celestial poles *(Astron.)*. The two points in which the earth's axis, produced indefinitely, cut the celestial sphere.

celestial sphere *(Astron.)*. An imaginary sphere, of indeterminate radius, of which the observer is the centre. On the surface of all stars, independent of their real distance, are points specified by two coordinates, referred to some chosen great circle of the sphere.

celestine *(Min.)*. Strontium sulphate, crystallizing in the orthorhombic system; occurs in association with rock-salt and gypsum; also in the sulphur deposits of Sicily, and in nodules of limestone.

Celite *(Chem.)*. TN for a form of **diatomite** used as an insulating material and filter aid.

cell *(Biol.)*. (1) The unit, consisting of nucleus and cytoplasm, of which plants and animals are composed; in the former surrounded by a non-living wall. (2) The whole bacterium or yeast, the *bacterial cell*. adj. *cellular*.

cell *(Elec.Eng.)*. (1) Chemical generator of emf. (2) Small item forming part of experimental assembly, e.g. Kerr cell, dielectric test cell, etc.

cell *(Maths.)*. Small volume unit in mathematical co-ordinate system.

cell *(Nuc.Eng.)*. (1) Unit of homogeneous reactivity in reactor core. (2) Small storage or work place for 'hot' radioactive preparations.

cell *(Zool.)*. One of the spaces into which the wing of an Insect is divided by the veins.

cell cavity *(Bot.)*. See **lumen**.

cell constant *(Chem.)*. The conversion factor relating the conductance of a conductivity cell to the conductivity of the liquid in it.

cell cycle *(Biol.)*. The series of events which describe the history of a dividing cell from one generation to another. It consists of 3 intermitotic stages, the G1, S and G2 phases of very variable length, together with the mitotic phase. DNA is duplicated during the S (*synthesis*) phase.

cell division *(Biol.)*. The formation of two daughter cells from one parent cell. See **amitosis**, **mitosis** and **meiosis**.

cell enlargement, extension *(Bot.)*. The growth in volume or length of a plant cell, produced from a meristem as it matures, involving *vacuolation* and the synthesis of protoplasmic and wall materials.

cell-free *(Biol.)*. Applied to biological phenomena like **translation** and **transcription** which can be made to occur in the laboratory in the absence of cells.

cell fusion *(Biol.)*. The merging of cells by fusion of their plasma membranes resulting in a bi- or multi-nucleate complex. Fusion can be induced by various agents like polyethylene glycol and is the crucial step in the formation of **hybridomas** during **monoclonal antibody** production.

cell genetics *(Biol.)*. The study of genetics, particularly the location of genes on chromosomes, by means of cells grown in culture.

cell inspection lamp *(Elec.Eng.)*. An electric filament lamp provided with narrow bulb so that it can be used for the examination of an accumulator by insertion between the plates.

cell line *(Biol.)*. Strictly a cell culture derived from a single progenitor cell and with therefore a homogeneous genetic constitution.

cell lineage *(Zool.)*. The developmental history of individual cells of an embryo during cell division following fertilization.

cell-mediated immunity *(Immun.)*. Specific immunity which depends on the presence of activated **T-lymphocytes** acting as cytotoxic cells and/or releasing **lymphokines** which activate monocytes and macrophages. Cell mediated immunity is responsible for protecting against intracellular microbes, but also for the rejection of **allografts** and for **delayed type hypersensitivity reactions**.

cellobiose, cellose *(Chem.)*. $C_{12}H_{22}O_{11}$, a disaccharide, obtained by complete hydrolysis of cellulose. G–β1→4–G (G=glucose). It is the repeating unit for cellulose.

cello foils *(Print.)*. See **vinyl foils**.

Cellosolve *(Plastics)*. Hydroxy-ether, 2-ethoxy-ethan 1-ol, $C_2H_5O.CH_2.CH_2OH$, a colourless liquid used as a solvent in the plastics industry. It is miscible with water, ethanol and ethoxyethane, and boils at 135.3°C.

cell plate *(Bot.)*. A thin partition, bounded by a membrane, growing centrifugally by the coalescence of vesicles across the equatorial plane of the telophase spindle to effect the division of the cytoplasm and to become the basis of the middle lamella of the new wall. Characteristic of cell division in larger plants and many algae. See also **cleavage, phragmoplast**.

cell tester *(Elec.Eng.)*. A portable voltmeter for checking the voltage of accumulator cells.

cell transformatiom *(Biol.)*. See **transformation (2)**.

cellular automaton *(Comp.)*. **Automaton** consisting of a number of cells, where the state of a given cell depends on the state of adjacent cells.

cellular concrete *(Build.)*. Concrete in the body of which bubbles of air are induced, either by chemical or mechanical means, in the process of manufacture, thereby producing a concrete of low unit weight.

cellular double bottom *(Ships)*. A common construction of the bottom of a ship, where an inner bottom extends throughout the length of the ship between the peak bulkheads and over the width of the ship. The space between the outer and inner bottom plating is divided into cells by transverse floors and fore and aft keelsons, some of which are oil and watertight so that the space may be used for fresh water, water ballast or oil fuel.

cellular fabric *(Textiles)*. The name applied to a woven or knitted fabric featuring an open, or cell-like, structure used mainly for shirting, blouse and underwear purposes.

cellular glass *(Glass)*. Glass foamed to a mass of separate air cells. This is dark in colour with a strong sulphide smell. It is a very moisture-resistant low-temperature insulation material of great value.

cellular horn *(Acous.)*. A horn for a high-frequency loudspeaker (tweeter), in which the path from the throat to the outer air is by a number of expanding channels of equal length, so that marked directivity, arising from the short wavelengths in relation to the width of the total opening, is not apparent.

cellular radio *(Telecomm.)*. Radio system linking mobile users (in vehicles or with hand-held sets) to the public telephone system. A nation-wide network of hundreds of UHF transmitters, all on separate frequencies (each transmitter and its service area being known as a *cell*) ensures full coverage. The mobile sets use frequency-searching and -hopping to maintain service when moving from one cell to another.

cellular silica *(Chem.)*. Inorganic silicates containing numerous air cells and low permeability to water vapour.

cellular slime moulds *(Bot.)*. See **Acrasiomycetes**.

cellular structure *(Eng.)*. See **network structure**.

cellular-type switchboard *(Elec.Eng.)*. A switchboard in which each switch with its associated apparatus is contained in a separate cell of fireproof material. Also called *cubicle-type switchboard*.

cellulase *(Bot.)*. An enzyme or mixture of enzymes capable of catalysing the hydrolysis of cellulose to cellobiose or glucose.

cellulitis *(Med.)*. A spreading infection of the subcutaneous tissues with pyogenic bacteria, often streptococcal.

celluloid *(Plastics)*. A thermoplastic made from nitrocellulose, camphor and ethanol; rel.d. 1.35–1.85, moulding temp. 29–49°C, breakdown strength 20 000 to 45 000 V/mm. Highly inflammable. It is elastic and very strong, and can be produced in very thin sheets. Its use has decreased with the introduction of non-inflammable thermoplastics.

cellulose *(Bot.)*. Occurs in **microfibrils** in the walls of most plant cells and since the secondary walls in the wood of trees commonly contain 50–60% cellulose, it is the most abundant organic molecular species on earth.

cellulose *(Chem.)*. $(C_6H_{10}O_5)n$, a complex polyose with many thousands of residues, forming cell walls in plants. The chief source of cellulose is wood, cotton and other fibrous materials (e.g. flax, hemp, nettle etc.). Pure cellulose is obtained by removing all incrustations of lignin resins and other inorganic and organic matter by treatment with alkali, acids, sodium sulphite etc. Cellulose is soluble in cuprammonium hydroxide (**Schweitzer's reagent**), ammoniacal copper carbonate, a solution of zinc oxide in conc. hydrochloric acid. It is the raw material for the manufacture of paper, rayon, cellulose lacquers, films. Cellulose can undergo many chemical transformations, e.g. strong acids transform it into **amyloid**; it can be hydrolysed and oxidized (*cellulose hydrates, hydrocelluloses, oxycelluloses*) and esterified (*cellulose acetates; cellulose nitrates; benzyl-cellulose; cellulose xanthate*). It contains the β1-4 glucoside linkage. See **cellobiose**.

cellulose acetate *(Textiles)*. See **acetate, triacetate**.

cellulose acetates *(Chem.)*. *Acetylcelluloses*. These are ethanoic acid esters of cellulose, obtained by the action of glacial ethanoic acid, ethanoic anhydride and sulphuric acid, upon cellulose. They are considerably less inflammable than cellulose nitrates, and are a raw material for films, windscreens, textile fibres, lacquers etc.

cellulose esters *(Chem.)*. Cellulose derivatives obtained by esterification with nitric acid, acetic acid etc.

cellulose lacquers *(Chem.)*. Lacquers prepared by dissolving nitrocellulose or acetylcellulose in a mixture of suitable solvents, with the admixture of resins, plasticizers and pigments.

cellulose nitrates *(Chem.)*. See **nitrocelluloses**.

cellulose paints *(Build.)*. Reversible coatings based on chemically treated cellulose compounds having wide application in the motor car industry but limited use in house painting and decorating. Their storage and use is governed by statutory regulations.

cellulose xanthate *(Chem.)*. $[C_6H_8O_3(ONa).OCS_2Na]n$, an acid salt of cellulose-dithiocarbonic acid, obtained by treating cellulose with concentrated caustic soda, with subsequent dissolution in carbon disulphide. The resulting product is called **viscose**, the intermediate for viscose rayon textile fibres.

cell unit *(Crystal.)*. See **unit cell**.

cell unit *(Elec.Eng.)*. A unit which forms the basis for an extended switchboard.

cell wall *(Bot.)*. A mechanically rigid layer outside the **plasma membrane** of plant cells, usually consisting of microfibrils of (a) cellulose in vascular plants, bryophytes and many algae, (b) xylan in some other algae and (c) chitin in fungi all in a matrix of other polysaccharides and other substances (e.g. lignin and suberin). Such a wall can withstand a substantial turgor pressure (say 2–20 bars).

Celotex *(Build.)*. TN for artificial building-board used for acoustic reflection, made from sugar-cane fibre (bagasse) compressed and baked.

celsian *(Min.)*. A barium feldspar; barium aluminium silicate. Found in association with some manganese ores.

Celsius scale *(Phys.)*. The SI name for *Centigrade scale*. The original Celsius scale of 1742 was marked zero at the boiling-point of water and 100 at the freezing point, the scale being inverted by Strömer in 1750. Temperatures on the International Practical Scale of Temperature are expressed in degrees Celsius. See however **Kelvin thermodynamic scale of temperature**.

cement *(Build.,Civ.Eng.)*. A material for uniting other materials or articles. It is generally plastic at the time of application, but hardens when in place. See **Portland cement**.

cement *(Geol.)*. The material which binds any loose sediment into a coherent rock. The commonest cements are ferruginous, calcareous, and siliceous.

cement *(Zool.)*. In Mammalian teeth, a layer resembling bone covering the dentine beyond the enamel.

cementation *(Civ.Eng.)*. See **grouting**.

cementation *(Eng.)*. Any process in which the surface of a metal is impregnated at high temperature by another substance. See **case-hardening**. Also *carbonization, carburization*.

cement copper *(Eng.)*. Impure copper, obtained when the metal is precipitated by means of iron from solutions resulting from leaching.

cemented carbides *(Eng.)*. *Sintered carbides*. See **carbide tools**.

cement fillet *(Build.)*. A substitute for lead flashings in the angles between, e.g. a chimney stack and roof, weatherproofing being provided by running in a band of cement mortar.

cement grout *(Civ.Eng.)*. A fluid cement mixture for filling crevices.

cement gun *(Civ.Eng.)*. An apparatus for spraying fine concrete or cement mortar by pneumatic pressure.

cementite *(Eng.)*. The iron carbide (Fe_3C) constituent of steel and cast-iron (particularly white cast-iron). Very hard and brittle.

cement joggle *(Build.)*. A key formed between adjacent stones in parapets etc., by running cement mortar into a square-section channel cut equally into each of the jointing faces, thereby preventing relative movement.

cement mortar *(Build.,Civ.Eng.)*. A hydraulic mortar composed of Portland cement (or other siliceous cement) and sand.

cement paints *(Build.)*. Powder materials based on Portland cement which are mixed with water immediately before use. Normal application is to exterior wall surfaces of brick, concrete, cement rendering etc.

cement rock *(Geol.)*. Argillaceous limestone containing >18% of clay.

cement-rubber-latex *(Build.)*. A flooring composed of cement, aggregate and an elastomer to give flexibility. See also **fleximer**.

Cenozoic *(Geol.)*. The youngest era of the Phanerozoic, covering approx. 65 million years ago to the present day. It includes the Tertiary and Quaternary periods.

censer mechanism *(Bot.)*. A means of seed liberation in which the seeds are shaken out of the fruits as the stem of the plant sways in the wind (e.g. in the poppy).

censor, censorship *(Behav.)*. In psychoanalytic theory, a powerful unconscious inhibitory mechanism in the mind, which prevents anything painful to the conscious aims of the individual from emerging into consciousness. It is responsible for the distortion, displacement and condensation present in dreams.

cent *(Nuc.Eng.)*. Unit of reactivity equal to one-hundredth of a **dollar**.

centering *(Civ.Eng.)*. See **centring**.

centi- *(Genrl.)*. Prefix meaning *one-hundredth*.

Centigrade scale *(Phys.)*. The most widely used method of graduating a thermometer. The fundamental interval of temperature between the freezing and boiling points of pure water at normal pressure is divided into 100 equal parts, each of which is a *Centigrade degree*, and the freezing point is made the zero of the scale. To convert a temperature on this scale to the Fahrenheit scale, multiply by 1.8 and add 32; for the Kelvin equivalent add 273.15. See also **Celsius scale**.

centimetre-gram-second unit *(Genrl.)*. See **CGS unit**.

centiMorgan *(Biol.)*. Measure of the distance between the loci of two genes on the same chromosome, obtained from the *cross-over frequency*. 1 *centiMorgan* = 1% *crossing-over*; 1 *Morgan* = 100 *centiMorgans* when summed over short distances between intervening loci.

centipedes *(Zool.)*. See **Chilopoda**.

centipoise *(Phys.)*. One-hundredth of a **poise**, the CGS unit of viscosity. Symbol cP; $1\ cP = 10^{-3}\ Nm^{-2}\ s$.

central angle *(Maths.)*. An angle bounded by two radii of a circle.

central cylinder *(Bot.)*. See **stele**.

central dogma *(Biol.)*. The postulate that genetic information resides in the nucleic acid and passes to the protein sequence, but cannot flow from protein to nucleic acid.

central force *(Phys.)*. A force is central with respect to an origin if a particle at an arbitrary point experiences a force directed along the line between the origin and the point. If the force, $f(r)$ is conservative it can be derived from a potential energy function $V(r)$, a function only of the distance r from the origin. $f(r) = -dV(r)/dr$. $V(r)$ is the **central potential**. See **angular momentum**.

central lubricating system *(Eng.)*. (1) An oil or grease lubricating system in which the lubricant is distributed by a single pump via pipes to a number of outlets, one actuation of the pump providing a discharge at every outlet. (2) A coolant distribution system in which a battery of machine tools is supplied with cleaned coolant by a single pump/filter unit.

central nervous system *(Zool.)*. The main ganglia of the nervous system with their associated nerve cords, consisting usually of a brain or cerebral ganglia and a dorsal or ventral nerve cord which may be double, together with associated ganglia. Abbrev. *CNS*.

central potential *(Phys.)*. A spherically symmetric potential in which the potential depends only on the *distance* from some centre; The orbital angular momentum is constant for a single particle moving in such a potential. In quantum mechanics, the Schrödinger equation can be solved for such a system; the hydrogen atom is an example in which an electron moves in the central coulomb potential provided by the nuclear charge.

central processing unit *(Comp.)*. See **central processor**.

central processor *(Comp.)*. Main part of a computer consisting of an **arithmetic logic unit** and a **control unit**. Also *central processing unit, CPU*.

central projection *(Maths.)*. One in which a geometrical configuration in one plane P_1 is projected on another plane P_2 by drawing straight lines from a point O (not in P_1 or P_2) through the various points in the configuration, so that these lines intersect P_2.

centre *(Civ.Eng.)*. A timber frame built as a temporary support during the construction of an arch or a dome.

centre *(Eng.)*. A conical support for workpieces, used in lathes and some other machine tools. Live centres, for headstocks, are usually left soft and machined to 60° included angle before use in the lathe. Dead centres, also normally at 60°, are heat treated or tipped. Antifriction centres are also used. Running centres rotate with the work.

centre *(Maths.)*. Of a conic: the point through which all diameters pass.

centre *(Surv.)*. Accurate vertical alignment of centre of rotation of survey instrument over, or under, a fixed point or station.

centre adjustment *(Surv.)*. On tripod-mounted survey instrument, device which allows instrument to be moved horizontally until exactly above (or underground, perhaps beneath) survey signal point.

centre arbor *(Horol.)*. The arbor in the train of a watch or clock which is planted in the centre of the plates.

centre-bit *(Build.)*. A wood-boring tool having a projecting central point and two side wings, one of which scribes the boundary of the hole to be cut, while the other removes the material.

centre-contact cap *(Elec.Eng.)*. A bayonet cap, fitted to an electric lamp, in which the outer wall forms one of the contacts, the other being the projection.

centre-contact holder *(Elec.Eng.)*. A lamp holder with a single centrally located contact, designed to be used with a centre-contact cap.

centre drill *(Eng.)*. Drill in which a short parallel portion leads to a coned section with an included angle of 60°. Used (1) to give accurate start to a twist drill and (2) to provide a conical recess to ends of workpieces to be machined between **centres** in a lathe or cylindrical grinder.

centre frequency *(Telecomm.)*. (1) Geometric or arithmetic midpoint of the cut-off frequencies of a filter. (2) Carrier frequency, when modulated symmetrically.

centre keelson *(Ships)*. See under **keelson**.

centre lathe *(Eng.)*. Machine tool for cutting metal in which a workpiece rotates and a rigidly held tool is applied so as to make a circular product. Includes ability to make flat surfaces by facing, and screw cutting, boring and drilling operations. The workpiece may be held in a **chuck** or between **centres**.

centreless grinding *(Eng.)*. A method of grinding cylindrical objects. The work is supported on a rest, between a pair of abrasive wheels revolving at different speeds in opposite directions, instead of between centres as in normal practice.

centre margin ring *(Print.)*. The bevelled strip around the circumference of the plate cylinder of a letterpress rotary press against which the plates are locked.

centre nailing *(Build.)*. A method of nailing slates on a roof; the nail is driven in any one slate just above the line of the head of the slate in the course below. Cf. **head nailing**.

centre note *(Print.)*. Notes placed between columns, as in bibles, to protect the small type and to economize in space.

centre of action *(Meteor.)*. A position occupied, more or less permanently, by an anticyclone or a depression, which largely determines the weather conditions over a wide area. The climate of Europe is dependent on the Siberian anticyclone and the Icelandic depression.

centre of anallatism *(Surv.)*. In a distance-measuring telescope, the point from which the distance to an observed staff is proportional to the staff intercept as seen between the upper and lower stadia lines of the diaphragm.

centre of buoyancy *(Ships)*. The centroid of the immersed portion of a floating body. In ship construction, a ship being symmetrical about a fore and aft plane, the centre of buoyancy lies in that plane. The vertical position is calculated from a succession of waterplane areas and the longitudinal position from a succession of transverse section areas.

centre of curvature *(Maths.)*. See **curvature**.

centre of gravity *(Phys.)*. The particles making up a body all experience gravitional forces. The resultant of these forces act through a single point, the *centre of gravity*. In a uniform gravitional field, such as that near the earth's surface, the centre of gravity coincides with the **centre of mass**.

centre of inversion *(Maths.)*. See **inversion**.

centre of lens *(Phys.)*. A point on the principal axis of a lens, through which passes any ray whose incident and emergent directions are parallel. Measurements of object and image distance, focal length, etc., are taken from this point.

centre of mass *(Aero.)*. Centroid of all distributed masses of a flight vehicle. Important design feature for determination of size of wings, layout, stability and payload. Also operationally in loading aircraft to safe limits for flight.

centre of mass *(Phys.)*. The point in an assembly of mass particles where the entire mass of the assembly may be regarded as being concentrated and where the resultant of the external forces may be regarded as acting for considerations not concerned with the rotation of the assembly. The point is defined vectorially by

$$\bar{\mathbf{r}} = \frac{\sum m\mathbf{r}}{\sum m},$$

where m is the mass of a particle at a position $\mathbf{r}$ and the summation extends over the whole assembly.

centre of oscillation *(Phys.)*. A point in a compound pendulum which, when the pendulum is at rest, is vertically below the point of suspension at a distance equal to the length of the equivalent simple pendulum (that is, the simple pendulum having the same period). If the pendulum is suspended at the centre of oscillation, its period is the same as before. Also called *centre of percussion*.

centre of pressure *(Aero.)*. The point at which the resultant of the aerodynamic forces (lift and drag) intersects the chord line of the aerofoil. Its distance behind the leading edge is usually given as a percentage of the chord length. Abbrev. *c.p.*

centre of pressure *(Phys.)*. The point in a surface immersed in a fluid at which the resultant pressure over the immersed area may be taken to act.

centre of symmetry *(Crystal.)*. A point within a crystal such that all straight lines that can be drawn through it pass through a pair of similar points, lying on opposite side of the centre of symmetry and at the same distance from it. Thus, faces and edges of the crystal occur in parallel pairs, on opposite sides of a centre of symmetry.

centre pinion *(Horol.)*. The first pinion in a watch or clock train, driven by the great wheel.

centre-point steering *(Autos.)*. The relative positioning of the steered wheel and the swivel axis so as to obtain coincidence between the point of intersection of the swivel axis with the road and the plane of the wheel.

centre punch *(Eng.)*. A punch with a conical point, used to mark or 'dot' the centres to be drilled, etc.

centre section *(Aero.)*. The central portion of a wing, to which the main planes are attached; in large aircraft it is often built into the fuselage and may incorporate the engine nacelles and the main landing gear.

centre spread *(Print.)*. A design occupying the area of two pages in the centre opening of a booklet or journal.

centre square *(Eng.)*. A device for marking the centres of circular objects and bars. The bar is placed in the centre of the square, which is bisected by a blade that serves as a guide for scribing a diametral line.

centre wheel *(Horol.)*. The wheel mounted on the arbor of the central pinion. It usually makes one turn per hour, so that any calculations relating to the train or number of vibrations made by the balance are taken from this wheel.

centre-zero instrument *(Elec.Eng.)*. An indicating instrument which has the zero at the centre of the scale and can therefore read both positive and negative values of the quantity indicated.

centric leaves *(Bot.)*. Cylindrical leaves, with the palisade tissues arranged uniformly around the periphery of the cylinder, e.g. Onion.

centrifugal *(Bot.)*. A developmental process, starting at the centre and working towards the outside.

centrifugal *(Zool.)*. See **efferent**.

centrifugal brake *(Eng.)*. An automatic brake used on cranes etc. in which excessive speed of the rope drum is checked by revolving brake shoes forced outwards into contact with a fixed drum by centrifugal force.

centrifugal casting *(Foundry)*. The casting of large pipes, cylinder liners etc. in a rotating mould of sand-lined or water-cooled steel.

centrifugal clutch *(Eng.)*. A type of clutch in which the friction surfaces are engaged automatically at a definite speed of the driving member, and thereafter maintained in contact, by the centrifugal force exerted by the weighted levers.

centrifugal compressor *(Eng.)*. One which passes entering air through a series of low-pressure, high volume fans via increasingly restricted chambers so that the emerging air attains the required higher pressure; also, small booster fans (superchargers).

centrifugal fan *(Eng.)*. A fan with an impeller of paddle-wheel form, in which the air enters axially at the centre and is discharged radially by centrifugal force. Also called *paddle wheel fan*.

centrifugal-flow compressor *(Aero.)*. A compressor in which a vaned rotor inspires air near the axis and throws it towards a peripheral diffuser, the pressure rising mainly through centrifugal forces. The maximum pressure ratio for a single stage is about four. Centrifugal compressors are universal for aero-engine superchargers and were widely used in the earlier turbojets.

centrifugal pulp cleaner *(Paper)*. A cleaning device for pulp stock which produces a high-speed rotation or rotary vortex motion causing dirt particles to separate out.

centrifugal pump *(Min.Ext.)*. Continuously acting pump with rotating impeller, used to accelerate water or suspensions through a fixed casing to peripheral discharge at the desired delivery height.

centrifugal starter *(Elec.Eng.)*. A device used with small induction motors; it consists of a centrifugally operated switch on the rotor, which automatically cuts out starting-resistance or performs some other operation as the motor runs up to speed.

centrifugal tension *(Eng.)*. The force per unit area of cross-section induced, in consequence of centrifugal force, in the material of a rotating rim, loop, or driving belt.

centrifuge *(Eng.)*. Rotating machine which uses centrifugal force to separate molecules from solution, particles and solids from liquids, and immiscible liquids from each other. Depends on differences in the relative densities of the substances to be separated. Used widely in science where forces up to 500 000 g may be obtained in an *ultracentrifuge*. In industry they are used in sugar and cream production, separating water from fuel and swarf from cutting oil etc.

centrifuge enrichment *(Phys.)*. The enrichment of uranium isotopes using a high-speed centrifuge. It uses the mass difference between ^{235}U and ^{238}U to effect the separation. The separation factor per stage is much greater than for gaseous diffusion enrichment.

centrifuge separation *(Nuc.Eng.)*. Technique of isotope separation applied to the enrichment of ^{235}U in uranium. Sometimes called *centrifugation*.

centring *(Eng.)*. (1) The marking of the centres of holes to be drilled in a piece of metal. (2) The adjusting of work in a lathe so that its axis coincides with the lathe axis.

centring, centering *(Civ.Eng.)*. The general term applied to centres used in constructional work.

centriole *(Biol.)*. A structure virtually identical with the basal body of cilia, which serves as the centre for the polymerization of the microtubules of specialized fibrillar structures, including sperm flagelli and the mitotic spindle. Centrioles are normally arranged in pairs at right angles to each other.

centripetal *(Bot.)*. A developmental process starting at the outside and working towards the centre.

centripetal *(Zool.)*. See **afferent**.

centripetal acceleration *(Phys.)*. The acceleration, directed towards the centre of curvature of the path, which is possessed by a body moving along a curved path with constant speed. Its value is v^2/R, where v is the speed and *R* the radius of curvature.

centripetal force, centrifugal force *(Phys.)*. A body is constrained to move along a curved path by a force, directed towards the centre of curvature, and deviating the body from a straight path. This force is called the *centripetal force*. It is equal and opposite to an inertial force directed away from the centre of curvature of the path, the *centrifugal force*. They are both equal in magnitude to the product of the mass and its **centripetal acceleration**.

centroid *(Maths.)*. See **centre of gravity**.

centrolecithal *(Zool.)*. Having the yolk in the centre.

centromere *(Biol.)*. Also *primary constriction*. In mitotic metaphase chromosomes it consists of a narrow region in which **chromatids** are joined. It has flanking **kinetochores** to which the microtubules of the spindle attach at mitosis.

centrosome *(Biol.)*. Minute, self-duplicating structure near the interphase nucleus, from which the fibres of the spindle radiate at mitosis. In animal cells it contains a **centriole**.

centrum *(Zool.)*. The basal portion of a vertebra which partly or entirely replaces the notochord, and from which arise the neural and haemal arches, transverse processes etc.

cepaceous *(Bot.)*. Smelling or tasting like onion or garlic.

cephalad *(Zool.)*. Situated near, facing towards, or passing to, the head region.

cephal-, cephalo- *(Genrl.)*. Prefix from Gk. *kephalē*, head.

cephalic *(Zool.)*. Pertaining to, or situated on or in, the head region.

cephalic index *(Med.)*. Ratio of maximum breadth of skull divided by maximum length, multiplied by 100.

cephalization *(Zool.)*. The specialization of the anterior end of a bilaterally symmetrical animal as the site of the mouth, the principal sense organs, and the principal ganglia of the central nervous system; the formation of the head.

cephalocele *(Med.)*. Protrusion of the membranes of the brain, with or without the substance of the brain, through a hole in the skull.

Cephalochordata *(Zool.)*. A subphylum of the Chordata having a persistent notochord, metameric muscles and gonads, a pharynx having a large number of gill-slits which are enclosed in an atrial cavity, and lacking paired fins, jaws, brain, and skeletal structures of bone and cartilage; marine sand living forms. Lancelets.

cephalometry *(Med.)*. The radiological or ultrasound measurement of the foetal head dimensions in utero.

Cephalopoda *(Zool.)*. A class of bilaterally symmetrical Mollusca in which the anterior part of the foot is modified into arms or tentacles, while the posterior part forms a funnel leading out from the mantle cavity, the mantle is undivided, and the shell is a single internal plate, or an external spiral structure, or absent. Squids, Octopods and Pearly Nautilus.

cephaloridine *(Med.)*. Cephalosporin C with the acetoxy group replaced by a pyridinium ion. Antibiotic used for treatment of severe sepsis.

cephalosporins *(Med.)*. Group of antibiotics derived from the fungus *Cephalosporium*, effective against a broad spectrum of organisms (**wide-spectrum antibiotic**). Common examples are *cefuroxime* and *cefotaxime*.

cephalothorax *(Zool.)*. In some Crustacea, a region of the body formed by the fusion of the head and thorax.

Cepheid parallax *(Astron.)*. See **period-luminosity law**.

Cepheid variable *(Astron.)*. A class of variable star with a period of 1 to 50 days, characterized by precise regularity. There is an exact correlation between the period and **luminosity**: the longer the period, the more luminous the star. This relationship means that the observed period of a Cepheid is an indicator of its distance, and for this reason they have been of immense importance for the determination of the distance scale of the universe. The prototype is beta Cephei.

ceramic capacitor *(Electronics)*. One using a high permittivity dielectric such as barium titanate to provide a high capacitance/unit volume.

ceramic filter *(Biol.)*. A deep filter, usually of ceramic, with fine pores in which small particles or bacteria become trapped. Also called *Pasteur filter*. Now largely superseded by **membrane filters**.

ceramic fuel *(Nuc.Eng.)*. Nuclear fuel with high resistance for temperature, e.g. uranium dioxide, uranium carbide.

ceramic insulator *(Elec.Eng.)*. An insulator made of ceramic material, e.g., porcelain; generally used for outdoor installations.

ceramics *(Chem.)*. The art and science of nonorganic nonmetallic materials. The term covers the purification of raw materials, the study and production of the chemical compounds concerned, their formation into components, the study of structure, constitution and properties. See **alumina, carbides** etc.

cerargyrite, horn silver *(Min.)*. Silver chloride. Also a group name for silver halides.

cerat- *(Genrl.)*. Same as *kerat-*.

cercal *(Zool.)*. Pertaining to the tail.

cercaria *(Zool.)*. The final larval stage of Trematoda which develops directly into the adult; usually characterized by the possession of a round or oval body, bearing eye-spots and a sucker, and a propelling tail.

cercus *(Zool.)*. In some Arthropoda, a multiarticulate sensory appendage at the end of the abdomen.

cere *(Zool.)*. In Birds, the soft skin covering the base of the upper beak.

cerebellar fossa *(Zool.)*. See **cerebral fossa**.

cerebellum *(Zool.)*. A dorsal thickening of the hind-brain in Vertebrates. adj. *cerebellar*.

cerebral *(Zool.)*. Pertaining to the brain; pertaining to the **cerebrum**.

cerebral abscess *(Med.)*. An abscess within the brain.

cerebral blood flow *(Med.)*. The blood flow to the brain.

cerebral flexure *(Zool.)*. The bend which develops between the axis of the forebrain and that of the hindbrain, in adult Craniata.

cerebral fossa *(Zool.)*. In Mammals, a concavity in the cranium corresponding to the cerebrum.

cerebral haemorrhage *(Med.)*. Bleeding within the substance of the brain. One of the main causes of a *stroke* or *cerebro-vascular accident*.

cerebral hemispheres *(Zool.)*. See **cerebrum**.

cerebral palsy *(Med.)*. Muscular paralysis or other dysfunction resulting from perinatal damage to the motor area of brain.

cerebral thrombosis *(Med.)*. Thrombosis or clotting within one of the vessels supplying the brain leading to a **stroke** or *cerebro-vascular accident*.

cerebral tumour *(Med.)*. A **tumour** within the cranial cavity.
cerebr-, cerebro- *(Genrl.)*. A prefix from L. *cerebrum*, brain.
cerebroside *(Biol.)*. The simplest glycolipid, consisting of N-acyl sphingosine with either a glucose or a a galactose residue.
cerebrospinal *(Zool.)*. Pertaining to the brain and spinal cord.
cerebrospinal fluid *(Med.)*. The clear colourless fluid which bathes the surfaces of the brain and spinal cord.
cerebrum *(Zool.)*. A pair of hollow vesicles or hemispheres forming the anterior and largest part of the brain of Vertebrates.
Cerenkov counter *(Nuc.Eng.)*. Radiation counter which operates through the detection of Cerenkov radiation.
Cerenkov radiation *(Phys.)*. Radiation emitted when a charged particle travels through a medium at a speed greater than the velocity of light in the medium. This occurs when the refractive index of the medium is high, i.e. much greater than unity, as for water.
ceriferous *(Bot.,Zool.)*. Wax-bearing, wax-producing.
cerium *(Chem.)*. A steel-grey metallic element, one of the 'rare earth' metals. Symbol Ce, at. no. 58. r.a.m. 140.12, rel.d. at 20°C 6.7, mp 623°C, electrical resistivity 0.78 microhm metres. When alloyed with iron and several rare elements, it is used as the sparking component in automatic lighters and other ignition devices. It is also a constituent (0.15%) in the aluminium base alloy ceralumin and is photosensitive in the ultraviolet region. It is also used on tracer bullets, for flashlight powders and in gas mantles. It is a getter for noble gases in vacuum apparatus. There are a number of isotopes which are fission products, ^{144}Ce, with half-life of 290 days, being a pure electron emitter of importance.
cermet *(Aero.,Eng.)*. Ceramic articles bonded with metal. Composite materials combining the hardness and high temperature characteristics of ceramics with the mechanical properties of metal, e.g. cemented carbides.
ceroma *(Zool.)*. A synonym for *cere*.
cerous *(Zool.)*. See **cere**.
Certificate of Airworthiness *(Aero.)*. Certification issued or required by the *CAA* confirming that a civil aircraft is *airworthy* in every respect to fly within the limitations of at least one of six categories. Abbrev. *C of A*. US *ATC* (Approved Type Certificate).
Certificate of Compliance *(Aero.)*. Certification that parts of an aircraft have been overhauled, repaired or inspected, etc. to comply with airworthiness requirements. Abbrev. *C of C*.
Certificate of Maintenance *(Aero.)*. Certification that an aircraft has been inspected and maintained in accordance with its maintenance schedule. Abbrev. *C of M*.
ceruminous glands *(Zool.)*. Modified sweat glands occurring in the external auditory meatus of Mammals and producing a waxy secretion.
cerussite *(Min.)*. Lead carbonate, crystallizing in the orthorhombic system. A common ore of lead.
cervical ganglia *(Zool.)*. Two pairs of sympathetic ganglia, anterior and posterior, situated in the neck of Craniata.
cervical smear *(Med.)*. The taking of a small sample of cells from the uterine cervix for the detection of cancer or the pre-cancerous stage.
cervicectomy *(Med.)*. Removal of the cervix uteri.
cervicitis *(Med.)*. Inflammation of the cervix uteri.
cervicum *(Zool.)*. In Insects, the neck or flexible intersegmental region between the head and the pro-thorax: in higher Vertebrates, the neck or narrow flexible region between the head and the trunk. adj. *cervical*. Pertaining to the neck or to the cervix uteri.
cervine *(Bot.)*. Dark-tawny.
cervix uteri *(Zool.)*. Neck of the uterus, situated partly above and partly in the vagina.
Cesarean section *(Med.)*. See **Caesarean section**.
cesium *(Chem.)*. See **caesium**.
cesspool *(Build.)*. (1) A small square wooden box, lined with lead, which serves as a cistern in parapet gutter at a point where roof water is discharged into a down pipe. (2) Underground pit (also *cesspit*) for the reception of sewage from houses not connected to a mains or community drainage system.
Cestoda *(Zool.)*. A class of Platyhelminthes, all the members of which are endoparasites; there is a tough cuticle; the alimentary canal is lacking; hooks and suckers for attachment occur at what is considered to be the anterior extremity. Tapeworms.
Cetacea *(Zool.)*. An order of large aquatic carnivorous Mammals; the forelimbs are fin-like, and hind-limbs lacking; there is a horizontal flattened tailfin; the skin is thick with little hair, there are two inguinal mammae flanking the vulva, and the neck is very short. Whales, Dolphins and Porpoises.
cetane *(Chem.)*. Normal **hexadecane**.
cetane number *(Chem.)*. Percentage of cetane in a mixture of cetane and 1-methylnaphthalene that has the same ignition factor as the fuel under test; measure of the ignition value of the fuel. Equivalent for diesel fuels of the octane number for petrol.
cetrimide *(Chem.)*. One of the quaternary ammonium **cationic detergents**, consisting very largely of alkylmethylammonium bromides. It is a powerful disinfectant for cleansing skin, wounds etc. TN *Cetavlon*.
cetyl alcohol *(Chem.)*. *Hexadecan-1-ol*. The palmitic ester is the chief constituent of **spermaceti**
Ceylon chrysolite *(Min.)*. See **chrysolite**.
ceylonite *(Min.)*. See **pleonaste**.
Ceylon peridot *(Min.)*. A misnomer for the yellowish-green variety of the mineral tourmaline, approaching olivine in colour; used as a semiprecious gemstone.
Ceylon satinwood *(For.)*. See **East Indian satinwood**.
Cf *(Chem.)*. The symbol for californium.
CFA *(Immun.)*. Abbrev. for *Complete Freunds Adjuvant*.
CFR engine *(Eng.)*. A specially designed petrol engine, standardized by the Co-operative Fuel Research Committee, in which the knock-proneness or detonating tendency of volatile fuels is determined under controlled conditions and specified as an *octane number*. See **detonation, octane number**.
cg limits *(Aero.)*. The forward and aft positions within which the resultant centre of gravity of an aircraft must lie if balance and control are to be maintained; cf. **loading and cg diagram**.
CGS unit *(Phys.)*. Abbrev. for *Centimetre-Gram-Second* unit, based on the centimetre, the gram and the second as the *fundamental units* of length, mass and time. For most purposes superseded by **SI units**.
chabazite *(Min.)*. A white or colourless hydrated silicate of aluminium and calcium, found in rhombohedral crystals and belonging to the zeolite group.
chaeta *(Zool.)*. In Invertebrates, a chitinous bristle, embedded in and secreted by an ectodermal pit.
chaetiferous *(Zool.)*. Also chaetigerous, chaetophorous. Bearing bristles.
Chaetognatha *(Zool.)*. A phylum of hermaphrodite Coelomata, having the body divided into three distinct regions, head, trunk, and tail; the head bears two groups of sickle-shaped setae; small transparent forms, of carnivorous habit, occur in the surface waters of the sea. Arrowworms.
chaetoplankton *(Bot.)*. Planktonic organisms bearing bristle-like outgrowths from the cells the function of which is taken to be to increase drag and reduce sedimentation rate.
Chaetopoda *(Zool.)*. A group of Annelida, the members of which are distinguished by the possession of conspicuous setae; it includes the **Polychaeta** and the **Oligochaeta**.
chaff *(Aero.)*. Radar reflective strip or particles dispensed from aircraft, missiles or guns to confuse radar. First used in WW2 and known as 'Window', some kinds can be cut to desired length in flight to cope with different radar frequencies.
Chagas' disease *(Med.)*. See **trypanosomiasis**.
chain *(Chem.)*. A series of linked atoms, generally in an organic molecule (catenation). Chains may consist of one kind of atom only (e.g. carbon chains), or of several kinds

of atoms (e.g. carbon-nitrogen chains). There are open-chain and closed-chain compounds, the latter being called ring or cyclic compounds.

chain *(Eng.)*. A series of interconnected links forming a flexible cable, used for sustaining a tensile load.

chain *(Maths.)*. Any series, members of which are related in a specific way.

chain *(Surv.)*. An instrument for the measurement of length. It consists of 100 pieces of straight iron or steel wire, looped together end-to-end, and fitted with brass swivel handles at both ends, the overall length being one chain. See **engineer's chain, Gunter's chain, link**.

chain barrel *(Eng.)*. A cylindrical barrel, sometimes grooved, on which surplus chain is wound, as in certain types of crane.

chain block *(Eng.)*. A lifting mechanism comprising chains and sheaves in combination, such as a differential chain block, which allows very heavy loads to be hoisted by hand.

chain bond *(Build.)*. The bonding together of a stone wall by a built-in chain or iron bar.

chain code *(Comp.)*. A sequence of 2^n or fewer binary digits arranged so that the pattern of n adjacent digits locates their position uniquely.

chain conveyor *(Eng.)*. Any type of conveyor in which endless chains are used to support slats, apron, pans, buckets, as distinct from the use of a simple band. See **apron conveyor, bucket conveyor**.

chain coupling *(Eng.)*. A shaft coupling allowing irregularities in shaft alignment and a small amount of end play, designed for easy disconnection of shafts, in which the torque is transmitted from a chain sprocket keyed to the driving shaft to a similar sprocket keyed to the driven shaft, via a length of duplex roller chain wrapped round the two sprockets.

chain grate stoker *(Eng.)*. A mechanical stoker for boilers, in which the furnace grate consists of an endless chain built up from steel links.

chaining *(Comp.)*. Method of processing a large program, segment by segment, often used where the program is too large to fit in the main memory all at once.

chain insulator *(Elec.Eng.)*. See **suspension insulator**.

chain lines *(Paper)*. The more widely spaced continuous lines in the watermark of a laid paper.

chain lockers *(Ships)*. The subdivisions within a ship's hull for the housing of the anchor cables. They are usually divided on the centre line to form separate compartments for port and starboard cables.

chain printer *(Comp.)*. **Line printer** in which characters are carried on a continuous chain between print hammers and paper.

chain pump *(Eng.)*. A method of raising water through small lifts by means of disks attached to an endless chain which passes upwards through a tube; a chain alone may be used.

chain reaction *(Chem.)*. Chemical or atomic process in which the products of the reaction assist in promoting the process itself such as ordinary fire or combustion, or atomic fission. Hence chain reactions are characterized by an 'induction period' of comparatively slow reaction rate, followed, as the chain-promoting products accumulate, by a vastly accelerated reaction rate.

chain survey *(Surv.)*. A survey in which lengths only are measured (by means of a chain) and no angular measurements are made.

chain terminator *(Biol.)*. See **stop codon**.

chain tongs *(Build.)*. A pipe grip formed by a chain hooked into a bar with a toothed projection.

chain wheel *(Eng.)*. A toothed disk which meshes with a (roller) chain to transmit motion.

chair *(Chem.)*. The conformation of a six-membered ring in which the atoms are alternately above and below their mean plane. It is the stable conformation for cyclohexane and for aldohexoses. Cf. **boat**.

chair *(Civ.Eng.)*. The cast-steel support which is spiked to the sleeper (US *tie*) and used to secure **bull-headed rail** in position.

chair *(Glass)*. (1) The 'chair' with long arms on which the glass-maker rolls his blowpipe whilst fashioning the ware. (2) The set of men who work together in the hand process of fabrication.

chalaza *(Bot.)*. The basal portion of the nucellus of an ovule.

chalaza *(Zool.)*. One of two spirally twisted spindlelike cords of dense albumen which connect the yolk to the shell membrane in a bird's egg.

chalazion *(Med.)*. A small nodule on the eyelid due to chronic inflammation of a sebaceous gland.

chalazion *(Vet.)*. A cystic swelling of the Meibomian gland of the eyelid, seen commonly in dogs.

chalazogamy *(Bot.)*. The entry of the pollen tube through the chalaza of the ovule. Cf. **porogamy**.

chalcanthite *(Min.)*. Hydrated copper sulphate, $CuSO_4.5H_2O$.

chalcedony *(Min.)*. A microcrystalline variety of quartz with abundant micropores. It occurs filling cavities in lavas, and in some sedimentary rocks; flint is a variety found in the Chalk.

chalcocite *(Min.)*. A greyish-black sulphide of copper, Cu_2S, an important ore of copper. Also called *copper glance, redruthite*.

chalcophile *(Geol.)*. Descriptive of elements which have a strong affinity for sulphur and are therefore more abundant in sulphide ore deposits than in other types of rock. Lead is an example.

chalcopyrite, copper pyrites *(Min.)*. Sulphide of copper and iron, crystallizing in the tetragonal system; the commonest ore of copper, occurring in mineral veins. The crystals are brassy yellow, often showing superficial tarnish or iridescence.

chalet *(Arch.)*. A type of country house, distinguished by having a steeply pitched roof, outside balconies, galleries and staircase; generally built of wood.

chalice *(Zool.)*. A flask-shaped gland consisting of a single cell, especially numerous in the epithelia of mucous membranes.

chalk *(Geol.)*. A white, fine-grained and soft limestone, consisting of finely divided calcium carbonate and minute organic remains. In N.W. Europe, it forms the upper half of the Cretaceous system.

chalk *(Med.)*. Employed in suitable mixtures as an antacid; also in the treatment of diarrhoea.

chalk gland *(Bot.)*. A secreting organ, present in some leaves around which a deposit of calcium carbonate accumulates as in many species of *Saxifraga*.

chalking *(Build.)*. A paint film defect caused by the disintegration of the binder resulting in a loose powdery deposit. This defect is normally attributable to atmospheric exposure or application of an underbound coating to a highly porous substrate. It can be removed by thorough rubbing down and repainting with a suitable coating.

chalking *(Print.)*. When a printed ink film has failed to dry and key correctly to the substrate and it can be removed by light rubbing. Caused by the ink vehicle being rapidly absorbed into the stock and leaving the pigments largely unbound at the surface.

chalk line *(Build.,Civ.Eng.)*. A length of well-chalked string used to mark straight lines on work by holding it taut in position close to the work and plucking it.

chalybite *(Min.)*. See **siderite**.

chamaephyte *(Bot.)*. Woody or herbaceous plant with perennating buds above but within 25 m of the soil surface (includes *cushion plants*). See **Raunkaier system**.

chambered level tube *(Surv.)*. A level tube fitted at one end with an air chamber. By tilting the tube, air may be added to or taken from the bubble, whose length (which tends to be shortened by rise of temperature) may thus be regulated to maintain the sensitivity of the instrument.

chamber process *(Chem.Eng.)*. A process for the manufacture of sulphuric acid, in which the reactions between the air, sulphur dioxide and nitric acid gases necessary to produce the sulphuric acid take place in large lead chambers.

chamfer *(Build.)*. The surface produced by bevelling an edge or corner.

chamfer plane *(Build.)*. A plane fitted with adjustable guides to facilitate the cutting of any desired chamfer.

chamomile, camomile *(Med.)*. Derived from the dried flower-heads of *Anthemis nobilis*; formerly a well-known stomachic and tonic, and sometimes used in powdered form in cosmetic preparations and shampoos.

chamosite *(Min.)*. A hydrous silicate of iron and aluminium occurring in oolitic and other bedded iron ores.

chance *(For.)*. Strictly, any unit of operation in the woods, with many and varied applications, of which the most familiar is *logging* or *cutting chance*. A logging or pulpwood operating unit.

chancel *(Arch.)*. The eastern part of a church, originally separated from the *nave* by a screen of latticework, so as to prevent general access thereto, though not to interrupt either sight or sound.

chancery *(Print.)*. A style of italic, the original being modelled on the handwriting used in the Papal chancery; the most often seen present-day example is *Bembo italic*.

chancre *(Med.)*. The hard swelling which constitutes the primary lesion in syphilis.

chancroid *(Med.)*. Nonsyphilitic ulceration of the genital organs due to venereally-contracted infection by gram negative bacillum *Haemophilus ducreyi*.

Chandrasekhar limit *(Astron.)*. The upper limit, 1.44 solar masses, for a **white dwarf** star.

change face *(Surv.)*. To rotate a theodolite telescope about its horizontal axis so as to change from 'face left' to 'face right', or vice versa. See **transit**.

change of state *(Phys.)*. A change from solid to liquid, solid to gas, liquid to gas, or vice versa.

change-over *(Image Tech.)*. The transference of projection from one machine to another at the end of one reel and the start of the next, without an apparent break in the sequence.

change-over contact *(Telecomm.)*. The group of contacts in a relay assembly, so arranged that, on operation, a moving contact separates from a back contact, is free during transit, and then makes contact with a front contact. See **make-before-break contact**.

change-over switch *(Elec.Eng.)*. A switch for changing a circuit from one system of connections to another.

change point *(Surv.)*. A staff station to which two sights are taken, the first a foresight from one set-up of the level, the second a backsight from the next set-up of the level. Sometimes called a *turning point*.

change-pole motor *(Elec.Eng.)*. An induction motor with a switch for changing the connections of the stator winding to give two alternative numbers of poles, so that the motor can run at either of two speeds.

change-speed motor *(Elec.Eng.)*. A motor which can be operated at two or more approximately constant speeds, e.g. a change-pole motor.

change wheels *(Eng.)*. The gear-wheels through which the lead screw of a screw-cutting lathe is driven from the mandril, the reduction ratio being varied by changing the wheels. See **screw-cutting lathe**.

changing bag *(Image Tech.)*. A light-tight bag to accommodate a camera and sensitive photographic material, so that the former can be loaded or unloaded with the latter in daylight.

channel *(Comp.)*. Physical path followed by data, particularly between a central processing unit and a peripheral device or, by extension, between the user's program and a file on backing store. See **bus**.

channel *(Electronics)*. The main current path between the **source** and **drain** electrodes in a field-effect transistor.

channel *(Eng.)*. A standard form of rolled-steel section, consisting of 3 sides at right-angles, in channel form. See **rolled-steel sections**.

channel *(Nuc.Eng.)*. Passage through reactor core for coolant, fuel rod or control rod.

channel *(Telecomm.)*. Any clear path along which signals and information can be sent. A radio channel may consist of a transmitter and distant receiver, tuned to operate on an assigned frequency, or channel; alternatively, a microwave radio system or a long distance underground cable or optical-fibre system may have total capacity of several thousand speech channels or several television channels. Sometimes a single amplifier may be described as a *channel*, or a stereophonic amplifier has l.h. or r.h. channels; a single television camera with control equipment is often called a *channel*.

channel capacity *(Telecomm.)*. The maximum possible information rate for a channel.

channel effect *(Electronics)*. In a transistor, bypassing the base component by leakage due to surface conduction.

channel gulley *(Build.)*. A channel, about 45 cm long on the sides of which grease from sink wastes are trapped before reaching the gulley proper. Allows periodic removal.

channelling effect *(Nuc.Eng.)*. In radiation, the greater transparency of an absorbing material with voids relative to similar homogeneous material. Expressed numerically by the ratio of the attenuation coefficients. Also used for escape of radiation through flaws in shielding of reactor, etc.

channel pipe *(Build.)*. An open drain of half or three-quarter circular section, used in inspection and intercepting chambers.

channel separation *(Telecomm.)*. (1) The frequency increment between assigned frequencies for radio communication or broadcasting; chosen with regard to like frequency variation of the transmitters, bandwidth of the transmissions, and the ability of receivers to discriminate between signals, all with a view to avoiding mutual interference between channels and services. (2) In a stereophonic sound system, the level of interference between *left* and *right* channels.

channel width *(Telecomm.)*. The width (in kHz, MHz or GHz) of any channel, which could be a radio channel or one within a **frequency-division multiplex** cable system, that is allotted to any particular form of transmission or service. Speech can be sent over a channel a few kHz wide, but television needs several MHz.

chantlate *(Build.)*. A projecting strip of wood fixed to the rafters at the eaves and supporting the normal roof covering; it serves to carry the drip clear of the wall.

chapel *(Print.)*. An association of journeyman printers (and, in certain districts, apprentices), who elect a 'father' to watch their interests.

Chaperon resistor *(Elec.Eng.)*. Wirewound resistor of low residual reactance.

chaplet *(Foundry)*. Iron support for core placed in moulding box in iron foundry.

chapters *(Horol.)*. The Roman figures used on the dials of watches and clocks to mark the hours.

character *(Biol.)*. In *Mendelian genetics*, the abnormality or variant caused by a gene. In *quantitative genetics*, whatever is measured for study, e.g. weight or yield.

character *(Comp.)*. Symbol which may be represented in a computer, such as letter, digit, space, punctuation mark. See **control character**, **alphanumeric**.

character code *(Comp.)*. The **binary code** used to represent a character in a computer. There are a number of standard codes, **ASCII**, the *American Standard Code for Information Interchange*, **IS07**, the *International Organisation for Standardisation 7 bit code*, **EBCDIC**, the *Extended Binary Coded Decimal Interchange Code*.

characteristic *(Electronics)*. Any measurable property (e.g. gain, loss, impedance) of a device, active or passive, taken under closely specified operating conditions (supply voltages, temperature, frequency etc.).

characteristic curve *(Electronics)*. A graph showing the variation of a particular characteristic of a device with other parameters e.g. for a transistor, the collector current against collector-emitter voltage, plotted for different values of base current.

characteristic curve *(Image Tech.)*. Graph plotting the relationship between the density of a developed emulsion and the logarithm of the exposure. Also called the *D Log E curve*, *H and D curve*.

characteristic curve *(Maths.)*. Given a 1-parameter family of surfaces, $\psi(x,y,x,a) = 0$, the limiting position of the curve of intersection of two adjacent surfaces of the family is given by

$$\begin{cases} \psi(x, y, z, a) = 0 \\ \dfrac{\partial\psi}{\partial a}(x, y, z, a) = 0, \end{cases}$$

for a given value of a. This curve is a characteristic curve of the family of surfaces. The locus of all the characteristic curves is the envelope of the family.

characteristic equation of a matrix *(Maths.)*. Only defined for a square matrix. The equation which results when the determinant of the matrix obtained by replacing every diagonal element a_{ii} of the given matrix by $a_{ii}x$, is equated to zero. Its roots are called either *latent roots* or *eigenvalues*.

characteristic equation of an ordinary differential equation *(Maths.)*. See **auxiliary equation**.

characteristic function of a set *(Maths.)*. That which is unity for all points of the set, but zero outside the set.

characteristic impedance *(Phys.)*. An impedance is presented to waves propagated through a continuous medium. The *characteristic impedance* Z_0 of the medium is defined as the ratio of exciting force (or voltage) to velocity (or current) at a point. (1) For acoustical plane waves $Z_0 = \rho v$ where ρ is the density of the medium and v is the velocity of the wave. (2) For waves in a *transmission line*, $Z_0 = Lc = \sqrt{L/C}$ where L and C are the inductance and capacity per unit length of the line and c is the velocity of electromagnetic waves. In free space, $Z_0 = \mu_0 c = 376.6\ \Omega$ where μ_0 is the permeability of free space.

characteristic of a logarithm *(Maths.)*. See **logarithm**.

characteristic points *(Maths.)*. (1) Of a 1-parameter family of curves in a plane, $\psi(x,y,a) = 0$, the characteristic points or points of contact with the envelope are given by the equations

$$\psi(x, y, a) = 0, \qquad \frac{\partial\psi}{\partial a}(x, y, a) = 0,$$

for a particular value of the parameter a. The envelope is found by eliminating a from these two equations. (2) Of a 2-parameter family of surfaces, $\psi(x,y,z,a,b)$, the characteristic points, as above, are defined by the equations

$$\psi(x, y, z, a, b) = 0,$$
$$\frac{\partial\psi}{\partial a}(x, y, z, a, b) = 0,$$
$$\frac{\partial\psi}{\partial b}(x, y, z, a, b) = 0,$$

for fixed a and b. The envelope is found by eliminating a and b from these three equations.

characteristic polynomial *(Maths.)*. The polynomial which is equated to zero in the **characteristic equation**.

characteristic radiation *(Phys.)*. That from an atom associated with electronic transitions between energy levels; the frequency of the radiation emitted is characteristic of the particular atom.

characteristic spectrum *(Phys.)*. Ordered arrangement in terms of frequency (or wavelength) of radiation (optical or X-ray) related to the atomic structure of the material giving rise to them.

characteristic velocity *(Space.)*. The change of velocity and the sum of changes of velocity (accelerations and decelerations) required to execute a space manoeuvre, e.g. transfer between orbits.

characteristic X-radiation *(Phys.)*. X-radiation consisting of discrete wavelengths which are characteristic of the emitting element. If arising from the absorption of X- or γ-radiation, may be called *fluorescence X-radiation*. See **characteristic radiation**.

character recognition *(Comp.)*. Optical or magnetic reading by a peripheral capable of directly absorbing information from diagrams or the printed or written page.

characters per second *(Comp.)*. Measure of the speed in which data, in coded form, may be transmitted between two points.

Charadriiformes *(Zool.)*. Order of wading and swimming birds found on sea coasts and inland waters. Most feed on animal life. Waders, Gulls and Auks.

Charales *(Bot.)*. Stoneworts. Small order in the Charophyceae of macroscopic fresh and brackish water algae with a distinct axis; anchored by rhizoids to the bottom, and with whorled branches. One very large cell runs the length of each internode. Oogamous.

charcoal *(Chem.)*. The residue from the destructive distillation of wood or animal matter with exclusion of air; contains carbon and inorganic matter.

charcoal *(Med.)*. *Activated charcoal* has value as an absorbent, especially in cases of alkaloid poisoning, and as a palliative for flatulent dyspepsia. Used also to adsorb toxic gases in masks.

charcoal blacking *(Foundry.)*. *Blacking* made from powdered charcoal, which is dusted over the surface of a mould to improve the smoothness of the casting in fine work.

charcoal iron *(Eng.)*. Pig-iron made in a blast-furnace using charcoal instead of coke. Sometimes also wrought-iron made from this. Obsolete.

charge *(Chem.)*. Electrical energy stored in chemical form in secondary cell.

charge *(Eng.)*. Amounts (ore, fuel and flux) charged into furnace at one loading, or proportions of these.

charge *(Glass.)*. (1) Synonym of *batch*. (2) The quantity of batch required to fill a *pot*.

charge *(Min.Ext.)*. Operation of placing explosive, primer, detonator and fuse in drill hole.

charge *(Nuc.Eng.)*. Fuel material in nuclear reactor.

charge *(Phys.)*. Quantity of unbalanced electricity in a body, i.e. excess or deficiency of electrons, giving the body negative or positive electrification, respectively. See **negative** and **positive**. That of an ion, one or more times that of an electron, of either sign.

charge-coupled device *(Electronics)*. A semiconductor device which relies on the short-term storage of minority carriers in spatially defined depletion zones on its surface. The charges thus stored can be moved about by the application of control voltages via metallic conductors to the storage points, in the manner of a shift register. Abbrev. *CCD*.

charged *(Phys.)*. Said of a capacitor when a working potential difference is applied to its electrodes. Said of a secondary cell or battery (accumulator) when it stores the maximum (rated) energy in chemical form, after passing the necessary ampere-hours of charge through it. Said of a conductor when it is held at an operating potential, e.g. traction conductors or rails, or mains generally.

charged-coupled device *(Comp.)*. Also *CCD*. A memory unit in which packets of electrically charged particles circulate continuously through cells printed onto a semiconductor. Single **chip** memories with a very large storage capacity can be made from CCDs.

charge (-discharge) machine *(Nuc.Eng.)*. A device for inserting or removing fuel in a nuclear reactor without allowing escape of radiation and, in some reactors, without shutting the reactor down. Also called *(Re)fuelling machine*.

charged system *(Textiles)*. Used in dry-cleaning. By adding a surface active agent a clear dispersion may be obtained when water is added to the organic liquid. The mixture has enhanced cleaning properties.

charge exchange *(Phys.)*. Exchange of charge between neutral atom and ion in a plasma. After its charge is neutralized, high-energy ions can normally excape from plasma – hence this process reduces plasma temperature.

charge face *(Nuc.Eng.)*. In a reactor, that face of the biological shield through which fuel is loaded.

charge independent *(Phys.)*. Said of nuclear forces between particles, the magnitude and sign of which do not depend on whether the particles are charged. See **nuclear force** and **short-range forces**.

charge indicator *(Elec.Eng.)*. See **potential indicator**.

charge-mass ratio *(Phys.)*. Ratio of electric charge to mass of particle, of great importance in physics of all particles and ions.

charging current *(Elec.Eng.)*. (1) Current passing through an accumulator during the conversion of electrical energy into stored chemical energy. (2) Impulse of current passing into a capacitor when a steady voltage is suddenly applied, the actual current being limited by the total resistance of the circuit. (3) Alternating current which flows through a capacitor when an alternating voltage is applied.

charging resistor *(Elec.Eng.)*. A resistance inserted in series with a switch to limit the rate of rise of current when making the circuit.

charging voltage *(Elec.Eng.)*. E.m.f. required to pass the correct charging current through an accumulator, about 2.5 V for each lead-acid cell.

Charles's law *(Phys.)*. The volume of a given mass of gas at constant pressure is directly proportional to the absolute thermodynamic temperature. This is also known as Gay-Lussac's law and is the equivalent of saying that all gases have the same coefficient of expansion at constant pressure. This is approximately true at low pressures and sufficiently high temperatures as the *ideal gas* behaviour is approached.

charm *(Phys.)*. A property characterizing *quarks* and so hadrons. The charm of leptons and gauge bosons is zero. Charm is conserved in strong and electromagnetic interactions between particles but not in weak interactions.

charnockite *(Geol.)*. A coarse-grained, dark granite rock, consisting of feldspars, blue quartz, and orthopyroxene; it occurs typically in Madras and is named after the founder of Calcutta, Job Charnock.

Charnoid direction *(Geol.)*. A NW-SE direction of folding in the rocks of central and eastern England. Exemplified by the Precambrian rocks of Charnwood Forest.

Charon *(Astron.)*. Pluto's only known satellite, of uncertain diameter, discovered photographically in 1978.

Charophyceae *(Bot.)*. Class of the green algae (Chlorophyta) characterized by motile cells (if produced) scaly or naked and asymmetric, with 2 flagella inserted laterally or subapically and associated and with a **multilayered structure**; no phycoplast; haplontic, the zygote being a resting stage. Includes unicellular, sarcinoid, filamentous and parenchymatous sorts. Predominantly freshwater. Examples include; *Klebsormidium*, *Spirogyra*, the **desmids**, *Coleochaeta* and *Chara*. (In earlier classification only the Charales were placed in the Charophyceae.)

Charpy test *(Eng.)*. A notched-bar impact test in which a notched specimen, fixed at both ends, is struck behind the notch by a striker carried on a pendulum. The energy absorbed in fracture is measured by the height to which the pendulum rises.

chart comparison unit *(Radar.)*. Type of display superimposed upon navigational chart.

Chartered Engineer *(Eng.)*. Abbrev. C.Eng. A style or title (U.K) which can only be used by persons registered with the Engineering Council, a body set up by Royal Charter in 1981.

Chartered Surveyor *(Build.)*. A style or title which may be used by a fellow or professional associate of the chartered surveyors professional organisation.

Chart-Pak *(Print.)*. Rules and border patterns on self-adhesive material.

chase *(Build.)*. (1) A trench dug to accommodate a drain-pipe. (2) A groove chiselled in the face of a wall to receive pipes, etc.

chase *(Print.)*. An iron or steel frame into which type is locked by means of wooden or mechanical wedges.

chase-mortising *(Build.)*. A method adopted to frame a timber in between two others already fixed, a sloping chase being cut to the bottom of the mortise so that the cross-piece may be got into position.

chaser *(Eng.)*. A cutter of which the edge is serrated to the profile of a screw thread, used for producing or accurately finishing screw threads. Single chasers may be mounted in lathe tool-posts; sets are used in die-heads, collapsible taps and other screw-cutting tools.

chasmocleistogamous *(Bot.)*. Producing both chasmogamous and cleistogamous flowers.

chasmogamous *(Bot.)*. Flowers which open normally to expose the reproductive organs. Cf. **cleistogamous**.

chassis *(Aero.,Electronics)*. Rigid base on which electronic units or other electrical components are mounted.

chassis *(Autos.)*. The main frame of a vehicle (as distinct from the body) to which the engine, steering and transmission gear, wheels etc. are attached. Vehicles in which a suitably reinforced steel body fulfills this function are sometimes called *chassis-less, monocoque* or *unitized*.

chassis *(Eng.)*. Generally, any major part or framework of an assembly, to which other parts are attached.

Chastek paralysis *(Vet.)*. *Paralytic disease of silver foxes*. A disease of fox, mink, and ferrets characterized by paralysis, convulsions and death, due to thiamin (*vitamin* B_1) deficiency. The deficiency is caused by feeding on raw fish, parts of which contain an anti-thiamin factor.

chatoyancy *(Min.)*. The characteristic optical effect shown by cat's eye and certain other minerals, due to the reflection of light from minute aligned tubular channels, fibres, or colloidal particles. When cut *en cabochon* such stones exhibit a narrow silvery band of light which changes its position as the gem is turned.

chatter *(Eng.)*. Vibration of a cutting tool or of the work in a machine; caused by insufficient rigidity of either, and results in noise and uneven finish.

chatter *(Telecomm.)*. Undesired rapid closing and opening of contacts on a relay, reducing their life and making switching uncertain.

Chatterton's compound *(Elec.Eng.)*. An adhesive insulating substance consisting largely of guttapercha; used as a cement or filling, especially in cable jointing.

Chebyshev filter *(Telecomm.)*. A *constant-k* filter in which a very rapid *cut-off*, or increase in attenuation with frequency is achieved at the expense of the evenness of the *insertion loss* within the passband. Named after a Russian mathematician, and can have various anglicized spellings. Cf. **Butterworth filter**.

Chebyshev inequality *(Maths.)*. If $a \geqq a_{r+1}$ and $b_r \geqq b_{r+1}$ for all r, then

$$\left(\sum_1^n a_r\right)\cdot\left(\sum_1^n b_r\right) < n \sum_1^n a_r b_r.$$

Chebyshev polynomials *(Maths.)*. The polynomials $T_n(x)$ in the expansion

$$\frac{1-h^2}{1-2xh+h^2} = \sum_{n=0}^{\infty} T_n(x).(2h)^n.$$

check digit *(Comp.)*. Redundant bit in stored word, used in self-checking procedure such as a **parity check**. If there is more than one check digit they form a *check number*.

checking *(Build.)*. A not very serious fault developed by paintwork to relieve minor stresses in the film, characterised by a network of fine cracks that does not usually go down very deep.

check-lock *(Build.)*. A device for locking in position the bolt of a door lock.

check nut *(Eng.)*. See **lock nut**.

check rail *(Civ.Eng.)*. A third rail laid on a curve alongside the inner rail and spaced a little from it, to safeguard rolling-stock against derailment due to excessive thrust on the outer rail. Also called *guard rail, rail guard, safeguard, safety rail, side rail*.

check receiver *(Telecomm.)*. Radio or television receiver for verifying quality or content of a programme. Also called *monitoring receiver*.

check throat *(Build.)*. A small groove cut in the face of a short step in the upper surface of a wooden window-sill, just behind the face of the sash. It serves to stop rain from driving up under the sash. Also *anti-capillary groove*.

check valve *(Eng.)*. A nonreturn valve, closed automatically by fluid pressure; fitted in a pipe to prevent return flow of the fluid pumped through it. See **clack, feed check valve**.

cheddite *(Chem.)*. Mixture of castor oil, ammonium perchlorate, and 2,4-dinitrotoluene; used as an explosive.
cheek *(Build.)*. One of the solid parts on each side of a mortise, or the removed side of a tenon.
cheek *(Zool.)*. In Trilobita, the pleural portion of the head; in Mammals, the side of the face below the eye, the fleshy lateral wall of the buccal cavity.
cheese *(Elec.Eng.)*. See **cable wax**.
cheese *(Textiles)*. Densely-wound cylindrical package of yarn made on various winding machines for warping and other purposes.
cheese cloth *(Textiles)*. A light-weight open cotton fabric used in cheese manufacture but sometimes adapted for shirts and blouses.
cheese-head screw *(Eng.)*. A screw with a cylindrical head, similar in shape to a round cheese, slotted for a screw-driver.
cheilectropion *(Med.)*. Turning outwards of the lip.
cheilitis *(Med.)*. Inflammation of the lip.
cheilognathus *(Med.)*. See **harelip**.
cheiropompholyx *(Med.)*. A skin disease in which vesicles filled with clear fluid suddenly appear on the hands and feet.
cheiropraxis *(Med.)*. A system of healing or alternative medicine, based on manipulation of the spinal column with the intention of restoring normal function to nerves.
chela *(Zool.)*. In Arthropoda, any chelate appendage. adjs. *cheliferous, cheliform*.
chelate *(Zool.)*. Of Arthropoda, having the penultimate joint of an appendage enlarged and modified so that it can be opposed to the distal joint like the blades of a pair of scissors to form a prehensile organ. Cf. **subchelate**.
chelating agent *(Med.)*. A chemical agent which combines with unwanted metal ions. Used to treat heavy metal poisoning, thus sodium calcium EDTA is a chelating agent promoting the excretion of lead.
chelicerae *(Zool.)*. In Arachnida, a pair of pre-oral appendages, which are usually chelate.
Chelicerata *(Zool.)*. Arthropod sub-phylum comprising animals with two major body regions, an anterior prosoma and a posterior ophistosoma and with the foremost appendages bearing chelae. Spiders, Scorpions, Horseshoe Crabs.
Chelonia *(Zool.)*. An order of Reptiles in which the body is encased in a horny capsule consisting of a dorsal carapace and a ventral plastron, the jaws are provided with horny beaks in place of teeth, and the lower temporal arcade alone is present. Tortoises and Turtles.
cheluviation *(Bot.)*. Leaching of iron and aluminium oxides from soils after the formation of soluble complexes with, especially, polyphenols from fresh litter of conifers or heath plants. See **podsol**.
chemautotroph *(Biol.)*. An organism which derives energy from the oxidation of inorganic compounds for the assimilation of simple materials, e.g. carbon dioxide and ammonia.
chemical affinity *(Chem.)*. See **affinity**.
chemical balance *(Chem.)*. An instrument used in chemistry for weighing, to a high degree of accuracy, the small amounts of material dealt with.
chemical binding effect *(Phys.)*. A variation in the cross-section of a nucleus for neutron bombardment depending on how the element is combined with others in a chemical compound.
chemical bond *(Chem.)*. The electric forces linking atoms in molecules or non-molecular solid phases. Three basic types of bond are usually distinguished: (a) Ionic or electrostatic bonding, in which valence electrons are lost or gained, and atoms which are oppositely charged are held together by coulombic forces. (b) Covalent bonding, in which valence electrons are associated with two nuclei, the resulting bond being described as polar if the atoms are of differing electronegativity. (c) Metallic bonding, in which valence electrons are shared over many nuclei, and electronic conduction occurs.
chemical closet *(Build.)*. A suitably shaped container for use in conjunction with special deodorizing and liquefying chemicals, when running water is not available.
chemical compound *(Chem.)*. A substance composed of two or more elements in definite proportions by weight, which are independent of its mode of preparation. Thus the ratio of oxygen to carbon in pure carbon monoxide is the same whether the gas is obtained by the oxidation of carbon, the reduction of carbon dioxide or by heating methanoic acid, ethandioic acid or potassium hexacyanoferrate (II) with concentrated sulphuric acid.
chemical constitution *(Chem.)*. See **constitution**.
chemical elements *(Chem.)*. See **Table of Chemical Elements** in Appendix.
chemical energy *(Chem.)*. The energy liberated in a chemical reaction. See **affinity**.
chemical engineering *(Genrl.)*. Design, construction and operation of plant and works in which matter undergoes change of state and composition.
chemical equation *(Chem.)*. A quantitative symbolic representation of the changes occurring in a chemical reaction, based on the requirement that matter is neither added nor removed during the reaction.
chemical finishing *(Textiles)*. The final preparation of bleached fabric during which the material is subjected to the action of various chemicals including resin forming compounds.
chemical fog *(Image Tech.)*. Overall **fog** in a photographic image caused by excessive development or chemical contamination or decomposition.
chemical hygrometer *(Meteor.)*. See **absorption hygrometer**.
chemical kinetics *(Chem.)*. The study of the rates of chemical reactions. See **order of reaction**.
chemical lead *(Eng.)*. Lead of purity exceeding 99.9%; suitable for the lining of vessels used to hold sulphuric acid and other chemicals.
chemically-formed rocks *(Geol.)*. Rocks formed by precipitation of materials from solution in water, e.g. calc-tufa, and various saline deposits.
chemical oxygen demand *(Chem.)*. A measure of the amount of potassium dichromate needed to oxidize reducing material in a water sample. It is expressed as the concentration of oxygen gas, in parts per million, chemically equivalent to the amount of dichromate consumed. It is generally higher than the **biological oxygen demand**, as some materials, such as cellulose, will react with dichromate but not with oxygen under biological conditions. Abbrev. *COD*.
chemical precipitation *(Build.)*. The process of assisting settlement of the solid matters in sewage by adding chemicals before admitting the sewage to the sedimentation tanks.
chemical pulp *(Paper)*. Pulp in which the fibres have been resolved by chemical, as distinct from mechanical means, with the removal of the greater part of the lignin and other non-cellulose material.
chemical reaction *(Chem.)*. A process in which at least one substance is changed into another.
chemical shift *(Phys.)*. A shift in position of a spectrum peak due to a small change in chemical environment. Observed in the *Mössbauer effect* and in *nuclear magnetic resonance*.
chemical symbol *(Chem.)*. A single capital letter, or a combination of a capital letter and a small one, which is used to represent either an atom or a mole of a chemical element; e.g. the symbol for sodium is Na, for sulphur is S etc.
chemical toning *(Image Tech.)*. The process of converting the silver image into, or replacing it by, a coloured substance by chemicals other than a dye.
chemical wood-pulp *(Paper)*. Pulp obtained from wood by the sulphite, sulphate, soda or other chemical process.
chemiluminescence *(Chem.)*. A process in which visible light is produced by a chemical reaction.
chemiosmosis *(Biol.)*. The mechanism underlying the formation of ATP synthesis by oxidative phosphorylation. The energy for ATP synthesis is derived from electrochemical gradients across the inner membrane of the mitochondrion. A similar mechanism operates during **photophosphorylation**.

chemisorption *(Chem.)*. Irreversible adsorption in which the absorbed surface is held on the substance by chemical forces.

chemistry *(Chem.)*. The study of the composition of substances and of the changes of composition which they undergo. The main branches of the subject are often considered to be **inorganic**, **organic** and **physical chemistry**.

chemokinesis *(Immun.)*. Random movement of cells such as leucocytes stimulated by substances in the environment.

chemonasty *(Bot.)*. A plant movement provoked, but not orientated, (see **nasty**) by a chemical stimulus.

chemoreceptor *(Zool.)*. A sensory nerve-ending, receiving chemical stimuli.

chemoreflex *(Med.)*. A reflex initiated by a chemical stimulus.

chemosis *(Med.)*. Oedema of the conjunctiva.

chemosphere *(Astron.)*. The atmospheric layers between 20 and 200 km.

chemostat *(Bot.)*. A culture vessel in which steady state growth is maintained by appropriate rates of harvest and the addition of the ingredients of the medium.

chemosynthesis *(Bot.)*. The use, as by some bacterium, of energy derived from chemical reactions (e.g. oxidation of sulphur or of ammonia) in the synthesis from inorganic molecules of their organic requirements. Cf. **photosynthesis**.

chemosynthetic autotroph *(Biol.)*. See **chemautotroph**.

chemotaxis *(Biol.,Immun.)*. Stimulation of movement by a cell or organism towards or away from substances producing a concentration gradient in the environment. E.g. the C5a complement peptide and leukotriene cause granulocytes to move into sites where they are released.

chemotaxomony *(Bot.)*. The use of chemical evidence (of both primary and secondary metabolism) in taxonomy.

chemotherapy *(Med.)*. Treatment of disease by chemical compounds selectively directed against invading organisms or abnormal cells.

chemotropism *(Bot.,Zool.)*. A differential growth movement or curvature of part of an organ in a direction related to the concentration gradient of a chemical.

chenodeoxycholic acid *(Med.)*. Used in certain patients to dissolve gall stones as an alternative to surgery.

Chenopodiaceae *(Bot.)*. Family of ca 1500 spp. of dicotyledonous flowering plants (superorder Caryophyllidae). Mostly herbs, temperate and subtropical, mostly in saline habitats. Flowers inconspicuous and wind pollinated. Some are C_4 plants. Includes sugar beet, mangelwurzels, beetroot, leaf beets (all species of *Beta*), spinach and quinoa.

chequer plate *(Eng.)*. Steel plate used for flooring; provided with a raised chequer pattern to give a secure foothold.

cheralite *(Min.)*. A radioactive mineral rich in thorium.

Cherenkov. *(Phys.)*. See **Cerenkov**.

Chernikeef log *(Ships)*. A device for measuring distance moved through the water. An impeller is lowered through the bottom of the ship to about 40 cm (18 in) below the hull. The rotations of the impeller are transmitted electrically to a distance recorder and combined with time to indicate speed.

chernozem *(Ecol.)*. Grassland soil formed in sub-humid cool to temperate areas, with humus at the surface with a blackish layer of mineral soil just beneath which grades downward to a lighter layer where lime has accumulated. Occurs under tall-grass communities such as Russian Steppes, North American Prairies and Argentinian Pampas.

cherry *(For.)*. Tree of the genus, *Prunus*, economically not regarded as a timber tree, though the wood has certain minor uses. It carves and turns well, and is used for picture frames, inlaying, tobacco pipes and walking-sticks.

cherry-picker *(Civ.Eng.)*. A working platform mounted on a two-limbed hydraulically or electrically operated raising mechanism, on the back of a truck; used for access to elevated inaccessible places, e.g. street lighting fittings.

chert *(Geol.)*. A siliceous rock consisting of cryptocrystalline silica, and sometimes including the remains of siliceous organisms such as sponges or radiolaria. It occurs as bedded masses, as well as concretions in limestone.

chestnut *(For.)*. A light- to dark-brown wood resembling oak; much used for fencing, posts and rails.

cheval-vapeur (CV), Pferdestärke (PS) *(Eng.)*. The metric unit of horse-power, equivalent to 75 kgf m/s, 735.5 W or 0.986 hp. Also *cheval* (ch), *force de cheval*.

Cheyne-Stokes breathing *(Med.)*. Waxing and waning or respiration in disease of the heart, kidney, or brain.

χ *(Phys.)*. Symbol for magnetic susceptibility.

chiasma *(Biol.)*. Point of contact between **chromatids** visible during meiosis and involved in **crossing over**. pl. *chiasmata*.

chiasma *(Zool.)*. A structure in the central nervous system, formed by the crossing over of the fibres from the right side to the left side and vice versa.

chiastolite *(Min.)*. A variety of andalusite characterized by cruciform inclusions of carbonaceous matter.

chickenpox *(Med.)*. *Varicella*. A common mild acute infectious disease in which papules, vesicles, and small pustules appear in successive crops, mainly on the trunk, face, upper arms, and thighs.

chicken wire *(Build.)*. Light-gauge wire netting.

chiffon *(Textiles)*. A very fine, soft woven dress material of silk or man-made fibres; the word is also used in other cloth descriptions to indicate lightness, e.g. chiffon velvet, chiffon taffeta.

chilblains *(Med.)*. See **erythema pernio**.

childhood psychosis *(Behav.)*. A group of childhood disorders characterized by disturbed social relationships, speech impairment, and bizarre motor behaviour. Three sub-groups are recognised: infant **autism**, late onset psychosis, e.g. **childhood schizophrenia** and disorders due to degeneration of the central nervous system.

childhood schizophrenia *(Behav.)*. A childhood disorder which manifests itself after a period of normal development, often in adolescence, when the child begins to show severe disturbances in social adjustment and reality contact. As a diagnostic category it is not distinguished from adult **schizophrenia**.

Child-Langmuir equation *(Electronics)*. In a thermionic diode, where the anode current is limited by the space-charge surrounding the cathode, this equation states that this current is proportional to the three-halves power of the anode-cathode voltage. $I = GV^{3/2}$ where I is the current, V the voltage, and G a constant depending on the physical form of the electrodes.

Chile nitre, Chile saltpetre *(Chem.)*. A commercial name for sodium nitrate (V), $NaNO_3$.

chill *(Foundry)*. An iron mould, or part of a mould, sometimes watercooled; used to accelerate cooling and give great hardness and density to the metal which comes into contact during cooling.

chill crystals *(Eng.)*. Small crystals formed by the rapid freezing of molten metal when it comes into contact with the surface of a cold metal mould.

chilled iron *(Eng.)*. Cast-iron cast in moulds constructed wholly or partly of metal, so that the surface of the casting is white and hard while the interior is grey.

Chilognatha *(Zool.)*. See **Diplopoda**.

Chilopoda *(Zool.)*. Class of Arthropoda having the trunk composed of numerous somites each bearing one pair of legs; the head bears a pair of uniflagellate antennae, a pair of mandibles, and two pairs of maxillae; the first body somite bears a pair of poison-claws; the genital opening is posterior; active carnivorous forms, some of considerable size and dangerous to Man; some are phosphorescent. Centipedes.

chime barrel *(Horol.)*. The cylinder on whose periphery short vertical pins lift the hammers in a chiming clock, the pins being so arranged as to give the required sequence of notes.

chimera *(Biol.)*. (1) In animals, an individual exhibiting two or more different genotypes in patches derived from two or more different embryos which have become fused, naturally or artificially, at an early stage to make a single embryo. Cf. **mosaic**. (2) A DNA molecule with sequences from more then one organism. adj. *chimeric*.

chimera *(Bot.)*. A plant or plant part in which there are two (rarely more) genetically different sorts of cell, a result of mutation or of grafting. The name is not applied to ordinary grafted plants where the stock differs from the scion but rather where cells of one sort have come to form a layer over a core of cells of the other throughout the shoot. See also **periclinal chimera, mericlinal chimera, sectorial chimera**.

chimney *(Min.Ext.)*. Either an ore shoot or more usually an especially rich steeply inclined part of the lode.

chimney bar *(Build.)*. An iron bar supporting an arch over a fireplace opening.

chimney bond *(Build.)*. A **stretching bond** which is generally used for the internal division walls of domestic chimney-stacks, as well as for outer walls.

chimney-breast *(Build.)*. The part of the chimney between the flue and the room.

chimney jambs *(Build.)*. The upright sides of the fireplace opening.

chimney lining *(Build.)*. The tile flues, or cement rendering, within a chimney space.

chimney shaft *(Build.)*. The part of a chimney projecting above the roof, or a chimney standing isolated like a factory chimney.

chimney stack *(Build.)*. The unit containing a number of flues together.

china clay *(Geol.)*. A clay consisting mainly of **kaolinite**, one of the most important raw materials of the ceramic industry. China clay is obtained from kaolinized granite, for example in S.W.England, and is separated from the other constituents of the granite (quartz and mica) by washing out with high pressure jets of water. Also called **kaolin, porcelain clay**.

china clay *(Med.)*. *Kaolin*. Used as an internal absorbent (e.g. of poisons), for poultices, dusting powders etc.

china stone *(Geol.)*. A kaolinitized granitic rock containing unaltered plagioclase. Also applied to certain limestones of exceptionally fine grain and smooth texture.

chine *(Aero.)*. The extreme outside longitudinal member of the planing bottom of a flying-boat hull, or of a seaplane float; runs approximately parallel with the keel. Also used on sharp fuselage edge of supersonic aircraft, e.g. SR 71.

Chinese binary *(Comp.)*. A form of coding for punched cards in which the card is read in columns rather than in rows.

chintz *(Textiles)*. A plain-woven cotton fabric that has been printed and then glazed by calendering. It may also be stiffened by the addition of starch.

chip *(Comp.)*. Popular name for an **integrated circuit**. The term derives from the method of manufacture, as each chip is made as part of a *wafer*, a flat sheet of silicon, impregnated with impurities in a pattern to form an array of transistors and resistors with the electrical interconnections made by depositing thin layers of gold or aluminium. Many copies of the integrated circuit are formed simultaneously and then have to be broken apart.

chip *(Eng.)*. See **swarf**.

chip-axe *(Build.)*. A small single-handed tool for chipping timber to rough size.

chip-board *(Build.)*. Boards made by glueing together chips of wood and pressed and cured to a given thickness. May have veneered faces. Also *particle board*.

chip-board *(Paper)*. A board, usually made from waste paper, used in box making.

chip breaker *(Eng.)*. A groove, step, or similar feature in one face of a (lathe) tool or cutter, designed to break long chips, or swarf, into small pieces which will more easily clear the tool and can be conveniently disposed of.

chip detector *(Aero.)*. Magnetic device fitted in the lubrication system of aero-engines to collect chips of steel from worn or broken parts.

chip log *(Ships)*. Quadranted piece of wood, weighted round its edge, attached to the log line. See **nautical log**.

chipping *(Eng.)*. The removing of surface defects from semifinished metal products by pneumatic chisels. Seams, laps and rokes are thus eliminated.

chipping chisel *(Build)*. See **cold chisel**.

chirality *(Chem.)*. Absence of symmetry involving reflection or inversion. A molecule is chiral if no stable conformation of it may be superimposed on its mirror image. Most chiral organic molecules can be described in terms of so-called chiral centres, in which an atom (usually carbon) has four distinct substituents. The chirality is specified in general by the **Cahn-Ingold-Prelog system**.

chirality *(Maths.)*. The property of being right-handed or left-handed, usually applied to co-ordinate axes.

chiropody *(Genrl.)*. The care and treatment of minor ailments of the feet.

Chiroptera *(Zool.)*. An order of aerial Mammals having the forelimbs specially modified for flight; mainly insectivorous or frugivorous nocturnal forms. Bats.

chiropterophilous *(Bot.)*. Pollinated by bats.

chirp radar *(Radar)*. A radar system using linear frequency-modulated pulses.

chisel *(Build)*. Steel tool for cutting wood, metal or stone; it consists of a shank whose end is bevelled to a cutting edge.

chiselled (boasted) ashlar *(Build.)*. A random tooled **ashlar** finished with a narrow chisel.

chi-squared distribution *(Stats.)*. The distribution of many quadratic forms in statistics, often encountered as the distribution of the sample variance and of a statistic measuring the agreement of a set of empirically observed frequencies with theoretically derived frequencies. The central chi-squared distribution is indexed by one parameter, the degrees of freedom.

chitin *(Zool.)*. A nitrogenous polysaccharide with the formula $(C_8H_{13}N_5)_n$ occurring as skeletal material in many Invertebrates particularly the cuticle of Arthropoda.

Chladni figure *(Acous.)*. The visual pattern produced by a fine powder on a vibrating plate. The powder particles accumulate at the nodes where the plate does not move.

Chlamydobacteriales *(Biol.)*. Chemosynthetic bacteria-like organisms, probably closely related to the true bacteria: filamentous: characterized by the deposition of ferric hydroxide in or on their sheaths. Occur in fresh water, particularly moorland bogs, where their oxidation of iron (II) to iron (III) results in the deposition of **bog iron ore**.

chlamydospore *(Bot.)*. A hyphal cell that becomes thick-walled, separates from the parent mycelium and functions as a spore.

chloanthite *(Min.)*. Nickel arsenide occurring in the cubic system. A valuable nickel ore, associated with smaltite.

chloragen, chloragogen cells *(Zool.)*. In Oligochaeta, yellowish flattened cells occurring on the outside of the alimentary canal, and concerned with nitrogenous excretion.

chloral *(Chem.)*. *Trichloroethanal*, $CCl_3.CHO$, bp97°C, a viscous liquid, of characteristic odour, obtained by the action of chlorine upon ethanol and subsequent distillation over sulphuric acid.

chloral hydrate *(Chem.)*. *Trichloroethanal hydrate*. $CCl_3.CH(OH)_2$, mp 57°C. bp (with decomposition) 97°C, large colourless crystals, soluble in water, a hypnotic and sedative. Obtained from **chloral** and water. One of the few compounds having 2 hydroxy groups attached to the same carbon atom.

chloramines *(Chem.)*. Compounds obtained by the action of hypochlorite solutions on compounds containing the NH or NH_2 groups. Important as disinfectants. Chloramine T, $CH_3.C_6H_4.SO_2.NClNa$, is the active constituent of an ointment employed as an antidote against *vesicant* war gases.

chloramphenicol *(Med.)*. Antibiotic from some species of *Streptomyces*; also obtained synthetically. Of therapeutic value for the treatment of e.g. typhoid, scrub-typhus, typhus, psittacosis.

chlorapatite *(Min.)*. Chlorophosphate of calcium, a

member of the apatite group of minerals. See **apatite, fluorapatite**.

chlorastrolite *(Min.)*. A fibrous green variety of *pumpellyite*; it occurs in rounded geodes in basic igneous rocks near Lake Superior. When cut *en cabochon*, it exhibits **chatoyancy**.

chlorates (V) *(Chem.)*. Salts of chloric acid. Powerful oxidizing agents. Explosive when ground or detonated in contact with organic matter.

chlorazide *(Chem.)*. N_3Cl; a colourless, highly explosive gas formed by the reaction between sodium chlorate (I) and sodium azide.

chlorazine *(Chem.)*. *2-Chloro-4,6-bisdiethylamino-1,3,5-triazine*, used as a herbicide.

chlordiazepoxide *(Med.)*. A mild tranquillizing drug. TN *Librium*.

Chlorella *(Bot.)*. Microscopic unicellular green algae (Chlorophyceae, Chlorococcales) reproducing by autospores, no sexual reproduction; easily grown in laboratory culture and used in biochemical studies.

chlorenchyma *(Bot.)*. Tissue composed of cells containing chloroplasts e.g. leaf mesophyll.

chlorhexidine *(Med.)*. A disinfectant used as a gluconate solution for skin cleansing, in creams, lozenges.

chloric (V) acid *(Chem.)*. $HClO_3$. A monobasic acid forming a series of salts, chlorates – where ClO_3 acts as an univalent radical.

chloride of lime *(Chem.)*. See **bleaching powder**.

chloride of silver cell *(Elec.Eng.)*. A primary cell having electrodes of zinc and silver and a depolarizer of silver chloride.

chlorides *(Chem.)*. Salts of hydrochloric acid. Many metals combine directly with chlorine to form chlorides.

chloridizing roasting *(Eng.)*. The roasting of sulphide ores and concentrates, mixed with sodium chloride, to convert the sulphides to chlorides.

chlorinated rubber paints *(Build.)*. Paints possessing good resistance to chemicals, acids, alkalis, moisture penetration and mould growths. Being derived from rubber they are extremely flexible. They have many applications in industrial, manufacturing and laboratory situations but their poor adhesion to bare metal means that a recommended primer must be used.

chlorination *(Build.)*. The addition of chlorine to sewage as a bactericide in the land treatment method.

chlorination *(Chem.)*. (1) The substitution or addition of chlorine in organic compounds. (2) The sterilization of water with chlorine, sodium hypochlorite, chloramine or bleaching powder.

chlorine *(Chem.)*. Element, symbol Cl, at.no. 17, r.a.m. 35.453, valencies 1−, 3+, 5+, 7+, mp −101.6°C, bp −34.6°C. The second halogen, chlorine is a greenish yellow gas, with an irritating smell and a destructive effect on the respiratory tract. Produced by the electrolysis of conc. brine, or by oxidation of hydrochloric acid. It is a powerful oxidizing agent, and is widely used, both pure and as *bleaching powder*, for bleaching and disinfecting. Chlorine is used in the organic chemicals industry to produce tetrachloromethane, trichloromethane, PVC, and many other solvents, plastics, disinfectants and anaesthetics.

chlorine number *(Paper)*. An assessment of the bleachability of a pulp by measuring its chlorine demand under specified conditions of test.

chlorine oxides *(Chem.)*. The (I) *oxide* (Cl_2O) and (IV) *oxide* (ClO_2) are gases similar to chlorine. The (I) oxide dissolves in water to give hypochlorous acid. The (IV) oxide is used as a bleach. The (VI) *oxide* (Cl_2O_6) and (VII) *oxide* (Cl_2O_7) are liquids. All are highly unstable, often explosive.

chlorite *(Min.)*. A group of minerals, typically green, somewhat resembling the micas, composed of hydrated magnesiun, iron and aluminosilicates. They occur as alteration products of igneous rocks, chlorite schists and in sediments.

chloritization *(Geol.)*. The replacement, by alteration, of ferro-magnesian minerals by chlorite.

chloritoid *(Min.)*. Hydrated iron, magnesium, aluminium silicate, crystallizing in the monoclinic or triclinic systems. Characteristic of low and medium grade regionally metamorphosed sedimentary rocks.

chloroacetic (ethanoic) acids *(Chem.)*. *Monochloroacetic acid*, $CH_2Cl.COOH$, mp 63°C, bp 189°C; formed by chlorination of acetic acid in the presence of acetic anhydride, sulphur or phosphorus. *Dichloroacetic acid*, $CHCl_2.COOH$, bp 191°C, formed by heating chloral hydrate with potassium cyanide. *Trichloroacetic acid*, $CCl_3.COOH$, mp 56°C, bp 198°C; formed by oxidizing chloral hydrate with nitric acid.

chloroauric (III) acid *(Chem.)*. $HAuCl_4$. A complex acid formed when gold (III) oxide (Au_2O_3) dissolves in hydrochloric acid. Forms a series of complex salts called *chloroaurates* (III).

chlorobutadiene (2-) *(Chem.)*. See **chloroprene**.

Chlorococcales *(Bot.)*. An order of the Chlorophyceae; the members are unicellular, but may form colonies of uninucleate, or multinucleate, cells, which never divide vegetatively; asexual reproduction by zoospores or autospores, and sexual reproduction by biflagellate gametes.

chlorocruorin *(Zool.)*. A green respiratory pigment of certain *Polychaeta*. Conjugated protein containing a prosthetic group similar to, but not identical with, reduced haematin.

chlorofibre *(Textiles)*. Fibres made from copolymers with vinyl chloride or vinylidene chloride predominating. Other constituent monomers include acrylonitrile. The material has markedly hydrophobic properties.

chloroform *(Chem.)*. *Trichloromethane*. $CHCl_3$, bp 62°C, rel.d. 1.49; a colourless liquid of a peculiar etherial odour, an anaesthetic, solvent for oils, resins, rubber and numerous other substances. It is prepared technically from ethanol and calcium chlorate (I).

chlorohydrins *(Chem.)*. A group of compounds with -Cl and -OH groups on adjacent carbon atoms, e.g. *ethylene chlorohydrin* or *1-hydroxy-2-chloroethane*, $CH_2Cl.$-CH_2OH.

chlorophaeite *(Min.)*. A mineral closely related to chlorite, dark green when fresh, but rapidly changing to brown, hence the names (Gk. *chloros*, yellowish-green; *phaios*, dun). Described from basic igneous rocks.

chlorophenol red *(Chem.)*. *Dichlorophenolsulphonphthalein*, an indicator used in acid-alkali titration, having a pH range of 4.8 to 6.4, over which it changes from yellow to red.

Chlorophyceae *(Bot.)*. Class of the green algae (Chlorophyta) characterized by: motile cells (if produced) radially symetrical with 2, 4 or many flagella inserted apically and with four, cruciate flagellar roots; with a phycoplast; haplontic, the zygote being a resting stage. Predominantly fresh water. Includes flagellate, coccoid motile and non-motile colonial (**coenobia**), sarcinoid, filamentous and parenchymatous sorts. Examples include; *Chlamydomonas*, *Volvox*, *Chlorella*, *Oedogonium*. See also **Chlorococcales**.

chlorophylls *(Bot.)*. Green pigments (variously substituted porphyrin rings with magnesium) involved in photosynthesis. Chlorophyll a is the primary photosynthetic pigment in all those organisms that release oxygen i.e. all plants and all algae including the blue-green algae. The other chlorophylls are *accessory pigments*; chlorophyll b in vascular plants, bryophytes and green algae; chlorophyll c in brown algae, diatoms, chrysophytes etc.

Chlorophyta *(Bot.)*. The green algae, a division of eukaryotic algae characterized by: chlorophyll a and b; irregularly stacked thylakoids; no chloroplastal endoplasmic reticulum; mitochondrial cristae; flagella not heterokont; storing starch in the chloroplasts. Contains the classes Chlorophyceae, Ulvophyceae and Charophyceae, and is obviously ancestral to land plants.

chloroplast *(Bot.)*. *Plastid*, one or more in a cell, containing the membranes, pigments and enzymes necessary for photosynthesis in a eukaryotic alga (of any colour) or green plant. See also **chloroplast-DNA, chloroplast ER, thylakoid, photosynthetic pigments, accessory pigments**.

chloroplast DNA *(Bot.)*. DNA present in chloroplasts, organized like that of prokaryotes and coding for some but not all chloroplast proteins.

chloroplast ER *(Bot.)*. An extension of the *endoplasmic reticulum* which encloses the chloroplast in some groups of eukaryotic algae e.g. the Heterokontophyta.

chloroplatinic (IV) acid *(Chem.)*. $H_2PtCl_6.6H_2O$. Formed when platinum is crystallized from a solution acidified with hydrochloric acid.

chloroprene *(Chem.)*. *2-Chlorobutan 1,2 : 3,4-diene*. Starting point for the manufacture of **neoprene**.

chloroquine *(Med.)*. Widely used synthetic antimalarial drug.

chlorosis *(Bot.)*. Deficiency of chlorophyll in a normally green part of a plant so that it appears yellow-green, yellow or white, as a result of mineral deficiency, inadequate light or infection.

chlorothiazide *(Med.)*. One of the thiazide class of diuretics, used in cases of oedema, hypertension, cirrhosis of the liver.

Chloroxone *(Chem.)*. TN for 2,4-dichlorophenoxyacetic (ethanoic) acid, also known as *2,4-D*, a selective weedkiller.

chlorpromazine *(Med.)*. A widely used tranquillizing drug, of great value in mental and behavioural disorders, also used as an anti-emetic, analgesic adjuvant, and for preoperational premedication.

chlorpropamide *(Med.)*. An antidiabetic drug of the sulphonyl urea type, for maturity onset diabetes.

chlorthalidone *(Med.)*. Thiazide-related diuretic with a long duration of action. Used to treat oedema, hypertension and diabetes insipidus.

choana *(Zool.)*. A funnel-shaped aperture; pl. *choanae*, the internal nares of Vertebrates.

Choanichthyes *(Zool.)*. See **Sarcopterygii**.

choanocyte *(Zool.)*. In Porifera, a flagellate cell, in which a collar surrounds the base of the flagellum.

Chobert rivet *(Eng.)*. A hollow rivet, used where only one side of the work is accessible. It comprises a headed tube having a concentric taper of which the smallest diameter is at the shank end. Drawing a wedge-headed mandrel through the bore clinches the rivet by upsetting the shank end on the inaccessible side of the structure. The bore may subsequently be plugged by a parallel sealing pin which also increases the shear strength of the rivet.

chock-to-chock *(Aero.)*. See **block time**.

choice point *(Behav.)*. The position in a T-maze or other type of maze or in any apparatus involved in discrimination training, when an animal can make only one of two or more alternative responses.

choke *(Autos.)*. (1) The venturi or throat in the air-passage of a **carburettor**. (2) A butterfly valve in the carburettor intake which reduces the air supply and so gives a rich mixture for starting purposes. US *strangler*.

choke *(Elec.Eng.)*. Colloquialism for inductor. Usually refers to coils of high self-inductance used to limit flow of an a.c. without power dissipation. Any d.c. flowing through the choke will be impeded only by the intrinsic resistance of the winding.

choke *(Radar)*. In a waveguide, a groove or discontinuity of such a shape and size as to prevent the passage of guided waves within a limited frequency range.

choke coupling *(Elec.Eng.)*. (1) Use of the impedance of a choke for coupling the successive stages of a multistage amplifier. (2) In a waveguide, coupling flanges with quarter-wavelength groove which breaks surface continuity at current node.

choke damp *(Min.Ext.)*. A term sometimes used for **black damp** (carbon dioxide). More correctly, any mixture of gases which causes choking or suffocation.

choke feed *(Elec.Eng.)*. The use of a high inductance path for the d.c. component of current to active device in an electrical circuit (e.g. a transistor). The a.c. signal is fed through a capacitor so there is separation of a.c. and d.c. components.

choke flange *(Radar)*. Type of waveguide coupling which obviates the need for metallic contact between the mating flanges, and yet offers no obstruction to the guided waves. One of the waveguide flanges has a slot formed in it of dimensions which prevent energy leakage within the desired frequency range.

choke-jointed *(Elec.Eng.)*. See **joint**.

choke modulation *(Elec.Eng.)*. System of anode modulation in which the modulating and modulated valves are coupled by means of a high-inductance choke included in the common anode-voltage supply lead.

cholaemia, cholemia *(Med.)*. Presence of bile pigments in the blood.

cholagogue *(Med.)*. Increasing evacuation of bile; a drug which does this.

cholangitis, cholangeitis *(Med.)*. Inflammation of the bile duct system.

cholecystectomy *(Med.)*. Excision of the gall bladder.

cholecystenterostomy *(Med.)*. An artificial opening made between the gall bladder and the upper part of the intestine.

cholecystitis *(Med.)*. Inflammation of the gall bladder.

cholecystography *(Radiol.)*. The X-ray investigation of the gall bladder previously filled with a substance opaque to X-rays. A *cholecystogram* is the X-ray photogaph obtained.

cholecystokinin *(Med.)*. A hormone liberated from the duodenal mucosa, causing contraction of the gall bladder and relaxation of the *sphincter of Oddi*.

cholecystostomy *(Med.)*. The surgical formation of an opening in the wall of the gall bladder.

choledochotomy *(Med.)*. Incision into one of the main bile ducts.

cholelithiasis *(Med.)*. Stones in the gall bladder and bile passages.

cholera *(Med.)*. An acute bacterial infection by *Vibrio cholerae* in Eastern countries; characterized by severe vomiting and diarrhoea, drying of the tissues and painful cramps; spread by infected food and water.

cholesteatoma *(Med.)*. A tumour in the brain, or in the middle ear, composed of cells and crystals of cholesterol.

cholesteraemia, cholesteremia *(Med.)*. See **cholesterolaemia**.

cholesterol *(Chem.)*. $C_{27}H_{45}OH$, a sterol of the alicyclic series, found in nerve tissues, gall stones, and in other tissues of the body. It is a white crystalline solid, mp148.5°C, soluble in organic solvents and fats. There are numerous stereoisomers known. Parent compound for many steroids.

cholesterolaemia, cholesterolemia *(Med.)*. Excess of cholesterol in the blood.

cholesterosis *(Med.)*. Diffuse deposits of cholesterol in the lining membranes of the gall bladder.

cholic acid *(Chem.)*. $C_{23}H_{39}O_3.COOH$, the product of hydrolysis of certain bile acids, is conjugated in the body, forming glycocholic acid with glycine and taurocholic acid with taurine.

choline *(Chem.,Biol.)*. *Ethylol-trimethyl-ammonia hydrate*, $OH.CH_2.CH_2.NMe_3.OH$, a strong base, present in the bile, brain, yolk of egg etc., combined with fatty acids or with glyceryl-phosphoric acid (**lecithin**). It is concerned in regulating the deposition of fat in the liver, and its acetyl ester is an important neurotransmitter. See **acetyl choline**.

cholinergic *(Med.)*. Activated by acetylcholine, e.g. transmission at preganglionic synapses of the autonomic system and post-ganglionic endings of the parasympathetic and some sympathetic nerve fibres.

choluria *(Med.)*. Bile pigments in the urine.

chomophyte *(Bot.)*. A plant growing on rock ledges littered with detritus, or in fissures and crevices where root hold is obtainable.

chondral *(Biol.)*. Pertaining to cartilage.

chondr-, chondrio-, chondro- *(Genrl.)*. Prefix from Gk. *chondros*, cartilage, grain.

Chondricthyes *(Zool.)*. Class of cartilaginous Fishes in which the skeleton may be calcified but not ossified. Teeth not fused to the jaw but serially replaced; fertilization internal. Rays, Dogfish, Sharks and Chimeras. Cf. **Osteichthyes**.

chondrification *(Biol.)*. Strictly, the formation of chondrin; hence, the development of cartilage. Also *chondrogenesis*.
chondrin *(Biol.)*. A firm, elastic, translucent, bluish-white substance of a gelatinous nature, which forms the ground-substance of cartilage.
chondrite *(Min.)*. A type of stony meteorite containing *chondrules* — nodule-like aggregates of minerals.
chondroblast *(Biol.)*. A cartilage cell which secretes the chondrin matrix.
chondroclast *(Biol.)*. A cartilage cell which destroys the cartilage matrix.
chondrocranium *(Zool.)*. The primary cranium of Craniata, formed by the fusion of the parachordals, auditory capsules and trabeculae.
chondrodite *(Min.)*. Hydrated magnesium silicate, crystallizing in the monoclinic system, and occurring in metamorphosed limestones.
chondroids *(Vet.)*. Compact lumps of dried pus commonly found in the exudate of inflamed guttural pouches of the horse.
chondroma *(Med.)*. A tumour composed of cartilage cells.
chondrosamine *(Chem.)*. $C_6H_{13}O_5N$, or *2-aminogalactose*, is the basis of *chondroitin*, which is the substance of cartilage and similar body tissues.
chondrosarcoma *(Med.)*. A malignant tumour composed of sarcoma cells and cartilage.
chondroskeleton *(Zool.)*. The cartilaginous part of the Vertebrate skeleton.
chondrule *(Min.)*. See under **chondrite**.
chop *(Build.)*. Movable jaw of a bench vice.
chopped wave *(Elec.Eng.)*. A travelling voltage wave which rises to a maximum and then rapidly falls to zero. Such a wave occurs on transmission lines when an ordinary voltage wave has caused an insulator flashover, thereby losing its tail.
chopper *(Elec.Eng.)*. (1) Electronic inverter circuit which converts d.c. to a.c. It operates by rapidly switching off and on (chopping) a direct voltage source to produce a square or rectangular waveform which varies between two discrete values, zero and the direct supply level. (2) Light interrupter used to produce a.c. output from a photocell.
chopper *(Nuc.Eng.)*. Device consisting of a rotating mechanical shutter made of a sandwich of aluminium and cadmium sheets which provides bursts of neutrons for a **time-of-flight spectrometer**. See also **neutron velocity selector, neutron spectrometer**.
chopper amplifier *(Elec.Eng.)*. See **chopper-stabilized amplifier**.
chopper disk *(Elec.Eng.)*. Rotating toothed or perforated disk which interrupts the light signal falling on a photocell at a desired frequency.
chopper-stabilized amplifier *(Elec.Eng.)*. Low-drift d.c. amplifier in which a semiconductor switch is used to convert a d.c., or low frequency a.c. signal into an a.c. square wave which can be more readily amplified. A chopper circuit, operating in synchronism with the a.c. may be used at the amplifier output as a rectifier. An equivalent input drift voltage of around 0.1 μV per °C. can be obtained.
chops *(Horol.)*. A clamp. The two flat pieces of metal, usually of brass, between which the end of a pendulum suspension spring is secured.
chord *(Aero.)*. See **chord line**.
chord *(Eng.)*. See **boom**.
chord *(Maths.)*. A straight line drawn between two points on a curve.
chorda *(Zool.)*. Any stringlike structure, e.g. the *chordae tendineae*, tendinous chords attaching the valves of the heart; also, the notochord.
chordacentra *(Zool.)*. Vertebral centra formed from the notochordal sheaths; cf. **archecentra**. adj. *chordacentrous*.
chordal thickness *(Eng.)*. Of a gear-wheel tooth, the length of the chord subtended by the tooth thickness arc.
Chordata *(Zool.)*. A phylum of the Metazoa containing those animals possessing a notochord.
chord line *(Aero.)*. A straight line joining the centres of curvature of the leading and trailing edges of any aerofoil section.
chord of contact *(Maths.)*. Of the conic: the line joining the points of contact of the tangents from an external point.
chordoma *(Med.)*. An invasive tumour arising from the remains of the notochord in the skull and the spinal column.
chordotomy *(Med.)*. The cutting of the nerve fibres in the spinal cord conveying the sensation of pain; done for the relief of severe pain.
chordotonal organs *(Zool.)*. In Insects, sense organs consisting of bundles of scolophores, sensitive to pressure, vibrations, and sound.
chorea *(Med.)*. Involuntary repetitive jerky movements of the body, particularly limbs and face. A manifestation of a number of neurological diseases including *Sydenham's chorea* which occurs in the course of acute rheumatic fever and *Huntingdon's chorea*, a genetically transmitted disease in which there is also intellectual and personality deterioration.
choria *(Behav.)*. A neurological disorder, characterized by spasmic, jerky and involuntary movements, particularly of the face, tongue, hands and arms.
choriocarcinoma *(Med.)*. A malignant tumour of the uterus composed of cells derived from the foetal chorion; it appears during or after pregnancy. Also known as *hydatiform mole, chorio-epithelioma*.
chorion *(Zool.)*. In higher Vertebrates, one of the foetal membranes, being the outer layer of the amniotic fold: in Insects, the hardened eggshell lying outside the vitelline membrane.
chorionic villus sampling *(Biol.)*. Method for diagnosing human foetal abnormalities in the 6th to 10th week of gestation, in which small pieces of foetally-derived chorionic villi are removed through the cervix; chromosomes in the tissue are examined for abnormalities, and DNA may be extracted and **probed** for disease-associated alleles. Abbrev. *CVS*.
chorioptic mange *(Vet.)*. A mange of horses and cattle due to mites of the genus *Chorioptes*. In horses, known as foot mange or itchy leg.
chorioretinitis *(Med.)*. Inflammation of the choroid and retina.
choroid *(Zool.)*. The vascular tunic of the Vertebrate eye, lying between the retina and the sclera. adj. *choroidal*.
choroiditis *(Med.)*. Inflammation of the choroid of the eye.
choroid plexus *(Zool.)*. In higher Craniata, the thickened, vascularized regions of the *pia mater* in immediate contact with the thin epithelial roofs of the diencephalon and medulla.
Christmas tree *(Min.Ext.)*. Casinghead assembly which controls oil flow. See **blowout preventer**.
Christoffel symbols *(Maths.)*. The Christoffel symbols of the first kind are defined as

$$\begin{bmatrix} r \\ pq \end{bmatrix} = \frac{1}{2}\left(\frac{\partial g_{pr}}{\partial x^q} + \frac{\partial g_{qr}}{\partial x^p} - \frac{\partial g_{pq}}{\partial x^r}\right).$$

Also denoted by

$$\begin{bmatrix} pq \\ r \end{bmatrix}, [pq, r]\ C^r_{pq}, \text{and by } \Gamma_{pqr}.$$

The Christoffel symbols of the second kind are defined as

$$\left\{\begin{matrix} r \\ pq \end{matrix}\right\} = g^{\alpha r}\begin{bmatrix} \alpha \\ pq \end{bmatrix}.$$

Also denoted by

$$\left\{\begin{matrix} pq \\ r \end{matrix}\right\}, \text{and by } \Gamma^r_{pq}.$$

g_{pq} and g^{pq} are components of the covariant and contravariant fundamental metric tensors.

chroma *(Image Tech.)*. See **chrominance**.

chroma *(Phys.)*. In the Munsell colour system a term to indicate degree of saturation; zero represents neutral grey, and, depending on the hue, the numbers 10 to 18 represent complete saturation.

chroma control *(Image Tech.)*. Control which adjusts colours in colour TV picture. Abbrev. for *chromaticity control*.

chromadizing *(Eng.)*. Treating aluminium or its alloys with chromic acid to improve paint adhesion.

chromaffinoma *(Med.)*. A benign or malignant tumour of chromaffin cells usually of the adrenal medulla. syn. *phaeocromocytoma*.

chroma-key *(Image Tech.)*. Video technique for the combination of images from two or more sources. See **CSO**.

chromates (VI) *(Chem.)*. The salts corresponding to chromium (VI) oxide. The normal chromates, M_2CrO_4, are generally yellow, while the dichromates (VI), $M_2Cr_2O_7$, are generally orange-red. Chromates are used as pigments and in tanning.

chromate treatment *(Eng.)*. Treating a metal with a hexavalent chromium compound to produce a *conversion coating*, so altering its surface properties.

chromatic aberration *(Image Tech.,Phys.)*. An enlargement of the focal spot caused (1) in a cathode ray tube, by the differences in the electron velocity distribution through the beam, and (2) in an optical lens system using white light, by the refractive index of the glass varying with the wavelength of the light, resulting in coloured fringes surrounding the image.

chromatic adaptation *(Bot.)*. Differences in amount or proportion of photosynthetic pigments in response to the amount and colour of light. It can be (1) phenotypic as in many Cyanophyceae or (2) constitutive as in the distribution of littoral algae (red lowest, green at top, brown between).

chromatic circle *(Build.)*. The range of colours in the visible spectrum displayed as wedges within a circle passing through yellow, yellow green, green, blue green, blue, blue purple, purple, red purple, red, red orange, orange, yellow orange.

chromatic colour *(Phys.)*. A colour which is not grey.

chromaticity *(Image Tech.,Phys.)*. Colour quality of light, as defined by its chromaticity coordinates, or alternatively by its purity (saturation) and dominant wavelength.

chromaticity diagram *(Image Tech.)*. Plane diagram in which one of the three chromaticity co-ordinates is plotted against another. Generally applied to the CIE (x,y) diagram in rectangular co-ordinates for colour television.

chromatics *(Phys.)*. The science of colours as affected by phenomena determined by their differing wavelengths.

chromatid *(Biol.)*. One of the two, thread-like structures joined at the **centromere**, which constitute a single **metaphase** chromosome, each containing a single, double-helical DNA molecule with associated protein.

chromatin *(Biol.)*. Network of more or less de-condensed DNA, and associated proteins and RNA, in the interphase nucleus forming higher-ordered structures in the nucleus of **eukaryotes**.

chromatin beads *(Biol.)*. See **nucleosomes**.

chromatography *(Chem.)*. Method of separating (often complex) mixtures. *Adsorption chromatography* depends on using solid adsorbents which have specific affinities for the adsolved substances. The mixture is introduced onto a column of the adsorbent, e.g. alumina, and the components eluted with a solvent or series of solvents and detected by physical or chemical methods. *Partition chromatography* applies the principle of **countercurrent distribution** to columns and involves the use of two immiscible solvent systems: one solvent system, the *stationary phase*, is supported on a suitable medium in a column and the mixture introduced in this system at the top of the column; the components are eluted by the other system, the *mobile phase*. See also **gas-liquid chromatography** and **paper chromatography**.

chromatophore *(Biol.)*. A cell containing pigment granules which may change its shape and colour effect on nervous or hormonal stimulation.

chrom-, chromo-, chromat-, chromato- *(Genrl.)*. Prefix from Gk. *chrōma, chrōmatos*, colour.

chrome alum *(Chem.)*. *Potassium chromium sulphate*, $K_2SO_4.Cr_2(SO_4)_3.24H_2O$, purple octahedral crystals obtained by the reduction of potassium dichromate (VI) solution acidified with sulphuric acid.

chrome brick *(Eng.)*. Brick incorporating chromite, used as a refractory lining in steelmaking furnaces.

chrome iron ore *(Min.)*. See **chromite**.

chromel *(Eng.)*. Alloy used in heating elements, based on nickel with about 10% chromium.

chrome spinel *(Min.)*. Another name for the mineral *picotite*, a member of the spinel group. Cf. **chromite**.

chromic oxide *(Chem.)*. Chromium (III) oxide, Cr_2O_3 an amphoteric oxide corresponding to chromium (III) salts, CrX_3, and to chromites (chromates (III)), $M_2Cr_2O_4$.

chromic salts *(Chem.)*. Salts in which chromium is in the (III) oxidation state. They are usually green or violet, and are not readily oxidized or reduced.

chromic (VI) acid *(Chem.)*. H_2CrO_4. An aqueous solution of chromic anhydride. Often applied to a solution of sodium dichromate in sulphuric acid, used for cleaning glassware. See **chromium (VI) oxide**.

chrominance *(Image Tech.,Phys.)*. (1) Colorimetric difference between any colour and a reference colour of equal luminance and specified chromaticity. The reference colour is generally a specified white, e.g. CIE illuminant C for artificial daylight. (2) In TV, the information which defines the colour (hue and saturation) of the image, as distinct from its brightness or **luminance**.

chrominance channel *(Image Tech.)*. Circuit carrying the chrominance signal in a colour TV system.

chrominance signal *(Image Tech.)*. See **chrominance** (2).

chrominance sub-carrier *(Image Tech.)*. A sub-carrier frequency in a TV signal which is modulated with the chrominance information.

chromite *(Min.)*. A double oxide of chromium and iron (generally also containing magnesium), used as a source of chromium and as a refractory for resisting high temperatures. It occurs as an accessory in some basic and ultrabasic rocks, and crystallizes in the cubic system as lustrous grey-black octahedra; also massive. Also called *chrome iron ore*.

chromium *(Chem.)*. Metallic element, symbol Cr, at.no. 24, r.a.m. 51.996, rel.d. at 20°C, 7.138, mp 1510°C, electrical resistivity at 20°C, 0.131 microhm metres. Obtained form chromite. Alloyed with nickel in heat-resisting alloys and with iron or iron and nickel in stainless and heat-resisting steels. Also used as a corrosion-resisting plating.

chromium (VI) oxide *(Chem.)*. CrO_3. Produced by the action of sulphuric acid on a concentrated solution of potassium dichromate (VI). It is deliquescent and a powerful oxidizing agent. See **chromic (VI) acid**.

chromoblast *(Zool.)*. An embryonic cell which will develop into a chromatophore.

chromocentre *(Biol.)*. Mass of localized interphase **chromatin**; usually refers to the fused, centromeric regions of dipteran salivary gland **polytene** chromosomes.

chromogen *(Chem.)*. A coloured compound containing a **chromophore**.

chromolithography *(Print.)*. The preparation of lithographic printing surfaces, whether for single or multicolour printing, by hand methods only, without a process camera as in **photolithography**. For multicoloured printing, an outline key drawing is made and transferred to the required number of stones or plates, on each of which the litho artist draws those portions which are required by the colour to be used when printing. See **autolithography**, **lithography**.

chromomere *(Biol.)*. One of the characteristic granules of compacted chromatin in serial array on a metaphase chromosome.

chromonema *(Biol.)*. Complete single thread of chromatin.

chromo-optometer *(Phys.)*. An optometer in which the

chromatic aberration of the eye is used to determine its refraction.

chromo paper *(Paper)*. Paper which is more heavily coated than art paper; used for chromolithography.

chromophil, chromophilic *(Biol.)*. Staining heavily in certain microscopical techniques.

chromophobe, chromophobic *(Biol.)*. Resisting stains, or staining with difficulty, in certain microscopical techniques.

chromophores *(Chem.)*. Characteristic groups which are responsible for the colour of dyestuffs. Such groups include: C = C, C = O, C = N, N = N, $O \leftarrow N_2O = O$ and others.

chromoplast *(Bot.)*. (1) A **plastid** containing a pigment, especially a yellow or orange plastid containing carotenoids. (2) A photosynthetic plastid, now usually called a chloroplast, of a non-green alga.

chromosomal aberration *(Biol.)*. Any visible abnormality in chromosome number or structure, including **trisomy** and **translocations**.

chromosomal chimera *(Bot.)*. That in which there are differences in number or morphology of the chromosomes.

chromosomally-enriched DNA library *(Biol.)*. **DNA library** made from DNA enriched for a specific chromosome; **flow-sorted** chromosomes are a possible source.

chromosome *(Biol.)*. In eukaryotes the deeply-staining rod-like structures seen in the nucleus at cell division, made up of a continuous thread of DNA which with its associated proteins (mainly *histones*) forms higher order structures called **nucleosomes** and has special regions, **centromere** and **telomere**. Normally constant in number for any species, there are 22 pairs and 2 sex chomosomes in the human. In micro-organisms the DNA is not associated with histones and does not form visible condensed structures.

chromosome arm *(Biol.)*. Part of a chromosome from the **centromere** to the end.

chromosome complement *(Biol.)*. The set of chromosomes characteristic of the nuclei of any one species of plant or animal. **Chromosome set** refers to the haploid set.

chromosome cores *(Biol.)*. Non-histone protein network left when **histones**, DNA and RNA are removed from mammalian metaphase chromosomes.

chromosome mapping *(Biol.)*. Assigning **genes** or **bands** to specific regions of the haploid chromosome complement; each chromosome is numbered in order of size, short arms are designated p, long arms q, the bands are numbered consecutively outward along each arm, and regions within a band also numbered, thus: 11p2.3 refers to the third region of band 2 on the short arm of chromosome 11. The ends are designated *ter*, as in 3pter, the end of the short arm of chromosome 3.

chromosome-mediated gene transfer *(Biol.)*. Transfer of genetic material into a cell by introducing foreign chromosomes. Techniques include **calcium phosphate precipitation** and **electroporation**.

chromosome set *(Biol.)*. The whole of the chromosomes present in the nucleus of a gamete, usually consisting of one each of the several kinds that may be present.

chromosome sorting *(Biol.)*. The ability to sort chromosomes by DNA content or size. See **fluorescence activated cell sorter**.

chromosphere *(Astron.)*. Part of the outer gaseous layers of the Sun, visible as a thin crescent of pinkish light in the few seconds of a total **eclipse** of the Sun. Located above the **photosphere** and below the **corona**, it has a temperature of 4 500 K.

chromous salts *(Chem.)*. Salts of chromium in the (II) oxidation state; they yield blue solutions with water and are strong reducing agents.

chromyl chloride *(Chem.)*. CrO_2Cl_2, a strongly oxidizing red liquid, bp 117°C.

chronaxie *(Med.)*. A time constant in nervous excitation, equal to the smallest time required for excitation of a nerve when the stimulus is an electrical current of twice the threshold intensity required for excitation when the stimulus is indefinitely prolonged.

chronic *(Med.)*. Said of a disease which is deep-seated or long-continued. Cf. **acute**.

chronic granulomatous disease *(Immun.)*. Inherited disease of male children characterized by recurrent abscesses and granuloma formation. The neutrophil leucocytes are deficient in a cytochrome component necessary for generating active oxygen free radicals for intracellular killing of microbes, even though these are ingested normally.

chronic respiratory disease of fowl *(Vet.)*. An infection of the respiratory tract by organisms of the genus *Mycoplasma*. Abbrev. *CRD*.

chronograph *(Horol.)*. (1) A watch with a centre seconds hand which can be caused to start, stop and fly back to zero by pressing the button or a push-piece on the side of the case. The chronograph mechanism is independent of the going train, so that the balance is not stopped when the seconds hand is stopped. See **stop watch**. (2) Any type of mechanism which gives a record of time intervals.

chronometer *(Horol.)*. A precision timekeeper. In UK and US the term denotes the very accurate timekeeper kept on board ship for navigational purposes, and fitted with the spring detent mechanism. On the Continent the term is also applied to any very accurate clock or watch which may be fitted with the spring detent or lever escapement.

chronometer clock *(Horol.)*. A clock fitted with a chronometer escapement.

chronometer escapement *(Horol.)*. The spring detent escapement. A highly detached escapement, capable of giving the most exacting performance. Impulse is given to the balance every alternate vibration. Locking is performed by a pallet on a spring detent, and unlocking by a discharging pallet carried on a roller on the balance staff. Also *detent escapement*.

chronoscope *(Horol.)*. Electronic instrument for precision measurement of very short time intervals.

chrysalis *(Zool.)*. The pupa of some Insects, especially Lepidoptera; the pupa-case.

chrysene *(Chem.)*. Colourless hydrocarbon found in the highest boiling fractions of coal tar. Exhibits red-violet fluorescence; mp 254°C, bp 448°C.

chrysoberyl *(Min.)*. Beryllium aluminate, crystallizing in the orthorhombic system. The crystals often have a stellate habit and are green to yellow in colour. See also **alexandrite**.

chrysoberyl cat's eye *(Min.)*. See **cymophane**.

chrysocolla *(Min.)*. A hydrated silicate of copper, often containing free silica and other impurities. It occurs in incrustations or thin seams, usually blue and amorphous.

chrysolite *(Min.)*. Sometimes applied to the whole of the olivine group of magnesium iron silicates but usually restricted to those richer in the magnesium component. In gemmology, an old name applied to several yellow and greenish-yellow stones.

Chrysophyceae *(Bot.)*. The golden-brown algae, a class of eukaryotic algae, in the division Heterokontophyta, ca 800 spp. Naked, scaly (silica) or walled. Flagellated unicellular and colonial, amoeboid, coccoid, palmelloid, simple and branched filamentous and a few thalloid types. Fresh water mostly. Prototrophs; osmotrophic and phagotrophic heterotrophs.

chrysoprase *(Min.)*. An apple-green variety of chalcedony; the pigmentation is probably due to the oxide of nickel.

chrysotherapy *(Med.)*. Treatment by injections of gold.

chrysotile *(Min.)*. A fibrous variety of serpentine forming the most valuable type of the asbestos of commerce. Also called *Canadian asbestos*.

Chrytridiomycetes *(Bot.)*. Class of the Mastigomycotina with posteriorly uniflagellate zoospores and gametes. Body unicellular or mycelial. Mostly aquatic and saprophytic, or parasitic on algae, fungi or plants. Includes *Synchytrium endobioticum* causing potato wart disease, and *Blastocladia*.

CHU, Chu *(Genrl.)*. Abbrev. for *Centigrade heat unit*, the same as the **pound-calorie**.

chuck *(Eng.)*. Device attached to the spindle of a machine tool for gripping the revolving work, cutting tool, or drill.
chucking machine *(Eng.)*. A machine tool in which the work is held and driven by a chuck, not supported on centres.
chuffs *(Build.)*. Bricks which have been rendered useless owing to the presence in them of cracks caused by rain falling on them while they were hot. Also called *shuffs*.
chumship *(Print.)*. The Scottish equivalent of **companionship**.
chunking *(Behav.)*. A process of reorganizing materials in memory which allows a number of items to be fitted into one larger unit, a form of *encoding*.
churn drill *(Min.Ext.)*. See **cable tool drilling**.
churning loss *(Autos.)*. In a gear box, the power wasted in fluid friction through the pumping action of the revolving gears in the oil.
chute *(Build.)*. A special tapered outlet pipe from a deep inspection chamber, employed to make rodding easier.
chute *(Min.Ext.)*. (1) An inclined trough for the transference of broken coal or ore. (2) An area of rich ore in an inclined vein or lode, generally of much greater vertical than lateral extent.
chute riffler *(Powder Tech.)*. Sampling device consisting of a V-shaped trough from which a series of chutes feed two receiving bins. Alternate chutes feed opposite bins. Particles have in theory an equal chance of being retained in either of the bins.
Chvostek's sign *(Med.)*. Twitching of the muscles of the face on tapping the facial nerve; a sign of tetany.
chyle *(Med.)*. Lymph, containing emulsified fat, from the lymphatic vessels draining the small intestine.
chyle *(Zool.)*. In Vertebrates, lymph containing the results of the digestive processes, and having a milky appearance due to the presence of emulsified fats and oils.
chylification, chylifaction *(Zool.)*. Formation of **chyle**.
chylomicron *(Biol.)*. The plasma lipoprotein with the lowest density. It transports dietary lipids from the intestine to the liver and adipose tissue.
chylomicrons *(Med.)*. Minute particles of emulsified fat present in the blood plasma, particularly after digestion of a fatty meal.
chyloperitoneum *(Med.)*. The presence of chyle in the peritoneal cavity as a result of obstruction of the abdominal lymphatics.
chylothorax *(Med.)*. The presence of chyle in the pleural cavity, due to injury to, or pressure on, the thoracic duct.
chyluria *(Med.)*. The presence of chyle in the urine.
chyme *(Zool.)*. In Vertebrates, the semifluid mass of partially digested food entering the small intestine from the stomach.
chymotrypsin *(Biol.)*. Peptidase of the mammalian digestive system which is specific for peptide bonds adjacent to amino acids with aromatic or bulky hydrophobic side chains.
cicatrix *(Med.,Biol.)*. The scar left after the healing of a wound; one which marks the previous attachment of an organ or structure, particularly in plants.
cicero *(Print.)*. See **Didot point system**.
Ciconiiformes *(Zool.)*. An order of Birds having a desmognathous palate and usually webbed feet; all are long-legged birds of aquatic habit, living mainly in marshes and nesting in colonies. They are powerful flyers and some migrate over long distances. Storks, Herons, Ibises, Spoonbills, and Flamingoes.
CIE *(Phys.)*. Abbrev. for *Commission Internationale d'Éclairage*. Formed to study problems of illumination.
CIE coordinates *(Phys.)*. Set of colour coordinates specifying proportions of three theoretical additive primary colours required to produce any hue. These theoretical primaries were established by the CIE and form basis of all comparative colour measurement. See **chromaticity diagram**.
CI engine *(Eng.)*. See **compression-ignition engine**.
cilia *(Biol.)*. (1) Fine hair-like protrusions of the cell surface which beat in unison to create currents of liquid over the cell surface or propel the cell through the medium. Each cilium has a complex and characteristic internal structure built around 9 peripheral pairs and one central pair of microtubules. (2) In Mammals, the eyelashes; in Birds, the barbicels of a feather. adjs. *ciliated, ciliate*.
ciliary *(Zool.)*. In general, pertaining to or resembling cilia; in Vertebrates, used of certain structures in connection with the eye, as the *ciliary ganglion, ciliary muscles, ciliary process*.
ciliate, ciliated *(Bot.)*. (1) Having a fringe of long hairs on the margin. (2) Having flagella.
ciliograde *(Zool.)*. Moving by the agency of cilia.
Ciliophora *(Zool.)*. A class of Protozoa, comprising forms which always possess cilia at some stage of the life-cycle, and usually have a meganucleus.
ciliospore *(Zool.)*. In Protozoa, a ciliated swarmspore. In Suctoria, a bud produced by asexual reproduction.
cilium *(Biol.)*. Sing. of *cilia*. A flagellum especially any one of the many short flagella of some protozoa.
Ciment Fondu *(Civ.Eng.)*. TN for a type of very rapid-hardening cement made by heating lime and alumina in an electric furnace to incipient fusion, and afterwards grinding to powder.
cimetidine *(Med.)*. Heals peptic ulcers by blocking the histamine$_2$ receptors (H_2 antagonist) which causes a reduced gastric acid output.
cinching *(Image Tech.)*. Tightening a roll of film by holding the centre and pulling the edge.
cinch marks *(Image Tech.)*. Abrasions or scratches along the length of the film caused by movement between surfaces in a roll.
cinchocaine *(Med.)*. A long-lasting local anaesthetic. The hydrochloride is used for spinal injection.
cinchona bases *(Chem.)*. Alkaloids present in cinchona bark, derivatives of quinoline.
cinchonine *(Chem.)*. $C_{19}H_{22}ON_2$, an alkaloid of the quinoline group, found in cinchona and cuprea barks; crystallizes in rhombic prisms from alcohol, mp 264°C. It behaves as a diacidic base and gives two series of salts.
cinchophen *(Med.)*. An analgesic formerly used for gout.
cinder pig *(Eng.)*. Pig iron made from a charge containing a considerable proportion of slag from puddling or reheating furnaces.
cinders *(Geol.)*. Volcanic *lapilli* composed mainly of dark glass and containing numerous vesicles (air or gas bubbles).
cine camera *(Image Tech.)*. A motion-picture camera, usually one for narrow-gauge film.
CinemaScope *(Image Tech.)*. TN for a system of **wide-screen** cinematography using **anamorphic lenses** with a horizontal compression/expansion of 2:1.
cine radiography *(Radiol.)*. The rapid sequence of X-ray films taken by a camera attached to an image intensifier.
Cinerama *(Image Tech.)*. TN for an early form of **wide-screen** cinematography shot with three adjacent cameras for presentation with three projectors on overlapping panels to form a continuous picture.
cingulum *(Zool.)*. Any girdle-shaped structure. In Annelida, the clitellum; in Rotifera, the outer post-oral ring of cilia; in Mammals, a tract of fibres connecting the hippocampal and callosal convolutions of the brain; in Mammals, a ridge surrounding the base of the crown of a tooth and serving to protect the gums from the hard parts of food.
cinnabar *(Min.)*. Mercury sulphide, HgS, occurring as red acicular crystals, or massive; the ore of mercury, worked extensively at Almadén, Spain, and elsewhere.
cinnamaldehyde *(Chem.)*. *3-phenylpropenal*, $C_6H_5CH{=}CH.CHO$, bp 246°C; an oil of aromatic odour; the chief constituent of cinnamon and cassia oils.
cinnamic acid *(Chem.)*. 3-Phenylpropenoic acid, $C_6H_5CH{=}CH\,COOH$, mp 133°C, bp 300°C; an unsaturated monobasic aromatic acid, prepared from benzal chloride by heating with sodium acetate.
cinnamon stone *(Min.)*. See **hessonite**.
cinquefoil *(Arch.)*. A 5-leaved ornament used in panellings, etc.
cinture *(Arch.)*. A plain ring or fillet round a column,

generally placed at the top and bottom to separate the shaft from capital and base.

CIP *(Min.Ext.)*. Abbrev. for *Carbon-In-Pulp*.

cipher tunnel *(Arch.)*. A false chimney built on to a house for symmetrical effect.

cipolin *(Build.)*. A white marble with green streaks; used for decorative purposes.

CIPW *(Geol.)*. A quantitative scheme of rock classification based on the comparison of norms; devised by four American petrologists, Cross, Iddings, Pirsson, and Washington.

circadian rhythm *(Biol.)*. A cyclical variation in the intensity of a metabolic or physiological process, or of some facet of behaviour, of ca. 24 hours.

circinate *(Bot.)*. Coiled, like a watch spring inwardly from base towards the apex. The leaf vernation in most ferns, some cycads and some seed-ferns.

circinate *(Med.)*. Rounded, circular.

circle *(Maths.)*. A plane curve which is the locus of a point which moves so that it is at a constant distance (the *radius*) from a fixed point (the *centre*). The length of the circumference of a circle is $2\pi r$, and its area πr^2, where r is the radius and π is equal to 3.141 593 (to 6 places). The cartesian equation of a circle, with centre (a,b) and radius k, is $(x-a)^2+(y-b)^2=k^2$.

circle coefficient *(Elec.Eng.)*. A term often used to denote the leakage factor of an induction motor.

circle diagram *(Elec.Eng.)*. Graphical representation of complex impedances at different points in a transmission system on an orthogonal network. Best known example is a *Smith chart*.

circle of confusion *(Image Tech.)*. Strictly *circle of least confusion*, the minimum image area of a point source of light formed by an optical system, the diameter of which determines the effective resolution. In practice, the maximum permissible diameter is affected by viewing conditions; thus it is taken as 0.01 in (0.25 mm) for a photographic print viewed directly but 0.001 in (0.025 mm) for a motion picture film image greatly magnified on projection.

circle of convergence *(Maths.)*. Circle within which a complex power series converges. Its radius is called the *radius of convergence*.

circle of curvature *(Maths.)*. See **curvature**.

circle of inversion *(Maths.)*. See **inversion**.

circling disease *(Vet.)*. See **listeriosis**.

circlip *(Eng.)*. A spring washer in the form of an incomplete circle, usually used as a retaining ring (e.g. for ball bearings). Internal or external types are fixed in a circular groove in a hole or shaft by temporary distortion (closing or opening the circular shape).

circuit *(Elec.Eng.)*. Arrangement of conductors and passive and active components forming a path, or paths, for electric current.

circuit *(Telecomm.)*. (1) Arrangement of one or more complete paths for electron flow. (2) Complete communication channel. (3) Assembly of electronic (or other) components having some specific function, e.g. amplifier, oscillator or gate. See **integrated circuit**, **printed circuit**.

circuital magnetization *(Phys.)*. See **solenoidal magnetization**.

circuit-breaker *(Elec.Eng.)*. Device for opening electric circuit under abnormal operating conditions, e.g. excessive current, heat, high ambient radiation level etc. Also *contact-breaker*. See **oil switch**, **air-blast switch**.

circuit cheater *(Elec.Eng.)*. One which, for test purposes, simulates a component or load. Cf. **dummy load**.

circuit diagram *(Elec.Eng.)*. Conventional representation of wiring system of electrical or electronic equipment.

circuit noise *(Electronics.)*. See **thermal noise**.

circuit parameters *(Elec.Eng.)*. Relevant values of physical constants associated with circuit elements.

circulant *(Maths.)*. A determinant in which each row is a cyclic permutation by one position of the previous row.

circular cone *(Maths.)*. See **cone (2)**.

circular error *(Horol.)*. The variation in the isochronism of a pendulum due to the path of the bob being that of a circular arc instead of a cycloid. For small amplitudes the error is small, and for precision clocks the amplitude is kept to about two degrees on either side of the line of suspension.

circular form tool *(Eng.)*. A ring-shaped profile cutter which is gashed to have a substantially radial surface. The curved surface is shaped, across its width, to correspond to the contour of the part to be produced, the cutting edge being formed by the junction of the radial and the curved surfaces.

circular functions *(Maths.)*. The *trigonometrial functions*, more particularly when defined with radian argument. Cf. **hyperbolic functions** and **elliptic functions**. All these functions are so named because of their association with the rectification of the similarly named curves.

circular knitting machine *(Textiles)*. A **weft-knitting machine** that produces fabric of circular cross-section in endless lengths.

circular level *(Surv.)*. Spirit-level with bubble housed under slightly concave glass.

circular magnetization *(Phys.)*. The magnetization of cylindrical magnetic material in such a way that the lines of force are circumferential.

circular measure *(Maths.)*. The expression of an angle in *radians*, one radian being the angle subtended at the centre of a circle by an arc of length equal to the radius. There are thus 2π, or approximately 6.283, radians in one complete revolution. 1 radian $=57.2958°$; 1 degree $=0.017\,453\,3$ radians.

circular mil *(Elec.Eng.)*. US unit for wire sizes, equal to area of wire 1 mil (=0.001 in =0.025 mm) diameter.

circular mitre *(Build.)*. A mitre formed between a curved and a straight piece.

circular pallets *(Horol.)*. **Pallets** equidistant from pallet-staff axis.

circular permutation *(Maths.)*. An arrangement of objects in a circle. There are $(n-1)!$ different circular permutations of n objects. See **permutations**.

circular pitch *(Eng.)*. The distance between corresponding points on adjacent teeth of a gear-wheel, measured along the **pitch circle**.

circular plane *(Build.)*. A plane adapted (though the use of shaped or flexible soles) for producing curved surfaces, either convex or concave.

circular point on a surface *(Maths.)*. A point at which the principal curvatures are equal.

circular points at infinity *(Maths.)*. Two conjugate imaginary points on the line at infinity on a plane which lie on all circles in the plane: denoted by I and J.

circular polarization *(Telecomm.)*. An electromagnetic wave for which either the electric or magnetic field vector describes a circle at the wave frequency; waves may have left or right handed circular polarization. Used widely in satellite communications.

circular saw *(Eng.)*. A steel disk carrying teeth on its periphery, used for sawing wood, metal or other materials; usually power driven.

circular shift *(Comp.)*. See **end-around shift**.

circular time base *(Electronics)*. Circuit for causing the spot on the screen of a CRT to traverse a circular path at constant angular velocity.

circular velocity *(Space)*. The horizontal velocity of body, required to keep it in a circular orbit, at a given altitude, about a planet. For a near Earth orbit, assuming no air drag, the circular velocity (V_c) is given by:

$$V_c=\sqrt{Rg}=7{\cdot}91 \text{ km/sec}$$

where R is the radius of the Earth and g the acceleration due to gravity. Also *orbital velocity*.

circular waveguide *(Telecomm.)*. A waveguide of circular cross-section (compared with conventional rectangular form). Can be used for very low-loss high-bandwidth communication links.

circulating current *(Elec.Eng.)*. That which flows round the loop of a complete circuit, as contrasted with *longitudinal current*, which flows along the two sides or legs of the same circuit, in parallel.

circulating-current protective system *(Elec.Eng.)*. A

form of Merz-Price protective system in which the current transformers at the two ends of the circuit to be protected are arranged to circulate current round the pilots, any difference in the currents in the two transformers passing through a relay.

circulating memory *(Comp.)*. Obsolete term for memory in which the impulses on a sonic delay line store are taken from the output, re-shaped, and re-inserted in the input. See **delay line**.

circulating pump *(Eng.)*. A pump, usually of centrifugal type, used to circulate cooling water or, more generally, any liquid. See **centrifugal pump**.

circulation *(Aero.)*. Used to describe the lift-producing airflow round an aerofoil, but strictly defined as the integral of the component of the fluid velocity along any closed path with respect to the distance round the path. See **super-circulation**.

circulation *(Maths.)*. Of a vector, the line integral of the vector along a closed path in the field of the vector.

circulation *(Med.)*. The continuous movement of the blood through the heart, arteries, capillaries, and veins.

circulation loss *(Min.Ext.)*. Of mud being pumped down a bore hole during drilling. It indicates the presence of porous or void conditions *downhole* and is potentially serious.

circulation of electrolyte *(Elec.Eng.)*. Movement of the electrolyte in an electroplating bath in order to ensure an even deposit.

circulator *(Radar)*. Device having a number of terminals, usually three, with internal circuits which ensure that energy entering one terminal flange flows out through the next in a particular direction. They may appear as waveguide, coaxial or stripline components.

circulatory integral *(Maths.)*. The line integral of a function over a closed curve.

circulatory system *(Zool.)*. A system of organs through which is maintained a constant flow of fluid, which facilitates the transport of materials between the different organs and parts of the body.

circumcentre of a triangle *(Maths.)*. The centre of the circumscribed circle. The point at which the perpendicular bisectors intersect.

circumcircle *(Maths.)*. See **circumscribed circle**.

circumcision *(Med.)*. Surgical removal of the prepuce or foreskin of the male or the labia minora in the female.

circumferential register adjustment *(Print.)*. The maintaining on rotary presses of one operation in relation to others in the direction of the web, manual control being superseded by electronic equipment.

circumferential register screws *(Print.)*. Adjustable screws, in place of margin bar, allowing slight movement of the plate around the cylinder on rotary presses.

circumnutation *(Bot.)*. The rotation of the tip of an elongating stem, so that it traces a helical curve in space.

circumpolar stars *(Astron.)*. Those stars which, for a given locality on the earth, do not rise and set but revolve about the elevated celestial pole, always above the horizon. To be circumpolar, a star's declination must exceed the co-latitude of the place in question.

circumscribed circle *(Maths.)*. Of a polygon: the circle which passes through all its vertices. Also called **circumcircle**.

cire perdue *(Eng.)*. See **investment casting**.

cirque, corrie *(Geol.)*. A semi-amphitheatre, or 'arm-chair-shaped' hollow, of large size, excavated in mountain country by, or under the influence of, ice.

cirrate, cirriferous *(Bot.,Zool.)*. Bearing cirri.

cirrhosis *(Med.)*. A disease of the liver in which there is increase of fibrous tissue and destruction of liver cells. Gk. *kirros*, orange-tawny.

Cirripedia *(Zool.)*. Subclass of marine Crustacea, generally of sessile habit when adult; the young are always free-swimming; the adult possesses an indistinctly segmented body which is partially hidden by a mantle containing calcareous shell plates; there are six pairs of biramous thoracic legs; attachment is by antennules; many species are parasitic. Barnacles.

cirrocumulus *(Meteor.)*. Thin, white patch, *sheet* or *layer* of *cloud* without shading, composed of very small elements in the form of grains, ripples etc., merged or separate, and more or less regularly arranged; most of the elements have an apparent width of less than one degree. Abbrev. *Cc*.

cirrose *(Bot.,Zool.)*. Curly, like a waved hair. Consisting of diverging filaments.

cirrostratus *(Meteor.)*. Transparent whitish *cloud veil* of fibrous (hair-like) or smooth appearance, totally or partly covering the sky, and generally producing halo phenomena. Abbrev. *Cs*.

cirrus *(Meteor.)*. Detached *clouds* in the form of white, delicate filaments of white or mostly white patches or narrow bands. These clouds have a fibrous (hair-like) appearance, or silky sheen, or both. Abbrev. *Ci*.

cirrus *(Zool.)*. In Protozoa, a stout conical vibratile process, formed by the union of cilia; in some Platyhelminthes, a copulatory organ formed by the protrusible end of the vas deferens; in Annelida, a filamentous tactile and respiratory appendage; in Cirripedia, a ramus of a thoracic appendage; in Insects, a hairlike structure on an appendage; in Crinoidea, a slender jointed filament arising from the stalk or from the centrodorsal ossicle and used for temporary attachment; in Fish, a barbel.

cirsoid aneurysm *(Med.)*. A mass of newly formed, tortuous, and dilated arteries.

cis- *(Chem.)*. A prefix indicating that geometrical isomer in which the 2 radicals are adjacent in a metal complex or on the same side of a double bond or alicyclic ring.

cisplatin *(Med.)*. An antitumour drug which has an alkylating action and is used in the treatment of ovarian carcinoma and testicular teratoma.

cissing *(Build.)*. A defect in paint, enamel, or varnish work due to poor adhesion; characterized by the appearance of pinholes, craters, and in serious cases the contraction of the wet paint to form blobs on the surface of the work.

cissoid *(Maths.)*. The inverse of a parabola in respect of its vertex. Its cartesian equation is $x(x^2+y^2)=by^2$.

cisternum *(Biol.)*. A compartment or vesicle, often flattened, formed within the cytoplasm by membranes of the **endoplasmic reticulum** or **Golgi apparatus**.

cistron *(Biol.)*. A *gene* defined as a stretch of DNA specifying one *polypeptide*.

cis x *(Maths.)*. Cos $x+i$ sin$x$, which equals e^{ix}.

citral *(Chem.)*. *Geranial*, $C_{10}H_{16}O$, an alkene-type terpene, formula $CH_2=CMe.CH_2.CH_2.CH_2.CMe=CH.$-CHO; bp 110°–112°C (12 mm); it occurs in the oil of lemons and oranges and in lemon-grass oil. Used as a flavouring and in perfumery.

citrates *(Chem.)*. The salts of citric acid.

citrene *(Chem.)*. See **limonene**.

citric acid *(Chem.)*. *2-hydoxypropane-1,2,3-tricarboxylic acid*. $C_6H_8O_7$, an important hydroxy-tricarboxylic acid, occurs in the free state in many fruits, especially lemons, but is now prepared commercially largely by fermentation with *Aspergillus*. Much used for flavouring effervescent drinks.

citric acid cycle *(Biol.)*. See **tricarboxylic acid cycle**.

citrine *(Min.)*. Not the true **topaz** of mineralogists but a yellow variety of quartz, which closely resembles it in colour but not in other physical characters; it is of much less value than topaz, and figures under a number of geographical names like *Spanish topaz*. Also *false topaz*. But see **Brazilian topaz**, the true mineral.

citronellal *(Chem.)*. Also know as rhodinal; $C_{10}H_{18}O$, an aldehyde forming the main consituent of citronella oil and lemon-grass oil. It is used in perfumery.

civery *(Arch.)*. One bay of a vaulted ceiling. Also spelt *severy*.

Civil Aviation Authority *(Aero.)*. An independent body which controls the technical, economic and safety regulations of British civil aviation. In 1972 it took over the relevant functions of the Department of Trade and Industry, and those of the Air Registration Board and the Air Transport Licensing Board. It is responsible for the civil side of the Joint National Air Traffic Control Services. Abbrev. *CAA*.

civil engineer *(Civ.Eng.)*. Someone engaged in the planning and construction of railways, roads, harbours, bridges etc.

civil twilight *(Astron.)*. The interval of time during which the sun is between the horizon and 6° below the horizon, morning and evening.

Cl *(Chem.)*. The symbol for chlorine.

clack, clack valve *(Eng.)*. A check valve admitting water from a feedpump to the boiler of a locomotive. A ball valve is used, the name *clack* being derived from the characteristic sound of the ball striking its seat.

clacking *(Vet.)*. *Forging*. An error of gait in the horse in which the toe of a hind foot strikes the sole or shoe of a fore foot.

cladding *(Build.)*. The material used for the external lining of a building.

cladding *(Min.Ext.)*. The material used for the lining of a mine shaft. Traditionally of timber battens or planks, now frequently of precast reinforced concrete slabs.

cladding *(Nuc.Eng.)*. Thin protective layer, usually metallic, of reactor fuel units to contain fission products and to prevent contact between fuel and coolant. See **can**.

cladding *(Phys.)*. The homogeneous dielectric material of lower refractive index surrounding the core of an *optical fibre*.

cladistics *(Bot.)*. Method of classifying organisms into groups (taxa) based on 'recency of common descent' as judged by the possession of shared derived (i.e. not primitive) characteristics.

clad metal *(Eng.)*. One with two or three layers bonded together to form a composite with e.g. a corrosion resistant layer formed over a stronger core by co-rolling, heavy plating, chemical deposition etc.

cladode *(Bot.)*. (1) Strictly a **phylloclade** of one internode. (2) More commonly, any phylloclade.

cladogram *(Bot.,Zool.)*. A branching diagram (dendrogram) reflecting the relationships between groups of organisms determined by the methods of **cladistics**.

cladophyll *(Bot.)*. Same as **phylloclade**.

Clairaut's differential equation *(Maths.)*. An equation of the form

$$y = xp + f(p),$$

where $$p = \frac{dy}{dx}.$$

The general solution is $y = cx + f(c)$.

There is also a singular solution given parametrically by

$$x = -f^1(p),\ y = -pf^1(p) + f(p)$$

Claisen condensation *(Chem.)*. An important synthetic reaction involving condensation between esters, or between esters and ketones, in the presence of sodium ethoxide. Used in the preparation of *ethyl acetoacetate*.

Claisen flask *(Chem.)*. A distillation flask used for vacuum distillations; it consists of a glass bulb with a neck for a thermometer, to which another neck with outlet tube is attached.

clamp *(Build)*. See **cramp**.

clamp *(Image Tech.)*. A circuit to hold part of a waveform signal at a specified level.

clamp *(Telecomm.)*. Circuit in which a waveform is adjusted and maintained at a definite level when recurring after intervals.

clamp allowance *(Print.)*. The portion of a plate outside the printing area which is held by the plate cylinder **clamp bars**.

clamp bar *(Print.)*. (1) A bar used to lock or unlock plates to a letterpress rotary or offset lithographic machine. (2) A bar to hold blankets in place on the pins of the impression cylinder of a letterpress rotary or the blanket cylinder of an offset press.

clamp connection *(Bot.)*. In some Basidiomycetes, a small connecting hypha across the septa between two adjacent cells of dikaryotic hypha. Formed as the two nuclei, lying one behind the other, in the terminal cell divide. Facilitates the maintenance of the dikaryon by allowing the second of the two daughter nuclei to pass the cell wall.

clamping diode *(Electronics)*. A **diode** used to clamp a voltage at some point in a circuit. See **catching diode**.

clamshell *(Aero.)*. (1) Cockpit canopy hinged at front and rear. (2) Hinged part of thrust reverser in gas turbine. (3) Hinged door of cargo aircraft.

clan *(Geol.)*. A suite of igneous rock-types closely related in chemical composition but differing in mode of occurrence, texture, and possibly in mineral contents.

clangbox *(Aero.)*. Deflector fitted to jet engine to divert gas flow for, e.g. V/STOL operation.

clap-board *(Build.)*. A form of weather-board which is tongued and rebated and frequently moulded, rather than being feather-edged.

clap-board gauge *(Build.)*. A gauge used in fixing clap-boarding, to ensure that each board is so set as to expose a parallel width.

clapper board *(Image Tech.)*. Board with clapstick and written details of the production (date, scene, take number etc.) photographed at the start of each shot.

clapper box *(Eng.)*. A slotted tool head carried on the saddle of a planing or a shaping machine. It carries a pivoted block to which the tool is clamped, thus allowing the tool to swing clear of the work on the return stroke of the table, or the head.

clappers *(Image Tech.)*. Pair of hinged arms closed sharply in view of the camera, the action and noise providing clear indications on the separate picture and sound records for subsequent synchronization. Also *clapstick*.

Clapp oscillator *(Telecomm.)*. Low-drift **Colpitts oscillator**.

clap-sill *(Hyd.Eng.)*. See **mitre sill**.

clarain *(Min.)*. Bands in coal, characterized by bright colour and silky lustre.

clarendon *(Print.)*. A heavy type-face. **Clarendon**

Clarke *(Min.)*. The average value of a chemical element in the Earth's crust.

Clark process *(Chem.)*. A process for effecting the partial softening of water by the addition of sufficient limewater to convert all the acid carbonates of lime and magnesium into the normal carbonates.

claspers *(Zool.)*. In Insects, an outer pair of gonapophyses; in male Selachian Fish, the inner narrow lobe of the pelvic fin, used in copulation; more generally, any organ used by the sexes for clasping one another during copulation.

clasp nail *(Build.)*. A square-section soft iron cut nail whose head has 2 pointed projections that sink into the wood, and the projecting end of the nail is bent over and punched into the back surface.

clasp nut *(Eng.)*. A nut split diametrally into halves, which may be closed so as to engage with a threaded shaft; used as a clutch between a lathe lead screw and the saddle.

class *(Biol.)*. In biometry, a group of organisms all falling within the same range, as indicated by the unit of measurement employed; in Zoology, one of the taxonomic groups into which a phylum is divided, ranking next above an order. In Botany the taxonomic rank below *division* and above *order*; the names end in -phyceae (algae), -mycetes (fungi) or -opsida (other plants).

class-A amplifier *(Telecomm.)*. An amplifier stage (valve or transistor) in which anode or collector current flows throughout the amplitude range of the applied signal.

class-AB amplifier *(Telecomm.)*. An amplifier stage (valve or transistor) in which anode or collector current flows for most, but not all, of the amplitude range of the applied signal. For small signals, operation is in **class-A**.

class-A, -B, -C etc. insulating materials *(Elec.Eng.)*. Classification of insulating materials according to the temperature which they may be expected to withstand.

class-B amplifier *(Telecomm.)*. An amplifier stage (valve or transistor) in which anode or collector current flows for approximately half of the amplitude range of the applied signal. Commonly a pair of such amplifiers is used in *push-pull* operation.

class-C amplifier *(Telecomm.)*. An amplifier stage (valve or transistor) in which anode or collector current flows during less than half of the amplitude variation of the applied signal.

class frequency *(Stats.)*. The number of observations in a set of data in a particular class interval.

Classical *(Arch.)*. The architecture pertaining to ancient Greece and Rome, the principles of which were revived throughout Europe during the **Renaissance**.

classical *(Phys.)*. Said of theories based on concepts established before relativity and quantum mechanics, i.e. largely in conformity with Newton's mechanics and Maxwell's electromagnetic theory. US *non-quantized.*

classical conditioning *(Behav.)*. A learning procedure involving the repeated pairing of two stimuli, one of which (the unconditioned stimulus or *UCS*) already elicits a response, the unconditioned response, and the other (the conditioned stimulus or *CS*), which does not. After one or more pairing the CS comes to elicit a response very similar to that initially elicited only by the UCS, even if the CS is now presented on its own. This learned response is referred to as the conditioned response.

classical flutter *(Aero.)*. See **flutter**.

classical scattering *(Phys.)*. See **Thomson scattering**.

classification *(Powder Tech.)*. Grading in accordance with particle size, shape, and density by fluid means.

classification of clouds *(Meteor.)*. An agreed classification such as that published by the *World Meteorological Organization.* There are 10 *cloud genera*, which can be divided into up to 14 *cloud species*, themselves divisible into one or more out of 9 *cloud varieties*. Agreed definitions of the 10 genera will be found elsewhere.

classification of ships *(Ships)*. Passenger ships are classified according to the nature of the voyages in which they are engaged. **Classification societies** assign ships to a class in the Register Book so long as they are built, equipped and maintained in accordance with the rules of the society.

classifier *(Min.Ext.)*. A machine for separating the product of ore-crushing plant into 2 portions consisting of particles of different sizes. In general, the finer particles are carried off by a stream of water, while the larger settle. The fine portion is known as the *overflow* or *slime*, the coarse as the *underflow* or *sand.*

class interval *(Stats.)*. A subset of the range of values of a variate.

clastic rocks *(Geol.)*. Rocks formed of fragments of pre-existing rocks.

clathrate *(Chem.)*. Form of compound in which one component is combined with another by the enclosure of one kind of molecule by the structure of another; e.g. rare gas in 1,4-dihydroxybenzene. See **molecular sieve**.

clathrin *(Biol.)*. A protein which is the major structural component of the proteinaceous layer of **coated pits** and **vesicles**. Clathrin takes the form of a triskelion and the association of triskelions into hexagons and pentagons forms the protein lattice of the coated pit and vesicle.

Claude process *(Eng.)*. A method of liquefying air in stages, the expanding gas being cooled by external work on the pistons.

claudication *(Med.)*. The action of limping. See **intermittent claudication**.

Clausius-Clapeyron equation *(Chem.)*. This shows the influence of pressure on the temperature at which a change of state occurs, and the variation of vapour pressure with temperature.

$$q = T\cdot\Delta v\cdot\frac{\mathrm{d}p}{\mathrm{d}T},$$

where q is the heat absorbed (latent heat), T is the absolute temperature, Δv is the change in volume.

Clausius' inequality *(Phys.)*. For any thermodynamic system undergoing a cyclic process

$$\oint\frac{dQ}{T}\leqslant 0$$

where dQ is an infinitesimal quantity of heat absorbed or liberated by the system at the temperature T kelvin. The equality is appropriate to a reversible process.

Clausius-Mosotti equation *(Phys.)*. One relating electrical polarizability to permittivity, principally for fluid dielectrics.

Clausius virial *(Phys.)*. A term in an expression for the calculation of the pressure of a gas based on the kinetic theory of gases. The term allows for the inclusion of the effect of forces between the molecules.

Claus process *(Chem.Eng.)*. A process for recovering elemental sulphur from H_2S, by **desulphurizing**. It is a 2-stage process as defined in the 2 chemical equations. The H_2S is divided into 2 streams. 1st step $2H_2S+3O_2=2H_2O+2SO_2$. 2nd step $2H_2S+SO_2=2H_2O+3S$. The second step is carried out at a high enough temperature to produce dry sulphur.

claustra *(Arch.)*. A panel pierced with geometrical designs giving relief to the concrete structures of the late 19th and early 20th centuries in France.

claustrophobia *(Med.)*. Fear of being in a confined space.

clavate *(Zool.)*. adj. from *clava*; shaped like a club, e.g. *clavate* antennae.

clave *(Zool.)*. A gradual swelling at the distal end of a structure, resembling a club.

clavicle *(Zool.)*. In Vertebrates, the collar bone, an anterior bone of the pectoral girdle. adj. *clavicular.*

claw *(Bot.)*. The narrow, elongated lower portion of a petal in some plants.

claw *(Build.)*. A small tool with a bent and split end, used for extracting tacks.

claw *(Zool.)*. A curved, sharp-pointed process at the distal extremity of a limb; a nail which tapers to a sharp point.

claw bolt *(Build.)*. A wrought-iron bolt with a long head flattened in a direction parallel to the bolt, and bent over at right angles near the end.

claw chisel *(Build.)*. A chisel, having a 2 in. (50 mm)-long serrated cutting edge, used for rough dressing building stone.

claw coupling, claw clutch *(Eng.)*. A shaft coupling in which flanges carried by each shaft engage through teeth cut in their opposing faces, one flange sliding axially to disengage the drive.

claw foot *(Med.)*. See **pes cavus**.

claw-hammer *(Build.)*. A hammer having a bent and split peen which may be used for extracting a nail by giving leverage under its head.

claw hand *(Med.)*. Clawlike position adopted by the hand when the muscles supplied by the ulnar nerve are paralysed.

claw ill *(Vet)*. See **foul in the foot**.

claws *(Image Tech.)*. Mechanism for advancing the film frame by frame in a camera or projector by intermittently pulling down the perforation holes.

clay *(Geol.)*. A fine-textured, sedimentary, or residual deposit. It consists of hydrated silicates of aluminium mixed with various impurities. Clay for use in the manufacture of pottery and bricks must be fine-grained and sufficiently plastic to be moulded when wet; it must retain its shape when dried, and sinter together, forming a hard coherent mass without losing its original shape, when heated to a sufficiently high temperature. In the mechanical analysis of soil, according to international classification, clay has a grain-size $< 1/256$ mm.

Clayden effect *(Image Tech.)*. Partial reversal of an image when a very brief exposure to an intensely bright source, such as an electric spark, is followed by a longer exposure at a much lower light level.

clay gun *(Eng.)*. Arrangement which shoots a ball of fireclay into a blast furnace's taphole.

claying *(Civ.Eng.)*. The operation of lining a blast-hole with clay to prevent the charge from getting damp.

clay ironstone *(Geol.)*. Nodular beds of clay and iron minerals, often associated with the Coal Measure rocks.

clay puddle *(Civ.Eng.)*. A plastic material produced by mixing clay thoroughly with about one-fifth of its weight of water. It is used in engineering construction to prevent the passage of water, e.g. for cores for earthen reservoir dams.

clay with flints *(Geol.)*. A stiff clay, containing unworn flints, which occurs as a residual deposit in chalk areas, but which is extensively mixed with other superficial deposits.

clean *(Print.)*. Said of a proof sheet containing no corrections.

cleaning eye *(Build.)*. See **access eye**.

clean room *(Space)*. A special facility for handling material, destined for space activities, in a sterile and dust-free environment.

clean up *(Elec.Eng.)*. (1) Improvement in vacuum which occurs in an electric discharge tube or vacuum lamp consequent upon absorption of the residual gases by the glass. (2) The removal of residual gas by a **getter**.

clear *(Comp.)*. To remove data from **memory** so that fresh data can be recorded.

clear *(Telecomm.)*. To take down all temporary connections, in the form of plugs inserted into jacks, on the termination of a call in a manual exchange.

clear air turbulence *(Meteor.)*. Turbulence in the free atmosphere that is not associated with cumulus or cumulonimbus clouds. It occurs mainly in the upper troposphere or lower stratosphere especially in the vicinity of jet streams.

clearance *(Eng.)*. (1) The distance between two objects, or between a moving and stationary part of a machine. (2) The angular backing-off given to a cutting tool in order that the heel shall clear the work.

clearance volume *(Eng.)*. In a reciprocating engine or compressor, the volume enclosed by the piston and the adjacent end of the cylinder, when the crank is on the top dead centre. See **compression ratio, cushion steam**.

clearing *(Print.)*. The disposal of letterpress formes after printing, the **furniture** and **quoins** being returned to stock and the type put aside, the machine-set to be remelted, and the hand-set distributed.

clearing *(Textiles)*. In dyeing or printing the removal of excess colouring material to improve the appearance of the fabric. Cutting out an imperfection in a yarn and knotting or splicing the resultant ends.

clearing agent *(Biol.)*. In microscopical technique, a liquid reagent which has the property of rendering objects immersed in it transparent and so capable of being examined by transmitted light.

clearing hole *(Eng.)*. A hole drilled slightly larger than the diameter of the bolt or screw which passes through it.

clearing manoeuvre *(Aero.)*. Alteration of aircraft attitude to give better view of other air traffic.

clear lamp *(Elec.Eng.)*. An electric filament lamp in which the bulb is made of clear glass.

clear span *(Build.)*. The horizontal distance between the inner extremities of the two bearings at the ends of a beam.

cleat *(Build.)*. A strip of wood or metal fixed to another for strengthening purposes, or as a locating piece to ensure that another part shall be in its correct position.

cleats *(Min.Ext.)*. The main cleavage planes or joint planes in a seam of coal.

cleat wiring *(Elec.Eng.)*. A system of wiring in which the wires are attached to the wall or other surface by cleats.

cleavage *(Bot.,Zool.)*. (1) Division of the cytoplasm by infurrowing of the plasmalemma. Cf. **cell plate**. (2) The series of mitotic divisions by which the fertilized ovum is transformed into a multicellular embryo.

cleavage *(Chem.)*. (1) The splitting of a crystal along certain planes parallel to certain actual or possible crystal faces, when subjected to tension. (2) The splitting up of a complex molecule, such as a protein, into simpler molecules, usually by hydrolysis and mediated by an enzyme.

cleavage *(Geol.)*. A property of rocks, such as slates, whereby they can be split into thin sheets. Cleavage is produced and oriented by the pressures that have affected rocks during consolidation and earth movements.

cleavage-nucleus *(Zool.)*. The nucleus of the fertilized ovum produced by the fusion of the male and female pronuclei: in parthenogenetic forms, the nucleus of the ovum.

cleavelandite *(Min.)*. A platy variety of **albite**.

cleaving-saw *(Build.)*. See **pit-saw**.

cleft palate *(Med.)*. A gap in the roof of the mouth as a result of congenital maldevelopment, with or without hare-lip.

cleidoic egg *(Zool.)*. Egg of a terrestrial animal with a protective shell.

cleidotomy *(Med.)*. The cutting of the clavicles when the shoulders of the foetus prevent delivery in difficult labour.

cleistocarp *(Bot.)*. Same as **cleistothecium**.

cleistogamy *(Bot.)*. The production of flowers, often inconspicuous, which do not open and in which self-pollination occurs. Cf. **chasmogamy**.

cleistothecium *(Bot.)*. A more or less globose **ascocarp** with no specialized opening.

Clemmensen reduction *(Chem.)*. Reduction of aldehydes and ketones, by heating with hydrochloric acid and zinc amalgam, to the corresponding hydrocarbons, e.g. propanone to propane.

clench nailing *(Build.)*. A method of nailing pieces together in which the end of the nail, after passing right through the last piece, is bent back and driven into this piece, so that it may not be drawn out.

clerestorey *(Arch.)*. The upper stage of the walls of a building, occurring, as in the case of a church above the projecting aisle roofs, and pierced with windows to admit light to the central portion of the building.

clerk of works *(Build.,Civ.Eng.)*. The official appointed by the employer to watch over the progress of any given building works, and to see that contractors comply with requirements for materials and labour.

cleveite *(Min.,Chem.)*. Variety of **pitchblende** containing uranium oxide and rare earths; often occluding substantial amounts of helium.

Cleveland Iron Ore *(Min.)*. An ironstone consisting of iron carbonate, which occurs in the Middle Lias rocks of North Yorkshire near Middlesbrough. The ironstone is oolitic and yields on the average 30% iron.

clevis *(Eng.)*. U-shaped component with holes drilled through the arms as a means of symmetrical attachment to another component. Used e.g. for attaching load carriers to overhead conveyors.

clevis joint *(Aero.)*. Fork and tongue joint secured by metal pin as used in joining solid rocket motor cases.

cliché *(Print.)*. International name for printing block; French origin.

click *(Acous.)*. Short impulse of sound, with wide frequency spectrum, with no perceptible concentration of energy-giving characterization.

click *(Horol.)*. Pawl or detainer used, in conjunction with a ratchet wheel, to permit rotation in 1 direction only.

clicker *(Print.)*. A compositor who receives copy and instructions from the overseer and distributes the work among his companions.

click spring *(Horol.)*. The spring which holds the click in the teeth of the ratchet wheel.

click stop *(Image Tech.)*. Aperture control which clicks at each reading, enabling it to be set without looking at the figures.

climacteric *(Biol.,Med.)*. A critical period of change in a living organism, e.g. the menopause.

climacteric *(Bot.)*. The period of ripening of some fruit e.g. apples, characterized by an increased rate of respiration.

climate *(Meteor.)*. The statistical ensemble of atmospheric conditions characteristic of a particular locality over a suitably long period (e.g., 30 years) including relevant parameters such as mean and extreme values, measures of variability, and descriptions of systematic seasonal variations. Aspects considered include temperature, humidity, rainfall, solar radiation, cloud, wind and atmospheric pressure.

climatic factor *(Bot.)*. A condition such as average rainfall, temperature, and so on, which plays a controlling part in determining the features of a plant community and/or the distribution and abundance of animals. Cf. **edaphic factor, biotic factor**.

climatic zones *(Meteor.)*. The earth may be divided into zones, approximating to zones of latitude, such that each zone possesses a distinct type of climate. Eight principal zones may be distinguished: a zone of tropical wet climate near the equator; 2 subtropical zones of steppe and desert climate; 2 zones of temperate rain climate; 1 incomplete zone of boreal climate with a great range of temperature in the northern hemisphere; and 2 polar caps of arctic snow climate.

climatology *(Meteor.)*. The study of climate and its causes.

climax *(Ecol.)*. The end-point in a succession of vegetation, when the community has reached an approximately steady state, in equilibrium with local conditions. See also **monoclimax theory, polyclimax theory, sere**.

climb cutting *(Eng.)*. The method of machining a surface with a multi-toothed cutter in which the workpiece moves in the same direction as the periphery of the cutter at the line of contact. It produces a better surface finish than can be obtained by cutting upwards but has the disadvantage that the workpiece tends to be drawn towards the cutter.

climbing form *(Civ.Eng.)*. A type of form sometimes used in the construction of reinforced concrete walls for buildings. The wall is built in horizontal sections, the climbing form being raised, after the pouring of each section, into a position convenient for the pouring of the next section.

clinch *(Eng.)*. To set or close a fastener, usually a rivet. Also called *clench*.

clinch nailing *(Build.)*. See **clench nailing**.

cline *(Ecol.)*. A quantitative gradation in the chacteristics of an animal or plant species across different parts of its range associated with changing ecological, geographic, or other factors, e.g., **ecocline, geocline**.

C-line *(Phys.)*. Fraunhofer line in spectrum of sun at 656.28 nm, arising from ionized hydrogen in its atmosphere.

clinic *(Med.)*. From the Greek *klinē*, a bed; *klinikos* was a doctor who visited patients in bed, and the word *clinic* has been used in English for both patient and doctor. Today its meanings include: the instruction of medicine in any of its branches at the bedside or in presence of hospital patients (whence *clinical medicine, clinical surgery, clinical psychology*, etc.): an institution, or a department of one, or a group of doctors, for treating patients or for diagnosis (*outpatients clinic, Health clinic*): a private hospital: by extension, any similar instructional and/or remedial meeting.

clinical psychology *(Behav.)*. A branch of psychology concerned with the application of research findings in the field of mental health.

clinical trial *(Med.)*. The method of testing the efficacy of a treatment or a hypothesis related to the cause of a disease, broadly divisible into two kinds. (1) *Retrospective trial*. This looks at existing records, patient habits, environmental situation etc., and attempts to make correlations between these and clinical outcome. Because past records may not be standardized and because it may be difficult to obtain a proper control group, this type of trial is at best indicative and is often considered unsatisfactory. (2) *Prospective trial*. In this kind there are two groups of patients, carefully matched for age, sex, clinical stage etc. One group (the control) will receive the current best available treatment or sometimes a *placebo*, while the other group (the experimental) will receive the new treatment. The clinical outcome is then assessed, preferably without the assessor knowing to which group the patient belongs. Statistical analysis is finally necessary to test whether the outcome in the experimental group differs significantly from that in the control.

clinker *(Eng.)*. Incombustible residue, consisting of fused ash, raked out from coal- or coke-fired furnaces; used for road-making and as aggregate for concrete. See **breeze**.

clinkers *(Build.)*. See **klinker brick**.

clinks *(Eng.)*. Internal cracks formed in steel by differential expansion of surface and interior during heating. The tendency for these to occur increases with the hardness and mass of the metal, and with the rate of heating.

clink-stone *(Geol.)*. See **phonolite**.

clinochlore *(Min.)*. A variety of chlorite.

clinograph *(Surv.)*. A form of adjustable set square, the two sides forming the right angle being fixed, while the third side is adjustable; it differs from the adjustable set square in that no scale is provided to show the angular position of the third side in relation to the other two. Used in surveying deep borehole to check departure from vertical.

clinohumite *(Min.)*. One of the four members of the humite group, magnesium silicate with hydroxide, occurring in metamorphosed limestones.

clinoid *(Med.)*. Bony processes of the sella turcica that encircle the pituitary gland's stalk.

clinometer *(Surv.)*. A hand instrument for the measurement of angles of slope.

clinopyroxene *(Min.)*. A general term for the monoclinic members of the pyroxene group of silicates.

clino-rhomboidal crystals *(Crystal.)*. See **triclinic system**.

clinostat *(Bot.)*. See **klinostat**.

clinozoisite *(Min.)*. A hydrous calcium aluminium silicate in the epidote group. Differs from zoisite in crystallizing in the monoclinic system.

clintonite *(Min.)*. Hydrated aluminium, magnesium, calcium silicate. One of the brittle micas; differs optically from xanthophyllite.

clioquinol *(Med.)*. Used as an internal amoebicide, also in creams and ointments (sometimes with hydrocortisone) for certain skin infections. Also called *iodochlorhydroxyquinoline*.

clip *(Image Tech.)*. A short continuous sequence selected from a motion picture or video production. Also *clipping*.

clipping *(Telecomm.)*. (1) Loss of initial or final speech sounds in telephony due to the operation of voice-operated or other switching devices. (2) Speech distortion caused by overloading, limiting or amplifier circuits, resulting in the cut off of amplitude peaks in louder passages. (3) Similar effects on extreme *black* or *white* or synchronizing pulse peaks in TV signals. (4) See **limiter**.

clipping circuit *(Telecomm.)*. One for removing peaks, tails or high-frequency ripple on leading and trailing edges and within pulses. See **limiter**.

clip screws *(Surv.)*. The screws by which the two verniers of the vertical circle of a theodolite may be adjusted so as to eliminate index error. Also called *antagonizing screws*.

clitellum *(Zool.)*. A special glandular region of the epidermis of *Oligochaeta* which secretes the cocoon and the albuminoid material which nourishes the embryo.

clitoridectomy *(Med.)*. Surgical removal of the clitoris.

clitoridotomy *(Med.)*. Circumcision of the female.

clitoris *(Zool.)*. In female Mammals, a small mass of erectile tissue, homologous with the glans penis of the male, situated just anterior to the vaginal aperture.

cloaca *(Zool.)*. Generally, a posterior invagination or chamber into which open the anus, the genital ducts, and the urinary ducts; in Urochordata, the median dorsal part of the atrium; in Holothuroidea, the wide posterior terminal part of the alimentary canal into which the respiratory trees open.

cloacitis *(Vet.)*. Inflammation of the cloaca.

Cloanthite *(Min.)*. See **chloanthite**.

clock *(Comp.)*. Electronic unit which synchronizes processes within a computer system by generating pulses at a constant rate. Cf. **real-time clock**.

clock *(Elec.Eng.)*. See **counter**.

clock *(Horol.)*. Timekeeper, other than chronometer or watch.

clock-driven behaviour *(Behav.)*. Rhythmic behaviour under the influence of an endogenous clock, found throughout the animal kingdom and involving many of the annual, lunar and daily rhythms of behaviour found in animals. Its presence must be determined by isolation and other experimental procedures. See **zeitgeber**.

clock gauge *(Eng.)*. A length-measuring instrument in which the linear displacement of an anvil is magnified and indicated by the deflection of a needle pointer rotating over a dial. Often used as part of a vertical comparator.

clock meter *(Elec.Eng.)*. An energy meter in which the current passing causes a change in the rate of a clock.

clock rate *(Comp.)*. Frequency at which the **clock** produces pulses, usually measured in megahertz.

clock track *(Comp.)*. Series of optical or magnetic marks on the relevant input medium giving information which locates the read areas.

clock-watch *(Horol.)*. A watch which strikes the hours.

clonal selection *(Immun.)*. Hypothesis proposed by Burnet which explains immune responses as due to selective stimulation by antigens of those lymphocytes bearing receptors capable of recognizing the antigens, from among a large population bearing a great variety of different receptors. The expansion of lymphocyte clones accounted for specific antibody production and immunological memory.

clone *(Biol.)*. (1) Organisms, cells or micro-organisms all derived from a single progenitor by asexual means. They have therefore an almost identical genotype. In plants it includes those derived by vegetative propagation such as grafting and taking cuttings. (2) Used loosely to describe the procedures by which a **vector** with an inserted DNA sequence makes multiple copies of itself and thus of the inserted sequence, i.e. *cloning*.

clonic phase *(Behav.)*. The third phase in a *grand mal seizure* in which the muscles contract and relax rhythmically while the body jerks in violent spasms.

close annealing *(Eng.)*. The operation of annealing metal products (e.g. sheets, strip and rod) in closed containers to avoid oxidation. Also called *pot (or box) annealing*.

close coupling *(Elec.Eng.)*. See **tight coupling**.

closed circuit *(Image Tech.)*. A system of video picture and sound transmission in which the camera and the display, even if remote, are directly linked, in contrast to broadcast methods.

closed circuit *(Phys.)*. One in which there is zero impedance to the flow of any current, the voltage dropping to zero.

closed-circuit grinding *(Chem.Eng.)*. Size reduction of solids done in several stages, the material after each stage being separated into coarser and finer fractions, the coarser being returned for further size reduction and the finer being passed on to a further stage, so that overgrinding is minimized.

closed-coil winding *(Elec.Eng.)*. An armature winding in which the complete winding forms a closed circuit.

closed community *(Bot.)*. A plant community which occupies the ground without leaving any closed spaces bare of vegetation.

closed-core transformer *(Elec.Eng.)*. A transformer in which the magnetic circuit is entirely of iron, i.e. with no air gaps.

closed cycle *(Eng.)*. Heat engine in which the working substance continuously circulates without replenishment.

closed-cycle control system *(Elec.Eng.)*. One in which the controller is worked by a change in the quantity being controlled, e.g. an automatic voltage regulator in which a field current is actuated by a deviation of the voltage from a desired value, the **reference voltage**.

closed diaphragm *(Acous.)*. Diaphragm or cone which is not directly open to the air, but communicates with the latter through a horn, which serves to match the high mechanical impedance of the diaphragm with the low radiation impedance of the outer air.

closed inequality *(Maths.)*. One which defines a closed set of points, e.g. $-1 \leqq x \leqq +1$.

closed interval *(Maths.)*. An interval which includes its end points. Cf. **open interval**.

closed-jet wind tunnel *(Aero.)*. Any wind tunnel in which the working section is enclosed by rigid walls.

closed loop *(Telecomm.)*. Description of any system in which part of the output is fed back to the input to effect a control or regulatory action; also, the performance of such a system when the feedback path is connected. Cf. **open loop**. See **feedback**.

closed-loop system *(Telecomm.)*. Any control system in which feedback is applied, and in which the controlled quantity is measured and compared with a standard representing the desired performance. Any deviation from the standard is fed back in to the control system in such a sense as to reduce the deviation.

closed magnetic circuit *(Elec.Eng.)*. The magnetic core of an inductor or transformer without air gap.

closed mitosis *(Bot.)*. Mitosis during which the nuclear envelope remains more or less intact, e.g. some algae and fungi. Ant. **open mitosis**.

closed pipe *(Acous.)*. An organ pipe which is closed at one end. The wavelength of its fundamental resonance is four times the length of the air column inside the pipe.

closed pore *(Powder Tech.)*. A cavity within a particle of powder which does not communicate with the surface of the powder.

closed set *(Maths.)*. A set of points which includes all its accumulation points. Complement of an **open set**.

closed slots *(Elec.Eng.)*. In the rotor or stator of an electric machine, slots receiving the armature winding, which are completely closed at the surface and therefore in the form of a tunnel. Often called *tunnel slots*.

closed stokehold *(Eng.)*. A ship's boiler room, closed to allow fans to maintain in it an air pressure slightly above atmospheric, so that forced draft may be provided to the furnaces.

closed subroutine *(Comp.)*. Separate calls to the **subroutine** using the same piece of code. Cf. **open-**.

closed traverse *(Surv.)*. A traverse in which the final line links up with the first line.

closed vascular bundle *(Bot.)*. A vascular bundle which does not include any cambium and which will not, therefore, form secondary tissues.

close file *(Comp.)*. Correct method of giving up access to a file on backing store; this allows operations hidden from a user, such as clearing of buffers, to be successfully completed. Cf. **open file**.

close-packed hexagonal structure *(Eng.)*. An arrangement of atoms in crystals which may be imitated by packing spheres; characteristic of many metals. The disposition of the atomic centres in space can be related to a system of hexagonal cells.

close plating *(Elec.Eng.)*. The application of thin sheet of metal by a process of soldering.

closer *(Build.)*. Any portion of a brick used, in constructing a wall, to close up the bond next to the brick of a course. Also *bat, glut*.

close sand *(Foundry)*. See under **open sand**.

close string *(Build.)*. See **housed string**.

close timbering *(Build.,Civ.Eng.)*. Trench or excavation lining in which boards have no space between them.

close-up *(Image Tech.)*. Shot taken from a short distance, typically one showing only the face or head of an actor.

closing error *(Surv.)*. In a closed traverse, designed to return to starting point, discrepancy revealed on plotting, which may be small enough to allow distribution through the series of measurements.

closing layer *(Bot.)*. One of the alternating layers of compact and loose suberized tissue in the lenticels of some species of plant.

closing membrane *(Bot.)*. Same as **pit membrane**.

closing-up *(Eng.)*. The operation of forming the head on a projecting rivet shank.

clostridium *(Biol.)*. An ovoid or spindle-shaped bacterium, specifically one of the anaerobic genus *Clostridium*, which contains several species pathogenic to man and animals: viz., *Cl. botulinum* (botulism), *Cl. chauvei* (black-leg), *Cl. tetani* (tetanus), *Cl. welchii* (gas gangrene).

clot *(Med.)*. The semisolid state of blood or of lymph when they coagulate. See also **thrombosis**, **embolism**.

cloth *(Textiles)*. Generic term embracing all fabrics woven, felted, knitted, or non-woven etc. using any material or man-made fibre or continuous filament or blends of these. Including apparel, furnishing or industrial *fabrics*.

clothing *(Build.)*. The covering (walls, roofing etc.) applied to the structural framework of a steel-framed building.

cloth joints *(Print.)*. See **joints**.

clothoid *(Maths.)*. See **Cornu's spiral**.

cloud *(Meteor.)*. A mass of water droplets or ice particles remaining more or less at constant altitude. Cloud is usually formed by condensation brought about by warm moist air which has risen by convection into cooler regions and has been cooled thereby, and by expansion, below its dew point.

cloud and collision warning system *(Aero.)*. A primary radar system with forward scanning which gives a VDU display of dangerous clouds and high ground at ranges sufficient to allow course to be altered for their avoidance. A second sytem is usually necessary to give short-range warning of the presence of other aircraft.

cloud chamber *(Phys.)*. A device for making visible the paths of charged particles moving through a gas. The gas, saturated with a vapour, is suddenly cooled by expansion and the vapour condenses preferentially on the ions produced by the passage of charged particles through the chamber. The tracks of the particles are revealed by the drops of liquid which, if suitably illuminated, can be photographed. See **bubble chamber**.

cloudiness *(Meteor.)*. The amount of sky covered by cloud irrespective of type. Estimated in eighths (*oktas*). Overcast, 8; cloudless, 0.

clouding *(Build.)*. See **blooming**.

cloudy swelling *(Med.)*. A mild degenerative change in cells in which the swollen cloudy appearance is due to the presence of small vacuoles containing fluid and indicating damage to the 'sodium pump' of the cell membrane.

clouring *(Build.)*. Chiselling or picking small indentations on a wall surface as a key for the finish.

clout nail *(Build.)*. A nail having a large, thin, flat head.

clove oil *(Chem.)*. An oil obtained from the flowers of *Eugenia aromatica*. It is a pale yellow, volatile liquid, of strong aromatic odour; bp 250°–260°C, rel.d. 1.048–1.070. Formerly much used in confectionery, microscopy and dentistry.

cloxacillin *(Med.)*. A **penicillin** which is resistant to penicillinase produced by *Staphylococcus* and therefore used to treat penicillin resistant staphylococcal infections.

clubbing of the fingers *(Med.)*. Increased convexity of the nails with loss of nail bed angle and thickening of soft tissue round the phalanges, associated with a number of diseases (carcinoma of the lung, chronic lung sepsis, chronic cyanotic heart disease, infective endocarditis and inflammatory bowel disease). Can be familial, cause uncertain.

club line *(Print.)*. A term used for the last (short) line of a paragraph. In common practice in the upmaking of a book, the occurrence of a club line at the top of the page is to be avoided. Also called break line.

club moss *(Bot.)*. In various usages, the Lycopsida, the orders Lycopodiales and Selaginellales, or just the Lycopodiaceae.

club-tooth escapement *(Horol.)*. The most widely used escapement for watches. See **escapement**.

clumps *(Print.)*. Metal spacing material usually 6 pt and 12 pt.

clunch *(Min.Ext.)*. (1) Sear earth below coal seam of marl or shale. (2) A tough fireclay.

Clupeiformes *(Zool.)*. An order of Osteichthyes mostly planktonic feeders containing some economically important species. Herrings and Anchovies.

Clusius column *(Nuc.Eng.)*. Device used for isotope separation by method of thermal diffusion, consisting of a long vertical cylinder with a hot wire up the axis.

cluster *(Chem.)*. A molecular compound containing four or more metal atoms bonded to one another without intervening ligands. Clusters have been prepared with more than forty metal atoms so bonded, and the internal structure of these approximates to that of metals themselves.

cluster *(Image Tech.)*. On a cinematograph projector, a group of magnetic heads for the reproduction of multi-channel sound from striped prints.

cluster *(Stats.)*. In statistical sampling a division of the population to be sampled into subsets, a sample of which will be taken.

cluster analysis *(Ecol.)*. Hierarchical classification technique, often used to reveal patterns of similarity among species lists from many sites.

cluster analysis *(Stats.)*. A statistical method of classifying observations into subsets, members of which satisfy some criterion of similarity.

cluster cup *(Bot.)*. The popular name for an *aecidium*.

clustered column *(Arch.)*. A column formed of several shafts bunched together.

cluster mill *(Eng.)*. Rolling mill in which two small diameter working rolls are each backed up by two larger rolls.

cluster variables *(Astron.)*. Short-period variable stars first observed in globular clusters; they are typical members of Population II. The periods are less than one day. See **RR Lyrae variables**.

clutch *(Build.,Civ.Eng.)*. A connecting bar used between adjacent flanges of I-section steel sheet piles to retain them in position. In section, it consists of a web part running between a pair of adjacent piles, with curved 'flanges' which slide over and secure the flanges of the piles.

clutch *(Eng.)*. A device by which two shafts or rotating members may be connected or disconnected, either while at rest or in relative motion.

clutch stop *(Autos.)*. A small brake arranged to act on the driven member of a clutch when it is fully withdrawn, to facilitate an upward gear change. With modern synchro gear-box this device is only used in heavy gear boxes. Also *inertia brake*.

clutter *(Radar)*. Unwanted echoes on a radar display, usually due to terrain in the immediate vicinity of the antenna, or to rain or sea.

Cm *(Chem.)*. The symbol for curium.

cm³ *(Genrl.)*. Abbrev. for *cubic centimetre*.

C-MAC *(Image Tech.)*. See MAC.

CMI *(Comp.)*. See **computer managed instruction**.

CMOS *(Comp.,Electronics)*. See **complementary metal-oxide silicon**.

CN *(Chem.)*. Abbrev. for *Coordination Number*.

cn *(Maths.)*. See **elliptical functions**.

cnemidium *(Zool.)*. In Birds, the lower part of the leg, bearing usually scales instead of feathers.

cnemis *(Zool.)*. The shin or tibia.

CNES *(Space)*. Abbrev. for *Centre National d'Etudes Spatiales*, the French national space agency.

C-network *(Telecomm.)*. One consisting of three impedance branches in series. The free ends are connected to one pair of terminals and the junctions to another.

Cnidaria *(Zool.)*. A phylum of Metazoa comprising forms which are of aquatic habit; they show radial or biradial symmetry; possess a single cavity in the body, the enteron, which has a mouth but no anus; and generally have only two germinal layers, the ectoderm and the endoderm, from one of which the germ cells are always developed. Polyps, Corals, Sea-anemones, Jelly-fish, and Hydra.

cnidoblast *(Zool.)*. a thread-cell or stinging-cell, containing a **nematocyst**; characteristic of the Cnidaria. Also *cnida*.

CNS *(Zool.)*. Abbrev. for *Central Nervous System*.

Co *(Chem.)*. The symbol for cobalt.

coacervation *(Chem.)*. The reversible aggregation of particles of an emulsoid into liquid droplets preceding flocculation.

coach bolt *(Eng.)*. One having a convex head with a square section beneath it which fits into a corresponding cavity in the material to be bolted and prevents the bolt from turning as the nut is screwed on.

coach screw *(Eng.)*. A large wood screw with a square head which is turned by a spanner; used in heavy timberwork.

co-adaptation *(Biol.)*. Correlated adaptation or change in two mutually dependent organisms.

coagulation *(Biol.)*. The irreversible setting of protoplasm on exposure to heat, extreme pH or chemicals. Cf. **denaturation**.

coagulation *(Chem.)*. The precipitation of colloids from solutions, particularly of proteins.

coagulum *(Med.)*. A clot.

coal *(Geol.)*. A general name for firm brittle *carbonaceous rocks*; derived from vegetable debris, but altered, particularly in respect of volatile constituents, by pressure, temperature and a variety of chemical processes. The various types are classified on basis of volatile content, calorific value, caking and coking properties.

coal ball *(Bot.)*. A calcareous nodule, usually containing abundant petrified plant remains found in some seams of coal.

coal-cutting machinery *(Min.Ext.)*. Mechanized systems used in colliery to detach coal from its face, and perhaps to gather and load it to transporting device.

coalescent *(Bot.,Zool.)*. Grown together, especially by union of walls.

Coal Measures *(Geol.)*. The uppermost division of the **Carboniferous System**, consisting of beds of coal interstratified with shales, sandstones, and limestones and conglomerates.

Coal Sack *(Astron.)*. A large obscuring dust cloud visible to the naked eye as a dark nebula in the Milky Way near the Southern Cross.

coal sizes *(Min.Ext.)*. Those officially (UK) recognized are: large coal, over 6 in; large cobbles 6 to 3 in; cobbles, 4 to 2 in; trebles, 3 to 2 in; doubles, 2 to 1 in; singles, 1 to $\frac{1}{2}$ in; peas $\frac{1}{2}$ to $\frac{1}{4}$ in; grains, $\frac{1}{4}$ to $\frac{1}{8}$ in.

coal-tar *(Chem.)*. The distillation products of the high- or low-temperature carbonization of coal. Coal-tar consists of hydrocarbon oils (benzene, toluene, xylene and higher homologues), phenols (carbolic acid, cresols, xylenols and higher homologues), and bases, such as pyridine, quinoline, pyrrole, and their derivatives.

coal tar paints *(Build.)*. Surface coatings based on combinations of coal tar pitch and epoxy resins, either single or two pack materials, which have excellent resistance to heat, moisture and chemical attack.

coal tar wood preservatives *(Build.)*. Cheap wood preservatives, e.g. creosote derived from coal tar distillation. Normally applied to rough hewn exterior timber such as garden fencing, telegraph poles etc. Subsequent decoration using oil paints is not recommended since bleeding will occur.

co-altitude *(Astron.)*. See **zenith distance**.

coal washery *(Min.Ext.)*. Cleaning plant where run-of-mine coal is processed to remove shale and pyrite, reduce ash and sort in sizes.

Coanda effect *(Aero.)*. Named after its discoverer, it is the tendency of a fluid jet to attach itself to a downstream surface roughly parallel to the jet axis. If this surface curves away from the jet the attached flow will follow it, deflecting from the original direction.

coarctation *(Med.)*. A narrowing or constriction especially of the aortic arch causing hypertension in upper part of body.

coarse aggregate *(Civ.Eng.)*. Gravel or crushed stone (forming a constituent part of concrete) which when dry will be retained on a sieve having 5 mm ($\frac{1}{4}$ in) diameter holes. See **aggregate**.

coarsening *(Eng.)*. Increase in the grain size of metals usually by *grain growth*.

coarse scanning *(Radar)*. Rapid scan carried out to determine approximate location of any target.

coarse screen *(Image Tech.,Print.)*. A term applied to half-tones for use on rough paper. The *screen* may be up to 33 lines to the centimetre (80 lines to the inch).

coarse stuff *(Build.)*. A mixture of lime mortar and hair formerly used as a first coat for plastering internal walls.

coastal reflection *(Telecomm.)*. Reflection of signal by a land mass so that the resultant received signal consists of direct and reflected waves. This can cause direction-finding errors.

coastal refraction *(Telecomm.)*. Refraction, towards the normal, of waves arriving from sea to land at their incidence with the shoreline, resulting in an appreciable error in radio direction finding for bearings making a small angle with the shoreline.

coasting *(Elec.Eng.)*. Running with motor supply cut off and the brakes not applied.

coasting of temperature *(Eng.)*. Rise above correct predetermined operating temperature when the fuel supply has been checked or shut-off; due to excessive thermal storage in the furnace brickwork through prior overheating.

coat *(Bot.)*. (1) See **integument** (ovule). (2) See **testa** (seed coat).

coated cathode *(Electronics)*. One sprayed or dipped with a compound having a lower work function than the base metal, in order to enhance electron emission, which may be thermionic or photo-emission.

coated fabric *(Textiles)*. A textile fabric to which is attached on one or both sides a layer of a coating material such as plastic (especially polyvinyl chloride) or rubber.

coated lens *(Image Tech.,Phys.)*. The amount of light reflected from a glass surface can be considerably reduced by coating it with a thin film of transparent material (blooming). The film has a thickness of a quarter of a wavelength and its refractive index is the geometric mean of that of air and glass. Used to increase the light transmission through an optical system by reducing internal reflections, thus also reducing **flare**.

coated pit *(Biol.)*. A small surface invagination of the plasma membrane distinguished by a thick proteinaceous layer composed largely of *clathrin* on its cytoplasmic side. The pits invaginate to form vesicles within the cytoplasm. See **receptor mediated endocytosis**.

coated vesicle *(Biol.)*. A cytoplasmic vesicle with a *clathrin* coat formed by the invagination of a coated pit.

coating *(Elec.Eng.)*. The metallic sheets or films forming the plates of a capacitor.

coating and wrapping *(Min.Ext.)*. The process of covering a pipeline with bitumen and winding protective paper around it. Often done by machine just before the welded pipe is lowered into the trench.

coating machine *(Paper)*. Any machine designed to apply a layer of a coating material to the surface(s) of a base paper.

coaxial antenna *(Telecomm.)*. Exposed $\lambda/4$ length of coaxial line, with reversed metal cover, acting as a dipole. λ = wavelength.

coaxial cable *(Electronics)*. One in which a solid inner conductor is surrounded by a tubular outer conductor, the dielectric separating them being (in the case of rigid coaxial cables) air, or for flexible cables, solid or expanded plastic, insulating beads, or discs. The electromagnetic field associated with the currents in the conductors is confined to the dielectric, and, at high frequencies, a high degree of screening is achieved. A correctly terminated cable, more than a few wavelengths long, will present a characteristic impedance to an imposed a.c., its value depending on relative dimension of the conductors and the permittivity of the dielectric. Used extensively to feed HF and VHF antennae, and for multiplex signals in long-distance and submarine telecommunications.

coaxial circles *(Maths.)*. Two families of circles having a common radical axis.

coaxial filter *(Telecomm.)*. One in which a section of coaxial line is fitted with re-entrant elements to provide the inductance and capacitance of a filter section.

coaxial line *(Telecomm.)*. A **transmission line** coaxial in form, e.g. **coaxial cable**.

coaxial-line resonator *(Telecomm.)*. One in which the standing waves are established in a coaxial line, short- or open-circuited at the end remote from the drive. Of very high Q, these are used for stabilizing oscillators, or for selective coupling between amplifying stages.

coaxial propellors *(Aero.)*. Two propellors mounted on

concentric shafts having independent drives and rotating in opposite directions.

coaxial relay *(Telecomm.)*. Switching device in which the coaxial circuit on both sides of the contact is maintained at its correct impedance level, thus avoiding wave reflection in the current path.

coaxial stub *(Telecomm.)*. Section of coaxial line which is short-circuited at one end and functions as a high impedance at quarter-wave resonance.

cob *(Build.)*. An unburnt brick.

cobalt *(Chem.)*. Hard, grey metallic element in the eighth group of the periodic system. It is magnetic below 1075°C, and can take a high polish. Symbol Co, at.no. 27, r.a.m. 58.9332, valency 2 or 3, mp 1490°C, elec. resist. at 20°C 0.0635 microhm metres. Tensile strength (commercial containing carbon) 450 MN/m^2. Similar properties to iron but harder; used extensively in alloys, and as a source for radiography or industrial irradiation.

cobaltammines *(Chem.)*. Complex compounds of cobalt salts with ammonia or its organic derivatives, e.g. *hexammino (III) compounds*, which contain the group $Co(NH_3)_6^{3+}$.

cobalt bloom *(Min.)*. See **erythrite**.

cobalt bomb *(Phys.)*. Theoretical nuclear weapon loaded with ^{59}Co. The long-life radioactive ^{60}Co, formed on explosion, would make the surrounding area uninhabitable. Also radioactive source comprising ^{60}Co in lead shield with shutter.

cobalt carbonyl *(Chem.)*. $Co_2(CO)_8$, used as a catalyst, esp. in the formation of aldehydes and diene polymerization.

cobalt (II) oxide *(Chem.)*. CoO. Used to produce a deep blue colour in glass, and, in small quantity, to counteract the green tinge in glass caused by the presence of iron.

cobaltite *(Min.)*. Sulphide and arsenide of cobalt, crystallizing in the cubic system; usually found massive and compact, with smaltite.

cobalt steels *(Eng.)*. Steels containing 5–12% cobalt, 14–20% tungsten, about 4% chromium, 1–2% vanadium, 0.8% carbon and a trace of molybdenum. They have great hardness and cutting power, but are brittle.

cobalt unit *(Radiol.)*. Source of radioactive cobalt (half-life 5.3 yrs) mounted in a protective capsule; operated like the **caesium unit**.

cobalt vitriol *(Min.)*. See **bieberite**.

cobble *(Geol.,Min.Ext.)*. (1) A rock fragment intermediate in size between a pebble and a boulder, dia > 64 mm < 256 mm. (2) See **coal sizes**.

cobblestone *(Civ.Eng.)*. A smallish roughly squared stone once used for paving purposes; superseded by setts.

cobb sizing test *(Paper)*. A method of measuring the amount of water taken up by a test piece in given time under the prescribed conditions of test. Expressed in g/m^2.

COBOL *(Comp.)*. Programming language used for business data processing. *COmmon Business Oriented Language.*

cocaine *(Med.)*. A coca-base alkaloid, the methyl ester of benzoyl-L-ecgonine. Used as a local anaesthetic, now mainly for eye, nose and throat surgery. A drug producing marked addiction.

coccidiomycosis *(Med.,Vet.)*. A fungal infection with *Coccidioides immitis* seen in South-Western US, Central and South America. Usually mild upper respiratory tract infection in man but may cause more severe infestation of lung. Causes a chronic infection of cattle, sheep, dogs, cats and certain rodents. Also *coccidioidomycosis*.

coccidiosis *(Vet.)*. A contagious infection of animals and birds by protozoa of the genera *Eimeria* and *Isospora*, usually affecting the intestinal epithelium and causing enteritis.

coccoid *(Bot.)*. Small, unicellular, walled, spherical and non-motile algae.

coccolith *(Bot.)*. Small (say 2–10 μm) calcified scale covering the cells of coccolithophorids (flagellated unicellular algae of the Haptophyceae). Abundant as fossils in chalk.

coccus *(Biol.)*. A spherical or near spherical bacterium with a diameter from 0.5–1.25 μm.

coccydynia *(Med.)*. Severe pain in the coccyx.

coccygectomy *(Med.)*. Excision of the coccyx.

coccyx *(Zool.)*. A bony structure in Primates and Amphibians, formed by the fusion of the caudal vertebrae; urostyle. pl. *coccyges*.

cochineal *(Chem.)*. The dried bodies of female insects of *Coccus cacti* plus the enclosed larvae. The colouring matter is carminic acid, $C_{17}H_{18}O_{10}$, soluble in water and ethanol.

cochlea *(Acous.,Zool.)*. In Mammals, the complex spirally coiled part of the inner ear which translates mechanical vibrations into nerve impulses.

cochlear potentials *(Acous.)*. Electrical potentials within the cochlear structures resulting from acoustic situation.

cochleate *(Bot.,Zool.)*. Spirally twisted, like the shell of a snail; cochleariform.

cock *(Build.)*. See **plug cock**.

cock *(Horol.)*. A carrier or bracket for a pivot.

cock-bead *(Build.)*. A bead projecting from the surface which it is decorating. Also *cocked bead*.

Cockcroft and Walton accelerator *(Elec.Eng.)*. High-voltage machine in which rectifiers charge capacitors in series, the discharge of these driving charged particles through an accelerating tube.

cocked bead *(Build.)*. See **cock-bead**.

cocket centring *(Civ.Eng.)*. Arch centring which leaves some head-room above the springing lines.

cocking rollers *(Print.)*. See **slewing rollers**.

cockle-stairs *(Arch.)*. Winding stairs.

cockling *(Paper)*. Local wrinkling of the surface of paper generally due to the release of dried-in strains as the result of moisture take-up.

cockling *(Textiles)*. A defect in a fabric which causes wrinkles usually arising from non-uniform shrinking. However, it may be used to produce an interesting irregularity in the appearance of a fabric.

cockpit *(Aero.)*. The compartment in which the pilot or pilots of an aircraft are seated. It is so called even where it forms a prolongation of the cabin. Also *flight deck*.

cockroach *(Print.)*. Typography set entirely in lower-case, without the use of any capitals.

cockscomb, cockscomb pyrites *(Min.)*. A twinned form of **marcasite**.

cockspur fastener *(Build.)*. A bronze or iron fastener for casement windows, used in conjunction with a stay bar and pin.

cock-up *(Print.)*. (1) A large initial that extends above the first line and ranges at the foot. (2) In line-casting machines when the 2-letter matrix is in the cock-up position the companion italic or bold letter is cast.

coconut oil *(Chem.)*. The oil obtained from the fruit of the coconut palm; a white waxy mass, mp 20°–28°C, rel.d. 0.912, saponification value 250–258, iodine value 8.9, acid value 5–50. Chief constituent lauric acid.

cocoon *(Textiles)*. Specifically, the envelope spun by the fully-grown silkworm round itself as a protective covering when entering the chrysalid state. If intended for silk manufacture, the cocoon is heated to destroy the pupa, and the thread is subsequently reeled.

cocoon *(Zool.)*. In Insects, a special envelope constructed by the larva for protection during the pupal stage; it consists either of silk or of extraneous matter bound together with silk.

co-current contact *(Chem.Eng.)*. The opposite of **counter-current contact**. At the start of the process, fresh A is in contact with fresh B, and at each succeeding stage progressively spent reagents are in contact, so that the effective driving force, and hence the overall economy, is lower.

cocuswood *(For.)*. A tropical American hardwood of the genus *Brya*, available only in the form of small logs. Typical uses include inlaying, brush backs, turnery, parquet, and musical instruments. Also called *Jamaica ebony, West Indian ebony*.

COD *(Chem.)*. Abbrev. for *Chemical Oxygen Demand*.

CODAN *(Telecomm.)*. Abbrev. for *Carrier Operated Device Anti-Noise*. Circuit for silencing a receiver in absence of a signal.

Coddington lens *(Phys.)*. A magnifying lens cut from a spherical piece of glass, having concentric spherical surfaces of equal radius at the ends, and a V-cut round the centre of its length to act as a stop at the centre of the sphere.

code *(Comp.)*. (1) See **binary code**. (2) A set of program instructions. (3) See **machine code**. See also **Universal Product Code, Binary Coded Decimal**.

codeine *(Med.)*. An alkaloid of the morphine group, the methyl derivative of *morphine*; a widely used analgesic.

coder *(Telecomm.)*. In a pulse-modulation system, the sampler that tests the signal at specified intervals.

codes of construction *(Chem.Eng.)*. Written procedures for design, selection of materials, and manufacturing and operating procedures for production of chemical plant items for onerous conditions. Many are internationally acceptable, e.g. *ASME* (American Society of Mechanical Engineers), *TEMA* (Tubular Exchange Manufacturers Association), *ABCM* (Association of British Chemical Manufacturers).

codes of practice *(Build.,Civ.Eng.)*. Recommendations drawn up by a regulatory authority describing what is regarded as good practice in particular types of work. May not be mandatory.

coding *(Comp.)*. Writing out a program for a computer.

coding *(Radar)*. Process of subdividing a relatively long pulse of transmitted power into a predetermined pattern of shorter pulses. The receiver uses **autocorrelation** or **matched filter** techniques to respond only to echoes bearing the transmitted code. This increases the radar's range, resolution and immunity to interference or jamming.

coding *(Telecomm.)*. Process of transforming information into code in accordance with definite rules. See **encode, pulse-amplitude modulation, pulse-code modulation, pulse-width modulation, ASCII**.

coding capacity *(Biol.)*. The number of different protein molecules which a DNA molecule could specify. Because of the presence of exons, reiterated sequences and other non-coding regions, it is not a very helpful concept to apply to whole chromosomes or the genomes of higher organisms.

coding sequence *(Biol.)*. That part of a nucleic acid molecule which can be transcribed and translated into polypeptide using the genetic code.

cod-liver oil *(Chem.)*. Oil obtained from the livers of cod-fish; a yellow or brown liquid, of characteristic odour, rich in vitamins A and D, rel.d. 0.992–0.930, saponification value 182–189, iodine value 141–159, acid value 204–207.

codominant *(Biol.)*. Describes a pair of alleles which both show their effects in heterozygotes, e.g. many blood group genes.

codominant *(Bot.)*. One of two or more species, which together dominate a plant community.

codon *(Biol.)*. A *triplet* of three consecutive bases in the DNA or in messenger RNA, which specifies (*codes for*) a particular *amino acid* for incorporation in a *polypeptide*. See also **genetic code**.

coefficient *(Maths.)*. In the algebraic expression or equation, a coefficient is the factor multiplying the variable (or quantity) under consideration. It may also be a number or parameter defining a certain characteristic, relationship etc., e.g. coefficient of correlation.

coefficient of absorption *(Phys.)*. See **absorption coefficient**.

coefficient of apparent expansion *(Phys.)*. The value of the coefficient of expansion of a liquid, which is obtained by means of a dilatometer if the expansion of the dilatometer is neglected. It is equal to the difference between the true coefficient of expansion of the liquid and the coefficient of cubical expansion of the dilatometer.

coefficient of compressibility *(Chem.)*. A measure of deviation of a gas from Boyle's law.

coefficient of dispersion *(Elec.Eng.)*. See **dispersion coefficient**.

coefficient of elasticity *(Phys.)*. See **elasticity**.

coefficient of equivalence *(Eng.)*. A factor used in converting amounts of aluminium, iron and manganese into equivalent amounts of zinc, in relation to their effect on the constitution of brass.

coefficient of expansion *(Phys.)*. The fractional expansion (i.e., the expansion of the unit length, area, or volume) per degree rise in temperature. Calling the coefficients of linear, superficial and cubical expansion of a substance α, β, and γ respectively, β is approximately twice, and γ three times, α.

coefficient of fineness of water plane *(Ships)*. The ratio of the actual area of the water plane to the area of the rectangle having length and breadth the same as the maximum dimensions of the water plane.

coefficient of friction *(Phys.)*. See **friction**.

coefficient of perception *(Phys.)*. A term used in connection with the effect of glare; equal to the reciprocal of Fechner's constant.

coefficient of performance *(Eng.)*. When applied to a heat pump it is the ratio of high temperature heat transfer to work input. In a refrigerator it is the ratio of low temperature heat transfer to work input. It may be greater then unity.

coefficient of reflection *(Phys.)*. See **reflection factor**.

coefficient of restitution *(Phys.)*. The ratio of the relative velocity of two elastic spheres after direct impact to that before impact. If a sphere is dropped from a height onto a fixed horizontal elastic plane, the coefficient of restitution is equal to the square root of the ratio of the height of rebound to the height from which the sphere was dropped. See **impact**.

coefficient of rigidity *(Phys.)*. See **elasticity of shear**.

coefficient of utilization *(Elec.Eng.)*. A term used in lighting calculations to denote the ratio of the useful light to the total output of the installation.

coefficient of variation *(Stats.)*. A dimensionless quantity measuring the relative dispersion of a set of observations, calculated as the ratio of the **standard deviation** to the mean of the data values (sometimes expressed as a percentage.

coefficient of viscosity *(Phys.)*. The value of the tangential force per unit area which is necessary to maintain unit relative velocity between two parallel planes unit distance apart in a fluid; symbol η. That is, if F is the tangential force on the area A and (dv/dz) is the velocity gradient perpendicular to the direction of flow, then $F = \eta A(dv/dz)$. For normal ranges of temperature, η for a liquid decreases with increase in temperature and is independent of the pressure. Unit of measurement is the **poise**, $10^{-1}\,Nm^{-2}s$ in SI units and 1 dyne $cm^{-2}s$ in CGS units.

coele-, -coele *(Zool.)*. Prefix and suffix derived from Gk. *koilia*, large cavity (of the belly).

Coelenterata *(Zool.)*. See **Cnidaria**.

coeliac *(Zool.)*. In Vertebrates, pertaining to the belly or abdomen.

coeliac disease *(Med.)*. A wasting disease of childhood in which failure to absorb fat from the intestines is associated with an excess of this substance in the faeces, due to a sensitivity to wheat gluten.

coelom *(Zool.)*. The secondary body cavity of animals, which is from its inception surrounded and separated from the primary body cavity by mesoderm. adj. *coelomic, coelomate*.

Coelomata *(Zool.)*. A group of Metazoa, including all those animals which possess a **coelom** at some stage of their life-history.

coelomere *(Zool.)*. In metameric animals, the portion of coelom contained within one somite.

coelomoduct *(Zool.)*. A duct of mesodermal origin, opening at one end into the coelom, at the other end to the exterior.

coelomostome *(Zool.)*. In Vertebrates, the ciliated funnel by which the nephrocoel opens into the splanchnocoel.

coelostat *(Astron.)*. An instrument consisting of a mirror (driven by clockwork) rotating about an axis in its own plane, and pointing to the pole of the heavens. It serves to reflect, continuously, the same region of the sky into the field of view of a fixed telescope.

coelozoic *(Zool.)*. Extracellular; living within one of the cavities of the body.

coenobium *(Bot.)*. An algal *colony* in which the number and arrangement of cells are initially determined and which grows only by enlargement of the cells. E.g. *Volvox, Pediastrum*.

coenocyte, coenocytia *(Bot.,Zool.)*. A multinucleate cell. Multinucleate syncytial tissues formed by the division of the nucleus without division of the cell, as striated muscle fibres and the trophoblast of the placenta.

coenogamete *(Bot.)*. A multinucleate gamete.

coenosarc *(Zool.)*. In Hydrozoa, the tubular common stem uniting the individual polyps of a hydroid colony.

coenosteum *(Zool.)*. In Corals and Hydrocorralinae, the common calcareous skeleton of the whole colony.

coenuriasis *(Vet.)*. *Sturdy, gid, turnsick*. A nervous disease of sheep and goats caused by invasion of the brain and spinal cord by 'coenurus cerebralis', the intermediate stage of the tapeworm *Taenia multiceps*. Controlled by routine worming of canines.

coenzyme *(Biol.)*. Small molecules which are essential in stoichiometric amounts for the activity of some enzymes. Their loose association with enzymes distinguish them from prosthetic groups which fulfill a similar role but are tightly bound to the enzyme.

coenzyme A *(Biol.)*. The coenzyme which acts as a carrier for acyl groups (A stands for acetylation). See **acyl-CoA**.

coenzyme Q *(Biol.)*. See **ubiquinone**.

coercimeter *(Phys.)*. Instrument for measurement of coercive force.

coercive force *(Phys.)*. Reverse magnetizing force required to bring magnetization to zero after a ferromagnetic material has been left with appreciable residual magnetism.

coercivity *(Phys.)*. Coercive force when the cyclic magnetization reaches saturation.

coesite *(Min.)*. A high-pressure variety of silica. Found in rocks subjected to the impact of large meteorites; but first made as a synthetic compound.

C of A *(Aero.)*. Abbrev. for *Certificate of Airworthiness*.

cofactor *(Maths.)*. Of an element a_{ij} in the determinant $|A|$ or in the square matrix A: its coefficient A_{ij} in the expansion of $|A|$. It equals $(-1)^{i+j}$ times the minor of a_{ij}. Also called *signed minor*.

coffer *(Arch.)*. A sunk panel in a ceiling or soffit.

coffer *(Hyd.Eng.)*. A canal lock-chamber.

coffer-dam *(Civ.Eng.)*. A temporary wall serving to exclude water from any site normally under water, so as to facilitate the laying of foundations or other similar work; usually formed by driving **sheet piling**.

coffer dam *(Ships)*. A short compartment, 1–1.6 m (3–5 ft) in length, separating oil-carrying compartments in an oil tanker from other compartments. Both bulkheads must be oil-tight and the space must be well ventilated.

coffering *(Min.Ext.)*. (1) The operation involved in the construction of dams (see **coffer-dam**) for impounding water. (2) Shaft lining impervious to normal water-pressure.

coffer work *(Build.)*. Stone-faced rubble-work.

coffin *(Nuc.Eng.)*. See **flask**.

C of M *(Aero.)*. Abbrev. for *Certificate of Maintenance*.

cog *(Build.)*. The solid middle part left between the two notches cut in the lower timber in a cogging joint.

cog *(Eng.)*. Wooden teeth, formerly used in gear wheels.

cogging *(Build.)*. A form of jointing used to connect one beam to another across which it is bearing. Notches as long as the top beam is wide are cut in the top surface of the lower beam opposite one another, so as to leave a solid middle part, and the upper beam has a transverse notch cut in it, to fit over this solid part. Also *caulking, cocking, corking*.

cogging *(Eng.)*. The operation of rolling or forging an ingot to reduce it to a bloom or billet.

cognition (animal) *(Behav.)*. See **cognitive ethology**.

cognition (human) *(Behav.)*. Pertaining to the intellect and the mental processes involved in the obtaining and processing of knowledge and information; cognitive psychology includes studies of perception, memory, concept formation and problem solving.

cognitive dissonance *(Behav.)*. In social psychology, the most influential of **balance theories**, originating with Festinger, that assumes that beliefs, attitudes and knowledge about related experiences must be consistent with one another. If they are not, the individual experiences an unpleasant state of dissonance and is motivated to reduce it by reinterpreting some aspect of their experience in a way that will maximize consistency (consonance).

cognitive ethology *(Behav.)*. A branch of ethology introduced by Griffin, concerned with the issues of whether or not conscious awareness, and/or intention, should be taken into account in explanations of animal behaviour.

cognitive map *(Behav.)*. A mental representation of physical space.

cognitive therapy *(Behav.)*. An approach to therapy which holds that emotional disorders are caused primarily by irrational but habitual forms of thinking. It is related to behaviour therapy because it regards such patterns as forms of behaviour and attempts to help the individual discover and change inappropriate thought patterns.

coherence *(Phys.)*. If two light waves are superposed so as to produce *interference* effects and there is a constant *phase* relation maintained between them, the waves are said to be coherent. Sources producing coherent light are necessary to produce observable interference effects and such sources can be formed by dividing the wave from one source into two parts. Coherence can be thought of in terms of both time and space and *lasers* are capable of producing light of great time and spatial coherence.

coherent *(Bot.)*. United, but so slightly that the coherent organs can be separated without very much tearing.

coherent oscillator *(Radar)*. One which is stabilized by being phase-locked to the transmitter of a radar for beating with the echo, and used with radar-following circuits.

coherent pulse *(Telecomm.)*. Said when individual trains of high frequency waves are all in the same phase.

coherent sources *(Telecomm.)*. Those between which definite phase relationships are maintained, so enabling meaningfull interference effects to occur.

coherent units *(Genrl.)*. A system of coherent units is one where no constants appear when units are derived from base units.

cohesion *(Bot.)*. The union of plant members of the same kind, as when petals are joined in a sympetalous corolla.

cohesion *(Phys.)*. The attraction between molecules of a liquid which enables drops and thin films to be formed. In gases the molecules are too far apart for cohesion to be appreciable (but see **Joule-Thomson effect**).

cohesion *(Powder Tech.)*. Attraction forces by which particles are held together to form a body, generally imparted by the introduction of a temporary binder, or by compaction. It is measured as green strength.

cohesion mechanism *(Bot.)*. Any mechanism in a plant, in particular one concerned with the dehiscence of a sporangium, which depends on the cohesive powers of water (i.e., that a mass of water resists disruption).

cohesion theory *(Bot.)*. The generally accepted explanation of water movement through the xylem, that the water is drawn though the vessels or tracheids under tension, the columns of water being maintained by the cohesion of the water molecules and their adhesion to the walls.

cohesive end *(Biol.)*. DNA in nature or when cut by many restriction enzymes often has short single-stranded sequences at their end. If the complementary sequence occurs elsewhere, the two ends will cohere. The basis of most sequence insertion procedures. syn. *sticky end*.

cohesive ends *(Biol.)*. See **sticky ends**.

cohesive soils *(Civ.Eng.)*. Soils possessing inherent strength due to the surface tension of capillary water.

cohort *(Zool.)*. A taxonomic group ranking above a superorder.

coil *(Elec.Eng.)*. Length of insulated conductor wound around a core which can be, for example, air (especially very high frequencies) or iron or ferrite. Because current passing in the coil will create a magnetic field which couples with the winding, the coil can be considered an **inductor**.

coil-and-wishbone *(Autos.)*. Type of independent front-wheel suspension in which a coil spring, usually with a telescopic hydraulic damper, is mounted upon a wish-bone-shaped frame, the steering swivels being mounted at the apex and top of the spring taking the weight of one corner of the chassis.

coiled-coil filament *(Elec.Eng.)*. A spiral filament for an electric lamp which is coiled into a further helix to reduce radiation losses and enable it to be run at a higher temperature.

coiler *(Textiles)*. See **can coiler**.

coil heating *(Build.)*. See **panel heating**.

coil ignition *(Autos.)*. See **battery-coil ignition**.

coil loading *(Elec.Eng.)*. The added inductance, in the form of coils, inserted at intervals along an extended line. Cf. **continuous loading**.

coil-side *(Elec.Eng.)*. That part of an armature coil lying in a single slot.

coil-span *(Elec.Eng.)*. The distance, measured round the armature periphery, between one side of an armature coil and the other; usually measured in electrical degrees or slots.

coil-span factor *(Elec.Eng.)*. A factor introduced into the equation giving the e.m.f. of an electric machine, to allow for the fact that the coils have a fractional pitch and therefore do not embrace the whole flux.

coil-winding machine *(Elec.Eng.)*. A machine for automatically, or semi-automatically, winding an electrical conductor coil to a given scheme.

coincidence detection *(Radiol.)*. The simultaneous detection of two annihilation **photons** emitted during positron decay.

coincidence gate *(Telecomm.)*. Electronic circuit producing an output pulse only when each of two (or more) input circuits receive pulses at the same instant, or within a prescribed time interval.

coincidence phenomenon *(Acous.)*. Equality of wave-length of two sound-carrying spheres. E.g. special form of interaction between the bending wave on a plate and sound waves in the surrounding medium, causing increased sound transmission. See **limiting frequency (1)**.

coincidence tuning *(Telecomm.)*. Tuning all stages to the midband frequency. Cf. **staggered tuning**.

coin-collector *(Elec.Eng.)*. A mechanical device attached to energy meters, in order that the consumer may close a switch after the insertion of a given number of coins.

coining *(Eng.)*. The impressing or repressing of a component in a die and tool set, in which all surfaces of the part are confined, to impart final shape and accuracy of dimensions. This does not generally imply intentional volume and density changes.

coir *(Textiles)*. The reddish-brown coarse fibre obtained from the coconuts of *Cocos nucifera*. The finer fibres are used for making mats, the coarser ones for brushes, and the shortest for filling mattresses and upholstery.

coition, coitus *(Med.)*. Sexual intercourse.

coitus interruptus *(Med.)*. A form of sexual intercourse in which the penis is withdrawn from the vagina before ejaculation occurs.

coke breeze *(Build.)*. The smaller grades of coke from coke ovens or gasworks, used in the manufacture of breeze concrete.

cokes *(Eng.)*. Originally, tin plates made from wrought-iron produced in a coke furnace. The term is now applied to plates with a thinner tin coating.

coking coals *(Eng.)*. Those with more than 15% volatile matter and 80% carbon, which can produce a crush-resistant coke.

col *(Meteor.)*. The region between two centres of high pressure, or anticyclones.

co-latitude *(Astron.)*. The complement of the latitude, terrestrial or celestial. On the celestial sphere it is, therefore, the angular distance between the celestial pole and the observer's zenith, and also the meridian altitude of the celestial equator above the observer's horizon.

Colby's bars *(Surv.)*. Compensated bimetallic bars 10 ft in length, arranged to show an unvarying length despite temperature change; used for base-line measurement in the Ordnance Survey.

colchicine *(Biol.,Chem.)*. $C_{22}H_{25}O_6N$, an alkaloid obtained from the root of the autumn crocus, *Colchicum autumnale*; pale yellow needles, mp 155°–157°C. Used to inhibit chromosome separation and so double the original number of chromosomes; much used in plant breeding for making *tetraploids*.

cold *(Med.)*. *Common cold, coryza.* An acute viral infectious catarrh of the nasal mucous membrane.

cold agglutinin *(Immun.)*. Antibody against red cells which causes agglutination at temperatures below body temperature but not at 37°C. Often reactive with the I antigen of red cells and induced by infection with *Mycoplasma pneumoniae*.

cold bend *(Eng.)*. A test of the ductility of a metal; it consists of bending a bar when cold through a certain specified angle.

cold-blooded *(Zool.)*. Having a bodily temperature which is dependent on the enviromental temperature; poikilothermal. Cf. **warm-blooded**.

cold cathode *(Electronics)*. Electrode from which electron emission results from high-potential gradient at the surface at normal temperatures.

cold-cathode discharge lamp *(Phys.)*. An electric discharge lamp in which the cathode is not heated, the electron emission being produced by a high-voltage gradient at the cathode surface.

cold chisel *(Eng.)*. A chisel for chipping or cutting away surplus metal; it is used with a hand hammer. Different forms of cutting edges (e.g. flat, cross-cut, half-round) are used for various purposes. See **cross-cut chisel, flat chisel**.

cold-drawing *(Eng.)*. Process of producing bar or wire by drawing through a steel die without heating the material. See **wire-drawing**.

cold front *(Meteor.)*. The leading edge of an advancing mass of cold air, often attended by line-squalls and heavy showers.

cold galvanizing *(Eng.)*. Coating of iron goods with zinc suspended in an organic liquid. This is evaporated leaving a zinc film on the article.

cold-heading *(Eng.)*. The process of forming the heads of bolts or rivets by upsetting the end of the bar without heating the material.

cold insulation mastic *(Build.)*. A type of coating prepared from bitumen solution, containing asbestos fibres and other fillers, which is used on cold surfaces where it is necessary to prevent entry of moisture and to provide protection.

cold junction *(Phys.)*. Junction of thermocouple wires with conductors leading to a thermoelectric pyrometer or other temperature indicator or recorder.

cold mirror *(Image Tech.)*. In a lamphouse system, a dichroic mirror reflecting visible light but transmitting the infra-red radiations which cause heating.

cold moulding *(Plastics)*. The use of resin, filler and accelerator to fill the mould, which then polymerizes to form the component.

cold pinch *(Min.Ext.)*. Emergency closing of a ruptured pipe line by flattening it with hydraulic pincers.

cold pool *(Meteor.)*. See **thickness chart**.

cold riveting *(Eng.)*. The process of closing a rivet without previous heating; confined to small rivets.

cold-rolled *(Eng.)*. Said of metal that has been rolled at a temperature close to atmospheric. The cold-rolling of metal sheets results in a smooth surface-finish.

cold saw *(Eng.)*. A metal-cutting circular saw for cold cutting. The teeth may be either integral with the disk or inserted.

cold-set ink *(Print.)*. Printing ink formulated to work in a warmed-up inking system and which sets on the cold stock.

cold sett, set, sate *(Eng.)*. A smith's tool similar to a short, stiff cold chisel; used for cutting bars etc. without heating. It is supported by a metal handle and struck with a sledge-hammer.

cold short *(Eng.)*. Metals brittle below their recrystallizing temperature.

cold shut *(Foundry)*. A casting imperfection due to metal entering the mould by different gates or sprues, cooling and failing to unite on meeting.

cold start *(Comp.)*. (1) Restart of a program following a stoppage in which the program has been lost or corrupted and thus must be re-loaded. (2) Restart of a computer which involves re-loading its operating system. See **bootstrap**. Cf. **warm start**.

cold store *(Electronics)*. General term for computer memories which, because they depend on superconductivity, have to be kept at temperatures close to absolute zero. See **superconducting memory**.

cold welding *(Eng.)*. The forcing together of like or unlike metals at ambient temperature, often in a shearing manner, so that normal oxide surface films are ruptured allowing such intimate metal contact that adhesion takes place.

cold-working *(Eng.)*. The operation of shaping metals at or near atmospheric temperature so as to produce *strain-hardening*. See also **work-hardening**.

Cole-Cole plot *(Phys.)*. Graph of real against imaginary part of complex permittivity, theoretically a semi-circle, from which the relaxation time of polar dielectrics can be determined.

colectomy *(Med.)*. Excision of the colon.

colemanite *(Min.)*. Hydrated calcium borate, crystallizing in the monoclinic system; occurs as nodules in clay found in California and elsewhere.

coleopter *(Aero.)*. Aircraft having an annular wing, the fuselage and engine lying on the centre line. Some French designs had VTOL capability.

Coleoptera *(Zool.)*. An order of Insecta, having the fore-wings or elytra thickened and chitinized, meeting in a straight line; the hind-wings, if present, are membranous; the mouthparts are adapted for biting. Beetles.

coleoptile *(Bot.)*. The sheath, probably a much modified first leaf, enclosing the epicotyl in the embryo of grasses and growing up, during germination, as far as the soil surface to protect the expanding leaves. It is a classical object for the study of auxin action and of phototropism.

coleorrhiza *(Bot.)*. The sheath enclosing the radicle of the embryo of grasses, through which the radicle grows at germination.

colibacillosis *(Vet.)*. Neonatal infection with *Escherichia coli*. Most common in calves under one month old and a septicaemic form effects calves from 3 to 6 days old. An enteric toxaemic form also effects calves under a week, while an enteric form affects those over this age. Infection may be complicated by a virus. Antisera and vaccines are available.

colic *(Med.,Vet.)*. Severe cramp-like spasmodic pain due to intermittent contractions of smooth muscle of the intestinal, renal or biliary tract.

colitis *(Med.)*. Inflammation of the colon.

collagen *(Biol.)*. Family of fibrous proteins abundant in the extracellular matrix, tendons and bones of animals.

collapse of lung *(Med.)*. An airless state of the lung, sometimes caused by obstruction of a bronchus, pneumothorax or after abdominal operation.

collapse therapy *(Med.)*. The treatment of lung disease by compression of the affected area, e.g. by injecting air between the layers of the pleura.

collapsible tap *(Eng.)*. A screw-cutting tap in which the chasers can be withdrawn radially, either for the purpose of producing a tapered thread or, in a more common type, to collapse the tap for quick withdrawal without risk of spoiling the thread produced.

collar *(Arch.)*. A band, either flat or slightly concave, plain or decorated, around a column.

collar *(Bot.)*. The junction between the stem and root of a plant, usually situated at soil level.

collar *(Eng.)*. A ring of rectangular section secured to a shaft to provide axial location with respect to a bearing: a similar ring formed integral with the shaft.

collar *(Min.Ext.)*. (1) Concrete platform from which shaft linings are suspended at top; entry end of drill hole. (2) Heavy components placed above the drill in a *drill string* to stabilize the drill and to smooth torsional shock.

collar *(Zool.)*. The rim of a choanocyte; in Hemichorda, a collarlike ridge posterior to the proboscis; in Gastropoda with a spiral shell, the collarlike fleshy mantle edge protruding beyond the lip of the shell; more generally, any collarlike structure.

collar beam *(Build.)*. The horizontal connecting beam of a collar-beam roof.

collar-beam roof *(Arch.)*. A roof composed of 2 rafters tied together by a horizontal beam connecting points about half-way up the rafters.

collar cell *(Zool.)*. See **choanocyte**.

collar-head screw *(Eng.)*. A screw in which the head is provided with an integral collar; used where fluid leakage may occur past the threads.

collaring *(Eng.)*. The term used to indicate that metal passing through a rolling mill follows one of the rolls so as to encircle it.

collars *(Eng.)*. In rolling mills, the sections of larger diameter separating the grooves in rolls used for the production of rectangular sections.

collate *(Print.)*. To check that the sections of a book are in correct order, after *gathering*.

collateral *(Zool.)*. (1) Running parallel or side by side. (2) Having a common ancestor several generations back.

collateral bud *(Bot.)*. An accessory bud located to the side of an axillary bud.

collateral bundle *(Bot.)*. A vascular bundle having phloem on one side only of the xylem, usually the abaxial side.

collating machine *(Print.)*. A rotary printing press which collects, in correct register and sequence, the components of e.g. multipart stationery, 'collating' being a misnomer.

collating sequence *(Comp.)*. Sequence of characters in order of their character codes.

collecting cell *(Bot.)*. A cell of the mesophyll of a leaf lying below and in contact with cells of the palisade from which it is presumed to collect photosynthetic products for transfer to the vascular tissue.

collecting cylinder *(Print.)*. On rotary presses, the cylinder for collecting sheets or sections before folding or delivery.

collecting electrode *(Elec.Eng.)*. See **passive electrode**.

collecting lens *(Image Tech.)*. In a multiple condensing lens, the component lens which is nearest the light source.

collecting power *(Phys.)*. The property of a lens to render parallel rays convergent or reduce the divergency of originally divergent rays.

collection *(Print.)*. Gathering the sections or pages of a web-fed rotary press.

collective dose equivalent *(Radiol.)*. The quantity obtained by multiplying the average effective dose equivalent by the number of persons exposed to a given source of radiation. Expressed as a sievert (*Sv*).

collective drive *(Elec.Eng.)*. The drive in an electric locomotive in which all the driving wheels are coupled and powered by a single motor. Cf. **individual drive**.

collective electron theory *(Phys.)*. Assumption that ferromagnetism arises from free electrons; Fermi-Dirac statistics identify the Curie point with the transition from ferro- to paramagnetic states.

collective fruit *(Bot.)*. A fruit derived from several flowers, as a mulberry.

collective model of the nucleus *(Phys.)*. Combines certain features of the *shell model* and *liquid-drop model*. It assumes that the nucleons move independently in a real potential but the potential is not the spherically symmetric potential of the shell model. It is instead a potential capable of undergoing deformation and this represents the collective motion of the nucleons as in the liquid-drop model.

collective pitch control *(Aero.)*. A helicopter control by which an equal variation is made in the blades of the rotor(s), independently of their azimuthal position, to give climb and descent.

collective unconscious *(Behav.)*. Jung: those aspects of unconscious mental life that represent the accumulated experiences of the human species, in contrast with the unconscious life of an individual based on personal experience.

collector *(Electronics)*. (1) Any electrode which collects electrons which have already completed and fulfilled their function, e.g. screen grid. (2) Outer section of a transistor which delivers a primary flow of carriers.

collector agent *(Min.Ext.)*. In froth-flotation, a chemical which is adsorbed by one of the minerals in an ore pulp, causing it to become hydrophobic and removable as a mineralized froth. Also *promoter*.

collector capacitance *(Electronics)*. The capacitance of the depletion layer forming the collector of a transistor.

collector current *(Electronics)*. The current which flows at the collector of a transistor on applying a suitable bias.

collector-current runaway *(Electronics)*. The continued increase of the collector current arising from an increase of temperature in the collector junction when the current grows. See **thermal runaway**.

collector efficiency *(Electronics)*. The ratio of the useful power output to the d.c. power input of a transistor.

collector junction *(Electronics)*. One biased in the high-resistance direction, current being controlled by **minority carriers**. The semiconductor junction between the collector and base electrodes of a transistor.

collector rings *(Elec.Eng.)*. See **slip rings**.

collector shoe *(Elec.Eng.)*. A metal shoe used on the vehicles of an electric traction system to maintain contact with the conductor-rail.

collector strip *(Elec.Eng.)*. See **contact strip**.

collect run *(Print.)*. A method of increasing the number of pages per copy on web-fed presses by *collection* before final fold and delivery.

College electros *(Print.)*. Electrotypes with a plastic backing which eliminates the need for 'slabbing' and results in a pliable plate; developed at the London College of Printing.

collenchyma *(Bot.)*. A mechanical tissue, typical of leaf veins, petioles and the outer cortex of stems, of more or less elongated cells with unevenly thickened non-lignified primary walls.

Colles' fracture *(Med.)*. Fracture of the lower part of the forearm above the wrist.

collet *(Eng.)*. An externally coned sleeve, slit in two or more planes for part of its length, and arranged to be closed by being drawn into an internally coned rigid sleeve, for the purpose of gripping articles or material.

collet *(Horol.)*. A circular flange or collar: the collar held friction-tight on a balance staff to which the inner end of the balance spring is pinned.

collet chuck *(Eng.)*. A collet mechanism used for holding work or a tool in a lathe, drilling machine, etc.

colleterial glands *(Zool.)*. One or two pairs of accessory reproductive glands, present in most female Insects. Their secretion forms the oötheca in Orthoptera, and a cement which fastens the eggs to the substratum in many other insects.

colliculus *(Zool.)*. A small prominence; as on the surface of the optic lobe of the brain, a rounded process of the arytaenoid cartilage.

colliding-beam experiment *(Phys.)*. A technique in high-energy physics whereby two beams of particles are made to collide head-on. A greater proportion of the energy of the incident particles is available for the creation of new particles in the collision than in a fixed target experiment of similar total energy.

colligative properties *(Chem.)*. Those properties of solutions which depend only on the concentration of dissolved particles, ions and molecules, and not on their nature. They include depression of freezing point, elevation of boiling point and osmotic pressure.

collimation *(Phys.)*. The process of aligning the various parts of an optical system. (The word is falsely derived from the Latin *collineare, -atum*, to bring together in a straight line).

collimation *(Radiol.)*. The limiting of a beam of radiation to the required dimensions.

collimation error *(Surv.)*. An error produced in levelling or in theodolite work when the line of collimation is out of its correct position; the latter, for the level, is parallel to the bubble line and perpendicular to the vertical axis of rotation of the instrument, and for the theodolite it is also perpendicular to the trunnion axis.

collimation system *(Surv.)*. In levelling along a series of instrument set-ups, transfer of **line of collimation** from back to fore sight, by change in reading on graduated staff. See **rise and fall system**.

collimator *(Phys.)*. A device for obtaining a parallel or near parallel beam of radiation or particles. An optical collimator consists of a source, usually a fine slit, at the principal focus of a converging lens or mirror. Penetrating radiation such as X-rays or γ-rays is collimated by a series of holes or slits in a highly absorbing material such as lead.

collimator *(Radiol.)*. The lens of a gamma camera imaging system which absorbs photons travelling in inappropriate directions and originating from parts of the body other than the region under examination.

collinear array *(Telecomm.)*. An antenna array consisting of a number of dipoles connected end-to-end and operated in phase. Maximum radiation is normal to the line of dipoles (cf. **end-fire array**). Groups of **Yagi** antennas may also be formed into collinear arrays, achieving very sharp directivity.

collinear transformation of a matrix *(Maths.)*. See **congruent transformation**.

collinear vectors *(Maths.)*. Two vectors which are parallel to the same line.

collineation *(Maths.)*. Analytical transformation having a one-to-one correspondence between points, collinear points being projected into collinear points.

Collins process *(Phys.)*. A process for the liquefaction of helium which combines the *Joule-Kelvin* expansion with the *Claude* process.

collision *(Phys.)*. An interaction between particles in which momentum is conserved. If also the kinetic energy of the particles is conserved, the collision is said to be *elastic*, if not then the collision is *inelastic*. With particles in nuclear physics, there is no contact unless there is *capture*. Collision then means a nearness of approach such that there is mutual interaction due to the forces associated with the particles.

collisional excitation *(Phys.)*. The transfer of energy when an atom is raised to an excited state by collision with another particle.

collision bulkhead *(Ships)*. A strong watertight bulkhead, not less than one-twentieth of the vessel's length from the fore end.

collision diameter *(Chem.)*. The distance of closest approach of the centres of two identical colliding molecules.

collision number *(Chem.)*. The frequency of collisions per unit concentration of molecules.

collodion *(Chem.,Med.)*. A cellulose tetranitrate, soluble in a mixture of ethanol and ethoxyethane (1:7); the solution is used for coating materials and, in medicine, for sealing wounds and dressings.

collodion process *(Image Tech.)*. Early photographic process using glass plates coated with iodized collodion (nitro-cellulose) sensitized with silver nitrate solution and exposed in the camera while still wet; hence known as the *wet plate* process.

colloid *(Chem.)*. From Gk. *kolla*, glue. Name originally given by Graham to amorphous solids, like gelatine and rubber, which spontaneously disperse in suitable solvents to form lyophilic sols. Contrasted with crystalloids on the one hand and with lyophobic sols on the other. The term currently denotes any colloidal system.

colloidal electrolyte *(Phys.)*. One formed from long-chain hydrocarbon compounds, end radicals of which can ionize, thus providing some properties of electrolytes.

colloidal fuel *(Eng.)*. A mixture of fuel oil and finely pulverized coal, which remains homogeneous in storage; calorific value high; used in oil-fired boilers as substitute for fuel oil alone.
colloidal graphite *(Eng.)*. Extremely fine dispersion of ground graphite in oil. Graphite lowers the surface tension of oil without lowering the viscosity; the oil spreads more easily, taking the graphite to rough surfaces where it can build up a smoothness.
colloidal movement *(Chem.)*. See **Brownian movement**.
colloidal mud *(Min.Ext.)*. Thixotropic mixture of finely divided clays with baryte and/or bentonite, used in the drilling of deep oil bores. See **drilling mud**.
colloidal state *(Chem.)*. A state of subdivision of matter in which the particle size varies from that of true 'molecular' solutions to that of coarse suspensions, the diameter of the particles lying between 1 nm and 100 nm. The particles are charged and can be subjected to electrophoresis, except at the **iso-electric point**. They are subject to Brownian movement and have a large amount of surface activity.
colloid goitre *(Med.)*. Abnormal enlargement of the thyroid gland due to accumulation in it of the viscid iodine-containing colloid.
colloid mill *(Chem.Eng.)*. Mill with very fine clearance between the grinding components, operating at high speed, and capable of reducing a given product to a particle size of 0.1 to 1 μm.
collophane *(Min.)*. Cryptocrystalline variety of *apatite*.
collotype *(Print.)*. A **planographic process** for printing of tone subjects in one or more colours without using the **half-tone process**. A film of dichromated gelatine is spread on a glass plate, printed down with a continuous tone negative, and developed with the use of glycerine and water. Dark parts of the subject become hardened and ink-accepting, light parts remain watery and ink-rejecting, and areas of intermediate tone accept ink to correspond with their variations in tone. During preparation, the surface develops a fine grain which provides a bite for the ink. Suitable for runs up to one thousand. Used for the finest facsimile reproductions of works of art.
colluviarium *(Civ.Eng.)*. An access opening in an aqueduct for maintenance and ventilation.
coloboma *(Med.)*. Congenital defect of development, especially of the lens, the iris, or of the retina.
coloenteritis *(Med.)*. Inflammation of the colon and small intestine. Also *enterocolitis*.
colon *(Zool.)*. In Insects, the wide posterior part of the hind-gut; the large intestine of Vertebrates. adj. *colonic*.
colonnade *(Arch.)*. A row of columns supporting an entablature.
colony *(Bot.)*. (1) The vegetative form of many species of algae in which the sister cells are connected in a group to function as a unit. In many sorts (e.g. *Synura*), the colonies, of no fixed number of cells, grow by division and reproduce asexually by fragmentation; but see also **coenobium**. (2) A fungal mycelium grown, e.g. on an agar plate, from one spore. (3) A bacterial colony similarly initiated and grown.
colony *(Zool.)*. A collection of individuals living together and in some degree interdependent, as a *colony* of polyps, a *colony* of social Insects: strictly, the members of a colony are in organic connection with one another.
colony stimulating factors *(Immun.)*. Substances made by a number of cells which cause haemopoietic stem cells to proliferate and differentiate into mature forms, appearing as colonies in tissue culture. There are separate factors for granulocytes and macrophages, for eosinophyl leucocytes and for erythrocytes. Abbrev. *CSF*.
colopexy, colopexia *(Med.)*. The anchoring of part of the colon by sewing it to the abdominal wall.
colophon *(Print.)*. Originally, a device and/or notice by the scribe or printer, placed at the end of a book before title-pages became customary. Replaced today by title page and imprint, although some publishers often append a colophon. In modern practice, a decorative device on the title-page or spine of a book.
colophony, colophonium *(Chem.)*. See **rosin**.
Colorado beetle *(Zool.)*. A black-and-yellow striped beetle (*Leptinotarsa decemlineata*), which feeds upon potato leaves, causing great destruction.
Colorado ruby *(Min.)*. An incorrect name for the fiery-red garnet (pyrope) crystals obtained from Colorado and certain other parts of US.
Colorado topaz *(Min.)*. True topaz of a brownish-yellow colour is obtained in Colorado, but quartz similarly coloured is sometimes sold under the same name.
colorimeter *(Phys.)*. An instrument used for the precise measurement of the hue, purity and brightness of a colour.
colorimetric analysis *(Chem.)*. Analysis of a solution by comparison of the colour produced by a reagent with that produced in a standard solution.
colorimetric purity *(Image Tech.)*. The ratio of the luminosity of a dominant hue to the total luminosity of a colour. See **saturation**.
colorimetry *(Image Tech.)*. The measurement of the spectral transmission and colour density of photographic images.
Color-key *(Print.)*. A method of producing proofs of colour separations using a film print of each colour mounted in register with the others.
colostomy *(Med.)*. A hole surgically made into the colon for the escape of faeces through the abdominal wall when the bowel below is obstructed.
colostrum *(Biol.)*. A clear fluid accumulated in the mammary glands in the latter part of pregnancy. It precedes the flow of milk at the onset of lactation and is an important source of maternal antibodies for the infant.
colostrum-corpuscles *(Zool.)*. Large cells containing fat-particles, which appear in the secretion of the mammary glands at the commencement of lactation.
colotomy *(Med.)*. An incision into the colon; (loosely) colostomy.
colour *(Phys.)*. Each *flavour* of quarks comes in three 'colours'; red, blue and green; and for antiquarks, three anticolours. All hadrons are 'colourless' *either* one red plus one blue plus one green quark *or* one quark and its oppositely coloured antiquark.
colour analyser *(Image Tech.)*. In film laboratory practice, a calibrated closed circuit TV system reproducing colour negative as a positive image to assess its required printing levels.
colour balance *(Image Tech.)*. See **balancing**.
colour bar *(Print.)*. A standard strip made up of solid and tone printed across the sheet. Used in conjunction with a reflection **densitometer** to obtain and maintain a standard ink film thickness and hence consistency of colour during printing.
colour bleeding *(Image Tech.)*. See **bleeding**.
colour blindness *(Med.)*. The lack of one or more of the spectral colour sensations of the eye. The commonest form, Daltonism, consists of an inability to distinguish between red and green. Even persons of normal sight may be colour blind to the indigo of the spectrum.
colour burst *(Image Tech.)*. Short sequence of colour sub-carrier frequency transmitted as a reference for the chrominance signal at the beginning of each line.
colour cast *(Image Tech.)*. Predominance of a particular colour affecting the whole image, resulting from unsatisfactory colour **balancing** in reproduction.
colour coder *(Image Tech.)*. Apparatus in colour TV to generate chrominance subcarrier and composite colour signal from the camera signals.
colour contamination *(Image Tech.)*. Error in colour reproduction caused by incomplete separation of primaries.
colour contrast *(Image Tech.)*. Visually, the subjective enhancement of one hue when seen in the surroundings of substantially complementary colours. Photographically, the **gamma** value of one of the component colour images.
colour coordinates *(Phys.)*. Set of numbers representing the location of a hue on a chromaticity diagram.

colour-corrected lens *(Image Tech.,Phys.)*. One in which chromatic aberration is much reduced.

colour correction *(Print.)*. When making sets of colour printing surfaces, it may be necessary to compensate for inherent faults of the printing inks available. Can be carried out by hand on the separation negatives or positives for any of the printing processes and on the actual plate for relief printing, by **colour masking** or electronically.

colour coupler *(Image Tech.)*. See **coupler**.

colour decoder *(Image Tech.)*. Circuit in a TV receiver which extracts, decodes and separates the 3 constituent colours.

colour developer *(Image Tech.)*. A processing solution in which a colour image is formed in association with the developed silver image.

colour difference signals *(Image Tech.)*. Signals for colour TV transmission obtained by subtracting the **luminance**, Y, from each of the three primary colour signals R, G and B, in the form B − Y, R − Y, which are then coded.

coloured cement *(Build.)*. Ordinary Portland cement into which selected pigments are introduced in the grinding process.

colour excess *(Astron.)*. The amount by which the colour index of a star exceeds the accepted value for its spectral class; used as a measure of absorption of starlight.

colour fastness *(Textiles)*. The resistance of a dyed textile to change of colour when exposed to specific agents such as water, light or rubbing.

colour fatigue *(Phys.)*. Changes in the sensation produced by a given colour, when the eye is fatigued by another or by the same colour.

colour filter *(Phys.)*. Film of material selectively absorbing certain wavelengths, and hence changing spectral distribution of transmitted radiation.

colour gate *(Image Tech.)*. Circuit in colour TV receiver which allows only primary colour signal, corresponding to excited phosphor, to reach modulation electrode of tube.

colour guides *(Print.)*. Term sometimes applied to **progressive proofs**.

colour index *(Astron.)*. The difference between the photographic and the visual magnitudes of a star, from which may be deduced the effective temperature.

colour index *(Geol.)*. A number which represents the percentage of dark-coloured heavy silicates in an igneous rock, and is thus a measure of its leucocratic, mesocratic, or melanocratic character.

colour index *(Image Tech.)*. A systematic arrangement of colours according to their hue, saturation and brightness for identification and reproduction. *The Colour Index* is a publication giving chemical details of commercially available dyestuffs and pigments.

colouring pigment *(Build.)*. See **stainer**.

colour intermediate *(Image Tech.)*. A masked **integral tripack** colour film intended for laboratory duplicating purposes, not for projection.

colour killer *(Image Tech.)*. Circuit rendering the chrominance channel of a colour TV receiver inoperative during the reception of monochrome signals.

colour-luminosity array *(Astron.)*. A variant of the **Hertzsprung-Russell diagram** in which the absolute magnitudes of stars are plotted as a function of their colours.

colour masking *(Image Tech.)*. In photographic colour reproduction, the use of additional images to compensate for the deficiencies of the dyes forming the principal colour records. These may be separate black-and-white images but the emulsions of modern colour negative films incorporate additional dye components for this purpose – *integral colour masking*. In the reproduction of colour film on television, colour masking is carried out electronically by matrixing the three colour signals.

colour masking *(Print.)*. Use of photographic masks in the process camera to compensate for inherent faults in the inks available for colour printing. Separate masks can be prepared, e.g. a low density positive of the cyan printer is used to modify the magenta printer negative, thus reducing the amount of magenta printing in blue areas; or an all-purpose masking colour film can be used, such as Multi-mask or Trimask. The problem has to some extent been overcome with modern electronic scanning techniques.

colour mixture curve *(Phys.)*. Representation of the specified three colours which match a given colour.

colour negative *(Image Tech.)*. A photographic image in which the tonal brightness values of the original subject are inverted, light to dark, and its colours are represented by substantially complementary hues.

colour phase alternation *(Image Tech.)*. Sequence of the colour signals in the video signal.

colour photographic sensitivity *(Image Tech.)*. Sensitivity of an emulsion to specified wavelength ranges.

colour picture signal *(Image Tech.)*. Monochrome video signal, plus a subcarrier conveying the colour information, which is transmitted with synchronizing signals.

colour positive *(Image Tech.)*. A photographic image in which the tonal values and colours are substantially similar to those of the original subject.

colour primaries *(Image Tech.,Print.)*. Set of colours, usually three, from which a colour picture can be reproduced: red, green and blue for an **additive process**, cyan, magenta and yellow for a **subtractive** one.

colour printing *(Print.)*. The reproduction of an original subject comprising 2 or more colours. Colour printing is achieved by any of the normal printing processes; each colour is printed separately, in a predetermined order, the superimposed impressions, if accurately registered, building up an image corresponding in colour to the original subject.

colour purity magnet *(Image Tech.)*. Magnet placed near to neck of colour picture tube to modify path of electron beam and thus improve purity of the displayed colour.

colour pyramid *(Image Tech.)*. See **colour triangle**.

colour reference signal *(Image Tech.)*. Continuous signal which determines the phase of the burst signal.

colour register *(Print.)*. The correct superimposing of two or more colours to achieve correct image fit.

colour saturation *(Image Tech.)*. See **saturation**.

colour screen *(Image Tech.)*. Either a filter or a mosaic of the primary colours.

colour separation *(Image Tech.,Print.)*. The production in the process camera of 3 separate negatives of the same subject through green, red, and blue filters to record the proportions of the respective colours in the image. The negatives can be used as a basis for colour prints or for half-tone colour blocks. The process can also now be done electronically. See **three-colour process**.

colour separation overlay *(Image Tech.)*. A system of video image combination in which the foreground action is shot against a uniform blue backing, these areas being replaced by a picture from another source. Also *chromakey*.

colours of thin films *(Phys.)*. When light is reflected from a thin film, such as a soap bubble or a layer of oil on water, coloured effects are seen which are due to optical interference between light reflected from the upper surface and that from the lower. The colours are not bright unless the film is less than about 10^{-3} mm thick.

colour specification *(Image Tech.)*. The description of a colour in a standard manner, so that it can be duplicated without comparison.

colour standards *(Image Tech.)*. A standard range of colours for reference purposes in making dyes or filters, or for composing colour patterns for colour photography.

colour subcarrier *(Image Tech.)*. Signal, conveying the colour informaton as a modulation, added to the monochrome video and synchronizing signals.

colour temperature *(Phys.)*. That temperature of a black body which radiates with the same dominant wavelengths as those apparent from a source being described. See **Planck's law**.

colour threshold *(Phys.)*. The luminance level below which colour differences are indiscernable.

colour transparency *(Image Tech.)*. A colour photograph to be viewed or projected with transmitted light.

colour triangle *(Phys.)*. That drawn on a chromaticity diagram, representing entire range of chromaticities obtainable from additive mixtures of three prescribed primaries, represented by the corners of the triangle.

colour under *(Image Tech.)*. Method of colour video recording in which **chroma** is recorded at a lower carrier frequency than **luminance**, thus reducing the band-width required but at the sacrifice of some colour quality.

colour vision *(Phys.)*. A normal eye can match any colour by an additive mixture of saturated red, green and blue. This has led to the Young-Helmholtz theory that the perception of colour depends on the combination of stimuli from three primary mechanisms in the eye. This theory has yet to be proved or satisfactorily replaced. See **trichromatic coefficients, colour**.

colour vision *(Zool.)*. The ability of animals to discriminate light of different wavelengths. Depending on the animal there may be specialized receptors (cones in vertebrates) which are preferentially sensitive to two, three or more wavelengths. Trichromatic vision is most common.

colpitis *(Med.)*. Inflammation of the vagina.

Colpitts oscillator *(Electronics)*. One in which a parallel tuned circuit is connected between grid and collector (or grid and anode in a valve circuit) with the capacitive part consisting of two series capacitors. The junction of these is at emitter (or cathode) potential and the positive feedback is via the capacitor leading back to the base (or grid).

colpocele *(Med.)*. A hernia into the vagina.

colpocystocele *(Med.)*. A hernia formed by protrusion of the bladder into the vagina.

colpocystotomy *(Med.)*. Incision of the bladder through the wall of the vagina.

colpoperineoplasty *(Med.)*. Repair of the vagina and perineum by plastic surgery.

colpoperineorrhaphy *(Med.)*. Sewing up of the torn vagina and perineum.

colpoptosis *(Med.)*. Prolapse of the vagina.

colporrhaphy *(Med.)*. Narrowing of the vagina by surgical operation.

colposcope *(Med.)*. An instrument for inspecting the vagina.

colpospasm *(Med.)*. Spasm of the vagina.

colpus *(Bot.)*. An elongated aperture in the wall of a pollen grain. Cf. **aperturate**.

columbite *(Min.)*. Niobate and tantalate of iron and manganese. When the Nb content exceeds that of Ta, the ore is called columbite. See also **tantalite**.

Columbus *(Space)*. The name given the European Space Agency's programme to ensure a European contribution to the International Space Station. It includes station-attached and free-flying space vehicles.

columella *(Bot.)*. A small column, especially: (1) The axial part of a root cap in some species, in which the cells are arranged in longitudinal files. (2) The sterile tissue in the centre of the sporangium of bryophytes and some fungi. (3) A radial rod in the wall of a spore or pollen grain.

columella *(Zool.)*. In Mammals, the central pillar of the cochlea; in lower Vertebrates, the auditory ossicle connecting the tympanum with the inner ear; in some lower Tetrapods, the epipterygoid; in spirally coiled gastropod shells, the central pillar; in the skeleton of some Corals, the central pillar. adj. *columellar*.

column *(Bot.)*. The central portion of the flower of an orchid (probably an outgrowth of the receptacle of the flower), bearing the anther, or anthers and the stigmas.

column *(Civ.Eng.,Eng.)*. A vertical pillar or shaft of cast-iron, forged steel, steel plate in box section, stone, timber etc., used to support a compressive load. See also **strut**.

column *(Zool.)*. In Crinoidea, the stalk; in Vertebrates, a bundle of nerve fibres running longitudinally in the spinal cord; the edge of the nasal septum; more generally, any columnar structure, as the vertebral column.

column analogy method *(Civ.Eng.)*. A method of analysing indeterminate frames of non-uniform section by comparing equilibrium conditions, at any point, with the stresses in an analogous eccentrically loaded column.

columnar crystals *(Eng.)*. Elongated crystals formed by growth taking place at right angles to the surface of the mould. The extent to which they grow before solidification is completed, by the formation of equi-axed crystals in the interior, is important.

columnar epithelium *(Zool.)*. A variety of epithelium consisting of prismatic columnar cells set closely side by side on a basement membrane, generally in a single layer.

columnar structure *(Geol.)*. A form of regular jointing, produced by contraction following crystallization and cooling in igneous rocks, especially those of basic composition. The columns are generally roughly perpendicular to the cooling surface.

column binary *(Comp.)*. See **Chinese binary**.

columns across *(Print.)*. Printing newspapers and magazines with the columns imposed across the plate cylinders.

columns around *(Print.)*. Printing newspapers and magazines with the columns imposed around the plate cylinders.

column vector *(Maths.)*. A matrix consisting of a single column.

colures *(Astron.)*. The great circles passing through (a) the poles of the celestial equator and ecliptic and through both solstitial points, and (b) the poles of the celestial equator and both equinoctial points, these two great circles being the solstitial and equinoctial colures respectively.

COM *(Comp.)*. Abbrev. for *Computer Output on Microfilm*. Technique for producing such output.

coma *(Astron.)*. The visible head of a comet.

coma *(Bot.)*. A tuft of hairs or leaves.

coma *(Electronics.)*. Of a CRT, plumelike distortion of spot arising from misalignment of focusing elements of gun.

coma *(Med.)*. A state of complete unconsciousness in which the patient is unable to respond to any external stimulation.

coma *(Phys.)*. An aberration of a lens or lens system whereby an off-axis point object is imaged as a small pear-shaped blob. The aberration is due to the power of the zones of the lens varying with distance from the axis.

comagmatic assemblage *(Geol.)*. Refers to igneous rocks with a common set of chemical, mineralogical and textural features, suggesting derivation from a common parent magma.

COMAL *(Comp.)*. Programming language which is an enhanced version of **BASIC**, with structured programming features.

comatose *(Med.)*. Being in a state of coma.

comb *(Arch.)*. The ridge of a roof.

comb *(Build.)*. A flat flexible wire- or rubber-toothed instrument used by the painter for graining surfaces.

comb *(Textiles)*. To prepare cotton and wool fibres for spinning by separating and straightening them and removing impurities and fibres below a specified length.

comb *(Zool.)*. In Ctenophora, a ctene; the framework of hexagonal wax cells produced by social Bees to shelter the young or for storing food.

combat rating *(Aero.)*. See **power rating**.

comb binding *(Print.)*. A style of loose-leaf binding in which the leaves are held together by a comb, usually of plastic, being passed through slots in the paper and then curved to form a tube.

combed yarns *(Textiles.)*. Highest quality yarns prepared from carded and combed fibres that have been mechanically straightened and freed from **neps** and short fibres.

Combescure transformation of a curve *(Maths.)*. A one-to-one transformation which maps one space curve on another so that the tangents at corresponding points are parallel.

comb filter *(Image Tech.,Telecomm.)*. Electronic circuit tuned to select a number of specific frequencies for transmission or rejection, leaving unaffected those between them.

combination *(Chem.)*. Formation of a compound.

combination chuck *(Eng.)*. A lathe chuck in which the jaws may be operated all together, as in a universal or **self-centring chuck**; or each operated separately for holding work of irregular shape, as in an **independent chuck**.

combination cylinder *(Build.)*. A household hotwater cylinder with a feed tank above it, all within the same outer case.

combination mill *(Eng.)*. A continuous rolling mill in which the shaping mills follow the roughing mill directly.

combinations *(Maths.)*. The different selections that can be formed from a given number of items, order within each group being immaterial. For n items, all different, taken r at a time, there are $n!/r!.(n-r)!$ combinations. Denoted by ^nC_r or $\binom{n}{r}$. Cf. **permutations**.

combination set *(Eng.)*. A fitter's instrument comprising a universal protractor, spirit level, rule and straight-edge, centre head, and square.

combination tones *(Acous.)*. Additional tones produced by a non-linear system when two or more tones are applied. The combination tones have frequencies which are sums (summation tones) and differences (difference tones) of the frequencies of the applied tones.

combined carbon *(Eng.)*. In cast-iron, the carbon present as iron-carbide as distinct from that present as graphite. See **graphitic carbon**.

combined half-tone and line *(Print.)*. A printing surface on which half-tone and line work are combined.

combined-impulse turbine *(Eng.)*. An *impulse turbine* in which the first stage consists of nozzles that direct the steam on to a wheel carrying two rows of moving blades, between which a row of fixed guide blades is interposed.

combined system *(Build.)*. A system of sewerage in which only one set of sewers is provided for the removal of both the sewage proper and also rain water. Cf. *separate system*.

combining weight *(Chem.)*. See **equivalent weight**.

combustion *(Chem.)*. Chemical union of oxygen with gas accompanied by the evolution of light and rapid production of heat (exothermic).

combustion chamber *(Aero.)*. The chamber in which combustion occurs: (1) the cylinder of a reciprocating engine; (2) the individual chambers or single annular chamber of a gas turbine; (3) the combustion zone of a ramjet duct; (4) the chamber, with a single venturi outlet, of a rocket.

combustion chamber *(Eng.)*. (1) In a boiler furnace, the space in which combustion of gaseous products from the fuel takes place. (2) In an internal-combustion engine, the space above the piston (when on its inner dead-centre) in which combustion occurs.

combustion control *(Eng.)*. The control, either by an attendant or by automatic devices, of the rate of combustion in a boiler furnace, in order to adjust it to the demand on the boiler.

combustion noise *(Acous.)*. Noise caused by combustion. It can be particularly loud if combustion takes place in an acoustic resonator (e.g. industrial burner, jet aircraft) where a feedback between the released heat and the sound waves can lead to **instabilities**.

combustion tube furnace *(Eng.)*. Laboratory appliance having one or more horizontal refractory tubes heated by gas or electricity; used chiefly for the estimation of carbon content of steels, temperatures 1100° to 1300°C.

come-and-go *(Print.)*. See **fore-and-aft**.

comedo *(Med.)*. A blackhead. A collection of cells, sebum, and bacteria, filling the dilated orifices of the sebaceous glands near hair follicles.

comet *(Astron.)*. A member of the solar system, of small mass, becoming visible as it approaches the sun, partly by reflected sunlight, partly by fluorescence excited by the solar radiation. A bright nucleus is often seen, and sometimes a tail. This points away from the sun, its gases and fine dust being repelled by radiation pressure and the solar wind.

comfort behaviour *(Behav.)*. Behaviours that have to do with body care, e.g. grooming, scratching, preening.

comma *(Acous.)*. The pitch error, not greater than 80:81.1, arising from tuning one note in various ways with natural ratios from a datum note.

commag *(Image Tech.)*. International code name for a picture film combined with a magnetic sound-track.

command guidance *(Aero.)*. The guidance of missiles or aircraft by electronic, optical or wire-borne signals from an external source controlled by human operator or automatically.

command language *(Comp.)*. The **language** used to communicate with the operating system.

Commelinidae *(Bot.)*. Subclass or superorder of monocotyledons. Almost all are terrestrial herbs, often of moist places, with the perianth differentiated into sepals and petals, or reduced and not petaloid, syncarpous and usually with a starchy endosperm. Contains ca 19000 spp. in 25 families including Cyperaceae, Gramineae, Bromeliaceae and Zingiberaceae.

commensalism *(Biol.)*. An external, mutually beneficial partnership between 2 organisms (*commensals*), one partner may gain more than the other. adj. *commensal*.

commensurable quantities *(Maths.)*. Quantities, each of which is an integral multiple of a common basic quantity or measure, e.g. the numbers 6 and 15 are commensurable but 6 and $\sqrt{3}$ are not.

comminuted *(Med.)*. Reduced to small fragments, e.g. *comminuted* fracture.

comminuted powder *(Powder Tech.)*. Material reduced to a powder by e.g. attrition, impact, crushing, grinding, abrasion, milling or chemical methods.

comminution *(Min.Ext.)*. Size-reduction by breaking, crushing, or grinding, e.g. in ore-dressing.

commissural bundle *(Bot.)*. A small vascular bundle interconnecting larger bundles.

commissure *(Bot.)*. A juncture or suture, especially a surface by which carpels are united.

commissure *(Zool.)*. A joint; a line of junction between two organs or structures; a bundle of nerve-fibres connecting two nerve-centres.

commode step *(Build.)*. A step having a riser curved to present a convex surface; used sometimes at and near the foot of a staircase.

common ashlar *(Build.)*. A block of stone which is pick- or hammer-dressed.

common-base connection *(Electronics)*. The operation of a transistor in which the signal is fed between base and emitter, the output being between collector and base with the latter earthed. Also *grounded base*.

common bricks *(Build.)*. A class of brick used in ordinary construction (especially in interior work) for filling in, and to make up the requisite thickness of heavy walls and piers. They usually have plain sides, are not neatly finished, and are much more absorbent and also much weaker than *engineering bricks*.

common bundle *(Bot.)*. A vascular bundle belonging in part to a stem and in part to a leaf.

common-collector connection *(Electronics)*. The operation of a transistor in which the collector is earthed, input is between collector and base, and output is between emitter and collector. Also known as *grounded collector*, or *emitter follower*, this circuit provides relatively high input impedance with low output impedance. Voltage gain is unity, whilst current gain depends on transistor characteristics.

common command language *(Comp.)*. A command language used to access a number of different **information retrieval systems**.

common dovetail *(Build.)*. An angle-joint between 2 members in which both show end grain.

common-emitter connection *(Electronics)*. The operation of a transistor in which the signal is fed between base and emitter with the latter earthed. The output is betwen emitter and collector. Also *grounded emitter*.

common-frequency broadcasting *(Telecomm.)*. The use of the same carrier frequency by 2 or more broadcast transmitters, sufficiently separated for their useful service areas not to overlap. Also *shared-channel broadcasting*.

common joist *(Build.)*. See **bridging joist**.

common lead *(Eng.)*. Lead of lower purity than chemical or corroding lead (about 99.85%).

common-mode failure *(Nuc.Eng.)*. Failure of two or more supposedly independent parts of a system (e.g. a reactor) from a common external cause or from interaction between the parts.

common-mode rejection ratio *(Electronics)*. For a differential amplifier, the ratio of the gain for a differential input to that for a common-mode input.

common-mode signal *(Electronics)*. A signal applied simultaneously to both inputs of a differential amplifier.

common rafter *(Build.)*. A subsidiary rafter carried on the purlins and supporting the roof covering. Also rafters of a common length and bevel in any roof. Also *intermediate rafter*.

common-rail injection *(Autos.)*. A fuel-injection system for multi-cylinder C.I. engines; an untimed pump maintains constant pressure in a pipe line (rail), from which branches deliver the oil to the mechanically operated injection valves.

common return *(Telecomm.)*. A single conductor which forms the return circuit for 2 or more otherwise separate circuits.

communication *(Behav.)*. No generally accepted definition; different researchers use the term according to their interest in: (1) goal-directed behaviour used to influence other individuals; (2) species-specific behaviour adapted through evolution for a signalling function.

communication, non verbal *(Behav.)*. Communication by means other than language in its spoken or written form.

communications satellite *(Space)*. An artificial satellite to aid global communications by the relay of data, voice and television. The satellite can be purely reflective, but usually fulfills a repeater role using an on-board **transponder**, operated by solar power. It may orbit the Earth at any altitude but a **geosynchronous** (and particularly a **geostationary**) orbit is normally used to provide continuous coverage over a large area. See **direct broadcast satellite**.

community *(Bot.)*. Any group of plants growing together under natural conditions and forming a recognizable sort of vegetation e.g. oak wood, blanket bog.

community *(Zool.)*. The animals inhabiting a restricted area, as field or pond; such a community is not necessarily stable.

community antenna television *(Telecomm.)*. Used in cable TV systems which cover a town or similar area and is fed from a single central antenna, which may take signals from existing channels or from satellites.

commutating field *(Elec.Eng.)*. The magnetic field under the compoles of a d.c. machine; it induces, in the conductors undergoing commutation, an e.m.f. in a direction to assist in the commutation process.

commutating machine *(Elec.Eng.)*. An electrical machine provided with a commutator.

commutating pole *(Elec.Eng.)*. See **compole**.

commutation factor *(Electronics)*. Product of rate of current decay and rate of voltage rise after a gas discharge, both expressed per microsecond.

commutation switch *(Telecomm.)*. That controlling the sequential switching operations required for multichannel pulse communication systems.

commutative *(Maths.)*. An operation is commutative if the order in which it is performed on two quantities does not affect the result, e.g. the operation of addition in arithmetic is commutative because a added to b equals b added to a. Cf. **associative, distributive, transitive**.

commutator *(Elec.Eng.)*. The part of a motor or generator armature through which electrical connections are made by rubbing brush contacts.

commutator *(Maths.)*. Of two elements a and b of a group: $aba^{-1}b^{-1}$ or $a^{-1}b^{-1}ab$ according as operations are written on the left or the right respectively. Of two functions a and b: $ab-ba$. In either case written $[a,b]$.

commutator bar *(Elec.Eng.)*. One of the copper bars forming part of a commutator. Also called a *commutator segment*.

commutator bush *(Elec.Eng.)*. See **commutator hub**.

commutator face *(Elec.Eng.)*. See **commutator surface**.

commutator grinder *(Elec.Eng.)*. A portable electric grinding equipment which can be mounted on a commutator machine to grind the commutator surface without removing the armature from the machine.

commutator hub *(Elec.Eng.)*. A metal structure used for supporting a commutator. Also called a *commutator bush, commutator shell, commutator sleeve*.

commutator losses *(Elec.Eng.)*. Losses occurring at the commutator of an electric machine; they include resistance loss in the segments, in the brushes, and at the contact surface, friction loss due to the brushes sliding on the commutator surface, loss due to sparking, and eddy-current loss in the segment.

commutator motor *(Elec.Eng.)*. An electric motor which embodies a commutator.

commutator ring *(Elec.Eng.)*. A ring, usually of cast-iron, made to fit into a dovetail in the commutator segments in order to clamp them firmly in position. The term is also used to denote the insulating rings which have to be placed between the metal of the above ring and the commutator segments.

commutator ripple *(Elec.Eng.)*. Small periodic variations in the voltage of a d.c. generator or rotary converter resulting from the fact that there can only be a finite number of commutator segments on the machine.

commutator segment *(Elec.Eng.)*. See **commutator bar**.

commutator surface *(Elec.Eng.)*. The smooth portion of a commutator upon which the current-collecting brushes slide. Also called the **commutator face**.

comopt *(Image Tech.)*. International code name for a picture film combined with an optical sound-track.

compact *(Eng.)*. That produced by confining a metal powder, with or without nonmetallic material, and compressing it in a die.

compact disk *(Acous.)*. Digitally encoded read-only disk, read by laser. Commonly used in domestic hifi systems for high-quality sound reproduction.

compacted graphite cast iron *(Eng.)*. Made by a method like that for **ductile cast iron** but without allowing the formation of completely spherulitic graphite nodules so that the graphite shape is between that of grey and ductile cast iron. Abbrev. *CG iron*. Also *vermicular iron*.

compaction *(Build.,Civ.Eng.)*. The process of consolidating soil or dry concrete by mechanical means, or of wet concrete by vibration.

compact space, set *(Maths.)*. A topological space, or set within one, is compact if every collection of open sets which covers it completely includes a finite subcollection which does so. A closed and bounded set in n-dimensional Euclidean space is compact.

compander *(Telecomm.)*. Device for compressing the volume range of the transmitted signal and re-expanding it at the receiver, thus increasing the signal/noise ratio (*COMpresser...exPANDER.*)

companion cell *(Bot.)*. A parenchyma cell, with dense cytoplasm and a conspicuous nucleus, in the phloem of an angiosperm, adjacent to and originating from the same mother cell as a sieve tube member. Cf. **albuminous cell**.

companionship *(Print.)*. A number of compositors working together on adjacent portions of the same job. Called in Scotland *chumship*.

comparative psychology *(Behav.)*. Refers to a field of animal study associated mostly with North American psychologists; ideally it stresses attention to the issues of development, motivation and causality, and relies on the method of cross-species comparison for making generalisations about the mechanisms underlying behaviour. Some comparative psychologists focus their studies on a single species for convenience, e.g. the white rat, and tend to focus on the area of learning.

comparator *(Comp.)*. Circuit which compares the magnitudes of two numbers.

comparator *(Phys.)*. (1) A form of apparatus used for the accurate comparison of standard of length. It has also been used for measuring the coefficients of expansion of metal bars. (2) A form of colorimeter.

comparator *(Telecomm.)*. Circuit which compares two sets of impulses, and acts on their matching or otherwise.
comparison lamp *(Phys.)*. A lamp used, when performing photometric tests, for making successive comparisons between the lamp under test and a standard lamp.
comparison prism *(Phys.)*. A small right-angled prism placed in front of a portion of the slit of a spectroscope or spectrograph for the purpose of reflecting light from a second source of light into the collimator, so that two spectra may be viewed simultaneously. See **comparison spectrum**.
comparison spectrum *(Phys.)*. A spectrum formed alongside the spectrum under investigation, for the purpose of measuring the wavelengths of unknown lines. It is desirable that the comparison spectrum should contain many standard lines of known wavelength. The spectrum of the iron arc is often used for this purpose. See **comparison prism**.
comparison surface *(Phys.)*. A surface illuminated by a standard lamp or a comparison lamp; used in photometry.
compartment *(Biol.)*. Region of an insect embryo within which all cells give rise to the same adult structure, e.g. leg, forewing.
compartment *(For.)*. Permanent administrative units of a plantation delineated by well-defined natural features.
compass *(Surv.)*. Instrument which indicates either (a) magnetic bearings by alignment of its needle on the magnetic poles, or (b) reference bearing by radio signal, or (c) change of orientation in relation to a gyro-maintained line.
compass brick *(Build.)*. A brick which tapers in at least one direction; specially useful for curved work, in arches, parts of furnaces etc.
compasses *(Eng.)*. An instrument for describing arcs, taking or marking distances, etc.; it consists essentially of two limbs hinged together at one end.
compass plane *(Build.)*. One of use on a curved surface, having a flexible metal sole which can be set concave or convex.
compass roof *(Arch.)*. A roof with rafters bent to the shape of an arc.
compass safe distance *(Aero.,Ships)*. The minimum distance at which equipment may safely be positioned from a direct-reading magnetic compass, or detector unit of a remote-indicating compass, without exceeding the values of maximum compass deviation change.
compass saw *(Build.)*. A narrow bladed saw for cutting small radius curves.
compass traverse *(Surv.)*. Rapid rough survey method in which the magnetic bearing of each line is measured.
compatible *(Comp.)*. Computer hardware is *compatible* if it can use the same software. Software is *upwardly compatible* if it can continue to be used when a system is improved.
compatible colour television *(Image Tech.)*. The technique of transmitting television pictures in colour by a combination of **luminance** and **chrominance signals**, compatibility with black-and-white television being preserved as the chrominance elements are virtually disregarded by the monochrome receiver.
compatible equations *(Maths.)*. See **consistent equations**.
compensated induction motor *(Elec.Eng.)*. An induction motor with a commutator winding on the rotor, in addition to the ordinary primary and secondary windings; this winding is connected to the circuit in such a way that the motor operates at unity or at a leading power factor.
compensated pendulum *(Phys.)*. A pendulum made of 2 materials which have different coefficients of expansion and are so chosen that the length of the pendulum remains constant when the temperature varies.
compensated semiconductor *(Electronics)*. Material in which there is a balanced relation between **donors** and **acceptors**, by which their opposing electrical effects are partially cancelled.
compensated series motor *(Elec.Eng.)*. The usual type of a.c. series motor, in which a compensating winding is fitted to neutralize the effect of armature reaction and so give a good powerfactor. Also called *neutralized series motor*.
compensated shunt box *(Elec.Eng.)*. A shunt box for use with a galvanometer, arranged so that on each step a resistance is put in series with the galvanometer, and the total resistance of galvanometer and shunt is not altered.
compensated voltmeter *(Elec.Eng.)*. A voltmeter arranged to indicate the voltage at the remote end of a feeder or other circuit, although connected at the sending end. A special winding compensates for the voltage drop in the feeder.
compensated wattmeter *(Elec.Eng.)*. A wattmeter in which there is an additional winding, arranged to compensate for the effect of the current flowing in the voltage circuit.
compensating coils *(Elec.Eng.)*. Current carrying coils to adjust distribution of magnetic flux.
compensating diaphragm *(Surv.)*. A fitment for a tacheometer which, by an adjustment to the stadia interval determined by the vertical angle, enables the horizontal component of a sloping sight to be deduced from the staff intercept.
compensating error *(Surv.)*. As opposed to systematic (biased) error in series of observations, one equally likely to be due to over- or under-measurement, and therefore reasonably likely to be conpensated by errors of opposite sign.
compensating field *(Elec.Eng.)*. A term sometimes used to indicate the field produced by a compensating winding or, occasionally, by a compole.
compensating filter *(Image Tech.)*. See **CC filter**.
compensating jet *(Autos.)*. An auxiliary petrol jet used in some carburettors to supplement the discharge from the main jet at low rates of air flow, and to keep the mixture strength constant. See **carburettor**.
compensating plate lock-up *(Print.)*. A plate lock-up with resilient means to compensate for imperfections in the bevel of plates.
compensating pole *(Elec.Eng.)*. See **compole**.
compensating roller *(Print.)*. See **jockey roller**.
compensating winding *(Elec.Eng.)*. A winding used on d.c. or a.c. commutator machines to neutralize the effect of armature reaction.
compensation *(Acous.)*. In a sound-reproducing system, adjustment of an actual frequency response to one specified.
compensation balance *(Horol.)*. A balance so constructed as to compensate for the changes of dimensions in the balance and the elastic properties of the balance spring, with changes of temperature. Actually, compensation cannot be complete over a range of temperature. A watch or chronometer adjusted for the extremes of the range will not be correct at the middle of the range. This gives rise to what is known as the 'middle-temperature error'.
compensation pendulum *(Horol.)*. A pendulum so constructed that the distance between the centre of oscillation and the point of suspension remains constant with changes of temperature. See **pendulum**.
compensation point *(Bot.)*. *CO_2 compensation point*. The light intensity at which, under specified conditions, photosynthesis and respiration just balance so that there is no net exchange of CO_2 nor O_2. For C_3 plants it is 50–70 ppm, for C_4 plants 0–10 ppm.
compensation theorem *(Phys.)*. That the change in current produced in a network by a small change in any impedance Z carrying a current I is the result of an apparent e.m.f. of $-I.\delta Z$.
compensation water *(Civ.Eng.)*. The water which has to be passed downstream from a reservoir to supply users who, prior to the construction of the dam, took their water directly from the stream.
compensator *(Image Tech.)*. A device such as a graduated filter used with an extreme **wide-angle** lens to improve uniformity of illumination of the image area.
compensator *(Phys.)*. (1) A glass plate used in various optical interferometers to achieve equality of optical path length. (2) A plate of variable thickness of optically active

quartz used to produce elliptically polarized light of a given orientation.

competition *(Biol.)*. The struggle between organisms for the necessities of life (water, light, etc.).

competitive exclusion principle *(Ecol.)*. The ecological 'law' that two species cannot occupy the same ecological niche or utilize the same limiting resource. One species always outcompetes the other.

compilation error *(Comp.)*. Error detected during compilation (e.g. a **syntax error**).

compiler *(Comp.)*. Program which translates a **high-level language** program into a computer's **machine code** or some other **low-level language**. Each high-level language instruction is changed into several machine-code instructions. It produces an independent program which is capable of being executed. n. *compilation*. See **execute**. Cf. **interpreter**, **assembler**.

complanate *(Bot.)*. Flattened, compressed.

complement *(Biol.)*. See **chromosome complement**, **chromosome set**.

complement *(Immun.)*. Name given to a system of proteins in the blood which act in series to produce a variety of biological effects, The system is triggered by antibodies when they become cross-linked through combination with antigens, irrespective of the antibody specificity. The *classical pathway* consists of 9 separate components, of which C1 to C5 are enzymes (specific esterases) in an inactive form. Combination of C1 with aggregated immunoglobulin Fc activates it; it activates in turn the next component by splitting off small peptides; this in turn acts on the next component *seriatim*. Results in a mechanism which amplifies the effects at each stage, but the activations are transient and regulated by inhibitory substances also present in blood. The most important products of the enzyme cascade are C3b and the peptides **C3a** and **C5a** which aid engulfment and killing of microbes by phagocytic cells and initiate inflammatory reactions. $C5^-$ acts on C6 and this causes the remaining components C6 to C9 to arrange themselves into tubular structures which become inserted in the lipid layer of the outer cell membrane, causing the cell to leak its contents. Many cells with attached antibody and some microbes can be killed this way. The *alternative pathway* of complement activation, which can be triggered by a variety of microbial polysaccharides and by some antigen-complexes, does not involve C1, C2, C4 but is initiated by combination with **properdin** and two factors, D and B, which activate C3 to C3b. The biological effects of either pathway are similar.

complementarity *(Phys.)*. In quantum mechanics the wave and particle models are complementary. There is a correspondence between particles of momentum p and energy E and the associated wavetrain of frequency $\nu = E/h$ and wavelength $\lambda = h/p$, where h is Planck's constant. A measurement proving the wave character of radiation on matter cannot prove the particle character in the same measurement, and conversely.

complementary *(Biol.)*. Relationships between single strands of DNA and RNA are complementary to each other if their sequences are related by the *base-pairing rules*, thus ATCG is complementary to the sequence TAGC, and can pair with it by hydrogen bonding.

complementary after-image *(Phys.)*. The subjective image, in complementary colours, that is experienced after visual fatigue induced by observation of a brightly coloured object.

complementary angles *(Maths.)*. Two angles whose sum is 90°. Each is said to be the complement of the other.

complementary colours *(Build.)*. Paint colours which lie opposite in the chromatic circle giving maximum contrast. Sometimes called *contrasting colours*.

complementary colours *(Phys.)*. Pairs of colours which combine to give spectral white.

complementary function *(Maths.)*. If $y = u + v$ is a solution of a differential equation, where u is a particular integral and v contains the full number of arbitrary constants, v is called the *complementary function*. The complementary function is the general solution of the auxiliary equation.

complementary genes *(Biol.)*. Two *non-allelic* genes which must both be present for the manifestation of a particular character.

complementary medicine *(Med.)*. Term used to denote systems of treatment not fully accepted by orthodox medical science. Homeopathy, osteopathy, cheiropraxis, acupuncture and herbal medicine are some of the best recognized examples.

complementary metal-oxide silicon *(Electronics)*. Used to describe logic arrays (*integrated circuits*), using **MOS** construction, which employ gates consisting of complementary pairs of p-channel and n-channel **FETs**. The switching signal is applied to both gates, so that only one FET is *on* at one time; significant current flows only during the switching operation, so that power consumption is greatly reduced. CMOS circuits are used where low heat dissipation and power consumption are important design criteria.

complementary symmetry *(Electronics)*. Shown by otherwise identical *p-n-p* and *n-p-n* transistors.

complementary transistors *(Electronics)*. A *n-p-n* and a *p-n-p* transistor pair used to produce a push-pull output using a common signal input.

complementation *(Biol.)*. The full or partial restoration of normal function when two recessive mutants, both deficient in that function, are combined in a double heterozygote. *Complementing* mutants are *non-allelic*, *non-complementing* are *allelic*.

complementation *(Comp.)*. Method of representing negative numbers. See **one's complement**, **two's complement**.

complement deficiency *(Immun.)*. Hereditary deficiencies of complement components are uncommon in man although strains of laboratory animals exist lacking C3, C4, C5 or C6. Absence of any single component is compatible with life but persons lacking the early acting components, especially with diminished C3, are unusually liable to bacterial infection and often show signs of immune complex disease. The genes for C4, C2 and Factor B lie within the **major histocompatibility complex**.

complement fixation *(Immun.)*. Term synonymous with activation but often applied to a system *in vitro* which detects complement by its capacity to cause lysis of red cells with antibody on the surface. Activation of complement by combination of antigen with antibody prior to adding the red cells diminishes the amount of complement available to lyse the red cells. This is a sensitive method for detecting the presence of antigen or antibody but has been superseded by other methods such as **ELISA**.

complement of a set *(Maths.)*. The complement of a set A is the set of all members of the universal set which are not members of A, and is sometimes denoted by A'.

complete combustion *(Eng.)*. Burning fuel without trace of unburnt gases in the products of combustion, usually accompanied by excess air in the flue products. Cf. **perfect combustion**.

complete differential *(Maths.)*. The complete differential df of a function $f(x,y,z)$ is

$$\frac{\partial f}{\partial x}\cdot dx + \frac{\partial f}{\partial y}\cdot dy + \frac{\partial f}{\partial z}\cdot dz.$$

Similar expressions apply for more or less variables. Also called *total differential*.

complete Freunds adjuvant *(Immun.)*. A water in oil emulsion with added heat-killed mycobacteria into which is incorporated an antigen for the purpose of immunization against it. This form of adjuvant is very effective for eliciting both T- and B-cell immunity, but it is not used in humans because it is liable to cause suppurating granulomas. Abbrev. *CFA*.

complete integral *(Maths.)*. See **complete primitive**.

complete metric space *(Maths.)*. If, for any distance ε, we can choose an N such that the distance between a_m and a_n is less than ε for all m and n greater than N, then the a_r are called a Cauchy sequence. A metric space in which all Cauchy sequences have limits is complete.

complete primitive *(Maths.)*. The solution of a differential equation containing the full number of arbitrary constants. Also called *general* or *complete solution, general*, or *complete integral*.

complete radiator *(Phys.)*. Same as **black body**.

complete reaction *(Chem.)*. A reaction which proceeds until one of the reactants has effectively disappeared.

complete set of functions *(Maths.)*. See **orthogonal functions**.

complete solution *(Maths.)*. See **complete primitive**.

complex *(Behav.)*. A term introduced by Jung to denote an emotionally toned constellation of mental factors formed by the attachment of instinctive emotions to objects or experiences in the environment, and always containing elements unacceptable to the self. It may be recognized in consciousness, but is usually repressed and unrecognized.

complex amplitude *(Acous.)*. Complex number with amplitude and phase information of a harmonic signal, e.g. sound pressure.

complex hyperbolic functions *(Phys.)*. Hyperbolic functions, with complex quantities as variables, which facilitate calculations of electric waves along transmission lines.

complex ion *(Chem.)*. See **co-ordination compound**.

complexity *(Comp.)*. See **computational complexity**.

complexity of DNA, RNA *(Biol.)*. A measure, obtained from renaturation kinetics, of the number of copies of a given sequence in, e.g. a genome or the mRNA in a cell.

complex number *(Maths.)*. A number z of the form $a+ib$, where $i=\sqrt{-1}$ and a and b are real numbers. The number a is called the *real part*, written Rz, and b the *imaginary part*, written Iz. E.g. if $z=3+4i$, $Rz=3$ and $Iz=4$. Cf. **modulus** and **argument**.

complexometric titration *(Chem.)*. Titration of a metal ion with a reagent, usually *EDTA*, which forms chelate complexes with the metal. The end point is accompanied by a sharp decrease in the concentration of metal ions, and is observed by a suitable indicator.

complexones *(Chem.)*. Collective term used to denote chelating reagents (usually organic) used in the analytical determination of metals, e.g. *EDTA*.

complex Poynting vector *(Elec.Eng.)*. See **Poynting vector**.

complex tissue *(Bot.)*. A tissue made up of cells or elements of more than one kind.

complex tone *(Acous.)*. Strictly, a musical note in which all the separate tones are exact multiples of a fundamental frequency, recognized as the pitch, even when the actual fundamental is absent, as in the lowest octave of the piano. Loosely, a mixed musical chord.

complex wave *(Phys.)*. One with a nonsinusoidal waveform which can be resolved into a fundamental with superimposed harmonics. See **Fourier principle**.

compliance *(Phys.)*. Mechanical compliance is the linear displacement produced by unit force in a vibrating system. It is the reciprocal of **stiffness** and is the mechanical analogue of capacitance.

complicate *(Bot.,Zool.)*. Folded together.

compo *(Build.)*. A cement mortar.

compole *(Elec.Eng.)*. An auxiliary pole, employed on commutator machines, which is placed between the main poles for the purposes of producing an auxiliary flux to assist commutation. Also called *commutating pole*, **compensating pole, interpole**.

component *(Elec.Eng.)*. A term sometimes used to denote one of the component parts into which a vector representing voltage, currents, or volt-amperes may be resolved; the component parts are usually in phase with, or in quadrature with, some reference vector. See **active voltage, active current, active volt-amperes, reactive voltage, reactive component of current, reactive volt-amperes**.

component *(Image Tech.)*. System in which the separate components of **chroma, luminance**, sound and synchronization for each line are digitally coded and transmitted in sequence with storage and restoration at the receiver.

component *(Phys.)*. The resolved part of a force (or velocity, acceleration, momentum, etc) in any particular direction, e.g. the component of a force **F** along a line making an angle θ with the line of action of **F** is **F** cos θ. See **resolution of forces**.

component of a vector *(Maths.)*. In general, one of two or more vectors whose sum is the given vector, but unless the contrary is stated the component in a specified direction is the vector in that direction, whose sum with one or more vectors in directions perpendicular to the specified direction is the given vector.

components *(Chem.)*. The individual chemical substances present in a system. See **phase rule**.

compose *(Print.)*. To assemble type matter for printing, either by hand or by type-setting machines.

Composertron *(Acous.)*. Complex magnetic-tape recording assembly, in which elements of recorded sound can be synthesized and reproduced as a composition.

composing frame *(Print.)*. A wooden or metal structure at which the compositor works. Originally providing accommodation on top for an upper and lower case with storage below for other cases, there is now a large variety of styles to suit the requirements of the work to be done, with sloping or flat tops, and storage for all kinds of typographic material in addition to cases.

composing machines *(Print.)*. The *Monotype* composes type matter in separate letters, which may be used for hand-setting or correcting; the *Linotype* composes in solid lines, or slugs, which must be reset when correcting. Both machines have a keyboard resembling that of a typewriter. See also **Intertype** and **filmsetting**.

composing rule *(Print.)*. A piece of brass or steel rule with a projecting nose piece, moved up as each line in the stick is completed by the compositor.

composing stick *(Print.)*. A metal or wooden 3-sided boxlike receptacle in which the compositor sets his type letter by letter. The width or measure can be altered as desired.

Compositae *(Bot.)*. *Asteraceae*. The daisy family, ca 25 000 spp. of dicotyledonous flowering plants (superorder Asteridae). The largest family of dicotyledons. Mostly herbs and shrubs, cosmopolitan. The inflorescence is a head (*capitulum*) made up of many small individual florets or florets surrounded by an involucre of bracts, the whole resembling, and functioning biologically as a single flower. The florets have a gamopetalous corolla that is usually **tubular** or **ligulate**, and the ovary is inferior and develops into a one-seeded indehiscent, dry fruit. The modified calyx or pappus is composed of hairs, scales or bristles, and often develops as a feathery parachute aiding wind dispersal. Includes relatively few economic plants, such as sunflower (for oil), lettuce, endive, chicory and a number of ornamentals, e.g. chrysanthemum and various daisies. The insecticide pyrethrum comes from the heads of a species of *Tanacetum*.

composite *(Textiles)*. A matrix such as cement or plastic reinforced by fibres.

composite balance *(Elec.Eng.)*. A modification of the Kelvin balance; the moving coils are of high resistance, enabling the instrument to be used as a wattmeter.

composite beam *(Eng.)*. A beam composed of two materials properly bonded together and having different moduli of elasticity, e.g. a reinforced concrete beam.

composite block *(Print.)*. (1) Combined half-tone and line block. (2) A block made up from 2 or more originals.

composite cable *(Elec.Eng.)*. Cable containing different purpose conductors inside a common sheath.

composite compact *(Eng.)*. A **compact** made by powder metallurgy which has several adherent layers of different alloys.

composite conductor *(Elec.Eng.)*. One in which strands of different metals are used in parallel.

composite material *(Eng.)*. Structural material made of two or more different materials e.g. cermet or carbon-fibre reinforced epoxy resin.

composite photography *(Image Tech.)*. General term for cinematographic **special effects** in which two or more separate shots are combined to give the effect of a single scene. See **back projection, front projection, travelling matte shot** etc.

composite resistor *(Elec.Eng.)*. One formed of solid rod of carbon compound.

composite structure *(Eng.)*. Any structure made by bonding two or more different materials, such as metal, plastic, composite material etc.

composite truss *(Build.)*. A roof truss formed of timber struts and steel or wrought-iron ties (apart from the main tie, which is usually of timber to simplify fixings).

composite yarn *(Textiles)*. Yarn made from a combination of staple fibres and continuous filaments.

composition *(Chem.)*. The nature of the elements present in a substance and the proportions in which they occur.

composition founts *(Print.)*. The smaller sizes of type, up to 14-point, as used for bookwork.

composition nails *(Build.)* Roofing nails made of a cast 60-40 copper-zinc alloy.

composition of atmosphere *(Chem.)*. Dry atmospheric air contains the following gases in the proportions (by weight) indicated: nitrogen, 75.5; oxygen, 23.14; argon, 1.3; carbon dioxide, 0.05; krypton, 0.028; xenon, 0.005; neon, 0.000 86; helium, 0.000 056. There are variable trace amounts of other gases incl. hydrogen and ozone. Water content, which varies greatly, is excluded from this analysis.

composition of forces *(Phys.)*. The process of finding the resultant of a number of forces, that is a single force which can replace the other forces and produce the same effect. See **parallelogram of forces**.

composition rollers *(Print.)*. (1) For letterpress printing, a mixture of glue, glycerine and molasses. (2) For lithographic printing, vegetable oils and rubber, vulcanized.

compositor *(Print.)*. A craftsman whose work consists of setting up type matter by hand, or correcting that set by machine. Skill and judgment in display work are part of his routine.

compost *(Bot.)*. (1) Rotted plant material and/or animal dung etc. used as a soil conditioner. (2) A medium in which plants (especially plants in pots) are grown, composed of one or more of sand, soil, grit, peat, perlite, vermiculite etc. with lime and fertilizers as necessary.

compound *(Bot.)*. Consisting of several parts: a leaf made up of several distinct leaflets; an inflorescence of which the axis is branched etc. Cf. **simple**.

compound *(Chem.)*. See **chemical compound**.

compound arch *(Arch.)*. An arch having an **archivolt** receding in steps, so as to give the appearance of a succession of receding arches of varying spans and rises.

compound brush *(Elec.Eng.)*. A type of brush used for collecting current from the commutator of an electric machine; the brush has alternate layers of copper and carbon so that the conductivity is greater longitudinally (i.e. in the direction of the main current flow) than laterally.

compound catenary construction *(Elec.Eng.)*. A construction used for supporting the overhead contact wire of an electric traction system; the contact wire is supported from an auxiliary catenary which, in turn, is supported from a main catenary, all 3 wires lying in the same plane.

compound curve *(Surv.)*. A curve composed of two arcs of different radii, having their centres on the same side of the curve, connecting two straights.

compound dredger *(Civ.Eng.)*. A type of dredger combining the suction or suction cutter apparatus with a bucket ladder.

compound engine *(Eng.)*. A development of the *simple* steam engine, the compound engine has two or more cylinders of different size, allowing the steam to expand over several stages and enabling more work to be done per unit mass of steam and thus give greater efficiency at the cost of increased complexity.

compound eyes *(Zool.)*. Paired eyes consisting of many facets or ommatidia, in most adult Arthropoda.

compound fault *(Geol.)*. A series of closely spaced parallel or subparallel faults.

compound filled apparatus *(Elec.Eng.)*. Electrical apparatus (e.g. bus-bars, potential transformers, switchgear) in which all live parts are enclosed in a metal casing filled with insulating compound.

compound generator *(Elec.Eng.)*. See **compound motor**.

compound girder *(Build.)*. A rolled-steel joist strengthened by additional plates riveted or welded to the flanges.

compounding *(Eng.)*. The principle, or the use of the principle, of expanding steam in two or more stages, either in reciprocating engines or steam-turbines.

compound lever *(Eng.)*. A series of levers for obtaining a large mechanical advantage, the short arm of one being connected to the long arm of the next; used in large weighing and testing machines.

compound magnet *(Elec.Eng.)*. A permanent magnet made up of several laminations.

compound microscope *(Phys.)*. See **microscope**.

compound modulation *(Telecomm.)*. Use of an already modulated wave as a further modulation envelope. Also called *double modulation*.

compound motor, generator *(Elec.Eng.)*. One which has both series and shunt field windings.

compound nucleus *(Phys.)*. In certain nuclear reactions, the bombarding particle forms a highly excited unstable *compound nucleus* with the target nucleus. This compound nucleus decays to complete the reaction.

compound pendulum *(Phys.)*. Any body capable of rotation about a fixed horizontal axis and in stable equilibrium under the action of gravity. If the centre of gravity is a distance h from the axis, and k is the *radius of gyration* about the horizontal axis through the centre of gravity, the period of small oscillations is

$$T = 2\pi\sqrt{\frac{h^2 + k^2}{hg}}.$$

compound pillar *(Build.)*. A pillar formed of a rolled-steel joist or channels strengthened by additional plates riveted or welded to the flanges.

compound press tool *(Eng.)*. A **press tool** which performs two or more operations at the same station at each stroke of the press.

compound reflex *(Zool.)*. A combination of several reflexes to form a definite coordination, either simultaneous or successive.

compound slide rest *(Eng.)*. Mounted on the upper face of the lathe cross-slide and carrying the tool post. Can be rotated or *set over* for cutting short internal or external tapers.

compound train *(Eng.)*. A train of gear-wheels in which intermediate shafts carry both large and small wheels, in order to obtain a large speed ratio in a small space.

compressed air *(Eng.)*. Air at higher than atmospheric pressure. It is used (often at about $600\,kN/m^2$) as a transmitter of energy where the use of electricity or an IC engine would be hazardous (e.g. in mining). The exhaust air may be used for cooling or ventilation.

compressed-air capacitor *(Elec.Eng.)*. An electric capacitor in which air at several atmospheres' pressure is used as the dielectric, on account of its high dielectric strength at these pressures.

compressed-air disease *(Med.)*. See **caisson disease**.

compressed-air inspirator *(Eng.)*. Injector used with pressure-air burners, by which a stream of compressed air is directed through a venturi throat to inspire additional combustion air.

compressed-air lamp *(Min.Ext.)*. An electric lamp for use in fiery mines; it is supplied from a small compressed-air-driven generator incorporated in the lamp-holder.

compressed-air tools *(Eng.)*. See **pneumatic tools**.

compressed-air wind tunnel *(Aero.)*. See **variable-density wind tunnel**.

compressibility *(Phys.)*. The reciprocal of the bulk modulus. See also **coefficient of compressibility**.

compressibility *(Powder Tech.)*. The property of a powder by which it accepts reduction in volume by pressure. It is measured as the ratio of the volume of loose powder to the volume of the compact, and is related to the pressure applied. This is sometimes referred to as the *compression ratio.*

compressibility drag *(Aero.).* The sharp increase of drag as air speed approaches the speed of sound and flow characteristics change from those of a viscous to those of a compressible fluid, causing the generation of shock waves.

compression cable *(Telecomm.).* See **pressure cable**.

compression fitting *(Build.).* See **capillary fitting**.

compression-ignition engine *(Eng.).* An internal-combustion engine in which ignition of the liquid fuel injected into the cylinder is performed by the heat of compression of the air charge. See **diesel engine**.

compression moulding *(Plastics).* The material is placed in a heated, hardened, polished steel container and forced down by means of a plunger at a pressure of ca. 20–35 MN/m^2.

compression plate lock-up *(Print.).* A method of locking plates to the cylinder by movable dogs which press the plate from the angled edge at the side towards the centre of the cylinder. Cf. **tension plate lock-up**.

compression ratio *(Eng.).* In an internal-combustion engine, the ratio of the total volume enclosed in the cylinder at the outer dead-centre to the volume at the end of compression; the ratio of swept volume, plus clearance volume, to clearance volume. See **clearance volume**.

compression ratio *(Powder Tech.).* See under **compressibility**.

compression rib *(Aero.).* See **rib**.

compression spring *(Eng.).* A helical spring with separated coils, or a conical coil spring, with plain, squared or ground ends, made of round, oblong, or square-section wire.

compression test *(Eng.).* A test in which specimens are subjected to an increasing compressive force, usually until they fail by cracking, buckling or disintegration. A stress-strain curve may be plotted to determine mechanical properties, as in a **tensile test**. Compression tests are often applied to materials of high compression but low tensile strength, such as concrete.

compression wood *(Bot.).* The **reaction wood** of conifers, as formed on the underside of horizontal branches, characterized by denser structure, higher lignin content and microfibrils in shallower helices than normal wood.

compressive shrinkage *(Textiles).* A process of forcing a fabric to shrink in length by subjecting it to compressive forces. This makes the fabric less prone to shrink in use.

compressor *(Aero.).* Compresses the air supply to a **gas turbine**.

compressor *(Eng.).* A reciprocating or rotary pump for raising the pressure of a gas.

compressor *(Telecomm.).* An electronic amplifier designed to reduce the dynamic range of speech, for transmission at an average higher level in the presence of interference.

compressor *(Zool.).* A muscle which by its contraction serves to compress some organ or structure.

compressor drum *(Aero.).* A cylinder composed of a series of rings or, more usually, disks wherein the blades of an axial compressor are mounted.

Compton absorption *(Phys.).* That part of the absorption of a beam of X-rays or γ-rays associated with Compton scattering processes. In general, it is greatest for medium-energy quanta and in absorbers of low atomic weight. At lower energies **photoelectric absorption** is more important, and at high energies **pair production** predominates.

Compton effect *(Phys.).* Elastic scattering of photons by electrons, i.e. scattering in which both momentum and energy are conserved. If λ_s and λ_i are respectively the wavelengths associated with scattered and incident photons, the Compton shift is given by

$$\lambda_s - \lambda_i = \lambda_o(1 - \cos\theta)$$

where θ is the angle between the directions of the incident and scattered photons and λ_0 is the *Compton wavelength* ($\lambda_0 = 0.002\,43$ nm) of the electron. The effect is only significant for incident X-ray and γ-ray photons.

Compton recoil electron *(Phys.).* An electron which has been set in motion following an interaction with a photon (**Compton effect**).

Compton scatter *(Radiol.).* A change in the direction of travel of a photon due to the interaction between the photon and the tissue. This is the major cause of loss of resolution in radionuclide imaging.

Compton's rule *(Chem.).* An empirical rule that the melting point of an element in kelvin is equal to half the product of the r.a.m. and the specific latent heat of fusion.

Compton wavelength *(Phys.).* Wavelength associated with the mass of any particle, such that $\lambda = h/mc$ where λ = wavelength, h = Planck's constant, m = rest mass and c = velocity of light.

compulsion *(Behav.).* An action which an individual may consider irrational but feels compelled to do.

computability *(Comp.,Maths.).* A property of **functions**. A function is computable if it can be proved that there exists a **Turing machine** to evaluate it at any given point.

computational complexity *(Comp.).* The study of the intrinsic difficulty of computing solutions to different types of mathematically posed problems.

computed tomography *(Radiol.).* Reconstruction of cross-sectional images of the body made by a rotating X-ray source and detector which move around the body and records the X-ray transmissions throughout the 360° rotation. A computer reconstructs the image in the **slice**. Also *CAT scanner, computer aided tomography.*

computer *(Genrl.).* (1) A device or set of devices that can store data and a program that operates on the data. A general purpose computer can be programmed to solve any reasonable problem expressed in logical and arithmetical terms. The first general purpose automatic computing machine called **analytical engine** was only partly built. The first fully operational general purpose computer, electromechanical and using binary digits, was probably the Z3, built in Germany by Konrad Zuse in 1941. Electronic **stored program** computers are **digital**. See also **analogue computer** or **hybrid computer**. (2) Widely used to mean an *electronic digital computer.* An electronic device or set of devices that can store **data** and **programs**, and **executes** the programs by performing simple **bit** operations at high speed. It normally has a **central processor, fast store, back-up store** and at least one **I/O device**. See **computer generations, desk-top computer, personal computer**. (3) Widely used to refer to a **computer system** which incorporates at least one *digital computer.* (4) An abstract model for a computing machine. See **Turing machine**. (5) A *calculator.*

computer aided design *(Comp.).* The use of the computer particularly with **high resolution graphics** in a wide range of design activities from the design of cars to the layout of **chips**. The designs can be modified and evaluated rapidly and precisely. Abbrev. *CAD.*

computer aided engineering *(Eng.).* The application of computers to manufacturing processes in which manual control of machine tools is replaced by automatic control resulting in increased accuracy and efficiency. Abbrev. *CAE.*

computer aided learning *(Comp.).* Encompasses the many ways in which a student or teacher in any field may make use of a computer. Also *computer assisted learning, computer augmented learning.* Abbrev. *CAL.*

computer aided machining *(Eng.).* A general term used to describe manufacturing processes which are computer controlled. The data from a **computer aided design** system (*CAD*) is used directly to produce the program needed for the machine control. Formerly developed as a separate system is now often integrated with CAD to form a complete design and manufacturing facility. Abbrev. *CAM.*

computer aided tomography *(Radiol.).* See **computed tomography**.

computer architecture *(Comp.).* The structure, behaviour and design of computers. See **computer generations, central processor, distributed computing**.

computer assisted instruction *(Comp.).* Use of the computer to provide educational exercises, e.g. *PLATO.*

Abbrev. *CAI.* Cf. **computer aided learning, computer-managed instruction.**

computer generations *(Comp.).* Convenient means of expressing stages in the advance of digital computer technology. See **first generation computer, second-, third-, fourth-, fifth-.**

computer graphics *(Comp.).* The automatic handling of diagrams, pictures and drawings. The **interactive** input and modification of drawings. See **joystick, light pen, mouse, graphics tablet, plottter, high-resolution graphics, raster graphics, vector graphics.**

computer-managed instruction *(Comp.).* Computer assistance to teachers for testing and keeping records.

computer science *(Comp.).* The study with the aid of computers, of computable processes and structures. It has many branches, for example: **Artificial Intelligence, Information Retrieval, computer architecture, software engineering, formal language theory, computability** and **complexity, cybernetics, communications, database management systems.**

computer system *(Comp.).* A linked system of processors and **I/O devices** with the **software** necessary to make them operate as a **computer.**

computer typesetting *(Print.).* The use of electronic equipment to process an unjustified input into an output of justified and hyphenated lines. The output can be a new tape, or disk to be used on a phototypesetting machine or it can be the final product in some filmsetting systems. Some equipment can accept a random input of items and produce a classified and alphabetically arranged output.

computer vision *(Comp.).* The objective of giving computers the power to 'see' and to interpret what they 'see'.

COMSAT *(Space).* Abbrev. for *Communications Satellite Corporation,* an organization which provides satellite services for international transmission of data.

concanavalin A *(Immun.).* A **lectin** derived from jack beans (*Canavalia ensiformis*) which binds to oligosaccharides present in the membrane glycoproteins on many cells. The lectin has 4 binding sites and so can cause cross-linking of the glycoproteins. it is a very effective polyclonal mitogen for T-cells, causing them to secrete **lymphokines.**

concave brick *(Build.).* A **compass brick.**

concave grating *(Phys.).* A diffraction grating ruled on the surface of a concave spherical mirror, made usually of speculum metal or glass. Such a grating needs no lenses for collimating or focusing the light. Largely on this account it is the most useful means of producing spectra for precise measurement. See **Rowland circle.**

concave lens *(Phys.).* A *divergent lens.*

concave mirror *(Phys.).* A curved surface, usually a portion of a sphere, the inner surface of which is polished. It is capable of forming real and virtual images, their positions being given by the equation $2/r = 1/l' + 1/l$, where r is the radius of curvature of the surface, l the distance of the object and l' the distance of the image from the mirror. (Cartesian **convention of signs.**)

concealed heating *(Build.).* See **panel heating.**

conceal/reveal *(Image Tech.).* Visual transition effect in which one picture appears to slide across another.

concentrate *(Min.Ext.).* The products of concentration operations in which a relatively high content of mineral has been obtained and which are ready for treatment by chemical methods.

concentrated load *(Build.).* A load which is regarded as acting through a point.

concentrating table *(Min.Ext.).* Supported deck, across or along which mineralized sands are washed or moved to produce differentiated products according to the gravitational response of particles of varying size and/or density. Stationary tables include strakes, sluices, buddles; moving tables include shaking tables (e.g. *Wilfley table*), vanners, and rockers.

concentration *(Chem.).* (1) Number of molecules or ions of a substance in a given volume, generally expressed as moles per cubic metre or cubic decimetre. (2) The process by which the concentration of a substance is increased, e.g. the evaporation of the solvent from a solution.

concentration *(Eng.).* Production of **concentrate.**

concentration cell *(Phys.).* One with similar electrodes in common electrolyte, the e.m.f. arising from differences in concentration at the electrodes.

concentration plant *(Min.Ext.). Concentrator mill, reduction works, washing, cleaning plant.* Buildings and installations in which ore is processed by physical, chemical and/or electrical methods to retain its valuable constituents and discard as tailings those of no commercial interest. See **mineral processing.**

concentration polarization *(Phys.).* A form of polarization occurring in an electrolytic cell, due to changes in the concentration of the electrolyte surrounding the electrode.

concentric *(Telecomm.).* Term replaced by *coaxial.*

concentric arch *(Arch.).* An arch laid in several courses whose curves have a common centre.

concentric chuck *(Eng.).* See **self-centring chuck.**

concentric plug-and-socket *(Elec.Eng.).* A type of plug-and-socket connection in which one contact is a central pin and the other is a ring concentric with it.

concentric vascular bundle *(Bot.).* A bundle in which a strand of xylem is completely surrounded by a sheath of phloem (*amphicribral*) or vice versa (*amphivasal*).

concentric winding *(Elec.Eng.).* An armature winding, used on a.c. machines, in which groups of concentric coils are used. Also used to denote the type of winding, used on transformers, in which the high-voltage winding is arranged concentrically with the low-voltage winding.

concentric wiring *(Elec.Eng.).* An interior wiring system in which the conductor consists of an insulated central core surrounded by a flexible metal sheath which forms the return lead.

conceptacle *(Bot.).* A flask-shaped cavity in a thallus, opening to the outside by a small pore, and containing reproductive structures (e.g. of *Fucus*).

conception *(Med.).* The fertilization of an ovum with a spermatozoon.

concertina fold *(Print.).* See **accordion fold.**

concert pitch *(Acous.).* The recognized pitch, i.e. frequency of the generated sound wave, to which musical instruments are tuned, so that they can play together. The exact value has varied considerably during musical history, but it has recently been internationally standardized so that A (above middle-C) becomes 440 hertz. Dance-bands are normally pitched a few hertz lower. Allowance must be made for the rise in temperature experienced in concert halls, which alters the pitch in ways peculiar to the different types of instrument.

concha *(Arch.).* The smooth concave surface of a vault.

concha *(Zool.).* In Vertebrates, the cavity of the outer ear; the outer or external ear; a shelf projecting inwards from the wall of the nasal cavity to increase the surface of the nasal epithelium.

conchiolin *(Zool.).* A horny substance forming the outer layer of the shell in Mollusca.

conchoid *(Maths.).* If, from a fixed point P, a line is drawn to meet any curve C in the point Q, and if A and B are any two points on the line PQ such that $AQ = BQ =$ a constant, then the locus of the points A and B is a conchoid with respect to the point P. The conchoid of a straight line not passing through P, first discussed by Nicomedes, 250 B.C., in connection with the trisection of an angle, and called the *conchoid of Nicomedes*, has either a node at P, a cusp at P or no double point at all, depending upon whether the perpendicular length from P to the line is greater than, equal to, or less than AQ respectively.

concolorate *(Zool.).* Having both sides the same colour.

concolor, concolorous *(Bot.,Zool.).* Uniform in colour.

conconavalin A *(Bot.).* A **lectin** from *Canavalia ensiformis* (Jack bean).

concrescence *(Bot.).* Union of originally distinct organs by the growth of the tissue beneath them.

concrete *(Build.,Civ.Eng.).* A mixture of cement, sand

and gravel or stone chips with water in varying proportions according to use.

concrete blocks *(Civ.Eng.)*. Solid or hollow pre-cast blocks of concrete used in the construction of buildings.

concrete mixer *(Civ.Eng.)*. An appliance in which the constituents of concrete are mixed mechanically.

concrete operations, stage, period *(Behav.)*. In Piagetian theory, the period between ages 8 and 11 years, when a child acquires the ability to think logically, but only in very concrete terms; the ability to reason abstractly is still very limited. This limitation is reflected in their inability to solve particular types of problems, e.g. those involving the mental operation of *reversibility*. See **concrete thinking, conservation**.

concrete paving slabs/flags *(Build.)*. Pre-cast concrete slabs for the top surface of pavements or paths. Often compressed in manufacture.

concrete thinking *(Behav.)*. A form of reasoning that is strongly tied to the immediate situation, or to very tangible and specific information, as opposed to abstract reasoning. See **concrete operations**.

concretion *(Geol.)*. Nodular or irregular concentration of siliceous, calcareous, or other materials, formed by localized deposition from solution in sedimentary rocks.

concretion *(Med.)*. Collection of organic matter with or without lime salts in bodily organs.

concussion *(Med.)*. A violent shaking or blow (especially of or to the head), or the condition resulting from it.

condensate *(Eng.)*. The liquid obtained as a result of removing from a vapour such portion of the latent heat of evaporation as it may contain.

condensation *(Chem.)*. The union of two or more molecules with the elimination of a simpler group, such as H_2O, NH_3 etc.

condensation *(Meteor.)*. The process of forming a liquid from its vapour. When moist air is cooled below its dew-point, water vapour condenses if there are extended surfaces or nuclei present. These nuclei may be dust particles or ions. Mist, fog and cloud are formed by nuclear condensation.

condensation gutter *(Build.)*. A small gutter provided at the curb of lantern lights to carry away condensed water formed on the interior surface of the glazing.

condensation sinking *(Build.)*. A groove cut in the bottom rails of skylights to carry away condensed water formed on the interior surface of the glazing.

condensation trails *(Meteor.)*. Artificial clouds caused by the passage of an aircraft due either to condensation following the reduction in pressure above the wing surfaces, or to condensation of water vapour contained in the engine exhaust gases. Also *contrails*.

condensed *(Bot.)*. An inflorescence with the flowers crowded together and nearly or quite sessile.

condensed *(Print.)*. See **elongated**.

condensed chromatin *(Biol.)*. See **heterochromatin**.

condensed nucleus *(Chem.)*. A ring system in which two rings have one or more (generally two) atoms in common, e.g. naphthalene, phenanthrene, quinoline.

condensed system *(Chem.)*. One in which there is no vapour phase. The effect of pressure is then practically negligible, and the *phase rule* may be written $P+F=C+1$.

condenser *(Chem.)*. Apparatus used for condensing vapours obtained in distillation. In laboratory practice usually a single tube, either freely exposed to air or contained in a jacket in which water circulates.

condenser *(Eng.)*. (1) A chamber into which the exhaust steam from a steam-engine or turbine is delivered, to be condensed by the circulation or the introduction of cooling water; in it a high degree of vacuum is maintained by an air pump. (2) The part of a refrigeration system in which the refrigerant is liquefied by transferring heat to the cooling medium, usually water or air.

condenser *(Phys.)*. A large lens or mirror used in an optical projecting system to collect light, radiated from the source, over a large solid angle, and to direct this light on to the object or transparency which is to be focused at a distance by a projection lens.

condenser bushing *(Elec.Eng.)*. A type of bushing used for terminals of high-voltage apparatus (e.g. transformers and switchgear) in which alternate layers of insulating material and metal foil form the insulation between the conductor and the outer casing; the metal foil serves to improve the voltage distribution. Also *capacitor bushing* or *terminal*.

condenser circulating pump *(Eng.)*. See **circulating pump**.

condenser spinning *(Textiles)*. Fibres are carded and the resultant web is divided into narrow strips which are rubbed together on oscillating rubber or leather aprons to form twistless **slubbings**. These are then spun into condenser yarns.

condenser terminal *(Elec.Eng.)*. **Capacitor terminal**. See **condenser bushing**.

condenser tissue *(Paper.)*. Thin rag paper used as a capacitor dielectric. Also *capacitor tissue*.

condenser tubes *(Eng.)*. The tubes through which the cooling water is circulated in a **surface condenser**, and on whose outer surfaces the steam is condensed.

condenser yarn *(Textiles)*. Yarns spun from clean soft waste material; suitable for cotton blankets, quiltings and towellings.

con-di nozzle *(Aero.,Eng.)*. See **convergent-divergent nozzle**.

conditional instability *(Meteor.)*. The condition of the atmosphere when the **temperature lapse rate** lies between the *dry* and **saturated adiabatic lapse rates** i.e. the atmosphere is stable for unsaturated air but unstable for saturated.

conditional instability of the second kind (CISK) *(Meteor.)*. A process whereby low-level **convergence** in the wind field produces convection and cumulus formation thereby releasing latent heat which enhances the convergence and increases convection; this 'positive feedback loop' may lead to the formation of a large-scale disturbance.

conditional jump *(Comp.)*. See **jump**.

conditional lethal *(Biol.)*. See **lethal**.

conditionally convergent series *(Maths.)*. A series which is convergent but not absolutely convergent, e.g. $1-\frac{1}{2}+\frac{1}{3}-\frac{1}{4}$ ... which converges to $\log_e 2$.

conditionally stable *(Telecomm.)*. Said of a system or amplifier which is stable for certain values of input signal and gain, but not others. See **Nyquist criterion**.

conditional probability *(Stats.)*. The probability of occurrence of an event given the occurrence of another *conditioning* event.

conditional probability distribution *(Stats.)*. The distribution of a random variable given the value of another (possibly associated) random variable or event.

conditioned reflex *(Behav.)*. Reflex action by an animal to a previously neutral stimulus as the result of **classical conditioning**.

conditioning *(Min.Ext.)*. In froth flotation, the treatment of mineral pulp with small additions of chemicals designed to develop specific aerophilic or aerophobic qualities on surfaces of different mineral species as a prelude to their separation.

conditioning *(Textiles)*. Allowing textile materials to absorb moisture until they are in equilibrium with the surrounding atmosphere. Samples are conditioned in a **standard atmosphere** before testing. For commercial transactions by weight a percentage moisture content is usually specified.

conditions of severity *(Elec.Eng.)*. A term used in connection with the testing of circuit-breakers to denote the conditions (e.g. power factor, rate of rise of restriking voltage etc.) obtaining in the circuit when the test is carried out.

conductance *(Phys.)*. The ratio of the current in the conductor to the potential difference between its ends; reciprocal of **resistance**. SI unit is *siemens*, abbrev. *S*. Also *reciprocal ohms* or *mhos*.

conductance ratio *(Elec.Eng.)*. That between equivalent conductance of a given solution and its value at infinite dilution.

conduct disorders *(Behav.)*. Childhood disorders involving antisocial behaviour.

conductimetric analysis *(Chem.)*. Volumetric analysis in which the end-point of a titration is determined by measurements of the conductance of the solution.

conducting tissue *(Bot.)*. Xylem and phloem in vascular plants; leptoids and hydroids in Bryophytes.

conduction angle *(Elec.Eng.)*. That part of a half cycle (expressed as an angle) for which a rectifier, or controlled rectifier (**thyristor** or **SCR**) conducts. By changing the conduction angle, the power delivered can be controlled.

conduction band *(Electronics.)*. In band theory of solids, band which is only partially filled, so that electrons can move freely in it, hence permitting conduction of current.

conduction by defect *(Electronics.)*. In a doped semiconductor, conduction by *holes* in the valency electron band.

conduction current *(Electronics.)*. That resulting from flow of electrons or ions in a conducting medium.

conduction electrons *(Electronics.)*. The electrons situated in the conduction band of a solid, which are free to move under the influence of an electric field.

conduction hole *(Electronics.)*. In a crystal lattice of semiconductor, conduction obtained by electrons filling holes in sequence, equivalent to a positive current.

conduction of heat *(Phys.)*. The transfer of heat from one portion of a medium to another, without visible motion of the medium, the heat energy being passed from molecule to molecule. See **thermal conductivity**.

conductivity *(Elec.Eng.)*. Inverse of **resistivity**. Units are Sm^{-1}, symbol σ.

conductivity *(Phys.)*. The ratio of the *current density* in a conductor to the electric field causing the current to flow. It is the conductance between opposite faces of a cube of the material of 1 metre edge. Reciprocal of *resistivity*. Unit $\Omega^{-1}m^{-1}$.

conductivity bridge *(Elec.Eng.)*. A form of Wheatstone bridge used for the comparison of low resistances.

conductivity cell *(Elec.Eng.)*. Any cell with electrodes for measuring conductivity of liquid or molten metals or salts.

conductivity modulation *(Electronics.)*. That effected in a semiconductor by varying a charge carrier density.

conductivity test *(Elec.Eng.)*. See **fall-of-potential test**.

conductor *(Build.)*. A pipe for the conveyance of rain water. Also called *leader*.

conductor *(Elec.Eng.)*. (1) A material which offers a low resistance to the passage of an electric current. (2) That part of an electric transmission, distribution or wiring system which actually carries the current.

conductor *(Min.Ext.)*. See **top casing**.

conductor *(Phys.)*. A material used for the transference of heat energy by conduction. All materials conduct heat to some extent, but it can be chanelled (like electricity) using materials of different conductivity.

conductor load *(Elec.Eng.)*. The total mechanical load to which an overhead electric conductor may be subjected, because of its own weight and that of any adhering matter such as snow or ice.

conductor rail *(Elec.Eng.)*. In some electric traction systems a bare rail laid alongside the running rails to conduct the current to or from the train. Also called *contact rail*.

conductor-rail insulator *(Elec.Eng.)*. An insulator used for supporting a conductor rail and for insulating it from the earth.

conductor-rail system *(Elec.Eng.)*. A system of electric traction in which current is collected from a *conductor rail* by *collector shoes*.

conduit *(Bot.)*. Functional element of the conducting system of the xylem: either a whole *vessel* or a single *tracheid*. Water moves within conduits though cell lumens and perforations in vessels and between conduits through pits (which may resist the spread of an **embolism**).

conduit *(Elec.Eng.)*. A trough or pipe containing electric wires or cables, in order to protect them against damage from external causes.

conduit *(Hyd.Eng.)*. A pipe or channel, usually large, for the conveyance of water.

conduit box *(Elec.Eng.)*. A box connected to the metal conduit used in some electric wiring schemes. The box forms a base to which fittings (e.g. switches or ceiling roses) may be attached, or it may take the place of bends, elbows, or tees, used to facilitate the installation of the wiring.

conduit fittings *(Elec.Eng.)*. A term applied to all the auxiliary items, such as boxes, elbows, etc., needed for the conduit system of wiring.

conduit system *(Elec.Eng.)*. (1) A system of wiring, used for domestic and other premises, in which the conductors are contained in a steel conduit. (2) A system of current collection used on some electric tramway systems; the conductor rail is laid beneath the roadway, and connection is made between it and the vehicle by means of a collector shoe passing through a slot in the road surface.

conduplicate *(Bot.)*. Folded longitudinally about the midrib, so that the two halves of the upper surface are brought together.

condyle *(Zool.)*. A smooth rounded protuberance, at the end of a bone, which fits into a socket on an adjacent bone, as the *condyle* of the lower jaw, the occipital *condyles*. Adjs. *condylar, condyloid.*

condylomata *(Med.)*. Inflammatory wartlike papules on the skin round the anus and external genitalia, especially in syphilis. sing. *condyloma.*

Condy's fluid *(Build.)*. A solution of sodium permanganate used as a brown stain for timber.

cone *(Bot.)*. See **strobilus**.

cone *(Eng.)*. A device used on top of blast furnaces to enable charge to be put in without permitting gas to escape. Also *bell.*

cone *(Maths.)*. (1) A (conical) surface generated by a line, one point (the *vertex*) of which is fixed and one point of which describes a fixed plane curve. Any line lying in the surface is called a *generator*. (2) A solid bounded by a conical surface and a plane (the *base*) which cuts the surface. If the fixed curve is a circle the cone is called a *circular cone*, and if also the vertex is perpendicularly over the centre of the circle the cone is called a *right circular cone*.

cone *(Textiles)*. A conical package of yarn wound on to a conical frame.

cone *(Zool.)*. Light sensitive structures in the retina of many Vertebrates which respond preferentially to particular wavelengths and thus provide the basis for colour vision. Cf. **rod**.

cone bearing *(Eng.)*. A shaft bearing consisting of a conical journal running in a correspondingly tapered bush, so acting as a combined journal and thrust bearing; used for some lathe spindles.

cone capacitor *(Elec.Eng.)*. A capacitor consisting of two conducting cones, one inside the other, separated by insulating material. By moving one cone relative to the other a variable capacitance can be obtained.

cone classifier *(Min.Ext.)*. Large inverted cone into which ore pulp is fed centrally from above. Coarser material settles to bottom discharge and finer overflows peripherally.

cone clutch *(Eng.)*. A friction clutch in which the driving and driven members consist of conical frusta. The externally coned (driven) member may be moved axially in or out of the internally coned member for engaging and disengaging the drive. See **clutch, friction clutch**.

cone diaphragm *(Acous.)*. One of paper, plastics or metal foil driven by a circular coil carrying speech currents near its apex; widely used for radiating sound in loudspeaking receivers. Also called *cone loudspeaker*.

cone drive, cone gear *(Eng.)*. A belt drive between two similar coned or tapered pulleys. A variable speed ratio is obtained by lateral movement (by means of a striker) of the belt along the pulleys.

cone-in-cone structure *(Geol.)*. Cones stacked inside one another, occurring in sedimentary rocks; usually of fibrous calcite, sometimes of other minerals.

cone loudspeaker *(Acous.)*. See **cone diaphragm**.

cone of silence *(Telecomm.)*. Cone-shaped space directly over the antenna of a radio-beacon transmitter in which signals are virtually undetectable.

cone pulley *(Eng.)*. A belt pulley stepped to give two or more diameters; used in conjunction with a similar pulley to obtain different speed ratios. See also **cone drive**.

cone sheets *(Geol.)*. Minor intrusions which occur as inwardly inclined sheets of igneous rock and have the form of segments of concentric cones.

confabulation *(Behav.)*. A tendency to fill in memory gaps with invented stories.

conference system *(Telecomm.)*. A telephone system used for conference between groups of persons at a distance. See **teleconferencing**.

confervoid *(Bot.)*. Consisting of delicate filaments.

confidence interval *(Stats.)*. An interval so constructed as to have a prescribed probability of containing the true value of an unknown parameter.

configuration *(Chem.)*. The spatial arrangement of atoms in a molecule, especially in one containing several asymmetric carbon atoms.

configuration control *(Nuc.Eng.)*. Control of reactivity of a reactor by alterations to the configuration of the fuel, reflector and moderator assembly.

confinement *(Nuc.Eng.)*. See **containment**.

conflict *(Behav.)*. *Animal behaviour*: refers to situations in which an individual animal appears motivated to engage in more than one activity, usually inferred from the simultaneous presence of causal factors that would normally impel the animal to behave in two or more incompatible ways. *Human behaviour*: refers to situations in which the tendency to respond in a particular way is inhibited either because of (1) environmental factors (e.g. fear of punishment), or (2) because of internal inhibitions (e.g. conscience). The study of conflict is a vast and diverse area, and includes *social conflict*, which focuses on conflicts between individuals or between groups of individuals.

confocal conics *(Maths.)*. Conics with the same foci.

confocal quadrics *(Maths.)*. Quadrics with the same foci.

conformable strata *(Geol.)*. An unbroken succession of strata. See also **nonsequence, unconformity**.

conformal-conjugate transformation *(Maths.)*. A conformal transformation in which a conjugate system on one surface corresponds to a conjugate system on the other.

conformal transformation *(Maths.)*. If the z-plane is mapped on the w-plane (or vice versa) so that both the magnitude and the sense of the rotation of the angles is the same in either plane, then the transformation is said to be *conformal*. Sometimes used in the sense of *isogonal transformation*.

conformational analysis *(Chem.)*. The study of the relative spatial arrangements of atoms in molecules, in particular in saturated organic molecules. Viewed along a C-C bond, the three other substituents of the nearer carbon atom are said to be *eclipsed* with respect to those of the further atom if they cover them. Rotating the nearer atom of the bond by 60° with respect to the further will give the more stable **staggered** conformation.

conforming *(Image Tech.)*. Finally selecting and matching original picture and sound components to an edited continuity, especially in videotape production.

confusion *(Image Tech.)*. *Circle of least confusion*. See **circle of confusion**.

congé *(Arch.)*. A small circular moulding, either concave or convex, at the junction of a column with its base.

congeneric *(Zool.)*. Belonging to the same genus.

congenic *(Immun.)*. Applied to inbred cells of animals which have been bred to be genetically identical except in respect to a single gene locus. See **recombinant inbred strains**.

congenital *(Zool.)*. Dating from or from before birth.

congenital deformity *(Med.)*. Malformation present at birth. Does not have to be a genetically determined defect, but may be due to environmental factors *in utero*, e.g. thalidomide.

congestion *(Med.)*. Pathological accumulation of blood in a part of the body.

congestion call meter *(Telecomm.)*. A meter which counts the number of calls made over the last choice of outlet in a grading scheme.

congestion traffic-unit meter *(Telecomm.)*. A meter which registers the traffic flow over the last choice of outlets in a grading scheme.

conglomerate *(Geol.)*. A coarse-grained clastic sedimentary rock composed of rounded or subrounded fragments larger than 2mm in size (e.g. pebbles, cobbles, boulders).

Congo red *(Chem.)*. A benzidene direct dyestuff (scarlet) produced by coupling diphenyl-bis-diazonium chloride with naphthionic acid; first of a long series of derivatives. Used as a chemical indicator for acid solutions in the pH range 3–5, (blue–red).

congruence *(Maths.)*. A statement that two numbers a and b are congruent with respect to a divisor n. Written $a \equiv b$ (mod n), and meaning that the remainders when a and b are divided by n are equal.

congruent *(Maths.)*. Alike in all relevant respects; capable of coincident superposition in Euclidean geometry.

congruent melting *(Eng.)*. Melting at a constant temperature or pressure of a material in which both phases retain the same composition.

congruent transformation *(Maths.)*. If a given matrix A is transformed into B by $B = P'AP$, where P is a nonsingular matrix and P' is its transpose, then B is said to be congruent to A.

conic *(Maths.)*. The curve in which a plane cuts a circular (but not necessarily right) cone. The curve is a parabola if the cutting plane is parallel to a generator, otherwise it is an ellipse or a hyperbola according to whether a plane parallel to the cutting plane but through the vertex of the cone wholly contains no generators or two generators respectively. Alternatively, the locus of a point which moves so that its distance from a fixed point (**focus**) is a constant e (**eccentricity**) times its distance from a fixed line (**directrix**). The conic is an ellipse, parabola or hyperbola according to whether e is less than, equal to, or greater than 1. Any second degree equation in two variables represents a conic. By a suitable choice of axes such an equation can be reduced to one of the following forms: (1)

(1) $\dfrac{x^2}{a^2} + \dfrac{y^2}{b^2} = 1$, an ellipse

(2) $\dfrac{x^2}{a^2} + \dfrac{y^2}{b^2} = 1$, a hyperbola

(3) $y^2 = 4ax$, a parabola

(4) $ax^2 + 2hxy + by^2 = 0$
a line pair if $h^2 > ab$
two coincident lines if $h^2 = ab$
a single point if $h^2 < ab$

Cf. **axes, diameter, focus** and **vertex**.

conical camber *(Aero.)*. An expression applied to the adjustment of the **camber** of a wing across the span to meet the variation of the upflow of the air from the fuselage side to the tip. Used on high-speed aircraft, the 'twist' is applied mainly to the leading edge. The name originates from the conic lofting process used in deriving the aerofoil sections.

conical drum *(Min.Ext.)*. Winding drum, to which hoisting rope of cage or skip is attached. This shape aids in smooth acceleration from rest to full speed where the rope reaches the flat central part of the compound drum (the full diameter between the cones).

conical horn *(Acous.)*. A horn in the form of a cone, the apex being truncated to form the throat.

conical pivot *(Horol.)*. (1) A pivot formed as a cone, which runs in a screw with a tapered hole, the angle of taper being greater than that of the conical pivot; used for the balance staff of alarm-clock movements, and for certain watches with pin-pallet escapement. (2) A form of pivot, used in English watches, which has a conical shoulder, the pivot itself being parallel.

conical projections *(Geol.)*. In these, the projection is on to an imaginary cone touching the sphere at a given standard parallel. There are equal-area and two standard (*Gall's projection*) versions, as in the cylindrical type. Best for middle latitudes.

conical refraction *(Phys.)*. An effect seen in *biaxial* crystals when light incident in a particular direction is spread on refraction into a hollow cone of rays. *Internal* and *external* conical refraction require different experimental conditions for observation.

conical scanning *(Radar.)*. Similar to **lobe switching**, but circular. The direction of maximum response generates a conical surface; used in missile guidance.

conical surface *(Maths.)*. See **cone**.

conidial *(Bot.)*. (1) Referring to, or pertaining to, a conidium. (2) Producing conidia.

conidiophore *(Bot.)*. A simple or branched hypha bearing one or more conidia.

conidiosporangium *(Bot.)*. A sporangium capable of direct germination, as well as producing zoospores.

conidium *(Bot.)*. An asexual fungal spore. Produced exogenously from a hyphal tip, never in a sporangium.

Coniferales *(Bot.)*. The conifers. Order of ca 600 spp. of gymnosperms (class Conferopsida). Also fossils from the Jurassic. Trees (mostly) and shrubs, with simple, usually small, leaves. Reproductive organs in unisexual cones. Siphonogamous. Dominating large areas of the earth (though reduced since the Cretaceous) and including many important *soft wood* trees for timber and pulp. Includes pines, spruces, cypresses, monkey puzzles.

Coniferopsida *(Bot.)*. Class of gymnosperms dating from the Carboniferous. Mostly substantial trees, with pycnoxylic wood, simple leaves, saccate pollen and flattened (platyspermic) seed. Includes Cordaitales (Carboniferous and Permian), Volztiales (Permian – Jurassic), and the Coniferales, Taxales (yews) and, in some classifications, the Ginkgoales.

coniferous *(Bot.)*. Cone-bearing; relating to a cone-bearing plant.

conifuge *(Powder Tech.)*. A conical type of centrifuge for the collection of aerosol particles.

coning and quartering *(Min.Ext.)*. Production of representative sample from a large pile of material such as ore, in which it is first formed into a cone by deposition centrally, and then reduced by removal, one shovelful at a time, into four separate piles drawn alternately from four peripheral opposed points. Two are then discarded and the process (perhaps with intermediate size reduction) is repeated until a manageable hand sample is obtained.

coning angle *(Aero.)*. Angle between longitudinal axis of blade of lifting rotor and the tip-path plane in helicopters.

Coniphora puteana *(Build.)*. Formerly *Coniphora cerebella*. One of several fungi which causes **wet rot** in timber.

conjugate acid and base *(Chem.)*. These are related, according to the **Brönsted-Lowry theory**, by the reversible exchange of a proton; thus Acid$\rightleftharpoons$Base + H^+.

conjugate algebraic numbers *(Maths.)*. Any numbers which are the roots of the same irreducible algebraic equation with rational coefficients.

conjugate angles *(Maths.)*. Two angles whose sum is equal to 360°. Each is said to be the *explement* of the other.

conjugate arcs *(Maths.)*. Two arcs whose sum is a complete circle.

conjugate axis of hyperbola *(Maths.)*. See **axes**.

conjugate Bertrand curves *(Maths.)*. Any two curves having the same principal normals. Also called *Bertrand curves* and *associate Bertrand curves*.

conjugate branches *(Phys.)*. Any two branches of an electrical network such that an e.m.f. in one branch produces no current in the conjugate branch.

conjugate complex number *(Maths.)*. The conjugate $\bar{z}$ of a complex number $z = x + iy$ is $x - iy$.

conjugate deviation *(Med.)*. The sustained deviation of the eyes in one direction as a result of a lesion in the brain.

conjugate diameters *(Maths.)*. Two diameters of a conic which are also conjugate lines with respect to the conic are conjugate diameters. Each bisects chords parallel to the other.

conjugate directions *(Maths.)*. At a point P on a surface, if Q is a point near P and Q tends to P, then the conjugate directions are the directions in the limiting case of: (a) the straight line between P and Q; and (b) the line of intersection of the tangent planes at P and Q. If these directions coincide, they are *self-conjugate* or *asymptotic directions*.

conjugate division *(Bot.)*. Simultaneous mitosis of a pair of associated nuclei in e.g. a dikaryotic cell.

conjugate double bonds *(Chem.)*. Di-alkene compounds with an arrangement of alternate single and double bonds between the carbon atoms, namely R.CH=CH.CH=CH.R. Additive reactions take place, inasmuch as atoms or radicals become attached to the two outside carbon atoms of the chain, thus creating a new alkene linkage in the centre.

conjugate dyadics *(Maths.)*. Two dyadics such that each can be obtained from the other by reversing the order of the factors in each dyad. If a dyadic is equal to its conjugate it is said to be *symmetric*; if it is equal to the negative of its conjugate it is *antisymmetric*.

conjugate elements of a determinant *(Maths.)*. Those which are interchanged when the rows and columns are interchanged, e.g. a_{ij} and a_{ji}.

conjugate elements of a group *(Maths.)*. Two elements A and B of a group are conjugate with respect to the group if there exists an element P in the group such that $B = P^{-1}AP$ or PAP^{-1}, according as operators are written on the right or left respectively. B is then referred to as the transform of A by P.

conjugate foci *(Phys.)*. Two points such that rays of light diverging from either of them are brought to a focus at the other. For a simple convergent lens an object and its real image are at *conjugate foci*.

conjugate functions *(Maths.)*. Two real functions which satisfy the Cauchy-Riemann equations.

conjugate harmonic functions *(Maths.)*. Two harmonic functions which satisfy the Cauchy-Riemann equations.

conjugate impedance *(Phys.)*. Two impedances are conjugate when their resistance components are equal and their reactances are equal in magnitude but opposite in sign.

conjugate lines (of a conic) *(Maths.)*. Two lines are said to be conjugate with respect to a conic if the pole of each lies on the other.

conjugate matrix *(Maths.)*. See **transpose of a matrix**.

conjugate planes *(Phys.)*. Planes perpendicular to the axis of an optical system such that any object point in one near the axis is imaged in the other.

conjugate point *(Maths.)*. See **double point**.

conjugate points (of a conic) *(Maths.)*. Two points are said to be conjugate with respect to a conic if each lies on the polar of the other.

conjugate solutions *(Chem.)*. If two liquids A and B are partially miscible, they will produce at equilibrium two conjugate solutions, the one of A in B, the other B in A.

conjugate subgroup *(Maths.)*. Each member of a set of subgroups, obtained by transforming a given subgroup of a group G by every element of the group G, is said to be conjugate to each of the other members. See **conjugate elements of a group**.

conjugate system of curves *(Maths.)*. A system consisting of two one-parameter families of curves on a surface, such that where a member of one family intersects a member of the other family, the directions of the tangents to the curves are conjugate directions.

conjugate triangles *(Maths.)*. Two triangles are conjugate with respect to a conic if the vertices of one are the poles of the sides of the other. Also called *reciprocal polar triangles*. Cf. **self-polar triangle**.

conjugation *(Biol.)*. In prokaryotes, the process by which genetic information is transferred from one bacterium to another.

conjugation *(Bot.)*. Form of sexual reproduction in some algae and fungi in which there is fusion of 2 non-flagellated gametes or protoplasts.

conjugation tube *(Bot.)*. A tubular outgrowth of a cell though which one non-flagellated gamete moves to fuse with another during conjugation.

conjunction *(Astron.)*. Term signifying that two heavenly bodies have the same apparent geocentric longitude. Applied to Venus and Mercury it is subdivided into *inferior conjunction* and *superior conjunction*, according as the planet is between the earth and sun, or the sun between the earth and planet, respectively.

conjunctiva *(Zool.)*. In Vertebrates, the modified epidermis of the front of the eye, covering the cornea externally and the inner side of the eyelid.

conjunctive tissue *(Bot.)*. Secondary tissue occupying the space between the vascular bundles where the secondary xylem does not form a solid cylinder.

conjunctivitis *(Med.)*. Inflammation of the conjunctiva.

conky *(For.)*. Applied to a log or tree bearing fruit bodies of wood-rotting fungi.

connate *(Bot.,Zool.)* Of plant or animals parts which are firmly joined, particularly of like parts.

connate water *(Geol.)*. Water trapped in a rock at the time of its deposition.

connected domain *(Maths.)*. A *simply-connected domain* is such that any closed curve in it and the region enclosed by the curve lie wholly within the domain. A *multiply-connected domain* has one or more 'holes'. A *doubly-connected domain* has one 'hole' and so on, an *n*tuply-connected domain having *n*-1 'holes'.

connected load *(Elec.Eng.)*. The sum of the rated inputs of all consumers' apparatus connected to an electric power supply system.

connecting-rod *(Eng.)*. In a reciprocating engine or pump, the rod connecting the piston or crosshead to the crank.

connecting-rod bolts *(Eng.)*. Bolts securing the outer half of a split big-end bearing of a connecting-rod to the rod itself. Sometimes known as *big-end bolts*.

connecting thread *(Bot.)*. See **plasmodesma**.

connection, connecting box *(Elec.Eng.)*. A box containing terminals to which are brought a number of conductors of a wiring or distribution system, to facilitate the making of connections between them.

connection machine *(Comp.)*. TN for a massively parallel computer. See **parallel processing, SISMD**.

connective *(Zool.)*. A bundle of nerve fibres uniting two nerve centres.

connective tissue *(Zool.)*. A group of animal tissues fulfilling mechanical functions, developed from the mesoderm and possessing a large quantity of nonliving intercellular matrix, which usually contains fibres; as bone, cartilage, and areolar tissue.

connective tissue diseases *(Med.)*. A term used to cover a number of diseases of uncertain aetiology including disseminated **lupus erythematosus, scleroderma** and **polyarteritis nodosa**.

connectivity *(Chem.)*. The way in which atoms are bound together in a molecule, i.e. the configurations of all the atoms. **Stereoisomers** have the same connectivity, while structural isomers do not.

connector *(Build.)*. See **timber connectors**.

connector bar *(Elec.Eng.)*. See **terminal bar**.

connivent *(Bot.)*. Converging and meeting at the tips.

conoscopic observation *(Crystal.)*. The investigation of the behaviour of doubly refracting crystal plates under convergent polarized light. Interference pictures obtained are important in explaining crystal optical phenomena.

consanguinity *(Geol.)*. Term applied to rocks having a similarity or community of origin, which is revealed by common peculiarities of mineral and chemical composition and often also of texture.

consciousness *(Behav.)*. Several related and general meanings: (1) Alert and capable of action; (2) Awareness of the environment, sentience; (3) Awareness of a person's thoughts and feelings; (4) The ability of an individual to perceive their own mental life; (5) States of mind of which one is aware, and which organize and co-ordinate one's activities, as opposed to subconscious factors organizing and guiding behaviour.

consensus sequence *(Biol.)*. A DNA sequence found with minor variations and similar function in widely divergent organisms.

consequent *(Maths.)*. (1) In a ratio $a : b$, the term b is the consequent; a is the **antecedent**. (2) In implication, (if p, then q), q is the consequent, p is the antecedent.

consequent drainage *(Geol.)*. A river system directly related to the geological structure of the area in which it occurs.

consequent pole *(Elec.Eng.)*. An effective pole not at the end of ferromagnetic material, e.g. at the ends of a diameter of a ring magnet; must occur in pairs. Used in the Broca galvanometer.

conservancy system *(Build.)*. Disposing of waste matter from buildings by earth closets and privies, without water.

conservation *(Behav.)*. A term in Piagetian theory, referring to the understanding that certain physical attributes such as quantity, volume and mass, do not change despite various transformations in their physical appearance. It is a knowledge gained between 8 and 11 years of age, during the periods of **concrete operations**.

conservation *(Ecol.)*. (1) Protection of natural ecosystems from the hand of man with the intention of preserving them as heritage or as a practical gene-bank. (2) Wise management of ecosystems, allowing exploitation at a level which does not impair the future capacity to produce.

conservation laws *(Chem.)*. Usually refers to the classical laws of the separate conservation of mass, of energy and of atomic species, which are sufficiently accurate for most chemical reactions.

conservation laws *(Phys.)*. In the interaction between particles certain quantities remain the same before and after the event. (1) Dynamical quantities such as mass-energy and momentum are conserved. (2) Intrinsic properties of charge and *baryon number* are conserved in nuclear interactions. In addition, in *strong* and *electromagnetic* interactions between elementary particles, the intrinsic properties of strangeness, charm, topness and bottomness are also conserved but not in *weak* interactions.

conservation of energy *(Phys.)*. The total energy of an isolated system is constant. Energy may be converted from one form to another, but is not created or destroyed. If the system has only conservative forces, then the total mechanical energy (kinetic and potential) is constant. (see **mass-energy equation**.)

conservation of matter *(Chem.)*. See **law of conservation of matter**.

conservation of momentum *(Phys.)*. The sum of the momenta in a closed system (i.e. one in which no influences act upon it from outside) is constant and is not affected by processes occurring within the system.

conservation of movement of the centre of gravity *(Phys.)*. The state of rest or of motion of the centre of gravity of a system can never be altered by the action of internal forces within the system.

conservative field of force *(Phys.)*. A field of force such that the work done in moving a particle from A to B is independent of the path followed.

conservative system *(Phys.)*. A system such that in any cycle of operations where the configuration of the system remains unchanged overall, the work done is zero.

conservator tank *(Elec.Eng.)*. A small tank, carried on the top cover of an oil-filled power transformer, to accommodate the change in oil volume with temperature as the load varies, and reduce the oil surface exposed to the air.

conservatory *(Arch.)*. A glazed building in which plants may be grown under controlled atmospheric conditions.

consistency *(Build.)*. The thickness, density or firmness of any thick liquid such as paint, or adhesive. Consistency may be measured using a *plastometer*.

consistency *(Paper.)*. The amount of bone dry fibre in pulp or stock, expressed as a percentage.

consistent equations *(Maths.)*. Equations having a common solution, e.g., the pair of equations $x+y=3$ and $x-y=1$ are consistent, but the pair $x+y=3$ and and $x+y=2$ are not. Sometimes called *compatible equations*.

console *(Build.)*. A bracket whose shelf is supported by volutes.

console *(Comp.)*. Special desk equipped with a keyboard and a VDU for communicating with a computer.

console *(Elec.Eng.)*. Type of **control panel**: central controlling desk in power station, process plant, computer, or reactor, where an operator can supervize operations and give instructions.

consolidation *(Geol.)*. Of strata, the drying, compacting, and induration of rocks, as a result of pressures operating after deposition.

consolidation of learning, memory *(Behav.)*. A process that is thought to follow a learning trial or experience and that continues for some time after the learning event, during which the memory for the event becomes stable and durable, presumably a physiological process involving a structural change in the brain.

consolidation pile *(Civ.Eng.)*. A pile driven (with others) into the ground to consolidate the soil and enable it to support heavier loads than would otherwise be possible.

consonance *(Acous.)*. The condition where two pure tones blend pleasingly.

conspecific *(Biol.)*. Relating to the same species. Often used as a noun.

constancy *(Ecol.)*. The percentage of sample plots in a plant community containing a particular species.

constant *(Maths.)*. In an algebraic expression or equation, a quantity (or parameter) which remains the same while the variables change. Constants assume different values for different initial and boundary conditions, but are constant for each particular set of conditions. An *absolute* constant, e.g. π is always the same.

constant-amplitude recording *(Acous.)*. Gramophone recording technique whereby the response of recording, i.e. amplitude of the track divided by the root of the applied power, is independent of frequency. Cf **constant-velocity recording**.

constantan *(Elec.Eng.)*. An alloy of about 40% nickel and 60% copper, having a high volume resistivity and almost negligible temperature coefficient; used as the resistance wire in resistance boxes, etc. Also called *Eureka*.

constant boiling mixture *(Chem.)*. An **azeotropic mixture**.

constant-current characteristic *(Electronics)*. A property of a circuit or transformer whereby current supplied to a load is independent of its impedance or any externally applied voltage. A transistor, with base-emitter voltage held constant, can meet this criterion when the load is connected to the collector.

constant-current modulation *(Elec.Eng.)*. See **choke modulation**.

constant-current motor *(Elec.Eng.)*. An electric motor designed to operate at a constant current from a constant-current generator.

constant-current system *(Elec.Eng.)*. A system of transmitting electric power in which all the equipment is connected in series and a constant current is passed round the circuit. Variations in power result in a variation of the voltage of the system, constant-current generators being used for the supply. See **series system**.

constant-current transformer *(Elec.Eng.)*. A transformer designed to maintain a constant secondary current within a specified working range, for all values of secondary impedance and all values of primary voltage.

constant-frequency oscillator *(Elec.Eng.)*. One in which special precautions are taken to ensure that the frequency remains constant under varying conditions of load, supply voltage, temperature, etc.

constant-k filter *(Telecomm.)*. Simple type formed from a **constant-k network** only.

constant-k network *(Telecomm.)*. Iterative network for which product of series and shunt impedance is frequency independent.

constant-level chart *(Meteor.)*. An upper air chart showing isobars for a particular level. Cf **constant-pressure chart**.

constant-level tube *(Surv.)*. A special form of level tube in which the volume ratio of bubble to liquid is fixed at such a value that decrease in length of the bubble due to expansion of the liquid is exactly counterbalanced by increase in length of the bubble due to diminished surface tension, so that the length of the bubble – and thus the sensitivity of the level tube – remains unaltered by rise in temperature.

constant-mesh gear-box *(Autos.)*. A gear-box in which the pairs of wheels providing the various speed ratios are always in mesh, the ratio being determined by the particular wheel which is coupled to the mainshaft by sliding dogs working on splines.

constant of integration *(Maths.)*. The arbitrary constant resulting from indefinite integration. It arises because the derivative of a constant is zero, and there is therefore no means of evaluating it without additional information.

constant of inversion *(Maths.)*. See **inversion**.

constant-power generator *(Elec.Eng.)*. An electric generator which, by variation of the generated voltage, gives a constant power output at varying currents.

constant-pressure chart *(Meteor.)*. An upper air chart showing contours of the geopotential height above sea level at which a particular pressure occurs. Cf. **constant-level chart**. See **thickness chart**.

constant-pressure cycle *(Autos.)*. See **diesel cycle**.

constant proportions *(Chem.)*. See **law of constant (definite) proportions**.

constant region *(Immun.)*. The carboxy-terminal half of the light or the heavy chain of an immunoglobulin molecule. Termed constant because the amino acid sequence is the same in all molecules of the same class or sub-class.

constant-resistance network *(Telecomm.)*. One in which iterative impedance in at least one direction is resistive and independent of the applied signal voltage.

constant-speed propeller *(Aero.)*. See **propeller**.

constant time-lag *(Elec.Eng.)*. See **definite time-lag**.

constant variable transmission *(Autos.)*. Form of automatic transmission using belt drive with variable diameter pulleys. The system is becoming more widespread with the development of a segmented steel driving belt.

constant velocity joint *(Autos.)*. Joint for transmitting drive, which does not show the cyclical velocity changes of the Hooke-type joint at higher angles. Used for the drive shafts of front-wheel drive cars.

constant-velocity recording *(Acous.)*. In gramophone disk recording the technique whereby the lateral r.m.s. velocity of the sinuous track is made proportional to the root of the electrical power applied to the recorder, irrespective of the frequency. This criterion is necessary for minimum wear and surface noise on the finished record, but has to be modified to constant-amplitude recording below about 250 Hz as a compromise with playing time.

constant-voltage system *(Elec.Eng.)*. The usual system of transmission of electric power, in which the voltage between the conductors is maintained approximately constant, and all apparatus is connected to the system in parallel across the conductors.

constant-voltage transformer *(Elec.Eng.)*. One with or without extra components which gives a constant voltage with varying load current, or with varying input voltage over a specified range.

constant-volume amplifier *(Telecomm.)*. See **vogad**.

constellation *(Astron.)*. A group of stars, not necessarily connected physically, to which have been given a pictorial configuration and a name (generally of Greek mythological origin) which persist in common use although of no scientific significance.

constipation *(Med.)*. A condition in which the faeces are abnormally dry and hard: retention of faeces in the bowel: infrequent evacuation.

constituent *(Eng.)*. Component of alloy or other compound or of a mixture. It may be present as an element or in chemical, physical or intermediate combination.

constituent *(Maths.)*. Of a matrix or determinant, one of the numbers in the array which forms the matrix or determinant. Also called an *element*.

constituents *(Chem.)*. All the substances present in a system.

constitution *(Chem.,Phys.)*. Structural distribution of atoms and/or ions composing a regularly co-ordinated substance. Includes percentage of each constituent and its regularity of occurrence through the material.

constitutional ash *(Min.Ext.)*. Ash resulting from combustion of coal, and derived from siliceous matter in the coal-forming plants. 'Fixed ash' as distinct from entrained impurity of 'free ash'.

constitutional formula *(Chem.)*. A formula which shows the arrangement of the atoms in a molecule.

constitutional water *(Chem.)*. See **water of hydration**.

constitution changes *(Eng.)*. Changes in solid alloys which involve the transformation of one constituent to another (as when pearlite is formed from austenite), or a change in the relative proportions of two constituents.

constitution diagram *(Eng.)*. *Phase diagram* which shows effect of temperature and composition on alloy in which two or more metals co-exist in solid solution and pass from one crystal lattice to another (e.g. face-centred, cubic, body-centred) and also change in general properties.

constitutive enzyme *(Bot.)*. An enzyme that is formed under all conditions of growth. Cf. **inducible enzyme**.

constitutive heterochromatin *(Biol.)*. **Heterochromatin** which is always condensed. **Satellite DNA** is found in these regions and coding sequences are apparently absent.

constraint *(Eng.)*. The property which distinguishes a mechanism from a kinematic chain. In a mechanism, the motion of one part is followed by a predetermined motion of the remainder of the mechanism.

constriction *(Biol.)*. Narrow, localized region in a chromosome, normally found at the centromere (*primary constriction*), and often also at other sites (*secondary constrictions*), including **nucleolar organizing region**.

constriction resistance *(Phys.)*. That across actual area of contact through which current passes from one metal to another, equal to $(\rho_1 + \rho_2)/4\alpha$, where α = radius of circular contact area and ρ_1 and ρ_2 = resistivities of the metals.

constrictive pericarditis *(Med.)*. Where chronic inflammation encircles the heart and prevents normal pump function.

constrictor *(Zool.)*. A muscle which by its contraction constricts or compresses a structure or organ.

constringence *(Phys.)*. Inverse of the dispersive power of a medium. Ratio of the mean refractive index diminished by unity to the difference of the refractive indices for red and violet light.

consumers *(Ecol.)*. In an ecosystem, the heterotrophic organisms, chiefly animals, which ingest either other organisms or particulate organic matter. Cf. **decomposers, producers**.

consummatory act, behaviour, phase *(Behav.)*. Historically in animal behaviour studies, the end phase of goal oriented behaviour, typified by a series of responses directed at that goal, often of a stereotyped nature; it follows the **appetitive phase**, a goal seeking phase of behaviour. The rigid distinction between appetitive and consummatory phases of a behaviour sequence has been largely abandoned, although the terms are still used descriptively.

contact *(Elec.Eng.)*. (1) That part of either of two conductors which is made to touch the other when it is desired to pass current from one to the other, as in a switch. (2) The juxtaposition of parts, usually of platinum or silver, which, when brought into contact, provide for the passage of a current and, when withdrawn, cause its cessation. For the former, suitable minimum pressure and areas of contact are necessary; for the latter, a sufficiently low recovery voltage, and rapidity of withdrawal of the contacts from each other.

contact angle *(Chem.)*. The angle between the liquid and the solid at the liquid-solid-gas interface. It is acute for wetting (e.g. water on glass) and obtuse for nonwetting (e.g. water on paraffin wax).

contact angle *(Min.Ext.)*. The angle (θ) between a bubble of air and the chemically clean, polished and horizontal surface of a specimen of mineral to which it clings, measured between that surface and the side of the bubble. This forms an index to the floatability of the species under prescribed conditions, in which chemicals are added to the water and change in angle observed.

contact bed *(Build.)*. A tank, filled with material such as broken clinker, used in the final oxidizing stage in sewage treatment, which consists in charging the filtering medium with the liquid sewage, allowing it to stand for a time, draining it off, and finally keeping the tank empty for a time. See **percolating filter**.

contact bounce, chatter *(Telecomm.)*. Intermittent opening and closing of relay contacts.

contact breaker *(Elec.Eng.)*. **Circuit breaker**.

contact e.m.f. *(Elec.Eng.)*. That which arises at the contact of dissimilar metals at the same temperature, or the same metal at different temperatures.

contact fingers *(Elec.Eng.)*. Contacts pressed by springs against the moving contacts of a drum-type controller.

contact flight *(Aero.)*. Navigation of an aircraft by the pilot observing the ground only.

contact herbicide *(Bot.)*. A herbicide which kills those plant parts that it comes into contact with, e.g. ioxynil, paraquat. Cf. **soil-acting herbicide, translocated herbicide**.

contact hypersensitivity *(Immun.)*. Hypersensitivity reaction provoked by application to the skin of substances which act as **haptens** or as antigens. Usually due to prior sensitization by the chemical, and may be immediate or delayed type.

contact inhibition *(Biol.)*. The inhibition of movement of some kinds of cells in tissue culture which occurs when they touch each other. See **transformation (2)**.

contact insecticide *(Chem.)*. One which kills on contact with insect surface (body, legs etc.); used against sucking insects (e.g. aphids, mosquitoes) which are not affected by insecticides acting only through the alimentary system; e.g. pyrethrins, rotenone, DDT.

contact ionization *(Electronics)*. Loss of electron by an easily ionized atom (e.g. caesium) when it comes into contact with the surface of a metal with affinity for electrons (e.g. tungsten).

contact jaw *(Elec.Eng.)*. (1) The clamping device of a resistance welding machine, which secures the parts to be welded and also conducts the current to them. (2) The fixed part of a switch, with which the moving blade makes contact in closing the circuit.

contact lens *(Phys.)*. A lens, usually of plastic material, worn in contact with the cornea instead of spectacles. Used to aid activity, or for cosmetic purposes.

contact maker *(Elec.Eng.)*. Any device used to make an electrical contact, especially a periodical contact, as in **sparking plugs**.

contact metal *(Telecomm.)*. That used for contacts on springs of relays, generally silver, platinum, tungsten, etc.

contact metamorphism *(Geol.)*. The alteration of rocks caused by their contact with, or proximity to, a body of igneous rock.

contact noise *(Elec.Eng.)*. Noise voltage arising across a contact, with or without, adsorbed gases, arising from differences in work function of contacting metal conductors, one of which may be a semiconductor.

contactor controller *(Elec.Eng.)*. A controller in which the various circuits are made and broken by means of contactors.

contactor starter *(Elec.Eng.)*. An electric motor starter in which the steps of resistance are cut out, or other operations are performed, by means of contactors.

contactor switching starter *(Elec.Eng.)*. A switching starter in which the switching operations are carried out by means of contactors.

contact potential *(Elec.Eng.)*. See **contact e.m.f.**

contact-potential barrier *(Electronics)*. Potential barrier formed at the junction between regions with different energy gap or carrier concentration.

contact pressure *(Telecomm.)*. The pressure between the contacts on relay springs, a minimum pressure being required for certain contact when the circuit is frequently broken.

contact print *(Image Tech.)*. A positive photographic or motion picture print made with the sensitive emulsion exposed in physical contact with the negative.

contact process *(Chem.Eng.)*. The most important process for making sulphuric acid. Sulphur dioxide is passed with dry air over a catalyst, usually vanadium (V) oxide, to produce sulphur (VI) oxide which is absorbed in sulphuric acid to which water is added, giving the product. If sulphur dioxide is produced from pyrites, special cleaning is required, usually by electrostatic precipitation, but it can be conveniently made by direct ignition of sulphur in dry air.

contact-radiation therapy *(Radiol.)*. Radiation from a very short distance, e.g. 20 mm, with voltages around 50 kV.

contact rail *(Elec.Eng.)*. See **conductor rail**.

contact resistance *(Elec.Eng.)*. Resistance at surface of contact between two conductors; mainly determined by constriction of current in the materials near the point of contact. Contact area is extended by pressure *increasing* this resistance.

contact scanning *(Eng.)*. Ultrasonic inspection procedures in which the ultrasonic head is acoustically coupled to the material being scanned.

contact screen *(Print.)*. A vignetted-dot screen on a film base used in the process camera in close contact with the emulsion, as an alternative to the glass half-tone screen; also for same-size screened work by contact.

contact segment *(Elec.Eng.)*. Contacts on certain types of motor starter, or other control equipment, which are segment-shaped, so that when in position they form a ring divided by radial gaps.

contact shoe *(Elec.Eng.)*. See **collector shoe**.

contact spring *(Elec.Eng.)*. The flexible metal holder of the contact in a relay. The stiffness of the holder determines the pressure between contacts for a given displacement.

contact strip *(Elec.Eng.)*. On a pantograph or bow type of current collector, the renewable metal or carbon strip that actually makes contact with the overhead wire of an electric traction system. Also called a *bow strip* when used on a bow collector.

contact stud *(Elec.Eng.)*. In the surface-contact system of electric traction, the studs in the roadway for making contact with the contact skate on an electric vehicle. The studs are only made alive when the car is actually passing over them.

contact vein *(Min.Ext.)*. A vein occurring along the line of contact of two different rock formations, one of which may be an igneous intrusion.

contact wire *(Elec.Eng.)*. The overhead conductor from which current is collected, by suitable forms of collector gear, for the vehicles of some electric traction systems.

contagion *(Med.)*. The communication of disease by direct contact between persons, or between an infected object and a person. adj. *contagious*.

contagious bovine pleuro-pneumonia *(Vet.)*. Lung plague. An acute, subacute, or chronic disease of cattle caused by *Mycoplasma mycoides*; characterized by fever, pneumonia, and pleurisy. Vaccines have been used experimentally.

contagious catarrh *(Vet.)*. See **infectious coryza**.

contagious distribution *(Ecol.)*. Pattern of distribution of plants and animals in which individuals occur closer together than would be expected on a random basis.

contagious equine abortion *(Vet.)*. A contagious form of abortion in horses due to infection of the placenta by *Salmonella abortivoequina*.

contagious equine metritis *(Vet.)*. Important cause of equine infertility due to a gram-negative microaerophilic coccobacillus. First reported in 1977, but now rare after intense screening and prohylaxis. Abbrev. *CEM*.

contagious ophthalmia *(Vet.)*. *Heather blindness*. A contagious disease of sheep, characterized by conjunctivitis and keratitis, caused by a rickettsial organism *Colesiota conjunctivae (Rickettsia conjunctivae)*.

contagious pustular dermatitis *(Vet.)*. *Malignant aphtha, orf.* A contagious virus disease of sheep and goats characterized by vesicle and pustule formation on the skin and mucous membranes, especially on lips, nose, and feet. Vaccination available.

containment *(Nuc.Eng.)*. In fusion, the use of shaped magnetic fields or of **inertial confinement** to contain a plasma (also called *confinement*). See also **magnetic confinement**.

contaminated rock *(Geol.)*. Igneous rock whose composition has been modified by the incorporation of other rock material.

contamination meter *(Radiol.)*. Particular design of Geiger-Müller circuit for indicating for civil defence purposes the degree of radio-active contamination in an area, especially for estimating the time for its safe occupation.

contiguity *(Behav.)*. The closeness in time of two events which is sometimes regarded as the condition leading to association, especially in **classical conditioning** procedures.

continental climate *(Meteor.)*. A type of climate found in continental areas not subject to maritime influences, and characterized by more pronounced extremes between summer and winter; the winters become colder to a greater degree than the summers become hotter; also relatively small rainfall and low humidities.

continental crust *(Geol.)*. That part of the earth's crust which underlies the continents and continental shelves. It is approximately 35 km thick in most regions but is thicker under mountainous areas. Sedimentary rocks predominate in its uppermost part and metamorphic rocks at depth, but the detailed composition of the lower crust in uncertain. Cf. **oceanic crust**.

continental deposit *(Geol.)*. A rock formed under subaerial conditions or in water not directly connected with the sea. See **aeolian deposits**, **glacial deposits**, **lacustrine deposits**.

continental drift *(Geol.)*. A hypothesis put forward by Wegener in 1912 to explain the distribution of the continents and oceans and the undoubted structural, geological and physical similarities which exist between continents. The continents were believed to have been formed from one large land mass and to have drifted apart. See **plate tectonics**.

continental shelf *(Geol.)*. The gently sloping offshore zone, extending usually to about 200 m depth.

continental slope *(Geol.)*. The relatively steep slope between the continental shelf and the more gentle rise from the abyssal plain.

contingency table *(Stats.)*. A table giving the frequency of observations cross-classified by variate values.

continued fraction *(Maths.)*. A terminating or infinite fraction of the form

$$a+\cfrac{b}{c+\cfrac{d}{e+\cfrac{f}{g+}}}\text{etc.,}$$

usually written $a+\frac{b}{c+}\frac{d}{e+}\frac{f}{g+}$, etc.

The values obtained by ending the fraction at a, at c, at e etc. are called *convergents*.

continuity *(Image Tech.)*. The co-ordination and matching in sequence of all the successive scenes making up a motion picture or television production and the records and documents so required.

continuity *(Phys.)*. The existence of an uninterrupted path for current in a circuit.

continuity *(Telecomm.)*. Supervision of succession of

items from various sources in making up a broadcast programme.

continuity-bond *(Elec.Eng.)*. A rail-bond used to maintain the continuity of the track- or conductor-rail circuit at junctions and crossings.

continuity-fitting *(Elec.Eng.)*. A device used in electric wiring installations for ensuring a continuous electric circuit between adjacent lengths of conduit.

continuous *(Maths.)*. A function $f(x)$ is continuous at a point x_0 if for every neighbourhood U_2 of $f(x_0)$, there exists a neighbourhood U_1 of x_0 such that $f(U_1)$ is contained in U_2.

continuous beam *(Eng.)*. A beam supported at a number of points and continuous over the supports, as distinct from a series of simple independent beams. Also *continuous girder*.

continuous brake *(Eng.)*. A brake system used on railway trains, in which operation at one point applies the brakes simultaneously throughout the train. See **air brake, electropneumatic brake, pneumatic brake, vacuum brake**.

continuous casting *(Eng.)*. Method in which the molten metal is added at the top of the mould while the externally solidified material is withdrawn from the bottom, the mould being cooled by water or air jets.

continuous control *(Telecomm.)*. System in which controller is supplied with continuously varying actuating signal. Cf. **on-off control**.

continuous creation *(Astron.)*. See **steady-state theory**.

continuous culture *(Bot.)*. A culture maintained at a steady state over a period, usually in a **chemostat**. Cf. **batch culture**.

continuous current *(Elec.Eng.)*. Earlier name for **direct current**. Now obsolete in UK and US but retained on Continent.

continuous diffusion *(Genrl.)*. Counterflow system of extracting sugar from beet whereby fresh beet slices (cosettes) are extracted by hot dilute sugar solution, and partially extracted slices are finally extracted with fresh hot water.

continuous-disk winding *(Elec.Eng.)*. A type of winding used for transformers; the whole winding is made from one continuous length of conductor instead of being split up.

continuous distillation *(Chem.Eng.)*. An arrangement by which a fresh distillation charge is continuously fed into the still in the same measure as the still charge is distilled off. The contrary process is known as *batch distillation*.

continuous electrode *(Elec.Eng.)*. A type of carbon electrode used in electric furnaces; the electrode is gradually fed forward as the lower part burns away, and the upper part is renewed by adding fresh material. The furnace can thus be worked continuously, without intervals for electrode renewal.

continuous extraction *(Chem.Eng.)*. Extraction of solids or liquids by the same solvent, which circulates through the extracted substance, evaporates, and is condensed again, and continues the same cycle over again; or by exhaustive extraction with solvents in counter-current arrangement.

continuous feeder *(Print.)*. An automatic sheet-feeder designed to permit continuous reloading without stopping the machine.

continuous filament yarn *(Textiles)*. Yarn composed of one or more unbroken filament. Man-made fibres are usually made as continuous filaments although these may be cut or broken into staple fibres for subsequent manufacture into fabric. Silk is a natural continuous filament. Cf. **staple fibres**.

continuous filter *(Build.)*. See **percolating filter**.

continuous furnace *(Eng.)*. Furnace in which the charge enters at one end, moves through continuously, and is discharged at the other.

continuous girder *(Eng.)*. See **continuous beam**.

continuous impost *(Build.)*. An impost which does not project from the general surfaces of the pier and arch.

continuous loading *(Elec.Eng.)*. Inductance loading with series inductions at intervals much less than a wavelength.

continuous mill *(Eng.)*. A rolling-mill consisting of a series of pairs of rolls in which the stock undergoes successive reductions as it passes from one end to the other end of the mill. See also **pull-over mill, reversing mill, three-high mill**.

continuous oscillations *(Telecomm.)*. Those which would occur in a tuned circuit of inductance, capacitance but no resistance: in practical circuits, the *damping* effect of resistance has to be overcome by injecting energy from an external source. Also *undamped oscillations*.

continuous printing *(Image Tech.)*. Contact printing in which negative film and print stock move continuously past an illuminated exposure aperture.

continuous processing machine *(Image Tech.)*. Equipment for processing photographic film or paper as a continuous strip passing through the successive chemical solutions and washing baths to the final drier.

continuous projector *(Image Tech.)*. A motion picture projector in which the film moves continuously and uniformly, the intermittent effect being obtained by optical means, such as a multi-faced rotating prism.

continuous rating *(Elec.Eng.)*. The maximum power dissipation which could be allowed to continue indefinitely. Cf. **intermittent rating**.

continuous reinforcement *(Behav.)*. *Schedules of reinforcement* in which every correct behaviour is reinforced.

continuous sections *(Print.)*. The normal arrangement of the sections for bookwork as distinct from **insetted**. Each section is made up of consecutive pages, its last page being followed by the first page of the next section.

continuous spectrum *(Phys.)*. One which shows continuous non-discrete changes of intensity with wavelengths or particle energy.

continuous stationery *(Comp.)*. Paper, perforated at page intervals and fan-folded to form a pack, used in **dot matrix** and **line printers**.

continuous variation *(Biol.)*. Variation of a character whose measurements do not fall into distinct classes, but take any value within certain limits.

continuous vent *(Build.)*. Extension of a vertical waste pipe above the point of entry of liquid wastes to a point above all windows, to provide ventilation.

continuous welded rail *(Civ.Eng.)*. Track which has been prewelded into lengths of up to 366 m (1200 ft) before being transported to the site, where successive lengths are welded together to produce unbroken track many miles long. The rail is hydraulically extended to the length which it would take up at a standard temperature (28°C in UK), so that it is in tension at any lower temperature.

continuum *(Ecol.)*. The pattern of overlapping populations in a large but definite community with component populations distributed along a gradient of e.g. altitude.

continuum *(Maths.)*. The real line (or any complete metric space). The term is little used now on its own, but certain combinations are important. The *cardinality of the continuum*, C, is the cardinal number of the points in the real line, which is greater than **aleph-0**. The *continuum hypothesis* is that there is no cardinal number between aleph-0 and C. This hypothesis cannot be proved or disproved from the basic axioms of set theory.

contorted *(Bot.)*. Petals in a flower bud which overlap a neighbour to one side and are overlapped on the other so that the whole appears twisted. See **aestivation**.

contour *(Surv.)*. The imaginary intersection line between the ground surface and any given level surface: a line connecting points on the ground surface which are at the same height above **datum**.

contour acuity *(Phys.)*. The power of the eye to distinguish a displacement between two sections of a line, as in reading a **vernier**.

contour fringes *(Phys.)*. Interference fringes formed by the reflection of light from the top and bottom surfaces of a thin film or wedge. The fringes correspond to optical thickness. Also called *Fizeau fringes*. See **Newton's rings**.

contour gradient *(Surv.)*. A line on the ground surface having a constant inclination to the horizontal.

contour interval *(Surv.)*. The vertical distance between adjacent contours in any particular case.

contraception *(Med.)*. The prevention of conception.
contraceptive *(Med.)*. Any agent which prevents the fertilization of the ovum with a spermatozoon. See **oral contraceptives**.
contract-demand tariff *(Elec.Eng.)*. A form of 2-part tariff in which the fixed charge is made proportional to the maximum kilowatt demand likely to be made.
contractile root *(Bot.)*. A root, some part of which shortens (by a change in shape of the inner cortical cells) so as to pull e.g. a herbacious plant closer to the ground or a bulb or corm deeper into the soil.
contractile tissue *(Zool.)*. A group of animal tissues which possess the property of contractility; more commonly spoken of as muscle.
contractile vacuole *(Zool.)*. In some Protozoa, a cavity, filled with fluid, which periodically collapses and expels its contents into the surrounding medium, so ridding the animal of surplus fluid.
contractility *(Zool.)*. The power of becoming reduced in length, exhibited by some cells and tissues, as muscle; the power of changing shape.
contraction *(Powder Tech.)*. The percentage shrinkage after heat treatment from the green dimensions.
contraction cavities *(Eng.)*. Porous zones in metal castings caused by contraction of cooling metal, bad mould design, interrupted pouring of molten metal or premature freezing of feeding head.
contraction coefficient *(Elec.Eng.)*. A coefficient used in making calculations on the magnetic circuit of an electric machine, to allow for the effect of the fringing of the flux in the air-gap due to open or semiclosed slots. The actual length of the gap is reduced by the coefficient in order to obtain an effective gap length, which is shorter than the actual value.
contraction in area *(Eng.)*. See **necking**.
contraction joint *(Civ.Eng.)*. An interruption of a structure specifically allowed for contraction in the time following building.
contraction ratio *(Aero.)*. The ratio of the maximum cross-sectional area of a wind tunnel to that of the working section.
contracture *(Med.)*. Distortion of shortening of a part, due to spasm or paralysis of muscles, or to the presence of scar tissue.
contracture *(Zool.)*. Muscular contraction which persists after the stimulus which caused it has ceased.
contrails *(Meteor.)*. See **condensation trails**.
contralateral *(Zool.)*. Pertaining to the opposite side of the body. Cf. **ipsilateral**.
Contran *(Comp.)*. *Control translator*. A programing language combining features of Fortran IV and Algol 60.
contraries *(Paper)*. Anything in the pulp or paper stock which is unwanted in the paper.
contrast *(Acous.)*. The relation, measured in decibels, between the maximum intensity level and the minimum useful intensity level in programme material such as speech or music.
contrast *(Image Tech.)*. (1) The difference in brightness between the lightest and darkest areas in a subject or its reproduction. (2) In a photographic image, the relation between the maximum and minimum densities. See **gamma**. (3) In lighting a subject, the relation between the **key** illumination and the **filler light** in the shadow areas.
contrast amplification *(Acous.)*. That in which dynamic contrast in sound reproduction is increased by electronic means, compensating imprecisely for contrast reduction which is necessary in most communication systems to avoid intrusion of noise.
contrast medium *(Med.)*. Substance injected into the bloodstream to increase the contrast in X-ray procedures; usually contains iodine. Widely used in diagnostic radiology.
contrast photometer *(Phys.)*. A class of photometer in which measurement is made by comparing the illumination produced on two adjacent surfaces by the lamp under test and by a standard lamp.
contrast range *(Image Tech.)*. See **dynamic range**.
contrate wheel *(Horol.)*. (1) A toothed wheel the teeth of which are formed at right-angles to the plane of the wheel; a wheel that transmits motion between two arbors at right angles. (2) The fourth wheel in a watch with the verge escapement.
contra wire *(Elec.Eng.)*. Same as *constantan wire*.
control *(Comp.)*. The selection, interpretation and sequencing of functions to be performed within a computing system. Control is held by the **control unit**.
control *(Elec.Eng.)*. General term for manual or automatic adjustment (usually by potentiometer, fader, or attenuator) of power level in a transmission within its dynamic range.
control *(Nuc.Eng.)*. Maintenance of power level of a reactor at desired setting by adjustments to the reactivity by control rods or other means.
control absorber *(Nuc.Eng.)*. See **control rod**.
control ampere-turns *(Elec.Eng.)*. Magnetomotive force applied to a magnetic amplifier.
control-board *(Elec.Eng.)*. A switchboard on which are mounted the operating handles, push-buttons, or other devices for operating switchgear situated remotely from the board. The board usually has mounted on it indicating instruments, key diagrams, and other accessory apparatus.
control character *(Comp.)*. A non-printing character which is treated as a signal to control operating functions. Cf. **alphanumeric**.
control characteristic *(Elec.Eng.)*. Curve connecting output quantity against control quantity under determined conditions in a magnetic amplifier.
control circuit *(Elec.Eng.)*. One which performs the function of control of the operation of a piece of equipment or of an electrical circuit.
control column *(Aero.)*. The lever supporting a hand-wheel or hand-grip by which the ailerons and elevator of an aircraft are operated. It may be a simple 'joystick', pivoted at the foot and rocking fore-and-aft and laterally; on military aircraft, usually fighters, it is often hinged halfway up for lateral movement; on transports it is usually either 'spectacle' or 'ram's horn' shape.
control-configured vehicle *(Aero.)*. One designed with artificial stability giving e.g. reduced wing size and control surfaces, enhanced manoeuvrability, reduced gust response and flutter suppression.
control current, voltage *(Elec.Eng.)*. One which, by its magnitude, direction or relative phase, determines the operation of an item of plant and/or electrical circuit.
control electrode *(Elec.Eng.)*. One, e.g. a grid, primary function of which is to control flow of electrons between two other electrodes, without taking appreciable power itself, control being by voltage which regulates electrostatic fields.
control hysteresis *(Elec.Eng.)*. Ambiguous control depending on previous conditions. Jump or snap action arising in electronic or magnetic amplifiers because of excessive positive feed-back which occurs under certain conditions of load.
control impedance *(Phys.)*. The electrical property of a device which controls power in one direction only, as a gas-filled relay.
controllable-pitch propeller *(Aero.)*. See **propeller**.
controlled air space *(Aero.)*. Areas and lanes clearly defined in 3 dimensions wherein no aircraft may fly unless it is under radio instructions from **air-traffic control**.
controlled carrier *(Telecomm.)*. Transmission in which the magnitude of the carrier is controlled by the signal, so that the depth of modulation is nearly independent of the magnitude of the signal.
controlled cooling *(Eng.)*. Methods of heat treatment in which the cooling cycle is accurately controlled so as to impart the desired properties.
controlled variable *(Eng.)*. Quantity or condition which is measured and controlled.
controller *(Elec.Eng.)*. An assembly of equipment for controlling the operation of electric apparatus.
control limit-switch *(Elec.Eng.)*. A limit-switch connected

in the control circuit of the motor whose operation is to be limited.

control-line *(Elec.Eng.)*. A train-line used on multiple-unit trains for connecting master controllers or contactor gear on the different coaches.

control magnet *(Elec.Eng.)*. A magnet used in electric indicating instruments to provide a force for controlling the movement of the moving system.

control panel *(Elec.Eng.)*. Panel containing full set of indicating devices and remote control units required for operation of industrial plant, reactor, chemical works, etc. Cf. **console**.

control program *(Comp.)*. See **monitor, operating system**.

control register *(Comp.)*. Computer **register** within the **control unit** and which stores a single control instruction.

control relay *(Elec.Eng.)*. See **relay types**.

control reversal *(Aero.)*. See **reversal of control**.

control rod *(Nuc.Eng.)*. Rod moved in and out of reactor core to vary reactivity. May be neutron-absorbing rod, e.g. boron or cadmium, or less often, a fuel rod. See **regulating rod, shim rod**.

control rod worth *(Nuc.Eng.)*. The change in reactivity of a reactor caused by the complete insertion or withdrawal of the control rod.

control (set) point *(Elec.Eng.)*. Value of controlled variable, departure from which causes a controller to operate in such a sense as to reduce the **error** and restore an intended steady state.

control total *(Comp.)*. The sum resulting from the addition of a specified field from each of a group of records, often used for checking purposes.

control turns, windings *(Elec.Eng.)*. Those wires on the core of a magnetic amplifier or transducer which carry the control current. US term *signal windings*.

control unit *(Comp.)*. Part of the **central processor** which supervizes the execution of instructions.

control voltage *(Elec.Eng.)*. See **control current**.

control word *(Comp.)*. One which transmits an operating instruction to a **central processor**, e.g. XEQ for execute. Cf. **control character**.

control zone *(Aero.)*. A volume of controlled airspace, precisely defined in plan and altitude, including airports, in which flight rules additional to those in a control area pertain. ICAO defines a specific upper limit.

conus *(Zool.)*. Any cone-shaped structure or organ.

conus arteriosus *(Zool.)*. In some lower Vertebrates, a valvular region of the truncus arteriosus, adjacent to the heart.

conus medullaris *(Zool.)*. The conical termination of the spinal cord.

convection *(Geol.)*. The very slow mass movement of subcrustal material; believed to be the mechanism that drives tectonic plates.

convection current *(Phys.)*. Current in which the charges are carried by moving masses appreciably heavier than electrons.

convection of heat *(Phys.)*. The transfer of heat in a fluid by the circulation flow due to temperature differences. The regions of higher temperature, being less dense, rise, while the regions of lower temperature move down to take their place. The convection currents so formed help to keep the temperature more uniform than if the fluid was stagnant.

convective transfer *(Astron.)*. The transfer of energy from one part of a star to another by convection.

convector *(Eng.)*. Heater which warms the air passing over it. Cf. **radiator**.

conventional signs *(Civ.Eng.,Surv.)*. Standard symbols, universally understood, used in the representation on maps and plans of features which would otherwise be difficult or impossible to represent.

convention of signs *(Phys.)*. In lens calculations all distances must be measured from some origin and to ensure consistency in the derivation and use of lens formulae, it is necessary to adopt a sign convention. More than one is in use, but in the *Cartesian* convention all distances are measured *from* the reflecting or refracting surface being considered, or *from* the principal planes in the case of a thick lens or lens system. Distances measured in the direction in which the incident light is travelling are given a positive sign, and those in the opposite direction a negative sign. Distances measured perpendicular to the axis e.g. size of image or object, are measured from the axis; above is positive, below is negative.

convergence *(Biol.)*. See **convergent evolution**.

convergence *(Image Tech.)*. In a colour TV display, alignment of the three electron or optical beams for correct image registration over the whole picture.

convergence (*Maths.*). (1) An infinite sequence $a_1, a_2, \ldots a_n, \ldots$ is said to be convergent to a if, as n tends to infinity, a_n tends to the limit a, i.e. $\lim_{n\to\infty} a_n = a$. (2) An infinite series $a_1 + a_2 + \ldots + a_n + \ldots$, is said to be convergent to the sum A, if the sum of the terms tends to the limit A, as the number of terms tends to infinity, i.e. $\lim_{n\to\infty} \sum a_n = A$. (3) An infinite product, $a_1, a_2 \ldots a_n \ldots$ is said to be convergent to α if, as n tends to infinity, the product tends to the limit α, i.e., $\lim_{n\to\infty} \prod a_n = \alpha$. (4) An infinite integral $\int_a^\infty f(t)dt$, is said to converge to I, if $f(t)$ is continuous for $t \geqq a$, and if $\lim_{x\to\infty} \int_a^x f(t)dt = I$.

convergence *(Meteor.)*. Negative divergence.

convergence coils, magnets *(Image Tech.)*. Used to ensure convergence of the electron beams in a triple gun colour TV tube.

convergence surface *(Electronics)*. Ideally, that generated by the point on which the electron beams in a multibeam CRT converge. In practice, this can only be regarded as a point to a first approximation.

convergence zone *(Geol.)*. The zone where moving plates collide and area is lost by shortening and crustal thickening or by subduction and destruction of the crust. The site of earthquakes, mountain building, trenches and vulcanism.

convergent *(Maths.)*. See **continued fraction**.

convergent-divergent nozzle *(Aero.,Eng.)*. A venturi type of nozzle in which the cross-section first decreases to a throat and then increases to the exit, such a form being necessary for efficient expansion of steam, as in the steam-turbine, or of other vapours or gases, as in the supersonic aero-engine, ramjet or rocket. Abbrev. *con-di nozzle*.

convergent evolution *(Biol.)*. The tendency of unrelated species to evolve similar structures, physiology or appearance due to the same selective pressures, e.g. the eyes of Vertebrates and Cephalopods or the succulent CAM plants of deserts.

convergent lens *(Phys.)*. A lens which increases the convergence or decreases the divergence of a beam of light. A simple lens is convergent if it is double-convex, plano-convex, or concavo-convex with the radius of curvature of the concave face greater than that of the convex.

convergent thinking *(Behav.)*. A type of ability which is sampled in standard intelligence tests. Rational, logical thought which works within certain confines, as in solving a mathematical problem. See **creativity, divergent thinking**.

converging-beam therapy *(Radiol.)*. A form of **cross-fire technique** in which a number of beams of radiation are used simultaneously to treat a particular region through different entry portals.

converging-field therapy *(Radiol.)*. Form of *moving field therapy* in which the source of radiation moves in a spiral path.

converse magnetostriction *(Phys.)*. The change of magnetic properties, e.g., induction, in a ferro-magnetic material when subjected to mechanial stress or pressure.

conversion *(Chem.)*. The fraction of some key reagent which has been used up, hence a measure of the rate or completeness of a chemical reaction.

conversion coating *(Eng.)*. A coating on a metal produced by chemical treatment e.g. chromatization and phosphating, which impart properties like corrosion resistance. See **passivation, activation**.

conversion coefficient *(Phys.)*. See **internal conversion**.

conversion detector *(Telecomm.)*. See **mixer (2)**.

conversion disorder *(Behav.)*. The loss or impairment of some motor or sensory function for which there is no known organic cause. Formerly *conversion hysteria*.

conversion efficiency *(Electronics)*. See **anode efficiency**.

conversion electron *(Electronics)*. One ejected from inner shell of atom when excited nucleus returns to ground state, and the energy released is given to orbital electron, instead of appearing as quantum.

conversion factor *(Phys.)*. Factor by which a quantity, expressed in one set of units, must be multiplied to convert it to another.

conversion gain, loss *(Telecomm.)*. Effective amplification, or loss, of a conversion detector, measured as the r output voltage of intermediate frequency to input voltage of signal frequency. Commonly expressed in decibels.

conversion mixer *(Telecomm.)*. Same as **frequency changer**.

conversion ratio *(Nuc.Eng.)*. Number of fissionable atoms e.g., ^{239}Pu produced per fissionable atom, ^{235}U destroyed in a reactor. Corresponding conversion gain is defined as $R-1$. Symbol R.

converter *(Comp.)*. A device for changing information coded in one form into the same information coded in another, e.g. analogue-digital converter.

converter *(Elec.Eng.)*. A circuit for changing a.c. to d.c. or vice versa. Rating can be a few watts to megawatts.

converter *(Telecomm.)*. US for **frequency changer**.

converter reactor *(Nuc.Eng.)*. One in which fertile material in reactor core is converted into fissile material different from the fuel material. Cf. **breeder reactor**.

convertible machine *(Print.)*. A multi-unit printing machine which can be mechanically altered to print either as a multi-colour press or as a **perfector** or in a combination of these operations.

converting *(Eng.)*. See **Bessemer process**.

converting *(Textiles)*. Producing a sliver of staple fibres from a continuous filament tow by cutting or breaking.

converting station *(Elec.Eng.)*. An electric power system substation containing one or more converters.

convertiplane *(Aero.)*. A VTOL aircraft which can take-off and land like a helicopter, but cruises like an aircraft; by swivelling the rotor(s) and/or wing to act as propellers, or by putting the rotor(s) into **autorotation** and using other means for propulsion.

convex lens *(Phys.)*. A **convergent lens**.

convex mirror *(Phys.)*. A portion of a sphere of which the outer face is a polished reflecting surface. Such a mirror forms diminished virtual images of all objects in front of it.

conveyor *(Eng.)*. Generally consists of a suitable tensioned endless belt made from hard wearing materials and arranged to run over rollers. Used to move materials in bulk from one point to another, including cross-country.

convolute *(Bot.)*. Coiled, folded, or rolled, so that one half is covered by the other.

convolute *(Zool.)*. Having one part twisted over or rolled over another part; twisted; as the cerebral lobes of the brain in higher Vertebrates, gastropod shells in which the outer whorls overlap the inner. n. *convolution*.

convolution *(Med.)*. Any elevation of the surface of the brain.

convolution integral *(Maths.)*. The integral $\int f(t)g(x-t)dt$, where the limits of the integral are variously defined. Also referred to as the *cross-correlation* between $f(x)$ and $g(x)$. When $f(x)=g(x)$ the integral is sometimes referred to as the *autocorrelation of $f(x)$* or simply the *convolution*.

convulsion *(Med.)*. Generalized involuntary spasm of the muscles normally under control of the will.

coolant *(Eng.)*. (1) A mixture of water, soda, oil, and soft-soap, used to cool and lubricate the work and cutting tool in machining operations. See **cutting compound**. (2) A fluid used as the cooling medium in the jackets of liquid-cooled IC engines; e.g. water, ethylene glycol (ethan 1.2-diol).

coolant, reactor *(Nuc.Eng.)*. The gas, liquid or liquid metal circulated through a reactor core to carry the heat generated in it by fission and radioactive decay to boilers or heat exchangers.

cooled-anode valve *(Electronics)*. Large thermionic valve in which special provisions are made for dissipating the heat generated at the anode, effected by circulating water, oil, or air around the anode, or by radiation from its surface.

cooling *(Nuc.Eng.)*. The decay of activity of irradiated nuclear fuel or highly radioactive waste before it is processed or disposed of.

cooling coil *(Eng.)*. Tubing which transfers heat from the material or space cooled to the primary refrigerant.

cooling curves *(Eng.)*. Curves obtained by plotting time against temperature for a metal cooling under constant conditions. The curves show the evolutions of heat which accompany solidification, polymorphic changes in pure metals, and various transformations in alloys.

cooling drag *(Aero.)*. That proportion of the total drag due to the flow of cooling air through and round the engine(s).

cooling duct *(Elec.Eng.)*. See **duct**.

cooling pond *(Eng.)*. An open pond in which water, heated through use in an industrial process, or after circulation through a steam-condenser, is, before re-use, allowed to cool through evaporation.

cooling tower *(Eng.)*. A tower of wood, concrete, etc., used to cool water after circulation through a condenser. The water is allowed to trickle down over wood slats, thus exposing a large surface to atmospheric cooling.

coomassie blue *(Biol.)*. TN for dye used as a sensitive stain to locate proteins after their fractionation by **electrophoresis** or **isoelectric focusing** in a gel matrix.

cooperation *(Ecol.)*. A category of interaction between two species where each has a beneficial effect on the other, increasing the size or growth rate of the population, but, unlike mutualism, not a necessary relationship. Termed **protocooperation** by some, since its basis is neither conscious nor intelligent, as in man.

Cooper pairs *(Phys.)*. In a *superconducting* material below its critical temperature, the electrons do not act independently but in dynamic pairs. The electrons are weakly bound together and form *Cooper pairs*. The BCS (Bardeen, Cooper, Schrieffer) theory uses this concept to give a detailed microscopic theory of superconductivity.

co-ordinate axes *(Maths.)*. See **axes**.

co-ordinate bond *(Chem.)*. See **covalent bond**.

co-ordinated transposition *(Telecomm.)*. The reduction of mutual inductive effects in multiline transmission systems (telephony or power) by periodically interchanging positions.

co-ordinate potentiometer *(Elec.Eng.)*. One in which two linear potentiometers carry a.c. currents 90° apart in phase, so that the resultant voltage between tappings can be adjusted in both phase and magnitude.

co-ordinates *(Maths.)*. Numbers which specify the position or orientation of a point or geometric configuration.

co-ordinating gap *(Elec.Eng.)*. A spark-gap, used in power transmission schemes, so arranged that it will break down at a voltage bearing a definite relation to the breakdown voltage of other apparatus in the system, thereby enabling surge voltages to be safely discharged to earth.

co-ordination compound *(Chem.)*. A compound generally described from the point of view of the central atom to which other atoms are bound or coordinated and are called ligands. Normally, the central atom is a (transition) metal ion, and the ligands are negatively charged or strongly polar groups.

co-ordination number *(Chem.)*. The number of atoms or groups (ligands) surrounding the central (nuclear) atom of a complex ion or molecule.

cop *(Textiles)*. Yarn package built on to mule spindle or ring tube.

copalite, copaline *(Min.)*. Synonym for **Highgate resin**.

cope *(Foundry)*. The upper half of a mould or moulding box.

Copepoda *(Zool.)*. Subclass of Crustacea, mainly of small size. Some are parasitic, others planktonic where they form an important food source for pelagic Fish like herring.

Copernican System *(Astron.)*. The heliocentric theory of planetary motion; called after Copernicus, who introduced it in 1543. It superseded the geocentric, or Ptolemaic System.

coping *(Build.)*. (1) A stone, brick, tile or concrete covering to the top of a wall exposed to the weather; it is designed to throw off the water, and is preferably wider than the wall, with *drips* cut in its projecting undersurfaces. (2) The operation of splitting stone by driving wedges into it.

coping *(Vet.)*. The operation of paring or cutting the beak or claws of a bird, particularly of hawks.

coping brick *(Build.)*. Specially shaped brick used for capping the exposed top of a wall; used sometimes with a creasing and sometimes without, in which latter case the brick is wider than the wall and has drips under its lower edges.

coping saw *(Build.)*. Small saw with narrow tensioned blade in a D-shaped bow, for cutting curves in wood up to about 15 mm thick (i.e. too thick for **fret-saw**).

coplanar vectors *(Maths.)*. Vectors are coplanar if they lie in the same plane.

copolymer *(Chem.)*. Polymer formed from the reaction of more than one species of monomer, e.g. poly(styrene)-butan 1,2 : 3,4-diene.

copper *(Eng.)*. Bright, reddish metallic element, symbol Cu, at.no. 29, r.a.m. 63.54, mp 1083°C resistivity at 0°C, 0.016 microhm metre. Native copper crystallizes in the cubic system. It often occurs in thin sheets or plates, filling narrow cracks or fissures. Copper is ductile, with high electrical and thermal conductivity, good resistance to corrosion; it has many uses, notably as an electrical conductor. Basis of brass, bronze, aluminium bronze and other alloys. Nickel-iron wires with a copper coating are frequently used for *lead-ins* through glass seals, forming vacuum-tight joints. ^{64}Cu is a mixed radiator, of half-life 13 h.

copperas *(Min.)*. Iron sulphate, $FeSO_4.7H_2O$. See **melanterite**. *White copperas* is goslarite, $ZnSO_4.7H_2O$.

copper brushes *(Elec.Eng.)*. Brushes occasionally used for electric commutator machines where high conductivity is required; they are made of copper strip, wire, or gauze.

copper-clad steel conductor *(Elec.Eng.)*. See **steel-cored copper conductor**.

copper factor *(Elec.Eng.)*. A term used in electric machine design to denote the ratio of the cross-sectional area of the copper in a winding to the total area of the winding, including insulating material and clearance space.

copper glance *(Min.)*. A popular name for *chalcocite*.

copper glazing *(Build.)*. Glazing formed of a number of individual panes separated by copper strips on the edges of which small flanges of copper have later been formed by deposition to retain the glass. Also *copperlite glazing*.

copper (II) sulphate *(Chem.)*. $CuSO_4$. Bluestone; a salt, soluble in water, used in copper-plating baths; formed by the action of sulphuric acid on copper; crystallizes as hydrous copper sulphate, $CuSO_4.5H_2O$, in deep-blue triclinic crystals. See **blue vitriol, chalcanthite**.

copper loss *(Elec.Eng.)*. The loss occurring in electric machinery or other apparatus due to the current flowing in the windings; it is proportional to the product of $(current)^2 \times resistance$.

copper loss *(Telecomm.)*. The power dissipated as heat in an antenna, or other oscillatory, circuit due to Joule effect, including that due to eddy currents in the conductors and nearby metallic objects.

copper nickel *(Min.)*. See **niccolite**.

copper nose *(Vet.)*. See **brown nose disease**.

copper pyrites *(Min.)*. See **chalcopyrite**.

copper-sheathed cable *(Elec.Eng.)*. See mineral-insulated cable.

coppersmith's hammer *(Eng.)*. A hammer having a long curved ball-pane head, used in dishing copper plates.

copper uranite *(Min.)*. See **torbernite**.

coppice *(Bot.,For.)*. (1) A traditional form of woodland in which trees are cut to near ground level every 10–15 years and allowed to grow again from the **stool**, to produce poles for firewood, charcoal and fencing (especially hurdle-making). (2) To cut such trees. In a *coppice-with-standard*, some trees are left to grow for several coppice cycles to form timber.

coprecipitation *(Phys.)*. The precipitation of a radioisotope with a similar substance, which precipitates with the same reagent, and which is added in order to assist the process.

coprodaeum *(Zool.)*. That part of the cloaca into which the anus opens, in Birds.

coprolalia *(Behav.)*. The utterance of filthy words by the insane.

coprolite *(Geol.)*. Fossilized excreta of animals. Generally composed largely of calcium phosphate.

coprophagous *(Zool.)*. Dung-eating.

coprophilia *(Behav.)*. Pleasure or gratification obtained from any dealing with faeces.

coprophilous, coprophilic *(Bot.)*. Growing on or in dung.

coprozoic *(Zool.)*. Living in dung, as some Protozoa.

copula *(Zool.)*. A structure which bridges a gap or joins two other structures, as the series of unpaired cartilages which unite successive gill arches in lower Vertebrates.

copulation *(Zool.)*. In Protozoa, a type of syngamy in which the gametes fuse completely; in higher animals, union in sexual intercourse.

copulation tube *(Bot.)*. See **conjugation tube**.

copy *(Comp.)*. The transfer of information from one store to another without changing the information in the original store.

copy *(Print.)*. Any matter supplied for setting or for reproduction by any of the printing processes.

copyholder *(Print.)*. (1) One who reads aloud from the copy as the proof-corrector follows the reading in his proof. (2) A contrivance for holding up sheets of copy on typesetting machines.

copying machine *(Eng.)*. A machine for producing numbers of similar objects by an engraving tool or end-cutter, which is guided automatically from a master pattern or template. See also **document copying**.

copy lens *(Image Tech.)*. A lens designed for optimum quality of image formation at comparatively short object distances and small degrees of magnification or reduction.

copy number *(Biol.)*. The number of genes or plasmid sequences per genome which a cell contains.

coquille *(Glass.)*. Glass in thin curved form used in the manufacture of sun glasses. The radius of curvature is usually 3½ in (9 cm). Similar glass of 7 in (18 cm) radius is called *micoquille*.

coquimbite *(Min.)*. Hydrated ferric sulphate, crystallizing in the hexagonal system, occurring in some ore deposits and also in volcanic fumaroles such as those of Vesuvius.

coquina *(Geol.)*. A limestone made up of coarse shell fragments, usually of molluscs.

Coraciiformes *(Zool.)*. An order of Birds, most of which are short-legged arboreal forms, nesting in holes and having nidicolous young. Mainly tropical and often brightly coloured. Kingfishers, Bee-eaters.

coracoid *(Zool.)*. In Vertebrates, a paired posterior ventral bone of the pectoral girdle, or the cartilage which gives rise to it.

CORAL *(Comp.)*. Programming language used for **on-line real-time** systems.

coral *(Zool.)*. The massive calcareous skeleton formed by certain species of *Anthozoa* and some *Hydrozoa*; the colonies of polyps forming this skeleton. adjs. *coralline, coralloid, coralliferous, corallaceous, coralliform*.

coral reef *(Geol.)*. A calcareous bank formed of the calcareous skeletons of corals and algae which live in colonies. The various formations of coral reefs are known as *atolls, barrier reefs*, and *fringing reefs*.

coral sand *(Geol.)*. A sand made up of calcium carbonate grains derived from eroded coral skeletons, often found in deep water on the seaward side of a coral reef.

corbeille *(Arch.)*. Carved work representing a basket, used as a form of decoration.

corbel *(Build.)*. Bricks or stones, frequently moulded, projecting from a wall to support a load.

corbelling *(Build.)*. Projecting courses of brick or stone forming a ledge used to support a load.

corbel-piece *(Build.)*. See **bolster**.

corbicula *(Zool.)*. The pollen basket of Bees, consisting of the dilated posterior tibia with its fringe of long hairs.

corbie-step gable *(Build.)*. A gable having a series of regular steps up each slope. Also called *crow-step gable*.

cord *(For.)*. A non-metric timber measure, 128 ft^3 (8 × 4 × 4 ft, about 3½ m^3).

cordate *(Bot.)*. Said of a leaf base which has the form of the indented end of a conventional heart.

corded way *(Build.)*. A sloping path formed with deep sloping steps separated by timber or stone risers.

cordierite *(Min.)*. Magnesium aluminium silicate with some iron, typically occurring in thermally metamorphosed rocks and in some gneisses. Often shows cyclic twinning. *Iolite* is a name often used for gem varieties and the mineral is also sometimes called *dichroite* from its strong blue to colourless dichroism.

cords *(Print.)*. Lengths of hemp across the back of a book, to which the sections are attached by sewing. See **bands**.

Cordtex *(Civ.Eng.)*. TN for a textile detonating fuse containing a core of pentaerythritol tetranitrate, used in the initiation of large explosive charges. Has a high velocity of detonation and will not inspire detonation unless directly connected to the actual explosive.

corduroy *(Textiles)*. Strong, hard-wearing cloth having a rounded or flattened cord or rib of weft pile running longitudinally; made entirely from cotton, or cotton warp and spun-rayon pile.

cordwood *(For.)*. Tree trunks of medium diameter sawn into uniform lengths.

core *(Aero.)*. Gas generator portion of a gas turbine engine which may be developed as a basis of several engines used on different types of aircraft.

core *(Build.)*. The material removed from a mortise.

core *(Civ.Eng.)*. (1) A watertight wall built within a dam or embankment as an absolute barrier to the passage of water. (2) Cylindrical sample of material obtained by driving a hollow-core drill into strata to ascertain the variation of composition.

core *(Elec.Eng.)*. (1) Region associated with a coil: may be air or a magnetic material to increase inductance. Typical materials are ferrites and punched laminations of soft iron. Construction may be as a complete magnetic circuit (divided to accommodate the coil) or simply a rod inserted into the coil. (2) Assembly consisting of the conductor and insulation of a cable but not including the external protective covering. An arrangement comprising many such assemblies is termed *multicore*.

core *(Foundry.)*. A solid mass of specially prepared sand or loam placed in a mould to provide a hole or cavity in the casting.

core *(Geol.)*. The central part of the Earth at a depth below 2900 km. It is believed to be composed of nickel and iron.

core *(Image Tech.)*. (1) A plastic cylinder on which film or magnetic tape is wound. (2) The inner part of the positive carbon of an arc lamp, impregnated with metallic salts to improve colour and brilliance.

core *(Min.Ext.)*. Cylindrical rock section cut by rotating hollow drill bit in prospecting, sampling, blasting.

core *(Nuc.Eng.)*. That part of a nuclear reactor which contains the fissile material, either dispersed or in cans.

core *(Phys.)*. In an atom, the nucleus and all complete shells of electrons. In the atoms of the alkali metals, the nucleus, together with all but the outermost of the planetary electrons, may be considered to be a core, around which the valency electron revolves in a manner analogous to the revolution of the single electron in the hydrogen atom around the nucleus. In this manner, the simple Bohr theory may be made to give an approximate representation of the alkali spectra. See also **atomic structure**.

core-balance protective system *(Elec.Eng.)*. An excess-current protective system for electric power systems, in which any leakage current to earth in a 3-phase circuit is made to produce a resultant flux in a magnetic circuit surrounding all 3 phases; this flux produces a current in a secondary winding on the magnetic circuit, which operates a relay controlling the appropriate circuit-breakers.

core bar *(Foundry.)*. (1) An iron bar on which cylindrical loam cores are built up. The bar is supported horizontally and rotated while a loam board is pressed against the core. (2) An iron rod for reinforcing a sand core.

coreboard *(Build.)*. See **blockboard**.

core box *(Foundry.)*. A wooden box shaped internally for moulding sand cores in the foundry.

CO_2 recorder *(Eng.)*. An instrument which analyses automatically the flue gas leaving a furnace, and records the percentage of carbon dioxide (CO_2) on a chart. See **exhaust-gas analyser**.

cored electrode *(Elec.Eng.)*. A metal electrode provided with a core of flux or other material; used in arc welding.

cored hole *(Eng.)*. A hole formed in a casting by the use of a core, as distinct from a hole that has been drilled.

cored solder *(Eng.)*. Hollow solder wire containing a flux paste, which allows flux and solder to be applied to the work simultaneously.

coreless induction furnace *(Elec.Eng.)*. A high-frequency induction furnace in which there is no iron magnetic circuit other than the charge in the furnace itself.

core losses *(Elec.Eng.)*. The losses occurring in electric machinery and equipment owing to hysteresis and eddy-current losses set up in the iron of the magnetic circuit, which are due to an alternating or varying flux.

coremium *(Bot.)*. (1) A ropelike strand of anastomosing hyphae. (2) A tightly packed group of erect conidiophores, somewhat resembling a sheaf of corn.

core oven *(Foundry.)*. A foundry oven used for drying and baking cores before insertion in a mould.

core plate *(Elec.Eng.)*. See **lamination**.

core plates *(Foundry.)*. Disks attached to a **core bar** to reinforce large cores.

core prints *(Foundry.)*. Projections attached to a pattern to provide recesses in the mould at points where cores are to be supported.

core register *(Foundry.)*. Corresponding flats or vees formed on cores and core prints, when correct angular location is necessary.

core sample *(Min.Ext.)*. Sample from a bore hole to give information on the rock formation at side or bottom. Usually a few inches in diameter.

core sand *(Foundry.)*. Moulding sand to which a binding material such as linseed oil has been added to obtain good cohesion and porosity after drying.

core-spun yarn *(Textiles)*. A yarn made with a core (usually of continuous filaments) surrounded by a sheath generally made of staple fibres.

core store *(Comp.)*. See **main memory**.

core-type induction furnace *(Elec.Eng.)*. An induction furnace in which there is an iron core to carry the magnetic flux.

core-type transformer *(Elec.Eng.)*. A transformer in which the windings surround the iron core, the former usually being cylindrical in shape.

coriaceous, corious *(Bot.,Zool.)*. Firm and tough, like leather in texture.

Coriolis acceleration *(Meteor.)*. The apparent tendency of a freely moving particle to swing to one side when its motion is referred to a set of axes that is itself rotating in space. The magnitude of the acceleration for a particle moving horizontally on the surface of the Earth is $2\Omega V\sin\phi$ where Ω is the angular magnitude of the angular velocity of rotation of the Earth, V is the speed of the particle relative to the Earth's surface, and ϕ is the latitude. The acceleration is perpendicular to the direction of V and is directed to the right in the northern hemisphere. For general three-dimensional motion the

Coriolis acceleration has some other small terms because the Earth's axis does not, normally, lie parallel to the local vertical; for meteorological purposes these additional terms are negligible.

Coriolis effect *(Phys.)*. In a rotating reference frame, Newton's second law of motion is not valid, but it can be made to apply if, in addition to the real forces acting on a body, a Coriolis force and a *centrifugal force* are introduced. The effect of the Coriolis force is to deviate a moving body perpendicular to its velocity. So a body freely falling towards the Earth is slightly deviated from a straight line and will fall to a point east of the point directly below its initial position. Coriolis forces explain the directions of the trade winds in equatorial regions and would effect astronauts in an artificial *g*-environment produced by rotation.

Coriolis force *(Meteor.)*. The force, acting on a given mass, which would produce the **Coriolis acceleration**.

Coriolis parameter *(Meteor.)*. The Coriolis parameter f is defined by $f = 2\Omega\sin\phi$ where Ω is the angular velocity of rotation of the earth, and ϕ is the latitude.

corium *(Zool.)*. The dermis of Vertebrates.

cork *(Bot.)*. *Phellem*. A tissue of dead cells with suberized cells which form a protective layer replacing the epidermis in older stems and roots of many seed plants. See **cork cambium, periderm**.

cork cambium *(Bot.)*. *Phellogen*. The layer of meristematic cells lying a little inside the suface of an older root or stem and forming cork on its outer surface and phelloderm internally. See **periderm**.

corkscrew rule *(Elec.Eng.)*. Rule relating the direction of the magnetic field to the current direction in a conductor. The corkscrew is driven in direction of current and sense of rotation of the handle gives direction of magnetic field.

corkscrew staircase *(Arch.)*. A helical staircase built about a solid central newel.

Corliss valve *(Eng.)*. A steam-engine admission and exhaust valve in the form of a ported cylinder which is given an oscillating rotary motion over the steam port by an eccentric driven wrist-plate.

corm *(Bot.)*. Organ of perennation and vegetative propagation in e.g. *Crocus* consisting of a short, usually erect and unicated, underground stem of one year's duration, next year's rising on top.

cormophyte *(Bot.)*. In former systems of classification, a plant of which the body is differentiated into roots, stems and leaves.

corn *(Bot.)*. In Britain, wheat as well as other cereals; in US, maize.

corn *(Med.)*. Localized overgrowth of the horny layer of the skin due to local irritation, the overgrowth being accentuated at the centre.

corn *(Vet.)*. A local inflammation due to bruising or compression of the keratogenous membrane of the posterior portion of the horse's foot; septic corn, an abscess localized to the sole of a bird's foot.

cornea *(Zool.)*. In Invertebrates, a transparent area of the cuticle covering the eye, or each facet of the eye; in Vertebrates, the transparent part of the outer coat of the eyeball in front of the eye. adj. *corneal*.

cornelian *(Min.)*. A synonym for *carnelian*.

corneous *(Bot.,Zool.)*. Resembling horn in texture.

corner *(Print.)*. The piece of leather covering each of the outer corners of a half-bound volume.

corner *(Telecomm.)*. See **bend**.

corner bead *(Build.)*. An **angle staff**.

corner cramp *(Build.)*. See **mitre cramp**.

corner-lap joint *(Build.)*. See **end-lap joint**.

corner reflector *(Radar.)*. Metal structure of three mutually perpendicular sheets, for returning signals.

corner tool *(Foundry.)*. A sleeking tool for finishing off the internal corners of a mould.

cornice *(Arch.)*. A projecting moulding decorating the top of a building, window etc.

corniculate, cornute *(Bot.,Zool.)*. (1) Shaped like a horn. (2) Bearing a horn or hornlike outgrowth.

Cornish boiler *(Eng.)*. A horizontal boiler with a cylindrical shell provided with a single longitudinal furnace tube or flue.

corn oil *(Chem.)*. A pale yellow oil obtained from maize; rel.d. 0.920–0.925, saponification value 188–193, iodine value 111–123, acid value 1.7–20.6. Used as a cooking oil. Also called *maize oil*.

cornstone *(Geol.)*. Concretionary limestone common in the Old Red Sandstone and Permo-Triassic rocks of Britain. Characteristic of soils formed in arid conditions.

cornua *(Zool.)*. Hornlike processes; as the posterior *cornua* of the hyoid. adj. *cornual, cornute*.

Cornu prism *(Phys.)*. A 60° prism formed of two 30° prisms cemented together, one being of right-handed and the other left-handed quartz, the optic axes of the two being parallel to the ray passing through the prism at minimum deviation, that is, parallel to the base. This device overcomes a defect of using a single 60° quartz prism in a spectrometer.

Cornu's spiral *(Maths.)*. A spiral with parametric cartesian equations

$$x = \int_0^s \cos \tfrac{1}{2}\theta^2 d\theta \text{ and } y = \int_0^s \sin \tfrac{1}{2}\theta^2 d\theta.$$

It has the property that its radius of curvature $\rho = 1/s$. Also called a *clothoid*.

corolla *(Bot.)*. Inner whorl of the **perianth** especially if different from the outer and then often brightly coloured, composed of the petals which may be free or fused to one another.

corona *(Arch.)*. The part of a **cornice** showing a broad projecting face and throated underneath to throw off the water.

corona *(Bot.)*. (1) A trumpetlike outgrowth from the perianth, as in the daffodil. (2) A ring of small leafy undergrowths from the petals, as in campion. (3) A crown of small cells on the Oögonium of Charophyta.

corona *(Meteor.)*. A system of coloured rings seen round the sun or moon when viewed through very thin cloud. They are caused by diffraction by water droplets. The diameter of the corona is inversely proportional to the size of the droplets.

corona *(Phys.)*. Phenomenon of air breakdown when electric stress at the surface of a conductor exceeds a certain value. At higher values, stress results in luminous discharge. See **critical voltage**.

corona *(Zool.)*. In Echinoidea, the shell or test; in Crinoidea, the disk and arms as opposed to the stalk; in Rotifera, the discoidal anterior end of the body; the head or upper surface of a structure or organ. adj. *coronal*.

coronagraph *(Astron.)*. A type of telescope designed by Lyot in 1930 for observing and photographing the solar corona, prominences etc. at any time.

corona radiata *(Zool.)*. A layer of cylindrical cells surrounding the developing ovum in Mammals.

coronary bypass *(Med.)*. Surgical bypassing of a blocked or narrow coronary artery.

coronary circulation *(Zool.)*. In Vertebrates, the system of blood vessels (coronary arteries) which supply the muscle of the heart-wall with blood.

coronary heart disease *(Med.)*. All forms of heart disorders arising from disease of the coronary arteries. Includes **angina** and **myocardial infarction**.

coronary sinus *(Med.)*. A channel opening into the right atrium draining blood from most of the cardiac veins.

coronary thrombosis *(Med.)*. Formation of a clot in one of the coronary arteries leading to obstruction of the artery and **infarction** of the area of the heart supplied by it.

corona voltmeter *(Elec.Eng.)*. Instrument for measuring high voltages by observing the conditions under which a corona discharge takes place on a specially designed wire.

coronene *(Chem.)*. $C_{24}H_{12}$. A yellow solid hydrocarbon, mp 430°C. Large planar molecule with seven benzene rings. ⇨

coronet *(Zool.)*. (1) The junction of the skin of the pastern with the horn of the hoof of a horse. (2) The knob at the base of the antler in deer.

coronoid *(Zool.)*. (1) In some Vertebrates, a membrane

bone on the upper side of the lower jaw. (2) More generally, beak-shaped.

co-routine *(Comp.)*. A kind of **subroutine** but whereas a subroutine can only be entered and left in predefined ways, a co-routine can be left and re-entered at any point.

corpora allata *(Zool.)*. In Insects, endocrine organs behind the brain which secrete **neotenin**, the juvenile hormone. In some species they are paired and laterally placed, but in others they fuse during development to form a single median structure, the *corpus allatum*.

corpora bigemina *(Zool.)*. In Vertebrates, the optic lobes of the brain.

corpora cardiaca *(Zool.)*. In Insects, paired neurohaemal organs lying behind the brain, and containing the nerve endings of the neurosecretory cells in the brain which produce several hormones including one involved in moulting. This hormone is released into the blood at the *corpora cardiaca.*

corpora cavernosa *(Zool.)*. In Mammals, a pair of masses of erectile tissue in the penis.

corpora geniculata *(Zool.)*. In the Vertebrate brain, paired protuberances lying below and behind the thalamus.

corpora lutea *(Biol.)*. See **corpus luteum**.

corpora pedunculata *(Zool.)*. In Insects, the *mushroom* or *stalked bodies*, which are the most conspicuous formations in the protocerebral lobes of the brain.

corpora quadrigemina *(Zool.)*. The optic lobes of the Mammalian brain, which are transversely divided.

cor pulmonale *(Med.)*. Hypertrophy and failure of the right side of the heart as a consequence of lung disease. In the UK usually as a consequence of chronic bronchitis and emphysema.

corpus *(Bot.)*. Inner core of cells, dividing in several planes and distinct from the more superficial *tunica* in a shoot apical meristem, giving rise to the inner tissues of the shoot. See **tunica-corpus concept**.

corpus adiposum *(Zool.)*. See **fat-body**.

corpus albicans *(Zool.)*. See **corpus mamillare**.

corpus callosum *(Zool.)*. In the brain of placental Mammals, a commissure connecting the cortical layers of the two lobes of the cerebrum.

corpuscle *(Zool.)*. A cell which lies freely in a fluid or solid matrix and is not in continuous contact with other cells.

corpuscular radiation *(Phys.)*. A stream of atomic or subatomic particles, which may be charged positively, e.g. α-particles, negatively, e.g. β-particles, or not at all, e.g. neutrons.

corpuscular theory of light *(Phys.)*. The view, held by Newton, that the emission of light consisted of the emission of material particles at very high velocity. Although this theory was discredited by observations of interference and diffraction phenomena, which could only be explained on the wave theory, there has been, to some extent, a return to the corpuscular idea in the conception of the photon.

corpus luteum *(Biol.)*. The endocrine structure developed in the ovary from a Graafian follicle after extrusion of the ovum, secreting progesterone; the yellow body.

corpus mamillare *(Zool.)*. In the brains of higher Vertebrates, a protuberance on the floor of the hypothalamic region in which the fornix terminates.

corpus spongiosum *(Zool.)*. In Mammals, one of the masses of erectile tissue composing the penis.

corpus striatum *(Zool.)*. In the Vertebrate brain, the basal ganglionic part of the wall of each cerebral hemisphere.

corrasion *(Geol.)*. The work of vertical or lateral cutting performed by a river by virtue of the abrasive power of its load. See **rivers, geological work of**.

correction for buoyancy *(Phys.)*. In precision weighing it is necessary to correct for the differences in the buoyancy of the air for the body being weighed and for the weights. The correction to be added to the value w, of the weights (in grams) is: $1.2w\,(1/D-1/\delta)$ mg, where D and δ are the densities of the body and the weights respectively in $\mathrm{g\,cm^{-3}}$.

correction of angles *(Surv.)*. The process of adjusting the observed angles in any triangle so that their sum shall equal 180°.

correlation *(Biol.)*. *Mutual relationship*, e.g. the condition of balance existing between the growth and development of various organs of a plant.

correlation *(Geol.)*. The linking together of strata of the same age occurring in separate outcrops.

correlation *(Maths.)*. A linear transformation which, in the plane, maps lines into points and points into lines and in space maps points into planes and planes into points.

correlation *(Stats.)*. The tendency for variation in one variate to be accompanied by linear variation in another.

correlation coefficient *(Stats.)*. A dimensionless quantity taking values in the range -1 to 1 measuring the degree of linear association between two variates. A value of -1 indicates a perfect negative linear relationship, 1 a perfect positive relationship.

correlation detection *(Telecomm.)*. Method of enhancing weak signals in noise by averaging the product of the received signal and a locally generated signal having some of the known or anticipated properties of the transmitted information. See **autocorrelation, matched filter**.

correspondence principle *(Phys.)*. The predictions of quantum and classical mechanics must correspond in the limit of very large quantum numbers.

corresponding angles *(Maths.)*. For a diagram consisting of two straight lines cut by a transversal, corresponding angles are angles on the same side of the transversal, each angle being above its appropriate line or each below it. If the two straight lines are parallel then the corresponding angles are equal.

corresponding states *(Phys.)*. Substances are said to be *corresponding states* when their pressures and temperatures are equal fractions of the critical values. A general form of *van der Waals' equation* may then be used which is applicable to all gases.

corridor *(Aero.)*. (1) Safe track for intruding aircraft. (2) Path through atmosphere of re-entering aerospace craft above which there is insufficient air density for lift control and below which kinetic heating is excessive.

corridor disease *(Vet.)*. A fatal disease of cattle in Africa due to infection by the protozoon *Theileria lawrencei*; transmitted by ticks.

corrie *(Geol.)*. See **cirque**.

corrosion *(Chem.)*. The slow wearing away of solids, especially metals, by chemical attack.

corrosion *(Geol.)*. (1) Petrologically, the modification of crystals formed early in the solidification of an igneous rock by the chemical action of the residual magma. (2) Geomorphologically, erosion by chemical processes.

corrosion embrittlement *(Eng.)*. Like the effects of severe rusting, the loss of ductility due to corrosion acting between the grains of the material. It is not always readily observed.

corrosion-fatigue *(Eng.)*. Acceleration of weakening of structure exposed to pulsed stress by chemical penetration and attack on metal components.

corrosion voltmeter *(Elec.Eng.)*. Instrument which locates and estimates corrosion of materials by measuring e.m.f. arising from electrochemical action between material and corrosive agent.

corrosive sublimate *(Chem.)*. *Mercuric chloride*.

corrugated board *(Paper)*. A layered packaging material produced by sticking a suitable liner to both sides of a fluted paper or papers.

corrugated iron *(Build.)*. Sheet-iron, usually *galvanized*, with corrugations for stiffening.

corrupt *(Comp.)*. To introduce errors into data or programs.

cortex *(Bot.)*. (1) Generally, the outer part of a thallus or an organ (cf. medulla). (2) Specifically, the tissue (often collenchyma and parenchyma), in a stem or root, between the epidermis and the vascular tissue (i.e. from hypodermis to endodermis inclusive).

cortex *(Zool.)*. The superficial or outer layers of an organ; cf. **medulla**.

cortical *(Bot.,Zool.)*. (1) Relating to bark. (2) Relating to the cortex.

cortical microtubules *(Bot.)*. Microtubules in the cytoplasm just below the plasmalemma in an interphase plant cell, commonly parallel to and perhaps controlling the shape of the developing cellulose microfibrils in the wall.

corticate *(Bot.)*. Possessing or producing a cortex.

corticolous *(Bot.)*. Living on the surface of bark.

corticosteroids *(Med.)*. Steroids secreted by the adrenal cortex or their synthetic analogues. *Glucocorticoids*, including cortisone and hydrocortisone, act on carbohydrate metabolism; *mineral corticoids* such as **aldosterone** have a primary role in maintaining fluid and electrolyte balance.

corticotrophic, corticotropic *(Med.)*. Having a stimulatory influence on the adrenal cortex; adrenocorticotrophic. See **ACTH**.

corticotrophin *(Med.)*. See **ACTH**.

cortisone *(Biochem.,Med.)*. A crystalline hormone isolated from the adrenal cortex.

Corti's organ *(Zool.)*. In Mammals, the modified epithelium forming the auditory apparatus of the ear, in which nerve fibres terminate.

corundum *(Min.)*. Oxide of aluminium, crystallizing in the trigonal system. It is next to diamond in hardness, and hence is used as an abrasive. The clear blue variety is *sapphire* and the clear red *ruby*. See also **white sapphire**.

corymb *(Bot.)*. A **racemose inflorescence** with the upper flower stalks shorter than the lower so that all the flowers are at approximately the same level.

Corynebacteriaceae *(Biol.)*. A family of bacteria belonging to the order *Eubacteriales*. Gram-positive rods; some pleomorphic species, mainly aerobes, occur in dairy products and the soil. Some pathogenic species e.g. *Erysipelothrix rhusiopathiae* (swine erysipelas). *Corynebacterium diphtheriae* (diphtheria).

coryza *(Med.)*. See **cold**.

coryza *(Vet.)*. See **infectious coryza, malignant catarrhal fever**.

cos Q *(Elec.Eng.)*. An expression often used to denote the power factor of a circuit, the power factor being equal to the cosine of the angle (ϕ) of the phase difference between the current and voltage in the circuit.

cos, cosine *(Maths.)*. See **trigonometrical functions**.

cosecant antenna *(Telecomm.)*. One comprising a surface so shaped that the radiation pattern is described by a cosecant curve over a wide angle; gives about the same signal strength for near and far sources. Used mainly in navigation radars; another variation is the *cosecant-squared antenna*.

cosec, cosecant *(Maths.)*. See **trigonometrical functions**.

coset *(Maths.)*. A collection of elements in a group formed by combining a fixed element of the group with each element of a subgroup separately under the group operation.

cosh *(Maths.)*. See **hyperbolic functions**.

cosine law *(Phys.)*. See **Lambert's cosine law**.

cosine potentiometer *(Elec.Eng.)*. Voltage divider in which the output of an applied direct voltage is proportional to the cosine of the angular displacement of a shaft.

cosmic abundance *(Astron.)*. The relative proportion of each atomic element found in the universe, determined from studies of the solar spectrum and the composition of the Earth, Moon and **meteorites**.

cosmic background radiation *(Astron.)*. See **microwave background**.

cosmic noise *(Telecomm.)*. Interference due to extraterrestial phenomena, e.g. sun spots.

cosmic rays *(Astron.)*. Highly penetrating rays from outer space. The *primary cosmic rays* which enter the earth's upper atmosphere consist mainly of protons with smaller amounts of helium and other heavier nuclei. Cosmic rays of low energy have their origin in the sun, those of high energy in galactic or extragalactic space, possibly as a result of supernova explosions. Collision with atmospheric particles results in *secondary cosmic rays* and particles of many kinds, including neutrons, mesons, and hyperons.

cosmic string *(Astron.)*. Hypothetical massive filaments of matter (10^{19} kg) predicted in **supersymmetry** theory as an important component of the very early universe.

cosmine *(Zool.)*. The dentinelike substance forming the outer layer of the cosmoid scales of Crossopterygii.

cosmogenic *(Phys.)*. Said of an isotope capable of being produced by the interaction of cosmic radiation with the atmosphere or the surface of the earth.

cosmogony *(Astron.)*. The science of the origins of stars, planets, and satellites. It deals with the genesis of the Galaxy and the solar system.

cosmoid scale *(Zool.)*. In Crossopterygii, the characteristic type of scale consisting of an outer layer of cosmine, coated externally with vitrodentine, a middle bony vascular layer and an inner isopedine layer. Cf. **ganoid scale**.

cosmological principle *(Astron.)*. The postulate, adopted generally in **cosmology**, that the universe is uniform, homogeneous, and isotropic, i.e., that it has the same appearance for all observers everywhere in the universe, and there is no preferred position.

cosmological redshift *(Astron.)*. A **redshift** in the spectrum of a **galaxy** that is caused by the recession velocity associated with the expansion of the universe.

cosmology *(Astron.)*. The study of the universe on the largest scales of length and time, particularly the propounding theories concerning its origin, nature, structure and evolution. A cosmology is any model said to represent the observed universe. Western cosmology is entirely scientific in its approach, and has produced two famous models, the **Big Bang** and the **steady-state theories**.

COSMOS *(Space)*. General term applied to Russian satellites used for a variety of missions: such as surveillance, atmospheric research, communications, solar wind studies, testing propulsion units and military purposes. COSMOS-1 was launched in March 1962 and by the end of 1986, 1810 objects bearing the name had been put in orbit. The series continues.

cosmotron *(Phys.)*. Large proton synchrotron using frequency modulation of an electric field; it accelerates protons to energies greater than 3 GeV.

cossyrite *(Min.)*. See **aenigmatite**.

costa *(Zool.)*. In Vertebrates, a rib; in Insects, one of the primary veins of the wing; in Ctenophora, one of the meridional rows of ctenes: more generally, any riblike structure. adj. *costal, costate*.

cost-benefit analysis *(Bot.)*. An assessment of the relative costs (in terms of the necessary investment of carbohydrate or nitrogen etc.) and benefits (in terms of enhanced photosynthesis, reduced losses to herbivores, increased probability of establishment of offspring etc.), and hence the likely selective advantage, of any observed or imagined morphological or physiological variation, such as hairier leaves, synthesis of novel toxin, larger seeds etc.

costeaning *(Min.Ext.)*. Prospecting by shallow pits or trenches designed to expose lode outcrop.

cot, cotangent *(Maths.)*. See **trigonometrical functions**.

Cot curve *(Biol.)*. A plot of concentration against time for the renaturation of DNA, gives a measure of the number of different sequences present.

coterminal angles *(Maths.)*. Two angles having the same vertex and the same initial line and whose terminal lines are coincident, e.g. 60° and 420°.

coterminous *(Zool.)*. Of similar distribution.

coth *(Maths.)*. See **hyperbolic functions**.

cotransport *(Biol.)*. The **active transport** of a solute that is driven by a concentration gradient of some other solute, usually an ion, e.g. the entry of amino acids into animal cells depends upon a sodium ion gradient across the **plasma membrane**.

co-trimoxazole *(Med.)*. Antibacterial drug, consisting of 5 parts sulphamethoxazole and 1 part of trimethoprim. Widely used to treat urinary tract infections.

cotter *(Eng.)*. A tapered wedge, usually of rectangular section, passing through a slot in one member and bearing against the end of a second encircling member whose axial position is to be fixed or adjustable.

cotter pin *(Eng.)*. A split-pin inserted in a hole in a cotter or other part, to prevent loosening under vibration.
cotter way *(Eng.)*. The slot cut in a rod to receive a **cotter**.
cotton *(Textiles)*. The seed hairs of the cotton plant of which there are many varieties (*Gossypium spp.*). The fibres vary in length according to variety and country of origin but on average are about 2–3 cm. Sea Island and Egyptian cottons are longer and are used for making high quality fine fabrics and sewing threads.
Cotton balance *(Elec.Eng.)*. An instrument for measuring the intensity of a magnetic field by finding the vertical force on a current-carrying wire placed at right angles to the field.
cotton ball *(Min.)*. See **ulexite**.
Cotton-Mouton effect *(Phys.)*. Effect occurring when a dielectric becomes double-refracting on being placed in a magnetic field H. The retardation δ of the ordinary over the extraordinary ray in traversing a distance l in the dielectric is given by $\delta = C_m \lambda l H^2$ where λ is the light wavelength and C_m is the Cotton-Mouton constant.
cottonseed oil *(Chem.)*. Oil from the seeds of *Gossypium heraceum*, a yellow, brown, or dark-red liquid, mp 34°–40°C. rel.d. 0.922–0.930, saponification value 191–196, iodine value 105–114, acid value 0. Used in manufacture of soaps, fats, margarine etc.
cottonwood *(For.)*. A North American species of poplar. Its considerable girth provides very wide boards. It is rated as moderately durable, and is rather difficult to impregnate with wood preservatives. It carves and turns well.
cotton wool *(Med.,Textiles)*. Loose cotton or vicose rayon fibres which have been bleached and pressed into a sheet; used as an absorbent or as a protective agent. Medicated cotton wool sometimes has a distinguishing colour to indicate its special property.
cotton-wool patches *(Med.)*. Areas of white exudate in the retina.
Cottrell precipitator *(Eng.)*. System used to remove dust from process gases electrostatically.
cotyledon *(Bot.)*. *Seed leaf*. The first leaf or one of the first leaves of the embryo of a seed plant; typically one in monocotyledons, two in dicotyledons, two to many in gymnosperms. In non-endospermic seed, e.g. pea, the cotyledons may act as storage organs. See also **hypogeal, epigeal, endosperm**.
cotyledonary placentation *(Zool.)*. Having the villi in patches, as Ruminants.
cotyloid *(Zool.)*. Cup-shaped; pertaining to the acetabular cavity. In Mammals, a small bone bounding part of the acetabular cavity.
cotype *(Zool.)*. An additional type specimen, being a brother or sister of the same brood as the type specimen.
couch *(Paper)*. To separate the newly formed wet sheet or web from the forming surface and transfer it to a felt.
couching *(Med.)*. Displacement of the lens in the treatment of cataract.
couch roll *(Paper)*. A roll which performs the action of couching.
coudé telescope *(Astron.)*. An arrangement by which the image in an equatorial telescope is formed, after an extra reflection, at a point on the polar axis. It is then viewed by a fixed eyepiece looking either down or up the polar axis. This type of mounting is much used for high dispersion spectroscopy with modern large telescopes. Also called *coudé mounting*
coulomb *(Phys.)*. SI unit of electric charge, defined as that charge which is transported when a current of one ampere flows for one second. Symbol C.
coulomb energy *(Phys.)*. Fraction of binding energy arising from simple electrostatic forces between electrons and ions.
coulomb force *(Phys.)*. Electrostatic attraction or repulsion between two charged particles.
coulomb potential *(Phys.)*. One calculated from Coulomb's inverse square law and from known values of electric charge. The term is used particularly in nuclear physics to indicate that component of the potential energy of a particle which varies with position as a consequence of an inverse square law of force of *Yukawa potential*.
coulomb scattering *(Phys.)*. Scattering of particles by action of coulomb force.
Coulomb's law *(Phys.)*. Fundamental law which states that the electric force of attraction or repulsion between two point charges is proportional to the product of the charges and inversely proportional to the square of the distance between them. The force also depends on the permittivity of the medium in which the charges are placed. In SI units, if Q_1 and Q_2 are the point charges a distance d apart, the force is

$$F = \frac{1}{4\pi\varepsilon}\frac{Q_1Q_2}{d^2}$$

where ε is the permittivity of the medium. The force is attractive for charges of opposite sign and repulsive for charges of the same sign.
Coulomb's law for magnetism *(Phys.)*. The force between two isolated point magnetic poles (theoretical abstractions) would be proportional to the product of their strengths and inversely proportional to the square of their distance apart times the permeablility of the medium between them.

$$F \propto \frac{M_1M_2}{\mu d^2},$$

where M_1 and M_2 are the strengths of the two poles, d is their distance apart and μ is the relative permeability of the medium.
coulometer *(Elec.Eng.)*. Voltameter or electrolytic cell, designed e.g. for use in measurement of the quantity of electricity passed.
Coulter counter *(Biol.)*. TN for a method of counting individual cells by pumping a suspension through an orifice and measuring the change in capacitance as each cell passes.
coumachlor *(Chem.)*. 3-(*a*-acetonyl-4-chlorobenzyl)-4-hydroxy-coumarin, an antiblood-coagulant type of rodenticide.
coumalic acid *(Chem.)*. *Pyrone-5-carboxylic acid*, formed by the action of concentrated sulphuric acid on malic acid; mp 206°C. Pyrone is *coumalin*.
coumaric acid *(Chem.)*. $HO.C_6H_4CH{=}CH.COOH$, *hydroxy-cinnamic acid*.
coumarin *(Chem.)*. Odoriferous principle of tonquin beans and woodruff, $C_9H_6O_2$, bp 200°C; used for scenting tobacco.
coumarone *(Chem.)*. The condensation product of a benzene nucleus with a furan ring. It is a very stable, inert compound, bp 169°C; found in coal-tar. Strong acids effect polymerization into para-coumarone and coumarone resins.
coumarone resins *(Chem.)*. Condensation and polymerization products obtained from *coumarone*; used for varnishes, in printing ink, and as plasticizers for moulding powders. They are neutral and acid- and alkali-resisting.
count *(Nuc.Eng.)*. Summation of photons or ionized particles, detected by a counting tube, which passes pulses to counting circuits.
count *(Textiles)*. See **count of yarn**.
countable set *(Maths.)*. See **denumerable set**.
count down *(Space)*. A sequence of events in the preparation for the firing of a launch system, denoted by counting time backwards towards zero where zero represents lift-off. The count starts some hours before launch and is finally reckoned in minutes and seconds.
counter *(Elec.Eng.)*. (1) Circuit in which a free-running oscillator of known frequency increments a numerical output at regular intervals. Can be used to determine the time between two events by indicating the number of counts which have occurred. (2) Circuit for registering the number of events which have occurred in an external piece of apparatus or circuit. (3) Part of an integrating electricity meter which indicates the number of revolu-

tions made by the spindle of the meter, this indication being proportional to the amount of energy which has passed through the circuit.

counter *(Eng.)*. An instrument for recording the number of operations performed by a machine, or the revolutions of a shaft.

counter *(Ships)*. A description applied to a form of ship's stern, implying an overhung portion of deck, abaft the stern post; hence the term 'under the counter'.

counter-arched *(Civ.Eng.)*. Said of a revetment having arches turned between counterforts.

counterboring *(Eng.)*. The operation of boring the end of a hole to a larger diameter.

counterbracing *(Eng.)*. The provision of two diagonal tie-rods in the panels of a frame girder or other structure. Also called *cross-bracing*.

counter-conditioning *(Behav.)*. A procedure for weakening a classically conditioned response by associating the stimuli that evoke it to a new response that is incompatible with it.

counter, counting tube *(Nuc.Eng.)*. Device for detecting ionizing radiation by electric discharge resulting from **Townsend avalanche** and operating in proportional or Geiger region.

countercurrent contact *(Chem.Eng.)*. In processes involving the transfer of heat or mass between two streams *A* and *B*, as in liquid extraction, the arrangement of flow so that at all stages the more spent *A* contacts the less spent *B*, thus ensuring a more even distribution and greater economy than with **co-current contact**.

countercurrent distribution *(Chem.)*. A repetitive distribution of a mixture of solutes between two immiscible solvents in a series of vessels in which the two solvent phases are in contact. The components are distributed in the vessels according to their partition coefficients.

countercurrent treatment *(Min.Ext.)*. Arrangement used in chemical extraction of values from ore, in washing rich liquor away from spent sands. The stripping liquid enters 'barren' at one end of a typical layout and the rich ore pulp at the other. They pass countercurrent through a series of vessels, the pulp emerging stripped after its final wash with the new 'barren' liquor and the liquor leaving at the far end, now rich with dissolved values and 'pregnant'.

counter efficiency *(Nuc.Eng.)*. Ratio of counts recorded by counter to number of incident particles or photons reaching detector. Counts may be lost due to (a) absorption in window, (b) passage through detector without initiating ionization. (c) passage through detector during dead time following previous count.

counter e.m.f. *(Elec.Eng.)*. See **back e.m.f.**

counter-flap hinge *(Build.)*. A hinge which is arranged, by the provision of separate centres of rotation for each leaf, so that it may fold back to back.

counter-floor *(Build.)*. An inferior floor laid as a base for a better surface (e.g. parquet). Also *sub-floor*.

counterflow jet condenser *(Eng.)*. A **jet condenser** in which the exhaust steam and air flow upwards to the airpump suction in the opposite direction to that of the descending spray of cooling water.

counterfort *(Civ.Eng.)*. A buttress giving lateral support to a retaining wall, to which it is bonded.

counterglow *(Astron.)*. See **zodiacal light**.

counterions *(Chem.)*. See **gegenions**.

counter lathing *(Build.)*. See **brandering**.

counter life *(Nuc.Eng.)*. The total number of counts a nuclear counter can be expected to make without serious deterioration of efficiency.

counterpoise *(Telecomm.)*. A network of conductors placed a short distance above the surface of the ground but insulated from it, and used for the earth connection of an antenna. It will have a large capacity to earth and serves to reduce greatly the earth current losses that would otherwise take place. Also known as a *capacity earth, artificial earth* or *counterpoise antenna*.

counter range *(Nuc.Eng.)*. See **start-up procedure**.

countershading *(Zool.)*. A type of protective coloration in which animals are darker on their dorsal surface than on their ventral surface, thus ensuring that illumination from above renders them evenly coloured and inconspicuous.

countershaft *(Eng.)*. An intermediate shaft interposed between driving and driven shafts in a belt drive, either to obtain a larger speed ratio or where direct connection is impossible.

countersinking *(Build.)*. The driving of the head of a screw or nail below the surface so that it may be hidden by a plug.

countersinking *(Eng.)*. The provision of a conical enlargement at the end of a hole to receive the head of a screw or rivet. See **counterboring**.

counter-stern *(Ships)*. A type of ship's stern construction. It is virtually an excrescence to the main hull, and is not waterborne.

countersunk head *(Eng.)*. A screw or rivet head with a conical base, allowing it to enter the countersunk workpiece so that the top surface of the head is substantially flush with the workpiece surface.

counter-transference *(Behav.)*. In psychoanalytic theory, the analyst's emotional response to the client, often involving personal and unconscious feeling projected onto the client. See **projection**.

counter-vault *(Civ.Eng.)*. An inverted arch.

countess *(Build.)*. A roofing slate, 20×10 in. (508×254 mm).

counting chain *(Phys.)*. A system for the detection and recording of ionizing radiation. Consisting essentially of a detector, linear amplifier, pulse height analyser and a device to display or record the counts.

counting machine *(Paper)*. A piece of equipment that records the number of sheets cut in a sheetcutter or stacked in a pile. Generally also capable of inserting a tab at the chosen sheet count.

count of yarn *(Textiles)*. A number which designates the size of a yarn. Historically this has been the number of lengths per unit mass with many local variations for the units (see **denier**). Now standardized on SI units of the **tex** system which uses the mass per unit length.

count ratemeter *(Nuc.Eng.)*. One which gives a continuous indication of the rate of count of ionizing radiation, e.g. for radiac survey.

country rock *(Min.Ext.)*. The valueless rock forming the walls of a reef or lode.

counts *(Radiol.)*. The disintegrations that a radionuclide detector records.

coupe *(For.)*. A felling area, usually one of an annual succession unless otherwise stated.

couple *(Eng.)*. A system of two equal but oppositely directed parallel forces. The perpendicular distance between the two forces is called its *arm* and any line perpendicular to the plane of the two forces its *axis*. The *moment* of a couple is the product of the magnitude of one of its two forces and its arm. A couple can be regarded as a single statical element (analogous to force); it is then uniquely specified by a vector along its axis having a magnitude equal to its moment. Couples so specified combine in accordance with the parallelogram law of addition of vectors.

couple-close roof *(Build.)*. A roof-form derived from the couple roof by connecting the lower ends of the two rafters together with a tie, so as to prevent spreading of the roof under load.

coupled flutter *(Aero.)*. See **flutter**.

coupled oscillator *(Electronics)*. Circuit in which positive feedback from output circuit to input circuit of an amplifier by mutual inductance is sufficient to initiate or maintain oscillation.

coupled rangefinder *(Image Tech.)*. A rangefinder co-ordinated with the focusing mechanism of a camera lens.

coupled switches *(Elec.Eng.)*. See **linked switches**.

coupled vibrations *(Phys.)*. Two or more vibrating systems connected in some manner so that energy can be transferred from one to another, execute *coupled vibrations*. The resultant complicated motion can be analysed in terms of a linear combination of the **normal modes** of the system.

coupled wheels *(Eng.)*. The wheels of a locomotive which are connected by coupling rods to distribute the driving effort over more than one pair of wheels.
couple roof *(Build.)*. A roof composed of two rafters not braced together.
coupler plug *(Elec.Eng.)*. A plug on a jumper cable, such as that used for making connection between the two coaches of an electric multiple-unit train.
couplers *(Image Tech.)*. Compounds included in photographic emulsions or processing solutions which form coloured dye images associated with the developed silver image by the oxidation products of the reduction.
coupler socket *(Elec.Eng.)*. A socket for receiving a coupler plug.
coupling *(Biol.)*. When two specified *non-allelic* genes are on the same chromosome, having come from the same parent, they are in *coupling*. Cf. **repulsion**.
coupling *(Build.)*. (1) A short collar screwed internally at each end to receive the ends of two pipes which are to be joined together. (2) A **capillary fitting** for the same purpose. Also *union*.
coupling *(Eng.)*. (1) A device for connecting two lengths of hose, etc. (2) A device for connecting two vehicles. (3) A connection between two co-axial shafts, conveying a drive from one to the other.
coupling *(Min.Ext.)*. Short tube internally threaded at both ends for joining two lengths of drilling tube. See also **slip joint**.
coupling capacitor *(Electronics)*. Any capacitor for coupling two circuits, e.g. an antenna to a transmitter or receiver, or one amplifying stage to another.
coupling coefficient, factor *(Telecomm.)*. Ratio of total effective positive (or negative) impedance common to two resonant circuits to geometric mean of total positive (or negative) reactances of two separate circuits.
coupling coil *(Telecomm.)*. One whose inductance is a small fraction of the total for circuit of which it forms a part; used for inductive transfer of energy to or from the circuit.
coupling element *(Telecomm.)*. The component through which energy is transferred in a coupled system.
coupling factor *(Telecomm.)*. See **coupling coefficient**.
coupling factors *(Biol.)*. Proteins of the mitochondrial inner membrane, essential for the *coupling* of the passage of electrons along the electron transport chain with the synthesis of ATP.
coupling loop *(Telecomm.)*. A loop placed in a waveguide at a position of maximum magnetic field strength in order to extract energy.
coupling probe *(Telecomm.)*. A probe placed in a waveguide at a position of maximum electric field strength in order to extract energy.
coupling resistance *(Telecomm.)*. Common resistance between two circuits for transference of energy from one circuit to the other.
coupling transformer *(Elec.Eng.)*. A transformer used as a coupling element.
coupon *(Eng.)*. An extra piece, attached to a forging or casting, from which a test piece can be prepared.
courbaril *(For.)*. A tall tree (*Hymenaea*) common to the American tropics. It yields *South American copal gum*, and a hardwood timber which is not easy to work. Typical uses are in furniture, cabinet-making and ship-building.
course *(Build.)*. A horizontal layer of bricks or building-stones running throughout the length and breadth of a wall. See also **cushion**.
course *(Ships)*. The angle between some datum line and the direction of the ship's head.
course *(Surv.)*. The known length and bearing of a survey line.
course *(Textiles)*. A row of loops across the width of a knitted fabric.
course correction *(Space)*. The firing or burning of a rocket motor, during a coasting flight, in a controlled direction and for a controlled duration to correct an error in course.
course density *(Textiles)*. The number of courses per cm measured along a **wale** of the knitted fabric.
course length *(Textiles)*. The length of yarn in one course of a knitted fabric.
coursing joint *(Build.)*. The mortar joint between adjacent courses of brick or stone. Also *bed joint*.
Courtelle *(Chem.)*. TN for synthetic fibres based on polycyanoethene.
courtship behaviour *(Behav.)*. Refers to a wide range of behaviours throughout the animal kingdom, often very conspicuous, leading to copulation and rearing of young.
covalency *(Chem.)*. The union of 2 or more atoms by the sharing of one or more pairs of electrons
covalent bond *(Chem.)*. A chemical bond in which two or more atoms are held together by the interaction of their nuclei with one or more pairs of electrons.
covalent radius *(Chem.)*. Half the internuclear separation of two bonded like atoms. For bonds between unlike atoms, approximate **bond lengths** may be derived from the sum of the covalent radii for the two bonded atoms.
cove *(Build.)*. A hollow cornice, usually large.
coved ceiling *(Arch.)*. A ceiling which is formed at the edges to give a hollow curve from wall to ceiling, instead of a sharp angle of intersection.
covellite, covelline *(Min.)*. Sulphide of copper, CuS. The colour is indigo-blue or darker. Also called *indigo copper*.
cover *(Build.)*. In coursed work, the hidden or covered width of a slate or tile.
cover *(Civ.Eng.)*. The thickness of concrete between the outer surface of any reinforcement and the nearest surface of the concrete. See **effective depth**.
cover *(Ecol.)*. The percentage of the ground surface covered by a plant species.
coverage, covering power *(Image Tech.)*. The area over which a lens can give a sharply focused image.
covered electrode *(Elec.Eng.)*. A metal electrode covered with a coating of flux; used in arc-welding.
cover flashing *(Build.)*. A separate flashing fastened into the upright surface and overlapping the flashing in the angle between the surfaces.
cover glass *(Image Tech.)*. Square of thin glass mounted to protect a photographic transparency when used as a **slide**.
covering power *(Image Tech.)*. See **coverage**.
cover iron *(Build.)*. See **back iron**.
cover paper *(Paper)*. A heavy paper or board, generally of distinctive appearance e.g. coloured and/or embossed and intended for use as the cover of booklets, pamphlets, menus etc.
cover slip *(Biol.)*. The thin slip of glass used for covering a specimen that is being observed under a microscope. Essential for all but the lowest magnifications.
cover stones *(Build.)*. Flat stones covering girders etc. and serving as a foundation for walls above.
coverts *(Zool.)*. See **tectrices**.
cover unit *(Print.)*. A separate printing unit coupled to the main press for producing the cover of a journal or paper.
coving *(Build.)*. (1) The upright splayed side of a fireplace opening. (2) The projection of upper storeys over lower.
co-volume *(Chem.)*. The correction term b in van der Waals' equation of state, denoting the effective volume of the gas molecules. It is approximately equal to four times the actual volume of the molecules.
cow-hocked *(Vet.)*. Said of horses whose hocks are abnormally close to each other.
cowl *(Build.)*. A cover, frequently louvred and either fixed or revolving, fitted to the top of a chimney to prevent down draught.
cowl flaps *(Aero.)*. See **gills**.
cowling *(Aero.)*. The whole or part of the streamlining covering of any aero-engine; in air-cooled engines, designed to assist cooling airflow.
Cowper's glands *(Zool.)*. In male Mammals, paired glands whose ducts open into the urethra near the base of the penis.
cow-pox *(Med.,Vet.)*. See **vaccinia**.

coxa *(Zool.)*. In Insects, the proximal joint of the leg. adj. *coxal*.

coxalgia *(Med.)*. Pain in the hip.

coxa valga *(Med.)*. A deformity of the hip in which the angle between the neck and the shaft of the femur exceeds 140°.

coxa vara *(Med.)*. A deformity of the hip in which the angle between the neck and the shaft of the femur is less than 120°.

c p *(Aero.)*. See **centre of pressure**.

CP *(Chem.)*. Abbrev. for *Chemically Pure*, a grade of chemical reagent for general laboratory use; less pure than *AR* (analytic reagent).

cP *(Phys.)*. Abbrev. for *centiPoise*.

CP *(Surv.)*. Abbrev. for *Change Point*.

CP filter *(Image Tech.)*. *Colour Printing filter*. One of a series of gelatine filters made with specific transmission values for the three primary colours, used for the correction of printing exposure by small increments of colour.

CPL *(Aero.)*. Abbrev. for *Commercial Pilot's Licence*.

C_3 plant *(Bot.)*. A plant in which CO_2 is fixed directly by ribulose 1,5 bisphosphate carboxylase oxygenase into 3-phosphoglyceric acid which is subsequently converted into sugars etc. by the Calvin cycle. Most plants are C_3 plants.

C_4 plant *(Bot.)*. A plant which fixes CO_2 in photosynthesis by the **Hatch-Slack pathway**. Identified C_4 plants (all are angiosperms including some crop plants such as maize and sugar cane) mostly grow in warm sunny places where they photosynthesize substantially faster than C_3 plants in which photosynthesis may be limited by CO_2. Most have **Krantz anatomy**. See **C_3 plant, CAM plant**.

CP/M *(Comp.)*. *Control Program Monitor*. TN given to a commonly-used **operating system** for microcomputers based on the Z80 microprocessor chip.

CPM *(Eng.)*. Abbrev. for *Critical Path Method*. See **critical path planning**.

cps *(Comp.)*. See **characters per second**.

c p s, CPS, c/s *(Phys.)*. Abbrevs. for *cycles per second*, superseded by hertz (Hz).

CPU *(Comp.)*. See **central processor**.

Cr *(Chem.)*. The symbol for chromium.

CR1, CR2, CR3 *(Immun.)*. Cell surface receptors for C3b and its decay products.

crab *(Image Tech.)*. Movement of a camera sideways, at right angles to its optical axis.

Crab Nebula *(Astron.)*. An expanding nebulosity in Taurus which represents the remains of the supernova of A.D. 1054. It is a powerful source of radio waves and of X-rays. The nebulosity arises from a faint star at the centre, which is a rapid pulsar with a period of 0.033 sec. Both the X-ray and optical radiation show the same pulse, the period of which is slowly increasing.

crabwood *(For.)*. Wood from a tree (*Carapa*) of Central and South America. It is inclined to warp and split during seasoning and is not resistant to insect attack, although it is resistant to wood-rotting fungi. Used for making veneers, roofing shingles and as a structural timber.

crack detector *(Elec.Eng.)*. An electromagnetic device for detecting flaws by the gathering of fine magnetic powder along the flaw lines in an iron specimen when magnetized, or by the reflection of ultrasonic waves.

cracked heels *(Vet.)*. *Grease*. A necrobacillosis of horses heels due to infection with *Fusiformis necrophorus*, the organism gaining entry through cuts and abrasions.

cracking *(Min.Ext.)*. Breaking down heavier crude oil fractions to lighter fractions by heat, pressure and the use of catalysts in the refinery.

crackled *(Glass)*. Said of glassware whose surface has been intentionally cracked by water immersion and partially healed by reheating before final shaping.

crack stopper *(Aero.)*. In structural design, a means of reducing the progression of potential cracks by placing adjacent components across the likely direction of the crack.

cradle *(Build.)*. (1) A frame of laths on which scrim is stretched to receive plaster in forming coved or other heavy cornices, etc. (2) See **cradle scaffold**.

cradle *(Elec.Eng.)*. An earthed metal net placed below a high-voltage overhead transmission line where it crosses a public highway, railway or telephone circuit; a conductor, if broken, falls on the net and is earthed without further damage. Not now in general use except in construction.

cradle *(Vet.)*. A frame encircling the neck of a horse; used as a means of restraint.

cradle scaffold *(Build.)*. A form of suspended scaffolding consisting of a strong framework fitted with guard rails and boards for the working platform, and slung from two fixed points or from a wire rope secured between two jibs. Also called *boat scaffold*.

cradling *(Build.)*. Rough timber fixings as grounds around steelwork.

cradling piece *(Build.)*. (1) A short timber fixed at each side of a fireplace hearth, between a chimney breast and trimmer, to support the ends of floorboards. (2) See **cradling**.

Crag *(Geol.)*. A local type of shelly and sandy rocks which have been deposited in relatively shallow water; found in the Pliocene of East Anglia.

crag-and-tail *(Geol.)*. A hill consisting partly of solid rock shaped by ice action, with a **tail** of morainic material banked against it on the lee-side.

cramp *(Build.)*. *Clamp*. Adjustable contrivance for holding parts of woodwork together while they are being assembled and glued.

cramp *(Med.)*. Painful spasm of muscle.

cramp-iron *(Build.)*. See **cramp**.

crampon, crampoon *(Build.)*. An appliance for holding stones or other heavy objects which are to be hoisted by crane. It consists of a pair of bars hinged together like scissors, the points of which are bent inwards for gripping the load, while the handles are connected by short lengths of chain to a common hoist-ring.

Crampton's muscle *(Zool.)*. In Birds, a muscle of the eye which by its contraction decreases the diameter of the eyeball and so aids the eye to focus objects near to it.

crane barge *(Min.Ext.)*. In off-shore drilling a special vessel for handling heavy loads in the supply and maintenance of drilling platforms.

crane magnet *(Elec.Eng.)*. A magnet (normally electromagnet) used for lifting, e.g. scrap metal.

crane motor *(Elec.Eng.)*. A motor specially designed for the operation of a crane or hoist. It should be very robust and have a high starting torque.

crane post *(Eng.)*. The vertical member of a jib crane, to the top of which the jib is connected by a tie-rod.

crane rating *(Elec.Eng.)*. A term sometimes employed to denote a method of specifying the rating of a motor for intermittent load, such as that of a crane. The maximum power and the load factor are stated.

cranial flexures *(Zool.)*. Flexures of the brain in relation to the main axis of the spinal cord, transitory in lower Vertebratres, permanent in higher Vertebrates. See **nuchal flexure, pontal flexure, primary flexure**.

Craniata *(Zool.)*. A subphylum of the Chordata, also called Vertebrata.

cranioclasis *(Med.)*. The instrumental crushing of the foetal skull in obstructed labour.

cranioclast *(Med.)*. An instrument for gripping and crushing the foetal skull in obstructed labour.

craniosacral system *(Zool.)*. See **parasympathetic nervous system**.

craniotomy *(Med.)*. Incision of the foetal skull and removal of its contents in obstructed labour.

cranium *(Zool.)*. That part of the skull which encloses and protects the brain; the brain-case. adj. *cranial*.

crank *(Eng.)*. An arm attached to a shaft, carrying at its outer end a pin parallel to the shaft; used either to give reciprocating motion to a member attached to the pin, or in order to transform such motion into rotary motion of the shaft.

crank-brace *(Build.)*. A brace having a bent handle by which it may be rotated.

crankcase *(Eng.)*. A boxlike casing, usually cast-iron or aluminium, which encloses the crankshaft and connecting-rods of some types of reciprocating engines, air-compressors etc.

crank effort *(Eng.)*. The effective force acting on the crank pin of an engine in a direction tangential to the circular path of the pin.

crank pin *(Eng.)*. The pin which is fitted into the web or arm of a crank, and to which a reciprocating member or connecting-rod is attached.

crankshaft *(Eng.)*. The main shaft of an engine or other machine which carries a crank or cranks for the attachment of connecting-rods.

crank throw *(Eng.)*. (1) The radial distance from the mainshaft to the pin of a crank, equal to one half the stroke of a reciprocating member attached to the pin. (2) The web or webs and pin of a crank.

crank web *(Eng.)*. The arm of a crank, usually of flat rectangular section.

crash line *(Print.)*. A mark inscribed on the bed of a letterpress machine and sometimes on the machine chase, to indicate that type must be kept outside of this mark to escape damage.

crash recorder *(Aero.)*. See **flight recorder**.

crassulacean acid metabolism, CAM *(Bot.)*. Form of photosynthesis characteristic of desert and some other succulent plants in which CO_2 is taken up (stomata open) during the night and fixed into malic acid from which it is released (stomata shut) during the day and then refixed in the normal way. Such CAM plants lose only say one tenth as much water in transpiration as a C_3 plant in fixing equivalent amounts of carbon. A few submerged aquatic plants (without stomata) also operate a form of CAM. See also **C_3 plant, C_4 plant, PEP carboxylase**.

Crassulaceae *(Bot.)*. Family of ca 1500 spp. of dicotyledonous flowering plants (superorder Rosidae). Most are leaf-succulent perennial CAM plants; widespread mostly in warm dry temperate regions. Of little economic importance other than as ornamentals e.g. *Sedum, Kalanchoe*.

crater *(Astron.)*. Circular depression on the surface of a planetary body. Those on Mercury, the Moon, and most of the natural satellites of planets have been formed by impacts with **meteorites** in the remote past. The Moon, Mars, and Io (one of Jupiter's satellites) have volcanic craters also.

crater *(Geol.)*. A more or less circular depression generally caused by volcanic activity, occasionally by meteoric activity.

crater lamp *(Electronics.)*. Discharge tube so designed that a concentrated light source arises in a crater in a solid cathode.

craton *(Geol.)*. A part of a continent that has attained crustal stability; typically Precambrian or Lower Palaeozoic in age.

crawl *(Print.)*. Continuous running of the press at very low speed. Cf. **inching**.

crawling *(Build.)*. A defect in paint or varnish work, characterized by the appearance of bare patches and paint ridges before drying.

crawling *(Elec.Eng.)*. A phenomenon sometimes observed with induction motors, the motor running up to about only one-seventh of full speed on account of the presence of a pronounced seventh harmonic in the field form. The phenomenon is also observed with other harmonics. Also called *balking*.

craze, crazing *(Build.)*. (1) Minute hair cracks which appear on the surface of pre-cast concrete work or artificial stone. (2) Fissuring of faulty coats of paint or varnish in irregular criss-cross cracks.

crazy chick disease *(Vet.)*. See **nutritional encephalomalacia**.

CRD *(Vet.)*. Abbrev. for *Chronic Respiratory Disease of fowl*.

C reactive protein *(Immun.)*. A plasma protein normally present in low amounts but increased greatly by trauma or infection, i.e. an *acute phase protein*. It was originally identified by its ability to bind a carbohydrate from *Streptococcus pneumoniae* containing phosphoryl choline groups (C-carbohydrate), but it can bind to nucleic acids, to some lipoproteins and can activate **complement**. Its biological function is unknown.

cream-laid *(Paper.)*. White writing-paper made with a laid water-mark.

cream of tartar *(Chem.)*. Commercial name for acid potassium tartrate.

cream-wove *(Paper.)*. White writing-paper, in the manufacture of which a wove dandy has been used.

crease-resist finish *(Textiles)*. A durable finish applied especially to cotton, linen or rayon fabrics to improve the capacity of the materials to resist and recover from creases formed in wear. The fabric is treated with precursors that form a resin within the fibres on curing.

creasing *(Build.)*. See **tile creasing**.

creatine phosphate *(Biol.)*. Phosphate ester of creatine which is a high energy compound capable of converting ADP to ATP. This capacity is exploited in the creatine phosphate of muscle as a short term source of energy during bursts of muscular activity.

creatinuria *(Med.)*. The presence of creatinine in the urine.

creativity *(Behav.)*. An area of study which attempts to explore the ability to construct original and viable products, ideas etc., and to go beyond conventional developments.

creatorrhoea, creatorrhea *(Med.)*. The abnormal presence of muscle fibres in the faeces.

credits *(Image Tech.)*. Titles at the beginning or end of a film or TV programme listing the cast, technicians and organizations concerned.

creel *(Textiles)*. The steel or wooden structure which holds the supply packages at spinning, winding, warping or other machines.

creep *(Build.)*. (1) The gradual alteration in length or size of a structural member or high tensile wire owing to inherent properties of the materials involved. Not to be confused with shrinkage or expansion. (2) The tendency for lead on a sloping roof to thicken near the lower edge and thin near the top.

creep *(Chem.)*. (1) The rise of a precipitate on the wet walls of a vessel. (2) The formation of crystals on the sides of a vessel above the surface of an evaporating liquid.

creep *(Eng.)*. Continuous deformation of metals under steady load. Exhibited by iron, nickel, copper and their alloys at elevated temperature, and by zinc, tin, lead and their alloys at room temperature. Sometimes taken to mean variable deformation.

creep *(Min.Ext.)*. Gradual rising of the floor in a coal-mine due to pressure. See **crush**.

creeping weasel *(Eng.)*. Instrument for measuring the internal details of long, narrow holes, e.g. the fuel channels in reactor cores.

creep limit *(Eng.)*. The maximum tensile stress which can be applied to a material at a given temperature without resulting in measurable creep.

creep strength *(Eng.)*. The ability of a material to resist deformation under constant stress, measured as the amount of creep induced by a constant stress acting for a given time and temperature.

creep tests *(Eng.)*. Methods for measuring the resistance of metals to creep. Time-extension curves under constant loads are determined.

cremaster *(Zool.)*. In the pupae of Lepidoptera, an organ of attachment developed from the tenth abdominal somite; in Mammals, a muscle of the spermatic cord; in Metatheria, a muscle whose contraction causes the expression of milk from the mammary gland.

cremorne bolt *(Build.)*. See **espagnolette**.

crenate *(Bot.)*. Leaf margins etc. having rounded teeth; scalloped.

creosote oil *(Chem.)*. A coal-tar fraction, boiling between 240°C and 270°C. The crude creosote oil is used as raw material for producing tar acids etc., or used direct as a germicide, insecticide, or disinfectant in various connections (e.g. soaps, sheep dips, impregnation of railway sleepers etc.).

crêpe *(Textiles)*. A woven fabric with a distinctive rough, crinkled appearance because of the special high-twisted yarns from which it is made. Similar fabrics are also made by warp- and weft-knitting.

crêpe de chine *(Textiles)*. A light crêpe fabric made from continuous filament yarns.

crêpe paper *(Paper)*. Crinkled paper produced by doctoring the moist web from a supporting cylinder, so increasing elongation in the machine direction. Used principally for packaging and industrial applications but also for decorative purposes.

crêpe rubber *(Chem.)*. Raw, unvulcanized sheet rubber, not chemically treated in any way.

crepitation *(Med.)*. (1) A crackling sensation felt by the observer on movement of a rheumatic joint. (2) The fine crackling noise made when two ends of a broken bone are rubbed together. (3) The fine crackling sounds heard over the chest in disease of the pleura or of the lungs. Also called *crepitus*.

crepitation *(Zool.)*. The explosive discharge of an acrid fluid by certain Beetles, which use this as a means of self-defence.

crepuscular *(Zool.)*. Active at twilight or in the hours preceding dawn.

crepuscular rays *(Meteor.)*. The radiating and coloured rays from the sun below the horizon, broken up and made apparent by clouds or mountains; also, the apparently diverging rays from the sun passing through irregular spaces between clouds.

crescent *(Horol.)*. The circular notch cut in the periphery of the roller of the lever escapement to allow the passing of the guard pin or safety finger. Also known as *passing hollow*.

crescent wing *(Aero.)*. A sweptback wing in which the angle of sweep and **thickness-chord ratio** are progressively reduced from root to tip so as to maintain an approximately constant **critical Mach number**.

cresol red *(Chem.)*. *2-cresolsulphonphthalein*. Indicator used in acid-alkali titrations, having a pH range of 7.2 to 8.8, over which it changes from yellow to red.

cresol resins *(Plastics.)*. Resins made from 1-methyl-3-hydroxybenzene and 1-methyl-4-hydroxybenzene and an aldehyde, similar in properties to the phenolics. The 2-compound reacts slowly, and is therefore likely to remain partly unchanged and act as a softener or plasticizer.

cresols *(Chem.)*. A technical name for the hydroxytoluenes, $CH_3.C_6H_4.OH$, monohydric phenols. There are three isomers, viz.: 2-cresol, mp 30°C, bp 191°C. 3-cresol, mp 4°C, bp 203°C. 4-cresol, mp 36°C, bp 202°C. Only 3- and 4-cresol form nitro-compounds with nitric acid, whereas the 2-cresol is oxidized. Important raw materials for plastics, expecially the 3-compound; also used for explosives, as intermediates for dyestuffs, and as antiseptics.

crest *(Build.,Civ.Eng.)*. The top of a slope or parapet: the ridge of a roof.

crest *(Eng.)*. Part of screw thread outline which connects adjacent flanks on the top of the ridge.

crest *(Geol.)*. The highest part of an anticlinal fold. A line drawn along the highest points of a particular bed is called the *crest line*, and a plane containing the crest lines of successive folds is called the *crest plane*.

crest *(Zool.)*. A ridge or elongate eminence, especially on a bone.

crest factor *(Elec.Eng.)*. See **peak factor**.

cresting *(Build.)*. Ornamental work along a ridge, cornice, or coping of a building. Also called *bratticing* or *brattishing*.

crest tile *(Build.)*. A purpose-made tile having a V-shape specially suiting it to location astride the ridge-line of a roof.

crest value *(Elec.Eng.)*. See **peak value**.

crest voltmeter *(Elec.Eng.)*. See **peak voltmeter**.

Cretaceous *(Geol.)*. The youngest period of the Mesozoic era covering an approx. time-span from 145–65 million years ago. The corresponding system of rocks, the **chalk** is its most striking deposit.

cretin *(Med.)*. One who is affected by **cretinism**. adjs. *cretinoid, cretinous*.

cretinism *(Med.)*. A congenital condition in which there is failure of mental and physical development, due to absence or insufficiency of the secretion of the thyroid gland.

cretonne *(Textiles)*. A heavy printed cotton fabric often used as a furnishing fabric.

Creutzfeldt-Jacob disease *(Med.)*. A rapidly progressive dementia usually beginning in middle age and now thought to be due to a transmissable agent or **slow virus** similar to **kuru**.

crevasse *(Geol.)*. A fissure, often deep and wide, in a glacier or ice-sheet.

crevice corrosion *(Eng.)*. In a liquid containing system, the increased corrosion found in crevices and cracks which are partly segregated from the main flow and where build up of ions and salts may occur.

CRI *(Image Tech.)*. *Colour Reversal Intermediate*, a duplicate colour negative printed directly from the original and processed by reversal.

crib *(Min.Ext.)*. (1) An interval from work underground for croust, bait, snack, downer, piece, chop, snap, bite or tiffin. (2) A job. (3) A form of timber support.

cribbing *(Civ.Eng.)*. An interior lining for a shaft, formed of framed timbers backed with boards; it is used to support the sides and keep back water.

crib-biting *(Vet.)*. A trait acquired by horses, in which some object is grasped with the incisor teeth and air swallowed; may lead to indigestion.

cribellum *(Zool.)*. In certain Spiders, a perforate oval plate, lying just in front of the anterior spinnerets, which produces a broad strip of silk composed of a number of threads.

cribriform *(Zool.)*. Perforate, sievelike; as the *cribriform plate*, a perforate cartilaginous element of the developing Vertebrate skull, which later gives rise to the ectethmoid.

cribrose *(Bot.)*. Pierced with many holes; resembling a sieve.

cribwork *(Civ.Eng.)*. Steel or timber cribs or boxes, sometimes filled with concrete and sunk below water-level to carry the foundations of bridges.

cricoid *(Zool.)*. Ring-shaped; as one of the cartilages of the larynx.

cri du chat syndrome *(Med.)*. A mental deficiency syndrome due to specific chromosomal aberration. Associated with **microcephaly**, widely separated eyes and a characteristic cry resembling that of a cat.

crimp *(Textiles)*. The waviness of a fibre, measured as the difference between the straightened and crimped fibre expressed as a percentage of the straightened length. The crimp of a yarn in a woven fabric is measured similarly by comparing the length of the fabric with the length of the yarn removed from it and straightened.

crimping *(Eng.)*. Pressing into small regular folds or ridges in: (1) the reduction of cross-section of a bar material by progressively corrugating it along its surface to give an increase in length; (2) bending or moulding to a required shape; (3) folding or bending sheet metal to provide stiffness.

crimp stability *(Textiles)*. The ability of **textured yarns** to recover after being extended.

crinanite *(Geol.)*. A basic igneous rock, consisting of intergrown crystals of feldspar, titanaugite, olivine, and analcite. Similar to *teschenite*.

Crinoidea *(Zool.)*. A class of Echinodermata with branched arms; the oral surface directed upwards; attached for part of the whole of their life by a stalk which springs from the aboral apex; suckerless tube feet; open ambulacral grooves; no madreporite, spines or pedicellariae. Most members extinct.

crisis *(Med.)*. (1) A painful paroxysm in tabes dorsalis. (2) The rapid fall of temperature marking the end of a fever.

crispate, crisped *(Bot.)*. Having a frizzled appearance.

crispening *(Image Tech.)*. Emphasis of edge effects to improve the visibility of video images.

crissum *(Zool.)*. In Birds, the region surrounding the cloaca or the feathers situated on that area. adj. *crissal*.

crista *(Biol.)*. An infolding of the inner membrane of a mitochondrion.
crista *(Zool.)*. A ridge or ridgelike structure; as the projection of the transverse crests of lophodont molars.
crista acustica *(Zool.)*. (1) A chordotonal apparatus forming part of the tympanal organ in Tetigoniidae and Gryllidae. (2) A patch of sensory cells in the ampulla, utricle, and saccule of the vertebrate ear.
cristate *(Bot.)*. Bearing a crest.
cristobalite *(Min.)*. The name is usually applied to the high-temperature form of SiO_2. It is found in volcanic rocks and is stable from 1470° to 1713°C, but exists at lower temperatures. Another, low temperature form is not stable above atmospheric pressure and is a constituent of *opal*.
crit *(Phys.)*. Abbrev. for *critical mass*.
crith *(Phys.)*. Unit of mass, that of 1 litre of hydrogen at s.t.p., i.e. 89.88 milligrams.
crithidial *(Zool.)*. Pertaining to, or resembling, the flagellate genus *Crithidia*; said of a stage in the life-cycle of some Trypanosomes.
critical angle *(Phys.)*. See **total internal reflection**.
critical angle *(Telecomm.)*. The angle of radiation of a radio wave which will just not be reflected from the ionosphere.
critical coefficient *(Chem.)*. Additive property, a measure of the molar volume. It is defined as the ratio of the critical temperature to the critical pressure.
critical corona voltage *(Elec.Eng.)*. The voltage at which a corona discharge just begins to take place around an electric conductor.
critical coupling *(Elec.Eng.)*. That between two circuits or systems tuned to the same frequency, which gives maximum energy transfer without overcoupling.
critical damping *(Phys.)*. Damping in an oscillatory electric circuit or in an oscillating mechanical system (such as the movement of an indicating instrument) just sufficient to prevent free oscillations from arising (ringing).
critical field *(Electronics)*. In the case of a **magnetron**, the smallest steady magnetic flux density (at a constant anode voltage) which would prevent an electron (assumed to be emitted at zero velocity from the cathode) from reaching the anode. It can also mean the magnetic field applied to a conductor below which the superconducting transition occurs at a given temperature. Also called *cut-off field*.
critical frequency *(Telecomm.)*. Frequency of a radio wave which is just sufficient to penetrate an ionized layer in the upper atmosphere. See **maximum usable frequency**, **MUF**.
critical humidity *(Chem.)*. That at which the equilibrium water vapour pressure of a substance is equal to the partial pressure of the water vapour in the atmosphere, so that it would neither lose nor gain water on exposure.
criticality *(Nuc.Eng.)*. State in nuclear reactor when multiplication factor for neutron flux reaches unity and external neutron supply is no longer required to maintain power level, i.e. chain reaction is self-sustaining. See **start-up procedure**.
critical Mach number, M_{CRIT} *(Aero.)*. M_{CRIT} is the **Mach number** at which the airflow over the aircraft first becomes locally supersonic. It can be as low as $M = 0.3$ in leading edge slat gaps during high incidence climbing but more usually between $M = 0.75$ and $M = 0.95$ for wings of decreasing thickness. It is generally the Mach number above which compressibility effects noticeably affect handling characteristics.
critical mass (crit) *(Nuc.Eng.)*. The minimum mass of fissionable material which can sustain a chain reaction.
critical path planning *(Eng.)*. A procedure used in planning a large programme of work. Using a digital computer, it determines the particular sequence of operations which must be followed to complete the overall programme in the minimum time and also determines which events have some 'float' or capacity to reprogramme without affecting the whole.
critical period *(Behav.)*. See **sensitive period**.
critical point *(Eng.)*. See **arrest points**.
critical point *(Phys.)*. The point on the **isothermal** for the critical temperature of a substance at which the pressure and volume have their critical values. At the critical point the densities (and other physical properties) of the liquid and gaseous states are identical.
critical point method *(Biol.)*. Technique for preparing tissue or **metaphase** chromosomes for electron microscopy, by **freeze-drying** at the critical point of water. Preserves structural features relatively well.
critical potential *(Phys.)*. A measure of the energy (in electron volts) required to ionize a given atom, or raise it to an excited state.
critical pressure *(Phys.)*. The pressure at which a gas may just be liquefied at its critical temperature.
critical range *(Eng.)*. The range of temperature in which the reversible change from austenite (stable at high temperature) to ferrite, pearlite and cementite (stable at low temperature) occurs. The upper limit varies with carbon content; the lower limit for slow heating and cooling is about 700°C.
critical rate *(Eng.)*. The rate of cooling required to prevent the formation of pearlite and to secure the formation of martensite in steel. With carbon steel, this means cooling in cold water, but it is reduced by the addition of other elements, hence oil- and air-hardening steels.
critical size *(Nuc.Eng.)*. The minimum size for a nuclear reactor core of given configuration.
critical solution temperature *(Chem.)*. The temperature above which two liquids are miscible in all proportions.
critical speed *(Aero.)*. (1) The speed during take-off at which it has to be abandoned if the aircraft is to stop in the available space. Cf. **accelerate-stop distance**. Also *decision speed*. Abbrev. V_1. (2) Rotational speed at which resonance or whirling may occur.
critical speed *(Eng.)*. The rotational speed of a shaft at which some periodic disturbing force coincides with the fundamental or some higher mode of the natural frequency of torsional or transverse vibration of the shaft and its attached masses.
critical state *(Phys.)*. The condition of a gas at its critical point, when it appears to hover between the liquid and gaseous states.
critical temperature *(Elec.Eng.)*. The temperature at which magnetic materials lose their magnetic properties; same as *Curie point (temperature)*.
critical temperature *(Eng.)*. See **transformation temperature**.
critical temperature *(Phys.)*. The temperature above which a given gas cannot be liquefied. See **liquefaction of gases**.
critical voltage *(Elec.Eng.)*. That which, when applied to a gas-discharge tube, just initiates discharge.
critical volume *(Phys.)*. The volume of unit mass (usually 1 mole) of a substance under critical conditions of temperature and pressure.
critical wavelength *(Phys.)*. Free space wavelength corresponding to the *critical frequency* for a waveguide.
crizzling *(Glass)*. Fine cracks in the surface of the glass, occasioned by local chilling during manufacture.
CR-law *(Elec.Eng.)*. Relates to exponential rise or decay of charge on capacitor in series with a resistor, and, by extension, to signal distortion on long submarine cables.
crochet *(Zool.)*. A hook which aids in locomotion, and is associated with the apex of the abdominal legs in Insect larvae.
crocidolite *(Min.)*. The blue asbestos of South Africa, a fibrous variety of the amphibole riebeckite. Long coarse flexible spinning fibre with a high resistance to acids. See also **Cape asbestos**, **tiger's eye**.
crocking *(Textiles)*. Testing the **colour fastness** of dyes when the fabric is rubbed. The apparatus used for carrying out the tests is called a *crockmeter*.
Crocodilia *(Zool.)*. An order of Reptiles having upper and lower temporal arcades, a hard palate, an immovable quadrate, loose abdominal ribs, socketed teeth; large

powerful amphibious forms. Crocodiles, Alligators, Caimans, Gavials. Also *Loricata*.

crocodiling *(Build.)*. A defect on a painted or varnished surface, characterized by the formation of ridges or cracks, sometimes going back to the surface, in irregular patches. Sometimes known as alligatoring.

crocoite, crocoisite *(Min.)*. Lead chromate, bright orange-red in colour.

croissant vitellogène *(Zool.)*. In the developing oöcyte, a crescentic area surrounding the archoplasm, in which the mitochondria are grouped.

Crookes dark space *(Phys.)*. A dark region separating the cathode from the luminous 'negative glow' in an electrical discharge in a gas at low pressure. The thickness of the Crookes dark space increases as the pressure is reduced. For air, it is about 1 cm thick at $10\,N/m^2$ pressure.

Crookes radiometer *(Phys.)*. A small mica 'paddlewheel' which rotates when placed in daylight in an evacuated glass vessel. Alternate faces of the mica vanes are blackened and the slight rise of temperature of the blackened surfaces caused by the radiation which they absorb warms the air in contact with them and increases the velocity of rebound of the molecules, the sum of whose impulse constitutes the driving pressure.

Crookes tube *(Phys.)*. Original gas-discharge tube, illustrating striated positive column, Faraday dark space, negative glow, Crookes dark space, and cathode glow.

crop *(Geol.)*. See **outcrop**.

crop *(Print.)*. To remove portions of copy, e.g. part of a photograph, so that it better fits the page design or focuses attention on the main subject involved.

crop *(Textiles)*. To cut the nap or pile of a fabric to a uniform length.

crop *(Zool.)*. See **proventriculus** (2).

crop bound *(Vet.)*. A term applied to birds suffering from impaction of the crop or ingluvies.

crop marks *(Print.)*. Lines drawn on an **overlay** or on a photograph to indicate where the image should be trimmed to remove unwanted copy.

cropped *(Print.)*. Said of the edges of a book which have been cut down to an extent that mars the appearance of the pages.

cropper *(Print.)*. A small platen printing machine.

cropping *(Eng.)*. The operation of cutting off the end or ends of an ingot to remove the pipe and other defects.

cropping *(Vet.)*. The operation of amputating a part of the comb or wattles of birds, or of the ears of dogs.

cross *(Biol.)*. The mating together of individuals of two different breeds, varieties, strains or genotypes. The progeny are *cross-bred*.

cross *(Build.)*. A special pipe-fitting having four branches mutually at right angles in one plane.

cross ampere-turns *(Elec.Eng.)*. The component of the armature ampere-turns which tends to produce a field at right angles to the main field.

cross arm *(Elec.Eng.)*. The horizontal crossmember attached to telegraph poles, or power transmission line towers, for supporting the insulators which carry the conductors.

cross association *(Print.)*. Placing in position across the machine the required strips or slit webs before finishing operations.

cross-axle *(Civ.Eng.)*. A driving axle having cranks mutually at right angles.

crossbar exchange *(Telecomm.)*. One with the following features: (1) **crossbar switches**; (2) common circuits which may be electro-mechanical or, more commonly, electronic, to select and establish the switched paths and to operate the switch contacts.

crossbars *(Print.)*. Book chases are fitted with two crossbars, a long and a short, which quarter the chase and facilitate the lock-up.

crossbar switch *(Telecomm.)*. One having multiple vertical and horizontal paths, with electromagnetically operated contacts for connecting any vertical with any horizontal path.

crossbar transformer *(Elec.Eng.)*. Coupling device between a coaxial cable and a waveguide, the latter having a short transverse rod, to the centre of which the central conductor is connected.

cross-bearing *(Surv.)*. Observation on survey point not in the immediate scheme of work, useful for checking purposes.

cross-bedding *(Geol.)*. Internally inclined planes in a rock related to the original direction of current flow.

cross-blast explosion pot *(Elec.Eng.)*. An explosion pot in which the pressure generated by the arc in the pot causes a stream of oil to be directed across the arc path at right angles to it.

cross-blast oil circuit-breaker *(Elec.Eng.)*. An oil circuit-breaker in which the pressure generated by the arc causes a stream of oil to be forced through ports placed opposite one pair of contacts, thereby cutting across the arc stream.

cross bombardment *(Phys.)*. A method of identification of radioactive nuclides through their production by differing reactions.

cross-bond *(Elec.Eng.)*. A rail-bond for connecting together the two rails of a track or the rails of adjacent tracks.

cross-bonding *(Elec.Eng.)*. The sheath of cable 1 is connected to that of cable 2 and farther on to cable 3. The total induced e.m.f. vanishes and there are no sheath-circuit eddies.

cross-bracing *(Eng.)*. See **counterbracing**.

cross-correlation *(Acous.)*. Multiplication of two signals and averaging over a time interval.

cross-coupling *(Elec.Eng.)*. Undesired transfer of interfering power from one circuit to another by induction, leakage, etc.

cross cut *(Min.Ext.)*. In metal mining, a level or tunnel driven through the country rock, generally from a shaft, to intersect a vein or lode.

cross-cut chisel *(Eng.)*. A cold chisel having a narrow cutting edge carried by a stiff shank of rectangular section; used for heavy cuts. See **cold chisel**.

cross-cut file *(Eng.)*. A file in which the cutting edges are formed by the intersection of two sets of teeth crossing each other.

cross-cut saw *(Build.)*. A saw designed for cutting timber across the grain.

cross direction *(Paper)*. The direction of a paper sheet or web, in the plane of the sheet, at a right angle to the machine direction.

cross dyeing *(Textiles)*. The further dyeing of a fabric made from a blend of fibres after one of the fibres has already been dyed.

crossed field tube *(Telecomm.)*. Any microwave beam tube in which the directions of the static magnetic field, the static electric field and the electron beam are mutually perpendicular, as required for converting the potential energy of the electron beam into radio-frequency energy. Examples include **magnetrons**, certain **backward wave tubes**. Cf. **travelling wave tubes** in which the kinetic energy of beam electrons is converted to radiofrequency energy.

crossed lens *(Phys.)*. A simple lens the radii of curvature of which have been chosen to give minimum spherical aberration for parallel incident rays. For a refractive index of 1.5, the radii should be in the ratio 1:6, the surface of smaller radius facing the incident light.

crossed Nicols *(Phys.)*. Two Nicol prisms arranged with their principal planes at right angle, in which position the plane-polarized light emerging from one Nicol is extinguished by the other.

crossette *(Build.)*. A projection formed on the flank of a voussoir at the top, giving it a bearing upon the adjacent voussoir on the side towards the springing.

cross fall *(Civ.Eng.)*. The difference in vertical height between the highest and lowest points on the cross-section of a road surface. Also, the average rate of fall from one side to the other, or from the crown to a side of a road.

cross-fertilization *(Bot.)*. The fertilization of the female gametes of one individual by the male gametes of another individual. Also called *allogamy*.

cross field *(Elec.Eng.)*. The component of the flux in an electric machine which is assumed to be produced by the cross ampere-turns.

cross-fire technique *(Radiol.)*. The irradiation of a deep-seated region in the body from several directions so as to reduce damage to surrounding tissues for a given dose to that region.

cross-frogs *(Civ.Eng.)*. See **crossings**.

cross front *(Image Tech.)*. A sliding front carrying the lens in field and technical cameras; used to avoid the consequence of tilting the axis of a camera away from normality with an object.

cross garnet *(Build.)*. A form of **strap hinge**.

cross girders *(Eng.)*. (1) Short girders acting as ties between two main girders. (2) The members which transmit the weight of the roadway to the main girders of a bridge.

cross-grained float *(Build.)*. A float made of a piece of cross-grained wood; used in finishing hard-setting plasterwork.

crosshair *(Surv.)*. A spider's thread fixed across the diaphragm of a level or theodolite.

cross-hatch pattern *(Image Tech.)*. A test pattern of vertical and horizontal lines used on TV picture tube.

crosshead *(Eng.)*. A reciprocating block, usually sliding between guides, forming the junction piece between the piston-rod and connecting-rod of an engine.

crosshead *(Print.)*. A heading or sub-heading centred on the measure.

crossings *(Civ.Eng.)*. The cast-steel railway track component which allows passage for the wheel-flanges at places where one line crosses another. Also *cross-frogs*.

cross-linking *(Chem.)*. The formation of side bonds between different chains in a polymer, thus increasing its rigidity.

cross-magnetizing *(Phys.)*. Effect of armature reaction on magnetic field of current generator.

cross matching *(Immun.)*. Procedure used in selecting blood for transfusion. The red cells to be transfused are mixed with the serum from the patient and if no agglutination occurs the red cells are suitable for transfusion, but an antiglobulin test may be necessary to detect *incomplete antibody*.

cross modulation *(Telecomm.)*. Impression of the envelope of one modulated carrier upon another carrier, due to nonlinearity in the medium transmitting both carriers. See **Luxemburg effect**.

cross neutralization *(Electronics)*. A method of neutralization in push-pull amplifiers. Each output is connected by a negative-feedback circuit to the other input.

Crossopterygii *(Zool.)*. A subclass of the class *Osteichthyes*, first known as fossils from the Middle Devonian period, and persisting to the present. They are of interest because the pectoral fins which are lobed and branched at their tips are attached to the girdle, an arrangement which could have led to the evolution of the tetrapod limb. Living forms include the *Coelocanths* and *Dipnoi* (lung-fish).

cross-over *(Acous.)*. In a twin loudspeaker system, the point in the frequency range above which the amplifier output is fed mainly to the treble speaker and below which mainly to the bass speaker.

cross-over *(Biol.)*. An exchange of segments of homologous chromosomes during **meiosis** whereby linked genes become recombined; also the product of such an exchange. The *cross-over frequency* is the proportion of gametes bearing a cross-over between two specified gene loci. It ranges from 0 for allelic genes to 50% for genes so far apart that there is always a cross-over between them. See also **chiasma**, **centiMorgan**.

cross-over *(Build.)*. A special pipe-fitting with its middle length cranked out for use in a *pass-over offset* when two pipes cross each other in a plane.

cross-over *(Civ.Eng.)*. On railways, a communicating track between two parallel lines, enabling rolling-stock to be transferred from one line to the other.

cross-over *(Electronics)*. In an electron-lens system, location where streams of electrons from the object pass through a very small area, substantially a point, before forming an image.

cross-over area *(Electronics)*. The point at which the electron beam comes to a focus inside the accelerating anode of a cathode-ray tube.

cross-over frequency *(Acous.)*. (1) The frequency in a two-channel loudspeaker system at which the high and low frequency units deliver equal acoustic power; alternatively it may be more generally applied to electric dividing networks when equal electric powers are delivered to each of adjacent frequency channels. (2) See **turn-over frequency**.

cross-over network *(Acous.)*. Same as *dividing network*.

cross-over site *(Biol.)*. The place in the genome where breakage and reunion of DNA strands occur during **recombination**.

cross-pane hammer *(Build.)*. A hammer with a pane consisting of a blunt chisel-like edge at right angles to the shaft.

cross-ply *(Autos.)*. Term applied to the older type of tyres with flexible tread and relatively stiff sidewalls. Cf. **radial-ply**.

cross pollination *(Bot.)*. The conveyance of pollen from an anther of one flower to the stigma of another, either on the same or on a different plant of the same, or related, species.

cross-product of a vector *(Maths.)*. See **vector product**.

cross protection *(Bot.)*. The protection offered by prior, systemic infection by one virus against infection by a second, related virus. Deliberate infection with a symptomless strain of tomato mosaic virus is used commercially to protect tomatoes from infection by other, more damaging strains. The phenomenon is also used experimentally to establish the relatedness of different isolates of viruses.

cross range *(Space)*. The distance either side of a nominal re-entry track which may be achieved by using the lifting properties of a re-entering space vehicle.

cross-ratio *(Maths.)*. (1) Of four numbers a,b,c,d, taken in the order a,b, c,d: the ratio

$$\frac{a-c}{a-d}\bigg/\frac{b-c}{b-d}.$$

Denoted by (ab,cd). (2) Of four points on a straight line A,B,C,D: the cross-ratio is

$$(AB, CD)=\frac{AB}{AD}\cdot\frac{CD}{CB}$$

(the usual convention as to signs being observed). (3) Of a pencil, consisting of four straight lines OA, OB, OC, OD: the cross-ratio is

$$O(AB, CD)=\frac{\sin AOB}{\sin AOD}\cdot\frac{\sin COD}{\sin COB}.$$

Any transversal (i.e. line not through O) cuts the pencil in four points having the same cross-ratio. Some writers associate the above fractions with the order a,c,b,d as opposed to the order a,b,c,d. Also called *anharmonic ratio*.

cross-section *(Eng.)*. The section of a body (e.g. a girder or moulding) at right angles to its length; a drawing showing such a section.

cross-section *(Maths.)*. The section made by a plane which cuts the axis of symmetry or the longest axis at right angles.

cross-section *(Phys.)*. In atomic or nuclear physics, the probability that a particular interaction will take place between particles. The value of the cross-section for any process will depend on the particles under bombardment and upon the nature and energy of the bombarding particles. Suppose N_0 particles per second are incident on a target area A containing N particles and N_0' of the incident particles produce a given reaction, then if $N_0' \ll N_0$, $N_0' = N_0 N\sigma/A$ where σ is the cross-section for the reaction; σ can be imagined as a disc of area σ surrounding each target particle. Measured in **barns**.

cross-sill *(Civ.Eng.)*. See **sleeper**.

cross-slide *(Eng.)*. That part of a planing machine or lathe on which the toolholder is mounted, and along which it may be moved at right angles to the bed of the machine.

cross-springer *(Build.)*. In a groined arch, the rib following the line of a groin.

cross-staff *(Surv.)*. An instrument for setting out right angles in the field. It consists of a frame or box having two pairs of vertical slits, giving two lines of sight mutually at right angles.

cross-talk *(Telecomm.)*. The interference caused by energy from one conversation invading another circuit by electrostatic or electromagnetic coupling. See **far-end-crosstalk, near-end-crosstalk**.

cross-talk meter *(Telecomm.)*. An arrangement for measuring the attenuation between circuits which are liable to permit cross-talk.

cross-tie *(Civ.Eng.)*. See **sleeper**.

cross-tongue *(Build.)*. A wooden tongue for a **ploughed-and-tongued joint**, cut so that the grain is at right angles to the grooves.

cross-tree *(Ships)*. A lateral formation on a ship's mast; its uses are for rigging to top masts, hooks, tackle etc. The term is derived from antique wooden ships.

cross wall construction *(Build.)*. A system where the main supports are the walls running back to front, supporting floors, roof and curtain walls.

crotonaldehyde *(Chem.)*. *But-2-enal.* $CH_3.CH=CH.CHO$, a liquid, of pungent odour, bp 105°C, an unsaturated aldehyde, obtained from ethanol by heating with dilute hydrochloric acid or with a solution of sodium acetate. As an intermediate product, **aldol** is formed.

crotonic acid *(Chem.)*. *But-2-enoic acid.* $CH_3.CH=COOH$, an olefinic monocarboxylic acid. There are two stereoisomers, viz., crotonic acid, mp 71°C, bp 180°C; and iso- or allo-crotonic acid, mp 15°C, bp 169°C. The first form is the *cis-*, the latter the *trans-*form. The crotonic acids are also isomers of methacrylic and vinylacetic acids.

crotonyl *(Chem.)*. The group $-CH.CH=CH.CH_3$ in organic compounds.

crotyl *(Chem.)*. The group $-CH_2CH=CHCH_3$ in organic compounds.

croup *(Med.)*. Hoarse croaking cough associated with inflammation of the larynx and trachea in children.

croup *(Vet.)*. The sacral region of the back of the horse.

crowbar *(Build.)*. A round iron bar, pointed at one end and flattened to a wedge shape at the other, used as a lever for moving heavy objects.

Crowe process *(Min.Ext.)*. Method of removing oxygen from cyanide solution before recovery of dissolved gold, in which liquor is exposed to vacuum as it flows over trays in a tower.

crown *(Bot.)*. A very short rootstock.

crown *(Build.,Civ.Eng.)*. The highest part of an arch. Also called the *vertex*.

crown *(For.)*. The upper branchy part of a tree above the bole.

crown *(Paper)*. A paper size in use in the UK, 385 × 505 mm.

crown *(Zool.)*. The part of a polyp bearing the mouth and tentacles; the distal part of a deer's horn; the grinding surface of a tooth; the disc and arms of a Crinoid; crest; head.

crown ether *(Chem.)*. A cyclic polyether, often of the form $(-O-CH_2-CH_2)_n$. They complex alkali metal ions strongly. See also **macrocycle**.

crown gall *(Bot.)*. Disease of dicotyledons, especially of fruit bushes and trees, caused by a soil bacterium *Agrobacterium tumefaciens* and characterized by the production of large, tumour-like galls. See also **Ti plasmid, opine**.

crown-gate *(Hyd.Eng.)*. A canal-lock headgate.

crown glass *(Glass)*. (1) Glass made in disk form by blowing and spinning, having a natural fire-finished surface but varying in thickness with slight convexity, giving a degree of distortion of vision and reflection. (2) Soda-lime-silica glass. See also **optical crown**.

crown octavo *(Print.)*. A book size, $7\frac{1}{4} \times 4\frac{7}{8}$ in (metric, 186 × 123 mm).

crown of thorns tuning *(Electronics)*. Tuning of cavity magnetrons involving changing inductance of cavities by the introduction of conducting rods along their axes.

crown quarto *(Print.)*. A book-size, $9\frac{3}{4} \times 7\frac{3}{8}$ in (metric, 246 × 186 mm).

crown-tile *(Build.)*. An ordinary flat tile. Also *plane-tile*.

crown wheel *(Eng.)*. The larger wheel of a bevel reduction gear. See **bevel gear**.

crown wheel *(Horol.)*. A contrate wheel with pointed teeth forming the escape wheel of a **verge escapement**.

crow-step gable *(Build.)*. See **corbie-step gable**.

crow steps *(Arch.)*. Steps on the coping of a gable, common in traditional Scottish and Dutch architecture.

croy *(Civ.Eng.)*. A protective barrier built out into a stream to prevent erosion of the bank.

crozier *(Bot.)*. The young ascus when it is bent in the form of a hook.

crozzle *(Build.)*. An excessively hard and misshapen brick which has been partially melted and overheated.

CRT *(Electronics)*. Abbrev. for *Cathode-Ray Tube*.

cruciate, cruciform *(Bot.)*. Having the form of, or arranged like, a cross.

crucible *(Chem.,Min.Ext.)*. A refractory vessel or pot in which metals are melted. In chemical analysis, smaller crucibles, made of porcelain, nickel or platinum, are used for igniting precipitates, fusing alkalies etc.

crucible furnace *(Eng.)*. A furnace, fired with coal, coke, oil, or gas, in which metal contained in crucibles is melted.

crucible steel *(Eng.)*. Steel made by melting blister bar or wrought-iron, charcoal and ferro-alloys in crucibles which hold about 100 lb. This was the first process to produce steel in a molten condition, hence product called *cast steel*. Mainly used for the manufacture of tool steels, but now largely replaced by the electric-furnace process.

crucible tongs *(Chem.,Eng.)*. Tongs used for handling crucibles.

Cruciferae *(Bot.)*. Family of ca 3000 spp. of dicotyledonous flowering plants (superorder Dilleniidae). Mostly herbs, rarely shrubs, cosmopolitan. The flowers characteristically have four sepals, four petals and six stamens, all free, and superior ovary of 2 fused carpels. Includes the genus *Brassica*, and a number of minor vegetable crops e.g. water-cress, and many ornamentals e.g. wallflower.

crude oil *(Chem.)*. See **petroleum**.

cruise control *(Autos.)*. System which automatically maintains a selected road speed.

cruise missile *(Aero.)*. Missile launched from a mobile platform, following a low altitude course and guided by an inertial guidance system which takes account of minute gravitational anomalies over the terrain on the way to the target.

cruiser stern *(Ships)*. Stern construction, integral with the main hull for strength and form and partially water-borne. It assists in manoeuvrability and wave formation and provides underdeck roominess. See **counter**.

crump *(Min.Ext.)*. Rock movement under stress due to underground mining, possibly violent.

crunode *(Maths.)*. See **double point**.

cruor *(Zool.)*. The coagulated blood of Vertebrates.

crura *(Zool.)*. See **crus**.

crural *(Zool.)*. Pertaining to or resembling a leg. See **crus**.

crureus *(Zool.)*. A leg muscle of higher Vertebrates.

crus *(Zool.)*. The zeugopodium of the hind-limb in Vertebrates; the shank; any organ resembling a leg or shank. pl. *crura*. adj. *crural*.

crush *(Min.Ext.)*. The broken condition of pillars of coal in a mine due to pressure of the strata. See **creep**.

crush breccia *(Geol.)*. A rock consisting of angular fragments, often recemented, which has resulted from the faulting or folding of pre-existing rocks. See also **crush conglomerate, fault breccia**.

crush conglomerate *(Geol.)*. A rock consisting of crushed and rolled fragments, often recemented; it has resulted from the folding or faulting of pre-existing rocks.

crusher *(Min.Ext.)*. Machine used in earlier stages of comminution of hard rock. Typically works on dry feed as it falls between advancing and receding breaking plates. Types include **jaw breaker** or *jaw crusher*, **gyrator, rolls, stamps**.

crushing *(Image Tech.)*. Loss of tonal gradation in the picture through reduced contrast at the extremes of the brightness range: *black crushing* in the shadows, *white crushing* in the highlights.

crushing *(Print.)*. See **smashing**.

crushing strip *(Print.)*. A strip sometimes required on fold rollers to increase the nip.

crushing test *(Civ.Eng.)*. A test of the suitability of stone to be used for roads or building purposes.

crush syndrome *(Med.)*. Where severe muscle injury results in the release of myoglobin and subsequent acute kidney failure.

crust *(Civ.Eng.)*. See **wearing course**.

crust *(Geol.)*. The outermost layer of the lithosphere consisting of relatively light rocks. Continental crust consists largely of granitic material; oceanic crust is largely basaltic.

Crustacea *(Zool.)*. A class of Arthropoda, mostly of aquatic habit and mode of respiration; the second and third somites bear antennae and the fourth a pair of mandibles. Shrimps, Prawns, Barnacles, Crabs, Lobsters, etc.

crust of the earth *(Geol.)*. See under **earth**.

crustose *(Bot.)*. Forming a crust; especially of lichens, having a crust-like thallus closely attached to, and virtually inseparable from the surface on which it is growing.

crutch *(Horol.)*. The lever or rod which transmits the impulse from the pallets to the pendulum rod. The end of the crutch may be in the form of a fork to embrace the pendulum rod, or else in the form of a pin which enters a slot in the pendulum rod.

crutching *(Vet.)*. The operation of removing the wool from around the tail and quarters of sheep as a preventive of **myiasis**.

cryogenic *(Phys.)*. Term applied to low-temperature substances and apparatus.

cryogenic gyro *(Phys.)*. One depending on electron spin in atoms at very low temperature.

cryogenics *(Phys.)*. The study of materials at very low temperatures.

cryoglobulin, cryoprecipitate *(Immun.)*. Precipitate which forms in serum at temperatures below about 10°C but goes into solution at body temperature. It is usually due to **rheumatoid factor** interacting with immunoglobulin, although constituents like heparin and fibrinogen may be included. Cryoglobulins occur in some proliferative B-cell disorders and in **systemic lupus erythematosus**. Their presence may cause restricted blood flow and vessel spasm in cold extremities.

cryolite *(Min.)*. Sodium aluminium fluoride, used in the manufacture of aluminium.

cryometer *(Phys.)*. A thermometer for measuring very low temperatures.

cryosar *(Electronics)*. Low-temperature germanium switch with on-off time of a few nano-seconds.

cryoscope *(Chem.)*. Instrument for the determination of freezing- or melting-points.

cryoscopic method *(Chem.)*. The determination of the rel. mol. mass of a substance by observing the lowering of the freezing-point of a suitable solvent.

cryostat *(Phys.)*. Low-temperature thermostat.

cryotherapy *(Med.)*. Medical treatment using the application of *extreme* cold.

cryotron *(Electronics)*. Miniature electronic switch, operating in liquid helium, consisting of a short wire wound with a very fine control wire. When a magnetic field is induced via the control wire, the main wire changes from super-conductive to resistive. Used as a memory, characterized by exceedingly short access time. See **superconductivity**.

crypt *(Zool.)*. A small cavity; a simple tubular gland.

cryptic coloration *(Zool.)*. Protective resemblance to some part of the environment or camouflage, from simple **countershading** to more subtle mimicry of, e.g. leaves or twigs. Cf. **aposematic coloration**.

crypto- *(Genrl.)*. Prefix from Gk. *kryptos*, hidden.

cryptobiosis *(Zool.)*. The state in which an animal's metabolic activities have come effectively, but reversibly, to a standstill. See **anabiosis**.

cryptogam *(Bot.)*. In earlier systems of classification, a plant without flowers or cones in which the method of reproduction was not apparent, i.e. algae, fungi, bryophytes and pteridophytes.

cryptometer *(Build.)*. An instrument used to determine the obliterating or hiding power, or the opacity, of paints and pigments.

Cryptophyceae *(Bot.)*. Small class of eukaryotic algae. Biflagellate unicells with **periplast**; chloroplasts with chlorophyll a and c and phycobilins, thylakoids paired, chloroplast ER and nucleomorph present, reserve polysaccharide starch stored between chloroplast and chloroplast ER. Fresh-water and marine.

cryptophyte *(Bot.)*. (1) A member of the **Cryptophyceae**. (2) Herb with perennating buds below soil (or water) surface, (includes geophyte, heliophyte, hydrophyte). See **Raunkaier system**.

cryptorchid *(Vet.,Zool.)*. An animal in which one or both testes have not descended from the abdominal cavity to the scrotum within a reasonable time.

cryptorchidectomy *(Vet.)*. Surgical removal of the testes from a cryptorchid animal.

cryptosystem *(Comp.)*. Abbrev. for *cryptographic system*. That which enables the transmission of secret and authenticated messages.

cryptozoic *(Zool.)*. Living in dark places, as in holes, caves, or under stones and tree-trunks.

crystal *(Bot.)*. Crystalline inclusion in a plant cell, usually of calcium oxalate. Types include the **druse** and the **raphide**.

crystal *(Electronics)*. Piezoelectric element, shaped from a crystal in relation to crystallographic axes, e.g. quartz, tourmaline, Rochelle salt, ammonium dihydrogen phosphate, to give **facets** to which electrodes are fixed or deposited for use as transducers or frequency standards.

crystal *(Genrl.)*. Solid substance showing some marked form of geometrical pattern, to which certain physical properties, angle and distance between planes, refractive index etc. can be attributed.

crystal *(Glass)*. See **crystal glass**.

crystal *(Horol.)*. The glass that covers the dial of a watch.

crystal *(Min.)*. Old name for *quartz*.

crystal anisotropy *(Crystal.)*. In general, directional variations of any physical property, e.g. elasticity, thermal conductivity etc., in crystalline materials. Leads to existence of favoured directions of magnetization, related to lattice structure in some ferromagnetic crystals.

crystal axes *(Crystal.)*. The axes of the natural coordinate system formed by the crystal lattice. These are perpendicular to the natural faces for many crystals. See **uniaxial, biaxial**.

crystal boundaries *(Crystal.)*. The surfaces of contact between adjacent crystals in a metal. Anything not soluble in the crystals tends to be situated at the crystal boundaries, but in the absence of this, the boundary between two similar crystals is simply the region where the orientation changes.

crystal counter *(Electronics)*. One in which an operating pulse is obtained from a crystal when made conducting by an ionizing particle or wave.

crystal cutter *(Electronics)*. A cutter used in gramophone recording, a piezoelectric crystal being the means of initiating the mechanical displacements of the stylus.

crystal detector *(Electronics)*. A **demodulator**, used primarily in microwave applications. The non-linear voltage-current characteristic of a point-contact diode is used to separate the signal information from the carrier frequency.

crystal diamagnetism *(Eng.)*. Property of negative susceptibility shown by silver, bismuth etc.

crystal diode *(Electronics)*. See **diode (2)**.

crystal dislocation *(Crystal.)*. Imperfect alignment between the lattices at the junctions of small blocks of ions ('mosaics') within the crystal. The resulting mobility and opportunity for realignment of the molecules is of importance in crystal growth, plastic flow, sintering etc.

crystal drive *(Electronics)*. System in which oscillations of low power are generated in a **piezoelectric crystal resonator**, being subsequently amplified to a level requisite for transmission.

crystal electrostriction *(Phys.)*. The dimensional changes of a dielectric crystal under an applied electric field. See **electrostriction, magnetostriction**.

crystal face *(Crystal.)*. One of the bounding surfaces of a crystal. In the case of small, undistorted crystals, each face is an optically plane surface. A *cleavage face* is the smooth surface resulting from cleavage; in such minerals as mica, the cleavage face may be almost a plane surface, diverging only by the thickness of a molecule.

crystal filter *(Electronics)*. Band-pass filter in which piezoelectric crystals provide very sharp frequency-discriminating elements, especially for group modulation of multichannels over coaxial lines.

crystal-gate receiver *(Telecomm.)*. Superhet receiver in which one (or more than one) piezoelectric crystal is included in the intermediate-frequency circuit, to obtain a high degree of selectivity.

crystal glass *(Glass)*. A colourless, highly transparent glass of high refractive index, which may be 'lead crystal'. (A somewhat misleading term since it denotes different things in different glassmaking districts.)

crystal goniometer *(Crystal.)*. Instrument for measuring angles between crystal faces.

crystal growing *(Electronics)*. Technique of forming semiconductors by extracting crystal slowly from molten state. Also called *crystal pulling*.

crystal indices *(Crystal.)*. See **Miller indices**.

crystal lattice *(Crystal.)*. Three-dimensional repeating array of points used to represent the structure of a crystal, and classified into fourteen groups by Bravais.

crystalline *(Geol.)*. Having a crystal structure.

crystalline cone *(Zool.)*. The outer refractive body of an ommatidium which acts as a light guide.

crystalline form *(Crystal.)*. The external geometrical shape of a crystal.

crystalline lens *(Zool.)*. The transparent refractive body of the eye in Vertebrates, Cephalopoda, etc. It is compressible by muscles and focuses images of objects emitting light onto the retina.

crystalline overgrowth *(Crystal.)*. The growth of one crystal round another, frequently observed with isomorphous substances. Cf. **cubic system**.

crystalline rocks *(Geol.)*. These consist wholly, or chiefly, of mineral crystals. They are usually formed by the solidification of molten rock, by metamorphic action, or by precipitation from solution.

crystalline solid *(Chem.)*. A solid in which the atoms or molecules are arranged in a regular manner, the values of certain physical properties depending on the direction in which they are measured. When formed freely, a crystalline mass is bounded by plane surfaces (faces) intersecting at definite angles. See **X-ray diffraction**.

crystalline style *(Zool.)*. In Bivalvia, a transparent rod-shaped mass secreted by a diverticulum of the intestine; composed of protein with an adsorbed amylolytic enzyme.

crystallinity *(Textiles)*. Fibres are formed from linear polymers which are oriented and show a high degree of order that varies for different compounds. The orderly three-dimensional arrangement of the molecules is revealed by the X-ray diffraction patterns that are obtained from them. These are akin to the patterns obtained from simple crystalline substances and are a measure of the crystallinity of the fibres.

crystallites *(Chem.)*. Very small, often imperfectly formed crystals.

crystallites *(Min.)*. Minute bodies occurring in glassy igneous rocks, and marking a stage in incipient crystallization.

crystallization *(Chem.)*. Slow formation of a crystal from melt or solution.

crystallized *(Build.)*. Said of an enamelled or varnished surface which presents the appearance of galvanized iron.

crystalloblastic texture *(Geol.)*. The texture of metamorphic rocks resulting from the growth of crystals in a solid medium.

crystallographic axes *(Crystal.)*. Lines of reference intersecting at the centre of a crystal. Crystal (or morphological) axes, usually three in number, by their relative lengths and attitude, determine the system to which a crystal belongs.

crystallographic notation *(Crystal.)*. A concise method of writing down the relation of any crystal face to certain axes of reference in the crystal.

crystallographic planes *(Crystal.)*. Any set of parallel and equally spaced planes that may be supposed to pass through the centres of atoms in crystals. As every plane must pass through atomic centres, and no centres must be situated between planes, the distance between successive planes in a set depends on their direction in relation to the arrangement of atomic centres.

crystallographic system *(Crystal.)*. Any of the major units of crystal classification embracing one or more symmetry classes.

crystallography *(Genrl.)*. Study of internal arrangements (ionic and molecular) and external morphology of crystal species, and their classification into types.

crystalloid *(Bot.)*. A crystal of protein in e.g. a cell of a storage organ.

crystal loudspeaker *(Acous.)*. See **piezoelectric loudspeaker**.

crystal microphone *(Acous.)*. See **piezoelectric microphone**,

crystal mixer *(Electronics)*. A **frequency changer** which uses the non-linear voltage-current characteristic of a point-contact diode to generate sum and difference frequencies. Used mainly in microwave and radar receivers to convert the incoming signal to a lower (intermediate) frequency for subsequent amplification and demodulation.

crystal momentum *(Phys.)*. The product of the Dirac constant, $\hbar$ and the wavevector q of a *phonon* in a crystal. For a *photon* the product $\hbar k$, where k is its wavevector, is the momentum it carries, as its frequency is proportional to the magnitude of k. This is not so for a phonon, so the crystal momentum is just a useful fiction used in the discussion of scattering processes. It is also called the *pseudo-momentum*.

crystal nuclei *(Chem.)*. The minute crystals whose formation is the beginning of crystallization.

crystal oscillator *(Electronics)*. Valve or transistor oscillator in which frequency is held within very close limits by rapid change of mechanical impedance (coupled piezoelectrically) when passing through resonance.

crystal pick-up *(Acous.)*. See **piezoelectric pick-up**.

crystal pulling *(Electronics)*. See **crystal growing**.

crystal rectifier *(Electronics)*. One which depends on differential conduction in semiconducting crystals, suitably 'doped', such as Ge or Si.

crystal sac *(Bot.)*. A cell almost filled with crystals of calcium oxalate.

crystal set *(Telecomm.)*. A simple radio receiver using only a crystal detector.

crystal spectrometer *(Phys.)*. An instrument that uses crystal lattice diffraction to analyse the wavelengths (and energies) of scattered radiation. The radiation can be neutrons, electrons, X-rays or γ-rays and the scattering can be elastic or inelastic. See **triple-axis neutron spectrometer**.

crystal structure *(Crystal.)*. This consists of the whole assemblage of rows and patterns of atoms, which have a definite arrangement in each crystal.

crystal structure *(Eng.)*. The arrangement in most pure metals may be imitated by packing spheres, and the same applies to many of the constituents of alloys. See **body-centred cubic, face-centred cubic, close-packed hexagonal structure**.

crystal systems *(Crystal.)*. A classification of crystals based on the intercepts made on the crystallographic axes by certain planes.

crystal texture *(Crystal.)*. The size and arrangement of the individual crystals in a crystalline mass.

crystal triode *(Telecomm.)*. Early name for *transistor*.

crystal violet *(Chem.)*. A dyestuff of the rosaniline series, hexamethyl-4- rosaniline.

Cs *(Chem.)*. The symbol for caesium.

c/s *(Phys.)*. Cycles per second. See **c.p.s., hertz**.

CSF *(Immun.)*. Abbrev. for *Colony Stimulating Factor*.

CSF *(Paper)*. Abbrev. for *Canadian Standard Freeness*.

CS gas *(Chem.)*. *Orthochlorobenzylidene malononitrile*. A potent tear-gas used for crowd dispersal. See **war gas**.

CSI *(Image Tech.)*. *Compact Source Iodide*. TN for a type of **metal halide** lamp.

CSNet *(Comp.)*. A US electronic mail network, linking users of **ARPANET** to other computer scientists.

CSO *(Image Tech.)*. Same as **colour separation overlay**.

C-spanner *(Eng.)*. One for turning large, narrow nuts, found particularly on machine tools. It is sickle-shaped with the end having a projection which fits into a peripheral notch in the nut. Also *sickle spanner*.

C-stage *(Plastics)*. Final stage in curing process of phenol formaldehyde resin, characterized by infusibility and insolubility in alcohol or acetone.

ctene *(Zool.)*. One of the comb-plates or locomotor organs of Ctenophora, consisting of a row of strong cilia of which the bases are fused.

ctenidium *(Zool.)*. Generally, any comblike structure; in aquatic Invertebrates a type of gill consisting of a central axis bearing a row of filaments on either side; in Insects, a row of spines resembling a comb.

ctenoid *(Zool.)*. Said of scales which have a comblike free border.

Ctenophora *(Zool.)*. A phylum of triploblastic animals showing biradial symmetry; they have a system of gastrovascular canals and typically eight meridional rows of swimming plates or ctenes, composed of fused cilia. Sea Acorns, Comb-Bearers.

CTR *(Nuc.Eng.)*. Abbrev. for *Controlled Thermonuclear Reactor*, or *Reaction*. See **fusion reactor**.

Cu *(Chem.)*. The symbol for copper.

Cuban 8 *(Aero.)*. Aerobatic manoeuvre in a vertical plane consisting of $\frac{3}{4}$ loop, $\frac{1}{2}$ roll, $\frac{3}{4}$ loop, and $\frac{1}{2}$ roll.

Cuban mahogany *(For.)*. Reddish hardwood from *Swietenia*, found in the West Indies and Florida.

cube *(Maths.)*. A solid with six square faces. A square parallelepiped. Its volume is the length of a side raised to third power, hence cubed.

cubical antenna *(Telecomm.)*. One consisting of two or more square loops, with sides one quarter-wavelength long, spaced by one-quarter wavelength; maximum radiation is along the axis of the loops. Also *cubical-quad antenna*.

cubical epithelium *(Zool.)*. A form of columnar epithelium in which the cells are short.

cubic close packing *(Chem.)*. That stacking of spheres formed by stacking close packed layers in the sequence ABCABC. The unit cell of such an arrangement is a face-centred cube, with four atoms per cell. This structure is adopted by many metals, e.g. Cu, Ag and Au. Abbrevs. *CCP, FCC*.

cubic equation *(Maths.)*. An algebraic equation of the third degree. The usual standard form is

$$x^3 + 3ax + b = 0,$$

whose solution can be expressed as

$$x = \sqrt[3]{p} - \frac{a}{\sqrt[3]{p}},$$

where $p = -b + \sqrt{b^2 + 4a^3}$, and where, to obtain all three solutions, all three cubic roots of p, including the complex roots, have to be used. Unlike the quadratic, this solution is of little practical value because complex cube roots can only be determined by solving cubic equations. The general equation $y^3 + uy^2 + vy + w = 0$ is reduced to standard form by the substitution $y = x - \frac{1}{3}u$.

cubicle-type switchboard *(Elec.Eng.)*. See **cellular-type switchboard**.

cubic system *(Crystal.)*. The crystal system which has the highest degree of symmetry; it embraces such forms as the cube and octahedron.

cubing *(Build.)*. An approximate method for estimating costs of buildings. The volume of a building is multiplied by a figure known from experience to represent a fair average figure for the cost of unit volume of such building.

cubital *(Zool.)*. See **secondary**.

cubital remiges *(Zool.)*. The primary quills connected with the ulna in Birds.

cuboid *(Maths.)*. A rectangular parallelepiped.

Cuboni test *(Vet.)*. A test for pregnancy in the mare, based on the chemical detection of oestrogens in the urine.

cucullate *(Bot.,Zool.)*. Hood-shaped.

cue marks *(Image Tech.)*. Dots or circles appearing in the corner of the frame near the end of a reel to warn the projectionist to prepare for a **change-over**.

cuesta *(Geol.)*. A hill or ridge with steep slope in one side and a gentle dip-slope on the other. syn. *escarpment*. Common term in US.

cuffing *(Med.)*. The accumulation of white cells round a blood vessel in certain infections of the nervous system.

cuirasse respirator *(Med.)*. A respirator which is attached, like armour, to the chest wall only, and assists respiration by fluctuations in its own internal pressure.

cullet *(Glass)*. Waste glass used with the 'batch' to improve the rate of melting and to save waste of materials.

culm *(Bot.)*. The stem especially the flowering stem, of grasses and sedges.

culm *(Geol.)*. The name given to the rocks of Carboniferous age in the south-west of England, consisting of fine-grained sandstones and shales, with occasional thin banks of crushed coal or 'culm'.

culmination *(Astron.)*. The highest or lowest altitude attained by a heavenly body as it crosses the meridian. *Upper culmination* indicates its meridian transit above the horizon, *lower culmination* its meridian transit below the horizon, or, in the case of a circumpolar, below the elevated pole.

cultivar *(Bot.)*. A subspecific rank used in classifying cultivated plants and indicated by the abbreviation cv. and/or by placing the name in single quotation marks; defined as an assemblage of cultivated plants which is clearly distinguished by any characters (morphological, physiological, cytological, chemical etc.), and which when reproduced (sexually or asexually, as appropriate) retains its distinguishing characters.

culture *(Biol.)*. A micro-organism, tissue or organ growing in or on a medium or other support; to cultivate such in this way.

culvert *(Civ.Eng.)*. Construction for the total enclosure of a drain or watercourse.

cumarone *(Chem.)*. See **coumarone**.

cumene *(Chem.)*. *Isopropyl (2 methylethyl)- benzene*, $C_6H_5(CH_3)_2$, bp 153°C.

cummingtonite *(Min.)*. Hydrated magnesium iron silicate, a monoclinic member of the amphibole group, occurring in metamorphic rocks. Differs from *grunerite* in having magnesium in excess of iron.

cumulative distribution *(Genrl.)*. In an assembly of particles, the fraction having less than a certain value of a common property, e.g. size or energy. Cf. **fractional distribution**.

cumulative distribution function *(Stats.)*. A function giving the probability that a corresponding continuous random variable takes a value less than or equal to the argument of the function.

cumulative dose *(Radiol.)*. Integrated radiation dose resulting from repeated exposure.

cumulative errors *(Civ.Eng.,Maths.)*. See **systematic errors**.

cumulative excitation *(Electronics)*. Successive absorption of energy by electrons in collision, leading to ionization. See **avalanche effect**.

cumulatively-compound machine *(Elec.Eng.)*. A compound-wound machine in which the series and shunt windings assist each other.

cumulonimbus *(Meteor.)*. Heavy and dense *cloud*, with a considerable vertical extent, in the form of a mountain or huge towers. At least part of its upper portion is usually smooth, or fibrous or striated, and nearly always flattened; this part often spreads out in the shape of an anvil or vast plume. Under the base of this cloud, which is often very dark, there are frequently low, ragged clouds either merged with it or not, and precipitation sometimes in the form of **virga**. Abbrev. *Cb*.

cumulus *(Meteor.)*. Detached *clouds*, generally dense and with sharp outlines, developing vertically in the form of rising mounds, domes or towers, of which the bulging upper part often resembles a cauliflower. The sunlit part of these clouds are mostly brilliant white; their base is relatively dark and nearly horizontal. Sometimes ragged. Abcumulus is ragged. Abbrev. *Cu*.

cumulus *(Zool.)*. The mass of cells surrounding the developing ovum in Mammals.

cumulus oōphorus *(Zool.)*. See **zona granulosa**.

cuneate, cuneal, cuneiform *(Bot.)*. Wedge-shaped.

Cunningham correction *(Powder Tech.)*. Modification of *Stoke's law* applied when particles of an aerosol are small compared to the mean free path of the molecules of the gas through which they are falling.

cup chuck *(Eng.)*. A lathe chuck in the form of a cup or bell screwed to the mandrel nose. The work is gripped by screws in the walls of the chuck. Also called *bell chuck*.

cupel *(Eng.)*. A thick-bottomed shallow dish made of bone ash; used in the cupellation of lead beads containing gold and silver, in the assay of these metals.

cupellation *(Eng.)*. The operation employed in recovering gold and silver from lead. It involves the melting of the lead containing these metals and its oxidation by means of an air-blast.

cupferron *(Chem.)*. Ammonium-nitroso-β-phenylhydrazine. Reagent used in the colorimetric detection and estimation of copper.

cup flow figure *(Plastics)*. The time in seconds taken by a mould of standard design to close completely under pressure, when loaded with a charge of moulding material.

cup head *(Eng.)*. A rivet or bolt head shaped like an inverted cup.

cupid's darts *(Min.)*. See **flèches d'amour**.

cup joint *(Build.)*. A joint formed between two lead pipes in the same line by opening out the end of one pipe to receive the tapered end of the other.

cup leather *(Eng.)*. A ring of leather moulded to U-section, used as a seal in hydraulic machinery. Now replaced in many applications by more durable materials like neoprene.

cupola *(Build.)*. A **lantern** constructed on top of a dome.

cupola *(Geol.)*. A dome-shaped offshoot rising from the top of a major intrusion.

cupola furnace *(Eng.)*. A shaft furnace used in melting pig-iron (with or without iron or steel scrap) for iron castings. The lining is firebrick. Metal, coke and flux (if used) are charged at top, and air is blown in near the bottom.

cupped wire *(Eng.)*. Wire in which internal cavities have been formed during drawings.

cuprammonia *(Chem.)*. A solvent for cellulose, prepared by adding ammonium chloride and then excess of caustic soda to a solution of a copper (II) salt, washing and pressing the precipitate, and dissolving it in strong ammonia.

cupric *(Chem.)*. *Copper (II)*. Containing divalent copper. Copper (II) salts are blue or green when hydrated and are stable.

cupriferous pyrites *(Min.)*. See **chalcopyrite**.

cuprite *(Min.)*. Oxide of copper, crystallizing in the cubic system. It is usually red in colour and often occurs associated with native copper; a common ore.

cupro-nickel *(Eng.)*. An alloy of copper and nickel; usually contains 15, 20, or 30% of nickel; is very ductile and has high resistance to corrosion.

cupro-uranite *(Min.)*. See **torbernite**.

cuprous *(Chem.)*. *Copper (I)*. Containing monovalent copper. Soluble copper (I) salts generally disproportionate to copper (0) and copper (II) salts.

cup shake *(For.)*. A shake between concentric layers. Also called *ring shake*.

cupula *(Zool.)*. Any domelike structure, e.g. the apex of the lungs, the apex of the cochlea.

cupule *(Bot.)*. One of a number of more or less cup-shaped organs, especially the structure that encloses the fruits of oak, beech, chestnut, birch etc., e.g. the acorn cup. (Cupuliferae).

cup wheel *(Eng.)*. An abrasive wheel in the form of a cylinder, used mainly for grinding. The cylindrical surfaces may be perpendicular or inclined to the end. Very shallow cup wheels are known as dish wheels or saucers.

curare *(Med.)*. South American native poison from the bark of species of *Strychnos* and *Chondodendron*.

curarine *(Med.)*. Paralysing toxic alkaloid ($C_{19}H_{26}ON_2$) extracted as *d*-tubocurarine chloride from crude curare; used in anaesthesia as a muscle relaxant.

curb *(Build.)*. (1) A wall-plate carrying a dome at the springings. Also *curb-plate*. (2) A frame of upstand round an opening in a floor or roof.

curb *(Civ.Eng.)*. A hollow timber or cast-iron cylinder used in sinking and lining a shaft or well, for which purpose it is laid over the site, and then, as earth is excavated from beneath it, the lining is built upon it and sinks with the curb. Also called *cutting curb, drum-curb*.

curb *(Min.Ext.)*. Framework fixed in rock of mine shaft to act as foundation for brick or timber lining.

curb *(Vet.)*. A swelling occurring just below the point of the hock of the horse, usually due to inflammation caused by sprain of the calcaneo-cuboid ligament.

curb pins *(Horol.)*. The two vertical pins attached to the index embracing the balance spring, near the point of attachment of the outer coil. By moving the pins nearer to, or farther away from, the point of attachment, a delicate regulation of the time of vibration of the balance is obtained.

curb-plate *(Build.)*. See **curb**.

curb-roof *(Build.)*. See **mansard roof**.

curettage *(Med.)*. The scraping of the walls of cavities (especially of the uterus) with a *curette* (or *curet*), a spoon-shaped instrument.

curie *(Phys.)*. Unit of radioactivity. 1 curie is defined as 3.700×10^{10} decays per second, roughly equal to the activity of 1 g of ^{226}Ra. Abbrev. *Ci*. Now replaced by the becquerel (*Bq*). $1\ \text{Bq} = 2.7 \times 10^{-11}$ Ci.

curie balance *(Phys.)*. A torsion balance for measuring the magnetic properties of nonferromagnetic materials by the force exerted on the specimen in a nonuniform magnetic field.

Curie point (temperature) *(Phys.)*. (1) Temperature above which a ferromagnetic material becomes paramagnetic. Also called *magnetic transition temperature*. (2) Temperature (*upper Curie point*) above which a ferroelectric material loses its polarization. (3) Temperature (*lower Curie point*) below which some ferroelectric materials lose their polarization.

Curies' law *(Phys.)*. For paramagnetic substances, the magnetic susceptibility is inversely proportional to the absolute temperature.

Curie-Weiss law *(Phys.)*. At the Curie temperature, θ, a ferromagnetic material becomes paramagnetic. Well above this temperature its paramagnetic susceptibility is $\chi = C/(T-\theta)$, where T is the absolute temperature and C is the Curie constant.

curine *(Chem.)*. $(C_{18}H_{19}O_3N)_2$, an alkaloid of the quinoline group, found in curare extract obtained from various *Strychnos* spp.

curing *(Chem.)*. A term applied usually to a fermentation or ageing process of natural products, e.g. rubber, tobacco etc.

curing *(Civ.Eng.)*. A method of reducing the contraction of concrete on setting; the surface is kept covered for a time with damp sacks, or with damp sawdust or sand, or sprayed with water.

curing *(Plastics)*. The chemical process undergone by a thermosetting plastic by which the hot liquid resin sets to a solid at the same temperature. Curing generally takes place during the moulding operation, and may require from 45 sec to 30 min for its completion.

curium *(Chem.)*. Manmade radioactive element, symbol Cm, at. no. 96, produced from americium. There are several long-lived isotopes (up to 1.7×10^7 years half-life), all α-emitters.

curl *(For.)*. A roughly hewn block of timber cut from a crotch and intended for cutting into veneers.

curl *(Maths.)*. For any small vector of any shape at any point in a vector field, the line integral of the vector V around its bounding edge will result in an orientation of the area for which the line integral is greatest. The amount of this maximum line integral, expressed per unit area, is called the curl of the vector field at the point and is given the vectorial sense of the positive normal drawn on the small exploring area when in the position giving the greatest integral.

curl *(Paper)*. A paper defect caused by unequal dimensions of the top and under sides of the sheet due to changes in the ambient moisture or temperature.

curled toe paralysis *(Vet.)*. A disease of chicks characterized by leg weakness and inward curling of the toes, associated with degenerative changes in the peripheral nerves; caused by a deficiency of riboflavin in the diet.

curly grain *(For.)*. A wavy pattern on the surface of worked timber due to the undulate course taken by the vessels and other elements of the wood.

current *(Genrl.)*. A flow of e.g. water, air etc.

current *(Phys.)*. Rate of flow of charge in a substance, solid, liquid or gas. Conventionally, it is opposite to the flow of (negative) electrons, this having been fixed before the nature of the electric current had been determined. Practical unit of current is the **ampere**.

current amplification *(Elec.Eng.)*. Ratio of output current to input current of an amplifier or photomultiplier, often expressed in decibels.

current antinode *(Elec.Eng.)*. A point of maximum current in a standing-wave system along a transmission line or aerial.

current attenuation *(Elec.Eng.)*. Ratio of output to input currents of a transducer, expressed in decibels.

current balance *(Elec.Eng.)*. A form of balance in which the force required to prevent the movement of one current-carrying coil in the magnetic field of a second coil carrying the same current is measured by means of a balancing mass. Cf. **magnetic balance**.

current bedding *(Geol.)*. See **cross-bedding**.

current-carrying capacity *(Elec.Eng.)*. The current which a cable can carry before the temperature rise exceeds a permissible value (usually 40°C). It depends on the size of the conductor, the thermal resistances of the cable, and surrounding medium.

current circuit *(Elec.Eng.)*. The electrical circuit associated with the current coil of a measuring instrument or relay.

current coil *(Elec.Eng.)*. A term frequently used with wattmeters, energy meters or similar devices, to denote the coil connected in series with the circuit and therefore carrying the main current.

current collector *(Elec.Eng.)*. The device used on the vehicles of an electric traction system for making contact with the overhead contact wire or the conductor-rail. See **bow, pantograph, plough, trolley system**.

current density *(Elec.Eng.)*. Current flowing per unit cross-sectional area of conductor or plasma, expressed in amperes per square metre.

current efficiency *(Elec.Eng.)*. The ratio of the mass of substance liberated in an electro-chemical process by a given current, to that which should theoretically be liberated according to Faraday's law.

current feed *(Telecomm.)*. Delivery of radio power to a current maximum (loop or antinode) in a resonating part of an antenna.

current feedback *(Elec.Eng.,Electronics)*. In amplifier circuits, a feedback voltage proportional to the load current. It may be applied in series or in shunt with the source of the input signal. See also **negative feedback, positive feedback, voltage feedback**.

current gain *(Electronics.)*. In a transistor, ratio of output current to input current. In common emitter configuration it may be as high as 100, whilst in common base not exceed unity.

current generator *(Phys.)*. Ideally, a current source of infinite impedance such that the current will be unaltered by any further impedance in its circuit. In practice, a generator whose impedance is much higher than that of its load.

current limiter *(Elec.Eng.)*. A component which sets an upper limit to the current which can be passed.

current margin *(Elec.Eng.)*. In a relay, difference between steady-state currents corresponding to values used for signalling and for just operating the relay.

current node *(Elec.Eng.)*. A point of zero electric current in a standing-wave system along a transmission line or aerial.

current regulator *(Elec.Eng.)*. Circuit employed to control the current supplied to a unit.

current saturation *(Elec.Eng.)*. Condition when anode current in triode valve has reached its maximum value.

current sensitivity *(Elec.Eng.)*. Magnitude of change in reading of a current-measuring instrument by a given change in current.

current transformer *(Elec.Eng.)*. (1) One designed to be connected in series with circuit, drawing predetermined current. Sometimes called *series transformer*. (2) Winding enclosing conductor of heavy alternating current; steps down the current in known ratio for measurement.

current, voltage resonance *(Elec.Eng.)*. Condition of a circuit when the magnitude of a current (or voltage) passed through the maximum as the frequency is changed through resonance; obs. *syntony* or *tuning* (Lodge).

current weigher *(Elec.Eng.)*. See **current balance**.

cursor *(Comp.)*. Character, often flashing on and off, which indicates the current display position on a **VDU**. See **addressable cursor**.

cursor *(Eng.)*. The adjustable fiducial part of a drawing or other instrument, with an engraved line on metal or a transparent window, both placed to reduce parallax error. See **vernier**.

cursorial *(Zool.)*. Adapted for running.

curtail step *(Build.)*. A step which is not only the lowest step in a flight but is also shaped at its outer end to the form of a scroll in plan.

curtain antenna *(Telecomm.)*. Large number of vertical radiators or reflectors in a plane.

curtain wall *(Build.)*. A thin wall whose weight is carried directly by the structural frame of the building, not by the wall below.

curtain walling *(Build.)*. Large-area prefabricated framed sections of lightweight material generally also pre-decorated on the exterior surface.

curtate cycloid *(Maths.)*. See **roulette**.

curtate trochoid *(Maths.)*. See **roulette**.

Curtis winding *(Elec.Eng.)*. The winding of low-capacitance and low-inductance resistors in which the wire is periodically reversed.

curvature *(Maths.)*. (1) Of a plane curve: the curvature at a point P on the curve measures the rate of change (at the point) in the angle ψ, which the tangent makes with a fixed axis, relative to the arc length s. It is thus defined as

$$K=\frac{d\psi}{ds}=\left(\frac{d^2y}{dx^2}\right)\Big/\left\{1+\left(\frac{dy}{dx}\right)^2\right\}^{3/2}.$$

K is the reciprocal of the radius of curvature, p, which is the radius of the circle which touches the curve (on the concave side) at the point in question. The circle is the *circle of curvature*, and its centre is the *centre of curvature* of the curve at the point. The circle of curvature is also called the *osculating* circle. (2) Of a space curve: the curvature at a point on a space curve is the rate of change of direction of the tangent with respect to the arc length, i.e.

$$K=\frac{1}{p}=\frac{d\theta}{ds},$$

where as before K is the curvature, p the radius of curvature, θ represents the change in direction of the tangent, and s is the length. This is called the *first curvature of a space curve*. The second curvature, or *torsion*, of the curve is the corresponding rate of change in the direction of the binormal, i.e.

$$\lambda=\frac{d\psi}{ds},$$

where ψ is the angle through which the binormal turns. $1/\lambda$ is the radius of torsion. Cf. **moving trihedral** and **osculating sphere**. (3) Of a surface: at any point P on the surface there is, in general, a single normal line. Planes through this line cut the surface in plane curves called *normal curvature at P* in that direction. The maximum and minimum values of the normal curvature at P are called the *principal curvatures at P*, and the directions in which they occur, which will be mutually perpendicular, are called the *principal directions at P*. The average of the principal curvatures at P is called the *average*, *mean* or *mean normal curvature at P*, and their product, ψ, the *total*, *total normal* or *Gaussian curvature at P*. The reciprocals of the normal and of the principal curvatures at P are called the *normal* and *principal radii of curvature at P* respectively.

curvature correction *(Civ.Eng.)*. A correction used in the calculation of quantities for earthworks following a curved line in plan; the quantities are taken out as if the line were straight, and a curvature correction made to account for the fact that it is not straight.

curvature of field *(Phys.)*. For an optical system, a planar object normal to the axis will not in general be imaged as a plane. Even in the absence of *spherical aberration*, *coma* and *astigmatism*, the image surface will be curved, the **Petzval curvature**.

curvature of spectrum lines *(Phys.)*. In a spectrum produced by a prism the lines are slightly convex towards the red end. Rays from the ends of the slit are inclined at a small angle to the plane at right angles to the refracting edge of the prism, and so suffer a slightly greater deviation than rays from the centre of the slit, appearing bent towards the violet end of the spectrum.

curve *(Eng.)*. An instrument used by the draughtsman for drawing curves other than circular arcs. It consists of a thin flat piece of transparent plastic or other material, having curved edges which are used as guides for the pencil.

curve *(Maths.)*. The locus of a point moving with one degree of freedom.

curve fitting *(Phys.)*. The process of finding the best algebraic function to describe a set of experimental measurements. Usually accomplished by using a least-squares process by which the parameters of the function are adjusted to minimize the sum of the squares of the deviations of the observations from the theoretical curve.

curve of pursuit *(Aero.)*. That path followed by a combat aircraft steering towards the present position of an adversary.

curve ranging *(Surv.)*. The operation of setting out on the ground points which lie on the line of a curve of given radius.

curvilinear asymptote *(Maths.)*. See **asymptote**.

curvilinear co-ordinates *(Maths.)*. (1) Of a point in space: three systems of surfaces may be defined by the parametric equation $x=x(u,v,w)$, $y=y(u,v,w)$, $z=z(u,v,w)$, and any point in space may be regarded as the intersection of three surfaces, one from each family. The parameters u,v,w, are then said to be the *curvilinear co-ordinates* of the point. If the three systems of curves are mutually orthogonal, the co-ordinates are *orthogonal* curvilinear co-ordinates. (2) Of a point on a surface: if a surface is given in parametric form $x=x(u,v)$, $y=y(u,v)$, and $z=z(u,v)$, the parameters u and v are the curvilinear co-ordinates of a point on the surface.

curvilinear distortion *(Image Tech.,Phys.)*. Curvature of lines which should be straight, as seen in the outer portion of the image from a stopped-down simple lens. See **barrel distortion, pincushion (pillow) distortion**.

Cushing's syndrome *(Med.)*. The concurrence of obesity, hairiness, linear atrophy of the skin, loss of sexual function, and curvature of the spine, due to a tumour in the pituitary or adrenal gland, causing oversecretion of corticosteroids.

cushion *(Build.)*. The capping stone of a pier.

cushion craft *(Ships)*. Name given to certain types of **hovercraft**.

cushion plant *(Bot.)*. Plant with many densely crowded upright shoots not more than a few centimetres high, forming a cushion-like mass on the ground; typical of alpine and arctic floras. Also *chamaephyte*.

cushion steam *(Eng.)*. The steam shut in the cylinder of a steam-engine after the closing of the exhaust valve.

cusp *(Bot.,Zool.)*. A sharp-pointed prominence, as on teeth.

cusp *(Maths.)*. See **double point**. adj. *cuspidate*.

cusps *(Astron.)*. The horns of the moon or of an inferior planet in the crescent phase.

customs plant *(Min.Ext.)*. A crushing or concentrating plant serving a group of mines on a contract basis. It buys ore according to valuable content and complexity of treatment, and relies for profit on sale of products.

cut *(Eng.)*. The thickness of the metal shaving removed by a cutting tool.

cut *(Hyd.Eng.)*. The water-way between the pontoons of a pontoon-bridge.

cut *(Image Tech.)*. (1) 'Cut!', instruction by the director to stop shooting. (2) In an edited film, an instantaneous change from one scene to another.

cut *(Min.Ext.)*. A petroleum fraction.

cut *(Nuc.Eng.)*. Proportion of input material to any stage of an isotope separation plant which forms useful product. Also *splitting ratio*.

cut *(Print.)*. A **block**.

cut-and-cover *(Civ.Eng.)*. A method often adopted in the construction of underground railways at only a moderate depth. A cutting is first excavated to accommodate the railway; it is then covered over to original ground-level by arching supported on side walls.

cut-and-fill *(Civ.Eng.)*. A term used to describe any cross-section of highway or railroad earthworks which is partly in cutting and partly in embankment.

cut-and-mitred-string *(Build.)*. A *cut string* which is mitred at the vertical parts of the notches in the upper surface, so that the end grain of the risers may be concealed.

cut-and-mitred valley *(Build.)*. A valley formed in a tiled roof by cutting one edge of the tiles on both sides of the valley so that they form a mitre, which is rendered watertight by lead soakers bonded in with the tiles.

cutaneous *(Zool.)*. Pertaining to the skin.

cut edges *(Print.)*. Said of the edges of a book when all three are clean cut as distinct from *trimmed edges*, where only the furthest projecting leaves at the tail are trimmed.

cut flush *(Print.)*. Cut after binding, so that the cover does not project.

cuticle *(Bot.)*. Layer of **cutin** on the outside of some plant cell walls especially the shoot epidermis where it forms a continuous layer which, with the **epicuticle**, has relatively low permeability to water and gases.

cuticle *(Textiles)*. Flat overlapping scales that lie on the surface of animal hair and wool. They cover the internal core or cortex.

cuticle *(Zool.)*. A nonliving layer secreted by and overlying the epidermis.

cuticular transpiration *(Bot.)*. The loss of water vapour from a plant through the cuticle.

cuticulin *(Zool.)*. The outermost layer of the insect epicuticle, consisting of lipoprotein.

cutin *(Bot.)*. A mixture of fatty substances especialy of cross-linked polyesters based on mostly C_{16} and C_{18} aliphatic acids and hydroxyacids in the **cuticle** of plants.

cutinization *(Bot.,Zool.)*. The formation of cutin; the deposition of cutin in a cell wall to form a cuticle.

cut-in notes *(Print.)*. Notes occupying a rectangular space, set into the text at the outer edge of a paragraph.

cutis *(Zool.)*. The dermis or deeper layer of the Vertebrate skin.

cut-off *(Eng.)*. The point in an engine cycle, expressed as percentage of stroke, at which the supply of steam, fuel oil, etc., is stopped.

cut-off *(Print.)*. A feature of reel-fed presses; the paper is cut after printing to a size determined by the cylinder periphery; a few models have a selection of cylinder sizes and a consequently variable cut-off.

cut-off current *(Electronics)*. The residual current flowing in a valve or transistor when the device is biased *off* in a specified way.

cut-off field *(Electronics)*. Same as *critical field*.

cut-off frequency *(Electronics)*. That above (or below) which gain, efficiency, or other desirable characteristic of a circuit or device is changing so rapidly that it is no longer useful, e.g. for an amplifier, cut-off frequency is commonly taken as that when gain is 3 dB less than the mid-band value. See **alpha cut-off, critical frequency**.

cut-off frequency *(Telecomm.)*. For any specified mode of propagation in a loss-less waveguide or other structure, the frequency at which the attenuation constant changes from zero to a positive value or vice versa. See **cut-off wavelength**.

cut-off knife *(Print.)*. A plain or serrated blade which severs each copy on a reel-fed rotary.

cut-off posture *(Behav.)*. In ethology, a term referring to postures that remove social stimuli (e.g. a potential mate, or opponent) from sight, and thus may serve to reduce the actor's arousal in a conflict situation.

cut-off rubbers *(Print.)*. The rubber strip set in a cylinder against which the **cut-off knife** presses when cutting the product into copies. Also called *cutting buffer, cutting strip*.

cut-off wavelength *(Telecomm.)*. The free-space wavelength corresponding to the lowest frequency at which a waveguide or other propagation structure can support a particular mode or field pattern.

cut off wheel *(Eng.)*. Thin abrasive wheel made of flexible material which is used to cut metals, concrete etc.

cut-out *(Elec.Eng.)*. Off-switch operated automatically if safe operating conditions are not maintained, e.g. water flow cut-out.

cut-out half-tone *(Print.)*. A half-tone from which the background is removed to give prominence to the subject.

cut-over *(Telecomm.)*. The rapid transfer of large numbers of subscribers' lines from one exchange to another, particularly from an electromagnetic (*Strowger*) to an electronic exchange.

cut-stone *(Build.)*. A stone hewn to shape with a chisel and mallet.

cut string *(Build.)*. A string whose upper surface is shaped to receive the threads and risers of the steps, while the lower surface is parallel to the slope of the stair. Also called an *open string*.

cutter *(Acous.)*. The sapphire or diamond point which removes the thread of lacquer in gramophone-disk recording.

cutter *(Eng.)*. Any tool used for severing, often more specifically a milling cutter. See **milling-machine**.

cutter-block *(Build.)*. The part holding the cutters in a wood-planing machine.

cutter dredge *(Civ.Eng.)*. Alluvial dredge which loosens material by means of powered cutting ring, draws it to a pump and delivers it for treatment aboard in adjacent plant.

cutter loader *(Min.Ext.)*. Coal-cutting machine which both severs the mineral and loads it on to a transporting device such as a face conveyor.

cutters *(Build.)*. Bricks which are made soft enough to be cut with a trowel to any shape required, and then rubbed to a smooth face and the correct shape. Also called *rubbers*.

cutting *(Bot.)*. A piece of a plant, usually shoot, root or leaf, which is cut off and induced to form adventitious roots and/or buds as a means of vegetative propagation. See also **rooting compound**.

cutting *(Civ.Eng.)*. An open excavation through a hill, for carrying a highway or railroad at a lower level than the surrounding ground.

cutting *(Image Tech.)*. The process of editing a film by the creative assembly of individual scenes of picture and sound to meet the director's intentions, resulting in a *cutting copy* or *work print*. Also, the actual assembly of the original negative to match.

cutting *(Vet.)*. See **brushing**.

cutting buffer *(Print.)*. See **cut-off rubbers**.

cutting compound *(Eng.)*. A mixture of water, oil, and soft soap, etc., used for lubricating and cooling the cutting tool in machining operations. See **coolant**.

cutting curb *(Civ.Eng.)*. See **curb**.

cutting cylinder *(Print.)*. The cylinder that holds knives to cut the web into separate copies. See **cut-off knife, cut-off rubbers**.

cutting disks *(Print.)*. See **slitters**.

cutting gauge *(Build.)*. A **marking gauge** fitted with a bevelled cutter in place of a pin. Used for cutting thin wood and for marking across the grain to obviate tearing.

cutting list *(Build.)*. A list giving dimensions, sometimes with diagrams of sections, of timber required for any given work.

cutting marks *(Print.)*. Short lines printed onto the sheet to indicate cutting, slitting or punching positions.

cutting speed *(Eng.)*. The speed of the work relative to the cutting tool in machining operations; usually expressed in feet or metres per minute.

cutting strip *(Print.)*. See **cut-off rubbers**.

cutting tools *(Eng.)*. Steel tools used for the machining of metals. See **broach, cutter, lathe tools, milling-cutter, planer tools, reamer, screwing die, shaper tools, slotting tools, tap, twist drill**.

cuttling *(Textiles)*. Operation of folding a fabric to make it convenient to handle; also known as *plaiting*.

cut-up trade (cut and sew) *(Textiles)*. Section of the knitting industry dealing with fabric made on a circular knitting-machine. The material is afterwards cut to shape from patterns, pieces being sewn together to form the final article.

cut-water *(Civ.Eng.)*. The angular edge of a bridge-pier, shaped to lessen the resistance it offers to the flow of water.

Cuvierian ducts *(Zool.)*. In lower Vertebrates, a pair of large venous trunks entering the heart from the sides.

cv *(Bot.)*. Abbrev. for *cultivar*.

c-value paradox *(Biol.)*. The paradox that some very similar animals and plants have unexpectedly large differences in the amount of their genomic DNA, e.g. amphibian genomes vary by over a 100-fold. Not simply an increase in the number of sequence copies per genome.

CVS *(Med.)*. See **chorionic villus sampling**.

CVT *(Autos.)*. See **constant variable transmission**.

CW radar *(Radar)*. *Continuous Wave radar*. One in which the transmitter emits a continuous radio-frequency signal; the receiving antenna is arranged so that a very small amount of the transmitted power enters it, along with the signal reflected from the target. Movement of the target causes Doppler frequency shift in the reflected signal and this difference can be detected at the output of a mixer. CW radar uses less bandwidth than conventional pulsed radar.

cyanamide process *(Chem.)*. The fixation of atmospheric nitrogen by heating calcium carbide (ethynide) in a stream of the gas. Calcium cyanamide, $CaCN_2$, is thus formed, and this, on treatment with water, a little sodium hydroxide, and steam under pressure, yields ammonia.

cyanates *(Chem.)*. Salts containing the monovalent acid radical CNO′.

cyanhydrins *(Chem.)*. A series of compounds formed by the addition of hydrogen cyanide to aldehydes and ketones. Their general formula is R′.C(OH)(CN).R″ and they are useful for the preparation of 2-hydroxy- acids. See **acetone cyanhydrin**. Also *cyanohydrins*.

cyanicide *(Min.Ext.)*. Any constituent in ore or chemical product made during treatment of gold-bearing minerals by cyanidation, which attacks or destroys the sodium or calcium cyanide used in the process.

cyanidation vat *(Eng.)*. A large tank, with a filter bottom, in which sands are treated with sodium cyanide solution to dissolve out gold.

cyanide hardening *(Eng.)*. Case-hardening in which the carbon content of the surface of the steel is increased by heating in a bath of molten sodium cyanide.

cyanides *(Chem.)*. (1) Salts of hydrocyanic acid. (2) See **nitriles**.

cyaniding *(Min.Ext.)*. The process of treating finely ground gold and silver ores with a weak solution of sodium cyanide, which readily dissolves these metals. The precious metals were formerly recovered by precipitation from solution with zinc. Currently a number of methods have proved more economical, including adsorption on resins and carbon. See **resin-in-pulp**, **carbon-in-pulp**.

cyanin *(Chem.)*. The colouring matter of the cornflower and the rose. It is an anthocyanin, and on hydrolysis yields cyanidin and two molecules of glucose.

cyanite *(Min.)*. See **kyanite**.

cyanizing *(Build.)*. See **kyanizing**.

cyanogen *(Chem.)*. A very poisonous, colourless gas with a smell of bitter almonds. It is soluble in 4 volumes of water, ammonium oxalate (ethandioate) being formed on standing. Its formula is C_2N_2, or $N{\equiv}C{-}C{\equiv}N$, and it somewhat resembles the halogens in its chemical behaviour.

cyanogenesis *(Bot.)*. The release from plant parts, usually after wounding, of hydrogen cyanide by cytoplasmic glycosidase action on a vacuolar glycoside containing e.g. mandelonitrile. Occurs in leaves of cherry laurel (*Prunus laurocerasus*), seeds of bitter almonds and fronds of bracken. Possibly a deterrent to herbivores.

cyanohydrins *(Chem.)*. See **cyanhydrins**.

Cyanophyceae *(Bot.)*. *Myxophyceae, Cyanobacteria*. Blue-green algae. Prokaryotic organisms with chlorophyll a and phycobilins. Reserve carbohydrate $\alpha, 1{\rightarrow}4$ glucan. Asexual reproduction by spores, division fragmentation or homogonia. Unicellular, colonial, or filamentous. Sorts with **heterocysts** fix nitrogen. Planktonic sorts may have **gas-vacuoles**. Occur in fresh- and salt-water (planktonic and benthic), in soils and as nitrogen-fixing symbionts in *Azolla* and the roots of Cycads and some flowering plants and in some lichens. See also **gliding**.

cyanosis *(Med.)*. Blueness of the skin and the mucous membranes due to insufficient oxygenation of the blood. May be peripheral due to poor circulation or central due to failure of oxygenation.

cyanotype *(Image Tech.)*. The ferroprussiate process, familiar as blue-printing; it depends on the light reduction of a ferric salt to a ferrous salt, with production of Prussian blue on development.

cyanuric acid *(Chem.)*. A tribasic, heterocyclic acid, having the formula $H_3C_3N_3O_3$. The trimer of cyanic acid which is too unstable to exist by itself, an aqueous solution being slowly converted to urea.

cyanuric dyes *(Chem.)*. Relatively new class of dyestuffs based on cyanuric chloride ($C_3N_3Cl_3$). Their importance lies in the ability to link chemically with the fabrics or fibres (particularly cellulosic fibres) being dyed. See **reactive dyes**.

cybernetics *(Comp.)*. The study of control and communications in complex electronic systems and in animals especially humans.

Cycadales *(Bot.)*. The cycads. Order of gymnosperms (Cycadopsida), widespread in the Mesozoic, now ca 65 spp. in Central America, S. Africa, SE Asia and Australia. Stems stout, unbranched, manoxylic; leaves large, pinnate, with haplocheilic stomata. Dioecious. Reproductive organs in large cones (except female *Cycas*), Zooidogamous. Radiospermic. The pith of 2 spp. is a minor source of sago.

Cycadopsida *(Bot.)*. Class of gymnosperms containing the superficially similar Cycadales and Cycadeoidales and a number of other orders. Probably not a natural group.

cyclamates *(Chem.)*. Derivatives of cyclohexyl-sulphamic acid, having 30 times the sweetening power of sucrose, much used in foods, drinks, and for dietary purposes; banned in some countries because of supposed health hazards.

cycle *(Genrl.)*. A series of occurrences in which conditions at the end of the series are the same as they were at the beginning. Usually, but not invariably a cycle of events is recurrent.

cycle *(Maths.)*. For a periodic quantity or function, the set of values that it assumes during a period.

cycle of erosion *(Geol.)*. The hypothetical course of development followed in landscape evolution; it consists of the major stages of youth, maturity, and old age.

cycle-time *(Comp.)*. The time interval between the start and restart of a particular hardware operation, e.g. *CPU cycle time*, *memory cycle time*.

cyclic *(Bot.)*. A flower having the parts arranged in whorls, rather than in spirals.

cyclic adenosine monophosphate, cAMP *(Biol.)*. A derivative of adenosine monophosphate in which the phosphate forms a ring involving the 3′ and 5′ hydroxyl groups of ribose. It is of major metabolic importance through its diverse effects on many enzymes.

cyclic compounds *(Chem.)*. Closed-chain or ring compounds consisting either of carbon atoms only (carbocyclic compounds), or of carbon atoms linked with one or more other atoms (heterocyclic compounds).

cyclic group *(Maths.)*. A group in which every element can be expressed as a power of a single element. Cyclic groups are *Abelian*, and those of the same order are *isomorphic*.

cyclic pitch control *(Aero.)*. Helicopter rotor control in which the blade angle is varied sinusoidally with the blade azimuth position, thereby giving a tilting effect and horizontal translation in any desired direction.

cyclic quadrilateral *(Maths.)*. A four-sided polygon whose vertices lie on a circle.

cyclic shift *(Comp.)*. See **end-around shift**.

cyclic test *(Min.Ext.)*. See **locked test**.

cyclitis *(Med.)*. Inflammation of the ciliary body of the eye.

cyclo- *(Chem.)*. Containing a closed carbon chain or ring.

cyclo- *(Genrl.)*. Prefix from Gk. *kyklos*, circle.

cycloalkanes *(Chem.)*. Hydrocarbons containing saturated carbon rings. Also called *cyclanes, polymethylenes*.

cyclobutane *(Chem.)*. $(CH_2)_4$. Alicyclic compound; a cycloalkane. Bp 11°C.

cyclogiro *(Aero.)*. An aircraft lifted and propelled by pivoted blades rotating parallel to roughly horizontal transverse axes.

cyclohexanamine *(Chem.)*. See **cyclohexylamine**.

cyclohexane *(Chem.)*. C_6H_{11}, mp 2°C, bp 81°C, rel.d. 0.78, a colourless liquid, of mild ethereal odour. Molecules generally adopt a **chair** conformation.

cyclohexanol *(Chem.)*. $C_6H_{11}.OH$, mp 15°C, bp 160°C, rel.d. 0.945, an oily, colourless liquid.

cyclohexanone *(Chem.)*. $C_6H_{10}O$, bp 154°–156°C rel.d. 0.945, a colourless liquid, of propanone-like odour, solvent for cellulose lacquers.

cyclohexylamine (cyclohexanamine) *(Chem.)*.

$C_6H_{11}NH_2$, colourless liquid, bp 134°C. A reduction product of aniline, its derivatives are used in the manufacture of plastics, etc.

cycloid *(Maths.)*. An arch-shaped curve with intrinsic equation $s = 4a \sin \psi$. Its parametric cartesian equations are $x = a(\theta + \sin\theta)$ and $y = a(1 - \cos\theta)$. See **roulette**.

cycloid *(Zool.)*. Evenly curved; said of scales which have an evenly curved free border.

cycloidal teeth *(Eng.)*. Gear teeth whose flank profiles consist of cycloidal curves.

cyclone *(Meteor.)*. (1) Same as *depression*. (2) A **tropical revolving storm** in the Arabian Sea, Bay of Bengal and South Indian Ocean.

cyclone *(Min.Ext.)*. Conical vessel used to classify dry powders or extract dust by centrifugal action. See **hydrocyclone**.

cyclonite *(Chem.)*. Hexogen, cyclotrimethylene trinitramine $(CH_2)_3(N.NO_2)_3$, a colourless, crystalline solid, mp 200°–202°C, odourless, tasteless, non-poisonous, soluble in acetone, prepared by oxidative nitration of hexamethylene tetramine. Used, generally with TNT, as an explosive.

cyclo-octadiene *(Chem.)*. C_8H_{12}, dimerization product of butadiene obtained by using **Ziegler catalysts**. Used as an intermediate in the preparation of nylon polymers from petrochemical sources.

cycloparaffins *(Chem.)*. Same as **cycloalkanes**.

cyclopean *(Build.)*. A name given to ancient dry-masonry works in which the stones are very large and are irregular in size.

cyclopentanal *(Chem.)*. C_5H_9CHO, *cyclic aliphatic alcohol*, bp 139°C, used as an intermediate in the preparation of perfumery and flavouring esters.

cyclopentane *(Chem.)*. A cycloalkane with the formula C_5H_{10}, bp 49°C. The ring of carbon atoms is nearly flat.

cyclopentanone *(Chem.)*. C_5H_8O, solvent for a wide range of synthetic polymers, particularly PVC.

cyclophon *(Electronics)*. Tube which uses the fundamental principle of electron-beam switching.

cyclophosphamide *(Immun.)*. A potent alkylating agent used as an anti-cancer drug, but also as an immunosuppressive agent which acts particularly on B-lymphocytes. Use is restricted by its toxicity to bone marrow and the bladder.

cyclophosphamide *(Med.)*. An alkylating drug which interferes with DNA synthesis and prevents cell replication. Used in the treatment of leukaemia and lymphoma.

cycloplegia *(Med.)*. Paralysis of the ciliary muscle.

cyclopropane *(Chem.)*. A cycloalkane with the formula C_3H_6. bp −33°C. As the C-C-C angles are constrained to be 60° in place of the normal tetrahedral angle, the molecule is very reactive.

cyclosilicates *(Min.)*. Silicate minerals whose atomic structure contains rings of SiO_4 groups, e.g. **beryl**.

cyclosis *(Biol.)*. The circulation of protoplasm within a cell.

cyclospondylous *(Zool.)*. Showing partial calcification of cartilaginous vertebral centra in the form of concentric rings.

cyclosporin A *(Immun.)*. A cyclic peptide used as an immunosuppressive agent which has a selective action on the generation of helper T-cells, which do not become functional while the drug is present. It is useful in preventing graft rejection but, because it produces renal damage easily, blood levels need repeated monitoring.

Cyclostomata *(Zool.)*. An order of the class Agnatha. Aquatic and gill-breathing, with a round suctorial mouth; buccal cavity contains a muscular tongue bearing horny teeth used to rasp the flesh from the prey; cartilaginous endoskeleton; no fins or limb girdles; slimy skin with no scales. Lampreys and Hagfish.

cyclostrophic wind *(Meteor.)*. The theoretical wind which, when blowing round circular isobars, represents a balance between the pressure gradient and the centrifugal force, the **Coriolis force** being neglected; it is a useful approximation only at low latitudes e.g. in tropical cyclones.

cyclothem *(Geol.)*. A series of beds formed during one sedimentary cycle. Particularly associated with coal-bearing rocks.

cyclotron *(Phys.)*. Machine in which positively charged particles are accelerated in a spiral path within *dees* in a vacuum between the poles of a magnet, energy being provided by a high frequency voltage across the dees. When the radius of the path reaches that of the dees, the particles are electrically deflected out of the cyclotron for use in nuclear experiments. See **betatron, synchrocyclotron, synchrotron, cyclotron frequency**.

cyclotron frequency *(Phys.)*. Any particle of charge q moving with a velocity v perpendicular to a magnetic field of flux density B, moves in a circular path. The number of revolutions around this path per second, the *cyclotron frequency*, is $f = Bq/2\pi m$, where m is the mass of the particle. The frequency is independent of the velocity, a result that is used in the *cyclotron* and the *magnetron*.

cyclotron resonance *(Phys.)*. The resonant coupling of electromagnetic power into a system of charged particles undergoing orbital movement in a uniform magnetic field. Used for the quantitative determination of the band parameters in semiconductors. See **Landau levels**.

cyclotron resonance heating *(Nuc.Eng.)*. Mode of heating of a plasma by resonant absorption of energy based on the waves induced in the plasma at the cyclotron frequency of electrons (abbrev. *ECRH*) or ions (abbrev. *ICRH*).

cyesis *(Med.)*. Pregnancy. See **pseudocyesis**.

Cygnus A *(Astron.)*. Strongest radio source in Cygnus, identified with a distant peculiar galaxy which is also an X-ray source.

cylinder *(Comp.)*. Name given to the set of tracks in a multi-disk pack which can be read without moving the read head. See **disk-pack**.

cylinder *(Eng.)*. The tubular chamber in which the piston of an engine or pump reciprocates; the internal diameter is called the *bore*, and the piston-travel the *stroke*.

cylinder *(Maths.)*. (1) A (cylindrical) surface generated by a line which moves parallel to a fixed line so as to cut a fixed plane curve. Any line lying in the surface is called a *generator*. (2) A solid bounded by a cylindrical surface and two parallel planes (the *bases*) which cut the surface. A cylindrical surface is named after its normal sections and a solid cylinder after its bases. The axis of a cylinder (if it has one) is its line of symmetry parallel to its generators or the line joining the midpoints of its bases. A circular cylinder is called a *right circular cylinder* if its bases are normal to its axis, otherwise it is called an *oblique circular cylinder*.

cylinder barrel *(Eng.)*. The wall of an engine cylinder, as distinct from the cylinder itself, which term includes the head or covers.

cylinder bearers *(Print.)*. At each end of impression, blanket and plate cylinders, to provide a datum for packing; may be integral with cylinder or be a removable band (bearer ring).

cylinder bit *(Build.)*. A steel drill with helical cutting edge, used for precise boring.

cylinder block *(Eng.)*. The largest part of an *IC engine* which is bored to receive the pistons and contains integrally cast cooling water channels made from cast iron or, more recently from lightweight aluminium alloy.

cylinder bore *(Eng.)*. See **cylinder**.

cylinder brakes *(Print.)*. Mechanism on a fast-running printing press which stops it quickly.

cylinder caisson *(Civ.Eng.)*. A caisson formed of hollow cylindrical cast-iron sections arranged one on top of another, so that there is always one above water-level, while the bottom one is a special cutting section. As excavation proceeds within the cylinder, the loaded sections sink, and when they have reached a sufficient depth, the cylinder is filled with concrete.

cylinder collection *(Print.)*. On rotary presses, the gathering of the required number of sections or sheets round a cylinder.

cylinder cover *(Eng.)*. The end cover of the cylinder of a reciprocating engine or compressor.

cylinder dressing *(Print.)*. The layers of board, paper, or other packing material required to produce the necessary impression.

cylinder-dried *(Paper)*. Paper which has been dried by being passed over heated cylinders.

cylinder escapement *(Horol.)*. A frictional-rest escapement in which the balance is mounted on a hollow cylinder, and a tooth of the escape wheel gives impulse to the balance by pressing against the lips of the cylinder, the action being that of a wedge. The escape wheel is locked by the tip of the tooth pressing against the outside or inside of the cylinder. The teeth of the escape wheel are mounted on 'stalks' and stand at right angles to the plane of the wheel. To admit the entry of the teeth into the cylinder, about one-half of the cylinder is cut away where the teeth enter. Also known as *horizontal escapement*.

cylinder head *(Eng.)*. Removable top part of an IC engine which, when in place, provides a gas-tight seal for the cylinders. Contains valves, valve ports, combustion chambers and cooling water channels.

cylinder mould machine *(Paper)*. A paper or board machine in which the forming unit comprises an endless wire cloth on the surface of a hollow metal cylinder situated in a vat supplied with stock. Also *vat machine*.

cylinder press *(Print.)*. A general term used to distinguish cylinder printing machines from hand presses, platens and rotary machines. See **stop-cylinder, single-revolution, two-revolution**.

cylinder wrench *(Build.)*. See **pipe wrench**.

cylindrical co-ordinates *(Maths.)*. Three numbers, r, θ and z, which represent the position of a point in space, the first two numbers r and θ representing, in polar co-ordinates, the position of the projection of the point on a reference plane, and the third z representing the height of the point above the reference plane. Related to rectangular cartesian by the equations $x = r \cos\theta$, $y = r \sin\theta$ and $z = z$.

cylindrical gauge *(Eng.)*. A length gauge of cylindrical form whose length and diameter are made to some standard size. See **gauge (3)**.

cylindrical grinding *(Eng.)*. The operation of accurately finishing cylindrical work by a high speed abrasive wheel. The work is rotated by the headstock of the machine and the wheel is automatically traversed along it under a copious flow of coolant.

cylindrical lens *(Image Tech.)*. Lens element having one or both surfaces of cylindrical curvature, used in **anamorphic** systems or to produce a line image.

cylindrical record *(Acous.)*. The Edison-type of gramophone record, in which the reproducing needle traverses a helical (spiral) record on its surface.

cylindrical rotor *(Elec.Eng.)*. A rotor of an electric machine in which the windings are placed in slots around the periphery, so that the surface is cylindrical.

cylindrical wave *(Phys.)*. One where equiphase surfaces form coaxial cylinders.

cylindrical winding *(Elec.Eng.)*. A type of winding used for core-type transformers; it consists of a single coil of one or more layers wound concentrically with the iron core; it is usually long compared with its diameter.

cyma *(Arch.)*. A much-used moulding showing a reverse curve in profile. Also called an *ogee*.

cyma reversa, inversa *(Arch.)*. A cyma which is convex at the top and concave at the bottom.

cyme *(Bot.)*. See **cymose inflorescence**.

cymene *(Chem.)*. $CH_3C_6H_4CH(CH_3)_2$, *1-(1-methylethyl)-4 methylbenzene*, bp 175°C.

cymophane *(Min.)*. A variety of the gem-mineral *chrysoberyl* which exhibits chatoyancy; sometimes known as *chrysoberyl cat's eye* or *Oriental cat's eye*.

cymose inflorescence, cyme *(Bot.)*. An inflorescence in which the main stem and each subsequent branch ends in a flower, with any further development of the inflorescence coming from a lateral branch or lateral branches arising below the flower. Cf. **racemose inflorescence**.

cynopodous *(Zool.)*. Having nonretractile claws, as dogs.

Cyperaceae *(Bot.)*. The sedge family, ca 4000 spp. of monocotyledonous flowering plants (superorder Commelinidae). Mainly rhizomatous, perennial, grasslike herbs, cosmopolitan especially in temperate and arctic region, often in wet habitats. The aerial stems are typically solid, triangular in section and bear grass-like leaves in three ranks, the flowers are inconspicuous and wind-pollinated. The leaves and stems of some are used for making hats, baskets, mats and paper (papyrus) and for thatching. Includes the large genus *Carex* (1000 spp.).

cypress knee *(Bot.)*. A vertical upgrowth from the roots of swamp cypress (*Taxidium*) apparently a **pneumatophore**.

Cypriniformes *(Zool.)*. An order of Osteichthyes almost entirely inhabiting fresh water, with over 3000 species. Characins, Loaches and Carp.

Cys *(Chem.)*. Symbol for **cysteine**.

cyst *(Zool.)*. A nonliving membrane enclosing a cell or cells; any bladderlike structure, as the gall bladder or the urinary bladder of Vertebrates; a sac containing the products of inflammation. adjs. *cystic, cystoid, cystiform*.

cysteine *(Chem.)*. *2-amino-3-mercaptopropanoic acid*. $HS.CH_2CH(NH_2).COOH$. The L- or R- form of this amino acid is found in proteins, often in its oxidized form, **cystine**. Symbol Cys, short form C. ⇨

cystic *(Zool.)*. Pertaining to the gall bladder; pertaining to the urinary bladder.

cystic adenoma *(Med.)*. An adenoma containing numerous cysts.

cystic duct *(Zool.)*. The duct from the gall bladder which meets the hepatic duct to form the common bile duct.

cysticercosis *(Med.)*. Infection with cysticerci.

cysticercus *(Zool.)*. Bladderworm; larval stage in many tapeworms, possessing a fluid-filled sac containing an invaginated scolex.

cystic fibrosis *(Med.)*. Autosomal recessive genetic disorder causing abnormal viscid mucous production throughout the body but particularly the lungs. Leads to recurrent severe chest infections. Elevated sodium ions in sweat is a diagnostic factor.

cysticolous *(Zool.)*. Cyst-inhabiting.

cystidium *(Bot.)*. A swollen, elongated, sterile hypha, occurring among the basidia of the hymenium of some Hymenomycetes, usually projecting beyond the surface of the hymenium.

cystine *(Chem.)*. The dimer resulting from the oxidation of cysteine. The resulting disulphide bridge is an important structural element in proteins, as it often connects groups otherwise distant in the protein chain.

cystitis *(Med.)*. Inflammation of the bladder.

cystocele *(Med.)*. Hernia of the bladder.

cystogenous *(Zool.)*. Cyst-forming; cyst-secreting.

cystography *(Radiol.)*. The radiological examination of the urinary bladder following the administration intravenously or through the urethra of a **contrast medium**.

cystolith *(Bot.)*. Mass of calcium carbonate within a plant cell, on a stalk-like projection from the cell wall.

cystoscope *(Med.)*. An instrument for inspecting the interior of the bladder.

cystostomy *(Med.)*. Formation of an opening in the bladder.

cystotomy *(Med.)*. Incision into the bladder.

cystozooid *(Zool.)*. In Cestoda, the bladder or tail portion of a bladderworm. Cf. **acanthozooid**.

cytase *(Biol.)*. A general term for an enzyme able to break down the β–1→4 link of cellulose.

cytochimera *(Bot.)*. See **chromosome chimera**.

cytochromes *(Biol.)*. Proteins of the electron transfer chain which can carry electrons by virtue of their haem **prosthetic groups**. Cytochromes b, c1 and c have the same prosthetic group as haemoglobin. Cytochromes a and a3 have the related haem A and together form the terminal complex of the chain, cytochrome oxidase.

cytogenesis *(Biol.)*. The formation and development of cells.

cytogenetic map *(Biol.)*. See **chromosome map**.

cytogenetics *(Biol.)*. Study of the chromosomal complement of cells, and of chromosomal abnormalities and their inheritance.

cytokinesis *(Biol.)*. The contraction of an equatorial belt of cytoplasm which brings about the separation of two daughter cells during cell division of animal tissues. In plants the division of the cytoplasm as distinct from the nucleus. See **cell plate, cleavage**.

cytokinin *(Bot.)*. Any of a group of plant **growth substances**, derivatives of adenine e.g. zeatin, synthesized especially in roots and promoting cell division and bud formation, delaying senescence and, sometimes, promoting flowering and breaking dormancy. Also the artificial analogues of the above.

cytology *(Biol.)*. The study of the structure and functions of cells.

cytolysis *(Biol.)*. Dissolution of cells.

cytophilic antibody *(Immun.)*. Antibodies which bind to **Fc receptors** on the cell membrane.

cytoplasm *(Biol.)*. That part of the cell outside the nucleus but inside the **cell wall** if it exists.

cytoplasmic inheritance *(Biol.)*. Inheritance of traits coded for by the chloroplast or mitochondrial genomes, maternal because of the inheritance of chloroplasts and mitochondria through the egg rather than the sperm or male cell.

cytoplasmic male sterility *(Bot.)*. Lack of functional pollen as a maternally inherited trait resulting from a defective mitochondrial genome. See **male sterility**. Cf. **cytoplasmic inheritance**.

cytorrhysis *(Bot.)*. Process in which a plant cell wall collapses inwardly following water loss as a result of the exposure of the cell to a solution of a macromolecular solute to which the cell wall is impermeable, of higher osmotic pressure than that of the cell contents. Turgor will be zero or possibly negative. Cf. **Plasmolysis**.

cytosine *(Chem.)*. *6-aminopurine-2-one*. One of the five major bases found in nucleic acids. It pairs with guanine in both DNA and RNA. See **DNA, genetic code**. ⇨

cytoskeleton *(Biol.)*. Structures composed of protein which serve as skeletal elements within the cell, e.g. **microtubules, microfilaments**.

cytosol *(Biol.)*. The soluble components of the cytoplasm.

cytotaxis *(Biol.)*. Rearrangement of cells as a result of stimulation.

cytotaxonomy *(Bot.)*. The use of studies of chromosome number, morphology and behaviour in taxonomy.

cytotoxic *(Immun.)*. Able to kill cells. Applies to cytotoxic T-lymphocytes, to killer cells and **natural killer cells**, and also to damage mediated by **complement**.

cytotoxic antibiotic *(Med.)*. Drugs used in the treatment of cancer derived from antibiotics and which mimic radiotherapy *(radiomimetic drugs)*. Common examples are *doxorubicin* and *bleomycin*.

cytotoxic drug *(Med.)*. Term used for drugs used in the treatment of cancer. They include **alkylating drugs, cytotoxic antibiotics, antimetabolites, vinca alkaloids**.

cytotoxin *(Biol.)*. A toxin having a destructive action on cells.

cytotrophoblast *(Zool.)*. The inner layer of the trophoblast; layer of Langhans.

D

d- *(Chem.)*. Abbrev. for *dextrorotatory.*

δ- *(Chem.)*. Substituted on the fourth carbon atom of a chain.

D *(Chem.)*. Symbol for deuterium.

D *(Phys.)*. Symbol for (1) angle of deviation, (2) electric flux density (displacement), (3) diffusion coefficient.

Δ *(Chem.)*. Prefixed symbol for a double bond beginning on the carbon atom indicated.

[d] *(Phys.)*. A line in the blue of the solar spectrum, having a wavelength of 437.8720 nm due to iron.

[D] *(Phys.)*. A group of 3 Fraunhofer lines in the yellow of the solar spectrum. $[D_1]$ and $[D_2]$, wavelengths 589.6357 and 589.0186 nm, are due to sodium, and $[D_3]$, wavelength 587.5618 nm, to helium.

D-A *(Image Tech.)*. Digital-to-Analogue, referring to the conversion of signals.

dabbing *(Build.)*. See **daubing**.

dacite *(Geol.)*. A volcanic rock intermediate in composition between andesite and rhyolite, the volcanic equivalent of a granodiorite.

Dacron *(Chem.)*. American equivalent of **Terylene** synthetic fibre.

dacryo-adenitis *(Med.)*. Inflammation of the lacrimal (tear) gland.

dacryocystitis *(Med.)*. Inflammation of the lacrimal sac.

dacryocystorhinostomy *(Med.)*. Formation of a direct opening between the tear sac and the nose.

dactyl *(Zool.)*. A digit. adj. *dactylar*.

dactylitis *(Med.)*. Inflammation of a finger or of a toe.

dado *(Arch.)*. Panelling applied to the lower half of the walls of a room, or alternatively, decoration to give a similar effect.

dado *(Build.)*. A border around the lower part of the wall of a room.

dado capping *(Build.)*. The name given to the dado rail when the dado occupies as much as two-thirds of the height of the room.

dado plane *(Build)*. Type of grooving plane with two projecting spurs, one on each side at the front, and an adjustable depth stop. Used for making grooves for shelving etc. The spurs cut across the grain and keep the plane in its correct path. Also *trenching plane*.

dado rail *(Build.)*. The moulding capping the dado in a room and separating it from the upper part of the walls. Also *surbase*.

daft lamb disease *(Vet.)*. *Border disease, hairy shaker disease, hypomyelinogenesis imperfecta.* Uncommon disease of sheep and goats caused by a togavirus. Ewe may abort or a poor-viability lamb born; the latter can have a long coat. Lambs can also show deformity and unco-ordinated gait due to chronic contraction of muscles.

daguerreotype *(Image Tech.)*. Early process using a silvered copper plate sensitized by fuming with iodine and bromine vapour and developed with mercury vapour.

dailies *(Image Tech.)*. The same as **rushes**.

daisy-wheel printer *(Comp.)*. Printer in which characters are arranged near the ends of the spokes of a rimless wheel (on the 'petals' of a 'daisy'). Daisy wheels are manually interchangeable to enable alternative character sets to be used.

dalapon *(Chem.)*. *2,2-Dichloropropanoic acid*, used as a weedkiller.

d'Alembert's principle *(Phys.)*. On a body in motion, the external forces are in equilibrium with the inertial forces.

d'Alembert's ratio test *(Maths.)*. A series of positive terms converges or diverges respectively according to whether the limit of the ratio of a term to its predecessor is less or greater than unity. When the limit is unity the test is inconclusive.

Dalitz pair *(Phys.)*. Electron-positron pair produced by the decay of a free neutral pion (instead of one of the two gamma quanta normally produced).

DALR *(Meteor.)*. See **dry adiabatic lapse rate**.

Dalradian Series *(Geol.)*. A very thick and variable succession of sedimentary and volcanic rocks which have suffered regional metamorphism. Occurring in the Scottish Highlands approximately between the Great Glen and the Highland Boundary fault. Referred to the Precambrian System.

Daltonism *(Med.)*. See **colour blindness**.

Dalton's atomic theory *(Chem.)*. States that matter consists ultimately of indivisible, discrete particles (atoms), and atoms of the same element are identical; chemical action takes place as a result of attraction between these atoms, which combine in simple proportions. It has since been found that atoms of the chemical elements are not the ultimate particles of matter, and that atoms of different mass can have the same chemical properties (**isotopes**). Nevertheless, this theory of 1808 is fundamental to chemistry. See **atomic structure**.

Dalton's law *(Chem.)*. See **law of multiple proportions**.

Dalton's law of partial pressures *(Chem.)*. The pressure of a gas in a mixture is equal to the pressure which it would exert if it occupied the same volume alone at the same temperature.

dam *(Civ.Eng.)*. An embankment or other construction made across the current of a stream.

dam *(Min.Ext.)*. (1) A retaining wall or bank for water or tailings. (2) An air-tight barrier to isolate underground workings which are on fire.

damask *(Textiles)*. (1) A figured fabric made with satin and sateen weaves, in which background and figure have a contrasting effect; used mainly for furnishing. (2) Linen cloth of damask texture, used for tablecloths and towellings; also a cotton cloth of similar nature, used for tablecloths; both fabrics are reversible.

damped balance *(Chem.)*. Chemical balance using magnetic or air dash pots to bring it quickly to rest.

damped oscillation *(Phys.)*. Oscillation which dies away from an initial maximum asymptotically to zero amplitude, usually with an exponential envelope; e.g. the note from a struck tuning fork. See **damping, critical damping, decay factor**, and cf. **continuous oscillations**.

dampener *(Eng.)*. A device attached to both inlet and outlet sides of large reciprocating machinery to reduce pulsations which, if not reduced, would cause either mechanical damage or noise nuisance or both.

dampening rollers *(Print.)*. Rollers on a lithographic printing machine which are kept moist by a water supply, and by means of which the plate is dampened before being rolled by the ink rollers.

damper *(Aero.)*. Widely used term applied to devices for the suppression of unfavourable characteristics or behaviour; e.g., *blade damper*, to prevent the hunting of a helicopter rotor; *flame damper*, to prevent visual detection at night of the exhaust of a military aircraft; *shimmy damper*, for the suppression of *shimmy*; *yaw damper*, suppresses directional oscillations in high-speed aircraft, while a *roll damper* does likewise laterally, in both cases the frequency of the disturbances being too high for the pilot to anticipate and correct manually.

damper *(Autos.)*. Frictional or hydraulic device attached between the chassis and axles to prevent spring rebound and damp out oscillation. Formerly inaccurately referred to as *shock absorber*.

damper *(Elec.Eng.)*. Energy-absorbing component often

used for reducing the transmission of oscillatory energy from a disturbing source. Also called *amortisseur, damper winding, damping grid, damping winding.*

damper *(Eng.)*. (1) An adjustable iron plate or shutter fitted across a boiler flue to regulate the draught. (2) A device for damping out torsional vibration in an engine crankshaft, the energy of vibration being dissipated frictionally within the damper. See **vibration dampers**. (3) Device for stiffening the steering of a motor cycle to obviate wheel wobble.

damping *(Acous.)*. Transfer of sound energy into heat. There are different mechanisms; structure-borne sound is damped e.g. by molecular displacement processes, and air-borne sound by friction on interfaces. See **Stokes layer**.

damping *(Aero.)*. The capability of an aircraft of suppressing or resisting harmonic excitation and/or flutter. *Internal damping* is intrinsic to the materials, while *structural damping* is the total effect of the built-up structure. See **resonance test**.

damping *(Eng.)*. The checking of a (vibratory) motion by friction, etc. The damping capacity of cast iron makes this metal useful for lathe beds and other machine tool castings required to be rigid.

damping *(Phys.)*. Extent of reduction of amplitude of oscillation in an oscillatory system, due to energy dissipation, e.g., friction and viscosity in mechanical system, and resistance in electrical system. With no supply of energy, the oscillation dies away at a rate depending on the **degree of damping**. The effect of damping is to increase slightly period of vibrations. It also diminishes sharpness of resonance for frequencies in the neighbourhood of natural frequency of vibrator. See **logarithmic decrement**.

damping down *(Eng.)*. The temporary stopping of a blast furnace by closing all apertures by which air could enter.

damping factor *(Phys.)*. See **decay factor**.

damping magnet *(Elec.Eng.)*. A permanent magnet used to produce damping by inducing eddy currents in a metal disk or other body.

damping-off *(Bot.)*. Collapse and death of seedlings around emergence due usually to fungal attack by *Pythium* and *Fusarium* spp. when conditions are unfavourable for the seedlings.

dam plate *(Eng.)*. Iron vertical plate holding the wall of refractory brick (the dam stone) which forms the fore-hearth of a blast furnace.

damp-proof course *(Build.)*. A layer of impervious material, as plastic or bituminous sheeting built into a wall 15 to 25 cm above ground-level, to prevent moisture from the foundations rising in the walls by capillary attraction. Vertical damp courses are also used at door and window openings. Also used in chimneys and parapet walls to prevent downward passage of moisture.

damp-proofing *(Build.)*. The process of coating a wall with a special preparation to prevent moisture from getting through.

danburite *(Min.)*. A rare accessory mineral, occurring in pegmatites as yellow orthorhombic crystals. Chemically, danburite is a calcium borosilicate. $CaB_2Si_2O_8$.

dancing roller *(Print.)*. See **jockey roller**.

dancing step *(Build.)*. A step intermediate between a flier and a winder, having its outer end narrower in plan than its inner end. Also *balanced step*. Helps to form a better shaped handrail.

D and K *(Textiles)*. *Damaged and Kept*, usually by the dyer and finisher of the fabric.

dandy roll *(Paper)*. A hollow cylinder covered with wire cloth situated on top of the paper machine wire so that the surface of the roll makes contact with the upper surface of the wet web. The wire cloth may be such that a wove or laid pattern is imparted to the paper or names or other designs secured to it to produce corresponding watermarks in the paper.

dangerous semicircle *(Meteor.)*. The right-hand half of the storm field in the northern hemisphere, the left-hand half in the southern hemisphere, when looking along the path in the direction a **tropical revolving storm** is travelling. Cf. **navigable semicircle**.

Daniell cell *(Chem.)*. Primary cell with zinc and copper electrodes, the zinc rod being inserted in sulphuric acid contained within a porous pot, which is itself immersed in a copper pot containing copper (II) sulphate solution.

dannemorite *(Min.)*. A rare manganese-rich monoclinic amphibole, the name being used for the manganese-rich, iron-rich end member and differing from **tirodite** in having more iron than magnesium.

Dano composting plant *(Build.)*. A method of composting the organic wastes in domestic refuse with sewage sludge by a process of fermentation and grinding, which reduces the materials to a moist granulated condition. The product is a valuable soil conditioner because of its high humus content.

DAP *(Comp.)*. Distributed array processor. See **array processor**.

daphnite *(Min.)*. A variety of chlorite very rich in iron and aluminium.

dapsone *(Med.)*. Sulphone drug widely used in the treatment of **leprosy**.

daraf *(Phys.)*. Unit of elastance, the reciprocal of capacitance in farads. (*Farad* backwards).

darby *(Build.)*. A **derby float**.

darcy *(Geol.)*. A unit used to express the **permeability coefficient** of a rock, for example in calculating the flow of oil, gas or water. More commonly used is the *millidarcy* (mD), one-thousandth of a darcy (D).

Darcy's law *(Powder Tech.)*. A permeability equation which states that the rate of flow of fluid through a porous medium is directly proportional to the pressure gradient causing the flow.

dark burn fatigue *(Phys.)*. Decrease of efficiency of a luminescent material during excitation.

dark current *(Image Tech.,Phys.)*. Residual current in a photocell, video camera tube etc, when there is no incident illumination. The current depends on temperature.

dark ground illumination *(Biol.)*. A method for the microscopic examination of living material, e.g. micro-organisms, tissue culture cells, by scattered light. A special condenser with a circular stop illuminates the specimen with a numerical aperture larger than that collected by the objective. Specimens appear luminous against a dark background.

dark nebulae *(Astron.)*. Obscuring clouds of dust and gases, common throughout the Milky Way, and also observed in other galaxies. See **Coal Sack**.

dark reactions *(Bot.)*. Those reactions in photosynthesis in which CO_2 is fixed and reduced. They depend on energy and reducing power from the **light reactions**. See also **Calvin cycle**.

dark red heat *(Eng.)*. Glow emitted by metal at temperatures between 550° and 630°C.

dark red silver ore *(Min.)*. See **pyrargyrite**.

dark resistance *(Electronics)*. Resistance of a selenium or other photocell in the dark.

dark-room camera *(Print.)*. A built-in process camera controlled completely from inside the dark room, with the copy board and its illumination equipment outside.

dark slide *(Image Tech.)*. The carrier for plates to be exposed in cameras, loaded in the dark room and uncovered, after attachment to the camera, by withdrawing a slide.

dark space *(Electronics)*. See **anode-, Aston-, Crookes-, Faraday-dark space**.

dark trace screen *(Electronics)*. Screen which yields a dark trace under electron-beam bombardment.

DARPA *(Aero.,Comp.)*. Abbrev. for *Defense Advanced Research Projects Agency*, US. See **ARPA**.

dart *(Horol.)*. See **safety finger**.

dart *(Zool.)*. Any dartlike structure, e.g. in certain Snails, a small pointed calcareous rod which is used as an incentive to copulation; in certain Nematoda, a pointed weapon used to obtain entrance to the host.

Darwinian theory *(Biol.)*. See **natural selection**.

dash pot *(Eng.)*. A device for damping-out vibration or for allowing rapid motion in one direction but only much slower motion in the opposite direction. It consists of a piston attached to the part to be controlled (fitted with a nonreturn valve if required) sliding in a cylinder containing liquid to impede motion.

dasypaedes *(Zool.)*. Birds which when hatched have a complete covering of down. Cf. **altrices**.

data *(Comp.)*. All the **operands** and results of computer operations directed by the detailed **instructions** comprising the **program**. A program can be *data* for another program, e.g., a **compiler** takes a program as data.

data bank *(Comp.)*. Collection of **databases** or large files of data.

data base *(Comp.)*. Collection of structured **data** independent of any particular application.

database management system *(Comp.)*. Software that handles the storage, retrieval and updating of data in a computer, often integrating data from a number of files. Also *DBMS*. See **data model**.

database typesetting *(Print.)*. The storing of information in a database for publications such as directories which can be periodically updated by computer processing and prepared for phototypesetting.

data capture *(Comp.)*. Collecting data for use in a particular computer process, e.g., for monitoring.

data compaction *(Comp.)*. Term often applied to **data compression** that involves only the removal of extraneous and unnecessary space and therefore is not reversible.

data compression *(Comp.)*. Altering the form of data to reduce its storage space.

data dictionary *(Comp.)*. Index of the contents of a set of files or a database. See **directory**.

Data Encryption Standard *(Comp.)*. An automatic method of data **encryption** designed by IBM and adopted as a standard.

data flow *(Comp.)*. An approach to the organization of complex algorithms and machines, in which operations are triggered by the arrival of data.

data flowchart *(Comp.)*. **Flowchart** used to describe a complete processing system, clerical operations and individual programs, but excluding details of such programs. Also *system flowchart*.

data handling *(Space)*. The management and flow of data to-and-from a space vehicle; the on-board subsystem might include data buses, commutators, computers, recorders, multiplexers, etc. whereas the ground segment uses equipment like de-multiplexers and display units to interpret the transmitted signal which is sent either directly or *via* a data relay satellite.

data-handling capacity, capability *(Telecomm.)*. The maximum amount of information which can be transmitted and received over a given channel or circuit.

data-handling system *(Comp.)*. Term, no longer widely used, for automatic or semi-automatic equipment for collecting, receiving, transmitting, and storing numerical data. It may be handled continuously (as analogue or position signals) or in discrete steps (as digital or binary signals). The system may also be able to perform calculations on the stored data.

data model *(Comp.)*. A structure for the arrangement of data which aids data retrieval. There are three models in general use, *a hierarchic model*, *a network model* and one giving a **relational data base**.

data preparation *(Comp.)*. Translation of data into machine readable form.

data processing *(Comp.)*. Traditional name given to business information processing. Abbrev. *DP*.

data protection *(Comp.)*. Safeguards to protect the integrity, privacy and security of data.

data reduction *(Comp.)*. The computerized repackaging of observational data to make it more concise and meaningful.

data retrieval *(Comp.)*. The search for and selection of data from a store.

data signalling rate *(Telecomm.)*. The aggregate rate at which binary digits, including any control bits, are transmitted over a channel or circuit, expressed in bits/second. Cf. **baud**.

data storage *(Comp.)*. See **memory capacity**.

data structure *(Comp.)*. Organized form in which grouped data items are held in the computer, such as **list**, **tree**, **table**, **string**.

data type *(Comp.)*. Most programming languages require a variable to be declared as a *data type*. Basic restrictions and assumptions will then control the use of the variable. See **character**, **tree**, **stack**, **queue**, **set**, **real**, **integer**, **Boolean**, **string**, **list**.

Datel *(Comp.)*. TN for data transmission facilities provided by British Telecom.

dative bond *(Chem.)*. See **covalent bond**.

datolite *(Min.)*. Hydrated calcium borosilicate occurring as a secondary product in amygdales and veins, usually as distinct prismatic white or colourless monoclinic crystals.

datum *(Aero.)*. *Datum level*, or *rigging datum*, is the horizontal plane of reference, in flying attitude, from which all vertical measurements of an aircraft are taken; *cg datum* is the point from which all mass moment arms are measured horizontally when establishing the centre of gravity and loading of an aircraft.

datum *(Eng.)*. A point, line or surface to which dimensions are referred on engineering drawings and from which measurements are taken in machining or other engineering operations.

datum *(Surv.)*. An assumed surface used as a reference surface for the measurement of reduced levels.

daubing *(Build.)*. (1) The operation of dressing a stone surface with a special hammer in order to cover it with small holes. (2) A rough-stone finish given to a wall by throwing a rough coating of plaster upon it. See **rough-cast**.

daughter *(Biol.)*. Offspring belonging to the first generation, whether male or female; as *daughter cell*, *daughter nucleus*.

daughter product *(Phys.)*. A nuclide that originates from the radioactive disintegration of another *parent* nuclide.

Davis apparatus *(Ships)*. A respiratory apparatus specially designed to permit escape from a pressure-equalizing chamber in a submarine. Oxygen is breathed from a chamber which, embracing the wearer, gives buoyancy and assists rise to the surface.

Davisson-Germer experiment *(Electronics)*. The first demonstration (1927) of wavelike diffraction patterns from electrons by passing them through a nickel crystal.

Davy lamp *(Min.Ext.)*. The name of the safety lamp invented by Sir Humphrey Davy in 1815.

day *(Astron.)*. See **apparent solar-**, **mean solar-**, **sidereal-**.

daylight *(Eng.)*. The distance between the bed surface and the bottom of the ram of a press.

daylight factor *(Elec.Eng.)*. The ratio of the illumination measured on a horizontal surface inside a building to that which obtains at the same time outside the building, due to an unobstructed hemisphere of sky. Occasionally called *window efficiency ratio*.

daylight lamp *(Phys.)*. A lamp giving light having a spectral distribution curve similar to that of ordinary daylight.

day-light size *(Build.)*. The distance between successive mullions in a window and between lintel and sill.

day-neutral plant *(Bot.)*. A plant in which flowering is not sensitive to day-length. Cf. **long-day plant**, **short-day plant**. See also **photoperiodism**.

dB *(Acous.,Telecomm.)*. Abbrev. for *decibel*.

dBA, dBB, dBC *(Acous.)*. Result of a **sound pressure level** measurement when the signal has been weighted with a frequency response of the A, B, or C curve. The dBA curve approximates the human ear and is therefore used most in noise control regulations.

dBm *(Telecomm.)*. A unit for expressing power level in decibels, relative to a reference level of one milliwatt.

DBMS *(Comp.)*. See **database management system**.

DBS *(Image Tech.)*. *Direct Broadcasting by Satellite*.

DBS *(Telecomm.)*. Abbrev. for *Direct Broadcast Satellite*.

d.c. *(Elec.Eng.)*. Abbrev. for **direct current**.

d.c. *(Print.)*. Abbrev. for *double column*; *double crown*.

d.c. amplifier *(Elec.Eng.)*. One which uses direct coupling between stages (i.e. no blocking capacitor) to amplify from zero frequency (d.c.) signals to signals of higher frequency.

d.c. balancer *(Elec.Eng.)*. The coupling and connecting of two or more similar direct-current machines, so that the conductors connected to the junction points of the machines are maintained at constant potentials.

d.c. bias *(Elec.Eng.)*. (1) In an electronic amplifier, the direct signal applied to an active component which sets the quiescent conditions for the device. Thereafter, an a.c. signal may be applied. (2) In a magnetic tape recorder, the addition of a polarizing direct current in the signal recording to stabilize magnetic saturation.

d.c. bridge *(Elec.Eng.)*. A four-arm null bridge energized by a d.c. supply. The prototype is the **Wheatstone bridge**, other examples the **metre bridge** and the **Post Office box**.

d.c. component *(Image Tech.)*. That part of the picture signal which determines the average or datum brightness of the reproduced picture.

d.c. converter *(Elec.Eng.)*. A converter which changes direct current from one voltage to another.

d.c. coupling *(Electronics)*. See **direct coupling**.

d.c./d.c. converter *(Elec.Eng.)*. A d.c. voltage transformer using an inverter and rectifier. Also called *d.c. transformer*.

DCF *(Build.)*. Abbrev. for *Deal-Cased Frame*.

d.c. generator *(Elec.Eng.)*. A rotary machine to convert mechanical into direct current power.

d.c. meter *(Elec.Eng.)*. One which responds only to d.c. component of a signal, e.g. moving coil instruments.

d.c. resistance *(Elec.Eng.)*. The resistance which a circuit offers to the flow of a direct current. Also called *true (ohmic) resistance*.

d.c. restoration *(Image Tech.)*. Re-insertion of a d.c. or very low frequency component which has been lost or reduced in transmission; in a TV receiver the use of a **clamp** to hold the level of the d.c. component.

d.c. testing of cables *(Elec.Eng.)*. The application of a d.c. voltage of 5 times the r.m.s. of the working a.c. voltage. Cables which have considerable tracking and are likely to break down in service are broken down by the d.c.; healthy cables are not affected.

d.c. transformer *(Elec.Eng.)*. (1) Device to measure large direct currents by means of associated magnetic field. (2) Colloq. for *d.c./d.c. converter*.

d.c. transmission *(Elec.Eng.)*. Method of connecting together different power generating systems for sending and receiving bulk quantities of electricity when a.c. is not attractive. Losses are lower, insulation is used more effectively, steady slate charging current is zero (important if cables are used for interconnection) and different power systems do not need to be synchronized. Disadvantage is cost of converter equipment at both sending and receiving ends.

d.c. transmission *(Image Tech.)*. Inclusion of d.c. or very low frequency in the transmitted video signal. If omitted, it has to be *restored* in relation to the **pedestal** in the receiver.

DDL *(Comp.)*. Data description languages. Used in a **DBMS**.

DDT *(Chem.)*. Abbrev. for a complex chemical mixture, in which *pp'*-dichlorodiphenyltrichloroethane predominates; a synthetic insecticide remarkable for high toxicity to insects at low rates of application. A **stomach insecticide** and **contact insecticide** with a very long persistence of activity from residual deposits, which have caused it to be banned in many countries.

deactivation *(Chem.)*. The return of an activated atom, molecule, or substance to the normal state. See **activation (2)**.

dead *(Acous.)*. An enclosure which has a period of reverberation much smaller than usual for its size and audition requirements. Applied to sets in motion-picture production.

dead *(Build.)*. Said of materials which have deteriorated.

dead angle *(Eng.)*. That period of crank angle of a steam-engine during which the engine will not start when the stop-valve is opened; due to the ports being closed by the slide-valve.

dead axle *(Eng.)*. An axle which does not rotate with the wheels carried by it. Cf **live axle**.

dead bank *(Eng.)*. A stoker-fired boiler furnace from which the coal feed is shut off, the fire being allowed to burn back as far as possible without going out entirely.

dead-beat compass *(Ships)*. A magnetic compass with a short period of oscillation and heavily damped so that it comes to rest very quickly.

dead-beat escapement *(Horol.)*. An escapement in which there is no 'recoil' to the escape wheel. The dead-beat action is obtained by making the locking faces of the pallets arcs of circles, struck from the pallet staff as centre. This escapement is the one used for regulators, and is capable of giving very accurate results.

dead burnt *(Eng.)*. Descriptive of such carbonates as limestone, dolomite, magnesite, when they have been so kilned that the associated clay is vitrified, part or all of the volatile matter removed and the slaking quality lowered.

dead-centre *(Eng.)*. (1) Either of the two points in the crankpin path of an engine at which the crank and connecting-rod are in line and the piston exerts no turning effort on the crank. See **inner (top) dead-centre**, **outer (bottom) dead-centre**. (2) A lathe centre. See **tail-stock** and **centre**.

dead-centre lathe *(Eng.)*. A small lathe (used in instrument-making) in which both centres are fixed, the work being revolved by a small pulley mounted on it.

dead coil *(Elec.Eng.)*. A coil in the winding of a machine which does not contribute any e.m.f. to the external circuit, because it is short-circuited or disconnected from the rest of the winding. See also **dummy coil**.

dead earth *(Elec.Eng.)*. A connection between a normally live conductor and earth by means of a path of very low resistance.

dead end *(Telecomm.)*. The unused portion of an inductance coil in an oscillatory circuit.

dead-ended feeder *(Elec.Eng.)*. See **independent feeder**.

dead end, leg *(Build.)*. The length of pipe between a closed end and the nearest connection to it, forming a 'dead' pocket in which there is no circulation.

dead-end tower *(Elec.Eng.)*. See **terminal tower**.

deadening *(Build.)*. (1) The operation of dealing with a surface, so as to give it a dead finish. (2) **Pugging**.

dead eye *(Eng.)*. (1) A sheaveless block used in setting up rigging. (2) A light type of bearing for supporting a spindle; it may consist merely of a hole in a sheet of metal or other material.

dead fingers *(Med.)*. A disease of pneumatic-drill operators, causing cyanosis, anaesthesia of finger-tips, and sometimes bone absorption.

dead finish *(Build.)*. A dull or rough finish particularly, in painting, a **flat finish**.

dead flue *(Build.)*. A flue which is bricked in at the bottom.

dead ground *(Min.Ext.)*. Ground devoid of values: ground not containing veins or lodes of valuable mineral: a barren portion of a coal seam. Also **deads**.

dead knot *(For.)*. A knot which is partially or wholly separated from the surrounding wood.

dead letters *(Print.)*. Letters remaining in the type case when setting has to stop because the supply of one *sort* has been exhausted.

dead load *(Civ.Eng.)*. The weight of a structure with finishings, fixtures and partitions. Cf. **live load**.

dead lock *(Build.)*. A lock the bolt of which is key operated from one or both sides as opposed to spring bolt or latch.

dead-man's handle *(Elec.Eng.)*. A form of handle commonly used on the controllers of electric vehicles; designed so that if the driver releases his pressure on the handle, owing to sudden illness or other causes, the current is cut off and the brakes applied. *Dead-man's pedal* is a similar foot-operated safety device.

dead matter *(Print.)*. Type which has been used and is awaiting distribution.

deadmen *(Civ.Eng.)*. The concrete, plate, or other anchorage for land ties.

dead oil *(Min.Ext.)*. Crude oil without dissolved gas.

dead points *(Eng.)*. See **dead-centre (1)**.

dead rise *(Aero.)*. At any cross-section of a flying-boat hull or seaplane float, the vertical distance between the keel and the chine.

dead roasting *(Eng.)*. Roasting carried out under conditions designed to reduce the sulphur content, or that of the other volatile matter, to the lowest possible value. Distinguished from **partial roasting** and **sulphating roasting**.

dead room *(Acous.)*. See **anechoic room**.

deads *(Min.Ext.)*. (1) Same as **dead ground**. (2) Waste (*back fill*) used to support roof.

dead segment *(Elec.Eng.)*. A commutator segment which is not connected either for accidental reasons or for a definite purpose, to the armature winding associated with the commutator.

dead shore *(Build.)*. A vertical post used to prop up temporarily any part of a building.

dead-smooth file *(Eng.)*. The smoothest grade of file ordinarily used, having 70 to 80 teeth to the inch for files of average length; used for finishing surfaces.

dead sounding *(Build.)*. **Pugging**.

dead spot *(Telecomm.)*. A region where the reception of radio transmissions over a particular frequency range is extremely weak.

dead time *(Phys.)*. Time after ionization during which a detector cannot record another particle. Reduced by a *quench* as in Geiger-Müller counters. When the dead time of a detector is variable a fixed electron dead time may be incorporated in subsequent circuits. Also called *insensitive time*.

dead-time correction *(Nuc.Eng.)*. Correction applied to the observed rate in a nuclear counter to allow for the probable number of events occurring during the dead time.

dead water *(Eng.)*. In a boiler or other plant, water not in proper circulation.

dead weight *(Ships)*. The difference, in tons, between a ship's displacement at load draught and light draught. It comprises cargo, bunkers, stores, fresh water etc.

dead-weight pressure-gauge *(Eng.)*. A device in which fluid pressure is measured by its application to the bottom of a vertical piston, the resulting upward force being then balanced by applying weights to the upper end; used for calibrating Bourdon gauges.

dead-weight safety valve *(Eng.)*. A safety valve in which the valve itself is loaded by a heavy metal weight; used for small valves and low pressures. See **safety valve**.

dead well *(Civ.Eng.)*. An **absorbing well**.

de-aerator *(Eng.)*. A vessel in which boiler feed water is heated under reduced pressure in order to remove dissolved air.

deaf aid *(Acous.)*. A device used by a person with hearing loss to improve audition of external sounds, either in the form of an acoustic amplifier (collector), or in the form of a microphone-receiver combination, with or without amplifier.

deafening *(Build.)*. **Pugging**.

deafness *(Med.)*. Lack of sensitivity of hearing in one or both ears, with consequent increase in the threshold of minimum audibility, measurement of which is useful in diagnosis.

deal *(For.)*. (1) A piece of timber of cross-section roughly 10 × 3 in (250 × 75 mm). (2) See **white deal**.

death *(Biol.)*. In a cell or an organism, complete and permanent cessation of the characteristic activities of living matter.

deathnium centre *(Electronics)*. Crystal lattice imperfection in semiconductor at which it is believed electron hole pairs are produced or recombine.

debacle, débâcle *(Meteor.)*. The breaking up of the surface ice of great rivers in spring.

débridement *(Med.)*. The removal of foreign matter and excision of infected and lacerated tissue from a wound.

debris flow *(Geol.)*. A mass movement involving rapid flow of various kinds of debris especially in mud. May be associated with earthquakes and volcanic eruptions (e.g. South America) or with excessive rainfall on unstable material (e.g. Aberfan disaster). See **mud flow**.

de Broglie wavelength *(Phys.)*. That associated with a particle by virtue of its motion, i.e. $\lambda = h/p$ where λ is the wavelength, h is Planck's constant and p the particle's relativistic momentum. Only for electrons and other elementary particles can the de Broglie wavelength be large enough to produce observable diffraction effects. See **electron diffraction**.

debug *(Comp.)*. To detect, locate and correct every **bug**. See **error, diagnostic**.

debunching *(Electronics)* Tendency for a beam of electrons or a velocity-modulated beam of electrons to spread because of mutual repulsion. See **buncher, bunching**.

deburring *(Textiles)*. Mechanical removal of dirt and vegetable matter from raw wool.

Debye and Scherrer method *(Crystal.)*. A method of X-ray crystal analysis applicable to powders of crystalline substances or aggregates of crystals.

Debye-Hückel theory *(Chem.)*. A theory of electrolytic conduction which assumes complete ionization and attributes deviations from ideal behaviour to inter-ionic attraction.

Debye length *(Phys.)*. Maximum distance at which coulomb fields of charged particles in a plasma may be expected to interact.

Debye temperature *(Phys.)*. The Debye theory of specific heats of solids can be usefully applied to many solids using only one fitting parameter, the Debye temperature.

Debye theory of specific heats of solids *(Phys.)*. Based on the assumption that the thermal vibrations of the atom of a solid can be presented by harmonic oscillators whose energies can be quantized. The oscillator frequencies are distributed, up to a maximum (cut-off) frequency, according to the normal modes of vibration of a *continuous* medium. For many substances the theory gives a satisfactory agreement with experiment over a wide range of temperature. See **Blackman theory of specific heats of solids**.

Debye unit *(Phys.)*. Unit of electric dipole moment equal to 3.34×10^{-30} C m or 10^{-18} e.s.u.

Debye-Waller factor *(Phys.)*. The intensities of coherently scattered X-rays or neutrons from a crystal are reduced by the thermal vibrations of the atoms. If it is assumed that the thermal vibrations are isotropic then the intensities are reduced by the Debye-Waller factor $\exp(-2B\sin\theta/\lambda^2)$ where 2θ is the radiation and B is the temperature factor and generally is in the range 2–3 K.

decade *(Phys.)*. Any ratio of 10:1. Specifically the interval between frequencies of this ratio.

decade box *(Elec.Eng.)*. A resistance (capacitance or inductance) box divided into sections so that each section has 10 switched positions and 10 times the value of the preceding section. The switches can therefore be set to any integral value within the range of the box.

decahydro-naphthalene. *(Chem.)*. See **decalin**.

decalcification *(Med.)*. The process of absorption of lime salts from bone.

decalcomania paper *(Paper)*. A transfer paper for conveying a design on to pottery, etc.

decalescence *(Eng.)*. The absorption of heat that occurs when iron or steel is heated through the arrest points. See **recalescence**.

decalin *(Chem.)*. *Decahydro-naphthalene*, $C_{10}H_{18}$, bp 190°C (103 kN/m^2), product of complete hydrogenation of naphthalene under pressure and in the presence of a catalyst.

decametric waves *(Telecomm.)*. Waveband from 10 to 100 m.

decant *(Chem.)*. To pour off the supernatant liquor when a suspension has settled.

Decapoda *(Zool.)*. (1) An order of Malacostraca with 3 pairs of thoracic limbs modified as maxillipeds, and 5 as walking legs. Shrimps, prawns, crabs, lobsters etc. (2) A suborder of Cephalopoda having 8 normal arms and 2 longer partially retractile arms; the suckers are pedunculate, there is a well-developed internal shell, and lateral fins are present; actively swimming forms, usually carnivorous. Squids and Cuttlefish.

decapsulation *(Med.)*. Removal of capsule or covering of an organ, especially of the kidney.
decarboxylase *(Bot.)*. An enzyme that catalyses the removal of CO_2 from its substrate.
decarburization *(Eng.)*. Removal of carbon from the surface of steel by heating in an atmosphere in which the concentration of decarburizing gases exceeds a certain value.
decatizing *(Textiles)*. Process for imparting a permanent finish to worsted and woollen fabrics by forcing steam through them while under tension, to improve their appearance and handle.
decay *(Radiol.)*. The process of spontaneous transformation of a radionuclide.
decay constant *(Phys.)*. See **disintegration constant**.
decay factor *(Phys.)*. That which expresses rate of decay of oscillations in damped oscillatory system, given by natural logarithm of ratio of two successive amplitude maxima divided by the time interval between them. Calculated from ratio of resistance coefficient to twice mass in a mechanical system, and ratio of resistance to twice inductance in an electrical system. Also called *damping factor*. See **logarithmic decrement**.
decay heat *(Nuc.Eng.)*. The heat produced by the radioactive decay of fission products in a reactor core. This continues to be produced even after the reactor is shut down.
decay law *(Phys.)*. If for a physical phenomenon, the rate of decrease of a quantity is proportional to the quantity at that time, then the decay law is an exponential, i.e. $a = a_0 e^{-\lambda t}$ where λ is the *decay constant* and a_0 is the value at time $t = 0$. E.g. activity in a radioactive substance, sound intensity in an enclosure, discharge of capacitor.
decay product *(Phys.)*. See **daughter product**.
decay time *(Phys.)*. That in which the amplitude of an exponentially decaying quantity reduces to e^{-1} (36.8%) of its original value.
Decca Navigator *(Aero.,Ships)*. TN for a navigation system of the radio position fixing type using continuous waves. The **fix**, given by the intersection of two hyperbolic position lines, is indicated on meters and can be plotted on a Decca chart or on a flight log which gives a continuous pictorial presentation of position. See **Dectra**.
decelerating electrode *(Electronics)*. One which is intended to reduce the velocity of electrons.
deceleration *(Phys.)*. Negative acceleration. The rate of diminution of velocity with time. Measured in metres (or feet) per second squared.
decerebrate *(Zool.)*. Lacking a cerebrum.
decerebrate rigidity *(Med.)*. A posture of extensor rigidity of trunk and limbs seen in man with disease or lesion of the brain-stem.
decerebrate tonus *(Zool.)*. A state of reflex tonic contraction of certain skeletal muscles following upon the separation of the cerebral hemispheres from the lower centres.
deci- *(Genrl.)*. Prefix with physical unit, meaning one-tenth.
deci-ampere balance *(Elec.Eng.)*. An ampere-balance having a range 0.1 to 10 amperes.
decibel *(Acous.)*. 10 times the logarithm to base 10 of an energy ratio. E.g. the sound pressure level is measured in decibels and defined as $10 \log p^2/p_0^2$, where p is the r.m.s sound pressure and p_0 is a reference sound pressure. For air-borne sound, p_0 is the *threshold of hearing*.
decibel *(Telecomm.)*. One tenth of a *bel*, signifying the ratio of two amounts of power; given by $n = 10 \log_{10} P_1/P_2$, where n is the ratio expressed in decibels, and P_1 and P_2 represent the power levels being compared. Used almost universally as a measure of response or performance in electronic and communication circuits. Under appropriate conditions, usually identical impedance in input and output circuits, the ratio may be expressed as $n = 20 \log_{10} V_1/V_2$ where V_1 and V_2 are the voltage levels of the signals being compared, or as the equivalent current ratio; the latter interpretation is prone to confusion.
decibel meter *(Acous.)*. Meter which has a scale calibrated in logarithmic steps and labelled with decibel units.
decidua *(Zool.)*. In Mammals, the modified mucous membrane lining the uterus at the point of contact with the placenta, which is torn away at parturition and then ejected; the afterbirth; the maternal part of the placenta.
deciduate *(Zool.)*. Said of Mammals in which the maternal part of the placenta comes away at birth. Cf. **indeciduate**.
deciduous *(Bot.)*. (1) Falling off, usually after a lengthy period of functioning. (2) Plants which shed leaves habitually before a cold period. Ant. *evergreen*.
deciduous *(Med.)*. Falling off; said of the first dentition teeth.
decimal fraction *(Maths.)*. A fraction having a power of ten as denominator. The denominator is not usually written but is indicated by the decimal point, a dot (on the Continent, a comma) placed between the unit figure and the numerator. The number of figures after the decimal point is equal to the power of 10 of the denominator. Thus 42.017 is equal to

$$42\tfrac{17}{1000}$$

Also *decimal, decimal number*.
decimal system *(Maths.)*. A number system whose base is 10 and in which every fraction is expressed as a decimal number. In this system each unit is in theory 10 times the next smaller one (in practice not necessarily so; cf. most decimal coinage systems). The **metric system** and **Système International** are decimal systems.
decimetre *(Genrl.)*. One-tenth of a **metre**.
decimetric waves *(Telecomm.)*. Waveband having range from 10 cm to 1 m.
decimolar calomel electrode *(Chem.)*. A calomel electrode containing 0.1 M potassium chloride solution.
decimo octavo *(Print.)*. See **18mo**.
decimo sexto *(Print.)*. See **16mo**.
decineper *(Phys.)*. Unit of voltage and current attenuation in lines and amplifiers, of magnitude one-tenth of the **neper**. Defined by $d = \log_e(x_1/x_2)$, where d is number of decinepers, x_1 and x_2 are currents (voltages or acoustic pressures). In properly terminated networks, 1 decineper equals 0.8686 dB. Abbrev. *dN*.
decision speed *(Aero.)*. See **critical speed**.
decitex *(Textiles)*. See **tex**.
deck *(Image Tech.)*. The assembly of transport mechanism and transducer heads for a disk or magnetic tape recording/reproducing system.
deck *(Ships)*. A platform which forms the top of one horizontal division of a ship and the bottom of that immediately above.
deck beam *(Ships)*. A stiffening member of a deck, which may be either transverse or longitudinal. It is supported at extremities by knee connections to frames or bulkheads or by supporting girders.
deck bridge *(Civ.Eng.)*. A bridge in which the track is carried by the upper stringer. Cf. **through bridge**.
deck crane *(Ships)*. A crane, either fixed or movable, mounted on deck for use in loading or discharging cargo.
deck houses *(Ships)*. See **top hamper**.
decking *(Build.)*. The platform supporting the derrick on a derrick tower gantry.
deckle edge *(Paper)*. A rough feather edge on the four sides of a hand-made sheet due to stock seeping beneath the deckle. Similar effects can be simulated artificially. Also the irregular edges of a web of paper before trimming.
decks *(Print.)*. On rotary presses, pairs of horizontal printing couples arranged one above the other.
deck stringer *(Ships)*. The main strength portion of a ship's deck, being that portion, on both sides, adjacent and attached to the shell plating. It comprises the stringer strake of plating and the stringer angle section forming such attachment.
declaration *(Comp.)*. Statement in a **high level language** that has the form of descriptive information rather than

an explicit instruction. Also *declarative statement*. See **data type**.

declared efficiency *(Elec.Eng.)*. The efficiency which the manufacturers of an electric machine or transformer declare it to have, under certain specified conditions.

declination *(Astron.)*. The angular distance of a heavenly body from the celestial equator measured positively northwards along the hour circle passing through the body.

declination *(Surv.)*. See **magnetic declination**.

declination circle *(Astron.)*. A graduated circle on the declination axis of an equatorial telescope which enables the telescope to be set to a given declination or the declination of a given star to be read.

declinimeter *(Elec.Eng.)*. An apparatus for determining the direction of a magnetic field with respect to astronomical or survey coordinates.

declutch *(Eng.)*. The action of separating the driving from the driven member of a clutch to prevent or interrupt the transmission of (rotary) motion.

DECnet *(Comp.)*. A widely used commercial network provided by Digital Equipment Corporation.

decoction *(Chem.)*. An extract of a substance or substances obtained by boiling.

decoder *(Telecomm.)*. (1) Any circuit which responds to a particular coded signal while rejecting others. (2) A circuit which converts coded information (e.g. *pulse-code-modulated* speech) into analogue information. (3) That part of a television receiver which separates the incoming colour information into the red, green and blue components necessary to operate the cathode-ray tube. (4) In a frequency-modulated broadcast receiver, the circuit which separates the stereophonic sound signal into left and right channels.

decollate *(Comp.)*. To separate the sheets of multipart continuous stationery.

decolorize *(Chem.)*. To remove the coloured material from a liquid by bleaching, precipitation or adsorption.

decolorizers *(Glass)*. Materials added to the batch for the express purpose of improving the appearance of the glass by hiding the yellow-green colour due to iron impurities.

decommissioning *(Nuc.Eng.)*. The permanent withdrawal from service of a nuclear facility and the subsequent operations to bring it to a safe and stable condition.

decompensation *(Med.)*. Failure of an organ to compensate for functional overload produced by a disease.

decomposers *(Ecol.)*. In an ecosystem, heterotrophic organisms, chiefly bacteria and fungi, which break down the complex compounds of dead protoplasm, absorbing some of the products of decomposition, but also releasing simple substances usable by producers. Cf. **consumers**, **producers**.

decomposition *(Chem.)*. The breaking down of a substance into simpler molecules or atoms.

decomposition voltage *(Elec.Eng.)*. The minimum voltage which will cause continuous electrolysis in an electrolytic cell.

decompound *(Bot.)*. Two or more times *compound*.

decompression *(Med.)*. Any procedure for relieving pressure or the effects of pressure.

decondensed chromatin *(Biol.)*. See **euchromatin**.

deconjugation *(Biol.)*. The separation of the paired chromosomes before the end of the prophase meiosis.

decontamination *(Nuc.Eng.)*. Removal of radioactivity from area, building, equipment or person to reduce exposure to radiation. More generally, removal or neutralization of bacteriological, chemical or other contamination.

decontamination factor *(Nuc.Eng.)*. Ratio of initial to final level of contamination for a given process.

decorative laminate *(Plastics)*. Laminates with a highly resistant, frequently decorative, surface based on melamine resins coated on resin impregnated paper.

decorticated *(Bot.)*. Deprived of bark; devoid of cortex.

decortication of the lung *(Med.)*. Operative removal from the lung of pleura thickened as a result of chronic inflammation.

decoupling *(Electronics)*. Reduction of a common impedance between parts of a circuit, e.g. by using a **by-pass capacitor**.

decoupling filter *(Electronics)*. Simple resistor-capacitor section(s), which decouple feedback circuits and so prevent oscillation or **motor-boating** (relaxation oscillation).

decrement *(Phys.)*. Ratio of successive amplitudes in a damped harmonic motion.

decrepitation *(Chem.)*. The crackling sound made when crystals are heated, caused by internal stresses and cracking.

decrepitation *(Powder Tech.)*. Breakdown in size of the particles of a powder due to internal forces, generally induced by heating.

decryption *(Comp.)*. The recovery of a plain message from an encrypted one.

Dectra *(Aero.,Ships)*. A radio position fixing system, based largely on **Decca Navigator** principles, designed to cover specific air route segments and trans-oceanic crossings. In addition to fix, location along a track and range information are given: hence, *Decca, Track and Range*.

decubitus *(Med.)*. The special or preferred posture in bed of a patient suffering from particular disease states. Patients with *pleuritic pain* lie on the affected side.

decubitus ulcer *(Med.)*. An ulcer or bed sore developing from prolonged immobility.

decumbent *(Bot.)*. A stem, lying flat, usually with a turned-up tip.

decurrent *(Bot.)*. (1) Running down, as when a leaf margin continues down the stem as a wing. (2) With several roughly equal branches, as in shrubs and in the crowns of some trees, especially when old. Cf. **excurrent**.

decussate *(Bot.)*. With leaves in opposite pairs, each pair being at right angles to the next. See **phyllotaxis**.

decussate texture *(Geol.)*. The random arrangement of prismatic or tabular crystals in a rock.

decussation *(Zool.)*. Crossing over of nerve-tracts with interchange of fibres.

Dedekind cut *(Maths.)*. A division of the rational numbers into 2 classes such that all numbers in one are greater than all numbers in the other; used to define irrational numbers.

dedendum *(Eng.)*. (1) Radial distance from the pitch circle of a gear-wheel to the bottom of the spaces between teeth. See **involute gear teeth**, **pitch diameter**. (2) Radial distance between the pitch and minor cylinders of an external screw thread. (3) Radial distance between the major and pitch cylinders of an internal thread.

dedicated computer *(Comp.)*. One which is permanently assigned to one application.

dedifferentiation *(Biol.)*. Changes in a differentiated tissue, leading to the reversion of cell types to a common indifferent form, such as the meristematic state.

dedolomitization *(Geol.)*. The recrystallization of a dolomite rock or dolomitic limestone consequent on contact metamorphism; essentially involving the breaking down of the dolomite into its two components, $CaCO_3$, and $MgCO_3$. The former merely recrystallizes into a coarse calcite mosaic; but the latter breaks down further into MgO and CO_2. See **forsterite-marble**.

deducted spaces *(Ships)*. Spaces deducted from the **gross tonnage** to obtain the **net register tonnage**. In general, deducted spaces are those spaces required to be used in the working of the ship and accommodating the crew.

de-emphasis *(Telecomm.)*. The use of an amplitude-frequency characteristic complementary to the one used for *pre-emphasis* prior to transmission and recording.

de-energize *(Elec.Eng.)*. To disconnect a circuit from its source of power.

deep bead *(Build.)*. Piece of timber covering the lower 3 in. or so of movement of the bottom sash in a window, so as to permit ventilation at the meeting rail while keeping bottom of the window to all intents and purposes shut.

deep drawing *(Eng.)*. The process of cold-working or drawing sheet or strip metal by means of dies into shapes involving considerable plastic distortion of the metal.

deep-dyeing fibres *(Textiles)*. Fibres whose chemical composition has been varied from the normal so that they take up more dyestuff in the dyeing process.

deep etch *(Print.)*. A lithographic process in which the image is very slightly etched into the surface of the plate, producing more stability than on a **surface** (or *albumen*) plate. A positive is used when printing down.

deeply-etched *(Print.)*. A half-tone block given a supplementary deepening etch to improve its printing and duplicating qualities.

deeps *(Surv.)*. See **lead-line**.

deep-sea deposits *(Geol.)*. Pelagic sediments accumulating in depths of more than 2000 m. They include, in order of increasing depth, calcareous oozes, siliceous oozes and red clay. Terrigenous material is absent.

deep-sea lead *(Surv.)*. A lead used for attachment to a lead-line measuring beyond 100 fathoms.

deep tank *(Ships)*. A large tank extending from the bottom up to the first deck and from side to side. May be used for liquid cargo or for water ballast.

deep therapy *(Radiol.)*. X-ray therapy of underlying tissues by **hard radiation** (usually produced at more than 180 kVp) passing through superficial layers.

deep-well pump *(Eng.)*. A centrifugal pump, generally electrically driven by a submerged motor built integrally with it, placed at the bottom of a deep bore hole for raising water.

deer-fly fever *(Med.)*. See **tularaemia**.

deerite *(Min.)*. A black, monoclinic hydrous silicate of iron and manganese.

dee(s) *(Electronics)*. Pair of hollow half-cylinders, i.e. D-shaped, in the vacuum of a cyclotron for accelerating charged particles in a spiral, a high-frequency voltage being applied to them in antiphase.

defaecation, defecation *(Med.)*. The ejection of faeces from the body.

default option *(Comp.)*. Specific alternative action to be taken automatically by the computer in the event of the omission of a definite instruction or action.

defect *(Crystal.)*. Lattice imperfection which may be due to the introduction of a minute proportion of a different element into a perfect lattice, e.g. indium into germanium crystal, to form an intrinsic semiconductor for a transistor. A 'point' defect is a 'vacancy' or an 'interstitial atom', while a 'line' defect relates to a dislocation in the lattice.

defect *(Eng.)*. Anything which can cause a product to be unable to fulfil its specified function. Officially classified as minor, major, serious and very serious, with the latter two potentially able to cause injury and severe injury and corresponding degrees of economic loss.

defective equation *(Maths.)*. An equation derived from another, but with fewer roots than the original.

defective virus *(Biol.)*. Virus unable to replicate without a *helper*, e.g. a **plasmid** providing a replicative function.

defect sintering *(Eng.)*. Sintering whereby particles are introduced as a fine dispersion in a sintered body, or by chemical action during sintering. Mobility in heat treatment, with or without working, enables the introduced atoms or molecules to migrate through defects giving marked modification of properties.

defect structure *(Crystal.)*. Intense localized misalignment or gap in the crystal lattice, due to migration of ions or to slight departures from stoichiometry. The resulting opportunity for mobility is important for semiconductors, catalysis, photography, rectifiers, corrosion etc. Cf. **dislocation**.

defence mechanism *(Behav.)*. In psychoanalytic theory, a collective term for a number of reactions that try to ward off or lessen anxiety, by a variety of means that seek to keep the source of anxiety out of consciousness (e.g. repression).

deferent *(Astron.)*. See **epicycle**.

defervescence *(Med.)*. The fall of temperature during the abatement of a fever: the period when this takes place.

defibrillator *(Med.)*. Electrical apparatus to arrest ventricular **fibrillation**.

deficiency *(Biol.)*. The absence, by loss or inactivation, of a gene or a part of a chromosome, that is normally present.

deficiency *(Maths.)*. The difference between the maximum possible number of double points on a curve and the actual number.

deficiency disease *(Bot.)*. That caused by a lack of essential minerals often due to their unavailability in the soil.

deficiency disease *(Med.)*. Any disease resulting from the deprivation of food substances (e.g. vitamins) necessary to good health.

deficient *(Build.)*. See **unstable**.

definite *(Bot.)*. (1) Sympodial growth. (2) A cymose inflorescence.

definite integral *(Maths.)*. If $F(x)$ equals

$$\int f(x)dx,$$

then

$$\int_a^b f(x)dx = F(b) - F(a)$$

is the *definite integral* over the range a to b.

definite proportions *(Chem.)*. See **law of constant (definite) proportions**.

definite time-lag *(Elec.Eng.)*. A time-lag fitted to relays or circuit breakers to delay their operation; it is quite independent of the magnitude of the current causing that operation. Also called *constant time-lag, fixed time-lag, independent time-lag*.

definition *(Acous.)*. Ill-defined quantity describing one aspect of the quality of concert halls and auditoria. Similar to clarity.

definition *(Image Tech.)*. The ability of an imaging system to reproduce fine detail, involving both its **resolution** and its reproduction of subject contrast. See **MTF**.

definition *(Phys.)*. See **resolving power of the eye**.

definitive *(Zool.)*. Final, complete: fully developed; defining or limiting.

deflagration *(Chem.)*. Sudden combustion, generally accompanied by a flame and a crackling sound.

deflation *(Geol.)*. The winnowing and transport of dry loose material, especially silt and clay, by wind.

deflecting electrodes *(Electronics)*. See **deflector plates**.

deflection *(Eng.)*. (1) The amount of bending or twisting of a structure or machine part under load. (2) The movement of the pointer or pen of an indicating or a recording instrument.

deflection angle *(Electronics)*. See **angle of deflection**.

deflection angle *(Surv.)*. The angle between one survey line and the prolongation of another survey line which meets it. See also **intersection angle**.

deflection coil *(Electronics)*. That which, when placed around the neck of a cathode-ray tube and energized with appropriate currents, achieves the desired deflection of the luminous spot on the screen.

deflection defocusing *(Electronics)*. Loss of focus of a CRT spot as deflection from the centre of the screen increases.

deflection sensitivity *(Electronics)*. (1) Ratio of displacement of spot or angle of an electron beam to the voltage producing it. (2) Ratio of displacement of spot or angle of beam to the magnetic field producing it, or the current in the deflecting coils. Also applies to galvanometer or other instrument for measuring voltage, current or quantity of electricity.

deflection yoke *(Electronics,Image.Tech.)*. An assembly (usually on a moulded plastic former) of specially shaped **deflection coils** placed around the neck of a cathode-ray tube in a television receiver.

deflectometer *(Eng.)*. A device for measuring the amount of bending suffered by a beam during a transverse test.

deflector *(Elec.Eng.)*. See **arc deflector**.

deflector *(Ships)*. An instrument used during the adjustment of magnetic compasses. It measures the strength of the field at the compass.

deflector coil(s) *(Electronics)*. Coil(s) so arranged that a current passing through produces a magnetic field which deflects the beam in a cathode-ray tube employing magnetic deflection. Usually applied around the neck of the tube.

deflector plates *(Electronics)*. Electrodes so arranged in a cathode-ray tube that the electrostatic field produced by a p.d. deflects the beam. Also *deflecting electrodes*. See **electrostatic deflection**.

deflexed *(Bot.)*. Bent sharply downwards.

deflocculate *(Chem.)*. To break up agglomerates and form a stable colloidal dispersion.

defoaming agent *(Chem.)*. Substance added to a boiling liquid to prevent or diminish foaming. Usually hydrophobic and of low surface tension, e.g. silicone oils.

deforestation *(Ecol.)*. Permanent removal of forest.

deformation *(Geol.)*. A general term used to describe the structural processes that may affect rocks after their formation. Includes folding and faulting.

deformation *(Maths.)*. See **homotopic mapping**.

deformation potential *(Electronics)*. Potential barrier formed in semiconducting materials by lattice deformation.

deformation ratio *(Maths.)*. See **magnification**.

degassing *(Electronics)*. Removal of last traces of gas from valve envelopes, by pumping or *gettering*, with or without heat, the former with eddy currents, electron bombardment, or simple baking.

degassing *(Min.Ext.)*. To remove entrained gas from drilling mud; important because the gas may seriously reduce the hydrostatic pressure available in the bore hole.

degaussing *(Image Tech.)*. Removal from colour CRT of spurious magnetization which could affect colour purity.

degaussing *(Phys.)*. Neutralization of the magnetization of a mass of magnetic material, e.g., a ship, by an encircling current.

degeneracy *(Phys.)*. Two or more quantum states are said to be degenerate if they have the same energy. If the energy level can be realised in *n* different ways the energy has an *n*-fold degeneracy.

degenerate *(Phys.)*. Two or more quantum states are said to be degenerate if they have the same energy. The *degeneracy* is the number of states having a given energy.

degenerate gas *(Electronics)*. (1) That which is so concentrated, e.g. electrons in the crystal lattice of a conductor, that the Maxwell-Boltzmann law is inapplicable. (2) Gas at very high temperature in which most of the electrons are stripped from the atoms. (3) An electron gas which is far below its Fermi temperature so that a large fraction of the electrons completely fills the lower energy levels and has to be excited out of these levels in order to take part in any physical processes.

degenerate semiconductor *(Electronics)*. One in which the conduction approaches that of a simple metal.

degeneration *(Biol.)*. Evolutionary retrogression; the process of returning from a higher or more complex state to a lower or simpler state.

degenerative disorders *(Behav.)*. See **dementia**.

deglutition *(Zool.)*. The act of swallowing.

degradation *(Phys.)*. Loss of energy of motion solely by collision. Deliberate slowing of neutrons in a reactor is **moderation**. In an isolated system, the **entropy** increases.

degreasing *(Textiles)*. Removal of natural fats, oils and waxes from textiles by extraction usually with an organic liquid.

degree *(Maths.)*. (1) Unit of angle, written °; thus 360° for a revolution, 90° for a right angle. (2) Degree of a polynomial is the highest power of the variable present, so that a quadrate has degree 2, for example. (3) Similar definitions to (2) apply in many areas.

degree *(Phys.)*. The unit of temperature difference. It is usually defined as a certain fraction of the fundamental interval, which for most thermometers is the difference in temperature between the freezing and boiling points of water. See **Celsius scale, centigrade scale, Fahrenheit scale, international practical temperature scale, Kelvin thermodynamic scale of temperature**.

degree of a curve *(Surv.)*. The angle subtended by a standard chord length of 100 ft at the centre of a curve.

degree of damping *(Phys.)*. The extent of the damping in an oscillatory system, expressed as a fraction or percentage of that which makes the system critically damped.

degree of dissociation *(Chem.)*. The fraction of the total number of molecules which are dissociated.

degree of ionization *(Chem.)*. The proportion of the molecules or 'ion-pairs' of a dissolved substance dissociated into charged particles or ions.

degrees of freedom *(Chem.)*. (1) The number of variables defining the state of a system (e.g. pressure, temperature) which may be fixed at will. See **phase rule**. (2) Number of independent capacities of a molecule for holding energy, translational, rotational and vibrational. A molecule may have a total of *3n* of these, where *n* is the number of atoms in the molecule.

de Haas-van Alphen effect *(Phys.)*. Oscillations in the magnetic moment of a metal as a function of $1/B$ where B is the magnetic induction. Interpretation of this effect gives important information about the shape of the *Fermi surface*.

dehiscence *(Bot.)*. The spontaneous opening at maturity of a fruit, anther, sporangium or other reproductive body, permitting the release of the enclosed seeds, spores etc. adj. *dehiscent*. Cf. **indehiscent**. More generally the act of splitting open as in **diapedesis**.

dehydration *(Chem.)*. (1) The removal of H_2O from a molecule by the action of heat, often in the presence of a catalyst, or by the action of a dehydrating agent, e.g. concentrated sulphuric acid. (2) The removal of water from crystals, tars, oils etc., by heating, distillation or by chemical action.

dehydration *(Med.)*. Excessive loss of water from the tissues of the body.

dehydrogenase *(Biol.)*. Enzymes which catalyse the oxidation of their substrate with the removal of hydrogen atoms.

de-icing *(Aero.)*. Method of protecting aircraft against icing by removing built-up ice before it assumes dangerous proportions. It may be based on pulsating pneumatic overshoes, chemical applications or intermittent operation of electrical heating elements. Cf. **anti-icing**.

Deimos *(Astron.)*. One of two natural satellites of **Mars**.

de-individuation *(Behav.)*. A term used by social psychologists to denote a loss of a sense of personal responsibility in conditions of relative anonymity, where a person cannot be identified as an individual but only as a member of a group. The individual engages in activities he or she would not normally do, presumably because of a weakening of internal controls (e.g. shame or guilt).

Deion circuit-breaker *(Elec.Eng.)*. TN for a circuit-breaker fitted with an arc-control device in which the arc takes place within a slot in a stack of insulated plates. The plates contain iron inserts or a magnet coil, so that the arc is blown magnetically towards the closed end of the slot, thereby coming into contact with cool oil which has a de-ionizing action and extinguishes the arc.

de-ionized water *(Chem.)*. Water from which ionic impurities have been removed by passing it through cation and anion exchange columns.

Deka-ampere-balance *(Elec.Eng.)*. An ampere-balance having a current range from 1 to 100 amperes.

del *(Maths.)*. In cartesian co-ordinates, the vector

delamination *(Biol.)*. The division of cells in a tissue, leading to the formation of layers.

De la Rue cell *(Elec.Eng.)*. See **chloride of silver cell**.

delay *(Acous.)*. Time shift which can be introduced into the transmission of a signal, by recording it magnetically on tape, disk, or wire, and reproducing it slightly later. Used in public-address systems, to give illusion of distance and to coalesce contributions from original source and reproducers.

delay circuit *(Telecomm.)*. One in which the output circuit is delayed by a specified time interval with respect to the input signal; used for phase adjustment or correction in radio-frequency circuits, or in digital (pulse) circuits for making signals from different sources coincide.

delay distortion *(Telecomm.)*. Change in wave form during transmission because of non-linearity of delay with frequency, which is $d\beta/d\omega$, where β = phase delay in radians, and $\omega = 2\pi \times$ frequency (Hz). See **envelope delay distortion**.

delayed action *(Elec.Eng.)*. Any arrangement which imposes an arbitrary delay in operation of e.g. a switch or circuit-breaker.

delayed automatic gain control *(Telecomm.)*. That which is operative above a threshold voltage signal in a radio receiver. Provides full amplification in radio receivers for very weak signals, and constant output from detector for signals above the threshold. Also called *quiet automatic volume control*. See also **automatic gain control**.

delayed critical *(Phys.)*. Assembly of fissile material critical only after release of delayed neutrons. Cf. **prompt critical**.

delayed drop *(Aero.)*. A live parachute descent in which the parachutist deliberately delays pulling the ripcord until after a descent of several thousand metres.

delayed neutrons *(Phys.)*. Those arising from fission but not released instantaneously. Fission neutrons are always **prompt neutrons**; those apparently delayed (up to seconds) arise from breakdown of fission products, not primary fission. Such delay eases control of reactors.

delayed opening *(Aero.)*. Delaying the opening of a parachute by an automatic device. In any flight above 40 000 ft (12 500 m), low temperature and pressure require that aircrew must reach lower altitude for survival as rapidly as possible, and it is usual to have a barostatic device to delay opening to a predetermined height, usually 15 000 ft (4500 m).

delayed type hypersensitivity *(Immun.)*. Hypersensitivity state mediated by T-lymphocytes. When the antigen is introduced locally, e.g. in the skin, a gradual local accumulation of T-cells and monocytes results. The visible reaction is reddening and local swelling, increasing for 24–48 hrs. and then subsiding sometimes leaving a small scar due to necrosis of blood vessels. Tuberculin testing is a good example. Frequently this and other types of hypersensitivty co-exist, and reactions are not clear cut.

delay element *(Comp.)*. Computer component which has the property that it stores a signal for a time interval before it transfers the signal to its output line. See **delay line**.

delay line *(Comp.)*. Column of mercury, a quartz plate, or length of nickel wire, in which impressed sonic signals travel at a finite speed and which, by the delay in travelling, can act as a **store**, the signals being constantly recirculated and abstracted as required. See **first generation computer**.

delay line *(Telecomm.)*. Real or artificial transmission line or network used to delay propagation of an electrical signal usually by a specified interval.

delay network *(Telecomm.)*. Artificial line of electrical networks, designed to give a specified phase delay in the transmission of currents over a specified frequency band.

delay period *(Eng.)*. The time or crank-angle interval between the passage of the spark and the resulting pressure rise in a petrol or gas engine, or between fuel injection and pressure rise in an oil engine.

Delbruck scattering *(Phys.)*. Elastic coherent scattering of gamma-rays in the coulomb field of a nucleus. The effect is small and so far has not been conclusively detected.

delessite *(Min.)*. An oxidized variety of chlorite, relatively rich in iron.

deleted neighbourhood *(Maths.)*. Of a point z_0; the set of all points z in the domain $|z - z_0| < \alpha$ (α a constant) excluding the point z_0.

deletion *(Biol.)*. Same as **deficiency**.

deletion mutation *(Biol.)*. That in which a base or bases are lost from the DNA. Cf. **base substitution**.

Delhi boil *(Med.)*. *Tropical sore; Baghdad boil.* Oriental sore resulting from infection of the skin with a protozoal parasite.

delinquency, delinquent *(Behav.)*. Conduct disorder, usually against the law, but including a range of antisocial, deviant or immoral behaviour. In Britain, young people are considered *juvenile delinquents* if they are under 17 years of age with criminal convictions.

deliquescence *(Chem.)*. The change undergone by certain substances which become damp and finally liquefy when exposed to the air, owing to the very low vapour pressure of their saturated solutions; e.g. calcium chloride.

delirium *(Med.)*. A profound disturbance of consciousness occurring in febrile and toxic states; characterized by restlessness, incoherent speech, excitement, delusions, illusions, and hallucinations.

delirium tremens *(Med.)*. An acute delirium in chronic alcoholism, characterized by insomnia, restlessness, terrifying hallucinations and illusions, and loss of orientation to time and place. Abbrev. *DT*.

delivery *(Eng.)*. (1) The discharge from a pump or compressor. (2) The withdrawal of a pattern from a mould.

delivery *(Print.)*. The mechanical arrangement for delivering sheets after printing, there being several designs.

Dellinger fade-out *(Telecomm.)*. Complete fade-out (which may last for minutes or hours) and inhibition of short-wave radio-communication because of the formation of a highly absorbing D-layer, lower than the regular E- and F-layers of the ionosphere, on the occasion of a burst of hydrogen particles from an eruption associated with a sun spot.

delph *(Hyd.Eng.)*. A drain behind a sea embankment, on the land side.

Delrin *(Plastics)*. TN for an acetal resin. Highly crystalline stable form of polymerized formaldehyde. Used as an engineering plastic for many applications.

delta *(Geol.)*. The more or less triangular area of river-borne sediment deposited at the mouth of rivers heavily charged with detritus. A delta is formed on a low-lying coastline, particularly in seas of low tidal range and little current action, and in areas where subsidence keeps pace with sediment deposition. The Nile Delta is a good example.

delta *(Space)*. The velocity change required to transform a particular trajectory into another.

delta connection *(Elec.Eng.)*. The connection of a 3-phase electrical system such that the 3-phasors representing system voltages form a closed triangle.

deltaic deposit *(Geol.)*. The accumulations of sand, silt, clay and organic matter, deposited as **topset, foreset** and **bottomset beds**. A good active example is the Mississipi delta and in the geological record, the Millstone grit in England.

delta impulse function *(Telecomm.)*. Infinitely narrow pulse of great amplitude, such that the product of its height and duration is unity.

delta iron *(Eng.)*. The polymorphic form of iron stable between 1403°C and the melting point (about 1532°C). The space lattice is the same as that of α-iron and different from that of γ-iron.

delta-matching transformer *(Telecomm.)*. A matching network between 2-wire transmission lines and half-wave antennae.

delta modulation *(Telecomm.)*. Form of **differential pulse-code modulation** in which only one bit for each sample is used.

delta network *(Telecomm.)*. One with three branches all in series.

delta-particle *(Phys.)*. Very short lived hyperon which decays almost instantaneously through the strong interaction.

delta-ray *(Phys.)*. Any particle ejected by recoil action from passage of ionizing particles, e.g. in a Wilson cloud chamber.

delta-ray spectrometer *(Nuc.Eng.)*. See **spectrometer** and **delta ray**.

delta voltages *(Elec.Eng.)*. The voltage between alternate lines in a delta connected system.

delta wave *(Med.).* The lower frequency brain waves (1–8 Hz) seen on an **electroencephalograph**.

delta wing *(Aero.).* A sweptback wing of substantially triangular planform, the trailing edge forming the base. It is longitudinally stable and does not require an auxiliary balancing aerofoil, although tail or nose planes are sometimes fitted to increase pitch control and trim so that landing flaps can be fitted.

deltoid *(Bot.,Zool.).* Having the form of an equilateral triangle: any triangular structure, as the deltoid muscle of the shoulder.

deltoid *(Maths.).* Recent name for Steiner's three-cusped hypocycloid.

delusions *(Behav.,Med.).* An irrational belief which an individual will defend with intensity, despite overwhelming evidence that it has no basis in reality, common among schizophrenic mental disorders.

delustrant *(Textiles).* Dense inorganic material, frequently titanium dioxide, added to a man-made fibre before it is extruded. In this way a range of fibres may be obtained with different lustres and opacities.

de luxe *(Print.).* A lavish style of bookwork design, characterized by generous margins.

demagnetization *(Elec.Eng.,Phys.).* (1) *Removal* of magnetization of ferromagnetic materials by the use of diminishing saturating alternating magnetizing forces. (2) *Reduction* of magnetic induction by the internal field of a magnet, arising from the distribution of the primary magnetization of the parts of the magnet. (3) *Removal* by heating above the Curie point. (4) *Reduction* by vibration.

demagnetization *(Min.Ext.).* In **dense media process** using ferro-silicon, passage of the fluid through an a.c. field to deflocculate the agglomerated solid.

demagnetization factor *(Elec.Eng.).* Diminution factor (N) applied to the intensity of magnetization (I) of a ferromagnetic material, to obtain the demagnetizing field (ΔH), i.e. $\Delta H = NI$. N depends primarily on the geometry of the body concerned.

demagnetizing ampere-turns *(Elec.Eng.).* See **back ampere-turns**.

demagnetizing coil *(Electronics).* One used to eliminate residual magnetization from a record or playback head of a tape recorder; powered by mains frequency a.c. Also *degaussing coil.*

demand *(Elec.Eng.).* See **maximum demand**.

demand factor *(Elec.Eng.).* Ratio of the maximum demand on a supply system to the total connected load.

demand indicator *(Elec.Eng.).* See **maximum-demand indicator**.

demand limiter *(Elec.Eng.).* See **current limiter**.

demand meter *(Elec.Eng.).* One reading or recording the loading on an electrical system.

demantoid *(Min.).* Bright-green variety of the garnet andradite, essentially silicate of calcium and iron.

deme *(Ecol.).* A local population of interbreeding organisms.

dementia *(Behav.).* Refers to degeneration of various functions governed by the central nervous system, including motor reactions, memory and learning capacity, problem solving etc. These functions normally decline with age, but several dementia syndromes result from pathological organic deterioration of the brain. See **senile dementia**.

dementia praecox *(Behav.).* Obsolete term for **schizophrenia**.

demersal *(Zool.).* Found in deep water or on the sea bottom; as Fish eggs which sink to the bottom, and midwater and bottom-living Fish as opposed to surface Fish (e.g. Herring) and Shellfish. Cf. **pelagic**.

demifacet *(Zool.).* One of the two half-facets formed when the articular surface for the reception of the capitular head of a rib is divided between the centra of two adjacent vertebrae.

demi-hunter *(Horol.).* See **half-hunter**.

demijohn *(Glass).* Narrow-necked wine or spirit container of more than 2 gal. capacity.

demineralization *(Chem.Eng.).* A process for cleaning water in which the anions and cations are removed separately by absorption in synthetic exchange materials, leaving the water free of dissolved salts. The removal cells are regenerated by treatment with alkali and acid.

demister *(Autos.).* Ducts arranged so that hot dry air is played on the interior of the windscreen to prevent condensation. Heat source now usually heat dissipated from the engine.

demodectic mange *(Vet.). Demodectic folliculitis.* Mange of animals caused by mites of the genus *Demodex*, which live in the hair follicles and sebaceous glands.

demodulation *(Telecomm.).* The inverse of **modulation**. Applied generally to the process of extracting the original information impressed on a **carrier**. In *amplitude modulation* the received signal is usually passed through non-linear circuits which generate sum and difference frequencies which allow the modulating signal and the carrier to be separated. In *frequency modulation* the signal is fed to a frequency sensitive circuit which generates an output proportional to the frequency shift of the carrier brought about by modulation. See **discriminator**. In pulse-coded systems the word demodulation is replaced by *decoding*. See **decoder**.

demodulation of an exalted carrier *(Telecomm.).* Same as *homodyne reception.*

demodulator *(Telecomm.).* A circuit or device which *demodulates*. See **detector**.

De Moivre's theorem *(Maths.).* That

$$(\cos\theta + i\sin\theta)^n = \cos n\theta + i\sin n\theta.$$

Expressed in terms of the exponential function, the theorem is $(e^{i}\theta)^n = e^{ni}\theta$.

demonstrator *(Aero.).* A new aircraft, engine or system constructed to prove its novel features prior to embarking on full development.

Demospongiae *(Zool.).* A class of *Porifera* usually distinguished by the possession of a skeleton composed of siliceous spicules, or of spongin. e.g., bath sponges.

demountable *(Phys.).* Said of X-ray tubes or thermionic valves when they can be taken apart for cleaning and filament replacement, and are continuously pumped during operation.

demulcent *(Med.).* Soothing; allaying irritation.

demulsification number *(Min.Ext.).* The resistance to emulsification by a lubricant when steam is passed through it; indicated in the minutes and half minutes required for the separation of a given volume of oil after emulsification.

demultiplexer *(Comp.).* Circuit that enables data from a single source to select one of several possible destinations. Cf. **multiplexer**.

demy octavo *(Print.).* A book size, $8\frac{1}{2} \times 5\frac{1}{2}$ in (metric 216×140 mm).

demy quarto *(Print.).* A book-size, $11 \times 8\frac{5}{8}$ in (metric 279×219 mm).

denary notation *(Comp.,Maths.).* Familiar system representing numbers in base ten with the digits 0, 1, 2,....9.

denaturant *(Phys.).* Isotope added to fissile material to render it unsuitable for military use, e.g. ^{238}U can be added to ^{233}U, but denaturing with ^{238}U necessarily produces plutonium which must be used or disposed of.

denaturation *(Biol.).* The destruction of the native conformation or state of a biological molecule by heat, extremes of pH, heavy metal ions, chaotropic agents etc., resulting in loss of biological activity. Specifically in DNA, the breakage of the hydrogen bonds maintaining the double-helical structure, a process which can be reversed by *renaturation* or annealing.

denatured alcohol *(Chem.).* Alcohol (ethanol) which according to law has been made unfit for human consumption by the admixture of nauseating or poisonous substances, e.g. methyl alcohol, pyridine, benzene.

dendr-, dendro- *(Bot.).* Gk. *dendra*, tree. Meaning tree, treelike or branching.

dendrite *(Crystal.).* A treelike crystal formation.

dendrite *(Eng.).* Metal crystals grow in the first instance by branches developing in certain directions from the nuclei. Secondary branches are later thrown out at periodic intervals by the primary ones and in this way a skeleton crystal, or *dendrite*, is formed. The interstices between the branches are finally filled with solid which in a pure metal is indistinguishable from the skeleton. In many alloys, however, the final structure consists of skeletons of one composition in a matrix of another.

dendrite *(Zool.)*. A branch of a **dendron**.

dendritic cell *(Immun.)*. A cell that has branching processes. The term is used in immunology to describe two distinct kinds of cells which have different functions. (1) *Follicular dendritic cells.* Present in **germinal centres** where they are in intimate contact with dividing B-cells. Antigen-antibody complexes become trapped at the surface of the dendritic processes and are retained there for long periods. They are involved in the generation of **B-memory cells**. (2) Cells with dendritic morphology which occur mainly in the thymus-dependent areas of lymph nodes and the spleen. Here they are often termed *interdigitating cells.* Very similar cells occur as Langerhans cells in the skin. Dendritic cells of this kind are not phagocytic and they typically express Ia. They are very effective accessory cells in stimulating T-lymphocytes.

dendritic markings *(Geol.)*. Treelike markings, usually quite superficial, occurring on joint-faces and other fractures in rocks, frequently consisting of oxide or manganese or of iron. Less frequently the appearance is due to the inclusion of a mineral of dendritic habit in another mineral or rock, e.g. chlorite in silica as in 'moss agate'.

dendritic ulcer *(Med.)*. A branching ulcer of the cornea, due to herpes of the cornea.

dendrochronology *(Ecol.)*. Science of reconstructing past climates from the information stored in tree trunks as annual radial increments of growth.

dendrogram *(Bot.)*. A branching diagram after the style of a family tree reflecting similarities or affinities of some sort. See also **cladogram, phenogram**.

dendrograph *(Bot.)*. An instrument which is used to measure the periodical swelling and shrinkage of tree trunks.

dendroid *(Bot.)*. (1) Tall, with an erect main trunk, as tree-ferns. (2) Freely branched.

dendron *(Zool.)*. The afferent or receptor process of a neuron. Often much branched. Cf. **axon**.

denervated *(Med.)*. Deprived of nerve supply.

dengue *(Med.)*. A tropical disease in which an influenza-like viral agent is transmitted by mosquito to man; characterized by severe pain in the joints and a rash. The condition is self-limiting.

denial *(Behav.)*. Refusal to acknowledge some unpleasant feature of the external world or some painful aspect of one's own experiences or emotions (a *defence mechanism*).

denier system *(Textiles)*. A system used in the 'counting' of silk and nylon yarns; designated by the weight in grams of 9 000 m of yarn. See **count of yarn, Tex**.

denim *(Textiles)*. A strong cotton twill fabric made from yarn-dyed warp (often blue) and undyed (sometimes even unbleached) weft yarn. Used for the manufacture of overalls, boiler suits and jeans.

denitrification *(Biol.)*. See **nitrate-reducing bacteria**.

denominational number system *(Maths.)*. A system of representing numbers by a combination of digits in which each digit contributes an amount dependent upon both its value and its position. Thus with a radix of *r*, the *n*-digit number $a_{nb1}\ a_{nb2} \ldots a_2\ a_1\ a_0$ represents

$$\sum_{s=0}^{n-1} a_s r^s,$$

where each digit a_s can have any of *r* values; e.g. in the decimal system $r = 10$, and the number 345 represents $3 \times 10^2 + 4 \times 10 + 5$. Cf. **non-denominational number system**.

denominator *(Maths.)*. See **division**.

dense *(Glass)*. Of optical glass, having a higher refractive index.

dense-media process *(Min.Ext.)*. **Heavy-media** or sink-float process. Dispersion of ferrosilicon or other heavy mineral in water separates lighter (floating) ore from heavier (sinking) ore.

dens epistrophei *(Zool.)*. See **odontoid process**.

dense set *(Maths.)*. A set of points is dense in itself if every point of the set is a limit point. A set is said to be everywhere dense in an interval, if every subinterval (no matter how small) contains points of the set. A subset X of a set Y is said to be dense in Y if the closure of X is Y.

densification *(Powder Tech.)*. All modes of increasing density, including the effect of sintering contraction.

densi-tensimeter *(Chem.)*. An apparatus for determining both the pressure and the density of a vapour.

Densithene *(Plastics)*. TN for polythene loaded with lead powder in the form of sheet, pipe etc. Used for radioactive shielding.

densitometer *(Image Tech.,Phys.)*. Any instrument for measuring the optical transmission or reflecting properties of a material, in particular the optical density (absorbance) of exposed and processed photographic images.

density *(Image Tech.)*. A measure of the light-stopping power of a transparent material; it is defined as the logarithm of the **opacity**, which is the reciprocal of the transmission ratio. See also **reflection density**.

density *(Phys.)*. The mass of unit volume of a substance, expressed in such units as kg/m^3, g/cm^3 or lb per cubic foot. See **relative density**.

density bottle *(Phys.)*. A thin glass bottle, accurately calibrated, used for the determination of the density of a liquid.

density change method *(Powder Tech.)*. Particle-size analysis technique which measures concentration changes within a sedimenting suspension by measuring the pressure exerted by a column of the suspension.

density-dependence *(Ecol.)*. The phenomenon whereby performance of organisms within a population is dependent on the extent of crowding.

density gradient centrifugation *(Biol.)*. An important method of physically separating DNA molecules according to their base composition. A solution of a highly soluble salt, like CsCl, can be centrifuged to an equilibrium gradient of density in which DNA of different densities will separate. DNA density is affected by base-composition, by the binding of small molecules or by DNA strandedness.

density of gases *(Chem.,Phys.)*. According to the **gas laws**, the density of a gas is directly proportional to the pressure and the rel. mol. mass and inversely proportional to the absolute temperature. At standard temperature and pressure the densities of gases range from $0.0899\ g/dm^3$ for hydrogen to $9.96\ g/dm^3$ for radon.

density of states *(Phys.)*. The number of electronic states per unit volume having energies in the range from E to $E + dE$; an important concept of the band theory of solids.

density range *(Print.)*. The range of density from shadow to highlight on a film negative or positive or on the printed image. When measured with a **densitometer** the range can be expressed numerically.

dent *(Textiles)*. Term denoting one wire in a loom reed; it also refers to the space between two wires, through which warp threads are drawn. The number of dents per cm is the sett of the cloth.

dental caries *(Med.)*. A disintegration of enamel and dentine of tooth, common where fluorine content of water is low. Probably due to acids formed by the action of bacteria on dietary carbohydrate.

dental formula *(Zool.)*. A formula used in describing the dentition of a Mammal to show the number and distribution of the different kinds of teeth in the jaws; thus a Bear has in the upper jaw 3 pairs of incisors, 1 pair of canines, 4 pairs of premolars, and 2 pairs of molars; and in the lower jaw 3 pairs of incisors, 1 pair of canines, 4 pairs of premolars, and 3 pairs of molars. This is expressed by the formula

$$\frac{3142}{3143}.$$

dentary *(Zool.)*. In Vertebrates, a membrane bone of the lower jaw which usually bears teeth. In Mammals, it forms the entire lower jaw.

dentate *(Bot.)*. Having a toothed margin.

dentelle *(Print.)*. A style of decoration of a toothlike or lacelike character; used in covers.

denticles *(Zool.)*. Any small toothlike structure: the placoid scales of Elasmobranchii.

denticulated *(Build.)*. A term applied to mouldings decorated with dentils.

dentigerous cyst *(Vet.)*. A cyst containing teeth; usually a teratomatous cyst on the malar bone of a horse.

dentil *(Build.)*. A projecting rectangular block forming one of a row of such blocks under the corona of a cornice.

dentine *(Med.,Zool.)*. A hard calcareous substance, allied to bone, of which teeth and placoid scales are mainly composed. Ivory is dentine. adj. *dentinal.*

dentirostral *(Zool.)*. Having a toothed or notched beak.

dentition *(Zool.)*. The kind, arrangement, and number of the teeth; the formation and growth of the teeth; a set of teeth, as the milk dentition.

denudation *(Geol.)*. The laying bare (L. *nudus*, naked) of the rocks by chemical and mechanical disintegration and the transportation of the resulting rock debris by wind or running water. Ultimately denudation results in the degradation of the hills to the existing base-level. The process is complementary to sedimentation, the amount of which in any given period is a measure of the denudation.

denuded quadrat *(Bot.)*. A square piece of ground, marked out permanently and cleared of all its vegetation, so that a study may be made of the manner in which the area is reoccupied by plants.

denumerable set *(Maths.)*. One that can be put into a one–one correspondence with the positive integers. Also *countable set, enumerable set, numerable set.*

deodar *(For.)*. Tree of the genus *Cedrus*, one of the most important Indian softwoods, probably second to teak in importance as native timber. It is not resistant to termite attack although it is resistant to powder-post beetle infestation. It has a distinctive unpleasant smell. In addition to its use for normal joinery purposes it also makes good structural timber.

deodorizing *(Chem.Eng.)*. A process extensively used in edible oil and fat refining, in which the oil or fat is held for several hours at high temperatures (varying with product, but 200°C is not uncommon), and low pressure ($\sim 10^4\,N/m^2$), during which time steam is blown through to remove the traces of odour-creating substances (usually free fatty acids of high rel. mol. mass).

deoxidation *(Eng.)*. The process of reduction or elimination of oxygen from molten metal before casting by adding elements with a high oxygen affinity, which form oxides that tend to rise to the surface.

deoxidized copper *(Eng.)*. Copper from which the oxygen remaining after poling has been removed by the addition of a deoxidizer, which, remaining in solid solution, lowers the conductivity below that of tough-pitch copper, but the product is more suitable for working operations.

deoxidizer *(Eng.)*. A substance which will eliminate or modify the effect of the presence of oxygen, particularly in metals. Also *deoxidant.*

departure time *(Aero.)*. The exact time at which an aircraft becomes airborne is an important factor in air traffic control; estimated time of departure (abbrev. *ETD*).

dependent functions *(Maths.)*. A set of functions such that one may be expressed in terms of the others.

dependent variable *(Maths.)*. A variable whose values are determined by one or more other (independent) variables.

depersonalization *(Behav.)*. A condition in which an individual experiences a range of feelings of unreality or remoteness in relation to the self, to the body or to other people, even extending to the feeling of being dead. Primarily a symptom in a range of neurotic syndromes.

depeter *(Build.)*. Plasterwork finished in imitation of tooled stone, small stones being pressed in with a board before the plaster sets. Also called *depreter.*

dephlogisticated air *(Chem.)*. The name given by Priestley to oxygen. The term is of historic interest only.

dephosphorization *(Eng.)*. Elimination, partial or complete, of phosphorus from steel, in basic steel-making processes. Accomplished by forming a slag rich in lime. See **acid process, basic process, Bessemer process, open-hearth process.**

depilate *(Med.)*. To remove the hair from.

depilatories *(Chem.)*. Compounds for removing or destroying hair; usually sulphide preparations.

depleted uranium *(Chem.)*. Sample of uranium having less than its natural content of ^{235}U.

depletion *(Phys.)*. Reduction in the proportion of a specific isotope in a given mixture.

depletion layer *(Electronics)*. In semiconductor materials, location where mobile electrons do not neutralize the combined charge of the donors and acceptors.

depletion-mode transistor *(Electronics)*. An insulated-gate field-effect transistor in which carriers are present in the **channel** when the gate-source voltage is zero. Channel conductivity is controlled by changing the magnitude and polarity of this voltage. See also **enhancement-mode transistor** and **pinch-off**.

depluming itch *(Vet.)*. A skin irritation of fowls due to infestation of the skin around the feather shafts by the mite *Cnemidocoptes gallinae.*

depolarization *(Elec.Eng.)*. Reduction of polarization, usually in electrolytes, but sometimes in dielectrics. In the former it may refer to removal of gas collected at plates of cell during charge or discharge.

depolarizing muscle relaxant *(Med.)*. Suxamethonium is the only commonly used drug in this class. It acts by mimicking the action of acetyl choline at the neuromuscular junction to produce a relatively short period of paralysis and relaxation. It is commonly used to facilitate the passage of an endotracheal tube to maintain airway patency at the start of a surgical operation.

depolymerization *(Chem.)*. The breakdown of a complex material into simple units, either with or without incorporation of any external reagent.

deposit *(Elec.Eng.)*. (1) The coating of metal deposited electrolytically upon any material. (2) The sediment which is sometimes found at the bottom of a secondary cell owing to gradual disintegration of the electrode material.

deposit *(Geol.)*. See under **deposition.**

deposition *(Geol.)*. The laying down or placing into position of sheets of sediment (often referred to as *deposits*) or of mineral veins and lodes. Synonymous with *sedimentation* in the former sense.

depreciation factor *(Elec.Eng.)*. A term commonly used in the design of floodlighting and similar installations to denote the ratio of the light output when the lighting equipment is clean to that when it is dirty (i.e. after having been in service for some time).

depressant *(Med.)*. Lowering functional activity; a medicine which lowers functional activity of the body.

depressed conductor-rail *(Civ.Eng.)*. A section of conductor-rail depressed below normal level where contact with the shoes is not desired.

depressing agent *(Min.Ext.)*. *Wetting agent.* One used in froth flotation to render selected fraction of pulp less likely to respond to aerating treatment.

depression *(Med.)*. A state of deep dejection and a pervasive feeling of helplessness, accompanied by apathy and a sense of personal worthlessness, resulting in retardation of thought and bodily functions. Endogenous depression is a spontaneous occurrence. Reactive depression is a response, sometimes exaggerated, to grief or other personal tragedy.

depression *(Meteor.)*. A cyclone. That distribution of atmospheric pressure in which the pressure decreases to a minimum at the centre. In the northern hemisphere, the winds circulate in a counter-clockwise direction, in such a system; in the southern hemisphere, in a clockwise direction. A depression usually brings stormy unsettled weather.

depression of freezing point *(Phys.)*. A solution freezes at a lower temperature than the pure solvent, the amount of the depression of the freezing point being proportional to the concentration of the solution, provided this is not too great. The depression produced by a 1% solution is called the *specific depression*, and is inversely proportional to the molecular weight of the solute. Hence the depression is proportional to the number of moles dissolved in unit weight of the solvent and is independent of the particular solute used.

depression of land *(Geol.)*. Depression relative to sea-level may be caused in many ways, including sedimentary consolidation, the superposition of large masses of ice, the migration of magma, or changes of chemical phase at depth. It is generally recognized by the marine transgression produced but is often difficult to distinguish from eustatic changes in sea-level. See **drowned valleys**.

depressor *(Zool.)*. A muscle which by its action lowers a part or organ; a reagent which, when introduced into a metabolic system, slows down the rate of metabolism.

depreter *(Build.)*. See **depeter**.

depth *(Horol.)*. The amount by which the teeth of a wheel intersect the teeth of the mating wheel or pinion.

depth *(Ships)*. The depth measured from the top of the keel to the top of some specified deck.

depth gauge *(Build.)*. Device clamped to a drill or bit to regulate the depth of the hole bored.

depthing tool *(Horol.)*. An instrument by means of which two wheels, or a wheel and pinion, can be mounted and their depth adjusted until it is correct, after which the distance apart of their centres can be transferred to the plates for the drilling of the pivot holes.

depth of field *(Image Tech.)*. The range of near and far distances from the camera within which the subject will be in reasonably sharp focus; it is determined by the focal length of the lens used and its **stop** as well as the actual focus setting.

depth of focus *(Image Tech.)*. Within a camera, the range of distances from the lens to the photo-sensitive surface within which the image is acceptably sharply defined for a given **circle of confusion**.

depth of fusion *(Eng.)*. The depth to which a new weld has extended into the underlying metal or a previous weld.

depth of modulation *(Telecomm.)*. See **modulation depth, modulation index**.

depth of penetration *(Elec.Eng.)*. Within a plane conductor the magnitude of an electromagnetic field and the associated current falls off exponentially from the surface. The depth of penetration or *skin depth* (δ metres) is normally considered as the depth for which the field magnitude is $1/e$ of its value at the surface. $\delta = (\pi f \sigma)^{-0.5}$ where f is the frequency, μ is the permeability and σ the conductivity.

deputy *(Min.Ext.)*. Man of responsibility in a coal mine. See also **fireman**.

dérailleur *(Eng.)*. Variable transmission gear mechanism whereby the driving chain may be 'derailed' from one sprocket wheel to another of different size, thus changing the driving ratio. Ten or more ratios are possible. Much used in racing bicycles. Cf. **epicyclic gear**.

de-rating *(Elec.Eng.)*. Deliberate reduction in the duty placed on components and equipment by the designer, to maintain an adequate margin of safety and improve reliability.

derby float *(Build.)*. A large trowel consisting of a flat board with two handles on the back.

Derbyshire neck *(Med.)*. See **goitre**.

Derbyshire spar *(Min.)*. A popular name for the mineral *fluorite* or *fluorspar*.

derivative *(Maths.)*. The derived function.

derivative feedback *(Telecomm.)*. Feedback signal in control system proportional to time derivative error.

derived fossils *(Geol.)*. Fossils eroded from an older sediment and redeposited in a younger sediment.

derived function *(Maths.)*. The derived function f' of the function f is defined by $f'(x)$ = the limit as h tends to zero of

$$\frac{f(x+h)-f(x)}{h}.$$

If $y = f(x)$, other notations for $f'(x)$ are

$$\frac{dy}{dx}, y', D_x y \text{ and } \frac{d}{dx} f(x).$$

The gradient of the curve $y = f(x)$ at the point (a,b) is $f'(a)$. Also called *differential coefficient* or *derivative*. Cf. **partial differential coefficient**.

derived units *(Phys.)*. Units derived from the fundamental units of a system by consideration of the dimensions of the quantity to be measured. See **dimensions**.

derm *(Zool.)*. See **dermis**.

dermal *(Bot.)*. Appertaining to the epidermis or other superficial layer of a plant member.

dermal *(Zool.)*. Pertaining to the skin; more strictly, pertaining to the dermis. Also *dermic*.

dermal branchiae *(Zool.)*. See **papulae**.

dermatitis *(Med.)*. Inflammation of the surface of the skin or epidermis.

dermatogen *(Bot.)*. A **histogen** precursor of the epidermis.

dermatographia, dermographia *(Med.)*. A sensitive condition of the skin in which light pressure with, for example, a pencil point will produce a reddish weal. A form of urticaria.

dermatology *(Med.)*. That branch of medical science which deals with the skin and its diseases.

dermatomyositis *(Med.)*. A disease, probably of **autoimmune** origin with inflammation and weakness of muscles, often with a purplish skin rash. Although some patients may have an underlying malignancy the majority respond to corticosteroids and immunosuppressive drugs.

Dermatophagoides pteronyssinus *(Immun.)*. A mite present in house dust. Antigens extracted from mites and their faeces are a common cause of allergy to house dust in W. European countries.

dermatophyte *(Bot.)*. A parasitic fungus which causes a skin desease in animals or man, e.g. ring worm, athlete's foot.

dermatosclerosis *(Med.)*. See **sclerodermia**.

dermis, derm *(Zool.)*. The inner layers of the integument, lying below the epidermis and consisting of mesodermal connective tissue. adjs. *dermal, dermic*.

dermoid *(Med.)*. A cyst of congenital origin containing such structures as hair, skin, and teeth; occurs usually in the ovary.

dermomuscular layer *(Zool.)*. A sheet of muscular tissue underlying the skin in lower Metazoa: it consists of longitudinal and, usually, circular layers.

derrick *(Build.,Civ.Eng.)*. An arrangement for hoisting materials, distinguished by having a boom stayed from a central post, which in turn is usually stayed in position by guys.

derricking jib crane *(Eng.)*. A jib crane in which the inclination of the jib, and hence the radius of action, can be varied by shortening or lengthening the tie-ropes between post and jib.

derris *(Chem.)*. An extract of the root of the *Derris* tree, of which rotenone is the chief toxic constituent; effective contact insecticide.

dertrotheca *(Zool.)*. In Birds, the horny covering of the maxilla. Also *dertrum*.

DES *(Comp.)*. See **data encryption standard**.

desalination *(Chem.)*. The production of fresh from sea water by one of several processes, including *distillation, electrodialysis* and reverse *osmosis*.

desaturation *(Image Tech.)*. The presence of grey in a colour; sometimes intentionally introduced in a colour reproduction process for artistic effect.

desaturation *(Phys.)*. Adding white light to a saturated colour (pure spectral wavelength) produces a desaturated or pale colour. Cf. **saturation**.

de-scaling *(Eng.)*. The process of (1) removing scale or metallic oxide from metallic surfaces by **pickling**; (2) removing scale from the inner surfaces of boiler plates and water tubes.

Descartes' rule of signs *(Maths.)*. No algebraic equation $f(x)=0$ can have more positive or negative roots respectively than there are changes of sign from + to − and from − to + in the polynomial $f(x)$ or $f(-x)$.

descending *(Zool.)*. Running from the anterior part of the body to the posterior part, or from the cephalic to the caudal region.

descending letters *(Print.)*. Letters the lower part of which is below the base line; e.g., *p, q, y*.

descending node *(Space)*. For Earth, the point at which a satellite crosses the equatorial plane travelling north to south.

descloizite *(Min.)*. Hydrated vanadium-lead-zinc ore with the general formula $PbZn(VO_4)OH$. Important source of vanadium. Occurs in the oxidation zone of lead-zinc deposits.

describer *(Civ.Eng.)*. The apparatus, either in signal cabins or for public use, which indicates movements, destinations etc. of trains.

de-seaming *(Eng.)*. Process of removing surface blemishes, or superficial slag inclusions from ingots or blooms.

desensitization *(Med.)*. The method of abolishing the sensitivity of a person to a protein by injecting small amounts of the same protein.

desensitization, systematic *(Behav.)*. A form of **behaviour therapy**, used especially in the treatment of **phobias**, in which fear is reduced by exposing the individual to the feared object in the presence of a stimulus that inhibits the fear; usually some form of relaxation is involved. See also **counter-conditioning**.

desensitize *(Print.)*. To treat a lithographic plate with a solution of gum arabic to promote the hydrophilic properties of the plate.

desensitizer *(Image Tech.)*. Substance which destroys the sensitivity of an emulsion without affecting the latent image.

desertification *(Ecol.)*. Formation of deserts from vegetated zones by the action of drought and/or increased populations of humans and herbivores.

desert rose *(Min.)*. A cluster of platy crystals, often including sand grains, formed in arid climates by evaporation. Typically barytes or gypsum. Also called *rock rose*.

desiccants *(Chem.)*. Substances of a hygroscopic nature, capable of absorbing moisture and therefore used as drying agents; e.g. anhydrous sodium sulphate, anhydrous calcium chloride, phosphorus pentoxide etc.

desiccation *(For.)*. See **seasoning**.

desiccator *(Chem.)*. Laboratory apparatus for drying substances; it consists of a glass bowl with ground-in lid, containing a drying agent, e.g. concentrated sulphuric acid or anhydrous calcium chloride; a tray for keeping glassware etc. in position is also provided, and if desired the desiccator can be evacuated.

designation marks *(Print.)*. An indication of the title of a book, printed in small letters in the same line with the **signature** on the first page of each section.

desilverization *(Eng.)*. The process of removing silver (and gold) from lead after softening. See **Parke's process**, **Pattinson's process**.

desizing *(Textiles)*. Removal of *size* from woven fabrics.

desk switchboard *(Elec.Eng.)*. A form of switchboard panel in which the operating switches, pilot lamps etc. are mounted on panels inclined to the horizontal like the surface of a desk.

desk top publishing *(Comp.)*. Software for use on a microcomputer which allows text to be manipulated with control of e.g. font, type size, microjustification, column organisation and the incorporation of graphics. This is done with **mouse** and **windows**, without the obvious control codes needed in normal typesetting. It will contain a *raster image processor* producing output compatible with a **page description language**. The resulting output should be able to drive a **laser printer** for proofing and short runs and a high-resolution typesetter for normal printing and publishing.

Deslandres equation *(Phys.)*. An empirical expression for the positions of the origins or heads in a band spectrum. $\nu = a + bn + cn^2$, ν being the wave number of the head, *a*, *b*, and *c* constants and *n* taking successive integral values.

de-sliming *(Min.Ext.)*. Removal of very fine particles from an ore pulp, or classification of it into relatively coarse and fine fractions.

desmids *(Bot.)*. Green algae, two families of the Charophyceae. Unicellular, the cells usually symmetrical about a median constriction, often elaborate in shape, and moving by the secretion of mucilage. Characteristic of rather acid fresh water habitats.

desmine *(Min.)*. See **stilbite**.

Desmodur *(Plastics)*. TN for isocyanates used to produce polyurethanes in conjunction with polyester resins.

desmognathous *(Zool.)*. In Birds, said of a type of palate in which the vomers are small or wanting and the maxillopalatines meet in the middle line; the palatines and pterygoids articulate with the basisphenoid rostrum.

desmosome *(Biol.)*. Strong intercellular junctions which bind cells together, either at discrete points at the surface, *spot desmosomes*, or as continuous bands around cells, *belt desmosomes*. The *hemidesmosome* is a similar structure binding an epithelial cell to the basal lamina. The two membranes of associated cells remain separated in the desmosome by an intercellular space which is traversed by protein filaments that extend into the adjacent cytoplasm of both cells.

desmotropism *(Chem.)*. A special case of **tautomerism** which consists in the change of position of a double bond, and in which both series of compounds can exist independently; e.g. keto and enol form of acetoacetic ester, malonic ester, phenyl-nitromethane etc.

desorption *(Chem.)*. Reverse process to **adsorption**. See **outgassing**.

de-spun antenna *(Space,Telecomm.)*. One used in a *spin-stabilized* communication satellite; the antenna rotates at a speed equal to and in the opposite sense to the body of the satellite, thereby continuing to point in the required direction, usually an antenna on Earth.

desquamation *(Med.)*. The shedding of the surface layer of the skin.

Destriau effect *(Phys.)*. A form of electroluminescence arising from localized regions of very intense electric field associated with impurity centres in the phosphor.

destructive distillation *(Chem.)*. The distillation of solid substances accompanied by their decomposition. The destructive distillation of coal results in the production of coke, tar products, ammonia, gas etc.

destructive read-out *(Comp.)*. Clearing of a storage location simultaneously with reading its contents. Abbrev. *DRO*.

destructor station *(Elec.Eng.)*. An electric generating station in which the fuel used consists chiefly of town or other refuse.

desulphurizing *(Chem.Eng.)*. The process of removing sulphur from hydrocarbons by chemical reactions. Many are known by proprietary names listed separately, e.g. **Appleby Frodingham**, **Claus**, **Houdry**.

desuperheater *(Eng.)*. A vessel in which superheated steam is brought into contact with a water spray in order to make saturated or less highly superheated steam. See **superheated steam**.

detachable key switch *(Elec.Eng.)*. A switch which can be operated only by a special key which is kept under supervisory control.

detached escapement *(Horol.)*. An escapement in which there is a minimum of interference with the free vibration of the pendulum or balance.

detail drawing *(Build.)*. A large-scale working drawing (usually of a part only) giving information which does not appear on small-scale drawings of the whole construction.

detail paper *(Paper)*. A translucent tracing paper, usually unoiled.

detector *(Nuc.Eng.)*. Device in which presence of radiation induces physical change which is observable. See **scintillation counter**, **proportional counter**, **Geiger Muller counter**, **germanium radiation detector**, **nuclear emulsion**, etc.

detector *(Telecomm.)*. A circuit which turns any carrier into a form which can be heard or displayed; if the carrier is unmodulated, a dc voltage is the output. A modulated carrier gives the modulating signal as an output; commonly called *demodulation*. See **linear detector**, **square-law detector**.

detector finger *(Print.)*. See **web break detector**.

detent *(Eng.,Horol.)*. A catch which, in being removed, initiates the motion of a machine. In the chronometer escapement of a clock, the detent carries a stone or jewel for locking the escape wheel.

detent escapement *(Horol.)*. See **chronometer escapement**.

detergents *(Chem.)*. Cleansing agents (solvents, or mixtures thereof, sulphonated oils, abrasives etc.) for removing dirt, paint etc. Commonly refers to soapless detergents, containing surfactants which do not precipitate in hard water; sometimes also a protease enzyme to achieve 'biological' cleaning and a whitening agent (see **fluorescent whitening agents**). See also **cationic detergents**, **non-ionic detergents**.

determinant *(Maths.)*. A square array of numbers representing the algebraic sum of the products of the numbers, one from each row and column, the sign of each product being determined by the number of interchanges required to restore the row suffices to their proper order, e.g.

$$\begin{matrix} a_1\, b_1\, c_1 \\ a_2\, b_2\, c_2 \\ a_3\, b_3\, c_3 \end{matrix} \quad = \quad \begin{matrix} a_1b_2c_3 - a_1b_3c_2 \\ -a_2b_1c_3 - a_2b_3c_1 \\ +a_3b_1c_2 - a_3b_2c_1. \end{matrix}$$

determinate *(Bot.)*. (1) Growth of a limited extent. (2) Sympodial growth. (3) Cymose inflorescence.

determinate *(Eng.)*. Said of a structure which is a **perfect frame**. Cf. **indeterminate**.

determinate cleavage *(Zool.)*. A type of cleavage in which each blastomere has a predetermined fate in the later embryo.

deterministic *(Comp.)*. Applies to a machine if the next state of the machine can be predicted from its present state and any new input. Cf **non-deterministic**.

de-tinning *(Eng.)*. Chlorine treatment to remove tin coating from metal scrap.

detonating fuse *(Min.Ext.)*. Fuse with core of TNT which burns at some 6 km/s and is itself set off by detonator. Varieties include **primacord**, **Cordtex**, and *Cordeau Detonnant*.

detonation *(Autos.)*. In a petrol engine, spontaneous combustion of part of the compressed charge after the passage of the spark; the accompanying knock. It is caused by the heating effect of the advancing flame front, which raises the gas remote from the plug to its spontaneous **ignition temperature**.

detonation *(Chem.)*. Decomposition of an explosive in which the rate of heat release is great enough for the explosion to be propagated through the explosive as a steep shock front, at velocities above 1 km/s and pressures above 10^9 N/m^2.

detonation meter *(Eng.)*. An instrument for measuring quantitatively the severity and frequency of detonation in a petrol-engine cylinder. See **bouncing-pin detonation meter**.

detorsion *(Zool.)*. In Gastropoda, partial or complete reversal of torsion, manifested by the untwisting of the visceral nerve loop and the altered position of the ctenidium and anus.

detrital mineral *(Geol.)*. A mineral grain derived mechanically from a parent rock. Typically such minerals are resistant to weathering; e.g. diamonds, gold and zircon. See **heavy minerals**.

detrition *(Geol.)*. The natural process of rubbing or wearing down strata by wind, running water or glaciers. The product of detrition is *detritus*.

detritovore *(Ecol.)*. Organism which eats detritus.

detritus *(Biol.)*. Organic material formed from decomposing organisms.

detritus *(Geol.)*. See **detrition**.

detritus chamber (pit) *(Build.)*. A tank through which crude sewage is first passed in order to allow the largest and heaviest of suspended matters to fall to the bottom, from which they can be removed.

detrusor *(Med.)*. The muscular coat of the urinary bladder (detrusor urinae); sometimes used for the outer of the three muscular coats of the bladder.

Dettol *(Chem.)*. TN for nontoxic and nonirritant germicide of which the active principle is chloroxylenol dissolved in a saponified mixture of aromatic oils, e.g. terpineol.

detumescence *(Med.)*. The reduction of a swelling.

detuning *(Telecomm.)*. Adjustment of a resonant circuit so that its resonant frequency does not coincide with that of the applied e.m.f.

Deuce *(Comp.)*. TN for a first-generation computer, descended from *pilot ACE*. (*Digital Electronic Universal Computing Engine*).

deuteranopic *(Phys.)*. Colour blind to green.

deuteration *(Chem.)*. The addition of or replacement by deuterium atoms in molecules.

deuterium *(Chem.)*. Isotope of element hydrogen having one neutron and one proton in nucleus. Symbol D, when required to be distinguished from natural hydrogen, which is both ^{1}H and ^{2}H. This heavy hydrogen is thus twice as heavy as ^{1}H, but similarly ionized in water. Used in isotopic 'labelling' experiments (e.g. mechanism of esterification). See **deuteron**.

deutero- *(Genrl.)*. Prefix, in general denoting second in order, derived from. Particularly, in chemistry, containing heavy hydrogen (**deuterium**).

Deuteromycotina *(Bot.)*. *Deuteromycetes, Fungi Imperfecti*. The imperfect fungi. Form subdivision or form class of the Eumycota or true fungi for which no sexual reproduction is known. Typically mycelial with simple septa. The affinities of many appear to be with the Ascomycotina. Include many saprophytes e.g. *Aspergillus*, causing food spoilage, *Penicillium*, a source of antibiotics and also plant parasites, e.g. *Fusarium*, *Verticillium* causing wilt diseases etc.

deuteron *(Chem.)*. Charged particle, D^+, the nucleus of **deuterium**, a stable but lightly-bound combination of one proton and one neutron. It is mainly used as a bombarding particle accelerated in cyclotrons.

deuterostoma *(Zool.)*. In development, a mouth which arises secondarily, as opposed to a mouth which arises by modification of the blastopore.

deuterotoky *(Zool.)*. Parthenogenesis leading to the production of both males and females.

deutocerebron, deutocerebrum *(Zool.)*. In higher Arthropoda, as Insects and Crustacea, the fused ganglia of the second somite of the head, forming part of the 'brain'.

developable surface *(Maths.)*. That which, without shrinking or stretching, may be rolled out on to a plane.

developed dyes *(Chem.)*. Dyes which are developed on the fibre by the interaction of the constituents which produce them. Dyeing with aniline black provides an example of a *developed dye*.

developer *(Image Tech.)*. Chemical solution which converts a **latent** photographic image to a visible one by reducing the exposed silver compounds to metallic silver; in colour photography this metallic silver may be associated with the formation of dyes. In addition to the **developing agent**, it generally contains an alkaline **accelerator**, a **restrainer** and a preservative.

developer streaks *(Image Tech.)*. Streaks, usually following areas of heavy image density, resulting from irregular chemical action by the solution in a processing machine. See **directional effects**.

developing agent *(Image Tech.)*. Chemical having the property of reducing light-exposed silver halide grains to metallic silver; for black-and-white processing the most common are *metol*, *hydroquinone* and *phenidone*.

development *(Image Tech.)*. The conversion of an

exposed **latent image** to a visible one by chemical reaction.

development *(Min.Ext.)*. Opening of ore body by access shafts, drives, crosscuts, raises and winzes for purpose of proving mineral value (ore reserve) and exploiting it.

deviance *(Stats.)*. A statistic measuring the degree of fit of a statistical model by means of comparison with the degree of fit of a more complete model.

deviation *(Elec.Eng.)*. The difference between the instantaneous quantity and the *index (or desired) value*, e.g. difference between the magnetic north and the setting of a compass needle due to a local magnetic disturbance, or between the actual value of the controlled variable and that to which the controlling mechanism is set in an automatic control system. Also called *error*.

deviation *(Min.Ext.)*. Of a drill hole departing from the vertical either by design or accidentally.

deviation *(Ships)*. The angle between the **magnetic meridian** and the **compass meridian**. A *deviation table* is normally prepared showing the deviation for various headings.

deviation *(Stats.)*. The difference between an observation and a fixed value. Statistical measures of dispersion are often based on the deviation of each value in a set of observations from their common mean.

deviation *(Telecomm.)*. In a **frequency modulation** system, the extent to which the carrier frequency moves from its unmodulated position when the modulating signal is applied.

deviation distortion *(Telecomm.)*. Consequence of any restriction of bandwidth or linearity of discrimination in the transmission and reception of a frequency-modulated signal.

deviation IQ *(Behav.)*. See **intelligence quotient**.

deviation ratio *(Telecomm.)*. In a system using **frequency modulation**, the ratio of the maximum possible frequency deviation of the carrier to the maximum frequency of the modulating signal.

device driver *(Comp.)*. See **driver**.

devil *(Meteor.)*. A small whirlwind due to strong convection, which, in the tropics, raises dust or sand in a column.

devil float *(Build.)*. A square float having four nails projecting from its working face; used to perform the scoring required in **devilling**.

devilling *(Build.)*. The operation of scoring the surface of a plaster coat to provide a key for another coat.

devitrification *(Geol.)*. Deferred crystallization which, in the case of glassy igneous rocks, converts obsidian and pitchstone into dull crypto-crystalline rocks (usually termed **felsites**) consisting of minute grains of quartz and feldspar. Such devitrified glasses give evidence of their originally vitreous nature by traces of perlitic and spherulitic textures.

devitrification *(Glass.)*. A physical process which causes a change from the glassy state to a minutely crystalline state and has to be avoided in manufacture. If the change is due to lapse of time, the glass becomes turbid and brittle.

Devonian *(Geol.)*. The oldest period of the Upper Palaeozoic, covering a time-span between app. 400 and 360 million years. The corresponding system of rocks.

dew *(Meteor.)*. The deposit of moisture on exposed surfaces which accumulates during calm, clear nights. The surfaces become cooled by radiation to a temperature below the dew-point, thus causing condensation from the moist air in contact with them.

Dewar flask *(Chem.)*. A silvered glass flask with double walls, the space between them being evacuated; used for storing, e.g. liquid air.

de-watering *(Civ.Eng.)*. The process of pumping water from the interior of a caisson, or of excluding water from an excavation, by the use of pipes sunk round the perimeter, with continuous pumping.

de-watering *(Min.Ext.)*. Partial or complete drainage by thickening, sedimentation, filtering or on screen as a process aid, to facilitate shipment or drying of product.

dew-blown *(Vet.)*. See **bloat**.

dew claw *(Zool.)*. In Dogs, the useless claw on the inner side of the limb (especially the hind limb) which represents the rudimentary first digit.

DEW line *(Radar.)*. Line of radar missile warning stations along 70th parallel of latitude from Alaska to Greenland. (*Distant Early-Warning line*).

dew-point *(Meteor.,Phys.)*. The temperature at which a given sample of moist air will become saturated and deposit dew, if in contact with the ground. Above ground, condensation into water droplets takes place.

dew-point hygrometer *(Meteor.)*. A type of hygrometer for determining the dew-point, i.e. the temperature of air when completely saturated. The relative humidity of the air can be ascertained by reference to vapour pressure tables.

dew pond *(Meteor.)*. A water-tight hollow, usually on elevated land, where the combined effects of rainfall and fog drip exceed that of evaporation. The effect of dew is negligible and the name is the result of an ancient misunderstanding.

dexiocardia *(Med.)*. See **dextrocardia**.

dexiotropic *(Zool.)*. Twisting in a spiral from left to right; spiral cleavage.

dextral *(Genrl.)*. See **dextrorse**.

dextral fault *(Geol.)*. A **tear fault** in which the rocks on one side of the fault appear to have moved to the right when viewed from across the fault. The opposite of a **sinistral fault**.

dextran *(Chem.)*. A polyglucose formed by micro-organisms in which the units are joined mainly by α—$1\rightarrow6$ links with variable amounts of crosslinking, via α—$1\rightarrow4$, α—$1\rightarrow3$, or α—$1\rightarrow2$. Hydrolysed by dextranases.

dextrin *(Chem.)*. *Starch gum*. A term for a group of intermediate products obtained in the transformation of starch into maltose and *d*-glucose. Dextrins are obtained by boiling starch alone or with dilute acids. They do not reduce **Fehling's solution**. Crystalline dextrins have been obtained by the action of *Bacillus macerans*.

dextrocardia *(Med.)*. A congenital anomaly with the heart situated in the right side of the chest often with similar transposition of **abdominal viscera**; the condition is then termed *situs inversus*.

dextrorotatory *(Phys.)*. Said of an optically active substance which rotates the plane of polarization in a clockwise direction when looking against the incoming light.

dextrorse *(Biol.)*. Helical, twisted or coiled in the sense of conventional (right-handed) screw thread or of a Z-helix. Cf. **sinistrorse**.

dextrose *(Chem.)*. See **glucose**.

DFS *(Comp.)*. See **disk filing system**.

DFVLR *(Space)*. Abbrev. for *Deutsche Forschung und Versuchanstalt für Luft und Raumfahrt*, the German centre for aerospace research.

DH *(Build.)*. Abbrev. for *Double-Hung*.

diabantite *(Min.)*. A variety of chlorite relatively rich in iron and poor in aluminium.

diabase *(Geol.)*. The American term for *dolerite*.

diabetes insipidus *(Med.)*. A condition in which there is an abnormal increase in the amount of urine excreted, as a result of disease of, or injury to, the posterior lobe of the pituitary gland, reducing the secretion of antidiuretic hormone.

diabetes mellitus *(Med.)*. A disorder of metabolism due to insulin lack in which there is an excess of sugar in the blood and urine with excessive thirst and loss of weight. In the young there is probably an auto-immune destruction of the islet cells of the pancreas leading to an absolute requirement for insulin therapy. In older patients (*maturity onset diabetes*) the insulin lack is only relative and diet or oral insulin stimulating drugs may provide sufficient control.

diabetic coma *(Med.)*. When diabetes becomes uncontrolled there is an accumulation of **ketones** causing metabolic acidosis. The blood sugar is very high and dehydration and loss of potassium also occur. The patient may become confused or stuporous, but this may be a late feature and the term coma is misleading. The condition is more often termed *diabetic keto-acidosis*.

diabetic keto-acidosis *(Med.)*. The build up of **ketones** resulting in **acidosis** in uncontrolled diabetes mellitus. There is a high blood sugar and dehydration.

diacetone alcohol *(Chem.)*. $CH_3.CO_2.CH_2.C(CH_3)_2.OH$,

diacetyl *(Chem.)*. *Butane 2,3-dione*, $CH_3.CO.CO.CH_3$, a yellow-green liquid, bp 87°C. It is the simplest diketone, and is obtained by the action of nitrous acid on methyl ethyl ketone. It occurs naturally in butter and is one of the constituents giving butter its characteristic flavour.

diachronism *(Geol.)*. The transgression across time planes by a geological formation. A bed of sand, when traced over a wide area, may be found to contain fossils of slightly different ages in different places, as, when deposited during a long-continued marine transgression, the bed becomes younger in the direction in which the sea was advancing. adj. *diachronous*.

diacoele *(Zool.)*. In Craniata, the third ventricle of the brain.

diacritical current *(Elec.Eng.)*. The current in a coil to produce a flux equal to half that required for saturation.

diacritical point *(Elec.Eng.)*. The point on the magnetizing curve of a sample of iron at which the intensity of magnetization has half its saturation value.

diadelphous *(Bot.)*. Stamens of a flower fused by their filaments into 2 groups.

diadochy *(Geol.,Min.)*. The replacement of one element by another in a crystal structure.

diagenesis *(Geol.)*. Those changes which take place in a sedimentary rock at low temperatures and pressures after its deposition. These include compaction, cementation, recrystallization, etc. Cf. **metamorphism**.

diagnosis *(Bot.)*. A formal description of a plant, having special reference to the characters which distinguish it from related species.

diagnosis *(Med.)*. The determination of the nature of a disordered state of the body or of the mind; the identification of a diseased state.

diagnostic *(Comp.)*. Aid in the debugging of programs or systems. See **syntax analysis, test data, trace**.

diagnostic characters *(Zool.)*. Characteristics by which one genus, species, family, or group can be differentiated from another.

diagonal *(Eng.)*. A tie or strut joining opposite corners of a rectangular panel in a framed structure.

diagonal *(Maths.)*. (1) Of a polygon: a straight line drawn between 2 nonadjacent vertices. (2) Of a polyhedron: a straight line drawn between two vertices which do not lie in the same face. (3) Of a square matrix or determinant matrix: the leading or principal diagonal is that running from the top left-hand corner to the bottom right-hand corner.

diagonal bond *(Build.)*. See **raking bond**.

diagonal eyepiece *(Surv.)*. One incorporating a right-angle prism for convenience in surveying steep lines with telescope or theodolite.

diagonal matrix *(Maths.)*. A square matrix in which all the elements except those in the leading diagonal are zero.

diagram *(Eng.)*. A curve which indicates the sequence of operations in a machine.

diagram *(Maths.)*. (1) An outline figure or scheme of lines, points, and spaces, designed to: (a) represent an object or area; (b) indicate the relation between parts; (c) show the value of quantities or forces.

dial *(Horol.)*. Graduated plate immediately behind the hands of a clock or other timekeeper, from which the time is read.

dial *(Min.Ext.)*. A large compass mounted on a tripod, used for surveying or mapping workings in coal-mines.

dial *(Telecomm.)*. A calling device operated by the rotation of a disk which, on its release, produces the pulses necessary to establish a connection in an automatic system.

dialdehydes *(Chem.)*. Compounds containing two aldehyde groups. The most important one is **glyoxal** (ethan 1,2-dial).

dial foot *(Horol.)*. (1) Circular pins, attached to the back of a dial, which enter corresponding holes in the pillar plate to ensure its correct location. (2) In some clocks, the dial feet are pinned to the pillar plate to hold the dial firm.

dial gauge *(Eng.)*. A sensitive measuring instrument in which small displacements of a plunger are indicated in 1/100 mm, or similar length units, by a pointer moving over a circular scale.

dialkanones *(Chem.)*. See **diketones**.

dialkenes *(Chem.)*. *Diolefins*. Hydrocarbons containing 2 double bonds in their molecules. They exist in 3 forms according to the disposition of the double bonds, i.e. adjacent as in the allenes; separated by 1 single bond in the conjugated compounds such as butadiene; or separated by 2 or more single bonds.

diallage *(Min.)*. An ill-defined, altered monoclinic pyroxene, in composition comparable with augite or diopside, with a lamellar structure; occurs typically in basic igneous rocks such as gabbro.

dialling *(Min.Ext.,Surv.)*. The process of running an underground traverse with a mining **dial**.

dialogite *(Min.)*. See **rhodochrosite**.

dial plate *(Horol.)*. See **bottom plate**.

dial switch *(Elec.Eng.)*. A multicontact switch in which the contacts are arranged in the arc of a circle, so that contact can be made by a radial moving arm.

dialypetalous *(Bot.)*. See **polypetalous**.

dialyser *(Chem.)*. Dialysis apparatus.

dialysis *(Chem.)*. The separation of a colloid from a substance in true solution by allowing the latter to diffuse through a semipermeable membrane. The process is used in the **artificial kidney**.

diamagnetism *(Phys.)*. Phenomenon in some materials in which the susceptibility is negative, i.e. the magnetization opposes the magnetizing force, the permeability being less then unity. It arises from the precession of spinning charges in a magnetic field. The susceptibility is generally one or two orders of magnitude weaker than typical paramagnetic susceptibility.

diameter *(Maths.)*. (1) Intercept made by the circumference on a straight line through the centre of a circle. (2) Straight line bisecting a family of parallel chords of a conic. Cf. **conjugate diameters**. (3) Straight line which passes through the centre of parallel sections of a quadric surface.

diameter of commutation *(Elec.Eng.)*. The diametral plane in which the coils of an armature winding that are undergoing commutation should be situated for perfect commutation.

diametral winding *(Elec.Eng.)*. An armature winding in which the number of slots is a multiple of the number of poles.

diametrical pitch *(Eng.)*. Of a gear-wheel, the number of teeth divided by the pitch diameter.

diametrical tappings *(Elec.Eng.)*. Tappings taken on a closed armature, which are diametrically opposite to each other, i.e. displaced from each other by 180 electrical degrees.

diametrical voltage *(Elec.Eng.)*. The voltage between opposite lines of a symmetrical 6-phase system, or the voltage between tappings on an armature winding which are diametrically opposite to each other, i.e. displaced from each other by 180°.

diametric system *(Crystal.)*. See **tetragonal system**.

diamines *(Chem.)*. Compounds containing two amino groups.

diaminoethanetetraacetic acid *(Chem.)*. $(HOOC.CH_2)_2N.CH_2.CH_2.N(CH_2COOH)_2$, usually known as EDTA. A tetrabasic acid whose anions can function as hexadentate ligands. It is much used in analysis and as a sequestrating agent for metal ions, particularly the group 2A ions, Mg^{2+} and Ca^{2+}.

diamino-pimelic acid *(Biol.)*. A cell wall constituent of some bacteria and blue-green algae, not known to occur in any other group.

diamond *(Min.)*. One of the crystalline forms of carbon; it crystallizes in the cubic system, rarely in cubes, commonly in forms resembling an octahedron, and less commonly in the tetrahedron. Curved faces are characteristic. It is the hardest mineral (10 in Mohs' scale); hence valuable as an abrasive, for arming rock-boring tools, etc.; its high dispersion and birefringence makes it valuable as a gemstone. Occurs in blue ground, in river gravels, and in shore sands. See **black diamond, bort, industrial diamonds, kimberlite**.

diamond *(Print.)*. An old type size, about $4\frac{1}{2}$-point.

diamond antenna *(Telecomm.)*. Same as *rhombic antenna*.

diamond bit *(Min.Ext.)*. Drilling bit in which industrial diamonds are set in the cutting portions of the bit. Cuts hard rock and wears longer, so increasing the time between replacing bits involving a **round trip**.

diamond die *(Eng.)*. A wire drawing die containing a diamond insert, for reduced wear.

diamond dust *(Min.Ext.)*. The hardest abrasive agent, used as loose powder or paste in polishing, or embedded in metal tool parts. e.g. *rock saws* and *drill bits*.

diamond mesh *(Build.)*. A form of **expanded metal** with a diamond-shaped network.

diamond saw *(Build.)*. (1) Stone-cutting circular saw used with diamond dust for cutting rock sections. (2) A band or frame saw with diamond-bonded cutting elements, used for cutting stone, concrete (e.g. in 'bump-cutting' in runways), plastics etc.

diamond-skin disease *(Vet.)*. See **swine erysipelas**.

diamond tool *(Eng.)*. A diamond of specified shape, mounted in a holder which is used for precision machining of non-ferrous metals and ceramics.

diamond wheel *(Eng.)*. A rotating wheel for a grinding machine in which small diamonds are embedded, and which is used for grinding and cutting very hard materials.

diamond-work *(Build.)*. A wall constructed of lozenge-shaped stones laid in courses.

dianysl-trichloroethane *(Chem.)*. See **methoxychlor**.

diapause *(Zool.)*. In Insects, a state, which may arise at any stage of the life-cycle, in which development is suspended and cannot be resumed, even in the presence of apparently favourable conditions, unless the diapause is first 'broken' by an appropriate environmental change.

diapedesis *(Zool.)*. In Porifera, the passage to the exterior of cells primarily occupying the interior of certain types of larva: in Vertebrates, the passage of blood leucocytes through the walls of blood vessels into the surrounding tissues.

diaper-work *(Build.)*. Paving constructed in a chequered pattern, composed of stones or tiles of different colours.

diaphoresis *(Med.)*. Perspiration.

diaphoretic *(Med.)*. Producing perspiration: a medicine which does this.

diaphragm *(Acous.)*. A vibrating membrane as in a loudspeaker, telephone and similar sound sources, also in receivers, e.g. the human ear-drum. Also *open-diaphragm loudspeaker, pleated-diaphragm loudspeaker*.

diaphragm *(Bot.)*. A plate of cells crossing a space. Especially a plate one cell thick with many intercellular pores through which air but not water may pass between intercellular air spaces in the submerged stems etc. of many aquatic plants.

diaphragm *(Build.)*. A web across a hollow terracotta block, forming compartments.

diaphragm *(Elec.Eng.)*. A sheet of perforated or porous material placed between the positive and negative plates of an accumulator cell.

diaphragm *(Image Tech.,Phys.)*. Same as **stop** (1).

diaphragm *(Surv.)*. A flanged brass ring which is held in place in a surveying telescope by means of four screws, and which receives the **reticule**.

diaphragm *(Zool.)*. Generally a transverse partition subdividing a cavity. In Mammals, the transverse partition of muscle and connective tissue which separates the thoracic cavity from the abdominal cavity; in Anura, a fan-shaped muscle passing from the ilia to the oesophagus and the base of the lungs; in some Arachnida, a transverse septum separating the cavity of the prosoma from that of the abdomen; in certain Polychaeta, a strongly developed transverse partition dividing the body cavity into two regions.

diaphragm cell *(Chem.Eng.)*. An electrolytic cell in which a porous diaphragm is used to separate the electrodes, thus permitting electrolysis of, principally, sodium chloride without recombination in the cell of electrolysis products.

diaphragm plate *(Build.)*. A connecting stiffener between the webs of a box girder.

diaphragm pump *(Eng.)*. A pump in which a flexible diaphragm set between two non-return valves replaces a piston or bucket, being clamped round the edge and attached at the centre to a reciprocating rod of short stroke.

diaphysectomy *(Med.)*. Excision of part of the shaft of a long bone.

diaphysis *(Zool.)*. The shaft of a long limb bone.

diaphysitis *(Med.)*. Inflammation of the diaphysis.

diapir *(Geol.)*. An intrusion of relatively light material into pre-existing rocks, doming the overlying cover. Applied especially to salt and igneous intrusions.

diapophyses *(Zool.)*. A pair of dorsal transverse processes of a vertebra, arising from the neural arch.

diapositive *(Image Tech.)*. A positive transparency on glass or film.

diapsid *(Zool.)*. Said of skulls in which the supra- and infra-temporal fossae are distinct.

diarrhoea, diarrhea *(Med.)*. The frequent evacuation of liquid faeces.

diarthrosis *(Zool.)*. A true (as opposed to a fixed) joint between two bones, in which there is great mobility; a cavity, filled with a fluid, generally exists between the two elements.

diaspore *(Min.)*. Alumina monohydrate, occurring in bauxite. A dimorph of **boehmite**.

diastase *(Biol.)*. An enzyme capable of converting starch into sugar. Produced during the germination of barley in the process of malting.

diastasis *(Med.)*. The separation, without fracture, of an epiphysis from the bone.

diastema *(Zool.)*. (1) An equatorial modification of protoplasm preceding cell-division. (2) A gap in a jaw where there are no teeth, as in Mammals lacking the canines.

diaster *(Biol.)*. In cell-division, a stage in which the daughter chromosomes are situated in two groups near the poles of the spindle, ready to form the daughter nuclei.

diastereoisomers *(Chem.)*. Stereoisomers which are not simple mirror images (enantiomers) of one another. For example, in a molecule with two chiral centres, the R,R- and S,S-forms are enantiomers, while the S,R- and R,S-forms are diastereoisomers.

diastole *(Bot.,Zool.)*. Rhythmical expansion, as of the heart, or of a contractile vacuole.

diastolic murmur *(Med.)*. A murmur heard over the heart during diastole usually indicative of valvular disease.

diastrophism *(Geol.)*. Large-scale deformation of the Earth's crust.

diathermanous *(Phys.)*. Relatively transparent to radiant heat.

diathermic surgery *(Med.)*. The use of an electric arc in preference to a knife. This has the advantage of sealing cuts and reducing bleeding.

diathermy *(Med.)*. Heating of tissues by high-frequency electric currents. In physiotherapy used to heat structures like joints and muscles under the skin. In surgery to cause coagulation and necrosis by localized application.

diathesis *(Med.)*. The constitutional state of the body which renders it liable to certain diseases.

diatomite, diatomaceous earth *(Min.)*. A siliceous deposit occurring as a whitish powder consisting essentially of the frustules of diatoms. It is resistant to heat and chemical action, and is used in fireproof cements, insulating materials, as an absorbent in the manufacture of explosives and as a filter. Also *infusorial earth, kieselguhr, tripolite.*

diatom ooze *(Min.).* A deep-sea deposit consisting essentially of the frustules of diatoms; widely distributed in high latitudes.

diatoms *(Bot.).* See **Bacillariophyceae**.

diatoni *(Build.).* Quoins having two dressed faces projecting from the wall.

diatropism *(Bot.).* A **tropism** in which a plant part becomes aligned at right angles to the source of the orientating stimulus; e.g. rhizomes are typical *diagravitropic*. Cf. **plagiotropic, orthotropic**.

diazepam *(Med.).* A **benzediazepine** drug used in the treatment of anxiety states. Often useful as adjunctive treatment in alcohol withdrawal.

diazo *(Print.).* See **dyeline**.

diazoamino compounds *(Chem.).* Pale yellow crystalline substances obtained by the action of a primary or secondary amine on a diazonium salt. Their general formula is R.N = N.NHR′. They do not form salts and most of them are easily transformed into the isomeric aminoazo compounds.

diazo compounds *(Chem.).* Compounds of the general formula R.N=N.R′, obtained by the action of excess nitrous acid on aromatic amines at temperatures below 10°C. They are important intermediates for dyestuffs.

diazomethane *(Chem.).* CH_2N_2, an aliphatic diazo compound, an odourless, yellow, poisonous gas, very reactive, used for introducing a methyl group into a molecule. It is prepared from nitroso-methyl-urethane by decomposition with alcoholic potassium hydroxide.

diazonium salts *(Chem.).* The acid salts of diazobenzene of the general formula (RN≡N+)Cl–, important intermediates for azo-dyestuffs. They are usually prepared only in aqueous solution by the action of nitrous acid on an aromatic amine at low temperatures in the presence of excess of acid. The –N=N– group can easily be replaced by hydrogen, hydroxyl, halogen etc., and the diazonium salts can thus be transformed into other benzene derivatives.

diazo process *(Image Tech.).* A document-copying process in which diazonium compounds in the paper are destroyed by ultra-violet light and the unexposed areas converted to a coloured dye. See **dyeline**.

diazotization *(Chem.).* The process of converting amino into diazo compounds.

dibasic acids *(Chem.).* Acids containing two replaceable hydrogen atoms in the molecule.

dibenzoyl *(Chem.).* See **benzil**.

dibenzoyl-peroxide *(Chem.).* $C_6H_5.CO.O.O.CO.C_6H_5$. Relatively stable organic peroxide used mainly as a catalyst in polymerization and polycondensation reactions.

dibenzyl *(Chem.).* Symmetrical diphenylethane, $C_6H_5.CH_2.CH_2.C_6H_5$. Mp 52°C.

dibenzyl group *(Chem.).* A synonym for the stilbene group, comprising compounds containing two benzene nuclei linked together by a chain of two or more carbon atoms.

diborane *(Chem.).* The simplest borane, B_2H_6, bp −92°C. It is an example of **electron-deficient** bonding.

dibranchiate *(Zool.).* Having 2 gills or ctenidia.

dice *(Electronics).* Small regular pieces of semi-conductor material for fabrication of devices.

dicentric *(Biol.).* Having two centromeres.

dicephalus *(Med.).* A developmental 'monstrosity' in which a foetus has 2 heads.

dichasial cyme *(Bot.).* See **dichasium**.

dichasium, dichasial cyme *(Bot.).* An inflorescence in which the main stem ends in a flower and bears, from near the flower, two lateral branches also ending in a flower. These branches may in turn bear further similar branches and so on. See **cymose inflorescence, monochasium**.

dichlamydeous *(Bot.).* Having distinct calyx and corolla. Cf. **perianth**.

dichlorodifluoromethane *(Chem.).* CCl_2F_2, *Freon 12*. Used as a refrigerant, solvent and in fire extinguishers, bp −30°C.

dichloroethylenes *(Chem.). Dichloroethenes*, $C_2H_2Cl_2$. Exist in *cis*-form, bp 48°C and *trans*-form, bp 60°C. Prepared industrially from tetrachloroethane. Used as source material for vinyl chloride, the momomer of PVC.

dichloromethane *(Chem.). Methylene chloride*, CH_2Cl_2. Colourless liquid; bp 41°C. Used widely as a solvent, e.g. in manufacture of cellulose acetate.

dichlorophen *(Chem.).* Organic compound based on **hexachlorophene** widely used as an antibacterial agent, particularly for water treatment.

dichlorophenoxyacetic acid *(Bot.). 2,4-D.* Synthetic **auxin** used as a selective herbicide and in media for tissue culture.

1,2-dichloropropane *(Chem.).* $CH_3.CHCl.CH_2Cl$, *propylene dichloride*. Bp 94°C. Solvent for oils, fats, waxes, and resins.

dichocephalous *(Zool.).* Said of ribs which have 2 heads, a tuberculum, and a capitulum. Cf. **holocephalous**.

dichogamy *(Bot.).* The maturation of the anthers at a different time from the stigmas in the same flower, *protandry* or *protogyny*. Cf. **homogamy**.

dichoptic *(Zool.).* Having the eyes of the two sides distinctly separated.

dichotomy *(Astron.).* The half-illuminated phase of a planet, as the moon at the quarters, and Mercury and Venus at greatest elongation.

dichotomy *(Bot.).* Bifurcation of an organ, by the division of an apical cell or meristem into 2 equal parts each growing into a branch. Common in algae, also *Selaginella*. Cf. **false dichotomy**.

dichroic *(Chem.).* Said of materials, such as solution of chlorophyll, which exhibit one colour by reflected light and another colour by transmitted light.

dichroic fog *(Image Tech.).* Fog which arises from the formation of an organic compound of silver; so-called because of its reddish coloration by transmitted light, and greenish coloration by reflected light.

dichroic mirror *(Phys.).* Colour-selective mirror, which reflects a particular band of spectral energy and transmits all others.

dichroism *(Crystal.).* The property possessed by some crystals (e.g. tourmaline) of absorbing the ordinary and extraordinary ray to different extents; this has the effect of giving to the crystal different colours according to the direction of the incident light.

dichroism *(Min.).* See **pleochroic haloes**.

dichroite *(Min.).* **Cordierite**.

dichromates (VI) *(Chem.).* See **chromates (VI)**.

dichromatism *(Med.).* Colour blindness in which power of accurate differentiation is retained for only 2 bands of colour in the spectrum.

Dicke radiometer *(Telecomm.).* Sensitive receiving circuit which detects weak signals in noise by modulating them at the input before they reach conventional demodulation circuits. The output of the demodulator is compared with a reference signal from the modulator; when they coincide, this indicates signal presence. Sometimes used for poise measurements in microwave systems.

dickite *(Min.).* A clay mineral of the kaolin (or kandite) group, a hydrated aluminosilicate. It generally occurs in hydrothermal deposits.

Dick test *(Immun.).* A test for immunity against the toxin of *Streptococcus pyogenes* which causes scarlet fever. A small amount of toxin injected into the skin causes an area of redness after 6 or more hours in persons who do not have antibodies against the toxin.

dicliny *(Bot.).* Having unisexual flowers, either male and female on different individual plants (dioecy) or both on one plant (monoecy). Cf. **hermaphrodite**. adj. *diclinous*.

dicophane *(Chem.).* See **DDT**.

Dicotyledones *(Bot.). Magnoliopsida.* The dicotyledons, or dicots, the larger of the two classes of **angiosperms** or flowering plants. Trees, shrubs and herbs of which characteristically the embryo has two cotyledons, the parts of the flowers are in twos or fives or multiples of these numbers, and the leaves commonly are netveined. Contains ca 165 000 spp. in 250 families usually divided among 6 subclasses or superorders: Magnoliidae,

Hamamelidae, Caryophyllidae, Dilleniidae, Rosidae and Asteridae. Cf. **monocotyledons**.

dicrotic *(Med.)*. Having a double beat or wave; an acceleration of the normal pulse found in fevers.

dicty-, dictyo- *(Genrl.)*. Prefix from Gk. *diktyon*, net.

dictyosome *(Bot.)*. *Golgi body*. An element of the **golgi apparatus**.

dictyostele *(Bot.)*. A type of **solenostele**, typical of many ferns e.g. *Dryopteris*, in which overlapping leaf gaps dissect the vascular cylinder into anastomosing strands (meristeles) each with xylem surrounded by phloem amd endodermis.

dicyclic *(Bot.)*. Having the perianth in two whorls.

didactyl *(Zool.)*. Having all the toes of the hind feet separate, as in many Marsupialia. Cf. **syndactyl**.

didelphic *(Med.)*. Pertaining to a double uterus.

Didot point system *(Print.)*. The continental point system based on a 12-point 'cicero', the equivalent of the British **pica** but measuring approximately 12.8 British points. In the Didot system the point measures 0.376 compared to the British point size of 0.351 mm.

didymous *(Bot.)*. Formed of two similar parts, partially attached; twinned.

didynamous *(Bot.)*. Having 2 long and 2 short stamens.

die *(Build.)*. (1) The body of a pedestal. (2) The enlarged part at either end of a baluster, where it comes into the coping or the plinth.

die *(Eng.)*. (1) A metal block used in stamping operations. It is pressed down on to a blank of sheet-metal, on which the pattern or contour of the die surface is reproduced. (2) The element complementary to the punch in press tool for piercing, blanking, etc. (3) An internally threaded steel block provided with cutting edges, for producing screw threads by hand or machine. (4) A tool made of very hard material, often tungsten, carbide or diamond, with a (bell-mouthed) hole, usually circular, used to reduce the product cross-section by plastic flow, in wire or tube drawing.

dieback *(Bot.)*. Necrosis of a shoot, starting at the apex and progressing proximally.

die box, die head *(Eng.)*. The holder into which screw dies are fitted in a screwing machine.

die case *(Print.)*. In the **Monotype system**, the frame in which the matrices of a fount are assembled.

diecasting *(Eng.)*. Casting of metals or plastics into permanent moulds, made of suitably resistant non-deforming metal. See **gravity-, pressure-**.

diecasting alloys *(Eng.)*. Alloys suitable for diecasting, which can be relied on for accuracy and resistance to corrosion when cast. Aluminium-base, copper-base, tin-base, zinc-base and lead-base alloys are those generally used.

die-fill ratio *(Powder Tech.)*. The ratio of uncompacted powder volume to the volume of the green compact.

dieldrin *(Chem.)*. A contact insecticide based on HEOD, a chlorinated naphthalene derivative. Typical use is moth-proofing of carpets and other furnishings. Highly persistent and therefore now less favoured.

dielectric *(Phys.)*. Substance, solid, liquid or gas, which can sustain a steady electric field, and hence an insulator. It can be used for cables, terminals, capacitors, etc.

dielectric absorption *(Phys.)*. Phenomenon in which the charging or discharging current of a dielectric does not die away according to the normal exponential law, due to absorbed energy in the medium.

dielectric amplifier *(Elec.Eng.)*. One which operates through a capacitor, the capacitance of which varies with applied voltage.

dielectric antenna *(Telecomm.)*. One in which required radiation field is principally produced from a noncon-ducting dielectric.

dielectric breakdown *(Elec.Eng.)*. Passage of large current through normally nonconducting medium at sufficiently intense field strengths accompanied by a relative reduction of resistance and, in solids, mechanical damage.

dielectric constant *(Elec.Eng.)*. See **permittivity**.

dielectric current *(Phys.)*. A changing electric field applied to a perfect dielectric gives rise to a displacement current. For a real dielectric there will also be a conduction current or absorption current giving rise to energy loss in the dielectric.

dielectric diode *(Elec.Eng.)*. A capacitor whose negative plate can emit electrons into, e.g. CdS crystals, so that current flows in one direction.

dielectric dispersion *(Elec.Eng.)*. Variation of permittiv-ity with frequency.

dielectric fatigue *(Elec.Eng.)*. Breakdown of a dielectric subjected to a repeatedly applied electric stress, insuffi-cient to break down dielectric if applied once or a few times.

dielectric guide *(Elec.Eng.)*. Possible transmission of very-high-frequency electromagnetic energy functionally realized in a dielectric channel, the permittivity of which differs from its surroundings.

dielectric heating *(Elec.Eng.)*. Radio frequency heating in which power is dissipated in a non-conducting medium through dielectric hysteresis. It is proportional to V^2fS/t, where V is applied voltage, f the frequency, S the area of the heated specimen and t its thickness. ϕ is the **loss factor** of the material (permittivity × power factor). This is the principle used in microwave ovens.

dielectric hysteresis *(Phys.)*. Phenomenon in which the polarization of a dielectric depends not only on the applied electric field but also on its previous variation. This leads to power loss with alternating electric fields.

dielectric lens *(Telecomm.)*. A lens made of dielectric material in such a form that it refracts radio waves in much the same way as a glass lens affects light. Used to shape the beam on microwave and radar antennae.

dielectric loss *(Phys.)*. Dissipation of power in a dielectric under alternating electric stress; $W = \omega CV^2\delta$, where W = power loss, V = r.m.s. voltage, C = capacitance, δ = power factor, $\omega = 2\pi \times$ frequency.

dielectric loss angle *(Phys.)*. Complement of **dielectric phase angle**.

dielectric phase angle *(Phys.)*. That between an applied electric field and the corresponding conduction current vector. The cosine of this angle is the power factor of the dielectric.

dielectric polarization *(Phys.)*. Phenomenon explained by formation of doublets (dipoles) of elements of dielectric under electric stress.

dielectric relaxation *(Phys.)*. Time delay, arising from dipole moments in a dielectric when an applied electric field varies.

dielectric strain *(Phys.)*. See **displacement**.

dielectric strength *(Elec.Eng.)*. Electric stress necessary to break down a dielectric. It is generally expressed in kV per mm of thickness. The stress, steady or alternating, is normally maintained for 1 minute when testing.

Diels-Alder reaction *(Chem.)*. An addition reaction across a pair of conjugated double bonds to form a ring. A typical Diels-Alder reagent is a compound with a double or triple bond activated by electronegative substituents, e.g. tetracyanoethene.

die lubricant *(Powder Tech.)*. Lubricant applied to reduce friction between the powder and the die walls during compaction; it is usually a solid such as a soap. It may be incorporated in the powder when interparticle friction is also reduced.

diencephalon *(Zool.)*. In Vertebrates, the posterior part of the fore-brain connecting the cerebral hemispheres with the midbrain.

diene *(Chem.)*. Organic compound containing two double bonds between carbon atoms in its structure.

diene synthesis *(Chem.)*. See **Diels-Alder reaction**.

die nut *(Eng.)*. Hardened steel nut, usually hexagonal in form with an internal screwed hole. Can be turned by spanner to rectify damage to existing threads.

diesel cycle *(Autos.)*. A compression-ignition engine cycle in which air is compressed, heat added at constant volume by the injection and ignition of fuel in the heated charge, expanded (so doing work on the piston), and the products exhausted, the cycle being completed in either two revolutions (4-stroke) or one (2-stroke). See **diesel**

engine, four-stroke cycle, two-stroke cycle. (Rudolph Diesel (1858–1913), German engineer.)

diesel-electric locomotive *(Eng.)*. A locomotive in which a diesel engine is coupled to an electric generator which powers the motors connected to the driving axles.

diesel engine *(Eng.)*. A compression-ignition engine in which the oil fuel is introduced into the heated compressed-air charge by means of a fuel pump which injects measured quantities of fuel to each cylinder in turn. Earlier models sprayed the fuel by means of an air-blast. See **compression-ignition engine.**

diesel generating station *(Elec.Eng.)*. A generating station in which the prime-mover is a **diesel engine.**

diesel-hydraulic locomotive *(Eng.)*. One whose prime mover is a diesel engine, the power being transmitted through an oil-filled **torque converter.**

diesel knock *(Autos.)*. See **knocking.**

diesel locomotive *(Eng.)*. A locomotive powered by a diesel or compression-ignition engine, as distinct from a **steam locomotive.** See **diesel-electric, diesel-hydraulic locomotive.**

die set *(Eng.)*. See **subpress.**

die sinking *(Eng.)*. The engraving of dies for coining, paper embossing and similar operations.

die stamping *(Print.)*. An **intaglio** method of printing requiring a steel die; used mainly for high-class stationery.

die-stock *(Eng.)*. A hand screw-cutting tool, consisting of a holder in which screwing dies can be secured; it is held and rotated by a pair of handles.

Dieterici's equation *(Chem.)*. The **van der Waals' equation** modified for the effect of molecules near the boundaries.

diethanolamine *(Chem.)*. $CH_2OH.CH_2NH.CH_2$.-CH_2OH. One of the **ethanolamines** used industrially as an intermediate in the preparation of emulsifying agents, corrosion inhibitors etc. Also used as a stripping agent for CO_2 and H_2S in gas streams.

diethyldithiocarbamic acid *(Chem.)*. $(C_2H_5)_2$.N.CS.SH. Colourless crystalline solid. Used as a reagent for detecting copper, with which it gives a characteristic brown colour. The zinc salt is used as an accelerator in vulcanization of rubber.

diethylene glycol *(Chem.)*. *2,2'-dihydroxyethoxy-ethane*, $(C_2H_4OH)_2O$. Colourless liquid; bp 245°C. Used as a solvent, e.g. for cellulose nitrate. Its monoethyl ether is known as *carbitol*, also used widely as a solvent. Derivatives (esters and ethers) used as plasticizers.

diethyl ether *(Chem.)*. *Ethoxyethane, ether.* $C_2H_5.O.C_2H_5$, mp −113°C, rel.d. 0.72, a mobile, very volatile liquid of ethereal odour, an anaesthetic and a solvent. It is prepared from ethanol and sulphuric acid.

difference *(Maths.)*. See **subtraction.**

difference engine *(Comp.)*. Name given to the very early mechanical computer designed by Charles Babbage and begun in 1823. See also **analytical engine.**

difference of phase *(Phys.)*. See **phase difference.**

difference of potential *(Phys.)*. See **magnetic difference of potential, potential difference.**

difference operators *(Maths.)*. The three main difference operators used in numerical analysis are defined as follows where u_r is a sequence of numbers, $r = 1,2,3 \ldots$ (1) $\Delta u_r = u_{r+1} - u_r$ (descending or forward difference). (2) $\delta u_{r+1/2} = u_{r+1} - u_r$ (central difference). (3) $\nabla u_{r+1} = u_{r+1} - u_r$ (ascending or backward difference). Higher order differences are denoted by indices as in the differential calculus. $\Delta^2 u_r$ is a second order difference.

difference threshold *(Behav.)*. The amount by which a given stimulus must be increased or decreased in order for a subject to perceive a *just notable difference*, JND.

difference tone *(Acous.)*. See **combination tone.**

differentiable function of a complex variable *(Maths.)*. For a function of a complex variable to be differentiable it must satisfy the Cauchy-Riemann equations, and each of the four partial derivatives must be continuous.

differential *(Electronics)*. A device or circuit whose operation depends on the difference of two opposing effects.

differential *(Maths.)*. An arbitrary increment dx of an independent variable x or, if $y = f(x)$, the increment dy of y defined by $dy = f'(x)dx$. See **complete differential, derived function.**

differential absorption ratio *(Radiol.)*. Ratio of concentration of radioisotope in different tissues or organs at a given time after the active material has been ingested or injected.

differential amplifier *(Electronics)*. One whose output is proportional to the difference between two inputs. Usually based on the balanced differential pair. Many linear integrated circuits are of this type.

differential analyser *(Comp.)*. An obsolete **analogue computer** designed to solve differential equations.

differential anode conductance *(Elec.Eng.)*. Reciprocal of **differential anode resistance.**

differential anode resistance *(Elec.Eng.)*. The slope of the anode voltage versus anode current curve of a multi-electrode valve, when taken with all other electrodes maintained at constant potentials with reference to the cathode. At high frequencies, resistance values are larger than for d.c. (see **skin effect**) and a separate value may be quoted. Also called *a.c. resistance, slope resistance.*

differential booster *(Elec.Eng.)*. A booster in which a series winding on the field is connected in opposition to the shunt winding.

differential calculus *(Maths.)*. A branch of mathematics dealing with continuously varying quantities; based on the differential coefficient, or derivative, of one quantity with respect to another of which it is a function.

differential capacitor *(Elec.Eng.)*. One with one set of moving plates and two sets of fixed plates so arranged that, as capacitance of the moving plates to one set of fixed plates is increased, that to the other set is decreased.

differential chain block *(Eng.)*. A lifting tackle in which a 2-diameter chain wheel carries a continuous chain. Rotation of the chain wheel by a hanging loop shortens a second loop, supporting the load pulley in such a way as to give a large mechanical advantage. Also called *differential pulley block.*

differential coefficient *(Maths.)*. See **derived function.**

differential cross-section *(Phys.)*. The ratio of the number of scattered particles per unit solid angle in a given direction to the number of incoming particles per unit area.

differential dyeing *(Textiles)*. Variation in fibres of the same class that lead to their dyeing differently when immersed in a dye-bath.

differential equation *(Maths.)*. An equation involving total or partial differential coefficients. Those not involving partial differential coefficients are *ordinary differential equations.*

differential flotation *(Eng.)*. Production of more than one valuable concentrate by a series of froth flotation treatments of prepared ore pulp.

differential gear *(Eng.)*. A gear permitting relative rotation of two shafts driven by a third. The driving shaft rotates a cage carrying planetary bevel wheels meshing with two bevel wheels on the driven shafts. The latter are independent, but the sum of their rotation rates is constant.

differential grinding *(Min.Ext.)*. Comminution so controlled as to develop differences in grindability of constituents of ore.

differential hardening *(Image Tech.)*. Hardening of a photographic emulsion in specific areas to form an image, either by the action of light on chemically treated gelatine (e.g. with bichromate) or by development products associated with the formation of a silver image (**tanning developer**).

differential ionization chamber *(Nuc.Eng.)*. Two-compartment system in which resultant ionization current recorded is the difference between the currents in the two chambers. One version *(compensated ion chamber)* may be used to distinguish between neutrons and gamma radiation.

differential iron test *(Elec.Eng.)*. An apparatus for iron testing consisting of two magnetic squares, one of the sample to be tested and the other of a standard material. The windings on the squares are connected to a differential wattmeter, so that there will be no deflection when the quality of the two specimens is the same.

differential leakage flux *(Elec.Eng.)*. A general term given to the leakage flux occurring in and around the air gap of an induction motor. See **zigzag leakage**.

differentially compound-wound machine *(Elec.Eng.)*. A compound-wound d.c. machine in which the magnetomotive forces of the two windings oppose one another.

differentially-wound motor *(Elec.Eng.)*. A d.c. motor with series and shunt windings on the field connected so that the series windings opposes the shunt winding, and therefore causes the speed of the motor to rise as load is put on the machine.

differential-mode signal *(Elec.Eng.)*. (1) That part of two signals, both measured to a common reference, by which they are different (as opposed to common-mode signal). Often used as the input to an amplifier when small signals have to be amplified in a high level of background interference. (2) In a balanced 3-terminal system, signal applied between the 2 underground terminals.

differential motion *(Eng.)*. A mechanical movement in which the velocity of a driven part is equal to the difference of the velocities of two parts connected to it.

differential permeability *(Elec.Eng.)*. Ratio of a small change in magnetic flux density of magnetic material to change in the magnetizing force producing it, i.e. slope of the magnetization loop at the point in question.

differential pressure gauge *(Eng.)*. A gauge, commonly of U-tube form, which measures the difference between two fluid pressures applied to it.

differential protective system *(Elec.Eng.)*. See **balanced protective system**.

differential pulse-code modulation *(Telecomm.)*. A version of **pulse-code modulation** in which the difference in value between a sample and its predecessor constitutes the transmitted information. For many types of signal fewer bits are needed; often used in satellite communication.

differential relay *(Elec.Eng.)*. See **relay types**.

differential resistance *(Elec.Eng.)*. Ratio of a small change in the voltage drop across a resistance to the change in current producing the drop, i.e. the slope of the voltage-current characteristic for the material.

differential stain *(Zool.)*. A stain which picks out details of structure by giving to them different colours or different shades of the same colour.

differential susceptibility *(Elec.Eng.)*. Ratio of a small change in the intensity of magnetization of a magnetic material to the change of magnetizing force producing it, i.e. the slope of the intensity-magnetization (hysteresis) loop.

differential thermal analysis *(Chem.)*. The detection and measurement of changes of state and heats of reaction, especially in solids and melts, by simultaneously heating two samples of identical heat capacities and noting the difference in temperature between them, which becomes very marked when one of the two samples passes through a transition temperature with evolution or absorption of heat but the other does not. Abbrev. *DTA*.

differential titration *(Chem.)*. Potentiometric titration in which the e.m.f. is noted between additions of small increments of titrant, the end point being where the e.m.f. changes most sharply.

differential winding *(Elec.Eng.)*. A winding in a compound motor which is in opposition to the action of another winding.

differentiating circuits *(Elec.Eng.)*. (1) Amplifier having a combination of resistive input and feedback inductance, or capacitive input and feedback resistance, such that the output is proportional to the rate-of-change (differential) of the input signal. (2) A passive circuit comprising either R and L or C and R, whose output is proportional to the rate-of-change of the input signal. This circuit does not produce as accurate a result as the active circuit described above. Used, for example, to detect sudden changes in otherwise steady wave form and to modify waveforms in digital circuits.

differentiating solvent *(Chem.)*. See **levelling solvent, non-aqueous solvents**.

differentiation *(Biol.)*. (1) The qualitative changes in morphology and physiology occurring in a cell, tissue or organ as it develops from a meristematic, primordial or unspecialized state into the mature or specialized state. (2) Removing excess stain from some of the structures in a microscope preparation so that the whole can be seen more clearly.

differentiation *(Geol.)*. The process of forming two or more rock types from a common magma.

differentiation *(Maths.)*. Operation of finding a differential coefficient.

differentiator *(Elec.Eng.)*. See **differentiating circuit**.

diffluence *(Meteor.)*. The spreading apart of streamlines.

diffraction *(Phys.)*. The phenomenon, observed when waves are obstructed by obstacles or apertures, of the disturbance spreading beyond the limits of the geometrical shadow of the object. The effect is marked when the size of the object is of the same order as the wavelength of the waves and accounts for the alternately light and dark bands, diffraction fringes, seen at the edge of the shadow when a point source of light is used. It is one factor that determines the propagation of radio waves over the curved surface of the earth and it also accounts for the audibility of sound around corners. See **Fraunhofer diffraction, Fresnel diffraction**.

diffraction analysis *(Crystal.)*. Analysis of the internal structure of crystals by utilizing the diffraction of X-rays caused by the regular atomic or ionic lattice of the crystal.

diffraction angle *(Phys.)*. That between the direction of an incident beam of light, sound, or electrons, and any resulting diffracted beam.

diffraction grating *(Phys.)*. One of the most useful optical devices for producing spectra. In one of its forms, the diffraction grating consists of a flat glass plate in the surface of which have been ruled with a diamond equidistant parallel straight lines, which may be as close as 1000 per millimetre. If a narrow source of light is viewed through a grating it is seen to be accompanied on each side by one or more spectra. These are produced by diffraction effects from the lines acting as a very large number of equally spaced parallel slits.

diffraction pattern *(Phys.)*. That formed by equal intensity contours as a result of diffraction effects, e.g. in optics or radio transmission.

diffractometer *(Phys.)*. An instrument used in the analysis of the atomic structure of matter by the diffraction of X-rays, neutrons or electrons by crystalline materials. A monochromatic beam of radiation is incident on a crystal mounted on a goniometer. The diffracted beams are detected and their intensities measured by a counting device. The orientation of the crystal and the position of the detector are usually computer-controlled.

diffuse density *(Image Tech.)*. The density of a photographic image when measured by diffuse light rather than specular. See **Callier effect**.

diffuse growth *(Bot.)*. Growth where cells divide throughout the tissue. Cf. **apical growth, intercalary** growth, **trichothallic growth**.

diffuse placentation *(Zool.)*. Having the villi developed in all parts of the placenta, except the poles, as in Lemurs, most Ungulates and Cetacea.

diffuse porous *(Bot.)*. Wood having the pores distributed evenly thoughout a growth ring or changing in diameter gradually across it, e.g. birch, evergreen oaks. Cf. **ring porous**.

diffuser *(Acous.)*. Irregular structure, pyramids and cylinders, to break up sound waves in rooms. See also **scatterer**.

diffuser *(Aero.)*. A means for converting the kinetic energy of a fluid into pressure energy; usually it takes the form of a duct which widens gradually in the direction of flow; also fixed vanes forming expanding passages in a compressor delivery to increase the pressure.

diffuser *(Eng.)*. A chamber surrounding the impeller of a centrifugal pump or compressor, in which part of the kinetic energy of the fluid is converted to pressure energy by a gradual increase in the cross-sectional area of flow.

diffuser *(Image Tech.)*. Translucent material in front of studio lamp to diffuse light and soften shadows.

diffuse reflection *(Phys.)*. Same as **nonspecular reflection**.

diffuse-reflection factor *(Phys.)*. The ratio of the luminous flux diffusely reflected from a surface to the total luminous flux incident upon the surface.

diffuse series *(Phys.)*. Series of optical spectrum lines observed in the spectra of alkali metals. Energy levels for which the orbital quantum number is two are designated *d-levels*.

diffuse sound *(Acous.)*. Sound which is reflected in all directions inside a volume.

diffuse tissue *(Bot.)*. A tissue consisting of cells which occur in the plant body singly or in small groups intermingled with tissues of distinct type.

diffuse transmittance *(Phys.)*. See **transmittance**.

diffusion *(Chem.)*. General transport of matter whereby molecules or ions mix through normal thermal agitation. Migration of ions may be directed and accelerated by electric fields, as in **dialysis**.

diffusion activation energy *(Phys.)*. The diffusion of interstitial atoms, lattice vacancies or impurities in a crystalline solid are temperature dependent. The diffusion coefficient D is given by $D = D_0 \exp(-E/kT)$, where D_0 is a constant, E is the *activation energy* of the process, T is the temperature and k is Boltzmann's constant.

diffusion area *(Nuc.Eng.)*. Term used in reactor diffusion theory. One-sixth of the mean square displacement (i.e. direct distance travelled irrespective of route) between point at which neutron becomes thermal and where it is captured.

diffusion attachment *(Image Tech.)*. Lens accessory for softening the outline of the image in a camera or enlarger, often a disk with a finely etched or engraved surface.

diffusion barrier *(Phys.)*. Porous partition for gaseous separation according to molecular weight and hydrodynamic velocities, especially for separation of isotopes. A fired but unglazed plate.

diffusion capacitance *(Phys.)*. The rate of change of injected charge with the applied voltage in a semiconductor diode.

diffusion coating *(Eng.)*. Methods by which an alloy or metal are allowed to diffuse into the surface of an underlying metal. They can involve heating and exposing the metal to a solution of the coating material.

diffusion coefficient *(Chem.)*. In the diffusion equation (*Fick's law*), the coefficient of proportionality between molecular flux and concentration gradient. Symbol D. Units, M^2/s, as for thermal diffusivity.

diffusion constant *(Phys.)*. The ratio of diffusion current density to the gradient of charge carrier concentration in a semiconductor.

diffusion current *(Chem.)*. In electrolysis, the maximum current at which a given bulk concentration of ionic species can be discharged, being limited by the rate of migration of the ions through the diffusion layer.

diffusion flame *(Eng.)*. Long luminous gas flame holding practically a constant rate of radiation for its designed length of travel, together with uniform precipitation of free carbon, diffusion occurring between adjacent strata of air and gas.

diffusion law *(Chem.)*. See **Fick's law, Graham's law**.

diffusion layer *(Chem.)*. In electrolysis, the layer of solution adjacent to the electrode, in which the concentration gradient of electrolyte occurs.

diffusion length *(Nuc.Eng.)*. Square root of **diffusion area**.

diffusion length *(Phys.)*. Average distance travelled by carriers in semiconductor between generation and recombination.

diffusion of solids *(Phys.)*. In semiconductors, the migration of atoms into pure elements to form surface alloy for providing minority carriers.

diffusion plant *(Nuc.Eng.)*. One used for isotope separation by **gaseous (molecular) diffusion** or **thermal diffusion**.

diffusion potential *(Elec.Eng.)*. The potential difference across the boundary of an electrical double-layer in a liquid.

diffusion pump *(Chem.Eng.)*. See **Gaede diffusion pump**.

diffusion theory *(Nuc.Eng.)*. Simplified neutron migration theory based on **Fick's law**. Less accurate than the more detailed **transport theory**.

diffusion transfer reversal *(Image Tech.)*. Process in which a direct positive image is formed from the material exposed in the camera, hence the basis of many instant photography systems. The exposed emulsion is developed by a viscous solution containing a silver halide solvent while in contact with a receiving layer on paper or transparent base. Unexposed silver halides become transferred to this receptor and form a positive image which can be separated on its new support.

diffusion welding *(Eng.)*. A method in which high temperature and pressure cause a permanent bond between two metallic surfaces without melting or large scale deformation. A solid metal filler may be sandwiched between the surfaces to aid the process.

diffusivity *(Phys.)*. A measure of the rate at which heat is diffused through a material. It is equal to the thermal conductivity divided by the product of the specific heat at constant pressure and the density. Units $m^2 s^{-1}$.

difluorophosphoric acid *(Chem.)*. HPO_2F_2. Formed by partial hydrolysis of phosphoryl fluoride, POF_3, with cold dilute alkali, or preferably by heating phosphoric acid with ammonium fluoride.

digametic *(Zool.)*. Having gametes of two different kinds.

digastric *(Zool.)*. Of muscles, having fleshy terminal portions joined by a tendinous portion, as the muscles which open the jaws in Mammals.

digenesis *(Zool.)*. (1) **Alternation of generations**. (2) The condition of having two hosts; said of parasites. adj. *digenetic*.

digenetic reproduction *(Zool.)*. See **sexual reproduction**.

digenite *(Min.)*. A cubic sulphide of copper, usually massive and associated with other copper ores. Composition probably near Cu_2S_5.

di George's syndrome *(Immun.)*. See **thymic hypoplasia**.

digester *(Paper)*. A vessel in which fibrous raw materials are heated under pressure with chemicals, in the first stages of papermaking.

digestion *(Min.Ext.)*. In chemical extraction of values from ore or concentrate, period during which material is exposed under stated conditions to attacking chemicals.

digestion *(Zool.)*. The process by which food material ingested by an organism is rendered soluble and assimilable by the action of enzymes. adj. *digestive*.

digestive gland *(Zool.)*. Gland(s) present in many Invertebrates and Protochordata, in which intracellular ingestion and absorption take place, as opposed to the alimentary canal proper.

digit *(Comp.)*. Discrete sign. See **bit, check digit, binary digit**.

digit *(Zool.)*. A finger or toe, one of the free distal segments of a pentadactyl limb.

digital *(Comp.)*. Representation of a numerical quantity by a number of discrete signals or by the presence or absence of signals in particular positions. See **bit**.

digital *(Telecomm.)*. Communications circuits in which the information is transmitted in the form of trains of pulses; speech and vision need to be converted into code before such transmission (see **pulse-code modulation**) whereas most data is already in suitable form. Advantages include immunity to noise and the possibility of electronic exchange switching.

digital computer *(Comp.)*. See **computer**.

digital differential analyser *(Comp.)*. An obsolete electronic computer for solution of differential equations by incremental means.

digital filter *(Telecomm.)*. One which passes, or rejects, pulsed or digital information whose pattern corresponds exactly with that laid down in the design of the filter circuit.

digitalis *(Med.)*. Dried leaves of the foxglove, *Digitalis purpurea*, whose steroid glycosides are used in heart disease.

digitalization *(Med.)*. Administration of digitalis to a patient with heart disease, in amounts sufficient to produce full therapeutic effect.

digital meter *(Elec.Eng.)*. One displaying the measured quantity as numerical value.
digital plotter *(Comp.)*. Graph plotter which receives digital input specifying the coordinates of the points to be plotted. Cf. **incremental plotter**.
digital subtraction angiography *(Radiol.)*. A radiological technique where an initial X-ray image is digitized and subtracted from another taken after the injection of **contrast medium**. As only the contrast in the blood vessels is added, high quality images of these blood vessels can be obtained after a small intravenous injection. Avoids catheterization of an artery for **angiography**.
digital watch, clock *(Horol.)*. A watch or clock which displays the time as a series of numbers, usually on the 24-hour system, instead of in the traditional swept-hands or *analogue* manner.
digitate *(Bot.)*. Palmate.
digitigrade *(Zool.)*. Walking on the ventral surfaces of the phalanges only, as terrestrial carnivores.
digitize *(Comp.)*. Convert an analogue signal to a digital signal. Also called *quantization*.
digitoxin *(Chem.)*. A vegetable glycoside isolated from the leaves of *Digitalis purpurea*.
digitule *(Zool.)*. Any small fingerlike process.
digonal *(Maths.)*. Of symmetry about an axis, such that a half-turn (180°) gives the same figure.
digoneutic *(Zool.)*. Producing offspring twice a year.
digoxin *(Med.)*. The commonest glycoside of digitalis used in the treatment of heart failure and **atrial fibrillation**.
dihedral *(Maths.)*. Angle between two planes, as measured in the plane normal to their line of intersection.
dihedral angle *(Aero.)*. Acute angle at which an aerofoil is inclined to the transverse plane of reference. A downward inclination is called *negative dihedral*, sometimes *anhedral*.
diheptal *(Electronics)*. Referring to 14 in number, e.g. pins on base of a tube or valve.
dihybrid *(Biol.)*. The product of a cross between parents differing in two characters determined by single genes; an individual heterozygous at two gene loci.
dihydroxyacetone *(Chem.)*. $CH_2OH.CO.CH_2OH$. The simplest *ketose*, used in sun tan lotion.
dikaryon *(Bot.)*. Fungal hypha or mycelium in which 2 nuclei of different genetic constitution (and different mating type) are present in each cell (or hyphal segment). adj. *dikaryotic*. See **dikaryophase**.
dikaryophase *(Bot.)*. That part of the life cycle of an ascomycete or basidiomycete in which the cells are dikaryotic, i.e. between **plasmogamy** and **karyogamy**.
dike *(Hyd.Eng.)*. See **dyke**.
diketen *(Chem.)*. Dimer of *ketene*. Bp 127°C. Useful intermediate in preparative organic chemistry.
diketones *(Chem.)*. *Dialkanones*. Compounds containing –CO– groups which, according to their position in the molecule, are termed 1,2-diketones –CO.CO–, or 1,3-diketones $-CO.CH_2.CO-$ etc.
dikkop *(Vet.)*. See **African horse-sickness**.
dilambdodont *(Zool.)*. Said of teeth in which the paracone and metacone are V-shaped, well separated and placed near the middle of the tooth, as in some *Insectivora*.
dilapidations *(Build.)*. A term applied to the damage done to premises during a period of tenancy.
dilatancy *(Chem.)*. Property shown by some colloidal systems, of thickening or solidifying under pressure or impact. Cf. **thixotropy**.
dilatometer *(Chem.)*. An apparatus for the determination of transition points of solids. It consists of a large bulb joined to a graduated capillary tube, and is filled with an inert liquid. The powdered solid is introduced, and the temperature at which there is a considerable change in volume on slow heating or cooling may be noted; alternatively, the temperature at which the volume shows no tendency to change with time may be found.
dilatometry *(Chem.)*. The determination of transition points by the observation of volume changes.
dilator *(Zool.)*. A muscle which, by its contraction, opens or widens an orifice. Cf. **sphincter**.
di litho *(Print.)*. See **direct lithography**.
Dilleniidae *(Bot.)*. Subclass or superorder of dicotyledons. Trees, shrubs and herbs, polypetalous or sympetalous, stamens (if numerous) developing centrifugally, mostly syncarpous, often with parietal placentation. Contains ca 24 000 spp. in 69 families including Malvaceae, Cruciferae and Ericaceae.
diluent *(Build.)*. A volatile liquid unable to act as a paint solvent on its own, but able to dilute a genuine solvent without disadvantage. It is used to cheapen the formulation.
diluent air *(Build.)*. Air admitted or induced into a flue to dilute the noxious effects of combustion.
dilution *(Chem.)*. (1) Decrease of concentration. (2) The volume of a solution which contains unit quantity of dissolved substance. The reciprocal of *concentration*.
dilution law *(Chem.)*. See **Ostwald's dilution law**.
dilution refrigerator *(Phys.)*. A device for producing very low temperature, down to 0.004 K, on a small sample. Uses the very low temperature properties of a mixture of ^3He and ^4He.
diluvium *(Geol.)*. An obsolete term for those accumulations of sand, gravel, etc. which, it was thought, could not be accounted for by normal stream and marine action. In this sense, the deposits resulting from the Deluge of Noah would be *diluvial*.
dimensional analysis *(Phys.)*. An equation representing a relationship between quantities which describes a physical phenomenon, must be homogeneous. If the quantities on each side of an equation are expressed in terms of the dimensions of the fundamental quantities of mass, length, time, current etc., then (1) the equation can be tested for homogeneity or (2) the way in which one quantity depends on the others can be predicted. Dimensional analysis gives no information about dimensionless quantities or pure numbers occurring in an equation.
dimensional stability *(Paper)*. The resistance offered by a paper to changes in its dimensions when ambient conditions alter.
dimensions *(Phys.)*. The dimensions of a **derived unit** used to express the measurement of a physical quantity are the *powers* to which the fundamental units are involved in the quantity; e.g., velocity has dimensions +1 in length and −1 in time or $[LT^{-1}]$, force has dimensions +1 in mass, +1 in length and −2 in time, $[MLT^{-2}]$.
dimension stone *(Build.)*. A term sometimes used for an *ashlar*.
dimer *(Chem.)*. Molecular species formed by the union of two like molecules, e.g. carboxylic acids in aprotic solvents: $2\ RCOOH \rightleftharpoons (RCOOH)_2$.
dimeric *(Chem.)*. See **dimer**.
dimerous *(Bot.)*. A flower with 2 members in a given whorl.
dimethacone *(Med.)*. An antifoaming agent sometimes added to antacids to relieve flatulence.
dimethylformamide *(Chem.)*. $(CH_3)_2N.CHO$. Solvent of wide industrial application, used in plastics manufacture, in gas separation processes, in the artificial fibre industry.
dimethyl glyoxime *(Chem.)*. Compound used in analysis as a specific and quantitative precipitant for palladium (weakly acid solution) and nickel (ammoniacal solution) with which it gives a brilliant red precipitate.
dimethyl hydrazine *(Chem.)*. See **hydrazine**.
dimethyl phthalate *(Chem.)*. *Dimethyl benzene 1,2-dicarboxylate*. High boiling-point ester, bp 280°C, widely used as an insect repellent, particularly against midges and mosquitoes.
dimethyl sulphate *(Chem.)*. $CH_3O.SO_2.OCH_3$, bp 187–188°C. Widely used as a methylating (i.e. introduction of a methyl (CH_3) group) agent in organic preparations. Manufactured from dimethyl ether and sulphur trioxide.
diminished stile *(Build.)*. A door-stile which is narrowed down for a part of its length, as, e.g. in the glazed portion of a sash door. Also *gunstock stile*.
diminishing courses *(Build.)*. See **graduated courses**.
diminishing pipe *(Build.)*. A tapered pipe length used to connect pipes of different diameters.

dimity *(Textiles.)*. Strong cotton fabric that appears striped because of a corded pattern; used chiefly for mattress coverings.

dimmer *(Elec.Eng.)*. A light intensity controller using thyristors or a variable resistor. Also called *dimming resistance*.

dimmer wheel *(Elec.Eng.)*. A handwheel for operating one or more dimmers; commonly used in stage-lighting equipment.

dimming resistance *(Elec.Eng.)*. See **dimmer**.

dimorphic *(Biol.)*. Organelles, organs, or individuals etc. existing in 2 distinct forms. Also *dimorphous*.

dimorphic, dimorphous *(Chem.)*. Capable of crystallizing in 2 different forms.

dimorphism *(Min.)*. Crystallization into 2 distinct forms of an element or compound, e.g. carbon as diamond and graphite; FeS_2 as pyrite and *marcasite*.

dimorphism *(Zool.)*. The condition of having 2 different forms, as animals which show marked differences between male and female (sexual dimorphism), animals which have 2 different kinds of offspring, and colonial animals in which the members of the colony are of 2 different kinds.

DIN *(Genrl.,Image Tech.)*. *Deutsche Institut für Normung*, the German national standards organization; in particular, their system of photographic **speed** rating with logarithmic increments.

Dinantian *(Geol.)*. The lower Carboniferous rocks of N.W. Europe, comprising Tournaisian and Visean Stages.

Dinas bricks *(Build.)*. Firebricks made almost entirely of sand with a small amount of lime.

dineutron *(Phys.)*. Assumed transient existence of a set of two neutrons in order to explain certain nuclear reactions.

dinging *(Build.)*. Rough plastering for walls, a single coat being put on with a trowel and brush.

dinitrocresol *(Chem.)*. See **DNOC**.

dinitrogen fixation *(Bot.)*. Same as **nitrogen fixation**.

dinky sheet *(Print.)*. A web much narrower than the full width of a rotary press.

dinocap *(Chem.)*. *2-(1-Methyl-heptyl)-4,6-dinitrophenyl crotonate*; used as a fungicide.

Dinoflagellata *(Zool.)*. See **Dinophyceae**.

Dinophyceae *(Bot.)*. The *dinoflagellates*. Mesokaryotic algae. Motile cells have 2 laterally inserted flagella, one lying in a transverse groove around the cell and helically coiled, the other lying in a longitudinal groove and posteriorly directed. Chloroplasts with chlorophyll a and c, and **peridinin**, thylakoids in threes, envelope of 3 membranes; store starch in the cytoplasm. Phototrophs include flagellate, colonial, coccoid, palmelloid and a few filamentous sorts. Marine (see **red tide**) and fresh water. Phagotrophic, parasitic and various symbiotic sorts occur.

diocoel *(Zool.)*. The lumen of the diencephalon.

diode *(Electronics)*. (1) Simplest electron tube, with a heated cathode and anode; used because of unidirectional and hence rectification properties. (2) Semiconductor device with similar properties, evolved from primitive crystal rectifiers for radio reception.

diode characteristic *(Electronics)*. A graph showing the current-voltage characteristics of a vacuum-tube or semiconductor diode. In particular it will show marked differences between currents in the forward and reverse directions, and, in the case of avalanche or Zener diodes, sudden increases in reverse current when the applied voltage reaches a critical value.

diode clipper *(Electronics)*. A limiting circuit using a diode.

diode isolation *(Electronics)*. Isolation of the circuit elements in a micro-electronic circuit by using the very high resistance of a reverse-biased *p-n* junction.

diode-pentode, diode-triode *(Electronics)*. Thermionic diode in same envelope as a pentode and triode, respectively.

diode-transistor logic *(Electronics)*. Logic circuitry in which arrays of diodes at the inputs perform logic functions prior to controlling the base current of a transistor which subsequently provides power gain for driving additional gates.

diode voltmeter *(Elec.Eng.)*. One in which measured voltage is rectified, amplified and displayed on a moving-coil meter.

dioecious *(Bot.,Zool.)*. Having the sexes in separate individuals. n. *dioecism*.

dioestrus *(Zool.)*. In female Mammals, the growth period following metoestrus.

diolefin *(Chem.)*. See **dialkene**.

diols *(Chem.)*. Dihydric alcohols, chiefly represented by the glycols in which the hydroxyl groups are attached to adjacent carbon atoms.

Dione *(Astron.)*. The fourth natural satellite of **Saturn**, 1100 km in diameter. It is heavily cratered.

Dione B *(Astron.)*. The twelfth natural satellite of **Saturn**, a tiny object just 20 km in diameter. It is orbitally associated with the much larger **Dione**.

Dionic *(Elec.Eng.)*. TN designating instruments for testing the purity of water by measuring its electrical conductivity.

diophantine equations *(Maths.)*. Indeterminate equations for which integral or rational solutions are required.

diopside *(Min.)*. A monoclinic pyroxene, typically occurring in metamorphosed limestones and dolomites, and composed of calcium magnesium silicate, $CaMgSi_2O_6$, usually with a little Fe.

dioptase *(Min.)*. A rare emerald green hydrated copper silicate.

dioptre *(Phys.)*. The unit of power of a lens. A convergent lens of 1 metre focal length is said to have a power of +1 dioptre. Generally, the power of a lens is the reciprocal of its focal length in metres, the power of a divergent lens being given a negative sign.

dioptre lens *(Image Tech.)*. A supplementary lens used in front of the main camera objective to bring close objects into focus; sometimes covering only part of the field of view to show both close and more distant objects (*split dioptre*). A *supplementary lens*.

dioptric mechanism *(Zool.)*. A mechanism, consisting of the cornea, aqueous humour, lens and vitreous humour, by which the images of external objects may be focused on the retina of the eye, in Chordata. An analogous mechanism in the ommatidia of Arthropods.

dioptric system *(Phys.)*. An optical system, which contains only refracting components.

diorite *(Geol.)*. A coarse-grained deep-seated (plutonic) igneous rock of intermediate composition, consisting essentially of plagioclase feldspar (typically near andesine in composition) and hornblende, with or without biotite in addition. Differs from granodiorite in the absence of quartz. See also **tonalite**.

dioxan *(Chem.)*. *1,4-Dioxycyclohexane*. Colourless liquid; mp 11°C, bp 101°C. Used as a solvent for waxes, resins, viscose etc.

dioxazine dyes *(Chem.)*. Range of dyestuffs which are relatively complex sulphonated compounds with an affinity for cellulose. Chiefly derived from chloranil by reaction with amines. The chief dyes in the series are light-fast in the blue range.

DIP *(Electronics)*. Abbrev. for *Dual In-line Package*.

dip *(Geol.)*. The angle that a plane makes with the horizontal. The dip is perpendicular to the **strike** of the structure.

dip *(Hyd.Eng.)*. Any departure from the regular slope at which a pipe is laid, when the slope is increased locally.

dip *(Phys.)*. The angle measured in a vertical plane between the earth's magnetic field at any point and the horizontal; also called *inclination*.

dip circle *(Phys.)*. An instrument consisting of a magnetic needle pivoted on a horizontal axis; accurate measurements of magnetic dip can be obtained with it.

dipentene *(Chem.)*. *dl-limonene*.

dip fault *(Geol.)*. A fault parallel to the direction of dip. See **strike fault**.

diphase *(Elec.Eng.)*. A term sometimes used in place of *2-phase*.

diphasic *(Zool.)*. Of certain parasites, having a life-cycle which includes a free active stage. Cf. **monophasic**.

diphenyl *(Chem.)*. *Phenylbenzene*, $C_6H_5.C_6H_5$; colourless; mp 71°C; bp 254–255°C; soluble in alcohol and ether. It occurs in coal-tar, and is prepared by heating iodobenzene to 220°C with finely divided copper.

diphenyl ether *(Chem.)*. $C_6H_5.O.C_6H_5$, *diphenyl oxide*, a liquid of pleasant odour, mp 28°C, bp 253°C, obtained from phenol by heating with $ZnCl_3$ or $AlCl_3$.

diphenylglyoxal *(Chem.)*. See **benzil**.

diphenylguanidine *(Chem.)*. Crystalline solid; mp 147°C. Used as an accelerator in vulcanization of rubber.

diphenylmethane *(Chem.)*. $(C_6H_5)_2CH_2$, colourless needles, mp 26°C. bp 262°C, obtained by the action of chloromethyl benzene on benzene in the presence of aluminium chloride.

diphtheria *(Med.)*. Infection, usually air-borne, with the bacillus *Corynebacterium diphtheriae*. Bacilli are confined to the throat, producing local necrosis ('pseudo-membrane'), but a powerful **exotoxin** causes damage especially to heart and nerves.

diphtheria toxin *(Immun.)*. Toxin made by *Corynebacterium diphtheriae* responsible for the damage caused by clinical infection. The toxin has one part of the molecule which attaches to a surface component on susceptible cells such as heart muscle and another which interferes with protein synthesis within them. Toxin is effectively neutralized by antitoxin, which is used to treat severe infections in unimmunized persons.

diphtheria toxoid *(Immun.)*. Toxin treated with formaldehyde so as to destroy toxicity but not alter its capacity to act as antigen. Used for active immunization against diphtheria. It is usually used after adsorption cnto aluminium hydroxide, which acts as an **adjuvant**, and in combination with tetanus toxoid and often with *Bordatella pertussis* vaccine.

diphycercal *(Zool.)*. Said of a type of tail-fin (found in Lung-fish, adult Lampreys, the young of all Fish, and many aquatic Urodela) in which the vertebral column runs horizontally, the fin being equally developed above and below it.

diphygenic *(Zool.)*. Having 2 different modes of development.

diphyletic *(Biol.)*. Of dual origin: descended from 2 distant ancestral groups.

diphyodont *(Zool.)*. Having 2 sets of teeth; a deciduous or milk dentition and a permanent dentition, as in Mammals.

diplegia *(Med.)*. Bilateral paralysis of like parts of the body.

diplex *(Telecomm.)*. The simultaneous transmission or reception of two signals using a specified common feature, e.g. a common carrier or antenna. Cf. **duplex**, **multiplex**.

diplexer *(Telecomm.)*. A means of coupling which permits two transmitters or receivers to operate with one aerial.

diplobiont *(Bot.)*. A plant which includes in its life-cycle at least 2 kinds of different individuals; if the species is dioecious, there are 3 kinds of individuals. adj. *diplobiontic*.

diploblastic *(Zool.)*. Having 2 primary germinal layers, namely, ectoderm and endoderm.

diplococcus *(Biol.)*. A coccus which divides by fission in one plane, the two individuals formed remaining paired.

diplogangliate, diploganglionate *(Zool.)*. Having paired ganglia.

diplohaplont *(Bot.)*. Organism in which there is an *alternation* of haploid and diploid *generations*. Cf. **haplont, diplont**.

diploid *(Biol.)*. Possessing two sets of chromosomes, one set coming from each parent. Most organisms are diploid. Cf. **haploid**.

diploidisation *(Bot.)*. The fusion within the vegetative mycelium of two haploid nuclei to give a diploid nucleus in some fungi.

diplonema *(Biol.)*. A stage in the meiotic division (**diplotene** stage) at which the chromosomes are clearly double.

diplont *(Bot.)*. Organisms in which only the zygote is diploid, meiosis occurring at its germination and the vegetative cells being haploid. Cf. **haplont, diplohaplont**.

diplophase *(Biol.)*. The period in the life-cycle of any organism when the nuclei are diploid. Cf. **haplophase**.

diplopia *(Med.)*. Double vision of objects.

Diplopoda *(Zool.)*. A class of Arthropoda having the trunk composed of numerous double somites, each with 2 pairs of legs; the head bears a pair of uniflagellate antennae, a pair of mandibles, and a gnathochilarium representing a pair of partially fused maxillae: the genital opening is in the third segment behind the head; vegetarian animals of retiring habits. Millipedes.

diplospondyly *(Zool.)*. The condition of having two vertebral centra, or a centrum and an inter-centrum, corresponding to a single myotome. adjs. *diplospondylic, diplospondylous*.

diplostemonous *(Bot.)*. Having twice as many stamens as there are petals, with the stamens in 2 whorls, the members of the outer whorl alternating with the petals.

diplotene *(Biol.)*. The fourth stage of meiotic prophase, intervening between pachytene and diakinesis, in which homologous chromosomes come together and there is condensation into tetrads.

diplozoic *(Zool.)*. Bilaterally symmetrical.

dip needle *(Phys.,Surv.)*. Dipping needle, inclinometer. Magnetic needle on horizontal pivot, which swings in vertical plane. When set in magnetic north-south plane, its inclination shows the *angle of dip*.

Dipnoi, Dipneusti *(Zool.)*. An order of Sarcopterygii, in which the air-bladder is adapted to function as a lung, and the dentition consists of large crushing plates. Lung-fish.

dipole *(Acous.)*. Radiator producing a sound field of two adjacent **monopoles** in antiphase. A localized fluctuating force is the prototype of a dipole. The directivity of a dipole has the shape of an eight.

dipole *(Phys.)*. Equal and opposite charges separated by a close distance constitute an electric dipole. A bar magnet or a coil carrying a steady current produce a magnetic dipole.

dipole antenna *(Telecomm.)*. Wire or rod antenna, half a wavelength long, and split at the centre for connection with a transmission line. Maximum radiation is at right angles to the axis, and dipoles also have maximum performance with waves polarized in the same plane as the axis. Also *half-wave dipole, doublet*. See **folded dipole**.

dipole molecule *(Phys.)*. One which has a permanent moment due to the permanent separation of the effective centres of the positive and negative charges.

dipole moment *(Phys.)*. See **electric dipole moment**, **magnetic dipole moment**.

dipping *(Build.)*. An industrial process of coating articles by immersion in paint, followed by withdrawal.

dipping *(Eng.)*. The immersion of pieces of material in a liquid bath for surface treatment such as pickling or galvanizing.

dipping *(Vet.)*. The process of immersing animals in a medicated bath, for the destruction of ectoparasites.

dipping needle *(Surv.)*. See **dip needle**.

dipping refractometer *(Chem.)*. A type of refractometer which is dipped into the liquid under examination.

diprotodont *(Zool.)*. Having the first pair of upper and lower incisor teeth large and adapted for cutting, the remaining incisor teeth being reduced or absent; pertaining to the Diprotodontia. Cf. **polyprotodont**.

dip slope *(Geol.)*. A land form developed in regions of gently inclined strata, particularly where hard and soft strata are interbedded. A long gentle sloping surface which coincides with the inclination of the strata below ground. See **cuesta**.

dip soldering *(Eng.)*. The method of soldering previously fluxed components by immersing them in a bath of molten solder. Ideal for bulky assemblies with complicated or multiple joints, and for fast automatic operation.

dipsomania *(Med.)*. The condition in which there is a recurring, temporary, and uncontrollable impulse to drink excessively.

dip stick *(Eng.)*. A rod inserted in a tank or sump to measure the depth of oil or other liquid.

Diptera *(Zool.)*. An order of Insects, having one pair of wings, the hinder pair being represented by a pair of club-shaped balancing organs or halteres; the mouth-parts are suctorial; the larva is legless and sluggish. Flies, Gnats, and Midges.

dipygus *(Med.)*. A developmental monstrosity in which a foetus has a double pelvis.

dipyre *(Min.)*. A member of the **scapolite** series, containing 20 to 50% of the meionite molecule.

Dirac's constant *(Phys.)*. *Planck's constant* (*h*) divided by 2π. Usually termed *h-bar*, and written $\hbar$. Unit in which *electron spin* is constant. See **Planck's law**.

Dirac's Theory *(Phys.)*. Using the same postulates as the **Schrödinger equation**, plus the requirement that quantum mechanics conform with the theory of relativity, an electron must have an inherent angular momentum and magnetic moment. (1928).

direct-acting pump *(Eng.)*. A steam-driven reciprocating pump in which the steam and water pistons are carried on opposite ends of a common rod.

direct-arc furnace *(Elec.Eng.)*. An electric arc furnace in which the arc is drawn between an electrode and the charge in the furnace.

direct broadcast satellite *(Telecomm.)*. High power *geostationary* communications satellite, usually having a specially designed antenna so that the **footprint** coincides with the region of the Earth's surface to which television programmes are to be beamed for direct reception by the viewer, rather than redistribution by cable or other means.

direct capacitance *(Phys.)*. That between two conductors, as if there are no other conductors.

direct chill casting *(Eng.)*. A method like **continuous casting** but for larger cross-sections in which the hollow mould is closed at the bottom by a platform. This is gradually lowered as the metal becomes solid on the outside and therefore able to contain the melt, platform, mould and metal being appropriately cooled.

direct-conversion reactor *(Elec.Eng.)*. One which converts thermal energy directly into electricity by means of thermoelectric elements (usually of silicon-germanium).

direct cooling *(Elec.Eng.)*. The cooling of transformer and machine windings by circulating the coolant through hollow conductors.

direct-coupled exciter *(Elec.Eng.)*. An exciter for a synchronous or other electric machine, which is mounted on the same shaft as the machine that it is exciting.

direct-coupled generator *(Elec.Eng.)*. A generator which is mechanically coupled to the machine which is driving it, i.e. not driven through gearing, a belt, etc.

direct coupling *(Electronics)*. Interstage coupling without the use of transformers or series capacitors, so that the d.c. component of the signal is retained; also called *d.c. coupling*.

direct current *(Elec.Eng.)*. Current which flows in one direction only, although it may have appreciable pulsations in its magnitude. Abbrevs. *dc, d.c.*

direct-current amplifier, -balancer *(Elec.Eng.)*. See **d.c. amplifier, d.c. balancer**.

direct data entry *(Comp.)*. Input of data directly to the computer using, normally, a **key-to-disk unit**. The data may be validated while held in a temporary file, before being written to the disk for subsequent processing.

direct-fired *(Eng.)*. Furnace in which the fuel is delivered into the heating chambers.

directing stimulus *(Behav.)*. Stimulus which, though not releasing a component of species-specific behaviour, is important in determining the direction of a response.

direct injection *(Aero.)*. The injection of metered fuel (for a spark-ignition engine) into the super-charger eye of the cylinders, which eliminates the freezing and poor high-altitude behaviour of carburettors.

direct-injection pump *(Aero.)*. A fuel-metering pump for injecting fuel direct to the individual cylinders.

direct interaction *(Phys.)*. A mechanism for describing how a nuclear reaction takes place. It assumes that the interaction between bombarding nucleus and target nucleus involves only a few nucleons near the surface of the nuclei. Cf. **compound nucleus**.

direction *(Maths.)*. The position of one point in space relative to another.

directional antenna *(Space,Telecomm.)*. One in which the transmitting and/or receiving properties are concentrated along certain directions, used in space over very long distances.

directional circuit-breaker *(Elec.Eng.)*. A circuit-breaker which operates when the current flowing through it is in the direction opposite to normal.

directional coupler *(Telecomm.)*. In a transmission line or a waveguide, a device which couples a secondary transmission path to a wave travelling in only one direction on the main path; there is no energy transfer for propagation in the other direction. The amount of energy coupled is usually only a small proportion, possibly 10 to 20 dB less than that in the main beam.

directional derivative *(Maths.)*. The rate of change of a function $F(x,y,z)$ with respect to arc length along a given curve (or in a given direction), i.e.

$$\frac{dF}{ds} = l\frac{\partial F}{\partial x} + m\frac{\partial F}{\partial y} + n\frac{\partial F}{\partial z},$$

where l,m,n are the direction cosines of the tangent to the curve at the point (x,y,z).

directional drilling *(Min.Ext.)*. The use of special down-hole drilling assemblies to turn a drill hole in the desired direction.

directional effects *(Image Tech.)*. Defects of non-conformity in a processed image caused by the action of depleted developer solution where there has been inadequate agitation in a continuous processing machine.

directional filter *(Telecomm.)*. A combination of filters, e.g. a high- and a low-pass filter or two different band-pass filters, to separate the bands of frequencies used for transmission in opposite directions in a **carrier system**.

directional gain *(Telecomm.)*. Ratio, expressed in decibels, of the response, generally along the axis where it is a maximum, to the mean spherical (or hemispherical with reflector or baffle) response, of an antenna loudspeaker or microphone. Also *directivity index*.

directional lighting fittings *(Elec.Eng.)*. Lighting fittings (often used in street lighting installations) which direct a high proportion of their light output towards a point on the roadway midway between adjacent lamp standards.

directional loudspeaker *(Acous.)*. A loudspeaker which radiates more strongly in one direction than in others. Normally the radiated sound power is directed in a beam. Often a combination of loudspeakers (array) is directional.

directional microphone *(Acous.)*. One which is directional in response. See also **acoustic telescope**.

directional receiver *(Telecomm.)*. Receiving system using a directional antenna for discrimination against noise and other transmissions.

directional relay *(Elec.Eng.)*. A relay whose operation depends on the direction of the current flowing through it.

directional transmitter *(Telecomm.)*. Transmitting system using a directional antenna, to minimize power requirements and to diminish effect of interference.

direction angles *(Maths.)*. The 3 angles which a line makes with the positive directions of the co-ordinate axes.

direction components *(Maths.)*. See **direction numbers**.

direction cosines *(Maths.)*. The cosines of the 3 direction angles of a line.

direction coupling *(Elec.Eng.)*. That afforded by a device in which power can be transmitted in specified paths and not in others.

direction-finding *(Telecomm.)*. Using a *direction-finder*. The principle and practice of determining a bearing by radio means, using a discriminating antenna system and a radio receiver, so that direction or bearing of a distant transmitter can be determined.

direction numbers *(Maths.)*. Of a line, any 3 numbers,

not all of which are zero, which are proportional to the direction cosines of the line. Also *direction components, direction ratios*.

direction of a curve *(Maths.)*. Direction of tangent to curve at the point.

direction ratios *(Maths.)*. See **direction numbers**.

direction switch *(Elec.Eng.)*. A switch which determines the direction of travel; used on electric lifts or similar equipment.

directive efficiency *(Telecomm.)*. Ratio of maximum to average radiation or response of an antenna; the **gain**, in dB, of antenna over a dipole being fed with the same power.

directive force *(Elec.Eng.)*. A term used to denote the couple which causes a pivoted magnetic needle to turn into a north and south direction.

directive gain *(Telecomm.)*. For a given direction, 4π times the ratio of the radiation intensity in that direction to the total power which is radiated by the aerial.

directivity *(Acous.,Telecomm.)*. Measurement, in dB, of the extent to which a directional loudspeaker, microphone or antenna concentrates its radiation or response in specified directions.

directivity angle *(Telecomm.)*. Angle of elevation of direction of maximum radiation or reception of electromagnetic wave by an antenna.

directivity factor *(Acous.)*. Non-dimensional quantity for loudspeakers and microphones which characterizes the strength of the directivity.

directivity index *(Telecomm.)*. Same as *directional gain*.

direct labour *(Build.)*. A mode whereby labour is employed directly by the client, as opposed to the usual method of working through independent architect, engineer, and surveyor.

direct laying *(Eng.)*. Cables are laid in a trench and covered with soil; planks, bricks, tiles, or concrete slabs are put over the cable as protection. Cables used to be armoured, but modern practice is merely to put a serving of bituminized paper or hessian over the *lead* sheath.

direct lighting *(Elec.Eng.)*. A system of lighting in which not less than 90% of the total light emitted is directed downwards, i.e. in the lower hemisphere.

direct lithography *(Print.)*. Lithographic printing whereby the plate prints directly on the paper, without first offsetting onto a blanket cylinder.

directly-heated cathode *(Electronics)*. Metallic (coated) wire heated to a temperature such that electrons are freely emitted. Also called *filament cathode*.

direct manipulation *(Comp.)*. An approach to the computer-aided restructuring of data by pointing and moving it rather than by entering coordinates or descriptions. See **mouse**.

direct metamorphosis *(Zool.)*. The incomplete metamorphosis undergone by exopterygote insects, in which a pupal stage is wanting.

director *(Med.)*. A grooved instrument for guiding a surgical knife.

director *(Telecomm.)*. (1) Free resonant dipole element in front of antenna array which assists the directivity of the array in the same direction. See **Yagi antenna**. (2) The apparatus which obtains a channel, through exchange junctions, to the required exchange. During dialling the trains of impulses are registered, and when the required exchange is found the numerical trains are passed to the required exchange, and operate the selectors to get the required subscriber.

director circle *(Maths.)*. See **orthoptic circle**.

director meter *(Telecomm.)*. In an automatic switching exchange, a meter which is attached to a director to total the number of times it is taken into operation.

director system *(Telecomm.)*. An automatic switching system for routing calls between exchanges. It uses a storage mechanism for the numerical impulse trains, while the code impulses are being translated and used to find a route over junctions to the required exchange, which, when found, receives the numerical trains and hence, by step-by-step mechanism, connects to the wanted subscriber.

directory *(Comp.)*. List of file names, together with information enabling the files to be retrieved from backing store by the operating system. See also **data dictionary**.

direct printing *(Print.)*. Method in which the print is made directly on the paper as in letterpress or photogravure, as distinct from the usual lithographic method of **offset printing**.

direct process *(Eng.)*. The method originally used for obtaining from ore a form of iron similar to wrought-iron in one operation, i.e. without first making pig-iron.

direct radiation *(Phys.)*. See **primary radiation**.

direct-reading instrument *(Elec.Eng.)*. An instrument in which the scale is calibrated in the actual quantity measured by the instrument, and which therefore does not require the use of a multiplying constant.

direct-recorded disk *(Acous.)*. One which, after a groove has been cut into its plastic surface, is immediately ready for **playback** (reproduction).

directrix *(Maths.)*. (1) Of a conic or quadric, the polar line or plane of a focus. See **conic** for alternative definition. (2) A curve of a ruled surface in general through which the generators pass.

direct rope haulage *(Min.Ext.)*. Engine plane. An ascending truck is partly balanced by a descending one, motive power being applied to the drum round which the haulage rope passes.

direct sound *(Acous.)*. The sound intensity arising from a source to a listener, as contrasted with the reverberant sound which has experienced reflections between the source and the listener.

direct stress *(Eng.)*. The stress produced at a section of a body by a load whose resultant passes through the centre of gravity of the section.

direct stroke *(Elec.Eng.)*. When a transmission line or other apparatus is struck by a lightning stroke, it is said to receive a *direct stroke*.

direct-suspension construction *(Elec.Eng.)*. A form of construction used for the overhead contact wire on electric traction systems; the contact wire is connected directly to the supports without catenary or messenger wires.

direct-switching starter *(Elec.Eng.)*. An electric motor starter arranged to switch the motor directly across the supply, without the insertion of any resistance or the performing of any other current-limiting operation.

direct-trip *(Elec.Eng.)*. A term used in connection with circuit-breakers, starters, or other similar devices, to indicate that the current which flows in the tripping coil is the main current in the circuit, not an auxiliary current obtained from a battery or other source.

direct vernier *(Surv.)*. A vernier in which n divisions on the vernier plate correspond in length to $(n-1)$ divisions on the main scale.

direct-vision prism *(Phys.)*. A compound prism with component prisms of two glasses having different dispersive powers and cemented together so that, in passing through the combination, light suffers dispersion but no deviation.

direct-vision spectroscope *(Phys.)*. A spectroscope employing a **direct-vision prism**. Such an instrument is usually in the form of a short straight tube with a slit at one end and an eyepiece at the other; it is used for rough qualitative examination of spectra.

direct vision viewfinder *(Image Tech.)*. A viewfinder in which the subject is viewed directly, not by reflection.

direct voice input *(Aero.)*. A means by which a pilot can command an aircraft to respond to his spoken instructions for such functions as change of radio frequency, flight performance and possibly weapon aiming and delivery.

direct wave, ray *(Telecomm.)*. That portion of the power radiated from an antenna which goes directly to the receiving antenna, without ionospheric reflection. Also *ground wave*.

dirigible *(Aero.)*. A navigable balloon or airship.

dirt *(Min.Ext.)*. Broken valueless mineral. Also *gangue*. In US *muck*.

DIR technology *(Image Tech.)*. *Developer Inhibitor Release*, couplers whose inclusion in photographic emulsions increase **border effects** during development, giving improved image sharpness.

dis *(Elec.Eng.)*. See **discontinuity**.

disaccharides *(Chem.)*. A group of carbohydrates considered to be derived from 2 molecules of a monosaccharide (either the same or different) by elimination of 1 molecule of H_2O; e.g. maltose is Gα—1→4G (G = glucose).

disadvantage factor *(Nuc.Eng.)*. Ratio of average neutron flux in reactor lattice to that within actual fuel element.

disappearing-filament pyrometer *(Eng.)*. An instrument used for estimating the temperature of a furnace by observing a glowing electric-lamp filament against an image of the interior of the furnace formed in a small telescope. The current in the filament is varied until it is no longer visible against the glowing background. From a previous calibration the required temperature is derived from the current value.

disarticulation *(Med.)*. Amputation of a bone through a joint.

disassembler *(Comp.)*. Program which translates from machine code to an assembly language, generally used to decipher existing machine code by generating the equivalent symbolic codes.

disazo dyes *(Chem.)*. Dyestuffs containing two azo groups of the type: $C_6H_5.N{=}N.C_6H_4.N{=}N.C_6H_4.OH$. These dyes are obtained by diazotizing an amino-derivative of azobenzene and then coupling it with a tertiary amine or with a phenol, or by coupling a diamine or dihydric phenol with 2 molecules of a diazonium salt.

disbudding *(Vet.)*. The destruction, usually by caustic chemicals or thermo-cautery, of the horn buds of calves, kids and lambs.

disc *(Genrl.)*. See **disk**.

discal *(Zool.)*. Pertaining to or resembling a disk or disklike structure; a wing-cell of various Insects.

discard *(Eng.)*. See **cropping**.

discharge *(Nuc.Eng.)*. Unloading of fuel from a reactor.

discharge *(Phys.)*. (1) The abstraction of energy from a cell by allowing current to flow through a load. (2) Reduction of the p.d. at the terminals (plates) of a capacitor to zero. (3) Flow of electric charge through gas or air due to ionization, e.g., lightning, or at reduced pressure, as in fluorescent tubes. See **field discharge**.

discharge bridge *(Elec.Eng.)*. Measurement of the ionization or discharge, in dielectrics or cables, depending on the amplification of the high-frequency components of the discharge.

discharge circuit *(Elec.Eng.)*. One arranged to discharge a capacitor or parasitic capacitance, either for circuit operational reasons or for safety.

discharge electrode *(Elec.Eng.)*. See **active electrode**.

discharge head *(Min.Ext.)*. Vertical distance between intake and delivery of pump, *plus* allowance for mechanical friction and other retarding resistances requiring provision of extra power.

discharge lamp *(Elec.Eng.)*. One in which luminous output arises from ionization in gaseous discharge. See **discharge (3)**.

discharge printing *(Textiles)*. The removal of a dye from a fabric to leave a white pattern.

discharger *(Elec.Eng.)*. (1) A device, such as a spark gap, which provides a path whereby a piece of electrical apparatus may be discharged. (2) An apparatus containing an electrically heated wire for firing explosives in blasting.

discharge rate *(Elec.Eng.)*. A term used in connection with the discharge of accumulators. An accumulator has a certain capacity at, e.g. a 1-hr discharge rate, when that capacity can be obtained if the accumulator is completely discharged in 1 hr. If the discharge rate is lower, i.e. the discharge takes more than 1 hr, the capacity obtainable will be higher.

discharge resistance *(Elec.Eng.)*. (1) A noninductive resistance placed in parallel with a circuit of high inductance (e.g. the field winding of an electric machine) in order to prevent a high voltage appearing across the circuit when the current in it is switched off. Also called *buffer resistance*. (2) Resistance placed in parallel with a capacitance or circuit with parasitic capacitance to provide a discharge path for stored charge, for reasons of circuit operation or for safety.

discharge tube *(Electronics)*. Any device in which conduction arises from ionization, initiated by electrons of sufficient energy.

discharge valve *(Eng.)*. A valve for controlling the rate of discharge of fluid from a pipe or centrifugal pump.

discharging arch *(Build.)*. An arch built in a wall to protect a space beneath from the weight above, and to allow access or discharge.

discharging pallet *(Horol.)*. The pallet mounted in the discharging roller of the chronometer escapement, which brings about the unlocking of the escape wheel by removing the locking pallet from one of its teeth.

discharging roller *(Horol.)*. The circular disk carrying the discharging pallet mounted on the balance staff in a chronometer escapement.

discharging tongs *(Elec.Eng.)*. A pair of metal tongs used for discharging capacitators before they are touched by hand.

discission *(Med.)*. An incision into a part; especially, needling of a cataract.

disclimax *(Ecol.)*. A stable community which is not the climatic or edaphic community for a particular place, but is maintained by man or his domestic animals, e.g. a desert produced by overgrazing, where the natural climax would be grassland. The name derives from *disturbance climax*.

Discolichenes *(Bot.)*. A group of lichens in which the fungus is a Discomycete.

discomposition effect *(Phys.)*. See **Wigner effect**.

Discomycetes *(Bot.)*. A class of fungi in the Ascomycotina in which the fruiting body (ascocarp) is usually an apothecium. Includes the Lecanorales (*lichen*-forming fungi), and many saprophytic and mycorrhizal sorts e.g. the morels and the truffles.

discone antenna *(Telecomm.)*. A biconical antenna, used for short-wave and VHF communication, having one cone flattened out to form a disk. The transmission line is connected between the centre of the disk and the apex of the cone. Its input impedance and radiation pattern remains constant over a wide frequency range and it gives an omnidirectional pattern in a horizontal plane when the axis of the cone is vertical.

disconformity *(Geol.)*. A break in the rock sequence in which there is no angular discordance of dip between the two sets of strata involved. Cf. **unconformity**.

disconnected set *(Maths.)*. One which can be divided into two sets having no points in common and neither containing an accumulation point of the other.

disconnection *(Elec.Eng.)*. See **discontinuity**.

disconnector *(Build.)*. See **interceptor**.

discontinued construction *(Build.)*. See **acoustic construction**.

discontinuity *(Elec.Eng.)*. A break, whether intentional or accidental, in the conductivity of an electrical circuit; also called *disconnection*, colloq. *dis*.

discontinuity *(Maths.)*. A point at which a function is not continuous.

discontinuous distribution *(Biol.)*. Isolated distribution of a species, as the Tapir, which is found in the Malay Peninsula and Sumatra and again in Central and South America.

discontinuous variation *(Biol.)*. A sudden change in otherwise smoothly varying characters in a group of organisms over, for example, a geographical range.

discovery well *(Min.Ext.)*. The first well to reveal oil in a new field or at a new level.

discriminant *(Maths.)*. (1) Of a polynomial equation $x^n + a_1x^{n-1} + \ldots + a_n = 0$: the product of the squares of all the differences of the roots taken in pairs. (2) Of a differential equation: the result of eliminating $p\left(=\dfrac{dy}{dx}\right)$ between the differential equation $F(x,y,p) = 0$ and

$$\frac{\partial}{\partial p}F(x, y, p) = 0$$

is the p-discriminant equation, which represents the p-discriminant locus. For a quadratic equation $ax^2+bx+c=0$, the discriminant is b^2-4ac. The roots are equal if $b^2-4ac=0$, real if $b^2-4ac>0$ and imaginary if $b^2-4ac<0$ (a,b,c being real). If the solution of the differential equation is $u(x,y,c)=0$ (c an arbitrary constant), the result of eliminating c between $u(x,y,c)=0$ and $\frac{\partial y}{\partial c}(x, y, c)=0$ gives the c-discriminant equation.

discriminant analysis *(Stats.)*. A method of assigning observations to groups on the basis of values of observations previously obtained from each group.

discriminating circuit-breakers *(Elec.Eng.)*. A term sometimes used to denote circuit-breakers which operate only when the current is in a given direction.

discriminating protective system *(Elec.Eng.)*. An excess-current protective system which causes to be disconnected only that portion of a power system upon which a fault has actually occurred.

discriminating satellite exchange *(Telecomm.)*. A small automatic exchange which can decide, without engaging its main exchange, whether or not it can complete a call arising from one of its subscribers.

discriminating selector *(Telecomm.)*. A selector which discriminates between calls which are to be completed locally and those which are to be completed through other exchanges, working through the absorption of a train of impulses.

discrimination *(Behav.)*. Animal behaviour: the ability to respond to different patterns of stimulation, often tested for by using a conditioning procedure (see **discrimination training**). Human: in social psychology, a term denoting behaviour towards people or groups of people based on their inclusion in a particular group (e.g. gender, race).

discrimination *(Telecomm.)*. The selection of a signal having a particular characteristic, e.g. frequency, amplitude, etc., by the elimination of all the other input signals at the discriminator.

discrimination training *(Behav.)*. Learning to respond to certain stimuli that are reinforced, and not to others that are not reinforced.

discriminator *(Telecomm.)*. (1) Circuit which rejects pulses below a certain amplitude level, and shapes the remainder to standard amplitude and profile. (2) Device which effects the routing and/or determines the fee units of a call originating at a satellite exchange in telephony.

discus proligerus *(Zool.)*. See **zona granulosa**.

dish *(Image Tech.)*. A directional antenna having a concave surface.

dish *(Telecomm.)*. Colloq. for *parabolic reflector*, which may be made of sheet metal or mesh; a form of microwave antenna used for point-to-point and satellite communication and for radio astronomy and satellite broadcast reception.

disharmony *(Zool.)*. See **hypertely**.

dished *(Eng.)*. Of wheels, especially steering wheels, having the hub inset on a different plane from the rim.

dished-out *(Build.)*. Said of the wooden framework or bracketing on which the laths and plastering are fixed in vaults, domes, coved ceilings, and the like.

dish wheel *(Eng.)*. See **cup wheel**.

disincrustant *(Eng.)*. See **anti-incrustator**.

disinfectant *(Chem.)*. Any preparation that destroys the causes of infection. The most powerful disinfectants are oxidizing agents and chlorinated phenols.

disinfection *(Med.)*. The destruction of pathogenic bacteria, usually with an antiseptic chemical or **disinfectant**.

disinfestation *(Med.)*. The destruction of insects, especially lice.

disintegrating mill *(Min.Ext.)*. A mill for reducing lump material to a granular product. It consists of fixed and rotating bars in close proximity, crushing being partly by direct impact and partly by interparticulate attrition. See also **beater mill**.

disintegration *(Phys.)*. A process in which a nucleus ejects one or more particles, applied especially, but not only, to spontaneous radioactive decay.

disintegration constant *(Phys.)*. The probability of radioactive decay of a given unstable nucleus per unit time. Statistically, it is the constant λ, expressing the exponential decay $\exp(-\lambda t)$ of activity of a quantity of this isotope with time. It is also the reciprocal of the mean life of an unstable nucleus. Also called *decay constant, transformation constant*.

disintegration energy *(Phys.)*. See **alpha-decay energy**.

disintegration of filament *(Elec.Eng.)*. The gradual breaking up of the filament of an electric filament lamp, owing to the projection of particles from the filament; the particles adhere to the inner surface of the bulb, causing blackening.

disjunct *(Bot.)*. Interrupted, disconnected, not continuous.

disjunct *(Zool.)*. Having deep constrictions between the different tagmata of the body.

disjunction *(Biol.)*. The separation during meiotic **anaphase** of the two members of each pair of homologous chromosomes.

disjunctor *(Bot.)*. A portion of wall material forming a link between the successive conidia in a chain, and serving as a weak plane where separation may occur.

disk *(Bot.)*. (1) A fleshy outgrowth from the receptacle of a flower, surrounding or surmounting the ovary and often secreting nectar. (2) The central part of a capitulum.

disk *(Comp.)*. See **magnetic disk, videodisk**.

disk *(Zool.)*. Any flattened circular structure.

disk-and-drum turbine *(Eng.)*. A steam-turbine comprising a high pressure impulse wheel, followed by intermediate and low-pressure reaction blading, mounted on a drum-shaped rotor. Also called *combination turbine* or *impulse-reaction turbine*.

disk area *(Aero.)*. The area of the circle described by the tips of the blades of a rotorcraft; similarly applied to propellers.

disk armature *(Elec.Eng.)*. One for a motor or generator wound to a large diameter on a short axle length.

disk brakes *(Aero.,Autos.)*. Type in which two or more pads close by caliper action on to a disk which is connected rigidly to the landing or car wheel-hub; more efficient than drum type, owing to greater heat dissipation.

disk camera *(Image Tech.)*. A camera in which the images are recorded on a small disk of photographic film, which is rotated through an appropriate angle after each exposure to present a new area.

disk capacitor *(Elec.Eng.)*. One in which the variation in capacitance is effected by the relative axial motion of disks.

disk centrifuge *(Powder Tech.)*. Apparatus for particle size analysis in which particles are sedimented in a rotating disk.

disk drive *(Comp.)*. Mechanism which causes magnetic disks to rotate between read/write heads.

Disk Filing System *(Comp.)*. Name given to the operating system in the BBC microcomputer. Abbrev. *DFS*.

disk filter *(Min.Ext.)*. *American filter*. Continuous heavy-duty vacuum filter in which separating membranes are disks, each revolving slowly through its separate compartment.

disk floret *(Bot.)*. (1) Usually in the Compositae one of the florets occupying the central part of the capitulum, whatever its morphology. (2) Sometimes, a *tubular floret*. Cf. **ray floret**.

disk formatting *(Comp.)*. The preparation of blank magnetic disk or tape for subsequent writing or reading by adding control information such as track and sector number. See **hard sectored-, soft sectored-**.

disk friction *(Eng.)*. The force resisting the rotation of a disk in a fluid. It is important in the design of centrifugal machinery as it decreases efficiency and causes a rise in the pressure of the fluid being pumped.

disk inking *(Print.)*. The least efficient arrangement for ink distribution, only suitable for the light *platen* *machine*. Ink is supplied to a slowly rotating disk from which it is picked up by the rollers.

disk loading *(Aero.)*. The lift, or upward thrust, of a rotor divided by the disk area.

disk operating system *(Comp.)*. See **DOS**.

disk pack *(Comp.)*. Set of magnetic disks fitted on a common spindle; each disk has its own set of read/write heads.

disk pile *(Civ.Eng.)*. A hollow pile having a wide flange at the foot with projecting radial ribs on it; used for piling in sand.

disk, plate clutch *(Eng.)*. A friction clutch in which the driving and driven members have flat circular or annular friction surfaces, and consist of one or a number of disks, running either dry or lubricated. See **single-plate clutch, multiple-disk clutch**.

disk record *(Acous.)*. The type of gramophone record, originally made of synthetic thermoplastic resin, and in which the reproducing needle follows a spiral groove, while the record is rotated at constant speed; devised by Berliner.

disk ruling *(Print.)*. Method by which ink is applied to the paper by disks instead of pens, permitting higher speeds on the ruling machine.

disk-seal tube *(Electronics)*. A valve constructed from metallic disks (which may be the electrodes) sandwiched with glass or ceramic insulating pillars. Characteristics include low inter-electrode capacitance, reduced lead inductances and high temperature ratings. Used in high-power applications at frequencies up to 2 GHz.

disk valve *(Eng.)*. A form of suction and delivery valve used in pumps and compressors; it consists of a light steel or fabric disk resting on a ported flat seating; steel valve disks are usually spring-loaded.

disk winding *(Elec.Eng.)*. A type of winding for medium and large transformers, in which the turns are made up into a number of annular disks.

dislocation *(Crystal.)*. A lattice imperfection in a crystal structure, classified according to type, e.g. edge dislocation, screw dislocation.

dislocation *(Med.)*. The displacement of one part from another; especially the abnormal separation of two bones at a joint.

disomic *(Biol.)*. Relating to two homologous chromosomes or genes.

dispensable circuit *(Elec.Eng.)*. a separate circuit used in wiring system to which is connected apparatus that can be cut out of circuit at times of heavy load.

dispensary *(Med.)*. A place where drugs, etc., are dispensed (i.e. are prepared for administration).

dispenser cathode *(Electronics)*. One which is not coated, but is continuously supplied with suitable emissive material from a separate cathode element.

dispermy *(Zool.)*. Penetration of an ovum by two spermatozoa.

dispersal *(Ecol.)*. The active or passive movement of individual plants or animals or their disseminules (seeds, spores, larvae, etc.) into or out of a population or population area. It includes emigration, immigration and migration. Should not be confused with *dispersion*.

dispersed phase *(Chem.)*. A substance in the colloidal state.

dispersion *(Ecol.)*. The distribution pattern in an animal or plant population, this being random, uniform (more regular than random), or clumped (see **aggregation**). Should not be confused with *dispersal* or with *distribution*, which refers to the species as a whole, although the dispersion of a population can be described as following a random, or **Poisson distribution**.

dispersion *(Phys.)*. The dependence of wave velocity on the frequency of wave motion; a property of the medium in which the wave is propagated. In the visible region of the electromagnetic spectrum, dispersion manifests itself as the variation of refractive index of a substance with wavelength (or colour) of the light. It is on account of its dispersion that a prism is able to form a spectrum. See **Hartmann dispersion formula, Cauchy's dispersion formula, anomalous dispersion**.

dispersion *(Stats.)*. The extent to which observations are dissimilar in value, often measured by standard deviation, range, etc.

dispersion coefficient *(Elec.Eng.)*. A term often used to denote the leakage factor of an induction motor.

dispersion curve *(Phys.)*. A plot of frequency against wavelength for a wave in a dispersive medium. See **phonon dispersion curve, acoustic branch, optic branch**.

dispersion forces *(Chem.)*. Weak intermolecular forces, corresponding to the term *a* in *van der Waal's equation*. See **London forces**.

dispersion hardening *(Eng.)*. See **precipitation hardening**.

dispersion medium *(Chem.)*. A substance in which another is colloidally dispersed.

dispersive medium *(Phys.)*. One in which the phase velocity of a wave is a function of frequency.

dispersive power *(Phys.)*. The ratio of the difference in the refractive indices of a medium for the red and violet to the mean refractive index diminished by unity. This may be written

$$\nu = \frac{n_V - n_R}{n - 1}.$$

dispersivity quotient *(Phys.)*. The variation of refractive index n with wavelength λ, $dn/d\lambda$.

displaced terranes *(Geol.)*. Internally consistent rock-masses within an orogenic area which are abruptly discontinuous with their surroundings. Sometimes called *suspect terranes*. See **terranes**.

displacement *(Aero.)*. The mass of the air displaced by the volume of gas in any lighter-than-air craft or water by a seaplane hull or float.

displacement *(Behav.)*. In psychoanalytic theory, a *defence mechanism* involving the transfer of emotional energy from an unacceptable object to a safer one, so that gratification of a need comes from a source that is personally or socially more acceptable. In dreams, for example, one image may be over-exaggerated and another, more central image, minimized in its affective quality.

displacement *(Eng.)*. (1) The volume of fluid displced by a pump plunger per stroke or per unit time. (2) The swept volume of a working cylinder.

displacement *(Hyd.Eng.)*. The weight of water displaced by a vessel. It is equal to the total weight of the vessel and contents. See **Archimedes' principle**.

displacement *(Phys.)*. Vector representing the electric flux in a medium and given by $\mathbf{D} = \varepsilon\mathbf{E}$, where ε is the permittivity and $\mathbf{E}$ is the electric field. Also called *dielectric strain, electric flux density*.

displacement activity *(Behav.)*. The performance of a behaviour pattern which is apparently irrelevant to the situation in which it occurs; common in conflict situations.

displacement current *(Phys.)*. Integral of the displacement current density through a surface. The time rate of change of the electric flux. Current postulated in a dielectric when electric stress or potential gradient is varied. Distinguished from a normal or conduction current in that it is not accompanied by motion of current carriers in the dielectric. Concept introduced by Maxwell for the completion of his electromagnetic equations.

displacement flux *(Phys.)*. Integral of the normal component of displacement over any surface in a dielectric. See **displacement**.

displacement law *(Phys.)*. Soddy and Fajans formulation that radiation of an α-particle *reduces* the atomic number by 2 and the mass number by 4, and that radiation of a β-particle *increases* the atomic number by 1, but does not change the mass number. It was later found that emission of positron *decreases* the atomic number by 1, but does not change the mass number. Gamma emission and isomeric transition change neither mass nor atomic number. Displacement laws are summarized as follows:

Type of disintegration	Change in atomic number	Change in mass number
alpha emission	−2	−4
beta electron emission	+1	0
beta positron emission	−1	0
beta electron capture	−1	0
isomeric transition	0	0
gamma emission	0	0

A change in atomic number means displacement in the periodic classification of the chemical elements; a change in mass number determines the radioactive series.

displacement pump *(Min.Ext.)*. One with pulsing action, produced by steam, compressed air or a plunger, causing non-return valves to prevent return flow of displaced liquid during the retracting phase of the pump cycle.

displacement series *(Chem.)*. See **electrochemical series**.

display *(Electronics)*. Mode of showing information on a cathode-ray tube screen, especially in computing, radar and navigation. See **scope**.

display behaviour *(Behav.)*. Species specific patterns of either sound or movement, often stereotyped in form, and which serve a great variety of communicative functions, e.g. in courtship or agonistic behaviour.

display work *(Print.)*. Displayed text setting (such as title pages, jobbing work, advertisements), distinguished from solid text composition.

disposable load *(Aero.)*. Maximum ramp weight minus *operating weight empty* (*OWE*); includes crew, fuel, oil and payload (civil) or armament (military).

disproportionation *(Chem.)*. A reaction in which a single compound is simultaneously oxidized and reduced, e.g. the spontaneous reaction in water of soluble copper (I) salts to form equal amounts of copper (0) and copper (II).

disruptive discharge *(Elec.Eng.)*. That of a capacitor when the discharge arises from breakdown (puncture) of the dielectric by an electric field which it cannot withstand.

disruptive strength *(Eng.)*. The stress at which a material fractures under tension.

disruptive voltage *(Elec.Eng.)*. That which is just sufficient to puncture the dielectric of a capacitor. A test voltage is normally applied for one minute. See also **breakdown voltage**.

dissecting *(Print.)*. The removal of type matter which is to be printed in a second colour, in order to impose it in another chase, position and spacing being carefully regulated.

dissecting aneurysm *(Med.)*. The leaking of blood through a tear in the inner wall of the **aorta** producing a cleavage in the layers of the vessel and tracking of blood in a 'false lumen' along the aorta and its vessel wall. Associated with a high mortality.

disseminated sclerosis *(Med.)*. See **multiple sclerosis**.

disseminated values *(Min.Ext.)*. Mode of occurrence in which small specks of concentrate are scattered evenly through the gangue mineral.

dissemination *(Bot.,Zool.)*. The spread or migration of species, usually by means of seeds, spores and larvae.

disseminule *(Bot.)*. A propagule.

dissimilar terms *(Maths.)*. Terms containing different powers of the same variable(s), or containing different variables, e.g. $xy,(xy)^2$; $3x$, $3z$.

dissipation *(Telecomm.)*. Loss or diminution, usually undesirable, of power, the lost power being converted into heat. Causes power loss in transmission lines etc. and can diminish cut-off sharpness in filters. In low frequency circuits, it is due largely to resistance and eddy-current losses. In high frequency circuits, resistance, radiation and dielectric losses all contribute. The heat if not removed by heat sinks, air or water cooling can damage components.

dissipation factor *(Elec.Eng.)*. The cotangent of the phase angle (δ) for an inductor or capacitor. For low loss components, the dissipation factor is approximately equal to the power factor,

$$\tan\delta \simeq \frac{\sigma}{\omega\varepsilon},$$

for a low loss dielectric, where σ is the electrical conductivity, ε the permittivity of the medium, and ω is $2\pi \times$ frequency.

dissipationless line *(Telecomm.)*. A hypothetical transmission line in which there is no energy loss. Also *lossless line*.

dissipation trails *(Meteor.)*. Lanes of clear atmosphere formed by the passage of an aircraft through a cloud.

dissipative network *(Telecomm.)*. One designed to absorb power, as contrasted with networks which attenuate power by impedance reflection. All networks dissipate to some slight extent, because neither capacitors nor inductors can be made entirely loss-free.

dissociation *(Chem.)*. The reversible or temporary breaking-down of a molecule into simpler molecules or atoms. See **Arrhenius theory of dissociation**.

dissociation constant *(Chem.)*. The equilibrium constant for a process considered to be a dissociation. Commonly it is applied to the dissociation of acids in water.

dissociation, dissociative disorder *(Behav.)*. An unconscious **defence mechanism** in which a group of psychological functions are separated from the remainder of the person's activities. In extreme cases this may result in a *dissociative disorder*, e.g. **amnesia**, **fugue**, **multiple personality**.

dissociation of gases *(Eng.)*. Chemical combustion reaction occurring at the highest temperature of the flame where carbon dioxide and water vapour tend to dissociate into CO, H_2 and O_2.

dissolution *(Chem.)*. The taking up of a substance by a liquid, with the formation of a homogeneous solution.

dissolve *(Image Tech.)*. Transition from one scene to another in which the whole image of the first gradually disappears as it is replaced by the second. Also *lap dissolve*.

dissolving pulp *(Paper)*. See **alpha pulp**.

dissonance *(Acous.)*. The playing of two or more musical terms simultaneously to produce an unpleasant effect on the listener.

dissymmetrical *(Genrl.)*. See **asymmetrical**.

dist *(Build.)*. Abbrev. for (1) *distemper*, (2) *distributed*.

distal *(Biol.)*. Far apart, widely spaced: pertaining to or situated at the outer end: farthest from the point of attachment. Cf. **proximal**.

distance *(Maths.)*. (1) The length of the line joining two points. (2) Angular, between two points: the angle between the two lines from the point of reference to the points in question. (3) The length of the segment between the two lines or two planes of the line perpendicular to both lines or both planes. (4) Distance from a line or plane to a point: the length of the perpendicular from the point to the line or plane.

distance block, distance piece *(Build.)*. A wooden or other block serving to separate 2 pieces by a desired distance.

distance control *(Elec.Eng.)*. See **remote control**.

distance mark *(Radar)*. Mark on the screen of a CRT to denote distance of target.

distance-measuring equipment *(Aero.)*. Airborne secondary-radar which indicates distance from a ground transponder beacon. Abbrev *DME*

distance meter *(Image Tech.)*. Same as *rangefinder*.

distance protection *(Elec.Eng.)*. See **impedance protective system**.

distance relay *(Elec.Eng.)*. See **impedance relay**.

distant-reading compass *(Eng.)*. Gyro flux-gate compass in which the indicator is remote from the sensing device.

distant-reading instrument *(Eng.)*. A recording or indicating instrument (such as a thermometer or pressure gauge) in which the reading is shown on a scale at some distance from the point of measurement. See **remote control**.

distemper *(Build.)*. A mixture of dry pigment with size, water and sometimes oil, used as a paint for internal walls and ceilings.

disthene *(Min.)*. A less commonly used name for the mineral **kyanite**.

distichiasis, distichia *(Med.)*. A condition in which there are two complete rows of eyelashes in one or both eyelids.

distichous *(Bot.)*. Leaves on a stem arranged in 2 diametrically opposite rows.

distillation *(Chem.)*. A process of evaporation and re-condensation used for separating liquids into various fractions according to their boiling points or boiling ranges. See also **molecular-**.

distillation flask *(Chem.)*. A laboratory apparatus, usually made of glass; it consists of a bulb with a neck for the insertion of a thermometer and a side tube attached to the neck, through which the vapours pass to the condenser.

distinct *(Bot.)*. Plant members not joined to one another.

distomiasis, distomatosis *(Vet.)*. Infection of the bile ducts by flukes or trematode worms.

distorted wave *(Elec.Eng.)*. A term often used in electrical engineering to denote a nonsinusoidal waveform of voltage or current.

distorting network *(Telecomm.)*. A network altering the response of part of a system, and anticipating the correction of response required to restore a signal waveform before actual distortion has occurred, e.g. owing to the inevitable frequency distortion in a line, or to minimize noise interference.

distortion *(Eng.)*. Any departure from the original shape because of applied stress or the release of residual stress in the material.

distortion *(Phys.)*. An aberration of a lens or lens system whereby a square object is imaged with either concave (**barrel distortion**) or convex (**pincushion distortion**) lines. The type and amount of distortion depend on the position of the lens stop.

distortion *(Telecomm.)*. Change of waveform, spectral content or pulse shape of any wave or signal due to any cause.

distortion factor *(Elec.Eng.)*. Ratio of the r.m.s. harmonic content to the total r.m.s. value of the distorted sine wave.

distortionless line *(Phys.)*. A transmission line with constants such that the attenuation (a minimum value) and the delay time are constant in magnitude with variation in frequency. The characteristic impedance is purely resistive. For such a line $LG = RC$, where R is resistance, L inductance, G leakage and C capacitance, all being distributed values per unit length. Also called *distortionless condition*.

distortion of field *(Elec.Eng.)*. A term commonly used in connection with electrical machines to denote the change in the distribution of flux in the air gap when the machine is put on load.

distraction display *(Behav.)*. Behaviour, especially of some female birds, which is sometimes similar to that of an injured individual (hence called *injury feigning*), generally a response to a predator threatening the eggs or young, and usually effective in diverting its attention.

distrails *(Meteor.)*. See **dissipation trails**.

distribute *(Print.)*. To put individual letters and spaces back into their proper compartments in the case after use. Machine-set matter is usually melted down to be used again in the caster.

distributed amplifier *(Telecomm.)*. Same as *transmission-line amplifier*.

distributed capacitance *(Phys.)*. (1) That distributed along a transmission line, which, with distributed resistance and/or inductance, reduces the velocity of transmission of signals. (2) That between the separate parts of a coil, lowering its inductance; represented by an equivalent lumped capacitor across the terminals, giving the same frequency of resonance.

distributed computing *(Comp.)*. The functional and geographical dispersion of computing power within a fully integrated system of processors and peripherals. It is a more economical and adaptable way to structure a very large computing system than to have one **mainframe**. See **front-end processor**.

distributed constants *(Phys.)*. Those applicable to real or artificial transmission lines and waveguides, because dimensions are comparable with the wavelength of transmitted energy.

distributed inductance *(Phys.)*. Said of a circuit which has an inductance distributed uniformly along it, e.g., a power transmission line, a loaded telephone circuit, or a travelling-wave valve or tube.

distributed winding *(Elec.Eng.)*. The winding of an electric machine which is spread uniformly over the stator or rotor surface.

distributing centre *(Elec.Eng.)*. In an electric power system, a point at which an incoming supply from a feeder is split up amongst a number of other feeders or distributors.

distributing main *(Elec.Eng.)*. See **distributor**.

distributing point *(Elec.Eng.)*. See **feeding point**.

distribution *(Autos.)*. The provision of the same quantity and quality of petrol-air mixture to each of the cylinders of a multi-cylinder engine by the carburettor and induction manifold.

distribution *(Ecol.)*. The occurrence of a species, considered from a geographical point of view, or with reference to altitude or other factors. Sometimes used as equivalent to *dispersal*. Should not be confused with *dispersion* which refers to individuals.

distribution *(Stats.)*. The partition of observations into intervals by value; the set of frequencies of observations in a set of intervals; a generic term for mathematical formulae giving probabilities related to values of random variables. See also **cumulative distribution function**.

distribution board *(Elec.Eng.)*. An insulating panel carrying terminals and/or fuses, for the distribution of power supplies to repeaters or telegraph circuits.

distribution cable *(Elec.Eng.)*. A communication cable extending from a feeder cable into a defined service area.

distribution coefficients *(Phys.)*. Chromaticity coordinates for spectral (monochromatic) radiations of equal power, i.e., for the component radiations forming an equal energy spectrum.

distribution factor *(Elec.Eng.)*. A factor used in the calculation of the e.m.f. generated in the winding of an a.c. machine, to allow for the fact that the e.m.f.s. in each of the individual coils are not in phase with one another. Also called *breadth coefficient, breadth factor*.

distribution factor *(Radiol.)*. A *modifying factor* used in calculating biological radiation doses, which allows for the nonuniform distribution of an internally-absorbed radioisotope.

distribution frame *(Telecomm.)*. A structure with large numbers of terminals, for arranging circuits in specified orders.

distribution-free methods *(Stats.)*. Methods of statistical analysis which under certain conditions do not depend on the probability distribution generating the observations.

distribution fuse-board *(Elec.Eng.)*. A distribution board having fuses in each of the separate circuits.

distribution law *(Chem.)*. The total energy in a given assembly of molecules is not distributed equally, but the number of molecules having an energy different from the median decreases as the energy difference increases, according to a statistical law.

distribution pillar *(Elec.Eng.)*. A structure in the form of a pillar, containing switches, fuses, etc., for interconnecting the distributing mains of an electric power system.

distribution reservoir *(Hyd.Eng.)*. See **service reservoir**.

distribution switchboard *(Elec.Eng.)*. A distribution board having a switch in each of the branch circuits.

distributive *(Maths.)*. An operation is distributive if the result of applying it to a sum of terms equals the sum of the results of applying it to the terms individually.

distributor *(Autos.)*. A device, geared to the cam-shaft, whereby H.T. current is transmitted in correct sequence to the sparking plugs.

distributor *(Elec.Eng.)*. The cable or overhead line

forming that part of an electric distribution system to which the consumers' circuits are connected. Also called a *distributing main.*

distributor rollers *(Print.)*. In a printing press, the rollers which distribute the ink, as distinct from the inking rollers which supply ink to the forme or plate.

district *(Min.Ext.)*. An underground section of a coal mine serviced by its own roads and ventilation ways: a section of a coal mine.

distrix *(Med.)*. Splitting of the ends of hairs.

disturbance *(Telecomm.)*. Any signal originating from a source other than the wanted transmitter, e.g. atmospherics, unwanted stations, noise in the receiver.

disuse atrophy *(Med.)*. Wasting of a part as a result of diminution or cessation of functional activity.

ditch canal *(Hyd.Eng.)*. See **level canal**.

ditching *(Aero.)*. Emergency alighting of a landplane on water.

dither *(Elec.Eng.)*. Small continuous signal supplied to servomotor operating hydraulic valve or similar device, and producing a continuous mechanical vibration which prevents sticking.

dithionic acid *(Chem.)*. $H_2S_2O_6$. Its salts are reducing agents and are called dithionates.

dithionous acid *(Chem.)*. $H_2S_2O_4$. Its salts are strong reducing agents and are called dithionites.

dithiothreitol *(Biol.)*. $(CHOH.CH_2SH)_2$. A mild reducing agent often used to reduce protein disulphide bonds.

ditrematous *(Zool.)*. Of hermaphrodite animals, having the male and female openings separate: of unisexual forms, having the genital opening separate from the anus.

Dittus-Boelter equation *(Eng.)*. An equation for the transfer of heat from tubes to viscous fluids flowing through them.

$$u \propto c\left(\frac{k\Delta s}{d}\right)^{1/3}\left(\frac{vd}{u/s}\right)^{1/12},$$

where u = film transfer factor, k = thermal conductivity, Δ = logarithmic mean temp. difference between tube and liquid, d = thickness of fluid stream, s = relative density of fluid, c = specific heat capacity of fluid, v = mean velocity of fluid in tube, and u/s = kinematic viscosity fluid.

diuresis *(Med.)*. The excretion of urine, especially in excess.

diuretic(s) *(Med.)*. Producing diuresis; class of drug which promotes sodium and water loss by the kidneys. Class is usually subdivided into **thiazide** and related diuretics and into the more potent **'loop' diuretics**.

diurnal *(Astron.)*. During a day. The term is used in astronomy and meteorology to indicate the variations of an element during an average day.

diurnal libration *(Astron.)*. The name given to the phenomenon by which, owing to the finite dimensions of both the earth and the moon, an observer can see rather more than half the moon's surface when his observations at different times or from different places on the earth are combined. The effect is one of **parallax**, the term *libration* being a misnomer in this case.

diurnal parallax *(Astron.)*. The change in the apparent position of a celestial object which results from the change in the observer's position caused by the Earth's daily rotation. Geometrically it is the angle subtended at the object by the Earth's radius. It is significant only for members of our solar system.

diurnal range *(Meteor.)*. The extent of the changes which occur during a day in a meteorological element such as atmospheric pressure or temperature.

diurnal variation *(Phys.)*. A variation of the earth's magnetic field as observed at a fixed station, which has a period of approximately 24 hours.

divalent *(Chem.)*. Capable of combining with 2 atoms of hydrogen or their equivalent. Also having an oxidation state of two.

divaricate *(Bot.,Zool.)*. Spreading widely apart, forked, divergent.

dive *(Aero.)*. A steep descent with or without power.

diver *(Min.Ext.)*. Small plummet adjusted to a desired relative density, so that it indicates the density of the fluid in which it is immersed by its up-and-down motion. Also *cartesian diver.*

dive-recovery flap *(Aero.)*. An **air brake** in the form of a flap to reduce the **limiting velocity** of an aircraft.

divergence *(Aero.)*. In aircraft stability, a disturbance that increases without oscillation; *lateral divergence* leads to a spin or an accelerating spiral descent; *longitudinal divergence* causes a nosedive, or a stall.

divergence *(Meteor.)*. If the components of the vector wind are (u,v,w) the divergence is defined as

$$\frac{\partial u}{\partial x}+\frac{\partial v}{\partial y}+\frac{\partial w}{\partial z}.$$

The horizontal divergence is defined as $\frac{\partial u}{\partial x}+\frac{\partial v}{\partial y}$ which is usually almost exactly compensated by $\frac{\partial w}{\partial z}$. Furthermore, the integrated divergence throughout a column of the atmosphere is almost zero i.e. is a small residual of larger positive and negative values; this is known as the *Dines compensation*. Hence strong negative values near the surface are matched by strong positive values at high levels.

divergence *(Phys.)*. Initiation of a chain-reaction in a reactor, in which slightly more neutrons are released than are absorbed and lost. The rate and extent of the divergence are normally controlled by neutron absorbing rods, e.g., of cadmium or hafnium.

divergence angle *(Electronics)*. Angle of spread of electron beam, arising from mutual repulsion or debunching.

divergence, divergent evolution *(Bot.)*. Where the same basic structure has evolved to give organs of different form and function. See also **homology**. Cf. **convergent evolution**.

divergence of a vector *(Maths.)*. The scalar product $\nabla.\mathbf{V}$ of the vector **V** with the vector operator ∇ (del). Written div **V**.

divergence speed *(Aero.)*. The lowest equivalent **air speed** at which **aeroelastic divergence** can occur.

divergent *(Bot.)*. Apices of organs wider apart than their bases.

divergent *(Phys.)*. Term applied to reactor or critical experiment when multiplication constant exceeds unity.

divergent junction *(Geol.)*. A zone where plates move apart and new crust and lithosphere are formed. Characterized by volcanism and earthquakes. e.g. mid-Atlantic ridge.

divergent lens *(Phys.)*. A lens which increases the divergence, or diminishes the convergence, of a beam of light passing through it. Such a lens will be double concave, plano-concave, or convexo-concave, the concave surface having the smaller radius of curvature.

divergent nozzle *(Eng.)*. A nozzle whose cross-section increases continuously from entry to exit; used, e.g. in compound impulse turbines.

divergent sequence, series *(Maths.)*. Definitions vary. Some writers count anything not convergent as divergent. Others use it as synonymous with *unbounded*, excluding finitely oscillating sequences such as $u_n=(-1)^n$. Still others confine it to sequences tending to $+\infty$ or $-\infty$, excluding infinitely oscillating sequences such as $u_n=n(-1)^n$.

divergent strabismus *(Med.)*. Squint in which the eyes diverge from each other.

divergent thinking *(Behav.)*. Thinking which is productive and original, involving the creation of a variety of ideas or solutions which tend to go beyond conventional categories (De Bono). See **convergent thinking**.

diversion cut *(Civ.Eng.)*. See **bye channel**.

diversity *(Ecol.)*. An index of the number of species in a defined area, often represented mathematically. *Alpha diversity* is on a local scale, *beta diversity* on a regional scale. See also **richness**.

diversity antenna *(Telecomm.)*. The antenna system of a diversity receiver.

diversity factor *(Elec.Eng.)*. The ratio of the arithmetical sum of the individual maximum demands of a number of consumers connected to an electric supply system, to the simultaneous maximum demand of the group.

diversity reception *(Telecomm.)*. System designed to reduce fading; several antennae, each connected to its own receiver, are spaced several wavelengths apart from one another, the demodulated outputs of the receivers being combined. Alternative systems use antennae orientated for oppositely polarized waves (*polarized diversity*), or independent transmission channels on neighbouring frequencies (*frequency diversity*).

divers' paralysis *(Med.)*. See **caisson disease**.

diverter *(Elec.Eng.)*. A low resistance connected in parallel with the series winding or the compole winding of a d.c. machine in order to divert some of the current from it, thereby varying the m.m.f. produced by the winding.

diverter relay *(Elec.Eng.)*. A relay employed with certain excess-current protective systems; it increases the stability of the protective system by putting resistance in parallel with the tripping relay in the case of a heavy fault.

diverticulitis *(Med.)*. Inflammation of diverticula in the colon.

diverticulosis *(Med.)*. The presence of diverticula in the colon.

diverticulum *(Med.,Zool.)*. (1) Saccular dilatation of a cavity or channel of the body. (2) Lateral outgrowth of the lumen of an organ. (3) Pouchlike protrusion of the mucous membrane of the colon through the weakened muscular wall. (4) A pouchlike side branch on the mycelium of some fungi. pl. *diverticula*.

divertor *(Nuc.Eng.)*. Trap used in thermonuclear device to divert magnetic impurity atoms from entering plasma, and fusion products from striking walls of chamber. Also *bundle divertor*.

divided bearing *(Eng.)*. See **split bearing**.

divided pitch *(Eng.)*. The axial distance between corresponding points on successive threads of a multiple-threaded screw.

divided touch *(Phys.)*. The magnetizing of a steel bar by stroking it with the opposite poles of two permanent magnets, these being drawn apart from the centre of the bar to the ends.

divided winding *(Elec.Eng.)*. A term proposed for that class of windings (for d.c. machines) usually called multiple or multiplex, in which there are two or more separate windings on the armature, joined in parallel by the brushes.

dividend *(Maths.)*. See **division**.

divider *(Elec.Eng.)*. Circuit which has an output which is a well defined fraction of a given input; can be constructed using resistors or capacitors. Also called *voltage divider* or *attenuator*.

dividers *(Eng.)*. Compasses used only for measuring or transferring distances, and not for describing arcs.

dividing box *(Elec.Eng.)*. A box for bringing out separately the cores of a multi-core cable. The insulation of the cable is hermetically sealed and the cores may be brought out either as bare or insulated conductors.

dividing engine *(Eng.)*. An instrument for marking or engraving accurate subdivisions on scales; it consists of a carriage adjusted by a micrometer screw and holding a marking tool.

dividing fillet *(Elec.Eng.)*. See **barrier**.

dividing head *(Eng.)*. See **indexing head**.

dividing network *(Acous.)*. A frequency-selective network which arranges for the input to be fed into the appropriate loudspeakers, usually two, covering high and low frequencies respectively. Also *loudspeaker, cross-over network*.

diving-bell *(Civ.Eng.)*. A water-tight working chamber, open at the bottom, which is lowered into water beneath which excavation or other works are to proceed. The interior is supplied with compressed air to maintain the water-level inside at a reasonable height, and thus leave free a space within which men may work.

divinity calf *(Print.)*. The name given to bindings in dark brown calfskin, with blind tooling; used chiefly for theological works.

division *(Bot.)*. Highest taxonomic rank used in the classification of plants (equivalent to the zoologist's **phylum**), ranking above *class*; the names end in -phyta or, for fungi, -mycota.

division *(Maths.)*. (1) For numbers, the operation of ascertaining how many times one number, the *divisor*, is contained in a second, the *dividend*. The result is called the *quotient*, and, if the divisor is not contained an integral number of times in the dividend, any number left over is called the *remainder*. Indicated either by the division sign, $\div$, or by a stroke or bar, in which case the expression as a whole is called a *fraction* and the dividend and divisor the *numerator* and *denominator* respectively. Fractions less than 1 are called *common* or *proper* or *vulgar fractions*, and those greater than 1, *improper fractions*. Colloquially, however, a *fraction* is less than 1. (2) For polynomials and other mathematical entities, the inverse operation to multiplication. When appropriate: nomenclature analogous to that outlined above is used.

division plate *(Eng.)*. A plate used for positioning the plunger of an indexing head; provided with several concentric rings of holes accurately dividing the circumference into various equal subdivisions.

division ring *(Maths.)*. A ring which, if zero is removed, is a group under multiplication, i.e. every non-zero element has an inverse. A commutative division ring is a field.

division wall *(Build.)*. A wall within a building or serving two houses. Also *party wall*.

divisor *(Maths.)*. See **division**.

dizygotic twins *(Biol.)*. Twins produced from two fertilized eggs. They may be the same or different sexes and are genetically equivalent to full sibs. syn. *fraternal twins*. Cf. **monozygotic twins**.

dl- *(Chem.)*. Containing equimolecular amounts of the dextrorotatory and the laevorotatory forms of a compound; racemic. Now usually written ($\pm$).

D-layer *(Telecomm.)*. The lowest region or layer of absorbing ionization, 55 to 95 km above the Earth. It impedes short-wave communications by absorbing some of the incident power, but it enhances long-wave communication.

d-levels *(Phys.)*. See **diffuse series**.

D lines *(Phys.)*. See **[D]**.

D log E curve *(Image Tech.)*. See **characteristic curve**.

DM *(Build.)*. Abbrev. for *Disconnecting Manhole*. See **interceptor**.

D max, D min *(Image Tech.)*. Abbreviations for the maximum and minimum densities of a photographic image.

DMDT *(Chem.)*. See **methoxychlor**.

DME *(Aero.)*. Abbrev. for *Distance-Measuring Equipment*.

D method, operator *(Maths.)*. Used in determining the solution of a linear differential equation with constant coefficients. D represents d/dx, and, under certain conditions, it can be manipulated by some of the procedures of ordinary algebra.

DMF *(Chem.)*. Abbrev. for *DiMethylFormamide*.

DML *(Comp.)*. *Data Manipulation Languages*. Used in **database management systems**.

dn *(Maths.)*. See **elliptic functions**.

DNA binding proteins *(Biol.)*. In prokaryotes, promoters, repressers etc. In eukaryotes, similar proteins, excluding the histones.

DNA, deoxyribose nucleic acid *(Biol.)*. In its double-stranded form the genetic material of most organisms and organelles, although phage and viral genomes may use single-stranded DNA, single-stranded RNA or double-stranded RNA. The two strands of DNA form a double-helix, the strands running in opposite directions, as determined by the sugar-phosphate 'backbone' of the molecule. The four bases project towards each other like the rungs of a ladder, with a purine always pairing with a pyrimidine, according to the *base-pairing rules*, in which thymidine pairs with adenine and cytosine with guanine. In its B molecular form the helix is 2.0 nm in diameter with a pitch of 3.4 nm (10 base pairs).

DNA library *(Biol.)*. Mixture of cloned DNA sequences derived from a single source, like a mouse or a chromosome, and containing ideally all, but in reality most, of the sequences from that source.

DNC *(Chem.)*. See **DNOC**.

DNOC *(Chem.)*. *2-Methyl-4,6-dinitro(1-hydroxybenzene)*, used as an insecticide and herbicide. Also *DNC, dinitrocresol.*

DNP *(Immun.)*. Abbrev. for *DiNitroPhenyl*, a commonly used **hapten**.

Doba's network *(Elec.Eng.)*. Shaping circuit used in pulse amplifiers where rise times of a few nanoseconds are required.

dobby *(Textiles)*. Mechanism over the top or at the side of a loom. Operated by punched cards it lifts and lowers the healds to move the warp threads in timed sequence to form the design in the cloth.

dobby fabric *(Textiles)*. A fabric made on a loom fitted with a **dobby**.

Dobson spectrophotometer *(Meteor.)*. An instrument used in the routine measurement of atmospheric ozone. It compares, using a photomultiplier and an optical wedge, the intensities of two wavelengths in the solar spectrum in the region of partial ozone absorption (0.30 to 0.33 μm), and from the result the total amount of ozone in a vertical column can be calculated. The instrument may be used to obtain the vertical distribution of ozone from the **umkehr effect**.

docking *(Space)*. The physical attaching of one space vehicle to another.

doctor *(Elec.Eng.)*. A device used in electroplating for depositing metal on imperfectly plated parts; it consists of an anode of the metal to be deposited, covered with a spongy material saturated with the plating material.

doctor *(Paper)*. A blade-like device resting at a shallow angle on the down-running surface of a roll or cylinder to remove unwanted material.

doctor *(Print.)*. On a photogravure machine, a blade which wipes superfluous ink from the non-printing area and from the walls between the cells of the cylinder.

doctor knife *(Textiles)*. A metallic blade set near the surface of a printing roller to remove excess colouring matter from the fabric.

doctor test *(Chem.)*. A test for sulphur in petroleum using a sodium plumbate (II) solution.

doctrine of specific nerve energies *(Behav.)*. The assertion that qualitative differences in sensory experience depend on which nerve is stimulated and not on the physical attributes of the stimulus.

documentation *(Comp.)*. Full description in words and diagrams accompanying a package, program or system.

document copying *(Print.)*. A variety of methods are in use based on **dyeline**, **dual spectrum**, electrostatic (**xerography**), **photocopying**, and **thermographic** principles, some able to produce lithographic masters for **small offset**.

document reader *(Comp.)*. Input device which reads marks and characters made in predetermined positions on special forms.

document retrieval *(Comp.)*. See **information retrieval**.

dodecagon *(Maths.)*. A 12-sided polygon.

dodecahedron *(Maths.)*. A 12-sided polyhedron.

dodecyl benzene *(Chem.)*. Important starting material for anionic detergents derived from petroleum. Based on the tetramer of propene.

dodging *(Image Tech.)*. Manipulation of the light projected through a negative by an enlarger to lighten or darken selected parts of the resultant print.

dodine *(Chem.)*. Dodecylguanidine, used as a fungicide.

doffer *(Textiles)*. (1) The operative who removes full cops or other yarn packages from a machine. (2) A fully automatic machine which performs this operation mechanically e.g. in spinning.

doffing tube *(Textiles)*. The tube of a **rotor spinning** unit through which the yarn passes on its way to the rotating collection package.

dog *(Build.,Eng.)*. A steel securing-piece used for fastening together 2 timbers, as in the process of shoring, for which purpose it is hooked at each end at right angles to the length, so that the hooked ends may be driven into the timbers. The term is also applied to a great variety of gripping implements, viz. a clutching attachment for withdrawing well-boring tools; a pawl; an adjustable stop used in machine tools; a spike for securing rails to sleepers; a lathe carrier; a circular clawed object to join members of a roof truss.

dog clutch *(Eng.)*. A clutch consisting of opposed flanges, or male and female members, provided with two or more projections and slots, one member being slidable axially for engaging and disengaging the drive. See **clutch**.

dog down *(Ships)*. To secure in position by pieces of bent-round iron, driven through holes in a cast-iron slab in such a manner as to be jammed.

Dogger *(Geol.)*. The middle epoch of the Jurassic period.

doggers *(Geol.)*. Flattened ovoid concretions, often of very large size, in some cases calcareous, in others ferruginous, occurring in sands or clays. They may be a metre or more in diameter.

dog-house *(Glass)*. A small extension of a glass furnace, into which the **batch** is fed.

dog-legged stair *(Build.)*. A stair having successive flights rising in opposite directions, and arranged without a well-hole.

dog-nail *(Build.)*. A large nail having a head projecting over one side.

dog's ear *(Build.)*. The corner of a sheet-lead tray, formed with a folding joint.

dog's tooth *(Build.)*. A string course in which bricks are so laid as to have one corner projecting.

dog-tooth spar *(Min.)*. A form of calcite in which the scalenohedron is dominant but combined with prism, giving a sharply pointed crystal like a canine tooth.

Doherty transmitter *(Telecomm.)*. One in which high efficiency of amplification of amplitude-modulated wave is obtained by two valves connected to the load, one directly, the other through a 90° retarding network.

Dolby *(Acous.,Image Tech.)*. TN for a noise-reduction system for magnetic and photographic sound recording and reproduction. Also for a system of stereophonic sound presentation in the cinema. Dolby B improves by 10 dB, Dolby C by 15 dB.

doldrums *(Meteor.)*. Regions of calm in equatorial oceans. Towards the solstices, these regions move about 5° from their mean positions, towards the north in June and towards the south in December.

dolerite *(Geol.)*. The general name for basic igneous rocks of medium grain- size, occurring as minor intrusions or in the central parts of thick lava flows; much quarried for road metal. Typical dolerite consists of plagioclase near labradorite in composition, pyroxene, usually augite, and iron ore, usually ilmenite, together with their alteration products.

Dolezalek quadrant electrometer *(Elec.Eng.)*. The original quadrant electrometer used for measurement of voltages and charges. It consisted of a suspended plate rotating over a metal box divided into four quadrants, two of which are earthed, the other two being charged by the current under measurement. (This is *heterostatic* operation. *Idiostatic* connection has the plate joined to one pair of quadrants.)

dolichocephalic *(Med.)*. *Long-headed*; said of a skull, the breadth of which is less than four-fifths the length.

dolichocolon *(Med.)*. An excessively long colon.

dolichol phosphate *(Biol.)*. A long chain unsaturated lipid with a terminal pyrophosphate found in the membranes of the **endoplasmic reticulum**. The core oligosaccharide for N-glycosylation of proteins is constructed on a dolichol phosphate molecule prior to its donation to the nascent polypeptide chain.

dolliform, dolioform *(Zool.)*. Barrel-shaped.

doll *(Civ.Eng.)*. A small arm or post carrying railway signalling apparatus, mounted on a gantry or bridge.

dollar *(Nuc.Eng.)*. US unit of reactivity corresponding to the prompt critical condition in a reactor, i.e. the reactivity contributed by the delayed neutrons. See also **cent**.

dollar spots *(Vet.)*. The skin lesions of **dourine**.

dolly *(Civ.Eng.)*. An object interposed between the monkey of a pile-driver and the head of a pile to prevent damage to the latter by blows from the former.

dolly *(Elec.Eng.)*. Operating member of a tumbler switch which projects through the outer casing.

dolly *(Eng.)*. (1) A heavy hammer-shaped tool for supporting the head of a rivet during the forming of a head on the other end. (2) A shaped block of lead used by panel beaters when hammering out dents.

dolly *(Image Tech.)*. Wheeled mobile mounting for a motion picture or television camera and its operator, allowing the action to be followed.

dolly truck *(Print.)*. See **reel bogie**.

dolly tub *(Min.Ext.)*. *Kieve*. A large wooden tub used for the final upgrading of valuable minerals separated by water concentration in ore dressing. See **tossing**.

dolomite *(Eng.)*. Calcined dolomite is used as a basic refractory for withstanding high temperatures and attack by basic slags in metallurgical furnaces. The name is also used to describe refractories made from magnesian limestone, which does not necessarily contain the mineral dolomite.

dolomite *(Min.)*. The double carbonate of calcium and magnesium, crystallizing in the rhombohedral class of the trigonal system, occurring as cream-coloured crystals or masses with a distinctive pearly lustre, hence the synonym *pearl spar*. A common mineral of sedimentary rocks and an important gangue mineral.

dolomitic limestone *(Geol.)*. A calcareous sedimentary rock containing calcite or aragonite in addition to the mineral dolomite. Cf. **dolostone**.

dolomitization *(Geol.)*. The process of replacement of calcium carbonate in a limestone or ooze by the double carbonate of calcium and magnesium (dolomite).

dolostone *(Geol.)*. A rock composed entirely of the mineral dolomite. The term is synonymous with *dolomite rock*, but avoids confusion between dolomite rock and the mineral dolomite.

dolphin *(Civ.Eng.)*. Permanent moorings for vessels, usually formed by driving a number of timber or steel box piles into the bed of the river or sea and strengthening them against impact or tug by some form of binding or fixing near their heads.

domain *(Immun.)*. Applied to immunoglobulins, it is the three-dimensional structure formed by a single homology region of an immunoglobulin heavy or light chain, i.e. V_L, C_L, V_H, CH1, CH2, CH3 or CH4.

domain *(Maths.)*. (1) a set of points. Also called a *region*. (2) Of a function or mapping, the set on which the function or mapping is defined, e.g. for a real function the domain is a subset of the set of real numbers, namely the set of those real numbers for which the function is defined.

domain *(Phys.)*. In ferroelectric or ferromagnetic material, region where there is saturated polarization, depending only on temperature. The transition layer between adjacent domains is the *Bloch wall*, and the average size of the domain depends on the constituents of the material and its heat treatment. Domains can be seen under the microscope when orientated by strong electric or magnetic fields.

domatium *(Bot.)*. A cavity in a plant inhabited by commensal mites or insects.

dome *(Arch.)*. A vault springing from a circular, or nearly circular, base.

dome *(Civ.Eng.)*. A domed cylinder attached to a locomotive boiler to act as a steam space and to house the regulator valve.

dome *(Crystal.)*. A crystal form consisting of two similar inclined faces meeting in a horizontal edge, thus resembling the roof of a house. The term is frequently incorrectly applied to a four-faced form which is really a prism lying on an edge.

dome *(Geol.)*. (1) An igneous intrusion with a domelike roof. (2) An anticlinal fold with the rock dipping outwards in all directions.

dome, apical dome *(Bot.)*. The part of the apical meristem distal to the first primordium.

dome nut *(Eng.)*. See **cap nut**.

dominance *(Crystal.)*. See under **habit**.

dominance hierarchy *(Behav.)*. An aspect of the social organization of various species, usually referring to aggressive interactions, in which certain individuals predictably dominate, or are dominated by, other members of the group.

dominant *(Biol.)*. Describes a gene (allele) which shows its effect in those individuals who received it from only one parent, i.e. in heterozygotes. Also describes a character due to a dominant gene. Cf. **recessive**.

dominant *(Bot.)*. The species which because of its number or size determines the character of a plant community or vegetation layer. Several species can be *co-dominant*.

dominant wavelength *(Phys.)*. The wavelength of monochromatic light which matches a specific colour when combined in suitable proportions with a reference standard light. On a chromaticity diagram it is determined by a straight line drawn from the achromatic point though the point representing the chromaticity of the colour of interest, to where the line intersects the pure colour perimeter.

Domin scale *(Ecol.)*. A ten-point scale used in estimating canopy cover. The scale is not linear.

donkey boiler *(Eng.)*. A small vertical auxiliary boiler for supplying steam-winches and other deck machinery on board ship when the main boilers are not in use.

donkey pump *(Eng.)*. A small steam reciprocating pump independent of the main propelling machinery of a ship; used for general duty.

donor *(Chem.)*. (1) That reactant in an induced reaction which reacts rapidly with the inductor. (2) That which 'gives', as in *proton-donor* or *electron-pair donor*.

donor *(Electronics)*. Impurity atoms which add to the conductivity of a semiconductor by contributing electrons to a nearly empty conduction band, so making *n*-type conduction possible. See also **acceptor, impurity**.

donor level *(Electronics)*. See **energy levels**.

donut *(Elec.Eng.)*. See **doughnut**.

door by-pass switch *(Elec.Eng.)*. See **gate by-pass switch**.

door case *(Build.)*. The frame into which a door fits to shut an opening.

door check *(Build.)*. (1) A device fitted to a door to prevent it from being slammed, and yet to ensure that it closes. (2) Scottish term for planted door stops round an unrebated door frame.

door cheeks *(Build.)*. The jambs of a door frame. Also *door posts*.

door closer *(Elec.Eng.)*. See **gate closer**.

door contact *(Elec.Eng.)*. A contact attached to a door or gate, so arranged that it closes a circuit when the door is opened or closed, and rings an alarm bell or gives some other signal.

door interlock *(Elec.Eng.)*. See **gate interlock**.

door-knob transformer *(Telecomm.)*. Device for coupling a coaxial cable to a waveguide; the inner of the coaxial cable is fed through a hole in the broad wall of the waveguide, and stops at half the depth of the waveguide. This protrusion has a spherical end or *doorknob* to improve impedance matching.

door operator *(Elec.Eng.)*. See **gate operator**.

door posts *(Build.)*. See **door cheeks**.

door stop *(Build.)*. A device fitted to a door, or to the floor near to a door, to hold it open.

door stops, checks *(Build.)*. Thin timbers round a door opening to stop the door as opposed to rebated jambs.

door strip *(Build.)*. A strip, often of flexible material, attached to a door to cover the space between the bottom of the door and the threshold. Also *weather strip, draught excluder*.

door switch *(Elec.Eng.)*. A switch mounted on a door so that the opening or closing of the door operates the switch. Also called a *gate switch* when used on electric lifts.

dopamine *(Med.)*. A **sympathomimetic** drug which acts on dopamine and adrenergic receptors to increase heart rate, cardiac output and blood pressure. Also important as a neural transmitter in the brain.

doped junction *(Electronics)*. One with a semiconductor crystal which has had impurity added during a melt.

doping *(Aero.)*. A chemical treatment with nitrocellulose or cellulose acetate dissolved in thinners, which is applied to fabric coverings, for the purposes of tautening, strengthening, weatherproofing, etc.

doping *(Electronics)*. Addition of known impurities to a semiconductor, to achieve the desired properties in diodes and transistors, i.e. **donor** in *n*-type and **acceptor** in *p*-type.

Doppler broadening *(Phys.)*. Frequency spread of radiation in single spectral lines, because of Maxwell distribution of velocities in the molecular radiators. This also broadens the resonance absorption curve for atoms or molecules excited by incident radiation.

Doppler effect *(Phys.)*. The apparent change of frequency (or wavelength) because of the relative motion of the source of radiation and the observer. For example, the change in frequency of sound heard when a train or aircraft is moving towards or away from an observer. For sound, the observed frequency is related to the true frequency by

$$f_o = \frac{V - V_o(+W)}{V - V_s(+W)} \cdot f_s$$

where V_s, V_o are velocities of source and observer, V is the phase velocity of the wave and W is the velocity of the wind. For electromagnetic waves, the *Lorentz transformation* is used to give

$$f_o = \sqrt{\frac{1 - v/c}{1 + v/c}} \cdot f_s$$

where V is the *relative* velocity of the source and observer and C is the velocity of light. In astronomy the measurement of the frequency shift of light received from distant galaxies, the **red shift**, enables their recession velocities to be found.

Doppler navigator *(Aero.)*. Automated dead reckoning by a device which measures true ground speed by the Doppler frequency shift of radio beam echoes from the ground and computes these with compass readings to give the aircraft's true track and position at any time. Entirely contained in the aircraft, this system cannot be affected by radio interference or hostile jamming.

Doppler radar *(Radar)*. Any means of detection by reflection of electromagnetic waves, which depends on measuremcent of change of frequency of a signal after reflection by a target having relative motion. See **CW radar, pulse Doppler radar**.

doran *(Radar)*. Doppler ranging system for tracking missiles (*D*oppler *ran*ge.)

doré silver *(Eng.)*. Silver bullion, i.e. ingots or bars, containing gold.

doric *(Print.)*. A type face. The same as sans-serif or gothic.

dormancy *(Biol.)*. A state of temporarily reduced but detectable metabolic activity, as in seeds.

dormant bolt *(Build.)*. A hidden bolt sliding in a mortise in a door; operated by turning a knob or by means of a key.

dormer *(Build.)*. A window projecting from a roof slope.

dormin *(Bot.)*. See **abscisic acid**.

Dorn effect *(Chem.)*. The production of a potential difference when particles suspended in a liquid migrate under the influence of mechanical forces e.g. gravity; the converse of **electrophoresis**.

dorsal *(Bot.)*. (1) The surface of flattened thalloid plants which faces away from the substrate. (2) More generally the *abaxial* surface which in a leaf is confusingly usually the lower surface. (The term is not consistently used.)

dorsal *(Med.)*. Said of the back of any part.

dorsal *(Zool.)*. Pertaining to that aspect of a bilaterally symmetrical animal which is normally turned away from the ground.

dorsal fins *(Aero.)*. Forward extensions along the top of the fuselage to increase effectiveness of the main fin in sideslip, especially in **asymmetric flight**.

dorsalgia *(Med.)*. Pain in the back.

dorsalis *(Zool.)*. An artery supplying the dorsal surface of an organ.

dorsal suture *(Bot.)*. The midrib of the carpel in cases where dehiscence occurs along it.

dorsal trace *(Bot.)*. The median vascular supply to a carpel.

dorsiferous *(Zool.)*. Said of animals which bear their young on their back.

dorsiflexion *(Med.)*. The bending towards the back of a part, e.g. flexion of the toes towards the shin.

dorsigrade *(Zool.)*. Walking with the backs of the digits on the ground, as Sloths.

dorsiventral *(Bot.)*. A flattened plant member having structural differences between its dorsal and ventral sides.

dorsum *(Zool.)*. In Anthozoa, the sulcular surface: in Arthropoda, the tergum or notum: in Vertebrates, the dorsal surface of the body or back.

Dortmund tank *(Build.)*. A deep tank, with conical bottom to which liquid sewage is supplied by a pipe reaching down nearly to the bottom. The resulting upward flow assists sedimentation of the sludge.

DOS *(Comp.)*. *Disk Operating System*. TN for a micro-computer operating system which handles files and programs stored on disk. Letters are often prefixed, e.g. *PC-DOS, MS-DOS*.

dose *(Med.)*. The prescribed quantity of a medicine or of a remedial agent.

dose *(Radiol.)*. General term for quantity of radiation. See **absorbed dose, dose equivalent, effective dose equivalent, collective dose equivalent, genetically significant dose**.

dose equivalent *(Radiol.)*. The quantity obtained by multiplying the absorbed dose by a factor to allow for the different effectiveness of the various ionizing radiations in causing harm to tissues. Unit Sievert (*Sv*).

dosemeter *(Radiol.)*. Instrument for measuring **dose** and used in radiation surveys, hospitals, laboratories and civil defence. It gives a measure of the radiation field and dosage experienced. Also *dosimeter*.

dose rate *(Radiol.)*. The **absorbed dose**, or other dose, received per unit time.

dose reduction factor *(Radiol.)*. A factor giving the reduction in radiation sensitivity for a cell or organism which results from some chemical protective agent.

dot gain *(Print.)*. An effect whereby **half-tone** dots are enlarged during image production at platemaking or in the printing operation, thus altering tonal values from the original film element. The amount of dot gain can be measured using a **densitometer** to determine whether the dot gain lies within acceptable limits.

dot-matrix printer *(Comp.)*. Printer in which characters are formed of ink dots from a rectangular matrix of printing positions, typically 9×7.

dots *(Build.)*. (1) Dabs of mortar or wooden blocks laid at intervals on a floor or roof surface to serve as a guide to the level of the screed. (2) Nails partially driven into a wall, so that the projecting lengths may serve as guides to thickness in laying on a coat of plaster, after which they are removed.

dot sequential *(Image Tech.)*. Colour TV system in which the three colour signals are sent in rapid succession for each point in a brief part of the line scan period.

dotting-on *(Build.)*. The process of adding together similar items when **taking off**.

double *(Bot.)*. (1) A flower having more than the normal number of petals, commonly by the conversion of stamens (or even carpels) to petals, i.e. *petalody*. (2) The capitulum of the Compositae having (in a superficially similar way) some or all of the tubular disc florets converted into ligulate florets. Ant. *single*.

double *(Build.)*. A slate size, 13×6 in. (330×252 mm).

double-acting engine *(Eng.)*. Any reciprocating engine in which the working fluid acts on each side of the piston alternately; most steam-engines and a few internal-combustion engines are so designed.

double-acting pump *(Eng.)*. A reciprocating pump in

which both sides of the piston act alternately, thus giving two delivery strokes per cycle.

double-action press *(Eng.)*. A press fitted with two slides, permitting multiple operations, such as blanking and drawing, to be performed.

double amplitude *(Phys.)*. The sum of the maximum values of the positive and negative half-waves of an alternating quantity. Also *peak-to-peak amplitude*.

double bar and yoke method *(Elec.Eng.)*. A ballistic method of magnetic testing, in which 2 test specimens are arranged parallel to each other and clamped to yokes at the ends to form a closed magnetic circuit. A correction for the effect of the yokes is made by altering their position on the bars and repeating the test.

double bar gauge *(Build.)*. Marking gauge in which the fence slides on 2 independently adjustable bars, each with marking pins.

double-bead *(Build.)*. Two side-by-side beads, separated by a quirk.

double-beam cathode-ray tube *(Electronics)*. One containing two complete sets of beam-forming and beam-deflecting electrodes operated from the same cathode, thus allowing two separate waveforms to appear on the screen simultaneously (*double-gun tube*). Also it may use a single gun with a beam-switching circuit, i.e. two inputs continuously interchanged, or a single beam tube in which the beam is split, and the two parts are separately deflected.

double-beat valve *(Eng.)*. A hollow cylindrical valve for controlling high-pressure fluids. The seatings at the two ends exposed to pressure are of only slightly different area, so that the valve is nearly balanced and easily operated.

double-bellied *(Arch.)*. A term applied to a baluster which has had both ends turned alike.

double beta decay *(Phys.)*. Energetically possible process involving the emission of two β-particles simultaneously. Not to be confused with **dual beta decay**.

doublebind (communication) *(Behav.)*. Mutually contradictory messages which set up a conflict with the individual who is simultaneously receiving the message that what is meant is the opposite of what is said. Believed by some to be a causative factor in schizophrenia, when a parent is consistently giving doublebind messages to a child.

double bond *(Chem.)*. Covalent bond involving the sharing of 2 pairs of electrons.

double-break *(Elec.Eng.)*. Said of switches or circuit-breakers in which the circuit is made or broken at 2 points in each pole or phase.

double bridge *(Elec.Eng.)*. Network for measuring low resistances. See **Kelvin bridge**.

double bridging *(Build.)*. Bridging in which 2 pairs of diagonal braces are used to connect adjacent floor joists at points dividing their length into equal parts.

double buffering *(Comp.)*. The use of two **buffers** so that one may be filled while the other is being emptied.

double case *(Print.)*. A case with accommodation for both upper-case and lower-case characters. Wrongly called a **half-case**.

double-catenary construction *(Elec.Eng.)*. A method of supporting the overhead contact wire of an electric traction system; the contact wire is suspended from two parallel catenary wires so that the three wires are in a triangular formation.

double cloth *(Textiles)*. Two distinct cloths woven and bound together simultaneously to obtain greater thickness without affecting the face texture, e.g. heavy overcoatings.

double-coated film *(Image Tech.)*. A film base with emulsions on both sides.

double-coil loudspeaker *(Acous.)*. Electrodynamic loudspeaker with two driving coils separated by a compliance, the coils driving one or two open cone diaphragms, thus operating more effectively over a wide range of frequencies.

double-cone loudspeaker *(Acous.)*. Large open coil-driven cone loudspeaker with a smaller free-edge cone fixed to the coil former, thus assisting radiation for high audio frequencies.

double contraction *(Foundry)*. Total shrinkage allowance necessary to add to the dimensions of a finished casting when making a wood pattern from which a metal pattern is to be cast.

double-cover butt joint *(Eng.)*. A butt joint with a cover plate on both sides of the main plates.

double-current furnace *(Elec.Eng.)*. A special form of electric furnace in which direct current is used for an electrolytic process and alternating current for heating, on the principle of the induction furnace.

double-current generator *(Elec.Eng.)*. An electric generator which can supply both alternating and direct current.

double-cylinder knitting machine *(Textiles)*. A **circular knitting machine** with two cylinders one above the other and having one set of double-ended needles that can knit in either cylinder. Often used for the manufacture of *hosiery*.

double decomposition *(Chem.)*. A reaction between two substances in which the atoms are rearranged to give two other substances. In general it may be written AB + CD = AC + BD.

double-delta connection *(Elec.Eng.)*. A method of connecting windings of a six-phase transformer, etc., so that they may be represented diagrammatically by two triangles.

double-diffused metal-oxide semiconductor *(Electronics)*. A metal-oxide semiconductor manufacturing process involving two stages of diffusion of impurities through a single **mask**, enabling depletion-mode or enhancement-mode transistors to be produced on a chip. The technique keeps **channels** short so that the devices are suitable for high-speed logic or microwave applications.

double-diffused transistor *(Electronics)*. One produced by using **double-diffused metal-oxide semiconductor** techniques.

double diffusion *(Immun.)*. Test for detecting antigens and antibodies in which the two are arranged to diffuse toward one another in a gel (usually agar or agarose). Where they meet lines of precipitation form. Since each antigen forms a separate line the method is used for analysing purity of preparations, and for detecting antigens which share common determinants.

double diode *(Electronics)*. Two diodes (usually vacuum-tube diodes) in a single package or envelope. May have a common cathode for full-wave rectification.

double-disk winding *(Elec.Eng.)*. A form of winding used for transformers, in which two disk coils are wound in such a way that, when placed side by side to form a single coil, the beginning and end of the complete coil are at the outside periphery.

double-dovetail key *(Build.)*. A piece of wood used to connect together 2 members lengthwise; shaped and fitted like a **slate cramp**.

doubled yarns *(Textiles)*. See **folded yarns**.

double earth fault *(Elec.Eng.)*. A fault on an electric power transmission system, caused by two phases going to earth simultaneously, either at the same point on the system or at different points

double embedding *(Biol.)*. A technique for embedding small objects, otherwise liable to distortion or disorientation, e.g. the specimen may be first orientated and embedded in celloidin and the small celloidin block embedded in hard paraffin wax.

double-ended boiler *(Eng.)*. A marine boiler of the shell type provided with a furnace at both ends, each with its independent tubes and uptake.

double-ended bolt *(Eng.)*. A bar screwed at each end for the reception of a nut.

double-entry compressor *(Aero.)*. A centrifugal compressor with double sided impeller so that air enters from both sides.

double exposure *(Image Tech.)*. The intentional exposure of two (or more) separate images on the same photographic record; they may be superimposed or separated by masking.

double-faced hammer *(Eng.)*. A hammer provided with a flat face at each end of the head.

double-faggoted iron *(Eng.)*. See **faggoted iron**.

double fertilization *(Bot.)*. The process, characteristic of angiosperms, in which 2 male nuclei enter the embryo sac. One fuses with the egg cell nucleus to form the zygote which develops into the embryo; the other typically fuses with the 2 polar nuclei to form a triploid nucleus from which the endosperm derives.

double Flemish bond *(Build.)*. A bond in which both exposed faces of a wall are laid in Flemish bond.

double floor *(Build.)*. A floor in which the bridging joists are supported at intervals by binding joists. Cf. **single floor, framed floor**.

double-flow turbine *(Eng.)*. A turbine in which the working fluid enters at the middle of the length of the casing and flows axially towards each end.

double frame *(Print.)*. A frame with accommodation for forty cases and a double-sized working top.

double-glazing *(Build.)*. (1) Hermetically sealed glazing where 2 sheets of glass enclose a 22 mm space, used where insulation and conservation of heat are important. (2) where sound insulation is the primary function the space should be about 175 mm wide.

double-helical gears *(Eng.)*. **Helical gears** in which two sets of teeth having opposite inclinations are cut on the same wheel, thus eliminating axial thrust.

double-hump effect *(Telecomm.)*. Property of two coupled resonant circuits, each separately resonant to the same frequency, of showing maximum response to two frequencies disposed about the common resonance frequency.

double-hung window *(Build.)*. A window having top and bottom sashes, each balanced by sash cord and weights so as to be capable of vertical movement in its own grooves.

double-image micrometer *(Biol.,Min.)*. A microscope attachment for the rapid and precise measurement of small objects, like cells or particles. The principle is that 2 images of the object are formed by means which allow the separation of the images to be varied and its magnitude read on a scale. By setting the 2 images edge-to-edge (a very precise setting) and then interchanging them, the difference of the scale-readings is a measure of the diameter of the object. The double image may be formed by birefringent crystals, prisms, an interferometer-like system or a vibrating mirror.

double-insulated *(Elec.Eng.)*. Descriptive of portable electric appliances (usually domestic) which have two completely separate sets of insulation between the current-carrying parts and any metal accessible to the user. An earth connection is, therefore, unnecessary.

double integral *(Maths.)*. A summation of the values of a function $f(x,y)$ over a specified region R of the x, y plane. Written

$$\iint_R f(x, y)\, dx\, dy.$$

double jersey *(Textiles)*. A range of weft-knitted fabrics (rib or interlock) made on fine gauge machines. The construction is usually chosen so that the fabric has reduced extensibility.

double junction *(Build.)*. A drainage or water-pipe fitting made with a branch on each side.

double-layer winding *(Elec.Eng.)*. An armature winding, always used for d.c. machines and frequently for a.c. machines, in which the coils are arranged in 2 layers, one above the other, in the slots.

double-length word *(Comp.)*. Hardware feature of many computers where two words (of, say, 16 bits) can be joined together and manipulated as a single (32 bit) word in the central processor.

double lock *(Hyd.Eng.)*. A construction consisting of 2 side-by-side lock chambers, across the same canal. They are interconnected through a sluice so that the amount of water lost in lockage is only half that which is lost by a single lock.

double magazine *(Image Tech.)*. A magazine holding two rolls of film feeding a camera or printer; one may be a **mask** or **matte** through which the other, the raw stock, is exposed.

double-margined door *(Build.)*. A door hinged as one leaf but having the appearance of a pair of folding doors.

double moding *(Electronics)*. Irregular switches of frequency by magnetron oscillator, due to changing mode.

double modulation *(Telecomm.)*. See **compound modulation**.

double oblique crystals *(Crystal.)*. See **triclinic system**.

double partition *(Build.)*. A partition having a cavity in which sliding doors may move.

double pica *(Print.)*. An old type size, about 22 points.

double-pipe exchanger *(Chem.Eng.)*. Heat exchanger formed from concentric pipes, the inner frequently having fins on the outside to increase the area. The essential flow pattern is that the 2 streams are always parallel.

double-pitch skylight *(Build.)*. A skylight having two differently sloped glazed surfaces.

double plating *(Print.)*. See **two set**.

double point *(Maths.)*. For the curve $f(x,y) = 0$: one at which $\frac{\partial f}{\partial x} = \frac{\partial f}{\partial y} = 0$ and $\frac{\partial^2 f}{\partial x^2}, \frac{\partial^2 f}{\partial x \partial y}$ and $\frac{\partial^2 f}{\partial y^2}$ are not all zero. A double point at which there are two real and distinct tangents, one real tangent, or no real tangents, is called a *node*, *cusp*, or *isolated point* respectively. A node is sometimes called a *crunode*, a cusp a *spinode*, and an isolated point an *acnode* or *conjugate point*. Nodes and cusps assume a variety of forms but their names have yet to be decided.

double-pole *(Elec.Eng.)*. Said of switches, circuit-breakers, etc., which can make or break a circuit on two poles simultaneously.

double precision *(Comp.)*. Operation using **words** of double normal **length**. See **double-length word**.

double printing *(Image Tech.)*. The process of exposing a positive emulsion in a printing machine successively with more than one negative resulting in a superposition of two positive images after development.

double quirk-head *(Build.)*. A bead sunk into a surface so as to leave a quirk on each side.

doubler *(Telecomm.)*. See **frequency doubler**.

doubler circuit *(Elec.Eng.)*. (1) A form of self-saturating magnetic amplifier having an a.c. output. (2) One driven so that a frequency can be filtered from the output which is double the frequency of the input.

double reception *(Telecomm.)*. Simultaneous reception of two signals on different wavelengths by two receivers connected to the same antenna.

double refraction *(Phys.)*. Division of an electromagnetic wave in an anisotropic medium into two components propagated with different velocities, depending on the direction of propagation. In uniaxial crystals the components are called the *ordinary* ray where the wavefronts are spherical and the *extraordinary* ray where the wavefronts are ellipsoidal. In **biaxial** crystals both wavefronts are ellipsoidal.

double-roller safety action *(Horol.)*. The safety action of a lever escapement in which two rollers are used, one carrying the impulse pin, the other being used for the guard finger.

double roof *(Build.)*. A roof in which the rafters are supported on purlins between walls.

double-row ball journal bearing *(Eng.)*. A rolling bearing which has two rows of caged balls running in separate tracks. There are *rigid*, *self-aligning*, and *angular contact* double-row ball journal bearings.

double-row radial engine *(Aero.)*. A **radial engine** where the cylinders are arranged in two planes, and operate on two crank pins, 180° apart.

doubles *(Min.Ext.)*. See **coal sizes**.

double salts *(Chem.)*. Compounds having two normal salts crystallizing together in definite molar ratios.

double scalp *(Vet.)*. A form of **osteodystrophia** of sheep thought to be caused by phosphorus deficiency; characterized by severe debility, anaemia, and weakening and thinning of the bones, notably the frontal bones.

double series *(Maths.)*. A series formed from all the terms of a 2-dimensional array which extends to infinity in both dimensions, i.e.,

$$\sum_{r,s=1}^{\infty} \alpha_{rs}.$$

double shrinkage *(Foundry)*. See **double contraction**.

double-sideband system *(Telecomm.)*. One which transmits both the sidebands produced in the modulation of a radio-frequency carrier wave. Cf. **single-sideband system**.

double-six array *(Telecomm.)*. A pair of six-element **Yagi antennae** placed side-by-side to improve directivity.

double spread *(Print.)*. Any pair of facing pages designed as one unit.

double squirrel-cage motor *(Elec.Eng.)*. A squirrel-cage motor with two cage windings on its rotor, one of high resistance and low reactance, and the other of high reactance and low resistance. The former carries most of the current at starting, and therefore gives a high starting torque, while the latter carries most of the current when running and results in a high efficiency.

double stars *(Astron.)*. A pair of stars appearing close together as seen by telescope. They may be at different distances (*optical double*) or physically connected. See **binary double**.

double-stream amplifier *(Electronics)*. A type of travelling-wave tube in which the operation depends upon the interaction of two electron beams of differing velocities.

double superhet receiver *(Telecomm.)*. Receiver in which amplification takes place at both high and low intermediate frequencies, thus requiring two frequency-changing stages.

doublet *(Image Tech.)*. A pair of simple lenses designed to be used together, so that optical distortion in one is balanced by reverse distortion in the other.

doublet *(Phys.)*. A pair of associated lines in a spectrum, such as the two D lines of sodium. The arc spectra of the alkali metals consist entirely of series of doublets.

doublet antenna *(Telecomm.)*. See **dipole antenna**.

double tenons *(Build.)*. Two parallel tenons on the end of thick material.

double-threaded screw *(Eng.)*. A screw having two threads, whose **pitch** is half the **lead**. Also called a *two-start thread*. See **multiple-threaded screw, divided pitch**.

double-throw switch *(Elec.Eng.)*. One which enables connections to be made with either of two sets of contacts.

double thrust-bearing *(Eng.)*. A thrust-bearing for taking axial thrust in either direction.

double-tone ink *(Print.)*. Ink for half-tone printing, containing a secondary pigment which spreads outward from each dot while the ink is drying to give a richer effect to the illustration.

double triode *(Electronics)*. Thermionic valve with two triode assemblies in the same envelope.

double-trolley system *(Elec.Eng.)*. A system of electric traction where, instead of the running rails, a second insulated contact wire is used for the return of negative current. It avoids trouble due to stray earth currents.

double-tuned circuit *(Elec.Eng.)*. Circuit containing 2 elements which may be tuned separately.

double wall *(Acous.)*. Wall of two plates with a layer of soft material between them. Above a certain frequency, the sound transmission of a double wall is much lower than that of a single wall of the same mass.

double-wall coffer-dam *(Civ.Eng.)*. A coffer-dam formed of a pair of parallel walls, separated by a mass of clay puddle, carried down to a watertight stratum and to a level below the bottom of the water.

double-webbed girder *(Eng.)*. A built-up box girder in which the top and bottom booms are united by two parallel webs.

double-wedge aerofoils *(Aero.)*. See **wedge aerofoil**.

double-width press *(Print.)*. A newspaper press with a width of four standard pages.

double window *(Build.)*. A window arranged with double sashes enclosing air, which acts as a sound and heat insulator. See **double-glazing**.

double-wire system *(Elec.Eng.)*. The usual system of electric wiring; it employs separate wires for the go and return conductors, instead of using the earth as a return.

doubling course *(Build.)*. A special course of slates laid at the eaves, to ensure that the lowest margin there has two thicknesses of slate throughout its depth. Also *double* or *eaves course*.

doubling piece *(Build.)*. A **tilting fillet**.

doubling time *(Nuc.Eng.)*. (1) Time required for the neutron flux in a reactor to double. (2) In a breeder reactor, time required for the amount of fissile material to double.

doubling, twisting frame *(Textiles)*. A machine in which yarns are *folded* or twisted together in a simple manner to give stronger uniform products or to produce a wide variety of fancy effects.

doublures *(Print.)*. The inside of book covers lined with silk or leather and specially decorated.

doubly-fed series (repulsion) motor *(Elec.Eng.)*. A single-phase series (or repulsion) motor in which the armature receives its power partly by conduction and partly by induction.

dough-moulding compound *(Plastics)*. Alkyd moulding materials in the form of dough or putty.

doughnut *(Elec.Eng.)*. (1) Anchor-ring shape used in circular-path accelerator tubes of glass or metal. (2) Traditional shape of a pile of annular laminations for magnetic testing, since there is no external field. (3) Traditional shape of loading coils on transmission lines, permitting exact balance in addition to inductance; also *donut, toroid*.

Douglas bag *(Med.)*. A specially constructed bag for the collection of air expired from the lung, from the analysis of which the oxygen consumption of the body can be determined.

Douglas fir *(For.)*. *Pseudotsuga menziesii*, the most important timber of the North American continent, and one of the best known softwoods in the world. The wood needs to be incised for preservative treatments by pressure processes; it is unaffected by powder-post beetle attack. Also called *British Columbian Pine, Oregon pine, red pine, false hemlock*.

Douglas's pouch *(Med.)*. In the female pelvis, the pouch of peritoneum between the rectum and the posterior wall of the uterus.

dourine *(Vet.)*. A contagious infection of breeding horses, characterized by inflammation of the external genital organs and paralysis of the hind limbs; due to *Trypanosoma equiperdum*, which is transmitted through coitus. Also *mal du coit*.

douzième *(Horol.)*. A unit of measurement used in the watch trade. See **ligne**.

douzième gauge *(Horol.)*. A form of calliper gauge on which the readings are given in douzièmes.

dovetail *(Build.)*. A joint formed between a flaring tenon, having a width diminishing towards the root, and a corresponding recess or mortise.

dovetail halving *(Build.)*. A form of **halving** in which the mating parts are cut to a dovetail shape.

dovetail hinge *(Build.)*. A hinge whose leaves increase in width outwards from the hinge joint.

dovetail key *(Build.)*. A batten, of dovetail-shaped section, which is driven into a corresponding groove cut across the back of adjacent boards in a panel, and serves to prevent warping.

dovetail key *(Eng.)*. A parallel key in which the part sunk in the boss or hub is of dovetail section, the portion on the shaft being of rectangular section. See **key**.

dovetail mitre *(Build.)*. See **secret dovetail**.

dovetail saw *(Build.)*. A saw similar to the tenon saw but of smaller size and having usually 12 teeth to the inch.

dowel *(Build.)*. A copper, slate, nonferrous metal, or stone pin sunk into opposing holes in the adjacent faces of two stones, when it is required to unite these more strongly than is possible with a mortar joint.

dowel *(Eng.)*. (1) A pin fixed in one part which, by accurately fitting in a hole in another attached part, locates the two, thus facilitating accurate reassembly. (2) A pin similarly used for locating divided patterns.

dowelling jig *(Build.)*. A device for directing the bit in drilling holes to receive dowels.

dowel plate *(Build.)*. A steel plate having a number of tapered holes in it, used as a pattern for main dowel pins.

dowel screw *(Build.)*. See **hand-rail screw**.

Dow-etch plates *(Print.)*. Magnesium alloy plates used in **powderless etching**.

down *(Comp.)*. A computer is *down* when it is not available for normal use.

downcast *(Min.Ext.)*. A contraction for *downcast shaft*, i.e. the shaft down which fresh air enters a mine. The fresh air may be sucked or forced down the shaft.

downcomer *(Build.)*. See **downpipe**.

downcomer *(Eng.)*. See **downtake**.

downdraught *(Autos.)*. Said of a carburettor in which the mixture is drawn downwards in the direction of gravity.

downdraught *(Meteor.)*. The downward draught of air occurring with the approach of a thunderstorm and due to evaporative cooling of descending air by heavy rain.

downer *(Print.)*. A sudden breaking of the web on a rotary machine, leading to **down time**.

down feathers *(Zool.)*. See **plumulae**.

downhole *(Min.Ext.)*. Describes the drills, measuring instruments and equipment used down the bore hole.

downlead *(Telecomm.)*. See **antenna downlead**.

down locks *(Aero.)*. See **up, down locks**.

downpipe *(Build.)*. A pipe (usually vertical) for conveying rain water from the gutter to the drain, or to an intermediate gulley. Also called *downcomer, downspout, fall pipe*.

downspout *(Build.)*. See **downpipe**.

Down's process *(Eng.)*. Electrolytic method of producing sodium metal and chlorine from fused salt at 600°C.

Down's syndrome *(Biol.,Med.)*. A form of mental retardation caused by a chromosomal abnormality, *trisomy 21*; the main features are moderate to severe retardation, a small round head, slanting eyes and minor abnormalities of hands and feet. These children also often have congenital heart lesions. Formerly *Mongolism*.

downtake *(Eng.)*. The pipe through which the blast-furnace gas is taken down outside the furnace from the top of the furnace to the duct catcher. Also *downcomer*.

downthrow *(Geol.)*. In a **fault**, the vertical displacement of the fractured strata. Indicated on geological maps by a tick attached to the fault line, with (where known) a figure alongside indicating the amount of downthrow.

down time *(Comp.)*. Period of time during which a computer is **down**.

down time *(Eng.)*. Time during which a machine, e.g. a computer or printing press (or series of machines), is idle because of adjustment, cleaning, reloading or other maintenance.

Downtonian *(Geol.)*. The lowest stage of the Old Red Sandstone facies of the Devonian System, named from its typical development around Downton Castle in the Welsh Borderlands.

downward modulation *(Telecomm.)*. See **negative modulation**.

downwash *(Aero.)*. The angle through which the airflow is deflected by the passage of an aerofoil measured parallel with the plane of symmetry.

downy mildew *(Bot.)*. One of several plant diseases of e.g. vines, onions, lettuce, caused by biotrophic fungi of the family Peronosporaceae. See **Oomycetes**.

Dow process *(Eng.)*. Extraction of magnesium from sea-water by precipitation as hydroxide with lime; also, electrolytic production of magnesium metal from fused chloride.

dowsing *(Genrl.)*. The process of locating underground water by the twitching of a twig (traditionally hazel or witch-hazel) held in the hand. The phenomenon, if it exists, continues to defy rational explanation, and is accordingly frequently scouted.

doxapram hydrochloride *(Med.)*. Centrally acting respiratory stimulant sometimes used in the treatment of severe respiratory failure.

DP *(Comp.)*. See **data processing**.

dracone *(Ships)*. A flexible sausage-shaped envelope of woven nylon fabric coated with synthetic rubber. Floats by reason of buoyancy of its cargo, oil or fresh water. It is towed.

draft *(Genrl.)*. See also **draught**.

draft *(Textiles)*. The drawing-out or attenuation of the **web** of textile fibres passing through a card, drawframe, speedframe, or spinning machine. Measured by the ratio of the linear density of input to output materials.

draft stop *(Build.)*. See **fire stop**.

draft tube *(Eng.)*. A discharge-pipe from a water-turbine to the tail race. It decreases the pressure at outlet and increases the turbine efficiency.

drag *(Aero.)*. Resistance to motion through a fluid. As applied to an aircraft in flight it is the component of the resultant force due to relative airflow measured along the drag axis, i.e. parallel to the direction of motion.

drag *(Build.)*. (1) A steel-toothed tool for dressing stone surfaces or for keying plasterwork. (2) A three pronged rake for mixing plaster.

drag *(Foundry)*. The bottom half of a moulding box or flask.

drag angle *(Eng.)*. The angle between the welding rod and the normal to the surface of the weld.

drag axis *(Aero.)*. A line through the centre of mass of an aircraft parallel with the relative airflow, the positive direction being downwind.

drag-bar *(Eng.)*. See **draw-bar**.

drag classifier *(Min.Ext.)*. Endless belt with transverse rakes, which moves upward through an inclined trough so as to drag settled material up and out while slow-settling sands overflow below as pulp arrives for continuous sorting into coarse and fine sands.

drag conveyor, drag chain conveyor *(Eng.)*. A conveyor in which an endless chain, having wide links carrying projections or wings, is dragged through a trough into which the material to be conveyed is fed.

drag-cup generator *(Elec.Eng.)*. A servo unit used to generate a feedback signal proportional to the time derivative of the error.

dragged work *(Build.)*. Stone-dressing done with a drag.

drag hinge *(Aero.)*. The pivot of a rotorcraft's blade which allows limited angular displacement in azimuth.

dragline excavator *(Civ.Eng.)*. A mechanical excavating appliance consisting of a steel scoop bucket which swings on chains from a movable jib; after biting into the material to be excavated, it is dragged towards the machine by means of a wire rope.

drag link *(Autos.)*. A link which conveys motion from the drop arm of a steering gear to the steering arm carried by a stub axle, which it connects through ball joints at its ends. Also called *steering rod*.

drag link *(Eng.)*. A rod by which the link motion of a steam-engine is moved for varying the cut-off. See **link motion**.

dragon-beam, dragging-beam *(Build.)*. The horizontal timber on which the foot of the hiprafter rests.

dragon's blood *(Build.,Print.)*. A resinous exudation from the fruit of palm trees and the stems of different species of *Dracaena*. It is a red, amorphous substance, mp 120°C, soluble in organic solvents. By destructive distillation the resin yields methylbenzene and phenylethene. It is used for colouring varnishes and lacquers; also, in photoengraving, to protect parts of a plate in the etching process.

drag struts *(Aero.)*. Structural members designed to brace an aerofoil against air loads in its own plane. Also, landing gear struts resisting the rearward component of impact loads.

drag wires *(Aero.)*. Streamlined wires or cables bracing an aerofoil against drag (rearward) loads. Applicable to biplanes and some early monoplanes.

drain *(Electronics)*. In a field-effect transistor, the region into which majority **carriers** flow from the **channel**; comparable with the collector in a conventional bipolar transistor.

drain *(Med.)*. Any piece of material, such as a plastic tube, used for directing away the discharges of a wound.

drainage *(Build.)*. The removal of any liquid by a system constructed for the purpose.

drainage *(Geol.)*. The removal of surplus meteoric waters by rivers and streams. The complicated network of rivers is related to the geological structure of a district, being determined by the existing rocks and superficial deposits in the case of youthful drainage systems; but in those that are mature, the courses may have been determined by strata subsequently removed.

drainage *(Med.)*. The action of draining discharges from wounds or infected areas.

drainage coil *(Elec.Eng.)*. A coil bridged between the legs of a communication pair, with its electrical midpoint, earthed, in order to prevent the accumulation of static charges on the conductors.

drain cock *(Eng.)*. A cock placed (a) at the lowest point of a vessel or space, for draining off liquid; (b) in an engine cylinder, for discharging condensed steam.

drainer *(Paper)*. A large compartmènt with perforated tiles or metal plates in the base for the purpose of allowing water to drain from stuff deposited in it.

drain holes *(Min.Ext.)*. Draining oil from strata by boring holes out from the main bore by the use of special **downhole** machinery such as **mud motor** driven drills.

drain plug *(Build.)*. A device for closing the outlet of drain pipes. See **bag plug, screw plug**.

drain tiles *(Civ.Eng.)*. Hollow clay or concrete tiles laid end to end without joints, to carry off surface or excess water.

drain-trap *(Build.)*. See **air-trap**.

Dralon *(Chem.)*. TN for German polyacrylonitrile staple fibre.

Draper effect *(Chem.)*. See **photochemical induction**.

draught *(Eng.)*. (1) The flow of air through a boiler furnace. (2) A measure of the degree of vacuum inducing air-flow through a boiler furnace.

draught *(Ships)*. The depth of water that a ship requires to float freely. More particularly; *draught forward* and *draught aft*, the depths of water required at the forward and after perpendiculars respectively. *Mean draught* is draught at midlength.

draught-bar *(Eng.)*. See **draw-bar**.

draught bead *(Build.)*. Same as **deep bead**.

draught excluder *(Build.)*. See **door strip**.

draught gauge *(Eng.)*. A sensitive vacuum gauge for indicating the draught in a boiler furnace or flue.

draughtsman, draftsman *(Eng.)*. One who makes engineering drawings (of models, articles to be made, electrical circuits, plans, etc.) from which prints, at one time blue-prints but now usually dye-lines, are made for actual use. See **tracing**. The draughtsman who makes such drawings frequently designs the details, the main design being laid down by an engineer or architect.

dravite *(Min.)*. Brown tourmaline, sometimes used as a gemstone.

draw *(Foundry)*. Internal cavity or spongy area occurring in a casting due tó inadequate supplies of molten metal during consolidation. See **drawing of patterns**.

draw *(Horol.)*. The action whereby one part is drawn into another. In a lever escapement, the locking faces of the pallets are formed at an angle relative to the teeth such that a tooth of the escape-wheel, when pressing against the locking face, tends to draw the pallet into the wheel.

draw *(Min.Ext.)*. (1) To allow ore to run from working places, stopes, through a chute into trucks. (2) To withdraw timber props or sprags from overhanging coal, so that it falls down ready for collection. (3) To collect broken coal in trucks.

draw *(Print.)*. When type has not been locked securely in the forme, the rollers *draw* it out during printing.

draw *(Textiles)*. The outward (drafting and twisting) and inward (winding) run of a **mule** spinning carriage.

drawability *(Eng.)*. A measure of the ability of a material to be *drawn*, as in forming a cup-like object from a flat metal blank.

draw-bar cradle *(Eng.)*. A closed frame or link for connecting the ends of the draw-bars of railway vehicles, so coupling them together.

draw-bar plate *(Eng.)*. On a locomotive frame, a heavy transverse plate through which the draw-bar is attached.

draw-bar pull *(Eng.)*. The tractive effort exerted, in given circumstances, by a locomotive or tractor.

draw-bore *(Build.)*. A hole drilled transversely through a mortise and tenon so that, when a pin is driven in, it will force the shoulders of the tenon down upon the abutment cheeks of the mortise.

draw-bridge *(Civ.Eng.)*. A general name for any type of bridge of which the span is capable of being moved bodily to allow of the passage of large vessels.

drawdown *(Min.Ext.)*. Fall of water level in natural reservoir. e.g. an **aquifer**, when rate of extraction exceeds rate of replenishment.

drawdown *(Print.)*. Method of checking the hue of a printing ink sample by scraping it down onto a standard paper alongside a standard sample to produce ink films with graduated density. Comparison of the two can give an indication of the correctness of hue, rate of absorption and drying ability.

draw, drawing rollers *(Print.)*. A pair of rollers, usually both driven, which control the web tension on rotary presses.

drawer-front dovetail *(Build.)*. See **lap dovetail**.

drawer-lock chisel *(Build.)*. A crank-shaped chisel with two edges, one being parallel to the shank and the other at right angles to it. Used for chopping drawer-lock recesses or other work in restricted positions.

draw-filing *(Eng.)*. The operation of finishing a filed surface by drawing the file along the work at right angles to the length of the file.

draw-gate *(Hyd.Eng.)*. A name given to the valve controlling a sluice.

draw-in box (pit) *(Elec.Eng.)*. A box or pit to enable cables to be drawn into, or removed from, a conduit or duct.

drawing *(Eng.)*. The operation of producing wire, or giving rods a good finish and accurate dimensions, by pulling through one or a series of tapered dies. See also **tempering**.

drawing-down *(Eng.)*. The operation of reducing the diameter of a bar, and increasing its length, by forging.

drawing fires *(Eng.)*. The operation of raking out fires from boiler furnaces when shutting down.

drawing-in *(Textiles)*. Drawing the warp yarns through the eyes of the loom heald preparatory to weaving.

drawing of patterns *(Foundry)*. The removal of a pattern from a mould; also termed lifting of patterns. It is facilitated by the taper or draught of the pattern, and by loosening the pattern by **rapping**.

drawing of tubes *(Eng.)*. See **tube drawing**.

drawing rollers *(Textiles)*. Pairs of steel rollers, each pair running at higher speed to draft the roving passing through. The front pair may comprise a fluted bottom roller with the top one covered with leather, rubber, or cork.

drawing staple fibres *(Textiles)*. (1) Running together and attenuation of a number of slivers, preparatory to making, slubbing and roving. (2) The operation in which worsted tops are reduced to a roving.

drawing synthetic continuous filaments *(Textiles)*. Hot or cold stretching of synthetic filaments during their manufacture in order to orient the molecules and so develop the tensile strength and elastic properties of the filaments.

drawing temper *(Eng.)*. The operation of tempering hardened steel by heating to some specific temperature and quenching to obtain some definite degree of hardness. See **tempering**.

draw-in pit *(Elec.Eng.)*. See **draw-in box**.

draw-in system *(Elec.Eng.)*. The system whereby cables are pulled into conduits or ducts of earthenware, concrete, or iron, from one man-hole to another.

draw knife *(Build.)*. A cutting blade with a handle at each end at right-angles to the blade; used for shaving wood.

draw-nail *(Foundry)*. A pointed steel rod driven into a pattern to act as a handle for withdrawing it from the mould. See **draw-screw**.

drawn on *(Print.)*. Said of a book cover which is attached by gluing down the back; if the endpapers are pasted down it is said to be *drawn on solid*.

drawn-wire filament *(Elec.Eng.)*. An incandescent lamp filament, made by a wire-drawing process as opposed to a squirting process.

draw-off valve *(Eng.)*. A **bib-valve**.

draw-out metal-clad switchgear *(Elec.Eng.)*. Metal-clad switchgear in which the switch itself can be isolated from the bus-bars for inspection and maintenance, by moving it away from the bars along suitable guides.

draw ratio *(Textiles)*. The ratio of the linear density of the undrawn yarn to that of the drawn yarn. See **drawing (synthetic continuous filaments)**.

draw-screw *(Foundry)*. A screwed rod provided with an eye at the end to act as a handle; screwed into a pattern for lifting it from the mould. See **draw-nail**.

draw sheet *(Print.)*. (1) The sheet drawn over the completed make-ready on a press before proceeding with the printing. (2) The top sheet of a **cylinder dressing**. (3) See **shim**.

draw stop *(Acous.)*. See **draw knob**.

draw works *(Min.Ext.)*. Surface gear of a drilling rig; it includes all the machinery needed to drive, place and extract the drilling machinery, from it the whole operation is controlled.

dream-interpretation *(Behav.)*. In psychoanalytic theory, a technique for understanding the individual's unconscious life by focusing on dream content and attempting to unravel its hidden meaning, which reflect unconscious wishes and conflicts. See **Freud's theory of dreams**.

dredge *(Min.Ext.)*. Barge or twin pontoons carrying chain of digging buckets, or suction pump, with over gear such as jigs, sluices, trommels, tailing stackers, manoeuvring anchor-lines, and power producer. Used to work alluvial deposits of cassiterite, gold, gemstones, etc.

dredger excavator *(Eng.)*. See **bucket-ladder excavator**.

D region *(Immun.)*. A short sequence of amino acids in the variable region of immunoglobulin heavy chains which is coded by one or another of several separate DNA **exons**. This contributes to antibody diversity (hence the use of *D*).

dreikanter *(Geol.)*. A wind-faceted pebble typically having three curved faces. Common in desert deposits. See **ventifact**.

dressed timber *(Build.)*. Timber which has been planed more or less to size.

dresser *(Eng.)*. (1) An iron block used in forging bent work on an anvil. (2) A mallet for flattening sheet-lead. (3) A tool for facing and grooving millstones, or for trueing grinding wheels.

dresser, dressing *(Build.)*. A boxwood tool for straightening lead piping and sheet-lead, and the operation of using it.

dressing *(Build.)*. The operation of smoothing the surface of stone or timber.

dressing *(Eng.)*. Fettling of castings, removal of flashes and runners.

dressing *(Med.)*. The application of sterile material, gauze, lint etc., to a wound or infected part; material so used.

dressing *(Min.Ext.)*. Grinding of worn crushing rolls, to restore cylindrical shape. Rock crushing and screening to required sizes. Preparation of amalgamation plates with liquid mercury for gold recovery.

dressing *(Print.)*. The operation of fitting furniture around the pages in a forme, preparatory to locking it up.

dressing iron *(Build.)*. See **break iron**.

dressings *(Build.)*. The mouldings, quoins, strings and like features in a room or building.

drier *(Eng.)*. Furnace used to de-water ore products without changing their composition.

driers *(Chem.)*. Substances accelerating the drying of vegetable oils, e.g. linseed oil in paints. The most important representatives of this group are the naphthenates, resinates and oleates of lead, manganese, cobalt, calcium and zinc.

drift *(Aero.)*. (1) The motion of an aircraft in a horizontal plane due to crosswind. (2) Slow unidirectional error of instrument or gyroscope.

drift *(Civ.Eng.)*. The direction in which a tunnel is driven.

drift *(Electronics,Telecomm.)*. Slow variation of any performance characteristic (gain, frequency, power output, noise level etc.) of any device or circuit. May be due to gradual self-heating of equipment, ambient temperature, or ageing. In particular the tendency of any tuned device (receiver, oscillator etc.) to move slowly away from its intended or selected frequency of operation.

drift *(Eng.)*. (1) A tapered steel bar used to draw rivet holes into line. (2) A brass or copper bar used as a punch.

drift *(Geol.)*. A general name for the superficial, as distinct from the solid, formations of the earth's crust. It includes typically the Glacial Drift, comprising all the varied deposits of boulder clay, outwash gravel, and sand of Quaternary age. Much of the drift is of fluvio-glacial origin.

drift *(Hyd.Eng.)*. The rate of flow of a current of water.

drift *(Min.Ext.)*. (1) A level or funnel pushed forward underground in a metal mine, for purposes of exploration or exploitation. The inner end of the drift is called a *dead end*. (2) A heading driven obliquely through a coal-seam. (3) A heading in a coal-mine for exploration or ventilation. (4) An inclined haulage road to the surface. (5) Deviation of borehole from planned course.

drift angle *(Aero.,Ships.)*. The angle between the planned course and the track. Also sometimes used for angle between heading and track made good.

drift chamber *(Phys.)*. A particle detector used in high-energy physics experiments whereby the tracks of charged particles in interactions may be recorded.

drift currents *(Meteor.)*. Ocean currents produced by prevailing winds.

drifter *(Min.Ext.)*. A cradle-mounted compressed-air rock drill, used when excavating tunnels (*drifts* or *cross-cuts*).

drifting *(Eng.)*. The process of bringing holes into line by hammering a drift through them.

drifting *(Min.Ext.)*. Tunnelling along the strike of a lode, horizontally or at a slight angle.

drifting test *(Eng.)*. A workshop test for ductility; a hole is drilled near the edge of a plate and opened by a conical drift until cracking occurs.

drift mobility *(Electronics)*. See **mobility**.

drift plug *(Build.)*. A wooden plug which is driven through the bore of a lead pipe in order to smooth out a kink.

drift sight *(Aero.)*. A navigational instrument for measuring drift angle.

drift space *(Electronics)*. In an electron tube which depends on the velocity-modulation of electron beams, the space which is free of externally applied alternating fields; here, relative repositioning of the electrodes takes place.

drift transistor *(Electronics)*. One in which resistivity increases continuously between emitter and collector junctions, improving high-frequency performance.

drift tube *(Electronics)*. A piece of metal tubing, held at a fixed potential, which forms the **drift space** in a microwave tube or linear accelerator.

drill *(Eng.)*. (1) A revolving tool with cutting edges at one end, and having flats or flutes for the release of chips; used for making cylindrical holes in metal. (2) Also the *drilling machine* which turns the drill.

drill *(Min.Ext.)*. (1) Hand drill, **auger**. (2) A compressed-air operated rock drill, jack-hammer, **pneumatic drill**. (3) Generally, the more elaborate equipment required in *power drilling*.

drill *(Textiles)*. A heavy woven fabric (often cotton) with diagonal lines on the surface.

drill bit *(Build.)*. A wood-boring bit with a cylindrical shank with twin helical grooves and cutting edges.

drill bit *(Min.Ext.)*. The actual cutting or boring tool in a *drill*. In rotary drilling it may be of the *drag* variety, with two or more cutting edges (usually hard-tipped against wear), **roller** (with rotating hard-toothed rollers), or

diamond, with cutting face (annular if core samples are required) containing suitably embedded **borts**. The assembly is attached to the bottom of the **drill pipe** and rotated with it. **Drilling mud** pumped down the drill pipe cools, lubricates and carries away the debris

drill bush *(Eng.)*. A hard sleeve inserted in a **drilling jig** to locate and guide a twist drill accurately, in repetition drilling.

drill chuck *(Eng.)*. A self-centring chuck usually having three jaws which are contracted on to the drill by the rotation of an internally coned sleeve encasing them.

drill collar *(Min.Ext.)*. Collar which attaches the drill to the drill string. In deep bores it can weigh 100 tons and is used to damp-out torsional stresses and stabilize the drill string.

drilled and strung *(Print.)*. A method of binding in which holes are drilled close to the back and the leaves then secured by thread or cord; superior to and more expensive than either **flat stitching** or **stabbing**.

drill-extractor *(Civ.Eng.)*. A tool used to remove from a boring a broken drill or one which has fallen free of the drilling apparatus.

drill feed *(Eng.)*. The hand- or power-operated mechanism by which a drill is fed into the work in a drilling machine.

drilling *(Min.Ext.)*. (1) The operation of tunnelling or stoping. (2) The operation of making short holes for blasting, or deep holes with a diamond drill for prospecting or exploration. Drilling may be percussive (repeated blows on the drilling tool) or rotary (circular grinding; see **drill bit**), or a combination of these. Cf. **boring**.

drilling jig *(Eng.)*. A device used in repetition drilling, which locates and firmly holds the workpiece accurately in relation to a **drill bush** or pattern of drill bushes accurately positioned in the jig.

drilling machine *(Eng.)*. A machine tool for drilling holes, consisting generally of a vertical standard, carrying a table for supporting the work and an arm provided with bearings for the drilling spindle. See **pillar drill**, **radial drill**, **sensitive drill**.

drilling mud *(Min.Ext.)*. Mixture of clays, water, density increasing agents like **barite** and sometimes thixotropic agents like *bentonite* pumped down through the drilling pipe and used to cool, lubricate and flush debris from the drilling assembly. It also helps to seal the bored rock and most importantly provides the hydrostatic pressure to contain the oil and gas when this is reached. **Mud motors** powered by the mud flow can be used for drilling, particularly in directions away from the main bore axis.

drill(ing) pipe *(Min.Ext.)*. The tube, joined by screwed **collars**, which connects the drilling platform to the drill bit and imparts rotary motion to the latter.

drilling platform *(Min.Ext.)*. Offshore platform which may be floating or fixed to the sea bed and from which over 50 wells may be bored radially and at various angles into the bearing strata.

Drinker respirator *(Med.)*. A respirator in which the whole body, excluding the head, is placed. It assists respiration by moving the chest. Commonly known as an *iron lung*.

driography *(Print.)*. A planographic printing process akin to lithography whereby the antipathy of the image and non-image areas on the plate is maintained using special plate coatings and inks. Usually relies on the properties of dissimilar silicones.

drip *(Build.,Civ.Eng.)*. A groove in the projecting under-surfaces of a coping brick or stone wider than the wall; designed to prevent water from passing from the coping to the wall. Also called *gorge* or *throat*.

drip-dry *(Textiles)*. Fabrics and garments that shed creases and wrinkles when hung out wet and allowed to dry. Some light ironing may still be required before wearing. Cotton fabrics require a special treatment with resin-forming compounds to confer drip-dry properties on them.

drip-feed lubricator *(Eng.)*. A small reservoir from which lubricating oil is supplied in drops to a bearing, sometimes through a glass tube to render the rate of feed visible. See **sight-feed lubricator**.

drip mould *(Build.)*. A projecting moulding arranged to throw off rainwater from the face of a wall.

dripping eave *(Build.)*. An eave which is not fitted with a gutter and which therefore allows the rain to flow over to a lower roof or to the ground.

drip-proof *(Elec.Eng.)*. Said of an electric machine or other electrical equipment which is protected by an enclosure whose openings for ventilation are covered with suitable cowls, or other devices, to prevent the ingress of moisture or dirt falling vertically.

drip-proof burner *(Eng.)*. Gas burner designed to prevent the choking of flame ports or nozzles by foreign matter that may drip or fall on to it.

drip sink *(Build.)*. A shallow sink at, or just higher than, floor level, to take the drip from a tap.

dripstone *(Build.)*. A projecting moulding built in above a doorway, window opening, etc., to deflect rain water.

drip tip *(Bot.)*. A marked elongation of the tip of the leaf, said to facilitate the shedding of rain from the surface of the leaf.

drive *(Behav.)*. A motivational concept used to describe changes in responsiveness to a consistent external stimulus as a function of varying *internal states*. Formerly a popular concept in both psychology and ethology, its usefulness is now considered problematic (e.g. *hunger* as a drive).

drive *(Elec.Eng.)*. Signal applied to the input of an amplifier, e.g. current to base of bipolar transistor and voltage to gate of field effect transistor.

drive *(Min.Ext.)*. Tunnel or level driven along or near a lode, vein or massive ore deposit as distinct from country rock. *Driving* is the process.

drive *(Telecomm.)*. That which controls a master resonator in an oscillator, e.g. **quartz crystal**, **resonating line**, **cavity**.

drive-in *(Image Tech.)*. Open-air cinema where spectators can view the programme from their parked cars, with individual loud speakers reproducing the sound.

driven elements *(Elec.Eng.)*. Those in an antenna which are fed by the transmitter, as compared with reflector, director, or parasitic elements.

driven roller *(Print.)*. A roller geared to the press drive and used to take the web from one section of the press to another, maintaining tension in the process.

driver *(Comp.)*. A piece of software that controls a device.

drive-reduction hypothesis *(Behav.)*. In learning theory, the idea that reinforcing stimuli must reduce some drive or need in an animal if learning is to occur.

driver plate *(Eng.)*. A disk which is screwed to the mandrel nose of a lathe, and carries a pin which engages with and drives a carrier attached to the work. Also *driving plate*.

driver stage *(Telecomm.)*. Amplifiers which drive the final stage in a radio transmitter.

driving axle *(Eng.)*. A vehicle axle through which the driving effort is transmitted to the wheels fixed to it. Also called a *live axle*.

driving chain *(Eng.)*. An endless chain consisting of steel links which engage with toothed wheels, so transmitting power from one shaft to another. See **roller chain**.

driving fit *(Eng.)*. A degree of fit between two mating pieces such that the inner member, being slightly larger than the outer, must be driven in by a hammer or press.

driving gear *(Eng.)*. Any system of shafts, gears, belts, chains, links, etc., by which power is transmitted to another system.

driving point impedance *(Elec.Eng.)*. The ratio of the e.m.f. at a particular point in a system to the current at that point.

driving side *(Eng.)*. The tension side of a driving belt; the side moving from the follower to the driving pulley.

driving wheel *(Eng.)*. (1) The first member of a train of gears. (2) The road wheels through which the tractive force is exerted in a locomotive or road vehicle.

drogue *(Aero.)*. A sea anchor used on seaplanes and flying-boats; it is a conical canvas sleeve, open at both ends, like a bottomless bucket. Used to check the way of the aircraft.

drogue parachute *(Aero.,Space)*. A small parachute used to (1) slow down a descending aircraft or spacecraft, (2) to extract a larger parachute or (3) extract cargo from a hold or wing mounting.

dromaeognathous *(Zool.)*. In Birds, said of the type of palate with large basipterygoid processes springing from the basisphenoid, and the vomers large and flat and connected posteriorly with the palatines, which do not articulate with the rostrum.

drone *(Aero.)*. Pilotless guided aircraft used as a target or for reconnaissance.

drone *(Zool.)*. In social Bees (Apidae), a male.

droop snoot *(Aero.)*. Cockpit section hinged on to main fuselage to provide downward visibility at low speeds. Colloq. for *droop nose*.

drop *(Bot.)*. Premature abscission of fruit especially (e.g. June drop) when half-grown or (pre-harvest drop) when almost mature.

drop *(Comp.)*. Digits are said to *drop-in* if they are recorded without a signal, and *drop-out* if not recorded from a signal.

drop *(Elec.Eng.)*. A term commonly used to denote **voltage drop**.

drop *(Horol.)*. The space moved by a tooth of the escape wheel when it is entirely free from contact with the pallets.

drop *(Min.Ext.)*. The vertical displacement in a down-throw fault: the amount by which the seam is lower on the other side of the fault. See **throw**.

drop *(Print.)*. To unlock a forme and remove the chase after printing. The type matter is then either distributed or tied up and stored.

drop arm *(Autos.)*. A lever attached to a horizontal spindle which receives rotary motion from the steering gear; used to transmit linear motion through attached steering rod or drag link to the arms carried by the stub axles.

drop-down curve *(Hyd.Eng.)*. The longitudinal profile of the water surface in the case of nonuniform flow in an open channel, when the water surface is not parallel to the invert, owing to the depth of water having been diminished by a sudden drop in the invert.

drop elbow, tee *(Build.)*. A small elbow or tee with ears for fixing to a support.

drop electrode *(Chem.)*. See **dropping mercury electrode**.

drop foot *(Med.)*. Dropping of the foot from its normal position, caused by paralysis of the muscles, due to injury or inflammation of the nerves supplying them.

drop forging *(Eng.)*. The process of shaping metal parts by forging between two dies, one fixed to the hammer and the other to the anvil of a hydraulic or mechanical hammer. The dies are expensive, and the process is used for the mass-production of parts such as connecting-rods, crankshafts etc.

drop gate *(Foundry)*. A pouring-gate or runner leading directly into the top of a mould.

drop hammer *(Eng.)*. A gravity-fall hammer or a double acting stamping hammer used to produce drop forgings by stamping hot metal between pairs of matching dies secured to the anvil block and to the top of the drop hammer respectively.

drop hammer test *(Paper)*. A means of measuring and recording photoelectrically the compression resistance of a box by allowing a standard weight to fall through a given distance on to the sample under the prescribed conditions of test.

drop-out *(Image Tech.)*. A brief loss of signal, especially in magnetic recording.

drop-out *(Telecomm.)*. Said of a relay when it de-operates, i.e. contacts revert to de-energized condition.

dropout halftone *(Print.)*. A **half tone** in which the highlight areas have no screen dot formation.

dropped beat *(Med.)*. Intermission of a regular pulse wave at the wrist, due to intermission of the heart beat or to an extrasystole.

dropped elbow *(Vet.)*. A condition in which there is inability to extend the forelimb, due to paralysis of the radial nerve.

dropped head *(Print.)*. The first page of a chapter, etc. which begins lower down than ordinary pages. As far as possible, the drop should be constant throughout the book.

dropper *(Elec.Eng.)*. In catenary constructions for electric traction systems, the fitting used for supporting the contact wire from the catenary wire.

dropping mercury electrode *(Chem.)*. A half-element consisting of mercury dropping in a fine stream through a solution. Used in polarography, a continuously renewed mercury surface being formed at the tip of a glass capillary, the accumulating impurities being swept away with the detaching drops of mercury.

dropping resistor *(Elec.Eng.)*. A resistor whose purpose is to reduce a given voltage by the voltage drop across the resistance itself.

drop-point slating *(Build.)*. A mode of laying asbestos slates so that one diagonal is horizontal.

drop siding *(Build.)*. **Weather-boarding** which is rebated and overlapped.

drop stamping *(Eng.)*. See **drop forging**.

dropsy *(Med.)*. See **oedema**.

drop tank *(Aero.)*. A fuel tank designed to be jettisoned in flight. Also *slipper tank*.

drop tracery *(Arch.)*. Tracery which lies partly below the springing of the arch which it decorates.

drop valve *(Eng.)*. A conical-seated valve used in some steam-engines; rapid operation by a trip-gear and return spring reduces wire-drawing losses.

drop wrist *(Med.)*. Limp flexion of the wrist from paralysis of the extensor muscles, as a result of neuritis or of injury to the nerve supplying them, e.g. in lead poisoning.

drosometer *(Meteor.)*. An instrument for measuring the amount of dew deposited.

Drosophila melanogaster *(Biol.)*. Common fruitfly, a dipteran used extensively for genetic experiments because its giant **salivary gland chromosomes** and other biological characteristics are very suitable for studies of chromosome organization and **gene mapping**.

dross *(Eng.)*. Metallic oxides that rise to the surface of molten metal in metallurgical processes.

dross *(Min.Ext.)*. Small coal, inferior or worthless.

drossing *(Eng.)*. Removal of scums, oxidized films and solidified metals from molten metals.

drought *(Meteor.)*. A marked deficiency of rain compared to that usually occurring at the place and season under consideration.

drove *(Build.)*. A broad-edged chisel for dressing stone.

drove *(Hyd.Eng.)*. A narrow channel used for irrigation.

drove work *(Build.)*. Stone dressing done with a boaster, leaving rows of parallel chisel marks on the slant across the face. Also called *boasted work*.

drowned *(Min.Ext.)*. Flooded, e.g. *drowned workings*.

drowned valleys *(Geol.)*. Literally, river valleys which have become drowned by a rise of sea level relative to the land. This may be due to actual depression of the land, sea level remaining stationary; or to a eustatic rise in sea level, as during the interglacial periods in the Pleistocene, when melting of the ice-caps took place. See also **ria** and **fiords**.

drowning pipe *(Build.)*. A cistern inlet pipe, reaching below the water surface and thereby reducing the noise of discharge. Also *silencer pipe*.

Drude law *(Phys.)*. A law relating the specific rotation for polarized light of an optically active material to the wavelength of the incident light.

$$\alpha = \frac{k}{\lambda^2 - \lambda_0^2},$$

where a = specific rotation, k = rotation constant for material, $\lambda_0{}^2$ = dispersion constant for material, λ = wavelength of incident light. The law does not apply near absorption bands.

drug *(Med.)*. Any substance, natural or synthetic, which has a physiological action on a living body, either when used for the treatment of disease or the alleviation of pain or for purposes of self-indulgence, leading in some cases to progressive addiction.

drug resistance *(Med.)*. The condition in which tissues become resistant after treatment with drugs, commonly found with many anti-tumour treatments. See **antibiotic resistance**.

drum *(Build.)*. (1) Any timber structure cylindrical in shape. (2) Any cylinder used as a form for bending wood to shape.

drum *(Image Tech.)*. (1) Heavy cylinder rotating in contact with the film in an optical sound recorder or reproducer to ensure uniform movement. (2) Assembly containing the rotating magnetic heads in a **helical-scan** videotape recorder.

drum *(Min.Ext.)*. Cylinder or cone, or compound of these, on and off which the winding rope is paid when moving cages or skips in a mine shaft.

drum armature *(Elec.Eng.)*. An armature for an electric machine, having on it a drum winding.

drum breaker starter *(Elec.Eng.)*. A drum starter in which a separate circuit-breaker is provided for interrupting the circuit.

drum controller *(Elec.Eng.)*. A controller in which the connections for performing the desired operation are made by means of fixed contact fingers, bearing on metallic contact strips mounted in the form of a rotating cylindrical drum.

drum-curb *(Civ.Eng.)*. See **curb**.

drum filter *(Min.Ext.)*. Thickened ore pulp is fed to a trough through which a cylindrical hollow drum rotates slowly. Vacuum draws liquid into pipes mounted internally, while solids (filter cake) are arrested on a permeable membrane wrapped round drum circumference, and are removed continuously before re-submergence.

drumlin *(Geol.)*. An Irish term, meaning a little hill, applied to accumulations of glacial drift moulded by the ice into small hog-backed hills, oval in plan, with the longer axes lying parallel to the direction of ice movement. Drumlins often occur in groups, giving the 'basket of eggs' topography which is seen in many parts of Britain and dates from the last glaciation.

Drumm accumulator *(Elec.Eng.)*. A special form of alkaline accumulator capable of high discharge rates; the positive plate contains nickel oxides and the negative plate is of zinc.

drum movement *(Horol.)*. A clock movement housed in a cylindrical metal case.

drum starter *(Elec.Eng.)*. A motor starter in which the necessary operations are carried out by fixed contact fingers bearing upon contact strips mounted in the form of a rotating cylindrical drum.

drum washer *(Paper)*. A large diameter cylinder with perforated surface rotating slowly in a vat supplied with uncleaned stuff so that a mat is created on the surface by drainage. The means are provided of subjecting the fibrous mat to sprays of clean water and removing the washings. Accepted stuff is generally removed by a doctor.

drum weir *(Civ.Eng.)*. A weir formed by a gate which can rotate about a horizontal axis and thereby control the flow.

drum winding *(Elec.Eng.)*. A winding for electrical machines in which the conductors are all placed under the outer surface of the armature core. It is the form of winding almost invariably used. Also called *barrel winding*.

drunken saw *(Build.)*. A circular saw revolving about an axis which is not absolutely at right angles to the plane of the saw, consequently cutting a wide kerf.

drunken thread *(Eng.)*. A screw in which the advance of the helix is irregular in every convolution.

drupe *(Bot.)*. A fleshy fruit with one or more seeds each surrounded by a stony layer (the endocarp), e.g plum.

drupel *(Bot.)*. A small drupe, usually one of a group forming an aggregate fruit, e.g. raspberry.

druse *(Bot.)*. A globose mass of crystals of calcium oxalate (ethandioate) around a central foundation of organic material; in some plant cells.

drusy *(Min.Ext.)*. Containing cavities often lined with crystals; said of mineralized lodes or veins. See **geodes**.

drusy cavities *(Geol.)*. See **geodes**.

dry adiabatic *(Meteor.)*. A curve on an aerological diagram representing the temperature changes of a parcel of dry air subjected to an adiabatic process.

dry adiabatic lapse rate *(Meteor.)*. The *temperature lapse rate* of dry air which is subjected to adiabatic ascent or descent. This lapse rate also applies to moist air which remains unsaturated. Its magnitude is 9.76°C per km. Abbrev. *DALR*.

dry area *(Build.)*. The 2 in or 3 in (50 or 75 mm) cavity in the wall below ground-level in basement walls built hollow; the purpose of the cavity is to keep the basement walls dry.

dry assay *(Chem.)*. The determination of a given constituent in ores, metallurgical residues and alloys, by methods which do not involve liquid means of separation. See also **wet assay, scorification, cupellation**.

dry-back boiler *(Eng.)*. A shell-type boiler with one or more furnaces passing to a chamber at the back, from which an upper bank of firetubes leads to the uptake at the front.

dry battery *(Elec.Eng.)*. A battery composed of **dry cells**.

dry blowing *(Min.Ext.)*. Manual or mechanical winnowing of finely divided sands to separate heavy from light particles, practised in arid regions.

dry bone, dry bone ore *(Min.)*. See **smithsonite**.

dry box *(Nuc.Eng.)*. Sealed box for handling material in low humidity atmosphere. Not synonymous with **glove box**.

dry brushing *(Textiles)*. The process of gently brushing a fabric to raise the fibres on the surface (e.g. with a teazle).

dry cell *(Elec.Eng.)*. A primary cell in which the contents are in the form of a paste. See **Leclanché cell**.

dry compass *(Ships)*. A mariner's magnetic compass which has no liquid in the bowl. Cf. **liquid compass**.

dry construction *(Build.)*. In building, the use of timber or plasterboard for partitions, lining of walls and ceilings, to eliminate the traditional use of plaster and the consequent drying out period. Also *dry lining*.

dry copper *(Eng.)*. Copper containing oxygen in excess of that required to give 'tough pitch'. Such metal is liable to be brittle in hot- and cold-working operations.

dry-core cable *(Telecomm.)*. A multicore lead-covered core for telephone or telegraph use, the separate conductors to which are covered with a continuous helix of ribbon-shaped paper. The paper provides the insulation after being dried with carbon dioxide, which is pumped through the cable and kept under pressure.

dry deposition *(Ecol.)*. Deposition of gaseous materials on natural surfaces involving turbulent transport. Applied to *pollutant gases*.

dry dock *(Civ.Eng.)*. A dock in which ships are repaired. Water is excluded by means of gates or caissons, after the dock has been emptied. Also called *graving dock*.

dry electrolytic capacitor *(Elec.Eng.)*. One in which the negative pole takes the form of a sticky paste, which is sufficiently conducting to maintain a gas and oxide film on the positive aluminium electrode.

dry flashover voltage *(Elec.Eng.)*. The breakdown voltage between electrodes in air of a clean dry insulator.

dry flue gas *(Eng.)*. The gaseous products of combustion from a boiler furnace, excluding water vapour. See **flue gas**.

dry fruit *(Bot.)*. A fruit in which the pericarp does not become fleshy at maturity.

dry ice *(Chem.)*. Solid (frozen) carbon dioxide, used in refrigeration (storage) and engineering. At ordinary atmospheric pressures it sublimes slowly.

dry indicator test *(Paper)*. One of the methods of determining the resistance of a paper to penetration by aqueous media. A powder that intensifies in colour when wet is sprinkled on a test sample floating on water and the time measured until an agreed degree of colour change has taken place.

drying cabinet *(Build.)*. A heated cabinet for drying or airing clothes. Commonly used in multistorey flats or where alternative drying facilities are not available, and heated by electricity, gas, or forced hot air.

drying cylinder *(Paper)*. A hollow metal cylinder, heated internally by live steam to remove moisture from a web of paper in contact with the greater part of its outer surface.

drying cylinder *(Textiles)*. A heated rotating hollow cylinder over which wet fabric is passed to dry it. Often a set of cylinders held closely together in a vertical stack is used with the fabric passing under and over each cylinder.

drying oils *(Build.)*. Vegetable or animal oils which harden by oxidation when exposed to air.

drying-out *(Elec.Eng.)*. (1) The process of heating the windings of electrical equipment to drive all moisture out of the insulation; usually done by passing current through the windings. (2) In electroplating, the process of removing moisture from a metal by passing it through hot water and then through sawdust or a current of hot air.

drying stove *(Foundry)*. A large stove or oven in which dry sand moulds and cores are dried. See **dry sand**.

dry joint *(Elec.Eng.)*. A faulty solder joint giving high-resistance contact due to residual oxide film.

dry laying *(Textiles)*. The formation of a **web** of fibres by carding or **air-laying** preparatory to the manufacture of a **non-woven fabric**.

dry liner *(Eng.)*. See **liner**.

dry mass *(Aero.)*. The mass of an aero-engine, including all essential accessories for its running and the drives for airframe accessories, without oil, fuel, or coolant.

dry moulding *(Foundry)*. The preparation of moulds in dry sand, as distinct from the use of greensand or **loam**.

dry mounting *(Image Tech.)*. Method of attaching a photographic print to a mounting card by sandwiching between them a sheet of tissue which becomes adhesive with the application of heat and pressure.

dryness fraction *(Eng.)*. The proportion, by weight, of dry steam in a mixture of steam and water, i.e. in wet steam.

dry offset *(Print.)*. Almost synonymous with **letterset printing** but chiefly used when relief plates are employed on conventional litho machines, the dampening rollers being unused.

dry pipe *(Eng.)*. A blanked-off and perforated steam-collecting pipe placed in the steam space of a boiler and leading to the stop-valve, for the purpose of excluding water resulting from *priming*. See **antipriming pipe**.

dry plate *(Image Tech.)*. Term used in the early days of photography to describe glass plates coated with a light-sensitive gelatine emulsion, in contrast to the former collodion *wet plates*.

dry-plate rectifier *(Elec.Eng.)*. See **metal rectifier**.

dry rot *(Bot.,Build.)*. (1) One of a number of plant diseases, e.g. of stored potatoes, in which a lesion dries out as it forms. (2) The rotting of timber by the fungus *Serpula (Merulius) lacrymans*, so that it becomes dry, light and friable, with a cracked appearance. Cf. **wet rot**.

dry run *(Comp.)*. Use of test data to check paths through a program or flowchart without the use of a computer. See **logical error**.

dry sand *(Foundry)*. A moulding sand possessing the requisite cohesion and strength when dried. It is moulded in a moist state, then dried in an oven, when a coherent and porous mould results.

Drysdale permeameter *(Elec.Eng.)*. An instrument for determining, by a ballistic method, the permeability of a sample of iron; a plug carrying a primary and secondary coil is inserted in an annular hole in the sample of material under test.

Drysdale potentiometer *(Elec.Eng.)*. An a.c. potentiometer of the polar type, comprising a phase-shifting transformer and resistive voltage divider. It is calibrated against a standard cell with a direct current and the alternating current is then set to the same value using an electrodynamic indicator.

dry spinning *(Textiles)*. The production of filaments by the extrusion of a solution of the polymer in a volatile liquid which is then removed by evaporation.

dry-spun flax *(Textiles)*. The coarse flax yarn obtained from a dry roving. See **wet-spinning**.

dry steam *(Eng.)*. Steam free from water, but unsuperheated. Often called *dry saturated steam*.

dry sump *(Autos.)*. An internal-combustion engine lubrication system in which the separate crankcase is kept dry by an oil scavenge pump, which returns the oil to a tank, from which it is delivered to the engine bearings by a pressure pump.

dry-sweating *(Vet.)*. See **anhidrosis**.

dry valley *(Geol.)*. A valley produced at some former period by running water, though at present streamless. This may be due to a fall of the water table, to river capture, or to climatic change.

dry weight *(Space)*. The weight of a system without fuel and consumables; for a launch vehicle it is the total launch weight minus that of the propellants and pressurizing gases.

d.s.c. *(Elec.Eng.)*. An abbrev. for *Double Silk-Covered wire*.

D slide-valve *(Eng.)*. A simple form of slide-valve, in section like a letter D, sliding on a flat face in which ports are cut. See **slide-valve**.

DTA *(Chem.)*. Abbrev. for *Differential Thermal Analysis*.

DTL *(Electronics)*. Abbrev. for *Diode-Transistor Logic*.

DTP *(Comp.)*. See **desk top publishing**.

DTR *(Image Tech.)*. Abbrev. for *Diffusion-Transfer Reversal*.

dual *(Maths.)*. In projective geometry, the dual of something is obtained by interchanging lines and points. For example, the dual of the statement 'Any two points are joined by a line' is 'Any two lines meet at a point' which is made true by including points at infinity.

dual beta decay *(Phys.)*. **Branching** where a radioactive nuclide may decay by either electron or positron emission. Not to be confused with **double beta decay**.

dual combustion cycle *(Eng.)*. An internal-combustion engine cycle sometimes taken as a standard of comparison for the compression-ignition engine, in which combustion occurs in two stages, i.e. partly at constant volume and partly at constant pressure.

dual-gate MOSFET *(Electronics)*. A *Metal-Oxide Semiconductor Field-Effect Transistor* having two separate gate electrodes. This imparts superior performance for applications such as mixers, amplifiers, and demodulators.

dual in-line package *(Electronics)*. A common integrated-circuit package having two parallel rows of connectors at right angles to the body, as required for insertion into pre-drilled holes in a printed-circuit board.

dual ion *(Chem.)*. See **zwitterion**.

duality *(Arch.,Civ.Eng.)*. The repetition of members in the same angular direction in a structure or building.

dualizing *(Maths.)*. See **polar reciprocation**.

dual modulation *(Telecomm.)*. The modulation of a carrier with two separate types of modulation (e.g. amplitude and frequency), each carrying different information.

dual spectrum *(Print.)*. A method of **document copying** in which the document, in contact with the copy paper, is heated by infrared and an image of the document is formed on the copy paper, which is then cascaded with a powder which adheres to the surface, is fused, and can be used for **small offset**.

dual track *(Acous.)*. Use of two tracks on a magnetic tape, so that recording and subsequent reproduction can proceed along one track and return along the other, thus obviating rewinding.

Duane and Hunt's law *(Phys.)*. The maximum photon energy in an X-ray spectrum is equal to the kinetic energy of the electrons producing the X-rays, so that the maximum frequency, as deduced from quantum mechanics, is eV/h, where V = applied voltage, e = electronic charge and h = Planck's constant.

dubbing *(Build.)*. The operation of filling in hollows in the surface of a wall with coarse stuff, as a preliminary to plastering.

dubbing *(Image Tech.)*. Re-recording to combine two or more sound records, including the replacement of original dialogue by another language. Also the transfer of a magnetic sound record to a photographic one.

Du Bois balance *(Elec.Eng.)*. An instrument used for measuring the permeability of iron or steel rods. The magnetic attraction across an air gap in a magnetic circuit, of which the sample forms a part, is balanced against the gravitational force due to a sliding weight on a beam.

Duchemin's formula *(Aero.)*. An expression giving the normal wind pressure on an inclined area in terms of that on a vertical area. It states that:

$$N = F\frac{2 \sin \alpha}{1 + \sin^2 \alpha}$$

where F = pressure of wind in N/m^2 of vertical surface; a = angle of the inclined surface with the horizontal; N = normal pressure in N/m^2 of inclined surface.

Duchenne-Erb paralysis *(Med.)*. A form of paralysis in which the arm can be neither abducted nor turned outwards nor raised nor flexed at the elbow, as a result of a lesion of the fifth and sixth cervical nerves in the brachial plexus.

Duchenne muscular dystrophy *(Med.)*. A common form of the inherited **muscular dystrophies**, affecting only male children and leading to progressive muscular weakness.

duchess *(Build.)*. A slate, 24 × 12 in. (610 × 305 mm).

duck *(Textiles)*. A plain, bleached cotton or linen cloth, used for tropical suitings. Heavier makes are used for sails, tents and conveyor belting. Similar to *canvas*.

duck board *(Build.)*. A board which has slats nailed across it at intervals and is used as steps in repair work on roofs, or for walking in excavations or valley gutters.

duck cholera *(Vet.)*. An infection of ducks by the bacterium responsible for **fowl cholera**.

duckfoot bend *(Build.)*. See **rest bend**.

duckfoot quotes *(Print.)*. The chevron-shaped quotation marks used by continental printers.

duck viral hepatitis *(Vet.)*. An acute and highly fatal virus disease of ducklings under 3 weeks old, characterized by liver cell necrosis and sudden death. Lower mortality after 4 weeks. Notifiable disease in UK for which an experimental live vaccine has been used.

duct *(Elec.Eng.)*. An air passage in the core or other parts of an electric machine along which cooling air may pass; also called *ventilating duct, cooling duct*.

duct *(Eng.)*. (1) A hole, pipe or channel carrying a fluid, e.g. for lubricating, heating or cooling. (2) A large sheet-metal tube or casing through which air is passed for forced-draught, ventilating or conditioning purposes.

duct *(Print.)*. A reservoir holding the ink in a printing machine. Usually the supply is regulated by a number of screws and by a ratchet.

duct *(Zool.)*. A tube formed of cells: a tubular aperture in a nonliving substance, through which gases and liquids or other substances (such as spermatozoa, ova, spores) may pass. Also *ductus*.

ducted cooling *(Aero.)*. A system in which air is constrained in ducts that convert its kinetic energy into pressure for more efficient cooling of an aero-engine or of its radiator.

ducted fan *(Aero.)*. A gas turbine aero-engine in which part of the power developed is harnessed to a fan mounted inside a duct. Also *turbofan*.

duct height *(Meteor.)*. Height above the earth's surface of the lower effective boundary of a tropospheric radio duct.

ductile cast iron *(Eng.)*. Cast iron in which the free graphite has been induced to form as nodules by adding cerium or magnesium in the molten state which gives a marked increase in ductility. Abbrev. *SG iron* for *Spherulitic Graphite cast iron*.

ductility *(Eng.)*. Ability of metals and alloys to retain strength and freedom from cracks when shape is altered. See **work-hardening**.

ductless glands *(Zool.)*. Masses of glandular tissue which lack ducts and discharge their products directly into the blood; as the lymph glands and the endocrine glands.

ductule *(Zool.)*. A duct with a very narrow lumen: a small duct; the fine terminal portion of a duct.

ductus *(Zool.)*. See **duct**.

ductus arteriosus *(Med.)*. A blood vessel important in foetal development linking the *pulmonary artery* to the *aorta*. It closes at birth. In some cases there is abnormal persistence known as *patent ductus arteriosus* where blood flows from the aorta to the pulmonary artery creating an abnormal shunt.

ductus caroticus *(Zool.)*. In some Vertebrates, a persistent connection between the systemic and carotid arches.

ductus Cuvieri *(Zool.)*. See **Cuvierian ducts**.

ductus ejaculatorius *(Zool.)*. In many Invertebrates, as the Platyhelminthes, a narrow muscular tube forming the lower part of the vas deferens and leading into the copulatory organ.

ductus endolymphaticus *(Zool.)*. In lower Vertebrates and the embryos of higher Vertebrates, the tube by which the internal ear communicates with the surrounding medium.

ductus pneumaticus *(Zool.)*. In physostomous Fish, a duct which connects the gullet with the air-bladder.

duct waveguide *(Phys.)*. Layer in the atmosphere which, because of its refractive properties, keeps electromagnetic radiated (or acoustic) energy within its confines. It is *surface* or *ground-based* when the surface of the earth is one confining plane.

duct width, thickness *(Meteor.)*. Difference in height between the upper and lower boundaries of a tropospheric radio duct.

Dufaycolor *(Image Tech.)*. Obsolete process of additive colour photography and cinematography, using black-and-white emulsion exposed through a three-colour **reseau** on the base.

duff *(Min.Ext.)*. Fine coal too low in calorific value for direct sale.

duffel, duffle *(Textiles)*. A heavy low quality woven woollen fabric raised on both sides. Short coats made from this fabric.

Dühring's rule *(Chem.Eng.)*. If the temperatures at which two chemically similar liquids have the same vapour pressure be plotted against each other, a straight line results, i.e. $t' = \alpha + \beta t$, where t' and t are the boiling points of the two liquids on the same scale of temperature and at the same pressure, and α and β are constants.

duke *(Paper)*. A notepaper size, 178 × 143 mm (7 × $5\frac{5}{8}$ in).

dulcin *(Chem.)*. *Sucrol*, *4-ethoxyphenylurea*. Colourless crystalline substance which is about 200 times as sweet as sugar.

dull *(Med.)*. Not resonant to percussion; said of certain regions of the body, especially the chest.

dull-emitter cathode *(Electronics)*. One from which electrons are emitted in large quantities at temperatures at which incandescence is barely visible. The emitting surface is the oxide of an alkaline-earth metal.

Dulong and Petit's law *(Chem.)*. See **law of Dulong and Petit**.

dulosis *(Zool.)*. Among Ants (*Formicoidea*), an extreme form of social parasitism in which the work of the colony of one species is done by captured 'slaves' of another species called *amazons*. Also *slavery, helotism*.

dumb buddle *(Min.Ext.)*. A buddle without revolving arms or sweeps, for concentrating tin ores.

dumb compass *(Ships)*. See **pelorus**.

dumb iron *(Autos.)*. Forgings attached to the front of the side-members of the frame, to carry the spring shackles and front cross-member, now found only on commercial vehicles.

dummy *(Build.)*. Mallet made by fastening a lump of lead to the end of a cane. Used for straightening out lead pipes.

dummy *(Print.)*. An unprinted volume, generally unbound, made up for the use of publishers in estimating their requirements. Thickness should be measured at the fore-edge and tail.

dummy antenna *(Telecomm.)*. Same as *artificial antenna.*

dummy coil *(Elec.Eng.)*. A coil put on to an armature to preserve mechanical balance and symmetry, but not electrically connected to the rest of the winding.

dummying *(Eng.)*. The preliminary rough-shaping of the heated metal before placing between the dies for drop-forging.

dummy load *(Elec.Eng.)*. One which matches a feeder or transmitter; so designed to absorb the full load without radiation, particularly for testing. Also called *antenna load.*

dummy piston *(Eng.)*. A disk placed on the shaft of a reaction turbine; to one side of it steam pressure is applied to balance the end thrust; sometimes called a *balance piston.*

dummy plates *(Print.)*. Blank, small diameter plates used to fill the cylinder of rotary presses for lock-up.

dummy variable *(Comp.)*. Identifier used for syntax reasons in a program, but which will be replaced by some other variable identifier when the program is executed; a common example is the use of a dummy variable in the definition of a subroutine parameter.

dump *(Comp.)*. To copy the contents of a file, or the contents of the immediate access store, to backing store or to an output device. The output is known as the dump, and may be used to ensure the **integrity** of the data or to assist in program error detection.

dump *(Min.Ext.)*. The heap of accumulated waste material from a metal mine, or of treated tailings from a mill or ore-dressing plant. Also called *tip.*

dump condenser *(Nuc.Eng.)*. Condenser which allows steam from heat exchangers to be bypassed from the turbines in a nuclear-power station, if the electrical load is suddenly taken off.

dumper *(Civ.Eng.)*. A wagon used, in the construction of earthworks, for conveying excavated material on site and dumping it where required.

dumpling *(Civ.Eng.)*. The soil remaining in the centre of an open excavation which is commenced by sinking a trench around the site; the dumpling is removed later.

dump valve *(Aero.)*. An automatic safety valve which drains the fuel manifold of a gas turbine when it stops, or when the fuel pressure fails. Also a large capacity valve to release residual pressure in any fluid system for emergency or operational reasons after landing, or to release all cabin pressure in an in-flight emergency.

dumpy level *(Surv.)*. A type of level in which the essential characteristic is the rigid connection of the telescope to the vertical spindle.

dune *(Geol.)*. An accumulation of sand formed in an area with a prevailing wind. The principal types are: *barchans*, crescent-shaped dunes which migrate in the direction of the point of the crescent; *seifs*, elongated ridges of sand aligned in the wind direction; *transverse dunes*, at right angles to the wind; and *whaleback dunes*, very large elongated dunes. Fossil sand dunes can be recognized, and they indicate desert conditions and wind conditions during past geological periods.

dune bedding *(Geol.)*. A large-scale form of cross-bedding characteristic of wind blown sand dunes.

dungannonite *(Geol.)*. A corundum-bearing diorite containing nepheline, originally described from Dungannon, Ontario.

dungaree *(Textiles)*. A cotton cloth, with a twill weave, made from coloured warp and weft yarns, generally used for men's overalls.

dunite *(Geol.)*. A coarse-grained, deep-seated igneous rock, almost monomineralic, consisting essentially of olivine, though chromite is an almost constant accessory. In several parts of the world (e.g. Bushveld Complex, S. Africa) it contains native platinum and related metals. Named from Mt. Dun, New Zealand.

dunkop *(Vet.)*. See **African horse-sickness**.

dunnage *(Ships)*. Loose wood laid in the hold to keep the cargo out of the bilge-water, or wedged between parts of the cargo to keep them steady.

Dunning process *(Image Tech.)*. Early form of black-and-white composite cinematography : a transparent yellow print of the background scene was used as a **mask** in the camera while shooting the foreground action illuminated by yellow light against a bright uniform blue-violet backing.

duode *(Acous.)*. Electrodynamic open diaphragm loudspeaker driven by eddy currents in a metal former, the **voice coil** being wound over a rubber compliance; the arrangement gives enhanced width of response with damping of diaphragm resonances.

duodecal *(Electronics)*. 12-contact tube base.

duodecimal system *(Maths.)*. A number system whose base is 12.

duodecimo *(Print.)*. See **12 mo**.

duodenal ulcer *(Med.)*. A **peptic ulcer** occurring in the 1st part of the duodenum.

duodenectomy *(Med.)*. Excision of the duodenum.

duodenitis *(Med.)*. Inflammation of the duodenum.

duodenocholecystostomy *(Med.)*. A communication, made by operation, between the duodenum and the gall bladder.

duodenojejunostomy *(Med.)*. A communication surgically made between the duodenum and the jejunum.

duodenum *(Zool.)*. In Vertebrates, the region of the small intestine immediately following the pylorus, distinguished usually by the structure of its walls. adj. *duodenal.*

duolateral coil *(Elec.Eng.)*. See **basket coil**.

duophase *(Elec.Eng.)*. Use of an inductor in output circuit of the active device in an amplifier to obtain a reversed-phase voltage for driving a push-pull stage.

duotone *(Print.)*. Two half-tone plates made from monochrome copy, the key plate at 45° and the tint at 75°, to produce two-tone effect.

dupe *(Image Tech.)*. Abbrev. for *duplicate negative*, prepared from the original camera negative for protection or to incorporate visual effects not originally photographed.

Duperry's lines *(Phys.)*. Lines on a magnetic map showing the direction in which a compass needle points, i.e. the direction of the magnetic meridian.

duplex *(Biol.)*. A double stranded part of a nucleic acid molecule.

duplex *(Comp.)*. Data transmission in both directions simultaneously. Also *full duplex*. Cf. **half duplex**.

duplex balance *(Telecomm.)*. Telegraph name for *line balance.*

duplex burner *(Aero.)*. A gas turbine fuel injector with alternative fuel inlets, but a single outlet nozzle.

duplex chain *(Eng.)*. A roller chain construction using 2 sets of rollers and 3 sets of link plates, used where the chain tension exceeds that which can be transmitted by a simple chain. Its use avoids the need for matching 2 simple chains run side by side.

duplexer *(Telecomm.,Radar)*. In radar, a system which takes advantage of the time delay between transmission of a pulse and reception of its echo to allow the use of the same aerial for transmission and reception. *Transmit-Receive* (TR), or sometimes TR and *Anti-Transmit-Receive* switches are used to isolate the delicate receiver during the high-power pulse transmission. More generally, in radio, any system or network allowing simultaneous transmission and reception on a single aerial, although separation is normally achieved by using different frequencies for transmission and reception.

duplex escapement *(Horol.)*. An escapement in which the escape wheel has two sets of teeth, one for giving impulse and one for locking. The locking teeth are in the plane of the wheel, and lock by pressing against the outside of a hollow cylinder on the axis of the balance staff. This cylinder has a notch to allow the tooth to escape for unlocking. The impulse teeth are raised above the plane of the escape wheel and give impulse to the balance by striking against a finger fixed to the balance staff; impulse is given every alternate vibration. Although capable of giving very satisfactory results, this escapement is now rarely met with in watches, as it is sensitive and liable to set.

duplex lathe *(Eng.)*. A lathe in which two cutting tools are used, one on each side of the work, either to avoid springing of the latter, or to increase the rate of working. See **multiple tool lathe**.

duplex paper *(Paper)*. A deprecated term. A paper or board comprising two noticeably different layers e.g. by reason of colour or furnish.

duplex processes *(Eng.)*. The combination of two alternative methods in performing one operation; as when steel making is carried out in two stages, first in the open hearth and second in the electric furnace.

duplex pump *(Eng.)*. A pump with two working cylinders side by side.

duplex set *(Print.)*. A combination of **barring motor** and main drive motor used to provide all speeds from **inching** and **crawl** to full speed.

duplex winding *(Elec.Eng.)*. A winding for d.c. machines in which there are two separate and distinct windings on the machine, the two being connected in parallel by the brushes.

duplicate feeder *(Elec.Eng.)*. A feeder forming an alternative path to that normally in use.

duplicate plates *(Print.)*. Stereotypes, electrotypes, and rubber and plastic plates made from original plates or from type formes.

duplicating *(Eng.)*. Use of special equipment for machining or forming an object which is a copy of two- or three-dimensional master.

duplication *(Biol.)*. Doubling of a gene or a larger segment of a chromosome, by a variety of mechanisms including unequal crossing-over and fusion between a chromosomal fragment and a whole chromosome of the same sort.

duplicator paper *(Paper)*. Paper intended for use in stencil duplicator machines.

duplicident *(Zool.)*. Having two pairs of incisor teeth in the upper jaw; as Hares and Rabbits.

Dupuit relation *(Powder Tech.)*. A relationship between the apparent linear velocity of a fluid flow through an isotropic powder bed and the actual velocity of flow which depends on the porosity of the powder bed.

Dupuytren's contraction *(Med.)*. Thickening and contraction of the fascia of the palm of the hand, with resulting flexion of the fingers, especially of the ring and little fingers.

durable press *(Textiles)*. A treatment applied to fabrics or garments to make them retain desired creases in normal wearing and washing. Compounds similar to those for *crease-resist finishes* are used.

durain *(Geol.)*. A separable constituent of dull coal; of firm, rather granular structure, sometimes containing many spores.

dura mater *(Zool.)*. A tough membrane lining the cerebro-spinal cavity in Vertebrates.

duramen *(Bot.)*. See **heart wood**.

Duranol dyes *(Chem.)*. Acetate (ethanoate) silk (viscose) dyes, derived from amino-anthraquinones.

durene *(Chem.)*. *1:2:4:5-tetramethylbenzene*, mp 79.3°C, bp 196°C.

duricrust *(Geol.)*. A hard crust formed in or on soil in a semi-arid environment. It is formed by the precipitation of soluble minerals from mineral waters, particularly during the dry season.

Durosier's murmur *(Med.)*. A murmur heard over the femoral artery during diastole of the heart; indicative of aortic valve incompetence.

dust *(Powder Tech.)*. Particulate material which is or has been airborne and which passes a 200 mesh B.S. rest sieve (76 μm).

dust chamber *(Eng.)*. Fume chamber in which dust is arrested as dry furnace gases are filtered, cyclozed or baffled.

dust core *(Phys.)*. Magnetic circuit embracing or threading a high-frequency coil, made of ferromagnetic particles compressed into an insulating matrix binder, thus obviating losses at high frequency because of eddy currents.

dust counter *(Meteor.)*. An instrument for counting the dust particles in a known volume of air.

dust counter *(Min.Ext.)*. Apparatus, usually portable, for collecting dust in mines, for display and check on working conditions underground.

dust cover *(Print.)*. See **jacket**.

dust explosion *(Eng.)*. An explosion resulting from the ignition of small concentrations of inflammable dust (e.g. coal dust or flour) in the air.

dust figure *(Elec.Eng.)*. See **Lichtenberg figure**.

dust monitor *(Nuc.Eng.)*. Instrument which separates airborne dust and tests for radioactive contamination.

dust panel *(Build.)*. A thin board fixed horizontally between two drawers.

dust-proof *(Elec.Eng.)*. Said of a piece of electrical apparatus which is constructed so as to exclude dust or textile flyings.

Dutch barn *(Arch.)*. A simple open structure, generally of mild steel, the roof support being formed of braced trusses, arch shaped, and carried by slender columns spaced normally at about 4 m intervals. The roof covering may be of corrugated iron or asbestos cement.

Dutch bond *(Build.)*. A bond differing from English bond only in the angle detail, the vertical joints of one stretching course being in line with the centre of the stretchers in the next stretching course.

Dutch elm disease *(For.)*. See **elm**.

Dutch gold *(Eng.)*. A cheap alternative to **gold-leaf**, consisting of copper-leaf, which, by exposure to the fumes from molten zinc, acquires a yellow colour.

dutchman *(Build.)*. A piece of wood driven into a gap left in a joint which has been badly cut.

Dutch process *(Chem.)*. A process of making **white lead** by corroding metallic lead in stacks where fermentation of tan or bark is taking place, in the presence of dilute ethanoic acid.

Dutch roll *(Aero.)*. Lateral oscillation of an aircraft, in particular an oscillation having a high ratio of rolling to yawing motion. Dutch roll can be countered by yaw dampers.

duty *(Elec.Eng.)*. The cycle of operations which an apparatus is called upon to perform whenever it is used, e.g. with a motor it is the starting, running for a given period, and stopping; or with a circuit-breaker, it may be closing and opening for a given number of times with given time intervals between; or the prescription of a process timer.

duty factor *(Elec.Eng.)*. (1) The ratio of the equivalent current taken by a motor or other apparatus running on a variable load to the full-load current of the motor (continuous rating). (2) Pulse signals giving ratio of pulse duration to space interval.

DVORAK keyboard *(Comp.)*. One laid out to minimize finger movement.

dwang *(Build.)*. (1) See **bridging**. (2) A mason's term for a crowbar.

dwarfism *(Image Tech.)*. Negative size-distortion in stereoscopic film caused by the camera lens separation being in excess of the normal interocular distance.

dwarf male *(Zool.)*. A male animal greatly reduced in size, and usually in complexity of internal structure also, in comparison with the female of the same species; such males may be free-living but are more usually carried by the female, to which they may be attached by a vascular connection in extreme cases, as some kinds of deep-sea Angler Fish.

dwarf rafter *(Build.)*. A **jack rafter**.

dwarf shoot *(Bot.)*. See **short shoot**.

dwarf star *(Astron.)*. The name given to a low-luminosity star of the **main sequence**. See also **White Dwarf**.

dwell *(Eng.)*. Of a cam, the angular period during which the cam follower is allowed to remain at a constant lift.

dwell *(Plastics)*. Term used to describe the pause in the application of pressure in a moulding press, which allows the escape of gas from the moulding material.

dwell *(Print.)*. The slight pause in the motion of a hand-press or platen when the impression is being made.

Dwight Lloyd machine *(Min.Ext.)*. A type of continuous *sintering* machine characterized by having air drawn down through the burning bed on a travelling grate into 'wind boxes' and used in roasting pyritic ore so that by segregating and recycling the gas streams, concentrations of sulphur dioxide sufficiently high for conversion to sulphuric acid are readily obtained.

Dy *(Chem.)*. The symbol for dysprosium.

dyad *(Biol.)*. Half of a tetrad group of chromosomes passing to 1 pole at meiosis.

dyad *(Maths.)*. A tensor formed from the product of two vectors. For the vectors A and B it is written AB where, unlike scalar and vector multiplication, no multiplication symbol separates the factors.

dyadic *(Maths.)*. The sum of two or more dyads. See **conjugate dyadics**.

Dycril plates *(Print.)*. See **photopolymer plates**.

dye *(Textiles)*. A coloured compound that has **substantivity** for the textile.

dye-coupling process *(Image Tech.)*. One using **couplers**.

dye-destruction *(Image Tech.)*. Colour-printing process in which the tripack layers incorporate dyes of the respective primary colours and the image is differentially bleached in development.

dye laser *(Phys.)*. A laser using an organic dye and excited by a separate laser. It can be tuned over a significant fraction of the visible spectrum by using a reflection grating as one of the cavity mirrors.

dyeline *(Print.)*. A **document-copying** method using **dyeline base paper**, printing down with ultra-violet light which neutralizes diazonium salts in non image areas, and processing with ammonium which develops the image. Sometimes called *diazo*.

dyeline base paper *(Paper)*. Paper with controlled chemical and physical properties to enable it to be coated satisfactorily with a diazo compound and thereafter used to make a dyeline print.

dyenin *(Biol.)*. The protein which forms columns of side arms along the peripheral **microtubules** of cilia and mediates the movement of the microtubules relative to each other, causing the cilia to bend. It has ATPase activity.

dyestuffs *(Chem.)*. Groups of aromatic compounds having the property of dyeing textile fibres, and containing characteristic groups essential to their qualification as dyes. The more important dyestuffs are classified as follows: (a) nitroso- and nitro-dyestuffs, (b) azo-dyes, (c) stilbene, pyrazole and thiazole dyestuffs, (d) di- and triphenylmethane dyes, (e) xanthene dyestuffs, (f) acridine and quinoline dyestuffs, (g) indamine and indophenol dyestuffs, (h) azines, oxazines and thiazines, (i) hydroxyketone dyestuffs, (j) sulphide dyes, (k) vat dyestuffs, indigo and indanthrenes, (l) reactive dyestuffs, (m) **acid dyes**, used in photography.

dye toning *(Image Tech.)*. The chemical process whereby a dye is made to replace the silver in a normal photographic image, or to adhere to it by mordanting.

dye transfer *(Image Tech.)*. Colour print process using three gelatine matrices taken from separation negatives and processed with a tanning developer. These are treated with dye of the appropriate colour and the images transferred in register on to paper.

dying shift *(Min.Ext.)*. Night (graveyard) shift.

Dykanol *(Elec.Eng.)*. TN for dielectric material used in paper capacitors.

dyke *(Geol.)*. A sheet-like body injected into the crust, with discordant contacts with the host rock. Dykes are most numerous in areas of volcanic activity. Characteristically they are a few metres wide but may extend for hundreds of kilometres.

dyke *(Hyd.Eng.)*. A wall or embankment of timber, stone, concrete, fascines or other material, built as training works for rivers, to confine flow rigidly within definite limits over the length treated.

dyke phase *(Geol.)*. That episode in a volcanic cycle characterized by the injection of minor intrusions, especially dykes. The dyke phase usually comes after the major intrusions, and is the last event in the cycle.

dyke swarm *(Geol.)*. A series of dykes of the same age, usually trending in a constant direction over a wide area. Occasionally, dykes may radiate outwards from a volcanic centre, as the Tertiary dyke swarm in Rum, Scotland; but usually they are parallel; e.g. the O.R.S. dyke swarm of S.W. Scotland, of which the trend is north-east to south-west.

dynamical stability *(Ships)*. The work done, usually in ft-tons or megajoules, when a vessel is heeled to a particular angle. See **heel**.

dynamical theory of X-ray and electron diffraction *(Phys.)*. A theoretical approach that takes into account the dynamical equilibrium between the incident and diffracted beams in a crystal, e.g. the effect of interference between the incident beam and multiply diffracted beams.

dynamic balance *(Aero.)*. The condition wherein centrifugal forces due to any rotating mass, e.g. a propeller, produce neither couple nor resultant force in the shaft and hence a reduction of vibration.

dynamic balancing *(Acous.)*. The technique of balancing the centrifugal forces in rotating machines so that there is no residual unbalance, and consequent vibration, to give rise to noise.

dynamic characteristic *(Elec.Eng.)*. Any characteristic curve obtained under normal working conditions for the device under question, e.g. the collector current (collector/emitter voltage relationship for a bipolar transistor when the effect of load impedance is included).

dynamic damper, detuner *(Eng.)*. Supplementary rotating mass driven through springs attached to a crankshaft at a point remote from the node to eliminate a troublesome critical speed.

dynamic electricity *(Elec.Eng.)*. A term sometimes used to denote electric currents, i.e. electric charges in motion, as opposed to static electricity, in which the charges are normally stationary. Also *current electricity*.

dynamic heating *(Aero.)*. Heat generated at the surface of a fast-moving body by the bringing to rest of the air molecules in the boundary layer either by direct impact or by viscosity.

dynamic isomerism *(Chem.)*. See **tautomerism**.

dynamic loudspeaker *(Acous.)*. Open diaphragm, driven by a *voice coil* on a former; intended to be used in a plane (Rice-Kellog) baffle, from the side of a box (box baffle), or at the neck of a large horn (flare).

dynamic memory *(Comp.)*. Needs to be periodically *refreshed*, i.e. read and rewritten every 2 ms or so, as stored charge tends to leak. Cf. **static-**. See also **volatile-**.

dynamic model *(Aero.)*. A free-flight aircraft model in which the dimensions, inertia and masses are such as to duplicate full-scale behaviour.

dynamic noise suppressor *(Acous.)*. One which automatically reduces the effective audio band-width, depending on the level of the required signal to that of the noise.

dynamic pressure *(Aero.)*. The pressure resulting from the instantaneous arresting of a fluid stream, the difference between total and static pressure.

dynamic psychology *(Behav.)*. Refers to a school of thought that assumes the primary importance of inner and subjective mental and emotional events in the explanation of behaviour. See **psychodynamics**.

dynamic range *(Image Tech.)*. The range between the maximum signal which can be satisfactorily handled and the minimum inherent noise level of the system.

dynamic range *(Telecomm.)*. The full range of signal levels, from highest to lowest, contained in any signal, transmission or recording; normally expressed in **decibels**. Needs to be assessed in most forms of electronic design.

dynamic resistance *(Elec.Eng.)*. The relationship between voltage and current at a given position on the non-linear static characteristic of a device in an electrical circuit, e.g. a diode. It may be regarded as the tangent to the characteristic curve at that point and is often assumed to be constant over a small range of voltage and current.

dynamics *(Maths.)*. That branch of applied mathematics which studies the way in which force produces motion.

dynamic sensitivity *(Elec.Eng.)*. The alternating component of the output of a photoelectric device divided by the alternating component of incident radiant flux.

dynamic viscosity *(Phys.,Eng.)*. See **coefficient of viscosity**.

dynamite *(Chem.)*. A mixture of nitroglycerine with

kieselguhr, wood pulp, starch flour etc., making the nitroglycerine safe to handle until detonated. The most common industrial high explosive.

dynamo *(Elec.Eng.)*. Electromagnetic machine which converts mechanical energy into a.c. or d.c. electrical supply. See **alternator**.

dynamo-electric amplifier *(Elec.Eng.)*. Low- and zero-frequency mechanically rotating armature in a controlled magnetic field; used in servo systems.

dynamometer *(Elec.Eng.)*. Instrument for measurement of supply torque. *Electric dynamometers* can be used for measurement of a.c. current, voltage, or power. See **Siemens-**.

dynamometer *(Eng.)*. A machine for measuring the brake horse-power of a prime-mover or electric motor. See **absorption-**, **electric-**, **transmission-**, **Froude brake**, **rope brake**.

dynamometer ammeter *(Elec.Eng.)*. An ammeter operating on the dynamometer principle, the fixed and moving coils being connected in series and carrying the current to be measured.

dynamometer voltmeter *(Elec.Eng.)*. A voltmeter operating on the dynamometer principle, the fixed and moving coils being connected in series, and in series with a high resistance across the voltage to be measured.

dynamometer wattmeter *(Elec.Eng.)*. A commonly used type of wattmeter operating on the dynamometer principle, the fixed coil being usually in the current circuit and the moving coil in the voltage circuit.

dynamothermal metamorphism *(Geol.)*. A regional metamorphism involving both heat and pressure.

dynamotor *(Elec.Eng.)*. An electric machine having two armature windings, one acting as a generator and the other as a motor, but only a single magnet frame. Also called a *rotary transformer*.

dynatron *(Electronics)*. Circuit in which steady-state oscillations are set up in a tuned circuit between screen and anode of a tetrode, the latter exhibiting negative resistance when the anode potential is below the potential of the screen.

dyne *(Phys.)*. The unit of force in the CGS system of units. A force of one dyne acting on a mass of 1 g, imparts to it an acceleration of 1 cm/s^2. Approximately 981 dynes are equal to 1 g weight. 10^5 dynes = 1 newton.

dynode *(Electronics)*. Intermediate electrode (between cathode and final anode) in photomultiplier or electron multiplier tube. Dynode electrons are those which emit secondary electrons and provide the amplification.

dynode chain *(Electronics)*. Resistance potential divider employed to supply increasing potentials to successive dynodes of an electron multiplier.

dys- *(Genrl.)*. Prefix from Gk. *dys-*, in English mis-, un-.

dysadaptation *(Med.)*. Marked reduction in rapidity of adaptation of the eye to suddenly reduced illumination, as in vitamin A deficiency.

dysarthria *(Med.)*. Difficult articulation of speech, due to a lesion in the brain.

dysbasia angiosclerotica *(Med.)*. Pain in the legs on walking, due to thickening of the arteries. See **intermittent claudication**.

dyschezia *(Med.)*. A form of constipation in which the faeces are retained in the rectum, as a result of blunting of a normal reflex, due to faulty habits.

dyscrasia *(Med.)*. Any disordered condition of the body, especially of the body fluids.

dyscrasite *(Min.)*. Silver ore consisting mainly of a silver antimonide, Ag_3Sb.

dysdiadokokinesia *(Med.)*. Inability to perform rapid alternate movements as a result of a lesion in the cerebellum.

dysentery *(Med.)*. A term formerly applied to any condition in which inflammation of the colon was associated with the frequent passage of bloody stools. Now confined to (1) *bacillary dysentery*, due to infection with *Bacterium dysenteriae*; (2) *amoebic dysentery*, the result of infection with the *Entamoeba histolytica*.

dysgenic *(Zool.)*. Causing, or tending towards, racial degeneration.

dysgraphia *(Med.)*. Inability to write, as a result of brain damage or other cause.

dyskinesia *(Med.)*. A term applied to any one of a number of conditions characterized by involuntary movements which follow a definite pattern, e.g. tics.

dyslalia *(Med.)*. Articulation difficulty due to defects in speech organs.

dyslexia *(Med.)*. *Word blindness*; great difficulty in learning to read, write, or spell, which is unrelated to intellectual competence and of unknown cause.

dysmelia *(Med.)*. Misshapen limbs.

dysmenorrhoea, dysmenorrhea *(Med.)*. Painful and difficult menstruation.

dysmetria *(Med.)*. Faulty estimation of distance in the performance of muscular movements, due to a lesion in the cerebellum.

dysostosis *(Med.)*. Defect in the normal ossification of cartilage.

dyspareunia *(Med.)*. Painful or difficult coitus.

dyspepsia *(Med.)*. Indigestion: any disturbance of digestion.

dysphagia *(Med.)*. Difficulty in swallowing.

dysphasia *(Med.)*. Disturbed utterance of speech due to a lesion in the brain.

dysphonia *(Med.)*. Difficulty of speaking, due to any affection of the vocal cords.

dysphoria *(Med.)*. Unease; absence of feeling of well-being.

dysplasia *(Med.)*. Abnormality of development.

dyspnea, dyspnoea *(Med.)*. Laboured or difficult respiration.

dysprosium *(Chem.)*. A metallic element, a member of the rare-earth group. Symbol Dy, at. no. 66, r.a.m. 162.5.

dyssynergia *(Med.)*. Incoordination of muscular movements, due to disease of the cerebellum.

dystectic mixture *(Chem.)*. A mixture with a maximum melting-point. Cf. **eutectic**

dystocia, dystokia *(Med.)*. Painful or difficult childbirth.

dystrophia adiposogenitalis *(Med.)*. *Fröhlich's syndrome*. A condition characterized by obesity, hairlessness of the body, and underdeveloped genital organs, due to disordered function of the pituitary gland.

dystrophia myotonica *(Med.)*. See **muscular dystrophy**.

dystrophic *(Ecol.)*. Said of a lake in which the water is rich in organic matter, such as humic acid, but this consists mainly of undecomposed plant fragments, and nutrient salts are sparse.

dystrophy *(Med.)*. A condition of impaired or imperfect nutrition, as in **muscular dystrophy**.

dysuria, dysury *(Med.)*. Painful or difficult passage of urine.

E

e *(Build.)*. Symbol for eccentricity of a load.
e *(Phys.)*. Symbol for the electron (e^-) or positron (e^+).
***e*, e, ε** *(Maths.)*. (1) A transcendental number, the base of natural (Napierian) logarithms, defined as the limiting value of $\left(1+\frac{1}{m}\right)^m$ as *m* approaches infinity; $e = 2.718\,281\,828\,5$ ... to ten places of decimals. (2) Eccentricity of a **conic**.
e *(Phys.)*. Symbol for the elementary charge; 1.6022×10^{-19}coulomb.
ε *(Chem.)*. Symbol for molar extinction coefficient.
ε *(Phys.)*. Symbol for (1) emissivity, (2) linear strain, (3) permittivity.
ε- *(Chem.)*. (1) Substituted on the fifth carbon atom. (2) *epi-*, i.e. containing a condensed double aromatic nucleus substituted in the 1.6 positions. (3) *epi-*, i.e. containing an intramolecular bridge.
η *(Chem.,Elec.Eng.)*. A symbol for electrolytic polarization, overvoltage.
η *(Phys.)*. Symbol for co-efficient of viscosity.
E *(Chem.Eng.)*. Symbol for eddy diffusivity.
E, EA, EAC *(Immun.)*. Abbrevs. for *Erythrocyte*, *Erythrocyte with Antibody on its surface*, and *Erythrocyte with Antibody and Complement*; the latter describing components which have become attached following activation (e.g. EAC 1423). Such red cells are used to detect Fc receptors or complement receptors on other cells.
E *(Phys.)*. Symbol for (1) potential difference, esp. electromotive force of voltaic cells, (2) (with subscript) single electrode potential, thus E_H, on the hydrogen scale; E_O, standard electrode potential.
E- *(Chem.)*. Prefix denoting 'on the opposite side' (Ger. *entgegen*), and roughly equivalent to *trans-*. See **Cahn-Ingold-Prelog system**.
E *(Phys.)*. Symbol for (1) electromotive force, (2) electric field strength, (3) energy, (4) illumination, (5) irradiance.
[E] *(Phys.)*. One of the Fraunhofer lines in the green of the solar spectrum. Its wavelength is 526.9723 nm, and it is due to iron.
Eagle mounting *(Phys.)*. A compact mounting of a concave diffraction grating based on the principle of the **Rowland circle**. The mounting suffers from less astigmatism than either the Rowland or **Abney mountings**, and is useful for studying higher order spectra.
EAN *(Comp.)*. *European Academic Network*. A communications network for the European research community.
EAN networks *(Comp.)*. A group of networks in Europe which share the same **protocol** for communication.
ear *(Build.)*. A **crossette**.
ear *(Eng.)*. A cast or forged projection integral with, or attached to, an object, for supporting it, or attaching another part to it pivotally; also called *lug*.
ear *(Zool.)*. Strictly, the sense-organ which receives auditory impressions; in Insects, various tympanic structures on the thorax or forelegs in some Birds and Mammals, a prominent tuft of feathers or hair close to the opening of the external auditory meatus; in Mammals, the pinna; more generally, any ear-like structure.
ear defenders *(Acous.)*. Plugs or muffs of various materials for insertion into, or fitting over, the ear to reduce reception of noise.
ear drum *(Med.)*. The outer termination of the aural mechanism of the ear, consisting of a membrane, in tension, for transferring the acoustic pressures applied from without to the ossicles for transmission to the inner ear.
earing *(Eng.)*. Excessive elongation along edges and folds of metals being shaped by deep-drawing or rolling.
Early Bird *(Telecomm.)*. The first (1964) **communications satellite** providing regular commercial telecommunications.
Early effect *(Electronics)*. Variation of junction capacitance and effective base thickness of transistor with the supply potentials.
early replicating regions *(Biol.)*. Parts of chromosomes which are replicated early in **S-phase**.
early-warning radar *(Radar)*. A system for the detection of approaching aircraft or missiles at greatest possible distances. See **BMEWS**.
early wood *(Bot.)*. *Spring wood*. The wood formed in the first part of a growth layer during the spring, typically with larger cells with thinner cell walls than the **late wood**.
ear muffs *(Acous.)*. (1) Pads of rubber or similar material which are placed on head telephone receivers to minimize discomfort during long use. (2) **Ear defenders**.
EARN *(Comp.)*. The *European Academic Research Network*. A European-wide network which forms part of **BITNET**. It was supported by funding from IBM Corporation.
earphone coupler *(Acous.)*. A suitably-shaped cavity with an incorporated microphone used for acoustical testing of earphones.
earphones *(Acous.)*. Electroacoustic transducers which transform electric signals into acoustic signals and are worn over the ear.
Earth *(Astron.)*. Third planet from the Sun and largest of the inner planets, with a mean equatorial radius 6378.17 km. Mass 5.977×10^{24} kg. Mean density 5.517. Earth has an *atmosphere* of oxygen and nitrogen, and an extensive **magnetosphere**. Internally there are three distinct regions. The *crust* is the outermost layer, about 30 km thick under the continents, consisting mainly of sedimentary rocks resting on ancient igneous rocks, principally basalt and granite. Beneath is the *mantle* reaching to nearly 3000 km, composed of silicate rocks. The central *core* is about 10 times denser than water and is probably composed of nickel and iron. In part at least the core is liquid, and this is the seat of the Earth's magnetic field. At the centre the temperature is 6400 K.
earth *(Phys.)*. Connection to main mass of earth by means of a conductor having a very low impedance.
earth *(Telecomm.)*. (1) A conducting path (intentional or otherwise) between an electrical circuit and earth or some conductor serving in its place. (2) A system of plates or wires buried in the ground to allow a path to earth for currents flowing in an antenna, used largely to improve the efficiency of broadcast and short-wave antenna.
earth closet *(Build.)*. A metal-lined receptacle, usually placed beneath a pierced seat, for receiving human excreta, the latter being deodorized by covering with dry earth, sometimes mixed with chemicals.
earth coil *(Elec.Eng.)*. A pivoted coil of large diameter for measuring the strength of the earth's magnetic field; this is done by suddenly changing the position of the coil in this field and observing the throw of a ballistic galvanometer connected to it. Also called an *earth indicator*.
earth colours *(Build.)*. Pigments prepared from natural earths, mostly oxides and silicates, e.g. ochres, umbers, and siennas.
earth continuity conductor *(Elec.Eng.)*. A third conductor (with line and neutral) in a mains distribution system, bonded to earth, provided for connection to any metal component not in the electrical circuit; for safety purposes. Abbrev. *ECC*.
earth currents *(Elec.Eng.)*. (1) Currents in the earth

which, by electromagnetic induction, cause irregular currents to flow in submarine cables and so interfere with the reception of the transmitted signals. (2) Direct currents in the earth, which are liable to cause corrosion of the lead sheaths of cables; they can be the earth-return currents of power systems. (3) Any current flowing in an item of equipment through that which is grounded or at earth potential.

earth dam *(Civ.Eng.)*. One built of local gravels, earth etc. with impervious clay core which rests on firm ground to avoid undermining.

earth detector *(Elec.Eng.)*. See **leakage indicator**.

earthed aerial *(Telecomm.)*. Marconi aerial, in which an elevated wire is earthed at its lower end.

earthed circuit *(Elec.Eng.)*. An electric circuit which is intentionally connected to earth at one or more points.

earthed concentric wiring system *(Elec.Eng.)*. A 2-wire system for wiring or general distribution, which uses twin-concentric conductors, the outer conductor being earthed.

earthed neutral *(Elec.Eng.)*. A neutral point of a polyphase system or piece of electrical apparatus which is connected to earth, either directly or through a low impedance.

earthed pole *(Elec.Eng.)*. The pole or line of an earthed circuit which is connected to earth.

earthed switch *(Elec.Eng.)*. A switch, used in wiring installations, in which provision is made for earthing all exposed metal parts. Also called a *Home-Office switch*.

earthed system *(Elec.Eng.)*. A system of electric supply in which one pole or the neutral point is earthed, either directly or through a low impedance, the former being known as a solidly earthed system.

earth electrode system *(Elec.Eng.)*. The totality of conductors, conduits, shields, and screens which are connected to the main earth by low-impedance conductors.

earth fault *(Elec.Eng.)*. An accidental connection between a live part of an electrical system and earth.

earth (ground) capacitance *(Phys.)*. The capacitance between an electrical circuit and earth (or conducting body connected to earth).

earth impedance *(Phys.)*. The impedance as normally measured, with all extraneous electromotive forces reduced to zero, between any point in a communicating system or a measuring circuit and earth.

earth inductor *(Elec.Eng.)*. See **earth coil**.

earthing autotransformer *(Elec.Eng.)*. See **earth reactor**.

earthing resistor *(Elec.Eng.)*. A resistance through which the neutral point of a supply system is earthed, in order to limit the current which flows on the occurrence of an earth fault. Also called *earthing resistance*.

earthing tyres *(Aero.)*. Tyres for aircraft having an electrically conductive surface in order to discharge static electricity upon landing.

earth lead *(Telecomm.)*. Connection between a radio transmitting or receiving apparatus and its earth.

earth leakage protection *(Elec.Eng.)*. Protection system, suitable for domestic use, in which imbalance between live and neutral currents is used to trip a circuit breaker, thus isolating the supply in the event of an earth fault. See **balanced protective system**.

earth-pillars *(Geol.)*. These occur where sediments consisting of relatively large and preferably flat stones, embedded in a soft, finer-grained matrix, are undergoing erosion, especially in regions of heavy rainfall. As the ground is progressively lowered the flat stones protect the softer material beneath them and are thus left standing on tall, acutely conical pillars.

earth plate *(Elec.Eng.)*. A metal plate buried in the earth for the purpose of providing an electrical connection between an electrical system and the earth.

earth potential *(Phys.)*. The electric potential of the earth; usually regarded as zero, so that all other potentials are referred to it.

earth pressure *(Civ.Eng.)*. The pressure exerted on a wall by earth which is retained, i.e. supported laterally by the wall.

earthquake *(Geol.)*. A shaking of the earth's crust caused, in most cases, by displacement along a fault. The place of maximum displacement is the *focus* (cf. **epicentre**). Although the amount of the displacement may be small, a matter of inches only, the destruction wrought at the surface may be very great, due in part to secondary causes, e.g. the severing of gas mains and water mains, as in the great San Francisco earthquake in 1906. See **seismology**.

earthquake intensity *(Geol.)*. The measurement of the effects of an earthquake at a particular place. One earthquake has one magnitude but different intensities. See **earthquake magnitude, Mercalli scale**.

earthquake magnitude *(Geol.)*. A measure of the strength of an earthquake as determined by instruments, expressed on a scale called the *Richter scale*. See **earthquake intensity**.

earth reactor *(Elec.Eng.)*. (1) A reactor connected between the neutral point of an a.c. supply system and earth, in order to limit the earth current which flows on the occurrence of an earth fault. (2) An arrangement of reactors or transformers, so connected to a polyphase system that a neutral point is artificially obtained. Also called *earthing autotransformer, neutral compensator, neutral autotransformer, neutralator*.

earth resistance *(Elec.Eng.)*. The resistance offered by the earth between two points of connection, and therefore forming a coupling between all circuits making use of the same current path in the earth.

earth-return circuit *(Elec.Eng.)*. One which comprises an insulated conductor between two points, the circuit being completed through the earth.

earth science *(Geol.)*. A term frequently used as a synonym for geology. Has also been used to include sciences which fall outside the scope of geology, e.g. meteorology.

earthshine *(Astron.)*. Close to new moon, the entire disc of the Moon is often bathed in a faint light, which is sunlight reflected from the Earth. Also called *ashen light*, or picturesquely *the old moon in the new moon's arms*.

earth's magnetic field *(Elec.Eng.)*. See **terrestrial magnetism**.

Earth station *(Telecomm.)*. A single transmitter and steerable antenna, or a site housing several of these, with the sole purpose of transmitting to or receiving from communication and direct broadcast satellites, and linking them to terrestrial communications networks.

earth system *(Telecomm.)*. See **earth**.

earth terminal *(Elec.Eng.)*. A terminal provided on the frame of a machine or piece of apparatus to make a connection to earth.

earth thermometer *(Meteor.)*. A thermometer used for measuring the temperature of the earth at depths up to a few metres. *Symon's earth thermometer* (the most commonly used) consists of a mercury thermometer, with its bulb embedded in paraffin wax, suspended in a steel tube.

earthware duct *(Elec.Eng.)*. A conduit made of earthenware, for carrying underground cable.

earthwork *(Civ.Eng.)*. A bank or cutting.

earthy *(Elec.Eng.)*. Said of (1) circuits when they are connected to earth, either directly (as for direct currents) or through a capacitor (with alternating currents); (2) any point in a communicating system (e.g. the midpoint of a shunting resistance across a balanced line) which is at earth potential, although not actually connected to earth through zero impedance; (3) points of a bridge when reduced to earth potential by a Wagner earth.

earthy cobalt *(Min.)*. A variety of wad containing up to about 40% of cobalt oxide. Also called *asbolite*.

EAS *(Aero.)*. Abbrev. for *Equivalent AirSpeed*.

easement curve *(Surv.)*. See **transition curve**.

easer *(Print.)*. A general term for additives mixed with printing ink to produce some particular result, e.g. quicker drying, reduction of tack.

easing *(Build.)*. The shaping of a curve so that there is no abrupt change in curvature, particularly in handrails.

easing centres *(Build.,Civ.Eng.)*. The process of gradually removing the centring from beneath a newly completed arch, thereby transferring its weight slowly to the arch abutments.

easing wedges *(Civ.Eng.)*. **Striking wedges**.

East Coast fever *(Vet.)*. A disease of cattle in Africa due to infection by the protozoan *Theileria parva*; characterized by fever, enlarged lymph glands, respiratory distress, diarrhoea, and loss of condition. Transmitted by the tick *Rhipicephalus appendiculatus*.

East Indian satinwood *(For.)*. A hardwood timber with a narrowly interlocked grain, and a fine and uniform texture. From *Chloroxylon swietenia*, a native of India and Ceylon. It is naturally durable when used in exposed positions, but is regarded as a cabinet wood.

easting *(Surv.)*. Departure, or displacement, along an east-west line of a point with reference to a true N-S datum.

easy-care *(Textiles)*. See **drip-dry**.

eave *(Build.)*. The lower part of a roof which projects beyond the face of the walls.

eave-board *(Build.)*. See **tilting fillet**.

eave-lead *(Build.)*. A lead gutter behind a parapet, around the edge of a roof.

eaves course *(Build.)*. See **doubling course**.

eaves fascia *(Build.)*. See **fascia (2)**.

eaves gutter *(Build.)*. A trough fixed beneath an eave to catch and carry away the rain flow from the roof. Also called *shuting*.

eaves plate *(Build.)*. A beam carried on piers or posts and supporting the feet of roof rafters in cases where there is no wall beneath.

eaves soffit *(Build.)*. The horizontal surface beneath a projecting eave.

ébauche *(Horol.)*. A partly finished watch movement, consisting of the dial plate, bridges and balance cock.

EBCDIC *(Comp.)*. See **character code**.

E-bend *(Telecomm.)*. A smooth bend in the axis direction of a waveguide, the axis being maintained in a plane parallel to the polarization direction.

Eberhard effect *(Image Tech.)*. Border effect in the developing of a heavy photographic image, showing higher density at the edges than in the centre.

EBM *(Aero.,Eng.)*. Abbrev. for *Electron Beam Machining*.

EBNA *(Immun.)*. *Epstein-Barr virus Nuclear Antigen*. Antigen detected in the nuclei of B-cells and tumour cells in conditions associated with infection by the Epstein Barr virus, such as *infectious mononucleosis, Burkitt's lymphoma* and *nasopharyngeal carcinoma*.

ebola disease *(Med.)*. A severe and often fatal viral disease causing one form of African haemorrhagic fever. The other is caused by the *Marburg virus* which is closely related.

ebonite *(Chem.)*. A hard insulating material of rubber which has been vulcanized, i.e. the latex molecules have been cross-linked through sulphur atoms.

ebony *(For.)*. Heavy hardwood from a tree of the genus *Diospyros*, a native of W. Africa, India and S.E. Asia. It is naturally resistant to most diseases and insects. Typical uses are for making brush backs, piano keys etc.

ebony Sindanyo *(Elec.Eng.)*. See **Sindanyo**.

EBR *(Image Tech.)*. *Electron Beam Recording*, a system for transferring a video picture to photographic motion picture film, the scanning modulated beam exposing the film frame in a vacuum enclosure.

EBU *(Genrl.)*. Abbrev. for *European Broadcasting Union*.

ebullator *(Phys.)*. A heated surface used to impart heat to a fluid in contact.

ebulliometer *(Phys.)*. A device which enables the true boiling point of a solution to be determined.

ebullioscopy *(Chem.)*. The determination of the molecular weight of a substance by observing the elevation of the boiling-point of a suitable solvent.

ebullition *(Phys.)*. See **boiling**.

eburnation *(Med.)*. Ivorylike hardening of bone which occurs at the joint surfaces in osteoarthritis.

EBW *(Aero.,Eng.)*. Abbrev. for *Electron Beam Welding*.

ECAC *(Aero.)*. Abbrev. for *European Civil Aviation Conference*.

ecad *(Bot.)*. A species with distinctive forms which depend simply on the environment rather than on genotypic differences. Cf. **ecotype**.

ECC *(Elec.Eng.)*. **Earth continuity conductor**.

eccentric *(Eng.)*. (1) Displaced with reference to a centre; not concentric. (2) A crank in which the pin diameter exceeds the stroke, resulting in a disk eccentric to the shaft; used as a crank, particularly for operating steam-engine valves, pump plungers etc.

eccentric angle *(Maths.)*. Of a point P on an ellipse, the angle θ, where $(a \cos \theta, b \sin \theta)$ are the parametric co-ordinates of P referred to the axes of the ellipse.

eccentric fitting *(Build.)*. A fitting in which the centre line is offset.

eccentric groove *(Acous.)*. See **locked groove**.

eccentricity *(Build.)*. The distance from the centre of application of a load or system of loads to the centroid of the section of the structural member to which it or they are connected. Symbol e.

eccentricity *(Maths.)*. See **conic**.

eccentric load *(Build.)*. A load which is carried by a structural member at a point other than the centroid of the section.

eccentric pole *(Elec.Eng.)*. A pole on an electric machine in which the pole face is not concentric with the armature but has a greater air gap at one pole tip than at the other to assist in neutralizing the effect of armature reaction.

eccentric sheave *(Eng.)*. The disk of an **eccentric**, often formed integral with the shaft.

eccentric station *(Surv.)*. One not physically occupied during triangulation, etc., but serving as a fixation point.

eccentric strap *(Eng.)*. A narrow split bearing, fitting on to an eccentric sheave and bolted to the end of a valve rod, etc.; corresponds to the 'big end' of a connecting-rod.

eccentric throw-out *(Eng.)*. A device for engaging the back gear of a lathe. The back gear shaft runs in eccentric-bored bearings, which are rotated to bring the gears in and out of mesh with those on the mandrel. See **back gear**.

ecchondroma *(Med.)*. A tumour composed of cartilage and growing from the surface of bone.

ecchondrosis *(Med.)*. An abnormal outgrowth of the joint cartilage, in chronic arthritis.

ecchymosis *(Med.)*. A large discoloured patch due to extravasion of blood under the skin.

Eccles-Jordan circuit *(Electronics)*. Original bistable multivibrator using two triodes or transistors. See **flip-flop**.

eccrine *(Med.)*. Said of a gland whose product is excreted from its cells.

ecdemic *(Zool.)*. Foreign; not *indigenous* or *endemic*.

ecdysis *(Zool.)*. The act of casting off the outer layers of the integument, as in *Arthropoda*.

ECFA *(Immun.)*. *Eosinophil Chemotactic Factor of Anaphylaxis*. A peptide released from mast cells which causes eosinophils to move into the site from the bloodstream. Local accumulation of eosinophil granulocytes takes place where type-I allergic reactions occur.

ECG *(Med.)*. Abbrev. for *ElectroCardioGram*. See **electrocardiograph**.

echelon grating *(Phys.)*. A form of interferometer resembling a flight of glass steps, light travelling through the instrument in a direction parallel to the treads of the steps. The number of interfering beams is therefore equal to the number of steps. Owing to the large path difference, $t(\mu - 1)$, where t is the thickness of a step and μ is the **index of refraction**, the order of interference and therefore the resolving power are high, making the instrument suitable for studying the fine structure of spectral lines.

echin, echino *(Bot.)*. Prefix meaning *spiny*.

echinococcosis *(Vet.)*. An infection of sheep, pigs and cattle, and sometimes man, by the intermediate hydatid stage of the tapeworm *Echinococcus granulosus*. The adult worm occurs in dogs and other carnivora.

echinococcus *(Zool.)*. A bladderworm possessing a well-developed bladder containing daughter bladders, each with numerous scolices.

Echinodermata *(Zool.)*. A phylum of radially symmetrical marine animals, having the bodywall strengthened by calcareous plates; there is a complex coelom; locomotion is usually carried out by the tube-feet, which are distensible finger-like protrusions of a part of the coelom known as the water-vascular system; the larva is bilaterally symmetrical and shows traces of metamerism. Starfish, Sea Urchins, Brittle Stars, Sea Cucumbers, and Sea Lilies.

Echinoidea *(Geol.)*. Fossil echinoids are found in strata ranging from the Lower Palaeozoic to the present. They are particularly important in the Jurassic (Clypeus Grit etc.) and Cretaceous, where, in the Chalk, they have proved useful indices of horizon, especially the various species of *Micraster* and *Holaster*.

Echinoidea *(Zool.)*. A class of Echinodermata having a globular, ovoid, or heart-shaped body which is rarely flattened; there are no arms; the tube-feet possess ampullae and occur on all surfaces, but not in grooves; the anus is aboral or lateral, the madreporite aboral; there is a well-developed skeleton; free-living forms. Sea Urchins.

echinus *(Arch.)*. An ornament in the shape of an egg carved on a moulding, etc.

Echiuroidea *(Zool.)*. A phylum of sedentary marine worm-like animals, in which nearly all trace of metamerism has been lost in the adult; the body is sac-shaped, and feeding is effected by an anterior non-retractile proboscis, bearing a ciliated groove leading to the mouth.

echo *(Acous.)*. Received acoustic wave, distinct from a directly received wave, because it has travelled a greater distance due to reflection.

echo *(Comp.)*. Data transmission in which data is returned to the point of origin for comparison with the original data.

echo *(Radar)*. Return signal in radar, whether from wanted object, or from side or back lobe radiation.

echo *(Telecomm.)*. The reception of a signal additional to, and later than, the desired signal; caused by reflection from hills, etc., or travel completely round the earth.

echo box *(Radar)*. Adjustable test resonator of high Q for returning a signal to the receiver from the transmitter.

echocardiography *(Med.)*. Examination of the structure and function of the heart using reflected pulsed ultrasound.

echo chamber *(Acous.)*. Same as *reverberation chamber*.

echo flutter *(Radar)*. A rapid sequence of reflected radar (or sound) pulses arising from one initial pulse.

echographia *(Med.)*. Ability to copy writing, associated with inability to express ideas in writing, due to a lesion in the brain.

echoic memory *(Behav.)*. Refers to the brief retention of auditory information, in an unprocessed or *echo* form; fades within 2–6 seconds. Cf. **iconic memory**.

echolalia *(Behav.,Med.)*. Aimless repetition of words heard without regard for their meaning, occurring in disease of the brain or in insanity; often seen in catatonic schizophrenia and autistic children.

echolocation *(Behav.)*. Means of locating objects in conditions of poor visibility; involves the production of high frequency sounds, and the detection of their echoes.

echopraxia, echopraxis *(Med.)*. Imitation by an insane person of postures or of movements of those near him; commonly present in the catatonic type of schizophrenia.

echo ranging sonar *(Acous.)*. Determination of distance and direction of objects, such as submarines, by the reception of the reflection of a sound pulse under water. See **asdic**.

echo sounding *(Acous.)*. Use of echoes of pressure waves sent down to the bottom of the sea and reflected, the delay between sending and receiving times giving a measure of the depth; used also to detect wrecks and shoals of fish.

echo studio *(Acous.)*. An enclosure of long reverberation time, used for the artificial introduction of an adjustable degree of reverberation in the main channel of a broadcast programme.

echo suppression *(Telecomm.)*. In telephone 2-way circuits, the attenuation of echo currents in one direction which is due to telephone currents in the other direction.

ECHO viruses *(Med.)*. *Entero Coxsackie Human Orphan viruses*, a group of Coxsackie viruses, often responsible for enteritis.

eckermannite *(Min.)*. A rare monoclinic alkali amphibole; a hydrous sodium lithium aluminium magnesium silicate.

eclampsia *(Med.)*. A term now restricted to the acute toxaemia occurring in pregnancy, parturition, or in the puerperium, associated with hypertension, convulsions and loss of consciousness. adj. *eclamptic*. See also **pre-eclampsia**.

E-class insulation *(Elec.Eng.)*. A class of insulating material to which is assigned a temperature of 120°C. See **class-A insulating material** etc.

eclipse *(Astron.)*. (1) The total or partial disappearance from view of an astronomical object when it passes directly behind another object. (2) The passage of a planet or satellite through the shadow cast by another body so that it is unable to shine as it normally does by reflected sunlight. An eclipse of the Sun, *solar eclipse* can only occur at new Moon, when the moon is directly between the Earth and the Sun. Although the Moon is much nearer the Earth than the Sun, a coincidence of nature makes them appear nearly the same size in our sky. A *total eclipse*, when the whole disc is obscured can last no more than 7.5 minutes and is often less. At this time the **chromosphere** and **corona** may be briefly seen by the eye. A *partial eclipse* of much longer duration occurs before, after and to each side of the *path of totality*. Sometimes the apparent size of the lunar disc is just too small for a total eclipse, and there is an *annular eclipse* in which a bright ring of sunlight surrounding the Moon is seen; **Baily's beads** may then be visible. A *lunar eclipse* occurs when the Moon passes into the shadow of the Earth, which can happen only at full Moon, the moon glowing with a dim coppery hue rather than disappearing from view. Moons and satellites of other bodies in the solar system are eclipsed when they pass through the shadow of their primary bodies. In **binary star** sytems it is also possible for one star to eclipse another, causing an *eclipsing binary star*.

eclipse year *(Astron.)*. The interval of time between two successive passages of the sun through the same node of the moon's orbit; it amounts to 346.620 03 days.

eclipsing binary *(Astron.)*. A binary whose orbital plane lies so nearly in the line of sight that the components pass in front of each other in the course of their mutual revolution.

ecliptic *(Astron.)*. The great circle in which the plane containing the centres of the earth and sun cuts the celestial sphere; hence, the apparent path of the sun's annual motion through the fixed stars. See **obliquity of the ecliptic**.

eclogite *(Geol.)*. A coarse-grained deep-seated metamorphic rock, consisting essentially of pink garnet, green pyroxene (some of which is often chrome-diopside) and (rarely) kyanite.

eclosion *(Zool.)*. The act of emergence from an egg or pupa case.

ECM *(Eng.)*. Abbrev. for *ElectroChemical Machining*.

ECM *(Telecomm.)*. Same as **electronic counter measures**.

ecocline *(Ecol.)*. A *cline* occurring across successive zones of an organism's habit.

ecological efficiency *(Ecol.)*. Ratios between the amount of energy flow at different points along a food chain, e.g., the *primary* or *photosynthetic efficiency* is the percentage of the total energy falling on the earth which is fixed by plants, this being approximately 1%.

ecological factor *(Bot.)*. Anything in the environment which affects the growth, development and distribution of plants, and therefore aids in determining the characters of a plant community.

ecological indicators *(Ecol.)*. Organisms whose presence in a particular area indicates the occurrence of a particular set of water and soil conditions, temperature zones, etc. Large species with relatively specific requirements are most useful in this way, and numerical relationships between species, populations, or whole communities are more reliable than a single species.

ecological niche *(Ecol.)*. The position or status of an organism within its community or ecosystem. This results from the organism's structural adaptations, physiological responses, and innate or learned behaviour. An organism's niche depends not only on where it lives but also on what it does.

ecological pyramids *(Ecol.)*. Diagrams in which the producer level forms the base and successive trophic levels the remaining tiers. They include the *pyramids of numbers, biomass* and *energy*.

ecological succession *(Ecol.)*. See **succession**.

ecology *(Genrl.)*. The scientific study of the interrelations between living organisms and their environment, including both the physical and biotic factors, and emphasizing both interspecific and intraspecific relations; the scientific study of the distribution and abundance of living organisms (i.e. exactly where they occur, and precisely how many there are), and any regular or irregular variations in distribution and abundance, followed by explanation of these phenomena in terms of the physical and biotic factors of the environment.

Econet *(Comp.)*. TN for a **local area network**.

econometrics *(Stats.)*. The application of statistical methods to economic phenomena.

economic geology *(Geol.)*. The study of geological bodies and materials that can be used profitably by man.

economic ratio *(Civ.Eng.)*. In reinforced concrete work, the ratio between steel reinforcement and concrete which allows the full strength of both to be developed at the same time.

economizer *(Eng.)*. A bank of tubes, placed across a boiler flue, through which the feed water is pumped, being heated by the otherwise waste heat of the flue gases.

Economo's disease *(Med.)*. *Von Economo's disease*. See **encephalitis lethargica**.

economy resistance *(Elec.Eng.)*. A resistance inserted into the circuit of a contactor coil or other electromagnetic device after its initial operation, in order to reduce the current to a value just sufficient to hold the device closed.

ecophysiology *(Ecol.)*. Branch of physiology concerned with how organisms are adapted to their natural environment.

Eco R1 *(Biol.)*. A type of **restriction enzyme** derived from a strain of *E. coli*, strain R, which cleaves double stranded DNA at a specific sequence.

ecospecies *(Bot.)*. An **ecotype** sufficiently distinct to be given a subspecific name.

ecosystem *(Ecol.)*. Conceptual view of a plant and animal community, emphasizing the interactions between living and non-living parts, and the flow of materials and energy between these parts. Ecosystems are usually represented as flow diagrams, showing the path of these flows between producers, consumers and decomposers.

ecotone *(Ecol.)*. A transitional zone between two habitats.

ecotype *(Bot.)*. A form, within a species, of which the distinguishing characteristics are, at least partly, genotypic rather than simply phenotypic. Cf. **ecad**.

ecotype *(Ecol.)*. A form or variety of any species possessing special inherited characteristics enabling it to succeed in a particular habitat.

ECRH *(Nuc.Eng.)*. *Electron Cyclotron Resonance Heating*. See **cyclotron resonance heating**.

ecru *(Textiles)*. Unbleached knitted fabrics.

ECT *(Med.)*. Abbrev. for *ElectroConvulsive Therapy*.

ectasia, ectasis *(Med.)*. Pathological dilation or distension of any structure of the body. adj. *ectatic*.

ectethmoid *(Zool.)*. One of a pair of cartilage bones of the Vertebrate skull, formed by ossification of the ethmoid plate.

ecthoraeum *(Zool.)*. The thread of a nematocyst.

ecthyma *(Med.)*. Local gangrene and ulceration of the skin as a result of infection, the ulcer being covered by a crust and the skin round it being inflamed.

ectoblast *(Zool.)*. See **epiblast**.

ectoderm *(Zool.)*. The outer layer of cells forming the wall of a gastrula: the tissues directly derived from this layer.

ectogenesis *(Zool.)*. Development outside the body.

ectogenous *(Zool.)*. Independent; self-supporting.

ectolecithal *(Zool.)*. Said of ova in which the yolk is deposited peripherally.

ectomorph *(Behav.)*. One of Sheldon's somatotyping classifications; ectomorphs are delicately built, not very muscular, and are withdrawn and intellectual. See **somatotype theory**.

-ectomy *(Genrl.)*. Suffix from Gk. *ektomē*, cutting out, used esp. in surgical terms.

ectomycorrhiza *(Bot.)*. See **ectotrophic mycorrhiza**.

ectoparasite *(Bot.)*. A parasite feeding on the internal tissues of the host, but having the greater part of its body and its reproductive structures on the surface.

ectoparasite *(Zool.)*. A parasite living on the surface of the host.

ectophloedal *(Bot.)*. Living on the outside of bark.

ectophloic *(Bot.)*. A stele or stem having phloem on the side of the xylem nearer the periphery of the organ, but not on the other side.

ectopia cordis *(Med.)*. Congenital displacement of the heart outside the thoracic cavity.

ectopia, ectopy *(Med.)*. Displacement from normal position. Adj. *ectopic*.

ectopia vesicae *(Med.)*. A congenital abnormality in which the anterior wall of the bladder is absent and the posterior wall opens on to the surface of the abdomen, the lower abdominal wall being also absent.

ectopic gestation *(Med.)*. Fertilization of the ovum and growth of the foetus outside the uterus.

ectoplasm *(Biol.)*. A layer of clear non-granular cytoplasm at the periphery of the cell adjacent to the **plasma membrane**.

Ectoprocta *(Zool.)*. A phylum of Metazoa with the anus outside the lophophore; with a coelomic body cavity and the lophophore retractable into a tentacle sheath and without excretory organs. *Bryozoa*.

ectopy *(Med.)*. See **ectopia**.

ectotrophic mycorrhiza *(Bot.)*. A mycorrhiza in which there is on the outside of the root a well-developed layer of fungal mycelium the hyphae of which interconnect with hyphae both within the root cortex and also ramifying through the soil. Most trees form an ectotrophic mycorrhizal association often with a basidiomycete.

ectozoon *(Zool.)*. See **ectoparasite**.

ectromelia *(Biol.)*. An infectious disease of mice, due to a virus.

ectropion, ectropium *(Med.)*. Eversion of the eyelid.

eczema *(Immun.)*. Itching, inflammatory skin condition in which papules, vesicles and pustules may be present together with oedema, scaling or exudation. Although the immediate cause may not be known underlying hypersensitivity to food (e.g. milk proteins) or an environmental allergen is often detectable in atopic persons. Allergens include chemical agents, plant poisons and materials used in trades.

eczematous conjunctivitis *(Med.)*. See **phlyctenular conjunctivitis**.

edaphic climax *(Bot.)*. A climax community of which the existence is determined by some property of the soil.

edaphic factor *(Bot.)*. Any property of the soil, physical or chemical, which influences plants growing on that soil.

eddy *(Genrl.)*. An interruption in the steady flow of a fluid, caused by an obstacle situated in the line of flow; the **vortex** so formed.

eddy-current brake *(Elec.Eng.)*. (1) A form of brake for the loading of motors during testing; it consists of a mass of metal rotating in front of permanent magnets so that heavy eddy currents are set up in it. (2) A form of brake, used on tramways, in which the retarding force is produced by the induction of eddy currents in the rail by an electromagnet on the vehicle.

eddy-current heating *(Elec.Eng.)*. See **induction heating**.

eddy currents *(Elec.Eng.)*. Those arising through varying electromotive forces consequent on varying magnetic fields, resulting in diminution of the latter and dissipation of power. They are one of the main causes of heating in motors, transformers, etc. (known also as *iron loss*, c.f. **copper loss**). To minimize them the iron cores of such machines are composed of many layers of thin iron sheet (laminations) which are insulated from one another to reduce the currents' ability to flow in the direction in which the field tries to induce them. This in turn slightly increases the reluctance of the magnetic circuit, thereby reducing the overall efficiency of the machine, hence a compromise must be struck. Used for mechanical damping and braking (as in electricity meters) and for induction heating as applied in case-hardening. Also called *Foucault current(s)*.

eddy-current speed indicator *(Elec.Eng.)*. A speed indicator consisting of a rotating disk and a spring-controlled magnetic needle; the latter is deflected as a result of eddy currents induced in the disk.

eddy-current testing *(Eng.)*. A non-destructive test using an electromagnetic field to induce eddy currents in a component. Changes of shape in the component, including internal cracks, are detected by variations in the signal produced in pick-up coils located nearby.

eddy diffusion *(Chem.Eng.)*. The migration and interchange of portions of a fluid as a result of their turbulent motion. Cf. **diffusion**.

eddy diffusion *(Meteor.)*. The transport of quantities such as heat and momentum by eddies in regions of the atmosphere which are in a state of turbulent motion. Eddy diffusion is roughly 10^5 times as effective as molecular diffusion which for meteorological purposes can normally be ignored.

eddy diffusivity *(Chem.Eng.)*. Exactly analogous, for eddy diffusion, to the **diffusivity** for molecular diffusion. Symbol *E*.

eddy flow *(Phys.)*. See **turbulent flow**.

Eddy's theorem *(Eng.)*. The bending moment at any point in an arch is equal to the product of the horizontal thrust at the abutment and the vertical distance between the line of action of this thrust and the given point in the arch.

edelopal *(Min.)*. A variety of *opal* with an exceptionally brilliant play of colours.

edema, edematous *(Med.)*. See **oedema, oedematous**.

edenite *(Min.)*. An end-member compositional variety in the hornblende group of monoclinic amphiboles: hydrated magnesium, calcium, and aluminium silicate.

Edentata *(Zool.)*. An order of primitive terrestrial Mammals characterized by the incomplete character of the dentition; the testes are abdominal; phytophagous or insectivorous forms. Sloths, Ant-eaters, Armadillos.

edentulous, edentate *(Zool.)*. Without teeth.

edge *(Aero.)*. See **leading edge, trailing edge**.

edge coal *(Min.Ext.)*. Highly inclined coal-seams.

edge effect *(Acous.)*. In acoustic absorption measurements, the variations which arise from the size, shape or division of the areas of material being tested in a reverberation room.

edge effect *(Elec.Eng.)*. Deviation from parallelism in fields at the edge of parallel plate capacitors, or between poles of permanent or electro-magnets, thus leading to nonuniformity of the field at the edges.

edge effect *(Image Tech.)*. See **border effect**. Also *fringe effect*.

edge-emitting diode *(Telecomm.)*. A **light-emitting diode** in which the radiation is parallel to the surface and perpendicular to the current flow; used as a light source in optical fibre communications.

edge filter *(Chem.Eng.)*. A type of filter in which a large number of disks are clamped on a perforated hollow shaft joining a cylinder. This is contained in another cylindrical vessel into which liquid is pumped and flows through the narrow spaces between the disks, the solids being trapped on the disk edges. Filtrate leaves via the perforations in the hollow shaft.

edge numbers *(Image Tech.)*. A series of numerals and letters printed or photographically exposed along the edge of a strip of motion picture film, usually at intervals of one foot (16 frames on 35 mm film), allowing individual frames to be identified and located.

edge plane *(Build.)*. One with the cutter at the extreme front for working corners.

edge planing *(Print.)*. Trimming and squaring the edges of printing plates, either original or duplicate, by hand- or power-operated tools.

edge rolled *(Print.)*. Said of a pattern on the edges of leather bindings, either blind or decorated, applied by a **roll**.

edge sealed *(Textiles)*. The edge of a fabric that has been made without the usual **selvedge** and has been cut and sealed (usually by the heating and melting of a thermoplastic fibre) to prevent fraying.

edge tones *(Acous.)*. Tones produced by the impact of an air jet on a sharply-edged dividing surface, as in an organ pipe mouth. See also **whistle**.

edge tool *(Eng.)*. A hand-worked, mallet-struck or machine-operated cutting tool with one or a regular pattern of cutting edges, i.e. excluding grinding wheels.

edge trimming plane *(Build.)*. One with a rebated sole, for squaring up edges.

edge water *(Geol.)*. That pressing inward upon the gas or oil in a natural reservoir.

edge winding *(Elec.Eng.)*. A form of winding frequently used for the field windings of salient-pole synchronous machines; it consists of copper strip wound on edge around the pole. Such a winding has good heat-dissipating properties.

edgewise instrument *(Elec.Eng.)*. A switchboard indicating instrument in which the pointer moves in a plane at right angles to the face of the switchboard. The end of the pointer is bent and moves over a narrow scale.

Edison accumulator *(Elec.Eng.)*. **Nickel-iron-alkaline accumulator**, or secondary cell.

Edison effect *(Electronics)*. The phenomenon of electrical conduction between an incandescent filament and an independent cold electrode contained in the same envelope, when the second electrode is made positive with reference to a part of the filament. Precursor of **Fleming diode**. Also *Richardson effect*.

Edison phonograph *(Acous.)*. The original type of gramophone, in which the records were registered on the surface of hollow cylindrical waxes.

Edison screw-cap *(Elec.Eng.)*. A lamp cap in which the outer wall forms one of the contacts, and which is in the form of a coarse screw for inserting into a corresponding socket. A central pin forms the other contact. Sizes in descending order include goliath, large, medium, small, miniature and lilliput.

Edison screw-holder *(Elec.Eng.)*. A holder for electric lamps with Edison screw-caps.

E-display *(Radar)*. Display in which target range and elevation are plotted as horizontal and vertical coordinates of the blip.

editing *(Image Tech.)*. The selection and arrangement in sequence of the individual scenes of a film or video production in accordance with the director's interpretation of the script. In film editing, the chosen sections are physically cut from *rush prints* and joined together but in video editing they are re-recorded in the required order.

editing terminal *(Print.)*. In phototypesetting the visual display unit (VDU) on which is displayed the results of keyboarding. Corrections and other editorial functions can be carried out before final phototypesetting.

edition *(Print.)*. A number of copies printed at one time, either as the original issue (*first edition*), or when the text has undergone some change, or the text has been partly or entirely reset, or the format has been altered.

edition binding *(Print.)*. The normal style of binding for hardcover books, highly mechanized, only occasionally sewn on tapes, with case usually cloth-covered and lettered on spine only. Also called *publisher's binding*. Cf. **bound book, library binding**.

editor *(Comp.)*. Program which enables the user to inspect and alter his program or data. See **screen editor**.

editor *(Image Tech.)*. (1) The creative craftsman who

works with the director to achieve the final assembly of a film or video production. (2) Equipment for the detailed examination of the action in a motion picture film on a small illuminated screen; an editing table may also include paths for the synchronization and reproduction of the associated magnetic sound records.

edriophthalmic *(Zool.)*. Having sessile eyes, as some Crustacea.

Edser and Butler's bands *(Phys.)*. Dark bands, having a constant frequency separation, which are seen in the spectrum of white light which has traversed a thin, parallel-sided plate of a transparent material, or a thin parallel-sided film of air between glass plates.

EDTA *(Biol.,Chem.)*. *Ethylene Diamine Tetra-Acetic (ethanoic) acid, diamino ethane tetracetic acid.* $CH_2N(CH_2COOH_2)$. A **chelating agent** frequently used to protect enzymes from inhibition by traces of metal ions and as an inhibitor of metal dependent proteases because of its ability to combine with metals. It is also used in special soaps to remove metallic contamination. Also *complexone*.

Edwards' roaster *(Min.Ext.)*. Long horizontal furnace through which sulphide minerals are rabbled counterwise to hot air, to remove part or all of the sulphur by ignition.

EEG *(Med.)*. Abbrev. for *ElectroEncephaloGraph* (or *-Gram*).

eel *(Zool.)*. Any fish of the Anguillidae, Muraenidae, or other family of the Anguilliformes. The name is extended to other fish of similar form, e.g. Sand Eel.

eel-grass *(Bot.)*. Species of *Zostera*, grass-like monocotyledons which grow in the sea mostly around or below the low-water mark, used for sound insulation and the correction of acoustic defects.

effective address *(Comp.)*. One which is obtained by modifying the address part of an **instruction** during processing.

effective antenna height *(Telecomm.)*. Height (in metres) which, when multiplied by the field strength (in volts per metre) incident upon the antenna, gives the e.m.f. (in volts) induced therein. It is less than the physical height and differs from the equivalent height in that it is also a function of the direction of arrival of the incident wave.

effective bandwidth *(Telecomm.)*. The bandwidth of an ideal (rectangular) band-pass filter which would pass the same proportion of the signal energy as the actual filter.

effective column length *(Build.)*. The column length which is used in finding the *ratio of slenderness* after taking into account the rigidity of the end fixings.

effective depth *(Civ.Eng.)*. The depth of a reinforced concrete beam as measured from the surface of the concrete on the compression side to the centre of gravity of the tensile reinforcement. See **cover**.

effective dose equivalent *(Radiol.)*. The quantity obtained by multiplying the dose equivalents to various tissues and organs by the risk weighting factor appropriate to each organ and summing the products. Unit Sievert (*Sv*).

effective energy *(Radiol.)*. The quantum energy (or wavelength) of a monochromatic beam of X-rays or γ-rays with the same penetrating power as a given heterogeneous beam. Its value depends upon the nature of the absorbing medium. Also known as *effective wavelength*.

effective half-life *(Phys.)*. The time required for the activity of a radioactive nuclide in the body to fall to half its original value as a result of both biological elimination and radioactive decay. Its value is given by

$$\frac{\tau(b\tfrac{1}{2}) \times \tau(r\tfrac{1}{2})}{\tau(b\tfrac{1}{2}) + \tau(r\tfrac{1}{2})},$$

where $\tau(b\frac{1}{2})$ and $\tau(r\frac{1}{2})$ are the biological and radioactive half-lives respectively.

effective heating surface *(Eng.)*. The total area of a boiler surface in contact with water on one side and with hot gases on the other.

effective mass *(Electronics)*. For electrons and/or holes in a semiconductor, effective mass is a parameter which may differ appreciably from the mass of a free electron, and which depends to some extent on the position of the particle in its energy band. This modifies the mobility and hence the resulting current.

effective particle density *(Powder Tech.)*. The mass of a particle divided by its volume, including opened and closed pores.

effective porosity *(Powder Tech.)*. That portion of the powder porosity which is readily accessible to a fluid moving through a powder compact.

effective radiated power *(Telecomm.)*. Actual maximum or unmodulated power delivered to a transmitting antenna multiplied by the factor of *gain* in a specified direction in the horizontal plane.

effective range *(Eng.)*. That part of the scale of an indicating instrument over which a reasonable precision may be expected.

effective resistance *(Elec.Eng.)*. The total a.c. resistance covering eddy-current losses, iron losses, dielectric and corona losses, transformed power, as well as conductor loss. For a sinusoidal current it is the component of the voltage in phase with current divided by that current. Measured in ohms.

effective sieve aperture size *(Powder Tech.)*. To allow for the size aperture distribution in a real sieve for accurate particle size analysis, sieves have to be calibrated. This is sometimes done by analysing a powder of known distribution. If A percentage by weight should be retained on a sieve of nominal aperture size a, then B percentage is retained on the actual sieve; then from the particle size distribution of the powder B, the size at which B percentage of the powder is greater than this size, is read off. The effective size of the sieve aperture is then stated to be B. The value of B will depend upon the size distribution of the powder used to calibrate the sieve. The magnitude of the quantity $a-B$ is termed the *aperture error of a sieve*.

effective span *(Build.)*. The horizontal distance between the centres of the two bearings at the ends of a beam.

effective temperature *(Astron.)*. The temperature which a given star would have if it were a perfect radiator, or a **black body**, with the same distribution of energy among the different wavelengths as the star itself.

effective value *(Genrl.)*. That of a simple parameter which has the same effect as a more complex one, e.g. r.m.s. value of ac = dc value for many purposes.

effective wavelength *(Phys.)*. See **dominant wavelength**.

effective wavelength *(Radiol.)*. See **effective energy**.

effector *(Zool.)*. A tissue-complex capable of effective response to the stimulus of a nervous impulse, e.g. a muscle or gland.

effector neurone *(Zool.)*. A motor neurone.

effects track *(Image Tech.)*. A sound track containing only the sound effects and noises required in a production, excluding speech and music; sometimes abbreviated *FX*.

effect threads *(Textiles)*. Yarns of striking appearance that are included in a fabric to attract attention.

efferent *(Zool.)*. Carrying outwards or away from; as the *efferent branchial vessels* in a Fish, which carry blood away from the gills, and *efferent nerves*, which carry impulses away from the central nervous system. Cf. **afferent**.

effervescence *(Chem.)*. The vigorous escape of small gas bubbles from a liquid, especially as a result of chemical action.

efficiency *(Genrl.)*. A non-dimensional measure of the performance of a piece of apparatus, e.g. an engine, obtained from the ratio of the output of a quantity, e.g. power, energy, to its input, often expressed as a percentage. The power efficiency of an *IC engine* is the ratio of the shaft- or brake-horsepower to the rate of intake of fuel, expressed in units of energy content per unit time. It must always be less than 100% which would imply perpetual motion. Not to be confused with efficacy, which takes account only of the output of the apparatus, and is not given an exact quantitative definition.

efficiency of impaction *(Powder Tech.)*. Ratio of the

cross-sectional area of the stream from which particles of a given size are removed, to the total cross-sectional area of the jet stream, both areas measured at the mouth of the jet.

efficiency of screening, numerical index of *(Min.Ext.)*. A quantity used to assess the efficiency of industrial screening procedures, i.e. sieving procedures. It is defined by the equation $E = F(D/B\text{-}C/A)$, where E = the numerical index of efficiency of screening, F = the percentage of the powder supply passed by the screen, A = the percentage of difficult oversize in the powder supply, B = the percentage of difficult undersize in the powder supply, C = the percentage of the difficult oversize passed by the screen, D = the percentage of the difficult undersize passed by the screen.

efficiency ratio *(Eng.)*. Of a heat engine, the ratio of the actual thermal efficiency to the ideal thermal efficiency corresponding to the cycle on which the engine is operating.

effleurage *(Med.)*. The action of lightly stroking in massage.

efflorescence *(Build.)*. Formation of a white crystalline deposit on the face of a wall; due to the drying out of salts in the mortar or stone.

efflorescence *(Chem.)*. The loss of water from a crystalline hydrate on exposure to air, shown by the formation of a powder on the crystal surface.

efflorescence *(Min.)*. A fine-grained crystalline deposit on the surface of a mineral or rock.

effluent *(Build.)*. Liquid sewage after having passed through any stage in its purification.

effluent monitor *(Nuc.Eng.)*. Instrument for measuring level of radioactivity in fluid effluent.

effluve *(Elec.Eng.)*. The corona discharge from an electrostatic machine or high-frequency generator, used to stimulate the human skin.

efflux *(Aero.)*. The mixture of combustion products and cooling air which forms the propulsive medium of any jet or rocket engine.

effort syndrome *(Med.)*. *Soldier's heart*. A condition in which the subject complains of palpitations, breathlessness and chest pain, often after exercise, in the absence of heart disease. Thought to be due to psychological factors.

effusiometer *(Chem.)*. An apparatus for comparing the relative molecular masses of gases by observing the relative times taken to stream out through a small hole.

effusion *(Chem.)*. The passage of a gas through a small aperture. See **Graham's Law**.

effusion *(Med.)*. An abnormal outpouring of fluid into the tissues or cavities of the body, as a result of infection or of obstruction to bloodvessels or lymphatics.

effusion *(Phys.)*. The flow of gases through larger holes than those to which diffusion is strictly applicable; see **Graham's law**. The rate of flow is approximately proportional to the square-root of the pressure difference.

EFP *(Image Tech.)*. *Electronic Field Production*, video programme production shooting outside a studio, usually with light-weight portable cameras.

EFT *(Comp.)*. See **electronic funds transfer**.

egest *(Zool.)*. To throw out, to expel; to defaecate, to excrete. n.pl. *egesta*.

egg *(Bot.,Zool.)*. See **ovum**.

egg and anchor *(Arch.)*. An ornament carved on a moulding, resembling eggs separated by vertical anchors.

egg and dart *(Arch.)*. Similar to **egg and anchor**, arrows taking the place of anchors.

egg apparatus *(Bot.)*. The egg and the two synergidae in the embryo sac of an angiosperm.

egg-bound *(Vet.)*. Said of the oviduct of birds when obstructed by an egg.

egg-box lens *(Telecomm.)*. Same as **slatted lens**.

egg-cell *(Zool.)*. The ovum, as distinct from any other cells associated with it.

egg-eating *(Vet.)*. A vice developed by individual birds, characterized by the eating of their own eggs or of those of other birds.

egg, egg cell *(Bot.)*. A nonflagellated, female gamete usually larger than the male gamete with which it fuses.

egg glair *(Build.)*. A substance produced by mixing the white of an egg with a half litre of lukewarm water. Egg glair is applied to painted surfaces and allowed to dry prior to gilding work. The glair, which removes any tack from the painted surface, is destroyed by the gold size applied to areas to be gilded. The surface must be washed with lukewarm water to remove excess glair when gilding work is complete.

egg nucleus *(Zool.)*. The female pronucleus.

egg-peritonitis *(Vet.)*. Septic peritonitis extending from an infected and obstructed oviduct of birds.

egg-shaped sewer *(Build.)*. A type of sewer section much used for fluctuating flows; the section resembles the longitudinal profile of an egg placed with the smaller radius at the bottom, thus increasing the velocity of flow at the bottom.

eggshell finish *(Build.)*. Paint or varnish which dries with a degree of sheen or lustre between matt and semi-gloss.

egg-shell finish *(Paper)*. A soft dull finish on paper.

egg sleeker *(Foundry)*. A moulder's sleeker with a spoon-shaped end; used for smoothing rounded corners in a mould.

egg tooth *(Zool.)*. A sharp projection at the tip of the upper beak of young birds and some mammals, by means of which they break open the egg-shell.

E-glass *(Glass)*. A glass containing not more than 1% of alkali (calculated as Na_2O) and used for the manufacture of glass fibre.

ego *(Behav.)*. Originally, a term in philosophy, denoting the existence of a sense of self; in psychoanalytic theory, the rational, reality oriented level of personality which develops in childhood, as the child gathers awareness of and comes to terms with, the nature of the social and physical environment; represents reason and common sense, in contrast to the **id**.

egocentrism *(Behav.)*. In Piaget's theory, a form of thinking most typically found in young children, in which the individual perceives and comprehends the world from a totally subjective point of view, being unable to differentiate between the objective and subjective components of experience.

egophony *(Med.)*. See **aegophony**.

ego psychology *(Behav.)*. Those Freudian theorists who emphasise ego processes, such as reality perception, learning and conscious control of behaviour, and who argue that the ego has its own energy and autonomous functions, not derived from the **id**; in contrast to *instinct theory*, which states that all mental energy is ultimately derived from the id.

EGT *(Aero.)*. Abbrev. for *Exhaust Gas Temperature*.

Egyptian *(Print.)*. See **slab serif**.

Egyptian blue *(Min.)*. An artificial mineral, calcium copper silicate, $CaCuSi_4O_{12}$ thought to be the pigment of ancient Egypt.

Egyptian jasper *(Min.)*. A variety of jasper occurring in rounded pieces scattered over the surface of the desert, chiefly between Cairo and the Red Sea; used as a broochstone and for other ornamental purposes. Typically shows colour zoning.

EHF *(Telecomm.)*. Abbrev. for *Extremely High Frequency*.

ehp *(Aero.)*. Abbrev. for *total Equivalent brake Horse-Power*. Also *tehp*.

EI *(Image Tech.)*. *Exposure Index*, generally taken as equivalent to the **ASA speed** rating.

eidetic imagery *(Behav.)*. The ability to reproduce on a dark screen, or when the eyes are closed, a vividly clear and detailed picture or visual memory-image of previously seen objects. Commonly present in children up to 14 years, and occasionally persisting into adult life.

eidograph *(Surv.)*. An instrument for reducing and enlarging plans.

Eidophor *(Image Tech.)*. TN for a projection TV system in which a beam of light from a powerful source passes through a liquid surface which is optically distorted by an electron beam modulated by the picture signal. From Gk. *eidos*, image.

Eiffel wind tunnel *(Aero.)*. An open jet, non-return flow wind tunnel.
eigenfrequency *(Acous.)*. Frequency of vibration of a system which vibrates freely. See also **acoustic resonance**.
eigenfunction *(Phys.)*. A solution of a wave equation which satisfies a set of boundary conditions. In quantum mechanics, a possible solution for the Schrödinger equation for a given system.
eigentones *(Phys.)*. The natural frequencies of vibration of a system.
eigenvalues *(Maths.)*. Of a matrix. See **characteristic equation of a matrix**.
eigenvalues *(Phys.)*. Possible values for a parameter of an equation for which the solutions will be compatible with the boundary conditions. In quantum mechanics, the energy eigenvalues for the **Schrödinger equation** are possible **energy levels** for the system.
eighteen-electron rule *(Chem.)*. The structures of most stable organometallic compounds of the transition elements, e.g. *ferrocene*, can be rationalized by showing that 18 valence electrons can be associated with each metal atom.
18mo *(Print.)*. The 18th of a sheet or a sheet folded to make eighteen leaves or thirty-six pages.
eight-millimetre (8mm) *(Image Tech.)*. (1) The narrowest gauge of motion picture film; the Super-8 format has now completely replaced the original Regular-8. (2) The narrowest gauge of videotape, used particularly in compact **camcorders**.
eight-to-pica leads *(Print.)*. Strips of metal, $1\frac{1}{2}$-point in thickness, used to space out lines of type. They are usually called thin leads, two being equivalent to a thick lead, and eight to a pica em.
Eikmeyer coil *(Elec.Eng.)*. The name given to the original type of former-wound armature coil which can be dropped straight into the slots of an electric machine.
Einstein-de Haas effect *(Phys.)*. When a magnetic field is applied to a body the precessional motion of the electrons produce a mechanical moment that is transferred to the body as a whole. See **Barnett effect**.
Einstein diffusion equation *(Chem.)*. As a result of Brownian motion, molecules or colloidal particles migrate an average distance δ in each small time interval τ. Hence the equation for the diffusion of a spherical particle of radius r through a fluid of viscosity η may be written

$$D=\frac{\delta^2}{2\tau}=\frac{RT}{6\pi\eta rN}$$

where D is the diffusion coefficient, N is Avogadro number, R is the gas constant and T the absolute termperature.
Einstein energy *(Phys.)*. See **mass-energy equation**.
Einstein equation for the specific heat of a solid *(Chem.)*. One mole of the solid consists of N molecules, each vibrating with frequency v in 3 dimensions. Hence, by quantum theory, the molar specific heat may be written:

$$C_v=\frac{3Rx^2e^x}{(e^x-1)^2}$$

where $x=hv/\mathrm{k}T$, R is gas constant, h Planck's constant, k Boltzmann's constant and T the absolute temperature.
Einstein law of photochemical equivalence *(Chem.)*. Each quantum of radiation absorbed in a photochemical process causes the decomposition of one molecule.
Einstein photoelectric equation *(Chem.)*. That which gives the energy of an electron, just ejected photoelectrically from a surface by a photon, i.e. $E=h\nu-\mathrm{w}$, where E = kinetic energy, h = Planck's constant, ν = frequency of photon, and w = work function.
Einstein shift *(Astron.)*. The **redshift** of spectral lines caused when electromagnetic radiation is emitted from an object with a significant gravitational field. Also termed *gravitational redshift*.
Einstein theory of specific heat of solids *(Phys.)*. Based on the assumption that the thermal vibrations of the atoms in a solid can be represented by harmonic oscillators of one frequency, whose energy is quantized. See **Debye theory**, **Blackman theory**.
einsteinum *(Chem.)*. Artificial element, symbol Es, at. no. 99, produced by bombardment in a cyclotron, but also recognized in H-bomb debris, it having been produced by beta decay of uranium which had captured a large number of neutrons. The longest lived isotope is ^{254}Es, with a half-life of greater than 2 years.
Einthoven galvanometer *(Elec.Eng.)*. A galvanometer in which the current is carried by a single current-carrying filament in a strong magnetic field, the deflection usually being magnified by a microscope. Also called a *string galvanometer*.
ejaculatory duct *(Zool.)*. See **ductus ejaculatorius**.
ejecta *(Geol.)*. Solid material thrown from a volcano or an impact crater. syn. *ejectamenta*.
ejection capsule *(Aero.)*. (1) A cockpit, cabin, or portion of either, in a high-altitude and/or high-speed military aircraft which can be fired clear in emergency and which after being slowed down, descends by parachute. (2) Container of recording instruments ejected and parachute recovered.
ejector *(Build.)*. An appliance used for raising sewage from a low-level sewer to a sewer at a higher elevation; worked by compressed air.
ejector *(Eng.)*. (1) A device for exhausting a fluid by entraining it by a high velocity steam or air jet, e.g. an **air ejector**. (2) A mechanism for removing a part or assembly from a machine at the end of an automatic sequence of operations, e.g. in assembling or machining.
ejector seat *(Aero.)*. A crew seat for high-speed aircraft which can be fired, usually by slow-burning cartridge, clear of the structure in emergency. Automatic releases for the occupant's safety harness and for parachute opening are usually incorporated. Also *ejection seat*.
eka- *(Chem.)*. A prefix, coined by Mendeleev, denoting the element occupying the next lower position in the same group in the periodic system; used in the naming of new elements and unstable radio-elements.
ektrodactylia *(Med.)*. Congenital absence of one or more fingers or toes.
elaeodochon *(Zool.)*. See **oil gland**.
elaeolite *(Min.)*. A massive form of the mineral nepheline, greenish-grey or (when weathered) red in colour.
elaiosome *(Bot.)*. An outgrowth from the surface of a seed, containing fatty or oily material (often attractive to ants) and serving in seed dispersal.
Elara *(Astron.)*. The seventh natural satellite of **Jupiter**.
Elasmobranchii *(Zool.)*. A subclass of Gnathostomata, highly developed usually predacious fishes with a cartilaginous skeleton and plate-like gills. Includes Sharks, Skates, Rays.
elastance *(Phys.)*. The reciprocal of the capacitance of a capacitor, so termed because of electromechanical analogy with a spring. Unit is the *daraf* (*farad* spelt backwards).
elastase *(Biol.)*. The proteolytic enzyme secreted by the pancreas which digests *elastin*.
elastic bitumen *(Min.)*. See **elaterite**.
elastic collision *(Phys.)*. A collision between two bodies in which, in addition to the total momentum being conserved, the total kinetic energy of the bodies is conserved.
elastic constants *(Phys.)*. Quantities used to describe the behaviour of a material when subjected to stress in one of three modes; longitudinal, shear or compression. These give rise to *Young's* (elongational), *rigidity* and *bulk* moduli of elasticity. They are expressed in units of stress ($\mathrm{N\,m^{-2}}$, $\mathrm{lbf\ ft^{-2}}$, $\mathrm{tonf\ in^{-2}}$). Also includes Poisson's ratio. Known also as the *elastic moduli* or *moduli of elasticity*.
elastic deformation *(Eng.)*. Changes in shape which are proportional to an applied force, returning to the initial state if the applied force is removed.
elastic fabric *(Textiles)*. Fabric containing rubber or **elastomeric yarns** that give it good elastic recovery and shape retaining properties.

elastic fatigue *(Eng.,Phys.)*. A temporary departure from perfect elasticity shown by some materials, which, after suffering elastic deformation, slowly return to their original form.

elastic fibres *(Zool.)*. See **yellow fibres**.

elastic fibrocartilage *(Zool.)*. See **yellow fibrocartilage**.

elasticity *(Phys.)*. The tendency of a body to return to its original size or shape, after having been stretched, compressed, or deformed. The ratio of the stress called into play in the body by the action of the deforming forces to the strain or change in dimensions or shape is called the *coefficient (modulus) of elasticity*. See the following definitions and **Hooke's law**.

elasticity of bulk *(Phys.)*. The elasticity for changes in the volume of a body caused by changes in the pressure acting on it. The bulk modulus is the ratio of the change in pressure to the fractional change in volume. See **elasticity**.

elasticity of elongation *(Phys.)*. The stress in this case is the stretching force per unit area of cross-section and the strain is the elongation per unit length. The modulus of elasticity of elongation is known as **Young's modulus**. See **elasticity**.

elasticity of gases *(Phys.)*. If the volume V of a gas is changed by δV when the pressure is changed by δp, the modulus of elasticity is given by

$$-V\frac{\delta p}{\delta V}.$$

This may be shown to be numerically equal to the pressure p for isothermal changes, and equal to γp for adiabatic changes, γ being the ratio of the specific heats of the gas.

elasticity of shear, rigidity *(Phys.)*. The elasticity of a body which has been pulled out of shape by a shearing force. The stress is equal to the tangential shearing force per unit area, and the strain is equal to the angle of shear, that is, the angle turned through by a straight line originally at right angles to the direction of the shearing force. See **elasticity**, **Poisson's ratio**.

elastic limit *(Eng.)*. The highest stress that can be applied to a metal without producing a measurable amount of plastic (i.e. permanent) deformation. Usually assumed to coincide with the limit of proportionality.

elastic limit *(Phys.)*. The limiting value of the deforming force beyond which a body does not return to its original shape or dimensions when the force is removed.

elastic medium *(Phys.)*. A medium which obeys **Hooke's law**. No medium is perfectly elastic, but many are sufficiently so to justify the making of calculations which assume perfect elasticity.

elastic moduli *(Eng.)*. *Elastic constants*. See **bulk modulus**, **modulus of rigidity**, **Young's modulus**.

elastic scattering *(Phys.)*. See **scattering**.

elastic strain *(Eng.)*. The strain or fractional deformation undergone by a material in the elastic state, i.e. a strain which disappears with the removal of the straining force.

elastic tissue *(Zool.)*. A form of connective tissue in which elastic fibres predominate.

elastin *(Biol.)*. A protein which is the major component of the elastic fibres of the **extracellular matrix**. The molecules provide an extensible network by being heavily cross-linked in a random coiled configuration. See **scleroproteins**.

elastivity *(Phys.)*. The reciprocal of the permittivity of a dielectric.

elastomer *(Chem.)*. A material, usually synthetic, having elastic properties akin to those of rubber.

elastomeric yarns *(Textiles)*. Yarns comprising filaments or staple fibres (e.g. polyurethane), which have good elastic properties.

elaterite *(Min.)*. A solid bitumen resembling dark-brown rubber. Elastic when fresh. Sometimes known as *mineral caoutchouc*.

E-layer *(Phys.)*. Most regular of the ionized regions in the ionosphere, which reflects waves from a transmitter back to earth. Its effective maximum density increases from zero before dawn to its greatest at noon, and decreases to zero after sunset, at heights varying between 110 and 120 km. There are at least two such layers. Also called *Heaviside layer* or *Kennelly-Heaviside layer*.

elbaite *(Min.)*. One of the three chief compositional varieties of tourmaline; a complex hydrated borosilicate of lithium, aluminium and sodium. Most of the gem varieties of tourmaline are elbaites.

elbow *(Build.)*. An arch stone whose lower bed is horizontal, while its upper bed is inclined towards the centre of the arch, to correspond with those of the voussoirs.

elbow *(Eng.)*. A short right-angle pipe connection, as distinct from a bend which is curved not angular.

elbow-board *(Build.)*. The window-board beneath a window, in the interior.

elbow linings *(Build.)*. The panelling at the sides of window recess, running from the floor to the level of the window-board. Cf. **jamb linings**.

Electra Complex *(Behav.)*. See **Oedipus and Electra Complexes**.

electret *(Elec.Eng.)*. Permanently polarized dielectric material, formed by cooling barium titanate from above a Curie point or waxes (e.g. carnauba) in a strong electric field.

electric *(Phys.)*. Said of any phenomena which depend essentially on a peculiarity of electric charges.

electrical *(Phys.)*. Descriptive of means related to, pertaining to, or associated with electricity, but not inherently functional.

electrical absorption *(Phys.)*. An effect in a dielectric whereby, after an initially charged capacitor has been once discharged, it is possible after a few minutes to obtain from it another discharge, usually smaller than the first.

electrical analogy *(Acous.)*. The correspondence between electrical and acoustical systems, which assists in applying to the latter procedures familiar in the former.

electrical bias *(Telecomm.)*. The use of a polarizing winding on a relay core, for adjusting the sensitivity of the relay to signal currents.

electrical conductivity *(Phys.)*. Ratio of current density to applied electric field. Expressed in siemens per metre (S/m) or $ohm^{-1}m^{-1}$ in SI units. Conductivity of metals at high temperatures varies as T^{-1}, where T is absolute temperature. At very low temperatures, variation is complicated but it increases rapidly (at one stage proportional to T^{-5}), until it is finally limited by material defects of structure.

electrical degrees *(Elec.Eng.)*. Angle, expressed in degrees, of phase difference of vectors, representing currents or voltages, arising in different parts of a circuit.

electrical discharge machining *(Eng.)*. *Spark erosion* technique in which metal is removed as sparks pass between a shaped electrode and the work. Can be used for machining irregular holes etc. Abbrev. *EDM*.

electrical dischargers *(Aero.)*. Devices for discharging static electricity, e.g. earthing tyres, static wick dischargers.

electrical double layer *(Chem.)*. The layer of adsorbed ions at the surface of a dispersed phase which gives rise to the **electrokinetic effects**.

electrical engineering *(Genrl.)*. That branch of engineering chiefly concerned with the design and construction of all electrical machinery and devices, power transmission etc.

electrical power distribution *(Space)*. The provision, conditioning and supply of electrical power to satisfy the needs of a spacecraft and its payload. Continuous sources of power may be the Sun (e.g. **solar cells**, thermal devices) or carried on board (e.g. **fuel cells**, radioisotopes). During certain mission phases of space-flight, such as launch and re-entry, an auxiliary power source (APU) must be used.

electrical prospecting *(Min.Ext.)*. Form of **geophysical prospecting** which identifies anomalous electrical effects associated with buried ore bodies. Most important techniques utilize resistivity or inverse conductivity and the inductive properties of ore minerals.

electrical reset *(Elec.Eng.)*. Restoration of a magnetic device, e.g. relay or circuit breaker, by auxiliary coils or relays.
electrical resistivity *(Phys.)*. Reciprocal of **electrical conductivity**. Units are ohm metres.
electrical resonance *(Elec.Eng.)*. Condition arising when a maximum of current or voltage occurs as the frequency of the electrical source is varied; also when the length of a transmission line approximates to multiples of a quarter-wavelength and the current or voltage becomes abnormally large.
electrical thread *(Elec.Eng.)*. A form of thread used on screwed steel conduit for electrical installation work.
electric-arc furnace *(Elec.Eng.)*. See **arc furnace**.
electric-arc welding *(Elec.Eng.)*. See **arc welding**.
electric axis *(Telecomm.)*. Direction in a crystal which gives the maximum conductivity to the passage of an electric current. The X-axis of a piezoelectric crystal.
electric balance *(Elec.Eng.)*. A name sometimes applied to a type of electrometer, to a current weigher (which establishes the absolute ampere), and to a Wheatstone bridge.
electric bell *(Elec.Eng.)*. A bell in which the hammer is operated electrically by means of a solenoid. A single stroke may be given, or, more commonly, a rapid succession of strokes may be maintained by means of a make-and-break contact on the solenoid.
electric braking *(Elec.Eng.)*. A method of braking for electrically driven vehicles; the motors are used as generators to return the braking energy to the supply, or to dissipate it as heat in resistances.
electric calamine *(Min.)*. See **hemimorphite**.
electric cautery *(Med.)*. Burning of parts of the human body for surgical purposes by means of electrically heated instruments. Also called *electrocautery*.
electric circuit *(Phys.)*. Series of conductors forming a partial, branched or complete path around which either a direct or alternating current can flow.
electric cleaner *(Elec.Eng.)*. In electroplating work, a cleansing solution in which the cleansing is accelerated by the passage of an electric current.
electric conduction *(Elec.Eng.)*. Transmission of energy by flow of charge along a conductor.
electric dipole *(Phys.)*. See **electric doublet**.
electric dipole moment *(Phys.)*. Product of the magnitude of the electric charges and the distance between it and its opposite charge in an electric dipole.
electric discharge *(Phys.)*. Same as **field discharge**.
electric-discharge lamp *(Elec.Eng.)*. A form of electric lamp in which the light is obtained from an electric discharge between two electrodes in an evacuated glass tube. Sometimes called a *gas-discharge lamp*.
electric double layer *(Elec.Eng.)*. A positive and negative layer distribution of electric charge very close together so that effectively the total charge is zero but the 2 layers form an assembly of dipoles, thus giving rise to an electric field.
electric doublet *(Phys.)*. System with a definite electric moment, mathematically equivalent to 2 equal charges of opposite sign at a very small distance apart.
electric dynamometer *(Eng.)*. An electric generator which is used for measuring power. The stator frame is capable of partial rotation in bearings concentric with those of the armature, and the torque is balanced and measured by hanging weights on an arm projecting from the frame.
electric, electrostatic component *(Elec.Eng.)*. That component of an electromagnetic wave which produces a force on an electric charge, and along the direction of which currents in a conductor exposed to the field are urged to flow.
electric eye *(Telecomm.)*. Miniature cathode-ray tube used, e.g. in a radio receiver to exhibit a pattern determined by the rectified output voltage obtained from the received carrier, thus assisting in tuning the receiver. Also used for balancing a.c. bridges. Also *cathode-ray tuning indicator, electron-ray indicator tube, magic eye*.
electric field *(Phys.)*. Region in which forces are exerted on any electric charge present.
electric field strength *(Phys.)*. The strength of an electric field is measured by the force exerted on a unit charge at a given point. Expressed in volts/metre. Symbol *E*.
electric flux *(Phys.)*. Surface integral of the electric field intensity normal to the surface. The electric flux is conceived as emanating from a positive charge and ending on a negative charge without loss. Symbol Ψ.
electric flux density *(Phys.)*. See **displacement**.
electric generator *(Elec.Eng.)*. See **generator**.
electric harmonic analyser *(Elec.Eng.)*. An electrical device for determining the magnitude of the harmonics in the wave shape of an alternating current or voltage. Also known as *spectrum analyser*.
electricity *(Elec.Eng.)*. The manifestation of a form of energy associated with static or dynamic electric charges.
electricity meter *(Elec.Eng.)*. See **integrating meter**.
electric lamp *(Elec.Eng.)*. A lamp in which an electric current is used as a source of energy for radiating light.
electric-light ophthalmia *(Med.)*. See **photophthalmia**.
electric locomotive *(Elec.Eng.)*. A locomotive in which the motive power is by electric motor, supplied either from batteries (battery locomotive), from a diesel engine/electrical generator set on the locomotive (diesel electric), from overhead contact wire or from a contact rail.
electric machine *(Elec.Eng.)*. See **electric motor, electromagnetic generator, electrostatic generator**.
electric moment *(Elec.Eng.)*. Product of the magnitude of either of 2 equal electric charges and the distance between their centres, with axis direction from the negative to the positive charge. See also **magnetic moment**.
electric motor *(Elec.Eng.)*. Any device for converting electrical energy into mechanical torque; occasionally called an *electromotor*.
electric organ *(Zool.)*. A mass of muscular or nervous tissue, modified for the production, storage, and discharge of electric energy; occurring in Fish.
electric oscillations *(Telecomm.)*. Electric currents which periodically reverse their direction of flow, at a frequency determined by the constants of a resonant circuit. See also **continuous oscillations, electronic oscillations**.
electric polarization *(Elec.Eng.)*. The dipole moment per unit volume of a dielectric.
electric potential *(Elec.Eng.)*. That measured by the energy of a unit positive charge at a point, expressed relative to **zero potential**.
electric propulsion *(Space)*. The use of electrostatic or electromagnetic fields to accelerate ions or plasma thereby producing propulsive thrust. See **ion propulsion**.
electric resistance welded tube *(Chem.Eng.)*. Much used in heat exchangers, it is made continuously by forming an accurately rolled strip over a mandrel and welding the edges electrically. Abbrev *ERW tube*.
electric shielding *(Elec.Eng.)*. See **Faraday cage**.
electric storm *(Meteor.)*. A meteorological disturbance in which the air becomes highly charged with static electricity. In the presence of clouds this leads to thunderstorms.
electric strength *(Elec.Eng.)*. Maximum voltage which can be applied to an insulator or insulating material without sparkover or breakdown taking place. The latter arises when the applied voltage gradient coincides with a breakdown strength at a temperature which is attained through normal heat dissipation.
electric susceptibility *(Elec.Eng.)*. The amount by which the relative permittivity of a dielectric exceeds unity, or the ratio of the polarization produced by unit field to the permittivity of free space.
electric traction *(Elec.Eng.)*. The operation of a railway or road vehicle by means of electric motors, which obtain their power from an overhead contact wire or from batteries mounted on the vehicle.
electric wind *(Elec.Eng.)*. Stream of air caused by the repulsion of charged particles from a sharply pointed portion of a charged conductor.
electrification *(Elec.Eng.)*. (1) Charging a network to a

high potential. (2) Conversion of any plant to operation by electricity, e.g. changeover of steam driven locomotives to electricity. (3) Charging a conductor by electric induction from another charged conductor. (4) Provision of a supply of electrical energy where none existed, e.g. rural electrification is the provision of electricity to consumers in country areas.

electroacoustics *(Acous.)*. The branch of technology dealing with the interchange of electric and acoustic energy, e.g. as in a transducer.

electroanalysis *(Chem.)*. Electrodeposition of an element or compound to determine its concentration in the electrolysed solution. See **conductimetric analysis, polarograph, potentiometer, voltameter**.

electroarteriograph *(Med.)*. Instrument for recording blood-flow rates.

electrobrightening *(Eng.)*. See **electrolytic polishing**.

electrocapillary effect *(Chem.)*. The decrease in interfacial tension, usually of mercury, caused by the mutual repulsion of adsorbed ions opposing the attractive force of interfacial tension. See also **capillary electrometer, electrical double layer**.

electrocapillary maximum *(Chem.)*. The potential at which a mercury surface in an electrolyte is charge-free and consequently has the maximum interfacial tension; about −0.28 volts.

electrocardiogram *(Med.)*. *ECG*. A record of the electrical activity of the heart.

electrocardiography *(Med.)*. The study of electric currents produced in cardiac muscular activity.

electrocataphoresis *(Phys.)*. See **electrophoresis**.

electrocautery *(Med.)*. See **electric cautery**.

electrochemical constant *(Chem.)*. See **faraday**.

electrochemical, electrode potential series *(Chem.)*. The classification of redox half-reactions, written as reductions, in order of decreasing reducing strength. Thus the combination of any half reaction with the reverse of one further down the series will give a spontaneous reaction. For reference, the half reaction $2H^+ + e^- = H_2$ is taken as having an energy of zero.

electrochemical equivalent *(Phys.)*. The mass of a substance deposited at the cathode of a voltameter per coulomb of electricity passing through it.

electrochemical machining *(Eng.)*. Removing material from a metal by anodic dissolution in a bath in which electrolyte is pumped rapidly through the gap between the shaped electrode and the stock. Abbrev. *ECM*.

electrochemistry *(Chem.)*. That branch of chemistry which deals with the electronic and electrical aspects of processes, usually, but not always, in a liquid phase.

electrochronograph *(Elec.Eng.)*. The combination of an electrically driven clock and an electromagnetic recorder for recording short time intervals.

electrocoagulation *(Med.)*. Coagulation of bodily tissues by high-frequency electric current.

electroconvulsive therapy *(Behav.)*. A form of therapy in which an electric current is passed through the brain, resulting in convulsive seizures, used primarily in the treatment of depression. Abbrev. *ECT*.

electroconvulsive therapy *(Med.)*. Passage of an electric current through the brain (usually under a light general anaesthetic) to produce a convulsion. Generally used to treat refractory **depression**.

electro copper glazing *(Build.)*. See **copper glazing**.

electrocution *(Med.)*. Death caused by an electric shock.

electrocyte *(Zool.)*. A cell, usually muscle but sometimes nerve, which is specialized to generate an electric discharge.

electrode *(Elec.Eng.)*. (1) Conductor whereby an electric current is lead into or out of a liquid (as in an electrolytic cell), or a gas (as in an electric discharge lamp or gas tube), or a vacuum (as in a valve).

electrode *(Electronics)*. In a semiconductor, emitter or collector of electrons or holes.

electrode admittance *(Elec.Eng.)*. Admittance measured between an electrode and earth when all potentials on electrodes are maintained constant.

electrode boiler *(Elec.Eng.)*. A boiler in which heat is produced by the passage of an electric current through the liquid to be heated.

electrode characteristic *(Elec.Eng.)*. Graph relating current in electronic device to potential of one electrode, that of all others being maintained constant.

electrode conductance *(Elec.Eng.)*. In-phase or real component of an electrode admittance.

electrode current *(Electronics)*. The net current flowing in a valve (or tube) from an electrode into the surrounding space.

electrode dark current *(Electronics)*. The current which flows in a camera tube or phototube when there is no radiation incident on the photocathode, given certain specified conditions of temperature and shielding from radiation. It limits the sensitivity of the device.

electrode dissipation *(Electronics)*. Power released at an electrode, usually an anode, because of electron ion impact. In large valves the temperature is held down by radiating fins, graphiting to increase radiation, or by water- or oil-cooling.

electrode efficiency *(Elec.Eng.)*. The ratio of the quantity of metal deposited in an electrolytic cell to the quantity which should theoretically be deposited according to Faraday's laws.

electrode holder *(Elec.Eng.)*. In electric arc-welding, a device used for holding the electrode and leading the current to it.

electrode impedance *(Elec.Eng.)*. Impedance (or its components) is the ratio of a small sinusoidal voltage on an electrode to the corresponding sinusoidal current, all other electrodes being maintained at constant potential.

electrodeposition *(Chem.)*. Deposition electrolytically of a substance on an electrode, as in electroplating or electroforming.

electrode potential series *(Chem.)*. See **electrochemical series**.

electrode resistance *(Elec.Eng.)*. Reciprocal of **electrode conductance**.

electrodermal effect *(Med.)*. Change in skin resistance – consequent upon emotional reactions.

electrodialysis *(Chem.)*. Removal of electrolytes from a colloidal solution by an electric field between electrodes in pure water outside the two dialysing membranes between which is contained the colloidal solution.

electrodisintegration *(Phys.)*. Disintegration of nucleus under electron bombardment.

electrodissolution *(Elec.Eng.)*. Dissolving a substance from an electrode by electrolysis.

electrodynamic instrument *(Elec.Eng.)*. An electrical measuring instrument which depends for its action on the electromagnetic force between 2 or more current-carrying coils.

electrodynamic loudspeaker *(Acous.)*. Loudspeaker, in which the radiating cone is driven by current in a coil which moves in a constant magnetic field.

electrodynamic microphone *(Acous.)*. The inverse of an **electrodynamic loudspeaker**.

electrodynamic wattmeter *(Elec.Eng.)*. One for low-frequency measurements. It depends on the torque exerted between currents carried by fixed and movable coils.

electroencephalogram *(Med.)*. *EEG*. A record of the electrical activity of the brain.

electroencephalograph *(Med.)*. Instrument for study of voltage waves associated with the brain; effectively comprises a sensitive detector (voltage or current), a d.c. amplifier of very good stability and an electronic recording system. Abbrev. *EEG*.

electroendosmosis *(Chem.)*. Movement of liquid, under an applied electric field, through a fine tube or membrane. Also *electro-osmosis, electrosmosis*.

electroextraction *(Eng.)*. The recovery of a metal from a solution of its salts by electrolysis, the metal depositing on the cathode.

electrofacing *(Elec.Eng.)*. The process of coating, by electrodeposition, a metal surface with a harder metal to render it more durable.

electrofluor *(Phys.)*. A transparent material which has the property of storing electrical energy and releasing it as visible light.

electrofluorescence *(Phys.)*. See **electroluminescence**.

electroformed sieves *(Powder Tech.)*. Fine mesh sieves formed by photoengraving and electroplating nickel.

electroforming *(Electronics)*. Electrodeposition of copper on stainless steel formers, to obtain components of waveguides with closely dimensioned sections.

electroforming *(Eng.)*. A primary process of forming metals, in which parts are produced by electrolytic deposition of metal on a conductive removable mould or matrix.

electrogenic pump *(Bot.)*. An ion-translocating pump which causes a net transfer of charge across a membrane and therefore an electrical potential difference across it.

electrogram *(Elec.Eng.)*. A chart, obtained from an electrograph, of variations in the atmospheric potential gradient.

electrograph *(Elec.Eng.)*. A recording electrometer for measuring the potential gradient in the atmosphere.

electrographite *(Powder Tech.)*. Carbon can be transformed into graphite by heat. Temperature in excess of 1500°C causes detectable graphite formation. Temperatures in the range 2200°C to 2800°C are commonly used.

electrohydraulic forming *(Eng.)*. The discharge of electrical energy across a small gap between electrodes immersed in water, which produces a shock wave. The wave travels through the water, until it hits and forces a metal workpiece into the particular shape of the die which surrounds it, thus, e.g. piercing holes, rock-crushing, etc. Sometimes called *explosive forming*.

electrokinetic effects *(Chem.)*. Phenomena due to the interaction of the relative motion with the potential between the two phases in a dispersed system. There are four: **electroendosmosis, electrophoresis, streaming potential** and **sedimentation potential**.

electrokinetic potential *(Chem.)*. Potential difference between surface of a solid particle immersed in aqueous or conducting liquid and the fully dissociated ionic concentration in the body of the liquid. Concept important in froth flotation, electrophoresis etc. Also called *zeta potential*.

electrokinetics *(Phys.)*. Science of electric charges in motion, without reference to the accompanying magnetic field.

electroluminescence *(Phys.)*. Luminescence produced by the application of an electric field to a dielectric phosphor. Also termed *electrofluorescence*. See **Gudden-Pohl effect**.

electrolysis *(Chem.)*. Chemical change, generally decomposition, effected by a flow of current through a solution of the chemical, or its molten state, based on ionization.

electrolysis *(Med.)*. The removal of hair by applying an electrically charged needle to the follicle.

electrolyte *(Chem.)*. Chemical, or its solution in water, which conducts current through ionization.

electrolyte *(Med.)*. A compound that in solution dissociates into ions. In clinical practice sodium, potassium and bicarbonate are the electrolytes of everyday concern.

electrolyte strength *(Chem.)*. Extent towards complete ionization in a dilute solution. When concentrated, the ions join in groups, as indicated by lowered mobility.

electrolytic capacitor *(Elec.Eng.)*. An electrolytic cell in which a very thin layer of nonconducting material has been deposited on one of the electrodes by an electric current. This is known as 'forming' the capacitor, the deposited layer providing the dielectric. Because of its thinness a larger capacitance is achieved in a smaller volume than in the normal construction of a capacitor. In the so-called *dry electrolytic capacitor* the dielectric layer is a gas which however is actually 'formed' from a moist paste within the capacitor.

electrolytic cell *(Chem.)*. An electrochemical cell in which an externally applied voltage causes a non-spontaneous change to occur, such as the breakdown of water into hydrogen and oxygen. Opposite of **galvanic cell**.

electrolytic copper *(Eng.)*. Copper refined by electrolysis. This gives metal of high purity (over 99.94% copper), and enables precious metals, such as gold and silver, to be recovered.

electrolytic corrosion *(Eng.)*. Corrosion produced by contact of two different metals when an electrolyte is present and current flows.

electrolytic depolarization *(Elec.Eng.)*. See **depolarization**.

electrolytic dissociation *(Chem.)*. The splitting-up (which is reversible) of substances into oppositely-charged ions.

electrolytic grinding *(Eng.)*. A metal bonded and diamond impregnated grinding wheel removes metal from the stock but in addition the abradant is an insulator between the wheel acting as anode and the stock which has electrolyte flowing over it. This allows **electrochemical machining** to occur.

electrolytic instrument *(Elec.Eng.)*. An instrument depending for its operation upon electrolytic action, e.g. an electrolytic meter.

electrolytic lead *(Eng.)*. Lead refined by the Betts Process; has purity of about 99.995–99.998% lead.

electrolytic lightning arrester *(Elec.Eng.)*. A lightning arrester consisting of a number of electrolytic cells in series; it breaks down, allowing the lightning stroke to discharge to earth, when the voltage across it exceeds about 400 volts/cell.

electrolytic machining *(Eng.)*. An electrochemical process based on the same principles as electroplating, except that the workpiece is the anode and the tool is the cathode, resulting in a deplating operation. Sometimes, e.g. in electrolytic grinding, combined with some abrasive action.

electrolytic meter *(Elec.Eng.)*. An integrating meter whose operation depends on electrolytic action.

electrolytic polarization *(Elec.Eng.)*. Change in the potential of an electrode when a current is passed through it. As current rises, polarization reduces the p.d. between the two electrodes of the system.

electrolytic polishing *(Eng.)*. By making the metal surface an anode and passing a current under certain conditions, there is a preferential solution so the microscopic irregularities vanish, leaving a smoother surface; also termed *anode brightening, anode polishing, electrobrightening, electropolishing*.

electrolytic refining *(Eng.)*. The method of producing pure metals, by making the impure metal the anode in an electrolytic cell and depositing a pure cathode. The impurities either remain undissolved at the anode or pass into solution in the electrolyte.

electrolytic tank *(Elec.Eng.)*. A device used to simulate field systems, e.g. the electrostatic field around the electrodes in a CRT. The tank is filled with a poorly conducting fluid in which is immersed a scale model in metal of the desired system. Appropriate voltages are applied between the various parts and the equipotentials are traced out using a suitable probe.

electrolytic wirebar *(Elec.Eng.)*. A bar of electrolytically refined copper of suitable dimensions for rolling to form wire.

electrolytic zinc *(Eng.)*. Zinc produced from its ores by roasting (to convert sulphide to oxide), solution of oxide in sulphuric acid, precipitation of impurities by adding zinc dust, and final electrolytic deposition of zinc on aluminium cathodes. Product has purity over 99.9%.

electromagnet *(Elec.Eng.)*. Soft iron core, embraced by a current-carrying coil, which exhibits appreciable magnetic effects only when current passes.

electromagnetic brake *(Elec.Eng.)*. A brake in which the braking force is produced by the friction between two surfaces pressed together by the action of a solenoid, or by magnetic attraction, the necessary flux being produced by an electromagnet.

electromagnetic clutch *(Autos.,Eng.)*. A friction clutch without pressure springs, operated by a solenoid connected to the dynamo. As the current increases the solenoid presses the plates together. A switch in the base of the gear lever breaks the circuit and releases the pressure for gear changes.

electromagnetic component *(Telecomm.)*. Strictly, that component of the combined field surrounding a transmitting antenna which represents the radiated energy.

electromagnetic control *(Elec.Eng.)*. A form of remote control for switchgear, etc., in which operation is effected by means of a solenoid.

electromagnetic damping *(Elec.Eng.)*. See **magnetic damping**.

electromagnetic deflection *(Electronics)*. Deflection of the beam in a cathode-ray tube by a magnetic field produced by a system of coils carrying currents, e.g. for scanning a TV image or for providing a time-base deflection.

electromagnetic field theory *(Phys.)*. Theory based on *Maxwell's field equations* and concerned with electromagnetic interactions. It predicts the existence of a wave comprising interdependent transverse waves of electric and magnetic fields, **electromagnetic waves**. The scope of the theory is enormous including as it does the operating principles of such electromagnetic devices as motors, television and radar.

electromagnetic focusing *(Electronics)*. The focusing of a beam of charged particles by magnetic fields associated with current-carrying coils. Used in cathode-ray tube and electron microscope. See **electron lens**.

electromagnetic generator *(Elec.Eng.)*. An electric generator which depends for its action on the induction of e.m.f.s in a circuit by a change in the magnetic flux linking with that circuit.

electromagnetic horn *(Telecomm.)*. See **horn antenna**.

electromagnetic induction *(Elec.Eng.)*. Transfer of electrical power from one circuit to another by varying the magnetic linkage. See **Faraday's law of induction**.

electromagnetic inertia *(Elec.Eng.)*. The energy required to stop or start a current in an inductive circuit. Inductance is analagous to mass in a mechanical system.

electromagnetic instruments *(Elec.Eng.)*. Electrical measuring instruments whose action depends on the electromagnetic forces set up between a current-carrying conductor and a magnetic field. See **moving-coil instrument, dynamometer**.

electromagnetic interaction *(Phys.)*. An interaction between charged elementary particles and mediated by **photons**; completed in about 10^{-18}s. Intermediate in strength between *strong* and *weak* interactions.

electromagnetic lens *(Electronics)*. One using current-carrying coils to focus electron beams.

electromagnetic loudspeaker *(Acous.)*. Device where a fluctuating magnetic field, induced by an electric current, excites a force in magnetic material connected to the loudspeaker cone.

electromagnetic microphone *(Acous.)*. Inverse of an **electromagnetic loudspeaker**,

electromagnetic mirror *(Elec.Eng.)*. A reflecting surface for electromagnetic waves.

electromagnetic pick-up *(Acous.)*. One in which the motion of the stylus, in following the recorded track, causes a fluctuation in the magnetic flux carried in any part of a magnetic circuit, with consequent electromotive forces in any coil embracing such magnetic circuit.

electromagnetic pole-piece *(Elec.Eng.)*. In a U-shaped core, the pole-pieces are attached at the free end and are often conical in shape to concentrate the magnetic field in the air gap.

electromagnetic prospecting *(Geol.)*: Method in which distortions produced in electric waves are observed and lead to pinpointing of geological anomalies.

electromagnetic pump *(Elec.Eng.)*. Pump designed to conduct fluids, e.g. liquid metals, and to maintain circulation without use of moving parts.

electromagnetic radiation *(Phys.)*. The emission and propagation of electromagnetic energy from a source including long (radio) waves, heat rays, light, X-rays and γ-rays.

electromagnetic reaction *(Elec.Eng.)*. Reaction between the anode and grid circuits of a valve obtained by electromagnetic coupling; also called *inductive reaction, magnetic reaction*.

electromagnetics *(Genrl.)*. See **electromagnetism**.

electromagnetic separation *(Min.Ext.)*. Removal of ferromagnetic objects from town refuse, or 'tramp iron' from bulk materials, as they travel along a conveyor, over a drum, or into a revolving screen, by setting up a magnetic field which diverts the ferromagnetic material from the rest. Concentration of ferromagnetic minerals from gangue.

electromagnetic separation *(Nuc.Eng.)*. Isotope separation by **electromagnetic focusing**, as in a mass spectrometer.

electromagnetic spectrum *(Phys.)*. See **electromagnetic wave**.

electromagnetic switch *(Elec.Eng.)*. A switch whose opening and closing is effected by means of electromagnets or solenoids.

electromagnetic theory *(Phys.)*. See **electromagnetic field theory**.

electromagnetic units *(Phys.)*. Any system of units based on assigning an arbitrary value to μ_0, the permeability of free space. $\mu_0 = 4\pi\lambda 10^{-7}\,\mathrm{Hm}^{-1}$ in the SI system; μ_0 is unity in the CGS electromagnetic system.

electromagnetic wave *(Phys.)*. A wave comprising two interdependent mutually perpendicular transverse waves of electric and magnetic fields. The velocity of propagation in free space for all such waves is that of the velocity of light, $2.997\,924\,58 \times 10^8\,\mathrm{ms}^{-1}$. The electromagnetic spectrum ranges from wavelengths of 10^{-15} m to 10^3 m, i.e. from γ-rays through X-rays, ultraviolet, visible light, infrared, microwave, short-, medium- and long-wave radio waves. Electromagnetic waves undergo reflection and refraction and exhibit interference and diffraction effects and can by polarized. The waves can be channelled by, for example, waveguides for microwaves or fibre optics for light. See **electromagnetic field theory**.

electromagnetism *(Elec.Eng.)*. Science of the properties of, and relations between, magnetism and electric currents. Also called *electromagnetics*.

electromechanical brake *(Elec.Eng.)*. A form of electric brake in which the braking force is obtained partly as the result of the attraction of two magnetized surfaces, and partly by mechanical means, as a result of the operation of a solenoid.

electromechanical counter *(Elec.Eng.)*. One which records mechanically the number of electric pulses fed to a *solenoid*.

electromechanical recorder *(Elec.Eng.)*. One which changes electrical signals into a mechanical motion of a similar form, and cuts or records the shape of the motion in an appropriate medium.

electrometallization *(Eng.)*. The electrodeposition of a metal on a nonconducting base, either for decorative purposes or to give a protective covering.

electrometallurgy *(Eng.)*. A term covering the various electrical processes for the industrial working of metals, e.g. electrodeposition, electrorefining and operations in electric furnaces.

electrometer *(Elec.Eng.)*. Fundamental instrument for measuring potential difference, depending on the attraction or repulsion of charges on plates or wires.

electrometer gauge *(Elec.Eng.)*. A small attracted-disk electrometer sometimes attached to the needle of a quadrant electrometer to determine whether the needle is sufficiently charged.

electrometric titration *(Chem.)*. See **potentiometric titration**.

electromotive force *(Elec.Eng.)*. Difference of potential produced by sources of electrical energy which can be used to drive currents through external circuits. Abbrev. *emf*, unit, *volt*.

electromotive series *(Chem.)*. See **electrochemical series**.

electromyography *(Med.)*. The study of electric currents set up in muscle fibres by bodily movement.

electron *(Phys.)*. A fundamental particle with negative electric charge of 1.602×10^{-19} coulombs and mass 9.109×10^{-31} kg. Electrons are a basic constituent of the atom; they are distributed around the nucleus in *shells* and the electronic structure is responsible for the chemical properties of the atom. Electrons also exist independently and are responsible for many electric effects in materials. Due to their small mass, the wave properties and relativistic effects of electrons are marked. The *positron*, the antiparticle of the electron, is an equivalent particle but with a positive charge. Either electrons or positrons may be emitted in β-decay. Electrons, muons, and neutrinos form a group of fundamental particles called **leptons**.

electron affinity *(Chem.)*. The energy required to remove an electron from a negatively charged ion to form a neutral atom.

electron affinity *(Electronics)*. (1) Tendency of certain substances, notably oxidizing agents, to capture an electron. (2) See **work function**.

electron attachment *(Phys.)*. Formation of negative ion by attachment of free electron to neutral atom or molecule.

electron beam *(Electronics)*. A stream of electrons moving with the same velocity and direction in neighbouring paths and usually emitted from a single source such as a cathode.

electron-beam analysis *(Eng.)*. Scanning a microbeam of electrons over a surface *in vacuo* and analysing the secondary emissions to determine the distribution of selected elements.

electron beam valve (tube) *(Electronics)*. One in which several electrodes control one or more electron beams.

electron-beam welding *(Eng.)*. Heating components to be welded by a concentrated beam of high-velocity electrons *in vacuo*.

electron binding energy *(Phys.)*. Same as **ionization potential**.

electron camera *(Image Tech.)*. Generic term for a device which converts an optical image into a corresponding electric current directly by electronic means, without the intervention of mechanical scanning.

electron capture *(Phys.)*. That of shell electron (K or L) by the nucleus of its own atom, decreasing the atomic number of the atom without change of mass. The capture is accompanied by the emission of *neutrino*.

electron charge/mass ratio *(Electronics)*. A fundamental physical constant, the mass being the rest mass of the electron: $e/m = 1.759 \times 10^{11}\,\mathrm{C\,kg^{-1}}$.

electron cloud *(Chem.)*. (1) The density of electrons in a volume of space, as the position and velocity of an electron cannot be simultaneously specified. (2) The nature of the valence electrons in a metal, where their non-attachment to specific nuclei gives rise to electronic conduction.

electron conduction *(Electronics)*. That which arises from the drift of free electrons in metallic conductors when an electric field is applied. See **n-type** and **p-type semiconductor**.

electron coupling *(Electronics)*. That between two circuits, due to an electron stream controlled by the one circuit influencing the other circuit. Such coupling tends to be unidirectional, the second circuit having little influence on the first.

electron-deficient *(Chem.)*. A substance which does not have enough valence electrons to form 'normal' chemical bonds. Usually, the term is used for the compounds of boron, but all metallic bonding is of this type.

electron density *(Phys.)*. The number of electrons per gram of a material. Approx. 3×10^{23} for most light elements. In an ionized gas the equivalent electron density is the product of the ionic density and the ratio of the mass of an electron to that of a gas ion.

electron device *(Electronics)*. One which depends on the conduction of electrons through a vacuum, gas or semiconductor.

electron diffraction *(Electronics)*. Investigation of crystal structure by the patterns obtained on a screen from electrons diffracted from the surface of crystals or as a result of transmission through thin metal films.

electron discharge *(Electronics)*. Current produced by the passage of electrons through otherwise empty space.

electron-discharge tube *(Electronics)*. Highly-evacuated tube containing two or more electrodes between which electrons pass.

electron dispersion curve *(Phys.)*. A curve showing the electron energy as a function of the wave vector under the influence of the periodic potential of a crystal lattice. Experiments and calculations which determine such curves give important information about the energy gaps, the electron velocities and the density of states.

electron drift *(Electronics)*. The actual transfer of electrons in a conductor as distinct from energy transfer arising from encounters between neighbouring electrons.

electronegative *(Phys.)*. Carrying a negative charge of electricity. Tending to form negative ions, i.e., having a relatively positive electrode potential.

electronegativity *(Chem.)*. The relative ability of an atom to retain or gain electrons. There are several definitions, and the term is not quantitative. It is, however, useful in predicting the strengths and the polarities of bonds, both of which are greater when there is a significant electronegativity difference between the atoms forming the bond. On the commonly used scale of Pauling, the range of values is from less than 1 (alkali metals) to 4 (fluorine).

electron-electron scattering *(Phys.)*. A possible process that contributes to the electrical resistivity of metals. Important at low temperatures in transition metals.

electron emission *(Electronics)*. The liberation of electrons from a surface.

electron gas *(Electronics)*. The 'atmosphere' of free electrons in vacuo, in a gas or in a conducting solid. The laws obeyed by an electron gas are governed by Fermi-Dirac statistics, unlike ordinary gases to which Maxwell-Boltzmann statistics apply.

electron gun *(Electronics)*. Assembly of electrodes in a cathode-ray tube which produces the electron beam, comprising a cathode from which electrons are emitted, an apertured anode, and one or more focusing diaphragms and cylinders.

electronic *(Genrl.)*. Pertaining to devices or systems which depend on the flow of electrons; the term covers most branches of electrical science other than electric power generation and distribution. Telecommunications, radar, and computers all use electronic components and techniques. Electronic engineering is a field which encompasses the application of electronic devices, as opposed to physical electronics which is the study of electronic phenomena in vacuum, in gases, or in solids.

electronic charge *(Phys.)*. The unit in which all nuclear charges are expressed. It is equal to 1.602×10^{-19} coulombs.

electronic configuration *(Chem.)*. The descriptions of the electrons of an atom or a molecule in terms of orbitals.

electronic control *(Elec.Eng.)*. Method of control which is based on the use of electronic circuits, suitable transducers and actuators where necessary. Modern examples often incorporate programmable units for both accuracy and flexibility. Applications are widespread, e.g. other circuits, instruments, machinery etc.

electronic counter measures *(Telecomm.)*. An offensive or defensive tactic using electronic systems and reflectors to impair the effectiveness of enemy guidance, surveillance or navigational equipment which depend on electromagnetic signals. Also *electronic warfare, EW*. Abbrev. *ECM*.

electronic engineering *(Electronics)*. See **electronic**.

electronic engraving *(Print.)*. The making of plates direct from the copy without the use of the camera or the etching bath. As the copy is scanned the plates are engraved. Some models produce sets of colour plates, applying the necessary **colour correction** electronically.

electronic flash *(Electronics)*. Battery or mains device which charges a capacitor, the latter discharging through a tube containing neon or xenon when triggered. Used for photography and stroboscopy.

electronic flash *(Image Tech.)*. A source of very brief illumination provided by a high-voltage discharge between electrodes in an envelope containing a rare gas, such as xenon. It may be used repeatedly.

electronic funds transfer *(Comp.)*. The use of computer systems to transfer credits and debits between cooperating organizations, such as banks or large companies. Abbrev *EFT*.

electronic ignition *(Autos.)*. Generic term for various systems which employ electronic circuitry, rather than the traditional coil and circuit breaker, to provide a timed high-voltage pulse to the sparking plugs. Cf. **coil ignition**.

electronic intelligence *(Telecomm.)*. Using airborne equipment, ground stations and surveillance satellites, to monitor and record enemy electromagnetic emanations and to reveal the nature and deployment of guidance, navigational and communication systems.

electronic keying *(Telecomm.)*. Production of telegraphic signals by all-electronic system.

electronic mail *(Comp.)*. Communications system in which a computer user can compose and address a letter at one terminal, which is then generated at another elsewhere, when the recipient *logs in*. Worldwide **networks** for electronic mail are in use. Abbrev. *email*.

electronic microphone *(Acous.)*. One in which the acoustic pressure is applied to an electrode of a valve.

electronic music *(Acous.)*. See **electrosonic music**.

electronic oscillations *(Telecomm.)*. Those generated by electrons moving between electrodes in an amplifying device, the frequency being determined by the transit time.

electronic pick-up *(Electronics)*. One in which external vibration affects the grid of a valve and thereby modulates the anode current. See **microphonic**.

electronic rectifier *(Elec.Eng.)*. See **rectifier**.

electronic register control *(Print.)*. Equipment to maintain lateral and lengthwise register on webfed printing and paper-converting machines.

electronic shutter *(Image Tech.)*. A camera shutter in which the rate of opening and closing is set by the charge of a capacitor through a transistor rather than by mechanical means.

electronic switch *(Elec.Eng.)*. Device for opening or closing electric circuit by electronic means.

electronic theory of valency *(Chem.)*. Valency forces arise from the transfer or sharing of the electrons in the outer shells of the atoms in a molecule. The two extremes are complete transfer (*ionic bond*) and close sharing (*covalent bond*), but there are intermediate degrees of bond strength and distance.

electronic traction control *(Autos.)*. System which adjusts the power provided to the driven wheels to prevent wheelspin by sensing excessive acceleration of the wheels.

electronic tuning *(Electronics)*. (1) Changing the operating frequency of an amplifier, oscillator, receiver, or other device, by altering a control voltage or signal rather than by mechanically changing any physical properties of tuned circuits. (2) Changing the operating frequency of a high-frequency valve, which depends on an electron beam for its operation, by altering the characteristics of the beam.

electronic voltmeter *(Elec.Eng.)*. One whose operation depends on the detection, measurement and amplifying properties of electronics. Normally the display is digital. See **digital meter**.

electronic wattmeter *(Elec.Eng.)*. One whose operation depends on the detection, measurement and amplifying properties of electronics. Normally the display is digital. See **digital meter**.

electron lens *(Electronics)*. A composite arrangement of magnetic coil and charged electrodes, to focus or divert electron beams in the manner of an optical lens.

electron mass *(Phys.)*. A result of relativity theory, that mass can be ascribed to kinetic energy, is that the effective mass (m) of the electron should vary with its velocity according to the experimentally confirmed expression:

$$m=\frac{m_0}{\sqrt{1-\left(\frac{v}{c}\right)^2}},$$

where m_0 is the mass for small velocities, c is the velocity of light, and v that of the electron.

electron micrograph *(Biol.)*. A photomicrograph of the image of an object, taken by substituting a photographic plate for the fluorescent viewing screen of an **electron microscope**.

electron microscope *(Biol.)*. Tube in which electrons emitted from the cathode are focused, by suitable magnetic and electrostatic fields, to form an enlarged image of the cathode on a fluorescent screen. By passing the electrons through an object, such as a virus, a vastly enlarged image can be obtained on a photographic plate. The instrument has very high resolving power compared with the optical microscope, due to the shorter wavelength associated with electron waves.

electron mirror *(Electronics)*. A 'reflecting' electrode in an electron tube, e.g. reflex klystron.

electron mobility *(Electronics)*. See **mobility**.

electron multiplier *(Electronics)*. Electron tube in which the anode is replaced by a series of auxiliary electrodes, maintained at successively increasing positive potentials up to the final anode. Electrons emitted from the cathode impinge on the first of the auxiliary electrodes, from which secondary electrons are ejected and travel to the next electrode, where the process is repeated. With suitable materials for the auxiliary electrodes, the number of secondary electrons emitted at each stage is greater than the number of incident electrons, so that very high overall amplification of the original tube current results. See **dynode**.

electron octet *(Chem.,Phys.)*. The (up to) eight valency electrons in an outer shell of an atom or molecule. Characterized by great stability, in so far as the complete shell round an atom makes it chemically inert, and round a molecule (by sharing) makes a stable chemical compound.

electron optics *(Phys.)*. Control of free electrons by curved electric and magnetic fields, leading to focusing and formation of images.

electron pair *(Chem.)*. Two valence electrons shared by adjacent nuclei, so forming a bond.

electron paramagnetic resonance *(Chem.)*. See **electron spin resonance**.

electron-phonon scattering *(Phys.)*. An important process that contributes to the cause of electrical resistivity; the electrons are scattered by the thermal vibrations of the crystal lattice.

electron probe analysis *(Chem.)*. A beam of electrons is focused on to a point on the surface of the sample, the elements being detected both qualitatively and quantitatively by their resultant X-ray spectra. An accurate (1%) non-destructive method needing only small quantities (micron size) of sample.

electron radius *(Phys.)*. The classical theoretical value is 2.82×10^{-15} m, but experimentally the effective value varies greatly with the interaction concerned.

electron runaway *(Electronics)*. Condition in a plasma when the electric fields are sufficiently large for an electron to gain more energy from the field than it loses, on average, in a collision. See **avalanche effect**.

electron scanning *(Image Tech.)*. Scanning or establishing a TV image by an electron beam in a TV camera tube or a CRT, normally using a rectangular raster, with horizontal lines.

electron sheath *(Electronics)*. Electron space charge around an anode, when the supply of electrons is greater than demanded by the anode circuit. See **electron cloud**.

electron shell *(Phys.)*. A grouping of electrons surrounding the nucleus of an atom. The *Pauli exclusion principle* governs the way in which electrons can fill the available *orbitals*. For a given principal quantum number n, there are n allowed values of l, the orbital angular momentum quantum number; for each value of l, there are $(2l+1)$ allowed values of m_l, the magnetic angular momentum quantum number; for each value of m_l, there are two values of m_s, the magnetic spin number. This makes a total of $2n^2$ orbitals for a given value of n, and, as the Pauli principle allows only one electron for each set of four quantum numbers n, l, m_l, m_s, this limited number of allowed orbitals makes up the electron shell for a given n. Inner filled shells are relatively inert and the chemical properties of the atom are determined by the electron arrangement in the outermost shell. See **K-**, **L-**, **M-...shells**.

electron spin *(Electronics)*. See **spin**.

electron spin resonance *(Phys.)*. A branch of microwave spectroscopy in which there is resonant absorption of radiation by a paramagnetic substance, possessing unpaired electrons, when the energy levels are split by the application of a strong magnetic field. The difference in energy levels is modified by the environment of the atoms. Information on impurity centres in crystals, the nature of the chemical bond and the effect of radiation damage can be found. Also *electron paramagnetic resonance*. Abbrev. *ESR*.

electron trap *(Electronics)*. An acceptor impurity in a semiconductor.

electron tube, valve *(Electronics)*. In US, any electronic device in which the electron conduction is in a vacuum or gas inside a gas-tight enclosure.

electron-volt *(Phys.)*. General unit of energy of moving particles, equal to the kinetic energy acquired by an electron losing one volt of potential, equal to 1.602×10^{-19} J; abbrev. *eV*.

electron volts *(Radiol.)*. The unit of energy for the photon. Used in radiology and expressed in thousands (kev) or millions (mev) of electron volts.

electro-optical effect *(Elec.Eng.)*. Same as *Kerr effect*. See **Pockel's effect**.

electro-osmosis *(Chem.)*. Same as *electro-endosmosis*.

electroparting *(Elec.Eng.)*. The electrolytic separation of two or more metals.

electrophonic effect *(Acous.)*. Sensation of hearing arising from the passage of an electric current of suitable magnitude and frequency through the body.

electrophonic music *(Acous.)*. Same as *electrosonic music*.

electrophoresis *(Chem.)*. Motion of charged particles under an electric field in a fluid, positive groups to the cathode and negative groups to the anode.

electrophorus *(Elec.Eng.)*. Simple electrostatic machine for repeatedly generating charges. A resinous plate, after rubbing, exhibits a positive charge, which displaces a charge through an insulated metal plate placed in partial contact. Earthing the upper surface of this plate leaves a net negative charge on the metal plate when it is removed, a process which can be repeated indefinitely.

electrophysiology *(Biol.)*. The study of electrical phenomena associated with living organisms, particularly nervous conduction.

electroplaque *(Zool.)*. Large, flat disc-shaped **electrocyte**. Usually stacked in series to produce a substantial voltage pulse.

electroplating *(Elec.Eng.)*. Deposition of one metal on another by electrolytic action on passing a current through a cell, for decoration or for protection from corrosion, etc. Metal is taken from the anode and deposited on the cathode, through a solution containing the metal as an ion.

electroplating bath *(Elec.Eng.)*. Tank in which objects to be electroplated are hung. It is filled with electrolyte at the correct temperature, with anodes of the metal to be deposited on articles which are made cathodes.

electroplating generator *(Elec.Eng.)*. A direct-current electric generator, specially designed for electroplating work; it gives a heavy current at a low voltage.

electropneumatic *(Elec.Eng.)*. Said of control system using both electrical and pneumatic elements.

electropneumatic brake *(Civ.Eng.)*. A type of brake which can be applied simultaneously throughout the length of the train.

electropneumatic contactor *(Elec.Eng.)*. A contactor operated by compressed air but controlled by electrically operated valves.

electropneumatic control *(Elec.Eng.)*. A form of remote control in which switches or other apparatus are operated by compressed air controlled by electrically operated valves; commonly used on electric trains.

electropneumatic signalling *(Elec.Eng.)*. A signalling system operated by compressed air, the valves which control the latter being operated electrically.

electropolishing *(Eng.)*. See **electrolytic polishing**.

electroporation *(Biol.)*. Method of introducing foreign DNA or chromosomes into cells by subjecting them to a brief voltage pulse, which transiently increases membrane permeability, allowing uptake of DNA or chromosomes from the surrounding buffer.

electropositive *(Phys.)*. Carrying a positive charge of electricity. Tending to form positive ions, i.e., having a relatively negative electrode potential.

electroradiescence *(Phys.)*. Emission of ultraviolet or infrared radiation from dielectric phosphors on the application of an electric field.

electroreceptor *(Zool.)*. Sense organ specialized for the detection of electric discharges. Found in a variety of Fish particularly Mormyrids, Gymnotids and some Elasmobranchs.

electrorefining *(Eng.)*. See **electrolytic refining**.

electroscope *(Elec.Eng.)*. Indicator and measurer of small electric charges, usually 2 gold leaves which diverge because of repulsion of like charges; with 1 gold leaf and a rigid brass plate, indication is more precise.

electroscopic powder *(Elec.Eng.)*. A mixture of finely divided materials which can acquire charges by rubbing together, so that, if dusted on to a plate, the different materials adhere to differently charged portions of the plate, forming a figure.

electrosmosis *(Chem.)*. See **electroendosmosis**.

electrosonic music *(Acous.)*. Music or other sounds produced by electronic means (e.g., by oscillators, photocells, generators or microprocessors), then combined electrically and reproduced through loudspeakers; also called *electronic music, electrophonic music*.

electrostatic accelerator *(Elec.Eng.)*. One which depends on the electrostatic field due to large d.c. potentials. Used, for example, for accelerating charged particles for certain experiments in physics.

electrostatic actuator *(Acous.)*. Apparatus used for absolute calibration of microphone through application of known electrostatic force.

electrostatic adhesion *(Elec.Eng.)*. That between two substances, or surfaces, due to electrostatic attraction between opposite charges.

electrostatic bonding *(Chem.)*. See **electrovalence**.

electrostatic charge *(Elec.Eng.)*. Electric charge at rest on the surface of an insulator or insulated body, and consequently leading to the establishment of an adjacent electrostatic field system.

electrostatic coupling *(Electronics)*. That between circuit components, one applying a signal to the next through a capacitor.

electrostatic deflection *(Electronics)*. Deflection of the beam of a cathode-ray tube by an electrostatic field produced by two plates between which the beam passes on its way to the fluorescent screen.

electrostatic field *(Elec.Eng.)*. Electric field associated with stationary electric charges.

electrostatic flocking *(Textiles)*. The application of a coloured flock directed by an electrostatic field on to a fabric pretreated with an adhesive. The fibres of the flock protrude from the surface of the fabric giving it a characteristically prickly feel. The products are often used as wall-hangings.

electrostatic focusing *(Electronics)*. Focusing in high-vacuum cathode-ray tubes by the electrostatic field produced by two or more electrodes maintained at suitable potentials.

electrostatic generator *(Elec.Eng.)*. One operating by electrostatic induction, e.g. **Wimshurst machine** and **Van de Graaff generator**. Also called *frictional machine, induction machine, influence machine, static machine*.

electrostatic induction *(Elec.Eng.)*. Movement and manifestation of charges in a conducting body by the proximity of charges in another body. Also the separation of charges in a dielectric by an electric field.

electrostatic instrument *(Elec.Eng.)*. An electrical measuring instrument depending for its action on electrostatic forces set up between charged bodies. See **electroscope, electrometer**.

electrostatic Kerr effect *(Elec.Eng.)*. Dispersion of the plane of polarization experienced by a beam of plane-polarized light on its passage through a transparent medium subjected to an electrostatic strain. The basis of action of some light-modulation systems.

electrostatic lens *(Electronics)*. An arrangement of tubes and diaphragms at different electric potentials.

electrostatic loudspeaker *(Acous.)*. Inverse of an **electrostatic microphone**.

electrostatic machine *(Elec.Eng.)*. See **electrostatic generator**.

electrostatic memory *(Electronics)*. A device in which the information is stored as electrostatic energy, e.g. storage tube. Also *electrostatic storage*.

electrostatic microphone *(Acous.)*. Microphone with a stretched or slack foil diaphragm, which is polarized at steady potential, or at a high-frequency voltage, both of which are modulated by variations in capacitance due to varying pressure.

electrostatic oscillograph *(Elec.Eng.)*. An oscillograph in which the moving element is actuated by electrostatic attraction or repulsion.

electrostatic precipitation *(Elec.Eng.)*. Use of an electrostatic field to precipitate solid (or liquid) particles in a gas, e.g. in dust removal.

electrostatic printing *(Print.)*. See **xerography**.

electrostatics *(Genrl.)*. Section of science of electricity which deals with the phenomena of electric charges substantially at rest.

electrostatic separator *(Elec.Eng.)*. Apparatus in which materials having different permittivities are deflected by different amounts when falling between charged electrodes, and therefore fall into different receptacles.

electrostatic shield *(Elec.Eng.,Telecomm.)*. Conducting shield surrounding instruments or other apparatus to prevent their being influenced by external electric fields, or between two circuits to prevent unwanted capacitance coupling.

electrostatic spraying *(Build.)*. A method of spray application whereby paint particles sprayed through a high voltage electric field becomes charged by electron absorption while the article to be painted is earthed so as to attract the paint particles. By this method overspray is eliminated since the attraction causes a *wrap-round effect* as paint is attracted to the sides of the article not facing the gun.

electrostatic storage *(Electronics)*. See **electrostatic memory**.

electrostatic voltmeter *(Elec.Eng.)*. One depending for its action upon the attraction or repulsion between charged bodies. Usual unit of calibration is *kilovolt*. See also **electrometer**.

electrostatic wattmeter *(Elec.Eng.)*. One which utilizes electrostatic forces to measure a.c. power at high voltages.

electrostriction *(Phys.)*. Change in the dimensions of a dielectric accompanying the application of an electric field.

electrosyntonic switch *(Elec.Eng.)*. A switch remotely controlled by means of a high-frequency current superimposed on the main circuit.

electrotaxis *(Zool.)*. See **galvanotaxis**.

electrotellurograph *(Elec.Eng.)*. An apparatus for the study of earth currents.

electrotherapy, electrotherapeutics *(Med.)*. The treatment of diseases involving the use of electricity.

electrothermic, electrothermal instrument *(Elec.Eng.)*. One depending on the Joulean heating of a current for its operation. See **bolometer**.

electrothermoluminescence *(Phys.)*. Changes in electroluminescent radiation resulting from changes of dielectric temperature. (Some dielectrics show a series of maxima and minima when heated.) The complementary arrangement of observing changes in thermoluminescent radiation when an electric field is applied is termed **thermoelectroluminescence**.

electrotint *(Print.)*. A printing block produced by drawing with varnish on a metal plate, and depositing metal electrically on the parts not covered by the varnish.

electrotonus *(Med.)*. The state of a nerve which is being subjected to a steady discharge of electricity.

electrotropism *(Zool.)*. See **galvanotaxis**.

electrotype *(Print.)*. A hard-wearing printing plate made by depositing a film of copper electrolytically on a mould taken from type or an original plate. The copper shell is backed with a lead alloy. Commonly abbreviated to *electro*.

electrovalence *(Chem.)*. Chemical bond in which an electron is transferred from one atom to another, the resulting ions being held together by electrostatic attraction.

electroviscosity *(Chem.)*. Minor change of viscosity when an electric field is applied to certain polar liquids.

electroweak theory *(Phys.)*. The **Weinberg and Salam theory** unifying electromagnetic and weak interactions between particles.

electrowinning *(Eng.)*. See **electroextraction**.

electrum *(Eng.)*. (1) An alloy of gold and silver (55–88% of gold) used for jewellery and ornaments. (2) Nickel-silver (copper 52%, nickel 26% and zinc 22%); it has the same uses as other nickel-silvers.

electuary *(Med.)*. A medicine consisting of the medicinal agent mixed with honey, syrup or jam.

Elektron alloys *(Eng.)*. TN for magnesium-based light alloys with up to 4.5% copper, up to 12% aluminium and perhaps some manganese and zinc.

element *(Chem.)*. Simple substance which cannot be resolved into simpler substances by normal chemical means. Because of the existence of **isotopes** of elements, an element cannot be regarded as a substance which has identical atoms, but as one which has atoms of the same atomic number.

element *(Elec.Eng.)*. A term often used to denote the resistance wire and former of a resistance type of electric heater.

element *(Electronics)*. Unit of an assembly, especially the detailed parts of electron tubes which affect operation or performance.

element *(Image Tech.,Phys.)*. Component of a lens, e.g. 6-element lens.

element *(Maths.)*. See **constituent**.

element *(Meteor.)*. See **meteorological-**.

elemental analysis *(Chem.)*. Quantitative analysis of a substance to determine the relative amounts of the elements that make it up.

elementary bodies *(Med.)*. Particles present in cells of the body in virus infections.

elementary colours *(Phys.)*. See **primary colours**.

elementary particle *(Phys.)*. Particle believed to be incapable of subdivision; the term **fundamental particle** is now more generally used.

element-former, carrier *(Elec.Eng.)*. A refractory substance upon which the heated wire of a resistance type of electric heater is wound.

elements of an orbit *(Astron.)*. The 6 data mathematically necessary to determine completely a planet's orbit and its position in it: (1) longitude of the ascending node, (2) inclination of the orbit, (3) longitude of perihelion, (4) semi-axis major, (5) eccentricity, (6) epoch, or date of planet's passing perihelion. Analogous elements are used in satellite and double star orbits.

elephantiasis *(Med.)*. Enlargement of the limbs, or of the scrotum, from thickening of the skin and stasis of lymph; due to obstruction of lymphatic channels, especially by filarial worms.

Elephantide pressboard *(Elec.Eng.)*. TN for pressboard with a large cotton content, used for insulation of transformers, armature and stator coils. It is specially suitable for use under oil.

eleutherodactyl *(Zool.)*. Having the hind toe free.

elevated duct *(Meteor.)*. Tropospheric radio duct which has both upper and lower effective boundaries elevated.
elevation *(Arch.,Build.)*. The view or representation of any given side of a building.
elevation *(Eng.)*. A view (e.g. side or end elevation) of a component or assembly drawn in projection on a vertical plane.
elevation *(Surv.)*. **Reduced level** (US).
elevation of boiling-point *(Chem.)*. The raising of the boiling-point of a liquid by substances in solution. May be used to determine molecular weights of solutes.
elevator *(Aero.)*. An aerodynamic surface, operated by fore-and-aft movement of the pilot's control column, governing motion in pitch.
elevator *(Eng.)*. (1) A type of **conveyor** for raising or lowering material which is temporarily carried in buckets or fingers attached to an endless chain or belt. (2) A **lift**.
elevator *(Zool.)*. A muscle which by its contraction raises a part of the body. Cf. **depressor**.
elevator chain *(Eng.)*. Chain used to carry a series of buckets or slats and to which the elevator drive is applied. *Cast chain* is used for slow to medium speeds, *precision (roller) chain* for higher speeds.
elevons *(Aero.)*. Hinged control surfaces on the wing trailing edge of tailless or delta aircraft which are moved in unison to act as elevators and differentially as ailerons.
elfin forest *(Bot.)*. See **Krummholz**.
elimination *(Chem.)*. The removal of a simple molecule (e.g. of water, ammonia etc.) from 2 or more molecules, or from different parts of the same molecule. See **condensation (2)**.
elimination filter *(Telecomm.)*. See **band-stop filter**.
ELINT *(Telecomm.)*. Abbrev. for *ELectronic INTelligence*.
Elinvar *(Eng.)*. A nickel-chromium steel alloy with variable proportions of manganese and tungsten. Used for watch hairsprings because of its constant elasticity at different temperatures.
ELISA *(Immun.)*. See **enzyme-linked immunosorbent assay**.
elixir *(Med.)*. A strong extract or tincture.
ell *(Build.)*. A short L-shaped connecting pipe.
ellipse *(Maths.)*. See **conic**.
ellipsoid *(Maths.)*. See **quadric**.
elliptical arch *(Build.)*. The arch formed to an elliptical curve, or sometimes to a curve which is not a true ellipse but a combination of circular arcs from three or five centres.
elliptical galaxies *(Astron.)*. A common type of galaxy of symmetrical form but having no spiral arms; the nearer elliptical galaxies have been resolved into stars, but contain no dust or gas.
elliptical orbit *(Space)*. The orbit of a space vehicle about a primary body in the shape of an ellipse. The primary centre of mass is one of the foci and the nearest and farthest points from it are the *pericentre* and *apocentre* respectively.
elliptical point on a surface *(Maths.)*. One at which the curvatures of all normal sections are of the same sign. Cf. **hyperbolic, parabolic** and **umbilical point on a surface**.
elliptical polarization *(Telecomm.)*. That in which the vector representing the wave varies as the radius of an ellipse while the vector rotates about a point. See **circular polarization**.
elliptic functions *(Maths.)*. In general, $f(z)$ is an elliptic function if it is doubly periodic, i.e., if there exist two numbers w_1 and w_2, $w_1 \neq w_2$, such that $f(z) = f(z + w_1) = f(z + w_2)$. The simplest elliptic functions are the Jacobian *sn*, *cn* and *dn* functions of u which are defined as follows:

$$\text{sn}\, u = x$$
$$\text{cn}\, u = \sqrt{1 - x^2}$$
$$\text{dn}\, u = \sqrt{1 - k^2 x^2}$$

where u is the elliptic integral of the 1st class. If x is replaced by sin φ, the following equation can be written from which the basic properties of the Jacobian *sn*, *cn* and *dn* functions follow:

$$\text{sn}\, u = \sin \varphi$$
$$\text{cn}\, u = \cos \varphi$$
$$\text{dn}\, u = \sqrt{1 - k^2 \sin^2 \varphi} = \Delta\varphi \text{ (say)}$$
$$\varphi = \text{am}\, u \text{ (say)}.$$

elliptic geometry *(Maths.)*. See **absolute** geometry.
elliptic integral *(Maths.)*. All integrals of the type $\supset \sqrt{P}\, dx$, where P is a fourth-degree polynomial, can be expressed in one of the following forms called respectively elliptic integrals of the 1st, 2nd and 3rd class:

$$F(k, x) = \int_0^x \frac{dt}{\sqrt{(1 - t^2)(1 - k^2 t^2)}}$$

$$E(k, x) = \int_0^x \sqrt{\frac{1 - k^2 t^2}{1 - t^2}}\, dt$$

$$\Pi(n, k, x) = \int_0^x \frac{dt}{(1 + nt^2)\sqrt{(1 - t^2)(1 - k^2 t^2)}}$$

k is called the modulus and may be taken $0 < k < 1$. Cf. **elliptic functions**.
elliptic polarization *(Phys.)*. This may be regarded as being produced by two mutually perpendicular plane-polarized components which are not in phase.
elliptic trammel *(Eng.)*. An instrument for drawing ellipses, consisting of a straight arm having a pencil point at one end and two adjustable studs (all three of which project at right angles to the arm), and a frame with two grooves crossing one another at right angles. If a stud is placed in each of the grooves, then, as the arm is rotated, the pencil point describes an ellipse.
elm *(For.)*. Tree of the genus *Ulmus*, a native of Northern Europe, usually felled when it is between 18–30 m in height; the tree may have a life as long as 140 years. It is noted for its durability under water, and is used for wagon making, coffins, agricultural implements, gymnasium equipment, pulley blocks, ship building etc. Susceptible to Dutch elm disease, caused by a fungus *Ceratocystis ulmi*) which, spread by bark beetles carrying infected spores, defoliates and kills the tree within weeks.
elmendorf tear tester *(Paper)*. A paper test apparatus utilizing the acceleration of a falling quadrant shaped pendulum, mounted on a frictionless bearing, to measure the force necessary to continue a tear, already made in the test piece, through a given length.
El Niño, El Niño southern oscillation *(Meteor.)*. El Niño – The Child – is the name originally given locally to a weak warm ocean current flowing south along the coast of Ecuador at Christmas time. The El Niño southern oscillation is the term now applied to a more intense, extensive and prolonged warming of the eastern tropical Pacific occurring every few years which is associated with major meteorological anomalies. Extreme cases have serious effects on fisheries, bird life and mainland weather. See also **southern oscillation**.
elongated *(Print.)*. A narrow form of type, often used in display work. It is commonly known as **condensed**.
elongation *(Astron.)*. The angular distance between the moon or planets and the sun. The planets Mercury and Venus have maximum elongations of about 28° and 48° respectively.
elongation *(Eng.)*. The percentage extension produced in a tensile test.
elongation ratio *(Powder Tech.)*. The ratio of the length of a particle to its breadth.
Elrod *(Print.)*. A machine for producing lead-alloy strip material, spacing, and rule, cut to size or in long lengths.
elution *(Min.Ext.)*. Washing of loaded ion-exchange resins to remove captured uranium ions or other seized elements in washing liquor. This liquor is the *eluant* and the enriched solution it becomes is the *eluate*. The resin (or **zeolite**) is regenerated (like rinsing water softener with brine).
elutriation *(Chem.Eng.)*. Process for separating into sized fractions finely divided particles in accordance with their

rate of gravitation relative to a rising stream of fluid. Used for biological molecules in a centrifugal *elutriator*.

elutriator, centrifugal *(Powder Tech.)*. Device for fractionating powder particles into sized fractions by means of a suspension undergoing centrifugal motion.

eluvial, eluvium gravels *(Geol.)*. Those gravels formed by the disintegration *in situ* of the rocks which contributed to their formation. Ant. *alluvial deposits*.

ELV *(Space)*. See **expendable launch vehicle**.

elvan *(Geol.)*. A term applied by Cornish miners to the dyke rocks associated with the Armorican granites of that county. Elvans are actually quartz-porphyries, microgranites, and other medium- to fine-textured dyke rocks of granitic composition.

elytra *(Zool.)*. In *Coleoptera*, the hardened, chitinized fore-wings which form horny sheaths to protect the hind-wings when the latter are not in use; in certain *Polychaeta*, plate-like modifications of the dorsal cirri, possibly for respiration. adjs. *elytroid, elytriform*.

e/m *(Phys.)*. The ratio of the electric charge to mass of an elementary particle. For slow moving electrons $e/m = 1.759 \times 10^{11}\,\mathrm{C\,kg^{-1}}$. This value decreases with increasing velocity because of the relativistic increase in mass. Also *specific charge*.

em *(Print.)*. The square of the body of any size of type; the 12-point em is the unit of measurement for spacing material and the dimensions of pages. 6 ems of 12-point = 1 inch approximately. See **em quad**.

emaciation *(Med.)*. Extreme wasting.

emagram *(Meteor.)*. An aerological diagram, the name being derived from *energy per unit mass diagram*. The axes, rectangular or oblique, are temperature and log(pressure).

emanations *(Chem.)*. Heavy isotopic inert gases resulting from decay of natural radioactive elements. They are radioisotopes 222, 220, 219 of element 86 or **radon**; these gases are short-lived and decay to other radioactive elements.

emarginate *(Bot.,Zool.)*. Notched; especially of a petal, with a small indentation at the apex.

emasculation *(Bot.)*. In plant breeding the surgical removal of the stamens from a flower, usually before pollen is shed, to prevent self-pollination when cross-pollination is planned.

emasculation *(Med.)*. Removal of testes, or of testes and penis.

emasculator *(Vet.)*. An instrument for castrating horses and bulls by crushing the spermatic cord.

embankment wall *(Civ.Eng.)*. A retaining wall from the top of which the supported earth normally rises at a slope.

embattlemented *(Arch.)*. A term applied to a building feature (such as a parapet) which is indented along the top like a battlement.

embedded column *(Arch.)*. A column which is partly built into the face of a wall.

embedded temperature detector *(Elec.Eng.)*. A resistance thermometer or thermocouple built into a machine or other piece of equipment during its construction, in order to be able to ascertain the temperature of a part which is inaccessible under working conditions.

embedding *(Biol.)*. The technique of embedding biological specimens in a supporting medium, such as paraffin wax or plastics like epoxy resin, in preparation for sectioning with a microtome.

embellishment *(Arch.)*. Ornamentation applied to building features.

embolectomy *(Med.)*. Removal of an embolus.

embolic gastrulation *(Zool.)*. Gastrulation by invagination.

embolism *(Bot.)*. The blockage of a xylem **conduit** by a bubble of air as a result of damage or following **cavitation**. See also **tylosis**.

embolism *(Med.)*. The blocking of a blood vessel usually by a blood clot or thrombus from a remote part of the circulation.

embolomerous *(Zool.)*. A type of vertebra consisting of a neural arch resting on two notochordal centra, an anterior intercentrum and a posterior pleurocentrum. Found in Labyrinthodontia.

embolus *(Med.)*. A clot or mass formed in one part of the circulation and impacted in another, to which it is carried by the blood stream.

emboly *(Zool.)*. Invagination; the condition of pushing in or growing in. adj. *embolic*.

embossed paper *(Paper)*. Paper, to the surface of which a pattern has been imparted by passing the sheet or web through the nip of suitable rolls in an embossing calender. The upper roll is of steel engraved with the appropriate design and the other, backing, roll is of compressible fibrous material.

embossing *(Print.)*. Producing a raised design on paper or board by the use of a die; if the design is unprinted it is called *blind embossing*.

embrasure *(Arch.)*. The splayed reveal of a window opening.

embrittlement *(Eng.)*. Some metals (e.g. steel, wrought iron, zinc or magnesium alloys) show reduced impact toughness while at subnormal temperatures; seasonal cracking of high-zinc brasses is a severe form of embrittlement. Absorption of atomic hydrogen, i.e. in electroplating or pickling, causes hydrogen embrittlement in iron and steel. Embrittlement of boiler plate, attributed to the action of sodium hydroxide, causes cracks along the boiler seams, from rivet to rivet, below the water line.

embrocation *(Med.)*. The action of applying or rubbing a medicated liquid into an injured part: the liquid so used.

embroidery *(Textiles)*. (1) Lace work consisting of a ground of net on which an ornamental design has been stitched. (2) Ornamental work done by needle or machine on a cloth, canvas, or other ground.

embryo *(Bot.)*. A plant at an early stage of development, e.g. within a seed.

embryo *(Med.)*. A developing organism. In man the term is restricted to the stages between two and eight weeks after conception.

embryo *(Zool.)*. An immature organism in the early stages of its development, before it emerges from the egg or from the uterus of the mother. adj. *embryonic*.

embryo culture *(Bot.)*. The aseptic culture on a suitable medium of an embryo excised at an early stage. The technique is useful in plant breeding in cases where, as in some hybrids, the embryos abort if left in the ovule. Cf. **ovule culture**.

embryogenesis *(Biol.)*. (1) The processes leading to the formation of the embryo. (2) Also production of embryoids in tissue culture.

embryogeny *(Bot.)*. See **embryogenesis**.

embryoid *(Bot.)*. Embryo-like structure, which may subsequently grow into a plantlet, produced in a tissue culture. See **micropropagation, anther culture, totipotency**.

embryology *(Biol.)*. The study of the formation and development of embryos.

embryoma *(Med.)*. A tumour formed of embryonic or foetal elements.

embryonic fission *(Zool.)*. See **polyembryony**.

embryonic tissue *(Bot.)*. Meristematic tissue.

embryophyte *(Bot.)*. A member of those plant groups in which an embryo, dependent at least at first on the parent plant, is formed i.e. the bryophytes and vascular plants.

embryo sac *(Bot.)*. The female gametophyte in angiosperms, formed within the ovule by the enlargement of the functional megaspore and containing usually 8 nuclei.

embryotomy *(Med.)*. The removal of the viscera or of the head of a foetus, in obstructed labour.

Emdecca *(Build.)*. Zinc sheets so treated as to resemble tiling.

emerald *(Min.)*. The brilliant green gemstone, a form of beryl; silicate of beryllium and aluminium, crystallizing in hexagonal prismatic forms, occurring in mica-schists, calcite veins and rarely in pegmatites.

emerald copper *(Min.)*. See **dioptase**.

emergence *(Bot.)*. (1) An outgrowth from a plant, derived from epidermal and cortical tissues which does not contain vascular tissue or develop into a stem or leaf. (2) The appearance above ground of germinating seedlings, *pre-emergence, post-emergence*.

emergence *(Geol.)*. The uplift of land relative to the sea.

emergence *(Zool.)*. An epidermal or subepidermal outgrowth. In Insects, the appearance of the imago from the cocoon, pupa-case, or pupal integument.

emergency diesel supply *(Nuc.Eng.)*. To provide essential electrical power to a nuclear power plant in the event of loss of grid.

emergency release-push *(Elec.Eng.)*. A switch fitted to an electric lift, to allow the car, in case of emergency, to be moved with the doors open.

emergency shutdown *(Nuc.Eng.)*. Rapid shutdown of a reactor to forestall or remedy a dangerous situation.

emergency shutdown system *(Nuc.Eng.)*. System of shutting down reactor if other methods fail e.g. injection of boron spheres which absorb neutrons strongly and quickly make reactor subcritical. See **secondary shutdown system**.

emergency stop *(Elec.Eng.)*. A switch installed in a lift-car, or other similar piece of equipment, by means of which the power to the operating motor can be cut off. Also called a *safety switch*.

emergency switch *(Elec.Eng.)*. A switch placed in a convenient position for cutting off the supply of electricity to a piece of apparatus or to a building, in case of emergency.

emergent ray point *(Radiol.)*. See **exit portal**.

emersed *(Bot.)*. Raised above or rising out of a surface, especially growing up out of water.

emersion *(Astron.)*. The exit of the moon, or other body, from the shadow which causes its eclipse.

Emerson enhancement effect *(Bot.)*. The more than additive effect on the rate of photosynthesis (in plants and algae) of illuminating simultaneously with far red light ($\lambda > 680$ nm) and light of shorter wavelength ($\lambda < 680$ nm), indicating the existence of the two **photosystems**, I and II.

emery *(Min.)*. A finely granular intimate admixture of corundum and either magnetite or haematite, occurring naturally in Greece and localities in Asia Minor, etc.; used extensively as an abrasive.

emery paper, emery cloth, emery buff *(Eng.)*. Paper, or more often cloth, surfaced with emery powder, held on by an adhesive solution; used for polishing and cleaning metal. See **emery**.

emery wheel *(Eng.)*. A **grinding wheel** in which the abrasive grain consists of emery powder, held by a suitable bonding material.

emesis *(Med.)*. The act of vomiting.

emetic *(Med.)*. Having the power to cause vomiting; a medicament which has this power.

emetine *(Chem.)*. $C_{29}H_{40}O_4N_2$ an alkaloid obtained from the roots of Brazilian ipecacuanha. It forms a white amorphous powder, mp 74°C, soluble in ethanol, ether or trichloromethane, slightly in water. It is used in medicine as an emetic; its principal use, however, is in the form of emetine bismuthous iodine or emetine hydrochloride, remedies for amoebic dysentery.

e.m.f., emf *(Elec.Eng.)*. Abbrevs. for *ElectroMotive Force*.

EMI *(Comp.)*. Abbrev. for *Electro-Magnetic Interference*, particularly from computers for which strict standards are now commonly set.

emigration *(Ecol.)*. A category of population dispersal covering one-way movement out of the population area. Cf. **immigration, migration**.

emissary *(Zool.)*. Passing out, as certain veins in Vertebrates which pass out through the cranial wall.

emission *(Phys.)*. Release of electrons from parent atoms on absorption of energy in excess of normal average. This can arise from (1) *thermal* (thermionic) agitation, as in valves, Coolidge X-ray tubes, cathode-ray tubes; (2) *secondary* emission of electrons, which are ejected by impact of higher energy primary electrons; (3) *photoelectric* release on absorption of quanta above a certain energy level; (4) *field* emission by actual stripping from parent atoms by high electric field.

emission current *(Electronics)*. The total electron flow from an emitting source.

emission efficiency *(Electronics)*. See **cathode efficiency**.

emission spectrum *(Phys.)*. Wavelength distribution of electromagnetic radiation emitted by self-luminous source.

emission tomography *(Radiol.)*. See **tomography, emission**.

emissive power *(Phys.)*. The total emissive power of a surface is the energy radiated at all wavelengths per unit area per unit time. It depends on the nature of the surface and on its temperature. See **emissivity** and **emission**.

emissivity *(Phys.)*. The ratio of emissive power of a surface at a given temperature to that of a black body at the same temperature and with the same surroundings. Values range from 1.0 for lampblack down to 0.02 for polished silver. See **Stefan-Boltzmann law**.

emitter *(Electronics)*. In a transistor, the region from which charge carriers, that are minority carriers in the base, are injected into the base.

emitter follower *(Electronics)*. See **common-collector connection**.

emitter junction *(Electronics)*. One biased in the low resistance direction, so as to inject minority carriers into the base region.

emmetropia *(Med.)*. The normal condition of the refractive system of the eye, in which parallel rays of light come to a focus on the retina, the eyes being at rest. adj. *emmetropic*.

empennage *(Aero.)*. See **tail unit**.

emphasizer *(Telecomm.)*. An audio-frequency circuit which selects and amplifies specific frequencies or frequency bands.

emphysema *(Med.)*. (1) The presence of air in the connective tissues. (2) The formation in the lung of bullae or spaces containing air, as a result of destruction of alveoli and rupture of weakened alveolar walls.

emphysematous chest *(Med.)*. The barrel-shaped, immobile chest which is the result of chronic bronchitis and emphysema.

empire cloth *(Elec.Eng.)*. An insulating fabric which is impregnated with linseed oil.

empirical formula *(Chem.)*. Formula deduced from the results of analysis which is merely the simplest expression of the ratio of the atoms in a substance. In molecular materials it may, or may not, show how many atoms of each element the molecule contains: e.g. methanal, CH_2O, ethanoic acid, $C_2H_4O_2$, and lactic acid, $C_3H_6O_3$, have the same percentage composition, and consequently, on analysis, they would all be found to have the same empirical formula.

empirical formula *(Genrl.)*. A formula founded on experience or experimental data only, not deduced in form from purely theoretical considerations.

empiricism *(Genrl.)*. The regular scientific procedure whereby scientific laws are induced by inductive reasoning from relevant observations. Critical phenomena are deduced from such laws for experimental observation, as a check on the assumptions or hypotheses inherent in the theory correlating such laws. Scientific procedure, described by empiricism, is not complete without the experimental checking of deductions from theory.

emplastrum *(Med.)*. A medicated plaster for external application.

emplectum *(Build.)*. An ancient form of masonry, showing a squared stone face, sometimes interrupted by courses of tiles at intervals.

empress *(Build.)*. A slate size, 26 × 16 in. (660 × 406 mm).

emprosthotonos *(Med.)*. Bending of the body forwards caused by spasm of the abdominal muscles, as in tetanus.

empty band *(Phys.)*. See **energy band**.

empyema *(Med.)*. Accumulation of pus in any cavity of the body, especially the pleural.

em quad *(Print.)*. A square quadrat of any size of type. Less than type height, it is used for spacing. Usually called a *mutton*, to distinguish it clearly from an *en* (or *nut*).

em rule *(Print.)*. The dash (—). A thin horizontal line 1 em of the type body in width. Apart from its uses as a mark of punctuation, it is often used to build up rules in tabular work. Also *em score, metal rule, mutton rule.*

emulsification *(Print.)*. A fault found in lithographic printing whereby fine droplets of water become dispersed in the ink on the press rollers during printing.

emulsified coolant *(Eng.)*. These are used as cutting media in (metal) machining. The 3 main types are: (*a*) an emulsion of water, a thick soap solution, and a mineral oil, (*b*) an emulsion of a mineral oil and soft soap or some other alkaline soap solution, and (*c*) an emulsion of a sulphurized or sulphonated oil neutralized and blended with a soluble oil.

emulsifier *(Chem.)*. An apparatus with a rotating, stirring or other device used for making emulsions.

emulsifying agents *(Chem.)*. Substances whose presence in small quantities stabilizes an emulsion, e.g. ammonium linoleate, certain benzene-sulphonic acids etc.

emulsion *(Build.)*. A preparation serving as a retarder.

emulsion *(Chem.)*. A colloidal suspension of one liquid in another, e.g. milk.

emulsion *(Image Tech.)*. A suspension of finely divided silver halide crystals in a medium such as gelatine which provides the light-sensitive coating on film, glass plates and paper and plastic supports. It may also contain sensitizing dyes and colour-forming **couplers**.

emulsion paint *(Build.)*. A water-thinnable paint made from a pigmented emulsion or dispersion of a resin (generally synthetic) in water. The resin may be polyvinyl acetate, polyvinyl chloride, an acrylic resin or the like.

emulsion technique *(Phys.)*. Study of nuclear particles, by means of tracks formed in photographic emulsion.

emunctory *(Med.)*. Conveying waste matter from the body; any organ or canal which does this.

en *(Chem.)*. Abbrev. for *1,2-diamino ethane, ethylene diamine* $(CH_2.NH_2)_2$ in complexes.

en *(Print.)*. A unit of measurement used in reckoning up composition. It is assumed that the average letter has the width of one *en* and that the average word, including the space following, has the width of six *ens*. See **en quad**.

enabling pulse *(Electronics)*. One which opens a **gate** which is normally closed.

enamel *(Build.)*. Glaze.

enamel *(Elec.Eng.)*. A finely ground oil paint containing resin. The oil is thickened by heating linseed oil with chinawood oil for some hours at 300°C. Unrivalled for insulating very fine wires.

enamel *(Zool.)*. The external calcified layer of a tooth, of epidermal origin and consisting of elongate hexagonal prisms, set vertically on the surface of the underlying dentine; enamel also occurs in certain scales.

enamel cell *(Zool.)*. See **ameloblast**.

enamel-insulated wire *(Elec.Eng.)*. Wire having an insulating covering of *enamel* used for winding small magnet coils, etc.

enamelled brick *(Build.)*. A brick having a glazed surface.

enamelled slate *(Elec.Eng.)*. Slate covered with a coating of hard enamel to render it suitable for the construction of switchboards.

enamel paint *(Build.)*. A high-grade paint prepared by careful grinding of pigment in an oily medium containing a proportion of resin, and the usual lesser ingredients. It may be anything from glossy to flat.

enanthem, enanthema *(Med.)*. An eruption on a mucous membrane.

enantiomerism *(Chem.)*. See **enantiomorphism**.

enantiomorphism *(Chem.)*. Mirror image isomerism. A classical example is that of the crystals of sodium ammonium tartrates which Pasteur showed to exist in mirror image forms. adj. *enantiomorphous*.

enargite *(Min.)*. Sulpharsenide of copper, often containing a little antimony.

enarthrosis *(Zool.)*. A ball-and-socket joint.

enation *(Bot.)*. (1) Generally an outgrowth as those on leaves caused by some viruses. (2) A non-vascularised, spine-like outgrowth from the axis of some primitive vascular plants e.g. *Zosterophyllum*. See **enation theory**.

enation theory *(Bot.)*. A theory that regards the *microphylls* of the *Lycopodiales* etc. as simple enations that have become vascularised and therefore different from megaphylls which are regarded as having evolved from branch systems.

encallow *(Build.)*. The mould which is first removed from the surface of the site where clay for brickmaking is to be obtained, and which is stored in a spoil bank for resurfacing later.

encapsulation *(Eng.)*. Provision of a tightly fitted envelope of material to protect, e.g. a metal during treatment or use in an environment with which it would otherwise react in a detrimental manner. Similarly, the coating of, e.g. an electronic component in a resin to protect it against the environment.

encase *(Build.)*. To surround or enclose with linings or other material.

encastré *(Build.)*. A term applied to a beam when the end of it is fixed.

encaustic painting *(Build.)*. A process of painting in which hot wax is used as a medium.

encaustic tile *(Build.)*. An ornamental coloured tile whose colours are produced by substances added to the clay before firing.

Enceladus *(Astron.)*. The second natural satellite of **Saturn**, 500 km in diameter.

encephalagia *(Med.)*. Pain inside the head.

encephalitis *(Med.)*. Inflammation of the brain substance usually by a viral agent. adj. *encephalitic*.

encephalitis lethargica *(Med.)*. *Sleepy sickness, Von Economo's disease*. An acute inflammation of the brain with a filterable virus; characterized by fever and disturbances of sleep, and followed by various persisting forms of nervous disorder (e.g. Parkinsonism), or by changes in character. Last UK outbreak in 1920s.

encephalitis periaxialis, encephalitis periaxialis diffusa *(Med.)*. *Schilder's disease*. A disease characterized by progressive destruction of the nerve fibres composing the central white matter of the brain, causing blindness, deafness, paralysis, and amentia.

encephalitogen *(Immun.)*. Substances present in extracts of brain which when administered with a potent adjuvant such as CFA cause experimental allergic encephalitis. The active material is myelin basic protein.

encephalocele *(Med.)*. Hernial protrusion of brain substance through a defect in the skull.

encephalogram *(Radiol.)*. X-ray plate produced in **encephalography**.

encephalography *(Radiol.)*. Radiography of the brain after its cavities and spaces have been filled with air, or dye, previously injected into the space round the spinal cord.

encephalomalacia *(Med.)*. Pathological softening of the brain.

encephalomyelitis *(Med.)*. Diffuse inflammation of the brain and the spinal cord.

encephalon *(Zool.)*. The brain.

encephalopathy *(Med.)*. A generalised disease of the brain often associated with toxic poisoning, e.g. *lead encephalopathy* is the brain disorder caused by lead poisoning; *hepatic encephalopathy* is the confusional state associated with severe liver disease.

encephalospinal *(Zool.)*. See **cerebrospinal**.

enceph-, encephalo- *(Genrl.)*. Prefix from Gk. *enkephalos*, brain.

enchondroma *(Med.)*. A tumour, often multiple, composed of cartilage and occurring in bones.

enclitic *(Med.)*. Having the planes of the foetal head inclined to those of the maternal pelvis.

enclosed fuse *(Elec.Eng.)*. A fuse in which the fuse wire is enclosed in a tube or other covering.

enclosed self-cooled machine *(Elec.Eng.)*. An electric machine which is enclosed in such a way as to prevent the circulation of air between the inside and the outside, but in which special provision is made for cooling the enclosed air by some attachment, e.g. an air cooler, forming part of the machine.

enclosed-ventilated *(Elec.Eng.)*. Said of electrical apparatus which is protected from ordinary mechanical damage by an enclosure, with openings for ventilation.

encode *(Telecomm.)*. (1) Conversion of information, by means of a code, in such a way that it can be subsequently re-converted to its original form. (2) Conversion of one system of communication to another, as from amplitude to pulse-code modulation.

encoding *(Behav.)*. In memory research, the process of transforming material into a form easily stored and retrieved. See **chunking**.

encoding altimeter *(Aero.)*. An **altimeter** designed for automatic reporting of altitude to **air-traffic control**. A special encoding disk within the instrument rotates in response to movement of the aneroid capsules, and transmits a signal which is amplified, fed to the aircraft's transponder and thence automatically to ATC.

encounter group *(Behav.)*. A general term for a range of group therapies with the general aim of increasing personal awareness and encouraging creative and open relations with others. Procedures vary widely, though all tend to emphasise free and candid expression of feeling and thought within the group.

encrinal limestone *(Geol.)*. A crinoidal limestone. syn. *encrinital limestone*.

encryption *(Comp.)*. Encoding data to make it incomprehensible to those without a decoder. See **decryption**.

encysted *(Med.)*. Enclosed in a cyst or a sac.

encystment *(Bot.)*. The formation of a walled nonmotile body from a swimming spore.

encystment *(Zool.)*. The formation by an organism of a protective capsule surrounding itself. Also *encystation*. v. *encyst*.

end *(Min.Ext.)*. Solid rock at end of underground passage.

end *(Textiles)*. (1) in spinning, an individual strand (2) in weaving, a warp thread (3) in finishing, each passage of a fibre through a machine (4) A length of finished fabric shorter than the normal **piece**.

endarch *(Bot.)*. A xylem strand having the protoxylem on the side nearer to the centre of the axis. Cf. **exarch**, **mesarch**.

end-around shift *(Comp.)*. One in which digits drop off one end of a word and return at the other. Also *circular shift, cyclic shift, ring shift, rotation shift*.

endarteritis *(Med.)*. Inflammation of the intima of an artery.

endarteritis obliterans *(Med.)*. Obliteration of the lumen of an artery, as a result of inflammatory thickening of the intima.

end-artery *(Med.)*. A terminal artery which does not anastomose with itself or with others.

end bell *(Elec.Eng.)*. A strong metal cover placed over the end-windings of the rotor of a high speed machine, e.g. a turbo-alternator, to prevent their displacement by centrifugal forces.

end bracket *(Elec.Eng.)*. An open structure fitted at the end of an electrical machine, for the purpose of carrying a bearing.

end cell *(Elec.Eng.)*. See **regulator cell**.

end connections *(Elec.Eng.)*. That part of an armature winding which does not lie in the slots, and which serves to join the ends of the active or slot portions of the conductors.

end correction *(Acous.)*. Ratio between two lengths in an open-ended resonating tube, where the pressure node is just outside the tube. The nominator is the distance between the tube end and the pressure node, and the denominator is the tube radius.

end-down *(Textiles)*. Broken yarn twisting on the rotating spindles in spinning. A warp thread broken during weaving.

endellionite *(Min.)*. Another name for *bournonite*.

endemic *(Ecol.,Med.)*. (1) A species or family confined to a particular region and thought to have originated there, e.g. *Primula scotica*, native only in the north of Scotland. (2) A disease permanently established in moderate or severe form in a defined area. Also *indigenous*.

endergonic *(Bot.)*. An energy-requiring biochemical reaction which cannot therefore proceed spontaneously. Cf. **exergonic**.

end-feed magazine *(Eng.)*. One of several types of magazine used for the automatic loading of centreless grinding machines. It is suitable for short, conical components.

end-fire array *(Radar,Telecomm.)*. *End-fire aerial array*. A linear array of radiators in which the maximum radiation is along the axis of the array; the antenna may be uni- or bi-directional. A **Yagi** array is an example, though in most end-fire arrays, each radiator is fed from a transmission line, the relative phase of each element being determined by its position along the line.

end fixing *(Build.)*. A term used in referring to the condition of the ends of a column, whether fixed or only partially so. See **effective length**.

end gauge *(Eng.)*. Gauge consisting of a metal block or cylinder the ends of which are made parallel within very small limits, the distance between such ends defining a specified dimension. See **limit gauge**.

end-hats *(Elec.Eng.)*. Metal fitments at the ends of the cathode in certain magnetrons to reduce space charge.

ending *(Textiles)*. A dyeing fault that results in the end of the fabric being a different shade from the other end or the bulk of the fabric.

end-labelling *(Biol.)*. The enzymatic attachment of a radioactive atom to the end of a DNA or RNA molecule.

end-lap joint *(Build.)*. A halving joint formed between the ends of two pieces of timber intersecting at an angle.

end leakage flux *(Elec.Eng.)*. Leakage flux associated with the end connections of an electric machine.

endless rope haulage *(Min.Ext.)*. A method of hauling trucks underground by attachment to a long loop of rope, guided by many pulleys along the roads or haulage ways, and actuated by a power-driven drum.

endless saw *(Eng.)*. See **bandsaw**.

end links *(Eng.)*. The links at either end of a chain; they are made slightly stronger than the remainder to allow for wear when attached to hooks, couplings, etc.

end matter *(Print.)*. The items which follow the main text of a book, i.e. appendices, notes, glossary, bibliography, index.

end measuring instruments *(Eng.)*. Measuring instruments (e.g. micrometers, callipers, gauges) which measure length by making contact with the ends of an object.

end mill *(Eng.)*. A **milling cutter** having radially disposed teeth on its circular cutting face; used for facing and grooving operations.

endo- *(Genrl.)*. Prefix from Gk. *endon*, within.

endo-aneurysmorrhaphy *(Med.)*. (1) Obliteration, by suture at either end, of an aneurysm of an artery. (2) Obliteration, by suture, of the aneurysmal sac, with reconstruction of the original arterial channel.

endobiotic *(Bot.)*. (1) Growing inside another plant. (2) Formed inside the host cell.

endoblast *(Zool.)*. See **hypoblast**.

endocardiac *(Zool.)*. Within the heart.

endocarditis *(Med.)*. Inflammation of the lining membrane of the heart, especially of that part covering the valves.

endocardium *(Zool.)*. In Vertebrates, the layer of endothelium lining the cavities of the heart.

endocarp *(Bot.)*. A differentiated innermost layer of a pericarp, usually woody in texture like the hard outside of a peach stone.

endocervicitis *(Med.)*. Inflammation of the mucous membrane lining the cervix uteri.

endochondral *(Zool.)*. Situated within or taking place within cartilage; as *endochondral ossification*, which begins within the cartilage and works outwards.

endocoelar *(Zool.)*. See **splanchnopleural**.

endocranium *(Zool.)*. In Insects, internal processes of the skeleton of the head, serving for muscle attachment.

endocrine *(Zool.)*. Internally secreting; said of certain glands, principally in Vertebrates, which pour their secretion into the blood, and so can affect distant organs or parts of the body. See **hormone**.

endocrinology *(Med.)*. The study of the internal secretory glands.

endocrinopathy *(Med.)*. Any disease due to disordered function of the endocrine glands.

endocuticle *(Zool.)*. The laminated inner layer of the insect cuticle.

endocytosis *(Biol.)*. The entry of material into the cell by the invagination of the **plasma membrane**. Material can enter in the fluid phase of the resulting vesicle or attached to its membrane.

endoderm *(Zool.)*. The inner layer of cells forming the wall of a gastrula and lining the archenteron: the tissues directly derived from this layer.

endodermis *(Bot.)*. (1) The innermost layer of the cortex in roots and stems, sheathing the stele, one cell thick. In roots typically and sometimes in stems there are no radial intercellular spaces and all anticlinal walls have a **casparian strip**. In some roots the cell walls become thickened later. See also **starch sheath**. (2) A similar layer, with casparian strip or the like, elsewhere, e.g. surrounding the vascular tissues in the pine leaf.

endo-ergic process *(Phys.)*. Nuclear process in which energy is consumed. Also *endo energetic*. The equivalent thermodynamic term is *endothermic*.

endo-exo configuration *(Chem.)*. Special case of *cis-trans* isomerism in ring compounds, e.g. borneol shows the *trans*- or *endo*-configuration while *iso*-borneol shows the *cis*- or *exo*-configuration.

end of file marker *(Comp.)*. Marker signalling the end of a file. Abbrev. *EOF*.

endogamy *(Bot.,Zool.)*. The production of a zygote by the fusion of gametes from 2 closely related parents, or from the same individual. Cf. **exogamy**. See also **autogamy, inbreeding**.

endogenous *(Behav.)*. A general term, used to suggest that the causes of some physical or psychological conditions are due to internal and probably physical factors.

endogenous *(Bot.)*. Processes or organs which originate or develop within the plant. Cf.**exogenous**.

endogenous *(Zool.)*. In higher animals said of metabolism which leads to the building of tissue, or to the replacement of loss by wear and tear. Cf. **exogenous**.

endogenous rhythm *(Bot.)*. A rhythm in movement or in a physiological process which depends on internal rather than external stimuli. It will often persist under constant environmental conditions and may show *entrainment*. See also **circadian rhythm**.

endoglycosidase *(Biol.)*. Enzyme which hydrolyses the internal glycosidic bonds of polysaccharides or oligosaccharides, converting them to disaccharides and terminal dextrins.

endolithic *(Bot.)*. Growing within rock like some algae do in limestone.

endolymph *(Zool.)*. In Vertebrates, the fluid which fills the cavity of the auditory vesicle and its outgrowths (semicircular canals etc.).

endolymphangial *(Zool.)*. Situated within a lymphatic vessel.

endolymphatic *(Zool.)*. Pertaining to the membranous labyrinth of the ear in Vertebrates, or to the fluid contained therein.

endometrial *(Med.)*. Pertaining to, or having the character of, endometrium.

endometrioma *(Med.)*. *Adenomyoma*. An endometrial tumour consisting of glandular elements and a cellular connective tissue, occurring in the regions where endometrium is normally absent.

endometriosis *(Med.)*. A condition in which fragments of tissue resembling **endometrium** occur in other tissue or organs. It is subject to the same menstrual changes as normal endometrium and causes severe pain and discomfort with menstruation.

endometritis *(Med.)*. Inflammation of the endometrium.

endometrium *(Med.)*. The mucous membrane lining the cavity of the uterus.

endomitosis *(Bot.)*. Process, occurring naturally in some differentiating cells or when induced by e.g. colchicine, in which the chromosomes divide without cell division giving double the original number of chromosomes in the cell. See also **restitution nucleus**.

endomorph *(Behav.)*. One of Sheldon's somatotyping classifications; endomorphs are soft and round; they are described as loving comfort, affection and are sociable. See **somatotype theory**.

endomorphy *(Behav.)*. Sheldon's *somatotype* which is characterized by general heaviness or stoutness of build. Said to be associated with more or less extraversive personality traits; sociability, hedonism etc. See **somatotype theory**.

endomyocardial fibrosis *(Med.)*. A form of **cardiomyopathy** common in East Africa which causes fibrosis of endocardium and myocardium leading to severely restricted function.

endomyocarditis *(Med.)*. Inflammation of the heart muscle and of the membrane lining the cavity of the heart.

endomysium *(Zool.)*. The intramuscular connective tissue which unites the fibres into bundles.

endoneurium *(Zool.)*. Delicate connective tissue between the nerve fibres of a funiculus.

endonuclease *(Biol.)*. An enzyme which cleaves in the middle of a polynucleotide chain.

endoparasite *(Bot.,Zool.)*. A parasite living inside the body of its host.

endopeptidase *(Biol.)*. An enzyme which cuts a polypeptide chain internally, not just removing terminal peptides.

endophlebitis *(Med.)*. Inflammation of the intima of a vein.

endophyte *(Bot.)*. A plant living inside another plant, but not necessarily parasitic on it.

endophytic *(Bot.)*. Within a plant, the hyphae of a symbiotic fungus.

endophytic mycorrhiza *(Bot.)*. Same as **endotrophic mycorrhiza**.

endopite *(Zool.)*. The inner ramus of a biramous arthropod appendage.

endoplasmic reticulum *(Biol.)*. A series of flattened membranous tubules and *cisternae* in the cytoplasm of eukaryotic cells which can either show a smooth profile, *smooth endoplasmic reticulum*, or be decorated with ribosomes, *rough endoplasmic reticulum*. It is the site of synthesis of lipids and some proteins. The membrane is continuous with that of the nuclear envelope. Abbrev. *ER*.

endopolyploid *(Bot.)*. The product of **endomitosis**.

Endoprocta *(Zool.)*. See **Entoprocta**.

Endopterygota *(Zool.)*. Subclass of the Insecta with complete metamorphosis and a larval form in which the wings develop internally. Also **holometabolous**; Cf. **Exopterygota**.

endoradiosonde *(Electronics,Med.)*. A miniature, battery powered, transmitter, encapsulated like a pill; designed to be swallowed by patients in order to transmit physiological data from their gastrointestinal tracts.

endorhachis *(Zool.)*. In Vertebrates, a layer of connective tissue which lines the canal of the vertebral column and the cavity of the skull.

endorphins *(Biol.,Med.)*. Peptides synthesised in the pituitary gland which have analgesic properties associated with their affinity for the opiate receptors in the brain.

endoscope *(Eng.,Med.)*. An instrument for inspecting and photographing (1) internal cavities of the body in medicine or (2) remote and inaccessible sites in industry. Fibre-optics are normally used to both illuminate and inspect the remote site from outside.

endoscopic embryology *(Bot.)*. The condition when the apex of the developing embryo points towards the base of the archegonium.

endoscopy *(Med.)*. Any technique for visual inspection of internal organs. Modern instruments are usually flexible fibre-optic devices with additional facilities for biopsy.

endoskeleton *(Zool.)*. In Craniata, the internal skeleton, formed of cartilage or cartilage bone. In Arthropoda, the endophragmal skeleton, i.e. hardened invaginations of the integument forming rigid processes for the attachment of muscles and the support of certain other organs.

endosome *(Biol.)*. A cytoplasmic vesicle derived from a **coated vesicle** by removal of the protein coat. See **receptor mediated endocytosis**.

endosperm *(Bot.)*. Tissue formed within the embryo-sac, usually triploid (cf. **double fertilization**), serving in the nutrition of the embryo and often increasing to form a storage tissue in the mature seed (e.g. cereals).

endospermic, endospermous *(Bot.)*. A mature seed having endosperm. syn. *albuminous*.

endospore *(Bot.)*. (1) The innermost layer of the wall of a spore: *intine*. (2) A very resistant thick-walled spore formed within a bacterial cell. (3) A spore produced within or by the division of the contents of the parent cell.

endosporic *(Bot.)*. A *gametophyte* developing within the spore, as in seed plants.

endostyle *(Zool.)*. In some Protochordata and in the larvae of Cyclostomata, a longitudinal ventral groove of the pharangeal wall, lined by ciliated and glandular epithelium. adj. *endostylar*.

endosymbiosis, endocytobiosis *(Bot.)*. **Symbiosis** in which one organism (prokaryote or eukaryote) lives inside the cells of another eukaryote, the association behaving typically as an organism.

endosymbiotic hypothesis *(Bot.)*. Hypothesis postulating that the chloroplasts, mitochondria and other plastids of most eukaryotes evolved from endosymbiotic prokaryotes, able to photosynthesize or respire aerobically. Some subsequent transfer of genes from endosymbiont to host must also have occurred.

endothecium *(Bot.)*. (1) The *fibrous layer* in the wall of an anther. (2) The inner tissues in the young sporophyte of bryophytes, giving rise to the sporogenous tissue and/or the columella. Cf. **amphithecium**.

endotheliochorial placenta *(Zool.)*. See **vasochorial placenta**.

endothelioma *(Med.)*. A tumour arising from the lining membrane of blood vessels or lymph channels. Also a tumour arising in the pleura, in the peritoneum, or in the meninges.

endothelium *(Zool.)*. Pavement epithelium occurring on internal surfaces, such as those of the serous membranes, blood-vessels, lymphatics.

endothermic *(Chem.)*. Accompanied by the absorption of heat. (ΔH positive.)

endotoxin *(Biol.)*. Toxin retained within, if not part of, the bacterial protoplasm, and only released on lysis or death of the bacterial cells. Examples of diseases caused by endotoxins are typhoid fever and cholera.

endotoxin *(Immun.)*. Synonym for lipopolysaccharides derived from Gram negative bacteria which are powerful activators of macrophages and cause release of **interleukin I** and **tumour necrosis factor**. Very small amounts are sufficient to cause fever and malaise in humans, and this accounts for the unpleasant effects of vaccines containing whole Gram negative bacteria such as TAB vaccine.

endotoxin shock *(Immun.)*. Syndrome following administration of endotoxin or systemic infection with endotoxin producing bacteria. Characterized by prostration, hypotension, fever and leucopaenia.

endotrophic mycorrhiza *(Bot.)*. A mycorrhiza in which the fungal hyphae grow between and within the cells of the root cortex and connect with hyphae ramifying though the soil but which do not form, like in *ectotrophic* mycorrhiza a thick mantle on the surface of the root. Vesicular-arbuscular **mycorrhizas** and the mycorrhizas of orchids and of the Ericaceae are endotrophic.

endozoic *(Bot.)*. (1) Living inside an animal. (2) Said of the method of seed dispersal in which seeds are swallowed by some animal and voided after having been carried for some distance.

end-papers *(Print.)*. Stout papers formed from a folded sheet which is firmly attached to the first and last sections of a volume at the fold. One half of each sheet is securely pasted to the inner side of the front and the back cover, the other half forming a fly-leaf.

end plate *(Zool.)*. A form of motor nerve-ending in muscle.

end-plate fins *(Aero.)*. Fins mounted at or near the tips of the tail plane or wing to increase its efficiency.

end-point *(Chem.)*. The point in a volumetric titration at which the amount of added reagent is equivalent to the solution titrated.

end product *(Phys.)*. The stable nuclide forming the final member of a radioactive decay series.

end-quench test *(Eng.)*. See **Jominy test**.

end sheet *(Elec.Eng.)*. A sheet of insulating material placed between the end section of an accumulator and the lining of the container.

end shield *(Elec.Eng.)*. A cover which wholly or partially encloses the end of an electric machine.

end speed *(Aero.)*. Naval term for the speed of an aircraft relative to its aircraft carrier at the moment of release from catapult or accelerator.

end spring *(Elec.Eng.)*. A small spring of hard lead placed in a lead-acid accumulator, between the end plates and the container to prevent the plates from spreading.

end stone *(Horol.)*. A flat jewel which acts as a bearing surface for the end of a pivot, e.g. the balance staff is provided with jewels, with through holes for the parallel portions of the pivots and end stones for the ends of the pivots.

endurance *(Aero.)*. The maximum time that an aircraft can continue to fly without refuelling, under any agreed conditions.

endurance limit *(Eng.)*. See **limiting range of stress, fatigue limit**.

endurance range *(Eng.)*. See **limiting range of stress**.

end-windings *(Elec.Eng.)*. See **armature end connections**.

end window counter *(Phys.)*. G-M counter designed so that radiation of low penetrating power can enter one end. (This is usually covered with a thin mica sheet.)

endysis *(Zool.)*. The formation of new layers of the integument following ecdysis.

enema *(Med.)*. Fluid injected into the rectum to promote evacuation of the bowels.

energetics *(Chem.)*. The abstract study of the energy relations of physical and chemical changes. See **thermodynamics**.

energy *(Phys.)*. The capacity of a body for doing work. Mechanical energy may be of two kinds: *potential energy*, by virtue of the position of the body, and *kinetic energy*, by virtue of its motion. Energy can take a wide variety of forms. Both mechanical and *electrical* energy can be converted into *heat* which is itself a form of energy. Electrical energy can be stored in a capacitor to be recovered on the discharge of the capacitor. *Elastic potential energy* is stored when a body is deformed or changes its configuration, e.g. in a compressed spring. All forms of wave motion have energy; in electromagnetic waves it is stored in the electric and magnetic fields. In any closed system, the total energy is constant – the *conservation of energy*. *Units of energy*: SI unit is the **joule** (symbol J) and is the work done by a force of 1 newton moving through a distance of 1 metre in the direction of the force. The CGS unit, the *erg* is equal to 10^{-7} joules and is the work done by a force of 1 dyne moving through 1 cm in the direction of the force. The foot-pound force (ft-lbf) of the British system equals 1.356 J. See **kinetic energy, potential energy, mechanical equivalent of heat, kilowatt-hour, electron-volt, Btu**.

energy balance *(Chem.Eng.)*. A **heat balance**.

energy balance *(Ecol.)*. A quantitative account of the exchanges of energy between organisms and their surroundings.

energy balance *(Eng.)*. A balance-sheet drawn up from an engine test, expressing the various forms of energy produced by the engine (e.g., B.H.P or output power, heat to cooling water, heat carried away in exhaust gases, and heat unaccounted for) as percentages of the heat supplied from the calorific value of the fuel.

energy band *(Phys.)*. In a solid the energy levels of the individual atoms combine to form bands of allowed energies separated by forbidden regions. The individual electrons are considered to belong to the crystal as a whole rather than to a particular atom. The energy bands are the consequence of the motion of the electron in the periodic potential of the crystal lattice. A solid for which a number of the bands are completely filled and the others empty, is an insulator provided the energy gaps are large. If one band is incompletely filled or the bands overlap then metallic conduction is possible. For semiconductors, there is a small energy gap between the filled and empty band and *intrinsic* conduction occurs when some electrons acquire sufficient energy to surmount the gap.

energy barrier *(Chem.)*. The minimum amount of free energy which must be attained by a chemical entity in order to undergo a given reaction. See **activation energy**.

energy component *(Elec.Eng.)*. See **active component**.

energy confinement time *(Nuc.Eng.)*. In fusion, the ratio of the total energy of a confined plasma to the rate of energy loss from it.

energy curve *(Image Tech.)*. The spectral distribution of radiated energy in a source of light, e.g. an arc, the ordinate being proportional to the energy contained in a specified small band of wavelengths.

energy density of sound *(Acous.)*. Sum of potential and kinetic sound energy per unit volume.

energy equivalent sound pressure level *(Acous.)*. Sound pressure level, where the squared sound pressure is averaged over a long time, typically more than 15 minutes. Used to characterize strongly fluctuating sound levels such as those of traffic noise.

energy gap *(Phys.)*. Range of forbidden energy levels between two permitted bands. See **energy band**.

energy levels *(Electronics)*. In semiconductors, a *donor level* is an intermediate level close to the conduction band; being filled at absolute zero, electrons in this level can acquire energies corresponding to the conduction level at other temperatures. An *acceptor level* is an intermediate level close to the normal band, but empty at absolute zero; electrons corresponding to the normal band can acquire energies corresponding to the intermediate level at other temperatures. See **Fermi characteristic-**. Electron energies in atoms are limited to a fixed range of values termed *permitted energy levels*, and represented by horizontal lines drawn against a vertical energy scale. See **eigenvalues, Schrödinger equation**.

energy management *(Aero.)*. Operational technique of minimizing energy loss by advanced automatic flight and engine monitoring and control systems. The actual method takes several forms but includes e.g. the measurement of an individual aircraft and engine performance in flight and adjustment of height and Mach number to suit the monitored conditions.

energy-mass equation *(Phys.)*. See **mass-energy equation**.

enfleurage *(Chem.)*. Process of cold extraction with fat, e.g. of essential oils from flowers. Used in perfumery.

eng *(For.)*. A very durable dark reddish-brown wood from Burma, much used for house-building and for parquet flooring. Also called *in*.

ENG *(Image Tech.)*. *Electronic News Gathering*, recording and sometimes transmitting TV news items direct from their location using very light-weight video equipment.

engaging speed *(Aero.)*. The relative speed of a carrier-borne aircraft to its ship at the moment when the **arrester gear** is engaged.

Engel process *(Plastics)*. Patented process for the production of large vessels.

engine *(Eng.)*. Generally, a machine in which energy is applied to do work: particularly, a machine for converting heat energy into mechanical work; loosely, a locomotive.

engine cylinder *(Eng.)*. That part of an engine in which the heat and pressure energy of the working fluid do work on the piston, and so are converted into mechanical energy. See **cylinder, cylinder barrel**.

engineer *(Eng.)*. One engaged in the science and art of engineering practice. The term is a wide one, but it is properly confined to one qualified to design and supervise the execution of mechanical, electrical, hydraulic, and other devices, public works etc. See **Chartered Engineer**.

engineering bricks *(Civ.Eng.)*. Bricks made of semitvitreous materials. They are dense and of high strength and low porosity. Manufactured chiefly in Staffordshire, Lancashire, Sussex, N. Wales, and Yorkshire. Used where severe loading conditions exist or exposure to damp, frost, acid or acid fumes etc.

engineering geology *(Geol.)*. Geology applied to engineering practice, especially in mining and civil engineering.

engineer's chain *(Surv.)*. A 100 ft measuring chain of 100 1 ft links.

engine friction *(Eng.)*. The frictional resistance to motion offered by the various working parts of an engine. See **friction horsepower**.

engine pit *(Eng.)*. (1) A hole in the floor of a garage to enable a man to work on the underside of a motor vehicle. (2) An engine sump or crank pit; the boxlike lower part of the crank-case. (3) A large pit for giving clearance to the flywheel of any large gas engine or winding engine.

engine plane *(Min.Ext.)*. In a coalmine, a roadway on which tubs, trucks, or trains are hauled by means of a rope worked by an engine.

engine pod *(Aero.)*. A complete turbojet power unit, including cowlings, supported on a pylon, usually under the wings of an aircraft, an installation method commonly adopted on most types of multi-engined high subsonic speed aircraft.

engine sized paper *(Paper)*. Paper made from stock to which appropriate chemicals have been added to confer resistance to penetration by aqueous liquids.

engine speed *(Eng.)*. In a turbine engine, the revolutions per minute of the main rotor assembly; in a reciprocating engine, those of the crankshaft.

englacial *(Geol.)*. Contained within the ice, e.g. *englacial stream*.

Engler distillation *(Chem.)*. The determination of the boiling range of petroleum distillates, carried out in a definite prescribed manner by distilling 100 cm^3 of the substance and taking the temperature after every 5 or 10 cm^3 of distillate has collected. The initial and final boiling points are also measured.

Engler flask *(Chem.)*. A 100 cm^3 distillation flask of definite prescribed proportions used for carrying out an **Engler distillation**.

English *(Print.)*. An old type size, approximately 14-point.

English bond *(Build.)*. The form of bond in which each course is alternately composed entirely of headers or of stretchers.

English cross bond *(Build.)*. A **Dutch bond**.

English lever *(Horol.)*. Type of watch with lever escapement having pointed teeth on the escape wheel.

English roof truss *(Eng.)*. A truss for roofs of large span in which the sloping upper and lower chords are symmetrical about the central vertical, and are connected by vertical and diagonal members.

engorgement *(Med.)*. Congestion of a tissue or organ with blood.

enhancement effect *(Bot.)*. Same as **Emerson enhancement effect**.

enhancement-mode transistor *(Electronics)*. A field-effect transistor in which there are no charge carriers in the channel when the gate-source voltage is zero. Channel conductivity increases when a gate-source voltage of appropriate polarity is applied.

enhancer *(Biol.)*. DNA sequence which can stimulate transcription of a gene while being at a distance from it.

enkephalins *(Biol.)*. Penta peptides isolated from the brain which have the same N-terminal amino acid sequences as **endorphins** and share their analgesic properties.

enkephalins *(Med.)*. Belong to the endorphins and thought to play a role in neurotransmission of painful stimuli.

enlarger *(Image Tech.)*. Apparatus for making photographic prints of larger size than the original camera negative by projecting an illuminated image through a lens on to the sensitized paper.

enlarging lens *(Image Tech.)*. Lens designed for use in a photographic enlarger, required to project at comparatively short distances and provide a flat distortion-free image from a flat original, the negative.

enneagon *(Maths.)*. More correct but less common name for *nonagon*.

enol form *(Chem.)*. Unsaturated alcohol form of a substance exhibiting keto-enol tautomerism, i.e. that form in which the mobile hydrogen atom is attached to the oxygen atom, and therefore has acidic properties. E.g., *ethyl aceto-acetate*.

enophthalmos, enophthalmus *(Med.)*. Abnormal retraction of the eye within the orbit.

enostosis *(Med.)*. A bony growth formed on the internal surface of a bone.

enprint *(Image Tech.)*. Standard size enlargement produced by D and P firms in lieu of contact print.

en quad *(Print.)*. A type space half an em wide. It is usually called a *nut*, to distinguish it clearly from an **em** (or *mutton*).

enriched uranium *(Phys.)*. Uranium in which the proportion of the fissile isotope ^{235}U has been increased above its natural abundance.

enrichment *(Arch.,Build.)*. Ornamentation applied to building features.

enrichment *(Biol.)*. A method of increasing the proportion of cells with a mutation which cannot be selected directly. Mutants unable to grow in a given medium, are tolerant of agents like penicillin which only kill growing cells.

enrichment *(Min.Ext.)*. Effect of superficial leaching of lode, whereby part of value is dissolved and redeposited in a lower enriched zone. See **secondary enrichment**.

enrichment *(Nuc.Eng.)*. (1) Raising the proportion of ^{235}U fissile nuclei above that for natural uranium in reactor fuel. (2) Raising the proportion of the desired isotope in an element above that present initially by isotope separation.

enrichment factor *(Phys.)*. (1) In the U.K., the abundance ratio of a product divided by that of the raw material. The *enrichment* is the enrichment factor less unity. (2) US **separation factor**.

enrockment *(Hyd.Eng.)*. The layer of stones placed over the face of a dyke or sea-embankment as a protection against the impact of the water.

en rule *(Print.)*. A dash (–) cast on an *en* body. Half the width of an *em* rule, it is often used to divide dates, express range, etc. Also *en score*.

ensemble *(Phys.)*. In statistical mechanics, a set of a very large number of systems, all dynamically identical with the system under consideration and differing in the initial condition.

ensiform *(Genrl.)*. Shaped like the blade of a sword.

ensiform process *(Zool.)*. See **xiphisternum**.

ensilage *(Genrl.)*. See **silage**.

enstatite *(Min.)*. An orthorhombic pyroxene, chemically magnesium silicate, $MgSiO_3$; it occurs as a rock-forming mineral and in meteorites.

enstrophy *(Meteor.)*. Half the square of the vorticity. It is conserved in two-dimensional, adiabatic, non-dissipative flow.

ENT *(Med.)*. Abbrev. for *Ear, Nose, and Throat* as a department of medical practice.

entablature *(Arch.,Eng.)*. (1) The whole of the parts immediately supported upon columns, consisting of an architrave, a frieze and a cornice. (2) An engine framework supported on columns.

entamoebiasis *(Med.)*. Infection with *Entamoeba histolytica*.

entasis *(Arch.)*. The slight swelling towards the middle of the length of a column to correct for the appearance of concavity in the outline of the column if it had a straight taper.

enteral *(Med.)*. Within, or by way of, the intestine.

enteral *(Zool.)*. Parasympathetic.

enterclose *(Arch.)*. A corridor separating two rooms.

enterectomy *(Med.)*. Removal of part of the bowel.

enter-, entero- *(Genrl.)*. Prefix from Gk. *enteron*, intestine.

enteric *(Med.)*. (1) Pertaining to the intestines. (2) Synonym for *enteric fever* (see **typhoid fever**).

entering edge *(Elec.Eng.)*. The edge of the brush of an electrical machine which is first met during revolution by a point on the commutator or slip ring. Also called *leading edge* or *toe of the brush*. Cf. **leaving edge**.

entering pallet *(Horol.)*. See **pallet**.

entering tap *(Eng.)*. See **taper tap**.

enteritis *(Med.)*. Inflammation of the small intestine.

entero-anastomosis *(Med.)*. The operative union of two parts of the intestine; the operation for doing this.

Enterobacteriaceae *(Biol.)*. A family of bacteria belonging to the order Eubacteriales. Gram-negative rods; aerobes; carbohydrate fermenters; many saprophytic. Includes some gut parasites of animals and some blights and soft rots of plants, e.g. *Salmonella typhi* (typhoid), *Escherichia coli* (some strains cause enteritis in calves and infants).

enterobiasis *(Med.)*. Infection by the thread- or pinworm, *Enterobius vermicularis*. Commonest intestinal parasitic infection in UK, causing anal irritation in children.

enterocele *(Med.)*. A hernia containing intestine.

enterocentesis *(Med.)*. Operative puncture of the intestine.

enterocentesis *(Vet.)*. The operation of puncturing the distended intestine of a horse suffering from colic, in order to liberate the gas.

enterocoel *(Zool.)*. A coelome formed by the fusion of coelomic sacs which have separated off from the archenteron. In Chaetognatha, Echinodermata, and Cephalochorda.

enterocolitis *(Med.)*. Inflammation of the small intestine and of the colon.

enterocolostomy *(Med.)*. A communication, made by operation, between the small intestine and colon.

enterocystocele *(Med.)*. Hernia containing intestine and bladder.

entero-enterostomy *(Med.)*. The operative formation of a communication between two separate parts of the small intestine; the operation for doing this.

enterogenous cyanosis *(Med.)*. A disorder characterized by chronic cyanosis and by the presence of methaemoglobin or sulphaemoglobin in the blood; due usually to taking drugs, especially aniline derivatives.

enterolith *(Med.)*. A concretion of organic matter and lime, bismuth, or magnesium salts, formed in the intestine.

enteron *(Zool.)*. The single body cavity of Coelenterata; it corresponds to the archenteron of a gastrula: in higher forms, the alimentary canal. See also **archenteron**. adj. *enteric*.

Enteropneusta *(Zool.)*. See **Hemichordata**.

enteroproctia *(Med.)*. The condition in which an artificial anus is formed from intestine.

enterostomy *(Med.)*. The surgical formation of an opening in the intestine, for the purpose of draining intestinal contents.

enterosympathetic *(Zool.)*. Said of that part of the autonomic nervous system which supplies the alimentary tract.

enterotomy *(Med.)*. Incision of the intestinal wall.

enterotoxaemias of sheep *(Vet.)*. A group of fatal toxaemic diseases of sheep due to alimentary infection and toxin production by the bacteria *Clostridium perfringens (Cl. welchii)*, types B, C or D. Vaccines widely used.

enthalpy *(Phys.)*. Thermodynamic property of a working substance defined as $H = U + PV$ where U is the internal energy, P the pressure and V the volume of a system. Associated with the study of heat of reaction, heat capacity and flow processes. SI unit is the joule.

enthalpy heat drop *(Eng.)*. The enthalpy drop which occurs during the adiabatic expansion of unit mass of steam or other vapour and is capable of transformation into work.

entire *(Bot.)*. Said of the margin of a flattened organ when it is continuous, being neither toothed nor lobed.

entire function *(Maths.)*. See **analytic function**.

ento- *(Genrl.)*. Prefix from Gk. *entos*, within.

entoderm *(Zool.)*. See **endoderm**.

entogastric *(Zool.)*. Within the stomach or enteron.

entomology *(Zool.)*. The branch of zoology which deals with the study of insects.

entomophagous *(Zool.)*. Feeding on insects.

entomophily *(Bot.)*. Pollination by insects. adj. *entomophilous.*

Entoprocta *(Zool.)*. A phylum of Metazoa in which the anus is inside the circlet of tentacles; mainly marine forms. Also *Endoprocta.*

entovarial *(Zool.)*. Within the ovary.

entozoic *(Bot.)*. Living inside an animal.

entozoon *(Zool.)*. An animal parasite living within the body of the host. adj. *entozoic.*

entrain *(Eng.)*. To suspend bubbles or particles in a moving fluid.

entrainer *(Chem.Eng.)*. See **azeotropic distillation**.

entrainment *(Behav.)*. The process whereby an endogenous clock driven rhythm is synchronised to the rhythm of environmental events. See **Zeitgeber**.

entrainment *(Chem.)*. Transport of small liquid particles in vapour, e.g. when drops of water are carried over in steam.

entrance lock *(Hyd.Eng.)*. A lock through which vessels must pass in entering or leaving a dock, on account of difference in level between the water impounded in the dock and that outside.

entrance pupil *(Phys.)*. The image of the *aperture stop* as formed by that part of an optical system nearer the object. It defines the cone of rays entering the system.

entresol *(Arch.)*. A **mezzanine** floor usually between the ground and first floors.

entropy *(Phys.)*. In thermal processes, a quantity which measures the extent to which the energy of a system is available for conversion to work. If a system undergoing an infinitesimal reversible change takes in a quantity of heat dQ at absolute temperature T, its entropy is increased by $dS = dQ/T$. The area under the absolute temperature-entropy graph for a reversible process represents the heat transferred in the process. For an adiabatic process, there is no heat transfer and the temperature-entropy graph is a straight line, the entropy remaining constant during the process. When a thermodynamic system is considered on the microscopic scale, equilibrium is associated with the distribution of molecules that has the greatest probability of occurring, i.e. the state with the greatest degree of disorder. *Statistical mechanics* interprets the increase in entropy in a closed system to a maximum at equilibrium as the consequence of the trend from a less probable to a more probable state. Any process in which no change in entropy occurs is said to be *isentropic*.

entropy *(Telecomm.)*. Deprecated term for *information rate.*

entropy of fusion *(Phys.)*. A measure of the increased randomness that accompanies the transition from solid to liquid or liquid to gas; equal to the latent heat divided by the absolute temperature. Also called *vaporization.*

entry portal *(Radiol.)*. Area through which a beam of radiation enters the body.

enucleate *(Bot.)*. (1) Lacking a nucleus. (2) To remove, e.g. by microdissection, the nucleus from a cell.

enucleation *(Med.)*. Removal of any tumour or globular swelling so that it comes out whole.

enucleation *(Zool.)*. The removal of the nucleus from a cell, e.g. by microdissection. adj. and v. *enucleate.*

enumerable set *(Maths.)*. See **denumerable set**.

enuresis *(Behav.,Med.)*. A lack of bladder control past the age when such control is normally achieved.

envelope *(Aero.)*. The gas-bag of a balloon, or of a nonrigid or semirigid airship.

envelope *(Maths.)*. Of a family of plane curves: the curve which touches every curve of the family.

envelope *(Paper)*. A container for a letter or similar flat document, generally made from paper by cutting a suitable blank shape which is then folded and glued. A flap is usually provided by which the contents may be sealed, either by tucking in or by use of an applied adhesive.

envelope *(Telecomm.)*. The modulation waveform within which the carrier of an amplitude-modulated signal is contained, i.e. the curve connecting the peaks of successive cycles of the carrier wave.

envelope delay *(Telecomm.)*. The time taken for the envelope of a signal to travel through a transmission system, without reference to the time taken by the individual components. Cf. **group delay**.

envelope-delay distortion *(Telecomm.)*. Distortion arising when the rate of change of phase shift with frequency for a transmission system is variable over the required frequency range.

envelope velocity *(Telecomm.)*. See **group velocity**.

environment *(Ecol.)*. (1) The physical and chemical surroundings of an object, e.g. the temperature and humidity, the physical structures, the gases. (2) When applied to human societies, the cultural, aesthetic and any other factors which contribute to the quality of life, are included in the definition.

environmental control *(Space)*. The provision and control of the environment of a space vehicle, or part of it, so that its payload (including man) can operate efficiently. It can involve control of temperature, humidity, atmosphere and contamination.

environmental pathway *(Nuc.Eng.)*. Route by which a radionuclide in the environment can reach man, e.g. from radioactivity in rain to grass, to cows, to milk, to people.

environmental variance *(Biol.)*. Variation of quantitative character due to non-genetic causes.

enzootic *(Vet.)*. Said of a disease prevalent in, and confined to, animals in a certain area; corresponding to *endemic* in man.

enzootic ataxia *(Vet.)*. See **swayback**.

enzootic bovine leukosis *(Vet.)*. Caused by an *oncorna* virus. Imported by infected animals. Tumour masses are found in lymph nodes and some other organs, e.g. spinal cord, heart and abomasum. Abbrev. *EBL.* Notifiable in UK where a slaughter policy exists.

enzootic marasmus *(Vet.)*. See **pine**.

enzootic ovine abortion *(Vet.)*. *Kebbing.* A contagious form of abortion in sheep caused by a *chlamydial* infection. Most common cause of ovine abortion in UK, but vaccine is available.

enzyme *(Biol.)*. A protein with catalytic activity, which is restricted to a limited set of reactions, defining the specificity of the enzyme. They are designated by the suffix *-ase*, frequently attached to the type of reaction catalysed. Thus enzymes catalysing hydrolytic reactions are termed *hydrolases.* Virtually all metabolic reactions are dependent on and controlled by enzymes.

enzyme-linked immunosorbent assay *(Immun.)*. *ELISA.* An assay method in which antigen or antibody are detected by means of an enzyme chemically coupled either to antibody specific for the antigen or to anti-Ig which in turn will bind to the specific antibody. Either the antigen or the antibody to be detected is attached to the surface of a small container or to plastic beads, and the specific antibody allowed to bind in turn. The amount bound is subsequently measured by addition of a substrate for an enzyme which develops a colour when hydrolysed. Commonly used enzymes are *horse radish peroxidase* or *alkaline phosphatase.*

Eocene *(Geol.)*. The oldest division of Tertiary rocks, now regarded as an epoch within the Paleogene system.

EOF *(Comp.)*. See **end of file marker**.

eolian deposits *(Geol.)*. See **aeolian deposits**.

eolith *(Geol.)*. Literally 'dawn stone'; a term applied to the oldest-known stone implements used by early man which occur in the Stone Bed at the base of the Crag in E. Anglia and in high-level gravel deposits elsewhere. The workmanship is crude, and some authorities question their human origin, thinking it likely that the chipping has been produced by natural causes.

Eolithic Period *(Geol.)*. The time of the primitive men who manufactured and used eoliths: the dawn of the Stone Age. Cf. **Palaeolithic Period, Neolithic Period**.

eon *(Geol.)*. A large part of geological time consisting of a number of **eras**. E.g. *Phanerozoic eon* which includes the **Palaeozoic**, **Mesozoic** and **Cenozoic** eras.

eosin *(Chem.)*. $C_{20}H_6Br_4O_5K_2$, the potassium salt of tetrabromo-fluorescein, a red dye.

eosinophil *(Biol.)*. Any cell whose protoplasmic granules readily stain red with the dye eosin, particularly a granulocyte in the blood and the oxyphil cells in the *pars anterior* of the pituitary gland.

eosinophilia *(Immun.)*. Increase in numbers of eosinophil leucocytes in the blood above the normal levels (up to $400/mm^3$ in humans). Usually associated with repeated immediate type hypersensitivity reactions. Unexplained very high levels of eosinophil leucocytes sometimes occur in subjects with eosinophilic cardiac myopathy and their granule contents cause necrosis of heart muscle cells.

eosinophilia *(Med.)*. A pathological excess of eosinophils in the blood.

eosinophil leucocyte *(Immun.)*. Polymorphonuclear leucocyte with large eosinophil granules in the cytoplasm containing cationic proteins. Other granules contain peroxidase. The cells have Fc receptors for IgE and for complement. The granules are secreted when the cells are activated (e.g. via these receptors). In parasite infections such as *schistosomiasis* in which IgE antibodies are made, eosinophils attach themselves to the parasites and the granule contents are secreted onto the parasites and kill them.

Eötvös balance *(Min.Ext.)*. Torsion balance sensitive to minute gravitational differences in land masses.

Eötvös equation *(Chem.)*. The molecular surface energy of a substance decreases linearly with temperature, becoming zero about 60°C below the critical point.

Eozoic *(Geol.)*. A term suggested for the Precambrian System, but little used. It means the 'dawn of life', and is comparable with *Palaeozoic*, *Mesozoic*, and *Cainozoic*.

eozoon *(Geol.)*. A banded structure found originally in certain Canadian Precambrian rocks and thought to be of organic origin; now known to be inorganic and a product of dedolomitization, consisting of alternating bands of calcite and serpentine replacing forsterite.

e-PAL *(Image Tech.)*. Enhanced PAL, a development of the **PAL** colour TV broadcast system with improved picture quality.

epapophysis *(Zool.)*. A median process of a vertebral neural arch.

eparterial *(Med.)*. Situated over an artery. *Eparterial bronchus*, the first division of the right bronchus, which passes to the upper lobe of the right lung.

epaxial *(Zool.)*. Above the axis, especially above the vertebral column, therefore, dorsal; as the upper of two blocks into which the myotomes of fish embryos become divided. Cf. **hypaxial**.

epaxonic *(Zool.)*. See **epaxial**.

epeirogenic earth movements *(Geol.)*. Continent-building movements, as distinct from mountain-building movements, involving the coastal plain and just-submerged 'continental platform' of the great land areas. Such movements include gentle uplift or depression, with gentle folding and the development of normal tensional faults.

epencephalon *(Zool.)*. See **cerebellum**.

ependyma *(Zool.)*. In Vertebrates, the layer of columnar ciliated epithelium, backed by neuroglia, which lines the central canal of the spinal cord and the ventricles of the brain. adj. *ependymal*.

ependymitis *(Med.)*. Inflammation of the ependyma.

ependymoma *(Med.)*. A tumour within the brain arising in or near the ventricles and containing ependyma-like cells.

Ephedra *(Bot.)*. Sea-grape, a genus of jointed, all but leafless, desert plants of the Gnetaceae. Source of **ephedrine**.

ephedrine *(Med.)*. An alkaloid isolated from *Ephedra vulgaris*. Used in medicine as a sympathomimetic agent for many conditions (asthma, allergies), and as an operational adjuvant.

ephemeral *(Bot.)*. A plant which completes its life cycle in a short period, weeks rather than months.

ephemeral fever *(Vet.)*. See **three-day sickness**.

ephemeris *(Astron.)*. A compilation, published at regular intervals, in which are tabulated the daily positions of the sun, moon, planets and certain stars, with other data necessary for the navigator and observational astronomer. See **Astronomical Ephemeris, Nautical Almanac**.

ephemeris time *(Astron.)*. Abbrev. *ET*. Uniform or Newtonian time, as used in the calculation of future positions of sun and planets. The normal measurement of time by observations of stars includes the irregularities due to the changes in the earth's rate of rotation. The difference between ephemeris time and universal time is adjusted to zero at an epoch in 1900; it amounted to about 40 sec in 1970. The *ephemeris second* is the fundamental unit of time adopted by the International Committee of Weights and Measures, its defined value being 1/31 556 925.9747 of the tropical year for 1900 January, 0 at 12^h ET.

Ephemeroptera *(Zool.)*. An order of Insects, in which the adults have large membranous forewings and reduced hindwings, and the abdomen bears two or three long caudal filaments; the adult life is very short and the mouthparts are reduced and functionless; the immature stages are active aquatic forms. May-flies.

epi- *(Chem.)*. (1) Containing a condensed double aromatic nucleus substituted in the 1,6-positions. (2) Containing an intramolecular bridge.

epibasal half *(Bot.)*. The anterior portion of an embryo.

epibiosis *(Ecol.)*. Relationship in which one organism lives on the surface of another without causing it harm. Plant epibionts are *epiphytes*, animal epibionts, *epizoites*.

epiblast *(Zool.)*. The outer germinal layer in the embryo of a metazoan animal, which gives rise to the ectoderm. Cf. **hypoblast**.

epiblem rhizodermis *(Bot.)*. The outermost cell layer (epidermis) of a root.

epiboly *(Zool.)*. Overgrowth; growth of one part or layer so as to cover another. adj. *epibolic*.

epicalyx *(Bot.)*. An extra, calyx-like structure below and close to the real calyx in some flowers e.g. many Rosaceae, including strawberry, and many Malvaceae.

epicanthus *(Med.)*. A semilunar fold of skin above, and sometimes covering, the inner angle of the eye; a normal feature of the Mongolian races.

epicardium *(Zool.)*. In Vertebrates, the serous membrane covering the heart: in *Urochordata*, diverticula of the pharynx, which grow out and surround the digestive viscera like a perivisceral cavity. adj. *epicardial*.

epicarp *(Bot.)*. The superficial layer of the pericarp, especially when it can be stripped off as a skin.

epicentre *(Geol.)*. That point on the surface of the earth lying immediately above the focus of an earthquake, or nuclear explosion.

epichlorhydrin *(Chem.)*. *1-Chloro-2,3-epoxypropane*, C_3H_5ClO, bp 117°C. A liquid derivative of glycerol formed by reaction with hydrogen chloride to give *dichlorohydrin*, which in turn is treated with concentrated potassium hydroxide solution. Used in the production of **epoxy resins**.

epicoele *(Zool.)*. In Craniata, the cerebellar ventricle or cavity of the cerebellum.

epicondyle *(Zool.)*. The proximal part of the condyle of the humerus or femur.

epicormic shoot *(Bot.)*. A shoot growing out, adventitiously or from a dormant bud, from a tree trunk or substantial woody branch.

epicotyl *(Bot.)*. Either (1) the part of the shoot of an embryo or seedling above the cotyledon(s), i.e. whole of the plumule, or, (2) the stem between the cotyledon(s) and the first leaf or leaves, i.e. the first internode of the plumule.

epicritic *(Zool.)*. Pertaining to sensitivity to slight tactile stimuli.

epicuticle *(Bot.)*. Layer of waxes including long-chain ($> C_{20}$) alkanes, alcohols, acids and esters on the surface of the cuticle. Cf. **bloom**.

epicuticle *(Zool.)*. The thin outermost layer of the insect cuticle consisting of lipid and protein. See **cuticulin**.

epicycle *(Astron.)*. The term applied in Ptolemaic or geocentric astronomy to a small circle, described uniformly by the sun, moon, or planet, the centre of that circle itself describing uniformly a larger circle (the *deferent*), concentric with the earth.

epicyclic gear *(Eng.)*. A system of gears, in which one or more wheels travel round the outside or inside of another wheel whose axis is fixed.

epicyclic train *(Eng.)*. A system of epicyclic gears, in which at least one wheel axis itself revolves about another fixed axis; used for giving a large reduction ratio in small compass, and for the gearboxes of some motor cars.

epicycloid *(Maths.)*. See **roulette**.

epidemic *(Med.)*. An outbreak of an infectious disease spreading widely among people at the same time in any region. Also used as adjective.

epidemic parotitis *(Med.)*. See **mumps**.

epidemic tremor *(Vet.)*. See **infectious avian encephalomyelitis**.

epidemiology *(Med.)*. The study of disease in the population, defining its incidence and prevalence, examining the role of external influences such as infection, diet or toxic substances and examining appropriate preventative or curative measures.

epidermal, epidermatic *(Bot.,Zool.)*. Relating to the epidermis.

epidermis *(Bot.)*. A layer, usually one cell thick, forming a skin over the young shoots and roots of plants, continuous except where perforated by stomata (on shoots) or over lateral or adventitious roots; often carries hairs. Eventually replaced by the periderm in woody plants.

epidermis *(Zool.)*. Those layers of the integument which are of ectodermal origin; the epithelium covering the body.

epidermoid cyst *(Med.)*. A *wen*. A cyst lined with epithelium, occurring in the scalp.

epidermolysis bullosa *(Med.)*. A congenital defect of the skin, in which the slightest blow produces a blister.

epidiascope *(Phys.)*. A projection lantern which may be used for transparencies or for opaque objects. See **episcope**.

epididymectomy *(Med.)*. Removal of the epididymis.

epididymis *(Zool.)*. In the male of Elasmobranchii, some Amphibians and Amniota, the greatly coiled anterior end of the Wolffian duct, which serves as an outlet for the spermatozoa.

epididymitis *(Med.)*. Inflammation of the epididymis.

epididymo-orchitis *(Med.)*. Inflammation of the epididymis and the testes.

epididymotomy *(Med.)*. Incision into the epididymis.

epidiorite *(Geol.)*. A metamorphosed gabbro or dolerite in which the original pyroxene has been replaced by fibrous amphibole. Other mineral changes have also taken step in the conversion, by dynamothermal metamorphism, of a basic igneous rock into a green schist.

epidosites *(Geol.)*. Metamorphic rocks composed of epidote and quartz. See **epidotization**.

epidote *(Min.)*. A hydrated aluminium iron silicate, occurring in many metamorphic rocks in lustrous yellow-green crystals, and as an alteration product in igneous rocks.

epidotization *(Geol.)*. A process of alteration, especially of basic igneous rocks in which the feldspar is albitized with the separation of epidote and zoisite representing the anorthite molecule of the original plagioclase.

epidural anaesthesia *(Med.)*. Loss of painful sensation in the lower part of the body produced by injecting an anaesthetic into the epidural space surrounding the spinal canal.

epigamic *(Zool.)*. Attractive to the opposite sex; as *epigamic colours*.

epigastric *(Zool.)*. Above or in front of the stomach; said of a vein in Birds, which represents the anterior part of the anterior abdominal vein of lower Vertebrates.

epigastrium *(Med.)*. The abdominal region between the umbilicus and sternum.

epigeal, epigaeous *(Bot.)*. (1) Germinating with the cotyledons appearing above the surface of the ground. (2) Living on the soil surface.

epigenesis *(Biol.)*. The theory, now universally accepted, that the development of an embryo consists of the gradual production and organization of parts, as opposed to the theory of *preformation*, which supposed that the future animal or plant was already present complete, although in miniature, in the fertilised egg.

epigenetic *(Min.)*. Ore deposits formed later than the rocks enclosing them. See **syngenetic**.

epiglottidectomy *(Med.)*. Excision of the epiglottis.

epiglottis *(Zool.)*. In Bryozoa, the epistome; in Insects, the epipharynx; in Mammals, a cartiliaginous flap which protects the glottis.

epignathous *(Zool.)*. Having the upper jaw longer than the lower jaw, as in Sperm Whales.

epignathus *(Med.)*. A foetal malformation in which the deformed remnants of one twin project through the mouth of the more developed twin.

epigynous *(Bot.)*. Having the calyx, corolla and stamens inserted on the top of the inferior ovary.

Epikote *(Plastics)*. TN for a range of **epoxy resins**, used for castings, encapsulation (potting) and surface coatings.

epilation *(Med.)*. Removal of the hair by the roots.

epilepsy *(Med.)*. A general term for a sudden disturbance of cerebral function accompanied by loss of consciousness, with or without convulsion. See **grand mal, petit mal, Jacksonian epilepsy**.

epileptiform *(Med.)*. Resembling epilepsy.

epileptogenic *(Med.)*. Exciting an attack of epilepsy.

epilimnion *(Ecol.)*. The warm upper layer of water in a lake. Cf. **hypolimnion**.

epilithic *(Bot.)*. Growing on a rock surface.

epiloia *(Med.)*. A condition characterized by feeble-mindedness, epileptic fits, sclerosis of the brain, and tumours in the skin and viscera; due to a defect in development. See also **tuberose sclerosis**.

epimenorrhoea *(Med.)*. Too frequent occurrence of menstrual periods.

epimerization *(Chem.)*. A type of asymmetric transformation in organic molecules, e.g. shown by the change of D-gluconic acid into an isomeric mixture of D-gluconic and D-mannonic acids.

Epimetheus *(Astron.)*. The eleventh natural satellite of **Saturn**, discovered in 1980.

epimorph *(Min.)*. A natural cast of a crystal.

epimysium *(Zool.)*. The investing connective-tissue coat of a muscle.

epinasty *(Bot.)*. **Nastic movement** in which the upper side of the base of an organ grows more than the lower, resulting in a downward bending of the organ as in the petals of an opening flower or the leaves of a tomato plant with water-logged roots. Cf. **hyponasty**.

epinephros *(Zool.)*. See **suprarenal body**.

epineural *(Zool.)*. In Echinodermata, lying above the radial nerve: in Vertebrates, lying above or arising from the neural arch of a vertebra.

epineurium *(Zool.)*. The connective tissue which invests a nerve trunk, uniting the different funiculi and joining the nerve to the surrounding and related structures.

epiparasite *(Zool.)*. See **ectoparasite**.

epipetalous *(Bot.)*. Attached to or inserted upon the petals, as stamens are in many flowers.

epipharynx *(Zool.)*. In Insects, the membranous roof of the mouth which in some forms is produced into a chitinized median fold, and in Diptera is associated with the labrum, to form a piercing organ; in Acarina, a forward projection of the anterior face of the pharynx. adj. *epipharyngeal*.

epiphloeodal, epiphloeodic *(Bot.)*. Growing on the surface of bark.

epiphora *(Med.)*. An overflow of the lacrimal secretion, due to obstruction of the channels which normally drain it.

epiphragm *(Zool.)*. In Gastropoda, a plate, mostly composed of calcium phosphate, with which the aperture of the shell is sealed during periods of dormancy.

epiphyllous *(Bot.)*. Growing upon, or attached to, the

upper surface of a leaf; sometimes, growing on any part of a leaf.

epiphysis *(Zool.)*. A separate terminal ossification of some bones, which only becomes united with the main bone at the attainment of maturity; the pineal body; in Echinoidea, one of the ossicles of Aristotle's lantern. adj. *epiphysial*.

epiphysis cerebri *(Zool.)*. See **pineal organ**.

epiphysitis *(Med.)*. Inflammation of the epiphysis of a bone.

epiphyte *(Ecol.)*. An organism which is attached to another with benefit to the former but not to the latter. See **nest epiphyte, tank epiphyte**.

epiphytotic *(Bot.)*. A widespread outbreak of disease among plants, by analogy with epidemic.

epipleura *(Zool.)*. In Coleoptera, the reflexed sides of the elytra; in Birds, the unicate process; in bony Fish, upper ribs formed from membrane bone; in Cephalochordata, horizontal shelves of membrane arising from the inner sides of the metapleural folds, and forming the floor of the atrial cavity.

epiplocele *(Med.)*. A hernia containing omentum (the **epiploon**) only.

epiploic foramen *(Zool.)*. See **Winslow's foramen**.

epiploon *(Zool.)*. In Mammals, a double fold of serous membrane connecting the colon and the stomach (the great omentum). adj. *epiploic*.

epipubic *(Zool.)*. In front of or above the pubis; pertaining to the epipubis.

episclera *(Med.)*. The connective tissue between the conjunctiva and the sclera.

episcleritis *(Med.)*. Inflammation of the episclera, sometimes involving the sclera.

episcope *(Phys.)*. A projection lantern which throws an enlarged image of a brilliantly illuminated opaque object onto a screen.

episematic *(Zool.)*. Serving for recognition; as *episematic colours*.

episepalous *(Bot.)*. (1) Borne on the sepals. (2) Placed opposite to the sepals.

episiostenosis *(Med.)*. Narrowing of the vulvar orifice.

episiotomy *(Med.)*. Cutting the vulvar orifice to facilitate delivery of the foetus.

episodic memory *(Behav.)*. Refers to personal memories, tied to a particular time and place. Cf. semantic memory.

episome *(Biol.)*. A self-replicating element able to grow independently of the host's chromosome, but also to integrate into it. Often termed a *plasmid* in prokaryotes.

epispadias *(Med.)*. Congenital defect in the anterior or dorsal wall of the urethra, commoner in the male than in the female.

epispore *(Bot.)*. The outermost layer of a spore wall, often consisting of a deposit forming ridges, spines, or other irregularities of the surface.

epistatic *(Biol.)*. Describes a gene, or character, whose effect overrides that of another gene with which it is not allelic; analogous to *dominant* applied to genes at different loci. More generally, *epistasis* exists when the effect of two or more non-allelic genes in combination is not the sum of their separate effects.

epistaxis *(Med.)*. Bleeding from the nose.

epistemics *(Behav.)*. The scientific study of the perceptual, intellectual and linguistic processes by which knowledge and understanding are acquired and communicated.

epistilbite *(Min.)*. A white or colourless zeolite; hydrated calcium aluminium silicate, crystallizing in the monoclinic system.

epistomatal, epistomatic *(Bot.)*. A leaf having stomata on the upper surface only. Cf. **amphistomatal, hypostomatal**.

epistropheus *(Zool.)*. The axis vertebra.

epistyle *(Arch.)*. See **architrave**.

epitaxial *(Electronics)*. Pertaining to a method of growing thin layers of semiconducting material onto an existing substrate, whilst ensuring that the crystalline orientation of the deposited layer is the same as that of the substrate. This is achieved by condensing silicon atoms onto a silicon substrate at 1200°C.

epitaxial transistor *(Electronics)*. One in which the collector consists of a high-resistivity epitaxial layer deposited on a low-resistivity substrate, the emitter and base regions being formed in or on this layer by diffusion techniques. This type of construction results in a very thin base region, important for effective high-frequency operation, and a relatively large collector, which ensures good heat dissipation.

epitaxy *(Crystal.)*. Unified crystal growth or deposition of one crystal layer on another.

epithalamus *(Zool.)*. In the Vertebrate brain, a dorsal zone of the thalamencephalon.

epithelioid *(Med.)*. Resembling epithelium.

epithelioma *(Med.)*. A malignant growth derived from epithelium.

epithelioma contagiosa *(Vet.)*. See **fowl pox**.

epithelium *(Bot.)*. A compact layer of cells often secretory, lining a cavity or covering a surface.

epithelium *(Zool.)*. A form of tissue characterized by the arrangement of the cells as an expansion covering a free surface, or as solid masses, by the presence of a basement membrane underlying the lowermost layer of cells, and by the small amount of intercellular matrix: the secretory substance of glands, the tissues lining the alimentary canal and blood-vessels etc. adjs. *epithelial, epitheliomorph*.

epithermal neutrons *(Nuc.Eng.)*. Neutrons having energies just above thermal, comparable to chemical bond energies.

epithermal neutrons *(Phys.)*. See **neutron**.

epithermal reactor *(Phys.)*. See **intermediate reactor**.

epitokous *(Zool.)*. Said of the heteronereid stage in Polychaeta.

epitope *(Immun.)*. Term used to describe an antigenic determinant in a molecule which is specifically recognised by an antibody combining site or by the antigen receptor of a T-cell.

epitrichium *(Zool.)*. A superficial layer of the epidermis in Mammals, which consists of greatly swollen cells and is found on parts of the body devoid of hair. adj. *epitrichial*.

epitrochlea *(Med.)*. The inner condyle, or bony eminence, on the inner aspect of the lower end of the humerus.

epitrochoid *(Maths.)*. See **roulette**.

epitropic fibre *(Textiles)*. A synthetic fibre whose surface contains particles (e.g. of carbon) that are electrically conducting.

epituberculosis *(Med.)*. Congestion and inflammation of the area surrounding a tuberculous focus, especially in the lung.

epixylous *(Bot.)*. Growing on wood.

epizoic *(Bot.)*. (1) Growing on a living animal, e.g. *epibiosis*. (2) Having the seeds or fruits dispersed by animals.

epizoon *(Zool.)*. An animal which lives on the skin of some other animal; it may be an **ectoparasite** or a **commensal**. adj. *epizoan*.

epizootic *(Vet.)*. Applied to a disease affecting a large number of animals simultaneously throughout a large area and spreading with great speed.

epizootic catarrhal fever *(Vet.)*. See **equine influenza**.

epizootic lymphangitis *(Vet.)*. A chronic contagious lymphangitis of horses, due to infection by *Histoplasma farciminosus*, a yeast.

epoch *(Astron.)*. The precise instant to which the data of an astronomical problem are referred; thus the elements of an orbit when referred to a specific epoch also implicitly define the obliquity of the ecliptic, the rate of precession and other conditions obtaining only at that instant.

epoch *(Geol.)*. A sub-division of a geological period. e.g. the **Wenlock** epoch of the **Silurian** period.

epoxy (epoxide) resins *(Chem.)*. Polymers derived from epichlorhydrin and bisphenol-A. Widely used as structural plastics, surface coatings and adhesives, and for encapsulating and embedding electronic components. Characterized by low shrinkage on polymerization, good adhesion, mechanical and electrical strength and chemical resistance.

EPR *(Chem.)*. Abbrev. for *Electron Paramagnetic Resonance.*

EPROM *(Comp.)*. See **erasable programmable read-only memory**.

epsomite, Epsom salts *(Min.)*. Hydrated magnesium sulphate, $MgSO_4.7H_2O$. Occurs as incrustations in mines, in colourless acicular to prismatic crystals. Many chemical and other uses.

Epstein-Barr virus *(Biol.)*. First virus to be implicated in human cancer because of its association with Burkitt's lymphoma.

epulis *(Med.)*. A tumour, innocent or malignant, of the gums, growing from the periosteum of the jaw.

EQ gate *(Comp.)*. See **equivalence gate**.

equal-area criterion *(Elec.Eng.)*. A term used in connection with the stability of electric power systems. The stability limit occurs when two areas on the power-angle diagram, governed by the load conditions obtaining, are equal.

equal energy source *(Phys.)*. An electromagnetic or acoustic source whose radiated energy is distributed equally over its whole frequency spectrum.

equal-, equi-tempered scale *(Acous.)*. See **tempered scale**.

equal falling particles *(Min.Ext.)*. Particles possessing equal **terminal velocities**; the underflow, oversize product of a classifier.

equalization *(Telecomm.)*. Electronically, the reduction of distortion by compensating networks which allow for the particular type of the distortion over the requisite band.

equalization of boundaries *(Surv.)*. See **give-and-take lines**.

equalizing bed *(Civ.Eng.)*. The bed of fine ballast or concrete laid immediately underneath a pipeline, e.g. in a trench, in cases where the bottom of the excavation is sound but uneven (as in rock, hard chalk etc.).

equalizing network *(Telecomm.)*. (1) Network, incorporating any inductance, capacitance or resistance, which is deliberately introduced into a transmission circuit to alter the response of the circuit in a desired way; particularly to equalize a response over a frequency range. (2) A similar arrangement incorporated in the coupling between stages in an amplifier.

equalizing pulse *(Image Tech.)*. A pulse used in television at twice the line frequency, which is applied immediately before and after the vertical synchronizing pulse. This is done to reduce any effect of line-frequency pulses on the interlace.

equalizing signals *(Image Tech.)*. Those added to ensure triggering at the exact time in a frame cycle.

equal-signal system *(Aero.,Telecomm.)*. One in which two signals are emitted for radio-range, an aircraft receiving equal signals only when on the indicated course.

equation *(Chem.)*. See **chemical equation**.

equation *(Maths.)*. (1) A sentence in which the verb is 'is equal to' and which involves a variable (e.g. x) on a set E; e.g. $x^2 + 6 = 5x$, where x is a variable on the set of integers. Those elements of E which turn the equation into a true statement, when they are put in place of x, constitute the solution set of the equation. For the equation $x^2 + 6 = 5x$ the solution set is [2,3]. If the equation is satisfied by every element of the set E, then the equation is said to be an *identity*; e.g., $\sin^2 x + \cos^2 x = 1$, where x is a variable on the set of real numbers, is an identity, for it is satisfied by every real number. (2) The equation of a curve or surface is an equation satisfied by the co-ordinates of every point on the curve or surface, but not satisfied by the co-ordinates of any point not on the curve or surface.

equation of maximum work *(Chem.)*. See **Gibbs-Helmholtz equation**.

equation of state *(Chem.)*. An equation relating the volume, pressure and temperature of a given system, e.g. van der Waals' equation.

equation of time *(Astron.)*. The difference between the right ascensions of the true and mean sun, and hence the difference between apparent and mean time. In the sense mean time minus apparent time, it has a maximum positive value of nearly $14\frac{1}{2}$ min in February, and a negative maximum of nearly $16\frac{1}{2}$ min in November, and vanishes 4 times a year.

equator *(Astron.)*. See **celestial-**, **terrestrial-**.

equatorial *(Astron.)*. The name given to an astronomical telescope which is so mounted that it revolves about a polar axis parallel to the earth's axis; when set on a star it will keep that star in the field of view continuously, without adjustment. It has two graduated circles reading *Right Ascension* and *Declination* respectively.

equatorial *(Biol.)*. Situated or taking place in the equatorial plane; as the *equatorial furrow* which precedes division of an ovum into upper and lower blastomeres, and the *equatorial plate*, which, during mitosis, is the assembly of chromosomes on the spindle in the *equatorial plane*.

equatorial horizontal parallax *(Astron.,Surv.)*. See **horizontal parallax**.

equi- *(Genrl.)*. Prefix from L. *aequus*, equal.

equiangular spiral *(Maths.)*. A spiral in which the angle between the tangent and the radius vector is constant. Also called a **logarithmic spiral**. Its polar equation is $\log r = \alpha\theta$.

equidistant locking pallets *(Horol.)*. **Pallets** in which locking corners of entering and exit pallets are equidistant from pallet-staff axis.

equilateral arch *(Build.)*. An arch in which the two springing points and the crown of the intrados form an equilateral triangle.

equilateral roof *(Build.)*. A pitched roof having rafters of a length equal to the span.

equilateral triangle *(Maths.)*. A triangle having 3 equal sides.

equilibration *(Behav.)*. In Piaget's theory, the motivational mechanism underlying intellectual development; contradictory explanations of perceived events produce a state of disequilibrium, and this acts as a motivation to reorganise thinking on a higher cognitive level.

equilibration *(Eng.)*. The production of balance or equilibrium; as in the provision of balance weights for a lift or cage.

equilibrium *(Chem.)*. The state reached in a reversible reaction when the reaction velocities in the two opposing directions are equal, so that the system has no further tendency to change.

equilibrium *(Phys.)*. The state of a body at rest or moving with constant velocity. A body on which forces are acting can be in equilibrium only if the resultant force is zero and the resultant torque is zero.

equilibrium constant *(Chem.)*. The ratio, at equilibrium, of the product of the active masses of the molecules on the right side of the equation representing a reversible reaction to that of the active masses of the molecules on the left side.

equilibrium diagram *(Eng.)*. See **constitution diagram**.

equilibrium moisture *(Chem.Eng.)*. The percentage water content of a solid material, when the vapour pressure of that water is equal to the partial pressure of the water vapour in the surrounding atmosphere.

equilibrium of floating bodies *(Phys.)*. For a body which floats, partly immersed in liquid, the weight of the body is equal to the weight of the fluid it displaces. The ratio of the volume of the body to the volume immersed is the ratio of the density of the fluid to that of the body. See **Archimedes' principle** and also **metacentre** and **centre of buoyancy** for the stability of the equilibrium.

equilibrium still *(Chem.Eng.)*. One designed to produce a boiling liquid mixture in complete phase equilibrium with its vapour, for the purposes of physicochemical measurement.

equilux spheres *(Phys.)*. Spherical surfaces which are concentric with a source of light so that the illumination is constant.

equine contagious catarrh *(Vet.)*. See **strangles**.

equine encephalomyelitis *(Vet.)*. An acute disease of horses, mules and donkeys, due to a virus which causes an encephalomyelitis; characterized by fever, nervous symptoms, and often death. Transmitted by mosquitoes, ticks and mites. Several strains of virus occur: eastern American, western American, Venezuelan, and Russian. Another form, **Borna disease** is caused by an unrelated virus.

equine infectious anemia *(Vet.)*. *Swamp fever*. An acute or chronic RNA virus disease of horses characterized by intermittent fever, weakness, emaciation and anaemia; spread by flies. Notifiable in UK.

equine influenza *(Vet.)*. *Epizootic catarrhal fever, shipping fever, stable pneumonia, pink eye, the cough.* Caused by myxoviruses, *A Equi 1* (Prague), *A Equi 2* (Miami). Highly contagious disease occurring as epidemics. Symptoms include; fever, cough, nasal discharge and depression. Vaccine available.

equine thrush *(Vet.)*. Inflammation of the frog of the horses's foot, attended with a fetid discharge.

equine viral arteritis *(Vet.)*. An acute, respiratory viral disease of horses, characterized by degenerative and inflammatory changes in the small arteries; the main symptoms are fever, respiratory difficulty, oedema of the legs, diarrhoea, and in pregnant mares, abortion. Modified live vaccines available in certain places.

equine virus abortion *(Vet.)*. See **equine virus rhinopneumonitis.**

equine virus rhinopneumonitis *(Vet.)*. *Equine Herpes Virus 1 infection, EHV1.* Two subtypes; Subtype 1 strains associated with respiratory, neurological and neonatal disease and abortion. Subtype 2 with mainly respiratory disease. Vaccination available.

equinoctial points *(Astron.)*. The two points, diametrically opposite each other, in which the celestial equator is cut by the ecliptic; called respectively the *First Point of Aries* and the *First Point of Libra*, from the signs of the Zodiac of which they are the beginning.

equinox *(Astron.)*. (1) Either of the two points on the **celestial sphere** where the **ecliptic** intersects the **celestial equator**. Physically they are the points at which the Sun, in its annual motion, crosses the celestial equator, the *vernal equinox* as the Sun crosses the south to north, and the *autumnal equinox* as it crosses from north to south. The vernal equinox is the zero point in celestial coordinate systems. (2) Either of the two instants of time at which the Sun crosses the celestial equator, being about March 21 and September 23.

equipartition of energy *(Chem.)*. The Maxwell-Boltzmann law, which states that the available energy in a closed system eventually distributes itself equally among the **degrees of freedom** present.

equipotent *(Zool.)*. See **totipotent.**

equipotential surface (region) *(Elec.Eng.)*. One where there is no difference of potential, and hence no electric field. See **Faraday cage.**

Equisetales *(Bot.)*. The horsetails. Order of ca 20 species of the genus *Equisetum*, cosmopolitan except for Australia and New Zealand. Also fossils from the Upper Carboniferous onwards. The sporophyte has roots and rhizomes, and an aerial stem which bears whorls of very small fused leaves and of branches. The stems are photosynthetic. The gametophytes are thalloid and photosynthetic.

equitant *(Bot.)*. Distichous leaves folded longitudinally and each overlapping the next. See **vernation.**

equitonic scale *(Acous.)*. The musical scale in which the main notes progress by whole tones, as contrasted with the Pythagorean diatonic, which uses both whole tones and half-tones.

equivalence class *(Maths.)*. See **equivalence relation.**

equivalence gate *(Comp.)*. A **gate** with two input signals. If both are 0, output is 1. If both are 1 output is 1. When incoming signals differ output is 0. Abbrev. *EQ gate.* Also *equivalence element.*

equivalence, photochemical *(Chem.)*. See **Einstein law of photochemical equivalence.**

equivalence relation *(Maths.)*. Relation which is reflexive, symmetric and transitive. It partitions a set into equivalence classes.

equivalent *(Chem.)*. See **equivalent weight.**

equivalent air speed *(Aero.)*. **Indicated air speed** corrected for position error (angle of incidence) and air compressibility. Abbrev. *E.A.S.*

equivalent circuit *(Elec.Eng.)*. One consisting of passive components (R, L and C) and ideal current (or voltage) sources, which behaves, as far as current and voltage at its terminals are concerned, exactly as some other circuit or component, e.g. a bipolar transistor may be represented by a combination of resistors, capacitors and a current generator.

equivalent conductance *(Chem.)*. Electrical conductance of a solution which contains 1 gram-equivalent weight of solute at a specified concentration, measured when placed between two plane parallel electrodes, 1 m apart. *Molar conductance* is more often used.

equivalent electrons *(Electronics)*. Those which occupy the same orbit in an atom, hence have the same principal and orbital quantum numbers.

equivalent focal length *(Phys.)*. The focal length of a thin lens which is equivalent to a thick lens in respect of the size of image it produces.

equivalent free-falling diameter *(Powder Tech.)*. The diameter of a sphere which has the same density and the same free-falling velocity in a given fluid as an observed particle of powder.

equivalent height *(Telecomm.)*. That of a perfect antenna, erected over a perfectly conducting ground, which, when carrying a uniformly distributed current equal to the maximum current in the actual antenna, radiates the same amount of power.

equivalent hiding power *(Powder Tech.)*. Of particles interrupting a light beam, in photo-sedimentation analysis, the weight of a given size, expressed as a fraction or a percentage of the weight of a standard size, which has the same hiding power as unit weight of the standard of the size when in suspension. The standard size is usually the largest particle present.

equivalent lens *(Phys.)*. A simple lens of the same *equivalent focal length* as an optical system which, when placed at the first principal plane of the system, produces an image identical to that of the system except for a shift along the axis of magnitude equal to the distance between the principal planes of the system.

equivalent network *(Telecomm.)*. One identical to another network either in general or at some specified frequency. The same input applied to each would produce outputs identical in both magnitude and phase generated across the same internal impedance.

equivalent points *(Phys.)*. The principal points of a lens that is used with the same medium on both sides. See **cardinal points.**

equivalent potential temperature *(Meteor.)*. The equivalent potential temperature of an air sample is the **equivalent temperature** of the sample when brought adiabatically to a pressure of 1000 mb. It is a conservative property for both dry and saturated adiabatic processes.

equivalent proportions *(Chem.)*. See **law of equivalent (or reciprocal) proportions.**

equivalent reactance *(Elec.Eng.)*. The value which the reactance of an equivalent circuit must have in order that it shall represent the system of magnetic or dielectric linkages present in the actual circuit.

equivalent resistance *(Elec.Eng.)*. The value which the resistance of an equivalent circuit must have in order that the loss in it shall represent the total loss occurring in the actual circuit.

equivalent simple pendulum *(Phys.)*. See **centre of oscillation.**

equivalent sine wave *(Elec.Eng.)*. One which has the same frequency and the same r.m.s. value as a given non-sinusoidal wave.

equivalent surface diameter *(Powder Tech.)*. The diameter of a sphere which has the same effective surface as that of an observed particle when determined under stated conditions.

equivalent temperature *(Meteor.)*. The equivalent temperature of a sample of moist air is the temperature which would be attained by condensing all the water vapour in the sample and using the latent heat thus released to raise the temperature of the sample.

equivalent T-networks *(Telecomm.)*. T- or pi-networks equivalent in electrical properties to sections of transmission line, provided these are short in comparison with the wavelength.

equivalent volume diameter *(Powder Tech.)*. The diameter of a sphere which has the same effective volume as that of an observed particle when determined under the same conditions.

equivalent weight *(Chem.)*. That quantity of one substance which reacts chemically with a given amount of a standard. In particular, the equivalent weight of an acid will react with one mole of hydroxide ions, while the equivalent weight of an oxidising agent will react with one mole of electrons.

equivalve *(Zool.)*. Said of bivalves which have the two halves of the shell of equal size.

ER *(Biol.)*. Abbrev. for *Endosplasmic Reticulum*.

Er *(Chem.)*. The symbol for erbium.

era *(Geol.)*. A geological-time unit within an eon. e.g. **Mesozoic** era.

erasable programmable read-only memory *(Electronics)*. *EPROM*. A read-only memory in which stored data can be erased, by ultraviolet light, or other means, and reprogrammed bit by bit with appropriate voltage pulses.

erase *(Comp.)*. To remove data in a store or register of a computer.

erase head *(Acous.)*. In magnetic tape recording, the head which saturates the tape with high-frequency magnetization, in order to remove any previous recording.

erasion *(Med.)*. Removal of all diseased structures from a joint by cutting and scraping.

erbium *(Chem.)*. A metallic element, a member of the rare-earth group. Symbol Er, at. no. 68, r.a.m. 167.26. Found in the same minerals as dysprosium (gadolinite, fergusonite, xenotime), and in euxenite.

erbium laser *(Phys.)*. Laser using erbium in YAG (*Yttrium-Aluminium-Garnet*) glass. It has the advantage of operating between 1.53 and 1.64 μm, a range in which there is a high attenuation in water. This feature is of particular importance in laser aplications to eye investigations, since a great deal of energy absorption will now occur in the cornea and aqueous humour before reaching the delicate retina.

ERC *(Image Tech.)*. Abbrev. for *Ever-Ready Case*.

erect *(Bot.)*. Set at right angles to the part from which it grows.

erect *(Zool.)*. See **erection**.

erectile tissue *(Med.)*. Tissue which contains baggy blood-spaces, which when distended with blood becomes *turgid*.

erecting prism *(Phys.)*. A right-angled prism used for erecting the image formed by an inverting projection system. The prism is used with its hypotenuse parallel to the beam of light incident on one of the other faces, which is totally reflected at the hypotenuse and emerges from the third face parallel to its original direction.

erecting shop *(Eng.)*. That part of an engineering works where finished parts are assembled or fitted together.

erection *(Build.)*. The assembly of the parts of a structure into their final positions.

erection *(Zool.)*. The *turgid* condition of certain animal tissues when distended with blood; an upright or raised condition of an organ or part. adj. *erect*.

erections *(Ships)*. See **top hamper**.

erector *(Image Tech.)*. A lens added to an optical viewing system to provide an image the right way up, rather than inverted.

erector *(Med.)*. A muscle which, by its contraction, assists in raising or erecting a part or organ.

erg *(Phys.)*. Unit of work or energy in CGS system. 1 erg of work is done when a force of 1 dyne moves its point of application 1 cm in the direction of the force. See **energy**.

ergastic substances *(Bot.)*. Non-protoplasmic substances; storage and waste products; starch, oil, crystals, tannins.

ergate *(Zool.)*. A sterile female ant or worker.

ergatogyne *(Zool.)*. An apterous queen ant.

ergatoid *(Zool.)*. Resembling a worker; said of sexually perfect but wingless adults of certain social Insects.

ergonomics *(Behav.)*. The application of various human studies to the area of work and leisure; includes anatomy, physiology and psychology.

ergonomy *(Med.)*. The physiological differentiation of functions.

ergosterol *(Chem.)*. A **sterol** which occurs in ergot, yeast and moulds. Traces of it are associated with cholesterol in animal tissues. On irradiation with ultraviolet light, **vitamin D_2** is produced.

ergotamine tartrate *(Med.)*. Ergot preparation which relieves migraine by constricting cranial arteries.

ergotism *(Med.)*. A condition characterized by extreme vasoconstriction leading to gangrene and convulsions; due to eating the grains of cereals which are infected by the ergot fungus *Claviceps purpurea*. Formerly known as *St. Anthony's Fire*.

Ericaceae *(Bot.)*. Family of ca 3000 spp. of dicotyledonous flowering plants (superorder Dilliniidae). Mostly shrubs, mostly calcifusge, cosmopolitan. Includes the heaths (*Erica* spp.) and heather (*Calluna*) and a number of ornamental plants e.g. Rhododendron.

ericaceous *(Bot.)*. Heatherlike.

ericeticolous *(Bot.)*. Growing on a heath.

Erichsen test *(Eng.)*. A test in which a piece of metal sheet is pressed into a cup by means of a plunger; used to estimate the suitability of sheet for pressing or drawing operations.

ericoid *(Bot.)*. Having very small tough leaves like those of heather.

erionite *(Min.)*. One of the less common zeolites; a hydrated aluminium silicate of sodium, potassium and calcium.

eriophorous *(Bot.)*. Having a thick cottony covering of hairs.

erlang *(Telecomm.)*. International unit of traffic flow in telephone calls.

Erlenmeyer flask *(Chem.)*. A conical glass flask with a flat bottom, widely used for titrating, as it is easily cleaned, stood, stoppered and swirled.

Ernie *(Comp.)*. A so-called 'computer' used to select winning numbers in the British Premium Bond lottery. Similar random number generators are used in the **Monte Carlo method**. (*Electronic Random Number Indicating Equipment*.)

erogenous zones *(Behav.)*. Sensitive areas of the body, the stimulation of which arouses sexual feelings and responses. They function as substitutes for the genital organs and are associated with stages of development in childhood.

eros *(Behav.)*. In psychoanalytic theory, the constructive life instinct, the urge for survival and procreation.

erosion *(Eng.)*. The removal of metal from components subject to fluid flow, particularly when the liquid contains solid particles.

erosion *(Geol.)*. The removal of the land surface by weathering, corrasion, corrosion, and transportation, under the influence of gravity, wind, and running water. Also, the eating away of the coastline by the action of the sea. Soil erosion may result from factors such as bad agricultural methods, excessive deforestation, overgrazing.

erratics *(Geol.)*. Stones, ranging in size from pebbles to large boulders, which were transported by ice, which, on melting, left them stranded far from their original source. They furnish valuable evidence of the former extent and movements of ice sheets.

error *(Comp.)*. Fault or mistake causing the failure of a computer program or system to produce expected results. See **compilation-**, **execution-**, **logical-**, **syntax-**.

error *(Elec.Eng.)*. See **deviation**.

error *(Eng.)*. In servo or other control systems, the difference between the actual value of a quantity arising from the process and the desired value in the controller.

error-correcting code *(Comp.)*. A code that by including extra **parity bits** can detect and correct certain types of errors that may occur during reading, writing and transmission of data.

error in indication *(Elec.Eng.)*. In indicating instruments, the difference between the indication of the instrument and the true value of the quantity being measured. It may be expressed as a percentage of the true value, and a positive value of the error means that the indication of the instrument is greater than the true value.

error message *(Comp.)*. Indication that an error has been detected.

error of closure *(Surv.)*. See **closing error**.

error signal *(Telecomm.)*. Feedback signal in control system representing deviation of controlled variable from set value.

Ertel potential vorticity *(Meteor.)*. A rigorous formulation by Ertel of the idea of **potential vorticity** for any compressible, thermodynamically active, inviscid fluid in adiabatic flow. If S is some conservative thermodynamic property of the fluid (e.g. the potential temperature), Ω is the angular velocity of the co-ordinate system, ρ is the density, and V is the velocity of the fluid relative to the co-ordinate system, then the Ertel potential vorticity Π is defined by

$$\Pi = \nabla S \cdot \left(\frac{2\Omega + \text{curl}\ \mathbf{V}}{\rho} \right)$$

Π is a conservative property for all individual fluid particles. Approximations to the Ertel potential vorticity are useful in dynamical studies of the general circulation of the atmosphere and in **numerical forecasting**.

erubescite *(Min.)*. See **bornite**.

erucic acid *(Chem)*. $CH_3[CH_2]_7CH{=}CH[CH_2]_{11}$·COOH. Mono-saturated fatty acid in the oleic acid series; mp 33°C. Occurs in rapeseed oil and mustard seed oil.

eructation *(Med.)*. A belching of gas from the stomach through the mouth.

erumpent *(Bot.)*. Developing at first beneath the surface of the substratum,then bursting out through the substratum and spreading somewhat.

eruption *(Med.)*. A breaking-out of a rash on the skin or on the mucous membranes; a rash.

eruptive rocks *(Geol.)*. A term sometimes used for all igneous rocks; but carefully compare *extrusive rocks*.

erysipelas *(Med.)*. A diffuse and spreading inflammation of skin and subcutaneous cellular tissue especially of face, neck, forearm and hands. The inflamed area being red, shiny and sharply demarcated; caused by *Streptococcus pyogenes*.

erythema *(Med.)*. A superficial redness of the skin, due to dilation of the capillaries.

erythema multiforme *(Med.)*. A skin disease in which raised red patches appear and reappear, especially on the upper part of the body, associated in severe cases with extensive skin necrosis and involvement of kidneys, lungs and gastrointestinal system.

erythema nodosum *(Med.)*. A skin disease in which red, painful, oval swellings appear, usually on the shins, associated with fever, joint pains, and sore throat; now believed to be due to a hypersensitive phase in certain infections, such as tuberculosis, sarcoidosis, and streptococci, or in drug hypersensitivity.

erythema pernio *(Med.)*. Painful red swellings of the extremities, known as *chilblains*.

erythraemia, erythremia *(Med.)*. An excess of red cells in the blood. Also *polycythaemia*.

erythrasma *(Med.)*. Infection of the horny layer of the skin with the fungus *Microsporon minutissimum*, giving rise to superficial, reddish-yellow patches.

erythrite *(Min.)*. Hydrated cobalt arsenate, occurring as pale reddish crystals or incrustations. Also called *cobalt bloom*.

erythritol *(Chem.)*. $CH_2OH.(CHOH)_2.CH_2OH$. A tetrasaccharide carbohydrate found in lichens.

erythroblast *(Zool.)*. A nucleated mesodermal embryonic cell, the cytoplasm of which contains haemoglobin, and which will later give rise to an erthrocyte. See also **megaloblast**.

erythroblastosis foetalis *(Immun.)*. *Haemolytic disease of the newborn*. Disease of the human foetus due to immunisation of the mother. Escape of foetal red cells into the maternal circulation during pregnancy (or during a previous pregnancy) elicits antibodies in the mother. If these are IgG, and can cross the placental membranes into the foetal circulation, they cause haemolysis of the foetal red cells. This may be sufficient to cause stillbirth or anaemia and severe jaundice with brain damage. *Rhesus antigens* are the commonest cause. See **rhesus blood group system**.

erythrocyte *(Zool.)*. One of the red blood corpuscles of Vertebrates; flattened oval, or circular disklike, cells (lacking a nucleus in Mammals), whose purpose is to carry oxygen in combination with the pigment haemoglobin in them, and to remove carbon dioxide.

erythrocytosis *(Med.)*. Excess in the number of red cells in the blood.

erythromelalgia *(Med.)*. A condition characterized by pain, redness, and swelling of the toes, feet, and hands, often associated with vascular disease.

erythromycin *(Med.)*. An antibiotic derived from *Streptomyces erythreus*, useful against some bacterial infections, particularly when the patient is allergic to penicillin.

erythropenia *(Med.)*. Diminution, below normal, of the number of red cells in the blood.

erythrophore *(Zool.)*. A chromatophore containing a reddish pigment.

erythropoiesis *(Med.)*. The formation of red blood cells.

erythropoietin *(Med.)*. A substance produced by the juxtaglomerular cells of the kidneys which stimulates the production of red blood cells by the bone marrow.

erythropsia *(Med.)*. The state in which objects appear red to the observer, e.g. snow blindness.

erythropterin *(Zool.)*. A red heterocyclic compound deposited as a pigment in the epidermal cells or the cavities of the scales and setae of many Insects.

erythrose *(Chem.)*. A tetrose. ➪

Es *(Chem.)*. Symbol for *einsteinium*.

ESA *(Space)*. Abbrev. for *European Space Agency*, formed in 1975 combining the activities of the European Space Research Organization (ESRO) and the European Launcher Development Organization (ELDO). ESA, an inter-governmental agency, co-ordinates European space activities and related technologies; in particular, it instigates and manages international space programs on behalf of its 13 member states.

Esaki diode *(Electronics)*. See **tunnel diode**.

escape *(Bot.)*. A plant growing outside a garden and derived from cultivated specimens, surviving but not well naturalised.

escape behaviour *(Behav.)*. Refers to defensive behaviour against a predator, often involves specialised evasive manoeuvres or structures (e.g. a special exit burrow).

escape boom *(Min.Ext.)*. Means of escape from an offshore platform. Pivoted chutes with a bouyant outer end are normally secured inboard. In emergency they are released and rotate outwards until the outer end floats and personnel can slide to safety.

escape conditioning *(Behav.)*. A procedure in which escape behaviour regularly terminates a negative reinforcement or aversive stimulus.

escapement *(Build.)*. The cut-away part above the mouth of a plane through which the shavings are voided.

escapement *(Eng.)*. A mechanism allowing one component at a time to move automatically from one machine to the next in a line.

escape, 'scape pinion *(Horol.)*. The pinion on the arbor of the escape wheel.

escape, 'scape wheel *(Horol.)*. The wheel of which the teeth act on the pallets.

escape velocity *(Astron.,Space)*. The minimum velocity necessary for an object to travel in a parabolic orbit about a massive primary body, and thus to escape its gravitational attraction. An object which attains this or any greater velocity will coast away from the primary.

For the surface of the Earth this velocity is 11.2 km s^{-1}, for the Moon 2.4 km s^{-1}, and for the Sun 617.7 km s^{-1}. It is given by the relation

$$V_e = \sqrt{2Rg} = 11{\cdot}2 \text{ km/sec}$$

where R is the radius and g the acceleration due to gravity. The escape velocity of a **black hole** exceeds the speed of light.

escarpment *(Geol.)*. A long cliff-like ridge developed by denudation where hard and soft inclined strata are interbedded, the outcrop of each hard rock forming an *escarpment*, such as those of the Chalk (Chiltern Hills, N. and S. Downs) and the Jurassic limestones (Cotswold Hills). Generally an escarpment consists of a short steep rise (the *scarp face*) and a long gentle slope (the *dip-slope*). Cf. **cuesta**.

eschar *(Med.)*. A dry slough produced by burning or by corrosives.

Eschka's reagent *(Chem.)*. Mixture of MgO and Na_2CO_3(2 : 1); used for estimation of the sulphur content of fuels.

escribed circle of a triangle *(Maths.)*. One which touches one side of a triangle externally and the extensions of the other two sides.

escutcheon *(Build.)*. A perforated plate around an opening, such as a keyhole plate or the plate to which a doorknob is attached.

esker *(Geol.)*. A long, narrow, steep-sided, sinuous ridge of poorly stratified sand and gravel deposited by a subglacial or englacial stream. Found in glaciated ares.

Esmarch's bandage *(Med.)*. A rubber bandage which, applied to a limb from below upwards, expels blood from the part.

esophagus *(Zool.)*. See **oesophagus**.

esophoria *(Med.)*. Latent internal squint, revealed in an apparently normal person by passing a screen before the eye.

espagnolette *(Build.)*. Fastening for a French window, having two long bolts which operate in slots at the top and bottom when the door handle is turned. Sometimes called *cremorne bolt*.

esparto *(Paper)*. A grass native to north Africa and southern Spain formerly widely used as the raw material for paper pulp because of the excellent printing qualities of the papers in which it was used.

ESPRIT *(Comp.)*. European Strategic Programme for Research in Information Technology. EEC initiative, begun 1982, to fund and encourage collaborative research in computing and information technology.

espundia *(Med.)*. An ulcerative infection of the skin, and of the mucous membranes of the nose and mouth, by the protozoal parasite *Leishmania braziliensis*; occurs in South America.

ESR *(Chem.)*. Abbrev. for *Electron Spin Resonance*.

essential element *(Bot.)*. An element without which a plant cannot grow and reproduce and which cannot be substituted by another element. See also **macronutrient, micronutrient, deficiency disease**.

essential minerals *(Geol.)*. Components present, by definition, in a rock, the absence of which would automatically change the name and classification of the rock. Cf. **accessory minerals**.

essential oils *(Bot.)*. Volatile **secondary metabolites** formed mainly in oil glands, rarely in ducts, mostly terpenoids also aliphatic and aromatic esters, phenolics and substituted benzene hydrocarbons, responsible for the odours of many aromatic plants (and steam-distilled as perfumes from some). Some appear to deter insects or herbivores, others to be **allelopathic**.

essential organs *(Bot.)*. The stamens and carpels of a flower.

Essex board *(Build.)*. A building-board made of layers of compressed wood-fibre material cemented with a fire-resisting cement.

essexite *(Geol.)*. A coarse-grained deep-seated igneous rock, essentially an alkali-gabbro, with preponderance of soda. Named from Essex Co., Mass.

Esson coefficient *(Elec.Eng.)*. See **specific torque coefficient**.

essonite *(Min.)*. Original spelling of **hessonite**.

establishment *(Print.)*. A workman who receives a weekly wage is said to be *on the establishment*, in contrast to one who does piecework. Usually abbreviated to *stab*.

establishment charges *(Build.,Eng.)*. See **on-costs**.

ester *(Chem.)*. Esters are derivatives of acids obtained by the exchange of the replaceable hydrogen for alkyl radicals. Many esters have a fruity smell and are used in artificial fruit essences; also used as solvents.

esterase *(Biol.)*. An enzyme which catalyses the hydrolysis of ester bonds.

esterification *(Chem.)*. The direct action of an acid on an alchol, resulting in the formation of esters. An equilibrium is reached between the quantities of acid and alcohol present and the quantities of ester and water formed. Catalysed by hydrogen ions.

ester value *(Chem.)*. The number of milligrams of potassium hydroxide required to saponify the fatty acid esters in one gram of a fat, wax, oil etc.; equal to the saponification value minus the acid value.

esthiomene *(Med.)*. A condition in which there are chronic hypertrophy and destructive ulceration of the external genitals of the female.

estradiol, estrogen *(Med.)*. Same as **oestradiol, oestrogen,** etc.

estuarine deposition *(Geol.)*. Sedimentation in the environment of an estuary. The deposits differ from those which form in a deltaic environment, chiefly in their closer relationship to the strata of the adjacent land, and are usually of finer grain and of more uniform composition. Both are characterized by brackish water and sediments which contain land-derived animal and plant remains.

estuarine muds *(Geol.)*. So-called *estuarine muds* are, in many cases, silts admixed with sufficient true clay to give them some degree of plasticity; they are characterized by a high content of decomposed organic matter.

estuary *(Geol.)*. An inlet of the sea at the mouth of a river; developed especially in areas which have recently been submerged by the sea, the lower end of the valley having been thus drowned. See **fiords** and cf.**delta**.

Et *(Chem.)*. A symbol for the ethyl radical C_2H_5—.

ETA *(Aero.)*. *Estimated Time of Arrival*, as forecast on a **flight plan**, for a civil aircraft or the time of arrival over a target for a military aircraft.

eta *(Nuc.Eng.)*. (η) In reactor theory, one of the factors in the **four factor formula** which represents average number of fission neutrons produced per neutron absorbed in the fuel.

etalon *(Phys.)*. An interferometer consisting of an air film enclosed between half-silvered plane parallel plates of glass or quartz having a fixed separation. It is used for studying the fine structure of spectral lines. See **Fabry and Pérot interferometer**.

Etard's reaction *(Chem.)*. The formation of aromatic aldehydes by oxidizing methylated derivatives and homologues of benzene with chromyl chloride, CrO_2Cl_2.

etchant *(Chem.)*. Chemical for removing copper from laminate during production of printed circuits.

etched (etch-) figures *(Crystal.)*. Small pits or depressions of geometrical design in the faces of crystals, due to the action of some solvent. The actual form of the figure depends upon the symmetry of the face concerned, and hence they provide invaluable evidence of the true symmetry in distorted crystals

etching *(Eng.)*. (1) Method of showing the structure of metals and alloys by attacking a highly polished surface with a reagent that has a differential effect on different crystals or different constituents. (2) Removing films from the surface of materials to facilitate the subsequent deposition of another coating e.g. paint.

etching *(Image Tech.)*. The process of (1) dissolving, with an acid, portions of a surface, such as copper or zinc sheet, where it is not protected with a resist; (2) soaking away gelatine differentially to form a relief image.

etching pits *(Eng.)*. Small cavities formed on the surface of metals during etching.

etching test *(Chem.)*. Test for the detection of fluorides. The substance under examination is heated with sulphuric acid in a lead vessel covered with a glass lid. If fluorides are present the glass will be etched owing to the action of hydrogen fluoride produced by the action of the acid on the fluoride.

ETD *(Aero.)*. *Estimated Time of Departure*. See **departure time**.

ethacrynic acid *(Med.)*. A **'loop' diuretic** used in the treatment of oedema and oliguria due to renal failure.

ethanal *(Chem.)*. See **acetaldehyde**.

ethanamide *(Chem.)*. See **acetamide**.

ethane *(Chem.)*. $H_3C.CH_3$, a colourless, odourless gas of the alkane series; the critical temperature is +34°C, the critical pressure is 50.2 atm, bp −84°C. The second member of the alkane series of hydrocarbons. Chemical properties similar to those of **methane**.

ethanoates *(Chem.)*. See **acetates**.

ethanoic acid *(Chem.)*. See acetic acid.

ethanol *(Chem.)*. The IUPAC name for *ethyl alcohol*. C_2H_5OH, mp −114°C, bp 78.4°C; a colourless liquid, of vinous odour, miscible with water and most organic solvents, rel. d. 0.789; formed by the hydrolysis of ethyl chloride or of ethyl hydrogen sulphate; it may be obtained by absorption of ethylene in fuming sulphuric acid at 160°C, followed by hydrolysis with water, by reduction of acetaldehyde, or by direct synthesis from ethylene and water at high temperatures in the presence of a catalyst. It is prepared technically by the alcoholic fermentation of sugar. It forms alcoholates with sodium and potassium.

ethanolamines *(Chem.)*. Amino derivatives of ethyl alcohol, *monoethanolamine*, $CH_2OH.CH_2.NH_2$; *diethanolamine*, $(CH_2.CH_2OH)_2NH$; *triethanolamine*, $(CH_2.CH_2OH)_3.N$; hygroscopic solids with strong ammoniacal smell. Used, in combination with fatty acids, to produce detergents and cosmetic products.

ethanoyl *(Chem.)*. Same as *acetyl*.

ethanoylation *(Chem.)*. See **acetylation**.

ethanoyl chloride *(Chem.)*. See **acetyl chloride**.

ethene *(Chem.)*. The IUPAC name for *ethylene*, $H_2C{=}CH_2$, mp −169°C, bp −103°C, a gas of the alkene series, contained in illuminating gas and in gases obtained from the cracking of petroleum. Used for synthetic purposes, e.g. polythene and ethylene oxide, and for maturing fruit in storage.

ethenoid resins *(Plastics)*. Resins made from compounds containing a double bond between 2 carbon atoms, i.e. the acrylic, vinyl and styrene groups of plastics.

ethephon *(Bot.)*. *2-chloroethylphosphoric acid*, which breaks down rapidly in water to yield *ethylene* and is used to promote controlled ripening of fruit.

ether *(Chem.)*. *Alkoxyalkane*. (1) Any compound of the type R—O—R', containing two identical, or different, alkyl groups united to an oxygen atom; they form a homologous series $C_nH_{2n+2}O$. (2) Specifically, **diethyl ether**.

ether *(Phys.)*. A hypothetical, non-material entity supposed to fill all space whether 'empty' or occupied by matter. The theory that electromagnetic waves need such a medium for propagation is no longer tenable.

Ethernet *(Comp.)*. A **local area network** designed on the principle that one computer wishing to communicate with another, broadcasts onto the network. Acknowledgement establishes the link.

ethidium bromide *(Biol.)*. 3.8.-diamino-5-ethyl-6-phenylphenanthridinium bromide. A fluorescent reagent which binds to double stranded DNA and RNA and is used for their detection after fractionation in a gel matrix. A mutagen.

Ethiopian region *(Zool.)*. One of the primary faunal regions into which the surface of the globe is divided; it includes all of Africa and Arabia south of the tropic of Cancer.

ethmo- *(Genrl.)*. Prefix from Gk. *ethmos*, sieve.

ethmohyostylic *(Zool.)*. In some Vertebrates, having the lower jaw suspended from the ethmoid region and the hyoid bar.

ethmoidalia *(Zool.)*. A set of cartilage bones (*ethmoids*) forming the anterior part of the brain-case in the Vertebrate skull.

ethmoidectomy *(Med.)*. Surgical removal of the ethmoid cells or of part of the ethmoid bone.

ethmoiditis *(Med.)*. Inflammation of the ethmoid cells.

ethmoturbinal *(Zool.)*. In Mammals, a paired bone or cartilage of the nose, on which are supported the folds of the olfactory mucous membrane.

ethogram *(Behav.)*. Refers to attempts to draw up a complete behavioural inventory for a particular species; the term has its origin in early instinct theories of animal behaviour and is rare in modern usage, because of the methodological complexities involved in categorising units of behaviour.

ethology *(Behav.)*. Describes an approach to the study of animal behaviour in which attempts to explain behaviour combine questions about its immediate causation, development, function and evolution.

ethoxyl group *(Chem.)*. The group $—O.C_3H_5$.

ethyl acetate *(Chem.)*. *Ethanoate*, $CH_3COOC_2H_5$, mp −82°C, bp 77°C; colourless liquid of fruity odour, used as a laquer solvent and in medicine.

ethyl aceto-acetate *(Chem.)*. $CH_3.CO.CH_2.COO.C_2H_5$. One of the best known examples of organic compounds existing in keto and enol forms. Widely used as a chemical intermediate, including the manufacture of pyrazolone dyes and mepacrine. Also *aceto-acetic ester*.

ethyl acrylate *(Chem.)*. $CH_2{=}C.COOC_2H_5$. Colourless liquid; bp 101°C. Used in the manufacture of plastics.

ethyl alcohol *(Chem.)*. See **ethanol**.

ethylamine *(Chem.)*. $C_2H_5.NH_2$, bp 19°C, a liquid or gas of ammoniacal odour, which dissolves in water, and forms salts; it dissolves $Al(OH)_3$.

ethylene *(Bot.)*. The gas *ethene*; a plant **growth substance** produced especially in wounded, diseased, ripening and senescent tissues, interacting with auxin and promoting e.g. fruit ripening, leaf abscission and epinasty. Ethylene and substances that release it (see **ethephon**) are used commercially to regulate the ripening of fruit, especially stored fruit.

ethylene *(Chem.)*. See **ethene**.

ethylene diamine tartrate *(Electronics)*. Chemical in crystal form, exhibiting marked piezoelectric phenomena; used in narrow-band carrier filters. Abbrev. *EDT*.

ethylene diamine tetra-acetic acid *(Chem.)*. See **EDTA**.

ethylene dichloride *(Chem.)*. 1,2-dichloroethane.

ethylene glycol *(Chem.)*. *Glycol*, $HO.CH_2.CH_2.OH$, bp 197.5°C, rel.d. 1.125, a colourless syrupy, hygroscopic liquid, miscible with water and ethanol. Prepared from ethylene dibromide or ethylene chlorhydrin by hydrolysing with caustic soda. Intermediate for glycol esters, which are solvents and plasticizers for lacquers; used in the textile industry, for printing-inks, foodstuffs, antifreezing mixtures, and for de-icing aeroplane wings.

ethylene oxide *(Chem.)*. C_2H_4O, bp 13.5°C, a mobile colourless liquid of ethereal odour, obtained by distilling 2-chloroethanol with concentrated potassium hydroxide. Manufactured directly from ethylene and oxygen in the presence of a catalyst. Very useful organic intermediate in the manufacture of solvents, detergents etc. Can also be used as a sterilizing medium in the gaseous state.

ethylene-propylene rubber *(Plastics)*. A type of synthetic rubber based on readily available petrochemical materials. Good all-round mechanical properties but not very resistant to organic solvents.

ethyl group *(Chem.)*. The monovalent radical $—C_2H_5$.

ethylidene *(Chem.)*. The organic group $CH_3.CH{:}$.

ethyl mercaptan *(Chem.)*. *Ethane thiol*, C_2H_5SH, a liquid, bp 37°C, of vile odour.

ethyne *(Chem.)*. See **acetylene**.

ethynide *(Chem.)*. See **acetylide**.

etiolation *(Bot.)*. The condition of a green plant which has not received sufficient light; the stems are weak, with abnormally long internodes, the leaves are small, yellowish or whitish, and the vascular strands are deficient in xylem. adj. *etiolated*.

etiology *(Med.)*. See **aetiology**.

etioplast *(Bot.)*. The form of **plastid** that develops in plants grown in the dark; colourless and containing a *prolamellar body*, rapidly converted to a chloroplast on illumination.

E-transformer *(Elec.Eng.)*. An electric sensing device in an automatic control system which gives an error voltage in response to linear motion. It consists of coils for detecting small displacements of magnetic armature, which affects balance of currents when off-centre.

Etruscan *(Arch.)*. An ancient Italian civilisation which immediately preceded and influenced that of Ancient Rome. It is noted for the massive scale of its fortifications, built from hewn rock, many of which stand today.

Ettinghausen effect *(Phys.)*. A difference in temperature established between the edges of a metal strip carrying a current longitudinally, when a magnetic field is applied perpendicular to the plane of the strip. Effect is very small and is analogous to **Hall effect**.

Eu *(Chem.)*. The symbol for europium.

eu- *(Genrl.)*. Prefix from Gk. *eu*, well, good.

Eubacteriales *(Biol.)*. One of the two main orders of the true bacteria, distinguished from the *Pseudomonadales* by the peritrichous flagella of the motile members. Spherical or rod-shaped cells, having no photosynthetic pigments; not acid fast; readily stained by aniline dyes.

euchromatin *(Biol.)*. Dispersed **chromatin** visible by microscopy of eukaryotic cell nuclei.

euclase *(Min.)*. A monoclinic mineral, occurring as prismatic, usually colourless, crystals; hydrated beryllium aluminium silicate.

Euclidean geometry, space *(Maths.)*. Ordinary, *flat*, geometry, with all measures of curvature zero.

eucrite *(Geol.)*. A coarse-grained, usually ophitic, deep-seated basic gabbro containing plagioclase near bytownite in composition, both ortho- and clino-pyroxenes, together with olivine. Eucrite is an important rock type in the Tertiary complexes of Scotland.

eucryptite *(Min.)*. A hexagonal lithium aluminium silicate, commonly found as an alteration product of spodumene.

eudialyte, eudialite *(Min.)*. A pinkish red complex hydrated sodium calcium iron zirconosilicate. It crystallizes in the rhombohedral system and occurs in some alkaline igneous rocks.

eudiometer *(Chem.)*. (1) A voltameter-like instrument in which quantity of electricity passing can be found from volume of gas produced by electrolysis. (2) Similar system to determine volume changes in a gas mixture due to combustion.

eugamic *(Zool.)*. Pertaining to the period of maturity.

eugenics *(Biol.)*. Study of the means whereby the characteristics of human populations might be improved by the application of genetics.

eugenol *(Chem.)*. $C_6H_3(OH)(OCH_3)(CH_2.CH{:}CH_2)$, a phenol homologue, the chief constituent of oil of cloves and cinnamon leaf oil, bp 252°C, rel. d. 1.07; used for manufacturing **vanillin**.

euglenoid movement *(Zool.)*. A type of movement undergone by many Protozoa, which possess a definite body form, by means of contractions of the protoplasm stretching the pellicle. Also called *metaboly*.

Euglenophyceae *(Bot.)*. Small group of eukaryotic algae. Chloroplasts with chlorophyll a and b, thylakoids in threes, chloroplast envelope of 3 membranes. Storage polysaccharide paramylon ($\beta, 1\rightarrow3$ glucan). Flagellated, usually 2 flagella, some times palmelloid. Phototrophs (often auxotrophic) and heterotrophs (both osmotrophic and phagotrophic). Fresh water and marine. See also **pellicle**.

euhedral crystals *(Geol.)*. See **idiomorphic crystals**.

eukaryote *(Biol.)*. A higher organism; literally those which have 'good nuclei' in their cells; that is animals, plants and fungi in contrast to the prokaryotic cells of bacteria and blue-green algae. Eukaryotes possess a nucleus bounded by a membranous nuclear envelope and have many cytoplasmic organelles. Their nuclear DNA is complexed with proteins to form chromosomes. Cf. **prokaryote**. adj. *eukaryotic*.

Eulerian angles *(Maths.)*. Three independent angles φ, θ and ψ which specify the direction of a line in space. The line specified can be visualized as follows. Starting with orthogonal cartesian axis $Oxyz$, first rotate the axes about Oz through an angle φ. Next rotate the displaced axes about the new position of Ox through an angle θ. Finally rotate the axes about the new position of Oz through an angle ψ and the resulting position of Ox is the line specified by the Eulerian angles φ, θ, ψ.

Euler's constant *(Maths.)*. The limit of

$$\sum_1^n \frac{1}{r} - \log_e n, \text{ as } n \rightarrow \infty.$$

Denoted by γ or C. Its value is 0.577 215 66...

Euler's formula *(Civ.Eng.)*. A formula giving the collapsing load for a long, thin column of given sizes. It only applies where the load passes down the centroid of the column. It states that

$$P = \frac{\pi^2 EI \text{ (min.)}}{L^2},$$

where P = the collapsing load, E = Young's modulus, I = the least moment of inertia, and L = the length of the pin-jointed column.

Euler's theorem *(Maths.)*. $K = A\cos^2\theta + B\sin^2\theta$, where K is the curvature of a normal section of a surface displaced by an angle θ from a principal direction, A is the curvature in that principal direction, and B is the other principal curvature.

eumerism *(Zool.)*. An aggregation of similar parts.

EUMETSAT *(Space)*. Abbrev. for *European Meteorological Satellite Organisation*, an inter-governmental agency which provides operational meteorological data for its member states.

Eumycota *(Bot.)*. Division containing the true fungi. Eukaryotic heterotrophic, walled organisms, typically with hyphae, or single cells (e.g. yeasts) or chains of cells. The walls usually contain chitin (chitosan in Zygomycotina) as a major constituent. Comprises the subdivisions; Mastigomycotina, Zygomycotina, Ascomycotina and Basidiomycotina.

eunuch *(Med.)*. A male who has no testes.

eunuchoid *(Med.)*. A male with testes, but no secondary male characteristics.

Euphausiacea *(Zool.)*. An order of shrimp-like Malacostraca which are filter-feeding and pelagic. They are a major source of food for some Whales. Krill.

Euphorbiaceae *(Bot.)*. Family of ca 7000 spp. of dicotyledonous flowering plants (superorder Rosidae). Trees, shrubs and herbs, mostly tropical, with unisexual flowers, and usually with latex. Includes *Hevea* the source of most natural rubber, manioc (*Manihot utilissima*, cassava or tapioca) and the very large genus *Euphorbia*, the spurges, (1600 spp.) the members of which include C_3 plants, C_4 plants and leafless, stem-succulent CAM plants. See also **cyathium**.

euphoria, euphory *(Med.)*. A feeling of well-being, not necessarily indicative of good health.

euphotic zone *(Ecol.)*. See **photic zone**.

euploid, euploidy *(Biol.)*. Having a chromosome complement consisting of one or more whole *chromosome sets* and, therefore, haploid, diploid or polyploid. Cf. **aneuploid**.

eupyrene *(Zool.)*. Spermatozoa which are normal, typical. Cf. **apyrene, oligopyrene**.

Euratom *(Nuc.Eng.)*. European Atomic Energy Community.

EURECA *(Space)*. An acronym for *European Retrievable Carrier*, a free-flying platform with a mission duration of six monthes, launched and recovered by the *Space Shuttle Orbiter*.

Eureka *(Elec.Eng.)*. See **constantan**.

eureka *(Radar)*. See **rebecca-eureka**.

Eurokom *(Comp.)*. A *teleconferencing* system for the participants of the **ESPRIT** program.

Euronet *(Comp.)*. A European information exchange **network**.

Europa *(Astron.)*. The second natural satellite of **Jupiter**, discovered by Galileo, and encased in a mantle of ice.

europium *(Chem.)*. Metallic element of the rare-earth group. Symbol Eu, at. no. 63, r.a.m. 151.96, valency 2 and 3. Contained in black monazite, gadolinite, samarskite, xenotime.

Eurovision *(Image Tech.)*. Exchange of TV programmes between member countries of the European Broadcasting Union.

Eurypterida *(Geol.)*. An order of Crustaceans ranging from Ordovician to Permian and represented by such types as *Eurypterus* and *Stylonurus*, the latter reaching almost 2 m in length.

eusporangium *(Bot.)*. A sporangium of a vascular plant in which the wall develops from superficial cells, and the sporogenous tissue from internal cells of the sporophyll or sporangiophore. The wall is usually more than one cell thick at maturity. Cf. **leptosporangiate**.

Eustachian tube *(Zool.)*. In land Vertebrates, a slender duct connecting the tympanic cavity with the pharynx.

Eustachian valve *(Zool.)*. In Mammals, a rudimentary valve separating the openings of the superior venae cavae from that of the inferior vena cava.

eustatic movements *(Geol.)*. Changes of sea level, constant over wide areas, due probably to alterations in the volumes of the seas. These may be due to variations in the extent of the polar ice-caps, large-scale crustal movements in ocean basins, or submarine volcanic activity.

eustele *(Bot.)*. A stele in which the primary vascular tissue is organised into discrete vascular bundles surrounding a pith; typical of dicotyledons and gymnosperms.

eustomatous *(Zool.)*. With a well-defined mouth or opening.

eustyle *(Arch.)*. A colonnade in which the space between the columns is equal to $2\frac{1}{4}$ times the lower diameter of the columns.

eutaxitic *(Geol.)*. Descriptive of the streaky banded structure of certain pyroclastic rocks, as contrasted with the smooth layered structure of flow-banded lavas.

eutectic *(Chem.)*. Relating to a mixture of two or more substances having a minimum melting point. Such a mixture behaves in some respects like a pure compound.

eutectic change *(Eng.)*. The transformation from the liquid to the solid state in a eutectic alloy. It involves the simultaneous crystallization of two constituents in a binary system and of three in a ternary system.

eutectic point *(Eng.,Min.)*. The point in the binary or ternary constitutional diagram indicating the composition of the eutectic alloy, or mixture of minerals, and the temperature at which it solidifies.

eutectic structure *(Eng.)*. The particular arrangement of the constituents in a eutectic alloy which arises from their simultaneous crystallization from the melt.

eutectic system *(Eng.)*. A binary or ternary alloy system in which one particular alloy solidifies at a constant temperature which is lower than the beginning of solidification in any other alloy.

eutectic welding *(Eng.)*. A metal welding process carried out at relatively low temperature, using the eutectic properties of the metals involved.

eutectoid *(Eng.)*. Similar to a eutectic except that it involves the simultaneous formation of two or three constituents from another solid constituent instead of from a melt. *Eutectoid point* and *eutectoid structure* have similar meanings to those given for *eutectic*.

eutectoid steel *(Eng.)*. Steel having the same composition as the eutectoid point in the iron-carbon system (0.87%C), and which therefore consists entirely of the eutectoid at temperatures below 710°C. See **pearlite**.

EUTELSAT *(Space)*. Abbrev. for *European Telecommunications Satellite Organisation*, a European intergovernmental agency which provides satellite communications for its participating countries.

eutexia *(Eng.)*. The property of being easily melted, i.e. at a minimum melting-point.

euthanasia *(Med.)*. Easy or painless death; the action of procuring this.

Eutheria *(Zool.)*. An infraclass of viviparous Mammals in which the young are born in an advanced stage of development; no marsupial pouch; an allantoic placenta occurs; the scrotal sac is behind the penis, the angle of the lower jaw is not inflexed and the palate is imperforate. The higher Mammals.

euthyroid *(Med.)*. Having a normal level of thyroid activity.

eutrophic *(Ecol.)*. Said of a type of lake in which the hypolimnion becomes depleted of oxygen during the summer by the decay of organic matter falling from the epilimnion. A eutrophic lake is usually shallow, with much primary productivity. Cf. **oligotrophic**.

euxenite *(Min.)*. Niobate, tantalate, and titanate of yttrium, erbium, cerium, thorium, and uranium, and valuable on this account. Commonly massive and brownish black in colour.

EV *(Image Tech.)*. Abbrev. for *Exposure Value*.

eV *(Phys.)*. Abbrev. for *electron-volt*.

EVA *(Space)*. Abbrev. for *Extra-Vehicular Activity*, i.e. operations performed outside the 'living environment' of a space vehicle. To accomplish EVA, it is necessary to wear a space (or pressure) suit provided with pressure control and life support systems.

evaginate *(Bot.,Zool.)*. Not having a sheath.

evagination *(Med.)*. The turning inside out of an organ.

evagination *(Zool.)*. Withdrawal from a sheath; the development of an outgrowth; eversion of a hollow ingrowth; an outgrowth; an everted hollow ingrowth. Cf. **invagination**.

evanescent mode *(Telecomm.)*. A waveguide propagation mode at a frequency below the **cut-off wavelength**. In this mode, the amplitude of the wave diminishes rapidly along the waveguide, but the phase does not change; applied in certain special waveguide filter designs.

evanescent waves *(Acous.)*. Non-decaying surface waves.

evaporation *(Eng.)*. The conversion of water into steam in a boiler.

evaporation *(Phys.)*. The conversion of a liquid into vapour, at temperatures below the boiling point. The rate of evaporation increases with rise of temperature, since it depends on the saturated vapour pressure of the liquid, which rises until it is equal to the atmospheric pressure at the boiling point. Evaporation is used to concentrate a solution.

evaporative capacity *(Eng.)*. The mass of steam at a stated temperature and pressure produced in one hour from unit area of heating surface from feed-water at a stated temperature, when steaming at the most economical rate. Measured in lb/ft^2hr or kg/m^2hr. Also called *evaporation rate*.

evaporative cooling *(Aero.,Chem.Eng.)*. The process of evaporating part of a liquid by supplying the necessary latent heat from the main bulk of liquid which is thus cooled. Used for some piston aero-engines in the 1930s, some current types of turbine and rocket components, and also for cooling purposes in cabin air conditioning systems. Cf. **sweat cooling**.

evaporator *(Chem.)*. A still designed to evaporate moisture or solvents to obtain the concentrate.

evaporimeter *(Meteor.)*. An instrument used for measuring the rate of natural evaporation.

evaporite *(Geol.)*. Sedimentary deposit of material previously in aqueous solution and concentrated by the evaporation of the solvent. Normally found as the result of evaporation in lagoons or shallow enclosed seas, e.g. salt and anhydrite.

evapotranspiration *(Ecol.)*. The total water loss from a particular area, being the sum of evaporation from the soil and transpiration from vegetation.

evection *(Astron.)*. The largest of the four principal periodic inequalities in the mathematical expression of the moon's orbital motion; due to the variable eccentricity of the moon's orbit, with a maximum value of 1° 16′ and a period of 31.81 days.

even-even nuclei *(Phys.)*. Those for which the numbers of protons and neutrons are both even and normally stable.

even function *(Maths.)*. f is an even function if $f(-x) = f(x)$. Cf. **odd function**.

evening star *(Astron.)*. The name given in popular language to a planet, generally Venus or Mercury, seen in the western sky at or just after sunset. Also used loosely to describe any planet which transits before midnight.

even-odd nuclei *(Phys.)*. Those with an even number of protons and an odd number of neutrons.

even parity *(Comp.)*. Binary representation in which the number of 1s is even. See **parity check**.

even pitch *(Eng.)*. In screw-cutting in the lathe, the thread cut is said to be of *even pitch* if its **pitch** is equal to, or an integral multiple of, the pitch of the lead screw.

even small caps *(Print.)*. Small capitals set up without capitals. Sometimes called LEVEL SMALL CAPS.

event horizon *(Astron.)*. The boundary of a **black hole**, inside this boundary no light can escape.

eventration *(Med.)*. Protrusion of the abdominal contents outside the abdomen, e.g. through the diaphragm into the thorax.

even working *(Print.)*. When the pages of a book occupy a number of complete *sections* (usually of 16 or 32 pages). If an oddment of 4 or 8 pages is required it is called an *uneven working*.

Everest theodolite *(Surv.)*. A form of theodolite differing from the transit in that reversal of the line of sight is effected by removing the telescope from its trunnion supports and turning it.

ever-ready case *(Image Tech.)*. A case from which the camera can operate without being removed. Usually the camera is screwed into the case, whose top and front can hinge down in one piece to expose lens and controls. Abbrev. *ERC*

evisceration *(Med.)*. *Disembowelment*; operative removal of thoracic and abdominal contents from the foetus in obstructed labour; operative removal of a structure (e.g. the eye) from its cavity.

evocation *(Bot.)*. *Flora evocation*. The initial event at the root apex in response to the arrival of the floral stimulus which commits the apex to the subsequent formation of flower primordia.

evolute *(Biol.)*. Having the margins rolled outwards.

evolute *(Maths.)*. Of a curve: locus of centre of curvature or envelope of normals. The original curve is the **involute** of the evolute.

evolution *(Biol.)*. Changes in the genetic composition of a population during successive generations. The gradual development of more complex organisms from simpler ones.

evolution *(Maths.)*. Raising a number to a power $1/n$, where n is a positive integer, i.e. finding a root. Cf. **involution** (1).

evolutionarily stable strategy *(Bot.)*. A strategy such that, if most members of a population adopt it, there is no *mutant* strategy that would give higher reproductive fitness.

evolutionary operation *(Min.Ext.)*. Introduction of sectionally controlled variants into a commercial process, during transfer of laboratory research into better production.

evulsion *(Med.)*. Plucking out by force.

EW *(Telecomm.)*. Abbrev. for *Electronic Warfare*; see **electronic counter-measures**.

E-wave *(Telecomm.)*. See **TM-wave**.

Ewing curve tracer *(Elec.Eng.)*. An instrument for throwing a curve representing the hysteresis loop of a sample of iron on to a screen. A mirror is deflected horizontally in proportion to the magnetizing force and vertically in proportion to the flux produced.

Ewing permeability bridge *(Elec.Eng.)*. A measuring device in which the flux produced in a sample of iron is balanced against that produced in a standard bar of the same dimensions. The magnetizing force on the bar under test is varied until balance is obtained, and from the value of the force so found the permeability can be estimated.

EWT *(Build.)*. Abbrev. for *ElseWhere Taken*.

exacerbation *(Med.)*. An increase in the severity of a disease, or of its manifestations.

exact equation *(Maths.)*. The differential equation

$$P(x, y)dx + Q(x, y)dy = 0, \text{ where } \frac{\partial P}{\partial y} = \frac{\partial Q}{\partial x}.$$

The primitive is $u(x, y) = c$, c being a constant, where

$$\frac{\partial u}{\partial x} = P \text{ and } \frac{\partial u}{\partial y} = Q.$$

exalbuminous *(Bot.)*. A seed, lacking endosperm when mature; non-endospermic.

exalted carrier *(Telecomm.)*. Addition of a synchronized carrier before demodulation, to improve linearity and to mitigate the effects of fading during transmission.

exanthema, exanthem *(Med.)*. An eruption on the surface of the body. pl. *exanthemata*. Used specifically for infectious diseases characterized by an exanthem.

exarch *(Bot.)*. Of a xylem strand, having the protoxylem of the side towards the outside of the axis (the normal condition in e.g. angiosperm roots.)

excavation *(Med.)*. The process of hollowing out; a part hollowed out.

excess-3 code *(Comp.)*. Binary coding of a decimal number, to which has been added 3. Complements are formed by 1 changed to 0, and 0 changed to 1, in all numerics.

excess air *(Eng.)*. The proportion of air that has to be supplied in excess of that theoretically required for complete combustion of a fuel, because of the imperfect conditions under which combustion takes place in practice.

excess code *(Comp.)*. One which increases decimal digits before conversion to binary digits.

excess conduction *(Electronics)*. That arising from excess electrons provided by donor impurities.

excess electron *(Electronics)*. One added to a semiconductor by, e.g. a donor impurity.

excess feed *(Print.)*. When the peripheral speed of a web-driving component is greater than the next ahead too much paper is drawn through; also called *making paper*. Cf. **insufficient feed**.

excess voltage *(Elec.Eng.)*. See **overvoltage protective device**.

exchange *(Phys.)*. (1) The interchange of one particle between two others (e.g., a pion between two nucleons), leading to establishment of exchange forces. (2) Possible interchange of state between two indistinguishable particles, which involves no change in the wave function of the system.

exchangeable disk pack *(Comp.)*. A **disk pack** which can be removed from the disk unit for storage and later use on the same or a compatible machine. See **Bernoulli disk**.

exchange force *(Phys.)*. A force acting between particles due to the exchange of some property. In quantum mechanics, such forces can arise when two particles interact. In the ion of the hydrogen molecule, the forces responsible for the binding can be regarded as the continual exchange of the single electron between the two protons. Exchange forces are an important concept in the understanding of nuclear forces. The strong force between nucleons is the exchange of a *pion* (π-meson) between the two interacting nucleons. See **particle exchange**.

excipient *(Med.)*. The inert ingredient in a medicine which takes up and holds together the other ingredients.

excision *(Med.)*. The action of cutting a part out or off; the surgical removal of a part.

excision repair *(Biol.)*. Enzymatic DNA repair process in which a mismatching DNA sequence is removed and the gap filled by synthesis of a new sequence complementary to the remaining strand.

excitable tissue *(Zool.)*. That which responds to stimulation, e.g. muscle or nervous tissue.
excitant *(Elec.Eng.)*. A term occasionally used to denote the electrolyte in a primary cell.
excitation *(Elec.Eng.)*. (1) Current in a coil which gives rise to a m.m.f. in magnetic circuit, especially in a generator or motor. (2) The m.m.f. itself. (3) The magnetizing current of a transformer.
excitation *(Phys.)*. Addition of energy to a system, such as an atom or nucleus, raising it above the **ground state**.
excitation *(Telecomm.)*. Signal which drives any amplifier stage in a transmitter or receiver.
excitation *(Zool.)*. The contraction of muscle resulting from nervous stimulation. cf. **inhibition**. adj. *excitatory*.
excitation anode *(Electronics)*. An auxiliary anode used to maintain the cathode spot of a mercury-pool cathode valve.
excitation loss *(Elec.Eng.)*. The ohmic loss (I^2R) in the field or exciting windings of an electric machine excited by direct current.
excitation potential *(Electronics)*. Potential required to raise orbital electron in atom from one energy level to another. Also called *resonance potential*. See also **ionization potential**.
excited atom *(Phys.)*. One with more energy than in the normal or ground state. The excess may be associated with the nucleus or an orbital electron.
excited ion *(Phys.)*. Ion resulting from the loss of a valence electron, and the transition of another valence electron to a higher energy level.
excited nucleus *(Phys.)*. A nucleus raised to an excited state with an excess of energy over its ground state. Nuclear reactions frequently leave the product nucleus in an excited state. It returns to its ground state with the emission of γ-rays.
exciter *(Elec.Eng.)*. A small machine for producing the current, usually d.c., necessary for supplying the exciting winding of a larger machine. It is frequently mounted on the shaft of the machine which it is exciting.
exciter field rheostat *(Elec.Eng.)*. A rheostat in the field of an exciter whereby the voltage of the exciter, and, therefore, the excitation on the main machine, can be controlled.
exciter lamp *(Image Tech.)*. In optical sound reproduction, a small incandescent lamp whose light is modulated by the passage of the photographic sound track on the film.
exciter set *(Elec.Eng.)*. An assembly of one or more exciters with a prime-mover of electric driving motor.
exciting circuit *(Elec.Eng.)*. The complete circuit through which flows the current for exciting an electric machine. It comprises the exciter, the windings of the main machine, and possibly a field rheostat and measuring instruments.
exciting coil *(Elec.Eng.)*. A coil on a field magnet, or any other electromagnet, which carries the current for producing the magnetic field.
exciting current *(Elec.Eng.)*. That drawn by a transformer, magnetic amplifier, or other electric machine under no load conditions.
exciting winding *(Elec.Eng.)*. The winding which produces the e.m.f. to set up the flux in an electric machine or other apparatus.
exciton *(Electronics)*. Bound hole-electron pair in semiconductor. These have a definite half-life during which they migrate through the crystal. Their eventual recombination energy releases as a photon or, less often, several phonons.
excitron *(Electronics)*. A single-anode steel-tank mercury-arc rectifier, with a means for maintaining a continuous cathode spot, even when current is not flowing to the anode. This improves the turn-on characteristics of the rectifier.
exclusion principle *(Phys.)*. See **Pauli exclusion principle**.
excoriation *(Med.)*. Superficial loss of skin.
excrescence *(Med.)*. Any abnormal outgrowth of tissue.
excreta *(Zool.)*. Poisonous or waste substances eliminated from a cell, tissue or organism. adjs. *excrete, excreted*. n. *excretion*.
ex. & ct. *(Build.)*. Abbrev. for *excavate and cart away*.
excurrent *(Bot.)*. (1) Running out, as when the midrib of a leaf is prolonged into a point. (2) With a single main axis or trunk and subordinate laterals, as in trees, e.g. pine, especially when young.
excurrent *(Zool.)*. Carrying an outgoing current; said of ducts, and, in certain, *Porifera*, of canals leading from the apopyles of the flagellated chambers to the exterior or to the paragaster.
excursion *(Nuc.Eng.)*. Rapid increase of reactor power above set operation level, either deliberately caused for experimental reasons or accidental.
ex-, e- *(Genrl.)*. Prefixes from L. *ex*, *e*, out of.
execute *(Comp.)*. Carry out instructions specified by the machine code version of a program. See **compiler**, **machine code instruction**.
execution error *(Comp.)*. Error detected during program execution (e.g. overflow, division by zero). Also called *run-time error*.
executive program *(Comp.)*. See **operating system**, **compiler**, **editor**.
exempted spaces *(Ships)*. Certain spaces in the ship which are not included in the **gross tonnage**, e.g. double bottoms used for water ballast, wheelhouse, galley, W.C.s etc.
exenteration *(Med.)*. *Disembowelment*; complete removal of the contents of a cavity.
exergonic *(Bot.)*. A biochemical reaction accompanied by the release of energy (strictly, with negative ΔG) and capable, therefore, of proceeding spontaneously.
Exeter hammer *(Build.)*. See **London hammer**.
exfoliation *(Bot.)*. The process of falling away in flakes, layers, or scales, as some bark in plants.
exfoliation *(Eng.)*. Lifting away the surface of a metal due to the formation of corrosion products beneath the surface, a common result of rusting.
exfoliation *(Geol.)*. The splitting off of thin folia or sheets of rock from surfaces exposed to the atmosphere, particularly in regions of wide temperature variation. It is one of the processes involved in spheroidal weathering.
exhalant *(Zool.)*. Emitting or carrying outwards a gas or fluid; as the *exhalant siphon* in some Mollusca.
exhaust *(Eng.)*. (1) The working fluid discharged from an engine cylinder after expansion. (2) That period of the cycle occupied by the discharge of the used fluid.
exhaust cone *(Aero.)*. In a turbojet or turboprop, the duct immediately behind the turbine and leading to the *jet pipe*, consisting of an inner conical unit behind the *turbine disk* and an outer unit of frustum form connecting the *turbine shroud* to the jet pipe.
exhaust-driven supercharger *(Aero.)*. A piston-engine supercharger driven by a turbine motivated by the exhaust gases; also *turbo-supercharger*.
exhaust fan *(Eng.)*. A fan used in artificial draught systems; placed in the smoke uptake of a boiler to draw air through the furnace and exhaust the flue gases.
exhaust gas *(Eng.)*. The gaseous exhaust products of an IC engine, containing in general CO_2, CO, O_2, N_2 and water-vapour.
exhaust-gas analyser *(Autos.)*. An instrument which records the mixture strength supplied to a petrol engine by automatic electrical measurement of the thermal conductivity of the exhaust gas. See also **CO_2 recorder**.
exhaustion dyeing *(Textiles)*. The process of dyeing in which the textile takes up substantially all the available dyestuff.
exhaustive methylation *(Chem.)*. The process of converting bases into their quaternary ammonium salts and subsequent distillation with alkalis, resulting in the formation of simpler unsaturated compounds which can be reduced to known saturated compounds. This method was of particular value for investigating the constitution of alkaloids and other complicated ring systems.
exhaust lap *(Eng.)*. Of a slide-valve, the distance moved by the valve from midposition on the port face, before

uncovering the steam port to exhaust; sometimes called *inside lap*.

exhaust line *(Eng.)*. The lower line of the enclosed area of an **indicator diagram** showing the back pressure on the piston during the exhaust stroke of an engine.

exhaust manifold *(Autos.)*. Branched pipe which channels burnt gases from the combustion chambers to the exhaust system.

exhaust pipe *(Eng.)*. The pipe through which the exhaust products of an engine are discharged.

exhaust port *(Eng.)*. In an engine cylinder, the port or opening through which a valve allows egress of the exhaust.

exhaust stator blades *(Aero.)*. An assembly of stator blades, usually in sections to allow for thermal expansion, mounted behind the turbine to remove residual swirl from the gases.

exhaust steam *(Eng.)*. See **live steam**.

exhaust stroke *(Eng.)*. In a reciprocating engine, the piston stroke during which the **exhaust** is ejected from the cylinder; sometimes called the *scavenging stroke*.

exhaust valve *(Eng.)*. The valve controlling the discharge of the exhuast gas in an internal-combustion engine. Risk of overheating presents a design problem, particularly in aircraft engines, which is met by the use of heat-resisting steels, facing with Stellite or a similar hard deposit.

exhaust velocity *(Space)*. The velocity at which a propellant gas leaves a rocket motor. It is related to the *specific impulse*, I_{sp} by the expression: $v_e = I_{sp}g$ where v_e is the exhaust velocity and g the acceleration due to gravity at the Earth's surface.

exhibitionism *(Behav.)*. Describes behaviour in which sexual gratification is obtained through displaying the genitals to members of the opposite sex; by extension all behaviour motivated by the pleasure of self-display.

exine *(Bot.)*. The outer part of the wall of a pollen grain or embryophyte spore, from the patterns of the surface of which it is often possible to identify the genus or even species of plant from which the pollen has come. Cf. **intine**. See also **sporopollenin, pollen analysis**.

exinguinal *(Zool.)*. In land Vertebrates, outside the groin.

exinite *(Geol.)*. A hydrogen-rich **maceral** which is found in coal.

exit *(Comp.)*. To transfer control from a **subprogram** back to the calling program or from the program entirely. There can be more than one exit in a program or subprogram.

exit pallet *(Horol.)*. See **pallet**.

exit portal *(Radiol.)*. The area through which a beam of radiation leaves the body. The centre of the exit portal is sometimes called the *emergent ray point*.

exit pupil *(Phys.)*. The image of the aperture stop as formed by that part of an optical system on the image side of the aperture. It defines the emergent cone of rays from the system. In a microscope or telescope it is usually the image of the objective formed by the eyepiece, and it is the position which should be occupied by the eye of the observer.

Exner function *(Meteor.)*. If p is the atmospheric pressure, p_0 a reference pressure, and γ is the ratio of the specific heats of a perfect gas, then the Exner function P of p is given by

$$P = (p/p_0)^{(\gamma-1)/\gamma}.$$

It is useful in studies of compressible adiabatic flow.

exobiology *(Biol.)*. The study of putative living systems which probably must exist elsewhere in the universe.

exocardiac *(Zool.)*. Outside the heart.

exocarp *(Bot.)*. See **epicarp**.

exoccipital *(Zool.)*. A paired lateral cartilage bone of the Vertebrate skull, forming the side-wall of the brain-case posteriorly.

exocoelar *(Zool.)*. See **somatopleural**.

exocoelom *(Zool.)*. The extraembryonic coelom of a developing Bird, Reptile or Elasmobranch.

exocrine *(Med.)*. A gland which secretes its hormone into a duct rather than direct into the blood stream as in endocrine glands.

exocrine *(Zool.)*. Said of glands the secretion of which is poured into some cavity of the body, or on to the external surface of the body by ducts. Cf. **endocrine**.

exocuticle *(Zool.)*. The layer of the insect cuticle, between the epicuticle and endocuticle, which becomes hardened and darkened in most Insects.

exocytosis *(Biol.)*. The exit of material from the cell by fusion of internal vesicles with the **plasma membrane** which either void their contents to the exterior or introduce new surface material into the plasma membrane.

exodermis *(Bot.)*. The outermost layer or layers of the cortex of some roots with more or less thickened and/or suberised cell walls; a specialised hypodermis.

exo-electron *(Electronics)*. One, emitted from the surface of a metal or semiconductor, which comes from a metastable trap with very low binding energy under conditions such that electrons in their ground state could not be emitted.

exo-ergic process *(Phys.)*. Nuclear process in which energy is liberated. Also *exo-energetic*. The equivalent thermodynamic term is *exothermic*.

exogamete *(Zool.)*. A gamete which unites with one from another parent.

exogamy *(Bot.,Zool.)*. The production of a zygote by the fusion of gametes from 2 unrelated parents. Cf. **endogamy, outbreeding, allogamy**.

exogenous *(Bot.)*. (1) Resulting from causes external to an organism. (2) Produced on the outside of another organ or developed from tissues at or near the surface (as leaf primordia are). (3) Growing by the addition of new layers at or near the surface. Cf. **endogenous**.

exogenous *(Zool.)*. In higher animals, said of metabolism which leads to the production of energy for activity. Cf. **endogenous**.

exomphalos *(Med.)*. A hernia formed by the protrusion of abdominal contents into the umbilicus.

exon *(Biol.)*. That part of the transcribed nuclear RNA of eukaryotes which forms the mRNA after the excision of the **introns**.

exonuclease *(Biol.)*. An enzyme which cleaves nucleic acids from a free end. It can thus digest a linear molecule by steps. Cf. **endonuclease**.

exophoria *(Med.)*. Latent external squint revealed in an apparently normal person by passing a screen before the eye.

exophthalmic goitre *(Med.)*. The protrusion of the eyes, or eye muscle disorder, associated with hyperthyroidism. Also *Graves disease*.

exopodite *(Zool.)*. The outer ramus of a crustacean appendage.

Exopterygota *(Zool.)*. A subclass of Insects in which wings occur, although sometimes secondarily lost; the change from young form to adult is gradual, the wings developing externally; the young form is usually a nymph.

EX OR *(Comp.,Maths.)*. See **logical operations**.

exoscopic embryology *(Bot.)*. The condition when the apex of the embryo is turned towards the neck of the archegonium.

exoskeleton *(Zool.)*. Hard supporting or protective structures that are external to and secreted by the ectoderm, e.g., in Vertebrates, scales, scutes, nails, and feathers, in Invertebrates, the carapace, sclerites, etc.

exosphere *(Astron.)*. Region of the Earth's atmosphere beyond about 500 km.

exospore *(Bot.)*. (1) The outermost layer of the wall of a spore; **exine**. (2) A spore formed by the extrusion of material from the parent cell.

exostosis *(Med.)*. A bony tumour growing outwards from a bone.

exothermic *(Chem.)*. Accompanied by the evolution of heat.

exotic *(Bot.,Zool.)*. Not native. *Ecdemic*.

exotoxin *(Biol.)*. A toxin released by a bacterium into the medium in which it grows. Frequently very toxic, e.g.

neurotoxins which destroy cells of the nervous system, haemolytic toxins which lyse red blood cells.

expanded *(Build.)*. Of cellular structure and therefore light in weight. Thus expanded concrete is *lightweight concrete*.

expanded metal *(Build.)*. A metal network formed by suitably stamping or cutting sheet metal and stretching it to form open meshes. It is used as a reinforcing medium in concrete construction, as lathing for plasterwork and various other purposes. Cf. **BRC fabric**.

expanded plastics *(Plastics)*. Foamed plastic materials, e.g. PVC, polystyrene, polyurethane, polythene etc., created by the introduction of pockets or cells of inert gas (air, carbon dioxide, nitrogen etc.) at some stage of manufacture. Used for heat insulation purposes or as core materials in 'sandwich' construction, because of their low densities; also, because of lightness, for packaging, and (as foam rubber) for upholstery cushioning, artificial sponges etc. Often highly inflammable.

expanded sweep *(Electronics)*. A technique for speeding-up the motion of the electron beam in an oscilloscope during a part of the sweep.

expander *(Acous.)*. Amplifying apparatus for automatically increasing the contrast in speech modulation, particularly after reception of speech which has had its contrast compressed by a **compressor**.

expander *(Elec.Eng.)*. An inert material, such as carbon or barium sulphate, added to the active material in accumulator plates to prevent shrinkage of the mixture.

expanding arbor *(Eng.)*. A lathe arbor expandable by blades or keys sliding in taper grooves, which allows work of various bore diameter to be supported and located.

expanding bit *(Build.)*. A boring-bit carrying a cutter on a radial arm, the position of the cutter being adjustable so that holes of different sizes may be cut.

expanding brake *(Eng.)*. A brake in which internal shoes are expanded to press against the drum, usually by a cam or toggle mechanism.

expanding cement *(Build.,Civ.Eng.)*. Cement containing a chemical agent which induces predetermined expansion to minimize the normal shrinkage which occurs in concrete during the setting and drying out process.

expanding mandrel *(Eng.)*. See under **mandrel** (1).

expanding metals *(Eng.)*. Metals or alloys, e.g. 2 parts antimony to 1 part bismuth, which expand in final stage of cooling from liquid; used in type-founding.

expanding plug *(Build.)*. A **bag plug**.

expanding reamer *(Eng.)*. A **reamer**, (1) partially slit longitudinally, and capable of slight adjustment in diameter by a coned internal plug, (2) machined with 6 or more grooves cut at a small angle to the axis along which hard steel blades, correspondingly tapered, can be moved by nuts at either end. The effective circumference of the blades is thus adjustable over a range of about 1.5 mm.

expanding universe *(Astron.)*. The view, based on the evidence of the red-shift, that the whole universe is expanding; supported by relativity theory, in which a static universe would be unstable.

expansion *(Electronics)*. A technique by which the effective amplification applied to a signal depends on the magnitude of the signal being larger for the bigger signals. See also **automatic volume expansion**.

expansion *(Eng.)*. (1) Increase in volume of working fluid in an engine cylinder. (2) Piston stroke during which such expansion occurs.

expansion *(Phys.)*. See **adiabatic-**, **coefficient of-**.

expansion circuit-breaker *(Elec.Eng.)*. A circuit-breaker in which arc extinction occurs as a result of the rapid cooling produced by the expansion of steam or of gases; these result from the arc which arises between the contacts in water or in a small quantity of oil.

expansion curve *(Eng.)*. The line on an **indicator diagram** which shows the pressure of the working fluid during the expansion stroke.

expansion engine *(Eng.)*. An engine which utilizes the working fluid expansively.

expansion gear *(Eng.)*. That part of a steam-engine valve gear through which the degree of expansion can be varied.

expansion joint *(Civ.Eng.)*. A joint arranged between two parts to allow them to expand with temperature rise, without distorting laterally, e.g. the gap left between successive lengths of rail or the joint made between successive sections of carriageway in road construction. See **continuous welded rail**.

expansion joint *(Eng.)*. A special pipe joint used in long pipelines to allow for expansion, e.g. a horseshoe bend, a corrugated pipe acting as a bellows, a sliding socket joint with a stuffing box.

expansion line *(Eng.)*. See **expansion curve**.

expansion of gases *(Phys.)*. All gases have very nearly the same coefficient of expansion, namely 0.003 66 per kelvin when kept at constant pressure. See **absolute temperature**, **gas laws**.

expansion pipe *(Build.)*. In a domestic system of heating, a pipe carried up from the hot-water tank to a point above the level of the cold-water tank, where its open end is bent over, so that, if the water boils, it may discharge any water or steam forced out.

expansion ratio *(Aero.)*. The ratio between the gas pressure in a rocket combustion chamber, or a jet pipe, and that at the outlet of the propelling nozzle.

expansion rollers *(Eng.)*. Rollers on which one end of a large girder or bridge is often carried, to allow of movement resulting from expansion; the other end of the girder etc. is fixed.

expansion tank *(Build.)*. In a hot-water system, the tank connected to, and above, the hot-water cylinder to allow of expansion of the water on heating; often the cold-water feed tank is so used. See **expansion pipe**.

expansion valve *(Eng.)*. An auxiliary valve working on the back of the main slide-valve of some steam-engines to provide an independent control of the point of cut-off.

expansive working *(Eng.)*. The use of a working fluid expansively in an engine; an essential feature of every efficient working cycle.

expectancy, principle of *(Behav.)*. Refers to a cognitive theory of animal and human learning which posits that the anticipation of events, and especially of rewards, is an important aspect of many learning events.

expectation *(Stats.)*. The average value resulting from an infinite series of repetitions of an experiment or observations of a random variable.

expectoration *(Med.)*. The coughing up of mucous or sputum from the air passages.

expedor phase advancer *(Elec.Eng.)*. A phase advancer which injects into the secondary circuit of an induction motor an e.m.f. which is a function of the secondary current. Cf. **susceptor phase advancer**.

expendable launch vehicle *(Space)*. A launch system which is made up of throw-away stages and has no recoverable parts. Abbrev. *ELV*.

experimental allergic encephalomyelitis *(Immun.)*. An auto-immune disease produced by injections into mice of proteins present in brain and spinal cord together with complete Freund's adjuvant. After a few days acute encephalomyelitis develops accompanied by demyelination and progressive paralysis. The main factor is T-cell sensitization against myelin basic protein but antibodies may also be involved. A similar condition has been described in humans following immunization against rabies using a crude brain-derived vaccine which is now obsolete.

experimental embryology *(Zool.)*. The experimental study of the physiology and mechanics of development.

experimental mean pitch *(Aero.)*. The distance of travel of a propeller along its own axis, while making one complete revolution, assuming the condition of its giving no thrust.

experimental petrology *(Geol.)*. A branch of petrology concerned with the laboratory study of rocks and minerals under different physical and chemical conditions.

expert system *(Comp.)*. Software designed to function like a specialist consultant. It usually has two parts, a base of organised expert knowledge, which can be easily expanded, and a set of rules for reaching conclusions. See **knowledge-based systems.**

expiration *(Zool.)*. The expulsion of air or water from the respiratory organs.

explant *(Bot.)*. A piece of tissue or an organ removed for experimental purposes, from a plant especially to start a tissue culture.

explantation *(Zool.)*. In experimental zoology, the culture, in an artificial medium, of a part or organ removed from a living individual, tissue-culture. n. *explant.*

explement *(Maths.)*. See **conjugate angles.**

expletive *(Build.)*. A stone used as a filling for a cavity.

explicit function *(Maths.)*. A variable x is an explicit function of y when x is directly expressed in terms of y. Cf. **implicit function**

exploding star *(Astron.)*. See **nova.**

exploitable girth *(For.)*. (1) The minimum girth at breast height at or above which trees are considered suitable for felling or for the purpose specified. (2) The girth down to which all portions of a bole or tree must be exploited as timber or fuel under a permit licence.

exploitation well *(Min.Ext.)*. Well drilled in a proved deposit.

exploratory behaviour *(Behav.)*. A form of **appetitive behaviour**, by which an animal gains information about its environment; some forms of exploratory behaviour are goal-linked; others seem motivated by a general curiosity mingled with mild fear and is not terminated by any apparent end goal.

Explorer *(Space)*. A series of American artificial satellites used to study the physics of space cosmic rays, temperatures, meteorites, etc; responsible for the discovery of the **Van Allen radiation belts**. Explorer I was the first Earth satellite launched by the US (January, 1958).

exploring brush *(Elec.Eng.)*. A small brush fitted to a d.c. machine for experimental purposes; it can be moved round the commutator in order to investigate the distribution of potential around it.

exploring coil *(Elec.Eng.)*. One used to measure magnetic field strengths by finding the change of magnetic linkages on removing the coil from a position in the field to a remote point out of the influence of the field. Also called *search coil.*

explosion *(Chem.)*. A rapid increase of pressure in a confined space. Explosions are generally caused by the occurrence of exothermic chemical reactions in which gases are produced in relatively large amounts. For nuclear explosion see **atomic bomb** and **hydrogen bomb.**

explosion pot *(Elec.Eng.)*. A strong metal container surrounding the contacts of an oil circuit-breaker; the high pressure set up inside the pot when an arc occurs assists in the extinction of the arc.

explosion-proof, flame-proof *(Elec.Eng.)*. Said of electrical apparatus so designed that an explosion of inflammable gas inside the enclosure will not ignite inflammable gas outside. Such apparatus is used in mines or other places having an explosive atmosphere.

explosion welding *(Eng.)*. Welding of two components made to hit each other at high speed due to a controlled explosion.

explosive *(Min.Ext.)*. There are two main classes, 'permitted' and 'non-permitted', i.e. those which are safe for use in coal-mines and those which are not. Ammonium nitrate mixtures are mostly used in coal-mines; nitroglycerine derivatives in metal-mines. ANFO (*Ammonium Nitrate and Fuel Oil*) is now widely used in hard rock mining. Explosives are used as propellants (*low explosives*) and for blasting (*high explosives*), in both civil and military applications.

explosive forming *(Eng.)*. One of a range of high-energy rate-forming processes by which parts are formed at a rapid rate by extremely high pressures. Low and high explosives are used in variations of the explosive forming process; with the former, known as the cartridge system, the expanding gas is confined; with the latter, the gas need not be confined and pressure of up to one million atmospheres may be attained. See **electrohydraulic forming.**

explosive fracturing *(Min.Ext.)*. The use of an explosive charge to crack strata and increase oil flow round a bore hole. Invented by Col. Roberts in Titusville, Pennsylvania as the *Roberts torpedo.*

explosive rivet *(Eng.)*. A type of **blind rivet** which is clinched or set by exploding, electrically, a small charge placed in the hollow end of the shank.

exponent *(Maths.)*. See **index.**

exponential baffle *(Acous.)*. A baffle approximating to a short section of an exponential horn.

exponential function *(Maths.)*. The function e^x. See *e* and **exponential series.**

exponential growth *(Biol.)*. A stage of growth occurring in populations of unicellular micro-organisms when the logarithm of the cell number increases linearly with time.

exponential horn *(Acous.)*. One for which the taper or flare follows an exponential law.

exponential reactor *(Nuc.Eng.)*. One with insufficient fuel to make it diverge, but excited by an external source of neutrons for the determination of its properties.

exponential series *(Maths.)*. The series

$$1+\frac{x}{1!}+\frac{x^2}{2!}+\frac{x^3}{3!}+\ldots+\frac{x^n}{n!}+\ldots$$

which converges to the value e^x for all values of x.

export *(Biol.)*. The process of transferring proteins across a membrane either into the medium or into another cellular compartment.

exposed pallets *(Horol.)*. Usually a clock movement in which the **pallets** are in front of the dial.

exposure *(Image Tech.)*. (1) The process of allowing light to fall on a photo-sensitive surface; numerically specified by the intensity of light I and the duration of the exposure, time T, $E = I \times T$, in candela-metre-seconds or lux-seconds. (2) In practice, exposure in a system is determined by the illumination and reflectivity of the subject, the light transmission of the lens, its **aperture**, and the length of time the shutter is open; each of these factors must be adjusted to suit the sensitivity of the material being exposed.

exposure *(Meteor.)*. The method by which an instrument is exposed to the elements. The exposure in meteorological stations is standardized so that records from different stations are comparable.

exposure dose *(Radiol.)*. See **dose.**

exposure learning *(Behav.)*. Changes in behaviour that result from an individual being exposed to an object or situation, under circumstances in which no consistent response apart from investigatory or exploratory behaviour is elicited by that situation, and with no obvious reward. See also **observational learning.**

exposure meter *(Image Tech.)*. Instrument, a combination of photocell and current meter, for measuring the light reflected from or falling on a subject (incident light) and expressing it in terms of exposure for a given stop or as an **exposure value**. Now often incorporated in the camera.

exposure value *(Image Tech.)*. An index combining both aperture and shutter-speed, expressed as a single figure which remains constant when both are altered, for example to change depth of focus, in such a way that the amount of light reaching the sensitized emulsion is unchanged. Some shutters have stop and speed rings coupled to give a constant EV when either is altered. Abbrev. *EV*

expression *(Textiles)*. Residual liquid left in a fabric after squeezing (e.g. on a mangle), calculated as a percentage of the dry fabric.

Expressionism *(Arch.)*. A concept of architectural design which prevailed in Germany at the beginning of the 20th century, based on symbolism frequently inspired by biological forms.

expression vector *(Biol.)*. A **vector** in genetic manipulation work, which is specially constructed so that a large amount of the protein product, coded by an inserted sequence, can be made.

expulsion fuse *(Elec.Eng.)*. An enclosed fuse-link in which the arc occurring when the link melts is extinguished by the lengthening of the break due to expulsion of part of the fusible material through a vent in the container.

expulsion gap *(Elec.Eng.)*. A special form of expulsion fuse connected in series with a gap and placed across insulator strings on an overhead transmission line; a voltage surge breaks down the gap and the resulting arc is quickly broken by the fuse, so that no interruption to the supply need take place.

exsanguination *(Med.)*. Severe loss of blood.

exserted *(Bot.)*. Protruding.

exstrophy, extrophy *(Med.)*. A turning inside out of a hollow organ (especially the bladder).

ex. sur. tr. & ct. *(Build.)*. Abbrev. for *excavate surface trenches and cart away*.

extended delivery *(Print.)*. The delivery of a machine lengthened by design to give the printed image on the sheet the maximum time to set before being covered by the next.

extender *(Build.,Plastics)*. (1) A substance added to paint as an adulterant or to give it body, e.g. barytes, china clay, French chalk, gypsum, whiting. (2) In synthetic resin adhesives, a substance (e.g. rye flour) added to reduce the cost of gluing or to adjust viscosity. (3) A noncompatible plasticizer used as an additive to increase the effectiveness of the compatible plasticizer in the manufacture of elastomers.

extender *(Image Tech.)*. See **teleconverter**.

extenders *(Print.)*. The parts of lower-case type characters which extend above or below the **x-height**, i.e. the *ascending* and *descending* part in, e.g. *b, f, p*.

extension flap *(Aero.)*. A landing flap which moves rearward as it is lowered so as to increase the wing area. See **Fowler flap**.

extension ore *(Min.Ext.)*. In assessment of reserves, ore which has not been measured and sampled, but is inferred as existing from geological reasoning supported by proved facts regarding adjacent ore.

extension spring *(Eng.)*. A helical spring with looped ends which stores energy when in the stretched condition.

extension tubes *(Image Tech.)*. Tubes which increase the distance between the lens and the focal plane of a camera for close-up work. Made in sets to give different degrees of enlargement.

extensometer *(Eng.)*. An instrument for measuring dimensional changes of a material by e.g. measuring changes in capacitance in a capacitor, one plate of which is attached to the material, while the other is fixed.

extensor *(Zool.)*. A muscle which by its contraction straightens a limb, or a part of the body. Cf. **flexor**.

exterior angle *(Maths.)*. The angle between any side produced and the adjacent side (not produced) of a polygon.

external angle *(Build.)*. A vertical or horizontal angle forming part of the projecting portion of a wall or feature of a building; also called *salient junction*. See also **arris**.

external characteristic *(Elec.Eng.)*. A curve showing the relation between the terminal voltage of an electric generator and the current delivered by it.

external circuit *(Elec.Eng.)*. The circuit to which current is supplied from a generator, a battery, or other source of electrical energy.

external compensation *(Chem.)*. Neutralization of optical activity by the mixture or loose molecular combination of equal quantities of two enantiomorphous molecules.

external conductor *(Elec.Eng.)*. The outer earthed conductor of an earthed concentric wiring system.

external digestion *(Zool.)*. A method of feeding, adopted by some Coelenterata, Turbellaria, Oligochaeta, Insects, and Araneida, in which digestive juices are poured on to food outside the body and imbibed when they have dissolved some or all of the food.

external firing *(Eng.)*. The practice of heating a boiler or pan by a furnace outside the shell; all modern boilers have internal furnaces and flues.

external indicator *(Chem.)*. An indicator to which are added drops of the solution in which the main reaction is taking place, away from the bulk of the solution.

external respiration *(Zool.)*. See **respiration, external**.

external screw-thread *(Eng.)*. A screw-thread cut on the outside of a cylindrical bar. Also called *a male thread*.

external secretion *(Zool.)*. A secretion which is discharged to the exterior, or to some cavity of the body communicating with the exterior. Cf. **internal secretion**.

exteroceptor *(Zool.)*. A sensory nerve-ending, receiving impressions from outside the body. Cf. *interoceptor*.

extinction *(Behav.)*. In **classical conditioning**, the weakening of the tendency of a conditioned stimulus to elicit a conditioned response by unreinforced presentations of the conditioned stimulus. In **operant conditioning**, a decline in the tendency to perform the operant response, as a result of the unreinforced occurrences of the response.

extinction coefficient *(Chem.)*. See **molar absorbance**.

extinction voltage, potential *(Electronics)*. (1) Lowest anode potential which sustains a discharge in a gas at low pressure. (2) In a gas-filled tube, the p.d. across the tube which will extinguish the arc.

extra- *(Bot.)*. Prefix meaning beyond, outside of, outwith.

extra bound *(Print.)*. A book completely bound by hand, expensive, using carefully chosen materials, sewn headbands, special end-papers, gold tooling, etc.

extracellular *(Bot.)*. Located or taking place outside a cell. Cf. **intracellular, intercellular**.

extracellular enzyme *(Biol.)*. One secreted out of the cell into the intercellular space or the lumen of the gut for example.

extracellular matrix *(Biol.)*. A non-cellular matrix of proteins and glycoproteins surrounding cells in some tissues. It can be extensive as in cartilage and connective tissue and calcified as in bone.

extra-chromosomal DNA *(Biol.)*. (1) Non-integrated viral DNA and other **episomes**. (2) DNA of cytoplasmic organelles, i.e. mitochondrial and chloroplastal DNA.

extrachromosomal inheritance *(Bot.)*. See **cytoplasmic inheritance**.

extraction *(Chem.)*. A process for dissolving certain constituents of a mixture by means of a liquid with solvent properties for one of the components only. Substances can be extracted from solids, e.g. grease from fabrics with petrol; or from liquids, e.g. extraction of an aqueous solution with ethoxyethane, the efficiency depending on the partition coefficient of the particular substance between the two solvents.

extraction fan *(Eng.)*. A fan used to extract foul air, fumes, suspended paint particles etc. from a working area.

extraction metallurgy *(Min.Ext.)*. First stage or stages of ore treatment, in which gangue minerals are discarded and valuable ones separated and prepared for working up into finished metals, rare earths or other saleable products. Characteristically, the methods used do not change the physical structure of these products save by comminution.

extraction thimble *(Chem.)*. A porous cylindrical cup containing the solid material to be extracted, usually by placing it under the hot reflux in a still.

extraction turbine *(Eng.)*. A steam-turbine from which steam for process work is tapped at a suitable stage in the expansion, the remainder expanding down to condenser pressure.

extractive distillation *(Chem.Eng.)*. A technique for improving, or achieving in cases impossible without it, distillation separation processes by the introduction of an additional substance which changes the system equilibrium. It is essentially different from **azeotropic distillation** in that the added substance is not distilled itself but is added as a liquid at some point, usually at or near top in the distillation column, and leaves at the base as liquid.

extract ventilator *(Build.)*. A cowl-like appliance fitted to the top of a ventilating shaft in a building to induce in it an up-draught.

extrados *(Build.,Civ.Eng.)*. The back or top surface of an arch. See **intrados**.

extradural *(Med.)*. Situated outside the dura mater.

extra-embryonic *(Zool.)*. In embryos developed from eggs containing a great deal of yolk, as those of Birds, pertaining to that part of the germinal area beyond the limits of the embryo.

extra-floral (extra-nuptial) nectary *(Bot.)*. A nectary occurring on or in some part of a plant other than a flower.

extra-galactic nebula *(Astron.)*. A **nebula** external to the Galaxy.

extra-heavy *(Build.)*. Pipe heavier than standard.

extra-high voltages *(Elec.Eng.)*. A term used in official regulations for voltages above 3.3 kV; but more commonly employed to denote voltages of the order of 100 kV or more.

extra-lateral rights *(Min.Ext.)*. See **apex law**.

extraneous ash *(Min.Ext.)*. In raw coal, the so called 'free dirt' or associated shale and enclosing beds.

extra-nuptial nectary *(Bot.)*. See **extra-floral nectary**.

extraordinary ray *(Phys.)*. See **double refraction**.

extrapolation *(Maths.)*. The estimation of the value of a function at a particular point from values of the function on one side only of the point. Cf. **interpolation**.

extras *(Build.,Civ.Eng.)*. All work the inclusion of which is not expressed or implied in the original contract price. Also *variations*.

extrasensory perception *(Behav.)*. Perception of phenomena without the use of the ordinary senses.

extrasystole *(Med.)*. A premature contraction of the heart interrupting the normal rhythm, the origin of the impulse to contraction being abnormally situated either in the ventricles or in the atrium.

extraterrestrial *(Astron.)*. Used to describe (hypothetical) intelligent life anywhere in the universe.

extra thirds *(Paper)*. A former size of cut card, 44.5 × 76 mm ($1\frac{3}{4} \times 3$ in).

extra-uterine *(Med.)*. Situated or happening outside the uterus.

extravasation *(Med.,Zool.)*. The abnormal escape of fluids, as blood or lymph, from the vessels which contain them. v. *extravasate*.

extravascular *(Med.)*. Placed or happening outside a blood vessel.

extraversion *(Med.)*. See **exstrophy**.

extraversion/introversion *(Behav.)*. In Jung's theory, a characterization of personality according to a tendency to focus on the external and objective aspects of experience or on internal and subjective ideas. Eysenck uses extraversion-introversion as a trait dimension along which individuals can be placed and uses the popular notion of extravert as *outgoing* and introvert as *withdrawn*.

extreme breadth *(Ships)*. The greatest breadth measured to the farthest out part of the structure on each side, including any rubbing strakes or other permanent attachments to the hull or **superstructure** but not to the **top hamper**.

extreme dimensions *(Ships)*. These dimensions are provided for general information, mainly for docking purposes. They are **length overall**, **extreme breadth** and **depth** or **summer draught**.

extreme pressure lubricant *(Eng.)*. A solid lubricant, such as graphite, or a liquid lubricant with additives which form oxide or sulphide coatings on metal surfaces exposed where, under very heavy loading, the liquid film is interrupted, thus mitigating the effects of dry friction.

extrinsic *(Crystal.)*. Said of electrical conduction properties arising from impurities in the crystal.

extrinsic *(Zool.)*. Said of appendicular muscles of Vertebrates which run from the trunk to the girdle, or the base of the limb. Cf. **intrinsic**.

extrinsic semiconductor *(Electronics)*. See **semiconductor**.

extrophy *(Med.)*. See **exstrophy**.

extrorse *(Bot.,Zool.)*. Directed or bent outwards; facing away from the axis; especially, of stamens opening towards the outside of the flower. Cf. **introrse**.

extrovert *(Zool.)*. An extrusible proboscis, found in certain aquatic animals. See **lophophore**.

extrusion *(Eng.)*. Operation of producing rods, tubes and various solid and hollow sections, by forcing suitable material through a die by means of a ram. Applied to numerous nonferrous metals, alloys and other substances, notably plastics (for which a screw-drive is frequently used). In addition to rods and tubes, extruded plastics include sheets, film and wire-coating.

extrusion *(Textiles)*. The process used in the manufacture of man-made fibres in which a viscous solution of the polymer or the molten polymer is forced through the fine holes of a **spinneret**.

extrusive rocks *(Geol.)*. Rocks formed by the consolidation of magma on the surface of the ground as distinct from *intrusive rocks* which consolidate below ground. Commonly referred to as *lava flows*; normally of fine grain or even glassy.

exudate *(Med.)*. The fluid which has formed in the tissues or the cavities of the body as a result of inflammation; it contains protein and many cells, and clots outside the body. n. *exudation*.

exudation pressure *(Bot.)*. See **root pressure**.

exudative diathesis of chicks *(Vet.)*. Subcutaneous oedema in chicks associated with excessive capillary permeability due to vitamin E deficiency.

exumbrella *(Zool.)*. The upper convex surface of a medusa. adj. *exumbrellar*.

exuviae *(Zool.)*. The layers of the integument cast off in ecdysis.

exuvial *(Zool.)*. Pertaining to, or facilitating, ecdysis.

eye *(Arch.)*. (1) The circular opening in the top of a dome. (2) A circular or oval window.

eye *(Build.)*. Of an axe or other tool, the hole or socket in the head for receiving the handle.

eye *(Eng.)*. (1) A loop formed at the end of a steel wire or bolt. See **eye bolt**. (2) The central inlet passage of the impeller of a centrifugal compressor or pump.

eye *(Glass)*. The hole in the centre (or elsewhere) of the floor of a pot furnace up which the combustible gases rise as flame to heat the furnace.

eye *(Meteor.)*. The central calm area of a cyclone or hurricane, which advances as an integral part of the disturbed system.

eye *(Zool.)*. The sense organ which receives visual impressions.

eye-and-object correction *(Surv.)*. A correction applied in precise work to the average angle of elevation read on the vertical circle, in order to compensate for the vertical axis of the theodolite not being truly vertical. The correction is

$$+\frac{\Sigma o - \Sigma e}{4}\cdot\theta,$$

where Σo = object-end reading of the altitude level, Σe = eye-end reading of the altitude level, and θ = angular value of 1 division of altitude level.

eye bolt *(Eng.)*. A bolt carrying an eye instead of the normal head; fitted to heavy machines and other parts for lifting purposes.

eye-cup *(Image Tech.)*. A soft rubber fitting for the eyepiece of a camera viewfinder.

eye-ground *(Med.)*. The fundus; that part of the cavity of the eyeball which can be seen through the pupil with an ophthalmoscope.

eyelids *(Aero.)*. Jet engine thrust-reverser nozzle deflectors so shaped and closed.

eyepiece *(Phys.)*. In an optical instrument, the lens or lens system to which the observer applies his eye in using the instrument.

eyepiece graticule *(Biol.,Min.)*. Grid incorporated in the eyepiece for measuring objects under the microscope. Special type used in particle-size analysis consists of a rectangular grid for selecting the particles and a series of graded circles for use in sizing the particles. Also called *micrometer eyepiece, ocular micrometer* (US).

eye spot *(Bot.,Zool.)*. *Stigma.* Usually orange or red spot composed of droplets containing carotenoids. Found in motile cells of many algae, phytoflagellates and lower animals, and presumed to act as a screen on one side of a photoreceptive area in the detection of the direction of illumination in phototaxis.

eye-stalk *(Zool.)*. A paired stalk arising close to the median line on the dorsal surface of the head of many Crustacea, bearing an eye.

Eyring formula *(Acous.)*. A formula proposed for the period of reverberation of an enclosure, taking into account that the sound waves lose energy every time they are reflected. The results are not identical to those of **Sabine's reverberation formula**.

F

f *(Chem.)*. A symbol for: (1) activity coefficient, for molar concentration; (2) partition function.

f_{hfb} *(Electronics)*. Symbol for common-base mode forward current gain cut-off frequency.

f_{hfe} *(Electronics)*. Symbol for common-emitter mode forward current gain cut-off frequency.

f_T *(Electronics)*. In transistors, symbol for transition-frequency; that at which the common-emitter forward current gain is reduced to unity.

f_α *(Electronics)*. (1) In transistors, symbol for *common emitter forward-current gain cut-off frequency*. Symbols f_{hfe}, f_{hfb}, f_T are more commonly used. (2) Abbrev. for *cut-off frequency*.

F *(Biol.)*. Symbol for *filial generation* in work on inheritance; usually distinguished by a subscript either F_1 or F1, first filial generation; F_2 or F2, second filial generation. Cf. **P**.

F *(Chem.)*. Symbol for fluorine.

F *(Elec.Eng.)*. Symbol for farad.

F *(Phys.)*. Symbol used, following a temperature (e.g. 41°F), to indicate the **Fahrenheit scale**.

F *(Build.)*. Abbrev. for *face* or *flat*.

F *(Phys.)*. Symbol for faraday.

[F] *(Phys.)*. A Fraunhofer line in the blue of the solar spectrum of wavelength 486.1527 nm. It is the second line in the Balmer hydrogen series, known also as Hβ.

F1 hybrid *(Biol.)*. Crop or strain variety, characterized by unusual vigour and uniformity, produced by crossing two selected inbred lines. Cf. **heterosis**, **cytoplasmic male sterility**, **nick**.

FAA *(Aero.)*. Abbrev. for *Federal Aviation Administration*, a US Government agency responsible for all aspects of US civil aviation. Cf. *CAA*.

Fabaceae *(Bot.)*. See **Leguminosae**.

Fab fragment *(Immun.)*. Fragment of immunoglobulin molecule obtained by hydrolysis with papain. It consists of one light chain linked to the N-terminal part of the adjacent heavy chain. Two Fab fragments are obtained from each molecule. Each contains one antigen-binding site, but none of the heavy chain Fc.

$F(ab')_2$ fragment *(Immun.)*. Fragment of immunoglobulin obtained by pepsin digestion. It contains both Fab fragments plus a short section of the hinge region of the heavy chain Fc. It behaves as a bivalent antibody but lacks properties associated with the Fc fragment (e.g. complement activation, placental transmission).

fabric *(Build.)*. Walls, floors, and roof of building.

fabric *(Geol.)*. The sum of all the textural and structural features of a rock.

fabric *(Textiles)*. A coherent assembly of fibres and/or yarns that is long and wide but relatively thin and strong; a *cloth*.

Fabry-Pérot inteferometer *(Phys.)*. An instrument in which multiple-beam circular **Haidinger fringes** are produced by the passage of monochromatic light through a pair of plane parallel half-silvered glass plates, one of which is fixed while the other can be moved by an accurately calibrated screw. In transmission the fringes appear as sharp bright fringes on a dark background. By observing the fringes as the separation of the plates is changed, the wavelength of the light can be determined.

façade *(Arch.)*. The front elevation of a building.

face *(Build.)*. (1) The front of a wall or building. (2) The exposed vertical surface of an arch.

face *(Crystal.)*. See **crystal-**.

face *(Eng.)*. The working surface of any part; as the sole of a carpenter's plane, the striking surface of a hammer, the surface of a slide-valve, or the surface of the steam chest on which it slides, the seating surface of a valve, the flank of a gear-tooth, etc.

face *(Min.Ext.)*. The exposed surface of coal or other mineral deposit in the working place where mining, winning, or getting is proceeding.

face-airing *(Min.Ext.)*. The operation of directing a ventilating current along the face of a working place; also called *flushing*.

face-centred cubic *(Chem.)*. A crystal lattice with a cubic unit cell, the centre of each face of which is identical in environment and orientation to its vertices. Abbrev. *FCC*. Specifically a common structure of metals, in which the unit cell contains 4 atoms, based on this lattice. See **cubic close packing**.

face chuck, face plate *(Eng.)*. A large disk which may be screwed to the mandrel of a lathe and is provided with slots and holes for securing work of a flat or irregular shape.

face edge *(Build.)*. See **working edge**.

face-hammer *(Build.)*. A hammer having a peen which is flat rather than pointed or edged.

face lathe *(Eng.)*. A lathe designed for work of large diameter but short length (e.g. large wheels or disks).

face left, face right *(Surv.)*. Expressions referring to the pointing of a theodolite telescope when the vertical circle is respectively *left* and *right* of the telescope, as seen from the eyepiece end.

face mark *(Build.)*. A distinguishing mark made on one face of a piece of wood to show that it was used as the basis for truing the other surfaces.

face mix *(Build.)*. A mixture of cement and stone dust used for facing concrete blocks in imitation of real stone.

face mould *(Build.)*. A templet used as a reference for shaping the face of wood, stone, etc.

face plate *(Electronics)*. That part of a cathode-ray tube which carries the phosphor screen.

face plate *(Eng.)*. (1) See **face chuck**. (2) A **surface plate**.

face-plate breaker controller *(Elec.Eng.)*. A face-plate controller having a separate contactor for breaking the circuit.

face-plate breaker starter *(Elec.Eng.)*. A face-plate starter having a separate interlocked contactor for breaking the circuit.

face-plate controller *(Elec.Eng.)*. See **face-plate starter**.

face-plate coupling *(Eng.)*. See **flange coupling**.

face-plate starter *(Elec.Eng.)*. An electric motor starter in which a contact lever moves over a number of contacts arranged upon a plane surface. Also called a *face-plate controller*.

face right *(Surv.)*. See **face left**.

face shovel *(Civ.Eng.)*. A mechanical mobile device for cutting into the vertical face of an excavation and depositing soil into vehicles for transporting it elsewhere.

face side *(Build.)*. A trued/faired side of a piece of wood bearing the face mark, used as the base for trueing and marking all other sides.

facet *(Arch.)*. A **facette**.

facet *(Crystal.)*. The flat side of a crystal.

facet *(Zool.)*. One of the corneal elements of a compound eye; a small articulatory surface.

facette *(Arch.)*. A projecting flat surface between adjacent flutes in a column. Also called a *listel*.

face-wall *(Build.)*. The front wall.

facial *(Zool.)*. Pertaining to or situated on the face; the seventh cranial nerve of Vertebrates, supplying the facial muscles and tongue of higher forms, the neuromast organs of the head and snout in lower forms, and the palate in both.

facies *(Geol.)*. The appearance or aspect of any rock; the sum total of its characteristics. Used of igneous, metamorphic and especially sedimentary rocks.

facies, stratigraphic *(Geol.)*. The sum of the rock and fossil features of a sedimentary rock. They include *lithofacies*: mineral composition, grain size, texture, colour, cross-bedding and other sedimentary features; and *biofacies*: the fossil plant and animal characteristics of the rock.

facilitated diffusion *(Biol.)*. The rapid permeation of solutes into cells by interaction of the solutes with specific carriers which facilitate their entry. Facilitated diffusion is distinguished from **active transport** by not allowing the entry of solutes against their concentration gradients.

facilitation *(Zool.)*. The augmented response of a nerve due to prestimulation; the activation of physiological and behavioural response resulting from nonspecific stimulation from a conspecific. See **social facilitation**.

facing *(Civ.Eng.)*. An outer covering applied to the exposed face of sea-walls, embankments, brick walls etc.

facing *(Eng.)*. (1) The operation of turning a flat face on the work in the lathe. (2) A raised machined surface to which another part is attached.

facing bond *(Build.)*. A general term for any bond consisting mainly of stretchers.

facing bricks *(Build.)*. A class of brick used for ordinary facing work; of better quality and appearance than common bricks, but not made to withstand heavy loads, as are engineering bricks.

facing paviors *(Build.)*. A class of hard-burnt bricks used as facing bricks in high-class work.

facing points *(Civ.Eng.)*. See **points**.

facing sand *(Eng.)*. Moulding sand with admixed coal dust, used near pattern in foundry flask to give the casting a smooth surface.

FACS *(Biol.)*. See **fluorescence activated cell sorter**.

facsimile *(Telecomm.)*. The scanning of any still graphic material to convert the image into electrical signals, for subsequent reconversion into a likeness of the original.

facsimile bandwidth *(Telecomm.)*. The frequency difference between the highest and lowest components necessary for the adequate transmission of the facsimile signals.

facsimile receiver *(Telecomm.)*. One for translating signals from a communication channel into a facsimile record of the original copy.

facsimile transmitter *(Telecomm.)*. Means for translating text, lines and half-tone copy into signals suitable for a communication channel.

factor *(Biol.)*. Obs. meaning *gene*.

factor analysis *(Behav.)*. A statistical method for studying the interrelations between various test elements or between different behaviours, in order to discover whether certain correlations allow an explanation in terms of one or more common factors.

factor B, factor D *(Immun.)*. Components involved in the alternative pathway of complement activation.

factor H, factor I *(Immun.)*. A glycoprotein and an enzyme respectively which together inactivate C3b and act as a damper on activated complement.

factorial *n* *(Maths.)*. The product of all whole numbers from a given number (n) down to 1; it is in fact the number of different ways of arranging n objects. Written $n!$ or $\underline{|n}$.

$$n! = n(n-1)(n-2)(n-3) \ldots 3.2.1.$$

Cf. **gamma function, Stirling's approximation, subfactorial *n***.

factor of merit *(Elec.Eng.)*. Of reflecting galvanometers, the deflection in millimetres, produced on a scale at a distance of 1 m by a current of 1 micro-ampere, the deflection being corrected for coil-resistance and time of swing.

factor of safety *(Build.,Eng.)*. The ratio, allowed for in design between the ultimate stress in a member or structure and the safe permissible stress in it. Abbrev. *FS*.

factory-fitting *(Elec.Eng.)*. An electric-light fitting in which the lamp is housed in a strong protecting glass globe. Also called a *mill-fitting*.

faculae *(Astron.)*. The name given to large bright areas of the photosphere of the sun. They can be seen most easily near sunspots and at the edge of the sun's disk; and are at a higher temperature than the average for the sun's surface.

facultative *(Biol.)*. Able but not obliged to function in the way specified; a facultative *anaerobe* can grow in and perhaps use free oxygen, but will also survive and perhaps grow in its absence. Cf. **obligate**.

facultative heterochromatin *(Biol.)*. Chromatin condensed in some cell types but not in others and which is not expressed when condensed despite containing **coding sequences**. Cf. **constitutive heterochromatin**.

facultative parasite *(Bot.)*. A parasite able to live saprophytically and be cultured on laboratory media. Cf. **obligate parasite**. See also **necrotroph**.

fade, fading *(Telecomm.)*. Variation of strength, sometimes periodic, of the signal received from a distant transmitter. In short-wave communication fading may be due to interference between the reflected and direct waves or to change in the properties of the ionosphere during the transmission. In microwave links, it may be caused by atmospheric absorption of the transmitted signal, or by refraction of the highly directional beams causing them to *miss* the receiving antenna.

fade-in *(Image Tech.)*. (1) Gradual appearance of a picture from uniform black. (2) Gradual increase in the level of a sound signal from silence or a very low intensity.

fade-out *(Image Tech.)*. (1) Gradual disappearance of a picture to uniform black. (2) Gradual decrease in the level of a sound signal to silence or a very low intensity.

fader *(Image Tech.)*. A control to vary audio or video signal level to produce fade effects.

fade shutter *(Image Tech.)*. Variable shutter on a motion picture camera or printer whose opening can be continuously altered while running.

fading *(Autos.)*. See **brake-fade**.

fading *(Build.)*. A paint film defect caused by ageing, weathering or exposure to sunlight or atmospheric pollution resulting in a loss of colour.

fading *(Image Tech.)*. Weakening of a photographic image as a result of age, exposure to light or chemical reactions. Black-and-white images may become pale brown or yellow, while in colour images the components may often fade to different extents, resulting in gross distortion of colour balance.

fading *(Telecomm.)*. See **fade**.

fading *(Textiles)*. The change in colour that sometimes takes place when a material is exposed to light or atmospheric fumes.

fading area *(Telecomm.)*. That in which fading is experienced in night reception of radio waves, between the primary and secondary areas surrounding a station transmitting on medium and long waves.

fadometer *(Chem.)*. An instrument used to determine the resistance of a dye or pigment to fading.

faecal pellets *(Geol.)*. Animal excrement, often in the form of rods or ovoid pellets, found in sedimentary rocks. See **coprolite**.

faeces *(Zool.)*. The indigestible residues remaining in the alimentary canal after digestion and absorption of food.

Fagaceae *(Bot.)*. Family of ca 100 spp. of dicotyledonous flowering plants (superorder Hamamelidae). Mostly trees, often dominant in broad-leaved forests of especially the N. hemisphere. Flowers typically in catkins. Includes oak, beech, chestnut and southern beech.

faggot *(Civ.Eng.)*. A bundle of brushwood. See **fascine**.

faggot *(Eng.)*. Made by forming a box with four long flat bars of wrought-iron and filling the interior with scrap and short lengths of bar.

faggoted iron *(Eng.)*. Wrought-iron bar made by heating a **faggot** to welding heat and rolling down to a solid bar. If the process is repeated *double-faggoted iron* is obtained.

fagopyrism *(Vet.)*. *Buckwheat rash*. A form of photosensitization affecting animals with white or lightly pigmented skin; due to ingestion of a fluorescent substance occurring in buckwheat (*Fagopyrum sagittatum*), with subsequent exposure to strong sunlight.

fagot, fagoted *(Genrl.)*. See **faggot, faggoted**.

Fahrenheit scale *(Phys.)*. The method of graduating a thermometer in which freezing point of water is marked 32° and boiling point 212°, the fundamental interval being therefore 180°. Fahrenheit is being widely replaced by the Celsius (centigrade) and Kelvin scales. To convert °F to °C subtract 32 and multiply by 5/9. For the **Rankine** equivalent add 459.67 to °F; this total multiplied by 5/9 gives the **Kelvin** equivalent.

FAI *(Build.)*. Abbrev. for *Fresh-Air Inlet.*

faïence *(Build.)*. Glazed terra-cotta blocks used as facings for buildings.

fail-operational *(Aero.)*. System designed so that it can continue to function after a single failure, warning being indicated.

fail safe *(Nuc.Eng.)*. Design in which power supply, control or structure is able to return to a safe condition in the event of failure or mal-operation, by automatic operation of protective devices or otherwise.

failure modes and effects analysis *(Aero.)*. Method used in the design of aircraft systems to assess the reliability of all components in all stages of flight, identify inadequate parts or system design and then rectify by replacement or redesign.

fair cutting *(Build.)*. The operation of cutting brickwork to the finished face of the work. Abbrev. *FC.*

fair ends *(Build.)*. Projecting masonry ends requiring to be dressed to a finished surface.

fairfieldite *(Min.)*. A hydrated phosphate of calcium and manganese, crystallizing in the triclinic system as prismatic crystals or fibrous aggregates.

fairing *(Aero.)*. A secondary structure added to any part of an aircraft to reduce drag by improving the streamlining.

fairing *(Ships)*. The process of ensuring that the lines of intersection of all planes with a true ship form are *fair;* the resulting lines are known as **quarter lines.**

Fair's graticules *(Powder Tech.)*. Types of eye-piece graticule marked with rectangles and circles for use in microscope methods of particle size analysis.

fairy ring *(Bot.)*. A ring, usually in grass, in which the plants near the periphery are green and healthy and those near the centre less so. It will persist and expand for many years and is associated with the mycelium of the fungus which forms fruiting bodies at the ring's periphery.

Fajans rule for ionic bonding *(Chem.)*. The conditions which favour the formation of ionic (as opposed to covalent) bonds are (*a*) large cation, (*b*) small anion, (*c*) small ionic charge, (*d*) the possession by the cation of an inert gas electronic structure.

Fajans-Soddy law of radioactive displacement *(Chem.)*. The atomic number of an element decreases by 2 upon emission of an α-particle, and increases by 1 upon emission of a β-particle.

falcate, falciform *(Bot.)*. Sickle-shaped.

falciform ligament *(Zool.)*. In higher Vertebrates, a peritoneal fold attaching the liver to the diaphragm.

Falconbridge process *(Eng.)*. Method of separating copper from nickel, in which matte is acid-leached to disssolve copper, after which residue is melted and refined electrolytically.

falcula *(Zool.)*. A sharp curved claw. adj. *falculate.*

fall *(Civ.Eng.)*. The inclination of rivers, streams, ditches, drains etc. quoted as a fall of so much in a given distance. See **cross fall.**

fall *(Eng.)*. A hoisting rope.

fall *(Min.Ext.)*. (1) The collapse of the roof of a level or tunnel, or of a flat working place or stall; the collapse of the hanging wall of an inclined working place or stope. (2) A mass of stone which has fallen from the roof or sides of an underground roadway, or from the roof of a working place.

fall bar *(Build.)*. The part of a latch which pivots on a plate screwed to the inner face of a door, and drops into a hook on the frame.

falling mould *(Build.)*. The development in elevation of the centre line of a handrail.

falling stile *(Build.)*. The shutting stile of a gate, especially of a gate so hung that the bottom of the shutting stile falls as the gate closes.

fall-of-potential test *(Elec.Eng.)*. A test for locating a fault in an insulated conductor; the voltage drop along a known length of the conductor is compared with the voltage drop between one end of the conductor and the fault. Also called **conductivity test, drop test.**

Fallopian tube *(Zool.)*. In Mammals, the anterior portion of the Müllerian duct; the oviduct.

Fallot's tetralogy *(Med.)*. A relatively common type of cyanotic congenital heart disease (blue baby) where there is an incomplete septum between left and right ventricles (hole in the heart), the pulmonary valve is small and stenotic, the aorta lies over both left and right ventricles leading to right ventricular overload and hypertrophy.

fall-out *(Ecol.)*. Particulate matter in the atmosphere, which is transported by natural turbulence, but which will eventually reach the ground by sedimentation or dry or wet deposition. Applied especially to material from nuclear explosions, when the fall-out is radioactive.

fall-out *(Phys.)*. Airborne radioactive contamination resulting from nuclear explosion, inadequately filtered reactor coolant, failure of reactor containment after an accident etc.

fall pipe *(Build.)*. See **downpipe.**

fall time *(Telecomm.)*. The decaying portion of a wave pocket or pulse; usually the time taken for the amplitude to decrease from 90% to 10% of the peak amplitude.

false amethyst *(Min.)*. An incorrect name given to a purple gemstone. Applied (wrongly) to purple fluorite, and sometimes to purple corundum. See **oriental amethyst.**

false amnion *(Zool.)*. See **chorion.**

false annual ring *(Bot.)*. A second ring of xylem formed in one season, following the defoliation of the tree by the attacks of insects or other accident; oaks are liable to this, as they may be completely stripped of leaves by the oak tortrix.

false bands *(Print.)*. Strips of board or leather glued across the spine of hollow-backed books before covering with leather. See **raised bands.**

false bearing *(Build.)*. A beam, such as a sill, when not supported under its entire length, is said to have a *false bearing.*

false bedding *(Geol.)*. See **cross bedding.**

false body *(Build.)*. A paint defect whereby a paint which appears to be full bodied undergoes a sharp drop in viscosity under the agitation or brushing but returns to its original condition whenever agitation stops. The speed of return to original consistency affects the flow of the applied coating and brush marks do not flow out.

false bottom *(Eng.)*. (1) A removable bottom placed in a vessel to facilitate cleaning; a casting placed in a grate to raise the fire bars and reduce the size of the fire. (2) Any secondary bottom plate or member used to reduce the volume of a container or to create a secondary container.

false colour *(Image Tech.)*. System of colour photography used for camouflage detection and aerial survey; a multi-layer colour film is used in which one layer is sensitized to infra-red, so that natural green subjects such as grass and foliage are reproduced in magenta.

false curvature *(Phys.)*. That of particle tracks (e.g., in cloud chambers, bubble chambers, spark chambers or photographic emulsions) which results from undetected interactions and not from an applied magnetic field.

false diamond *(Min.)*. Several natural minerals are sometimes completely colourless and, when cut and polished, make brilliant gems. These include zircon, white sapphire, and white topaz. All three, however, are birefringent and can be easily distinguished from true diamond by optical and other tests.

false ellipse *(Build.)*. An approximate ellipse, composed of circular arcs.

false fruit *(Bot.)*. A fruit formed from other parts of the flower in addition to the carpels e.g. the strawberry where the receptacle becomes fleshy.

false header *(Build.)*. A half-length brick, sometimes used in Flemish bond.

false hemlock *(For.)*. See **Douglas fir.**

false key *(Eng.)*. A circular key for attaching a hub to a shaft; it is driven into a hole which is parallel with the shaft axis and has been drilled half in the hub and half in the shaft.

false pile *(Civ.Eng.)*. A length added to the top of a pile which has been driven.

false pregnancy *(Med.,Zool.)*. See **pseudocyesis**.

false ribs *(Zool.)*. In higher Vertebrates, ribs which do not reach the sternum.

false ruby *(Min.)*. Some species of garnet (*Cape ruby*) and some species of spinel (*balas ruby* and *ruby spinel*) possess the colour of ruby, but have neither the chemical composition nor the physical attributes of true ruby.

false septum *(Bot.)*. A **replum**.

false tissue *(Bot.)*. See **pseudoparenchyma**.

false topaz *(Min.)*. A name wrongly applied to yellow quartz. See **citrine**.

false-twist *(Textiles)*. A process of twisting a yarn so that although nearly all the yarn is twisted the total resultant twist is zero. Some of the yarn is twisted in one direction and the rest in the opposite direction. This may be achieved by twisting a yarn at some intermediate position along its length whereas in spinning during which real twist is inserted a free end of the yarn is twisted. See **textured yarn**.

falsework *(Civ.Eng.)*. The temporary work known as **centring**; scaffolding, or other temporary supports used in construction.

false-zero test *(Elec.Eng.)*. A test, made on a bridge or potentiometer, in which a balance is obtained, not with zero galvanometer reading, but with some definite value caused by a constant extraneous current.

falx *(Zool.)*. Any sickle-shaped structure. adjs. *falciform, falcate*.

falx cerebri *(Zool.)*. A strong fold of the dura mater, lying in the longitudinal fissure between the two cerebral hemispheres.

family *(Biol.)*. A group of similar genera of taxonomic rank below **order** and above **genus**; with plants, the names usually end in -aceae.

family *(Phys.)*. The group of radioactive nuclides which form a decay series.

family therapy *(Behav.)*. Psychotherapy which regards the family as a unit and as the object of therapy, rather than its individual members, so that roles and attitudes within the family can be explored and changed.

fan *(Aero.)*. Rotating bladed device for moving air in ducts or in e.g. wind tunnels. See **turbofan**. Cf. **propeller**.

fan *(Eng.)*. (1) A device for delivering or exhausting large volumes of air or gas with only a low pressure increase. It consists either of a rotating paddle-wheel or an airscrew. (2) A small vane to keep the wheel of a wind pump at right angles to the wind.

fan *(Geol.)*. (1) A detrital cone found at the foot of mountains and also in the deep sea (submarine fan). (2) Fan cleavage; an axial-plane cleavage in which the cleavage planes fan out.

fan *(Horol.)*. A wheel whose velocity is regulated by air resistance; used in some clocks.

fan antenna *(Telecomm.)*. Antenna in which a number of vertically inclined wires are arranged in a fanwise formation, the apex being at the bottom.

fan characteristic *(Eng.)*. A graph showing the relation between pressure and delivery, used as a basis for fan selection. The characteristic is determined by the shape of the fan blades.

Fanconi's anaemia *(Biol.)*. Human clinical syndrome in which spontaneous rearrangements preferentially involving non-homologous chromosomes occur at an increased rate.

Fanconi's syndrome *(Med.)*. A kidney disease where the renal tubules are unable to conserve aminoacids, phosphates, glucose, bicarbonate and water. Patients present with **acidosis** or with **rickets**. It may occur as a genetically determined disease or may complicate **amyloidosis**, multiple myeloma or poisoning with heavy metals.

fan cooling *(Autos.)*. The use of an engine-driven fan to induce a greater airflow through the radiator at low speeds than would result from the forward motion of the vehicle.

fancy yarns *(Textiles)*. Yarns made for decorative purposes. The ornamentation of the thread may be due to a variety of reasons, such as (*a*) colour; (*b*) the combination of threads of different types; (*c*) the production of thick and thin places; (*d*) the production of curls, loops, slubs etc. at suitable intervals. The majority of these fancy yarns are *folded yarns*, two or more threads being combined in some special way to produce the desired effect.

fan dipole *(Telecomm.)*. A **dipole** antenna consisting of two triangular sheets of metal, with the feeder connected between them. Used for VHF and UHF, with the advantage of a broad operating wavelength band.

fan drift *(Min.Ext.)*. Ventilating passage along which air is moved by means of a fan.

fang *(Build.)*. The part of an iron railing which is embedded in the wall.

fang *(Zool.)*. The grooved or perforate poison-tooth of a venomous serpent: one of the cuspidate teeth of carnivorous animals, especially the canine or carnassial.

fang bolt *(Eng.)*. A bolt having a nut which carries pointed teeth for gripping the wood through which the bolt passes, so preventing the nut from rotating when the bolt is tightened.

fanglomerate *(Geol.)*. A conglomerate formed by lithification of a fan.

fan-guard *(Build.)*. A protective parapet formed of boarding secured around the platforms of builders' stagings or gantries, when the platforms are to be used for receiving and distributing materials.

fan marker beacon *(Aero.)*. A form of marker beacon radiating a vertical fan-shaped pattern.

fan shaft *(Min.Ext.)*. Mine shaft or pit at the top of which a ventilating fan is placed.

fantail burner *(Eng.)*. A pulverized-coal burner which discharges the fuel and primary air vertically downwards into the furnace in a thin flat stream to meet heated secondary air which is discharged horizontally from the walls.

Fantasound *(Image Tech.)*. Obsolete form of stereo-sound presentation in the cinema, using multi-channel tracks on a separate film run in synchronism with the picture.

fantasy, phantasy *(Behav.)*. *General*, sequences of private mental images, sometimes in anticipation of possible events, but sometimes irrational and referring to extremely unlikely possibilities. In *psychoanalytic theory*, an imaginary episode, operating on either a conscious or unconscious level, in which the subject is a central figure and which fulfils a conscious or unconscious wish.

fan vaulting *(Arch.)*. Tracing rising from a capital or a corbel, and diverging like the folds of a fan on the surface of a vault.

farad *(Elec.Eng.)*. The practical and absolute SI unit of electrostatic capacitance, defined as that which, when charged by a p.d. of one volt, carries a charge of one coulomb. Equal to 10^{-9} electromagnetic units and 9×10^{11} electrostatic units. Symbol F. This unit is in practice too large, and the subdivisions, *microfarad* (μF), *nanofarad* (nF), and *picofarad* (pF), are in more general use.

faraday *(Phys.)*. Quantity of electric charge carried by one mole of singly charged ions, i.e. 9.6487×10^4 coulombs. Symbol *F*.

Faraday cage *(Elec.Eng.)*. An arrangement of conductors, or conducting sheet or conducting mesh, bonded together so that they provide an **electrostatic shield**, but connected in such a way that induced currents cannot circulate. Used, for example, to provide an equipotential screen around equipment and/or personnel to enable them to work on live high voltage equipment. See **Faraday shield**.

Faraday dark space *(Phys.)*. Dark region in a gas-discharge column between the negative glow and the positive column.

Faraday disk *(Elec.Eng.)*. Rotating disk in the gap of an electromagnet, so that a low e.m.f. is generated across a radius. Used, e.g. to generate a calculated electromotive force to balance against the drop across a resistance due to a steady current, thus establishing the latter in absolute terms from the calculation of the mutual inductance between the disk and the exciting air-cored coil.

Faraday effect *(Phys.)*. Rotation of the plane of polarization of plane-polarized light when passed through an isotropic transparent medium placed in a strong magnetic field, the light being passed in a direction in which the field has a component. If l is the length of path traversed, H the strength of the field in the direction of propagation and θ is the angle of rotation, then $\theta = ClH$ where C is Verdet's constant. Effect is also exhibited by a plane-polarized microwave passing through a ferrite.

Faraday shield *(Elec.Eng.)*. See **Faraday cage**.

Faraday's ice-pail experiment *(Phys.)*. Classical experiment which consists in lowering a charged body into a metal pail connected to an electroscope, in order to show that charges reside only on the outside surface of conductors.

Faraday's law of induction *(Phys.)*. The e.m.f. induced in any circuit is proportional to the rate of change of magnetic flux linked with the circuit. Principle used in every practical electrical machine. *Maxwell's field equations* involve a more general mathematical statement of this law.

Faraday's laws of electrolysis *(Phys.)*. (1) The amount of chemical change produced by a current is proportional to the quantity of electricity passed. (2) The amounts of different substances liberated or deposited by a given quantity of electricity are proportional to the chemical equivalent weights of those substances.

faradic currents *(Med.)*. Induced currents obtained from secondary winding of an induction coil and formerly used for curative purposes.

faradmeter *(Elec.Eng.)*. Generic name for direct-reading capacitance meters. Typically, they use the mains voltage in series with an a.c. milliammeter.

farcy *(Med.,Vet.)*. See **glanders**.

far-end cross-talk *(Telecomm.)*. Cross-talk heard by a listener, and caused by a speaker at the distant end of the parallelism.

far field *(Acous.)*. Sound field a long distance from the source. Every sound source has a *far field* and a *near field*. E.g. a monopole source has a far field which decays as $1/r$ (r is the distance from the source) and a near field which decays as $1/r^2$ so that the far field dominates at a large distance.

farina *(Genrl.)*. Generally, ground corn, meal, starch etc.

farinose *(Bot.)*. Covered with a mealy powder.

-farious *(Genrl.)*. A suffix meaning *arranged in so many rows*.

farmer's lung *(Immun.)*. Respiratory disease due to hypersensitivity to spores of a thermophilic bacterium, *Micropolyspora faeni*, present in the dust of mouldy hay. Mainly occurs among farmers in damp areas. Characterized by attacks of breathlessness coming on some hours after inhaling the dust. A combination of Type III and Type IV hypersensitivity reactions are probably involved. After several seasons can result in pulmonary fibrosis.

Farmer's reducer *(Image Tech.)*. A reducing bath for photographic images made by the addition of potassium ferricyanide to hypo.

far point *(Phys.)*. Object point conjugate to the retina when accommodation is completely relaxed; at infinity in emmetropia, between infinity and the eye in myopia, and behind the eye in hyperopia.

far-red light *(Bot.)*. In the context of plant responses mediated by *phytochrome*, light of wavelength around 730 nm.

Farror's process *(Eng.)*. Case-hardening by mixture of ammonium chloride, manganese dioxide and potassium ferrocyanide.

fascia *(Arch.,Build.)*. (1) A wide flat member in an entablature. (2) A board embellishing a gutter around a building. (3) The broad flat surface over a shop front or below a cornice.

fascia *(Autos.)*. The instrument board of an automobile.

fascia *(Zool.)*. Any bandlike structure; especially the connective-tissue bands which separate muscles, nerves, etc. adj. *fascial*.

fasciation *(Bot.)*. An abnormal condition, resulting from damage, infection or mutation, in which a shoot (or other organ) grows broad and flattened, resembling several shoots fused laterally.

fascicle *(Bot.)*. A vascular *bundle*.

fascicular cambium *(Bot.)*. The flat strand of cambium between xylem and phloem in a vascular bundle.

fasciculus *(Zool.)*. A small bundle, as of muscle or nerve fibres.

fasciitis, fascitis *(Med.)*. Inflammation of fascia.

fascine *(Civ.Eng.)*. A bundle of brushwood used to help make a foundation on marshy ground, or to make a wall to protect a shore against erosion by sea or river, or to accumulate sand and silt on the bed of an estuary.

fascine buildings *(Build.)*. A building constructed with logs and boards.

fasciola *(Zool.)*. A narrow band of colour; a delicate lamina in the Vertebrate brain.

Fasciola hepatica *(Zool.)*. (Trematoda, Malacocotylea). The *liver fluke*. The adult lives in the liver of sheep, and the secondary host, entered by the *miracidium* larva, and in which the *redia* and *cercaria* larvae develop, is a water snail, usually of the genus *Limnaea* (Gastropoda, Pulmonata).

fascioliasis *(Med.,Vet.)*. Infection of Man and other animals with the liver fluke *Fasciola hepatica*.

fasciotomy *(Med.)*. Surgical incision of fascia.

fassaite *(Min.)*. A monoclinic pyroxene rich in aluminium, calcium, and magnesium, and poor in sodium; found in metamorphosed limestones and dolomites.

fast *(Acous.)*. Measuring mode of a **sound-level meter** with a time constant of 0.125 s.

fast *(Build.)*. Said of colours which are not affected by the conditions of use (i.e. light, heat, chemical action, damp etc.) to which they are subjected. n. *fastness*.

fast *(Image Tech.)*. Contributing to reduction of time of exposure; said of an emulsion or lens.

fast *(Min.Ext.)*. (1) A heading or working place which is driven in the solid coal, in advance of the open places, said to be in the *fast*. (2) A hole in coal which has had insufficient explosive used in it, or which has required undercutting. (3) In shaft sinking, a hard stratum under poorly consolidated ground, on which a **wedging crib** can be laid.

fast-acting relay *(Telecomm.)*. A relay designed to act with minimum delay after the application of voltage, usually by increasing the resistance of the circuit in comparison with the inductance, and by minimizing moving masses.

fast coupling *(Eng.)*. A coupling which permanently connects two shafts.

fast effect *(Phys.)*. See **fast fission**.

fastener *(Eng.)*. An article designed to fasten together two or more other articles, usually in the form of a shaft passing through the articles to be fastened, e.g. nails, screws, rivets, pins.

fast fission *(Phys.)*. ^{238}U has a fission threshold for neutrons of energy about 1 MeV, and the fission cross-section increases rapidly with energy. Fission of this isotope by fast neutrons may cause a substantial increase in the reactivity of a thermal reactor (*fast effect*).

fast fission factor *(Nuc.Eng.)*. Ratio of the total number of fast neutrons produced by fissions due to neutrons of all energies (fast and thermal) to the number resulting from thermal/neutron fissions. Symbol ε.

fast Fourier transform *(Acous.)*. See **real time analyser**.

fast head *(Eng.)*. The fixed headstock of a lathe.

fastigiate *(Bot.)*. Having the branches more or less erect and parallel, e.g. Lombardy poplar.

fastigium *(Med.)*. The highest point of temperature in a fever.

fast-needle surveying *(Surv.)*. See **fixed-needle surveying**.

fastness *(Textiles)*. The ability of a colour to remain unchanged when exposed to a specified agency including light, rubbing and washing.

fast neutrons *(Phys.)*. See **neutron**.

fast pulley *(Eng.)*. A pulley fixed to a shaft by a key or set bolt, as distinct from a **loose pulley** which can revolve freely on the shaft.

fast reaction *(Phys.)*. Nuclear reaction involving strong interaction and occurring in a time of the order of 10^{-23} seconds. Due to **strong interaction** forces.

fast reactor *(Nuc.Eng.)*. One without a moderator in which chain reaction is maintained almost entirely by fast fission. It may also be a **breeder reactor**.

fast sheet *(Build.)*. See **stand sheet**.

fast store *(Comp.)*. Computer memory with a very fast **access time**. See **main memory**.

fast-time constant circuits *(Telecomm.)*. Those for which the circuit parameters (particularly resistance and capacitance) permit a very rapid response to a step signal.

fast to light *(Build.)*. Paints or pigments so described will not fade on exposure to light.

fat *(Build.)*. Part of a cement mortar mix containing a higher proportion of cement than the rest. This comes to the surface before the mixture has set.

fat *(Chem.)*. See **fats**.

fat *(Zool.)*. See **adipose tissue**.

fata morgana *(Meteor.)*. A complicated mirage caused by the existence of several layers of varying refractive index, resulting in multiple images, possibly elongated. Especially characteristic of the Strait of Messina and Arctic regions.

fat board *(Build.)*. A board on which the brick-layer collects the fat during the process of pointing.

fat-body *(Zool.)*. In Insects, a mesodermal tissue of fatty appearance, the cells of which contain reserves of fat and other materials and play an important part in the metabolism of the animal; in Amphibians, highly vascular masses of fatty tissue associated with the gonads.

fat coals *(Min.Ext.)*. Coals which contain plenty of volatile matter (gas-forming constituents).

fat edges *(Build.)*. A defect in paintwork, characterized by the formation of ripples at edges and in angles; due to excess of paint.

fat face *(Print.)*. Heavy types with hairline serifs and main strokes at least half as wide as the height of the letters, *Ultra Bodoni* being a well- known example.

father file *(Comp.)*. See **grandfather, father, son files**.

father of the chapel *(Print.)*. A person elected by the associated employees of a printing department to represent them and to watch their interests.

fathom *(For.)*. A timber measure $= 216\,\text{ft}^3$, $= 6 \times 6 \times 6\,\text{ft}$.

fathom *(Genrl.)*. A unit of measurement. Generally, a nautical measurement of depth = 6 ft.

fathom *(Min.Ext.)*. In general mining, the volume of a 6 ft cube; in gold mining, often a volume 6 ft by 6 ft by the thickness of the reef; in lead mining, sometimes a volume 6 ft by 6 ft by 2 ft. It is the unit of performance of a rock drill - 'fathoms per shift'.

fathometer *(Acous.)*. An ultrasonic depth-finding device.

fatigue *(Zool.)*. The condition of an excitable cell or tissue which, as a result of activity, is less ready to respond to further stimulation until it has had time to recover.

fatigue limit *(Eng.)*. The upper limit of the range of stress that a metal can withstand indefinitely. If this limit is exceeded, failure will eventually occur.

fatigue of metals *(Eng.)*. The phenomenon of the failure of metals under the repeated application of a cycle of stress. Factors involved include amplitude, average severity, rate of cyclic stress and temperature effect. **Notch brittleness** commences at a scratch or blemish.

fatigue test *(Eng.)*. A test made on a material to determine the range of alternating stress to which it may be subjected without risk of ultimate failure. By subjecting a series of specimens to different ranges of stress, while the mean stress is constant, a stress-number curve is obtained.

fatigue-testing machine *(Eng.)*. A machine for subjecting a test piece to rapidly alternating or fluctuating stress, in order to determine its fatigue limit. See **Haigh fatigue-testing machine**.

fat lime *(Build.)*. **Lime** made by burning a pure, or very nearly pure, limestone, such as chalk.

fat matter *(Print.)*. A composing-room term for easily set portions of the work in hand.

fat-necrosis *(Med.)*. The splitting of fat, due to the escape of a fat-splitting enzyme from the pancreas into the abdominal cavity, with death of the fat-containing cells so affected.

fats *(Chem.)*. An important group of naturally occurring substances consisting of the glycerides of higher fatty acids, e.g. palmitic acid, stearic acid, oleic acid. Essential componenet of the human diet, repairing the wastage of human fat, and, by breakdown and oxidation, providing some of the required energy.

fat splitting *(Chem.)*. Term used to describe the hydrolysis of animal and vegetable fats into glycerol and fatty acids. Can be effected in a number of ways but chiefly by using strong alkalis (as in soapmaking) or inorganic acids.

fattening *(Build.)*. Thickening of varnish in the can, especially the appearance of gelatinous bits. Also *curdling, livering*.

fatty acids *(Chem.)*. A term for the whole group of saturated and unsaturated monobasic aliphatic carboxylic acids. The lower members of the series are liquids of pungent odour and corrosive action, soluble in water; the intermediate members are oily liquids of unpleasant smell, slightly soluble in water. The higher members from C_{10} upwards, are mainly solids, insoluble in water, but soluble in ethanol and in ethoxyethane.

fatty acids *(Med.)*. Monobasic organic acids with the general formula R.COOH. Saturated fatty acids with no double bonds are linked with the development of **atheroma**. In contrast the polyunsaturated fatty acids, linoleic, linolenic and arachadonic are termed essential fatty acids as they must be included in the diet and may have a preventitive role against atheroma and are required for the synthesis of prostaglandins.

fatty degeneration *(Med.)*. Degeneration of the cell substance, accompanied by the appearance in it of droplets of fat, due to the action of poisons or to lack of oxygen.

faucet *(Build.)*. (1) A small tap or cock. (2) The enlarged or socket end of a pipe at a *spigot-and-socket joint*.

faucet ear *(Build.)*. See **ear**.

faujasite *(Min.)*. One of the less common zeolites; a hydrated silicate of sodium, calcium, and aluminium. It exhibits a wider range of molecular absorption than any other zeolite.

fault *(Geol.,Min.Ext.)*. A fracture in rocks along which some displacement (the *throw* of the fault) has taken place. The displacement may vary from a few millimetres to thousands of metres. Movement along faults is the common cause of earthquakes.

fault breccia *(Geol.)*. A fragmental rock of breccia type resulting from shattering during the development of a fault.

fault current *(Elec.Eng.)*. That caused by defects in electrical circuit or device, such as short-circuit in system. The peak value of current is the accepted measure.

fault-finding *(Elec.Eng.)*. General description of locating and diagnosing faults, according to a pre-arranged schedule, generally arranged in a chart or table, with or without special instruments. U.S. *trouble-shooting*.

fault rate *(Elec.Eng.)*. See **reliability**.

fault resistance *(Elec.Eng.)*. A term sometimes used to denote insulation resistance, but more commonly the resistance of an actual fault, e.g. an arc between a conductor and earth.

fault tolerance *(Comp.)*. The ability of a system to execute specific tasks correctly regardless of failures and errors. See **redundancy**.

fault tolerant *(Elec.Eng.)*. Method of design and construction of an electrical circuit to make it highly reliable. The circuit configuration is arranged in such a way that strategic-components failure does not mean circuit operation is completely lost.

fauna *(Zool.)*. A collective term denoting the animals occurring in a particular region or period. pl. *faunas* or *faunae*. adj. *faunal*.

faunal region *(Zool.)*. An area of the earth's surface characterized by the presence of certain species of animals.

faveolate, favose *(Bot.,Zool.)*. Resembling a honeycomb in appearance. *Favous*, said of a surface pitted like a honeycomb. *Favus*, a hexagonal pit or plate.

favism *(Med.)*. Haemolytic red cell destruction in people of Mediterranean origin with glucose-6-phosphate dehydrogenase deficiency after ingesting the broad bean, *Vicia faba*.

favus *(Med.)*. A contagious skin disease, especially of the scalp, due to infection with the fungus *Achorion schönleinii*.

fax *(Telecomm.)*. Abbrev. for *facsimile*.

fayalite *(Min.)*. A silicate of iron, Fe_2SiO_4, crystallizing in the orthorhombic system; a common constituent of slags but occurring also in igneous rocks, chiefly of acid composition, including pitchstone, obsidian, quartz-porphyry, rhyolite, and also in ferrogabbro.

faying face, surface *(Eng.)*. That part of a surface of wood or metal specially prepared to fit an adjoining part.

FBR *(Nuc.Eng.)*. Fast breeder reactor. See **breeder reactor, fast reactor**.

FC *(Build.)*. Abbrev. for *Fair Cutting*.

FCC *(Chem.)*. Abbrev. for *Face-Centred Cubic*.

F-centre *(Phys.)*. An electron trapped at a negative ion vacancy in an ionic crystal. *F*-centres can be formed by the release of electrons by irradiation with X-rays or by producing stoichiometric excess of anions in the crystal. F^1-centres consist of *F*-centres with a further electron trapped in the same vacancy. F_A-centres are *F*-centres modified by one neighbouring cation being different from the cations of the lattice. *F*-centre aggregates can be formed by arrays of nearest neighbour centres. *F*-centres give rise to broad optical absorption bands.

Fc fragment *(Immun.)*. Fragment of immunoglobulin obtained by papain hydrolysis representing the C-terminal halves of the two heavy chains linked by disulphide bonds. It has no antibody activity but contains the sites involved in complement activation and placental transmission, and some of the Gm allotype markers.

F-class insulation *(Elec.Eng.)*. A class of insulating material to which is assigned a temperature of 155°C. See **class-A, -B, -C**, etc., **insulating materials**.

Fc receptor *(Immun.)*. Receptor present on the plasma membranes of cells which bind the Fc fragment of immunoglobulin. Neutrophils, mononuclear phagocytes, eosinophils, B-lymphocytes have receptors for Fc of IgG. These may differ for different IgG subclasses. Mast cells, basophil leucocytes and eosinophils have receptors for Fc of IgE.

fd *(Build.)*. Abbrev. for *framed*.

F-diagram *(Chem.Eng.)*. The cumulative residence time distribution in a continuous flow system, plotted on dimensionless co-ordinates. Important in assessing the performance of chemical reactors, kilns etc.

F-display *(Radar)*. Type of radar display, used with directional antenna, in which the target appears as a bright spot which is off-centre when the aim is incorrect.

FDS law *(Phys.)*. Fermi-Dirac-Sommerfeld law, which gives the algebraic number of a quantized system of particles which have velocities within a small range.

Fe *(Chem.)*. The symbol for *iron*.

fear *(Behav.)*. In animal behaviour, a motivational state aroused by specific stimuli and which normally gives rise to avoidance, defensive or escape behaviour.

feather *(Build.)*. A thin strip of wood fitted into a groove on each edge of adjacent butted boards.

feather *(Eng.)*. (1) A rectangular key sunk into a shaft to permit a wheel to slide axially, while preventing relative rotation. (2) Iron slips for reducing the friction between a wedge and an object to be split.

Feather analysis *(Phys.)*. An approximate method of determining the range of β-rays forming part of a combined β-γ spectrum, by comparison of the absorption curve with that for a pure β-emitter.

feather eating *(Vet.)*. A habit, acquired by birds, characterized by pecking, plucking, or eating their own plumage or that of other birds. The habit may develop into **cannibalism**.

feather-edge brick *(Build.)*. A brick similar to a compass brick, used especially for arches.

feather-edged coping *(Build.)*. A coping-stone sloping in one direction on its top surface. Also known as *splayed coping*.

feathering *(Paper)*. The irregular edge of an ink line due to the paper being insufficiently sized.

feathering hinge *(Aero.)*. A pivot for a rotorcraft blade which allows the angle of incidence to change during rotation.

feathering paddles *(Eng.)*. Paddle-wheels so controlled that the floats enter and leave the water at right angles to the surface.

feathering pitch *(Aero.)*. The blade angle of a propeller giving minimum drag when the engine is stopped.

feathering propeller *(Aero.)*. See **propeller**.

feathering pump *(Aero.)*. A pump for supplying the necessary hydraulic pressure to turn the blades of a *feathering propeller* to and from the feather position.

feather joint *(Build.)*. See **ploughed-and-tongued joint**.

feather ore *(Min.)*. A plumose or acicular form of the sulphide of lead and antimony, Also *jamesonite*. The name is also used for *stibnite*.

feathers *(Zool.)*. Epidermal out-growths forming the body covering of Birds; distinguished from scales and hair, to which they are closely allied, by their complex structure, and by the possession of a vascular core which at first projects from the surface.

feather tongue *(Build.)*. A wooden tongue for a **ploughed-and-tongued joint**, so that the grain is across the grooves.

febrifuge *(Med.)*. Against fever; a remedy which reduces fever.

febrile *(Med.)*. Pertaining to, produced by, or affected with fever.

Fechner colours *(Phys.)*. The visual sensations of colour which are induced by intermittent achromatic stimuli.

Fechner law *(Med.)*. See **Weber-Fechner law**.

fecundity *(Ecol.)*. The number of young produced by a species or individual.

feebly hydraulic lime *(Build.)*. **Lime** made by burning a limestone containing 5–12% clay.

feed *(Eng.)*. (1) The rate at which the cutting tool is advanced. (2) Fluid pumped into a vessel, e.g. feed-water to a boiler. (3) Mechanism for adavancing material or components into a machine for processing.

feed *(Min.Ext.)*. Forward motion of drill or cutter.

feed *(Telecomm.)*. To offer a programme or signal at some point in a communication network.

feedback *(Acous.)*. Phenomenon in which part of an output signal is fed back into the input of the system. If the feedback signal is in phase with the primary input signal, the system can become unstable (positive feedback). This often occurs in electroacoustic systems in which microphone and loudspeaker are in the same room. Negative feedback occurs if the feedback signal decreases the input signal.

feedback *(Telecomm.)*. Transfer of some output energy of an amplifier to its input, so as to modify its characteristics. *Current* (or *voltage*) *feedback* is when feedback signal depends on current (or voltage) in output respectively. See **negative feedback, positive feedback**.

feedback admittance *(Telecomm.)*. The short-circuit transfer admittance from the output to the input terminals of any circuit, filter, electronic device or combination of all three.

feedback characteristic *(Electronics)*. See **transistor characteristics**.

feedback circuit *(Telecomm.)*. See **feedback path**.

feedback control loop *(Telecomm.)*. A closed transmission path including an active transducer, forward and feedback paths and one or more mixing points. The system is such that a given relation is maintained between the input and output signals of the loop.

feedback control system *(Telecomm.)*. US for **closed loop system**.

feedback factor *(Telecomm.)*. Defined $m = 1 - \beta A$; m is the feedback factor, β is the **feedback ratio** and A is the open-loop gain of the amplifier.

feedback inhibition *(Biol.)*. The ability of a later product of a chain of biochemical reactions to lower the synthesis of an earlier metabolite, e.g. by binding with its **promoter**.

feedback path *(Telecomm.)*. That from the loop output signal to the feedback signal in a feedback control loop.

feedback ratio *(Telecomm.)*. A property of the *feedback path* which determines the amount of feedback applied to the input of an amplifier. Defined by $\beta = e_f/e_o$ where β is the feedback ratio, e_o is the signal voltage at the amplifier output and e_f is the signal fed back to the input via the *feedback path*.

feedback signal *(Telecomm.)*. That which is responsive in an automatic controller to the value of the controlled variable.

feedback transducer *(Telecomm.)*. One which generates a signal, generally electrical, depending on quantity to be controlled, e.g. for rotation potentiometer, synchro or tacho, giving proportional derivative or integral signals respectively.

feedback windings *(Elec.Eng.)*. Those control windings in a saturable reactor to which are made the feedback connections.

feed-check valve *(Eng.)*. Nonreturn valve in the delivery pipe between feed-water pump and boiler.

feeder *(Elec.Eng.)*. (1) An overhead or underground cable, of large current-carrying capacity, used in the transmission of electric power; it serves to interconnect generating stations, sub-stations, and feeding points, without intermediate connections. (2) In electrical circuits, the lines running from the main switchboard to the branch panels in an installation.

feeder *(Foundry)*. The runner or riser hole of a mould, containing sufficient molten metal to feed the casting and so compensate for contraction of the solidifying metal.

feeder *(Hyd.Eng.)*. A natural or artificial channel supplying water to a reservoir or canal.

feeder *(Min.Ext.)*. A mechanical appliance for supplying broken rock or crushed ore, at a predetermined rate, to some form of crusher or concentrator.

feeder *(Telecomm.)*. Conductor, or system of conductors, connecting the radiating portion of an antenna to the transmitter or receiver. It may be a balanced pair, a quad, coaxial or waveguide.

feeder bus-bars *(Elec.Eng.)*. In a generating station or main substation, bus-bars to which the outgoing feeders are connected.

feeder ear *(Elec.Eng.)*. A type of ear for attaching an overhead contact wire of a tramway system to the supporting wire; it serves also to lead current to the contact wire.

feeder head *(Eng.)*. See **hot top**.

feeder mains *(Elec.Eng.)*. See **feeder** (1).

feeder panel *(Elec.Eng.)*. A switchboard panel on which are mounted the switchgear and instruments for controlling one or more feeders.

feeder pillar *(Elec.Eng.)*. A pillar containing switches, links, and fuses, for connecting the feeders of an electric power distributing system with the distributors.

feed finger *(Eng.)*. A rod used in association with a collet to push or pull the bar of material forward in a capstan lathe or automatic until the bar touches a stop, thus governing the feed length.

feeding *(Build.)*. The thickening of paint due to a reaction in its constituents. This defect often occurs when non compatible materials are used when mixing or stirring paints. Paints affected by feeding lose opacity and attempts to reduce to a working consistency using solvents often result in further thickening.

feeding head *(Foundry)*. Same as *feeder*.

feeding point *(Elec.Eng.)*. The junction point between a feeder and a distribution system. Also called a *distributing point*.

feeding rod *(Foundry)*. A heated iron rod inserted in the feeder of a mould, and worked with a pumping motion to assist feeding during the cooling of the molten metal.

feeding-up *(Build.)*. A thickening of paints (and sometimes also said of varnishes) in the can, sufficient to make them unsatisfactory for use.

feed mechanism *(Elec.Eng.)*. The mechanism which causes the carbons of an arc lamp to move gradually towards the arc at the speed necessary to compensate for the rate at which they burn away, thereby keeping the arc length constant.

feed pipe *(Eng.)*. The pipe carrying feed-water from the feed pump to a boiler.

feed reel *(Image Tech.)*. The reel of film which is being unwound as the film is taken off to pass through the gate in a camera, printer, or projector.

feed rollers *(Print.)*. Driven rollers which convey a web into a printing couple. Cf. **idler** or **idling roller**.

feed screw *(Eng.)*. A screw used for supplying motion to the feed mechanism of a machine tool.

feedthrough *(Elec.Eng.)*. A conductor used to connect patterns on opposite sides of the board of a printed circuit, or an insulated conductor for connection between two sides of a metal earthing screen.

feed-water *(Eng.)*. The water, previously treated to remove air and impurities, which is supplied to a boiler for evaporation.

feed-water heater *(Eng.)*. An arrangement for heating boiler feed-water by means of steam which has done work in an engine or turbine. It is similar in principle to a steam condenser of either the surface or the jet type.

feel *(Textiles)*. Term describing the physical character of a cloth when handled.

feeler *(Textiles)*. Mechanical or electrical device, used with automatic weft replenishing motion on a loom to determine when the pirn change is necessary.

feeler gauge *(Eng.)*. A thin strip of metal of known and accurate thickness, usually one of a set, used to measure the distance between surfaces or temporarily placed between working parts while setting them an accurate distance apart.

feeler switch *(Elec.Eng.)*. A switch sometimes forming part of the equipment of an auto-reclose circuit-breaker; after the circuit-breaker has opened on a fault, the feeler switch closes a test circuit to determine whether or not the fault has cleared itself, and, if not, it prevents the circuit-breaker from reclosing.

feet *(Print.)*. The underside of the type, grooved in **foundry type**, but not in **Monotype**. When type is badly locked up, it is said to be *off its feet*.

feet-switch *(Elec.Eng.)*. See **tropical switch**.

Fehling's solution *(Chem.)*. A solution of cupric sulphate and potassium sodium tartrate (Rochelle salt) in alkali, used as an oxidizing agent. It is an important analytical reagent for aldehydes, glucose, fructose etc., which reduce it to cuprous oxide.

feint *(Print.)*. The pale, edge to edge, horizontal ruling in account books and notebooks.

feldspar, felspar *(Min.)*. A most important group of rock-forming silicates of aluminium, together with sodium, potassium, calcium, or (rarely) barium, crystallizing in closely similar forms in the monoclinic and triclinic systems. The chief members are *orthoclase* and *microcline* (potassium feldspar); *albite* (sodium feldspar); and the *plagioclases* (sodium-calcium feldspar). The form *felspar*, though still commonly used, perpetuates a false derivation from the German *fels* (rock); actually it is from the Swedish *feldt* (field).

feldspathic sandstone *(Geol.)*. See **arkose**.

feldspathoids *(Min.)*. A group of rock-forming minerals chemically related to the feldspars, but undersaturated with regard to silica content, and therefore incapable of free existence in the presence of magmatic silica. The chief members of the group are **haüyne, leucite, nepheline, nosean**, and **sodalite**.

Felici balance *(Elec.Eng.)*. An a.c. electrical measuring bridge for determining mutual inductance between windings.

Felici generator *(Elec.Eng.)*. A modern form of electrostatic high-voltage generator developed in France and comparable with a Van de Graaff generator.

feline distemper *(Vet.)*. See **feline enteritis**.

feline enteritis *(Vet.)*. *Panleucopenia, feline distemper*. Highly contagious and often fatal parvovirus disease of cats characterized by vomiting, dehydration, abdominal pain, anorexia and death. Abortion and foetal resorption can occur in queens. Kittens can be born with *cerebellar ataxia*. Vaccination widely used.

feline infectious anemia *(Vet.)*. Caused by *Haemobartonella felis*. Symptons include inappetance, weakness, anemia, jaundice, wasting and splenomegaly. Parasite is associated with red blood cells.

feline influenza *(Vet.)*. *Cat flu*. Complex of conditions with several pathogens involved and numerous secondary invaders. The two most commonly encountered agents are *Feline rhinotracheitis virus (FVR)* and *Feline calcivirus (FCV)*. Vaccines available.

feline leukemia *(Vet.)*. Characterized by fever, weakness, inappetance, wasting and anemia. Caused by leukemia virus. Mainly abdominal and thymic forms. Experimental vaccines have been used.

feline pneumonitis *(Vet.)*. Caused by a cat-adapted *Chlamydia psittaci*. The symptoms are mucopurulent conjunctivitis, sneezing and a mild nasal discharge. No vaccine is yet available.

feline viral rhinotracheitis *(Vet.)*. An acute viral infection of the upper respiratory tract of the cat. See **feline influenza**. Vaccines avaialable.

fell *(Textiles)*. The edge of the cloth in a loom, where the **picks** of weft are beaten up by the reed.

fellmongering *(Textiles)*. Obtaining wool from the skins of slaughtered sheep.

felloe *(Build.)*. (1) The outer part of the framing for a centre. (2) A segment of the rim of a wooden wheel, about which a tyre is usually shrunk. The term is sometimes applied to the whole rim.

felon, fellon *(Vet.)*. Suppurative arthritis of cattle; commonly associated with mastitis.

felsite *(Geol.)*. An 'omnibus term' for fine-grained igneous rocks of acid composition, occurring as lavas or minor intrusions, and characterized by felsitic texture; a fine patchy mosaic of quartz and feldspar, resulting from the devitrification of an originally glassy matrix.

felspar *(Min.)*. See **feldspar**.

felt *(Build.)*. A fibrous material, treated so as to be rendered watertight, used as underlining for roofs; also as an overlining for roofs when underlaid with asphalt or an asphalt compound.

felt *(Paper)*. A woven blanket or a synthetic fabric in the form of an endless band to give support to the web at various points on the paper machine e.g. at a wet press, MG dryer or drying cylinders.

felt *(Textiles)*. (1) A densely-matted non-woven fabric containing wool or hair that has passed through a felting process of heat, steam, and pressure. (2) Heavily-milled woven fabric with a matted fibrous surface.

felting *(Paper)*. The natural action by which fibres adhere in paper-making.

felting *(Textiles)*. The formation of a felt during processing (e.g. by **milling**) or during wear; a property especially of woollen fabrics resulting from the structure of the **cuticle** of the fibres.

female *(Bot.)*. (1) The larger and less motile gamete, the egg. (2) Gametophytes, and their reproductive structures which produce eggs but not male gametes. (3) Sporophytes and their reproductive structures which produce the megaspores and hence seeds. (4) Individual seed plants or flowers which have functional carpels but not functional stamens. Cf. **male, hermaphrodite**.

female *(Eng.)*. See **male and female**.

female *(Zool.)*. An individual whose gonads produce ova.

female gauge *(Eng.)*. See **ring gauge**.

female pronucleus *(Biol.)*. The nucleus remaining in the ovum after maturation.

female thread *(Eng.)*. See **internal screw-thread**.

femerell *(Build.)*. A roof lantern having louvres for ventilation.

femic constituents *(Geol.)*. Those minerals which are contrasted with the salic constituents in determining the systematic position of a rock in the *C.I.P.W.* scheme of classification. Note that these are the *calculated* components of the 'norm'; the corresponding *actual* minerals in the 'mode' are said to be *mafic*, i.e. rich in magnesium and iron.

femur *(Arch.)*. See **meros**.

femur *(Zool.)*. The proximal region of the hind limb in land Vertebrates: the bone supporting that region: the third joint of the leg in Insects, Myriapoda, and some Arachnida. adj. *femoral*.

fen *(Ecol.)*. Vegetation developed naturally on waterlogged land, forming a peat that is neutral or alkaline from the dead parts of tall grasses, sedges and herbs. See **carr**. Cf. **bog**.

fence *(Build.)*. (1) An adjustable grinding edge or plate directing or limiting the movement of one piece with respect to another. (2) An attachment to a plane (cf. **fillister**) which keeps it at a fixed distance from the edge of the work, used especially on rebate and grooving planes.

fence *(Eng.)*. (1) A guard or stop to limit motion. (2) A guide for material, as in a circular saw or planing machine.

fenchel wet expansion tester *(Paper)*. Test apparatus in which a strip of paper is held vertically between two clamps under low tension and immersed in water. The resultant increase in length (wet expansion) is indicated by a needle connected to one of the clamps.

fenchone *(Chem.)*. A dicyclic ketone. Optical isomers occur in fennel and lavender oils (dextrorotatory form), and in thuja oil (laevorotatory). Crystalline solid; mp 5°C and bp 192°C.

fender *(Elec.Eng.)*. A metal cover attached to the end of the frame of an electric machine in such a way as to prevent accidental contact with live or moving parts. It does not carry a bearing. Also called a *protection cap*.

fender wall *(Build.)*. A dwarf brick wall supporting the hearth of a ground-floor fireplace.

fenestra *(Arch.)*. A window or other opening in the outer walls of a building. pl. *fenestrae*.

fenestra *(Zool.)*. An aperture in a bone or cartilage, or an opening between two or more bones.

fenestra metotica *(Zool.)*. In a typical chondrocranium, an opening behind the auditory capsule, through which pass the ninth and tenth cranial nerves and the internal jugular vein.

fenestra ovalis *(Zool.)*. In Vertebrates in which the middle-ear is developed, the upper of two openings in the cochlea. Cf. **fenestra rotunda**.

fenestra pro-otica *(Zool.)*. In a typical chondrocranium, an opening in front of the auditory capsule, through which pass the fifth, sixth, and seventh cranial nerves.

fenestra rotunda *(Zool.)*. In Vertebrates in which the middle-ear is developed, the lower of two openings in the cochlea. Cf. **fenestra ovalis**.

fenestrate, fenestrated *(Bot.,Zool.)*. With window-like perforations; perforated or having translucent spots.

fenestration *(Build.)*. (1) The arrangement of window and other openings in the outer walls of a building. (2) The controlling of light emission into a room or building.

fenestration *(Med.)*. A surgical operation to improve hearing which involves a new 'window' being opened to the inner ear.

fenestra tympani *(Zool.)*. See **fenestra rotunda**.

fenestra vestibuli *(Zool.)*. See **fenestra ovalis**.

fenoprop *(Chem.)*. *2-(2,4,5-trichlorophenoxy)propanoic acid*, used as a weedkiller. Also *silvex, 2,4,5-TP*.

fent *(Textiles)*. A damaged piece of cloth cut from a length, or a short piece of material; usually sold by weight to wholesalers and market traders.

fentanyl *(Med.)*. A powerful analgesic resembling morphine in its action.

FEP *(Plastics)*. *Fluorinated Ethene Propene*. Plastic with many of the properties of *PTFE*, having very good chemical resistance; it is unaffected by moisture and has a wide temperature range of application from −260° to +200°C.

feral *(Zool.)*. Of a domesticated animal which has reverted to the wild.

ferberite *(Min.)*. Iron tungstate, the end member of the wolframite group of minerals, the series from $FeWO_4$ to $MnWO_4$.

fergusite *(Geol.)*. An alkaline **syenite** containing large crystals of pseudoleucite in a matrix of aegirine-augite, olivine, apatite, sanidine, and iron oxides.

fergusonite *(Min.)*. A rare mineral occurring in pegmatites; it is a niobate and tantalate of yttrium, with small amounts of other elements.

Fermat's last theorem *(Maths.)*. No integral values of x, y and z can be found to satisfy the equation $x^n + y^n = z^n$ if n is an integer greater than 2. This theorem is still unproved.

Fermat's principle of least time *(Phys.)*. Principle stating that the path of a ray, e.g., of light, from one point to another (including refractions and reflections) will be that taking the least time.

Fermat's spiral *(Maths.)*. See **parabolic spiral**.

fermentation *(Bot.)*. *Anaerobic respiration* of, especiallly, sugars to ethanol, lactic acid, butyric acid etc.

fermentation *(Chem.)*. A slow decomposition process of organic substances induced by micro-organisms, or by complex nitrogenous organic substances (**enzymes**) of plant or animal origin; usually accompanied by evolution of heat and gas. Important fermentation processes are the alcoholic fermentation of sugar and starch, and the lactic fermentation.

fermi *(Phys.)*. Femtometre (fm). A very small length unit, i.e., 10^{-15} m. Used in nuclear physics, being of the order of the radius of the proton, 1.2 fm.

Fermi age *(Nuc.Eng.)*. Slowing down area for neutrons calculated from *Fermi age theory* which assumes that neutrons, on being slowed down, lose energy continuously in an infinite homogeneous medium. Has the dimensions of length^2. Also called *neutron age*. See **age theory**.

Fermi characteristic energy level *(Electronics)*. See **Fermi level**.

Fermi constant *(Phys.)*. A universal constant which indicates the coupling between a nucleon and a lepton field. Its value is 1.4×10^{-50} J/m^3, and it is important in β-decay theory.

Fermi decay *(Electronics)*. Theory of ejection of electrons as β-particles.

Fermi-Dirac distribution curve *(Electronics)*. A function, ranging from unity to zero, specifying the probability that an electron in a semiconductor will occupy certain quantum states when thermal equilibrium exists. The energy level at which the value of the function is 0.5 is called the **Fermi level**.

Fermi-Dirac gas *(Phys.)*. An assembly of particles which obey Fermi-Dirac statistics and the Pauli exclusion principle. For an extremely dense Fermi gas, such as electrons in a metal, all energy levels up to a value E_F, the Fermi energy, are occupied at absolute zero.

Fermi-Dirac-Sommerfeld law *(Phys.)*. See **FDS law**.

Fermi-Dirac statistics *(Phys.)*. Statistical mechanics laws obeyed by system of particles whose wave function changes sign when two particles are interchanged, i.e. the Pauli exclusion principle applies.

Fermi level *(Electronics)*. The energy level at which there is a 0.5 probability of finding an electron; it depends on the distribution of energy levels and the number of electrons available. In semiconductors, the number of electrons is relatively small and the Fermi level is affected by donor and acceptor impurities.

fermion *(Phys.)*. A particle which obeys Fermi-Dirac statistics. Fermions have total spin angular momentum of $(n+\frac{1}{2})\hbar$ where $n = 0,1,2,..$ and $\hbar$ is the Dirac constant. Baryons and leptons are fermions and are subject to the Pauli exclusion Principle.

Fermi plot *(Phys.)*. See **Kurie plot**.

Fermi potential *(Phys.)*. The equivalence of the energy of the Fermi level as an electric potential.

Fermi selection rules *(Phys.)*. See **nuclear selection rules**.

Fermi surface *(Phys.)*. A constant energy surface in k-space which encloses all occupied electron states at absolute zero in a crystal.

Fermi temperature *(Phys.)*. The degeneracy temperature of a Fermi-Dirac gas which is defined by E_F/k, where E_F is the energy of the Fermi level and k is Boltzmann's constant. This temperature is of the order of tens of thousands of kelvins for the free electrons in a metal.

fermium *(Chem.)*. Manmade element. Symbol Fm, at.no. 100. Principal isotope ^{257}Fm, half-life 95 days.

ferns *(Bot.)*. Pteridophytes of the class *Filicopsida*.

ferractor *(Phys.)*. Magnetic amplifier with ferrite core.

Ferranti effect *(Elec.Eng.)*. The rise in voltage which takes place at the end of a long transmission line when the load is thrown off; it is due to the charging current flowing through the inductance of the line.

Ferranti-Hawkins protective system *(Elec.Eng.)*. A discriminative protective system for feeders; core balance transformers are placed at each end, with their secondary windings connected to each other through pilot wires.

Ferranti meter *(Elec.Eng.)*. A name often given to the mercury-motor type of supply meter invented by Ferranti.

ferredoxin *(Bot.)*. Non-haem iron-sulphur protein, component of the electron transport system in photosynthesis.

Ferrel cell *(Meteor.)*. A mid-latitude mean atmospheric circulation cell proposed by Ferrel in the 19th century in which air flows poleward and eastward near the surface and equatorward and westward at higher levels. This disagrees with reality. The term is now sometimes used to describe a mid-latitude circulation identifiable in mean meridional wind patterns.

ferric [iron (III)] chloride *(Chem.)*. $FeCl_3$. Brown solid. Deliquescent, soluble in ethanol. Uses: coagulant in sewage and industrial wastes, mordant, photo-engraving etching of copper, cholorination and condensation catalyst, disinfectant, pharmaceutical, analytic reagent.

ferric [iron (III)] oxide *(Chem.)*. Fe_2O_3. The common red oxide of iron. Used in metallurgy, pigments, polishing and theatrical rouge, gas purification, and as a catalyst.

ferric [iron (III)] sulphate *(Chem.)*. $Fe_2(SO_4)_3$. A yellowish-white powder which dissolves slowly in water. Uses: pigments, water purification, dyeing, disinfectant, medicine.

ferricyanide *(Chem.)*. The complex ion $Fe(CN)_6^{3-}$. Also *[hexacyano ferrate (III)]*.

ferri-, ferro-. *(Chem.)*. Denoting trivalent and divalent iron respectively.

ferri-, ferro- *(Genrl.)*. Prefixes from L. *ferrum*, iron.

ferrimagnetism *(Phys.)*. Phenomenon in some magnetically ordered materials in which there is incomplete cancellation of the *anti-ferromagnetically* arranged spins giving a net magnetic moment; observed in ferrites and similar materials.

ferrimolybdite *(Min.)*. Hydrated molybdate of iron. Most so-called molybdite is ferrimolybdite. It occurs as a yellowish alteration product of molybdenite.

ferrite *(Eng.)*. A ceramic iron oxide compound having ferromagnetic properties with a general formula MFe_2O_4,where M is generally a metal such as cobalt, nickel, zinc.

ferrite bead *(Telecomm.)*. Small element of ferrite material used particularly for threading on to wire of transmission line to increase the series inductance, but also once used for computer memory as *ferrite-bead memory*.

ferrite core *(Phys.)*. A magnetic core, usually in the form of a small toroid, made of ferrite material such as nickel ferrite, nickel-cobalt ferrite, manganese-magnesium ferrite, yttrium-iron garnet, etc. These materials have high resistance and make eddy-current losses very low at high frequencies.

ferrite-core memory *(Comp.)*. Memory widely used in **second generation computers**. It consisted of tiny rings of magnetic material through which several wires were threaded. Each core could be magnetized in either direction to store a **bit**.

ferrite-rod antenna *(Telecomm.)*. Small reception antenna, using ferrite rod to accept electro-magnetic energy, output being from an embracing coil. Also called *loopstick antenna*.

ferritin *(Biol.)*. Protein which functions as an iron store in the liver. As the central iron core is visible in the electron microscope ferritin can be used as a tag for the localisation of proteins in electron microscopy.

ferro- *(Genrl.)*. Prefix. See **ferri-**.

ferro-actinolite *(Min.)*. An end-member compositional variety in the monoclinic amphiboles; essentially $Ca_2Fe_5Si_8O_{22}(OH,F)_2$,but the name is applied to a member of the actinolite series with more than 80% of this molecule.

ferrochromium *(Eng.)*. A **master alloy** of iron and chromium (60–72% chromium) used in making additions of chromium to steel and cast-iron.

ferrocyanide *(Chem.)*. The complex ion $Fe(CN)_6{}^{3-}$. Also *[hexacyano ferrate (II)]*.

ferrodynamometer *(Elec.Eng.)*. Any **dynamometer** which incorporates ferromagnetic material to enhance the torque.

ferro-edenite *(Min.)*. An end-member compositional variety in the monoclinic amphiboles; a hydrous sodium, iron, calcium and aluminium silicate.

ferroelectric materials *(Phys.)*. Dielectric materials (usually ceramics) with domain structure, which exhibit spontaneous electric polarization. Analogous to ferromagnetic materials (see **ferromagnetism**). They have relative permittivities of up to 10^5, and show dielectric hysteresis. Rochelle salt was first to be discovered. Others include barium titanite, and potassium dihydrogen phosphate.

ferrogedrite *(Min.)*. See **gedrite**.

ferrohastingsite *(Min.)*. A compositional variety in the hornblende group of monoclinic amphiboles.

ferromagnetic amplifier *(Elec.Eng.)*. A paramagnetic amplifier which depends on the non-linearity in ferroresonance phenomena at high RF power levels.

ferromagnetic resonance *(Elec.Eng.)*. A special case of paramagnetic resonance, exhibited by ferromagnetic materials and often termed *ferroresonance*. Explained by simultaneous existence of two different pseudo-stable states for the magnetic material B-H curve, each associated with a different magnetisation current for the material. Oscillation between these two states leads to large currents in associated circuitry.

ferromagnetism *(Phys.)*. Phenomenon in some magnetically ordered materials in which there is a bulk magnetic moment and the magnetization is large. The electron spins of the atoms in microscopic regions, *domains*, are aligned. In the presence of an external magnetic field the domains oriented favourably with respect to the field grow at the expense of the others and the magnetization of the domains tends to align with the field. Above the *Curie temperature*, the thermal motion is sufficient to offset the aligning force and the material becomes *paramagnetic*. Certain elements (iron, nickel, cobalt), and alloys with other elements (titanium, aluminium) exhibit permeabilities up to 10^4 (*ferromagnetic materials*). Some show marked hysteresis and are used for permanent magnets, magnetic amplifiers etc.

ferromanganese *(Eng.)*. A **master alloy** of iron and manganese, used in making additions of manganese to steel or cast-iron.

ferrometer *(Elec.Eng.)*. An a.c. instrument for measuring the magnetization of a ferromagnet.

ferromolybdenum *(Eng.)*. A **master alloy** of iron and molybdenum (55–65% molybdenum) used in adding molybdenum to steel and cast-iron.

ferronickel *(Eng.)*. Alloy of iron and nickel containing more than 30% nickel. Lower nickel alloys are known as nickel steel. See **Elinvar**, **Invar**, **Mumetal**, **permalloy**.

ferroprussiate paper *(Paper)*. See **blueprint paper**.

ferroresonance *(Elec.Eng.)*. See **ferromagnetic resonance**.

ferrosilicon *(Eng.)*. A **master alloy** of iron and silicon, used in making additions of silicon to steel and cast-iron. When containing 15% silicon, widely used in dense media processes.

ferrospinel *(Eng.)*. A crystalline material which has the equivalent function to that of the *M* in a ferrite. See **ferrite**.

ferrotype *(Image Tech.)*. A wet-collodion positive on a plate of darkened metal. Formerly favoured by street-photographers owing to immediate availability of the print. Also *melanotype*, *tin-type*.

ferrous [iron(II)] oxide *(Chem.)*. FeO. Black oxide of iron.

ferrous [iron(II)] sulphate *(Chem.)*. $FeSO_4.7H_2O$. Also mineral *copperas*.

ferroxyl indicator *(Chem.)*. A little potassium hexacyanoferrate (III) and phenolphthalein, together with a corroding solution, e.g. of sodium chloride, made into a jelly with agar. It is used to show the presence of anodic and cathodic areas in an apparently uniform piece of iron, by turning blue and pink respectively.

ferruginous deposits *(Geol.)*. Sedimentary rocks containing sufficient iron to justify exploitation as iron ore. The iron is present, in different cases, in silicate, carbonate, or oxide form, occurring as the minerals chamosite, thuringite, siderite, haematite, limonite, etc. The ferruginous material may have formed contemporaneously with the accompanying sediment, if any, or may have been introduced later.

ferrule *(Build.)*. The brass ring round the handle of a chisel, or similar tool, at the end where the tang enters or at the striking end to prevent splitting.

ferrule *(Eng.)*. (1) A short length of tube. (2) A circular gland nut used for making a joint. (3) A slotted metal tube into the ends of which the conductors of a joint are inserted. The whole is soldered solid. When the conductors are oval, the ferrule is in two parts to allow for the fact that the major axes of the oval sections may not coincide.

ferrule *(Horol.)*. A small grooved pulley around which is wrapped the string of a bow, to give rotary motion.

fertile *(Bot.)*. Able to produce asexual spores and/or sexual gametes.

fertile *(Phys.)*. An isotope in a nuclear reactor which can be converted by the capture of a neutron into *fissile* isotope, e.g. ^{238}U is converted by series of reactions into ^{239}Pu.

fertile flower *(Bot.)*. (1) A flower with functional carpels and/or stamens. (2) Sometimes, a female flower.

fertilisin *(Zool.)*. A substance which is present in the cortex of an ovum which increases sperm motility.

fertility *(Ecol.)*. The number or percentage of eggs produced by a species or individual which develop into living young. Cf. **fecundity**.

fertilization *(Biol.)*. The union of 2 sexually differentiated gametes to form a zygote.

fertilization cone *(Zool.)*. A conical projection of protoplasm arising from the surface of an ovum containing many microtubules which are thought to facilitate the entry of the sperm.

fertilization tube *(Bot.)*. See **conjugation tube**.

Fery spectrograph *(Phys.)*. A spectrograph in which the only optical element is a back-reflecting prism with cylindrically curved faces. Considerable astigmatism is experienced with this instrument.

Fessenden oscillator *(Acous.)*. A low-frequency underwater sound source of the moving-coil type.

festination *(Med.)*. Involuntary quick walking with short steps, occurring in certain diseases of the nervous system, e.g. in **Parkinson's disease**.

festoon dryer *(Paper)*. A means of drying coated paper by the circulation of hot air while the web is in the form of slowly travelling loops carried on wooden rods.

FET *(Electronics)*. Abbrev. for *Field-Effect Transistor*.

fetch *(Civ.Eng.)*. The distance of the nearest coast in the direction of the strongest and most prevalent winds on a harbour site.

fetch *(Meteor.)*. The length of the traverse of an airstream of fairly uniform direction across a sea or ocean area.

fetch/execute cycle *(Comp.)*. See **instruction cycle**.

fetishism *(Behav.)*. Sexual gratification in which an object, or some part of the body, is the main source of sexual arousal, to the exclusion of the person as a whole.

fetlock *(Vet.)*. The metacarpophalangeal and metatarsophalangeal regions of horses.

fettler *(Textiles)*. An operative who clears away the fibrous waste and dirt from card cylinders, doffers, and rollers etc.

fettling *(Eng.)*. 'Making good', e.g. of ore hearth and walls of an open-hearth, reverberatory or cupola furnace, where erosion or chemical action has damaged the refractory lining.

fever *(Med.)*. The complex reaction of the body to infection, associated with a rise in temperature. Less accurately, a rise of the temperature of the body above normal.

Feynman diagram *(Phys.)*. One which shows the contributions to the rate of a elementary particle reaction. A powerful method of finding the physical properties of a system of interacting particles.

f-factor *(Radiol.)*. The ratio of absorbed dose to exposure dose for a given material and X-ray energy.

FFT *(Acous.)*. See **fast Fourier transform**.

FHP *(Eng.)*. Abbrev. for *Friction HorsePower*.

Fibonacci numbers, sequence, series *(Maths.)*. Sequence of integers, where each is the sum of the two preceding it. 1,1,2,3,5,8,13,21,...

fibre *(Bot.)*. An elongated sclerenchyma cell, typically tapering at both ends, with a thick secondary wall containing steeply helical or longitudinal cellulose microfibrils, lignified or not and with or without a living protoplast at maturity. Bundles of such fibres constitute some economically important textile fibres, e.g. flax, jute, hemp, sisal.

fibre *(Eng.)*. Term for arrangement of the constituents of metals parallel to the direction of working. It is applied to the elongation of the crystals in severely cold-worked metals, to the elongation and stringing out of the inclusions in hot-worked metal, and to preferred orientations.

fibre *(Textiles)*. Any type of vegetable, animal, regenerated, synthetic, or mineral fibre or filament from which yarns and fabrics are manufactured, by spinning and weaving, or knitting, felting, bonding etc. A fibre is long in relation to its thickness and is fine and flexible.

fibre-board *(Build.)*. Building or insulating board made from fibrous material such as wood pulp, waste paper, and other waste vegetable fibre. May be homogeneous, bitumen-bonded or laminated. See **hardboard, insulating board**.

fibre brushes *(Build.)*. Cheap paint brushes in which the filling consists of vegetable fibre. Normally limited in use to work with alkaline materials such as paint removers, cement paints etc.

fibre bundle *(Phys.)*. A bundle of optical fibres used in endoscopes for the inspection of body cavities. An incoherent bundle is used for illumination. A coherent bundle, in which the relative positions of the individual fibres are maintained, is used for the transmissoin of images. See **fibre optics**.

fibre camera *(Phys.)*. Instrument for measuring the X-ray diffraction pattern of fibrous materials.

fibre, man-made *(Textiles)*. A regenerated or synthetic fibre that has been manufactured, as distant from a natural fibre. Man-made fibres are obtained as continuous filaments which may be cut or broken into **staple fibres**.

fibre metallurgy *(Eng.)*. The metallurgy of the manufacture of the fibres and products made from metallic fibres by sintering.

fibre optics *(Telecomm.)*. See **optical fibre**.

fibre-optics gyro *(Aero.)*. Instrument for measuring angular rotation by passing two beams of coherent light in both directions round a closed loop of optical fibre. Rotation affects the phase shift at the output of the two beams. The loops are often of triangular shape, each side being between 20 and 200 mm long. The 'gyro' is fixed to the aircraft, has no rotating parts and is strictly not a gyro, but is so called because it provides equivalent information.

fibre, regenerated *(Textiles)*. A fibre that has been manufactured from a naturally-occurring linear polymer by dissolving it in a suitable solvent and extruding the resulting viscous solution through a **spinneret**. The polymer is regenerated as continuous filaments by a subsequent evaporation or precipitation process.

fibre, synthetic *(Textiles)*. A fibre that has been manufactured by synthesis of a suitable linear polymer followed by extrusion of the molten material through a **spinneret**.

fibre-tracheid *(Bot.)*. A cell of the secondary xylem intermediate between a **libriform fibre** and a tracheid.

fibril *(Bot.)*. A small fibre.

fibril, fibrilla *(Zool.)*. Any minute threadlike structure, such as the longitudinal contractile elements of a muscle fibre. adj. *fibrillar, -ate*.

fibrillation *(Med.)*. (1) Twitching of individual muscle fibres, or bundles of fibres, in certain nervous diseases. (2) Incoordinate contraction of individual muscle fibres of the heart, giving rise to an irregular and inefficient action of the heart; i.e. *atrial fibrillation, ventricular fibrillation*.

fibrillation *(Textiles)*. The process by which a film (e.g. of polypropylene) is converted into fibres. The film is deliberately stretched in order to orientate the molecules. Further manipulation such as twisting splits the film longitudinally into a yarn comprising an interconnected mass of fibres.

fibrin *(Biol.)*. Protein which forms the fibrin blood clot. The monomer spontaneously associates into long insoluble fibres which are subsequently stabilised by covalent bonds between the polypeptide side chains.

fibrin *(Med.)*. An insoluble protein precipitated from blood which forms a network of fibres during the process of clotting. Its immediate precursor is the soluble protein *fibrinogen*.

fibrinogen *(Biol.)*. The soluble precursor of **fibrin**. it is converted to fibrin under the influence of thrombin by proteolytic cleavage of terminal peptides.

fibrinolysis *(Med.)*. Enzymatic breakdown of **fibrin**. A variety of compounds can be used as fibrinolytic drugs in the treatment of disease where **thrombus** or clotting takes place.

fibrino-purulent *(Med.)*. Containing fibrin and pus.

fibro-adenoma *(Med.)*. An adenoma in which there is an overgrowth of fibrous tissue.

fibroblast *(Med.)*. The main cell of connective tissue.

fibroblasts *(Zool.)*. Flattened connective-tissue cells of irregular form, believed to be responsible for the secretion of the white fibres; lamellar cells, fibrocytes.

fibrocartilage *(Zool.)*. A form of cartilage which has white or yellow fibres embedded in the matrix.

fibroid *(Med.)*. Resembling fibrous tissue; a fibromyoma, usually applied to muscular and fibrous growths in the uterus.

fibroin *(Textiles)*. The main protein constituent of the silk fibre.

fibrolite *(Min.)*. A variety of the aluminium silicate **sillimanite** occurring as felted aggregates of exceedingly thin fibrous crystals; also used when the mineral is cut as a gemstone.

fibroma *(Med.)*. A tumour composed of fibrous tissue.

fibromyectomy *(Med.)*. Removal of a fibromyoma.

fibromyositis *(Med.)*. Inflammation of fibrous tissue in muscle and in the muscle fibres adjacent to it.

fibrose *(Med.)*. To form fibrous tissue.

fibrosis *(Med.)*. The formation of fibrous tissue as a result of injury or inflammation of a part, or of interference with its blood supply.

fibrositis *(Med.)*. Inflammation (often presumed rheumatic) of fibrous tissue.

fibrotic *(Med.)*. Pertaining to fibrosis.

fibrous concrete *(Build.)*. Concrete in which fibrous aggregate, such as asbestos, sawdust, etc., is incorporated either as alternative or additional to the sand and gravel.

fibrous layer *(Bot.)*. In the wall of the anther in angiosperms, a layer of cells below the epidermis with uneven cell walls probably responsible, because of the way they shrink on drying, for the opening of the mature anther; also called *endothecium*.

fibrous plaster *(Build.)*. (1) Prepared plaster slabs formed of canvas stretched across a wooden frame and coated with a thin layer of gypsum plaster. (2) Any decorative plasterwork reinforced with hessian or similar.

fibrous root system *(Bot.)*. Root system composed of many roots of roughly equal thickness and length, as in grasses. Cf. tap root system.

fibrous tissue *(Zool.)*. A form of connective-tissue consisting mainly of bundles of white fibres; any tissue containing a large number of fibres.

fibrovascular bundle *(Bot.)*. A vascular bundle accompanied, usually on its outer side, by a strand of sclerenchyma.

fibula *(Build.)*. A bent iron bar, used to fasten together adjacent stones.

fibula *(Zool.)*. In Tetrapoda, the posterior of the two bones in the middle division of the hind-limb.

fibulare *(Zool.)*. A bone of the proximal row of the tarsus, in line with the fibula, in Tetrapoda.

Fick principle *(Med.)*. A method introduced by the German physiologist A.E. Fick for the measurement of cardiac output. Flow or cardiac output is the oxygen uptake of either an organ or the whole body, divided by the oxygen extraction of the tissue being examined.

Fick's law of diffusion *(Chem.)*. The rate of diffusion in a given direction is proportional to the negative of the concentration gradient, i.e.

$$\text{molar flux} = -D\frac{\partial C}{\partial x},$$

where D is the diffusion coefficient.

ficoll-hypaque *(Immun.)*. A proprietary mixture of a large polysaccharide and a dense synthetic organic molecule used in radiography, but the density can also be adjusted to allow separation of different cells by centrifugation. Frequently used for separating mononuclear cells and granulocytes from blood.

fidelity *(Ecol.)*. The degree of restriction of a species to a particular situation, a species with high fidelity having a strong preference for a particular type of community.

fidelity *(Telecomm.)*. The exactness of reproduction of the input signal at the output end of a system transmitting information.

Fido Net *(Comp.)*. A co-operative network for personal computer users. Most nodes are in the US centred around St Louis but have begun to spread to Europe and other parts of the world.

fiducial *(Surv.)*. Said of a line or point established accurately as a basis of reference.

fiducial points *(Elec.Eng.)*. Points on the scale of an indicating instrument located by direct calibration, as contrasted with the intervening points, which are inserted by interpolation or subdivision.

fiducial temperature *(Meteor.)*. The temperature at which a sensitive barometer reads correctly, the maker's calibration holding for latitude 45° at the temperature 285 K (12°C) only.

field *(Comp.)*. Predetermined section of a record.

field *(Image Tech.)*. A single complete scanning of the picture from top to bottom; in an **interlaced scanning** system two successive fields complete a **frame**.

field *(Maths.)*. See **ring**.

field *(Phys.)*. (1) The interaction between bodies or particles is explained in terms of fields. For example, the *potential energy* of a body may depend on its position and then is represented by a *scalar field* with magnitude only. Other physical quantities carry direction as well as magnitude and they are represented by *vector fields*, e.g. electric, magnetic, gravitational fields. (2) Space in which there are electromagnetic oscillations associated with a radiator; the *induction* field which represents the interchange of energy between the radiator and space is within a few wavelengths of the radiator; *radiation field* represents the energy lost from the radiator to space. The region where components radiated by antenna elements are parallel is the *Fraunhofer region* and , where not, is the *Fresnel region*. The latter will exist between the antenna and the Fraunhofer region and is usually taken to extend a distance $2D^2/\lambda$, where λ is the wavelength of the radiation and D is the aerial aperture in a given aspect.

field ampere-turns *(Elec.Eng.)*. The ampere-turns producing the magnetic field of an electric machine.

field blanking *(Image Tech.)*. The period between successive **fields** during which the picture information is suppressed and **sync pulses** transmitted. Also known as the *vertical blanking interval*.

field-breaking resistance *(Elec.Eng.)*. See **field-discharge resistance**.

field-breaking switch *(Elec.Eng.)*. See **field-discharge switch**.

field capacity *(Bot.)*. Amount of water held in a given soil by capillarity against drainage by gravity to a water potential or suction of about −0.05 bars.

field coil *(Elec.Eng.)*. The coil which carries the current for producing the m.m.f. to set up the flux in an electric machine; occasionally called a *field spool*. See also **magnetizing coil**.

field control *(Elec.Eng.)*. The adjustment of the field current of a generator or motor to control the voltage or speed respectively.

field copper *(Elec.Eng.)*. A term used in the design of electrical machines to denote the total quantity of copper used in the field windings of a machine.

field current *(Elec.Eng.)*. The current in the field winding of an electric machine.

field density *(Phys.)*. The number of lines of force passing normally through unit area of an electric or magnetic field.

field discharge *(Phys.)*. The passage of electricity through a gas as a result of ionization of the gas; it takes the form of a brush discharge, an arc, or a spark. Also *electric discharge*.

field-discharge resistance *(Elec.Eng.)*. A discharge resistance used for connecting across the terminal of the shunt or separately excited field winding of an electric machine, to prevent high induced voltages when interrupting the field circuit. Also called a *field-breaking resistance*.

field-discharge switch *(Elec.Eng.)*. A switch for controlling the field circuit of a generator. It is provided with special contacts so that a discharge resistance is connected across the winding at the moment of breaking the field circuit. Also called a *field-breaking switch*.

field-diverter rheostat *(Elec.Eng.)*. A rheostat connected in parallel with the series field winding or the compole winding of a d.c. machine to give control of the m.m.f. produced by the winding, independently of the current flowing through the main circuit.

field drain *(Civ.Eng.)*. See **sub-soil drain**.

fielded panel *(Build.)*. A panel which is moulded, sunk, raised, or divided into smaller panels.

field-effect transistor *(Electronics)*. Abbrev. *FET*. One consisting of a conducting channel formed by a strip of *n* or *p-type* semiconductor adjacent to a gate (consisting of *p* or *n-type* material. The voltage applied to the gate affects the thickness of the depletion layer, thereby controlling the resistivity of the channel. The input impedance at the gate is very high and the FET's mode of operation resembles a triode valve. FETs can be of the J-FET or MOSFET variety. See also **depletion-mode transistor** and **enhancement-mode transistor**.

field emission *(Electronics)*. That arising, at normal temperature, through a high-voltage gradient causing an intense electric field at a metallic surface and stripping electrons from surface atoms.

field-emission microscope *(Phys.)*. One in which the positions of the atoms in a surface are made visible by means of the electric field emitted on making the surface the positive electrode in a high-voltage discharge tube containing argon at very low pressure. When an argon atom passes over a charged surface atom, it is stripped of an electron, and thus is drawn toward the negative electrode, where it hits a fluorescent screen in a position corresponding to that of the surface atom.

field enhancement *(Phys.)*. Local increase in electrical field strength due to convex curvature of an electrode, or proximity of another electrode.

field frequency *(Image Tech.)*. The number of fields scanned per second, currently 60 Hz for US broadcast practice and 50 Hz for European.

field intensity *(Phys.)*. Same as *field strength*.

field lens *(Phys.)*. Lens placed in or near the plane of an image to ensure that the light to the outer parts of the image is directed into the subsequent lenses of the system, and thus uniform illumination over the field of view is ensured.

field magnet *(Elec.Eng.)*. The permanent or electromagnet which provides m.m.f. for setting up the flux in an electric machine. See **rotating-field magnet**.

field of force *(Phys.)*. Principle of *action at a distance*, i.e. mechanical forces experienced by an electric charge, a magnet, or a mass, at a distance from an independent electric charge, magnet or mass, because of fields established by these and described by uniform laws.

field of view *(Phys.)*. The area over which the image is visible in the eyepiece of an optical instrument. It is usually limited by a circular stop in the focal plane of the eye-lens. See **sagittal field, tangential field**.

field oscillator *(Electronics)*. Same as **frame oscillator**.

field resistance *(Bot.)*. Resistance shown by a plant to natural infection in the field which is more dependent on the environment and the nature of pathogen and vector than is inoculation in laboratory or greenhouse.

field rheostat *(Elec.Eng.)*. A variable resistance (rheostat) connected in series or parallel with the field winding of an electrical machine for the purpose of varying the current in the winding. Also called a *field regulator*.

field rivet *(Eng.)*. A rivet which is put in when the work is on the site. Also called *site rivet*.

field sequential *(Image Tech.)*. A colour TV system in which successive fields are scanned in the three primary colours.

Field's siphon flush tank *(Build.)*. See **flushing tank**.

field star *(Astron.)*. An individual star which is not a member of any *cluster* or *association*. They are numerous on all astronomical photographs.

field strength *(Phys.)*. Vector representing the quotient of a force and the *charge* (or *pole*) in an electric (or magnetic) field, with the direction of the force; also called *field intensity*.

field strength *(Telecomm.)*. The measure of the intensity of an electric, magnetic or electromagnetic field, most commonly expressed in volts per metre, mV/m or μV/m. In microwave engineering, field strength is often identified with the power flux density, measured in w/m^2. Field strength is inversely proportional to distance from an antenna.

field strength meter *(Telecomm.)*. A calibrated radio receiver and antenna system for measuring **field strength**.

field suppressor *(Elec.Eng.)*. An arrangement for automatically reducing the field current of a generator when a short-circuit or other fault occurs on the machine or its adjacent connections.

field sync pulse *(Image Tech.)*. A signal transmitted during the **field blanking** interval to synchronize the time-base of the receiver with the transmitter.

field theory *(Phys.)*. As yet unverified attempt to link the properties of all fields (electric, magnetic, gravitational, nuclear) into a unified system.

field tube *(Eng.)*. A special form of boiler tube, consisting of an outer tube which is closed at its lower end and contains a second concentric tube down which the water passes to return up the annular space between the two.

field winding *(Elec.Eng.)*. The winding placed on the field magnets of an electric machine and producing the m.m.f. necessary to set up the exciting flux.

fiery mine *(Min.Ext.)*. One in which there is a possibility of explosion from gas or coal-dust.

FIFO *(Comp.)*. See **queue**.

fifth freedom traffic *(Aero.)*. Passengers or freight carried between two countries by an airline of a third country.

fifth generation computer *(Comp.)*. A forward looking generic term to apply to computer systems predicted for the early 1990s. Will have very fast processing with **VLSI** based on **logic programming**, providing access to large knowledge bases through novel human/machine interfaces. See **computer generations**.

figure of loss *(Elec.Eng.)*. A term occasionally used in connection with transformers to denote the energy loss per unit mass of material (iron or copper).

figure of merit *(Elec.Eng.)*. General parameter which describes the quality of performance of an instrument, a circuit or other item associated with a given system, e.g. the voltage gain of an amplifier or the bandwidth of a filter.

filament *(Bot.)*. (1) Generally a chain (unbranched or branched) of cells joined end on end. (2) A fungal hypha. (3) The stalk of a stamen in angiosperms.

filament *(Elec.Eng.)*. A fine wire of high resistance, which is heated to incandescence by the passage of an electric current. In an electric filament-lamp it acts as the source of light, and in thermionic tubes it acts as an emitter of electrons.

filament *(Electronics)*. See **heater**. (Historically, this term remains from when the source of electrons was simply a fine wire heated to incandescence by a current.)

filament *(Textiles)*. See **continuous filament yarn**.

filament *(Zool.)*. Any fine threadlike structure; the axis of a down-feather.

filamentation *(Textiles)*. Mechanical damage to continuous filaments in a yarn or fabric giving the material a fuzzy appearance.

filament getter *(Electronics)*. One which adsorbs gas readily when hot, and is used for this purpose in a high-vacuum assembly.

filament lamp *(Phys.)*. An electric lamp in which a filament in a glass bulb, evacuated or filled with inert gas, is raised to incandescence by the passage through it of an electric current.

filariasis *(Med.,Vet.)*. Infestation with nematode worms of the family *Filariidae*, which inhabit the lymphatic channels, often causing elephantiasis.

filar micrometer *(Biol.,Min.)*. Modified eyepiece graticule with movable scale or movable cross-hair.

file *(Comp.)*. General term for named set of data items stored in machine readable form. See **master file, transaction-**.

file card *(Eng.)*. A short wire brush used to clean swarf from files.

file server *(Comp.)*. The file store in a distributed computing environment.

filet net *(Textiles)*. Any type of square ground-mesh of lace; a woven net.

Filicales *(Bot.)*. *Polypodiales*. The leptosporangiate ferns. Order of ca 9000 extant species of Filicopsida. (also fossils from the Carboniferous onwards). Sporophyte may have long, creeping, horizontal rhizomes with fronds at intervals, short more or less upright stems with a rosette of fronds, or grow into *tree-ferns*. Leaves are circinate in bud and usually pinnately compound. Many are poisonous, some carcinogenic, the expanding fronds (croziers) of a few spp. are eaten; some are cultivated as ornamentals.

Filicopsida *(Bot.)*. *Pteropsida, Polypodiopsida*. The ferns, A class of Pteridophytes from the Devonian onwards. The sporophyte usually has roots, a rhizome or stem and spirally arranged often pinnately-compound leaves (fronds, *megaphylls*) with circinate venation. (The paleozoic sorts show little distinction of stem and leaf.) Sporangia typically borne marginally or abaxially on leaves, mostly homosporous. Spermatozoids multiflagellate. Includes Marattiales (eusporangiate), Filicales and Salviniales (floating, heterosporous aquatica).

filiform *(Bot.,Zool.)*. Threadlike, e.g. *filiform antenna*.

fill *(Build.,Civ.Eng.)*. Rock or soil dumped to bring a site to the required level. After dumping, the fill has to be progressively consolidated.

filled band *(Chem.)*. An energy-level band in which there are no vacancies. Its electrons do not contribute to valence or conduction processes.

filler *(Build.)*. (1) A material used to fill in the pores of, or any holes in, wood, plaster, etc., which is to be painted, varnished, or otherwise decorated. Supplied in several forms: (a) powder to be mixed with water; (b) ready mixed, oil, water or resin based; (c) two pack materials; (d) cellulose materials. (2) An **extender**.

filler *(Civ.Eng.)*. A finely divided substance added to bituminous material for road surfacing, in order to reduce it to a suitable consistency.

filler *(Paper)*. See **loading**.

filler *(Plastics)*. An inert solid substance added to a synthetic resin either for economy (e.g. wood-flour, clay) or to modify its properties (e.g. mica, aluminium or other metal powder).

filler joist floor *(Build.)*. A type of floor in which the principal supporting members are steel joists, the gaps between the joists being filled with concrete.

filler light *(Image Tech.)*. Lighting directed into the shadow areas of a scene in order to avoid excessive contrast.

filler metal *(Eng.)*. The metal required to be added at the weld in welding processes in which the fusion temperature is reached. In metal-electrode welding, the electrode, usually fluxed and coated, is melted down to provide the filler metal.

filler rod *(Elec.Eng.)*. See **welding rod**.

fillet *(Aero.)*. A fairing at the intersection of 2 surfaces intended to improve the airflow by reducing breakaway and turbulence.

fillet *(Arch.)*. (1) A flat and narrow surface separating or strengthening curved mouldings. (2) A **listel**. See **facette**.

fillet *(Build.)*. A thin strip of wood fixed into the angle between 2 surfaces, to a wall as a shelf support, to a floor as a door stop etc.

fillet *(Eng.)*. (1) A narrow strip of metal raised above the general level of a surface. (2) A radius provided, for increased strength, at the intersection of 2 surfaces, particularly in a casting.

fillet *(Plastics)*. Term used to describe a rounded internal corner in a plastic article, to avoid a possible weakness arising from an abrupt change in cross-section.

fillet *(Print.)*. A band or line of gold leaf, or a plain band or line, on a cover.

filleting *(Build.)*. See **cement fillet, fillet**.

fillet weld *(Eng.)*. A weld at the junction of 2 parts, e.g. plates, at right angles to each other, in which a fillet of welding metal is laid down in the angle created by the intersection of the surfaces of the parts.

filling *(Min.Ext.)*. (1) The loading of tubs or trucks with coal, ore, or waste. (2) The filling-up of worked-out areas in a metal-mine.

filling *(Textiles)*. (1) Insoluble materials such as China clay added with starch or gum to increase the weight of fabrics or to alter their appearance. (2) Used in Canada and US for *weft yarns*.

filling-in *(Build.)*. The operation of building in the middle part of a wall, between the face and the back.

filling-in *(Print.)*. The spreading of ink on the printing image causing halftone dots to join up or the bowls of letters to be filled with ink.

filling pile *(Hyd.Eng.)*. A pile serving to retain the sheeting of a coffer-dam.

filling post *(Build.)*. A middle post in a timber frame.

fillister *(Build.)*. (1) A rebate in the edge of a sash bar, to receive the glass and putty. (2) A rabbeting plane having a movable stop to regulate depth or cut.

fillister-head screw *(Eng.)*. A cheese-head screw with a slightly convex upper head surface.

film *(Chem.)*. (1) Any thin layer of substance, e.g. that which carries a light-sensitive emulsion for photography, that which carries iron (III) oxide particles in a matrix for sound recording. (2) Thin layer of material deposited, formed or adsorbed on another, down to monomolecular dimensions, e.g. electroplated films, oxide on aluminium, sputtered depositions on glass or microcomponents.

film *(Image Tech.)*. (1) A photographic emulsion carried on a thin flexible transparent support; motion picture film is manufactured in long continuous strips with perforation holes along one or both edges. (2) The craft and techniques of cinematography. (3) A completed motion picture production.

film badge *(Radiol.)*. Small photographic film used as radiation monitor and dosimeter. Normally worn on lapel, wrist or finger and sometimes partly covered by cadmium and tin screens so that exposure to neutrons, and to beta and gamma rays, can be estimated separately.

film circuits *(Electronics)*. See **thin-film circuits**.

film coefficient *(Chem.Eng.)*. The constant of proportionality between the flux (e.g. of mass or heat) and the difference in driving force (e.g. concentration or temperature) across a film. Symbol *h*.

film foils *(Print.)*. Blocking foils with a viscose film substrate. These give better release and finer detail than **paper foils**.

film glue *(Build.)*. Resin-impregnated paper, or phenolic resin film, used in making resin-bonded plywood.

film pack *(Image Tech.)*. A package of cut sheets of photographic film which can be successively exposed in a camera and individually removed for processing.

film recording *(Acous.)*. Process of recording sound on a sound-track on the edge of cinematograph film, for synchronous reproduction with the picture.

filmsetting *(Print.)*. See **phototypesetting**.

film sizing *(Min.Ext.)*. Concentration of finely divided heavy mineral by gently sloped surfaces which may be plane, riffled, or vibrated. See **buddle, Wilfley table**.

film speed *(Image Tech.)*. (1) The sensitivity of a photographic material as rated by a standard method of exposure determination, such as *ASA* or *DIN*. (2) The rate at which film passes through a motion picture camera or projector, specified in frames per second or in feet or metres per minute.

film strip *(Image Tech.)*. A strip of film, usually 35 mm gauge, containing a series of still pictures for individual projection, often used for educational illustration and demonstrations.

film theory *(Chem.Eng.)*. Mass transfer between two well-stirred fluid phases, or between a solid and a well-stirred fluid, takes place by eddy diffusion in the bulk fluid phase, combined with molecular diffusion through a stagnant film at the interface, the latter being the slower process.

filoplumes *(Zool.)*. Delicate hairlike contour feathers with a long axis and few barbs, devoid of locking apparatus at the distal end.

filopodium *(Biol.)*. A long filamentous spike containing actin filaments found at the surface of some cells.

filter *(Chem.)*. A sheet of material with pore sizes within a defined range and used to separate particles or macromolecules from a suspension or solution. See **filter paper, membrane filter**.

filter *(Chem.Eng.)*. Equipment used to separate liquids and suspended solids, either for recovery of solid, classification of liquid, or both simultaneously. The liquid flows through the pores in a cloth, wire mesh, or granular bed, and thus the solid is sieved out.

filter *(Electronics)*. A transmission network used in electronic circuits, or an optical device in optical communication systems, for the selective enhancement or reduction of specified components of an input signal. Filtering is achieved by selectively attenuating those components of the input signal which are undesired, relative to those which it is desired to enhance. A filter may consist of inductances and capacitances, resistor and capacitors and gyrators (to transform specified capacitors into inductances); it may contain amplifying stages (see **active filter**), or it may rely on resonances in piezoelectric, ceramic, or magnetic materials. *Digital filters* depend on the action of externally-manipulated gates to selectively block or pass certain of the pulses making up a digital signal.

filter *(Image Tech.)*. A device, usually consisting of a glass plate or a sheet of gelatin, interposed across a beam of light for the purpose of altering the relative intensity of the different component wavelengths in the beam.

filter aid *(Chem.Eng.)*. Diatom earth added to solutions, or precoated on to filter before use, to aid separation of difficult solids, usually colloidal, which would otherwise pass through or choke the filter.

filter attenuation *(Telecomm.)*. The loss of signal power in its passage through a filter due to absorption, reflection, or radiation. It is usually given in decibels.

filter attenuation band *(Telecomm.)*. See **stop band**.

filter bed *(Build.)*. A **contact bed** or any similar bed used for filtering purposes.

filter cake *(Chem.)*. The layer of precipitate which builds up on the cloth of a filter press.

filter circuit *(Phys.)*. One, usually composed of **reactors** arranged in resonant circuits, designed to accept certain desired frequencies and to reject all others. Often used to remove noise from a signal, e.g., between a radio transmitter and its antenna. Also *filter network*.

filter cut *(Image Tech.)*. The wavelength at which the relative absorption for light of different wavelengths changes rapidly.

filtered equations *(Meteor.)*. Modified forms of the equations of motion, especially of derived forms such as the **vorticity equation**, which exclude certain solutions such as fast moving sound and gravity waves that are irrelevant to the types of atmospheric motion producing phenomena of meteorological interest. Such filtering is effected by use of the **hydrostatic approximation** and the judicious use of the **geostrophic approximation** and the **balance equation**.

filtered model *(Meteor.)*. A numerical forecast model which makes use of the **filtered equations**. Such models use much less computer time than those based on the **primitive equations**.

filter factor *(Image Tech.)*. The number of times a given exposure must be increased because of the presence of a filter which absorbs light and reduces the effective exposure of a lens system.

filter feeders *(Ecol.)*. In benthic and planktonic communitites, detritus feeders which remove particles from the water. In benthic communities they usually predominate on sandy bottoms. Cf. **deposit feeders**.

filtering basin *(Hyd.Eng.)*. A tank through which water passes on its way from the reservoir to the mains, and in which it is subjected to a process of filtration.

filter overlap *(Image Tech.)*. The band of wavelengths transmitted in common by filters of different (usually broad) spectral absorption.

filter paper *(Chem.)*. Paper, consisting of pure cellulose, which is used for separating solids from liquids by filtration. Filter paper for quantitative purposes is treated with acids to remove all or most inorganic substances, and has a definite ash content.

filter press *(Chem.)*. An apparatus used for filtrations; it consists of a set of frames covered with filter cloths into which the mixture which is to be filtered is pumped.

filter-press action *(Geol.)*. A differentiation process involving the mechanical separation of the still liquid portion of a magma from the crystal mesh. The effective agent is pressure operating during crystallization.

filter pulp *(Paper)*. Rag fibre made into convenient slubs or cakes at the paper mill.

filter transmission band *(Telecomm.)*. See **pass band**.

filtrate *(Chem.)*. The liquid freed from solid matter after having passed through a filter.

filtration *(Chem.)*. The separation of solids from liquids by passing the mixture through a suitable medium, e.g. filter paper, cloth, glass wool, which retains the solid matter on its surface and allows the liquid to pass through.

filtration *(Radiol.)*. Removal of longer wavelengths in a composite beam of X-rays by the interposition of thin metal, e.g. copper or aluminium.

fimbria *(Zool.)*. Any fringing or fringe-like structure; the delicate processes fringing the internal opening of the oviduct in Mammals; the ridge of fibres running along the anterior edge of the hippocampus in Mammals; the processes fringing the openings of the siphons in Molluscs.

fimbriate, fimbriated *(Bot.,Zool.)*. Having a fringed margin.

fimbriocele *(Med.)*. Hernia containing the fimbriae of the Fallopian tube.

fimicolous *(Bot.)*. Growing on or in dung.

fin *(Aero.)*. A fixed vertical surface, usually at the tail, which gives directional stability to a fixed wing aircraft in motion and to which a rudder is usually attached. In an airship, any fixed stabilizing surface. See **stabilizer**.

fin *(Eng.)*. (1) One of several thin projecting strips of metal formed integral with an air-cooled engine cylinder or a pump body or gear-box to increase the cooling area. (2) A thin projecting edge on a casting or stamping, formed by metal extruded between the halves of the die; any similar projection.

fin *(Zool.)*. In Fish, some Cephalopoda, and other aquatic forms, a muscular fold of integument used for locomotion or balancing; supported in the case of Fish by internal skeletal elements.

final approach *(Aero.)*. The part of the landing procedure from the time when the aircraft turns into line with the runway until the flare-out is started. Colloquially *finals*.

final limit-switch *(Elec.Eng.)*. A limit-switch used on an electric lift; it is arranged to operate in case of failure of the ordinary limit-switch. Also called an *ultimate limit-switch*.

final selector *(Telecomm.)*. In a **Strowger exchange**, the selector which is operated by two trains of impulses so that the wipers finally rest on contacts which are connected to the required subscriber's line.

finder *(Astron.)*. A small auxiliary telescope of low power fixed parallel to the optical axis of a large telescope for the purpose of finding the required object and setting it in the centre of the field; also used in stellar photography for guiding during an exposure.

finder *(Image Tech.)*. See **view finder**.

finder *(Telecomm.)*. In a **Strowger exchange**, a uniselector which automatically hunts to find the line of a subscriber, when he lifts his telephone receiver, in order to connect selectors to his circuit. Also *line finder*.

fine aggregate *(Civ.Eng.)*. Sand or the screenings of gravel or crushed stone.

fine boring *(Eng.)*. A high-precision final machining process for bores of internal diameters between 5 mm and 750 mm, using a single-point cutter in a very accurate and rugged machine tool in which the cutting stresses, and consequently the clamping distortion, are kept particularly low.

fine etching *(Print.)*. The finishing stages in the making of a half-tone block to achieve the required contrast and range of tones.

fine gold *(Eng.)*. Pure 24-carat gold.

fine-grain developer *(Image Tech.)*. Developer solution which minimizes clumping of the silver grains by partial development only, but at the cost of a lower image density.

fine-grained *(Ecol.)*. Especially small-scale pattern of environmental or resource heterogeneity. Often used in relation to the foraging activities of animals.

fine machining *(Eng.)*. A family of precision finishing processes, including fine boring, milling, grinding, honing, lapping, diamond turning, and superfinishing, using particularly accurate machine tools with special provision to eliminate vibration and, often, cemented carbide or diamond cutters to remove small amounts of material in order to attain high accuracy and excellent surface finish.

fineness *(Chem.)*. The state of subdivision of a substance.

fineness *(Eng.)*. The purity of a gold or silver alloy; stated as the number of parts per thousand that are gold (or silver).

fineness modulus *(Civ.Eng.)*. A numeral indicating the fineness of an aggregate, as determined by ascertaining the percentage residue, by weight or volume, remaining on each of a series of nine sieves with apertures ranging from 37.5 mm (1.5 in) to 0.15 mm (0.06 in), summing and dividing by 100.

fineness-of-grind gauge *(Powder Tech.)*. Device to provide control tests for the presence of large particles of powder in a slurry. It generally comprises a stainless steel block with parallel grooves of gradual decrease of depth from an inch to zero, along which the slurry is distributed by means of a smooth stroke with a rigid blade. When the groove depth is equal to the largest particle present, the smooth surface of the slurry film is marked by score lines.

fineness ratio *(Aero.)*. The ratio of the length to the maximum diameter of a streamlined body, or flying boat planing bottom.

fine papers *(Paper)*. Papers of high quality, used for graphic purposes.

fines *(Powder Tech.)*. That portion of a powder composed of particles under a specified size.

fine screen *(Image Tech.,Print.)*. The term for half-tones suitable for art paper, usually 52 but also 59 lines per centimetre.

fine silt *(Geol.)*. See **silt grade**.

fine structure *(Electronics)*. Splitting of optical spectrum lines into multiplets due to interaction between spin and orbital angular momenta of electrons in the emitting atoms.

fine stuff *(Build.)*. Fine type of plaster, usually composed of lime and plaster of Paris, used for the finishing coat.

fingering *(Textiles)*. Combed, soft-twisted, worsted yarn of the type generally used for hand knitting; usually 2, 3 or 4 ply.

finger plate *(Build.)*. A plate fixed on the side of the meeting stile of a door, near the lock, to prevent damage to the paintwork by finger-marks.

finger stop *(Eng.)*. A sliding stop in a press tool, used to locate and position the material to be processed in relation to the tool, usually at the commencement of a production run.

finger-type contact *(Elec.Eng.)*. A type of contact which, as usually fitted to drum-type controllers, is in the form of a finger which is pressed against the contact surface by means of a spring.

finial *(Build.)*. A term applied to an ornament placed at the summit of a gable, pillar or spire.

fining *(Glass)*. See **founding**.

fining coat *(Build.)*. See **setting coat**.

fining-off *(Build.)*. The operation of applying the setting coat.

fining-upwards cycle *(Geol.)*. A sedimentary cycle in which coarse-grained material grades up into finer-grained material. e.g. turbidite deposits.

finishing *(Print.)*. The lettering and ornamentation of a bound volume by the finisher; the term does not apply to 'cased' work.

finishing coat *(Build.)*. See **setting coat**.

finishing cut *(Eng.)*. A fine cut taken to finish the surface of a machined workpiece.

finishing stove *(Print.)*. A small gas or electric stove on which the finisher heats the tools required for his work.

finishing tool *(Eng.)*. A lathe or planer tool, generally square-ended and cutting on a wide face. Used for taking the final or finishing cut.

fink truss *(Eng.)*. See **French truss**.

fin rays *(Zool.)*. In Fish, the distal skeletal elements which support the fins; in Cephalochorda, the rods of connective tissue supporting the dorsal and vertral fins.

fiords, fjords *(Geol.)*. Narrow winding inlets of the sea bounded by mountain slopes; formed by the drowning of steep-sided valleys, deeply excavated by glacial action; in many cases a rock-bar partially blocks the entrance and impedes navigation.

FIR *(Aero.)*. Abbrev. for *Flight Information Region.*

fir *(For.)*. Trees of the genus *Abies*, giving a valuable structural softwood; also used for parts of musical instruments and in shipbuilding. The species *A. balsamea*, found in Canada and the Eastern US, yields **Canada balsam** and is used for pulp.

fireback boiler *(Build.)*. The boiler fitted in a kitchen stove or at the back of any fire.

fireball *(Astron.)*. See **bolide**.

fire bank *(Min.Ext.)*. A slack or rubbish heap or dump, at surface on a colliery, which becomes fired by spontaneous combustion.

fire-bar *(Elec.Eng.)*. A heating element fitted to a high-temperature electric radiator.

fire barriers *(Build.)*. Fire-resisting doors, enclosed staircases, and similar obstructions to the spread of fire in a building. See **fire stop**.

fire bars *(Eng.)*. Cast-iron bars forming a grate on which fuel is burnt, as in domestic fires, boiler furnaces, etc.

fire-box *(Eng.)*. That part of a locomotive-type boiler containing the fire; the grate is at the bottom, the walls and top being surrounded by water. See **locomotive boiler**.

firebrick arch *(Eng.)*. An arch built at the end of a boiler furnace, either to deflect the burning gases or to assist the combustion of volatile products. Also *flue bridge*.

fire cement *(Build.)*. See **refractory cement**.

fireclay *(Geol.,Eng.)*. Clay consisting of minerals predominantly of SiO_2 and Al_2O_3, low in Fe_2O_3, CaO, MgO etc. Those clays which soften only at high temperatures are used widely as refractories in metallurgical and other furnaces. Fireclays occur abundantly in the Carboniferous System, as 'seat earths' underneath the coal-seams.

fire cracks *(Build.)*. Fine cracks which appear in a plastered surface, due to unequal contractions between the different coats.

fired *(Electronics)*. Said of gas tubes when discharging, particularly TR pulse tubes during the transmission condition.

fire damp *(Min.Ext.)*. The combustible gas contained naturally in coal; chiefly a mixture of methane and other hydrocarbons; forms explosive mixtures with air.

fire-damp cap *(Min.Ext.)*. Blue flame which forms over the flame of a safety-lamp when sufficient fire-damp is present in colliery workings.

firedog *(Build.)*. See **andiron**.

fire door *(Build.)*. (1) A fire-resisting door of wood, metal, or both. (2) The door of a boiler furnace.

fire extinguisher *(Genrl.)*. Any of several portable apparatus for emergency use against fire. In general, these depend upon the ejection (by rapid chemical reaction or compressed gas, CO_2 or nitrogen) of the fire-inhibiting medium. The latter may be: water or an alkaline solution; a rapidly evolved chemical foam; CO_2 gas and 'snow'; tetrachloromethane, CCl_4; chlorobromomethane, CH_3ClBr, or bromochlorodifluoromethane, CF_3BrCl, both heavy smothering gases; a suitable hydrogen carbonate in dry powder form.

firefly *(Zool.)*. Beetles, family Lampyridae, with light-producing organs whose main function is in attracting a mate.

fire foam *(Genrl.)*. Mixture of foaming but non-inflammable substances used to seal off oxygen and to extinguish fire without use of water.

fire load *(Build.)*. The heat in MJ/m^2 or Btu/ft^2 of floor area of a building if destroyed by fire (including contents).

fireman *(Min.Ext.)*. (1) In a metal-mine, a miner whose duty it is to explode the charges of explosive used in headings and working places. (2) In a coal-mine, an official responsible for safety conditions underground.

fire opal *(Min.)*. A variety of opal (cryptocrystalline silica) characterized by a brilliant orange-flame colour. Particularly good specimens, prized as gemstones, are of Mexican origin.

fire polishing *(Glass)*. The polishing of glassware, decorated with a pressed pattern, by holding it in a **glory-hole**.

fireproof aggregates *(Build.)*. Materials such as crushed firebricks, fused clinkers, slag etc., incorporated in concrete to render it fire-resisting.

fire refining *(Eng.)*. The refining of blister copper by oxidizing the impurities in a reverberatory furnace and removing the excess oxygen by poling. May be used as an alternative to electrolytic refining, and is in any case carried out as a preliminary to this.

fire retardant adhesive *(Eng.)*. Heavy duty adhesives intended to fasten lagging around hot surfaces. These are usually alkyd based, but may contain chlorinated paraffins and antimony oxide to render the film noncombustible.

fire retardant paints *(Build.)*. Surface coatings which reduce the degree of flame spread over a surface and afford protection to the underlying substrate. These coatings act as a buffer coat between a flammable substrate and any source of ignition. Some rely solely upon their incombustible nature while those of the intumescent type swell on heating to form an insulating barrier.

fire ring *(Eng.)*. A top piston-ring of a special heat-resisting design, used in some 2-stroke oil engines.

fire sand *(Eng.)*. Refractory oxide or carbide suitable for lining furnaces.

fire stink *(Min.Ext.)*. The smell given off underground when a fire is imminent, e.g. in the **gob**; also, smell of sulphuretted hydrogen from decomposing pyrite.

fire-stone *(Geol.)*. A stone or rock capable of withstanding a considerable amount of heat without injury. The term has been used with reference to certain Cretaceous and Jurassic sandstones employed in the manufacture of glass furnaces.

fire stop *(Build.)*. An obstruction across an air passage in a building to prevent flames from spreading further. Also *draft stop*.

fire top-centre *(Autos.)*. The top dead-centre of an internal combustion engine, when the piston is about to make its power stroke.

fire trap *(Image Tech.)*. A pair of rollers through which 35 mm film passed on leaving and entering the magazines of a cinema projector, intended to limit the spread of fire from the gate when highly inflammable nitrate film was in use; now obsolete.

fire-tube boiler *(Eng.)*. A boiler in which the hot furnace gases, on their way to the chimney, pass through tubes in the water space, as opposed to a **water-tube boiler**. See **marine boiler**, **locomotive boiler**.

firewall *(Aero.)*. See **bulkhead**.

firing *(Electronics)*. Establishment of discharge through gas tube.

firing *(Eng.)*. (1) The process of adding fuel to a boiler furnace. (2) The ignition of an explosive mixture, as in a petrol or gas engine cylinder. (3) Excessive heating of a bearing.

firing *(Vet.)*. The application of thermocautery to the tissues of animals.

firing angle *(Electronics)*. Phase angle (of an applied voltage) at which a thyristor (or a gas filled valve, e.g. thyratron) starts to conduct.

firing key *(Elec.Eng.)*. A key which fires a charge of explosive by completing the electric circuit to a fuse.

firing order *(Autos.)*. The sequence in which the cylinders of a multi-cylinder internal combustion engine fire, e.g. 1, 3, 4, 2 for a 4-cylinder engine.

firing power *(Electronics)*. The minimum RF power required to start a discharge in a switching tube for a specified ignitor current.

firing stroke *(Autos.)*. The power or expansion stroke of an internal combustion engine.

firing time *(Electronics,Radar)*. The interval between applying a d.c. voltage to the trigger electrode of a thyristor or switching tube and the beginning of conduction. In radar the time required to establish an RF discharge in a switching tube (*transmit-receive* or *anti-transmit-receive*) after the application of RF power.

firing tools *(Eng.)*. Implements (e.g. shovel, rakes, and slicers or slicing bars) used in firing a boiler furnace by hand.

firmer chisel *(Build.)*. A woodcutting chisel, usually thin in relation to its width (1/16 to 2 in or 3 to 50 mm). Stouter than a paring chisel but less robust than a mortise chisel.

firmer gouge *(Build.)*. Standard type of **gouge** with the bevel on the outside (cf. **scribing gouge**). Used for cutting grooves and recesses.

firmware *(Comp.)*. See **hardwired logic**.

firn *(Geol.)*. See **névé**.

firring *(Build.)*. Timber strips of constant width but varying depth, which are nailed to the wood bearers to flat roofs as a basis for roof boarding, to which they give a suitable fall. Also spelt *furring*.

first detector *(Telecomm.)*. See **mixer**.

first generation computer *(Comp.)*. An early machine built around 1951 which used electronic **valves** and **mercury delay lines** in the processor and electrostatic and **magnetic drums** and **magnetic tape** for on-line storage. In 1949 the first operational **stored program** computer, the Manchester Mark 1, was completed. Other important development machines were the Cambridge EDSAC, NPL Pilot ACE, Washington SEAC, BINAC, the very large commercial UNIVAC, EDVAC, Princeton's IAS, WHIRLWIND 1. Many of these were logically and functionally equivalent to a present day **microcomputer**. See **computer generations**.

first-in, first-out *(Comp.)*. See **queue**.

first-order reaction *(Chem.)*. One in which the rate of reaction is proportional to the concentration of a single reactant, i.e.

$$\frac{dc}{dt} = -kc,$$

where c is the concentration of the reagent.

First Point of Aries *(Astron.)*. The point in which the ecliptic intersects the celestial equator, crossing it from south to north; the origin from which both right ascension and celestial longitude are measured. See **equinoctial points**, **equinox**.

First Point of Libra *(Astron.)*. See **equinoctial points**.

first runnings *(Chem.)*. The first fraction collected from a **fractional distillation** process, usually containing low boiling impurities.

first ventricle *(Zool.)*. In Vertebrates, the cavity of the left lobe of the cerebrum.

first weight *(Min.Ext.)*. The first indications of roof pressure which occur after the removal of coal from a seam.

'fir tree' roots *(Aero.)*. A certain type of fixing adopted for turbine blades, the outline form of the root resembling that of a fir tree.

FIS *(Aero.)*. Abbrev. for *Flight Information Service*.

Fischer reagent for water *(Chem.)*. The sample is titrated with a methanol solution of iodine, sulphur (IV) oxide and pyridine, free iodine not appearing until after the end point, according to the reaction $H_2O + I_2 + SO_2 + CH_3OH \rightarrow 2HI + CH_2HSO_4$.

Fischer-Tropsch process *(Chem.)*. Method of obtaining fuel oil from coal, natural gas etc. Cf. **hydrogenation**. The 'synthesis gas', hydrogen and carbon monoxide in proportional volumes 2:1, is passed, at atmospheric or slightly higher pressure and temperature up to 200°C, through contact ovens containing circulating water with an iron or cobalt catalyst. The gases are washed out and the resultant oil contains alkanes and alkenes, from the lower members up to solid waxes; fractionation yields petrols, diesel oils etc.

fish-beam *(Civ.Eng.)*. A beam which is *fish-bellied*.

fish bellied *(Eng.)*. Said of (1) steel girders with a convex lower edge; (2) long straight-edges, which are convex upward. Such a form results in greater resistance to bending.

fishbone antenna *(Telecomm.)*. An **end-fire array** consisting of dipoles spaced along a twin transmission line, in the same pattern as the bone structure of many fishes.

fished joint *(Civ.Eng.)*. A **butt joint** between rails or beams, in which fish-plated or cover straps are fitted on both sides of the joint and bolted together.

fish-eye lens *(Image Tech.)*. Camera lens covering an extremely wide angle, up to at least 180°, but with considerable barrel distortion and fore-shortening of perspective.

fish glue *(Genrl.)*. (1) *Isinglass.* (2) Any glue prepared from the skins of fish (esp. sole, plaice), fish-bladders and offal.

fishing *(Eng.)*. Recovering tools dropped from drilling tackle during deep rock drilling operations.

fishing tool *(Min.Ext.)*. A tool attached to a drill string and designed to catch and retrieve components lost down-hole. Also *overshot tool.*

fish-paper *(Elec.Eng.)*. A flexible insulating material, usually of varnished cambric, for separating coil windings, etc.

fish-plate *(Civ.Eng.)*. A steel or wood cover-plate, fitted one to each side of a fished joint between successive lengths of beam or rail. Also called *fish-bar, fish piece, shin, splice piece.*

fish-wire *(Elec.Eng.)*. A thin wire drawn into a conduit for electric cables or wires during construction, and subsequently used for drawing in the cables or wires themselves.

fissile *(Phys.)*. Capable of nuclear fission, i.e. breakdown into lighter elements of certain heavy isotopes (^{232}U, ^{235}U, ^{239}Pu) when these capture neutrons of suitable energy. Also *fissionable.* See **reactor**.

fission *(Bot.)*. Asexual reproduction of some unicellular organisms in which the cell divides into two more or less equal parts as in the fission yeast *Schizosaccharomyces.* Also *binary fission.* Cf. **budding**.

fission *(Phys.)*. The spontaneous or induced disintegration of a heavy atomic nucleus into two or more lighter fragments. The energy released in the process is referred to as *nuclear energy.* See **binding energy of the nucleus**.

fissionable *(Phys.)*. See **fissile**.

fission bomb *(Phys.)*. Same as **atomic bomb**.

fission chain *(Phys.)*. Atoms formed by uranium or plutonium fission have too high a neutron-proton ratio for stability. This is corrected either by neutron emission (the delayed neutrons) or more usually by the emission of a series of beta-particles, so forming a short radioactive decay chain.

fission chamber *(Nuc.Eng.)*. Ionization chamber lined with a thin layer of uranium. This can experience fission by slow neutrons, which are thereby counted by the consequent ionization.

fission neutrons *(Phys.)*. Those released by nuclear fission, having a continuous spectrum of energy with a maximum of about 10^6 eV.

fission parameter *(Phys.)*. The square root of the atomic number of a fissile element divided by its relative atomic mass.

fission poisons *(Nuc.Eng.)*. Fission products with abnormally high thermal neutron absorption cross-sections, which reduce the reactivity of nuclear reactors. Principally ^{135}Xe and ^{149}Sm.

fission products *(Phys.)*. Atoms, often radioactive, resulting from nuclear fission. They have masses of roughly half of that of the fissioning nucleus, e.g. ^{90}Sr, a major contributor to radiation in *fall-out* from nuclear explosions.

fission spectrum *(Phys.)*. The energy distribution of neutrons released by nuclear fission.

fission yield *(Phys.)*. The percentage of fissions for which one of the products has a specific mass number. Fission yield curves show two peaks of approximately 6% for mass numbers of about 97 and 138. The probability of fission dividing into equal mass products falls to about 0.01%.

fissiped *(Zool.)*. Having free digits. Cf. **pinnatiped**.

fissure *(Geol.,Min.Ext.)*. A cleft in rock determined in the first instance by a fracture, a joint plane, or fault, subsequently widened by solution or erosion; may be open, or filled in with superficial deposits, often minerals of pneumatolytic and hydatogenetic provenance. See also **grike**.

fissure *(Med.)*. (1) Any normal cleft or groove in organs of the body. (2) Linear ulceration of the anus, usually the result of constipation.

fissure eruption *(Geol.)*. Throwing-out of lava and (rarely) volcanic 'ashes' from a fissure, which may be many kilometrs in length. Typically there is no explosive violence, but a quiet welling-out of very fluid lava. Recent examples are known from Iceland.

fistula *(Build.)*. An ancient name for a water-pipe.

fistula *(Med.)*. An epithelial lined track connecting two hollow viscera. adj. *fistulous.*

fistulous withers *(Vet.)*. Abscess and fistula formation in the withers of the horse; *Brucella abortus* infection has been found in some cases, and in others a nematode worm, *Onchocerca cervicalis,* has been found in the affected region.

fit *(Eng.)*. The dimensional relationship between mating parts. Limits of tolerances for shafts and holes to result in fits of various qualities, e.g. clearance, transition and interference fits are laid down by national Standard Specifications.

fit *(Med.)*. A sudden attack of disturbed function of the sensory or of the motor parts of the brain, with or without loss of consciousness. See also **epilepsy**.

FITC *(Immun.)*. Abbrev. for *Fluorescein IsoThioCyanate* used for making antibodies fluorescent for use in fluorescent antibody techniques.

fitch *(Build.)*. A small, long-handled, hog's-hair brush, used for fine finishing work. Made with chisel, filibert or round tapered tips.

fitness *(Ecol.)*. (1) In natural history, the degree to which an organism is adapted to its environment and can therefore survive the struggle for existence. (2) In evolutionary ecology, the extent to which an individual passes on its genes to the next generation. Cf. **fitness, Darwinian, fitness, inclusive**.

fitness, Darwinian *(Biol.)*. The number of offspring of an individual, or the number relative to the mean (*relative fitness*), that live to reproduce in the next generation; effectively, the number of genes passed on.

fitness, inclusive *(Biol.)*. Same as *Darwinian fitness,* but including those genes that the individual shares with its relatives and are passed on by them.

fitter *(Eng.)*. A mechanic who assembles finished parts in an engineering workshop.

fitter's bench *(Eng.)*. A heavy wooden bench provided with a vice and a drawer for tools.

fitter's hammer *(Eng.)*. A hand hammer having a flat striking face, and either a straight, cross, or ball pane.

Fittig's synthesis *(Chem.)*. The synthesis of benzene hydrocarbon homologues by the action of metallic sodium on a mixture of a brominated benzene hydrocarbon and bromo- or iodo-alkane in a solution of dry ether.

fitting *(Elec.Eng.)*. A device used for supporting or containing a lamp, together with its holder, and its shade or reflector; part of the equipment of an electric-light installation.

fitting *(Eng.)*. Hand or bench work involved in the assembly of finished parts by a fitter.

fittings *(Eng.)*. (1) Small auxiliary parts of an engine or machine. (2) Boiler accessories, as valves, gauges etc.

fitting shop *(Eng.)*. The department of an engineering workshop where finished parts are assembled. See **erecting shop**.

FitzGerald-Lorentz contraction *(Phys.)*. The contraction in dimensions (or time scale) of a body moving through the ether with a velocity approaching that of light, relative to the frame of reference (Lorentz frame) from which measurements are made. Also *Lorentz contraction.*

five-centred arch *(Build.)*. An arch having the form of a false ellipse struck from 5 centres.

five-unit code *(Telecomm.)*. The Baudot code, as used for machine transmission of telegraphic signals in synchronous and start-stop systems.

fix *(Aero.)*. The exact geographical position of an aircraft, as determined by terrestrial or celestial observation, or by radio cross-bearing. Cf. **pinpoint**.

fix *(Surv.)*. Point accurately established on plan, perhaps by observations for latitude and longitude.

fixation *(Behav.)*. In psychoanalytic theory, a defence mechanism caused by acute anxiety and frustration during development; the individual is temporarily or permanently halted at a particular psychosexual stage of growth and this is reflected in an unevenness of personality development (e.g. fixation at the oral stage may result in a compulsive eating or talking disorder).

fixation *(Biol.)*. (1) See **carbon fixation, nitrogen fixation, phosphate fixation**. (2) In microscopy, preparative treatment, especially before embedding and sectioning, aimed to stabilize (or fix) structures against subsequent treatments; e.g. for electron microscopy, with glutaraldehyde which crosslinks proteins and osmium tetroxide which stabilises lipids, or, for light microsopy, with mixtures containing ethanol, acetic acid, chromium trioxide, mercuric ions etc. which may precipitate proteins.

fixation *(Zool.)*. The action of certain muscles which prevent disturbance of the equilibrium or position of the body or limbs; the process of attachment of a free-swimming animal to a substratum, on the commencement of a temporary or permanent sessile existence.

fixation of nitrogen *(Chem.)*. For commercial purposes (fertilizers etc.) the **Haber process** is now the most important method of nitrogen fixation.

fixed action pattern *(Behav.)*. Abbrev. *FAP*. Species specific movements, recognised by their relatively high degree of stereotype, although they may show variation in their orientation aspects. Originally all FAPs were assumed to be innate, but it is now recognised that such patterns are often influenced by environmental factors during development.

fixed beam *(Build.)*. A beam with fixed ends.

fixed carbon *(Eng.)*. Residual carbon in coke after removal of hydrocarbons by distillation in inert atmosphere.

fixed-charge collector *(Elec.Eng.)*. A device, attached to prepayment meters, which is arranged so that the insertion of coins corresponding to a fixed charge (rental or hire purchase charge) allows a consumer to close a switch to receive a supply, the switch being opened automatically after a predetermined time.

fixed contact *(Elec.Eng.)*. The contact of a switch or fuse which is permanently fixed to the circuit terminal.

fixed eccentric *(Eng.)*. An eccentric which is permanently keyed to a shaft, not capable of angular movement, as is a *loose eccentric*.

fixed end *(Build.)*. The term applied, in theory, to the end of a beam when it is built in or otherwise secured so that the tangent at the end to the curve taken up by the beam when it is deflecting under applied loading remains fixed.

fixed expansion *(Eng.)*. A steam engine in which the cut-off cannot be altered and which thus works with a constant expansion ratio.

fixed head disk unit *(Comp.)*. One where a separate read/write head is positioned over each track on each surface; this reduces access time, at increased cost.

fixed interval schedule *(Behav.)*. See **interval schedule of reinforcement**.

fixed-length record *(Comp.)*. One where the number of available bits (or characters) is predetermined. Cf. **variable-length records**.

fixed-loop aerial *(Aero.)*. A loop aerial, used with a homing receiver, which is fixed in relation to the aircraft's centreline.

fixed-needle surveying *(Surv.)*. Traverse with magnetic compass locked, instrument being used like theodolite for measuring azimuth angles, except where there is no nearby iron or steel to affect bearing, when compass needle may be used to give a check reading.

fixed-pitch propeller *(Aero.)*. See **propeller**.

fixed point notation *(Comp.)*. Numbers are expressed as a set of digits with the decimal point in position (e.g. 142.687). The position of the decimal point is generally maintained by the program or programmer. The magnitude of fixed point numbers is limited by the construction of the computer, but operations are generally very fast and preferred for most commercial data processing work. See **floating-point notation**.

fixed points *(Phys.)*. Temperature which can be accurately reproduced and used to define a temperature scale and for the calibration of thermometers. The temperature of pure melting ice and that of steam from pure boiling water at one atmosphere pressure define the Celsius and Fahrenheit scales. The *International Practical Temperature Scale* defined ten fixed points ranging from the triple point of hydrogen (13.81 K) to the freezing point of gold (1337.58 K). See **triple point, Kelvin thermodynamic scale of temperature**.

fixed pulley *(Eng.)*. A pulley keyed to its shaft. See **fast pulley**.

fixed ratio schedule *(Behav.)*. See **ratio schedule of reinforcement**.

fixed sash *(Build.)*. (1) A **stand sheet**. (2) A sash permanently fixed in a solid frame. Also *dead light*.

fixed time-lag *(Elec.Eng.)*. See **definite time-lag**.

fixed-trip *(Elec.Eng.)*. A term applied to certain forms of circuit-breaker or motor starter to indicate that the tripping mechanism cannot operate while the breaker or starter is actually being closed. Also called *fixed-handle circuit-breaker*. Cf. **free-trip**.

fixed-type metal-clad switchgear *(Elec.Eng.)*. Metal-clad switchgear in which all parts are permanently fixed, no provision being made for easy removal of any part for inspection or maintenance purposes.

fixing *(Image Tech.)*. Process for removing unreduced silver halides after development of an emulsion.

fixing block *(Build.)*. A block of material, having the shape of a brick, which can be built into the surface of a wall to provide a substance to which joinery, such as window frames, may be nailed. Fixing blocks are made of porous concrete, of coke breeze, or of a special brick made with a mixture of sawdust which burns away in the kiln to leave a porous brick material. See also **wood brick**.

fixing fillet, pad *(Build.)*. (1) A **slip**. (2) A strip fixed to a surface to support something such as a shelf.

fixing plugs *(Build.)*. Plugs made of plastic, metal or other material used in conjunction with screws or bolts to provide fixings to walls or concrete surfaces.

fixings *(Build.)*. Supports, such as grounds and plugs, for securing joinery in position.

fixture *(Build.)*. An attachment to a building.

fixture *(Eng.)*. A device used in the manufacture of (interchangeable) parts to locate and hold the work without guiding the cutting tool.

Fizeau fringes *(Phys.)*. See **contour fringes**.

fjords *(Geol.)*. See **fiords**.

flabellate, flabelliform *(Bot.)*. Shaped like a fan.

flaccid *(Bot.)*. A cell or cells in a tissue which are not *turgid* but limp. Turgor pressure is zero as a result of water loss. See also **plasmolysis**.

flag *(Comp.)*. Indicator which can be set or unset to indicate a condition in a program or a set of data. See **sentinel, carry flag, EOF, overflow flag, separator**.

flag *(Geol.)*. Natural flagstones are sedimentary rocks of any composition which can be readily separated, on account of their distinct stratification, into large slabs. They are often fine-grained sandstones interbedded with shaly or micaceous partings along which they can be split.

flag *(Image Tech.)*. A small opaque shield used in studio lighting to provide a local shadow. Also *french flag*.

flag *(Paper)*. A coloured slip of paper inserted in a reel to indicate the position of a join or break in the web.

flag alarm *(Elec.Eng.)*. See **flag indicator**.

flagellar root *(Bot.)*. A group of microtubular structures lying under the plasmalemma close to the basal body in flagellated cells.

Flagellata *(Zool.)*. See **Mastigophora**.

flagellate *(Bot.,Zool.)*. (1) Having flagella. (2) An organism in which the body form is a unicell or colony of cells with flagella. (3) Bearing a long thread-like appendage. (4) A member of the Mastigophora.

flagellin *(Biol.)*. The protein monomer which is assembled into a helical array to form the filament of a bacterial flagellum.

flagellum *(Biol.)*. Long hair-like extension from the cell surface whose movement is used for locomotion. In sperm and protozoa its internal structure is very similar to that of cilia, but bacterial flagella have a unique structure and a rotatory action.

flagellum *(Zool.)*. (1) A threadlike extension of the protoplasm of a cell or of a Protozoan. (2) In other Arthropoda, a filiform extension to an appendage, e.g. a crustacean limb, an insect antenna. (3) In Gastropoda, one of the male genitalia.

flag indicator *(Elec.Eng.)*. In navigational and other instruments, a semaphore-type signal which warns viewer when instrument readings are unreliable. Also called *alarm flag, flag alarm.*

flag leaf *(Bot.)*. The top-most leaf on a culm, just below the ear, in such cereal grasses as wheat and barley.

flail chest *(Med.)*. A term used to describe paradoxical movement of the chest wall during respiration which can occur after major injury to the chest with multiple fractured ribs.

flail joint *(Med.)*. A joint in which there is, as a result of disease or of operation, excessive mobility.

flaking *(Build.)*. (1) The breaking away of surface plaster due to non-adhesion with the undercoat or to free lime or impurities in the basic surface. (2) A defect in paintwork, the paint film breaking away in greater or lesser areas from the surface it was covering.

flame *(Chem.)*. A region in which chemical interaction between gases occurs, accompanied by the evolution of light and heat.

flame arc *(Elec.Eng.)*. An electric arc maintained between carbons containing certain metallic salts, which give a colour to the arc flame.

flame-arc lamp *(Elec.Eng.)*. An arc lamp using flame carbons.

flame blow-off factor *(Eng.)*. Relation between the velocity of combustible mixture and the rate of flame propagation, the latter varying appreciably with gases of different composition. See also **burner firing block, flame retention, piloted head** and **tunnel burners.**

flame carbons *(Elec.Eng.)*. Carbon electrodes containing certain metallic salts, which have the effect of colouring an arc maintained between them. See **flame-cored carbon.**

flame cell *(Zool.)*. See **solenocyte.**

flame cleaning *(Build.)*. A method of preparing steelwork in which an oxy-acetylene flame is passed over the surface to loosen mill scale and rust. The rust and scale are then removed by wire brushing and the surface primed while still warm.

flame-cored carbon *(Elec.Eng.)*. A carbon electrode having a central core of a material designed to colour the flame of an arc drawn from it.

flame cutting *(Eng.)*. Cutting of ferrous metals by oxidation, using a stream of oxygen from a blow pipe or torch on metal preheated to about 800°C by fuel gas jets in the cutting torch.

flame damper *(Aero.)*. See **damper.**

flame failure *(Eng.)*. The accidental extinction of a burner flame, e.g. in an oil-fired boiler, which has to be detected in order to stop or otherwise control the supply of further fuel, to prevent explosion or damage.

flame-failure control *(Eng.)*. Direct-flame thermostat with interconnected relay valve, which provides a constantly burning pilot flame for igniting the main gas burners, and automatically shuts off the gas supply to the main burner in the event of the pilot flame becoming extinguished.

flame hardening *(Eng.)*. The use of an intense flame from e.g. an oxy-acetylene burner for local heating of the surface of a hardenable ferrous alloy which is then immediately cooled.

flame ionization gauge *(Chem.)*. Used in gas chromatography, as a very sensitive detector for the separate fractions in the effluent carrier gas, by burning it in a hydrogen flame, the electrical conductivity of which is measured by inserting two electrodes in the flame with an applied voltage of several hundred volts.

flame lamp *(Elec.Eng.)*. A filament lamp having the bulb in the form of a flame.

flame plates *(Eng.)*. Those plates of a boiler firebox subjected to the maximum furnace temperature.

flame-proof *(Elec.Eng.)*. See **explosion-proof.**

flame retention *(Eng.)*. Ability to retain a stable flame with gas burners at all rates of gas flow, irrespective of adverse combustion conditions. See also **burner firing block, flame blow-off factor, piloted head, tunnel burners.**

flame spectrum *(Phys.)*. That imparted to non-luminous flame by substances volatilized in it.

flame temperature *(Eng.)*. The temperature at the hottest spot of a flame.

flame test *(Chem.)*. The detection of the presence of an element in a substance by the colouration imparted to a Bunsen flame.

flame trap *(Autos.)*. A gauze or grid of wire, or coiled corrugated sheet, placed in the air intake to a carburettor to prevent the emission of flame from a 'pop-back'.

flame trap *(Eng.)*. Device inserted in pipelines carrying air-gas mixture of a combustible or self-burning nature to arrest the flame in the event of a flash-back (or backfire) occurring at the burner.

flame tube *(Aero.)*. The perforated inner tubular 'can' of a gas turbine combustion chamber in which the actual burning occurs; cf. **combustion chamber.**

flange *(Eng.)*. (1) A projecting rim, as the rim of a wheel which runs on rails. (2) The top or bottom members of a rolled I-beam. (3) A disk shaped rim formed on the ends of pipes and shafts, for coupling them together; or on an engine cylinder, for attaching the covers. See also **flanged rail.**

flange coupling *(Eng.)*. A shaft coupling consisting of 2 accurately faced flanges keyed to their respective shafts and bolted together.

flanged beam, -girder *(Eng.)*. A rolled-steel joist of 1-section.

flanged nut *(Eng.)*. A nut having a flange or washer formed integral with it. See **collar-head screw.**

flanged pipes *(Eng.)*. Pipes provided either with integral or attached flanges for connecting them together by means of bolts.

flanged rail *(Civ.Eng.)*. A rail section of inverted T shape, now almost universally used in which the flange is at the bottom and the end of the stem of the T at the top. The latter part is enlarged locally to form the head of the rail. Also called *flat-bottomed rail.*

flanged seam *(Eng.)*. A joint made by flanging the ends of furnace tubes and bolting them together between a pair of steel rings.

flange joint *(Eng.)*. Any joint between pipes, made by bolting together a pair of flanged ends.

flange protection *(Elec.Eng.)*. The rendering of electrical apparatus flame-proof by providing all joints with very wide flanges.

flank *(Build.)*. A roof valley.

flank *(Eng.)*. (1) That part of a gear-tooth profile which lies inside the pitch-line or circle. (2) The working face of a cam. (3) The straight side connecting the crest and the root of a screw-thread.

flank angle *(Eng.)*. An angle between the flanks of a thread and a plane perpendicular to the axis, measured in an axial plane.

flank dispersion *(Elec.Eng.)*. See **end leakage flux.**

flanking transmission *(Acous.)*. In architectural acoustics, sound transmission by the flanking walls of two rooms. When the wall between the two rooms has a high transmission loss (e.g. double wall), flanking transmission limits the sound isolation of the two rooms.

flanking window *(Build.)*. A window located beside an external door.

flanks *(Build.,Civ.Eng.)*. (1) The parts of the intrados of an arch near to the abutments. Also called *haunches.* (2) The side surfaces of a building stone or ashlar, when it is built into a wall.

flank wall *(Build.)*. A side wall.

flannel *(Textiles)*. A soft all-wool fabric, the weave being either plain or twill. The cloth is pre-shrunk and lightly raised.

flannelette *(Textiles)*. A cotton fabric of plain or twill weave, raised on both sides and used for pyjamas, nightdresses, sheets, and working shirts.

flanning *(Build.)*. The internal splay of a window jamb, or of a fireplace.

flap *(Aero.)*. Any surface attached to the wing, usually to the trailing edge, which can be adjusted in flight, either automatically or through controls, to alter the lift as a whole; primarily on fixed-wing aircraft, but occasionally on rotor systems.

flap *(Med.)*. An area of tissue partly separated by the knife from the surface of the body, in connection with amputation of a limb or for the purpose of grafting skin.

flap angle *(Aero.)*. The angle between the chord of the flap, when lowered or extended, and the wing chord.

flap attenuator *(Telecomm.)*. One consisting of a strip of absorbing material which is introduced through a non-radiating slot in a waveguide.

flapping angle *(Aero.)*. The angle between the tip-path plane of a helicopter rotor and the plane normal to the hub axis.

flapping hinge *(Aero.)*. The pivot which permits the blade of a helicopter to rise and fall within limits, i.e. variation of zenithal angle in relation to the rotor head.

flap setting *(Aero.)*. The flap angle for a particular condition of flight, e.g. take-off, approach, landing.

flap tile *(Build.)*. A purpose made tile, shaped so as to fit over a hip or valley line, or to catch water.

flap trap *(Build.)*. A type of antiflood valve, in which back flow is prevented by a hinged metal flap fitted in an intercepting chamber, to allow flow in one direction only.

flap valve *(Build.,Eng.)*. (1) A sheet of flexible material like polythene hinged about one edge, allowing one way flow of air through ventilators. Formerly made of mica. See also **flap trap**. (2) A similar non-return valve for liquids, made of metal, faced with rubber or an O-ring, used for low pressures.

flare *(Acous.)*. The prominent part of the opening of a horn, bell, or trumpet attached to a loud speaking unit.

flare *(Astron.)*. An energetic outburst in the lower atmosphere of the **Sun**.

flare *(Image Tech.)*. Scattering of light within an optical system, such as a camera lens; it produces unwanted exposure, unrelated to the required image, and may appear as a patch of light or as an overall haze, reducing the effective tonal range. See **flare factor**.

flare factor *(Image Tech.)*. Ratio between the luminance scale of the subject and that of the camera image.

flare gas *(Min.Ext.)*. See **flare stack**.

flare header *(Build.)*. A brick which has been burnt to a darker colour at one end, so that it may be used with others in facing-work to vary the effect.

flare-out *(Aero.)*. Controlled approach path of aircraft immediately prior to landing.

flaring *(Acous.)*. A term applied to the end of a pipe, etc., when it is shaped out so as to be of increasing diameter towards the end. See also **flare** (Acous.).

flaser structure *(Geol.)*. A streaky, patchy structure in a dynamically metamorphosed rock. Flaser gabbro, for example, shows an apparent crude flow structure round unaltered granular lenses.

flash *(Eng.)*. A thin fin of metal formed at the sides of a forging where some of the metal is forced between the faces of the forging dies. By extension, a similar extrusion in other (e.g. moulded) materials.

flash *(Plastics)*. The excess material forced out of a mould during moulding. Sometimes referred to as a *fin*, if the mould is in two halves.

flashback voltage *(Electronics)*. The inverse peak voltage in a gas tube at which ionization occurs.

flash boiler *(Eng.)*. A steam boiler consisting of a long coil of steel tube, usually heated by oil burners, in which water is evaporated as it is pumped through by the feed pump. See **steam car**.

flashbulb *(Image Tech.)*. A source of very brief illumination by the electrical ignition of metallic wire (magnesium or zirconium) in an oxygen-filled bulb; it can only be used once but multi-bulb arrays are made for operation in rapid succession.

flashbulb memory *(Behav.)*. Clear recollections that people sometimes have of the events surrounding a dramatic event, e.g. the assassination of a head of state.

flash-butt welding *(Elec.Eng.)*. See **resistance-flash welding**.

flash colours *(Zool.)*. See **startle colours**.

flash cube *(Image Tech.)*. A group of four flashbulbs in one mount, individually fired by a mechanical striker igniting a primer.

flash distillation *(Chem.)*. The spraying of a liquid mixture into a heated chamber of lower pressure, in order to drive off some of the more volatile constituents. Used in the petroleum industry.

flash drying *(Eng.)*. Removal of moisture as stream of small particles falls through current of hot gas.

flashed glass *(Glass)*. A term sometimes applied to glass coloured by the application of a thin layer of densely coloured glass to a thicker, colourless, base layer.

flasher *(Elec.Eng.)*. (1) An oscillator, or thermally operated switch and heater, arranged to switch a lamp on and off repetitively. (2) The lamp itself.

flash gun *(Image Tech.)*. An assembly of flash bulb or flash tube in a reflector complete with its power source, mounted on or connected to the camera for synchronization.

flashing *(Build.)*. (1) Method of brick burning in which the air supply is periodically stopped in order that the colouring of the bricks shall be irregular. (2) Glossy patches or streaks on flat-finished surfaces, usually attributable to poor technique.

flashing *(Hyd.Eng.)*. The process of passing a boat across any point in a river where there is a sudden fall; effected by constructing a convergent passage from the high to the low level, shutting it by a sluice gate to allow the water to pond up, and then opening the gate and allowing the boat to be carried through the sluice way by the artificially deepened water.

flashing board *(Build.)*. A board to which flashings are secured.

flashing compound *(Build.)*. Thick nondrying materials used for filling in crevices, e.g. between insulation blocks. Essential properties are impermeability and elasticity.

flashing light *(Ships)*. A navigation mark identified during darkness by a distinctive pattern of flashes of light. See **alternating light**, **occulting light**.

flashlight photography *(Image Tech.)*. Photography where rapid subject movement must be frozen, or where inadequate available light is supplemented, using a very brief source of illumination. Formerly, magnesium powder was ignited but current practice employs **flashbulbs** or **electronic flash**, usually synchronized with the camera shutter.

flashover *(Elec.Eng.)*. An electric discharge over the surface of an insulator.

flashover test *(Elec.Eng.)*. A test applied to electrical apparatus to determine the voltage at which a flashover occurs between any two parts, or between a part and earth. Also called *sparkover test*.

flashover voltage *(Elec.Eng.)*. The highest value of a voltage impulse which just produces flashover.

flash photolysis *(Chem.)*. Photolysis induced by light flashes of short duration but high intensity, e.g. from a laser.

flash point *(Build.)*. The temperature at which material gives off a vapour which will ignite on exposure to a flame.

flash point *(Min.Ext.)*. The temperature at which a liquid, heated in a Cleveland cup (open test) or in a Pensky-Martens apparatus (closed test), gives off sufficient vapour to flash momentarily on the application of a small flame. The *fire point* is ascertained by continuing the test.

flash radiography *(Radiol.)*. High-intensity, short duration X-ray exposure.

flash roasting *(Eng.)*. The roasting of finely-ground concentrates by introducing them into a large combustion chamber in which the sulphur is burned off as they fall.

flash spectrum *(Astron.)*. A phenomenon seen at the first instant of totality in a solar eclipse; the dark lines of the Fraunhofer spectrum formed in the chromosphere flash out into bright emission lines as soon as the central light of the sun is cut off.

flash suppressor *(Elec.Eng.)*. A device for preventing

flashovers on the commutators of d.c. generators; it consists of an automatically operated switch for short-circuiting certain points in the winding, thereby reducing the voltage to zero before a flashover has had time to develop.

flash-synchronized *(Image Tech.)*. Said of shutters which, when released, can close an electrical circuit for firing a flash gun.

flash test *(Elec.Eng.)*. A test applied to electrical equipment for testing its insulation strength; it consists of the application of a voltage of about twice the working voltage, for a period of not more than about one minute.

flash tube *(Image Tech.)*. Discharge tube used for **electronic flash**.

flash welding *(Eng.)*. An electric welding process similar to butt welding, in which the parts are first brought into very light contact. A high voltage starts a flashing action between the two surfaces, which continues while sufficient forging pressure is applied to the parts to complete the weld.

flask *(Foundry)*. A moulding box of wood, cast-iron, or pressed steel, for holding the sand mould in which a casting is made; it may be in several sections. See **cope**, **drag**.

flask *(Nuc.Eng.)*. Lead case for storing or transporting multicurie radioactive sources or a container for the transport of irradiated nuclear fuel. Also called *cask*, *coffin*.

flat *(Elec.Eng.)*. A term used to denote a point on the surface of a commutator where the bars are lower than normal, due to wear or displacement.

flat *(Image Tech.)*. The unit panel from which studio sets are constructed.

flat arch *(Build.)*. An arch whose intrados has no curvature and whose voussoirs (laid in parallel courses) are arranged to radiate to a centre. It is used over a doorway, fireplace, and window-openings to relieve the pressure on the beam or lintel below it. Also called a *jack arch* and *straight arch*.

flatback stope *(Min.Ext.)*. A stope, in **overhand stoping**, worked upwards into a lode more or less parallel to level.

flat band *(Build.)*. A square and plain impost stone.

flat bed *(Print.)*. A general name distinguishing a cylinder machine with a horizontal flat bed from a *rotary machine* or from a *platen machine*.

flat-bottomed rail *(Civ.Eng.)*. See **flanged rail**.

flat chisel *(Eng.)*. A cold chisel having a relatively broad cutting edge, used in chipping flat surfaces.

flat compounded *(Elec.Eng.)*. Said of a compound-wound generator the series winding of which has been so designed that the voltage remains constant at all loads between no-load and full-load. Also *level-compounded*.

flat file *(Comp.)*. See **relational database**.

flat finish *(Build.)*. Surface coating showing no gloss, sheen or lustre.

flat foot *(Med.)*. See **pes planus**.

flat four *(Aero.)*. Four cylinder, horizontally opposed piston engine.

flat gouge *(Build.)*. A gouge having a cutting edge shaped to a large radius of curvature.

flat joint *(Build.)*. The type of mortar joint made in **flat pointing**.

flat-joint jointed *(Build.)*. A flat joint which has had a narrow groove struck along the middle of its face by means of a jointer.

flat keel *(Ships)*. See under **keelson**.

flat knitting machine *(Textiles)*. A **weft-knitting machine** with straight needle beds.

flat lead *(Build.)*. Sheet-lead.

flat lighting *(Image Tech.)*. General diffuse lighting, minimising the contrast between high-light and shadow areas of the scene.

flat of keel *(Ships)*. The portion of a ship's form actually coinciding with the baseline in a transverse plane.

flat pointing *(Build.)*. The method of pointing, used for uncovered internal wall surfaces, in which the stopping is formed into a smooth flat joint in the plane of the wall.

flat random noise *(Acous.)*. See **white noise**.

flat-rate tariff *(Elec.Eng.)*. A method of charging for electrical energy in which only one single charge is made, e.g. a fixed price per unit consumed.

flat region *(Nuc.Eng.)*. Portion of reactor core over which neutron flux (and hence power level) is approximately uniform.

flat roof *(Build.)*. A roof surface laid nearly horizontal, i.e. having a fall of only about 1 in 80, provided for drainage.

flats *(Eng.)*. (1) Iron or steel bars of rectangular section. (2) The sides of a (hexagonal) unit.

flat spot *(Autos.)*. In a carburettor, a point during increase of air flow (resulting from increased throttle opening or speed) at which the air-fuel ratio becomes so weak as to prevent good acceleration.

flat stitching *(Print.)*. The stitching of a book close to the back with wire which passes through from the first page to the last, as distinct from **saddle stitching**. Cf. **stabbing**, **drilled and strung**.

flattener *(Glass)*. One who takes a cylindrical piece of glass like a wide tube, cracked longitudinally, and, after heating it to softening in a furnace, flattens it out to form a sheet. An old process only used for making special types of sheet.

flattening material *(Nuc.Eng.)*. Neutron absorber or depleted fuel rod used in centre of reactor core to give larger flat region.

flatter *(Eng.)*. (1) Smith's tool resembling a flat-faced hammer, which is placed on forged work and struck by a sledge hammer. (2) Draw-plate for producing flat wire such as hair springs.

flatting mill *(Eng.)*. Rolling mill which produces strip metal or sheet.

flatting varnish *(Build.)*. An oil varnish containing resin used as a basis for the final coat of varnish after having been rubbed down with abrasives to give a flatter surface.

flat-top antenna *(Telecomm.)*. One in which the uppermost wires are horizontal; also *roof antenna*.

flat tuning *(Telecomm.)*. A broadband receiver or amplifier, unable to discriminate between closely-spaced signals having different frequencies.

flat twin cable *(Elec.Eng.)*. Cable for wiring work in which two conductors are laid side by side (but not twisted together) and surrounded by a suitable covering or sheath.

flatulence *(Med.)*. Air or gas in the intestinal tract.

flatus *(Med.)*. Gas or air accumulated in the stomach or intestines.

flat yarn *(Textiles)*. (1) Man-made continuous filaments that have not been twisted or *textured*. (2) A straw-like filament.

flaunching *(Build.)*. The slope given to the top surface of a chimney to throw off the rain.

flavescent *(Bot.)*. Becoming yellow; yellowish.

flavone *(Chem.)*. A yellow plant pigment; the phenyl derivative of chromone, parent substance of a number of natural vegetable dyes.

flavones *(Chem.)*. Yellow pigments occurring widely in plants, derivatives of **flavone**.

flavonoids *(Bot.)*. A large group of **secondary metabolites** of bryophytes and vascular plants, based on 2-phenylbenzopyran, often as glycosides. Some sorts are pigments and others may be *phytoalexins*, insecticides or of no known function. Many are significant in chemotaxonomy.

flavoproteins *(Biol.)*. Proteins which serve as electron carriers in the electron transport chain by virtue of their prosthetic group, *flavine adenine dinucleotide*.

flavour *(Phys.)*. An index which denotes different types of quarks. The six types are: up(u), down(d), strange(s), charm(c), bottom (or beauty) (b), top (or truth) (t). Also *flavor*.

flax *(Textiles)*. Plants of *Linum usitatissimum* grown in temperate regions and used for the fibres obtained from their stems and for their seeds (linseed oil and animal food). The cellulosic fibres are prepared by **retting** followed by cleaning (including **hackling** and **scutching**).

flax tow *(Textiles)*. The short flax fibres removed during cleaning.

F-layer *(Phys.)*. Upper ionized layer in the ionosphere resulting from ultraviolet radiation from the sun and capable of reflecting radio waves back to earth at frequencies up to 50 MHz. At a regular height of 300 km during the night, it falls to about 200 km during the day. During some seasons, this remains as the F_1 layer while an extra F_2 layer rises to a maximum of 400 km at noon. Considerable variations are possible during particle bombardment from the sun, the layer rising to great heights or vanishing. It is also known as the *Appleton layer*.

fleaking *(Build.)*. Thatching a roof with reeds.

fleam *(Build.)*. Angle of rake between the cutting edge of a sawtooth and the plane of the blade.

fleam-tooth *(Build.)*. A sawtooth having the shape of an isosceles triangle.

flèche *(Arch.)*. A slender spire, particularly a timber one, springing from a roof ridge.

flèches d'amour *(Min.)*. Acicular, hairlike crystals of rutile, a crystalline form of the oxide of titanium, TiO_2, embedded in quartz. Used as a semiprecious gemstone. Also called *love arrows* (the literal translation), *cupid's darts*, or *Venus' hair stone*.

fleece wool *(Textiles)*. Wool obtained by shearing the living sheep.

fleecy fabric *(Textiles)*. (1) Term in the hosiery trade for fabrics having at the back a thick yarn which is brushed to raise a pile. (2) Any woven apparel cloth raised on one or both sides and made to resemble a fleece.

fleeting tetanus *(Vet.)*. See **transit tetany**.

Fleming diode *(Electronics)*. The 1904 detector of radio signals, with an incandescent filament and a separate anode.

Flemish bond *(Build.)*. A bond consisting of alternate headers and stretchers in every course, each header being placed in the middle of the stretchers in the courses above and below.

Flemish garden-wall bond *(Build.)*. A bond in which each course consists of three stretchers alternating with a header, each header being placed in the middle of the stretchers in the courses above and below.

Fletcher-Munson curves *(Acous.)*. Equal-loudness curves for aural perception, measured just outside the ear, extending from 20 to 20 000 Hz, and from the threshold of hearing to the threshold of pain. They are the basis of the **phon** scale.

fletton *(Build.)*. A well-known type of brick of a mottled pink and yellow colour and having sharp arrises and a deep frog; made chiefly around Peterborough and Bedford.

F-levels *(Phys.)*. See **fundamental series**.

flexible bands *(Print.)*. Those used in attaching the sections of a volume together where the bands are not let into the back of the sections, the sewing thread passing completely round each band. See **raised bands**.

flexible cable *(Elec.Eng.)*. A cable containing one or more cores of such cross-section and fine stranding as to make the whole quite flexible.

flexible cord *(Elec.Eng.)*. A flexible cable of small cross-section, consisting usually of a large number of fine wire strands, surrounded by plastic or by rubber insulation and braiding. Used for connections to portable domestic apparatus, pendant lamps etc. See **twin flexible cord**.

flexible coupling *(Eng.)*. A coupling used to connect two shafts in which perfectly rigid alignment is impossible; the drive is commonly transmitted from one flange to another through a resilient member, such as a steel spring, or a rubber disk or bushes.

flexible manufacturing system *(Eng.)*. An arrangement of computer controlled machine tools which can automatically carry out a variety of machining operations on a range of components. Abbrev. *FMS*.

flexible resistor *(Elec.Eng.)*. One resembling a flexible cable.

flexible roller bearing *(Eng.)*. An antifriction bearing containing hollow cylindrical rollers made by winding strip steel into helical form. The hollow construction permits greater deflection under load. Also called *Hyatt roller bearing*.

flexible support *(Elec.Eng.)*. A support for an overhead transmission line, which is designed to be flexible in a direction along the line, but rigid in a direction at right angles to the line.

flexible suspension *(Elec.Eng.)*. A method of suspending the contact wire of a traction system so that it has a certain amount of lateral and vertical movement relative to the fixed supports.

flexible wiring *(Elec.Eng.)*. The use of flexible cables in wiring an interior installation.

fleximer *(Build.)*. See **cement-rubber-latex**.

flexographic printing *(Print.)*. Usually reel-fed rotary, using relief rubber plates and spirit- or water-based inks which dry quickly, leaving no odour; much used for food wrappers.

flexor *(Zool.)*. A muscle which by its contraction bends a limb or a part of the body. Cf. **extensor**.

flexuose, flexuous *(Bot.)*. A zig-zag, wavy stem.

flexural rigidity *(Phys.)*. A measure of the resistance of a beam to bending.

flexural strength *(Build.)*. Cross-breaking or transverse strength. A measure of the resistance to fracture on transverse loading, i.e. the load applied centrally across a simple beam of the material under test.

flexure *(Build.)*. The bending of a member, e.g. under load.

flex-wing *(Aero.)*. A collapsible single surface fabric wing of delta planform investigated first for the return of space vehicles as gliders; later applied to army low-performance tactical aircraft of collapsible type; now for **microlights** and hang gliders.

flicker *(Phys.)*. Visual perception of fluctuation of brightness at frequencies lower than that covered by persistence of vision. Threshold of flicker depends on brightness and angle from optic axis. *Chromaticity flicker* arises from variations in **chromaticity** only.

flicker effect *(Electronics)*. Irregular emission of electrons from a thermionic cathode due to spontaneous changes in condition of emitting surface; resulting in an electronic noise.

flicker-fusion frequency *(Phys.)*. The rate of successive intermittent light stimuli just necessary to produce complete fusion, and thereby the same effect as continuous lighting.

flicker photometer *(Phys.)*. A photometer in which a screen is illuminated alternately and in quick succession by the lamp under test and a standard lamp, thus producing a flickering effect. When the illumination from the two sources of light is equal, the flickering effect disappears.

flicker shutter *(Image Tech.)*. Rotating shutter in a motion picture projector interrupting the light at least 48 times a second to minimise the perception of flicker in the projected picture.

flick roll *(Aero.)*. See **roll**.

flier *(Build.)*. See **flying-shore**.

flight *(Eng.)*. (1) The helical element in a worm or screw conveyor, usually fabricated from sheet metal, which may be hollow to allow a heating or cooling liquid to pass through it. (2) The helical element in a vibratory bowl feeder over which components pass while being unscrambled and orientated.

flight control *(Aero.)*. Control of vehicle, e.g. spacecraft, missile, or module, so that it attains its target, taking all conditions and corrections into account. Generally done by computer, controlled by signals representing actual and intended path.

flight deck *(Aero.)*. (1) Upper part of an aircraft carrier. (2) Crew compartment of a large aircraft.

flight director *(Aero.)*. An aircraft instrument (i.e. blind) flying system in which the dials indicate what the pilot must do to achieve the correct flight path as well as the actual attitude of the aircraft. The dial display is usually an **artificial horizon** with a spot or pointer which must be centred. The equipment can be coupled to a radio- or **gyro-compass** to bring the aircraft on to a desired heading, preset on the compass, or it can be coupled to the **instrument landing system** to receive these signals. In all cases, the flight director can be made automatic by switching its signals into the autopilot.

flight engineer *(Aero.)*. A member of the flying crew of an aircraft responsible for engineering duties. i.e. management of the engines, fuel consumption, power systems, etc.

flight envelope *(Aero.)*. Plot of altitude vs speed defining performance limits within which an aircraft and/or its equipment can operate.

flight fine pitch *(Aero.)*. The minimum blade angle, held by a removable stop, which the propeller of a turboprop engine can reach while in the air, and which provides braking drag for the landing approach.

flight information *(Aero.)*. A *flight-information centre* provides a *flight-information service* of weather and navigational information within a specified **flight-information region**.

flight-information region *(Aero.)*. An airspace of defined dimensions within which information on air-traffic flow is provided according to the types of airspace therein. Abbrev. *FIR*.

flight-information service *(Aero.)*. One giving advice and information to assist in the safe, efficient conduct of flights. Abbrev. *FIS*.

flight level *(Aero.)*. **Air traffic control** instructions specify heights at which controlled aircraft must fly and these are given in units of 100 ft (30 m) for altitudes of 3000 or 5000 ft (900 or 1500 m) and above.

flight Mach number *(Aero.)*. Ratio of true air speed of an aircraft to speed of sound under identical atmospheric conditions.

flight management system *(Aero.)*. Computer controlled automatic flight control system allowing the pilot to select specific modes of operation. These could include: standard instrument departure; autothrottle; standard terminal arrival system; Mach hold.

flight path *(Aero.)*. The path in space of the centre of mass of an aircraft or projectile. (Its *track* is the horizontal projection of this path).

flight plan *(Aero.)*. A legal document filed with *air traffic control* by a pilot before or during a flight (by radio), which states his destination, proposed course, altitude, speed, ETA and alternative airfield(s) in the event of bad weather, fuel shortage etc.

flight recorder *(Aero.)*. A device which records data on the functioning of an aircraft and its systems on tape or wire. (1) The *Flight Data Recorder* (*FDR*) should be in a crashproof, floatable box which may be ejected in case of an accident and is usually fitted with a homing radio beacon and flashlight. (2) The *Cockpit Voice Recorder* (*CVR*) stores all speech between crew on the flight deck and between crew and ground ATC. (3) The *Maintenance Data Recorder* (e.g. *AIDS*) receives data from hundreds of inputs from engineering systems. Popular name, *black box* (frequently a misnomer).

flight time *(Aero.)*. See **block time**.

Flinders bar *(Ships)*. A bar of soft iron attached to the magnetic compass binnacle to correct for disturbing effects of that part of the ship's field due to vertical soft iron in the fore and aft line.

flint *(Geol.)*. Concretions of silica, sometimes tabular, but usually irregular in form, particularly abundant on the bedding planes of the Upper Chalk. See also **chert**, **paramudras**.

flint glass *(Glass)*. Originally lead glass; the good quality silica needed to ensure freedom from colour was obtained from crushed flints. Now colourless glass other than flat glass.

flint gravel *(Geol.)*. A deposit of gravel in which the component pebbles are dominantly of flint, e.g. the Tertiary and fluvioglacial gravels in S.E. England.

flip-flop *(Comp.)*. Circuit which can be in either of two states. The receipt of a pulse will reverse the state. It is used to construct **delay elements** and as a one-bit storage device. Also *bistable multivibrator*, *trigger circuit*.

FLIR *(Aero.)*. See **forward-looking infra red**.

flirt *(Horol.)*. A device for bringing about the sudden movement of mechanism.

flitch *(For.)*. A piece of timber of greater size than 4 × 12 in, intended for reconversion.

flitch beam *(Build.)*. A built-up beam formed with an iron plate between 2 timber beams.

flit-plug *(Elec.Eng.)*. A detachable connecting-box for coupling cables.

float *(Aero.)*. (1) The distance travelled by an aircraft between flattening-out and landing. (2) A watertight buoyancy unit which is of combined streamline and hydrodynamic form to reduce air and water resistance; *main floats* are the principal hydrodynamic support of float-planes, while *wing-tip floats*, often retractable, give lateral stability to flying boats.

float *(Autos.)*. A small buoyant cylinder of thin brass, steel or proofed cork, placed in the float chamber of a **carburettor** for actuating a valve controlling the petrol supply from the main tank.

float *(Build.)*. (1) A plasterer's trowel. (2) A polishing block used by marble-workers. Cf. **float stone**. (3) A flat piece of wood with a handle on one side used for **floating**.

float *(Eng.)*. A single-cut file (or *float-cut file*), i.e. a file having only one set of parallel teeth, as distinct from a *cross-cut file*.

float *(For.)*. A measure of timber, equalling 18 loads.

float *(Min.Ext.)*. (1) Values so fine that they float on the surface of the water when crushed or washed, e.g. *float gold*. (2) Surfacial deposit of rock or mineral detached from the main dyke or vein. (3) Term for blocks of bedrock within soil or superficial deposits encountered during prospecting or drilling, e.g. **erratics**.

float *(Textiles)*. (1) In a woven fabric the length of yarn between adjacent intersections. (2) A defect in a fabric caused by a thread passing over other threads with which it is designed to interweave (3) A pattern thread in a lace.

float bowl *(Autos.)*. US for **float chamber**.

float chamber *(Autos.)*. In a **carburettor**, the petrol reservoir from which the jets are supplied, and in which the fuel level is maintained constant by means of a float-controlled valve.

float-cut file *(Eng.)*. See **float**.

floated coat *(Build.)*. A coat of plaster smoothed with a float.

float glass *(Glass)*. A transparent glass, the two surfaces of which are flat, parallel and fire polished so that they give clear, undistorted vision and reflection, manufactured by floating hot glass in ribbon form on a heated liquid of greater density than the glass, e.g. molten tin.

floating *(Build.)*. (1) The second of three coats applied in plastering, applied with a **float** to bring the coat level with the **screeds**. (2) A paint film defect caused when pigments separate and rise to the surface during drying, resulting in a patchy finish.

floating address *(Comp.)*. See **symbolic address**.

floating anchor *(Ships)*. See **sea anchor**.

floating balance *(Horol.)*. Modern type of clock escapement with balance wheel suspended from a cylindrical hairspring.

floating battery *(Elec.Eng.)*. An electrical supply system in which a storage battery and electrical generator are connected in parallel to share load, so that the former carries the whole load if the generator fails.

floating bay *(Build.)*. An area between screeds, which is to be filled in with plaster.

floating-carrier wave *(Telecomm.)*. Transmission in which the carrier is reduced or even suppressed to zero during intervals of no modulation by signals, to economize in power or to avoid interference with reception.

floating compass *(Genrl.)*. An early type of primitive compass, probably Chinese. The needle was attached to a float of wood or straw and floated in a bowl of water. The directions were marked inside the bottom of the bowl. See **liquid compass**.

floating dam *(Hyd.Eng.)*. A **caisson**.

floating floor *(Acous.)*. Common type of floor to reduce sound transmission. The load-carrying floor is covered with a layer of fibrous material or soft foam, and a hard plate of about 40 mm thickness is put over the foam.

floating gudgeon pin *(Eng.)*. One free to revolve in both the connecting-rod and the piston bosses.

floating harbour *(Hyd.Eng.)*. A breakwater formed of booms fastened together and anchored, to afford protection to vessels behind it.

floating kidney *(Med.)*. See **nephroptosis**.

floating-point notation *(Comp.)*. Numbers are expressed as a fractional value (mantissa) followed by an integer exponent of the base, e.g. 0.34791×10^3. This enables a large range of numbers to be represented with a fixed number of digits. The advantage of the extra range is gained at the expense of processing time and precision. Cf. **fixed-point notation**.

floating potential *(Electronics)*. That appearing on an isolated electrode when all other potentials on electrodes are held constant.

floating ribs *(Zool.)*. See **false ribs**.

floating rule *(Build.)*. A long straight edge used to form flat surfaces in plaster or cement work.

floating temperature control *(Eng.)*. The use, in a furnace, of an automatic temperature controller which functions by electrically operated valves.

float seaplane *(Aero.)*. An aircraft of the sea-plane type, in which the water support consists of floats in place of the main undercarriage, and sometimes at the tail and wing tips. It may be of the *single-* or *twin-float type*.

float stone *(Build.)*. A shaped iron block which is rubbed over curved brickwork, such as cylindrical backs, in order to remove marks left on the surface by rough dressing.

float stone *(Min.)*. A coarse, porous, friable variety of impure silica, which on account of its porosity, floats on water until saturated.

float switch *(Elec.Eng.)*. A switch operated by a float in a tank or reservoir, and usually controlling the motor of a pump.

flocculation *(Chem.)*. The coalescence of a finely divided precipitate into larger particles.

flocculation *(Immun.)*. Formation of floccules in a precipitin test or of agglutinated bacteria in an agglutination test for flagellar antigens of *Salmonella* species.

flocculation *(Min.Ext.)*. Coagulation of ore particles by use of reagents which promote formation of flocs, as a preliminary to settlement and removal of excess water by thickening and/or filtration.

flocculent *(Chem.)*. Existing in the form of cloud-like tufts or flocs.

flocculi *(Astron.)*. See **plages**.

floccus *(Zool.)*. In Birds, the downy covering of the young forms of certain species; in Mammals, the tuft of hair at the end of the tail; more generally, a tuft.

flock *(Behav.)*. A group of birds that remains together because of social attraction between individuals, rather than, e.g. because of a shared interest in some environmental feature (an *aggregation*). See **schooling** and **herding**.

flock *(Textiles)*. (1) Waste fibres produced in the processes of finishing woollen cloths, used for bedding and upholstery purposes. (2) Short-cut or ground wool, cotton or man-made fibres for spraying on to adhesive coated backings for furniture and upholstery purposes. See also **electrostatic flocking**.

flogger *(Build.)*. A brush with a long bristle filling used in graining to create the effect of pores in certain timbers.

flogging *(Build.)*. The operation of rough-dressing a timber to shape, when the material is removed in large pieces.

flong *(Paper,Print.)*. A board made from papier-mâché used for making moulds from which stereo plates are cast.

flood *(Image Tech.)*. *Floodlight*. A studio light source to give general illumination over a wide angle.

floodable length *(Ships)*. The maximum length of that portion of a ship, centred at a given point, which can be flooded without submerging the **margin of safety line**.

flood basalts *(Geol.)*. Widespread plateau basalts originating from fissure eruptions.

flood fencing *(Hyd.Eng.)*. Fencing which is so anchored as (*a*) to enable it to withstand the force of flood waters, or (*b*) to permit it to hinge over when the water rises sufficiently.

flood flanking *(Hyd.Eng.)*. The constructing of an embankment by depositing stiff moist clay in separate small loads, so that each shall unite so far as is possible with the others, while the crevices left when the clay has dried out are filled with **sludging**.

flooding *(Behav.)*. A conditioning technique in which extinction of avoidance and fear responses is achieved by confronting the anxiety-producing stimulus, often using only imaginal exposure and verbal description by the therapist.

flooding *(Chem.)*. The condition of a fractionating column in which the upward flow of vapour has become great enough to prevent the downward flow of liquid, giving poor performance.

flooding *(Med.)*. Copious bleeding from the uterus.

floodlighting *(Elec.Eng.)*. The lighting of a large area or surface by means of light from projectors situated at some distance from the surface.

floodlight projector *(Elec.Eng.)*. The housing and support for a lamp used in a floodlighting scheme; it is designed with a reflector which directs the light from the lamp into a suitable beam.

flood plain *(Geol.)*. A plain of stratified alluvium bordering a stream and covered when the stream floods.

flood track *(Image Tech.)*. A photographic sound track with full uniform exposure across its complete width.

floor *(Foundry)*. The bed of sand constituting the floor of a foundry; in it large castings are often made.

floor *(Min.Ext.)*. The upper surface of the stratum underlying a coal-seam.

floor *(Ships)*. Transverse vertical plate connected to the shell plating and to the inner bottom and extending from side to side.

floor contact *(Elec.Eng.)*. A switch contact which is attached to the floor of an automatic electric lift and is operated by the passenger stepping into the lift; it is usually arranged to prevent the lift from being operated from any of the landings.

floor cramp *(Build.)*. A cramp for closing up the joints of floor-boarding when it is being nailed in position.

floor guide *(Build.)*. A groove formed in a floor surface to receive a sliding door or partition and direct its movement.

flooring saw *(Build.)*. One with a curve towards the toe and extra teeth on the back above the toe. Used for cutting floorboards in order to raise them.

floor joist *(Build.)*. A **bridging joist**.

floor line *(Build.)*. A mark made at the lower end of a door-post, or other finishing, to indicate the level of the floor when the finishing is in position.

floor plan *(Arch.)*. A plan drawn for any given floor of a building, normally showing the dimensions of the rooms, etc.

floor sand *(Foundry)*. Foundry sand in which new and used moulding sand and coal dust are mixed.

floor stop *(Build.)*. A *door-stop* projecting from the floor near a door.

floor-switch *(Elec.Eng.)*. The *landing switch* of a lift or elevator.

flop damper *(Build.)*. A damper which stays under its own weight in the open or shut position.

flopgate *(Min.Ext.)*. Diverting gate which directs moving material into alternative routes. Can be worked by remote control.

flop-over *(Image Tech.)*. A visual effect in which the picture is reversed from left to right.

floppy disk *(Comp.)*. Lightweight, flexible magnetic disk which behaves as if rigid when rotated rapidly. It is robust and light enough to send through the post. Also called *diskette, floppy*.

FLOPS *(Comp.)*. *FLOating Point operations per Second.* A measure of speed of a processor. Also *megaflops*. Cf. **MIPS, LIPS**.

flora *(Bot.)*. (1) The plants of a particular region habitat or epoch. (2) A catalogue or description of such plants.

floral diagram *(Bot.)*. A conventional plan of the arrangement of the parts of a flower as seen in cross-section.

floral envelope *(Bot.)*. The calyx and corolla, or perianth.

floral formula *(Bot.)*. A representation of the structure of a flower by means of the letters P (perianth) or K (calyx) and C (corolla), A (androecium) and G (gynoecium), of figures (numbers of parts), of parentheses (connation), of horizontal brackets (adnation) and of a line above or below the G (superior or inferior ovary), thus; P 3+3 A 3+3 *G(3)* is the tulip.

floral leaf *(Bot.)*. (1) A bract or bracteole. (2) A sepal or petal.

floral mechanism *(Bot.)*. The arrangement of the flower parts to ensure either cross- or self-pollination.

Florence flask *(Chem.)*. A long-necked round flask with a flat bottom.

Florentine arch *(Arch.)*. An arch having a semi-circular intrados and a pointed extrados, giving greater strength at the crown.

Florentine blind *(Build.)*. An outside roller blind, similar to the Italian blind but having side pieces.

flore pleno *(Bot.)*. With *double flowers*. Abbrev. *fl pl.*

floret *(Bot.)*. An individual flower in a crowded inflorescence. In grasses, typically consists of an ovary (with 2 styles) and 3 stamens enclosed by 2 bracts (lemma and palea).

floriated *(Arch.)*. Said of an elaborately ornamented building style, particularly with flowers and leaves featured.

florigen *(Bot.)*. A hypothetical plant hormone inferred to be induced in leaves, to move to the shoot apices and there to cause the initiation of flowers.

flos ferri *(Min.)*. A stalactite form of the orthorhombic carbonate of calcium, *aragonite*, some of the masses resembling delicate coralline growths; deposited from hot springs.

flotation *(Min.Ext.)*. See **froth flotation**.

flour *(Build.,Civ.Eng.)*. The fine dust incidentally formed in crushing material to be used as an aggregate.

flow box *(Paper)*. A compartment immediately before the machine wire or other forming unit, supplied with stock and designed to ensure uniform mixing within the stock and the means to control its flow on to the wire and even distribution.

flowchart *(Comp.)*. Diagrammatic representation of the operations involved in an **algorithm** or automated system. Flow lines indicate the sequence of operations or the flow of data, and special standard symbols are used to represent particular operations. See **data flowchart**, and **program flowchart**.

flow counter *(Nuc.Eng.)*. See **gas-flow counter**.

flow cytometry *(Biol.)*. Sorting cells or chromosomes, to which a fluorochrome has been stoichiometrically bound, on the basis of size. They are passed through a beam of light and directed into one of two flasks depending on how much light they emit or scatter.

flower *(Bot.)*. The reproductive structure in angiosperms consisting of a shoot axis bearing, as lateral members traditionally interpreted as modified leaves, one or more of sepals and petals, or tepals, stamens and carpels.

flowering plants *(Bot.)*. The angiosperms, comprising the monocotyledons and the dicotyledons.

flowers *(Print.)*. Small type ornaments, copied from early designs, used for building up fancy borders, etc.

flowers of sulphur *(Chem.)*. Finely divided sulphur obtained by allowing sulphur vapour to fall on a cold surface.

flow forming *(Eng.)*. A metal-forming operation, in which thick blanks of aluminium, copper, brass, mild steel or titanium are made to flow plastically by rolling them under pressure in the direction of roller travel to result in components, often conical, having a wall thickness much less than the original blank thickness. Also *flospinning*.

flow improver *(Min.Ext.)*. Chemicals added to oil passing through a pipeline which reduce frictional losses.

flow-line production *(Eng.)*. A system of mass production, and certain types of batch production, in which machines are arranged in flow lines to enable components to progress from operation to operation in correct sequence.

flow lines *(Eng.)*. Lines which appear on the surface of iron and steel when stressed to the yield point. They arise from the fact that all parts of a given sample do not yield at the same time; the lines are traces on the surface of the planes along which yielding first occurs.

flowmeter *(Phys.)*. Device for measuring, or giving an output signal proportional to, the rate of flow of a fluid in a pipe.

flow noise *(Acous.)*. Acoustic signal caused by a flow process, e.g. siren, ventilator noise, jet noise.

flow-off *(Foundry)*. A channel cut from a riser to allow metal to escape when it has reached a pre-determined height.

flow pipe *(Build.)*. The pipe conveying hot water from boiler to cylinder in a domestic hot-water system.

flow process *(Glass)*. See **gob process**.

flow sheet *(Min.Ext.)*. A diagram showing the sequence of operations used in a process of production with a given plan, e.g. the extraction and refining of metals.

flow-sorted chromosomes *(Biol.)*. Those sorted on the basis of size by **flow cytometry**.

flow-structure *(Geol.)*. A banding, often contorted, resulting from flow movements in a viscous magma, adjacent bands differing in colour and/or degree of crystallization. It is also shown by the alignment of phenocrysts or of minute crystals and crystallites, in the groundmass of lavas and, more rarely, minor intrusions. Also loosely used of metamorphic rocks.

fl. pl. *(Bot.)*. Abbrev. for *flore pleno*.

fluctuation *(Med.)*. The palpable undulation of fluid in any cavity or abnormal swelling of the body.

fluctuation noise *(Acous.)*. Noise produced in the output circuit of an amplifier by shot and flicker effects.

fludrocortisone acetate *(Med.)*. Mineral corticoid used in replacement therapy for adrenocortical insufficiency.

flue *(Eng.)*. A passage or channel through which the products of combustion of a domestic fire, boiler etc., are taken to the chimney.

flue bridge *(Eng.)*. See **firebrick arch**.

flue gas *(Eng.)*. The gaseous products of combustion from a boiler furnace, consisting chiefly of CO_2, CO, O_2, N_2 and water vapour, whose analysis is used as a check on the furnace efficiency. See **CO_2** recorder.

flue-gas temperature *(Eng.)*. Temperature of flue gases at the point in the flue where it leaves the furnace.

flue gathering *(Build.)*. See **gathering**.

flueing soffit *(Build.)*. A flush soffit under a geometrical stair.

flue lining *(Build.)*. Flexible stainless steel, fireclay or fire-resistant concrete pipe, arranged within a flue passage to protect the walls.

fluff *(Paper)*. Fibres or other débris removed from the surface of the paper during printing and converting operations.

fluffing *(Print.)*. A tendency with soft-sized paper for fibres to be detached from the surface and edges, producing difficulties in printing, particularly if by lithography.

fluid *(Genrl.)*. A substance which flows. It differs from a solid in that it can offer no permanent resistance to change of shape. See **gas, liquid**.

fluid flywheel *(Eng.)*. A device for transmitting power through the medium of the change in momentum of a fluid, usually oil. Similar in principle to a Froude brake in which the stator is released and forms the driven member.

fluidics *(Eng.,Phys.)*. The science of liquid flow in tubes etc. which strongly simulates electron flow in conductors and conducting plasma. The interaction of streams of fluid can thus be used for the control of instruments or industrial processes without the use of moving parts.

fluid inclusion *(Min.)*. See **inclusion**.

fluidity *(Phys.)*. The inverse of **viscosity**.

fluidization *(Chem.Eng.)*. A technique whereby gas or

vapour is passed through solids so that the mixture behaves as a liquid, and of special significance when the solid is a catalyst to induce reactions in the fluidizing medium.

fluidized bed *(Chem.Eng.)*. If a fluid is passed upward through a bed of solids with a velocity high enough for the particles to separate from one another and become freely supported in the fluid, the bed is said to be fluidized. Then the total fluid frictional force on the particles is equal to the effective weight of the bed. Fluidized beds are used in chemical industry because of the intimate contact between solid and gas, the high rates of heat transfer and the uniform temperatures within the bed, and the high heat transfer coefficients from the bed to the walls of the containing vessel.

fluidized-bed reactor *(Nuc.Eng.)*. One in which the active material is supported in a finely divided form by a gas or liquid.

fluid lubrication *(Eng.)*. A state of perfect lubrication in which the bearing surfaces are completely separated by a fluid or viscous oil film which is induced and sustained by the relative motion of the surfaces.

fluid-mosaic *(Bot.)*. Model of cell membrane structure postulating a bimolecular leaflet of lipid, interspersed with various specific proteins which lie at the surfaces of or in the membrane, the lipid being fluid and the proteins able to move laterally.

fluid needle *(Build.)*. The fluid control in a spray gun. Operated by the trigger, it seats into the fluid tip which meters and directs the fluid into the air stream.

fluing *(Build.)*. A term applied to window jambs which are splayed. See **splayed jambs**.

fluing arch *(Build.)*. See **splaying arch**.

fluke *(Zool.)*. (1) A semipopular name for worms belonging to the group Trematoda. (2) The tail of a Cetacean.

flukes *(Med.)*. Flat worms of the Trematoda class. Blood flukes are responsible for **schistosomiasis** and liver flukes for **fascioliasis**.

flukes *(Ships)*. The flattened and curving points terminating the arms of an anchor.

flume *(Build.,Civ.Eng.)*. A metal chute used for the distribution of concrete from a placing plant.

flume *(Min.Ext.)*. A flat-bottomed timber trough, or other open channel, generally nowadays formed in concrete, for the conveyance of water, e.g. to ore-washing plant, or as a by-pass.

flume stabilizer *(Ships)*. A roll stabilization system using passive fluid tanks fitted athwartships, and having specifically designed nozzles to cause a phase difference of 90° between the movement of the liquid in the tank and the roll of the ship.

Fluon *(Chem.)*. TN for **polytetrafluoroethene**.

fluorapatite *(Min.)*. The commonest form of **apatite**.

fluorene *(Chem.)*. *Diphenylenemethane*, $(C_6H_4)_2CH_2$; colourless fluorescent plates; mp 113°C, bp 295°C; contained in coal-tar; produced by leading diphenylmethane through red-hot tubes. ⇨

fluorescein *(Chem.)*. $C_{20}H_{12}O_5$, *(1,3-dihydroxy-benzene) phthalein*, red crystals which dissolve in alkalis with a red colour and green fluorescence.

fluorescein isothiocyanate *(Biol.)*. Fluorochrome commonly conjugated with antibodies for use in **indirect immunofluorescence**.

fluorescence *(Phys.)*. Emission of radiation, generally light, from a material during illumination by radiation of usually higher frequency, or from the impact of electrons. See **phosphor**.

fluorescence activated cell sorter *(Biol.)*. Instrument in which cells or chromosomes in a suitable medium have their fluorescence measured as they pass down a fine tube. They are then ejected from an aperture which causes the stream to break up into fine droplets which pass between electrostatic deflector plates. Depending on the amount or nature of the fluorescence, the droplet containing the cell or chromosome is given an electrical charge which causes it to be diverted into an appropriate reservoir. The sorter can also be used to measure the amount of fluorescence and hence e.g. DNA content of the cells.

fluorescence microscopy *(Biol.)*. Light microscopy in which the specimen is irradiated at wavelengths which will excite the natural or artificially introduced fluorochromes. An optical filter absorbs the exciting wavelengths but transmits the fluorescent image which can be studied normally.

fluorescent brightener *(Textiles)*. See **fluorescent whitening agents**.

fluorescent lamp *(Elec.Eng.)*. A mercury-vapour electric-discharge lamp having the inside of the bulb or tube coated with fluorescent material so that ultraviolet radiation from the discharge is converted to light of an acceptable colour.

fluorescent penetrant inspection *(Eng.)*. The use of a fluorescent dye which will penetrate any minute crack in a component. After immersion the component is wiped dry and any subsequent seepage from fissures is detected by irradiation at the exciting wavelength.

fluorescent screen *(Electronics)*. One coated with a layer of luminescent material so that it fluoresces when excited, e.g. by X-rays or cathode rays. See **fluoroscopy**, **phosphor (2)**.

fluorescent whitening agents *(Chem.)*. Special dyes widely used to 'whiten' textiles, paper etc., and sometimes incorporated in detergents. Their effect is based on ability to convert invisible ultraviolet light into visible blue light, giving fabrics greater uniformity of reflectance over the visible part of the spectrum. Also *optical bleaches*, *optical whites*.

fluorescent yield *(Phys.)*. Probability of a specific excited atom emitting a photon in preference to an Auger electron.

fluoridation *(Chem.,Med.)*. The addition of inorganic fluorides (usually *sodium fluoride*) to water supplies with the intention of preventing dental **caries**. The amount added is usually about 1 part/million.

fluorimeter *(Radiol.)*. One used for measuring the intensity of fluorescent radiation.

fluorimetry *(Min.Ext.)*. Analytical method in which fluorescence induced by ultraviolet light or X-rays is measured. Also *fluoremetry*.

fluorinated ethene propene *(Plastics)*. See **FEP**.

fluorination *(Chem.)*. The replacement of atoms (usually hydrogen atoms) in molecules by fluorine. Can be carried out catalytically using fluorine in the vapour phase or by using fluorine 'carriers' such as cobaltic fluoride or silver fluoride in solution with increased temperature.

fluorine *(Chem.)*. Pale greenish yellow gas, the most electronegative (non-metallic) of the elements and the first of the halogens. Chemically highly corrosive and never found free. Discovered by Scheele in 1771 and isolated by Moissan in 1886. Symbol F, at.no. 9, r.a.m. 18.9984, valency 1, mp −223°C, bp −187°C, density 1.696 g/dm^3 at s.t.p. Used in separating the isotopes of uranium (see **uranium hexafluoride**) Combines with carbon to form inert polymers with low coefficient of friction, e.g. Teflon. See **fluorocarbons**.

fluorite, fluorspar *(Min.)*. Calcium fluoride, CaF_2, crystallizing in the cubic system, commonly in simple cubes. Occasionally colourless, yellow, green, but typically purple; the coloured varieties may fluoresce strongly in ultraviolet light. See **Blue John**.

fluoroboric acid *(Chem.)*. A complex monobasic acid formed by the combination of hydrogen fluoride and boron (III) fluoride. Salts called *fluoroborates*.

fluorocarbons *(Chem.)*. Hydrocarbons in which some or all of the hydrogen atoms have been replaced by fluorine. The fluorinated derivatives of methane are widely used as refrigerating agents and propellants for aerosols. See also **Freons** and **polytetrafluoroethene**.

fluorochrome *(Chem.)*. A molecule or chemical group which fluoresces on irradiation. Can be made to bind to a specific site and thus localize it.

fluorography *(Radiol.)*. The photography of fluoroscopic images.

fluorophore *(Chem.)*. A group of atoms which give a molecule fluorescent properties.
fluoroscope *(Radiol.)*. Measurement system for examining fluorescence optically. Fluorescent screen assembly used in fluoroscopy.
fluoroscopy *(Eng.)*. Examination of objects by observing their X-ray shadow on a fluorescent screen; used to examine contents of luggage, packages without unwrapping, quality of welding etc.
fluorosis *(Med.)*. Chronic poisoning with fluorine.
fluorspar *(Min.)*. See **fluorite**.
Fluothane *(Chem.)*. See **halothane**.
flush *(Bot.)*. Area of land, fed by ground-water which may be oligotrophic or eutropic and, in the latter, may enrich the soil locally.
flush *(Build.)*. In the same plane.
flush bead *(Build.)*. A sunk bead, finished so as to be level with the surface which it decorates.
flush boards *(Print.)*. A method of binding in which boards are drawn on and trimmed with the book. The covers are then flush with the page edges.
flush-bolt *(Build.)*. A sliding bolt sunk into the side or edge of a door so as to be flush with the surface.
flushing *(Build.)*. A crushing of the edges of a stone at a **hollow bed**, due to excessive pressure upon them.
flushing *(Hyd.Eng.)*. The process of cleansing a sewer or other space by suddenly passing through it a quantity of water.
flushing *(Min.Ext.)*. The operation of clearing off accumulation of fire-damp or noxious gases underground by means of air currents. See **face-airing**.
flushing of ewes *(Vet.)*. The practice of increasing the nutritional plane of ewes a few weeks before they are served by the ram, with the object of increasing fertility.
flushing tank *(Build.)*. A tank used to accumulate the water for flushing a drain or sewer which is not laid at a self-cleansing gradient. The discharge is often effected automatically by a siphoning device.
flush joint *(Build.)*. The type of mortar joint made in **flat pointing**.
flush panel *(Build.)*. A panel whose surface is in line with the faces of the stiles.
flush-plate *(Elec.Eng.)*. See **switch plate**.
flush soffit *(Build.)*. The continuous surface formed under any ceiling or stair etc.
flush-switch *(Elec.Eng.)*. A switch which can be mounted flush with the wall. Also called *panel-switch*, *recessed switch*.
flush valve *(Build.)*. A valve operating the flushing system for a sanitary appliance.
flute *(Build.)*. A long vertical groove, usually circular in form, in the surface of a column or other member.
flute cast *(Geol.)*. A hollow eroded by turbulent flow and subsequently filled by sediment. More properly called a *mould*, it is common in **turbidite** deposits, and can be used to determine the direction of flow of the depositing currents.
fluting *(Build.)*. See **flute**.
fluting *(Eng.)*. Parallel channels or grooves, longitudinal or helical, cut in a cylindrical object such as a tap or reamer.
fluting plane *(Build.)*. A plane for cutting round bottomed grooves.
flutter *(Acous.)*. Rapid fluctuation of frequency or amplitude.
flutter *(Aero.)*. Sustained oscillation, usually on wing, fin or tail, caused by interaction of aerodynamic forces, elastic reactions and inertia, which rapidly break the structure. *Asymmetrical flutter* occurs where the port and starboard sides of the aircraft simultaneously undergo unequal displacements in opposite directions, as opposed to *symmetrical flutter*, where the displacements and their direction are the same; *classical* or *coupled flutter* is due solely to the inertial, aerodynamic or elastic coupling of two or more degrees of freedom.
flutter *(Image Tech.)*. Variation in brightness of a reproduced picture, arising from additional radio reflection from a moving object, e.g. an aircraft.
flutter *(Med.)*. An abnormality of cardiac rhythm, in which the atrium of the heart contracts regularly at a greatly increased frequency (between 180 and 400 beats/min), the ventricles contracting at a slower rate.
flutter echo *(Acous.)*. See **multiple echo**.
flutter speed *(Aero.)*. The lowest **equivalent air speed** at which flutter can occur.
fluviatile *(Bot.,Zool.)*. Of streams and rivers, *fluvial*.
fluviatile deposits *(Geol.)*. Sand and gravel deposited in the bed of a river.
fluviomarine *(Zool.)*. Able to live in rivers and in the sea, as the Salmon.
fluvioterrestrial *(Zool.)*. Found in rivers and on their banks, as the Otter.
flux *(Chem.)*. A substance added to a solid to increase its fusibility. In soldering, welding etc. added to the molten metal to dissolve infusible oxide films which prevent adhesion.
flux *(Eng.)*. Material added to a furnace charge, which combines with those constituents not wanted in the final product and issues as a separate slag.
flux *(Maths.)*. Through a surface in a vector field: the integral over the surface of the product of the elementary area by the normal component of the vector, i.e. $\supset$ F.ds.
flux *(Phys.)*. The rate of flow of mass, volume, or energy per unit cross-section normal to the direction of flow. See also, **electric-**, **magnetic-**.
flux density *(Phys.)*. The number of photons (or particles) passing through unit area normal to the beam, or the energy of the radiation passing through this area.
flux gate *(Elec.Eng.)*. Magnetic reproducing head in which magnetic flux (due to flux leakage from signals recorded on magnetic tape) is modulated by high-frequency saturating magnetic flux in another part of the magnetic circuit.
flux-gate compass *(Eng.)*. Device in which the balance of currents in windings is affected by the earth's magnetic field.
flux guidance *(Elec.Eng.)*. Directing the electric or magnetic flux in high-frequency heating by shaped electrodes or magnetic materials, respectively.
flux link, linkage *(Elec.Eng.)*. Conservative flux across a surface bounded by a conducting turn. For a coil, *flux linkage* is the integration of the flux with individual turns.
fluxmeter *(Elec.Eng.)*. Electrical instrument for measuring total quantity of magnetic flux linked with a circuit; it consists of a search coil placed in the magnetic field under investigation, and a ballistic galvanometer, or an uncontrolled moving-coil element (Grassot) or a semi-conductor probe generating Hall voltage. See also **gaussmeter**.
flux quantization *(Phys.)*. A magnetic effect in superconductors. A ring of material is placed in a uniform magnetic field and then cooled below its critical temperature so that it becomes superconducting. When the external field is removed it is found that the ring has trapped the field in its hole. If the flux of the trapped field is measured as a function of the strength of the applied field, it is found to be quantized in steps of $(h/2e)$ where h is Planck's constant and e is the electronic charge. This shows superconductivity to be a quantum effect.
fly *(Horol.)*. An air brake; a fan with two or four blades, used in clocks to maintain uniformity between the blows of the hammers when striking or chiming.
fly *(Print.)*. A term indicating a sheet of paper folded once to make 4 pages, with the printing, if any, on the first page only.
fly *(Textiles)*. Fine short fibres that escape into the atmosphere especially in the carding of cotton.
fly ash *(Build.)*. A fine ash from the pulverized fuel burned in power stations, used in brick-making and as a partial substitute for cement in concrete.
flyback *(Image Tech.,Radar)*. The return of the scanning beam to its starting point at the completion of a radar trace or a line of a TV picture, the line being blanked out during the process. Also called *retrace*.
flyback action *(Horol.)*. In a chronograph or stop-watch, that part of the action which causes the hands to fly back to zero when the button is pressed.

fly before buy *(Aero.)*. Process of procuring new military aircraft by flying a prototype prior to giving the production order. An alternative is to order 'off the drawing board', which shortens delivery time at the risk of inadequate performance or delays in fixing inadequacies found during flight test.

fly-blown *(Vet.)*. Affected by **myiasis**.

fly-by *(Space)*. A type of space mission where the spacecraft passes close to the target but does not rendezvous with it, orbit around it or land on it.

fly-by-light *(Aero.)*. Flight control system in which the signalling is performed by coherent light beams travelling in optic fibres.

fly-by-wire *(Aero.)*. Flight control system using electric/electronic signalling.

fly cutter *(Eng.)*. A single point tool used in milling machine to produce a flat surface.

fly disease *(Vet.)*. See **nagana**.

flyer *(Build.)*. **Flying shore**.

fly hand *(Print.)*. A rotary press assistant who removes the printed product from the delivery or the conveyor.

flying-boat *(Aero.)*. A seaplane wherein the main body or hull provides water support.

flying bond *(Build.)*. See **monk bond**.

flying buttress *(Arch.)*. An arched buttress giving support to the foot of another arch. Also called *arc-boutant, arch(ed) buttress*.

flying deck *(Ships)*. See **hurricane deck**.

flying levels *(Surv.)*. Back-sight and fore-sight readings taken between any 2 points, without reference to bench marks, when only the difference of level of the points is required.

flying paster *(Print.)*. See **automatic reel change**.

flying scaffold *(Build.)*. A *suspended scaffold, flying shore*. A support, independent from the ground, between two buildings. Used usually as temporary support after the removal of one building from a row.

flying-shore *(Build.)*. A horizontal baulk of timber or scaffolding rig used to provide temporary support between two opposite walls, usually not more than about 9 m apart. Also *flier, flyer*.

flying speed *(Aero.)*. The *maximum flying speed* is the highest attainable speed in level flight, under specified conditions and corrected to standard atmosphere, and the *minimum flying speed* is the lowest speed at which level flight can be maintained.

flying-spot microscope *(Biol.)*. A type of light microscope in which the object is scanned in two dimensions by a light spot formed by the diminished image of a cathode ray tube placed at the eyepiece plane of a compound microscope. All the transmitted energy can be collected by a photomultiplier and an image, suitable for electronic analysis, reconstructed using the timing circuits driving the CRT. The **scanning electron microscope** is an analogous instrument.

flying spot scanner *(Image Tech.)*. Device used in **telecine** equipment for producing TV picture signals from photographic transparencies or motion picture film; the subject is scanned with a spot of light generated on a CRT and the modulated transmitted light is received by a photocell.

flying tail *(Aero.)*. See **all-moving tail**.

flying tuck *(Print.)*. A gear-driven rotating folding blade mounted on a cylinder on web-fed presses.

fly leaf *(Print.)*. A blank leaf at the beginning and at the end of a bound volume. It may be part of an end-paper.

fly nut *(Eng.)*. See **wing nut**.

fly pinion *(Horol.)*. The pinion on the arbor of which a *fly* is mounted.

fly press *(Eng.)*. A hand-operated press for punching holes, making driving fits, etc.; it consists of a bed supporting a vertical frame through which a square-threaded screw is fitted. The screw is turned by a cross-piece terminating in one or two heavy steel balls, for giving additional impetus to the descent of the die attached to the bottom end.

fly rail *(Build.)*. A flap attached to a table-frame by means of a vertical hinge; it swings out to provide support for a folding leaf.

flysch *(Geol.)*. Sediments derived from the Alpine orogeny. More generally applied to almost any **turbidite** deposits, derived from developing, large-scale, fold structures.

fly shuttle *(Textiles)*. Mechanism, invented by John Kay (Bury, Lancs) in 1733, for propelling the shuttle across the loom. It superseded hand-shuttling.

flywheel *(Eng.)*. A heavy wheel attached to a shaft (e.g. an engine crankshaft) either to reduce the speed fluctuation resulting from uneven torque, or to store up kinetic energy to be used in driving a punch, shears, etc., during a short interval.

flywheel effect *(Image Tech.)*. Maintenance of synchronization in a TV receiver by internal circuits which can operate for a short time in the absence of transmitted sychronizing signals, to assist under adverse reception conditions.

fly wire *(Build.)*. A fine woven wire mesh used to cover the joint between adjacent pieces of building-board.

Fm *(Chem.)*. Symbol for fermium.

FM *(Genrl.)*. Abbrev. for *Frequency Modulation*.

FMEA *(Aero.)*. See **failure modes and effects analysis**.

FMS *(Eng.)*. See **flexible manufacturing system**.

f-number *(Image Tech.)*. The relative **aperture** of a lens, particularly when stopped down, representing its light transmission; it expresses the diameter of the lens diaphragm as a fraction of its focal length, e.g. f/8, also written f:8 or f8. See **aperture, stop**.

foam *(Chem.)*. See **froth**; also *fire foam*.

foambacked fabric *(Textiles)*. Dress and furnishing fabrics bonded on the back to polyether or polyester foams by adhesive or flame treatment.

foamed plastics *(Plastics)*. See **expanded plastics**.

foamed slag *(Build.)*. Blast furnace slag aerated while still molten. Used for building blocks and for acoustic and thermal insulation.

foam plug *(Min.Ext.)*. Mass of foam generated and blown into underground workings to seal off a fire or keep out oxygen, where a fire risk exists.

foam separation *(Chem.)*. Removal of solutes or ions from a liquid by bubbling air through in the presence of surface active agents which tend to be adsorbed on to the bubbles. Cf. **froth flotation**, for larger particles.

focal length of a lens *(Phys.)*. The distance measured along the principal axis, between the principal focus and the second principal point. In a thin lens both principal points may be taken to coincide with the centre of the lens. See **back focus, equivalent focal length**; also **convention of signs**.

focal plane *(Image Tech.,Phys.)*. The plane, at right angles to the principal axis of a lens or lens system, in which the image of a particular object is formed. The principal focal plane passes through the principal focus, and contains the images of objects at infinity. It is the normal location of the sensitive surface of a film or plate, or a ground-glass focusing screen.

focal-plane shutter *(Image Tech.)*. Camera shutter in the form of a blind with a slot, which is pulled rapidly across, and as close as practicable to, the film or plate, exposure time being varied by adjusting the width of the slot. Called *self capping* because the slot is closed during retensioning.

focal point *(Phys.)*. The focal spot formed on the axis of a lens or curved mirror by a parallel beam of incident radiation. In its general form, this definition includes acoustic lenses, electron lenses, and lenses or mirrors designed for use with radio waves, infrared or ultraviolet radiation.

focal spot *(Phys.)*. A spot on to which a beam of light or charged particles converges. See also **X-ray-**.

focometer *(Phys.)*. An instrument for measuring the focal length of a lens.

focus *(Geol.)*. See **earthquake**.

focus *(Maths.)*. Of a conic: a point such that the two lines of every pair of conjugate lines through it are mutually perpendicular. The ellipse and hyperbola each have two real and two imaginary foci, and the parabola has one real focus. See **conic** for alternative definition. Of a quadric: a point not on the quadric, such that the three planes of every set of three mutually conjugate planes through it are mutually perpendicular.

focus *(Phys.)*. A point to which rays converge after having passed through an optical system, or a point from which such rays appear to diverge. In the first case the focus is said to be *real*; in the second case, *virtual*. The *principal focus* is the focus for a beam of light rays parallel to the principal axis of a lens or spherical mirror.

focusing *(Image Tech.)*. Adjustment of the axial position of a camera lens to form a sharp image in the required plane, such as the film or plate in photography or the camera tube target in TV.

focusing *(Phys.)*. The convergence to a point of (a) beams of electromagnetic radiation, (b) charged particle beams, (c) sound or ultrasonic beams.

focusing coil, electrode *(Elec.Eng.)*. One used to focus a charged particle beam by magnetic or electrostatic field control.

focusing screen *(Image Tech.)*. Screen, usually of ground glass, located in the place of a film or plate, or on the top of a reflex camera, for adjusting the focusing of the lens before exposure.

focus-skin distance *(Radiol.)*. The distance from the focus of an X-ray tube to the surface of incidence on a patient, usually measured along the beam axis. Abbrev. *FSD*.

foetal membranes *(Zool.)*. In Reptiles, Birds and Mammals, outgrowths from the embryo, or the extra-embryonic tissue, which surround and protect the foetus and facilitate respiration. See **allantois, amnion, chorion**.

foetus *(Zool.)*. A young mammal within the uterus of the mother or in oviparous animals the young within the egg, from the commencement of organogeny until birth. adj. *foetal*.

fog *(Image Tech.)*. An overall density in a photographic record not related to the exposed image. It may be caused by an unwanted exposure to light or radiations such as X-rays, by incorrect chemical processing or by protracted and unsuitable storage.

fog *(Meteor.)*. Minute water droplets with radii in the range 1 to 10 μm suspended in the atmosphere and reducing visibility to below 1 km (1100 yd in Britain).

fogbow *(Meteor.)*. A bow seen opposite the sun in fog. The bow is similar to the rainbow, but the colours are faint, or even absent, owing to the smallness of the drops, which causes diffraction scattering of the light.

fog fever *(Vet.)*. *Atypical interstitial pneumonia*. An acute respiratory distress syndrome of cattle which usually comes on within two weeks of introduction to lush pasture. Most common in suckler herds where morbidity may reach 50%.

fog levels *(Image Tech.)*. The minimum density of a processed photographic image in an unexposed area.

fog signal *(Civ.Eng.)*. A detonating cap which is placed on a rail before the passage of a train, so that the detonation occurring when a wheel passes over it shall serve as a signal to the driver in bad visibility.

fog-type insulator *(Elec.Eng.)*. A type of overhead-line insulator having long leakage distances; specially designed for areas in which fog is prevalent.

föhn wind *(Meteor.)*. A warm, dry wind which blows to the lee of a mountain range. It is prevalent on the northern slopes of the Alps.

foid *(Min.)*. A term meaning **feldspathoid** used by international agreement on rock classification.

foldback DNA *(Biol.)*. Sequence complementarity which allows a single-stranded molecule to form secondary structure. **Hairpin** DNA is one form with a minimal loop at its end.

folded dipole *(Telecomm.)*. A **dipole** antenna with a separate, parallel rod or wire connected from one tip to another of the original dipole; done to increase the impedance at the feed point, for matching purposes.

folded horn *(Acous.)*. An acoustic horn which is turned back in itself to reduce necessary space.

folded scale *(Maths.)*. On a slide-rule, scale labelled CF or DF calibrated same as principal scales but displaced about half scale length. Used to avoid resetting rule.

folded yarns *(Textiles)*. Yarns formed from two or more single yarns twisted together for strength or special appearance. The products are known as 2-, 3-, 4- etc. fold or ply yarns. (See also **doubling frame**.

folder *(Print.)*. The section of a web-fed press where the webs are **associated**, folded, cut, delivered, and sometimes stitched.

folding *(Geol.)*. Folding (*bending*) of strata is usually the result of compression that causes the formation of the geological structures known as anticlines, synclines, monoclines, isoclines, etc. The amplitude (i.e., vertical distance from crest to trough) of a fold ranges from a centimetre to thousands of metres.

folding blade *(Print.)*. A metal strip which thrusts the web or webs into jaws or rollers to produce a fold. Also *tucker, tucking blade*.

folding cylinder *(Print.)*. The cylinder which holds the folder blade or jaw for making a fold on web-fed presses.

folding jaws *(Print.)*. The gripping section on a **nip and tuck folder**.

folding plates *(Print.)*. (1) Adjustable parts on a folding machine into which the paper travels and, being stopped, receives a **buckle fold**. (2) Large sized illustrations which require folding before inclusion in the book.

folding rollers *(Print.)*. Driven rollers on folding machines, the paper being folded as it passes between them.

folding shutters *(Build.)*. See **boxing shutters**.

folding strength *(Paper)*. The number of double folds needed to break a test strip of the paper, under prescribed conditions.

folding wedges *(Build.,Civ.Eng.)*. **Striking wedges** used for tightening and easing shoring and centring, and in some joint construction in joinery.

fold-out *(Print.)*. A large leaf in a book which must be *folded out* when the book is being used.

foliaceous, foliose *(Bot.)*. (1) Flat and leaflike. (2) Bearing leaves.

foliar feeding *(Bot.)*. Supplying mineral elements in solution by watering them onto the foliage, useful where soil conditions prevent uptake through the roots as in *lime-induced chorosis*.

foliar gap, trace *(Bot.)*. See **leaf gap, leaf trace**.

foliation *(Geol.)*. The arrangement of minerals normally possessing a platy habit (such as the micas, chlorites, and talc) in folia and leaves, lying with their principal faces and cleavages in parallel planes; such arrangement is due to development under great pressure during regional metamorphism.

folic acid *(Chem.)*. See **vitamin B**.

folio *(Print.)*. (1) A sheet of paper folded in half. (2) A book made up of sheets folded once, so having 4 pages to the sheet. (3) The number of a page.

foliose *(Bot.)*. (1) Leafy, having the body differentiated into stem and leaves. Cf. **thallose**. (2) Lichens with a flattened thallus rather loosely attached to the substrate. Cf. **crustose, fruticose**.

folium of Descartes *(Maths.)*. The curve defined by the cartesian equation, $x^3+y^3=3axy$.

folk weave *(Textiles)*. A loosely woven rough fabric made from coarse yarns including coloured ones.

follicle *(Bot.)*. A dry, dehiscent, many-seeded fruit formed from one carpel, dehiscing along one side only, as in *Delphinium*, where there is a group of follicles.

follicle *(Immun.)*. Spherical accumulations of lymphocytes (largely B-cells) present in lymphoid tissues which become enlarged after antigenic stimulation to become secondary follicles or **germinal centres**.

follicle *(Zool.)*. Any small saclike structure, such as the pit surrounding a hair-root. adjs. *follicular, folliculose*. See **Graafian follicle**.

follicle-stimulating hormone *(Med.)*. A gonadotrophic mucoprotein hormone secreted by the basophil cells of the pars anterior of the pituitary gland, which stimulates growth of the Graafian follicles of the ovary and spermatogenesis. Abbrev. *FSH*.

follicular mange *(Vet.)*. See **demodectic mange**.
folliculitis *(Med.)*. Inflammation of a follicle, especially of the ovary.
folliculoma *(Med.)*. A tumour arising from cells in the Graafian follicle of the ovary.
follower *(Civ.Eng.)*. An intermediate length of timber which transmits the blow from the monkey to the pile; used when driving below water-level.
follower *(Eng.)*. That which follows the profile of a **cam**.
follower *(Horol.)*. The driven wheel of a pair of wheels engaging with each other. In clocks and watches the wheels are the drivers and the pinions the followers. In synchronous electric clocks the wheels are the followers.
follower *(Surv.)*. A chainman who has charge of the rear end of a chain and is responsible for lining-in the leader at each chain's length.
follow focus *(Image Tech.)*. Adjusting lens focus during action at varying distances to maintain a sharp image.
following dirt *(Min.Ext.)*. Following *stone*. A thin bed of loose shale above coal; a parting between the top of a coal-seam and the roof. See **pug**.
following response *(Behav.)*. Refers to the fact that newly hatched presocial birds will follow a moving object during a fairly brief period soon after hatching; under natural conditions this would be the parents or siblings. See **imprinting**.
follow-rest *(Eng.)*. A supporting member attached to the rear of the saddle of a lathe to steady the workpiece or bar material against the cutting pressure. See **steady**.
folly *(Arch.)*. A structure, usually tower or sham ruin built in a **Gothic** or **Classical** style, the purpose of which was to enhance a view. Follies were built generally towards the end of the eighteenth century, when landscaping became very popular. See **Picturesque**.
fomes *(Med.)*. Any infected object other than food. pl. *fomites*. Fomites such as clothing, bedding etc. may convey infection from one person to another.
font *(Print.)*. See **fount**.
fontanelle *(Zool.)*. In Craniata, a gap or space in the roof of the cranium.
food allergy *(Med.)*. The presence of **allergens** in food that provoke hypersensitivity reactions. Infants may be allergic to bovine milk and many other allergies have been described. *Food idiosyncracy* describes adverse effects occurring after certain foods without an immunological basis.
food body *(Bot.)*. A soft mass of cells, containing oil and other nutrient substances, attached to the outside of the seed coat; it is eaten by ants, which drag the seed along, leave it when they have eaten the food body, and so assist dispersal.
food chain *(Ecol.)*. Sequence of organisms dependent on each other for food. The number of links in the chain is usually only four or five. See also **food webs**.
food groove *(Zool.)*. A groove along which food is passed to the mouth; median and ventral in Branchiopoda; along the edge of each ctenidium in Bivalvia.
food poisoning *(Med.)*. Usually used to describe acute **gastro-enteritis** caused by bacterial or chemical contamination of food.
food pollen *(Bot.)*. Pollen formed by some flowers, which attracts insects; it may be incapable of bringing about fertilization and may be formed in special anthers. Insects seeking food pollen help in conveying good pollen to other flowers.
food vacuole *(Zool.)*. In the cytoplasm of some *Protozoa*, a space surrounding a food-particle and filled with fluid.
food webs *(Ecol.)*. The interlocking patterns formed by a series of interconnected food chains.
foolscap *(Paper)*. A superseded size of writing and printing paper $13\frac{1}{2} \times 17$ in. (US, 13×16 in.).
foolscap octavo *(Print.)*. A book-size, $6\frac{1}{2} \times 4\frac{1}{8}$ in.
fool's gold *(Min.)*. See **pyrite**.
foot *(Bot.)*. The basal part of an embryo, a developing sporophyte or a spore-producing body, embedded in the parental tissues and serving to absorb nutrients.
foot *(Zool.)*. A locomotor appendage; in Crustacea, any appendage used for swimming or walking; in Arachnida, Myriapoda, and Insects, the tarsus; in Echinodermata, the podia (see **podium**); in Mollusca, a median ventral muscular mass, used for fixation or locomotion; in land Vertebrates, the podium of the hind limb, or of all limbs in Tetrapoda.
footage number *(Image Tech.)*. See **edge number**.
foot-and-mouth disease *(Vet.)*. An acute febrile contagious disease of cloven-footed animals, due to infection by a virus; characterized by a vesicular eruption on the mucous membrane and skin, especially in the mouth and in the clefts of the feet. Also *aphthous fever*. Controlled by slaughter in UK and by vaccines where it is endemic.
foot block *(Build.)*. An **architrave block**.
foot-board *(Civ.Eng.)*. See **foot-plate**.
foot bolt *(Build.)*. A robust form of tower bolt, fixed near the foot of a door in a vertical position.
footbridge *(Civ.Eng.)*. A bridge for the use of pedestrians only.
footing *(Build.,Civ.Eng.)*. The lower part of a column or wall, standing immediately upon the foundation; usually enlarged locally in order to distribute the load over a greater area.
footing *(Elec.Eng.)*. The foundation which is set in the ground to support a tower of an overhead transmission line.
footing resistance *(Elec.Eng.)*. The ohmic resistance between a transmission-line tower and the earth.
foot irons *(Build.)*. Shaped iron bars which can be built into the joints of a manhole wall, leaving projecting steps for use by workmen descending the manhole. Also called *step irons*.
foot mange *(Vet.)*. See **chorioptic mange**.
Footner process *(Build.)*. A process which uses phosphate pickling to remove scale from steelwork prior to painting.
foot-plate *(Build.)*. A **hammer-beam**.
foot-plate *(Civ.Eng.)*. The platform on which the driver and fireman of a steam locomotive stand.
foot-pound (force) *(Phys.)*. Unit of work in the ft-lb-sec system of units. The work done in raising a mass of one pound through a vertical distance of one foot against gravity, i.e. 1.3558 J.
footprint *(Telecomm.)*. The service area, or the outline of the area on the Earth's surface within which a communication or direct broadcast satellite gives satisfactory results.
foot rot *(Bot.)*. One of a number of plant diseases, caused by a variety of fungi, in which the primary symptom is the death of the roots and stems around or below soil level.
foot rot *(Vet.)*. A suppurative infection between the horn and the sensitive corium of the hoof of the sheep, caused by *Fusiformis nodosus* and possibly other bacteria; causes lameness. Vaccines available.
foot run *(Build.)*. A term meaning foot of length, as in speaking of a loading or of a price *per foot run*.
foot screws *(Surv.)*. See **plate screws**.
foot-stall *(Build.)*. The base of a pillar.
foot-stick *(Print.)*. The name given to a **side-stick** when used at the foot of a page.
footstone *(Build.)*. The lowest coping-stone over a gable.
foot switch *(Elec.Eng.)*. A switch arranged for operation by the foot.
foot thumper *(Aero.)*. Stall warning device that vibrates the rudder pedals when the detector senses that a stall is imminent.
foot-ton (force) *(Phys.)*. 2240 foot-pounds (force).
foot valve *(Eng.)*. (1) The nonreturn or suction valve fitted at the bottom of a pump barrel, or in the valve chest of a pump. (2) A nonreturn valve at the inlet end of a suction pipe.
footwall *(Geol.,Min.Ext.)*. The lower wall of country rock in contact with a vein or lode. The upper wall is the *hanging wall*.
footway *(Min.Ext.)*. A colliery shaft in which ladders are used for descending and ascending.
forage mites *(Vet.)*. Acari of the family Acaridae, which commonly infest the skin of animals and birds.

foraging *(Behav.)*. Behaviour that involves searching for, capturing and consuming food.

foramen *(Zool.)*. An opening or perforation, especially in a chitinous, cartilaginous, or bony skeletal structure.

foramen lacerum *(Zool.)*. An opening of the Vertebrate skull in the side of the brain-case, which is situated between the alisphenoid and the orbitosphenoid, and through which pass the third, fourth, fifth ophthalmic, and sixth cranial nerves.

foramen magnum *(Zool.)*. The main opening at the back of the Vertebrate skull, by which the spinal cord issues from the brain-case.

foramen triosseum *(Zool.)*. A large aperture between the three bones of the pectoral girdle in Birds.

Foraminifera *(Zool.)*. An order of Sarcodina, the members of which have numerous fine anastomosing pseudopodia and a shell which is usually calcereous; the ectoplasm is sometimes vacuolated.

forb *(Bot.)*. Any herb other than a grass.

forbidden clone *(Immun.)*. Clones of lymphocytes reactive with 'self' antigens. According to the clonal selection theory they should have been eliminated during the early maturation of lymphocytes, and hence forbidden. The term is of historical interest, and is sometimes used, but the actual mechanisms by which 'self-reactivity' is avoided are more complicated.

forbidden land *(Phys.)*. The gap between two bands of allowed energy levels in a crystalline solid; see **energy band, energy gap**.

forbidden lines *(Astron.,Phys.)*. Spectral lines which cannot be reproduced under laboratory conditions. Such lines correspond to transitions from a metastable state, and occur in extremely rarefied gases, e.g., in the solar corona, and in gaseous nebulae.

forbidden transition *(Phys.)*. Transition of electrons between energy states, which, according to Pauli selection rules, have a very low probability in relation to those which have a high probability, and are *allowed*.

force *(Phys.)*. That which, when acting on a body which is free to move, produces an acceleration in the motion of the body, measured by rate of change of momentum of body. The unit of force is that which produces unit acceleration in unit mass. See **newton**; also **dyne, poundal**. Extended to denote loosely any operating agency. Electromotive force, magnetomotive force, magnetizing force, etc., are strictly misnomers.

forced-circulation boilers *(Eng.)*. Steam boilers in which water and steam are continuously circulated over the heating surface by pumps (as opposed to natural circulation systems) in order to increase the steaming capacity. See **Löffler boiler**.

forced commutation *(Elec.Eng.)*. (1) The usual process of commutation in which the change of direction of the current in the coil actually undergoing commutation is assisted by flux from a commutating pole. (2) Method of commutating conducting devices in an electronic converter whereby circuits containing inductance and capacitance are arranged to drive the current through the conducting devices to zero. Ant. *natural* commutation.

forced development *(Image Tech.)*. Processing with increased time or temperature to increase the image density for under-exposed film.

forced draught *(Elec.Eng.)*. Said of electrical apparatus cooled by ventilating air supplied under pressure from some external source.

forced draught *(Eng.)*. An air supply to a furnace driven or induced by fans or steam jets (as opposed to the *natural* draught created by a chimney) in order to obtain a high rate of combustion. See **balanced draught, closed stokehold, induced draught**.

forced-draught furnace *(Eng.)*. A furnace, but more particularly a boiler furnace, arranged to work under **forced draught**.

forced-flow boilers *(Eng.)*. See **forced-circulation boilers**.

force diagram *(Eng.)*. A diagram in which the internal forces in a framed structure, assumed pin-jointed, are shown to scale by lines drawn parallel to the members themselves. Also called a *reciprocal* or *stress diagram*.

forced lubrication *(Eng.)*. The lubrication of an engine or machine by oil under pressure. See **force feed, full force feed**.

forced movements *(Med.)*. See **tropism**.

forced oscillations *(Phys.)*. Oscillatory currents whose frequency is determined by factors other than the constants of the circuit in which they are flowing, e.g., those flowing in a resonant circuit coupled to a fixed frequency oscillator. See **forced vibrations**.

force drying *(Build.)*. Accelerated drying through moderate heat, generally below about 82°C.

forced vibrations *(Phys.)*. Vibrations which result from the application of a periodic force to a body capable of vibrating. The amplitude of forced vibrations becomes very great when resonance occurs, that is, when the frequency of the applied force equals the natural frequency of the vibrator, particularly if the damping is small.

force feed *(Eng.)*. Lubrication of an engine by forcing oil to main bearings and through the hollow crankshaft to the big-end bearings.

force on a moving charge *(Phys.)*. If a charge q is moving with a velocity $\mathbf{v}$ in a magnetic field of intensity $\mathbf{B}$, then the force on the charge is $\mathbf{F}=q(\mathbf{v}\times\mathbf{B})$. If the moving charges are within a conductor, the force of a short length $\mathbf{l}$ is $\mathbf{F}=i(\mathbf{l}\times\mathbf{B})$, where i is the current. See **Lorentz force**.

forceps *(Med.)*. (1) A pincerlike instrument with two blades, for holding, seizing or extracting objects. *Obstetrical forceps* have large blades, which, applied to the foetal head, aid delivery. (2) That part of either of the two ends of the *corpus callosum* of the brain which diverges into the adjacent brain tissue on each side.

forceps *(Zool.)*. In Dermaptera, the pincer-shaped cerci; in Arachnida and Crustacea, the opposable distal joints of the chelae; in Echinodermata, the distal opposable jaws of pedicellariae. adj. *forcipulate*.

force pump *(Build.)*. An air pump used to clean out gas and other service pipes by blowing air through them.

force pump *(Eng.)*. A pump which delivers liquid under a pressure greater than its suction pressure. It consists of a barrel fitted with a solid plunger, and a valve chest with suction and delivery valve.

Ford cup *(Build.)*. A device used to ascertain the viscosity of paints. The time taken for paint to flow from the cup though an accurately machined orifice in its base is measured in seconds.

fore-and-aft *(Print.)*. The first and last sections of the book are printed and folded in one sheet, the first page being head to head with the last page; the second section is similarly combined with the second last section, and so on. Used for the production of large paperback editions; there are fewer sheets to handle and the economy of two-on working is maintained throughout all printing and binding operations. Also *come-and-go*.

forearc basin *(Geol.)*. A sedimentary basin developed in the gap between a volcanic arc and its **subduction zone**.

forebody strake *(Aero.)*. Low aspect-ratio extension of the wing at the root along sides of the forebody. These create powerful vortices during high incidence flight, thereby improving handling and increasing lift.

fore-brain *(Zool.)*. In Vertebrates, that part of the brain which is derived from the first or anterior brain-vesicle of the embryo, comprising the olfactory lobes, the cerebral hemispheres, and the thalamencephalon; the first or anterior brain-vesicle itself.

forecast *(Meteor.)*. A statement of the anticipated weather conditions in a given region, for periods of from 1 hour to 30 days in advance, the longer term being less reliable; made from a study of current synoptic charts, or by carrying out a **numerical forecast**.

fore-edge *(Print.)*. The outside margin of a book page; the edge opposite to the back; the outer edge of a volume. Cf. **head, tail**.

fore-edge painting *(Print.)*. Pictures painted on the fanned-out edges of a book, unseen when the book is closed, as an additional decoration to an already elaborate binding.

foreground/background processing *(Comp.)*. Method of

organising a **time-sharing** computer system so that while main tasks may claim the use of the computer when required, other less pressing tasks utilize the remaining time.

fore-gut *(Zool.)*. That part of the alimentary canal of an animal which is derived from the anterior ectodermal invagination or stomodaeum of the embryo.

forehand welding *(Eng.)*. That in which the palm of the torch or electrode hand faces the direction of travel so that the metal ahead of the weld position is preheated.

forehearth *(Eng.)*. Bay in front of a furnace into which molten products can be run.

Forensic Medicine *(Med.)*. Any aspect of medicine which has civil or statutory legal implications. Most simply *legal medicine*.

fore peak *(Ships.)*. The spaces forward of the **collision bulkhead**. Lower part frequently used as freshwater tank and upper part may be used as storeroom.

foreplane *(Aero.)*. Horizontal aerofoil mounted on front fuselage for pitch control. In a tail-first or *canard* configuration it replaces the function of the tail plane, in a delta wing design it may assist slow speed behaviour. Never used for roll control, a foreplane may be fixed or retractable and have **slats**, **flaps** or **elevators**.

forepoling *(Min.Ext.)*. A method of progressing through loosely consolidated ground by driving poles forward over frames.

foreset beds *(Geol.)*. Gently inclined cross-bedded units progressively covering **bottomset beds** and covered in turn by **topset beds**. Foreset beds form the major bulk of a delta.

fore sight *(Surv.)*. The levelling staff reading as taken forward to a station which has not been passed by the instrument. It transfers collimation line from back-sight station to fore-sight station. See also **back observation, intermediate sight**.

forest *(Ecol.)*. (1) Natural vegetation in which the dominant species are trees, with crowns that touch each other to form a continuous canopy. (2) Almost any vegetation with trees, including plantations. (3) In Britain, areas which were formerly with trees and used for hunting but which may now be treeless.

forfex *(Zool.)*. A pair of pincers, as of an earwig.

forge *(Eng.)*. Open hearth or furnace with forced draught; place where metal is heated and shaped by hammering.

forge pigs *(Eng.)*. Pig-iron suitable for the manufacture of wrought-iron.

forge tests *(Eng.)*. Rough workshop tests made to check the malleability and ductility of iron and steel.

forging *(Eng.)*. The operation of shaping (hot) malleable metals by means of hammers or presses. It includes hand-hammer, steam-hammer, press and drop forging.

forging *(Vet.)*. See **clacking**.

forging machines *(Eng.)*. Power hammers and presses used for forging and drop forging. See **drop forging**.

fork *(Horol.)*. In lever escapements, the end of the lever which receives the impulse pin.

forked lightning *(Meteor.)*. A popular name given to a lightning stroke; the name derives from the branching of the stroke channel which is commonly observed.

forked tenon *(Build.)*. A joint formed by a slot mortise astride a tenon cut across the length of a member.

fork-lift truck *(Genrl.)*. A vehicle with power operated prongs which can be raised or lowered at will, for loading, transporting and unloading goods; chiefly used in factories and warehouses. Loads are usually stacked upon stands or pallets with sufficient ground clearance for the prongs to be inserted beneath them.

form *(Civ.Eng.)*. See **mould**.

form *(Crystal.)*. A complete assemblage of crystal faces similar in all respects as determined by the symmetry of a particular class of crystal structure. Thus the *cube*, consisting of 6 similar square faces, and the *octahedron*, consisting of 8 faces, each an equilateral triangle, are crystal *forms*. The number of faces in a form ranges from 1 (the pedion) to 48 (the haxakis-octahedron). A natural crystal may consist of one form or many.

formability *(Eng.)*. See **drawability**.

formaldehyde *(Chem.)*. *Methanal*. H.CHO, bp −21°C, a gas of pungent odour, readily soluble in water, and usually used in aqueous solution. Formaldehyde easily polymerizes to *paraformaldehyde* or metaformaldehyde. It is produced by oxidation of methanol, or by the oxidation of ethene in the presence of a catalyst. It forms with ammonia *hexamethylene-tetramine*. It is a disinfectant and is of great importance in plastics manufacture.

formaldehyde (methanal) resins *(Chem.)*. Synthetic resins which are condensation products of methanal with hydroxybenzenes, urea etc. See also **Bakelite, Delrin**.

formalin *(Chem.)*. The TN for a commercial 40% aqueous methanal solution.

formal language theory *(Comp.)*. A subject in computer science concerned with the specification and use of artificial languages.

formal operations *(Behav.)*. In Piagetian theory, the fourth and final major stage of cognitive development, occurring during adolescence; it reflects a transition from logic bound to the real and concrete to a logic capable of dealing with abstract events. See **concrete operations**.

formal specification *(Comp.)*. (1) Subject concerned with techniques and methods of specifying a program using a **formal language**. (2) A program specification in a formal language.

formant *(Acous.)*. Envelope of the frequency pattern arising in the vocal cords, which determines the distribution of energy in unvoiced sounds and the reinforcement of harmonics in voiced sounds. It is the formant which leads to recognition of speech sounds by aural perception.

format *(Comp.)*. See **disk formatting**.

format *(Print.)*. The general size, quality of paper, type face and binding.

formates, methanoates *(Chem.)*. The salts of **formic acid**.

formation *(Geol.)*. A stratigraphical rock unit used as a basis for rock mapping. A formation has a recognizable lithological identity and divisable into *members* or combined into *groups*. It has been casually used in Britain but more precisely in North America.

formation *(Paper)*. The pattern of the fibres in the paper when viewed by transmitted light.

formative time *(Electronics)*. The time interval between the first Townsend discharge in a given gap and the formation there of a self-maintaining glow discharge.

formatting *(Print.)*. Term used to describe the translating of type specifications into format or command codes prior to phototypesetting.

form drag *(Aero.)*. The difference obtained when the *induced drag*, i.e. the fraction of the total drag induced by lift, is subtracted from the **pressure drag**. See **drag**.

forme *(Print.)*. Type matter assembled and locked up in a chase ready for printing.

formed plate *(Elec.Eng.)*. A type of plate used in lead-acid accumulators; made by electrolytically converting the substance of which the plate is made into active material.

former *(Aero.)*. A structural member of a fuselage, nacelle, hull or float to which the skin is attached and having the primary purpose of preserving form or shape. It generally carries structural loads. Cf. **frame**.

former *(Elec.Eng.)*. A tool for giving a coil or winding the correct shape; it sometimes consists of a frame upon which the wire can be wound, the frame afterwards being removed.

forme rollers *(Print.)*. Rollers which supply ink to a forme or plate, as distinct from **distributor rollers**.

former-wound coil *(Elec.Eng.)*. An armature coil built to the correct shape by means of a *former*, it being then dropped into the slots on the armature.

form factor *(Phys.)*. The ratio of the effective value of an alternating quantity to its average value over a half-period.

form genus *(Bot.)*. See **form taxon**.

form grinding *(Eng.)*. See **profile grinding**.

formic, methanoic acid *(Chem.)*. HCOOH, a colourless liquid, of pungent odour, corrosive, mp 9°C, bp 101°C. Prepared by passing carbon monoxide and steam at 200°–300°C under pressure over a catalyst.

forming *(Eng.)*. Changing the shape of a metal component without in general altering its thickness. Cf. **drawing**.
forming cutter *(Eng.)*. See **form tool**.
forming rollers *(Print.)*. See **bending rollers**.
formol titration *(Chem.)*. A method of estimating volumetrically the amount of amino acids present in a solution. It is based upon the fact that amino acids and their derivatives possess both a carboxyl and an amino group which neutralize each other, and that by the addition of methanal the amino group is converted into a methylene derivative without basic properties, by which reaction it becomes possible to titrate subsequently the carboxyl in the usual manner.
formol toxoid *(Immun.)*. Any toxoid prepared by formaldehyde treatment of a toxin.
formoxyl *(Chem.)*. The organic radical O.CHO-.
form taxon *(Bot.)*. An artificial taxon, e.g. a *form genus*, intended to provide a name for morphologically similar but possibly unrelated organisms or parts of organisms when it is not possible to determine their correct taxonomic position. Thus an *organ genus* in palaeobotany provides names for fossil leaves, spores, pieces of cuticle etc. where it is not (yet) possible to reconstruct the whole plant. Again in Deuteromycetes, it provides names for fungi with no, or no known, sexual reproduction.
form tool, forming cutter *(Eng.)*. Any cutting tool which produces a desired contour on the work-piece by being merely fed into the work, the cutting edge having a profile similar to, but not necessarily identical with, the shape produced. See **chaser**.
formula *(Chem.)*. The representation of the types and relative numbers of atoms in a compound, or the actual number of atoms in a molecule of a compound. It uses chemical symbols and subscripts, e.g.: H_2SO_4 or C_6H_6.
formula *(Genrl.)*. A fixed rule or set form.
formula *(Maths.)*. A rule expressed in algebraic symbols.
formula *(Med.)*. A prescription or specification.
formwork *(Civ.Eng.)*. **Shuttering**.
formyl *(Chem.)*. The organic radical OCH-.
fornacite, furnacite *(Min.)*. A basic chromarsenate of copper and lead, crystallizing in the monoclinic system.
fornix *(Zool.)*. In the brains of higher Vertebrates, a tract of fibres connecting the posterior part of the cerebrum with the hypothalamus.
Forssman antigen, Forssman antibody. *(Immun.)*. A glycolipid antigen present on tissue cells of many species, and on the red cells of some species such as sheep. Forssman antibody may be elicited by immunization with the antigen, but often present as a natural antibody in humans. Sheep red cells coated with Forssman antibody are used in complement fixation tests.
forsterite *(Min.)*. An end-member of the olivine group of minerals, crystallizing in the orthorhombic system. Chemically forsterite is a silicate of magnesium, Mg_2SiO_4. It typically occurs in metamorphosed impure dolomitic limestones.
forsterite-marble, ophicalcite *(Geol.)*. A characteristic product of the contact metamorphism of magnesian (dolomitic) limestones containing silica of organic or inorganic origin.
Forstner bit *(Build.)*. A patent centre-bit with a small centre-point and a circular rim, for sinking blind holes in timber.
FORTH *(Comp.)*. Programming language using **reverse Polish notation**, with applications in control.
Fortin's barometer *(Meteor.)*. A type of mercury barometer suited for accurate readings of the pressure of the atmosphere. The zero of the scale is indicated by a pointer inside the mercury cistern, the bottom of which is flexible and may be moved by an adjusting screw until the mercury surface just touches the pointer.
FORTRAN *(Comp.)*. Programming language widely used for scientific programming; *FORmula TRANslation*.
forward-bias *(Electronics)*. Said of a semiconductor diode, p-n junction, or the emitter-base junction of a transistor, when the polarity of the applied e.m.f. is such as to allow substantial conduction to take place.
forward eccentric *(Eng.)*. On a steam-engine having link motion reverse gear, the eccentric which drives the valve when the engine is going ahead. See **link motion**.
forwarding *(Print.)*. The operations entailed in bookbinding, until a book has been placed in its covers. See **finishing**.
forward lead *(Elec.Eng.)*. See **forward shift**.
forward-looking infra red *(Aero.)*. Sensor systems in the nose of aircraft or guided weapon for target detection and vehicle guidance.
forward path *(Telecomm.)*. The transmission path from the loop-activating signal to the loop-output signal in a feedback control loop.
forward perpendicular *(Ships)*. The forward side of a ship's stem post when this is truly perpendicular to the longitudinal base line; but in cases when the stem post is 'raked', i.e. angled to the baseline, it is the perpendicular intersecting the forward side of the stem post at the summer load water line. Abbrev. *FP*.
forward scatter *(Phys.)*. Scattering of particles through an angle of less than 90° to the original direction of the beam. Cf. **back scatter**.
forward scatter *(Telecomm.)*. Multiple transmission on centimetric waves, using reflection down and forwards from ionization in troposphere; range about 150 km. See **tropospheric scatter**.
forward shift *(Elec.Eng.)*. A movement of the brushes of a commutator machine around the commutator, from the neutral position, and in the same direction as that of rotation. Also called *forward lead*.
forward sweep *(Aero.)*. See **sweep**.
forward transfer function *(Telecomm.)*. That of the forward path of a feedback control loop.
forward voltage *(Elec.Eng.)*. The voltage of the polarity which produces the larger current in an electrical system.
forward wave *(Electronics)*. One whose group velocity is in the same direction as the motion of an electron stream in a travelling-wave tube. See **backward wave**.
fossa *(Zool.)*. A ditchlike or pitlike depression, as the *glenoid fossa*.
fossa rhomboidalis *(Zool.)*. See **fourth ventricle**.
fossette *(Zool.)*. In general, a small pit or depression: in some Arthropoda, the socket which receives the base of the antennule.
fossil *(Geol.)*. The relic or trace of some plant or animal which has been preserved by natural processes in rocks of the past.
fossil meal *(Build.)*. A diatomaceous earth used in the manufacture of light-weight and heat resistant, often hollow, blocks.
fossil zone *(Geol.)*. The stratigraphical horizon characterized by a *zone fossil*. See **zone**.
fossorial *(Zool.)*. Adapted for digging.
Foster-Seeley discriminator *(Elec.Eng.)*. One for demodulating FM transmission using a balanced pair of diodes. To these are applied voltages which are sum and difference of limiter signal voltage and half transformer coupled voltage, diode outputs being differenced.
Föttinger coupling, transmitter *(Eng.)*. A hydraulic coupling, gear or clutch for transmitting power from, e.g. an engine to a ship's propeller; it consists essentially of an outward-flow water turbine driving an inward-flow turbine, within a common casing.
Föttinger speed transformer *(Eng.)*. A hydraulic reduction gear formerly used in marine propulsion, comprising a centrifugal pump and turbine runner in a single unit, giving a speed reduction of 5:1.
Foucault currents *(Elec.Eng.)*. See **eddy currents**.
Foucault knife-edge test *(Phys.)*. A method of testing for the form and optical quality of a lens or mirror by placing a knife-edge in the focus of a point source and observing the pattern of light and shade as seen in the lens when an observer places his eye immediately behind the knife-edge.
Foucault's measurement of the velocity of light *(Phys.)*. One of the first successful attempts to obtain an accurate result for this important constant. Foucault, in 1862, made use of a rapidly rotating mirror sending light to a distant fixed concave mirror which reflected it back.

Measurement of the displacement of the reflected image gave a value of 2.986×10^8 m/s for the velocity of light *in vacuo*.

Foucault's pendulum *(Astron.)*. An instrument devised by Foucault in 1851 to demonstrate the rotation of the earth; it consists of a heavy metal ball suspended by a very long fine wire. The plane of oscillation slowly changes through 15° sin (latitude) per sidereal hour.

foul air flue *(Build.)*. A ventilating flue through which vitiated air from a room is drawn.

foul anchor *(Ships)*. An anchor which has become entangled in its own cable or with some obstruction on the bottom.

foulard *(Textiles)*. A light-weight dress fabric with a printed pattern, made either of silk or man-made filament yarns or of high quality Sea Island or Egyptian cotton.

foul berth *(Ships)*. A situation where anchored ships may collide when swinging round with the change of tidal stream.

foul clay *(Build.)*. A brick earth composed of silica and alumina combined with only a small percentage of lime, magnesia, soda or other salts. Such a clay lacks sufficient fluxing material to fuse its constituents at furnace temperature, and is improved by the addition of sand or loam, lime or ashes. Also called *plastic clay*, *pure clay*, *strong clay*.

fouling *(Eng.)*. (1) Coming into accidental contact with. (2) Deposition or incrustation of foreign matter on a surface, as of carbon in an engine cylinder, or marine growth on the bottom of a ship or on structures subject to the action of sea-water.

fouling point *(Civ.Eng.)*. The location before the meeting of two tracks where the loading-gauge outlines come into contact.

foul in the foot *(Vet.)*. *Infectious pododermatitis, claw ill.* Suppuration and necrosis of the interdigital tissues of the foot in cattle due to infection by the bacterium *Sphaerophorus necrophorus* (*Fusiformis necrophorus*); causes lameness.

fouls *(Min.Ext.)*. The cutting-out of portions of the coal-seam by 'wash outs'; barren ground.

foul solution *(Min.Ext.)*. In cyanide process, one contaminated by salts or metals other than gold and silver, and which is not fit to be recirculated in the process.

foul water *(Build.)*. Water contaminated by soil, waste or trade effluent.

foundation cylinder *(Eng.)*. A large steel or iron cylinder sunk into the ground to provide a solid foundation for bridge piers, etc., in soft ground.

foundation piles *(Civ.Eng.)*. Piles driven into the ground to provide an unyielding support for a structure.

foundation ring *(Eng.)*. In a locomotive boiler, a rectangular iron ring of rectangular section, to which the lower edges of the inner and outer plates of the fire box are secured.

founded *(Civ.Eng.)*. Said of a caisson which has been sunk to a firm level.

founder's type *(Print.)*. Founts of type produced by the specialist typefoundries, as distinct from that produced by the printer himself, in hard type metal; used mainly for display lines.

founding *(Glass)*. The melting operation in which molten glass is made almost free from undissolved gases. Also *fining*, *plaining*, *refining*.

foundry *(Eng.)*. A workshop in which metal objects are made by casting in moulds.

foundry *(Print.)*. That department of printing establishment where work in connection with duplicate plates is carried out.

foundry chase *(Print.)*. A chase with type-high rims and sometimes with built-in locking-up device, used when taking a mould of type matter.

foundry ladle *(Foundry)*. A steel ladle lined with fireclay; used for transporting molten metal from a foundry cupola to the moulds. Small ladles are carried by hand, large ones by a truck or crane. See **hand shank**.

foundry pig-iron *(Eng.)*. Bars of cast-iron up to 1 m in length and 1.5 cm in diameter, as bought by an iron foundry.

foundry pit *(Foundry)*. A large hole in the floor of a foundry, which serves the purpose of a moulding box for very large or deep castings.

foundry sand *(Foundry)*. Silica-based sand with either enough natural clay or special additions (oil, molasses, fullers' earth etc.) to give it some cohesive strength when used in moulding.

foundry stove *(Foundry)*. A large oven for drying moulds and cores, heated either externally by hot gases or internally by a fire-basket.

fountain *(Print.)*. The ink reservoir on a printing press. In **lithography** that part of the machine which holds the dampening or fountain solution.

fountain effect *(Phys.)*. Elegant experiment in superfluidity demonstrating the behaviour of liquid helium II.

fountain solution *(Print.)*. In lithographic printing a solution of water or alcohol, acid, buffer and gum arabic which when fed to the plate by the machine dampers prevents the non-image areas from taking ink.

fount, font *(Print.)*. A complete set of type of the same face and size, containing proportionate quantities of the individual characters.

fount management system *(Print.)*. Character-generating software which controls the format of typefaces produced by a *non-impact printer*.

four-bar chain *(Eng.)*. A common and simple versatile mechanism comprising an assemblage of four rigid members, which are capable of relative motion.

four-centred arch *(Arch.)*. A pointed arch struck from 4 centres.

fourchette *(Med.)*. The posterior or junction of the labia minora.

four-colour press *(Print.)*. Sheet-fed and web-fed models are available for the main printing processes. Four colours are the minimum for full-colour reproduction but extra units are often added for additional printing or varnishing.

four-colour reproduction *(Print.)*. Printing by any of the main printing processes with standard yellow, magenta, cyan, and black inks, not always in that order, from sets of colour plates or cylinders, to produce a reproduction of a coloured subject. See **colour correction, three-colour process**.

fourdrinier *(Paper)*. The standard type of paper making machine characterized by a wire part of which the upper, forming, surface is horizontal or nearly so.

four factor formula *(Nuc.Eng.)*. That giving the multiplication factor of an infinite thermal reactor as the product of **fast fission factor, resonance escape probability, thermal utilization factor** and the number of neutrons absorbed per neutron absorbed in the fuel (η).

Fourier analysis *(Phys.,Maths.)*. The determination of the harmonic components of a complex waveform (i.e. the terms of a Fourier series that represents the waveform) either mathematically or by a wave-analyser device.

Fourier half-range series *(Maths.)*. Fourier series with only sine or cosine terms. Valid for x between 0 and π.

Fourier integral *(Phys.)*. The limiting form of the *Fourier series* when the period of the waveform becomes infinitely long. It is the Fourier representation of a non-repeated waveform, i.e. pulses, wave-packets, and wave trains of limited extent.

Fourier number *(Phys.)*. A dimensionless parameter used for studying heat flow problems. It is defined by $\lambda t/C_p\rho l^2$ where λ is the thermal conductivity, t is the time, C_p the specific heat at constant pressure, ρ the density and l a linear dimension.

Fourier optics *(Phys.)*. The application of Fourier analysis and the use of Fourier transforms to problems in optics, in particular to image formation. The *Fraunhofer diffraction* pattern is the Fourier transform of the distribution of amplitude of light across the diffracting object. The distribution of amplitude in the Fraunhofer pattern is modified by the optical system and the image formed is the transform of this modified distribution. The same principle is used in X-ray crystal structure analysis where an 'image' of the atomic arrangement is constructed mathematically from the X-ray diffraction pattern.

Fourier principle *(Phys.)*. That which shows all repeating waveforms can be resolved into sine wave components consisting of a fundamental and a series of harmonics at multiples of this frequency. It can be extended to prove that non-repeating waveforms occupy a continuous frequency spectrum.

Fourier series *(Maths.)*. The series

$$\tfrac{1}{2}a_0 + \sum_{n=1}^{\infty} (a_n \cos nx + b_n \sin nx)$$

where

$$a_n = \frac{1}{\pi}\int_{-\pi}^{+\pi} f(x) \cos nx \, dx$$

and

$$b_n = \frac{1}{\pi}\int_{-\pi}^{+\pi} f(x) \sin nx \, dx,$$

which, subject only to certain very general restrictions upon $f(x)$, converges to $f(x)$ in the interval $(-\pi, +\pi)$. Fourier series are used in the determination of the component frequencies of vibrating systems. See **Fourier analysis**.

Fourier transform *(Maths.,Phys.)*. A mathematical relation between the energy in a transient and that in a continuous energy spectrum of adjacent component frequencies. The Fourier transform $F(u)$ of the function $f(x)$ is defined by

$$F(u) = \int_{-\infty}^{+\infty} e^{-iut} f(t) \, dt.$$

Some writers use e^{+iut} instead of e^{-iut}.

Fourier transform spectroscopy *(Phys.)*. The production of a spectrum by taking the **Fourier transform** of a two-beam interference pattern.

four-jaw chuck *(Eng.)*. A chuck used in a lathe or in certain other types of machine tools, comprising four jaws disposed at right angles to each other. Usually, the jaws act independently of each other. Used for holding rectangular or irregularly contoured workpieces, or for revolving circular workpieces eccentrically.

four-part vault *(Arch.)*. A vault formed at the intersection of 2 barrel vaults.

four-phase system *(Elec.Eng.)*. A name sometimes given to a 2-phase system in which the midpoints of the 2 phases are connected to form a neutral point.

four-stroke cycle *(Autos.)*. An engine cycle completed in 4 piston strokes (i.e. in 2 crankshaft revolutions), consisting of suction or induction, compression, expansion or power stroke, and exhaust. See **diesel cycle, Otto cycle.**

four-terminal resistor *(Elec.Eng.)*. Laboratory standard fitted with current and potential terminals, arranged so that measurement of the potential drop across the resistor is not affected by contact resistances at the terminals.

fourth generation computer *(Comp.)*. Term somewhat imprecisely applied to a computer system in use in 1987; physically small and neat using low-cost **chip** technology with **LSI**. Processors and computers are being linked up to form systems and **networks** rather than a computer. **Computer graphics** is blossoming. Typical machines so far might be Amdahl 470, Intel 8748. See **distributed computing, microcomputer, computer generations, workstation.**

fourth pinion *(Horol.)*. The pinion on which the fourth wheel is mounted.

fourth rail *(Elec.Eng.)*. A conductor-rail on an electric traction system. When there are two running rails and two conductor-rails, the fourth rail generally carries the return current, instead of its being allowed to return along the running rails.

fourth-rail insulator *(Elec.Eng.)*. An insulator for supporting a fourth rail in an electric-traction system.

fourth ventricle *(Zool.)*. In Vertebrates, the cavity of the hind-brain.

fourth wheel *(Horol.)*. The wheel in a watch which drives the escape pinion. If the train is suitable for a seconds hand (i.e. the fourth wheel makes one turn per minute), the hand is carried on an extension of the fourth wheel arbor.

fourth wire *(Elec.Eng.)*. A name sometimes given to the neutral wire in a 3-phase, 4-wire distribution system.

four-way tool post *(Eng.)*. A supporting and clamping mechanism for cutting tools used in a lathe or certain other machine tools, in which up to 4 tools may be mounted simultaneously but in various positions, the required tool being brought into the operative position by indexing the tool post appropriately.

four-wire circuit *(Telecomm.)*. One in which information is transmitted in one direction on one path and in the opposite direction on the other path. Need not necessarily consist of four separate wires; electronic means may be used to reproduce the effect of a physical four-wire circuit.

four-wire repeater *(Telecomm.)*. A repeater for insertion into a 4-wire telephone circuit, in which the 2 amplifiers, 1 for amplifying in each direction, are kept separate.

four-wire system *(Elec.Eng.)*. A system of distribution of electric power requiring 4 wires. In a 3-phase system, the 4 wires are connected to the 3 line terminals of the supply transformer and the neutral point; and in the 2-phase system, the wires are connected to the ends of the 2 transformer windings.

fovea *(Zool.)*. A small pit or depression. adjs. *foveate, foveolar, foveolate.*

fovea centralis *(Zool.)*. In the Vertebrate eye, the areas of greatest visual resolution, seen as a small depression at the centre of the **macula lutea**.

foveola *(Zool.)*. Same as *fovea.*

fowl cholera *(Vet.)*. An acute and usually fatal septicaemia of domestic fowl and other birds, caused by the bacterium *Pasteurella multocida* (*P. aviseptica*). Characterized by dejection, loss of appetite, raised temperature and diarrhoea. Mild and chronic forms of the disease also occur.

fowl coryza *(Vet.)*. See **infectious coryza**.

Fowler flap *(Aero.)*. A high-lift trailing-edge flap that slides backward as it moves downward, thereby increasing the wing area, also leaving a slot between its leading edge and the wing when fully extended.

Fowler position *(Med.)*. The semisitting position in which the patient may be placed in bed after an abdominal operation with head elevated and knees drawn up so that the pelvis is the lowest part of the body.

fowl paralysis *(Vet.)*. *Marek's disease, neuro-lymphomatosis.* Chickens are the only important natural host. It is one of the most ubiquitous avian infections, caused by a herpes virus with three serotypes. Paralysis sometime seen but bird usually shows depression prior to death. Various peripheral nerves become enlarged and lymphoid tumours are observed in various organs. Controlled by vaccination.

fowl pest *(Vet.)*. A term usually embracing both **fowl plague** and **Newcastle disease**.

fowl plague *(Vet.)*. An acute, contagious virus disease of chickens and other domestic and wild birds; the main symptoms are high temperature, oedema of the head, nasal discharge, and rapid death.

fowl pox *(Vet.)*. *Avian diphtheria.* Caused by large DNA virus. A world-wide and relatively slowly spreading infection of chickens and turkeys. Lesions consist of nodules in the skin which progress to scabs, and of diphtheria membranes in the respiratory and upper digestive tracts.

fowl typhoid *(Vet.)*. A contagious septicaemic disease of chickens and other domesticated birds caused by *Salmo-*

nella gallinarum. Vaccination using a *rough strain* of *S. gallinarum*.

fox encephalitis *(Vet.)*. *Epizootic fox encephalitis, infectious encephalitis of foxes*. An acute infection characterized by severe nervous symptoms, caused by the virus of **infectious canine hepatitis**.

fox marks *(Print.)*. Brown stains on the pages of books that have been allowed to get damp.

foxtail wedging *(Build.)*. The tightening-up of a tenon in a blind mortise by inserting small wedges in saw-cuts in the end of the tenon before inserting the latter in the mortise. The operation of driving the tenon into position then forces the wedges into the saw-cuts and spreads the fibres of the tenon, giving a secure hold resisting withdrawal. Also *fox wedging*.

foyaite *(Geol.)*. A widely distributed variety of nepheline-syenite, described originally from the Foya Hills in Portugal. Typically it contains about equal amounts of nepheline and potassium feldspar, associated with a subordinate amount of coloured mineral such as aegirine.

fp *(Phys.)*. Abbrev. for *Freezing-Point*.

FP *(Ships)*. Abbrev. for *Forward Perpendicular*.

FPS *(Genrl.)*. The system of measuring in *feet, pounds*, and *seconds*.

Fr *(Chem.)*. The symbol for francium.

fracking *(Min.Ext.)*. Forcing liquid out into the strata round a well bottom to split it. It contains sand or other material which remains in the fissures and prevents them closing. See **proppants, tertiary production**.

fractal *(Maths.)*. Formally, a set such that two standard definitions of its dimension give different answers. A curve is one-dimensional according to one definition but a curve with arbitrarily small irregularities may have a larger dimension by the other definition, and it is then called a fractal curve. For example, it is estimated that the coastline of the UK can be regarded as a fractal curve whose larger dimension is about 1.25 that of the smaller.

fraction *(Maths.)*. See **division**.

fractional crystallization *(Chem.)*. The separation of substances by the repeated partial crystallization of a solution.

fractional crystallization *(Geol.)*. Separation of a cooling magma into parts by crystallization of different minerals at successively lower temperatures.

fractional distillation *(Chem.)*. Distillation process for the separation of the various components of liquid mixtures. An effective separation can only be achieved by the use of **fractionating columns** attached to the still.

fractional distribution *(Chem.)*. In an assembly of particles having different values of some common property such as size or energy, the fraction of particles in each range of values is called the fractional distribution in that range. Cf. **cumulative distribution**.

fractional pitch *(Eng.)*. Of a screw-thread cut in the lathe, a pitch not an integral multiple or submultiple of the pitch of the lead screw. See **even pitch**.

fractionating column *(Chem.)*. A vertical tube or column attached to a still and usually filled with packing, e.g. Raschig rings, or intersected with bubble plates. An internal reflux takes place, resulting in a gradual separation between the high- and the low-boiling fractions inside the column, whereby the fractions with the lowest boiling-point distil over. The efficiency of the column depends on its length or on the number of bubble plates used, and also on the ability of the packing to promote contact between the vapour and liquid phases.

fractionation *(Chem.)*. See **fractional distillation**.

fractionation *(Radiol.)*. A system of treatment commonly used in radiotherapy in which doses are given daily or at longer intervals over a period of 3 to 6 weeks.

fraction I protein *(Bot.)*. See **ribulose 1,6-bisphosphate carboxylase oxygenase**.

fractography *(Eng.)*. The description of a fracture face by means of photographs at various magnification.

fracture *(Med.)*. Breaking of a bone. Fractures may be *simple* (broken bone only), *compound* (external wound communicating with fracture), *complicated* (additional injury, e.g. to internal organs, blood vessels etc.), *comminuted* (bone broken in several or many parts), *fissured* (bone cracked, e.g. skull), **impacted, greenstick**.

fracture *(Min.)*. The broken surface of a mineral as distinct from its cleavage. The fracture is described, in different cases, as conchoidal (shell-like), platy or flat, smooth, hackly (like that of cast-iron) or earthy. Thus calcite has a perfect rhombohedral cleavage, but conchoidal fracture.

fracture cleavage *(Geol.)*. A set of closely-spaced, parallel joints. Common in shallowly deformed metamorphic rocks.

fracture test *(Eng.)*. Deliberate breakage of a metal part to allow the examination of the fracture face to determine grain size, case depth etc.

fragile-X syndrome *(Biol.)*. Hereditary human condition involving mental subnormality in which a portion of the X-chromosome tends to show breaks in appropriately prepared lymphocytes.

fragmentation *(Biol.)*. (1) See **amitosis**. (2) The breaking-off of a portion from the main body of a chromosome.

fragmentation *(Bot.)*. Form of vegetative or asexual reproduction in which pieces of the parent become detached and grow into new individuals, especially in many filamentous algae.

fraktur *(Print.)*. A **black letter** type peculiar to Germany.

framboesia, frambesia *(Med.)*. **Yaws**.

frame *(Aero.)*. A transverse structural member of a fuselage, hull, nacelle or float, which follows the periphery and supports the skin, or the skin-stiffening structure. Cf. **frame**, and see **spar**.

frame *(Build.)*. See **framework**.

frame *(Comp.)*. One complete picture on a television or VDU screen, consisting of 625 lines; a new frame is transmitted every 1/25th second, as two interlaced fields: the odd-numbered lines followed by the even-numbered lines.

frame *(Image Tech.)*. (1) The individual unit picture on a strip of motion picture film. (2) The complete picture produced by a pair of interlaced fields in TV.

frame *(Print.)*. The compositor's workplace, providing storage for cases of type and provision for mounting them on top.

frame antenna *(Telecomm.)*. One comprising a loop of one or more turns of conductor wound on a frame, its plane being oriented in the direction of the incoming waves, or, in transmission, for the direction of maximum radiation. Also called *coil antenna, loop antenna*.

frame counter *(Image Tech.)*. (1) Automatic indicator in a camera showing the number of frames of film which have been exposed. (2) A measurer for processed motion picture film, showing the length in feet and frames which has passed over a **sprocket**.

framed *(Build.)*. Said of work assembled with mortise and tenon joints.

framed and braced door *(Build.)*. A boarded door secured in a frame consisting of two stiles, and top, middle, and bottom rails, with diagonal braces between.

framed floor *(Build.)*. A floor in which the bridging joists are supported at intervals by binding joists, which in turn are supported at intervals by girders. Cf. **double floor**.

framed grounds *(Build.)*. Grounds used in good work around openings such as door openings, the head being tenoned into the posts on each side.

frame direction-finding system *(Telecomm.)*. Simple type of direction-finder using a loop, preferably screened to obviate *antenna effect*, the polar response of which is a figure of eight; the loop is rotated until the received signal vanishes, when the axis of frame is in line with direction of arrival of wave. Also *loop direction-finding system*.

framed, ledged, and braced door *(Build.)*. A boarded door secured in a frame consisting of two stiles and a top-rail, and braced on one side with middle and bottom rails and diagonal braces.

frame finder *(Image Tech.)*. Viewfinder comprising a wire frame and a peep hole separated by a distance equivalent to the focal length of the lens.

frame frequency *(Image Tech.)*. The number of complete interlaced pictures, **frames**, scanned per second; currently

30 Hz for American and Japanese broadcast practice and 25 Hz for Europe.

frame grid *(Electronics)*. Said of a rugged high-performance thermionic valve in which the grid is held very rigid and close to the cathode surface by winding it on stiff rods.

frame level *(Build.)*. A mason's level.

frame line *(Image Tech.)*. The thin black line dividing the frames in the positive projection print of a motion picture.

frame noise *(Image Tech.)*. Noise caused in an optical sound reproducer if the slit is displaced so as to scan the edge of the picture area, including the frame lines.

frame oscillator *(Electronics)*. That which generates frame-scanning voltage or current. Also *field oscillator*.

frames *(Build.,Civ.Eng.)*. (1) The centring used in concrete construction. (2) The built-up superstructure in any suitable structural material to form the skeleton of a building. (3) The surrounding timber or metal members of a door or window.

frame saw *(Build.)*. A thin-bladed saw, which is held taut in a special frame. Also called *span saw*.

frame slip *(Image Tech.)*. Lack of exact synchronization of the vertical scanning and the incoming signal whereby the reproduced video picture progresses vertically.

frame store *(Image Tech.)*. Device for storing the information of two successive interlaced **fields**; it can then release this in sequential (non-interlaced) form.

frame-sync pulse *(Image Tech.)*. A pulse transmitted at the end of each complete frame-scanning operation, to synchronize the framing oscillator at the receiver with that at the transmitter.

frame turner *(Ships)*. A tradesman engaged in turning and bevelling ships' frames, when red hot, to the shape of the ship's form.

frame-type switchboard *(Elec.Eng.)*. See **skeleton-type switchboard**.

frame weir *(Civ.Eng.)*. A type of movable weir consisting of a wooden barrier supported against iron frames placed at intervals across a river, and capable of being lowered on to the bed of the river in flood-time, or of being entirely removed.

framework *(Build.)*. The supporting skeleton of a structure.

framing *(Build.)*. The operation of assembling into final position the members of a structure.

framing *(Image Tech.)*. Vertical adjustment of film in the gate of a projector to centre the picture correctly in the aperture.

francium *(Chem.)*. The heaviest alkali metal. Symbol Fr, at.no. 87. No stable isotopes exist. ^{223}Fr of half-life 22 min. is most important.

Franck-Condon principle *(Phys.)*. An electronic transition takes place so fast that a vibrating molecule does not change its internuclear distance appreciably during the transition. Applied to the interpretation of molecular spectra.

Franck-Hertz experiment *(Phys.)*. An early experiment that showed the existence of energy levels in an atom by measuring the excitation potentials by a study of the impact of electrons with varying energies in atoms in a discharge tube.

franking *(Build.)*. The operation of notching a sash-bar to make a mitre joint with a transverse bar.

franklinite *(Min.)*. Iron-zinc spinel, occurring rarely, as at the type-locality, Franklin, New Jersey.

Frasch process *(Min.Ext.)*. Method of mining elemental sulphur, by drilling into deposit and flushing it out by hot compressed air as a foam of low density. The sulphur is melted by superheated water.

Frasnian *(Geol.)*. The lower stage of the Upper Devonian rocks of Europe.

frass *(Zool.)*. Faeces; excrement.

frater *(Arch.)*. A refectory; sometimes applied in error to a monastic common room or to a chapter house. Also *frater house*, *fratry*.

Fraunhofer diffraction *(Phys.)*. The field transmitted through an aperture in an absorbing screen when the field pattern is at large distances from the screen, the incoming and outgoing wavefronts' approach being planar (to within a fraction of a wavelength) over the extent of the aperture. Cf. **Fresnel diffraction**. See **field**.

Fraunhofer lines *(Astron.)*. Sharp, narrow, absorption lines in the spectrum of the Sun, 25 000 of which are now identified. The most prominent lines are due to the presence of calcium, hydrogen, sodium and magnesium. Most of the absorption occurs in cool layers of the atmosphere, immediately above the incandescent **photosphere**. See **[A]**, **[B]**, **[C]** etc.

Fraunhofer region *(Phys.)*. See **field**.

frazil ice *(Meteor.)*. Ice, in the form of small spikes and plates, formed in rapidly flowing streams, where the formation of large slabs is inhibited.

free *(Bot.)*. Not joined to another member.

free *(Phys.)*. Said of a transducer when it is not *loaded*, e.g., the input has a *free impedance*.

free-air anomaly *(Geol.)*. A gravity anomaly which had been corrected for the height of the measured station above datum. Cf. **Bouguer anomaly**.

free-air dose *(Radiol.)*. A dose of radiation, measured in air, from which secondary radiation (apart from that arising from air, or associated with the source) is excluded.

free association *(Behav.)*. A psychoanalytic technique for probing unconscious ideas and feelings; the individual verbalizes whatever thought comes to mind, without structuring or censoring their content.

free atmosphere *(Meteor.)*. The atmosphere above the friction layer where motion is determined primarily by the large-scale pressure field.

free atom *(Chem.)*. Unattached atom assumed to exist during reactions. See **free radicals**.

free balloon *(Aero.)*. Any balloon floating freely in the air, not propelled or guided by any power or mechanism, either within itself or from the ground.

free beaten stuff *(Paper)*. Lightly beaten stuff with minimum hydration. The resultant paper is low in strength, bulky, porous and opaque. Such stock permits easy drainage of water on the machine wire.

freeboard *(Ships)*. An assignment made by law to prevent overloading of a ship; calculated from statutory tables based on the vessel's form. Permanent markings are made on the ship's side to indicate the depth to which a ship may be loaded, and severe penalties are incurred by any overloading.

freeboard deck *(Ships)*. Uppermost complete deck having permanent means of closing all openings.

free cell formation *(Bot.)*. Free nuclear division followed by delimitation of separated cells without all parental cytoplasm being used, as in ascospore formation.

free cementite *(Eng.)*. Iron carbide in cast-iron or steel not associated with the ferrite in pearlite.

free central placentation *(Bot.)*. Placentation in which the placentas develop on a central column or projection which arises from the base of a unilocular ovary and is not connected by septa to the ovary wall, as in many Caryophyllaceae, Primulaceae.

free charge *(Elec.Eng.)*. An electrostatic charge which is not bound by an equal or greater charge of opposite polarity.

free-cutting *(Eng.)*. See **free machining**.

free-cutting brass *(Eng.)*. α–β Brass containing about 2–3% of lead, to improve the machining properties. Used for engraving and screw machine work.

freedom *(Eng.)*. A body free to move in space is said to have 6 degrees of freedom, 3 of translation and 3 of rotation.

free electron theory *(Electronics)*. Early theory of metallic conduction based on concept that outer valence electrons, which do not form crystal bonds, are free to migrate through crystal, so forming **electron gas**. Now superseded by **energy band** theory.

free end *(Build.)*. The end of a beam which is not fixed or built in.

free energy *(Chem.)*. The capacity of a system to perform work, a change in free energy being measured by the maximum work obtainable from a given process. See **Gibbs free energy, Helmholtz free energy**.

free fall *(Space)*. The motion of an unpropelled body in a gravitational field. In orbit beyond the earth's atmosphere, free fall produces near-weightlessness. See **microgravity**.

free-falling velocity *(Powder Tech.)*. The velocity of fall of a particle of powder through a still fluid, at which the effective weight of the particle is balanced by the drag exerted by the fluid on the particle.

free ferrite *(Eng.)*. Ferrite in steel or cast-iron not associated with the cementite in pearlite.

free field *(Acous.)*. Sound field which is radiated directly from a source, without being reflected elsewhere.

free-field emission *(Phys.)*. That from an emitter when the electric gradient at a surface is zero.

free-flight wind tunnel *(Aero.)*. (1) A wind tunnel, usually of up-draught type, wherein the model is not mounted on a support, but flies freely. (2) One in which three pilots each control one axis of a free-flying model, as used by NASA in the large Ames 80 ft tunnel. (3) Ballistic shape fired into airflow of a wind tunnel for shock wave or re-entry experiments.

free floating anxiety *(Behav.)*. See **generalized anxiety disorder**.

free-handle *(Elec.Eng.)*. See **free-trip**.

free haul *(Civ.Eng.)*. See **overhaul**.

free-hearth electric furnace *(Elec.Eng.)*. A direct-arc furnace in which one electrode forms a part of the bottom of the hearth.

free impedance *(Phys.)*. That at input terminals of a transducer when its load impedance is zero.

freeing port *(Ships)*. An opening in the bulwark to free the deck from large quantities of water which may come on board in heavy weather. Usually with a hinged plate fitted to prevent water coming in.

freelay *(Print.)*. See **Multifont type tray**.

free machining *(Eng.)*. An alloy with additions made to make it easier to machine and so reduce machining time and power consumption, e.g. the addition of lead to steel.

freemartin *(Zool.)*. In cattle, a sterile female intersex occurring as co-twin with a normal bull-calf.

free milling *(Min.Ext.)*. Descriptive term for gold or silver ore, which contains its metal in amalgamable state.

free-needle surveying *(Surv.)*. Traverse work done with a compass, the magnetic bearing of each line being read in turn. Cf. **fixed-needle surveying**.

free oscillations *(Telecomm.)*. (1) Oscillatory currents whose frequency is determined by constants of the circuit in which they are flowing, e.g. those resulting from the discharge of a capacitance through an inductance. (2) Mechanical oscillations governed solely by natural elastic properties of vibrating body.

free path *(Phys.)*. See **mean free path**.

free pendulum clock *(Horol.)*. Type of **impulse clock** in which a free pendulum swings in a vacuum chamber, receiving impulses from a gravity lever released at regular intervals by an electromagnetically operated slave pendulum.

free-piston engine *(Eng.)*. A prime mover of the internal-combustion type, in which a power piston acts directly on a compressor piston working in the opposite direction. May be compounded, as an alternative to a rotary compressor, with a gas turbine.

free pole *(Elec.Eng.)*. A magnet pole which is imagined, for theoretical purposes, to exist separately from its corresponding opposite pole.

free radicals *(Chem.)*. Groups of atoms in particular combinations capable of free existence under special conditions, usually for only very short periods (sometimes only microseconds). Because they contain unpaired electrons, they are paramagnetic, and this fact has been used in determining the degree of dissociation of compounds into free radicals. The existence of free radicals such as methyl ($CH_3\cdot$) and ethyl ($C_2H_5\cdot$) has been known for many years.

free recall *(Behav.)*. A test of memory in which the subject is required to produce as many learned items of the learned task as possible, without regard for the order in which it was first learned.

free-running circuit *(Electronics)*. Running without external synchronization.

free-running frequency *(Telecomm.)*. That of an oscillator, otherwise uncontrolled, when not locked to a synchronizing signal, e.g. a time-base generator in a television receiver.

free-running speed *(Elec.Eng.)*. The speed which a vehicle or train will attain when propelled by a constant tractive effort, i.e. the speed at which the applied tractive effort exactly equals the forces resisting motion. Also called *balancing speed*.

free-settling *(Min.Ext.)*. In classification of finely ground mineral into equal-settling fractions, use of conditions in which particles can fall freely through the environmental fluid, as opposed to the packed conditions of mineral settling.

free space *(Bot.)*. The *apoplast*, especially in estimates of the fraction it forms (the apparent free space) of a tissue.

free space impedance *(Phys.)*. For electromagnetic waves the characteristic impedance of a medium is given by the square root of the ratio of permeability to permittivity (or 4π times this in unrationalized systems of units). This gives a free space value of 376.6 ohms.

free-sprung *(Horol.)*. A watch is said to be *free-sprung* when no index and curb pins are available for the correction of its rate. The balance and spring are so proportioned and adjusted as to give the best possible performance under all conditions. Chronometers and chronometer watches are always free-sprung.

freestone *(Build.)*. A building-stone which can be worked with a chisel without tending to split into definite layers.

free surface *(Ships)*. A free surface exists when any compartment in the ship is partly filled with liquid.

free-trip *(Elec.Eng.)*. Said of certain types of circuit-breaker or motor starter in which the tripping mechanism is independent of the closing mechanism, and will therefore allow the switch to trip while the latter is being operated. Also *free-handle*. Cf. **fixed-trip**.

free turbine *(Aero.)*. A power take-off turbine mounted behind the main turbine/compressor rotor assembly and either driving a long shaft inside the main rotor to a gearbox at the front of the engine, or a short shaft to a gearbox at the rear of the engine. It can also be a separate unit fed by a remotely produced gas supply.

free vibrations *(Phys.)*. The vibrations which occur at the natural frequency of a body when it has been displaced from its position of rest and allowed to vibrate freely without the application of any periodic force.

free vortex flow *(Aero.)*. A vortex persisting away from a solid surface as in a natural tornado. A bound vortex is one attached to the body creating e.g. a wing tip vortex.

free-wheel *(Autos.)*. A 1-way clutch, usually depending on the wedging action of rollers, placed in the transmission line, so as to transmit torque only when the engine is driving.

freeze-drying *(Biol.)*. A method of fixing tissues by freezing sufficiently rapidly as to inhibit the formation of ice crystals, e.g. by immersion in isopentane cooled to −190°C, followed by dehydration in vacuo.

freeze-etch, freeze-fracture *(Bot.)*. A technique for specimen preparation for electron microscopy in which tissue is frozen (strictly vitrified) by very rapid cooling to below *minus* 100°C, fractured and a surface replica made (for microscopical examination) either immediately (freeze-fracture) or after allowing some water to sublime (freeze-etch). See **carbon replica technique**.

freeze frame *(Image Tech.)*. Repetition of a single frame to hold the action stationary.

freeze sinking *(Min.Ext.)*. Shaft sinking, or penetration of water-logged strata, in which refrigerated brine is circulated to freeze the ground and make establishment of an imperviously lined shaft possible.

freeze-substitution *(Bot.)*. Specimen preparation for electron microscopy in which tissue is rapidly 'frozen' (see **freeze-etch**) and then the ice dissolved out at low temperatures with suitable solvents (often containing a fixative) before embedding in the normal way.

freezing *(Phys.)*. The conversion of a liquid into the solid form. This process occurs at a definite temperature for each substance, this temperature being known as the *freezing-point*. The freezing of a liquid invariably involves the extraction of heat from it, known as *latent heat of fusion*. See **latent heat, depression of freezing-point**.

freezing mixture *(Chem.)*. A mixture of two substances, generally of ice and a salt, or solid carbon dioxide with an alcohol, used to produce a temperature below 0°C.

freezing-point *(Eng.)*. The temperature at which a metal solidifies. Pure metals, eutectics and some intermediate constituents freeze at constant temperature; alloys generally solidify over a range.

freezing-point *(Phys.)*. The temperature at which a liquid solidifies, which is the same as that at which the solid melts (the *melting-point*). The freezing-point of water is used as the lower fixed point in graduating a thermometer. Its temperature is defined as 0°C (273.15 K). Abbrev. *f.p.*. See **triple point**; also **water** and **depression of freezing-point**.

freezing-point method *(Chem.)*. See **cryoscopic method**.

freight car *(Civ.Eng.)*. US for a goods wagon.

fremitus *(Med.)*. Palpable vibration, especially of the chest wall during speech or coughing; variations in intensity are of diagnostic value.

Fremont test *(Eng.)*. A type of impact test in which a beam specimen notched with a rectangular groove is broken by a falling weight.

Frémy's salt *(Chem.)*. Potassium hydrogen fluoride, acid potassium fluoride $K^+(HF_2{}^-)$.

French bit *(Build.)*. A boring tool having a flat blade, shaped at the two cutting edges in continuous curves, from the point to and beyond a place of maximum diameter; used in a lathe-head for drilling hard wood.

French chalk *(Min.)*. The mineral, talc, ground into a state of fine subdivision, its softness and its perfect cleavage contributing to its special properties when used as a filler or dry lubricant.

French curve *(Eng.)*. A drawing instrument used to guide the pen or pencil in drawing curved lines. It consists of a thin flat sheet of transparent plastic or other material cut to curved profiles at the edges.

French door *(Arch.)*. **French window**.

French fliers *(Build.)*. Steps in an **open newel stair** with **quarter-space landings**.

French fold *(Print.)*. A sheet of paper folded twice to make 8 pages, but left uncut, and printed only on pages 1, 4, 5, and 8.

French gold *(Eng.)*. *Oroide*. A copper-based alloy with about 16.5% zinc, 0.5% tin and 0.3% iron.

French joint *(Print.)*. One in which the book cover is given a wide space between board and spine to enable the board to hinge back freely. Also called *French groove*.

frenchman *(Build.)*. A joint-trimming tool, used for pointing.

French moult *(Vet.)*. A defective development of the first plumage of birds leading to the shedding of the wing and tail primaries; particularly observed in aviary-bred budgerigars.

French pitch *(Phys.)*. One for musical instruments. A is 435 Hz at 15°C. See **concert pitch**.

French polish *(Build.)*. A solution of shellac dissolved in methylated spirit. Applied to wood surfaces to produce a high polish on them.

French roof *(Build.)*. A **mansard roof**.

French sewing *(Print.)*. (1) The normal method of present-day machine-sewing, i.e. without tapes. (2) A method of hand-sewing at the edge of the bench without using a frame.

French stuc *(Build.)*. Plasterwork finished to present a surface resembling that of stonework.

French truss *(Eng.)*. A symmetrical roof truss for large spans, composed of a pair of braced isosceles triangles based on the sloping sides of the upper chord, their apices being jointed by a horizontal tie. Also called *Belgian truss, Fink truss*.

French window *(Arch.)*. A glazed casement, serving as both window and door.

Frenet's formulae *(Maths.)*.

$$\mathbf{t}' = \qquad + \frac{1}{\rho}\mathbf{n}$$

$$\mathbf{n}' = -\frac{1}{\rho}\mathbf{t} \qquad + \frac{1}{\tau}\mathbf{b}$$

$$\mathbf{b}' = \qquad -\frac{1}{\tau}\mathbf{n}$$

where **t**, **n** and **b** are unit vectors along the tangent, principal normal and binormal respectively at a point on a space curve, ρ and τ are the radii of curvature and torsion respectively, and a dash (′) denotes differentiation with respect to arc length. Also called *Serret-Frenet formulae*.

Frenkel defect *(Crystal.)*. Disorder in the crystal lattice, due to some of the ions (usually the cations) having entered interstitial positions, leaving a corresponding number of normal lattice sites vacant. Likely to occur if one ion (in practice the cation) is much smaller than the other, e.g. in silver chloride and bromide.

frenotomy, fraenotomy *(Med.)*. Cutting of the frenum of the tongue for tongue-tie.

Freons *(Chem.)*. TN for compounds consisting of ethane or methane with some or all of the hydrogen substituted by fluorine, or by fluorine and chlorine. Used as refrigerants, in fire-extinguishers, as aerosol fluids for insecticides, etc., because of their low bp and chemical inertness, and for insulating atmospheres in electrical apparatus because of their high breakdown strengths. Principle ones are: Freon 11 trichlorofluoromethane CCl_3F, Freon 12 dichlorodifluoromethane CCl_2F_2, Freon 21 dichlorofluoromethane $CHCl_2F$, Freon 114, dichlorotetrafluoroethane $CClF_2.CClF_2$, Freon 142, 1-chloro-1 :1-difluoroethane CH_2CClF_2.

frequency *(Ecol.)*. The number of any given species in an area, or the percentage of sample squares which contain the species.

frequency *(Phys.)*. Rate of repetition of a periodic disturbance, measured in hertz (cycles per second). Also called *periodicity*.

frequency allocation *(Telecomm.)*. Frequency on which a transmitter has to operate, within specified tolerance. Bands of frequencies for specified services are allocated by international agreement.

frequency band *(Telecomm.)*. Interval in the frequency spectrum occupied by a modulated signal. In sinusoidal amplitude modulation, it is twice the maximum modulation frequency, but it is much greater in frequency or pulse modulation. See **frequency allocation**.

frequency bridge *(Elec.Eng.)*. An a.c. measuring bridge whose balancing condition is a function of the supply frequency, e.g. *Robinson bridge*.

frequency changer *(Elec.Eng.)*. Circuit (or machine) designed to receive an input at one frequency and to convert this to an output at a different frequency.

frequency changer *(Telecomm.)*. *Mixer, frequency converter*. See **superhet receiver**.

frequency converter *(Telecomm.)*. Same as **frequency changer**.

frequency counter *(Electronics)*. A device which produces an alphanumeric read-out of the frequency of an incoming signal. Frequencies up to 100 MHz can be measured, the accuracy being limited only by the internal reference oscillator or frequency source. Microwave frequencies can be accommodated by means of external frequency converters and oscillators. See **frequency divider**.

frequency demultiplication *(Telecomm.)*. See **frequency division**.

frequency departure *(Telecomm.)*. Discrepancy between actual and nominal carrier frequencies of a transmitter. Formerly termed *frequency deviation*.

frequency deviation *(Telecomm.)*. (1) In frequency modulation, maximum departure of the radiated frequency from mean quiescent frequency (carrier). (2) Greatest deviation allowable in operation of frequency modulation. In broadcast systems within the range 88 to 108 MHz, the maximum deviation is ±75 kHz. (3) See **frequency departure**.

frequency-discriminating filter *(Telecomm.)*. See **filter**.

frequency discriminator *(Telecomm.)*. Circuit the output from which is proportional to frequency or phase change in a carrier from condition of no frequency or phase modulation.

frequency distortion *(Acous.)*. In sound reproduction, variation in the response to different notes solely because of frequency discrimination in the circuit or channel. Generally plotted as a decibel response on a logarithmic frequency base.

frequency distribution *(Stats.)*. See **distribution**.

frequency diversity *(Telecomm.)*. See **diversity reception**.

frequency divider *(Electronics)*. A digital circuit, an essential part of a **frequency counter**, which by responding to individual cycles of an incoming signal and feeding the resultant pulses to an array of gating circuits, can produce subharmonics of the input frequency.

frequency division *(Telecomm.)*. Dividing a frequency by hormonic locking oscillators or stepping pulse circuits. Specific integral division alone can be obtained; an arbitrary collection of frequencies cannot be divided, except by recording and reproducing at a lower speed; also called *frequency demultiplication*.

frequency-division multiplex *(Telecomm.)*. A method of *multiplex* transmission in which individual speech or information channels are modulated to separate channels and then transmitted simultaneously over a cable or microwave link. Single-sideband suppressed carrier methods are normally employed. See **group, supergroup**.

frequency doubler *(Telecomm.)*. Frequency multiplier in which the output current or voltage has twice the frequency of the input. Achieved by a combination of non-linear and tuned circuits.

frequency doubling *(Telecomm.)*. Introduction of marked double-frequency components through lack of polarization in an electromagnetic or electrostatic transducer, in which the operating forces are proportional to the square of the operating currents and voltages respectively.

frequency drift *(Telecomm.)*. Change in frequency of oscillation because of internal (ageing, change of characteristic) or external (variation in supply voltages or ambient temperature) causes. Also *oscillator drift*.

frequency factor *(Chem.)*. The pre-exponential factor in Arrhenius' equation, expressing the frequency of successful collisions between the reactant molecules.

frequency meter *(Electronics)*. A device which compares an unknown frequency with known standard, which may be a standard-frequency transmission derived from an **atomic clock**, the output of a crystal-controlled or other precision oscillator, or the comparison may be made with the resonant frequency of a tuned circuit or resonant line or structure. See **wavemeter, absolute wavemeter, heterodyne wavemeter, frequency counter**.

frequency-modulated cyclotron *(Elec.Eng.)*. One in which frequency of the voltage applied to the **dees** is varied, so as to keep synchronous orbiting of accelerated particles when their mass increases through relativity effect at the high velocities attained.

frequency modulation *(Phys.)*. Variation of frequency of a transmitted wave in accordance with impressed modulation, the amplitude remaining constant. Abbrev. *FM*.

frequency-modulation receiver *(Electronics, Telecomm.)*. One incorporating a frequency demodulator. See **frequency discriminator, Foster-Seeley discriminator, ratio detector**.

frequency monitor *(Elec.Eng.)*. Nationally or internationally operated equipment to ascertain whether or not a transmitter is operating within its assigned channel.

frequency multiplier *(Electronics)*. Non-linear circuit in which the output circuit is tuned to select a harmonic of the input signal. Transistor or varactor diode circuits may be used.

frequency of gyration *(Electronics)*. That of electrons about a line indicating direction of magnetic field in ionosphere.

frequency of infinite attenuation *(Telecomm.)*. A frequency at which a filter inserted in a communication channel provides a maximum attenuation theoretically infinite with loss-free inductances and capacitances. Such large attenuation is generally provided by an antiresonant series arm, or by an acceptance resonant shunt arm.

frequency of penetration *(Phys.)*. That of a wave which just fails to be reflected by ionospheric layer.

frequency overlap *(Telecomm.)*. Common parts of frequency bands used, e.g. for the regular video signal and the chrominance signal in colour TV.

frequency pulling *(Telecomm.)*. Change in oscillator frequency resulting from variation of load impedance.

frequency relay *(Elec.Eng.)*. See **relay types**.

frequency response *(Phys.)*. If constant input power is applied to a **transducer** through a range of discrete frequencies, its frequency response over the given range is the **envelope** of the output powers at each of those frequencies. The response may either be measured absolutely in watts against hertz; by implication, e.g. volts or intensity against frequency; or proportionately, e.g. **decibels** below peak output response against frequency. A *flat* or *level response* therefore indicates equal response to all frequencies within the stated range, e.g. for audio equipment an equal response to within, say, 1 dB for the range 20 Hz to 20 kHz.

frequency selectivity *(Telecomm.)*. See **selectivity**.

frequency-shift keying *(Telecomm.)*. In radio telegraphy, altering the carrier by mark-and-space keying.

frequency-shift transmission *(Telecomm.)*. A form of modulation used in communication systems in which the carrier is caused to shift between two frequencies denoting respectively *on* and *off* pulse.

frequency stabilization *(Telecomm.)*. Prevention of changes produced in frequency of oscillation of a self-oscillating circuit by changes in supply voltage, load impedance, valve parameters, etc. Achieved by resonating crystals, tuned cavities or transmission lines.

frequency standard *(Telecomm.)*. (1) Reference oscillator of very high frequency stability; may be quartz-crystal controlled, although atomic beam standards provide the ultimate reference. (2) Special transmissions, often with precision time codes added, which can be received worldwide and used as standards.

frequency swing *(Telecomm.)*. Extreme difference between maximum and minimum instantaneous frequencies radiated by a transmitter.

frequency synthesizer *(Telecomm.)*. Source of signals of precisely-defined frequency; they may be sine, square or pulse waveform and frequencies may range from zero (dc) to microwave. The output is derived from one or more precision crystal-controlled oscillators, working with multipliers, dividers and mixers.

frequency table *(Stats.)*. A table classifying a set of observations by the number of occurrences of particular values or types.

frequency tolerance *(Telecomm.)*. Extent to which frequency of the carrier of a transmission is permitted to deviate from its allocation.

frequency transformer *(Elec.Eng.)*. See **frequency changer**.

frequency translation *(Telecomm.)*. Shifting all frequencies in a transmission by the same amount (not through zero).

frequency tripler *(Electronics)*. See **frequency multiplier**.

fresco *(Build.)*. A method of painting on plastered walls with lime-fast colours while the plaster is still wet.

fresh-water allowance *(Ships)*. The difference between the **freeboard** in sea water of 1025, and in fresh water of 1000 oz/ft^3 or g/dm^3.

fresh-water sediments *(Geol.)*. Sediments which are accumulating or have accumulated in fresh-water, i.e. river, lake or glaciofluvial environments.

fresnel *(Phys.)*. A unit of optical frequency, equal to 10^{12}Hz = 1THz (terahertz).

Fresnel-Arago laws *(Phys.)*. These concern the condition of interference of polarized light: (1) two rays of light emanating from the same polarized beam, and polarized in the same plane, interfere in the same way as ordinary light; (2) two rays of light emanating from the same polarized beam and polarized at right angles to each other will interfere only if they are brought into the same plane of polarization; (3) two rays of light polarized at right angles and emanating from ordinary light will not interfere if brought into the same plane of polarization.

Fresnel diffraction *(Phys.)*. The study of the diffracted field at a distance from an aperture in an absorbing screen, the distance being large compared with the wavelength but not so large that the curvature of the wavefronts can be neglected. Cf. **Fraunhofer diffraction**. See **field**.

Fresnel ellipsoid *(Phys.)*. A method of representing the doubly refracting properties of a crystal, used in crystal optics.

Fresnel lens *(Image Tech.)*. A lens having a surface of stepped concentric circles, thinner and flatter than a conventional lens of equivalent focal length; used in viewfinders and as a condenser in studio **spot lights**.

Fresnel region *(Phys.)*. See **field**.

Fresnel's bi-prism *(Phys.)*. An isosceles prism having an angle of nearly 180°, used for producing interference fringes from the two refracted images of an illuminated slit.

Fresnel's mirrors *(Phys.)*. Two plane mirrors inclined at an angle of a little less than 180°, used for producing interference fringes from the two reflected images of an illuminated slit.

Fresnel's reflection formula *(Phys.)*. A formula giving the fraction of the intensity of unpolarized incident light reflected at the surface of a transparent medium. The fraction equals

$$\frac{1}{2}\left\{\frac{\sin^2(i-r)}{\sin^2(i+r)}+\frac{\tan^2(i-r)}{\tan^2(i+r)}\right\},$$

where i and r are the angles of incidence and refraction respectively.

Fresnel's rhomb *(Phys.)*. A glass rhomb which is used for obtaining circularly-polarized light from plane-polarized light by total internal reflection. The rhomb is so constructed that two such reflections at an angle of 54° are obtained, each of which introduces a phase difference of one-eighth of a period between the two components obtained from the incident plane-polarized light.

Fresnel zones *(Phys.)*. Zones into which a wave-front is divided, according to the phase of the radiation reaching any point from it. The radiation from any one zone will reach the given point half a period out of phase with the adjacent zones.

fret *(Print.)*. A continuous border design of interlaced bands or fillets, tooled on a book cover or used for typographic decoration.

fret-saw *(Build.)*. One with a very thin and narrow blade (usually 5 in or 125 mm) kept in tension by an elongated metal bow 12–20 in (300–500 mm) in length. Used for cutting narrow curves in thin wood. There is also a mechanical type.

fretted lead *(Build.)*. Lead section used for joining glass panes. See **came**.

fretting corrosion *(Eng.)*. Corrosion due to slight movements of unprotected metal surfaces, left in contact either in a corroding atmosphere or under heavy stress.

fret-work *(Build.)*. (1) A mode of glazing in which diamond-shaped panes (*quarrels*) are connected together by leaden canes to form a window. (2) Panels with holes cut through to form designs.

Freudian slip *(Behav.)*. An error in speech, which Freud believed revealed unconscious ideas or wishes.

Freud's theory of dreams *(Behav.)*. Holds that the dream is a disguised transformation of unconscious wishes. The *manifest content* of the dream refers to the images we remember upon waking, and is a transformation of repressed wishes and ideas, the *latent dream content*; dream interpretation consists of working out the nature and meaning of the transformation.

Freundlich's adsorption isotherm *(Chem.)*. The concentration c of adsorbate is related to the equilibrium partial pressure p as $c = ap^b$, where a and b are empirical constants.

Freysinnet *(Civ.Eng.)*. A method used for the stretching and anchoring of high tensile wires in a post-tensioned system.

friability test *(Civ.Eng.)*. A test for determining the suitability (that is, its resistance to crushing) of any given stone for use in asphalt work. A sample is heated for 15 min in a sand bath at 175°C, and should then not disintegrate on receiving a blow from a hammer.

friable *(Min.Ext.)*. Of ore, easily fractured or crumbled during transport or comminution.

friar *(Print.)*. An area of print with too little ink. See **monk**.

friction *(Med.)*. (1) The sound produced by the rubbing together of two inflamed surfaces, as in pleurisy or pericarditis. (2) Rubbing of a part, as in massage.

friction *(Phys.)*. The resistance to motion which is called into play when it is attempted to slide one surface over another with which it is in contact. The frictional force opposing the motion is equal to the moving force up to a value known as the *limiting friction*. Any increase in the moving force will then cause slipping. *Static friction* is the value of the limiting friction just before slipping occurs. *Kinetic friction* is the value of the limiting friction after slipping has occurred. This is slightly less than the static friction. The *coefficient of friction* is the ratio of the limiting friction to the normal reaction between the sliding surfaces. It is constant for a given pair of surfaces.

frictional damper *(Eng.)*. A device consisting of a supplementary mass frictionally driven from a crankshaft at a point remote from a node, which dissipates vibrational energy in heat.

frictional electricity *(Phys.)*. Static electricity produced by rubbing bodies or materials together, e.g. an ebonite rod with fur.

frictional machine *(Elec.Eng.)*. See **electrostatic generator**.

frictional-rest escapement *(Horol.)*. See **escapement**.

friction and windage loss *(Elec.Eng.)*. Losses in an electrical machine due to friction of sliding parts (see **friction loss**) and also to air resistance. These losses are frequently considered together in designing and testing electrical machinery.

friction calendering *(Textiles)*. Passing fabric between two bowls of a calender designed so that one of the bowls is highly polished and rotates faster that the other. This produces a glaze on the fabric surfaces.

friction clutch *(Eng.)*. A device for connecting or disconnecting two shafts which are in line, in any relative position, through the friction of two surfaces in contact. It consists of a pair of opposed members, between which the drive is transmitted through the friction of their contact surfaces, and which may be separated by a lever system.

friction compensation *(Elec.Eng.)*. A small torque, additional to the main torque, provided in a motor-type integrating meter to compensate for the effect of friction of the moving parts.

friction drive *(Eng.)*. A drive in which one wheel causes rotation of a second wheel with which it is pressed into contact, through the agency of the friction forces at the contact surfaces.

friction feed *(Comp.)*. Mechanism for advancing paper by gripping it between rollers.

friction gear *(Eng.)*. A gear in which power is transmitted from one shaft to another through the tangential friction set up between a pair of wheels pressed into rolling contact. One of the contacting surfaces is usually fabric-faced. Suitable only for small powers.

friction glazing *(Paper)*. A method of glazing paper in which one of the calender rolls revolves at a greater peripheral speed than that of the others. A very high polish is obtained.

friction horsepower *(Eng.)*. That part of the gross or

indicated horsepower developed in an engine cylinder which is absorbed in frictional losses; the difference between the indicated and the brake horsepower.

friction layer *(Meteor.)*. The atmospheric layer, extending to a height of about 600 m, in which the influence of surface friction is appreciable. Sometimes referred to as the 'planetary boundary layer'.

friction loss *(Elec.Eng.)*. The power absorbed in the bearings, commutator, or slip-ring surfaces, or at any other sliding contacts of an electric machine.

friction pile *(Build.)*. A pile which supports its load only by the friction on its sides.

friction rollers *(Eng.)*. See **antifriction bearing**.

friction spinning *(Textiles)*. A method for converting staple fibres into a yarn by feeding a sliver on to a rotating perforating roller through which air is being sucked. Another roller is set near so that the yarn is formed from the rapidly twisting fibres at the nip of the rollers.

friction twisting *(Textiles)*. Method of texturing yarns.

friction welding *(Eng.)*. Welding in which the necessary heat is produced frictionally, e.g. by rotation, and forcing the parts together.

Friedel and Crafts' synthesis *(Chem.)*. The synthesis of alkyl substituted benzene hydrocarbons and aromatic ketones, by the action of halogenoalkanes or acyl halides on aromatic hydrocarbons in the presence of anhydrous aluminium chloride.

Friedreich's ataxia *(Med.)*. An autosomal recessive heriditary disease of the central nervous system which usually occurs in childhood. It is often associated with skeletal discontinuities and about half have cardiomyopathy.

friendly numbers *(Maths.)*. Pairs of numbers each of which is the sum of the factors of the other, including unity, e.g. 220 and 284. Also called *amicable numbers*. Cf. **perfect number**.

frieze *(Arch.,Build.)*. (1) The middle part of an entablature, between the architrave and the cornice. (2) The decorated upper part of a wall, below the cornice.

frieze panel *(Build.)*. An upper panel in a 6-panel door.

frieze rail *(Build.)*. The rail next to the top rail in a 6-panel door.

frilling *(Image Tech.)*. Crinkling and separation of the emulsion layer from the glass of a photographic plate, usually as a result of too high a temperature in processing.

fringe area *(Image Tech.)*. In radio or TV broadcasting, regions remote from a transmitter in which good reception is uncertain.

fringe effect *(Image Tech.)*. A faint light line on the low-density side of the boundary between a lightly and a heavily exposed area on a developed emulsion. Also called *edge effect*. See **Mackie lines**.

fringe medicine *(Med.)*. See **alternative medicine**.

fringes *(Image Tech.)*. Coloured edges visible in an image caused by **chromatic aberration** or as a result of imperfect registration of the colour components.

fringing *(Elec.Eng.)*. The spreading of flux lines in a given field configuration (conduction, electric or magnetic) e.g. at the edges of a parallel plate capacitor or the edges of an air gap in a magnetic circuit.

fringing coefficient *(Elec.Eng.)*. A coefficient used in making magnetic circuit calculations to allow for the effect of fringing of the flux.

frisket *(Print.)*. (1) On a hand press, a thin iron frame covered with paper, to hold the sheet to the tympan, prevent it from being soiled, and strip it from the forme after printing. (2) On a platen machine, the adjustable metal fingers which strip the sheet from the forme after printing.

frisking *(Nuc.Eng.)*. Searching for radioactive radiation by contamination meter, usually a portable ionization chamber.

fritting *(Eng.)*. Condition in fire assaying in which the powdered ore, flux and other reagents are in a pasty condition a little below melting point.

frog *(Build.)*. Of a plane, the surface against which the blade rests. It determines the **pitch**.

frog *(Civ.Eng.)*. The point of intersection of the inner rails, where a train or tram crosses from one set of rails to another. The frog is in the form of a V. See **crossing**, **turnout**.

frog *(Vet.)*. A V-shaped band of horn passing from each heel to the centre of the sole of a horse's foot.

Fröhlich's syndrome *(Med.)*. See **dystrophia adiposogenitalis**.

Froin's syndrome *(Med.)*. The presence of yellow cerebrospinal fluid, which has a high content of protein but no cells, below the site of obstruction (e.g. by a tumour) of the spinal cord.

frond *(Bot.)*. (1) A large compound leaf especially of a fern, cycad or palm. (2) A more or less leaf-like thallus of lichen, liverwort or alga. (The term is imprecise in either usage.)

frons *(Zool.)*. In Insects, an unpaired sclerite of the front of the head; in higher Vertebrates, the front of the head above the eyes. adj. *frontal*.

front *(Build.)*. The sole face of a plane.

front *(Meteor.)*. (1) Surface of separation of two air masses. (2) Line of intersection of the surface of separation of two air masses with another surface or with the ground. If warm air replaces cold, it is a warm front; if cold replaces warm, a cold front.

frontage line *(Build.)*. The **building line**.

frontal *(Zool.)*. (1) A paired dorsal membrane bone of the Vertebrate skull, lying between the orbits. (2) Pertaining to the frons.

frontal lobes *(Zool.)*. The front part of the cortex. It is thought to be involved in immediate memory.

frontal plane *(Zool.)*. The median horizontal longitudinal plane of an animal.

frontal sinuses *(Zool.)*. Air cavities, connected with the nasal chambers, extending into the frontal bones, in Mammals.

front clearance *(Eng.)*. In a single-point cutting tool, the clearance of the edge below the point; expressed as the angle between this edge as seen in side view on the tool shank and a line through the point and perpendicular to the tool shank base.

front end *(Image Tech.)*. General term for all the operations from original photography through editing and sound mixing up to the first show print.

front-end processor *(Comp.)*. Small computer which receives data from a number of input devices, organizes it and transmits it to a more powerful computer for processing.

front hearth *(Build.)*. The part of the hearth extending beyond the chimney breast.

frontispiece *(Print.)*. An illustration facing the title-page of a book.

frontogenesis, frontolysis *(Meteor.)*. Respectively, the intensification or realization of a front, and its weakening or disappearance.

front porch *(Image Tech.)*. A short period of **black level** signal transmitted between the end of the picture information and the horizontal **sync pulse**.

front projection *(Image Tech.)*. Projection of a picture from the same side of the screen as the audience. In particular, the system of composite cinematography also known as **reflex projection**.

front-to-back ratio *(Telecomm.)*. Ratio of effectiveness of a directional antenna or microphone, etc., in the forward and reverse directions.

frontwall cell *(Electronics)*. Semiconductor cell in which light passes through a conducting layer to the active layer, which is separated from the base metal by a semiconductor.

frost *(Meteor.)*. A *frost* is said to occur when the air temperature falls below the freezing-point of water (0°C or 32°F). See **hoar frost**.

frosted lamp *(Elec.Eng.)*. A filament-lamp the bulb of which is etched or sand-blasted to break up any direct rays of light from the filament.

frost hollow *(Meteor.)*. A hollow in hilly ground into

which cold air drains, stagnates and becomes exceptionally cold owing to radiation at night.

frost-point *(Meteor.)*. The temperature at which air becomes saturated with respect to ice if cooled at constant pressure.

froth *(Chem.,Min.Ext.)*. *Foam*. A gas-liquid continuum in which bubbles of gas are contained in a much smaller volume of liquid, which is expanded to form bubble walls. The system is stabilized by oil, soaps or emulsifying agents which form a binding network in the bubble walls.

frother *(Eng.)*. A substance used to promote the formation of a foam in the flotation process.

froth flotation *(Min.Ext.)*. Process in dominant use for concentrating values from low-grade ores. After fine grinding, chemicals are added to a pulp (ore and water) to develop differences in surface tension between the various mineral species present. The pulp is then copiously aerated, and the preferred (*aerophilic*) species clings to bubbles and floats as a mineralized froth, which is skimmed off.

Froude brake *(Eng.)*. An absorption dynamometer consisting of a rotor inside a casing, itself free to rotate, the space between the two being filled with water. The energy is dissipated in eddy formation and heat, the torque absorbed being measured by the torque necessary to prevent rotation of the casing.

Froude's transition curve *(Surv.)*. A transition curve the equation to which is that of a cubic parabola, the offset y from the straight produced being given by

$$y=\frac{x^3}{6lr},$$

where x = distance from tangent point, l = length of transition, r = radius of the circular arc.

frozen bearing *(Eng.)*. A seized bearing. See **seizure**.

frozen equilibrium *(Chem.)*. The state of a solid at low temperature, which is prevented from attaining the theoretically possible thermodynamic equilibrium, because its molecular motion has become too slow. Cf. **Nernst theory**.

fructification *(Bot.)*. Any seed- or spore-bearing structure like the large spore-bearing structures of many fungi.

L-fructose *(Chem.)*. *Fruit-sugar, laevulose*, $C_6H_{12}O_6$, mp 95°C. A **ketohexose** is prepared by heating inulin with dilute acids, and is always found together with d-glucose in sweet fruit juices. See **sucrose**.

frue vanner *(Min.Ext.)*. Endless rubber belt which is driven slowly upslope while finely ground ore is washed gently downslope. Belt is given a side shake to aid distribution, and wash water is so adjusted that heavy material stays on belt, while light gangue is washed down to bottom end of pulley system round which belt circulates.

frugivorous *(Zool.)*. Fruit-eating.

fruit *(Bot.)*. (1) The structure that develops from the ovary of an angiosperm as the seeds mature, with (false fruit) or without (true fruit) associated structures. (2) Sometimes, any of the various structures associated with the mature seed of a gymnosperm, especially e.g. the fleshy cone scales of the juniper 'berry'. (3) A fructification.

fruiting body, fruit body *(Bot.)*. See **fructification**.

frusemide *(Med.)*. A **'loop' diuretic** used in the treatment of pulmonary oedema and heart failure.

frustration *(Behav.)*. *Animal behaviour*: a motivational state assumed to occur in situations in which an animal's actions do not lead to an expected outcome (e.g. a food reward). Often the initial response to a frustrating situation is to increase the persistence and intensity of behaviour; continued frustration often leads to **redirected behaviour**. *Human behaviour*: refers to either the prevention of activity that is directed towards a goal, or to the psychological state that results from being prevented from reaching a goal.

frustule *(Bot.)*. The silicified cell wall of a diatom (Bacillariophyceae) consisting of two halves which fit one (the hypotheca) inside the other (the epitheca) like the two halves of a petri-dish (together, in some usages, with the cell itself.)

frustum of a cone *(Maths.)*. A cone truncated by a plane parallel to its base.

frutescent *(Bot.)*. Shrubby.

fruticose *(Bot.)*. Shrubby.

frying arc *(Elec.Eng.)*. A *hissing arc*.

F-scale *(Behav.)*. A scale constructed to measure the **authoritarian personality**; contains questions designed to measure proposed aspects of authoritarianism, e.g. submission to authority, admiration for power etc.

FSD *(Genrl.)*. Abbrev. for *Full-Scale Deflection.*

FSH *(Med.)*. Abbrev. for *Follicle-Stimulating Hormone.*

FSK *(Telecomm.)*. Abbrev. for *Frequency-Shift Keying.*

f state *(Electronics)*. That of an orbital electron when the orbit has angular momentum of three *Dirac units*.

fuchsin *(Chem.)*. Magenta, the hydrochloride of rosaniline, a basic triphenylmethane dyestuff, dark green crystals, dissolving in water to form a purple-red solution. Used as a disinfectant, especially in certain skin infections.

fuchsite *(Min.)*. A green variety of muscovite (white mica) in which chromium replaces some of the aluminium.

fucivorous *(Zool.)*. Seaweed-eating.

fucoxanthin *(Bot.)*. *xanthophyll*; a major **accessory pigment** in most members of the Heterokontophyta and responsible for their brownish colours (e.g. the brown algae).

fucoxanthin *(Chem.)*. $C_{40}H_{56}O_6$ or $C_{40}H_{60}O_6$; the main carotenoid found in brown algae. Brown crystalline solid; mp 168°C.

fudge *(Print.)*. A space reserved in a newspaper for late news. Also known as *stop press*.

fuel accumulator *(Aero.)*. A reservoir which augments the fuel supply when the critical fuel pressure is reached during the starting cycle of a gas turbine.

fuel assembly *(Nuc.Eng.)*. A group of nuclear fuel elements forming a single unit for purposes of charging or discharging a reactor. The term includes bundles, clusters, stringers etc.

fuel cell *(Chem.)*. A galvanic cell in which the oxidation of a fuel (e.g. methanol) is utilized to produce electricity.

fuel-cooled oil cooler *(Aero.)*. A compact oil cooler for high-performance gas turbines in which heat is transferred to fuel passing in the counter bores of the device, instead of to air.

fuel cut-off *(Aero.)*. A device which shuts off the fuel supply of an aero-engine; also *slow-running cut-out*.

fuel cycle *(Nuc.Eng.)*. The stages involved in the supply and use of fuel in nuclear power generation. The main steps are mining, milling, extraction, purification, enrichment (if required), fuel fabrication, irradiation in the reactor, cooling, reprocessing, recycling and waste management and disposal.

fuel element *(Nuc.Eng.)*. Unit of nuclear fuel which may consist of a single cartridge, or a cluster of fuel **pins**.

fuel grade *(Aero.)*. The quality of piston aero-engine fuel as expressed by its **knock rating**.

fuel injection *(Autos.)*. A method of operating a spark-ignition engine by injecting liquid fuel directly into the induction pipe or cylinder during the suction-stroke, thus dispensing with a carburettor; the standard method in diesels, it avoids freezing in aero engines.

fuel injectors *(Autos.)*. See **injectors**.

fuel jettison *(Aero.)*. Apparatus for the rapid emergency discharge of fuel in flight.

fuelling machine/refuelling machine *(Nuc.Eng.)*. See **charge (-discharge) machine**.

fuel manifold *(Aero.)*. The main pipe, or gallery, with a series of branch pipes, which distributes fuel to the burners of a gas turbine.

fuel, nuclear fuel *(Nuc.Eng.)*. Fissile material inserted in or passed through a reactor; the source of the chain reaction of neutrons, and so of the energy released.

fuel oils *(Chem.)*. Oils obtained as residues in the distillation of petroleum; used, either alone or mixed with other oils, for domestic heating and for furnace firing (particularly marine furnaces); also as fuel for I.C. engines.

fuel rating *(Nuc.Eng.)*. The ratio of total energy released to initial weight of heavy atoms (U, Th, Pu) for reactor fuel. Usually expressed in megawatts per tonne. US term *specific power*.

fuel reprocessing *(Nuc.Eng.)*. The processing of nuclear fuel after use to remove fission products etc. and to recover fissile and fertile materials for further use.

fuel rod *(Nuc.Eng.)*. Unit of nuclear fuel in rod form for use in a reactor. Short rods are sometimes termed *slugs*.

fuel tanks *(Aero.)*. These may be of many forms, for which the names vary. The *main tanks* are normally all those carried permanently and are usually composed either of flexible, self-sealing bags, or cells in wing or fuselage, or are integral with the wing structure; *auxiliary tanks* can be mounted additionally to increase range. See **drop tank**.

fuel trimmer *(Aero.)*. A variable-datum device for resetting in flight the automatic fuel regulation, by *barostat*, of a gas turbine to meet changes in ambient temperature.

fugacity *(Chem.)*. The tendency of a gas to expand or escape; substituted for pressure in the thermodynamic equations of a real gas. Analogous to **activity**. See **ideal gas**.

fugitive *(Geol.)*. Decriptive of the dissolved volatile constituents of magma, which are commonly lost by evaporation when the magma is erupted as lava, and which are partly responsible for metasomatic alteration when magma is intruded.

fugitive colours *(Build.)*. Colours which fade on exposure to sunlight.

fugitometer *(Chem.)*. An apparatus for testing the fastness of dyed materials to light.

fugue *(Behav.)*. A condition, related to amnesia, in which the individual takes a sudden and unexpected trip from home, assuming a new identity and forgetting the past, in an attempt to escape from overwhelming stress.

fulcrum *(Phys.)*. The point of support, or pivot, of a lever.

fulguration *(Med.)*. The destruction of tissue by means of electric sparks. Used in the treatment of some malignant tumours.

fulgurites *(Min.)*. *Lightning tubes*. Tubular bodies, branching or irregularly rod-like, produced by lightning in loose unconsolidated sand; caused by the vitrification of the sand grains forming silica glass. Although of very narrow cross-section, some specimens have been found to exceed 6 m in length. Also *lechatelierite*.

fuliginous *(Bot.)*. Soot-coloured.

full *(Eng.)*. A term signifying slightly larger than the specified dimension. Cf. **bare**.

full adder *(Comp.)*. Logic circuit which adds a pair of corresponding bits of two numbers expressed in binary form and any carry from a previous stage, producing a sum and a new carry. Also *three-input adder*. Cf. **half-adder**.

full annealing *(Eng.)*. Of steel, heating above the critical range, followed by slow cooling, as distinguished from (*a*) annealing below the critical range, and (*b*) normalizing, which involves air-cooling.

full aperture *(Image Tech.)*. A frame size occupying the full width of 35 mm film between the perforations.

full bound *(Print.)*. *Whole bound*. Said of a volume the sides and back of which are covered with leather or cloth.

full-centre arch *(Arch.)*. A semicircular arch or vault.

fuller's earth *(Geol.)*. A nonplastic clay consisting essentially of the mineral montmorillonite, and similar in this respect to bentonite. Used originally in 'fulling', i.e. absorbing fats from wool, hence the name. The Fullers' Earth of English stratigraphy is a small division of the Jurassic System in the S. Cotswolds.

full-force feed *(Eng.)*. An engine lubrication system in which oil is forced to main bearings, connecting-rod big-end bearings, and thence, by drilled holes or attached pipes, to the gudgeon pins and cylinder walls.

full gear *(Eng.)*. Of a steam-engine valve gear, the position giving maximum valve travel and cut-off for full power.

full hard *(Eng.)*. The stage in the tempering of some ferrous and non-ferrous alloys just below that at which the metal cannot be formed by bending.

full-load *(Elec.Eng.)*. The normal rated output of an electric machine or transformer.

full-load *(Eng.)*. The normal maximum load under which an engine or machine is designed to operate.

full moon *(Astron.)*. The instant when the geocentric longitudes of the sun and moon differ by 180°; the moon is then opposite the sun, and therefore fully illuminated.

full out *(Print.)*. An instruction to set the matter with no indention.

full-pitch winding *(Elec.Eng.)*. An armature winding in which the span of the coils is equal to a pole pitch.

full-plate watch *(Horol.)*. A watch in which the top plate is circular and the balance is mounted above the plate.

full radiator *(Phys.)*. See **black body**.

full-satellite exchange *(Telecomm.)*. A small automatic telephone exchange which is entirely dependent for completion of calls on its parent or main exchange.

full shroud *(Eng.)*. A gear-wheel in which the shrouding extends up to the tips of the teeth.

full thread *(Eng.)*. A screw-thread cut to the theoretically correct depth.

full-wave rectification *(Elec.Eng.)*. That in which current flows, during both half-cycles of the alternating voltage, through similar rectifying devices alternately, e.g. in a double diode or bridge rectifier.

fully fashioned *(Textiles)*. Knitted fabrics and garments that are shaped by increasing or decreasing the number of wales. This ensures that the garment fits more closely.

fulminates *(Chem.)*. Compounds containing the ion CNO^-, which explode under slight shock or on heat detonators, e.g. mercury fulminate, $Hg(CNO)^-$.

fumaric acid *(Chem.)*. *Ethane 1,2-dicarboxylic acid*, $HOOC.CH=CH.OOH$, small prisms which do not melt, but sublime at about 200°C, with the formation of maleic anhydride. Fumaric acid being the *trans*-form. Used in polyester resins.

fumaroles *(Geol.)*. Small vents on the flanks of a volcanic cone, or in the crater itself, from which gaseous products emanate.

fume *(Chem.,Powder Tech.)*. Cloud of airborne particles, generally visible, of low volatility and less than a micrometre in size, arising from condensation of vapours or from chemical reaction.

fume cupboard *(Chem.)*. A glass chamber or cupboard where laboratory operations involving obnoxious fumes are carried out under forced ventilation.

fumigants *(Chem.)*. Substances which, when volatilized, are capable of destroying vermin, insects, bacteria, moulds, or which act as disinfectants. Examples are hydrogen cyanide and ethylene oxide for vermin and insects, and formaldehyde (disinfectant).

fumigation *(Med.)*. Disinfection by means of gas or vapour.

fuming liquids *(Chem.)*. Liquids which give off vapours which unite with water to form a mixture or compound with a lower vapour pressure than water.

fuming (Nordhausen) sulphuric acid *(Chem.)*. A solution of sulphur (VI) oxide in concentrated sulphuric acid.

function *(Biol.)*. The normal activity of a biological structure, as *digestive function* or *ribosomal function*.

function *(Comp.)*. See **operation, subroutine**.

function *(Maths.)*. A *function*, f, (or *mapping*, f) from a set X to a set Y, denoted by $f{:}X \rightarrow Y$, associates with each element x of X a single element y of Y, denoted by $f(x)$.

functional *(Biol.)*. Carrying out normal activities; active (as opposed to *passive*).

functional disease *(Med.)*. Term used to describe symptoms for which it is believed there is no organic basis.

Functionalism *(Arch.)*. A principle of design whereby the architect is committed to the basic functional requirements of a building, regardless of style or aesthetics.

functional psychosis *(Behav.)*. A severe mental disorder which cannot be attributed to any certain physical pathology.

function chamber *(Build.)*. A closed chamber, generally of brick and concrete, inserted in a sewer system for accepting the inflow of one or more sewers and allowing for the discharge thereof.

function generator *(Comp.)*. Element in an analogue computer capable of generating voltage wave approximately following any desired, single-valued, continuous algebraic function of one variable.

function generator *(Telecomm.)*. Signal generator with range of alternative non-sinusoidal output waveforms.

function switch *(Comp.)*. A network with a number of input and output lines, connected so that the output signals give the information in a different code from that of the input.

functor *(Maths.)*. A mapping between categories which preserves their structure.

fundamental colours *(Phys.)*. See **primary additive colours**.

fundamental component *(Telecomm.)*. The harmonic component of an alternating wave which has the lowest frequency and which usually represents the major portion of the wave.

fundamental crystal *(Electronics)*. A crystal which is designed to vibrate at the lowest order of a given mode.

fundamental dynamical units *(Phys.)*. The basic equations of dynamics are such as to be the same for any system of fundamental units. Unit force acting on unit mass produces unit acceleration; unit force moved through unit distance does unit work; unit work done in unit time is unit power. Four systems are, or have been, in general use, the SI system now being the only one employed in scientific work.

fundamental theorem of algebra *(Maths.)*. Every polynomial in one variable is zero for some complex value of the variable, i.e. every polynomial equation has a solution.

fundamental theorem of arithmetic *(Maths.)*. Every positive integer has a unique factorization into prime numbers.

fundamental wavelength *(Telecomm.)*. Wavelength in free space corresponding to a fundamental frequency.

fundus *(Med.)*. That part of the cavity of the eyeball which can be seen through the pupil with an ophthalmoscope.

fungi *(Bot.)*. Heterotrophic, eukaryotic organisms reproducing by spores, and their allies. See **Eumycota, Myxomycota**.

fungible *(Min.Ext.)*. Oil products which are interchangeable and can therefore be mixed during transport. Makes it difficult to trace the origins of a given sample.

fungicidal paints *(Build.)*. Various forms of liquid materials from antiseptic washes to wood preservatives and stains which contain substances which destroy fungi.

fungicide *(Bot.)*. A substance that kills fungal spores and/or mycelium.

Fungi Imperfecti *(Bot.)*. Older name for Deuteromycotina.

fungistatic *(Bot.)*. Preventing the growth of a fungus without killing it.

funicle *(Bot.)*. The stalk by which the ovule (and seed) is attached to the placenta in angiosperms.

funicular railway *(Civ.Eng.)*. A form of **cable railway**.

funiculus *(Zool.)*. In some Invertebrates (as Bryozoa), thickened strands of mesoderm attaching the digestive organs to the body-wall; more generally, any small cord, as a tract of nerve fibres in the central nervous system. adj. *funicular*.

	System				Dimensions
	ft. lb. sec.	Gravitational	CGS	SI (MKSA)	
length	foot (ft.)	foot	centimetre (cm)	metre (m)	L
mass	pound (lb.)	slug	gram (g)	kilogram (kg)	M
time	second (s)	second	second	second	T
velocity	ft./s	ft./s	cm/s	m/s	LT^{-1}
acceleration	ft./s^2	ft./s^2	cm/s^2	m/s^2	LT^{-2}
force	poundal (pdl.)	pound force (lbf.)	dyne	newton (N)	MLT^{-2}
work	ft. pdl.	ft. lbf.	erg	joule (J)	ML^2T^{-2}
power	ft. pdl./s	ft. lbf./s	erg/s	watt (W)	ML^2T^{-3}

Notes: (1) There is no name for the unit of power except in the SI system. It is possible to express power in the ft.lb.sec system and in the gravitational system by the horse-power (550 ft-lbf s^{-1}) and in the CGS system by the watt (10^7 erg s^{-1}). (2) The unit of force (lbf) in the gravitational system is also known as the pound weight (lb.wt).

fundamental frequency *(Telecomm.)*. In a steady periodic oscillation, a frequency which divides into all components in the waveform. See **harmonic**.

fundamental frequency of antenna *(Telecomm.)*. Lowest frequency at which antenna is resonant when not loaded.

fundamental interval *(Phys.)*. The number of degrees between the two fixed points on a thermometer scale.

fundamental metric tensor *(Maths.)*. See **metric (1)**.

fundamental mode *(Telecomm.)*. (1) In an antenna, the pattern of current and voltage distribution at the fundamental frequency, usually showing a maximum at one end of a radiator and a minimum at the other. (2) In a waveguide or resonant cavity, the field distribution at frequencies (or corresponding free-space wavelengths) between the lowest which that structure can support and the first harmonic of that frequency.

fundamental particle *(Phys.)*. A particle that is incapable of subdivision. There are believed to be three kinds of such particle; *leptons*, *quarks* and *gauge bosons*.

fundamental series *(Phys.)*. Series of optical spectrum lines observed in the spectra of alkali metals. Energy levels for which the orbital quantum number is three are designated *f-levels*.

funnel *(Zool.)*. A modified part of the foot in Cephalopoda, protruding from the mantle cavity and acting as an exhalent channel. In Annelida, a nephrostome.

funnelling *(Meteor.)*. The strengthening of a wind blowing along a valley, especially when the valley narrows.

fur *(Zool.)*. In Mammals, the thick undercoat of short, soft, silky hairs.

fural *(Chem.)*. See **furfural**.

furan *(Chem.)*. A heterocyclic ring compound, like pyrole but with oxygen in place of –NH. ⇨

furan group *(Chem.)*. A group of heterocyclic compounds derived from furan, C_4H_4O, a compound containing a ring of 4 carbon atoms and 1 oxygen atom.

furan resins *(Chem.)*. A group of plastics derived from the partial polymerization of furfuryl alcohol, or from condensation of furfuryl alcohol with either furfural or methanal, or of furfural with ketones, and used widely as baked plastic coatings on metal, as adhesives and as resin binders for stoneware.

furca *(Zool.)*. Any forked structure; in Vertebrates, a divergence of nerve-fibres; in some *Arthropods*, a pair of divergent processes at the end of the abdomen.

furcula *(Zool.)*. In Collembola, the leaping apparatus, consisting of a pair of partially fused appendages arising from the fourth abdominal somite; in Birds, the partially fused clavicles, the wishbone; more generally, any forked structure.

furfural *(Chem.)*. Fural or furfuraldehyde, $C_4H_3O.CHO$, a colourless liquid, bp 162°C, obtained by distilling pentoses with diluted hydrochloric acid. Used as a solvent, particularly for the selective extraction of crude rosin, also as raw material for synthetic resins. Used in petroleum refining for the selective extraction of impurities.

6-furfurylaminopurine *(Bot.)*. See **kinetin**.

furlong *(Genrl.)*. A distance of 10 Gunter's chains, i.e. 220 yd or one-eighth of a mile.

furnace atmosphere *(Eng.)*. Three main classes: (1) *oxidizing*, produced when air volumes are in excess of fuel requirements; (2) *neutral*, when air to fuel ratios are perfectly proportioned; (3) *reducing*, due to deficiency of combustion air. See also **protective-**.

furnace brazing *(Eng.)*. A high-production method of copper-brazing steel, without flux, in a reducing atmosphere, or of brazing steels, copper and copper alloys with brasses or silver-brazing alloys, in continuous or in batch furnaces.

furnace clinker *(Build.)*. The final residue from the combustion of coke or coal which has been burnt and reburnt in order to consume the maximum of combustible matter in it. It is useful as an aggregate in the manufacture of concrete.

furnace linings *(Eng.)*. The interior portions of metallurgical furnaces which are in contact with hot gases and the charge, and must therefore be constructed of materials resistant to heat, abrasion, chemical action etc. See **refractories**.

furnish *(Paper)*. The ingredients from which paper is manufactured.

furniture *(Build.)*. A general name for all metal fittings for doors, windows, etc.

furniture *(Print.)*. Lengths of wood, plastic or metal, less than type height, used in a forme for making margins. etc. They are made to standard widths and lengths.

furring *(Build.)*. See **firring**.

furrowed *(Build.)*. A term applied to margin-drafted ashlars having parallel vertical grooves cut in the face.

furrowing *(Bot.)*. Cell division by intucking the plasmalemma to pinch the cell into two. Cf. **cell plate**.

furuncle *(Med.)*. See **boil**.

furunculosis *(Med.)*. The condition of having several boils.

fusain *(Min.)*. Mineral charcoal, the soiling constituent of coal, occurring chiefly as patches or wedges. It consists of plant remains from which the volatiles have been eliminated.

fuse *(Elec.Eng.)*. A device used for protecting electrical apparatus against the effect of excess current; it consists of a piece of fusible metal, which is connected in the circuit to be protected, and which melts and interrupts the circuit when an excess current flows. The term fuse also includes the necessary mounting and cover (if any).

fuse *(Min.Ext.)*. A thin waterproof canvas length of tube containing gunpowder arranged to burn at a given speed for setting off charges of explosive.

fuse-board *(Elec.Eng.)*. See **distribution-**.

fuse-box *(Elec.Eng.)*. Term sometimes used for a distribution fuse-board enclosed in a box.

fuse-carrier *(Elec.Eng.)*. A carrier for holding a fuse-link; arranged to be easily inserted between fixed contacts, so that a replacement of the fuse-link can be quickly carried out. Also called a *fuse-holder*.

fused junction *(Electronics)*. See **alloy junction**.

fused ring *(Chem.)*. See **condensed nucleus**.

fused silica *(Glass)*. See **vitreous silica**.

fusee *(Horol.)*. A spirally grooved pulley of gradually increasing diameter, used to equalize the pull of the mainspring as it runs down by increasing the leverage on a chain or gut-line wound round the fusee and the mainspring barrel.

fuse-element, fuse-link, fuse *(Elec.Eng.)*. The essential part of a fusible cut-out.

fuse-holder *(Elec.Eng.)*. See **fuse-carrier**.

fuselage *(Aero.)*. The name generally applied to the main structural body of a heavier-than-air craft, other than the hull of a flying-boat or amphibian.

fuse-link *(Elec.Eng.)*. See **fuse-element**.

fusel oil *(Chem.)*. Mainly optically inactive 3-methylbutan-1-ol, $(CH_3)_2.CH.CH_2.CH_2OH$, accompanied by active amyl alcohol, usually occurring in the products of alcoholic fermentation.

fuse rating *(Elec.Eng.)*. The maximum current a fuse will carry continuously and/or (less frequently) the minimum current at which it can be relied upon to blow.

fuse-switch *(Elec.Eng.)*. A **switch-fuse**.

fuse tongs *(Elec.Eng.)*. Tongs with insulating handles, used for withdrawing or replacing fuses on high-voltage circuits.

fusible alloys, metals *(Eng.)*. Alloys of bismuth, lead and tin (and sometimes cadmium or mercury) which melt in the 47–248°C temperature range; used as solders and for safety devices in fire extinguishers, boilers etc.

fusible cut-out *(Elec.Eng.)*. See **fuse**.

fusible plug *(Eng.)*. A plug containing a metal of low melting-point used, for example, in the crown of a boiler fire box to prevent serious overheating of the plates if the water-level falls below them.

fusiform *(Bot.)*. Spindle-shaped; elongated, broadest in the middle and tapering towards both ends.

fusiform *(Bot.,Zool.)*. Elongated and tapering towards each end; shaped like a spindle.

fusiform initials *(Bot.)*. More or less elongated initial cells, in the cambium, giving rise to all the cells of the secondary xylem and phloem except for the ray cells. Cf. **ray initials**.

fusing factor *(Elec.Eng.)*. The minimum current required to blow a fuse, expressed as a ratio to the rated current.

fusing point *(Eng.)*. See **melting point**.

fusion *(Phys.)*. (1) The process of forming new atomic nuclei by the fusion of lighter ones; principally the formation of helium nuclei by the fusion of hydrogen and its isotopes. The energy released in the process is referred to as *nuclear energy* or *fusion energy*. See **binding energy of the nucleus**. (2) The conversion of a solid into a liquid state; the reverse of *freezing*. Fusion of a substance takes place at a definite temperature, the melting point, and is accompanied by the absorption of latent heat of fusion.

fusion bomb *(Phys.)*. Same as **hydrogen bomb**.

fusion cones *(Eng.)*. See **Seger cones**.

fusion drilling *(Min.Ext.)*. Method of hard-rock boring with a paraffin-oxygen jet which melts the rock, the slag being decrepitated and flushed out by a water spray.

fusion energy *(Phys.)*. Energy released by the process of nuclear *fusion* usually in the formation of helium from lighter nuclei.The energy released in stars is by fusion processes; see **carbon cycle, proton-proton chain**. Fusion is the source of energy in the hydrogen bomb.

fusion-fission hybrid reactor *(Nuc.Eng.)*. Proposed reactor system in which neutrons from fusion produce fissile material from a U or Th blanket and electricity. See **blanket, breeder reactor**.

fusion reactor *(Nuc.Eng.)*. Reactor in which **nuclear fusion** is used to produce useful energy. Much research is going into the development of such a machine.

fusion welding *(Elec.Eng.)*. A process of welding metals in which the weld is carried out solely by the melting of the metals to be joined, without any mechanical pressure.

fust *(Arch.)*. The shaft of a column.

fustian *(Textiles)*. A term including a number of hard-wearing fabrics usually of cotton but differing widely in structure and appearance, but all heavily wefted; they are used for clothing and furnishings. See **corduroy, moleskin, velveteen**.

G

g *(Aero.)*. See **load factor**.
g *(Chem.)*. Abbrev. for *gram(me)*.
g *(Chem.)*. Symbol for osmotic coefficient.
g *(Genrl.)*. A symbol for specified efficiency.
g *(Phys.)*. Symbol for acceleration due to gravity.
γ *(Chem.)*. Symbol for: (1) substituted on the carbon atom of a chain next but two to the functional group; (2) substituted on one of the central carbon atoms of an anthracene nucleus; (3) substituted on the carbon atom next but two to the hetero-atom in a heterocyclic compound; (4) a stereoisomer of a sugar.
γ *(Maths.)*. See **Euler's constant**.
γ *(Phys.)*. Symbol for (1) ratio of specific heats of a gas, (2) surface tension, (3) propagation coefficient, (4) Gruneisen constant, (5) molar activity coefficient, (6) coefficient of cubic thermal expansion, (7) the greatest refractive index in a biaxial crystal, (8) electrical conductivity.
G *(Build.)*. Abbrev. for *gulley*.
G *(Elec.Eng.)*. See **G-value**.
G *(Genrl.)*. Symbol for giga, i.e. 10^9.
G *(Chem.)*. Symbol for: (1) thermodynamic potential: (2) Gibbs function; (3) free energy.
G *(Phys.)*. Symbol for (1) the constant of **gravitation**, (2) shear modulus, rigidity, (3) conductance.
Γ *(Maths.)*. Symbol for gamma function.
Γ *(Phys.)*. Symbol for surface concentration excess.
[G] *(Phys.)*. A pair of Fraunhofer lines in the deep blue of the solar spectrum. One, of wavelength 430.8081 nm, is due to iron; the other, of wavelength 430.7907 nm, is due to calcium.
G1 *(Bot.)*. Period of growth in the **cell cycle** between the end of cell division and the beginning of DNA synthesis.
G-11 *(Chem.)*. See **hexachlorophene**.
G2 *(Bot.)*. As for G1 but between the end of DNA synthesis and the beginning of the next division.
Ga *(Chem.)*. The symbol for gallium.
gab *(Build.)*. A pointed tool for working hard stone.
GABA *(Med.)*. Abbrev. for *Gamma-Amino Butyric Acid*.
gabbart scaffold *(Build.)*. Scaffolding in which sawn timbers are used instead of round poles.
gabbro *(Geol.)*. The name of a rock clan, and also of a specific igneous rock type. The rock gabbro is a coarse grained plutonite, consisting essentially of plagioclase, near labradorite in composition, and clinopyroxene, with or without olivine in addition. The gabbro clan includes also norite, eucrite, troctolite, kentallenite, etc.
gaberdine *(Textiles)*. A firm twill fabric (e.g. with worsted warp and cotton weft), with the warp predominating on the surface; used for dress and suiting cloths and light showerproof overcoatings. All-cotton gaberdine is also used for similar purposes.
gabers scaffold *(Build.)*. A **gabbart scaffold**.
gabion *(Civ.Eng.)*. A long wicker or wire basket, containing earth or stones, deposited with others to serve the same purposes as **fascines**.
gable *(Build.)*. The triangular part of an external wall at the end of a ridged roof.
gable board *(Build.)*. A **barge board**.
gable shoulder *(Build.)*. The projecting masonry or brickwork supporting the foot of a gable.
gable springer *(Build.)*. The concrete, brick, or tile corbel supporting the gable shoulder.
gable tiles *(Build.)*. Purpose-made arris tiles to cover the intersection between gable and roof.
gaboon *(For.)*. Mahogany-like wood from a tree of the genus *Aucoumea*, found in parts of central and western Africa. Planed surfaces of the wood are silky. Used for veneer and plywood manufacture, furniture and fittings.
Gabriel synthesis *(Chem.)*. A reaction used in organic syntheses for the preparation of pure primary amines. Based on the use of potassium phthalimide (benzene 1,2-dicarboximide) and halogenoalkanes.
gadding *(Vet.)*. The excited behaviour of cattle when irritated by gad-flies.
gad-fly *(Vet.)*. A fly of the genus *Hypoderma*, the larva of which parasitizes cattle and is known as a *warble*.
Gadiformes *(Zool.)*. An order of mainly marine and often deep-water Osteichthyes with elongated body and long dorsal and anal fins. Some are economically important. Cod, Haddock and Grenadiers.
gadolinite *(Min.)*. Silicate of beryllium, iron, and yttrium, often with cerium; occurs in pegmatite.
gadolinium *(Chem.)*. A rare metallic element; trivalent; a member of the rare earth group. Symbol Gd, at.no. 64, r.a.m. 157.25. Only known in combination; obtained from the same sources as europium.
Gaede diffusion pump *(Chem.Eng.)*. Pump using mercury vapour, which entrains molecules of gas from a low pressure established by a backing pump. Oil of low vapour-pressure (apiezon) is a modern alternative.
Gaede molecular pump *(Chem.Eng.)*. Rotary pump which ejects molecules of gas by imparting a drift velocity to their random motion.
gaffer *(Image Tech.)*. Senior lighting electrician on a film or TV unit.
gag *(Med.)*. To retch; also a device for keeping the mouth open for surgical procedures.
gagger *(Foundry)*. See **lifter**.
gahnite *(Min.)*. A mineral belonging to the spinel group; occurs as grey octahedral cubic crystals. Also known as *zinc-spinel* (see **spinel**) the composition being zinc aluminate, $ZnAl_2O_4$.
Gaia *(Ecol.)*. Theory proposed by J.E. Lovelock in 1979 concerning the role of biota in maintaining a climatic homeostasis.
gain *(Telecomm.)*. (1) In electric systems, generally provided by insertion of an amplifier into a transmission circuit, or by matching impedances by a loss-free transformer. Measured in decibels or nepers, and defined as the increase in power level in the load, i.e. the ratio of the actual power delivered to that which would be delivered if source were correctly matched, without loss, to the load, on the absence of the amplifier. (2) In a directional antenna, ratio (expressed in decibels) of voltage produced at the receiver terminals by a signal arriving from the direction of maximum sensitivity of the antenna, to that produced by the same signal in an omnidirectional reference antenna (generally a half-wave dipole). In a transmitting antenna, ratio of the field strength produced at a point along the line of maximum radiation by a given power radiated from antenna, to that produced at the same point by the same power from an omnidirectional antenna.
gain-bandwidth product *(Telecomm.)*. A *figure of merit* rating for amplifiers, or transmission paths incorporating amplifiers, based on the product of gain and bandwidth as measured under specified conditions.
gain control *(Telecomm.)*. Means for varying the degree of amplification of an amplifier, often a simple potentiometer. See **automatic-**.
gaiting *(Textiles)*. The operation of preparing a loom for weaving by placing the warp in position.
Gal *(Chem.)*. Abbrev. for *Galactose*.
galactagogue, galactagog *(Med.)*. Promoting the secretion of milk (Greek, *gala*, gen. *galaktos*); any medicine which does this.
galactans *(Chem.)*. The anhydrides of galactose. They

comprise many gums, agar, and fruit pectins, and occur in algae, lichens, mosses.

galactic circle *(Astron.)*. The great circle of the celestial sphere in which the latter is cut by the galactic plane: hence the primary circle to which the galactic coordinates are referred.

galactic cluster *(Astron.)*. Alternative name for **open cluster**.

galactic coordinate *(Astron.)*. Two spherical coordinates referred to the galactic plane; the origin of galactic longitude is at R.A. $17^h42.4^m$, dec. −28°55′ (1950); galactic latitude is measured positively from the galactic plane towards the north galactic pole.

galactic halo *(Astron.)*. An almost spherical aggregation of stars, gas and dust, which is concentric with the Galaxy. It contains stars of Population II, and is responsible for much of the background of radio emission from the sky. Similar halos surround other galaxies.

galactic noise *(Telecomm.)*. That arriving from outer space, similar to electronic circuit noise, but apparently arising from sources in galaxies.

galactic plane *(Astron.)*. The plane passing as nearly as possible through the centre of the belt known as the Milky Way or Galaxy.

galactic rotation *(Astron.)*. The rotation of the Galaxy about its centre. All stars, gas and dust within the Galaxy share this rotation, which is fastest towards the centre. In the Sun's vicinity the velocity due to galactic rotation is about 250 km/sec. The Sun takes about 250 million years to complete one orbit round the galaxy.

galactobolic *(Zool.)*. Refers to the action of neurohypophysial peptides which contract mammary myoepithelium and so eject milk.

galactocele *(Med.)*. A cystic swelling in the breast, due to retention of milk as the result of a blockage of a milk duct.

galactophoritis *(Med.)*. Inflammation of the milk ducts.

galactophorous *(Zool.)*. See **lactiferous**.

galactopoiesis *(Med.)*. Increase in milk secretion.

galactorrhoea, galactorrhea *(Med.)*. Excessive secretion of milk by the breast, causing it to overflow through the nipple.

galactosaemia *(Med.)*. An inborn error of metabolism. Infants with the defect are unable to metabolize **galactose** to glucose because of the absence of the enzyme galactose-1-phosphate utidyl transferase. Untreated survivors are mentally retarded.

galactose *(Chem.)*. A hexose of the formula $CH_2OH.(CHOH)_4.CHO$; thin needles; mp 166°C; dextrorotatory. It is formed together with D-glucose by the hydrolysis of milk-sugar with dilute acids. Stereoisomeric with glucose, which it strongly resembles in properties. Present in certain gums and seaweeds as a polysaccharide *galactan* and as a normal constituent of milk. ⇨

galactosis *(Zool.)*. See **lactation**.

galactotrophic, galactotropic *(Med.)*. Stimulating the secretion of milk by the mammary gland.

galatin dynamite *(Min.Ext.)*. High explosive containing nitroglycerine, sodium nitrate, collodion cotton, and such inert fillers as wood meal and sodium carbonate.

galaxite *(Min.)*. A rare manganese aluminium spinel, formula $MnAl_2O_4$.

Galaxy *(Astron.)*. (1) The name given to the belt of faint stars which encircles the heavens and which is known as the Milky Way. (2) The name is also used for the entire system of dust, gases and stars within which the sun moves; now known to have the typical spiral structure. Still more generally the term signifies any **extra-galactic nebula** each being a vast collection of stars, dust and gas. Probably about a thousand million galaxies are observable; they show a marked tendency to form clusters. (3) Automatic star-plate analyser for measuring brightness and position of high density photographic images of portions of the Galaxy. (*General Automatic Luminosity And XY measuring engine*).

gale *(Meteor.)*. A wind having a speed of 34 knots (63 km/h) or more, at a height of 33 ft (10 m) above the ground. On the Beaufort scale, a gale is a wind of force 8.

galeate, galeiform *(Bot.,Zool.)*. Shaped like a helmet or a hood.

galena *(Min.)*. Lead sulphide, PbS; commonest ore of lead, occurring as grey cubic crystals, often associated with zinc blende, in mineralized veins. Also *lead glance*.

galet *(Build.)*. See **spall**.

Galilean binoculars *(Phys.)*. Those in which the objectives are the usual doublet telescope objectives and the eyepieces are negative lenses.

Galilean transformation *(Phys.)*. In classical kinematics, the space and time coordinates of an event in one frame of reference as seen by an observer in another frame moving with a constant velocity relative to it. See **Lorentz transformation**.

gall *(Bot.)*. An abnormal localized swelling or outgrowth, usually of characteristic shape which follows an attack by a parasite or pest.

gall *(Med.)*. See **bile**.

gall *(Vet.)*. An injury of the skin of animals due to the pressure of harness.

gallamine *(Med.)*. A muscle relaxant used in anaesthesia.

gall bladder *(Zool.)*. In Vertebrates, a lateral diverticulum of the bile-duct in which the bile is stored.

gallery *(Arch.)*. An elevated floor projecting beyond the walls of a building and supported on pillars, brackets, or otherwise, so as to command a view upon the main floor, as at a theatre etc. Sometimes cantilevered to eliminate obstruction of view by pillars.

gallery *(Elec.Eng.)*. A device for attaching to a lampholder to provide a support for a glass shade or reflector which is too large or heavy to be supported by the shade-carrier ring.

gallery *(Min.Ext.)*. A tunnel or passage in a mine.

gallery furnace *(Eng.)*. Type used in distillation of mercury from its ores.

gallet *(Build.)*. A splinter of stone.

galleting *(Build.)*. See **garreting**.

galley *(Print.)*. A steel tray open at one end, in which type matter is held after setting. Corrections and deletions are more easily made to type in galley form than in page form, and are generally marked on the galley proof or slip, itself commonly referred to as a *galley*.

galley press *(Print.)*. A proofing press which allows for the thickness of the galley.

galley proof *(Print.)*. A proof taken while the type is on a galley and used for checking and pasting-up.

galley rack *(Print.)*. A rack for storing galleys of type.

gallic acid *(Chem.)*. $C_6H_2(OH)_3COOH$, *3,4,5-trihydroxybenzoic acid*; crystallizes with 1 H_2O; thin needles; decomposes at about 200°C into CO_2 and pyrogallol (3,4,5-trihydroxybenzene). Occurs in nut-galls, tea, divi-divi and other plants; it is obtained from tannin by hydrolysis.

Galliformes *(Zool.)*. An order of ground-living Birds with feet well adapted for running. Game birds (which seek their food, berries, seeds, buds, and insects, on the ground). Brush-turkeys, Curassows, Turkeys, Pheasants, Partridge, Grouse, and Quail.

gallium *(Chem.)*. A metallic element in the third group of the periodic system. Symbol Ga, at.no. 31, r.a.m. 69.72, rel.d. 5.9, oxidation state +3, mp 30.15°C, bp 2000°C. Used in fusible alloys and high temperature thermometry. Gallium arsenide it is an important semiconductor.

gallium arsenide *(Chem.)*. Semiconductor in which gallium and arsenic are combined in near **stoichiometric** proportions. It has sphalerite (zinc blende) structure.

gallon *(Genrl.)*. Liquid measure. One imperial gallon is the volume occupied by 10 lb avoirdupois of water. 1 Imp. gallon = 4.54609 litres, = 6/5 US gallon. US gallon = 3.785 43 litres, = 5/6 Imp. gallon.

galloon *(Arch.)*. Decorated work for a band or moulding, to which is applied a row of small round balls.

gallop rhythm *(Med.)*. Heard when listening to the heart beat when there is an added heart sound. Usually indicates heart failure.

Galloway boiler *(Eng.)*. A cylindrical boiler of the Lancashire type, in which the two furnace tubes unite, at a short distance from the grates, in to a single arched oval flue, crossed by water tubes.

Galloway tubes *(Eng.)*. The inclined water tubes which cross the flue of a Galloway boiler to assist circulation and increase the heating surface.

gall-sickness *(Vet.)*. See **anaplasmosis**.

gallstones *(Med.)*. *Biliary calculi.* Pathological concretions in the gall bladder and bile passages. They have not a uniform composition but some constituents may be preponderant, e.g. cholesterol, or calcium carbonate and phosphate.

galvanic cell *(Chem.)*. An electrochemical cell from which energy is drawn. Cf. **electrolytic cell**.

galvanic corrosion *(Eng.)*. Corrosion resulting from the current flow between two dissimilar metals in contact with an electrolyte.

galvanic current *(Med.)*. Steady unidirectional current for therapeutic use.

galvanic skin response *(Behav.)*. A change in the electrical resistance of the skin, recorded by a polygraph, widely used as an index of autonomic reaction. Abbrev. *GSR*.

galvanized iron *(Build.)*. Iron which has been subjected to **galvanizing**, e.g. zinc coating, widely used, especially in corrugated form, *corrugated iron*, for minor roofing purposes, e.g. on wooden buildings, etc. Also for nails, bolts, etc., where moisture may produce corrosion.

galvanizing *(Eng.)*. The coating of steel or iron with zinc, generally by immersion in a bath of zinc, covered with a flux, at a temperature of 425–500°C. The zinc may alternatively be electrodeposited from cold sulphate solutions. The zinc is capable of protecting the iron from atmospheric corrosion even when the coating is scratched, since the zinc is preferentially attacked by carbonic acid, forming a protective coating of basic zinc carbonates.

galvanoluminescence *(Elec.Eng.)*. Feeble light emitted from the anode in some electrolytic cells.

galvanomagnetic effect *(Elec.Eng.)*. See **Hall effect**.

galvanometer *(Elec.Eng.)*. Current-measuring device depending on forces on the sides of a current-carrying coil normal to magnetic fields in gaps. In a moving-magnet instrument, the suspended coil is replaced by an astatic magnet system which is magnetically shielded for very sensitive work.

galvanometer constant *(Elec.Eng.)*. A number by which the scale reading of a galvanometer must be multiplied in order to give a reading of current in amperes or other suitable units.

galvanometer shunt *(Elec.Eng.)*. A shunt connected in parallel with a galvanometer to reduce its sensitivity. See **universal shunt**.

galvanoscope *(Elec.Eng.)*. A term sometimes used to denote and instrument for detecting, but not measuring, an electric current.

galvanotaxis *(Zool.)*. Tendency of organisms to grow or move into a particular orientation relative to an electric current passing through the surrounding medium. Also *galvanotropism*.

gambrel roof *(Build.)*. See **mansard roof**.

games paddle *(Comp.)*. Name for a hand-held device used for computer games.

games, theory of *(Behav.)*. A mathematical formalization of decision and strategic processes involving the probabilities and values of various outcomes of action-choices for the decision-makers.

gametangium *(Bot.)*. Any cell or organ within which gametes are formed, e.g. antheridium, oogonium, archogonium.

gametes *(Biol.)*. Reproductive cells which will unite in pairs to produce zygotes; germ cells. adj. *gametal*.

gametogenesis *(Biol.)*. The formation of gametes from gametocytes: *gametogeny*.

gametophore *(Bot.)*. The branch of a moss which bears the sex organs.

gametophyte *(Bot.)*. The characteristically haploid generation in the life-cycle of a plant showing **alternation of generations**, producing the gametes, forming the zygote which germinates to give the **sporophyte**.

gam-, gamo- *(Genrl.)*. Prefix from Gk. *gamos*, marriage, union.

gamma *(Image Tech.)*. A measure of the contrast in image reproduction. In a photographic system it is the increment of image density produced by a given increment of log. exposure, $\gamma = \Delta D/\Delta \log E$, generally derived from the **characteristic curve**. In television, overall gamma similarly relates the logarithmic increments of receiver screen luminance and those of the brightness of the original scene.

gamma *(Phys.)*. See **microgram**.

gamma-amino butyric acid *(Med.)*. An inhibitory *neurotransmitter* in the brain. A malfunction of it may exist in **Huntingdon's chorea**.

gamma-BHC *(Chem.)*. Same as *Gammexane*.

gamma brass *(Eng.)*. The *γ-constituent in brass* is hard and brittle and is stable between 60 and 68% of zinc at room temperature. γ-Brass is an alloy consisting of this constituent.

gamma camera *(Radiol.)*. See **scintillation camera**.

gamma correction *(Image Tech.)*. Non-linear amplification of the picture signal applied on transmission to compensate for preceding tonal distortions and to obtain the desired contrast in the received image.

gamma detector *(Radiol.)*. A radiation detector specially designed to record or monitor gamma-radiation.

gamma function *(Maths.)*. The function $\Gamma(x)$ defined by Euler as $\int_0^\infty e^{-t}t^{x-1}dt$ and by Weierstrass by the equation

$$\frac{1}{\Gamma(x)} = xe^{\gamma x}\prod_1^\infty\left(1+\frac{x}{n}\right)e^{-x/n}$$

where γ is Euler's constant. Its main properties are

$$\Gamma(1+x) = x\Gamma(x)$$

$$2^{2x-1}\Gamma(x)\Gamma(x+\tfrac{1}{2}) = \sqrt{\pi}\,\Gamma(2x)$$

$$\Gamma(x)\Gamma(1-x) = \frac{\pi}{\sin \pi x}$$

$$\Gamma(\tfrac{1}{2}) = \sqrt{\pi}$$

$$\Gamma(n+1) = n!.$$

gamma globulin *(Immun.)*. Describes the serum proteins which on electrophoresis have the lowest anodic mobility at neutral pH. These are mainly immunoglobulins, and the term was used to describe immunoglobulins until more specific means of distinguishing them were developed.

gamma infinity *(Image Tech.)*. The maximum γ obtainable with prolonged development.

gamma iron *(Eng.)*. The polymorphic form of iron stable between 906 and 1403°C. It has a face-centred cubic lattice and is nonmagnetic. Its range of stability is lowered by carbon, nickel and manganese, and it is the basis of the solid solutions known as *austenite*.

gamma-radiation *(Phys.)*. Electromagnetic radiation of high quantum energy emitted after nuclear reactions or by radioactive atoms when nucleus is left in excited state afer emission of α- or β-particle.

gamma-ray astronomy *(Astron.)*. The study of radiation from celestial sources at wavelengths shorter than 0.01 nm. Gamma-rays have been detected from the *gamma-ray background*, a few energetic galaxies and **quasars**, and from certain highly-evolved stars.

gamma-ray capsule *(Nuc.Eng.)*. Usually metal, sealed and of sufficient thickness to reduce γ-ray transmission to a safe value.

gamma-ray energy *(Phys.)*. Energy of a gamma-ray photon given by $h\nu$ where ν is the frequency and h is Planck's constant. The energy may be determined by diffraction by a crystal or by the maximum energy of photoelectrons ejected by the γ-rays. The depth of penetration into a material is determined by the energy.

gamma-ray photon *(Phys.)*. A quantum of gamma-radiation energy given by $h\nu$ where ν is the frequency and h is Planck's constant.

gamma-rays *(Phys.)*. See **gamma-radiation**. In medicine this is the commonest form of radionuclide emission in patient imaging.

gamma-ray source *(Radiol.)*. A quantity of matter emitting γ-radiation in a form convenient for radiology.

gamma-ray spectrometer *(Nuc.Eng.)*. Instrument for investigation of energy distribution of γ-ray quanta. Usually a scintillation or germanium counter followed by a **pulse-height analyser**.

Gammexane *(Chem.)*. TN applied to the γ isomer of **benzene hexachloride**; a synthetic stomach and contact insecticide of great toxicity to a wide range of pests.

gamocyte *(Zool.)*. In Protozoa, a phase developing from a trophozoite and giving rise to gametes.

gamone *(Bot.)*. Any chemical substance released by a gamete or hypha that is attractive to another appropriate gamete or hypha in sexual reproduction, e.g. malic acid in ferns, called *sirenin*.

gamopetalous *(Bot.)*. Having a corolla consisting of a number of petals united by their edges. Cf. **polypetalous**.

gamophyllous *(Bot.)*. Having the perianth members united by their edges. Cf. **polyphyllous**.

Gamow-Teller selection rules *(Phys.)*. See **nuclear selection rules**.

gamut *(Phys.)*. The range of chromaticities available through the addition of three colours.

gang *(Min.Ext.)*. A train or *journey* of tubs or trucks.

gang boarding *(Build.)*. A board with battens nailed across to form steps; used as a gangway during building operations.

ganged capacitor *(Telecomm.)*. Assemblage of two or more variable capacitors mechanically coupled to the same control mechanism.

ganging *(Elec.Eng.)*. Mechanical coupling of the movements of variable circuit elements for simultaneous control.

ganging oscillator *(Elec.Eng.)*. One giving a constant output, whose frequency can be rapidly varied over a wide range; used for testing accuracy of adjustment of ganged circuits over their tuning range.

ganglion *(Med.)*. Aggregation of nerve cells.

ganglion *(Zool.)*. A plexiform collection of nerve fibre terminations and nerve cells. pl. *ganglia*.

ganglionectomy, gangliectomy *(Med.)*. Surgical removal of a nerve ganglion, or of a ganglion arising from a tendon sheath.

ganglioneuroma *(Med.)*. A tumour composed of ganglion nerve cells, nerve fibre, and fine fibrous tissue, usually arising in connection with sympathetic nerves, e.g. in the medulla of the suprarenal gland.

ganglion impar *(Zool.)*. The unpaired, most posterior of the abdominal sympathetic ganglia.

ganglionitis *(Med.)*. Inflammation of a nerve ganglion.

ganglioside *(Biol.)*. Glycolipids derived from cerebroside by the addition of complex oligosaccharide chains.

gang milling *(Eng.)*. The use of several milling cutters on one spindle to produce a surface with a required profile or to mill the face and sides of the work at one operation.

gang mould *(Build.,Civ.Eng.)*. A mould in which a number of similar concrete units may be cast simultaneously.

gang punch *(Comp.)*. To punch (the same information) in a number of cards.

gangrene *(Med.)*. Death of a part of the body, associated with putrefaction; due to infection or to cutting off of the blood supply.

gangrenous coryza *(Vet.)*. See **malignant catarrhal fever**.

gang saw *(For.)*. An arrangement of parallel saws secured in a frame to operate simultaneously in sawing a log into strips.

gang switches *(Elec.Eng.)*. A number of switches mechanically connected together so that they can all be operated simultaneously.

gang tool *(Eng.)*. A tool holder having a number of cutters; used in lathes and planes, each tool cutting a little deeper than the one ahead of it.

gangue *(Eng.)*. The portion of an ore which contains no metal; valueless minerals in a lode.

gangway *(Build.)*. Rough planks laid to provide a footway for the passage of workmen on a site.

gangway *(Min.Ext.)*. Main haulage road, or level underground.

gannister *(Geol.)*. A particularly pure and even-grained siliceous grit or loosely cemented quartzite, occurring in the Upper Carboniferous of northern England, and used in the manufacture of silica-bricks.

ganoid *(Zool.)*. Formed of, or containing, ganoin. Said of fish scales of rhomboidal form, composed of an outer layer of ganoin and an inner layer of isopedin; (fish) having these scales.

ganoin *(Zool.)*. A calcareous substance secreted by the dermis and forming the superficial layer of certain fish scales; it was formerly supposed to be homologous with enamel.

gantry *(Build.)*. A temporary erection having a working platform used as a base for building operations or for the support of cranes, scaffolding, or materials.

gantry *(Space)*. The servicing tower beside a rocket on its launching-pad.

Gantt chart *(Eng.)*. Graphic construction chart which shows each operation and its connection and timing as part of the overall scheme.

Ganymede *(Astron.)*. The third natural satellite of **Jupiter**, discovered by Galileo. It is 5200 km in diameter, making it the largest moon in the Solar System, and larger than Mercury. It is the brightest of the Galilean satellites.

gap *(Acous.)*. Air gap in magnetic circuit of recording or erasing head in tape-recorder; allows signal to interact with oxide film.

gap *(Aero.)*. The distance from the leading edge of a biplane's upper wing to the point of its projection on to the chord line of a lower wing.

gap *(Comp.)*. Digits which separate signals for data or program. Also *space*.

gap *(Electronics)*. Range of energy levels between the lowest of conduction electrons and the highest of valence electrons.

gap *(Phys.)*. Space between discharge electrodes.

GAPAN *(Aero.)*. Abbrev. for *Guild of Air Pilots and Air Navigators*.

gap arrester *(Elec.Eng.)*. A lightning arrester consisting essentially of a small air gap connected between the circuit to be protected and earth; the gap breaks down on the occurrence of a lightning surge, and discharges the surge to earth. See **multigap arrester**.

gap bed *(Eng.)*. A lathe bed having a gap near the headstock, to permit of turning large flat work of greater radius than the centre height.

gap breakdown *(Phys.)*. The cumulative ionization of the gas between electrodes, leading to a breakdown of insulation and a **Townsend avalanche**.

gap bridge *(Eng.)*. A bridge casting of the same cross-section as the bed in a gap-bed lathe, and used to close the gap.

gape *(Min.Ext.)*. Aperture below which a crushing machine can receive and work on entering ore.

gape *(Zool.)*. The width of the mouth when the jaws are open.

gapes *(Vet.)*. See **syngamiasis**.

gap factor *(Elec.Eng.)*. Ratio of energy gain in electron-volts traversing a gap across which an accelerating field acts, to the actual voltage across the gap.

gap filler *(Aero.)*. Radar to supplement long range surveillance radar.

gap junction *(Biol.)*. Junction between cells which allows direct communication between cells by molecules which can diffuse through pores in the junction. The flow is controlled by the opening or closing of these pores.

gap lathe *(Eng.)*. A lathe with a **gap bed**.

gap length *(Acous.)*. The distance between adjacent

surfaces of the poles in a longitudinal magnetic recording system.

gap window *(Arch.)*. A long and narrow one.

garboard strake *(Ships)*. The first strake or line of plating attached to the keel on either side.

garden city *(Arch.)*. An independent community established on the outskirts of cities from the end of the nineteenth century, in an attempt to integrate industry with the pastoral life of the village.

garden-wall bonds *(Build.)*. Forms of bond with an increased number of stretchers used largely for building low boundary walls of single brick thickness, when the load to be carried is that of the wall only and it is desired to show a fair face on both sides of the wall.

gargle *(Acous.)*. A **wow** which has fluctuation changes ranging between about 30 and 200 Hz.

gargoyle *(Build.)*. A grotesquely-shaped spout projecting from the upper part of a building, to carry away the rainwater.

gargoylism *(Med.)*. Former name for *Hurler's syndrome*.

garnet *(Min.)*. A group of cubic minerals which are silicates of di- and tri-valent metals and occur typically in metamorphic rocks e.g. garnetiferous schists. Some species are of value as gems, rivalling ruby in colour. See also **andradite, grossular, melanite, pyrope, spessartine, uvarovite.**

garnet hinge *(Build.)*. A form of *strap hinge*.

garneting *(Build.)*. See **garreting.**

garnet paper *(Build.)*. Type of sandpaper having powdered garnet as the abrasive coating.

garnett machine *(Textiles)*. A strong carding machine consisting of rollers fitted with saw teeth; used for breaking down waste materials to fibres for re-use.

garnierite *(Min.)*. A bright green nickeliferous serpentine, hydrated nickel magnesium silicate. It occurs in serpentinite as a decomposition product of olivine, and in other deposits, and is an important ore of nickel.

garnish bolt *(Build.)*. A bolt whose head is chamfered or faceted.

garreting *(Build.)*. A term applied to the process of inserting small stone splinters in the joints of coarse masonry. Also *galleting, garneting*.

garret window *(Build.)*. A skylight of which the glazing lies along the slope of the roof.

garter spring *(Eng.)*. An endless band formed by connecting the two ends of a long helical spring; used to exert a uniform radial force on any circular piece round which it is stretched, as in a *carbon gland*.

garth *(Arch.)*. An enclosed area attached to a building and surrounded usually by a cloister.

gas *(Min.Ext.)*. Explosive mixture of combustible gasses with air, particularly *methane* and *carbon monoxide*. Also used for accumulations of combustion products, e.g. *carbon dioxide*. See **afterdamp, black damp, choke damp, fire damp, white damp.**

gas *(Phys.)*. (1) A state of matter in which the molecules move freely, thereby causing the matter to expand indefinitely, occupying the total volume of any vessel in which it is contained. (2) The term is sometimes reserved for a gas at a temperature above the critical value. Also defined as a definitely compressible fluid. See **gas laws**, and references at **gases.**

gas amplification *(Phys.)*. Increase in sensitivity of a Geiger or proportional counter compared with a corresponding ionization chamber.

gas analysis *(Chem.)*. The quantitative analysis of gases by absorption. A measured quantity of gas, 100 cm^3, is brought into intimate contact with the various reagents, and the reduction in volume is measured after each absorption process. Carbon dioxide is absorbed in a concentrated potassium hydroxide solution, oxygen in alkaline pyrogallol solution, carbon monoxide in ammoniacal cuprous chloride solution, unsaturated compounds by absorption in bromine water etc., hydrogen by combustion with a measured quantity of air over palladium asbestos. Nitrogen is estimated by difference.

gas-and-pressure-air burner *(Eng.)*. Industrial gas burner designed to operate with low pressure gas and with air under pressure from fans and compressors.

gas-bag *(Aero.)*. Any separate gas-containing unit of a rigid airship.

gas-blast circuit-breaker *(Elec.Eng.)*. A high power circuit breaker in which a blast of gas is directed across the contacts at the instant of separation in order to extinguish the arc. See **air-blast switch.**

gas cap *(Min.Ext.)*. Gas accumulated above oil reservoir or salt dome, encountered when unsuccessfully drilling for oil or natural gas.

gas carbon *(Chem.)*. A hard dense deposit of almost pure carbon which slowly collects on the inside of a coal-gas retort.

gas carburizing *(Eng.)*. The introduction of carbon into the surface layers of mild steel by heating in a reducing atmosphere of gas high in carbon, usually hydrocarbons or hydrocarbons and carbon monoxide.

gas cell *(Chem.)*. A galvanic cell in which at least one of the reactants is a gas.

gas checking *(Build.)*. A paintwork defect characterized by the appearance of fine wrinkles. Appears particularly when paints based on undercooked oils dry in the presence of coal-gas combustion products.

gas chromatography *(Chem.)*. See **gas-liquid chromatography.**

gas colic *(Vet.)*. Colic associated with **tympanites.**

gas concrete *(Civ.Eng.)*. Lightweight concrete in which bubbles of gas are generated when a metallic additive (powdered aluminium or zinc) reacts chemically with the water and cement in the concrete.

gas conditioning *(Eng.)*. See **protective furnace atmosphere.**

gas constant *(Phys.)*. The constant of proportionality R in the equation of state for 1 mole of an ideal gas, $pV = RT$, where p is the pressure, V the volume and T the absolute temperature. $R = 8.314\,J\,K^{-1}\,mol^{-1}$. Also *molar gas constant*.

gas-cooled reactor *(Nuc.Eng.)*. One in which cooling medium is gaseous, usually carbon dioxide, air or helium.

gas counter *(Nuc.Eng.)*. (1) A gas-filled counter operating in the proportional or Geiger-Müller modes. (2) Geiger counter into which radioactive gases can be introduced.

gas counting *(Nuc.Eng.)*. That of radioactive materials in gaseous form. The natural radioactive gases (radon isotopes) and carbon dioxide (^{14}C) are common examples. See also **gas flow counter.**

gas-discharge lamp *(Phys.)*. See **electric-discharge lamp.**

gas-discharge tube *(Electronics)*. Generally, any tube in which an electric discharge takes place through a gas. Specially, a tube comprising a hot or cold cathode, with or without a control electrode (grid) for initiating the discharge, and with gas at an appreciable pressure. See **gas-filled relay, glow tube, grid-glow tube, ignitron, mercury-arc rectifier, thyratron.**

gas drain *(Min.Ext.)*. A tunnel or borehole for conducting gas away from old workings.

gas-electric generating set *(Elec.Eng.)*. A gas- or petrol-driven electric generating set.

gas electrode *(Phys.)*. One which holds gas by adsorption or absorption, so that it becomes effective as an electrode in contact with an electrolyte.

gas engine *(Autos.)*. One in which gaseous fuel is mixed with air to form a combustible mixture in the cylinder and fired by spark ignition. Used for stationary power and may operate on 2- or 4-stroke cycle.

gaseous diffusion *(Nuc.Eng.)*. Isotope separation process based on principle of molecular diffusion.

gaseous diffusion enrichment *(Phys.)*. The enrichment of uranium isotopes using gaseous UF_6 passing through a porous barrier. The separation factor is 1.0043 and this means more than 100 stages are required to produce fuel sufficiently enriched for a commercial nuclear reactor.

gaseous discharge *(Phys.)*. Flow of charge arising from ionization of low-pressure gas between electrodes, initiation being by electrons of sufficent energy released from hot or cold cathodes. Various gases give characteristic spectral colours, e.g. mercury, sodium vapour, neon, hydrogen etc.

gaseous exchange *(Biol.)*. See **respiration**.

gases *(Phys.)*. See **density of-**, **expansion of-**, **kinetic theory of-**, **liquefaction of-**.

gas evolution *(Eng.)*. The liberation of gas bubbles during the solidification of metals. It may be due to the fact that the solubility of a gas is less in the solid than in the molten metal, as when hydrogen is evolved by aluminium and its alloys, or to the promotion of a gas-forming reaction, as when iron oxide and carbon in molten steel react to form carbon monoxide. See also **blowholes**, **unsoundness**.

gas exchange *(Bot.)*. The uptake and output of gases, especially of carbon dioxide and oxygen in photosynthesis and respiration.

gas-filled cable *(Telecomm.)*. An impregnated paper-insulated power cable in which gas (nitrogen) at a high pressure is admitted within the lead sheath to minimize ionization.

gas-filled filament lamp *(Elec.Eng.)*. One in which the bulb contains an inert gas.

gas-filled photocell *(Electronics)*. One in which anode and photocathode are enclosed in atmosphere of gas at low pressure. It is more sensitive than the corresponding high-vacuum cell because of formation of positive ions by collision of the photoelectrons with the gas molecules.

gas-filled relay *(Electronics)*. Thermionic tube, usually of the mercury-vapour type, when used as a relay; a thyratron.

gas-flow counter *(Nuc.Eng.)*. (1) Counter tube through which gas is passed to measure its radioactivity, (2) One used to measure low intensity α or β sources. These are introduced into the interior of the counter, and to prevent the ingress of air, the counting gas flows through it at a pressure slightly above atmospheric. Also *flow counter*.

gas gangrene *(Med.)*. Rapidly spreading infection of a wound with gas-forming anaerobic bacteria, causing progressive gangrene of the infected part.

gas generator *(Aero.)*. The high-pressure compressor/combustion/turbine section of a gas turbine which supplies a high-energy gas flow for turbines which drive propellers, fans or compressors.

gas generator *(Chem.)*. Chemical plant for producing gas from coal, e.g. water gas, by alternating combustion of coal and reduction of steam.

gas gland *(Zool.)*. A structure in the wall of the air-bladder in certain Fish which is capable of secreting gas into the bladder. See **rete mirabile**.

gas governor *(Eng.)*. Device in which diaphragms are used to maintain a constant pressure supply to an appliance.

gash *(Elec.Eng.)*. *Guanidine Aluminium Sulphate Hexahydrate*, a ferroelectric compound with an almost square hysteresis loop suitable for constructing a binary cell.

gas-impregnated cable *(Telecomm.)*. See **gas filled cable**.

gasket *(Eng.)*. (1) A layer of packing material placed between contact surfaces or parts needing a sealed joint. It can consist of thin copper sheets, compressed rubber bonding asbestos, etc. Use of asbestos has however gone out of favour. Used between cylinder blocks and heads, etc. (2) Jointing or packing material, such as cotton rope impregnated with graphite grease; used for packing stuffing-boxes on pumps, etc.

gasket, gaskin *(Build.)*. Hemp or cotton yarn wound round the spigot end of a pipe at a joint, and rammed into the socket of the mating pipe to form a tight joint.

gas laws *(Phys.)*. Boyle's law, Charles's law and the pressure law which are combined in the equation $pV = RT$, where p is the pressure, V the volume, T the absolute temperature and R the gas constant for 1 mole. A gas which obeys the gas laws perfectly is known as an *ideal* or *perfect* gas. Cf. **van der Waals' equation**.

gas lift *(Min.Ext.)*. Method of pumping oil from the bottom of a well by releasing compressed liquid gas there. On vaporization it lifts and entrains the oil.

gas lime *(Chem.)*. The spent lime from gasworks after it has been used for the absorption of hydrogen sulphide and carbon dioxide in the gas purifiers.

gas-liquid chromatography *(Chem.)*. A form of partition or adsorption chromatography in which the mobile phase is a gas and the stationary phase a liquid. Solid and liquid samples are vaporized before introduction on to the column. The use of very sensitive detectors has enabled this form of chromatography to be applied to submicrogram amounts of material. Abbrev. *GLC*.

gas mantle, incandescent mantle *(Eng.)*. A small dome-shaped structure of knitted or woven ramie or rayon, impregnated with a solution of the nitrates of cerium and thorium, then dried, and the textile fabric burned off.

gas maser *(Electronics)*. One in which the interaction takes place between molecules of gas and the microwave signal.

gas mask *(Chem.)*. A device for protection against poisonous gases, which are absorbed by activated charcoal or by other reactive substances, e.g. soda-lime. The choice of the absorbing material depends on the nature of the gas to be counteracted.

gasoline engine *(Eng.)*. US term for *petrol engine*.

Gasparcolor *(Image Tech.)*. Obsolete process of subtractive colour printing for stills and motion pictures, using an integral tripack with pre-coloured emulsion layers and processed by a dye-destruction (bleach-out) method.

gas-pipe tongs *(Eng.)*. A wrench used for turning pipes when screwing them into, or out of, coupling pieces.

gas pliers *(Eng.)*. Stout pliers with narrow jaws, the gripping faces of which are concave and serrated, to provide a secure grip.

gas-pressure cable *(Elec.Eng.)*. See **pressure-cable**.

gas-pressure regulator *(Phys.)*. Diaphragm operated valve or other device actuated by gas pressure and balanced to produce a constant outlet pressure irrespective of fluctuating initial pressures.

gas producer *(Aero.)*. A turbo-compressor of which the power output is in the form of the gas energy in the efflux, sometimes mixed with air from an auxiliary compressor. Essentially, the gas producer is mounted remotely from the point of utilization of its energy, e.g. helicopter **rotor-tip jets** or a **free turbine**.

gas pump *(Autos.)*. See **petrol pump**.

gas pump *(Eng.)*. See **Humphrey gas pump**.

gas regulator *(Eng.)*. (1) Same as *gas-pressure regulator*. (2) The throttle valve of a gas engine.

gas scrubber *(Chem.)*. An apparatus for the purification of gas from tarry matter and water soluble impurities, such as ammonia.

gasserectomy *(Med.)*. Excision of the Gasserian ganglion, sometimes performed for the relief of trigeminal neuralgia (tic douloureux).

Gasserian ganglion *(Zool.)*. In Vertebrates, a large ganglion of the fifth cranial nerve, near its origin.

gas show *(Geol.)*. A surface indication of the escape of natural gas from underground reservoirs; of importance in oilfield exploration.

gassing *(Aero.)*. See **inflation**.

gassing *(Elec.Eng.)*. (1) The evolution of gas which occurs in an accumulator towards the end of its charging period. (2) Evolution of gas from metal (and other) surfaces in high vacuum equipment.

gassing *(Textiles)*. The process of passing yarns or fabrics through a gas flame or over electrically heated elements to remove protruding fibres and make the material smoother.

gassing of copper *(Eng.)*. A process which denotes the brittleness produced when copper containing oxide is heated in an atmosphere containing hydrogen. The hydrogen diffuses into the metal and combines with oxygen, forming steam which cannot diffuse out. A high steam pressure is built up at the crystal boundaries and the cohesion is diminished.

gas stocks and dies *(Eng.)*. Stocks and dies (see **die (2)**, **die-stock**) for cutting screw-threads on gas barrel or piping.

gas sweetening *(Min.Ext.)*. The removal of hydrogen

sulphide and carbon dioxide from natural gas, then called *sour gas*.

gas tar *(Chem.)*. Coal-tar condensed from coal-gas, consisting mainly of hydrocarbons. Distillation of tar provides many substances, e.g. ammoniacal liquor, 'benzole', naphtha, and creosote oils, with a residue of pitch. Dehydrated, it is known as 'road tar', and used as a binder in road-making.

gas temperature *(Aero.)*. The temperature of the gas stream resulting from the combustion of fuel and air within a turbine engine. For engine performance monitoring, the temperature may be measured at either of two points signified by the abbrevs. *JPT* and *TGT*. *JPT* (jet pipe temperature) is the measured temperature of the gas stream in the exhaust system, usually at a point behind the turbine. *TGT* (turbine gas temperature) is the measured temperature of the gas stream between turbine stages. *EGT* (exhaust gas temperature), frequently used, is synonymous with *JPT*.

gaster *(Zool.)*. The abdomen proper in Hymenoptera, being the region posterior to the first abdominal segment which, in many members, is constricted.

gastero-, gastr-, gastro- *(Genrl.)*. Prefixes from Gk. *gastēr*, gen. *gastros*, stomach.

Gasteromycetes *(Bot.)*. The puffballs, earth-stars and stinkhorn fungi, a class of Basidiomycotina in which the hymenium is enclosed until after the spores have matured. Most are saprophytic in soil.

gas thread *(Eng.)*. See **British Standard pipe thread**.

gas transport *(Zool.)*. The transport by the blood of oxygen from the site of *external respiration* to cells where it is needed for aerobic respiration (see **respiration, aerobic**), this frequently involving a *respiratory pigment*, and the transport away from the respiring cells of any carbon dioxide produced.

gastrectomy *(Med.)*. Removal of the whole, or part, of the stomach.

gastric *(Zool.)*. Pertaining to, or in the region of, the stomach.

gastric juice *(Med.)*. Human gastric juice consists principally of water (99.44%), free HCl (0.02%), and small quantities of NaCl, KCl, $CaCl_2$, $Ca_3(PO_4)_2$, $FePO_4$, $Mg_3(PO_4)_2$ and organic matter including enzymes.

gastrin *(Biol.)*. Polypeptide hormone secreted by specialized cells in the stomach mucosa which stimulates the secretion of acid and pepsin by other cells in the mucosa.

gastritis *(Med.)*. Inflammation of the mucous membrane of the stomach.

gastrocele *(Med.)*. Hernia of the stomach.

gastrocentrous *(Zool.)*. Said of vertebrae in which the centrum is composed largely of the pleurocentrum and the intercentrum is reduced or absent; in all *Amniota*.

gastrocnemius *(Zool.)*. In land Vertebrates, a muscle of the shank.

gastrocoel *(Zool.)*. See **archenteron**.

gastrocolic *(Med.)*. Pertaining to, or connected with, the stomach and the colon.

gastroduodenal *(Med.)*. Pertaining to, or connected with, the stomach and the duodenum.

gastroduodenitis *(Med.)*. Inflammation of the mucous membrane of the stomach and the duodenum.

gastroduodenostomy *(Med.)*. A communication between the stomach and the duodenum, made by operation.

gastro-enteritis *(Med.)*. Inflammation of the mucous membrane of the stomach and the intestines.

gastroenterostomy *(Med.)*. A communication between the stomach and the small intestine, made by operation.

gastrogastrostomy *(Med.)*. A communication, made by operation, between the upper and lower parts of the stomach when these are pathologically separated by a stricture.

gastrojejunal *(Med.)*. Pertaining to, or connected with, the stomach and the jejunum.

gastrojejunostomy *(Med.)*. A communication between the stomach and the jejunum, made by operation.

gastrolienal *(Med.)*. Pertaining to the stomach and the spleen.

gastromyotomy *(Med.)*. Incision of the muscle of the stomach round a gastric ulcer.

gastropexy *(Med.)*. The operation of suturing the stomach to the abdominal wall for the treatment of gastroptosis.

Gastropoda *(Zool.)*. A class of Mollusca with a distinct head bearing tentacles and eyes, a flattened foot, a visceral hump which undergoes torsion to various degrees and is often coiled, always bilaterally symmetrical to some extent, with the shell usually in one piece. Snails, Slugs, Whelks, etc. Also *Gasteropoda*.

gastroptosis *(Med.)*. Abnormal downward displacement of the stomach in the abdominal cavity.

gastroscope *(Med.)*. An instrument passed through the gullet to inspect the interior of the stomach and by means of which a photograph of the lining of the stomach may be taken or pieces of tissue sampled for histological diagnosis. See **endoscope**.

gastrostaxis *(Med.)*. Bleeding or oozing of blood from the stomach, the mucous membrane of which is intact.

gastrostomy *(Med.)*. The operative formation of an opening into the stomach, through which food may be passed when the normal channels are obstructed.

gastrotomy *(Med.)*. Incision of the stomach wall.

Gastrotricha *(Zool.)*. A class of phylum Aschelminthes. Small, free-living worm-like aquatic animals with ciliated tracts on the cuticle which has scales and bristles.

gastrovascular *(Zool.)*. Combining digestive and circulatory functions, as the canal system of Ctenophora and the coelenteron of Cnidaria.

gastrula *(Zool.)*. In development, the double walled stage of the embryo which succeeds the blastula.

gastrulation *(Zool.)*. The process of formation of a gastrula from a blastula during development.

gas tube *(Phys.)*. Since it is so far impossible to obtain a perfect vacuum or even one approaching outer space, all tubes and valves are gas tubes; in a so-called gas tube, the pressure of residual gas is sufficiently high to influence the operation. See **gas-filled photocell, thyratron**.

gas tungsten-arc welding *(Eng.)*. Welding in which a tungsten electrode forms the arc but does not contribute to the weld, and the weld puddle and arc are protected by an inert gas.

gas turbine *(Eng.)*. A simple, high-speed machine used for converting heat energy into mechanical work, in which stationary nozzles discharge jets of expanded gas (usually products of combustion) against the blades of a turbine wheel. Used in stationary power and other plants, locomotives, marine (especially naval) craft, jet aero engines, and experimentally in road vehicles. See **by-pass turbojet, gas producer, shaft turbine, ducted fan, turbojet, turboprop**.

gas-turbine plant *(Eng.)*. A power plant comprising at least a gas turbine, an air compressor driven by the gas turbine, and a combustion chamber in which liquid (usually) fuel is burnt to form the working medium of the turbine. Intercooling or other thermodynamic modifications may be added to increase the thermal efficiency of the plant.

gas vacuole *(Bot.)*. Structure in the cells of some planktonic blue-green algae which provide buoyancy and are composed of many small, more or less cylindrical, gas-filled vesicles.

gas welding *(Eng.)*. Any metal welding process in which gases are used in a combination to obtain a hot flame. The most commonly used gas welding process employs the oxy-acetylene combination which develops a flame temperature of 3200°C. Some plastics, especially polythene, may be fusion-jointed by a form of low temperature gas welding without flame.

gas well *(Geol.)*. A deep boring, generally in an oil-field, which yields natural gas rather than oil. See **natural gas**.

gate *(Comp.)*. A circuit which controls the flow of binary signals. The components of a computer processor and memory can all be constructed from combinations of very simple gates, each with a combination of input signals and a single output. See **AND gate, OR-, NOT-, Equivalence-, XOR-, NAND-, NOR-**.

gate *(Electronics)*. Electronic circuit which passes impressed signals when permitted by another independent source of similar signals.

gate *(Foundry)*. In a mould, the channel or channels through which the molten metal is led from the runner, down-gate, or pouring-gate to the mould cavity. Also called *geat, git, sprue*.

gate *(Hyd.Eng.)*. A movable barrier for stopping or regulating the flow in a channel.

gate *(Image Tech.)*. The aperture and associated mechanism at which film is exposed in a motion picture camera, printer or projector.

gate *(Plastics)*. Term used to denote the small restricted space in an injection mould between the mould cavity and the passage carrying the plastic moulding material.

gate array *(Electronics)*. An integrated logic circuit consisting of a two-dimensional array of logic cells, each consisting of one or more **gates**. The final details of metallization and interconnection determine the performance of the chip, according to the required application.

gate by-pass switch *(Elec.Eng.)*. A switch fitted in an electric lift-car so that the operator can render the gate or door locks inoperative in cases of emergency.

gate-chambers *(Hyd.Eng.)*. Recesses in the side walls of a lock to accommodate the gates when open.

gate-change gear *(Autos.)*. A multi-speed gearbox in which the control lever moves sideways as well as backwards and forwards in a *gate*. This may be in the form of a simple H, or a more complex pattern depending on the number of ratios. Moving the lever into each arm of the gate selects one ratio.

gate closer *(Elec.Eng.)*. A device for automatically closing the gates of a lift-car. Also called *door closer*.

gate current *(Elec.Eng.)*. (1) Current flowing in the gate-cathode circuit of a **thyristor**. (2) Current flowing in the gate-source circuit of a field effect transistor (normally very small). (3) a.c. or d.c. pulse which saturates the core of a reactor. Also termed *gate-drive* in (1) and (2).

gated-beam tube *(Electronics)*. A valve in which a flat electron beam passes through a slotted plate; beam deflection causes very rapid cut-off of current. Used in some switching and FM detector circuits.

gate detector *(Electronics)*. One whose operation is controlled by an external gating signal.

gated throttle *(Aero.)*. A supercharged aero-engine throttle quadrant with restricting stop(s) to prevent the throttle being wrongly used. See **boost control**.

gate-end box *(Elec.Eng.)*. A cable box used in mining work for making a joint between one of the main supply cables and a trailing cable for use near the coal face.

gate interlock *(Elec.Eng.)*. A combination of a gate or door switch with an automatic gate or door lock.

gate leakage current *(Electronics)*. The d.c. gate current which flows in an FET under normal operating conditions.

gate operator *(Elec.Eng.)*. A power-operated device for closing the gates of a lift-car.

gate stick *(Foundry)*. A stick placed vertically in the cope while it is rammed up; on removal it provides the gate or runner passage into the mould. Also called *runner stick*.

gate switch *(Elec.Eng.)*. See **door switch**.

gate valve *(Eng.)*. One which can be moved across the line of flow in a pipe line.

gate voltage *(Electronics)*. The control voltage for an electronic gate. The voltage applied to the 'gate' electrode of a field-effect transistor.

gateway *(Comp.)*. Device connecting one communicating **network** with another.

gateway *(Min.Ext.)*. A road through the worked-out area (goaf) for haulage in longwall working of coal. Road connecting coal working with main haulage. Also *gate road*.

gate winding *(Elec.Eng.)*. That used to obtain gating action in a magnetic amplifier.

gather *(Glass)*. A charge of glass picked up by the gatherer.

gatherer *(Glass)*. A person who gathers a charge of glass on a blowpipe or gathering iron for the purpose of forming it into ware or feeding a charge to a machine for that purpose.

gathering *(Build.)*. The contracting portion of the chimney passage to the flue, situated a short distance above the source of heat.

gathering *(Print.)*. Collecting and arranging in proper sequence the folded sections forming a volume.

gathering line *(Min.Ext.)*. Small bore pipes which collect oil or gas from peripheral wells and take them to central distributing station.

gathering motor *(Min.Ext.)*. Light electric loco used to move loaded coal trucks from filling points to main haulage system.

gathering pallet *(Horol.)*. In the striking mechanism of a clock or repeater watch, a revolving finger which lifts the rack one tooth for each blow struck; a single-toothed wheel.

gating *(Phys.)*. Selection of part of a wave on account of time or magnitude. Operation of a circuit when one wave allows another to pass during specific intervals.

gating *(Textiles)*. See **gaiting**.

Gattermann reactions *(Chem.)*. Reactions used in organic syntheses for preparing aromatic aldehydes from hydocarbons. Based on passing a mixture of carbon monoxide and hydrogen chloride into the hydrocarbon in the presence of cuprous chloride and aluminium chloride.

gauche *(Chem.)*. A form of **staggered conformation** in which the substituents being considered make a dihedral angle of $\pm 60°$ to one another. The most stable conformation of hydrogen peroxide has the hydrogen atoms in this relationship.

Gaucher's disease *(Med.)*. A rare familial disorder of lipid metabolism resulting in the accumulation of abnormal **gluco cerebrosides** in the *reticuloendothelial system*. The condition usually presents in childhood with anaemia and enlargement of liver and spleen.

gauge *(Build.)*. Device for marking lines parallel with an edge. See **marking and cutting gauge**.

gauge *(Civ.Eng.)*. Distance between the inside edges of the rails of a permanent way.

gauge *(Eng.)*. (1) An object or instrument for the measurement of dimensions, pressure, volume, etc. See **pressure gauge, water gauge**. (2) An accurately dimensioned piece of metal for checking the dimensions of work or less precisely made gauges. See **limit gauge, master gauge, plug gauge, ring gauge**. (3) A tool used for measuring lengths, as a **micrometer gauge**. (4) The diameter of wires and rods. See **Birmingham Wire Gauge, Brown & Sharpe Wire Gauge**.

gauge *(Image Tech.)*. The width of motion picture film or magnetic tape; for film standard sizes are 70 mm, 35 mm, 16 mm and 8 mm while for tape they are 2″, 1″, $\frac{3}{4}$″, $\frac{1}{2}$″, $\frac{1}{4}$″ and 8 mm.

gauge *(Textiles)*. Relates to the fineness of a knitted fabric and generally indicates the number of needles per cm. in warp or weft knitting machines.

gauge bosons *(Phys.)*. A type of particle that mediates the interaction between two fundamental particles. There are four types: *photons* for electromagnetic interactions, *gluons* for strong interactions, *intermediate vector bosons* for weak interactions and *gravitons* for gravitational interactions.

gauge box *(Civ.Eng.)*. A box which measures a known quantity of material, such as cement, sand, or coarse aggregate or similar substance, for testing or making mixtures. Also called *batch box*.

gauge cocks *(Eng.)*. Small cocks fitted to pressure vessels to which pressure gauges are attached or which carry liquid level gauge glasses.

gauge-concussion *(Civ.Eng.)*. The lateral outward impact of the wheel flanges against the rails due to centrifugal force.

gauged arch *(Build.)*. An arch built from special bricks cut with a bricklayer's saw and rubbed to exact shape on a stone.

gauged mortar *(Build.)*. Mortar made of cement, lime, and sand, to proportions suitable for the bricks, blocks, or other material used.

gauge door *(Min.Ext.)*. A door underground for controlling the supply of air to part of the mine.

gauged stuff *(Build.)*. A stiff plaster used for cornices, mouldings, etc.; made with lime putty to which plaster of Paris is added to hasten setting. Also called *putty and plaster*.

gauge glass *(Eng.)*. The tube fitted vertically between a pair of gauge fittings and used to indicate the liquid level in a tank or boiler. Such glasses are usually protected by flat glass panels or perforated metal.

gauge number *(Eng.)*. An (arbitrary) number denoting the gauge or thickness of sheet metal or the diameter of wire, rod, or twist drills in one of many gauge number systems, e.g. British Standard Wire Gauge (SWG), Birmingham Wire Gauge (BWG), Brown & Sharpe Wire Gauge in US. See also **letter sizes**.

gauge pot *(Build.)*. A small receptacle for cement grout, facilitating pouring of small quantities.

gauge pressure *(Eng.)*. Of a fluid, the pressure as shown by a pressure gauge, i.e. the amount by which the pressure exceeds the atmospheric pressure, the sum of the two giving the absolute pressure.

gauge rod *(Build.)*. A rod used in laying graduated courses of slates.

gauge theory *(Phys.)*. Theories in particle physics which attempt to describe the various types of interaction between fundamental particles. *Quantum electrodynamics* describes relativistic quantum fields. The *Weinberg and Salam* theory unifies the weak and electromagnetic interactions. *Quantum chromodynamics* is designed to explain the binding together of *quarks* to form *hadrons*. Grand unified theories are gauge theories which set out to unify the three interactions, electromagnetic, weak and strong.

gauging-board *(Build.)*. A platform on which mortar or concrete may be mixed.

gaul *(Build.)*. A hollow spot in the **setting**.

Gause's principle *(Ecol.)*. The idea that closely related organisms do not co-exist in the same niche, except briefly. See also **competitive exclusion principle**.

gauss *(Phys.)*. CGS electromagnetic unit of magnetic flux density; equal to 1 maxwell cm^{-2}, each unit magnetic pole terminating 4π lines. Now replaced by the SI unit of magnetic flux density, the tesla (T). $1T = 10^4$ gauss.

Gauss' convergence test *(Maths.)*. If, for a series of positive terms Σa_n

$$n\left(\frac{a_n}{a_{n+1}}-1\right)=\sigma+\left(\frac{1}{n^\delta}\right),\quad \delta>0,$$

then Σa_n is convergent if $\sigma > 1$ and divergent if $\sigma \leqq 1$. This test is an extension of Raabe's test.

Gauss' differential equation *(Maths.)*. The equation

$$x(1-x)\frac{d^2y}{dx^2}+\{c-(a+b+1)x\}\frac{dy}{dx}-aby=0.$$

It is satisfied by the hypergeometric function $F(a;b;c;x)$. Also called *hypergeometric equation*.

Gauss eyepiece *(Phys.)*. A form of eyepiece used in optical instruments, such as spectrometers and refractometers, to facilitate setting the axis of the telescope at right angles to a plane-reflecting surface. Light enters the side of the eyepiece and is reflected down the telescope tube by a piece of unsilvered glass, being then reflected back into the eyepiece by the plane surface.

Gaussian curvature *(Maths.)*. See **curvature (3)**.

Gaussian distribution *(Stats.)*. See **normal distribution**.

Gaussian noise *(Acous.)*. See **white noise**.

Gaussian optics *(Phys.)*. Refers to simple optical theory which does not consider the aberration of lenses. Practically, it applies only to paraxial rays.

Gaussian points *(Phys.)*. Same as **principal points of a lens**.

Gaussian response *(Phys.)*. Response, e.g., of an amplifier, for a transient impulse, which, when differentiated, matches the Gaussian distribution curve.

Gaussian units *(Phys.)*. Formerly widely used system of electric units where quantities associated with electric field are measured in e.s.u. and those associated with magnetic field in e.m.u. This involves introducing a constant *c* (the free space velocity of electromagnetic waves) into Maxwell's field equations.

Gaussian well *(Phys.)*. A particular form of *potential well* used to describe the distribution of potential energy of a nuclear particle in the field of the nucleus or other nuclear particle.

Gauss' laws of electrostatics (and magnetostatics) *(Elec.Eng.)*. The surface integral of the normal component of electric displacement (or magnetic flux) over any closed surface in a dielectric is equal to the total electric charge enclosed (or to zero in the magnetic case). Differential forms of these laws comprise two of **Maxwell's field equations**; see also **Poisson's equation**.

gaussmeter *(Elec.Eng.)*. Instrument measuring magnetic flux density. This term is most widely used in US.

Gauss' theorem *(Maths.)*. $\int_S \frac{\partial\phi}{\partial n}\,dS = \int_V \nabla^2\phi\,dV$, where the surface S is the boundary of the volume V and n is the normal direction to S. Cf. **Green's theorem**.

gauze *(Textiles)*. A light weight woven fabric of open texture.

gavage *(Vet.)*. Forced feeding of birds being fattened for meat production.

gavel *(Build.)*. A mallet used for setting stones.

gavelock *(Build.)*. An iron crowbar. Also *gablock*.

Gay-Lussac's law *(Chem.)*. *Charles' law*. Of volumes: when gases react, they do so in volumes which bear a simple ratio to one another and to the volumes of the resulting substances in the gaseous state, all volumes being measured at the same temperature and pressure.

gay-lussite *(Min.)*. A rare grey hydrated carbonate of sodium and calcium, occurring in lacustrine deposits.

gazebo *(Arch.)*. A summerhouse resembling a temple in form and commanding a wide view.

G-banding *(Biol.)*. See **banding techniques**.

GCA *(Aero.,Radar)*. Abbrev. for *Ground Controlled Approach*.

GCI *(Aero.,Radar)*. Abbrev. for *Ground Controlled Interception*.

G cramp *(Build.)*. One in the shape of a G, with a screw passing through one end. The shoe is sometimes swivelled to enable the cramp to be used on tapered surfaces.

Gd *(Chem.)*. The symbol for gadolinium.

G-display *(Radar)*. Similar to **F-display** but indicating increasing or diminishing range of target by increasing or diminishing lateral extension of the spot.

Ge *(Chem.)*. The symbol for germanium.

geanticline *(Geol.)*. A regional upwarping of the crust of the earth. Cf. **geosyncline**.

gear *(Eng.)*. (1) Any system of moving parts transmitting motion, e.g. levers, gear-wheels etc. (2) A set of tools for performing some particular work. (3) A mechanism built to perform some special purpose, e.g. steering gear, valve gear. (4) The position of the links of a steam-engine valve motion, as astern gear, mid-gear etc. (5) The actual gear ratio in use, or the gear-wheels involved in transmitting that ratio, in an automobile gearbox, as *first gear*, *third gear* etc.

gearbox *(Eng.)*. Casing containing a **gear train**. The term commonly stands also for the casing including its gear train, particularly when applied to gearboxes used with engines or with machine tools.

gear cluster *(Eng.)*. A set of gear-wheels integral with, or permanently attached to, a shaft, as on the lay shaft of an automobile gearbox.

gear cutters *(Eng.)*. Milling cutters, hobs etc. having the requisite tooth form for cutting teeth on gear-wheels.

geared lathe *(Eng.)*. A lathe provided with a *back gear* or a multi-speed gearbox between the driving motor and the head.

geared locomotive *(Elec.Eng.)*. An electric locomotive in which the motors drive the axles through reduction gears.

geared turbo-generator *(Elec.Eng.)*. An electric generator driven through a reduction gear from a steam-turbine, the object being to enable both machines to operate at their most economical speeds.

gearing *(Eng.)*. Any set of gear-wheels transmitting motion. See **gear**.

gearing-down *(Eng.)*. A reduction in speed between a driving and a driven wheel or unit e.g. between the engine of an automobile and the road wheels.

gearing-up *(Eng.)*. Raising the speed of a driven unit above that of its driver by the use of gears; opposite of *gearing-down*.

gearless locomotive *(Elec.Eng.)*. An electric locomotive with a **gearless motor**.

gearless motor *(Elec.Eng.)*. A traction motor mounted directly on the driving axle of an electric locomotive.

gear lever *(Eng.)*. A lever used to move gearwheels relative to each other to change gear. In a motor car, this lever acts on the gear-wheels indirectly through **selector forks**.

gear marks *(Print.)*. Slurred streaks or bands across the printed sheet or web caused by uneven rotation of cylinder.

gear pump *(Eng.)*. A small pump consisting of a pair of gear-wheels in mesh, enclosed in a casing, the fluid being carried round from the suction to the delivery side in the tooth spaces; used for lubrication systems, etc.

gear-tooth forming *(Eng.)*. A family of engineering processes, including casting, plastic moulding, stamping from sheet metal for watch and clock gears, form-cutting, gear shaping, hobbing and other methods of gear-tooth generating.

gear train *(Eng.)*. Two or more **gear-wheels**, transmitting motion from one shaft to another. With external spur or bevel gears, the velocity ratio is inversely proportional to the number of gear teeth.

gear-wheel *(Eng.)*. A toothed wheel used in conjunction with another, or with a rack, to transmit motion.

gedrite *(Min.)*. An orthorhombic amphibole, containing more aluminium and less silicon than anthophyllite. The iron-aluminium end-member has been called *ferrogedrite*. Gedrite occurrences are restricted to metamorphic and metasomatic rocks.

gegenions *(Chem.)*. The simple ions, of opposite sign to the colloidal ions, produced by the dissociation of a colloidal electrolyte. Also called *counterions*.

gegenschein *(Astron.)*. (*Ger.* counter-glow.). A term applied to a faint illumination of the sky sometimes seen in the ecliptic, diametrically opposite the sun, and connected with the zodiacal light.

gehlenite *(Min.)*. The calcium magnesium end member of the melilite group of minerals.

Geiger characteristic *(Phys.)*. Plot of recorded count rate against operating potential for a Geiger or proportional counter detecting a beam of radiation of constant intensity.

Geiger-Müller counter *(Nuc.Eng.)*. One for ionizing radiations, with a tube carrying a high voltage wire in an atmosphere containing argon plus halogen or organic vapour at low pressure, and an electronic circuit which quenches the discharge and passes on an impulse to record the event. Also *Geiger counter*, *G-M counter*. See **Townsend avalanche**.

Geiger-Müller tube *(Nuc.Eng.)*. The detector of a G-M counter, i.e. without associated electronic circuits.

Geiger-Nuttall relationship *(Phys.)*. An empirical rule relating the half-life T of radioactive materials emitting α-particles to the range R of the particles emitted, i.e. $\log (1/T) = a \log R + b$ where a and b are constants.

Geiger region *(Nuc.Eng.)*. That part of the characteristic of a counting tube, where the charge becomes independent of the nature of the ray intercepted. Also called *Geiger plateau*, since in this region the efficiency of counting varies only slowly with voltage on the tube.

Geiger threshold *(Nuc.Eng.)*. Lowest applied potential for which Geiger tube will operate in Geiger region.

Geissler pump *(Chem.)*. A glass vacuum pump which operates from the water supply.

geitonogamy *(Bot.)*. Fertilization involving pollen and ovules from different flowers on the same individual plant (*ramet*) or from the same clone (*genet*); a form of *allogamy* (1). See **self-pollination**, **self-fertilization**, **inbreeding**.

gel *(Biol.)*. The inert matrix in which polynucleotides and polypeptides can be separated by electrophoresis.

gel *(Chem.)*. The apparently solid, often jellylike, material formed from a colloidal solution on standing. A gel offers little resistance to liquid diffusion and may contain as little as 0.5% of solid matter. Some gels, e.g. gelatin, may contain as much as 90% water, yet in their properties are more like solids than liquids.

gel *(Image Tech.)*. A coloured gelatine filter for light sources.

gelatine *(Build.)*. Refined animal tissues of skin, tendon, bones and horn which form a strong jelly when mixed with boiling water and allowed to cool. Used as a mordant in glass gilding.

gelatin(e) *(Chem.)*. A colourless, odourless and tasteless glue, prepared from albuminous substances, e.g. bones and hides. Used for foodstuffs, photographic films, glues etc.

gelatin filter *(Image Tech.)*. See **filter**.

gelation *(Plastics)*. The process whereby plasticized PVC compounds by the application of heat undergo an irreversible change to soft, rubbery thermoplastic materials. The resultant soft solids are referred to as gels.

gel diffusion tests *(Immun.)*. Precipitin tests in which antigen and antibody are placed separately in a gel medium (commonly agar) and allowed to diffuse towards one another and to form lines of precipitate where they meet in suitable proportions.

gelignite *(Chem.,Min.Ext.)*. Explosive used for blasting, composed of a mixture of nitroglycerine (60%), guncotton (5%), woodpulp (10%) and potassium nitrate (25%).

gemel window *(Arch.)*. A 2-bay window.

gem gravels *(Geol.)*. Sediments of the gravel grade containing appreciable amounts of gem minerals, and formed by the disintegration and transportation of pre-existing rocks, in which the gem minerals originated. They are really placers of a special type, in which the heavy minerals are not gold or tin, but such minerals as garnets, rubies, sapphires, etc. As most of the gem minerals are heavy and chemically stable, they remain near the point of origin, while the lighter constituents of the parent rocks are washed away, a natural concentration of the valuable components resulting.

gemma *(Bot.)*. (1) A multicellular structure for vegetative reproduction in algae, pteridophytes and, especially, bryophytes. (2) Same as **chlamydospore**.

gemma *(Zool.)*. A bud that will give rise to a new individual. pl. *gemmae*.

gemma cup *(Bot.)*. Cup-like structure in which gemmae are borne in some mosses and liverworts.

gemmation *(Bot.,Zool.)*. Budding; gemma-formation.

gemmiferous, gemmiparous *(Zool.)*. Producing gemmae.

gemmule *(Zool.)*. In fresh water Porifera, an aggregation of embryonic cells within a resistant case, which is formed at the onset of hard conditions when the rest of the colony dies down, and which gives rise to a new colony when conditions have once more become favourable.

gender identity *(Behav.)*. The individual's sense of being a man or a woman.

gender role *(Behav.)*. The set of attitudes and behaviour a given culture considers appropriate for each sex.

gene *(Biol.)*. The hereditary determinant of a specified difference between individuals. Can refer either to a particular *allele* or to all alleles at a particular *locus*. Molecular analysis has shown that a specific sequence or parts of a sequence can be identified with the classical gene. See also **cistron**.

genealogy *(Bot.,Zool.)*. The study of the development of plants and animals from earlier forms.

gene bank *(Bot.)*. A collection of plants or, more often of seeds, cell cultures, frozen pollen etc. of species of known or potential use to man, and especially of landraces that may contain genes of use in crop breeding.

genecology *(Ecol.)*. The branch of ecology which seeks

genetic explanations of the patterns of distribution of plants and animals in time and space.

gene dosage *(Biol.)*. Number of copies of a gene in a given cell or individual organism.

gene-for-gene concept *(Bot.)*. The concept that there are corresponding genes for virulence and resistance in pathogen and host respectively.

gene frequency *(Biol.)*. The proportion of all representatives of a particular gene in a population that contain the specified allele.

gene mapping *(Biol.)*. See **mapping**.

gene number *(Biol.)*. The total number of different **coding sequences** which are transcribed into RNA in an individual or species.

general aviation *(Aero.)*. Private, agricultural and survey aviation.

general integral *(Maths.)*. See **complete primitive**.

generalization *(Behav.)*. In learning, when the learned behaviour occurs in situations similar to, but not identical with, that in which learning first occurred.

generalized anxiety disorder *(Behav.)*. A chronic state of diffuse, unfocused anxiety, in the absence of specific symptoms such as are found in phobic reactions, and without any associated specific stimuli or objects.

generalist *(Ecol.)*. Organism with many food sources, or able to live in many habitats. See also **specialist**.

general lighting *(Elec.Eng.)*. A system of lighting employing fittings which emit the light in approximately equal amounts in an upward and a downward direction.

general paresis *(Behav.,Med.)*. A psychosis of organic origin (produced by syphilitic infection), in which there is a progressive deterioration of cognitive or motor functions, culminating in death.

general purpose foils *(Print.)*. Blocking foils suitable for marking paper, bookcloths, etc. but not normally used on thermoplastics.

general sexual dysfunction *(Behav.)*. In women, the absence or weakness of the physiological changes normally accompanying the excitement phase of sexual response.

general solution *(Maths.)*. See **complete primitive**.

general theory of relativity *(Phys.)*. Einstein's extension of the special theory of relativity to deal with accelerating frames of reference. Based on the *principle of equivalence*, it shows that the laws of physics must be in such a form that it is impossible to distinguish between a uniform gravitational field and an accelerated frame of reference.

generating circle *(Eng.)*. Any circle in which a point on the circumference is used to trace out a curve when the circle rolls along a straight line or curve.

generating function *(Maths.)*. A function which can be regarded as summarizing a sequence of functions which become apparent when the generating function is expanded as a series; e.g. (1) The function $(q+pt)_n$ generates the binomial probability distribution when expanded in powers of t. The coefficient of t_x in the expansion of $(q+pt)_n$ is the probability $\binom{n}{x}p^x q^{n-x}$ of x successes out of n trials. (2) The function $(1-2xh+h^2)^{-1/2}$, when expanded in powers of h, generates the Legendre polynomials $P_n(x)$.

generating line *(Eng.)*. A straight or curved line rotated about some axis to generate a surface.

generating set *(Elec.Eng.)*. An electric generator, together with the prime mover which drives it.

generating station *(Elec.Eng.)*. A building containing the necessary equipment for generating electrical energy.

generation *(Biol.)*. The individuals of a species which are separated from a common ancestor by the same number of broods in the direct line of descent.

generation rate *(Electronics)*. Rate of production of *electron-hole pairs* in semiconductors.

generation time *(Phys.)*. Average life of fission neutron before absorption by fissile nucleus.

generative cell *(Bot.)*. A cell of the male gametophyte in pollen grain or pollen tube. In gymnosperms it divides to give the sterile (stalk) cell and the spermatogenous (body) cell. In angiosperms it divides to give the two male gametes. See **double fertilization**.

generator *(Elec.Eng.)*. Electrostatic or electromagnetic device for conversion of mechanical into electrical energy. Also known as *electric generator*.

generator *(Maths.)*. See **cone (1)**, **cylinder (1)**, **ruled surface**.

generator bus-bars *(Elec.Eng.)*. Bus-bars in a generating station to which all the generators can be connected.

generator-field control *(Elec.Eng.)*. See **variable voltage control**.

generator panel *(Elec.Eng.)*. A panel of a switchboard upon which are mounted all the switches, instruments, and other apparatus necessary for controlling a generator.

generator potential *(Zool.)*. An electrical potential arising in a sensory neuron as a result of a sensory stimulus.

genesis *(Biol.)*. The origin, formation, or development of some biological entity. adj. *genetic*.

genet *(Bot.)*. A genetic individual, the product of one zygote; either an individual plant grown from a sexually-produced seed or all the individuals produced by vegetative reproduction from one such plant. Cf. **ramet**.

gene therapy *(Biol.)*. Colloquial term for the substitution of a functional for a defective gene as a treatment for a *genetic disease*.

genetically significant dose *(Radiol.)*. The dose that, if given to every member of a population prior to conception of children, would produce the same genetic or hereditary harm as the actual doses received by the various individuals.

genetic code *(Biol.)*. The rules which relate the four bases of the DNA or RNA with the 20 amino acids found in proteins. There are 64 possible different 3-base sequences (triplets) using all permutations of the four bases. One triplet uniquely specifies one amino acid, but each amino acid can be coded by up to 6 different triplet sequences. The code is therefore *degenerate*. Triplets also act as **stop codons**. The evidence suggests that the code is universal applying from the simplest to the most evolutionarily advanced organism, although minor variations have been found.

genetic correlation *(Biol.)*. Measure of the extent to which the variation of two different quantitative characters is caused by the same genes.

genetic drift *(Biol.)*. The process by which *gene frequencies* are changed by the chances of random sampling in small populations.

genetic engineering *(Biol.)*. Colloq. for **genetic manipulation**.

genetic equilibrium *(Biol.)*. The situation reached when the frequencies of genes and of genotypes in a population remain constant from generation to generation.

genetic manipulation *(Biol.)*. Term for the procedures with which it is now possible to combine DNA sequences from widely different organisms *in vitro*, often with great precision.

genetics *(Biol.)*. The study of heredity; of how differences between individuals are transmitted from one generation to the next; and of how the information in the genes is used in the development and functioning of the adult organism.

genetic spiral *(Bot.)*. In phyllotaxis, a hypothetical line though the centres of successive leaf primordia at the shoot apex.

genetic variance *(Biol.)*. Measure of the variation between individuals of a population due to differences between their *genotypes*.

Geneva movement *(Eng.)*. Intermittent movement from a continuous motion by a wheel with slots.

Geneva stop work *(Horol.)*. See **stop work**.

-gen, -gene *(Genrl.)*. Suffix meaning *generating, producing*.

gen-, geno- *(Genrl.)*. Prefix from Gk. *genos*, race, descent.

genial *(Zool.)*. Pertaining to the chin.

genicular *(Zool.)*. Pertaining to, or situated in, the region of the knee.

geniculate *(Bot.,Zool.)*. Bent rather suddenly, like the leg at the knee; as *geniculate antennae*.

geniculate ganglion *(Zool.)*. In Craniata, the ganglion from which the seventh cranial nerve arises.

geniohyoglossus *(Zool.)*. The muscle which moves the tongue in Vertebrates.

genioplasty *(Med.)*. Plastic surgery of the chin.

genital atrium *(Zool.)*. In Platyhelminthes and some Mollusca, a cavity into which open the male and female genital ducts.

genitalia, genitals *(Zool.)*. The gonads and their ducts and all associated accessory organs.

genital stage *(Behav.)*. In psychoanalytic theory, the final phase of psychosexual development in which sexual pleasure involves not only gratification of one's own impulses but attention to and pleasure in, the social and physical pleasure of one's mate. Freud was referring to heterosexual patterns, and to sexual pleasure derived mainly through the genitalia.

genito-urinary *(Med.)*. Pertaining to the genital and the urinary organs.

genlock *(Image Tech.)*. Device for synchronizing locally generated pulse signals with those transmitted from a distant source.

Gennari's band, fibres, line *(Med.)*. A layer of nerve fibres in the cerebral cortex.

genome *(Biol.)*. The totality of the DNA sequences of an organism or organelle.

genomic DNA *(Biol.)*. Nuclear chromosomal DNA.

genomic library *(Biol.)*. **DNA library** derived from a whole, single **genome**.

genotype *(Biol.)*. The particular alleles at specified loci present in an individual; the genetic constitution. adj. *genotypic*. Cf. **phenotype**.

gentian violet *(Chem.)*. Mixture of the three dyes, methyl rosaniline, methyl violet and crystal violet, which is antiseptic and bactericidal. Used as a fungicide and anthelmintic.

gentiobiose *(Chem.)*. $C_{12}H_{22}O_{11}$. A disaccharide based on glucose. A reducing sugar which occurs in combination in **amygdalin**. Contains G.β1→6.G (G = glucose unit).

genu *(Zool.)*. A kneelike structure, i.e. a bend in a nerve tract; more particularly, part of the corpus callosum in Mammals.

genu recurvatum *(Med.)*. The condition in which there is hyperextension of the knee-joint.

genus *(Biol.)*. A taxonomic rank of closely related forms, which is further subdivided into species and therefore below **family** and above **species**. pl. *genera*. adj. *generic*.

genu valgum *(Med.)*. *Knock-knee*. The angle between the femur and the tibia is so altered that the leg deviates laterally from the midline.

genu varum *(Med.)*. *Bow leg*. The reverse of genu valgum, the altered angle between the femur and the tibia being such that the legs bow outwards at the knee.

genys *(Zool.)*. In Vertebrates, the lower jaw.

geobiotic *(Zool.)*. Terrestrial; living on dry land.

geobotanical indicator *(Min.Ext.)*. See **geobotanical surveying**.

geobotanical surveying *(Min.Ext.)*. Form of **geochemical prospecting**. (1) Identification and systematic surveying of distibution of metallophile plant species, e.g. *calamine violet* associated with zinc anomalies in central Europe (*geobotanical indicators*). (2) Identification and systematic surveying of pathological conditions in plants caused by metal toxaemia. (3) Systematic sampling of vegetation to identify anomalous concentrations of metals in plant tissues.

geocarpy *(Bot.)*. The ripening of fruits underground, the young fruits being pushed into the soil by a post-fertilization curvature of the stalk.

geocentric *(Astron.)*. The term applied to any system or mathematical construction which has as its point of reference the centre of the earth.

geocentric altitude *(Surv.)*. The **true altitude** of a heavenly body as corrected for **geocentric parallax**.

geocentric latitude *(Astron.)*. See under **latitude and longitude** (*terrestrial*).

geocentric parallax *(Astron.)*. The apparent change of position of a heavenly body due to a shift of the observer by the rotation of the earth; hence only observed in bodies (e.g. the moon and sun) sufficiently close for the earth's radius to subtend a measurable angle when seen from the body. Also *diurnal parallax*.

geocentric parallax *(Surv.)*. The correction which must be applied to the altitude of a heavenly body in the solar system as observed, in order to give the altitude corrected to the earth's centre. Its value is given by $p = +P.\cos\alpha$, where p = geocentric parallax, P = **horizontal parallax**, and α = observed altitude corrected for refraction.

geochemical prospecting *(Min.Ext.)*. Application of **geochemistry** to mineral exploration by the systematic analysis of bedrock, soil, stream, river and ground water, stream gravels and vegetation for the purposes of identifying anomalous concentrations of particular elements of economic interest, or elements commonly associated with such ore bodies. See **geobotanical surveying, soil sampling, stream sampling, tracers**.

geochemistry *(Chem.)*. The study of the chemical composition of the earth's crust.

geochronology *(Geol.)*. Study of time with respect to the history of the earth, primarily through the use of *absolute* and *relative age-dating* methods.

geocline *(Ecol.)*. A **cline** occurring across topographic or spatial features of an organism's range.

geode *(Geol.,Min.Ext.)*. Hollow, rounded rock, mineral nodule or concretion, often lined with crystals which have grown inwards. Also *drusy*.

geodesic *(Maths.)*. The shortest path between two points on any surface.

geodesic structures *(Civ.Eng.)*. Structures consisting of a large number of few identical parts and therefore simple to erect; and whose pressure is loadshed throughout the structure, so that the larger it is, the greater its strength.

geodesy *(Surv.)*. Branch of surveying concerned with extensive areas, in which, to obtain accuracy, allowance must be made for the curvature of the earth's surface. Also *geodetic surveying*.

geodetic construction *(Aero.)*. A redundant space frame whose members follow diagonal geodesic curves to form a lattice structure, such that compression loads induced in any member are braced by tension loads in crossing members. Does not need stress-carrying covering.

geodetic surveying *(Surv.)*. See **geodesy**.

geognosy *(Geol.)*. An old term for absolute knowledge of the earth, as distinct from geology, which includes various theoretical aspects.

geographical latitude *(Astron.)*. See under **latitude and longitude** (*terrestrial*).

geographical mile *(Genrl.)*. The length of one minute of latitude, a distance varying with the latitude, and having a mean value of 6076.8 ft. (1852.2 m). US, one minute of longitude at the equator, i.e. 6087.1 ft. (1855.3 m).

geographical race *(Zool.)*. A collection of individuals within a species, which differ constantly in some slight respects from the normal characters of the species, but not sufficiently to cause them to be classified as a separate species, and which are peculiar to a particular area.

geoid *(Surv.)*. A gravitational equipotential surface at approximately mean sea level used as the datum level for gravity surveying.

geo-isotherms *(Geol.)*. Lines or surfaces of equal temperature within the Earth.

geological column *(Geol.)*. A diagram that shows the subdivisions of part or all of geological time or the stratigraphical sequence in a particular area.

geological time *(Geol.)*. The time extending from the end of the Formative Period of earth history to the beginning of the Historical Period. It is conveniently divided into Periods, each being the time of formation of one of the Systems into which the Stratigraphical Column is divided.

geology *(Geol.)*. The study of the planet Earth. It embraces mineralogy, petrology, geophysics, geochemistry, physical geology, palaeontology and stratigraphy. It increasingly involves the use of the chemical, physical, mathematical and biological sciences. See **earth science**.

geomagnetic effect *(Phys.)*. The effect of the earth's

magnetic field on cosmic rays by which positively charged particles are deflected towards the east.

geometrical attenuation *(Phys.)*. Reduction in intensity of radiation on account of the distribution of energy in space, e.g. due to inverse-square law, or progression area along the axis of a horn.

geometrical cross-section *(Phys.)*. Area subtended by a particle or nucleus. This does not usually resemble the interaction cross-section.

geometrical isomerism *(Chem.)*. A form of stereoisomerism in which the difference arises because of hindered rotation about a double bond or a bond that is part of a ring. Thus but-2-ene has two isomers, depending on whether the methyl groups are on the same (cis) or opposite (trans) sides of the double bond. In the **Cahn-Ingold-Prelog system**, these are called Z- and E- respectively.

geometrical optics *(Phys.)*. The study of optical problems based on the conception of light rays. See **physical optics**.

geometrical stair *(Build.)*. A stair arranged about a well-hole and curved between the successive flights.

geometric capacitance *(Phys.)*. That of an isolated conductor *in vacuo*, uninfluenced by dielectric material, and depending only on shape.

geometric distortion *(Image Tech.)*. Any departure in an image from the representation of the correct form and perspective of the original subject. See **barrel distortion**, **keystone distortion**, **pincushion distortion**.

geometric mean *(Maths.)*. Of n positive numbers a_r: the nth root of their product, i.e.

$$\left(\prod a_r\right)^{1/n}.$$

geometry *(Maths.)*. A branch of mathematics concerned generally with the properties of lines, curves and surfaces. Usually divided into *pure*, *algebraic* and *differential geometry* in accordance with the mathematical techniques utilized.

geometry factor *(Phys.)*. $1/(4\pi)$ of the solid angle subtended by the window or sensitive volume of a radiation detector at the source.

geophagous *(Zool.)*. Earth-eating.

geophilous *(Zool.)*. Living on, or in, the soil.

geophones *(Min.Ext.)*. Array of sound detectors used to collect information in seismic surveys from planned explosions.

geophysical prospecting *(Min.Ext.)*. Prospecting by measuring, directly or indirectly, differences in the density, electrical resistance, magnetic properties, elastic properties and radioactivity of the earth's crust.

geophysics *(Genrl.)*. Study of physical properties of the earth; it makes use of the data available in geodesy, seismology, meteorology and oceanography, as well as that relating to atmospheric electricity, terrestrial magnetism and tidal phenomena.

geophyte *(Bot.)*. Herbs with perennating buds below the soil surface. It includes plants with bulbs, corms, rhizomes etc. See **Raunkiaer system**.

geopotential height *(Meteor.)*. The height of a point in the atmosphere expressed in units (geopotential metres) proportional to the geopotential at that height. One geopotential metre is numerically equal to (9.8/g) of a geometric metre, where g is the local acceleration of gravity. All reported measured heights of pressure levels are given in geopotential metres thus obviating the necessity for considering variations of gravity when making dynamical calculations.

George *(Aero.)*. Colloq. for *automatic pilot*.

Georgian glass *(Glass)*. A reinforced fire-resisting glass.

geostationary *(Space)*. Of an orbit lying above the equator, in which an artificial satellite moves at the same speed as the Earth rotates, thus maintaining position above the Earth's surface.

geostrophic approximation *(Meteor.)*. Use of the *geostrophic wind* as an approximation to the actual wind, either in operational forecasting or as a replacement for certain terms in the equations of motions.

geostrophic force *(Meteor.)*. A virtual force used to account for the change in direction of the wind relative to the Earth's surface, arising from the Earth's rotation and the **Coriolis force**.

geostrophic wind *(Meteor.)*. The theoretical wind arising from the **pressure gradient** force and the **geostrophic force**.

geosynchronous *(Space)*. Same as **geostationary**.

geosyncline *(Geol.)*. A major elongated downwarp of the Earth's crust, usually hundreds of kilometres long and filled with sediments and lavas many kilometres in thickness. The rocks are generally deformed and metamorphosed later.

geotaxis *(Biol.)*. The locomotory response of a motile organism or cell to the stimulus of gravity. adj. *geotactic*. Cf. **geotropism**.

geotechnical process *(Civ.Eng.)*. Means employed to alter the properties of soil for construction purposes. These include compaction and vibration, soil stabilization by the addition of cement, bitumen emulsions etc., the latter sometimes carried out by injection.

geotextile *(Textiles)*. A textile material used in civil engineering. Strong fabrics made of synthetic fibres are frequently used in road-making and for the stabilization of embankments.

geothermal gradient *(Geol.)*. The rate at which the temperature of the earth's crust increases with depth.

geothermal power *(Geol.)*. Power generated by using the heat energy of crustal rocks. Active volcanic areas have traditionally been a source, but more recently deep boreholes in areas with a high geothermal gradient have shown economic potential.

geotropism *(Biol.)*. The reaction of a plant member or sessile animal to the stimulus of gravity, shown by a growth curvature; cells become more elongated on one side than the other, tending to bring the axis of the affected part into a particular relation to the force of gravity. adj. *geotropic*. Cf. **geotaxis**.

geotropism *(Bot.)*. See **gravitropism**.

geranial *(Chem.)*. See **citral**.

geraniol *(Chem.)*. $C_{10}H_{18}O$. A terpene alcohol forming a constituent of many of the esters used in perfumery.

geriatrics *(Med.)*. The specialized medical care of the elderly and aged.

germ *(Zool.)*. The primitive rudiment which will develop into a complete individual, as a fertilized egg or a newly formed bud; a unicellular micro-organism.

germanium *(Chem.)*. A metalloid element in the fourth group of the periodic system. Greyish-white in appearance. Symbol Ge, at.no. 32, r.a.m. 72.59, rel.d. 5.47, mp about 958°C. Germanium occurs in a few minerals including coal. The chief mineral sources are **argyrodite**, also certain zinc and copper concentrates, and to some extent the flue dusts of gasworks. The main use of the metal, which has exceptional properties as a semiconductor, is in the manufacture of solid rectifiers or diodes in microwave detectors, and, in a highly pure state, in transistors.

germanium diode *(Electronics)*. A semiconductor diode that uses a pellet of germanium as the rectifying element. It has a lower forward voltage than a silicon diode, but its characteristics are less predictable and drift with temperature.

germanium radiation detector *(Nuc.Eng.)*. A semiconductor detector with relatively large sensitive volume for γ-ray spectrometry. It has a much higher resolution but (in general) less sensitivity than a scintillation spectrometer.

germanium rectifier *(Elec.Eng.)*. A *p-n* junction diode often with high current rating. It requires a lower forward voltage than a silicon diode, but it has higher reverse leakage current.

germanium rectifier *(Elec.Eng.,Electronics)*. A *p-n* junction diode. It requires a lower forward voltage than a silicon diode, but it has higher reverse leakage current.

German lapis *(Min.)*. See **Swiss lapis**.

German measles *(Med.)*. See **rubella**.

German siding *(Build.)*. Weather-boards finished with a hollow curve along the outside of the top edge, and rebated along the inside of the lower edge.

German silver *(Eng.)*. A series of alloys containing copper, zinc and nickel within the limits; copper, 25–50%; zinc, 10–35%; and nickel 5–35%. Also *nickel silver*.

germ band *(Zool.)*. In Insects, a ventral plate of cells, produced in the egg by cleavage, which later gives rise to the embryo.

germ cells *(Zool.)*. In Metazoa, the reproductive cells. Gametes; spermatozoa and ova, or the cells which give rise to them.

germinal aperture, germinal pore *(Bot.)*. See **germ pore**.

germinal cells *(Zool.)*. See **germ cells**.

germinal centre *(Immun.)*. An aggregation of lymphocytes, mainly B-cells with numerous blast forms undergoing cell division, which develops from a primary follicle in response to antigenic stimulation. A prominent feature is the presence of follicular dendritic cells bearing antigen-antibody complexes closely entangled with the B-cells, and of tingible body macrophages (so called because they contain remnants of the nuclei of dead lymphocytes). Germinal centres are thought to be sites at which B-memory cells are produced with receptors which recognize antigens in the complexes. The cell death which occurs there may represent B-cells which have undergone non-viable mutations or which have receptors which bind *idiotypic antibodies*.

germinal disk *(Zool.)*. The flattened circular region at the top of a megalecithal ovum, in which cleavage takes place.

germinal epithelium *(Zool.)*. A layer of columnar epithelium which covers the stroma of the ovary in Vertebrates.

germinal layers *(Zool.)*. See **germ layers**.

germination *(Bot.)*. The beginning of growth in a spore, seed, zygote etc., especially following a dormant period.

germ layers *(Zool.)*. The three primary cell-layers in the development of Metazoa, i.e. ectoderm, mesoderm, and endoderm.

germ line *(Biol.)*. The cells whose descendants give rise to the *gametes*.

germ nucleus *(Zool.)*. See **pronucleus**.

germ pore *(Bot.)*. A thin area in the wall of a spore or pollen grain though which the germ or pollen tube emerges at germination.

germ tube *(Bot.)*. The hypha or other tubular outgrowth that emerges from a spore at germination (e.g. pollen tube).

grontic *(Zool.)*. Pertaining to the senescent period in the life history of an individual.

gerontology *(Med.)*. The scientific study of the processes of aging.

gersdorffite *(Min.)*. Metallic grey sulphide-arsenide of nickel, occurring as cubic crystals or in granular or massive forms.

gesso *(Build.)*. A pasty mixture of whiting, prepared with size or glue, applied to a surface as a basis for painting or gilding.

Gestalt *(Behav.)*. A German word meaning *whole*; Gestalt psychology is an approach to perception and other cognitive skills, which stresses the need to understand the underlying organization of these functions, and believes that to dissect experience into its 'constituent bits' is to lose its essential meaning.

gestalt therapy *(Behav.)*. A therapeutic approach, developed by Fritz Perls, which focuses on the present manifestations of past conflict, often using role playing and other *acting out* techniques in order to help individuals gain insight into themselves and their behaviour. Not to be confused with the work of the Gestalt school of psychology, which focuses primarily on cognitive functions.

gestation *(Zool.)*. In Mammals, the act of retaining and nourishing the young in the uterus; pregnancy.

get *(Min.Ext.)*. To win or mine.

getter *(Electronics)*. Material (K, Na, Mg, Ca, Sr, or Ba), used, when evaporated by high frequency induction currents, for *cleaning* the vacuum of valves, after sealing on the pump line during manufacture.

gettering discharge *(Electronics)*. That used to assist the **getter** in the *clean up* of vacuum in valves, through ionization of remaining gas molecules.

GeV *(Phys.)*. Abbrev. for *giga-electron-volt*; unit of particle energy, 10^9 electron volts, 1.602×10^{-10} J. US, sometimes *BeV*.

geyser *(Build.)*. A water-heating appliance providing supplies of hot water rapidly for domestic purposes, the source of heat being gas or electricity.

geyser *(Geol.)*. A volcano in miniature, from which hot water and steam are erupted periodically instead of lava and ashes, during the waning phase of volcanic activity. Named from the Great Geyser in Iceland, though the most familiar example is probably 'Old Faithful' in the Yellowstone Park, Wyoming. The eruptive force is the sudden expansion which takes place when locally heated water, raised to a temperature above boiling point, flashes in to steam. Until the moment of eruption, this had been prevented by the pressure of the overlying column of water in the pipe of the geyser, which is usually terminated upwards by a sinter crater. Also *gusher*.

geyserite *(Min.)*. See **sinter**.

G-gas *(Phys.)*. Gaseous mixture (based on helium and isobutane) used in low-energy β-counting (e.g., of tritium) by gas-flow proportional counter.

ghaut *(Build.)*. A landing-stage on a riverside.

ghost *(Eng.)*. In steel, a band in which the carbon content is less than that in the adjacent metal and which therefore consists mainly of ferrite. Also *ghost line*.

ghost *(Image Tech.)*. Vertical streaks on highlights in a projected picture, arising from incorrect phasing of the rotary shutter with respect to the moving film.

ghost crystal *(Min.)*. A crystal within which may be seen an early stage of growth, outlined by a thin deposit of dust or other mineral deposit.

ghost image *(Phys.)*. The image arising from a mirror when the rays have experienced reflection within the glass between the surface and the silvering.

ghost line *(Eng.)*. See **ghost**.

giant *(Min.Ext.)*. *Hydraulic giant*. See **monitor**.

giant cells *(Zool.)*. Cells of unusual size, as the myeloplaxes of bone-marrow; certain cells of the excitable region of the cerebrum; certain cells sometimes found in lymph glands; large multinucleate cells of the thymus and of the spleen. **Osteoclasts**.

giant fibres *(Zool.)*. In many Invertebrates (e.g. Annelida, Cephalopoda and some Arthropoda) enlarged motor axons in the ventral nerve cord which transmit impulses very rapidly and initiate escape behaviour.

giantism *(Med.)*. Uniformly excessive growth of the body, due to overactivity of the anterior lobe of the pituitary gland. Also *gigantism*.

giant powder *(Min.Ext.)*. Dynamite.

giant source *(Med.)*. Large source of radioactivity, e.g. 150 000 curies of ^{60}Co, used for industrial sterilization of packed food or chemical processing (e.g. cross-linking of polymers).

giant star *(Astron.)*. A star which is more luminous than the main sequence stars of the same spectral class. Smaller groups of subgiants and supergiants are recognized.

giardiasis *(Med.)*. Infestation of the intestinal tract with the flagellate protozoon *Giardia lamblia*, sometimes causing severe diarrhoea.

gib *(Eng.)*. (1) Metal piece used to transmit the thrust of wedge or cotter, as in some connecting-rod bearings. (2) A brass working surface let in to the working surface of a steam-engine cross-head. (3) Tapered or parallel strip in bearings for reciprocating slides, used to fit or to clamp the slide in the guide.

gibberellic acid *(Bot.)*. A **gibberellin** obtained from cultures of the fungus *Gibberella fuijikuroi* used commercially e.g. to promote rapid, even malting of barley.

gibberellin *(Bot.)*. Any of a large group of terpenoid plant *growth substances*, synthesized mainly in young leaves and promoting stem elongation and, sometimes, flowering, germination and the utilization of reserves as in germinating barley grains. See also **gibberellic acid**.

gibbous *(Astron.)*. The word applied to the phase of the moon, or of a planet, when it is between either quadrature and opposition, and appears less than a circular disk but greater than a half disk.

gibbous *(Bot.)*. (1) Swollen, especially if swollen more on one side than another. (2) With a pouch.

Gibbs' adsorption theorem *(Chem.)*. Solutes which lower the surface tension of a solvent tend to be concentrated at the surface, and conversely.

Gibbs-Duhem equation *(Chem.)*. For binary solutions at constant pressure and temperature, the chemical potentials (μ_1,μ_2) vary with the mole fractions (x_1,x_2) of the two components as follows:

$$\left(\frac{\partial \mu_1}{\partial \ln x_1}\right)_{T,P} = \left(\frac{\partial \mu_2}{\partial \ln x_2}\right)_{T,P}$$

Gibbs' free energy *(Chem.)*. The difference of the enthalpy and the product of the entropy and the temperature of a system. Usual symbol G. A calculated negative change in G indicates a spontaneous process in a closed system at constant pressure.

Gibbs-Helmholtz equation *(Chem.)*. An equation of thermodynamics,

$$F - \Delta F = -\Delta U + T\cdot\frac{d(-\Delta F)}{dT}$$

where $-\Delta F=$ decrease in free energy; $-\Delta U=$ decrease in intrinsic energy; $T=$ the absolute temperature. It is applied to the voltaic cell in the form:

$$yFE = -q + yFT\cdot\frac{dE}{dT}$$

where $Y=$ number of gram equivalents of chemical change, $G=$ faraday, $E=$ e.m.f. of the cell, and $-q=$ heat of reaction.

gibbsite *(Min.)*. Hydroxide of aluminium, $Al(OH)_3$, occurring as minute mica-like crystals, concretional masses, or incrustations. An important constituent of bauxite. Also called *hydrargillite*.

Gibbs-Konowalow rule *(Chem.)*. For the phase equilibrium of binary solutions. At constant pressure the equilibrium temperature is a maximum or minimum when the compositions of the two phases are identical, and vice versa, e.g. in eutectics or azeotropes. The corresponding statements hold for pressure at constant temperature.

Gibbs' phase rule *(Chem.)*. See **phase rule**.

gib-headed key *(Eng.)*. A key for securing a wheel, etc., to a shaft, having a head formed at right angles to its length.

giblet check *(Build.)*. An exterior rebate for a door which opens outwards.

gid *(Vet.)*. See **coenuriasis**.

Gies' biuret reagent *(Chem.)*. A reagent for testing the presence of proteins by the biuret reaction; it consists of a solution of 10% KOH and 0.075% copper (II) sulphate.

Giffard's injector *(Eng.)*. The original steam **injector**.

giga- *(Genrl.)*. Prefix used to denote 10^9 times, e.g. a gigawatt is 10^9 watts.

giga-electron-volt *(Phys.)*. See **GeV**.

gigantism *(Bot.)*. Abnormal increase in size, often associated with polyploidy.

gigantism *(Med.)*. See **giantism**.

giggering *(Print.)*. A method of producing lines on the back of a volume by means of a catgut cord.

gig stick *(Build.)*. See **radius rod**.

gilder's cushion *(Build.)*. A pad comprising a flat wooden board padded with felt covered with tightly stretched chamois leather having a parchment screen at one end. A gilder uses the pad as a base on which to manipulate and cut loose leaf metals when gilding.

gilder's knife *(Build.)*. A long, thin, blunt edged knife used for cutting loose leaf gold on a gilder's cushion.

gilder's mop *(Build.)*. A soft brush with a round, bushy camel hair filling used to press gold leaf into recesses and enrichments.

gilder's tip *(Build.)*. A small brush comprising a single line of badger hair set between two thin sheets of thin card. It is used to pick up loose gold leaf from the gilder's cushion and transfer it to the surface being gilded.

gilder's wheel *(Build.)*. A device used to apply lines from a roll of ribbon gold to a prepared surface.

gilding *(Build.)*. The application of gold leaf by means of an adhesive, normally gold size, as mordant.

gilding metal *(Eng.)*. Copper-zinc alloy containing zinc up to 10%.

gill *(Bot.)*. One of the vertical plates, bearing the hymenium, on the undersurface of the cap of the fruiting body of the mushroom and other agarics.

gill *(Zool.)*. A membranous respiratory outgrowth of aquatic animals, usually in the form of thin lamellae or branched filamentous structures.

gill arch *(Zool.)*. In Fish, the incomplete jointed skeletal ring supporting a single pair of gill slits; one segment of the branchial basket.

gill bars *(Zool.)*. See **gill rods**.

gill basket *(Zool.)*. In Fish and Cyclostomes, the skeletal ring supporting a single pair of gill slits; one segment of the branchial basket.

gill book *(Zool.)*. The booklike respiratory lamellae of Xiphosura borne by the opisthosoma, of which they represent the appendages. Also called *book gill*.

gill clefts *(Zool.)*. See **gill slits**.

gill cover *(Zool.)*. See **operculum**.

gillion *(Genrl.)*. Rarely 10^9; preferable term *giga-* or *G*.

gill pouch *(Zool.)*. One of the pouchlike gill slits of Cyclostomes and Fish.

gill rakers *(Zool.)*. In some Fish, small processes of the branchial arches, which strain the water passing out via the gill slits and prevent the escape of food particles.

gill rods *(Zool.)*. In Cephalochorda, skeletal bars which support the pharynx. Also called *gill bars*.

gills *(Aero.)*. Controllable flaps which vary the outlet area of an aircooled engine cowling or of a radiator; also *cooling gills, cowl flaps, radiator flaps*.

gill slits *(Zool.)*. In Chordata, the openings leading from the pharynx to the exterior, on the walls of which the gills are situated. Also called *brachial clefts, gill clefts*.

gilsonite *(Min.)*. See **uintaite**.

gimbal mount *(Eng.)*. One giving rotational freedom about two perpendicular axes, as used for gyroscope and nautical compass.

gimbals, gymbals *(Horol.)*. Self-aligning bearings for supporting a chronometer in its box. Used to ensure that the chronometer is kept level, irrespective of the ship's motion.

gimmick *(Elec.Eng.)*. Colloquialism for small capacitor formed by twisting insulated wires.

gin *(Eng.,Min.Ext.)*. (1) A hand hoist which consists of a chain or rope barrel supported in bearings and turned by a crank. (2) A portable tripod carrying lifting tackle.

gin *(Textiles)*. A machine that is used to remove cotton fibres from the seeds: the process is known as ginning.

ginger-beer plant *(Bot.)*. A symbiotic association of a yeast and a bacterium, which ferments a sugary liquid containing oil of ginger, giving ginger beer. Often known popularly as *Californian bees*, and by similar names.

gingival *(Med.,Zool.)*. In Mammals, pertaining to the gums.

gingivectomy *(Med.)*. The cutting back of inflammed or excess gum.

gingivitis *(Med.)*. Inflammation of the gums.

ginglymus *(Zool.)*. An articulation which allows motion to take place in one plane only; a hinge joint. adj. *ginglymoid*.

Ginkgoales *(Bot.)*. Order of gymnosperms, abundant world-wide in the Mesozoic, with one extant species, *Ginkgo biloba*, a Chinese tree with small, fan-shaped leaves. Zooidogamous.

Giorgi system *(Phys.)*. System of units proposed in 1904 and later adopted as the MKSA system, the common *practical units* of ohm, volt, ampere, etc., becoming identified with *absolute units* of the same entities. See **SI units**.

GIOTTO *(Space)*. Name of the **ESA** spacecraft which flew through the coma of Halley's comet and approached within 600 km of the nucleus. The **fly-by** took place at a relative velocity of 68 km/sec in March 1986. The encounter was supported by data from two Japanese probes (*Suisei* and *Sakigake*), two USSR spacecraft (*Vega 1* and *2*) and the US International Cometary Explorer (*ICE*). Coordination was performed by the truly international *Inter-Agency Consultative Group* (IACG).

GIP *(Paper)*. See **glazed imitation parchment**.

girder *(Bot.)*. Usually longitudinal strand of mechanical tissue, often T or I shaped in cross section, giving strength and, especially, stiffness to a plant part as in a grassleaf.

girder *(Eng.)*. A beam, usually steel, to bridge an open space. Girders may be rolled sections, built up from plates or of lattice construction. See also **continuous beam**.

girder bridge *(Civ.Eng.)*. A bridge in which the loads are sustained by beams (generally compound) resting across the bridge supports.

girdle *(For.)*. A continuous incision which is made all round a bole, cutting through at least bark and cambium, generally with the object of killing the tree.

girdle *(Zool.)*. In Vertebrates, the internal skeleton to which the paired appendages are attached, consisting typically of a U-shaped structure of cartilage or bone with the free ends facing dorsally. In *Amphineura*, the part of the mantle which surrounds the shells

girth *(For.)*. See **exploitable-, mid-, mid-timber-, quarter-, top-**.

girt strip *(Build.)*. See **ribbon strip**.

gismondine *(Min.)*. A rare zeolite; a hydrated calcium aluminium silicate which occurs in cavities in basaltic lavas.

give-and-take lines *(Surv.)*. Straight lines drawn on a plan of any area having irregular boundaries, each line following the trend of a part of the boundary so that any small piece that it cuts off the area is balanced by an equal piece added by it.

gizzard *(Zool.)*. See **proventriculus**.

glabrescent, glabrate *(Bot.)*. (1) Almost hairless. (2) Becoming hairless.

glabrous *(Bot.,Zool.)*. Having a smooth hairless surface.

glacial acetic acid *(Chem.)*. Pure concentrated acetic (ethanoic) acid. Owing to its comparatively high mp (16.6°C) it solidifies easily, forming icelike crystals.

glacial action *(Geol.)*. All processes relating to the action of glacier ice, comprising: (*a*) the grinding, scouring, plucking, and polishing effected by ice, armed with rock fragments frozen into it; and (*b*) the accumulation of rock debris resulting from these processes.

glacial deposits *(Geol.)*. These include spreads of boulder clay, sheets of sand and gravel occurring as outwash fans, outwash deltas, and kames; also deposits of special topographical form, such as drumlins and eskers.

glacial erosion *(Geol.)*. The removal of rock materials by the action of glaciers and associated meltwater streams. Includes grinding, scouring, plucking, grooving and polishing by rock fragments contained in the ice.

glacial phosphoric acid *(Chem.)*. See **metaphosphoric acid**.

glacial sands *(Geol.)*. These cover extensive areas in advance of sheets of boulder clay, and together with glacial (largely fluvioglacial) gravels, represent the outwash from ice-sheets.

glaciation *(Geol.)*. Embraces both the processes and products arising from the presence of ice masses on the Earth. The effects are most obvious on land but there is increasing evidence that the shallow sea floors too were affected. Glaciation, traditionally connected with the Pleistocene Period, is now known from older geological periods including the Permo-Carboniferous and the Precambrian.

glacier *(Geol.)*. A large mass of ice. Three kinds can be recognized (*a*) valley glaciers, (*b*) Piedmont glaciers which overflow from valleys and coalesce on the lower ground, and (*c*) large continental ice-sheets (e.g. Greenland) and smaller ice-caps (e.g. Iceland).

glacier lake *(Geol.)*. See **lake**.

glacis *(Civ.Eng.)*. An inclined bank.

Gladstone and Dale Law *(Phys.)*. Law relating the refracting index of a gas and its density as it is changed by variations in pressure and temperature:

$$\frac{n-1}{\rho} = \text{constant}$$

where n is the refractive index and ρ the density.

glair *(Print.)*. A preparation made from white of egg and vinegar, used as the adhesive for gold leaf in gold finishing and blocking.

glance *(Min.)*. Opaque mineral with a resinous or shining lustre.

glancing angle *(Phys.)*. The complement of the **angle of incidence**.

gland *(Bot.,Zool.)*. A structure at or near the surface of a plant or a single or an aggregate of epithelial cells in animals, specialized for the elaboration of a secretion useful to the organism or of an excretory product. adj. *glandular*.

gland *(Eng.)*. (1) A device for preventing leakage at a point where a rotating or reciprocating shaft emerges from a vessel containing a fluid under pressure. (2) A sleeve or nut used to compress the packing in a **stuffing-box**.

gland bolts *(Eng.)*. Bolts for holding and tightening down a gland.

gland cell *(Zool.)*. A unicellular gland, consisting of a single goblet-shaped epithelial cell producing a secretion, usually mucus.

glanders *(Med.)*. A contagious bacterial disease of horses, mules, and asses, due to infection by *Actinobacillus mallei* (*Malleomyces mallei*); inflammatory nodules occur in the respiratory passages and lungs and also in other parts of the body. Infection of the lymphatics under the skin is known as *farcy*. Glanders is communicable to man.

glandular epithelium *(Zool.)*. Epithelial tissue specialized for the production of secretions.

glandular fever *(Med.)*. See **infectious mononucleosis**.

glandular tissue *(Zool.)*. See **glandular epithelium**.

glans *(Bot.)*. A nut.

glans *(Zool.)*. A glandular structure.

glans penis *(Zool.)*. A dilatation of the extremity of the mammalian penis.

glare *(Phys.)*. The visual discomfort experienced by observers in the presence of a visible source of light. The term can also refer to the visual disability produced by the presence of visible sources in the field of view when these sources do not assist the viewing process.

glarimeter *(Paper)*. An instrument for measuring the gloss of a paper surface based on light reflectance.

glass *(Genrl.)*. A hard, amorphous, brittle substance, made by fusing together one or more of the oxides of silicon, boron, or phosphorus, with certain basic oxides (e.g. sodium, magnesium, calcium, potassium), and cooling the product rapidly to prevent crystallization or devitrification. The melting point varies between 800°C and 950°C. The tensile strength of glass resides almost entirely in the outer skin; if this is scratched or corroded, the glass is much more easily broken. See also **natural glass**.

glass blocks *(Arch.,Build.)*. Hollow blocks made of glass usually with a patterned surface. Used where translucence and decorative effect are required. They also provide insulation against heat and sound. Sometimes called *glass bricks*.

glass-bulb rectifier *(Elec.Eng.)*. A mercury-arc rectifier in which the arc takes place within a glass bulb. Cf. **steel-tank rectifier**.

glassed *(Build.)*. A term applied to stones such as granite and marble which are highly polished by being held against a revolving disk covered with felt.

Glasser's disease *(Vet.)*. *Infectious polyarthritis of pigs. Hemophilus suis, H. parasuis* and *Mycoplasma* spp. are all implicated. Sudden-onset fever, tachypnoea and dyspnoea are common findings, usually with anorexia and lameness with all joints swollen and painful.

glass fibre *(Textiles)*. Glass melted and then drawn out through special bushings into fine fibres which may be spun into threads and woven into tapes and cloths. The fibres may also be formed into pads and quiltings for thermal insulation. Plastics reinforced with glass fibres are used for making some car and boat bodies.

glass-fibre paper *(Paper)*. That formed by using glass fibres as part or the whole of the **furnish**.

glass gilding *(Build.)*. The application of loose leaf gold to the reverse side of glass using isinglass or gelatine as a mordant.

glassification *(Nuc.Eng.)*. See **vitrification**.

glassine *(Paper)*. Transparent glazed greaseproof paper. Produced by long beating of the stock and high glazing of the finished paper.

glasspaper *(Paper)*. Paper coated with glue on which is sprinkled broken glass of a definite grain size; used for rubbing down surfaces. Cf. **sandpaper**.

glass shot *(Image Tech.)*. A composite shot in which live action is photographed through the clear portions of a glass plate on which the remainder of the scene has been painted.

glass tile *(Build.)*. A small glass sheet in a roof, bonded in with slates, plain tiles, or pantiles, to admit light within the roof space.

glass wool *(Build.,Chem.)*. A felt of fine glass fibres; used as a relatively inert filter, packing or insulation.

glauberite *(Min.)*. Monoclinic sulphate of sodium and calcium, occurring with rock salt, anhydrite, etc., in saline deposits.

Glauber salt *(Min.)*. Properly termed *mirabilite* (hydrated sodium sulphate, $Na_2SO_4.10H_2O$). A monoclinic mineral formed in salt lakes, deposited by hot springs, or resulting from the action of volcanic gases on sea water.

glaucescent *(Bot.)*. (1) Somewhat glaucous. (2) Becoming glaucous.

glaucoma *(Med.)*. An eye condition in which, from various causes, the intra-ocular pressure rises, leading to damage to optic nerve fibres and causing a quarter of all blindness after the age of 45.

glauconite *(Min.)*. Hydrated silicate of potassium, iron, and aluminium, a green mineral occurring almost exclusively in marine sediments, particularly in greensands. It is generally found in rounded fine-grained aggregates of ill-formed platelets although it has a mica structure. The manner of its formation is somewhat uncertain.

glaucophane *(Min.)*. A blue monoclinic amphibole occurring in crystalline schists formed at high pressures. A hydrated sodium iron magnesium silicate, the name is used for an end-member compositional variety of amphibole.

glaucous *(Bot.)*. Greyish- or bluish-green; covered with a bloom as a plum is.

glaze *(Build.)*. (1) Brilliant glass-like surface given to tiles, bricks etc. (2) Transparent or semi-transparent varnish coating used in graining, marbling or colour glazing.

glazed brick *(Build.)*. A brick having a glassy finish to the surface produced by spraying it with special surface preparations before firing.

glazed door *(Build.)*. A door fitted with glass panels.

glazed frost *(Meteor.)*. A smooth layer of ice which is occasionally formed when rain falls and the temperature of the air and the ground is below freezing-point.

glazed imitation parchment *(Paper)*. A highly glazed packaging paper, generally bleached and frequently opacified, suitable for waxing or coating for use in food wrapping applications.

glazed morocco *(Print.)*. Goatskin crushed and polished by rolling.

glazier *(Build.)*. A workman who cuts panes of glass to size and fits them in position.

glazier's putty *(Build.)*. A mixture of whiting and linseed oil, sometimes including white lead, forming a plastic substance for sealing panes of glass into frames.

glazing *(Build.)*. The operation of fitting panes of glass into sashes.

glazing *(Image Tech.)*. Forming a shiny surface on photographic prints by heating them in contact with a highly polished surface, e.g. chromium plate.

glazing bead *(Build.)*. A bead nailed, instead of putty, to secure a pane.

gleba *(Bot.)*. The spore-bearing tissue enclosed within the peridium of the fructification of *Gasteromycetes*, and in truffles.

gleet *(Med.)*. Chronic discharge from the urethra as a result of gonococcal infection.

glenoid *(Zool.)*. Socket-shaped; any socket-shaped structure; as the cavity of the pectoral girdle which receives the basal element of the skeleton of the fore-limb.

glenoid fossa *(Zool.)*. In Mammals, a hollow beneath the zygomatic process.

gley, glei soil *(Bot.)*. A soil that is permanently or periodically waterlogged and, therefore anaerobic, characterized by its blue-grey colours (due to ferrous iron) often mottled with orange-red (ferric iron).

glia *(Zool.)*. Same as *neuroglia*. Gk. *gliā*, meaning glue.

glibenclamide *(Med.)*. A **sulphonylurea** used in the treatment of maturity onset diabetes.

glide path *(Aero.)*. The approach slope (usually $3\frac{1}{2}°$ or $5°$) along which large aircraft are assumed to come in for a landing. The term is imprecise because such aircraft do not glide, but are brought in with a considerable amount of power.

glide path beacon *(Aero.)*. A directional radio beacon, associated with an ILS, which provides an aircraft, during approach and landing, with indications of its vertical position relative to the desired approach path.

glide path landing beam *(Aero.)*. Radio signal pattern from a radio beacon, which aids the landing of an aircraft during bad visibility.

glider *(Aero.)*. A heavier than air craft not power driven within itself, although it may be towed by a power driven aircraft. Cf. **sailplane**.

gliding *(Aero.)*. (1) Flying a heavier-than-air craft without assistance from its engine, either in a spiral or as an approach glide before flattening-out antecedent to landing. (2) The sport of flying *gliders* which are catapulted into the air, or launched by accelerating with a winch, or towed by car, or towed to height by an aircraft.

gliding *(Bot.)*. Form of movement in algae in which cells or filaments, without flagella, move against a solid surface.

gliding angle *(Aero.)*. The angle between the flight path of an aircraft in a glide and the horizontal.

gliding growth *(Bot.)*. Same as **sliding growth**.

gliding planes *(Crystal.)*. In minerals, planes of molecular weakness along which movement can take place without actual fracture. Thus calcite crystals or cleavage masses can be distorted by pressure and pressed into quite thin plates without actual breakage. See **slip planes**.

glimmerite *(Geol.)*. An ultrabasic rock composed mainly of mica, usually of the biotite variety.

glint *(Radar)*. Pulse-to-pulse variation in the amplitude of a reflected signal; may be due to reflection from different surfaces of a rapidly moving target, or from propellors or rotor blades.

glioblastoma multiforme *(Med.)*. A very malignant tumour of the central nervous system.

glioma *(Med.)*. A general term applied to tumours arising from nueroglial nervous tissue in the brain and, more rarely, in the spinal cord.

gliomatosis *(Med.)*. Diffuse overgrowth of neuroglia in the brain or in the spinal cord.

glissette *(Maths.)*. The locus of any point, or envelope of any line, moving with a first curve which slides against two fixed curves. Many curves can be regarded as either glissettes or roulettes, e.g. the astroid.

Gln *(Chem.)*. Symbol for **glutamine**.

global variable *(Comp.)*. One which is available for the whole of a program, including any subprograms.

globe photometer *(Phys.)*. See **Ulbricht sphere photometer**.

globigerina ooze *(Geol.)*. A deep-sea deposit covering a large part of the ocean floor (one quarter of the surface of the globe); it consists chiefly of the minute calcareous shells of the foraminifer, *Globigerina*.

globoid *(Bot.)*. A rounded inclusion of phytin in a protein body.

globular cementite *(Eng.)*. In steel, cementite occurring in the form of globules instead of in lamellae (as in pearlite) or as envelopes round the crystal boundaries (as in hyper-eutectoid steel). Produced by very slow cooling, or by heating to between 600 and 700°C.

globular cluster *(Astron.)*. Densely packed family of stars arranged characteristically as a compact sphere of stars. They contain tens of thousands to millions of stars, all thought to have been formed at the same time. Over 100 are known as members of our Galaxy, distributed through the **galactic halo**.

globular pearlite *(Eng.)*. See **granular pearlite**.

globule *(Astron.)*. Small dark nebula composed of opaque gas and dust, representing an early stage of star formation.

globulites *(Geol.)*. Crystallites (i.e. incipient crystals) of minute size and spherical shape occurring in natural glasses such as pitchstones.

globus *(Zool.)*. Any globe-shaped structure; as the *globus pallidus* of the Mammalian brain. adj. *globate*.

globus hystericus *(Med.)*. The sensation as of a lump in the throat experienced in hysteria.

glochidiate *(Bot.)*. Bearing barbed hairs or bristles.

glomera carotica *(Med.)*. See **carotid bodies**.

glomerate *(Bot.)*. Collected into heads.

glomerulitis *(Med.)*. Inflammation of the glomeruli of the kidney.

glomerulonephritis *(Immun.,Med.)*. Kidney diseases in which the major lesion is in the glomeruli. The capillary walls contain deposits of IgG and often of activated complement components. These are due to deposition of immune complexes from the circulation, or to antibodies against some component present in the glomerular capillary wall. The glomeruli are infiltrated with inflammatory cells. This condition is reversible, but if it becomes chronic gradual obliteration of the glomeruli occurs resulting in renal insufficiency.

glomerulus *(Zool.)*. A capillary blood-plexus, as in the Vertebrate kidney; a nestlike mass of interlacing nerve-fibrils in the olfactory lobe of the brain. adj. *glomerular*.

glory *(Meteor.)*. A small system of coloured rings surrounding the shadow of the observer's head, cast by the sun on a bank of mist, as in the *spectre of Brocken*. The *glory* is produced by diffraction caused by the water droplets in the mist.

glory-hole *(Glass,Eng.)*. (1) A subsidiary furnace, in which articles may be reheated during manufacture. (2) An opening exposing the hot interior of a furnace.

glory-hole *(Min.Ext.)*. Combination of open pit mining with underground tunnel through which spoil is removed after gravitating down.

glossa *(Zool.)*. In Vertebrates, the tongue; any tongue-like structure. adj. *glossate, glossal*.

glossectomy *(Med.)*. Removal of the tongue.

gloss-, glosso- *(Genrl.)*. Prefix from Gk. *glōssa*, tongue.

glossitis *(Med.)*. Inflammation of the tongue.

glossodynia *(Med.)*. Pain in the tongue.

glossopharyngeal *(Zool.)*. Pertaining to the tongue and the pharynx; the ninth cranial nerve of Vertebrates, running to the first gill-cleft in lower forms, to the tongue and the gullet in higher forms.

glossoplegia *(Med.)*. Paralysis of the tongue.

glossospasm *(Med.)*. Spasm of the muscles of the tongue.

gloss paint *(Build.)*. Paint to which varnish is added as an ingredient in the manufacturing process; characterized by a glossy finish.

glottis *(Zool.)*. In higher Vertebrates, the opening from the pharynx into the trachea.

glove box *(Nuc.Eng.)*. An enclosure in which radioactive or toxic material can be handled by use of special rubber gloves attached to the sides of the box thus preventing contamination of the operator. Normally operated at a slightly reduced pressure so that any leakage is inward.

Glover tower *(Chem.Eng.)*. A tower of a sulphuric acid plant used to recover the nitrogen oxides from the Gay-Lussac tower, to cool the gases from the burners, to concentrate the acid trickling down the tower, to partly oxidize the gases from the sulphur burners, and to introduce the necessary nitric acid into the chambers by running nitric acid down the tower along with the nitrated acid from the Gay-Lussac tower.

glow discharge *(Electronics)*. Visible discharge near a cathode when the potential drop is slightly higher than the ionization potential of the gas; consists of luminous spectral bands.

glow plug *(Aero.)*. In a gas turbine, an electrical igniting plug which can be switched on to ensure automatic re-lighting when the flame is unstable, e.g. under icing conditions. Also used in diesel engines.

glow potential *(Electronics)*. Potential which initiates sufficient ionization to produce a gas discharge between electrodes, but is below sparking potential.

glow switch *(Electronics)*. Tube in which a glow discharge thermally closes a contact, starting fluorescent tubes.

glow tube *(Electronics)*. Cold-cathode, gas-filled diode, with no space-current control, the colour of glow depending on contained gas.

Glu *(Chem.)*. Abbrev., depending on the context, for *glucose* or *glutamic acid*.

glucagon *(Med.)*. A hormone released by the alpha cells of the **islets of Langerhans** which raises the blood sugar by stimulating the breakdown of *glycogen* in the liver to *glucose*.

glucans *(Chem.)*. The condensation polymers glucose, e.g. cellulose, starch, dextrin, glycogen etc.

gluco- *(Genrl.)*. See **glyc-**.

glucocorticoid *(Med.)*. Any of the **corticosteroids** which act as carbohydrate, fat and protein metabolism.

d-gluconic acid *(Chem.)*. $CH_2OH(CHOH)_4COOH$, an oxidation product (at C-1) of d-glucose.

glucophore *(Chem.)*. A group of atoms which causes sweetness of taste.

glucose *(Biol.,Chem.)*. The commonest **aldohexose**, with the formula $C_6H_{12}O_6$ and the major source of energy in animals. Starch and cellulose are condensation polymers of glucose. Also *dextrose, grape sugar*. ⇨

glucose-6-phosphate dehydrogenase *(Med.)*. An enzyme that catalyses the dehydrogenation of glucose-6-phosphate to 6-phosphogluconolactone, an important pathway in carbohydrate metabolism. Deficiency may cause the development of *haemolytic anaemia*.

glucosuria *(Med.)*. See **glycosuria**.

glucuronic acid *(Chem.)*. $CHO.(CHOH)_4.COOH$. It can be prepared by reduction of the lactone of saccharic acid, and occurs in small amounts in the urine. It forms glycosides or esters with phenols or aromatic acids which are removed from the body in the form (6-CH_2OH of glucose→6-COOH).

glue *(Build.)*. A substance used as an adhesive agent between surfaces to be united. Glue is obtained from various sources, e.g. bones, gelatin, starch, resins etc. See **adhesive**.

glueline *(Elec.Eng.)*. High-frequency heating technique for drying glue films in woodwork construction, by applying electric field in line with the film, with specially shaped electrodes. The film should have a high loss factor compared with medium to be 'glued'.

glue size *(Build.)*. Refined animal material available in granular or powder form which is soaked in cold water before melting with boiling water and allowed to cool and gel. Used to pretreat the surface to equalize porosity before paperhanging.

glue sniffing *(Med.)*. Sniffing solvents produces intoxication and is a practice that adolescents adopt as a form of drug abuse.

glume *(Bot.)*. (1) One of the two bracts at the base of each **spikelet** in the inflorescence of a grass. (Old name: *sterile glume*.) (2) The bract subtending the flower in the

Cyperaceae (sedges and allies). (3) *Flowering glume* is an old name for **lemma**.

gluon *(Phys.)*. A *gauge boson* that mediates the strong interaction between quarks (and antiquarks). According to quantum chromodynamic theory, there should be eight different gluons each with zero mass and zero charge.

glut *(Build.)*. See **closer**.

glutamic acid *(Chem.)*. *2-amino-pentane-1,6-dioic acid*, HOOC.CH_2.CH_2.CH(NH_2).COOH. The L- or S- form is a constituent of proteins, and is classed as acidic as it has two carboxylic acid groups. Symbol Glu, short form E. ⇨

glutamine *(Chem.)*. The 6-amide of glutamic acid. Symbol Gln, short form Q. ⇨

glutaraldehyde *(Chem.)*. CHO.$(CH_2)_3$CHO, an oil, soluble in water and volatile in steam; used in tanning leather, esp. for clothing, as it imparts resistance to perspiration. Also a disinfectant.

gluteal *(Zool.)*. Pertaining to the buttocks.

gluten *(Bot.,Med.)*. The reserve protein of wheat grain, a mixture of gliadin and glutenin. Gluten intolerance is important in the development of **coeliac disease**.

gluteus *(Zool.)*. In land Vertebrates, a retractor and elevator muscle of the hind-limb.

Gly *(Chem.)*. Symbol for **glycine**.

glycerides *(Chem.)*. A term for glycerine esters, the most important of which are the **fats**.

glycerine *(Chem.)*. *Glycerol, propan-1,2,3-triol*, CH_2OH.-CHOH.CH_2OH, a syrupy hygroscopic liquid, mp 17°C. bp 290°C, obtained by the hydrolysis of oils and fats, or by the alcoholic fermentation of glucose in the presence of sodium sulphite solution, which reacts with the aldehydes formed, thus liberating a larger amount of glycerine. It is also prepared synthetically from propylene by chlorination and hydrolysis. Glycerine is a trihydric alcohol, forming alcoholates, esters and numerous derivatives. Colourless. It is a raw material for **alkyd resins**, **nitroglycerine**, printing inks, foodstuff preparations etc. Also *glycerin*.

glycerine litharge cement *(Chem.)*. A mixture of litharge (lead (II) oxide) and glycerine which rapidly sets to a hard mass.

glycerol *(Chem.)*. See **glycerine**.

glycerol-phthalic resins *(Plastics)*. See **alkyd resins**.

glyceryl trinitrate *(Med.)*. Short acting vasodilator which reduces venous return to the heart, heart work and oxygen consumption. Main use is in the treatment of **angina pectoris**.

glycine *(Chem.)*. *Aminoethanoic acid*, the simplest of the naturally occurring amino acids. Symbol Gly, short form G. ⇨

glycocalyx *(Biol.)*. The carbohydrate layer on the outer surface of animal cells which is composed of the oligosaccharide termini of the membrane glycoproteins and glycolipids.

glycocoll *(Chem.)*. *Aminoethanoic acid.*

glycogen *(Bot.)*. Storage polysaccharide, α, 1→4 linked with frequent α, 1→6 branches, existing as small granules in blue-green algae and bacteria and in the cytoplasm of eumycete fungi and animals but not in plants or eukaryotic algae.

glycol *(Chem.)*. See **ethylene glycol**.

glycolipid *(Biol.)*. A glycosylated lipid. See **cerebroside**, **ganglioside**.

glycols *(Chem.)*. Dihydric alcohols, of the general formula $C_nH_{2n}(OH)_2$, viscous liquids with a sweet taste or crystalline substances. They give all the alcohol reactions and, having two hydroxyl groups in the molecule, they can also form mixed compounds, e.g. ester-alcohols.

glycolysis *(Biol.)*. The sequence of reactions which converts glucose to pyruvate with the concomitant net synthesis of 2 molecules of ATP. Under aerobic conditions it is the prelude to the complete oxidation of glucose *via* the **tricarboxylic cycle**.

glycon *(Med.)*. A carbohydrate storage substance found in liver and muscles and broken down to glucose when energy is needed.

glycophyte *(Bot.)*. A plant which will grow only in soils containing little sodium chloride or other sodium salt. Most plants are glycophytes. Cf. **halophyte**.

glycoprotein *(Biol.)*. A protein with covalently linked sugar residues. The sugars may be bound to OH side chains of the polypeptide *(O-linked)* or to the amide nitrogen of asparagine side chains *(N-linked)*.

glycosides *(Chem.)*. A group of compounds, derived from the other monosaccharides in the same way as glucosides are derived from glucose, including glucosides as a subclass.

glycosuria, glucosuria *(Med.)*. The presence of sugar in the urine.

glycosyltransferase *(Biol.)*. An enzyme which catalyses the transfer of sugars onto a protein or another sugar side chain. These enzymes bring about the glycosylation of glycoproteins.

glycuronic acid *(Chem.)*. See **glucuronic acid**.

glyoxal *(Chem.)*. *Ethan 1,2-dial.* CHO.CHO, a dialdehyde, a yellow liquid, mp 15°C, bp 51°C (776 mm), forming green vapours, but it is not stable, and polymerizes to insoluble paraglyoxal, $(CHO.CHO)_3$.

glyoxalic acid *(Chem.)*. *Ethan-1-al-2-oic acid*, CHO.-COOH + H_2O or $CH(OH)_2$.COOH, an aldehyde monobasic acid occurring in unripe fruit; rhombic prisms, soluble in water, volatile in steam, which can be obtained by the oxidation of ethanol with nitric acid, or by the hydrolysis of dichloroacetic or dibromoacetic acid.

glyoxalines *(Chem.)*. See **imidazoles**.

glyoxylate cycle *(Bot.)*. Series of reactions resulting in the formation of succinate from acetyl CoA, enabling carbohydrates to be made from fatty acids as in germinating oil-storing seeds. See **glyoxysome**.

glyoxylic acid *(Chem.)*. See **glyoxalic acid**.

glyoxysome, glyoxisome *(Biol.)*. Cytoplasmic organelles of plant cells similar to **peroxisomes** but also containing enzymes of the *glyoxylate cycle*, a cyclic metabolic pathway that generates succinate from acetate. They are abundantly present in e.g. the endosperm or cotyledons of oil rich seeds.

glyph *(Arch.)*. A short upright flute. See **triglyph**.

glyptal resins *(Plastics)*. See **alkyd resins**.

Gm allotype *(Immun.)*. An allotypic determinant (recognizable by specific antibodies) due to amino acid substitutions at various positions in the constant region of human IgG heavy chains. At least 25 different allotypes are known, useful in genetic studies.

G-M counter *(Nuc.Eng.)*. See **Geiger-Müller counter**.

gmelinite *(Min.)*. A pseudohexagonal zeolite, white or pink in colour and rhombohedral in form, resembling chabazite. Chemically, it is hydrated silicate of aluminium, sodium and calcium.

Gmelin test *(Chem.)*. A test for the presence of bile pigments; based upon the formation of various coloured oxidation products in treatment with concentrated nitric acid.

GMT *(Genrl.)*. Abbrev. for *Greenwich Mean Time*.

gnathic *(Zool.)*. Pertaining to the jaws.

gnathites *(Zool.)*. Mouth-parts, especially those of Insects.

gnathobase *(Zool.)*. In Arthropoda, a masticatory process on the innerside of the first joint of an appendage.

gnathopod *(Zool.)*. In Arthropoda, any appendage modified to assist in mastication.

Gnathostomata *(Zool.)*. A superclass of Chordata, including all Vertebrate animals with upper and lower jaws; comprises a wide range of animals, from Fishes to Tetrapods. Cf. **Agnatha**.

gnathostomatous *(Zool.)*. Having the mouth provided with jaws.

gnathotheca *(Zool.)*. The horny part of the lower beak of Birds.

gneiss *(Geol.)*. A metamorphic rock of coarse grain size, characterized by a mineral banding, in which the light minerals (quartz and feldspar) are separated from the dark ones (mica and/or hornblende). The layers of dark minerals are foliated, while the light bands are granulitic. See also **metamorphism**.

gneissose texture *(Geol.)*. A rock texture in which foliated and granulose (granulitic) bands alternate.

Gnetopsida *(Bot.)*. Class of the gymnosperms of ca 80 spp. Mostly trees, shrubs, lianas and switch plants. Leaves reticulately veined or scales. Xylem with vessels. Reproductive structures organized into compound strobili. The micropyle projects as a long tube. Siphonogamous. Three genera, *Gnetum*, *Ephedra* and *Welwitchia* (the last a curious turnip-like plant with 2 strap-shaped leaves, of the Namibian desert.)

gnomon *(Maths.)*. The remainder of a parallelogram after a similar parallelogram has been removed from one corner.

gnomon *(Surv.)*. An early instrument for determination of time and latitude, involving the measurement of the shadow of an upright rod as cast by the sun. The pointer of a sundial.

gnotobiotic *(Biol.)*. Of a known (defined) environment for living organisms, as a sterile culture inoculated with one or a few strains of bacteria.

gnotobiotic *(Immun.)*. Describes an environment in which animals can be reared in which all the living microbes are known. This avoids antigenic stimulation by casual infections. If all living microbes are absent the environment is said to be *germ-free*.

go *(Build.)*. The **going**.

goaf *(Min.Ext.)*. See **gob**.

goal-directed behaviour *(Behav.)*. Implies that the individual has some model of the goal situation, and that discrepancies between the current and goal situation are used to guide behaviour; conscious awareness of the goal or complex cognitive abilities are not necessarily associated with goal-directed behaviour, which probably occurs in many species.

goat pox *(Vet.)*. An epidemic disease of goats due to infection by a virus; characterized by fever and a papulovesicular eruption of the skin and mucous membranes.

gob *(Glass)*. (1) A measured portion of molten glass as fed to machines making glass articles. (2) A lump of hot glass gathered on a punty or blowing iron.

gob *(Min.Ext.)*. The space left by the extraction of a coal-seam, into which waste is packed, also loose waste. Also *goaf*.

gob fire *(Min.Ext.)*. A fire occurring in a worked-out area, due to ignition of timber or broken coal left in the gob.

gob heading, gob road *(Min.Ext.)*. A roadway driven through the gob after the filling has settled.

goblet cell *(Zool.)*. A goblet- or flask-shaped epithelial gland cell, occurring usually in columnar epithelium.

gobo *(Image Tech.)*. An opaque mask on a stand used in the studio to provide local shadow areas or shield the camera lens.

gob process *(Glass)*. One for making hollow ware, in which glass is delivered by an automatic feeder in the form of soft lumps of suitable shape to a forming unit. Also *flow process, gravity process*.

gob stink *(Min.Ext.)*. A smell indicating spontaneous combustion or a fire in the gob.

godets *(Textiles)*. (1) Small rollers used to regulate the speed of extruded filaments during man-made fibre production. (2) An insert of material used to shape a garment being assembled from fabric.

go-devil *(Min.Ext.)*. Cylindrical plug with brushes, scrapers and rollers on its periphery and able to move under the oil pressure through a pipeline and clean it. Also *pig, rabbit*.

goethite *(Min.)*. Orthorhombic hydrated oxide of iron with composition FeO.OH. Dimorphous with lepidocrocite.

Goetz size separator *(Powder Tech.)*. Instrument for the size classification of airborne particles, comprising a high-speed rotor with a helical channel along which the particles move in laminar flow and are deposited by centrifugal force on a removable envelope surrounding the rotor.

going *(Build.)*. The horizontal interval between consecutive risers in a stair.

going-barrel *(Horol.)*. A barrel in which the winding takes place from the arbor. Power is transmitted direct from such a barrel to the train by teeth on the barrel.

going fusee *(Horol.)*. A fusee with maintaining power.

going light *(Vet.)*. A term popularly applied to emaciation of animals.

going rod *(Build.)*. A rod used for setting out the going of the steps in a flight.

goitre *(Med.)*. Morbid enlargement of the thyroid gland. See also **Basedow's disease**.

goitrogenous *(Med.)*. Producing, or tending to produce, goitre.

goitrous *(Med.)*. Affected with, or pertaining to, goitre.

Golay cell *(Phys.)*. Pneumatic cell used as detector of heat radiation, e.g., in infrared spectrometer.

gold *(Chem.)*. A heavy, yellow, metallic element, occurring in the free state in nature. Symbol Au, at.no. 79, r.a.m. 196.967, rel.d. at 20°C 19.3, mp 1062°C, electrical resistivity about 0.02 microhm metres. Most of the metal is retained in gold reserves but some is used in jewellery, dentistry, and for decorating pottery and china. In coinage and jewellery, the gold is alloyed with varying amounts of copper and silver. *White gold* is usually an alloy with nickel, but as used in dentistry this alloy contains platinum or palladium.

gold amalgam *(Min.)*. A variety of native gold containing approximately 60% of mercury.

gold blocking *(Print.)*. The process of pressing a design upon gold-leaf spread out on the cover of a book, the tools or dies, which are heated, leaving the desired impression. Also carried out by machine, gold-foil being fed from a spool.

gold cushion *(Print.)*. A small board, covered with rough calfskin, which is padded with a soft material. The gold-leaf required for gold blocking is placed on the cushion ready for use.

golden beryl *(Min.)*. A clear yellow variety of the mineral beryl, prized as a gemstone. Heliodor is a variety from S.W. Africa. Golden beryl has been used as a name for *chrysoberyl*.

golden number *(Astron.)*. A term derived originally from medieval church calendars, and still used to signify the place of a given year in the Metonic cycle of 19 years.

golden section *(Arch.)*. A proportion thought to be ideal by Renaissance theorists. It is defined as a line cut in such a way that the smaller part is to the greater part as the greater part is to the whole.

gold filled *(Eng.)*. The agreed term for a coating of gold over a base metal which should be stamped with a fraction giving the proportion by weight of the gold present and the karat fineness of the gold alloy.

gold grains *(Radiol.)*. Small lengths of activated gold wire (half-life 2.70 days) having similar advantages to **radon seeds** and used in a similar manner.

gold leaf *(Build.)*. 23 to 24 carat gold which is beaten into leaf form available in books of 25 sheets 82 mm^2 in either loose or transfer form.

gold-leaf *(Eng.)*. Pure gold beaten out into extremely thin sheets, so that it may be applied to surfaces which are to be gilded.

gold-leaf electrometer *(Elec.Eng.)*. **Gold-leaf electroscope**, modified for the measurement of very small currents by observing the rate of movement of the gold leaf through a microscope. The **personal dosimeter** is a development of this instrument.

gold-leaf electroscope *(Elec.Eng.)*. A device for detecting small electric charges which are applied to a piece of thin gold foil usually attached at upper end to a metal electrode. Mutual repulsion between the foil and the similarly charged plate electrode leads to the former being displaced.

gold paints *(Build.)*. Paints made of bronze powders mixed with varnish.

gold rug, gold rubber *(Print.)*. A piece of flannel cloth or soft rubber used for wiping off surplus gold after gilding. It is sold to a refiner after a period of use.

Goldschmidt process *(Chem.)*. (1) See **aluminothermic process**. (2) Detinning of coated iron by use of chlorine.

gold-size *(Build.)*. Type of **size** used as a basis to secure gold-leaf on to surfaces which are to be gilded, and for other purposes. Formulated with specific drying times, e.g. 1 hr, 4 hr and 24 hr.

Gold slide *(Meteor.)*. A slide, named after its designer, which is attached to a marine mercury barometer to make the corrections for index error, latitude, height and temperature mechanically.

gold spring *(Horol.)*. The delicate spring forming part of the detent escapement. One end is anchored to the detent, and the other end rests against the detent horn and projects just beyond it. On one vibration of the balance, the discharging pallet presses against the end of the gold spring and moves the whole detent, bringing about unlocking of the escape wheel. On the return vibration, the discharging pallet merely lifts the gold spring without causing movement of the detent.

gold toning *(Image Tech.)*. A metallic toning process in which the original silver image becomes associated with a gold deposit.

gold tooling *(Print.)*. The decorating of book covers by hand, using gold-leaf and heated tools.

golf-ball printer *(Comp.)*. Uses a moving spherical print head, which is easily removable, enabling many different character sets to be used.

Golgi apparatus *(Biol.)*. A cytoplasmic organelle consisting of a stack of plate-like **cisternae** often close to the nucleus. It is the site of protein glycosylation.

golgi body *(Bot.)*. See **golgi apparatus**.

Golgi's organs *(Zool.)*. In Vertebrates, a type of sensory nerve-ending occurring in tendons near the point of attachment of muscle fibres, stretch receptors.

goliath Edison screw-cap *(Elec.Eng.)*. An Edison screw-cap having a screw thread about 1.5 in diameter, with 4 threads/in.; used with large metal-filament lamps.

gomphosis *(Zool.)*. A type of articulation in which a conical process fits into a cavity; as the roots of teeth into their sockets.

gonad *(Zool.)*. A mass of tissue arising from the primordial germ cells and within which the spermatozoa or ova are formed; a sex gland, ovary or testis. adj. *gonadal*.

gonadotrophic, gonadotropic *(Med.)*. Pertaining to gonadotrophins, gonad-stimulating hormones.

gonal *(Zool.)*. Forming or giving rise to a gonad: as the *gonal ridge*.

Gondwanaland *(Geol.)*. The hypothetical continent of the southern hemisphere which broke up and drifted apart to form bits of the present continents of S. America, Africa, India, Australia and Antarctica.

gone to bed *(Print.)*. A newspaper term meaning that the newspaper has been made up and sent to press.

gong *(Horol.)*. Rectangular steel strip bent into the form of a spiral which, after blueing, provides a deep note when struck; used in clocks when striking the hours.

gon-, gono- *(Genrl.)*. Prefix from Gk. **gonos**, seed, offspring.

goni- *(Genrl.)*. Prefix from Gk. *gōnia*, angle; *gony*, knee. Not to be confused with the prefix **gon-, gono-**.

gonidium *(Bot.)*. (1) A cell which gives rise to an asexual daughter colony in *Volvox*. (2) Any algal cell in the thallus of a lichen.

goniometer *(Eng.)*. A device for measuring angles or for direction finding.

gonioscope *(Med.)*. An instrument used in ophthalmology to measure, and to inspect the anterior chamber of the eye (i.e. the region between cornea and iris).

gonitis *(Vet.)*. Inflammation of the stifle joint of animals.

gonnardite *(Min.)*. A rare zeolite; a hydrated sodium, calcium, aluminium silicate.

gonoblast *(Zool.)*. A reproductive cell.

gonochorism *(Zool.)*. Sex determination.

gonochoristic *(Zool.)*. Having separate sexes.

gonococcus *(Med.)*. A Gram-negative diplococcus, the causative agent of **gonorrhoea**.

gonocoel *(Zool.)*. That portion of the coelom, the walls of which give rise to the gonads: hence, the cavity of the gonads.

gonoduct *(Zool.)*. A duct conveying genital products to the exterior; a duct leading from a gonad to the exterior.

gonopods *(Zool.)*. The external organs of reproduction in Insects, associated with the 8th and 9th abdominal segments in females and with the 9th only in males. In Chilopoda, a pair of modified appendages borne on the 17th (genital) body segment.

gonopore *(Zool.)*. The aperture by which the reproductive elements leave the body.

gonorrhoea, gonorrhea *(Med.)*. A contagious infection of the mucous membrane of the genital tract with the gonococcus, contracted usually through sexual intercourse.

gonosome *(Zool.)*. In colonial animals, all the individuals concerned with reproduction.

go, not-go gauge *(Eng.)*. A **limit gauge** of which one section is above and another below the specified dimensional limits of the part to be gauged. If the 'go' section accepts the part and the 'not-go' section interferes with the part, the latter is accurate within the limits specified.

Gooch crucible *(Chem.)*. A filter used in laboratories, which consists usually of a small porcelain cup, the bottom of which is perforated with numerous small holes and covered with a thin layer of washed asbestos fibres, which act as a filtering medium. Now mainly superseded by sintered glass crucibles. Also *Gooch filter*.

good colour *(Print.)*. A term which indicates consistent distribution of ink throughout a print job.

Good Pasture's syndrome *(Med.)*. A **vasculitis** of the small blood vessels of the kidney and lung. Probably an abnormal immune reaction.

googol *(Maths.)*. A name for the number 10^{100}.

gooseberry stone *(Min.)*. The garnet **grossular**, which was named for the resemblance of this green variety, both in form and colour, to the gooseberry which has the botanical name *Ribes grossularia*.

goose flesh *(Med.)*. See **horripilation**.

go-out *(Hyd.Eng.)*. A sluice in an embankment impounding tidal waters, which can pass through it when the tide is out.

Gordon's formula *(Civ.Eng.)*. An empirical formula giving the collapsing load for a given column. It states that

$$P = \frac{fc.A}{1 + c\left(\frac{l}{d}\right)^2}$$

where P = the collapsing load, fc = safe compressive stresses for very short lengths of the material, A = area of cross-section, l = length of the pin-jointed column, d = the least breadth or diameter of the cross-section, c = a constant for the material and the shape of cross-section.

gore *(Aero.)*. One of the sectorlike sections of the canopy of a parachute.

gorge *(Build.)*. A **drip**.

gorge *(Geol.)*. A general term for all steep sided, relatively narrow valleys, e.g. canyons, overflow channels etc.

goslarite *(Min.)*. Hydrated zinc sulphate, a rare mineral precipitated from water seeping through the walls of lead mines; formed by the decomposition of sphalerite. See also **white vitriol**.

gossan *(Min.Ext.)*. The leached, oxidized material found in surface exposures of an ore deposit; represents the residue left after **secondary enrichment** of a mineral vein or lode. Often stained brown by iron oxides and rich in quartz, gossans are an indication of mineral deposits below the surface, although they may not be of any value themselves.

Gothic *(Arch.)*. The style of architecture, mainly ecclesiastical, prevalent in Western Europe from the late 12th century. It is characterized by pointed arches, ribbed vaults, cruciform plans, flying buttresses and glazed **clerestoreys**.

Gothic *(Print.)*. Originally a term applied to black letter, or Old English type, it is now used to include bold sanserif faces.

Gothic Revival *(Arch.)*. An architectural movement of the late 18th and the 19th centuries to revive the **Gothic** style. Since it had originated as a Christian form of architecture it was appropriate that it should be adopted in the construction of the many new parish churches being built at that time.

gothic wing *(Aero.)*. A very low **aspect ratio** wing with the double curvature plan form similar to a gothic arch of the perpendicular period, used for supersonic aircraft to combine low **wave drag** with **separated lift**. See **ogee wing**.

Gothlandian *(Geol.)*. See **Silurian**.

Göttingen wind tunnel *(Aero.)*. A return flow wind tunnel in which the working section is open. See **open-jet wind tunnel**.

Gott's method *(Elec.Eng.)*. A bridge method of finding the capacitance of a cable. The cable is made to form one arm of the bridge, the others being a standard capacitor and two sections of a slide wire.

gouache *(Build.)*. Opaque colours mixed with water, honey, and gum, applied in impasto style.

gouge *(Build.)*. Similar to a chisel, but having a curved blade and a cutting edge capable of forming a rounded groove.

gouge *(Med.)*. A hollow chisel for removing and cutting bone.

gouge *(Print.)*. A hand tool used to form curved lines.

gouge slip *(Build.)*. A shaped piece of oilstone on which the concave side of the cutting edge of a gouge may be rubbed to sharpen it.

goundou *(Med.)*. *Dog nose, gros nez.* Symmetrical bony overgrowth at the sides of the nose, thought to be a late sequel of yaws.

gout *(Med.)*. A disorder of metabolism in which there is an excess of uric acid in the blood; this is deposited, as sodium biurate, in the joints, bones, ligaments, and cartilages. In acute gout there is a sudden very painful swelling of the joint, usually of the big toe.

Gouy layer *(Min.Ext.)*. Diffuse layer of counterions surrounding charged lattices at surface of particle immersed in liquid.

governor *(Elec.Eng.)*. Speed regulator on rotating machine, e.g. turbo-alternator set.

governor *(Eng.)*. A device for controlling the fuel or steam supply to an engine in accordance with the power demand, so that the speed is kept constant under all conditions of loading.

GPI *(Med.)*. See **paresis**.

Graafian follicle *(Zool.)*. A vesicle, containing an ovum surrounded by a layer of epithelial tissue, which occurs in the ovary of Mammals.

grabbing crane *(Civ.Eng.)*. An excavator consisting of a crane carrying a large grab or bucket in the form of a pair of half-scoops, so hinged as to scoop or dig into the earth as they are lifted.

grab-dredger *(Civ.Eng.)*. A dredging appliance consisting of a grab or grab-bucket suspended from the jib-head of a crane, which does the necessary raising and lowering. Also called a *grapple dredger*.

graben *(Geol.)*. An elongated downthrown block bounded by faults along its length. Cf. **horst**.

grab, grab-bucket *(Civ.Eng.)*. A steel bucket or cage made of two halves hinged together, so that they dig out and enclose part of the material on which they rest; used in mechanical excavators and dredgers. See **grabbing crane, grab-dredger**.

gracilis *(Zool.)*. A thigh muscle of land Vertebrates.

grade *(Civ.Eng.)*. See **gradient**.

grade *(Eng.)*. A measure (expressed by a letter of the alphabet) of the hardness of a grinding wheel as a whole, due to the strength of the bond holding the abrasive particles in place.

graded bedding *(Geol.)*. Bedding which shows a sorting effect with the coarser material at the base progressively changing upwards to finer sediment at the top. Occasionally such grading may be *reversed*.

graded brush *(Elec.Eng.)*. A brush for collecting current from the commutator of an electrical machine; made up of layers of different materials, or of material which has different values of lateral and longitudinal resistance.

graded filter *(Image Tech.)*. A camera filter having variations of density, colour or diffusion across its surface to modify selected areas of the resultant image.

graded-index fibre *(Telecomm.)*. A fibre in an **optical-fibre** cable, in which the refractive index of the glass decreases gradually or in several small steps from the central core to the periphery. See **stepped-index fibre, monomode fibre, multimode fibre**.

grade of service *(Telecomm.)*. The proportion of calls in the busy hour which must fail to be completed through insufficiency of apparatus.

grade pegs *(Surv.)*. Pegs driven into the ground as references, to establish gradients in constructional work. Better, *gradient pegs*.

grader *(Civ.Eng.)*. A power-operated machine provided with a blade for shaping excavated surfaces to the desired shape of slope.

gradient *(Civ.Eng.)*. The degree of slope, e.g. of a highway or railway. US *grade*.

gradient *(Maths.)*. Of a plane curve at a given point: the slope of the tangent at the point. Of a scalar function $f(x,y,z)$: the vector $\nabla f(x,y,z)$ where ∇ is the vector operator *nabla*. Written grad f.

gradient *(Phys.)*. The rate of change of a quantity with distance, e.g., the temperature *gradient* in a metal bar is the rate of change of temperature along the bar.

gradient *(Surv.)*. The ratio of the difference in elevation between two given points and the horizontal distance between them, or the distance for unit rise or fall. Also *incline*.

gradient analysis *(Ecol.)*. Method of ordination in which vegetation samples or stands are plotted on axes representing environmental variables.

gradient of reinforcement *(Behav.)*. The curve which describes the declining effectiveness of reinforcement as the interval between the response and reinforcement increases.

gradient pegs (posts). *(Surv.)*. **Grade pegs**.

gradient post *(Civ.Eng.)*. A short upright post fixed at the side of a railway at a point of change of gradient. An arm on each side of the post indicates whether the track rises or falls, the figure on it giving the distance to be travelled to rise or fall one of the same unit of distance.

gradient wind *(Meteor.)*. A theoretical wind which, when blowing along curved isobars with no tangential acceleration, represents a balance between the pressure gradient, the Coriolis force and the centrifugal force. It is less than the **geostrophic wind** round a **depression**, and greater round an **anticyclone**.

grading *(Build.)*. (1) The proportions of the different sizes of stone or sand used in mixing concrete. (2) The selection of these proportions.

grading *(Civ.Eng.)*. The operation of preparing a surface to follow a given gradient.

grading *(Image Tech.)*. The process of selecting optimum printing values of colour and intensity for the successive scenes of an assembled motion picture negative.

grading *(Telecomm.)*. (1) The scheme of connecting trunks or outlets so that a group of selectors is given access to individual trunks, while larger groups of selectors share trunks when all the individual trunks are found to be in use. (2) An arrangement of trunks connected to the banks of selectors by the method of grading. See **grade of service**.

grading coefficient *(Elec.Eng.)*. A figure denoting the ratio of the lower to the upper limit of current for motor starters.

grading group *(Telecomm.)*. The group of selectors which are concerned in one grading scheme.

grading instrument *(Surv.)*. A general name for any instrument of the **gradiometer** class.

grading shield *(Elec.Eng.)*. A circular conductor placed concentric with a string of suspension insulators on an overhead transmission line to equalize the potential across the individual insulator units. Also called an *arcing shield*.

grading, size *(Chem.)*. See **size-grading**.

gradiometer *(Phys.)*. Magnetometer for measurement of gradient of magnetic field.

gradiometer *(Surv.)*. An instrument for setting out long uniform gradients; it consists of a level that may be elevated or depressed, by known amounts, by means of a vertical tangent screw.

graduated circle *(Surv.)*. A circular plate, marked off in degrees and parts of degrees, used on surveying instruments as a basis for the measurement of horizontal or vertical angles. See **horizontal circle, vertical circle**.

graduated courses *(Build.)*. Courses of slates laid so that the gauge diminishes from eaves to ridge.

graduated vessels *(Chem.)*. Vessels which are used for measuring liquids and are adapted to measure or deliver definite volumes of liquid.

Graff's 'C' stain *(Paper)*. A microscopical stain for assisting the identification of paper, prepared from aluminium, calcium and zinc chlorides, together with iodine and potassium iodide.

graft *(Med.)*. Any transplanted organ or tissue.

graft chimera *(Bot.)*. A more or less stable **periclinal chimera** in which the skin and core (typically) derive from different species. Graft chimeras arise, rarely, from the junction of stock and scion in interspecific grafts, can be propagated only vegetatively and exhibit characteristics between the two species. Graft chimeras are indicated by a + sign before the specific or generic name (intra- and inter- generic chimaeras respectively). The most famous is probably + *Laburnocytisus adami* which consists of a skin of *Cytisus purpureus* over a core of *Laburnum anagyroides*.

graft hybrid *(Bot.)*. In some usages, *graft chimera*, in others, *burdo*.

grafting *(Bot.)*. The placing for propagation or experiment of a piece of one plant (scion) on to a piece, usually with roots, of another (stock) so that the tissues may unite and growth follow.

grafting *(Build.)*. The operation of lengthening a timber by jointing another piece on to it.

graft-versus-host disease *(Med.)*. *GVH*. A **graft-versus-host reaction** of transfused or transplanted lymphocytes against *host* antigens. Major complication of **bone marrow grafting**.

graft-versus-host reaction *(Immun.)*. Occurs when a tissue graft (notably bone marrow) contains T-lymphocytes which can respond to antigens present in the recipient which are not identical with those of the donor. If the recipient is unable to suppress the donor lymphocytes, because of immunological immaturity or to immunosuppression due to X-irradiation or to drugs, they cause severe damage to the recipient which begins in the skin and gut and may progress to death. Abbrev. *GVH*.

Graham escapement *(Horol.)*. The dead-beat clock escapement.

grahamite *(Min.)*. A member of the asphaltite group.

Graham's law *(Chem.)*. The velocity of effusion of a gas is inversely proportional to the square root of its density.

grain *(Bot.)*. (1) See **caryopsis**. (2) The pattern on the surface of worked wood due to variations in the size, shape, arrangement and composition of the cells forming the wood.

grain *(Chem.)*. Unit of mass. See **apothecaries' weight**.

grain *(Geol.,Min.)*. (1) See **rift and grain**. (2) Average size of mineral crystals composing a rock. Direction in which it tends to split. Of a metal, texture of broken surface (smooth-, rough- etc.).

grain *(Image Tech.)*. The component particles of black silver which make up a photographic image after development and their aggregation into clumps; colour images from which silver has been removed may still show something of the original grouping. *Graininess* is a subjective term describing the visual appearance of a picture, whereas *granularity* is an objective measurement using a micro-densitometer to examine the image structure.

grain growth *(Eng.)*. Associated with recrystallization, this refers to an increase in the average grain size resulting from some crystals absorbing adjacent ones.

graining *(Build.)*. The operation of brushing, combing, or otherwise marking a painted surface while the paint is still wet, in order to produce an imitation of the grain of wood.

graining boards *(Print.)*. Boards or metal plates with parallel lines in relief, running diagonally. Used to produce a diced effect on covers.

graining comb *(Build.)*. See **comb**.

graining plates *(Print.)*. The aluminium or zinc plates for lithography require a grained surface, which will retain moisture and also hold the image areas or 'work'. Plates can be used several times and regrained after each use in a graining machine, where the plate is covered with balls of porcelain, glass, or steel, and, after adding water and an abrasive such as carborundum, is vibrated mechanically to produce the required grain.

grain refining *(Eng.)*. Production of small closely knit grains, resulting in improved mechanical properties. Particularly with aluminium alloys it is achieved by small additions to melts of substances such as boron which cause fine nucleation on casting. For ferrous metallurgy, see **grain-size control**.

grains *(Min.Ext.)*. See **coal sizes**.

grain size *(Geol.,Eng.)*. *Particle size*. The average size of the grains or crystals in a sample of metal or rock.

grain-size analysis *(Powder Tech.)*. US term for particle-size analysis. It is also the literal translation for the German phrase used to specify particle-size analysis. Confusion arises because grain-size analysis is also used to describe procedures for measuring the size of crystallites in a cast or sintered metal.

grain-size control *(Eng.)*. Specifically, control of the rate at which the austenite grains grow when steel is heated above the critical range; the control is effected by the addition of aluminium before casting.

gramicidin *(Biol.)*. A heterogeneous group of ionophores. Thus gramicidin A is an open chain polypeptide while gramicidin S is a cyclic peptide.

graminacious, gramineous *(Bot.)*. Relating to grasses.

Graminae *(Bot.)*. *Poaceae*. The grass family, ca 9000 spp. of monocotyledonous flowering plants (superorder Commelinidae). Annual or perennial herbs, sometimes woody (the bamboos), often tufted or rhizomatous; cosmopolitan and represented in most habitats of earth. The aerial stems are usually hollow and bear leaves, sheathing the stem at the base and with flat, long and narrow blades, in two ranks. The flowers are relatively inconspicuous and wind pollinated (see **floret** and **spikelet**). Many tropical grasses are C_4 plants, including maize. The grasses are extremely important economically for food (all the cereals — rice, wheat, oats, maize etc. and sugar cane), for fodder, and also for some constructional and furniture-making materials (bamboos) and in lawns, sportsfields etc.

graminicolous *(Bot.)*. Living on grasses; esp. of parasitic fungi.

graminivorous *(Zool.)*. Grass-eating.

grammatite *(Min.)*. Syn. for *tremolite*.

gram(me) *(Phys.)*. The unit of *mass* in the CGS system. It was originally intended to be the mass of $1 cm^3$ of water at 4°C but was later defined as one-thousandth of the mass of the International Prototype Kilogramme, a cylinder of platinum-iridium kept at Sèvres.

gram(me)-atom *(Chem.)*. The quantity of an element whose mass in grams is equal to its relative atomic mass. A **mole** of atoms.

gram(me) calorie *(Genrl.)*. See **calorie**.

gram(me)-ion *(Phys.)*. Mass in grams of an ion, numerically equal to that of the molecules or atoms constituting the ion.

gram(me)-molecular volume *(Chem.)*. See **molar volume**.

gram(me)-molecule *(Chem.)*. See **mole**.

gram(me)-röntgen *(Phys.)*. The real conversion of energy when one röntgen is delivered to 1 g of air (approx. 8.4 μJ). A convenient multiple is the **meg**

gram(me)s per square metre *(Paper)*. The mass in

grams of one square metre of the paper or board under the prescribed conditions of test. An expression of the basis weight or substance (now preferred term) of paper.

Gram-negative bacteria *(Biol.)*. Those bacteria which fail to stain with Gram's reaction. The reaction depends on the complexity of the cell wall and has for long determined a major division between bacterial species. Cf. **Gram-positive bacteria**.

gramophone *(Acous.)*. An instrument for reproducing sound, using a stylus in contact with a spiral groove on a revolving disk. In U.S.A. this term is proprietary: the general equivalent is *phonograph*.

gramophone audiometer *(Acous.)*. A quick method of testing the hearing of a large number of subjects, who are required to write down numbers perceived at diminishing intensities through head-telephones, the sounds being obtained from a gramophone record.

gramophone record *(Acous.)*. See **disk record**.

Gram-positive bacteria *(Biol.)*. The comparative simplicity of the cell-wall of some bacterial species allows them to be stained by Gram's procedure. See **Gram-negative bacteria**.

grandfather, father, son files *(Comp.)*. Three most recent versions of a file which is periodically updated, retained for security purposes.

grand mal *(Med.)*. General convulsive epileptic seizure, with loss of consciousness. See also **epilepsy**, **petit mal**.

grand period of growth *(Bot.)*. The period in the life of a plant, or of any of its parts, during which growth begins slowly, gradually rises to a maximum, gradually falls off, and comes to an end, even if external conditions remain constant.

grand swell *(Acous.)*. A swell-like balanced pedal for bringing in, as it is depressed, all the stops in an organ in a graded series.

Grand Unified Theory *(Astron.,Phys.)*. A theory to describe elementary particles in which it is postulated that the *strong* and *weak* nuclear forces, together with the *electromagnetic force* merge into a single force field. The theory finds a particular application in attempts to describe the very early universe. Abbrev. *GUT*. See **strong interaction**, **weak interaction**.

granite *(Geol.)*. A coarse-grained igneous rock containing megascopic quartz, averaging 25%, much feldspar (orthoclase, microcline, sodic plagioclase), and mica or other coloured minerals. In the wide sense, granite includes alkali-granites, adamellites, and granodiorites, while the *granite clan* includes the medium- and fine-grained equivalents of these rock types.

granite-aplite *(Geol.)*. See **aplite**.

granite-porphyry *(Geol.)*. Porphyritic microgranite, a rock of granitic composition but with a groundmass of medium grain size in which larger crystals (phenocrysts) are embedded.

granite series *(Geol.)*. A series relating the different types of granitic rock with respect to their time and place of formation in an orogenic belt, starting with early-formed deep-seated autochthonous granites and ending with post-tectonic high-level plutons.

granitic finish *(Build.)*. A surface finish, resembling granite, given to cement work by the use of a suitable face mix.

granitic texture *(Geol.)*. See **granitoid texture**.

granitization *(Geol.)*. A metamorphic process by which rocks can be changed into granite *in situ*.

granitoid texture *(Geol.)*. A rock fabric in which the minerals have no crystal form and occur in shapeless interlocking grains. Such rocks are in the coarse grain-size group. Also called *xenomorphic granular texture*.

granoblastic texture *(Geol.)*. An arrangement of equigranular mineral grains in a rock of metamorphic origin similar to that of a normal granite but produced by recrystallization in the solid and not by crystallization from a molten condition. The grains show no preferred orientation.

granodiorite *(Geol.)*. An igneous rock of coarse grain size, containing abundant quartz and at least twice the amount of plagioclase over orthoclase, in addition to coloured minerals such as hornblende and biotite. Cf. **diorite**.

granolithic *(Build.)*. A rendering of cement and fine granite chippings, used as a covering for concrete floors, on which it is floated in a layer 1 to 2 in. (25 to 50 mm) in thickness. Used because of its hard-wearing properties.

granophyre *(Geol.)*. An igneous rock of medium grain size, in which quartz and feldspar are intergrown as in graphic granite.

granular *(Powder Tech.)*. Said of particles having an approximately equidimensional but irregular shape.

granular, globular pearlite *(Eng.)*. Pearlite in which the cementite occurs as globules instead of as lamellae. Produced by very slow cooling through the critical range, or by subsequent heating just below the critical range.

granularity *(Image Tech.)*. See **grain**.

granulation *(Astron.)*. The mottled appearance of the Sun's **photosphere** when viewed at very high resolution.

granulation tissue *(Med.)*. A new formation of vascular connective tissue which grows to fill up the gap of a wound or ulcer; when healing is completed, a white scar is left.

granule *(Geol.)*. A rock or mineral with a grain size between 2 and 4 mm.

granulite *(Geol.)*. A granular-textured metamorphic rock, a product of regional metamorphism.

granulitic texture *(Geol.)*. The texture of a granulite, sometimes referred to as *granulose* or *granoblastic*, is an arrangement of shapeless interlocking mineral grains resembling the granitic texture but developed in metamorphic rocks. Fewer than 10% of the grains have a preferred orientation.

granulitization *(Geol.)*. The process in regional metamorphism of reducing the components of a solid rock to grains. If the reduction of the size of the particles goes farther, mylonite is produced.

granulocyte *(Immun.)*. A general term describing polymorphonuclear leucocytes.

granulocytopenia *(Med.)*. An abnormal diminution in the number of granulocytes in the blood.

granuloma *(Immun.)*. A localized accumulation of macrophages around the site of some continuing stimulus, such as a persisting antigen which causes delayed type hypersensitivity and the release of lymphokines chemotactic to monocytes (which turn into macrophages). The macrophages may become tightly compressed and their cell membranes often fuse together so as to form multinucleated giant cells.

granuloma annulare *(Med.)*. A condition in which rings of white cellular nodules appear on the back of the hands, and occasionally elsewhere.

granuloma inguinale *(Med.)*. *Ulcerating granuloma, granuloma venereum*. A chronic disease occurring in the tropics, in which ulcerating nodules appear on the genital organs, the perineum and the groin.

granulomatous *(Med.)*. Of the nature of, or resembling, granuloma.

granum *(Bot.)*. A stack, rather like a pile of coins, of ca 5–50 **thylakoids** (or disks) forming one of say 40–60 such bodies in the **stroma** of a chloroplast in vascular plants, bryophytes and some green algae.

grape-sugar *(Chem.)*. **glucose**.

graph *(Maths.)*. A drawing depicting the relationship between two or more variables. The relationship may be known mathematically, e.g. the graph of the equation $x^2 + y^2 = 1$ is a circle, or it may be unknown, e.g. the graph of a hospital patient's temperature.

graphecon *(Radar)*. A double-ended storage tube used for the integration and storage of radar information and as a translating medium.

graphical display unit *(Comp.)*. Output device, incorporating a cathode ray tube, on which line drawings and text can be displayed. Used in conjunction with a **light pen** to input or re-position data.

graphical methods *(Eng.)*. The name given to those methods in which items, such as forces in structures, are determined by drawing diagrams to scale.

graphic formula *(Chem.)*. A formula in which every atom is represented by the appropriate symbol, valency bonds being indicated by dashes; e.g. H—O—H, the graphic formula for water.

graphic granite *(Geol.)*. Granite of pegmatitic facies, in which quartz and alkali-feldspar are intergrown in such a manner that the quartz simulates runic characters. Also called *runite*.

graphics *(Comp.)*. See **computer graphics, raster graphics, vector graphics**.

graphics tablet *(Comp.)*. Input device where the movement of a pen over a sensitive pad is translated into digital signals giving the pen's position.

graphic statics *(Eng.)*. A method of finding the stresses in a framed structure, in which the magnitude and direction of the forces are represented by lines drawn to a common scale.

graphic texture *(Geol.)*. A rock texture in which one mineral intimately intergrown with another occurs in a form simulating ancient writing, especially runic characters; produced by simultaneous crystallization of two minerals present in eutectic proportions. See **graphic granite**.

graphite *(Chem.,Min.)*. One of the two naturally occurring forms of crystalline carbon, the other being diamond. It occurs as black, soft masses and, rarely, as shiny crystals (of flaky structure and apparently hexagonal) in igneous rocks; in larger quantities in schists particularly in metamorphosed carbonaceous clays and shales, and in marbles; also in contact metamorphosed coals and in meteorites. Graphite has numerous applications in trade and industry now much overshadowing its use in 'lead' pencils. Much graphite is now produced artificially in electric furnaces using petroleum as a starting material. Also called *black lead, plumbago*. See **colloidal graphite**.

graphite brush *(Elec.Eng.)*. A brush, made of graphite, for collecting the current from the commutator of an electric machine. It has a higher conductivity and better lubricating properties than an ordinary carbon brush.

graphite paint *(Build.)*. Paint consisting of fine graphite and bitumen solution, with or without oil, and supplied chiefly for the protection of hot wet surfaces such as the insides of industrial boilers.

graphite reactor *(Nuc.Eng.)*. One in which fission is produced principally or substantially by slow neutrons moderated by graphite. See **moderator**.

graphite resistance *(Elec.Eng.)*. A resistance unit consisting of a rod of graphite, which has a high ohmic value; also a variable resistance made up of piles of graphitized disks of cloth under a variable pressure.

graphitic acid *(Chem.)*. Graphite which has been treated with nitric acid and potassium (V) chlorate for a prolonged period. It is an *intercalation compound*.

graphitic carbon *(Eng.)*. In cast-iron, carbon occurring as graphite instead of as cementite.

graphitization *(Chem.)*. The transformation of amorphous carbon to graphite brought about by heat. It results in a volume change due to the alteration in atomic lattice layer spacing. It is reversible under bombardment by high-energy neutrons and other particles. See **Wigner effect**.

grapnel, grappel *(Eng.)*. Any device used to grapple with an object which is obscured to view, such as a submarine object. Grapnels generally take the form of grapnel hooks having several flukes.

grapnel, grappel *(Min.Ext.)*. An extracting tool used in boring operations.

graptolite *(Geol.)*. An extinct group of animals, represented by their abundant fossil remains in Palaeozoic rocks, and used for dating Ordovician and Silurian sediments.

grass *(Bot.)*. See **Gramineae**.

grass *(Radar)*. Irregular deflection from the time-base of a radar display, arising from electrical interference or noise.

grass disease *(Vet.)*. See **bovine hypomagnesaemia**.

grassland *(Ecol.)*. Natural, semi-natural or farm vegetation in which the dominant species are grasses. Major grasslands of the world include pampas, prairies and savannas.

Grassot fluxmeter *(Elec.Eng.)*. See **fluxmeter**.

grass sickness *(Vet.)*. A disease of unknown cause affecting the horse, often occurring soon after the horse is put on to grass, associated with dysfunction of the bowels. Acute cases are characterized by colic and more chronic cases by emaciation. The essential pathological change is believed to be degeneration of the neurones in the ganglia of the autonomous nervous system.

grass staggers *(Vet.)*. See **bovine hypomagnesaemia**.

grass table *(Build.)*. A **ground table**.

grass tetany *(Vet.)*. See **bovine hypomagnesaemia**.

grate *(Eng.)*. That part of a furnace which supports the fuel. It consists of fire bars or bricks so spaced as to admit the necessary air. See **fire bars**.

grate area *(Eng.)*. The area of the grate in a furnace burning solid fuel; for a boiler furnace, a measure of the evaporative capacity of the boiler.

graticule *(Surv.)*. See **reticule**.

grating *(Phys.)*. An arrangement of alternate reflecting and nonreflecting elements, e.g. wire screens or closely spaced lines ruled on a flat (or concave) reflecting surface, which, through diffraction of the incident radiation analyses this into its frequency spectrum. An *optical grating* can contain a thousand lines or more per cm. A **standing-wave** system of high-frequency sound waves with their alternate compressive and rarefied regions can give rise to a diffraction grating in liquids and solids. With a *criss-cross* system of waves, a 3-dimensional grating is obtainable.

grating spectrum *(Phys.)*. An optical spectrum produced by a **diffraction grating**. Cf. **prismatic spectrum**.

Grätz rectifier *(Elec.Eng.)*. Type of bridge rectifier using 6 rectifying elements for 3-phase supply.

gravel *(Build.)*. A natural mixture of rounded rock fragments and generally sand used in the manufacture of concrete.

gravel *(Geol.)*. The name of the aggregate consisting dominantly of pebbles, though usually a considerable amount of sand is intercalated. In the Stratigraphical Column, gravels of different ages and origins occur abundantly, e.g. in S.E. England, where they consist chiefly of well-rounded flint pebbles originally derived from the Chalk. These gravels are mainly of fluviatile and fluvioglacial origin, but marine gravels are also common in the littoral zone. The indurated equivalent of gravel is conglomerate. See **particle size**.

gravel *(Med.,Vet.)*. Small calculi in the kidneys, ureters or urinary bladder.

gravel board, gravel plank *(Build.)*. A long board standing on its edge at the bottom of a wooden fence, so that the upright boards of the fence do not have to reach down to the ground.

graveolent *(Bot.)*. Having a strong rank odour.

Graves disease *(Med.)*. Excessive secretion of thyroid hormone causing **thyrotoxicosis** with loss of weight, sweating, rapid pulse and prominent eyes (exophthalmic goitre).

graveyard *(Nuc.Eng.)*. See **burial site**.

gravid *(Med.,Zool.)*. Pregnant; carrying eggs or young.

gravimetric analysis *(Chem.)*. The chemical analysis of materials by the separation of the constituents and their estimation by weight.

graving dock *(Civ.Eng.)*. See **dry dock**.

gravipause *(Space)*. That point or border in space in which the gravitational force of one body is matched by the counter-gravity of another; also *neutral point*.

graviperception *(Bot.)*. The perception of gravity by plants.

gravitation *(Phys.)*. The name given to that force of nature which manifests itself as a mutual attraction between masses, and whose mathematical expression was first given by Newton, in the law which states: 'Any two particles of matter attract one another with a force directly proportional to the product of their masses and inversely proportional to the square of the distance between them.' This may be expressed by the equation:

$$F = G\frac{m_1 m_2}{d^2},$$

where F is the force of gravitational attraction between bodies of mass m_1 and m_2, separated by a distance D. G is the constant of gravitation and has the value $6.670 \times 10^{-11}\,\mathrm{N\,m^2\,kg^{-2}}$.

gravitational astronomy *(Astron.)*. See **celestial mechanics**.

gravitational differentiation *(Geol.)*. The production of igneous rocks of contrasted types by the early separation of denser crystals such as olivine, pyroxenes, etc. which become concentrated in the basal parts of intrusions. The ultramafic rocks such as peridotites and picrites may have been formed in this way.

gravitational field *(Phys.)*. That region of space in which at all points a gravitational force would be exerted on a test particle.

graviton *(Phys.)*. A *gauge boson* that is the agent for gravitational interactions between particles. The graviton has not been detected but is believed to have zero mass and zero charge. It bears the same relationship to the gravitational field as the photon does to the electromagnetic field. Gravitational interactions between particles are so weak that no quantified effects have been observed.

gravitropism, geotropism *(Bot.)*. A **tropism** in which a plant part comes orientated with respect to the direction of gravity.

gravity assist *(Space)*. Use of a body to provide an energy boost to a spacecraft on a trajectory planned to pass close to the body for this purpose.

gravity clock *(Horol.)*. One actuated by its own weight, e.g. by moving down a racked pillar.

gravity-controlled instruments *(Elec.Eng.)*. Electrical measuring instruments in which the controlling torque is provided solely by the action of gravity.

gravity conveyor *(Eng.)*. A conveyor in which the weight of the articles handled is sufficient to effect their transport from a higher to a lower point; as when they are allowed to slide down an inclined runway.

gravity dam *(Civ.Eng.)*. A dam which is prevented by its own weight from overturning.

gravity diecasting *(Eng.)*. A process by means of which castings of various alloys are made in steel or cast-iron moulds, the molten metal being poured by hand. See **diecasting, pressure diecasting**.

gravity drop hammer *(Eng.)*. A type of machine hammer used for drop forging, in which the impact pressure is obtained from the kinetic energy of the falling ram and die. Also *board hammer*.

gravity escapement *(Horol.)*. An escapement in which a constant impulse is given to the pendulum by a weight which is lifted and which always falls though a constant distance. The best-known gravity escapement is the *double three-legged*, in which pivoted arms are lifted a constant height to give, on falling, a constant impulse to the pendulum. Used in Big Ben.

gravity feed gun *(Build.)*. A spray gun for use with gravity feed containers having the fluid connection on the upper gun body.

gravity feed tanks *(Aero.,Autos.)*. Fuel tanks, situated above the point of delivery to the engine, which feed the engine solely by gravity.

gravity plane *(Min.Ext.)*. An inclined plane on which the descending full trucks pull up the ascending empty ones.

gravity process *(Glass)*. See **gob process**.

gravity roller conveyor *(Eng.)*. A fabricated framework, usually in longitudinal sections and slightly inclined, carrying freely rotating tubular rollers on which a load may be moved by the force of gravity acting on it.

gravity scale *(Eng.)*. A measure of certain characteristics of diesel fuel oil, related to the density of the oil by an arbitrary formula.

gravity separation *(Min.Ext.)*. Use of differences between relative densities of roughly-sized grains of mineral to promote settlement of denser species while less-dense grains are washed away. See **buddles, dense media process, Humphrey spirals, hydrocyclones, jig, shaking table, sluice, vanner**.

gravity stabilization *(Space)*. A way of stabilizing the orientation of a spacecraft with reference to a primary body (such as Earth) by using the gravity gradient, in which the long axis of the spacecraft is directed toward the centre of the body.

gravity stamp *(Min.Ext.)*. Set of 5 heavy pestles, lifting about 10 in at 90 times a minute by a cam and allowed to fall on ore spread in mortar-box, for crushing purposes. Also *California stamp*.

gravity tectonics *(Geol.)*. Processes of rock deformation and folding which are activated by gravity applied over considerable periods of geological time.

gravity transport *(Geol.)*. The movement of material under the influence of gravity. It includes downhill movement of weathering products, and movement of unweathered material in landslides.

gravity water system *(Phys.)*. A system in which flow occurs under the natural pressure due to gravity.

gravity waves *(Phys.)*. Liquid surface waves controlled by gravity and not surface tension.

gravure *(Print.)*. Abbrev. for *photogravure*.

Grawitz's tumour *(Med.)*. See **hypernephroma**.

gray *(Radiol.)*. SI unit of absorbed dose. Symbol Gy. See **absorbed dose**.

Gray-King Test *(Min.Ext.)*. Test of coking quality of coal under prescribed conditions of heating to 600°C.

graywacke *(Geol.)*. See **greywacke**.

grazing angle *(Phys.)*. A very small **glancing angle**.

grease *(Vet.)*. *Greasy heels, sore-heels*. A chronic seborrheic dermatitis associated with the caudal aspect of the pastern and fetlock. Usually associated with poor management.

grease cup *(Eng.)*. A lubricating device consisting of an externally threaded cylindrical cup. Screwing down an internally threaded cap forces grease into the bearing on which the grease cup is fixed.

grease gun *(Eng.)*. A device for forcing grease into bearings under high pressure. It consists of a cylinder from which the grease is delivered by hand pressure on the piston, intensified by a second plunger which forms the delivery pipe and which is pressed against a nipple screwed into the bearing.

greaseproof paper *(Paper)*. Any paper that in its natural state or as the result of coating or other treatment resists penetration by oils or greases. *Greaseproof*; a quality of paper possessing grease resistant characteristics brought about by heavy beating of a suitable fibrous furnish.

grease-spot photometer *(Phys.)*. A simple means of comparing the intensities of two light sources. A screen of white paper, rendered partially translucent by a spot of grease, is illuminated normally by the two sources, one on each side. The position of the screen is adjusted until the grease spot is indistinguishable from its surround, when the illuminations on the two sides may be assumed to be equal. Also called *Bunsen photometer*. See **photometer**.

grease table *(Min.Ext.)*. Sloping table anointed with petroleum jelly, over which diamondiferous concentrate is washed, the diamond adhering strongly while the gangue is worked away.

grease trap *(Build.)*. A trapped gulley receiving sink wastes, and specially designed to prevent obstruction of gulley or drain by congealed fatty matter.

greasy heels *(Vet.)*. See **grease**.

greasy pig disease *(Vet.)*. *Marmite disease, exudative epidermitis. Staphylococcus hyicus* infection of piglets. Skin becomes covered with a greasy brown exudate which dries and then peels off. Mortality high if left untreated.

great circle *(Maths.)*. The intersection of a sphere by a plane passing through its centre. The shortest distance between two points on the surface of a sphere is along the great circle passing them.

Great Ice Age *(Geol.)*. See **Pleistocene Period**.

great primer *(Print.)*. An old type size, approximately 18-point.

green *(Civ.Eng.)*. Colloq. for concrete in the hardening stage, after pouring and before setting.

green *(Powder Tech.)*. Describes the unheat-treated condition of preforms held together by the cohesive forces resulting from compaction alone or with the assistance of a temporary binder.

green algae *(Bot.)*. The *Chlorophyta*.

greenalite *(Min.)*. A septechlorite related to the chlorites chemically and to the serpentines structurally. A hydrous, iron silicate of composition $Fe_6Si_4O_{10}(OH)_3$.

green bricks *(Build.)*. Moulded clay shapes which after undergoing a burning process will become bricks.

Greenburg-Smith impinger *(Powder Tech.)*. Device for sampling gas streams, consisting of a tall, flat-bottomed flask containing a vertical tube narrowing to a jet at the lower end, which is immersed in dust-free water. The gas stream is sucked through the jet and any dust particles are collected in the water.

green carbonate of copper *(Min.)*. See **malachite**.

green flash *(Astron.)*. A phenomenon sometimes seen in clear atmospheres at the instant when the upper rim of the sun finally disappears below the horizon as a bright blob of light; green is the last apparent colour from the sun, because the more greatly refracted blue is dispersed.

green glands *(Zool.)*. The antennal excretory glands of decapod Crustacea.

greenheart *(For.)*. A very strong yellowish-green timber from South America; largely used for piles and underwater work because of its considerable resistance to the attack of the *Teredo*, a wood-boring mollusc. Used for fishing rods.

greenhouse effect *(Astron.)*. Phenomenon by which thermal radiation from the Sun is trapped within the atmospheric environment of a planet, causing a higher surface temperature. The effect is most pronounced for **Venus** and the **Earth**.

greenhouse effect *(Geol.)*. The heating of the atmosphere by the absorption of infrared energy re-emitted by the Earth from incoming solar energy.

greenhouse effect *(Meteor.)*. The trapping by water vapour and carbon dioxide of long-wave radiation emitted by the Earth which leads to the temperature at the Earth's surface being considerably higher than would otherwise be the case.

Greenland spar *(Min.)*. See **cryolite**.

green manure *(Bot.)*. A method of increasing soil organic matter by planting a crop on temporarily free land and ploughing it in while still green.

greenockite *(Min.)*. Cadmium sulphide occurring as small yellow hexagonal crystals in cavities in altered basic lavas.

green sand *(Foundry)*. Highly siliceous sand, with some alumina, dampened and used with admixed coal or charcoal in moulding.

greensand *(Geol.)*. A sand or sandstone with a greenish colour, due to the presence of the mineral glauconite.

green sand casting *(Foundry)*. One made in foundry sand not strengthened by kiln-drying before pouring.

Green's theorem *(Maths.)*. (1) If C is the boundary curve of plane region R, then

$$\int_C (\mathrm{P}dx + \varphi dy) = \int_R \left(\frac{\partial \varphi}{\partial x} - \frac{\partial \mathrm{P}}{\partial y}\right) dx\, dy.$$

Also known as *Gauss Theorem* or *Ostrogradski's theorem*. It is a case of *Stokes' Theorem*. (2) If S is the boundary surface of volume V, then $\int_S \mathbf{F} \cdot d\mathbf{S} = \int_V \operatorname{div} \mathbf{F}\, dV$. Also known as *Green's lemma*, *Gauss theorem*, *Ostrogradski's theorem* or the *divergence theorem*. (3) With S and V as above,

$$\int_V (\mathrm{P}\nabla^2\varphi + \varphi\nabla^2\mathrm{P})\, dV = \int_S \left(\mathrm{P}\frac{\partial \varphi}{\partial n} - \varphi\frac{\partial \mathrm{P}}{\partial n}\right) dS,$$

where n is in the direction normal to S.
Cf. **Stokes' theorem, Gauss' theorem**.

greenstick fracture *(Med.)*. Fracture of rickety bones, which break like a green stick.

greenstone *(Geol.)*. An omnibus term lacking precision and applied indiscriminately to basic and intermediate igneous rocks in which much chlorite has been produced as a result of metamorphism.

green vitriol *(Min.)*. **Melanterite**.

Greenwich Mean Time *(Astron.)*. Abbrev. *GMT*. *Mean solar time* referred to the zero meridian of longitude, i.e. that through Greenwich. GMT is the basis for scientific and navigational purposes. See **universal time**.

gregaria phase *(Zool.)*. In locusts (Orthoptera), a phase characterized by high activity and gregarious tendencies, differing morphologically from the **solitaria phase** with which, under natural conditions, it alternates.

Gregorian calendar *(Genrl.)*. The name commonly given to the civil calendar now used in most countries of the world. It is the Julian calendar as reformed by decree of Pope Gregory XIII in 1582. The Gregorian reform omitted certain leap years, and brought the length of the year on which the calendar is based nearer to the true astronomical value.

Gregorian telescope *(Astron.)*. A form of reflecting telescope, very similar in principal to the Cassegrainian, in which the large mirror is pierced at the centre and the light is reflected back into an eye-piece in the centre by a small concave mirror on the principal axis and a little outside the focus.

Gregory formula *(Maths.)*. Obtained from the Newton interpolation formula, used for numerical integration. It takes the form

$$\int_a^b f(x)dx = \frac{h}{2}\left(y_0 + 2y_1 + 2y_2 + \ldots + 2y_{n-1} + y_n\right)$$
$$-\frac{h}{12}\left(\Delta y_{n-1} - \Delta y_0\right) - \frac{h}{24}\left(\Delta^2 y_{n-2} + \Delta^2 y_0\right)$$
$$-\frac{19h}{720}\left(\Delta^3 y_{n-3} - \Delta^3 y_0\right) - \frac{3h}{160}\left(\Delta^4 y_{n-4} + \Delta^4 y_0\right) - \ldots$$

where h is the common interval between successive values of x, and $\Delta^m y_{n-m}$ represent finite differences.

greige *(Textiles)*. See **grey**.

greisen *(Geol.)*. A rock composed essentially of mica and quartz, resulting from the alteration of a granite by percolating solutions. Greisens often contain small amounts of fluorite, topaz, tourmaline, cassiterite, and other relatively uncommon minerals, and may be associated with mineral deposits, as in Cornwall. See **pneumatolysis**.

greisenization *(Geol.)*. The process by which granite is converted to greisen. Greisenization is a common type of wall rock alteration in areas where granite is traversed by hydrothermal veins. Sometimes *greisening*.

grenz rays *(Radiol.)*. X-rays produced by electron beams accelerated through potentials of 25 kV or less. These are generated in many types of electronic equipment; low penetrating power.

grey *(Image Tech.)*. Neutral in colour, having no perceptible hue and zero saturation; in colour reproduction, a balance of the component colours to give this subjective impression.

grey *(Textiles)*. Fabrics in the state they leave the loom or knitting machine before any scouring or bleaching has been carried out.

grey body *(Phys.)*. A body which behaves as a black body in respect of the absorbed fraction of the incident radiation, $E_a = E_i(1 - A)$ where E_a is the absorbed fraction, E_i incident energy, and A albedo, and complies with the **Stefan-Boltzmann law**.

grey copper ore *(Min.)*. See **tetrahedrite**.

grey iron *(Eng.)*. Pig- or cast-iron in which nearly all the carbon not included in pearlite is present as graphite carbon. See **mottled iron, white iron**.

grey key image *(Image Tech.)*. A neutral image printed in shades of black and white used to supplement the primary colour images in colour reproduction, extending the density range and adding visibility.

grey matter *(Zool.)*. The centrally-situated area of the central nervous system, mainly composed of cell bodies. Cf. **white matter**.

grey scale *(Image Tech.)*. A series of shades of grey ranging from white to black used as a test object in reproduction processes.

greywacke *(Geol.)*. A sandstone containing silt, clay, and rock fragments in addition to quartz grains. It is much more poorly sorted than other types of sandstone, and often occurs on beds which show gradation in grain size from fine at the top to coarse at the bottom. See **turbidite**.

greywethers *(Geol.)*. Grey-coloured rounded blocks of sandstone or quartzite left as residual boulders on the surface of the ground when less resistant material was denuded. From a distance they resemble sheep grazing. See **sarsen**.

grid *(Build.)*. The plan layout of the structure of a given building.

grid *(Elec.Eng.)*. The electrical power generation, transmission and, to a lesser extent, distribution system. See also **supergrid, accumulator grid, damper resistance grid**.

grid *(Electronics)*. Control electrode having an open structure (e.g. mesh) allowing the passage of electrons; in an electron gun it may be a hole in a plate.

grid *(Eng.)*. A grating made up of a number of parallel bars, such as that required to prevent foreign matter from entering a pump intake.

grid *(Surv.)*. A network of lines superimposed upon a map and forming squares for referencing, the basis of the network being that each line in it is at a known distance either east or north of a selected origin. In ordnance surveying, triangulation system covering large area.

grid bias *(Electronics)*. The d.c. negative voltage applied to the control grid of a thermionic valve.

grid-bias battery *(Elec.Eng.)*. Once widely used to provide a potential difference for the grid polarization of valves. US, *C-battery*.

grid capacitor *(Elec.Eng.)*. A capacitor between the grid and the remainder of the grid circuit.

grid circuit *(Elec.Eng.)*. The circuit connected between the grid and the cathode of a thermionic valve.

grid conductance *(Phys.)*. The in-phase component of the grid input admittance, due to grid current, Miller effect, etc.

grid control *(Electronics)*. That provided by voltage on grid of a thyratron or mercury-arc rectifier; at a sufficient positive voltage, anode current flows, grid loses control until deionization is effected after loss of anode voltage.

grid-controlled mercury-arc rectifier *(Elec.Eng.)*. One in which initiation of arc in each cycle is determined by phase of voltage on a grid for each anode.

grid dip meter *(Elec.Eng.)*. Wavemeter in which absorption of energy by tuned circuit at resonance is indicated by decrease of valve grid current.

grid drive *(Elec.Eng.)*. Voltage or power required to drive a valve when delivering a specified load.

grid-glow tube *(Electronics)*. Cold-cathode gas-discharge tube in which glow is triggered by a grid.

grid-iron pendulum *(Horol.)*. A compensation pendulum with 5 parallel rods of iron and 4 of brass, the total length of each metal being in inverse ratio of its coefficient of linear expansion.

grid modulation *(Electronics)*. See **suppressor-grid modulation**.

grid navigation *(Aero.,Ships)*. A navigational system in which a grid instead of true North is used for the measurement of angles and for heading reference. The grid is of parallel lines referenced to the 180° meridian which is taken as Grid North. Used principally in polar route flying and, as the system is independent of convergence of meridians, headings remain the same from departure to destination.

grid neutralization *(Elec.Eng.)*. The method of neutralization of an amplifier through an inverting network in the grid circuit, which provides the requisite phase shift of 180°. Also called *phase inverter*.

grid pool tube *(Electronics)*. A tube having a mercury-pool cathode, one or more anodes, and a control electrode or *grid* which controls the onset of current flow in each cycle. The *excitron* and *ignitron* are examples.

grid ratio *(Radiol.)*. In grid therapy, the ratio of the total area of holes to that of the grid. Also ratio of lead strips to radiolucent plastic used in radiography to reduce scattered radiation effect on film.

grid resistance *(Elec.Eng.)*. A resistance unit for heavy current work, e.g. starters for railway motors; it is made up of a number of resistance grids placed side by side and mounted in a metal frame.

grid therapy *(Radiol.)*. A method of treatment in radiotherapy, in which radiation is given through a grid of holes in a suitable absorber, e.g. lead, rubber. By careful positioning, a greater depth dose can be given, and skin breakdown reduced to a minimum.

grief *(Med.)*. Deep sense of sorrow and suffering caused by bereavement. In excessive prolongation it may develop into overt *depressive illness*.

grief stem *(Min.Ext.)*. See **kelly**.

Griffin mill *(Min.Ext.)*. Pendulum mill in which a hanging roller bears against a stationary bowl as it rotates, crushing passing ore.

Grignard reagents *(Chem.)*. Mixed organometallic compounds prepared by dissolving dry magnesium ribbon or filings in an absolutely dry ethereal solution of an alkyl bromide or iodide. The value of Grignard reagents in the preparation of secondary or tertiary alcohols, and the alkenes, so easily produced therefrom, and in the synthesis of alkyl and aryl derivatives from many halogen compounds, can hardly be overestimated.

grike *(Geol.)*. A fissure in limestone rock caused by the solvent action of rainwater. See **karst**.

grill *(Build.)*. A layer of joists in a grillage foundation.

grillage foundation *(Build.)*. A type of foundation often used at the base of a column. It consists of one, two, or more tiers of steel beams superimposed on a layer of concrete, adjacent tiers being placed at right angles to each other, while all tiers are encased in concrete.

grille *(Build.)*. A plain or ornamental openwork of wood or metal, used as a protecting screen or grating.

grindability *(Min.Ext.)*. Empirical assessment of response of ore to pulverizing forces, applied under specified conditions.

grinder *(Paper)*. Equipment for producing mechanical woodpulp by holding logs of wood against the surface of a revolving natural or synthetic grindstone kept wet by water showers.

grinder's rot *(Med.)*. Lung disease caused by inhalation of metallic particles by steel-grinders.

grinding *(Min.Ext.)*. Comminution of minerals by dry or, more usually, wet methods, mainly in rod, ball, or pebble mills.

grinding-in *(Eng.)*. The process of obtaining a pressure-tight seal between a conical-faced valve and its seating by grinding the two together with an abrasive mixture such as silicon carbide and oil.

grinding machine *(Eng.)*. A machine tool in which flat, cylindrical, or other surfaces are produced by the abrasive action of a high-speed grinding wheel. See **centreless grinding, cylindrical grinding, profile grinding, surface grinding machine, thread grinding**.

grinding medium *(Eng.)*. The solid charges (balls, pebbles, rods, etc.) used in suitable mills for grinding certain materials, e.g. cement, pigments etc., to a fine powder.

grinding slip *(Build.)*. See **gouge slip**.

grinding wheel *(Eng.)*. An abrasive wheel for cutting and finishing metal and other materials. It is composed of an abrasive powder, such as silicon carbide or emery, held together by a bond or binding agent, which may be either a vitrified material or a softer material, such as shellac or rubber.

grinning through *(Build.)*. A paintwork defect characterized by the fact that the **hiding power** of the paint is insufficient to obscure completely the surface underneath.

grip *(Build.)*. (1) A small channel cut to carry away rain water during construction of foundations. (2) Small channel across the road-side to conduct surface water to a drain. Also *grippers*.

grip *(Image Tech.)*. Member of a camera crew moving equipment or camera mountings.

gripper edge *(Print.)*. See **grippers**.

gripper-feed mechanism *(Eng.)*. A type of strip-feeding mechanism used in presswork, which advances the strip material during its forward stroke, and then returns to its initial position with grippers open, and thus inoperative.

grippers *(Print.)*. Attachments which grip the edge (*gripper edge*) of a sheet of paper when it is fed into the printing machine.

gripper-shuttle *(Textiles)*. A projectile that grips the weft thread and inserts it into the fabric being manufactured on a weaving machine.

grisaille *(Arch.)*. Decorative painting in monochromatic colours to give the impression of design in relief.

griseofulvin *(Med.)*. An oral antibiotic which is excreted in the cells of the skin, thus destroying any cutaneous fungus infections.

grit *(Eng.)*. A measure indicating the sizes of the abrasive particles in a grinding wheel, usually expressed by a figure denoting the number of meshes per linear inch in a sieve through which the particles will pass completely.

grit *(Geol.)*. Siliceous sediment, loose or indurated, the component grains being angular. Contrast *sand* and *sandstone*, in which the grains are rounded.

grit *(Powder Tech.)*. Hard particles, usually mineral, of natural or industrial origin, retained on a 200 mesh B.S. test sieve (76 μm).

grit blasting *(Eng.)*. A process used in preparation for metal spraying, which cleans the surface and gives it the roughness required to retain the sprayed metal particles.

grit cell *(Bot.)*. A **stone cell** occurring in a leaf or in the flesh of a fruit e.g. pear.

grit chamber *(Build.)*. A **detritus chamber**.

grizzle bricks *(Build.)*. Bricks which are underburnt and of bad shape. They are soft inside and unsuitable for good work but are often used for the inside of walls. Also called *place bricks, samel bricks*.

grizzly *(Min.Ext.)*. The set of parallel bars or grating used for the coarse screening of ores, rocks etc.

grog *(Build.)*. Bricks or waste from a clayworks broken down and added to clay to be used for brick manufacture.

groin *(Arch.)*. The line of junction of the 2 constituent arches in a **groined arch**.

groined arch *(Arch.)*. An arch which is intersected by other arches cutting across it transversely.

groin rib *(Arch.)*. A projecting member following the line of a groin.

grommet *(Eng.)*. A ring or collar used to line a sharp-edged hole through which a cable or similar material passes.

grooming *(Behav.)*. Refers to maintenance of and attention to all aspects of the body surface; it can be performed by individuals to their own bodies, or between members of a dyad or larger social group.

groove *(Print.)*. The separation between the board and the spine of the book cover, permitting free opening.

grooving *(Eng.)*. (1) Cracking of the plates of steam boilers at points where stresses are set up by the differential expansion of hot and colder parts. (2) Producing a rectangular, V-shaped or similar groove or channel by milling, shaping, grinding etc. to provide e.g. a location for a reciprocating slide or as a lubricant reservoir.

grooving plane *(Build.)*. One with a narrow, interchangeable blade and adjustable fence, for cutting grooves.

grooving saw *(Build.)*. A circular saw which may be of the drunken type, used for cutting grooves.

gros nez *(Med.)*. See **goundou**.

gross register(ed) tonnage *(Ships)*. See **gross tonnage**.

gross tonnage *(Ships)*. The cubic capacity ($100\,ft^3 = 1$ ton) of all spaces below the freeboard deck and permanently closed-in spaces above that deck, except *exempted spaces*.

grossular *(Min.)*. An end-member of the garnet group, the composition being represented by $Ca_3Al_2Si_3O_{12}$; formed in the contact-metamorphism of impure limestone. Commonly contains some iron, and is greenish, brownish, or pinkish. Also *gooseberry stone*. See also **hydrogrossular**.

gross weight *(Aero.)*. See **weight**.

grosswetterlage *(Meteor.)*. The sea-level pressure distribution averaged over a period during which the essential characteristics of the atmospheric circulation over a large region remain nearly unchanged.

gross wing area *(Aero.)*. The full area of the wing, including that covered by the fuselage and any nacelles.

grotesque *(Print.)*. The name given to early sanserif types and still used, especially the contraction *grot*.

Grotthus-Draper law *(Chem.)*. Only such energy of electromagnetic radiation as is absorbed by tissue is effective in chemical action following ionization in that tissue.

ground *(Min.Ext.)*. The mineralized deposit and rocks in which it occurs, e.g. payground, payable reef; barren ground, rock without value.

ground *(Telecomm.)*. US for *earth*.

ground *(Textiles)*. (1) In lace manufacture, the mesh which forms a foundation for a pattern. (2) The base fabric to which is secured the figuring, threads, e.g. pile or loops in carpet or terry cloths.

ground absorption *(Telecomm.)*. The energy loss in radio-wave propagation due to absorption in the ground.

ground air *(Build.)*. In land used for sewage treatment, the air contained in the upper layers in the subsoil; it has a variable composition, including carbon dioxide, ammonia, and other gases resulting from oxidization of organic matters, and may be noxious.

ground auger *(Build.)*. An auger specially adapted for boring holes in the ground, for artesian-wells etc.

ground capacitance *(Phys.)*. See **earth capacitance**.

ground clutter *(Radar)*. The effect of unwanted ground-return signals on the screen pattern of a radar indicator.

ground coat *(Build.)*. A base coat of paint on which further treatments such as glazing, graining or marbling will be worked.

ground control *(Radar)*. Control of an aircraft or guided missile in flight from the ground.

ground-controlled approach *(Aero.,Radar)*. Aircraft landing system in which information is transmitted by a *ground controller* from a ground radar installation at end of runway to a pilot intending to land. Also called *talk-down*. Abbrev. *GCA*.

ground-controlled interception *(Aero.,Radar)*. Radar system whereby aircraft are directed on to an interception course by a station on the ground. Abbrev. *GCI*.

grounded base *(Electronics)*. See **common-base connection**.

grounded-cathode amplifier *(Electronics)*. One with cathode at zero alternating potential, drive on the grid and power taken from the anode; the normal and original use of a triode valve.

grounded circuit *(Elec.Eng.)*. A circuit which is deliberately connected to earth at one point or more, for safety or testing.

grounded collector *(Electronics)*. See **common-collector connection**.

grounded emitter *(Electronics)*. See **common-emitter connection**.

grounded-grid amplifier *(Electronics)*. One with grid at zero alternating voltage, drive between cathode and earth, output being taken from the anode; there is no anode-grid feedback.

ground engineer *(Aero.)*. An individual, selected by the licensing authorities, who has power to certify the safety for flight of an aircraft, or certain specified parts of it. Term now superseded by *licensed aircraft engineer*.

ground fine pitch *(Aero.)*. A very flat blade angle on a turboprop propellor, which gives extra braking drag and low propellor resistance when starting the engine. Colloq. *disking*.

ground frost *(Meteor.)*. A temperature of 0°C (32°F) or below, on a horizontal thermometer in contact with the shorn grass tips on a turf surface.

ground joist *(Build.)*. A horizontal timber supported off the ground at a basement or ground-floor level.

ground-level *(Surv.)*. The **reduced level** of a survey station with reference to an official bench mark.

ground loop *(Aero.)*. An uncontrollabe and violent swerve or turn by an aircraft while taxiing, landing or taking-off.

groundmass *(Geol.)*. In igneous rocks which have crystallized in two stages, the groundmass is the finer-grained portion, in which the phenocrysts are embedded. It may consist wholly of minute crystals, wholly of glass, or partly of both.

ground meristem *(Bot.)*. Partly differentiated meristematic tissue (**primary meristem**) derived from the apical meristem and giving rise to ground tissue.

ground mould *(Civ.Eng.)*. A timber piece or frame used as a template to bring earthworks such as embankments to the required form.

ground noise *(Acous.)*. See **background noise**.

ground plan *(Build.,Civ.Eng.)*. A drawing showing a plan view of the foundations for a building or of the layout of rooms etc. on the ground floor.

ground plate *(Build.)*. The bottom horizontal timber to which the frame of a building is secured.

groundplot *(Aero.)*. Method of calculating the position of an aircraft by relating groundspeed and time on course to starting position.

ground-position indicator *(Aero.)*. An instrument which continuously displays the dead-reckoning position of an aircraft.

ground ray *(Telecomm.)*. Same as *direct ray*.

ground reflection *(Radar)*. The wave in radar transmission which strikes the target after reflection from the earth.

ground resonance *(Aero.)*. A sympathetic response between the dynamic frequency of a rotorcraft's rotor and the natural frequency of the alighting gear which causes rapidly-increasing oscillations.

ground return *(Radar)*. The aggregate sum of the radar echoes received after reflection from the earth's surface. May include **clutter**.

grounds *(Build.)*. Strips of wood which are nailed to a wall or partition (fixing plugs being used when necessary) as a basis for the direct attachment of joinery.

ground safety lock *(Aero.)*. See **retraction lock**.

ground sill *(Build.)*. A **sleeper**.

ground sills *(Hyd. Eng.)*. Underwater walls built at intervals across the bed of a channel to prevent excessive scour of the bed or to increase the width of flow.

ground sluicing *(Min.Ext.)*. Bulk concentration of heavy minerals *in situ*, by causing a stream of water to flow over unconsolidated alluvial ground with just enough force to flush away the lighter, less valuable sands leaving the heavier ones to be removed for further treatment.

groundspeed *(Aero.)*. Speed of aircraft or missile relative to the ground and not to the surrounding medium. Cf. **air speed**.

ground state *(Phys.)*. State of nuclear system, atoms, etc., when at their lowest energy, i.e., not *excited*. Also *normal state*.

ground support equipment *(Aero.)*. All the handling facilities employed to service aircraft on the airport, e.g. tractors, steps, fuelling tankers, food and cleaning supplies. Also weapon trolleys and installation check-out gear for military aircraft.

ground system *(Telecomm.)*. See **earth system**.

ground table *(Build.)*. The course of stones at the foundation of a building. Also *grass table*.

ground tissue *(Bot.)*. Tissue other than vascular tissue, epidermis and periderm, mostly parenchma and collenchyma of e.g. pith and cortex.

ground water *(Build.)*. Water naturally contained in, and saturating, the subsoil.

ground water *(Geol.)*. Water occupying space in rocks. It may be *juvenile*, having arisen from a deep magmatic source, or *meteoric*, the result of rain percolating into the ground.

ground wave *(Phys.)*. See **wave**.

groundwood *(Paper)*. Pulp produced in a grinder.

group *(Chem.)*. (1) A vertical column of the periodic system, containing elements of similar properties. (2) Metallic radicals which are precipitated together during the initial separation in qualitative analysis. (3) A number of atoms which occur together in several compounds.

group *(Geol.)*. A stratigraphic rock unit consisting of two or more **formations**.

group *(Maths.)*. A set together with an operation * is said to be a group if it satisfies the four following conditions: (1) The set G is *closed* with respect to *; i.e. for all elements x,y in G, $x*y$ is an element of G. (2) The operation * in G is *associative*; i.e. for all elements x, y, z in G, $(x*y)*z = x*(y*z)$. (3) The set G has an *identity element* with respect to *; i.e. there is an element in G such that, for all elements x in G, $x*e = x = e*x$. (4) Each element in G has an *inverse* with respect to *; i.e. for each element x in G there is an element x' in G such that $x*x' = e = x'*x$. A group $(G;*)$ is said to be an *abelian group* if the operation * in G is commutative; i.e. if, for all elements x, y in G, $x*y = y*x$. The real numbers with addition, the non-zero real numbers with multiplication, and the three-dimensional vectors with vector addition are abelian groups.

group *(Telecomm.)*. An assembly of telephone speech channels, 48 kHz wide, comprising 12 channels of 4 kHz each, in a **frequency-division multiplex** system. Cf. **supergroup**.

group automatic operation *(Elec.Eng.)*. A method of automatic control sometimes employed with electric lifts; the pressing of a landing push-button calls the next lift.

group delay *(Telecomm.)*. For any signal travelling through a network or along a transmission line, the derivative of phase with respect to frequency. Where the medium is *non-dispersive*, group delay is equal to *envelope display*; in real situations, the phase response of a device or transmission path nearly always varies with frequency, so that group delay becomes a complex quantity.

group mixer, fader *(Telecomm.)*. See **mixer**.

group modulation *(Telecomm.)*. Use of one carrier for transmitting a group of communication channels, with demodulation on reception and ultimate separation. Carrier side frequencies may represent different groups.

group reaction *(Chem.)*. The reaction by which members of a **group** are precipitated.

group selection *(Ecol.)*. A form of natural selection proposed to explain the evolution of behaviour which appears to be for the long-term good of a group or species, rather than for the immediate advantage of the individual.

group selector *(Telecomm.)*. In a **Strowger exchange**, a selector which is first operated by a train of impulses, so that the wipers are lifted to a desired level of bank contacts, and then hunts by rotating the wipers over the contacts until a free outlet is found.

group theory *(Nuc.Eng.)*. Approximate method for the study of neutron diffusion in a reactor core, in which neutrons are divided into a number of velocity groups in which they are retained before transfer to the next group. See **one-**, **two-**, and **multigroup theory**.

group therapy *(Behav.,Med.)*. Any form of psychotherapy involving several persons at the same time, such as a small group of patients with similar psychological or physical problems who discuss their difficulties under the chairmanship of, e.g. a doctor. They thus learn from the experiences of others and teach by their own.

group velocity *(Phys.)*. The velocity of energy propagation for a wave in a dispersive medium. Given by

$$v_g = \left(\frac{d\omega}{dk}\right) = v - \lambda\frac{dv}{d\lambda}$$

where $k = 2\pi/\lambda$, $\omega = 2\pi \times$ frequency, v is the phase velocity and λ is the wavelength.

grouse disease *(Vet.)*. A popular term for the specific infection of the intestines of grouse by the nematode worm *Trichostrongylus pergracilis*.

grout *(Civ.Eng.)*. See **cement grout**.

grouting *(Build.,Civ.Eng.)*. The process of injecting cement grout for strengthening purposes. Also called *cementation*.

grouting *(Eng.)*. Setting a machine foot or base to a required level by filling the space between it and the supporting floor or foundation with cement grout.

groutnick *(Build.)*. A groove cut in a masonry joint to give access to grout.

growing *(Electronics)*. The production of semiconductor crystals by slow crystallization from the molten state.

growing point *(Bot.)*. Apical meristem.

grown diffusion transistor *(Electronics)*. A junction transistor in which the junctions are formed by the diffusion of impurities.

grown-junction *(Electronics)*. One having junctions produced by changing the types and amounts of acceptor and donor impurities that are added during **growing**.

growth *(Biol.)*. An irreversible change in an organism accompanied by the utilization of material, and resulting in increased volume, dry weight or protein content; increase in population or colony size of a culture of micro-organisms.

growth *(Eng.)*. The permanent increase in size which occurs in cast iron which has been kept for long periods at 480°C, due to changes in the carbides.

growth *(Nuc.Eng.)*. (1) Elongation of fuel rods in reactor under irradiation. (2) Build-up of artificial radioactivity in a material under irradiation, or of activity of a daughter product as a result of decay of the parent.

growth *(Phys.)*. Build-up of artificial radioactivity in a material under irradiation, or of activity of a daughter product as a result of decay of the parent.

growth curvature *(Bot.)*. A curvature in an elongated plant organ, caused by one side growing more than the other.

growth form *(Bot.)*. See **life form**.

growth hormones *(Biol.)*. Substances having growth-promoting properties, e.g. pituitary growth hormone; in plants, *auxins*.

growth inhibitor *(Bot.)*. A substance that inhibits plant growth at low concentrations, especially an endogenous substance, the best characterized of which is **abscisic acid**.

growth in soft agar *(Biol.)*. The ability of cells in culture to grow in a low-concentration gel of agar. It is one of the properties which distinguish a **transformed cell** from normal cells which can only grow in culture in contact with a substrate.

growth movement *(Bot.)*. Movement of a plant part brought about by differential growth. Cf. **turgor movement**.

growth regulator *(Bot.)*. A **growth substance**, especially one of the synthetic types.

growth retardant *(Bot.)*. A synthetic substance used to retard the growth of a plant, for example to stop the sprouting of stored onions or to restrict the height of grain crops e.g. maleic hydrazide.

growth ring *(Bot.)*. A recognizable increment of wood (secondary xylem) in a cross-section of a stem; most commonly an **annual ring**, but under some conditions more than one (or no) growth ring is formed within one year. See **early wood**, **late wood**.

growth room *(Bot.)*. A room in which plants are grown under controlled artificial lighting, photoperiod, temperature etc.

growth substance *(Bot.)*. One of number of substances (sometimes called *hormones*) formed in plants, or a synthetic analogue thereof, which have specific effects on plant growth or development at low concentrations (say $>10\,\mu M$). *Abscisic acid, auxins, cytokinins, ethylene, gibberellins.*

groyne *(Hyd.Eng.)*. Structures of timber, fascine or rubble used to control water scour. They can confine a river channel or be built out to sea on a beach.

GR-S *(Plastics)*. Abbrev. for *Government Rubber-Styrene*, the chief synthetic rubber developed in US during World War II. Based on styrene and butadiene. See **Buna**.

grub axe *(Build.)*. A tool for digging up roots; it has a broad chisel-shaped point on one side and a flat adze-like blade on the other.

grub saw *(Build.)*. A hand-saw for cutting marble, having a steel blade stiffened along the back with wooden strips.

grub screw *(Eng.)*. A small headless screw used without a nut to secure a collar or similar part to a shaft.

grummet *(Build.)*. Hemp and red-lead putty mixed as a jointing material for water-tightness.

Grüneisen's relation *(Phys.)*. The coefficient of volume expansion of a solid β, is given by $\beta = \gamma\kappa C$, where κ is the compressibilty, C the specific heat capacity of unit volume and γ is Grüneisen's constant which for most materials lies between 1 and 2 and is practically independent of temperature.

grunerite *(Min.)*. A monoclinic calcium-free amphibole; a hydrated silicate of iron and magnesium, differing from cummingtonite in having Fe > Mg. Typically found in metamorphosed iron-rich sediments.

gryke *(Geol.)*. See **grike**.

GSE *(Aero.)*. See **ground support equipment**.

GSR *(Behav.)*. Abbrev. for *Galvanic Skin Response*.

G-string *(Phys.)*. Colloquiallism for single-wire transmission line loaded with dielectric so that surface-wave propagation can be employed. (Named after Dr Groubau).

g-suit *(Aero.)*. See **anti-g suit**.

g-tolerance *(Space)*. The tolerance of an object or person to a given value of g-force.

guag *(Min.Ext.)*. The space left after the mineral has been extracted. Also *gunis*.

guaiacol *(Chem.)*. $HO.C_6H_4.OCH_3$, the monomethyl ether of *catechol* (*1-methoxybenzene 2-hydroxybenzene*), found in beechwood tar; a very unstable compound, with strong reducing properties. Used in medicine and in veterinary practice as an expectorant.

guanidine *(Chem.)*. $HN{:}C(NH_2)_2$, imido-urea or the amidine of amidocarbonic acid; a crystalline compound, easily soluble in water, strongly basic.

guanine *(Chem.)*. Purine base which occurs in DNA and RNA, pairing with cytosine. See **genetic code**. ⇨

guanophore *(Zool.)*. See **xanthophore**.

guard *(Build.,Eng.)*. A protection on a scaffold to prevent persons from falling; a fence or other safety device on a machine, to prevent injury to the operator or others from gears, cutting tools etc.

guard *(Telecomm.)*. (1) Signal which prevents accidental operation by spurious signals or avoids ambiguity. (2) The wire accompanying a speaking pair through an automatic telephone exchange; it is earthed while the speaking pair is being used by subscribers, thus indicating that the speaking pair cannot be engaged by any other circuit. Also *guard wire*.

guard band *(Image Tech.,Telecomm.)*. (1) In radio and TV transmission, an additional frequency band on each side of the allocated band to reduce interference from adjacent channels. (2) In magnetic recording, an unused space between tracks to avoid cross-talk.

guard cell *(Bot.)*. One of a pair of specialized epidermal cells which surround and by increase and decrease in their turgor, open and close a stoma, thus regulating **gas exchange**.

guard circle *(Acous.)*. Inner groove on disk recording which protects stylus from being carried to centre of turntable.

guard cradle *(Elec.Eng.)*. A network of wires serving the same purpose as a **guard wire**. Also called *guard net*.

guard magnet *(Min.Ext.)*. Strong magnet, usually suspended above moving stream of lump ore, to remove steel from broken drills etc. (**tramp iron**), which might damage the crushing machine.

guard net *(Elec.Eng.)*. See **guard cradle**.

guard pin *(Horol.)*. See **safety finger**.

guard rail *(Civ.Eng.)*. See **check rail**.

guard ring *(Elec.Eng.)*. Auxiliary electrode used to avoid distortion of electric (or heat) field pattern in working part of a system as a result of the **edge effect**, or to bypass leakage current through insulator to earth in an ionization chamber.

guard-ring capacitor *(Elec.Eng.)*. A standard capacitor consisting of circular parallel plates with a concentric ring maintained at the same potential as one of the plates to minimize the edge effect.

guards *(Print.)*. (1) Narrow strips of paper or linen projecting between sections in a book, for the attachment of plates, maps, etc. (2) Strips of paper sewed into a book to compensate for the thickness of inserted plates.

guard wire *(Elec.Eng.)*. An earth wire used on an overhead transmission line; it is arranged in such a position that, should a conductor break, it will immediately be earthed by contact with the wire. See **Price's guard-wire**.

guard wire *(Telecomm.)*. See **guard**.

gubernaculum *(Zool.)*. In Mammals, the cord supporting the testes, in the scrotal sac: in Hydrozoa, an ectodermal strand supporting the gonophore in the gonotheca: in Mastigophora, a posterior flagellum used in steering. adj. *gubernacular*.

Gudden-Pohl effect *(Phys.)*. A form of electro-luminescence which follows metastable excitation of a phosphor by ultraviolet light.

Guddermanian *(Maths.)*. A curious function defined by

$$\mathrm{gd}\,x = \int_0^x \operatorname{sech}\theta\, d\theta.$$

It then follows that if $u = \mathrm{gd}\,x$, then $\sin u = \tanh x$ and $\cos u = \operatorname{sech} x$, and that

$$x = \int_0^u \sec\theta\, d\theta.$$

gudgeon *(Build.)*. The wrought-iron pin which is fastened to a gate-post or door frame, and about which the leaf of a strap-hinge turns.

gudgeon *(Eng.)*. A pivot at the end of a beam or axle on which a bell, wheel etc. works.

gudgeon pin *(Autos.)*. The pin connecting the piston with the bearing of the little end of the connecting-rod. Also called *piston pin*. See also **floating-**.

Guerin process *(Eng.)*. Used in presswork to cut or form sheet metal by placing it between a die made of a cheap material and a thick rubber pad which adapts itself to the die while under pressure.

guest *(Zool.)*. An animal living and/or breeding in the nest of another animal, as a Myrmecophile in an Ants' nest.

guidance *(Aero.)*. See **guided missile**.

guide bars *(Eng.)*. (1) Bars with flat or cylindrical surfaces provided to guide the crosshead of a steam-engine, and so avoid lateral thrust on the piston rod. Also called *motion bars, slide bars*. (2) Any bars used as guides to control one machine element in relation to another.

guide bead *(Build.)*. A bead fixed to the inside of a cased frame as a guide for the sliding sash. Also *inner bead, parting bead*.

guided missile *(Aero.)*. *Guided weapon*, UK. Strategic and tactical unmanned weapon which is guided to its target; propulsion is usually by **rocket**, **ramjet** or simplified short-life **turbojet**. Guided missiles are broadly divided into categories using the initials of the launch and target media (Air, Surface or Underwater) followed by M for missile, e.g. ASM or SSM. In addition there are the ICBM (*Inter-Continental Ballistic Missiles*) and IRBM (*Intermediate Range Ballistic Missiles*) which, as their names imply, are guided on only part of their journeys and end in a ballistic descent through the atmosphere. Guidance systems vary greatly: *direct command guidance* is control entirely from the launcher by radio or wire signals; *radar command guidance* is radio-signal guidance from a lock-on radar/computer system on the launcher; *beam rider* is a missile which follows a radar beam directly from launcher to target; *semi-active homing* is the radar 'illumination' of the target, on the reflection from which the missile 'homes'; *collision course homing* is similar, but employs an offset missile aerial to bring it onto a converging course with the target; *fully-active homing* is self-contained, the missile generating its own radar signals and carrying a lock-on or collision-course computer device; *passive homing* is by a sensitive detector on to infrared, heat, sound, static electricity, magnetic or other wave emissions from the target. Long-range guidance devices include the *celestial*, wherein the missile's automatic astronavigation equipment is given a preflight programme relating its course to the target and to a fixed star(s) on which its tracking telescope is focused; and the *inertial* system, which depends upon a precise knowledge of the target, so that the course can be planned by the variation of inertial forces, which are followed by the internal device which can measure and correct minute variations in gravitational, and other, forces. The latter is the only known system completely independent of jamming or other external interference. See **stand-off bomb, cruise missile**.

guided wave *(Phys.)*. Electromagnetic or acoustic wave is constrained within certain boundaries as in a wave guide.

guide field *(Elec.Eng.)*. That component of field particles in a cyclotron or betatron which maintains particles in their intended path.

guide mill *(Eng.)*. A rolling-mill equipped with guides to ensure that the stock enters the mill at the correct point and angle.

guide pulley *(Eng.)*. A loose pulley used to guide a driving-belt past an obstruction or to divert its direction. Also called *idler pulley*.

guide rail *(Civ.Eng.)*. A **check rail**.

guides *(Min.Ext.)*. Timbers, ropes, or steel rails at sides of shaft used to steady the cage or skip.

guide track *(Image Tech.)*. Sound track, usually speech or dialogue, recorded during picture shooting under adverse recording conditions, to serve as a guide for subsequent replacement in the studio.

guide-vanes *(Aero.)*. A general term for aerofoils which guide the airflow in a duct, also *cascades*. See **impeller-intake-, nozzle-, toroidal-intake-**.

guide wavelength *(Phys.)*. Wavelength in a guide operated above the cut-off frequency. $1/\lambda^2_g = 1/\lambda^2_0 - 1/\lambda^2_{0c}$, where λ_g is the guide wavelength, λ_0 the wavelength in the unbounded medium at the same frequency, and λ_{0c} the wavelength in the unbounded medium at the cut-off frequency for the mode in question.

guild *(Ecol.)*. A division or category of a plant species from one area, made on the basis of similar *phenology* and *morphology*. See also **synusia**. Also, of an animal community, a group of species within the same trophic level which exploits a common resource, in a similar manner.

Guillain Barre syndrome *(Med.)*. An acute or subacute disorder of peripheral nerves causing progressive muscular weakness which may even paralyse the muscles of respiration. Often thought to be related to a recent viral illness but **aetiology** is not certain. Provided the patient is kept alive during acute phase full recovery is usual.

Guillemin effect *(Phys.)*. The tendency of a bent magnetostrictive rod to straighten in a longitudinal magnetic field.

Guillemin line *(Telecomm.)*. A network designed to produce a nearly square pulse with a steep rise and fall.

guillotine *(Med.)*. An instrument for cutting off tonsils.

guillotine *(Paper)*. A machine having a heavy steel blade, used for cutting or trimming stacks of paper, or for trimming books.

guinea-pig paralysis *(Vet.)*. A viral infection of cavies characterized by a diffuse meningo-myelo-encephalitis.

gula *(Zool.)*. In Vertebrates, the upper part of the throat.

gular *(Zool.)*. In some Fish, a bone developed between the rami of the lower jaw: in Chelonia, an anterior unpaired element of the plastron.

gulching *(Min.Ext.)*. The noise which generally precedes a fall or settlement of overlying strata.

Guldberg and Waage's law *(Chem.)*. See **law of mass action**.

Guldin's theorems *(Maths.)*. See **Pappus' theorems**.

gullet *(Build.)*. A depression cut in the face of a saw in front of each tooth, alternately on each side of the blade.

gullet *(Civ.Eng.)*. A narrow trench dug the full depth of a proposed cutting (in the case of large cuttings). A track is laid along the bottom of this trench, and wagons carry away the earth as the trench is widened into the full cutting.

gullet *(Zool.)*. The oesophagus; in Protozoa the cytopharynx.

gulleting *(Civ.Eng.)*. The process of excavating road or railway cuttings on a series of steps worked simultaneously. See also **gullet**.

gullet saw *(Build.)*. A saw with gullets cut in front of each tooth. Also called *brier-tooth saw*.

gulley *(Build.)*. A fitting installed at the upper end of a drain, to receive the discharge from rainwater or waste pipes.

gulley trap *(Build.)*. A device installed at a gulley to imprison foul air within the drain pipe. Also called *yard trap*.

gulose *(Chem.)*. A monosaccharide belonging to the group of aldohexoses. ⇨

gum *(Bot.)*. Viscid plant secretion exuding naturally or on wounding, soluble or swelling in water. Mostly complex, branching polysaccharides. Some e.g. gum arabic, alginic acid, agar, are economically important.

gum *(For.)*. See **American red gum, eucalyptus**.

gum arabic *(Chem.)*. A fine, yellow or white powder, soluble in water, rel.d. 1.355. It is obtained from certain varieties of acacia, the world's main supply coming from the Sudan and Senegal. Used in pharmacy for making emulsions and pills; also in glues and pastes. Also called *acacia gum, Senegal gum*.

gum-boil *(Med.)*. A small abcess on the palate, associated with a carious tooth, or the result of infection following upon local injury.

gum dichromate process *(Image Tech.)*. The use of gum as a vehicle for pigments and dichromate on printing papers; the exposed image being developed by water.

gum-lac *(Zool.)*. An inferior type of lac, containing much wax, produced by some lac Insects (Hemiptera) in Madagascar.

gumma *(Med.)*. A mass of cellular granulation tissue, due to syphilitic infection in the late or tertiary stage. adj. *gummatous*.

gummed paper *(Paper)*. Paper coated with a moisture activated adhesive, dextrin, gum arabic etc.

gumming up *(Print.)*. Application of film of gum arabic to the surface of a lithographic plate, thinning down and drying it evenly, to protect and enhance the hydrophilic properties on the non-image areas.

gummosis *(Bot.)*. A pathological condition of plants characterized by the conspicuous secretion of gums.

gums *(Chem.)*. Nonvolatile, colloidal plant products which either dissolve or swell up in contact with water. On hydrolysis, they yield certain complex organic acids in addition to pentoses and hexoses.

gums *(Zool.)*. In higher Vertebrates, the thick tissue masses surrounding the bases of the teeth.

gum streaks *(Print.)*. Streaks across a lithographic plate caused through improper gumming-up of the plate. If the gum is too thick and not thinned down, it can attack the plate image and cause it to lose ink receptivity.

gun *(Build.)*. See **spray gun**.

gun *(Electronics)*. Assemblage of electrodes, comprising cathode, anode, focusing and modulating electrodes, from which the electron beam is emitted before being subjected to deflecting fields in a *CRT*.

gun *(Image Tech.)*. Electron gun used to excite the phosphor in a TV tube.

guncotton *(Chem.)*. A **nitrocellulose** (cellulose hexanitrate) with a high nitrogen content. It burns readily and explodes when struck or strongly heated. Used for explosives.

gun current *(Electronics)*. In electron beam tubes, the total electron current flowing to the anode, only part of which forms the beam current.

gun drill *(Eng.)*. A trepanning drill or a centre-cut drill used for deep-hole drilling.

gunis, gunnis, gunnice, gunnies *(Min.Ext.)*. See **guag**.

gunite *(Civ.Eng.)*. A finely graded cement concrete (a mixture of cement and sand), which is sprayed into position under air pressure to produce a dense, impervious adherent layer; used to line or repair existing works, or to build up dense concrete in inaccessible areas of operation.

gunmetal *(Eng.)*. A copper-tin alloy (i.e. bronze) either Admiralty gunmetal (copper 88%, tin 10%, zinc 2%) or copper 88%, tin 8% and zinc 4%. Lead and nickel are frequently added, and the alloys are used as cast where resistance to corrosion or wear is required, e.g. in bearings, steam-pipe fittings, gears.

Gunn effect *(Electronics)*. A pair of electrodes connected across a small piece of N-type gallium arsenide can, when the applied voltage exceeds a certain level, demonstrate a negative resistance effect, due to the electron velocity actually reducing when the electric field exceeds about 3 kV/cm. Installed in a suitable tuned circuit or cavity, a Gunn-effect diode can operate as a oscillator from 500 Hz to well over 50 GHz.

gunning *(Build.)*. The forcible application of refractory, sound insulating or corrosion resistant linings with a gun, operated usually by compressed air.

gun perforation *(Min.Ext.)*. See **perforation**.

gunstock stile *(Build.)*. See **diminished stile**.

Gunter's chain *(Surv.)*. A chain having overall length of 66 ft (=20.116 m). One acre equals 10 sq. chains; and 1 chain is 1/10th of a furlong.

Guppy *(Aero.)*. Aircraft modified by substituting a very large diameter section for a major part of the fuselage. Nose and tail hinge sideways for carrying large indivisible loads as in Boeing 377. *Airbus* major assemblies are delivered this way. Also *Super Guppy* and *Pregnant Guppy*.

gurley *(Paper)*. The reading, generally expressed in seconds, obtained from testing a paper for air resistance utilising the gurley densometer.

gurley densometer *(Paper)*. A test instrument for measuring the air resistance of a sample of paper. It consists of two concentric upright cylinders, the outer being closed at the bottom and containing oil, the inner open at both ends and having a clamping device at the top. After raising the inner cylinder and placing the sample in position it is allowed to fall under its own weight, its progress being retarded by the entrapped air which passes through the sample at a rate dependent on the resistance of the paper. The time in seconds is recorded for the passage of a given volume of air e.g. 100 ml.

gusher *(Geol.)*. See **geyser**.

gusset, gusset plate *(Eng.)*. A bracket or stay, cast or built up from plate and angle, used to strengthen a joint between 2 plates which meet at a joint, as the junction of a boiler shell with the front and back plates, or between connecting members of a structure.

gusset piece *(Build.)*. A piece of timber covering the triangular end-gap between the roof slope and the horizontal gutter boarding behind a chimney stack.

gustatory calyculus *(Zool.)*. See **taste-bud**.

GUT *(Astron.,Phys.)*. Generic acronym for **Grand Unified Theory**.

gut *(Zool.)*. The alimentary canal.

Gutenberg discontinuity *(Geol.)*. The seismic-velocity discontinuity separating the mantle of the Earth from the core at a depth of approximately 2900 km.

gut oedema *(Vet.)*. See **bowel oedema disease**.

gutta *(Zool.)*. A patch of colour or other marking, resembling a small drop, on the surface of an animal. pl. *guttae*. adj. *guttulate*.

guttae *(Arch.)*. An ornament in the form of a line of truncated cones used to decorate entablatures or hollow mouldings.

gutta-percha *(Telecomm.)*. The coagulated latex of *Iso-*

nandra (or *Palaquium*) *gutta* and other trees. Having a high resistivity and permittivity, and being waterproof, it was especially suitable without lead sheath, for submarine cables before the introduction of plastics.

guttation *(Bot.)*. The exudation of drops of water (containing some solutes) from an uninjured part of a plant, commonly under conditions of high humidity, typically from **hydathodes** at tips or margins of leaves.

gutter *(Build.,Civ.Eng.)*. A channel along the side of a road, or around the eaves of a building, to collect and carry away rainwater.

gutter *(Print.)*. (1) The spacing material in a forme between the fore-edges of the pages. (2) The space between two fore-edges on the printed sheet and including any trims.

gutter bearer *(Build.)*. A timber about $2 \times 1\frac{1}{2}$ in (50 × 38 mm), carrying gutter boarding.

gutter bed *(Build.)*. A lead sheet fixed behind the eaves gutter and over the tilting fillet to prevent overflow from the gutter from soaking into the wall.

gutter boards *(Build.)*. See **snow boards**.

gutter bolt *(Build.)*. A securing bolt between the spigot and the socket ends at a joint in a gutter.

gut-tie *(Vet.)*. Strangulation of a loop of intestine which has herniated through a rupture in the peritoneal covering of the right spermatic cord of castrated cattle.

guttural *(Zool.)*. Pertaining to the throat.

guttural pouch *(Vet.)*. A diverticulum of the Eustachian tube of the horse.

gutturoliths *(Vet.)*. See **chondroids**.

Gutzeit test *(Chem.)*. A method of determining arsenic, by adding metallic zinc and hydrochloric acid. The evolved gases darken mercury (II) salts.

guy *(Civ.Eng.)*. A rope holding a structure in position.

guy *(Telecomm.)*. Thin tension support for antenna mast or similar structure.

guy derrick *(Build.,Civ.Eng.)*. A crane operating from a mast held upright by guy-ropes.

guying *(Civ.Eng.)*. The operation of holding or adjusting a structure in position by means of guy-ropes.

G-value *(Chem.)*. A constant in radiation chemistry denoting the number of molecules reacting as a result of the absorpton of 100 eV radiation energy.

GVH *(Immun.)*. See **graft-versus-host reaction**.

gymno- *(Genrl.)*. Prefix from Gk. *gymnos*, naked.

gymnocyte *(Biol.)*. A cell without a cell-wall.

Gymnomycota *(Bot.)*. See **Myxomycota**.

gymnosperms *(Bot.)*. Group (classified as a division Gymnospermophyta or a class Gymnospermae, or regarded as polyphyletic) containing those **seed plants** in which the ovules are not enclosed in carpels, the pollen typically germinating on the surface of the ovule. There is no double fertilization and the xylem is vessel-less (except in the Gnetales). Contains ca 700 extant spp., often classified as 3 classes: Cycadopsida, Coniferopsida and Gnetopsida.

gynaecium *(Bot.)*. See **gynoecium**.

gynaecology, gynecology *(Med.)*. That branch of medical science which deals with the functions and diseases peculiar to women.

gynaecomastia, gynecomastia *(Med.)*. Abnormal enlargement of the male breast.

gynandrism *(Zool.)*. See **hermaphrodite**.

gynandromorph *(Zool.)*. An animal exhibiting male and female characters.

gynandromorphism *(Zool.)*. The occurrence of secondary sexual character of both sexes in the same individual.

gynandrous *(Bot.)*. Having the stamens and style united.

gyn-, gyno-, gynaeco- *(Genrl.)*. Prefix from Gk. *gyne*, gen. *gynaikos*, woman.

gynobasic *(Bot.)*. A style which appears, because of the infolding of the ovary wall, to be inserted at the base of the ovary, e.g. Labiatae.

gynodioecious *(Bot.)*. A species having some individual plants with hermaphrodite flowers only and others with female flowers only. Cf. **dioecious**.

gynoecium *(Bot.)*. The female part of a flower, consisting of one or more carpels. Cf. **androecium**.

gynomonoecius *(Bot.)*. Species having all the plants bearing both female and hermaphrodite flowers. Cf. **monoecious**.

gynophore *(Bot.)*. An elongation of the receptacle between androecium and gynoecium.

gynospore *(Bot.)*. Same as **megaspore** in heterosporous plants, Cf. **androspore**.

gypsum *(Min.)*. Crystalline hydrated calcium sulphate, $CaSO_4.2H_2O$. Occurs massive as *alabaster*, fibrous as *satin spar*, and as clear, colourless, monoclinic crystals known as *selenite*. Used in making plaster of Paris, plaster and plaster-board used in building.

gypsum plate *(Phys.)*. A thin plate of gypsum used in the determination of the sign of the birefringence of crystals in a polarizing microscope.

gyration, radius of *(Maths.)*. See **moment of inertia**.

gyrator *(Elec.Eng.)*. Electronic component which does not obey reciprocity law. Frequently based on the Faraday effect in ferrites.

gyrator *(Electronics)*. A circuit which can perform impedance inversion; the equivalent impedance at one port is inversely proportional to the impedance connected at the other port. With suitable phase alteration, a capacitance connected at the output can be made to appear as an inductance at the input. Used in integrated circuits where inductance cannot be realized on account of bulk.

gyratory *(Min.Ext.)*. A form of rock-breaker, in which an inner cone gyrates in a larger outer hollow cone.

gyro *(Eng.)*. See **gyroscope**.

gyrocompass *(Eng.)*. A gyroscope, electrically rotated, controlled and damped either by gravity or electrically so that the spin axis settles in the meridian.

gyrodyne *(Aero.)*. A form of rotorcraft in which the rotor is power-driven for take-off, climb, hovering and landing, but is in **autorotation** for cruising flight, there usually being small wings further to unload the rotor.

gyro-frequency *(Electronics)*. See **frequency of gyration**.

gyro horizon *(Aero.)*. An instrument which employs a gyro with a vertical spin axis, so arranged that it displays the attitude of the aircraft about its pitch and bank axes, referenced against an artificial horizon. It is normally electrically, but sometimes pneumatically, operated.

gyrolite *(Min.)*. A hydrated calcium silicate, formula $Ca_2Si_3O_7(OH)_2.H_2O$. Often occurs in amygdales with apophyllite.

gyromagnetic compass *(Aero.)*. A magnetic compass in which direction is measured by gyroscopic stabilization.

gyromagnetic effect *(Phys.)*. See **Barnett effect**.

gyromagnetic ratio *(Phys.)*. The ratio γ of the magnetic moment of a system to its angular momentum. For orbiting electrons $\gamma = e/2m$ where e is the electronic charge and m is the mass of the electron. γ for electron spin is twice this value.

gyroplane *(Aero.)*. Rotorcraft with unpowered rotor(s) on a vertical axis. Also *autogyro*.

gyroscope *(Eng.)*. Spinning body in a **gimbal mount** or similar, which resists torques altering the alignment of the spin axis, and in which **precession** or **nutation** replace the direct response of static bodies to such applied torques. Used as a compass, or as a controlling device for servos which reduce the misalignment and thus correct the course or e.g. stabilize a ship. Also *gyro, gyrostat*.

gyrostat *(Eng.)*. See **gyroscope**.

Gyrosyn *(Aero.,Ships)*. TN for a remote-indicating compass system employing a directional *gyroscope* which is monitored by and *synchronized* with signals from an element fixed in azimuth and designed to sense its angular displacement from the earth's magnetic meridian by **fluxgate**. The element is located at some remote point, e.g. wing tips, away from extraneous magnetic influences.

gyrotron *(Nuc.Eng.)*. In fusion studies, very high-frequency power generator for microwave heating at the electron cyclotron resonance.

gyrus *(Med.,Zool.)*. A ridge between two grooves: a convolution of the surface of the cerebrum.

G-Y signal *(Image Tech.)*. Component of colour TV signal which, when combined with the **Y signal** produces the green chrominance signal.

Gzelian *(Geol.)*. The youngest epoch of the Pennsylvanian period.

H

h *(Genrl.)*. Symbol for height.

h *(Phys.)*. Symbol for (1) Planck's constant. See **Planck's law**. (2) specific enthalpy.

H *(Chem.)*. The symbol for *hydrogen*.

H *(Phys.)*. Symbol for henry.

H_α, H_β, H_γ, etc. *(Phys.)*. The lines of the Balmer series in the hydrogen spectrum. Their wavelengths are: H_α, 656.299; H_β, 486.152; H_γ, 434.067; H_δ 410.194 nm. The series continues into the ultraviolet, where about 20 more lines are observable.

H *(Elec.Eng.)* Symbol for magnetic field strength.

H *(Phys.)*. Symbol for enthalpy.

[H] *(Phys.)*. One member of the strongest pair (H and K) of Fraunhofer lines in the solar spectrum, almost at the limit of visibility in the extreme violet. Their wavelengths are [H], 396.8625 nm; [K], 393.3825 nm; and the lines are due to ionized calcium.

haar *(Meteor.)*. Local term for a wet sea-fog advancing in summer from the North Sea upon the shores of England and Scotland.

Haas effect *(Acous.)*. Phenomenon associated with a long-delayed echo which has been applied to reinforcement systems in auditoria.

habenula *(Zool.)*. A strap-like structure; in particular, a nerve-centre of the diencephalon.

Haber process *(Chem.)*. Currently the most important process of fixing nitrogen, in which the nitrogen is made to combine with hydrogen under influence of high temperatures (400°–500°C), high pressure (2×10^7 N/m^2), and catalyst of finely-divided iron from iron(III) oxides, in large continuous enclaves. Many variants operate at different pressures according to the catalyst. The product ammonia may be dissolved in water or condensed, and unreacted gases recycled.

habit *(Behav.)*. As a general term, and in learning theory, refers to learned patterns of behaviour which are very consistent and predictable in particular situations. Its specific usage varies in different branches of psychology, e.g. referring to compulsive behaviour in clinical psychology, to a particular cognitive style in cognitive psychology.

habit *(Crystal.)*. A term used to cover the varying development of the crystal forms possessed by any one mineral. Thus calcite may occur as crystals showing the faces of the hexagonal prism, basal pinacoid, scalenohedron, and rhombohedron. According to the relative development or *dominance* of one or other of these forms, the habit may be prismatic, tabular, scalenohedral, or rhombohedral.

habit *(Zool.)*. The established normal behaviour of an animal species.

habitat *(Biol.)*. The normal locality inhabited by a plant or animal, particularly in relation to the effect of its environmental factors.

habit spasm *(Med.)*. *Tic*. A repeated, rapidly performed, involuntary, and coordinated movement, occurring in a nervous person.

habituated culture *(Bot.)*. A **tissue culture** that has developed an ability to synthesize auxin since its isolation and can now grow without having it added.

habituation *(Behav.)*. An aspect of learning in which there is a decrease in responsiveness as a consequence of repeated stimulation; the habituated response reappears if the stimulus is withheld for a long time.

hachure *(Surv.)*. *Hatching*. The use of lines to shade a plan and indicate hills and valleys.

hack *(Build.)*. A long parallel bank, about 6 in (150 mm) high, of brick, rubbish, and ashes, on which bricks are laid in the course of manufacture, when it is intended to dry them in the open.

hack-barrow *(Build.)*. A barrow used to carry green bricks to the hack for drying.

hack-cap *(Build.)*. A small timber structure erected to provide cover for a hack.

hacker *(Comp.)*. Colloq. for person who writes programs of a somewhat routine nature. Also those who obtain unauthorized access through information **networks** to computer systems.

hacking *(Build.)*. (1) The operation of piling up green bricks on a hack to dry. (2) The process of making a surface rough, in order to provide a key for plasterwork. (3) A course of stones in a rubble wall, the course being composed partly of single stones of the full height of the course and partly of shallower stones arranged two to the height of the course.

hacking-out knife *(Build.)*. A knife used to remove old putty from sash rebates before reglazing.

hackling *(Textiles)*. Process of combing scutched flax in the hackling machine, in order to divide and parallelize the long fibres, and to remove the short ones and impurities.

hackmanite *(Min.)*. A fluorescent variety of sodalite, showing on freshly fractured surfaces a pink colour which fades on exposure to light, but which returns if kept in the dark or subjected to X-rays or ultraviolet light. See **tenebrescence**.

hack-saw *(Eng.)*. (1) A mechanic's hand-saw used for cutting metal. It consists of a steel frame, across which is stretched a narrow saw-blade of hardened steel. (2) A larger saw, similar to the above, but power-driven.

Hackworth valve gear *(Eng.)*. A radial gear in which an eccentric opposite the crank operates a link whose other end slides along an inclined guide, the valve rod being pivoted to a point on the link.

hade *(Geol.)*. The angle of inclination of a fault-plane, measured from the vertical.

Hadley cell *(Meteor.)*. A meridional circulation of the atmosphere consisting of low-level equatorward movement of air from about 30° to the equator, rising air near the equator, poleward flow aloft, and descending motion near 30°. It was suggested by Hadley in the 18th century, and is at least partially confirmed by observation.

hadrom, hadrome *(Bot.)*. (1) The conducting elements and associated parenchyma in xylem tissue, cf. **leptom**. (2) The *hydroids* of mosses.

hadron *(Phys.)*. An elementary particle that interacts *strongly* with other particles. Hadrons include *baryons* and *mesons*.

Haeckel's law *(Biol.)*. See **recapitulation theory**.

haem *(Med.)*. A complex iron containing organic compound which, in combination with the protein globin, forms **haemoglobin** and which gives it its oxygen carrying capacity.

haemad *(Zool.)*. Situated on the same side of the vertebral column as the heart.

haemagglutinin *(Immun.)*. A substance which agglutinates red blood cells. This may be a specific antibody, or a **lectin**, or a component of certain viruses (e.g. influenza or measles) by which they bind to cell surfaces.

haemal arch *(Zool.)*. A skeletal structure arising ventrally from a vertebral centrum, which encloses the caudal blood vessels.

haemal canal *(Zool.)*. The space enclosed by the centrum and the haemal arch of a vertebra, through which pass the caudal blood vessels.

haemal, haematal, haemic *(Zool.)*. Pertaining to the blood or to blood vessels.

haemal ridges *(Zool.)*. See **haemapophyses**.

haemal spine *(Zool.)*. The median ventral vertebral spine formed by the fusion of the haemapophyses, below the haemal canal.

haemal system *(Zool.)*. The system of vessels and channels in which the blood circulates.

haemangioma *(Med.)*. *Angioma*. A tumour composed of blood vessels irregularly disposed and of varying size.

haemapoiesis, haemopoiesis *(Med.,Zool.)*. *Haematopoiesis, haematosis*. The process of forming new blood.

haemapophyses *(Zool.)*. A pair of plates arising ventrally from the vertebral centrum, and meeting below the haemal canal to form the haemal arch and spine. Also called *ridges*.

haemarthrosis *(Med.)*. A joint containing blood which has effused into it.

haematemesis *(Med.)*. The vomiting of blood, or of blood-stained contents of the stomach.

haematinic *(Med.)*. Pertaining to the blood.

haematite *(Min.)*. See **hematite**.

haematobium *(Zool.)*. An organism living in blood. adj. *haematobic*.

haematoblast *(Zool.)*. A primitive blood cell, which may develop into an erythrocyte or a leucocyte: a blood platelet.

haematocele *(Med.)*. An effusion of blood localized in the form of a cyst in a cavity of the body.

haematochrome *(Bot.)*. A red or orange pigment accumulated in the cells of some green algae usually when nitrogen starved, as *Chlamydomonas nivalis*, responsible for 'red snow'.

haematocolops *(Med.)*. Accumulation of menstrual blood in the vagina, due to an imperforate hymen.

haematocolpometra *(Med.)*. Accumulation of menstrual blood in the vagina and uterine cavity.

haematocrit *(Biol.)*. A graduated capillary tube of uniform bore in which whole blood is centrifuged, to determine the ratio, by volume, of blood cells to plasma.

haematogenesis *(Zool.)*. See **haemopoiesis**.

haematogenous *(Zool.)*. Having origin in the blood.

haematologist *(Med.)*. One who specializes in the study of the blood and its diseases.

haematoma *(Med.)*. A swelling composed of blood effused into connective tissue.

haematophagous *(Zool.)*. Feeding on blood.

haematozoon *(Zool.)*. An animal living parasitically in the blood.

haematuria *(Med.)*. Presence of blood in the urine.

haemic *(Zool.)*. See under **haemal**.

haemin *(Chem.)*. The hydrochloride of haematin, $C_{34}H_{32}N_4O_4FeCl$, brown crystals. Its molecule contains four pyrrole radicals.

haemochromatosis *(Med.)*. *Bronzed diabetes*. A disease in which the iron-containing pigment haemosiderin is deposited in excess in the organs of the body, giving rise to cirrhosis of liver, enlargement of spleen, diabetes, skin pigmentation.

haemocoel *(Zool.)*. A secondarily formed body cavity derived from the blood vessels which replaces the coelom in Arthropods and Molluscs.

haemocyanin *(Zool.)*. A blue respiratory pigment, containing copper, in the blood of *Crustacea* and *Mollusca*. It has respiratory functions similar to haemoglobin.

haemocytes *(Zool.)*. The corpuscles found floating in haemolymph.

haemocytoblast *(Immun.)*. A stem cell in bone marrow or in other haemopoietic tissues.

haemocytolysis *(Zool.)*. See **haemolysis**.

haemocytometer *(Biol.)*. An apparatus consisting of a special glass slide with a grid of lines engraved on the bottom of a shallow rectangular trough so that if a coverslip is placed over the trough the grid demarcates known volumes. Cells from a well-mixed suspension are introduced into the space and the number in the grid squares counted under the microscope. Used for blood counts, mitotic counts etc.

haemodialysis *(Med.)*. The restoration of diffusable chemical constituents of the blood towards normal by passing blood across a cellulose membrane which has on the other side a fluid containing **electrolytes** at the desired concentration. Principally used to restore body chemistry in patients with kidney failure, *artificial kidney*.

haemoglobinaemia *(Med.)*. The abnormal presence of haemoglobin in the blood, as a result of destruction of red blood cells.

haemoglobin, hemoglobin *(Biol.)*. The red pigment of blood whose major function is the transport of oxygen from the lungs to the tissues. It is a protein of 4 polypeptide chains each bearing a haem prosthetic group which serves as an oxygen binding site.

haemoglobinometer *(Med.)*. An instrument for measuring the percentage of haemoglobin in the blood.

haemoglobinuria *(Med.)*. Presence of haemoglobin in the urine, as a result of excessive destruction of red blood cells.

haemolymph *(Zool.)*. The watery fluid containing leucocytes, believed to represent blood, found in the haemocoelic body-cavity of certain Invertebrates.

haemolysin *(Immun.)*. (1) Antibody capable of lysing red cells in the presence of complement. (2) A bacterial toxin which lyses red cells.

haemolysis *(Biol.)*. The lysis of red blood cells.

haemolytic anaemia *(Immun.)*. Anaemia due to an abnormal increase in the rate of destruction of circulating erythrocytes. This can result from the presence of antibodies against the erythrocytes (e.g. against the Rhesus antigen as in **erythroblastosis foetalis**, or to autoantibodies); or from over-activity of mononuclear phagocytes in association with grossly enlarged spleen (hypersplenism); or from metabolic abnormalities of the erythrocytes such as glucose-6-phosophatase deficiency aggravated by some drugs.

haemolytic disease of newborn animals *(Vet.)*. Seen mainly in fowl and piglets, it is the result of an immune reaction between antigens on the **red blood cells** of the neonate and colostral antibody from the dam. Intravascular haemolysis occurs giving rise to weakness, dyspnoea, anaemia and jaundice.

haemolytic disease of the newborn *(Med.)*. See **erythroblastosis foetalis** for the immunological background. The consequent anaemia and jaundice can occur when a *Rhesus negative* mother carries a *Rhesus positive* child and can be prevented by immunizing the mother so that maternal antibodies are not produced.

haemolytic plaque assay *(Immun.)*. A method used to detect and enumerate individual cells secreting antibody in vitro. Sheep red cells (treated when necessary so as to bind the antibody) are mixed with a cell suspension to be assayed in a thin layer of agarose, and incubated so that they can secrete antibody which diffuses onto the red cells. Complement is then added and cells secreting antibody are revealed by the presence of an area of haemolysis around them.

haemopericardium *(Med.)*. The presence of blood in the pericardial sac.

haemophilia *(Med.)*. A hereditary bleeding tendency, with deficiency of a normal blood protein (factor VIII, anti-haemophilic globulin) preventing normal clotting. The classic **sex-linked** disorder. Clinically indistinguishable from *Christmas disease* where there is a factor IX deficiency and is sometimes called *h*.B.

haemopneumothorax *(Med.)*. The presence of blood and air in the pleural cavity.

haemoptysis *(Med.)*. The expectoration of blood, or of blood-stained sputum.

haemorrhage *(Med.)*. Bleeding; escape of blood from a ruptured blood vessel.

haemorrhagic disease *(Vet.)*. See **redwater**.

haemorrhagic septicaemia *(Vet.)*. *Bovine pasteurellosis, shipping fever, shipping pneumonia*. A bacterial disease of cattle caused by *Pasteurella multocida* and characterized by high fever, pneumonia and oedematous swelling of the skin. A similar disease occurs in sheep. *Pasteurella* vaccines available.

haemorrhoid *(Med.)*. *Pile* (usually pl.). Varicose dilation of the haemorrhoidal veins at the lower end of the rectum and the anus.

haemosiderosis *(Med.)*. Deposition, in the tissues of the body, of the iron-containing pigment *haemosiderin*, after excessive destruction of red blood cells. See also **haemochromatosis**.

haemostasis *(Med.)*. The arrest of bleeding.

haemostatic *(Med.)*. Arresting or checking bleeding; an agent which does this.

haemotropic *(Zool.)*. Affecting blood.

hafnium *(Chem.)*. A metallic element in the fourth group of the periodic system. Symbol Hf, at. no. 72, r.a.m. 178.49 rel.d. 12.1, mp about 2000°C. It occurs in zirconium minerals, where its chemical similarity but relatively high neutron absorption makes it a troublesome impurity in zirconium metal for nuclear engineering. Used to prevent recrystallization of tungsten filaments.

haft *(Build.)*. A tool handle.

hagatalite *(Min.)*. A variety of **zircon** containing an appreciable quantity of the rare earth elements.

ha-ha *(Build.)*. A fence sunk in a ditch below ground-level so as to give an uninterrupted view.

Haidinger fringes *(Phys.)*. Optical interference fringes produced by transmission and reflection from two parallel partly reflecting surfaces, e.g. a plate of optical glass. The fringes are produced by division of amplitude of the wave front and are circular fringes formed at infinity (cf. **contour fringes**). Used extensively in interferometry, e.g. Fabry and Pérot, and Michelson interferometers.

Haigh fatigue-testing machine *(Eng.)*. A machine for testing the resistance of materials to fatigue under alternating direct stress; the specimen is loaded by means of a powerful electromagnet and excited by an alternating current.

hail, hailstones *(Meteor.)*. Precipitation in the form of hard pellets of ice, which often fall from cumulo-nimbus clouds and accompany thunderstorms. They are formed when raindrops are swept up by strong air-currents into regions where the temperature is below freezing-point. In falling, the hailstone grows by condensation from the warm moist air which it encounters.

hail stage *(Meteor.)*. That part of the condensation process taking place at a temperature of 0°C so that water vapour condenses to water liquid which then freezes.

hair *(Bot.)*. See **trichome**.

hair *(Med.,Zool.)*. A slender, elongate structure, mostly composed of **keratins**, arising by proliferation of cells from the Malpighian layer of the epidermis in Mammals; more generally, any threadlike growth of the epidermis.

hair *(Textiles)*. Animal fibre other than silk or sheep's wool. However, the soft shorter fibres from certain animals may also be called wool but qualified by the animal's name e.g. angora wool.

hair cloth *(Textiles)*. A material generally composed of coarse hair and cotton-yarn; used as a stiffening for coats and in upholstery.

haircord *(Textiles)*. Cotton fabrics of light weight, in which cords are produced by running two warp threads together at frequent intervals.

haircord carpet *(Textiles)*. A hard carpet made from animal hair.

hair follicle *(Zool.)*. See **follicle**.

hair hygrometer *(Meteor.)*. A form of **hygrometer** which is controlled by the varying length of a human hair with humidity. It is not an absolute instrument, but it can be used at temperatures below freezing-point, and it can be made self-recording.

hairline *(Phys.)*. A fine straight line in the optical system of an instrument; used for the positive location of an image or correlation on to a measuring scale.

hairpin loop *(Biol.)*. A short double-stranded region made possible in a single-stranded nucleic acid molecule because of the complementarity of neighbouring sequences. Common in tRNA.

hair plates *(Zool.)*. Groups of articulated sensory hairs occurring near the joints of the appendages in insects and acting as proprioceptors.

hair space *(Print.)*. The thinnest space between words, about 1-point wide.

hairspring *(Horol.)*. The balance spring.

hair stippler *(Build.)*. A decorative brush having a bristle filling set in rows in ebonite secured to a rectangular wooden base. Used in colour blending and colour glazing to equalize colour distribution and eliminate brush marks.

hake *(Build.)*. A **hack** built to dry tiles in the course of their manufacture.

halation *(Electronics)*. In a CRT, glow surrounding spot on phosphor arising from internal reflection within the thickness of the glass. Also called *halo*.

halation *(Image Tech.)*. Unwanted exposure surrounding the image of a bright object caused by light reflected from the rear surface of the film base or plate; it is reduced by a light-absorbing backing layer. See **anti-halation**.

Haldane apparatus *(Chem.)*. An apparatus for the analysis of air; used for the analysis of mine gases.

half-adder *(Comp.)*. Logic circuit which adds a pair of binary bits producing a sum and a carry. Also *two-input adder*. Cf. **full-adder**.

half-anchor ear *(Elec.Eng.)*. An anchor ear to which only one anchoring wire is attached.

half-bed, half-joint *(Build.)*. In pricing the labour charge for stonework, each horizontal surface on a stone is spoken of as a *half-bed*, as it contributes one-half to the cost of preparing each bed-joint; similarly, *half-joint* refers to the vertical jointing surfaces.

half-blind dovetail *(Build.)*. See **lap dovetail**.

half-bound *(Print.)*. Said of a book having its back, a portion of the sides, and the corners bound in one material (originally leather) and the remainder of the sides in some other material (e.g. cloth or paper).

half-brick wall *(Build.)*. A wall built entirely of stretchers and therefore 4.5 in thick.

half-case *(Print.)*. (1) A type case of the usual width, but half the length, for holding display type or special letters. (2) See **double case**.

half-cell *(Elec.Eng.)*. See **single-electrode system**.

half-closed slot *(Elec.Eng.)*. See **semiclosed slot**.

half-column *(Arch.)*. An **embedded column** of which half projects.

half-deflection method *(Elec.Eng.)*. A method of finding the internal resistance of a cell when the value is known to be high. A second cell, a galvanometer, and a resistance are connected in series with the cell under test, and the value of the resistance required to give a galvanometer deflection of half the value obtained with the cell alone is found.

half duplex *(Comp.)*. Data transmission in both directions but in only one direction at any one time.

half-element *(Elec.Eng.)*. See **single-electrode system**.

half-frame camera *(Image Tech.)*. A camera which takes pictures half the normal size for the film used, in particular, half the size of the standard 36 × 24 mm still picture on 35 mm film.

half-hour rating *(Elec.Eng.)*. A form of rating for electric machinery supplying an intermittent load. It indicates that the machine delivers the specified rating for a period of half an hour without exceeding the specified temperature rises. Cf. **one-hour rating**.

half-hunter *(Horol.)*. A form of watch case in which the glass occupies one-half of its hinged cover. Also *demi hunter*.

half-lap joint *(Build.)*. The name of the joint formed by the process of **halving**.

half-lattice girder *(Build.,Civ.Eng.)*. See **Warren girder**.

half-life *(Phys.)*. (1) Time in which half of the atoms of a given quantity of radioactive nuclide undergo at least one disintegration. Also *half-value period*. The half-life T is related to the *decay constant* λ by $T = (\ln 2)/\lambda$. (2) See biological **half-life, effective half-life**.

half-line block *(Print.)*. One in which the strength of the lines is reduced to produce a pleasing effect by using in the process camera a coarse cross-line screen or a single-line screen either vertical, horizontal, or diagonal.

half-measure *(Print.)*. Type matter set to half the usual page or column width to accommodate an illustration.

half-normal bend *(Elec.Eng.)*. A bend serving to connect two lengths of the conduit used in electrical installation work which are at an angle of 135°.

half-period zones *(Phys.)*. A conception, due to Huyghens, whereby an optical wavefront is considered to be divided into a number of concentric annular zones, so that, at a given point in front of the wave, the illumination from each zone is half a period out of phase with that from its neighbour. The use of half-period zones facilitates the study of diffraction problems.

half plate *(Image Tech.)*. A standard format, $4\frac{3}{4} \times 6\frac{1}{2}$ in.

half-power *(Acous.)*. Condition of a resonant system, electrical, mechanical, acoustical, etc. when amplitude response is reduced to $1/\sqrt{2}$ of maximum, i.e. by 3 dB.

half-power point *(Telecomm.)*. Useful reference point(s) in the performance characteristics of any device, e.g. amplifier, antenna or transistor, whose performance varies with frequency; it is the frequency or frequencies at which the gain or response falls 3 dB below a mean value.

half-principal *(Build.)*. A short rafter which does not reach the ridge of a roof.

half-residence time *(Phys.)*. Time in which half the radioactive debris deposited in the stratosphere by a nuclear explosion would be carried down to the troposphere.

half-roll *(Aero.)*. See **roll**.

half-round chisel *(Eng.)*. A **cold chisel** having a small half-round cutting edge; used for chipping semicircular grooves such as oil-ways.

half-round file *(Eng.)*. A file whose cross-section has one flat and one convex face.

half-round screws *(Eng.)*. See **button-headed screws**.

half-sawn *(Build.)*. Said of a stone face as left from the saw.

half-secret dovetail *(Build.)*. See **lap dovetail**.

half-section *(Eng.)*. On an engineering drawing, a sectional view of a symmetrical workpiece or assembly, terminating at an axis of symmetry.

half-section *(Telecomm.)*. See **section**.

half-sheet work *(Print.)*. Method in which one forme or plate is used to print on each side of the paper which is split after printing, thus producing two identical copies, each requiring half the sheet of paper. See also **work-and-turn**.

half-shroud *(Eng.)*. A gear-wheel shroud extending only up to half the tooth height. See **full shroud, shroud**.

half-silvered *(Phys.)*. A metallic film deposited e.g. by anode sputtering onto glass or other surface which reflects a substantial proportion of incident light.

half-socket pipe *(Build.)*. A drain-pipe having a socket for the lower half only.

half space *(Build.)*. (1) A landing at the end of a flight of steps. (2) A raised floor in a window bay.

half-stuff *(Paper)*. Raw materials which have been converted into pulp but not yet beaten.

half-supply voltage principle *(Elec.Eng.)*. If the collector-emitter voltage of a power stage transistor is less than half the supply voltage, the circuit will be inherently safe from thermal runaway, as the thermal loop gain will be less than unity.

half-thickness *(Phys.)*. See **half-value thickness**.

half-timbering *(Build.)*. An early mode of house-building in which the foundations and principal members were of stout timber, and walls were formed by filling the spaces between members with plaster.

half-title *(Print.)*. The title of a book printed on the leaf preceding the title page.

half-tone *(Telecomm.)*. **Facsimile systems** which can accept a continuous range of contrast.

half-tone block *(Print.)*. A screened plate mounted on a base of type height and used for the reproduction of continuous tone copy, such as photographs.

half-tone process *(Image Tech.,Print.)*. The reproduction of a subject containing a range of tones by photographing through a cross-line screen which translates these tones into dots of varying size. Used for all the usual printing methods, relief (letterpress), planographic (lithography), intaglio (gravure) and stencil (screen process).

half-twist bit *(Build.)*. Bit shaped like a gimlet, used for making screw-holes.

half-uncials *(Print.)*. An early style of lettering in which square forms of letter are mixed with rounded forms; a few type faces are available which copy this style.

half-value layer *(Phys.)*. See **half-value thickness**.

half-value period, half-period *(Phys.)*. See **half-life** (1).

half-value thickness *(Phys.)*. The thickness of a specified substance which must be placed in the path of a beam of radiation in order to reduce the transmitted intensity by one-half. Also *half-thickness, half-value layer*.

half-wave antenna *(Telecomm.)*. An antenna whose overall length is one half-wavelength. The voltage distribution is from a maximum at one end to a minimum in the middle and a maximum at the other.

half-wave plate *(Phys.)*. A plate of doubly refracting, uniaxial crystal cut parallel to the optic axis, of such thickness that, if light is transmitted normally through it, a phase difference of half a period is introduced between the ordinary and extraordinary waves. A half-wave plate is used in Laurent's polarimeter.

half-wave rectification *(Telecomm.)*. Rectification in which current flows only during the positive (or negative) half-cycles of the alternating voltage. The commonest form of detection of amplitude modulated radio signals.

half-wave rectifier *(Elec.Eng.)*. One in which there is conduction for part of one-half of the applied alternating cycle of voltage.

half-wave suppressor coil *(Telecomm.)*. An inductance coil inserted at half-wavelength intervals along an antenna wire; used in some forms of directional antenna to suppress radiation in reverse phase from alternate half-wavelength sections of the wire.

half-wave transmission *(Elec.Eng.)*. A method of transmission of electrical energy in which the natural period of oscillation of the transmission line is equal to four times the frequency of the transmitted currency.

half-width *(Phys.)*. A measure of sharpness on any function $y = f(x)$ which has a maximum value y_m at x_0 and also falls off steeply on either side of the maximum. The half-width is the difference between x_0 and the value of x for which $y = y_m/2$. Used particularly to measure the width of spectral lines or of a response curve.

halides *(Chem.)*. Fluorides, chlorides, bromides, iodides and astatides.

hali-, halo- *(Genrl.)*. Prefix from Gk. *hals*, salt.

haliplankton *(Ecol.)*. The plankton of the seas.

halite *(Min.)*. *Common or rock salt*. The naturally occurring form of sodium chloride, crystallizing in the cubic system; forming deposits of considerable thickness in close association with anhydrite and gypsum especially in the Permian and Triassic rocks. The salt being pumped out as brine or mined. Salt domes, with which oil or gas may be associated, occur in many parts of the world.

halitosis *(Med.)*. Offensive breath.

Hallade recorder *(Eng.)*. An instrument for recording vibration of rolling-stock due to track irregularities etc. in planes parallel, transverse and perpendicular relative to the track.

Hall coefficient *(Electronics)*. The coefficient of proportionality (R_H) in the relation $E_H = R_H(J \times B)$ where E_H is the resulting transverse electric field, J is the current density and B is magnetic flux density. The magnitude of R_H depends on the number of electrons available, and the sign of R_H indicates the sign of the majority charge carrier.

Hall effect *(Electronics)*. Disturbance of the lines of current flow in a conductor due to the application of a magnetic field, leading to an electric potential gradient transverse to direction of current flow.

Hall mobility *(Electronics)*. Mobility (mean drift velocity in unit field) of current carriers in a semiconductor as calculated from the product of the Hall coefficient and the conductivity.

halloysite *(Min.)*. One of the clay minerals, a hydrated form of kaolinite and member of the kandite group; consists of hydrated aluminium silicate.

Hall probe *(Phys.)*. A small probe which uses the **Hall effect** to compare magnetic fields.

Hall process *(Eng.)*. A process for the extraction of

aluminium by the electrolysis of a fused solution of alumina in cryolite at a temperature of approximately 1000°C. The aluminium is molten at this temperature and settles to the bottom of the bath, from which it is drawn off. Contains up to 99.8% aluminium.

hallucination *(Behav.)*. A perceptual experience that occurs in the absence of any appropriate external stimulus.

hallucinogen *(Med.)*. A drug or chemical which induces **hallucinations** (e.g. cannabis, lysergic acid (LSD) or mescaline).

hallux *(Zool.)*. In land Vertebrates, the first digit of the hind-limb.

hallux flexus *(Med.)*. A late stage of hallux rigidus, the big toe being rigidly flexed on the sole.

hallux rigidus *(Med.)*. Rigid stiffness of the big toe, due to osteoarthritis of the joint between the toe and the foot.

hallux valgus *(Med.)*. A deformity of the big toe, in which it turns towards and comes to lie above the toe next to it; usually associated with bunion.

hallux varus *(Med.)*. A rare deformity of the big toe, in which it diverges from the toe next to it.

halo *(Astron.)*. Spherical region space surrounding our Galaxy (and others), and permeated by very hot transparent gas.

halo *(Electronics)*. Same as **halation**.

halo *(Meteor.)*. A bright ring or system of rings often seen surrounding the sun or moon. The *large halo*, of radius 22°, is due to light refracted at minimum deviation by ice crystals in high cirrostratus clouds. See also **corona**.

halo *(Min.Ext.)*. Product of diffusion in rock surrounding an ore deposit of traces of mineral or element being sought, identified by geochemical tests. The halo may be the only surface indication of an ore deposit at depth. See **geochemical prospecting**.

halobiotic *(Zool.)*. Living in salt water, especially in the sea.

halochromism *(Chem.)*. The formation of coloured salts from colourless organic bases by the addition of acids.

halo effect *(Behav.)*. In the perception of other people, the tendency to generalize an impression of one characteristic of a personality to other, unrelated aspects of the personality.

halogen *(Chem.)*. One of the seventh group of elements in the periodic table, for which there is one electron vacancy in the outer energy level, viz., F., Cl, Br, I, At. The main oxidation state is -1.

halogenation *(Chem.)*. The introduction of *halogen* atoms into an organic molecule by substitution or addition.

halogen-quench Geiger tube *(Nuc.Eng.)*. Low-voltage tube for which halogen gas (normally bromine) absorbs residual electrons after a current pulse, and so quenches the discharge in preparation for a subsequent count.

haloid acids *(Chem.)*. A group consisting of hydrogen fluoride, hydrogen chloride, hydrogen bromide and hydrogen iodide.

halolimnic *(Zool.)*. Originally marine but secondarily adapted to fresh water.

haloperidol *(Med.)*. Drug belonging to the butyrophenones which as a class all have anti-psychotic actions. Haloperidol is used in the treatment of schizophrenia.

halophile *(Ecol.)*. A freshwater species capable of surviving in salt water.

halophilic bacteria *(Biol.)*. Salt-tolerant bacteria occurring in the surface layers of the sea, where they are important in the nitrogen, carbon, sulphur and phosphorus cycles. Many are pigmented or phosphorescent and if present in quantity may colour the surface water. Some halophilic bacteria, e.g. halobacterium, are able to live in salted meats.

halophyte *(Bot.)*. A plant able to grow where the soil is rich in sodium chloride or other sodium salts. Cf. **glycophyte**.

halophytic vegetation *(Bot.)*. A population of halophytes, e.g. on mangrove swamp, salt marsh or alkaline soil.

halosere *(Bot.)*. A *sere* that starts on land emerging from the sea, e.g. a salt marsh.

Halothane *(Chem.)*. $CF_3CHClBr$. (*1-chloro-1-bromo-2,2,2 trifluoroethane*). TN for a nonexplosive anaesthetic, now widely used, containing 1 atom of chlorine in its molecular structure. Also *Fluothane*.

halotrichite *(Min.)*. Hydrated sulphate of iron and aluminium, occurring rarely as yellowish fibrous silky colourless crystals. *Iron alum*.

halteres *(Zool.)*. A pair of capitate threads which take the place of the hind wings in *Diptera*, and assist the insect to maintain its equilibrium while flying; balancers.

halving *(Build.)*. A method of jointing (e.g. two timbers); it consists in cutting away half the thickness from the face of one, and the remaining half from the back of the other, so that when the two pieces are put together the outer surfaces are flush.

Hamamelidae *(Bot.)*. Subclass or superorder of dicotyledons. Mostly woody plants, the perianth poorly developed or none, flowers often unisexual, often borne in catkins, often wind pollinated. Contains ca 3400 spp. in 23 families including Betulaceae and Fagaceae. (Includes the Amentiferae).

hamartoma *(Med.)*. A benign tumour due to excess growth of one of the cellular elements, often vascular, in normal tissue.

hambergite *(Min.)*. Basic beryllium borate, crystallizing in the orthorhombic system.

Hamilton's principle *(Maths.)*. The motion of a system from time t_1 to t_2 is such that the line integral $\int_{t_1}^{t_2}(T+W)dt$ is stationary, where T is the kinetic energy of the system and $W=\sum_i F_i.r_i$, where F_i are the external forces acting at points r_i. This principle can be taken as a basic postulate in place of Newton's laws. For systems to which Newton's laws can be applied it gives the same results, but it is important because it can also be applied to other systems.

hammer-axe *(Build.)*. A tool with a double head-axe at one side and hammer at the other of the handle.

hammer beam *(Build.)*. A short cantilever beam projecting into a room or hall from the springing level of the roof, strengthened by a curved strut underneath, and carrying a hammer-beam roof.

hammer-beam roof *(Build.)*. A type of timber roof existing in various forms, all affording good headroom beneath. It consists essentially of arched ribs, supported on hammer beams at their feet, and carrying the principal rafters, strengthened sometimes by a collar-beam and/or struts.

hammer blow *(Eng.)*. (1) The alternating force between a steam locomotive's driving-wheels and the rails, due to the centrifugal force of the balance weights used to balance the reciprocating masses. (2) The noise due to a pressure wave travelling along a pipe in which the flow of a liquid has been suddenly impeded.

hammer break *(Elec.Eng.)*. A name given to the electromagnetic trembler device used on an electric bell or the primary of an induction coil.

hammer-dressed *(Build.)*. A term applied to stone surfaces left with a rough finish produced by the hammer.

hammer-drill *(Min.Ext.)*. A compressed-air rock drill in which the piston is not attached to the steel or borer but moves freely.

hammer-drive screw *(Eng.)*. A self-tapping screw with a very long pitch, akin to a nail, which can be driven by a hammer or inserted by a press into a plain hole in a relatively soft material.

hammer-headed *(Build.)*. A term applied to masons' chisels intended to be struck by a hammer rather than by a mallet.

hammerman *(Eng.)*. (1) The operator of a power-driven hammer. (2) A smith's mate.

hammer mill *(Min.Ext.)*. See **impact crusher**.

hammer scale *(Eng.)*. The scale of iron oxide which forms on work when it is heated for forging.

hammer test *(Eng.)*. Drop test for impact strength of large metal parts, e.g. rails. Weight is dropped from increasing heights until specified deflection is produced.

hammer toe *(Med.)*. A deformity of any toe, especially

the second, in which the toe, flexed on itself, is, at its junction with the foot, bent towards the instep.

hammer track *(Phys.)*. Highly characteristic track resembling hammer, formed by decay of ^{8}Li nucleus into two α-particles emitted in opposite directions at right angles to the lithium track.

Hammond organ *(Acous.)*. See **organ**.

hance *(Arch.)*. That part of the intrados, close to the springing of an elliptical or many-centred arch, which forms the arc of smaller radius.

Hancock jig *(Min.Ext.)*. One in which ore is jigged up and down with some throw forward in a tank of water, the heavy mineral stratifying down and being separately removed.

hand *(Textiles)*. Used in the US for **handle**.

hand brace *(Build.)*. See **brace**.

hand-cut overlay *(Print.)*. Plies of paper cut out in the highlights and built up in the dark tones and included in the *make-ready* to improve the printing of half-tone blocks. Cf. **mechanical overlay**.

H and D (Hurter and Driffield) curve *(Image Tech.)*. See **characteristic curve**.

hand feed *(Eng.)*. The hand operation of the feed mechanism of a machine tool. See **feed**.

hand hole *(Eng.)*. A small hole, closed by a removable cover, in the side of a pressure vessel or tank; it provides means of access for the hand to the inside of the vessel.

hand ladle *(Foundry)*. A small foundry casting ladle supported by a long handle of steel bar. See also **hand shank**.

handle *(Textiles)*. The subjective reaction obtained by feeling a fabric and assessing its roughness, harshness, flexibility, softness etc.

hand lead *(Surv.)*. Lead plummet used at sea, by attachment to a lead-line measuring within 100 fathoms (183 m).

hand letters *(Print.)*. Letters formed of brass and mounted on a handle, with the finisher impresses the title on the back or side of a bound volume.

handle-type fuse *(Elec.Eng.)*. A fuse in which the carrier containing the fuse-link is provided with a handle to facilitate withdrawal and replacement.

hand level *(Surv.)*. Small and light sighting tube above which a spirit level and mirror are arranged, so that the bubble can be seen while sighting on a station.

handmade paper *(Paper)*. Paper made in single sheets by dipping a mould into a vat containing stock so that the requisite amount of stock is picked up and, by skilful shaking, is disturbed and formed into the sheet. The wet sheets are couched on to felts, pressed to remove water, dried and, if necessary, tub sized. Handmade papers are characterized by their permanence and durability, appearance of quality and excellent properties for water-colour painting.

hand monitor *(Radiol.)*. Radiation monitor designed to measure radioactive contamination on the hands of an operator, or to be held in the hand.

hand mould *(Paper)*. Wooden frame, accompanied by a pair of deckles, covered with a wove or laid pattern woven wire on which a sheet of hand-made paper is formed.

hand press *(Print.)*. A press operated by hand; now used chiefly for proof-pulling.

hand-rail bolt *(Build.)*. A rod which is threaded and fitted with a nut at both ends; used to draw together the mating surfaces of a butt joint, such as that between adjacent lengths of hand-rail. The rod passes though holes in the members, and at one end the nut, a square one, is housed in a square mortise which prevents it from turning. At the other end the nut is circular, with notches cut in its periphery, so it can be turned within its mortise.

hand-rail plane *(Build.)*. A plane having a specially shaped sole and cutting-iron, adapting it to the finishing of the top surface of a hand-rail.

hand-rail punch *(Build.)*. A small tool which is inserted in the notches in the periphery of the circular nut of a **hand-rail bolt** and is used to tighten it up.

hand-rail screw *(Build.)*. A small rod, taper-threaded at each end, used to connect adjacent lengths of hand-rail as an alternative to the **hand-rail bolt**. Also a *dowel screw*.

hand reset *(Elec.Eng.)*. Restoration of a magnetic device, e.g. relay or circuit-breaker, by a manual operation.

hand rest *(Eng.)*. A support, shaped like a letter T, on which a turner rests a hand tool, during wood-turning or metal-spinning in a lathe.

hand roller *(Print.)*. A roller used for inking type-matters on the hand press, preparatory to pulling a proof.

hand-rope operation *(Elec.Eng.)*. A method of control of electric lifts; a rope is passed through the car and attached to control equipment mounted at the top of the shaft.

Hand-Schuller-Christian disease *(Med.)*. A lipoid granuloma of unknown aetiology, affecting mainly the skull which may lead to exophthalmus and diabetes insipidus. The granulomas contain cholesterol.

handscrew *(Build.)*. A wooden clamp consisting of two parallel bars connected by two tightening screws; used to hold parts together while a glued joint dries, or to secure work in process of being formed.

handset *(Telecomm.)*. A rigid combination of microphone and receiver, in a form convenient for holding simultaneously to mouth and ear.

handshaking *(Comp.)*. The **protocol** which enables two devices to establish (or break) communications.

hand shank *(Eng.)*. A **foundry ladle** supported at the centre of a long iron bar, formed into a pair of handles at one end for control during pouring; carried by two men.

hand specimen *(Min.Ext.)*. Fragment of ore deposit chosen for special study or test as displaying one or more characteristic minerals.

hand tools *(Eng.)*. All tools used by fitters when doing hand work at the bench, as hammers, files, scrapers, etc.

hand wheel *(Horol.)*. A grooved pulley provided with a cranked handle and mounted on a universal form of vice, used for driving a lathe or other tool by hand.

hand winding *(Elec.Eng.)*. The process of winding a machine by the insertion of the coils, turn by turn, into the slots; used in cases where it is not convenient to use former-wound coils.

hangar *(Aero.)*. A special construction for the accommodation of aircraft.

hanger *(Build.)*. A bracket for the support of a gutter at the eaves.

hanger *(Elec.Eng.)*. (1) Plates of glass or other material standing on edge in an accumulator cell, and supporting the accumulator-plates by means of their lugs. (2) A fitting used for supporting the overhead contact wire of a traction system from a transverse wire or structure.

hanger *(Eng.)*. (1) A bracket, usually of cast-iron, bolted to a wall or to the underside of a girder, to hold a bearing for supporting overhead shafting. (2) A bracket, usually of steel strip, used to support a pipe from a roof.

hangfire *(Min.Ext.)*. Unexpected delay or failure of explosive charge to detonate, thus creating a dangerous situation.

hang glider *(Aero.)*. Original manned glider as used by Lilienthal in Germany in 1890s. Revived in the 1960s as the **Rogallo wing**, employing flexible wing surfaces and now a major class of ultra-light aircraft. Both flexible and fixed hard wings are now used.

hanging *(Eng.)*. *Hang-up*. In the blast furnace, adhesion of partly melted charge to walls, thus upsetting smooth working.

hanging battens *(Elec.Eng.)*. A suspended row of lamps used in stage lighting.

hanging buttress *(Build.)*. A buttress carried upon a corbel at its base.

hanging drop preparation *(Biol.)*. A preparation for the microscope in which the specimen, in a drop of medium on the undersurface of a coverslip, is suspended over a hallow-ground slide, to which it is sealed to prevent evaporation.

hanging figures *(Print.)*. Figures normally supplied with *old style* founts; 1, 2, 0 conform to the *x-height*, 6 and 8 have the height of *ascenders*, and 3, 4, 5, 7 and 9 hang below the *base line*. Also called *old style figures*.

hanging indentation *(Print.)*. Layout in which the first line of the paragraph is set *full-out* and the succeeding lines are indented one em or more, as employed in this dictionary.

hanging post *(Build.)*. See **hingeing post**.

hangings *(Build.)*. A term applied to materials such as wallpaper, used as wall coverings.

hanging sash *(Build.)*. A sash arranged to slide in vertical grooves, and counterweighted so as to be balanced in all positions.

hanging steps *(Build.)*. Steps which are built into a wall at one end and are unsupported at the other end. Also known as *cantilevered steps*.

hanging stile *(Build.)*. That stile of a door to which the hinges are secured.

hanging valley *(Geol.)*. A tributary valley not graded to the main valley. It is a product of large-scale glaciation and due to the glacial overdeepening of the main valley relative to the hanging valley. The two valleys are connected by rapids or waterfalls.

hanging wall *(Min.Ext.)*. Rock above the miner's head, usually the country rock above the deposit being worked.

hang-over *(Telecomm.)*. (1) The delay in restoration of speech-activated switches, as in the **vodas**, to ensure the non-clipping of weak final consonants of words. (2) Excessive prolongation of any *on-off* type of signal or current or voltage pulse.

hang-up *(Eng.)*. See **hanging**.

hank *(Textiles)*. A general term for a reeled length of yarn.

Hankel functions of the first and second kind *(Maths.)*. See **Bessel functions** (of the third kind).

hapanthous *(Bot.)*. See **hapaxanthic**.

hapaxanthic *(Bot.)*. Flowering and fruiting once and then dying. See also **monocarpic**.

haplo- *(Genrl.)*. Prefix from Gk. *haploos*, single, simple.

haplobiont *(Bot.)*. A plant which has only one kind of individual or form in its life-history. adj. *haplobiontic*.

haplodiploidy *(Zool.)*. A means of sex determination where females develop from fertilized eggs and are therefore diploid and the males from unfertilized eggs and are haploid. e.g. Honeybees.

haplodont *(Zool.)*. Having molars with simple crowns.

haploid *(Biol.)*. Of the reduced number of chromosomes characteristic of the germ cells of a species, equal to half the number in the somatic cells. Cf. **diploid**.

haploidization *(Bot.)*. In the **parasexual cycle** of fungi, the progressive loss of chromosomes from the diploid set by occasional non-disjunction until stable haploid nuclei are formed.

haplont *(Bot.)*. Organism in which only the gametes are haploid, meiosis occurring at their formation and the vegetative cells being diploid. Cf. **diplont**.

haplophase *(Biol.)*. The period in the life-cycle of any organism when the nuclei are haploid. Cf. **diplophase**.

haplostele *(Bot.)*. A protostele in which the solid central core of xylem is circular in cross-section.

haplostemonous *(Bot.)*. Having a single whorl of stamens. Cf. **diplostemonous**.

haplotype *(Biol.)*. A particular set of *alleles* at several very closely linked loci.

haploxylic *(Bot.)*. Said of a leaf containing one vascular strand.

hapten *(Immun.)*. A substance which can combine with antibody but cannot itself initiate an immune response unless it is attached to a carrier molecule. Most haptens are small molecules (e.g. dinitrophenyl). Haptens are often conjugated chemically to carrier proteins for experimental purposes, since they provide easily recognized antigenic determinants.

hapteron *(Bot.)*. A holdfast, i.e. a unicellular or multicellular organ attaching a plant to the substrate.

haptonema *(Bot.)*. Appendage arising between the flagella of the motile cells of the Haptophyceae sometimes serving for temporary attachment to a surface.

Haptophyceae *(Bot.)*. *Prymnesiophyceae*. Class of eukaryotic algae, flagellated and palmelloid sorts (often interconvertible). Motile cells usually with 2 equal, smooth, flagella and a **haptonema**. Some sorts or stages bear **coccoliths**. Mostly phototrophic, some heterotrophic (both osmotrophic and phagotrophic). Mostly marine.

haptotropism *(Bot.)*. *Thigmotropism*. A **tropism** like that of a tendril coiling round its support in which differential growth is determined by touch.

hard *(Electronics)*. Adjective, synonymous with *high-vacuum*, which differentiates thermionic vacuum valves from gas-discharge tubes.

hard *(Glass)*. Having a relatively high softening point.

hard *(Hyd.Eng.)*. A layer of gravel or similar materials put down on swampy or sodden ground to provide a way for passage on foot.

hard and soft acids and bases *(Chem.)*. A terminology used with **Lewis** acids and bases to indicate non-polarizable (hard) or polarizable (soft). In general, hard-hard and soft-soft interactions are stronger than hard-soft ones.

hard bast *(Bot.)*. Sclerenchyma present in phloem.

hardboard *(Build.)*. Fibre-board that has been compressed in drying, giving a material of greater density than **insulating board**.

hard bronze *(Eng.)*. Copper-based alloy used for tough or dense castings; based of 88% copper plus tin with either some lead or zinc.

hard copy *(Comp.)*. Computer output printed on paper.

hard core *(Build.,Civ.Eng.)*. Lumps of broken brick, hard natural stone etc. used to form the basis of a foundation for road or paving or floors to a building.

hard disk *(Comp.)*. Rigid magnetic disk. It normally allows a higher recording density than a floppy disk thus providing more storage for the same physical dimensions. May be stacked as *platters*. See **disk pack**.

hard drawn *(Eng.)*. Term applied to wire or tube which has been greatly reduced in cross-section without annealing.

hardenability *(Eng.)*. The response of a metal to quenching for the purpose of hardening, measured, most commonly, by the **Jominy (end quench) test**.

hardener *(Image Tech.)*. Chemical (formalin, acrolein, chrome alum etc.) added to the fixing bath to toughen the emulsion of a film.

hardener *(Plastics)*. An **accelerator**.

hardening *(Bot.)*. Increasing resistance to cold as temperatures are gradually lowered either naturally or as the result of horticultural practice. Analogous hardening to drought, heat, wind etc. occurs.

hardening *(Eng.)*. The process of making steel hard by cooling from above the critical range at a rate that prevents the formation of ferrite and pearlite and results in the formation of martensite. May involve cooling in water, oil or air, according to composition and size of article.

hardening media *(Eng.)*. Liquids into which steel is plunged in hardening. They include cold water containing sodium chloride or hydroxide to increase the cooling power.

hardening of oils *(Chem.)*. The hydrogenation of oils in the presence of a catalyst, usually finely divided nickel, in which the unsaturated acids are transformed into saturated acids, with the result that the glycerides of the unsaturated acids become hard. This process is of great importance for the foodstuffs industries, e.g. margarine is prepared in this way.

Harder's glands *(Zool.)*. In most of the higher Vertebrates, an accumulation of small glands near the inner angle of the eye, closely resembling the lacrimal gland.

hard-facing *(Eng.)*. (1) The application of a surface layer of hard material to impart, in particular, wear resistance. (2) A surface so formed. The composition is generally of high melting-point metals, carbides etc., applied by powder, wire or plasma arc spraying, or by welding.

hard glass *(Chem.)*. Borosilicate glass, whose hardness is principally due to boron compounds. Resistant to heat and to chemical action.

hard-gloss paint *(Build.)*. A popular class of paint that dries hard with a high gloss. It always contains some hard resin in the medium.

hard head *(Eng.)*. Alloy of tin with iron and arsenic left after refining of tin.

hard heading *(Min.Ext.)*. Sandstone or other hard rock met with in making headings or tunnels in a coal mine.

Hardinge mill *(Min.Ext.)*. Widely used grinding mill, made in three sections: a flattish cone at the feed end, a cylindrical drum centrally and a steep cone; the assembly being hung horizontally between trunnions.

hard lead *(Eng.)*. All antimonial lead; metal in which the high degree of malleability characteristic of pure lead is destroyed by the presence of impurities, of which antimony is the most common.

hardmetal *(Eng.)*. Sintered tungsten carbide. Used for the working tip of high-speed cutting tools. See **sintering**.

hard metals *(Eng.)*. Metallic compounds with high melting-points; typified by refractory carbides of the transition metals (4th to 6th groups of the periodic system), notably tungsten, tantalum, titanium and niobium carbides.

hardness *(Electronics)*. Degree of vacuum in an evacuated space, especially of a thermionic valve or X-ray tube. Also penetrating power of X-rays, which is proportional to frequency.

hardness *(Eng.)*. Signifies, in general, resistance to cutting, indentation and/or abrasion. It is actually measured by determining the resistance to indentation, as in **Brinell, Rockwell, Vickers** and **scleroscope hardness tests**. The value of hardness obtained by the different methods are to some extent related to each other and to the ultimate tensile stress of non-brittle metals.

hardness *(Min.)*. The resistance which a mineral offers to abrasion. The absolute hardness is measured with the aid of a sclerometer. The comparative hardness is expressed in terms of Mohs' scale, and is determined by testing against ten standard minerals: (1) talc, (2) gypsum, (3) calcite, (4) fluorite, (5) apatite, (6) orthoclase, (7) quartz, (8) topaz, (9) corundum, (10) diamond. Thus a mineral with 'hardness 5' will scratch or abrade fluorite, but will be scratched by orthoclase. Hardness varies on different faces of a crystal, and in some cases (e.g. kyanite) in different directions on any one face.

hard packing *(Print.)*. Hard paper employed to cover the cylinder of a printing press when printing on hard, smooth papers from engravings, etc,; used in order to obtain a sharp impression. Also necessary when printing from plastic or rubber plates.

hard pad *(Vet.)*. A name for the **hyperkeratosis** of dog's pads due to infection of the animal with canine distemper.

hard palate *(Zool.)*. In Mammals, the anterior part of the roof of the buccal cavity, consisting of the horizontal palatine plates of the maxillary and palatine bones covered with mucous membrane.

hard pan *(Min.)*. Lateritic gravels near or at surface of alluvial strata which have been consolidated by weathering and deposition of iron oxide or clay.

hard plaster *(Build.)*. Term usually applied to hard-setting forms of gypsum plaster, e.g. **Keene's cement**.

hard plating *(Eng.)*. Chromium plating deposited in appreciable thickness directly on to the base metal, that is, without a preliminary deposit of copper or nickel. The coating is porous, but offers resistance to corrosion and to wear.

hard radiation *(Radiol.)*. Qualitatively, the more penetrating types of X-, beta-, and gamma-rays.

hard-rock geology *(Geol.)*. An informal term for the geology of igneous and metamorphic rocks. Cf. **soft-rock geology**.

hard-rock mining *(Min.Ext.)*. Term used to distinguish between deposits soft enough to be detached by mechanical excavator and those which must first be loosened by blasting.

hard-rock phosphate *(Geol.)*. A phosphatic deposit resulting from the leaching of calcium carbonate out of a phosphatic limestone, leaving a phosphatic residue. Applied specifically to the phosphate deposits of Florida which have this origin.

hard-sectored disk *(Comp.)*. One which is formatted partially or wholly by markers which are put on the disk when it is made.

hard soaps *(Chem.)*. See **soaps**.

hard solder *(Eng.)*. One containing more than 0.3% of carbon.

hard stocks *(Build.)*. Bricks which are sound but have been overburnt and are not of good shape and colour.

hard twist *(Textiles)*. A yarn with more than the standard amount of twist, inserted to secure the desired effects in particular fabrics, e.g. crêpes, voiles, gaberdines.

hardware *(Comp.)*. General term for all mechanical, electric and electronic components of a **computer system** including **peripherals**. Cf. **software**.

hardware *(Space)*. The physical material (e.g. structure, electrical harness, computers) produced for space systems as opposed to non-tangible aspects such as computer software and operating procedures.

hard waste *(Textiles)*. Waste from a single or folded twisted yarns from cop ends and waste made during winding, warping, reeling and weaving.

hard water *(Chem.)*. Water having magnesium and calcium ions in solution and offering difficulty in making a soap lather. See **permanent hardness, temporary hardness**.

hardwired logic *(Comp.)*. Permanent circuitry, often integrated circuit elements and their interconnections. Sometimes called *firmware*. See **logic array**.

hardwood *(For.)*. Dense, close-grained wood from deciduous dicotyledonous trees (oak, beech, ash, teak etc.). Cf. **softwood**.

Hardy and Schulze 'law' *(Min.Ext.)*. The efficiency of an ion used as a coagulating agent is roughly proportional to its state of oxidation.

Hardy-Weinberg law *(Biol.)*. The gene frequencies in a large population remain constant from generation to generation if mating is at random and there is no selection, migration or mutation. If two alleles A and a are segregating at a locus, and each has a frequency of p and q respectively, then the frequencies of the genotypes AA, Aa, and aa are p^2, $2pq$ and q^2 respectively.

harelip *(Med.)*. A congenital cleft in the upper lip, often associated with cleft palate.

Hare's apparatus *(Phys.)*. A simple apparatus for comparing the densities of two liquids. An inverted U-tube has its limbs dipping each into one of the liquids. By suction at the top of the U-tube, the liquids rise in the limbs. The densities are inversely proportional to the heights the liquids rise above their reservoirs.

Harker diagram *(Geol.)*. A variation diagram in which chemical analyses of rocks are plotted to show their relationships. The constituents are plotted as ordinates against the silica content as abscissa.

H-armature *(Elec.Eng.)*. See **shuttle armature**.

harmonic *(Phys.)*. Sinusoidal component of repetitive complex waveform with frequency which is an exact multiple of basic repetition frequency (the fundamental). The full set of harmonics forms a Fourier series which completely represents the original complex wave. In acoustics, harmonics are often termed overtones, and these are counted in order of frequency above, but excluding, the lowest of the detectable frequencies in the note; the label of the harmonic is always its frequency divided by the fundamental. The nth overtone is the $(n + 1)^{th}$ harmonic.

harmonic absorber *(Phys.)*. Arrangement for removing harmonics in current or voltage waveforms, using tuned circuits or a wave filter.

harmonic analysis *(Phys.)*. Process of measuring or calculating the relative amplitudes of all the significant harmonic components present in a given complex waveform. The result is frequently presented in the form of a Fourier series, e.g. $A = A_0 \sin\omega t + A_1 \sin(2\omega t + \varphi_1) + A_2 \sin(3\omega t + \varphi_2) + ...$etc., where ω = pulsatance and φ = phase angle.

harmonic antenna *(Telecomm.)*. One whose overall length is an integral number (greater than one) of quarter-wavelengths.

harmonic components *(Phys.)*. Any term (but the first) in a Fourier series which represents a complex wave.

harmonic conjugate *(Maths.)*. Two of four collinear points are harmonic conjugates of the other two if the cross-ratio of the four points is harmonic (i.e. has the value minus one). Also used of a pencil of four concurrent lines.

harmonic distortion *(Phys.)*. Production of harmonic components from a pure sine wave signal as a result of non-linearity in the response of a transducer or amplifier.

harmonic drive *(Telecomm.)*. See **harmonic excitation**.

harmonic excitation *(Telecomm.)*. (1) Excitation of an antenna on one of its harmonic modes. (2) Excitation of a transmitter from a harmonic of the master oscillator. Also *harmonic drive*.

harmonic filter *(Telecomm.)*. One which separates harmonics from fundamental in the feed to an antenna. Also *harmonic suppressor*.

harmonic function *(Maths.)*. A function which satisfies Laplace's equation.

harmonic generator *(Phys.)*. Waveform generator with controlled fundamental frequency, producing a very large number of appreciable-amplitude odd and even harmonic components which provide a series of reference frequencies for measurement or calibration. See **multivibrator**.

harmonic interference *(Telecomm.)*. That caused by harmonic radiation from a transmitter and outside the specified channel of radio communication.

harmonic means *(Maths.)*. (1) Numbers inserted between two numbers such that the resulting sequence is a harmonic progression. (2) The *harmonic mean* of n numbers a_r is the reciprocal of the arithmetic mean of their reciprocals, i.e.

$$n\Big/\sum_{1}^{n}\frac{1}{a_r}.$$

harmonic progression *(Maths.)*. The numbers, a, b, c, d,... form a harmonic progression if their respective reciprocals form an arithmetical progression.

harmonic ratio *(Maths.)*. If the cross-ratio of four collinear points equals -1, it is harmonic.

harmonic series *(Phys.)*. One in which each basic frequency of the series is an integral multiple of a fundamental frequency.

harmonic suppressor *(Telecomm.)*. Same as *harmonic filter*.

harmonic wave *(Phys.)*. A wave whose profile is a pure sine curve; also *sinusoidal wave*.

harmotome *(Min.)*. A member of the zeolite group, hydrated silicate of aluminium and barium, crystallizing on the monoclinic system, though the symmetry approaches that of the orthorhombic system. Best known by reason of the distinctive cruciform twin groups that are not uncommon. Occurs in mineralized veins.

harness *(Aero.)*. (1) The entire system of engine ignition leads, particularly those which are screened to prevent electromagnetic interference with radio equipment. (2) The parallel combination of leads interconnecting the thermocouple probes of a turbine engine exhaust gas temperature indicating system. See **gas temperature**. (3) Prefabricated electrical connections for any electrical or electronic system. (4) Straps by which aircrew are held in their seats.

Harris process *(Eng.)*. A method of softening lead. Arsenic, antimony and tin are oxidized by adding sodium nitrate and lead oxide, and the oxides formed are caused to react with sodium hydroxide and chloride to form arsenates, antimonates and stannates.

hartite *(Min.)*. A naturally occurring hydrocarbon-compound, $C_{20}H_{34}$, crystallizing in the triclinic system.

hartley *(Telecomm.)*. Unit of **information content**.

Hartley oscillator *(Electronics)*. An oscillator circuit incorporating a parallel-tuned resonant circuit with a tapping point on the inductance. Feedback is provided by the section of inductance connected between grid and cathode (in valve circuit) or between source and gate (in an FET) or between emitter and base (in a transistor).

Hartley principle *(Telecomm.)*. General statement that amount of information which can be transmitted through a channel is the product of frequency bandwidth and time during which it is open, whether time division is used or not. See **information content**.

Hartmann dispersion formula *(Phys.)*. An empirical expression for the variation of refractive index n of material with the wavelength of light λ;

$$n = n_0 + \frac{c}{(\lambda - \lambda_0)^a},$$

where n_0, c, λ_0 are constant for a given material. For glass a is about 1.2.

Hartman oscillator *(Acous.)*. A device, consisting basically of a conical nozzle, supersonic gas jet and cylindrical cavity (resonator), used for generating high-intensity ultra-sound in fluids or gases.

Hartman test *(Phys.)*. Test for aberration of a lens, in which a diaphragm containing a number of small apertures is placed in front of the lens and the course of the rays is recorded by photographing the pencils of light in planes on either side of the focus.

Hartnell governor *(Eng.)*. An engine governor in which the vertical arms of two or more bell-crank levers support heavy balls, the horizontal arms carrying rollers which abut against the central spring-loaded sleeve operating the engine-governing mechanism.

Hartree equation *(Electronics)*. One relating flux density in travelling-wave magnetron to minimum anode potential required for oscillation in any given mode. Graphs representing this relationship for different mode numbers form a *Hartree diagram*.

Harvard classification *(Astron.)*. A method of classifying stellar apectra, employed by the compilers of the Draper Catalogue of the Harvard Observatory and now in universal use. See **spectral types**.

harvest mite *(Vet.)*. The popular name of the parasitic larval stage of the mite *Trombicula autumnalis*, which occurs in the skin of animals and man. The nymphal and adult stages of the mite are free-living.

harvest moon *(Astron.)*. The name given in popular language to the full moon occurring nearest to the autumnal equinox, at which time the moon rises on several successive nights at almost the same hour. This retarded rising, due to the small inclination of the moon's path to the horizon, is most noticeable at the time of the full moon, although it occurs for some phase of the moon each month.

harvest spider *(Zool.)*. Common name for Arachnids of the order *Opiliones*. Also called *harvestmen*.

Harvey process *(Eng.)*. Toughening treatment for alloy steels, involving superficial carburization, followed by heating to a high temperature and quenching with water.

harzburgite *(Geol.)*. An ultrabasic igneous rock belonging to the peridotite group. It consists almost entirely of olivine and orthopyroxene, usually with a little chromite, magnetite and diopside.

Harz jig *(Min.Ext.)*. Concentrating appliance in which water is pulsed through a submerged fixed screen, across which suitably sized ore moves. Heaviest particles gravitate down and through and lighter ones overflow.

Hashimoto's disease *(Med.)*. *Auto-immune thyroiditis*. See **Hashimoto thyroiditis**.

Hashimoto thyroiditis *(Immun.)*. A disease of the thyroid gland, characterized by chronic inflammatory changes due to infiltration by lymphocytes, plasma cells and macrophages. The gland becomes enlarged and hard. Auto-antibodies against thyroglobulin and other thyroid antigens are usually present in the blood.

hashing *(Comp.)*. Generating **hash totals** to associate with items of data. The numbers are used as index numbers or storage addresses for the data.

hashish *(Bot.)*. See **cannabis**.

hash total *(Comp.)*. A meaningless number generated from a coded data item or its key. See **hashing**.

hasp *(Build.)*. A fastening device in which a slotted plate fits over a staple and is secured to it by means of a padlock or peg.

hastate *(Bot.)*. Shaped like an arrow head with narrow basal lobes pointing outwards (i.e. halberd-shaped).

hastingsite *(Min.)*. A monoclinic amphibole, hydrated sodium calcium iron aluminosilicate. The name is used for an end-member compositional variety of amphibole.

hatch coaming *(Ships)*. The strengthened frame surrounding a **hatchway**. Its functions are to replace the strength lost by cutting the beams and deck plating, protect the opening and provide support for the hatch cover.

hatchet *(Build.)*. A small axe used for splitting or rough-dressing timber.

Hatch-Slack pathway *(Bot.)*. Method of CO_2 fixation in C_4 plants. Typically, in the mesophyll cells of a leaf with Kranz anatomy, CO_2 is fixed into the 4-carbon malic acid which is transported to the bundle sheath cells and there decarboxylated. The CO_2 thus released is refixed by **RUBISCO** in the normal way to sugars while the three-carbon acid remaining is returned to the mesophyll cells. The whole cycle acts as a concentrating and scavenging mechanism for CO_2. See **PEP carboxylase**.

hatchway *(Ships)*. An opening in the deck to allow cargo to be shipped into and removed from the hold.

haulage level *(Min.Ext.)*. Underground tramming road in, or parallel to, strike of the ore deposit, usually in footwall. Broken ore gravitates, or is moved, to ore chutes and drawn to trucks in this level.

haul distance *(Civ.Eng.)*. The distance, at any particular time, that excavated material from e.g. a cutting has to be carried before deposition in order to form an embankment. The *haul* is the sum of the products of each load by its haul distance.

haunch *(Build.)*. The part forming a stub tenon, left near the root of a haunched tenon. Also called a *hauncheon.*

haunched tenon *(Build.)*. A tenon from the width of which a part has been cut away, leaving a haunch near its root.

hauncheon *(Build.)*. See **haunch**.

haunches *(Build.,Civ.Eng.)*. See **flanks**.

haunching *(Build.)*. A mortise cut to receive the haunch of a haunched tenon.

Hausdorff space *(Maths.)*. A topological space in which any 2 points can be separated into 2 disjoint open sets.

hausmannite *(Min.)*. A blackish-brown crystalline form of manganese oxide, occurring with other manganese ores. Crystallizes in the tetragonal system but is often found massive.

haustellate *(Zool.)*. Mouthparts modified for sucking as in many Insects.

haustellum *(Zool.)*. In *Diptera*, the distal expanded portion of the proboscis. adj. *haustellate.*

haustorium *(Bot.)*. An outgrowth from a parasite which penetrates a tissue or cell of its host and acts as an organ for absorbing nutrients.

haüyne *(Min.)*. A feldspathoid, crystallizing in the cubic system, consisting essentially of silicate of aluminium, sodium, and calcium, with sodium sulphate; occurs as small blue crystals chiefly in phonolites and related rock types.

Haversian canals *(Zool.)*. Small channels pervading compact bone and containing blood vessels.

Haversian lamellae *(Zool.)*. In compact bone, and concentrically arranged lamellae which surround a Haversian canal.

Haversian spaces *(Zool.)*. In the development of bone, irregular spaces formed by the internal absorption of the original cartilage bone.

Haversian system *(Zool.)*. In compact bone, a Haversian canal with surrounding lamellae.

haversine *(Maths.)*. Half of the **versine**, i.e. $\frac{1}{2}(1 - \cos\theta)$.

hawk *(Build.)*. A small square board, with handle underneath, used to carry plaster or mortar.

hawk's eye *(Min.)*. A dark-blue form of silicified crocidolite found in Griqualand West; when cut *en cabochon*, it is used as a semi-precious gemstone. Cf. **tiger's eye**.

hawser, hawse-pipe *(Ships)*. A tubular casting fitted to the bows of a ship, through which the anchor chain or cable passes.

Hawthorne effect *(Behav.)*. The observation that experimental subjects who are aware that they are part of an experiment often perform better than totally naive subjects. See **demand characteristics**.

Hay bridge *(Elec.Eng.)*. An a.c. bridge quite widely used for the measurement of inductance.

hay fever *(Immun.,Med.)*. *Allergic rhinitis.* Acute nasal catarrh and conjunctivitis in atopic subjects, caused by inhalation of allergens (usually pollens). Due to an immediate hypersensitivity (Type 1) reaction resulting from combination of cell–IgE antibody with the causative allergen. There is a tendency for this type of sensitization to be inherited.

haze *(Meteor)*. A suspension of solid particles of dust and smoke etc. reducing visibility above 1 km.

Hazeltine neutralization *(Elec.Eng.)*. A method of anode neutralization used in single-stage power valve amplifiers.

Hazen and Williams' formula *(Eng.)*. An empirical formula relating the flow in pipes and channels to the hydraulic radius and the slope, using a coefficient which varies with the surface roughness.

Hazen's uniformity coefficient *(Powder Tech.)*. A measure of the range of particle sizes present in a given powder. It is defined by the equation $H = A/B$, where A is the 60th percentile of the cumulative percentage undersize by weight of the powder, and B is the 10th.

Hb *(Chem.)*. A symbol for a molecule of haemoglobin minus the iron atom.

ħ *(Phys.)*. *H-bar.* The symbol for Dirac's constant.

H-beam *(Eng.)*. A steel beam with a section shaped like the letter H, the cross-piece of which is relatively long. Also called *H-girder, I-beam, rolled steel joist.*

H-class insulation *(Elec.Eng.)*. A class of insulating material to which is assigned a temperature of 180°C. Other letters, which see, denote different temperatures.

HCP *(Chem.)*. Abbrev. for *Hexagonal Closet Packing.*

HDD *(Aero.)*. See **head down display**.

H-display *(Radar)*. Modified **B-display** to include angle of elevation. The target appears as two adjacent bright spots and the slope of the line joining these is proportional to the sine of the angle of elevation.

HDTV *(Image Tech.)*. *High Definition Television*, in general systems using 1000 or more scanning lines per frame.

He *(Chem.)*. The symbol for *helium.*

head *(Arch.)*. The capital of a column.

head *(Bot.)*. A dense inflorescence of sessile flowers, usually a **capitulum**.

head *(Build.)*. See **lintel**.

head *(Elec.Eng.)*. Recording and reproducing unit for magnetic tape, containing exciting coils, a laminated core, in ring form with a minute gap. Flux leakage across this gap enters the tape and magnetizes it longitudinally. See **flux gate**.

head *(Eng.)*. (1) Any part having the shape or position of a head, e.g. the *head* of a bolt. (2) Any part or principal part analogous to a head, e.g. the head of a hammer or a lathe.

head *(Geol.)*. A superficial deposit consisting of angular fragments of rock, originating from the breaking up of rock by alternate freezing and thawing of its contained water, followed by downhill movement. Head is found in valleys in periglacial regions, i.e. those formerly near the edge of an ice sheet. In south-eastern England, head is also known as *coombe rock.*

head *(Hyd.Eng.)*. A distance representing the height above a datum which would give unit mass of a fluid in a conduit the potential energy equal to the sum of its actual potential energy, its kinetic energy and its pressure energy. See **Bernoulli Law**.

head *(Image Tech.)*. (1) General term for the central mechanism of a motion picture projector or printer. See also **sound head**. (2) The mounting for a camera on a tripod. (3) The start of a roll of film or magnetic tape;

head end. (4) The top of a motion picture frame. (5) In magnetic reproduction and recording, the device converting magnetic variations to electrical signals and vice versa.

head *(Min.Ext.)*. (1) An advance main roadway driven in solid coal. (2) The top portion of a seam in the coal face. (3) The difference in air pressure producing ventilation. (4) The whole falling unit in a stamp battery, or merely the weight at the end of the stem.

head *(Print.)*. The top edge of a volume: the top margin of a page. See **fore-edge**, **tail**.

head amplifier *(Electronics)*. A pre-amplifier which has to be placed physically adjacent to the signal source in order to provide gain before the signal is lost though transmission lines or cables or before it is swamped by interference. Used especially with receiving aerials and sensitive microphones.

headband *(Print.)*. A decorative band of silk or other material at the head of a book, between the back and the cover.

head-bay *(Hyd.Eng.)*. The part of a canal lock immediately above the head-gates.

head-box *(Paper)*. See **flow box**.

head cap *(Print.)*. The leather at the head and tail of the spine folded over in a curve, the *head band* if present, being left visible.

head crash *(Comp.)*. Dramatic and expensive descent of the read/write head onto the surface of a magnetic disk, caused by mechanical malfunction.

head down display *(Aero.)*. Usually a visual display mounted inside the cockpit to supplement the **head up display**.

header *(Build.)*. A whole brick which has been laid so that its length is at right angles to the face of the wall.

header *(Elec.Eng.,Electronics)*. Base for a relay, or transistor can, with hermetically sealed insulated leads.

header *(Eng.)*. A box or manifold supplying fluid to a number of tubes or passages, or connecting them in parallel.

header block *(Comp.)*. Data placed at the beginning of a file which identifies the file and may describe its structure.

header joist *(Build.)*. See **trimmer**.

headframe *(Min.Ext.)*. The steel or timber frame at the top of a shaft, which carries the sheave or pulley for the hoisting rope, and serves such other purposes as, e.g. acting as transfer station for hoisted ore, or as loading station for man and materials.

head-gates *(Hyd.Eng.)*. The gates at the high-level end of a lock.

headgear *(Min.Ext.)*. See **headframe**.

head grit *(Vet.)*. See **yellowses**.

heading *(Build.)*. A **heading course**.

heading *(Civ.Eng.)*. A relatively small passage driven in the line of an intended tunnel, the latter being afterwards formed by enlarging the former.

heading *(Min.Ext.)*. Passage-way through solid coal.

heading bond *(Build.)*. The form of bond in which every brick is laid as a header, each 4.5 in (112 mm) face breaking joint above and below; used for footings and corbellings but not for walling.

heading chisel *(Build.)*. See **mortise chisel**.

heading course *(Build.)*. An external or visible course of bricks which is made up entirely of headers.

heading indicator *(Aero.)*. The development of supersonic fighters, which climb at extremely steep angles, made it essential to have an even more comprehensive instrument than the *attitude indicator*. A sphere enables compass heading to be included, thereby giving complete 360° presentation about all three axes.

heading joint *(Build.)*. A joint between the ends of boards abutting against each other.

headings *(Min.Ext.)*. *Head tin.* Concentrate settling nearest to the entry point of a concentrating device such as a **sluice** or **buddle**.

headline *(Print.)*. The line of type placed at the top of the page, giving either the title of the book or the chapter heading.

head motion *(Min.Ext.)*. Vibrator, a sturdy device which gives reciprocating movement to shaking tables, used in gravity methods of concentration.

head moulding *(Build.)*. A moulding situated above an aperture.

head nailing *(Build.)*. The method of nailing slates on a roof in which nails are driven in the slates near their heads or higher edges. Cf. **centre nailing**.

head race *(Hyd.Eng.)*. A channel conveying water to a hydraulically operated machine.

head-rail *(Build.)*. The horizontal member of a door-case.

headroom *(Build.)*. The uninterrupted height within a building on any floor, or within a staircase, tunnel, doorway etc.

heads *(Build.)*. A term applied to the tiles forming a course around the eaves.

headstock *(Eng.)*. In general, a device for supporting the end or head of a member or part; e.g. (1) the part of a lathe that carries the spindle, (2) the part of a planing machine that supports the cutter or cutters, (3) the supports for the gudgeons of a wheel, (4) the movable head of some measuring machines.

head tree *(Build.)*. A timber block placed on the top of a post, so as to provide increased bearing-surface.

head-up display *(Aero.)*. The projection of instrument information on to the windscreen or a sloping glass screen in the manner of the **reflector sight**, so that the pilot can keep a continual lookout and receive flight data at the same time. Used originally with fighter interception and attack radar, later adapted for blind approach, landing, and general flight data. Abbrev. *HUD*.

head valve *(Eng.)*. The delivery valve of a pump, as distinct from the suction or **foot valve**.

head wall *(Build.)*. A wall built in the same plane as the face of a bridge arch.

headway *(Build.)*. Same as **headroom**.

headworks *(Eng.)*. In a hydroelectric scheme, a dam forming a reservoir or a low weir across a river or stream, which possesses the necessary intakes and control gear to divert the water into an aqueduct.

heald *(Textiles)*. Part of the loom mechanism used to raise and lower the warp. Comprises eyes formed of coated twine or wire through which warp ends are drawn.

healing *(Build.)*. The operation of covering a roof with tiles, lead etc.

health physics *(Radiol.)*. Branch of radiology concerned with health hazards associated with ionizing radiations, and protection measures required to minimize these. Personnel employed for this work are *health physicists* or *radiological safety officers*.

heap leaching *(Min.Ext.)*. This, perhaps aided by heap roasting, is the dissolution of copper from oxidized ore by solvation with sulphuric acid. The resulting liquor is run over scrap iron to precipitate out its copper.

heap sampling *(Min.Ext.)*. See **quartering**.

hearing *(Acous.)*. The subjective appreciation of externally applied sounds.

hearing loss *(Acous.)*. At any frequency, the hearing loss of a partially deaf ear is the difference in decibels between the threshold of hearing and that of a normal ear.

heart *(Zool.)*. A hollow organ, with muscular walls, which by its rhythmic contractions pumps the blood through the vessels and cavities of the circulatory system.

heart attack *(Med.)*. The common term for a *myocardial infarction*.

heart-block *(Med.)*. The condition in which a lesion of the special tissue that conducts the contraction impulse from the atrium to the ventricle prevents the spread of the wave of contraction, usually leads to a slow pulse.

heart bond *(Build.)*. A form of bond having no through-stones, headers consisting of a pair of stones meeting in the middle of the wall, the joint between them being covered by another header stone.

heart-burn *(Med.)*. A burning sensation in the mid-line of the chest, usually associated with *dyspepsia*.

heart cam *(Horol.)*. A cam in the form of a heart, used in stop watches and chronographs to bring the recording hand instantly back to zero, on pressing the button. Also called *heart piece*.

heart failure *(Med.)*. Condition in which the heart fails to maintain an adequate flow of blood to all the body tissues, for whatever reason.

hearth *(Eng.)*. Floor of reverberatory, open hearth, cupola or blast furnace, made of refractory material able to support the charge and collect molten products for periodic removal.

hearting *(Build.)*. The operation of building the inner part of a wall, between its facings.

heart sounds *(Med.)*. Sounds generated by the heart and heard by **auscultation** over the left side of the chest. The first sound coincides with the closure of the *mitral* and *tricuspid* valves, the second sound with closure of the aortic and pulmonary valves. Added sound may be heard in diseases of the heart.

heart transplant *(Med.)*. When a human donor heart is implanted into a patient whose heart disease can be controlled by no other means. The donor heart takes on the pumping function for the recipient.

heartwater *(Vet.)*. A disease of cattle, sheep, and goats, occurring in parts of Africa, caused by *Cowdria ruminantium (Rickettsia ruminantium)*, and transmitted by ticks of the genus *Amblyomma*. Characterized by fever, nervous symptoms, and often death.

heart wood *(Bot.)*. *Duramen*. The inner, older layers of wood in the trunk or branch of a tree or shrub. Compared to the surrounding sap wood, it is denser and darker, the cell walls are impregnated with resins, phenolics etc. and it no longer functions in storage or conduction. It is more prone to rot in the tree but more durable as timber.

heat *(Phys.)*. Heat is energy in the process of transfer between a system and its surroundings as a result of temperature differences. However, the term is still used also to refer to the energy contained in a sample of matter; also for **temperature**, e.g., forging or welding **heat**. For some of the chief branches in the study of heat see **calorimetry**, **heat units**, **internal energy**, **latent heat**, **mechanical equivalent of heat**, **radiant heat**, **radiation**, **specific heat capacity**, **temperature**, **thermal conductivity**, **thermometry**.

heat *(Zool.)*. The period of sexual excitation.

heat balance *(Chem.Eng.)*. A statement which relates for a chemical process or factory all sources of heat and all uses. The allowance of heats of reaction, solution and dilution must be included as well as conventional sources for fuel, steam etc. If done graphically, it is usually called a *Sankey diagram*. Also called *energy balance*.

heat balance *(Eng.)*. Evaluation of operating efficiency of a furnace or other appliance, the total heat input being apportioned as to heat in the work, heat stored in brickwork, or refractory materials, loss by conduction, radiation, unburnt gases in waste products, sensible heat in dry flue gases, and latent heat of water vapour, thus determining the quantity and percentage of heat usefully applied and the sources of heat losses.

heat coil *(Elec.Eng.)*. A small protective resistance coil, which heats on prolonged excess current in telegraph lines, melts a small quantity of solder, and disconnects the line, so preventing possible damage to apparatus.

heat density *(Eng.)*. Weight and pressure of live gases in heating chambers of industrial furnaces, upon which the rate of heat transfer depends.

heat detector *(Eng.)*. An indirect-acting thermostat for operation in conjunction with a gas-flow control valve, and for controlling working temperatures in furnaces and heating appliances up to about 1000°C.

heat drop *(Eng.)*. Colloq. transfer of heat energy, but often used for **enthalpy heat drop**.

heater *(Electronics)*. Conductor carrying current for heating a cathode, generally enclosed by the latter.

heater box *(Print.)*. The heated part of a blocking press. This is usually electrically heated and thermostatically controlled.

heater platen *(Print.)*. A grooved platen sliding into the heater box of a blocking press and holding the die.

heater transformer *(Electronics)*. In a piece of equipment utilizing thermionic valves, a mains transformer giving a low-voltage output which is used to feed the valve heaters.

heat exchange *(Chem.Eng.)*. The process of using two streams of fluid for heating or cooling one or the other either for conservation of heat or for the purpose of adjusting process streams to correct processing temperatures.

heat exhaustion *(Med.)*. Less severe than **heat-stroke** and usually occurring after several days of salt and water depletion due to heat exposure.

heat filter *(Image Tech.)*. In an illuminating system, a filter which reduces or reflects infra-red radiation in the beam and hence its heating effect.

heat-flow measurement *(Geol.)*. The measurement of the amount of heat leaving the Earth.

heat flux *(Chem.Eng.)*. The total flow of heat in *heat exchange*, in appropriate units of time and area.

heath *(Ecol.)*. Vegetation type consisting of evergreen woody shrubs growing on acid soil. In northern Europe the species are largely members of the *Ericaceae*, but the term heath is often used more widely to cover dwarf shrub communities in other parts of the world.

heather blindness *(Vet.)*. See **contagious ophthalmia**.

heating curves *(Eng.)*. Curves obtained by plotting time against temperature for a metal heating under constant conditions. The curves show the absorption of heat which accompanies melting and arrest points marking polymorphic changes in pure metals and various transformations in alloys.

heating depth *(Elec.Eng.)*. Thickness of skin of material which is effectively heated by dielectric or eddy-current induction heating, or radiation.

heating element *(Elec.Eng.)*. The heating-resistor, together with its former, of an electric heater, electric oven, or other device, in which heat is produced by the passage of an electric current through a resistance.

heating inductor *(Elec.Eng.)*. Conductor, usually water-cooled, for inducing eddy currents in a charge, workpiece or load. Also called *applicator*, *work coil*.

heating limit *(Elec.Eng.)*. See **thermal limit**.

heating muff *(Aero.)*. A device for providing hot air, consisting of a chamber surrounding an exhaust pipe or jet pipe.

heating resistor *(Elec.Eng.)*. The wire or other suitable material used as the source of heat in an electric heater.

heating time *(Electronics)*. Time required after switching on, before a valve cathode or other piece of equipment reaches normal operating temperature.

heat-insulating concrete *(Eng.)*. High-alumina cement and a lightweight aggregate, e.g. kieselguhr, diatomic earth, or vermiculite, to reduce heat transfer through furnace walls, etc.

heat insulation *(Build.)*. The property, possessed in varying degrees by different materials, of impeding the transmission of heat.

heat liberation rate *(Eng.)*. A measure of steam boiler performance, expressed in kJ/lh (=26.8 Btu/ft^{2}h).

heat of formation *(Chem.)*. The net quantity of heat evolved or absorbed during the formation of one mole of a substance from its component elements in their standard states.

heat of solution *(Chem.)*. The quantity of heat evolved or absorbed when one mole of a substance is dissolved in a large volume of a solvent.

heat pipe *(Space)*. A means of cooling where heat is transferred along a tube from a heat source to a heat sink of small temperature difference. Heat transfer is effected by a liquid which vaporizes at a desired temperature.

heat pump *(Eng.)*. Machine operating on a reversed heat engine cycle to produce a heating effect. Energy from a low temperature source, e.g. earth, lake or river is absorbed by the working fluid, which is mechanically compressed, resulting in a temperature increase. The high temperature energy is transferred in a heat exchanger.

heat regenerator *(Chem.Eng.)*. A matrix of metal which is alternately heated by the waste gases from an industrial process and used to heat up the incoming gases.

heat-resisting alloy *(Eng.)*. Alloys developed to withstand high stresses at very high temperatures as in the fan blades of aeroengines.

heat-resisting paints *(Build.)*. A range of paints formulated to withstand the effects of exposure to high temperatures. Synthetic resins are employed, being mixed with pigments unaffected by heat, producing paints classified as *withstanding* temperatures: (1) Up to 93°C (200°F) includes alkyd resin paints pigmented with titanium oxide, synthetic reds, monastral blue, yellow oxides etc. (2) Between 93°C and 260°C (500°F) includes acrylic resin based paints pigmented with titanium oxide and suitable colours. (3) Between 260°C and 540°C (1000°F) based on silicone resins.

heat run *(Elec.Eng.)*. A test in which an electric machine or other apparatus is operated at a specified load for a long period to ascertain the temperature which it reaches.

heat-set ink *(Print.)*. Printing ink formulated to set by heating the surface of the stock.

heat setting *(Textiles)*. Stabilizing fibres, yarns, or fabrics by means of heating under controlled conditions. Thus a fabric may be heat set when it is held flat under tension while being extended in length and breadth. The resultant fabric will tend to retain its flatness in use. Pleats may also be heat set so that they remain clearly defined in a garment.

heat sink *(Elec.Eng.)*. Usually metal plate specially designed (e.g. with matt black fins) to conduct (and radiate) heat from an electrical component, e.g. a transistor.

heat spot *(Zool.)*. An area of the skin sensitive to heat owing to the presence of certain nerve-endings beneath the skin.

heat stroke *(Med.)*. *Heat hyperpyrexia*. The combination of coma, convulsions, a high temperature and other symptoms, as a result of exposure to excessive heat.

heat transfer *(Phys.)*. See **conduction of heat, convection of heat, radiation**.

heat transfer coefficient *(Chem.Eng.)*. The rate of heat transfer q between two phases may be expressed as

$$q = hA(T_1 - T_2),$$

where A is the area of the phase boundary, $(T_1 - T_2)$ is their difference in temperature, and the heat transfer coefficient h depends on the physical properties and relative motions of the two phases. See **Newton's law of cooling**.

heat transfer salt *(Chem.Eng.)*. Molten salts used as a heating or quenching medium. Usually mixtures of sodium or potassium nitrate or nitrite, range 200°C to 600°C. Abbrev. *HTS*.

heat treatment *(Eng.)*. Generally, any heating operation performed on a solid metal, e.g. heating for hot-working, or annealing after cold-working. Particularly, the thermal treatment of steel by normalizing, hardening, tempering etc.; used also in connection with precipitation-hardening alloys, such as those of aluminium.

heat units *(Phys.)*. See **joule, calorie, British thermal unit**; also *calorific value*.

heave *(Geol.Min.Ext.)*. (1) The horizontal distance separating parts of a faulted seam, bed, vein or lode, measured normal to the fault plane. (2) Rising of the floor of a mine.

heavier-than-air aircraft *(Aero.)*. See **aerodyne**.

Heaviside-Campbell bridge *(Elec.Eng.)*. An electrical network for comparing mutual and self inductances.

Heaviside layer *(Phys.)*. See **E-layer**.

Heaviside-Lorentz units *(Phys.)*. Rationalized CGS Gaussian system of units, for which corresponding electric and magnetic laws are always similar in form. See **rationalized units**.

Heaviside unit function *(Telecomm.)*. Step change in a magnitude, with an infinite rate of change, required in pulse analysis and transient response of circuits.

heavy-aggregate concrete *(Nuc.Eng.)*. That containing very dense aggregate material in place of some or all of the usual gravel, so increasing its gamma-ray absorption coefficient. Also used for appropriate general purposes (e.g. sea walls). See **loaded concrete**.

heavy chemicals *(Chem.)*. Those basic chemicals which are manufactured in large quantities, e.g. sodium hydroxide, chlorine, nitric acid, sulphuric acid.

heavy ground *(Min.Ext.)*. Unstable roof rock requiring special care and support.

heavy (H) chain *(Immun.)*. Polypeptide chain in immunoglobulins which together with the light chain makes up the complete molecule. Each heavy chain consists of a variable region and a constant region composed of three or four domains, depending on the class. See **immunoglobulin**.

heavy hydrogen *(Chem.)*. Same as **deuterium**.

heavy liquids *(Min.Ext.)*. Liquids, organic or solutions of heavy salts, of relative densities adjustable in the range 1.4 to 4.0, used to separate ore constituents into relatively heavy (sink) and light (float) fractions with fair precision and to carry out specific gravity tests on minerals. They include tetrachloromethane, saturated calcium chloride solution, bromoform, methyl iodide and Clerici's solution.

heavy media separation *(Min.Ext.)*. Dense media separation. Method of upgrading ore by feeding it into liquid slurry of intermediate density, the heavier fraction sinking to one discharge arrangement and the light ore overflowing. Used to remove shale from coal (*sink-float process*) and to reject waste from ore.

heavy metal *(Biol.)*. (1) In electron microscopy, metal of high atomic number used to introduce electron density into a biological specimen by staining, negative staining or shadowing. (2) In plant nutrition, metals of moderate to high atomic number e.g. Cu, Zn, Ni, Pb, present in soils due to an outcrop or mine spoil, preventing growth except for a few tolerant species and ecotypes.

heavy metal replacement *(Crystal.)*. See **isomorphous replacement**.

heavy mineral *(Geol.)*. A detrital mineral from a sedimentary rock having a higher then normal specific gravity. Commonly applied to minerals which sink in bromoform (density 2.9).

heavy particle *(Phys.)*. See **hyperon**.

heavy spar *(Min.)*. See **barytes**.

heavy water *(Chem.)*. Deuterium oxide, or water containing a substantial proportion of deuterium atoms (D_2O or HDO).

heavy-water reactor *(Nuc.Eng.)*. One using **heavy water** as moderator, e.g. **CANDU** and **SGHWR** types.

hebephrenia *(Behav.)*. *Hebephrenic schizophrenia, disorganized schizophrenia*. A type of schizophrenia characterized by incoherence of thought, odd and childlike behaviour, and inappropriate emotional expression, occurring about puberty.

Heberden's nodes *(Med.)*. Small bony knobs occurring on the bones of fingers of the old.

hebetude *(Med.)*. Lethargy and mental dullness, with impairment of the special senses.

hectare *(Surv.)*. Metric unit of area, equal to 100 *ares*. Abbrev. *ha*. 1 ha = 2.47 acres; 100 ha = 1 km^2, = 0.386 $mile^2$.

hecto- *(Genrl.)*. Metric prefix meaning × 100.

hectobar *(Eng.)*. See **bar**.

hectocotylized arm *(Zool.)*. In some male *Cephalopoda*, one of the tentacles modified for the purpose of transferring sperm to the female.

hectometric waves *(Telecomm.)*. See **medium frequency**.

hectorite *(Min.)*. A rare lithium-bearing mineral in the montmorillonite group of clay minerals.

heddle *(Textiles)*. See **heald**.

hedenbergite *(Min.)*. An important calcium-iron pyroxene, $CaFeSi_2O_6$, occurring as black crystals, and also as a component molecule in many of the rock-forming clinopyroxenes.

Hedley's dial *(Surv.)*. A form of compass adapted for taking inclined sights; it consists of a pair of sighting vanes (or a telescope capable of rotation about a horizontal axis), carrying with them a vertical graduated arc moving over a fixed reference mark.

heel *(Build.)*. The back end of a plane.

heel *(Ships)*. An angle of transverse inclination arising from a force external to the ship, e.g. wind. Cf. **list**.

heeling error *(Ships)*. **Deviation** at a magnetic compass occurring when a vessel heels. It can be eliminated by a group of vertical permanent magnets.

heel-post *(Hyd.Eng.)*. The vertical post at one side of a lock-gate, about which the lock-gate swings. Also called a *quoin-post*.

heel strap *(Build.)*. A strap fastening the foot of a principal rafter to the tie-beam on a timber truss. The strap is threaded at the ends to take a cover plate so that the joint is completely encircled.

Heenan dynamatic dynamometer *(Eng.)*. An eddy-current dynanometer in which the steel rotor has the shape of a gear-wheel and rotates in a cylindrical stator, with smooth bore, surrounded by d.c. excited field coils.

Heenan hydraulic torque meter *(Eng.)*. An arrangement of input and output shafts with vanes forming a concentric annular working compartment through which oil is forced to transmit torque, the differential pressure across the vanes being a measure of the torque.

Hegman gauge *(Powder Tech.)*. A **fineness-of-grind gauge** used widely in the British paint industry. The groove depth from four-thousandths of an inch (0.1 mm) to zero is marked off arbitrarily into a linear scale of 8 units. The fitness of a paint pigment is often quoted by reference to the *Hegman gauge number* at which scoring occurs.

Hehner's test *(Chem.)*. A test for the presence of formaldehyde in milk. It is based upon the appearance of a blue or violet ring when the milk is mixed with a dilute ferric chloride solution, and concentrated sulphuric acid is added to form a layer beneath the milk.

height board *(Build.)*. A gauge for the treads and risers of a timber staircase.

height control *(Image Tech.)*. The means of varying the amplitude of the vertical deflection of a *CRO*.

height of a transfer unit *(Chem.Eng.)*. A measure of the separating efficiency of packed columns for mass transfer operations. It is the height of packed column used for liquid-liquid extraction in which one theoretical transfer unit takes place. Abbrev. *HTU*.

height of instrument *(Surv.)*. (1) In levelling or trigonometrical survey work, the vertical distance of the plane of collimation of the level, or the horizontal axis of the theodolite above datum. (2) In tacheometry, the vertical distance of the horizontal axis of the instrument above ground-level. Abbrev. *H.I.*.

Heisenberg principle *(Phys.)*. See **uncertainty principle**.

Heising modulation *(Electronics)*. Constant-current modulation, arising from one valve driven by signal and another valve driven by carrier, having their anodes fed through the same inductor; the modulated carrier is taken from the anode circuit by capacitive or inductive coupling.

held water *(Geol.)*. That kept above the natural water table through capillary force. US *free water*.

helianthine *(Chem.)*. $(CH_3)_2N.C_6H_4.N{=}N.C_6H_4.SO_3H$, *4-dimethylamino-azobenzene-4-sulphonic acid*, a chrysoidine dye. The sodium salt of this acid is methyl orange, used as an indicator.

heli-arc welding *(Eng.)*. A welding process in which helium is used to shield the weld area from contamination by atmospheric oxygen and nitrogen.

helical antenna *(Telecomm.)*. An antenna in the form of a helix; when the circumference is one wavelength, maximum radiation is along its axis. Used in VHF and UHF bands.

helical coil model *(Biol.)*. Model of **metaphase** chromosome organization which envisages that the primary DNA helix is packed by secondary and higher orders of coiling.

helical gears *(Eng.)*. Gear-wheels in which the teeth are not parallel with the wheel axis, but *helical* (i.e. parts of a helix described on the wheel face), being therefore set at an angle with the axis. See **double-helical gears**.

helical hinge *(Build.)*. A type of hinge used for hanging swing-doors which have to open both ways.

helical rising, helical setting *(Astron.)*. The rising or setting of a star or planet, simultaneously with the rising or setting of the sun. It was much observed in ancient times as a basis for a solar calendar for agricultural purposes.

helical scan *(Image Tech.)*. Describing videotape recording and reproducing systems in which the tape path follows a helix around a rotating drum containing the magnetic heads.

helical spring *(Eng.)*. A spring formed by winding wire into a helix along the surface of a cylinder; sometimes erroneously termed a *spiral spring*.

helical thickening *(Bot.)*. Secondary wall deposited in the form of a helix in the tracheids and vessel elements of xylem.

helical tip speed *(Aero.)*. The resultant velocity of the tip of a propeller blade, which is a combination of the linear speed of rotation and the flight speed.

helical waveguide *(Telecomm.)*. A **circular waveguide** consisting of closely-spaced turns of fine copper wire, clad in a jacket which absorbs any radiation which might otherwise escape.

helicoid *(Biol.)*. Coiled like a flat spring.

helicoid *(Maths.)*. A surface generated by a screw motion of a curve about a fixed line (axis); i.e. the curve rotates about the axis and moves in a direction of the axis in such a way that the ratio of the rate of rotation and rate of translation is constant. This constant multiplied by 2π is called the *pitch* of the helicoid. See **right helicoid**.

helicopter *(Aero.)*. A **rotorcraft** whose main rotor(s) are power driven and rotate about a vertical axis, and which is thus capable of vertical take-off and landing.

helicotrema *(Zool.)*. An opening at the apex of the cochlea, by which the scala vestibuli communicates with the scala tympani.

helictic *(Geol.)*. Descriptive of the S-shaped trails of inclusions found in the minerals of some metamorphic rocks, especially abundant in garnet and staurolite. These inclusion trails are often continuous with the mineral alignments of surrounding crystals and may help to indicate the crystallization history of the rock.

helio- *(Genrl.)*. Prefix from Gk. *hēlios*, sun.

heliocentric parallax *(Astron.)*. See **annual parallax**.

heliodor *(Min.)*. A beautiful variety of clear yellow beryl.

heliograph *(Surv.)*. An instrument similar to the heliostat but fitted with a spring device by which it can be made to flash long or short flashes.

heliometer *(Astron.)*. An instrument for determining the sun's diameter and for measuring the angular distance between two celestial objects in close proximity. It consists of a telescope with its object glass divided along a diameter, the two halves being movable, so that a superposition of the images enables a value of the angular separation to be deduced from a reading of the micrometer.

heliophyte *(Bot.)*. See **sunplant**.

heliostat *(Astron.)*. An instrument designed on the same principle as the coelostat, but with certain modifications that make it more suitable for reflecting the image of the sun than for use on a larger region of the sky; hence used, in conjunction with a fixed instrument, especially for photographic and spectroscopic study of the sun.

heliostat *(Surv.)*. An instrument used to reflect the sun's rays in a continuous beam, so as to serve as a signal enabling a station to be sighted over long distances.

heliotaxis, heliotropism *(Biol.)*. Response or reaction of an organism to the stimulus of the sun's rays. adj. *heliotactic, heliotropic*.

heliotherapy *(Med.)*. The treatment of disease by the exposure of the body to the sun's rays.

heliotrope *(Min.)*. See **bloodstone**.

heliotrope *(Surv.)*. A form of **heliograph**.

heliotropin *(Chem.)*. See **piperonal**.

heliotropism *(Biol.)*. See **heliotaxis**. More particularly response by growth curvature.

Helipot *(Elec.Eng.)*. TN for precision multiturn potential divider, resettable to within 0.1%.

helium *(Chem.)*. Inert gaseous element, symbol He, at. no.

2, r.a.m. 4.0026, with extremely stable nucleus identical to α-particle. It liquefies at temperatures below 4K, and undergoes a phase change to a form known as *liquid helium II* at 2.2K. The latter form has many unusual properties believed to be due to a substantial proportion of the molecules existing in the lowest possible quantum energy state; see **superfluid**. Liquid helium is the standard coolant for devices working at cryogenic temperatures.

helium diving-bell *(Civ.Eng.)*. A **diving-bell** in which the nitrogen in the compressed-air is replaced by helium, thus reducing tendency to the *bends* and permitting effective operation at greater depths than normal.

helium-neon laser *(Phys.)*. One using a mixture of helium and neon, energized electrically. Its output is on the visible region at 632.8 nm and can have a continuous power of 1 W or a pulsed output with peak powers up to 100 W. Also *He-Ne laser*.

helium stars *(Astron.)*. Those stars, of spectral type B in the Harvard classification, whose spectrum shows only dark lines, in which those due to the element helium predominate.

helix *(Genrl.)*. A line, thread, wire, or other structure curved into a shape such as it would assume if wound in a single layer round a cylinder; a form like a screw-thread.

helix *(Maths.)*. A spiral curve lying on a cone or cylinder and cutting the generators at a constant angle.

helix *(Phys.)*. Spiral path as followed by charged particle in magnetic field.

helix *(Telecomm.)*. (1) See **helical antenna**. (2) The *slow-wave structure* in a travelling wave tube.

hell-box *(Print.)*. A container for broken type and, in the type-case, a compartment to hold also wrong founts for correct distribution later.

Hellesen cell *(Elec.Eng.)*. A dry cell with zinc and carbon electrodes and a depolarizer of manganese dioxide.

Helmert's formula *(Phys.)*. An empirical formula giving the value of g, the acceleration due to terrestrial gravity, for a given latitude and altitude: $g = 9.806\,16 - 0.025\,928 \cos 2\lambda + 0.000\,069 \cos^2 2\lambda - 0.000\,003\,086 H$, where λ is the latitude and H is the height in metres above sea-level. g in ms^{-2}.

helmet *(Civ.Eng.)*. A cast-iron **dolly** used at the head of a reinforced concrete pile, the two being separated by a sand cushion.

Helmholtz coils *(Elec.Eng.)*. Pair of identical compact coaxial coils separated by a distance equal to their radius. These give a uniform magnetic field over a relatively large volume at a position midway between them.

Helmholtz double layer *(Chem.)*. Electrical double layer. This assumes an interphase between a relatively insoluble solid and an ambient ionized liquid, in which oppositely charged ions tend to concentrate in layers. See also **electrokinetic potential**.

Helmholtz free energy *(Chem.)*. Similar to **Gibbs free energy** but with internal energy substituted for enthalpy. Symbol A or F. A negative change in A is indicative of a spontaneous change in a closed system at constant volume.

Helmholtz galvanometer *(Elec.Eng.)*. A type of **tangent galvanometer** in which an approximately uniform field is produced by having two coils parallel to each other, a few centimetres apart.

Helmholtz resonance *(Acous.)*. The type of acoustic resonance arising in a **Helmholtz resonator**.

Helmholtz resonator *(Acous.)*. An air-filled cavity with an opening. The resonance frequency depends on the stiffness of the cavity and the mass of air which oscillates in the opening.

Helmholtz's theorem *(Elec.Eng.)*. See **Thévenin's theorem**.

Helminthes *(Zool.)*. A name formerly used in classification to denote a large group of worm-like Invertebrates now split up into *Platyhelminthes*, *Nematoda*, and smaller groups.

helminthiasis *(Med.)*. Infestation of the body with parasitic worms.

helophyte *(Bot.)*. Marsh plant with perennating buds below the surface of the marsh. See **Raunkiaer system**.

helper T-lymphocyte *(Immun.)*. Often termed *helper T-cell*. A thymus derived lymphocyte which cooperates with B-lymphocytes to enable them to produce antibody when stimulated by antigen or by some polyclonal mitogens. Helper T-cells release lymphokines causing differentiation and growth of B-cells. They also influence the generation of cytotoxic and suppressor T-cells.

helve *(Build.)*. The handle of an axe, hatchet or similar chopping tool.

helve hammer *(Eng.)*. A power hammer used for plating, etc., in which the tup is secured to one end of a helve or beam, which is supported on a pivot and actuated by a crank or cam mechanism.

HEM *(Phys.)*. Abbrev. for *hybrid electromagnetic wave*.

hematite, haematite *(Min.)*. Oxide of iron, Fe_2O_3, crystallizing in the trigonal system. It occurs in a number of different forms: kidney iron-ore massive; specular iron-ore in groups of beautiful, lustrous, rhombohedral crystals as, for example, from Elba; bedded ores of sedimentary origin, as in the Pre-Cambrian throughout the world; and as a cement and pigment in sandstones.

hemeralopia *(Med.)*. *Day-blindness*; objects sometimes being better seen in a dull light. The term is wrongly used also to mean night-blindness.

hem-, hema-, hemo- *(Genrl.)*. US for *haem-*, *haema-*, *haemo-*.

hemi- *(Genrl.)*. Prefix from Gk. *hemi*, half.

hemianaesthesia, hemianesthesia *(Med.)*. Loss of sensibility to touch on one side of the body; usually connotes also loss of sensibility to pain and temperature.

hemianalgesia *(Med.)*. Loss of sensibility to pain on one side of the body.

hemianopia, hemianopsia *(Med.)*. Loss of half the field of vision.

Hemiascomycetes *(Bot.)*. A class of fungi in the Ascomycotina in which no ascocarps are formed. Mostly unicellular or with poorly developed mycelium. Includes the yeasts, *Saccharomyces* and *Schizosaccharomyces*, and some plant parasites causing e.g. peach leaf curl.

hemiataxy, hemiataxia *(Med.)*. Loss of coordination of the muscles of one side of the body.

hemiatrophy *(Med.)*. Wasting of muscles of one side of the body, or of one half of a part of the body.

hemiballism *(Med.)*. Involuntary violent twitching and jerking affecting one side of the body. Condition due to a hypothalamic lesion.

hemibranch *(Zool.)*. The single row of gill lamellae or filaments, borne by each face of a gill arch in Fish: a gill arch with respiratory lamellae or filaments on one face only.

hemicelluloses *(Bot.)*. A group of polysaccharides in the matrix of plant cell walls; homo- and hetero-polymers, linear and branched, of xylose, glucose and other sugars.

Hemichorda *(Zool.)*. A subphylum of Chordata, lacking any bony or cartilaginous skeletal structures; without tail or atrium; having a reduced notochord in the pre-oral region, and a superficial central nervous system; the three primary coelomic cavitites persist in the adult. Also *Hemichordata, Protochordata*.

hemichorea *(Med.)*. Chorea on one side of the body.

hemicolloid *(Chem.)*. A particle up to 2.5×10^{-8} m in length; 20–100 molecules.

hemicrania *(Med.)*. See **migraine**.

hemicryptophyte *(Bot.)*. Plant with buds at soil level. See **Raunkiaer system**.

hemicrystalline rocks *(Geol.)*. Those rocks of igneous origin which contain some interstitial glass, in addition to crystalline minerals. Cf. **holocrystalline rocks**.

hemicyclic *(Bot.)*. A flower with some parts inserted in helices and some in whorls.

hemignathous *(Zool.)*. Having jaws of unequal length.

hemihydrate plaster *(Build.)*. See **plaster of Paris**.

Hemimetabola *(Zool.)*. See **Exopterygota**.

hemimetabolic *(Zool.)*. Having an incomplete or partial metamorphosis.

hemimorphism *(Min.)*. The development of polar symmetry in minerals, in consequence of which different forms are exhibited at the ends of bi-terminated crystals. Hemimorphite shows this character in a marked degree.

hemimorphite *(Min.)*. An orthorhombic hydrated silicate of zinc; one of the best minerals for demonstrating polar symmetry, the two ends being distinctly dissimilar. US, *calamine* or *electric calamine*.

hemiparasite *(Bot.)*. See **partial parasite**.

hemipenes *(Zool.)*. In Squamata, the paired eversible copulatory organs.

hemiplegia *(Med.)*. Paralysis of one side of the body.

Hemiptera *(Zool.)*. An order of Insecta, having two pairs of wings of variable character; the mouth-parts are symmetrical and adapted for piercing and sucking; in some forms, the females are wingless; many are ectoparasitic, others feed on plant juices. Bugs, Cicadas, Aphids, Plant Lice, Scale Insects, Leaf Hoppers, White Flies, Black Flies, Green Flies, Cochineal Insects.

hemisection *(Med.)*. The cutting through of half of a part, e.g. of the spinal cord.

hemisphere *(Genrl.)*. The half of a sphere, obtained by cutting it by a plane passing through the centre. As applied to the earth, the term usually refers to the *Northern* or the *Southern hemisphere*, the division being by the equatorial plane.

hemisphere *(Zool.)*. One of the cerebral hemispheres (see **cerebrum**).

hemithyroidectomy *(Med.)*. Removal of one half of the thyroid gland.

hemizygous *(Biol.)*. Having only one representative of a gene or a chromosome, as are male mammals which have only one *X-chromosome*.

hemoglobin *(Zool.)*. See **haemoglobin**.

hemorrhage *(Med.)*. See **haemorrhage**.

hemorrhoid *(Med.)*. See **haemorrhoid**.

hemp *(Textiles)*. The bast fibre of the hemp plant, *Cannabis sativa*, generally used for making string and ropes. Certain other fibres such as manila and sisal are sometimes incorrectly called hemps.

Hempel burette *(Chem.)*. Used for measuring the volume of a gas, e.g. in gas analysis.

hempel quoins *(Print.)*. The simplest type of mechanical quoin, consisting of two separate and identical cast-iron wedges with a rack on the inner side, placed to oppose each other, and operated by turning a ribbed key which fits in the rack of each, causing them to slide and thus increase or decrease their combined width.

hench *(Build.)*. The narrow side of a chimney shaft.

He-Ne laser *(Phys.)*. See **helium-neon laser**.

henequen *(Textiles)*. Obtained from *Agava fourcroydes* and is similar to **sisal**.

Henle's loop *(Zool.)*. The loop formed by a uriniferous tubule when it enters the medulla of the Mammalian kidney and turns round to pass upwards again to the cortex.

Henoch-Schönlein purpura *(Med.)*. A disease characterized by purpura, urticaria, swollen joints, and abdominal pain, thought to be caused by an immune mechanism. In adults **glomerulonephritis** may predominate.

Henrici's notation *(Eng.)*. See **Bow's notation**.

henry *(Phys.)*. SI unit of self and mutual inductance. A circuit has an inductance of 1 henry if an e.m.f. of 1 volt is induced in the circuit by a current variation of 1 ampere per second. *Or*, a coil has a self-inductance of 1 henry when the magnetic flux linked with it is 1 weber per ampere. The mutual inductance of two circuits is 1 henry when the flux linked with one circuit is 1 weber per ampere of current in the other. Symbol H.

Henry's law *(Chem.)*. The amount of a gas absorbed by a given volume of a liquid at a given temperature is directly proportional to the pressure of the gas.

Henry Williams fishplate *(Eng.)*. A drop-forged fishplate for insulated rail joints in which the insulating fibre is not subject to mechanical wear, the load on the joint being transmitted by side flanges secured together by bolts.

HEOD *(Chem.)*. See **dieldrin**.

heparin *(Med.)*. A short acting anticoagulant used to prevent thrombus and clot formation. Can only be given parenterally and usually to initiate treatment of deep venous thrombosis and pulmonary embolism until anticoagulation by oral drugs can be established. Also used in low doses to prevent deep-vein thrombosis formation during surgery.

hepatectomy *(Med.)*. Removal of part of the liver.

hepat-, hepato- *(Genrl.)*. Prefix from Gk. *hēpar*, gen. *hēpatos*, the liver.

hepatic *(Bot.)*. A liverwort. See **Hepaticopsida**.

hepatic *(Med.)*. Pertaining to the liver.

hepatic artery *(Zool.)*. In Craniata, a branch of the coeliac artery which conveys arterial blood to the liver.

hepatic duct *(Zool.)*. In Craniata, a duct conveying the bile from the liver and discharging into the intestine as the common bile duct.

Hepaticopsida, Hepaticae *(Bot.)*. Class of the Bryophyta containing ca 10 000 spp. The gametophyte is thalloid or leafy with unicellular rhizoids and the capsule (sporophyte) without a columella.

hepatic portal system *(Zool.)*. In Craniata, the part of the vascular system which conveys blood to the liver; it consists of the **hepatic portal vein** and the **hepatic artery**.

hepatic portal vein *(Zool.)*. In Craniata, the vein which conveys blood from the alimentary canal to the liver.

hepatitis *(Med.)*. Inflammation of the liver. See **infective hepatitis**.

hepatitis contagiosa canis *(Vet.)*. See **infectious canine hepatitis**.

hepatization *(Med.)*. Pathological change of tissue so that it becomes liverlike in consistency; as of the lung in pneumonia.

hepatogenous *(Med.)*. Having origin in the liver.

hepatolenticular degeneration *(Med.)*. Degeneration of the liver and the lenticular nucleus (part of the brain). See **Wilson's disease**.

hepatolith *(Med.)*. A gallstone present in the liver.

hepatoma *(Med.)*. A tumour composed of liver tissue.

hepatomegaly, hepatomegalia *(Med.)*. Enlargement of the liver.

hepatopancreas *(Zool.)*. In many Invertebrates (as Mollusca, Arthropoda, Brachiopoda), a glandular diverticulum of the mesenteron, frequently paired, consisting of a mass of branching tubules, and believed to carry out the functions proper to the liver and pancreas of higher Vertebrates.

hepatopexy *(Med.)*. Fixation of the liver by suturing it to the abdominal wall.

hepatoportal system *(Zool.)*. See **hepatic portal system**.

hepatorrhexis *(Med.)*. Rupture of the liver.

hepatotomy *(Med.)*. Incision of the liver.

hepta- *(Chem.)*. Containing seven atoms, groups etc.

heptane *(Chem.)*. C_7H_{16}, an alkane hydrocarbon, a colourless liquid, bp 98°C, rel.d. 0.68. There are nine isomers. The foregoing properties relate to *normal* heptane, which is a constituent of petrol and resembles hexane in its chemical behaviour.

heptavalent, septavalent *(Chem.)*. Capable of combining with seven hydrogen atoms or their equivalent. Having an oxidation or co-ordination number of seven.

heptoses *(Chem.)*. A subgroup of the monosaccharides containing seven oxygen atoms, of the general formula $HO.CH_2.(CHOH)_5.CHO$.

herb *(Bot.)*. (1) A plant that does not develop persistent woody tissues above ground. (2) A plant that is used for medicinal purposes or for flavour.

herbaceous *(Bot.)*. A soft and green plant organ or a plant without persistent woody tissues above ground.

herbaceous perennial *(Bot.)*. A perennial plant with a perennating structure at or below ground level and producing aerial shoots that die at the end of the growing season.

herbarium *(Bot.)*. A collection of dried plants: by extension, the place where such a collection is kept.

hercogamy *(Bot.)*. Physical arrangement of anthers and stigma so that pollen is not transferred from one to the other in the absence of an insect visit.

Hercynian orogeny *(Geol.)*. The late Palaezoic orogeny in Europe.

hercynite *(Min.)*. See under **spinel**.

herding *(Behav.)*. The control of the movements of animals by individuals of the same or different species.

hereditary *(Biol.)*. Inherited; capable of being inherited; passed on or capable of being passed on from one generation to another.

hereditary angioneurotic oedema *(Immun.)*. Disease characterized by recurrent episodes of transient oedema of skin and mucous membranes, and due to absence or functional inactivity of *Cl^--esterase inhibitor*. Inherited as a Mendelian dominant. See **Cl^-**-inhibitor.

hereditary ataxia *(Med.)*. A group of inherited central nervous system diseases causing **ataxia**. *Fredreich's ataxia* is the best known, presenting in late childhood with poor co-ordination and the development of sensory and motor loss. About half have **cardiomyopathy**.

hereditary haemorrhagic telangiectasia *(Med.)*. A genetic disease with multiple vascular anomalies (telangiectasia) of the skin and mucous membranes. The condition is often complicated by internal bleeding. Also *Osler-Rendu-Weber syndrome*.

heredity *(Biol.)*. The relation between successive generations, by which characters persist.

Hereford disease *(Vet.)*. See **bovine hypomagnesaemia**.

heritability *(Biol.)*. Measure of the degree to which the variation of a character is inherited. Ranges from 0 to 100%. It is the proportion of *additive genetic variance* in the total *phenotypic variance*. Usually symbolized by H^2.

hermaphrodite *(Behav.)*. A person whose reproductive organs are anatomically ambiguous, so that they are not exclusively male or female.

hermaphrodite *(Bot.)*. (1) An individual organism which produces both male and female gametes. (2) Flowers with male and female organs functional in the one flower. Also *monoclinus*. Cf. **unisexual, dioecious**.

hermaphrodite *(Zool.)*. Having both male and female reproductive organs in one individual. n. *hermaphroditism*.

HERMES *(Space)*. A manned spaceplane for servicing European elements of the international space station.

hernia *(Med.)*. Protrusion of a viscus, or part of a viscus, through an opening or weak spot or defective area in the cavity containing it; especially of an abdominal viscus.

herniorrhaphy *(Med.)*. Surgical repair of a hernia.

herniotomy *(Med.)*. Cutting operation for hernia.

Hero's, Heron's formula *(Maths.)*. The area of a triangle with sides *a*, *b*, and *c*, is given by $\sqrt{s(s-a)(s-b)(s-c)}$ where $2s = a+b+c$.

Héroult process *(Eng.)*. An electrolytic process for the manufacture of aluminium from a solution of bauxite in fused cryolite.

herpes simplex *(Med.)*. A virus causing a number of clinical disorders ranging from the common eruption of vesicles round the mouth in a febrile illness, through the disabling *genital herpes* to the life threatening infection of the brain, *encephalitis*.

herpes zoster *(Med.)*. *Shingles*. A painful eruption of crops of firm vesicles along the course of a nerve, the posterior root ganglia of which are inflamed. Due to reactivation of chicken pox virus which has lain dormant in nerve roots and may be associated with depressed immunity.

herring-bone ashlar *(Build.)*. Blocks of stone tooled in grooves of herring-bone design.

herring-bone bond *(Build.)*. A form of **raking bond** in which the bricks are laid with rake in opposite directions from the centre of the wall, to form a herring-bone pattern. This bond is also used for brick pavings, and has the advantage of making effective bond in the middle.

herring-bone gear *(Eng.)*. See **double-helical gears**.

Herschel formula *(Eng.)*. A hydraulic formula used to calculate the rate of flow over a drowned nappe.

hertz *(Phys.)*. SI unit of frequency, indicating number of cycles per second (c/s). Symbol *Hz*.

Hertz antenna *(Telecomm.)*. Original half-wave dipole, fed at the centre.

Hertzian dipole *(Elec.Eng.)*. Pair of opposite and varying charges, close together, with an electric moment. Also called *Hertzian doublet*.

Hertzian oscillator *(Telecomm.)*. Idealized system envisaged by Hertz, comprising two point charges of opposite sign and separated by an infinitesimal distance, whose electric moment varies harmonically with time.

Hertzian waves *(Phys.)*. Electromagnetic waves from, e.g. 10^4 to 10^{10} Hz, used for communication through space, covering the range from very low to ultra-high frequencies, i.e., from audio reproduction, through radio broadcasting and television to radar.

Hertzsprung-Russell diagram *(Astron.)*. A graphical representation of the correlation between the **spectral type** and **luminosity** for a sample of stars, first described in 1913 by H.N. Russell. The diagram achieves its astrophysical significance when the plotted points are restricted to a particular sample (such as all the members of a **star cluster**) or a particular type (such as one class of **variable star**). It then effectively shows the correlation of surface temperature against luminosity. Points are not distributed randomly on the H-R diagram: normal stars which are burning hydrogen form a broad band known as the *main sequence*. Evolved stars are clumped together as **red giants**. Stars which have consumed their fuel supplies are grouped together as **white dwarfs**. It is an important task for the theory of **stellar evolution** to explain the main features of the H-R diagram.

Herzberg stain *(Paper)*. A general-purpose stain for identifying paper fibres under the microscope, consisting of a mixture of zinc chloride, iodine and potassium iodide.

hessian *(Maths.)*. The determinant obtained by replacing the *n* functions u_i in the Jacobian by the *n* partial derivatives of a single function *u*.

hessian *(Textiles)*. A strong plain-weave jute fabric, used for packing material, sacks, in tarpaulin manufacture, and as a furnishing fabric and wall covering.

hessite *(Min.)*. Silver telluride, a metallic grey pseudo-cubic mineral occurring in silver ores in various parts of the world.

hessonite *(Min.)*. *Cinnamon stone*. A variety of garnet containing a preponderance of the grossular molecule, and characterized by a pleasing reddish-brown colour.

Hess's law *(Chem.)*. The net heat evolved or absorbed in any chemical change depends only on the initial and final states, being independent of the stages by which the final state is reached.

heter-, hetero- *(Genrl.)*. Prefix from Gk. *heteros*, other, different.

hetero-agglutination *(Biol.)*. (1) The adhesion of spermatozoa to one another by the action of a substance produced by the ova of another species. (2) The adhesion of erythrocytes to one another when blood of different groups is mixed. Cf. **is**

heteroauxin *(Bot.)*. Old name for indole acetic acid, auxin.

heteroblastic *(Bot.)*. (1) Having a marked morphological difference between the first-formed structures, like leaves on a seedling, and those formed later. See also **juvenile phase**. Cf. **homoblastic**. (2) A plant whose seeds vary in the conditions they require for germination.

heteroblastic *(Zool.)*. Showing indirect development.

heterocercal *(Zool.)*. Said of a type of tail-fin, found in adult Sharks, Rays, Sturgeons, and many other primitive Fish, in which the vertebral column bends abruptly upward and enters the epichordal lobe, which is larger than the hypochordal lobe.

heterochlamydeous *(Bot.)*. Having a distinct calyx and corolla.

heterochromatin *(Biol.)*. Relatively dense chromatin visible by microscopy in eukaryotic cell nuclei. Generally contains DNA sequences inactive in transcription. See **facultative-, constitutive-**.

heterocoelous *(Zool.)*. Said of vertebral centra in which the anterior end is convex in vertical section, concave in horizontal section, while the posterior end has these outlines reversed.

heterocotylized arm *(Zool.)*. See **hectocotylized arm**.

heterocyclic compounds *(Chem.)*. Cyclic or ring compounds containing carbon atoms and other atoms, e.g. O, N, S, as part of the ring.

heterodactylous *(Zool.)*. Of Birds, having the first and second toes directed backwards, the third and fourth forwards, as the Trogons.

heterodesmic structure *(Crystal.)*. One including two or more types of crystal bonding.

heterodont *(Zool.)*. Of teeth, having different forms adapted to different functions.

heterodromous *(Bot.)*. Having asymmetric flowers of two kinds, one the mirror image of the other and sometimes associated in pairs.

heteroduplex DNA *(Biol.)*. A duplex DNA made by **renaturing** DNA strands from different sources. Depending on conditions stable heteroduplexes can form with varying proportions of single stranded regions. These can be 'mapped' in the electron microscope.

heterodyne *(Telecomm.)*. Combination of two sinusoidal RF waves in a nonlinear device resulting in sum and difference frequencies. The latter is the *heterodyne frequency*, and will produce an audio-frequency *beat note* when the original two sine waves are sufficiently close in frequency.

heterodyne (beat-frequency) wavemeter *(Telecomm.)*. One in which a continuously variable oscillator is adjusted to give zero beat with an unknown frequency, the value of which then coincides with the calibration of the oscillator.

heterodyne conversion *(Telecomm.)*. Change in the frequency of a modulated carrier wave produced by heterodyning it with a second unmodulated signal. The sum and difference frequencies will carry the original modulation signal and either of these can be isolated for subsequent amplification. The frequency changing stage of a superhet radio receiver employs this principle, an oscillator being tuned to a fixed amount away from the signal frequency so that the difference frequency (*intermediate frequency signal*) remains constant for all incoming signals.

heterodyne-frequency meter *(Telecomm.)*. See **heterodyne wavemeter**.

heterodyne interference *(Telecomm.)*. That arising from simultaneous reception of two stations the difference between whose carrier frequencies is an audible frequency.

heterodyne whistle *(Telecomm.)*. See **heterodyne interference**.

heteroecious *(Bot.,Zool.)*. A parasite which requires two, usually unrelated, host species to complete different stages of its life cycle. Cf. **autoecious**.

heterogamete *(Biol.)*. See **anisogamete**.

heterogametic sex *(Biol.)*. The sex that is heterozygous for the *sex-determining chromosome*, the male in mammals, the female in birds; its gametes determine the sex of the progeny. Cf. **homogametic**.

heterogamous *(Bot.)*. Producing unlike gametes or flowers.

heterogeneous *(Chem.)*. Said of a system consisting of more than one phase.

heterogeneous radiation *(Phys.)*. Radiation comprising a range of wavelengths or particle energies.

heterogeneous reactor *(Nuc.Eng.)*. One in which the fuel is present as rods spread in an array or lattice within (but separate from) the moderator. Cf. **homogeneous reactor**.

heterogenesis *(Zool.)*. See **metagenesis, abiogenesis**. adj. *heterogenetic*.

heterogenous summation *(Behav.)*. Refers to the fact that where a response is influenced by stimulus characters acting through more than one sensory modality, their effects may supplement each other, e.g. parts of the stimulus presented successively may produce the same response as when they are presented simultaneously.

heterogeny *(Zool.)*. Cyclic reproduction in which several broods of parthenogenetic individuals alternate with one or more broods of sexual forms.

heterogony *(Zool.)*. Reproduction by both parthenogenesis and amphigony.

heterokaryon *(Biol.)*. **Somatic cell hybrid** containing separate nuclei from different species. Also *heterokaryote*.

heterokaryosis *(Bot.)*. Coexistence of genetically different nuclei in a common cytoplasm, especially in a fungal hypha.

heterokont, heterokontan *(Bot.)*. Having flagella of unequal length or different type.

Heterokontophyta *(Bot.)*. A division of eukaryotic algae, characterized by: chlorophyll a and, usually, c; mostly with fucoxanthin; triplet thylakoids; chloroplast ER; mitochondrial microvilli; **heterokont** flagellation; storing $\beta, 1 \rightarrow 3$ glucans in the cytoplasm. Contains the classes: Xanthophyceae, Chrysophyceae, Bacillariophyceae, Raphidophyceae and Paeophyceae.

heterolecithal *(Zool.)*. With unequally distributed yolk.

heteromastigote *(Zool.)*. Having one or more anterior flagella directed forwards and a posterior flagellum directed backwards.

heteromerous *(Bot.)*. A lichen thallus with the algal cells confined to a distinct layer, with pure fungus above and below. Cf. **homoiomerous**.

heterometabolic *(Zool.)*. Having incomplete metamorphosis.

heterometry *(Chem.)*. A process of titration in which precipitation is plotted as an optical density curve.

heteromorphic, heteromorphous *(Bot.)*. (1) Having more then one form. (2) A species with **alternation of generations** in which the latter differ, often considerably, in form.

heteromorphous rocks *(Geol.)*. Rocks of closely similar chemical composition, but containing different mineral assemblages.

heteronomous *(Zool.)*. Subject to different laws, especially of growth and specialization. Cf. **autonomous**.

heterophil antigen *(Immun.)*. Antigens occurring on cells of many different animal species, plants and bacteria which are sufficiently similar to elicit antibodies which cross react extensively, e.g. *Forssman antigen*.

heterophoria *(Med.)*. Latent squint revealed by passing a screen before each eye. See also **esophoria** and **exophoria**.

heterophylly *(Bot.)*. An individual plant with two or more different forms of leaf as in the submerged, floating and aerial leaves of many water plants.

heteroplasma *(Zool.)*. In tissue culture, a medium prepared with plasma from an animal of a different species from that from which the tissue was taken. Cf. **autoplasma, homoplasma**.

heteroplastic *(Zool.)*. In experimental zoology, said of a graft which is transplanted to a site different from its point of origin, e.g. epithelial cells of cornea to a skin site.

heteroplasty *(Med.)*. The operation of grafting on one person body-tissue removed from another.

heteropolar *(Chem.)*. Having an unequal distribution of charge, as in covalent bonds between unlike atoms.

heteropolar generator *(Elec.Eng.)*. An electromagnetic generator of the usual type, i.e. one in which the conductors pass alternate north and south poles, or in which alternate poles pass the conductors.

heteropolar liquids *(Chem.)*. Compounds such as alcohols, amines and organic acids which contain molecules that have localized associated polar groups.

heteropycnosis *(Biol.)*. Differential stainability of bands of chromosomes.

heteroscedastic *(Stats.)*. Having unequal variances.

heterosexual *(Behav.)*. Sexual interest directed at members of the opposite sex.

heterosis *(Biol.)*. The difference between the mean of a quantitative character in a crossbred generation and the mean of the two parental strains. Also *heterozygous advantage*. See **hybrid vigour**.

heterospory *(Bot.)*. The production of more than one type of spore, typically megaspores and microspores. adj. *heterosporous*. Cf. **homospory**.

heterostyly *(Bot.)*. The condition in which individuals of a species have style and thus stigma lengths falling into 2 or more distinct classes. This causes the anthers to be placed low in flowers with high stigmas and vice versa. Heterostyly appears to promote crosspollination. Cf. **homostyly, illegitimate pollination, pin, thrum**.

heterothallism *(Bot.)*. The condition in which there are 2 (or more) mating types with sexual reproduction only successful between individuals of a different type. Also *self incompatibility*. adj. *heterothallic*. Cf. **homothallism**.

heterotopia *(Med.)*. Displacement of a group of cells of an organ from their normal position during the course of development.

heterotrichous *(Zool.)*. Having cilia or flagella of two or more different kinds.

heterotrophic *(Bot.)*. Organisms which require carbon in organic form, as do all animals, fungi, some algae, parasitic plants and most bacteria. Cf. **autotrophic**.

heterotypic *(Zool.)*. Differing from the normal condition. Cf. **homotypic**.

heterotypic division *(Biol.)*. The first (reductional) of two nuclear divisions in meiosis in which the number of chromosomes is halved. The second (equational) is a *homotypic division*.

heterozygosis *(Biol.)*. The condition of being *heterozygous*.

heterozygosity *(Biol.)*. The proportion of individuals in a population that are *heterozygous* at a specified locus, or at a number of loci averaged.

heterozygote *(Biol.)*. An individual with two different alleles at a particular locus, the individual having been formed from the union of gametes carrying different alleles. adj. *heterozygous*. Cf. **homozygote**.

HETP *(Chem.Eng.)*. *Height Equivalent to a Theoretical Plate*. A measure of the separation efficiency of a distillation column, i.e. the height of packing in a packed distillation column which behaves as a **theoretical plate**.

heulandite *(Min.)*. One of the best-known zeolites, often beautifully crystalline, occurring as coffin-shaped monoclinic crystals in cavities in basic igneous rocks. In composition similar to hydrated calcium sodium aluminium silicate.

heuristic *(Comp.)*. Describes an approach based on common sense rules and trial and error rather than on comprehensive theory.

heuristic program *(Comp.)*. One which attempts to improve its own performance as a result of learning from previous actions within the program.

Hewlett disk insulator *(Elec.Eng.)*. A disk-form of suspension-type insulator.

hewn stone *(Build.)*. Blocks of hammer-dressed stone.

hex- *(Chem.)*. Colloquialism for *uranium* (VI) *fluoride*, the compound used in the separation of uranium isotopes by gaseous diffusion.

hex *(Comp.)*. See **hexadecimal notation**.

hex- *(Genrl.)*. Prefix from Gk. *hex*, six.

hexa- *(Chem.)*. Containing 6 atoms, groups etc.

hexachlorophane *(Chem.)*. *2,2'*-methylene bis-(3,4,6 trichloro-hydroxybenzene). White powder. Mp approx 160°C. Widely used bactericide in soaps, deodorants and other toilet products. Sometimes referred to as *G-11*.

Hexactinellida *(Zool.)*. A class of Porifera, usually distinguished by the possession of a siliceous skeleton composed of triaxial spicules, and large thimble-shaped flagellated chambers.

hexadecane *(Chem.)*. An alkane hydrocarbon ($C_{10}H_{34}$) found in petroleum, especially that showing a straight chain structure.

hexadecimal notation *(Comp.,Maths.)*. System of representing numbers in base sixteen with the digits 0,1,2,...9,A,B,C,D,E,F.

hexafluorophosphoric acid *(Chem.)*. HPF_6. Produced by the action of strong hydrofluoric acid on difluorophosphoric acid. Also referred to as *phosphorofluoric acid*.

hexagon *(Maths.)*. A six-sided polygon.

hexagonal closet packing *(Chem.)*. The stacking of close-packed layers of spheres in an ABAB sequence. Several metals have this structure, including Mg.

hexagonal packing *(Crystal.)*. System in which many metals crystallize, thus achieving minimum volume. Each lattice point has twelve equidistant neighbours in such a cell construction.

hexagonal system *(Crystal.)*. A crystal system in which three equal coplanar axes intersect at an angle of 60°, and a fourth, perpendicular to the others, is of a different length.

hexagon dresser *(Eng.)*. A metal disk tool used for dressing grinding wheels.

hexagon voltage *(Elec.Eng.)*. The voltage between 2 lines, adjacent as regards phase-sequence, of a 6-phase system.

hexahydrobenzene *(Chem.)*. See **cyclohexane**.

hexahydrocresol *(Chem.)*. See **methylcyclohexanol**.

hexahydrophenol *(Chem.)*. **Cyclohexanol**.

hexahydropyridine *(Chem.)*. **Piperidine**.

hexamerous *(Bot.)*. Having parts in sixes.

hexametaphosphates *(Chem.)*. Salts of hexa-metaphosphoric acid, $H_6(PO_3)_6$, a polymer of metaphosphoric acid. Cf. **Calgon**.

hexamethylene *(Chem.)*. **Cyclohexane**.

hexamethylenediamine *(Chem.)*. *(1,6-diaminohexane)* $H_2N(CH_2)_6NH_2$. Important historically as a constituent material of the first nylon (nylon 6.6), which was a condensation polymer of hexamethylenediamine and **adipic acid**.

hexamethylenetetramine *(Chem.)*. *Hexamine*, $(CH_2)_6N_4$, a condensation product of methanal with ammonia, a crystalline substance with antiseptic and diuretic properties. Used in the production of *cyclonite*, a highly efficient explosive.

hexamitiasis *(Vet.)*. A disease of turkeys due to infection by the flagellate protozoon *Hexamita meleagridis*, which causes enteritis.

hexane *(Chem.)*. C_6H_{14}. There are five compounds with this formula: normal hexane, a colourless liquid, of ethereal odour, bp 69°C, rel.d. 0.66, is an important constituent of petrol and of solvent petroleum ether or ligroin.

hexapod *(Zool.)*. Having 6 legs.

Hexapoda *(Zool.)*. See **Insecta**.

hexarch *(Bot.)*. Having 6 strands of protoxylem.

hexastyle *(Arch.)*. A portico formed of 6 columns in front.

hexavalent, sexavalent *(Chem.)*. Capable of combining with 6 hydrogen atoms or their equivalent. Having an oxidation or co-ordination number of six.

hexobarbitone sodium *(Chem.)*. The monosodium derivative of 5-Δ'-cyclohexenyl-

hexogen *(Chem.)*. **Cyclonite**.

hexoses *(Chem.)*. A subgroup of the monosaccharides containing 6 oxygen atoms, of the general formula $HO.CH_2.(CHOH)_4.CHO$ and $HO.CH_2.(CHOH)_3.CO.CH_2OH$. The first formula signifies an *aldohexose*, the second a *ketohexose*. See appendix.

hexphase *(Elec.Eng.)*. A term sometimes used instead of *6-phase*.

Heyland diagram *(Elec.Eng.)*. A particular application of the circle diagram of an a.c. circuit to represent the behaviour of an induction motor.

HF *(Aero.)*. Abbrev. for *High Frequency*. Radio transmissions between 3 000 and 30 000 kHz.

Hf *(Chem.)*. The symbol for *hafnium*.

h_{fb} *(Electronics)*. In transistors, the symbol for common-base mode current gain. See **transistor parameters**.

h_{fe} *(Electronics)*. In transistors, the symbol for common-emitter mode current gain. See **transistor parameters**.

Hg *(Chem.)*. The symbol for *mercury*.

H-girder *(Build.)*. See **H-beam**.

H-hinge *(Build.)*. A hinge which when opened has the shape of the letter H. Also a *parliament hinge*.

H-2 histocompatibility system *(Immun.)*. The major histocompatibility system in the mouse. H-2 genes determine the major histocompatibility antigens on the surface of somatic cells and also the immune response (Ir) genes. The antigens in a given strain of mice are controlled by alleles within the H-2 locus. The H-2 system corresponds closely to the **HLA system** of humans, and much of our understanding of the latter was derived from studies in mice.

HI *(Surv.)*. Abbrev. for *Height of Instrument*.

HI Arc *(Image Tech.)*. High Intensity Arc, a high current carbon arc.

hiatus *(Geol.)*. A break or gap in the stratigraphical record, because of non-deposition or erosion.

Hibbert standard *(Elec.Eng.)*. A standard of magnetic

flux linkage suitable for fluxmeter or galvanometer calibration. It comprises a stabilized magnet producing a radial field in an annular gap, through which a cylinder carrying a multiturn coil can be dropped.

hibernation *(Zool.)*. The condition of partial or complete torpor into which some animals relapse during the winter season. v. *hibernate*.

hiccup *(Med.)*. Sudden spasm of the diaphragm followed immediately by closure of the glottis.

hickie *(Print.)*. A blemish on a printed image which appears as a spot surrounded by a halo, caused by a small fragment of paper adhering to the plate as a result of **fluffing** or **picking**.

hick joint *(Build.)*. A **flat joint** formed in fine mortar when pointing, after the old mortar has been raked out of the joints.

hickory *(For.)*. The product of a hardwood tree (*Carya*) common to the eastern US. It bends well and may be used for making boats, chairs, tool handles etc.

Hicks hydrometer *(Elec.Eng.)*. A form of hydrometer used for finding the relative density of the electrolyte in an accumulator, to determine the state of its charge; the hydrometer consists of a glass tube containing a number of coloured beads, which float at different relative densities.

hiddenite *(Min.)*. See **spodumene**.

hiding power *(Build.)*. The power of a paint to obscure a black-and-white contrast; generally expressed as the number of square feet per gallon or square metres per litre of paint. Also *obliterating power, hiding power*.

hidrosis *(Zool.)*. Formation and excretion of sweat.

hi-fi *(Acous.)*. Abbrev. for *high fidelity*.

high alumina cement *(Eng.)*. Made by fusing a mixture of bauxite and chalk or limestone and grinding the resultant clinker. It hardens much more rapidly (24 hr) than Portland cement and has great heat and chemical resistance.

high aspect ratios *(Aero.)*. See **aspect ratio**.

high brass *(Eng.)*. Common brass of 65/35 copper-zinc alloying, as distinct from deep-drawing brass with 66–70% copper.

high by-pass ratio *(Aero.)*. Applied to a **turbofan** in which the **air mass flow** ejected directly as propulsive thrust by the fan is more than twice the quantity passed internally through the *gas generator* section.

high-carbon steel *(Eng.)*. One such as spring steel, with between ½% and 1½% carbon.

high-conductivity copper *(Eng.)*. Metal of high purity, having an electrical conductivity not much below that of the international standard, which is a resistance of 0.153 28 ohms for a wire 1 m in length and weighing 1 g.

high definition *(Image Tech.)*. In current TV practice, the term is applied to systems using 1000 or more scanning lines to make up the picture.

high-definition developer *(Image Tech.)*. One which increases the contrast at boundaries between light and dark tones, the light boundary being enhanced by bromide from the heavily exposed part of the image and the dark boundary by comparatively fresh developer from the lightly exposed part.

high endothelial venule *(Immun.)*. Specialized venules in the thymus dependent area of lymph nodes characterized by prominent high endothelial lining cells. Recirculation of lymphocytes from blood to lymph takes place through the walls of these venules.

high-energy ignition *(Aero.)*. A gas-turbine ignition system using a very high voltage discharge.

high energy phosphate compounds *(Biol.)*. Phosphate compounds with a high negative free energy of hydrolysis. Endergonic metabolic reactions are driven by coupling them with the exergonic hydrolysis of these phosphate esters, the most common example being the hydrolysis of ATP.

high-energy physics *(Phys.)*. Synonymous with *particle physics*.

high energy-rate forging, forming *(Eng.)*. Methods in which the ram is accelerated to very high velocities by the release of compressed gas, usually to complete the operation in one blow. Abbrev. *HERF*. Also *high velocity forging*.

high energy rate forming *(Eng.)*. Any of a recently developed family of processes, in which metal parts are rapidly compacted, forged, extruded, by the application of extremely high pressures.

higher-order conditioning *(Behav.)*. A form of conditioning in which a conditioned stimulus from earlier training serves as an unconditioned stimulus.

high explosive *(Min.Ext.)*. One in which the active agent is in chemical combination and is readily detonated by application of shock. Nitrated cotton, nitroglycerine and ammonium nitrate are widely used, diluted to required explosive strength by inert fillers such as kieselguhr or wood pulp. See also **gelignite, trinitrotoluene**.

high-fidelity *(Acous.)*. Said of sound reproduction of exceptionally high quality. Abbrev. *hi-fi*.

high-fidelity amplifier *(Acous.)*. One in which the input signal is reproduced with a very high degree of accuracy.

Highfield booster *(Elec.Eng.)*. An automatic battery booster consisting of a generator, a motor and an exciter. Automatic regulation is carried out by balancing the exciter voltage against that of the battery.

high-flux reactor *(Nuc.Eng.)*. One designed to operate with a greater neutron flux than normal for testing materials for radiation effects and experiments requiring intense beams of neutrons. Also *materials testing reactor*.

high-frequency amplification *(Telecomm.)*. That at frequencies used for radio transmission. In a receiver, any amplification which takes place before detection, frequency conversion, or demodulation.

high-frequency capacitance microphone *(Acous.)*. One which uses audio variation of capacitance to vary the frequency of an oscillator, or response of a tuned circuit.

high-frequency heating *(Phys.)*. Heating (induction or dielectric) in which the frequency of the current is above mains frequency; from rotary generators up to ~3000 Hz and from electronic generators 1-100 MHz. Also *radio heating, microwave heating*.

high-frequency induction furnace *(Eng.)*. Essentially an air transformer, in which the primary is a water-cooled spiral of copper tubing, and the secondary the metal being melted. Currents at a frequency above about 500 Hz are used to induce eddy currents in the charge, thereby setting up enough heat in it to cause melting. Used in melting steel and other metals.

high-frequency repeater distribution frame *(Telecomm.)*. In carrier telephony, a frame providing flexibility primarily in the connecting of apparatus to external cables on which 12-circuit groups are transmitted. It also provides for the interconnection of apparatus used for the assembly of a number of 12-circuit groups before transmission to line.

high-frequency resistance *(Telecomm.)*. That of a conductor or circuit as measured at high frequency, greater than that measured with dc because of **skin effect**.

high-frequency transformer *(Telecomm.)*. One designed to operate at high frequencies, taking into account self-capacitance, usually with band-pass response.

high-frequency welding *(Elec.Eng.)*. Welding by RF heating. See **seam welding** (2).

Highgate resin *(Min.)*. A popular name for fossil gum-resin occurring in the Tertiary London Clay at Highgate in North London. See **copalite**.

high grading *(Min.Ext.)*. Selective mining, in which sub-grade ore is abandoned unworked. Also, theft of valuable concentrates or specimens such as nuggets of gold.

high-intensity separation *(Min.Ext.)*. Dry concentration of small particles of mineral in accordance with their relative ability to retain ionic charge after passing through an ionizing field.

high key *(Image Tech.)*. Describing a scene containing mainly light tones well illuminated without large areas of strong shadow.

Highland Boundary Fault *(Geol.)*. One of the most important dislocations in the British Isles, extending from the Clyde to Stonehaven and separating the Highlands of Scotland from the Midland Valley.

high-lead bronze *(Eng.)*. Soft matrix metal used for bearings, of copper/tin/lead alloys in approximate proportions 80, 10 and 10.

high-level language *(Comp.)*. Problem-orientated programming language in which each instruction may be equivalent to several **machine-code instructions**.

high-level (low-level) modulation *(Telecomm.)*. Conditions where modulation of a carrier for transmission takes place at high level for direct coupling to the radiating system, *or* is at lower level than this, with subsequent push-pull or straight amplification. Also known as *high (low) power modulation*.

high-level RF signal *(Radar)*. An RF signal having sufficient power to fire a switching tube.

high level waste *(Nuc.Eng.)*. Nuclear waste requiring continuous cooling to remove the heat produced by radioactive decay.

highlight *(Image Tech.)*. An area of maximum brightness in a scene and its reproduction in a photographic or TV image.

high-opacity foils *(Print.)*. A type of blocking-foil especially suitable for marking undressed bookcloths and deep-grained materials. These are only made in white and pastel shades and have a considerable weight of pigment to obliterate the surface completely.

high-pass (low-pass) filter *(Telecomm.)*. One which freely passes signals of all frequencies above, or below, a reference value known as the cut-off frequency, f_c. (*N.B.* Beyond f_c attenuation only rises slowly and seldom approaches complete cut-off as implied by this name.)

high-power modulation *(Telecomm.)*. See **high-level modulation**.

high-pressure compressor *(Aero.)*. In a gas-turbine engine with two or more compressors in series, the last is the high-pressure one. In a dual-flow turbojet this feeds the combustion chamber(s) only. Abbrev. *HP compressor*.

high-pressure cylinder *(Eng.)*. The cylinder of a compound or multiple-expansion steam-engine in which the steam is first expanded.

high-pressure hose *(Min.Ext.)*. *Armoured hose*, reinforced with circumferentially embedded wire, and hence able to withstand moderately high pressure and rough usage.

high-pressure turbine *(Aero.)*. The first turbine after the combustion chamber in a gas-turbine engine with two or more turbines in series. Abbrev. *HP turbine*.

high-pressure turbine stage *(Aero.)*. The first stage in a **multistage turbine**. Abbrev. *HP stage*.

high recombination rate contact *(Electronics)*. The contact region between a metal and semiconductor (or between semiconductors) in which the densities of charge carriers are maintained effectively independent of the current density.

high-resistance joint *(Elec.Eng.)*. See **dry joint**.

high-resistance voltmeter *(Elec.Eng.)*. One drawing negligible current and typically having a resistance in excess of 1000 ohms per volt.

high resolution graphics *(Comp.)*. Term generally applied to graphical display units capable of fine definition by plotting around 300 or more distinct points in the width of a domestic TV screen.

high spaces *(Print.)*. The normal height of spacing is about 0.75 in (19 mm) but higher spaces are more convenient when pages are to be stereotyped, and can also be used as a mount for original plates.

high-speed circuit-breaker *(Elec.Eng.)*. A circuit-breaker in which special devices are used to ensure very rapid operation; used particularly on d.c. traction systems.

high-speed steam-engine *(Eng.)*. A vertical steam-engine, generally compound, using a piston valve, or valves, whose moving parts are totally enclosed and pressure-lubricated.

high-speed steel *(Eng.)*. A hard steel used for metal-cutting tools. It retains its hardness at a low red heat, and hence the tools can be used in lathes etc., operated at high speeds. It usually contains 12 to 18% tungsten, up to 5% chromium, 0.4 to 0.7% carbon, and small amounts of other elements (vanadium, molybdenum etc.).

high-speed wind tunnel *(Aero.)*. A high **subsonic** wind tunnel in which compressibility effects can be studied.

high spot *(Radiol.)*. A small volume so situated that the dose therein is significantly above the general dose level in the region treated.

high-stop filter *(Telecomm.)*. Same as **low-pass filter**.

high-strength brass *(Eng.)*. A type of brass based on the 60% copper-40% zinc composition, to which manganese, iron and aluminium are added to increase the strength. **Manganese bronze** denotes a variety in which manganese is the principal addition, but most varieties now contain all 3 elements.

high-temperature reactor *(Nuc.Eng.)*. One designed to attain core temperatures above 660°C. Usually uses coated uranium dioxide or carbide pellets, cooled by helium gas.

high-tension *(Elec.Eng.)*. See **high-voltage**.

high-tension battery *(Elec.Eng.)*. One supplying power for the anode current of valves. US term *B-battery*. Also called *anode battery*.

high-tension ignition *(Elec.Eng.)*. An ignition system for internal-combustion engines which employs a spark from a high-tension magneto or an induction coil.

high-tension magneto *(Elec.Eng.)*. The usual form of magneto used for producing the high-voltage spark required for the ignition of an internal-combustion engine.

high-tension separation *(Min.Ext.)*. Electrostatic separation, in which small particles of dry ore fall through a high-voltage d.c. field, and are deflected from gravitational drop or otherwise separated in accordance with the electric charge they gather and retain.

high-vacuum *(Electronics)*. A system so completely evacuated that the effect of ionization on its subsequent operation may be neglected. See also **hard**.

high-voltage *(Elec.Eng.)*. Legally, any voltage above 650 volts. In batteries, etc., often called *high-tension*. See **high-tension battery**.

high-voltage test *(Elec.Eng.)*. The application of a voltage greater than working voltage to a machine, transformer, or other piece of electrical apparatus to test the adequacy of the insulation.

highway *(Comp.)*. British term, now dying out, for **bus**.

high-wing monoplane *(Aero.)*. An aircraft with the wing mounted on or near the top of the fuselage.

Hi-k capacitor *(Elec.Eng.)*. One in which the dielectric of barium and strontium titanates has permittivities above 1000.

Hilbert transformer *(Elec.Eng.)*. A device for obtaining a phase shift of 90°. It consists of a delay line, fed from a travelling-wave source and terminated by a negligibly small resistor, the p.d. across this forming the 0° output signal. The 90° signal is obtained by integrating the voltage along the line with a weighting function inversely proportional to the distance from the termination.

Hildebrand electrode *(Chem.)*. See **hydrogen electrode**.

hill-climbing *(Telecomm.)*. Continuous or periodic adjustment of self-regulating adaptive control systems to achieve an optimum result.

hillebrandite *(Min.)*. *Dicalcium silicate hydrate*. Occurs as white fibrous aggregates in impure thermally metamorphosed limestones and in boiler scale.

Hill reaction *(Bot.)*. The light-driven transport of electrons from water to some acceptor other then CO_2 (e.g. ferricyanide) with the production of oxygen, by isolated chloroplasts or chloroplast-containing cells.

Hilt's Law *(Geol.)*. An expression of the observation that the more deeply buried a coal seam, the higher is the rank of its coal.

hilum *(Bot.)*. A scar or mark, especially: (1) on the testa where the stalk was attached to the seed; (2) the central part of a starch grain around which the starch is deposited.

hilum *(Zool.)*. A small depression in the surface of an organ, which usually marks the point of entry or exit of blood-vessels, lymphatics, or an efferent duct. See *hilus*.

Himalia *(Astron.)*. The sixth natural satellite of **Jupiter**. Its diameter is about 100 km.

HiMAT *(Aero.)*. Abbrev. for *Highly Manoeuvrable Aircraft Technology*.

hind-brain *(Zool.)*. In Vertebrates, that part of the brain which is derived from the third or posterior brain-vesicle of the embryo, comprising the cerebellum and the medulla oblongata: the posterior brain-vesicle itself.

hindered settling *(Min.Ext.)*. Hydraulic classification of sand-sized particles in accordance with their ability to gravitate through a column of similar material expanded by a rising current of water.

hindered settling *(Powder Tech.)*. Settling of solids in a suspension of a concentration greater than 15% by volume, in which the predominant physical process is the draining of the fluid out of a thick slurry. No particle segregation by size occurs, a clear boundary being formed between the supernatant fluid and the settling suspension.

hind-gut *(Zool.)*. That part of the alimentary canal of an animal which is derived from the posterior ectodermal invagination or proctodaeum of the embryo.

hindrance *(Phys.)*. Impedances (0 for zero, and 1 for infinite) used in theoretical manipulation of switching.

hinge *(Zool.)*. The flexible joint between the 2 valves of the shell in a bivalve Invertebrate, such as a pelecypod Mollusc or a Brachiopod: any similar structure: a joint permitting of movement in one plane only.

hinge-bound door *(Build.)*. A door which will not close easily or fully owing to the hinges being too deeply sunk.

hinge fault *(Geol.)*. A fault along which the displacement increases from zero at one end to a maximum at the other end.

hingeing post *(Build.)*. The post from which a gate is hung. Also called *swinging post*.

hinge ligament *(Zool.)*. The tough uncalcified elastic membrane which connects the 2 valves of a bivalve shell.

hinge moment *(Aero.)*. The moment of the aerodynamic forces about the hinge axis of a control surface, which increases with speed, necessitating **aerodynamic balance**.

hinge region *(Immun.)*. A flexible region of immunoglobulin heavy chains near the junction of the Fab and Fc portions. This flexibility allows the angle between the arms bearing the antigen combining sites to vary widely and so accommodate different dispositions of the antigen.

hip *(Arch.)*. The salient angle formed by the intersection of 2 inclined roof slopes.

hip hook *(Build.)*. A strap of wrought-iron fixed at the foot of a hip rafter and bent into the form of a scroll, as a support for the hip tiles.

hip iron *(Build.)*. See **hip hook**.

hip knob *(Build.)*. A finial surmounting the peak of a gable or a hipped roof.

hipped end *(Build.)*. The triangular portion of roof covering the sloping end of a hipped roof.

hipped roof *(Build.)*. A pitched roof having sloping ends at the gable ends.

hippocampus *(Zool.)*. In the Vertebrate brain, a tract of nervous matter running back from the olfactory lobe to the posterior end of the cerebrum. adj. *hippocampal*.

hippuric acid *(Chem.)*. *Benzoyl-aminoethanoic acid*, $C_6H_5CO.NH.CH_2.COOH$, rhombic crystals, mp 187°C, occurring in the urine of many animals, particularly herbivores.

hippus *(Med.)*. Rhythmical alternate contraction and dilatation of the pupil of the eye.

hip rafter *(Build.)*. The rafter at the hip of roof where the two slopes meet. It supports the top ends of the jack rafters.

hip replacement *(Med.)*. Operative replacement of hip joint particularly for severe osteo-arthritis.

hip roll *(Build.)*. A timber of circular section with a vee cut out along its length, so as to adapt it for sitting astride the hip of a roof.

hip tile *(Build.)*. A form of arris-tile laid across the hip of a roof.

Hirschsprung's disease *(Med.)*. A condition occurring in children in which there is great hypertrophy and dilatation of the colon. Also *megacolon*.

hirsute *(Bot.,Zool.)*. Hairy; having a covering of stiffish hair or hairlike feathers.

hirsuties *(Med.)*. Excessive hairiness.

hirudin *(Zool.)*. An anticoagulin, present in the salivary secretion of the leech, which prevents blood clotting by inhibiting the action of thrombin on fibrinogen.

Hirudinea *(Zool.)*. A class of Annelida the members of which are ectoparasitic on a great variety of aquatic and terrestrial animals; they possess anterior and posterior suckers, and most of them lack setae; hermaphrodite animals with median genital openings; the development is direct. Leeches.

His *(Chem.)*. Symbol for **histidine**.

hispid *(Bot.,Zool.)*. Coarsely and stiffly hairy as in many Boraginaceae; having a covering of stiffish hair or hair-like feathers.

hiss *(Telecomm.)*. See **noise**.

His's bundle *(Med.)*. See **bundle of His**.

histamine *(Med.)*. A base $C_5H_9N_3$, 2-imidazolyl-4 (or 5) ethylamine. Formed *in vivo* by the decarboxylation of **histidine**. Released during allergic reactions, e.g. hay fever; large releases cause the contraction of nearly all smooth muscle and dilatation of capillaries, which cause a fall of arterial blood pressure and shock.

histidine *(Chem.)*. *2-Amino-3-imidazolepropanoic acid*. The L- or S-isomer is a constituent of proteins and a precursor of **histamine**. Symbol His, short form H. ⇨

histiocyte *(Immun.)*. A macrophage found within the tissues in contrast to those found in the blood (monocytes).

histioma, histoma *(Med.)*. Any tumour derived from fully developed tissue, such as fibrous tissue, cartilage, muscle, blood vessels.

histochemistry *(Zool.)*. The chemistry of living tissues.

histocompatibility antigen *(Immun.)*. Genetically determined antigens present on the surface of nucleated cells, including blood leucocytes. Coded for by *MHC genes*. They are responsible for the differences between genetically non-identical individuals which cause rejection of homografts. See **major histocompatibility complex**.

histocompatibility testing *(Immun.)*. Tests whereby donor and recipient are matched as closely as possible prior to tissue grafting in humans.

histogen *(Bot.)*. One of three meristems (**dermatogen**, **periblem**, **plerome**) at the shoot or root tip that give rise exclusively to particular tissues in that organ (epidermis, cortex, stele + pith, respectively). (The concept of discrete histogens has been replaced more recently by the **tunica-corpus concept**).

histogenesis *(Zool.)*. Formation of new tissues.

histogram *(Stats.)*. A graphical representation of class frequencies as rectangles against class interval, the value of frequency being proportional to the area of the corresponding rectangle.

histology *(Zool.)*. The study of the minute structure of tissues and organs.

histolysis *(Bot.)*. The breakdown, and sometimes liquefaction, of a cell or tissue.

histolysis *(Zool.)*. Dissolution and destruction of tissues.

histoma *(Med.)*. See **histioma**.

histones *(Biol.)*. Basic proteins involved in the packaging of DNA in the eukaryotic nucleus to form **chromatin**, which is packed into **nucleosomes**, the first level of chromosome organization above the DNA helix. There are five types of histone molecule, four of which have been highly conserved in sequence throughout eukaryotic evolution.

histoplasmosis *(Med.,Vet.)*. A disease of animals and man due to infection by the fungal organism *Histoplasma capsulatum*. Affects the lungs in man, being relatively common in the US.

Historical Geology *(Geol.)*. That major branch of geology that is concerned with the evolution of the Earth and its environment from its origins to the present day.

history of geology *(Geol.)*. The study of the history or development of geological knowledge.

histozoic *(Zool.)*. Living in the tissues of the body, amongst the cells.

hit *(Comp.)*. Occurrence of a successful search. If a file is searched for a particular item and several occurrences are found these are called hits.

hit-and-miss ventilator *(Build.)*. A ventilating device consisting of a slotted plate over which may be moved another slotted plate, so that the openings for access of air may be restricted as required.

hitch *(Min.Ext.)*. (1) A fault of minor importance, usually not exceeding the thickness of a seam. (2) Ledge cut in rock face to hold mine timber in place.

Hittorf dark space *(Phys.)*. See **Crookes dark space**.

HIV *(Immun.)*. Abbrev. for *Human Immunodeficiency Virus*.

HLA-A, HLA-B, HLA-C *(Immun.)*. Histocompatibility antigens, each coded by different loci in the MHC genes, and for which there are numerous allelic products at each locus. These antigens belong to Class I and have similar structures, consisting of a membrane bound polypeptide to which is attached non-covalently β2 microglobulin. Class I antigens are present on almost all nucleated cells. Cytotoxic T-cells recognize antigens in association with Class I molecules.

HLA-D *(Immun.)*. Histocompatibility antigens coded for by separate loci in the MHC gene complex. There are three distinct loci — DP, DR and DQ — each of which has several alleles. These antigens belong to Class II, and are dimers composed of two different membrane bound polypeptide chains. HLA-D antigens are normally absent from most nucleated cells but present on B-lymphocytes, dendritic (interdigitating) cells, and macrophages stimulated by **interferon**. T-helper cells recognize antigens in association with Class II molecules.

HLB *(Chem.)*. Term for *Hydrophilic-Lipophilic Balance* in emulsifying agents.

HMI *(Image Tech.)*. TN for a type of metal halide lamp.

H-network *(Telecomm.)*. Symmetrical section of circuit, with one shunt branch and four series branches.

Hn RNA, heterogeneous nuclear RNA *(Biol.)*. Population of RNA molecules in the nucleus including the precursors of mature messenger RNA, which are eventually found in the cytoplasm.

HNW *(Build.)*. An abbrev. for *Head, Nut, and Washer*.

Ho *(Chem.)*. The symbol for *holmium*.

hoarding *(Behav.)*. Refers to the storing of food or other items in the animal's home or territory; occurs in small mammals and in some bird species.

hoarding, hoard *(Build.)*. A close-boarded fence of temporary character erected around a building site.

hoar frost *(Meteor.)*. A deposit of ice crystals formed on objects, especially during cold clear nights when the dew-point is below freezing-point. The conditions favouring the formation of hoar-frost are similar to those which produce *dew*.

hob *(Eng.)*. (1) A hardened master punch, used in die sinking, which is a duplicate of the part to be produced by the die. It is pressed into an unheated die blank to make the die impression. (2) A gear-cutting tool resembling a milling cutter or a worm gear, whose thread is interrupted by grooves so as to form cutting faces. Also *hobbing cutter*.

hobbing machine *(Eng.)*. A machine for cutting teeth on gear blanks, for the production of spur, helical, and worm gears by means of a hobbing cutter.

hobbles *(Vet.)*. An apparatus applied to the legs of a horse for casting.

Höchstäter cable *(Elec.Eng.)*. A high-voltage multi-core cable in which a thin metallized sheath is placed over the insulation of each core, in order to control the distribution of electric stress in the dielectric and ensure that it is purely radial.

hock *(Zool.)*. The tarsal joint of a Mammal.

hoctonspheres *(Chem.Eng.)*. Storage vessels for low boiling-point liquids stored under pressure at temperatures at which they would at normal pressure vaporize. The vessels are made from segments specially pressed before welding so that the finished vessel is truly spherical. Diameters up to 10 m are common.

hod *(Build.)*. A 3-sided container, supported on a long handle, used for carrying bricks and mortar on the site.

Hodgkin's disease *(Immun.)*. One of a group of diseases named lymphomas which involve lymphoid tissues. It is characterized by destruction of the normal architecture of lymph nodes and replacement with reticular cells, lymphocytes, neutrophil and eosinophil leucocytes, and an unusual kind of giant cell with two nuclei. This is accompanied by deficient cell-mediated immunity although antibody formation is normal.

hodograph *(Phys.)*. A curve used to determine the acceleration of a particle moving with known velocity along a curved path. The hodograph is drawn through the ends of vectors drawn from a point to represent the velocity of the particle at successive instants.

hodoscope *(Nuc.Eng.)*. Apparatus (e.g. an array of radiation detectors) which is used for tracing paths of charged particles in a magnetic field.

Hoechst 33258 *(Biol.)*. TN for a DNA-specific stain, used in chromosome **banding**.

Hofmann degradation *(Chem.)*. A process used in organic chemistry to determine the structure of amines. It involves exhaustive methylation, i.e. the use of excess of a methylating agent, such as methyl iodide. Good for alkaloids.

Hofmann's reaction *(Chem.)*. A method of preparing primary amines from the amides of acids by the action of bromine and then of caustic soda. The number of carbon atoms in the chain should not be more than six, and the resulting amine has one carbon atom less then the amide from which it has been prepared.

Hofmeister series *(Chem.)*. The simple anions and cations arranged in the order of their ability to coagulate solutions of lyophilic colloids.

hogback *(Geol.)*. A ridge with a sharp summit and steep slopes on both sides, usually $>20°$.

hog-back girder *(Build.,Civ.Eng.)*. Girder which curves along its top edge to be convex upwards.

hog cholera *(Vet.)*. See **swine fever**.

hog-frame *(Build.)*. A term applied to some forms of truss which are shaped so as to bulge on the upper side.

hogging *(Build.)*. A mixture of gravel and clay, used for paving. Also *hoggin*.

hogging *(Ships)*. This occurs when the middle of the ship is supported in the crest of the wave while the ends are in troughs or when the ends are more heavily loaded than the middle. If the ship actually bends she is said to be hogged. Cf. **sagging**.

hog-pit *(Paper)*. The pit below the couch of the paper making machine, equipped with an agitator, in which backwater, edge trims (and the whole web at times of a break) are collected for pumping to an earlier part of the system for re-use.

hohlraum *(Acous.)*. German for *cavity*.

Hohmann orbit *(Space)*. A space trajectory tangential to, or *osculating*, two co-planar planetary orbits at its **perihelion** and **aphelion** respectively: it is also the most energetically economical transfer orbit.

hoist *(Min.Ext.)*. An engine with a drum, used for winding up a load from a shaft or in an underground passage such as a winze.

hoisting motor *(Elec.Eng.)*. See **lift motor**.

hol- *(Genrl.)*. Prefix. See **holo-**.

Holarctic region *(Zool.)*. One of the primary faunal regions into which the surface of the globe is divided. It includes North America to the edge of the Mexican plateau, Europe, Asia (except Iran, Afghanistan, India south of the Himalayas, the Malay peninsula), Africa north of the Sahara, and the Arctic islands.

hold *(Electronics)*. The maintenance, in charge-storage tubes, of the equilibrium potential by means of electron bombardment.

hold *(Image Tech.,Telecomm.)*. Synchronization control, in which oscillator frequency of the receiver is adjusted to that of incoming synchronizing pulses.

hold *(Ships)*. A compartment within a ship's hull for the carriage of cargo. Below the lowermost deck it is termed *hold*; above this, *'tween decks*. For identification, the holds are numbered from the fore end of the ship.

holdback *(Nuc.Eng.)*. Agent for reducing an effect, e.g. a large quantity of inactive isotope reduces the coprecipitation or absorption of a radioactive isotope of the same element.

hold-down roller *(Image Tech.)*. Small roller pressing the edges of motion picture film against a sprocket to ensure that the teeth fully engage the perforations. Also *pad roller*.

Holden permeability bridge *(Elec.Eng.)*. A permeability bridge in which the standard bar and the bar under test carry magnetizing coils, and are connected by yokes to form a closed magnetic circuit. The magnetizing currents are varied until there is no magnetic leakage between the yokes.

holder *(Elec.Eng.)*. See **lampholder**.

holderbat *(Build.)*. A metal collar formed in two half-round parts, capable of being clamped togther around a rain-water, soil, or waste pipe, and having a projecting leg on one part for fixing to a wall.

Holder's inequality *(Maths.)*. If a_r and b_r are positive, nonproportional, sets, and if $\alpha + \beta = 1$, then (1)

$$(1)\ \sum_1^n a_r^\alpha b_r^\beta < \left(\sum_1^n a_r\right)^\alpha \left(\sum_1^n b_r\right)^\beta,$$

if α and β are positive and (2)

$$(2)\ \sum_1^n a_r^\alpha b_r^\beta < \left(\sum_1^n a_r\right)^\alpha \left(\sum_1^n b_r\right)^\beta.$$

if $\alpha > 1$ or if $\alpha < 0$.

holdfast *(Bot.)*. Any single-celled or multicellular organ other than a root, which attaches a plant (especially an alga) to the substrate.

holdfast *(Build.)*. A device for holding down work on a bench, comprising a main pillar which passes through a hole in the bench, and an adjustable clamp with a shoe for placing on the work.

hold frame *(Image Tech.)*. See **freeze frame**.

holding altitude *(Aero.)*. The height at which a controlled aircraft may be required to remain at a **holding point**.

holding anode *(Electronics)*. Auxiliary d.c. anode in a mercury-arc rectifier for maintaining an arc.

holding beam *(Electronics)*. Widely-spread beam of electrons used to regenerate charges retained on the dielectric surface of a storage tube or electrostatic memory.

holding pattern *(Aero.)*. A specified flight track, e.g. *orbit* or figure-of-eight, which an aircraft may be required to maintain about a holding point.

holding point *(Aero.)*. An identifiable point, such as a radio beacon, in the vicinity of which an aircraft under **air-traffic control** may be instructed to remain.

holding-up *(Eng.)*. The action of pressing a heavy hammer against the head of a rivet while closing or forming the head on the shank.

hold-on coil *(Elec.Eng.)*. An electromagnet which holds the moving arm of a motor starter or other similar device, in the 'on' position; if the current in the coil is reduced or interrupted, the arm returns to the 'off' position under the action of a spring.

hold-up *(Chem.Eng.)*. In any process plant, the amount of material which must always be present in the various reactors etc. to ensure satisfactory operation.

hole *(Electronics)*. Vacancy in a normally filled energy band, *either* as result of electron being elevated by thermal energy to the conduction band, and so producing a hole-electron pair; *or* as a result of one of the crystal lattice sites being occupied by an acceptor impurity atom. Such vacancies are mobile and contribute to electric current in the same manner as positive carriers and mathematically are equivalent to positrons. Also called *negative-ion vacancy*.

hole control *(Min.Ext.)*. Altering the composition of the **drilling mud**, drill pressure and rate to accommodate the changes in the rock formation which it penetrates.

hole current *(Electronics)*. That part of the current in a semiconductor due to the migration of holes.

hole density *(Electronics)*. The density of the holes in a semiconductor in a band which is otherwise full.

hole injection *(Electronics)*. Holes can be *emitted* in *n*-type semiconductor by applying a metallic point to its surface.

hole mobility *(Electronics)*. See **mobility**.

hole theory of liquids *(Chem.)*. Interpretation of the fluidity of liquids by regarding them as disordered crystal lattices with mobile vacancies or holes.

hole trap *(Electronics)*. An impurity in a semiconductor which can release electrons to the conduction or valence bands and so trap a 'hole'.

holiday *(Build.)*. A greater or lesser part of the surface accidentally missed during painting.

holing *(Build.)*. The operation of piercing slates to receive nails.

holland *(Textiles)*. A glazed cotton or linen fabric used principally for window-blinds and **interlinings**.

hollander beater *(Paper)*. Horizontal trough with a dividing wall parallel to its longer side which stops short of the ends to provide a continuous channel around which stuff may circulate under the propulsion of the rotating beater roll. The surface of the latter is fitted with metal bars parallel to the roll shaft which impart a rubbing and cutting action on the fibres in the narrow gap between them and similar fixed bars beneath the roll. The extent of this action largely controls the properties of the finished paper. Now generally superseded by the **refiner**, which performs a similar function.

hollow back *(Print.)*. Type of binding in which the spine of the book is not pasted or glued down, leaving a space between the back of the sections and the leather or cloth when the book is open. Also called *open back*.

hollow bed *(Build.)*. A bed joint in which, owing to the surfaces of the stones not being plane, there is contact only at the outer edges.

hollow-cathode tube *(Electronics)*. A gas-discharge lamp in which the glowing plasma forms only inside a small tubular cathode, giving, under appropriate current and gas pressure conditions, an intense light with high spectral purity.

hollow fusee *(Horol.)*. A fusee with its top pivot sunk into the body to reduce the height of the movement.

hollow mandrel lathes *(Eng.)*. The term formerly applied to lathes capable of having bar stock fed through the mandrel for repetition work.

hollow newel *(Build.)*. The well-hole of a winding stair.

hollow pinion *(Horol.)*. A pinion drilled throughout its length.

hollow punch *(Eng.)*. A hollow cylindrical tool tapered on its outside diameter to form a cutting edge and used to punch circular washers from sheets of soft materials, e.g. rubber, leather etc.

hollow quoin *(Hyd.Eng.)*. A quoin accommodating the heel-post of a lock-gate in a vertical recess.

hollow roll *(Build.)*. A joint between the edges of two lead sheets on the flat, made by turning up each edge at right angles to the flat surface, bringing the two turned-up parts together, and shaping them over to form a roll.

hollows *(Foundry)*. Fillets, or curves of small radius, uniting two surfaces intersecting at an angle; added to a pattern to give strength to the casting and facilitate withdrawal of the pattern from the mould.

hollows *(Print.)*. Strips of strong paper, etc. attached to the boards and to the back of a book to strengthen the spine. Also called *back lining*.

hollow spindle spinning *(Textiles)*. A specialized method of yarn formation in which a core of untwisted staple fibres is wrapped with a binder thread as it passes through a rotating hollow spindle.

hollow walls *(Build.)*. Same as **Cavity walls**.

holmium *(Chem.)*. A metallic element, a member of the rare earth group. Symbol Ho, at.no. 67, r.a.m. 164.930. It occurs in euxenite, samarskite, gadolinite and xenotine.

holmquistite *(Min.)*. A rare lithium-bearing calcium-free variety of orthorhombic amphibole.

holoaxial *(Crystal.)*. A term applied to those classes of crystals characterized by axes of symmetry only; such crystals are not symmetrical about planes of symmetry.

holobenthic *(Zool.)*. Passing the whole of the life-cycle in the depths of the sea.

holoblastic *(Zool.)*. Said of ova which exhibit total cleavage.

holobranch *(Zool.)*. In Fish, a branchial arch carrying two rows of respiratory lamellae or filaments, one on the posterior and one on the anterior face.

holocarpic *(Bot.)*. Having the whole thallus transformed at maturity into a sporangium or a sorus of sporangia.

Holocene *(Geol.)*. The younger epoch of the **Quaternary** period.

holocentric chromosome *(Biol.)*. One lacking a localized **centromere**, and along whose full length spindle **microtubules** attach. Found in some plants, protozoa and certain classes of insect.

holocephalous *(Zool.)*. Said of single-headed ribs.

holocrystalline rocks *(Geol.)*. Those igneous rocks in which all the components are crystalline; glass is absent. Cf. **hemicrystalline rocks**.

hologamy *(Bot.)*. The condition of having gametes which resemble in size and form the ordinary cells of the species; union of such cells.

hologram *(Phys.)*. See **holography**.

holographic interferometry *(Eng.)*. This involves double exposure **holography** which results in a series of light and dark bands appearing on the image of the object being examined. The bands give information about deformations of the object's surface.

holography *(Phys.)*. A method of recording and reconstructing the wave front emanating from an illuminated object. Coherent light from a laser is split in two; one is a reference beam and the other illuminates the object. The waves scattered by the object and the reference beam are recombined to form an interference pattern on a photographic plate, the *hologram*; this records both the amplitude and phase of the scattered light. When the hologram is itself illuminated by light from a laser or other point source, two images are produced, one is virtual but the other is real and can be viewed directly. So a 3-dimensional image of the object can be produced.

holohedral *(Crystal.)*. Term applied when a crystal is complete, showing all possible faces and angles.

holo-, hol- *(Genrl.)*. Prefix from Gk. *holos*, whole.

holomastigote *(Zool.)*. Having numberous flagella scattered evenly over the body.

Holometabola *(Zool.)*. See **Endopterygota**.

holometabolic *(Zool.)*. Showing a complete metamorphosis, as members of the *Endopterygote*. Cf. **hemimetabolic**. n. *holometabolism*.

holomorphic function *(Maths.)*. See **analytic function**.

holophytic *(Bot.)*. Living by photosynthesis, like most plants which are not parasitic, saprophytic or phagocytic etc. Cf. **holozoic**.

holostyly *(Zool.)*. A type of jaw suspension in which the upper jaw fuses with the cranium, the hyoid arch playing no part in the suspension; characteristic of *Holocephali*.

Holothuroidea *(Zool.)*. A class of *Echinodermata* having a sausage-shaped body without arms; the tube feet possess ampullae and may occur on all surfaces; the anus is aboral, the madreporite internal; the skeleton is reduced to small ossicles embedded in the soft integument; free-living mud-feeders. See Cucumbers.

holotrichous *(Zool.)*. Bearing cilia of uniform length over the whole surface of the body.

holotype *(Bot.)*. The one specimen designated as the nomenclatural type in a published description of the species.

holozoic *(Bot.,Zool.)*. Living by ingestion or phagocytosis and digestion, like most animals; not *osmotrophic*. n. *holozoon*. Cf. **holophytic**.

homeomorphic *(Maths.)*. Topological spaces T_1 and T_2 are said to be homeomorphic, or topologically equivalent, if there is a mapping f from T_1 on to T_2 which is one-to-one and such that both f and its inverse f^{-1} are continuous. Such a mapping f is said to be a *homeomorphism*. In effect, two geometrical figures are homeomorphic if each can be transformed into the other by a continuous deformation, e.g. the circle, the square and the triangle are homeomorphic, i.e. topologically equivalent.

homeopathy *(Med.)*. A system of medicine, founded by Dr Samuel Hahnemann (1755–1843), the basic principle of which is *Similia similibus curantur* (let likes be cured by likes). Experimental observations led Hahnemann to conclude that (*a*) a disease is characterized by a definite symptom complex, (*b*) it can be effectively treated by the drug which produces in a healthy individual the most similar symptom complex, and (*c*) for a given disease, the proven drug is best administered in extreme dilution.

homeostasis *(Biol.)*. The tendency for the internal environment of an organism to be maintained constant.

homeostasis *(Ecol.)*. Tendency of plant and animal populations to remain constant as a result of density-dependent mechanisms operating on birth rate, survival or death rate.

homeotic mutants *(Biol.)*. Mutants in *Drosophila* which effect large-scale changes in development, e.g. the substitution of a leg for a wing.

homeotypic division *(Biol.)*. See **mitosis**.

homer *(Telecomm.)*. Any arrangement which provides signals or fields which can be used to guide a vehicle to a specified destination, e.g. a *homing aid* for aircraft, a *leader cable* for ships, or *guidance* system for missiles.

home range *(Ecol.)*. A definite area to which individuals, pairs, or family groups of many types of animals restrict their activities.

homespun *(Textiles)*. A coarse tweed hand-woven from handspun wool.

homing *(Telecomm.)*. The operation of a selector in returning to a predetermined normal position following the release of the connection.

homing aid *(Aero.)*. Any system designed to guide an aircraft to an airfield or aircraft carrier.

homing behaviour *(Behav.)*. Refers to navigational behaviour in a number of species, ranging from returning home after a daily foraging or other excursion, to the more complex navigational task involved in large migrations.

homo- *(Genrl.)*. Prefix from Gk. *homos*, same.

homoblastic *(Bot.)*. A species in which the first formed leaves in a seedling or shoot are very like those formed later. Cf. **heteroblastic**.

homoblastic *(Zool.)*. Showing direct embryonic development: originating from similar cells.

homocentric *(Phys.)*. Term applied when rays are either parallel or pass through one focus.

homocercal *(Zool.)*. Said of a type of tail-fin, found in all the adults of the higher Fish, in which the vertebral column bends abruptly upwards and enters the epichordal lobe, which is equal in size to the hypochordal lobe.

homochlamydeous *(Bot.)*. Having a perianth consisting of members all of the same kind, not distinguishable into sepals and petals.

homocyclic *(Chem.)*. Containing a ring composed entirely of atoms of the same kind.

homocystinuria *(Med.)*. A metabolic defect resulting in the excretion of homocystine (the oxidized form of cysteine, an essential amino acid) in the urine.

homodesmic structure *(Crystal.)*. Crystal form with only one type of bond (either ionic or covalent).

homodont *(Zool.)*. Said of teeth which all have the same characteristics.

homodyne reception *(Telecomm.)*. That using an oscillating valve adjusted to, or locked with, an incoming carrier, to enhance its magnitude and improve demodulation. Also *demodulation of an exalted carrier*.

homoeomerism *(Zool.)*. In metameric animals, the condition of having all the somites alike; also *heteromerism*. adj. *homoeomeric*.

homoeopathy *(Med.)*. See **homeopathy**.

homoeosis *(Zool.)*. In metameric animals, the assumption by a merome of the characters of the corresponding merome of another somite. Cf. **heterosis**.

homogametic *(Zool.)*. Having all the gametes alike.
homogametic sex *(Biol.)*. The sex that is *homozygous* for the *sex-determining chromosomes*. Cf. **heterogametic sex**.
homogamy *(Bot.)*. The simultaneous maturation of the anthers and stigmas in a flower. Cf. **dichogamy, protandry, protogyny**.
homogamy *(Zool.)*. Inbreeding, usually due to isolation.
homogeneous *(Chem.)*. Said of a system consisting of only one phase, i.e. a system in which the chemical composition and physical state of any physically small portion are the same as those of any other portion.
homogeneous co-ordinates *(Maths.)*. A system of co-ordinates in which any multiple of the co-ordinates of a point or line represents the same point, e.g. in line-co-ordinates the line (a,b,c) is the same as the line (ka,kb,kc).
homogeneous function *(Maths.)*. An algebraic function such that the sum of the indices occurring in each term is constant, e.g. $x^3+2x^2y+y^3$.
homogeneous ionization chamber *(Nuc.Eng.)*. One in which both walls and gas have similar atomic composition, and hence similar energy absorption per unit mass.
homogeneous light *(Phys.)*. The same as *monochromatic light*.
homogeneous radiation *(Phys.)*. That of constant wavelength (monochromatic), or constant particle energy.
homogeneous reactor *(Nuc.Eng.)*. One in which the fuel and moderator are finely divided and mixed (or the fuel may be dissolved in a liquid moderator) so as to produce an effectively homogeneous core material. See **slurry reactor**.
homogenesis *(Zool.)*. The type of reproduction in which the offspring resemble the parents.
homogenizer *(Phys.)*. A device in which coarse and polydisperse emulsions are transformed into nearly monodisperse systems. The liquid is subjected to an energetic shear.
homogeny *(Zool.)*. Individuals or of parts which are **homologous**. adj. *homogenous*.
homograft *(Immun.)*. Graft of tissue from one individual to another of the same species.
homoiohydric *(Bot.)*. Plants able to regulate water loss and to remain hydrated for some time (hours, days, or years) when the external water supply is restricted, e.g. most terrestrial vascular plants, few of which however can survive desiccation. Cf. **poikilohydric**.
homoiomerous *(Bot.)*. A lichen thallus which has an even distribution of algal cells through its thickness. Cf. **heteromerous**.
homoioplastic, homoplastic *(Zool.)*. In experimental zoology, said of a graft which is transplanted to a site identical with its point of origin, e.g. a skin graft to a skin site.
homoiothermal *(Zool.)*. See **warm-blooded**.
homokaryon *(Biol.)*. **Somatic cell hybrid** containing separate nuclei from the same species.
homologous *(Bot.,Zool.)*. Of the same essential nature, and of common descent.
homologous alternation of generations *(Bot.)*. See **homologous theory of alternation** and **isomorphic alternation of generations**. Cf. **antithetic theory of alternation.**
homologous chromosomes *(Biol.)*. Chromosomes that pair with each other during synapsis at *meiosis*, so that one member of each pair is carried by every gamete.
homologous organs *(Bot.)*. Organs which are equivalent morphologically and of common evolutionary origin but which may be similar or dissimilar in appearance or function.
homologous series *(Chem.)*. A series of organic compounds, each member of which differs from the next by the insertion of a $—CH_2—$ group in the molecule. Such a series may be represented by a general formula and shows a gradual and regular change of properties with increasing molecular weight.
homologous theory of alternation *(Bot.)*. The hypothesis that the sporophyte is of a similar nature to the gametophyte and thus that vascular plants evolved from algae with an **isomorphic alternation of generation**. Cf. **antithetic theory of alternation**.
homologous variation *(Bot.)*. The occurrence of similar variations in related species.
homology *(Biol.)*. Of DNA sequences or peptide sequences, the degree of similarity.
homology *(Bot.,Zool.)*. Morphological equivalence, common evolutionary origin. n. *homologue*. Cf. **analogy**.
homology group *(Maths.)*. Important group used in algebraic topology.
homomorphic *(Biol.)*. Said of chromosome pairs which have the same form and size.
homomorphous *(Bot.,Zool.)*. Alike in form.
homoplasma *(Zool.)*. In tissue culture, a medium prepared with plasma from another animal of the same species as that from which the tissue was taken. Cf. **autoplasma, heteroplasma**.
homoplastic *(Bot.,Zool.)*. Of the same structure and manner of development but not descended from a common source. n. *homoplasty*.
homopolar *(Chem.)*. Having an equal distribution of charge, as in a covalent bond between like atoms.
homopolar generator *(Elec.Eng.)*. Low-voltage d.c. generator based on Faraday disk principle which produces ripple-free output without commutation.
homopolar magnet *(Elec.Eng.)*. One with concentric pole pieces.
homopolar molecule *(Elec.Eng.)*. One without effective electric dipole moment.
homopolymer *(Biol.)*. DNA or RNA strand whose nucleotides are all of the same kind. Usually made enzymatically from a single nucleotide precursor.
homoscedastic *(Stats.)*. Having the same variance (applied to sets of observations).
homosexuality *(Behav.)*. Sexual interest directed at members of one's own sex.
homospory *(Bot.)*. A species which produces only one type of a spore. adj. *homosporous*. Cf. **heterospory**.
homostyly *(Bot.)*. The condition in which all the styles are the same length. Cf. **heterostyly**.
homotaxis *(Geol.)*. A term introduced by Huxley in 1862 to indicate that strata or sequences of strata in different areas shared the same fossil characteristics are not necessarily the same age. A faunal assemblage may originate in locality A, be gradually dispersed or migrate to locality B and eventually reach locality C. The strata accumulating at these three localities are *homotaxia* although not necessarily contemporaneous.
homothallism *(Bot.)*. The condition in which successful fertilization can take place between any two gametes from the same organism. Also *self compatibility*. adj. *homothallic*. Cf. **heterothallic**. It is analogous to self compatibility in flowering plants.
homothermous *(Zool.)*. See **warm-blooded**.
homotopic mapping *(Maths.)*. Two continuous mappings f and g of a topological space A into a topological space B are said to be homotopic if there is a function $F(x,t)$, representing a continuous mapping of $A \times I$ into B (I being the unit interval), for which $F(x,0) = f(x)$, $F(x,1) = g(x)$, for all x in A and for $0 \leqq t \leqq 1$. f and g are also said to be continuously deformed into each other.
homotypic *(Zool.)*. Conforming to the normal condition. Cf. **heterotypic**.
homozygosis *(Biol.)*. The condition of being *homozygous*.
homozygote *(Biol.)*. An individual whose two genes at a particular locus are the same allele, the individual having been formed by the union of gametes carrying the same allele. adj. *homozygous*. Cf. **heterozygote**.
homunculus *(Biol.,Med.)*. A dwarf of normal proportions; a mannikin or little man created by the imagination; a miniature human form believed by animalculists to exist in the spermatozoon.
hone, honestone, whetstone *(Geol.)*. Term applied to fine-textured even-grained indurated sedimentary rocks which may be used as oilstones for imparting a keen edge to cutting tools. Honestone has been largely replaced now by emery and silicon carbide products.
honeycomb *(Aero.)*. A gridwork across the duct of a wind tunnel to straighten the airflow; also *straighteners*.

honeycomb *(Textiles)*. Fabric with the threads forming ridges and hollows to give a cell-like appearance. Generally woven from coarse soft yarns in compact structures and used for towels and bedspreads.

honeycomb bag *(Zool.)*. See **reticulum**.

honeycomb base *(Print.)*. A metal base for printing plates providing a means of both securing them and moving them into register. The base is drilled with a pattern of holes in which register hooks are placed, there being suitable holes for any position of the plate.

honeycomb coil *(Telecomm.)*. One in which wire is wound in a zig-zag formation around a circular former. The adjacent layers are staggered, so that the wires cross each other obliquely to reduce capacitance effects between turns.

honeycomb slating *(Build.)*. Similar to **drop-point slating**; the tiles, however, have their bottom corners removed.

honeycomb structure *(Aero.)*. Lightweight, very rigid, material for aircraft skin or floors, usually made from thin light-alloy plates with a bonded foil interlayer of generally honeycomb-like form. For high supersonic speeds, a heat-resistant material is made by brazing together stainless-steel or nickel-alloy skins and honeycomb core.

honeycomb wall *(Build.)*. A wall built so as to leave regular spaces, usually entirely with stretchers. See also **sleeper wall**.

honey dew *(Zool.)*. A sweet substance secreted by certain *Hemiptera*; emitted through the anus.

honey guide, nectar guide *(Bot.)*. Lines or dots (sometimes visible only in UV photographs), on the perianth, that direct a pollinating insect to the nectar in a flower.

honing *(Eng.)*. The process of finishing cylinder bores etc. to a very high degree of accuracy by the abrasive action of stone or silicon carbide slips held in a head having both a rotatory and axial motion.

honing machine *(Eng.)*. A partly hand-operated or wholly automatic machine for *honing*.

hood *(Build.)*. (1) A cowl for a chimney.

hood jettison *(Aero.)*. Mechanism, often operated by explosive bolts or cartridge, for releasing the pilot's canopy in flight. Confined to military aircraft.

hood mould *(Build.)*. A projecting moulding above a door or window opening.

hoof *(Zool.)*. In Ungulates, a horny proliferation of the epidermis, enclosing the ends of the digits.

hook bolt *(Build.)*. A galvanized-iron bolt formed out of a rod which is bent at one end into a hook serving as the head, and threaded at the other to take a nut; used for fixing corrugated sheeting.

hook-down *(Print.)*. See **hook-up**.

hooked joint *(Build.)*. A form of joint used between the meeting edges of a door and its case when an airtight or dustproof joint is necessary; the meeting edges on the door have a projection on them fitting into a corresponding recess in the case.

hooked plates *(Print.)*. Single-leaf illustrations sewn in with the *section*, having been printed on paper large enough to allow a narrow fold down the back edge.

Hooke's joint *(Eng.)*. A common form of **universal joint**, comprising two forks arranged at right angles and coupled by a cross-piece.

Hooke's law *(Phys.)*. For an elastic material the *strain* is proportional to the applied *stress*. The value of the stress at which a material ceases to obey Hooke's law is the *limit of proportionality*. See **elasticity**.

hook-out blind *(Build.)*. An outside roller blind fitted with metal side arms and bottom rail so that the blind may be supported away from the window.

hook rebate *(Build.)*. The S-shaped rebate formed on the meeting edges of a **hooked joint**.

hook transistor *(Telecomm.)*. A junction transistor which uses an extra *p-n* junction to act as a trap for holes, thus increasing the current amplification.

hook-up *(Min.Ext.)*. In pilot plant testing, flexible assembly of machines into continuous flow line before final treatment.

hook-up *(Print.)*. The end of a line turned over and bracketed in the line above or below (*hook-down*). Often used in setting up poetry.

hook-up *(Telecomm.)*. A temporary communication channel.

hookworm disease *(Med.)*. See **ankylostomiasis**.

hookworms *(Med.,Zool.)*. Parasitic strongyloid nematodes with hooklike organs on the mouth for attachment to the host. Humans are attacked chiefly by the genus *Ankylostoma*, which penetrates the bare feet and induces a form of anaemia. See also **ankylostomiasis**.

Hoopes process *(Eng.)*. A process for the refining of aluminium electrolytically to a purity of 99.99%. Metal made by the Hall process is alloyed with 33% of copper and made the anode in a nonaqueous electrolytic bath composed of alumina and fluorides of barium, sodium and aluminium. When current is passed between the hearth and carbon electrodes on top of the bath, aluminium dissolves from the anode alloy and pure metal accumulates at the cathodes.

hooping *(Civ.Eng.)*. Reinforcing bars for concrete, bent either to a circular or helical shape.

hoop iron *(Build.)*. Thin strip-iron for securing barrels, and also for various purposes in the building trades, e.g. as reinforcement in brick walls, also for the packaging of bricks or building blocks to enable them to be lifted in quantity.

hoop stress *(Eng.)*. The stress in a tube under pressure, acting tangentially to the perimeter of a transverse section.

hoose *(Vet.)*. See **husk**.

hop *(Telecomm.)*. Distance along earth's surface between successive reflections of a radio wave from an ionized region; also called a *skip*.

'Hope sapphire' *(Min.)*. Synthetic stone having the composition of spinel and blue colour which turns purple in artificial light. First produced in the attempts to synthesize sapphire.

Hopkinson test *(Elec.Eng.)*. A method of testing two similar d.c. machines on full load without requiring a large consumption of power from the supply; one machine fed from the supply drives the other as a generator, which returns power to the supply.

Hopkinson-Thring torsion meter *(Eng.)*. A torsion meter in which two mirrors are mounted on a flange fixed to one end of the shaft under test, one mirror being fixed and the other so mounted that it tilts as the shaft twists. The separation of light beams reflected by the mirrors is a measure of the shaft twist.

hopper *(Build.)*. A draught-preventer at the side of a **hopper light**.

hopper *(Min.Ext.)*. A container or surge bin for broken ore, used to hold small amounts.

hopper crystal *(Crystal.,Min.)*. A crystal which has grown faster along its edges than in the centres of its faces, so that the faces appear to be recessed. This type of skeletal crystallization is often shown by rock-salt.

hopper dredger *(Civ.Eng.)*. The type of bucket-ladder or suction dredger which not only dredges material from below but has hopper compartments fitted with flap-doors which allow the material to be discharged after the vessel has moved to the place of deposit.

hopperfeed *(Eng.)*. A machine used for unscrambling identical components from bulk, orienting them, and delivering them in an orderly arrangement to an assembling machine or other mass-production process.

hopper light *(Build.)*. A window-sash arranged to open inwards about hinges on its lower edge.

hopper window *(Build.)*. A hopper light fitted at the sides with hoppers.

Hoppus foot *(For.)*. A largely obsolete unit of volume obtained by using the square of the quarter girth for trees or logs instead of the true sectional area; only applied to timber in the round.

hopsack (matt) weave *(Textiles)*. A development of plain weave in which two or more **ends** and **picks** weave as one.

hordeolum *(Med.)*. See **sty**.

horizon *(Astron.)*. That great circle, of which the zenith and the nadir are the poles, in which the plane tangent to the earth's surface, considered spherical at the point where the observer stands, cuts the celestial sphere.

horizon *(Behav.)*. The more or less coloured visual impression experienced subjectively by blind persons.

horizon *(Bot.)*. A layer in a soil distinguishable from others by colour, hardness, inclusions or other visible or tangible properties.

horizon *(Geol.)*. The surface separating two beds of rock. It has no thickness, and is more frequently used in the sense of a thin bed or time-plane with a characteristic lithofacies or biofacies, persistent over a wide area.

horizon *(Surv.)*. A plane perpendicular to the direction of gravity shown by a plumbline at the point of observation.

horizon glass *(Surv.)*. See **sextant**.

horizon sensor *(Electronics)*. Sensor providing a stable vertical reference level for missiles and depending on the use of a thermistor to detect the thermal discontinuity between earth and space.

horizontal antenna *(Telecomm.)*. One comprising a system of one or more horizontal conductors, radiating or responding to horizontally-polarized waves.

horizontal axis *(Surv.)*. See **trunnion axis**.

horizontal blanking *(Image Tech.)*. The elimination of the horizontal trace in a CRT during flyback.

horizontal circle *(Surv.)*. The graduated circular plate used for the measurement of horizontal angles by theodolite.

horizontal component *(Elec.Eng.)*. Component of Earth's magnetic field which acts (i.e. exerts a force on a unit pole placed in it) in a horizontal direction.

horizontal draw-out metal-clad switchgear *(Elec.Eng.)*. Metal-clad switchgear in which the switch itself can be isolated by removing it along suitable guides in a horizontal direction from the fixed portion of the panel.

horizontal engine *(Eng.)*. Any engine in which the cylinders are horizontal.

horizontal escapement *(Horol.)*. See **cylinder escapement**.

horizontal parallax *(Astron.,Surv.)*. The value of the geocentric parallax for a heavenly body in the solar system when the body is on the observer's horizon. In astronomy, the *equatorial horizontal parallax* of a planet or of the moon is the angle subtended at the centre of that body by the equatorial radius of the earth. See also **solar parallax**.

horizontal polarization *(Phys.)*. That of an electromagnetic wave when the alternating electric field is horizontal.

horizontal polarization *(Telecomm.)*. Transmission of radio waves in such a way that the electric lines of force are horizontal and the magnetic vertical; transmitting and receiving **dipoles** are mounted horizontally to handle signals polarized in this way. Cf. **vertical polarization**.

horizontal resolution *(Image Tech.)*. The number of separate picture elements which can be resolved along each horizontal scanning line in a television picture or facsimile reproduction.

horizontal sheeting *(Civ.Eng.)*. Long horizontal poling boards placed on each side of a trench excavated in bad ground and strutted apart. Short vertical walings are introduced when the trench has been sunk about 1 m.

horizontal stabilizer *(Aero.)*. Tail plane. See **stabilizer**.

hormone *(Biol.)*. A substance released by an *endocrine gland* into the bloodstream which carries it to remote sites in the body. There are many kinds with a wide range of activating and repressing functions; their synthetic derivatives are often used therapeutically.

hormone *(Bot.)*. Substance in plants which at low concentrations (say $< 10\,\mu M$) regulates some process in cells other than the ones in which it is made, e.g. abscisic acid, auxins, cytokinins, ethylene, gibberellins and, in some usages, their synthetic analogues. See also **growth substance**.

horn *(Acous.)*. Tube of continuously varying cross-section used in launching or receiving of radiation, e.g. acoustic horn, electromagnetic horn. Horns are best classified by their geometric or other shapes which include: *compound, conical, corner, exponential, folded, logarithmic, pyramidal, re-entrant, sectoral, tractrix*.

horn *(Eng.)*. Any projecting part, such as the two jaws of a horn-plate carrying an axle-box.

horn *(Geol.)*. A steep-sided mountain peak formed by the coalescence of three or more cirques.

horn *(Zool.)*. Keratin; one of the pointed or branched hard projections borne on the head in many Mammals; any conical or cylindrical projection of the head resembling a horn; in some Birds, a tuft of feathers on the head; in some *Gastropoda*, a tentacle; in some Fish, a spine. adjs. *horned, horny*.

horn antenna *(Telecomm.)*. A microwave antenna consisting of a flared-out section of a waveguide which may be rectangular, square or circular. Maximum radiation is along the central axis of a straight horn; other types may be curved, folded or bifurcated.

horn arrester *(Elec.Eng.)*. A lightning arrester consisting of a horn gap which arcs over on the occurrence of a lightning surge, but which rapidly extinguishes the arc on account of the special shape of the electrodes.

horn balance *(Aero.)*. An **aerodynamic balance** consisting of an extension forward of the hinge line at the tip of a control surface; it may be *shielded* (i.e. screened by a surface in front) or *unshielded*.

hornbeam *(For.)*. Tree of the genus *Carpinus*, yielding a moderately durable wood whose uses are confined to indoor conditions such as inlaying, marquetry and the making of small novelties. It can also be dyed black to simulate ebony.

hornblende *(Min.)*. Important members of the amphibole group of rock-forming minerals. They are of complex composition, essentially silicate of calcium, magnesium, and iron, with smaller amounts of sodium, potassium, hydroxyl, and fluorine; crystallizes in the monoclinic system; occurs as black crystals in many different types of igneous and metamorphic rocks, including hornblende-granite, syenite, diorite, andesite, etc., and hornblende-schist and amphibolite.

hornblende-gneiss *(Geol.)*. A coarse-grained metamorphic rock, containing hornblende as the dominant coloured constituent, together with feldspar and quartz, the texture being that typical of the gneisses. Differs from hornblende-schist in grain size and texture only.

hornblende-schist *(Geol.)*. A type of green schist, formed from basic igneous rocks by regional metamorphism, and consisting essentially of sodic plagioclase, hornblende, and sphene, frequently with magnetite and epidote. See also **glaucophane**.

hornblendite *(Geol.)*. An igneous rock composed almost entirely of hornblende.

horn-break fuse *(Elec.Eng.)*. A fuse fitted with arcing horns to assist in the rapid extinction of any arc which may be formed.

horn centre *(Eng.)*. A small transparent disk, originally of horn, used by draughtsmen to provide a substance into which the point of the compasses may be placed, in lieu of the paper beneath, at points which are to be much used as centres for describing arcs.

Horner's syndrome *(Med.)*. The combination of small pupil (*miiosis*), sunken eye (*erophthalmos*) and drooping of upper eyelid (*phosis*), due to paralysis of the sympathetic nerve in the region of the neck.

horn feed *(Radar)*. The feed to an antenna in the form of a horn.

hornfels *(Geol.)*. A fine-grained rock which has been partly or completely recrystallized by contact metamorphism.

horn gap *(Elec.Eng.)*. A spark gap of gradually increasing length, such that an arc struck across it gets longer and finally extinguishes itself.

horn gate *(Foundry)*. Horn-shaped ingates or sprues, radiating from the bottom of a runner, which supply several small moulds made in the same moulding box. See **ingate**.

horn lead *(Min.)*. Sometimes applied to the mineral **phosgenite**.

horn loudspeaker *(Acous.)*. A loudspeaker in which the radiating device is acoustically coupled to the air by means of a horn.

horns *(Build.)*. The ends of the head in a door or window frame when these project beyond the outer surfaces of the posts.

horn silver *(Min.)*. See **cerargyrite**.

hornwort *(Bot.)*. See **Anthocerotopsida**.

horology *(Genrl.)*. The science of time measurement, or of the construction of timepieces.

horripilation *(Med.)*. Erection of the hairs on the skin, giving rise to *goose flesh*.

horse *(Build.)*. (1) One of the **strings** supporting the treads and risers of a stair. (2) A trestle for supporting a board or timber while it is being sawn.

horsehead *(Min.Ext.)*. The curved part at the end of the arm of an oil well pump; it keeps the cable attached between the arm and the pump rods running at a fixed radius.

Horsehead Nebula *(Astron.)*. A famous dark nebula in the constellation Orion; it bears a plausible resemblance to the silhouette of a horse's head.

horse latitudes *(Meteor.)*. See **trade-winds**.

horse path *(Hyd.Eng.)*. A canal towing-path.

horse power *(Eng.)*. The mechanical engineering unit of power, equivalent to working at the rate of 33 000 ft lbf/min, or 42.41 B.t.u./min, or 745.70 watts. Abbrev. *h.p.* Also *cheval-vapeur*. Replaced in the current S.I. system of units by the watt.

horse pox *(Vet.)*. A contagious viral infection of equines. It is characterized by a papulo-vesicular eruption of the skin and mucous membranes.

horseradish peroxidase *(Biol.)*. Enzyme commonly conjugated with antibodies for use in *immunoassays*. Catalyses deposit of dye at the site of binding. Abbrev. *HRP*.

horseshoe curve *(Surv.)*. A curve whose arc subtends an angle of more than 180° at the centre, so that the **intersection point** lies on the same side of the curve as the centre.

horseshoe filament *(Elec.Eng.)*. An electric lamp filament in the shape of a single half-turn.

horseshoe kidney *(Med.)*. Congenital fusion of corresponding poles of two kidneys.

horseshoe magnet *(Elec.Eng.)*. Traditional form of an electro- or permanent magnet, as used in many instruments, e.g. meters, magnetrons.

horsing-up *(Build.)*. A term applied to the building-up of the mould used in running cornices etc.

horst *(Geol.)*. An elongated uplifted block bounded by faults along its length. Cf. **graben**.

horst faults *(Geol.)*. Two parallel normal faults heading outwards and throwing in opposite directions, the resulting structure being termed a *horst*.

hose *(Textiles)*. A tubular woven fabric made from flax or hemp for conveying liquids under pressure (e.g. a fire hose). May also be a fibre-reinforced plastic tube.

hose bit *(Build.)*. A type of **shell bit** with a projecting tip which withdraws the waste.

hose-proof *(Elec.Eng.)*. Said of a type of enclosure for electrical apparatus so constructed as to exclude water when the apparatus is washed down with a hose.

hosiery *(Textiles)*. Knitted articles for covering the feet and legs. Formerly used to include all types of knitted fabrics and garments.

hospital bus-bars *(Elec.Eng.)*. A set of bus-bars provided in a power station or substation for temporary or emergency purposes.

hospital switch *(Elec.Eng.)*. (1) A switch used on tramway or railway controllers to cut a faulty motor out of circuit. (2) Any switch for changing a circuit over to an emergency supply in case of failure of the main supply.

host *(Biol.)*. An organism which supports another organism (parasite) at its own expense, in molecular biology that in which a plasmid or virus can replicate.

host *(Phys.)*. Essential crystal, base material, or matrix of a luminescent material.

host range *(Biol.)*. The range of species or strains of bacteria which will support the replication of a plasmid or virus.

host rock *(Min.Ext.)*. See **country rock**.

hot *(Phys.)*. Charged to a dangerously high potential.

hot *(Radiol.)*. Colloquial reference to high levels of radioactivity; hence *hot laboratory*, a designated area for handling radioactive substances where extra precautions against irradiation of staff are taken.

hot-air engine *(Eng.)*. One in which the working fluid, air, is alternately heated and cooled by a furnace and regenerator. See **Stirling engine**.

hot-air gun *(Build.)*. An electric power tool which generates a stream of hot air which, when directed on a paint film, softens it sufficiently to facilitate its removal.

hot-air heater *(Build.)*. A heater which supplies warm air through gratings in the floor or openings in the walls.

hot-blast stoves *(Eng.)*. Large stoves, filled with a brick chequerwork, used for pre-heating the air blown into the blast-furnace. Also called *Cowper stoves*.

hot cathode *(Electronics)*. One in which the electrons are produced thermionically.

hot-cathode discharge lamp *(Phys.)*. A discharge lamp employing a heated cathode to increase its efficiency, improve the starting, and reduce the voltage drop across the tube.

hot-cathode rectifier *(Electronics)*. One with emitting cathode heated independently of the rectified current, e.g. a mercury thyratron.

hot crack *(Eng.)*. Crack formed during cooling by the stresses set up by the solidification of one or other of the components of an alloy or metal.

hot-die steels *(Eng.)*. Shock-resistant alloys used in high-temperature forging. In one type, from 3 to 10% chromium is used, and in another, 8 to 10% of tungsten, as main alloying constituents.

hot dip *(Eng.)*. Galvanizing of iron by immersion in molten zinc or spelter.

hot-drawn *(Eng.)*. Term describing metal wire, rod or tubing which has been produced by pulling it heated through a constricting orifice.

hot electron *(Electronics)*. An electron in excess of the thermal-equilibrium number and, for metals, having energy greater than the Fermi level. In semiconductors, the energy must be a definite amount above the edge of the conduction band. Hot electrons (or holes) can be generated by photo-excitation, tunnelling, minority-carrier injection or Schottky emission over a forward-biased p-n junction, or by abnormal electric fields in non-conductors.

hot-fluid injection *(Min.Ext.)*. Pumping steam, hot water or gas into the formation to increase flow of low specific gravity oil. Also *hot footing*.

hot galvanizing *(Eng.)*. The process of producing a corrosion-resistant zinc coating on articles by immersion in molten zinc or spelter, as distinct from *electroplating* with zinc.

hot-ground pulp *(Paper)*. Mechanically ground wood pulp in which the minimum of water is used, so allowing the temperature to rise by friction.

hot insulation mastics *(Build.)*. 'Breathing' mastics prepared by emulsifying bitumen in water and filling with asbestos floats, etc. Although the film formed is black and waterproof, it permits the passage of water vapour and allows the lagging around a hot surface to dry out.

HOTOL *(Space)*. Abbrev. for *Horizontal Take-Off and Landing*, a proposed single-stage-to-orbit launch vehicle burning oxygen and hydrogen propellants; a revolutionary engine uses atmospheric oxygen below Mach 5.

hot press *(Print.)*. See **blocking press**.

hot-pressed *(Paper)*. Originally paper finished by glazing between hot metal plates.

hot pressing *(Powder Tech.)*. Making a compact by application of heat as well as pressure. The combined effect is general softening and/or the liquefaction of a phase to allow sintering. Far greater amounts of pressure and heat would be required separately.

hot-press stamping *(Print.)*. See **blocking**.

hot shoe *(Image Tech.)*. Camera **accessory shoe** with electrical connections for flash sychronization.

hot-short, red-short *(Eng.)*. Said of metals that tend to be brittle at temperatures at which hot-working operations are performed.

hot spot *(Autos.)*. Part of the wall surface of the induction manifold of a petrol-engine on which the mixture impinges; heated by exhaust gases or coolant water to assist vaporization and distribution.

hot spot *(Biol.)*. Region of a chromosome peculiarly susceptible to mutation or recombination.

hot spot *(Electronics)*. A small region of an electrode in a valve (or CRO screen) which has a temperature above the average.

hot spot *(Image Tech.)*. Local area of excessive brightness in a projected image.

hot top *(Eng.)*. Feeder head, containing a reservoir of molten metal drawn on by a cooling ingot as it solidifies, thus avoiding porosity. See **ingot, ingot mould**.

hot well *(Eng.)*. The tank or pipes into which the condensate from a steam-engine or turbine condenser is pumped, and from which it is returned by the feed pump to the boiler.

hot-wire *(Elec.Eng.)*. Said of an electrical indicating instrument whose operation depends on the thermal expansion of, or change in resistance of a wire or strip when it carries a current. See **hot-wire ammeter**, etc.

hot-wire ammeter *(Elec.Eng.)*. An ammeter operating on the *hot-wire* principle; of use chiefly for very high frequencies.

hot-wire anemometer *(Meteor.)*. An instrument which measures wind velocities by using their cooling effect on a wire carrying an electric current, the resistance of the wire being used as an indication of the velocity.

hot-wire arc lamp *(Elec.Eng.)*. A form of clutch arc lamp in which the clutch controlling the movement of the carbons is operated by the expansion of a wire, according to the current passing through it.

hot-wire detector *(Telecomm.)*. A fine wire which is heated by the passage of high-frequency currents, producing a change in its d.c. resistance. See **bolometer**.

hot-wire microphone *(Acous.)*. One in which a d.c. signal through a hot wire is modulated by resistance variations consequent upon the cooling effect of an incident sound wave.

hot-wire oscillograph *(Elec.Eng.)*. An oscillograph in which a moving mirror is supported by an arrangement of wires carrying the current to be measured, and which is deflected as a result of the thermal expansion of these wires.

hot-wire voltmeter *(Elec.Eng.)*. A voltmeter operating on the *hot-wire* principle, the current heating the wire which expands and moves the needle.

hot-wire wattmeter *(Elec.Eng.)*. A wattmeter in which the deflection is indicated by means of a mirror mounted on an arrangement of fine wires carrying currents proportional to the voltage, and which is deflected as a result of the thermal expansion of the wires.

hot-working *(Eng.)*. The process of shaping metals by rolling, extrusion, forging etc., at elevated temperatures. The hot-working range varies from metal to metal, but it is, in general, a range in which recrystallization proceeds concurrently with the working, so that no strain-hardening occurs.

Houdry process *(Chem.)*. Catalytic cracking of petroleum, using activated aluminium hydrosilicate.

hour angle *(Astron.)*. The angle, generally measured in hours, minutes and seconds of time, which the hour circle of a heavenly body makes with the observer's meridian at the celestial pole; it is measured positively westwards from the meridian from 0 to 24 hr.

hour circle *(Astron.)*. (1) The great circle passing through the celestial poles and a heavenly body, cutting the celestial equator at 90°. (2) The graduated circle of an equatorial telescope which reads sidereal time and right ascension.

hour counter *(Elec.Eng.)*. See **time meter**.

hour-glass stomach *(Med.)*. Constriction of the middle part of the stomach, due either to spasm of stomach muscle or to the formation of scar tissue in connection with a gastric ulcer, the constriction in the latter case being permanent.

hour-meter *(Elec.Eng.)*. See **time-meter**.

hour rack *(Horol.)*. The toothed quadrant in a striking clock, one tooth of which is picked up by the gathering pallet for each hour struck.

hour wheel *(Horol.)*. The wheel in the motion work which carries the hour hand.

house corrections *(Print.)*. The correcting of mistakes made by the compositor when setting by hand or machine, as distinct from author's corrections made by him after checking the proofs.

housed joint *(Build.)*. A fitted joint, such as a tenon in its mortise.

housed string *(Build.)*. A string which has its upper and lower edges parallel to the slope of the stair, and houses, in grooves specially cut in the inner side, the ends of the steps. Also a *close string*.

housekeeping *(Comp.)*. Colloquialism for allocation decisions and record keeping within a computer system.

housekeeping data *(Space)*. A term used to denote information on the working of a space system, subsystem or component (as opposed to scientific data *per se*).

housekeeping gene *(Biol.)*. Term for genes specifying those functions common to all cells and their metabolism, in contrast to those needed only in differentiated cells.

housemaid's knee *(Med.)*. Inflammation of the bursa in front of the patella of the knee.

house mites *(Med.)*. Various species of arthropods, *Dermatophagoides*, thought to play an important allergenic role in house dust induced **asthma**.

house service meter *(Elec.Eng.)*. An integrating meter for measuring the electrical energy consumption of a domestic installation.

house style *(Print.)*. See **style of the house**.

housing *(Build.)*. A method of jointing two timbers in which the whole of the end of one is fitted into a corresponding blind mortise cut in the other.

housing *(Elec.Eng.)*. Containment of apparatus to prevent damage in handling or operation.

hovercraft *(Ships)*. A craft which can hover over or move across water or land surfaces while being held off the surfaces by a cushion of air. The cushion is produced either by pumping air into a plenum chamber under the craft or by ejecting air downwards and inwards through a peripheral ring of nozzles. Propulsion can be by tilting the craft, or by jet, or air propeller, or, over water, by water propeller, or, over land, by low-pressure tyres or tracks.

hovership *(Ships)*. A **hovercraft** intended for use over water only but capable of coming in over a beach to land.

Hovmüller diagram *(Meteor.)*. A diagram, with one axis representing time and the other longitude, on which are shown isopleths of an atmospheric variable such as pressure or *thickness*, usually averaged over a band of latitude. It demonstrates, very effectively, the movement over weeks or months of large-scale atmospheric features.

Howard protective system *(Elec.Eng.)*. A form of earth-leakage protection sometimes applied to a.c. machines and equipment. It consists of a current transformer connected between the frame of the machine and earth; if a current flows through the transformer a relay is operated, which opens the main circuit-breaker.

howieite *(Min.)*. A black triclinic hydrated silicate of sodium, manganese and iron.

howl *(Acous.)*. A high-pitched audio-tone due to unwanted acoustic (or electrical) feedback.

howl *(Aero.)*. See **screeching**.

howler *(Telecomm.)*. A device which uses acoustic feedback between a telephone transmitter and a telephone receiver to maintain a continuous oscillation and so provides suitable currents for testing telephonic apparatus.

Hoyt's metal *(Eng.)*. A tin base (91.5%) *white metal*, containing also antimony (3.4%), lead (0.25%), copper (4.3%) and nickel (0.55%).

h.p. *(Eng.)*. Abbrev. for *horsepower*.

h-parameters *(Electronics)*. See **transistor parameters**.

H-plane *(Phys.)*. The plane containing the magnetic vector **H** in electromagnetic waves, and containing the direction of maximum radiation. The electric vector is normal to it.

H-radar *(Radar)*. Navigation system in which an aircraft interrogates two ground stations for distance.

HRP *(Biol.)*. See **horseradish peroxidase**.

H-section *(Telecomm.)*. An electrical network derived from the T-section, in which half of each series arm is placed in the other leg of the circuit, making the section balanced.

HTO *(Chem.)*. Water in which an appreciable proportion of ordinary hydrogen is replaced by tritium.

HTR *(Nuc.Eng.)*. See **High-temperature reactor**.

HTU *(Chem.Eng.)*. Abbrev. for *Height of a Transfer Unit.*

H-type pole *(Elec.Eng.)*. A type of wooden support for overhead transmission lines, being two vertical poles centrally braced.

hub *(Surv.)*. See **change point**.

Hubble classification *(Astron.)*. A scheme for the classification of galaxies according to morphology, the principle types being **elliptical** and **spiral**.

Hubble constant *(Astron.)*. A measure of the rate at which the expansion of the universe varies with distance. It is expressed in the units km/sec per megaparsec, and has the value 50–100 km/sec/Mps. It is determined from the observed recession velocities or **redshifts** of distant **galaxies**.

Hubble diagram *(Astron.)*. A plot of redshifts of galaxies against their inferred distances.

Hubble law *(Astron.)*. Relationship first described by E. Hubble in 1925, which states that the recession velocity (or **redshift**) of a distant galaxy is directly proportional to the distance from the observer. The **Hubble constant** of proportionality is a measure of the expansion of the universe.

Hubble space telescope *(Space)*. A 2.5 m dia. optical telescope carried by satellite above the Earth's atmosphere; it will permit seeing up to 7 times deeper into space than is possible with Earth-bound instruments.

hübnerite, huebnerite *(Min.)*. Manganese tungstate, the end member of the wolframite group of minerals, the series from $MnWO_4$ to $FeWO_4$.

huckaback *(Textiles)*. A woven linen or cotton cloth with a rough surface, used for towels and glass-cloths. Different from *terry-towelling fabric* which is a pile fabric. See **terry fabric**.

HUD *(Aero.)*. Abbrev. for *Head-Up Display*.

hue *(Phys.)*. The perception of colour which discriminates different colours as a result of their wavelengths. Hue, chroma (degree of saturation) and value (brightness) specify a colour on the Munsell scale.

huebnerite *(Min.)*. See **hübnerite**.

hue sensibility *(Phys.)*. The ability of the eye to distinguish small differences of colour.

Huff separator *(Min.Ext.)*. High-tension or electrostatic separator used to concentrate small particles of dry ore.

hull *(Aero.)*. The main body (structural, flotation and cargo-carrying) of a flying-boat or boat amphibian.

hull *(Ships)*. A term used in its widest sense to signify the ship itself exclusive of masts, funnels and **top hamper**, but including **superstructure**. In a more restricted sense it means the shell of the ship. Also used to distinguish between ship and machinery.

hullite *(Min.)*. See **chlorophaeite**.

hum *(Acous.)*. Objectionable low-frequency components induced from power mains into sound reproduction, caused by inadequate smoothing of rectified power supplies, induction into transformers and chokes, unbalanced capacitances, or leakages from cathode heaters.

human immunodeficiency virus *(Immun.)*. The retrovirus responsible for **acquired immunodeficiency syndrome**, several types of which have been recognized. Abbrev. *HIV*, with a number added to denote the type.

hum bars *(Image Tech.)*. Horizontal band in the picture, often moving vertically, caused by AC interference, usually at mains frequency.

hum-bucking coil *(Elec.Eng.)*. See **bucking coil**.

humectant *(Chem.)*. Any hygroscopic substance, e.g. CuO, conc. H_2SO_4; applied to substances added to products to keep them moist, e.g. glycerol in tobacco.

humeral *(Zool.)*. In Vertebrates, pertaining to the region of the shoulder; in Insects, pertaining to the anterior basal angle of the wing; in *Chelonia*, one of the horny plates of the plastron.

humerus *(Zool.)*. The bone supporting the proximal region of the fore-limb in land Vertebrates. adj. *humeral*.

humic acids *(Chem.)*. Complex organic acids occurring in the soil and in bituminous substances formed by the decomposition of dead vegetable matter.

humicole, humicolous *(Bot.)*. Growing on soil or on humus.

humidifier *(Build.)*. An apparatus for maintaining desired humidity conditions in the air supplied to a building.

humidity *(Meteor.)*. See **absolute-**, **relative-**, **specific-**.

humidity mixing ratio *(Meteor.)*. The ratio of the mass of water vapour in a sample of moist air to the mass of dry air with which it is associated.

humification *(Bot.)*. The formation of humus during the decomposition of organic matter in soils.

humite *(Min.)*. An orthorhombic magnesium silicate, also containing magnesium hydroxide. Found in impure marbles. The humite group also includes chondrodite, clinohumite, and norbergite.

hummer screen *(Min.Ext.)*. Type used to grade smallish minerals, using a.c. to provide vibration by solenoid action.

hum note *(Acous.)*. The pitch of the note of the sound from a bell which persists after the strike note has died away.

humoral immunity *(Immun.)*. Specific immunity attributable to antibodies as opposed to cell-mediated immunity.

humoralism, humorism *(Med.)*. The doctrine that diseases arise from some change in the humours or fluids of the body.

humour, humor *(Zool.)*. A fluid; as the *aqueous humour* of the eye.

Humphrey gas pump *(Eng.)*. A large water-pump in which periodic gas explosions are made to act directly on an oscillating column of water, thereby effecting a pumping cycle; used in waterworks.

Humphreys spiral *(Min.Ext.)*. Spiral sluice which combines separation of mineral sands by simple gravitational drag with mild centrifugal action as the pulp cycles downward.

hump speed *(Aero.)*. The speed, on the water, at which the water resistance of the floats or boat body of a seaplane or flying-boat is a maximum. After this is past the craft begins to plane over the water.

humulene *(Chem.)*. $C_{15}H_{24}$, a sesquiterpene found in oil of hops.

humus *(Bot.)*. Amorphous, black or dark-brown material, a mixture of macromolecules based on benzene carboxylic and phenolic acids, often complexed with clays, which results from the decomposition of organic matter in soils. See also **mull, mor, moder**.

humus plant *(Bot.)*. A flowering plant, often with little chlorophyll, depending for much of its nutrition on a mycorrhizal association with a fungus growing in a rich humus.

Hund rule *(Chem.)*. Transition metals which have incomplete inner electron shells have large magnetic moments, as the electron spins in the incomplete shell are self-aligning.

Hund's rule *(Chem.)*. For electrons of otherwise equal energy, spins are aligned parallel as much as possible. This results in the strongly paramagnetic properties of many transition metal compounds.

Hungarian cat's-eye *(Min.)*. An inferior greenish cat's-eye obtained in the Fichtelgebirge in Bavaria. No such stone occurs in Hungary.

hung sash *(Build.)*. See **hanging sash**.

hunting *(Aero.)*. (1) An uncontrolled oscillation, of approximately constant amplitude, about the flight path of an aircraft. (2) The angular oscillation of a rotorcraft's blade about its drag hinge. (3) The oscillation of instrument needles.

hunting *(Autos.)*. Irregular running of a gasoline engine at idling speed, usually caused by over-lean mixture.

hunting *(Eng.)*. The tendency of rotating mechanisms which should run at constant speed to pulsate above and below that speed, due to shortcomings in the governor or other speed control systems. Also *cycling, oscillation.*

hunting *(Telecomm.)*. The operation of a selector, or other similar device, to establish connection with an idle circuit of a group.

Huntington mill *(Min.Ext.)*. Wet-grinding mill in which 4 cylindrical mullers, hung inside a steel tub, bear outward as they rotate, thus grinding the passing ore.

Huntington's chorea *(Med.)*. Hereditary **chorea** occurring in adults; associated with progressive mental deterioration and involuntary movements.

hunting tooth *(Eng.)*. An extra tooth added to a gear-wheel in order that its teeth shall not be an integral multiple of those in the pinion.

Huronian *(Geol.)*. A major division of the Proterozoic of the Canadian Shield, typically exposed on the northern shores of Lake Huron.

hurricane *(Meteor.)*. (1) A wind of force 12 on the *Beaufort Scale.* A mean wind speed of 75 mph (120 km/h). (2) A *tropical revolving storm* in the North Atlantic, eastern North Pacific and the western South Pacific.

hurricane deck *(Ships)*. A term, not normally in use, for a superstructure deck. Sometimes termed *flying deck.* It is independent of the ship from the point of view of strength.

hurter *(Build.)*. A cast-iron, timber, stone, or concrete block, which is so placed as to protect a quoin from damage from passing vehicles.

Hurter and Driffield (H and D) curve *(Image Tech.)*. See **characteristic curve**.

hushing, hush *(Min.Ext.)*. A washing away of the surface soil to lay bare the rock formation for prospecting.

husk *(Vet.)*. *Hoose, parasitic bronchitis, lungworm infection, dictyocauliasis.* Infection of cattle with *Dictyocaulus viviparus.* Affects young stock. Adult worms infest the bronchi and bronchioles resulting in varying degrees of respiratory disease. Vaccine available.

hutch *(Min.Ext.)*. (1) A small train or wagon. (2) A basket for coal. (3) A compartment of a jig used for washing ores. The concentrate which passes through a jig screen is called the *hutchwork.*

Hutchinson's teeth *(Med.)*. Narrowing and notching of the permanent incisor teeth, occurring in congenital syphilis.

Huygens' eyepiece *(Phys.)*. A combination of two plano-convex lenses placed with their plane sides towards the observer, at a distance apart equal to half the sum of their focal lengths, which are in the ratio of three to one, the shorter focus lens being nearer the observer. Huygens' eyepiece is often used in microscopes, but is not suited for use with cross-wires or an eyepiece scale.

Huygens' principle *(Phys.)*. The assumption that every element of a wavefront acts as a source of so-called secondary waves. The principle can be applied to the propagation of any wave motion, but it is more frequently used in optics than in acoustics.

H-wave *(Telecomm.)*. See **TE-wave**.

hyacinth, jacinth *(Min.)*. The reddish-brown variety of transparent zircon, used as a gemstone. The name has also been used for a brownish grossular from Ceylon.

hyades *(Astron.)*. The name of a star cluster, of the 'open' type, situated in the constellation Taurus; visible to the naked eye.

hyal-, hyalo- *(Genrl.)*. Prefix from Gk. *hyalos*, clearstone, glass.

hyaline *(Bot.,Zool.)*. Translucent and colourless; without fibres or granules, e.g., *hyaline cartilage.*

hyaline membrane disease *(Med.)*. One of the causes of the *respiratory distress* syndrome of the newborn where **surfactant** has not yet developed sufficently to maintain normal *alveolar* opening. Cyanosis and difficulty in breathing result.

hyalite *(Min.)*. A colourless transparent variety of *opal*, occurring as globular concretions and crusts. Also called *Müller's glass.*

hyaloid *(Zool.)*. Clear, transparent; as the *hyaloid membrane* of the eye which envelops the vitreous humour.

hyalophane *(Min.)*. Feldspar containing barium, with up to 30% $BaAl_2Si_2O_8$.

hyalopilitic texture *(Geol.)*. A texture of andesitic volcanic rocks in which the groundmass consists of small microlites of feldspar embedded in glass.

hyaloplasm *(Bot.)*. (1) The ground substance or matrix of the cytoplasm, between the organelles. (2) Cytoplasm.

Hyatt roller bearing *(Eng.)*. See **flexible roller bearing**.

hybrid *(Biol.)*. Offspring of a *cross* between two different strains, varieties, races or species. (n. or adj.)

hybrid-π *(Electronics)*. The most complex **equivalent circuit** used for analysis of the characteristics of transistors operating at high frequencies on the common-emitter configuration.

hybrid antibodies *(Immun.)*. Antibody molecules in which the two antigen combining sites are of different specificities. This can be achieved by recombining half molecules of two specific antibodies in vitro, or may also occur when cells from two different antibody producing cell lines are fused to make a **hybridoma**. A distinct kind of hybrid antibody is made from the variable region of a monoclonal antibody from one species joined to the constant region of immunoglobulin of another species. This can be achieved by insertion of hybrid genes into a myeloma cell line, with the aim of producing antibodies with the general attributes of human Ig but with a desired antigen binding capacity borrowed from a mouse hybridoma.

hybrid coil, transformer *(Telecomm.)*. In telephone circuits a coil, comprising four equal windings and an additional winding, used for the separation of incoming and outgoing currents in a two-wire repeater.

hybrid computer *(Comp.)*. One that combines a digital processor with a number of analogue units.

hybrid electromagnetic wave *(Phys.)*. One having longitudinal components of both the electric and magnetic field vectors. Abbrev. *HEM.*

hybrid integrated circuit *(Electronics)*. A complete circuit formed by combining different types of individual integrated circuits and, in some cases, a number of discrete components.

hybridization *(Biol.)*. (1) The formation of a new organism by normal sexual processes or more recently by protoplast fusion. (2) When DNA or RNA molecules from two different sources are annealed or renatured together, they will form hybrid molecules, whose stability is some measure of their sequence relatedness.

hybrid junction *(Telecomm.)*. A waveguide transducer which is connected to four branches of a circuit and designed so that these branches conjugate in pairs.

hybridoma *(Immun.)*. (1) B-cell hybridoma. A cell line obtained by the fusion of a myeloma cell line, which is able to grow indefinitely in culture, with a normal antibody secreting B-cell. The resulting cell line has the properties of both partners, and continues to secrete the antibody product of the normal B-cell. By choosing a myeloma which has ceased to make its own immunoglobulin product but has retained the machinery for doing this, the hybridoma secretes only the normal B-cell antibody. Since the cell line is cloned the antibody is monoclonal. (2) T-cell hybridoma. A cell line obtained by fusion of a T-lymphoma cell line with a normal T-lymphocyte. The resulting hybridoma may retain the properties of the lymphoma but acquires the antigen recognition specificity of the normal T-cell. Such hybridomas provide a source of homogeneous T-cell antigen receptors, and are also useful for recognizing specific antigens which stimulate T-cells such as histocompatibility antigens.

hybrid ring *(Telecomm.)*. A microwave junction, realized in waveguide, co-axial or stripline form, which has the properties of a **hybrid junction**. Also *rat race.*

hybrid rocks *(Geol.)*. Rocks which originate by interaction between a body of magma and its wall-rock or roof-rock, which may be another igneous, or sedimentary, or metamorphic rock.

hybrid set *(Telecomm.)*. In telephony, two or more coils or transformers connected together to form a hybrid junction. See **hybrid**.

hybrid sterility *(Bot.)*. The lack of fertility of some interspecific hybrids which results from lack of homology between chromosome sets and thus abnormal segregation, which may be overcome by chromosome doubling. See **amphidiploid**.

hybrid tee *(Telecomm.)*. Four-way waveguide junction arranged so that the power entering one port always divides into the two adjacent ports, but never reaches an opposite port. Sometimes used for a waveguide **balanced mixer**.

hybrid vigour *(Biol.)*. The increase in desirable qualities such as growth rate, yield, fertility, often exhibited by *hybrids*; i.e. favourable *heterosis*.

hydathode *(Bot.)*. Structure through which water is exuded from uninjured plants.

hydatid cyst *(Zool.)*. A large sac or vesicle containing a clear watery fluid and encysted immature larval *Cestoda*.

hydatidiform mole *(Med.)*. An affection of the chorionic villi (vascular tufts of the foetal part of the placenta) whereby they become greatly enlarged, resembling a bunch of grapes.

hydragogue *(Med.)*. Having the property of removing water; a purgative drug which produces watery evacuations.

hydralazine *(Med.)*. Vasodilator drug used to treat hypertension.

hydramnios *(Med.)*. Excess of fluid in the amniotic sac of the foetus.

hydranth *(Zool.)*. In *Hydrozoa*, a nutritive polyp of a hydroid colony.

hydrargillite *(Min.)*. See **gibbsite**.

hydrargyrism *(Med.)*. The state of being poisoned by mercury and its compounds.

hydrarthrosis *(Med.)*. Swelling of the joint, due to the accumulation in it of clear fluid.

hydrated electron *(Chem.)*. Very reactive free electron released in aqueous solutions by the action of ionizing radiation.

hydrated ion *(Chem.)*. Ion surrounded by molecules of water which it holds in a degree of orientation. Hydronium ion. Solvated H-ion of formula $[H_2O \rightarrow H]^+$ or H_3O^+.

hydrate of lime, hydrated lime *(Build.)*. See **caustic lime**.

hydrates *(Chem.)*. Salts which contain water of crystallization. See also **water of hydration**.

hydration *(Geol.)*. The addition of water to anhydrous minerals, the water being of atmospheric or magmatic origin. Thus anhydrite, by hydration, is converted into gypsum, and feldspars into zeolites.

hydration *(Paper)*. The result in the fibre of prolonged or heavy beating/refining whereby the stuff is said to be 'wet' i.e. water does not drain very readily from it on the machine wire.

hydraulic *(Genrl.)*. See **hydraulics**.

hydraulic accumulator *(Aero.,Eng.)*. A weight-loaded or pneumatic device for storing liquid at constant pressure, to steady the pump load in a system in which the demand is intermittent. In aircraft hydraulic systems an accumulator also provides fluid under pressure for operating components in an emergency, e.g. failure of an engine-driven pump.

hydraulic air compressor *(Min.Ext.)*. Arrangements in which water falling to the bottom of a shaft entrains air which is released in a tunnel at depth, while the water rises to a lower discharge level.

hydraulic amplifier *(Eng.)*. A power amplifier employed in some servomechanisms and control system, in which power amplification is obtained by the control of the flow of a high-pressure liquid by a valve mechanism.

hydraulic blasting *(Min.Ext.)*. In fiery mines, rock breaking by means of a ram-operated device acting on a **hydraulic cartridge**.

hydraulic brake *(Eng.)*. (1) An absorption dynamometer. See **Froude brake**. (2) A motor-vehicle brake applied by small pistons operated by oil under pressure supplied from a pedal-operated master cylinder.

hydraulic capacity *(Bot.)*. The amount of water held by a soil between **field capacity** and the **permanent wilting point**, a measure of available water.

hydraulic cartridge *(Civ.Eng.)*. An apparatus for splitting rock, mass concrete etc.; it consists of a long cylindrical body which has numerous pistons projecting from one side and moving in a direction at right angles to the body (under hydraulic pressure from within the body), which is placed in a hole drilled to take it.

hydraulic cement *(Build.,Civ.Eng.)*. A cement which will harden under water.

hydraulic classifier *(Min.Ext.)*. Device in which vertically flowing column of water is used to carry up and on light and small mineral particles while heavy and large ones sink.

hydraulic control *(Min.Ext.)*. Use of fluid in sensing mechanism to actuate a signalling or a correcting device, in response to pressure changes.

hydraulic coupling *(Eng.)*. A traction coupling used for automobiles or for diesel engines and electric motors, in which each half-coupling contains a number of radial vanes and rotation causes a vortex in the hydraulic fluid between them, power being dissipated by the slip.

hydraulic cyclone elutriator *(Powder Tech.)*. *Hydrocyclone* fitted with an apex container which serves as a return flow device for recycling the fine particles in suspension to give very efficient fractionation.

hydraulic engineering *(Eng.)*. That branch of engineering chiefly concerned in the design and production of hydraulic machinery, pumping plants, pipe-lines etc.

hydraulic fill *(Eng.)*. An embankment or other fill in which the materials are deposited in place by a flowing stream of water, gravity and velocity control being used to bring about selected deposition.

hydraulic fracturing *(Min.Ext.)*. Method of increasing oil flow from less permeable strata by forcing liquid into it under very high pressure. See **fracking**.

hydraulic glue *(Chem.)*. A glue which is able partially to resist the action of moisture.

hydraulic intensifier *(Eng.)*. A device for obtaining a supply of high-pressure liquid from a larger flow at a lower pressure.

hydraulicity *(Build.,Civ.Eng.)*. The property of a lime, cement, or mortar which enables it to set under water or in situations where access of air is not possible.

hydraulic jack *(Eng.)*. A **jack** in which the lifting head is carried on a plunger working in a cylinder, to which oil or water is supplied under pressure from a pump.

hydraulic leather *(Genrl.)*. A flexible leather prepared by being heavily treated, after tanning, with cod oil and then stoved; while hot, it may be shaped to requirements.

hydraulic lift *(Eng.)*. A lift or elevator operated either directly by a long vertical ram, working in a cylinder to which liquid is admitted under pressure, or by a shorter ram through ropes. See **jigger**.

hydraulic mining, hydraulicking *(Min.Ext.)*. The operation of breaking down and working a bank of gravel, alluvium, poorly consolidated or decomposed bed-rock by high-pressure water jets as in mining gold, tin *placers* or china clay deposits.

hydraulic mortar *(Build.,Civ.Eng.)*. A mortar which will harden under water.

hydraulic motor *(Eng.)*. A multicylinder reciprocating machine, generally of radial type, driven by water or oil.

hydraulic packing *(Eng.)*. L- or U-section rings providing a self-tightening packing under fluid pressure; used on rams and piston-rods of hydraulic machines. See **U-leather**.

hydraulic press *(Eng.)*. An upstroke, downstroke or horizontal press, with one or more rams, working at approximately constant pressure for deep drawing and extruding operations or at progressively increasing pressure for baling, plastics moulding etc.

hydraulic ram *(Eng.)*. (1) The plunger of a hydraulic press. (2) A device whereby the pressure head produced when a moving column of water is brought to rest is caused to deliver some of the water under pressure.

hydraulic reservoir *(Aero.)*. In an aircraft hydraulic system, the header tank which holds the fluid; not to be confused with the *accumulator*, which is a pressure vessel wherein hydraulic energy is stored.

hydraulic riveter *(Eng.)*. A machine for closing rivets by hydraulic power.

hydraulics *(Genrl.)*. The science relating to the flow of fluids. adj. *hydraulic*.

hydraulic squeezer *(Foundry)*. See **squeezer**.

hydraulic stowing *(Min.Ext.)*. The filling of worked-out portions of a mine with water-borne waste material. The water drains off and is pumped to surface.

hydraulic test *(Eng.)*. A test for pressure-tightness and strength applied to pressure vessels, pipe lines, instruments, etc.

hydraulic torque converter *(Eng.)*. A variable speed mechanism similar to a hydraulic coupling, in which a decrease in secondary speed is accompanied by an increase in torque.

hydrazides *(Chem.)*. The mono-acyl derivatives of hydrazine.

hydrazine *(Chem.)*. A fuming strongly basic liquid, $H_2N.NH_2$, bp 113°C. A powerful reducing agent, it (also its derivatives, in particular **dimethyl hydrazine**) is used as a high-energy propellant in rockets.

hydrazine hydrate *(Chem.)*. $N_2H_4.H_2O$. Diacid base. Attacks glass, rubber, and cork. Used in some liquid rocket fuels and as an oxygen scavenger in boiler water treatment. Derivatives are widely used as blowing agents in the production of expanded plastics and rubbers.

hydrazo compounds *(Chem.)*. Symmetric derivatives of hydrazine, colourless, crystalline, neutral substances, obtained by the reduction of azo compounds.

hydrazoic acid *(Chem.)*. N_3H, $HN^-{-}N^+{\equiv}N$. The aqueous solution is a strong monobasic acid and forms azides with many common metals. Used in Schmidt reaction for converting aromatic carboxylic acids into primary amines.

hydrazones *(Chem.)*. The condensation products of aldehydes and ketones with hydrazine, water being eliminated from the two molecules.

hydr-, hydro- *(Genrl.)*. Prefix from Gk. *hydōr*, gen. *hydatos*, water.

hydrides *(Chem.)*. Compounds formed by the union of hydrogen with other elements. Those of the nonmetals are generally molecular liquids or gases, certain of which dissolve in water (oxygen hydride) to form acid (e.g. hydrogen chloride) or alkaline (e.g. ammonia) solutions. The alkali and alkaline earth hydrides are crystalline, saltlike compounds, in which hydrogen behaves as the electronegative element. They contain H^--ions and, when electrolysed, give hydrogen at the anode. Transition elements give alloy or interstitial hybrids.

hydriodic acid *(Chem.)*. HI. An aqueous solution of hydrogen iodide. Forms salts called *iodides*. Easily oxidized.

hydro *(Genrl.)*. Prefix. See **hydr-**.

hydroa *(Med.)*. A skin disease in which groups of vesicles appear on reddened patches in the skin, associated with intense itching.

hydro-acoustics *(Acous.)*. Branch of acoustics concerned with radiation, propagation measurement etc. of sound in water.

hydroborons *(Chem.)*. See **boranes**.

hydrobromic acid *(Chem.)*. HBr. An aqueous solution of hydrogen bromide.

hydrocarbons *(Chem.)*. A general term for organic compounds which contain only carbon and hydrogen. They are divided into saturated and unsaturated hydrocarbons, aliphatic (alkane or fatty) and aromatic (benzene) hydrocarbons.

hydrocarbons *(Geol.)*. See **native-**.

hydrocele *(Med.)*. A swelling in the scrotum due to an effusion of fluid into the sac (*tunica vaginalis*) which invests the testis.

hydrocelluloses *(Chem.)*. Products obtained from cellulose by treatment with cold concentrated acids. They still retain the fibrous structure of cellulose, but are less hygroscopic.

hydrocephalus *(Med.)*. An abnormal accumulation of cerebrospinal fluid in the cavities (ventricles) of the brain, distending them and stretching and thinning the brain tissue over them.

hydrocerussite *(Min.)*. A colourless hydrated basic carbonate of lead occurring as an encrustation on native lead, galena, cerussite, and other lead minerals.

hydrochloric acid *(Chem.)*. HCl. An aqueous solution of hydrogen chloride gas. Dissolves many metals, forming chlorides and liberating hydrogen. Used extensively in industry for numerous purposes, e.g. for the manufacture of chlorine, pickling, tinning, soldering etc.

hydrocoel *(Zool.)*. In *Echinodermata*, the **water-vascular system**.

hydrocyanic acid *(Chem.)*. An aqueous solution of **hydrogen cyanide**. Dilute solution called *prussic acid*. Monobasic. Forms cyanides. Very poisonous.

hydrocyclone *(Min.Ext.)*. A small cyclone extractor for removing suspended matter from a flowing liquid by means of the centrifugal forces set up when the liquid is made to flow through a tight conical vortex. Used to separate solids in mineral pulp into coarse and fine fractions. Fluent stream enters tangentially to cylindrical section and coarser sands gravitate down steep-sided conical section to controlled apical discharge. Bulk of pulp, containing finer particles, overflows through a pipe inserted in the central vortex. Classification into fractions is aided by the centrifugal force with which the pulp is delivered to the appliance, which can handle large tonnages.

hydrodynamic lubrication *(Eng.)*. Thick-film lubrication in which the relatively moving surfaces are separated by a substantial distance and the load is supported by the hydrodynamic film pressure.

hydrodynamic power transmission *(Eng.)*. A power-transmission system which, in general, employs a hydraulic coupling or a torque converter or a reaction coupling.

hydrodynamic process *(Eng.)*. A process for shallow forming and embossing operations, in which high-pressure water presses the blank against a female die, there being no solid punch to conform to the die contour. Similar processes involve explosive charges.

hydrodynamics *(Eng.)*. That branch of dynamics which studies the motion produced in fluids by applied forces.

hydroelectric generating (power) station *(Elec.Eng.)*. An electric generating station in which the generators are driven by water turbines.

hydroelectric generating set *(Elec.Eng.)*. An electric generator driven by a water turbine.

hydrofining *(Chem.)*. See **hydroforming**.

hydrofluoric acid *(Chem.)*. Aqueous solution of hydrogen fluoride. Dissolves many metals, with evolution of hydrogen. Etches glass owing to combination with the silica of the glass to form silicon fluoride, hence it is stored in, e.g. polythene or gutta-percha vessels. See **etching test**.

hydrofoil *(Aero.)*. An immersed aerofoil-like surface to facilitate the take-off of a seaplane by increasing the hydrodynamic lift.

hydrofoil *(Ships)*. A fast, light craft fitted with winglike structures (foils) on struts under the hull. These may be extendable and adjustable. Propelled by propeller in water or air, or by jet. Foils may act entirely or partly submerged with hull lifted clear of the water at speed. Steered by water rudder. Cf. **hydroplane**.

hydroforming *(Chem.)*. The catalytic refining or reforming of petroleum oils by hydrogen at high temperatures and pressures; also called **hydrofining**.

hydroforming *(Eng.)*. A hydraulic forming process in which the shape is produced by forcing the material by means of a punch against a flexible bag partly filled with hydraulic fluid and acting as a die.

hydrofuge *(Zool.)*. Water-repelling; said of certain hairs possessed by some aquatic insects and used for retaining a film of air.

hydrogel *(Chem.)*. A gel the liquid constituent of which is water.

hydrogen *(Chem.)*. The least dense element, forming diatomic molecules H_2. Symbol H, at. no. 1, r.a.m. 1.007 97, valency 1. It is a colourless, odourless, diatomic gas, water being formed when it is burnt; mp −259.14°C, bp −252.7°C, density 0.089 88 g/dm^3 at stp. It is widely distributed as water, occurs in many minerals, e.g. petroleum, and in living matter. Hydrogen is manufactured chiefly as a by-product of electrolysing caustic soda, water gas and gas cracking. It is used in the Haber process for the fixation of nitrogen, and in the hardening of fats (e.g. in the manufacture of margarine); also for hydrogenation of oils, manufacture of hydrochloric acid, filling small balloons, and as a reducing agent for organic synthesis and metallurgy, oxyhydrogen and atomic hydrogen welding flames. It is of great importance in the moderation (slowing down) of neutrons as hydrogen atoms are the only ones of similar mass to a neutron and are therefore capable of absorbing an appreciable proportion of the neutron energy on collision. Isotopes, with 1 proton and 1 electron are:

Name	Atomic symbol	Relative atomic mass	Atomic number	Neutrons
protium	^{1}H	1·007825	1	0
deuterium	D,^{2}H	2·01410	2	1
tritium	T,^{3}H	3·0221	3	2

hydrogenation *(Chem.)*. Chemical reactions involving addition of hydrogen, present as a gas, to a substance, in the presence of a catalyst. Important processes are: the hydrogenation of coal; the hydrogenation of fats and oils; the hydrogenation of naphthalene and other substances. See **Fischer-Tropsch process**.

hydrogen bacteria *(Biol.)*. Chemosynthetic bacteria which obtain the energy required for carbon dioxide assimilation from the oxidation of hydrogen to water. Organic compounds may also be oxidized by these species under suitable conditions.

hydrogen bomb *(Phys.)*. Uses the nuclear fusion process to release vast amounts of energy. Extremely high temperatures are required for the process to occur and these temperatures are obtained by an *atomic bomb* around which the fusion material is arranged. Lithium deuteride can initiate a number of fusion processes involving the hydrogen isotopes, deuterium and tritium. These reactions also produce high-energy neutrons capable of causing *fission* in a surrounding layer of the most abundant isotope of uranium, ^{238}U so that further energy is released.

hydrogen bond *(Chem.)*. A strong inter- or intramolecular force resulting from the interaction of a hydrogen atom bonded to an electronegative atom with a lone-pair of electrons on another electronegative atom. This usually results in the two electronegative atoms being well within the sum of their van der Waals radii of one another. Hydrogen bonding is important in associated liquids, particularly water, and is responsible for much of the tertiary structure of proteins see, e.g. **alpha helix**.

hydrogen-bonding *(Biol.)*. Hydrogen atoms in the groups -O-H and -N-H can form non-covalent bonds with a nitrogen or oxygen atom. The bonding is ionic and weak, but hydrogen bonding between purines and pyrimidines contribute to the stability of the DNA double helix. The planar stacking of the bases and cooperation between adjacent hydrogen bonds making for a very stable and rigid structure.

hydrogen bromide *(Chem.)*. HBr. Hydrogen bromide gas can be made by direct combination of the two elements, particularly in the presence of a catalyst. Closely resembles hydrogen chloride, but is less stable.

hydrogen chloride *(Chem.)*. HCl. A colourless gas which dissolves in water to form hydrochloric acid. Produced by the action of concentrated sulphuric acid on chlorides, on the industrial scale as a by-product of the chlorination of hydrocarbons, and directly from the elements H_2 and Cl_2, both of which are formed as a by-product of caustic soda manufacture. Uses: to produce hydrochloric acid; chlorination of unsaturated organic compounds, e.g. chloroethene in polymerization, isomerization and alkylation, as a catalyst.

hydrogen cooling *(Elec.Eng.)*. A method of cooling rotating electric machines; the machine is totally enclosed and runs in an atmosphere of hydrogen.

hydrogen cyanide *(Chem.)*. HCN. Highly poisonous liquid, mp −13.4°C, bp 26°C, permittivity 95, dissociation constant 1.3×10^{-9} at 18°C. Dissolves in water to form hydrocyanic or prussic acid. Faint odour of bitter almonds. Uses: fumigant; medicine; organic synthesis, e.g. acrylic resins; analytical reagent.

hydrogen electrode *(Chem.)*. For pH measurement, a platinum-black electrode covered with hydrogen bubbles. Although rarely used in practice, it defines the **hydrogen scale** of electrode potentials. Also *Hildebrand electrode*.

hydrogen embrittlement *(Eng.)*. Effect produced on metal by sorption of hydrogen during pickling or plating operations.

hydrogen fluoride *(Chem.)*. HF. A liquid which fumes strongly in air. Dissolves in water to form hydrofluoric acid. Produced by the action of sulphuric acid on fluorides. Uses: catalyst in organic reactions; preparation of uranium; fluorides and hydrofluoric acid.

hydrogen I, II *(Astron.)*. The hydrogen of interstellar space, known in two clearly defined states: H I (*neutral hydrogen*) and H II (*ionized hydrogen*). The H I regions, confined mainly to the spiral arms of the galaxy, emit no visible light, and are detected solely by their emission of the 21-cm radio line. The H II regions, found in the gaseous nebulae, emit both visible and radio radiations.

hydrogen iodide *(Chem.)*. HI. A heavy colourless gas, formed by the direct combination of hydrogen and iodine; fumes strongly in air. Usually made by the decomposition of phosphorus (III) iodide by the action of water. Mp −50°C, bp −35°C. See **hydriodic acid**.

hydrogen ion *(Chem.)*. An atom of hydrogen carrying a positive charge, i.e. a proton; in aqueous solution, hydrogen ions are hydrated, H_3O^+.

hydrogen ion concentration *(Chem.)*. See **pH value**.

hydrogenous *(Phys.)*. Said of a substance rich in hydrogen and therefore suitable for use as a moderator of neurons in a nuclear reactor.

hydrogen oxide *(Chem.)*. Water.

hydrogen peroxide *(Chem.)*. H_2O_2. A viscous liquid with strong oxidizing properties. Powerful bleaching agent; and, as its decomposition products are water and oxygen, it is much used as a disinfectant. The strength of an aqueous solution is represented commercially by the number of volumes of oxygen which 100 cm^3 of the solution will give on decomposition. Recent interest in high concentration (ca. 90%) hydrogen peroxide has centred on its use as both a technical and a military oxidizing agent, especially in rocket fuels as an *oxidant*. Prepared via the anodic oxidation of hydrogen sulphates to peroxydisulphate followed by hydrolysis and steam distillation.

hydrogen phosphide *(Chem.)*. PH_3. See **phosphine** (1).

hydrogen scale *(Chem.)*. A system of relative values of electrode potentials, based on that for hydrogen gas, at standard pressure, against hydrogen ions at unit **activity (2)**, as zero. See **hydrogen electrode**.

hydrogen sulphide *(Chem.)*. H_2S. May be prepared by direct combination of the 2 elements or by the action of dilute hydrochloric or sulphuric acid on iron sulphide. It is readily decomposed. Reacts with bases forming sulphides and with some metals to produce metal sulphides and liberate hydrogen. Poisonous, with a characteristic smell of rotten eggs.

hydrogen thyratron *(Electronics)*. A *thyratron* containing hydrogen, used in radar transmitters to provide high peak currents at high voltage with very rapid switch-on and recovery time.

hydrogeology *(Geol.)*. The study of the geological aspects of the Earth's water.

hydrographical surveying *(Surv.)*. A branch of surveying dealing with bodies of water at the coastline and in harbours, estuaries and rivers.

hydrography *(Genrl.)*. The study, determination, and publication of the conditions of seas, rivers and lakes, viz., surveying and charting of coasts, rivers, estuaries and harbours, and supplying particulars of depth, bottom, tides, currents etc.

hydrogrossular *(Min.)*. A hydrated variety of garnet, close to grossular in composition but with some of the silica replaced by water. A massive variety from South Africa is sometimes called 'Transvaal Jade'.

hydrohematite, turgite *(Min.)*. $Fe_2O_3.nH_2O$. Probably a mixture of the two minerals *hematite* and *goethite*, the former being in excess. It is fibrous and red in the mass, with an orange tint when powdered.

hydroid *(Bot.)*. Water-conducting cell in the stems of some moss gametophytes, e.g. *Polytrichum*.

hydroid *(Zool.)*. An individual of the asexual stage in Coelenterata which show alternation of generations.

hydrolapse *(Meteor.)*. The rate of decrease with height of atmospheric water vapour as measured by humidity mixing ratio, dew point or other suitable quantity.

Hydrolastic *(Autos.)*. Said of a proprietary type of suspension in which front and rear suspensions are interconnected and compensated hydraulically.

hydrological cycle *(Meteor.)*. The evaporation and condensation of water on a world scale.

hydrology *(Genrl.)*. The study of water, including rain, snow and water on the earth's surface, covering its properties, distribution, utilization, etc.

hydrolysis *(Chem.)*. (1) The formation of an acid and a base from a salt by interaction with water; it is caused by the ionic dissociation of water. (2) The decomposition of organic compounds by interaction with water, either in the cold or on heating, alone or in the presence of acids or alkalis; e.g. esters form alcohols and acids; oligo- and polysaccharides on boiling with dilute acids yield monosaccharides.

hydromagnesite *(Min.)*. Hydrated magnesium hydroxide and carbonate, occurring as whitish amorphous masses, or rarely as monoclinic crystals in serpentines.

hydromagnetic *(Elec.Eng.)*. Pertaining to the behaviour of a plasma in a magnetic field. See also **magnetohydrodynamics**.

Hydromedusae *(Zool.)*. See **Hydrozoa**.

hydromelia *(Med.)*. Dilatation of the central canal of the spinal cord.

hydrometallurgy *(Min.Ext.)*. Extraction of metals or their salts from crude or partly concentrated ores by means of aqueous chemical solutions; also electrochemical treatment including electrolysis or ion exchange.

hydrometeor *(Meteor.)*. A generic term for all products of the condensation or sublimation of atmospheric water vapour, including: ensembles of falling particles which may either reach the Earth's surface (rain and snow) or evaporate during their fall (virga); ensembles of particles suspended in the air (cloud and fog); particles lifted from the Earth's surface (blowing or drifting snow and spray); particles deposited on the ground or on exposed objects (dew, hoar frost etc.).

hydrometer *(Phys.)*. An instrument by which the relative density of a liquid may be determined by measuring the length of the stem of the hydrometer immersed, when it floats in the liquid with its stem vertical.

hydronephrosis *(Med.)*. Distension of the kidney with urine held up as a result of obstruction elsewhere in the urinary tract.

hydronium ion *(Chem.)*. See **hydroxonium ion**.

hydropericardium *(Med.)*. Collection of clear fluid in the pericardial sac.

hydroperitoneum *(Med.)*. Ascites. Accumulation of clear fluid in the abdominal cavity.

hydroperoxides *(Chem.)*. Intermediate compounds formed during the oxidation of unsaturated organic substances, e.g. fatty oils, such as linseed oil. They contain the group -OOH.

hydrophane *(Min.)*. A variety of opal which, when dry, is almost opaque, with a pearly lustre, but becomes transparent when soaked with water, as implied in the name.

hydrophilic colloid *(Chem.)*. A colloid which readily forms a solution in water.

hydrophily *(Bot.)*. (1) Living in water. (2) Pollination by water.

hydrophobia *(Med.)*. Literally fear of water, a symptom of rabies. Used synonymously with *rabies*.

hydrophobic cement *(Build.,Civ.Eng.)*. Cement into which a special agent is introduced during the grinding process to reduce the deterioration which occurs in other cements when exposed to damp.

hydrophobic colloid *(Chem.)*. A colloid which forms a solution in water only with difficulty.

hydrophone *(Acous.)*. Electroacoustic transducer used to detect sounds or ultrasonic waves transmitted through water. Also *subaqueous microphone*.

hydrophyte *(Bot.)*. (1) A plant with leaves partly or wholly submerged in water. (2) Water plant with perennating buds at the bottom of the water (a sort of **cryptophyte**). See **Raunkaier system**.

hydroplane *(Ships)*. (1) A powered boat which skims the surface of the water; cf. **hydrofoil**. (2) A planing surface which enables a submarine to submerge.

hydroponics *(Bot.)*. Technique of growing plants without soil for experimental purposes and sometimes crops. The roots can be in either a nutrient solution or in an inert medium percolated by such a solution, *water culture*, *sand culture*. See **nutrient film technique**.

hydropote *(Bot.)*. A gland-like structure found on the submerged surfaces of the leaves of many water plants.

hydrops folliculi *(Med.)*. An ovarian cyst formed by the accummulation of clear fluid in a Graafian follicle.

hydropyle *(Zool.)*. A modified area of the serosal cuticle of the developing egg of some *Orthoptera* for the uptake or loss of water.

hydroquinone *(Chem.,Image Tech.)*. *1,4-dihydroxybenzene*. It is a strong reducing agent and is extensively used as a developer in photography. Also called *quinol*.

hydrosalpinx *(Med.)*. Accumulation of clear fluid in a Fallopian tube which has become shut off as a result of inflammation.

hydrosere *(Bot.)*. A sere beginning with submerged soil, as at the margin of a lake.

hydroskis *(Aero.)*. Hydrofoils, proportioned like skis and usually retractable, fitted to seaplanes without a **planing bottom** as the sole source of hydrodynamic lift. They are also fitted to aircraft landing gears to make them amphibious (*pantobase*), in which case a minimum taxiing speed is necessary to keep the aircraft above water.

hydrosol *(Chem.)*. A colloidal solution in water.

hydrostatic approximation *(Meteor.)*. The assumption that the atmosphere is in hydrostatic equilibrium in the vertical which is equivalent to ignoring the vertical components of acceleration and **Coriolis force** in the equations of motion. The approximation is valid for atmospheric disturbances on horizontal scales of not less than about 10 km.

hydrostatic extrusion *(Eng.)*. Form of extrusion in which the metal to be shaped is preshaped to fit the die which forms the lower end of a high-pressure container. The container is filled with the pressure-transmitting liquid, and pressure is built up in the liquid by a plunger until the metal is forced through the die.

hydrostatic joint *(Build.)*. A joint of the spigot-and-socket type, formed in a large water-main by forcing sheet-lead into the socket to ensure security against hydraulic pressure.

hydrostatic pressing *(Powder Tech.)*. The application of liquid pressure directly to a preform which has been sealed, e.g. in a plastic bag. It is characterized by equal pressure in all directions maintaining preform shape to reduced scale.

hydrostatics *(Genrl.)*. Branch of statics which studies the forces arising from the presence of fluids.

hydrostatic skeleton *(Zool.)*. A body form maintained by muscles acting upon a fluid-filled cavity, usually the coelom as in many soft-bodied invertebrates.

hydrostatic test *(Build.)*. A test to detect leakage in a drain. The latter is plugged at the outlet end and filled with water; any fall of level of the water indicates leakage.

hydrostatic valve *(Eng.)*. Apparatus which tends to maintain an underwater body (e.g. a moving torpedo) at the desired depth.

hydrosulphides *(Chem.)*. Salts formed by the action of hydrogen sulphide on some of the hydroxides. They contain the ion HS^-.

hydrosulphuric acid *(Chem.)*. An aqueous solution of hydrogen sulphide.

hydrotaxis *(Biol.)*. Response or reaction of an organism to the stimulus of moisture. adj. *hydrotactic*.

hydrotherapy *(Med.)*. Treatment of disease by water, externally or internally.

hydrothermal metamorphism *(Geol.)*. That kind of change in the mineral composition and texture of a rock which was effected by heated water.

hydrothorax *(Med.)*. Clear fluid in the pleural cavity formed by transudation from blood vessels.

hydrotropism *(Bot.)*. A tropism in which the orientating stimulus is water.

hydrovane *(Aero.)*. See **hydrofoil, hydroskis, sponson**.

hydroxides *(Chem.)*. Compounds of the basic oxides with water. The term *hydroxide* (a contraction of *hydrated oxide*) is applied to compounds that contain the -OH or hydroxyl group.

hydroxonium ion *(Chem.)*. The hydrogen ion, normally present in hydrated form as H_3O^+. Also *hydronium ion*.

hydroxyapatite *(Min.)*. The hydroxyl-bearing variety of the phosphate apatite.

hydroxyl *(Chem.)*. -OH. A monovalent group consisting of a hydrogen atom and an oxygen atom linked together.

hydroxylamine *(Chem.)*. *Hydroxy-ammonia*, NH_2OH, rather explosive, deliquescent colourless crystals which may be obtained by the reduction of nitric oxide, ethyl nitrate, or nitric acid under suitable conditions; mp 33°C, bp 58°C at 3 kN/m^2. Its aqueous solution is alkaline, its salts are powerful reducing agents.

hydroxylamines *(Chem.)*. Derivatives of hydroxylamine, NH_2OH, in which the hydrogen has been exchanged for alkyl radicals.

hydroxyproline *(Biol.,Chem.)*. *4-hydroxy-pyrrole-2-carboxylic acid*. An imino-acid formed by the post-translational hydroxylation of proline residues within protein molecules, common in cell-wall proteins.

hydrozincite, zinc bloom *(Min.)*. A monoclinic hydroxide and carbonate of zinc. It is an uncommon ore, occurring with smithsonite as an alteration product of sphalerite in the oxide zone of some lodes.

Hydrozoa *(Zool.)*. A class of *Cnidaria*, in which alternation of generations typically occurs; the hydroid phase is usually colonial, and gives rise to the medusoid phase by budding.

hyetograph *(Meteor.)*. An instrument which collects, measures and records the fall of rain (Greek *hyetos*, rain).

hygristor *(Elec.Eng.)*. Resistance element sensitive to ambient humidity.

hygro- *(Genrl.)*. Prefix from Gk. *hygros*, wet, moist.

hygrochastic *(Bot.)*. A plant movement caused by the absorption of water.

hygroma *(Vet.)*. A fluid-filled swelling, usually associated with a joint, due to distension of a synovial sac.

hygrometer *(Phys.)*. Instrument for measuring or giving output signal proportional to, atmospheric humidity. Electrical hygrometers make use of **hygristors**.

hygrometry *(Meteor.)*. The measurement of the hygrometric state, or **relative humidity**, of the atmosphere.

hygrophyte *(Bot.)*. A plant of wet or waterlogged soils.

hygroscopic *(Chem.)*. Tending to absorb moisture; in the case of solids, without liquefaction.

hygroscopic movement *(Bot.)*. *Imbibitional movement. Hygrometric movement.* That caused by changes in moisture content of unevenly thickened cell walls e.g. of *elators* or the awns of the grains of some grasses.

hygrostat *(Chem.)*. Apparatus which produces constant humidity.

hylophagous *(Zool.)*. Wood-eating.

hymen *(Zool.)*. In Mammals, a fold of mucous membrane which partly occludes the opening of the vagina in young forms.

hymenitis *(Med.)*. Inflammation of the hymen.

hymenium *(Bot.)*. The fertile layer containing the asci or basidia in Ascomycetes and Basidiomycetes.

Hymenomycetes *(Bot.)*. The mushrooms, toadstools, agarics and bracket fungi, a class of the Basidiomycotina in which the hymenium is exposed at the time of spore formation. Many are saprophytic in soil, dung, leaf litter etc., some are parasitic e.g. honey fungus *Armillaria mellea*, many form ectotrophic mycorrhizas with trees.

hymenophore *(Bot.)*. Any fungal structure which bears a hymenium.

Hymenoptera *(Zool.)*. An order of Endopterygota having usually two almost equal pairs of transparent wings, which are frequently connected during flight by a series of hooks on the hindwing; mandibles always occur but the mouthparts are often suctorial; the adults are usually of diurnal habit; the larvae show great variation of form and habit. Saw-flies, Gall-flies, Ichneumons, Ants, Bees, and Wasps.

hymenotomy *(Med.)*. Cutting of the hymen.

hyo- *(Genrl.)*. Prefix from Gk. *hyoeidēs*. U-shaped.

hyoid *(Zool.)*. In higher Vertebrates, a skeletal apparatus lying at the base of the tongue, derived from the hyoid arch of the embryo.

hyoid arch *(Zool.)*. The second pair of visceral arches in lower Vertebrates, lying between the mandibular arch and the first branchial arch.

hyoideus *(Zool.)*. In Vertebrates, the post-trematic branch of the facial (seventh cranial) nerve, which runs to the mucosa of the mouth and to the muscles of the hyoid region, and, in aquatic forms, to the neuromast organs of the region below and behind the orbit.

hyomandibular *(Zool.)*. In Craniata, the dorsal element of a hyoid arch.

hyomandibular nerve *(Zool.)*. In Craniata, the post-trematic branch of the seventh cranial nerve; it divides into anterior and posterior branches.

hyoscine *(Med.)*. A coca-base alkaloid, present in *Datura meteloides*. It is the scopoline ester of tropic acid. Has a sedative effect on the central nervous system and the hydrobromide is used to induce *twilight sleep*; once used to obtain criminal confessions. Also called *scopolamine*.

hyoscyamine *(Chem.)*. $C_{17}H_{23}O_3N$, a coca base alkaloid, optically active, stereoisomeric with atropine, forming colourless needles or plates, mp 109°C; it can be prepared from *Datura stramonium*.

hyostyly *(Zool.)*. A type of jaw suspension, found in most Fish, in which the upper jaw is attached to the cranium anteriorly by a ligament, posteriorly by the hyomandibular. adj. *hyostylic*.

hypabyssal rocks *(Geol.)*. Literally, igneous rocks that are not quite abyssal (i.e. *deep-seated, plutonic*), occurring as minor intrusions. The three main divisions, based on mode of occurrence, are *plutonic, hypabyssal* and *volcanic*.

hypaethral *(Arch.)*. Said of a building without a roof, or with an opening in its roof.

hypalgesia *(Med.)*. Diminished sensitivity to pain.

hypanthium *(Bot.)*. The flat or concave receptacle of a perigynous flower.

hypapophyses *(Zool.)*. Paired ventral processes of the vertebrae of many higher *Craniata*.

hypaxial *(Zool.)*. Below the axis, especially below the vertebral column, therefore ventral; as the lower of two blocks into which the myotomes of fish embryos become divided.

hyper- *(Genrl.)*. Prefix from Gk. *hyper*, above.

hyperacidity *(Med.)*. Excessive acidity, especially of the stomach juices.

hyperacusis *(Med.)*. Abnormally increased acuity of hearing.

hyperadrenalism *(Med.)*. Abnormally increased activity of the adrenal gland.

hyperaemia, hyperemia *(Med.)*. Congestion, or excess of blood, in a part of the body.

hyperaesthesia *(Med.)*. Lowered threshold to a given stimulus, e.g. excessive sensitivity to painful or tactile stimuli.

hyperalgesia *(Med.)*. Heightened sensitivity to painful stimuli.

hyperaphia *(Med.)*. Excessive sensitivity to touch.

hyperbaric chamber *(Med.)*. A chamber containing oxygen at high pressures, in which patients are placed for decompression, to undergo radiotherapy for certain tumours or treatment for certain forms of poisoning.

hyperbaric chamber *(Med.,Radiol.)*. A chamber containing oxygen at high pressures, in which patients are placed to undergo radiotherapy or treatment for certain forms of poisoning.

hyperbarism *(Space)*. An agitated bodily condition when the pressure within the body tissues, fluids and cavities is countered by a greater external pressure, such as may happen in a sudden fall from a high altitude.

hyperbilirubinaemia, hyperbilirubinemia *(Med.)*. Excess of the bile pigment **bilirubin** in the blood.

hyperbola *(Maths.)*. See **conic**.

hyperbolic *(Telecomm.)*. Said of any system of navigation which depends on difference in cycles and fractions between locked waves from two or more stations, e.g. **Decca**.

hyperbolic functions *(Maths.)*. A set of six functions, analogous to the trigonometrical functions sin, cos, tan, etc. The hyperbolic sine and cosine are written *sinh* and *cosh*: $\sinh x = \frac{1}{2}(e^x - e^{-x})$; $\cosh x = \frac{1}{2}(e^x + e^{-x})$. The other four functions, *tanh*, *cosech*, *sech* and *cotanh* may be derived from sinh and cosh by the same rules as apply to the trigonometrical forms. In electrical communication they are useful in calculations involving the transmission of currents along wires and in filters.

hyperbolic geometry *(Maths.)*. See **absolute** geometry.

hyperbolic paraboloid *(Maths.)*. A saddle-shaped surface generated by the equation $x^2/a - y^2/b = z$.

hyperbolic point on a surface *(Maths.)*. One at which the curvatures of the normal sections are not all of the same sign. The directions which separate normal sections with positive curvature from those with negative curvature are called *asymptotic directions*. Cf. **elliptical, parabolic**, and **umbilical points on a surface**.

hyperbolic spiral *(Maths.)*. A spiral with polar equation $r\theta = a^2$.

hyperboloid *(Maths.)*. See **quadric**.

hypercalcaemia *(Med.)*. Rise in the calcium content of the blood beyond normal limits.

hypercapnia *(Med.)*. Excess of carbon dioxide in the lungs or the blood.

hypercharge *(Phys.)*. See **strangeness**.

hyperchlorhydria *(Med.)*. Increased secretion of hydrochloric acid by the acid-secreting cells of the stomach.

hypercholesterolaemia *(Med.)*. Increase of cholesterol in the blood beyond normal limits.

hyperdactyly *(Zool.)*. The condition of having more than the normal number (five) of digits, as in Cetacea.

hyperdiploidy *(Biol.)*. The condition where the full chromosome complement is present, as well as portion of one chromosome which has been translocated.

hyperemesis *(Med.)*. Excessive vomiting.

hyperemesis gravidarum *(Med.)*. Continued vomiting during pregnancy.

hyperemia *(Med.)*. See **hyperaemia**.

hyper-eutectoid steel *(Eng.)*. Steel with more carbon than is contained in pearlite. In carbon steels, one containing more than 0.9% carbon.

hyperfine structure *(Phys.)*. Splitting of spectral lines into two or more closely spaced components due to the interaction between the orbital magnetic moment and the nuclear magnetic moment of the atom. *Also*, the effect of nuclear mass on the energy levels in isotopically related atoms. Both hyperfine effects are of the same order of magnitude.

hyperfocal distance *(Image Tech.)*. The distance in front of a lens focused at infinity beyond which all objects are acceptably in focus. If a lens is focused on the hyperfocal distance, the **depth of field** extends from half that distance to infinity.

hypergammaglobulinaemia *(Immun.)*. Term used to describe clinical conditions in which the concentration of immunoglobulins in the blood exceeds normal limits. May result from continuous antigenic stimulation in chronic infections, from autoimmune diseases, or from abnormal proliferation of B-cells as occurs in *Waldenstrom macroglobulinaemia* or in *myelomatosis*. The term came into use before immunoglobulins were identified and accepted.

hypergeometric equation *(Maths.)*. See **Gauss' differential equation**.

hypergeometric function *(Maths.)*. The function $F(a;b;c;x)$ which is the sum of the *hypergeometric series*

$$1 + \frac{ab}{1c}x + \frac{a(a+1)b(b+1)}{1.2.c(c+1)}x^2 + \frac{a(a+1)(a+2)b(b+1)(b+2)}{1.2.3.c(c+1)(c+2)}x^3 + \dots$$

which converges either if $|x| < 1$ or if $x = 1$ and $a + b < c$.

hypergeometric series *(Maths.)*. See **hypergeometric function**.

hyperglycaemia, hyperglycemia *(Med.)*. An increase in the sugar content of the blood beyond normal limits. See also **diabetes mellitus**.

hypergolic *(Space)*. A rocket propellant mixture (fuel and oxidizer) which ignites spontaneously upon mixing.

hyperhidrosis, hyperidrosis *(Med.)*. Excessive perspiration.

hypericism *(Vet.)*. A form of photosensitization occurring in sheep and cattle following the ingestion of the St. John's Wort plant (*Hypericum perforatum*).

hyperinosis *(Med.)*. Excess of fibrin in the blood; opp. *hypinosis*.

hyperinsulinism *(Med.)*. A condition in which the blood sugar falls below normal limits, due to oversecretion of insulin by the pancreas; usually associated with pancreatic tumours, which provide the excess insulin.

Hyperion *(Astron.)*. The seventh natural satellite of **Saturn**.

hyperkeratosis *(Med.)*. Overgrowth of the horny layer of the skin.

hyperkinesia *(Med.)*. Excessive motility of a person, or of muscles.

hyperkinetic state, hyperactive state *(Behav.)*. In children, characterized by motor restlessness, poor attention, and excitability; often associated with learning difficulties.

hypermetamorphic *(Zool.)*. Of Insects, passing through two or more sharply distinct larval instars. n. *hypermetamorphosis*.

hypermetropia *(Med.)*. *Long-sightedness*. An abnormal condition of the eyes in which parallel rays of light come to a focus behind the retina instead of on it, the eyes being at rest. Also *hyperopia*.

hypermnesia *(Med.)*. Exceptional power of memory.

hypernephroma *(Med.)*. A tumour occurring in the kidney, resembling adrenal tissue but now known to be a carcinoma of the renal tubular cells.

hyperon *(Phys.)*. Elementary particles with masses greater than that of a neutron and less that of a deuteron and having a life-time of the order of 10^{-10} s.

hyperopia *(Med.)*. See **hypermetropia**.

hyper-osmotic *(Biol.)*. A solution of greater osmolarity than that in a cell suspended in the solution. Water consequently passes out of the cell.

hyperparasite *(Bot.)*. An organism parasitic on a parasite.

hyperparasitism *(Zool.)*. The condition of being parasitic on a parasite. n. *hyperparasite*.

hyperphalangy *(Zool.)*. The condition of having more than the normal number of phalanges, as in Whales.

hyperpharyngeal *(Zool.)*. Above the pharynx as the *hyperpharyngeal band* in *Tunicata*.

hyperpituitarism *(Med.)*. Overactivity of the pituitary gland; any condition due to over-activity of the pituitary gland, e.g. acromegaly, gigantism.

hyperplasia *(Bot.)*. Abnormal, usually pathological, enlargement of an organ or tissue by an increase in the numbers of cells. Cf. **hypertrophy**, **hypoplasia**.

hyperplasia *(Med.,Zool.)*. Excessive multiplication of cells of the body; an overgrowth of tissue due to increase in the number of tissue elements: generally, overgrowth. adj. *hyperplastic*.

hyperploid *(Biol.)*. Having a chromosome number slightly exceeding an exact multiple of the haploid number.

hyperpnea, hyperpnoea *(Med.)*. Increase in the depth and frequency of respiration; over-ventilation of the lungs.

hyperpyrexia *(Med.)*. A degree of body temperature greatly above normal (e.g. >41°C or 105°F). See **heat-stroke**.

hypersensitivity *(Bot.)*. Condition in which a pathogen kills its host cells so quickly that the spread of infection is prevented.

hypersensitization *(Image Tech.)*. Any method by which the effective speed of a photographic emulsion can be increased before exposure, for example by chemical solutions or vapours or by pre-exposure uniformly to very low light levels.

hypersonic *(Aero.,Space)*. Velocities of **Mach number** 5 or more. *Hypersonic flow* is the behaviour of a fluid at such speeds, e.g. in a shock tube.

hypersound *(Acous.)*. Sound of a frequency over 10^9 Hz; not audible.

hypersthene *(Min.)*. An important rock-forming silicate of magnesium and iron, $(Mg,Fe)SiO_3$, crystallizing in the orthorhombic system; an essential constituent of norite, hypersthene-pyroxenite, hypersthenite, hypersthene-andesite, and charnockite. Strictly, an ortho-pyroxene containing 50–70% of the enstatite molecule.

hypersthenic *(Med.)*. Having increased strength or tonicity. *Hypersthenic gastric diathesis*. The constitutional disposition in which the stomach is short and overactive in secretion and movement.

hypersthenite *(Geol.)*. A coarse-grained igneous rock, consisting essentially of only one component, hypersthene, together with small quantities of accessory minerals.

hyperstomatal *(Bot.)*. Having stomata only on the upper surface. Cf. **amphistomatal**, **hypostomatal**.

hypertelorism *(Med.)*. The condition of excessive width between two organs or parts, particularly related to the eyes.

hypertely *(Zool.)*. The progressive attainment of disproportionate size, either by a part or by an individual.

hypertension *(Behav.)*. *High blood pressure*. A chronic elevation of blood pressure due to constricted arteries; often a consequence of intense and persistent stress.

hypertension *(Med.)*. Increase in blood-pressure above the normal which, if prolonged, can lead to heart or renal failure, strokes and myocardial infarction. The majority of cases have no clear cause but a minority may be due to endocrine or renal causes.

hyperthyroidism *(Med.)*. Abnormally high rate of secretion of thyroid hormone by the thyroid gland, resulting in an increase in the basal metabolic rate, with symptoms of tiredness, anxiety, heat intolerance, palpitations and weight loss.

hypertonic *(Biol.)*. See **hyper-osmotic**.

hypertonic *(Chem.)*. Having a higher osmotic pressure than a standard, e.g. that of blood, or of the sap of cells which are being tested for their osmotic properties.

hypertonus *(Med.)*. A state of excessive muscular tone; opp. *hypotonus*.

hypertrichiasis, hypertrichosis *(Med.)*. Abnormal overgrowth of hair; excessive hairiness.

hypertrophic obstructive cardiomyopathy *(Med.)*. A relatively common *cardiomyopathy* where there is abnormal thickening of the muscle in the left ventricle below the *aortic* valve often causing obstruction to ejection of blood during *systole*.

hypertrophic pyloric stenosis *(Med.)*. A disorder in children in which there is hypertrophy of the muscle in the pyloric region of the stomach, leading to obstruction to the passage of food into the small intestine, and vomiting.

hypertrophy *(Bot.,Zool.)*. Abnormal, usually pathological, enlargement of an organ or cell. Cf. **hyperplasia**.

hypervitaminosis *(Med.)*. The condition arising when too much of any vitamin (especially vitamin D) has been taken by a person.

hypha *(Bot.)*. (1) The mycelium of a fungus which is a branching, filamentous structure with apical growth; the tubular cytoplasm contains the nuclei and may be divided by septa. (2) Elongated tubular cell in the thallus of some algae.

hyphopodium *(Bot.)*. A more-or-less lobed outgrowth from a hypha, often serving to attach an epiphytic fungus to a leaf.

hypidiomorphic, subhedral *(Geol.)*. A term referring to the texture of igneous rocks in which some of the component minerals show crystal faces, the others occurring in irregular grains. Cf. **idiomorphic crystals**.

hypnagogic imagery *(Behav.)*. Hallucinatory type of phenomena or fantasy occurring in the drowsy state just before falling asleep.

hypnagogic state *(Behav.)*. The transitional state of consciousness while falling asleep.

hypnophrenosis *(Med.)*. Any type of disturbance of sleep.

hypnosis *(Behav.)*. A temporary, trance-like state, induced by certain verbal or non-verbal procedures (hypnotism), characterized by heightened suggestibility both during and after hypnoses. See **post-hypnotic suggestion**.

hypnospore *(Bot.)*. A thick-walled, non-motile, resting spore. See also **aplanospore**.

hypnotic *(Med.)*. Of the nature of, or pertaining to, hypnosis; a drug which induces sleep.

hypo *(Chem.)*. A colloquial abbrev. for *sodium thiosulphate*, the normal fixing solution for silver halide emulsions, the unreduced silver being removed by the hypo, which forms complexes with Ag^+.

hypo *(Chem.,Image Tech.)*. A colloquial abbrev. for *sodium thiosulphate*, the normal fixing solution for silver halide emulsions, the unreduced silver being removed by the hypo.

hypo- *(Genrl.)*. Prefix from Gk. *hypo*, under.

hypoacidity *(Med.)*. A deficiency of acid, especially in the gastric juice.

hypoadrenalism *(Med.)*. The condition in which the activity of the adrenal glands is below normal.

hypoblast *(Zool.)*. The innermost germinal layer in the embryo of a metazoan animal, giving rise to the endoderm and sometimes also to the mesoderm. Cf. **epiblast**.

hypobranchial *(Zool.)*. The lowermost element of a branchial arch.

hypobranchial space *(Zool.)*. The space below the gills in Decapoda.

hypocalcaemia, hypocalcemia *(Med.)*. A calcium content of the blood below normal limits.

hypocaust *(Build.)*. A hollow space beneath the floor of a room or bath, serving as a flue for the hot gases from a furnace which, in circulating, give warmth to the room or bath.

hypocercal *(Zool.)*. Said of a type of caudal fin, found in Anaspida and Pteraspida, in which the vertebral column bends downwards and enters the hypochordal lobe which is larger than the epichordal lobe.

hypochlorhydria *(Med.)*. Diminished secretion of hydrochloric acid by the acid-secreting cells of the mucous membrane of the stomach.

hypochlorites *(Chem.)*. See **hypochlorous acid**.

hypochlorous acid *(Chem.)*. HClO. An aqueous solution of chlorine (I) oxide. Monobasic acid which forms salts called *hypochlorites (chlorates* (I)). Weak acid, easily decomposed.

hypochlorous anhydride *(Chem.)*. Same as *chlorine (I) oxide* (see under **chlorine oxides**.

hypochondriasis *(Behav.,Med.)*. A disorder characterized by excessive preoccupation with bodily functions and sensations with the false belief that the latter indicate bodily disease; associated with a number of psychiatric syndromes.

hypocone *(Zool.)*. A fourth cusp arising on the cingulum on the postero-internal side of an upper molar tooth, producing a quadritubercular pattern.

hypocotyl *(Bot.)*. The part of the axis of the embryo between the radicle and the cotyledon(s), and the region of the seedling which derives from it.

hypocycloid *(Maths.)*. See **roulette**.

hypodermic *(Med.)*. Under the skin; a medical agent injected under the skin.

hypodermis, hypoderm *(Bot.)*. A layer, one or more cells thick immediately below the epidermis and differing morphologically from the underlying ground tissue.

hypodermis, hypoderm *(Zool.)*. In Arthropoda and other Invertebrates with a distinct cuticle, the epithelial cell-layer underlying the cuticle, by which the cuticle is secreted. adj. *hypodermal*.

hypodermoclysis *(Med.)*. The injection of fluid (e.g. salt solution) under the skin.

hypo eliminator *(Image Tech.)*. Solution used for removing all trace of fixative after washing prints, thus ensuring permanence. Usually hydrogen peroxide and ammonia.

hypo-eutectoid steel *(Eng.)*. Steel with less carbon than is contained in pearlite, i.e. the iron-cementite eutectoid. In carbon steels, one containing less than 0.9% carbon.

hypogammaglobulinaemia *(Immun.)*. Condition in which the concentration of immunoglobulins in the blood is much lower than normal. In the infantile sex-linked form there is a maturation defect in B-cells. A late acquired form has various causes, one of which is excessive activity of suppressor T-cells. A clinically similar condition may result from displacement of lymphoid tissues by lymphoma or leukaemic cells. There is greatly increased susceptibility to bacterial, but not to viral infections. Treatment is by regular administration of immunoglobulin concentrates prepared from the pooled blood of normal persons which contain antibodies against most environmental pathogenic microbes.

hypogastrium *(Med.)*. The lower median part of the abdomen; cf. **epigastrium**. adj. *hypogastric*.

hypogeal, hypogaeous *(Bot.)*. (1) Living beneath the surface of the ground. (2) Germinating with the cotyledons remaining in the soil.

hypogene *(Geol.)*. Said of rocks formed, or agencies at work, under the earth's surface.

hypoglossal *(Zool.)*. Underneath the tongue; the 12th cranial nerve of Vertebrates, running to the muscles of the tongue.

hypoglottis *(Zool.)*. In Vertebrates, the under part of the tongue; in *Coleoptera*, part of the labium.

hypoglycaemia, hypoglycemia *(Med.)*. A reduction in the level of sugar (glucose), in the blood.

hypognathous *(Zool.)*. Having the under jaw protruding beyond the upper jaw; having the mouth-parts directed downwards.

hypogonadism *(Med.)*. The condition in which there is a deficiency of the internal secretion of the gonads.

hypogynous *(Bot.)*. Flower having the stamens inserted on the convex receptacles close beside or beneath the base of the ovary, e.g. buttercup (*Ranunculus*). See **superior**. Cf. **perigynous, epigynous**.

hypohidrosis *(Med.)*. Abnormal diminution in the secretion of sweat.

hypohyal *(Zool.)*. The lowermost element of a hyoid arch.

hypoid bevel gear *(Eng.)*. A bevel gear in which the axes of the driving and driven shafts are at right angles but not in the same plane, resulting in some sliding action between the teeth; used in the back-axle drive of some automobiles.

hypolimnion *(Ecol.)*. The cold lower layer of water in a lake. Cf. **epilimnion**.

hypomania *(Med.)*. *Simple mania*. A condition characterized by mental excitement in the absence of mental confusion or of symptoms of insanity.

hypomenorrhea, hypomenorrhoea *(Med.)*. The condition in which the interval between 2 menstrual periods is increased to between 35 and 42 days.

hyponasty *(Bot.)*. The greater growth of the lower side which causes the upward bending of an organ. See also **nastic movement**. Cf. **epinasty**.

hyponome *(Zool.)*. In *Cephalopoda*, the funnel by which water escapes from the mantle cavity.

hypo-osmotic *(Biol.)*. A solution of lower osmolarity than that in a cell suspended in the solution. Water consequently passes into the cell.

hypopharyngeal *(Zool.)*. Below the pharynx, as the *hypopharyngeal* groove of *Cephalochorda*.

hypophloedal *(Bot.)*. Growing just within the surface of bark.

hypophosphoric acid *(Chem.)*. H_2PO_3, or $H_4P_2O_6$. Obtained by the slow oxidation of phosphorus in moist air. Stable at ordinary temperatures. Hydrolyzed by mineral acids, forming a mixture of phosphoric and phosphorous acids.

hypophosphorous acid *(Chem.)*. H_3PO_2. Feeble monobasic acid, which forms a series of salts called *hypophosphites* (phosphates (I)), oxidized to phosphates by oxidizing agents.

hypophysectomy *(Med.)*. Removal of the pituitary gland.

hypophysis *(Bot.)*. The suspensor cell closest to the embryo.

hypophysis *(Zool.)*. A downwardly growing structure; in Cephalochorda, the olfactory pit; in Vertebrates, the pituitary body. adj. *hypophysial*.

hypopituitarism *(Med.)*. A general term for any condition caused by diminished activity of the pituitary gland; characterized usually by obesity and imperfect sexual development.

hypoplasia *(Bot.,Zool.)*. Abnormal, usually pathological, under-development of a tissue because the cells are smaller or fewer. adj. *hypoplastic*. Cf. **hyperplasia, hypertrophy**.

hypoploid *(Biol.)*. Having a chromosome number a little less than some exact multiple of the haploid number.

hypopteronosis cystica *(Vet.)*. An inherited disease of budgerigars and canaries characterized by the formation of dermal cysts containing immature feathers.

hypopyon *(Med.)*. A collection of pus in the anterior chamber of the eye, between the iris and the cornea.

hyposensitization *(Immun.)*. Administration of a graded series of doses of an allergen to atopic subjects suffering from immediate type hypersensitivity to it. This must be done with great care to avoid anaphylactic reactions. The aim is to increase the level of specific IgG antibodies and/or to diminish the level of IgE antibodies so that subsequent contact with the allergen under natural conditions causes less severe reactions.

hypospadias *(Med.)*. A congenital deficiency in the floor of the urethra.

hypostasis *(Med.)*. Sediment or deposit. Passive hyperaemia in a dependent part owing to sluggishness of the circulation.

hypostatic *(Biol.)*. Opposite of *epistatic*; analogous to *recessive* applied to genes at different loci.

hypostoma, hypostome *(Zool.)*. In some Cnidaria, the raised oral cone: in Insects, the labrum: in Crustacea, the lower lip or fold forming the posterior margin of the mouth: in some Acarina, the lower lip formed by the fusion of the pedipalpal coxae.

hypostomatal *(Bot.)*. Leaf etc. with stomata only on the lower surface. Cf. **amphistomatal, epistomatal**.

hypostomatous *(Zool.)*. Having the mouth placed on the lower side of the head, as Sharks.

hypostyle hall *(Arch.)*. A hall having columns to support the roof.

hyposulphuric acid *(Chem.)*. *Dithionic acid*, $H_2S_2O_6$.

hyposulphurous acid *(Chem.)*. *Dithionous acid*, $H_2S_2O_4$.

hypotarsus *(Zool.)*. In Birds, the fibulare.

hypotension *(Med.)*. Low blood-pressure.

hypotenusal allowance *(Surv.)*. The distance added to each chain length, when chaining along sloping ground, in order to give a length whose horizontal projection shall be exactly one chain. For the 100-link chain the hypotenusal allowance is $100(\sec\theta - 1)$, where θ is the angle of slope of the ground from the horizontal.

hypotenuse *(Maths.)*. The side opposite the right angle of a right-angled triangle.

hypothalamus *(Zool.)*. In the Vertebrate brain, the ventral zone of the thalamencephalon. In Man the part of the brain which makes up the floor and part of the lateral walls of the third ventricle. The mammillary bodies, tuber cinerium, infundibulum, neurohypophysis and the optic chiasma are also part of the hypothalamus.

hypothermia *(Med.)*. Subnormal body temperature; occurs in the very young, the aged or in coma; used therapeutically for heart and other surgery.

hypothesis *(Genrl.)*. A prediction based on theory, an educated guess derived from various assumptions, which can be tested using a range of methods, but is most often associated with experimental procedure; a proposition put forward for proof or discussion.

hypothetical exchange *(Telecomm.)*. A telephone exchange which, until a new exchange is constructed, is made up from parts of existing exchanges, the subscribers being numbered according to the system of the new exchange.

hypothyroidism *(Med.)*. The condition accompanying the diminished secretion of the thyroid gland. See also **cretinism** and **myxoedema**.

hypotonic *(Biol.)*. See **hypo-osmotic**.

hypotonic *(Chem.)*. Having a lower osmotic pressure than a standard, e.g. that of blood, or of the sap of cells which are being tested for their osmotic properties.

hypotonus *(Med.)*. See **hypertonus**.

hypotrichous *(Zool.)*. Having cilia principally on the lower surface of the body.

hypotrochoid *(Maths.)*. See **roulette**.

hypovitaminosis *(Med.)*. The condition resulting from deficiency of a vitamin in the diet.

hypoxanthine *(Chem.)*. *6-hydroxypurine*. Formed by the breakdown of nucleoproteins by enzymatic action.

hypoxia *(Med.)*. Lack of oxygen supply.

hypso- *(Genrl.)*. Prefix from Gk. *hypsos*, height.

hypsochrome *(Chem.)*. A radical which shifts the absorption spectrum of a compound towards the violet end of the spectrum. Cf. **bathochrome**.

hypsodont *(Zool.)*. Of a Mammalian tooth with a high crown and deep socket. Cf. **brachyodont**.

hypsoflore *(Chem.)*. A radical which tends to shift the fluorescent spectrum of a compound toward shorter wavelengths. Cf. **bathoflore**.

hypsometer *(Phys.)*. An instrument used for determining the boiling-point of water, either with a view to ascertaining altitude, by calculating the pressure, or for correcting the upper fixed point of the thermometer used.

hypsophyll *(Bot.)*. A non-foliage leaf inserted high on a shoot, e.g. a floral bract. Cf. **cataphyll**.

Hyracoidea *(Zool.)*. An order of small Eutherian Mammals having 4 digits on the fore limb and 3 on the hind limb, pointed incisor teeth with persistent pulps, lophodont grinding teeth, no scrotal sac and 6 mammae; terrestrial African forms. Dassies, Hyraxes.

hy-rib *(Build.)*. A proprietary form of metal lath used for ceilings, walls, surroundings of columns etc. to which plaster is to be applied. Also used as reinforcement for concrete slabs carrying minor loads.

hysteranthous *(Bot.)*. Said of leaves which develop after the plant has flowered.

hysterectomy *(Med.)*. Surgical removal of the uterus.

hysteresis *(Phys.)*. The retardation or lagging of an effect behind the cause of the effect, e.g. dielectric hysteresis, magnetic hysteresis etc.

hysteresis coefficient *(Elec.Eng.)*. See **Steinmetz coefficient**.

hysteresis error *(Elec.Eng.)*. Instruments or control systems may show nonreversibility similar to hysteresis. The maximum difference between the readings or settings obtainable for a given value of the independent variable is the *hysteresis error*.

hysteresis heat *(Elec.Eng.)*. That arising from hysteresis loss, in contrast with that from ohmic loss associated with eddy currents.

hysteresis loop *(Elec.Eng.)*. A closed figure formed by plotting magnetic flux density **B** in a magnetizable material against magnetizing field H when the latter is taken through a complete cycle of increasing and decreasing values. The area of this loop measures the energy dissipated during a cycle of magnetization. Also called *B/H loop*. Similar effects occur with applied mechanical or electrical stresses, and this is by analogy known as *mechanical* or *electrical hysteresis*. Dielectric materials with appreciable hysteresis loss are termed *ferroelectric materials*.

hysteresis loss *(Elec.Eng.)*. Energy loss in taking unit quantity of material once round a hysteresis loop. It can arise in a dielectric material subjected to a varying electric field or in a magnetic material in a varying magnetic field.

hysteresis motor *(Elec.Eng.)*. A synchronous motor which starts by reason of the hysteresis losses induced in its steel secondary by the revolving field of the primary.

hysteresis tester *(Elec.Eng.)*. A device, invented by Ewing, for making a direct measurement of magnetic hysteresis in samples of iron or steel.

hyster-, hystero- *(Genrl.)*. Prefix from Gk. *hysteria*, womb or *hysteros*, later.

hysteria *(Med.)*. A physical disability with no apparent organic cause. It is a form of *psychoneurosis* which can present with sensory or motor dysfunction, i.e. blindness, paralysis, loss of speech or even fits. Direct confrontation often leads to rapid recovery of the physical presentation but may then re-present with another physical disability.

hysterocolpectomy *(Med.)*. Removal of the vagina (or part of it) and the uterus.

hysteropexy *(Med.)*. The fixation of a displaced uterus surgically.

hysterotomy *(Med.)*. Incision of the uterus.

Hz *(Phys.)*. Symbol for **hertz**.

I

I *(Chem.)*. A symbol for *van't Hoff's factor*.

I *(Maths.)*. The symbol used by mathematicians to represent an imaginary number whose square equals minus one, i.e. $i^2 = -1$. See **j**.

I- *(Chem.)*. An abbrev. for: (1) *optically inactive*; (2) *iso-*, i.e. containing a branched hydrocarbon chain.

I *(Chem.)*. The symbol for *iodine*.

I *(Chem.)*. A symbol for *ionic strength*.

I *(Eng.)*. Symbol for moment of inertia.

I *(Phys.)*. Symbol for (1) electric current, (2) luminous intensity.

IA *(Phys.)*. Abbrev. for *International Ångström*.

IAA *(Bot.)*. Abbrev. for *Indole-3-Acetic Acid*.

Ia antigens *(Immun.)*. Class II histocompatibility antigens with functions similar to human **HLA-D antigens** on mouse B-cells, macrophages and accessory cells.

IACS *(Eng.)*. Abbrev. for *International Annealed Copper Standard*.

IAEA *(Nuc.Eng.)*. *International Atomic Energy Agency*. Autonomous intergovernmental body concerned with the promotion of nuclear energy and to ensure as far as possible that it is not used to further military objectives.

Iapetus *(Astron.)*. The eighth natural satellite of **Saturn**.

IAR *(Comp.)*. See **program counter**.

IAS *(Aero.)*. Abbrev. for *Indicated Air Speed*.

IATA *(Aero.)*. Abbrev. for *International Air Transport Association*.

Iatrochemistry *(Med.)*. The study of chemical phenomena in order to obtain results of medical value; practised in the 16th century. Modern equivalents **chemotherapy** or **pharmacology**.

Iatrogenic disease *(Med.)*. A disease produced by a doctor, usually occurring as a side-effect of pharmacological agents.

IBA *(Bot.)*. Abbrev. for *Indole-3-Butyric Acid*.

I-beam *(Build.,Civ.Eng.)*. See **H-beam**.

ibuprofen *(Med.)*. A **non-steroidal anti-inflammatory drug** used in the treatment of musculo-skeletal pain and **rheumatoid arthritis**.

IC *(Comp.)*. See **integrated circuit**.

ICAO *(Aero.)*. Abbrev. for *International Civil Aviation Organization*.

Icarus *(Astron.)*. A minor planet, discovered in 1949, with a small eccentric orbit approaching the sun to within 30 million kilometres, inside the orbit of Mercury. In June 1968 Icarus passed within four million miles of the earth.

Ice *(Meteor.)*. Ice is formed when water is cooled below its freezing-point. It is a transparent crystalline solid of rel. d. 0.916 and specific heat capacity 0.50. Because water attains its maximum density at 4°C, ice is formed on the surface of ponds and lakes during frosts, and thickens downwards.

Ice action *(Geol.)*. The work and effects of ice on the earth's surface. See **glacial erosion, glaciation, glacier**.

Ice age *(Geol.)*. A period when glacial ice spread over regions which were normally ice-free. *The Ice Age* is a synonym of the Pleistocene epoch.

Iceberg *(Meteor.)*. A large mass of ice, floating in the sea, which has broken away from a glacier or ice barrier. Icebergs are carried by ocean currents for great distances, often reaching latitudes of 40° to 50° before having completely melted. Approximately one-tenth of an iceberg shows above the surface.

Iceblink *(Meteor.)*. A whitish glare in the sky over ice which is too distant to be visible.

Ice-breaker *(Civ.Eng.)*. (1) Protection on the upstream side of a bridge pier. (2) A projecting pier so arranged in relation to a harbour entrance that floating ice is kept outside. (3) A vessel specially equipped for clearing a passage through ice-bound waters.

Ice colours *(Chem.)*. Dyestuffs produced on the cotton fibre direct, by the interaction of a second component with a solution of a diazo-salt cooled with ice.

Ice contact slope *(Geol.)*. The steep slope of material originally deposited at an ice front and in contact with it.

Ice guard *(Aero.)*. A wire-mesh screen fitted to a piston aero-engine intake so that ice will form on it and not inside the intake; a *gapped ice guard* is mounted ahead of the intake so that air can pass round it, while a *gapless ice guard* is inside the intake and an alternative air path comes into use when it ices up.

Iceland agate *(Min.)*. A name quite erroneously applied to the natural glass *obsidian*.

Iceland spar *(Min.)*. A very pure transparent and crystalline form of calcite, first brought from Iceland. It has perfect cleavage, is noted for its double refraction, and hence is used in construction of the Nicol prism.

IC engine *(Autos.)*. Abbrev. for *Internal-Combustion Engine*.

Ice pellets *(Meteor.)*. Precipitation of transparent or translucent pellets of ice with diameters of 5 mm or less.

Ichor *(Geol.)*. The name applied by Sederholm to highly penetrating granitic liquids, charged with magmatic vapours (emanations), which he believed to operate in palingenesis.

Ichor *(Med.)*. A thin, watery discharge from a wound or a sore. adj. *ichorous*.

Ichthyic *(Zool.)*. Pertaining to, or resembling, Fish.

Ichthy-, Ichthyo- *(Genrl.)*. Prefix from Gk. *ichthys*, fish.

Ichthyopterygium *(Zool.)*. A paddle-like fin, or limb, used for swimming, e.g. pectoral or pelvic fin of Fish.

Ichthyosis *(Med.)*. *Xerodermia*. A disease characterized by dryness and roughness of the skin, resembling fish scales, due to lack of secretion of the sweat and the sebaceous glands.

Ichthyosis *(Vet.)*. Hardening of the skin which develops cracks which become filled with dirt and thereby suppurate. Seen congenitally in calves also over the elbows and hocks of dogs.

Iconic memory *(Behav.)*. A transient visual trace that fades rapidly after removal of the stimulus. Cf. **echoic memory**.

Icosahedron *(Maths.)*. A twenty-faced **polyhedron**. The faces of a *regular icosahedron* are identical equilateral triangles.

Icositetrahedron *(Min.)*. A sold figure having 24 trapezoidal faces, and belonging to the cubic system. Exemplified by some garnets.

ICRH *(Nuc.Eng.)*. *Ion Cyclotron Resonance Heating*. See **cyclotron resonance heating**.

ICSH *(Med.)*. Abbrev. for *Interstitial Cell Stimulating Hormone*. See **luteinizing hormone**.

Icteric *(Med.)*. Of the nature of, or affected with, jaundice.

Icterus *(Med.)*. **Jaundice**.

Ictus *(Med.)*. A stroke or sudden attack.

ICW *(Telecomm.)*. Abbrev. for *Interrupted Continuous Waves*.

Id *(Behav.)*. A term originally introduced by Groddeck and later used by Freud to denote the sum total of the primitive instinctual forces in an individual. It subserves the pleasure-pain principle, in which the activities of the organism are concerned with the immediate increase of pleasurable and reduction of painful stimuli. It is dominated by blind impulsive wishing and forms a major part of the unconscious mind.

Iddingsite *(Min.)*. An alteration product of olivine consisting of goethite, quartz, montmorillonite group clay materials, and chlorite.

ideal *(Maths.)*. A subset of a ring which is a subgroup with respect to addition and which contains all products of its elements with any element of the ring.

ideal crystal *(Crystal.)*. One in which there are no imperfections or alien atoms.

ideal gas *(Chem.)*. Gas with molecules of negligible size and exerting no intermolecular forces. Such a gas is a theoretical abstraction which would obey the ideal gas law under all conditions. The behaviour of real gases becomes increasingly close to that of an ideal gas as the pressure on them is reduced. Also called *perfect gas*.

ideal transducer *(Elec.Eng.)*. Any transducer which converts without loss all the power supplied to it.

ideal transformer *(Elec.Eng.)*. A hypothetical transformer corresponding to one with a coefficient of coupling of unity.

ideas of reference *(Behav.)*. A characteristic of some mental disorders, notably schizophrenia, in which the individual perceives irrelevant and independent environmental and social events as relating to himself or herself ('people are looking at me').

idempotent *(Maths.)*. If S is a set and $*$ is an operation in S, then an element e of S is said to be an idempotent with respect to $*$ if $e*e=e$. For the usual addition of real numbers 0 is the only idempotent; for the usual multiplication of real numbers 0 and 1 are the only idempotents. Each group has one and only one idempotent, namely the identity element. There is no restriction on the number of idempotents to be found in a semigroup.

identification *(Behav.)*. In psychoanalytic theory, the way in which an individual incorporates (introjects) the values, standards, sexual orientation, and mannerisms of the same sex parent, as part of the development of the **super-ego**. It can also be used to describe the influence of any relevant and powerful figure for the internalisation of external norms.

identification dimensions *(Ships)*. Same as **registered dimensions**.

identifier *(Comp.)*. Name or label chosen by the programmer.

identity *(Maths.)*. (1) See **equation**. (2) If S is a set with an operation $*$, and e is an element in the set S such that, for all elements x in the set S we have $x*e=x=e*x$, then e is said to be an identity element, or neutral element, with respect to the operation $*$ in S. E.g. in the set of real numbers, 0 is an identity element with respect to addition, and 1 is an identity element with respect to multiplication.

identity mapping *(Maths.)*. The identity mapping, or identity function, on a set S is the mapping i_s from S on to S defined by $i_s(x)=x$, for all elements x in S.

idio- *(Genrl.)*. Prefix from Gk. *idios*, peculiar, distinct.

idioblast *(Bot.)*. A cell of clearly different properties to the others in the tissue, as a stone cell in pear fruit.

idioblast *(Geol.)*. A crystal which grew in metamorphic rock and is bounded by its own crystal faces. Cf. **idiomorphic crystals**. adj. *idioblastic*. See also **porphyroblastic**.

idioglossia *(Med.)*. The wrong use of consonants by a child, making speech unintelligible.

idiogram *(Biol.)*. *Karyogram*. A diagram (or photomontage) of the chromosome complement of a cell, conventionally arranged to show the general morphology including relative sizes, positions of centromeres etc.

idiomorphic crystals *(Geol.)*. Igneous rock minerals which are bounded by the crystal faces peculiar to the species. Cf. **allotriomorphic** (anhedral) and **hypidiomorphic** (subhedral).

idiopathy *(Med.)*. Any morbid condition arising spontaneously, having no known origin. adj. *idiopathic*.

idiosyncrasy *(Med.)*. Individual hypersensitivity to a drug or food but not explained by altered immunity.

idiot *(Med.)*. A term no longer used but it formerly described a person so defective in mind from birth as to be unable to protect himself against ordinary physical dangers. Now defined as *mentally severely handicapped*.

idiothermous *(Zool.)*. See **warm-blooded**.

idiotope *(Immun.)*. Antigenic determinant on immunoglobulin molecules characteristic of the product of a single clone or a small minority of clones, and associated with or part of the antigen binding site.

idiot savant *(Behav.)*. A child who, despite generally diminished skills, shows astonishing proficiency in one isolated skill ('foolish wise one').

idiot tape *(Print.)*. A continuous unjustified tape, containing only signals for new paragraphs, which must be processed into a new justified tape before it can control a typesetting or filmsetting machine. See **computer typesetting**.

idiotype *(Immun.)*. Set of one or more idiotopes by which a clone of immunoglobulin-forming cells can be distinguished from other clones. Some idiotypes appear to be unique to an individual animal; others are found in many members of the same animal species.

idioventricular *(Med.)*. Pertaining to the ventricle of the heart alone.

I-display *(Radar)*. Display representing the target as a full circle when the antenna is pointed directly at it, the radius being in proportion to the range.

idler *(Eng.)*. See **idle wheel**.

idler *(Horol.)*. See **intermediate wheel**.

idler *(Print.)*. A free-running roller on a web-fed press. Also *idling roller*.

idler pulley *(Eng.)*. See **guide pulley**.

idle wheel *(Eng.)*. A wheel interposed in a gear train, either solely to reverse the direction of rotation or also to modify the spacing of centres, without affecting the ratio of the drive. Also *idler*.

idle wire *(Elec.Eng.)*. The part of the armature winding of an electric machine which does not actually cut the lines of force, i.e. that part comprising the end connections.

idling *(Autos.)*. The slow rate of revolution of an automobile or aero engine, when the throttle pedal or lever is in the closed position.

idling adjustment *(Autos.)*. A setting of the slow-running jet and throttle position of a carburettor, so as to give regular **idling**.

idling roller *(Print.)*. See **idler**.

idocrase *(Min.)*. See **vesuvianite**.

idose *(Chem.)*. A monosaccharide belonging to the group of **aldohexoses**. ⇨

idoxuridine *(Med.)*. Antiviral agent effective against herpes viruses. Used *topically* in the treatment of *herpetic* lesions.

IF *(Telecomm.)*. Abbrev. for *Intermediate Frequency*.

iff *(Maths.)*. Abbrev. form of 'if and only if'.

IFF *(Radar)*. Abbrev. for WWII system whereby vessels and aircraft carried **transponders** capable of indicating to a 'friendly' radar system that they were not hostile. *IFF* is used now as a general description of such radar identification systems. See **ATCRBS**.

IFIP *(Comp.)*. Abbrev. for *International Federation for Information Processing*.

IFR *(Aero.)*. Abbrev. for *Instrument Flight Rules*.

IFRB *(Telecomm.)*. Abbrev. for *International Frequency Registration Board*.

Ig *(Immun.)*. General abbrev. for **immunoglobulin**, under which there is information on structure and genetic control.

IgA *(Immun.)*. The major immunoglobulin present in mucosal secretions, where it appears as a dimer linked by a peptide made separately by epithelial cells, which assists to transport it across the mucosa. In the blood IgA is present in monomeric and polymerised forms. It is mainly synthesised by B-cells in the lymphoid tissues of the gut and respiratory tract, and IgA antibodies constitute the main humoral defence mechanism against microbes on mucosal surfaces.

IgD *(Immun.)*. Immunoglobin present only in very low concentrates in blood, but present at the surface of B-lymphocytes where it probably functions as an antigen receptor.

IgE *(Immun.)*. Immunoglobulin the Fc region of which binds very strongly to a receptor on the surface of mast cells and basophils. Normally present in blood only in very low concentrations, but the concentration is increased in atopic subjects, and in infection by several helminth parasites.

IgG *(Immun.)*. The major immunoglobulin in humans and most species from amphibians upwards (but not in fish). IgG fixes complement and crosses the placenta (i.e. can be passed from mother to foetus). Although mainly present in body fluids some is also present at mucosal surfaces. In humans there are four varieties (sub-classes) differing from each other in respect of parts of their heavy chains and the number of disulphide bonds which link the heavy chains together. This confers different biological properties, the functions of which are not fully understood. The sub-classes are termed IgG1, IgG2, IgG3, and IgG4. Similar subclasses are present in mice and rats but not in rabbits.

IgM *(Immun.)*. High molecular weight immunoglobulin, consisting in humans of five basic units (of two light and two heavy chains) arranged as a pentamer and joined together by disulphide bonds and a small link peptide (J chain). A monomeric form is present on the cell surface of B-lymphocytes, where it acts as the earliest form of antigen receptor. In some species the form present in the blood may be a monomer, in others a tetramer and in others a hexamer. IgM activates complement very effectively but does not cross the human placenta. IgM antibodies are the first to be synthesized and released after a primary antigenic stimulation, and are the predominant form made in response to many bacterial capsular polysaccharides. Since their pentameric structure enables them to attach firmly to antigenic sites situated close together on a surface, they provide effective protection against many bacteria.

igneous complex *(Geol.)*. A group of rocks, occurring within a comparatively small area, which differ in type but are related by similar chemical or mineralogical peculiarities. This indicates derivation from a common source.

igneous intrusion *(Geol.)*. A mass of igneous rock which crystallized before the magma reached the earth's surface; including **dykes, sills, stocks, bosses** and **batholiths**.

igneous (magmatic) cycle *(Geol.)*. The sequence of events usually followed in igneous activity; it consists of an eruptive phase, a plutonic phase, and a phase of minor intrusion.

igneous rocks *(Geol.)*. Rock masses generally accepted as being formed by the solidification of magma injected into the earth's crust, or extruded on its surface.

ignimbrite *(Geol.)*. A pyroclastic rock consisting originally of lava droplets and glass fragments which were so hot at the time of deposition that they were welded together. Also *welded tuffs*.

ignite *(Chem.)*. To heat a gaseous mixture to the temperature at which combustion occurs, e.g. by means of an electric spark.

igniter *(Civ.Eng.)*. A blasting fuse or other contrivance used to fire an explosive charge.

igniter plug *(Aero.)*. An electrical-discharge unit for lighting up gas turbines.

ignition *(Elec.Eng.,Autos.)*. The firing of an explosive mixture of gases, vapours or other substances by means e.g. of an electric spark.

ignition advance *(Autos.)*. The crank angle before top dead-centre, at which the spark is timed to pass in a petrol or gas engine. See **ignition timing, angle of advance**.

ignition coil *(Autos.)*. An induction coil for converting the low-tension current supplied by the battery into the high-tension current required by the sparking-plugs.

ignition lag *(Autos.)*. Of a combustible mixture in an engine cylinder, the time interval between the passage of the spark and the resulting pressure rise due to combustion.

ignition rating *(Elec.Eng.)*. A special rating, in *ampere hours* employed for accumulators used for supplying ignition systems; it is generally twice the continuous rating at a low discharge rate.

ignition system *(Autos.)*. The arrangement for providing the high-tension voltage required for ignition. See **electronic ignition, ignition coil, magneto ignition**.

ignition temperature *(Eng.)*. In the combustion of gases, the temperature at which the heat loss due to conduction, radiation, etc., is more than counterbalanced by the rate at which heat is developed by the combustion reaction.

ignition temperature *(Nuc.Eng.)*. In fusion, the point at which alpha-particle heating can sustain the fusion reaction.

ignition timing *(Autos.)*. The crank angle relative to top dead-centre at which the spark passes in a petrol or gas engine. See **angle of advance**.

ignition voltage *(Elec.Eng.)*. That required to start discharge in a gas tube.

ignitor *(Electronics)*. See **pilot electrode**.

ignitor drop *(Electronics)*. The voltage drop between cathode and anode of the ignitor discharge in a switching tube.

ignitron *(Electronics)*. Mercury-arc rectifier with *ignitor*, which is an electrode which can dip into the cool mercury pool and draw an arc to start the ionization.

IGV *(Aero.)*. Abbrev. for *Inlet Guide Vanes*.

IHP *(Eng.)*. Abbrev. for *Indicated HorsePower*.

ijolite *(Geol.)*. A coarse-grained igneous rock, consisting of nepheline, aegirine-augite, with usually melanite garnet as a prominent accessory, occurring in nepheline-syenite complexes in the Kola Peninsula, White Sea, the Transvaal, and elsewhere.

IKBS *(Comp.)*. Abbrev. for *Intelligent Knowledge-Based Systems*. See **knowledge-based systems**.

IL-1, IL-2 *(Immun.)*. See **interleukin 1, 2**.

Ile *(Chem.)*. Symbol for **isoleucine**.

ileitis *(Med.)*. Inflammation of the ileum.

ileocolitis *(Med.)*. Inflammation of the ileum and the colon.

ileocolostomy *(Med.)*. The making of a communication between the ileum and the colon by operation.

ileostomy *(Med.)*. An artificial opening in the ileum, made surgically.

ileum *(Zool.)*. In Vertebrates, the posterior part of the small intestine.

ileus *(Med.)*. Colic due to obstruction in the intestine; obstruction of the intestine. See **paralytic ileus**.

Ilgner system *(Elec.Eng.)*. See **Ward-Leonard-Ilgner system**.

iliac region *(Zool.)*. The dorsal region of the pelvic girdle in Vertebrates.

iliac veins *(Zool.)*. In Fish, the paired veins from the pelvic fins, draining into the lateral veins.

ilio- *(Zool.)*. A prefix which refers to that part of the pelvic girdle of a Vertebrate known as the *ilium*; used in the construction of compound terms, e.g. *iliofemoral*, pertaining to the ilium and the femur.

ilium *(Zool.)*. A dorsal cartilage bone of the pelvic girdle in Vertebrates. adj. *iliac*.

Ilkovic equation *(Chem.)*. In polarography, the equation expresses the current to the **dropping-mercury electrode** as, $i = AnCD^{1/2}m^{2/3}\tau^{1/6}$, where i = diffusion current, A = numerical constant, n = ionic charge, D = diffusivity, m = rate of flow of mercury, τ = lifetime of a drop.

ill-conditioned *(Surv.)*. A term used in **triangulation** to describe triangles of such a shape that the distortion resulting from errors made in measurement and in plotting may be great, the criterion often used being that no angle in a triangle should be less than 30°.

illegitimate pollination *(Bot.)*. The transfer of pollen in a way the floral structure appears to discourage, e.g. from the anther of one **pin** flower to the stigma of another.

illegitimate recombination *(Biol.)*. Recombination between species whose DNA shows little or no homology, facilitated by duplicate sequences casually present.

illite *(Min.)*. A monoclinic clay mineral, a hydrated silicate of potassium and aluminium. It is the dominant clay mineral in shales and mudstones. The illite group somewhat resembles muscovite.

illuminance *(Phys.)*. See **illumination**.

illuminated diagram *(Elec.Eng.)*. A circuit diagram on a switchboard, or a track diagram in a railway signal box, so arranged that lamps behind the diagram illuminate any part of the circuit which is alive or any part of the track upon which a train is standing.

illuminated-dial instrument *(Elec.Eng.)*. An electric measuring instrument for switchboard use, having an illuminated scale of translucent glass.

illumination *(Phys.)*. The quantity of light or luminous flux falling on unit area of a surface. Illumination is inversely proportional to the square of the distance of the surface from the source of light, and proportional to the cosine of the angle made by the normal to the surface with the direction of the light rays. The unit of illumination is the *lux*, which is an illumination of 1 lumen/m^2. Symbol *E*. Also *illuminance*.

illusion *(Behav.)*. A surprising perceptual experience, due to the fact that some aspect of the relation between the physical stimulus and the individual's perception of it, violates normal expectation.

ilmenite *(Min.)*. An oxide of iron and titanium, crystallizing in the trigonal system; a wide-spread accessory mineral in igneous and metamorphic rocks, especially in those of basic composition. A common mineral in detrital sediments, often becoming concentrated in beach sand.

ILS *(Aero.)*. Abbrev. for *Instrument Landing System*.

ILT *(Vet.)*. See **infectious laryngotracheitis**.

ilvaite *(Min.)*. Silicate of iron, calcium and manganese. It crystallizes in the orthorhombic system.

image *(Phys.)*. Optical images may be of two kinds, real or virtual. A *real image* is one which is formed by the convergence of rays which have passed through the image-forming device (usually a lens) and can be thrown on to a screen, as in the camera and the optical projector. A *virtual image* is one from which rays appear to diverge. It cannot be projected on to a screen or a sensitive emulsion.

image *(Print.)*. The general term to describe the subject on negatives, positives, and plates, at any stage of preparation.

image admittance *(Phys.)*. The reciprocal of *image impedance*.

image charge *(Phys.)*. Hypothetical charge used in electrostatic theory as a substitute for a conducting (equipotential) surface. The charge must not modify the field distribution at any point outside this surface.

image converter tube *(Phys.)*. One in which an optical image applied to a photoemissive surface produces a corresponding image on a luminescent surface.

image curvature *(Phys.)*. Concerned with the aberrations of the image in an electron microscope. Provided curvature of the field is the only aberration present, a sharp image is formed on a curved surface tangential to the image plane on the axis.

image-dissection camera *(Image Tech.)*. High-speed camera having, in place of a normal lens, an array of parallel light-guides made of thin glass fibres embedded in an opaque matrix, which transmit light by internal reflection on to successive portions of a moving plate.

image force *(Phys.)*. That force on an electric (or magnetic) charge between itself and its image induced in a neighbouring body.

image frequency *(Telecomm.)*. In a **superhet** receiver, that which differs from the *local oscillator* frequency by an amount equal to the *intermediate frequency*, and is on the opposite side of the local oscillator frequency from that of the desired signal; a direct result of a mixer's ability to produce sum and difference frequencies. Unless the mixer is preceded by tuned stages, to provide *image rejection*, images can cause serious interference with reception.

image impedance *(Phys.)*. A network is image operated when the input generator impedance equals the network input impedance, and the load impedance equals the output impedance of the network. If Z_{sc} is the short-circuit impedance and Z_{oc} the open-circuit impedance for the network, then the image impedance Z_o is given by $Z_o^2 = Z_{sc} . Z_{oc}$. See **iterative impedance**.

image intensifier *(Radiol.)*. An electronic device screen for enhancing brightness of an image in fluoroscopy, at the same time reducing patient dose. See **intensifying screen**.

image interference *(Telecomm.)*. That produced by any signal received with a frequency at or near the image frequency.

image orthicon *(Image Tech.)*. Improved version of the **orthicon** camera tube in which the electron image from a continuous photo-emissive target surface was focused on a storage photo-mosaic for scanning. Now obsolete.

image phase constant *(Telecomm.)*. In a filter section, terminated in both directions, unreal or imaginary part of the (image) transfer constant; phase delay of the section in radians.

image processing *(Comp.)*. Techniques for filtering, storing and retrieving images.

image rejection ratio *(Telecomm.)*. The ratio of the image frequency signal input at the aerial to the desired signal input, in a superhet receiver, for identical outputs.

image response *(Telecomm.)*. Unwanted response of a superhet receiver to the image frequency.

image transfer *(Image Tech.)*. Process in which the unexposed areas of a photographic emulsion are used to form an image in another layer; used in **instant photography** and document copying. See also **diffusion transfer reversal**.

image tube *(Electronics)*. One in which an optically focused image on a photo-emissive plate releases electrons which are focused on a phosphor by electric or magnetic means. Used in X-ray intensifiers, infrared telescopes, electron telescopes, and microscopes.

imaginal bud (disk) *(Zool.)*. One of a number of masses of formative cells which are the principal agents in the development of the external organs of the imago, during the metamorphosis of the *Endopterygota*.

imaginary axis *(Maths.)*. See **Argand diagram**.

imaginary circle *(Maths.)*. The name given to a circle defined analytically so as to have an imaginary radius, e.g., the circle defined by $(x-a)^2+(y-b)^2+r^2=0$.

imaginary number *(Maths.)*. The product of a real number *x* and *i*, where $i^2+1=0$. A complex number in which the real part is zero.

imaginary part *(Maths.)*. See **complex number**.

imaging system *(Space)*. An instrument with its supporting **hardware** and **software** for obtaining remote images of the Earth and objects in space. The types of image may be optical, microwave or obtained by using some other selected part of the electromagnetic spectrum. The images may be recorded, e.g. on film or digitally and recovered later or transmitted in real-time to an Earth receiving station.

imago *(Zool.)*. The form assumed by an Insect after its last ecdysis, when it has become fully mature; final instar. adj. *imaginal*.

IMAX *(Image Tech.)*. TN for a system of very large **wide-screen** motion-picture presentation, using film 70 mm wide running horizontally with a frame size 70 × 46 mm. Special measures are needed to maintain a flat focal plane.

imbalance *(Med.)*. A lack of balance, as between the ocular muscles, or between the activities of the endocrines, or between parts of the involuntary nervous system.

imbecile *(Med.)*. A term formerly used for a person with moderately severe mental subnormality.

imbibition *(Bot.)*. (1) Uptake of water where the driving force is a difference of **matric potential** rather than osmotic. (2) Uptake of water and swelling by seeds, the first step in germination.

imbibition *(Chem.)*. The absorption or adsorption of a liquid by a solid or a gel, accompanied by swelling of the latter.

imbibition *(Image Tech.)*. The transfer of dye from a matrix to an absorbing surface, such as a gelatine layer, to produce a dye image. Also known as *dye transfer*.

imbibition matrix *(Image Tech.)*. A relief or differentially hardened image capable of selectively absorbing dye and transferring it to another receptive layer.

imbricate *(Bot.,Zool.)*. Organs such as scales, leaves or petals in buds which overlap like the tiles on a roof. See also **aestivation, vernation**. Cf. **valvate**.

imbricated *(Build.)*. Said of slates or tiles which are laid so as to overlap.

imbricate structure *(Geol.)*. (1) A structure produced by thrust faulting leading to the development of numerous small faults and rock slices arranged in parallel like a pack of fallen cards. (2) A sedimentary structure in which pebbles with a flat surface are stacked in the same direction, dipping upcurrent.

IMC *(Aero.)*. *Instrument Meteorological Conditions*, wherein aircraft must conform to **instrument flight rules**.

IMEP *(Eng.)*. Abbrev. for *Indicated Mean Effective Pressure*.

Imhoff tank *(Build.)*. A form of settling tank to which sewage is passed, the solid matter being exposed to a fermentation process, with the production of methane gas and an inoffensive sludge which can be easily dried.

imidazoles *(Chem.)*. *Glyoxalines*; heterocyclic, compounds produced by substitution in a five-membered ring containing two nitrogen atoms on either side of a carbon atom. Benzimidazoles are formed by the condensation of *ortho*-diamines with organic acids, and contain a condensed benzene nucleus. ⇨

imides *(Chem.)*. Organic compounds containing the group —CO.NH.CO—, derived from acid anhydrides.

imino group *(Chem.)*. The group

$$\begin{matrix} R \\ R' \end{matrix} > NH.$$

Imino compounds are secondary amines obtained by the substitution of two hydrogen atoms in ammonia by alkyl radicals.

imitation *(Behav.)*. See **observational learning**.

imitation parchment *(Paper)*. A wood-pulp paper to which some degree of transparency, and grease resistance has been imparted by prolonged beating of the pulp.

IMMARSAT *(Space)*. Abbrev. for *International Maritime Satellite Organisation*, concerned with ship-to-ship and ship-to-shore communications *via* satellite.

immature cotton *(Textiles)*. Cotton picked before it is fully mature. The fibres are not properly formed and the yarn made from them is generally weaker and inferior.

immediate access store *(Comp.)*. See **main memory**.

immediate address *(Comp.)*. The address field itself is used to hold data which is needed for the operation.

immediate hypersensitivity *(Immun.)*. Antibody mediated hypersensitivity characteristically due to release of histamine and other vasoactive substances. Also *type 1 reaction*.

immediate mode *(Comp.)*. The use, for immediate execution, of a program language statement outside the program.

immersed liquid-quenched fuse *(Elec.Eng.)*. A fuse in which liquid is used for extinguishing the arc, the fuse-link being totally immersed in the liquid.

immersed pump *(Aero.)*. An electrical pump mounted inside a fuel tank.

immersible apparatus *(Elec.Eng.)*. Electrical aparatus designed to operate continuously under water.

immersion *(Astron.)*. The entry of the moon, or other body, into the shadow which causes its eclipse.

immersion foot *(Med.)*. A cold injury to the foot produced by cold wet exposure causing vasoconstriction resulting in anoxic tissue damage.

immersion heater *(Elec.Eng.)*. An electric heater designed for heating water or other liquids by direct immersion in the liquid.

immigration *(Ecol.)*. A category of population dispersal covering one-way movement into the population area. Cf. **emigration, migration**.

immiscibility *(Chem.)*. The property of two or more liquids of not mixing and of forming more than one phase when brought together.

immittance *(Phys.)*. Combined term covering **impedance** and **admittance**.

immobilised culture *(Biol.)*. The use of plastic foam, beads or sheets to let cells grow on the surfaces and interstices in close contact with the circulating medium. Combines the advantages of surface and suspension culture.

immune *(Bot.)*. Not susceptible to infection.

immune *(Med.)*. Protected against any particular infection; one who is in this state, i.e. has **immunity**. v. *immunize*, to make immune against infection. n. *immunization*.

immune adherence *(Immun.)*. Adherence of antigen-antibody complexes or antibody coated microbes which contain bound C3b or C4b to complement receptors on erythrocytes or platelets of some species and on macrophages and polymorphs.

immune bodies *(Med.)*. See **antibodies**.

immune complex *(Immun.)*. Macromolecular complex formed by antigen and antibody linked by their combining sites. The size of the complex depends upon the ratio of antigen to antibody. At **optimal proportions** the complexes come out of solution as precipitates, but in antigen excess they may remain in solution. Such complexes if present in the blood stream may be deposited in the walls of small blood vessels or in renal glomeruli, and there activate complement. This elicits an inflammatory reaction (Type 3) and is the cause of *immune complex diseases*.

immunity *(Immun.)*. A state of not being susceptible to the invasive or pathogenic effects of potentially infective microbes or to the effects of potentially toxic antigenic substances. Due to natural or non-specific mechanisms or to specific acquired immunity, or to both.

immunization *(Immun.)*. Administration either of antigen to produce active immunity or of antibody to confer passive immunity, and thereby to protect against the harmful effects of antigenic substances or microbes.

immunoblot *(Biol.)*. Technique used to detect the presence of a polypeptide which is antigenic to the specific antibody used in the assay. A protein mixture is denatured and separated on the basis of size by **SDS gel electrophoresis**, then blotted on to a sheet of nitrocellulose, preserving the pattern due to size. The proteins in the pattern are reacted with the specific antibody, then with an **AP-** or **HRP-**conjugated second antibody against the first which after development, produces visible staining at the site of binding. Also *western blot*.

immunofluorescence *(Immun.)*. Technique in which antigen or antibody is conjugated to a fluorescent dye and then allowed to react with the corresponding antibody or antigen in a tissue section or a cell suspension. This enables the location of antibodies or antigens in or on cells to be determined by fluorescence microscopy.

immunogen *(Immun.)*. A substance that stimulates humoral and/or cell mediated specific immunity when introduced into the body.

immunoglobulin *(Immun.)*. A family of proteins all of which have a similar basic structure, made up of **light chains** and **heavy chains** linked together by disulphide bonds so as to form a Y shaped molecule with two flexible arms. The detailed shape of the ends of the arms, the *variable region*, varies from one molecule to another within the family, so providing a wide range of antigen binding sites. The structure of the rest of the molecule is relatively constant from one molecule to another, and generally similar, but members are divided into immunoglobulin classes and sub-classes determined by the amino acid sequences of their heavy chains. (See **IgA**, **IgD**, etc.). All antibodies are immunoglobulins, and the same antigen binding sites can be present on the arms of immunoglobulins belonging to any of the classes.

immunoglobulin genes *(Immun.)*. Each polypeptide chain is coded for by several genes, each having a number of alleles, which are spliced together before transcription. Heavy chains are coded by V (variable region), D (diversity) and J (joining) genes which are linked to genes for the C (constant) region of one or other class. Light chains are coded by a different set of V and J genes joined to genes for one or other of the light chain constant regions. Splicing together of the V, D and J genes prior to joining the C region genes appears to take place more or less at random during the early stages of B-cell development. This results in an enormous variety of possible immunoglobulins, of which any individual B-cell can make one.

immunological memory *(Immun.)*. Describes the fact that antibody and cell mediated responses occur more rapidly and are quantitatively greater after a second exposure to antigen (provided that the interval is more than a few days and less than several years). Due to the stimulation on first exposure to the antigen of increased numbers of B- and T-cells capable of recognising the antigen, and which have reverted to a resting long lived form ('memory cells'). These are available to respond on second exposure to the antigen.

immunological tolerance *(Immun.)*. A state in which an animal fails to respond to an antigen capable of inducing humoral or cell mediated immunity, which has been induced by prior administration of the same antigen by a route or in a form such as either to eliminate potentially responsive lymphocytes or to induce specific suppressor T-cells. Can result from administration of antigens in very early life (before the immune system is fully developed).

immunosorbent *(Immun.)*. Use of an insoluble preparation of an antigen to bind specific antibodies from a mixture, so that the antibodies can later be eluted in pure form.

immunosuppression *(Immun.)*. Artificial suppression of immune response by the use of drugs which interfere with lymphocyte growth (antimetabolites), or by irradiation or by antibodies against lymphocytes. A state of immunosuppression can also exist as a result of infections (such as by **human immunodeficiency virus**, HIV or cytomegalovirus) which damage lymphocytes.

immunosuppressive *(Med.)*. Applied to drugs which lessen the body's rejection of, e.g. transplanted tissue and organs, or decrease the response to **auto-immune disease**.

immunotoxin *(Immun.)*. Molecule toxic to cells coupled to antibody specific for some antigen present at the surface of a particular type of cell, whereby the toxin can be selectively directed to cells of that type.

impact *(Phys.)*. For the direct impact of two elastic bodies, the ratio of the relative velocity after impact to that before impact is constant and is called the *coefficient of restitution* for the materials of which the bodies are composed. This constant has the value 0.95 for glass/glass and 0.2 for lead/lead, the values for most other solids lying between these two figures.

impact accelerometer *(Aero.)*. An **accelerometer** which measures the deceleration of an aircraft while landing.

impact crusher *(Min.Ext.)*. Machine in which soft rock is crushed by swift blows struck by rotating bars or plates. The material may break against other pieces of rock or against casing plates surrounding the rotating hammers.

impacted *(Med.)*. Firmly fixed, pressed closely in; said of a tooth which has failed to erupt, or of a fracture in which the broken bones are firmly wedged together.

impacter forging hammer *(Eng.)*. A horizontal forging machine in which two opposed cylinders propel the dies until they collide on the forging, which is worked equally on both sides.

impact extrusion *(Eng.)*. A fast, cold-working process for producing tubular components by one blow with a punch on a slug of material placed on the bottom of a die, so that the material squirts up around the punch into the die clearance.

impact ionization *(Electronics)*. The loss of orbital electrons by an atom of a crystal lattice which has experienced a high-energy collision.

impaction sampler *(Powder Tech.)*. Device for sampling particles of dust and spray droplets from a gaseous stream, which is forced through a jet and directed to a horizontal surface, usually a microscope slide smeared with grease, on which particles are collected by inertial impaction. Also *jet impactor*.

impact parameter *(Phys.)*. The distance at which two particles which collide would have passed if no interaction had occurred between them.

impact printer *(Comp.)*. Any **printer** where the standard typewriter action of 'character pressing on paper through inked ribbon' is used.

impact strength *(Eng.)*. Measured resistance of metal specimen to impact loading applied in Izod or Charpy tests.

impact test *(Eng.)*. Usually means a **notched-bar test** but it may also mean an Izod, Charpy or Fremont test performed on un-notched specimens, or a test in which the suddenly applied load is in tension instead of bending.

IMPATT diode *(Electronics)*. *IMPact Avalanche Transit-Time diode*; a microwave diode which exhibits a negative resistance characteristic due to avalanche breakdown and charge carrier transit time in chips made of gallium arsenide or silicon. When linked to a wave guide or resonant cavity or similar structure it can be used as a microwave oscillator or amplifier.

impedance *(Phys.)*. (1) Complex ratio of sinusoidal voltage to current in an electric circuit or component. Its real part is the **resistance** (dissipative or wattful impedance) and its imaginary part **reactance** (non-dissipative or wattless), which may be positive or negative according to whether the phase of the current lags or leads on that of the voltage. Resistance, reactance and impedance are all measured in ohms. Expressed symbolically, the impedance $Z = R + jX$ where R = resistance, X = reactance and $j = \sqrt{-1}$. Sometimes called *apparent resistance*. (2) Component offering electrical impedance. This is preferably termed *impedor*.

impedance bond *(Elec.Eng.)*. A special rail-bond of high reactance and low resistance designed to allow the passage of d.c. traction current but not the a.c. used for signalling purposes.

impedance circle *(Elec.Eng.)*. Locus of the end of the impedance vector in an Argand (R,X) diagram of a system, e.g. drawn to show the variation of input impedance of an improperly terminated line with frequency.

impedance coupling *(Elec.Eng.)*. The coupling of two circuits by means of a tuned circuit or an impedance.

impedance drop (rise) *(Elec.Eng.)*. A drop or rise in the voltage at the terminals of a circuit, caused by current passing through the impedance of the circuit.

impedance factor *(Elec.Eng.)*. The ratio of the impedance of a circuit to its resistance.

impedance matching *(Elec.Eng.)*. See **matching**.

impedance matching stub *(Elec.Eng.)*. See **matching stub**.

impedance matching transformer *(Elec.Eng.)*. See **matching transformer**.

impedance protective system *(Elec.Eng.)*. A discriminative protective equipment in which discrimination is secured by a measurement of the impedance between the point of installation of the relays (impedance relays) and the point of fault. Also called *distance protection*.

impedance relay *(Elec.Eng.)*. A relay, used in discriminative protective gear, whose operation depends on a measurement of the impedance of the circuit beyond the point of installation of the relay; if this falls below a certain value, when a fault occurs, the relay operates. Also called a *distance relay*.

impedance rise *(Elec.Eng.)*. See **impedance drop**.

impedance transforming filter *(Elec.Eng.)*. A filter network which has differing image impedances, and which can therefore act as a transformer over a band of frequencies.

impedance triangle *(Elec.Eng.)*. The right-angled triangle formed by the vectors representing the resistance drop, the reactance drop, and the impedance drop of a circuit carrying an alternating current.

impedance voltage *(Elec.Eng.)*. The voltage produced as a result of a current flowing through an impedance.

impedometer *(Elec.Eng.)*. Device for measuring impedances in waveguides.

impedor *(Phys.)*. Physical realization of an impedance. An inductor, capacitor, or resistor, or any combination of these.

impeller *(Aero.)*. The rotating member of a **centrifugal-flow compressor** or **supercharger**.

impeller *(Eng.)*. The rotating member of a centrifugal pump or blower, which imparts kinetic energy to the fluid.

impeller-intake guide-vanes *(Aero.)*. The curved extension of the vanes of a centrifugal impeller which extend into the intake eye or throat, and which thereby give the airflow initial rotation.

imperfect *(Build.)*. Said of a structural framework which has either more or fewer members than it would require to be determinate.

imperfect dielectric *(Phys.)*. One in which there is a loss element resulting in part of the electric energy of the applied field being used to heat the medium.

imperfect flower *(Bot.)*. A flower in which either the stamens or the carpels are lacking or, if present, nonfunctional.

imperfect fungi *(Bot.)*. The Deuteromycotina.

imperfect stage *(Bot.)*. Stage in the life cycle of e.g. a fungus in which the organism can only reproduce asexually if at all.

imperforate *(Med.)*. Not perforated; closed abnormally.

imperforate *(Zool.)*. Lacking apertures, especially of shells; said of Gastropod shells which have a solid columella.

imperial *(Build.)*. (1) A slate size, 33 × 24 in. (2) A domed roof shaped to a point at the top.

Imperial Standard Wire Gauge *(Eng.)*. See Standard Wire Gauge.

impermeable *(Chem.,Geol.)*. Not permitting the passage of liquids or gases.

impervious *(Build.)*. Said of materials which have the property of satisfactorily resisting the passage of water.

impetiginous *(Med.)*. Resembling, or of the nature of, impetigo.

impetigo *(Med.)*. A contagious skin disease, chiefly of the face and hands, due to infection with pus-forming bacteria (*Staphylococcus aureum*).

implant *(Biol.)*. A graft of an organ or tissue to an abnormal position.

implant *(Radiol.)*. The radioctive material, in an appropriate container, which is to be imbedded in a tissue for therapeutic use, e.g. needle or seed.

implantation *(Zool.)*. In Mammals, the process by which the blastocyst becomes attached to the wall of the uterus.

implementation *(Comp.)*. The various steps involved in producing a functioning system from a design.

implicit function *(Maths.)*. A variable x is an implicit function of y when x and y are connected by a relation which is not explicit. See **explicit function**.

imploding linear system *(Nuc.Eng.)*. Fusion device in which a cylindrical plasma is formed by the implosion of material lining the reactor vessel.

implosion *(Eng.)*. Mechanical collapse of a hollow structure, e.g. a cathode-ray tube.

implosive therapy *(Behav.)*. See **flooding**.

imposing stone *(Print.)*. A heavy iron-topped table on which the type matter is locked up preparatory to printing. In the early days of printing, level stone-topped tables were used.

imposition *(Print.)*. (1) The process of assembling letterpress pages of type in their proper order on the stone, arranging appropriate furniture or spacing material, and locking the whole into a chase. After imposition the unit is known as a **forme**, and from it a book section or signature is printed. (2) Assembly of film elements for platemaking in a predetermined pattern so that when the work is printed and folded the pages will run in the correct sequence. (3) A plan showing the arrangement of pages to suit the printing and folding operations and to give the desired page sequence.

impost *(Build.)*. The top member of a pier or pillar from which an arch springs.

impregnation *(Powder Tech.)*. The partial or complete filling of the pores of a powder product with an organic material, glass, salt or metal, to make it impervious or impart to it secondary properties. Vacuum, pressure and capillary forces may be employed. It may include stoving to produce setting. See **infiltration**.

impregnation *(Zool.)*. The passage of spermatozoa from the body of the male into the body of the female.

impression *(Print.)*. (1) All copies of a book printed at one time from the same type or plates. (2) The pressure applied to a type forme by the cylinder or platen.

impression cylinder *(Print.)*. The cylinder which presses the stock against the printing surface, which may be flat or cylindrical, and either letterpress, gravure or lithographic.

impression formation *(Behav.)*. A traditional area of research in social psychology, referring to the issue of how information about other individuals is integrated into a unified impression, often on the basis of very little information.

imprint *(Print.)*. The name of the publisher and/or printer which must appear on certain items, particularly books, periodicals and election literature.

imprinting *(Behav.)*. An aspect of learning in some species, through which attachment to the important parental figure develops; it involves a narrowing of the stimuli effective for several filial responses to those first encountered during a short period after birth. See **response, sensitive period**.

improper fraction *(Maths.)*. See **division**.

improving *(Eng.)*. See **softening**.

impsonite *(Min.)*. A member of the asphaltite group.

impulse *(Horol.)*. The force or blow imparted to the pendulum or balance by the escape wheel through the escapement.

impulse *(Phys.)*. When two bodies collide, over the period of impact there is a large reaction between them. Such a force can only be measured by its time integral ($\int F\,dt$) which is defined as the impulse of the force, and which equals the change of momentum produced in either body.

impulse *(Telecomm.)*. Obsolete term for *pulse*.

impulse circuit *(Telecomm.)*. In an automatic switching exchange, a source of machine-generated impulse trains for operating step-by-step switches, controlled by relays.

impulse circuit-breaker *(Elec.Eng.)*. A circuit-breaker, requiring only a small quantity of oil, in which the arc is extinguished by a mechanically produced flow of oil across the contacts.

impulse clock *(Horol.)*. One in which a master pendulum clock controls the operation of a number of slave clocks. The vibrations of the pendulum of the master clock are maintained by an electromagnetically reset gravity arm falling on to a pallet on the pendulum rod. When the gravity arm falls for impulse it also closes the circuit to the slave clocks, giving impulse to them. The **free pendulum clock** is also of the impulse type.

impulse excitation *(Electronics)*. Maintenance of oscillatory current in a tuned circuit by pulses synchronous with free oscillations, or at a sub-multiple frequency.

impulse flashover voltage *(Elec.Eng.)*. The value of the impulse voltage which just causes flashover of an insulator or other apparatus.

impulse frequency *(Telecomm.)*. The number of impulses per second in the impulse trains used in dialling and operating selectors.

impulse function *(Telecomm.)*. See **delta impulse function**.

impulse generator *(Elec.Eng.)*. Circuit providing single, or a continuous series of, pulses, generally by capacitor discharge and shaping, e.g. by the charging of capacitors in parallel and the discharging of them in series. Also called *surge generator*.

impulse inertia *(Elec.Eng.)*. That property of an insulator by which the voltage required to cause disruptive discharge varies inversely with its time of application.

impulse machine *(Telecomm.)*. A machine which generates accurately timed pulses for operating selector switches.

impulse period *(Telecomm.)*. The time between identical phases of a train of impulses: the time between the start of one impulse and the start of the next.

impulse pin *(Horol.)*. The vertical pin in the roller of the lever escapement which receives the impulse from the pallets, via the notch in the lever. It also effects the unlocking, on the reverse vibration.

impulse plane *(Horol.)*. That part of the pallet on which a tooth of the escape wheel acts when giving impulse.

impulse ratio *(Elec.Eng.)*. The ratio between the breakdown voltage of an insulator or piece of insulating material when subjected to an impulse voltage to the breakdown when subjected to a normal-frequency (50 or 60 hz) voltage.

impulse ratio *(Telecomm.)*. The ratio of the time during an impulse to the total time of impulse plus interval before another impulse.

impulse-reaction turbine *(Eng.)*. See **disk-and-drum turbine**.

impulse repeater *(Telecomm.)*. A relay mechanism for repeating impulses from one circuit to another.

impulse starter *(Aero.)*. A mechanism in a magneto which delays the rotor against a spring so that, when released, there is a strong and retarded spark to help starting.

impulse turbine *(Eng.)*. A steam turbine in which steam is expanded in nozzles and directed on blades carried by a rotor, in one or more stages, there being no change in pressure as the steam passes the blade-ring.

impulse voltage *(Elec.Eng.)*. A transient voltage lasting only for a few microseconds; very frequently used in high-voltage testing of electrical apparatus in order to simulate voltage due to lightning strokes or other similar causes.

impulse wheel *(Eng.)*. The wheel of an **impulse turbine**. Also used to denote one of the two principal types of turbines, in which the whole available head is tranformed into kinetic energy before reaching the wheel. See also **Pelton wheel**.

impulsive sound *(Acous.)*. Short sharp sound, the energy spectrum of which spreads over a wide frequency range.

impurity *(Electronics)*. Small proportion of *foreign matter*, e.g. indium, arsenic, gallium, lead, lithium, tin etc. added to a pure semiconductor, e.g. silicon or germanium to obtain the required type of conduction and conductivity for diodes, transistors and other solid-state devices. The impurity in the crystal lattice may add to, or subtract from the average densities of free electrons and holes in the semiconductor. See **acceptor, carriers, donor**.

impurity levels *(Electronics)*. Abnormal energy levels arising from slight impurities, resulting in conduction in semiconductors.

In *(Chem.)*. The symbol for *indium*.

in *(For.)*. See **eng**.

in- *(Genrl.)*. Prefix from Latin, meaning either 'in(to)' or 'not'.

inactivation *(Chem.)*. The destruction of the activity of a catalyst, serum etc.

inanition *(Med.)*. Exhaustion and wasting of the body from lack of food.

inband *(Build.)*. A header stone.

inband rybat *(Build.)*. A header stone laid to form the jamb of an opening.

inbred *(Biol.)*. The condition of the offspring produced by *inbreeding*. See also **inbreeding coefficient**.

inbred line *(Biol.)*. A strain which has been *inbred* over many generations and whose **inbreeding coefficient** is nearly 100%. All members are genetically identical and *homozygous* at all loci, or very nearly so. Cf. **isogenic**.

inbreeding *(Biol.)*. The mating together of individuals that are related by descent. The offspring are *inbred* to an extent depending on the degree of relationship. (See **inbreeding coefficient**). Inbreeding produces *homozygosis*.

inbreeding coefficient *(Biol.)*. A measure of the degree to which an individual is *inbred*. Ranges from 0 when the parents are unrelated to 100% when the parents for many generations back have been related. Usually symbolized by *F*.

inbreeding depression *(Biol.)*. The reduction of desirable characters such as growth rate, yield, fertility, consequent on the *homozygosis* produced by *inbreeding*, especially in those which normally outbreed.

inbye *(Min.Ext.)*. The direction from a haulage way to a working face.

incandescence *(Phys.)*. The emission of light by a substance because of its high temperature, e.g., a glowing electric-lamp filament. In the case of solids and liquids, there is a relation between the colour of the light and the temperature. Cf. **luminescence**.

incandescent lamp *(Phys.)*. A lamp in which light is produced by heating some substance to a white or red heat, e.g. a filament lamp.

incentive learning *(Behav.)*. A motivational concept that refers to the expectation of rewards or punishments from the environment. The high or low incentive value of a goal is reflected in the amount of energy the organism will expend to obtain it.

incept *(Bot.)*. The rudiment of an organ.

incertae sedis *(Bot.)*. Of uncertain taxonomic position.

incertum *(Build.)*. An early form of masonry work in which squared stones were used as a facing, with rubble filling as a backing.

incest *(Med.)*. Sexual intercourse between close relatives i.e. brother and sister or father and daughter.

incest taboo *(Behav.)*. A strong negative social sanction which forbids sexual relations between members of the same immediate family, found in all human societies.

inching *(Eng.)*. Very slow, closely controlled step-by-step movement of a usually fast moving machine, e.g. (Print.) turning the press by small amounts at a time, using an electrical inching button switch. Cf. **crawl**.

inching starter *(Elec.Eng.)*. An electric-motor starter in which provision is made for inching the motor, i.e. running it very slowly for such purposes as the threading of the paper in a printing press.

inch-penny weight *(Min.Ext.)*. In valuation of gold ore, the width of the lode or reef measured normal to the enclosing rock, multiplied by the assay value in penny weights per ton.

inch-tool *(Build.)*. A steel chisel having a cutting edge 1 in (25 mm) wide, used by the mason for dressing stone.

inch trim moment *(Ships)*. Same as *moment to change trim one inch*. Abbrev. *ITM*.

incidence *(Med.)*. The frequency with which new cases of a given disease presents in a particular period for a given population. Different from **prevalence**.

incidence, angle of *(Aero.)*. See **angle of incidence**.

incidental learning *(Behav.)*. Learning without trying to learn. Cf **intentional learning**.

incident beam *(Phys.)*. Any wave or particle beam the path of which intercepts a surface of discontinuity.

incipient plasmolysis *(Bot.)*. The state of a plant cell in which the **turgor** pressure is zero but the protoplast is in contact with the cell wall all round. Loss of water will result in **plasmolysis**; uptake of water will generate turgor.

incise *(Arch.)*. To cut in; to carve.

incised meander *(Geol.)*. An entrenched bend of a river, which results from renewed down-cutting at a period of rejuvenation.

incisiform *(Zool.)*. Shaped like an incisor tooth.

incision *(Med.)*. A surgical cut.

incisors *(Zool.)*. The front teeth of Mammals; they have a single root, are adapted for cutting, and are the only teeth borne by the premaxillae in the upper jaw.

incisura *(Med.)*. A cut or notch. Various notches in the body are thus designated.

inclination *(Phys.)*. See **dip**.

inclination factor *(Phys.)*. Used in Fresnel's theory of the propagation of light waves, where the disturbance at a point, due to the contributions from the secondary waves, is assumed to depend on the angle θ between the normal to the primary wavefront and the direction to the point. The term containing $f(\theta)$ is the *inclination factor* or *obliquity factor*.

incline *(Surv.)*. See **gradient**.

inclined-catenary construction *(Elec.Eng.)*. A catenary construction for the overhead contact wire of an electric traction system; in it, the catenary wire is not placed vertically above the contact wire.

inclined plane *(Phys.)*. For a smooth plane inclined at an angle θ to the horizontal, the force parallel to the plane required just to move a mass up the plane is *mg* sinθ. The inclined plane may therefore be regarded as a machine having a velocity ratio of cosecθ.

inclined shore *(Build.)*. See **raking shore**.

inclining experiment *(Hyd.Eng.)*. A practical method of determining the metacentric height and the height of the centre of gravity of a floating vessel; accomplished by observing the angle of heel of the vessel resulting from a measured transverse movement of a known weight across the deck.

inclinometer *(Phys.,Surv.)*. See **dip needle**.

inclusion *(Eng.)*. A particle of alien material retained in a solid metal. Such inclusions are generally oxides, sulphides or silicates of one or other of the component metals of the alloy, but may also be particles of refractory materials picked up from the furnaces or ladle lining.

inclusion *(Min.)*. A foreign body (gas, liquid, glass or mineral) enclosed by a mineral. Fluid inclusions (e.g. liquid carbon dioxide) may be used to study the genesis of the minerals in which they occur. See also **xenolith**.

inclusion bodies *(Med.)*. Particulate bodies found in the cells of tissue infected with a virus.

inclusive fitness *(Ecol.)*. The number of copies of an individual's genes passed on to the next generation plus an additional number passed on by relatives as a result of the individual's behaviour towards those relatives. Allied to the concept of **kin selection**.

incoherent *(Phys.)*. Said of radiation of the same frequency emitted from discrete sources with random phase relationships. All light sources except the *laser* emit incoherent radiation.

incoming feeder *(Elec.Eng.)*. A feeder in a sub-station through which power is received.

incompatibility *(Bot.)*. (1) The consistent failure of fertilization or hyphal fusion between particular combinations of individual plants, algae, fungi etc. See also **self-incompatibility**. (2) Interaction between stock and scion resulting in the failure of the graft either (a) immediately, leading to the rapid death of at least the scion, or (b) after some years of apparently successful growth, resulting in fracture across the graft union. (3) The relationship between a plant and a pathogen to which the plant is not susceptible.

incompatibility *(Med.)*. Mismatch which may be due to immunological, chemical or physical factors. Commonly applied to blood transfusion of mismatched or incompatible blood groups.

incompatible behaviours *(Behav.)*. Behaviour patterns which cannot occur simultaneously, because of reciprocal inhibition in the case of reflexes for example, but also due to psychological factors, such as the limits of attention.

incompetence *(Med.)*. Inability to perform proper function; said especially of diseased valves of the heart which allow the blood to pass in the wrong direction, e.g. *aortic incompetence*, *mitral incompetence*.

incomplete flower *(Bot.)*. A flower in which the calyx and corolla (or one of these) are lacking.

incomplete metamorphosis *(Zool.)*. In Insects, a more or less gradual change from the immature to the mature state, a pupal stage being absent and the young forms resembling the parents, except in the absence of wings and mature sexual organs.

incomplete reaction *(Chem.)*. A reversible reaction which is allowed to reach equilibrium, a mixture of reactants and reaction products being obtained.

incompressible volume *(Chem.)*. See **co-volume**.

inconels *(Eng.)*. Nickel-based heat-resistant alloys containing some 13% of chromium, 6% iron, and a little manganese, silicon, or copper. *Inconel X* is much used in gas-turbine blades.

incontinence *(Med.)*. Inability to retain voluntarily natural excretions of the body (e.g. faeces and urine); lack of self-control.

incoordination *(Med.)*. Inability to combine muscular movements in the proper performance of an action, the component muscle groups working independently instead of together.

incremental hysteresis loss *(Elec.Eng.)*. A small pulsation of the magnetic field about a fixed value leading to a small hysteresis loop on the boundary of a full loop.

incremental induction *(Elec.Eng.)*. The difference between the maximum and minimum value of a magnetic induction at a point in the polarized material, when subjected to a small cycle of magnetization; cf. **incremental hysteresis loss**.

incremental iron losses *(Elec.Eng.)*. A term sometimes used to denote iron losses occurring in an a.c. machine due to frequencies higher than the fundamental, e.g. tooth pulsation losses.

incremental permeability *(Elec.Eng.)*. The gradient of the curve relating flux density to magnetizing force (the *B/H curve*). This represents the effective permeability for a small alternating field superimposed on a larger steady field.

incremental plotter *(Comp.)*. Graph plotter which receives input data specifying increments to its current position, rather than data specifying co-ordinates. Cf. **digital plotter**.

incremental resistance *(Elec.Eng.)*. The small signal resistance for a component or network, $r = \Delta V/\Delta I$.

incrustation *(Build.)*. A term applied to a wall facing which is of different material from that forming the rest of the wall.

incubation *(Behav.)*. Behaviour which maintains the eggs of birds and other species in a fairly stable thermal and gaseous environment; most bird species accomplish it by sitting on the eggs, but other methods are also used by various species of birds and insects, e.g. mound-building, sunning.

incubation *(Med.)*. The period intervening between the infection of a host by bacteria or viruses and the appearance of the first symptoms.

incubous *(Bot.)*. Said of the leaf of a liverwort when its upper border (the border towards the apex of the stem) overlaps the lower border of the next leaf above it and on the same side of the stem.

incudectomy *(Med.)*. Removal of the incus by operation.

incurrent *(Zool.)*. Carrying an ingoing current; said of ducts, and, in certain *Porifera*, of canals leading from the exterior to the spongocoel.

incus *(Zool.)*. In Mammals, an ear ossicle, derived from the quadrate; more generally, any anvil-shaped structure. pl. *incudes*.

indanthrene *(Chem.)*. $C_{28}H_{14}O_4N_2$. *N-dihydro-1,2,2′,1′-anthraquinone-azine*, an anthraquinone vat dyestuff, a dark-blue powder, practically insoluble in water and organic solvents. For dyeing purposes, indanthrene is reduced by sodium hydrosulphite to the water-soluble salt of the dihydro derivative, and re-oxidized to indanthrene by exposure to air. TNs *Indanthrone*, *Caledon blue* and *Duranthrene*.

indeciduate *(Zool.)*. Said of Mammals in which the maternal part of the placenta does not come away at birth.

indefinite *(Bot.)*. (1) Numerous but not fixed in number. (2) Monopodiol growth. (3) A racemose inflorescence.

indefinite integral *(Maths.)*. See **integral**.

indehiscent *(Bot.)*. An organ which does not open spontaneously to release the seeds, spores etc. Cf. **dehiscence**.

indene *(Chem.)*. An aromatic double-ring liquid hydrocarbon (C_9H_8) occurring in coal-tar, bp 182°C. Usually contains coumarone.

indent *(Build.)*. A notch made in a timber.

indent *(Print.)*. To commence a line with a blank space, which in bookwork paragraphs may be 1, 1½, or 2 ems, according to the width of the line.

indentation test *(Build.,Civ.Eng.)*. A test for a paving, roofing, or roadmaking asphalt, in which a steady load is applied, under constant temperature conditions, to the asphalt surface, through the sector of a wheel resting upon it, the amount of indentation being measured after a fixed time.

indented bar *(Civ.Eng.)*. A special type of reinforcing bar used in reinforced concrete work to provide a mechanical bond and having for its full length a series of depressions and ridges all round.

indenter *(Civ.Eng.)*. A roller having projections from its curved surface, so that, when it is rolled over newly laid asphalt paving, indentations shall be left in the latter surface to render it non-skid. Also called *branding iron*, *crimper*.

independent axle-drive *(Elec.Eng.)*. See **individual axle-drive**.

independent chuck *(Eng.)*. A lathe chuck in which each of the jaws is moved independently by a key; used for work of irregular shape, or when very accurate centring is necessary.

independent equations *(Maths.)*. A set of equations none of which can be deduced from a combination of any of the others.

independent feeder *(Elec.Eng.)*. A feeder in an electric-power distribution system which is used solely for supply to a substation or a feeding point, and not as an inter-connector. Also called *dead-ended feeder*, *radial feeder*.

independent particle model (of a nucleus) *(Phys.)*. Model in which each nucleon is assumed to act quite separately in a common field to which they all contribute.

independent seconds watch *(Horol.)*. A watch having an independent train for driving the seconds hand.

independent suspension *(Autos.)*. A springing system in which the wheels are not connected by an axle beam, but are mounted separately on the chassis through the medium of springs and guide links, so as to be capable of independent vertical movement.

independent time-lag *(Elec.Eng.)*. See **definite time-lag**.

independent trip *(Elec.Eng.)*. A tripping device for a circuit-breaker, starter, or similar apparatus, in which the current operating the device is independent of the current flowing in the circuit to which the device is connected.

independent variable *(Maths.)*. See **dependent variable**.

indestructibility of matter *(Chem.)*. See **law of conservation of matter**.

indeterminacy principle *(Phys.)*. See **uncertainty principle**.

indeterminate *(Eng.)*. Said of a structure which is **redundant**. Cf. **determinate**.

indeterminate equations *(Maths.)*. Equations which by reason of certain relations between the coefficients, or from insufficient data, have an infinite number of solutions.

index *(Horol.)*. The regulating lever by means of which the rate of a watch may be adjusted. The lever is usually carried on the balance cock, and its short end carries the curb pins, the long end moving over a scale which indicates the amount of movement given to the curb pins. One end of the scale is marked *A* (advance) and the other *R* (retard). Hence, if the watch is losing, the index is moved towards *A*.

index *(Maths.)*. The small number written to the right of and above a number or term to indicate how many times that number or term has to be multiplied by itself, e.g. x in a^x, $(a+b)^2$. Note: $a^0=1$, $a^{\frac{1}{2}}=\sqrt{a}$, $a^{x/y}=\sqrt[y]{a^x}$, $a^{-1}=\frac{1}{a}$, $a^x=e^{x\log a}$. Also called *exponent*.

index case *(Med.)*. The first or original case of a disease. A term used in the **epidemiology** of infectious disease. In genetics synonymous with the *proband* or *propositus*.

indexed address *(Comp.)*. The **direct-** or **indirect address** is modified by the addition of a number held in an **index register**.

indexed sequential access *(Comp.)*. Process of storing or retrieving data directly, but only after reading an index to locate the address of that item. See **direct access**.

index error *(Surv.)*. Difference between the horizontal or vertical angular reading of theodolite and the true line of collimation, with regard to concentric centring of the azimuth and plate circle and accurate engraving of the reading lines.

index fossil *(Geol.)*. A fossil species which characterizes a particular geological **horizon**. It tends to be abundant, with a narrow time range and a wide geographical spread.

indexing head *(Eng.)*. A machine-tool attachment for rotating the work through any required angle, so that faces can be machined, holes drilled etc. in definite angular relationship.

index mineral *(Geol.)*. One whose appearance marks a particular grade of metamorphism in progressive regional metamorphism.

index of refraction *(Phys.)*. See **refractive index**.

Indian hemp *(Bot.)*. See **cannabis**.

Indian ink *(Genrl.)*. Ink in a solid form made from lamp-black mixed with parchment size or fish glue. Rubbed down in water it produces an intensely black permanent ink, used for line and wash-drawings etc.

Indian topaz *(Min.)*. See **citrine**; also a misnomer for yellow corundum.

India paper *(Paper)*. A thin, strong, opaque rag paper, made for Bibles and other books where many pages are required in a small compass.

India-rubber *(Chem.)*. See **rubber**.

indicated air speed *(Aero.)*. The reading of an air-speed indicator which when corrected for instrument errors, reads low by a factor equal to the square root of the relative air density as the latter falls with altitude. Abbrev. *IAS*.

indicated (horse-)power *(Eng.)*. Of a reciprocating engine, the (horse-)power developed by the pressure-volume changes of the working agent within the cylinder; it exceeds the useful or brake (horse-)power at the crankshaft by the power lost in friction and pumping. Abbrev. *IHP*.

indicated mean effective pressure *(Eng.)*. The average pressure exerted by the working fluid in an engine cylinder throughout the cycle, equal to the mean height of the indicator diagram in kN/m^2 or lbf/in^2. Abbrev. *IMEP*.

indicated ore *(Min.Ext.)*. Proved limits of deposit, in the light of known geology of mine and economic factors.

indicated thermal efficiency *(Eng.)*. The ratio between the indicated power output of an engine and the rate of supply of energy in the steam or fuel.

indicating instrument *(Eng.)*. One in which the immediate value only of the measured quantity is visually indicated.

indication *(Eng.)*. A sign on inspection which indicates an imperfection of the material.

indicator *(Bot.)*. (1) The presence of a species which gives an indication of features of the habitat, or method of land management by growing well or badly, e.g. the stinging nettle which indicates a high level of available phosphorus in the soil. (2) Plants which react to a particular pathogen or environmental factor with obvious symptoms and may, therefore, be used to identify that pathogen or factor.

indicator *(Chem.)*. (1) A substance whose colour varies with the acidity or alkalinity of the solution in which it is dissolved. (2) Any substance used to indicate the completion of a chemical reaction, generally by a change in colour.

indicator *(Elec.Eng.)*. Same as *annunciator*.

indicator *(Eng.)*. An instrument for obtaining a diagram of the pressure-volume or pressure-time changes in an engine or compressor cylinder during the working cycle.

indicator card *(Eng.)*. A chart on which the trace of an **indicator** is recorded, producing an **indicator diagram**.

indicator diagram *(Eng.)*. A graphical representation of the pressure and volume changes undergone by a fluid, while performing a work-cycle in the cylinder of an engine or compressor, the area representing, to scale, the work done during the cycle. See **indicated mean effective pressure**, **light-spring diagram**.

indicator gate *(Electronics)*. A step or pulse signal applied to a CRT to control its sensitivity in order to highlight a certain part of the display.

indicator range *(Chem.)*. The range of pH-values within which an indicator (1) changes colour.

indicator species analysis *(Ecol.)*. Multivariate statistical technique to enable classification of vegetation on the basis of the presence or absence of key species.

indicator tube *(Electronics)*. Miniature CRT in which size or shape of target glow varies with input signal.

indicator vein *(Min.Ext.)*. In prospecting, one associated with the lode or vein being traced, thus guiding the search.

indices of crystal faces *(Crystal.)*. See **Miller indices**.

indicial admittance *(Telecomm.)*. Transient current response of a circuit to the application of a **step function** of one volt, using Heaviside operational calculus.

indicial response *(Telecomm.)*. Output waveform from a system when a step pulse of unit magnitude is applied to the output.

indicolite, indigolite *(Min.)*. A blue (either pale or bluish-black) variety of tourmaline.

indigenous *(Zool.)*. Native; not imported.

indigestion *(Med.)*. A condition, marked by pain and discomfort, in which the normal digestive functions are impeded. Also *dyspepsia*.

indigo *(Chem.)*. $C_{16}H_{10}N_2O_2$, a dye occurring in a number of plants, especially in species of *Indigofera*, in the form of a glucoside. It is an indole derivative. Indigo is a very important blue vat dyestuff, and can be synthesized in various ways.

indigo copper *(Min.)*. See **covellite**.

indigolite *(Min.)*. See **indicolite**.

indirect address *(Comp.)*. The address specified in the instruction is that of a location which in turn contains the required address.

indirect-arc furnace *(Elec.Eng.)*. An electric-arc furnace in which the arc is struck between two electrodes mounted above the charge, the latter being heated chiefly by radiation.

indirect cylinder *(Build.)*. Hot-water system in which the boiler water only heats the supply in the cylinder, itself fed from an independent second circuit.

indirect-fired furnace *(Eng.)*. One in which the combustion chamber is separate from the one in which the charge is heated.

indirect heating *(Eng.)*. A system of heating by convection, as opposed to *direct heating* by radiation.

indirect immunofluorescence *(Biol.)*. Technique in which a specific antibody is first bound to its antigen. A fluorochrome visible under **fluorescence microscopy** and conjugated to a second antibody specific to the first is then used to detect the presence of the first antibody and therefore the original antigen.

indirect lighting *(Elec.Eng.)*. A system of lighting in which more than 90% of the total light flux from the fittings is emitted in the upper hemisphere.

indirectly-heated cathode *(Electronics)*. One with an internal heater, highly insulated from the cathode on a surrounding ceramic cylinder.

indirectly-heated valve *(Electronics)*. A valve using an **indirectly heated cathode**.

indirect metamorphosis *(Zool.)*. The complex change characterizing the life cycles of Endopterygota; the young are larvae and the imago is preceded by a pupal instar.

indirect wave, ray *(Telecomm.)*. See **ionospheric wave**.

indium *(Chem.)*. A silvery metallic element in the third group of the periodic system. Symbol In, at. no. 49, r.a.m. 114.82, mp 155°C, bp 2100°C, rel. density 7.28 at 13°C, electrical resistivity 9×10^{-8} ohm metres. Found in traces in zinc ores. The metal is soft and marks paper like lead; it forms compounds with carbon compounds. It has a large cross-section for slow neutrons and so is readily activated. Also used in manufacture of transistors and as bonding material for acoustic transducers.

individual *(Zool.)*. A single member of a species; a single zooid of a colony of Coelenterata or Polyzoa; a single unit or specimen.

individual distance *(Behav.)*. A spatial relationship between members of a flock (or other social group, e.g. a school or herd) which is maintained through the two conflicting tendencies of social attraction and aggression/avoidance.

individual drive *(Elec.Eng.)*. A system used for the electric operation of factories, in which each machine is driven by a separate electric motor. See also **individual axle-drive**.

individual (independent) axle-drive *(Elec.Eng.)*. A term applied to the arrangement of an electric locomotive in which each driving axle is driven by a separate motor.

indol-3-butyric acid *(Bot.)*. A synthetic plant growth regulator with auxin-like activity, used especially in **rooting compounds**.

indole *(Chem.)*. *Benzpyrrole*. C_8H_7N, colourless plates, mp 52°C, bp (decomposition) 245°C, volatile in steam. Indole forms the basis of the indigo molecule and results from the condensation of a benzene nucleus with a pyrrole ring. ⇨

indole-3-acetic acid *(Bot.)*. *Auxin, heteroauxin, IAA*. The commonest naturally-occurring plant growth substance of the **auxin** type.

indolent *(Med.)*. Causing little or no pain, e.g. *indolent* ulcer.

indomethacin *(Med.)*. A **non-steroidal anti-inflammatory drug** used in the treatment of **rheumatoid arthritis** and **gout**.

induced charge *(Phys.)*. That produced on a conductor as a result of a charge on a neighbouring conductor.

induced current *(Phys.)*. That which flows in a circuit as a result of induced e.m.f.

induced dipole moment *(Phys.)*. Induced moment of an atom or molecule which results from the application of an electric or magnetic field.

induced drag *(Aero.)*. The portion of the **drag** of an aircraft attributable to the derivation of lift.

induced draught *(Eng.)*. A forced draught system used for boiler furnaces, in which a fan placed in the uptake induces an air-flow through the furnace. See **balanced draught, extraction fan, fan**.

induced e.m.f. *(Phys.)*. That which appears in a circuit as a result of changes in the interlinkages of magnetic flux with part of the circuit. The e.m.f. in the secondary of a transformer. Discovered by Faraday in 1831. See **Faraday's law of induction**.

induced moving-magnet instrument *(Elec.Eng.)*. An instrument whose operation depends on the force exerted by the resultant of the fields produced by a fixed coil carrying a current, and a permanent magnet fixed at an angle thereto, on a movable piece of magnetic material.

induced polarization *(Phys.)*. That which is not permanent in a dielectric, but arises from applied fields.

induced radioactivity *(Phys.)*. That induced in non-radioactive elements by neutrons in a reactor, or protons or deuterons in a cyclotron or linear accelerator. X-rays or gamma-rays do not induce radioactivity unless the gamma-ray energy is exceptionally high.

induced reaction *(Chem.)*. A chemical reaction which is accelerated by the simultaneous occurrence in the same system of a second, rapid reaction.

inducer *(Biol.)*. An agent which increases the transcription of specific genes.

inducible enzyme *(Bot.)*. An enzyme that is formed only in response to an inducing agent, often its substrate. Cf. **constitutive enzyme**.

inductance *(Phys.)*. (1) That property of an element or circuit which, when carrying a current, is characterized by the formation of a magnetic field and the storage of magnetic energy. (2) The magnitude of such capability. See **inductance coefficient, self-inductance, mutual inductance**.

inductance-capacitance filter *(Elec.Eng.)*. A circuit comprising inductors and capacitors connected in such a way as to give a frequency-sensitive output when connected to an input containing different frequency components. The filters may be designed to allow through low frequency components, high frequency components or a band of frequency components between two set limits. These are termed *low-pass, high-pass* and *band-pass* respectively. Often abbreviated to *I-C filter*.

inductance coefficient *(Elec.Eng.)*. Property of a component (inductor) in a circuit whereby back e.m.f. arises because of the rate of change of a current. It is 1 *henry* when 1 volt is generated by a rate of change of 1 ampere per second.

inductance coil *(Elec.Eng.)*. Coil, with or without an iron circuit, for adding inductance to a circuit. Also called *inductor*.

inductance coupling *(Elec.Eng.)*. That between 2 circuits whereby a changing current in one induces a current in the other, proportional to the rate of change of the first current.

inductance factor *(Elec.Eng.)*. A term sometimes used to denote the ratio of the reactive current to the total current in an a.c. circuit, i.e. the *sine* of the angle of lag.

induction *(Chem.)*. Change in the electronic configuration and hence reactivity of one group in a molecule upon addition of a neighbouring polar group.

induction *(Maths.)*. See **mathematical induction**.

induction *(Zool.)*. The production of a definite condition by the action of an external factor.

induction balance *(Elec.Eng.)*. An electrical network, i.e. a bridge, to measure inductance.

induction coil *(Elec.Eng.)*. Transformer for producing high-voltage pulses in the secondary winding, obtained from interrupted d.c. in the primary, as for a petrol engine. The original Ruhmkorff induction coil was magnetically open-circuit and self-interrupting, like a buzzer or relay; used for early discharges in gas tubes.

induction compass *(Phys.)*. One which indicates the direction of the earth's magnetic field by a rotating coil, in which an e.m.f. is induced.

induction field *(Phys.)*. See **field**.

induction flame damper *(Aero.)*. See **flame trap**.

induction furnace *(Elec.Eng.)*. Application of induction heating in which the metal to be melted forms the secondary of a transformer.

induction generator *(Elec.Eng.)*. An electric generator similar in construction and operation to an induction motor; in order to generate, it must be driven above synchronous speed and must be excited from the a.c. supply into which it is delivering power.

induction hardening *(Eng.)*. Using high-frequency induction to heat the part for hardening. The method is rapid and lends itself to restriction of the hot zone.

induction heating *(Elec.Eng.)*. That arising from eddy currents in conducting material, e.g. solder, profiles of gear-wheels, conductor coils around vessels for heating liquids, etc. Generated with a high-frequency source, usually oscillators of high power, operating at 10^6 to 10^7 Hz. Also *eddy-current heating*.

induction, inlet manifold *(Autos.)*. In a multi-cylinder petrol-engine, the branched pipe which leads the mixture from the carburettor to the combustion chambers.

induction instrument *(Elec.Eng.)*. An electrical measuring instrument in which the pointer is moved as the result of the interaction between an alternating flux produced by the quantity to be measured, and currents induced by this flux in a disk.

induction lamp *(Elec.Eng.)*. See **neon-**.

induction machine *(Elec.Eng.)*. See **induction motor**, **induction generator**.

induction meter *(Elec.Eng.)*. The most common type of a.c. integrating meter; it is a motor meter in which the torque is produced as the result of the interaction between an alternating flux and currents induced in a disk by this flux.

induction motor *(Elec.Eng.)*. An a.c. motor in which currents in the primary winding (connected to the supply) set up a flux which causes currents to be induced in the secondary winding (usually the rotor); these currents interact with the flux to produce rotation. Also called *asynchronous motor, nonsynchronous motor*.

induction motor-generator *(Elec.Eng.)*. A motor-generator set driven by an induction motor.

induction period *(Chem.)*. The interval of time between the initiation of a chemical reaction and its observable occurrence.

induction port, -valve, *(Autos.)*. The port, valve, etc., through which the charge is induced into the cylinder during the suction stroke. Also *inlet port, -valve*.

induction regulator *(Elec.Eng.)*. A voltage regulator having a winding connected in series with the supply; voltages are induced in this winding from a primary winding connected across the supply, and regulation of the voltage is carried out by varying the relative position of the two windings.

induction relay *(Elec.Eng.)*. A relay, for use in an electrical circuit, in which the contacts are closed as the result of the interaction between an alternating flux and currents induced in a disk by this flux.

induction stroke *(Autos.)*. The suction stroke, charging stroke, or intake stroke, during which the working charge or air is induced into the cylinder of an engine.

inductive *(Elec.Eng.)*. Said of an electric circuit or piece of apparatus which possesses self or mutual inductance, which tends to prevent current changes. Always present to some extent, but may often be neglected.

inductive circuit *(Elec.Eng.)*. One in which effects arising from inductances are not negligible, the back e.m.f. tending to oppose a change in current, leading to sparking or arcing at contacts which attempt to open the circuit.

inductive drop *(Elec.Eng.)*. Voltage drop produced in an a.c. circuit owing to its self or mutual inductance.

inductive load *(Elec.Eng.)*. Terminating impedance which is markedly inductive, taking current lagging in phase on the source e.m.f., e.g. electrodynamic loudspeaker or motor. Also *lagging load*.

inductive neutralization *(Elec.Eng.)*. An amplifier in which the feedback of the self capacitance of the circuit elements is balanced by the equal and opposite susceptance of an inductor.

inductive pick-off *(Elec.Eng.)*. One in which changes in reluctance of a laminated path alter a current or generate an e.m.f. in a winding.

inductive reactance *(Elec.Eng.)*. That which relates the current through an inductance to the voltage appearing across it. Of magnitude WL, where W is angular frequency of supply and L is inductance in **henry** and associated phase angle such that current lags voltage by 90°.

inductive reaction *(Elec.Eng.)*. Same as **electromagnetic reaction**.

inductive resistor *(Elec.Eng.)*. Wirewound resistor having appreciable inductance at frequencies in use.

inductor *(Chem.)*. A substance which accelerates a slow reaction between two or more substances by reacting rapidly with one of the reactants.

inductor *(Elec.Eng.)*. Any circuit component whose inductance cannot be treated as negligible.

inductor generator *(Elec.Eng.)*. An electric generator in which the field and armature windings are fixed relative to each other, the necessary changes of flux to produce the e.m.f. being produced by rotating masses of magnetic material.

inductor loudspeaker *(Acous.)*. See **electromagnetic loudspeaker**.

indumentum *(Bot.,Zool.)*. (1) The hairy covering of a plant. (2) A covering of hair and feathers.

indurated *(Eng.)*. Hardened, made hard. n. *induration*.

induration *(Geol.)*. The process of hardening a soft sediment by heat, pressure and cementation. cf. **diagenisis**.

induration *(Med.)*. Hardness. Often used to describe tumours.

indusium *(Zool.)*. In some Insects, a third embryonic envelope lying between the chorion and the amnion in the early stages of development of the egg; a cerebral convolution of the brain in higher Vertebrates; an insect larva case. adjs. *indusiate, indusiform*.

industrial diamond *(Min.)*. Small diamonds, not of gemstone quality, e.g. **black diamond** and **bort**; used to cut rock in borehole drilling, and in abrasive grinding. Now synthesized on a considerable scale by subjecting carbon to ultra-high pressures and temperature.

industrial frequency *(Elec.Eng.)*. A term used to denote the frequency of the alternating current used for ordinary industrial and domestic purposes, usually 50 or 60 hz. Also *mains* or *power frequency*.

industrialized building *(Build.)*. See **system building**.

industrial melanism *(Ecol.)*. **Melanism** which has developed as a response to blackening of trees, etc., by industrial pollution. This favours melanic forms, especially among moths which rest on trees during the day.

inelastic collision *(Phys.)*. In atomic or nuclear physics, a collision in which there is a change in the total energies of the particles concerned resulting from the excitation or de-excitation of one or both of the particles. See **collision**.

inelastic scattering *(Phys.)*. See **scattering**.

inequality *(Astron.)*. The term used to signify any departure from uniformity in orbital motion; it may be (a) *periodic*, that is, completing a full cycle within a specific time and then repeating it; or (b) *secular*, that is, increasing steadily in magnitude with time.

inequality *(Maths.)*. A statement as to which is the larger or smaller of two quantities. The statement that a is larger than b is written $a>b$, and the consequential statement that b is smaller than a, $b<a$. The inequality sign, $>$ or $<$, can be coupled with the equality sign, $=$, and is then written $\geqq$ or $\leqq$.

inequipotent *(Zool.)*. Possessing different potentialities for development and differentiation.

inequivalve *(Zool.)*. Having the 2 valves of the shell unequal.

inert *(Chem.)*. Not readily changed by chemical means.

inert anode *(Ships)*. An anode of platinized titanium, used in **cathodic protection**. Requires an impressed direct current. Long-lasting. Cf. **virtually inert anode**.

inert gases *(Chem.)*. See **noble gases**.

inertia *(Image Tech.)*. A factor used in some systems of photographic speed rating, obtained by extrapolating the linear portion of the **characteristic curve** to indicate the nominal exposure for zero density.

inertia *(Phys.)*. The property of a body, proportional to its mass, which opposes a change in the motion of the body. See **inertial forces, inertial reference frame**.

inertia governor *(Eng.)*. A shaft type of centrifugal governor using an eccentrically pivoted weighted arm, which responds rapidly to speed fluctuations by reason of its inertia, and in such a way as to suppress them.

inertial confinement *(Nuc.Eng.)*. In fusion studies, short-term plasma confinement arising from inertial resistance to outward forces (mainly by the compression and heating of deuterium or mixed deuterium-tritium pellets by a powerful laser). See **containment, magnetic confinement, inertial fusion system**.

inertial damping *(Phys.)*. That which depends on the acceleration of a system, and not velocity.

inertial force *(Phys.)*. Newton's laws of motion do not apply in accelerating and rotating frames of reference. Newton's laws may still be used in these frames if *inertial forces* are introduced to preserve the second law of motion. In the case of rotating frames these forces are the centrifugal and Coriolis forces.

inertial fusion system *(Nuc.Eng.)*. System in which small capsules (*pellets*) containing deuterium and tritium are injected into a reaction chamber and ignited by high energy laser or ion beams. See **magnetic confinement fusion system**.

inertial guidance *(Aero.)*. Navigation of a aircraft, spacecraft or missile by measuring the inertial forces during flight and comparing them with a program held on board. It is not subject to outside interference.

inertial impaction *(Powder Tech.)*. Method of collecting small particles of dust and droplets from a fluid stream by allowing them to impinge upon an interposed deflecting surface.

inertial reference frames *(Phys.)*. In mechanics, a reference frame in which Newton's first law of motion is valid.

inertia switch *(Elec.Eng.)*. One operated by an abrupt change in its velocity, as for some meters, to avoid overloading.

inertinite *(Geol.)*. A carbon-rich **maceral** found in coal.

inert metal *(Phys.)*. Alloy (usually Ti-Zr) for which scattering of neutrons by nuclei is negligible.

I neutrons *(Phys.)*. Those possessing such energy as to undergo resonance absorption by iodine.

infanticide *(Med.)*. Killing of an infant, particularly the killing of a newborn infant by its mother.

infantile paralysis *(Med.)*. See **poliomyelitis**.

infantilism *(Med.)*. A disturbance of growth, the persistence of infantile characters being associated with general retardation of development.

infarct *(Med.)*. That part of an organ which has had its blood supply cut off, the area so deprived undergoing aseptic necrosis.

infarction *(Med.)*. The formation of an infarct; the infarct itself.

infection *(Med.)*. The invasion of body tissue by living micro-organisms, with the consequent production in it of morbid change; a diseased condition caused by such invasion; the infecting micro-organism itself.

infectious anemia of horses *(Vet.)*. *Swamp fever*. A viral disease of equines spreading by biting insects and other mechanical vectors. Symptoms include pyrexia, depression, oedema and anemia. Virus found in all tissues and persist in **white blood cells** for life. Vaccines available.

infectious avian bronchitis *(Vet.)*. An acute, highly contagious respiratory disease of chickens, caused by a virus and associated with inflammation of the respiratory tract, especially the trachea and bronchi; the main symptoms are nasal discharge, gasping, rales, and coughing.

infectious avian encephalomyelitis *(Vet.)*. *Epidemic tremor*. An encephalomyelitis of young chicks, caused by a virus, and characterized by muscular inco-ordination, muscular tremor, and death.

infectious bovine rhinotracheitis *(Vet.)*. Common herpes virus infection of cattle. Most common signs are associated with the upper respiratory tract, with reproductive, nervous and alimentary symptoms also. Vaccines available.

infectious bulbar paralysis *(Vet.)*. See **Aujesky's disease**.

infectious canine hepatitis *(Vet.)*. *Rubarth's disease*. Caused by canine adenovirus, type 1. Acute, contagious and often fatal disease of the dog characterized by fever, diarrhoea, vomiting melaena, abdominal pain, jaundice and nervous signs. Corneal oedema (blue eye) sometimes during recovery. Vaccines available.

infectious coryza *(Vet.)*. *Contagious catarrh, fowl coryza, roup*. An acute, contagious bacterial infection of the upper respiratory tract of domestic fowl.

infectious hepatitis *(Med.)*. See **infective hepatitis**.

infectious icterohaemoglobinuria *(Vet.)*. See **red-water**.

infectious jaundice *(Med.)*. Same as *leptospirosis icterohaemorrhagica*. See **leptospirosis**.

infectious keratitis *(Vet.)*. See **infectious ophthalmia**.

infectious laryngotracheitis *(Vet.)*. A highly contagious, often fatal, virus infection of the respiratory tract of the chicken. Signs include coughing and sneezing. Abbrev. *ILT*. Vaccination available.

infectious mononucleosis *(Med.)*. *Glandular fever*. An acute viral infectious disease characterized by slight sore throat, enlargement of glands in the neck, and an increase in the white (mononuclear) cells of the blood. The causative virus is the *Epstein-Barr virus*. Common in adolescents and usually mild but jaundice may occur in some and convalescence may also be greatly prolonged.

infectious ophthalmia *(Vet.)*. *New Forest disease, infectious bovine kerato conjunctivitis, pink eye*. A contagious form of conjunctivitis and keratitis occurring in cattle. *Moraxella bovis* with or without *Neisseria* spp. are implicated.

infectious parotitis *(Med.)*. See **mumps**.

infectious pig paralysis *(Vet.)*. See **Teschen disease**.

infectious pododermatitis *(Vet.)*. See **foul in the foot**.

infectious sinusitis of turkeys *(Vet.)*. *Big head disease of turkeys*. A disease of turkeys caused by infection of the infraorbital sinuses of the head by organisms of the genus *Mycoplasma*; characterized by swelling of the face and a discharge from the eyes and nostrils.

infectious synovitis *(Vet.)*. A disease of chickens caused by infection by organisms of the genus *Mycoplasma* resulting in exudative synovitis.

infective endocarditis *(Med.)*. A serious infection of the endocardium overlying the heart valves, particularly if they are diseased or altered. Formerly termed *sub-acute bacterial endocarditis* as the majority were caused by *streptococcus viridans*. Because normal valves can be attacked by a wide variety of micro-organisms, the term *infective endocarditis* has been adopted.

infective, infectious hepatitis *(Med.)*. Term applied to viral infection of the liver, causing jaundice. *Hepatitis A* is usually food-borne with an incubation period of 1 month with recovery the rule. *Hepatitis B* is more severe and is usually transmitted by infected blood or needles or instruments contaminated by blood. It is now recognised to be sexually transmitted also. The incubation period is 3–6 months with a mortality rate of 5–20%. The virus may remain in the blood in the so-called *carrier state*.

inferior *(Bot.)*. An ovary having perianth and stamens inserted round the top i.e. *epigynous*. The ovary appears to be sunk into and fused with the receptacle. Cf. **superior**.

inferior *(Zool.)*. Lower; under; situated beneath, e.g. the *inferior* rectus muscle of the eyeball.

inferior conjunction *(Astron.)*. See **conjunction**.

inferior figures, letters *(Print.)*. Small figures or letters set below the general level of the line; as in chemical formulae, e.g. C_6H_3.

inferiority complex *(Behav.)*. A concept, first proposed by Adler, referring to the repressed and powerful conviction of inferiority, whose basis lies in the universal experience of infantile helplessness and dependancy; these feelings become repressed during development, and are a powerful dynamic force in determining adult personality and often, character disorders.

inferior planets *(Astron.)*. See **planet**.

inferior vena cava *(Zool.)*. See **postcaval vein**.

inferred-zero instrument *(Elec.Eng.)*. See **suppressed-zero instrument**.

infertility *(Med.)*. Unable to produce offspring.

infestation *(Med.)*. The condition of being occupied or invaded by parasites, usually parasites other than bacteria.

infilling *(Build.)*. Material, such as hard core, used for making up levels, e.g. under floors.

infiltration *(Med.)*. (1) The accumulation of abnormal substances (or of normal constituents in excess) in cells of the body. (2) The gradual spread of infection in an organ (e.g. tuberculous *infiltration* of the lung).

infiltration *(Powder Tech.)*. Impregnation using capillary forces to soak up the impregnant.

infimum *(Maths.)*. Greatest lower bound.

infinite attenuation *(Phys.)*. The property of some filters of providing a theoretically infinite attenuation for one or more specified frequencies against which strong discrimination is required.

infinite line *(Phys.)*. A transmission line which is infinitely long, or finite but terminated with its characteristic impedance, and along which there is uniform attenuation and phase delay.

infinite loop *(Comp.)*. **Loop** from which there is no exit, other than by terminating the run.

infinite set *(Maths.)*. A set that can be put into a one-one correspondence with part of itself, e.g. the positive integers, which can be put into a one-one correspondence with the positive even integers.

infinitesimal *(Maths.)*. A vanishingly small part of a quantity, which, although it retains the dimensions or other qualities of the quantity, is negligible in magnitude compared with the quantity.

infinity *(Maths.)*. That which is larger than any quantified concept. For many purposes it may be considered as the reciprocal of zero, and minus infinity equated to plus infinity. Written ∞.

infinity plug *(Elec.Eng.)*. A plug in a resistance box which, when withdrawn, breaks the circuit, i.e. introduces an infinite resistance.

infix *(Comp.)*. Form of algebraic notation in which the operators are placed between the operands (e.g. $A+B$, X and Y). Cf. **postfix**.

inflammation *(Med.)*. The reaction of living tissue to injury or to infection, the affected part becoming red, hot, painful, and swollen, due to hyperaemia, exudation of lymph, and escape into the tissue of blood cells.

inflatable aircraft *(Aero.)*. A small low-performance aircraft for military use, in which the aerofoil surfaces (and sometimes the fuselage) are inflated so that it can be compactly packed for transport.

inflation *(Aero.)*. The process of filling an airship or balloon with gas. Sometimes called *gassing*.

inflationary universe *(Astron.)*. Model of the very early universe (10^{-44} seconds after the **Big Bang**) in which the universe expands momentarily much faster than the speed of light. This model is able to account for the flatness of spacetime and the isotropy of the observed universe.

inflected arch *(Civ.Eng.)*. See **inverted arch**.

inflexion *(Maths.)*. See **point of inflexion on a curve**.

inflorescence *(Bot.)*. (1) Flowering branch (or portion of the shoot above the last stem leaves) including its branches, bracts and flowers. See also **racemose, cymost, mixed, capitulum, umbel, panicle**. (2) In bryophytes, a group of antheridia or archegonia and associated structures.

influence line *(Build.,Eng.)*. An *influence line* for a structure is a curve, the ordinate to which at any point represents the value of some variable (such as the bending moment) at another particular point in the structure, due to the presence of a unit load at the point where the ordinate is taken.

influence machine *(Elec.Eng.)*. See **electrostatic generator**.

influenza *(Med.)*. An air-borne respiratory virus infection, causing epidemics which are often world-wide, but whose severity varies with the (constantly-changing) virus type.

influenza *(Vet.)*. See **equine influenza, swine influenza**.

information *(Comp.)*. Meaning given to data by the way in which it is interpreted.

information *(Telecomm.)*. (1) In telecommunications, any intelligence (code speech, images etc.) which can be communicated to a remote destination by electrical or electromagnetic means. (2) More generally, knowledge or intelligence unknown to the receiver before its receipt.

information content *(Telecomm.)*. In *information theory*, a measure of the information conveyed by the occurrence of an allowable symbol in a transmitted message; defined as the negative of the probability that this particular symbol might be sent. If the base of the logarithm chosen is 2, then information content is measured in *hartleys*. One hartley equals $\log_2 10$, or 3.323 **bits**.

information processing *(Comp.)*. The organization, manipulation and distribution of information. Central to almost every use of the computer and almost synonymous with *computing*.

information rate *(Telecomm.)*. The number of symbols transmitted per second multiplied by the average **information content** per symbol. See **bit**.

information retrieval *(Comp.)*. Abbreviated expression for *information storage and retrieval*. A major branch of **computer science** concerned with the use of computers to structure and retrieve information from large information stores. *Data retrieval* is the special case where the nature of the information stored is fully expressed and retrieval involves perfect matching. The more difficult situation occurs when, as in *document retrieval*, the nature of the stored information cannot be fully summarized and the request for information does not fully anticipate the nature of the response. Abbrev. *IR*. See **keywords**.

information storage and retrieval *(Comp.)*. See **information retrieval**.

information technology *(Comp.)*. The application to **information processing** of current technologies from computing, **telecommunications** and **microelectronics**. Abbrev. *IT*.

information theory *(Telecomm.)*. Mathematical study of the information rate, channel capacity, noise and other factors affecting information transmission and reception. Initially applied to electrical communications, now applied universally to business systems and other areas concerned with information in its broadest sense and its flow through networks. See **Shannon's theorem**.

infra- *(Genrl.)*. Prefix from L. *infra*, below.

infrablack *(Image Tech.)*. Amplitude in a TV signal beyond the black level of the picture.

infracostal *(Med.)*. Beneath the ribs.

infradyne *(Telecomm.)*. Supersonic heterodyne receiver in which the intermediate frequency is higher than that of the incoming signal.

inframarginal *(Zool.).* Below the margin; a marginal structure; in *Chelonia*, one of certain plates of the carapace lying below the marginals; in *Asteroidea*, one series of ossicles situated on the lower margin of each ray.

infraorbital foramen *(Zool.).* In Mammals, a foramen on the outer surface of each maxilla for the passage of the second division of the fifth cranial nerve.

infraorbital glands *(Zool.).* In Mammals, one of the four pairs of salivary glands.

infrared astronomy *(Astron.).* The study of radiation from celestial objects in the wavelength range 800 nm–1 mm. Absorption by water vapour in our atmosphere poses severe difficulties, some of which are overcome at high altitude observatories such as Mauna Kea in Hawaii at 4000 m. There are many thousands of infrared sources in our Galaxy, principally cool giant stars, nascent stars and the Galactic centre itself.

infra red countermeasures *(Aero.).* Means of deceiving missiles guided by infra-red sensors by (1) flares deployed as decoys from aircraft or (2) in design by reducing heat output from jet exhaust or shielding hot parts.

infrared detection *(Phys.).* Rays detected and registered photographically with special dyes: photosensitively with a special Cs—O—Ag surface; by photoconduction of lead sulphide and telluride; and, in absolute terms, by bolometer, thermistor, thermocouple or Golay detector.

infrared maser *(Phys.).* One which radiates or detects signals of mm wavelengths. See also **laser**.

infrared photography *(Image Tech.).* Photography using materials specially sensitized to infra-red radiation; applications include photography without visible light, haze-penetration, camouflage and forgery detection and medical records. See also **false colour**.

infrared radiation *(Phys.).* Electromagnetic radiation in the wavelength range from 0.75 to 1000 μm approximately; i.e. between the visible and microwave regions of the spectrum. The *near* infrared is from 0.75 to 1.5 μm, the *intermediate* from 1.5 to 20 μm and the *far* from 20 to 1000 μm.

infrared spectrometer *(Phys.).* An instrument similar to an optical spectrometer but employing non-visual detection and designed for use with infrared radiation. The infrared spectrum of a molecule gives information as to the functional groups present in the molecule and is very useful in the identification of unknown compounds.

infrasound *(Acous.).* Sound of frequencies below the usual audible limit, viz. 20 Hz.

infundibulum *(Zool.).* A funnel-shaped structure; in Vertebrates, a ventral outgrowth of the brain; a pulmonary vesicle; in Cephalopoda, the siphon; in Ctenophora, the flattened gastric cavity. adj. *infundibular*.

infusible *(Eng.). Refractory*; not rendered liquid under specified conditions of pressure, temperature or chemical attack.

infusion *(Chem.).* Solution of the soluble constituents of vegetable matter, obtained by steeping the vegetable matter in liquid, often hot and sometimes under pressure.

infusorial earth *(Min.).* See **tripolite**.

ingate *(Foundry).* The channel, or channels, by which the molten metal is led from the runner hole into the interior of a mould.

ingestion *(Zool.).* The act of swallowing or engulfing food material (*ingesta*) so that it passes into the body. v. *ingest*.

ingluvies *(Zool.).* An oesophageal dilatation of Birds; the crop.

ingluvitis *(Vet.).* Inflammation of the crop, or ingluvies, of Birds.

ingo, ingoing *(Build.).* See **reveal**.

ingold cutter *(Horol.).* A special cutter used to correct inaccuracies in the teeth of a wheel.

ingot *(Eng.).* A metal casting of a shape suitable for subsequent hot working, e.g. rolling.

ingot iron *(Eng.).* Iron of comparatively high purity, produced, in the same way as steel, in the open-hearth furnace, but under conditions that keep down the carbon, manganese and silicon content.

ingot mould *(Eng.).* The mould or container in which molten metal is cast and allowed to solidify to form an ingot.

ingot stripper *(Eng.).* Mechanism for extracting ingots from ingot moulds.

ingravescent *(Med.).* Gradually increasing in severity.

ingroup/outgroup *(Behav.).* A set of related concepts used in studies of intergroup relations; a person's ingroup is the group the individual perceives as the group they belong to; and the **outgroup** is the group to which one does not belong, and feels no allegiance to.

inguinal *(Zool.).* Pertaining to, or in the region of, the groin, e.g. the *inguinal canal* through which the testes descend in male mammals.

inguinodynia *(Med.).* Pain in the groin.

inhalant *(Zool.).* Pertaining to, or adapted for, the action of drawing in a gas or liquid, e.g. the *inhalant* siphon in some *Mollusca*.

inhalation *(Med.).* The act of breathing in, or taking into the lungs; any medicinal agent breathed into the lungs.

inherent ash *(Min.Ext.).* Noncombustible material intimately bound in the original coal-forming vegetation, as distinct from 'dirt' from extraneous sources.

inherent filtration *(Radiol.).* That introduced by the wall of the X-ray tube as distinct from added primary or secondary filters.

inherent floatability *(Min.Ext.).* Natural tendency of some mineral species to repel water and to become part of the 'float' without preliminary conditioning in the froth-flotation process.

inherent regulation *(Elec.Eng.).* The change in voltage at the output terminals of an electric machine (a.c. or d.c. generator, or a converter) when the load is applied or removed, all other conditions remaining constant. The change in secondary voltage of a transformer when the load is changed between zero and full load.

inherited error *(Comp.).* That in one stage of a multistage calculation which is carried over as an initial condition to a subsequent stage.

inhibited oil *(Elec.Eng.).* Transformer or switch oil which includes an antioxidant to delay the onset of sludge and acid formation.

inhibition *(Zool.).* The stopping or deceleration of a metabolic process; cf. **excitation**. adj. *inhibitory*.

inhibitor *(Bot.).* A substance which reduces or prevents some metabolic or physiological process. Also *growth inhibitor*.

inhibitor *(Chem.).* Additive which retards or prevents an undesirable reaction, e.g. phosphates, which prevent corrosion by the glycols in **antifreeze** solutions; antioxidants in rubber.

inhibitory *(Zool.).* Said of a nerve, whose action upon other nerves tends to render them less liable to stimulation.

inhibitory phase *(Chem.).* The protective colloid in a lyophobic sol.

inhour *(Nuc.Eng.).* Unit of reactivity equal to reciprocal of period of nuclear reactor in hours, e.g. a reactivity of two inhours will result in a **period** of half an hour.

inion *(Med.).* The external bony protuberence on the occiput, at the back of the skull.

initial *(Bot.).* (1) A dividing cell in a meristem; one of the daughters or its progeny adding to the tissues of the plant, the other remaining in the meristem and repeating the process. (2) Cell in the earliest stages of specialization.

initial conditions *(Comp.,Maths.).* Starting values of the parameters of an equation which are used in calculating numerical solutions.

initial consonant articulation *(Telecomm.).* See **articulation**.

initialize *(Comp.).* To set counters or variables to some starting value (often zero).

initial permeability *(Elec.Eng.).* Limiting value of differential permeability when magnetization tends to zero. Applicable to small alternations of magnetization when there is no magnetic bias.

initial stability *(Ships).* The moment of the couple tending to return the vessel to her equilibrium position when her

angle of heel is small. Depends directly on **metacentric height** and is frequently expressed as such.

initiator *(Chem.)*. The substance or molecule which starts a chain reaction.

initiator codon *(Biol.)*. The sequence AUG or rarely GUG which always specify the first amino acid of a protein.

injected *(Bot.)*. Having the intercellular spaces filled with water or other fluid.

injection *(Autos.)*. The process of spraying fuel into the inlet manifold or cylinder of an internal combustion engine by means of an injection pump.

injection *(Geol.)*. The emplacement of fluid rock matter in crevices, joints, or fissures found in rocks. See **intrusions**.

injection carburettor *(Aero.)*. A pressure carburettor in which the fuel delivery to the jets is maintained by pressure instead of by a float chamber. It is unaffected by negative *g* in aerobatics or severe atmospheric turbulence.

injection complex *(Geol.)*. An assemblage of rocks, partly igneous, partly sedimentary or metamorphic, the former in intricate intrusive relationship to the latter, occurring in zones of intense regional metamorphism.

injection condenser *(Eng.)*. See **jet condenser**.

injection efficiency *(Electronics)*. The fraction of the current flowing across the emitter junction in a transistor which is due to the minority carriers.

injection lag *(Autos.)*. In a compression-ignition engine, the time interval between the beginning of the delivery stroke of the fuel injection pump and the beginning of injection into the engine cylinder.

injection moulding *(Plastics)*. A method for the fabrication of thermoplastic materials. The viscous resin is squirted, by means of a plunger, out of a heated cylinder into a water-chilled mould, where it is cooled before removal. Method used also with thermosetting moulding powders.

injection pump *(Autos.)*. Unit of a diesel engine, or petrol-injection system, which injects measured quantities of fuel to each cylinder in turn on the compression stroke. See **common-rail injection, fuel injection, jerk-pump**.

injection string *(Min.Ext.)*. Pipe run in addition to the *production string* in the bore hole to allow the passage of additives or of drilling mud to stop the well. Also *kill string*.

injection valve *(Autos.)*. See **injectors, atomizer**.

injector *(Eng.)*. A device by which a stream of fluid, as steam, is expanded to increase its kinetic energy, and caused to entrain a current of a second fluid, as water, so delivering it against a pressure equal to, or greater than, that of the steam; the steam injector is commonly used for feeding boilers. See also **Giffard's injector**.

injectors *(Autos.)*. Plugs with valved nozzles through which fuel is metered to the combustion chambers in diesel- or fuel-injection engines. Connected with the fuel-pump by a capillary tube. In diesels, the injectors are generally screwed into the cylinder head (corresponding to the sparking-plug in a conventional petrol engine); in petrol-injection engines, they are located beside the inlet valves.

ink *(Genrl.)*. **Writing ink** usually consists of a fluid extract of tannin with the addition of solutions of iron salts. The semi-solid ink used in ball-point pens consists of highly-concentrated dyes in a non-volatile solvent. **Coloured writing inks** are prepared by dissolving suitable dyes in water. **Marking inks** are made from solutions of silver or copper compounds, aniline being sometimes added. See also **Indian ink, printing ink**.

ink *(Zool.)*. See **ink sac**.

inkblot test *(Behav.)*. See **Rorschach test**.

ink-jet printer *(Comp.)*. Produces characters by using a fine jet of quick drying ink.

ink jet printing *(Print.)*. Process whereby charged ink particles are projected in a continuous stream (or singly) and then deflected by an electrostatic field using computer control to create text and graphics on the printing substrate.

ink misting *(Print.)*. Ink is sometimes reduced to a fine mist, particularly on fast running rotary presses, by the friction between two surfaces rotating in opposite directions, *long inks* being more susceptible than *short inks*.

ink pump *(Print.)*. On rotary presses : (1) a special type of pump-operated ink supply system, (2) a circulating pump to agitate liquid inks, particularly for photogravure, and (3) a pump to maintain ink supply to the ink ducts.

ink rail *(Print.)*. Part of a pump-operated ink duct which extends across the machine to apply ink to the system.

ink sac *(Zool.)*. In some *Cephalopoda*, a large gland, opening into the alimentary canal near the anus, which secretes a dark-brown pigment (sepia).

ink table, slab *(Print.)*. A flat surface on a printing machine, part of the ink distribution system; also a flat surface used with a hand roller.

inlay *(Build.)*. Pieces of metal, ivory etc., inserted ornamentally (inlaid) in furniture, etc.

inlay *(Image Tech.)*. Combination of two video images using a separate signal to **key** one into the other.

inlet guide vanes *(Aero.)*. Radially positioned aerofoils in the annular air intake of axial-flow compressors which direct the airflow on to the first stage at the most efficient angle. The vanes are often rotatable about their mounting axes so that different entry air speeds can be accomodated; they are then called variable inlet guide vanes. Abbrev. *IGV*.

inlet manifold *(Autos.)*. See **induction manifold**.

inlet port, valve *(Autos.)*. See **induction port, valve**.

inlier *(Geol.)*. An outcrop of older rocks surrounded by those of younger age.

in-line assembly *(Electronics)*. Assembly by machine, in which an array of heads inserts components into a printed circuit board.

in-line engine *(Autos.)*. A multicylinder engine, consisting of a bank of cylinders mounted in line along a common crankcase.

INMARSAT *(Telecomm.)*. The *INternational MARitime SATellite organisation*. Responsible for maritime communications, distress and safety services via communications satellites.

innate *(Behav.)*. Refers to behaviour which normally occurs in all members of a species despite natural variation in environmental influences. The term is sometimes used to imply genetic determinism, but it is now accepted that the development of innate behaviour is complex, and often involves both endogenous and environmental factors.

innate *(Bot.)*. (1) Sunken into or originating within the thallus. (2) An anther joined by its base to the filament.

innate capacity for increase *(Ecol.)*. See **r**. Abbrev. r_{max}.

innate releasing mechanism (IRM) *(Behav.)*. In classical ethological models of motivation, a term referring to the notion that there is a specific relation between a particular stimulus and a particular response, such that a specific external stimulus (see **sign stimulus**) releases the motivational energy underlying the discharge of a specific response. The term is currently used in a general descriptive fashion, without the early implications about underlying mechanisms, which are more complex than was initially supposed. Abbrev. *IRM*.

inner bead *(Build.)*. See **guide bead**.

inner conductor *(Elec.Eng.)*. (1) See **internal conductor**. (2) The neutral conductor of a 3-wire system.

inner dead-centre *(Eng.)*. Of a reciprocating engine or pump, the piston position at the beginning of the outstroke, i.e. when the crank-pin is nearest to the cylinder. Also called *top dead-centre*.

inner ear *(Med.)*. Structure encased in bone and filled with fluid, including the cochlea, the semi-circular balancing canals and the vestibule.

inner forme *(Print.)*. In **sheet-work** printing the imposition containing the second and second-last pages of the printed sheet, the *outer forme* containing the first and last pages and being printed on the other side.

inner glume *(Bot.)*. Same as **palea**.

inner marker beacon *(Aero.)*. A vertically-directed radio beam which marks the airport boundary in a beam-approach landing system, such as the **instrument landing system**.

innervation *(Zool.)*. The distribution of nerves to an organ.

innings *(Civ.Eng.)*. Lands reclaimed from the sea.

innocent *(Med.)*. Not malignant; not cancerous. See **tumour**. A heart murmur of no pathological significance.

innominate *(Zool.)*. Without a name, e.g. the *innominate artery* of some Mammals, which leads from the aortic arch to give rise to the carotid artery and the subclavian artery, the *innominate vein* of Cetacea, Edentata, Carnivora and Primates, which leads across from the jugular-subclavian trunk of one side to that of the other, the *innominate bone*, which is the lateral half of the pelvic girdle.

inoculation *(Bot.)*. The placing of cells, spores etc. on or in a potential host, soil or culture medium.

inoculation *(Chem.)*. The introduction of a small crystal into a super-saturated solution or supercooled liquid in order to initiate crystallization.

inoculation *(Eng.)*. Modification of crystallizing habit or grain refinement for imparting alloy qualities to molten metal in furnace or ladle, by addition of small quantities of other metals, de-oxidants etc.

inoculation *(Med.)*. The introduction into an experimental animal, by various routes, of infected material or of pathogenic bacteria: the injection of a vaccine into a person for protection against subsequent infection by the organisms contained in the vaccine, but also used loosely in immunology for the introduction of a substance into the body, usually by parenteral injection.

inoculum *(Med.)*. The material used in inoculation.

inorganic chemistry *(Chem.)*. The study of the chemical elements and their compounds, other than the compounds of carbon; however, the oxides and sulphides of carbon and the metallic carbides are generally included in inorganic chemistry.

inosilicates *(Min.)*. Those silicate minerals having an atomic structure in which the SiO_4 groups are linked together in chains, e.g. **pyroxenes**.

inositol *(Chem.)*. $C_6H_6(OH)_6$. *Cyclohexanehexol, hexahydroxycyclohexane.*

in parallel *(Elec.Eng.)*. See **parallel**.

in phase *(Elec.Eng.)*. See under **phase**.

in-phase component *(Elec.Eng.)*. See **active component**.

in-phase loss *(Elec.Eng.)*. See **ohmic loss**.

in-pile test *(Nuc.Eng.)*. One in which the effects of irradiation are measured while the specimen is subjected to radiation and neutrons in a reactor.

input *(Comp.)*. (1) Information entering a device, data about to be processed. (2) To enter information, to read in data.

input capacitance *(Phys.)*. The effective capacitance between the input terminals of a network. See **Miller effect**.

input characteristic *(Electronics)*. See **transistor characteristics**.

input device *(Comp.)*. **Peripheral** which can accept data, presented in the appropriate machine readable form, decode it and transmit it as electrical pulses to the **central processing unit**.

input gap *(Electronics)*. See **buncher**.

input impedance *(Elec.Eng.,Electronics)*. The small signal impedance measured between the input terminals of a network.

input/output processor *(Comp.)*. Processor which supervises input-output operations. See **front-end processing**.

input signal *(Elec.Eng.)*. That connected to the input terminals of any instrument or system (usually electronic).

input transformer *(Elec.Eng.)*. One used for isolating a circuit from any d.c. voltage in the applied signal and/or to provide a change in voltage. Also used to match the impedance of an input signal to that of the circuit to maximise power transfer.

input voltage *(Elec.Eng.,Electronics)*. The voltage applied between the input terminals of a network.

inquartation *(Eng.)*. *Quartation.* Removal of silver from gold-silver bullion. The proportion of silver to gold must be raised by fusion to at least three to one, the silver being then dissolved in nitric acid. The silver can also be dissolved in concentrated sulphuric acid or converted to chloride by bubbling chlorine through the molten bullion. Also *parting of bullion.*

inquiline *(Zool.)*. A guest animal living in the nest of another animal, or making use of the food provided for itself or its offspring by another animal.

insanity *(Behav.)*. A medical legal term used in the defence of individuals who have committed crimes while their capacity for rational thought and behaviour was seriously impaired.

inscribed circle of a triangle *(Maths.)*. The circle touching all three sides of a triangle internally. Its centre is the intersection of the angle bisectors.

Insecta *(Zool.)*. A class of mainly terrestrial mandibulate Arthropoda which breathe by tracheae; they possess uniramous appendages; the head is distinct from the thorax and bears one pair of antennae; there are 3 pairs of similar legs attached to the thorax, which may also bear wings; the body is sharply divided into head, thorax, and abdomen. Also *Hexapoda.*

insecticides *(Chem.)*. Natural (e.g. derris, pyrethrins) or synthetic substances for destroying insects. The widely used synthetic compounds are broadly classified according to chemical composition, viz. chlorinated (e.g. **DDT**), organophosphates (e.g. **malathion**), carbamates, dinitrophenols. Some of these have properties now considered undesirable like persistence in *food chains.* Newer compounds like the modified pyrethrins have greater stability than the natural substance and still retain their effectiveness. See **contact-, fumigants, stomach-**

Insectivora *(Zool.)*. An order of small, mainly terrestrial Mammals having numerous sharp teeth, tuberculate molars, well-developed collar-bones, and plantigrade unguiculate pentadactyl feet; insectivorous. Shrews, Moles, and Hedgehogs. Sometimes divided into two orders, Lipotyphla and Menotyphla.

inselberg *(Geol.)*. A steep-sided knob or hill arising from a plain; often found in the semi-arid regions of tropical countries.

insemination *(Zool.)*. The approach and entry of the spermatozoon to the ovum, followed by the fusion of the male and female pronuclei.

insert *(Eng.)*. In casting, a small metal part which is fitted in the mould or die in such a manner that the metal flows around it to cast it in position. It may provide a hard metal wearing surface, or sintered metal oil-retaining bearing surface etc.

insert *(Image Tech.)*. A single shot, such as the close-up of a newspaper headline, photographed separately and inserted in a sequence.

insert *(Plastics)*. A metal core inserted into a plastic article during the (injection) moulding process.

insert *(Print.)*. See **inset**.

inserted tooth cutter *(Eng.)*. A milling cutter in which replaceable teeth, of expensive material, are inserted into slots in the cutter body.

insertion *(Biol.)*. A stretch of chromosome that has been inserted into another.

insertion *(Bot.)*. (1) The place where one plant member grows out of another, or is attached to another. (2) The manner of attachment.

insertion *(Zool.)*. The point or area of attachment of a muscle, mesentery or other organ.

insertion element *(Biol.)*. DNA sequence a few hundred bases long which can be naturally inserted into genomic DNA. Does not code for mRNA but can inactivate a coding sequence by its presence.

insertion gain, loss *(Elec.Eng.)*. That gain or loss in decibels when a transformer, filter or impedance matching transducer or network is inserted into a circuit.

insertion head *(Eng.)*. An automatic mechanism, which may incorporate cutting, forming or fastening tools, for placing components into a partly completed assembly.

insessorial *(Zool.)*. Adapted for perching.

inset, insert *(Print.)*. (1) An extra leaf or section loosely inserted in a book. (2) One folded sheet put inside another; an *insetted* (or *inserted*) *book* is one having its section placed one within the other.

Insetter *(Print.)*. Electronically-controlled equipment for registering pre-printed reels into rotary presses to be incorporated in the final product, overprinted if required; with controlled **automatic reel change**; half-width or full-width reels.

inside callipers *(Eng.)*. **Callipers** with the points turned outwards, used for taking inside dimensions.

inside crank *(Eng.)*. A crank, with two webs, placed between bearings, as distinct from an outside or overhung crank.

inside cylinders *(Eng.)*. In a steam locomotive, those cylinders which are fixed inside the frame.

inside lap *(Eng.)*. See **exhaust lap**.

inside spider *(Acous.)*. A flexible device inserted within a voice coil so as to centre the coil accurately between the pole pieces of a dynamic loudspeaker.

insight learning *(Behav.)*. A form of intelligent activity that involves the apprehension of relations between the elements of a problem; it often appears to occur suddenly, as a function of cognitive rather than behavioural effort, and for this reason is often contrasted with a more passive trial and error mode of learning or problem solving.

in-situ hybridization *(Biol.)*. The identification by **autoradiography** in a fixed cell or section of the place where a radioactive or similar **probe** binds. Can identify a complementary sequence or an antigen.

insolation *(Med.)*. *Sunstroke*. It results from exposure of the head and the neck to the direct rays of the sun, and is characterized by high fever, headache, and mental excitement, followed by unconsciousness.

insolation *(Meteor.)*. The radiation received from the sun. This depends on the position of the earth in its orbit, the thickness and transparency of the atmosphere, the inclination of the intercepting surface to the sun's rays, and the **solar constant**.

insoluble *(Chem.)*. Incapable of being dissolved. Most 'insoluble' salts have a definite, though very limited, solubility.

insoluble electrode *(Elec.Eng.)*. Non-ionizing cathode or anode.

insomnia *(Med.)*. A persistent inability to sleep.

inspection chamber *(Build.)*. A pit formed at regular points in the length of a drain or sewer to give access for purposes of inspection, testing and the removal of obstructions.

inspection fitting *(Eng.)*. A bend, elbow, or tee used in a conduit wiring system, which is fitted with a removable cover to facilitate the drawing-in of the wires and subsequent inspection.

inspection gauges *(Eng.)*. Gauges used by the inspection department of a works for testing the accuracy of finished parts.

inspection junction *(Build.)*. A drain-pipe with a vertical branch to ground level to allow access for inspection.

inspection plug *(Elec.Eng.)*. A plug fitted in the cover of an accumulator, through which the electrolyte can be inserted and its level observed.

inspiration *(Zool.)*. The drawing-in of air or water to the respiratory organs.

inspirator *(Eng.)*. The injector of a pressure gas burner, combined with a venturi mixing tube, primary combustion air being entrained by the projection of a gas stream into the injector throat.

inspissation *(Med.)*. The thickening of pus as a result of the removal of fluid.

instability *(Acous.)*. Vibration with an exponentially growing amplitude. It occurs if there is some sort of positive **feedback**; e.g. thermoacoustic instability, screech.

instability *(Aero.)*. An aircraft possess *instability* when any disturbance of its steady motion tends to increase, unless it is overcome by a movement of the controls by the pilot.

instability *(Nuc.Eng.)*. In a *plasma*, various instabilities of which the principal are: bending of plasma, *kink instability*, and bead-like instability, *sausage instability*, in both of which the magnetic field change magnifies the variations; acceleration-driven *Rayleigh-Taylor* instability when the magnetic fields and plasma behave like a light fluid in contact with and accelerated against a heavier one, causing spiky irregularities; *flute instability* in mirror machines.

instability *(Phys.)*. See **unstable equilibrium**.

instability *(Telecomm.)*. Tendency for a circuit to break into unwanted oscillation.

instantaneous automatic gain control *(Radar)*. A rapid action system in radar for reducing the clutter.

instantaneous carrying-current *(Elec.Eng.)*. Of switches, circuit-breakers, and similar apparatus, the maximum value of current which the apparatus can carry instantaneously.

instantaneous centre *(Eng.)*. The imaginary point, not necessarily in the body, about which a body having general motion may be considered to be rotating at a given instant.

instantaneous frequency *(Telecomm.)*. That calculated from the instantaneous rate of change of angle of a waveform on a time base. In any oscillation, the rate of change of phase divided by 2π.

instantaneous fuse *(Min.Ext.)*. Rapid-burning, as distinct from slow fuse. Ignition proceeds at a few kilometres per second, but is slower than that of **detonating fuse**.

instantaneous power *(Elec.Eng.)*. For a circuit or component, the product of the instantaneous voltage and the instantaneous current. This may not be zero even for a nondissipative (wattless) system on account of stored energy, although in this case its time integral must be zero.

instantaneous specific heat capacity *(Phys.)*. Specific heat capacity at any one temperature level; *true s.h.c.* to distinguish from *mean s.h.c.*.

instantaneous value *(Genrl.)*. A term used to indicate the value of a varying quantity at a particular instant. More correctly it is the average value of that quantity over an infinitesimally small time interval.

instantaneous velocity of reaction *(Chem.)*. Rate of reaction, measured by the change in concentration of some key reagent. For instance, in the first order reaction, $A \rightarrow B$, the *ivr* at any time t is given by $dC_A/dt = -dC_B/dt = -k.C_A$, where C_A, C_B are the concentrations of reagent and product, respectively, and k is the 'velocity constant' of the reaction. Similarly, for the second order reaction, $A+B \rightarrow C$, the *ivr* is $dC_A/dt = dC_B/dt = -dC_C/dt = -k.C_A.C_B$.

instant photography *(Image Tech.)*. A system which rapidly produces a positive picture from the material originally exposed, sometimes by viscous processing within the camera itself.

instar *(Zool.)*. The form assumed by an Insect during a particular stadium.

in step *(Elec.Eng.)*. See **step**.

instinct *(Behav.)*. Refers to a variety of related notions about the causes of behaviour, and its usage has undergone many historical changes, referring sometimes to the hypothetical motivational forces that impel behaviour (e.g. Descartes), to reflexive behaviour (Darwin), to the irrational aspects of human behaviour (Freud), and most recently, in classical ethology, to the explanation of innate behaviour patterns. Among modern thinkers it is generally agreed that the concept has little descriptive or explanatory value, and tends to be avoided, although the term still crops up in the popular literature. See **innate**.

institution *(Behav.)*. A type of relationship or pattern of relationships characteristic of a given society; institutionalized practices of customs within a society may be social, economic, legal etc. See **institutionalization**.

institutionalization *(Behav.)*. A syndrome of apathy and withdrawal resulting from long periods in any understimulating institution (e.g. a mental hospital).

Instron tensile strength tester *(Paper)*. A precise instrument for measuring tensile strength, in which the rate of strain on the paper sample is constant.

instruction *(Comp.)*. (1) A **machine-code instruction**. (2) A statement in a **high level language**. See also **microinstruction, macroinstruction**.

instruction address register *(Comp.)*. See **program counter**.

instruction cycle *(Comp.)*. Complete process of fetching one **instruction** from store decoding and executing it by the **central processor**. Also *fetch/execute cycle*. See **machine code instruction**.

instruction number *(Comp.)*. One used to label a specific statement in a program so that the user and the machine can refer back to it subsequently. Such numbers do not indicate the order in which instructions must be carried out, or the number of the address at which the instruction will be stored. Also *instruction number, line number*.

instruction set *(Comp.)*. Complete collection of instruments available in a **particular machine** code or **assembly language**.

instrument *(Elec.Eng.)*. The term for any item of electrical or electronic equipment which is designed to carry out a specific function or set of functions. Examples are found in a wide range of applications and include; in an electrical network, e.g. a voltmeter, in an engineering system, e.g. a displacement measuring unit or in medicine, e.g. an ECG unit.

instrumental sensitivity *(Elec.Eng.)*. The ratio of the magnitude of the response to that of the quantity being measured, e.g. divisions per milliamp.

instrument approach *(Aero.)*. Aircraft approach made using **instrument landing system**.

instrument decoder *(Comp.)*. Part of the **central processor** which decodes **machine-code instructions**.

instrument flight rules *(Aero.)*. The regulations governing flying in bad visibility under strict flight control. Abbrev. *IFR*.

instrument landing system *(Aero.)*. A localizer for direction in the horizontal plane and glideslope in the vertical plane, usually inclined at 3° to the horizontal. Two markers are used for linear guidance. Abbrev. *ILS*.

instrument range *(Nuc.Eng.)*. The intermediate range of reaction rate in a nuclear reactor, when the neutron flux can be measured by permanently installed control instruments, e.g. ion chambers. See **start-up procedure**.

instrument shunt *(Elec.Eng.)*. A resistance of appropriate value connected across the terminals of a measuring instrument to extend its range.

instrument transformer *(Elec.Eng.)*. One specially designed to maintain a certain relationship in phase and magnitude between the primary and secondary voltages or currents.

insufficient feed *(Print.)*. When too little paper is drawn through the press, there is an increase in tension between units or components. Cf. **excess feed**.

insufflation *(Med.)*. (1) The action of blowing gas, air, vapour, or powder into a cavity of the body, e.g. the lungs. (2) The powder, etc., used in insufflation.

insufflator *(Med.)*. An instrument used for insufflation.

insulant *(Build.)*. See **insulation** (2).

insulated *(Arch.)*. Said of any building or column which stands detached from other buildings.

insulated bolt *(Elec.Eng.)*. A bolt having a layer of insulating material around its shank.

insulated clip *(Elec.Eng.)*. A clip, incorporating an insulated eye, used for supporting flexible electrical connections.

insulated eye *(Elec.Eng.)*. An eye for supporting flexible connections; it has an insulating bush to prevent these from making contact with the metal.

insulated hanger *(Elec.Eng.)*. A hanger for the contact wire of an electric traction system, which insulates the contact wire from the main supporting system.

insulated hook *(Elec.Eng.)*. A hook terminating in an insulated eye through which flexible electric connections may be passed and supported.

insulated metal roofing *(Build.)*. Roofing panels made up of light-gauge aluminium or sheet steel, corrugated for strength, and with a top covering of insulation board and bituminous felt. The ceiling underlining is also frequently prefixed.

insulated neutral *(Elec.Eng.)*. A term used to denote: (1) the neutral point of a star-connected generator or transformer when it is not connected to earth directly or through a low impedance; (2) the middle wire of a three-wire distribution system when the wire is an insulated cable.

insulated pliers *(Elec.Eng.)*. Pliers having the handles covered with insulating material; used by electricians in order to avoid electric shocks when working on live conductors.

insulated-return system *(Elec.Eng.)*. A system of supply for an electric traction system, in which both the outgoing and the return conductors are insulated from earth.

insulated screw-eye *(Elec.Eng.)*. A screw terminating in an insulated eye through which flexible electric connections may be passed.

insulated system *(Elec.Eng.)*. A system of electric supply in which each of the conductors is insulated from earth for its normal voltage.

insulated wire *(Elec.Eng.)*. A solid conductor insulated throughout its length.

insulating beads *(Elec.Eng.)*. Beads of glass or similar material strung over a bare conductor to provide an insulating and heat-resisting covering which is also flexible.

insulating board *(Build.)*. Fibre board of low density, about 20 lb/ft^3 or 300 kg/m^3, used for thermal insulation and acoustical control.

insulating compound *(Elec.Eng.)*. An insulating material which is liquid at fairly low temperatures so that it can be poured into joint-boxes of cables and other similar pieces of apparatus and then allowed to solidify.

insulating materials, classification *(Elec.Eng.)*. See **class-A, -B, -C** etc.

insulating oils *(Elec.Eng.)*. Special types of oil having good insulating properties; used for oil-immersed transformers, circuit-breakers etc.

insulating tape *(Elec.Eng.)*. Tape impregnated with insulating compounds, frequently adhesive; used for covering joints in wires etc.

insulation *(Genrl.)*. (1) Any means for confining as far as possible a transmissible phenomenon (e.g. electricity, heat, sound, vibration) to a particular channel or location in order to obviate or minimize loss, damage, or annoyance. (2) Any material (also *insulant*) or means suitable for such a purpose in given conditions, e.g. dry air suitably enclosed, polystyrene and polyurethane foam slab, glass fibre, rubber, porcelain, mica, asbestos, hydrated magnesium carbonate, cork, kapok, crumpled aluminium foil etc. See also **lagging, shielding**.

insulation resistance, insulance *(Elec.Eng.)*. The resistance between two conductors, or between a conductor and earth, when they are separated only by insulating material. See **fault resistance**.

insulation test *(Elec.Eng.)*. A test made to determine the insulation resistance of a piece of apparatus or of a system of electric conductors.

insulation tester *(Elec.Eng.)*. An instrument for measuring **insulation resistance**.

insulator *(Elec.Eng.)*. Unit of insulating material specifically designed to give insulation and mechanical support, e.g. for telegraph lines, overhead traction and transmission lines, plates and terminals of capacitors.

insulator *(Phys.)*. Materials having a high electrical resistivity in range 10^6–10^{15} Ωm. The electrons in the material are not free to move under the influence of an electric field.

insulator arcing horn *(Elec.Eng.)*. A metal projection placed at the upper and lower ends of a suspension-type or other insulator, in order to deflect an arc away from the insulator surface.

insulator cap *(Elec.Eng.)*. A metal cap placed over the top of a suspension insulator, which serves to attach it to the next insulator.

insulator pin *(Elec.Eng.)*. The central metal support of a pin insulator, or the metal projection on the under side of a suspension insulator, serving to attach it to the cap of the next unit in the string.

insulator rating number *(Elec.Eng.)*. The voltage (in kilovolts) used in the 'thirty-seconds rain test'.

insulator strength *(Elec.Eng.)*. The maximum mechanical and/or the maximum electrical stress, which can be applied.

insulin *(Biol.)*. Polypeptide hormone produced by the β islet cells of the pancreas. It has many diverse effects but is best known for its ability to reduce blood sugar levels. Its absence results in the disease *diabetes mellitus* which is treated by administration of the hormone.

insulin *(Med.)*. A protein hormone which is produced by the islets of Langerhans of the pancreas. Widely used in the treatment of **diabetes mellitus**, insulin injection results promptly in a decline in blood glucose concentration and an increase in formation of products derived from glucose. Insulin was the first protein to have its structure completely established and consists of two peptide chains joined by two disulphide bridges. Insulin was formerly entirely produced from pancreatic extracts from pig and oxen, but can now be made from a **genetic engineered** strain of the bacterium, *Escherischia coli.*

intaglio *(Glass)*. A form of decoration in which the depth of cut is intermediate between deep cutting and engraving.

intaglio *(Print.)*. A printing process, such as photogravure, in which the printing image is cut or etched into the surface of the plate or cylinder, normally in a characteristic cell formation. The thickness of ink transferred to the paper during printing varies with the depth of the cell.

intarsia *(Textiles)*. A weft-knitted fabric having designs in two or more colours.

integer *(Comp.,Maths.)*. The set of integers is the set of positive and negative whole numbers with zero; i.e. the set $(\ldots -4,-3,-2,-1,0,1,2,3,4,\ldots)$. Cf. **real number**; see **data type**.

integral *(Maths.)*. The function which, when differentiated, equals the given function. If $\varphi(x)$ is the derivative of $f(x)$, then $f(x)$ is the integral of $\varphi(x)$>. More precisely, $f(x)$ is the *indefinite integral* of $\varphi(x)$. Also used generally as the adjective from *integer*.

integral calculus *(Maths.)*. The inverse of the **differential calculus**, its chief concern being to find the value of a function of a variable when its differential coefficient is known. This process, termed *integration*, is used in the solution of such problems as finding the area enclosed by a given curve, the length of a curve, or the volume enclosed by a given surface.

integral colour masking *(Image Tech.)*. See **colour masking**.

integral convergence test *(Maths.)*. The series of positive terms $\sum_1^n f(r)$ where $f(r)$ is a monotonic decreasing function, and the integral $\int_1^n f(x)\,dx$, converge and diverge together.

integral domain *(Maths.)*. A ring in which the product of 2 non-zero elements is non-zero.

integral dose *(Radiol.)*. See **dose**.

integral function *(Maths.)*. See **analytic function**.

integral stiffeners *(Aero.)*. The stiffening ridges left when an aircraft skin panel is machined from a solid billet.

integral tripack *(Image Tech.)*. Colour film or paper containing three distinct emulsion layers, each sensitized to a specific spectral range, which can be processed to produce superimposed colour separation images. Also termed *monopack*.

integrand *(Maths.)*. A function upon which integration is to be effected.

integrated circuit *(Comp.)*. Very small solid-state circuit consisting of interconnected semiconductor devices like transistors, capacitors and resistors printed into a single silicon **chip**. It cannot be subdivided without destroying its intended function. See also **SSI**, **MSI**, **LSI**, **VLSI**.

integrated virus *(Biol.)*. Viral DNA sequence which has *integrated* into the host's chromosomes at one or more sites.

integrating ammeter *(Elec.Eng.)*. Also -voltmeter, -wattmeter etc. Instrument which measures the time integral of the named quantity.

integrating circuit, network *(Elec.Eng.)*. (1) An amplifier with feedback capacitance and input resistance whose output is proportional to the integral of the input signal. (2) A network comprising passive components, which will perform the same mathematical function. This technique is not as accurate as that described in (1).

integrating factor *(Maths.)*. A multiplying factor which enables a differential equation to be transformed into an exact equation.

integrating frequency meter *(Elec.Eng.)*. A meter which sums the total number of cycles of an a.c. supply in a given time. Also called a *master frequency meter.*

integrating meter *(Elec.Eng.)*. An electrical instrument which sums up the value of the quantity measured with respect to time.

integrating motor *(Elec.Eng.)*. Permanent magnet d.c. motor, angular rotation of which is integration of current in armature.

integrating photometer *(Phys.)*. One which measures the total luminous radiation emitted in all directions.

integration *(Biol.)*. The insertion of one DNA sequence into another by recombination.

integration *(Maths.)*. See **integral calculus**.

integrator *(Comp.)*. Device which performs the mathematical operation of integration, usually with reference to time.

integrator *(Telecomm.)*. Any device which integrates a signal over a period of time.

integrity *(Comp.)*. A property of data which has retained its accuracy and correctness during processing and transmission. Cf. **corrupt**.

integument *(Bot.)*. A layer or several layers of tissue of an ovule surrounding and often more or less fused with the **nucellus** which develops into the testa.

integument *(Zool.)*. A covering layer of tissue; especially the skin and its derivatives. adj. *integumented.*

intelligence quotient *(Behav.)*. Abbrev. *IQ*. The score on an intelligence test. In early versions it was obtained by multiplying by 100 the ratio of *mental age* to *chronological age* to find out whether a child's mental age was ahead or behind its chronological age; in newer versions it is read off a table constructed from norms for individuals of the same age. The *deviation IQ*.

intelligent terminal *(Comp.)*. Terminal within which a certain amount of computing can be done without contacting a central computer.

intelligibility *(Acous.,Telecomm.)*. Percentage of correctly understood syllables, words or sentences. Important for speech transmission.

INTELSAT *(Space,Telecomm.)*. The *INTernational TELecommunications SATellite organisation.* Body responsible for the design, construction, development, operations and maintenance of the world-wide satellite communications system. Organized by government telecommunication authorities, who have shares in the organisation in proportion to their use of it.

intensification *(Image Tech.)*. Increasing the density of a processed image, usually by chemical addition of another metal, such as chromium or mercury, to the developed silver.

intensifier electrode *(Electronics)*. One which provides post-deflection acceleration in the electron beam of a cathode-ray tube.

intensifying screen *(Radiol.)*. (1) Layer or screen of fluorescent material adjacent to a photographic surface, so that registration by incident X-rays is augmented by local fluorescence. (2) Thin layer of lead which performs a similar function for high-energy X-rays or gamma-rays, as a result of ionization produced by secondary electrons.

intensitometer *(Radiol.)*. Instrument for measuring intensities of X-rays during exposures.

intensity level *(Acous.)*. Level of power per unit area as expressed in decibels above an arbitrary zero level, e.g. sound intensity level.

intensity modulation *(Image Tech.)*. Modulation of the beam current of a CRT to vary the spot luminosity. Also *Z-modulation*.

intensity of field *(Elec.Eng.)*. The vector quantity by which an electric or magnetic field at a point is measured. Precisely, it is the same number of units of intensity as the force in newtons on a unit charge (unit electric charge or unit fictitious pole) placed at the point in the electric or magnetic field respectively. Also *field strength*.

intensity of magnetization *(Phys.)*. Vector of magnetic moment of an element of a substance divided by the volume of that element.

intensity of radiation *(Phys.)*. Energy flux, i.e., of photons or particles, per unit area normal to the direction of propagation.

intensity of sound *(Acous.)*. The magnitude of a sound wave, measured in terms of the power transmitted, in watts, through unit area normal to the direction of propagation.

intensity (of wave) *(Elec.Eng.)*. The energy carried by any sinusoidal disturbance is proportional to the square of the wave amplitude. See **r.m.s. value**.

intensive reflector *(Elec.Eng.)*. A reflector for incandescent lamps, of such a shape as to produce an intense illumination at the point where it is required.

intentional learning *(Behav.)*. Learning when informed that there will be a later test of learning.

intention movement *(Behav.)*. The incomplete phases of behaviour patterns that often occur in conflict situations; many are assumed to provide the phylogenetic basis for the evolution of threat and courtship displays.

intention tremor *(Med.)*. Tremor of the arms on carrying out a voluntary movement, indicative of disease of the nervous system.

interaction *(Phys.)*. (1) Transfer of energy between two particles. (2) Interchange of energy between particles and a wave motion. (3) Between waves, see **interference**.

interaction gap *(Electronics)*. The space in a microwave tube in which the electron beam interacts with the wave system.

interaction space *(Electronics)*. That in a vacuum tube where an electron beam interacts with an alternating electromagnetic field.

interactive *(Image Tech.)*. Describing a video programme whose presentation of its contents can be varied by action on the part of the viewer.

interactive computing *(Comp.)*. A conversational mode of communication between computer and user. Input is commonly via a **keyboard** or a **mouse** and both input and output may be displayed on a VDU. See **prompt, multi-access, log in/out, light pen, teletypewriter**.

interambulacrum *(Zool.)*. In Echinodermata, especially Echinoidea, the region intervening between two ambulacral areas.

interbase current *(Electronics)*. The current flowing between the two base connectors in junction-type tetrode transistors.

inter-block gap *(Comp.)*. Space separating two recorded portions of data on backing store (e.g. to allow a tape to stop or to accelerate to reading speed).

interbranchial septa *(Zool.)*. The stout fibrous partitions separating the branchial chambers in Fish.

intercalare *(Zool.)*. A cartilage or ossification lying between the basiventrals, or between the basidorsals of the vertebral column.

intercalary *(Bot.)*. Placed between other bodies or the ends of a stem, filament, hypha etc.

intercalary *(Zool.)*. See **intercalate**.

intercalary meristem *(Bot.)*. A meristem located somewhere along the length of a plant member and which divides to give *intercalary growth*, as at the base of a grass leaf.

intercalate *(Zool.)*. To add, to insert, as in *intercalated somite*. adj. *intercalary*.

intercalating dyes *(Biol.)*. Chemical compounds with a high affinity for DNA whose molecules *intercalate* between the planar base-pairs of the DNA helix. They are used for visualizing DNA, changing its density and to induce breakage. Often mutagenic and carcinogenic.

intercavitary X-ray therapy *(Radiol.)*. X-ray therapy in which the appropriate part of suitable X-ray apparatus is placed in a body cavity.

intercellular *(Bot.,Zool.)*. Between cells. Cf. **intracellular, extracellular**.

intercellular spaces *(Bot.)*. The interconnecting spaces between cells in a tissue, air filled in vascular plants, and providing for gas exchange.

intercentra *(Zool.)*. See **hypapophyses**.

interceptor, intercepting trap *(Build.)*. A trap fitted in the length of a house drain, close to its connection to the sewer, which provides a water seal against foul gases rising up into the drain. Also called a *disconnector*.

interchange *(Biol.)*. The mutual transfer of parts between two chromosomes.

interchondral *(Zool.)*. Said of certain ligaments and articulations between the costal cartilages.

interclavicle *(Zool.)*. In Vertebrates, a bone lying between the clavicles, forming part of the pectoral girdle.

intercolumniation *(Arch.)*. The distance between the columns in a colonnade, in terms of the lower diameter of the columns as a unit.

interconnected star connection *(Elec.Eng.)*. See **zigzag connection**.

interconnecting *(Telecomm.)*. The commoning of outlets for the bank multiples of selectors on different shelves, when there is an insufficiency of outlets for full availability. See **grading**.

interconnector, interconnecting feeder *(Elec.Eng.)*. A feeder which serves to interconnect two substations or generating stations, and along which energy may flow in either direction.

intercooler *(Aero.)*. (1) A heat-exchanger on the delivery side of a supercharger which cools the charge heated by compression. (2) A secondary heat exchanger used in a cabin air conditioning system for cooling the charge air from the compressor to the turbine of a **bootstrap cold-air unit**.

intercooler *(Autos.)*. A form of heat exchanger which lowers the temperature of inlet air to improve volumetric efficiency in an engine with a **turbocharger**.

intercooler *(Eng.)*. A cooler, generally consisting of water-cooled tubes, interposed between successive cylinders or stages of a multi-stage compressor or blower, to reduce the work of compression.

INTERCOSMOS *(Space)*. A consortium of Soviet allied nations which carries out co-operative efforts in space matters.

intercostal *(Zool.)*. Between the ribs.

intercrystalline failure *(Eng.)*. Refers to metal fractures that follow the crystal boundaries instead of passing through the crystals, as in the usual transcrystalline fracture. It is frequently due to combined effect of stress and chemical action, but may be produced by stress alone when the conditions permit a certain amount of recrystallization under working conditions.

interdentil *(Build.)*. The space between successive *dentils*.

interdependent functions *(Maths.)*. See **dependent functions**.

interdigital cyst *(Vet.)*. An abscess occurring between the digits of the paw of the dog due to bacterial infection.

interdigital structure *(Electronics)*. One in which the effective path length between a pair of terminal electrodes is increased by an interlocking-finger design, which may be three-dimensional or formed by metallization of an insulating surface. Used for transistors, capacitors and integrated-circuit devices, for converting microwave signals into surface acoustic waves and as a slow-wave structure or part of a filter.

interdorsal *(Zool.)*. An intercalary element lying between adjacent basidorsals of the vertebral column.

inter-electrode capacitance *(Electronics)*. That of any pair of electrodes in a valve, other electrodes being earthed.

interface *(Aero.,Space)*. Relationship between parts of a system or subsystem which ensures that their eventual meeting will be harmonious; the interface may be physical (e.g. mechanical, thermal, electrical) or non-physical (e.g. software, organizational) and all conditions of the eventual union are controlled as part of the system documentation.

interface *(Comp.)*. Hardware and associated software needed to enable one device to communicate with another.

interface *(Electronics)*. A shared boundary. It may be a piece of hardware used between two pieces of equipment, a portion of computer storage accessed by two or more programs, or simply a surface that forms the boundary between two different materials.

interface *(Min.Ext.)*. Sharp contact boundary between two phases, either or both of which may be solid, liquid or gaseous. Differs from interphase in lacking a diffuse transition zone.

interfacial film *(Min.Ext.)*. Special state developed in emulsification, in which oriented molecules of the emulsifying agent are loosely aggregated in such a way as to surround and enclose droplets of one phase of the emulsified mixture.

interfacial surface tension *(Phys.)*. The *surface tension* at the surface separating two non-miscible liquids.

interfasicular cambium *(Bot.)*. Vascular cambium between the vascular bundles of the stem and joined to the cambium in the bundles to make a complete cylinder.

interfasicular region *(Bot.)*. Tissue between the vascular bundles of a stem. Also *medullary ray, pith ray*.

interference *(Aero.)*. Mutual aerodynamic interactions between solid bodies in airflow, the drag of the combined bodies exceeding that of their separate drags by the interference drag. Thus the lift of the lower wing of a biplane is reduced by the flow under the upper wing.

interference *(Biol.)*. The usually negative effect that the presence of one **chiasma** has on the probability of a second occurring in its vicinity.

interference *(Phys.)*. Interaction between two or more waves of the same frequency emitted from coherent sources. The wavefronts are combined, according to the *principle of superposition*, and the resulting variation in the disturbances produced by the waves is the interference pattern. See **interference fringes**.

interference *(Telecomm.)*. Any desired energy that tends to interfere with the reception of desired signals. May be *man-made*, e.g. from electrical machinery, power lines or radio-frequency heating systems, or due to natural phenomena, especially atmospheric electricity. In crowded wavebands, or abnormal propagation conditions, interference may be from other transmitters.

interference colours *(Phys.)*. See **colours of thin films**.

interference factor *(Telecomm.)*. See **telephone interference factor**.

interference fading *(Telecomm.)*. Fading of signals because of interference among the components of the signals which have taken slightly different paths to the receiver.

interference figure *(Crystal.)*. More or less symmetrical pattern of concentric rings or lemniscates, cut by a black cross or hyperbolae, exhibited by a section of anisotropic mineral when viewed in convergent light between crossed Nicol prisms. See also **uniaxial** and **biaxial**. Sometimes called *directions image*.

interference filter *(Phys.)*. If a beam of light is incident normally on a Fabry-Pérot etalon with a plate spacing d of between 2×10^{-7} and 6×10^{-7}, then there will be only one wavelength in the visible region for which there is an inteference maximum ($n\lambda = 2d$, where λ is the wavelength and $n = 0,1,2,...$). The highly reflecting surfaces of the etalon ensure that only a narrow band of wavelengths around λ is transmitted. Interference filters can also be made by successively evaporating dielectric and silvered films of suitable thickness on a glass plate.

interference filter *(Telecomm.)*. Means of reducing interference, e.g. a tuned rejector circuit for a single steady transmission, or a band-pass filter to reduce the accepted band of frequencies to the minimum. Also *interference trap*.

interference fit *(Eng.)*. A negative fit, necessitating force sufficient to cause expansion in one mating part, or contraction in the other, during assembly.

interference fringes *(Phys.)*. Alternate light and dark bands formed when two beams of monochromatic light having a constant phase relation, overlap and illuminate the same portion of a screen. The method of producing fringes is either by division of wavefront, see **Fresnel's biprism**, **Lloyd's mirror**, or by division of amplitude, see **Haidinger fringes**, **Fabry-Pérot interferometer**, **Fizeau fringes**, **Newton's rings**.

interference microscope *(Biol.)*. An instrument for the measurement of the optical retardation of a microscopic specimen, e.g. a living cell, which by conversion gives its total dry mass.

interference pattern *(Phys.)*. See **interference**.

interference trap *(Telecomm.)*. Same as *interference filter*.

interfering *(Vet.)*. An injury inflicted by a horse's foot on the opposite leg during progression.

interferometer *(Eng.)*. Instrument in which an acoustic, optical, or microwave interference pattern of 'fringes' is formed and used to make precision measurements, mainly of wavelength.

interferons *(Immun.,Med.)*. A group of proteins with antiviral activity originally identified as products released by cells in response to virus infection. When taken up by other cells the replication of virus within them is inhibited, by a mechanism which blocks the translation of viral messenger RNA. Interferons have since been found to be released from cells by other agents also, including immunological stimuli. α-interferons are released from leucocytes and β-interferons from fibroblasts in response to viral infection. γ-interferon is distinct and is released by activated macrophages and by T-lymphocytes and is regarded as a **lymphokine**. It can up-regulate the expression of MHC class II antigens and down-regulate the expression of MHC class I antigens, and activates **NK cells** to increase their cytocidal capacity. It also increases the sensitivity of B-cells to growth and differentiation factors. Interferons have been used as anti-cancer agents in some clinical trials.

interfluve *(Geol.)*. A ridge separating two parallel valleys.

intergalactic medium *(Astron.)*. General term for any material which might exist isolated in space far from any galaxy. Theoretical considerations suggest that an enormous quantity of undetected material exists, the total mass of which could far exceed the visible galaxies.

interglacial stage *(Geol.)*. A period of milder climate between two glacial stages.

intergranular corrosion *(Eng.)*. Corrosion in a polycrystalline mass of metal, taking place preferentially at the boundaries between the crystal grains. This leads to disintegration of the metallic mass before the bulk of the metal has been attacked by the corrosive agent.

intergranular texture *(Geol.)*. A texture characteristic of holocrystalline basalts and doleritic rocks, due to the aggregation of augite grains between feldspar laths arranged in a network.

interhalogen compound *(Chem.)*. Compound of two members of the halogen family, e.g. iodine monochloride ICl, iodine heptafluoride IF_7.

interkinesis *(Biol.)*. See **interphase**.

interlaced fencing *(Build.)*. See **interwoven fencing**.

interlaced scanning *(Image Tech.)*. A scanning system in which the lines of successive **fields** are displaced to form an interlaced pattern of alternate lines.

interlay *(Print.)*. Paper inserted between a printing plate and its mount in order to raise the plate to type height. Its thickness may be varied to produce increased or decreased impression where necessary.

interleukin *(Immun.)*. A term used to describe products of macrophages and T-lymphocytes which influence the differentiation of themselves or of other cells. Although hard to obtain pure under natural conditions, the genes responsible have been cloned and their products synthesized by recombinant DNA techniques. Abbrev. *IL*.

interleukin-1 *(Immun.)*. Secreted by activated macrophages. Causes fever by acting as a pyrogen at the thermoregulation centre in the brain, and the synthesis and release of **acute phase proteins** by liver cells. It is also required for activation of T-lymphocytes to develop receptors for interleukin-2 and to secrete interleukin-2. It has a direct mitogenic activity for thymocytes. There are two forms of IL-I termed alpha and beta which differ partly in respect of their amino acid sequences, but they have similar biological activities.

interleukin-2 *(Immun.)*. A growth factor for T-lymphocytes made by T-lymphocytes which have been activated by IL-I and a mitogenic agent such as concanavalin A or a specific antigen. It binds to specific receptorson the same or other T-cells and is thus able to maintain a population of T-cells in continuous growth and differential.

interleukin-3 *(Immun.)*. A factor made by activated T-lymphocytes which acts to stimulate growth and differentiation of the progenitors of all haemopoietic cells.

interleukin-4 *(Immun.)*. A factor made by activated T-lymphocytes which causes resting B-lymphocytes to divide.

interlining *(Textiles)*. A fabric placed in a garment between the lining and outer layer to act as a stiffener or to add bulk.

interlobar *(Med.)*. Situated or happening between 2 lobes, especially between 2 lobes of the lung.

interlock *(Elec.Eng.)*. Arrangement of controls in which those intended to be operated later are disconnected until the preliminary settings are correct, e.g. it is normally impossible to apply the anode voltage to an X-ray tube before the cooling water is flowing.

interlock *(Image Tech.)*. Maintaining synchronism between a motion picture camera and a sound recorder by electrical control.

interlock *(Min.Ext.)*. Arrangement of switch-gear by which the controlling source prevents premature loading, starting or continuance during partial malfunction where a series of operations is so interlinked as to require smooth on-line operation.

interlock *(Nuc.Eng.)*. A mechanical and or electrical device to prevent hazardous operation of a reactor, e.g. to prevent withdrawal of control rods before coolant flow has been established.

interlock *(Textiles)*. A double-faced weft-knitted fabric made of two rib fabrics joined by interlocking loops. Although originally made from cotton for underwear the fabrics are now knitted from various fibres and also used for outerwear.

intermediate *(Chem.)*. (1) A general term for any chemical compound which is manufactured from a substance (see **primary**) obtained from raw materials, and which serves as a starting material for the synthesis of another product. (2) A short-lived species in a complex reaction.

intermediate *(Image Tech.)*. General term for a print of a motion picture film made on colour duplicating stock having integral masking. The detailed usage is: *interdupe*, a duplicate colour negative derived from an interpositive; *interneg*, a duplicate colour negative from a reversal original; *interpositive, interpos*, a colour master positive printed from an original negative.

intermediate constituent *(Eng.)*. A constituent of alloys that is formed when atoms of two metals combine in certain proportions to form crystals with a different structure from that of either of the metals. The proportions of the two kinds of atoms may be indicated by formulae, e.g. CuZn; hence these constituents are also known as *intermetallic compounds*.

intermediate coupling *(Phys.)*. Coupling between spin and orbital angular momenta of valence electrons with characteristics between those of **j.j. coupling** and those of **Russell-Saunders coupling**.

intermediate filaments *(Biol.)*. Cytoplasmic filaments of *intermediate* diameter from 7 to 11 nm. There are several different subclasses, e.g. *desmin, keratin, lamin, vimentin*, each with a characteristic cellular distribution. They are all constructed of proteins possessing a rod-like structure built from four α-helical domains.

intermediate frequency *(Telecomm.)*. Output carrier frequency of a frequency changer (first detector) in a superhet receiver, adjusted to coincide with the centre of the frequency pass band of the intermediate amplifier. Abbrev. *IF*.

intermediate-frequency amplifier *(Telecomm.)*. That, in a superhet receiver, which is between the first (frequency changer) and second (final) demodulator, and which provides the main gain and band pass of the receiver.

intermediate-frequency oscillator *(Telecomm.)*. That which, in heterodyne reception, is combined with the output of the intermediate amplifier for demodulation in the second (final) detector. See **heterodyne conversion**.

intermediate-frequency response ratio *(Telecomm.)*. The ratio in decibels of an input signal at the intermediate frequency to one at the required signal frequency which would produce a corresponding output.

intermediate-frequency strip *(Telecomm.)*. An intermediate-frequency amplifier. Abbrev. *IF strip*.

intermediate-frequency transformer *(Telecomm.)*. One specially designed for coupling components and providing selectivity in an intermediate-frequency amplifier, the pass band being determined by antiresonance of the windings and coefficient of coupling.

intermediate host *(Zool.)*. In the life-history of a parasite, a secondary host which is occupied by the young forms, or by a resting stage between the adult stages in the primary host.

intermediate igneous rocks *(Geol.)*. Igneous rocks containing from 55% to 66% silica, and essentially intermediate in composition between the acid (granitic) and basic (gabbroic or basaltic) rocks. See **syenite, syenodiorite, diorite**.

intermediate level waste *(Nuc.Eng.)*. Nuclear waste not included in the categories **high level waste** or **low level waste**.

intermediate neutrons *(Phys.)*. See **neutron**.

intermediate phase *(Eng.)*. A homogeneous phase in an alloy with a composition range different from the pure components of the system.

intermediate pressure compressor *(Aero)*. The section of the axial compressor of a turbofan or by-pass turbojet between the *LP* and *HP* sections. It may be on a shaft of its own with a separate turbine, or it may be mounted on either of the other shafts.

intermediate rafter *(Build.,Civ.Eng.)*. See **common rafter**.

intermediate reactor *(Nuc.Eng.)*. One designed so that the majority of fissions will be produced by the absorption of intermediate **neutrons**. Also *epithermal reactor*.

intermediate sight *(Surv.)*. Reading on levelling staff, held at point not to be occupied by level other than back- or fore-sight observation from a levelling point between these last.

intermediate switch *(Elec.Eng.)*. A switch for controlling a circuit where more than two positions of control are required; it is connected between the two-way switches which must also be used in such a scheme.

intermediate vector boson *(Phys.)*. A *gauge boson* that mediates weak interactions between particles. There are three intermediate vector bosons, W^+, W^- and Z^0 and all have been observed. See **Weinberg and Salam's theory**.

intermediate wheel *(Horol.)*. *Idle wheel*. A wheel used to connect two other toothed wheels which are so far apart that they cannot mesh unless they are of excessive size.

intermedium *(Zool.)*. A small bone of the proximal row of the basipodium, lying between the tibiale and fibulare, or between the radiate and ulnare.

intermenstrual *(Med.)*. Occurring between two menstrual periods.

intermetallic compounds *(Eng.)*. See **intermediate constituents**.

intermingled yarn *(Textiles)*. A continuous filament yarn in which the constituent filaments are entangled by passing a turbulent air stream through the yarn.

intermission *(Med.)*. Temporary cessation, as of fever or of the normal pulse.

intermittent *(Image Tech.)*. The mechanism of a motion-picture camera or projector by which the film is advanced frame-by-frame.

intermittent claudication *(Med.)*. Pain in the legs on walking caused by inadequate bood supply to the leg muscles; **atherosclerosis** is responsible for arterial narrowing.

intermittent control *(Telecomm.)*. Control system in which controlled variable is monitored periodically, an intermittent correcting signal thus being supplied to the controller.

intermittent duty *(Telecomm.)*. The conditions of use for a component operated at its intermittent rating.

intermittent earth *(Elec.Eng.)*. An accidental earth connection which is present intermittently.

intermittent filtration *(Build.)*. The **land treatment** process of sewage purification, in which the land is drained artificially by ordinary earthenware or perforated plastic pipes. Cf. **broad irrigation**.

intermittent jet *(Aero.)*. See **pulse jet**.

intermittent rating *(Elec.Eng.)*. The specified power handling capacity of a component or instrument under specified condition of continuous usage. See **continuous rating, half-hour rating, one-hour rating**.

intermittent reinforcement *(Behav.)*. Schedules of reinforcement in which only some responses are reinforced; it can be based on *ratio* or *interval reinforcement*.

intermodulation *(Telecomm.)*. Undesired modulation of all frequencies with each other in passing through a nonlinear element in a transmission path.

intermodulation distortion *(Acous.)*. Amplitude distortion in which the intermodulation products are of greater importance than the harmonic products, as in AF amplifiers for high-quality speech or music.

intermolecular forces *(Phys.)*. Term referring to the forces binding one molecule to another. They are very much weaker than the bonding forces holding together the atoms of a molecule. See **van der Waals' forces**.

intermontane basin *(Geol.)*. A basin between mountain ranges often associated with a graben, e.g. Midland Valley of Scotland.

internal capacitance *(Phys.)*. Same as **inter-electrode capacitance**.

internal characteristic *(Elec.Eng.)*. A curve showing the relation between the load on an electric generator and the internal e.m.f.

internal-combustion engine *(Eng.)*. An engine in which combustion of a fuel takes place within the cylinder, and the products of combustion form the working medium during the power stroke. See **compression-ignition engine, diesel engine, gas engine, petrol engine**.

internal compensation *(Chem.)*. Neutralization of optical activity within the molecule by the combination of two enantiomorphous groups, e.g. in meso-tartaric acid.

internal conductor *(Elec.Eng.)*. The inner conductor of a concentric cable. Also called *inner conductor*.

internal conversion *(Phys.)*. Nuclear transition where energy released is given to orbital electron which is usually ejected from the atom (conversion electron), instead of appearing in the form of a γ-ray photon. The **conversion coefficient** for a given transition is given by the ratio of conversion electrons to photons.

internal e.m.f. *(Elec.Eng.)*. The e.m.f. generated in an electric machine; voltage appearing at the terminals is the internal e.m.f. minus any voltage drop caused by the current in the machine.

internal energy *(Chem.)*. The store of energy possessed by a material system. It is not usually possible to determine its absolute magnitude, but changes in its value can be measured. Changes in the internal energy of a system depend only upon the initial and final conditions, and are therefore independent of the paths of change.

internal energy *(Phys.)*. For a thermodynamic system, the difference between the heat absorbed by the system and the external work done by the system, is the change in its internal energy. (The *first law of thermodynamics*). The internal energy takes the form of the kinetic energy of the constituent molecules and their potential energies due to the molecular interactions. The internal energy is manifest as the temperature of the system, latent heat, as shown by a change of state, or the repulsive forces between molecules, seen as expansion. Symbol U; SI unit is the joule.

internal-expanding brake *(Eng.)*. See **expanding brake**.

internal flue *(Eng.)*. A furnace tube, or fire tube, running through the water space of a boiler.

internal focusing telescope *(Surv.)*. One in which focusing is effected by the movement of an internal concave lens fitted between the object glass and the eyepiece, both of which are fixed.

internal friction *(Eng.)*. The reaction which opposes flow in liquids and gases and may thus be said to be the phenomenon underlying **viscosity**.

internal gear *(Eng.)*. A spur gear in which teeth, formed on the inner circumference of an annular wheel, mesh with the external teeth of a smaller pinion. Both wheels revolve in the same direction.

internal grinding *(Eng.)*. The grinding of internal cylindrical surfaces by an abrasive wheel, which is either traversed along the revolving work or (in addition) given a planetary motion, the work being fixed.

internal image *(Immun.)*. Since an antibody made in response to an epitope of an antigen has a binding site (idiotope) complementary to the epitope, antibodies raised against the idiotope of the first antibody will have binding sites which have a similar shape to the epitope on the original antigen, i.e. have an internal image of that antigen. Anti-idiotypic antibodies have been shown to elicit antibodies against an antigen in an animal which was never exposed to the antigen itself.

internal impedance *(Phys.)*. The **output impedance** of an amplifier or generator.

internal indicator *(Chem.)*. An indicator which is dissolved in the solution in which the main reaction takes place.

internally fired boiler *(Eng.)*. A boiler whose fire box or furnace is inside the boiler and surrounded by water, as in the Lancashire, marine, and locomotive types.

internal mix spray gun *(Build.)*. A spray gun in which the fluid and atomising air are mixed within the air cap. Normally limited in use to work with small compressors and low pressure work.

internal pair production *(Phys.)*. Production of electron positron pair in the coulomb field of a nucleus. For transitions where the excitation energy released exceeds 1.02 MeV, this process will be competitive with both internal conversion and γ-ray emission. It occurs most readily in nuclei of low atomic number when the excitation energy is several MeV.

internal phloem *(Bot.)*. *Intraxylary phloem*. Primary phloem on the centric side of the xylem as in the Solanaceae.

internal resistance *(Elec.Eng.)*. That of any voltage source resulting in a drop in terminal voltage when current is drawn.

internal respiration *(Zool.)*. See **respiration, internal**.

internal screw-thread *(Eng.)*. A screw-thread cut on the inside of a cylindrical surface, as distinct from an **external screw-thread**. Also called a *female thread*.

internal secretion *(Zool.)*. A secretion which is poured into the blood vessels, or into the canal of the spinal cord; a hormone.

internal store *(Comp.)*. See **main memory**.

internal stress *(Eng.)*. Residual stress existing between different parts of metal products, as a result of the differential effects of heating, cooling or working operations, or of constitutional changes in the solid metal.

internal voltages *(Electronics)*. Those, such as **contact potential** or **work function**, which add an effect to the external voltages applied to an active device.

internasal septum *(Zool.)*. See **mesethmoid**.

International Air Transport Association *(Aero.)*. Association founded in Havana (1945) for the promotion of safe, regular and economic air transport, to foster air commerce and collaboration among operators, and to co-operate with **ICAO** and other international organizations. Abbrev. *IATA*.

international ångström *(Phys.)*. A unit which, although very nearly equal to the ångström unit (10^{-10} m), is defined in a different way. It is such that the red cadmium line at 15°C and 760 mm Hg pressure would have a wavelength of 6438.4696 I Å. Formerly the reference standard for metrology and spectroscopy. Abbrev. *IÅ*.

International Annealed Copper Standard *(Eng.)*. Abbrev. *IACS*. Standard reference for the conductivity of copper and its alloys. %IACS equals 1724.1/nΩ, where nΩ is the resistivity of the alloy per metre.

International Atomic Energy Agency *(Phys.)*. Organization for promoting the peaceful uses of nuclear power. Abbrev. *IAEA*.

international candle *(Phys.)*. Former unit of luminous intensity stated in terms of a point source, which was expressed in an equivalence of a standard lamp burning under specified conditions. Replaced by **candela**.

International Civil Aviation Organization *(Aero.)*. An intergovernmental co-ordinating body, which has its headquarters at Montreal, Canada, for the regulation and control of civil aviation and the co-ordination of **airworthiness** requirements and other safety measures on a worldwide basis. Abbrev. *ICAO*.

international electrical units *(Phys.)*. Units (amp, volt, ohm, watt) for expressing magnitudes of electrical quantities, adopted internationally until 1947; replaced first by **MKSA**, later by **SI units**.

International Frequency Registration Board *(Telecomm.)*. Committee responsible for reviewing and allocating radio frequencies used throughout the world, in order to ensure efficient use of channel space and avoid interference.

international paper sizes *(Paper)*. See **ISO paper sizes**.

international practical temperature scale *(Phys.)*. A practical scale of temperature which is defined to conform as closely as possible with the thermodynamic scale. Various fixed points were defined initially using the gas thermometer, and intermediate temperatures are measured with a stated form of thermometer according to the temperature range involved. The majority of temperature measurements of this scale are now made with platinum resistance thermometers.

international screw-thread *(Eng.)*. A metric system on which the pitch of the thread is related to the diameter, the thread having a rounded root and flat crest.

International Standard Atmosphere *(Aero.)*. A standard fixed atmosphere, adopted internationally, used for comparing aircraft performance; mean sea-level temperature 15°C at 1013.2 millibars, lapse rate 6.5°C/km altitude up to 11 km (ISA tropopause), above which the temperature is assumed constant at −56.5°C in the stratosphere. Abbrev. *ISA*.

International Style *(Arch.)*. An American term, originally introduced as the title of a book to complement the first International Exhibition of Modern Architecture which took place in New York in 1932, but now widely accepted as a name for the architectural style which has prevailed since the 1920s and which has formed the basis of architectural development to the present day.

International system of units *(Genrl.)*. See **SI units**.

International Telecommunication Union *(Telecomm.)*. Specialised agency of the United Nations, established to promote international collaboration in telecommunications with a view to improving the efficiency of world-wide services. It has three permanent committees, the **CCIR**, the **CCITT** and the **IFRB**. Abbrev. *ITU*.

interneuron *(Zool.)*. A nerve cell within the central nervous system which is neither sensory or motor but communicates between other nerve cells.

internode *(Bot.)*. Between two successive nodes of a stem or hypha.

internode *(Zool.)*. The myelinated part of a nerve between two adjacent nodes of Ranvier.

internuncial *(Biol.)*. Interconnecting, e.g. neurone of central nervous system interposed between afferent and efferent neurones of a reflex arc.

interoceptor *(Zool.)*. A sensory nerve-ending, specialized for the reception of impressions from within the body. Cf. **exteroceptor**.

interopercular *(Zool.)*. In Fish, a ventral membrane bone supporting the operculum.

interparietal *(Zool.)*. A median dorsal membrane bone of the Vertebrate skull, situated between the parietals and the supraoccipital.

interpenetration twins *(Min.)*. Two or more crystals united in a regular fashion, according to a fixed plan (the *twin law*), the individual crystals appearing to have grown through one another. Cf. **juxtaposition twins**.

interphase *(Biol.)*. The period of the **cell cycle** between mitoses.

interphase *(Min.Ext.)*. Transition zone between two phases in a system (solid/liquid, liquid/liquid, liquid/gas). In solid/fluid system; zone of shear through which the physical or chemical qualities of the contacting surfaces migrate toward one another.

interphase transformer *(Elec.Eng.)*. A centre-tapped, iron-cored reactor, connected between the two star-joints of the supply transformer for a double three-phase mercury-arc rectifier. The centre tap forms the d.c. negative terminal. Also called *interphase reactor, absorption inductor*, and *phase equalizer*.

interplane struts *(Aero.)*. In a multiplane structure, those struts, either vertical or inclined, connecting the spars of any pair of planes, one above the other.

interplanetary matter *(Astron.)*. Material in the solar system other than the planets and their satellites. This matter includes the streams of charged particles from the **solar wind**, the resulting tenuous gas being termed the *interplanetary medium*. Interplanetary matter also includes dust, **meteorites** and **comets**.

interplanetary space *(Space)*. That region of space generally outside the influence of planetary bodies which is occupied by a tenuous gas. It consists of charged and neutral particles of gas (mainly hydrogen), dust and streams and clouds of particles (protons, electrons, etc.) ejected by the Sun. A weak magnetic field pervades the whole region.

interpolation *(Maths.)*. The estimation of the value of a function at a particular point from values of the function on either side of the point. Cf. **extrapolation**.

interpolation of contours *(Surv.)*. The joining of points derived from spot levelling or in conformity with the existing mapped contour lines, assuming that the recorded data disclose all abrupt changes in ground level and configuration.

interpolator *(Comp.)*. Apparatus for comparing cards in separate packs and merging them. See **merge**.

interpole *(Elec.Eng.)*. See **compole**.

interpole motor *(Elec.Eng.)*. An electric motor fitted with **compoles**.

interpositional growth *(Bot.)*. See **intrusive growth**.

interpreter *(Comp.)*. Program which translates and executes a source program one statement at a time. Cf. **compiler**.

inter-renal body *(Zool.)*. In selachian Fish, a ductless gland which lies between the kidneys and corresponds to the cortex of the suprarenal gland of higher Vertebrates.

interrogation *(Telecomm.)*. Transmission of a pulse to a **transponder** (pulse repeater).

interrupt *(Comp.)*. Signal which causes a break in the execution of the current routine. Control passes to another routine in such a way that later the original routine may be resumed. See **packet switching**, **time sharing**.

interrupted continuous waves *(Telecomm.)*. Electromagnetic waves generated by an oscillator the output from which is interrupted periodically at an audible frequency, so that the received signal is directly audible after detection. Abbrev. *ICW*. Also called *chopped continuous waves*.

interruptedly pinnate *(Bot.)*. A pinnate leaf having pairs of large and small leaflets alternating along the rachis.

interrupter *(Elec.Eng.)*. A device which interrupts periodically the flow of a continuous current, such as the mechanical 'make-and-break' of an induction coil.

interscapular *(Med.)*. Between the two shoulder blades.

intersection *(Maths.)*. The intersection of sets *A* and *B* is the set of elements *x* such that *x* is in *A* and *x* is also in *B*.

intersection *(Surv.)*. Plane-table surveying in which the plane table is set up consecutively at each end of a measured base line. Rays are drawn on paper at each set-up to show the direction of the point that it is required to fix on the plan, the intersection of these rays giving the position of the point.

intersection angle *(Surv.)*. The angle of deflection as measured at the intersection point between the straights of a railway or highway curve.

intersection point *(Surv.)*. The point at which the straights of a railway or highway curve would meet if produced.

intersegmental membrane *(Zool.)*. The flexible infolded portion of the cuticle between adjacent definitive segments to allow the freedom of movement of the body in *Arthropoda*.

intersertal texture *(Geol.)*. The texture characterized by the occurrence of interstitial glass between divergent laths of feldspar in basaltic rocks.

inter-setting *(Print.)*. The introduction on a rotary press of a pre-printed web in register.

intersex *(Biol.)*. An individual which exhibits characters intermediate between those of the male and female of the same species. Often due to a chromosomal abnormality.

intersheath *(Elec.Eng.)*. Cylindrical electrodes in the interior of a cable dielectric, used for the purpose of keeping the variation of stress at a minimum. The intersheaths must be kept at certain potentials in order to achieve this purpose. See **stress**.

interspecific *(Biol.)*. Said of an event such as a cross between individuals from separate species. Cf. **intraspecific**.

interstation interference *(Telecomm.)*. That from another transmitter on the same or an adjacent wavelength, as distinct from atmospheric interference.

interstation noise suppression *(Telecomm.)*. Suppression of radio receiver output when no carrier is received.

interstellar hydrogen *(Astron.)*. The **interstellar medium** is predominantly hydrogen. Cool hydrogen clouds *HI regions* are diffuse clouds of atomic hydrogen detected by the radio emission at a frequency of 1420.4 Mhz ('21 centimetre line') associated with a **forbidden transition** between two ground state energy levels. Ionized hydrogen, or *HII regions*, exist as discrete clouds often associated with star formation or other sources of ultraviolet radiation.

interstellar medium *(Astron.)*. The gaseous and dusty matter pervading interstellar space and amounting to one-tenth of the mass of the Galaxy. Where cold gas and dust conglomerate, regions of star formation occur. The medium comprises hydrogen, helium, **interstellar molecules** and dust. It is replenished by stellar winds and the ejecta of **nova** outbursts and **supernova** explosions.

interstellar molecule *(Astron.)*. More than fifty species of molecule are found within the gas nebulae of the **interstellar medium**, particularly in cold dense clouds. The most common molecules include: CO, H_2O, NH_3, HCHO, CH_4 and CH_3OH.

interstice *(Chem.)*. Space between atoms in a lattice where other atoms can be located, e.g. in close-packed metallic lattices.

interstitial *(Zool.)*. Occurring in the interstices between other structures; as the *interstitial* cells of *Cnidaria*, which are small rounded embryonic cells occurring in the interstices between the columnar cells forming the ectoderm and endoderm.

interstitial cell stimulating hormone *(Med.)*. Abbrev. *ICSH*. See **luteinizing hormone**.

interstitial compounds *(Eng.)*. Metalloids in which small atoms of nonmetallic elements (H,B,C,N,) occupy positions in the interstices of metal lattices. In general, the structure of the metal is preserved, though somewhat distorted. They are often (especially those of Group IV and V metals) characterized by exceptionally high-melting points and hardness, by chemical inertness and by metallic lustre and conductivity.

intertrack bond *(Elec.Eng.)*. A conductor for connecting electrically the rails of separate tracks on electric railways or tramways to reduce the total resistance of the return path.

intertrigo *(Med.)*. Excoriation of the skin in the skin folds, usually due to excessive moisture.

intertrochanteric *(Med.)*. Situated between the two trochanters of the upper part of the femur.

intertropical convergence zone (ITCZ) *(Meteor.)*. A narrow, low-latitude zone in which air masses originating in the northern and southern hemispheres converge. Over the Atlantic and Pacific oceans it is the boundary between the northeast and southeast trade winds. The mean position is somewhat north of the equator but over the continents the range of movement is considerable. It is a zone of generally cloudy, showery weather. Sometimes called the *intertropical front (ITF)*.

intertype *(Print.)*. A typesetting machine, using circulating matrices, similar in most respects to the **Linotype**. See also **composing machines**.

interval *(Acous.)*. Ratio of the frequencies of two sounds; or in some cases its logarithm. See **semitone**.

interval *(Maths.)*. See **closed-**, **open-**.

interval schedules of reinforcement *(Behav.)*. Schedules in which reinforcement is given for the first response made after a certain period of time has passed. On fixed interval schedules, the period of time is always the same; on variable interval schedules the period of time fluctuates. Both types are forms of *partial* or *intermittent* reinforcement. Cf. **continuous reinforcement**.

intervening sequence *(Biol.)*. See **intron**.

interventional radiology *(Radiol.)*. A term used to describe radiological procedures undertaken for therapeutic rather than diagnostic purposes, e.g. **angioplasty**, where a balloon on a catheter is dilated in a narrowed artery.

intervertebral *(Zool.)*. Between the vertebrae; said of the fibro-cartilage disks between the vertebral centra in Crocodiles, Birds, and Mammals.

interwoven fencing *(Build.)*. Solid wood, or metal, fencing built up of very thin boards interlaced. Also called *interlaced fencing* or *wovenboard*.

intestine *(Zool.)*. In Vertebrates, that part of the alimentary canal leading from the stomach to the anus; in Invertebrates, that part of the alimentary canal which corresponds to the Vertebrate intestine, or was thought by the early investigators so to correspond. adj. *intestinal*.

intine *(Bot.)*. The inner part of the wall of a pollen grain or vascular plant spore. Cf. **exine**.

intoxication *(Med.)*. The state of being poisoned.

intracapsular *(Med.)*. Situated within a capsule, especially within the ligamentous joint capsule enveloping the head and neck of the femur.

intracavitary *(Radiol.)*. Applied within cavities of the body; said, e.g. of radium applied in the cavity of the uterus, also of irradiation of part of the body through natural or artificial body cavities.

intracellular *(Biol.)*. Within the cell.

intracellular enzyme *(Biol.)*. One secreted and functioning within the cell. Cf. **extracellular enzyme**.

intracerebral *(Med.)*. Situated in the substance of the brain.

intracervical *(Med.)*. Situated in, or applied to, the canal of the cervix uteri (the lowest part or neck of the uterus).

intracranial *(Med.)*. Situated within the skull.

intradermal *(Med.)*. Situated in, or introduced into, the skin.

intrados *(Build.)*. The under surface of an arch. Also *soffit*. See **extrados**.

intrafusal *(Zool.)*. A muscle fibre contained within a muscle spindle. Concerned with proprioceptive reflexes.

intrahepatic *(Med.)*. Situated or occurring in the substance of the liver.

intramammary *(Med.)*. Situated or occurring in the breast.

intramedullary *(Med.)*. In the substance of the medulla oblongata (the brain stem); situated or occurring in the substance of the spinal cord.

intranuclear forces *(Phys.)*. Those operative between nucleons at close range comprising **short range forces** and **coulomb forces**. According to the hypothesis of charge independence, the former are always the same for two nucleons of corresponding angular momentum and spin regardless of whether these are protons or neutrons. See **isotopic spin**.

intra-ocular *(Med.)*. Situated within the eyeball.

intraperitoneal *(Med.)*. Situated in, or introduced into, the peritoneal cavity.

intrapleural *(Zool.)*. Within the thoracic cavity.

intraspecific *(Zool.)*. Pertaining to interactions within members of a species. Cf. **interspecific**.

intrathecal *(Med.)*. Within the sheath of membranes investing the spinal cord.

intratracheal *(Med.)*. Within, or introduced into, the trachea.

intratubal *(Med.)*. Situated within a Fallopian tube.

intra-uterine *(Med.)*. Situated within, or developing within, the uterus.

intra-uterine device *(Med.)*. Contraceptive device of relatively high efficacy, inserted in the uterus; consists usually of a spiral or a loop-shaped copper-coated filament. Its action, though not completey understood, may be physiological. Abbrevs. *IUD, IUCD*.

intravenous *(Med.)*. Within, or introduced into, a vein.

intra-vitam staining *(Biol.)*. The artificial staining of living cells. Usually by injection of the stain into a circulatory system.

intraxylary phloem *(Bot.)*. See **internal phloem**.

intrazonal soil *(Bot.)*. Well developed soil in which local conditions have modified the influence of the regional climate. Cf. **zonal soil, azonal soil**.

intrinsic *(Zool.)*. Said of appendicular muscles of Vertebrates which lie within the limb itself, and originate either from the girdle or from the limb-bones. Cf. **extrinsic**.

intrinsic angular momentum *(Phys.)*. The total spin of atom, nucleus, or particle as an idealized point, or arising from orbital motion. When quantized, the former and latter are, respectively,

$$\frac{1}{2}\frac{h}{2\pi} \quad \text{and} \quad \frac{h}{2\pi},$$

where h = Planck's constant.

intrinsic conduction *(Electronics)*. That in a semiconductor when electrons are raised from a filled band into the conduction band by thermal energy, so producing hole-electron pairs. It increases rapidly with rising temperature.

intrinsic crystal *(Crystal.)*. One whose photoelectric properties do not depend on impurities.

intrinsic equation *(Maths.)*. Of a curve: an equation connecting the arc length s of any point on the curve with the gradient ψ at the point. Such an equation is independent of any co-ordinate system and is an inherent property of the curve.

intrinsic factor *(Med.)*. A glycoprotein secreted from the parietal cells of the gastric mucosa which is essential for the absorption of *cyanocobalamin*. Absence of intrinsic factor causes **pernicious anaemia**.

intrinsic impedence *(Phys.)*. Wave impedence depending on the medium alone, i.e., $Z_0{}^2 = \mu/\varepsilon$, where μ is the absolute permeability and ε the absolute permittivity of the medium.

intrinsic induction *(Phys.)*. Synonymous with **intensity of magnetization**. (*N.B.* In unrationalized units, it is 4π times this quantity.).

intrinsic mobility *(Electronics)*. The mobility of electrons in an intrinsic semiconductor.

intrinsic semiconductor *(Electronics)*. See **semiconductor**.

introgression *(Ecol.)*. *Introgressive hybridization*. Incorporation of genes of one species into the gene pool of another *via* an interspecific hybrid.

introitus *(Med.)*. Entry to a cavity, e.g. vagina.

introjection *(Behav.)*. In social psychology, the internalization of the norms and values of one's social group, so that the individual comes to be guided by a sense of what is appropriate or inappropriate, rather than by external rewards and punishments. In psychoanalytic theory, the term is synonomous with **identification**.

intromission *(Med.)*. The insertion of one part into another, especially of the penis into the vagina.

intromittent *(Zool.)*. Adapted for insertion, as the copulatory organs of some male animals.

intron *(Biol.)*. Genes in eukaryotes are organized in such a way that while the whole sequence is transcribed, only part of it forms the messenger RNA. *Introns* or *intervening sequences*, which are often long, are excised during the maturation of the RNA. Cf. **exons**, the part which is expressed in the protein product.

introrse *(Bot.,Zool.)*. Directed or bent inwards. Facing towards the axis; especially stamens opening towards the centre of the flower. Cf. **extrorse**.

introvert *(Behav.)*. In Jungian theory, a person whose libido is inwardly directed resulting in the shunning of interpersonal relations and absorption in egocentric thoughts. Generally, a person more interested in his thoughts and feelings than in the external world and social activity. n. *introversion*. See **extraversion**.

introvert *(Zool.)*. A structure or part of the body which may be involuted, as the proboscis of a Nemertinean worm, or turned inside-out.

intrusion *(Geol.)*. The process of emplacement of a magma into pre-existing rock. Also used in the sense of injection of a plastic sediment, e.g. a salt dome.

intrusive growth *(Bot.)*. Type of tissue growth in which an elongating cell grows by insertion between other cells. Cf. **symplastic growth**.

intubation *(Med.)*. The introduction of a tube, especially through the larynx into the trachea.

intumescence *(Chem.)*. The swelling of crystals on heating, often with the violent escape of moisture.

intumescence *(Med.)*. The process of swelling; a swelling.

intumescent paints *(Build.)*. Fire retardant paints which upon exposure to heat swell to form an insulating barrier which protects the underlying surface.

intussusception *(Bot.)*. New material inserted into the thickness of an existing cell wall. Cf. **apposition**.

intussusception *(Med.)*. The pushing down, or invagination, of one part of the intestine into the part below it. Usually occurs in children causing intestinal obstruction.

inulin *(Med.)*. A polysaccharide obtained from the tubers of the dahlia, used in assessing renal function.

invagination *(Zool.)*. Insertion into a sheath; the development of a hollow ingrowth; the pushing-in of one side of the blastula in embolic gastrulation. Cf. **evagination**. adj. *invaginate*.

inv allotypes *(Immun.)*. Allotypic antigenic determinants on the constant region of the kappa chain of human immunoglobulins.

invar *(Eng.)*. TN for iron-nickel alloy. Composed of 36% nickel, 63.8% iron and 0.2% carbon. Coefficient of thermal expansion is very small. Used for measuring-tapes, tuning-forks, pendulums and in instruments.

invariable plane *(Astron.)*. A certain plane which remains absolutely unchanged by any mutual action between the planets in the solar system; defined by the condition that the total angular momentum of the system about the normal to this plane is constant. The plane is inclined at 1°35′ to the ecliptic.

invariant *(Chem.)*. Possessing no **degrees of freedom** (1).

invariant *(Maths.)*. Any characteristic of a system which is unchanged by transformations of the type under consideration.

inventory *(Nuc.Eng.)*. Total quantity of fissile material in a reactor.

inverse *(Maths.)*. If S is a set containing an **identity** element e with respect to an operation * in S, then the element x of S is said to have an inverse x' in S, if $x*x' = e = x'x$. For the usual addition of real numbers, the inverse of x is written $-x$ and is called the negative of x. For the usual multiplication of real numbers, the inverse of the non-zero real number x is written $1/x$ or x^{-1} and is called the reciprocal of x. Note that

$$x + (-x) = 0 = (-x) + x,$$

$$xx^{-1} = 1 = x^{-1}x.$$

inverse current *(Electronics)*. Same as *reverse current.*

inverse hyperbolic function *(Maths.)*. If y is a hyperbolic function of x, then x is the inverse hyperbolic function of y, e.g. if $y = \sinh x$, the inverse hyperbolic function is $x = \text{arc sinh } y$ or $x = \sinh^{-1}y$, where arc sinh y or $\sinh^{-1}y$ is the number whose sinh is x.

inverse networks *(Elec.Eng.)*. A pair of two-terminal networks whose impedances are such that their product is independent of frequency.

inverse of a matrix *(Maths.)*. For a square matrix, the quotient of the adjoint of the matrix and its determinant provided the determinant is not zero. The inverse of the square matrix A has the same order as A and is denoted by

$$AA^{-1} = I = A^{-1}A,$$

where I is the unit matrix having the same order as A.

inverse of a one-to-one mapping *(Maths.)*. If f is a one-to-one mapping from set S on to set T, the inverse of f is the mapping g from T to S such that $g(b(x)) = x$. g is written f^{-1}.

inverse power factor *(Elec.Eng.)*. A term sometimes used to denote sec $\phi(= 1/\cos\phi)$.

inverse segregation *(Eng.)*. A type of segregation in which the content of impurities, inclusions and alloying elements in metals tends to decrease from the surface to the centre. See also **segregation, normal segregation.**

inverse-speed motor *(Elec.Eng.)*. See **series-characteristic motor.**

inverse square law *(Phys.)*. The intensity of a field of radiation is inversely proportional to the square of the distance from the source. Applies to any system with spherical wavefront and negligible energy absorption.

inverse time-lag *(Elec.Eng.)*. A time-lag which is approximately proportional to the inverse of the current causing its operation. Also called *inverse time-element, inverse time-limit.*

inverse trigonometrical function *(Maths.)*. If y is a trigonometrical ratio of the angle x, then x is the inverse trigonometrical function of y, e.g. if $y = \sin x$, the inverse trigonometrical function is $x = \text{arc sin } y$ or $\sin^{-1}y$, where arc sin$y$ or $\sin^{-1}y$ is the angle whose **sine** is y.

inverse voltage *(Elec.Eng.)*. That generated across a rectifying element during the half-cycle when no current flows. Its maximum value is the *peak inverse voltage.*

inversion *(Biol.)*. A stretch of chromosome that has been turned round so that the order of the nucleotides in the DNA is reversed.

inversion *(Chem.)*. The formation of a laevorotatory solution of fructose and glucose by the hydrolysis of a dextrorotatory solution of sucrose (cane-sugar).

inversion *(Maths.)*. The operation of deriving from a first set of points P a second set of points P' by the rule $OP.OP' = a^2$ where O is a fixed point (*the centre of inversion*), a is a constant (*the radius of inversion*), and P and P' lie on a straight line through O and on the same side of it.

inversion *(Meteor.)*. Reversal of the usual temperature gradient in the atmosphere, the temperature increasing with height. Inversions are of frequent occurrence near the ground on clear nights and in anticyclones, often causing dense smoke fogs over cities.

inversion of relief *(Geol.)*. A condition whereby synclinal ridges are separated by anticlinal depressions.

invert *(Build.,Civ.Eng.)*. The lowest part of the inner surface of the cross-section of a non-vertical drain or sewer at any given point, and related to the datum level.

invertase *(Biol.)*. A plant enzyme which hydrolizes cane-sugar. Also *sucrase.*

Invertebrata *(Zool.)*. A collective name for all animals other than those in the phylum Chordata; that is, all those animals that do not exhibit the characteristics of vertebrates, namely possession of a notochord or vertebral column, ventral heart, eyes etc. Some animals near the chordate boundary line, e.g. Hemichorda and Urochorda, are sometimes regarded as invertebrates. The term is not used as a scientific classification.

inverted arch *(Civ.Eng.)*. An arch having the crown below the line of the springings, e.g. the floor of a tunnel, in order to distribute the pressure on the walls over a greater area. Also called an *inflected arch.*

inverted-brush contact *(Elec.Eng.)*. A laminated switch-contact in which the laminations are carried on the fixed, instead of on the moving, contact.

inverted engine *(Aero.)*. An in-line engine having its cylinders below the crankshaft. Adopted in certain types of aircraft to improve the forward view of the pilot.

inverted-L antenna *(Telecomm.)*. One comprising a vertical wire joined to one end of a horizontal conductor.

inverted loop *(Aero.)*. An *aerobatic* manoeuvre consisting of a complete revolution in the vertical plane with the upper surface of the aircraft outside, which is started from the inverted position. Also *outside loop.*

inverted machine *(Elec.Eng.)*. Any electric machine in which the usual arrangement of the stator and rotor windings is inverted, e.g. an induction motor in which power is supplied to the rotor, the stator winding being short-circuited.

inverted rectifier *(Elec.Eng.)*. (1) Circuit arranged to convert d.c. to a.c. (2) Amplifier which inverts polarity. Same as *inverter.*

inverted rotary converter *(Elec.Eng.)*. A rotary converter which is used to convert from d.c. to a.c.

inverted speech *(Telecomm.)*. Inversion of the order of speech frequencies before modulation for privacy; effected by modulating with a carrier just above the maximum speech frequency, then discarding this carrier and its upper sideband.

inverted telephoto lens *(Image Tech.)*. Lens having a short focal length and comparatively long back focus, thus allowing space for a prism system between the rear of the lens and the sensitive surface.

inverted-V antenna *(Telecomm.)*. Two wires, several quarter-wavelengths long, joined at the top, one lower end being terminated by a resistance, the other end by the transmitter or receiver; the direction of maximum response is horizontal and in the plane of the wires.

inverter *(Elec.Eng.)*. A d.c.-a.c. converter using e.g. a transistor or thyristor oscillator.

inverter *(Telecomm.)*. Arrangement of modulators and filters for inverting speech or music for privacy. Also *speech inverter.*

invertible *(Maths.)*. Having an inverse.

invert soap *(Chem.)*. Term sometimes used to denote certain **cationic detergents**, e.g. cetyl pyridinium bromide.

invert sugar *(Chem.)*. The product obtained by the hydrolysis of cane-sugar with acids; it is a mixture of equal parts of d-fructose and d-glucose. Most fruits contain invert sugar, and honey averages over 70%.

investment *(Zool.)*. The outer layers of an organ or part, or of an animal.

investment casting *(Eng.)*. Forming a mould round a pattern whose shape can be destroyed to allow its removal. Patterns of complex shapes can be used which would otherwise be impossible to draw from the mould. Patterns may be of wax, plastic, frozen mercury etc. The ancient lost wax or *cire perdue* process is a good example.

invisible glass *(Phys.)*. Glass which has its surface curved, or has been coated with molecular thickness material, to eliminate surface reflections.

in vitro *(Biol.)*. Literally 'in glass'. Used to describe the experimental reproduction of biological processes in the more easily defined environment of the culture vessel or plate; cf. **in vivo.**

in vitro transcription *(Biol.)*. Use of a laboratory medium without the presence of cells to obtain specific mRNA production from a DNA sequence. Also called *cell-free transcription.*

in vitro translation *(Biol.)*. The use of pure mRNA, ribosomes, factors, enzymes and precursors to obtain a specific protein product without the presence of cells. The procedure can be coupled to *in vitro transcription*. Also called *cell-free translation.*

in vivo *(Biol.)*. Used to describe biological processes occurring within the living organism or cell; cf. **in vitro**.

involucre *(Bot.)*. (1) Bracts forming a calyx-like structure close to the base of a usually condensed inflorescence, as in the Compositae. (2) A sheath or ring of leaves surrounding a group of archegonia or antheridia in bryophytes.

involucrum *(Med.)*. Sheath of new bone formed round bone which has died as the result of infection of the bone.

involuntary muscle *(Zool.)*. See **unstriated muscle**.

involute *(Bot.)*. The margins of a leaf rolled towards the *adaxial surface*. Cf. **revolute**. See also **vernation**.

involute *(Maths.)*. See **evolute**. Every curve has many involutes. They are traced out by the different points on a piece of string unwinding from the curve.

involute *(Zool.)*. Tightly coiled; said of Gastropod shells.

involute gear teeth *(Eng.)*. Gear teeth whose flank profile consists of an involute curve given by the locus of the end of a taut string uncoiled from a base circle; the commonest form of tooth.

involution *(Maths.)*. (1) Raising a number to a positive integral power. Cf. **evolution**. (2) An **automorphism**, which, when applied twice, gives the identity mapping.

Io *(Astron.)*. The first natural satellite of **Jupiter**, discovered by Galileo. It is characterized by very active volcanoes, and stimulates radio emission in the ionosphere of **Jupiter**.

iodates (V) *(Chem.)*. Salts of iodic acid, HIO_3.

iodazide *(Chem.)*. N_3I. An iodine azide.

IO device *(Comp.)*. Abbrev. for *Input/Output device*. Peripheral which may be used as an **input device** or as an **output device**.

iodic anhydride *(Chem.)*. See **iodine oxides**.

iodic (V) acid *(Chem.)*. HIO_3. Formed by the direct oxidation of iodine with nitric acid. White crystalline solid. Soluble in water. Forms iodates (V).

iodides *(Chem.)*. Salts of hydriodic acid. Most metallic iodides liberate free iodine and leave behind the metal or metallic oxide when heated. See **hydriodic acid**.

iodination *(Chem.)*. (1) The substitution or addition of iodine atoms in or to organic compounds. (2) The addition of iodine to table salt in districts where thyroid deficiencies are prevalent. See **goitre**.

iodine *(Chem.)*. Nonmetallic element in the seventh group of the periodic system, one of the halogens. Symbol I, at. no. 53, r.a.m. 126.9044, valencies 1, 3, 5, 7. It forms blackish scales with a violet lustre and a characteristic smell; mp 113.5°C, bp 184.35°C, rel.d. 4.95 at 20°C. It is widely but sparingly distributed as iodides, and is a constituent of the thyroid gland. The important commercial sources are crude Chile saltpetre (caliche) and certain seaweeds; it is used in organic synthesis.

iodine *(Med.)*. Powerful disinfectant once widely used; a constituent, as iodide, of **contrast media**. See also **radio iodine**.

iodine monochloride *(Chem.)*. ICl. Brown-red crystals, mp 13.9°C. Prepared by reaction of chlorine with iodine. Used as an iodinating agent in analytical and organic chemistry.

iodine oxides *(Chem.)*. Iodine has four oxides with the empirical formulae, I_2O_4, I_4O_9, I_2O_5, IO_4. They differ in marked degree from those of the other halogens. The (V) oxide, I_2O_5 (**iodic anhydride**) has acidic properties; stable, with water yields iodic acid.

iodine pentafluoride *(Chem.)*. IF_5. Colourless liquid, formed with incandescence by the direct combination of fluorine and iodine, mp −8°C, bp 97°C.

iodine trichloride *(Chem.)*. ICl_3, yellow powder, a powerful disinfectant.

iodine value *(Chem.)*. The number of grams of iodine absorbed by 100 grams of a fat or oil. It gives an indication of the amount of unsaturated acids present in fats and oils.

iodism *(Med.)*. The condition resulting from overdosage of, or sensitivity to, iodine; characterized by running at the eyes and the nose, salivation, and skin eruptions.

iodo compounds *(Chem.)*. Compounds containing the covalent iodine radical —I, e.g. iodobenzene.

iodoform *(Chem.)*. Tri-iodomethane. CHI_3, yellow hexagonal plates, of peculiar, saffron odour, mp 119°C, volatile in steam, mildly antiseptic. It is prepared by warming alcohol with iodine and alkali; or by an electrolytic method, in which a current is passed through a solution containing KI, Na_2CO_3, and ethanol, the temperature being about 65°C. This reaction is the basis of the *iodoform test* for compounds containing the group $-CO.CH_3$ or $-CHOH.CH_3$.

iodogorgoic acid *(Chem.)*. *3,5-di-iodotyrosine*. A constituent of the thyroid hormone.

iodometric *(Chem.)*. Measured by iodine, e.g. in a titration.

iodophilic bacteria *(Biol.)*. Bacteria which stain blue with iodine, revealing the presence of starchlike compounds. They occur in large numbers in the rumen where they are associated with cellulose decomposition.

iodopsin *(Phys.)*. See **visual violet**.

iodoso compound *(Chem.)*. Compound containing the iodoso radical —IO.

iodoxy compounds *(Chem.)*. Compound containing the radical $-IO_2$.

iodyl ion *(Chem.)*. The monoxide of trivalent electropositive iodine: $(IO)^+$. Forms salts, e.g. iodyl sulphate $(IO)_2SO_4$, which are hydrolysed by water to give iodine, iodic (V) acid, and the acid corresponding to the anion.

iolite *(Min.)*. See **cordierite**.

ion *(Phys.)*. Strictly, any atom or molecule which has resultant electric charge due to loss or gain of valence electrons. Free electrons are sometimes loosely classified as *negative ions*. Ionic crystals are formed of ionized atoms and in solution exhibit ionic conduction. In gases, ions are normally molecular and cases of double or treble ionization may be encountered. When almost completely ionized, gases form a fourth state of matter, known as a **plasma**. Since matter is electrically neutral, ions are normally produced in pairs.

ion beam *(Phys.)*. A beam of ions moving in the same direction with similar speeds, especially when produced by some form of accelerating machine or mass spectrograph. Also *ionic beam*.

ion burn *(Electronics)*. Damage to the phosphor of a magnetic-deflection CRT because of relatively minute deflection of heavy negative ions by the magnetic control which deflects the electron beam.

ion cluster *(Chem.)*. Group of molecules loosely bound (by electrostatic forces) to a charged ion in a gas.

ion concentration *(Chem.)*. That expressed in moles per unit volume for a particular ion. Also *ionic concentration*. See **pH value**.

ion concentration *(Phys.)*. The number of ions of either sign, or of ion pairs, per unit volume. Also *ionization density*.

ion cyclotron frequency *(Phys.)*. An ion in a uniform magnetic field, perpendicular to its motion, follows a circular path similar to that in a *cyclotron*. The critical frequency depends on the magnitude of the field and on the charge and mass of the ion.

ion engine *(Electronics)*. A device for propulsion or attitude control of spacecraft or satellites, esp. in outer space. Ions and electrons are expelled at high velocity from *combustion chamber*. Outside, they recombine, thus preventing any space-charge effect which would counteract thrust.

ion exchange *(Chem.)*. Abbrev. *IX*. Use of zeolites, artificial resins or specially treated coal to capture anions or cations from solution. Used in water-softening, separating isotopes, de-salting sea water, and wide range of industrial processes such as chemical extraction of uranium from its ores. Instead of solid particles through which liquid runs, immiscible liquid-liquid systems may be employed to transfer ions from one phase to another.

ion exchange *(Med.)*. Interchange of ions of similar charge between an insoluble solid and a solution brought in contact with it. Administration of ion exchange resins orally or rectally is used to lower potassium concentration in patients with renal failure.

ion-exchange capacity *(Min.Ext.)*. Electrical charge surplus to that uniting the framework of the ion-exchange resin or other exchange vehicle.

ion-exchange liquids *(Min.Ext.)*. Those immiscible with water (e.g. kerosene) which are rendered active by addition of suitable chemicals, such as tri-lauryl amine. These liquids can be used in solvent extraction processes in place of solid ion-exchange resins.

ion-exchange resins *(Chem.)*. Term applied to a variety of materials, usually organic, which have the capacity of exchanging the ions in solutions passed through them. Different varieties of resin are used dependent on the nature (cationic or anionic) of the ions to be exchanged. Many of the resins in present-day use are based on polystyrene networks cross-linked with divinyl benzene.

ion flotation *(Chem.)*. Removal of ions or gels from water by adding a surface active agent which forms complexes. These are floated as a scum by the use of air bubbles. See **froth flotation**, **Gibbs' adsorption theorem**.

ionic *(Phys.)*. Appertaining to or associated with gaseous or electrolytic ions. *N.B. Ion* is frequently used interchangeably with *ionic* as an adjective, e.g., in *ion(ic)* conduction.

ionic beam *(Phys.)*. See **ion beam**.

ionic bombardment *(Electronics)*. The impact on the cathode of a gas-filled electron tube of the positive ions created by ionization of the gas. The bombardment may cause electrons to be ejected from the cathode.

ionic bond *(Chem.)*. Coulomb force between ion-pairs in molecule or ionic crystal. These bonds usually dissociate in solution.

ionic concentration *(Chem.)*. See **ion concentration**.

ionic conduction *(Chem.)*. That which arises from the movement of ions in a gas or electrolytic solution. In solids, it refers to electrical conductivity of an ionic crystal, arising from movement of positive and negative ions under an applied electric field. It is a diffusion process and hence very temperature-dependent.

ionic conductivity *(Chem.)*. An additive property, symbol l_i; the equivalent **conductivity** of the ion i at infinite dilution. Thus for KCl, $\Lambda_\infty = l_{K^+} + l_{Cl^-}$.

ionic conductor *(Chem.)*. One in which conduction is predominantly by ions, rather than by electrons and holes.

ionic crystal *(Crystal.)*. Lattice held together by the electric forces between ions, as in a crystalline chemical compound.

ionic current *(Electronics)*. Current carried by positively charged ions in a gas at low pressure.

ionic-heated cathode *(Electronics)*. Hot cathode that is heated primarily by ionic bombardment of the emitting surface.

ionic migration *(Chem.)*. Transport of ion-bearing particle to an electrode oppositely charged with electricity.

ionic potential *(Electronics)*. Ratio of an ionic charge to its effective radius, regarded as a capacitor.

ionic product *(Chem.)*. The product of the activities (see **activity (2)**) of the ions into which a pure liquid dissociates. For water, these ions are H_3O^+ and OH^-.

ionic radius *(Crystal.)*. Approximate limiting radius of ions in crystals, ranging, for common metals (including carbon), from a fraction to about one nanometre.

ionic strength *(Chem.)*. Half the sum of the terms obtained by multiplying the **activity** (2) of each ion in a solution by the square of its valency; it is a measure of the intensity of the electrical field existing in a solution.

ionic theory *(Chem.)*. The theory that substances whose solution conduct an electric current undergo electrolytic dissociation on dissolution. This assumption explains both the laws of electrolysis and the abnormal colligative properties, such as osmotic pressure, of electrolyte solutions.

ion implantation *(Electronics)*. Technique used to produce semiconductor devices. Impurities are introduced by firing high energy ions at the substrate material.

ionization *(Phys.)*. Formation of ions by separating atoms, molecules, or radicals, or by adding or subtracting electrons from atoms by strong electric fields in a gas, or by weakening the electric attractions in a liquid, particularly water.

ionization by collision *(Phys.)*. The removal of one or more electrons from an atom by its collision with another particle such as another electron or an α-particle. Very prominent in electrical discharge through a rarefied gas.

ionization chamber *(Nuc.Eng.)*. Instrument used in study of ionized gases and/or ionizing radiations. It comprises a gas-filled enclosure with parallel plate or coaxial electrodes, in which ionization of the gas occurs. For fairly large applied voltages, the current through the chamber is dependent only upon the rate of ion production. For very large voltages, additional ionization by collision enhances the current. The system is then known as a **proportional counter** or **Geiger-Müller counter**.

ionization constant *(Chem.)*. The ratio of the product of the activities (see **activity (2)**) of the ions produced from a given substance to the activity of the undissociated molecules of that substance. See **dissociation constant**.

ionization cross-section *(Phys.)*. Effective geometrical cross-section offered by an atom or molecule to an ionizing collision.

ionization current *(Phys.)*. (1) That passing through ionization chamber. (2) Current passed by an ionization gauge, when used for measuring low gas pressures.

ionization density *(Phys.)*. See **ion concentration**.

ionization gauge *(Electronics)*. Vacuum gauge formed by small thermionic triode attached to a chamber in which it is desired to measure residual gas pressure. Current is passed from anode to cathode, the grid being made negative; grid current is measured, giving degree of vacuum.

ionization manometer *(Nuc.Eng.)*. See **ionization gauge**.

ionization potential *(Phys.)*. Energy, in electron-volts (eV), required to detach an electron from a neutral atom. For hydrogen, the value is 13.6 eV. Atoms, other than hydrogen, may lose more than one electron and can be multiply ionized. Also **electron binding energy**, **radiation potential**, **photoelectron spectroscopy**.

ionization temperature *(Phys.,Astron.)*. A critical temperature, different for different elements, at which the constituent electrons of an atom will become dissociated from the nucleus; hence a factor in deducing stellar temperatures from observed spectral lines indicating any known stage of ionization.

ionization time *(Electronics)*. Delay between the application of ionizing conditions and the onset of ionization, depending on temperature and other factors, e.g. in a mercury-pool rectifier.

ionized *(Chem.,Phys.)*. (1) Electrolytically dissociated. (2) Converted into an ion by the loss or gain of an electron.

ionized atom *(Phys.)*. One with a resultant charge arising from capture or loss of electrons; an *ion* in gas or liquid.

ionizing collision *(Phys.)*. Interaction between atoms or elementary particles in which an ion pair is produced.

ionizing energy *(Phys.)*. That required to produce an ion pair in a gas under specified conditions. Measured in eV. For air it is about 32 eV.

ionizing event *(Phys.)*. Any interaction which leads to the production of ions.

ionizing particle *(Phys.)*. Charged particle which produces considerable ionization on passing through a medium. Neutrons, neutrinos, and photons are not ionizing particles although they may produce some ions.

ionizing radiation *(Phys.)*. Any electromagnetic or particulate radiation which produces ion pairs when passing through a medium.

ionizing voltage *(Electronics)*. See **starting voltage**.

ion migration *(Chem.)*. The movement of ions in an

electrolyte or semiconductor due to applying a voltage across electrodes.

ion mobility *(Chem.)*. **Ion velocity** in unit electric field (1 volt per metre).

ionogenic *(Chem.)*. Forming ions, e.g. electrolytes.

ionophone *(Acous.)*. See **cathodophone**.

ionophore *(Biol.)*. A compound which enhances the permeation of biological membranes by specific ions, acting either as specific ion carriers or by creating ion channels across the membrane. E.g., *gramicidin, valinomycin.*

ionosphere *(Meteor.)*. That part of the earth's atmosphere (the upper atmosphere) in which an appreciable concentration of ions and free electrons normally exist. This shows daily and seasonal variations. See **E-layer, F-layer, hop**.

ionospheric control points *(Telecomm.)*. Points in the ionosphere distant 2000 km and 1000 km from each ground terminal, used respectively to control transmission by way of the F_2- and *E-layers*.

ionospheric disturbance *(Meteor.)*. An abnormal variation of the ion density in part of the ionosphere, commonly produced by solar flares. Has a marked effect on radio communication. See **ionospheric storm**.

ionospheric forecast (prediction) *(Meteor.)*. The forecasting of ionospheric conditions relevant to communication.

ionospheric regions *(Meteor.)*. These are: D region, between 90 and 150 km; F region, over 150 km; all above surface of earth. Internal effective layers are labelled *E*, *sporadic E*, E_2, F, F_1, $F_{1.5}$, F_2.

ionospheric storm *(Telecomm.)*. Turbulence in parts of the ionosphere, probably connected with sunspot activity, causing dramatic changes in its reflective properties and sometimes totally disrupting short-wave communications.

ionospheric wave, ray *(Telecomm.)*. Radiation reflected from an upper ionized region between transmitter and receiver. Also *indirect-*, *reflected-* or *sky-wave*.

ionotropy *(Chem.)*. The reversible interconversion of certain organic isomers by migration of part of the molecules as an ion; e.g. hydrogen ion (**prototropy**).

ion pair *(Phys.)*. Positive and negative ions produced together by transfer of electron from one atom or molecule to another.

ion propulsion *(Space)*. Method of rocket propulsion in which charged particles (e.g. lithium or caesium ions) are accelerated by an electrostatic field, giving a small thrust but a high specific impulse; the thrust-to-weight ratio is large so that the main use is for station keeping.

ion source *(Phys.)*. Device for releasing ions as these are required in an accelerating machine. It usually consists of a minute jet of gas or vapour of the required compound which is ionized by bombardment with an electron beam.

ion spot *(Electronics)*. Deformation of target, cathode or screen by ion bombardment in a camera tube or CRT. In a camera, a spurious signal results; in a CRT, *ion burn* becomes apparent.

iontophoresis *(Chem.)*. Migration of ions into body tissue through electric currents. See **electrophoresis**.

ion trap *(Electronics)*. Means of preventing *ion burn* in a magnetic-deflection CRT. The electron beam is deflected through a large angle to reach the screen, so that the heavier negative ions fall elsewhere.

ion velocity *(Chem.)*. Velocity of translation of drift of ions under the influence of an electric field, in a gas or electrolyte. See also **ion mobility**.

ion yield *(Phys.)*. The average number of ion pairs produced by each incident particle or photon.

IOP *(Comp.)*. See **input/output processor**.

IPA *(Bot.)*. Abbrev. for *IsoPentenyl Adenine*.

i.p.h. *(Print.)*. Abbrev. for *impressions per hour*.

ipsilateral *(Zool.)*. Pertaining to the same side of the body. Cf. **contralateral**.

ipso- *(Chem.)*. Prefix meaning 'itself'. Used to describe the atom of a substituent group through which that group is attached, e.g. the carbon atom of a phenyl ring attached to phosphorus in triphenylphosphine.

IPT thermometers *(Chem.)*. Thermometers conforming to the standards laid down by the *Institute of Petroleum Technologists*.

IR *(Aero.)*. Abbrev. for *Infra Red*.

IR *(Chem.)*. Abbrev. for *InfraRed* (spectroscopy).

IR *(Comp.)*. See **information retrieval**.

IRCM *(Aero.)*. See **infra red countermeasures**.

IR drop *(Elec.Eng.)*. The voltage drop due to a current flowing through a resistance.

I region *(Immun.)*. Region in the murine major histocompatibility complex which contains genes coding for Class II histocompatibility antigens.

Ir gene *(Immun.)*. *Immune response gene*. Found in the I region, and so called because in inbred strains of mice the ability to respond to certain simple peptides depends upon which Class II antigens are expressed on their cells.

iridalgia *(Med.)*. Pain in the iris of the eye.

iridectomy *(Med.)*. Excision of part of the iris of the eye.

iridescence *(Phys.)*. The production of fine colours on a surface; due to the interference of light reflected from the front and back of a very thin film.

iridescent clouds *(Meteor.)*. High clouds which show colours, generally delicate pink and green, in irregular patches. It is thought that the effect is caused by the diffraction of sunlight by supercooled water droplets.

irid-, irido- *(Genrl.)*. Prefix from Gk. *iris*, gen. *iridos*, rainbow.

iridium *(Chem.)*. A brittle, steel-grey metallic element of the platinum family. Symbol *Ir*, at. no. 77, r.a.m. 192.2, rel.d. at 20°C 22.4, mp 2375°C, electrical resistivity 6×10^{-8} ohm metres. Alloyed with platinum or osmium to form hard, corrosion-resisting alloys, used for pen points, watch and compass bearings, crucibles, standards of length. The radioactive isotope ^{192}Ir is a medium-energy gamma-emitter used for industrial radiography.

iridochoroiditis *(Med.)*. Inflammation of the iris and of the choroid of the eye.

iridocoloboma *(Med.)*. Congenital absence of part of the iris, a gap or fissure being present in it.

iridocyclitis *(Med.)*. Inflammation of the iris and of the ciliary body of the eye.

iridocyte *(Zool.)*. A reflecting cell containing guanin, found in the integument of Fish and of certain Cephalopods, to which it gives an iridescent appearance.

iridodialysis *(Med.)*. Separation of the iris from its attachment to the ciliary body of the eye.

iridokeratitis *(Med.)*. Inflammation of the iris and of the cornea.

iridoplegia *(Med.)*. Paralysis of the sphincter, or circular muscle, of the iris.

iridosmine *(Min.)*. An ore of iridium and osmium, a natural alloy, with Os greater than 35%. Crystallizes in the hexagonal system.

iridotomy *(Med.)*. Surgical cutting of the iris.

iris *(Elec.Eng.)*. Apertured diaphragm across a waveguide, for introducing specific impedances.

iris *(Image Tech.)*. Adjustable circular aperture used in conjunction with camera lens for control of exposure.

iris *(Min.)*. A form of quartz showing chromatic reflections of light from fractures, often produced artificially by suddenly cooling a heated crystal. Also called *rainbow quartz*.

iris *(Zool.)*. In the Vertebrate eye, that part of the choroid, lying in front of the lens, which takes the form of a circular curtain with a central opening. adj. *iridial*.

irisation *(Phys.)*. Same as **iridescence**.

iris wipe *(Image Tech.)*. A transitional wipe effect in which the image boundary is in the form of a circle, increasing (*iris-in*) or decreasing (*iris-out*).

iritis *(Med.)*. Inflammation of the iris.

I^2R loss *(Elec.Eng.)*. The power loss caused by the flow of a current *I* through a resistance *R*.

IRM *(Behav.)*. Abbrev. for *Innate Releasing Mechanism*.

iroko *(For.)*. A general utility timber from *Chlorophora excelsa*, a tropical African tree, used for structural work, shipbuilding, cabinet work and furniture.

iron *(Chem.)*. A metallic element in the eighth group of the periodic system. It exists in three forms; see **alpha-,**

gamma-, delta-. Symbol Fe, at. no. 26, r.a.m. 55.847, rel. d. at 20°C 7.86, mp 1525°C, bp 2800°C, electrical resistivity 9.8×10^{-8} ohm metres. As basis metal in steel and cast-iron, it is the most widely used of all metals. See **iron ores**.

iron *(Print.)*. The type-founder's unit when supplying sorts, being a measure of approximately 25 ems.

iron alum *(Min.)*. See **halotrichite**.

iron arc *(Phys.)*. An arc between iron electrodes, used for obtaining light containing standardized lines, for spectrometer and spectrograph calibrations.

iron bacteria *(Biol.,Min.)*. Filamentous bacteria, which can convert iron oxide to iron hydroxide, deposited on their sheaths. Important in formation of **bog iron ore**.

ironbark *(For.)*. *Eucalyptus leucoxylon*, yielding wood which varies in colour from greyish-brown to reddish-brown. The sapwood is susceptible to powder-post beetle attack, whilst the heart-wood is impermeable to treatment with wood preservatives. It is not an easy timber to work.

iron-clad electromagnet *(Elec.Eng.)*. One in which the return path for the flux is formed by an iron covering surrounding the winding.

iron-clad switchgear *(Elec.Eng.)*. See **metal-clad switchgear**.

iron deficiency anaemia *(Med.)*. **Anaemia** due to either poor intake or loss of iron. Common in menstruating women or when there is blood loss from gastro-intestinal disease.

iron dust core *(Elec.Eng.)*. One used in a high-frequency transformer or inductor to minimize eddy-current losses. It consists of minute magnetic particles bonded in an insulating matrix.

iron-glance *(Min.)*. From the German *Eisenglanz*, a name sometimes applied to specular iron-ore (haematite).

iron (III) oxide *(Chem.)*. See **ferric oxide**.

iron (II) oxide *(Chem.)*. See **ferrous oxide**.

iron loss *(Elec.Eng.)*. The power loss due to hysteresis and eddy currents in the iron of magnetic material transformers or electrical machinery.

iron meteorites *(Geol.)*. One of the two main categories of meteorites, the other being the *stony meteorites*. They are composed of iron and of iron-nickel alloy, with only a small proportion of silicate or sulphide minerals.

iron-nickel accumulator *(Elec.Eng.)*. See **nickel-iron-alkaline accumulator**.

iron-olivine *(Min.)*. See **fayalite**.

iron ores *(Geol.)*. Rocks or deposits containing iron-rich compounds in workable amounts; they may be primary or secondary; they may occur as irregular masses, as lodes or veins, or interbedded with sedimentary strata. See **chamosite**, **goethite**, **haematite**, **limonite**, **magnetite**, **siderite** (Min. (1)).

iron pan *(Geol.)*. A hard layer often found in sands and gravels; caused by the precipitation of iron salts from percolating waters. It is formed a short distance below the soil surface.

iron pattern *(Foundry)*. A pattern made of cast-iron; used when a large number of castings are required from it, and long life is necessary. See **double contraction**.

iron pentacarbonyl *(Chem.)*. $Fe(CO)_5$. Formed at ordinary temperatures when carbon monoxide is passed over finely divided iron. A liquid which readily decomposes.

iron pyrites *(Min.)*. See **pyrite**.

iron spinel *(Min.)*. See **spinel**.

ironstone *(Geol.)*. An iron-rich sedimentary rock, found in nodules, layers, or beds.

ironwood *(For.)*. Hardwood trees from the genus, *Mesua*, found in the tropics. It is naturally durable but of limited use.

irradiance *(Phys.)*. See **radiant-flux density**.

irradiation *(Image Tech.)*. Image spread, arising from scatter within the emulsion from the silver halide grains. See **halation**.

irradiation *(Radiol.)*. Exposure of a body to X-rays, gamma-rays or other ionizing radiations.

irradiation swelling *(Nuc.Eng.)*. Changes in density and volume of materials due to neutron irradiation.

irrational number *(Maths.)*. A real number which cannot be expressed as the ratio of two integers e.g. $\sqrt{2}$. Cf. **surd**.

irregular *(Biol.)*. (1) Asymmetric, not arranged in an even line or circle. (2) Not divisible into halves by an indefinite number of longitudinal planes.

irregular-coursed *(Build.)*. Said of rubble walling built up in courses of different heights.

irregular galaxies *(Astron.)*. Small galaxies showing no symmetry; they contain little dust or gas. See **Magellanic Clouds**.

irregular variables *(Astron.)*. See **variable stars**.

irreversibility *(Phys.)*. Physical systems have a tendency to change spontaneously from one state to another but not to change in the reverse direction. *Entropy* provides an indication of irreversibility.

irreversible colloid *(Chem.)*. See **lyophobic colloid**.

irreversible controls *(Aero.)*. A flying control system, hydraulically or electrically operated, wherein there is no feed-back of aerodynamic forces from the control surfaces.

irreversible reaction *(Chem.)*. A reaction which takes place in one direction only, and therefore proceeds to completion.

irrigation *(Build.)*. The method of sewage disposal by **land treatment**.

irritability *(Biol.)*. A property of living matter, namely, the ability to receive and respond to external stimuli.

irritant *(Biol.)*. Any external stimulus which produces an active response in a living organism.

irrotational field *(Elec.Eng.)*. A field in which the *field circulation* is everywhere zero.

ISA *(Aero.)*. Abbrev. for *International Standard Atmosphere*.

isallobar *(Meteor.)*. The contour line on a weather chart, signifying the location of equal changes of pressure over a specified period.

isallobaric high and low *(Meteor.)*. Centres, respectively, of rising and falling **barometric tendency**.

isallobaric wind *(Meteor.)*. Theoretical component of the wind arising from the spatial non-uniformity of local rates of change of pressure.

ISAS *(Space)*. Abbrev. for *Institute of Space and Aeronautical Science* of the University of Tokyo, mainly reponsible for the Japanese scientific satellites.

ischaemia, ischemia *(Med.)*. Deficiency of blood flow to part of the body, causing inadequate tissue perfusion with oxygen. adj. *ischaemic*.

ischium *(Zool.)*. A posterior bone of the pelvic girdle in *Tetrapods*. adjs. *ischial, ischiadic*.

isenthalpic *(Phys.)*. Of a process carried out at constant enthalpy, or heat function H.

isentropic *(Phys.)*. See **entropy**.

Isherwood system *(Ships)*. A method of ship construction of which the dominant feature is longitudinal framing. Named after its originator.

I signal *(Image Tech.)*. In the NTSC colour TV system, that corresponding to the wideband axis of the chrominance signal.

isinglass *(Build.)*. Gelatine dissolved in hot water before being stained and used as a **mordant** in gilding on glass.

isinglass *(Chem.)*. Fish glue. A white solid amorphous mass, prepared from fish bladders; chief constituent, gelatin. It has strong adhesive properties. Used in various food preparations, as an adhesive, and in the fining of beers, wines etc.

island arc *(Geol.)*. A chain of volcanic islands formed at a convergent plate boundary. Deep ocean trenches occur on the convex side and deep basins on the concave side. e.g. Japan.

island universe *(Astron.)*. The name applied to an extra-galactic nebula.

iso- *(Chem.)*. A prefix indicating: (1) the presence of a branched carbon chain in the molecule; (2) an isomeric compound.

iso- *(Genrl.)*. Prefix from Greek *isos*, equal.

ISO7 *(Comp.)*. See **character code**.

iso-agglutination *(Biol.,Zool.)*. (1) The adhesion of spermatozoa to one another by the action of some substance produced by the ova of the same species. (2) The adhesion of erythrocytes to one another within the same blood-group. Cf. **hetero-agglutination**.

iso-antigen *(Immun.)*. Antigen carried by an individual which is capable of eliciting an immune response in genetically different individuals of the same species but not in the individual bearing it. Hence iso-immunization.

isobar *(Chem.)*. A curve relating qualities measured at the same pressure.

isobar *(Meteor.)*. A line drawn on a map through places having the same atmospheric pressure at a given time.

isobar *(Phys.)*. One of a set of nuclides having the same total of protons and neutrons with the same *mass number* and approximately the same *atomic mass*; e.g. the isotopes of hydrogen and helium, $_1{}^3H$ and $_2{}^3He$.

isobaric spin *(Phys.)*. See **isotopic spin**.

isobarometric, isobaric charts *(Meteor.)*. Maps on which isobars are drawn. See **isobar**.

isobases *(Geol.)*. Lines drawn through places where equal depression of the land mass took place in Glacial times, as a result of the weight of the ice load.

isobilateral *(Bot.)*. Divisible into symmetrical halves along two distinct planes; especially of a leaf which has palisade towards both faces, *unifacial*.

isobutane *(Chem.)*. An isomeric form of **butane** having the structure $(CH_3)_2CH.CH_3$, as compared with $CH_3CH_2.CH_2CH_3$ for normal butane. Found in natural gas, and produced by cracking petroleum. Used as a refrigerant.

isobutyl alcohol *(Chem.)*. *2-methylpropan-1-ol.* $(CH_3)_2CH.CH_2OH$, bp 107°C, partly miscible with water; formed during sugar fermentation.

isoceral *(Zool.)*. Said of a type of secondarily symmetrical tail-fin (in Fish) in which the areas of the fin above and below the vertebral column are equal.

isochore *(Chem.)*. A curve relating quantities measured under conditions in which the volume remains constant.

isochromatic *(Phys.)*. Interference fringe of uniform hue observed with white light source; especially in photoelastic strain analysis where it joins points of equal phase retardation.

isochrone *(Telecomm.)*. Hyperbola (or ellipse) on a chart or map, along which there is a constant difference (or sum) in time of arrival of signals from two stations at ends of a baseline.

isochronism *(Horol.)*. For a clock pendulum, isochronism implies that the time of vibration should be the same whatever the amplitude (see **circular error**), and for the balance that the time of vibration should be the same whatever the arc of vibration. In practice, the balance approaches nearer to true isochronism than does the pendulum.

isochronism *(Phys.)*. Regular periodicity, as the swinging of a pendulum. adj. *isochronous*.

isoclinal fold *(Geol.)*. A fold in which both limbs dip in the same direction. See **folding**.

isocline *(Phys.)*. Line on a map, joining points where the angle of dip (or inclination) of the earth's magnetic field is the same.

isoclinic *(Aero.)*. A wing designed to maintain a constant angle of incidence even when subject to dynamic loads.

isocoria *(Med.)*. Equality in the size of the pupils of the eye.

isocount contours *(Radiol.)*. The curves formed by the intersection of a series of **isocount surfaces** with a specified surface.

isocount surface *(Radiol.)*. A surface on which the counting rate is everywhere the same.

isocyanides *(Chem.)*. Isonitriles or carbylamines. Isocyano-compounds, R-N≡C. They are colourless liquids, only slightly soluble in water with a feebly alkaline reaction, having a nauseous odour, obtained by the action of trichloromethane and alcoholic potash on primary amines. They are very stable towards alkali, form additive compounds with halogens, HCl, H_2S etc., and can be hydrolysed into methanoic acid and a primary amine, having one carbon atom less than the original compound.

isocyclic compounds *(Chem.)*. **Carbocyclic compounds**.

isodactylous *(Zool.)*. Having all the digits of a limb the same size.

isodesmic structure *(Crystal.)*. Crystal structure with equal lattice bonding in all directions, and no distinct internal groups.

isodiametric, isodiametrical *(Genrl.)*. Of the same length vertically and horizontally.

isodiapheres *(Phys.)*. Two or more nuclides having the same difference between the number of neutrons and the number of protons.

isodimorphous *(Chem.)*. Existing in two isomorphous crystalline forms.

isodisperse *(Chem.)*. Dispersible in solutions having the same pH value.

isodomon *(Build.)*. An ancient form of masonry in which the facing consisted of squared stones laid in courses of equal height, and the filling of coursed stones of smaller size.

isodont *(Zool.)*. Having all the teeth similar in size and form.

isodose chart *(Radiol.)*. A graphical representation of a number of isodose contours in a given plane, which usually contains the central ray.

isodose curve, contour *(Radiol.)*. The curve obtained at the intersection of a particular **isodose surface** with a given plane.

isodose surface *(Radiol.)*. A surface on which the dose received is everywhere the same.

isodulcite *(Chem.)*. See **rhamnose**.

isodynamic lines *(Phys.)*. Lines on a magnetic map which pass through points having equal strengths of the earth's field.

iso-electric focusing *(Biol.)*. A technique for separating proteins according to their **iso-electric points**. The mixture of proteins is placed in a pH gradient established by **ampholines** in an electric field across a liquid or a gel matrix: the proteins move in the electric field until they reach their iso-electric points in the pH gradient where, having lost net charge, they are focused.

iso-electric point *(Chem.)*. Hydrogen ion concentration in solutions, at which dipolar ions are at a maximum. The point also coincides with minimum viscosity and conductivity. At this pH value, the charge on a colloid is zero and the ionization of an ampholyte is at a minimum. It has a definite value for each amino acid and protein.

iso-electronic *(Electronics)*. Said of similar electron patterns, as in valency electrons of atoms.

iso-enzymes *(Biol.)*. See *isozymes*.

isogamy *(Bot.)*. Sexual fusion of similar gametes. Cf. **anisogamy, oogamy**.

isogenetic *(Zool.)*. Having a similar origin.

isogenic *(Biol.)*. Describes a strain in which all individuals are genetically identical but not necessarily homozygous. Cf. **inbred line**.

isogonal transformation *(Maths.)*. A transformation from the *z*-plane to the *w*-plane, in which corresponding curves intersect at the same angle in each plane.

isogonic line *(Phys.)*. Line on a map joining points of equal magnetic declination, i.e., corresponding variations from true north.

isograd *(Geol.)*. A line joining points where metamorphic rocks have attained the same facies, by being subjected to the same temperature and pressure.

isogrivs *(Aero.,Ships)*. Lines drawn on a **grid navigation** system chart joining points at which convergency of meridians and magnetic variation are equal .

isohel *(Meteor.)*. A line drawn on a map through places having equal amounts of sunshine.

isohydric *(Chem.)*. Having the same pH value, or concentration of hydrogen ions.

isohyet *(Meteor.)*. A line drawn on a map through places having equal amounts of rainfall.

isokinetic sample *(Powder Tech.)*. Particles from a fluid stream or aerosol entering the mouth of the sampling device at the linear velocity at which the stream was flowing at that point prior to the insertion of the device.

isokont, isokonton *(Bot.)*. Having two or more flagella of equal length or identical morphology as the whiplash flagella of *Chlamydomonas*.

isolate *(Bot.)*. To establish a pure culture of a micro-organism.

isolated essential singularity *(Maths.)*. See **pole**.

isolated phase switchgear *(Elec.Eng.)*. Switchgear in which all the apparatus associated with each phase is segregated in separate cubicles or on separate floors of the switch-house.

isolated point *(Maths.)*. (1) A point of a set in the neighbourhood of which there is no other point. (2) See **double point**.

isolating mechanism *(Bot.)*. Anything which prevents the exchange of genetic material between two populations. It can include geography, physiology or behaviour.

isolation *(Acous.)*. Prevention of sound transmission by walls, mufflers, resilient mounts etc.

isolation diode *(Telecomm.)*. One used to block signals in one direction but to pass them in the other.

isolation transformer *(Elec.Eng.)*. One used to isolate electrical equipment from its power supply.

isolator *(Telecomm.)*. Passive device for insertion into VHF or microwave transmission lines (waveguide, coaxial or stripline) with low loss in the forward direction and high attenuation in the reverse. Used to isolate oscillators, transmitters etc. from a mismatched or reflective load. See **gyrator**.

isolecithal *(Zool.)*. Said of ova which have yolk distributed evenly through the protoplasm.

isoleucine *(Chem.)*. An essential amino acid, $CH_3CH_2CH(CH_3)CH(NH_2)COOH$. Symbol Ile, short form I. ⇨

isolux *(Phys.)*. Locus, line or surface where the light intensity is constant. Also called *isophot*.

isomagnetic lines *(Phys.)*. Lines connecting places at which a property of the earth's magnetic field is a constant.

isomastigote *(Zool.)*. Having two or four flagella of equal length.

isomerism *(Chem.)*. The existence of more than one substance having a given molecular composition and rel. mol. mass but differing in constitution or structure. (See also **optical isomerism**.) The compounds themselves are called *isomers* or *isomerides* (Gr. 'composed of equal parts'). Isobutane and butane have the same formula, C_4H_{10}, but their atoms are placed differently; one type of alkane molecule, $C_{40}H_{82}$, has over 50^{12} possible isomers. Isomerism is frequently met with among organic compounds and complex inorganic salts.

isomerism *(Phys.)*. The existence of nuclides which have the same atomic number and the same mass number, but are distinguishable by their energy states; that having the lowest energy is stable, the others having varying life-times. If the life-times are measurable the nuclides are said to be in *isomeric states* and undergo *isomeric transitions* to the ground state.

isomerized rubber *(Chem.)*. Rubber in which the molecules have been rearranged by heating in solution in the presence of suitable catalysts. It is more soluble and more compatible with other varnish constituents than ordinary rubber while maintaining good chemical resistance.

isomerous *(Bot.)*. Equal numbers as in the parts in two whorls of a flower.

isomers *(Chem.)*. See **isomerism**.

isomer separation *(Phys.)*. The chemical separation of the lower energy member of a pair of nuclear isomers.

isometric contraction *(Zool.)*. The type of contraction involved when a muscle produces tension but is held so that it cannot change its length.

isometric projection *(Arch.)*. A type of axonometric drawing in which all horizontal lines are drawn at 30° to the horizontal plane of projection; the result is a three dimensional drawing which gives equal emphasis to all three planes.

isometric system *(Crystal.)*. The cubic system.

isomorphic *(Bot.)*. Morphologically similar.

isomorphic alternation of generations *(Bot.)*. Alternation of generations which are morphologically alike. Also *homologous alternation of generations*. Cf. **heteromorphic alternation of generations**.

isomorphic groups *(Maths.)*. The groups $(G_1; *)$ and $(G_2; \circ)$ are said to be isomorphic to one another if there exists a one-to-one mapping f from G_1 on to G_2, and for all elements x,y in G_1 we have $f(x*y) = f(x) \circ f(y)$.

isomorphism *(Crystal.)*. The name given to the phenomenon whereby two or more minerals, which are closely similar in their chemical constitution, crystallize in the same class of the same system of symmetry, and develop very similar forms. adjs. *isomorphic, isomorphous*.

isomorphism *(Maths.)*. A one-to-one mapping from one algebraic system on to another which shows the systems to have the same abstract structure.

isomorphous replacement *(Crystal.)*. Replacing atoms at a given position in a crystal structure by others, usually those of a heavy metal. The determination of crystal structure of, particularly complex molecules is made much more difficult by the investigator's inability to determine the phase relations of the diffraction pattern. Heavy metal replacement is an important method of overcoming the problem.

isoniazid *(Med.)*. Drug used in the treatment of tuberculosis.

isonitriles *(Chem.)*. **Isocyanides**.

isonome *(Bot.)*. Line on a map joining points of equal abundance of a given species of plant.

iso-osmotic *(Biol.)*. A solution with the same osmolarity as that in a cell suspended in the solution. There is no net flux of water across the membrane.

isopach *(Geol.)*. A line drawn through points of equal thickness on a rock unit.

isopedin *(Zool.)*. The thin layer of bone forming the inner layer of the cosmoid scales of *Choanichthyes*.

isopentenyladenine *(Bot.)*. A natural cytokinin.

isopiestic *(Chem.)*. Having equal pressure in the system or conditions described.

isopleth *(Maths.)*. See **nomogram**.

Isopoda *(Zool.)*. An order of Malacostraca, in which the carapace is absent, the eyes are sessile or borne on immovable stalks; the body is depressed and the legs used for walking; they show great variety of form, size, and habit; some are terrestrial, plant-feeders, or ant-guests, others are marine, free-living, and feeding on seaweeds or ectoparasitic on fish. Woodlice, etc.

isopodous *(Zool.)*. Having the legs all alike.

isoprenaline *(Med.)*. A drug mimicking the action of sympathetic stimulation by adrenaline to produce tachycardia and increase the output of the heart. See **sympathomimetics**.

isoprene *(Chem.)*. $CH_2=C(CH_3).CH=CH_2$, an alkadiene, a colourless liquid, bp 37°C, obtained by the destructive distillation of rubber, by dehydrogenation of 2-methylbutane, from propene, or by several other methods. It is used in the synthesis of rubber. The *isoprene unit* is the basis of **terpenes**. See **butyl rubber**.

isopropyl alcohol *(Chem.)*. *Propan-2-ol*. $(CH_3)_2CHOH$, a colourless liquid, miscible with water, bp 81°C.

isopropyl benzene *(Chem.)*. See **cumene**.

isopropyl group *(Chem.)*. The monovalent radical, $(CH_3)_2CH—$.

Isoptera *(Zool.)*. An order of social Exopterygota living in large communities which occupy nests excavated in the soil or built up from mud and wood; different polymorphic forms or castes occur in each species; the mouthparts are adapted for biting and both pairs of wings, if present, are membranous and can be shed by means of a basal suture; exclusively herbivorous. White Ants, Termites.

isopycnic *(Meteor.)*. A line on a chart joining points of equal atmospheric density.

isoquinoline *(Chem.)*. Occurs in coal tar. It is an isomer of quinoline and a condensation product of a benzene ring with a pyridine ring. It forms colourless crystals, mp 23°C, bp 240°C. ⇨

isosceles triangle *(Maths.)*. A triangle having two equal sides.

isoseismal line *(Geol.)*. A line drawn on a map through places recording the same intensity of earthquake shocks. See **earthquake**.

ISO sizes *(Paper)*. A series of trimmed, international, metric paper sizes based on a width to length ratio of 1:1.414. The next smaller size in the series is produced by halving the longer dimension e.g. AO is 841 × 1189 mm, A1, 594 × 841 mm, A2, 420 × 594 mm etc. The range includes the A, B and C series of sizes.

isosorbide *(Med.)*. Long-acting nitrate vasodilator used in the treatment and prophylaxis of angina pectoris. Available as its dinitrate or its active metabolite, mononitrate.

isospin *(Phys.)*. Contraction of **isotopic spin**.

isostasy *(Geol.)*. The process whereby areas of crust tend to float in conditions of near equilibrium on the plastic mantle.

isostemonous *(Bot.)*. With stamens in one whorl and equal in number to petals.

isostere *(Meteor.)*. A line on a chart joining points of equal atmospheric **specific volume**.

isosteric *(Chem.)*. Consisting of molecules of similar size and shape.

isotach *(Meteor.)*. A line on a chart joining points of equal wind speed.

isotactic *(Plastics)*. Term denoting linear-substituted hydrocarbon polymers in which the substituent groups all lie on the same side of the carbon chain. See also **atactic** and **syndiotactic**.

isotaxy *(Chem.)*. Polymerization in which the monomers show stereochemical regularlity of structure. adj. *isotactic*. Cf. **syndyotaxy**.

isoteniscope *(Chem.)*. An instrument for the static measurement of vapour pressure by observing the change of level of a liquid in a U-tube.

isotherm *(Meteor.)*. A line drawn on a chart joining points of equal temperature.

isothermal *(Phys.)*. (1) Occurring at constant temperature. (2) A curve relating quantities measured at constant temperature.

isothermal change *(Phys.)*. A change in the volume and pressure of a substance which takes place at constant temperature. For gases, Boyle's law applies to isothermal changes.

isothermal efficiency *(Eng.)*. Of a compressor, the ratio of the work required to compress a gas isothermally to the work actually done by the compressor.

isothermal lines, curves *(Phys.)*. Curves obtained by plotting pressure against volume for a gas kept at constant temperature. For a gas sufficiently above its critical temperature for Boyle's law to be obeyed, such curves are rectangular hyperbolas.

isothermal process *(Eng.)*. A physical process, particularly one involving the compression and expansion of a gas, which takes place without temperature change.

isothermal transformation *(Eng.)*. Change in phase which occurs in a metal or alloy at constant temperature after cooling or heating through the equilibrium temperature.

isotones *(Phys.)*. Nuclei with the same neutron number but different atomic numbers (i.e., those lying in a vertical column of a *Segrè chart*).

isotonic *(Biol.)*. See **iso-osmotic**.

isotonic *(Chem.)*. Having the same osmotic pressure, e.g. as that of blood, or of the sap of cells which are being tested for their osmotic properties.

isotonic contraction *(Zool.)*. The type of contraction involved when a muscle shortens while maintaining a constant tension.

isotope *(Phys.)*. One of a set of chemically identical species of atom which have the same **atomic number** but different **mass numbers**. A few elements have only one natural isotope, but all elements have artificially produced radio-isotopes. (Gk. *isos*, same; *topos*, place).

isotope geology *(Geol.)*. The study of the relative abundances of radioactive and stable isotopes in rocks to determine radiometric ages and conditions of formation.

isotope separation *(Phys.)*. Process of altering the relative abundance of isotopes in a mixture. The separation may be virtually complete as in a mass spectograph, or may give slight enrichment only as in each stage of a diffusion plant.

isotope structure *(Phys.)*. Hyperfine structure of spectrum lines resulting from mixture of isotopes in source material. The wavelength difference is termed the *isotope shift*.

isotope therapy *(Radiol.)*. Radiotherapy by means of radioisotopes.

isotopic abundance *(Phys.)*. See **abundance ratio**.

isotopic dilution *(Radiol.)*. The mixing of a particular nuclide with one or more of its isotopes.

isotopic dilution analysis *(Phys.)*. A method of determining the amount of an element in a specimen by observing the change in isotopic composition produced by the addition of a known amount of radioactive allobar.

isotopic number *(Phys.)*. See **neutron excess**.

isotopic spin *(Phys.)*. Also called *isobaric spin*, *isospin*, *i-spin*. A quantum number assigned to members of a group of elementary particles differing only in electric charge; the particle groups are known as *multiplets*. Thus it is convenient to regard protons and neutrons as two manifestations of the nucleon, with isotopic spin either parallel or anti-parallel to some preferred direction, i.e. they have isotopic spin $+\frac{1}{2}$ and $-\frac{1}{2}$. The nucleon is then a doublet. This can be extended to all baryons and mesons. For example, the triplet π-meson consists of three pions. The small mass differences between the members of a multiplet is associated with their differing charges. The number of members of a multiplet set is $2I + 1$ where I is the isotopic spin, 0 for a singlet, $\frac{1}{2}$ for a doublet, 1 for a triplet, etc. The justification for the classification of particles is that all the members of a multiplet respond identically to strong nuclear interactions, the charges affecting only electromagnetic interactions. Isotopic spin is conserved in all strong interactions and never changes by more than one in a weak interaction. This classification is introduced by *analogy* with the spin or intrinsic angular momentum of atomic spectroscopy; isotopic spin has nothing to do with the nuclear spin of the particles.

isotopic symbols *(Chem.)*. Numerals attached to the symbol for a chemical element, with the following meanings; *upper left*, mass number of atom; *lower left*, nuclear charge of atom; *lower right*, number of atoms in molecule, e.g. ${}_1^2H_2$, ${}_{12}^{24}Mg$.

isotron *(Phys.)*. A device for the separation of isotopes. Pulses from a source of ions are synchronized with a deflecting field. The ions undergo deflections according to their mass.

isotropic *(Phys.)*. Said of a medium, the physical properties of which, e.g., magnetic susceptibility or elastic constants, do not vary with direction.

isotropic dielectric *(Elec.Eng.)*. One in which the electrical properties are independent of the direction of the applied electric field.

isotropic radiator *(Telecomm.)*. An idealised antenna which sends out energy equally in all directions; virtually impossible to realise in practice. Cf. **omnidirectional antenna**.

isotropic source *(Electronics)*. Theoretical source which radiates all its electromagnetic energy equally in all directions.

isozyme *(Biol.)*. *Isoenzyme*. Electrophoretically distinct forms of an enzyme with identical activities, usually coded by different genes.

ISRO *(Space)*. Abbrev. for *Indian Space Research Organisation*, which oversees all Indian space activities.

isthmus *(Zool.)*. A neck connecting two expanded portions of an organ; as the constriction connecting the mid-brain and the hind-brain of Vertebrates.

IT *(Comp.)*. See **information technology**.

itacolumite *(Geol.)*. A micaceous sandstone with loosely-interlocking grains, which enable the rock to bend when cut into thin slabs.

Italian asbestos *(Min.)*. A name often given to tremolite asbestos to distinguish it from Canadian or chrysotile asbestos.

Italian blind *(Build.)*. An outside roller-blind similar to the hook-out blind but having the side arms attached to the blind and capable of sliding up and down on side rods. Also called a **canalette blind**.

Italian roof *(Arch.)*. See **hipped roof**.

italic, italics *(Print.)*. A sloping style of type, thus *italic*.

itchy leg *(Vet.)*. See **chorioptic mange**.

iter *(Zool.)*. A canal or duct, as the reduced ventricle of the mid-brain in higher Vertebrates.

iterated fission expectation *(Phys.)*. Limiting value, after a long time, of the number of fissions per generation in the chain reaction initiated by a specified neutron to which this term applies.

iteration *(Comp.)*. To obtain a result by repeatedly performing the same sequence of steps until a specified condition is satisfied. See **loop**.

iterative impedance *(Phys.)*. The **input impedance** of a four-terminal network or transducer when the output is terminated with the same impedance, or when an infinite series of identical such networks are cascaded. See **image impedance**.

iteroparous *(Zool.)*. Reproducing on two or more occasions during a lifetime.

ITM *(Ships)*. Abbrev. for *Inch Trim Moment*. Same as *moment to change trim one inch*.

ITU *(Telecomm.)*. Abbrev. for *International Telecommunications Union*.

IUPAC *(Chem.)*. Abbrev. for the *International Union of Pure and Applied Chemistry*, a body responsible, among other things, for the standardization of chemical nomenclature, which it alters frequently.

ivory *(Zool.)*. The dentine of teeth, especially the type of dentine composing the tusks of elephants.

ivory board *(Paper)*. Genuine ivory board is formed from high-quality papers by starch-pasting two or more together.

ivorywood *(For.)*. Rare Australian hardwood from *Siphonadendron*, prized for engraving, turnery, mirror frames, inlaying etc.

IVP, IVU *(Med.)*. Abbrev. for *IntraVenous Pyelography, IntraVenous Urography*, i.e. the demonstration by X-ray of the renal tract after the intravenous injection of a radio-opaque contrast medium.

IW *(Chem.)*. An abbrev. for *Isotopic Weight*.

IX *(Chem.,Min.Ext.)*. Abbrev. for *Ion eXchange*.

Izod test *(Eng.)*. A notched-bar impact test in which a notched specimen held in a vice is struck in the end by a striker carried on a pendulum; the energy absorbed in fracture is then calculated from the height to which the pendulum rises as it continues its swing.

Izod value *(Eng.)*. The energy absorbed in fracturing a standard specimen in an Izod pendulum impact-testing machine.

J

j *(Elec.Eng.)*. The symbol *j* is used by electrical engineers in place of the mathematician's *i*. Its main use is that in circuits carrying sinusoidal current of angular frequency ω, any inductance *L* and any capacitance *C* can be replaced by reactances $j\omega L$ and

$$\frac{1}{j\omega C}$$

respectively and Ohm's law can then be used. Also referred to as the *90°* operator.

J *(Chem.)*. In names of dyestuffs, a symbol for yellow.

J *(Phys.)*. Symbol for joule, the unit of energy, work, quantity of heat.

J *(Eng.)*. Symbol for polar moment of inertia.

J *(Phys.)*. Symbol for (1) electric current density, (2) magnetic polarization.

jacaranda *(For.)*. See **rosewood**.

jacinth *(Min.)*. See **hyacinth**.

jack *(Eng.)*. A portable lifting machine for raising heavy weights through a short distance, consisting either of a screw raised by a nut rotated by hand gear and a long lever, or a small hydraulic ram. See **hydraulic-**.

jack *(Telecomm.)*. Socket whose connections are short-circuited until a jack plug is inserted. A *break-jack* is one which breaks the normal circuit on inserting plug, while *branch jack* is one which does not.

jack arch *(Build.)*. A **flat arch**.

jackbit *(Min.Ext.)*. Detachable cutting end fitted to shank of miner's rock drill, used to drill short blast-holes. Also called *rip-bit*.

jackblock *(Build.)*. A method of system building in which the roof and floor slabs are cast on top of each other and hydraulically jacked up to their respective levels, walls being built as and when required.

jack box *(Elec.Eng.)*. One containing switches or connections for changing circuits.

jacket *(Eng.)*. An outer casing or cover constructed round a cylinder or pipe, the space being filled with a fluid for either cooling or heating the contents, or with insulating material for keeping the contents at substantially constant temperature, e.g. the water *jackets* of an I.C. engine.

jacket *(Nuc.Eng.)*. See **can**.

jacket *(Print.)*. The wrapper, or dust cover, in which a book is enclosed. Bookjackets are usually artistically designed, and executed in colour, as their purpose is to enhance the appeal of the volume as well as to protect it from dust.

jackhammer *(Min.Ext.)*. A hand-held compressed-air hammer drill for rock-drilling.

jack plane *(Build.)*. A bench plane about 16 in long, used for bringing the work down to approximate size, prior to finishing with a trying or smoothing plane.

jack rafter *(Build.)*. A short rafter connecting a hip-rafter and the eaves, or a valley-rafter and the ridge.

jack shaft *(Elec.Eng.)*. An intermediate shaft used in locomotives having collective drive; the jack shaft is geared to the motor shaft and carries cranks which drive the coupling rods on the driving wheels.

Jacksonian epilepsy *(Med.)*. A convulsion of a limited group of muscles spreading gradually from one group to the other, usually without loss of consciousness; the result of a lesion (e.g. tumour) of the brain.

Jacobian *(Maths.)*. Of *n* functions u_i each of *n* variables x_j, the determinant whose *i,j*th element is $\frac{\partial u_i}{\partial x_j}$. Written

$$\frac{\partial(u_1, u_2, u_3 \dots u_n)}{\partial(x_1, x_2, x_3 \dots x_n)}.$$

Jacobian elliptic functions *(Maths.)*. See **elliptic functions**.

jacobsite *(Min.)*. An oxide of manganese and iron, often with considerable replacement of manganese by magnesium; crystallizes in the cubic system (usually in the form of distorted octahedra). A spinel.

Jacob's ladder *(Eng.)*. Vertical belt conveyor with cups or buckets.

Jacobson's glands *(Zool.)*. In some Vertebrates, nasal glands the secretion of which moistens the olfactory epithelium.

Jacobson's organ *(Zool.)*. In some Vertebrates, an accessory olfactory organ developed in connection with the roof of the mouth.

jacot tool *(Horol.)*. A tool used by watchmakers for polishing and burnishing pivots.

jacquard *(Textiles)*. A device, frequently incorporating punched cards or punched continuous strip, used to produce patterned fabrics during weaving, warp-knitting, weft-knitting and lace making. Named after the French inventor, Joseph-Marie Jacquard, 1752-1834.

Jacquet's method *(Eng.)*. Final polishing of metal surfaces by electrolysis.

jactitation *(Med.)*. Restless tossing of a patient severely ill; a twitching or convulsion of muscle or of a limb.

jacupirangite *(Geol.)*. A nepheline-bearing pyroxenite consisting of titanaugite, biotite, iron ores, and nepheline, the last being subordinate to the mafic minerals.

jade *(Min.)*. A general term loosely used to include various mineral substances of tough texture and green colour used for ornamental purposes. It properly embraces **nephrite** and **jadeite** but is sometimes misapplied to green varieties of minerals such as **amazonstone**, **bowenite**, **hydrogrossular**, **quartz** and **vesuvianite**.

jadeite *(Min.)*. A monoclinic member of the pyroxene group; sodium aluminium silicate. Usually white, grey or mauve, it occurs only in metamorphic rocks, and is the rare form of jade (*Chinese Jade*).

jad, jud *(Min.Ext.)*. Deep groove cut into bed to detach block of natural stone. To undercut. A *jadder* is a stonecutter and his working tool is a *jadding pick*.

jag-bolt *(Eng.)*. See **rag-bolt**.

jail fever *(Med.)*. See **typhus**.

jalousies *(Build.)*. Hanging or sliding wooden sun-shutters giving external protection to a window, and allowing for ventilation through louvres or holes cut in the shutters themselves. Also called *Venetian shutters*.

jamaicin *(Chem.)*. See **berberine**.

jamb *(Build.)*. The side of an aperture.

jamb linings *(Build.)*. The panelling at the sides of a window recess, running from the floor to the level of the window head. Cf. **elbow linings**.

jamb post *(Build.)*. An upright member on one side of a doorway opening.

jamb stone *(Build.)*. A stone forming one of the upright sides of an aperture in a wall.

James-Lange theory of emotions *(Behav.)*. A theory that emotion is the subjective experience of one's own bodily reactions in the presence of certain arousing stimuli; the stimuli cause certain physiological responses, and the awareness of these responses causes emotion.

jamesonite *(Min.)*. See **feather ore**.

Jamin interferometer *(Phys.)*. A form of interferometer in which two interfering beams of light pursue parallel paths a few centimetres apart. The instrument is used to measure the refractive index of a gas, by observing the fringe shift when one of the light beams traverses a tube filled with the gas, while the other traverses a vacuum.

jamming *(Telecomm.)*. Deliberate inteference of transmission on one carrier by transmission on or near the same frequency, with wobble or noise modulation.

JANET *(Comp.)*. *Joint Academic NETwork*. Dating back to the 1970's, it links universities and research institutions in UK.

jansky *(Astron.)*. A unit in radio astronomy to measure the power received at the telescope from a cosmic radio source. 1 jansky (Jy) = 10^{-26} W m^{-2} Hz^{-1} sr^{-1}.

J-antenna *(Telecomm.)*. Half-wave antenna fed and matched at the end by a quarter-wavelength line.

Janus *(Astron.)*. The tenth natural satellite of **Saturn**, discovered in 1980.

Janus *(Telecomm.)*. Transmitting or receiving antenna which can be switched between opposite directions. Used for airborne Doppler navigation systems.

japan *(Build.)*. A black, glossy, paint based on asphaltum and drying oils. Also *black japan*.

Japan camphor *(Chem.)*. See **camphor**.

Japanese paper *(Paper)*. Japanese hand-made paper prepared from mulberry bark. The surface is similar to that of **Japanese vellum**.

Japanese vellum *(Paper)*. An expensive handmade paper. Prepared from the inner bark of the mulberry tree, thicker than **Japanese paper**.

japanners' gold-size *(Build.)*. See **gold-size**.

japanning *(Build.)*. The process of finishing an article with japans, especially the **stoving** of japans.

Japan wax *(Chem.)*. A natural wax obtained from sumach, mp 50°C. It has a high content of palmitin.

jargon aphasia *(Med.)*. Rapid unintelligible utterance, due to a lesion in the brain.

jargons, jargoons *(Min.)*. A name given in the gem trade to the zircons (chiefly colourless, smoky or of golden-yellow colour) from Ceylon. They resemble diamonds in lustre but are less valuable. See also **hyacinth**.

jarosite *(Min.)*. A hydrous sulphate of iron and potassium crystallizing in the trigonal system; a secondary mineral in ferruginous ores.

jarrah *(For.)*. A dense resistant wood from Australia. It is of a deep-red colour, and is used for construction, piles, heavy framing and railway sleepers.

jar-ramming machine *(Foundry)*. See **jolt-ramming machine**.

jaspé *(Textiles)*. (1) Plain woven fabric with a shaded appearance resulting from a warp-thread colour pattern. Used mainly for bedspreads and curtains. (2) Yarn made from two chemically different continuous filament yarns (e.g. nylon and polyester) *textured* together and then dyed in such a way that only one component is coloured.

jasper *(Min.)*. An impure opaque chalcedonic silica, commonly red owing to the presence of iron oxides.

JATO *(Aero.)*. Abbrev. for *Jet-Assisted Take-Off*.

jaundice *(Med.)*. *Icterus*. Yellow coloration of the skin and other tissues of the body, by excess of bile pigment present in the blood and the lymph. May be caused by excessive breakdown of blood (*haemolytic jaundice*), by failure of the liver to transport the pigments (*hepatic jaundice*) or failure to excrete the pigment through the biliary system (*obstructive jaundice*).

Javel water, eau de Javelle *(Chem.)*. A mixture of potassium chloride and hypochlorite in solution. Chiefly used for bleaching and disinfecting.

jaw *(Eng.)*. (1) One of a pair or group of members between which an object is held, crushed, or cut, as the *jaws* of a vice or chuck. (2) One of several members attached to an object, to locate it by embracing another object.

jaw breaker *(Min.Ext.)*. *Jaw crusher, Blake crusher, alligator*. Heavy-duty rock-breaking machine with fixed vertical, and inclined swing jaw, between which large lumps of ore are crushed.

jaws *(Print.)*. See **folding jaws**.

jaws *(Zool.)*. In gnathostomatous Vertebrates, the skeletal framework of the mouth enclosed by flesh or horny sheaths, assisting in the opening and closing of the mouth, and usually furnished with teeth of horny plates to facilitate seizure of the prey or mastication; in Invertebrates, any similar structures placed at the anterior end of the alimentary tract.

J chain *(Immun.)*. A polypeptide chain with a high content of the amino acid cysteine, which enables it to form disulphide bonds. It helps to link together IgA molecules into polymeric forms and to hold IgM in pentameric form.

JCL *(Comp.)*. See **job control language**.

J-display *(Radar)*. A modified **A-display** with circular time base.

jean *(Textiles)*. Strong woven twilled fabric, used for overalls or casual wear. See **denim**.

jedding axe *(Build.)*. An axe having one flat face and one pointed peen.

jejunectomy *(Med.)*. Excision of part of the jejunum.

jejunitis *(Med.)*. Inflammation of the jejunum.

jejunocolostomy *(Med.)*. The formation, by operation, of a communication between the jejunum and the colon.

jejunoctomy *(Med.)*. Incision of the jejunum.

jejunojejunostomy *(Med.)*. The formation, by anastomosis, of a communication between two parts of the jejunum, thus short-circuiting the part in between.

jejunostomy *(Med.)*. The operative formation of an opening into the jejunum.

jejunum *(Zool.)*. In Mammals, that part of the small intestine which intervenes between the duodenum and the ileum.

jelly *(Image Tech.)*. See **gel**.

jelutong *(For.)*. Lightwood Malayan hardwood from the genus *Dyera*. It is almost white in colour and takes paints better than polishing techniques. The tree also yields a resinous type of rubber.

jemmy, jimmy *(Build.)*. A small crowbar.

Jeppesen chart *(Aero.)*. Airway charts, airport maps and information, named after Ebroy Jeppesen who built up the basic format from 1926 to 1940.

jerk *(Med.)*. A sudden and brief contraction of a group of muscles. Often used to test a **reflex**, e.g. knee jerk.

jerkin head *(Build.)*. The end of a pitched roof which is hipped, but not down to the level of the feet of the main rafters, thus leaving a half-gable. Also called **shread head**.

jerk-pump *(Autos.)*. A timed fuel-injection pump in which a cam-driven plunger overruns a spill port, thus causing the abrupt pressure rise necessary to initiate injection through the atomizer.

jerks *(Print.)*. Violent intermittent pulls of paper through a web-fed printing press or folder due usually to incorrect or worn drives.

jersey fabric *(Textiles)*. The general name for knitted fabrics supplied in lengths.

jet *(Genrl.)*. A fluid stream issuing from an orifice or nozzle; a small nozzle, as the *jet* of a carburettor.

jet *(Min.)*. A hard coal-black variety of lignite, sometimes exhibiting the structure of coniferous wood; worked for jewellery in the last century.

JET *(Nuc.Eng.)*. *Joint European Torus*, at Culham, UK. Large **tokamak** experiment designed to use deuterium and tritium to produce energy by a fusion process.

jet coefficient *(Aero.)*. The basic nondimensional thrust-lift relationship of the **jet flap**;

$$C_J = J/\tfrac{1}{2}e^{V^2 S}$$

where J = jet thrust, e = air density, V = speed and S = wing area.

jet condenser *(Eng.)*. One in which exhaust steam is condensed by jets of cooling water introduced into the steam space.

jet deflection *(Aero.)*. A jet-propulsion system in which the thrust can be directed downwards to assist take-off and landing.

jet drilling *(Min.Ext.)*. See **fusion drilling**.

jet dyeing *(Textiles)*. A machine for dyeing delicate fabrics, or garments, in which the material is gently circulated by the dye liquor being pumped at high velocity through jets or a narrow throat.

jet flap *(Aero.)*. A high-lift flight system in which (a) the whole efflux of turbojet engines is ejected downward from a spanwise slot at the wing trailing edge, or (b) a large surplus efflux from turboprop engines is so ejected, with the propellers providing a relative airflow over the wing. The downward ejection of the jet forms a barrier to the passage of air under the wing and induces more air to flow across the upper surface, so giving very high lift coefficients of the order of 10 and even higher. See alse **NGTE rigid rotor**.

jet impactor *(Powder Tech.)*. See **impaction sampler**.

jet lag *(Aero.)*. Delayed bodily effects felt after long flight by fast jet aircraft.

jet loom *(Textiles)*. High-speed machine in which the weft is propelled through the **shed** by a high pressure jet of air or water.

jet mill *(Chem.Eng.)*. A mill in which particles are pulverized to micrometre size by the collisions occurring among them when they are swept into a small jet of gas at sonic velocity.

jet noise *(Aero.)*. The noise of jet efflux varies as the eighth power of its velocity.

jet nozzle process *(Nuc.Eng.)*. Process whereby isotope separation, based on the mass dependence of centrifugal force, is obtained by the fast flow of uranium hexafluoride in a curved duct.

jet pipe shroud *(Aero.)*. A covering of heat-insulating material, usually layers of bright foil, round a jet pipe.

jet propulsion *(Aero.)*. Propulsion by reaction from the expulsion of a high-velocity jet of the fluid in which the machine is moving. It has been used for the propulsion of small ships by pumping in water and ejecting it at increased velocity, but the principal application is to aircraft. See **pulse-jet**, **ramjet**, **reaction propulsion**, **rocket propulsion**, **turbojet**.

jet pump *(Min.Ext.)*. Hydraulic elevator, in which a jet of high-pressure water rises in a pipe immersed in a sump containing the water, sands, or gravels which are to be entrained and pumped. Inefficient but cheap and effective where surplus hydraulic power exists.

jet shales *(Geol.)*. Shales containing 'jet rock', found in the Upper Lias of the Whitby district of England.

jet spinning *(Textiles)*. See **air jet spinning**.

jet stream *(Meteor.)*. A fairly well-defined core of strong wind, perhaps 200–300 miles (320–480 km) wide with wind speeds up to perhaps 200 mph (320 km/h) occurring in the vicinity of the **tropopause**.

jet-textured yarn *(Textiles)*. See **textured yarn**.

jetties *(Hyd.Eng.)*. See **groynes**.

jetting-out *(Arch.)*. The projection of e.g., a corbel from the face of a wall.

jewel *(Horol.)*. Natural ruby or sapphire, or synthetic stone, used for pivot bearings, also for the pallets and impulse pin. Owing to the high polish that can be obtained with such stones, combined with a hardness surface they provide wearing surfaces which have a long life and which cause little friction.

jewelled *(Horol.)*. Fitted with jewels. In watches (except in the lowest grades), the balance staff is always jewelled with two through holes and two end-stones. The pallet staff is also jewelled, but only the higher grades of watches have end-stones fitted. In watches with the club-tooth escapement, the pallets and impulse pin are invariably jewels. A '15-jewel' watch has the following jewels: balance staff 4, pallet staff 2, escape wheel 2, fourth wheel 2, third 2, pallets 2, impulse pin 1. Watches may have as many as 23 jewels. For platform escapements the holes are generally jewelled. In precision clocks, all the holes may be jewelled, and often the acting faces of the pallets are formed of inset jewels.

J exon *(Immun.)*. See **J gene**.

JFET *(Electronics)*. Abbrev. for *Junction Field-Effect Transistor*. See **field-effect transistor**.

J gene *(Immun.)*. *J exon.* A short sequence of DNA coding for part of the hypervariable region of immunoglobulin light or heavy chains near to the site of joining to the constant region. There are several possible J exons any of which may be used. The gene for the beta chain of the T-cell antigen receptor also includes a different J exon which has similar variability.

jib *(Eng.)*. The boom of a crane or derrick.

jib barrow *(Eng.)*. A wheelbarrow consisting of a platform without sides; used in foundries and workshops.

jib crane *(Eng.)*. An inclined arm or jib attached to the foot of a rotatable vertical post and supported by a tie-rod connecting the upper ends of the two. The load rope or chain runs from a winch on the post, and over a pulley at the end of the jib.

jib door *(Build.)*. A door which carries and continues the general decoration of the wall.

jig *(Eng.)*. A device used in the manufacture of (interchangeable) parts to locate and hold the work and to guide the cutting tool.

jig *(Min.Ext.)*. Device for concentrating ore according to relative density of its constituent minerals.

jig borer *(Eng.)*. A vertical-spindle machine for accurately boring and locating holes, having a horizontal table which can be precisely positioned by transverse and longitudinal feed motions.

jigger *(Eng.)*. A hydraulic lift or elevator in which a short-stroke hydraulic ram operates the lift through a system of ropes and pulleys in order to increase the travel.

jig, jigger *(Textiles)*. Machines with two rollers commonly used for dyeing. The open-width fabric passes repeatedly from one roller to the other and back again while immersed in a bath of the appropriate solution. The machines are also used for scouring and bleaching fabrics.

jig saw *(Build.)*. A mechanical saw with a short narrow reciprocating blade which cuts on the up stroke; used for curved as well as straight cuts.

jim-crow *(Eng.)*. (1) A rail-bending device, operated by hand or by hydraulic power. (2) A crowbar fitted with a claw. (3) A swivelling tool-head used on a planing machine, cutting during each stroke of the table.

jimmy *(Build.)*. See **jemmy**.

jitter *(Telecomm.)*. Small rapid irregularities in a waveform arising from fluctuations in supply voltages, components etc.

j.j. coupling *(Phys.)*. Extreme form of coupling between the orbital electrons of atoms. Electrons showing individual spin-orbital coupling also interact with each other. See also **intermediate coupling**, **Russell-Saunders coupling**.

JND *(Behav.)*. Abbrev. for *Just Noticeable Difference*.

job *(Comp.)*. Normal term for a complete item of work performed by a computer system.

jobbing chases *(Print.)*. Plain chases without bars, ranging from card sizes upward to folio.

jobbing founts *(Print.)*. Founts of type used for display purposes.

jobbing machines *(Print.)*. The class of machines, usually platens, used for printing commercial or jobbing work.

jobbing work *(Print.)*. Small printed matter such as handbills, billheads, cards, etc.

job case *(Print.)*. A case with accommodation for capitals and small letters, suitable for **jobbing founts**.

job control *(Comp.)*. One function of the **operating system** as it controls the provision of internal resources and the flow of each job through the computer system.

job control language *(Comp.)*. Special language used to identify a **job** and describe its requirements to the **operating system**. Abbrev. *JCL*.

job queue *(Comp.)*. Queue of jobs waiting to be serviced within a **multiprogramming** or **batch processing** system. See **background job**.

jockey roller *(Print.)*. On web-fed machines, a roller, usually the first to be traversed by the web, arranged to compensate any uneven tension as the reel unwinds.

jog *(Phys.)*. A discontinuity in an edge **dislocation** in a crystal. The dislocation will be made up of many sections of varying lengths lying on neighbouring slip planes and joined together by jogs.

jogger *(Print.)*. (1) A rapidly-vibrating inclined tray in which sheets of paper are placed to be jogged up to two adjacent edges prior to cutting. (2) An adustable fitment at the delivery of the printing machine which straightens the sheets as they are delivered, giving a neat delivery pile.

joggle *(Build.)*. See **stub tenon**.

joggle *(Eng.)*. (1) A small projection on a piece of metal fitting into a corresponding recess in another piece, to prevent lateral movement. (2) A lap joint in which one plate is slightly cranked so as to allow the inner edges of the two plates to form a continuous surface.

joggle *(Ships)*. A sharp distortion in a plate, angle or other section, made purposely to permit over-riding of contacting members. It reduces the amount of steel packing.

joggle joint *(Build.)*. A connection between adjacent ashlars in which **joggles** are used.

joggle work *(Build.)*. Coursed masonry in which slipping between the stones is prevented by the insertion of **joggles**.

johannsenite *(Min.)*. A silicate of calcium, manganese and iron. It is a member of the pyroxene group, crystallizing in the monoclinic system.

Johne's disease *(Vet.)*. *Paratuberculosis*. A chronic disease of cattle, sheep, and goats caused by infection by *Mycobacterium paratuberculosis (M. johnei)* which causes a chronic enteritis affecting the small intestine, caecum and colon; the main symptoms are diarrhoea and emaciation.

Johnson concentrator *(Min.Ext.)*. Machine used to arrest heavy auriferous material flowing in ore pulp. An inclined cylindrical shell rotates slowly, metallic particles being caught in rubber-grooved linings at periphery, lifted and separately discharged.

Johnson noise *(Genrl.)*. Same as **thermal noise**.

Johnston's organ *(Zool.)*. In Insects, a sensory structure situated within the second antennal joint, mechanoreceptive in function.

joiner *(Image Tech.)*. See **splicer**.

joiner's chisel *(Build.)*. A **paring chisel**.

joinery *(Genrl.)*. (1) The craft of working timber to form the finishings of a building, as distinct from **carpentry**. (2) The material worked in this way. n. *joiner*.

joint *(Elec.Eng.)*. (1) permanent connection between two lengths of cable. Waveguide connection which may be *butt-jointed* (intermetallic contact maintained) or *choke-jointed* (when a half-wavelength short-circuited line is used to provide an effective contact at the guide walls). (2) Term given to the contact formed when either two conductors, or a conductor and a device are connected together.

joint *(Foundry)*. The parting plane in the sand round a rammed mould, to enable the pattern to be withdrawn. It is covered with **parting sand** before the cope or top half is rammed.

joint *(Geol.)*. An actual or potential fracture in a rock, in which there is no displacement. See **columnar structure, rift and grain**.

joint *(Min.Ext.)*. A length of drilling pipe or casing, usually 20 to 30 feet long.

joint *(Radar)*. Permanent or semi-permanent connection between two lengths of waveguide. It may consist of plain flanges with direct metallic contact and no discontinuity in the waveguide walls or it may be a **choke flange**.

joint chair *(Civ.Eng.)*. A type of chair used at the joint between successive lengths of rail and providing support for the ends of both lengths.

joint efficiency *(Eng.)*. The ratio, expressed in per cent, of the strength of an analogous section of solid plate.

jointer *(Build.)*. A tool used by bricklayers to form the mortar joint between the courses of bricks in pointing.

jointer plane *(Build.)*. See **jointing plane**.

joint fastening *(Civ.Eng.)*. A fish-plate or other means of fastening together the adjacent ends of successive lengths of rail.

joint hinge *(Build.)*. A **strap hinge**.

joint-ill *(Vet.)*. *Navel-ill, pyosepticaemia*. A disease of young foals, calves, lambs, and piglets, caused by a variety of bacteria, and characterized by abscess formation in the umbilicus, pyaemia, abscesses in various organs, and arthritis affecting notably the leg joints.

jointing *(Build.)*. The operation of making and/or finishing the joints between bricks, stones, timbers, pipes etc.

jointing *(Eng.)*. Material used for making a pressure-tight joint between two surfaces; e.g. asbestos sheet, corrugated steel rings, vulcanized rubber etc. See **gasket**.

jointing *(Geol.)*. See **joint**.

jointing plane *(Build.)*. A bench plane, similar to the jack plane but larger (30 in, 750 mm long), used for truing the edges of timbers which are to be accurately fitted together. Also *jointer plane, shooting plane*.

jointing rule *(Build.)*. A straightedge about 6 ft (2 m) long, used as a guide when *pencilling*, i.e. painting the mortar joints of bricks to accentuate them.

jointless flooring *(Build.)*. See **magnesite flooring**.

joint-mouse *(Med.)*. A hard body, often a piece of cartilage, loose in the joint cavity; found especially in the joints of those suffering from osteoarthritis.

joints *(Print.)*. The lateral projections formed on each side of a volume in the process of backing; the cover hinges along it. A linen strip pasted down the fold of the endpaper is called a *cloth joint*.

joist *(Build.)*. A horizontal beam of timber, reinforced or prestressed concrete, or steel, used with others as a support for a floor and/or ceiling.

Jolly balance *(Chem.)*. A spring balance used to measure density by weighing in air and water.

Jolly's apparatus *(Chem.)*. Apparatus for the volumetric analysis of air.

jolt-, jar-ramming machine *(Foundry)*. A moulding machine (see **machine moulding**) in which the box, pattern, and sand are repeatedly lifted by a table operated by air pressure, and allowed to drop by gravity, the resulting jolt or jar packing or ramming the sand in an efficient manner. Also called *jolt-ram machine*.

jolt-squeeze machine *(Foundry)*. A moulding machine (see **machine moulding**) used for deep patterns; in it, jolting is used to pack the sand on to the pattern followed by squeezing from the top to complete the ramming.

Jominy (end-quench) test *(Eng.)*. A test for determining the relative hardenability of steels, in which one end of a heated cylindrical specimen is quenched, the resulting hardness decreasing towards the unquenched end.

Joosten process *(Min.Ext.)*. Use of chemical reaction between solutions of calcium chloride and sodium silicate to consolidate running soils or gravels when tunnelling. A water-resistant gel is formed.

Jordanon species *(Bot.)*. *Microspecies*. One of a number of true-breeding, morphologically slightly different, lines within a complex of largely inbreeding plants. Cf. **Linnaean species**.

jordan refiner *(Paper)*. A type of conical refiner in which a tapering rotor revolves in a hollow shell, both being equipped with longitudinal metal bars so that stuff passing through the gap between them is abraded and cut.

Josephson effects *(Electronics,Phys.)*. Two effects which can occur when, at very low temperatures, two *superconductors* are separated by a narrow insulating gap. By *tunnelling* through the gap a direct current can pass from one superconductor to another without an applied potential. Also, when a potential difference V is established between the superconductors there is an alternating current across the gap of frequency $\nu = 2Ve/h$, where e is the charge on the electron and h is Planck's constant. Applications include ultra high-speed switching of logic circuits, memory cells and parametric amplifiers operating up to 300 GHz; the Josephson effect is being widely adopted as the basis of the **standard volt**.

Joshi effect *(Phys.)*. Change of current in a gas because of light irradiation.

joule *(Phys.)*. SI unit of work, energy and heat. 1 joule is the work done when a force of 1 newton moves its point of application 1 metre in the direction of the forces. Abbrev. J. $1\,\text{erg} = 10^{-7}\,\text{J}$, $1\,\text{kWh} = 3.6 \times 10^6\,\text{J}$, $1\,\text{eV} = 1.602 \times 10^{-19}\,\text{J}$, $1\,\text{calorie} = 4.18\,\text{J}$, $1\,\text{Btu} = 1055\,\text{J}$.

Joule effect *(Elec.Eng.)*. Production of heat solely arising from current flow in a conductor. See **Joule's law**.

Joule effect *(Phys.)*. Slight increase in the length of an iron core when longitudinally magnetized. See also **magnetostriction**.

Joule magnetostriction *(Phys.)*. That for which length increases with increasing longitudinal magnetic field. Also *positive magnetostriction*.

Joule meter *(Elec.Eng.)*. An integrating watt-meter whose scale is calibrated in joules.

Joule's equivalent *(Phys.)*. See **mechanical equivalent of heat**.

Joule's law *(Chem.)*. (1) The internal energy of a given mass of gas is a function of temperature alone; it is independent of the pressure and volume of the gas. (2) The molar heat capacity of a solid compound is equal to the sum of the atomic heat capacities of its component elements in the solid state.

Joule's law *(Elec.Eng.)*. Heat liberated is $H = I^2Rt$ joules, where I = current in amperes, R = resistance in ohms, t = time in seconds. This is the basis of all electrical heating, wanted or unwanted. With high frequency a.c. R is an **effective resistance** and I may be confined to a thin skin of the conductor.

Joule-Thomson effect *(Phys.)*. When a gas is subjected to an *adiabatic expansion* through a porous plug or similar device, the temperature of the gas generally decreases. This effect is due to energy being used to overcome the cohesion of the molecules of the gas. The liquefaction of gases by the *Lindė process* depends on this effect. Also *Joule-Kelvin effect*.

journal *(Eng.)*. That part of a shaft which is in contact with, and supported by, a bearing.

journal file *(Comp.)*. Permanent record of every interaction with the computer **operating system**.

joystick *(Aero.)*. Colloq. for *control column*.

joystick *(Comp.)*. Convenient form of analogue-to-digital converter where input is the movement of a control lever in two dimensions.

Joy's valve-gear *(Eng.)*. A steam-engine valve-gear of the radial type used on some locomotives; in it, motion is taken entirely from a point on the connecting-rod.

JP- *(Aero.)*. Nomenclature for jet fuels: JP-1, *Avtur*, original jet fuel, now Jet A-1, NATO F35; JP-4, *Avtag*, wide range distillate, now Jet-B, NATO F40; JP-5, *Avcat*, dense, high flash kerosene, NATO F44; JP-7 is for high speed aircraft, e.g. SR-71; JP-10 is special high density fuel for missiles.

J/Ψ particle *(Phys.)*. A *meson* with zero strangeness and zero *charm*, but with a very high mass, 3.1 GeV, and an exceptionally long life. Discovered in 1974 at Brookhaven, by high-energy proton beam experiments (J-particle) and independently at the Stanford Linear Accelerator Centre by electron-positron colliding beam experiments (ψ-particle). The existence of this particle necessitated the postulation of the charm and the anticharm quarks: the J/ψ particle is composed of one charm and one anticharm quark.

JPT *(Aero.)*. See **gas temperature**.

JT60 *(Nuc.Eng.)*. Large tokamak experiment at the Japan Atomic Energy Research Institute. See **tokamak**.

jugal *(Zool.)*. A paired membrane bone of the zygoma of the Vertebrate skull, lying between the squamosal and the maxilla.

jugular *(Zool.)*. Pertaining to the throat or neck region, e.g., a *jugular* vein.

jugular nerve *(Zool.)*. The posterior branch of the hyomandibulár component of the 7th cranial nerve in Vertebrates; it carries the visceromotor component of the hyomandibular branch.

Julian calendar *(Astron.)*. The system of reckoning years and months for civil purposes, based on a tropical year of 365.25 days; instituted by Julius Caesar in 45 BC and still the basis of our calendar, although modified and improved by the Gregorian reform.

Julian date *(Astron.)*. The number of days which have elapsed since 12.00 GMT on 1 January 4713 BC. This consecutive numbering of days gives a calendar independent of month and year which is used for analysing periodic phenomena, especially in astronomy. This system devised in 1582 by J.Julius Scaliger has no connection whatsoever with the Julian calendar.

jump *(Comp.)*. Departure from the normal sequential execution of program steps. May be conditional on the result of a test or unconditional. Also *branch*.

jump cut *(Image Tech.)*. In film editing, the intentional deletion of part of the continuous action within a scene.

jumper *(Build.)*. A **through-stone**.

jumper *(Civ.Eng.)*. A pointed steel rod which is repeatedly dropped on the same spot from a suitable height (being turned slightly between blows), and which, by pulverizing the earth, forms a bore-hole.

jumper *(Elec.Eng.)*. (1) A short section of overhead transmission line conductor serving to form an electrical connection between two sections of a line. (2) Multi-core flexible cable making connection between the coaches of a multiple-unit train.

jumper *(Horol.)*. A click in the form of a wedge which causes a star wheel to jump forward one space.

jumper *(Min.Ext.)*. The borer, steel, or bit for a compressed-air rock drill.

jumper *(Telecomm.)*. Length of wire used in telephony to re-arrange permanent circuit connections.

jumper cable *(Elec.Eng.)*. A cable for making electrical connection between two sections of conductor-rail in an electric traction system.

jumper field *(Telecomm.)*. The cross-connection or translation field in a director. Any space devoted to jumpers, i.e. temporary connections, especially within a distribution frame. Also *cross-connection field*.

jumper-top blast pipe *(Eng.)*. A locomotive **blast pipe** in which the back pressure, and hence the draught, are automatically limited by the lifting of an annular valve, which increases the nozzle area.

jumper wire *(Telecomm.)*. See **jumper**.

jumping-figure watch *(Horol.)*. A watch in which the time is indicated by figures on disks jumping into position in windows in the watch dial. Also *digital watch*.

jumping-up *(Eng.)*. The operation of thickening the end of a metal rod by heating and hammering it in an endwise direction. Also called *upsetting*.

jump joint *(Eng.)*. A butt joint made by **jumping-up** the ends of the two pieces before welding them together.

junction *(Electronics)*. Area of contact between semiconductor material having different electrical properties.

junction *(Telecomm.)*. Union or division of waveguides in either *H-* or *E-planes*, *tee* or *wye*, *tapered* to broaden the frequency response, or *hybrid* to direct flow of wave energy.

junction chamber *(Build.)*. A closed chamber, generally of brick and concrete, inserted in a sewer system for accepting the inflow of one or more sewers and allowing for the discharge thereof.

junction circuit *(Telecomm.)*. One directly connecting two exchanges situated at a distance apart less than that specified for a trunk circuit.

junction coupling *(Elec.Eng.)*. In coaxial-line cavity resonators, coupling by direct connection to the coaxial conductor.

junction diode *(Electronics)*. One formed by the junction of *n-* and *p-* type semiconductors, which exhibits rectifying properties as a result of the potential barrier built up across the junction by the diffusion of electrons from the *n*-type material to the *p*-type. Applied voltages, in the sense that they neutralize this potential barrier, produce much larger currents than those that accentuate it.

junction field-effect transistor *(Electronics)*. An *FET* in which a bar of one sort of semiconductor material has junctions of the opposite sort diffused on both sides and connected together to form the **gate**. A terminal at one end of the bar becomes the **source** and the other is the **drain**. See also **MOSFET**.

junction rectifier *(Electronics)*. One formed by a *p-n* junction by *holes* being carried into the *n*-type semiconductor.

junction transistor *(Electronics)*. One in which the base receives minority carriers through *junctions* on each side. In normal operation, the emitter is forward biased. Signal currents entering the base tend to neutralize reverse bias in the collector junction, thus controlling the relatively large current flowing in collector circuit. See **transistor**.

juniper *(For.)*. A conifer, *Juniperus*, yielding an essential oil used mechanically; its fruits are used to flavour gin, and its wood for veneers, pencils etc.

junk ring *(Eng.)*. A metal ring attached to a steam-engine piston for confining soft packing materials; or for similarly holding a cast-iron piston-ring in position.

Jupiter *(Astron.)*. The fifth planet from the Sun, orbiting at a mean distance of 7.783×10^8 km and a sidereal period of 11.86 years. Its mass, 1.90×10^{27} kg is 318 times that of the Earth. The relative density is 1.33. Composition is 82% hydrogen, 17% helium, and 1% all heavier elements. Rotation period is 9 h 50 min at the equator, but five times longer at the poles. This differential rotation of Jupiter's cloudy atmosphere leads to a richly coloured banded structure, the most prominent feature of which is the *Great Red Spot*, a long-lived storm in the atmosphere which has been observed since the 17th. century. Due to rapid rotation Jupiter is oblate, the equatorial diameter being 6.4% larger than the polar diameter. There are 14 natural satellites, the four largest of which, the *Galilean satellites* Io, Europa, Ganymede and Callisto are visible with the simplest telescope. Beneath the atmosphere, the hydrogen is compressed so enormously that it becomes a metallic conductor, and this is the source of Jupiter's very strong magnetic field. Within the associated **magnetosphere** intense bursts of radio emission are detected.

Jurassic *(Geol.)*. The middle period of the Mesozoic era covering an app. time-span from 215–145 million years ago. Named after the type area, the Jura mountains. The corresponding system of rocks.

jury strut *(Aero.)*. A strut giving temporary support to a structure. Usually required for folding-wing biplanes, sometimes for naval monoplanes with folding parts.

justification *(Comp.,Print.)*. Process of arranging text so that the first (or last) characters of each line occur at the same position in print or on a VDU.

justification *(Print.)*. The correct spacing of words to a given measure of line.

just noticeable difference *(Behav.)*. See **difference threshold**. Abbrev. *JND*.

just scale, just temperament *(Acous.)*. The same as *natural scale*.

jute *(Textiles)*. Strong, brownish, bast fibre from the Asian plants *Corchorus olitorius* and *C. capsularus*. Used in cordage, canvas, hessian, and carpet backings.

juvenile *(Bot.)*. A structure characteristic of the *juvenile phase*.

juvenile hormone *(Zool.)*. See **neotenin**.

juvenile phase *(Bot.)*. Phase before flowering in woody plants; differ in many attributes including the ease of rooting from cuttings; usually transient but can be maintained as in some cultivated ornamental plants including many conifers.

juvenile water *(Geol.)*. Water derived from magma, as opposed to meteoric water (derived from rain or snow), or connate water (trapped in sediments at the time of deposition).

juxtaglomerular *(Med.)*. Close to the renal glomerulus, e.g. *juxtaglomerular apparatus*, a small group of cells located on the afferent arteriole of the glomerulus of the kidney, secreting **renin**.

juxtaposition twins *(Min.)*. Two (or more) crystals united regularly in accordance with a 'twin law', on a plane (the 'composition plane') which is a possible crystal face of the mineral. Cf. **interpenetration twins**.

K

k *(Chem.)*. A symbol for the *velocity constant* of a chemical reaction.

k *(Chem.Eng.)*. Symbol for *mass transfer coefficient*.

k *(Phys.)*. Symbol for (1) Boltzmann constant, (2) radius of gyration.

κ- *(Chem.)*. A symbol for: (1) *cata-*, i.e. containing a condensed double aromatic nucleus substituted in the 1,7 positions; (2) substitution on the tenth carbon atom; (3) *electrolytic conductivity*.

κ- *(Phys.)*. Symbol for (1) compressibility, (2) magnetic susceptibility.

K *(Chem.)*. The symbol for *potassium*.

K *(Phys.)*. The symbol for kelvin.

K *(Chem.)*. A symbol for *equilibrium constant*; K_s, solubility product.

K *(Chem.Eng.)*. See ***k***.

K *(Eng.)*. Symbol for **stress-intensity factor**.

K *(Phys.)*. Symbol for bulk modulus.

[K] *(Phys.)*. A very strong Fraunhofer line in the extreme violet of the solar spectrum. See **[H]**.

K-acid *(Chem.)*. *1,8-Aminonaphthol-4,6-disulphonic acid*, an intermediate for dyestuffs.

kaersutite *(Min.)*. A hydrated silicate of calcium, sodium, magnesium, iron, titanium, and aluminium; a member of the amphibole group crystallizing in the monoclinic system. It occurs in somewhat alkaline igneous rocks.

kainite *(Min.)*. Hydrated sulphate of magnesium, with potassium chloride, which crystallizes in the monoclinic system. It usually occurs in salt deposits.

Kainozoic *(Geol.)*. See **Cenozoic**.

kaiwekite *(Geol.)*. A volcanic rock containing phenocrysts of olivine, titanaugite, barkevikite and anorthoclase, probably a hybrid between basalt and trachyte.

kala-azar *(Med.)*. A disease due to infection with the protozon *Leishmania donovani*; characterized by enlargement of the liver and spleen, anaemia, wasting, and fever. Also *leishmaniasis*.

Kalanite *(Telecomm.)*. TN for a hard insulating material, not affected by oil, used for spreaders in cable joints.

kaliophilite *(Min.)*. Silicate of potassium and aluminium, which crystallizes in the hexagonal system. It has a similar composition to kalsilite and is probably metastable at all temperatures at atmospheric pressure.

kallitron *(Telecomm.)*. A periodic combination of two triode valves for obtaining negative resistance.

kalsilite *(Min.)*. A potassium-aluminium silicate, $KAlSiO_4$; it is related to nepheline and crystallizes in the hexagonal system.

kamacite *(Min.)*. A variety of nickeliferous iron, found in meteorites; it usually contains about 5.5% nickel. Metallurgically, alpha-iron.

kame *(Geol.)*. A mound of gravel and sand which was formed by the deposition of the sediment from a stream as it ran from beneath a glacier. Kames are thus often found on the outwash plain of glaciers.

kandite *(Min.)*. A collective term for the kaolin minerals or members of the kaolinite group. These include kaolinite, dickite, nacrite, anauxite, halloysite, metahalloysite, and allophane.

kanthals *(Eng.)*. Alloys with high electrical resistivity, used as heating elements in furnaces. General composition iron 67%, chromium 25%, aluminium 5%, cobalt 3%.

kaolin *(Geol.)*. Clay consisting of the mineral kaolinite; china clay.

kaolinite *(Min.)*. A finely crystalline form of hydrated aluminium silicate, $(OH)_4Al_2Si_2O_5$, occurring as minute monoclinic flaky crystals with a perfect basal cleavage; resulting chiefly from the alteration of feldspars under conditions of hydrothermal or pneumatolytic metamorphism, or by weathering. The kaolinite group of clay minerals includes the polymorphs **dickite** and **nacrite**. See **kandite**.

kaolinization *(Geol.)*. The process by which the feldspars in a rock such as granite are converted to kaolinite.

kaon *(Phys.)*. See under **meson**.

Kaplan water turbine *(Eng.)*. A propellor-type *water turbine* in which the pitch of the blades can be varied in accordance with the load, resulting in high efficiency over a large load-range.

kapok *(Textiles)*. The seed hairs of the kapok tree, *Ceiba pentranda*. They are light and fluffy and in loose form are used as an insulating or flotation material e.g. in life jackets. They are not spun or converted into fabrics commercially.

Kaposi's sarcoma *(Med.)*. A malignant vascular skin tumour that appears in two forms. The first is a slowly progressive lesion on toes or legs of elderly men of Mediterranean origin. The second is a much more invasive disseminated form and is common in children and young men in Central Africa. It is now seen as a common feature in *AIDS* or **acquired immune deficiency syndrome**.

kappa chain *(Immun.)*. One of the two types of **light chain** of immunoglobulins, the other being the lambda chain. An individual immunoglobulin molecule bears either two kappa or two lambda chains. The proportion of molecules bearing each chain varies in different species. In humans about 60% are of the kappa and 40% of the lambda type.

kappa number *(Paper)*. A bleachability test for pulps measured by the number of mls of 0.1 N potassium permanganate solution consumed by 1 g of dry pulp.

Kapp phase advancer, Kapp vibrator *(Elec.Eng.)*. A form of phase advancer for use with slip-ring induction motors. It consists of a small armature connected to each phase of the rotor circuit and allowed to oscillate freely in a d.c. field so that it has a leading e.m.f. induced in it.

kapur *(For.)*. Wood from *Dryoblanops*, found in Malaysia and Indonesia which also yields Borneo camphor (*borneol*). When in contact with iron the wood develops bad stains and eventually the metal corrodes.

karat *(Genrl.)*. See **carat**.

Karbate *(Chem.Eng.)*. TN for a form of extremely dense carbon having high enough strength to permit its being made into special shapes, e.g. tubes, pumps, and possessing corrosion resistance and good heat conductivity.

Karman vortex street *(Acous.)*. Regular vortex pattern behind an obstacle in a flow where vortices are generated and travel away from the object. The frequency of vortex generation is determined by the **Strouhal number**.

karri *(For.)*. A dense deep-red wood from Australia, similar but not so resistant as **jarrah**, used as a structural timber.

karst *(Geol.)*. Any uneven limestone topography, characterized by joints enlarged into criss-cross fissures (*grikes*) and pitted with depressions resulting from the collapse of roofs of underground caverns. It is formed by the action of percolating waters and underground streams.

kary-, karyo- *(Genrl.)*. Prefix from Gk. *karyon*, nucleus. Also **cary-**, **caryo-**.

karyogamy *(Bot.)*. The fusion of the two gametes to form a zygote. It usually follows cytoplasmic fusion but may as in some fungi be followed by a prolonged binucleate stage, *dikaryophase*. See **plasmogamy**.

karyogram *(Bot.)*. See **idiogram**.

karyon *(Biol.)*. The cell-nucleus, only used in compound words. From Gk. for nut or kernel.

karyotype *(Biol.)*. The appearance, number, and arrangement of the chromosomes in the cells of an individual.

Kasimovian *(Geol.)*. An epoch of the Pennsylvanian period.

Kaspar-Hauser experiments *(Behav.)*. Experiments in which animals are reared in complete isolation from other animals of their own or other species.

kata- *(Genrl.)*. Prefix from Gk. *kata*, down. Also *cata-*.

katabatic wind *(Meteor.)*. A local wind flowing down a slope cooled by loss of heat through radiation at night. It is caused by the difference in density between cold air in contact with the ground and the warmer air at corresponding levels in the *free atmosphere*. It is often quite strong, especially when channelled down a narrow valley, and can reach gale-force at the edges of the Antarctic and Greenland ice-caps.

katadromous *(Zool.)*. Of Fish, migrating to water of greater density than that of the normal habitat to spawn, as the Freshwater Eel which migrates from fresh to salt water to spawn. Cf. **anadromous**. Also *catadromous*.

kata-front *(Meteor.)*. A situation at a front, warm or cold, where the warm air is sinking relative to the **frontal zone**.

kataklastic *(Geol.)*. See **cataclasis**.

kataphorite *(Min.)*. See **katophorite**.

kataplexy *(Zool.)*. The state of imitation of death, adopted by some animals when alarmed.

katathermometer *(Min.Ext.)*. Instrument used in mine ventilation survey to assess cooling effect of air current. Thermometer bulb is first exposed dry, then when covered with wetted gauze, and time taken for temperature to fall from 100° to 95°F (38° to 35°C) is observed.

Katayama disease *(Med.)*. A disease due to invasion of the body by the blood fluke *Schistosoma Japonicum*, characterized by urticaria, painful enlargement of liver and spleen, bronchitis, diarrhoea, loss of appetite, and fever lasting a few days to several weeks.

katharometer *(Chem.)*. An instrument for the analysis of gases by means of measurements of thermal conductivity.

kathetron *(Electronics)*. See **cathetron**.

katophorite, kataphorite, cataphorite *(Min.)*. Hydrated silicate of sodium, calcium, magnesium, iron and aluminium; a member of the amphibole group. It crystallizes in the monoclinic system and occurs in basic alkaline igneous rocks.

kauri-butanol value *(Chem.)*. Term applied to solvents to indicate their dissolving powers for resins. Based on natural kauri gum as the standards.

kauri gum *(Chem.)*. A gum found in New Zealand, used for varnishes and linoleum cements. It is the resinous exudation of the kauri pine (*Agathis australis*), a tree whose timber is of value for general joinery and decorative purposes.

Kawasaki's disease *(Med.)*. *Mucocutaneous lymphnode syndrome*. A disease occurring in children under 5 years which causes a skin rash and glandular enlargement. In some it produces severe dilatation of the coronary arteries with thrombus formation leading to occlusion.

Kaye disk centrifuge *(Powder Tech.)*. A transparent disk centrifuge with an entry port coaxial with the axis of revolution. The suspension of the powder under test is injected into the centrifuge while it is running at the final speed. Concentration changes are measured optically.

Kayser-Fleischer ring *(Med.)*. A ring of brownish-yellow pigmentation found in the cornea of patients suffering from **Wilson's disease**.

Kazanian *(Geol.)*. A stratigraphical stage in the Permian rocks of Russia and eastern Europe.

K-band *(Telecomm.,Radar)*. US designation of microwave band between 12 and 40 GHz, widely adopted in the absence of internationally agreed standards. Subdivided into Ku-band (*K-under*), 12–19 GHz, K-band, 18–26 GHz and Ka (*K-above*), 26–40 GHz; replacing approximately the UK designations of *J-*, *K-* and *Q-bands*.

K-capture *(Phys.)*. Absorption of an electron from the innermost (K) shell of an atom into its nucleus. An alternative to ejection of a *positron* from the nucleus of a radioisotope. Also *K-electron capture*. See also **X-rays**.

K-cell *(Immun.)*. A cell generally resembling a lymphocyte which bears Fc receptors by which it can bind to other cells to which are attached antibodies against antigens on the cell surface (thus exposing the Fc portions). K-cells induce lesions in the membrane of the target cell which kill it. This is an example of antibody-dependent cell-mediated cytotoxicity.

kCi *(Phys.)*. Abbrev. for *kiloCurie*, i.e. radioactivity equivalent to 1000 curies.

K-display *(Radar)*. A form of **A-display** produced with a lobe-switching antenna. Each lobe produces its own peak and the antenna is directly on target when both parts of the resulting double peak have the same height.

keatite *(Min.)*. A high-pressure synthetic form of silica.

kebbing *(Vet.)*. See **enzootic ovine abortion**.

kedge anchor *(Ships)*. A small anchor used for steadying and warping purposes.

keel block *(Eng.)*. A standard casting shaped like a ship's keel which is used to provide a test specimen for steel or other alloys subject to high shrinkage.

keelson *(Aero.)*. A longitudinal structural member inside the bottom of the hull of a flying-boat. It forms part of the main framework, connecting the transverse members and bulkheads.

keelson *(Ships)*. (1) A term descriptive of the longitudinal strengthening members of a ship, which form the shell-plating stiffeners. A *flat keel* is the lower horizontal member of the ship's backbone; a *centre keelson* is the vertical member thereof; a *bar keel* is similar to the latter, external to the hull; *side keelsons* are vertical members, off the ship's centre line. (2) The wrought-iron saddles or standards which support cylindrical boilers of the Scotch marine type. Sometimes called *boiler cradles*.

Keene's cement *(Build.)*. A quick-setting hard plaster, made by soaking plaster of Paris in a solution of alum or borax and cream of tartar.

keep *(Eng.)*. See **keeper**.

keep down *(Print.)*. A typographic instruction to avoid the use of capital initials so far as possible.

keeper *(Build.)*. (1) The part of a Norfolk latch limiting the travel of the fall bar. (2) The socket fitted on a door jamb to house the bolt of the lock in the shut position.

keeper *(Elec.Eng.)*. See **armature** (2).

keeper *(Eng.)*. The lower part of the bearing in a railway-truck axle-box, which limits the downward movement of the box due to track irregularities.

keeper *(Phys.)*. Bar used to close the magnetic circuit of a permanent magnet when not in use, thereby conserving its strength. Also *keep*.

keep standing *(Print.)*. An instruction to keep the printing surface intact, particularly to keep the type locked up in chase, usually in anticipation of a reprint.

keep up *(Print.)*. (1) A typographic instruction to use capital initials in preference to small letters. (2) A synonym for *keep standing*.

Keewatin Group *(Geol.)*. A series of basic pillow lavas associated with sedimentary iron ores (worked in the 'Iron Ranges'); forms part of the Precambrian succession in the Canadian Shield.

kefir *(Chem.)*. A fermentation product of milk in which the lactose has undergone both alcoholic and lactic fermentation simultaneously.

keilhauite *(Min.)*. A variety of sphene containing more than 10% of the rare earths.

K-electron capture *(Phys.)*. See **K-capture**.

Kel-F *(Plastics)*. TN (US) for *polytrifluorochloroethene*.

Keller furnace *(Elec.Eng.)*. A form of electric furnace for iron smelting, in which part of the heat is produced by the passage of the current through charge and part by arcs between the electrodes and the charge.

Kelling's test *(Chem.)*. A test for the detection of lactic acid in gastric juice, based upon the colouring effect produced by the addition of a few drops of a very dilute neutral iron (III) chloride solution.

kelly *(Min.Ext.)*. The topmost **joint** of a drill string, attached below to the next drill joint and above to the swivel and mudhose connection. It is a heavy tubular part of square and sometimes other cross-section externally which fits into the corresponding hole of the **kelly bushing**, itself fixed to the **rotary table**. To add a fresh pipe to the drill string, it has to be raised by the **draw works** until the kelly and bushing are clear of the rotary table. The string is then locked by wedge-shaped *slips* to the table and the kelly joint unscrewed and parked in the *rat hole*. After a new pipe has been attached the kelly is rescrewed and the whole string lowered so that drilling can be restarted.

kelly bushing *(Min.Ext.)*. The relaceable bearing with a square hole in which the **kelly** slides and which is attached to the **rotary table** during drilling.

keloid *(Med.)*. A dense new growth of skin occurring in skin that has been injured.

kelp *(Bot.)*. (1) A general name for large seaweeds, especially *Laminaria* and allies. (2) The ashes from burning such seaweed, formerly a source of soda and iodine.

kelvin *(Phys.)*. The SI unit of thermodynamic temperature. It is 1/273.16 of the temperature of the **triple point** of water above **absolute zero**. The temperature interval of 1 kelvin (K) equals that of 1°C (degree Celsius). See **Kelvin thermodynamic scale of temperature**.

Kelvin ampere-balance *(Elec.Eng.)*. Laboratory instrument for measuring current; in it the calculated force between two coils carrying the current to be measured is balanced by the force of gravity on a weight sliding along a beam.

Kelvin bridge *(Elec.Eng.)*. An electrical network involving two sets of ratio arms used for accurate measurement of low resistances.

Kelvin compass *(Ships)*. A form of ship's compass having a very light card and a number of short parallel needles held by silk cords, as well as other special features. Also called *Thomson's compass*.

Kelvin effect *(Elec.Eng.)*. Skin effect, whereby varying currents in a conductor tend to concentrate near the surface. Important in radio circuits, and in eddy-current (high-frequency) heating.

Kelvin electrometer *(Elec.Eng.)*. Same as **quadrant electrometer**.

Kelvin's law *(Elec.Eng.)*. A principle regarding the transmission of electrical energy. It states that the most economical size of conductor to use for a line is that for which the annual cost of the losses is equal to the annual interest and depreciation on that part of the capital cost of the conductor which is proportional to its cross-sectional area.

Kelvin thermodynamic scale of temperature *(Phys.)*. A scale of temperature based on the thermodynamic principle of the performance of a reversible heat engine. The scale cannot have negative values so *absolute zero* is a well defined thermodynamic temperature. The temperature of the **triple point** of water is assigned the value 273.16 K. The temperature interval corresponds to that of the Celsius scale so that the freezing point of water (0°C) is 273.15 K. Unit is the **kelvin**. Symbol K.

Kelvin-Varley slide *(Elec.Eng.)*. Constant resistance decade voltage divider of the type used in vernier potentiometers. It consists of a resistor of $2r$ ohms shunting two adjacent units in a series array of eleven resistors, each r ohms. The $2r$ resistor may similarly be divided into eleven equal parts, and shunted if additional decade(s) are required.

kelyphitic rim *(Geol.)*. A shell of one mineral enclosing another in an igneous rock, produced by the reaction of the enclosed mineral with the other constituents of the rock. Kelyphitic rims are most common in basic and ultrabasic rocks.

Kennelly-Heaviside layer *(Phys.)*. See **E-layer**.

kentallenite *(Geol.)*. A coarse-grained, basic igneous rock, named for the type locality, Kentallen, Argyllshire; it consists essentially of olivine, augite, and biotite, with subordinate quantities of plagioclase and orthoclase in approximately equal amounts.

Kent claw hammer *(Build.)*. One with thick, slightly-curved claw and hexagonal head.

kentledge *(Civ.Eng.)*. Scrap iron, rails, heavy stones etc. used as loading on a structure (e.g. upon the top section in sinking a cylinder caisson), or as a counterbalance for a crane.

kenyte *(Geol.)*. A fine-grained igneous rock, occurring as lava flows on Mt. Kenya, E. Africa, and in the Antartic; essentially an olivine-bearing phonolite with phenocrysts of anorthoclase.

Kepler's laws of planetary motion *(Astron.)*. (1) The planets describe ellipses with the sun at a focus. (2) The line from the sun to any planet describes equal areas in equal times. (3) The squares of the periodic times of the planets are proportional to the cubes of their mean distances from the sun.

keratectasia *(Med.)*. Large bulging of part of the cornea.

keratectomy *(Med.)*. Excision of part of the cornea.

keratin *(Biol.)*. The class of **intermediate filament** which is characteristic of epithelial cells. α-keratin is a major component of skin; β-keratin exists as a β-sheet and is a major component of *silk*.

keratitis *(Med.)*. Inflammation of the cornea.

keratocele *(Med.)*. Protrusion of the innermost layer (Descemet's membrane) of the cornea through a corneal ulcer.

keratoconus *(Med.)*. Conical cornea. Cone-shaped deformity of the cornea owing to a weakness and thinness of the centre.

keratodermia blennorrhagica *(Med.)*. Red patches on the skin which become hard, dry, yellow, and raised above the skin, occurring in gonorrhoea.

keratogenous *(Zool.)*. See **keratin**.

keratoma *(Med.)*. A tumour of the skin in which overgrowth of the horny layer predominates.

keratomalacia *(Med.)*. A disease in which the cornea first becomes dry and lustreless and then softens; associated with deficiency of vitamin A in the diet.

keratophyre *(Geol.)*. A fine-grained igneous rock of intermediate composition. It is essentially a soda-trachyte, containing albite-oligoclase or anorthoclase in a cryptocrystalline groundmass. The pyroxenes, when present, are often altered to chlorite or epidote.

keratoplasty *(Med.)*. The grafting of a new cornea on to an eye whose cornea has become opaque.

keratoscleritis *(Med.)*. Inflammation of the cornea and of the sclera of the eye.

keratosis *(Med.)*. Overgrowth of the horny layer of the skin.

keratotomy *(Med.)*. Incision of the cornea.

kerf *(Build.)*. The cut made by a saw.

kerf *(Eng.)*. The part of the original material which was removed by cutting, pressing etc.

kerf *(Min.Ext.)*. In coal winning, undercut made by coal-cutting machine to depth of 1 m or more. Also called *kirve*.

kermesite *(Min.)*. Oxysulphide of antimony, which crystallizes in the monoclinic system. It is a secondary mineral occurring as the alteration product of stibnite. Also called *pyrostibite*.

kern *(Print.)*. The portion of some type letters which projects beyond the body and rests on the body of the preceding or following letter. e.g. the tail of an italic *f*.

kernel *(Maths.)*. The kernel of a mapping is the set of elements mapped to the identity (0 or 1 according to context).

kernicterus *(Med.)*. Damage to the brain in infants caused by the passage of unconjugated bilirubin across the blood-brain barrier. Usually a complication of **haemolytic disease of the newborn**.

Kernig's sign *(Med.)*. A sign of meningitis. When the patient is lying on his back with the thigh bent at right angles to the body, the leg cannot be bent straight at the knee.

kernite *(Min.)*. Hydrated sodium borate, $Na_2B_4O_7.4H_2O$.

kerogen *(Geol.)*. Fossilized, insoluble, organic material present in a sediment, that yields petroleum products on distillation.

Kerr cell *(Elec.Eng.)*. A light modulator consisting of a liquid cell between crossed polaroids. The light transmission is modulated by an applied electric field. See **Kerr effect**.

Kerr effect *(Elec.Eng.)*. Double refraction produced in certain transparent dielectrics by the application of an electric field. Also *electro-optical effect*.

Kerr effect *(Phys.)*. Plane-polarized light at normal incidence on the polished surface of a magnetized material when reflected becomes elliptically polarized. Also *magneto-optical effect*.

kersantite *(Geol.)*. A mica-lamprophyre, named from the type locality Kersanton, near Brest; it consists essentially of biotite and plagioclase feldspar. See **minette**.

ketenes *(Chem.)*. Compounds of the general formula $R_2C=C=O$. The ketene series may be considered homologues of carbon monoxide. The first member of the series is ketene, $CH_2=CO$, is readily obtainable by passing propanone vapours through a red-hot glass tube filled with broken tile. Propanone is then decomposed into ketene and methane. Ketenes form acids or acid derivatives on adding water, alcohols, ammonia, amines etc. to one of their double bonds. They are liable to autoxidation; they react with other unsaturated compounds, forming 4-membered rings. They are very unstable and polymerize easily.

ketoacidosis, ketosis *(Med.)*. The excessive formation in the body (due to incomplete oxidation of fats) of ketone or acetone bodies (aceto-acetic acid and β-hydroxybutyric acid) which are accompanied by ketonaemia and ketonuria; occurs, e.g. in diabetes.

keto-enolic tautomerism *(Chem.)*. The formation by certain compounds of two series of derivatives, based upon their ketonic or enolic constitution. The enol-form is produced from the keto-form by the migration of a hydrogen atom, which forms a hydroxyl group with the ketone oxygen, accompanied by a change in the position of the double bond. Thus:

$$\underset{\text{keto-form}}{-CH_2-CO-} \rightleftarrows \underset{\text{enol-form}}{-CH=C(OH)-}$$

keto form *(Chem.)*. That form of a substance exhibiting keto-enol tautomerism which has the properties of a ketone, e.g. ethyl aceto-acetate.

ketogenic *(Med.)*. Capable of producing ketone bodies, e.g. *ketogenic diet*.

ketohexoses *(Chem.)*. $HO.CH_2.(CHOH)_3.CO.CH_2OH$, a group of carbohydrates, isomers of *aldohexoses*. The formula contains three asymmetric carbon atoms, and there are eight stereoisomers, viz. D- and L-fructose, -sorbose, -tagatose and -allulose. They reduce an alkaline copper solution. Ketohexoses can be oxidized to acids containing fewer carbon atoms in the molecule.

ketonaemia, ketonemia *(Med.)*. the presence of ketone bodies in the blood. See **ketoacidosis**.

ketones *(Chem.)*. Compounds containing a carbonyl group, -CO-, in the molecule attached to two hydrocarbon radicals. The general formula is

$$R-\overset{\overset{\displaystyle O}{\|}}{C}-R'$$

Ketones are formed by the oxidation of secondary alcohols, by the dry distillation of the calcium or barium salt of an organic acid, by the catalytic condensation of acids or esters, by synthesis with Grignard reagents, and by the action of CO on sodium alkyls. Ketones are very reactive substances, forming additive compounds, e.g. with sodium hydrogen sulphite etc. Important derivatives are the oximes, obtained by the action of hydroxylamine. Hydrogen reduces ketones to secondary alcohols. The simplest ketone is **acetone** (propanone).

ketonuria *(Med.)*. The presence of ketone bodies in the urine. See **ketoacidosis**.

ketoses *(Chem.)*. A general term for monosaccharides with a ketonic constitution. They always form mixtures of acids on oxidation, containing a smaller number of carbon atoms than the original ketose.

ketoximes *(Chem.)*. The reaction products of ketones with hydroxylamine, containing the oximino group =N.OH attached to the carbon atom.

kettle *(Eng.)*. An open-top vessel used in carrying out metallurgical operations on low melting-point metals, e.g. in drossing and desilverizing lead.

kettle *(Geol.)*. A steep-sided basin in glacial drift, often a lake or swamp, and derived from the melting of a block of stagnant ice.

kettlestitch *(Print.)*. The stitch which is made at the head and tail of each section on a book to interlock the sections.

Keuper *(Geol.)*. The upper series of rocks assigned to the Triassic System in N.W. Europe, lying above the Muschelkalk.

keV *(Phys.)*. Abbrev. for kilo-electron-volt; unit of particle energy, 10^3 **electron volts**.

kevel *(Build.)*. A hammer, edged at one end and pointed at the other, used for breaking and rough-hewing stone.

Kew Certificates *(Horol.)*. Certificates of performance of watches and chronometers issued by the National Physical Laboratory, Teddington. (This work was originally undertaken at Kew, hence the name.) For an 'absolutely perfect' watch, 100 marks are awarded, made up as follows: 40 for a complete absence of variation of daily rate; 40 for absolute freedom from change of rate with change of position; and 20 for perfect compensation for effects of temperature.

Keweenawan *(Geol.)*. Conglomerates, arkoses, red sandstones and shales of desert origin, associated with great thicknesses of basic lavas and intrusives. This is the youngest of the Precambrian divisions in the Canadian Shield.

Kew-pattern barometer, compensated scale barometer *(Meteor.)*. The Adie barometer. Specially graduated so that error arising from changes in the free level in the cistern is obviated.

Kew-pattern magnetometer *(Elec.Eng.)*. A delicate type of reflecting magnetometer having a photographic recording arrangement for keeping record of changes in the earth's field.

key *(Build.)*. In any surface to be plastered, the roughening, lathing, or other preliminary process undertaken in order to give a grip to the coat of plaster and so enable it to adhere more satisfactorily.

key *(Comp.)*. (Of a data item). Part of the data description used to determine its location in store. See **associative memory, hashing**.

key *(Ecol.)*. Set of instructions devised to enable an unknown organism to be identified by a student on the basis of critical characteristics.

key *(Eng.)*. A piece inserted between a shaft and a hub to prevent relative rotation. It fits in a key-way, parallel with the shaft axis, in one or both members, the commonest form being the parallel key, of rectangular section.

key *(For.)*. *Hinge*. In tree felling, the portion of a tree bole deliberately left uncut until the last moment to lessen the risk of splitting.

key *(Image Tech.)*. By analogy with music, the character of the tonal range of subject matter and brightness in a picture as a subjective whole — **high key, low key**.

key *(Print.)*. The name applied to a small tool used for securing the bands while a book is being sewn.

key *(Telecomm.)*. Hand-operated spring-loaded switch used for telegraphy transmissions.

keyboard *(Comp.)*. **Input device** with an array of keys to be pressed. See **teletypewriter, ASCII keyboard, QUERTY-, Dvorak-, numeric-**.

key boss *(Eng.)*. A local thickening up of a boss or hub at the point at which a key-way is cut, to compensate for loss of strength due to the cut.

key chuck *(Eng.)*. A jaw chuck whose jaws are adjusted by screws turned by a key or spanner. See **self-centring chuck**.

key click *(Telecomm.)*. Click produced by **ringing** in key circuit as contact is opened or closed. Normally minimized by use of key-click filter. Also *key chirp*.

key course *(Civ.Eng.)*. The course of stones in an arch corresponding to the keystone.

key drawing *(Print.)*. In lithography and line-colour block making, an outline drawing which serves as a guide in the making of the separate colour plates.

key drop *(Build.)*. A guard plate covering a keyhole and falling into position by its own weight.

keyed pointing *(Build.)*. Pointing which is finished with lines or grooves struck on the flat joint. See **flat pointing**.

keyer *(Telecomm.)*. A device for changing the output of a transmitter from one frequency (or amplitude) to another according to the intelligence transmitted.

key fossil *(Geol.)*. See **index fossil**.

Key-Gaskel syndrome *(Vet.)*. Feline dysautonomia of unknown aetiology. Symptoms include dilated pupils, dehydration, constipation, regurgitation and sometimes loss of anal sphincter tone.

keyhole limpet haemocyanin *(Immun.)*. A large copper-containing protein from a particular kind of limpet. Haemocyanins normally function as oxygen carrying molecules, but are widely used as immunogens in immunology since they are likely to be completely foreign to mammals. Abbrev. *KLH*.

keyhole saw *(Build.)*. One with stiff, narrow blade 6–10 in (150–250 mm) long, for internal, curved and small cuts. Also *padsaw*.

keying *(Eng.)*. The process of fitting a key to the key-ways in a shaft and boss.

keying *(Image Tech.)*. A video switching effect which creates a space within a picture into which another image is inserted.

keying *(Print.)*. Use of keyboard to input coded copy for eventual typesetting by mechanical means in single letter (**Monotype**), or slug form (**Linotype**), or via a **phototypesetter**.

keying wave *(Telecomm.)*. See **marking wave**.

key light *(Image Tech.)*. The principal lighting of the main subject in a scene.

key plan *(Build.,Eng.)*. A small-scale plan showing the relative disposition of a number of items in a scheme.

key plate *(Build.)*. An **escutcheon**.

key print *(Image Tech.)*. See **grey key image**.

key seating *(Eng.)*. A key-way, or the surface on to which a key is bedded.

key-seating machine *(Eng.)*. A machine tool for milling key-ways in shafts, etc., by means of an end mill, the work being supported on a table at right angles to the axis of the spindle. Feed is obtained by an automatic traverse of either the tool or the table.

keystone *(Civ.Eng.)*. The central voussoir at the crown of an arch.

keystone distortion *(Image Tech.)*. Distortion of an optical or electronic image in which a rectangle is reproduced as a trapezium with the vertical sides converging. Generally a result of the beam axis not meeting the screen at right angles.

key-to-disk *(Comp.)*. Input device for accepting data from a keyboard and writing directly to magnetic disk.

key-to-tape unit *(Comp.)*. Input device for accepting data from a keyboard and writing it directly onto tape.

key way *(Eng.)*. A longitudinal slot cut in a shaft or hub to receive a **key**.

key-way tool *(Eng.)*. A slotting machine tool used for the vertical cutting of key-ways, equal in width to that of the key-way. See **slotting machine**.

keywords *(Comp.)*. The most informative or significant words in a piece of text. These are some of the elements stored in most **information retrieval** systems. Also *index terms*.

kick *(Build.)*. See **frog**.

kick *(Min.Ext.)*. Sudden increase in pressure during well drilling, which if not controlled quickly can cause a **blowout**. See **kill line**.

kick-back *(Autos.)*. Shocks felt at the steering wheel, due to the *reversibility* of many steering devices.

kick copy *(Print.)*. On a web-fed printing press the printed and folded product is counted off in batches, each batch indicated by a displaced copy.

kicker *(Print.)*. The mechanism on a web-fed printing press which displaces one copy of the product to indicate the intervals between batches. See **quire spacing**.

Kick's law *(Min.Ext.)*. Assumes that the energy required for subdivision of a definite amount of material is the same for the same fractional reduction in average size of the individual particles, i.e. $E = k_k \log_e d_1/drt_2$, where E is the energy used in crushing, k_k is a constant, depending on the characteristics of the material and method of operation of the crusher, and d_1 and d_2 are the average linear dimensions before and after crushing.

kicksorter *(Telecomm.)*. See **pulse height analyser (1)**.

kick stage *(Space)*. A propulsive stage used to provide an additional velocity increment required to put a spacecraft on a given trajectory.

kid *(Hyd.Eng.)*. A bundle of brushes serving as a **groyne**.

kidney *(Zool.)*. A paired organ for the excretion of nitrogenous waste products in Vertebrates.

kidney machine *(Med.)*. See **artificial kidney**.

kidney ore *(Min.)*. A form of the mineral haematite, oxide of iron, Fe_2O_3, which occurs in reniform masses, hence the name (Latin *ren*, kidney).

kidney piece *(Horol.)*. A cam, shaped like a kidney, used in perpetual calendar work to give the equation of time.

kidney stone *(Min.)*. A name given to **nephrite**, which was once supposed to be efficacious in diseases of the kidney (Gk. *nephros*, kidney).

kidney stones *(Med.)*. Hard deposits formed in the kidney. The composition varies, and kidney stones have been found to consist of uric acid and urates, calcium oxalate, calcium and magnesium phosphate, silica and alumina, cystine, xanthine, fibrin, cholesterol and fatty acids. Passage of the stones down the ureter may cause severe pain (*renal colic*).

kidney worm disease *(Vet.)*. In pigs the parasite the causative parasite if *Stephanurus dentalus*. The main target for infestation is the kidney with occasional spinal canal involvement. The intermediate host is the earthworm. In dogs *Dictyophyma renale* is the parasite,

kier *(Textiles)*. A large steel vessel in which yarn and cloth are boiled with alkaline liquors for scouring and bleaching. Now frequently replaced by continuous processing machinery.

kieselguhr *(Min.)*. See **diatomite**.

kieserite *(Min.)*. Hydrated magnesium sulphate which crystallizes in the monoclinic system; found in large amounts in some salt deposits.

kieve *(Min.Ext.)*. See **dolly tub**.

kieving *(Min.Ext.)*. See **tossing**.

kill *(Print.)*. In printing, an editorial instruction to delete entirely some item in preparation, derived from the use of the word as an instruction to distribute type.

kill, choke line *(Min.Ext.)*. Small bore pipe lines connected through the **blowout preventer** stack; they allow denser mud to be pumped into a borehole which has been shut out because of the danger of a **blowout**.

killed steel *(Eng.)*. Steel that has been *killed*, i.e. fully deoxidized before casting, by the addition of manganese, silicon and sometimes aluminium. There is practically no evolution of gas from the reaction between carbon and iron-oxide during solidification. Sound ingots are obtained. See also **rimming steel**.

killer *(Phys.)*. See **poison**.

kill string *(Min.Ext.)*. See **injection string**.

kiln *(Min.Ext.)*. Furnace used for: drying ore; driving off carbon dioxide from limestone; roasting sulphide ores or concentrates to remove sulphur as dioxide; reducing iron (II) ores to magnetic state in reducing atmosphere.

kilo- *(Genrl.)*. Prefix for denoting 1000; used in the metric system. E.g. 1 kilogram = 1000 grams.

kilocalorie *(Phys.)*. See **calorie**.

kilocurie source *(Radiol.)*. Giant radioactive source, usually in form of ^{60}Co.

kilocycles per second *(Phys.)*. See **kilohertz**.

kilo-electron-volt *(Phys.)*. See **keV**.

kilogram(me) *(Genrl.)*. Unit of mass in the MKSA (SI) system, being the mass of the *International prototype kilogram*, a cylinder of platinum-iridium alloy kept at Sèvres.

kilohertz *(Phys.)*. One thousand **hertz** or cycles per second. A multiple of the SI unit of frequency. Abbrev. kHz.

kilometric waves *(Telecomm.)*. Those with wavelengths between 1000 and 10 000 m.

kiloparsec *(Astron.)*. See **parsec**.

kilotex *(Textiles)*. See **tex**.

kiloton *(Phys.)*. Unit of explosive power for nuclear weapons equal to that of 10^3 tons of TNT.

kilovar *(Elec.Eng.)*. A unit of reactive volt-amperes equal to 1000 VArs. Abbrev. **kVAr**.

kilovolt-ampere *(Elec.Eng.)*. A commonly used unit for expressing the rating of a.c. electrical machinery and for other purposes; it is equal to 1000 volt-amperes. Abbrev. **kVA**.

kilowatt *(Elec.Eng.)*. A unit of power equal to 1000 watts and approximately equal to 1.34 h.p. Abbrev. **kW**.

kilowatt-hour *(Elec.Eng.)*. The commonly used unit of electrical energy, equal to 1000 watt-hours or 3.6 MJ. Often called simply a *unit*; abbrev. *kWh*. See **Board of Trade Unit**.

Kimball tag *(Comp.)*. Small punched card attached to merchandise which is detached when goods are sold, to provide machine-readable sales data.

Kimberley horse disease *(Vet.)*. *Walk-about disease*. A disease of horses in Australia due to poisoning caused by eating whitewood (*Atalaya hemiglauca*).

kimberlite *(Geol.)*. A type of mica-peridotite, occurring in volcanic pipes in South Africa and elsewhere, and containing xenoliths of many types of ultramafic rocks, and diamonds.

kinaesthesis *(Behav.)*. A general term for sensory feedback from muscles, tendons and joints, which inform the individual of the movements of the body or limbs, and the position of the body in space.

kinaesthetic *(Zool.)*. Pertaining to the perception of muscular effort.

kinase *(Biol.)*. An enzyme which catalyses the phosphorylation of its substrate by ATP. Thus protein kinases phosphorylate proteins and hexose kinases phosphorylate hexose.

kinematical theory of X-ray diffraction *(Phys.)*. A treatment which does not take account of the attenuation of the incident beam as it passes through the crystal nor the interference between the incident beam and multiply diffracted beams; a theory which can be applied to very thin or very small crystals.

kinematic chain *(Eng.)*. A number of links connected to one another to allow motion to take place in combination. It becomes a mechanism when so constructed as to allow constrained relative motion between its links.

kinematics *(Maths.)*. That branch of applied mathematics which studies the way in which velocities and accelerations of various parts of a moving system are related.

kinematic viscosity *(Phys.,Eng.)*. The coefficient of viscosity of a fluid divided by its density. Symbol ν. Thus $\nu = \eta/\rho$. Unit in the CGS system is the stokes ($cm^2\,s^{-1}$); in SI $m^2\,s^{-1}$.

kinesalgia *(Med.)*. Feeling of pain on movement.

kinescope *(Image Tech.)*. US term for the picture tube in a receiver.

kinescoping *(Image Tech.)*. US term for film recording of a TV programme. Also *telerecording*.

kinesin *(Biol.)*. A protein of wide distribution in eukaryotes which is responsible for the movement of organelles to which it is attached along microtubules.

kinesis *(Behav.)*. A simple response to environmental stimuli in which the animal's response is proportional to the intensity of stimulation; it involves a change in speed of movement or rate of turning. Unlike a **taxis**, however, the animal's body is not oriented to the stimulus although the effected movements often produce a change of position relative to it. Cf. **taxis**.

kinetic energy *(Phys.)*. Energy arising from motion. For a particle of mass m moving with a velocity v it is $\frac{1}{2}mv^2$, and for a body of mass M, moment of inertia I_g, velocity of centre of gravity v_g and angular velocity ω, it is

$$\tfrac{1}{2}Mv_g^2 + \tfrac{1}{2}I_g\omega^2.$$

kinetic friction *(Phys.)*. See **friction**.

kinetic heating *(Aero.)*. See **dynamic heating**.

kinetic pressure *(Aero.)*. See **dynamic pressure**.

kinetics *(Med.)*. The study of the rates at which chemical reactions and biological processes proceed.

kinetic theory of gases *(Phys.)*. A theory which accounts for the bulk properties of gases in terms of the motion of the molecules of the gas. In its simplest form the gas molecules are conceived as elastic spheres whose bombardment of the walls of the containing vessel causes the pressure exerted by the gas. If it is assumed that the size of the molecules is small compared with their mean spacing and that the molecules do not exert forces on each other except on collision, then the theory gives a simple explanation of the **gas laws** and yields useful results concerning gaseous viscosity and thermal conductivity.

kinetin *(Bot.)*. *6-furfurylaminopurine*. Synthetic plant growth regulator of the cytokinin type.

kinetochore *(Biol.)*. Paired structures within the **centromeric** region of metaphase chromosomes, to which spindle **microtubules** attach. They lie on each side of the **primary constriction**, and when viewed with the electron microscope appear as a trilaminar plate with microtubules entering at regular intervals.

kinetodesma *(Zool.)*. See **kinety**.

kinetosome *(Bot.)*. See **basal body**.

kinetosomes *(Zool.)*. See **kinety**.

kinety *(Zool.)*. A unit of structure in the Protozoa comprising the *kinetosomes* (the basal granules of the cilia and flagella) and the *kinetodesma* (a fine strand running from the kinetosomes). In Flagellata the line of cell division is parallel to the *kinetia* (symmetrical division), and in Ciliata the plane of cleavage cuts across the *kinetia* (percentien division).

king closer *(Build.)*. A three-quarter brick used to maintain the bond of the surface.

kingdom *(Bot.)*. Higher taxonomic rank; composed of number of **divisions**.

king pile *(Civ.Eng.)*. A pile driven down the centre of a wide trench to enable two short struts (butting on opposite sides of the pile) to be used, instead of one long one, for keeping the poling boards of opposite sides of the trench in position.

king pin *(Autos.)*. The pin by which a stub axle is articulated to an axle-beam or steering head; it is inclined to the vertical to provide caster action. Also called *swivel-pin*. For light vehicles, the king pin is now usually replaced by a pair of ball joints.

king post *(Build.)*. A vertical timber tie connecting the ridge and the tie-beam of a roof, shaped at its lower end to afford bearing to two struts supporting the middle points of the rafters. Also *broach-post*, *joggle-piece*, *joggle-post*, *king-piece*, *middle-post*.

king rod *(Build.)*. A vertical steel rod connecting the ridge and tie-beam of a couple-close roof, to prevent sagging of the tie-beam when it is required to support ceiling loads.

king's evil *(Med.)*. An old name for **scrofula**.

Kingston valve *(Eng.)*. A sea-valve fitted to a ship's side for the purpose of admitting water to circulating pumps, or flooding or blowing out ballast tanks.

kingswood *(For.)*. See **Brazilian kingswood**.

king tower *(Build.)*. In a derrick tower gantry, that one of the three timber towers through which the weight of the derrick itself is transmitted directly to the foundation below.

kinin *(Bot.)*. See **cytokinin**.

kinins *(Med.)*. A class of vasoactive peptides that are associated with local regulation of blood flow, e.g. bradykinin.

kink *(Electronics)*. An abrupt change or reversal in the slope of a *characteristic curve*, e.g. the change from forward to reverse bias in a semiconductor diode, the sudden increase of reverse current at a certain voltage in a Zener diode, or the region where a negative-resistance device shows increasing current with decreasing applied voltage.

kink instability *(Phys.)*. See **instability**.

kino *(For.)*. A red resinous exudate, rich in tannins, from various trees, notably *Pterocarpus spp.* Used in tanning. Also *gum kino*.

kin selection *(Ecol.)*. Natural selection for behaviour that lowers an individual's own chance of survival but raises that of a relative.

Kipp's apparatus *(Chem.)*. A generator for hydrogen sulphide, carbon dioxide, or other gases, by reacting a solid with a liquid. By opening or closing the gas outlet, gas pressure automatically fills or empties the solids container with liquid.

Kirchhoff's diffraction theory *(Phys.)*. A mathematical description of diffraction based on *Huygens' principle*.

Kirchhoff's equation *(Phys.)*. The rate of change of the heat ΔH of a process with temperature, carried out at constant pressure, is given by:

$$\left(\frac{\partial \Delta H}{\partial T}\right)_p = \Delta\left(\frac{\partial H}{\partial T}\right)_p = \Delta C_p,$$

where ΔC_p is the change in the heat capacity at constant pressure for the same process. Similarly, for a process carried out at constant volume:

$$\left(\frac{\partial \Delta U}{\partial T}\right)_V = \Delta\left(\frac{\partial U}{\partial T}\right)_V = \Delta C_V.$$

Kirchhoff's law *(Phys.)*. The ratio of the coefficient of absorption to the coefficient of emission is the same for all substances and depends only on the temperature. The law holds for the total emission and also for the emission of any particular frequency.

Kirchhoff's laws *(Elec.Eng.)*. Generalized extensions of *Ohm's law* employed in network analysis. They may be summarized as: (1) $\Sigma i = 0$ at any junction (2) $\Sigma E = \Sigma iZ$ round any closed path. E = e.m.f., i = current, Z = complex impedance.

Kirkendall effect *(Eng.)*. If a piece of a pure metal is placed in contact with a piece of an alloy of that metal and the whole heated, the region of alloy will diffuse into the pure metal.

kish *(Eng.)*. Solid graphite which has separated from, and floats on the top of, a molten bath of cast-iron or pig-iron which is high in carbon.

kiss impression *(Print.)*. The lightest possible impression on paper so that the type or blocks do not press into the paper, requiring careful **make-ready**; also required when printing rubber or plastic plates to avoid distortion.

kiss of life *(Med.)*. The 'mouth-to-mouth' method of resuscitation in which the rescuer places his mouth over the patient's and inflates the latter's lungs by breathing into them.

kitchen midden *(Geol.)*. The dump of waste material, largely shells and bones associated with ashes, marking the site of a kitchen in a settlement of early man.

kite *(Aero.)*. Any aerodyne anchored or towed by a line, not mechanically or power-driven. Derives its lift from the aerodynamic forces of the relative wind.

kite winder *(Build.)*. A **winder** used at the angle of a change of direction in a stair, and having consequently the shape of a kite in plan.

Kjeldahl flask *(Chem.)*. A glass flask with a round bottom and a long wide neck. Used in Kjeldahl's method for the estimation of nitrogen.

Kjeldahl's method *(Chem.)*. A method for the quantitative estimation of nitrogen in organic compounds, based on the conversion of the organic nitrogen into ammonium sulphate, and subsequent distillation of ammonia after the solution has been made alkaline. The ammonia which distils over can be titrated.

Kjellin furnace *(Elec.Eng.)*. A form of electric induction furnace having an iron core; used for melting ferrous metals.

Klein bottle *(Maths.)*. A one-sided surface obtained by pulling the narrow end of a tapering cylinder through the wall of the cylinder and then stretching the narrow end and joining it to the larger end. Cf. **Möbius strip**.

Kleine-Levin syndrome *(Med.)*. A rare disorder of young men with somnolence, excessive eating and perverted sexual behaviour.

Klein-Gordon equation *(Phys.)*. That describing the motion of a spinless charged particle in an electromagnetic field.

Klein-Nishina formula *(Phys.)*. Theoretical expression for the cross-section of free electrons for the scattering of photons. See **Compton effect**.

kleptomania *(Med.)*. Morbid condition where there is an uncontrollable desire to steal.

KLH *(Immun.)*. Abbrev. for *Keyhole Limpet Haemocyanin*.

Klieg lights *(Image Tech.)*. Large and extremely powerful light sources necessary in the early days of studio cinematography; prolonged exposure to their high intensity and U-V radiation caused inflammation known as *Klieg eyes*.

Klinefelter's syndrome *(Med.)*. Impaired gonadal development in male with one or more extra X chromosomes, it results in underdeveloped testes, mixed (male and female) secondary sex characteristics, sterility and sometimes, mild mental retardation.

K-lines *(Phys.)*. Characteristic X-ray frequencies from the atoms due to excitation of electrons in the K-shell. Denoted by K_α, K_β, K_γ; the lines are all doublets.

klinker brick *(Build.)*. A very hard type of brick much used in Holland and Germany, principally for paving purposes.

klinostat, clinostat *(Bot.)*. An apparatus to rotate a plant, e.g. slowly about a horizontal axis in order to cancel out at least some of the effects of gravity in investigations of gravitropism etc.

klippe *(Geol.)*. An erosional remnant of a thrust sheet. It is essentially an outlier lying on a thrust plane. pl. *klippen*.

Klippel-Feil syndrome *(Med.)*. A congenital shortening of the neck that may be associated with **syringomyelia**.

K- ,L- ,M- , ... shells *(Phys.)*. Imaginary spherical shells surrounding the nucleus of a many-electron atom in which groups of electrons are arranged according to the *Pauli exclusion principle*. Starting with the innermost shell the shells are called K, L, M, ...Q corresponding to the principal quantum numbers n of 1,2,3,....7. Each shell contains $2n^2$ electrons. See entries on individual shells, **K-shell**, etc. See also **atomic structure**.

klystron *(Electronics)*. General name for class of UHF and microwave electron tubes (amplifiers, oscillators, frequency multipliers, cascade amplifiers, etc.), in which electrons in a stream have their velocities varied (velocity modulation) by a high-frequency field; and subsequently impart energy to it or to other high-frequency fields, contained in a resonator. See **rhumbatron, reflex klystron**.

klystron frequency multiplier *(Electronics)*. A two-cavity klystron whose output cavity is tuned to an intergral multiple of the fundamental frequency.

knapping hammer *(Build.)*. A hammer used for breaking and shaping stones and flints.

knapsack abutments *(Civ.Eng.)*. Abutments provided with an additional integral slab behind them to reduce the overturning pressure.

knebelite *(Min.)*. A silicate of manganese and ferrous iron, in the olivine series. It crystallizes in the orthorhombic system and occurs as the result of metamorphism and metasomatism.

knee *(Build.)*. A sudden rise in a handrail when it is convex upwards. Cf. **ramp**.

knee *(Build.,Eng.)*. An elbow pipe (see **elbow**).

knee *(Electronics)*. The region of maximum curvature on a characteristic curve.

knee *(Zool.)*. In land Vertebrates, the joint between the femur and the crus.

knee brace *(Build.)*. A stiffening member fixed across the inside of an angle in a framework, particularly at the angle between roof and wall in a building frame.

knee gall *(Vet.)*. *Knee throughpin*. A distention of the carpal synovial sheath at the back of the carpal joint of the horse.

knee jerk *(Med.)*. A normal reflex extension of the leg at the knee joint, obtained by tapping the tendon below the patella.

kneeler *(Build.)*. The return of the dripstone at the spring of an arch.

knee rafter *(Build.)*. A rafter having its lower end bent downwards.

knee roof *(Arch.)*. A **mansard roof**.

knee thoroughpin *(Vet.)*. See **knee-gall**.

knib *(Print.)*. The projecting portion of a setting rule, by which the compositor lifts it out of the composing stick.

knickpoint *(Geol.)*. A change of slope in the longitudinal profile of a stream as a result of **rejuvenation**.

knife cheeks *(Print.)*. Spring-loaded gripping edges to hold the web while it is being severed.

knife edge *(Eng.)*. Suppport for a balance beam or similar instrument member, usually in the form of a hardened steel wedge, the apex of which gives line support.

knife fold *(Print.)*. A fold made by pushing the paper into the folding rollers with a blunt knife, a necessary arrangement when the fold is at right angles to the leading edge of the paper, or to the previous fold. Cf. **buckle-fold**.

knife switch *(Elec.Eng.)*. An electric circuit switch in which the moving element consists of a flat blade which engages with fixed contacts.

knife tool *(Eng.)*. A lathe tool having a straight lateral cutting edge, used for turning right up to a shoulder or corner.

Knight shift *(Phys.)*. Shift in nuclear magnetic resonance frequency in metals from that of the same isotope in chemical compounds in the same magnetic field. It is due to the paramagnetism of the conduction electrons.

knitting *(Textiles)*. The rapid process of making a fabric from yarn by the formation of intermeshing loops. See **warp-knitting** and **weft-knitting**.

knitting machine *(Textiles)*. A machine for making fabrics by **warp-knitting** or **weft-knitting**.

knitwear *(Textiles)*. A general term used for all knitted outer garments except stockings and socks.

knobbing *(Build.)*. The operation of breaking projecting pieces off stones in the quarry.

knock *(Eng.)*. See **knocking**.

knocked up toe of dogs *(Vet.)*. Synovitis and fibrositis affecting the first or second interphalangeal joint of the toe of the dog; caused by sprain and usually met with in racing greyhounds.

knocker-out *(Eng.)*. The horns of a planing machine against which the tappets strike to reverse the motion of the table.

knocking-down iron *(Print.)*. A piece of iron, fixed to the lying press, on which joints are beaten out and lacings flattened.

knocking, knock *(Autos.)*. The characteristic metallic noise, often called 'pinking', resulting from **detonation** in a petrol engine, or 'diesel knock', due to the rapid pressure rise during combustion in a diesel engine.

knocking, knock *(Eng.)*. A periodic noise made by a worn bearing in a reciprocating engine, due to reversal of the load on the shaft or pin.

knockings *(Build.)*. The stone pieces, smaller than spalls, knocked off in the process of chiselling or hammering.

knockings *(Min.Ext.)*. See **riddle**.

knock knee *(Med.)*. See **genu valgum**.

knock rating *(Autos.)*. The measurement of the **anti-knock value** of a volatile liquid fuel.

knoll *(Geol.)*. See **reef knolls**.

knot *(Aero.,Ships)*. Speed of 1 nautical mph (1.15 mph or 1.85 km/h) used in navigation and meteorology.

knot *(Bot.)*. (1) A node in a grass stem. (2) A hard and often resinous inclusion in timber, formed from the base of a branch which became buried in secondary wood as the trunk thickened.

knotter *(Paper)*. An appliance for the removal of knots, undigested particles or similar unwanted matter from paper pulp.

knotting *(Build.)*. A solution of shellac in spirit used for covering knots in wood, prior to painting, to prevent exudation of resin.

knowledge-based systems *(Comp.)*. Software systems which aim to store and effectively utilize large amounts of specialist knowledge.

knuckle *(Build.)*. The parts of a hinge receiving the pin.

knuckle joint *(Eng.)*. A hinged joint between 2 rods, in which the ends are formed into an eye and a fork respectively and united by a pin.

knuckle-joint press *(Eng.)*. A heavy power press incorporating a knuckle-joint mechanism for actuating the slide, It has a high mechanical advantage near the bottom end of its stroke, which is relatively short, and is therefore used for coining, sizing, and embossing work.

knuckle pins *(Autos.)*. See **wrist pins**.

knuckling *(Vet.)*. A state of abnormal flexion of the fetlock joint of the horse, due to a variety of causes, and associated with an inability or lack of desire to extend the joint fully.

Knudsen flow *(Chem.Eng.)*. The flow of low pressure gas through a tube whose diameter is much smaller than the mean free path of the molecules of the gas. Under these conditions, the macroscopic concept of viscosity needs to be modified since the resistance to motion is due primarily to molecular collisions with the passage walls. Also *molecular streaming*.

knurling tools *(Eng.)*. Small, hard steel rollers, serrated or cross-hatched on their peripheries, mounted on a pin carried by a holder. They are pressed against circular work in the lathe, to knurl or roughen a surface required to give a grip to the fingers, as the head of a 'milled' screw.

kobellite *(Min.)*. A lead, bismuth, antimony sulphide, crystallizing in the orthorhombic system.

Koch resistance *(Chem.)*. The resistance of a vacuum photocell or phototube when its active surface is irradiated with light.

Koch's postulates *(Med.)*. Four criteria laid down by Koch to show that a disease was caused by a micro-organism: (1) organism must be observed in all cases of the disease; (2) it must be isolated and grown in pure culture; (3) the culture must be capable of reproducing the disease when inoculated into a suitable experimental animal; (4) the organism must be recovered from the experimental disease.

Kodak relief plate *(Print.)*. TN for an easily processed relief plate suitable for either direct or 'dry offset' printing with considerable wearing capacity.

Kohler's disease *(Med.)*. Aseptic necrosis of the navicular bone in the tarsus; a self-limiting disease.

Kohlrausch's law *(Chem.)*. The contribution from each ion of an electrolytic solution to the total electrical conductance is independent of the nature of the other ion.

koilonychia *(Med.)*. Spoon-shaped depression of the finger nails.

konimeter *(Powder Tech.)*. Type of impaction sampler in which dirty air, sucked through a round orifice, impinges upon a circular glass plate smeared with grease, which collects the particles. In mining, for example, it is used to check dust content of mine atmosphere.

Kooman's array *(Telecomm.)*. Original name for the category of **broadside antennae** consisting of rows and columns of dipoles. The rows are half-wavelength apart and fed by twin-wire feeders, twisted between each row. This ensures that the dipoles are all fed in the same phase.

Koplik's spots *(Med.)*. Bluish-white specks on the inner side of the cheek, occurring in measles 2 to 3 days before the rash appears.

Köppen classification *(Meteor.)*. A system of classifying climate by groups of letters. The general climatic type is shown by a letter or letter-pair in capitals, and further detail is given by additional lower-case letters. E.g. the climate of the British Isles is coded as Cf where C indicates 'warm or temperate rainy' and f is 'moist, no marked dry season'.

Kopp's law *(Chem.)*. See **Joule's law**.

Korndorfer starter *(Elec.Eng.)*. A variant of the **auto-transformer starter** for 3-phase induction motors which involves no interruption of the supply of power after the starting cycle has been initiated. Used with high-powered and high-voltage machines.

kornerupine *(Min.)*. A magnesium aluminium borosilicate, crystallizing in the orthorhombic system. Rare, but collected, and cut as a gemstone.

Korotkoff sounds *(Med.)*. Sounds heard on auscultation over the brachial artery when a pneumatic cuff is deflated. Korotkoff's first sound corresponds to **systolic** blood pressure and his fifth sound when the sounds disappear is taken as the **diastolic** blood pressure.

Korsakoff's syndrome, Korsakoff's psychosis *(Med.)*. Mental confusion associated with confabulation and disorientation, usually due to chronic alcoholism.

Korsakoff's syndrome, psychosis *(Behav.)*. An irreversible nutritional deficiency, caused by chronic alcoholism and related malnutrition; characterised by severe **anterograde amnesia** and **confabulation**.

Kovar *(Eng.)*. TN for an alloy of Ni-Co-Fe for glass-to-metal seals over working ranges of temperature, when temperature coefficients of expansion coincide.

Kozeny-Carman equation *(Powder Tech.)*. A permeability equation used to calculate the surface area of a powder from the permeability of a packed powder bed to the flow of fluid through it.

Kr *(Chem.)*. The symbol for *krypton*.

Kraemer-Sarnow test *(Build.,Civ.Eng.)*. A test for the determination of the melting point of a bitumen for use in building or roadmaking. The melting point is obtained as the temperature at which superincumbent mercury falls through a standard sample of the bitumen, heated under standard conditions.

Kramer control *(Elec.Eng.)*. A form of speed and power factor control for large induction motors in which the slip energy is supplied to a rotary converter, the resulting d.c. power being used to drive a motor either mounted on the main motor shaft or driving an alternator and returning power to the supply.

Krantz anatomy *(Bot.)*. The leaf anatomy of most C_4 plants, characterised by the bundle sheath cells being large and having conspicuous chloroplasts. The mesophyll cells have less conspicuous choloroplasts.

kraton *(Geol.)*. See **craton**.

kraurosis *(Med.)*. Atrophy of the vulva, associated with narrowing of the vaginal orifice.

Kroll's process *(Eng.)*. Reduction to metal of tetrachloride of titanium or zirconium *in vacuo* or by reaction with magnesium in a neutral atmosphere.

Kronecker delta *(Maths.)*. A function δ_{ij} of two variables i and j, defined to have the value zero unless $i=j$ when its value is unity.

Kronig-Penny model *(Phys.)*. A relatively simple model for a one-dimensional lattice from which the essential features of the behaviour of electrons in a periodic potential may be illustrated. See **band theory of solids**.

Kruger flap *(Aero.)*. Leading edge flap which is hinged at its top edge so it normally lies flush with the underwing surface and swings down and forward to increase circulation and lift for low speed flight.

Krukenberg's tumour *(Med.)*. A solid tumour appearing in each ovary and believed to be always secondary to cancer of the stomach.

Krummholz *(Bot.)*. Stunted, wind-trimmed trees between the *timber-line* and the *tree-line* on mountains.

krypton *(Chem.)*. A zero valent element, one of the noble gases. Symbol Kr, at.no. 36, r.a.m. 83.80, mp −169°C, bp −151.7°C, density 3.743 g/dm^3 at s.t.p. It is a colourless and odourless monatomic gas, and constitutes about 1-millionth by volume of the atmosphere, from which it is obtained by liquefaction. It is used in certain gas-filled electric lamps. It forms few compounds, e.g. KrF_4.

K-shell *(Phys.)*. The innermost **electron shell** in an atom corresponding to a principal quantum number of unity. The shell can contain two electrons.

k-space *(Phys.)*. Symbol for *momentum space* or *wave vector space*. This is an important concept in semiconductor energy band theory. See **Brillouin zone**.

K-strategist *(Bot.)*. Organism which assigns relatively little of its resources to reproduction; characteristic of stable, saturated communities; has low fecundity and high competitive ability. Cf. **R-strategist**.

Kuchemann tip *(Aero.)*. Low drag wingtip shape developed by Kuchemann at RAE for transonic aircraft. It has a large radius curve in plan finishing with a corner at the trailing edge.

kulaite *(Geol.)*. An amphibole-bearing nepheline basalt.

Kümmell's disease *(Med.)*. Delayed collapse or crumbling of one or more spinal vertebrae after injury to the spine.

Kummer's convergence test *(Maths.)*. If, for a divergent series of positive terms $\Sigma(v_n)^{-1}$ and a series of positive terms Σu_n, the expression $v_n \cdot \dfrac{u_n}{u_{n+1}} - v_{n+1} \to l$, then the series Σu_n converges if $l>0$, and diverges if $l<0$.

Kundt constant *(Phys.)*. The ratio of **Verdet's constant** to magnetic susceptibility.

Kundt's rule *(Phys.)*. The refractive index of a medium on the shorter wavelength side of an absorption band is abnormally low, and on the other side abnormally high. This gives rise to **anomalous dispersion**.

Kundt's tube *(Acous.)*. Transparent tube in which standing sound waves are established, indicated by lycopodium powder, which accumulates at the nodes. Used for measuring sound velocities. Also see **acoustic streaming**.

kunzite *(Min.)*. See **spodumene**.

kupfernickel *(Min.)*. See **niccolite**.

Kupffer cell *(Immun.)*. The form of macrophages which line the blood sinuses of the liver. They are phagocytic and remove foreign particles from the blood. They have Fc receptors and can engulf blood cells which have become coated with antibody.

Kurie plot *(Phys.)*. One used for determining the energy limit of a β-ray spectrum from the intercept of a straight line graph. Prepared by plotting a function of the observed intensity against energy, the intercept on the axis being the energy limit for the spectrum. Also *Fermi plot*.

kurtosis *(Stats.)*. The degree to which a distribution is sharply peaked at its centre.

kuru *(Med.)*. A rapidly fatal brain degeneration formerly found in primitive people in New Guinea. It is thought to have been transmitted by the cannibalistic custom of eating human brain, thus transmitting a **slow virus**. Of interest because a slow virus is also responsible for **Creuzfeldt-Jacob disease**.

Kussmaul breathing *(Med.)*. Deep sighing breathing often known as air-hunger seen in severe metabolic **acidosis**.

kV *(Phys.)*. Abbrev. for *kilovolt*.

kVA *(Elec.Eng.)*. Abbrev. for *kilovolt-ampere*.

kVAr *(Elec.Eng.)*. Abbrev. for *kilovar*.

kVp *(Radiol.)*. Abbrev. for *kiloVolts, peak*. Voltage applied across an X-ray tube, hence designating the maximum energy of emitted X-ray photons.

kW *(Elec.Eng.)*. Abbrev. for *kilowatt*.

kwashiorkor *(Med.)*. A disease of the tropics due to protein and calorie deficiency in children. There is gross loss of weight, oedema, and pigmentation. Also known as *protein-energy malnutrition*.

kWh *(Elec.Eng.)*. Abbrev. for *kilowatt-hour*.

kyanite, cyanite *(Min.)*. A silicate of aluminium which crystallizes in the triclinic system. It usually occurs as long bladed crystals, blue in colour, in metamorphic rocks. See also **disthene**.

kyanizing *(Build.)*. The process of impregnating timber with a solution of corrosive sublimate as a preservative. Also called *cyanizing*.

kylite *(Geol.)*. An olivine rich theralite.

kymography *(Radiol.)*. A method of recording in a single radiograph the excursions of moving organs in the body, by use of a *kymograph*, which records physiological muscular waves, pulse beats or respirations.

kyphoscoliosis *(Med.)*. A deformity of the spine in which dorsal convexity is increased, the spine being also bent laterally.

kyphosis *(Med.)*. A deformity of the spine in which the dorsal convexity is increased.

L

l *(Chem.)*. An abbrev. for *litre*. (1 dm^3 is now used officially).

l *(Chem.)*. (1) A symbol for: (a) specific latent heat per gramme; (b) mean free path of molecules; (c) (with subcript) equivalent ionic conductance, 'mobility'. (2) An abbrev. for *Laevorotatory*.

λ *(Phys.)*. Symbol for (1) wavelength, (2) linear coefficient of thermal expansion, (3) mean free path, (4) radioactive decay constant, (5) thermal conductivity.

L *(Chem.)*. Symbol for *molar latent heat*.

L *(Phys.)*. Symbol for (1) angular momentum, (2) luminance, (3) self inductance.

Λ *(Chem.)*. A symbol for *molar conductance*; Λ_0, at infinite dilution.

La *(Chem.)*. The symbol for *lanthanum*.

label *(Arch.)*. A projecting moulding above a door or window opening.

label-corbel table *(Build.)*. A **dripstone** supported by a corbel.

labelled atom *(Min.Ext.)*. One which has been made radioactive, in a compound introduced into a flowline in order to trace progress.

labelled atom *(Phys.)*. That atomic position in a molecular formula which is occupied by an isotopic tracer. See **radioactive tracer**.

labelled molecules *(Phys.)*. Molecules containing a radioactive isotopic tracer.

labelling theory *(Behav.)*. The view that no actions are inherently deviant or abnormal, and that the label *mental illness* applied to certain behaviours and to individuals, acts as a self-fulfilling prophecy; once applied, the individual's self expectations, and the expectations of others result in behaviour associated with the label, and so perpetuate the condition.

labellum *(Bot.)*. (1) The posterior petal of the flower of an orchid and often the most conspicuous part of the flower. (2) The lower lip of the corolla forming a landing platform for pollinating insects in a number of families, e.g. Labiatae.

labellum *(Zool.)*. A spoon shaped lobe at the apex of the glassa in Bees; (pl.) in certain Diptera, a pair of fleshy lobes into which the proboscis (**labium**) is expanded distally. pl. *labella*.

labia *(Med.)*. *Lips*. Usually applied to lesser and greater pudendal lips of female external genitalia.

labia *(Zool.)*. Any structures resembling lips; as the lips of the vulva in Primates. See also the *sing*. form *labium*. adj. *labial*.

labia majora *(Zool.)*. In Mammals, the two prominent folds which form the outer lips of the vulva.

labia minora *(Zool.)*. In Mammals, the small inner lips of the vulva; nymphae.

labiate *(Bot.)*. (1) With a lip or lips, as the corolla of various Labiatae. (2) A species of the Labiatae.

labile *(Chem.)*. Unstable, liable to change. Usually a kinetic, not a thermodynamic term.

labile *(Electronics)*. Said of an oscillator which can be synchronized from a remote source.

labioglossopharyngeal *(Med.)*. Pertaining to the lips, tongue and pharynx.

labium *(Zool.)*. In Insects, the lower lip, formed by the fusion of the second maxillae; in the shells of Gastropoda, the inner or columellar lip of the margin of the aperture. See also the pl. form *labia*.

laboratory sand-bath *(Phys.)*. Gas heated steel bath containing sand in which crucibles or other vessels are indirectly heated.

labradorescence *(Min.)*. A brilliant play of colours shown by some labradorite feldspars. Probably due to a fine intergrowth of two phases.

labradorite *(Min.)*. A plagioclase feldspar containing 50 to 70% of the anorthite molecule; occurs in basic igneous rocks; characterized by a beautiful play of colours in some specimens, due to schiller structure.

La Brea Sandstone *(Geol.)*. A sandstone which succeeds the oil bearing sands of Miocene age in Trinidad; it contains the well known Pitch Lake, 104 acres (42 ha).

labrum *(Zool.)*. In Insects and Crustacea, the platelike upper lip; in the shells of Gastropoda, the outer lip or right side of the margin of the aperture. adj. *labral*.

labyrinth *(Acous.)*. (1) Folded path for loading the rear of loudspeaker diaphragms and microphone ribbons. (2) Inner ear containing cochlea and semi-circular canals.

labyrinth *(Nuc.Eng.)*. See **radiation trap (2)**

labyrinth *(Zool.)*. Any convolute tubular structure; especially, the bony tubular cavity of the internal ear in Vertebrates, or the membranous tube lying within it.

labyrinthitis *(Med.)*. Inflammation of the labyrinth of the ear.

labyrinthodont *(Zool.)*. Having the dentine of the teeth folded in a complex manner.

labyrinth packing *(Eng.)*. A type of gland, with radial or axial clearance, used, e.g. in steam engines and turbines.

labyrithectomy *(Med.)*. Surgical removal of the labyrinth of the ear.

lac *(Zool.)*. A resinous substance, an excretion product of certain Coccid insects (Gascardia, Tachardia etc.) in certain jungle trees; used in the manufacture of shellac. Chief source, India. See **shellac**, for which the name is frequently loosely used.

laccolith *(Geol.)*. A concordant intrusion with domed top and flat base, the magma having been instrumental in causing the up-arching of the 'roof'. Cf. **phacolith**.

lace *(Textiles)*. A fine open-work decorative fabric comprising a ground net on which patterns are formed by looping, twisting, or knitting. The process may be done by hand using a needle or bobbin or by machine.

laced valley *(Build.)*. A valley formed in a tiled roof by interlacing tile-and-a-half tiles across a valley board.

lace machine *(Textiles)*. There are several different machines developed for the manufacture of lace but particularly important are the Leavers machines. In this the threads are in narrow brass bobbins mounted in carriages and the pattern threads and the warp threads are manipulated by guide bars controlled by a jacquard to produce the desired pattern.

lachrymal *(Zool.)*. See **lacrimal**.

lacing course *(Build.)*. A brickwork bond-course built into rubble or flint walls.

lacing-in *(Print.)*. The operation of attaching the boards to a book by lacing the ends of the bands through holes made in the boards to receive them.

laciniate *(Bot.)*. Deeply cut into narrow segments; fringed.

lacquer *(Build.)*. A solution of film-forming substances in volatile solvents, e.g. a spirit lacquer or varnish consists of shellac or other gums dissolved in methylated spirit. Lacquers may be pigmented or clear, and dry by the evaporation of solvents only, not by oxidation. (In US lacquers include those drying by oxidation.) The most important lacquers are the **cellulose lacquers**.

lacquering *(Image Tech.)*. Coating processed motion picture film with a transparent protective layer to minimize scratches and abrasions in subsequent handling.

lacrimal *(Zool.)*. Pertaining to, or situated near, the tear gland; in some Vertebrates, a paired lateral membrane bone of the orbital region of the skull, in close proximity to the tear gland.

lacrimal duct *(Zool.)*. In most of the higher Vertebrates, a duct leading from the inner angle of the eye into the cavity of the nose; it serves to drain off the secretion of the lacrimal gland from the surface of the eye.

lacrimal gland *(Zool.)*. In most of the higher Vertebrates, a gland situated at the outer angle of the eye; it secretes a watery substance which washes the surface of the eye and keeps it free from dust.

lacrimation *(Med.)*. Shedding of tears.

lactation *(Zool.)*. The production of milk from the mammary glands of Mammalia.

lactation tetany *(Vet.)*. See **bovine hypomagnesaemia**.

lacteals *(Zool.)*. In Vertebrates, lymphatics, in the region of the alimentary canal, in which the lymph has a milky appearance, due to minute fat globules in suspension.

lactic *(Zool.)*. Pertaining to milk.

lactic acid *(Med.)*. The dextrorotatory isomer D-lactic acid is produced by anaerobic metabolism of glucose. Pathological accumulations of lactic acid occur where there is poor tissue perfusion or where the liver is unable to metabolise lactic acid.

lactic acids *(Chem.)*. $CH_3.CH(OH).COOH$ and $CH_2(OH).CH_2.COOH$, i.e. *hydroxy-propanoic acids*. There are two isomers, viz., 2-hydroxy-propionic acid, and 3-hydroxy-propionic acid; only the former, known as *lactic acid* or *fermentation lactic acid*, is of importance. Lactic acid is a synonym for racemic 2-hydroxy-propanoic acid, $CH_3.CH(OH).COOH$, hygroscopic crystals, mp18°C, bp83°C at 1 mm, rel.d. 1.248. It is produced commercially by the lactic fermentation of sugars with *Bacillus acidi lactici* or by heating glucose with caustic potash solution of a certain concentration.

lactiferous *(Bot.)*. Containing latex.

lactiferous *(Zool.)*. Milk-producing, milk-carrying.

Lactobacillaceae *(Biol.)*. A family of bacteria belonging to the order *Eubacteriales*. Gram-positive cocci or rods; carbohydrate fermenters are anaerobic or grow best at low oxygen tensions. Includes several pathogenic organisms, e.g. *Streptococcus pyogenes* (scarlet fever, tonsillitis, erysipelas, puerperal fever), *Diplococcus pneumoniae* (one cause of pnuemonia).

lactogenic hormone *(Med.)*. See **prolactin**.

lactones *(Chem.)*. Intramolecular esters of hydroxycarboxylic acids. Most lactones have 5 or 6 membered rings (γ- or δ-lactones) but others are known.

lactose *(Chem.)*. $C_{12}H_{22}O_{11}+H_2O$, rhombic prisms which become anhydrous at 140°C, mp 201°–202°C with decomposition; when hydrolysed it forms D-galactose and D-glucose. It reduces Fehling's solution and shows **mutarotation**. It occurs in milk and is not so sweet as glucose.

lactosuria *(Med.)*. The presence of lactose in the urine.

lacuna *(Bot.)*. A cavity, gap, space or depression such as a space in a tissue caused by the breakdown of the protoxylem, a **leaf gap** or sometimes an intercellular space.

lacuna *(Zool.)*. A small cavity or space, as one of the cell-containing cavities in bone. adjs. *lacunar, lacunose*.

lacustrine *(Geol.)*. Related to a **lake**.

LAD *(Behav.)*. Abbrev. for *Language Acquisition Device*.

ladder *(Textiles)*. In a knitted structure, especially stockings, a defect caused by the breaking of stitches resulting in the thread reverting in long runs in the wale direction to its original linear form.

ladder network *(Telecomm.)*. One which comprises a number of filter sections, all alike, in series, acting as a transmission line, with attenuation and delay properties.

ladies *(Build.)*. Slates 16 × 8 in. (406 × 203 mm.)

ladle *(Eng.)*. An open-topped vessel lined with refractory material; used for conveying molten metal from the furnace to the mould or from one furnace to another.

ladle addition *(Eng.)*. Addition of alloying metal (e.g. nickel shot) to molten iron in ladle before casting, to give special quality to product.

Laënnec's cirrhosis *(Med.)*. Term formerly used for multilobular cirrhosis of the liver, in which degeneration of liver cells is associated with areas of fibrosis enclosing many regenerating lobules of the liver. See **cirrhosis**.

laevo- *(Genrl.)*. Prefix from L. *laevus*, left.

laevorotatory *(Phys.)*. Said of an optically active substance which rotates the plane of polarization in an anticlockwise direction when looking against the oncoming light.

laevulose *(Chem.)*. $C_6H_{12}O_6$. Fruit sugar or d(—)-*fructose*.

lag *(Min.Ext.)*. To protect a shaft or level from falling rock by lining it with timber (*lagging*).

lag *(Telecomm.)*. Any time delay between an initiating action and a desired effect; e.g. a trigger pulse to a circuit may be followed by a lag before the circuit operates, or the image on a TV camera tube may persist for a number of frames, impairing the depiction of rapid movement.

lagena *(Zool.)*. In higher Vertebrates, a pocket lined by sensory epithelium and developed from the posterior side of the sacculus, which becomes transformed in Mammals into the scala media or canal of the cochlea.

lagging *(Biol.)*. Slow movement towards the poles of the spindle by one or more chromosomes in a dividing nucleus, with the result that these chromosomes do not become incorporated into a daughter nucleus.

lagging *(Build.,Civ.Eng.)*. (1) The wooden boards nailed across the framework of a centre, to form the immediate supporting surface for the arch. (2) Material generally used in floors and roofs to eliminate sound and loss of heat. See **insulation**.

lagging *(Eng.)*. (1) The process of covering a vessel or pipe with a non-conducting material to prevent either loss or gain of heat. (2) The non-conducting material itself. See **insulation**.

lagging current *(Elec.Eng.)*. An alternating current which reaches its maximum value at a later instant in the cycle than the voltage which is producing it.

lagging load *(Elec.Eng.)*. See **inductive load**.

lagging phase *(Elec.Eng.)*. A term used in connection with measuring equipment on 3-phase circuits to denote a phase whose voltage is lagging behind that of one of the other phases by approximately 120°. Also used, particularly in connection with the 2-wattmeter method of 3-phase power measurement, to denote the phase in which the current at unity power factor lags behind the voltage applied to the meter in which that current is flowing.

Lagomorpha *(Zool.)*. An order of Eutheria with two pairs of upper, and one of lower, incisor teeth which grow throughout life; there are no canines, and there is a wide diastema between the incisors and the cheek teeth into which the cheeks can be tucked to separate the front part of the mouth from the hind during gnawing. Rabbits and Hares.

lagoon *(Geol.)*. A shallow stretch of sea water close to the sea but partly or completely separated from it by a low strip of land. Often associated with coral reefs.

lagopodous *(Zool.)*. Having feet covered by hairs or feathers.

Lagrange's dynamical equations *(Maths.)*. The equations

$$\frac{d}{dt}\left(\frac{\partial T}{\partial \dot{q}_i}\right)-\frac{\partial T}{\partial q_i}=Q_i, \quad i=1,2,3,\text{etc.},$$

where T is the kinetic energy of a system, q_i the various co-ordinates which define the position of the system and Q_i the so called generalized forces, that is, $Q_i q \delta q_i$ is the work done by the external forces during a small displacement δq_i. These equations are characterized by being independent of any reactions internal to the system.

Lagrangian point *(Astron.)*. A point in the orbital plane of two objects in orbit around their common centre of gravity where a third particle of negligible mass can remain in equilibrium.

Lagrangian points *(Space)*. The five points associated with a binary system (particularly Earth-Moon), where the combined gravitational forces are zero (also called *libration points*).

lahar *(Geol.)*. A mudflow of volcanoclastic material on the slopes of a volcano.

LAI *(Bot.)*. Abbrev. for *Leaf Area Index*.

laid-dry *(Build.)*. Said of bricks or blocks which have been laid without mortar.

laid-in moulding *(Build.)*. A moulding cut out of a separate strip of wood of the required section, and sunk in a special groove in the surface which it is intended to decorate.

laid-on moulding *(Build.)*. A **planted moulding**.

laid paper *(Paper)*. Writing or printing paper watermarked with a pattern of spaced parallel lines (chain lines) generally disposed in the machine direction and usually accompanied by more closely spaced parallel lines at right angles (laid lines).

laitance *(Build.)*. (1) The milky scum from grout or mortar, squeezed out when tesserae or tiles are pressed into place. (2) The milky scum formed on trowelled cement, concrete or rendering.

lake *(Geol.)*. A body of water lying on the surface of a continent, and unconnected (except indirectly by rivers) with the ocean. Lakes may be *fresh water lakes*, provided with an outlet to the sea: or *salt lakes*, occurring in the lowest parts of basins of inland drainage, with no connection with the sea. Lakes act as natural settling tanks, in which sediment carried down by rivers is deposited, containing the shells of molluscs, etc. The lakes of former geological periods may thus be recognized by the nature of the sediments deposited in them and the fossils they contain. Lakes occur plentifully in glaciated areas, occupying hollows scooped out by the ice, and depressions lying behind barriers of morainic material.

lakes *(Chem.)*. Pigments formed by the interaction of dyestuffs and 'bases' or 'carriers', which are generally metallic salts, oxides or hydroxides. The formation of insoluble lakes in fibres, which are being dyed, is known as *mordanting*, the hydroxides of aluminium, chromium, and iron generally being employed as *mordants*.

lalling, lallation *(Med.)*. Babbling speech of infants; lack of precision in the articulatory mechanism of the mouth.

Lamarckism *(Biol.)*. A theory, now discredited, that evolutionary change takes place by the *inheritance of acquired characters*; i.e. that characters acquired during the lifetime of an individual (e.g. an athlete's strong muscles) are transmitted to its offspring.

lambda chain *(Immun.)*. One of the two types of light chain of immunoglobulins, the other being the **kappa chain**.

lambda leak *(Phys.)*. The leakage of liquid helium II through holes so small that normal liquids cannot pass.

lambda-particle *(Phys.)*. Hyperon with hypercharge 0 and isotopic spin 1.

lambda phage *(Biol.)*. A well-studied temperate phage which can either grow in synchrony with its host (*E.coli*) in its **lysogenic** phase or go into a *lytic* phase, when its genome is replicated many times by a *rolling circle* mechanism. An important vector in genetic manipulation procedures.

lambda point *(Phys.)*. (1) Transition temperature of helium I to helium II. (2) Temperatures characteristic of second order phase change, e.g. ferromagnetic Curie point.

lamb dysentery *(Vet.)*. An acute and fatal toxaemic disease of newborn lambs caused by an intestinal infection by the bacterium *Clostridium perfringens (Cl. welchii)*, Type B. The main symptom is diarrhoea due to enteritis. Vaccination widely used.

lambert *(Phys.)*. Unit of luminance or surface brightness of a diffuse reflector emitting $1\,\text{lumen}\,\text{cm}^{-2}$ or $10^4/\pi\,\text{cd}\,\text{m}^{-2}$. The millilambert is used for low illuminations. See **lumen**.

Lambert's cosine law *(Phys.)*. The energy emitted from a perfectly diffusing surface in any direction is proportional to the cosine of the angle which that direction makes with the normal.

Lambert's law *(Phys.)*. The illumination of a surface on which the light falls normally from a point source is inversely proportional to the square of the distance of the surface from the source.

lambing sickness (ewe) *(Vet.)*. See **milk fever**.

Lamb shift *(Phys.)*. A small difference in the $2S\frac{1}{2}$ and $2P\frac{1}{2}$ energy levels in the hydrogen atom, which, according to the Dirac theory, should be the same. The effect is explained by the theory of *quantum electrodynamics*.

lambswool roller *(Build.)*. A tool consisting of handle, frame and sheepskin fabric sleeve used for application of paints. The sleeves are made in short, medium and long pile to suit the appropriate paint and surface.

lamé *(Textiles)*. Fabric with conspicuous decorative metallic threads.

Lamé formula *(Eng.)*. A formula for calculating the stresses in thick (hydraulic) cylinders under elastic deformation.

lamella *(Bot.)*. (1) A thin plate or layer of cells. (2) A gill of an agaric.

lamella *(Zool.)*. A structure resembling a thin plate. adjs. *lamellar, lamellate*.

lamellar magnetization *(Elec.Eng.)*. Magnetization of a sheet or plate distributed in such a way that the whole of the front of the sheet forms one pole and the whole of the back forms the other.

lamellibranch *(Zool.)*. Having platelike gills, as members of the group, Bivalvia.

lamell-, lamelli- *(Genrl.)*. Prefix from L. *lamella*, thin plate.

lamina *(Biol.)*. The network of lamin proteins which lie on the inner surface of the nuclear membrane.

lamina *(Bot.,Zool.)*. (1) Expanded blade-like part of a leaf, as distinct from petiole and leaf-base. (2) Any flattened platelike structure.

lamina *(Elec.Eng.)*. Thin sheet steel.

lamina *(Geol.)*. See **lamination**.

lamina propria *(Immun.)*. Layer of connective tissue which supports the epithelium of the digestive tract and with it forms the mucous membrane. The lamina propria is the site at which accumulate lymphocytes and plasma cells, mast cells and accessory cells in immunological reactions involving the gut.

laminar flow *(Phys.)*. A type of fluid flow in which adjacent layers do not mix except on the molecular scale. See **streamline flow**, **viscous flow**, **boundary layer**.

laminarin *(Bot.)*. The storage polysaccharide of the brown algae, a $\beta, 1\rightarrow 3$ glucan.

laminate *(Eng.)*. A structural member made from two or more components bonded together like two sheets separated by a honeycomb.

laminate *(Plastics)*. See **laminated plastics**.

laminated arch *(Build.)*. An arch formed of successive thicknesses of planking, which are bent into shape and secured together with fastenings or glue.

laminated bending *(Eng.)*. The practice of bending several layers of material and at the same time joining them along their surfaces in contact to form a unit.

laminated brush *(Elec.Eng.)*. A brush for an electric machine, made up of a number of layers insulated from one another, so that the resistance is greater across the brush than along its length (i.e. in the direction of normal current flow).

laminated-brush switch *(Elec.Eng.)*. A switch in which one or both of the contacts are laminated. See **laminated contact**.

laminated conductor *(Elec.Eng.)*. A conductor commonly used for armature windings of large machines or for heavy-current bus-bars; it is made up of a number of thin strips, in order to reduce eddy currents in the conductor or to make it more flexible.

laminated contact *(Elec.Eng.)*. A switch contact made up of a number of laminations arranged so that each lamination can be pressed into contact with the opposite surface, thereby giving a large area of contact and also a wiping action. Also called *brush contact*.

laminated core *(Elec.Eng.)*. A core built up of laminations; this is the usual type of core for an electric machine or transformer.

laminated glass. *(Glass)*. See **safety glass**.

laminated magnet *(Elec.Eng.)*. (1) A permanent magnet built up of magnetized strips to obtain a high intensity of magnetization. (2) An electromagnet for a.c. circuits, having a laminated core to reduce eddy currents.

laminated paper *(Paper)*. Product formed by bonding the whole of the surface of a sheet of paper to another paper or sheet material such as metal foil, plastics film etc.

laminated plastics *(Plastics)*. Superimposed layers of a synthetic resin-impregnated or coated filler which have been bonded together, usually by means of heat and pressure, to form a single piece. Also *laminate*.

laminated pole *(Elec.Eng.)*. A pole for the field windings of an electric machine, having the core built up of laminations to reduce eddy currents caused by flux pulsations in the air gap.

laminated pole-shoe *(Elec.Eng.)*. A field magnet pole for an electric machine having the pole-shoe built up of laminations to reduce eddy currents in it caused by flux pulsations in the air gap.

laminated record *(Acous.)*. A gramophone record in which the surface material differs from that in the inside, or core, in being finer grained and therefore freer from surface noise. The superior material is carried on a fine sheet of paper, being pressed on to the hot core in the press.

laminated spring *(Eng.)*. A flat or curved spring consisting of thin plates or leaves superimposed, acting independently, and forming a beam or cantilever of uniform strength.

laminated yoke *(Elec.Eng.)*. A yoke for an electric machine, built up of laminations; used in some forms of a.c. motors.

lamina terminalis *(Zool.)*. In Craniata, the anterior termination of the spinal cord which lies at the anterior end of the diencephalon.

lamination *(Elec.Eng.)*. A sheet steel stamping shaped so that a number of them can be built up to form the magnetic circuit of an electric machine, transformer, or other piece of apparatus. Also called *core plate, punching, stamping*.

lamination *(Geol.)*. Stratification on a fine scale, each thin stratum, or *lamina*, frequently being a millimetre or less in thickness. Typically exhibited by shales and fine-grained sandstones.

laminboard *(Build.)*. See **coreboard**.

laminectomy *(Med.)*. Surgical removal of the posterior arch or arches of one or more spinal vertebrae.

laminin *(Biol.)*. A large fibrous protein which is a major component of the basal lamina.

laminitis *(Vet.)*. Inflammation of the sensitive laminae of the hoof of the horse and ox.

laminography *(Radiol.)*. See **tomography**.

lamins *(Biol.)*. A group of three proteins, lamin A, B and C, of the **intermediate filament** type which forms the nuclear lamina.

Lamont's law *(Elec.Eng.)*. That the permeability of steel, at any flux density, is proportional to the difference between that flux density and the saturation value.

lampas *(Vet.)*. A swelling of the palatal mucous membrane behind the upper incisor teeth of horses.

lampblack *(Chem.)*. The soot (and resulting pigment) obtained when substances rich in carbon (e.g. mineral oil, turpentine, tar etc.) are burnt in a limited supply of air so as to burn with a smoky flame. The pigment is black with a blue undertone, containing 80–85% carbon and a small percentage of oily material. See also **carbon black**.

lampbrush chromosome *(Biol.)*. Chromosome of the first meiotic prophase observed in many **eukaryotes**, which has a characteristic appearance due to the orderly series of lateral loops of chromatin, arranged in pairs on either side of the chromosome axis.

lamp cap *(Elec.Eng.)*. The cap of brass, plastic, or other material, at the base of a filament or electric discharge lamp, which contains the terminals, and also serves to support the lamp in the holder; also called *lamp base*. See **bayonet cap, Edison screw-cap**.

lampholder *(Elec.Eng.)*. A device for supporting an electric lamp; fitted with contacts connected to the source of supply; also called a *holder* or *lamp socket*. See **backplate lampholder, bayonet holder, Edison screw-holder**.

lampholder plug *(Elec.Eng.)*. A device for connecting a flexible electric cord, instead of a lamp, to an ordinary lampholder. Also called *plug adaptor*.

lamphouse *(Image Tech.)*. That part of a projector, printer or enlarger which contains the source of illumination.

lamping *(Min.Ext.)*. Use of ultraviolet light to detect fluorescent minerals either when prospecting or in checking concentrates during ore treatment.

lamp man *(Min.Ext.)*. Colliery surface worker in charge of miners' lamps. Works in lamp room or cabin and controls repairs, recharge, issue, lighting, etc. of portable lamps.

lamp resistance *(Elec.Eng.)*. A resistance consisting of one or more electric filament lamps.

lamprophyres *(Geol.)*. Igneous rocks usually occurring as dykes intimately related to larger intrusive bodies; characterized by abnormally high contents of coloured silicates, such as biotite, hornblende, and augite, and a correspondingly small amount of feldspar, some being feldspar-free. See also **minette, monchiquite**.

lamp socket *(Elec.Eng.)*. See **lampholder**.

lamp working *(Glass)*. Making articles, usually from glass tubing or rod, with the aid of an oxy-gas or air-gas flame.

lamziekte *(Vet.)*. A form of botulism occurring in cattle in South Africa due to the ingestion, in phosphorus deficient areas, of bones contaminated with *Clostridium botulinum*, Type D.

LAN *(Comp.)*. See **local area network**.

lanarkite *(Min.)*. A rare monoclinic sulphate of lead, occurring with anglesite and leadhillite (into which it easily alters), as at Leadhills, Lanarkshire, Scotland.

lanate *(Bot.,Zool.)*. Woolly.

Lancashire boiler *(Eng.)*. A cylindrical steam boiler having two longitudinal furnace tubes containing internal grates at the front. After leaving the tubes the gases pass to the front along a bottom flue, and return to the chimney along side or wing flues.

lance *(Build.)*. A sharp scribing part of a cutting tool, serving to cut through the grain in advance and on each side of the cutting tool proper.

lanceolate *(Bot.)*. Lance-shaped; much longer than broad and tapering at both ends.

lancet arch *(Arch.)*. A sharply pointed arch, of greater rise than an equilateral arch of the same span.

lancet window *(Arch.)*. A tall narrow window surmounted by a lancet arch.

lancewood *(For.)*. Durable straight grained wood chiefly from *Oxandra lanceolatea*, a native of tropical America.

lancinating *(Med.)*. Of pain, acute, shooting, piercing, cutting.

lancing *(Eng.)*. A line cut made in a press which does not remove metal but only separates it.

land and sea breezes *(Meteor.)*. Winds occurring at the coast during fine summer weather especially when the general pressure gradient is small. During the day, when the land is warmer than the sea, the air over the land becomes warmer and less dense than that over the sea and a local circulation is created with air flowing from land to sea at high levels and from sea to land near the surface. At night, conditions are reversed. The sea breeze can penetrate many miles inland and can become strong at low latitudes.

Landau levels *(Phys.)*. *Conduction electrons* of a solid in a magnetic field will describe complete orbits if $\omega\tau > 1$ where ω is the *cyclotron frequency* and τ is the time between scattering events. The electron density of states will be altered and the allowed energy levels, the *Landau levels*, will differ by $\hbar\omega$, where $\hbar$ is the Dirac constant. See **de Haas-van Alphen effect**.

Landau theory *(Chem.)*. (1) Theory for calculating diamagnetic susceptibility produced by free conduction electrons. (2) That explaining the anomalous properties of *liquid helium II* in terms of a mixture of normal and superfluids; see **helium, superfluid**.

Land camera *(Image Tech.)*. See **Polaroid camera**.

Landé splitting factor *(Phys.)*. Factor employed in the calculation of the splitting of atomic energy levels by a magnetic field. See **Zeeman effect**.

landing *(Min.Ext.)*. Stage in hoisting shaft at which cages are loaded or discharged.

landing area *(Aero.)*. That part of the movement area of an unpaved airfield intended primarily for take off and landing.

landing beacon *(Aero.,Telecomm.)*. A transmitter used to produce a landing beam.

landing beam *(Aero.,Telecomm.)*. Field from a transmitter along which an aircraft approaches a landing field during blind landing. See **instrument landing system**.

landing direction indicator *(Aero.)*. A device indicating the direction in which landings and take offs are required to be made, usually a T, toward the cross bar of which the aircraft is headed.

landing gear *(Aero.)*. That part of an aircraft which provides for its support and movement on the ground and also for absorbing the shock on landing. It comprises main support assemblies incorporating single or multi-wheel arrangements, and also auxiliary supporting assemblies such as nose-wheels, tail-wheels or skids. See also **bogie-**, **drag struts**, **oleo**.

landing ground *(Aero.)*. In air transport, any piece of ground that has been prepared for landing of aircraft as required; not necessarily a fully equipped airfield. An *emergency landing ground* is any area of land that has been surveyed and indicated to pilots as being suitable for forced or emergency landings. See **air strip**.

landing parachute *(Aero.)*. See **brake parachute**.

landing procedure *(Aero.)*. The final approach manoeuvres, which begins when the aircraft is in line with the axis of the runway, either for the landing, or upon the reciprocal for the purpose of a procedural turn, and ends when the actual landing is made, or *overshoot* action has to be taken.

landing speed *(Aero.)*. The minimum airspeed, with or without engine power, at which an aircraft normally alights.

landing wires *(Aero.)*. Wires or cables which support the wing structure of a biplane when on the ground and also the negative loads in flight. Also *anti-lift wires*.

landrace *(Bot.)*. Ancient or primitive cultivar of a crop plant.

Landsberger apparatus *(Chem.)*. An apparatus for the determination of the boiling point of a solution by using the vapour of the solvent to heat the solution. Prevents errors due to superheating.

landscape *(Print.)*. A term which indicates that the breadth of the page is greater than the depth: e.g. 138 × 216mm, the convention being to state the upright edge first. Also applied to a similarly shaped full page book illustration or map imposed at right angles to the main text.

landscape lens *(Image Tech.)*. Photographic objective of meniscus form, generally with the concave surface towards the object and with the diaphragm on the object side of the lens.

landscape marble *(Geol.)*. A type of limestone containing markings resembling miniature trees, etc.; when polished, the surface has the appearance of a sepia drawing.

landslip *(Geol.)*. The sudden sliding of masses of rock, soil, or other superficial deposits from higher to lower levels, on steep slopes. Landslips on a very large scale occur in mountainous districts as a consequence of earthquake shocks, stripping the valley sides bare of all loose material. In other regions landslips occur particularly where permeable rocks, lying on impermeable shales or clays, dip seawards or towards deep valleys. The clays hold up water, becoming lubricated thereby, and the overlying strata, fractured by joints, tend to slip downhill, a movement that is facilitated on the coast by marine erosion.

land treatment *(Build.)*. The final or oxidizing stage in sewage treatment, in which the liquid sewage is distributed over an area of land, through which it filters to underdrains. If the land will not permit of easy filtering, the sewage is applied to one plot of land by irrigation, and is then passed on to a second, third, and fourth plot, before final discharge into a stream.

Langerhans cell *(Immun.)*. A cell with dendritic shape present in the epidermis and characterised by the presence in the cytoplasm of Birbeck granules, i.e. small racquet shaped objects detected by electron microscopy, probably composed of some undigested material. Langerhans cells are the form taken in passage through the skin by the type of accessory cell which includes interdigitating or **dendritic cells**. MHC Class II antigens are strongly expressed, and these cells are very effective at antigen presentation to T-lymphocytes.

Langerhans cells *(Zool.)*. Spindle-shaped cells in the centre of each acinus of the pancreas.

Langerhans islets *(Zool.)*. Irregular masses of hyaline epithelium cells, unfurnished with ducts, occurring in the Vertebrate pancreas; they are responsible for the elaboration of the hormone insulin.

Langevin theory *(Phys.)*. (1) A classical expression for diamagnetic susceptibility produced by orbital electrons of atoms. (The quantum mechanical equivalent of this was derived subsequently by Pauli.) (2) An expression for the resultant effect of atomic magnetic moments which enters into explanations of both **paramagnetism** and **ferromagnetism**.

langite *(Min.)*. A very rare ore of copper occurring in Cornwall, blue to greenish-blue in colour; essentially hydrated copper sulphate, crystallizing in the orthorhombic system.

Lang lay *(Eng.)*. A method of making wire ropes in which the wires composing the strands, and the strands themselves, are laid in the same direction of twist.

Langmuir adsorption isotherm *(Chem.)*. The fraction θ of the adsorbent surface which is covered by molecules of adsorbed gas is given by $\theta = bp(1+bp)$ where p is the gas pressure, and b is a constant.

Langmuir dark space *(Electronics)*. Nonglow region surrounding a negative electrode placed in the luminous positive column of a gas discharge.

Langmuir frequency *(Electronics)*. The natural frequency of oscillation for electrons in a plasma.

Langmuir law *(Electronics)*. See **Child-Langmuir equation**.

Langmuir probe *(Electronics)*. Electrode(s) introduced into gas discharge tube to study potential distribution along the discharge.

Langmuir's theory *(Chem.)*. (1) The assumption that the extranuclear electrons in an atom are arranged in shells corresponding to the periods of the periodic system. The chemical properties of the elements are explained by supposing that a complete shell is the most stable structure. (2) The theory that adsorbed atoms and molecules are held to a surface by residual forces of a chemical nature.

Langmuir trough *(Min.Ext.)*. Apparatus in which a rectangular tank is used to measure surface tension of a liquid.

language *(Comp.)*. See **natural-**, **programming-**, **formal-**, **assembly-**, **high-level-**, **low-level-**.

language acquisition device *(Behav.)*. According to Chomsky, a hypothetical brain structure that enables an individual to learn the rules of grammar on hearing spoken language. Abbrev. *LAD*.

laniary *(Zool.)*. Adapted for tearing, as a canine tooth.

lanolin *(Chem.)*. Fat from wool (*adeps lanae*), a yellowish viscous mass of waxlike consistency, very resistant to acids and alkalis; it emulsifies easily with water and is used for making ointments. It consists of the palmitate, oleate and stearate of cholesterol.

lansfordite *(Min.)*. Hydrated magnesium carbonate, crystallizing in the monoclinic system Lansfordite occurs in some coal mines but is not stable on exposure to the atmosphere and becomes dehydrated.

lantern *(Arch.)*. An erection on the top of a roof, projecting above the general roof level, and usually having glazed sides to admit light, as well as openings for ventilation.

lantern wheel, pinion *(Horol.)*. A form of pinion consisting of two circular brass disks connected by cylindrical pins, the pins acting as the leaves in an ordinary pinion. Lantern pinions are very satisfactory as followers but should not be used as drivers. Used extensively in cheap clocks, alarm clocks with pin pallet escapement and in some turret clocks. The lantern wheel is obsolete in general engineering.

lanthanide contraction *(Chem.)*. The peculiar characteristic of the lanthanides that the ionic radius decreases as the atomic number increases, because of the increasing pull of the nuclear charge on the unchanging number of electrons in the two outer shells. Thus the elements after lanthanum, e.g. platinum, are very dense and have chemical properties very similar to their higher homologues, e.g. palladium.

lanthanide series *(Chem.)*. Name for the rare earth elements at. nos. 57-71, after lanthanum, the first of the series. Cf. **actinides**. In both series an incomplete f-shell is filling.

lanthanum *(Chem.)*. A metallic element in the third group of the periodic system, belonging to the rare earths group. Symbol La, at. no. 57, r.a.m. 138.91, mp 810°C.

lanthanum glass *(Glass)*. Optical glass used for high quality photographic lenses etc. with high refractive index and low dispersion.

lanuginose *(Bot.,Zool.)*. Woolly.

lanugo *(Zool.)*. In Mammals, prenatal hair.

lap *(Build.)*. The length of overlap (6 to 10 cm, 2½ to 4 in.) of a slate over the slate next but one below it, in centre nailed work; or that of a slate over the nail securing the slate next but one below it, in head nailed work.

lap *(Eng.)*. (1) A surface defect on rolled or forged steel. It is caused by folding a fin to the surface and squeezing it in; as welding does not occur, a seam appears on the surface. (2) Polishing cloth impregnated with diamond dust or other abrasive, used in polishing rock specimens, gemmology etc. (3) In mining, one coil of rope on the mine hoisting drum.

lap *(Glass)*. (1) A rotating disk or other tool for grinding or polishing glass. (2) A square piece of material, usually rubber, to protect the hands when handling glass.

lap *(Textiles)*. (1) A sheet of fibres or fabric wrapped round a core (e.g. of wood) for transfer to the next process e.g. the rolled sheet of fibres collected from the opening machine ready for carding. (2) The undesired build-up of fibres on a roller e.g. after a yarn break in spinning. (3) The length of fabric between folds when it is being plaited into a vertical pile.

laparoscopy *(Med.)*. The insertion of a rigid or flexible device to inspect the abdominal cavity. Simple surgical operations, e.g. tubal ligation may be carried out using a laparoscope.

laparotomy *(Med.)*. Cutting into the abdominal cavity. *Exploratory laparotomy*, the operation of cutting into the abdominal cavity so that direct examination of abdominal organs may be made.

lap dovetail *(Build.)*. An angle joint between two members, in which only one shows end grain, a sufficient thickness of wood having been left on this member, in cutting the joint, to cover the end grain of the other member. Also *drawer front dovetail*, from one of its common uses.

lapel microphone *(Acous.)*. Small microphone which does not impede vision of the speaker.

lapidicolous *(Zool.)*. Living under stones.

lapilli *(Geol.)*. Small rounded pieces of lava whirled from a volcanic vent during explosive eruptions; lapilli are thus similar to volcanic bombs but smaller in size, usually about the size of walnuts. sing. *lapillus*.

lapis lazuli *(Min.)*. Original sapphire of ancients, a beautiful blue stone used extensively for ornamental purposes. It consists of the deep blue feldspathoid **lazurite**, usually together with calcite and spangles of pyrite.

lap joint *(Build.)*. A joint between two pieces of timber, formed by laying one over the other for a certain length and fastening the two together with metal straps passing around the timbers, or with bolts passing through them.

lap joint *(Eng.)*. A plate joint in which one member overlaps the other, the two being riveted or welded along the seam single, double, or treble.

Laplace linear equation *(Maths.)*. A differential equation of the form

$$(a_0+b_0x)\frac{d^ny}{dx^n}+(a_1+b_1x)\frac{d^{n-1}y}{dx^{n-1}}+\dots+(a_n+b_nx)y=0.$$

Laplace's equation *(Maths.)*. The equation $\nabla^2V=0$, where ∇ is the vector operator nabla,

$$\text{i.e. } \frac{\partial^2V}{\partial x^2}+\frac{\partial^2V}{\partial y^2}+\frac{\partial^2V}{\partial z^2}=0.$$

It is satisfied by electric and gravitational potential functions. Cf. **Poisson's equation**.

Laplace transform *(Maths.)*. The Laplace transform $F(p)$ of the function $f(t)$ is defined by

$$F(p)=\int_0^\infty e^{-pt}f(t)dt.$$

In the *two sided* Laplace transform the range of integration is from $-\infty$ to $+\infty$.

La Pointe picker *(Min,Ext.)*. Small belt conveyor on which ore is so displayed that radioactive pieces are removed as they pass a Geiger-Müller counter.

lapping *(Elec.Eng.)*. The final abrasive polishing of a quartz crystal to adjust its operating frequency. Also, smoothing of surface of crystalline semiconductors.

lapping *(Eng.)*. The finishing of spindles, bored holes etc. to fine limits, by the use of laps of lead, brass, etc.

lapping *(Image Tech.)*. Rubbing one surface against another, generally with an abrasive such as rouge, so that the softer takes up the contour of the harder, e.g. in polishing a lens or optical flat.

lapping machine *(Eng.)*. A machine tool for finishing the bores of cylinders, etc., to fine limits by the use of revolving circular laps supplied with an abrasive powder suspended in the coolant.

lapse rate *(Meteor.)*. The rate of fall of a quantity with increase in height.

lap winding *(Elec.Eng.)*. A form of 2-layer winding for electric machines in which each coil is connected in series with the one adjacent to it. Cf. **wave winding**.

Laramide orogeny *(Geol.)*. In the narrow sense, that mountain building movement associated with the production of the Rocky Mountains between late Cretaceous and Palaocene times. More broadly sometimes embraces all orogenies that took place during that span of time.

lardalite *(Geol.)*. See **laurdalite**.

large-area foils *(Print.)*. Blocking foils for display card printing. These are capable of filling in very large areas and yet of giving fine detail for half-tone work.

large calorie *(Phys.)*. See **calorie**.

large crown octavo *(Print.)*. A nonstandard book size, larger than crown 8vo and smaller than demy 8vo, varying slightly from publisher to publisher.

large intestine *(Zool.)*. See **colon**.

large panel construction *(Build.)*. See **system building**.

large scale integration *(Comp.)*. See **LSI**.

large scale integration *(Electronics)*. Production of an integrated circuit with more than 100 gates on a single chip of semiconductor. Abbrev. *LSI*.

larmier *(Build.)*. A corona placed over a door or window opening to serve as a dripstone.

Larmor frequency *(Electronics)*. The angular frequency of precession for the spin vector of an electron acted on by an external magnetic field.

Larmor precession *(Electronics)*. The precessional motion of the orbit of a charged particle when subjected to a magnetic field. Precession occurs about the direction of the field.

Larmor radius *(Phys.)*. That of the circular or helical path followed by a charged particle in a uniform magnetic field.

larnite *(Min.)*. A rare calcium silicate, Ca_2SiO_4, formed at very high temperatures in contact metamorphosed limestone.

larry *(Build.)*. A tool having a curved steel blade fixed to the end of a long handle, to which it is bent normally; used for mixing mortar, or for mixing hair with coarse stuff to form a plaster.

larrying *(Build.)*. The process of pouring a mass of mortar upon the wall and working it into the joints; sometimes used in building large masses of brickwork.

larva *(Zool.)*. The young stage of an animal if it differs appreciably in form from the adult; a free-living embryo.

larvikite *(Geol.)*. See **laurvikite**.
larviparous *(Zool.)*. Giving birth to offspring which have already reached the larva stage.
larvivorous *(Zool.)*. Larva-eating.
laryngectomy *(Med.)*. Surgical removal of the larynx.
laryngismus *(Med.)*. Spasm of the larynx.
laryngitis *(Med.)*. Inflammation of the larynx.
laryngofissure *(Med.)*. *Thyrotomy*. Surgical exposure of the larynx by dividing the thyroid cartilage (Adam's apple) in the midline.
laryngology *(Med.)*. That branch of medical science which treats diseases of the larynx and adjacent parts of the upper respiratory tract. n. *laryngologist*.
laryngopharyngitis *(Med.)*. Inflammation of the larynx and the pharynx.
laryngophone *(Acous.)*. See **throat microphone**.
laryngoscope *(Med.)*. An instrument used for viewing the larynx.
laryngostenosis *(Med.)*. Pathological narrowing of the larynx.
laryngostomy *(Med.)*. The surgical formation of an opening in the larynx.
laryngotomy *(Med.)*. The operation of cutting into the larynx.
laryngotracheal chamber *(Zool.)*. In Amphibians, a small chamber into which the lungs open anteriorly, and which communicates with the buccal cavity by the glottis.
larynx *(Acous.)*. Organ in the throat of man and vertebrates which with the lungs forms the voice.
laser *(Phys.)*. *Light Amplification by Stimulated Emission of Radiation*. A source of intense monochromatic light in the ultraviolet, visible or infrared region of the spectrum. It operates by producing a large population of atoms with their electrons in a certain high energy level. By *stimulated emission*, transitions to a lower level are induced, the emitted photons travelling in the same direction as the stimulating photons. If the beam of inducing light is produced by reflection from mirrors or *Brewster's windows* at the ends of a resonant cavity, the emitted radiation from all stimulated atoms is in phase, and the output is a very narrow beam of coherent monochromatic light. Solids, liquids and gases have been used as lasing materials. See **He-Ne laser, carbon-dioxide laser, maser, Q-switching, mode-locking**.
laser-beam cutting *(Eng.)*. Using the intense narrow beam of radiation from a laser to cut often complex shapes in sheet or plate. Good finish and high precision can be achieved.
laser-beam machining *(Eng.)*. The use of a focused beam of high-intensity radiation from a laser to vapourise and so machine material at the point of focus. Cf. **laser beam cutting**.
laser compression *(Nuc.Eng.)*. See **inertial confinement**.
laser enrichment *(Phys.)*. The enrichment of uranium isotopes using a powerful laser to ionize atoms of the selected isotope of molecules bearing the uranium isotope. The separation can then be done by chemical or physical means.
laser fusion *(Phys.)*. The initiation of the nuclear fusion process by directing energy from a laser beam on to the fusion fuel contained in pellets. The laser both heats and compresses the material.
laser fusion reactor *(Nuc.Eng.)*. A proposed type of fusion reactor in which pellets of deuterium and tritium are contained in a suitable vessel and bombarded by a pulsed laser beam. See **magnetic confinement fusion system, inertial fusion system**.
laser gyro *(Eng.)*. An integrating rate gyroscope which combines the properties of the optical oscillator, the laser and general relativity. It differs from conventional gyros in the absence of a spinning mass and so its performance is not affected by accelerations. It has a low power consumption and is not subject to prolonged starting time, used in missile guidance systems. Modern versions can measure rates as low as 0.1 degree per hour.
laser level *(Surv.)*. An instrument for accurate single-handed levelling. A laser beam, narrowly confined in the vertical dimension, is continuously rotated around an accurately vertical axis. The staff has a cursor which is able to sense the laser beam and indicate its height which is then read off.
laser printer *(Comp.)*. Uses a laser beam to form characters on paper.
laser threshold *(Phys.)*. The minimum pumping power (or energy) required to operate a laser.
Laservision *(Image Tech.)*. TN for a **video-disc** system read by a laser beam.
lashing *(Min.Ext.)*. A South African term for removing broken rock after blasting. Canadian term, *mucking* or *mucking out*.
lash-up *(Telecomm.)*. The temporary connection of apparatus for experimental or emergency use.
lassa fever *(Med.)*. A viral haemorrhagic fever which occurs sporadically in rural West Africa. The animal reservoir is a wild rodent, *Mastomys natalensis*. Causes a severe illness with shock and had a high mortality.
Lassaigne's test *(Chem.)*. Test for the presence of nitrogen in an organic substance. The sample is heated with metallic sodium in a test tube; the product is placed in water and filtered. Iron (II) sulphate is added and the mixture boiled, cooled, and sodium cyanide and iron (III) chloride added. If nitrogen is present the characteristic *Prussian blue* colour is observed.
last-in, first-out *(Comp.)*. See **stack**.
last-subscriber release *(Telecomm.)*. The release of automatic switching plant when the last of both subscribers has replaced his receiver and opened his loop. Also *last party release*.
latching *(Surv.)*. See **dialling**.
latching *(Telecomm.)*. Arrangement whereby a circuit is held in position, e.g. in read-out equipment, until previous operating circuits are ready to change this circuit. Also called *locking*.
latch needle *(Textiles)*. See **needle** (machine knitting).
late-choice call meter *(Telecomm.)*. A traffic meter so connected as to record the number of calls carried by a late-choice trunk of a grading.
latency *(Behav.)*. General term for the interval before some reaction; also, the dormancy of a particular behaviour or response.
latency *(Comp.)*. Delay, in digital computers, between the initiation of the call for data and the start of the transfer. Latency forms part of the total **access time**.
latency period, stage *(Behav.)*. According to Freud, the fourth psychosexual stage of development, in which sexuality lies essentially dormant, occurring roughly between the ages of six and the onset of puberty.
latensification *(Image Tech.)*. Increasing the **latent image** before development, usually by a long uniform exposure to light of very low intensity, but chemical methods can be used.
latent *(Zool.)*. In a resting condition or state of arrested development, but capable of becoming active or undergoing further development when conditions become suitable; said also of hidden characteristics which may become evident under the right circumstances.
latent content *(Behav.)*. In psychoanalytic theory, the unconscious material or hidden meaning of a dream that is being expressed in a disguised fashion through symbols contained in the dream. See **manifest dream content**.
latent heat *(Phys.)*. More correctly, **specific latent heat**. The heat which is required to change the state of unit mass of a substance from solid to liquid, or from liquid to gas, without change of temperature. Most substances have a latent heat of fusion and a latent heat of vaporization. The specific latent heat is the difference in *enthalpies* of the substance in its two states.
latent image *(Image Tech.)*. The invisible image formed in a sensitive emulsion by molecular ionisation in the silver halide grains affected by exposure to radiation; it is made visible by the process of **development**.
latent instability *(Meteor.)*. A type of **conditional instability** of the atmosphere which exists only if a rising parcel of air reaches a critical level.
latent learning *(Behav.)*. Learning the characteristics of a situation which is not manifested by any immediate overt

behaviour, but which is revealed when the individual is later placed in a situation requiring previously acquired information.

latent magnetization *(Phys.)*. The property possessed by certain feebly magnetic metals (e.g. manganese and chromium) of forming strongly magnetic alloys or compounds.

latent neutrons *(Phys.)*. In reactor theory, the delayed neutrons due from (but not yet emitted by) fission products.

latent period *(Bot.)*. (1) Period of time between stimulation and the first signs of a response. (2) Period of time between infection and the appearance of the symptoms of disease.

latent period *(Radiol.)*. That between exposure to radiation and its effect.

latent roots of a matrix *(Maths.)*. See **characteristic equation of a matrix**.

later *(Build.)*. A brick or tile.

lateral *(Bot.)*. Arising from a parent axis or the structure so formed. Cf. **leader**.

lateral *(Genrl.)*. Situated on or at, or pertaining to, a side.

lateral axis *(Aero.)*. The crosswise axis of an aircraft, particularly that passing through its centre of mass, parallel to the line joining the wing tips.

lateral canal *(Hyd.Eng.)*. A separate navigational canal constructed to follow the lie of a river which does not lend itself to canalization.

lateral instability *(Aero.)*. A condition wherein an aircraft suffers increasing oscillation after a rolling disturbance. Also *rolling instability*.

lateralization, laterality *(Behav.)*. The organization of brain functions such that each of the cerebral hemispheres control different psychological functions, particularly in verbal and visual spatial skills. In most right handers, the left hemisphere is specialised for language functions, while the right hemisphere is better at various visual and spatial tasks.

lateral line *(Zool.)*. In Fish, a line of neuromast organs running along the side of the body. In Nematoda, a paired lateral concentration of hypodermis containing the excretory canal.

lateral load *(Eng.)*. A force acting on a structure or a structural member in a transverse direction, e.g. wind forces on a bridge or building at right angles to its length, which trusses and girders are not primarily designed to withstand.

lateral meristem *(Bot.)*. A meristem lying parallel to the sides of the axis, e.g. cambium.

lateral moraine *(Geol.)*. A low ridge formed at the side of a valley glacier. It is composed of material derived from glacial plucking and abrasion or material that has fallen on to the ice.

lateral plate *(Zool.)*. In Craniata, a ventral portion of each mesoderm band which surrounds the mesenteron in embryo; in Insects, the paired lateral region of the germ band of the embryo which becomes separated from the opposite member of its pair by the middle plate.

lateral recording *(Acous.)*. That in which the cutting stylus removes a thread (*swarf*) from the surface of a blank disk, the modulation being realized as a lateral (radial) deviation as the spiral is transversed, in contrast with the obsolete method of *vertical recording*. Cf. **stereophonic recording**.

lateral shift *(Geol.)*. The displacement of outcrops in a horizontal sense, as a consequence of faulting. Cf. **throw**.

lateral stability *(Aero.)*. Stability of an aircraft's motions out of the plane of symmetry, i.e. side slipping, rolling and yawing.

lateral traverse *(Eng.)*. The longitudinal play given to locomotive trailing axles to permit of taking sharp curves.

lateral velocity *(Aero.)*. The rate of **sideslip** of an aircraft, that is, the component of velocity which is resolved in the direction of the lateral axis.

laterigrade *(Zool.)*. Moving sideways, as some Crabs.

laterite *(Geol.)*. A residual clay formed under tropical climatic conditions by the weathering of igneous rocks, usually of basic composition. Consists chiefly of hydroxides of iron and aluminium. See **bauxite**.

laterization *(Geol.)*. The process whereby rocks are converted into laterite. Essentially, the process involves the abstraction of silica from the silicates. See **laterite**.

laterosphenoid *(Zool.)*. The so called 'alisphenoid' of Fish, Reptiles, and Birds (representing an ossification of the wall of the chondrocranium), as distinct from the alisphenoid of Mammals (developed from the splanchnocranium).

latewood *(Bot.)*. The wood formed in the later part of a growth ring, usually with smaller cells with thicker walls than the **early wood**.

latex *(Bot.)*. Fluid, often milky, exuded from *laticifers* when many plants are cut. It consists of a watery solution containing many different substances including terpenoids which form rubber, alkaloids such as the opium alkaloids, sugar, starch etc.

latex *(Chem.)*. (1) A milky viscous fluid extruded when rubber trees (e.g. *Hevea Brasiliensis*) are tapped. It is a colloidal system of caoutchouc dispersed in an aqueous medium, rel.d. 0.99, which forms rubber by coagulation. The coagulation of latex can be prevented by the addition of ammonia or formaldehyde. Latex may be vulcanized directly, the product being known as *vultex*. (2) In synthetic rubber manufacture, the process stream in which the polymerized product is produced.

latex *(Paper)*. An aqueous emulsion of synthetic rubber-like compounds used to increase the flexibility and durability of paper e.g. for bookbinding papers or base material for imitation leather. It may be added to the stock or used as an impregnant. Also extensively employed as the binder in mineral coating applications.

lath *(Build.)*. See **lathing**.

lath *(Min.)*. A term commonly applied to a lathlike crystal.

lathe *(Eng.)*. A machine tool for producing cylindrical work, facing, boring, and screw cutting. It consists generally of a bed carrying a head stock and tail stock, by which the work is driven and supported, and a saddle carrying the slide rest by which the tool is held and traversed.

lathe bed *(Eng.)*. That part of a lathe forming the support for the head stock, tail stock, and carriage. It consists of a rigid cast box section girder, supported on legs, its upper face being planed and scraped to provide true working surfaces, or 'ways'.

lathe carrier *(Eng.)*. A clamp consisting of a shank which is formed into an eye at one end and provided with a set screw. It is attached to work supported between centres and driven by the engagement of the driver plate pin with the shank or 'tail' of the carrier, which may be straight or bent.

lathe tools *(Eng.)*. Turning tools with edges of various shape (round nosed, side, etc.), and cutting angles, varying with the material worked on, formed by giving clearance to the front of the cutting edge and rake to the top of the tool. See **finishing tool, knife tool, roughing tool, side tool**.

lathe work *(Eng.)*. Any work ordinarily performed in the lathe, such as all classes of turning, boring, and screw cutting.

lathing *(Build.)*. Material fixed to surfaces to provide a basis for plaster. Formerly soft wood strips (*laths*) 3–4 ft long were used; nowadays steel meshing or perforated steel sheet is usual.

lathyrism *(Med.)*. A disease characterized by stiffness and paralysis of the legs, due to poisoning with certain kinds of chick pea.

lati- *(Bot.)*. Prefix meaning broad.

laticifer *(Bot.)*. A cell or vessel containing **latex**, present in tissues of many plants.

latitude *(Image Tech.)*. The range of exposure permissible, or range of density usefully obtainable in a photographic emulsion. The range of exposure obtainable with the linear portion of the gamma curve of an emulsion.

latitude *(Surv.)*. *Northing*. Distance of a point north (if positive) or south (negative) from point of origin of survey. This, together with departure (easting), locates the point in a rectangular grid orientated on true meridian.

latitude and longitude, celestial *(Astron.)*. Spherical co-ordinates referred to the ecliptic and its poles. *Celestial latitude* is the angular distance of a body north or south of the ecliptic. *Celestial longitude* is the arc of the ecliptic intercepted between the latitude circle and the First Point of Aries, and is measured positively eastwards from 0° to 360°.

latterkin *(Build.)*. Hardwood suitably shaped for clearing grooves in **cames**.

lattice *(Crystal.)*. A regular space arrangement of points as for the sites of atoms in a crystal.

lattice *(Maths.)*. A partially ordered set in which any two elements have a least upper bound and a greatest lower bound.

lattice *(Nuc.Eng.)*. Regular geometrical pattern of discrete bodies of fissionable and non-fissionable material in a nuclear reactor. The arrangement is subcritical or just critical if it is desired to study the properties of the system.

lattice bars *(Eng.)*. The diagonal bracing of struts and ties in an **open frame girder** or **lattice girder**.

lattice bridge *(Eng.)*. A bridge of lattice girders.

lattice coil *(Elec.Eng.)*. A form of coil used for the armature winding of electric machines, which is arranged so that the end connections cross over one another in a regular pattern, giving a lattice appearance.

lattice coil *(Telecomm.)*. An inductance coil in which the turns are wound so as to cross each other obliquely, to reduce the self capacitance. See **honeycomb coil**.

lattice diagram *(Elec.Eng.)*. A diagram for simplifying the calculation of travelling waves on a transmission line when there are a large number of successive reflections.

lattice dynamics *(Phys.)*. The study of the excitations a crystal lattice can experience and their consequences for the thermal, optical and electrical properties of solids.

lattice energy *(Crystal.)*. Energy required to separate the ions of a crystal from each other to an infinite distance.

lattice filter *(Telecomm.)*. One or more lattice networks acting as a wave filter.

lattice girder *(Eng.)*. A girder formed of upper and lower horizontal members connected by an open web of diagonal crossing members, used in structures such as bridges and large cranes.

lattice hypothesis *(Immun.)*. An hypothesis proposed to explain how aggregates are formed when antibodies combine with a soluble antigen, the size and composition of which varies with the ratio of the two components. Divalent antibody molecules combine with antigenic determinants on a polyvalent antigen to form a lattice. When the ratio of the two is such that all the antigen molecules are attached to one another by antibody molecules a macromolecular lattice is formed which comes out of solution. This ratio is known as the optimal proportion. In antigen excess the complexes will be smaller and may remain in solution.

lattice network *(Telecomm.)*. One formed by two pairs of identical arms on opposite sides of a square, the input terminals being across one diagonal and the output terminals across the other. Also *bridge network*, *lattice section*.

lattice structure *(Crystal.)*. One of three types of crystal structure: (1) ionic, with symmetrically arranged ions and good conducting power; (2) molecular, covalent, usually volatile and nonconducting; (3) layer, with large ions each associated with two small ones, forming laminae weakly held by nonpolar forces.

lattice vibration *(Crystal.)*. That of atoms or molecules in a crystal due to thermal energy.

lattice water *(Chem.)*. Water of crystallization, which is present in stoichiometric proportions and occupies definite lattice positions, but is in excess of that with which the ions could be co-ordinated. This water apparently fills in holes in the crystal lattice, as with **clathrate** compounds.

lattice winding *(Elec.Eng.)*. A winding, made up of lattice coils, for electric machines; always used for d.c. machines and frequently for a.c. machines.

lattice window *(Arch.)*. A window in which diamond shaped panes are supported in a leaden frame consisting of diagonally intersecting cames, the longer axes of the diamonds being vertical.

laudanum *(Med.)*. Tincture of opium.

Laue pattern *(Crystal.)*. Pattern of spots produced on photographic film when a heterogeneous X-ray beam is passed through a thin crystal, which acts like an optical grating. Used in the analysis of crystal structure.

laughing gas *(Med.)*. Popular name for anaesthetic gas, nitrous oxide.

laumontite *(Min.)*. A zeolite consisting essentially of hydrated silicate of calcium and aluminium, crystallizing in the monoclinic system; occurs in cavities in igneous rocks and in veins in schists and slates.

launching *(Telecomm.)*. Said of the operation of transmitting a signal from a conducting circuit into a waveguide.

launch pad *(Space)*. Special area from which a launch system is fired; it contains all the necessary support facilities such as a servicing tower, cooling water, safety equipment and flame deflectors.

launch system *(Space)*. An assemblage of propulsive devices (stages) capable of accelerating a space vehicle to a velocity needed to achieve a particular space trajectory. Examples are: Titan and Space Shuttle (US); Proton (USSR); Ariane (Europe); H-1 (Japan) and Long March (China). Also *launcher, launch vehicle*.

launch window *(Space)*. The time slot within which a spacecraft must be launched to best achieve its given mission trajectory; a launch window may become available a number of times during a particular *launch opportunity*.

launder *(Min.Ext.)*. Inclined trough for conveying water or crushed ore and water (pulp).

Lauraceae *(Bot.)*. Family of ca 2500 spp. of dicotyledonous flowering plants (superorder Magnoliidae). Mostly woody, tropical and subtropical. Includes avocado, cinnamon, camphor, bay laurel and some timber trees.

Laurasia *(Geol.)*. The palaeogeographic supercontinent of the northern hemisphere corresponding to Gondwanaland in the southern hemisphere.

laurdalite, lardalite *(Geol.)*. A coarse-grained soda-syenite from S. Norway; it resembles laurvikite but contains nepheline (elaeolite) as an essential constituent.

Laurentian Granites *(Geol.)*. Precambrian granitic intrusives in the Canadian Shield.

Laurent's expansion *(Maths.)*. If the function $f(z)$ is analytic in the annulus formed by two circles C_1 and C_2, of radii r_1 and r_2, respectively, then within this annulus:

$$f(z) = \sum_1^{\infty} a_n(z - z_0)^n + \sum_1^{\infty} b_n(z - z_0)^{-n},$$

where $$a_n = \frac{1}{2\pi i}\int_{C_1} \frac{f(z)dz}{(z - z_0)^{n+1}}$$

$$b_n = \frac{1}{2\pi i}\int_{C_2} (z - z_0)^{n-1} f(z)dz.$$

lauric acid *(Chem.)*. *Dodecanoic acid*, $CH_3.(CH_3)_{10}.COOH$. Crystalline solid; mp 44°C. Occurs as glycerides in milk, laurel oil, palm oil etc.

laurvikite, larvikite *(Geol.)*. A soda syenite from S. Norway, very popular as an ornamental stone when cut and polished; widely used for facing buildings, the distinctive feature being a fine blue colour, produced by schiller structure in the anorthoclase feldspars. Syn. *blue granite*.

lauryl alcohol *(Chem.)*. *Dodecan-1-ol*, $CH_3.(CH_2)_{10}.CH_2OH$. Insoluble in water, crystalline solid; mp 24°C. Used in manufacture of detergents.

lautarite *(Min.)*. Monoclinic iodate of calcium, occurring rarely in caliche in Chile.

lava *(Geol.)*. The molten rock material that issues from a volcanic vent or fissure and consolidates on the surface of the ground (*subaerial lava*), or on the floor of the sea (*submarine lava*). Chemically, lava varies widely in composition; it may be in the condition of glass, or a holocrystalline rock. See **volcano**; also *basalt, obsidian, pillow structure, pumice*.

lava flows *(Geol.)*. See **extrusive rocks**.

lavage *(Med.)*. Irrigation or washing out of a cavity, such as the stomach or the bowel, e.g. gastric *lavage*.

Lavalier microphone *(Acous.)*. Microphone worn on a cord round the neck. It has a frequency response such as to compensate for the change in directivity produced by that of the mouth.

law *(Genrl.)*. A scientific law is a rule or generalization which describes specified natural phenomena within the limits of experimental observation. An apparent exception to a law tests the validity of the law under the specified conditions. A true scientific law admits of no exception. A law is of no scientific value unless it can be related to other laws comprehending relevant phenomena.

law calf *(Print.)*. Calf leather with a rough surface and light in colour; used for account book bindings, etc.

lawn *(Textiles)*. Fine plain cloth made of flax or cotton yarns. See **organdie**.

law of conservation of matter *(Chem.)*. Matter is neither created nor destroyed during any physical or chemical change. See **relativity**.

law of constant (definite) proportions *(Chem.)*. Every pure substance always contains the same elements combined in the same proportions by weight.

law of Dulong and Petit *(Chem.)*. The atomic heat capacities of solid elements are constant and approximately equal to 26 (when the specific heat capacity is in joules). Certain elements of low atomic mass and high melting point have, however, much lower atomic heat capacities at ordinary temperatures.

law of effect *(Behav.)*. Thorndike's formulation of the importance of reward in learning, which states that the tendency of a stimulus to evoke a response is strengthened if the response is followed by a satisfactory or pleasant consequence, and is weakened if the response is followed by an annoying or unpleasant consequence.

law of equivalent (reciprocal) proportions *(Chem.)*. The proportions in which two elements separately combine with the same weight of a third element are also the proportions in which the first two elements combine together.

law of Guldberg and Waage *(Chem.)*. See **law of mass action**.

law of isomorphism *(Chem.)*. See **Mitscherlich's law of isomorphism**.

law of mass action *(Chem.)*. The velocity of a homogeneous chemical reaction is proportional to the concentrations of the reacting substances.

law of multiple proportions *(Chem.)*. When two elements combine to form more than one compound, the amounts of one of them which combine with a fixed amount of the other exhibit a simple multiple relation.

law of octaves *(Chem.)*. The relationship observed by Newlands (1863) which arranges the elements in order of atomic weight and in groups of 8 (octaves) with recurring similarity of properties. See **periodic system**.

law of partial pressures *(Chem.,Phys.)*. See **Dalton's law of partial pressures**.

law of photochemical equivalence *(Chem.)*. See **Einstein law of photochemical equivalence**.

law of rational indices *(Crystal.)*. A fundamental law of crystallography which states, in the simplest terms, that in any natural crystal the indices may be expressed as small whole numbers.

law of reciprocal proportions *(Chem.)*. See **law of equivalent proportions**.

law of volumes *(Chem.)*. See **Gay-Lussac's law**.

lawrencium *(Chem.)*. Transuranic element, symbol Lr, at.no. 103. Its only known isotope has a short half-life of 8 seconds.

laws of reflection *(Phys.)*. When a ray of light is reflected at a surface, the reflected ray is found to lie in the plane containing the incident ray and the normal to the surface at the point of incidence. The angle of reflection equals the angle of incidence.

Lawson criterion *(Nuc.Eng.)*. The minimum physical conditions of plasma temperature (T), plasma density (n) and confinement time (τ) needed for the production of net power in fusion. It is expressed as $n_\tau > f(T)$ where $f(T)$ has a minimum value of around 2×10^{20} sec/m^3 for the D-T reaction.

lawsonite *(Min.)*. An orthorhombic hydrated silicate of aluminium and calcium. It occurs in low grade regionally metamorphosed schists, particularly in glaucophane schists.

laxatives *(Med.)*. An agent to promote evacuation of the bowel.

lay *(Print.)*. See **lay edges**.

lay *(Telecomm.)*. The axial length of one turn of the helix formed by the core (in a telephone cable) or a strand of a conductor (in a power cable). See also **lay ratio**.

lay *(Textiles)*. Fabrics set out in piles of identical length ready for cutting out for making into garments.

lay barge *(Min.Ext.)*. Vessel for the storage, welding and laying of pipelines underwater.

lay boy *(Paper)*. Equipment at the end of a sheet cutter to collect the cut sheets and stack them in neat piles with aligned edges.

lay cords *(Print.)*. Cords with which a book is tied, to prevent the covers from warping while drying.

lay edges *(Print.)*. The edges of a sheet of paper which are laid against the guides or lays in a printing or folding machine. The front edge (or *gripper edge*) is laid to the *front lays* and the side of the sheet to the *side lay*.

layer *(Build.)*. (1) A course in a wall. (2) A bed of mortar between courses.

layer board *(Build.)*. See **lear board**.

layered igneous rocks *(Geol.)*. Igneous rocks which display layers of differing mineral and chemical composition. e.g. Bushveld complex in South Africa, the Skaergaard Complex of East Greenland.

layered map *(Surv.)*. See **relief map**.

layering *(Bot.)*. (1) See **stratification**. (2) A method of artificial propagation in which stems are pegged down and covered with soil until they root, when they can be detached from the parent plant.

layering *(Geol.)*. The high-temperature sedimentation feature of igneous rocks.

layer lattice *(Crystal.)*. Concentration of bonded atoms in parallel planes, with weaker and nonpolar bonding between successive planes. This gives marked cleavage, e.g. in graphite, mica.

layer line *(Crystal.)*. One joining a series of spots on X-ray rotating crystal diffraction photograph. Its position enables the crystal lattice spacing parallel to the axis of rotation to be determined.

layer-on *(Print.)*. A printer's assistant who feeds the sheets to the machine, an occupation largely superseded by mechanical feeders.

layers *(Phys.)*. Ionized regions in space which vary vertically and affect radio propagation. See **E-layer, F-layer**.

laying *(Build.)*. The first coat of plaster in 2-coat work.

laying in *(Build.)*. A term generally meant to mean the covering of a surface with a material, e.g. paint stripper, texture, goldsize etc. before further working.

laying off *(Build.)*. The finishing strokes in the brush painting process. These strokes should eliminate or minimise brushmarks.

laying-off *(Ships)*. The process of transferring the design form to full scale, for the purpose of *fairing* and ultimately of fabrication of details.

laying-out *(Eng.)*. The marking-out of material, especially plate work, full size, for cutting and drilling.

laying press *(Print.)*. See **lying press**.

laying the bearings *(Acous.)*. In tuning fixed pitch instruments, such as the piano or organ, the technique of tuning the 12 semitones of a central octave. All other notes are then tuned by unison or octaves.

lay light *(Build.)*. A window or sash, fixed horizontally in a ceiling, to admit light to a room.

lay marks *(Print.)*. Marks printed on the sheet to indicate the lay edges of the sheet during printing. Used as a guide for subsequent operations to ensure uniformity.

lay of the case *(Print.)*. The position of the characters in the type case.

layout *(Print.)*. (1) The general appearance of a printed page. (2) The art and practice of arranging display (e.g. advertising) matter to the best advantage. (3) An outline and accurate drawing showing all the information required to assemble a printed job, including the relative positions of the text and illustrations, dimensions, folds, trims, etc.

lay panel *(Build.)*. A long panel of small height formed in a panelled wall above a doorway, or all round the room immediately below the cornice.

lay ratio *(Telecomm.)*. The ratio of the **lay** to the mean diameter of the helix.

Layrub universal joint *(Eng.)*. A flexible joint depending on the deflection of a group of rubber bushes. Frequently used where the angle between shafts is not great, say up to 10° on either side of the straight line.

lay shaft *(Eng.)*. An auxiliary geared shaft, e.g. a secondary shaft running alongside the main shaft of an automobile gearbox, to and from which the drive is transferred by gear wheels of varying ratio.

lazulite *(Min.)*. A deep sky blue, strongly pleochroic mineral, crystallizing in the monoclinic system. In composition essentially a hydrated phosphate of aluminium, magnesium, and iron, with a little calcium. Found in aluminous high grade metamorphic rocks and in granite pegmatites.

lazurite *(Min.)*. An ultramarine blue mineral occurring in cubic crystals or shapeless masses; it consists of silicate of sodium and aluminium, with some calcium and sulphur, and is considered to be a sulphide bearing variety of haüyne. A constituent of **lapis lazuli**.

lb *(Genrl.)*. Abbrev. for *pound.*

L-band *(Telecomm.)*. A radar and microwave frequency band, generally accepted as being 1 to 2 GHz; not frequently referred to because of lack of international conformity in its use.

lb.s.t. *(Aero.)*. Abbrev. for *static thrust* in pounds.

lc *(Print.)*. Abbrev. for *lower case.*

L-capture *(Phys.)*. Absorption of an electron from the L-shell into the nucleus, giving rise to X-rays of characteristic wavelength depending on atomic number of the element. See also **K-capture, X-rays**.

LC coupling *(Electronics)*. Inductor output load of an amplifier circuit is connected through a capacitor to the input of another curcuit.

LCD *(Comp.)*. See **liquid crystal display**.

L-C filter *(Elec.Eng.)*. Property of a component (inductor) in a circuit whereby back e.m.f. arises because of rate of change of current. It is 1 *henry* when 1 volt is generated by rate of change of 1 ampere per second. Often abbreviated to *inductance.*

LCN *(Aero.)*. Abbrev. for *Load Classsification Number.*

LD 50 *(Bot.)*. The dose of a toxic substance that, administered in a named way, will kill 50% of a large number of individuals of a given species.

L-display *(Radar)*. A radar display in which the target appears as two horizontal pulses, left and right from a central vertical time base, varying in amplitude according to accuracy of aim.

leached zone *(Min.Ext.)*. See **gossan, secondary enrichment**.

leaching *(Bot.)*. The removal of substances from soils by percolating water. Cf. **cheluviation, flush, lessivage, podsol.**

leaching *(Min.Ext.)*. The extraction of a soluble metallic compound from an ore by dissolving in a solvent e.g. cyanide, sulphuric acid. The metal is subsequently precipitated or adsorbed from the solution.

lead *(Build.)*. The leaden came of a lattice window.

lead *(Chem.)*. A metallic element in the fourth group of the periodic system. Symbol Pb, at.no. 82, r.a.m. 207.19, valency 2 or 4, mp 327.4°C, bp 1750°C, rel.d. at 20°C 11.35. Specific electrical resistivity 20.65×10^{-8} ohm metres. Occurs chiefly as **galena**. The naturally occurring stable element consists of four isotopes: Pb 204 (1.5%), Pb 206 (23.6%), Pb 207 (22.6%), and Pb 208 (52.3%). It is used as shielding in X-ray and nuclear work because of its relative cheapness, high density and nuclear properties. Other principal uses: in storage batteries, ammunition, foil and as a constituent of bearing metals, solder and type metal. Lead can be hardened by the addition of arsenic or antimony. It occurs very rarely in the native form, and then appears to have been formed by fusion of some simple lead ore accidentally incorporated in lava.

lead *(Civ.Eng.)*. (1) See **haul distance**. (2) In railway track, the distance from the nose of the **crossing** to the nose of the **switch**.

lead *(Elec.Eng.)*. (1) A term often used to denote an electric wire or cable. (2) See **backward shift, brush shift, forward shift**.

lead *(Eng.)*. Distance between consecutive contours, on the same helix of a screw thread measured parallel to the axis of the screw; it is the axial distance a nut would advance in one complete revolution, equal to pitch in single start threads.

lead *(Surv.)*. The leaden sinker secured at one end of a **lead-line**. See also **leads**.

lead-acid (lead) accumulator *(Elec.Eng.)*. A secondary cell consisting of lead electrodes, the positive one covered with lead dioxide, dipping into sulphuric acid solution. Its e.m.f. is about 2 volts; very widely used.

lead age *(Min.)*. The age of a mineral or rock calculated from the ratios of its radiogenic and non-radiogenic lead isotopes.

lead and leave edges *(Print.)*. On web fed presses the first and last edges of the product to be printed.

lead azide *(Chem.)*. PbN_6. Explosive like most azides. Sometimes used, instead of mercury fulminate, as a detonator for TNT.

lead burning *(Build.)*. The process of welding together two pieces of lead, thus forming a joint without the use of solder.

lead (II) carbonate *(Chem.)*. $PbCO_3$. Occurs in nature as cerussite. At about 200°C it decomposes into the (II) oxide and carbon dioxide. Readily reduced to metal by CO.

lead (II) chromate (IV) *(Chem.)*. $PbCrO_4$.Precipitated when potassium chromate (VI) is added to the solution of a lead salt. Used as pigments, called *chrome yellows.* The colour may be varied by varying the conditions under which the precipitation is made.

lead disilicate *(Chem.)*. Obtained by fusing lead (II) oxide and silica together. As *lead frit,* it is used as a ready means of incorporating lead oxide in the making of lead glazes.

lead dot *(Build.)*. A lead peg or dowel used to fasten sheet lead to the upper surface of a coping or cornice, for which purpose it is run into a mortise in the stone.

leaded *(Eng.)*. Term descriptive of copper, bronze, brass, steel, nickel and phosphor alloys to which from 1% to 4% of lead has been added, mainly to improve machinability.

leaded lights *(Build.)*. A window formed of (usually) diamond shaped panes of glass connected together by leaden cames.

lead equivalent *(Radiol.)*. Absorbing power of radiation screen expressed in terms of the thickness of lead which would be equally effective for the same radiation.

leader *(Bot.)*. The younger part of the main stem or a main branch of woody plants. Its branches are called *laterals.* See also **spur**.

leader *(Build.)*. See **conductor**.

leader *(Image Tech.)*. A strip at the beginning (*head leader*) or end (*tail leader*) of a reel of film or tape for protection and identification.

leader *(Min.Ext.)*. A thin mineralized vein parallel to or otherwise related to the main ore-carrying vein, so aiding its discovery.

leader *(Print.)*. A group of dots (...) on one em set together to form dotted lines, or spaced at intervals to guide the eye on contents pages etc.

leader *(Surv.)*. The chainman who has charge of the forward end of a chain. He is directed into line by the follower.

leader-hook *(Build.)*. A device, such as a **holderbat**, for securing a rainwater pipe to a wall.

lead frit *(Chem.)*. See **lead disilicate**.

lead glance *(Min.)*. See **galena**.

lead glass *(Glass)*. Glass containing lead oxide. The amount may vary from 3–4% to 50% or more in special cases. ('English Lead Crystal', used for tableware, contains 33–34%.) Used as transparent radiation shield, esp. for X-rays.

lead grip *(Elec.Eng.)*. A bonding device for providing continuity of a lead sheathed cable.

leadhillite *(Min.)*. Hydrated carbonate and sulphate of lead, so called from its occurrence with other ores of lead at Leadhills (Scotland).

lead (II) hydroxide *(Chem.)*. $Pb(OH)_2$. Dissolves in excess of alkali hydroxides to form plumbites (plumbates (II)).

lead-in *(Acous.)*. Unmodulated groove at the start of a recording on a disk, so that the stylus falls into the groove correctly before the start of the modulation. The corresponding final groove after modulation ends is the *lead-out*.

leading current *(Elec.Eng.)*. An alternating current whose phase is in advance of that of the applied e.m.f. creating the current. For a pure capacitance in circuit, the phase of the current is $\pi/2$ radians in advance of that of the applied voltage. See **angle of lead, phase angle**.

leading edge *(Aero.)*. The edge of a streamline body or aerofoil which is forward in normal motion; structurally, the member that constitutes that part of the body or aerofoil.

leading edge *(Elec.Eng.)*. A term used in connection with the brushes of electric machines. See **entering edge**.

leading edge *(Telecomm.)*. Rising amplitude portion of a pulse signal. See **rise time, trailing edge**.

leading-edge flap *(Aero.)*. A hinged portion of the wing leading edge, usually on fast aircraft, which can be lowered to increase the camber and so reduce the stalling speed. Colloq. *droops*.

leading-in wire *(Elec.Eng.)*. The wire conducting the current from the cap contacts of an electric filament or discharge lamp to the filament itself or to the electrodes.

leading load *(Elec.Eng.)*. See **capacitative load**.

leading note *(Acous.)*. The note one semitone below the tonic or key note of the normal musical scale; essential in combining harmonic frequencies.

leading-out *(Print.)*. The process of inserting *leads* between lines of type matter, in order to open them out, thus presenting more white space between the printed lines. Also used in **phototypesetting** to indicate the placing of space between lines of text.

leading-out wire *(Elec.Eng.)*. The flexible insulated wire which is attached to the more delicate insulated wire used for the windings of transformers, etc. It is sufficiently robust for connecting to terminals.

leading phase *(Elec.Eng.)*. (1) A term used in connection with measuring equipment on 3-phase circuits to denote a phase whose voltage is leading upon that of one of the other phases by approximately 120°. (2) Particularly in connection with the two-wattmeter method of 3-phase power measurement, the phase in which the current at unity power factor leads upon the voltage applied to the meter in which that current is flowing.

leading pole horn *(Elec.Eng.)*. The portion of the pole-shoe of an electric machine which is first met by a point on the armature or stator surface as the machine revolves. Hence also *leading pole tip*.

leading ramp *(Elec.Eng.)*. The end of a conductor rail at which the collector shoe of an electric train first makes contact.

lead line *(Surv.)*. A line with which soundings are taken. The depth of water is indicated on the line by 'marks', or by knots in the line, indicating fathoms; fathoms not indicated on the line (e.g. between 7 and 10) are **deeps**.

lead network *(Electronics)*. One which provides a signal proportional to rate of change, i.e. time differential or derivative, of error signal.

lead of the web *(Print.)*. The reel of paper is torn diagonally to thread it through the press; when the press is started a wrongly threaded path is called a *lost lead* or *wrong lead*.

lead-out *(Acous.)*. See **lead-in**.

lead (II) oxide *(Chem.)*. PbO. An oxide of lead, varying in colour from pale yellow to brown depending on the method of manufacture. An intermediate product in the manufacture of red lead. See also **litharge** and **massicot**.

lead (IV) oxide *(Chem.)*. PbO_2. A strong oxidizing agent. Industrial application very limited. Present, in certain conditions, in accumulators or electrical storage batteries as a chocolate brown powder.

lead oxychloride *(Chem.)*. $PbCl_2.PbO$. See **Cassel's yellow**.

lead paint *(Build.)*. Paints containing lead. Because of toxicity their preparation, storage and application is usually governed by regulation.

lead paint regulations *(Build.)*. Detailed regulations which govern the use of lead paints, which are toxic.

lead plug *(Build.)*. A cast lead connecting piece binding together adjacent stones in a course; formed by running molten lead into suitably cut channels in the jointing faces.

lead poisoning *(Med.)*. Chronic poisoning due to inhalation, ingestion, and skin absorption of lead. Recognized as a hazard, both for young children (formerly through sucking lead or lead painted articles or toys), and also industrially. Characterized by anaemia, constipation, severe abdominal pain, and perhaps ultimately renal damage. Lesser degrees now recognised and soft water delivered by lead pipes is a potential hazard.

lead protection *(Radiol.)*. Protection provided by metallic lead against ionizing radiation. Joined with other substances to provide further protection, e.g. lead glass, lead rubber.

lead rubber *(Radiol.)*. Rubber containing high properties of lead compounds. Used as flexible protective material for, e.g. gloves and aprons.

leads *(Eng.)*. Lengths of thin lead wire inserted between a very large journal and the bearing cap during assembly to test the clearance.

leads *(Print.)*. Strips of lead placed between lines ot type. Usual thicknesses: 1-, $1\frac{1}{2}$-, 2-, and 3-point (*thick*), 6-point (*nonpareil*), 12-point (*pica*). Others are also used e.g. 4-point.

lead sulphate *(Chem.)*. $PbSO_4$. Formed as a white precipitate when sulphuric acid is added to a solution of a lead salt.

lead (II) sulphide *(Chem.)*. Found in nature as *galena*. Black precipitate formed when lead sulphate is reduced by carbon and when hydrogen sulphide is passed through a solution of a lead salt.

lead tetraethyl *(Chem.)*. $Pb(C_2H_5)_4$. A colourless liquid, obtained by the action of a zinc or magnesium ethyl halide on lead chloride. Used in motor spirit to increase the **antiknock value**.

lead tree *(Chem.)*. Form, in shape of tree, which lead takes after electrodeposition from simple salts.

lead zirconate titanate *(Chem.)*. A piezoelectric ceramic with a higher Curie temperature than barium titanate (IV).

leaf *(Bot.)*. (1) In modern vascular plants a lateral organ of limited growth which develops from a primordium at a shoot apex. In angiosperms a leaf typically has a bud in its axil. Most leaves are more or less flat and green and photosynthetic in function; modified leaves include bud scales, bulb scales, many sorts of spine and tendril, bracts and probably sepals and petals and possibly stamens and carpels. Cf. **sporophyll**. (2) In bryophytes, similar but usually smaller and thinner (mostly 1 cell thick) structures. See **phyllid**.

leaf area index *(Bot.)*. The ratio of the total area of leaves of a plant or crop to the area of soil available to it.

leaf filter *(Chem.Eng.)*. A type of filter in which large thin frames with a pipe connection through the frame are covered in cloth or metal gauze and placed inside a closed cylindrical vessel. Liquid to be filtered is forced under pressure into cylinder and clear filtrate leaves via frame's pipe connections (Kelly and Sweetland are this type). Similarly ore pulp is filtered by use of vacuum or pressure.

leaf gap *(Bot.)*. Region of parenchyma in the vascular cylinder of a stem above a leaf trace.

leafing *(Build.)*. The property of the small metallic flakes used in aluminium, bronze, and similar paints, of floating flat at the surface of the paint, thus increasing the metallic lustre. The overlapping effect of the flakes forms a barrier coat resistant to permeation.

leaflet *(Bot.)*. One of the leaf-like units which together make up the lamina of a compound leaf.

leaf mosaic *(Bot.)*. See **mosaic**.

leaf scar *(Bot.)*. The scar, usually covered by a thin protective layer, left on a stem following the abscission of a leaf.

leaf sheath *(Bot.)*. The sheath surrounding the stem at the base of a leaf in grasses.

leaf spring *(Eng.)*. A machine component comprising one or a group of relatively thin, flexible, or resilient strips, reacting as a spring to forces applied to a main surface.

leaf succulent *(Bot.)*. A plant with succulent photosynthetic leaves. Many are CAM plants e.g. *Aloe*. Cf. **stem succulent**.

leaf trace *(Bot.)*. A vascular bundle in a stem from its junction with another bundle of the stele to the base of the leaf; a leaf may have one trace or more.

leak *(Elec.Eng.)*. (1) A path between electrically isolated parts of a circuit, or of a component, which has reduced resistance and can cause small unwanted currents to flow. (2) A high-valued resistor deliberately placed in a circuit to permit the controlled discharge of electrically charged components, usually associated with the discharge of a capacitor.

leakage *(Nuc.Eng.)*. Net loss of particles from a region or across a boundary e.g. neutrons from the core of a reactor, often split up for calculation purposes into *fast non-leakage probability* (P_f); *resonance non-leakage probability* (P_r); *thermal non-leakage probability* (P_t).

leakage coefficient *(Elec.Eng.)*. The ratio of the total flux in the magnetic circuit of an electric machine or transformer to the useful flux which actually links with the armature of secondary winding. Also called *leakage factor*.

leakage conductance *(Elec.Eng.)*. In electrical circuits, the leakage current expressed by the reciprocal of insulation resistance of the circuit.

leakage current *(Elec.Eng.)*. Small unwanted current flowing through a component such as a capacitor or a reverse biased diode.

leakage factor *(Elec.Eng.)*. See **leakage coefficient**.

leakage flux *(Elec.Eng.)*. (1) That which, in any type of electric machine or transformer, does not intercept all the turns of the winding intended to enclose it. (2) That which crosses the air gap of a magnet other than through the intended pole faces. Often called *lost flux*.

leakage indicator *(Elec.Eng.)*. An instrument for measuring or detecting a leakage of current from an electric system to earth. Also called an *earth detector*.

leakage protective system *(Elec.Eng.)*. A protective system which operates as a result of leakage of current from electrical apparatus to earth.

leakage reactance *(Elec.Eng.)*. That, in a transformer, which arises from flux in one winding not entirely linking another winding; measured by the inductance of one winding when the other is short circuited. Leakage inductance, taken with the effective self-capacitance of a winding, determines the range of frequencies effectively passed by the transformer.

leak detector *(Nuc.Eng.)*. Syn. for **burst-can detector**.

leak detector *(Phys.)*. Device for indicating points of ingress of gas into a high vacuum system.

lean burn engine *(Autos.)*. Engine employing **stratified charge combustion** to burn unusually weak (non-stoichiometric) mixtures to reduce unwanted exhaust emissions.

lean ore *(Min.Ext.)*. Ore of marginal value; low grade ore.

lean-to roof *(Arch.)*. A roof having only one slope.

leap-frog test *(Comp.)*. Program used to test the internal operation of a computer.

leaping weir *(Civ.Eng.)*. A special arrangement whereby flood flows may be diverted from a channel into which normal flows would ordinarily pass, the water having to go over a weir set at such a height that flood flows leap beyond the channel to an overflow. Also called a *separating weir*.

leap second *(Astron.)*. A periodic adjustment of time signal emissions to maintain synchronism with co-ordinated universal time (UTC). A positive leap second may be inserted, or a negative one omitted, at the end of December or June.

leap years *(Astron.)*. Those years in which an extra day, February 29, is added to the civil calendar to allow for the fractional part of a tropical year of 365.2422 days. Since the Gregorian reform of the Julian calendar, the leap years are those whose number is divisible by 4, except centennial years unless these are divisible by 400.

lear board *(Build.)*. A board carrying a lead gutter. Also called a *layer board*.

learned helplessness *(Behav.)*. Described by Seligman, a condition characterized by a general sense of powerlessness, which has its origins in a traumatic and inescapable event, but which persists in situations where escape or avoidance is possible. It specifically refers to effects created in the laboratory in animals exposed to uncontrollable events, but is thought by some to be one of the factors underlying depression in humans.

learning set *(Behav.)*. Refers to the increased ability to solve a particular kind of problem as a consequence of previous experience with similar kinds of problems.

learning theory *(Behav.)*. A range of theories which, unlike **depth psychology**, explain behaviour in terms of learning and conditioning. Largely the offspring of *behaviourism*, it takes as its starting point the association of various stimuli and responses. See **behaviour therapy, social learning theory, S-R theory**.

lease rods *(Textiles)*. Two rods, across a warp sheet, which separate the yarns, equalize the tension and keep them a the correct height.

least action, principle of *(Phys.)*. See **principle of least action**.

least distance of distinct vision *(Biol.)*. For a normal eye it is assumed that nothing is gained by bringing an object to be inspected nearer than 25 cm, owing to the strain imposed on the ciliary muscles if the eye attempts to focus for a shorter distance.

least energy principle *(Phys.)*. That a system is only in stable equilibrium under those conditions for which its potential energy is a minimum.

least time, principle of *(Phys.)*. See **Fermat's principle of least time**.

leat *(Min.Ext.)*. Ditch following contour line used to conduct water to working place.

leather *(Genrl.)*. A universal commodity made by **tanning** and other treatment of the hides or skins of a great variety of creatures (mostly domesticated animals, but including also the whale, seal, shark, crocodile, snake, kangaroo, camel, ostrich etc.). *Artificial* or *imitation leathers* are commonly based on cellulose nitrate, PVC or polyurethane, suitably treated to simulate leather.

leatherboard *(Paper)*. Board made largely or wholly from leather scraps, generally on an intermittent board machine and containing latex as a binder and to impart flexibility.

leathercloth *(Textiles)*. A woven or knitted fabric coated on one side with a polymer (e.g. rubber, cellulose derivative, polyvinyl chloride) which is embossed to simulate leather.

leather hollows *(Eng.)*. Strips of leather used by pattern makers to form the fillets in wood patterning.

Leavers machine *(Textiles)*. See **lace machines**.

leaving edge *(Elec.Eng.)*. The edge of the brush of an electric machine which is last met during revolution by a point on the commutator or slip ring. Also called *heel, back, trailing edge.*

Leber's disease *(Med.)*. Hereditary optic atrophy; hereditary optic neuritis. A hereditary *sex linked* disease in which there is gradual loss of sight due to an affection of the optic nerve behind the eyeball.

Lebesgue integral *(Maths.)*. An advanced concept based upon the theory of sets whereby functions, not ordinarily integrable, can be said to be integrated, e.g. the function which is one when x is rational and zero elsewhere, is not ordinarily integrable but it does have a Lebesgue integral. Where both an ordinary integral (i.e. that of Riemann) and a Lebesgue integral exist they are equal.

Leblanc connection *(Elec.Eng.)*. Method of connecting transformers for linking a 3-phase to a 2-phase system.

Leblanc phase advancer *(Elec.Eng.)*. A d.c. armature with 3 sets of brushes per pole pair on the commutator; the brushes are connected to the 3 slip rings of the induction motor, and the advancer is driven from the motor shaft at an appropriate speed, causing it to take a leading current and improve the motor power factor.

Leblanc process *(Chem.)*. A process, formerly of great importance but now obsolete, for the manufacture of sodium carbonate and intermediate products from common salt, coal, limestone, and sulphuric acid.

LE cell *(Immun.)*. See **lupus erythematosus cell**.

Le Chatelier-Braun principle *(Chem.)*. If any change of conditions is imposed on a system at equilibrium, then the system will alter in such a way as to counteract the imposed change. This principle is of extremely wide application.

lechatelierite *(Min.)*. A name sometimes applied to naturally fused vitreous silica, such as that which occurs as fulgurites (*lightning tubes*).

Le Chatelier test *(Build.)*. A simple method for checking the soundness of cement by checking its expansion.

Lecher wires *(Elec.Eng.)*. Two insulated parallel stretched wires tunable by means of sliding short-circuiting copper strip. The wires form a microwave electromagnetic transmission line which may be used as a tuned circuit, as an impedance matching device or for the measurement of wavelengths. Also called *parallel wire resonator*.

lecithin *(Biol.)*. See **phosphatidyl choline**.

lecith-, lecitho- *(Genrl.)*. Prefix from Gk. *lekithos*, yolk of egg.

lecithocoel *(Zool.)*. The segmentation cavity of a holoblastic egg.

Leclanché cell *(Elec.Eng.)*. A primary cell, good for intermittent use, has a positive electrode of carbon surrounded by a mixture of manganese dioxide and powdered carbon in a porous pot. The pot and the negative zinc electrode stand in a jar containing ammonium chloride solution. The e.m.f. is approximately 1.4 volt. The *dry cell* is a particular form of Leclanché.

lectin *(Immun.)*. Proteins derived usually from plants that bind specifically to sugars or to oligosaccharides. Since similar sugar residues are present in glycoproteins or glycolipids on the surface of many cells, lectins cause agglutination of these cells. Some lectins bind specifically to certain cell types or even to cells at a particular stage of differentiation, and are used to identify them or to isolate glycoproteins extracted from them. Several act as polyclonal mitogens for T- and/or B-lymphocytes (e.g. *Concanavalin A* from Jack beans, *Phytohaemagglutinin* from red kidney beans and *Pokeweed mitogen* from *Phytolacca americana*).

LED *(Comp.)*. See **Light-Emitting Diode**.

Leda *(Astron.)*. The thirteenth natural satellite of **Jupiter**, perhaps only 10 km in diameter, discovered in 1979.

ledge *(Build.)*. One of the battens across the back of a batten door.

ledged-and-braced door *(Build.)*. A door similar to a batten door, but framed diagonally with braces across the back, between the battens.

ledged door *(Build.)*. See **batten door**.

ledgement *(Build.)*. A horizontal line of mouldings, or a string course.

ledger *(Build.)*. A horizontal pole or member, lashed or otherwise fastened across the standards in a scaffold or in a trench.

ledger board *(Build.)*. See **ribbon strip**.

Leduc effect *(Electronics)*. A magnetic field applied at right angles to the direction of a temperature gradient in an electrical conductor, will produce a temperature difference at right angles to the direction of both the temperature gradient and the magnetic field.

LEED *(Phys.)*. *Low-Energy Electron Diffraction.* Used to study the structure of surfaces using electrons of energy in the range 10–500 eV.

lee wave *(Meteor.)*. Stationary wave set up in an air stream to the lee of a hill or range over which air is flowing. Special conditions of atmospheric stability and vertical wind-structure are required. Lee waves are sometimes of large amplitude and can be dangerous to aircraft.

left-handed engine *(Aero.)*. An aero engine in which the propeller shaft rotates counter clockwise with the engine between the observer and the propellor.

left-hand propeller *(Aero.)*. See **propeller**.

left-hand thread *(Eng.)*. A screw thread cut in the opposite direction to the normal right hand. Viewed in elevation, the external thread is inclined upwards from right to left. Used when a normal thread would tend to unscrew.

left-hand tools *(Eng.)*. Lathe side tools with the cutting edge on the right, thus cutting from left to right.

leg *(Bot.)*. In horticulure a single short trunk from which branches arise, as of a fruit bush that is not managed as a **stool**.

leg *(Telecomm.)*. One side of a loop circuit, i.e. either the *go* or *return* of an electrical circuit.

legend *(Print.)*. Descriptive matter accompanying an illustration; a caption.

Legendre's differential equation *(Maths.)*. The equation

$$(1-x^2)\frac{d^2y}{dx^2}-2x\frac{dy}{dx}+n(n+1)y=0.$$

It is satisfied by the Legendre polynomial $P_n(x)$.

Legendre's polynomials *(Maths.)*. The polynomials $P_n(x)$ in the expansion

$$(1-2xh+h^2)^{-\frac{1}{2}}=\sum_{n=0}^{\infty}P_n(x).h^n.$$

In particular, $P_0(x)=1$, $P_1(x)=x$, $P_2(x)=\frac{1}{2}(3x^2-1)$, and $P_3(x)=\frac{1}{2}(5x^3-3x)$. Cf. **Chebyshev polynomials**.

leghaemoglobin *(Bot.)*. A protein, similar to haemoglobin, in the nitrogen-fixing root nodules of leguminous plants, where it is involved in the maintenance of a low concentration of oxygen. See **nitrogenase**.

legionnaire's disease *(Med.)*. An *atypical pneumonia* caused by *Legionella pneumophilia.* Infection by droplet inhalation from air conditioners and showers. In severe cases there is also renal, hepatic and neurological involvement.

legume *(Bot.)*. (1) A fruit from one carpel, dehiscent along top and bottom, e.g. pea pod. (2) A member of the Leguminosae. adj. *leguminous.*

Leguminosae *(Bot.)*. *Fabaceae.* The pea family, ca 17 000 spp. of dicotyledonous flowering plants (superorder Rosidae). Trees, shrubs and herbs, cosmopolitan. The flowers have five free petals and a superior ovary of one carpel. The fruit is characteristically a legume or pod. There are three subfamilies (sometimes ranked as families): Mimosoideae, Caesalpinioideae and Papilionoideae, the last with more or less butterfly-shaped or 'pea' flowers. They form root-nodules with symbiotic, nitrogen-fixing bacteria, *Rhizobium* spp. Although the seeds of many are poisonous (especially uncooked), the Papilionoideae includes extremely important crops e.g. the protein-rich seeds and pods of many sorts of peas and beans, lentils and groundnuts, and forage crops e.g. alfalfa.

lehr *(Glass)*. An enclosed oven or furnace used for annealing or other forms of heat treatment. It is a kind of tunnel down which the glass, hot from the forming process, is sent to cool slowly, so that strain is removed, and cooling takes place without additional strain being introduced. Lehrs may be of the open type (in which the flame comes in contact with the ware) or of the muffle type. Also *lear, leer, lier*.

Leica *(Image Tech.)*. TN of the first miniature still camera using 35 mm film, giving its name to the corresponding standard frame size of 36 × 24 mm.

leiomyoma *(Med.)*. A tumour composed of unstriped muscle fibres.

leishmaniasis, leishmaniosis *(Med.)*. A term applied to a group of diseases caused by infection with protozoal parasites of the genus *Leishmania*. See also **kala-azar**.

Lemberg's stain test *(Geol.)*. A black iron sulphide stain, used to distinguish between calcite and dolomite in limestones.

lemma *(Bot.)*. The lower of the two bracts enclosing a floret of a grass. Cf. **palea**.

lemma *(Maths.)*. A subsidiary theorem. The proof of a complicated theorem is generally made more clear by proving any subsidiary facts in a series of lemmas.

lemniscate of Bernoulli *(Maths.)*. A figure-eight-shaped curve with polar equation $r^2 = a^2\cos 2\theta$. Cf. **ovals of Cassini**.

length *(Comp.)*. The number of **bits** in a **word**, or the number of columns or spaces in a **field**.

length between perpendiculars *(Ships)*. The length on the summer load waterline from the fore side of the stem to the after side of the stern post or, in a vessel without a stern post, to the centre of the rudder stock.

length-fast, length-slow *(Min.)*. In optically birefringent minerals, terms used to denote whether the fast or the slow vibration direction is aligned parallel or nearly parallel to the length of prismatic crystals.

length fold collection *(Print.)*. Bringing together, on rotary presses, folded products from a series of formers sometimes mounted one above another. See **balloon former**.

length of lay *(Telecomm.)*. See **lay**.

length overall *(Ships)*. The length from the foremost point of the structure, including any figurehead or bowsprit, to the aftermost point of the ship.

Lennard-Jones potential *(Chem.)*. Potential due to molecular interaction, also called the *6-12 potential* because the attractive potential varies as the inverse 6th power of the intermolecular distance, and the repulsive potential as the inverse 12th.

leno fabric *(Textiles)*. A woven fabric in which the warp threads are made to cross one another between **picks**. Light-weight fabrics of this kind are known as gauzes; heavier qualities of cotton are used as blankets.

lens *(Electronics)*. (1) Device for focusing radiation; see **lens antenna**. (2) An arrangement of coils, permanent magnets or electrodes which can produce magnetic electrical fields capable of focusing electron beams. Used particularly in electron microscopes and accelerators.

lens *(Phys.)*. A portion of a homogeneous transparent medium bounded by spherical surfaces. Each of these surfaces may be convex, concave, or plane. If in passing through the lens a beam of light becomes more convergent or less divergent, the lens is said to be *convergent* or *convex*. If the opposite happens, the lens is said to be *divergent* or *concave*. See **chromatic aberration, focal length of a lens, image, lens formula, spherical aberration, thick lens**.

lens *(Zool.)*. In Arthropoda, the cornea of an ocellus or compound eye; in Craniata and Cephalopoda, a structure immediately behind the iris which serves to focus light on to the retina. See also **crystalline lens**.

lens antenna *(Telecomm.)*. A microwave antenna in which a focusing arrangement is placed in front of the radiator of energy in order to concentrate it into a beam of predetermined form and direction. The lens may consist of a system of metal slats or pieces of shaped dielectric material, or a lattice of metal spheres, the intention of all designs is to introduce selective phase delays over different paths through the lens in order to shape the beam.

lens barrel *(Image Tech.)*. The metal tube in which one or more lenses are mounted.

lens cap *(Image Tech.)*. A temporary light-tight protective covering for the external end of a camera lens; removed during exposure and focusing.

lens formula *(Phys.)*. The equation giving the relation between the image and object distances, l' and l and the focal length f of lens:

$$\frac{1}{f} = \frac{1}{l'} - \frac{1}{l}$$

in the cartesian **convention of signs** for a thin lens in air.

lens grinding *(Glass)*. The process of grinding pieces of flat sheet-glass (or pressed blanks) to the correct form of the lens. Cast iron 'tools' of the correct curvature, supplied with a slurry of abrasive and water, are used.

lens hood *(Image Tech.)*. Tubular or rectangular shade in front of the camera to exclude stray light from entering the lens; it may also include mountings for filters and fixed **mattes**. See also **matte box**.

lens mount *(Image Tech.)*. The metal tubular fitting housing the elements of a camera lens, incorporating the focus adjustment and sometimes the iris diaphragm; it is usually attached to the camera body by a standard screw or bayonet lock.

lentic *(Ecol.)*. Associated with standing water; inhabiting ponds, swamps, etc.

lenticel *(Bot.)*. A small patch of the periderm in which intercellular spaces are present allowing some gas exchange between the internal tissues and the atmosphere.

lenticle *(Geol.)*. A mass of lenslike (lenticular) form. The term may refer to masses of clay in sand, or vice versa, and, in metamorphic rocks, to enclosures of one rock type in another.

lenticonus *(Med.)*. Abnormal curvature of the lens of the eye, in which the surface becomes conical instead of spherical.

lenticular *(Bot.,Zool.)*. Shaped like a double convex lens.

lenticular *(Min.)*. Said of a mineral or rock of this particular shape, embedded in a matrix of a different kind.

lenticular bob *(Horol.)*. A pendulum bob whose cross section corresponds to that of a double convex lens; the usual form of bob for household clocks.

lenticular girder bridge *(Eng.)*. A type of girder bridge composed of an arch whose thrust is taken by a suspension system hanging below it, the anchorages of the latter system being provided by the arch thrust.

lenticular process *(Image Tech.)*. Obsolete process of colour photography and cinematography in which a black-and-white emulsion was exposed through a base embossed with minute lenses or ribs, using a banded tri-colour filter on the camera lens. After reversal processing, the image was projected through a similar filter to produce an *additive* colour picture.

lentigo *(Med.)*. Freckles. adjs. *lentiginose, lentiginous*.

Lentz valve gear *(Eng.)*. A locomotive valve gear in which the steam is admitted and exhausted through two pairs of *poppet valves*, spring controlled and operated from a camshaft rotating at engine speed.

Lenz's law *(Elec.Eng.)*. An induced e.m.f. will tend to cause a current to flow in such a direction as to oppose the cause of the induced e.m.f.

Leonids *(Astron.)*. A swarm of meteors whose orbit round the sun is crossed by the earth's orbit at a point corresponding to about November 17, when a display of more than average numbers is to be expected; the radiant point is in the constellation Leo.

leontiasis ossea *(Med.)*. A rare condition characterized by diffuse hypertrophy of the bones of the skull.

lepido- *(Genrl.)*. Prefix from Gk. *lepis*, gen. *lepidos*, a scale.

lepidocrocite *(Min.)*. An orthorhombic hydrated oxide of iron (FeO.OH) occurring as scaly brownish-red crystals in iron ores. It is often one of the constituents of limonite, together with its dimorph goethite.

lepidolite *(Min.)*. A lithium bearing mica crystallizing in the monoclinic system as pink to purple scaly aggregates. It is essentially a hydrated silicate of potassium, lithium, and aluminium, often also containing fluorine, and is the most common lithium bearing mineral. It occurs almost entirely in pegmatites. Also known as *lithia mica*.

lepidomelane *(Min.)*. A variety of biotite, rich in iron, which occurs commonly in igneous rocks.

Lepidoptera *(Zool.)*. An order of Endopterygota, having two pairs of large and nearly equal wings densely clothed in scales; the mouth parts are suctorial, the mandibles being absent and the maxillae forming a tubular proboscis; the larva or caterpillar is active and usually herbivorous, with biting mouth parts. Butterflies and Moths.

lepidote *(Bot.)*. With a covering of scale-like hairs.

lepospondylous *(Zool.)*. Said of vertebral centra in which there is a skeletal ring constricting the notochord in the intervertebral region, with an expansion between each pair of adjacent centra.

leproma *(Med.)*. A nodular lesion of leprosy.

lepromin test *(Immun.)*. Test in which killed *Mycobacterium leprae* organisms are injected into the skin of subjects with leprosy. Those with lepromatous leprosy do not react, whereas those with the tuberculoid form of leprosy show a tuberculin type response. Not diagnostic of leprosy but an aid to classification and prognosis.

leprosy *(Med.)*. Infection with the bacillus *Mycobacterium leprae*; a highly chronic and moderately contagious disease. *Lepromatous* and *neural* forms are recognized, the one with nodular and destructive lesions chiefly of the face and nose, the other, even more chronic, involving chiefly the nerves.

leptite *(Geol.)*. An even-grained metamorphic rock composed mainly of quartz and feldspars. Approximately synonymous with *granulite* and *leptynite*.

lepto- *(Genrl.)*. Prefix from Gk. *leptos*, slender.

leptocercal, leptocercous *(Zool.)*. Having a long slender tail.

leptodactylous *(Zool.)*. Having slender digits.

leptodermatous *(Zool.)*. Thin skinned.

leptomeninges *(Med.)*. The two innermost membranes, the arachnoid and the pia mater, investing the brain and the spinal cord.

leptomeningitis *(Med.)*. Inflammation of the leptomeninges.

leptom, leptome *(Bot.)*. The sieve elements and associated parenchyma cells (but not sclerenchyma cells) of the phloem. Cf. **hadrom**.

lepton *(Phys.)*. A fundamental particle that does not interact strongly with other particles. There are several different types of lepton; the electron, the negative muon, the tau-minus particle and their associated neutrinos. See **antilepton**.

leptonema *(Biol.)*. See **leptotene**.

lepton number *(Phys.)*. The number of **leptons** less the number of corresponding antiparticles taking part in a process. The number appears to be conserved in any process.

lepton-quark symmetry *(Phys.)*. The number of *pairs* of leptons (Charge $-e$ and charge 0) is equal to the number of *pairs* of quarks (charge $+2e/3$ and charge $-e/3$). The number of types of leptons is thus equal to the number of types of quarks. Required by the theory unifying electromagnetic and weak interactions between particles. See **Weinberg and Salam's theory**.

leptospirosis *(Med.)*. Disease caused by infection with one of three leptospira, causing malaise, fever and jaundice.

leptospirosis *(Vet.)*. A disease of animals or man caused by infection by an organism of the genus *Leptospira*. Vaccines can be used.

leptosporangium *(Bot.)*. A sporangium characteristic of the Filicales to which most modern ferns belong. Cf. **eusporangium**.

leptotene *(Biol.)*. The first stage of meiotic prophase, in which the chromatin thread acquires definite polarity.

leptynite *(Geol.)*. See **leptite**.

lesbianism *(Med.)*. Homosexuality between females.

lesion *(Med.)*. Any wound or morbid change anywhere in the body.

Leslie matrix model *(Ecol.)*. Specific deterministic model to predict the age structure of a population given the age structure at some past time and the age-specific survival and fecundity rates. Proposed by P.H. Leslie in 1945.

lessivage *(Bot.)*. The process in which clay particles are washed downwards in a soil by percolating water. Cf. **leaching, cheluviation**.

lethal *(Biol.)*. Causing death. (1) Of an environmental factor fatal to an organism. (2) Of a genetic factor causing death often, as in bacteria, only in a particular medium or environment, then a *conditional lethal*.

lethargic encephalitis *(Med.)*. See **encephalitis lethargica**.

lethargy of neutrons *(Nuc.Eng.)*. The natural logarithm of the ratio of initial and actual energies of neutrons during the moderation process. The lethargy change per collision is defined similarly in terms of the energy values before and after the collision. Also called *logarithmic energy decrement*.

let-off motion *(Textiles)*. Mechanism behind the machine to regulate the tension and delivery of the sheet of warp threads coming from the beam to the loom or warp-knitting machine.

Letraset *(Print.)*. TN for a **transfer lettering system**.

letterpress *(Print.)*. (1) A term applied to printing from relief surfaces (type or blocks), as distinct from lithography, intaglio etc. See **printing**. (2) The reading matter in a book, apart from illustrations, etc.

letterset printing *(Print.)*. See under **offset printing**.

letter sizes *(Eng.)*. A series of drill sizes in which each letter of the alphabet represents a diameter, increasing in irregular increments from A = 0.234 in (5.94 mm) to Z = 0.413 in (10.49 mm). Below A, *gauge numbers*, from 1 (0.228 in, 5.77 mm) to 80 (0.0135 in, 0.343 mm), are used in the reverse direction.

letting down *(Aero.)*. The reduction of altitude from cruising height to that required for the approach to landing.

letting down *(Eng.)*. The process of tempering hardened steel by heating until the desired temperature, as indicated by colour, is reached, and then quenching.

Leu *(Chem.)*. Symbol for **leucine**.

leucine *(Chem.)*. *2-amino-4-methylpentanoic acid*, $(CH_3)_2CH.CH_2.CH(NH_2).COOH$. The L- or S-isomer is a constituent of proteins. Symbol Leu, short form L. ⇨

leucite *(Min.)*. A silicate of potassium and aluminium, related in chemical composition to orthoclase, but containing less silica. At ordinary temperatures, it is tetragonal (pseudocubic), but on heating to about 625°C it becomes cubic. Occurs in igneous rocks, particularly potassium rich, silica poor lavas of Tertiary and Recent age, as e.g. at Vesuvius.

leucitophyre *(Geol.)*. A fine-grained igneous rock, commonly occurring as a lava, carrying phenocrysts of leucite and other minerals in a matrix essentially trachytic; a well known example comes from Rieden in the Eifel.

leuco- *(Genrl.)*. Prefix from Gk. *leukos*, white.

leuco-bases, leuco-compounds *(Chem.)*. Colourless compounds formed by the reduction of dyes, which when oxidized are converted back into dyes.

leucoblast *(Zool.)*. A cell which will give rise to a leucocyte.

leucocratic *(Geol.)*. A term used to denote a light colour in igneous rocks, due to a high content of felsic minerals, and a correspondingly small amount of dark, heavy silicates.

leucocyte *(Zool.)*. A white blood-corpuscle; one of the colourless amoeboid cells occurring in suspension in the blood-plasma of many animals.

leucocythaemia *(Med.)*. Obsolete term for *leukaemia*.

leucocytolysis *(Med.)*. The breaking down of white cells of the blood.

leucocytosis *(Med.)*. An increase in the number of leucocytes in the blood.

leucodermia, leucoderma *(Med.)*. *Vitiligo; melanodermia.* A condition in which white patches, surrounded by a pigmented area, appear in the skin.

leuco-erythroblastic anaemia *(Med.)*. A form of leukaemia in which immature forms of red and white cell series are found in the peripheral blood.

leucopenia, leucocytopenia *(Med.)*. Abnormal diminution in the numbers of white cells in the blood.

leucoplakia *(Med.)*. The stage of a chronically inflamed area at which the surface becomes hard, white, and smooth. On the tongue, it is usually due to syphilitic infection.

leucoplast *(Bot.)*. A colourless **plastid**.

leucopoiesis *(Med.)*. The production of white cells of the blood.

leucorrhoea, leucorrhea *(Med.)*. A whitish discharge from the vagina.

leucosapphire *(Min.)*. See **white sapphire**.

leucotomy *(Med.)*. Surgical scission of the association fibres between the frontal lobes of the brain and the thalamus, occasionally performed to relieve cases of severe schizophrenia and manic depression. Also called *prefrontal lobotomy*.

leucoxene *(Min.)*. An opaque whitish mineral formed as a decomposition product of ilmenite. Normally consists of finely crystalline rutile, but may be composed of finely divided brookite.

leukaemia, leukemia *(Med.)*. Progressive uncontrolled overproduction of any one of the types of white cell of the blood, with suppression of other blood cells and infiltration of organs such as the spleen and liver. Cancers of the bloodcell forming tissues. Occurs in acute and chronic **myeloid**, **lymphatic**, and **monocytic** varieties.

leuko- *(Genrl.)*. A variant spelling of **leuco-**.

leukotrienes *(Immun.,Med.)*. Pharmacologically active substances related to **prostaglandins** and generated from arachidonic acid by the action of lipoxygenases. A series of hydroxyicosatetraenoic acids, of which leukotriene B_4 is chemotactic for granulocytes; leukotrienes C_4, D_4 and E_4 (which contain cysteine) have the properties of a 'slow reacting substance of anaphylaxis' (*SRS-A*), i.e. they cause contraction of some types of smooth muscle, especially bronchial muscle, and increase vascular permeability. Leukotrienes may be released from platelets and various leucocytes when damaged, and SRS-A activity is generated as a result of combination of antigen with IgE antibody on mast cells.

levator *(Zool.)*. See **elevator**.

levee *(Geol.)*. A long low ridge built up on either side of a stream on its flood plain. It consists of relatively coarse sand and silt deposited by the stream when it overflows its banks.

levee *(Hyd.Eng.)*. A well consolidated bank of earth or spoil, having a central core forming an impervious connection with the natural ground; used as a form of training works to control river flow and prevent flooding of adjacent country. In America the name also signifies a quay, landing place, or pier.

level *(Acous.)*. Logarithmic ratio of two energies or two field quantities, where the nominator is the measured quantity and the denominator a reference quantity.

level *(Build.)*. A *spirit level*.

level *(Civ.Eng.)*. (1) To reduce a cut or fill surface to an approximately horizontal plane. (2) The state whereby any given surface is at a tangent to the perfect earth's perimeter at any given point. (3) A ditch or channel for drainage, especially in flat country.

level *(Min.Ext.)*. An approximately horizontal tunnel in a mine, generally marking a working horizon or level of exploitation, and either in or parallel to the ore body.

level *(Phys.)*. Possible energy value of electron or nuclear particle.

level *(Surv.)*. An instrument for determining the difference in height between two points. See also **level tube**.

level *(Telecomm.)*. (1) The difference between a measured and an arbitrarily defined reference quantity, usually expressed as the logarithm of the ratio of the quantities. In telecommunications, a common reference level is 1 mW, and power levels are expressed in decibels relative to this (dBm). (2) A specified point on an amplitude scale applied to a signal waveform, e.g. *reference white* or *black* in a television transmission. (3) A single bank of contacts, e.g. in a stepping relay.

level canal *(Hyd.Eng.)*. A canal which is level throughout. Also called a *ditch canal*.

level-compounded *(Elec.Eng.)*. See **flat-compounded**.

level indicator *(Telecomm.)*. Voltage indicator on a transmission line, calibrated to indicate decibels in relation to a zero power level; also called *power-level indicator*. See **volume unit**.

levelling *(Surv.)*. The operation of finding the difference of elevation between two points.

levelling *(Textiles)*. Process that allows a freshly dyed material to become uniformly coloured.

levelling agent *(Chem.)*. Agent added to a dye bath to produce uniform precipitation of the dye on to the fibre.

levelling bulb *(Chem.)*. In a gas pipette, the pressure may be equalized to that of the atmosphere by connecting with flexible tubing the liquid in the pipette to that in an open bulb, the level of which is adjusted till the heights of the two liquid interfaces (and hence their hydrostatic pressures) are equal.

levelling solvent *(Chem.)*. One which has a high dielectric constant and high polarity, so that most electrolytes appear strong in solution.

levelling staff *(Surv.)*. Light extensible system of wooden rods graduated in feet and tenths, used to transfer **line of collimation** of surveyor's level from back sight, on which it is first held vertically, to fore sight, its second vertical station. As the line of collimation is horizontal, change of height is thus measured. See **target rod**.

level-luffing crane *(Eng.)*. A jib crane in which, during derricking or luffing, the load is caused to move radially in a horizontal path, with consequent power saving.

level setting *(Telecomm.)*. Provision for adjusting the base voltage for an irregular waveform, e.g. in TV scanning circuit voltages and signals. See **clamp**.

level small caps *(Print.)*. See **even small caps**.

level trier *(Surv.)*. An apparatus for measuring the angular value of a division on a level tube; it consists of a beam, hinged about a horizontal axis at one end and capable of being moved up or down at the other end by means of a micrometer screw, which records the inclination corresponding to a given number of divisions of movement of the bubble.

level tube, spirit level *(Surv.)*. A specially shaped glass tube nearly filled with spirit, so as to leave a 'bubble' of air and spirit vapour, which always rises to the highest part of the tube. The level tube is used to test whether a surface to which it is applied is horizontal. It is an essential feature of many forms of surveying instrument. See **dumpy level**, **tilting level**.

lever *(Horol.)*. The pivoted arm which carries the pallets in the lever escapement.

lever *(Phys.)*. One of the simplest machines. It may be considered as a rigid beam pivoted at a point called the *fulcrum*, a load being applied at one point in the beam and an effort, sufficient to balance the load, at another. Three classes of lever may be distinguished: (a) fulcrum between effort and load; (b) effort between fulcrum and load; (c) load between fulcrum and effort. See **machine**, **mechanical advantage**, **velocity ratio**.

lever escapement *(Horol.)*. The most important type of watch escapement. The impulse from the escape wheel is transmitted to the balance by the equivalent of two levers – a pivoted lever carrying the pallets, and the roller carrying the impulse pin. In the *English* lever, the escape wheel is planted at right angles to the line joining the pallet staff and balance staff centres. In the *Swiss* lever or *straight-line* escapement the escape wheel, pallet staff and balance staff centres are in a straight line. Strictly speaking, the pin-pallet escapement is a lever escapement, but it is not generally referred to as such.

lever key *(Telecomm.)*. A hand-operated key for telephone switchboards; operated by a small lever, which opens and closes one or more spring contacts. May be locking or nonlocking.

lever safety valve *(Eng.)*. A safety valve in which the valve is held on its seating by a long lever, loaded by a weight at the other end; a form of **dead-weight safety valve**.

lever-type brush-holder *(Elec.Eng.)*. A type of brush-holder in which the brush is held at the end of an arm pivoted about the brush spindle.

lever-type starter *(Elec.Eng.)*. See **face-plate starter**.

lever-wind *(Image Tech.)*. Device on a camera which transports the film and sets the shutter in one movement.

levigation *(Min.Ext.)*. Use of sedimentation or elutriation to separate finely ground particles into fractions according to their movement through a separating fluid. Also used in connection with wet grinding.

levitation melting *(Eng.)*. A process in which the melt is suspended in an electromagnetic field while being heated by induction. It is therefore not contaminated by the material of a container while molten.

levitron *(Nuc.Eng.)*. Toroidal fusion system in which the poloidal magnetic field is provided by a current flowing in a solid ring situated at the circular axis of the torus. The ring may be supercooled so can be levitated indefinitely, avoiding the use of material supports.

levodopa *(Med.)*. The amino acid precursor of **dopamine**; by replenishing dopamine in the basal ganglia it improves mobility in *Parkinsonism*. Usually administered with an extra-cerebral dopa-decarboxylase inhibitor (carbidopa) to prevent peripheral effects of *dopamine*.

levyne, levynite *(Min.)*. A zeolite; hydrated silicate of calcium and aluminium crystallizing in the trigonal system.

lew *(Build.)*. A light covering or roof of straw used to protect bricks on **hacks** during the drying period.

lewis *(Build.)*. A truncated steel wedge or dovetail made in 3 pieces, with the larger end downwards and fitting into a similarly shaped hole in the top of a block of masonry; it then provides, by its attached hoist ring, a means of lifting the stone.

Lewis acids and bases *(Chem.)*. The concept that defines any substance donating an electron pair as a base, and any substance accepting an electron pair as an acid. Conventional acids and bases fit this definition, as do complex-forming reactions. Thus in the reaction $AlCl_3 + Cl^- = AlCl_4^-$, $AlCl_3$ is classed as an acid, and Cl^- as a base.

lewis bolt *(Civ.Eng.)*. A foundation bolt with a tapered and jagged head, which is securely fixed into a hole in the anchoring masonry by having molten lead run round it; also used in concrete foundations.

Lewis formula *(Eng.)*. A formula used to calculate the strength of gear teeth.

lewisite *(Chem.)*. $ClCH{=}CH.AsCl_2$. *2-chloro-ethenyl dichloro arsine*. Dark coloured oily liquid (colourless when pure) with a strong smell of geraniums. Vesicant having lachrymatory and nose irritant action. Used as poison gas.

lewisson *(Build.)*. A **lewis**.

Lewis's theory *(Chem.)*. The assumption that atoms can combine by sharing electrons, thus completing their shells without ionization.

Lewis structure *(Chem.)*. A possible structure for a molecule in which all electrons are specifically associated with one or two atoms. Many structures can only be described by a mixture of two or more Lewis structures, the best known example being benzene. Also *resonance structure*.

lexical analysis *(Comp.)*. Stage during the **compilation** of a program, in which standard components of a statement, such as PRINT, IF etc. are replaced by internal codes (*tokens*) which identify their meaning.

Leydig's duct *(Zool.)*. See **Wolffian duct**.

L-forms *(Biol.)*. Morphological variants developed from large bodies by prolonged exposure to various treatments. Consist of colonies of filterable bodies with a cytoplasmic matrix. Frequently revert to the normal form. If blood or serum is supplied in their nutrient medium may breed true for some generations.

LH *(Med.)*. Abbrev. for *Luteinizing Hormone*.

L-head *(Autos.)*. A petrol engine cylinder head carrying the inlet and exhaust valves in a pocket at one side; resembles an inverted L.

lherzolite *(Geol.)*. An ultramafic plutonic rock, a peridotite, consisting essentially of olivine, with both ortho- and clino-pyroxene; named from Lake Lherz in the Pyrenees.

L'Hospital's rule *(Maths.)*. A rule for evaluating certain limits. If $f(a)=0$ and $g(a)=0$, and if n is the lowest integer for which $f^n(a)$ and $g^n(a)$ are not both zero, then

$$\lim_{x \to a} \frac{f(x)}{g(x)} = \frac{f^n(a)}{g^n(a)}.$$

Li *(Chem.)*. The symbol for *lithium*.

liana, liane *(Bot.)*. A climbing plant, especially a woody climber of tropical forests.

Lias *(Geol.)*. The oldest epoch of the Jurassic period.

Lias Clay *(Geol.)*. A thick bed of clay found in the Lower Jurassic rocks of Britain. See **Blue Lias**.

liberation *(Min.Ext.)*. First stage in ore treatment, in which comminution is used to detach valuable minerals from gangue.

libethenite *(Min.)*. An orthorhombic hydrated phosphate of copper, occurring rarely as olive-green crystals in the oxide zone of metalliferous lodes.

libido *(Behav.)*. According to Freud, a motivating force associated with instinctual drives towards survival, pleasure and the avoidance of pain, and which manifests itself in wish fulfilment.

libollite *(Min.)*. A pitchlike member of the asphaltite group.

library *(Comp.)*. Collection of **programs**, **subroutines**, **macroinstructions** and sections of code available in a computer system for a user to **call** or incorporate in a program.

library binding *(Print.)*. Stronger than the usual **edition binding**; sewn on tapes which are inserted in **split boards**.

library program *(Comp.)*. One of a collection of programs on backing store available to computer users. Also *library routine*, *library sub-routine*.

libration in latitude *(Astron.)*. A phenomenon by which, owing to the moon's axis of rotation not being perpendicular to its orbital plane, an observer on the earth sees alternately more of the north and south regions of the lunar surface, and so, in a complete period, more than a hemisphere.

libration in longitude *(Astron.)*. A phenomenon by which, owing to the uniform rotation of the moon on its axis combined with its non-uniform orbital motion, an observer on the earth sees more, now on the east and now on the west, of the lunar surface than an exact hemisphere.

librations *(Astron.)*. Apparent oscillations of the moon (or other body). The actual physical librations (due to changes in the moon's rate of rotation) are very small; the other librations are librations in latitude or in longitude, and diurnal libration.

libriform fibre *(Bot.)*. Fibre in the **xylem**, usually the longest cell type in the tissue with thick walls and slit-like pits.

licensed aircraft engineer *(Aero.)*. An engineer licensed by the airworthiness authority (in the UK the *Civil Aviation Authority*) to certify that an aircraft and/or component complies with current regulations.

lichen *(Bot.)*. A symbiont association of a fungus with an alga forming a macroscopic body, a thallus. Lichen-forming fungi are mostly Ascomycotina (see Discomycetes), a few are Basidiomycotina. The alga is usually Chlorophyceae, occasionally Cyanophyceae, rarely one of each. Lichens are mostly very sensitive to air pollution (especially by SO_2) and the species present can be used as an index of pollution. Estimates suggest 20 000 spp.

lichen *(Build.)*. Commonly found on masonry and must be removed with fungicidal wash prior to painting.

lichen *(Med.)*. A term for papular skin eruption. *Lichen planus* has a papular patchy eruption with itch and a blue-violet colour.

Lichtenberg figure *(Elec.Eng.)*. A figure appearing on a photographic plate or on a plate coated with fine dust when the plate is placed between electrodes and a high voltage is applied between them.

lid *(Meteor.)*. Temperature inversion in the atmosphere which prevents the mixing of the air above and below the inversion region.

lidar *(Meteor.)*. Device for detection and observation of distant cloud patterns by measuring the degree of back scatter in a pulsed laser beam. (*LIght Detection And Ranging*.)

lidocaine *(Med.)*. See **lignocaine**.

Lie algebra *(Phys.)*. That dealing with groups of quantities subject to relationships which reduce the number that are independent. These are known as *special unitary groups*, and groups SU (2) and SU (3) have proved very valuable in elucidating relationships between fundamental particles.

Lieberkühn's crypts *(Zool.)*. Simple tubular glands occurring in the mucous membrane of the small intestine in Vertebrates.

Liebermann-Storch test *(Paper)*. A test for the detection of rosin in paper by extracting a sample with acetic anhydride, cooling and adding a drop of concentrated sulphuric acid which produces a red colour if rosin is present.

Liebermann test for phenols *(Chem.)*. Colour changes observed on treating the phenol with sulphuric acid and sodium nitrite (nitrate (III)), then pouring the solution into excess of aqueous alkali.

Liebig condenser *(Chem.)*. The ordinary water cooled glass condenser used in laboratory distillations.

lien *(Zool.)*. See **spleen**.

lienal *(Zool.)*. Pertaining to the spleen.

lienogastric *(Zool.)*. Pertaining to, or leading to, the spleen and the stomach, as the *lienogastric artery* and *vein* in Vertebrates.

lier *(Glass)*. See **lehr**.

Liesegang rings *(Chem.)*. The stratification, under certain conditions, of precipitates formed in gels by allowing one reactant to diffuse into the other.

life-cycle *(Biol.)*. The various stages through which an organism passes, from fertilized ovum to the fertilized ovum of the next generation.

lifeform *(Bot.)*. The overall morphology of a plant categorising it as an annual herb, a shrub, a succulent etc. See also **Raunkiaer system**.

life support *(Space)*. The provision of the necessary conditions of health and comfort to support a human in space, either during the occupation of a space vehicle or during *extra-vehicular activity* (EVA).

life table *(Ecol.)*. Tabulated display of the mortality schedule of a population, first devised by insurance companies and now used by ecologists in the study of wild populations.

lifetime *(Phys.)*. The mean period between the birth and death of a charge carrier in a semi-conductor. See also **half-life, mean life**.

LIFO *(Comp.)*. See **stack**.

lift *(Aero.)*. *Aerodynamic lift* is the component of the aerodynamic forces supporting an aircraft in flight, along the lift axis, due solely to relative airflow. Lift force acts at right angles to the direction of the undisturbed airflow relative to the aircraft.

lift *(Build.)*. *Elevator*. An enclosed platform or car moving in a well to carry persons or goods up and down. Cf. **paternoster**.

lift *(Image Tech.)*. See **pedestal**.

lift axis *(Aero.)*. See **axis**.

lift bridge *(Civ.Eng.)*. A type of movable bridge which is capable of being lifted bodily through a sufficient vertical distance to allow of the passage of a vessel beneath.

lift coefficient *(Aero.)*. A nondimensional number representing the aerodynamic lift on a body.

$$C_L = \frac{L}{1/2\,\rho V^2 S},$$

where L = lift, ρ = air density, V = air speed, S = wing area.

lift engine *(Aero.)*. An engine used on *VTOL* or *STOL* aircraft, having the primary purpose of providing lifting force. It is shut down in normal flight and the intake usually closed by a sealing flap.

lifter *(Foundry)*. An L- or Z-shaped bar of cast or wrought iron, used for supporting the sand in a cope, the upper end being hooked on to a box bar.

lifter *(Min.Ext.)*. (1) Projecting rib or wave in lining plates of ball or rod mill. (2) Perforated plate in drum washer or heavy media machine which aids tumbling action or removes floating fraction of ore being treated.

lift gate *(Build.)*. A gate which opens by bodily vertical movement, as distinct from one swinging about an axis at one end.

lifting *(Build.)*. The result of applying a paint on top of a coating which is not fully dry. This can also happen when subsequent coats of paint contain excessively strong solvents.

lifting blocks *(Eng.)*. A lifting machine consisting of a continuous rope passing round pulleys mounted in blocks, whereby an effort applied at the free end of the rope lifts a larger weight attached to the lower block. See **mechanical advantage**.

lifting magnet *(Elec.Eng.)*. A large electromagnet used, instead of a hook, on cranes or hoists, when lifting iron and steel.

lifting of patterns *(Foundry)*. See **drawing of patterns**.

lifting piece *(Horol.)*. In the rack-striking work of a striking clock, a cranked lever which carries the warning piece at one end and lifts the rack hook just before the hour.

lifting screw *(Foundry)*. An iron rod screwed into a pattern to withdraw it from the mould.

lifting the offsets *(Ships)*. The process of measuring the ship's form as 'laid off' on buttocks, waterlines and sections. These offsets are the permanent record and are used to reproduce the ship's form, initially, for design work and ordering material and, subsequently, in cases of repair or alterations.

lifting truck *(Eng.)*. A truck with three, or four, wheels, drawn by a handle which can be raised and lowered to lift a loaded platform standing on feet. The lift is effected either by leverage or by hydraulic mechanism.

lift-lock *(Hyd.Eng.)*. A canal lock serving to lift a vessel from one reach of water to another.

lift motor *(Elec.Eng.)*. A motor, sometimes having special characteristics, used for operating an electric lift. Also called *hoisting motor*.

lift-off *(Aero.)*. The speed at which a pilot pulls back on the control column to make an aircraft leave the ground. It is a carefully defined value for large aircraft depending upon the weight, runway surface, gradient, altitude and ambient temperature. It is one of the functions established from **WAT curves**. Colloquially *unstick*.

lift-off *(Space)*. The point at which the vertical thrust of a space craft exceeds the force due to local gravity and it begins to rise.

lift-valve *(Eng.)*. Any valve consisting of a disk, ball, plate etc., which lifts or is lifted vertically to allow the passage of a fluid.

ligament *(Zool.)*. A bundle of fibrous tissue joining two or more bones or cartilages.

ligand field theory *(Chem.)*. An essentially ionic interpretation of bonding in transition metal compounds, in which the spectroscopic and magnetic properties are rationalised in terms of the distortion of the non-bonding d-electrons of the metal by the electric field of the ligands.

ligands *(Chem.)*. In a complex ion, the ions, atoms or molecules, surrounding the central (nuclear) atom, e.g. $(CN)^-$ in $Fe(CN)_6^{4-}$.

ligase *(Biol.)*. An enzyme which seals nicks in one strand of a duplex DNA, much used to seal the gaps formed when one DNA sequence is artificially inserted into another.

ligasoid *(Phys.)*. A colloidal system in which the continuous phase is gaseous and the dispersed phase is liquid.

ligate *(Med.)*. To tie with a ligature. n. *ligation*.

ligation *(Biol.)*. Joining two linear nucleic acid molecules together by a phosphodiester bond. Such linear molecules often have **sticky ends** which facilitate ligation.

ligature *(Med.)*. A piece of thread, silk, wire, catgut or any other material, for tying round blood vessels etc.: to tie with thread etc.

ligature *(Print.)*. Two or more letters cast together on one body, e.g. fi, ffi, ffl.

light *(Build.)*. (1) A term applied to any glazed opening admitting light to a building. (2) A single division of a window.

light *(Glass)*. Of optical glass, having a lower refractive index.

light *(Phys.)*. *Electromagnetic radiation* capable of inducing visual sensation, with wavelengths between about 400 and 800 nm. See also **illumination, velocity of light**.

light-adapted *(Phys.)*. See **adaptation**.

light aircraft *(Aero.)*. One having a maximum take-off weight less than 12 500 lb (5670 kg).

light alloys *(Eng.)*. See **aluminium alloys**.

light-centre length *(Elec.Eng.)*. The distance from the geometrical centre of the filament of an electric filament lamp to the contact plate or plates at the end of the lamp cap remote from the bulb. With automobile headlight lamps the measurement is taken from the bulb side instead of from the remote side of the pin.

light chain *(Immun.)*. A polypeptide chain present in all immunoglobulin molecules. In most molecules two identical light chains are linked to two heavy chains by disulphide bonds. There are either of two types of light chain, **kappa** and **lambda**, but never both on the same molecule. Light chains have a N-terminal variable region which forms part of the antigen combining site and a C-terminal constant region.

light change points *(Image Tech.)*. The series of pre-set steps of exposure used in a motion-picture film printer; abbrev. *LCP*. (US *printer lights*).

light-curve *(Astron.)*. The line obtained by plotting, on a graph, the apparent change of brightness of a variable star, against the observed times.

light distribution curve *(Phys.)*. A graph showing the relation between the luminous intensity of a light source and the angle of emission.

light efficiency *(Phys.)*. Measured, e.g. in an electric lamp, by the ratio (total luminous flux/total power input), and expressed in lumens per watt.

light-emitting diode *(Electronics)*. Semiconductor diode which radiates in the visible region. Used in *alphanumeric* displays and as an indicator or warning lamp. Abbrev. *LED*.

lighter-than-air craft *(Aero.)*. See **aerostat**.

light face *(Print.)*. A lighter weight of type face, 'medium' being the usual normal in the **type family**.

light fastness *(Build.)*. The ability of paint pigments or dyes to retain their colour when exposed to light.

light filter *(Image Tech.)*. Filter used in photography etc. to change or control the total (or relative) energy distribution from the source. It consists of a homogeneous optical medium (sometimes of a specific thickness as in interference filters) with characteristic light absorption regions. Can be graded for special effects.

light flux *(Phys.)*. The measure of the quantity of light passing through an area, e.g., through a lens system. Light flux is measured in lumens. Illumination is light flux per unit area.

light fog *(Image Tech.)*. See **fog**.

lighting contrast *(Image Tech.)*. The ratio between the level of the **key light** plus filler to the **filler light** alone.

light meter *(Image Tech.)*. Device for measuring the intensity of incident or reflected light, usually comprising a photo-sensitive cell and a meter; when this measurement is expressed as the corresponding camera lens aperture and exposure time for a given photographic sensitivity it becomes an **exposure meter**.

lightness *(Image Tech.)*. The degree of illumination of a surface, measured in lumens per square metre.

lightning *(Meteor.)*. Luminous discharge of electric charges between clouds, and between cloud and earth (or sea). A path is found by the *leader stroke*, the main discharge following along this ionized path, with possible repetition. See **thunderstorm**.

lightning conductor *(Elec.Eng.)*. A metal strip connected to earth at its lower end, and its upper end terminated in one or more sharp points where it is attached to the highest part of a building. By electrostatic induction it will tend to neutralize a charged cloud in its neighbourhood and the discharge will pass indirectly to earth through the conductor. Also called *lightning rod*.

lightning protector *(Elec.Eng.)*. See **lightning arrester**.

lightning rod *(Elec.Eng.)*. See **lightning conductor**.

lightning (surge) arrestor *(Elec.Eng.)*. A device for the protection of apparatus from damage by a lightning discharge or other accidental electrical-surge. A surge arrestor is effectively a shunt device whose impedance is sensitive to voltage. At normal operating voltage, impedance is high, but, under conditions of overvoltage, the impedance reduces and the device protects by providing a bypass path for the surge.

lightning tubes *(Min.)*. See **fulgurites**.

light oils *(Chem.)*. A term for oils with a boiling range of about 100°–210°C, obtained from the distillation of coal-tar.

light pen *(Comp.)*. **Input device** used in conjunction with a **graphical display unit**. The graphical display unit's hardware, together with special software, senses the position of the pen and relays this information to the **central processing unit**.

light quanta *(Phys.)*. When light interacts with matter, the energy appears to be concentrated in discrete packets called *photons*. The energy of each photon $E = h\nu$ where ν is the frequency and h is Planck's constant. See **quantum, X-rays, gamma radiation**.

light railway *(Civ.Eng.)*. One of standard or narrow guage, subject to severe restrictions on speed, load etc., and thereby relieved of many of the operating requirements (signalling etc.) applied to main line railways.

light reactions *(Bot.)*. Those reactions in **photosynthesis** in which light is absorbed, transferred between pigments, and used to generate the ATP and reduced NADP used in the **dark reactions** for CO_2 fixation. See **photosystems I and II, photosynthetic pigments**.

light red silver ore *(Min.)*. See **proustite**.

light relay *(Telecomm.)*. Relay operating circuit using a photoelectric detector.

light resistance *(Electronics)*. The resistance, when exposed to light, of a photocell of the photoconductive type. Cf. **dark resistance**.

light sensitive *(Phys.)*. Said of thin surfaces of which the electrical resistance, emission of electrons, or generation of a current, depends on incidence of light.

light spring diagram *(Eng.)*. An indicator diagram taken by a piston or diaphragm type engine indicator, using a specially weak control spring or diaphragm in order to reproduce the low pressure part of the diagram to a large scale.

light table *(Print.)*. A ground glass surface, illuminated from below, there being a variety of sizes and styles, for use in the graphic reproduction processes when working with negatives, positives, and proofs. Also *shiner*.

light trap *(Image Tech.)*. General term for any mechanical arrangement which permits movement while excluding light, such as the doors and partitions for operational access to a dark-room (*light-lock*) or the rollers and lips over which film passes from and into a magazine.

light trap *(Zool.)*. A device for catching night-flying moths and/or other insects attracted by light.

light-up *(Aero.)*. The period during the starting of a

turbojet or *turboprop* engine when the fuel/air mixture has been ignited.

light valve *(Image Tech.)*. Device for rapidly varying light transmission and hence the exposure of motion picture film, as in photographic sound recording or in printing from an assembled negative.

light water *(Chem.)*. Normal water (H_2O) as distinct from **heavy water**.

light-water reactor *(Nuc.Eng.)*. A reactor using ordinary water as moderator and coolant. See **boiling-water reactor, pressurized-water reactor**. Abbrev. *LWR*.

light watt *(Phys.)*. The photometric radiation equivalent, the radiation being measured in lumens or watts. The ratio of lumens to watts depends on the wavelength. At the wavelength of greatest sensitivity of the eye $\lambda = 555$ nm, the value of the light watt is 682 lumens.

lightweight aggregate *(Build.,Civ.Eng.)*. Aggregate which is used in the manufacture of lightweight concrete. Normally clinker ash or clays or other materials which have been fired to reduce their carbon content and weight.

lightweight concrete *(Build.)*. Concrete of low density (20–90 lb/ft^3 or 300–1400 kg/m^3) made by using special aggregates or air entraining processes. Used particularly where lightness and insulation are required rather than load resisting properties.

lightwood *(For.)*. Coniferous wood having an abnormally high resin content.

light-year *(Astron.)*. An astronomical measure of distance, being the distance travelled by light in space during a year, which is approximately 9.46×10^{12} km (5.88×10^{12} miles).

ligne *(Horol.)*. A unit used in the measurement of watch movements. It is equal to 2.256 mm. The *twelfth*, or *douzième* (0.188 mm), is the unit used for the height or thickness of a movement. There are 12 twelfths in a ligne. Ladies' wrist watches vary from 3 to 8 lignes; gents' wrist watches from 10 to 13, pocket watches from 17 to 19, and deck watches from 20 to 32. Sometimes spelt *line*. Symbol ‴.

lignicole, lignicolous *(Bot.)*. Growing on or in wood, or on trees.

lignicole, lignicolous *(Zool.)*. Living on or in wood, as certain species of Termites.

lignin *(Bot.)*. A complex cross-linked polymer based on variously substituted p-hydroxyphenyl propane units; a constituent with cellulose etc. of the cell walls of xylem tracheids and vessels and of many sclerenchyma and some parenchyma cells etc. where its major function appears to be to impart rigidity to an otherwise flexible wall.

lignite *(Geol.)*. A brownish, black coal intermediate between peat and subbituminous coal. It is commoner than coal in Mesozoic and tertiary deposits. Commonly known as *brown coal*.

lignivorous *(Zool.)*. Wood-eating.

lignocaine *(Med.)*. Local anaesthetic agent which is also used to treat ventricular arrhythmias of the heart.

ligno-celluloses *(Chem.)*. Compounds of lignin and cellulose found in wood and other fibrous materials.

ligroin *(Chem.)*. A term for a petroleum fraction with a boiling range of from about 90° to 120°C.

ligulate *(Bot.)*. Strap-shaped.

Liliaceae *(Bot.)*. Family of ca 3500 spp. of monocotyledonous flowering plants (superorder Liliidae). Mostly herbs, often with bulbs, corms or rhizomes, cosmopolitan. The flowers are usually showy, with 6 perianth members, 6 stamens and a superior ovary of three carpels. Includes *Allium*, the onions, garlic, leeks, *Colchicum*, the source of **colchicine**, and many ornamental plants.

Liliidae *(Bot.)*. Subclass or superorder of monocotyledons. Mostly herbs, sepals and petals both usually petaloid, mostly syncarpous, ovary superior or more often inferior. Contains ca 28 000 spp. in 17 families including Liliaceae, Amaryllidaceae, Iridaceae and Orchidaceae.

Liliopsida *(Bot.)*. See **Monocotyledones**.

lim *(Maths.)*. Abbrev. for *limit*.

limaciform *(Zool.)*. Sluglike.

limaçon *(Maths.)*. A heart shaped curve with an inner loop at its vertex. An epitrochoid in which the rolling circle equals the fixed circle. Polar equation $r = 2a(k + \cos\theta)$.

limb *(Astron.,Surv.)*. Term applied to the edge or rim of a heavenly body having a visible disk; used specially of the sun and moon.

limb *(Bot.)*. (1) The lamina of a leaf. (2) The widened upper part of a petal. (3) The upper, often spreading, part of a sympetalous corolla.

limb *(Geol.)*. The side of a fold.

limb *(Surv.)*. Lower horizontal plate of theodolite.

limb *(Zool.)*. A jointed appendage, as a leg.

limb darkening *(Astron.)*. The apparent darkening of the limb of the sun due to the absorption of light in the deeper layers of the solar atmosphere near the edge of the disk.

limberneck *(Vet.)*. **Botulism** of Birds.

limbic *(Med.)*. Marginal, bordering; e.g. lobe of cerebral cortex.

limbous *(Zool.)*. Overlapping.

limbric *(Textiles)*. Plain-weave cotton cloth of light to medium weight in which the weft predominates.

limburgite *(Geol.)*. An ultramafic, fine grained igneous rock occurring in lava flows, similar to the dyke rock monchiquite, but having interstitial glass between the dominant olivine and augite crystals. Typically, limburgite is feldspar-free; but it does occur.

lime *(Chem.)*. A substance produced by heating limestone to 825°C or more, as a result of which the carbonic acid and moisture are driven off. Lime, which is much used in the building, chemical, metallurgical, agricultural and other industries, may be classified as *high calcium*, *magnesian*, or *dolomitic* depending on the composition. Unslaked lime is commonly known as *caustic lime*, also *anhydrous lime, burnt lime, quicklime*.

lime *(For.)*. A moderately lightweight hardwood, from the genus, *Tilia*, a native of W. Europe. Typical uses include carving, inlaying, marquetry work and cabinet making.

lime bag *(Foundry)*. A bag of powdered lime used for testing the fit of joints. Lime is sprinkled on the parting face, and the cope is lowered and lifted; if the lime adheres to the top face, the joint is good.

lime-induced chlorosis *(Bot.)*. Chlorosis of young leaves induced in (relatively) calcifuge plants by growth on calcareous or over-limed soils (i.e. soils of too high pH), caused apparently by a disturbance in iron metabolism and curable horticulturally by the application of e.g. chelated iron *sequestrene* to leaves or roots.

lime juice *(Med.)*. Described by James Lind (1716-94) as a cure for scurvy.

lime light *(Eng.)*. Intense white light obtained by heating a cylinder of lime in an oxyhydrogen flame.

lime mortar *(Build.,Civ.Eng.)*. A mortar composed of lime and sand, with the addition sometimes of other material, such as crushed bricks, ground slag, or coke. It is not suitable for use under water.

limen *(Med.)*. Smallest difference in pitch (frequency) or intensity of a sound or colour which can be perceived by the senses.

lime paste *(Build.)*. Slaked lime.

lime powder *(Build.)*. The material produced as a result of subjecting quicklime to the process of air-slaking.

lime-silicate rocks *(Geol.)*. These result from the contact (high-temperature) metamorphism of limestones containing silica in detrital grains, nodules of flint or chert, or siliceous skeletons, the silica combining with the lime to form such silicates as lime-garnet, anorthite, wollastonite.

lime slurry *(Min.Ext.)*. Thick aqueous suspension of finely-ground slaked lime, used to control alkalinity of ore pulps in flotation and cyanide process.

lime-soda process *(Chem.)*. Standard water-softening process, carried out either hot or cold; uses lime and soda ash to reduce the hardness of the treated water by precipitating the dissolved calcium and magnesium salts as insoluble calcium carbonate and magnesium hydroxide respectively. Often used in conjunction with **zeolite process**. Also used in the preparation of caustic soda, by mixing slaked lime with soda and filtering off the precipitated calcium carbonate. $Ca(OH)_2 + Na_2CO_3 = 2NaOH + CaCo_3$.

limestone *(Geol.)*. A sedimentary rock consisting of more than 50% by weight of calcium carbonate. Mineralogically, limestone consists of either aragonite or calcite, although the former is abundant only in Tertiary and Recent limestones. Dolomite may also be present, in which case the rock is called *dolostone*. Limestones may be *organic*, formed from the calcareous skeletal remains of living organisms, *chemically precipitated*, or *detrital*, formed of fragments from pre-existing limestones. The majority are organic, and can be classified according to their texture and the nature of the organisms whose skeletons are incorporated, e.g. oölitic limestone, shelly limestone, algal limestone, crinoidal limestone.

limewash *(Build.)*. A mixture which is prepared by slaking lump lime with about one-third of its weight of water, and then adding sufficient water to make a 'milk'; used as wall covering in cases where a frequent application is necessary. An improved, more water-resistant type embodies tallow.

lime water *(Chem.)*. Saturated calcium hydroxide solution.

limicolous *(Zool.)*. Living in mud.

limit *(Maths.)*. A sequence of numbers u_r has a *limit* L if for every **neighbourhood** U of L there exists m such that for all $n \geqq m$ u_n is in U.

limited stability *(Telecomm.)*. A property of a servomechanism or a communication system which is characterized by stability only when the input signal falls within a particular range.

limiter *(Nuc.Eng.)*. An aperture which defines the boundary of a plasma and protects the vacuum vessel from damage by contact with hot plasma.

limiter *(Telecomm.)*. Any transducer in which the output, above a threshold or critical value of the input, does not vary, e.g. a shunt-polarized diode between resistors. Particularly applied to the circuit in a frequency-modulation receiver in which all traces of amplitude modulation in the signal have to be removed before final demodulation.

limit gauge *(Eng.)*. A gauge used for verifying that a part has been made to within specified dimensional limits. Limit gauges consist, for example, of a pair of plug gauges on the same bar, one of which should just enter a hole ('go') and the other just not enter ('not go').

limit gauging *(Eng.)*. A method of measurement which ensures that pieces intended to fit together shall do so within certain specified limits of clearance, and that similar pieces shall be interchangeable.

limiting conductivity *(Chem.)*. The molar conductivity of a substance at infinite dilution, i.e. when completely ionized.

limiting density *(Chem.)*. The relative density of a gas at vanishingly low pressures.

limiting factor *(Ecol.)*. The variable, or combination of variables, which limits the rate of growth of an organism or a population, or which limits the rate of a physiological process.

limiting frequency *(Acous.)*. (1) Frequency at which the wavelength of an air-borne sound wave coincides with the wavelength of a bending wave of a wall or plate. Important in sound transmission. (2) Name for the highest or lowest frequency transmitted by an electro-acoustic system or by a wave guide.

limiting friction *(Phys.)*. See **friction**.

limiting gradient *(Civ.Eng.)*. **Ruling gradient**.

limiting Mach number *(Aero.)*. The maximum permissible *flight Mach number* at which any particular aircraft may be flown, either because of the **buffet boundary** or for structural strength limitations.

limiting range of stress *(Eng.)*. The greatest range of stress (mean stress zero) that a metal can withstand for an indefinite number of cycles without failure. If exceeded, the metal fractures after a certain number of cycles, which decreases as the range of stress increases. Also called *endurance range*; half this range is the *fatigue limit* or *endurance limit*.

limiting velocity *(Aero.)*. The steady speed reached by an aircraft when flown straight, the angle to the horizontal, power output, altitude and atmospheric conditions all specified. See **max level speed, terminal velocity.**

limit load *(Aero.)*. The maximum load anticipated from a particular condition of flight and used as a basis when designing an aircraft structure.

limit of proportionality *(Eng.)*. The point on a stress-strain curve at which the strain ceases to be proportional to the stress. Its position varies with the sensitivity of the extensometer used in measuring the strain.

limit point *(Maths.)*. See **accumulation point**.

limit state *(Civ.Eng.)*. The state at which a structure has become unfit for use. The main limit states are *limit state of collapse* and *serviceability limit state*.

limit switch *(Elec.Eng.)*. A switch fitted to electric lifts, travelling cranes, etc., to cut off the power supply if the lift-car or moving carriage travels beyond a certain specified limit.

limivorous *(Zool.)*. Mud-eating; as certain aquatic Invertebrates which swallow mud to extract from it the nutritious organic matter that it contains.

limnobiotic *(Zool.)*. Living in fresh water.

limnology *(Geol.)*. The study of lakes.

limnophilous *(Zool.)*. Living in marshes, especially freshwater marshes.

limonene *(Chem.)*. *Hesperidene, citrene, carvene*. The oil of the orange peel consists almost entirely of +-limonene, bp 175°C. −-Limonene is present in the oil of fir cones. These compounds, which are monocylic terpenes, are also called *1,8-menthadiene*.

limonite *(Min.)*. Although originally thought to be a definite hydrated oxide of iron, now known to consist mainly of cryptocrystalline goethite or lepidocrocite along with adsorbed water; some haematite may also be present. It is a common alteration product of most iron-bearing minerals and also the chief constituent of bog iron ore.

limp *(Print.)*. Said of a book having nonrigid sides; described as *limp cloth, limp leather*, according to the covering material.

limpet washer *(Build.)*. A form of washer used in fixing corrugated sheeting, for which purpose it is shaped on one side to conform to the curve of a corrugation.

linarite *(Min.)*. Hydrated sulphate of lead and copper, found in the oxide zone of metalliferous lodes; a deep-blue mineral resembling azurite, also crystallizing in the monoclinic system.

linch pin *(Eng.)*. A pin placed in a transverse hole on the outside of the axles of a vehicle, to retain a wheel; at the top it has a projection on one side, to prevent it from passing through the hole.

Lincoln index *(Ecol.)*. See **mark and recapture**.

Lindeck potentiometer *(Elec.Eng.)*. One which differs from most potentiometers in using a fixed resistance variable current to obtain balance.

Lindé process *(Chem.)*. A process for the liquefaction of air and for the manufacture of oxygen and nitrogen from liquid air.

Lindé sieve *(Chem.)*. See **molecular sieve**.

line *(Acous.)*. Said of a microphone or loudspeaker when it is considerably extended in one direction, for directivity in a normal plane.

line *(Build.)*. A cord stretched as a guide to the bricklayer for level and direction of succeeding courses; also for the setting out of foundations.

line *(Elec.Eng.)*. (1) Direction of an electric or magnetic field, as a line of flow or force. See **line of flux** (or *force*). (2) Term used to describe overhead conductor in a power transmission system, normally of steel-core aluminium.

line *(Genrl.)*. The 12th part of an inch.

line *(Horol.)*. (1) The cord or gut supporting the weight or weights of a weight-driven clock. (2) See **ligne**.

line *(Image Tech.)*. Single scan in a facsimile or television picture transmission system.

line *(Maths.)*. In a plane: the shortest distance between two points. On a sphere: a portion of a great circle.
line *(Radiol.)*. Single frequency of radiation as in a luminous, X-ray, or neutron spectrum.
line *(Telecomm.)*. Transmission line, coaxial, balanced pair, or earth return, for electric power, signals or modulation currents.
linea alba *(Med.)*. The tendinous line which extends down the front of the belly, from the lower end of the chest to the pubic bone, and gives attachment to abdominal muscles.
lineament *(Geol.)*. A long feature on the Earth's surface, often more clearly visible by satellite photography. It may be structural or volcanic and can be related to **plate tectonics**.
line amplifier *(Telecomm.)*. In broadcasting, that which supplies power, at a specified level to the line, either to a control centre or to a transmitter.
line amplitude *(Image Tech.)*. The amplitude of the voltage generated by the line-scanning generator, or the length of the line on the screen produced thereby.
linea nigra *(Med.)*. Pigmented linea alba, occurring in pregnant women.
linear *(Bot.)*. (1) A leaf having parallel sides, and at least four to five times as long as broad. (2) A tetrad of pollen grains in a single row.
linear *(Phys.,Eng.)*. Said of any device or motion where the effect is exactly proportional to the cause, as rotation and progression of a screw, current and voltage in a wire resistor (Ohm's law) at constant temperature, output versus input of a modulator (or demodulator). See **ultralinear**.
linear absorption coefficient *(Phys.)*. See **absorption coefficient**.
linear accelerator *(Electronics)*. Large device for accelerating electrons or positive ions up to nearly the velocity of light. The particles are accelerated through loaded waveguides by high-frequency pulses or oscillations of the correct phase.
linear amplifier *(Electronics)*. One for which the output power is directly proportional to the input level.
linear array *(Telecomm.)*. See **collinear array**.
linear-ball bearing *(Eng.)*. The linear analogue of the radial ball bearing, in which a column of balls rolls in a straight track between the bearing and a hardened and ground shaft, recirculating in a space formed parallel to the track but out of contact with the shaft. Commonly 5 or 6 tracks are arranged round the bearing, constraining it radially. Cf. **recirculating-ball thread**.
linear detector *(Telecomm.)*. One in which the demodulated signal voltage is directly proportional to the changes in input carrier amplitude or frequency, for amplitude or frequency modulation respectively.
linear differential equation of order n *(Maths.)*. An equation of the form

$$q_0(x)\frac{d^n y}{dx^n} + q_1(x)\frac{d^{n-1}y}{dx^{n-1}} + \dots + q_{n-1}(x)\frac{dy}{dx} + q_n(x)y = f(x).$$

linear distortion *(Telecomm.)*. That which results in the nonlinear response of a system, such as an amplifier, to the envelope of a varying signal, such as speech, without distorting (within acoustic perception) the detailed waveform.
linear energy transfer *(Phys.)*. The linear rate of energy dissipation by particulate or EM radiation while penetrating absorbing media. Abbrev. *LET*.
linearity *(Telecomm.)*. The condition in which the change in value of one quantity is directly proportional to that of another quantity; said especially of amplifiers, operating in the range where this gives minimum distortion, before the process of *limiting* sets in as the output approaches the maximum available.
linearity control *(Electronics)*. See **strobe** (2).
linearly dependent *(Maths.)*. n quantities $u_1, u_2 \dots u_n$ are linearly dependent if it is possible to find n numbers $\lambda_1, \lambda_2, \dots \lambda_n$ (at least one of which is not zero) such that $\lambda_1 u_1 + \lambda_2 u_2 + \dots + \lambda_n u_n = 0$. Otherwise the u_n are independent.
linear modulation *(Telecomm.)*. Modulation in which the change in the modulated characteristic of the carrier signal is proportional to the level of the modulating signal.
linear momentum *(Phys.)*. See **momentum**.
linear motor *(Elec.Eng.)*. A particular form of induction motor in which the stator and rotor are linear and parallel instead of cylindrical and coaxial.
linear network *(Telecomm.)*. One with electrical elements which are constant in magnitude with varying current.
linear programming *(Eng.)*. General term implying that, in a complicated process of operations on an article, the various steps have been mutually planned for maximum economy.
linear rectifier *(Elec.Eng.)*. One in which the output current is strictly proportional to the envelope of the applied alternating voltage.
linear resistor *(Elec.Eng.)*. One which 'obeys' Ohm's law, i.e. under certain conditions the current is always proportional to voltage. Also *ohmic resistor*.
linear scan *(Electronics)*. The sweeping of a cathode spot across a screen at constant velocity using a deflecting sawtooth waveform.
linear stopping power *(Phys.)*. See **stopping power**.
linear superpolymer *(Chem.)*. Polymer in which the molecules are essentially in the form of long chains with an average molecular weight greater than 10 000.
linear sweep *(Electronics)*. The use of a sawtooth waveform to obtain a linear scan.
linear time-base oscillator (generator) *(Electronics)*. The circuit for producing the deflecting voltage (or current) of the linear time base. See **linear scan, sweep circuit**.
linear transformation *(Maths.)*. A transformation between two sets of n variables represented by n equations in which the variables of both sets occur linearly. Thus when $n = 1$, the equation would be $ax + bxx' + cx'd = 0$. Cf. **bilinear transformation**.
line at infinity *(Maths.)*. The line joining the circular points at infinity.
lineation *(Geol.)*. Any one-dimensional structure in a rock.
line-blanking *(Image Tech.)*. The reduction of the amplitude of the television video signal to below the black level at the end of each line period. This allows the transmission of the line-synchronizing pulses.
line block *(Eng.)*. A sliding block pivoted to the end of the valve rod, and working in the slotted link of a **link motion**.
line block *(Print.)*. Relief block surface produced from a line drawing and incapable of printing with tonal gradation.
line-breaker *(Elec.Eng.)*. A contactor on an electric vehicle, arranged for closing or interrupting the main current circuit.
line broadening *(Astron.)*. A term used for the increase in width of the lines of a stellar spectrum due to rotation of the star, turbulence in the stellar atmosphere, or the *Stark* or *Zeeman effects*.
line-casting machines *(Print.)*. **Linotype** and **Intertype** machines.
line choking coil *(Elec.Eng.)*. An inductor included in an electric power supply circuit in order to protect plant connected to the line from the effect of high-frequency or steep-fronted surges. Also called *screening protector*.
line colour *(Print.)*. A colour-picture produced by superimposed impressions of two or more line illustrations printed in different colours. Varied tones are obtainable by the use of stipples.
line co-ordinates *(Maths.)*. A co-ordinate system in which each set of co-ordinates specifies a line, usually by reference to cartesian axes. Thus the line co-ordinates (a,b,c) could specify a line whose ordinary cartesian equation is $ax + by + c = 0$. In a line co-ordinate system a point is specified by a linear equation.

line coupling *(Telecomm.)*. Transfer of energy between resonant (tank) circuits in a transmitter, using a short length of line with small inductive coupling at each end. Also called *link coupling*.

line defect *(Crystal.)*. See **defect**.

line distortion *(Telecomm.)*. That arising in the frequency content or phase distribution in a transmitted signal, as a result of the propagation constant of the line.

line drop *(Telecomm.)*. The potential drop between any two points on a transmission line due to resistance, leakage, or reactance.

line equalizer *(Telecomm.)*. Device which compensates for attenuation and/or phase delay for transmission of signals along a line over a band of frequencies. Also *lumped loading, phase compensation, phase equalization*.

line finder *(Telecomm.)*. See **finder**.

line flyback *(Electronics)*. (1) The time interval corresponding to the steeper portion of a sawtooth wave. (2) The return time of the image spot from its deflected position to its starting point.

line focus *(Electronics)*. CRT in which electron beam meets screen along a line and not at a point. (When unintended this is due to astigmatism.)

line frequency *(Image Tech.)*. The number of lines scanned per second; for example, with 625-line pictures at 25 frames per second, the line frequency is 15 625 Hz.

line-frequency generator *(Image Tech.)*. The generator of the voltage or current which causes the scanning spot to traverse each line of the image.

line hold *(Image Tech.)*. Time-based control in a TV receiver to synchronize with the incoming signal.

line impedance *(Telecomm.)*. The impedance measured across the terminals of a transmission line; when the line is correctly matched to its load, this equals the *characteristic impedance* of that particular type of line.

line integral *(Maths.)*. Of a function along a path: the limiting sum along the path of the product of the value of the function and an element of length of the path.

linen *(Textiles)*. Yarns, fabrics and articles made from flax fibres.

line of action *(Phys.)*. The line along which a force acts.

line of apsides *(Astron.)*. See **apse line**.

line of centres *(Horol.)*. A line passing through two or more centres: the line joining the centres of two or more wheels; in a lever escapement, the line joining the balance and pallet staffs.

line of collimation *(Surv.)*. In a surveying telescope, the imaginary line passing through the optical centre of the object glass and the intersection of the cross-hairs in the diaphragm.

line of flux (force) *(Elec.Eng.,Phys.)*. A line drawn in a magnetic (or electric) field so that its direction at every point gives the direction of magnetic (or electric) flux (or force) at that point.

line of sight *(Surv.)*. Alternative term for *line of collimation*.

line of sight *(Telecomm.)*. Said of a transmission system when there has to be a straight line between transmitting and receiving antennae, as in UHF and radar.

line-of-sight velocity *(Astron.)*. The velocity at which a celestial body approaches, or recedes from, the earth. It is measured by Doppler shift of the spectral lines emitted by the body as observed on the earth. Also *radial velocity*.

line oscillator *(Electronics)*. One which has its frequency stabilized either by a resonant low-loss (high-Q) coaxial line, or by a resistance-capacitance ladder which gives the necessary delay (phase shift) in a feedback loop. Also *phase-shift oscillator*.

line pad *(Telecomm.)*. A resistance-attenuation network which is inserted between the programme amplifier and the transmission line to the broadcasting transmitter. Its purpose is to isolate electrically the amplifier output from the variations of impedance of the line.

line printer *(Comp.)*. Prints a complete line of characters at one time and hence is generally faster than a character printer, such as a **daisy wheel printer**.

line profile *(Phys.)*. A graph showing the fine structure of a spectral line, the intensity of the line, measured with a microphotometer, being plotted as a function of wavelength.

liner *(Eng.)*. A separate sleeve placed within an engine cylinder to form a renewable and more durable rubbing surface; in I.C. engines, termed *dry* if in continuous contact with the cylinder wall, and *wet* if supported only at the ends and surrounded by cooling water.

line ranger *(Surv.)*. An instrument for locating an intermediate point in line with two distant signals. It consists of two reflecting surfaces so arranged as to bring images of the two signals into coincidence when the instrument is in line with the signals.

line reflection *(Telecomm.)*. The reflection of some signal energy at a discontinuity in a transmission line.

liner-off *(Ships)*. A tradesman engaged in shipbuilding, whose function it is to 'mark off' by fair lines, using battens to enable plate workers and others to prepare material to fit *in situ*.

line scanning *(Image Tech.)*. A method of scanning in which the scanning spot repeatedly traverses the field of the image in a series of straight lines.

line screen process *(Image Tech.)*. A colour photographic process in which the screen takes the form of lines ruled on the emulsion.

line-sequential *(Image Tech.)*. Said of colour TV system in which successive scanning lines generate images in each of the 3 primary colours.

line shafting *(Eng.)*. Overhead shafting formerly used in factories to transmit power from a prime mover to many individual machines by pulley wheels and belts.

line slip *(Image Tech.)*. An apparent horizontal movement of part (or all) of a reproduced screen picture due to lack of synchronism between the line frequencies of the signal and the scanning system.

lines of curvature *(Maths.)*. Curves on a surface that are tangential to the principal directions at every point of the surface.

line spectrum *(Phys.)*. A spectrum consisting of relatively sharp lines, as distinct from a **band spectrum** or a **continuous spectrum**. Line spectra originate in the atoms of incandescent gases or vapours. See **Bohr theory, spectrum**.

line squall *(Meteor.)*. A system of squalls occurring simultaneously along a line, sometimes hundreds of miles long, which advances across the country. It is characterized by an arch or line of low dark cloud and a sudden drop in temperature and rise in pressure. Thunderstorms and heavy rain or hail often accompany these phenomena.

line stabilization *(Telecomm.)*. Dependence of an oscillator on a section of transmission line for stabilization of its frequency of oscillation; e.g. a quarter-wave line acts as a rejector circuit of very high-Q, thus giving a highly critical change in phase at the resonant frequency.

line standard *(Eng.)*. A standard of length consisting of a metal bar near whose extremities are engraved fine lines, the standard length being the distance between these lines measured under specified conditions.

line stretcher *(Telecomm.)*. A section of waveguide or coaxial line tuned by adjusting its physical length.

line switch *(Telecomm.)*. A small uniselector which immediately hunts to seize a free A-digit selector when the subscriber lifts his receiver.

line synchronization *(Image Tech.)*. Synchronization of the line-scanning generator at the receiver with that at the transmitter so that the scanning spots at the two ends keep in step throughout each line.

line-up *(Telecomm.)*. Adjustment of a number of circuits in series so that they function in the desired manner.

line voltage *(Elec.Eng.)*. See **voltage between lines**.

line width *(Telecomm.)*. The wavelength spread (or energy spread) of radiation which is normally characterized by a single value. The spread is defined by the separation between the points having half the maximum intensity of the line. In TV, the reciprocal of the number of lines per unit length in the direction of line progression.

lingering period *(Phys.)*. The time interval during which an electron remains in its orbit of highest excitation before jumping to the energy level of a lower orbit.

lingua *(Zool.)*. Any tonguelike structure; in Insects, the hypopharynx; in *Acarina*, the floor of the mouth.

lingual *(Zool.)*. In Arthropods, pertaining to the lingua; in Molluscs, pertaining to the radula; in Vertebrates, pertaining to the tongue.

linguidental *(Med.)*. Of the tongue and teeth.

lingulate *(Bot.)*. Tongue-shaped.

linguogingival *(Med.)*. Of the tongue and the gums.

lining *(Hyd.Eng.)*. A layer of clay puddle covering the sides of a canal, making them watertight.

lining *(Print.)*. (1) The operation of pasting a strip of strong paper down the back of a book after backing. (2) A strip of linen fixed down the middle of a section for strengthening purposes.

lining *(Textiles)*. A fabric attached to the inside of a garment to make it more acceptable to the wearer (e.g. more comfortable, more durable).

lining figures *(Print.)*. Said of figures which have the height of the capital letters of a fount, as distinct from *hanging figures*; e.g. 1,6,7. Also called *ranging figures*.

lining fitch *(Build.)*. A hog hair brush cut at a slant and used in conjunction with a wooden straight edge for the purpose of painting lines; made in various sizes.

lining-papers *(Print.)*. Synonym for **end-papers**.

lining-up *(Eng.)*. The operation of arranging the bearings of an engine crankshaft etc. in perfect alignment.

lining wheel *(Build.)*. A small device consisting of a metal cylinder and a range of serrated wheels used to paint lines of various widths. An alternative to a lining fitch.

link *(Eng.)*. (1) Any connecting piece in a machine, pivoted at the ends. (2) The curved slotted member of a **link motion**.

link *(Surv.)*. The one-hundredth part of a chain. In the Gunter's chain, 1 link = 7.92 in.; in the engineer's chain, 1 link = 1 ft.

link *(Telecomm.)*. A circuit or outlet between one rank of selectors and the next in order of operation, or between such selectors and a manual position. In US, a *trunk*.

linkage *(Biol.)*. The tendency of genes, or characters, to be inherited together because the genes are on the same chromosome.

linkage *(Chem.)*. A chemical bond, particularly a covalent bond in an organic molecule.

linkage *(Elec.Eng.)*. The product of the total number of lines of magnetic flux and the number of turns in the coil (or circuit) through which they pass.

linkage group *(Biol.)*. All the genes known from their linkage to be on the same chromosome.

linkage map *(Biol.)*. A diagram showing the positions of genes on a chromosome or set of chromosomes.

link coupling *(Telecomm.)*. See **line coupling**.

linked list *(Comp.)*. List where each item contains both data and a pointer to the next item.

linked numbering scheme *(Telecomm.)*. One in which a range of numbers (including the code numbers in a director automatic area) is distributed between the subscribers on several exchanges in a given area.

linked switches *(Elec.Eng.)*. Switches mechanically linked, so that they operate together or in a definite sequence. Also called *coupled switches*.

linker *(Comp.)*. A program which links programs together.

link mechanism *(Eng.)*. A system of rigid members joined together with constraints so that motion can be amplified or can be changed in direction.

link motion *(Eng.)*. A valve motion, invented by Stephenson, for reversing and controlling the cut-off of a steam-engine. It consists of a pair of eccentrics, set for ahead and reverse rotation, connected to the ends of a slotted link carrying a block attached to the valve rod. Variation of the link position (known as *linking-up*) makes either eccentric effective, and also varies the cut-off.

link resonance *(Telecomm.)*. That resulting in a repeated network when the sections are coupled together.

lin-log receiver *(Telecomm.)*. A radio receiver which gives a linear amplitude response for small signals but logarithmic for larger signals.

Linnaean (Linnean) system *(Biol.)*. The system of classification and of **binomial nomenclature** established by the Swedish naturalist Linnaeus.

Linnaean species *(Bot.)*. A species defined broadly, often including many varieties. Cf. **Jordanon**.

linocut *(Print.)*. A printing surface cut by hand in linoleum, giving a bold broad effect; not suitable for fine detail or long runs.

linoleic acid *(Chem.)*. Unsaturated fatty acid, $CH_3.(CH_2.CH\ CH)_3.(CH_2)_7.COOH$. Occurs as glycerides in various vegetable oils, especially linseed oil to which, by forming solid oxidation products on exposure to air, it imparts the 'drying' quality responsible for its utility in paints etc.

linoleum *(Build.)*. A floor-covering material, now largely superseded by *vinyl floor coverings*, made by impregnating a foundation of hessian fabric or waterproofed felt with a mixture of oxidized linseed oil (linoxyn), resins (e.g. kauri gum) and fillers. Abbrev. *lino*.

Linotype composing machine *(Print.)*. See **composing machines**.

linseed oil *(Chem.)*. An oil obtained from the seeds of flax (*Linum usitatissimum*). It contains solid and liquid glycerides of oleic and other unsaturated acids. Its iodine value is 160–200, which puts it into the class of drying oils. It is easily oxidized and polymerized for forming elastic films. Once much used for the mixing of paints and varnishes and for the manufacture of linoleum.

lint *(Textiles)*. (1) The main seed hairs of the cotton plant. (2) A plain-woven cotton fabric raised on one side to make it highly absorbent; used, after sterilization, as a wound dressing.

lintel *(Build.)*. A beam across the top of an aperture. Also called the *head*.

linters, cotton linters *(Textiles)*. Short fuzzy fibres remaining on cotton seeds after substantial removal of the longer fibres (*lint*); removed before the seeds are crushed. Linters are used extensively in the manufacture of cellulose nitrate and acetate etc.

lintol *(Build.)*. Erroneous spelling for *lintel*.

Linville truss *(Eng.)*. See **Whipple-Murphy truss**.

lionism *(Med.)*. A lion-like appearance, often characteristic of lepromatous *leprosy*.

lip *(Horol.)*. The edge of the cylinder of the cylinder escapement which receives impulse from the escape wheel.

lipaemia *(Med.)*. Excess of fat in the blood.

liparite *(Geol.)*. Syn. for **rhyolite**.

lipase *(Biol.)*. Enzyme which cleaves the hydrocarbon chains from lipids.

lipectomy *(Med.)*. Surgical removal of fatty tissue.

lipid body *(Bot.)*. A cytoplasmic inclusion, especially in the storage tissue of oil- and fat-rich seeds, typically spherical, 0.1-5 μm diameter, bounded by what appears to be half a unit membrane, and containing lipids.

lipids, lipoids *(Chem.)*. Generic terms for oils, fats, waxes and related products found in living tissues.

lip microphone *(Acous.)*. One constructed in the form of a box shaped to fit the face around the mouth. This reduces extraneous noises as at sporting events, etc.

lipochromes *(Chem.)*. Pigments of butter fat.

lipogenous *(Zool.)*. Fat-producing.

lipoma *(Med.)*. A tumour composed of cells containing fat.

lipoplast *(Bot.)*. A lipid body.

lipopolysaccharide *(Immun.)*. *LPS*. Term commonly used to refer to bacterial lipopolysaccharides which constitute the O-antigens of Gram negative bacilli, especially enterobacteria. Chemically they consist of various antigenically specific polysaccharides linked to lipid A. The latter contains rhamnose linked to galactosamine diphosphate which is esterified with myristic acid, and is responsible for the endotoxin properties common to all LPS. They are also polyclonal mitogens for B-lymphocytes.

lipoprotein *(Biol.)*. Complex of protein and lipid in varying proportions, classified according to their increasing density into **chylomicrons**, *very low density lipoproteins (VLDL), low density lipoproteins (LDL)* and *high density lipoproteins (HDL)*. They transport lipids in the plasma between various organs in the body, particularly the gut, liver and adipose tissues.

lipoproteins *(Med.)*. Soluble complex of fat and protein which serve to transport fat in the blood. Classified on density. Abnormal distribution associated with coronary artery disease.

liposome *(Biol.,Immun.)*. Spherical shell formed when mixtures of phospholipids with or without cholesterol are dispersed in aqueous solutions. Liposomes are made up of one or several concentric phospholipid bilayers within which other molecules can be incorporated. They simulate many permeability properties of membranes and are used for the administration of certain drugs.

lipotropic *(Med.)*. Having an effect on fat metabolism by accelerating fat removal or decreasing fat depositon, e.g. of liver fat.

Lippmann process *(Image Tech.)*. Early method of colour photography using a transparent silver halide emulsion backed with mercury. When viewed at an angle the reflected image appears in natural colours.

LIPS *(Comp.)*. Abbrev. for *Logical Inferences Per Second*. A measure of speed of a processor. Cf. **MIPS, FLOPS**.

lip seal *(Eng.)*. An oil-retaining shaft seal, often used adjacent to a bearing, in which an annular rubber sealing element having an aperture slightly smaller than the shaft diameter is deformed into the shape of a sealing lip, the sealing pressure being due partly to the deformation stress and partly to the force supplied by an additional **garter spring**.

lip sync *(Image Tech.)*. Precise synchronization in simultaneous picture and sound recording, in contrast to **wild shooting**.

lipuria *(Med.)*. The presence of fat in the urine.

liquation *(Eng.)*. Partial melting of an alloy due to heterogeneity of composition.

liquefaction *(Geol.)*. The change in packing of the grains of a water-filled sediment, turning it into a fluid mass which can then flow.

liquefaction of gases *(Phys.)*. To liquefy a gas, it must be cooled below its critical temperature and, in some cases, compressed. For the so-called 'permanent' gases, oxygen, nitrogen, hydrogen, and helium, having very low critical temperatures, the problem of liquefaction becomes one of obtaining low temperatures. This is done mainly by allowing the compresed gas to expand through a nozzle, cooling occurring by the Joule-Thomson effect.

liquefaction temperature *(Phys.)*. The temperature at which a gas changes state to liquid. Physically the same as the boiling point.

liquefied petroleum gases *(Eng.)*. These fall within three main categories, viz. propane, butane and pentanes. The two sources are natural gas wells and oil refinery separation. Abbrev. *LPG*.

liquid *(Genrl.)*. A state of matter between a solid and a gas, in which the shape of a given mass depends on the containing vessel, the volume being independent. A liquid is practically incompressible.

liquid compass *(Ships)*. A magnetic compass fitted in a bowl containing a suitable liquid. The weight of the card on the pivot is reduced and oscillations of the card are damped out.

liquid counter *(Phys.)*. A counter for measuring the radioactivity of a liquid, usually designed to measure β-, as well as γ-rays.

liquid crystal display *(Electronics)*. A digital or alphanumeric display consisting of two sheets of glass separated by a sealed-in liquid crystal material. Both sheets have a thin transparent coating of conducting material, with the viewing side etched into character-forming segments with leads going to the edges of the display. Voltages applied between the front and back coatings disrupt the orderly arrangement of the molecules sufficiently to darken the liquid and form visible characters. Power consumption is negligible, and edge-lighting can be provided for use in darkness.

liquid crystals *(Chem.)*. Certain pure liquids which are turbid and, like crystals, anisotropic over a definite range of temperature above their freezing-points. See **liquid crystal display**.

liquid-drop model *(Phys.)*. A model of the atomic nucleus using the analogy of a liquid drop in which the various concepts of surface tension, heat of evaporation, etc. are employed. A semi-empirical mass formula can be developed from the model which describes the masses of many stable and unstable nuclei in terms of a few parameters.

liquid-flow counter *(Phys.)*. One for continuous monitoring of radioactivity in flowing liquids.

liquid helium II *(Chem.)*. See under **helium**.

liquid honing *(Eng.)*. Honing by a jet of liquid containing abrasives.

liquid-liquid extraction *(Chem.Eng.)*. Process, both batch and continuous, whereby two nonmixing liquids are brought together to transfer soluble substance from one to the other for useful recovery of these soluble substances.

liquid-metal reactor *(Nuc.Eng.)*. (1) Normally used of a reactor designed for liquid-metal (usually sodium) cooling. (2) Occasionally used to indicate a liquid-metal fuelled reactor.

liquid oxygen *(Chem.)*. See **oxygen**.

liquid paraffin *(Chem.)*. A liquid form of *petroleum jelly*, colourless and tasteless, used as a mild laxative. A mixture of alkanes with more than twelve carbon atoms to the molecule.

liquid-penetrant inspection *(Eng.)*. See **fluorescent-penetrant inspection**.

liquid-phase sintering *(Eng.)*. Sintering in which a small proportion of the material becomes liquid. It may or may not speed up the sintering process by solution transfer of the phases forming the matrix.

liquid-quenched fuse *(Elec.Eng.)*. A fuse in which a liquid is used for quenching the arc. See **semi-immersed-, immersed-**.

liquid resistance *(Elec.Eng.)*. A resistance consisting of a liquid of low conductivity, the current being let to and from the liquid by means of suitable electrodes.

liquid rheostat *(Elec.Eng.)*. One in which a liquid column is used as the resistive element, the terminals being attached to suitable metal plates, one of which is usually movable, thus providing a continuous variation. Only used where current control need not be too precise.

liquid starter *(Elec.Eng.)*. A liquid rheostat arranged to operate as a motor starter.

liquidus *(Chem.)*. A line in a constitutional diagram indicating the temperatures at which solidification of one phase or constituent begins or melting is completed. See **solidus** and **solidification range**.

liquor amnii *(Med.)*. The clear fluid in the amniotic cavity, in which the embryo is suspended.

liquor ratio *(Textiles)*. Used in dyeing or finishing to express the ratio of the weight of liquor used to the weight of material being treated.

L-iron *(Eng.)*. A structural member of wrought-iron or rolled steel, having an L-shaped cross-section. Also called **angle iron**.

lisle *(Textiles)*. Long-staple, highly-twisted folded cotton hosiery yarn, gassed and often mercerized to produce a lustrous effect.

LISP *(Comp.)*. A general purpose programming language in which the expressions are represented as lists. *LISt Processing*.

Lissajous' curves, figures *(Maths.,Phys.)*. Plane curves formed by the composition of two sinusoidal waveforms in perpendicular directions. The parametric cartesian equations are: $x = a\sin(\omega t + \alpha)$, $y = b\sin(\Omega t + \beta)$. The curves embrace a great variety of forms. If the frequencies are commensurable they consist, in general, of a plurality of loops determined by the ratio of the frequencies, and this property is extensively exploited to compare the frequencies of two sinusoidal voltages by applying them to the plates of a cathode-ray tube.

lissencephalous *(Zool.)*. Having smooth cerebral hemispheres.

list *(Comp.)*. (1) A linearly ordered data structure. See **stack, queue, array, linked list**. (2) A **data type**, a list of **atoms** or smaller lists. See **LISP**. (3) To print or display a **listing**.

list *(Ships)*. An angle of transverse inclination arising from unsymmetrical distribution of internal weight. Cf. **heel**.

listel *(Arch.)*. See **facette**.

listening key *(Telecomm.)*. The lever key which the operator throws, to put her head-set on to a cord circuit and speak to a subscriber.

listerellosis *(Vet.)*. See **listeriosis**.

listeriosis *(Vet.)*. *Listerellosis.* A bacterial infection of animals and birds by *Listeria monocytogenes* (*Listerella monocytogenes*). The infection occurs in cattle and sheep (*circling disease*) and more rarely in pigs, and causes a febrile meningo-encephalitis. In poultry, and sometimes in young animals, the disease occurs as a septicaemia.

listing *(Build.)*. (1) A narrow edge of a board. (2) The operation of removing the sappy edge of a board.

listing *(Comp.)*. The sequences of program statements or data in printed form or on a screen.

listing *(Textiles)*. (1) In a dyed fabric, a variation in shade between the selvedges and the centre. (2) selvedge.

literal *(Print.)*. Casual error of composition, such as one character substituted for another, worn letters, transpositions, turned letters, etc.

litharge *(Build.,Chem.)*. Lead (II) oxide, used in paint-mixing as a drier; used also in the rubber and electrical accumulator industries.

litharge cement *(Chem.)*. See **glycerine litharge cement**.

lithia *(Chem.)*. *Lithium oxide*, Li_2O.

lithia emerald *(Min.)*. Misnomer for *hiddenite*.

lithia mica *(Min.)*. See **lepidolite**.

lithiasis *(Med.)*. The formation of calculi in the body. The condition in which an excess of uric acid and urates is excreted in the urine; the gouty diathesis.

lithic arenite *(Geol.)*. A sediment of sand-sized particles many of which are rock fragments rather than mineral grains.

lithification *(Geol.)*. Processes which convert an unconsolidated sediment into a **sedimentary rock**.

lithiophilite *(Min.)*. Orthorhombic phosphate of lithium and manganese, forming with triphylite a continuously variable series.

lithite *(Zool.)*. See **statolith**.

lithium *(Chem.)*. An element symbol Li, r.a.m. 6.939, at. no. 3, mp 186°C, bp 1360°C, rel.d. 0.585. It is the least dense solid, chemically resembling sodium but less active. It is used in alloys and in the production of tritium; also as a basis for lubricant grease with high resistance to moisture and extremes of temperature; and as an ingredient of high-energy fuels. Its salts are used medically as an anti-depressant.

lithium 12-hydroxy stearate *(Chem.)*. A lithium 'soap' widely use in high-performance greases as the main thickening agent. Helps to confer high water-resistance and good low-temperature performance.

lithium aluminium hydride *(Chem.)*. *Lithium tetrahydroaluminate*, $LiAlH_4$. Important reducing agent widely used in organic chemistry because of its effectiveness in 'difficult' reductions, e.g. the reduction of carboxylic acid groups (COOH) to primary alcohol groups (CH_2OH).

lithium carbonate *(Med.)*. Salts of lithium, used as a tranquilizer, and prophylactically and therapeutically in acute mania.

lithium drifted silicon detector *(Phys.)*. An energy-sensitive solid-state detector for ionizing radiation. Lithium is thermally diffused into almost pure but p-type silicon crystal. Used for low energy X-ray and γ-ray spectroscopy. The crystal is kept at liquid nitrogen temperature to reduce thermal noise and to avoid the redistribution of lithium.

lithium-drift germanium detector *(Nuc.Eng.)*. One type of **germanium radiation detector**. There is also an *intrinsic germanium detector* which does not contain lithium.

lithium hydride *(Chem.)*. LiH. Formed when lithium unites with hydrogen at a red heat, it is a strong reducing agent, used as a hydrogen carrier, e.g. for small balloons at sea.

lithium hydroxide *(Chem.)*. LiOH. A strong base, but not so hygroscopic as sodium hydroxide. Used in the form of a coarse, free flowing powder as a carbon dioxide absorber, e.g. in submarines. Also to make lithium stearates etc. in driers and lubricants.

litho *(Print.)*. A common abbreviation for *lithography*.

lithocyst *(Zool.)*. See **statocyst**.

lithodomous *(Zool.)*. Living in rocks.

lithogenous *(Zool.)*. Rock-building, as certain Corals.

lithographic paper *(Paper)*. High machine-finished or supercalendered paper, made so that any stretch occurs the narrow way of the sheet.

lithographic stone *(Geol.)*. A compact, porous fine-grained limestone, often dolomitic, employed in lithography. Pale creamy-yellow in colour, but occasionally grey. Fair samples may be obtained from the Jurassic rocks in Britain, but the finest material comes from Solenhofen and Pappenheim in Bavaria. See also **Solenhofen stone**.

lithographic varnish, lithographic oil, litho oil *(Print.)*. *Heat-bodied* oil used as binder for lithographic inks.

lithography *(Geol.)*. The systematic description of rocks, more especially sedimentary rocks. See **petrology**.

lithography *(Print.)*. Originally the art of printing from stone (*lithographic stone*), but now applied to planographic printing processes depending on the mutual repulsion of water and greasy ink. Damp rollers pass over the surface, followed by inking rollers. The design, which is greasy, repels the water but retains the ink (also greasy), which is transferred to the paper. In modern practice, a metal or paper plate is commonly used in place of stone, the plate being attached to a cylinder for rotary printing, either sheet-fed or web-fed. Most lithography is offset, the impression from the plate being transferred to rubber-covered blanket cylinder and thence to the stock which may be paper or board or other flexible sheets including metal. See also **chromolithography, photolithography, offset printing, small offset**.

litholapaxy *(Med.)*. The operation of crushing a stone in the bladder, followed by the washing out of the crushed fragments.

lithology *(Geol.)*. The character of a rock expressed in terms of its mineral composition, its structure, the grain-size and arrangement of its component parts; that is, all those visible characters that in the aggregate impart individuality to the rock. The term is most commonly applied to sedimentary rocks in hand specimen and outcrop.

lithophagous *(Zool.)*. Stone-eating; said of graminivorous Birds which take small stones into the gizzard, to aid mastication; also of certain Molluscs which tunnel in rock.

lithophile *(Geol.)*. Synonymous with *oxyphile*.

lithophyte *(Bot.)*. A plant growing on rocks or stones, especially an alga attached to a rock or stone.

lithosphere *(Geol.)*. The outer, rigid shell of the Earth lying over the **asthenosphere**. It includes the *crust, continents*, and *plates*. See **continental crust, oceanic crust**.

lithotomous *(Zool.)*. Stone-boring, as certain Molluscs.

lithotomy *(Med.)*. Cutting into the bladder or ureter for the removal of a stone or calculus.

lithotripsy *(Med.)*. A technique using ultrasound for destroying kidney stones.

lithotrite *(Med.)*. An instrument, with special blades, adapted for crushing stones in the bladder.

lithotrity *(Med.)*. The operation of crushing stones in the bladder. Cf. **litholapaxy, lithotripsy**.

lith process *(Image Tech.)*. Use of high-contrast emulsions and developers to give black-and-white images of high maximum density without half-tone gradation, particularly for photo-mechanical reproduction.

litmus *(Chem.)*. A material of organic origin used as an indicator; its colour changes to red for acids, and to blue for alkalis at pH 7. See also **indicator (1)**.

lit-par-lit injection *(Geol.)*. The injection of fluid or molten material, mostly granitic, along bedding, cleavage or schistosity planes in a rock. The product is an alternation of apparently igneous and non-igneous material, known as a *migmatite*. Some of the rocks previously considered to have been formed by the process are now thought to be the products of partial melting, the igneous layers representing the fraction of the rock which became liquid.

litre *(Genrl.)*. Unit of volume equal to 1 cubic decimetre.

litre-atmosphere *(Chem.)*. A former unit of work equal to 101.325 joules.

litter *(Bot.)*. More or less undecomposed fallen leaves and other plant residues at the soil surface.

little-end *(Autos.)*. That part of the connecting rod which is attached to the gudgeon pin.

Little's disease *(Med.)*. Spastic paralysis mainly affecting the lower half of the body, producing the classical 'spasm' child due to aortic infarction of the white matter of brain. Strongly associated with prematurity, better management of which is reducing this form of cerebral palsy.

littoral *(Bot.)*. (1) The area between high and low tide marks. (2) The shallower water of lakes where light reaches the bottom and where rooted plants may grow.

littoral zone *(Zool.)*. Faunal zone bounded by the continental shelf, i.e., down to approximately 200 m.

Littrow grating mounting *(Phys.)*. The mounting of a plane diffraction grating such that a plane mirror reflects back light transmitted by the grating so that it passes through the grating a second time. By rotating the mirror the spectrum can be scanned.

Littrow prism spectrograph *(Phys.)*. A spectrograph in which the same lens serves both to collimate the light and to focus the spectrum on the photographic plate, the light being reflected back through the prism by a plane mirror behind the prism.

litz(endraht) wire *(Elec.Eng.)*. Multiple-stranded wire, each strand being separately insulated so as to reduce the relative weighting of the **skin effect**, i.e. concentration of high-frequency currents in the surface. Much used in compact low-loss coils in filters and in high-frequency tuning circuits.

live *(Acous.)*. Said of an enclosure, or motion-picture set, which is not rendered dead by the presence of sound-absorbing areas, and in which the reverberation is normal or above normal.

live *(Elec.Eng.)*. Connected to a voltage source.

live *(Image Tech.)*. Direct transmission of sound or television without recording. A *live insert* is that part of an otherwise recorded transmission which is live.

live axle *(Eng.)*. A revolving axle to which the road-wheels are rigidly attached, as distinct from a fixed or **dead axle**.

live centre *(Eng.)*. In machine tools, a centre which rotates with the workpiece.

livedo *(Med.)*. Reddish mottling of the skin.

live load *(Eng.)*. A moving load or a variable force on a structure; e.g. that imposed by traffic movement over a bridge, as distinct from a dead weight or load, such as that due to the weight of the bridge.

liver *(Zool.)*. In Invertebrates, the digestive gland or **hepatopancreas**; in Vertebrates, a large mass of glandular tissue arising as a diverticulum of the gut, which secretes the bile and plays an important part in the storage and synthesis of metabolites.

live rail *(Elec.Eng.)*. The supply rail of a 3- or 4-rail electric traction system, usually a few hundred volts d.c. above or below earth potential.

liver flukes *(Med.,Zool.)*. A group of trematode parasites, especially *Fasciola hepatica* and *Clonorchis sinensis*, infecting man via various species of water Snail as intermediate hosts, causing damage to the liver and surrounding organs.

livering *(Build.)*. The condition when a paint in bulk becomes jellylike or tough. Also *curdling, livering up*.

live ring *(Eng.)*. A large roller-bearing, used for supporting turntable and revolving cranes.

live room *(Acous.)*. One which has a longer period of reverberation than the optimum for the conditions of performance and listening.

liver opal *(Min.)*. A form of opaline silica, in colour resembling liver.

liver rot *(Vet.)*. See **distomiasis**.

liverworts *(Bot.)*. See **Hepaticopsida**.

live steam *(Eng.)*. Steam supplied direct from a boiler, as distinct from exhaust steam or steam which has been partly expanded.

lixiviation *(Eng.)*. See **leaching**.

lizardite *(Min.)*. A mineral of the **serpentine** group.

Ljungstrom regenerative air heater *(Eng.)*. A heat exchanger largely used in power stations. It consists essentially of a large, vertical, slowly rotating drum filled with honeycomb arrangement of very thin steel sheet, the flue gases passing upwards through one half of the honeycomb while the air passes downwards through the other.

llama fibre *(Textiles)*. The hair of the llama (*Lama glama*) of South America.

Llandeilo *(Geol.)*. An epoch of the Ordovician period.

Llandovery *(Geol.)*. The oldest epoch of the Silurian period.

Llanvirn *(Geol.)*. An epoch of the Ordovician period.

L-lines *(Phys.)*. Characteristic X-ray frequencies from atoms due to the excitation of electrons from the L-shell. Denoted by L_α, L_β, ...; the lines are multiplets.

L_+, L_0, L_R dose of toxin *(Immun.)*. Used in standardizing diptheria toxin and antitoxin. Describes the quantity of toxin which, when mixed with one standard unit of antitoxin, will respectively kill a guinea pig under standardized conditions, or when injected subcutaneously produces no observable reaction, or when inoculated into the skin produces a minimal lesion.

Lloyd Morgan's canon *(Behav.)*. The proposal that one should never explain behaviour in terms of a higher mental function if a more simple process will explain it.

Lloyd's mirror *(Phys.)*. A device for producing interference fringes. A slit, illuminated by monochromatic light, is placed parallel to and just in front of the plane of a plane mirror or piece of unsilvered glass. Interference occurs between direct light from the slit and that reflected from the mirror. See **interference fringes**.

Lloyd's rules *(Ships)*. A set of rules laid down by the Classification Society, Lloyd's Register of Shipping, governing the construction of steel ships and their machinery.

LLTV *(Aero.)*. Abbrev. for *Low Light TeleVision*.

lm *(Phys.)*. Abbrev. for *lumen*.

LMFBR *(Nuc.Eng.)*. *Liquid Metal cooled Fast Breeder Reactor*. See **breeder reactor, fast reactor**.

LMR *(Nuc.Eng.)*. See **liquid metal reactor**.

L-network, -attenuator, -filter *(Telecomm.)*. Half an unbalanced T-network. See **L-section**.

lo *(Aero.)*. Low level military flight, usually at less than 200 ft.

load *(Comp.)*. To transfer a program and/or data into **main memory** from a **backing store**.

load *(Elec.Eng.)*. (1) Electrical impedance to which the output of source is connected. The source may be e.g. an electrical generator, a transformer, an amplifier or even a single transistor. (2) Total demand for electrical power on a supply, which may be e.g. an electrical generator or an amplifier. (3) Material placed between electrodes for the inductance of heat-through dielectric loss by means of high-frequency electric fields.

load *(Phys.)*. (1) The weight supported by a structure. (2) Mechanical force applied to a body.

load *(Telecomm.)*. (1) Termination of an amplifier or line which absorbs the transmitted power, which is a maximum when this load **matches** the output impedance.

(2) The actual power drawn by or received by a terminating impedance.

load capacitor *(Elec.Eng.)*. That which tunes and maximizes the power to a load in induction or dielectric heating.

load cast *(Geol.)*. A *sole mark* appearing as an irregular bulge on the underside of a sedimentary rock. It lacks evidence of current direction. See **flute cast**.

load cell *(Eng.)*. A load detecting and measuring element utilizing electrical or hydraulic effects which are remotely indicated or recorded.

load characteristic *(Elec.Eng.)*. *Instantaneous* voltage-current characteristic at the output of a generator or amplifier under loading conditions. Also called *operating characteristic*.

load classification number *(Aero.)*. A number defining the load-carrying capacity of the paved areas of an airport without cracking or permanent deflection. Abbrev. *LCN*.

load coil *(Elec.Eng.)*. The coil in an induction heater used to carry the alternating current which induces the heating current in the specimen or charge.

load curve *(Elec.Eng.)*. A curve whose ordinates represent the load on a system or piece of apparatus, and whose abscissae represent time of day, month, or year, so that the curve indicates the value of the load at any time.

load curve *(Eng.)*. See **influence line**.

load despatcher *(Elec.Eng.)*. An engineer who is responsible for the distribution of load over a large interconnected power system.

load displacement *(Ships)*. A ship's displacement at load draught, i.e. the draught to the centre of the freeboard disk marking, which is set off to the summer freeboard.

load draught *(Ships)*. The draught when loaded to the minimum **freeboard** permitted for the place and season.

load dumping *(Elec.Eng.)*. The automatic disconnection of circuits connected to an electrical supply. This is done either to maintain system frequency, if there is a loss in supply or to prevent overloading of the supply equipment. See **load management**, **load governing**.

loaded antenna *(Telecomm.)*. An antenna in which series inductance has been added to increase its natural wavelength.

loaded concrete *(Nuc.Eng.)*. Concrete used for shielding nuclear reactors, loaded with elements of high atomic number, e.g. lead, iron shot, barytes.

loaded impedance *(Telecomm.)*. That of the input of a transducer or other device when output load is connected.

load efficiency *(Electronics)*. Ratio of the useful power delivered by the output stage to a specified load, and the dc input power to the stage.

loader *(Comp.)*. Program which copies an **object program** held on **backing store** into **main store**.

loader *(Min.Ext.)*. A mechanical shovel or other device for loading trucks underground.

load-extension curve *(Eng.)*. A curve, plotted from the results obtained in a tensile test, showing the relations between the applied load and the extension produced.

load factor *(Aero.)*. (1) In relation to the structure, it is the ratio of an external load to the weight of an aircraft. Loads may be centrifugal and aerodynamic due to manoeuvring, to gravity, to ground or water reaction; usually expressed as g, e.g. 7 g is a load seven times the weight of the aircraft. (2) In aircraft operations, the actual *payload* on a particular flight as a percentage of the maximum permissible payload.

load factor *(Elec.Eng.)*. The ratio of the average load to peak load over a period. See **plant load factor**.

load governing *(Elec.Eng.)*. The automatic maintenance of supply frequency from a rotating machine by connecting and disconnecting load as the input power increases and decreases respectively. This is often preferred to conventional governing in small plant because the equipment is cheaper.

load impedance *(Telecomm.)*. Impedance of any device which accepts power from a source, e.g. amplifiers, antenna, loudspeakers, magnetizing coils.

loading *(Build.)*. See **on-costs**.

loading *(Elec.Eng.)*. An inductance added to a line for the purpose of improving its transmission characteristics throughout a specified frequency-band. See **coil-, continuous-**.

loading *(Min.Ext.)*. Adsorption by resins of dissolved ionized substances in ion-exchange process. In uranium technology, the loading factor is the mass of U_3O_8 adsorbed by unit volume of the resin.

loading *(Nuc.Eng.)*. The introduction of fuel into a reactor.

loading *(Paper)*. White mineral matter in finely divided form added to the stock to improve opacity, dimensional stability and reduce costs. In some circumstances it may assist in achieving a smooth finish. Certain loadings such as titanium dioxide are expensive but have special properties such as preventing undue loss of opacity when paper is waxed.

loading *(Textiles)*. Increasing the weight of fabrics by use of starch, size, china clay, etc.

loading and cg diagram *(Aero.)*. A diagram, usually comprising a side elevation of the aircraft concerned, with a scale and the location of all items of removable equipment, payload and fuel, which is used to adjust the weight of the aircraft so that its resultant lies within the forward and aft **cg limits**.

loading and unloading machine *(Nuc.Eng.)*. System for introducing and withdrawing fuel elements from a reactor, with safety provision for personnel. See **charge, discharge machine**.

loading capacity *(Min.Ext.)*. In ion exchange, saturation limit of resin.

loading coil *(Elec.Eng.)*. The coil inserted, in series with a line's conductors, at regular intervals. Also called *Pupin coil*.

loading gauge *(Civ.Eng.)*. (1) The limiting dimensions governing height, width etc., of rolling stock to ensure that adequate clearance is obtained for passage under bridges and through tunnels. (2) A shaped bar suspended over a railway track, at the correct height and position, to check compliance of trucks passing underneath with the above limiting dimensions.

load leads *(Elec.Eng.)*. The connections or transmission lines between the power source of an induction (or dielectric) heater and the load coil or applicator.

load-levelling relay *(Elec.Eng.)*. A relay used in connection with certain, temporarily dispensable, forms of apparatus such as storage water-heaters. It automatically switches them off when the demand on the system exceeds a certain value.

load line *(Electronics)*. On a set of output characteristic curves for an amplifying device, a line representing the load, straight if entirely resistive, elliptical if reactive.

load lines *(Ships)*. A group of lines marked on the outside of both sides of a ship to mark the minimum **freeboard** permitted in different parts of the world and seasons. Sometimes called *plimsoll mark*.

load management *(Elec.Eng.)*. The connection and disconnection of circuits fed from an electrical supply to maintain the average load constant, or within preset limits. This is done for reasons of stable operation of a power system and for economy.

load matching *(Telecomm.)*. Adjusting circuit conditions to meet requirements for maximum energy transfer to load.

load-rate prepayment meter *(Elec.Eng.)*. A form of prepayment meter in which the charge per unit is changed whenever the load exceeds a certain predetermined value.

loadstone *(Min.)*. See **lodestone**.

loam *(Bot.)*. A soil that is a mixture of sand, silt and clay particles, a desirable texture for horticultural and agricultural uses.

loam *(Build.)*. Earth composed of clay and sand. Also called *mild clay, sandy clay*.

loam *(Foundry)*. A clayey sand milled with water to a thin plastic paste, from which moulds are built up on a backing of soft brick; generally swept or strickled to shape without the use of a pattern.

loam *(Geol.)*. A rich friable soil containing more or less equal amounts of sand, silt, clay and usually organic matter.

loam board *(Foundry)*. See **strickle board**.

loam bricks *(Foundry)*. Cakes of loam, or soft building-bricks built up with loam, which form a solid but porous support for the loam forming the wall of the mould. See **loam**.

loan *(Paper)*. High quality bond paper, originally and occasionally made from a rag furnish and tub sized, of durable character intended for documents which resist repeated handling and are required to last.

lobar pneumonia *(Med.)*. Inflammation of one or more lobes of the lung, the affected lobes becoming solid; due usually to infection with the pneumococcus. See also **pneumonia**.

lobe *(Autos.)*. A rounded projection, usually on a cam.

lobe *(Med.)*. (1) A natural division, formed by a fissure, of an organ, e.g. the brain, liver, lungs. See **olfactory lobes**, **optic lobes**. (2) The soft lower part of the external ear.

lobe *(Telecomm.)*. Enhanced response of an antenna in the horizontal or vertical plane, as indicated by a lobe or loop in its radiation pattern. The **beam** effect arises from a *major lobe*, generally intended to be along the forward axis.

lobe *(Zool.)*. A rounded or flap-like projection. adjs. *lobate, lobed, lobose, lobulate.*

lobed *(Bot.)*. Leaf, petal etc. divided into curved or rounded parts connected to each other by an undivided centre.

lobeline *(Chem.)*. $C_{22}H_{27}O_2N$, a piperidine alkaloid obtained from *Lobelia inflata*. Forms broad needles, mp 130–131°C. It is monoacidic, and is remarkable for yielding acetophenone when heated with water. Used as a smoking deterrent.

lobe switching *(Radar)*. Method of determining the precise direction of a target without resorting to impossibly narrow antenna beams. While the antenna is turning, its beam is switched periodically to the left and right of the dead-ahead position; when equal signals are received in both positions, the antenna is accurately aimed.

lobotomy *(Behav.)*. See **pre-frontal lobotomy**.

lobotomy *(Med.)*. See **leucotomy**.

lobule, lobulus *(Zool.)*. A small lobe; one of the polyhedral cell masses forming the liver in Vertebrates. adjs. *lobular, lobulate.*

LOCA *(Nuc.Eng.)*. *Loss-Of-Coolant-Accident*. The conditions which might arise in the event of loss of primary or secondary coolant in a reactor.

local action *(Elec.Eng.)*. Deterioration of battery due to currents flowing to and from the same electrode.

local area network *(Comp.,Telecomm.)*. High bandwidth (allowing high data transfer rate) computer network, operating over a small area, such as an office or group of offices. Abbrev. *LAN*.

local attraction *(Phys.)*. See **magnetic anomaly**.

local carrier *(Telecomm.)*. In suppressed carrier systems, demodulation with an adequate carrier wave inserted before demodulation.

local exchange *(Telecomm.)*. The exchange to which a given subscriber has a direct line. Sometimes called *local central office*.

local group *(Astron.)*. The name of the family (or cluster) of **galaxies** to which our **Milky Way** belongs. There are two dozen members within 5 million light years or so. The **Andromeda Nebula** is another prominent member of the group.

localization *(Behav.)*. An ability of the sense organs of an animal to determine the source of a stimulus in space. Many sense organs can detect direction, but probably only the eyes and ears or lateral line system, can judge distance. See also **echolocation**.

localized vector *(Maths.)*. See **bound vector**.

localizer beacon *(Aero.)*. A directional radio beacon associated with the **ILS**, which provides an aircraft during approach and landing with an indication of its lateral position relative to the runway in use.

local junction circuit *(Telecomm.)*. A junction between exchanges in the local call area, i.e. where connections are made by local codes as opposed to **SDD**.

local Mach number *(Aero.)*. The ratio of the velocity of the airflow over a part of a body in flight to the local speed of sound. Usually it is concerned with a part of greater curvature, where the airflow accelerates momentarily, thereby increasing the Mach number above that of the body as a whole, e.g. over the wing or canopy.

local oscillations *(Telecomm.)*. Those generated within the apparatus which uses them.

local oscillator *(Telecomm.)*. That in a **superhet** receiver which is mixed with the incoming signal, in the *mixer* or *frequency changer* to produce an output of sum- and difference-frequencies. The *difference* signal will generally be quite low and it is this which is the *intermediate frequency*.

local time *(Astron.)*. Applied to any of the 3 systems of time reckoning, sidereal, mean solar, or apparent solar time, it signifies the hour angle of the point of reference in question measured from the local meridian of the observer. The local times of a given instant at 2 places differ by the amount of their difference in longitude expressed in time, the local time at a place east of another being the greater.

local variable *(Comp.)*. One whose use is restricted to a particular subprogram.

local vent *(Build.)*. A connection enabling foul air in a room or plumbing fixture to escape to the outer air.

location *(Comp.)*. A position in a **memory** which holds a **word** or part of a word.

location *(Eng.)*. A geometrical feature, e.g. a projection or a recess, by which one machine component or article may be correctly positioned by engagement in or with a corresponding feature in another, for the correct spatial relationship of machine or other components.

location *(Image Tech.)*. Place of motion picture or video production other than a permanently established studio.

locator *(Acous.)*. See **sound locator**.

locator beacon *(Aero.)*. The 'homing' beacon on an airfield used by the pilot until he picks up the localizer signals of the **ILS**.

lochia *(Med.)*. The normal discharge from the vagina during the first week or two after childbirth.

lock *(Horol.)*. The stopping of the escape wheel. Also **locking**.

lock *(Hyd.Eng.)*. A communicating channel, having gates at each end, between the higher and lower reaches of a canal. It is used to transfer a vessel from one reach to the other.

lock *(Print.)*. A safety device in most control boxes to prevent the press being moved under power; the press can be made free only by cancelling the lock; may also be used during the run to prevent increase of speed.

lockage *(Hyd.Eng.)*. Water lost, i.e. transferred from a higher to a lower level, in the operation of passing a vessel through a lock.

lock-bay *(Hyd.Eng.)*. The water space enclosed in a **lock-chamber**.

lock-chamber *(Hyd.Eng.)*. The space between the head-gates and tail-gates of a lock.

locked *(Elec.Eng.)*. Said of an oscillator when it is held to a specific frequency by an external source.

locked-coil conductor *(Elec.Eng.)*. A form of stranded conductor in which the outer wires are so shaped that they are prevented from having any radial movement.

locked (eccentric) groove *(Acous.)*. Finishing groove on the surface of a gramophone record, motion of the needle in this groove operating *stopping* or *record-changing* mechanisms.

locked oscillator detector *(Telecomm.)*. Form of FM detector using a pentode valve which incorporates a self-oscillating tuned circuit connected to the suppressor grid. The FM input signal is applied to a resonant tuned circuit which is connected to the control grid. The tuned circuit of the suppressor grid locks to the frequency of the incoming signal and leads to an AF anode current proportional to the FM input; this circuit does not respond to amplitude-modulated signals, including noise, so that a separate limiting stage is unnecessary.

locked test *(Min.Ext.)*. In preliminary tests on unknown ores, retention of a potentially troublesome fraction of the test product for addition to a new batch of ore, to ascertain whether in continuous work such a build-up would be upsetting. Also called *cyclic test*.

lock gate *(Hyd.Eng.)*. A pair of doors at one end of a lock, serving, in conjunction with a similar pair at the other end, to enclose water within the lock-chamber.

lock-in *(Electronics)*. Generally, to synchronize one oscillator with another, as in a homodyne or frequency doubler. One oscillator must be free-running and capable of being pulled. See **phase-locked loop**.

lock-in amplifier *(Telecomm.)*. Synchronous amplifier, sensitive to variation of signal of its own frequency.

locking *(Horol.)*. See **lock**.

locking *(Image Tech.,Telecomm.)*. (1) See **latching**. (2) The control of frequency of an oscillating circuit by means of an applied signal of constant frequency, e.g. controlling the time base of a TV receiver by the incoming signal.

locking angle *(Horol.)*. *The angle of lock*, measured from the pallet centre, through which the pallets have to move before unlocking can take place.

locking face *(Horol.)*. The portion of the pallet upon which the teeth of the escape wheel drop for locking.

locking key *(Telecomm.)*. A hand-operated telephone key, which, when operated, remains in its operated position until released by the hand.

locking plate *(Horol.)*. A circular plate around the periphery of which notches are cut, the distance between the notches regulating the number of hours struck.

locking relay *(Telecomm.)*. A telephone relay which, when operated, remains in its operated condition when the operating current ceases, either by closing a winding which carries a sustaining current or, more rarely, by mechanical means.

lockjaw *(Med.)*. See **tetanus**.

locknit *(Textiles)*. A fabric produced by **warp-knitting**.

lock nut *(Eng.)*. (1) An auxiliary nut used in conjunction with another, to prevent it from loosening under vibration. (2) Any special type of nut designed to prevent accidental loosening.

lock paddle *(Hyd.Eng.)*. A sluice through which water is passed to fill an empty lock-chamber.

lock rail *(Build.)*. (1) The door-rail which is level with the lock. (2) The front member of a piano, running along the keyboard, which usually contains the lock.

lockrand *(Build.)*. A course of stones laid as bondstones.

lock seam *(Eng.)*. A type of joint, not absolutely tight, produced in the manufacture of sheet metal drums, cans, etc., by simply folding, interlocking and pressing together the longitudinal edge areas of the product.

lock sill *(Hyd.Eng.)*. See **mitre sill**.

lock stile *(Build.)*. That stile of a door in or on which the lock is fastened.

locks, up and/or down *(Aero.)*. See **up and/or down locks**.

lock up *(Print.)*. To secure type matter firmly in the *chase*, ready for printing.

lock-woven mesh *(Civ.Eng.)*. A mechanically woven fabric, made of steel wires crossing at right-angles and secured at the intersections, used in reinforced concrete construction.

locomotive *(Eng.)*. A vehicle driven by oil, electricity or coal, for rolling stock on a railway.

locomotive boiler *(Eng.)*. The type of boiler used on steam locomotives; it consists of an internal fire-box at one end of the horizontal cylindrical shell, from which the hot gases are led through fire-tubes passing through the water space into the smoke-box at the front of the boiler. See **fire-tube boiler, fire-box**.

locomotor ataxia (ataxy) *(Med.)*. See **tabes dorsalis**.

locular, loculatous *(Bot.)*. Divided into compartments by septa.

locule, loculus *(Bot.)*. A cavity, especially one within a sporangium or an ovary, containing the spores or ovules repectively.

locus *(Biol.)*. The position on a chromosome occupied by a specified gene and its alleles. pl. *loci*.

locust *(Zool.)*. One of several kinds of winged insects of the family *Acridiidae*, akin to Grasshoppers, highly destructive to vegetation.

lode *(Civ.Eng.)*. An artificial dyke.

lode *(Geol.,Min.Ext.)*. (1) A mineral deposit composed of a zone of veins. (2) Steeply inclined fissure of non-alluvial mineral enclosed by walls of country rock of different origin. Of magmatic or hydatogenetic origin.

loden *(Textiles)*. A coarse-milled woollen fabric used for jackets and coats because of its weather resisting properties.

lodestone, loadstone *(Min.)*. Iron oxide. A form of magnetite exhibiting polarity, behaving, when freely suspended, as a magnet.

Lodge-Cottrell detarrer *(Chem.Eng.)*. Plant consisting of a series of pipes with high-tension electrodes for removal of tar particles from gases.

Lodge-Cottrell precipitator *(Chem.Eng.)*. Plant for carrying out *electrostatic precipitation* of dust from gases.

lodicules *(Bot.)*. Small scales inserted below the stamens in the florets of most grasses; when the floret is mature the lodicules swell forcing apart lemma and palea and allowing stamens and stigmata to grow out.

loess *(Geol.)*. A clay, consisting of fine rock-flour (mainly quartz), originating in arid regions and transported by wind. Vast accumulations cover large areas of China, eastern Europe and N. and S. America.

Löffler boiler *(Eng.)*. A high-pressure boiler employing forced circulation, by pumping steam through small-diameter tubes. Part of the high-temperature steam is returned to the water drum, to produce the saturated steam supply to the pump.

Lofton-Merritt stain *(Paper)*. A stain consisting of malachite green and basic fuchsin for distinguishing between unbleached sulphite and sulphate wood pulp fibres, under the microscope.

log *(Maths.)*. Abbrev. of *logarithm*.

log *(Ships)*. See **nautical log**.

logagraphia *(Med.)*. Loss of the ability to express ideas in writing.

logarithm *(Maths.)*. The logarithm of a number N to a given base b is the power to which the base must be raised to produce the number. It is written as $\log_b N$, or as $\log N$ if the base is implied by the context. There are two systems of logarithms in common use: (a) *common* or *Briggs'* logarithms, with base 10. If N is any number, it can be written in the form $N = 10^n \times M$ (where n is an integer and M is between 1 and 10) so that $\log N = n + \log M$. n is called the *characteristic* of the logarithm and $\log M$, which is obtained from tables, the *mantissa*. (b) *natural* or *Napierian* or *hyperbolic* logarithms with base $e (= 2.718...)$. To distinguish them, a logarithm to the base 10 is often written $\log x$; a logarithm to the base e being $\ln x$. However, in mathematical usage $\log x$ always means base e. The following conversions are useful: $\log_e x = \log_{10} x \times 2.30259$; $\log_{10} x = \log_e x \times 0.43429$.

logarithmic amplifier *(Acous.)*. One with an output which is related logarithmically to the applied signal amplitude, as in decibel meters or recorders.

logarithmic array *(Radar,Telecomm.)*. Tapered **end-fire array** designed to operate over a wide range of frequency.

logarithmic damping ratio *(Elec.Eng.)*. In a plane wave, the logarithmic ratio of the amplitudes at two adjacent maxima or minima.

logarithmic decrement *(Phys.)*. Logarithm to base e of the ratio of the amplitude of diminishing successive oscillations.

logarithmic function *(Maths.)*. The logarithmic function $\log x$ is defined by the equation

$$\log x = \int_1^x \frac{dt}{t}.$$

It equals $\log_e x$. See e and **exponential function**.

logarithmic horn *(Acous.)*. The horn of a loudspeaker

which is of *exponential* form. Also a metal horn for microwave work.

logarithmic mean temperature difference *(Chem.Eng.)*. Universally used relationship in heat exchange calculations. If temperature difference at one end of exchanger is T_1, and at other t_2, then the mean temperature at which heat exchange is calculated to occur is the log mean which is

$$\frac{T_1 - t_2}{\log_e \dfrac{T_1}{t_2}},$$

and this is less than the arithmetic mean. In complex exchange conditions this may have to be corrected further to allow for special factors and is then usually said to be *weighted*.

logarithmic resistor, capacitor, potentiometer *(Elec. Eng.)*. A variable form of resistor, etc. for which the movement of a control is directly or inversely proportional to the fractional change of resistance, etc. Such characteristics are used to counteract opposite characteristics in amplifiers, etc.

logarithmic spiral *(Maths.)*. See **equiangular spiral**.

logatom *(Acous.)*. Artificial syllable without meaning of the form consonant – vowel – consonant, used in articulation testing.

log-dec *(Phys.)*. Abbrev. for *LOGarithmic DECrement*.

logger *(For.)*. One who works in the felling and extraction of timber. Also *bucheron* (Can.), *lumberjack* (US).

logger *(Telecomm.)*. Arrangement of electronic devices for obtaining an output indication which is proportional to the logarithm of the input amplitude or intensity. Required in modulation and noise meters.

loggia *(Arch.)*. A covered gallery or portico built into, or projecting from, the face of a building, and bounded by a colonnade on its open side.

logging wheels *(For.)*. A pair of wheels 2–3.5 m (7–12 ft) in diameter for transporting logs, which are slung below the axle. Also *katydid*.

logic *(Comp.)*. See **mathematical logic, symbolic logic, logical operation**.

logical design *(Comp.)*. The basic planning and synthesizing of a network of **logic elements** to carry out a particular function. See **machine architecture**.

logical error *(Comp.)*. Mistake in the design of a program (e.g. a branch in the wrong statement).

logical operation *(Comp.)*. A nonarithmetical operation performed on binary variables called *Boolean* variables, which can take the values TRUE or FALSE. The three basic operators are **AND, OR** and **NOT**. All other logical operators can be expressed in terms of these three; see **NAND, NOR, XOR**. All logical operations can be performed by a computer using 1 for TRUE and 0 for FALSE. See **gate**.

logical shift *(Comp.)*. One where bits shifted from the end of the location are lost, and zeros are shifted in at the opposite end.

logic array *(Comp.)*. An integrated circuit consisting of an array of **logic gates** that are interconnected in a specified way. See **uncommitted-, programmable-**.

logic element *(Comp.)*. **Gate** or combination of gates. Also *logic unit*.

logic programming *(Comp.)*. Style of programming based on reasoning in formal logic exemplified by resolution theorem proving which underlies **PROLOG**.

logic unit *(Comp.)*. See **logic element**.

log in (out) *(Comp.)*. Method by which the remote terminal user enters (leaves) a multi-access computer system.

logistic equation *(Ecol.)*. Equation describing the typical increase of a population towards an asymptotic value:

$$\frac{dN}{dt} = rN\left(\frac{K - N}{K}\right)$$

where N is the population size, t is time, r is the innate capacity for increase and K is the maximum or asymptotic value of N.

LOGO *(Comp.)*. Programming language with list processing features, widely known for its **turtle** graphics.

logotype *(Print.)*. A word, or several letters, cast as one piece of type; also, a block or line illustration of a trade mark.

log-periodic *(Telecomm.)*. A tapered **end-fire array**, which may consist of dipoles, folded dipoles or monopoles, fed in such a way that the maximum radiation is along the major axis of the array. At any given frequency, only one element and a few immediate neighbours are in operation; hence, the tapered form of the elements gives the antenna a wide frequency range.

log washer *(Min.Ext.)*. Trough or tank set at a slope, in which ore is tumbled with water by means of one or more box girders with projecting arms set helically, so that as they rotate coarse lump is cleaned and delivered upslope while mud and sand overflows downslope.

logwood *(For.)*. The heartwood of *Haematoxylon campechianum*, having a distinctive sweet taste and a smell resembling that of violets. It yields a black dye, and also *haematoxylin*, a stain once much used in microscopy.

loktal base *(Electronics)*. A valve base with a centre pin to lock the base securely in an appropriate socket. It has 8 pins which extend directly through the glass envelope of the valve.

löllingite, loellingite *(Min.)*. Arsenide of iron, $FeAs_2$, occurring as steel-grey crystals, prismatic in habit, belonging to the orthorhombic system.

lomasome *(Bot.)*. A compact mass of membranes in the cytoplasm adjacent to a cell wall, apparently invaginated from the plasmalemma, of obscure function.

lomentum *(Bot.)*. A fruit, usually elongated, which develops constrictions as it matures, finally breaking across these into 1-seeded portions. adj. *lomentose*.

London forces *(Chem.)*. Forces arising from the mutual perturbations of the electron clouds of two atoms or molecules, the forces varying as the inverse sixth power of the distance between the molecules.

London hammer *(Build.)*. Light type of cross-pane hammer. Also *Exeter hammer*.

London plane *(For.)*. *Platanus acerifolia*, yielding a useful general-purpose timber. Also called *sycamore* (chiefly US), but not to be confused with the true sycamore, *Acer pseudoplatanus*.

London screwdriver *(Build.)*. One with flat blade and tapered end, usually favoured for large sizes.

London-shrunk *(Textiles)*. Term in the woollen and worsted trades to indicate that a fabric has been preshrunk. Dry fabric is allowed to dry naturally without tension.

lone-pair *(Chem.)*. Pair of valency electrons unshared by another atom. Such lone pairs are responsible for the formation of co-ordination compounds.

long *(Glass)*. Slow-setting.

long-and-short work *(Build.)*. A mode of laying quoins and of forming door and window jambs in rubble walling, the stones being alternately laid horizontally and set up on end; the latter stones are usually longer than the former.

long-bodied type *(Print.)*. Type case on a larger body to avoid leading, e.g. 9-point on 11-point will have the appearance of 9-point leaded 2-point. Also called *bastard fount*.

long column *(Civ.Eng.)*. A slender column which fails by bending rather than by crushing. It typically has a length 20 to 30 times its diameter.

long-day plant *(Bot.)*. A plant that will flower better under conditions of long days and short nights e.g. spinach, *Spinacia oleracea*. Cf. **short-day plant, day-neutral plant**. See **photoperiodism**.

long descenders *(Print.)*. The length of the descenders (j,g, etc.) is a feature of type design, long descenders being usually considered to confer elegance. Certain typefaces are supplied with alternative long descenders to improve their appearance for bookwork, among them being Monotype Baskerville, Plantin, Times.

longeron *(Aero.)*. Main longitudinal member of a fuselage or nacelle.

long float *(Build.)*. A plasterer's trowel so long as to need two men to handle it.

longi- *(Genrl.)*. Prefix from L, *longus*, long.

longicorn *(Zool.)*. Having elongate antennae, as some Beetles.

long inks, short inks *(Print.)*. An empirical method of comparison between inks of differing viscosity by determining the length of the ink thread when drawn upwards with an ink knife to the point at which it breaks.

longipennate *(Zool.)*. Having elongate wings or feathers.

longirostral *(Zool.)*. Having a long beak or rostrum.

longitude *(Astron.)*. See **latitude and longitude**.

longitude *(Surv.)*. The perpendicular distance of the midpoint of a survey line from the reference meridian.

longitudinal *(Aero.)*. A girder that runs fore and aft on the outside of a rigid airship frame. Longitudinals connect the outer rings of the transverse frames.

longitudinal axis *(Aero.)*. See **axis**.

longitudinal current *(Elec.Eng.)*. See **circulating current**.

longitudinal dune *(Geol.)*. A long, narrow sand-dune parallel to the direction of the prevailing wind.

longitudinal frame *(Ships)*. A stiffening member of a ship's hull disposed longitudinally, as opposed to a transverse frame. It is supported at ends by either bulkheads or web frames, disposed transversely.

longitudinal heating *(Elec.Eng.)*. Dielectric heating in which electrodes apply a high-frequency electric field parallel to lamination. See **glueline**.

longitudinal instability *(Aero.)*. The tendency of an aircraft's motion in the plane of symmetry to depart from a steady state, i.e. to pitch up or down, to rise or fall or to vary in horizontal speed.

longitudinal joint *(Build.)*. A joint used to secure two pieces of timber together in the direction of their length. See **splice, scarf**.

longitudinal magnetization *(Acous.)*. That of magnetic recording medium along an axis parallel to the direction of motion.

longitudinal magnification *(Phys.)*. The ratio of the length of the image to the length of the object in a lens system when the object is small and lies along the axis of the system.

longitudinal metacentre *(Ships)*. The **metacentre** obtained by producing an angle of **trim** about a transverse axis by an external force. Cf. **transverse metacentre**.

longitudinal oscillation *(Aero.)*. A periodic variation of speed, height and angle of pitch. See also **phugoid oscillation**

longitudinal stability *(Aero.)*. See **stability**.

longitudinal valve *(Zool.)*. In Amphibians, a large, flaplike valve which traverses the *conus arteriosus* longitudinally and obliquely.

longitudinal wave *(Acous.)*. Propagating sound wave in which the motions of the relevant particles are in line with the direction of translation of energy.

long letters *(Print.)*. Letters with the long accent added (ā, ē, ī etc.).

long-oil *(Build.)*. Term applied to varnishes, etc., containing a high proportion, i.e. more than about 60%, of oil.

long-period variables *(Astron.)*. See **variable stars**.

long-persistence screen *(Electronics)*. CRT screen coated with long afterglow phosphor (up to several seconds).

longprimer *(Print.)*. An old type-size, approximately 10-point.

long range *(Aero.)*. Aircraft, ship or missile capable of covering great distances without re-fuelling.

long saw *(Build.)*. A **pit-saw**.

long shoot *(Bot.)*. A shoot of mostly woody plants with relatively long internodes, that extends the canopy. Long shoots typically bear few or no flowers and in some species bear no foliage leaves but only scales (e.g. pines) or spines (e.g. *Berberis*). Cf. **short shoot, growth bud**.

longshore current *(Geol.)*. A current which flows parallel to the shore-line.

longshore drift *(Geol.)*. The movement of material along the shore by a **longshore current**.

long shot *(Image Tech.)*. A scene photographed from a distance, showing the general setting of the action.

long-shunt compound winding *(Elec.Eng.)*. A field winding arrangement for a compound-wound d.c. machine in which the shunt winding is connected across the external terminals, i.e. across the armature and the series winding.

long-sightedness *(Med.)*. See **hypermetropia**.

long superstructure *(Ships)*. A superstructure which is sufficiently long to be included in calculations of the ship's main strength, and not simply as an excrescence.

long-tail pair *(Electronics)*. A two-valve or two transistor circuit with a resistor common to both cathode or emitter connections. The common resistor provides strong negative feedback, and the circuit is inherently stable. Antiphase outputs can be obtained from each anode or emitter, and if both inputs are energised, differential amplification can take place.

long-term memory *(Behav.)*. According to the 3-store model of memory, a memory system that keeps memories for long periods, has a very large capacity, and stores items in an organised and processed form. See **short term memory, sensory store**.

long tom *(Min.Ext.)*. Portable sluice in which rough concentrates made during treatment of alluvial sands or gravels are sometimes worked up to a better grade.

long ton *(Genrl.)*. A unit of mass, 2240 lb. See **ton**.

longwall coal-cutting machine *(Min.Ext.)*. A machine which severs coal in mechanized mining.

longwall working *(Min.Ext.)*. Method of mining bedded deposits, notably coal, in which the whole seam is removed, leaving no pillars. If advancing, work goes outward from shaft and roads must be maintained through excavated areas. If retreating, ground behind workers can be allowed to cave if surface rights are not thus affected.

long waves *(Telecomm.)*. Low frequency radio waves of wavelength > 1000 m (frequency < 300 KHz).

look-through *(Paper)*. Term for the appearance of the paper by transmitted light..

loom *(Textiles)*. A machine for weaving cloth, in which two sets of threads, **warp** and **weft**, are interlaced. Conventional models employ shuttles to carry the weft between the warp threads. Later types have small carriers or rapiers, or project the weft by jets of air or water (**jet looms**). Modern highly engineered models are called *weaving machines*.

loom efficiency *(Textiles)*. The average number of **picks** per minute inserted by a loom expressed as a percentage of what is theoretically possible if the machine were running continuously.

looming *(Meteor.)*. The vague enlarged appearance of objects seen through a mist or fog, particularly at sea.

looming *(Phys.)*. A particular form of mirage in which the images of objects below the horizon appear in a distorted form.

looming *(Textiles)*. Drawing the threads of the warp through the eyes of the heald shaft and the reed, in the order arranged for the predetermined pattern.

looming response *(Behav.)*. Refers to the fact that many animals, including humans, respond to a sudden increase in stimulus size by ducking or turning away from it.

loop *(Aero.)*. An aircraft manoeuvre consisting of a complete revolution about a lateral axis, with the normally upper surface of the machine on the inside of the path of the loop. See also **inverted loop**.

loop *(Comp.)*. Sequence of instructions which is executed repeatedly until some specified condition is satisfied. See **nested loop, infinite loop**.

loop *(Electronics)*. Feedback control system.

loop *(Image Tech.)*. (1) The length of slack film between two driven points, as between the intermittent mechanism and a sprocket in a motion picture camera or projector. (2) A length of film or tape joined end to end, for use as a continuous strip.

loop *(Phys.)*. (1) Closed graphical relationship, e.g.

hysteresis loop. (2) Same as **antinode** of displacement in standing waves.

loop actuating signal *(Electronics)*. The combined input and feedback signals in a closed loop system.

loop antenna *(Telecomm.)*. See **frame antenna**.

loop cable *(Elec.Eng.)*. See **twin cable**.

loop coupling *(Elec.Eng.)*. Small loop connected between conductor and shield of a coaxial transmission line, for collecting energy from one of the series of resonators in a magnetron or a rhumbatron in a klystron valve.

loop dialling *(Telecomm.)*. See **loop-disconnect pulsing**.

loop difference signal *(Electronics)*. Output signal at point in feedback loop produced by input signal applied at the same point.

loop-disconnect pulsing *(Telecomm.)*. The normal subscriber dialling, whereby trains of impulses are set up by interrupting the loop current from the exchange. Also *loop dialling*.

loop diuretics *(Med.)*. Group of drugs which inhibit resorption from the *loop of Henle* in the renal tubule. They are powerful diuretics and include *frusemide, bumetanide* and *ethacrynic acid*.

looped filament *(Elec.Eng.)*. A filament arranged in the form of a large helix of one or more turns; usually employed for carbon filament lamps.

loop feedback signal *(Electronics)*. The part of the loop output signal fed back to the input to produce the **loop actuating signal**.

loop gain *(Electronics)*. Product gain of amplifier and feedback circuit; if greater than unity, the system is liable to sustain oscillations. See **Nyquist criterion**.

loop galvanometer *(Elec.Eng.)*. A sensitive galvanometer in which the moving element is a U-shaped current-carrying loop of aluminium foil, the two sides of the loop being in magnetic fields of opposite direction.

looping-in *(Elec.Eng.)*. A term used in wiring work to denote a method of avoiding the use of a tee-joint by carrying the conductor to and from the point to be supplied.

looping mill *(Eng.)*. A rolling mill in which the product from one stand is fed into another in the opposite direction, usually above the first.

looping roller *(Print.)*. A roller on a web-fed press which moves to provide an intermittent flow to the web.

loop input signal *(Electronics)*. An external signal applied to a feedback control loop in control systems.

loop output signal *(Electronics)*. The extraction of the controlling signal from a feedback control.

loop pile *(Textiles)*. Fabric made with loops protruding from the surface of a firm ground. See **moquette**.

loop-raised fabric *(Textiles)*. A warp-knitted fabric in which some of the loops are plucked by wires in a subsequent raising process.

loopstick antenna *(Telecomm.)*. Same as *ferrite-rod antenna*.

loop test *(Elec.Eng.)*. A method of test used for locating faults in electric cables. The faulty conductor is made to form part of a closed circuit or loop, an adjacent sound conductor usually forming another part. See **Allen's-, Varley-**.

loop yarn *(Textiles)*. A fancy yarn formed by wrapping an effect thread round a twisted core so that loops protrude from the yarn surface.

loose butt hinge *(Build.)*. A butt hinge in which one leaf may be lifted from the other, enabling, e.g. a door to be easily removed.

loose centres *(Eng.)*. Heads similar to the tailstock of a lathe; used for supporting work on the table of a planing machine etc. so that it may be rotated.

loose coupling *(Elec.Eng.)*. See **weak coupling**.

loose coupling *(Eng.)*. A shaft coupling capable of instant disconnection, as distinct from a **fast coupling**.

loose eccentric *(Eng.)*. An eccentric used on small reversing steam-engines. It rides freely on the shaft but is located by either of two stops on the shaft, which position and drive it for ahead- and reverse-running respectively.

loose gland *(Eng.)*. A ring used in making an **expansion joint** between hot-water pipes. It slides on the spigot and compresses a rubber ring against the socket, to which it is bolted.

loose-leaf *(Print.)*. A binding system in which separate leaves are held together within a cover by means of a spring, ring, spiral, or other device, by unlocking which a leaf may be removed or inserted at any point.

loose leaf gold *(Build.)*. Real gold beaten into very thin leaves and supplied in small books with tissue between each leaf. Produces a better lustre than transfer leaf. Used in **glass gilding**.

loose needle survey *(Surv.)*. Traverse with miner's dial, in which the magnetic bearing is read at each station.

loose piece *(Eng.)*. Part of a foundry pattern which has to be withdrawn from the mould after the main pattern, through the cavity formed by the latter.

loose pulley *(Eng.)*. A pulley mounted freely on a shaft; generally used in conjunction with a *fast pulley* to provide means of starting and stopping the shaft by shifting the driving belt from one to the other.

loparite *(Min.)*. A titanate of sodium, calcium and the rare earth elements; a variety of **perovskite**.

loph *(Zool.)*. A crest connecting the cusps of a molar tooth.

lophobranchiate *(Zool.)*. Having tuftlike, crestlike, or lobelike gills.

lophodont *(Zool.)*. Of Mammals, having cheek teeth with transverse ridges on the grinding surface.

lophophore *(Zool.)*. A ciliated tentacle used for food gathering; found in some aquatic invertebrates.

lordoscoliosis *(Med.)*. A deformity of the spine in which *lordosis* is associated with lateral curvature of the spine.

lordosis *(Med.)*. Deformity of the spine in which there is an increase of the forward convexity of the lower half of the spinal column.

lore *(Zool.)*. In Birds, the space between the beak and the eye. adj. *loral*.

Lorentz contraction *(Phys.)*. See **FitzGerald-Lorentz contraction**.

Lorentz force *(Phys.)*. The force experienced by a point charge q moving with a velocity $\mathbf{v}$ in a field of magnetic induction $\mathbf{B}$ and in an electric field $\mathbf{E}$. The force $\mathbf{F}$ is given by

$$\mathbf{F} = q[(\mathbf{v} \times \mathbf{B}) + \mathbf{E}]$$

Lorentz-Lorenz equation *(Chem.)*. The equation by which molecular refraction is defined:

$$[R] = \frac{n^2-1}{n^2+2} \cdot \frac{M}{\rho}$$

where $[R]$ is the molecular refraction, n is the refractive index, M is the molecular mass, ρ is the density.

Lorentz transformations *(Phys.)*. The relations between the co-ordinates of space and time of the same event as measured in two inertial frames of reference moving with a uniform velocity relative to one another. They are derived from the postulates of the special theory of **relativity**.

Lorenz apparatus *(Elec.Eng.)*. A delicate apparatus for the absolute determination of the value of a resistance. A metal disk is rotated in an accurately known uniform field, and the e.m.f. produced between its centre and its periphery is balanced against the drop of potential caused by the field-producing current in the resistance to be measured.

Loschmidt number *(Chem.)*. (1) Number of molecules in unit volume of an ideal gas at standard temperature and pressure, i.e. $2.687 \times 10^{25}\,m^{-3}$. (2) Name sometimes given, esp. on the Continent, to *Avogadro number*.

lõss *(Geol.)*. See **loess**.

loss *(Telecomm.)*. Opposite of *gain*, the diminution of power, as in a transformer or line. Loss is realized as a standard in a resistive *pad (attenuator)*, which introduces a known loss into a measuring circuit.

loss angle *(Elec.Eng.)*. The difference between 90° and the angle of lead of current over voltage in a capacitor.

Lossev effect *(Electronics)*. The radiation due to the

recombination of charge carriers injected into a *p-i-n* or *p-n* junction biased in the forward direction.

loss factor *(Acous.)*. Quantity to describe the damping of structure-borne sound in materials or structures.

loss factor *(Telecomm.)*. (1) The ratio of average power loss in a circuit or device to the power loss under peak loading. (2) The **power factor** of a material multiplied by its **dielectric constant**. This varies with frequency and governs the amount of heat generated in a material. See **lossy**.

lossless line *(Telecomm.)*. See **dissipationless line**.

loss of charge method *(Elec.Eng.)*. A method of measuring very high resistances. The resistance is placed across the terminals of a charged capacitor and the rate at which the charge leaks away is observed.

loss of vend *(Min.Ext.)*. Difference in weight between raw (run-of-mine) coal and saleable product leaving washery.

lossy *(Elec.Eng.,Telecomm.)*. Of a material or apparatus which dissipates energy, e.g. dielectric material or transmission line with a high attenuation. The attenuation loss will be in *decibels* (dB) while the rate of loss in the dielectric is proportional to its *loss factor*, i.e. to the product of power factor and relative permittivity.

lost flux *(Elec.Eng.)*. See **leakage flux**.

lost lead *(Print.)*. See **lead (of the web)**.

lost wax casting *(Eng.)*. See **investment casting**.

Lotka's equations *(Ecol.)*. Mathematical relationships between (1) the populations of two species living together and competing for food or space and (2) the situation where one is the predator and the other the prey.

loud-hailer *(Acous.)*. See **megaphone**.

loudness *(Acous.)*. The intensity or volume of a sound as perceived subjectively by a human ear.

loudness contour *(Acous.)*. Line drawn on the audition diagram of the average ear which indicates the intensities of sounds that appear to the ear to be equally loud.

loudness level *(Acous.)*. That of a specified sound is the intensity of the *reftone*, 1000 Hz, on the *phon* scale, which is adjusted to equal, in apparent loudness, the specified sound. The adjustment of equality is made either subjectively or objectively as in special sound-level meters.

loudspeaker *(Acous.)*. An electro-acoustic **transducer** which accepts transmission currents and radiates (by horn or by diaphragm) corresponding sound waves. See **electrodynamic-, electromagnetic-, piezoelectric-, electrostatic-, magnetostriction-**.

loudspeaker dividing network *(Acous.)*. See **dividing network**.

loudspeaker microphone *(Acous.)*. A microphone and dynamic loudspeaker combined which is useful for intercommunication systems.

loudspeaker response *(Acous.)*. Response measured under specified conditions over a frequency range, at a specified direction and distance.

loughlinite *(Min.)*. A variety of sepiolite (see **meerschaum**) containing sodium.

louping ill *(Med.)*. An encephalomyelitis of sheep and cattle. Occasionally may cause human infection.

louping ill *(Vet.)*. An encephalomyelitis affecting principally sheep and cattle, transmitted by ticks and caused by a virus, and characterized by acute fever and nervous symptoms. Vaccines available.

louvre, louver *(Build.)*. A system of covering a space (e.g. window, air vent) with sloping slats of glass, metal, wood, etc., so as to allow ventilation while protecting against rain, wind, etc. The effect is also used architecturally, in indirect lighting, and purely decoratively in furniture design.

love arrows *(Min.)*. See **flèches d'amour**.

Lovibond comparator *(Chem.)*. Term applied to a range of instruments used in the estimation of concentrations of chemical compounds in solution by comparison of colour intensities with known standards.

Lovibond tintometer *(Chem.)*. Instrument for measuring the colours of liquids or solids by transmitted or reflected light and relating the colours to standards. The colour is matched against combinations of glass slides of three basic colours.

low *(Meteor.)*. A region of low pressure, or a *depression*.

low *(Print.)*. Type or blocks which are below the level of the forme surface are said to be *low*, and must be adjusted by underlaying or interlaying.

low-angle plane *(Build.)*. One with the iron set at 12° level uppermost, for use on end grain.

low aspect ratios *(Aero.)*. See **aspect ratio**.

low-carbon steel *(Eng.)*. Arbitrarily, steel containing from 0.04% to 0.25% of carbon. See **mild steel**.

löweite *(Min.)*. A hydrated sulphate of sodium and magnesium, crystallizing in the trigonal system, occurring in salt deposits.

lower bound *(Maths.)*. See **bounds of a function**.

lower case *(Print.)*. The small letters as distinct from the capitals, the name deriving from the arrangement (now largely superseded) of the small letters in a lower case with the capitals in an *upper case*. Abbrev. *l.c.*

lower culmination *(Astron.)*. See **culmination**.

lower deck *(Ships)*. A deck below the weather deck. The term has no legal definition status, and is usually applied to a partial deck which acts simply as a platform and contributes nothing to main longitudinal strength.

lowering wedges *(Civ.Eng.)*. See **striking wedges**.

lower mean hemispherical candle-power *(Phys.)*. See **mean hemispherical candle-power**.

lower-pitch limit *(Acous.)*. The minimum frequency for a sinusoidal sound wave which produces a sensation of pitch.

lower quartile *(Stats.)*. The argument of the cumulative distribution function corresponding to a probability of 0.25; (of a sample) the value below which occur a quarter of the observations in the ordered set of observations.

lower transit *(Astron.)*. Another name for *lower culmination*.

lowest common multiple *(Maths.)*. *LCM*. Of a set of numbers, the smallest number which is exactly divisible by all numbers of the set.

low fidelity *(Acous.)*. Opposite of *high fidelity*.

low frequency *(Telecomm.)*. Vague term widely used to indicate audiofrequencies as distinct from radio frequencies, but more correctly implying radio frequencies between 30 and 300 kHz.

low frequency amplifier *(Telecomm.)*. An amplifier for audiofrequency signals.

low hysteresis steel *(Eng.)*. Steel with from 2.5 to 4.0% silicon, a high permeability and electric resistance, but low loss through hysteresis.

low key *(Image Tech.)*. Describing a scene containing mostly dark tones with substantial areas of shadow, often used to create a dramatic or sinister mood.

low-level language *(Comp.)*. Machine-orientated programming language in which each program instruction corresponds to a single **machine-code instruction**. See **assembly language**.

low-level modulation *(Telecomm.)*. See **high-level modulation**.

low level waste *(Nuc.Eng.)*. In general, those radioactive wastes which, because of their low activity, do not require shielding during normal handling or transport.

low light television *(Aero.)*. Sensor system, capable of operating at dawn and dusk, smaller and cheaper than radar.

low-melting-point alloys *(Eng.)*. See **fusible alloys**.

low-pass filter *(Telecomm.)*. See **high-pass filter**.

low-power modulation *(Telecomm.)*. Same as *low-level modulation*.

low-pressure compressor, stage, turbine *(Aero.)*. Stages of an **axial flow turbine**.

low-pressure cylinder *(Eng.)*. The largest cylinder of a multiple-expansion steam-engine (e.g. the third of a triple-expansion engine), in which the steam is finally expanded.

low red heat *(Eng.)*. A temperature between 550° and 700°C.

low-resolution graphics *(Comp.)*. Term generally applied to graphical display units where simple pictures can be built up by plotting relatively large blocks of colour or by using special graphics characters. Cf. **high-resolution graphics**.

lowry *(Civ.Eng.)*. An open form of box-car.

low-stop filter *(Telecomm.)*. Same as **high pass filter**.

low temperature carbonization *(Chem.Eng.)*. Processing of bituminous coal at temperatures up to 760°C to produce smokeless fuel, gas and by-product oil.

low tension *(Telecomm.)*. Term loosely applied to the currents and voltages associated with the filament or heater circuits of a thermionic valve or tube.

low-tension detonator *(Elec.Eng.)*. The usual form of detonator, in which a charge is fired by heating a wire by an electric current.

low-tension ignition *(Elec.Eng.)*. Electric ignition of the charge in the cylinder of an IC engine by the interruption of a current-carrying circuit inside the cylinder, no special means being used to produce a high-voltage spark as in high-tension ignition.

low-tension magneto *(Elec.Eng.)*. A magneto for producing the current impulses necessary in a low-tension ignition system.

low-velocity scanning *(Electronics)*. Scanning of target by electron beam under conditions such that the secondary emission ratio is less than one.

low voltage *(Elec.Eng.)*. Legally, any voltage not exceeding 250 volts d.c. or r.m.s.

low-volt release *(Elec.Eng.)*. See **under-voltage release**.

low-water alarm *(Eng.)*. An arrangement for indicating that the water-level in a boiler is dangerously low. See **low-water valve**.

low-water valve *(Eng.)*. A boiler safety valve which is opened by a float in the water if the level of the latter falls dangerously low; generally fitted to stationary boilers such as the Lancashire.

low-wing monoplane *(Aero.)*. A monoplane wherein the main planes are mounted at or near the bottom of the fuselage.

lox *(Chem.)*. Abbrev. for *Liquid OXygen*.

LP compressor, stage, turbine *(Aero.)*. Abbrev. for *Low Pressure-*.

LPG *(Chem.)*. *Liquefied Petroleum Gases* such as propane and butane, used for fuels.

LPS *(Immun.)*. Abbrev. for *LipoPolySaccharide*.

L-rest *(Eng.)*. A lathe rest used in hand turning, shaped like an inverted L.

L-ring *(Elec.Eng.)*. A method of tuning a magnetron.

l-s coupling *(Phys.)*. See **Russell-Saunders coupling**.

LSD *(Med.)*. See **lysergic acid diethylamide**.

L-section *(Telecomm.)*. Section or half-section of a filter, having shunt and series arms.

L-shell *(Phys.)*. The **electron shell** in an atom corresponding to a principal quantum number of two. The shell can contain up to 8 electrons.

LSI *(Comp.)*. Abbrev. for *Large Scale Integration* which refers to **chips** containing many thousands of logic **gates**. See also **VLSI**.

Lu *(Chem.)*. The symbol for *lutetium*.

lubricants *(Eng.)*. Compounds (solid, plastic or liquid) entrained between two sliding surfaces. 'Wetting' lubricants adhere strongly to one or both such surfaces while leaving the intervening film fairly nonviscous. Solids include graphite, molybdenite, talc. Plastics include fatty acids and soaps, sulphur-treated bitumen, and residues from petroleum distillation. Liquids include oils from animal, vegetable or mineral sources. See **boundary lubrication, fluid lubrication**.

Lucas theory *(Chem.)*. Theory that substituent groups attract or repel electrons from neighbouring groups and thus alter their reactivity, e.g. in benzene, chlorine attracts electrons and de-activates the benzene ring, hydroxyl and amine lend electrons and activate.

lucid dreaming *(Behav.)*. A dreaming state in which the dreamer feels awake and in conscious control over dream events.

luciferase *(Zool.)*. An oxidizing enzyme which occurs in the luminous organs of certain animals (e.g. Firefly) and acts on luciferin to produce luminosity. Also *photogenin*.

luciferin *(Zool.)*. A proteinlike substance which occurs in the luminous organs of certain animals and is oxidized by the action of luciferase. Also *photophelein*.

Lucite *(Plastics)* US for polymethyl methacrylate resin. See **Perspex**.

Lüders lines *(Eng.)*. See **flow lines**.

Ludlow *(Print.)*. A machine which produces slugs from hand-set matrices, used mainly for headings and display lines for newspapers etc.

Ludwig's angina *(Med.)*. Term applied to infection of the mouth spreading along the parapharangeal spaces to produce a hard swelling of the neck, with difficulty in swallowing and fever; usually caused by a β-haemolytic streptococcal infection.

lues *(Med.)*. A plague or pestilence. The term is now synonymous with *syphilis*. adj. *luetic*.

luffer-boarding *(Build.)*. Sloping slats arranged as in a **louvre**.

luffing-jib crane *(Eng.)*. A common form of jib crane, in which the jib is hinged at its lower end to the crane structure, so as to permit of alteration in its radius of action. See **derricking jib crane, level-luffing crane**.

lug *(Elec.Eng.)*. On an accumulator plate, a projection to which the electrical connection is made. See **terminal-**.

lug *(Eng.)*. See **ear**.

lug sill *(Build.)*. A sill which is of greater length than the distance between the jambs of the opening, so that its ends have to be built into the wall.

lumbago *(Med.)*. A term applied to pain of the muscles and ligaments in the lumbar region or lower part of the back.

lumbar *(Zool.)*. Pertaining to, or situated near, the lower or posterior part of the back; as the *lumbar* vertebrae.

lumbar puncture *(Med.)*. The process of inserting a needle into the lumbar subarachnoid space to obtain a specimen of cerebrospinal fluid.

lumber *(For.)*. The term employed in the United States and Canada for sawn wood of all descriptions.

lumberjack *(For.)*. US term for *logger*.

lumen *(Bot.)*. The space enclosed by the walls of a cell, especially if the protoplast has disappeared.

lumen *(Phys.)*. Unit of **luminous flux**, being the amount of light emitted in unit solid angle by a small source of output one **candela**. In other words, the lumen is the amount of light which falls on unit area when the surface area is at unit distance from a source of one candela. Abbrev. *lm*.

lumen *(Zool.)*. The central cavity of a duct or tubular organ.

lumen-hour *(Phys.)*. Quantity of light emitted by a 1-lumen lamp operating for 1 hour.

lumenmeter *(Phys.)*. An integrating photometer in which the total luminous flux is integrated in a matt white diffusing enclosure and measured through an opening.

luminaire *(Elec.Eng.)*. The British Standard term for any sort of electric-light fitting, whether for a tungsten or a fluorescent lamp.

luminaire *(Image Tech.)*. General term for a studio light source with its mounting and controls.

luminance *(Phys.)*. Measure of brightness of a surface, e.g., candela per square metre of the surface radiating normally. Symbol *L*.

luminance channel *(Image Tech.)*. In a colour TV system, any circuit path intended to carry the **luminance signal**.

luminance flicker *(Image Tech.)*. Perceptible rapid periodic variations in brightness, more obvious at higher levels of picture brightness; the eye is much less sensitive to chrominance flicker.

luminance signal *(Image Tech.)*. Signal controlling the luminance of a colour TV picture, its point-by-point image brightness, as distinct from its **chrominance**; known as the *Y-signal*. Sometimes abbreviated to *luma* by analogy with *chroma*.

luminescence *(Phys.)*. Emission of light (other than from thermal energy causes) such as *bioluminescence* etc. Thermal luminescence in excess of that arising from temperature occurs in certain minerals. See **fluorescence, phosphorescence**.

luminescent centres *(Chem.)*. Activator atoms, excited by free electrons in a crystal lattice and giving rise to electroluminescence.

luminophore *(Chem.)*. (1) A substance which emits light at room temperature. (2) A group of atoms which can make a compound luminescent.

luminosity *(Astron.)*. The intrinsic or absolute amount of energy radiated per second from a celestial object. Its units in astronomy are usually absolute **magnitude**, rather than watts, largely for historical reasons.

luminosity *(Phys.)*. The visual perception of the brightness of an area. The density of luminous intensity on a particular direction.

luminosity coefficients *(Phys.)*. Multipliers of the *tristimulus values* of a colour, so that the total represents the luminance of the colour according to its subjective assessment by the eye.

luminosity curve *(Phys.)*. That which gives the relative effectiveness of perception or of sensitivity of vision in terms of wavelength. See **photopic-**, **scotopic-**.

luminosity factor *(Phys.)*. The ratio of the total luminous flux emitted by a light source at a given wavelength to the total energy emitted.

luminous efficiency *(Phys.)*. The ratio of luminous flux of lamp to total radiated energy flux; usually expressed in lumens per watt for electric lamps. Not to be confused with **overall luminous efficiency**.

luminous flame *(Chem.)*. One containing glowing particles of solid carbon due to incomplete combustion.

luminous flux *(Phys.)*. The flux emitted within unit solid angle of one steradian by a point source of uniform intensity of one candela. Unit of measurement is the *lumen*.

luminous flux density *(Phys.)*. The quantity of luminous flux, passing through a normal unit area, weighted according to an internationally accepted scale of differential visual sensitivity.

luminous gas flame *(Eng.)*. Thermal breakdown of the hydrocarbons into carbon and hydrogen arising from lack of primary combustion air, the carbon being heated to incandescence.

luminous intensity *(Phys.)*. See **candela**.

luminous paints *(Build.)*. Special paints with the ability to glow in the dark.

luminous paints *(Chem.)*. Paints which glow in the dark. Based on the salts, e.g. sulphide, silicate, phosphate of, e.g. zinc, cadmium, calcium, with traces of heavy metals, e.g. manganese. Phosphorescent paints glow for longer or shorter periods after exposure to light. Fluorescent paints glow continuously under the action of radioactive additives.

luminous sensitivity *(Phys.)*. In a photoconductive cell, it is the ratio (output current/incident luminous flux), at a constant electrode voltage.

luminous sulphides *(Chem.)*. See **luminous paint**.

Lummer-Brodhun photometer *(Phys.)*. A form of contrast photometer in which, by an arrangement of prisms, the surfaces illuminated respectively by the standard source and the source under test are next to one another, thus enabling an easy comparison to be made.

Lummer-Gehrcke interferometer *(Phys.)*. A very accurately worked, plane, parallel-sided glass plate, so arranged that light is internally reflected in a zig-zag path through the plate. The rays which emerge into the air at each reflection are in a condition to produce interference fringes of a very high order. The instrument is used for studying the fine structure of spectral lines. See **interference fringes**.

lump *(Textiles)*. Piece of cloth taken straight from the loom; usually larger than normal.

lumped constant *(Electronics)*. A single constant which is electrically equivalent to the total of that type of distributed constant existing in a coil or circuit.

lumped constant *(Phys.)*. The electric properties of a component may be taken as *lumped* or *concentrated* when its dimensions are small in comparison with the wavelength of propagation of currents in it. Cf. **distributed constants**. See **primary constants**.

lumped impedance *(Electronics)*. An impedance concentrated in a single component rather than being distributed throughout the length of a transmission line.

lumped loading *(Telecomm.)*. See **line equalizer**.

lumped parameters *(Phys.)*. In analysis of electrical circuits, the circuit parameters which, under certain frequency conditions, may be regarded as behaving as localized units of inductance, resistance, capacitance etc.

lump lime *(Build.)*. The **caustic lime** produced by burning limestone in a kiln.

lumpy jaw *(Vet.)*. See **actinomycosis**.

lumpy skin disease *(Vet.)*. An acute virus disease of cattle in Africa, characterized by inflammatory nodules in the skin and generalized lymphadenitis.

lumpy withers *(Vet.)*. Small swellings on the withers of the horse caused by irritation of the subcutaneous bursae by the collar.

lumpy wool *(Vet.)*. See **mycotic dermatitis of sheep**.

lunar *(Zool.)*. In tetrapods, the middle member of the three proximal carpals in the fore-limb, corresponding to part of the astragalus in the hind-limb.

lunar bows *(Meteor.)*. Bows of a similar nature to *rainbows* but produced by moonlight.

lunar month *(Astron.)*. See **synodic month**.

lunate, lunulate *(Genrl.)*. Crescent-shaped; shaped like the new moon.

lunation *(Astron.)*. See **synodic month**.

lune *(Maths.)*. The portion of the surface of a sphere intercepted by two great circles. The remaining portion of a circle when a part is removed by an intersecting circle.

lunette *(Arch.)*. (1) A semicircular window or pediment over a doorway. (2) A small arched opening in the curved side of a vault.

lung *(Zool.)*. The respiratory organ in air-breathing Vertebrates. The lungs arise as a diverticulum from the ventral side of the pharynx; they consist of two vascular sacs filled with constantly renewed air.

lung book *(Zool.)*. An organ of respiration in some Arachnida (Scorpions, Spiders), consisting of an air-filled cavity opening on the ventral surface of the body; it contains a large number of thin vascular lamellae, arranged like the leaves of a book.

Lunge nitrometer *(Chem.)*. Apparatus devised for the determination of oxides of nitrogen by absorption in a gas burette. May be used for other analytical processes which involve the measurement of a gas.

lung plague *(Vet.)*. See **contagious bovine pleuro-pneumonia**.

lungworm disease *(Vet.)*. See **husk**.

lunitidal interval *(Genrl.)*. The time interval between the moon's transit and the next high-water at a place.

lunule, lunula *(Zool.)*. A crescentic mark. Specifically, the white crescent at the root of the nail. adj. *lunular*.

lupinosis *(Vet.)*. Poisoning of sheep, goats, cattle, and horses, by plants of the genus *Lupinus*.

lupus erythematosus *(Med.)*. A connective tissue disease thought to be due to an **auto-immune** reaction. Occurs in two main forms: *discoid*, where the skin of the nose and cheeks is raised and reddened giving a butterfly appearance; *systemic*, where tissue throughout the body is involved, causing kidney disease, painful joints, anaemia amd mental abnormality.

lupus erythematosus cell *(Immun.)*. A neutrophil leucocyte which contains in its cytoplasm homogeneous masses of phagocytosed nuclear material derived from other dead neutrophils. This occurs when anti-nuclear antibodies are present in the blood which opsonize the released nucleoproteins. LE cells are formed in vitro in blood drawn from subjects with systemic lupus erythematosus, and their presence is of diagnostic significance.

lupus vulgaris *(Med.)*. Tuberculosis of the skin particularly affecting the face.

lustre *(Build.)*. The degree of sheen on a surface or coating.

lustre *(Min.)*. This depends upon the quality and amount of light that is reflected from the surface of a mineral. The highest degree of lustre in opaque minerals is *splendent*, the comparable term for transparent minerals being *adamantine* (i.e. the lustre of diamond). *Metallic* and *vitreous* indicate less brilliant lustre, while *silky, pearly, resinous,* and *dull* are self-explanatory terms covering other degrees of lustre.

lute *(Build.)*. A straightedge for levelling off clay in a brick mould by removing the excess.

lute *(Foundry)*. Fireclay mixture use to seal cracks in e.g. a crucible.

luteal *(Zool.)*. Pertaining to, or resembling, the *corpus luteum*.

lutein cells *(Zool.)*. Yellowish coloured cells occurring in the corpus luteum of the ovary; they contain fat-soluble substances, developed from either the follicular cells or the theca interna.

luteinizing hormone *(Med.)*. Abbrev. *LH*. A gonadotrophic mucoprotein hormone, secreted by the basophil cells of the anterior lobe of the pituitary gland, which causes growth of the corpus luteum of the ovary and also stimulates activity of the interstitial cells of the testis; also called *interstitial cell stimulating hormone*.

luteinoma *(Med.)*. A tumour occurring in the ovary, composed of cells resembling those of the corpus luteum.

luteotrophic hormone *(Med.)*. See **prolactin**.

lutetium *(Chem.)*. A metallic element, a member of the rare earth group. Symbol Lu, at. no. 71, r.a.m. 174.97. It occurs in bloomstrandite, gadolinite, polycrase and xenotime. Formerly called *cassiopeium*.

luthern *(Arch.)*. A vertical window set in a roof.

lutidines *(Chem.)*. Dimethyl-pyridines, which all occur in bone-oil and in coal-tar; general formula, $C_5(CH_3)_2H_3N$.

lutite *(Geol.)*. A consolidated rock composed of silt and/or clay. e.g. **shale, mudstone**.

lux *(Phys.)*. Unit of illuminance or illumination in SI system, 1 lm/m^2. Abbrev. *lx*.

luxation *(Med.)*. Dislocation.

Luxemburg effect *(Telecomm.)*. Cross-modulation of radio transmissions during ionospheric propagation. Caused by nonlinearity in motion of the ions.

Lw *(Chem.)*. The symbol for *lawrencium*.

LWR *(Nuc.Eng.)*. See **light water reactor**.

lych (lich) gate *(Arch.)*. A roofed gateway entrance to a churchyard.

Lycopsida *(Bot.)*. The Lycopods, a class of Pteridophyta dating from the Early Devonian onwards. The sporophyte usually has roots, stems and leaves (**microphylls**), is protostelic and has lateral sporangia. Includes the Protolepidodendrales, Lycopodiales, Lepidodendrales, Isoetales, Selaginellales.

lyddite *(Chem.)*. See **picric acid**.

Lydian stone, lydite *(Min.)*. *Touchstone*. Black flinty jasper, also other silicified fine-grained rocks. The name touchstone has reference to the use of lydite as a streak plate for gold: the colour left after rubbing the metal across it indicates to the experienced eye the amount of alloy.

lye *(Chem.)*. Strong solution of sodium or potassium hydroxide.

lying, laying press *(Print.)*. A small portable screw-press in which books are held firmly during various operations.

lying panel *(Build.)*. A door panel whose width is greater than its height.

Lyman series *(Phys.)*. One of the hydrogen series occurring in the extreme ultraviolet region of the spectrum. The series may be represented by the formula

$$\nu = N\left(\frac{1}{1^2} - \frac{1}{n^2}\right),$$

$n = 2, 3, 4$...(see **Balmer series**), the series limit being at wave number $N = 109\,678$, which corresponds to wavelength, 91.26 nm. The leading line, called *Lyman alpha*, has a wavelength of 121.57 nm, and is important in upper atmosphere research, as it is emitted strongly by the sun.

lymph *(Zool.)*. A colourless circulating fluid occurring in the lymphatic vessels of Vertebrates and closely resembling blood plasma in composition. adj. *lymphatic*.

lymphadenitis *(Med.)*. Inflammation of lymph glands.

lymphadenoid *(Med.)*. Resembling the structure of a lymphatic gland.

lymphangiectasis *(Med.)*. Dilatation and distension of lymphatic vessels, due usually to obstruction.

lymphangiology *(Med.)*. The anatomy of the lymphatics.

lymphangioma *(Med.)*. A nodular tumour consisting of lymphatic channels.

lymphangitis *(Med.)*. Inflammation of a lymphatic vessel or vessels.

lymphangitis *(Vet.)*. See **epizootic-, ulcerative-**.

lymphatic leukaemia *(Med.)*. See **leukaemia**.

lymphatic system *(Zool.)*. In Vertebrates, a system of vessels pervading the body, in which the lymph circulates and which communicates with the venous system; lymph glands and lymph hearts are found on its course.

lymph gland *(Zool.)*. An aggregation of reticular connective tissue, crowded with lymphocytes, surrounded with a fibrous capsule, and provided with afferent and efferent lymph vessels.

lymph heart *(Zool.)*. A contractile portion of a lymph vessel, which assists the circulation of the lymph and forces the lymph back into the veins.

lymph-, lympho- *(Genrl.)*. Prefix from L. *lympha*, water.

lymphocyte function associated molecules *(Immun.)*. *LFA-1, CD2, LAF-3*. Factors present on the surface of leucocytes which are involved in their ability to adhere to other cells and to exert other functions such as killing or phagocytosis. Their nature is at present ill defined, but their absence in certain rare subjects is accompanied by defective immunity.

lymphocytes *(Immun.)*. Spherical cells with a large round nucleus (often slightly indented) and very scanty cytoplasm. Diameter varies between 7 and 12 µm. They are actively mobile. The term is essentially morphological and is used to refer to cells responsible for development of specific immunity, B-lymphocytes being associated with humoral and T-lymphocytes with cellular immunity. Lymphocytes are the predominant constituents of **lymphoid tissues**. A normal adult human contains about 2×10^{12} lymphocytes.

lymphocytopenia *(Med.)*. Diminution, below normal, of the number of lymphocytes in the blood.

lymphocytosis *(Med.)*. Increase in the number of lymphocytes in the blood, as in certain infections.

lymphogenous *(Zool.)*. Lymph-producing.

lymphogranuloma inguinale *(Med.)*. *Poradenitis venerea*. A virus venereal infection, characterized by enlargement of the glands in the groin; common in the tropics.

lymphoid tissues *(Immun.)*. Body tissues in which the predominant cells are lymphocytes, i.e. lymph nodes, spleen, thymus, Peyer's patches, pharyngeal tonsils, adenoids, and in birds the caecal tonsils and bursa of Fabricius.

lymphokine *(Immun.)*. Generic name for proteins (other than antibodies or surface receptors) which are released by lymphocytes, stimulated by antigens or by other means, which act on other cells involved in the immune response. The term includes interferons, interleukins, lymphotoxins, migration inhibition factor *inter alia*.

lymphoma *(Immun.,Med.)*. Neoplastic disease of lymphoid tissues and sometimes involving bone marrow, in which the neoplastic cells originate from lymphocytes or mononuclear phagocytes. Includes Hodgkin's disease, reticulosarcoma, giant follicular lymphoma, lymphatic leukaemia and Burkitt's lymphoma. Clinically divided by histological appearance into two main groups, *Hodgkin's* and *non-Hodgkin's lymphoma*, with the former carrying the more favourable prognosis.

lymphosarcoma *(Med.)*. A term used to describe one of the lymphomas with an intermediate grade of malignancy.

lymphotoxin *(Immun.)*. A lymphokine which is cytolytic for non-lymphocytic cells.

lymph sinuses *(Zool.)*. Part of the **lymphatic system** in some lower Vertebrates, consisting of spaces surrounding the blood vessels and communicating with the coelom by means of small apertures.

lymph vessels *(Zool.)*. See **lymphatic system**.

lyocytosis *(Zool.)*. Histolysis by the action of enzymes secreted outside the tissue, as in Insect metamorphosis.

lyolysis *(Chem.)*. The formation of an acid and a base from a salt by interaction with the solvent; i.e. the chemical reaction which opposes neutralization.

lyophilic colloid *(Chem.)*. A colloid (*lyophilic*, solvent-loving) which is readily dispersed in a suitable medium, and may be redispersed after coagulation.

lyophobic colloid *(Chem.)*. A colloid (*lyophobic*, solvent-hating) which is dispersed only with difficulty, yielding an unstable solution which cannot be reformed after coagulation.

lyosorption *(Chem.)*. The adsorption of a liquid on a solid surface, especially of solvent on suspended particles.

Lyot filter *(Astron.)*. See **polarizing monochromator**.

lyotropic *(Phys.)*. Concentration determines the phase of lyotropic materials.

lyotropic series *(Chem.)*. Ions, radicals, or salts arranged in order of magnitude of their influence on various colloidal, physiological and catalytic phenomena, an influence exerted by them as a result of the interaction of ions with the solvent. Cf.**Hofmeister series**.

lyra *(Zool.)*. The psalterium in the brain of Mammals; any lyre-shaped structure, as the lyre pattern on a bone.

lyrate *(Bot.)*. Shaped like a lyre. A pinnately lobed leaf with a large terminal and small lateral lobes.

lyriform organs *(Zool.)*. Patches, consisting of well-innervated ridges of chitin, on the legs, palpi, chelicerae, and body of various *Arachnida*; mechanoreceptive in function.

Lys *(Chem.)*. Symbol for **lysine**.

lyse *(Chem.,Biol.)*. To cause to undergo *lysis*.

lysergic acid diethylamide *(Med.)*. Lysergide. Compound which, when taken in minute quantities, produces hallucinations and thought processes outside the normal range. The results sometimes resemble schizophrenia. Abbrev. *LSD*.

Lysholm grid *(Radiol.)*. A type of grid interposed between the patient and film in diagnostic radiography in order to minimize the effect of scattered radiation.

Lysholm-Smith torque converter *(Eng.)*. A variable-ratio hydraulic gear of the Föttinger type, but in which multistage turbine blading gives high efficiency of transmission over a wide range: used in road and rail vehicles.

lysigenic, lysigenous *(Bot.,Zool.)*. Said of a space formed by the breakdown and dissolution of cells. Also *lysogenous*. Cf. **rexigenous, schizogenous**.

lysin *(Zool.)*. A substance which will cause dissolution of cells or bacteria.

lysine *(Chem.)*. *2,6-diaminohexanoic acid*, $H_2N.(CH_2)_4.CH(NH_2).COOH$. A 'basic' amino acid, as it contains two amino groups. The L- or S-isomer is a constituent of proteins. Symbol Lys, short form K. ⇨

lysis *(Biol.)*. Decomposition or splitting of cells or molecules, e.g. hydrolysis, lysin.

lysis *(Med.)*. Destruction of cells or tissues by various pathological processes, e.g. autolysis, necrobiosis.

Lysithea *(Astron.)*. The tenth natural satellite of **Jupiter**.

lysogeny *(Biol.)*. That part of the life-cycle of a temperate phage in which it replicates in synchrony with its host. Cf. **lytic cycle**.

Lysol *(Chem.)*. TN for a solution of cresols in soft soap. It is a well-known disinfectant.

lysosome *(Biol.)*. A vesicular cytoplasmic organelle containing hydrolytic enzymes that degrade those cellular constituents which become incorporated into the vesicle.

lysozyme *(Biol.)*. Enzyme present in egg-white, tears, nasal secretions, on the skin and in monocytes and granulocytes. It lyses certain bacteria, chiefly Gram-positive cocci, by splitting the muramic acid-β(1-4)-N-acetylglucosamine linkage in their cell walls. The first *enzyme* to have its three-dimensional structure determined.

lyssa *(Zool.)*. See **lytta**.

lytic cycle *(Biol.)*. That part of the life cycle of a temperate phage in which it multiplies rapidly, destroying its host and releasing many copies into the medium.

lytic infection *(Biol.)*. Infection of a bacterium by a phage which replicates uncontrollably, destroying its host and eventually releasing many copies into the medium.

lytta *(Zool.)*. In *Carnivora*, a rod of cartilage or fibrous tissue embedded in the mass of the tongue. Also *lyssa*.

M

m *(Genrl.)*. Abbrev. for *milli-*.
m *(Phys.)*. Symbol for (1) electromagnetic moment, (2) mass, (3) mass of electron, (4) molality.
m- *(Chem.)*. An abbrev. for: (1) *meta-*, i.e., containing a benzene nucleus substituted in the 1,3 positions; (2) *meso-* (1).
μ *(Chem.)*. A symbol for: (1) chemical potential; (2) dipole moment.
μ *(Electronics)*. Symbol for amplification factor.
μ *(Phys.)*. The symbol for (1) index of refraction, (2) magnetic permeability, (3) micron (Obs.), (4) the prefix *micro-*.
μ- *(Chem.)*. A symbol signifying: (1) *meso-* (2); (2) *meso-* (3); (3) a bridging ligand.
M *(Chem.)*. A general symbol for a metal or an electropositive radical. Abbrev. for *mega-*, i.e. 10^6.
M *(Phys.)*. Symbol for magnetization per unit volume.
M *(Chem.)*. Symbol for *rel. molecular mass.*
M *(Phys.)*. Symbol for (1) luminous emittance, (2) moment of force, (3) mutual inductance.
maar *(Geol.)*. The explosion vent of a volcano.
MAC *(Image Tech.)*. *Multiplexed Analogue Components*, a system of colour TV signal coding of the picture information; the form of audio signal coding is denoted by a prefix; e.g. *C-MAC, D2-MAC.*
macadamized road *(Civ.Eng.)*. A road whose surface is formed with broken stones of fairly uniform size rolled into a 15–25 cm (6–10 in) layer, with gravel to fill the interstices. See **tarmacadam**.
Macardle's disease *(Med.)*. An inborn error of metabolism leading to abnormal glycogen storage in the muscle and causing muscle pain and stiffness.
maceral *(Geol.)*. The organic material which comprises coal. It includes **exinite**, **inertinite** and **vitrinite**.
maceration *(Zool.)*. The process of soaking a specimen in a reagent in order to destroy some parts of it and to isolate other parts.
Mach angle *(Aero.)*. In supersonic flow, the angle between the **shock wave** and the airflow, or line of flight, of the body. The cosecant of this angle is equal to the Mach number.
Mach cone *(Aero.)*. In supersonic flow, the conical **shock wave** formed by the nose of a body, whether stationary in a **wind tunnel** or in free flight through the air.
machinability *(Eng.)*. The ease with which a metal or alloy can be machined.
machine *(Phys.)*. A device for overcoming a resistance at one point by the application of a force at some other point. Typical simple machines are the inclined plane, the lever, the pulley, and the screw. See also **mechanical advantage, velocity ratio**.
machine architecture *(Comp.)*. See **computer architecture**.
machine chase *(Print.)*. A chase specially designed for a particular letterpress machine and supplied with it.
machine code *(Comp.)*. Programming language in which each **machine-code instruction** can be recognized and executed by the computer without any intermediate translation.
machine-code instruction *(Comp.)*. A **word** which contains several codes; the *operation code* which identifies a particular, elementary operation, and others which are address codes for the data and the result. The instruction is linked to a series of hardware operations either directly or through **microprograms**. See **instruction cycle, instruction set, address calculation.**
machine direction *(Paper)*. The direction in the plane of a sheet or web of paper corresponding with the direction of travel of the paper machine. The direction at right angles to this is the *cross direction.*
machine finished *(Paper)*. Paper on which the requisite degree of surface finish has been attained by use of a calender or calenders forming part of the paper making machine.
machine glazed *(Paper)*. Paper or board with a characteristic high finish on one side, produced by causing the damp web to adhere to the surface of a large diameter, highly polished, steam heated drying cylinder (MG or Yankee dryer). As water is evaporated the paper surface assumes the polish of the cylinder.
machine instruction *(Comp.)*. See **machine code instruction.**
machine language *(Comp.)*. See **machine code**.
machine moulding *(Foundry)*. The process of making moulds and cores by mechanical means, as, for example, by replacing hand-ramming by power-squeezing of the sand, or by the jolting of the box on a vibrating table.
machine proof *(Print.)*. A proof taken on a printing machine as distinct from a proofing press, and implying a high standard in the result.
machine revise *(Print.)*. A proof taken on the printing machine and submitted for a final check before printing.
machine ringing *(Telecomm.)*. The normal ringing current which is placed on a subscriber's line to attract his attention. The currents are generated by a machine and interrupted by a suitable commutator.
machine riveting *(Eng.)*. Clenching rivets by the use of mechanical or compressed-air hammers or hydraulic riveters. See also **hydraulic riveter, pneumatic riveter**.
machine room *(Print.)*. That department of a printing establishment where the actual process of printing (by machine) is carried out.
machine tools *(Eng.)*. Any power-driven, non-portable machine designed primarily for shaping and sizing (metal) parts. See **drilling machine, key-way tool, lathe, milling machine, planing machine, shaping machine**.
machine-washable *(Textiles)*. A textile article that maybe washed in a domestic washing machine in aqueous detergent solutions at controlled temperatures suitable for the particular article. The conditions may be specified on a *care label* attached to the article.
machine wire *(Paper)*. Originally a woven mesh of phosphor bronze for use on a fourdrinier or cylinder mould machine for forming the sheet. Stainless steel is now used for machine wires and the term is retained for plastics cloths serving the same purpose.
machine word *(Comp.)*. See **word**.
machining allowance *(Eng.)*. The material provided beyond the finished contours on a casting, forging, or roughly prepared component, which is subsequently removed in machining to size.
machiolation *(Arch.)*. A projecting gallery on a fortification which had a pierced floor through which boiling oil was poured on the enemy.
machmeter *(Aero.)*. A pilot's instrument for measuring **flight Mach number**.
Mach number *(Aero.)*. The ratio of the speed of a body, or of the flow of a fluid, to the speed of sound in the same medium. At *Mach* 1, speed is *sonic*; below *Mach* 1, it is *subsonic*; above *Mach* 1, it is *supersonic*, creating a *Mach* (or *shock*) *wave. Hypersonic* conditions in air are reached at Mach numbers exceeding 5. See **velocity of sound**.
Mach principle *(Genrl.)*. That scientific laws are descriptions of nature, are based on observation and alone can provide deductions which can be tested by experiment and/or observation.
mackerel sky *(Meteor.)*. Cirrocumulus or altocumulus cloud arranged in regular patterns suggesting the markings on mackerel.

Mackie line *(Image Tech.)*. A boundary effect locally increasing the difference between areas of high and low image density, used to advantage in **high definition** developers.

mackle *(Print.)*. A defective printed sheet, having a blurred appearance due to incorrect impression. adj. *mackled.*

Maclaurin's series *(Maths.)*. Modification of *Taylor's series*, putting $a = 0$.

macle *(Min.)*. The French term for a **twin crystal**; in the diamond industry, more commonly used than *twin*, especially for twinned octahedra.

MacLeod's equation *(Chem.)*. The surface tension of a liquid is given as the difference between the densities of the liquid and its vapour raised to the fourth power, multiplied by a constant.

Macnaughten rules *(Med.)*. Rules formulated by the House of Lords where a defence of insanity must show the accused was labouring under such a defect of reason from disease of the mind, as not to know the nature or quality of the act he was doing; or if he did know it, he did not know that what he was doing was wrong.

Maco template *(Build.)*. Adjustable template consisting of a clamp which holds a stack of thin brass strips. When the strips are pressed edgewise against a moulding or curved surface, they conform with its shape and can be clamped in position to give an accurate replica.

Macpherson strut suspension *(Autos.)*. Widely used system for independent front suspension, in which a single lower **wishbone** locates the wheel in the fore and aft plane, and springing and damping are accommodated by a combined coilspring/damper unit attached between wheel hub and chassis.

macr-, macro- *(Genrl.)*. Prefix from Gk. *makros*, large.

macro *(Comp.)*. See **macro instruction**.

macro-axis *(Crystal.)*. The long axis in orthorhombic and triclinic crystals.

macrocephaly, macrocephalia *(Med.)*. Abnormal largeness of the head due to excess fluid, resulting in mental retardation. Cf. **hydrocephalus**.

macrocheilia *(Med.)*. Abnormal increase in the size of the lips.

macrocheiria *(Med.)*. Abnormally large hands.

macrocode *(Comp.)*. See **macroinstruction**.

macrocycle *(Chem.)*. A large ring compound, usually heterocyclic, and often containing repeating units of the form —$CH_2.CH_2.X$—, where X is commonly O, S or NH. See **crown ether**.

macrocyte *(Med.)*. An abnormally large red cell in the blood.

macrocytosis *(Med.)*. The presence of many abnormally large red cells in the blood.

macrodactyly, macrodactylia *(Med.)*. Congenital hypertrophy of a finger or fingers.

macrogamete *(Zool.)*. The larger of a pair of conjugating gametes, generally considered to be the female gamete.

macroglia *(Zool.)*. A general term for neuroglial cells, including astroglia, oligodendroglia and ependyma.

macroglobulin *(Immun.)*. Any globulin with a molecular weight above about 400 000 but usually applied to IgM (mol.wt. 900 000) or to a protease inhibitory molecule alpha-2-macroglobulin (mol.wt. 820 000). Macroglobulinaemia refers to markedly raised levels of IgM in the blood, due to an abnormally proliferating clone of neoplastic plasma cells, as in **Waldenstrom's macroglobulinaemia**.

macroglossia *(Med.)*. Abnormal enlargement of the tongue.

macrogols *(Med.)*. Polyethylene glycols, used as ointment bases.

macrograph *(Image Tech.)*. A photograph reproducing an object much larger than its actual size.

macro-instruction *(Comp.)*. An instruction in a programming language which causes several other instructions in the same language to be carried out and which is automatically replaced by them in the program.

macro lens *(Image Tech.)*. A camera lens designed for extreme close-up photography, down to a subject distance of a few centimetres.

macromere *(Zool.)*. In a segmenting ovum, one of the large cells which are formed in the lower or vegetable hemisphere.

macromolecular crystals *(Crystal.)*. Many proteins because of their regular 3-dimensional structures will form crystals under suitable conditions which can then be analysed by **X-ray crystallography** to determine their fine structure down to the level of 2×10^{-10} m. The interrelations of the amino-acid side chains and the particular conformation of the active site of an enzyme can, for example, be analysed.

macromolecule *(Chem.)*. Term applied to a very large molecule such as *haemoglobin* (containing about 10 000 atoms) and sometimes to a polymer of high molecular mass.

macronucleus *(Zool.)*. In *Ciliophora*, the larger of the two nuclei which is composed of vegetative chromatin. Cf. **micronucleus**.

macronutrient *(Ecol.)*. Element required in relatively large quantities by living organisms; for plants they are H, C, O, N, K, Ca, Mg, P, S. Cf. **micronutrient**. See also **essential element**.

macrophage *(Immun.)*. Cell of the **mononuclear phagocyte system**. Derived from blood monocytes which migrate into tissues and differentiate there. They are actively phagocytic, and ingest particulate materials including microbes. They contain lysosomes and an oxidative microbicidal system which becomes functional when the macrophages are activated by ingestion of degradable particles or by lymphokines such as interferon. They contain and secrete large amounts of lysozyme. Macrophages are motile, although some remain in the same site (e.g. as Kupffer cells in liver sinusoids) for long periods. They have Fc and complement receptors which are involved in attachment and engulfment of antibody-coated materials. The microbicidal activity is much increased by the action of lymphokines secreted by T-lymphocytes. They adhere firmly to glass or plastic surfaces, and are often separated in vitro by such means. In tissues at the site of a delayed hypersensitivity reaction they may adhere to one another, and alter their morphology to resemble epithelial cells or fuse together to form multinucleated giant cells. Macrophages are important accessory cells for presentation of antigens to T-lymphocytes. Although most resting macrophages express little or no Class II histocompatibility antigens they are induced to do so by interferon. Antigens ingested by them do not become wholly degraded, and peptide fragments are returned to the cell surface where they are associated with the Class II antigen, and can be recognised by antigen receptors of T-cells.

macrophagous *(Zool.)*. Feeding on relatively large particles of food. Cf. **microphagous**.

macrophotography *(Image Tech.)*. General term for photography at extremely close range, so that the image size is as large or larger than the actual subject.

macrophyll *(Bot.)*. A large leaf. Also *megaphyll.*

macrophyric *(Geol.)*. A textural term for medium to fine-grained igneous rocks containing phenocrysts > 2 mm long. Cf. **microphyric**.

macro processing language *(Comp.)*. A language for processing **macroinstructions**.

macroscopic *(Genrl.)*. Visible to the naked eye.

macroscopic state *(Chem.)*. One described in terms of the overall statistical behaviour of the discrete elements from which it is formed. Cf. **microscopic state**.

macrosection *(Eng.)*. A section of metal, ceramic etc., mounted, cut, polished and etched as necessary to exhibit the **macrostructure**.

macrosmatic *(Zool.)*. Having a highly developed sense of smell.

macrosome *(Zool.)*. A large protoplasmic granule or globule.

macrosplanchnic *(Zool.)*. Having a large body and short legs; as a Tick.

macrospore *(Bot.)*. See **megaspore**.

macrosporophyll *(Bot.)*. *Megasporophyll*. Leaf-like structure bearing mega-sporangia.

macrostoma *(Med.)*. Abnormal width of the mouth due to a defect in development.

macrostructure *(Eng.)*. Specifically, the structure of a metal as seen by the naked eye or at low magnification on a ground or polished surface or on one which has been subsequently etched. Generally, the structure of any body visible to the naked eye. Cf. **microstructure**.

macrotous *(Zool.)*. Having large ears.

macula acustica *(Zool.)*. Patches of sensory epithelium in the utriculus, sacculus and lagena of the Dogfish and the utricle and saccule of Mammals. In the latter they are organs of static or tonic balance consisting of supporting cells and hair cells, the hairs being embedded in a gelatinous membrane containing crystals of calcium carbonate, i.e., an otolith.

macula lutea *(Zool.)*. See **yellow spot**.

macula, macule *(Bot.,Zool.)*. A blotch or spot of colour; a small tubercle; a small shallow pit.

macula, macule *(Med.)*. A small discoloured spot on the skin, not raised above the surface; in particular a small yellow area seen on examination of the retina. adjs. *macular, maculate*.

MAD *(Aero.)*. Abbrev. for (1) *Mutual Assured Destruction*., (2) *Magnetic Anomaly Detection*.

Madagascar aquamarine *(Min.)*. A strongly dichroic variety of the blue beryl obtained, as gemstone material, from the Malagasy Republic.

Madagascar topaz *(Min.)*. See **citrine**.

made end-papers *(Print.)*. A term for any of the several varieties of specially assembled end-papers, having cloth joints, doublures, etc. as used in hand binding.

made ground *(Build.)*. Ground formed by filling in natural or artificial pits with hardcore or rubbish.

Madeira topaz *(Min.)*. A form of *Spanish topaz*. See **citrine**.

Madelung's constant *(Chem.)*. A constant representing the sum of the mutual potential coulombic attraction energy of all the ions in a lattice, in the equation for the lattice energy of an ionic crystal.

MADGE *(Aero.)*. Abbrev. for *Microwave Aircraft Digital Guidance Equipment*.

mad itch *(Vet.)*. See **Aujesky's disease**.

madreporite *(Zool.)*. In Echinodermata, a calcareous plate with a grooved surface, perforated by numerous fine canaliculi and situated in an interambulacral position, through which water passed to the axial sinus. adj. *madreporic*.

Madura foot *(Med.)*. *Mycetoma*. A disease, endemic in India and occurring elsewhere, in which nodular, ulcerated swellings appear on the foot, due to infection with a fungus.

Mae West *(Aero.)*. Personal lifejacket designed for airmen, inflated by releasing compressed CO_2.

mafic *(Min.)*. A mnemonic term for the ferromagnesian and other nonfelsic minerals actually present in an igneous rock.

magamp *(Acous.)*. Abbrev. for *magnetic amplifier*.

magazine *(Image Tech.)*. A removable light-tight container for feeding film to and from a motion picture camera, printer or processing machine.

magazine *(Print.)*. The unit on the Linotype and Intertype slug-casting machines, and on the Fotosetter and Fotomatic filmsetting machines, which houses the matrices, a channel for each sort.

magazine arc-lamp *(Elec.Eng.)*. A form of electric arc-lamp having a number of carbons which are automatically brought into operation as the others burn away, so that the lamp can burn for long periods without attention.

magazine projector *(Image Tech.)*. One capable of holding a number of slides, which are fed automatically for projection.

magazine reel stand *(Print.)*. A unit on a web-fed press which supports the reel in running position with one or more reels in the ready position.

Magellanic Clouds *(Astron.)*. Two dwarf galaxies, satellites of the **Milky Way**, visible as cloudy patches in the night sky in the southern hemisphere. First recorded by F. Magellan in 1519. They are about 180 000 light years away and contain a few thousand million stars. They are of immense astrophysical importance because individual stars within them can be studied and yet they are essentially all at the same distance from us. This removes a great source of uncertainty compared with the situation within our own Galaxy, where actual distances to individual stars are hard to determine.

Magendie's foramen *(Zool.)*. In Vertebrates, an aperture in the roof of the fourth ventricle of the brain, through which the cerebrospinal fluid communicates with the fluid in the spaces enclosed by the meningeal membranes.

magenta *(Chem.)*. See **fuchsin**.

maggot *(Zool.)*. An acephalous, apodous, eruciform larva such as that of certain Diptera.

maghemite *(Min.)*. An iron oxide with the crystal structure of magnetite but the composition of hematite. A cation-deficient spinel, produced by the oxidation of magnetite.

magic eye *(Telecomm.)*. A simple cathode-ray tube in which a metal fin is employed as a control electrode. The opening or closing of a fluorescent pattern is the indication of balance. Used as a detector for a.c. bridges and as a tuning indicator in radio receivers. Also *tuning indicator*.

magic number *(Phys.)*. Extra stable atomic nuclei exist which have a *magic number*, 2, 8, 20, 28, 50, 82, or 126, of either protons or neutrons.

magic-tee *(Radar)*. A four-port waveguide junction, a combination of an E-plane and an H-plane junction, having the properties of a **hybrid junction**.

magistral *(Eng.)*. Powdered roasted copper pyrite used in amalgamation of silver ores in Mexican process.

magma *(Geol.)*. Molten rock, including dissolved water and other gases. It is formed by melting at depth and rises either to the surface, as lava, or to whatever level it can reach before crystallizing again, in which case it forms an igneous intrusion.

magmatic cycle *(Geol.)*. See **igneous cycle**.

magnesia *(Chem.)*. See **magnesium oxide**.

magnesia alba *(Chem.)*. Commercial basic magnesium carbonate.

magnesia alum *(Min.)*. See **pickeringite**.

magnesia cement *(Build.)*. See **Sorel's cement**.

magnesia glass *(Glass)*. Glass containing usually 3–4% of magnesium oxide. Electric-lamp bulbs have been mainly made from this type of glass since fully automatic methods of production were adopted.

magnesia mixture *(Chem.)*. A mixture of magnesium chloride, ammonium chloride and ammonia solution used in chemical analysis for the estimation of phosphates.

magnesian spinel *(Min.)*. See **spinel**.

magnesite *(Eng.)*. Carbonate of magnesium, crystallizing in the trigonal system. Magnesite is a basic refractory used in open-hearth and other high-temperature furnaces; it is resistant to attack by basic slag. It is obtained from natural deposits (mostly magnesium carbonate, $MgCO_3$) which is calcined at high temperature to drive off moisture and carbon dioxide, before being used as a refractory.

magnesite flooring *(Build.)*. A composition of cement, magnesium compounds, sawdust and sand to form a hard continuous screed for floor surfaces. Also called *jointless flooring*.

magnesium *(Chem.)*. A light metallic element in the second group of the periodic system. Symbol Mg, at.no. 12, r.a.m. 24.312, mp 651°C, rel.d. 1.74, electrical resistivity 42×10^{-8} ohm metres, bp 1120°C at 1 atm, specific latent heat of fusion 377 kJ/kg. Found in nature only as compounds. The metal is a brilliant white in colour, and magnesium ribbon burns in air, giving an intense white light, rich in ultraviolet rays. It is used as a deoxidizer for copper, brass and nickel alloys. A basis metal in strong light alloys which are used in aircraft and automobile construction and for reciprocating parts.

New alloys with zirconium and thorium are used in aircraft construction.

magnesium carbonate *(Chem.)*. $MgCO_3$. See **magnesite**.

magnesium orthodisilicate *(Chem.)*. Occurs in nature as **serpentine**.

magnesium oxide *(Chem.)*. MgO. Obtained by igniting the metal in air. In the form of calcined magnesite and dolomite, it is used as a refractory material. See also **periclase**.

magnesium oxychloride cement *(Build.)*. Used in **magnesite flooring**.

magneson *(Chem.)*. *4-(4-nitrophenolazo)resorcinol*: used as a reagent for detection and determination of magnesium, with which it forms a characteristic blue colour in alkaline solution.

magnet *(Phys.)*. A mass of iron or other material which possesses the property of attracting or repelling other masses of iron, and which also exerts a force on a current-carrying conductor placed in its vicinity. See **electromagnet, permanent magnet**.

magnet coil *(Elec.Eng.)*. See **magnetizing coil**.

magnet core *(Phys.)*. The iron core within the coil of an electromagnet.

magnetic *(Phys.)*. Said of all phenomena essentially depending on magnetism. See **diamagnetism, paramagnetism**.

magnetic alloys *(Eng.)*. Generally, any alloy exhibiting ferromagnetism, e.g. and most importantly, **silicon iron**, but also iron-nickel alloys, which may contain small amounts of any of a number of other elements (e.g. copper, chromium, molybdenum, vanadium etc.), and iron-cobalt alloys.

magnetic amplifier *(Acous.)*. One in which the saturable properties of magnetic material are utilized to modulate an exciting alternating current, using an applied signal as *bias*; the signal output when rectified becomes a magnification of the input signal.

magnetic annealing *(Eng.)*. Heat treatment of magnetic alloy in a magnetic field, used to increase its permeability.

magnetic anomaly *(Geol.)*. The value of the local magnetic field remaining after the subtraction of the dipole portion of the Earth's field. The local deviation measured in magnetic prospecting.

magnetic armature *(Elec.Eng.)*. Ferromagnetic element, the position of which is controlled by external magnetic fields.

magnetic axis *(Phys.)*. A line through the effective centres of the poles of a magnet.

magnetic balance *(Elec.Eng.)*. A form of fluxmeter in which the force required to prevent the movement of a current-carrying coil in a magnetic field is measured. Cf. **current balance**.

magnetic bearing *(Surv.)*. The horizontal angle between a survey line and magnetic north.

magnetic bias *(Acous.)*. A steady magnetic field added to the signal field in magnetic recording, to improve linearity of relationship between applied field and magnetic remanence in recording medium.

magnetic blowout *(Elec.Eng.)*. A special magnet coil fitted to circuit-breakers or other similar apparatus in order to produce a magnetic field at the point of opening the circuit, so that any arc which is formed is deflected in such a direction as to be lengthened or brought into contact with a cool surface and is rapidly extinguished. Also called *blowout coil*.

magnetic bottle *(Nuc.Eng.)*. The containment of a plasma during thermonuclear experiments by applying a specific pattern of magnetic fields.

magnetic braking *(Elec.Eng.)*. A method of braking a moving system in which a brake is applied and released by an electromagnet.

magnetic bubble *(Phys.)*. Single-crystal rare-earth garnet sheet can be grown on a substrate with a preferred direction of magnetization perpendicular to the plate. Very small round domains (bubbles) can be formed by applying a magnetic field. See **magnetic bubble memory**.

magnetic bubble memory *(Comp.,Electronics)*. Potentially cheap and reliable *chip* memory in which the presence or absence of a *magnetic bubble* in a localized region of a thin magnetic film designates 0 or 1. Bubbles representing stored data can be moved and manipulated by current pulses in thin-film or conducting loops placed on the surface to produce localized magnetic fields. Very compact high-capacity memories can be realized by this method.

magnetic card *(Comp.)*. One with a suitable surface which can be magnetized in selective areas so that data can be stored.

magnetic character reading *(Print.)*. Scanning and interpretation of characters printed with magnetic ink on documents, e.g. cheques; special lettering is necessary.

magnetic chuck *(Eng.)*. A chuck having a surface in which alternate steel elements, separated by insulating material, are polarized by electromagnets, so as to hold light flat work securely on the table of a grinding machine or other machine tool.

magnetic circuit *(Elec.Eng.)*. Complete path, perhaps divided, for magnetic flux, excited by a permanent magnet or electromagnet. The range of reluctance is not so great as resistance in conductors and insulators, so that leakage of the magnetic flux into adjacent non-magnetic material, especially air, is significant.

magnetic clutch *(Autos.)*. (1) A friction clutch without pressure springs in which a solenoid pulls the plates together. (2) A clutch in which ferromagnetic powder takes up the drive when drawn into an annular gap by a solenoid. Also *Smith's coupling*.

magnetic component *(Elec.Eng.)*. The magnetic field associated with an electromagnetic wave. The magnetic and electric fields are related through the intrinsic impedance of the medium,

$$E/H = \eta = \sqrt{\mu/\varepsilon}.$$

magnetic confinement *(Nuc.Eng.)*. In fusion research, the use of shaped magnetic fields to confine a plasma.

magnetic confinement fusion system *(Nuc.Eng.)*. Fusion system in which plasma is confined, i.e. not allowed to come in contact with the walls of the chamber, by use of specially shaped magnetic fields. See **inertial fusion system**.

magnetic controller *(Elec.Eng.)*. Unit in control system operated by magnetic field applied to magnetic armature.

magnetic core *(Comp.)*. See **ferrite-core memory**.

magnetic coupling *(Elec.Eng.)*. The magnetic flux linkage between one circuit and another.

magnetic creeping *(Elec.Eng.)*. A gradual increase in the intensity of magnetization of a piece of magnetic material, after a continued application of the magnetizing force.

magnetic cutter *(Acous.)*. A cutter used in recording in which the motions of the recording stylus are operated by magnetic fields.

magnetic damping *(Elec.Eng.)*. Damping of motion of a conductor by induced eddy currents in it when moving across a magnetic field, particularly applicable to moving parts of instruments and electricity integrating meters.

magnetic declination *(Surv.)*. The angular deviation of a magnetic compass, uninfluenced by local causes, from the true north and south. The declination varies at different points on the earth's surface and at different times of the year. Also called *magnetic deviation*.

magnetic deflection *(Electronics)*. That of an electron beam in a cathode-ray tube, caused by a magnetic field established by current in coils where the beam emerges from the electron gun which forms the beam, the deflection being at right angles to the direction of the field.

magnetic deviation *(Surv.)*. See **magnetic declination**.

magnetic difference of potential *(Elec.Eng.)*. A difference in the magnetic conditions at two points which gives rise to a magnetic flux between the points.

magnetic dip *(Geol.)*. See **dip**.

magnetic dipole moment *(Phys.)*. A small current loop gives rise to a magnetic field characteristic of a magnetic *dipole*. The magnitude of the magnetic dipole moment of such a loop is $\mu = iA$, where a is the current and A the area of the loop. An orbiting electron in an atom will have an *orbital* magnetic dipole moment and also have a *spin* magnetic dipole moment due to its intrinsic spin. Nuclei of atoms with non-zero spin also have magnetic dipole moments. See **Bohr magneton**.

magnetic discontinuity *(Elec.Eng.)*. An air gap, or a layer of non-magnetic material, in a magnetic circuit.

magnetic disk *(Comp.)*. Storage device consisting of a flat rotatable circular plate, coated on both surfaces with a magnetic material. Data is written to and read from a set of concentric circular tracks.

magnetic-disk memory *(Acous.)*. See **disk memory**.

magnetic displacement *(Elec.Eng.)*. An alternative name for *magnetic induction*.

magnetic domain *(Eng.)*. Aggregated ferromagnetic group of atoms, usually well below one micrometre in diameter, forming part of a system of such groups with no preferred orientation.

magnetic doublet radiator *(Elec.Eng.)*. A hypothetical radiator consisting of two equal and opposite varying magnetic poles whose distant field is equivalent to that of a small loop aerial.

magnetic drum *(Comp.)*. Device used for **backing store** which consists of a cylinder driven very uniformly at high speed, carrying a layer of magnetic material on which data is stored and read.

magnetic elongation *(Elec.Eng.)*. The slight increase in length of a wire of magnetic material when it is magnetized. See **magnetostriction**.

magnetic energy *(Elec.Eng.)*. Product of flux density and field strength for points on the demagnetization curve of a permanent magnetic material, measuring the energy established in the magnetic circuit. Normally required to be a maximum for the amount of magnetic material used.

magnetic epoch *(Geol.)*. A long period of geological time in which the Earth's magnetic field was essentially of one polarity.

magnetic escapement *(Horol.)*. One comprising a small horseshoe magnet on the end of a vibrating spring, which actuates the teeth of the escape wheel as they come under the influence of its field.

magnetic events *(Geol.)*. Geologically short periods during **magnetic epochs** when the magnetic field had a reversed polarity.

magnetic ferrites *(Elec.Eng.)*. Those having magnetic properties and at the same time possessing good electrical insulating properties by reason of their ceramic structure.

magnetic field *(Elec.Eng.)*. Modification of space, so that forces appear on magnetic poles or magnets. Associated with electric currents and the motions of electrons in atoms.

magnetic-field intensity *(Elec.Eng.)*. The magnitude of the field strength vector in a medium (i.e. the magnetic strain produced by neighbouring magnetic elements or current-carrying conductors). The SI unit is the ampere per metre and the CGS unit is the oersted. Also called *magnetic-field strength, magnetic intensity, magnetizing force*.

magnetic-field strength *(Elec.Eng.)*. See **magnetic-field intensity**.

magnetic film *(Image Tech.)*. A full-width magnetic coating on a film base, slit and perforated to regular 35 mm or 16 mm motion picture standards.

magnetic flux *(Elec.Eng.)*. The surface integral of the product of the permeability of the medium and the magnetic-field intensity normal to the surface. The magnetic flux is conceived, for theoretical purposes, as starting from a positive fictitious north-pole and ending on a fictitious south-pole, without loss. When associated with electric currents, a complete circuit, the *magnetic circuit*, is envisaged, the quantity of magnetic flux being sustained by a magneto-motive force, m.m.f. (coexistent with ampere-turns linked with the said circuit). Permanent magnetism is explained similarly in terms of molecular m.m.f.s (associated with orbiting electrons) acting in the medium. Measured in *maxwells* (CGS) or *webers* (SI).

magnetic flux density *(Elec.Eng.)*. Product of magnetic-field strength (or intensity), H, and permeability of a material. Expressed in tesla (old unit, weber per square metre), has the symbol B. CGS unit is the gauss.

magnetic focusing *(Electronics)*. That of an electron beam by applied magnetic fields, e.g. in CRT or electron microscope.

magnetic forming *(Eng.)*. A fast, accurate production process for swaging, expanding, embossing, blanking etc., in which a permanent or expendable coil is moved by electromagnetism to act on the workpiece.

magnetic gate *(Elec.Eng.)*. A gate circuit used in magnetic amplifiers.

magnetic head *(Acous.)*. Recording, reproducing or erasing head in magnetic recorder.

magnetic hysteresis *(Elec.Eng.)*. Nondefinitive value of magnetic induction in ferromagnetic medium for given magnetic-field intensity, depending upon magnetic history of sample.

magnetic hysteresis loop *(Phys.)*. See **hysteresis loop**.

magnetic hysteresis loss *(Elec.Eng.)*. The energy expended in taking a piece of magnetic material through a complete cycle of magnetization. The magnitude of the loss per cycle is proportional to the area of the magnetic hysteresis loop.

magnetic induction *(Elec.Eng.)*. Induced magnetization in magnetic material, either by saturation, by coil-excitation in a magnetic circuit, or by the primitive method of stroking with another magnet.

magnetic ink *(Print.)*. Ink containing particles of ferromagnetic material, used for printing data so that magnetic character recognition is possible.

magnetic ink character recognition *(Comp.)*. Machine recognition of stylized characters printed in magnetic ink.

magnetic intensity *(Elec.Eng.)*. See **magnetic-field intensity**.

magnetic iron-ore *(Min.)*. See **magnetite**.

magnetic lag *(Elec.Eng.)*. The time required for the magnetic induction to adjust to a change in the applied magnetic field.

magnetic leakage *(Elec.Eng.)*. That part of the magnetic flux in a system which is useless for the purpose in hand and may be a nuisance in affecting nearby apparatus.

magnetic lens *(Phys.)*. The counterpart of an optical lens — in which a magnet or system of magnets is used to produce a field which acts on a beam of electrons in a similar way to a glass lens on light rays. Comprises current-carrying solenoids of suitable design.

magnetic levitation *(Phys.)*. A method of opposing the force of gravity using the mutual repulsion between two like magnetic poles.

magnetic link *(Elec.Eng.)*. A small piece of magnet steel placed in the immediate vicinity of a conductor carrying a heavy surge current, e.g. a transmission line tower carrying a lightning stroke current. The magnetization of the link affords a means of estimating the value of the current.

magnetic map *(Phys.)*. A map showing the distribution of the earth's magnetic field.

magnetic memory *(Comp.)*. See **backing store**.

magnetic mirror *(Nuc.Eng.)*. Device based on the principle that ions moving in a magnetic field towards a region of considerably higher magnetic field strength are reflected. This principle is used in **mirror machines**. Mirrors may be *simple, minimum* β (more efficient) or *tandem* (solenoid plugged at both ends by minimum β filters).

magnetic modulator *(Elec.Eng.)*. One using a magnetic circuit as the modulating element.

magnetic moment *(Phys.)*. Vector such that its product with the magnetic induction gives the torque on a magnet in a homogeneous magnetic field. Also called *moment of a magnet*.

magnetic monopole *(Phys.)*. Single isolated magnetic pole. Theoretically predicted but not yet observed. Monopoles would have to be quantized, the quantum g related to the quantum of electic charge e by $ge = h$ where h is Planck's constant.

Magnetic North *(Phys.)*. The direction in which the north pole of a pivoted magnet will point. It differs from the Geographical North by an angle called the *magnetic declination*.

magnetic oxide of iron *(Min.)*. See **magnetite**.

magnetic oxides *(Eng.)*. The iron oxides which are ferromagnetic and which, suitably fabricated from the powder form, provide efficient permanent magnets.

magnetic particle clutch *(Eng.)*. A form of hydraulic coupling in which the fluid is a suspension of magnetizable particles, the viscosity of the fluid, and thus the degree of slip of the clutch, being variable by varying the intensity of magnetization.

magnetic particle inspection *(Eng.)*. A rapid, non-destructive test for fatigue cracks and other surface and subsurface defects in steel and other magnetic materials, in which the workpiece is magnetized so that local flux-leakage fields are formed at the cracks or other discontinuities and shown up by the concentration at these fields of a magnetic powder applied to the workpiece.

magnetic pendulum *(Elec.Eng.)*. Suspended magnet executing torsional oscillations in any horizontal magnetic field. It forms the basis of the horizontal component magnetometer.

magnetic polarization *(Phys.)*. The magnetic polarization **J** is related to the magnetic field **H** and the magnetic induction **B** by

$$\mathbf{B} = \mu_0 \mathbf{H} + \mathbf{J}$$

where μ_0 is the permeability of free space.

magnetic polarization *(Chem.)*. The production of optical activity by placing an inactive substance in a magnetic field.

magnetic pole *(Elec.Eng.)*. A convenient conception, which cannot exist, deduced from the experimental indication of the direction of the magnetic field arising from a permanent magnet. If the latter is long in comparison with its cross-section and the ends are provided with soft-iron balls, the direction of the magnetic field, as indicated by iron-filings, appears to radiate from the centres of such spheres, called *poles*. Experimentally, such poles appear as magnetic charges, from which are deduced the magnitude of magnetic poles, magnetic-field strength, magnetic flux, magnetic potential, and electrical units.

magnetic potential *(Phys.)*. A continuous mathematical function the value of which at any point is equal to the potential energy (relative to infinity) of a theoretical unit north-seeking magnetic pole placed at that point.

magnetic potentiometer *(Elec.Eng.)*. A flexible solenoid (wound on a nonmagnetic base) used with a ballistic galvanometer to explore the distribution of magnetic potential in a field, etc.

magnetic pressure *(Nuc.Eng.)*. The pressure which a magnetic field is capable of exerting on a plasma.

magnetic printing *(Print.)*. Printing ink containing magnetic elements and capable of being read both visually and by computer.

magnetic printing *(Telecomm.)*. Transfer of recorded signal from one magnetic recording medium or element to another.

magnetic prospecting *(Min.Ext.)*. Form of **geophysical prospecting** which measures either the distortion of the Earth's magnetic field by local accumulation of ferromagnetic materials, e.g. magnetite, pyrrhotite, chromite or ilmenite, or the magnetic susceptibility of rocks. Methods include the use of a simple **dip needle**, instrumental **magnetometer** in field studies or of airborne magnetometers (*aeromagnetic survey*).

magnetic pumping *(Phys.)*. Use of radiofrequency currents in coils over bulges in the tube of a stellarator to modulate the steady axial field and provide heat to the plasma. This process is most efficient when there is resonance between the RF signals and vibrations of the molecules of plasma. The process is then called *resonance heating*.

magnetic pyrites *(Min.)*. See **pyrrhotite**.

magnetic quantum numbers *(Phys.)*. Those associated with the quantized orientation of the orbital and spin angular moments with respect to an applied magnetic fields.

magnetic reaction *(Elec.Eng.)*. Same as **electromagnetic reaction**.

magnetic recording *(Acous.)*. (1) The magnetic tape, from which the recorded signal may be reproduced. (2) The process of preparing a magnetic recording.

magnetic resonance imaging *(Radiol.)*. The use of nuclear magnetic resonance of protons to produce proton density maps or images of the human body.

magnetic reversal *(Geol.)*. A change of the Earth's magnetic field, to the opposite polarity.

magnetic rigidity *(Nuc.Eng.)*. A measure of the momentum of a particle. It is given by the product of the magnetic intensity perpendicular to the path of the particle and the resultant radius of curvature of this path.

magnetic rotation *(Phys.)*. See **Faraday effect**.

magnetic saturation *(Phys.)*. The limiting value of the magnetic induction in a medium when its *magnetization* is complete and perfect.

magnetic screen *(Elec.Eng.)*. A screen of a high magnetic permeability material such as Mu-metal, used to surround certain electrical and electronic components, in order to protect them from the effect of external magnetic fields.

magnetic separator *(Phys.)*. A device for separating, by means of an electromagnet, any magnetic particles in a mixture from the remainder of the mixture, e.g. for separating iron filings from brass filings.

magnetic shell *(Elec.Eng.)*. A magnetized body of dimensions identical to a current-carrying coil, which may be considered instead of the latter when considering the forces acting in a system.

magnetic shield *(Elec.Eng.)*. Surface of magnetic material which reduces the effect on one side of a magnetic field on the other side. A substantially complete shield is used to protect a.c. indicating instruments from errors arising from external alternating magnetic fields.

magnetic shift register *(Elec.Eng.)*. One in which the pattern of settings of a row of magnetic cores is shifted one step along the row by each fresh pulse.

magnetic shunt *(Elec.Eng.)*. A piece of magnetic material in parallel with a portion of a magnetic circuit, so arranged as to vary the amount of magnetic flux in that portion of the circuit.

magnetic slot-wedge *(Elec.Eng.)*. A slot-wedge of magnetic material which gives the same effect as a closed slot.

Magnetic South *(Phys.)*. The direction in which the South pole of a pivoted magnet will point. It differs from the Geographical South by an angle called the *magnetic declination*.

magnetic spectrometer *(Nuc.Eng.)*. One in which the distribution of energies among a beam of charged particles is investigated by means of magnetic focusing techniques.

magnetic stability *(Phys.)*. A term used to denote the power of permanent magnets to retain their magnetism in spite of the influence of external magnetic fields, vibration, etc.

magnetic storm *(Meteor.)*. Magnetic disturbance in the earth, causing spurious currents in submarine cables; probably arises from variation in particle emission from the sun, which affects the ionosphere.

magnetic surface wave *(Radar)*. Magnetic wave propagated along the surface of a ferromagnetic garnet substrate. Used in microwave delay lines, filters, etc. Abbrev. *MSW*.

magnetic susceptibility *(Elec.Eng.)*. The amount by which the relative permeability of a medium differs from unity, positive for a paramagnetic medium, but negative for a diamagnetic one. Equals intensity of magnetization ÷ applied field.

magnetic suspension *(Elec.Eng.)*. Use of a magnet to assist in the support of e.g. a vertical shaft in a meter, thereby relieving the jewelled bearings of some of the weight.

magnetic suspension *(Horol.)*. Type of balance-wheel assembly embodying a pair of complementary ring magnets or a magnet above the balance staff to reduce friction.

magnetic tape *(Telecomm.)*. Flexible plastic tape, typically 6 to 50 mm wide, coated on one side with dispersed magnetic material, in which signals are registered for subsequent reproduction. Used for storing television images, sound or computer data.

magnetic tape encoder *(Comp.)*. Input device which accepts data from a keyboard and writes it directly to magnetic tape.

magnetic tape reader *(Comp.)*. One which has a multiple head which transforms the pattern of registered magnetic signals into electrical pulse signals.

magnetic track *(Image Tech.)*. A magnetic stripe on cinematograph film for recording and reproducing sound.

magnetic transition temperature *(Phys.)*. See **Curie point** (1).

magnetic transmission *(Autos.)*. See **magnetic clutch**.

magnetic trap/bottle *(Nuc.Eng.)*. Configuration of magnetic fields which will confine plasma for times long enough for it to react. See **magnetic confinement**.

magnetic tube *(Phys.)*. See **trapping region**.

magnetic tuning *(Elec.Eng.)*. Control of very-high-frequency oscillator by varying magnetization of a rod of ferrite in the frequency-determining cavity by an externally applied steady field.

magnetic units *(Phys.)*. Units for electric and magnetic measurements in which μ_0, the permeability of free space, is taken as unity. Now replaced by SI units in which μ_0 is given the value $4\pi \times 10^{-7}\,\mathrm{Hm}^{-1}$.

magnetic variables *(Astron.)*. Stars in which strong variable magnetic fields have been detected by the Zeeman effect.

magnetic variations *(Meteor.)*. Both diurnal and annual variations of the magnetic elements (dip, declination etc.) occur, the former having by far the greater range. In the northern hemisphere, the declination moves to the west during the morning and then gradually back, the extreme range being nowhere more than 1°. The dip varies by a few minutes during the day. It is thought that these effects are caused by varying electric currents in the ionized upper atmosphere.

magnetic wire *(Acous.)*. A wire of magnetic material used in recording.

magnetism *(Phys.)*. Science covering magnetic fields and their effect on materials, due to unbalanced spin of electrons in atoms. See **Coulomb's law for magnetism, paramagnetism, diamagnetism, ferrimagnetism, ferromagnetism, antiferromagnetism**.

magnetite, magnetic iron-ore *(Min.)*. An oxide of iron, crystallizing in the cubic system. It has the power of being attracted by a magnet, but it has no power to attract particles of iron to itself, except in the form of lodestone.

magnetization *(Phys.)*. Orientation from randomness of saturated *domains* in a ferromagnetic material. The magnetization per unit volume **M** is related to the field strength **H** and the magnetic induction **B** by

$$\mathbf{B} = \mu_0(\mathbf{H} + \mathbf{M})$$

tesla, where μ_0 is the permeability of free space. SI unit is ampere per metre.

magnetization curve *(Elec.Eng.)*. A curve showing ferromagnetic characteristic of a material. The magnetic flux density, B, is usually plotted as a function of the magnetic field intensity H, since H is proportional to the current in the exciting coil. Also called *B/H curve*. It also denotes the relation between the total m.m.f. and the total flux in a magnetic circuit, such as that of an electric machine. Since m.m.f. is proportional to the e.m.f. generated in a machine, a curve between exciting current and e.m.f. is also called a *magnetization curve*.

magnetize *(Phys.)*. (1) To induce magnetization in ferromagnetic material by direct or impulsive current in a coil. (2) To apply alternating voltage to a transformer or choke having ferromagnetic material, generally laminated, in its core.

magnetizing coil *(Elec.Eng.)*. A current-carrying coil used to magnetize an electromagnet, such as the field coil of an electric generator or motor. Also called *field coil, magnet coil*.

magnetizing current *(Elec.Eng.)*. (1) Current (direct or impulsive) in a coil for the magnetization of ferromagnetic material in its core. (2) The a.c. taken by the primary of a transformer, apart from a load current in the secondary.

magnetizing roast *(Eng.)*. Reduction of weakly magnetic iron ore or concentrate by heating in a reducing atmosphere to convert it to a more strongly ferromagnetic compound.

magneto *(Elec.Eng.)*. A small permanent-magnet electric generator capable of producing periodic high-voltage impulses; used for providing the ignition of internal-combustion engines, firing of explosives etc.

magneto-caloric effect *(Phys.)*. The reversible heating and cooling of a medium when the magnetization is changed. Also called *thermomagnetic effect*.

magnetochemistry *(Chem.)*. The study of the relation of magnetic properties to chemical structure. Particularly, the extent of paramagnetism in transition metal compounds may be related to the type of ligand bonded to the metal.

magnetoelectric *(Phys.)*. Of certain materials, e.g., chromium oxide, the property of becoming magnetized when placed in an electric field. Conversely, they are electrically-polarized when placed in a magnetic field. May be used for measuring pulse electric or magnetic fields.

magneto generator *(Elec.Eng.)*. An electric generator in which the existing flux is obtained from permanent magnets.

magnetohydrodynamic generator *(Elec.Eng.)*. Abbrev. *MHD generator*. See **magnetoplasmadynamic generator**.

magnetohydrodynamic instability *(Nuc.Eng.)*. See **instability**.

magnetohydrodynamics *(Phys.)*. The study of the motions of an electrically conducting fluid in the presence of a magnetic field. The motion of the fluid gives rise to induced electric currents which interact with the magnetic field which in turn modifies the motion. The phenomenon has applications both to magnetic fields in space and to the possibility of generating electricity. If the free electrons in a plasma or high velocity flame are subjected to a strong magnetic field, then the electrons will constitute a current flowing between two electrodes in a flame. Abbrev. *MHD*.

magneto ignition *(Elec.Eng.)*. An ignition system for IC engines in which the voltage necessary to produce the spark is generated by a magneto.

magneto-ionic *(Phys.)*. Said of components of an electromagnetic wave passing through an ionized region and divided into ordinary and extraordinary waves by magnetic field of the earth.

magnetometer *(Surv.)*. Any instrument for measurement either of the absolute value of a magnetic field intensity, or of one component of this, e.g. horizontal component magnetometer for earth's magnetic field. See also **proton precessional-**.

magnetomotive force *(Phys.)*. Line integral of the magnetic field intensity round a closed path. Abbrev. *m.m.f.*

magneton *(Phys.)*. See **Bohr-**.

magneto-optical effect *(Phys.)*. See **Kerr effect**.

magneto-optic rotation *(Chem.)*. See **magnetic polarization**.

magnetophone *(Acous.)*. Any recording device involving the magnetization of a medium, e.g. magnetic tape.

magnetoplasmadynamic generator *(Elec.Eng.)*. A device which produces electrical energy from an electrically-conducting gas (plasma) flowing through a transverse magnetic field. Abbrev. *MPD generator*. Also called *magnetohydrodynamic generator*.

magnetoresistance *(Elec.Eng.)*. The resistivity of a

magnetic material in a magnetic field which depends on the direction of the current with reference to the field. If parallel to one another, the resistivity increases, but if mutually perpendicular, it decreases.

magnetosphere *(Astron.)*. The asymmetrically shaped volume round the Earth and other magnetic planets in which charged particles are subject to the planet's magnetic field rather than the Sun's. Its radius is least towards the Sun and greatest away from it.

magnetostatics *(Phys.)*. Study of steady-state magnetic fields.

magnetostriction *(Phys.)*. Phenomenon of elastic deformation of certain ferromagnetic materials, e.g., nickel, on the application of magnetizing force. Used in ultrasonic transducers and, formerly, as memory in computers.

magnetostriction loudspeaker *(Acous.)*. Loudspeaker based on the magnetostriction effect, in which the length of materials varies according to an impressed magnetic field. The diaphragm is driven by a nickel or rare earth rod in which magnetostrictive changes are induced by an embracing coil.

magnetostriction microphone *(Acous.)*. Inverse of a **magnetostrictive loudspeaker**.

magnetostriction transducer *(Elec.Eng.)*. Any device employing the property of magnetostriction to convert electrical to mechanical oscillations, e.g. by using a rod clamped at its centre and passing a.c. through a coil wound around the rod. Also *magnetostrictor*.

magnetostrictive filter *(Elec.Eng.)*. A filter network which utilizes magnetostrictive elements, bars or rods, with their energizing coils.

magnetostrictive oscillation *(Phys.)*. One based on the principle of the alteration of dimensions of a bar of magnetic material when the magnetic flux through it is changed. Nickel contracts with an increasing applied magnetic field but iron expands in weak fields and contracts in strong fields. The mechanical oscillatory system, e.g. a bar clamped at its centre, can be coupled magnetically to an amplifier to maintain the oscillation, which may be in the audiofrequency range.

magnetostrictive reaction *(Phys.)*. The inverse magnetostrictive effect, i.e. a change in magnetization under applied stresses.

magnetostrictor *(Elec.Eng.)*. See **magnetostriction transducer**.

magnet pole *(Elec.Eng.)*. See **pole piece**.

magnetron *(Electronics)*. A two-electrode valve in which the flow of electrons from a large central cathode to a cylindrical anode is controlled by crossed electric and magnetic fields; the electrons gyrate in the axial magnetic field, their energy being collected in a series of slot resonators in the face of the anode. Magnetrons, used mainly as oscillators, can produce pulsed output power at microwave frequencies, with high peak power ratings. Used in microwave and radar transmitters, and microwave cookers. See **cavity magnetron**, **crossed-field tube**.

magnetron amplifier *(Electronics)*. A high-power microwave amplifier using a magnetron in such a configuration that it is not self-oscillatory.

magnetron critical field *(Electronics)*. That which would just prevent an electron emitted from the cathode with zero energy from reaching the anode at a given value of anode voltage.

magnetron critical voltage *(Electronics)*. The voltage which would just enable an electron emitted from the cathode with zero velocity to reach the anode at a given value of magnetic-flux density.

magnetron modes *(Electronics)*. Different frequencies of oscillation corresponding to different field configurations and selected by interconnecting cavities in various ways to control their phase differences.

magnet steel *(Eng.)*. A steel from which permanent magnets are made. It must have a high remanence and coercive force. Steels for this purpose may contain considerable percentages of cobalt (up to 35%) as well as nickel, aluminium, copper etc. See also **permanent magnet**.

magnettor *(Elec.Eng.)*. A second-harmonic type of magnetic modulator which uses a saturable reactor to amplify d.c., or low alternating-frequency, signals.

magnet yoke *(Elec.Eng.)*. Sometimes applied to the whole of the magnetic circuit of an electromagnetic (or transformer, etc.), but strictly speaking, should refer only to the part which does not carry the windings.

magnification *(Maths.)*. In a conformal mapping, given by $w = f(z)$, the magnification varies from point to point but is the same in all directions at a given point. The linear magnification at a point z_0 is given by $|f'(z_0)|$ and the surface magnification by $|f'(z_0)|^2$.

magnification *(Phys.)*. The ratio of the size of the image to that of the object in an optical system where the object is plane and lies perpendicular to the axis of the system. Also *transverse magnification*. See **longitudinal magnification**, **magnifying power**.

magnification factor *(Phys.)*. See *Q*.

magnifier *(Radiol.)*. Any thermionic amplifier, especially one used for the amplification of audiofrequencies.

magnifying power *(Phys.)*. The ratio of the apparent size of the image of an object formed by an optical instrument to that of the object seen by the naked eye. For a microscope, it is necessary to assume that the object would be examined by the naked eye at the least distance of distinct vision, 25 cm. Unless otherwise stated, the *linear* magnification is assumed to be indicated. See also **longitudinal magnification**.

Magnistor *(Elec.Eng.)*. TN for type of toroidal reactor, which is used for specific circuit applications, e.g. in gating, switching, etc.

magnitude *(Astron.)*. A measure of the apparent or absolute brightness of a celestial object, first used by Hipparchos in about 120 BC, who referred to the brightest stars as first magnitude and the dimmest as sixth. This was put on a scientific basis by Pogson in 1854, who showed that equal magnitude steps were in logarithmic progression: a magnitude difference of one unit corresponds to a brightness ratio of 2.512; five magnitudes to 100 in actual brightnesses. The zero of the magnitude scale is essentially arbitrary. The *apparent magnitude* is the brightness measured at the Earth and it depends on distance and **luminosity**. More useful physically is the *absolute magnitude*, in which the observed apparent magnitude is corrected to that which it would have if placed at an (arbitrary) distance of 10 parsecs (32.6 light years) from the Earth. Properties of stars and galaxies can only be properly compared by means of absolute magnitudes. The observed magnitude will depend on the technique used to measure the brightness: *visual magnitude* refers to observation by eye; *photographic magnitude* is measured from the blackening of photographic emulsion; *photoelectric magnitude* is determined photoelectrically; *bolometric magnitude* is the integrated brightness across the whole electromagnetic spectrum.

magnolia metal *(Eng.)*. A lead-base alloy, containing 78–84% lead; remainder is mainly antimony, but small amounts of iron and tin are present. Used for bearings.

Magnoliidae *(Bot.)*. Subclass or superorder of dicotyledons. Trees, shrubs and herbs characterized by a well-developed perianth, numerous centripetal stamens and an apocarpous gynoecium. Contains ca 11 000 spp. in 36 families including the Magnoliaceae and the Ranunculaceae. More or less synonymous with *ranalian complex*; generally regarded as the most primitive of extant angiosperms.

Magnoliophyta *(Bot.)*. In some classifications a division containing the **angiosperms**. Subdivided into two classes, the Magnoliopsida (see **Dicotyledones**) and the Liliopsida (see **Monocotyledones**.)

Magnoliopsida *(Bot.)*. See **Dicotyledones**.

magnon *(Phys.)*. Quantum of spin wave energy in magnetic material.

Magnox *(Eng.)*. Group of magnesium alloys for canning uranium reactor fuel elements. Best known are Magnox B and Magnox A 12. The latter is Mg with 0.8% Al and 0.01% Be.

Magnus effect *(Phys.)*. Force experienced by a spinning ball or cylinder in a fluid. The effect is responsible for the swerving of golf and tennis balls when hit with a slice.

mag-opt *(Image Tech.)*. International code name for a print of a motion picture film having both magnetic and photographic (*optical*) sound tracks.

Magslip *(Elec.Eng.)*. TN for a synchro system for remote control or indication.

mahlstick *(Build.)*. A slender stick padded at one end with cloth or leather; used as a support for the hand guiding the brush. Also spelt **maulstick**.

mahogany *(For.)*. Reddish-brown hardwood from *Swietenia macrophylla*, found in Central America. Also called *baywood*. See **African-**, **Brazilian-**, **Cuban-**, **Uganda-**.

main airway *(Min.Ext.)*. In colliery, one directly connected with point of entry to mine.

main and tail *(Min.Ext.)*. Single-track underground rope haulage by means of rope to draw out the full wagons and the tail rope to draw back the empties.

main anode *(Electronics)*. That carrying the load current in mercury-arc rectifiers which have an independent **excitation anode**.

main beam *(Build.)*. In floor construction, a beam transmitting loads direct to the columns.

main circuit *(Elec.Eng.)*. See **current circuit**.

main contacts *(Elec.Eng.)*. The contacts of a switch which normally carry the current; cf. **arcing contacts**, which carry the current at the instant when the circuit is being interrupted.

main deck *(Ships)*. Usually means the principal deck or the upper of two decks or the second deck from the top when there are more than two decks, excluding the decks covering deck houses or erections. Frequently used to name a deck in a passenger liner.

main distribution frame *(Telecomm.)*. A frame for rearranging the incoming lines to a telephone exchange into the numerical order required in the exchange. Abbrev. *MDF*.

main exchange *(Telecomm.)*. One which has other exchanges, such as satellites, dependent on it for extension from them to other exchanges.

main field *(Elec.Eng.)*. The chief exciting field in an electric machine, as opposed to an auxiliary field, such as that produced by the compoles.

main float *(Aero.)*. The two single, or one central, float(s) which give bouyancy to a seaplane or amphibian.

mainframe *(Comp.)*. Computer with a variety of peripheral devices, a large amount of backing store and a fast central processing unit. The term came in with third generation computers and is generally used in comparison with smaller or subordinate computers.

main line *(Min.Ext.)*. Pipeline connecting producing areas to a refinery. Also *trunk line*.

main memory *(Comp.)*. Computer storage which holds programs and data during execution. It is described in terms of **locations** each of which has an **address**. It is fast, read/write, **random access** storage largely built from **chips**. Also referred to as *internal store*, *immediate access store*, *IAS*, *primary store*. Also *core store* whether or not it uses **ferrite core memory**. See **RAM**, **MOS**, **cache memory**, **charged-coupled device**, **bubble memory**, **thin film memory**, **access time**, **paging**. Cf. **backing store**.

main plane *(Aero.)*. The principal supporting surface, or wing, of an aircraft or glider, which can be divided into centre, inner, outer, and/or wing-tip sections.

main rope *(Min.Ext.)*. See **main and tail**.

main rotor *(Aero.)*. (1) The principal assembly or assemblies of rotating blades which provide lift to a rotorcraft. (2) The assembly of compressor(s) and turbine(s) forming the rotating parts of a gas turbine engine.

mains *(Elec.Eng.)*. Source of electrical power; normally the electricity supply system.

main sequence *(Astron.)*. In the **Hertzsprung-Russell diagram**, the vast majority of stars lie in a broad band known as the main sequence, running diagonally from high temperature, high luminosity stars to low temperature, low luminosity stars, in a smooth progression. A star spends most of its life on this main sequence, and throughout that time it converts hydrogen to helium. The position on main sequence depends mostly on mass, the more massive and luminous stars being located higher on the sequence. Once the hydrogen in the core is consumed, the star evolves away from the main sequence, becoming first a **red giant**. The Sun is a main sequence star.

mains frequency *(Elec.Eng.)*. Electricity a.c. supply frequency; 50 Hz in UK, 60 Hz in US, often 400 Hz in aircraft or ships. Also *power frequency*.

mainspring *(Horol.)*. The spring in a watch or spring-driven clock which provides the motive power.

mainspring hook *(Horol.)*. The means by which a mainspring is attached to its barrel.

mainspring winder *(Horol.)*. A tool for coiling a mainspring prior to its insertion or withdrawal from the barrel.

main store *(Comp.)*. See **main memory**.

mains transformer *(Electronics)*. See **power transformer**.

maintained tuning fork *(Acous.)*. One associated with an electronic oscillator so that the latter supplies energy continuously to maintain the fork in steady oscillation. The frequency of the oscilliation is substantially that of the free fork, and provides a method for establishing frequencies with great accuracy. Also *tuning-fork oscillator*.

maintaining power *(Horol.)*. A device which permits of power being transmitted to the train while winding is in progress.

maintaining voltage *(Elec.Eng.)*. That which just maintains ionization and discharge.

Main tanks *(Aero.)*. See under **fuel tanks**.

main tie *(Build.)*. The lower tensional members of a roof-truss, connecting the feet of the principal rafters.

main title *(Image Tech.)*. The section of a film or video programme giving the name of the production and the associated credits.

main transformer *(Elec.Eng.)*. A term sometimes used in connection with the Scott transformer which is connected across 2 of the phases on the 3-phase side.

main wheel *(Horol.)*. The first wheel in the train; the great wheel. In a going-barrel, it is integral with the barrel.

maisonette *(Arch.,Build.)*. A two-storey self-contained dwelling incorporated in a building of three or more storeys.

maize oil *(Chem.)*. See **corn oil**.

major axis *(Maths.)*. Of an ellipse: see **axes**.

major depression *(Behav.)*. A type of **affective disorder** characterized by major depressive episodes occurring without intervening manic episodes.

major histocompatibility complex *(Immun.)*. *MHC*. The collection of genes coding for the major histocompatibility antigens. See **HLA system**.

majority carrier *(Electronics)*. In a semiconductor, the electrons or *holes*, whichever carry most of the measured current. Cf. **minority carrier**.

majority emitter *(Electronics)*. Electrode releasing majority carriers into a region of semiconductor.

majority voting system *(Aero.)*. In a redundant electrical or computer system a means whereby signals from all channels (>3) are continually monitored; any discrepancy in a single channel is recognized and 'voted out' of circuit so that the system can continue to function.

major lobe *(Telecomm.)*. That containing the direction of highest sensitivity or maximum radiation for any form of **polar diagram**.

majuscule *(Print.)*. A capital letter as distinct from the small letter or miniscule.

make *(Telecomm.)*. The operation, partial or complete, of a telephone relay, when current is passed through its windings.

make-before-break contact *(Telecomm.)*. The group of contacts in a relay assembly so arranged that the one which moves makes contact with a front contact before it separates from a back contact, and so is never free, whether the relay is operated or not.

make-contact *(Telecomm.)*. A pair of contacts in a relay assembly which are brought together on operation of the relay and so close a circuit.

make even *(Print.)*. To arrange type so that the last word of a portion or 'take' of copy ends in a full line.

make-ready *(Print.)*. See **making-ready**.

make-up *(Print.)*. The arrangement of type-matter and blocks into pages.

making-capacity *(Elec.Eng.)*. A term used in connection with switchgear rating to denote the capability of a switch to make a circuit under certain specified conditions.

making-current *(Elec.Eng.)*. A term used to denote the maximum peak of current which occurs at the instant of closing the switch.

making good *(Build.)*. The repair of defective surfaces.

making paper *(Print.)*. See **excess feed**.

making-ready, make-ready *(Print.)*. Operations involved in preparing a printing machine to run. Includes machine adjustments, setting ink levels, fitting and adjusting plates, or forme, adjusting impression, etc.

making-up *(Textiles)*. (1) Examination, packing, and ticketing of fabrics before dispatch from the mills. (2) Factory manufacture of garments from the cut-out pieces of fabric.

Maksutov telescope *(Astron.)*. Optical telescope in which the image-forming surfaces are spherical, and therefore easy to make. A deeply curved meniscus lens corrects for aberrations in the primary mirror. Design published by Maksutov in 1944.

malabsorption *(Med.)*. Failure of adequate intestinal absorption of one or more groups of nutrients: occurs in **coeliac disease**, **sprue** and chronic pancreatitis.

malachite *(Min.)*. Basic copper carbonate $CuCO_3Cu(OH)_2$, crystallizing in the monoclinic system. It is a common ore of copper, and occurs typically in green botryoidal masses in the oxidation zone of copper deposits.

malachite green *(Chem.)*. A triphenylmethane dyestuff of the rosaniline group. Obtained by the condensation of benzaldehyde with dimethylaminobenzene in the presence of $ZnCl_2$, HCl, or H_2SO_4, and by oxidation of the resulting leuco-base with PbO_2.

malacia *(Med.)*. Pathological softening of any organ or tissue.

malacology *(Zool.)*. The study of Molluscs.

malacophily *(Bot.)*. Pollination by snails.

malacoplakia *(Med.)*. The occurrence of soft, rounded, pale plaques in the wall of the bladder, in chronic inflammation of the bladder.

Malacostraca *(Zool.)*. Largest subclass within the Crustacea. The body clearly divided into head, thorax and abdomen often with a carapace and possessing 20 segments. Crabs, Lobsters, Shrimps.

malacostracous *(Zool.)*. Having a soft shell.

malar *(Zool.)*. Pertaining to the **mala**; pertaining to, or situated in, the cheek region of Vertebrates; the **jugal**.

malaria *(Med.)*. Infection with *Plasmodium*, a genus of protozoa that live in red cells of the blood. The recurrent fever occurs in *quartan*, *tertian* or *subtertian* (malignant) forms according to the species of Plasmodium. Common in most tropical areas; transmitted by anopheline mosquitoes.

malathion *(Chem.)*. *S-[1,2-di(ethoxycarbonyl)ethyl]dimethyl phosphorothiolothionate*, used as an insecticide.

mal de caderas *(Vet.)*. A chronic infectious disease of horses in South America; due to *Trypanosoma equinum*, and characterized by weakness of the hindquarters.

mal du coït *(Vet.)*. See **dourine**.

male *(Bot.)*. (1) The smaller and motile gamete; the antherozoid or spermatozoid. (2) A gametophyte which produces male but not female gametes. (3) A sporophytes which produces microspores or pollen and hence, (4) individual seed plants or flowers which have functional stamens but not functional carpels. Cf. **hermaphrodite**.

male *(Zool.)*. An individual of which the gonads produce spermatozoa or some corresponding form of gamete. Cf. *female*

male and female *(Elec.Eng.,Telecomm.)*. Applied similarly to a plug and its complementary socket.

male and female *(Eng.)*. Trade terms applied to inner and outer members respectively of pipe fittings, thread pieces, etc. See also **external screw-thread**, **internal screw-thread**.

maleic acid *(Chem.)*. *cis-Butenedioic acid*. HOOC.CH:CH.COOH, mp 130°C, with decomposition into its anhydride and water, readily soluble in water. It has the cis-configuration, whereas its isomer, **fumaric acid**, has the *trans*-configuration. Important in polyester resins.

maleic anhydride *(Chem.)*. *cis-Butenedioic anhydride*. The anhydride of *maleic acid*. Important in establishing the structure of organic compounds containing conjugated double bonds. Also important industrially, mainly as an intermediate. Manufactured by passing a mixture of benzene vapour and air over heated vanadium pentoxide.

maleic hydrazide *(Bot.)*. See **growth retardant**.

male pronucleus *(Biol.)*. Nucleus of the spermatozoon.

male sterility *(Bot.)*. Condition in which viable pollen is not formed. Used by plant breeders to ensure cross pollination especially in the production of **F1 hybrid** seed. See **cytoplasmic male sterility**.

male thread *(Eng.)*. See **external screw-thread**.

malic acid *(Chem.)*. $HOOC.CH_2.CH(OH).COOH$, hygroscopic needles, mp 100°C., found in unripe fruit; it also occurs in wines. When attacked by certain ferments, butanoic, lactic and propanoic acids are produced. Heat causes the loss of a molecule of water, producing maleic and fumaric acids. It has been synthesized by various methods.

malignant *(Med.)*. Tending to go from bad to worse; especially, cancerous (see **tumour**).

malignant aphtha *(Vet.)*. See **contagious pustular dermatitis**.

malignant catarrhal fever *(Vet.)*. Uncommon disease of cattle caused by a herpes virus. Not apparently contagious between cattle, but sheep and deer may act as carriers. There is pyrexia, anorexia, depression, enlarged superficial lymph nodes, oculo-nasal discharges, keratitis and profuse salivation. Mortality 100%.

malignant oedema *(Vet.)*. An acute toxaemia of cattle, sheep, pigs, and horses due to wound infection by *Clostridium septicum* or *Clostridium novyi* (*Cl. oedematiens*). Vaccines available.

malignant stomatitis *(Vet.)*. See **calf diphtheria**.

malignite *(Geol.)*. An alkaline igneous rock, a *melanocratic* variety of nepheline syenite.

mall *(Build.)*. See **beetle**.

malleability *(Eng.)*. Property of metals and alloys which affects their alteration by hammering, rolling, extrusion. Temperature, as it affects crystallization, may alter resistance.

malleable cast-iron *(Eng.)*. A variety of cast-iron which is cast white, and then annealed at about 850°C to remove carbon (*white-heart process*) or to convert the cementite to rosettes of graphite (*black-heart process*). Distinguished from grey and white cast-iron by exhibiting some elongation and reduction in area in tensile test. See also **pearlitic iron**.

malleable iron *(Eng.)*. Now usually means malleable cast-iron, but the term is sometimes applied to wrought-iron.

malleable nickel *(Eng.)*. Nickel obtained by remelting and deoxidizing electrolytic nickel and casting into ingot moulds. Can be rolled into sheet and used in equipment for handling food, for coinage, condensers and other purposes where resistance to corrosion, particularly by organic acids, is required.

mallein *(Vet.)*. A concentrated filtrate of broth cultures of *Actinobacillus mallei* (*Malleomyces mallei*) which have been killed by heat; used as an inoculum for the diagnosis of glanders in horses.

mallenders and sallenders *(Vet.)*. Psoriasis affecting the skin at the flexures of the carpus (knee) joint (*mallenders*) and tarsus (hock) joint (*sallenders*) in the horse.

malleolar *(Zool.)*. Pertaining to, or situated near, the malleolus; in Ungulata, the reduced fibula.

malleolus *(Zool.)*. A process of the lower end of the tibia or fibula.

mallet *(Build.)*. A wooden hammer, or one made of raw-hide or rubber.

mallet-finger *(Med.).* Permanent flexion of the end-joint of a finger or thumb.

malleus *(Zool.).* In Mammals, one of the ear ossicles; in Rotifera, one of the masticatory ossicles of the mastax; more generally, any hammer-shaped structure.

Mallophaga *(Zool.).* An order of the Psocopteroidea (Paraneoptera); ectoparasites, usually of Birds, with reduced eyes, flattened form, tarsal claws and biting mouthparts, e.g. biting lice.

malm *(Build.).* An artificial imitation of natural marl made by mixing clay and chalk in a wash mill; the product is used as a clay for the manufacture of bricks. Also called *washed clay.*

Malm *(Geol.).* The youngest epoch of the Jurassic period.

malm rubber *(Build.).* A soft form of malm brick, capable of being cut or rubbed to special shapes.

malonic acid *(Chem.).* *Propanedioic acid.* HOOC.-CH_2.COOH, soluble in water, alcohol, ether; mp 132°C; it decomposes at a slightly higher temperature, giving ethanoic acid; it occurs in beetroot as its calcium salt and can be obtained from malic acid by oxidation with chromic acid. *Malonic ester*, or diethyl malonate, is a liquid of aromatic odour, bp 198°C, $CH_2(CO.OC_2H_5)_2$. The hydrogen of the methylene group is replaceable by sodium which in turn can be exchanged for an alkyl group. In this way, malonic ester is important for the synthesis of higher dibasic acids. Numerous derivatives of *malonylurea*, $C_4H_4O_3N_2$, barbituric acid, are used as hypnotics and anaesthetics.

Malpighian body (corpuscle) *(Zool.).* In the Vertebrate kidney, the expanded end of a uriniferous tubule surrounding a glomerulus of convoluted capillaries: in the Vertebrate spleen, one of the globular or cylindrical masses of lymphoid tissue which envelops the smaller arteries.

Malpighian cell *(Bot.).* A macrosclereid in the epidermis of the testa of a leguminous seed.

Malpighian layer *(Zool.).* The innermost layer of the epidermis of most Chordates, containing polygonal cells which continually proliferate, dividing by mitosis. Also known as *rete Malpighii, stratum germinativum.*

Malpighian tubes *(Zool.).* In Insects, some *Arachnida* and Myriapods, tubular glands of excretory function opening into the alimentary canal, near the junction of the mid-gut and hind-gut.

malpresentation *(Med.).* Abnormal posture of the foetus during birth.

Malta fever *(Med.).* See **brucellosis**.

maltese cross *(Image Tech.).* Mechanism providing intermittent frame-by-frame movement in a motion picture film projector; see **Geneva movement**.

malthenes *(Chem.).* Such constituents of asphaltic bitumen as are soluble in carbon disulphide and petroleum spirit. See **asphaltenes**, **carbenes**.

maltobiose *(Chem.).* See **maltose**.

maltose *(Chem.).* 4-O-(α-D-glucopyranosyl)-D-glucopyranose, maltobiose, $C_{12}H_{22}O_{11}$, a disaccharide, a white crystalline mass, dextrorotatory, formed by the action of diastase upon starch during the germination of cereals. It reduces Fehling's solution, and when hydrolysed is converted into D-glucose. In cereal chemistry, the *maltose figure* indicates the natural sugar content and diastatic activity of flour or meal. The unit of starch.

malt-sugar *(Chem.).* See **maltose**.

Malus' law *(Phys.).* For a plane-polarized beam of light incident on a polarizer, $I = I_0\cos^2\theta$, where I = intensity of transmitted beam; I_0 = intensity of incident beam; θ = angle between plane of vibrations of beam and plane of vibrations which are transmitted by the polarizer.

Malvaceae *(Bot.).* Family of ca 1000 spp. of dicotyledonous flowering plants (superorder Dilleniidae). Trees, shrubs and herbs, cosmopolitan except in very cold regions. The flowers have 5 sepals (sometimes with an epicalyx), 5 free petals, many stamens with the filaments united at the base and 5 fused carpels. Includes cotton, okra and some ornamental plants e.g. hollyhock.

mamilla *(Zool.).* A nipple.

mamillary body *(Zool.).* See **corpus mamillare**.

mamma *(Meteor.).* Clouds with rounded protuberances on their lower surfaces, like udders. They often occur below thunder-clouds.

mamma *(Zool.).* In female Mammals, the milk gland; the *breast.* adj. *mammary.*

Mammalia *(Zool.).* A class of Vertebrates. The skin is covered by hair (except in aquatic forms) and contains sweat glands and sebaceous glands; they are homoiothermous; the young are born alive (except in the Prototheria) and are initially nourished by milk; respiration is by lungs and a diaphragm is present; the circulation is double, and only the left systemic arch is present; dentition heterodont and diphyodont; there is a double occipital condyle; the lower jaw articulates with the squamosal; the long bones and vertebrae have three centres of ossification; there is an external ear and three auditory ossicles in the middle ear; large cerebral hemispheres.

mammogenic *(Med.).* Promoting growth of the duct and alveolar systems of the mammary gland, e.g. applied to hormones.

mammography *(Radiol.).* The radiological examination of the breast.

mamu *(Phys.).* Abbrev. for *millimass unit.*

Man *(Zool.).* The human race, all living races being included in the genus *Homo*, suborder Anthropoidea of the Primates. Man's distinguishing features include elaboration of the brain and behaviour, including communication by facial gestures and speech; the erect posture; the structure of the limbs (including the opposable thumb) and skull; the dentition with small canines; the long period of postnatal development associated with parental care.

Manchester yellow *(Chem.).* See **Martius yellow**.

mandelic acid *(Chem.).* *Phenyl-glycollic acid*; C_6H_5.CH(OH).COOH, glistening crystals; mp 133°C; soluble in water. It occurs naturally in the form of its glycoside, **amygdalin**, and can be synthesized by the hydrolysis of benzaldehyde cyanhydrin. Mandelic acid possesses an asymmetric carbon atom, and exists in a (+) and a (−) form and as the racemic compound.

mandible *(Zool.).* In Vertebrates, the lower jaw; in Arthropoda, a masticatory appendage of the oral somite; in Polychaeta and Cephalopoda, one of a pair of chitinous jaws lying within the buccal cavity. adj. *mandibular, mandibulate.*

mandibular disease *(Vet.).* *Shovel beak.* A malformation, accompanied by bacterial infection, affecting the beak of young chicks; associated with feeding dry meal.

mandibular glands *(Zool.).* In some Insects, glands opening near the articulation of the mandibles; in some *Lepidopterous larvae*, they function as salivary glands, the true salivary glands secreting silk; also present in the hive Bee and other adult *Hymenoptera.*

mandrel *(Eng.).* *Arbor* US. Accurately turned rod over which metal is forged, drawn or shaped during working so as to create or preserve desired axial cavity. A tapered mandrel is also used for holding and locating a bored component so that external diameters can be machined true to the bore. Also *mandril.*

mandrel *(Glass).* (1) A refractory tube used in the manufacture of glass tubing or rod. (2) A former used in making lamp-blown articles from tubing and in making precision-bore tubing.

mandrel *(Horol.).* A special faceplate lathe, usually hand driven, for turning or recessing watch plates or similar operations.

manganates *(Chem.).* See **permanganates**.

manganepidote *(Min.).* See **piedmontite**.

manganese *(Chem.).* A hard, brittle metallic element, in the seventh group of the periodic system, which exists in 4 polymorphic forms, α, β, γ and δ, and the first two of which have complicated crystal structures. It is brilliant white in colour, with reddish tinge. Symbol Mn, at. no. 25, r.a.m. 54.94, valency 2,3,4,6,7, rel.d. at 20°C 7.39, mp 1220°C, bp 1900°C, electrical resistivity 5.0×10^{-8} ohm metres, hardness in Mohs' scale 6. The principal manganese ores are *pyrolusite, hausmannite, psilomelane*;

pure manganese is obtained electrolytically. Manganese is mainly used in steel manufacture, as a deoxidizing and desulphurizing agent.

manganese alloys *(Eng.)*. Manganese is not used as the basis of alloys, but is a common constituent in those based on other metals. It is present in all steel and cast-iron, and in larger amount in special varieties of these, e.g. manganese steel, silico-manganese steel; also in many varieties of brass, in aluminium-bronze, and aluminium and nickel base alloys.

manganese bronze *(Eng.)*. Originally an alpha-beta brass containing about 1% of manganese; the term is now applied generally to **high-strength brass**, with up to 4% manganese.

manganese dioxide *(Chem.)*. *Manganese (IV) oxide*, MnO_2. Black solid, insoluble in water; occurs in **pyrolusite**. Basic and (mainly) acidic. Forms manganites. Oxidizing agent; uses: depolarizer in Leclanché cells, decolorizing oxidant for green ferrous ion in glass, laboratory reagent.

manganese epidote *(Min.)*. See **piedmontite**.

manganese garnet *(Min.)*. See **spessartine**.

manganese heptoxide *(Chem.)*. *Manganese (VII) oxide*, Mn_2O_7. A heavy, dark coloured oil, prepared by adding concentrated sulphuric acid to potassium manganate (VII). Unstable, explosive, strong oxidizing agent. Acidic. Forms manganates (VII).

manganese nodules *(Geol.)*. Irregular small concretions with high concentrations of Mn (< 30%) and Fe, Cu, Ni minerals. They are most common on the deep ocean floors, notably the N. Pacific where there is little sedimentation. See **todorokite, birnessite**.

manganese spar *(Min.)*. See **rhodochrosite**.

manganese steel *(Eng.)*. A term sometimes applied to any steel containing more manganese than is usually present in carbon steel (i.e. 0.3–0.8%), but generally to *austenitic manganese steel*, which contains 11–14%. This steel is resistant to shock and wear. Used for railway crossings, rock crusher parts etc.

manganin *(Eng.)*. A copper-base alloy, containing 13–18% of manganese and 1.5–4% of nickel. The electrical resistivity is high (about 38×10^{-8} ohm metres), and its temperature coefficient low; it is therefore suitable for resistors.

manganite *(Min.)*. A grey or black hydrated oxide of manganese crystallizing in the monoclinic system (*pseudo-orthorhombic*). It is a minor ore of manganese, and also occurs in deep-sea **manganese nodules**.

mangano-manganic oxide *(Chem.)*. *Manganese (II,III) oxide*. Mn_3O_4. A *spinel*.

manganophyllite *(Min.)*. A phlogopite or biotite mica containing appreciable manganese.

manganosite *(Min.)*. An oxide of manganese (MnO) which crystallizes in the cubic system.

manganous oxide *(Chem.)*. *Manganese (II) oxide*. MnO. Basic. Forms manganese (II) salts. Occurs naturally as *manganosite*.

mange *(Vet.)*. Inflammation of the skin of animals due to infection by certain species of acarids or mites. See **chorioptic-, demodectic-, notoedric-, otodectic-, psoroptic-, sarcoptic-**.

mangle *(Textiles)*. A machine consisting of two or more rollers running in contact with each other. Wet fabric is passed through the nip between the rollers to have as much liquid as possible removed.

man help *(Build.)*. An adjustable handle attached to a brush which is used to reach awkward or inaccessible areas.

mania *(Med.)*. Mental abnormality where the mood is of extreme elation with speed of thought giving a flight of ideas. Patient has little or no insight and business or family affairs can be seriously damaged.

manic-depressive *(Behav.)*. See **bipolar disorder**.

manic-depressive psychosis *(Med.)*. Where periods of **mania** and **depression** can alternate. Sometimes known as *bipolar affective psychosis*.

manifest dream content *(Behav.)*. See **Freud's theory of dreams**.

manifold *(Eng.)*. A pipe or chamber with several openings. See **exhaust-, induction-**.

manifold *(Maths.)*. A **Hausdorff space** with a countable basis, which is a union of open sets each of which can be mapped homeomorphically to an open set in some fixed n-dimensional Euclidean space.

manifold bank *(Paper)*. A thin bank paper, generally around 30 g/m^2.

manifold pressure *(Aero.)*. The absolute pressure in the induction manifold of a reciprocating aero-engine which, indicated by a cockpit gauge, is used together with r.p.m. settings to control engine power output and fuel consumption. See also **supercharger**.

manifold pressure gauge *(Aero.)*. See **boost gauge**.

manila *(Textiles)*. A rough fibre obtained from the leaves of *Musa textilis*, a banana-like plant grown in the Philippines.

manilla paper *(Paper)*. Paper generally made from unbleached chemical woodpulp, with admixtures of waste paper and/or mechanical woodpulp and buff/brown in colour (unless dyed). MG varieties are generally intended for conversion into envelopes. Solid or pasted varieties are generally used for files or folders.

manipulator *(Nuc.Eng.)*. Remote handling device used, e.g. with radioactive materials.

man lock *(Civ.Eng.)*. An air-lock enabling workmen to pass into, and out of, spaces filled with compressed air.

man-made fibre *(Textiles)*. See **fibre, man-made**.

manna *(Zool.)*. See **honey dew**.

mannans *(Chem.)*. The condensation polymers of mannose.

manned space flight *(Space)*. Refers to a manned presence on-board a spacecraft; the additional flexibility provided must be reconciled with the design impacts to safeguard the well-being of the individual. Pioneering efforts in manned space flight involved the Russian *Vostok* and American *Mercury* projects. More sophisticated systems are associated with Earth-orbiting space stations.

Mannerism *(Arch.)*. A recently coined term to describe the predominantly Italian architectural style which attempted to expand upon the classical ideals of the **Renaissance** (1530–1600). The Mannerist phase is often referred to as *Proto-Baroque*, when the departure from the ancient Roman precedents, both in plan forms and decorative practices, became apparent.

Mannesmann process *(Eng.)*. A process for making seamless metal tubing from a solid bar of metal by the action of 2 eccentrically mounted rolls which simultaneously rotate the bar and force it over a mandrel.

Mannheim process *(Chem.)*. In manufacture of sulphuric acid, catalysis of SO_2, in two stages, first with iron (III) oxide and then with platinized asbestos.

mannitol *(Chem.)*. $HO.CH_2.(CH.OH)_4.CH_2OH$; a hexahydric alcohol, fine needles or rhombic prisms, soluble in water and in hot alcohol, mp 166°C. It is found in many plants, and is an important diuretic.

mannitol *(Med.)*. Osmotic diuretic used to treat cerebral oedema.

mannosans *(Chem.)*. See **mannans**.

mannose *(Chem.)*. An aldohexose. D-mannose can be obtained by the oxidation of **mannitol**, and is a stereoisomer of d-glucose. ⇨

manocryometer *(Chem.)*. A combination manometer and cryoscope for measuring the effect of pressure on melting point.

manometer *(Phys.)*. An instrument used to measure the pressure of a gas or liquid. The simplest form consists of a U-tube containing a liquid (water, oil or mercury), one limb being connected to the enclosure whose pressure is to be measured, while the other limb is either open to the atmosphere or is closed or otherwise (e.g., by float) connected to a registering or recording mechanism. There are, however, numerous more sophisticated variations (e.g. balanced hollow ring, bell chamber, diaphragm, and metallic bellows). See also **pressure gauge**.

manoscopy *(Chem.)*. The measurement of gas densities.

manostat *(Eng.)*. Device for keeping the pressure in an enclosure at a constant level.

manoxylic wood *(Bot.)*. Secondary xylem in gymnosperms with very wide parenchymatous rays between the small groups of conducting cells, as in cycads. Cf. **pycnoxylic wood**.

man-riding *(Min.Ext.)*. Said of equipment on oil rigs which is used by personnel and is built to a higher standard of safety than material handling systems.

mansard roof *(Build.)*. A double-sloped pitched roof rising steeply from the eaves, and having a summit of flatter slope on both sides of the ridge. Also called *curb roof, French roof, gambrel roof.*

mantel tree *(Build.)*. The lintel of a fireplace.

mantissa *(Maths.)*. The positive, fractional part of a logarithm. See **logarithm**.

mantle *(Geol.)*. That part of the Earth between **crust** and **core**, ranging from depths of app. 40–2900 km.

mantle *(Zool.)*. In Urochordata, the true body-wall lying below the test and enclosing the atrial cavity; in Mollusca, Brachiopoda, and Cirripedia, a soft fold of integument which encloses the trunk and which is responsible for the secretion of the shell or carapace.

mantle cavity *(Zool.)*. In Urochorda, the atrial cavity: in Mollusca, Brachiopoda and Cirripedia, the space enclosed between the mantle and the trunk.

mantled gneiss dome *(Geol.)*. A dome-like structure consisting of granite at the centre, surrounded by gneiss, and intruded into low-grade regionally metamorphosed rocks. Such structures are found in Precambrian shield areas.

Mantoux test *(Immun.)*. A form of **tuberculin test** commonly used in humans to indicate present or past infection with *Mycobacterium tuberculosis* (or previous immunization with BCG). A tuberculin preparation, usually PPD, is injected intracutaneously. A positive test indicates that delayed hypersensitivity is present.

manubrium *(Zool.)*. Any handle-like structure; the anterior sternebra in Mammals: part of the malleus of the ear in Mammals; the pendant oral portion of a medusa.

manufacturing gauge *(Eng.)*. A gauge used by machine operators in the production of parts, as distinct from an inspection or master gauge.

manus *(Zool.)*. The podium of the fore-limb in land Vertebrates.

map *(Biol.)*. Depending on the level of organization, it can refer to the ordering of genes in an eukaryotic chromosome or to the determination of the arrangement of DNA sequences in a gene or cluster of genes. Thus *gene mapping* or *sequence mapping*.

map *(Maths.)*. Alternative for *mapping*.

map comparison unit *(Radar)*. See **chart comparison unit**.

maple *(For.)*. Various species of the genus *Acer*, found in Europe, Asia and America, used for cabinet work, flooring, musical instruments, panelling etc.

maple syrup, maple sugar *(Chem.)*. Syrup and sugar made from sap tapped from the maple tree. The delicate flavour is probably due to the small protein content.

map matching *(Aero.)*. Navigation by auto-correlation of terrain with data stored in aircraft, missile or RPV, often in the form of film. See **tercom**.

map measurer *(Surv.)*. An instrument used to find the length of a route on a map. It consists of a small wheel which is made to roll over the route, in so doing actuating a needle which records the distance traversed. Also called *opisometer*.

mapping *(Biol.)*. The determination of the positions and relative distances of genes on chromosomes by means of their **linkage**.

mapping *(Maths.)*. In modern mathematical literature generally regarded as equivalent to *function*.

MAR *(Chem.)*. An abbrev. for *MicroAnalytical Reagent*, a standard of purity which indicates that a reagent is suitable for use in *microanalysis*.

MAR *(Comp.)*. See **memory address register**.

maraging *(Eng.)*. Heat treatment used to harden alloy steels involving precipitation of intermetallic compounds in a carbon-free martensite. These include nickel-iron martensites with high toughness and resistance to shock and saline corrosion.

marangoni convection *(Space)*. The flow resulting from gradients in surface tension giving rise to the transfer of heat and mass; it is particularly relevant to **microgravity** conditions when gravity-induced convection is absent.

marasmus *(Med.)*. Progressive wasting, especially in infants. adj. *marasmic*.

marble *(Geol.)*. The term strictly applies to a granular crystalline metamorphosed limestone, but in a loose sense it includes any calcareous or other rock of similar hardness that can be polished for decorative purposes.

marble bones *(Med.,Vet.)*. See **osteopetrosis, osteopetrosis gallinarum**.

marblewood *(For.)*. A straight grained, fine and evenly textured hardwood with figuring of dark black bands on a grey-brown background, from *Diaspyros kurzii*, a native of the Andaman and Nicobar Islands.

marbling *(Print.)*. The operation of decorating book-edges, etc. with a variegated 'marble' effect. Carried out in a trough containing gum (made from carragheen moss), on the surface of which pigments (containing ox-gall) are worked in intricate patterns.

Marburg disease *(Med.)*. Together with **Ebola disease** are known as *African haemorrhagic fever* and caused by a very similar virus infection. Mortality rate is high.

marcasite *(Min.)*. (1) A disulphide of iron which crystallizes in the orthorhombic system. It resembles iron pyrites, but has a rather lower density, is less stable, and is paler in colour when in a fresh condition. (2) In the gemstone trade, *marcasite* is either pyrites, polished steel (widely used in ornamental jewellery in the form of small 'brilliants'), or even white metal.

marcescent *(Bot.)*. Withered but remaining attached to the plant.

marchioness *(Build.)*. A slate, 22 × 12 in. (558 × 305 mm).

Marconi antenna *(Telecomm.)*. Original simple vertical wire, fed between base end and earth.

Marconi-Franklin beam array *(Telecomm.)*. A *broadside* array consisting of a curtain of dipoles, mounted end-to-end and fed in phase; often with a second reflecting curtain. Gives a sharp beam with little upward radiation, for long-distance short wave communication.

marcus *(Build.)*. A large hammer with an iron head.

mare *(Geol.)*. A lunar plain filled with mafic volcanic rocks. See **maria**.

marekanite *(Geol.)*. A rhyolitic perlite broken down into more or less rounded pebbles, named from the type locality, Marekana river, Eastern Siberia.

Marek's disease *(Vet.)*. See **fowl paralysis**.

Marezzo marble *(Build.)*. An artificial marble made with Keene's cement.

Marfan's syndrome *(Med.)*. A connective tissue disorder which affects the heart valves and aortic tissue and has various skeletal abnormalities including long fingers, *arachnodactyly*, high arched palate and dislocation of lens.

Marform process *(Eng.)*. A proprietary production process for forming and deep-drawing irregularly shaped sheet metal parts, in which a confined rubber pad is used on the movable platen of a press and a rigid punch on the fixed platen. The unformed portion of the blank is subjected to sufficient pressure during the progress of the operation to prevent wrinkling.

margaric acid *(Chem.)*. *Heptadecanoic acid*, $CH_3(CH_2)_{15}COOH$. A long-chain saturated 'synthetic' fatty acid. It has an odd number of carbon atoms in the molecule in contrast with the even number of almost all the naturally-occurring fatty acids.

margarine *(Chem.)*. A butter substitute made from various mixtures of animal and vegetable fats, suitably treated by heating and cooling, churning with milk, colouring, and adding concentrates of vitamins A and D.

margarite *(Geol.)*. An aggregate of minute spherelike crystallites, arranged like beads, found as a texture in glassy igneous rocks.

margarite *(Min.)*. Hydrated silicate of calcium and aluminium, crystallizing in the monoclinic system, often as lamellae with a pearly lustre; one of the so-called **brittle micas**.

margin *(Build.)*. The flat surface of stiles or rails in panelled framing.

margin *(Telecomm.)*. In an amplifier or system the number of decibels increase in gain before oscillation occurs.

marginal *(Bot.)*. A species or community occurring at the boundary between two distinct habitats.

marginal bars *(Build.)*. Glazing bars so arranged as to divide the glazed opening into a large central part bordered by narrow panes at the edges.

marginal meristem *(Bot.)*. Meristem along margin of a leaf primordium giving rise to the tissues of the blade.

marginal ore *(Min.Ext.)*. Ore which, at current market value of products from its excavation and processing, just repays the cost of its treatment.

marginal testing *(Electronics)*. A form of test where the operation of a piece of equipment is tested with its operating conditions altered to decrease the safety margin against faults, e.g. an amplifier may be required to give a certain minimum gain with reduced supply voltage.

margin draft *(Build.)*. A smooth face round a joint in ashlar work.

margin lights *(Build.)*. Narrow panes of glass near the edges of a sash.

margin of safety line *(Ships)*. A line drawn 3 in below the upper surface of the bulkhead deck at the side. A passenger ship is subdivided into watertight compartments in such a way that this line is not submerged if two adjacent watertight compartments are flooded.

margin trowel *(Build.)*. A box-shaped float for finishing internal angles.

maria *(Astron.)*. The Latin designation of the so-called 'seas' on the lunar surface, named before the modern telescope showed their dark areas to be dry planes. Since 1959 spacecraft probes and landings (manned and unmanned) have provided much detailed information but their origin is still problematical. The sing. is *Mare* (e.g. *Mare Imbrium*), but there is a marked tendency to delatinize this picturesque terminology (e.g. Sea of Tranquility, Sea of Fertility, Ocean of Storms, Sea of Moscow, etc.). See **mascons, Moon**.

marialite *(Min.)*. Silicate of aluminium and sodium with sodium chloride, crystallizing in the tetragonal system. It is one of the end-members in the isomorphous series of the scapolite group.

marihuana *(Bot.)*. See **cannabis**.

marine boiler *(Eng.)*. A cylindrical boiler, of large diameter and short length, provided with two or more furnaces in flue tubes leading to combustion chambers, surrounded by water, at the back. The gases pass through banks of fire tubes to the smoke-box or uptakes at the boiler front. Also called *Scotch boiler*. See also **dry-back boiler, Yarrow boiler**.

marine chronometer *(Horol.)*. A specially mounted chronometer for use on board ship in the determination of longitude.

marine coatings *(Build.)*. Paints or varnishes specially formulated to withstand exposure to salt water, sea air, marine growths, extreme changes in climatic conditions etc.

marine compass *(Genrl.)*. See **floating compass, gyro compass**.

marine engineering *(Ships)*. That branch of mechanical engineering concerned with the design and production of propelling machinery and auxiliary equipment for use in ships.

marine engines *(Eng.)*. Engines used for ship propulsion.

marine erosion *(Geol.)*. The processes, both physical and chemical, which are responsible for the wearing-away and destruction of coastlines.

marine glue *(Build.)*. A form of glue resisting the action of water.

marine screw propeller *(Ships)*. A boss, carrying 2, 3 or 4 blades of helical form, which produces the thrust to drive a ship by giving momentum to the column of water which it displaces in an astern direction.

marine surveying *(Surv.)*. Hydrographical surveying undertaken in tidal waters.

Mariotte's law *(Phys.)*. Same as *Boyle's law*.

mark *(Telecomm.)*. Departure (positive or negative) from a neutral or no-signal state (space) in accordance with a code, using equal intervals of time. See **mark/space ratio**.

mark and recapture *(Ecol.)*. Method of animal census, in which a marked sample of animals is returned to the population, allowed to mix with unmarked individuals, and then the proportion of the marked individuals in a second sample is used to derive an estimate of the population size. Also called *Lincoln index*.

Markarian galaxy *(Astron.)*. Galaxy with excessive ultraviolet emission.

marker *(Aero.)*. See **airport-**.

marker *(Radar)*. Pip on a radar display for calibration of range and direction.

marker *(Telecomm.)*. Device used in a common control system which tests and selects incoming and outgoing circuits and causes switching apparatus to connect them together.

marker antenna *(Aero.)*. One giving a beam of radiation for marking air routes, often vertically for blind or instrument landing.

marker beacon *(Aero.)*. A radio beacon in aviation which radiates a signal to define an area above the beacon. See **fan-**, **inner-**, **middle-** , **outer-**, **Z-**, **glide-path beacon-**, **localizer beacon-**, **locator beacon-**, **radar beacon-**, **VHF rotating talking-beacon-** and **track** guide.

marker horizon *(Geol.)*. A layer of rock in a sedimentary sequence which, because of its distinctive appearance or fossil content, is easily recognized over large areas, facilitating correlation of strata.

marker light *(Civ.Eng.)*. An indicating light on a signal post, to indicate the position or aspect of the main signal should its light have failed.

marker pulses *(Telecomm.)*. (1) Pulses superimposed on a CRT display for timing purposes. (2) Pulses used to synchronize transmitter and receiver, e.g. in time-division multiplex.

marker system *(Telecomm.)*. A system of automatic switching in which the setting up of a speech path is controlled by a marker from information supplied by a register.

markfieldite *(Geol.)*. A variety of diorite with a porphyritic structure and a granophyric groundmass.

marking blue *(Eng.)*. (1) A mixture of spirit and blue dye applied to metal objects before scratching dimensions onto work. (2) Used for bedding two surfaces together. See also **reddle**.

marking gauge *(Build.)*. A tool for marking lines on the work parallel to one edge of it. It consists of a wooden bar having a projecting steel marking pin near one end and a sliding block adjustable for position along the bar.

marking, keying wave *(Telecomm.)*. In **frequency-shift keying** a signal slightly different in frequency from the spacing signal, which corresponds to the mark of the signal code.

marking knife *(Build.)*. A small steel tool having a chisel edge at one end and pointed at the other. It is used for setting out fine work.

marking-out *(Build.,Surv.)*. Setting out boundaries and levels for a proposed piece of work.

marking-out *(Eng.)*. Setting out centre lines and other dimensional marks on material, as a guide for subsequent machinine operations. See also **laying-out**.

marking-up *(Print.)*. Adding typographic instructions to the copy or the layout as a guide to setting.

mark of reference *(Print.)*. A sign which directs the reader to a footnote. The commonest marks of reference are * asterisk, † dagger, ‡ double dagger, § section, ‖ parallel, ¶ paragraph, in order of use. **Superior figures** are now more commonly used.

mark sense reader *(Comp.)*. Input device which reads special forms (or cards), usually by electrically sensing the marks made in predetermined positions. Also *mark sense device*.

mark/space ratio *(Telecomm.)*. The ratio of the time occupied by a *mark* to that occupied by the *space* in between marks in a telecommunication channel or recording system using pulsed rather than continuous signals.

marl *(Build.)*. A brick earth which contains a high percentage of carbonate of lime; it is the best clay for making bricks without addition of other substances. Also called *calcareous clay*.

marl *(Geol.)*. A general term for a very fine-grained rock, either clay or loam, with a variable admixture of calcium carbonate.

marl yarn *(Textiles)*. Folded yarn made from single yarns of two or more colours. May be continuous filament or woollen spun single yarn.

marmite disease *(Vet.)*. See **greasy pig disease**.

marmoration *(Build.)*. A marble casing for a building.

marmoratum, marmoretum *(Build.)*. A cement containing pulverized marble and lime.

marmorization *(Geol.)*. The recrystallization of a limestone by heat to give a marble.

marquise *(Arch.)*. A projecting canopy over the entrance to a building.

married print *(Image Tech.)*. A positive print of a motion picture with both picture and sound track images correctly synchronized for projection.

marrow *(Zool.)*. The vascular connective tissue which occupies the central cavities of the long bones in most Vertebrates, and also the spaces in certain types of cancellated bone.

marrying *(Image Tech.)*. The synchronization and make-up of the separate picture and sound track negatives for printing together.

Mars *(Astron.)*. Fourth planet in the solar system, and distinctly red even to the unaided eye. It moves on an eccentric orbit at a mean distance of 1.52 AU and a period of 686.98 days. Its diameter is half and its mass 0.107 that of the Earth. Atmospheric pressure is 0.007 compared to the Earth, and the composition is 95% carbon dioxide with small quantities of nitrogen and other gases. Its remarkable surface features were revealed by the Mariner (1965, 1969, 1971) and Viking (1976) probes. There are numerous large extinct volcanoes, huge canyons and a great many impact craters. There are two permanent ice caps of frozen carbon dioxide. It has two tiny natural satellites: Phobos (20–30 km) and Deimos (10–15 km).

marsh *(Ecol.)*. Vegetation which is seasonally water-logged and non-peat forming.

Marshall valve gear *(Eng.)*. A radial gear of the *Hackworth* type, in which the straight guide is replaced by a curved slot to correct inequalities in steam distribution.

Marsupiala *(Zool.)*. The single order included in the Mammalian group Metatheria, and having the characteristics of the subclass, e.g., Opossum, Tasmanian wolf, Marsupial Mole, Kangaroo, Koala bear.

marsupium *(Zool.)*. A pouchlike structure occupied by the immature young of an animal during the later stages of development; as the abdominal pouch of metatherian Mammals. adj. *marsupial*.

Martello Tower *(Arch.)*. Low round tower for guns, built as a coastal defence in England from 1793.

martempering *(Eng.)*. See **austempering**.

martensite *(Eng.)*. A constituent formed in steel when it is cooled at a rate sufficiently rapid to suppress the change from austenite to pearlite. Results from the decomposition of austenite at low temperatures. Consists of a solid solution of carbon in α-iron, and is responsible for the hardness of quenched steel.

Martin's cement *(Build.)*. A quick-setting hard plaster, made by soaking plaster of Paris in a solution of potassium carbonate.

martite *(Min.)*. A variety of **hematite** (Fe_2O_3) occurring in dodecahedral or octahedral crystals believed to be pseudomorphous after magnetite, and in part perhaps after pyrite.

martius yellow *(Chem.)*. Salts of dinitro-α-naphthol, used as a pigment and for colouring soap; poisonous. Also known as *Manchester yellow*.

Martonite *(Chem.)*. A tear gas, a mixture (4:1) of monobromopropanone and monochloropropanone.

marver *(Glass)*. A flat cast-iron or stone (marble) block upon which glass is rolled during the hand method of working.

mascon *(Astron.)*. *Mass concentrations*. Regions of high gravity occurring within certain **maria** of the Moon. Their origin is still conjectural.

maser *(Phys.)*. *Microwave Amplification by Stimulated Emission of Radiation*. A *microwave* oscillator that operates on the same principle as the **laser**. Maser oscillations produce coherent monochromatic radiation in a very narrow beam. Less noise is generated than in other kinds of microwave oscillators. Materials used are generally solid-state, but masers have been made using gases and liquids.

maser relaxation *(Phys.)*. The process by which excited molecules in the higher energy state revert spontaneously or under external stimulation to their equilibrium or ground state. Maser action arises when energy is released in this process to a stimulating microwave field which is thereby reinforced.

mash seam welding *(Eng.)*. An electrical resistance welding process in which the slightly overlapping edges of the workpiece are forged together during welding by broad-faced, flat electrodes.

mask *(Electronics)*. A thin sheet of metal or other coating containing a selected open area, to shield the surface of a semiconductor substrate selectively during deposition or other stages of integrated circuit manufacture. A series of different masks and processes is used to build up a complex circuit structure.

mask *(Image Tech.)*. Opaque material used to determine the size and shape of a photographic image in either the camera or print.

mask *(Print.)*. (1) Material which blocks off portions of an illustration in order to protect it. (2) Opaque material used to protect specified areas of a printing plate during exposure. (3) Material placed in position over a film assembly to prevent the marked areas from being exposed on the printing plate, the technique being used to create the separate plates in multi-colour printing when only one film assembly carrying all the information required for the work is used.

mask *(Zool.)*. A prehensile structure of the nymphs of certain Dragon-flies (Odonata), formed by the labium.

maskelynite *(Min.)*. A glass which occurs in colourless isotropic grains in meteorites and has the composition of a plagioclase. It probably represents re-fused feldspar.

masking *(Acous.)*. Loss of sensitivity of the ear for specified sounds in the presence of other louder sounds.

masking *(Comp.)*. An operation that selects certain of the bits in a register for subsequent processing. A register of equal length holds a bit pattern called a *mask* with each bit set to '1' where a corresponding bit is to be selected and '0' otherwise.

masking *(Image Tech.)*. See **colour masking**.

masking frame *(Image Tech.)*. Device for holding printing paper in an enlarger, comprising a baseplate to which is hinged a frame forming a border for the print, with moveable metal strips for adjusting horizontal and vertical margins.

masking paper *(Paper)*. A paper, generally semi-bleached and crêped, coated on one side with a self-adhesive compound. Usually sold in coil form and may be peeled from its support after use.

masochism *(Behav.)*. Sexual gratification through having pain inflicted on oneself.

masonry cement *(Build.)*. A cement incorporating inert material and an air entrainer to increase plasticity in mortar. Specially useful in minimizing possibility of damage in cold weather work.

mason's joint *(Build.)*. Pointing finished with a projecting vee.

mason's mitre *(Build.)*. The name given to an effect

similar to the **mitre** but produced (particularly in stonework) by shaping the intersection out of the solid.

mason's putty *(Build.)*. A mixture of Portland cement, lime putty, and stone dust, usually in the proportions 2:5:7, with water. Used for making fine joints, especially in ashlar work.

mason's scaffold *(Build.)*. A form of scaffold used in the erection of stone walls, when it is not convenient to leave holes for the support of one end of the putlogs; an inner set of standards and ledgers is used to provide this support.

mason's stop *(Build.)*. Same as *mason's mitre*.

mason's trap *(Build.)*. A form of trap in which a stone slab on edge dips below the water surface.

mass *(Phys.)*. The quantity of matter in a body. Two principle properties are concerned with mass: (1) The *inertial mass*, that is the mass as a measure of resistance of a body to changes in its motion; (2) The *gravitational mass*, that is the mass as a measure of the attraction of one body to another. The general theory of relativity shows inertial and gravitational mass are equivalent. By a suitable choice of units they can be made numerically equal for a given body, which has been confirmed to a high degree of precision by experiment. Mass is always conserved. Cf. **rest mass**. See **relativistic mass equation**, **mass-energy equation**.

mass absorption coefficient *(Phys.)*. See **absorption coefficient**.

mass action *(Chem.)*. See **law of mass action**.

mass balance *(Aero.)*. A weight or mass attached ahead of the hinge line of an aircraft control surface to give static balance with no moment about the hinge, and to reduce to zero inertial coupling due to displacement of the control surface. Mass balancing is a precaution against control surface **flutter**.

mass concrete *(Civ.Eng.)*. Concrete which is placed without reinforcement, and employing large-sized coarse aggregate to reduce the mortar content. Also called *bulk concrete*.

mass control *(Acous.)*. Said of mechanical systems, particularly those generating sound waves, when the mass of the system is so large that the compliance and resistance of the system are ineffective in controlling motion.

mass decrement, mass defect *(Phys.)*. Both terms may be used for (1) the binding energy of a nucleus expressed in mass units, and (2) the measured mass of an isotope less its mass number. The B.S.I. recommend the use of *mass decrement* for (2) above. In US, it is usually used for (1) and *mass excess* for (2). See also **packing fraction**.

mass effect *(Eng.)*. The tendency for hardened steel to decrease in hardness from the surface to the centre, as a result of the variation in cooling throughout the section. Becomes less marked as the rate of cooling required for hardening decreases, i.e. as the content of alloying elements increases.

mass-energy equation *(Phys.)*. $E = mc^2$. Confirmed deduction from Einstein's special theory of relativity that all energy has mass. If a body gains energy E, its inertia is increased by the amount of mass $m = E/c^2$, where c is the speed of light. Derived from the assumption that all conservation laws must hold equally in all frames of reference and using the principle of conservation of *momentum*, of *energy* and of *mass*. See **rest-mass energy**.

masseter *(Zool.)*. An elevator muscle of the lower jaw in higher Vertebrates. adj. *masseteric*.

mass excess *(Phys.)*. See **mass decrement**.

mass-flow hypothesis *(Bot.)*. Hypothesis proposing that translocation in the **phloem** results from a continuous flow of water and solutes, especially sugars, through the sieve tubes from source to sink, the flow being driven by a hydrostatic pressure difference caused by the osmotic movement of water following the loading of solutes into the sieve tubes at the source end and their removal (unloading) at the sink end.

mass-haul curve *(Civ.Eng.)*. A curve used in the design of earthworks involving cuttings and embankments, the abscissae representing distance along the centre line, and the ordinates the excess of cutting over filling, i.e. the material requiring to be hauled to another position.

massicot *(Min.)*. Yellow lead oxide. A rare mineral of secondary origin, associated with galena.

mass law *(Acous.)*. Law describing the sound transmission through walls. The transmission coefficient is approximately proportional to the inverse of the mass per unit area and to the inverse of the frequency, i.e. the **transmission loss** increases by approximately 6 dB when mass or frequency are doubled. There are many exceptions.

mass-luminosity law *(Astron.)*. A relationship between the mass and absolute magnitude of stars, the most massive stars being the brightest; applicable to all stars except the white dwarfs.

mass number *(Phys.)*. Total of *protons* and *neutrons* in a nucleus, each being taken as a unit of mass. Also *nuclear, nucleon, number*.

mass of electron *(Phys.)*. See **electron mass**.

mass ratio *(Aero.,Space)*. The ratio of the fully-fuelled mass of a rocket-propelled vehicle or stage to that when all fuel has been consumed.

mass resistivity *(Elec.Eng.)*. The product of the volume resistivity and the density of a given material at a given temperature.

mass spectrograph *(Phys.)*. A vacuum system in which positive rays of various charged atoms are deflected through electric and magnetic fields so as to indicate, in order, the charge-to-mass ratios on a photographic plate, thus measuring the atomic masses of isotopes with precision. System used for the first separation for analysis of the isotopes of uranium.

mass spectrometer *(Phys.)*. A *mass spectograph* in which the charged particles are detected electrically instead of photographically.

mass spectrum *(Phys.)*. See **spectrum**.

mass stopping power *(Phys.)*. See **stopping power**.

mass storage device *(Comp.)*. Term used to denote a large-capacity **backing store**, such as a **magnetic drum**.

mass transfer *(Chem.Eng.)*. The transport of molecules by convection or diffusion, as in the operations of extraction, distillation and absorption.

mass transfer coefficient *(Chem.Eng.)*. The molecular flux per unit driving force. Symbol k, K.

mass unit *(Phys.)*. See **atomic mass unit**.

mass wasting *(Geol.)*. Downhill gravity movement of rock material, e.g. *landslides*, *rockfall*.

mast *(Bot.)*. Fruit of beech, oak and other forest trees.

mast *(Eng.)*. A slender vertical structure which is not self-supporting and requires to be held in position by guy-ropes. Cf. *pylon*.

mast cell *(Immun.)*. Cell with basophil cytoplasmic granules similar to but smaller than those of basophil leucocytes in the blood. The granules contain mainly histamine, serotonin and heparin. Mast cells bind the Fc region of IgE antibodies. Reaction with an antigen (or the action of anaphylotoxin) causes extrusion of the granule contents, and the release of various newly formed pharmacologically active substances. These include leukotrienes, **platelet activating factor** and a peptide which is chemotactic for eosinophil leucocytes. There are two populations of mast cells. One is normally present in connective tissues, in the neighbourhood of small blood vessels. The other, *mucosal mast cells*, is induced by IL-3 secreted by T-cells. They have sparse granules and are more like lymphocytes, and become prominent in the mucosa of the gut in intestinal nematode infections.

mastectomy *(Med.)*. Surgical removal of the breast.

master *(Acous.)*. In gramophone-record manufacture, the metal (negative) disk obtained by plating the original lacquer surface on which a recording has been cut.

master *(Eng.)*. (1) The term applied to special tools, gauges, etc., used for checking the accuracy of others used in routine work. (2) The chief or key member of a system, as the *master cylinder* of a hydraulic brake mechanism.

master *(Image Tech.)*. (1) The original 16 mm reversal film exposed in the camera, after processing. (2) A positive print made for protection or duplication, not for normal projection.

master alloy *(Eng.)*. An alloy enriched in the required components which can be added to a melt to bring it to the correct composition.

master clock *(Comp.)*. See **clock**.

master clock *(Elec.Eng.)*. A clock for use with certain electrical time-keeping equipment. It sends out impulses at predetermined time-intervals for the operation of other clocks or similar equipment which are thereby kept in synchronization.

master connecting-rod *(Aero.)*. The main member of the master and articulated assembly of a radial aero-engine. It incorporates the crank-pin bearing, and the *articulated rods* of the other cylinders oscillate on it by means of wrist pins.

master controller *(Elec.Eng.)*. A controller for electrical equipment, the operation of which energizes or de-energizes the contactors which perform the actual switching operations which may be remote from the operator.

master cylinder *(Eng.)*. The container holding the fluid which operates the mechanism of hydraulic brakes or clutch.

master file *(Comp.)*. The principal source of information for a job. Cf. **transaction file**.

master frequency meter *(Elec.Eng.)*. See **integrating frequency meter**.

master gain control *(Acous.)*. (1) In a broadcasting studio, the attenuator connected between the programme input and main amplifiers in order to regulate the gain as desired. (2) In a stereo amplifier, the control which adjusts simultaneously the gains of both channels to equalize them.

master gauge *(Eng.)*. A standard or reference gauge made to specially fine limits; used for checking the accuracy of inspection gauges.

master group *(Telecomm.)*. In a **frequency-division multiplex** system, an assembly of 10 **supergroups**, formed by remodulation of super-groups onto that number of new carriers. (In some systems just 5 super-groups constitute a mastergroup.) A master group will then contain 600 (or 300) voice channels.

master oscillator *(Telecomm.)*. One, often of low power, which establishes the frequency of transmission of a radio transmitter.

master station *(Telecomm.)*. (1) Transmitting station from which one or more *satellite* stations receive a programme for rebroadcasting. (2) In a radio-navigation system depending on a chain of synchronized transmitters, e.g. *loran*, the main transmitter to which the signals from all the others are referred.

master switch *(Elec.Eng.)*. A switch for controlling the effect of a number of other switches or contactors; e.g. if the master switch is open, none of the other switches is operative.

master tap *(Eng.)*. A substandard screw-tap, sometimes used after the plug-tap when great accuracy is required.

master telephone transmission reference system *(Telecomm.)*. High quality transmission system provided with means for calibrating its transmitting, attenuating and receiving components in terms of absolute units. Used for comparison of quality between different systems. Located at CCIT laboratories in Geneva.

mastic *(Build.)*. A term applied to preparations used for bedding and pointing window frames, bedding wood-block flooring and repairing flat roofs.

mastic *(Chem.)*. A pale-yellow resin from the bark of *Pistacia lentiscus*, used in the preparation of fine varnishes.

mastic asphalt *(Build.)*. A mixture of bitumen with stone chippings or sand, used for roofing, roads, paving and damp-proof courses.

mastication *(Zool.)*. The act of reducing solid food to a fine state of subdivision or to a pulp.

masticator *(Chem.)*. An apparatus consisting of 2 revolving and heated cylinders studded with teeth or knives; used for converting rubber into a homogeneous mass.

masticatory *(Zool.)*. Pertaining to the trituration of food by the mandibles, teeth, or gnathobases, prior to swallowing.

Mastigomycotina *(Bot.)*. Subdivision or class containing those Eumycota or true fungi in which the spores and/or gametes are motile. Includes the Chytridiomycetes and, in some schemes, the Oomycetes.

Mastigophora *(Zool.)*. A class of Protozoa which possess one or more flagella in the principal phase; may be amoeboid but usually have a pellicle or cuticle; often parasitic but rarely intracellular; have no meganucleus; reproduction mostly by longitudinal binary fission; nutrition may be holophytic, saprophytic, or holozoic. Also *Flagellata*.

mastitis *(Med.,Vet.)*. Inflammation of the mammary gland.

mastodynia *(Med.)*. Pain in the breast.

mastoid *(Bot.,Zool.)*. Resembling a nipple; as a posterior process of the otic capsule in the Mammalian skull.

mastoidectomy *(Med.)*. Excision of the (infected) air-cells of the mastoid bone.

mastoiditis *(Med.)*. Inflammation of the air-cells of the mastoid bone.

mat *(Print.)*. The commonly-used contraction for **matrix**.

match-boarding, match-lining *(Build.)*. Same as **matched boards**.

matched boards *(Build.)*. Boards specially cut at the edges to enable close joints to be made, either by tongue and groove or by rebated edge.

matched filter *(Electronics)*. One used to maximize the signal-to-noise ratio in order to detect weak echoes or those where jamming is present. The filter has an impulse response which corresponds precisely to the spectrum of the transmitted pulse(s), which may be frequency modulated or coded. See **coding, pulse compression**.

matched load *(Telecomm.)*. One with impedance equalling the characteristic impedance of a source, line or waveguide, so that there is no reflection of power, and the received power is the maximum possible. Also *matched termination*. See **termination**.

matching *(Build.)*. See **matched boards**.

matching *(Radar)*. Said of the insertion of *matching sections* into radio-frequency transmission lines, with the aim of minimizing power reflections at a **mismatch**. A matching section may consist of specifically chosen lengths of waveguide, stripline or coaxial cable having a different impedance from the main system and connected in series so as to cause impedance transformation; alternatively, a *matching stub* may be connected in parallel with the transmission path.

matching *(Telecomm.)*. Adjusting a load impedance to match the source impedance with a transformer or network, so that maximum power is received, i.e. so that there is no reflection loss due to *mismatch*. The principle applies to many physical systems, e.g. non-reflecting optical surfaces, use of a horn in loudspeakers to match impedance of vibrator to that of air, matching load on electrical transmission line. Reactance in the load impedance can be neutralized or *tuned-out* by an equal reactance of opposite sign.

matching stub *(Telecomm.)*. Short or open circuited stub line attached to main line to neutralize reactive component of load and so improve matching.

matching transformer *(Telecomm.)*. One expressly inserted into a communication circuit to avoid reflection losses when the load and source impedances differ.

matchwood *(For.)*. Billets suitable for manufacturing into matches.

materialization *(Phys.)*. Reverse of Einstein energy released with annihilation of mass. A common example is by *pair production* (electron-positron) from gamma rays.

materials handling *(Eng.)*. The process, which includes mechanical handling, of transporting and positioning raw materials, semifinished and finished products in connection with industrial operations, by conveyors, cranes, trucks, hopperfeeds etc.

materials testing reactor *(Nuc.Eng.)*. See **high-flux reactor**.

maternal effect *(Biol.)*. A *non-heritable* influence of a mother on characters in her offspring, e.g. through her milk supply.

maternal immunity *(Immun.)*. Passive immunity acquired by the newborn animal from its mother. In humans and other primates this is mainly by active transport of IgG antibodies across the placenta. In species which have thicker placentas, such as ungulates, antibody is not transferred in utero but is acquired from the colostrum, and is absorbed intact from the gut during the first few days of life. The young of birds acquire maternal immunity from antibody in the egg yolk.

mathematical induction *(Maths.)*. An argument for verifying identities when the variable is limited to positive integral values. The two steps of the argument are (1) by assuming the correctness of the identity when the variable equals n, to deduce the corresponding identity for when the variable equals $n+1$, (2) to verify that the identity is correct when the variable has a low value such as 1 or 2. Step (1) shows that if the identity is ever correct it is correct for all subsequent values of the variable, and step (2) shows that the identity is correct for some low value of the variable.

mathematical logic *(Maths.)*. The application of mathematical techniques to logic in an attempt both to deduce new propositions by formal manipulations and to detect any underlying inconsistencies. Its study, by many eminent mathematicians and philosophers, as a means of clarifying the basic concepts of mathematics, has revealed a number of paradoxes, several of which have yet to be resolved. Also called *symbolic logic*. Cf. *Boolean algebra*.

mathematics *(Genrl.)*. May be defined as the study of the logical consequences of sets of axioms. Pure mathematics, roughly speaking, comprises those branches studied for their own sake or their relation to other branches. The most important of these are algebra, analysis and topology. The term *applied mathematics* is usually restricted to applications in physics. Applications in other fields, e.g. economics, mainly statistical, are sometimes referred to as *applicable mathematics*.

mating *(Build.)*. Said of surfaces or pieces which come into contact or interlock.

mating type *(Bot.)*. A group of individuals, within a species, which cannot breed among themselves but which are able to breed with individuals of other such groups. See also **plus strain, minus strain, incompatibility**.

matric potential *(Bot.)*. That component of the **water potential** due to the interaction of the water with colloids and to capillary forces (surface tension). Often important component in soils and cell walls. Symbol ψ.

matrix *(Bot.)*. More or less continuous matter in which something is embedded, e.g. the non-cellulosic substances of the cell wall in which the cellulose microfibrils lie in vascular plant cell walls. adj. *matrix*.

matrix *(Build.)*. The lime or cement constituting the cementing material that binds together the aggregate in a mortar or concrete.

matrix *(Image Tech.)*. A network of electronic circuits to combine several signal sources in a specified mathematical arrangement, for example, to transform colour co-ordinates in a colour TV system.

matrix *(Maths.)*. A system of elements e.g. real or complex numbers arranged in a square or rectangular formation, e.g.

$$\begin{bmatrix} a & b & c \\ d & e & f \\ g & h & j \end{bmatrix}$$

matrix *(Powder Tech.)*. The phase or phases which form the continuous skeleton of a powder body, thus forming the cells in which constituents imparting particular qualities may be held.

matrix *(Print.)*. The mould from which type is cast, produced from an impression with a punch: also the mould made from a relief surface in stereotyping and electrotyping. Usually contracted to *mat*.

matrix *(Zool.)*. The intercellular ground-substance of connective tissues.

matrixing *(Image Tech.)*. In colour TV, performing a colour coordinate transformation by computation with electrical or optical methods.

matroclinous *(Biol.)*. Exhibiting the characteristics of the female parent more prominently than those of the male parent. Cf. **patroclinous**.

matromorphic *(Biol.)*. Resembling the mother.

matte *(Image Tech.)*. An opaque mask determining the image area exposed in a motion picture camera or printer. It may be a physical object in front of the camera lens or a strip of film with silhouette images of high density, used in special effects work for image combination.

matte *(Min.Ext.)*. Fusion product consisting of mixed sulphides produced in the smelting of sulphide ores. In the smelting of copper, for example, a slag containing the gangue oxides and a matte consisting of copper and iron sulphides are produced. The copper is subsequently obtained by blowing air through the matte, to oxidize the iron and sulphur.

matte box *(Image Tech.)*. An extension in front of the lens of a motion picture camera in which filters and mattes can be mounted.

matter *(Med.)*. See **pus**.

matter *(Phys.)*. The substances of which the physical universe is composed. Matter is characterized by gravitational properties (on earth by weight) and by its indestructibility under normal conditions.

matte shot *(Image Tech.)*. A special effects shot involving the combination of images by the use of **mattes**.

matt finish *(Build.)*. A surface of finish which has no sheen or gloss.

Matthiessen hypothesis *(Elec.Eng.)*. That the total electrical resistivity of a metal may be equated to the sum of the various resistivities due to the different sources of scattering of free electrons; this also applies to thermal resistivity.

matt, matte *(Genrl.)*. Smooth but dull; tending to diffuse light; said, e.g of a surface painted or varnished so as to be dull or flat.

mattress *(Civ.Eng.)*. (1) Sheet expanded metal for reinforcement of concrete roads. (2) A rectangular mesh of reinforcing rods used normally in reinforced concrete foundations.

Matura diamonds *(Min.)*. Colourless (fired) zircons from Ceylon, which on account of their brilliancy are useful as gemstones. A misleading name.

maturation *(Bot.,Zool.)*. Final stages in the development of the germ cells; more generally, the process of becoming adult or fully developed.

maturation divisions *(Biol.)*. The divisions by which the germ cells are produced from the primary spermatocyte oocyte, during which the number of chromosomes is reduced from the diploid to the haploid number. See **meiosis**.

maturation of behaviour *(Behav.)*. Refers to environmentally stable characteristics, that are irreversible during development despite variation in environmental influences, and which develop with little or no practice.

maturity *(Geol.)*. (1) A stage of the geomorphological cycle characterized by maximum relief and well-developed drainage. (2) The ultimate stage in the development of a sediment, characterized by stable minerals and rounded grains.

MATV *(Image Tech.)*. *Master Antenna Television*, in which a number of individual receivers are fed from one communal antenna.

maul *(Build.)*. See **beetle**.

maulstick *(Build.)*. See **mahlstick**.

Maunder diagram *(Astron.)*. See butterfly diagram.

mauveine *(Chem.)*. *Perkin's mauve*, the first synthetic dyestuff.

max gross *(Aero.)*. See **maximum weight**.

maxilla *(Zool.)*. In Vertebrates, the upper jaw: a bone of the upper jaw: in *Anthropoda*, an appendage lying close behind the mouth and modified in connection with feeding. pl. *maxillae*. adj. *maxillary, maxilliferous, maxilliform*.

maxillary *(Zool.)*. Pertaining to a maxilla: pertaining to the upper jaw: a paired membrane bone of the Vertebrate skull which forms the posterior part of the upper jaw.

maxilliped *(Zool.)*. In *Arthropoda*, especially *Crustacea*, an appendage behind the mouth, adapted to assist in the transferance of food to the mouth.

maximally-flat *(Telecomm.)*. Said of amplifiers so designed that the circuit elements are transformed from filter sections incorporating stray admittances.

maximum and minimum thermometer *(Meteor.)*. An instrument for recording the maximum and minimum temperatures of the air between 2 inspections, usually a period of 24 hr. A type widely used is *Six's thermometer*.

maximum continuous rating *(Aero.)*. See **power rating**.

maximum demand *(Elec.Eng.)*. The maximum load taken by an electrical installation during a given period. It may be expressed in kW, kVA, or amperes.

maximum-demand indicator *(Elec.Eng.)*. An instrument for indicating the maximum demand which has occurred on a cicuit within a given period.

maximum-demand tariff *(Elec.Eng.)*. A form of charging for electrical energy in which a fixed charge is made, depending on the consumer's maximum demand, together with a charge for each unit (kWh) consumed.

maximum equivalent conductance *(Elec.Eng.)*. The value of the equivalent conductance of an electrolyic solution at infinite dilution with its own solvent.

maximum flying speed *(Aero.)*. See **flying speed**.

maximum landing weight *(Aero.)*. See **weight**.

maximum permissible concentration *(Radiol.)*. The recommended upper limit for the dose which may be received during a specified period by a person exposed to ionizing radiation. Also called *permissible dose*.

maximum permissible dose rate, flux *(Radiol.)*. That dose rate or flux which, if continued throughout the exposure time, would lead to the absorption of the maximum permissible dose.

maximum permissible level *(Radiol.)*. A phrase used loosely to indicate *maximum permissible concentration, dose* or *dose rate*.

maximum point on a curve *(Maths.)*. A peak on a curve. For the curve $y=f(x)$, the point where $x=a$ is a maximum if $f(a+h)-f(a)$ is negative for all values of h sufficiently small.

maximum-reading accelerometer *(Aero.)*. See **accelerometer**.

maximum safe air-speed indicator *(Aero.)*. A pilot's *air-speed indicator* with an additional pointer showing the **indicated air speed** corresponding to the aircraft's **limiting Mach number** and also having a mark on the dial for the maximum permissible air speed.

maximum take-off rating *(Aero.)*. See **power rating**.

maximum tensile stress *(Eng.)*. See **ultimate tensile stress**.

maximum traction truck *(Elec.Eng.)*. A special form of bogie or truck often used on trams, and arranged so that the greater part of the weight comes on the driving wheels, thereby enabling the maximum tractive effort to be obtained.

maximum usable frequency *(Telecomm.)*. That which is effective for long-distance communication, as predicted from diurnal and seasonal ionospheric observation. Varies on an eleven year cycle. Abbrev. *MUF*.

maximum value *(Elec.Eng.)*. See **peak value**.

maximum weight *(Aero.)*. See under **weight**. Also called *max take-off weight*, colloq. *max gross*.

max level speed *(Aero.)*. The maximum velocity of a power-driven aircraft at full power without assistance from gravity; the altitude should always be specified.

max take-off weight *(Aero.)*. See **weight**.

maxwell *(Elec.Eng.)*. The CGS unit of magnetic flux, the MKSA (or SI) unit being the *weber*. One maxwell = 10^{-8} weber.

Maxwell-Boltzmann distribution law *(Phys.)*. A law expressing the energy distribution among the molecules of a gas in thermal equilibrium.

Maxwell bridge *(Elec.Eng.)*. An early form of a.c. bridge which can be used for the measurement of both inductance and capacitance.

Maxwell experiment *(Image Tech.)*. The pioneer demonstration of 3-colour additive synthesis, using 3 black-and-white negatives.

Maxwellian viewing system *(Phys.)*. In some photometers, spectrophotometers, colorimeters, etc., an arrangement in which the field of view is observed by placing the eye at the focus of a lens, instead of using an eyepiece.

Maxwell primaries *(Image Tech.)*. The colours red, green and blue-violet, used in Maxwell's experiment.

Maxwell's circuital theorems *(Phys.)*. Generalized forms of Faraday's law of induction and Ampère's Law (modified to incorporate the concept of displacement current). Two of *Maxwell's field equations* are direct developments of the circuital theorems.

Maxwell's circulating current *(Elec.Eng.)*. A mesh or cyclic current inserted in closed loops in a complex network for analytical purposes.

Maxwell's demon *(Chem.)*. Imaginary creature who, by opening and shutting a tiny door between two volumes of gases, could in principle concentrate slower (i.e. colder) molecules in one and faster (i.e. hotter) molecules in the other, thus reversing the normal tendency toward increased disorder or entropy and breaking the second law of thermodynamics.

Maxwell's distribution law *(Phys.)*. The distribution of numbers of gas molecules which have given speeds, or kinetic energies in a gas of uniform temperature. The law can be deduced from the *kinetic theory of gases*.

Maxwell's equations *(Telecomm.)*. Those fundamental laws of theoretical physics which govern the behaviour of electromagnetic waves in all practical situations. Used to analyse the propagation of radio waves in free space, at all sorts of boundaries and in all guided-wave structures or transmission lines.

Maxwell's field equations *(Phys.)*. Mathematical formulations of the laws of Gauss, Faraday and Ampère from which the theory of electromagnetic waves can be conveniently derived.

$$\operatorname{div} B=0;\quad \operatorname{curl} H=\frac{\partial D}{\partial t}+j;$$

$$\operatorname{div} D=\rho;\quad \operatorname{curl} E=-\frac{\partial B}{\partial t}.$$

Maxwell's rule *(Elec.Eng.)*. A law stating that every part of an electric circuit is acted upon by a force tending to move it in such a direction as to enclose the maximum magnetic flux.

Maxwell's theorem *(Eng.)*. See **reciprocal theorem**.

Maxwell's thermodynamic relations *(Phys.)*. Four mathematical identities relating the pressure, the volume, the entropy and the thermodynamic temperature for a system in equilibrium. They are expressed in the form of partial derivatives relating the quantities.

mayday *(Telecomm.)*. Verbal international radio-telephone distress call or signal, corresponding to SOS in telegraphy. Corruption of Fr. *m'aidez*.

May's graticule *(Powder Tech.)*. An eyepiece graticule used in microscopic methods of particle-size analysis, having a rectangular grid for selecting particles, and a series of eight circles for sizing particles. The size of the circle diameters increases by $\sqrt{2}$ progression, and a series of parallel lines is superimposed on the rectangular grid.

maze *(Behav.)*. An apparatus consisting of a series of pathways in a more or less complicated configuration, beginning with a starting box, possibly including blind alleys, and ending in a goal box which generally contains a reward, this not being visible from the starting box. The simplest mazes are the T- and Y-mazes.

maze *(Nuc.Eng.)*. See **radiation trap** (2).
MBR *(Comp.)*. See **memory buffer register**.
McBurney's point *(Med.)*. A point situated on a line joining the umbilicus to the bony prominence of the hip-bone at the upper end of the groin, and 38 mm from the latter; a point of maximum tenderness in appendicitis.
McCabe-Thiele diagram *(Chem.Eng.)*. A graphical method, based on vapour-liquid equilibrium properties, for establishing the theoretical number of separation stages in a continuous distillation process.
McColl protective system *(Elec.Eng.)*. A form of protective system used on electric power networks; it operates on the balanced principle embodying biased beam relays.
McLeod gauge *(Chem.)*. Vacuum pressure gauge in which a sample of low pressure gas is compressed in a known ratio until its pressure can be measured reliably. Used for calibrating direct-reading gauges.
MC, mc *(Genrl.)*. Abbrev. for *Metric Carat*. See **carat**.
McNally tube *(Electronics)*. A reflex klystron capable of being tuned electronically to a wide range of frequencies.
MCPA, MCP *(Chem.)*. *4-chloro-2-methyl-phenoxyacetic acid*: used as a selective weedkiller. Also called *methoxone*.
MCPB *(Chem.)*. *4-(4-chloro-2-methylphenoxy)butanoic acid*, used as a weedkiller.
McQuaid-Ehn test *(Eng.)*. A method of showing grain size after heating a ferrous alloy to the austenitic temperature range. Grain sizes are classified from one (the finest) to eight.
M_{CRIT} *(Aero.)*. See **critical Mach number**.
Md *(Chem.)*. The symbol for *mendelevium*.
m-derived section, network, filter *(Telecomm.)*. A *T* or *pi* network section so designed that when two or more sections are joined to form a filter, their impedances match at all frequencies, although the individual sections may have different resonant frequencies.
MDF *(Telecomm.)*. Abbrev. for *Main Distribution Frame*.
M display *(Radar)*. Modified form of *A display* in which range is determined by moving an adjustable pedestal signal along the baseline until it coincides with the target signal; range is read off the control which moves the pedestal.
Me *(Chem.)*. A symbol for the methyl radical $-CH_3$.
M & E *(Image Tech.)*. Same as **music and effects**.
mean *(Stats.)*. See **expectation**; (of a sample) the arithmetic mean.
mean calorie *(Phys.)*. See **calorie**.
mean chord *(Aero.)*. See **standard mean chord**.
mean curvature *(Maths.)*. See **curvature**.
mean daily motion *(Astron.)*. The angle through which a celestial body would move in the course of 1 day if its motion in the orbit were uniform. It is obtained by dividing 360° by the period of revolution.
meander *(Geol.)*. Sharp sinuous curves in a stream particularly in the mature part of its course. The meanders are accentuated by continuing erosion on the convex side and deposition on the concave side of the stream course.
mean draught *(Ships)*. Half of the sum of the forward and after draughts of a vessel; differs slightly from draught at half length.
mean effective pressure *(Eng.)*. See **brake-, indicated-**.
mean establishment *(Surv.)*. The average value of the lunitidal interval at a place.
mean free path *(Acous.)*. Average distance travelled by a sound wave in an enclosure between wall reflections; required for establishing a formula for reverberation calculations.
mean free path *(Phys.)*. The mean distance travelled by a molecule in a gas between collisions. It is dependent on the molecular cross-section $\pi\sigma^2$ so that

$$\lambda = \frac{1}{\sqrt{2\pi}\, n\sigma^2}$$

where n is the number of molecules per unit volume. According to the kinetic theory of gases it is related to the viscosity η by $\lambda = k\eta/\rho u$ where ρ = density, u = mean molecular velocity and k is a constant between $\frac{1}{3}$ and $\frac{1}{2}$ depending on the approximation in the theory.
mean free time *(Phys.)*. Average time between collisions of electrons with impurity atoms in semiconductors; also of intermolecular collision of gas molecules.
mean hemispherical candle-power *(Phys.)*. The average value of the candle-power in all directions above or below a horizontal plane passing through the source; called the *upper* or *lower* mean hemispherical candle-power according as the candle-power is measured above or below the horizontal plane through the source.
mean horizontal candle-power *(Phys.)*. The average value of the candle-power of a light source in all directions in a horizontal plane through the source.
mean lethal dose *(Radiol.)*. The single dose of whole body irradiation which will cause death, within a certain period, to 50% of those receiving it. Abbrev. *MLD*.
mean life *(Phys.)*. (1) The average time during which an atom or other system exists in a particular form, e.g. for a thermal neutron it will be the average time interval between the instant at which it becomes thermal and the instant of its disappearance in the reactor by leakage or by absorption. Mean life = 1.443 × *half-life*. Also *average life*. (2) The mean time between birth and death of a charge carrier in a semiconductor, a particle (e.g., an ion, a pion), etc.
mean noon *(Astron.)*. The instant at which the mean sun crosses the meridian at upper culmination at any place; unless otherwise specified, the meridian of Greenwich is generally meant.
mean normal curvature *(Maths.)*. See **curvature**.
mean place *(Astron.)*. The position of a star freed from the effects of precession, nutation, and aberration, and of parallax, proper motion, and orbital motion where appreciable. These corrections can be computed for any future date, and when applied to the mean place give the apparent place.
mean residence time *(Phys.)*. Mean period during which radioactive debris from nuclear weapon tests remains in stratosphere.
mean sea level *(Surv.)*. In UK the ordnance survey datum level, determined at Newlyn, Cornwall.
means, inequalities between *(Maths.)*. $A \geq G \geq H$, where A is the arithmetic, G the geometric and H the harmonic mean of n positive numbers.
mean solar day *(Astron)*. The interval, perfectly constant, between two successive transits of the mean sun across the meridian.
mean solar time *(Astron.)*. Time as measured by the hour angle of mean sun. When referred to the meridian of Greenwich it is called Greenwich Mean Time. Before 1925 this began at noon but, by international agreement, is now counted from midnight; it is thus the hour angle of mean sun plus 12 hr, and is identical with *universal time*.
mean-spherical candle power *(Phys.)*. The average value of the candle-power of a light source taken in all directions.
mean-spherical response *(Acous.)*. That of a microphone or loudspeaker taken over a complete sphere, the radius of which is large in comparison with the size of the apparatus. For a loudspeaker, this response (total response) determines the total output of sound power, and therefore, in conjunction with the acoustic properties of an enclosure, the average reverberation intensity in the enclosure. For a microphone, this response is substantially equal to the response for reverberant sound. See **reverberation response, total response**.
mean-square error *(Stats.)*. The expectation of the square of the difference between an estimate of a parameter and its true value, taken with respect to the sampling distribution of the estimate.
mean stress *(Eng.)*. The midpoint of a range of stress. When it is zero, the upper and lower limits of the range have the same value but are in tension and compression respectively.
mean sun *(Astron.)*. A fictitious reference point which has a constant rate of motion and is used in timekeeping in preference to the non-uniform motion of the real sun.

The mean sun is imagined to follow a circular orbit along the celestial equator and is used to measure **mean solar time**.

mean-zonal candle-power *(Phys.)*. The average value of the candle-power of a light source taken in a given zone, the angular limits of the zone being stated.

measles *(Med.)*. *Morbilli*. An acute infectious fever caused by infection with a virus; characterized by catarrh of the respiratory passages, conjunctivitis, **Koplik's spots** and a distinctive rash.

measles of beef *(Vet.)*. Infection of beef by the cyst stage, 'Cysticercus bovis', of the human tapeworm *Taenia saginata*.

measles of pork *(Vet.)*. Infection of pork by the cyst stage, 'Cysticercus cellulosae', of the human tapeworm *Taenia solium*. Hence *measly pork*.

measure *(Print.)*. The width of the type-matter on a page or column, measured in ems.

measured ore *(Min.Ext.)*. Proved quantity and assay grade of ore deposit as ascertained by competent measurement of exposures and an adequate sampling campaign.

measuring chain *(Build.,Surv.)*. See **chain**.

measuring frame *(Build.)*. A wooden box without top or bottom used as a measure for aggregates in mixing concrete.

measuring instrument *(Elec.Eng.)*. A device serving to indicate or record one or more of the electrical conditions in an electrical circuit. Literally, the term also includes integrating meters, but is not generally used in this connection.

measuring tape *(Build.,Surv.)*. See **tape**.

measuring wheel *(Surv.)*. See **perambulator**.

meatotomy *(Med.)*. Incision of the urinary meatus to widen it.

meatus *(Zool.)*. A duct or channel, as the external auditory *meatus* leading from the external ear to the tympanum.

mechanical advantage *(Phys.)*. The ratio of the resistance (or load) to the applied force (or effort) in a **machine**.

mechanical analogue *(Eng.)*. That which can be drawn between mechanical and electrical systems obeying corresponding equations, e.g. mechanical and electrical resonators.

mechanical bond *(Civ.Eng.)*. The use of reinforcing bars with special ribs or after shaping e.g. as bends or hooks, to supplement the bond between concrete and steel.

mechanical characteristic *(Elec.Eng.)*. See **speed-torque characteristic**.

mechanical depolarization *(Elec.Eng.)*. Dissipation, by mechanical means, of the hydrogen bubbles causing polarization of an electrolytic cell.

mechanical deposits *(Geol.)*. Those of sediments which owe their accumulation to mechanical or physical processes.

mechanical efficiency *(Eng.)*. Of an engine, the ratio of the brake or useful power to the indicated power developed in the cylinders, i.e. the efficiency of the engine regarded as a machine. In other types of machines, the mechanical efficiency similarly accounts for the friction losses.

mechanical engineering *(Eng.)*. That branch of engineering concerned primarily with the design and production of all mechanical contrivances or machines, including prime movers, vehicles, machine tools, and production machines.

mechanical equivalent of heat *(Phys.)*. Originally conceived as a conversion coefficient between mechanical work and heat (4.186 joules = 1 calorie) thereby denying the identity of the concepts. Now recognized simply as the specific heat capacity of water, 4.186 kJ/kg.

mechanical equivalent of light *(Phys.)*. The ratio of the radiant flux, in watts, to the luminous flux, in lumens, at the wavelength for which the **relative visibility factor** is a maximum. Its value is about 0.0015 watts per lumen.

mechanical filter *(Image Tech.)*. An arrangement of springs and masses interposed in the drive of a camera or projector, to smooth out variations in the required constant speed.

mechanical finishing *(Textiles)*. A finish obtained by mechanical means such as raising or calendering as distinct from **chemical finishing**.

mechanical impedance *(Phys.)*. In a mechanical vibrating system, the ratio of the force and the velocity it produces. It consists of the *real* part, the mechanical resistance, which represents the power lost in the system, and the reactance which depends on the mass and **compliance** of the system.

mechanical overlay *(Print.)*. An **overlay** prepared without handcutting of which there are several varieties.

mechanical paper *(Paper)*. Paper composed substantially of mechanical woodpulp.

mechanical plating *(Eng.)*. Placing a metallic powder on a surface by hammering during e.g. a tumbling process.

mechanical refrigerator *(Eng.)*. A plant comprising a compressor for raising the pressure of the refrigerant, a condenser for removing the latent heat, a regulating valve for lowering its pressure and temperature by throttling, and an evaporator in which it absorbs heat at a low temperature.

mechanical resonance *(Phys.)*. In a mechanical vibrating system, the enhanced response to a driving force as the frequency of this force is increased through a resonant frequency at which the inertial reactance balances the reactance due to the compliance of the system.

mechanical scanning *(Image Tech.)*. Systems of **scanning** used in early forms of TV, both for the original scene and the displayed image; these included disks with a spiral of holes or lenses (**Nipkow disk**) and **mirror drums** with a series of reflecting facets or prisms.

mechanical stipple, tint *(Print.)*. An aid to the making of line blocks and litho plates. The stipple is chosen from a large variety and applied either to the drawing, to the negative made from it or to the block or plate during preparation. When used in line-colour work, a range of tones can be obtained.

mechanical stoker *(Eng.)*. *Automatic stoker*. A device for stoking or firing a steam boiler. It receives fuel continuously by gravity, carries it progressively through the furnace, and deposits or discharges the ash. See **chain grate stoker, overfeed stoker, underfeed stoker**.

mechanical tissue *(Bot.)*. Tissues such as collenchyma and sclerenchyma whose primary function is the support of the plant body.

mechanical woodpulp *(Paper)*. Pulp produced from wood entirely, or almost entirely, by mechanical means e.g. by grinding logs (groundwood) or by passing wood chips through a refiner (refiner mechanical pulp).

mechanical working *(Eng.)*. Rolling, spinning, pressing, hammering etc. a metal to change its shape.

mechanics *(Phys.)*. The study of forces on bodies and of the motions they produce. See **dynamics**, **kinematics**, **statics**.

mechanomotive force *(Phys.)*. The r.m.s. value of an alternating mechanical force, in newtons, developed in a transducer.

mechanoreceptor *(Med.)*. A sense organ specialized to respond to mechanical stimuli, such as pressure or deformation.

Meckel's diverticulum *(Med.)*. A diverticular outgrowth from the lower end of the small intestine, as a result of the persistence in the adult of the vitelline or yolk-sac duct of the embryo.

meconium *(Med.)*. The green coloured faecal material passed by the newborn infant.

meconium *(Zool.)*. In certain Insects, liquid expelled from the anus immediately after the emergence of the imago; also the first faeces of a new-born child or animal.

media *(Comp.)*. Collective name for materials (tape, disk, paper, cards, etc.) used to hold data.

mediad *(Zool.)*. Situated near, or tending towards, the median axis.

median *(Stats.)*. The central value in a set of observations ordered by value, dividing the ordered set into two equal parts; the argument of the cumulative distribution function of a random variable corresponding to a probability of one half.

mediastinitis *(Med.)*. Inflammation of the tissues of the mediastinum.

mediastinotomy *(Med.)*. The surgical exposure of the mediastinal region in the chest. Now used extensively in cardiac surgery.

mediastinum *(Zool.)*. In higher vertebrates, the mesentery-like membrane which separates the pleural cavities of the two sides ventrally; in Mammals, a mass of fibrous tissue representing an internal prolongation of the capsule of the testis.

medical model *(Behav.)*. The conceptualization of mental disorders as a group of diseases analogous to physical diseases.

medi-, medio- *(Genrl.)*. Prefix from L. *medius*, middle.

Mediterranean fever *(Med.)*. Recurrent *polyserositis*.

Mediterranean fever *(Vet.)*. A disease of cattle and water buffalo in Mediterranean countries due to infection by the protozoon *Theileria annulata*; characterized by fever, ocunasal discharge, anaemia, and diarrhoea, and transmitted by ticks.

medium *(Biol.)*. Substance in which an organism or part exists naturally, e.g. *aqueous medium*, or has been placed experimentally. See also **culture medium, mountant**.

medium *(Build.)*. A liquid or a semiliquid vehicle, such as water, oil, spirit, wax, which makes pigment and other components of paint workable. See **vehicle**.

medium Edison screw-cap *(Elec.Eng.)*. An **Edison screw-cap** having a diameter of approximately 1 in. and approximately 7 threads per inch. Abbrev. *MES*.

medium frequency *(Telecomm.)*. Between 300 KHz and 3 MHz; sometimes known as *hectometric waves*.

medium octavo *(Print.)*. A book-size, $9 \times 5\frac{3}{4}$ in. untrimmed.

medium scale integration *(Comp.)*. See **MSI**.

medium screen *(Image Tech.,Print.)*. The term for half-tones suitable for semi-smooth paper, usually either 100 or 120 lines per inch.

medium voltage *(Elec.Eng.)*. Legally, a voltage which is over 250 volts and not greater than 650 volts.

medium waves *(Telecomm.)*. Vague description of the range of wavelengths from about 200 to 1000 m; used generally for short- and medium-range broadcasting.

medulla *(Bot.)*. (1) Central part of an organ or a thallus. (2) Same as pith. adj. *medullary*. Cf. **cortex**.

medulla *(Textiles)*. Hollow cellular central portion of some animal fibres which is surrounded by the cortex.

medulla *(Zool.)*. The central portion of an organ or tissue, as the *medulla* of the Mammalian kidney; bone-marrow. adj. *medullary*.

medullablastoma *(Med.)*. A malignant and rapidly growing tumour occurring in the cerebellum.

medulla oblongata *(Zool.)*. The hind brain in Vertebates, excluding the cerebellum.

medullary bundle *(Bot.)*. Vascular bundle in the pith.

medullary canal *(Zool.)*. The cavity of the central nervous system in Vertebrates; the central cavity of a shaft-bone.

medullary folds *(Zool.)*. In a developing Vertebrate, the lateral folds of a medullary plate, by the upgrowth and union of which the tubular central nervous system is formed.

medullary plate *(Zool.)*. In a developing Vertebrate, the dorsal platelike area of ectoderm which will later give rise to the central nervous system.

medullary ray *(Bot.)*. *Primary ray*. (1) The interfasicular region. (2) Ray stretching from pith to cortex and deriving from the interfasicular region.

medullary sheath *(Bot.)*. The peripheral layers of cells of the pith. The cells are usually small, sometimes thick-walled, and sometimes more or less lignified.

medullary sheath *(Zool.)*. A layer of white fatty substance (myelin) which, in Vertebrates, surrounds the axons of the central nervous system and acts as an insulating coat.

medullated nerve fibres *(Zool.)*. Axons of the central nervous system which are provided with a **medullary sheath**.

medullated protostele *(Bot.)*. **Protostele** with a central core (medulla) of nonvascular tissue.

medullate, medullated *(Bot.)*. Having a pith.

medusa *(Zool.)*. In metagenetic Coelenterata, a free-swimming sexual individual.

Meehanite *(Eng.)*. High-silicon cast-iron produced by inoculation with calcium silicide. Resistant to acidic corrosion.

meerschaum *(Min.)*. A hydrated silicate of magnesium. It is claylike, and is shown microscopically to be a mixture of a fibrous mineral called parasepiolite and an amorphous mineral β-sepiolite. It is used for making pipes, and was formerly used in Morocco as a soap. Also called *sepiolite*.

meeting post *(Hyd.Eng.)*. The vertical post at the outer side of a lock-gate, which is chamfered so as to fit against the corresponding edge of the other gate of a pair when the gates are shut. Also *mitre post*.

meeting rail *(Build.)*. The top rail of the lower sash, or the bottom rail of the upper sash, of a double-hung window.

meeting stile *(Build.)*. See **shutting stile**.

mega- *(Genrl.)*. Prefix denoting 1 million, or 10^6, e.g. a frequency of 1 *megahertz* is equal to 10^6 Hz; *megawatt* $= 10^6$ watts, *megavolt* $= 10^6$ volts.

megabit *(Comp.)*. Unit of million *bits*, used in large store calculations.

megacolon *(Med.)*. Abnormally large colon.

mega-electron-volt *(Phys.)*. See **Mev**.

megagamete *(Zool.)*. See **macrogamete**.

mega-gramme-röntgen *(Phys.)*. See **gram(me)-röntgen**.

megahertz *(Telecomm.)*. The unit of frequency in which there are one million complete cycles of alternation per second. Used in preference to wavelength when the latter attribute of a wave or oscillation is very short. Abbrev. *MHz*.

megalecithal *(Zool.)*. Said of eggs which contain a large quantity of yolk. Cf. **microlecithal**.

megaloblast *(Zool.)*. An embryonic cell which has a large spherical nucleus and of which the cytoplasm contains haemoglobin, which will later give rise to *erythroblasts* by mitotic division within the blood vessels.

megaloblastic anaemia *(Med.)*. A form of anaemia in which megaloblasts are found in the bone marrow, and often macrocytes in the peripheral blood. The usual cause is a deficiency in Vitamin B_{12} or folic acid.

megamere *(Zool.)*. See **macromere**.

meganucleus *(Zool.)*. See **macronucleus**.

megaparsec *(Astron.)*. Unit used in defining distance of extragalactic objects. $1\,\text{Mpc} = 10^6$ parsec, $= 3.26 \times 10^6$ light years. See **parsec**.

megaphanerophyte *(Bot.)*. A tree over 30 m high.

megaphone *(Acous.)*. A horn to direct the voice. Can include microphone, amplifier and sound reproducer. Then called a *loudhailer*.

megaphyll *(Bot.)*. (1) Leaf, typically associated with a **leaf gap** and often relatively large and with a branching system of veins, assumed to have evolved from a branch system or directly from a **telome** truss, and supposed to be the sort of leaf possessed by seed plants and ferns. (2) Any large leaf. adj. *megaphyllous*. Cf. **microphyll**.

megaripple *(Geol.)*. A sandwave with a wavelength of < 1 m.

megascopic *(Genrl.)*. Visible to the naked eye

megasporangium *(Bot.)*. A sporangium which contains megaspores. It corresponds to ovule or nucellus of seed plant. See also **heterospory**.

megaspore *(Bot.)*. In a heterosporous species, a **meiospore** potentially developing into a female gametophyte.

megasporophyll *(Bot.)*. A leaf-like structure (or a structure thought to be homologous to a leaf) bearing megasporangia. See **sporophyll**. Cf. **microsporophyll**.

megaton *(Phys.)*. Explosive force equivalent to 1 000 000 tons of TNT. Used as a unit for classifying nuclear weapons.

megaureter *(Med.)*. Enormous dilation of a ureter with no demonstrable abnormality.

megavoltage therapy *(Radiol.)*. See **supervoltage therapy**.
megawatt days per tonne *(Nuc.Eng.)*. Unit for energy output from reactor fuel; a measure of burn-up. Expressed as MWd/tonne.
meglip *(Build.)*. A preparation of mastic varnish which has been diluted with linseed oil; used as a medium in the mixing of paints for fine work, e.g. **graining**.
megohmmeter *(Elec.Eng.)*. Portable apparatus for indicating values of high resistance, containing a circuit excited by a battery, transistor converter, or d.c. generator. Not to be used for measuring insulation resistance of all types of capacitor.
Meibomian glands *(Zool.)*. In Mammals, sebaceous glands on the inner surface of the eyelids, between the tarsi and conjunctiva. Also called *tarsal glands*.
meiomerous *(Bot.,Zool.)*. Having a small number of parts. n. *meiomery*.
meionite *(Min.)*. Silicate of aluminium and calcium, together with calcium carbonate, which crystallizes in the tetragonal system. It is an end-member of the isomorphous series forming the scapolite group.
meiosis *(Biol.)*. A process of cell division by which the chromosomes are reduced from the diploid to the haploid number. adj. *meiotic*.
meiospore *(Bot.)*. Spore, commonly one of four, containing a nucleus formed by meiosis and therefore haploid. The spores of bryophytes and vascular plants.
meiotic arrest *(Biol.)*. The waiting at a particular stage of meiosis by the oocyte until some stimulus is received, usually the entry of the sperm.
Meissner effect *(Phys.)*. A superconducting material, cooled below its critical temperature in the presence of an applied magnetic field, expels all magnetic flux from its interior. If the field is applied *after* the substance has been cooled below its critical temperature, the magnetic flux is excluded from the superconductor. These are the Meissner effects and show that the superconductor is perfectly diamagnetic. See **superconducting levitation**, **superconductivity**.
Meissner's corpuscules *(Zool.)*. In Vertebrates, a type of sensory nerve found in the skin, in which the nerve breaks up into numerous branches which surround a core of large cells in a connective-tissue capsule. Probably sensitive to touch.
MEK *(Chem.)*. Abbrev. for *Methyl Ethyl Ketone* (butan-2-one).
Meker burner *(Chem.)*. A Bunsen burner using a wire mesh over its outlet as a stabilizer to support a wide, hot flame.
mel *(Acous.)*. Unit of subjective pitch in sound, a pitch of 131 mels being associated with a simple tone of frequency 131 Hz.
melaconite *(Min.)*. Cupric oxide crystallizing in the monoclinic system. It is a black earthy material found as an oxidation product in copper veins, and represents a massive variety of **tenorite**.
melaena *(Med.)*. The passage of black, pitchlike faeces due to the admixture of altered blood, the result of haemorrhage in the alimentary tract.
melamine-formaldehyde methanal resin *(Plastics)*. A synthetic resin derived from the reaction of melamine with methanal or its polymers. It is thermosetting and much used for moulding and laminating. Also *MF resin*.
Melanex *(Plastics)*. Thin plastic foil used as light shield over scintillation detectors for α-particles.
melange *(Geol.)*. A heterogeneous mixture of rock materials mappable as a rock unit. Mostly of tectonic origin but some are caused by large-scale slumping of sediments.
melange *(Textiles)*. Yarns made from colour-printed top or slivers.
melanin *(Chem.)*. A dark-brown or black pigment occurring in hair and skin. Soluble only in alkali and is formed by the oxidation of tyrosine.
melanism *(Biol.,Ecol.)*. The abnormal situation, caused by over-production of melanin, where a proportion of the individuals in an animal population are black or melanic. See **industrial melanism**.
melanite *(Min.)*. A dark-brown or black variety of andradite garnet containing appreciable titanium.
melan-, melano-, *(Genrl.)*. Prefix from Gk. *melas*, gen. *melanos*, black
melanoblast *(Zool.)*. A special connective tissue cell containing melanin.
melanocratic *(Geol.)*. A term applied to rocks which are abnormally rich in dark and heavy ferro-magnesian minerals (to the extent of 60% or more). See also **leucocratic**.
melanodermia, melanoderma *(Med.)*. See **leucodermia**.
melanoglossia *(Med.)*. Black hairy tongue. An overgrowth of the papillae of the tongue, which are stained as the result either of bacterial action or of chemical action on certain food substances.
melanoma *(Med.)*. A tumour, of variable malignancy, arising from cells in the skin and retina that produce melanin. Common in people of N. European origin in sunny climates.
melanophore *(Zool.)*. A chromatophore containing black pigment.
melanosis *(Med.)*. The abnormal deposit of the pigment melanin in the tissues of the body.
melanosporous *(Bot.)*. Having black spores.
melanterite *(Min.)*. Hydrated ferrous sulphate which crystallizes in the monoclinic system. It usually results from the decomposition of iron pyrites or marcasite. Also called *copperas*.
melanuria *(Med.)*. The presence in the urine of the pigment melanin.
Melastomaceae *(Bot.)*. Family of ca 3000 spp. of dicotyledonous flowering plants (superorder Rosidae). Mostly shrubs and small trees, mostly tropical, important in the South American flora. Of little economic value.
melded fabric *(Textiles)*. A fabric made from **bicomponent fibres** in which the outer component melts at a lower temperature than the core. The cohesion is obtained by subjecting the pressed fibre mass to a temperature at which partial fusion occurs.
melena *(Med.)*. See **melaena**.
melibiose *(Chem.)*. Naturally occurring di-saccharide based on glucose and α-galactose. (*D-galactosyl-α1→6-D-glucose*).
melildosis *(Vet.)*. A glanders-like disease of wild rodents in the tropics. It occasionally effects horses, sheep, goats, and man. Caused by *Pfeifferella pseudomallei*.
melilite *(Min.)*. Calcium magnesium aluminium silicate; crystallizing in the tetragonal system, and occurring in alkaline igneous and contact metamorphosed rocks, and slags.
Melinex *(Plastics)*. TN for polythene terephthalate (e.g. Terylene) in good film form. Very strong film with extremely good transparency and electric properties. Similar to *Mylar*.
melliphagus, mellivorous *(Zool.)*. Honey-eating.
mellitic acid *(Chem.)*. Made by oxidizing carbon with strong nitric acid. Condensation with 1,3-ol:hydroxybenzene and aminophenols produces phthalein and rhodamine dyestuffs respectively.
meltback transistor *(Electronics)*. A junction transistor in which the junction is formed by allowing the molten doped semi-conductor to solidify.
melting band *(Meteor.)*. A bright horizontal band often observed in vertical cross-section (RHI) **weather radar** displays. It is due to strong reflections from snowflakes which become covered with a film of water as they fall through the 0°C level and begin to melt.
melting point *(Chem.)*. The temperature at which a solid begins to liquefy. Pure metals, eutectics, and some intermediate constituents melt at a constant temperature. Alloys generally melt over a range. Abbrev. mp. Also *fusing point*.
melting point test *(Build.,Civ.,Eng.)*. A test for the determination of the melting point of a bitumen for use in building or roadmaking. Also *fusing point test*.
melting temperature of DNA *(Biol.)*. Heating or raising the pH of DNA causes it to *melt* or separate into single strands. This can be detected by the measurement of

absorbance in the ultraviolet, because the **absorptivity** of single-stranded DNA is some 40% higher than that of its parent duplex. The melting temperature also gives information about base composition because GC pairs are more thermally stable than AT.

melton *(Textiles)*. A strong heavily milled and cropped fabric used for making overoats. Made from pure wool or cotton warp and woollen weft and having a felt-like appearance.

melt-spinning *(Textiles)*. The formation of continuous filaments by extrusion of molten polymer. A subsequent drawing process is necessary to develop the desired mechanical properties, including strength, in the filaments.

member *(Bot.)*. Any part of a plant considered from the standpoint of morphology.

member *(Build.)*. (1) A constituent part of a structural framework. (2) A division of a moulding.

member *(Zool.)*. An organ of the body, especially an appendage.

membrana *(Zool)*. A thin layer or film of tissue; a membrane.

membrana tectoria *(Zool.)*. A soft fibrillated membrane overlying Corti's organ.

membrana tympani *(Zool.)*. A thin fibrous membrane forming the tympanum or ear-drum.

membrane *(Biol.)*. (1) A thin sheet-like structure, often fibrous, connecting other structures or covering or lining a part. (2) *Cell membrane, unit membrane*; a sheet (say 10 nm thick), composed characteristically of a bimolecular leaflet of lipid and protein, enclosing a cell, an organelle or a vacuole. See also **tonoplast, pit membrane**.

membrane filter *(Chem.)*. Thin layer filter made by fusing cellulose ester fibres or by β-bombardment of thin plastic sheets, so that they are perforated by tiny uniform channels. Also *molecular filter*.

membranella *(Zool.)*. In *Ciliophora*, an undulating membrane formed of two or three rows of cilia.

memory *(Comp.)*. See **store**.

memory address register *(Comp.)*. A central processor register which holds the **address** of the next store location to be referenced. Abbrev. *MAR*.

memory buffer register *(Comp.)*. A central processor register which acts as a **buffer** for all data transfers to and from main store. Abbrev. *MBR*.

memory capacity *(Comp.)*. The amount of information, usually expressed by the number of **words**, which can be retained by a **memory**. It may also be expressed in **bits**. Also *data storage*. See also **access to store**.

memory location *(Comp.)*. See **store location**.

memory register *(Comp.)*. A **register** in the storage of a computer

memory span *(Behav.)*. The number of items a person can recall after just one presentation.

memory trace *(Behav.)*. The inferred change in the nervous system that persists after learning or experiencing something.

menarche *(Med.)*. The first menstruation.

Mendeleev's law *(Phys.)*. 'Law of Octaves'. If elements are listed according to increasing atomic weight, their properties vary but show a general similarity at each period of eight true rises.

Mendeleev's table *(Chem.)*. See **periodic system**.

mendelevium *(Chem.)*. Manmade element, symbol Md, at. no. 101, principal isotope ^{258}Md, half-life 54 days. Named after Mendeleev, Russian chemist, associated with the *periodic table*.

Mendelian character *(Biol.)*. A character determined by a single gene and inherited according to *Mendel's laws*.

Mendelian genetics *(Biol.)*. The genetics of characters determined by single genes with effects large enough to be easily recognizable.

Mendel's laws *(Biol.)*. Laws dealing with the mechanism of inheritance. 1st law (law of segregation) states (in modern terminology) that the two alleles received one from each parent are segregated in gamete formation, so that each gamete receives one or the other with equal probability. This results in various characteristic ratios in the progeny depending on the parental genotypes and dominance. 2nd law (law of recombination) states that two characters determined by two unlinked genes are recombined at random in gamete formation, so that they segregate independently of each other, each according to the 1st law. This results in various dihybrid ratios.

mending *(Print.)*. A corrected piece which is inserted in a printing plate.

mending *(Textiles)*. Making good any imperfection in cloth caused during weaving by yarn breakages, etc. or knots. Some faults are made good by hand sewing.

Ménière's disease *(Med.)*. *Labyrinthine vertigo*. A disorder characterized by attacks of dizziness, buzzing noises in the ears, and progressive deafness; due to chronic disease of the labyrinth of the ear.

menilite *(Min.)*. An alternative and more attractive name for **liver opal**; it is a grey or brown variety of that mineral.

meninges *(Zool.)*. In Vertebrates, envelopes of connective tissue surrounding the brain and spinal cord. sing. *meninx* (Gr. membrane).

meningioma *(Med.)*. A tumour of the meninges of the brain, and, more rarely, of the spinal cord.

meningism, meningismus *(Med.)*. The presence of the symptoms of meningitis in conditions in which the meninges are neither diseased or inflamed.

meningitis *(Med.)*. Inflammation of the **meninges** characterized by severe headache, **photophobia** and neck stiffness.

meningocele *(Med.)*. The hernial protrusion of the meninges through some defective part of the skull or the spinal column.

meningo-encephalitis *(Med.)*. Inflammation of the meninges and of the brain substance.

meningo-encephalocele *(Med.)*. A hernia of the meninges and brain through some defect in the skull.

meningomyelitis *(Med.)*. Inflammation of the meninges and of the spinal cord.

meningomyelocele *(Med.)*. Hernial protrusion of meninges and spinal cord through a defect in the spinal column.

meningovascular *(Med.)*. Pertaining to, or affecting, the meninges and the blood vessels (especially of the nervous system).

meniscus *(Phys.)*. The departure from a flat surface where a liquid meets a solid, due to surface tension. The effect can be seen clearly through the wall of a glass tube. The surface of water in air rises up the wall of the tube, while that of mercury is depressed.

meniscus *(Zool.)*. A small interarticular plate of fibrocartilage which prevents violent concussion between two bones; as the intervertebral disks of mammals.

meniscus lens *(Phys.)*. One which is convex on one face and concave on the other, the radius of curvature of the concave face being greater than that of the convex face.

meniscus telescope *(Astron.)*. A compact instrument, developed by Marksutov in 1941, in which the spherical aberration of a concave spherical mirror is corrected by a meniscus lens. It differs from the Schmidt type in having a correcting plate with 2 spherical surfaces.

menopause *(Med.)*. The natural cessation of menstruation in women.

menorrhagia *(Med.)*. Excessive loss of blood owing to increased discharge during menstruation.

mensa *(Zool.)*. The biting surface of a tooth.

menstruation *(Zool.)*. The periodical discharge from the uterus occurring from puberty to menopause. Also *xenomenia*.

mental age *(Behav.)*. A score devised by Binet to represent a child's test performance; it indicates the average age (50 percent of that age group) of those children who pass the same items as the person whose mental age is being computed, expressed in terms of years and months. See **intelligence quotient**.

mental retardation *(Behav.)*. A condition characterized by a very low intellectual ability and a serious inability to cope with the environment.

mental set *(Behav.)*. The tendency to view new problems in the same fashion as old problems; also a predisposition to perceive or think in one way rather than another.

menthol *(Med.)*. A camphor compound: The (−)isomer is the chief constituent of peppermint oil. It is used as an antiseptic, and local analgesic.

mentum *(Zool.)*. In higher Vertebrates, the chin; in some Gastropoda, a projection between the head and foot; in Insects, the distal sclerite forming the basal portion of the labium, situated between the submentum and the prementum. adj. *mental*.

menu *(Comp.)*. Displays option choices on a screen for an interactive program.

mepacrine *(Med.)*. A yellow bitter powder, formerly widely used in the prophylaxis and treatment of malaria; replaced by **chloroquine**.

meprobamate *(Med.)*. Tranquillizing drug, used in motion sickness and alcoholism.

meralgia paraesthetica *(Med.)*. An affection of the nerve supplying the skin of the front and outer part of the thigh; characterized by pain, tingling, and/or numbness.

meranti *(For.)*. A Malayan hardwood, from the genus *Shorea*. Three varieties are recognized, red, yellow and white.

Mercalli scale *(Geol.)*. A scale of intensity used to measure earthquake shocks. The various observable movements are graded from 1 (very weak) to 12 (total destruction).

mercaptans *(Chem.)*. *Thio-alcohols*. General formula, R.SH. They form salts with sodium, potassium and mercury, and are formed by warming alkyl halides or sulphates with potassium hydrosulphide in concentrated alcoholic or aqueous solution. Ethyl mercaptan, (ethanethiol) C_2H_5SH; is a liquid of nauseous odour; bp 36°C; it is readily oxidized to ethyl disulphide $(C_2H_5)_2S_2$, by exposure to the air. It is an intermediate for *sulphonal* and for rubber accelerators.

mercaptides *(Chem.)*. The salts of mercaptans.

Mercator's projection *(Geol.)*. A derivative of the simple *cylindrical projection*. The meridian scale is adjusted to coincide with the latitudinal scale at any given point, allowing true representation of shape but not of area, regions towards the poles being exaggerated in size. Long used by navigators, for straight lines drawn on it represent a constant bearing. A transverse form of Mercator is sometimes used in large-scale maps of limited E.-W. extent.

mercerization *(Textiles)*. A process which greatly increases the lustre of cotton yarns and fabrics. It consists of treating the material under tension with aqueous sodium hydroxide (caustic soda) which causes swelling and results in increased strength and dye absorption properties. Invented by John Mercer in 1844.

merchant convertor *(Textiles)*. A supplier of finished fabric who buys loom-state material and commissions its further processing.

merchant iron *(Eng.)*. Bar-iron made by repiling and rerolling puddled bar. All wrought-iron is treated in this way before being used for manufacture of chains, hooks etc.

mercurial pendulum *(Horol.)*. A compensation pendulum in which mercury is used as the compensating medium.

mercuric (mercury (II)) chloride *(Chem.)*. *Corrosive sublimate*. $HgCl_2$. Prepared by sublimation from a mixture of mercury (II) sulphate and sodium chloride

mercuric (mercury (II)) iodide *(Chem.)*. HgI_2. See also **Nessler's solution**.

mercurous (mercury (I)) chloride *(Chem.)*. See **calomel**.

Mercury *(Astron.)*. The planet closest to the sun, with a diameter 4880 km, orbital period 88 days and rotation period 59 days. The surface is heavily cratered and there is no true atmosphere, although traces of helium and argon from the **solar wind** or rocks are present. Its relative density of 5.5 is almost as large as the Earth's and may indicate an iron core. Daytime temperatures reach 450°C, falling to −180°C at night.

mercury *(Chem.)*. A white metallic element which is liquid at atmospheric temperature. Chemical symbol Hg, at. no. 80, r.a.m. 200.59, rel.d. at 20°C 13.596, mp −38.9°C, bp 356.7°C, electrical resistivity 95.8×10^{-8} ohm metres. A solvent for most metals, the products being called *amalgams*. Its chief uses are in the manufacture of batteries, drugs, chemicals, fulminate and vermilion. Used as metal in mercury-vapour lamps, arc rectifiers, power-control switches, and in many scientific and electrical instruments. Also called *quicksilver*. ^{198}Hg is a mercury isotope made from gold in a reactor, for use in a quartz mercury-arc tube, light from which has an exceptionally sharp green line, because of even mass number of a single isotope. This considerably improves comparisons of end-gauges in interferometers.

Mercury *(Space)*. Name given to the project which resulted in the first US manned orbital flight by John Glenn, Feb. 20 1962.

mercury arc *(Electronics)*. An electrical discharge though ionized mercury vapour, giving off a brilliant blue-green light containing strong ultraviolet radiation. Used in hot- and cold-cathode tubes, for power switching and for illlumination in fluorescent tubes, the ultraviolet radiation is used to excite visible radiation from a phosphor coating on the tube wall. See also **mercury-vapour tube**, **mercury-vapour lamp**.

mercury-arc converter *(Elec.Eng.)*. A *converter* making use of the properties of the *mercury-arc rectifier*.

mercury-arc rectifier *(Elec.Eng.)*. One in which rectification arises from the differential migration of electrons and heavy mercury ions in a plasma, formed by evaporation from a *hot spot* on a cathode pool of mercury. This vaporization has to be started by withdrawing an electrode from this pool, thus generating an arc. Used for a.c./d.c. conversion for railways, trams etc. before the introduction of silicon rectifiers. See **thyratron**.

mercury barometer *(Meteor.,Phys.)*. An instrument used for measuring the pressure of the atmosphere in terms of the height of a column of mercury which exerts an equal pressure. In its simplest form, it consists of a vertical glass tube about 80 cm long, closed at the top and having its lower open end immersed in mercury in a dish. The tube contains no air, the space above the mercury column being known as a *Torricellian vacuum*.

mercury cell *(Chem.)*. (1) Electrolytic cell with mercury cathode. (2) Dry cell employing mercury electrode, e.m.f. ca. 1.3 volts; e.g. *Mallory battery*, *Reuben-Mallory cell*.

mercury delay line *(Telecomm.)*. One in which mercury is used as the medium for sound transmission, conversion from and to electrical energy being through suitable transducers at the ends of the mercury column.

mercury discharge lamp *(Electronics)*. See **mercury arc**.

mercury fulminate *(Chem.)*. A crystalline solid, $Hg(ONC)_2$, prepared from ethanol, mercury and nitric acid; used as an initiator in detonators and percussion caps.

mercury intrusion method *(Chem.)*. Finding the distribution of sizes of capillary pores in a body by forcing in mercury, the radius being found from the pressure, and the percentage from the volume of mercury absorbed at each pressure.

mercury-pool cathode *(Electronics)*. Cold cathode in valve where arc discharge releases electrons from surface of mercury pool, e.g. in **mercury-arc rectifier**.

mercury seal *(Chem.)*. A device which ensures that the place of entry of a stirrer into a piece of apparatus is gas-tight, while allowing the free rotation of the stirrer.

mercury switch *(Elec.Eng.)*. A switch in which the fixed contacts consist of mercury cups into which the moving contacts dip, or in which the mercury is contained in a tube which is made to tilt, thereby causing the mercury to bridge the contacts.

mercury-vapour cycle *(Eng.)*. The use of mercury in a closed loop for external combustion turbines. It has certain advantages over water, but is not widely used.

mercury-vapour lamp *(Electronics)*. Quartz tube containing a mercury arc; used to provide ultraviolet rays for therapeutic and cosmetic treatment.

mercury-vapour pump *(Chem.Eng.)*. See **Gaede diffusion pump**.

mercury-vapour rectifier *(Elec.Eng.)*. See **mercury-arc rectifier**.

mercury-vapour tube *(Electronics)*. Generally, any device in which an electric discharge takes place through mercury vapour. Specifically, a triode valve with mercury vapour, which is ionized by the passage of electrons, and reduces the space charge and the anode potential necessary to maintain a given current. The grid is effective in controlling the start of the discharge. See **thyratron**.

merge *(Comp.)*. Combine two or more ordered files into a single ordered file.

mericlinal chimaera *(Bot.)*. A chimaera in which one component does not completely surround the other; an incomplete *periclinal chimaera*.

meridian *(Astron.)*. That great circle passing through the poles of the celestial sphere which cuts the observer's horizon in the north and south points, and also passes through his zenith. Also *meridian of longitude*.

meridian circle *(Astron.)*. A telescope mounted on a horizontal axis lying due east and west, so that the instrument itself moves in the meridian plane. It is used to determine the times at which stars cross the meridian, and is equipped with a graduated circle for deducing declinations. Also called *transit circle*.

meridional *(Zool.)*. Extending from pole to pole; as *meridional furrow* in a segmenting egg.

merino wool *(Textiles)*. Wool of fine quality from merino sheep or any other fine wove. The name is also used in the woollen trade for wool fibre recovered from fine woollen and worsted clothing rags.

Merioneth *(Geol.)*. The youngest epoch of the Cambrian period.

meristele *(Bot.)*. A strand of vascular tissue, enclosed in a sheath of endodermis, forming part of a dictyostele.

meristem *(Bot.)*. A group of actively dividing cells including **initials** and their undifferentiated derivatives. See also **apical meristem**, **intercalary meristem**, **lateral meristem**.

meristem culture *(Bot.)*. *Shoot tip culture*. The aseptic culture on a suitable medium of excised shoot apical meristems including one or a few leaf primordia. Used for **micropropagation** and to obtain virus-free plants from virus-infected stocks because meristems are normally virus-free. See also **tissue culture**.

meristic *(Zool.)*. Segmented; divided up into parts; pertaining to the number of parts, as in *meristic variation*. See also **merome**.

meristic variation *(Bot.,Zool.)*. Variation in the number of organs or parts; as in the number of body somites of a metameric animal.

merlons *(Arch.)*. The projecting parts of a battlement.

mermaid's purse *(Zool.)*. A popular name applied to the horny purselike capsule in which the eggs of certain Selachian Fish (Sharks, Dogfish, Skates, Rays) are enclosed.

mero- *(Genrl.)*. Prefix from Gk. *meros*, part.

meroblastic *(Zool.)*. Said of a type of ovum in which cleavage is restricted to a part of the ovum, i.e. is incomplete, usually due to the large amount of yolk. Cf. **holoblastic**.

merogamy *(Bot.)*. Having gametes smaller than the vegetative cells and produced by a special division; union of such gametes.

merogenesis *(Zool.)*. Segmentation, formation of parts.

merogony *(Zool.)*. See **schizogony**.

merome *(Zool.)*. A body somite or segment of a metameric animal.

meromorphic function *(Maths.)*. See **analytic function**.

meroplankton *(Ecol.)*. Organisms which spend only part of their life history in the plankton. Cf. **holoplankton**.

meros *(Arch.)*. The surfaces between the channels in a triglyph. Also called *femur*.

merosmia *(Med.)*. Temporary deficiency of the sense of smell which may be organic in nature (e.g. sinusitis) or a result of hysteria.

merosthenic *(Zool.)*. Having the hind limbs exceptionally well developed, as Frogs, Kangaroos.

merozoite *(Zool.)*. In *Protozoa*, a young trophozoite which results from the division of the schizont.

Merulius lacrymans *(Build.)*. See **Serpula lacrymans**.

merycism *(Med.)*. *Rumination*. The return, after a meal, of gastric contents to the mouth; they are then chewed and swallowed once more.

Merz-Hunter protective system *(Elec.Eng.)*. See **split-conductor protection**.

Merz-Price protective system *(Elec.Eng.)*. A form of balanced protective system for electric-power networks, in which the current entering a section of the network is balanced against that leaving it. If a fault occurs on the section this balance is upset, and a relay is caused to operate and trip circuit-breakers to clear the faulty section from the network.

Merz slit *(Phys.)*. A variable-width bilateral slit for spectrographs. Characterized by the fixed position of the centre of the slit.

mesa *(Electronics)*. Type of transistor in which one electrode is made very much smaller than the other, to control bulk resistance. Also, by selective etching, the base and emitter are raised above the region of the collector.

mesa *(Geol.)*. A flat-topped, steep-sided tableland. Small mesas are called *buttes*.

mesarch *(Bot.)*. A strand of xylem having the protoxylem in the centre with metaxylem developing centripetally and centrifugally.

mesa transistor *(Electronics)*. A transistor using the *mesa* method of construction; the semiconductor wafer is etched down in steps so that the base and emitter regions appear as plateaus above the collector. Connections are terminated at the edges of the material, instead of coming to the surface, as in *planar* construction.

mesaxonic foot *(Zool.)*. A foot in which the skeletal axis passes down the third digit, as in *Perissodactyla*. Cf. **paraxonic foot**.

mescaline *(Med.)*. A hallucinogenic drug derived from the Mexican cactus, mescal.

mesectoderm *(Zool.)*. Parenchymatous tissue formed from ectoderm cells which have migrated inwards.

mesencephalon *(Zool.)*. The mid-brain of Vertebrates.

mesenchyma *(Zool.)*. Parenchyma; embryonic mesodermal tissue of spongy appearance. adj. *mesenchymatous*.

mesenchyme *(Zool.)*. Mesodermal tissue, comprising cells which migrate from ectoderm, or endoderm, or mesothelium into the blastocoele. Cf. *mesothelium*.

mesenteric *(Zool.)*. Pertaining to the mesenteron; pertaining to a mesentery.

mesenteric caeca *(Zool.)*. Digestive diverticula of the mesenteron in many Invertebrates (e.g. in Arachnida, Crustacea, Echinodermata, Insects).

mesenteron *(Zool.)*. See **mid-gut**.

mesentery *(Zool.)*. In Coelenterata, a vertical fold of the body wall projecting into the enteron. More generally, a fold of tissue supporting part of the viscera. adj. *mesenterial, mesenteric*.

mesethmoid *(Zool.)*. A median cartilage bone of the Vertebrate skull, formed by ossification of the ethmoid plate. Also called *internasal septum*.

mesh *(Build.,Civ.Eng.)*. *Expanded metal* used as a reinforcement for concrete.

mesh *(Telecomm.)*. A complete electrical path (including capacitors) in the component branches of a complex network. Also called *loop*.

mesh connection *(Elec.Eng.)*. A method of connecting the windings of an a.c. electric machine; the windings are connected in series so that they may be represented diagrammatically by a polygon. The **delta connection** is a particular example of this method.

mesh network *(Telecomm.)*. One formed from a number of impedances in series.

mesh voltage *(Elec.Eng.)*. The voltage between any two lines of a symmetrical polyphase system which are consecutive as regards phase sequence. Called *delta voltage* in a three-phase system, and *hexagon voltage* in a six-phase system.

mesiad *(Zool.)*. Situated near, or tending towards, the median plane.

mesial, mesian *(Zool.)*. In the median vertical or longitudinal plane.

mesic atom *(Phys.)*. Short-lived atom in which a negative *muon* has displaced a normal electron.

mesitylene *(Chem.)*. $C_6H_3(CH_3)_3$, *1,3,5-trimethyl-benzene*; a colourless liquid, occurring in coaltar; bp 164°C.

mesityl oxide *(Chem.)*. $(CH_3)_2C=CH.CO.CH_3$; a colourless liquid of peppermint-like odour; bp 122°C. Obtained from propanone by an **aldol condensation** type of reaction induced by ammonia.

mesmerism *(Behav.)*. An early name for *hypnosis*.

mes-, meso- *(Genrl.)*. Prefix from Gk. *mesos*, middle.

meso- *(Chem.)*. (1) Optically inactive by intra-molecular compensation. Abbrev. *m*-. (2) Substituted on a carbon atom situated between 2 hetero-atoms in a ring. (3) Substituted on a carbon atom forming part of an intramolecular bridge.

meso- *(Genrl.)*. Prefix. See **mes-**.

mesobenthos *(Zool.)*. Fauna and flora of the sea-floor, at depths ranging from 100 to 500 fathoms (200 to 1000 m).

mesoblast *(Zool.)*. The mesodermal or third germinal layer of an embryo, lying between the endoderm and ectoderm. adj. *mesoblastic*.

mesoblastic somites *(Zool.)*. In developing metameric animals, segmentally arranged blocks of mesoderm, the forerunners of the somites.

mesocarp *(Bot.)*. The middle, often fleshy, layer of the pericarp.

mesocoele *(Zool.)*. In Vertebrates, the cavity of the mid-brain; mid-ventricle; Sylvian aqueduct.

mesocolloid *(Chem.)*. A particle whose dimensions are 25 to 250 nm containing 100–1000 molecules.

mesoderm *(Zool.)*. See **mesoblast**.

mesogaster *(Zool.)*. In Vertebrates, the portion of the dorsal mesentery which supports the stomach.

mesogloea *(Zool.)*. In Coelenterata, a structureless layer of gelatinous material intervening between the ectoderm and the endoderm.

mesokaryotic *(Bot.)*. *Dinokaryotic*. Having a nucleus with a nuclear envelope and with chromosomes, but having very little histone associated with the DNA of the chromosomes which remain condensed throughout interphase; the condition in the Dinophyseae. Cf. **prokaryotic, eukaryotic**.

mesolecithal *(Zool.)*. A type of egg of medium size containing a moderate amount of yolk which is strongly concentrated in one hemisphere, this being the lower one in eggs floating in water. Found in Frogs, Urodela, Lungfishes, Lower Actinopterygii and Lampreys. Cf. **oligolecithal, telolecithal**.

mesolite *(Min.)*. A zeolite intermediate in composition between **natrolite** and **scolecite**. Crystallizes in the monoclinic system, and occurs in amygdaloidal basalts and similar rocks.

Mesolithic *(Geol.)*. The middle division of the Stone Age. cf. **Palaeolithic, Neolithic**.

mesomerism *(Chem.)*. (1) See **desmotropism**. (2) See **resonance**.

mesometrium *(Zool.)*. The mesentery which supports the uterus and related structures.

mesomorph *(Behav.)*. One of Sheldon's somatotyping classifications; mesomorphs are hard, rectangular and well muscled; delight in physical activity, love adventure and power, and are indifferent to people. See **somatotype theory**.

mesomorphous *(Chem.)*. Existing in a state of aggregation midway between the true crystalline state and the completely irregular amorphous state. See also **liquid crystals**.

meson *(Phys.)*. A hadron with a baryon number of 0. Mesons generally have masses intemediate between those of electrons and nucleons and can have negative, zero or positive charges. Mesons are *bosons* and may be created or annihilated freely. There are three groups of mesons: π-mesons (pions), K-mesons (kaons) and η-mesons.

mesonephric duct *(Zool.)*. See **Wolffian duct**.

mesonephros *(Zool.)*. Part of the kidney of Vertebrates, arising later in development than, and posterior to, the *pronephros*, and discharging into the **Wolffian duct**; beomes the functional kidney in adult anamniotes. adj. *mesonephric*.

meson field *(Phys.)*. That which is concerned with the interchange of protons and neutrons in the nucleus of an atom, mesons transferring the energy.

mesopause *(Meteor.)*. The top of the mesosphere at about 80 to 85 km.

mesophilic bacteria *(Biol.)*. Bacteria which grow best at temperatures of 20–45°C.

mesophyll *(Bot.)*. The ground tissue of a leaf, located between the upper and lower epidermis and typically differentiated as chlorenchyma. See also **palisade** mesophyll, **spongy mesophyll**.

mesophyte *(Bot.)*. A plant adapted to habitats that are neither very wet nor dry. Cf. **hydrophyte, xerophyte**.

mesorchium *(Zool.)*. In Vertebrates, the mesentery supporting the testis.

mesosphere *(Meteor.)*. Region of the atmosphere lying between the **stratopause** and **mesopause** (50–85 km), in which temperature generally decreases with height.

mesosternum *(Zool.)*. In Insects, the sternum of the mesothorax; in Vertebrates, the middle part of the sternum, connected with the ribs; the gladiolus.

mesotarsal *(Zool.)*. In Insects, the tarsus of the second walking leg; in land Vertebrates, the ankle joint or joint between the proximal and distal rows of tarsals.

mesothelium *(Zool.)*. Mesodermal tissue comprising cells which form the wall of the cavity known as the coelom. Cf. **mesenchyme**.

mesothorax *(Zool.)*. The second of the 3 somites composing the thorax in Insects. adj. *mesothoracic*.

mesotrochal *(Zool.)*. Having an equatorial band of cilia.

mesovarium *(Zool.)*. In Vertebrates, the mesentery supporting the ovary.

Mesozoic *(Geol.)*. The era embracing the Triassic, Jurassic, and Cretaceous periods.

message switching *(Comp.)*. Method using sequentially organized switching mechanisms, of batching and storing messages for economical transmission on shared network lines.

messenger wire *(Elec.Eng.)*. A strong suspension wire for holding aerial cables. See **bearer cable**.

Messier catalogue *(Astron.)*. A listing of 108 galaxies, star clusters and nebulae drawn up by the French comet hunter Charles Messier in 1770. Objects are designated M1, M2, and so on. Many of these alphanumeric names are still widely used in Astronomy.

messuage *(Build.)*. A dwelling-house and its adjacent land and buildings.

mestome *(Bot.)*. Conducting tissue, with associated parenchyma, but without mechanical tissue.

mestome sheath, mestom sheath *(Bot.)*. An endodermis-like bundle sheath; the inner of two sheaths round the vascular bundle of some grasses, e.g. wheat.

Met *(Chem.)*. Symbol for **methionine**.

meta- *(Chem.)*. See ***m*-**.

meta- *(Genrl.)*. Prefix from Gk. *meta*, after.

meta- *(Chem.)*. (1) Derived from an acid anhydride by combination with one molecule of water. (2) A polymer of ... (3) A derivation of ... (4) The 1-3 relationship of substituents on a benzene ring, e.g. m-cresol.

meta-aldehyde *(Chem.)*. See **metaldehyde**.

metabolism *(Biol.)*. The sum-total of the chemical and physical changes constantly taking place in living matter. adj. *metabolic*.

metabolite *(Zool.)*. A substance involved in metabolism, being either synthesized during metabolism or taken in from the environment.

metaboly *(Bot.)*. The power possessed by some cells of altering their external form, e.g. as in Euglenida.

metaboric acid *(Chem.)*. See **boric acid**.

metacarpal, metacarpale *(Zool.)*. One of the bones composing the **metacarpus** in Vertebrates.

metacarpus *(Zool.)*. In land Vertebrates, the region of the fore-limb between the digits and the carpus.

metacentre *(Phys.)*. If a vertical line is drawn through the centre of gravity of a body floating in equilibrium in a liquid, and if the body is displaced slightly from its equilibrium position, a second vertical line is drawn through the centre of buoyancy (centre of gravity of the displaced liquid), the two lines meet at a point called the *metacentre*. According to whether this is above or below the centre of gravity of the body, the equilibrium is stable or unstable.

metacentric *(Ships)*. The measure of stability of a vessel at small angles of heel, indicative of its behaviour when rolling.

metacentric height *(Phys.)*. The distance between the centre of gravity of a floating body and its *metacentre*.

metachronal rhythm *(Biol.)*. The rhythmic beat of cilia on a cell surface in which the beat of adjacent cilia is slightly out-of-phase. Consequently it is seen as a series of waves passing over the ciliary surface.

metachrosis *(Zool.)*. The ability, shown by some animals (as the Chameleon) to change colour by expansion or contraction of chromatophores.

metacoele *(Zool.)*. In *Craniata*, the cavity of the hind-brain; the fourth ventricle.

metadiscoidal placentation *(Zool.)*. Having the villi at first scattered and then restricted to a disk as in Primates.

metadyne *(Elec.Eng.)*. See **amplidyne**.

metagenesis *(Zool.)*. See **alternation of generations**.

metal *(Civ.Eng.)*. See **road metal**.

metal-arc welding *(Elec.Eng.)*. A type of electric welding in which the electrodes are of metal, and melt during the welding process to form filler metal for the weld.

metal-clad switchgear *(Elec.Eng.)*. A type of switchgear in which each part is completely surrounded by an earthed metal casing. Cf. **metal-enclosed switchgear**.

metaldehyde *(Chem.)*. *Meta-aldehyde*. $(CH_3CHO)_4$, long glistening needles which sublime at 115°C with partial decomposition into ethanal. Acetaldehyde is polymerized to metaldehyde by the action of acids at temperatures below 0°C. Sometimes used as a portable fuel, *meta-fuel*.

metal detector *(Eng.)*. An instrument, widely used in industrial production, for detecting the presence of embedded stray metal parts in food products. It is usually incorporated in a conveyor line and gives visible or audible warning or automatically stops the line. Also used for detecting buried metal.

metal electrode *(Elec.Eng.)*. A form of electrode used in **metal-arc welding**.

metal-enclosed switchgear *(Elec.Eng.)*. A type of switchgear in which the whole equipment is enclosed in an earthed metal casing. Cf. **metal-clad switchgear**.

metal feeder *(Print.)*. A device which maintains a supply of metal in typesetting machines from an ingot fed at the required rate into the metal pot.

metal filament *(Elec.Eng.)*. A fine metal conductor heated to incandescence to provide illumination or to act as a source of electrons in a vacuum tube.

metal-filament lamp *(Elec.Eng.)*. A filament lamp in which the light is produced by raising a fine wire or filament of metal to white heat.

metal-film resistor *(Elec.Eng.)*. One formed by coating a high-temperature insulator with a very thin layer of metallic film, e.g. by vacuum deposition.

metal-halide lamp *(Image Tech.)*. A compact-source enclosed mercury arc lamp with metal halide additions, for a.c. operation. Abbrev. *MH lamp*.

metal inert-gas welding *(Eng.)*. Arc welding with a metal electrode shielded by an inert gas such as argon or carbon dioxide.

metal insulator *(Telecomm.)*. Waveguide or transmission line an odd number of quarter-wavelengths long, which has a high impedance. Used as a support or anchor without normal insulators at very high frequencies.

metal lathing *(Build.)*. **Expanded metal** used to cover surfaces to provide a basis for plaster.

metallic bond *(Chem.)*. In metals, the valence electrons are not even approximately localized in discrete covalent bonds, but are delocalized and interact with an indefinite number of atomic nuclei. This gives rise both to the opacity and lustre of metals and to their electrical conductivity.

metallic conduction *(Electronics)*. That which describes the movement of electrons which are freely moved by an electric field within a body of metal.

metallic-film resistor *(Elec.Eng.)*. One formed by coating a high-temperature insulator, such as mica, ceramic, Pyrex glass or quartz, with a metallic film.

metallic lens *(Telecomm.)*. One with slats or louvres which give varying retardation to a passing electro-magnetic wave, so that it is controlled or focused. Can also be used for sound waves.

metallic lustre *(Min.)*. A degree of lustre exhibited by certain opaque minerals, comparable with that of polished steel.

metallic packing *(Eng.)*. A **packing** consisting of a number of rings of soft metal, or a helix of metallic yarn encircling the piston rod and pressed into contact therewith by a gland nut.

metallization *(Chem.)*. (1) The vacuum deposition of thin metal films on to glass or plastic, for decorative or electrical purposes. (2) The conversion of a substance, e.g. selenium, into a metallic form.

metallized yarn *(Textiles)*. An effect yarn containing some metallic components which may be separate filaments or fibres made of metal or a thin metal strip (e.g. of anodized aluminium) which is protected by a transparent film (e.g. of cellulose acetate).

metallography *(Eng.)*. The study of metals and their alloys with the aid of various procedures e.g. microscopy, X-ray diffraction etc.

metalloid *(Chem.)*. An element having both metallic and non-metallic properties, e.g. arsenic.

metallo-organic compounds *(Chem.)*. See **organo-metallic compounds**.

metallurgical balance sheet *(Min.Ext.)*. Report in equation form of the products from treatment of a known tonnage of ore of specified assay value, yielding a known weight of concentrate and tailing, to which the head value can be attributed for economic and technical control.

metallurgy *(Eng.)*. The science and technology of metals and their alloys including methods of extraction and use.

metal matrix composite *(Aero.)*. Usually a refractory metal reinforced by a different fibre, e.g. silica reinforced aluminium or boron-titanium.

metal-oxide semiconductor transistor *(Electronics)*. An active semiconductor device in which a conducting channel is induced in the region between the electrodes by applying a voltage to an insulated electrode placed on the surface in this region. It is self-isolating by virtue of its construction, and so can be fabricated in a smaller area than a **bipolar transistor**.

metal pattern *(Foundry)*. A **pattern** made in cast-iron, brass, or light alloy, in order to ensure durability and permanence of form when a large number of castings are required, as in the case of repetition work on moulding machines.

metal (powder) spraying *(Eng.)*. A method of applying protective metal coatings or building up worn parts by spraying molten metal from a gun. The coating metal is supplied as wire or powder, melted by flame, and blown out of the gun as finely divided particles.

metal rectifier *(Elec.Eng.)*. A form of rectifier making use of the rectifying property of a layer of oxide on a metal disk, e.g. copper oxide on a copper disk. A number of such disks can be connected in series or in parallel to give high-voltage or high-current rectifiers. Generally superseded by semi-conductor rectifying devices. Also called *dry plate rectifier*.

metal rule *(Print.)*. See **rule, em rule**.

metals *(Civ.Eng.)*. The rails of a railway.

metal spinning *(Eng.)*. The shaping of thin sheet-metal disks into cup-shaped forms by the lateral pressure of a steel roller or a stick on the revolving disk, which is gradually pressed into contact with a former on the lathe faceplate.

metal trim *(Build.)*. Architraves and other finishings made out of pressed metal sheeting clipped or screwed in position around door or window openings.

metal valley *(Build.)*. A gutter, lined with lead, zinc or copper, between two roof-slopes.

metamere *(Zool.)*. See **merome**.

metameric match *(Image Tech.)*. A subjective match of two colours whose actual spectral composition differs.

metameric segmentation *(Zool.)*. See **metamerism**.

metamerism *(Textiles)*. A marked change in colour of material subjected to different lighting. Thus two such fabrics may match in daylight but appear different when examined in artificial light.

metamerism *(Zool.)*. Repetition of parts along the long axis of an animal. adj. *metameric*.

metamict *(Min.)*. Mineral which has been exposed to natural radioactivity so that its crystalline structure breaks down to a glassy amorphous state, e.g. zircon.

metamorphic facies *(Geol.)*. All those rocks that have reached chemical equilibrium under the same pressure-temperature range of metamorphism.

metamorphism *(Geol.)*. Change in the mineralogical and structural characteristics of a rock as a consequence of heat and/or pressure.

metamorphosis *(Zool.)*. Pronounced change of form and structure taking place within a comparatively short time, as the changes undergone by an animal in passing from the larval to the adult stage. adj. *metamorphic*.

metanephric duct *(Zool.)*. Ureter of amniote Vertebrates.

metanephridia *(Zool.)*. Nephridia which open into the coelom, the open end (nephrostome) being ciliated. Cf. **protonephridium**.

metanephros *(Zool.)*. In amniote Vertebrates, part of the kidney arising later in development than, and posterior to, the mesonephros; becomes the functional kidney, with a special *metanephric duct*. adj. *metanephric*.

metanilic acid *(Chem.)*. $C_6H_4(NH_2)(SO_3H)$, *meta-aminobenzene-sulphonic acid*; an intermediate for dyestuffs.

metaphloem *(Bot.)*. The later-formed *primary phloem*, especially that which matures after the organ has ceased to elongate. Cf. **protophloem**.

metaphosphoric acid *(Chem.)*. HPO_3. Formed as a viscous solid when phosphorus (V) oxide is left exposed to the air.

metaphysis *(Med.)*. The end of the shaft (diaphysis) of a long bone where it joins the epiphysis.

metaplasia *(Zool.)*. Tissue transformation, as in the ossification of cartilage.

metaplasis *(Zool.)*. The period of maturity in the life-cycle of an individual.

metapodium *(Zool.)*. (1) In Vertebrates, the second podial region; metacarpus or metatarsus; palm or instep. (2) In Insects, that portion of the abdomen posterior to the *petiole*. (3) In Gastropoda, the posterior part of the foot. adj. *matapodial*.

metapophysis *(Zool.)*. In some Mammals, a process of the vertebrae above the prezygapophysis which strengthens the articulation.

metaraminol *(Med.)*. A **sympathomimetic** drug given to produce vasoconstriction and to increase blood pressure.

metarchon *(Chem.)*. An agent which, without being toxic, so changes the behaviour of a pest that its persistence is diminished, e.g. a confusing sex attractant.

metasitism *(Zool.)*. Cannibalism.

metasoma *(Zool.)*. In Arachnida, the posterior part of the abdomen, or hindermost tagma of the body, which is always devoid of appendages. adj. *metasomatic*.

metasomatism *(Geol.)*. Change in the bulk chemical composition of a rock by the introduction of liquid or gaseous material from elsewhere.

metastable state *(Chem.)*. State which is apparently stable, often because of the slowness with which equilibrium is attained; said, for example, of a supersaturated solution.

metastasic *(Phys.)*. Said of electrons which move from one shell to another or are absorbed from a shell into the nucleus.

metastasis *(Med.)*. The transfer, by lymphatic channels or blood vessels, of diseased tissue (especially cells of malignant tumours) from one part of the body to another; the diseased area arising from such transfer.

metastasis *(Zool.)*. Transference of a function from one part or organ to another; metabolism.

metastasize *(Med.)*. To form metastases.

metatarsalgia *(Med.)*. A painful neuralgic condition of the foot, felt in the ball of the foot and often spreading thence up the leg.

metatarsal, metatarsale *(Zool.)*. One of the bones composing the **metatarsus** in Vertebrates.

metatarsus *(Zool.)*. In Insects, the first joint of the tarsus when it is markedly enlarged; in land Vertebrates, the region of the hind-limb between the digits and the tarsus.

Metatheria *(Zool.)*. A subclass of viviparous Mammals in which the newly born young are carried in an abdominal pouch which encloses the teats of the mammary glands; an allantoic placenta is usually lacking; the scrotal sac is in front of the penis, the angle of the lower jaw is inflexed, and the palate shows vacuities. Contains only one order, the Marsupiala.

metathorax *(Zool.)*. The third or most posterior of the 3 somites composing the thorax in Insects. adj. *metathoracic*.

metaxenia *(Bot.)*. Any effect that may be exerted by pollen on the tissues of the female organs.

metaxylem *(Bot.)*. The later-formed *primary xylem*, especially that which matures after the organ has ceased to elongate; commonly with reticulately thickened or pitted walls. Cf. **protoxylem**.

Metazoa *(Zool.)*. A subkingdom of the animal kingdom, comprising multicellular animals having two or more tissue layers, never possessing choanocytes, usually having a nervous system and enteric cavity, and always showing a high degree of coordination between the different cells composing the body. Cf. **Protozoa**, **Parazoa**.

metecdysis *(Zool.)*. The period after a moult in *Arthropoda* when the new cuticle is hardening and the animal is returning to normal in its physiological condition.

metencephalon *(Zool.)*. The anterior portion of the hind-brain in Vertebrates, developing into the cerebellum, and, in Mammals, the **pons** (part of the **medulla oblongata**). Cf. **myelencephalon**.

meteor *(Astron.)*. A 'shooting star'. A small body which enters the Earth's atmosphere from interplanetary space and becomes incandescent by friction, flashing across the sky and generally ceasing to be visible before it falls to Earth. See also **bolide**.

meteor craters *(Astron.)*. Circular unnatural craters of which Meteor Crater in Arizona is best known; believed to be caused by the impact of meteorites.

meteoric shower *(Astron.)*. A display of meteors in which the number seen per hour greatly exceeds the average. It occurs when Earth crosses the orbit of a meteor swarm and the swarm itself is in the neighbourhood of the point of section of the two orbits.

meteoric water *(Geol.)*. Ground-water of recent atmospheric origin.

meteorism *(Med.,Vet.)*. Excessive accumulation of gas in the intestines. See also **tympanites**.

meteorite *(Astron.)*. Mineral aggregates of cosmic origin which reach the Earth from interplanetary space; cf. **meteor** and **bolide**. See **achondrite, aerolite, chondrite, iron meteorite, pallasite, siderite, stony meteorite**.

meteorites *(Astron.,Min.)*. Mineral aggregates of cosmic origin which reach the earth from interplanetary space; cf. **meteor** and **bolide**. See **achondrite, aerolites, chondrite, iron meteorites, pallasite, siderite, stony meteorites**.

meteorological satellite *(Meteor.)*. An artificial satellite orbiting the Earth, either in the equatorial plane at such a distance (36 000 km) that it is stationary relative to the Earths's surface (*geostationary*), or in a plane inclined at a small angle to the Earth's axis which precesses slowly in space so that the phase of the orbit is constant relative to the Sun (*sun-synchronous* or *polar orbiting*). Sun-synchronous satellites orbit at heights of about 1000 km and cross the equator about 15 times a day in each direction, always at the same local times. Meteorological satellites carry television cameras and radiometers which transmit cloud pictures and measurements of terrestrial and solar radiation in various wavebands. These yield estimates of the vertical temperature structure of the atmosphere.

meteorology *(Genrl.)*. The study of the earth's atmosphere in its relation to weather and climate.

meteor stream *(Astron.)*. Streams of dust revolving about the Sun, whose intersection by the Earth causes meteor shows. Some night-time showers have orbits similar to those of known comets; day-time showers, detected by radio-echo methods, have smaller orbits, similar to those of minor planets.

meter *(Elec.Eng.)*. A general term for any electrical measuring instrument, but usually confined to integrating meters.

meter *(Genrl.)*. US spelling of **metre**.

meter protection circuit *(Elec.Eng.)*. One designed to avoid transient overloads damaging a meter. It may employ a gas-discharge tube which breaks down at a dangerous voltage or a Zener diode or similar device.

methadone *(Med.)*. Methadone hydrochloride. An analgesic with similar but less marked properties, both good and bad, than morphine.

methaemoglobin *(Chem.)*. A compound of haemoglobin and oxygen, more stable than oxyhaemoglobin, obtained by the action of oxidizing agents on blood, e.g. nitirites and chlorates.

methaemoglobinaemia *(Med.)*. Haemoglobin, modified by drugs or hereditary defect to form methaemoglobin, is incapable of carrying oxygen. Patients are blue and have shortness of breath.

methaemoglobinuria *(Med.)*. The presence of methaemoglobin in the urine.

methanal *(Chem.)*. See **formaldehyde**.

methane *(Chem.)*. CH_4. The simplest **alkane**, a gas, mp $-186°C$, bp $-164°C$; occurs naturally in oil-wells and as marsh gas. *Fire-damp* is a mixture of methane and air; coal-gas contains a large proportion of methane. It can be synthesized from its elements, and prepared by various methods, as by catalytic reduction of CO or CO_2, or by passing CO and H_2O over heated metal oxides, or by the action of water on aluminium carbide (methanide).

methanides *(Chem.)*. Carbides, such as aluminium and beryllium carbides, which give methane when decomposed by water.

methanoic acid *(Chem.)*. See **formic acid**.

methanol *(Chem.)*. CH_3OH; a colourless liquid, bp 66°C, rel.d. 0.8. It may be produced by the destructive distillation of wood; is nowadays synthesized from CO and H_2 in the presence of catalysts. It is an important intermediate for numerous chemicals, and is used as a solvent and for denaturing ethanol. Also called *methyl* or *wood alcohol*.

methanol *(Med.)*. *Methyl alcohol*; when taken it is poisonous, damaging the optic nerve to produce blindness and also producing a severe *metabolic acidosis*.

methene *(Chem.)*. See **methylene**.

methine *(Chem.)*. The trivalent radical $CH\equiv$.

methine dyes *(Chem.)*. Group of dyestuffs important in the development of photography. Consist mainly of dyes containing two quinoline or benzthiazole groups joined by conjugated aliphatic chains.

methionine *(Chem.)*. *2-amino-5-thio-hexanoic acid*, $CH_3.S.(CH_2)_2.CH(NH_2).COOH$. The L- or S- isomer is an essential amino acid for protein synthesis. Symbol Met, short form M. ⇨

method study *(Eng.)*. A branch of work study, concerned with determining the best production methods and the corresponding equipment for making an article in the desired quantities at optimum quality and cost.

methoxone *(Chem.)*. See MCPA.

methoxychlor *(Chem.)*. *1,1,1-Trichloro-2,2-di-(4-methoxyphenyl)ethane*, used as an insecticide; also known as *methoxy-DDT*, *DMDT* and *dianysl-trichloroethane*.

methoxyl group *(Chem.)*. The monovalent radical $-OCH_3$. In certain compounds it can be estimated analytically by *Zeisel's method*.

Methylal *(Chem.)*. $(CH_3O)_2CH_2$, a colourless liquid; bp 42°C, widely used as a solvent.

methyl alcohol *(Chem.)*. *Methanol*.

methylamines *(Chem.)*. *Mono-*, $CH_3.NH_2$; *di-* $(CH_3)_2NH$; and *tri-*, $(CH_2)_2N$, may be regarded as ammonia in which one or more of the hydrogen atoms is replaced by the methyl group. Ammoniacal liquids with fishy smell, occurring in herring brine. Used in manufacturing drugs, dyes etc.

methylated spirit *(Chem.)*. See **denatured alcohol**.

methylation *(Biol.)*. The addition of a methyl ($-CH_3$) group to a nucleic acid base, usually cytosine or adenine. In bacteria this can protect the site against cleavage by the appropriate restriction enzyme. In eukaryotes the transcription of certain genes is inhibited by DNA methylation, although the function is not fully understood.

methylation *(Chem.)*. The introduction of methyl (CH_3) groups into organic compounds. Carried out by using methylating agents such as diazomethane or dimethyl sulphate.

methylbenzene *(Chem.)*. See **toluene**.

methyl bromide *(Chem.)*. CH_3Br. *Bromomethane*. Bp 4°C, rel.d. (0°C) 1.732. Organic liquid widely used as a fire extinguishing medium.

methyl cellulose *(Chem.)*. An **ether** prepared from **cellulose** which gives highly viscous solutions in water. It is used in distempers and other water paints, foodstuffs, cosmetics etc.

methyl cellulose *(Med.)*. A bulking agent used to increase faecal mass and relieve constipation.

methyl chloride *(Chem.)*. CH_3Cl. Chloromethane. Bp $-23.7°C$, rel.d. (0°C) 0.952. Used as local anaesthetic and refrigerant.

methylcyclohexanol (IV) *(Chem.)*. The methyl derivative of *cyclohexanol*, actually a mixture of the 2-, 3- and 4-isomers. Used widely as a solvent for fats, waxes, resins and in lacquers. Soaps prepared from it are used as detergents. Also *hexahydrocresol*, *sextol*.

methyldopa *(Med.)*. Centrally and peripherally active drug used to treat hypertension. Use becoming less common as newer drugs are introduced.

methylene *(Chem.)*. The group $—CH_2—$.

methylene blue *(Chem.)*. A thiazine dyestuff prepared by oxidizing dimethylaniline in the presence of sodium thiosulphate and zinc chloride. It is fixed to cotton with the aid of tannin.

methylene chloride *(Chem.)*. *Dichloromethane*. CH_2Cl_2. Organic solvent widely used in paint strippers. Bp 41°C. By-product in the manufacture of trichloromethane.

methylene iodide *(Chem.)*. *Di-iodomethane*. CH_2I_2. High relative density (3.32) organic liquid. Used in flotation processes for ore separation and in the density determination of minerals.

methyl ethyl ketone *(Chem.)*. *Butan-2-one*. $CH_3.CO.C_2H_5$; a colourless liquid of ethereal odour, bp 81°C; prepared by the oxidation of butan-2-ol; an important solvent. Its peroxide is widely used as a catalyst for curing polyester resins. Abbrev. *MEK*.

methyl group *(Chem.)*. Monovalent radical $CH_3—$.

methyl iodide *(Chem.)*. CH_3I. *Iodomethane*. Bp 42.5°C. Prepared by adding iodine to methanol and red phosphorus. Useful reagent for the preparation of methyl ethers by reaction with primary alcohols in the presence of silver (I) oxide.

methyl methacrylate resins *(Plastics)*. See **acrylic resins**.

methylol group *(Chem.)*. The group $.CH_2OH$. Found in primary alcohols and glycols.

methyl orange *(Chem.)*. The sodium salt of helianthine, $(CH_2)_2N.C_6H_4.N=N.C_6H_4.SO_2O^-Na^+$. It is a chrysoidine dye, and is used as an indicator in volumetric analysis.

methyl-pyridines *(Chem.)*. See **picolines**.

methyl-rubber *(Chem.)*. The polymerization product of 2,3-dimethylbutadiene, $CH_2=C(CH_3).C(CH_3)=CH_2$, one of the first synthetic rubbers, much inferior to the natural product. It oxidizes easily, and can be vulcanized only by the addition of organic catalysts.

methyl salicylate *(Chem.)*. The main constituent of oil of wintergreen, with a characteristic aromatic odour. Used as an ointment or liniment for treatment of rheumatism.

methyl sulphate *(Chem.)*. $(CH_3)_2SO_4$; a colourless syrupy oil, very poisonous, bp 188°C; used for introducing the methyl group into phenols, alcohols and amines.

methyl violet *(Chem.)*. A triphenylmethane dyestuff consisting of a mixture of the hydrochlorides of tetra-, penta- and hexamethyl-pararosaniline.

methysergide *(Med.)*. Antiserotinergic drug occasionally used to prevent migraine but can cause adverse effects including retroperitoneal fibrosis.

Metis *(Astron.)*. A tiny natural satellite of **Jupiter**, discovered in 1979 by Voyager 2 mission.

metoecious *(Bot.,Zool.)*. Heteroecious.

metoestrus *(Zool.)*. In Mammals, the recuperation period after oestrus.

metol *(Image Tech.)*. See **4-methylaminophenol**.

Metonic cycle *(Astron.)*. A period of 19 years, which is very nearly equal to 235 synodic months, this relationship having been introduced in Greece in 433 BC by the astronomer Meton; its effect is that after a full cycle the phases of the Moon recur on the same days of the year.

metope *(Arch.)*. A slab or tablet (generally of marble and ornamental) between the triglyphs in a Doric frieze.

metoxenous *(Bot.,Zool.)*. Heteroecious.

metre *(Phys.)*. The Système Internationale (SI) fundamental unit of length. The metre is defined (1983) in terms of the velocity of light. The metre is the length of path travelled by light in vacuum during a time interval of 1/299 792 458 of a **second**. Originally intended to represent 10^{-7} of the distance on the earth's surface between the North Pole and the Equator, formerly it has been defined in terms of a line on a platinum bar and later (1960) in terms of the wavelength from ^{86}Kr.

metre bridge *(Elec.Eng.)*. A Wheatstone bridge in which one metre of resistance wire, usually straight, with a sliding contact is used to form two variable ratio arms in a Wheatstone bridge network.

metre-candle *(Phys.)*. See **lux**.

metre-kilogram(me)-second-ampere *(Phys.)*. See **MKSA**.

metric *(Maths.)*. (1) A differential expression of distance in a generalized vector space, i.e. $ds^2 = g_{\alpha\beta} dx\alpha dx\beta$. The coefficient $g_{\alpha\beta}$ forms a tensor known as the *fundamental metric tensor*. (2) Let M be any set, and let d be a mapping from the set of ordered pairs (x,y), where x and y belong to M, into the field of real numbers. Then d is called a metric in M if it satisfies the following conditions: For all x,y,z, in M, (i) $d(x,x)=0$; (ii) $d(x,y) \geqq 0$; (iii) $d(x,y)=0$ implies $x=y$; (iv) $d(x,y)=d(y,x)$; (v) $d(x,y)+d(y,z) \geqq d(x,z)$. The idea of metric is a generalization of the familiar notion of distance. For any given metric d, the value of $d(x,y)$ is called the distance between x and y. See **metric space**.

metric quad demy *(Paper)*. A size of paper for the production of demy quarto or demy octavo books.

metric quad royal *(Paper)*. A size of paper for the production of royal quarto or royal octavo books. 102 × 127 cm.

metric screw thread *(Eng.)*. A standard screw thread in which the diameter and pitch are specified in millimetres with a 60° angle.

metric space *(Maths.)*. A set M together with a metric d is called a metric space. See **metric**.

metric system *(Genrl.)*. A system of weights and measures based on the principle that each quantity should have one unit whose multiples and sub-multiples are all derived by multiplying or dividing by powers of ten. This simplifies conversion, and eliminates completely the complicated tables of weights and measures found in the traditional British system. Originally introduced in France, it is the basis for the now universal *Système International* (*SI*).

metric trait *(Biol.)*. Same as **quantitative character**.

metritis *(Med.)*. Inflammation of the **uterus**.

metrology *(Genrl.)*. The science of measuring.

metronidazole *(Med.)*. Antomicrobial drug with high activity against anaerobic bacteria and protozoa. Often used to treat surgical and gynaecological sepsis.

metrorrhagia *(Med.)*. Bleeding from the uterus between menstrual periods.

metrostaxis *(Med.)*. See **metrorrhagia**.

-metry *(Chem.)*. A suffix denoting a method of analysis or measurement, e.g. acidimetry, iodimetry, nephelometry.

metyrapone *(Med.)*. Competitive inhibitor for 11β-hydroxylation in the synthesis of cortisol by the adrenal. Used to suppress cortisol output in some tumours of the adrenal gland and in the assessment of adrenal-pituitary interaction.

MeV *(Phys.)*. Abbrev. for mega-electron-volts. Unit of particle energy. 10^6 electron-volts.

mevalonic acid *(Chem.)*. $CH_2OH.CH_2.C(OH)(CH_3).CH_2.COOH$. 3,5-Dihydroxy-3-methylpentanoic acid, a precursor in the biosynthesis of terpenes and steroids.

Mexican onyx *(Min.)*. A translucent, veined and particoloured aragonite found in Mexico and in the southwestern US.

mezzanine *(Build.)*. An intermediate floor constructed between 2 other floors in a building.

mezzotint *(Print.)*. An intaglio process in which printing is done from a copper plate, grained by rocking a semicircular toothed knife over the surface, the lighter tones being produced by scraping or burnishing away the grain to reduce the ink-holding capacity.

MF *(Aero.)*. Abbrev. for *Medium Frequency*. Frequencies from 300 to 3000 kHz. .

MF *(Paper)*. Abbrev. for *Machine Finished*.

MF keypad *(Telecomm.)*. Keypad, incorporated into a telephone and capable of producing *multifrequency* signals for operating electronic exchange equipment. Low-speed data (< 600 baud) can also be transmitted from an MF keypad.

MF resin *(Plastics)*. Abbrev. for *Melamine-Formaldehyde resin*.

MF switching *(Telecomm.)*. Operation of electronic telephone exchange equipment by **multifrequency** tones, rather than by electrical impulses.

MG *(Build.)*. Abbrev. for *Make Good*.

Mg *(Chem.)*. The symbol for *magnesium*.

MG *(Paper)*. Abbrev. for *Machine Glazed*.

mg *(Phys.)*. Abbrev. for *milligram*.

MG machine *(Paper)*. A paper machine incorporating a Yankee or MG drying cylinder in the drying section to produce machine glazed paper.

Mg point *(Glass)*. See **transformation points**.

MHC *(Immun.)*. Abbrev. for *Major Histocompatibility Complex*.

MHC restriction *(Immun.)*. The recognition by T-lymphocytes (cytotoxic or helper cells) of foreign antigen on the surface of another cell, e.g. a virus-infected cell or an accessory cell, only occurs when the antigen is associated with *self antigens* of the major histocompatibility complex. Cytotoxic T-lymphocytes usually respond to foreign antigen in association with Class I MHC antigens, whereas helper T-lymphocytes respond to foreign antigen in association with Class II MHC antigens. This was deduced from observations in vitro, and is important for understanding how T-lymphocytes function. It has no practical relevance to their function in any normal individual, all of whose cells are 'self', but may be relevant when chimerism is produced by bone marrow grafting.

MHD generator *(Elec.Eng.)*. Abbrev. for *MagnetoHydroDynamic generator*.

mho *(Elec.Eng.)*. Name for the reciprocal of the ohm in the CGS system. See **siemens**.

mianserin *(Med.)*. Anti-depressive drug reputed to have fewer cholinergic effects and fewer cardiovascular adverse affects. Does not interfere in the re-uptake of amines into the neurone.

miarolitic structure *(Geol.)*. A structure found in an igneous rock, consisting of irregularly shaped cavities into which the constituent minerals may project as perfectly terminated crystals.

mica *(Min.)*. A group of silicates which crystallize in the monoclinic system; they have similar chemical compositions and highly perfect basal cleavage. Mica is one of the best electrical insulators.

micaceous iron-ore *(Min.)*. A variety of specular **haematite** (Fe_2O_3) which is foliated or which simulates mica in the flakiness of its habit.

micaceous sandstone *(Geol.)*. A sandstone containing conspicuous flakes of mica.

mica cone *(Elec.Eng.)*. See **mica V-ring**.

micafolium *(Elec.Eng.)*. A composite insulating material consisting of a paper backing covered with mica flakes and varnish. Much used for insulating wire, machine coils, etc.

micanite *(Elec.Eng.)*. Mica splittings bonded by varnish or shellac into a large sheet; mechanically weak at high temperatures.

mica-schist *(Geol.)*. Schist composed essentially of micas and quartz, the foliation being mainly due to the parallel disposition of the mica flakes. See also **schist**.

mica V-ring *(Elec.Eng.)*. A ring of V-shaped cross-section made of a mica compound and used to insulate a metal V-ring from the bars of the commutator which it supports.

micelle *(Bot.)*. Crystalline region, inferred from X-ray diffraction data, within a microfibril of cellulose or similar structure.

micelle *(Chem.)*. A colloidal sized aggregate of molecules, such as those formed by surface active agents.

Michell bearing *(Eng.)*. A thrust or journal bearing in which pivoted pads support the thrust collar or journal in such a way that they tilt slightly under the wedging action of the lubricant induced between the surfaces by their relative motion. The fluid lubrication conditions thus produced result in a very low friction coefficient and power loss in the bearing.

Michelson interferometer *(Phys.)*. An early inteferometer designed for the investigation of the fine structure of spectrum lines and for the evaluation of the standard metre in wavelengths of light. The principle of operation is similar to that of the **Fabry-Pérot interferometer**, that is, using circular *Haidinger fringes*.

Michelson-Morley experiment *(Phys.)*. An attempt to detect and measure the relative velocity of the earth and the *ether* by observation of the shift of interference fringes produced in an interferometer whose orientation could be changed by 90°. No shift was detected and so no relative motion measured. This result can be explained in terms of the principle of the constancy of the speed of light. See **principle of relativity**.

MICR *(Comp.)*. See **magnetic ink character recognition**.

micro- *(Genrl.)*. Prefix from Gk. *mikros*, small. When used of units it indicates the basic unit $\times 10^{-6}$, e.g. 1 microampere (μA) $= 10^{-6}$ ampere. Symbol μ.

micro *(Geol.)*. Applied to names of rocks, it indicates the medium-grained form, e.g. *microdiorite*, *microsyenite*, *microtonalite*.

microaerophile *(Bot.)*. An organism which grows well at low oxygen concentrations.

microammeter *(Elec.Eng.)*. A most sensitive form of robust current-measuring instrument.

microanalysis *(Chem.)*. A special technique of both qualitative and quantitative analysis, by means of which very small amounts of substances may be analysed.

microanalytical reagent *(Chem.)*. See **MAR**.

microbalance *(Chem.)*. Sensitive to one microgram, for use in microanalysis; may be either beam or quartz fibre.

microbar *(Phys.)*. Unit of pressure $= 10^{-6}$ bar $= 10^{-1} Nm^{-2} = 1 dyne\,cm^{-2}$

microbe *(Biol.)*. An organism which can only be seen under the microscope.

microbiological mining *(Min.Ext.)*. Also *biomining*. (1) Use of natural or **genetically engineered** strains of bacteria to enhance or induce acid leaching of metals from ores (*bacterial leaching*), either in situ within an ore deposit, or to promote leaching of metals from mine waste. (2) Use of bacteria to recover useful or toxic metals from natural drainage waters, mine drainage or waste water from tips (*bacterial recovery*).

microbody *(Bot.)*. Cytoplasmic organelle of eukaryotes up to ca 1.5 μm diameter, bounded by a single membrane. Types include the **glyoxysome** and **peroxisome**.

microbore *(Build.)*. Pump assisted central heating system in which the hot-water pipes have diameters of 6, 8 or 10 mm.

microburst *(Aero.)*. Dangerous vertical gust having a core of ~1.5 miles (2.5 km) diameter in which downward velocities of 4000 ft/m (20 m/s) can occur down to low altitude.

microcanonical assembly *(Phys.)*. In statistical thermodynamics, designates an assembly which consists of a large number of systems each having the same energy.

microcephaly, microcephalia *(Med.)*. Abnormally small size of the head.

microchemistry *(Chem.)*. Preparation and analysis of very small samples, usually less than ten milligrams but, in radioactive work, down to a few atoms.

microcircuit isolation *(Elec.Eng.)*. The electrical insulation of circuit elements from the electrically conducting silicon wafer. The two main techniques are **oxide isolation** and **diode isolation**.

microcircuits *(Elec.Eng.)*. Those with components formed in one unit of semiconductor crystal.

microclimate *(Ecol.)*. The climate in small places, e.g. very close to organisms or in specific habitats such as nests or on bare ground.

microcline *(Min.)*. A silicate of potassium and aluminium which crystallizes in the triclinic system. A feldspar, it resembles orthoclase, but is distinguished by its optical and other physical characters. See also **potassium feldspar**.

Micrococcaceae *(Biol.)*. A family of bacteria belonging to the order *Eubacteriales*. Gram-positive cocci; includes free living, parasitic, saprophytic and pathogenic species, e.g. *Staphylococcus aureus* (one cause of food poisoning).

microcode *(Comp.)*. Software equivalent of a **micro instruction**, written to extend the machine-code instruction of the computer without the addition of further *hardwired logic*.

microcomputer *(Comp.)*. Computer based on a **microprocessor**. Generally this is a cheap and relatively slow computer with a limited access store, a simple instruction set and only elementary backing store (e.g. cassette tapes, floppy disks). See **personal computer**.

microcosmic salt *(Chem.)*. *Sodium ammonium hydrogen phosphate*, $NaNH_4HPO_4.4H_2O$. Present in urine.

microcrystalline texture *(Geol.)*. A term applied to a rock or groundmass in which the individual crystals can be seen only under the microscope.

microdensitometer *(Image Tech.)*. A densitometer capable of measuring the density variations of extremely small areas, down to the size of individual grains in a photographic image.

microdissection *(Biol.)*. A technique for small-scale dissection, e.g. on living cells, using **micro-manipulators** and viewing the object through a microscope.

microelectronics *(Electronics)*. The technology of constructing and utilizing complex electronic circuits and devices in extremely small packages by using integrated-circuit manufacturing techniques.

microencapsulation *(Paper)*. A process whereby a substance in a state of extreme comminution is enclosed in sealing capsules from which the material is released by impact, solution, heat or other means. See **NCR**.

micro-environment *(Ecol.)*. The environment of small areas in contrast to large areas, with particular reference to the conditions experienced by individual organisms and their parts (e.g. leaves, etc.).

μF *(Elec.Eng.)*. Abbrev. for *microfarad*.

microfarad *(Elec.Eng.)*. A unit of capacitance equal to one-millionth of a farad; more convenient for use than the farad. Abbrev. *μF*.

microfelsitic texture *(Geol.)*. A term applied to the cryptocrystalline texture seen, under the microscope, in the groundmass of quartz-felsites and similar rocks; due to the devitrification of an originally glassy matrix.

microfibril *(Bot.)*. Fine fibril, 5–30 nm wide, in a **cell wall**;

of cellulose in vascular plants and some algae but of other polysaccharide (e.g. xylans) in other algae. Cf. **matrix**.

microfiche *(Comp.)*. Output medium consisting of large-capacity microfilm sheets which may be randomly accessed by a special optical reader/magnifier. Also used in non-computer applications and then usually A6 size with images on 60–98 pages reduced 20–25 times.

microfilament *(Biol.)*. A filament of F actin, about 6 nm in diameter, found in the cytoplasm usually in long bundles and involved in localized cell contractions and streaming. Cf. **microtubule**. See **stress fibres**.

microfilaria *(Zool.)*. The early larval stage of certain parasitic Nematoda.

microfilm *(Image Tech.)*. (1) A black-and-white film of extremely high resolution and fine grain, available in long rolls especially for **microphotography**. (2) A record produced by **microphotography**.

microfossils *(Geol.)*. Fossils or fossil fragments too small to be studied without using a microscope.

microgamete *(Biol.)*. The smaller of a pair of conjugating gametes, generally considered to be the male gamete.

microgametocyte *(Zool.)*. In Protozoa, a stage developing from a trophozoite and giving rise to to male gametes.

microgap switch *(Elec.Eng.)*. A switch, used on low-power, low-voltage circuits, which relies for arc extinction on the lateral spread of the arc stream by mutual repulsion between contacts separated by thousandths of an inch.

microglia *(Zool.)*. A small type of neuroglia cell (occurring more frequently in grey matter than in white matter) having an irregular body and freely branching processes; can be phagocytic.

microglobulin *(Immun.)*. Any small globulin. Used in respect of Bence-Jones protein in urine or of β2 **microglobulin**.

microgram(me) *(Genrl.)*. Unit of mass equal to one millionth of a gram (10^{-9} kg). Sometimes known as *gamma*.

microgranite *(Geol.)*. A medium-grained, microcrystalline, acid igneous rock having the same mineral composition and texture as a granite.

micrographic texture *(Geol.)*. A distinctive rock texture in which the simultaneous crystallization of quartz and feldspar has led to the former occurring as apparently isolated fragments, resembling runic hieroglyphs, set in a continuous matrix of feldspar.

microgravity *(Space)*. The condition of near-weightlessness induced by **free-fall** or unpowered space flight; it is characterized by the virtual absence of gravity-induced convection, hydrostatic pressure and sedimentation. The term also refers to the scientific discipline which is concerned with the evaluation of processes in a near-zero *g* environment, particularly those of *Fluid Physics, Life* and *Material Sciences*.

microgroove records *(Acous.)*. Long-playing records with as many as 200–300 grooves/in., and played at rotational speeds of 33.33 or 45 rev/min.

microgyria *(Med.)*. Abnormal smallness of the convolutions of the brain.

micro-incineration *(Biol.)*. A technique for examining the distribution of minerals in slide preparations of tissue-sections or cells. The organic material is vaporized by heat and the nature and position of the mineral ash determined by microscopic examination.

microlecithal *(Zool.)*. Said of eggs containing very little yolk. Cf. **megalecithal**.

microlight *(Aero.)*. Aircraft whose empty weight does not exceed 330 lb (150 kg). In US *ultralight* is used for weights up to 254 lb (115 kg).

microlite *(Geol.)*. A general term for minute crystals of tabular or prismatic habit found in microcrystalline rocks. These give a reaction with polarized light.

microlux *(Phys.)*. A unit for very weak illuminations, equal to one-millionth of a lux.

micromanipulator *(Biol.)*. An instrument used to handle cells seen in a microscope, e.g. to remove a nucleus or inject RNA. The fine movements are controlled indirectly by pneumatic, mechanical or other means.

micromazia *(Med.)*. Failure of the female breast to develop after puberty.

micromere *(Zool.)*. In a segmenting ovum, one of the small cells which are formed in the upper or animal hemisphere.

micromesh sieves *(Powder Tech.)*. See **electro-formed sieves**.

micrometeorite *(Astron.)*. A particle of cosmic dust with mass <10 microgram and diameter <0.1mm. Such particles do not burn up in the Earth's atmosphere, but drift down to the surface. **Comets** are probably abundant sources of new micrometeorites.

micrometeorites *(Space)*. Extremely small particles, typically of mass less than 10^{-6} gm, which are present in space; these hyper-velocity particles represent a hazard for space-flight and must be protected against.

micrometer *(Astron.)*. Measures small angular separations in the telescope. It consists of 3 frameworks carrying spider-webs close to the image plane, one is fixed and the others are each adjustable by micrometer heads, by which the separation is read, with a graduated circle giving the angular relation of a double star. See also **micrometer gauge**.

micrometer eyepiece *(Biol.)*. See **eyepiece graticule**.

micrometer gauge *(Eng.)*. A U-shaped length gauge in which the gap between the measuring faces is adjustable by an accurate screw whose end forms one face. The gap is read off a scale uncovered by a thimble carried by the screw, and by a circular scale which is engraved n the thimble. Commonly *micrometer*.

micrometer theodolite *(Surv.)*. A theodolite equipped with micrometers instead of the usual verniers for reading the horizontal and vertical circles.

micrometre *(Phys.)*. One-millionth of a metre. Symbol *μm*. Also called *micron* in past.

micromicro- *(Genrl.)*. Prefix for 1-millionth-millionth, or 10^{-12}: replaced in SI by *pico-* (p).

μμF *(Elec.Eng.)*. Abbrev. for *micromicrofarad*.

micromodule *(Elec.Eng.)*. Sometimes said of circuits or components formed from the same crystal of material, e.g. germanium. An *integrated circuit*.

micron *(Genrl.)*. Obs. measure of length equal to one millionth of a metre, symbol μ. Replaced in SI by *micrometre*, symbol μm.

micronized coal *(Eng.)*. Pulverised coal in which > 80% will pass through a 40 μm sieve.

micronucleus *(Zool.)*. In Ciliopora, the smaller of the two nuclei which is involved with sexual reproduction. Cf. **macronucleus**.

micronutrient *(Ecol.)*. A *trace element* required in relatively small quantities by living organisms; for plants the micronutrients include Fe, B, Mn, Zn, Cu, Mo, Cl. Cf. **macronutrient**. See also **essential element**.

micropalaeontology *(Geol.)*. The study of **microfossils**.

microperthite *(Min.)*. A feldspar which consists of intergrowths of potassium feldspar and albite in a microscopic scale.

microphage *(Zool.)*. A small phagocytic cell in blood or lymph, chiefly the polymorphonuclear leucocytes (*neutrophils*). adj. *microphagocytic*.

microphagous *(Zool.)*. Feeding on small particles of food. Cf. **macrophagous**.

microphanerophyte *(Bot.)*. A **phanerophyte**, 2–8 m high.

microphone *(Acous.)*. An acousto-electric transducer, essential in all sound-reproducing systems. The fluctuating pressure in the sound wave is applied to a mechanical system, such as a ribbon or diaphragm. The motion of which generates an electromotive force, or modulates a current or voltage. See also **carbon-, directional-, electromagnetic-, hot-wire-, lapel-, moving-coil-, omnidirectional-, Olson-, pressure-, pressure-gradient-,**

microphone response *(Acous.)*. That measured over the operating frequency range in a particular direction, or averaged over all directions. The characteristic response is usually given by the ratio of the open circuit voltage generated by the microphone to the sound pressure (N/m^2) existing in the free progressive wave before introducing the microphone.

microphonic *(Electronics)*. Said of a component which responds to acoustic vibrations and/or knocks.

microphonic noise *(Acous.)*. That in the output of a valve related to mechanical vibration of the electrode system. Also called *microphonicity*.

microphotography *(Image Tech.)*. Photography of normal sized objects, especially documents, plans and graphic materials, as greatly reduced images of small area which must be examined by magnification or enlarged projection.

microphyll *(Bot.)*. (1) A small leaf, not associated with a leaf gap, assumed to have evolved from an **enation** and supposed to be the sort of leaf possessed by lycopods. (2) Any very small leaf, e.g. of heather or *Tamarix*. adj. *microphyllous*. Cf. **megaphyll**.

Microphyllophyta *(Bot.)*. Division of the plant kingdom, here treated as the class Lycopsida.

microphyric *(Geol.)*. A textural term descriptive of medium- to fine-grained igneous rocks containing phenocrysts < 2 mm long. Cf. **macrophyric**.

Micropodiformes *(Zool.)*. An order of Birds with a short humerus and long distal segments to the wings; Swifts and Hummingbirds.

micropodous *(Zool.)*. Having the foot, or feet, small or vestigial.

microporosity *(Eng.)*. Minute cavities generally found in heavy sections usually due to lack of efficient feeding or chilling; particularly found in magnesium, when they are highly coloured.

micro-porous coatings *(Build.)*. Paints, stains or clear coatings which allow a surface to breathe. Primarily intended for use on timbers, these coatings do not trap moisture unlike conventional paint systems.

microprism *(Image Tech.)*. Focusing device which breaks the image up into dots when it is not in focus.

microprocessor *(Comp.,Electronics)*. A computer **central processing unit** (CPU) that is contained on one or a few integrated-circuit chips, using large-scale integration (LSI) technology. A microprocessor can be used as part of an automatic control system or as the main element of a *microcomputer*.

microprogram *(Comp.)*. A sequence of **micro-instructions** used as part of the design of a processor which enables **control** logic to be established by **software**. Microprogrammed control is not quite as fast but much more flexible than control by **hardware**.

micropropagation *(Bot.)*. The use of small pieces of tissue, such as meristem grown in culture, to obtain large numbers of individuals. Needs sterile precautions and the proper nutrients.

micropsia *(Med.)*. The condition in which objects appear to the observer smaller than they actually are; may be due to retinal disease.

micropterous *(Zool.)*. Having small or reduced fins or wings.

micropyle *(Bot.)*. (1) A tiny opening in the integument at the apex of an ovule, through which the pollen tube usually enters. (2) The corresponding opening in the testa of the seed. Also called *foramen*.

micropyle *(Zool.)*. An aperture in the chorion of an Insect egg through which a spermatozoon may gain admittance.

microradiography *(Radiol.)*. Exposure of small thin objects to **soft X-rays**, with registration on a fine-grain emulsion and subsequent enlargement up to 100 times. Also used to signify the optical reproduction of an image formed, e.g. by an electron microscope.

microscope *(Phys.)*. An instrument used for obtaining magnified images of small objects. The *simple microscope* is a convex lens of short focal length, used to form a virtual image of an object placed just inside its principal focus. The *compound microscope* consists of two short-focus convex lenses, the objective and the eyepiece mounted at opposite ends of a tube. For most microscopes, the magnifying power is roughly equal to $450/f_o f_e$, where f_o and f_e are the focal lengths of objective and eyepiece in centimetres. See also **electron-**, **ultraviolet-microscope**.

microscope count method *(Powder Tech.)*. Technique for measuring the average particle diameter of a powder by microscope examination of a weighed drop of suspension.

microscopic state *(Chem.)*. One in which condition of each individual atom has been fully specified. Cf. **macroscopic state**.

microscopic stress *(Eng.)*. Those set up at a level of the grain size of an metal due to heat treatment etc.

microsection *(Eng.)*. A section of metal, ceramic etc., mounted, cut, polished and etched as necessary to exhibit the microstructure.

microsmatic *(Zool.)*. Having a poorly developed sense of smell.

microsomes *(Biol.)*. The membranous pellet obtained by centrifugation of a cell homogenate after removal of the mitochondria and nuclei. It was originally believed to represent a cell organelle but is now known to consist of fragments of endoplasmic reticulum, plasma membrane, Golgi apparatus etc.

microspecies *(Bot.)*. See **Jordanon**.

microspherulitic texture *(Geol.)*. A texture in which spherulites on a microscopic scale are distributed through the groundmass of an igneous rock.

microsplanchnic *(Zool.)*. Having a small body and long legs, as a Harvestman.

microsporangium *(Bot.)*. The structure within which **microspores** are formed. In seed plants, the pollen sac.

microspore *(Bot.)*. In heterosporous species a **meiospore** able to develop into a male gametophyte; in seed plants, the pollen grain.

microspore *(Zool.)*. A small swarm-spore or anisogamete of Sarcodina.

microsporocyte *(Bot.)*. A cell which divides to give microspores, i.e. a microspore mother cell.

microsporophyll *(Bot.)*. A leaf-like structure bearing microsporangia. Cf. **megasporophyll**. See also **sporophyll**.

microsporophyte *(Bot.)*. A typically diploid cell in a microsporangium, which divides by meiosis to give four microspores.

microstoma, microstomia *(Med.)*. An abnormally small mouth, due either to developmental defect or to contraction of scar tissue.

microstrip *(Telecomm.)*. A microwave transmission line consisting basically of a dielectric sheet carrying a conducting strip on one side and an earthed conducting plane on the other. The strip with its image in the plane (see **image charge**) forms a parallel strip transmission line.

microstructure *(Eng.)*. A term referring to the size, shape and arrangement with respect to each other (as seen under the microscope) of the constituents present in a metal or alloy, too small to be seen by the naked eye.

microswitch *(Elec.Eng.)*. A switch operated by very small movements of a lever, the circuit being made or broken by spring-loaded contacts.

microsyn *(Elec.Eng.)*. An a.c. transducer employed in measuring angular rotation and used as output device in precision gyroscopes and servomechanisms. Its construction is similar to that of an **E-transformer**.

microtome *(Biol.)*. An instrument for cutting thin sections of specimens, especially sections 1–10 μm thick for light microscopy. See also **ultramicrotome**.

microtubule *(Biol.)*. Tubular structure about 24 nm in diameter formed by the aggregation of **tubulin** dimers and small amounts of associated proteins in a helical array. They function as skeletal components within cells and may be arranged into complex structures such as cilia and mitotic spindles.

microtubule-organizing centre *(Bot.)*. Cytoplasmic site where microtubules are formed and organized; associated with or including centromeres, basal bodies, centrioles etc. Abbrev. *MTOC*.

microvilli *(Biol.)*. Finger-like protrusions of the cell surface. Their length and abundance varies characteristically between cells. When present in very high density, as on the apical surface of epithelial cells of the small intestine, they form a *brush border*.

microvillus *(Bot.)*. Finger-like infolding of the inner membrane of a mitochondria, as in the brown algae, diatoms etc. See **Heterokontophyta**. Cf. **crista**.

microwave background *(Astron.)*. Discovered in 1963, a weak radio signal which is detectable in every direction with almost identical intensity. It has a temperature of 2.7 K. The slight asymmetry is due to the motion of our Galaxy relative to this radiation. The radiation is the relic of the early hot phase in the **Big Bang** universe.

microwave heating *(Phys.)*. Heating (induction or dielectric) of materials in which the current frequency is in the range 0.3×10^{12} to 10^9 Hz. Extensively used in domestic microwave ovens.

microwave resonance *(Elec.Eng.)*. One between microwave signals and atoms or molecules of medium.

microwave resonator *(Telecomm.)*. Effective tuned circuit for microwave signal. Usually a cavity resonator but tuned lines are also used.

microwaves *(Telecomm.)*. Those electromagnetic wavelengths between 1 mm and 30 cm, from 300 GHz to 1 Ghz in frequency, thus bridging gap between normal radio waves and heat waves.

microwave spectrometer *(Phys.)*. An instrument designed to separate a complex microwave signal into its various components and to measure the frequency of each; analogous to an *optical spectrometer*. See **spectrometer**.

microwave spectroscopy *(Phys.)*. The study of atomic and/or molecular resonances in the microwave spectrum.

microwave spectrum *(Phys.)*. The part of the electromagnetic spectrum corresponding to microwave frequencies.

micturition *(Zool.)*. In Mammals, the passing to the exterior of the contents of the urinary bladder.

mid-brain *(Zool.)*. In Vertebrates, that part of the brain which is derived from the second or middle brain-vesicle of the embryo; the second or middle-brain vesicle itself dorsally comprises the tectum, including optic lobes (*corpora quadrigemina* in Mammals), and laterally the tegmentum and ventrally the *crura cerebri* (cerebral peduncles).

middle conductor *(Elec.Eng.)*. Same as *neutral conductor*.

middle ear *(Zool.)*. In Tetrapoda excluding Urodela, Gymnophiona and snakes, the cavity containing the auditory ossicles.

middle lamella *(Bot.)*. Layer of intercellular material, mostly pectic substances, developing from the cell plate and cementing together the primary walls of contiguous cells.

middle marker beacon *(Aero.)*. A marker beacon associated with the **ILS**, used to define the second predetermined point during a beam approach.

middle oils *(Chem.)*. Carbolic oils, obtained from coal-tar distillation. Their boiling range is from about 210°C to 240°C.

middle rail *(Build.)*. The rail next above the bottom rail in doors, framing, and panelling.

middle shore *(Build.)*. An inclined shore placed between the bottom and top shores in a set of raking shores.

middle temperature error *(Horol.)*. The error in the time of vibration of a compensation balance due to the fact that compensation for the dimensional changes of the balance and the elastic properties of the spring is not complete over a range of temperature. A watch or chronometer regulated to be correct at the extremes of the temperature range over which it is to be used will show an error at temperatures between the extremes.

middle third *(Civ.Eng.)*. The middle part of a brickwork or masonry structure, such as an arch or dam. It is equal in width, at any section, to one-third the width of the section, and is centrally disposed. The importance of the middle third is that, providing the line of resultant pressure lies wholly within it, no tensile forces come into play.

middle wire *(Elec.Eng.)*. Same as *neutral conductor*.

middlings *(Min.Ext.)*. In ore dressing, an intermediate product left after the removal of clean concentrates and rejected tailings. It consists typically of interlocked particles of desired mineral species and gangue, or of by-product minerals not responding to the treatment used. See also **riddle**.

mid-feather *(Build.)*. (1) See **cross-tongue**. (2) See **parting-slip**.

mid-gut *(Zool.)*. That part of the alimentary canal of an animal which is derived from the archenteron of the embryo.

midland tariff *(Elec.Eng.)*. A name sometimes given to that form of tariff for electrical energy in which a fixed charge per year per kVA of maximum demand is made, together with a charge per kWh.

mid- (middle) gear *(Eng.)*. The position of a steam-engine link motion or valve gear when the valve travel is minimal.

mid-ocean ridge *(Geol.)*. A major, largely submarine, mountain range where two **plates** are being pulled apart and new volcanic **lithosphere** is being created.

mid-point protective system *(Elec.Eng.)*. A method of balanced protection used for protecting generators against faults between turns by balancing the voltage of one half of the winding against that of the other.

mid-rib *(Bot.)*. (1) The main vein or nerve of a leaf. (2) Thickened region down the middle of a thallus.

mid-riff *(Zool.)*. See **diaphragm**.

mid space *(Print.)*. A type space cast four to the em.

mid-wing monoplane *(Aero.)*. A monoplane wherein the main planes are located approximately midway between the top and bottom of the fuselage.

Miehle *(Print.)*. The first successful **two-revolution** printing machine.

MIF *(Immun.)*. Abbrev. for *Migration Inhibition Factor*.

MIG *(Eng.)*. See **metal inert gas welding**.

migmatite *(Geol.)*. A rock with both igneous and metamorphic characteristics; generally consisting of a host metamorphic rock injected by granitic material.

migraine *(Med.)*. Hemicrania or paroxysmal headache. A condition in which recurring headaches are often associated with vomiting and disturbances of vision.

migration *(Chem.)*. Movement of ions under influence of electric field and against the viscous resistance of the solvent. Measured by observation in thin tubes.

migration *(Ecol.)*. Long-distance animal movement, often involving large populations and often seasonal.

migration area *(Phys.)*. One-sixth of the mean square distance covered by a neutron between creation and capture. Its square root is the *migration length*.

migration area/length *(Nuc.Eng.)*. One sixth of the mean square distance covered by a neutron between creation and capture. Its square root is called *migration length*. See **slowing down area, diffusion area**.

migration inhibition factor *(Immun.)*. *MIF*. Lymphokine which acts on macrophages so as to increase their adhesiveness. When released by T-lymphocytes in vivo (e.g. in the peritoneal cavity) it causes macrophages to clump together. In vitro it inhibits their migration (e.g. out of a capillary tube), this can be used as a semi-quantitative means of detecting the existence of delayed type hypersensitivity.

migratory cell *(Zool.)*. See **amoebocyte**.

Mikulicz's disease *(Med.)*. A chronic inflammation of salivary and lacrimal glands; seen in *sarcoidosis*, *lymphoma*, *leukaemia* and *Sjögren's disease*.

mil *(Genrl.)*. A unit of length equal to 10^{-3} in., used in measurement of small thicknesses, i.e. thin sheets. Also colloq. *thou*.

MIL-1553 B *(Aero.)*. Standard requirement for airborne digital databus. Originally US but now international.

milanese fabric *(Textiles)*. A warp-knitted fabric made with a warp containing twice as many threads as there are **wales** in the fabric with resultant reinforcement of the fabric.

Milankovitch theory of climatic change *(Meteor.)*. The theory that large oscillations in climate are related to changes in solar radiation received by the Earth as a result of: (1) the variations in eccentricity of the Earth's orbit (periods of 10^5 and 4.10^5 years); (2) variations in the obliquity of tilt of the Earth's axis (period 4.10^4 years); (3) the precession of the equinoxes (period 2.10^4

years). There is evidence for the two shorter cycles in data obtained from analysis of deep-sea bottom cores.

milarite *(Min.)*. A hydrated silicate of aluminium, beryllium, calcium and potassium, crystallizing in the hexagonal system.

mild clay *(Build.)*. See **loam**.

mildew *(Bot.)*. A plant disease in which fungal mycelium is visible on the surface of the host, e.g. downy mildew, powdery mildew. Also *mould*.

mild steel *(Eng.)*. Hot-rolled steel containing approximately 0.04% of carbon. See **low-carbon steel**.

mile *(Genrl.)*. A unit of length commonly used for distance measurement in the British Commonwealth and the US. A *statute mile* = 1760 yd, = 1609.34 m. See **geographical-**, **nautical-**.

miliaria *(Med.)*. *Prickly heat*. Inflammation of the sweat glands, accompanied by intense irritation of the skin.

miliary *(Med.)*. Like a millet seed; said of lesions which are small and multiple, e.g. *miliary tuberculosis*.

miliary tuberculosis *(Med.)*. A form of tuberculosis in which small tuberculous lesions are found in various organs of the body, especially in the meninges and in the lungs.

milk fever *(Vet.)*. *Parturient paresis, parturient fever (cow), lambing sickness (ewe), parturient eclampsia (sow and bitch)*. A metabolic disease of unknown cause affecting the parturient cow, ewe, goat, sow, and bitch; characterized by hypocalcaemia and muscular weakness and incoordination, tetany, loss of consciousness, and death. The disease also occurs during pregnancy in the ewe and during lactation in the goat.

milk glands *(Zool.)*. The mammary glands of a female Mammal: in viviparous Tsetse flies, special uterine glands by which the larva is nourished until it is ready to pupate.

milk-sugar *(Chem.)*. Lactobiose or **lactose**.

milk teeth *(Zool.)*. In diphyodont Mammals, the first or deciduous dentition.

milk tetany *(Vet.)*. See **calf tetany**.

Milky Way *(Astron.)*. See **Galaxy**.

mill *(Comp.)*. Obsolete term for the arithmetical unit of a computer.

mill *(Eng.)*. Generally (1) a machine for grinding or crushing, as a *flour mill, paint mill*, etc.; (2) a factory fitted with machinery for manufacturing, as a cotton mill, saw-mill, etc.

mill *(Min.Ext.)*. In UK, a crushing and grinding plant. In US, the whole equipment for comminuting and concentrating an ore.

millboards *(Paper)*. Boards, usually very dense, manufactured on an intermittent board machine from a variety of furnishes.

milled *(Eng.)*. Having the edge grooved or fluted, as a coin or the head of an adjusting screw. See **knurling tools**.

milled cloth *(Textiles)*. Woven or knitted wool or woollen fabric in which the milling action causes felting. The cloth has a fibrous surface, the threads and structure being almost completely hidden. The process involves the application of pressure and friction to the cloth, while it is wet, e.g. with soapy water.

milled lead *(Build.)*. Sheet-lead formed from cast slabs by a rolling process.

millefiori glass *(Glass)*. Glassware in which a large number of sections of glass rods of various colours form a pattern and are fused together or set in a clear glass matrix.

Miller bridge *(Elec.Eng.)*. A particular form used to measure amplification factors of valves.

Miller circuit *(Electronics)*. Amplifier in which negative feedback from output to input is regulated by a capacitor.

Miller effect *(Electronics)*. The change in effective input impedance of an amplifier due to unwanted shunt voltage feedback, which makes the input impedance a function of the voltage gain. In particular the increase in effective input capacitance for a valve or transistor used as a voltage amplifier.

Miller indices *(Crystal.)*. Integers which determine the orientation of a crystal plane in relation to the three crystallographic axes. The reciprocals of the intercepts of the plane on the axes (in terms of lattice constants) are reduced to the smallest integers in ratio. Also called *crystal indices*.

millerite, capillary pyrite *(Min.)*. Nickel sulphide, crystallizing in the trigonal system. It usually occurs in very slender crystals and often in delicately radiating groups.

Miller process *(Min.Ext.)*. Purification of bullion by removal of base metals as chlorides. Chlorine gas is bubbled through the molten metal.

millers' disease *(Vet.)*. See **osteodystrophia fibrosa**.

millet-seed sandstone *(Geol.)*. A sandstone consisting essentially of small spheroidal grains of quartz; typical of deposits accumulated under desert conditions.

mill-fitting *(Elec.Eng.)*. See **factory-fitting**.

mill-head *(Min.Ext.)*. Ore accepted for processing after removal of waste rock and detritus. Also, *assay grade* of such ore.

milli- *(Genrl.)*. Prefix from L. *mille*, thousand. When attached to units, it denotes the basic unit $\times 10^{-3}$. Symbol *m*.

milliammeter *(Phys.)*. An ammeter calibrated and scaled in milliamperes; used for measuring currents up to about 1 ampere.

millibar *(Meteor.)*. See **bar**.

millicurie *(Phys.)*. One thousandth of a **curie**.

millidarcy *(Geol.)*. See **darcy**.

Millikan oil-drop experiment *(Phys.)*. First experiment to determine the value of the electronic charge. The vertical motion of very small charged oil-drops in the electric field between two horizontal parallel plates was measured, and the charge deduced.

millilambert *(Phys.)*. A unit of brightness equal to 0.001 lambert; more convenient magnitude than the lambert.

millilitre *(Chem.)*. A unit of volume, one thousandth of a litre, essentially equivalent to 1 cm^3 or 10^{-6} m^3.

millilux *(Phys.)*. Unit of illumination equal to one-thousandth of a lux.

millimass unit *(Phys.)*. Equal to 0.001 of atomic mass unit. Abbrev, *mu*.

millimetre *(Genrl.)*. The thousandth part of a metre.

millimetre pitch *(Eng.)*. See **metric screw-thread**.

millimicron *(Elec.Eng.)*. Obsolete term for nanometre, 10^{-9} m. Used in the measurement of wavelengths of e.g. light.

milling *(Eng.)*. A machine process in which metal is removed by a revolving multiple-tooth cutter, to produce flat or profiled surfaces, grooves and slots. See also **milling cutter**, **milling machine**.

milling *(Min.Ext.)*. *Comminution; dressing*; removing valueless material and harmful constituents from an ore, in order to render marketing more profitable.

milling *(Textiles)*. A process in the finishing of woollen fabrics, carried out in a milling-machine by the agency of soap, alkali, or acid (depending on the nature of the fabric and the dye), pressure, and friction. See **milled cloth**.

milling cutter *(Eng.)*. A hardened steel disk or cylinder on which cutting-teeth are formed by slots or grooves on the periphery and faces, or into which separate teeth are inserted; used in the milling-machine for grooving, slotting, surfacing etc. See **end mill**, **milling machine**.

milling grade, milling width *(Min.Ext.)*. *Milling grade* denotes ore sufficiently rich to repay cost of processing; *milling width* that of the lode which will determine tonnage sent daily from mine to mill.

milling machine *(Eng.)*. A machine tool in which a horizontal arbor or a vertical spindle carries a rotating multi-tooth cutter, the work being supported and fed by an adjustable and power-driven horizontal table. See also **milling cutter**.

Millington reverberation formula *(Acous.)*. A modified formula for calculating the period of reverberation time, taking into account the random disposition of the reflecting and acoustically absorbing surfaces in an enclosure. Not applicable if one partial surface has the absorption coefficient 1 (e.g. open window).

million-electron-volt *(Phys.)*. See **MeV**.
millipede *(Zool.)*. Any myriapod of the class Diplopoda, vegetarian cylindrical animals with many joints most of which bear two pairs of legs.
Millipore filter *(Chem.)*. TN for a type of **membrane filter**.
milliradian *(Genrl.)*. 10^{-3} radian.
mill join *(Print.)*. An adhesion between the ends of two similar webs made at paper mill, the joints being sometimes marked for the printer's guidance.
Millon's reaction *(Chem.)*. A test for proteins, based on the formation of a pink or dark-red precipitate of coagulated proteins on heating with a solution of mercuric nitrate containing some nitrous acid.
mill race *(Eng.)*. Channel or **flume** by which water is led to a mill wheel or waterwheel.
mill scale *(Build.,Eng.)*. A thin flaky layer of blue/black iron oxide found on new hot rolled steel. Best removed by abrasive blasting before painting.
mill tail *(Eng.)*. The channel conveying water away from a mill wheel.
mill wheel *(Eng.)*. A water-wheel driving the machinery in a mill.
MILNET *(Comp.)*. A long distance US military communication network which was built using the results of the **ARPANET** research.
Milroy's disease *(Med.)*. *Hereditary oedema*. A disorder in which persistent oedema (swelling) of the legs occurs in members of the same family in successive generations.
Mil-Spec *(Aero.)*. *Military Specification* issued in the US, which lay down basic requirements to be observed by design teams in the development of aircraft. Abbrev. also *MS*.
milt *(Zool.)*. The spleen; in Fish, the testis or spermatozoa; to fertilize the eggs.
Mimas *(Astron.)*. The natural satellite orbiting closest to **Saturn**, 400 km in diameter.
MIMD *(Comp.)*. *Multiple Instruction stream, Multiple Data stream*. Term which describes the architecture of the **processor**. See **multiprocessor**.
mimetic diagram *(Elec.Eng.)*. In a control room of a large process plant or an electrical network, the animated diagram which indicates to the controller the state of operations, by coloured lights, recorders, or indicating instruments.
mimetite *(Min.)*. A chloride-arsenate-phosphate of lead with As>P; cf. **pyromorphite**. It crystallizes in the hexagonal system, often in barrel-shaped forms, and is usually found in lead deposits which have undergone a secondary alteration.
mimicry *(Zool.)*. The adoption by one species of the colour, habits, sounds, or structure of another species. adjs. *mimic, mimetic*.
minaret *(Arch.)*. A lofty slender tower rising from a mosque or similar building and surrounded by a gallery near the top.
mine *(Min.Ext.)*. (1) Subterranean excavation made in connection with exploitation of, or search for, minerals of economic interest. Terms *quarry*, *pit*, and *opencast* are reserved for workings open to daylight. (2) Term (N. England) for any coal seam irrespective of thickness or grade.
mine detector *(Eng.)*. An electronic device for the detection of buried explosive mines (or buried metal) depending on the change in the electromagnetic coupling between coils in the search head.
mineral *(Min.)*. A naturally-occurring substance of more or less definite chemical composition and physical properties. It has a characteristic atomic structure frequently expressed in the crystalline form or other properties.
mineral caoutchouc *(Min.)*. See **elaterite**.
mineral deposit *(Geol.)*. A naturally occurring body containing minerals of economic value.
mineral dressing *(Min.Ext.)*. See **mineral processing**.
mineral flax *(Build.)*. A fibrized form of asbestos much used in the manufacture of asbestos-cement sheeting.
mineral-insulated cable *(Elec.Eng.)*. One in which the conductor runs in an earthed copper sheath filled with magnesium oxide, which makes it fireproof and able to withstand excess loads. Also *copper-sheathed cable*.
mineralization *(Bot.)*. The decomposition in soils of organic matter by micro-organisms with the release of the mineral elements (N, P, K, S etc.) as inorganic ions.
mineralization *(Geol.)*. The process by which mineral(s) are introduced into a rock.
mineral nutrient *(Bot.)*. An essential element, other than C, H or O, normally obtained as an inorganic ion taken up e.g. through the roots of a land plant; e.g. N, P, K, Ca. See **macronutrient**, **micronutrient**.
mineralogy *(Genrl.)*. The scientific study of minerals.
mineral oils *(Chem.)*. Petroleum and other hydrocarbon oils obtained from mineral sources. Cf. *vegetable oils*.
mineral processing, dressing *(Min.Ext.)*. Crushing, grinding, sizing, classification or separation of ore into waste and value by chemical, electrical, magnetic, gravity, and physico-chemical methods. First-stage extraction metallury. Also *ore dressing*, *beneficiation*, *préparation mécanique* (*French*).
mineral vein *(Min.Ext.)*. A fissure or crack in a rock which has been subsequently lined or filled with minerals. See also **lode**.
mineral wool *(Eng.)*. See **rock wool**.
miner's dip needle *(Min.Ext.)*. A portable form of **dip needle** used for indicating the presence of magnetic ores.
miner's lamp *(Min.Ext.)*. A portable lamp specially designed to be of robust construction and adequate safety for use in mines.
minette *(Geol.)*. (1) A lamprophyre composed essentially of biotite and orthoclase, occurring in dykes associated with major granitic intrusions. (2) Jurassic ironstones of Briey and Lorraine, and is still so used.
miniature camera *(Image Tech.)*. A term sometimes applied to any small camera, but normally used for those taking 35 mm film.
miniature Edison screw-cap *(Elec.Eng.)*. An Edison screw-cap for electric filament lamps, in which the screw-thread has a diameter of about 0.4 in and about 14 threads per inch.
miniature (subminiature) valve *(Electronics)*. One in which all the dimensions are reduced to very small values, to keep down the interelectrode capacitances and the electron transit time, making it suitable for HF, VHF and UHF applications.
minicomputer *(Comp.)*. Computer whose size and capabilities lie between those of a **mainframe** and a **microcomputer**. The term referred originally to a range of third generation computer, cheaper and less well-equipped than contemporary mainframe machines. With changes in technology the term is becoming vaguer.
minimal area *(Ecol.)*. Concept used in sampling vegetation: the minimum area which must be searched in order to find (nearly) all of the species. See **species/area curve**.
minimum access program *(Comp.)*. Program routine involving minimum **access time**.
minimum blowing current *(Elec.Eng.)*. The minimum current which will cause melting of a fuse link under certain specified conditions.
minimum burner pressure valve *(Aero.)*. A device which maintains a safe minimum pressure at the burners of a gas turbine when it is idling.
minimum deviation *(Phys.)*. See **angle of minimum deviation**.
minimum discernible signal *(Telecomm.)*. Smallest input power to any unit which just produces a discernible change in output level.
minimum flying speed *(Aero.)*. The minimum speed at which an aircraft has sufficient lift to support itself in level flight in standard atmosphere. There is a close relationship with the weight, which affects the **wing loading** and the term must be stated with the weight (and altitude if *ISA*, sea level is not implied) and *TAS* specified.
minimum ionization *(Phys.)*. The smallest possible value of the specific ionization that a charged particle can produce in passing through a given substance. It occurs

for particles having velocities = 0.95 c, where c = velocity of light.

minimum pause *(Telecomm.)*. The interval of lost time which is necessarily introduced into the operation of a dial to ensure that the selectors have time to complete their hunting.

minimum point on a curve *(Maths.)*. A trough on a curve. For the curve $y = f(x)$, the point where $x = a$ is a minimum if $f(a+h) - f(a)$ is positive for all values of h sufficiently small.

minimum sampling frequency *(Telecomm.)*. Lowest sampling rate which can provide an accurate reproduction of the signal in a pulse-code-modulation system. Equal to twice the maximum signal frequency.

minimum two-part prepayment meter *(Elec.Eng.)*. A 2-part prepayment meter in which the time element is arranged to collect a charge based upon a minimum yearly consumption as well as the usual fixed charge.

minimum wavelength *(Phys.)*. The shortest wavelength emitted in an X-ray spectrum. It is determined by the maximum voltage applied to the X-ray tube. The emitted X-ray photon acquires all the energy of the electron accelerated by the voltage v, so the minimum wavelength

$$\lambda_m = \frac{hc}{eV} = \frac{1.2396 \times 10^{-6}}{V} m.$$

where h is Planck's constant, e is the electronic charge and c is the speed of light.

mining dial *(Min.Ext.)*. See **dial**.

mining engineering *(Genrl.)*. That branch of engineering chiefly concerned with the sinking and equipment of mineshafts and workings, and all operations incidental to the winning and preparation of minerals.

minion *(Print.)*. An old type size, approximately 7-point.

miniscule *(Print.)*. A lower-case, or small, letter.

minitrack *(Telecomm.)*. Phase-comparison angle tracking system, used for tracking satellites.

Minkowski diagram *(Phys.)*. A space-time diagram used to represent the positions and times of events relative to an inertial reference frame.

Minkowski's inequality *(Maths.)*.

$$\left(\sum_1^n |a_r + b_r|^\alpha\right)^{1/\alpha} \leq \left(\sum_1^n |a_r|^\alpha\right)^{1/\alpha} \cdot \left(\sum_1^n |b_r|^\alpha\right)^{1/\alpha},$$

if $\alpha \geqq 1$.

minnesotaite *(Min.)*. The iron-bearing equivalent of talc. A major constituent of the siliceous iron-ores of the Lake Superior region.

minnikin *(Print.)*. The smallest of the old type sizes, approximately 3-point.

minor *(Maths.)*. A sub-determinant, i.e. a determinant contained in another determinant of higher *order*. The minor of a particular element of a determinant is the determinant of the elements which remain when the row and column containing the element are deleted.

minor axis *(Maths.)*. Of an ellipse: see **axes**.

minor exchange *(Telecomm.)*. One directly connected to its group centre.

minor intrusions *(Geol.)*. Igneous intrusions of relatively small size, compared with **plutonic (major) intrusions**. They comprise dykes, sills, veins, and small laccoliths. The injection of the minor intrusions constitutes the dyke phase of an volcanic cycle.

minority carrier *(Electronics)*. In a semiconductor, the electrons or **holes** which carry the lesser degree of measured current. See **majority carrier**.

minor planet *(Astron.)*. Term used generally in professional astronomy for **asteroid**.

minuend *(Maths.)*. See **subtraction**.

minus colour *(Image Tech.)*. The complementary to a given colour, that is, the remainder when that colour is subtracted from white light. Thus, *minus red* is blue-green, *cyan*. Hence, the equivalent term *subtractive colour*.

minus strain *(Bot.)*. One of the two, arbitrarily designated, mating types of a heterothallic species. Cf. **plus strain**. See also **heterothallism**.

minute *(Genrl.)*. (1) A 60th part of an hour of time. (2) A 60th part of an angular degree. (3) A 60th part of the lower diameter of a column.

minute pinion, nut *(Horol.)*. The pinion in the motion work that drives the hour wheel.

minute wheel *(Horol.)*. The wheel in the motion work driven by the cannon pinion.

minverite *(Geol.)*. A basic intrusive rock, in essentials a dolerite, containing a brown, soda-rich hornblende; named from the type-locality, St Minver, Cornwall.

Miocene *(Geol.)*. An epoch of the Neogene sub-period of the Cenozoic era.

miosis *(Med.)*. Contraction of the pupil of the eye.

MIPS *(Comp.)*. *Millions of Instructions Per Second*. The number of million instructions that can be executed by a computer in a second. It will vary from 1–2 MIPS in a good personal computer to hundreds of MIPS in a supercomputer. Cf. **FLOPS**.

Mir *(Space)*. Advanced USSR space station, developed from experience gained with **Salyut**; it has the capability for considerable growth by adding further modules.

mirabilite *(Min.)*. See **Glauber salt**.

miracidium *(Zool.)*. The ciliated first-stage larva of a Trematode.

mirage *(Meteor.)*. An effect caused by total reflection of light at the upper surface of shallow layers of hot air in contact with the ground, the appearance being that of pools of water in which are seen inverted images of more distant objects. Other types of mirage are seen in polar region, where there is a dense, cold layer of air near the ground. See **fata morgana**.

mirage *(Telecomm.)*. Radio signal from which transmission path includes reflection from layer of rarified air, in analogous manner to the above.

Mira Stars *(Astron.)*. Long-period variable stars named after Mira Ceti; more than 3000 are known, with periods from 2 months to 2 years, and all are red giant stars.

mire *(Ecol.)*. Synonymous with **bog**.

mired *(Image Tech.)*. *MIcro-REciprocal Degree*, a measure of colour temperature being one-million divided by the value in Kelvins.

mirror *(Phys.)*. A highly-polished reflecting surface capable of reflecting light rays without appreciable diffusion. The commonest forms are plane, spherical (convex and concave) and paraboloidal (usually concave). The materials used are glass silvered on the back or front, speculum metal, or stainless steel.

mirror arc *(Image Tech.)*. A projector lamp-house system in which the source of light, a carbon or xenon arc, is at the focus of a parabolic mirror which concentrates the illumination on the gate.

mirror drum *(Image Tech.)*. A rotating cylinder or cone with a number of mirrors on its circumference to reflect a beam of light in **mechanical scanning**; still of possible application in projection TV systems.

mirror finish *(Eng.)*. A very smooth, lustrous surface finish produced, for example, on stainless steels and other metals by electrolytic polishing or lapping, and on electroplated surfaces by mechanical polishing.

mirror galvanometer *(Elec.Eng.)*. A galvanometer having a mirror attached to the moving part, so that the deflection can be observed by directing a beam of light on to the mirror and observing the movement of the reflection of this over a suitable scale. Also called *reflecting galvanometer*.

mirror lens *(Image Tech.)*. A compact **telephoto** system employing two concave reflecting surfaces to form the image; because a diaphragm cannot be included, exposure must be controlled by neutral density filters or by shutter time.

mirror machine *(Nuc.Eng.)*. Fusion machine using the **magnetic mirror** principle to trap high-energy ions in a plasma.

mirror nuclides *(Phys.)*. Those with the same number of nucleons, but with proton and neutron numbers interchanged.

mirror reflector *(Telecomm.)*. Surface or set of metal rods which reflects a wave geometrically.

mirror shutter *(Image Tech.)*. In a motion picture camera, a shutter having one surface of its opaque blades as mirrors, which reflect light from the lens into a viewfinder system when the shutter is closed.

mirror symmetry *(Phys.)*. See **parity**.

miscarriage *(Med.)*. Expulsion of the foetus before the 28th week of pregnancy. Loosely, abortion.

miscibility *(Chem.)*. The property enabling two or more liquids to dissolve when brought together and thus form one phase.

miscibility gap *(Chem.)*. The region of composition and temperature in which two liquids form two layers or phases when brought together.

MISD *(Comp.)*. *Multiple Instruction stream, Single Data stream.* Term which describes the architecture of the **processor**. See **pipelining**.

miser *(Build.)*. A large *auger* used for boring holes in the ground in wet situations.

misfire *(Elec.Eng.)*. Failure to establish, during an intended conducting period, a discharge or arc in a gas-discharge tube or a mercury-pool rectifier.

misfiring *(Autos.)*. The failure of the compressed charge to fire normally, generally due either to ignition failure or to an over-rich or weak mixture.

mismatch *(Telecomm.)*. Load impedance of incorrect value for maximum energy transfer. See **matching**.

mispickel *(Min.)*. See **arsenopyrite**.

missile *(Aero.,Space)*. There are two basic types of missile, in the current sense, *guided* and *ballistic*. The former is controlled from its launch until it hits its target; the latter, always of long-range, surface-to-surface type, is controlled into a precision ballistic path so that its course cannot be deflected by counter-measures. See **guided missile**.

mission *(Space)*. Succession of events which must happen to achieve the objectives stipulated; it includes everything which must be done from conception to the delivery of the results (more loosely, the actual flight of the spacecraft).

mission adaptive wing *(Aero.)*. A wing whose section profile is automatically adjusted to suit different flight conditions, e.g. Mach number, lift and altitude.

mission control centre *(Space)*. Room or building in which are assembled the means necessary to visualize and control a space system so that its mission objectives can be achieved.

mission specialist *(Space)*. Member of the crew of the **Space Shuttle** whose responsibilities are concerned with mission aspects, such as the control of the Orbiter's resources to a payload, the handling of payload equipment and the performance of the experiments in orbit.

Mississippian *(Geol.)*. A period of the Upper Palaeozoic era, lying between the Devonian and Pennsylvanian. It covers an app. time-span from 360–330 million years and is the North American equivalent of the *Lower Carboniferous* in Europe. The corresponding system of rocks.

mist *(Chem.)*. A suspension, often colloidal, of a liquid in a gas.

mist *(Meteor.)*. A suspension of water droplets (radii less than 1 μm) reducing the visibility to not less than 1 km. See **fog**.

mist coat *(Build.)*. A highly atomized thin coat of paint applied by spray. Can be applied prior to the application of a 'full' coat or is often used as a finishing process on spraying cellulose.

mitochondrion *(Biol.)*. Thread-like membranous cytoplasmic organelle whose major function is the generation of ATP. This takes place on infoldings of the inner membrane of the mitochondrion known as *cristae*.

mitogen *(Immun.)*. Any agent which induces mitosis in cells. In immunology often used to refer to substances such as lectins or lipopolysaccharides which cause a wide range of T- or B-lymphocytes to undergo cell division.

mitosis *(Biol.)*. The normal process of somatic cell division, in which each of the daughter cells is provided with a chromosome set identical to that of the parent cell. adj. *mitotic*. Cf. **amitosis**, **meiosis**.

mitospore *(Bot.)*. Spore formed by mitosis and hence having the same number of chromosomes as the parent.

mitotic crossing-over *(Bot.)*. The exchange of genetic material between homologous chromosomes during mitosis, resulting in genetic **recombination**. See **crossing-over**.

mitotic index *(Biol.)*. The proportion in any tissue of dividing cells, usually expressed as per thousand cells.

mitral *(Med.)*. Pertaining to, or affecting, the mitral valve, or valves, of the heart.

mitral *(Zool.)*. Mitre-shaped or *mitriform*; as the *mitral valve*, guarding the left auriculo-ventricular aperture of the heart in higher Vertebrates, or the *mitral layer*; of the olfactory bulb, composed of mitre-shaped cells.

mitral stenosis *(Med.)*. Narrowing of the communication between the left atrium and the left ventricle of the heart, as a result of disease of the mitral valves.

mitral valve *(Zool.)*. See **bicuspid valve**.

mitre *(Build.)*. A joint between two pieces at an angle to one another, each jointing surface being cut at an angle.

mitre block *(Build.)*. A block of wood rebated along one edge and having saw-cuts in the part above the rebate, with the kerfs inclined at 45° to the face of the rebate so as to guide the saw when cutting mouldings for a mitred joint.

mitre board *(Build.)*. See **mitre shoot**.

mitre box *(Build.)*. An open-ended box having saw-cuts in the sides at 45° to the length of the box; used like the mitre block but capable of taking deeper mouldings.

mitre cramp, clamp *(Build.)*. For holding together temporarily the two parts of a mitre joint.

mitre-cut piston-ring *(Eng.)*. A piston-ring in which the ends are mitred at the joint, as distinct from stepped or square ends.

mitre dovetail *(Build.)*. See **secret dovetail**.

mitred valley *(Build.)*. See **cut-and-mitred valley**.

mitre saw *(Build.)*. See **tenon saw**.

mitre-saw cut *(Build.)*. A device, such as a mitre-block or box, for keeping the saw at the required angle to the work when cutting mouldings for a mitre joint. Also *mitre-sawing board*.

mitre shoot *(Build.)*. A block of wood rebated along one edge as a guide for a jointing plane, and having a pair of wood strips fixed to the top face of the part above the rebate, at 45° to the face of the rebate, so as to hold the mitre face of a moulding at the right angle to the plane while it is being shot. Also called *mitre board*.

mitre sill *(Hyd.Eng.)*. The raised part of the bed of a canal lock against which the parts of the gates abut in closing. Also called *clap-sill, lock-sill*.

mitre square *(Build.)*. A tool similar to the bevel, having the blade at 45° to the stock.

mitre wheels *(Eng.)*. See **bevel gear**.

mitriform *(Zool.)*. See **mitral**.

Mitscherlich's law of isomorphism *(Chem.)*. Salts having similar crystalline forms have similar chemical constitutions.

mixed *(Zool.)*. Said of nerve trunks containing motor and sensory fibres.

mixed bud *(Bot.)*. A bud containing young foliage leaves and also the rudiments of flowers or of inflorescences.

mixed coupling *(Telecomm.)*. Simultaneous inductive and capacitance coupling between two resonant circuits.

mixed crystal *(Crystal.)*. A crystal in which certain atoms of one element are replaced by those of another.

mixed-flow (American) water turbine *(Eng.)*. An inward-flow reaction turbine in which the runner vanes are so curved as to be acted on by the water as it enters radially and as it leaves axially.

mixed forme base *(Print.)*. Precisely-machined units of aluminium alloy, in a selection of point sizes, which are assembled together with type to form a base for moumting an individual plate; in heights to suit *original* and *duplicate* plates.

mixed highs *(Image Tech.)*. In some colour TV systems, high frequency components of the picture, representing fine detail, are transmitted or recorded and reproduced as a mixed (**monochrome**) signal to conserve the band-width necessary.

mixed inflorescence *(Bot.)*. An inflorescence in which some of the branching is racemose and some is cymose.

mixed melting point *(Chem.)*. Technique used in the identification of chemical compounds, particularly organic, whereby a sample of known identity and melting point is mixed with a purified unknown sample and the melting point determined. Generally the mp of the known sample is lowered if the two samples are not identical. See **molecular depression of freezing point**.

mixed-pressure turbine *(Eng.)*. A steam turbine operated from two or more sources of steam at different pressures, the low-pressure supply, from, for example, the exhaust of other engines, being admitted at the appropriate pressure stage.

mixed service *(Telecomm.)*. Service provided by a PBX to the main exchange for a number of extension lines only.

mixer *(Build.)*. See **concrete mixer**.

mixer *(Telecomm.)*. (1) Collection of variable attenuators, hand controlled, which allows the combination of several transmissions to be independently adjusted from zero to maximum. A *group mixer* or *fader* deals with several groups of transmissions. Hence *mixer*, the person who operates mixers. (2) Frequency conversion stage in a **superhet receiver**. See **frequency changer**.

mixer-settlers *(Chem.Eng.)*. A countercurrent *liquid-liquid extractor*, consisting of a series of tanks in which the two liquids are alternately dispersed in one another (*mixers*) and separated by gravity (*settlers*).

mixing *(Image Tech.)*. General term for the combination of two or more picture or sound signals. In particular, the dubbing of several sound recordings into a single mixed track.

mixing *(Phys.)*. In gaseous isotope separation, the process of reducing the concentration gradient for the lighter isotope close to the diffusion barrier.

mixing *(Textiles)*. Mechanically blending staple fibres, especially wools or cottons of similar staple and colour, to obtain the most suitable material for spinning uniform yarns economically.

mixing efficiency *(Phys.)*. A measure of the effectiveness of the mixing process in isotope separation.

mixing length *(Meteor.)*. The average distance travelled by an eddy which is transporting heat, momentum or water vapour in the atmosphere.

mixotrophic *(Zool.)*. Combining two or more fundamental methods of nutrition, as certain *Mastigophora* which combine holophytic with saprophytic nutrition, or as a partial parasite.

mixture *(Autos.)*. The combined inflammable gas and air constituting the explosive charge.

mixture control *(Aero.)*. An auxiliary control fitted to a carburettor to allow of variation of mixture strength with altitude. May be manually operated or automatic.

mizzonite *(Min.)*. One of the series of minerals forming the scapolite group, consisting of a mixture of the meionite and marialite molecules. Mizzonite includes those minerals with 50–80% of the meionite end-member molecule. Found in metamorphosed limestones and in some altered basic igneous rocks.

MKSA *(Phys.)*. *Metre-Kilogram(me)-Second-Ampere* system of units, adopted by the International Electrotechnical Commission, in place of all other systems of units. See also **SI units**.

ml *(Genrl.)*. Abbrev. for *millilitre*.

MLC mixed lymphocyte culture *(Immun.)*. Culture together of lymphocytes from two individuals for 3 to 5 days, at the end of which the number of dividing cells is measured, usually by incorporation of ^{3}H-thymidine. This provides a measure of the extent to which the histocompatibility antigens of the two differ, and is used to assess the suitability for tissue transplantation.

MLD *(Radiol.)*. Abbrev. for *Mean Lethal Dose*.

***M*-lines** *(Phys.)*. Characteristic X-ray frequencies from atoms due to the excitation of electrons from the *M*-shell. Only developed in atoms of high atomic number.

M-loop *(Image Tech.)*. Tape path employed in **VHS** videotape recorders, giving rather more than 180° wrap on the drum; from the M-shaped tape guide in the mechanism.

mm *(Genrl.)*. Abbrev. for *millimetre*.

MMA *(Chem.)*. *MonoMethyl Aniline, N-methyl-aminobenzene*, C_6H_5.NH.CH_3. Used occasionally as an **anti-knock substance**, as an alternative to lead tetraethyl.

mmf *(Elec.Eng.)*. Obsolete abbreviation for *MicroMicroFarad*. Now replaced by *picofarad, pF*.

m.m.f. *(Phys.)*. Abbrev. for *MagnetoMotive Force*.

MMM (3M) overlay *(Print.)*. A special material in which an impression of the entire letterpress forme is taken and which, after heat treatment, becomes a **mechanical overlay**.

Mn *(Chem.)*. The symbol for *manganese*.

M_{ne} *(Aero.)*. Abbrev. for the maximum permissible indicated **Mach number**: a safety limitation, the suffix means 'never exceed' because of strength or handling considerations. The symbol is used mainly in operational instructions.

mnemonics *(Behav.)*. Rules of learning that improve recall (e.g. rhyming).

M_{no} *(Aero.)*. Abbrev. for *Normal Operating Mach number*, usually of a jet airliner, the term being used mainly in flight operations instructions for flight levels above 7600 m.

mo *(Build.)*. An abbrev. for *MOulded*.

Mo *(Chem.)*. The symbol for *molybdenum*.

mobbing *(Behav.)*. A form of harassment directed at predators by potential prey.

mobile belt *(Geol.)*. A long zone of the Earth's crust associated with igneous activity and deformation. Traditionally associated with geosynclinal development. See **plate tectonics**.

mobile element *(Biol.)*. DNA sequence capable of excising itself from a chromosome and then re-integrating itself, or its copies, into different sites in the chromosomes.

mobile phase *(Chem.)*. See **chromatography**.

mobility *(Electronics)*. (1) The freedom of particles to move, either in random motion or under the influence of fields or forces. (2) The average drift velocity of charge carriers per unit electric field in a homogeneous semiconductor. Mobility affects conductivity and Hall effect; the mobility of holes and electrons differs greatly.

Möbius strip *(Maths.)*. The one-sided surface formed by joining together the two ends of a long rectangular strip, one end being twisted through 180° before the join is made.

Möbius transformation *(Maths.)*. See **bilinear transformation**.

Mocha stone *(Min.)*. See **moss agate**.

mock moons *(Meteor.)*. Lunar images similar to **mock suns**. Also called *paraselenae*.

mock suns *(Meteor.)*. Images of the sun, not usually very well defined, seen towards sunset at the same altitude as the sun and 22° from it on each side. They are portions of the 22° ice *halo* formed by ice crystals which, for some reasons, are arranged with their axes vertical. Also called *parhelia*.

modacrylic fibre *(Textiles)*. Fibres made from synthetic linear polymers containing 35–85% of acrylonitrile groups.

modal fibre *(Textiles)*. Regenerated cellulose (viscose rayon) of high tenacity and high wet strength.

modal interval *(Stats.)*. The class interval corresponding to the largest frequency in a tabulation of observations into class intervals.

modality *(Zool.)*. A category of sensation, e.g. touch, smell or sight.

modal position *(Telecomm.)*. The position assumed by a telephone handset when its earpiece is in close contact with the ear of a person whose head has dimensions of *modal value*, itself derived from a representative sample.

modal value *(Stats.)*. See **mode**.

mode *(Geol.)*. The actual mineral composition of a rock expressed quantitatively in percentages. Cf. **norm**.

mode *(Phys.)*. (1) One of several electromagnetic wave frequencies which a given oscillator may generate, or to which a given resonator may respond, e.g., magnetron modes, tuned line modes. In a waveguide, the mode gives the number of half-period field variations parallel to the transverse axes of the guide. Similarly, for a cavity resonator, the half-period variations parallel to all three axes must be specified. In all cases, different modes will be characterized by different field configurations. (2) Similarly, one of several frequencies of mechanical vibration which a body may execute or with which it may respond to a forcing signal. (3) A well-defined distribution of the radiation amplitude in a cavity which results in the corresponding distribution pattern in the laser output beam. In a multimodal system the beam will tend to diverge.

mode *(Space)*. Situation or method of performing a specified task.

mode *(Stats.)*. The most frequent value in a set of observations; the value of a random variable at which the corresponding probability density function is a maximum.

mode *(Telecomm.)*. In optical fibres, the manner in which light rays travel inside the fibre. There are a variety of paths because the light can be reflected internally at a variety of angles. See **monomode fibre, multimode fibre**.

mode dispersion *(Telecomm.)*. In optical fibre communications, distortion or *smearing* of individual pulsed components of digital signals, caused by the different modes of propagation of the light inside the fibre arriving at the receiver at different times. Pulses need to be detected and re-generated before this leads to distortion and/or errors.

mode jump *(Telecomm.)*. Switch of a **magnetron** or similar microwave generator from one mode to another; also *mode shift*.

modelling *(Behav.)*. In behaviour therapy, the learning of a new behaviour by imitation of a model, and usually overtly or covertly reinforcing the desired behaviour.

mode-locking *(Phys.)*. A technique for producing laser pulses of extremely short duration. Laser cavities have modes with frequency spacing $c/2L$ where c is the speed of light and L is the length of the cavity. Oscillations can occur in any mode as long as its frequency is within the natural line width of the laser transition. If a property of the cavity is modulated at a frequency of $c/2L$, then all the modes become coherently coupled. A train of extremely short pulses is emitted where the time duration is roughly the inverse of the line width. See **laser**.

modem *(Comp.,Telecomm.)*. Device which converts between digital **bits** and analogue electrical impulses which can be transmitted as a frequency modulated tone over telephone circuits. *MOdulator/DEModulator*.

mode number(s) *(Electronics)*. These indicate the mode in which devices capable of operating with more than one field configuration are actually being used, e.g. in a cavity resonator, the mode numbers indicate the number of half wavelengths in the field pattern parallel to the three axes; in a magnetron, the mode number gives the number of cycles through which the phase shifts in one circuit of the anode; and in a klystron, it gives the number of cycles of the field which occur while an electron is in the field-free drift space.

moder *(Bot.)*. Form of humus intermediate between **mull** and **mor**.

moderation *(Nuc.Eng.)*. The slowing down of neutrons in a reactor to thermal energies.

moderation *(Phys.)*. See **degradation**.

moderator *(Nuc.Eng.)*. Material such as water, heavy water, graphite used to slow down neutrons in a reactor. See **lethargy of neutrons**.

moderator control *(Nuc.Eng.)*. Control of a reactor by varying the position or quantity of the moderator.

modern face *(Print.)*. A style of type with contrasting thick and thin strokes, serifs at right angles, curves thickened etc. See **type**.

mode separation *(Telecomm.)*. The frequency difference between operation of a microwave tube in adjacent modes. See **mode jump**.

mode shift *(Telecomm.)*. See **mode jump**.

modified refractive index *(Meteor.)*. Sum of the refractive index of the atmosphere at a given height and the ratio of the height to the radius of the earth.

modifier *(Biol.)*. A gene which influences the effect of another.

modifier *(Comp.)*. A code element used to alter the address of an operand. See **address calculation**.

modifier *(Min.Ext.)*. Modifying agent used in froth flotation to increase *either* the wettability *or* the water-repelling quality of one or more of the minerals being treated.

modiolus *(Zool.)*. The conical central pillar of the cochlea.

modular *(Electronics)*. Form of construction in which units, often with differing functions, are therefore quickly interchangeable.

modular design *(Arch.)*. A design on a grid of fixed dimensions, generally to facilitate the prefabricated manufacture of building components.

modular programming *(Comp.)*. An approach to programming in which separate logical tasks are programmed separately and joined later. See **structured programming, segmentation**.

modular ratio *(Civ.Eng.)*. The ratio betweeen Young's modulus for steel and that for the concrete in any given case of reinforced concrete.

modulated amplifier *(Telecomm.)*. An amplifier stage during which modulation of the signal is carried out. Also *modulated stage*.

modulated carrier, wave *(Telecomm.)*. A frequency which can be transmitted or received through space or a transmission circuit, with a superposed information signal, which, by itself, could not be effectively transmitted and received.

modulated continuous wave *(Telecomm.)*. Transmission in which a carrier is modulated by a tone and interrupted by keying. Abbrev. *MCW*.

modulated stage *(Telecomm.)*. See **modulated amplifier**.

modulating electrode *(Electronics)*. That of a thermionic valve to which a voltage is applied to control the size of the beam current.

modulation *(Acous.)*. (1) Change of amplitude or frequency of a carrier signal of given frequency. (2) Changing from one key to another in music. The continual change from one fundamental frequency to another in speech.

modulation *(Telecomm.)*. The process of impressing information (code, speech, video, data etc.) onto a higher frequency **carrier**. See **frequency modulation, delta modulation, phase modulation, frequency-shift transmission, dual modulation, pulse code-, pulse width-** and **pulse-amplitude modulation**.

modulation capability *(Telecomm.)*. The maximum percentage modulation which can be used without exceeding a specified distortion level.

modulation condition *(Telecomm.)*. The condition of voltages and currents in an amplifier for a modulated signal when the carrier is steadily modulated to a stated degree, e.g. 100%.

modulation depth *(Telecomm.)*. Factor indicating extent of amplitude modulation of a carrier. *Difference* ÷ *sum* of peak and trough values of a modulated wave. Often expressed as a percentage.

modulation distortion *(Telecomm.)*. Any distortion in the transmission of a signal introduced during the process of modulating that signal onto the carrier.

modulation frequency *(Telecomm.)*. One impressed upon a carrier wave in a modulator.

modulation index *(Telecomm.)*. In a *frequency modulation* system and in the case of a simple sinusoidal modulating signal, the ratio of the frequency deviation to the frequency of the modulating signal.

modulation pattern *(Telecomm.)*. That on an oscilloscope when the amplitude-modulated wave is connected to the

Y-deflection system and the modulation signal to the X-deflection plates. The result is a trapezoidal pattern which enables the modulation depth to be measured.

modulation rate *(Telecomm.)*. The reciprocal of the shortest time interval between successive significant instances of the modulating signal. If the interval is in seconds, the modulation rate is given in **baud**.

modulation transformer *(Telecomm.)*. One which applies the modulating signal to the carrier-wave amplifier in a transmitter.

modulator *(Acous.)*. Circuit in an electronic organ which changes the pitch of the notes.

modulator *(Telecomm.)*. Any circuit unit which modulates a radio carrier, at *high level* directly for transmission, or at *low level* for amplification of the modulated carrier before transmission.

module *(Arch.)*. The radius of the lower end of the shaft of a column.

module *(Build.)*. Unit of size used in the standardized planning of buildings and design of components.

module *(Eng.)*. Of a gear-wheel, the pitch diameter divided by the number of teeth. The reciprocal of **diametrical pitch**.

module *(Maths.)*. A left module M over a ring R is a set of elements with an additional operation defined under which they form a commutative group, and a multiplication defined on pairs $r \times m$, where r is in R and m in M, such that multiplication is associative, and distributive over addition.

module *(Space)*. A separate, and separable, compartment of a space vehicle.

modulus *(Eng.)*. Various moduli determine the deflection of a material under stress. Each is the ratio of stress to strain, which is a constant for a given material up to the elastic limit stress. See **elasticity** and **bulk modulus**.

modulus *(Genrl.)*. Constant for units conversion between systems.

modulus *(Maths.)*. The absolute value of a number. The modulus of the complex number $z = x + iy$ is

$$|z| = +\sqrt{x^2 + y^2}$$

value. Sometimes called *absolute value norm*. See also **amplitude, complex number**.

modulus of elasticity *(Phys.)*. For a substance, the ratio of stress to strain within the elastic range, i.e., where Hooke's law is obeyed. Measured in units of stress, e.g. GN/m^2. See also **elasticity**.

modulus of rigidity *(Eng.)*. See **elasticity of shear**.

modulus of rupture *(Eng.)*. A measure of the ultimate strength of the breaking load per unit area of a specimen, as determined from a torsion or a bending test.

Moebius process *(Eng.)*. An electrolytic process for parting gold-silver bullion. The electrolyte is silver nitrate. Bullion forms the anode, silver passes into solution and is deposited on the cathode. Gold remains on the anode.

moellon *(Build.)*. A rubble filling between the facing walls of a structure, sometimes laid in mortar.

Moerner's test *(Chem.)*. A test for the presence of tyrosine, based on the appearance of a green colour when the solution is heated with a mixture of formalin and sulphuric acid.

mofette *(Geol.)*. A volcanic opening through which emanations of carbon dioxide, nitrogen, and oxygen pass. It marks the last phase of volcanic activity.

mogas *(Aero.)*. Abbrev. for *MOtor GASoline*, 91 to 93 octane.

mohair *(Textiles)*. The long fine hair from the angora goat, *Capra hircus aegagrus*.

Mohr balance *(Chem.)*. Balance used to determine density by weighing a solid when suspended in air and in a liquid.

Mohr-Coulomb theory *(Min.Ext.)*. The resistance of rock to crushing is due to internal friction *plus* cohesion of bonding materials.

Mohr's salt *(Chem.)*. *Ammonium iron (II) sulphate*. $(NH_4)_2SO_4.FeSO_4.6H_2O$.

Mohs' scale of hardness *(Min.)*. A scale introduced by Mohs to measure the hardness of minerals. See **hardness**.

moil *(Glass)*. (1) Glass left on a *punty* or blowing iron after the gather has been cut off or after a piece of ware has been blown and severed. (2) Glass originally in contact with the blowing mechanism or head, which becomes **cullet** after the desired article is severed from it.

moiré effect *(Image Tech.)*. Pattern formed by interference or combination between two sets of regular divisions, for instance between two line-screens in printing or between the TV **raster** and a striped object within the scene.

moiré fibre *(Textiles)*. A ribbed fabric in which the yarns have been partially flattened by heat and pressure in calendering. This gives rise to the optical interference effect commonly know as *watered silk*.

moiré fringe *(Phys.)*. A set of dark fringes produced when two ruled gratings or uniform patterns are superimposed. The separation D of the moiré fringes is equal to d/θ where d is the line spacing of the grating and θ is the angle of intersection of the gratings. Moiré fringes can be used to measure the displacement of one ruled pattern with respect to the other to a high degree of precision.

moiré pattern *(Print.)*. A regular patterned effect formed by superimposing two or more sets of lines or dots of different pitch, or at certain angles; a defect to be avoided, especially in half-tone reproduction and in half-tone 4-colour process work.

moisture content *(Textiles)*. See **regain**.

moisture expansion *(Build.,Civ.Eng.)*. Increase in the volume of a material from absorption of moisture. Also *bulking*.

mol *(Chem.)*. See **mole**.

molality *(Chem.)*. The concentration of a solution expressed as the number of moles of dissolved substance per kilogram of solvent.

molal specific heat capacity *(Phys.)*. The *specific heat capacity* of 1 mole of an element or compound. Also called *volumetric heat* (for gases).

molar absorbance *(Chem.)*. The **absorbance** of a solution with a concentration of $1\,mol\,dm^{-3}$ measured in a cell of a thickness of 1 cm.

molar conductance *(Chem.)*. The conductance which a solution would have if measured in a cell large enough to contain one mole of solute between electrodes 1 cm apart.

molar conductivity *(Phys.)*. The electrical conductivity of an electrolyte having 1 mole of solute in $1\,dm^3$ of solution. Measured in $S\,cm^2/mol$.

molar heat *(Chem.)*. See **molar heat capacity**.

molar heat capacity *(Chem.)*. The heat required to raise the temperature of a substance by 1 K. The symbol for that measured at constant volume is C_V and for that at constant pressure C_P.

molarity *(Chem.)*. The concentration of a solution expressed as the number of moles of dissolved substance per dm^3 of solution.

molars *(Zool.)*. The posterior grinding or cheek teeth of Mammals which are not represented in the milk dentition.

molar surface energy *(Chem.)*. The surface energy of a sphere containing 1 mole of liquid; equal to $\gamma V^{2/3}$, where γ is the surface tension and V the molar volume. It is zero near the critical point and its temperature coefficient is often a colligative property. See **Eötvös equation**.

molar volume *(Chem.)*. The volume occupied by one mole of a substance under specified conditions. That of an ideal gas at s.t.p. is $2.2414 \times 10^{-2}\,m^3\,mol^{-1}$.

molasse *(Geol.)*. Sediments produced by erosion of mountain ranges following the final stages of an *orogeny*. Sandstones and other detrital rocks are the dominant products of this type of sedimentation. Cf. **flysch**.

molasses *(Chem.)*. Residual sugar syrups from which no crystalline sugar can be obtained by simple means. An important raw material of ethyl and other alcohols.

moldavite *(Min.)*. A type of **tektite**.

mold, molding *(Foundry)*. A variant spelling of *mould*, *moulding* etc.

mole *(Chem.)*. The amount of substance that contains as many entities (atom, molecules, ions, electrons, photons etc.) as there are atoms in 12 g of ^{12}C. It replaces in SI the older terms *gram-atom, gram-molecule* etc., and for any chemical compound will correspond to a mass equal to the relative molecular mass in grams. Abbrev. *mol.* See **Avogadro number**.

mole *(Civ.Eng.)*. A breakwater or masonry pier.

mole *(Med.)*. (1) **Naevus**. (2) A haemorrhagic mass formed in the Fallopian tube as a result of bleeding into the sac enclosing the embryo.

molecular *(Chem.)*. (1) Pertaining to a molecule or molecules. (2) Pertaining to 1 mole.

molecular association *(Chem.)*. The relatively loose binding together of the molecules of a liquid or vapour in groups of two or more.

molecular beam *(Chem.)*. Directed stream of un-ionized molecules issuing from a source and depending only on their thermal energy.

molecular biology *(Biol.)*. The study of the structure and function of macromolecules in living cells. Notably successful in explaining the structure of proteins, and the role of DNA as the genetic material, but with the ultimate aim of explaining as completely as possible in molecular terms the biology of cells and organisms. It is not primarily concerned with metabolic pathways or with the chemistry of natural products.

molecular distillation *(Chem.)*. The distillation, in a vacuum, of a labile material in which the temperature and time of heating are minimized by providing a condensation surface at such a distance from the evaporating surface that molecules can reach the condenser without intermolecular collision.

molecular electronics *(Electronics)*. The technique of growing solid-state crystals so as to form transistors, diodes, resistors, in a single mass, for microelectronic devices.

molecular elevation of boiling point *(Chem.)*. The rise in the boiling-point of a liquid which would be produced by the dissolution of 1 mole of a substance in 1 kg of the solvent, if the same laws held as in dilute solution.

molecular filter *(Chem.)*. Alternative name for **membrane filter**.

molecular formula *(Chem.)*. A representation of the atomic composition of a molecule. When no structure is indicated, atoms are usually given in the order C, H, other elements alphabetically. Functional groups may be written separately, thus sulphanilamide may be written $C_6H_8N_2O_2S$ or $H_2N.C_6H_4.SO_2NH_2$.

molecular genetics *(Biol.)*. The study and manipulation of the molecular basis of heredity.

molecular heat *(Chem.)*. See **molar heat capacity**.

molecular models *(Crystal.)*. Three dimensional models of the structures of many molecules including complex molecules like proteins and DNA have been used as an aid both in determining their structure and understanding their function. *Space filling models* use truncated spheres of diameters corresponding to the atomic radius which can be fitted together to give the proper bond angles. *Stick models* show only the positions of a special repeating feature in the structure such as the *alpha carbon* of a peptide. These positions can be represented by a marker on a rod (*stick*) which can be set to the appropriate three-dimensional co-ordinates. These mechanical models are becoming redundant as computer modelling has become widely available.

molecular orbital *(Chem.)*. A wave-function defining the energy of an electron in a molecule. Molecular orbitals are often constructed from linear combinations of the atomic orbitals of the constituent atoms.

molecular refraction *(Chem.)*. See **Lorentz-Lorenz equation**.

molecular rotation *(Chem.)*. One-hundredth of the product of the specific rotation and the relative molecular mass of an optically-active compound.

molecular sieve *(Chem.)*. Framework compound, usually a synthetic zeolite, used to absorb or separate molecules. The molecules are trapped in 'cages', the sizes of which can be selected to suit solvent.

molecular stopping power *(Phys.)*. The energy loss per molecule per unit area normal to the motion of the particle in travelling unit distance. It is *approximately* equal to the sum of the atomic stopping powers of the constituent atoms.

molecular streaming *(Chem.Eng.)*. See **Knudsen flow**.

molecular structure *(Chem.)*. The way in which atoms are linked together in a molecule.

molecular volume *(Chem.)*. The volume occupied by 1 mole of a substance in gaseous form at standard temperature and pressure (approx. $2.2414 \times 10^{-2}\,m^3$).

molecular weight *(Chem.)*. Better *relative molecular mass*. (1) The mass of a molecule of a substance referred to that of an atom of ^{12}C taken as 12.000. (2) The sum of the relative atomic masses of the constituent atoms of a molecule.

molecule *(Chem.)*. An atom or a finite group of atoms which is capable of independent existence and has properties characteristic of the substance of which it is the unit. Molecular substances are those which have discrete molecules, such as water, benzene or haemoglobin. Diamond, sodium chloride and zeolites are examples of non-molecular substances.

mole fraction *(Chem.)*. Fraction, of the total number of molecules in a phase, represented by a given component.

moler *(Build.)*. A diatomaceous earth used in the manufacture of light-weight temperature resistant components.

moleskin *(Textiles)*. A heavy fustian type of cotton fabric, with smooth face and twill back; used for working clothes.

Mollier diagram *(Chem.Eng.)*. A diagram, for any substance relating total heat and entropy, in which heat and work quantities are represented by line segments, not areas. Used to calculate the efficiency of steam engines or refrigeration cycles.

mollites ossium *(Med.)*. See **osteomalacia**.

Mollusca *(Zool.)*. Unsegmented coelomate Invertebrates with a head (usually well developed), a ventral muscular foot and a dorsal visceral hump; the skin over the visceral hump (the *mantle*) often secretes a largely calcareous shell and encloses a mantle cavity into which open the anus and kidneys and in which are the **ctenidia**, originally used for gaseous exchange. There is usually a **radula**; some have haemocyanin as a respiratory pigment; well-developed blood and nervous systems; often with larva of the **trochophore** type. Chitons, Slugs, Snails, Mussels, Whelks, Limpets, Squids, Cuttlefish, and Octopods.

molybdates (VI) *(Chem.)*. Salts of molybdic acid.

molybdenite *(Min.)*. Disulphide of molybdenum, crystallizing in the hexagonal system. It is the most common ore of molybdenum, and occurs in lustrous lead-grey crystals in small amounts in granites and associated rocks.

molybdenosis *(Vet.)*. *Teart*. A disease of cattle and sheep caused by an excess of molybdenum in the diet; characterized by chronic diarrhoea and emaciation. Occurs where high levels of molybdenum are present in the soil.

molybdenum *(Chem.)*. A metallic element in the sixth group of the periodic system. Symbol Mo, at.no. 42, r.a.m. 95.94, rel.d. at 20°C 10.2, hardness 147 (Brinell), mp 2625°C, bp 3200°C, resistivity *c.* 5×10^{-8} ohm metres. Its physical properties are similar to those of iron, its chemical properties to those of a nonmetal. Used in the form of wire for filament supports, hooks etc. in electric lamps and radio valves, for electrodes of mercury-vapour lamps, and for winding electric resistance furnaces. Is added to a number of types of alloy steels, certain types of Permalloy and Stellite. It seals well to Pyrex, spot-welds to iron and steel.

molybdenum blues *(Chem.)*. A variety of reducing agents (including *molybdenum* (III)) convert molybdates into colloidal molybdenum blues, the compositions of which approach $Mo_8O_{23}xH_2O$.

molybdenum (VI) oxide *(Chem.)*. MoO_3. Behaves as an acid anhydride, forming molybdic acid. The essential starting-point in the manufacture of molybdenum metal, and most other compounds of molybdenum.

molybdic acid *(Chem.)*. H_2MoO_4. See also **molybdenum (VI) oxide**.

molybdite *(Min.)*. Strictly, molybdenum oxide. Much so-called molybdite is ferrimolybdite, a hydrated ferric molybdate which crystallizes in the orthorhombic system. It is commonly impure and occurs in small amounts as an oxidation product of molybdenite. Also called *molybdic ochre*.

moment *(Aero.)*. See **hinge-**, **pitching-**, **rolling-**, **yawing-**.

moment *(Elec.Eng.)*. See **electric-**.

moment *(Phys.)*. (1) Of a couple, see **couple**. (2) Of a force or vector about a point, the product of the force or vector and the perpendicular distance of the point from its line of action. In vector notation, $\mathbf{r} \wedge \mathbf{F}$, where **r** is the position vector of the point, and **F** is the force or vector. (3) Of a force or vector about a line, the product of the component of the force or vector parallel to the line and its perpendicular distance from the line.

moment *(Stats.)*. The expected value of the nth power of a random variable, where *n* is an integer indexing the set of moments.

moment distribution *(Civ.Eng.)*. A method of analysing the forces in a continuous structure by adjusting between the imposed loads, spans and sectional properties of the supporting members concerned.

moment of a magnet *(Elec.Eng.)*. See **magnetic moment**.

moment of inertia *(Phys.)*. Of a body about an axis: the sum Σmr^2 taken over all particles of the body where *m* is the mass of a particle and *r* its perpendicular distance from the specified axis. When expressed in the form Mk^2, where *M* is the total mass of the body ($M = \Sigma m$), *k* is called the *radius of gyration* about the specified axis. Also used erroneously for *second moment of area*.

moment to change trim one inch *(Ships)*. The moment, taken about the *centre of flotation* which will change the *trim* by 1 in. Expressed in foot-tons.

momentum *(Phys.)*. A dynamical quantity, conserved within a closed system. A body of mass *M* and whose centre of gravity *G* has a velocity v has a *linear momentum* of *M*v. It has an *angular momentum* about a point *O* defined as the moment of the linear momentum about *O*. About *G* this reduces to $I\omega$ where *I* is the moment of inertia about *G* and ω the angular velocity of the body.

momentum wheel *(Space)*. A flywheel, part of an attitude control system, which stores momentum by spinning; three wheels with their axes at right angles can serve to stabilize a satellite's attitude. Also *inertia wheel*.

monacid *(Chem.)*. Containing one hydroxyl group, replaceable by an acid radical, with the formation of a salt.

monad *(Bot.)*. (1) A flagellated unicellular organism. (2) A single pollen grain, not united with others. Cf. **tetrad**.

monadelphous *(Bot.)*. Having all the stamens in the flower joined together by their filaments e.g. many Leguminosae.

monandrous *(Bot.)*. (1) Having 1 antheridium. (2) Having 1 stamen.

monarch *(Bot.)*. Having a single strand of protoxylem in the stele.

monaural *(Acous.)*. Pertaining to use of one ear instead of two (cf. *binaural*). In sound-recording, the (inaccurate) opposite of *stereophonic*. See **monophonic**.

monazite *(Min.)*. Monoclinic phosphate of the rare earth metals, $CePO_4$, containing cerium as the principal metallic constituent, and also some thorium. Monazite is exploited from beach sands, where it may be relatively abundant; one of the principal sources of rare earths and thorium.

monchiquite *(Geol.)*. An alkaline lamprophyre with phenocrysts of olivine, pyroxene and usually mica or amphibole, in a glassy groundmass.

Mönckeberg's sclerosis, degeneration *(Med.)*. Degeneration of the middle coat of medium-sized arteries in old people, characterized by the deposit of calcium in it.

Monday morning disease *(Vet.)*. See **sporadic lymphangitis**.

Mond gas *(Chem.)*. The gas produced by passing air and a large excess of steam over coal-slack at about 650°C. See also **semiwater gas**.

Mond process *(Eng.)*. A process used by Mond Nickel Co. in extracting nickel from a matte consisting of copper-nickel sulphides. The matte is roasted to obtain oxides, the copper is leached out with H_2SO_4, the nickel oxide is reduced to nickel with hydrogen, then the nickel is caused to combine with CO to form a carbonyl, which is decomposed by heating.

Monel metal *(Eng.)*. A nickel-base alloy containing nickel 68%, copper 29%, and iron, manganese, silicon, and carbon 3%. Has high strength (about 500 MN/m^2), good elongation (about 45%), and high resistance to corrosion. Used for condenser tubes, propellers, pump fittings, turbine blades, and for chemical and food-handling plant.

mongolism *(Med.)*. See **Down's syndrome**.

mongrel *(Bot.,Zool.)*. The offspring of a cross between varieties or races of a species.

moniliasis *(Med.)*. Infection with any of the species of the fungus *Monilia (Candida)*. See **thrush**.

moniliasis *(Vet.)*. *Candidiasis*. Infection of the mucosa of the mouth and other parts of the digestive tract of birds and animals by yeastlike organisms of the genus *Monilia*.

monimostyly *(Zool.)*. In Vertebrates, the condition of having the quadrate immovably united to the squamosal; cf. **streptostyly**.

monitor *(Comp.)*. A supervisory program which controls sequencing and **time sharing** procedures in a computer system. Also *executive program, supervisor program*. See also **operating system**.

monitor *(Image Tech.)*. A video display screen for critical picture presentation, not usually provided with RF reception circuits to act as a TV receiver.

monitor *(Min.Ext.)*. In hydraulic mining, high-pressure jet of water used to break down loosely consolidated ground in open-cast work. Also *giant*. In ore treatment, a device which checks part of the process and sounds a signal, makes a record, or initiates compensating adjustment if the detail monitored requires it.

monitor *(Radiol.)*. Ionization chamber or other radiation detector arranged to give a continuous indication of intensity of radiation, as in radiation laboratories, radiation protection from fallout contamination, industrial operations or X-ray exposure.

monitor *(Telecomm.)*. An arrangement for reproducing and checking any transmission without interfering with the regular transmission.

monitoring *(Radiol.)*. Periodic or continuous determination of the amount of specified substances, e.g. toxic materials or radioactive contamination, present in a region or a person or of a flow rate, pressure etc., in a (continuous) process, as safety measures.

monitoring loudspeaker *(Acous.)*. One of high quality and exactly matching others, for verifying quality of programme transmission before radiation or recording. Located in a monitoring room adjacent to film or broadcasting studio.

monitoring receiver *(Telecomm.)*. See **check receiver**.

monitoring station *(Telecomm.)*. National service for verifying frequencies of various transmitters within their allocated frequency bands; also for listening to international broadcasts and other traffic for the purposes of compiling bulletins and for security.

monitor position *(Telecomm.)*. The special position at any exchange at which operators deal with queries and complaints.

monk *(Print.)*. An area of printing with too much ink. See **friar**.

monk bond *(Build.)*. A modification of the *Flemish bond*, each course consisting of two stretchers and one header alternately. Also *flying bond*.

monkey tail *(Build.)*. A vertical scroll at one end of a hand-rail.

monkey-tail bolt *(Build.)*. A long-handled bolt for the top of a door, capable of operation from the floor.

mon-, mono- *(Chem.)*. Containing one atom, group etc., e.g. monobasic.

mon-, mono- *(Genrl.)*. Prefix from Gr. *monos*, alone, single.

mono *(Acous.)*. Abbrev. for *monophonic*.

monoamine oxidase *(Biol.)*. An enzyme involved in the inactivation of catecholamine neurotransmitters. Inhibitors of monoamine oxidase are used as antidepressants.

monoamine-oxidase inhibitors *(Med.)*. Class of drugs used to treat very severe depressive illness. Little used because of its dangerous interactions with foods containing *tyramine* and with **sympathomimetics**.

monobasic *(Chem.)*. Containing 1 hydrogen atom replaceable by a metal with the formation of a salt.

monobath *(Image Tech.)*. Solution which develops and fixes an emulsion in one stage.

monobloc *(Autos.)*. The integral casting of all the cylinders of an engine, i.e. in the same cylinder block.

monocable *(Civ.Eng.)*. The type of **aerial ropeway** in which a single endless rope is used both to support and to move the loads.

monocardian *(Zool.)*. Having a completely undivided heart.

monocarpellary *(Bot.)*. Having, or consisting of, a single carpel.

monocarpic *(Bot.)*. Dying at the end of its first fruiting season, as do annuals, biennials and some perennials. See also **hapaxanthic**. Cf. **polycarpic**.

monocerous *(Zool.)*. Having a single horn.

monochasium, monochasial cyme *(Bot.)*. An inflorescence in which the main stem ends in a flower and bears, below the flower, a lateral branch which itself ends in a flower. This branch may in turn bear further similar branches. Also *cymose inflorescence*. Cf. **dichasium**.

monochlamydeous *(Bot.)*. Having a perianth of one whorl of members.

monochlamydeous chimera *(Bot.)*. A periclinal chimera in which one component is present, at least at the shoot apex, as a single superficial layer of cells.

monochord *(Acous.)*. A primitive apparatus with a string and wooden resonator for demonstrating the properties of single stretched wires and the sounds generated thereby.

monochromatic *(Build.)*. A term used in colour meaning tints and shades of one hue or colour, e.g. blue used along with dark and/or light blue.

monochromatic *(Phys.)*. By extension from *monochromatic light*, any form of oscillation or radiation characterized by a unique or very narrow band of frequency.

monochromatic filter. *(Image Tech.)*. A filter which transmits light of a single wavelength, or, in practice, a very narrow band of wavelengths.

monochromatic light *(Phys.)*. Light containing radiation of a single wavelength only. No source emits truly monochromatic light, but a very narrow band of wavelengths can be obtained, e.g. the cadmium red spectral line, wavelength 643.8 nm with a *half-width* of 0.0013 nm. Light from some lasers have extremely narrow line widths.

monochromatic radiation *(Phys.)*. Electromagnetic radiation (originally visible) of one single frequency component. By extension, a beam of particulate radiation comprising particles all of the same type and energy. *Homogeneous* or *monoenergic* is preferable in this sense.

monochromator *(Phys.)*. Device for converting heterogeneous radiation (electromagnetic or particulate) into a homogeneous beam by absorption, refraction or diffraction processes.

monochrome *(Image Tech.)*. Picture reproduction in one colour only, but generally used to describe a black-and-white image.

monochrome receiver *(Image Tech.)*. TV receiver which reproduces a black-and-white transmission, or a colour transmission in black-and-white. See **compatible colour television**.

monochrome signal *(Image Tech.)*. That part of a TV signal controlling luminance only, not including the **chrominance signal**. In compatible systems, this signal provides the black-and-white picture.

monoclimax theory *(Ecol.)*. Idea, proposed by F.E. Clements in 1916, that all successional sequences in a region lead to a single climax vegetation of a type determined by the climate. Cf. **polyclimax theory**.

monocline *(Geol.)*. A fold with one limb which dips steeply; the beds, however, soon approximate to horizontality on either side of this flexure.

monoclinic system *(Crystal.)*. The style of crystal architecture in which the 3 crystal axes are of unequal lengths, having 1 of their intersections oblique and the other 2 at right angles. Also called *oblique system*.

monoclinous *(Bot.)*. Hermaphrodite flowers. Cf. **diclinous**.

monoclonal antibody *(Immun.)*. Antibody produced by a single clone of cells or a cell line derived from a single cell. Such antibodies are all identical and have unique amino acid sequences. Commonly used to refer to antibody secreted by a hybridoma cell line, but can also refer to the immunoglobulin produced in vivo by a B-cell clone such as a plasmacytoma if this has identifiable antibody properties.

monocolpate *(Bot.)*. A pollen grain having a single colpus.

monocoque *(Aero.)*. A fuselage or nacelle in which all structural loads are carried by the skin. In a *semimonocoque*, loads are shared between skin and framework, which provides local reinforcement for openings, mountings, etc. See **stressed skin construction**.

monocoque *(Autos.)*. See **chassis**.

Monocotyledones *(Bot.)*. *Liliopsida*. The monocotyledons, or monocots, the smaller of the two classes of **angiosperm** or flowering plants. Mostly herbs of which the embryo has characteristically one cotyledon, the parts of the flower are in threes or sixes and the leaves often have the main veins parallel; very few are woody, fewer still have secondary thickening. Contains ca. 55 000 spp. in 60 families divided among four subclasses or superorders; Alismatidae, Arecidae, Commelinidae and Liliidae.

monocotyledonous *(Bot.)*. (1) Belonging to the Monocotyledones. (2) An embryo or seed having one cotyledon.

monocule *(Zool.)*. An animal possessing a single eye, as the Water Flea *Daphnia*. adj. *monocular*.

monoculture *(Ecol.)*. A culture, crop or plantation with only one species.

monocyclic *(Bot.)*. Stamens or other floral parts in a single whorl.

monocyte *(Immun.)*. A large motile phagocytic cell with an indented nucleus present in normal blood, where it is the blood representative of the **mononuclear phagocyte system**. Monocytes are derived from promonocytes in the bone marrow. They remain in the blood for only a short time and then migrate into the tissues where they become macrophages.

monodactylous *(Zool.)*. Having only a single digit.

monodisperse system *(Chem.)*. A colloidal dispersion having particles all of effectively the same size.

monodont *(Zool.)*. Having a single persistent tooth, as the male Narwhal.

monoecious *(Bot.)*. (1) Flowering plants having separate male and female flowers on each individual plant. (2) Moss gametophytes, algae etc., producing male and female gametes on the same individual.

monoecious *(Zool.)*. See **hermaphrodite**.

monoenergic *(Phys.)*. See **monochromatic radiation**.

monoestrous *(Zool.)*. Exhibiting only one oestrous cycle during the breeding season. Cf. **polyoestrous**.

monofilament *(Plastics)*. A single filament of indefinite length. Used for ropes, surgical sutures etc., and in finer gauge, for textiles.

monogenetic *(Zool.)*. Multiplying by asexual reproduction; showing a direct life-history; of parasites, having a single host.

monogenic function *(Maths.)*. See **analytic function**.

monogerm *(Bot.)*. Varieties of sugar beet etc. in which each small dry fruit, sown like a seed, produces a single seedling rather than the group of seedlings typically produced by such fruits.

monogony *(Zool.)*. Asexual reproduction.

monohydric alcohols *(Chem.)*. Alcohols containing one hydroxyl group only.
monoid *(Maths.)*. Semigroup with identity; equivalently, category with only one object.
monolayer *(Chem.)*. See **monomolecular layer**.
monolayer culture *(Biol.)*. A **tissue culture** technique whereby thin sheets of cells are grown, on glass, in a nutrient medium.
monolete *(Bot.)*. A pollen grain or other spore with a single linear scar or aperture.
monolith *(Build.)*. A single detached column or block of stone.
monolithic *(Build.,Civ.Eng.)*. The term applied to a structure made of a continuous mass of material.
monolithic integrated circuit *(Electronics)*. Electronic circuit formed by diffusion or ion-implantation on a single crystal of semiconductor, usually silicon.
monomer *(Chem.)*. A substance considered as a single unit, in contrast with **dimers** and **polymers**. In particular, a molecule from which a polymer may be built, either by *addition* or *condensation polymerization*.
monomineralic rocks *(Geol.)*. Rocks consisting essentially of one mineral, e.g. dunite and anorthosite.
monomode fibre *(Telecomm.)*. A **stepped index** optical fibre where the diameter of the inner core (of higher refractive index) is comparable with the wavelength of light; this results in there being only one mode of light propagation; **mode dispersion** is thus eliminated.
monomolecular layer, monolayer *(Chem.)*. A film of a substance 1 molecule in thickness.
monomolecular reaction *(Chem.)*. A reaction in which only one species is involved in forming the activated complex of the reaction.
monomorphic *(Biol.)*. Showing little change of form during its life-history; structures with the same form or appearance. Cf. **polymorphic**.
monomorphous *(Crystal.)*. Existing in only one crystalline form.
mononuclear phagocyte system *(Immun.)*. A classification of phagocytic cells of which the typical mature form is the macrophage. They are all derived from bone marrow promonocytes, and share in varying degrees the ability to ingest and digest particulate materials, and to express the receptors characteristic of macrophages. However their appearance differs according to the tissue in which they are situated. Macrophages in the alveoli of the lung, the peritoneal cavity, and moving in tissues and in the lymph are free to migrate. Others such as Kupffer cells, tissue histiocytes, osteoclasts and astroglia remain in situ for long periods of time.
monopack *(Image Tech.)*. See **integral tripack**.
monophagous *(Zool.)*. Feeding on one kind of food only; as *Sporozoa*, living always in the same cell, or phytophagous Insects, with only one food-plant. Also *monotrophic*.
monophasic *(Zool.)*. Having an abbreviated life-cycle, without a free active stage; said of certain Trypanosomes. Cf. **diphasic**.
monophonic *(Acous.)*. Single-channel sound reproduction, recreating the acoustic source, as compared with *stereophonic* which uses two or more identical channels for auditory perspective. See **monoaural**.
Monophoto *(Print.)*. A major **filmsetting** system developed from the Monotype hot-metal system, the caster being replaced by a photographic unit.
monophyletic *(Biol.)*. Of a group of species which are descended from a single ancestral species. See **polyphyly**.
monophyletic group, monophyly *(Bot.)*. (1) Generally a group that is descended from a common ancestor. (2) In cladistics a group comprising a common ancestor and all of its descendents. Cf. **polyphyletic group, paraphyletic group**.
monophyodont *(Zool.)*. Having only a single set of teeth, the permanent dentition. Cf. **polyphyodont**.
monoplane *(Aero.)*. A heavier-than-air aircraft, either an aircraft or glider, having one main supporting surface.
monoplegia *(Med.)*. Paralysis of an arm or of a leg.
monoploid *(Biol.)*. True **haploid**.
monopodial growth *(Bot.)*. *Indefinite growth, indeterminate growth*. Pattern of growth in which a shoot continues to grow indefinitely and bears lateral shoots which behave similarly, e.g. in pines and other conifers. See also **racemose inflorescence**. Cf. **sympodial growth**.
monopod platform *(Min.Ext.)*. Drilling or production rig with one central leg, used in arctic conditions because of risk of ice damage to conventional designs.
monopole *(Acous.)*. Spherical radiator whose surface moves inwards and outwards with the same phase and amplitude everywhere. Any sound source which produces an equivalent sound field is also called a monopole, e.g. any small source generating volume flow.
monopropellant *(Space)*. Single propellant which produces propulsive energy as the result of a chemical reaction, usually induced by the presence of a catalyst. Cf. **bipropellant**.
monopulse *(Radar)*. A radar system with an antenna system with two or more overlapping lobes in its radiation pattern. From the transmission of a single pulse and analysis of error signals due to the target being off-axis in one or more lobes, detailed information about direction can be obtained. Used in many gun-control and missile guidance systems.
monorail *(Civ.Eng.)*. A railway system in which carriages are suspended from, and run along, a single continuous elevated rail.
monorail *(Image Tech.)*. Said of technical cameras having the focusing movement of the bellows accommodated on a single rail, obviating the need for a baseboard.
monosaccharides *(Chem.)*. The simplest group of carbohydrates, classified into tetroses, pentoses, hexoses etc. according to the number of carbon atoms, and into aldoses and ketoses depending on whether they contain a potential aldehyde or ketone group.
monosodium glutamate *(Chem.)*. *Sodium hydrogen glutamate*, $HOOC.CH(NH_2)CH_2CH_2COONa$, a derivative of glutamic acid, used as condiment or flavouring.
monosome *(Biol.)*. The unpaired accessory or X-chromosome. See **sex determination**.
monosomy *(Bot.)*. Condition in which a particular chromosome is represented once only in an otherwise diploid complement; a sort of **aneuploidy**.
monospermy *(Zool.)*. Fertilization of an ovum by a single spermatozoon.
monosporous *(Bot.)*. (1) Containing 1 spore. (2) Derived from 1 spore.
monostable *(Electronics)*. Of a circuit or system, fully stable in one state only but *metastable* in another to which it can be driven for a fixed period by an input pulse.
monostichous *(Bot.)*. Forming 1 row.
monosymmetric system *(Crystal.)*. See **monoclinic system**.
monotocous *(Zool.)*. Producing a single offspring at a birth.
monotonic *(Maths.)*. Having the property of either never increasing or never decreasing. Applied to a function or sequence.
monotonic *(Telecomm.)*. Said of transient response when it increases continuously with time.
Monotremata *(Zool.)*. The single order included in the Mammalian group Prototheria and having the characteristics of the subclass, e.g. Spiny Ant-eater (*Tachyglossus*), Duck-billed Platypus (*Ornithorhynchus*).
monotrophic *(Zool.)*. See **monophagous**.
monotropic *(Chem.)*. Existing in only one stable crystalline form, the other forms being unstable under all conditions.
Monotype *(Print.)*. See **composing machines**.
Monotype set system *(Print.)*. To make the best interpretation of a type of design, the em quad for each fount is allotted a set, there being $\frac{1}{4}$-point intervals between each. There are a few 'normal set' founts in which the width of the em quad is the same as the depth of the body, but in the majority of founts the set is narrower, e.g. 10 pt Times, $9\frac{3}{4}$ set. See also **Monotype unit system**.
Monotype unit system *(Print.)*. Each character of a fount

is allotted an appropriate number of units out of the 18 which represent the em quad of the 'set'. See also **Monotype set system**. An arrangement of this kind is essential for all tape-operated type-setting and filmsetting systems.

monotypic *(Bot.,Zool.)*. A taxonomic group containing only one subordinate member, a family with a single genus or a genus with a single species.

monovalent *(Chem.)*. Capable of combining with 1 atom of hydrogen or its equivalent, having an oxidation number or co-ordination number of one.

monozygotic twins *(Biol.)*. Twins produced by the splitting in two of a single fertilized egg, or of the early embryo derived from it. syn. *identical twins*. They are always of the same sex and are genetically equivalent to a single individual. Cf. **dizygotic twins**.

Monro's foramen *(Zool.)*. A narrow canal connecting the first or second ventricle with the third ventricle in the brain of Vertebrates.

monsoon *(Meteor.)*. Originally winds prevailing in the Indian Ocean, which blow S.W. from April to October and N.E. from October to April; now generally winds which blow in opposite directions at different seasons of the year. Similar in origin to land and sea breezes, but on a much larger scale, both in space and time. Particularly well developed over southern and eastern Asia, where the wet summer monsoon from the S.W. is the outstanding feature of the climate.

mons pubis *(Med.)*. A convex formation of subcutaneous fatty tissue over the pubic symphysis.

monster *(Biol.)*. An abnormal form of a species.

montage *(Image Tech.)*. (1) A film sequence containing rapidly changing multiple images, often superimposed and dissolving together, to convey an integrated visual impression. (2) Composite photograph made from the juxtaposition of cut-up photographs arranged in a pattern.

montasite *(Min.)*. An asbestiform variety of the amphibole grunerite. It differs from **amosite** in having less harsh and more silky fibres.

montebrasite *(Min.)*. A variety of **amblygonite** in which the amount of hydroxyl exceeds that of fluorine.

Monte-Carlo methods *(Stats.)*. Procedures employed to obtain numerical solutions to mathematical problems by means of random sampling.

month *(Astron.)*. See **anomalistic-**, **sidereal-**, **synodic-**, **tropical-**.

monticellite *(Min.)*. A silicate of calcium and magnesium which crystallizes in the orthorhombic system. It occurs in metamorphosed dolomitic limestones and, more rarely, in some ultrabasic igneous rocks.

montmorillonite *(Min.)*. A hydrated silicate of aluminium, one of the important clay minerals and the chief constituent of bentonite and fuller's earth. The montmorillonite group of clay minerals are also collectively termed *smectites*.

monzonite *(Geol.)*. A coarse-grained igneous rock of intermediate composition, characterized by approximately equal amounts of orthoclase and plagioclase (near andesine in composition) together with coloured silicates in variety. Named from Monzoni in the Tyrol. Often referred to as *syenodiorite*.

Moon *(Astron.)*. The Earth's only natural satellite and closest neighbour in space. Its average distance is 382 000 km. Its diameter of 3476 km is 1/4 that of Earth. The relative density is also lower, 3.3. Consequently the Moon has a mass only 0.012 that of the Earth. The Moon orbits the Earth in 27.32 days (relative to the stars), passing through the familiar cycle of *lunar phases*. The interval between successive new moons is 29.53 days, the *lunar month*. The Moon's rotation is in resonance with its orbit, so the same face is always presented to the Earth apart from small effects of **libration**. The lunar orbit is inclined at about 5° to the **ecliptic**, so the Moon's path is close to the **zodiac**. Even to the unaided eye the Moon has a range of features: Darker *maria* and lighter mountainous regions are easy to see. The whole surface is heavily cratered; in parts the density is so great they overlap; larger ones have a central mountain. All are believed to have been created by impact with **meteorites**. There is no water, atmosphere or magnetic field. The 386 kg of rock and dust retrieved in the Apollo missions have shown that the age of the Moon is about 4650 million years. Until 3300 million years ago it was geologically active, and the maria date from that time. The internal structure is quite unlike the Earth. There is a thick crust 30–100 km thick; it is immobile and overlays a solid mantle.

moonstone *(Min.)*. A variety of alkali feldspar or sometimes plagioclase, which possesses a bluish pearly opalescence attributed to lamellar micro- or cryptoperthitic intergrowth. It is used as a gemstone.

Moore lamp *(Elec.Eng.)*. A type of electric discharge lamp, usually in the form of a long tube, using a gas other than mercury or sodium vapour.

mooring tower *(Aero.)*. A permanent tower or mast for the mooring of airships. Provided with facilities for the transference of passengers and freight and arrangements for replenishing ballast, gas, and fuel.

moor, moorland *(Ecol.)*. Vegetation composed of ericaceous plants and certain grasses, growing on acid soil.

mopboard *(Build.)*. See **skirting board**.

mop-stick hand-rail *(Build.)*. A timber hand-rail having a circular section flattened on the under side.

moquette *(Textiles)*. A heavy warp pile fabric used for upholstery. The pile may be cut or left as uncut loops.

mor *(Ecol.)*. Crumbly layer of partially-decomposed plant-like litter occurring under some coniferous trees and woody shrubs, in which the pH is acid and the soil fauna poor. See also **mull**, **moder**.

moraine *(Geol.)*. Material laid down by moving ice. Moraines are found in all areas which have been glaciated, and are of several types. Terminal moraines are irregular ridges of material marking the farthest extent of the ice and representing debris pushed along in front of the ice. Lateral moraines are found along the sides of present-day and former glaciers. Medial moraines form by the combination of the two lateral moraines when two glaciers join. Ground moraine is an irregular sheet of **till** laid down beneath an ice sheet.

morbid *(Med.)*. Diseased: pertaining to, or of the nature of, disease.

morbidity *(Med.)*. The state of being diseased: the sick-rate in a community.

morbilli *(Med.)*. See **measles**.

mordant *(Textiles)*. A compound, frequently a metallic salt or oxide, applied to a fabric to form a stable complex with a dyestuff. The complex has superior fastness on the fabric compared with the dyestuff itself.

mordanting *(Image Tech.)*. Adding to a silver image, or replacing a silver image by, a substance having the requisite affinity for a specified dye, which silver in itself has not.

mordants *(Build.)*. Preparations applied to surfaces to assist paint or gold-leaf to adhere thereto.

mordants *(Chem.)*. Substances, chiefly the weakly basic hydroxides of aluminium, chromium, and iron, which combine with and fix a dyestuff on the fibre in cases where the fibre cannot be dyed direct. The product formed by the action of a dye on a mordant is called a *lake*.

mordenite *(Min.)*. A hydrated sodium, potassium, calcium, aluminium silicate. A zeolite crystallizing in the orthorhombic system and occurring in amygdales in igneous rocks and as a hydration product of volcanic glass.

more hug *(Print.)*. An increase in the wrap of the paper round the cylinder(s) of web-fed presses.

Morgagni's ventricle *(Zool.)*. In the higher Mammals, a paired pocket of the larynx, anterior to the vocal cords, and acting in the Anthropoidea as a resonator.

morganite *(Min.)*. A pink or rose-coloured variety of beryl, used as a gemstone.

morion *(Min.)*. A variety of smoky-quartz which is almost black in colour.

morning star *(Astron.)*. Popularly, a planet, generally Venus or Mercury, seen in the eastern sky at or about sunrise; also, loosely, any planet which transits after midnight.

morph *(Ecol.)*. A specific form or shape of an organism, singled out for attention.

morphactins *(Bot.)*. Group of synthetic compounds, based on fluorene-carboxylic acid which have a variety of effects, mostly inhibitory, on plant growth and development.

morphallaxis *(Zool.)*. Change of form during regeneration of parts, as the development of an antenna in certain Crustacea to replace an eye; gradual growth or development.

morphine *(Med.)*. The principal alkaloid present in opium. Characterized by containing a phenanthrene nucleus in addition to a nitrogen ring. Extensively used as a hypnotic to obtain relief from pain.

morph-, morpho-, -morph *(Genrl.)*. Prefix and suffix from Gk *morphē*, form.

morpho- *(Genrl.)*. See **morph-**.

morphoea, morphea *(Med.)*. *Localized sclerodermia.* A disease in which thickened patches appear on the skin of the trunk.

morphogenesis *(Zool.)*. The origin and development of a part, organ, or organism. adj. *morphogenetic.*

morpholine *(Chem.)*. A 6-membered heterocyclic compound, C_4H_9NO. It is a liquid, bp 128°C with strong basic properties. Miscible with water and used as a solvent in organic reactions.

morphology *(Bot.,Zool.)*. (1) The study of the structure and forms of organisms, as opposed to the study of their functions. (2) By extension, the nature of a member. adj. *morphological.*

morphosis *(Zool.)*. The development of structural characteristics; tissue formation. adj. *morphotic.*

Morquio-Brailsford disease *(Med.)*. An hereditary disease characterized by dwarfism, kyphosis and skeletal defects in the hip joint.

Morse code *(Telecomm.)*. System used in signalling or telegraphy, which consists of various combinations of dots and dashes.

Morse equation *(Phys.)*. An equation which relates the potential energy of a diatomic molecule to the internuclear distance.

Morse key *(Telecomm.)*. A hand-operated device that opens and closes contacts which modulate currents with coded telegraph signals. See also **bug key**.

Morse taper *(Eng.)*. A system of matching tapered shanks and sockets used for holding tools in lathes, drills etc. There are six sizes, with tapers approx. 3°.

mortality *(Ecol.)*. The death of individuals in a population. Cf. **natality**.

mortar *(Build.,Civ.Eng.)*. A pasty substance formed normally by the mixing of cement, sand and water, or cement, lime sand and water in varying proportions. Used normally for the binding of brickwork or masonry. It is **thixotropic** and its working life can be extended by addition of soap solutions. Hardens on setting and forms the bond between the bricks or stones.

mortar *(Chem.)*. A bowl, made of porcelain, glass or agate, in which solids are ground up with a pestle.

mortar board *(Build.)*. See **hawk**.

mortar mill *(Build.)*. Am appliance in which the ingredients of a mortar are crushed mechanically by two rollers running on the ends of a horizontal bar rotating about a central axis, the rollers running around a shallow pan containing the ingredients. Also called *pan-mill mixer*.

mortar structure *(Geol.)*. A cataclastic structure resulting from dynamic metamorphism in which small grains produced by granulation occupy the interstices between larger grains.

mortise-and-tenon joint *(Build.)*. A forming joint between a **mortise** and a **tenon**.

mortise bolt *(Build.)*. A bolt which is housed in a mortise in a door so as to be flush with its edge.

mortise chisel *(Build.)*. A more robust type of chisel than the *firmer*, for use in cutting mortises and therefore designed to withstand blows from a mallet. Also *framing chisel, heading chisel.*

mortise gauge *(Build.)*. A tool similar to the **marking gauge** but having an additional marking pin, which is adjustable for position along the bar and allows parallel lines to be set out in marking tenons and mortises. Also *counter gauge.*

mortise joint *(Build.)*. See **mortise-and-tenon joint**.

mortise lock *(Build.)*. A lock sunk into a mortise in the edge of a door.

mortise, mortice *(Build.)*. A rectangular hole cut in one member of a framework to receive a corresponding projection on the mating member.

mortuary fat *(Med.)*. See **adipocere**.

morula *(Zool.)*. A solid spherical mass of cells resulting from the cleavage of an ovum.

MOS *(Comp.,Electronics)*. Abbrev. for *Metal-Oxide Semiconductor*; a fabrication method utilizing a metal-insulator-semiconductor sandwich construction, the insulating layer being an oxide of the substrate material; in silicon chip, the insulator is silicon dioxide (SiO_2). Used widely in FETs, transistors, capacitors and resistors, and integrated circuits. See **CMOS**.

mosaic *(Biol.)*. An individual displaying the effects of different alleles or genes in different parts of the body, but derived from a single embryo, e.g. a tortoiseshell cat. Cf. **chimaera**.

mosaic *(Bot.)*. (1) The disposition in three dimensions of the leaves of a shoot or a plant, resulting from phyllotaxis and leaf morphology, which appears to maximize interception of light while minimizing mutual shading. (2) A patchy variation of the normal green colour of leaves; usually the result of infection by a virus. (3) A plant or part of a plant with an irregular distribution of cells of different genetic constitution, resulting from e.g. the mutation of an unstable gene or the sorting out of dissimilar plastids.

mosaic *(Build.)*. Inlaid work on plaster or stone, formed with small cubes, *tesserae*, or irregular-shaped fragments of marble, glazed pottery or glass.

mosaic *(Chem.)*. Structure of crystalline solids due to dislocations, consisting of an irregular matrix or mosaic of otherwise perfect crystallites.

mosaic *(Image Tech.)*. (1) See **photo-mosaic**. (2) Video effect in which the picture is represented by a number of small square or rectangular elements, also known as **pixillation, tile**.

mosaic *(Zool.)*. Animal chimaera.

mosaic development *(Zool.)*. Development when the eventual fate of cells of the developing embryo is determined, e.g. on formation of the blastomeres in the spirally segmented eggs of some Invertebrates.

mosaic egg *(Zool.)*. Egg which shows mosaic development, i.e., where areas of future functional development are determined in the early stages of cleavage.

mosaic gold *(Chem.)*. Complex tin (IV) sulphide obtained by heating dry tin amalgam, ammonium chloride and sulphur in a retort. Sometimes used as a pigment.

mosaicism *(Med.)*. A condition descriptive of individuals who possess cell populations with mixed numbers of chromosomes, e.g. some cells have 45, others 47, making a mosaic of 45/47.

mosaic screen *(Image Tech.)*. A screen made up of a continuous pattern of very small dots or lines of the three primary colours, red, green and blue, used in early forms of additive colour photography, either as a separate layer or forming an integral part of the support for a black-and-white emulsion.

mosaic structure *(Eng.)*. Discontinuous structure of a compound or metal consisting of minute domains, each bounded by its discontinuity lattice at the interface with other domains of the like composition.

Moscicki capacitator *(Elec.Eng.)*. A capacitator on the principle of the Leyden jar; sometimes used on transmission lines to act as a protective device against effects of high-frequency surges.

Moscovian *(Geol.)*. An epoch of the Pennsylvanian period.

Moseley's law *(Phys.)*. For one of the series of characteristic lines in the X-ray spectrum of atoms, the square root of the frequency of the lines is directly proportional to the *atomic number* of the element. This result stresses the importance of atomic number rather than atomic weight.

MOSFET *(Electronics)*. Abbrev. for *Metal-Oxide Semiconductor Field-Effect Transistor*; an insulated-gate FET, the insulator between the metal gate electrodes and the **channel** being silicon dioxide.

mosquito *(Zool.)*. The family Culicidae of the Diptera; have piercing probosces and suck blood, often transmitting diseases (e.g. malaria, yellow fever, elephantiasis) while doing so; larvae and pupae aquatic, e.g. *Anopheles, Culex, Aedes*.

moss *(Bot.)*. See **Bryopsida**.

moss agate *(Min.)*. A variegated cryptocrystalline silica containing visible impurities, as manganese oxide, in mosslike or dendritic form. Also called *Mocha stone*.

Mössbauer effect *(Phys.)*. When an atomic nucleus emits a γ-ray photon it must recoil to conserve linear momentum. Consequently there is a change of frequency of the radiation due to the movement of the source (*Doppler effect*). If the atom is firmly bound in a crystal lattice so that it may not recoil, the momentum is taken up by the whole lattice; an effect much used in the study of the structure of solids.

most economical range *(Aero.)*. The range obtainable when the aircraft is flown at the height, air speed and engine conditions which give the lowest fuel consumption for the aircraft weight and the wind conditions prevailing.

most significant bit *(Comp.)*. Bit with the greatest place value in a word. Abbrev. *MSB*.

mother *(Acous.)*. Metal positive which is made from the master in gramophone-record manufacture.

mother board *(Comp.)*. **Printed circuit board** which holds the principal components in a microcomputer system.

mother cell *(Bot.,Zool.)*. A cell which divides to give daughter cells; the term is applied particularly to cells which divide to give spores, pollen grains, gametes and blood corpuscles.

mother liquor *(Chem.)*. The solution remaining after a solute has been crystallized out.

mother of emerald *(Min.)*. A variety of *prase*, a leek-green quartz owing its colour to included fibres of actinolite; thought at one time to be the mother-rock of emerald.

mother set *(Print.)*. A set of printing plates (e.g. of a standard reference work) kept solely for the purpose of electro- or stereotyping further sets therefrom, as required. Not to be used for printing.

motional impedance *(Telecomm.)*. In an electro-mechanical transducer, e.g. a telephone receiver or relay, that part of the input electrical impedance due to the motion of the mechanism; the difference between the input electrical impedance when the mechanical system is allowed to oscillate and the same impedance when the mechanical system is stopped from moving, or blocked.

motion bars *(Eng.)*. See **guide bars**.

motion compensation *(Min.Ext.)*. Automatic machinery which maintains a constant downhole pressure on the drill bit bored from a floating platform.

motion picture camera *(Image Tech.)*. A camera for the exposure of cinematograph film, intermittently, one frame at a time, to record movement.

motion work *(Horol.)*. The auxiliary train of wheels, normally under the dial, which gives the correct relative motion to the hour and minute hands.

motivation *(Behav.)*. As a general term, and used mostly with reference to human behaviour, it refers to the desire to act in certain ways in order to achieve a goal; these may be transitory impulses or more persistent intentions. Motives may or may not be conscious and/or rational. In animal behaviour, it refers to a set of problems characterized by variations in responsiveness to a constant environmental situation, where the explanation assumes some reversible internal change in the individual, excluding changes due to learning, maturation or fatigue.

motoneuron(e) *(Biol.)*. A motor neuron(e).

motor *(Bot.,Zool.)*. Pertaining to movement; as nerves which convey movement-initiating impulses to the muscles from the central nervous system.

motor *(Eng.)*. A machine used to transform power into mechanical form from some other form.

motor areas *(Zool.)*. Nerve-centres of the brain concerned with the initiation and coordination of movement.

motor-boating *(Telecomm.)*. Very-low-frequency relaxation oscillation in an amplifier, arising from inadequate decoupling of common sources of current supply.

motor bogie *(Elec.Eng.)*. A bogie or truck on a railway locomotive or motor-coach which carries one or more electric motors.

motor cell *(Bot.)*. One of a number of cells which together can expand or contract and so cause movement in a plant member.

motor end plates *(Zool.)*. The special end-organ in which a motor nerve terminates in a striated muscle.

motor generator *(Elec.Eng.)*. A converter consisting of a motor connected to a supply of one voltage, frequency, or number of phases, and a generator providing output power to a system of different voltage, frequency, or number of phases, the motor and generator being mechanically connected.

motor habits *(Behav.)*. Repetitive, nonfunctional patterns of motor behaviour that seem to occur in response to stress (e.g. nail biting, thumb sucking). See also **stereotyping**.

motorized fuel valve *(Eng.)*. See **adjustable-port proportioning valve**.

motor meter *(Elec.Eng.)*. An integrating meter embodying a motor whose speed is proportional to the power flowing in the circuit to which it is connected, so that the number of revolutions made by the spindle is proportional to the energy consumed by the circuit. See **induction meter**.

motor-operated switch (circuit-breaker) *(Elec.Eng.)*. A large switch or circuit-breaker which is closed by means of an electric motor. Cf. *solenoid-operated switch*.

motor starter *(Elec.Eng.)*. A device for operating the necessary circuits for starting and accelerating to full speed an electric motor, but not for controlling its speed when running. See **controller**.

motor system *(Bot.)*. Tissues and structures concerned in the movements of plant members, as in pulvini.

motte *(Arch.)*. The steep mound, natural or man made, on which mediaeval castles were often built within an open fortified space, or **bailey**; thus the term *motte and bailey*.

mottled iron *(Eng.)*. Pig-iron in which most of the carbon is combined with iron in the form of cementite (Fe_3C) but in which there is also a small amount of graphite. The fractured pig has a white crystalline fracture with clusters of dark spots, indicating the presence of graphite.

mottler *(Build.)*. A small brush used in graining to create the effect of mottling or highlighting and shading areas or the timber. Short haired with a rectangular handle.

mottle yarn *(Textiles)*. Same as **marl yarn**.

mottramite *(Min.)*. Descloizite in which the zinc is almost entirely replaced by copper.

MOU *(Aero.)*. Abbrev. for *Memorandum Of Understanding*.

mould *(Bot.)*. A fungus, especially one that produces a visible mycelium on the surface of its host or substrate.

mould *(Build.)*. Zinc sheet or thin board cut to a given profile; used in running cornices etc.

mould *(Civ.Eng.)*. A temporary construction to support setting concrete in position. Also *form*.

mould *(Geol.)*. The impression of an original shape. Usually refers to fossils, but may be minerals or sedimentary structures. Cf. **casts**.

mould *(Print.)*. An impression of the type or blocks made in flong or plastic, and used for making stereotypes, electrotypes, or rubber or plastic duplicates.

moulded breadth *(Ships)*. The breadth over the frames of a ship, i.e. heel of frame to heel of frame. It is the breadth termed *B* by Lloyd's Register, and it is the line fairest in the moulding loft. Measured at the widest part of the hull.

moulded depth *(Ships)*. The depth of a ship from the top of the keel to the top of the beam at side, measured at midlength; referred to as Lloyd's *D*.

moulded dimensions *(Ships)*. The dimensions used in the calculation and tabulation of the scantlings (sizes) of the various members of the structure. They are *length between perpendiculars, moulded breadth, moulded depth.*

moulded-intake belt course *(Build.)*. An **intake belt course** shaped along the projecting corner to a more or less ornamental profile.

moulding *(Build.)*. A more or less ornamental band projecting from the surface of a wall or other surface.

moulding *(Foundry)*. See **foundry, machine-, moulding sands, pipe-, plate-.**

moulding *(Plastics)*. The moulding powder is weighed carefully into small containers or preformed in pill machines to pellets. The material is then placed in a steel die, heated to about 160°C, and subjected to pressure of 20–35 MN/m^2. Wood moulds are used to make articles from heated thermo-plastic sheets. See also **blow-, cold-, compression-, injection-, rotational-, transfer.**

moulding box *(Foundry)*. See **flask.**

moulding cutter *(Build.)*. A specially shaped cutting tool which, when revolving about its own axis, is capable of cutting a desired moulding profile.

moulding machines *(foundry)*. See **machine moulding.**

moulding plane *(Build.)*. Plane with a shaped sole and blade for cutting mouldings. There are two types: *English* (held at an angle) and *Continental* (held upright).

moulding powder *(Plastics)*. The finely ground mixture of binder, accelerator, colouring matter, filler and lubricant converted under pressure into the final moulding.

mouldings *(Build.)*. Strips of wood cut to a given cross-sectional profile and applied to surfaces required to be decorated.

moulding sands *(Foundry)*. Siliceous sands (containing clay or aluminium silicate as a binding agent) possessing naturally or by blending the qualities of fineness, plasticity, adhesiveness, strength, permeability, and refractiveness. See **dry sand, green sand, loam.**

mould machine *(Paper)*. See **cylinder mould machine.**

mould oil *(Build.,Civ.Eng.)*. A substance applied to shuttering to prevent adherence of the concrete.

moult *(Zool.)*. See **ecdysis.**

mount *(Image Tech.)*. Card or other backing surface to which a photograph is permanently attached for display.

mount *(Print.)*. The base on which a letterpress plate is mounted to bring it to type height.

mount *(Telecomm.)*. That part of a switching tube or cavity which enables it to be connected to a waveguide.

mountain cork *(Min.)*. A variety of asbestos which consists of thick interlaced fibres. It is light and will float, and is of a white or grey colour.

mountain leather *(Min.)*. A variety of asbestos which consists of thin flexible sheets made of interlaced fibres.

mountain wood *(Min.)*. A compact fibrous variety of asbestos looking like dry wood.

mountants *(Image Tech.)*. Special adhesives for fixing prints on mounts, free from chemicals which might attack the silver or other image during the course of time.

mouse *(Comp.)*. Input device, hemispherical in appearance. Moving the device around on a flat surface causes a cursor to move around the display screen.

mouse roller *(Print.)*. A small extra roller used to obtain better distribution of the ink on a machine.

mouth *(Acous.)*. See **flare.**

mouth *(Build.)*. The slot in the sole of a plane though which the cutting iron projects.

mouth parts *(Zool.)*. In Arthropoda, the appendages associated with the mouth.

movable types *(Print.)*. Single types, as distinguished from Linotype slugs or from blocks.

move *(Glass)*. A fixed number of articles to be made for a given rate of pay by a **chair.**

movement *(Horol.)*. The mechanism of a clock or watch, not including the case or dial.

movement area *(Aero.)*. That part of an airport reserved for the take-off, landing and movement of aircraft.

Movietone *(Image Tech.)*. TN for one of the first systems for recording sound on film, using a **variable density track.** The reduced size of the picture image allowing room for the track became known as the *Movietone frame*.

moving-coil galvanometer *(Elec.Eng.)*. A galvanometer depending for its action on the movement of a current-carrying coil in a magnetic field. Cf. **moving-magnet galvanometer.**

moving-coil instrument *(Elec.Eng.)*. An electrical measuring instrument depending for its action upon the force on a current-carrying coil in the field of a permanent magnet.

moving-coil loudspeaker *(Acous.)*. See **electrodynamic loudspeaker.**

moving-coil microphone *(Acous.)*. See **electrodynamic microphone.**

moving-coil regulator *(Elec.Eng.)*. A type of voltage regulator, for use on a.c. circuits, in which a short-circuited coil is made to move up and down the iron core of a specially arranged autotransformer.

moving-coil transformer *(Elec.Eng.)*. A type of transformer, occasionally used in constant-current systems, in which one coil is made to move relatively to the other for regulating purposes.

moving-coil voltmeter *(Elec.Eng.)*. One constructed like a galvanometer and used for d.c. measurements.

moving-conductor microphone *(Acous.)*. See **ribbon microphone.**

moving-field therapy *(Radiol.)*. A form of crossfire technique in which there is relative movement between the beam of radiation and the patient, so that the entry portal of the beam is constantly changing. See **converging-field therapy.**

moving form *(Civ.Eng.)*. See **climbing form.**

moving-iron *(Elec.Eng.)*. Descriptive of one-time drive in microphones and loudspeakers, which depends on the motion of magnetic material which is part of a magnetic circuit.

moving-iron voltmeter *(Elec.Eng.)*. One used for a.c. measurements, depending on attraction of a soft iron vane into the magnetic field due to current.

moving load *(Civ.Eng.)*. A variable loading on a structure, consisting of the pedestrians or vehicles passing over it. Also *live load.*

moving-magnet galvanometer *(Elec.Eng.)*. A galvanometer depending for its action on the movement of a small permanent magnet in the magnetic field produced by the current to be measures. Cf. **moving-coil galvanometer.**

moving-target indicator *(Radar)*. A device for restricting the display of information to moving targets. Abbrev. *MTI.*

moving trihedral *(Maths.)*. Three mutually orthogonal unit vectors t, n and b along the tangent, principal normal and binormal respectively at a point P on a space curve. The three planes perpendicular to the three vectors are called respectively the *normal, tangent* and *osculating* planes. The triad is determined by two limits, i.e. (1) the limiting position of a chord PQ as Q approaches P, which determines t, and (2) the limiting position of a plane PQR as Q and R approach P, which determines the osculating plane. Cf. **curvature.** $t = n \wedge b$, $n = b \wedge t$.

Mpc *(Astron.)*. Abbrev. for *Megaparsec.*

MPD *(Elec.Eng.)*. Abbrev. for *Magneto-PlasmaDynamic.*

MPD *(Radiol.)*. Abbrev. for *Maximum Permissible Dose.*

mph *(Genrl.)*. Abbrev. for *miles per hour.*

m.p., mp *(Chem.)*. Abbrev. for *melting point.*

MQ *(Image Tech.)*. An abbrev. for *Metol-Quinol* or *metol-hydroquinone* developers. See **4-methyl-aminophenol.**

mRNA, messenger RNA *(Biol.)*. The RNA whose sequence of nucleotide triplets determines the sequence of a polypeptide.

MS *(Aero.)*. See **Mil-Spec.**

M-shell *(Phys.)*. The **electron shell** in an atom corresponding to a principal quantum number of three. The shell can contain up to 18 electrons.

MSI *(Comp.)*. Medium scale integration refers to a **chip** with not more than a few hundred logic **gates**.

MSW *(Radar)*. Abbrev. for *Magnetic Surface Wave*.

MTBF *(Aero.)*. Abbrev. for *Mean Time Between Failures*.

M-test of Weierstrass for uniform convergence *(Maths.)*. Both the series $\Sigma u_n(x)$ and the product $\Pi(1+u_n(x))$ are uniformly convergent in the interval (a,b) if it is possible to find a convergent series of positive constants ΣM_n such that $|u_n(x)| \leqq M_n$, for all values of x in the interval (a,b).

MTF *(Image Tech.)*. *Modulation Transfer Function*, a method of expressing the performance of a lens, photographic emulsion or complete image reproduction system by its degree of modulation (image contrast) for varying unit image size (frequency). This gives a better evaluation of subjective **sharpness** than simple **resolution** values.

MTI *(Radar)*. Abbrev. for *Moving-Target Indicator*.

MTOC *(Bot.)*. Abbrev. for *MicroTubule-Organizing Centre*.

mucic acid *(Chem.)*. HOOC.$(CHOH)_4$COOH, an acid obtained by the oxidation of galactose.

mucigen *(Zool.)*. A substance occurring as granules or globules in chalice cells and later extruded as mucin.

mucilages *(Chem.)*. Complex organic compounds related to the polysaccharides, of vegetable origin, and having gluelike properties.

mucilaginous *(Bot.,Zool.)*. Pertaining to, containing, or resembling mucilage or mucin.

mucinogen *(Zool.)*. A substance producing or being the precursor of mucin, e.g. granules of mucous gland cells.

mucins *(Chem.)*. A group of glycoproteins occurring in mucus, saliva and other secretions. Widely distributed in nature, the polypeptide chains are densely glycosylated.

muck *(Min.Ext.)*. See **dirt**.

mucocele *(Med.)*. A localized accumulation of mucous secretion, in, e.g. a hollow organ, the outlet of which is blocked.

mucomembranous colic *(Med.)*. A condition in which constipation is associated with abdominal pain and the passing in the stools of membranes or casts of mucus.

mucopolysaccharides *(Chem.)*. Heteropolysaccharides each containing a hexosamine in its characteristic repeating disaccharide unit.

mucoproteins *(Chem.)*. Conjugated proteins with carbohydrate side-chains, which may include hexoses, hexosamines or glucuronic acids.

mucopurulent *(Med.)*. Consisting of mucus and pus.

mucosa *(Zool.)*. See **mucous membrane**.

mucosal disease *(Vet.)*. *Bovine viral diarrhoea (BVD)*. Common togavirus infection of cattle, of which acute and sub-acute signs may be seen. Symptoms include fever, anorexia, diarrheoa, ulceration of the abdominal tract and oculo-nasal discharge. Mortality can be high.

mucosanguineous *(Med.)*. Consisting of mucus and blood.

mucous glands *(Zool.)*. Glands secreting or producing mucus.

mucous membrane *(Zool.)*. A tissue layer found lining various tubular cavities of the body (as the gut, uterus, trachea, etc.). It is composed of a layer of epithelium containing numerous unicellular mucous glands and an underlying layer of areolar and lymphoid tissue, separated by a basement membrane. Also *mucosa*.

muco-viscidosis *(Med.)*. See **cystic fibrosis**.

mucro *(Bot.)*. A short, sharp, terminal point.

mucronate *(Bot.)*. Terminated by a short point or *mucro*.

mucus *(Zool.)*. The viscous slimy fluid secreted by the mucous glands. adjs. *mucous, mucoid, muciform*.

mud *(Geol.)*. A fine-grained unconsolidated rock, of the clay grade, often with a high percentage of water present. It may consist of several minerals.

mud *(Min.Ext.)*. See **drilling mud**.

mud acid *(Min.Ext.)*. Inorganic acids and chemicals used to promote flow in oil wells by acid treatment.

mud column *(Min.Ext.)*. The column of mud from the top to the bottom of a bore hole.

mud drum *(Eng.)*. A vessel placed at the lowest part of a steam-boiler: a similar plant to intercept and retain insoluble matter or sludge, as the lowest drum of a **water-tube boiler**.

mud fever *(Vet.)*. See **grease**.

mud flow *(Geol.)*. Mass movement of fine-grained material in a highly fluid state. When abundant coarse material is also carried it is known as a *debris flow*.

mud hole *(Eng.)*. A *hand hole* in a mud drum, or in the bottom of a boiler, for the removal of scale and sludge.

mudline *(Min.Ext.)*. Separating line between clear overflow water and settled slurry in dewatering or thickening plant.

mud motor *(Min.Ext.)*. Hydraulic motor situated downhole and actuated by the pressure of the **drilling mud** forced down the bore hole. Can be used for the main drilling, when the drill pipe does not rotate or to actuate subsiduary drills, reamers etc. See **drain holes**.

mud pipe *(Min.Ext.)*. An outer casing which is run down through semi-liquid material found during drilling until it rests on firmer strata. It protects the casing proper which contains the **drilling mud**.

mud pump *(Min.Ext.)*. One used in drilling of deep holes (e.g. oil wells) to force thixotropic mud to bottom of hole and flush out rock chips.

mudstone *(Geol.)*. An argillaceous sedimentary rock characterized by the absence of obvious stratification. Cf. **shale**.

mud volcano *(Geol.)*. A conical hill formed by the accumulation of fine mud which is emitted together with various gases, from an orifice in the ground. Generally derived from a volcanic hot spring but sedimentary mud volcanoes can also be produced by earthquake activity, which generates the extrusion of liquid mud.

MUF *(Telecomm.)*. Abbrev. for *Maximum Usable Frequency*.

muffle furnace *(Eng.)*. A furnace in which heat is applied to the outside of a refractory chamber containing the charge.

muffler *(Autos.)*. US for *silencer*.

mugearite *(Geol.)*. A dark, finely-crystalline, basic igneous rock which contains oligoclase, orthoclase, and usually olivine in greater amount than augite. Occurs typically at Mugeary in Skye.

mulch *(Bot.)*. In horticulture, matter placed on the soil surface in order e.g. to suppress the growth of weed seedlings, or to reduce temperature fluctuations at the soil surface; often organic, e.g. peat, shredded bark, sometimes inorganic, e.g. pebbles.

mule *(Textiles)*. Machine, intermittent in action, which first drafts and twists the yarn on the outward run, then on the inward run winds it on a spindle. It can spin very fine counts but there are also woollen mules for heavy yarns and condenser mules for waste yarns. (Invented by Samuel Crompton, Bolton, ca. 1770.) Now only used for wool spinning.

mull *(Ecol.)*. Loose crumbly layer of soil occurring in some deciduous woodlands, in which leaf litter is being broken down rapidly at near-neutral pH in the presence of a rich fauna including earthworms. Cf. **mor, moder**.

Mullen burst test *(Paper)*. The **burst test** when performed on a Mullen instrument.

muller *(Min.Ext.)*. Pestle, dragstone, iron wearing plate or shoe used to crush and/or abrade rock entrained between it and a baseplate over which it is moved.

Müllerian duct *(Zool.)*. A duct which arises close beside the oviduct, or which, by the actual longitudinal division of the archinephric duct, in many female Vertebrates becomes the oviduct.

Müllerian mimicry *(Zool.)*. Resemblance in colour between two animals, both benefiting by the resemblance. Cf. **Batesian mimicry**.

Müller's glass *(Min.)*. See **hyalite**.

Müller's muscle *(Zool.)*. The circular ciliary muscle of the Vertebrate eye.

mullion *(Build.)*. A vertical member of a window frame separating adjacent panes. Also called *monial, munnion*.

mullion structure *(Geol.)*. A linear structure found in severely folded sedimentary and metamorphic rocks in which the harder beds form elongated fluted columns. Mullions are parallel to fold axes.

mullite *(Min.)*. A silicate of aluminium, rather similar to sillimanite but with formula close to $Al_6Si_2O_{13}$. It occurs in contact-altered argillaceous rocks.

mulluscum contagiosum *(Med.)*. A contagious condition in which small, white, waxy nodules appear on the skin; believed to be due to infection with a virus.

multi-access *(Comp.)*. System which allows several users to have apparently simultaneous access to a computer. Used mainly for **interactive** computing on a **time-sharing** basis.

multiarticulate *(Zool.)*. Many-jointed.

multiaxial *(Bot.)*. Having a main axis consisting of several more or less equal files of cells with or without subordinate branches. Cf. **uniaxial**.

multi-break switch, circuit-breaker *(Elec.Eng.)*. A switch or circuit-breaker in which the circuit is broken at two or more points in series on each pole or phase.

multicavity magnetron *(Telecomm.)*. One with many cavities cut in the inner face of the solid cylindrical anode, with the mouths of the cavities facing the cathode. Sizes of alternate cavities may differ to assist in *mode separation*. See **magnetron**.

multicellular *(Biol.)*. Consisting of a number of cells.

multicellular voltmeter *(Elec.Eng.)*. A form of electrostatic voltmeter in which a number of moving vanes mounted upon the spindle are drawn into the spaces between a corresponding number of fixed vanes.

multi-centred bonding *(Chem.)*. The bonding characteristic of **electron deficient** compounds. In such compounds, 'bonds' can only be described that involve three or more atoms at one time.

multichannel *(Telecomm.)*. Any system which divides the frequency spectrum of a signal into a number of bands which are separately transmitted, with subsequent recombination.

multichip integrated circuit *(Electronics)*. An electronic circuit comprising two or more semi-conductor wafers which contain single elements (or simple circuits). These are interconnected to produce a more complex circuit and are encapsulated within a single pack.

multicipital *(Zool.)*. Many-headed.

multi-coloured paints *(Build.)*. Special paints comprising different colours, different media within the one container. These decorative finishes are usually sprayed but brush applied types are available.

multicuspidate *(Zool.)*. Teeth with many cusps.

multi-electrode valve *(Electronics)*. One comprising two or more complete electrode systems having independent electron streams in the same envelope. There is sometimes a common electrode, e.g. double-diode.

multi-exchange system *(Telecomm.)*. A group of local exchanges in an exchange area.

multifactorial *(Biol.)*. Determined by many genes and non-genetic factors. See also **polygenic**.

multifee metering *(Telecomm.)*. Successive operations of a subscriber's meter to correspond with the multi-fee chargeable for the call.

multifilament lamp *(Elec.Eng.)*. An electric filament-lamp having more than one filament in the same bulb, so that failure of one filament will not cause the lamp to be extinguished.

Multifont type tray *(Print.)*. The first British adaptation of the American *Rob Roy* System, using 108 small interchangeable plastic boxes, one for each sort, and arranging them on trays which can be stored in a galley rack, thus taking up much less room than the ordinary type case, either when in use or in store. Several colours of box are used, to distinguish one fount, or range of sorts, from another. *Compacto* is a similar system; *Freelay* is a development of it, using three sizes of box to house an appropriate quantity of each character, and suitable for the larger sizes of type.

multifrequency *(Telecomm.)*. Said of a signal, mostly used for *multifrequency signalling* or data transmission, consisting of several superimposed audiotones. According to CCITT specifications *1* consists of 697 and 1209 Hz, *2* is 697 and 1336 Hz, *3* is 687 and 1477 Hz, *4* is 770 and 1209 Hz etc.

multifrequency generator *(Telecomm.)*. The multifrequency inductor generator which is used for the multichannel voice-frequency telegraph system operated by teleprinter and transmitting over normal telephone lines.

multigap discharger *(Elec.Eng.)*. A discharger for use in a spark-type radio transmitter, making use of a number of gaps in series.

multigravida *(Med.)*. A woman who has been pregnant more than once.

multigroup theory *(Nuc.Eng.)*. Theoretical reactor model in which presence of several energy groups among the neutrons is taken into account.

multilayered structure *(Bot.)*. Single broad band of microtubules forming a *flagellar root* in the motile cells and gametes of the Charophyceae, bryophytes and vascular plants.

multilayer film *(Image Tech.)*. A photographic material comprising several distinct emulsion coatings and filter layers, especially in colour photography. See **tripack**.

multilayer winding *(Elec.Eng.)*. A type of cylindrical winding, used chiefly for transformers, in which several layers of wire are wound one over the other with layers of insulation between.

multilobal *(Textiles)*. A man-made fibre or filament which is extruded so that the cross-section has several rounded lobes. More precisely named according to the number of lobes e.g. trilobal.

multilocular *(Bot.)*. Having a number of compartments or loculi.

multimode fibre *(Telecomm.)*. An optical fibre (usually a **graded index** one) with a core diameter sufficiently larger than the wavelength of light to allow propagation of light energy in a large number of different modes.

multinet growth *(Bot.)*. Steady state pattern postulated for the growth of the cell walls of some elongating cells in which the cellulose microfibrils, deposited more or less transversely on the cytoplasmic surface of the wall, become passively reorientated by the growth of the cell so as to lie more nearly longitudinally in the older, outer parts of the wall.

multinucleate *(Biol.)*. Having many nuclei. Also *polynucleate*.

multiparous *(Zool.)*. Bearing many offspring at a birth.

multipath reception *(Telecomm.)*. Signals from the transmitter arriving by two or more paths; one direct the other(s) by reflection from buildings or other obstructions. Can lead to distortion in radio and *ghost* images in television.

multiple *(Maths.)*. If an integer a is a divisor of an integer b, then b is said to be a multiple of a. Formally, if a, b and x are integers such that $ax = b$, then b is a multiple of a and a is a divisor of b.

multiple-circuit winding *(Elec.Eng.)*. See **lap winding**.

multiple decay, multiple disintegration *(Phys.)*. Branching in a radioactive decay series.

multiple-disk clutch *(Eng.)*. A friction clutch similar in principle to the **single-plate clutch**, but in which a smaller diameter is obtained by using a number of disks, alternately splined to the driving and driven members, loaded by springs, and usually run in oil. See **friction clutch**.

multiple duct *(Elec.Eng.)*. A cable duct having a number of tunnels for the reception of several cables.

multiple echo *(Acous.,Telecomm.)*. Perception of a number of distinct repetitions of a signal, because of reflections with different delays of separate waves following various paths between the source and observer.

multiple effect evaporation *(Chem.Eng.)*. A system of evaporation which the hot vapour produced in one vessel instead of being condensed by cooling water is condensed in the calandria of another evaporator to produce further evaporation. To achieve this, the pressure in second evaporator must be artificially lowered. Can, according to conditions, work as double, triple, or quadruple effect.

multiple-expansion engine *(Eng.)*. An engine in which

the expansion of the steam or other working fluid is divided into two or more stages, which are performed successively in cylinders of increasing size. See **compound engine**, **triple-expansion engine**, **quadruple-expansion engine**.

multiple feeder *(Elec.Eng.)*. A feeder consisting of a number of cables connected in parallel: used where a single cable to carry the load would be prohibitively large.

multiple fission *(Zool.)*. A method of multiplication found in Protozoa, in which the nucleus divides repeatedly without corresponding division of the cytoplasm, which subsequently divides into an equal number of parts leaving usually a residium of cytoplasm. Cf. **binary fission**.

multiple fruit *(Bot.)*. A fruit formed from the maturing ovaries of a group of flowers, often with their receptacles and their floral parts, e.g. pineapple, fig.

multiple-hearth furnace *(Eng.)*. A type of roasting furnace consisting of a number of hearths (six or more). The charge enters on the top hearth and passes downwards from hearth to hearth, being rabbled by rotating arms from centre to circumference and circumference to centre of alternate hearths.

multiple-hop transmission *(Telecomm.)*. Transmission which uses multiple reflection of the sky wave by the ground and ionosphere.

multiple intrusions *(Geol.)*. Minor intrusions formed by several successive injections of approximately the same magma.

multiple modulation *(Telecomm.)*. The use of a modulated wave for modulating a further independent carrier of much higher frequency.

multiple neuritis *(Med.)*. See **polyneuritis**.

multiple-operator welding-unit *(Elec.Eng.)*. An electric-arc welding generator or transformer designed to supply current to two or more welding arcs operating in parallel.

multiple personality disorder *(Behav.)*. An extreme form of dissociative reaction in which 2 or more complete personalities, each well developed and distinct, are found in one individual.

multiple point *(Maths.)*. A point on the curve $f(x,y)$ at which the first non-zero derivative of f is of order n, is a multiple point of order n. Cf. **double point**.

multiple proportions *(Chem.)*. See **law of multiple proportions**.

multiple-retort underfeed stoker *(Eng.)*. A number of underfed inclined retorts arranged side by side with tuyères between, resulting in a fuel bed the full width of the furnace walls. See **single-retort underfeed stoker**.

multiple sclerosis *(Med.)*. A chronic progressive disease in which patches of thickening occur throughout the central nervous system leading to progressive paralysis. Also *disseminated sclerosis*.

multiple-spindle drilling machine *(Eng.)*. A drilling machine having two or more vertical spindles for simultaneous operation.

multiple star *(Astron.)*. A system in which three or more stars, united by gravitational forces revolve about their common centre of gravity.

multiple-switch starter (controller) *(Elec.Eng.)*. A starter (or controller) for an electric machine in which the steps of resistance are cut out, or other operations performed, by hand-operated switches.

multiple-threaded screw *(Eng.)*. A screw of coarse pitch in which two or more threads are used to reduce the size of thread and maintain adequate core strength. Also called *multistart thread*. See also **divided pitch**, **multistart worm**.

multiple-tool lathe *(Eng.)*. A heavy lathe having two large tool posts, one on either side of the work, each carrying several tools operating simultaneously on different parts of the work.

multiplets *(Phys.)*. (1) Optical spectrum lines showing fine structure with several components, i.e. triplet or more complex structures. Due to spin-orbit interactions in the atom. (2) See **isotopic spin**.

multiple-tuned antenna *(Telecomm.)*. A transmitting antenna system comprising an extensive horizontal 'roof' with a number of spaced vertical leads, each connected to earth through appropriate tuning circuits, and all tuned to the same frequency, the connection to the transmitter being made through one of them.

multiple-twin cable *(Telecomm.)*. A cable in which there are numbers of cores, each comprising two pairs twisted together.

multiple-unit control *(Elec.Eng.)*. The method of control by which a number of motors operating in parallel can be controlled from any one of a number of points; used on multiple-unit trains.

multiple-unit train *(Elec.Eng.)*. An electric or diesel train consisting of a number of motor-coaches and trailers, all controlled from one driving position at the front or rear of the train.

multiple valve *(Electronics)*. See **multi-electrode valve**.

multiplex *(Telecomm.)*. Use of one channel for several messages by **time-division multiplex** or **frequency division**.

multiplexor *(Comp.)*. An I/O device that routes data from several sources to a common destination. Cf. **demultiplexor**.

multiplex transmission *(Telecomm.)*. Transmission in which two or more signals modulate the carrier wave.

multiplicand *(Maths.)*. See **multiplication**.

multiplication *(Electronics)*. See **secondary emission**.

multiplication *(Maths.)*. The process of finding the *product* of two quantities which are called the *multiplicand* and the *multiplier*. Denoted by the multiplication sign ×, or by a dot, or by the mere juxtaposition of two quantities. Some branches of mathematics use special signs, e.g. the vector product of **a** and **b** is sometimes indicated by $\mathbf{a} \wedge \mathbf{b}$. Numbers are multiplied in accordance with rules of arithmetic, but multiplication of other mathematical entities has to be defined specifically for the entities concerned.

multiplication constant *(Phys.)*. The ratio of the average number of neutrons produced by fission in one neutron lifetime to the total number of neutrons absorbed or leaking out in the same interval. Also called *reproduction constant* and denoted by k.

multiplier *(Electronics)*. See **electron multiplier**.

multiplier *(Maths.)*. See **multiplication**.

multiply-connected domain *(Maths.)*. See **connected domain**.

multiplying camera *(Image Tech.)*. A camera for taking a number of small exposures on one negative, using a deflecting mirror or a number of lenses which can be traversed.

multiplying constant *(Surv.)*. A factor in the computation of distance by tacheometric methods (see **tacheometer**). It is that constant value for the particular instrument by which the staff intercept must be multiplied in order to give the distance of the staff from the focus of the object glass. If the distance from the centre of the instrument is required, it is necessary to add (see **additive constant**) the distance between the focus of the object glass and the centre of the instrument.

multipolar *(Zool.)*. Said of nerve cells having many axons. Cf. **bipolar**, **unipolar**.

multipole moments *(Phys.)*. These are magnetic and electric, and are measures of the charge, current and magnet (via intrinsic spin) distributions in a given state. These **static** multipole moments determine the interaction of the system with weak external fields. There are also *transition* multipole moments which determine radiative transitions between two states.

multiposition *(Telecomm.)*. Said of a system in which output or final control can take on 3 or more preset values.

multiprocessor *(Comp.)*. A linked set of central processors which allows **parallel processing**. See **MIMD**.

multiprogramming *(Comp.)*. The capacity of a central processor to handle more than one program at a time. See **time sharing**.

multirow radial engine *(Aero.)*. A radial *aero-engine* with two or more rows of cylinders.

multi-sensor *(Aero.)*. The use of more than one sensor to obtain information.
multiseriate *(Bot.)*. In several rows; vascular ray several cells wide.
multispeed supercharger *(Aero.)*. A gear-driven **supercharger** in which a clutch system allows engagement of different ratios to suit changes in altitude.
multispindle automatic machine *(Eng.)*. A fully-automatic, high-speed, special-purpose lathe primarily for bar work, in which the work sequence is divided so that a portion of it is performed at each of the several spindle stations, one part being completed each time the tools are withdrawn and the spindles indexed.
multistage *(Electronics)*. Said of a tube in which electrons are progressively accelerated by anode rings at increasing potentials, also of an amplifier with transistors or valves in series, or cascade.
multistage *(Space)*. Said of a space-vehicle having successive rocket-firing stages, each capable of being jettisoned after use.
multistage compressor *(Aero.)*. A gas turbine compressor with more than one stage; each row of blades in an **axial-flow compressor** is a stage, each **impeller** is a stage in a **centrifugal-flow compressor**. In practice, all axial compressors are multistage, while almost all centrifugal compressors are single stage. Occasionally an initial axial stage is combined with a centrifugal delivery stage.
multistage supercharger *(Aero.)*. A supercharger with more than one impeller in series.
multistage turbine *(Aero.)*. A turbine with two or more disks joined and driving one shaft.
multistart thread *(Eng.)*. See **multiple-threaded screw**.
multistart worm *(Eng.)*. A worm in which two or more helical threads are used to obtain a larger pitch and hence a higher velocity ratio of the drive.
multitone *(Acous.)*. Generator, thermionic or mechanical, which produces a spectrum of currents, i.e. with a large number of components, equally spaced in frequency.
multituberculate *(Zool.)*. Said of tuberculate teeth with many cusps: having many small projections.
multi-turn current transformer *(Elec.Eng.)*. A current transformer in which there are several turns on the primary winding. Cf. **bar-type current transformer**.
multivalent *(Bot.)*. A group of three or more partly homologous chromosomes held together during the prophase of the first division of meiosis. Cf. **bivalent**, **univalent**.
multivariate analysis *(Stats.)*. Statistical analysis of several measurements of different characteristics on each unit of observation.
multivibrator *(Electronics)*. A *relaxation oscillator* circuit consisting of two active elements (transistors, valves etc.) connected so that when one is conducting, the other is not. An *astable vibrator* switches spontaneously between these two states, the frequency of switching depending on the time constants of the coupling elements, or on the frequency of an external synchronizing voltage. A *monostable vibrator* can be triggered into an unstable state by an external signal, remaining in that state for a period determined by the time constants. The *flip-flop*, or bistable multivibrator can remain indefinitely in either state, external triggering being required to change from one state to the other.
multivoltine *(Zool.)*. Having more than one brood in a year. Cf. **univoltine**, **bivoltine**.
multiwire antenna *(Telecomm.)*. An antenna consisting of a number of horizontal wires in parallel.
mu-meson *(Phys.)*. An elementary particle once thought to be a *meson* but now known to be a *lepton*. See **muon**.
Mumetal *(Eng.)*. TN of high-permeability, low-saturation magnetic alloy of about 80% nickel, requiring special heat-treatment to achieve special low-loss properties. Useful in nonpolarized transformer cores, a.c. instruments, small relays. Also for shielding devices from external field as in CROs.
Mummery's plexus *(Zool.)*. A network of fine nerve-fibrils lying between the odontoblasts and the dentine in a tooth.
mumps *(Med.)*. Epidemic or infectious parotitis. An acute infectious disease characterized by a painful swelling of the parotid gland; caused by a virus.
mundic *(Min.)*. See **pyrite**.
mungo *(Textiles)*. A low grade of waste fibres recovered by pulling down old *hard*-woven wool rags, including tailors' cuttings and old felt. Cleaned and spun again as woollen yarns and used for weaving into lower grade fabrics. Cf. **shoddy**.
Munsell colour system *(Build.)*. A system of colour notation devised by Albert Munsell which breaks colour into 3 attributes: (1) *hue* colour, red, blue, green etc; (2) *value*, the lightness or darkness of a colour; (3) *chroma*, strength or saturation of a colour.
Munsell scale *(Phys.)*. A scale of chromaticity values giving approximately equal magnitude changes in visual hue.
munting, muntin *(Build.)*. The vertical framing piece separating the panels of a door.
Muntz metal *(Eng.)*. Alpha-beta brass, 60% copper and 40% zinc. Stronger than alpha brass and used for castings and hot-worked (rolled, stamped or extruded) products.
muon *(Phys.)*. Fundamental particle with a rest mass equivalent to 106 MeV; it is one of the *leptons* and has a negative charge and a half-life of about 2 μs. Decays to electron, neutrino and antineutrino. It participates only in *weak* interactions. The *antimuon* has a positive charge and decays to positron, neutrino and antineutrino.
muramic acid *(Chem.)*. Occurs in bacterial cell walls and consists of N-acetylglucosamine joined by an ethereal linkage to d(—)-lactic acid.
muramyl dipeptide *(Immun.)*. *N-acetyl-muramyl-L-alanyl-D-isoglutamine*. This is the simplest structural unit of bacterial peptidoglycans, which can mimic the adjuvant effects of mycobacteria in complete Freund's adjuvant. It causes release of IL-1 from macrophages, and is sometimes incorporated in a vaccine to increase its immunogenicity.
murexide *(Chem.)*. The dye ammonium purpurate, used as an indicator for calcium in **complexometric titration** using EDTA.
Murex process *(Min.Ext.)*. Magnetic separation of desired mineral from pulp, in which the ore is mixed with oil and magnetite and then pulped in water. Magnetite clings selectively to the desired mineral.
Murgatroyd belt *(Glass)*. A circumferential band in the side wall of a glass container, extending from the bottom upwards for about 20 mm. In this band, maximum stresses due to thermal shock occur.
muriatic acid *(Chem.)*. See **hydrochloric acid**.
muricate *(Bot.)*. Having a surface roughened by short, sharp points.
murmur *(Med.)*. An irregular sound which follows, accompanies, or replaces the normal heart sounds and often indicates disease of the valves of the heart; similar sound heard over blood-vessels.
Murphree efficiency *(Chem.Eng.)*. In the performance of a single plate in a **fractionating column**, the ratio of the actual enrichment of the vapour leaving the plate, to the enrichment which would have occurred if there had been sufficient time for the vapour to reach equilibrium with the liquid leaving the plate.
Musci *(Bot.)*. The mosses; Bryopsida.
muscle *(Zool.)*. A tissue whose cells can contract; a definitive mass of such tissue. See **cardiac-**, **striated-**, **unstriated-**, **voluntary-**.
muscovite *(Geol.,Min.)*. The common or white mica; a hydrous silicate of aluminium and potassium, crystallizing in the monoclinic system. It occurs in many geological environments but deposits of economic importance are found in granitic pegmatites. Used as an insulator and a lubricant.
Muscovy glass *(Min.)*. Formerly a popular name for **muscovite**.
muscular dystrophy *(Med.)*. Hereditary diseases marked by hypertrophy or wasting of pelvic and limb muscles.

See **Duchenne muscular dystrophy, fascioscapulohumeral muscular dystrophy**.

musculature *(Zool.)*. The disposition and arrangement of the muscles in the body of an animal.

musculocutaneous *(Zool.)*. Pertaining to the muscles and the skin.

mush *(Aero.)*. Condition of flight at the stall when the aircraft tends to maintain **angle of attack** while losing height rather than the sharp nose down pitch which is more common.

mush *(Telecomm.)*. Noise and distortion in reception due to the interaction of waves from two or more radio transmitters.

mushroom *(Bot.)*. Common name for the fruiting body of an **agaric**, especially of species of *Agaricus*.

mushroom bodies *(Zool.)*. Paired nerve centres of the protocerebrum in Insects, regarded as the principal association areas. Also known as *corpora pedunculata*.

mushroom construction *(Civ.Eng.)*. Reinforced concrete construction composed only of columns and floor slabs, the columns having a spread at their head to counteract shear forces.

mushroom follower *(Eng.)*. A cam follower in the form of a mushroom, i.e. with a domed surface, as distinct from a roller-type follower.

mushroom valve *(Autos.)*. See **poppet valve**.

mush winding *(Electronics)*. A type of winding used for a.c. motors and generators in which the conductors are placed one by one into partially closed and lined slots, the end connections being subsequently insulated separately.

musical echo *(Acous.)*. One in which the interval between the reception of successive echoes is so small that the impulses received appear to have the quality of a musical tone.

music and effects *(Image Tech.)*. A sound track containing recordings of both these but without speech or dialogue. Abbrev. *M & E*.

music synthesizer *(Comp.)*. Output device which on receipt of digital signals generates sounds similar to musical notes.

muskeg *(Bot.)*. Mossy growth found in high latitudes, usually as deep and treacherous swamp, bog or marsh. Unstable peat formation.

musk glands *(Zool.)*. In some Vertebrates, glands, associated with the genitalia, the secretion of which has an odour of musk.

muslin *(Textiles)*. Lightweight, plain-woven cloths of open texture and soft finish, bleached, and dyed; used as dress fabrics. There are also unbleached butter, cheese, or meat etc. muslins for wrapping purposes.

mustard gas *(Chem.)*. $(CH_2Cl.CH_2)_2S$, *dichlorodiethyl sulphide*; a poison gas manufactured from ethene and S_2Cl_2. See also **nitrogen mustards**.

mustine *(Chem.)*. See **nitrogen mustards**.

mutagen *(Biol.)*. A substance which causes *mutation*.

mutant *(Biol.)*. An individual which displays the result of a *mutation*. A gene which has undergone *mutation*.

mutarotation *(Chem.)*. The change with time of the optical activity of a freshly prepared solution of an active substance. The sugars are the best-known class to exhibit the phenomenon. See **anomerism**.

mutation *(Biol.)*. A change, spontaneous or induced, that converts one allele into another (*point mutation*). More generally, any change of a gene or of chromosomal structure or number. A *somatic mutation* is one occurring in a somatic cell and not in the germ line.

mutation rate *(Biol.)*. The frequency, per gamete, of mutations of a particular gene or a class of genes. Sometimes, *esp.* in micro-organisms, the number per unit of time (esp. generation time).

mute *(Image Tech.)*. A negative or positive film containing the picture record only, without any associated sound record.

muticate, muticous *(Bot.)*. Without a point, mucro or awn. Unarmed.

muting *(Electronics)*. Suppression of the output of a receiver or amplifier unless there is sufficient signal-to-noise ratio to comfortably ensure intelligibility.

muting circuit, switch *(Telecomm.)*. Arrangement for attenuating amplifier output unless paralysed by a useful incoming carrier; used with automatic tuning devices.

mutton *(Print.)*. See **em quad**.

mutton cloth *(Textiles)*. A plain circularly knitted cotton fabric often used in the unbleached state as a cleaning cloth.

mutton rule *(Print.)*. See **em rule**.

mutual capacitance *(Phys.)*. Capacitance calculated from the displacement current flowing between two conducting bodies when all adjacent conducting bodies are earthed.

mutual conductance *(Electronics)*. Transconductance specifically applied to a thermionic valve. Differential change in a space or anode current divided by differential change of grid potential which causes it. Colloquially termed *slope* of a valve, measuring the effectiveness of the valve as an amplifier in normal circuits. Expressed in milliamperes per volt, and denoted by *gm*.

mutual coupling *(Elec.Eng.)*. See **transformer coupling**.

mutual impedance *(Elec.Eng.)*. See **transfer impedance**.

mutual inductance *(Elec.Eng.)*. Generation of e.m.f. in one system of conductors by a variation of current in another system linked to the first by magnetic flux. The unit is the henry, when the rate of change of current of one ampere per second induces an e.m.f. of one volt.

mutual inductor *(Elec.Eng.)*. A component consisting of two coils designed to have a definite mutual inductance (fixed or variable) between them.

mutualism *(Biol.)*. Any association between 2 organisms which is beneficial to both and injurious to neither. See **symbiosis**.

muzzle *(Zool.)*. See **rhinarium**.

MWE, MWe *(Nuc.Eng.)*. Abbrev. for *MegaWatt Electric*; unit for electric power generated by a nuclear reactor.

myalgia *(Med.)*. The sensation of pain in muscle.

myarian *(Zool.)*. Based on musculature, as a system of classification: pertaining to the musculature.

myasthenia *(Med.)*. Muscular weakness.

myasthenia gravis *(Immun.)*. Disease characterized by increasing muscular weakness on exercise caused by faulty transmission at the neuromuscular junction. Autoantibodies are present in the blood against antigens of the acetylcholine receptor on the post synaptic membrane, and are the putative cause of the symptoms by interfering with or damaging this receptor. One form is associated with the presence of a thymoma, and thymectomy may reverse the condition.

Mycalex *(Phys.)*. TN for mica bonded with glass. It is hard, and can be drilled, sawn and polished; has a low power factor at high frequencies, and is a very good insulating material at all frequencies.

mycelium *(Bot.)*. A mass of branching hyphae; the vegetative body (thallus) of most true fungi. adj. *mycelial*.

mycetocytes *(Zool.)*. Cells containing symbiotic micro-organisms occurring in the **pseudovitellus**.

mycetoma *(Med.)*. See **Madura foot**.

mycetome *(Bot.,Zool.)*. Pseudovitellus: special organ in some species of Insects, Ticks, and Mites, inhabited by intracellular symbionts, e.g. Bacteria, Fungi, Rickettsias.

mycetophagous *(Zool.)*. Fungus-eating.

mycobiont *(Bot.)*. The fungal partner in a symbiosis, e.g. in a lichen.

mycology *(Bot.)*. The study of fungi.

myco-, myc-, myceto- *(Genrl.)*. Prefix from Gk. *mykēs*, gen. *mykētos*, fungus.

mycophthorous *(Bot.)*. A fungus parasitic on another fungus.

mycoplasma *(Med.)*. The smallest free-living organism known. *Mycoplasma pneumoniae* is an important cause of atypical pneumonia.

Mycoplasmatales *(Biol.)*. Organisms of this group are associated with pleuropneumonia in cattle and contagious agalactia in sheep: very variable in form, consisting of cocci, rods etc., and a cytoplasmic matrix: very similar to the L-forms of true bacteria: nonpathogenic morphologically similar organisms have also been found in animals and man.

mycoplasmosis *(Vet.)*. Infection of animals and birds by organisms of the genus *Mycoplasma*, e.g. chronic respiratory disease of fowl, contagious agalactia of sheep and goats, contagious pleuropneumonia of cattle.

mycorrhiza *(Bot.)*. A symbiotic association between a fungus and the roots, or other structures of a plant. See also **endotrophic mycorrhiza, ectotrophic mycorrhiza, vesicular-arbuscular mycorrhiza.**

mycosis *(Med.,Vet.)*. Any disease of man or animals caused by a fungus.

mycosis fungoides *(Med.)*. A chronic, and usually fatal, disease in which multiple funguslike tumours appear in the skin. Associated with lymphoma.

mycotic dermatitis of sheep *(Vet.)*. *Lumpy wool.* A dermatitis of sheep due to infection of the skin by the fungus *Dermatophilus dermatonomus.*

mycotoxin *(Bot.)*. A substance produced by a fungus and toxic to other organisms, especially to man and animals, e.g. the **aflatoxins**.

mycotrophic plant *(Bot.)*. A plant which lives in symbiosis with a fungus, i.e. has mycorrhizas.

mydriasis *(Med.)*. Extreme dilation of the pupil of the eye.

mydriatic *(Med.)*. Producing dilation of the pupil of the eye: any drug which does this.

mydriatic alkaloids *(Med.)*. Alkaloids which cause dilation of the pupil of the eye, e.g. atropine.

myelencephalon *(Zool.)*. The part of the hind-brain in Vertebrates forming part of the *medulla oblongata*. Cf. **metencephalon**.

myelin *(Zool.)*. A white fatty substance which forms the medullary sheath of nerve fibres.

myelination *(Zool.)*. Formation of a myelin sheath.

myelin sheath *(Zool.)*. See **medullary sheath**.

myelitis *(Med.)*. Inflammation of the spinal cord.

myel-, myelo- *(Genrl.)*. Prefix from Gk. *myelos*, marrow.

myelocele *(Med.)*. A condition in which the spinal cord protrudes on to the surface of the body; due to a defect of the spinal vertebrae.

myelocoel *(Zool.)*. The central canal of the spinal cord.

myelocyte *(Zool.)*. A bone-marrow cell; a large amoeboid cell found in the marrow of the long bones of some higher Vertebrates, which give rise, by division, to leucocytes.

myelography *(Radiol.)*. The radiological examination of the space between the theca and the spinal cord, following the injection of air or other contrast media.

myeloid cell *(Immun.)*. Cells derived from stem cells in the bone marrow which mature to form the granular leucocytes (granulocytes) of blood. They differentiate into different forms — neutrophils, eosinophils or basophils — under the influence of **colony stimulating factors**.

myeloma *(Immun.)*. Synonym for **plasmacytoma**. Hence myeloma protein.

myeloma *(Med.)*. A tumour composed of plasma cells, appearing usually in the marrow of several bones causing skeletal pain. These cells secrete abnormal immunoglobulins which are deposited in the kidney causing kidney failure.

myelomalacia *(Med.)*. Pathological softening of the spinal cord.

myelomatosis *(Med.)*. The occurrence of myelomata in several bones. See **myeloma**.

myelomeningocele *(Med.)*. Protrusion of the spinal cord on spinal membranes; due to a defect in the spinal column.

myelopathy *(Med.)*. Any disease affecting spinal cord or myeloid tissues.

myeloplast *(Zool.)*. A leucocyte of bone marrow.

myeloplax *(Zool.)*. A giant cell of bone marrow and other blood-forming organs, sometimes multinucleate, which gives rise to the blood-platelets. Also called *megakaryocyte*.

myenteric *(Zool.)*. Pertaining to the muscles of the gut; as a sympathetic nerve plexus controlling their movements.

myiasis *(Vet.)*. Parasitism of the tissues of animals by the larvae of flies of the suborder *Cyclorrhapha*.

Mylar *(Plastics)*. TN for US polyethylene terepthalate polyester film. Very strong with extremely good transparency and electrical properties. UK, *Melinex*.

mylonite *(Geol.)*. A banded, chertlike, cataclastic rock produced by the shearing and granulation of rocks associated with intense folding and thrusting.

mylonitization *(Geol.)*. The process by which rocks are granulated and pulverized and formed into mylonite.

myoblast *(Zool.)*. An embryonic muscle cell which will develop into a muscle fibre.

myocardial *(Med.)*. Pertaining to, or affecting, the myocardium.

myocardial infarction *(Med.)*. The infarction or death of heart muscle usually as a result of a **coronary thrombosis**. Colloq. *heart attack*.

myocarditis *(Med.)*. Inflammation of the muscle of the heart.

myocardium *(Zool.)*. In Vertebrates, the muscular wall of the heart.

myoclonia congenita *(Vet.)*. *Trembling, congenital tremors.* Six types of tremor recognized. Piglets show no tremor when asleep, slight tremor when lying down and marked tremor when standing. Aetiology varies in all six types.

myoclonus *(Med.)*. (1) Paramyoclonus multiplex. A condition in which there occur sudden shocklike contractions of muscles, often associated with epilepsy and progressive mental deterioration. (2) A sudden shocklike contraction of a muscle.

myocoel *(Zool.)*. The coelomic space within a myotome.

myocomma *(Zool.)*. A partition of connective tissue between 2 adjacent myomeres. Also *myoseptum*.

myocyte *(Zool.)*. A muscle cell: a deep contractile layer of the ectoplasm of certain Protozoa.

myo-edema *(Med.)*. See **myoidema**.

myo-epithelial *(Zool.)*. A term used to describe the epithelial cells of Coelenterata which are provided with taillike contractile outgrowths at the base.

myofibril *(Biol.)*. The contractile filament consisting of actin, myosin and associated proteins within muscle cells.

myogenic *(Zool.)*. Said of contraction arising in a muscle independent of nervous stimuli. Cf. **neurogenic**.

myoglobin *(Biol.)*. A haem protein related to haemoglobin but consisting of only one polypeptide chain. It is present in large amounts in the muscles of mammals such as whales and seals, where it acts as an oxygen store during diving. It was the first protein to have its 3-dimensional structure worked out.

myohypertrophy *(Med.)*. Increase in the size of muscle fibres.

myoidema, myo-edema *(Med.)*. *Mounding*. A localized swelling of wasting muscle obtained when the muscle is lightly struck.

myolemma *(Zool.)*. See **sarcolemma**.

myology *(Zool.)*. The study of muscles.

myoma *(Zool.)*. A tumour composed of unstriped (leiomyoma) or striped (rhabdomyoma) muscle fibres.

myomectomy *(Med.)*. Surgical removal of a myoma, especially of a fibromyoma of the uterus.

myomere *(Zool.)*. In metameric animals, the voluntary muscles of a single somite.

myometrium *(Biol.)*. The muscular coat of the uterus.

myo-, my- *(Genrl.)*. Prefix from Gk. *mys*, gen. *myos*, muscle.

myoneme *(Zool.)*. In Protozoa, a contractile fibril of the ectoplasm.

myoneural *(Zool.)*. Pertaining to muscle and nerve, as the junction of a muscle and a nerve.

myopathy *(Med.)*. Any one of a number of conditions in which there is progressive wasting of skeletal muscles from no known cause.

myophily, myiophily *(Bot.)*. Pollination by flies or other Diptera.

myopia *(Med.)*. *Short-sightedness.* A condition of the eye in which, with the eye at rest, parallel rays of light come to a focus in front of the retina. adj. *myopic*.

myosarcoma *(Med.)*. A malignant tumour composed of muscle cells and sarcoma cells.

myoseptum *(Zool.)*. See **myocomma**.

myosin *(Biol.)*. Large highly asymmetric protein found originally in association with actin in muscle but now known to have a wide cellular distribution. The myosin molecules aggregate into fibres which, in muscle, interdigitate between the actin fibres. The relative movement of the two sets of fibres provides the molecular basis of *muscular contraction*.

myosis *(Med.)*. See **miosis**.

myositis *(Med.)*. Inflammation of striped muscle.

myositis ossificans progressiva *(Med.)*. A condition in which there is progressive ossification of the muscles of the body.

myotasis *(Med.)*. Muscular tension. adj. *myotatic*.

myotome *(Zool.)*. A muscle merome; one of the metameric series of muscle masses in a developing segmented animal.

myotonia atrophica *(Med.)*. *Dystrophia myotonica*. A disease characterized by wasting of certain groups of muscles, difficulty in relaxing muscles after muscular effort, and general debility.

myotonia congenita *(Med.)*. *Thomsen's disease*. A rare and congenital malady characterized by extreme slowness in relaxation of muscles after voluntary effort.

myriametric waves *(Telecomm.)*. Waves corresponding to VLF, longer than 10 km.

myriapod *(Zool.)*. A general term denoting arthropods with many similar segments and comprising the classes Chilopoda and Myriapoda (Centipedes and Millepedes).

myringitis *(Med.)*. Inflammation of the drum of the ear.

myringoscope *(Med.)*. An instrument for viewing the drum of the ear.

myringotomy *(Med.)*. Incision of the drum of the ear.

myristic acid *(Chem.)*. *Tetradecanoic acid*, $CH_3.(CH_3)_{12}.COOH$. Crystalline solid; mp 58°C. Found (as glycerides) in milk and various vegetable oils.

myrmecochory *(Bot.)*. Distribution of seeds or other reproductive bodies by ants.

myrmecophagous *(Zool.)*. Feeding on ants.

myrmecophily *(Bot.)*. (1) Symbiosis with ants, e.g. Acacia. (2) Pollination by ants.

myrmekite *(Min.)*. An intergrowth of plagioclase and vermicular quartz.

Myrtaceae *(Bot.)*. Family of ca 3000 spp. of dicotyledonous flowering plants (superorder Rosidae). Woody plants, tropical and sub-tropical and especially Australian. Includes some sources of spices, e.g. cloves, allspice, and the genus Eucalyptus, trees important in the hardwood forests of Australia.

mysophobia *(Med.)*. Morbid fear of being contaminated.

myxamoeba *(Bot.)*. The amoeboid stage of a slime mould. See **Myxomycota**.

myxo- *(Genrl.)*. Prefix from Gk. *myxa*, mucus, slime.

Myxobacteriales *(Biol.)*. A group of bacteria characterized by the absence of a rigid cell wall, gliding movements and the production of slime. Some genera produce characteristic fruiting bodies.

myxoedema *(Med.)*. A condition due to deficiency of thyroid secretion; characterized by loss of hair, increased thickness and dryness of the skin, increase in weight, slowing of mental processes, and diminution of metabolism.

myxoma *(Med.)*. A term applied to a tumour composed of a clear jellylike substance and star-shaped cells.

myxomatosis *(Vet.)*. A highly contagious and fatal virus disease of rabbits characterized by tumourlike proliferation of myxomatous tissue especially beneath the skin of the head and body. Virus lethality has progressively decreased. Vaccine available.

Myxomycetes *(Bot.)*. The true (or acellular) slime moulds. Class of slime moulds (Myxomycota) first feeding phagotrophically as individual **myxamoebae** which can interconvert with flagellated swarm cells, either of which fuse in pairs. The zygote develops as a multinucleate **plasmodium**, which is phagotrophic and eventually forms a fruiting body liberating spores. E.g. *Physarum*.

Myxomycota *(Bot.)*. *Gymnomycota*. The slime moulds. Wall-less heterotrophic organisms, usually classified with the fungi, phagotrophic on bacteria or parasitic within plant cells. Dispersed by zoospores and/or small, walled, wind-blown spores. Includes Acrasiomycetes, Mysomycetes and Plasmodiophoromycetes.

Myxophyceae *(Bot.)*. Old name for blue-green algae; Cyanophyceae.

myxoviruses *(Biol.)*. RNA containing viruses pathological to Vertebrates, including the influenza virus and related species. Many have considerable genetic variability which results in continual changes in their antigenic status and consequent difficulty in producing an effective vaccine.

N

n *(Genrl.)*. Symbol for nano-.
n *(Chem.)*. Symbol for amount of substance.
n *(Phys.)*. Symbol for neutron.
n- *(Chem.)*. An abbrev. for *normal*, i.e. containing an unbranched carbon chain in the molecule.
ν *(Phys.)*. Symbol for (1) frequency, (2) neutrino, (3) Poisson's ratio, (4) kinematic viscosity.
N *(Chem.)*. Symbol for nitrogen.
N *(Eng.)*. A symbol often used for modulus of rigidity.
N *(Phys.)*. Symbol for newton.
N *(Chem.)*. Symbol for: (1) Avogadro number; (2) number of molecules.
N *(Elec.Eng.)*. Symbol for number of turns.
N *(Phys.)*. Symbol for neutron number.
N- *(Chem.)*. Symbol indicating substitution on the nitrogen atom.
N_A *(Phys.)*. Symbol for Avogadro number.
Na *(Chem.)*. Symbol for sodium.
NA *(Eng.)*. An abbrev. for *Neutral Axis.*
NA *(Phys.)*. Abbrev. for *Numerical Aperture.*
NAA *(Bot.)*. Abbrev. for *NaphthaleneAcetic Acid.*
nab *(Build.)*. The keeper part of a door-lock.
nabla *(Maths.)*. See **del**.
nacelle *(Aero.)*. A small streamlined body on an aircraft, distinct from the fuselage, housing engine(s), special equipment or crew.
nacre, nacreous layer *(Zool.)*. The iridescent calcareous substance, mostly calcium carbonate, composing the inner layer of a Molluscan shell, which is formed by the cells of the whole of the mantle. Mother of pearl.
nacreous *(Min.)*. A term applied to the lustre of certain minerals, usually on crystal faces parallel to a good cleavage, the lustre resembling that of pearls.
nacreous clouds *(Meteor.)*. Clouds composed of ice crystals in 'mother of pearl' formations, found at a height of 25 to 30 km. They may be **wave clouds**.
nacrite *(Min.)*. A species of clay mineral, identical in composition with kaolinite, from which it differs in certain optical characters and in atomic structure.
NAD *(Biol.)*. See **nicotinamide adenine dinucleotide**.
nadir *(Astron.)*. That pole of the horizon vertically below the observer; hence, the point on the celestial sphere diametrically opposite the **zenith**.
NADP *(Biol.)*. See **nicotinamide adenine dinucleotide phosphate**.
naevus *(Med.)*. Birthmark; mole. (1) A pigmented tumour in the skin. (2) A patch or swelling in the skin composed of small dilated blood vessels.
nagana *(Vet.)*. *Fly disease, tsetse fly disease.* A name embracing a group of diseases of animals in Africa, due to infection by trypanosomes.
nagging piece *(Build.)*. A horizontal timber between the studs in a partition.
Nagra *(Image Tech.)*. TN for a range of magnetic tape sound recording equipment, much used in association with motion picture cameras.
nail *(Zool.)*. In higher Mammals, a horny plate of epidermal origin taking the place of a claw at the end of a digit.
nail punch *(Build.)*. A small steel rod tapering at one end almost to a point; used to transmit the blow from a hammer to drive a nail in so that its head is beneath the surface. Also *nail set.*
NaK *(Eng.)*. Acronym for sodium (Na) and potassium (K) alloy used as coolant for liquid metal reactor. It is molten at room temperature and below.
naked *(Bot.)*. Lacking a structure or organ.
naked-light mine *(Min.Ext.)*. Non-fiery mine, where safety lamps are not required.
name table *(Comp.)*. See **symbol table**.
Namurian *(Geol.)*. A stratigraphical stage of the Carboniferous in Europe.
NAND *(Comp.)*. A logical operator where (*p* NAND *q*) written ($\overline{p \cdot q}$) takes the value FALSE if both *p* and *q* are TRUE, otherwise (*p* NAND *q*) is TRUE. See **logical operation**.
NAND gate *(Comp.)*. **Gate** with output signal 0 when all input signals are 1, otherwise output is 1. A NAND gate can be made from a single transistor and all other gates can be constructed from combinations of NAND gates. Also *NAND element.*
nanism *(Med.)*. The condition of being a dwarf (Latin *nanus*); dwarfism.
nano- *(Genrl.)*. Prefix for 10^{-9}, i.e. equivalent to millimicro or one thousand millionth. Symbol *n.*
nanophanerophyte *(Bot.)*. Phanerophyte, with buds 25 cm to 2 m above soil level.
nanoplankton *(Zool.)*. Plankton of microscopic size from 0.2 to 20 µm.
nap *(Textiles)*. A fluffy surface on fabrics produced by the finishing process of raising. *Napping* is raising by means of a revolving cylinder covered with stiff wire brushes, or rollers covered by teazles.
napalm *(Chem.)*. A gel of inflammable hydrocarbon oils with soaps, used in warfare because of its cheapness, ease of application and ability to stick to the target while burning.
naphtha *(Build.)*. A highly volatile solvent distilled from coal tar or petroleum.
naphtha *(Chem.)*. A mixture of light hydrocarbons which may be of coal tar, petroleum or shale oil origin. *Coal tar naphtha* (boiling range approx. 80–170°C) is characterized by the predominant aromatic nature of its hydrocarbons. Generally *petroleum naphtha* is a cut between gasoline and kerosine with a boiling range 120–180°C, but much wider naphtha cuts may be taken for special purposes, e.g. feedstock for high temperature cracking for chemical manufacture. The hydrocarbons in petroleum naphthas are predominantly aliphatic.
naphthalene *(Chem.)*. $C_{10}H_8$; consists of 2 condensed benzene rings. Glistening plates, insoluble in water, slightly soluble in cold ethanol and ligroin, readily soluble in hot ethanol and ethoxyethane; mp 80°C, bp 218°C; sublimes easily and is volatile in steam. It occurs in the coal tar fraction boiling between 180°C and 200°C. It forms an additive compound with picric acid (trinitrophenol). Naphthalene is more reactive than benzene, and substitution occurs in the first instance in the *alpha* position. It is an important raw material for numerous derivatives, many of which play a rôle in the manufacture of dyestuffs. ⇨
naphthalene acetic acid *(Bot.)*. A synthetic plant growth regulator with **auxin**-like activity, used especially in **rooting compounds**.
naphthalene derivatives *(Chem.)*. Substitution products of naphthalene.
naphthenates *(Chem.)*. Metal salts of the naphthenic acids which occur naturally in crude petroleum. Mainly used as wood and textile fungicides, e.g. copper naphthenate, and also as paint driers.
naphthenes *(Chem.)*. Cycloalkanes; polymethylene hydrocarbons. Many occur in petroleum.
naphthionic acid *(Chem.)*. *1,4-Naphthylamine-monosulphonic acid*, $H_2N.C_{10}H_6.SO_3H$, obtained by the sulphonation of 1-naphthylamine. Intermediate for azo-dyes.
naphthoic acids *(Chem.)*. $C_{10}H_7$.COOH, *naphthalene carboxylic acids.* There are two isomers, of which the 1-naphthoic acid crystallizes in fine needles, mp 160°C.

On distillation with lime, they are decomposed into naphthalene and CO_2.

naphthols *(Chem.)*. $C_{10}H_7OH$. There is a 1-naphthol, mp 95°C, bp 282°C, and a 2-naphthol, mp 122°C, bp 288°C. Both are present in coal tar and can be prepared from the respective naphthalene sulphonic acids or by diazotizing the naphthylamines. 1-Naphthol is prepared on a large scale by heating α-naphthylamine under pressure with sulphuric acid. They have a phenolic character. 2-Naphthol is an antiseptic and can be used as a test for primary amines.

naphthylamines *(Chem.)*. $C_{10}H_7.NH_2$. 1-Naphthylamine forms colourless prisms or needles, mp 50°C, bp 300°C, soluble in ethanol. It is of unpleasant odour, sublimes readily, and turns brown on exposure to the air. 2-Naphthylamine, odourless, mp 112°C, bp 294°C, forms colourless plates. Originally used as an anti-oxidant in rubber manufacture, but found to be a cause of bladder cancer.

Napierian logarithm *(Maths.)*. See **logarithm**.

Napier's analogies *(Maths.)*. Formulae for solving spherical triangles.

Napier's compasses *(Eng.)*. A form of compasses having a needle point and a pencil holder pivoted at the end of one limb, and a needle point and a pen pivoted at the end of the other, both limbs being jointed for safe carrying.

napoleonite *(Geol.)*. A diorite containing spheroidal structures, about 2.5 cm in diameter, which consist of alternating shells essentially of hornblende and feldspars.

nappe *(Geol.)*. A major structure of mountain chains such as the Alps, consisting essentially of a great recumbent fold with both limbs lying approximately horizontally. It is produced by a combination of compressional earth movements and sliding under gravity, resulting in translation of the folded strata over considerable horizontal distances.

nappe *(Maths.)*. A sheet; one of the two sheets on either side of the vertex forming a cone.

nap roller *(Print.)*. A hand roller with a rough surface used for lithographic transfer work.

narcissism *(Behav.)*. (1) An erotic preoccupation with one's own body. (2) In psychoanalytic theory, investment of libidinal energy in the self, the developmental basis of self regard and not necessarily pathological.

narcolepsy *(Behav.)*. A sleep disturbance characterized by uncontrollable sleepiness and the tendency to fall asleep in inappropriate or dangerous circumstances.

narcolepsy *(Med.)*. Periodic attacks of an uncontrollable urge to sleep.

narcosis *(Med.)*. A state of unconsciousness produced by a drug; the production of a narcotic state.

narcotic *(Med.)*. Tending to induce sleep or unconsciousness; a drug which does this.

narcotine *(Chem.)*. $C_{22}H_{23}O_7N$, an alkaloid of the *iso*quinoline series; occurs in opium; forms colourless needles; mp 176°C.

narcotize *(Med.)*. To subject to the influence of a narcotic.

nares *(Zool.)*. Nostrils; nasal openings; as the internal or posterior *nares* to the pharynx, the external or anterior *nares* to the exterior. adj. *narial, nariform.*

narrow-base tower *(Elec.Eng.)*. A tower for overhead transmission lines having a base sufficiently small to be supported on a single foundation. Cf. **broad-base tower**.

narrow-cut filter *(Image Tech.)*. A colour filter which transmits only a very limited part of the spectrum.

narrow gauge *(Civ.Eng.)*. A railway gauge less than the standard 4 ft 8.5 in (1.435 m).

narrow gauge film *(Image Tech.)*. Motion picture film of 16 mm or 8 mm width.

NASA *(Space)*. Abbrev. *National Aeronautics and Space Administration* responsible for civil space activities in the US, both research and development.

nasal *(Zool.)*. Pertaining to the nose; a paired dorsal membrane bone covering the olfactory region of the Vertebrate skull.

nasal sinusitis *(Med.)*. See **sinusitis**.

NASDA *(Space)*. Abbrev. for *National Space Development Agency*, the Japanese space agency mainly responsible for applications of space activities and launch systems.

nasolacrimal canal *(Zool.)*. A passage through the skull of Mammals, passing from the orbit to the nasal cavity, and through which the tear duct passes.

nasopalatine duct *(Zool.)*. In some Reptiles and Mammals, a duct piercing the secondary palate and connecting the vomeronasal organs with the mouth.

nasopharyngeal duct *(Zool.)*. The posterior part of the original vault of the Vertebrate mouth which, in Mammals, due to the development of a secondary palate, carries air from the nasal cavity to the pharynx.

nasopharyngitis *(Med.)*. Inflammation of the nasopharynx.

nasosinusitis *(Med.)*. Inflammation of the air-containing bony cavities in communication with the nose.

nasoturbinal *(Zool.)*. In Vertebrates, a paired bone or cartilage of the nose which supports the folds of the olfactory mucous membrane.

nastic movement, nasty *(Bot.)*. A plant movement in response to but not orientated by an external stimulus, e.g. the opening or closing of some flowers in response to increasing or decreasing temperature. See also **epinasty**, **hyponasty**, **photonasty**, **thermonasty**. Cf. **tropism**, **taxis**.

natal *(Med.,Zool.)*. (1) Pertaining to birth. (2) Pertaining to the buttocks.

natality *(Ecol.)*. The inherent ability of a population to increase. *Maximum* or *absolute natality* is the theoretical maximum production of new individuals under ideal conditions. *Ecological* or *realized natality* is the population increase occurring under specific environmental conditions. Cf. **mortality**.

natatory, natatorial *(Zool.)*. Adapted for swimming.

nates *(Med.)*. The buttocks.

National Bureau of Standards *(Genrl.)*. US federal department set up in 1901 to promulgate standards of weights and measures and generally investigate and establish data in all branches of physical and industrial sciences. Abbrev. *NBS*.

National Nature Reserve *(Ecol.)*. In Britain, extensive area set aside for nature conservation, under the auspices of the *Nature Conservancy Council*.

National Physical Laboratory *(Genrl.)*. UK authority for establishing basic units of mass, length, time, resistance, frequency, radioactivity etc. Founded by the Royal Society in 1900; now government controlled and engaged in a very wide range of research. Abbrev. *NPL*.

native *(Min.Ext.)*. Said of naturally occurring metal; e.g. *native* gold, *native* copper.

native hydrocarbons *(Geol.)*. A series of compounds of hydrogen and carbon formed by the decomposition of plant and animal remains, including the several types of coal, mineral oil, petroleum, paraffin, the fossil resins, and the solid bitumens occurring in rocks. Many which have been allotted specific names are actually mixtures. By the loss of the more volatile constituents as natural gas, the liquid hydrocarbons are gradually converted into the solid bitumens, such as *ozocerite*. See also **asphalt**, **bitumen**, **coal**, **mineral oils**.

nativism *(Behav.)*. In philosophy, the position that humans are born with some innate knowledge; in the study of perception, the view that some important abilities are innate.

natrojarosite *(Min.)*. Hydrated sulphate of sodium and iron crystallizing in the trigonal system.

natrolite *(Min.)*. Hydrated silicate of sodium and aluminium crystallizing in the orthorhombic system. A sodium-zeolite. It usually occurs in slender or acicular crystals, and is found in cavities in basaltic rocks and as an alteration product of nepheline or plagioclase.

natron *(Min.)*. Hydrated sodium carbonate, occurring in soda-lake deposits.

natural abundance *(Phys.)*. See **abundance**.

natural aging *(Eng.)*. The changes which take place in a material after it has cooled subsequent to heat treatment. May take months or years as in some cast irons.

natural antibody *(Immun.)*. Antibody present in the blood of normal individuals not known to have been immunized against the relevant antigen, e.g. in humans antibodies against antigens of the **ABO blood group system** or Forssman antibodies. They are generally induced by organisms in the gut owing to shared antigenic determinants. Animals which are reared under germ-free conditions do not develop such antibodies, but do so if the gut is colonized by enterobacteria.

natural background *(Phys.)*. In the detection of nuclear radiation, the radiation due to *natural radioactivity* and to cosmic rays, enhanced by contamination and fall-out.

natural cement *(Build.)*. A cement similar to a hydraulic lime, made from a natural earth, with but little preparation.

natural classification *(Bot.)*. One based on many characters, and likely to have a predictive value.

natural draught *(Eng.)*. The draught or air flow through a furnace induced by a chimney and dependent on its height and the temperature difference between the ascending gases and the atmosphere.

natural evaporation *(Meteor.)*. The evaporation that takes place at the surface of ponds, rivers etc. which are exposed to the weather; it depends on solar radiation, strength of wind and relative humidity.

natural frequency *(Phys.)*. See **normal frequency, normal mode.**

natural frequency *(Telecomm.)*. That of free oscillations in a system. See **natural period.**

natural frequency of antenna *(Telecomm.)*. The lowest frequency at which an unloaded antenna system is resonant.

natural gas *(Geol.)*. Any gas found in the earth's crust, including gases generated during volcanic activity (see **pneumatolysis, solfatara**). The term, however, is particularly applied to natural hydrocarbon gases which are associated with the production of petroleum. These gases are principally methane and ethane, sometimes with propane, butane, nitrogen, CO_2 and sulphur compounds (notably H_2S). The gas is found both above the petroleum and dissolved in it, but many very large gas fields are known which produce little or no petroleum. Natural gas has largely replaced **town gas** in many countries where it occurs abundantly (US, Canada, Algeria, W. Europe); since 1967 important finds in the North Sea have provided supplies for the British gas-grid system.

natural glass *(Geol.)*. Magma of any composition is liable to occur in the glassy condition if cooled sufficiently rapidly. Acid (i.e. granitic) glass is commoner than basic (i.e. basaltic) glass; the former is represented among igneous rocks by pumice, obsidian, and pitchstone; the latter by tachylite. Natural quartz glass occurs in masses lying on the surface of certain sandy deserts (e.g. Libyan Desert); while both clay rocks and sandstones are locally fused by basic intrusions. See also **buchite, fulgurite, tektites.**

natural immunity *(Immun.)*. Immunity conferred before birth and not *acquired* subsequently by exposure to antigens from the environment.

naturalized *(Ecol.)*. Introduced from another region but growing, reproducing and maintaining itself in competition with the native vegetation.

natural killer cell *(Immun.)*. *NK cell*. A lymphoid cell which kills a range of tumour cell targets in the absence of prior immunization and without evident antigen specificity. Morphologically it is a lymphocyte with large granules visible under the light microscope or after staining with azure dyes. NK cells are activated by interferon, and may also be important in killing some virally infected cells in vivo. Killing results from binding of the NK cell to its target and the insertion of granules into it. The granules contain proteases and **perforin**. Killer cells appear to use a similar mechanism. The relationship of these cells to one another is uncertain. Their origin is uncertain, but may be common with T-lymphocytes, since T-cells during prolonged culture in vitro sometimes acquire NK cell characteristics.

natural language *(Comp.)*. Any naturally evolved human language.

natural load *(Elec.Eng.)*. A resistive load impedance, numerically equal to the characteristic impedance of the transmission line which it terminates, or the power which the line would transmit if it were so loaded.

natural magnet. *(Phys.)*. See **lodestone.**

natural modes *(Telecomm.)*. See **normal modes.**

natural number *(Maths.)*. Term used to describe any element of the set [1,2,3...] or the set [0,1,2,3,...]. The inclusion of zero is a matter of definition.

natural order of colour *(Build.)*. The arrangement of colours of maximum purity in a circle to show natural progression from lightest to darkest tone, e.g. yellow is lightest, blue-purple is darkest. If this order is reversed discordant colours are produced.

natural period *(Telecomm.)*. Time of one cycle of oscillation arising from free oscillation, depending on inertia and elastance of a system. Reciprocal of natural frequency.

natural radioactivity *(Phys.)*. That which is found in nature. Such radioactivity indicates that the isotopes involved have a half-life comparable with the age of the earth or result from the decay of such isotopes. Most such nuclides can be grouped in one of three **radioactive series.**

natural resonance *(Telecomm.)*. Response of system to signal with period equal to its own natural period.

natural scale *(Acous.)*. The musical scale in which the frequencies of the notes within the octave are proportional to 24, 27, 30, 32, 36, 40, 45, and 48, and which can be realized in continuously variable pitch instruments, such as the human voice and stringed instruments, but not in keyboard instruments, which use the **tempered scale.** Also *just scale, just temperament.*

natural scale *(Surv.)*. A term applied to a section drawn with equal vertical and horizontal scales.

natural selection *(Biol.)*. An evolutionary theory which postulates the survival of the best adapted forms, with the inheritance of those characteristics wherein their fitness lies, and which arise as random variations due to mutation; it was first propounded by Charles Darwin, and is often referred to as *Darwinism* or the *Darwinian Theory.*

natural slope *(Civ.Eng.)*. The maximum angle at which soil in cutting or bank will stand without slipping.

natural uranium *(Chem.)*. That with its natural isotopic abundance, not depleted by the removal of ^{235}U.

natural-uranium reactor *(Nuc.Eng.)*. One in which natural, i.e. unenriched, uranium is the chief fissionable material.

natural wavelength of antenna *(Telecomm.)*. The free space wavelength corresponding to the natural frequency of an antenna.

nauplius *(Zool.)*. The typical first larval form of Crustacea; egg-shaped, unsegmented, and having three pairs of appendages and a median eye; found in some members of every class of Crustacea, but often passed over, becoming an entirely embryonic stage.

Nauta mixer *(Chem.Eng.)*. A proprietary mixer comprising a stationary cone, with point vertically downwards, within which a screw arm rotates on its own axis parallel to the conical surface. The lower end of the screw is fixed in a universal joint at its lower end, and to an arm at its upper end, by means of which the whole screw, while rotating, is moved round the interior surface of the cone. Much used for mixing small quantities of one constituent with large quantities of another and mostly used on dry solids.

Nautical Almanac *(Astron.)*. An astronomical ephemeris published annually in advance, for navigators and astronomers. First published in 1767, it is now called the *Astronomical Ephemeris*. An abridged version, for the use of navigators, is given the original title *The Nautical Almanac.*

nautical log *(Ships)*. A device for estimating the speed of a vessel. In the old fashioned log, a line divided into equal spaces (knots) runs freely off a reel and is attached to a *chip log*, which is stationary in the water as the vessel travels. Time is measured by a log glass. The modern *patent* (or *taffrail*) *log* mechanically indicates the rate of travel by means of a submerged fly or rotator, whose revolutions are conveyed to a register on the rail of the vessel by a braided hemp line secured to the rotator.

nautical mile *(Genrl.)*. One-sixtieth of a degree of latitude, a distance varying with latitude. The *UK nautical mile* is 6080 ft (1853.18 m), differing slightly from the *international nautical mile*, 1852 m.

nautical twilight *(Astron.)*. The interval of time during which the sun is between 6° and 12° below the horizon, morning and evening. See **astronomical twilight**, **civil twilight**.

Nautiloidea *(Zool.)*. A sub-class of the Cephalopoda, having a wide central siphuncle and a planospiral chambered shell. Abundant from early Cambrian to late Cretaceous, but now represented by one genus, *Nautilius*, which lives in tropical seas. All chambers except the terminal living chamber contain gas which buoys up the heavy shell.

naval brass *(Eng.)*. See **Tobin bronze**.

nave *(Arch.)*. The middle or main body of a basilica, rising above the *aisles*: the main part of a church, generally west of the crossing, including or excluding its aisles.

navel *(Textiles)*. Part of the spinning head through which the yarn is withdrawn in **rotor spinning**.

navel *(Zool.)*. In Mammals, the point of attachment of the umbilical cord to the body of the foetus.

navel-ill *(Vet.)*. See **joint-ill**.

navicular bone *(Zool.)*. In Mammals, one of the tarsal bones, also known as the *centrale* or *scaphoid*.

navicular disease *(Vet.)*. Chronic osteitis of the navicular bone in the foot of the horse. Vasodilators used in treatment.

navigable semicircle *(Meteor.)*. The left hand half of the storm field in the northern hemisphere, the right hand half in the southern hemisphere, when looking along the path in the direction a *tropical revolving storm* is travelling. Cf. **dangerous semicircle**.

navigation *(Behav.)*. In biology refers to complex forms of long distance orientation by animals.

navigation *(Genrl.)*. The art of directing a vessel or vehicle by terrestrial or stellar observation, or by radio and radar signals.

navigation *(Hyd.Eng.)*. A name frequently given to a canalized river the flow of which is more or less under artificial control.

navigational system *(Telecomm.)*. Any system of obtaining bearings and/or ranges for navigational purposes by radio techniques.

navigation flame float *(Aero.)*. A pyrotechnic device, dropped from an aircraft, which burns with a flame while floating on the water. Used to determine the drift of the aircraft at night.

navigation lights *(Aero.)*. Aircraft navigation lights consist of red, green and white lamps located in the port wing tip, starboard wing tip and tail respectively.

navigation smoke float *(Aero.)*. A pyrotechnic device, dropped from an aircraft, which emits smoke while floating on the water. Used for ascertaining the direction of the wind or the drift of the aircraft.

Nb *(Chem.)*. The symbol for *niobium*.

N-bands *(Biol.)*. See **banding techniques**.

NCR *(Paper)*. *No Carbon Required*. TN for stationery, especially for office machine use, in which simultaneous duplicate copies are obtained without the use of carbon paper. Microcapsules containing dyes on the verso of the 'top' copy are fractured by the impact of the writing medium (pen, typewriter, etc.), the image being then transferred to the under copy or copies, the receiving surface of which has a coating of attapulgite with a starch or latex binder. See **microencapsulation**.

Nd *(Chem.)*. The symbol for *neodymium*.

NDB *(Aero.)*. Abbrev. for *Non-Directional Beacon*. See **beacon**.

N-display *(Radar)*. A radar **K-display** in which the target produces two breaks on the horizontal time base. Direction is proportional to the relative amplitude of the breaks, and range is indicated by a calibrated control which moves a pedestal signal to coincide with the breaks.

NDT *(Eng.)*. See **non-destructive testing**.

Ne *(Chem.)*. The symbol for *neon*.

neanic *(Zool.)*. Said of the adolescent period in the life-history of an individual.

neap tides *(Astron.)*. High tides occurring at the moon's first or third quarter, when the sun's tidal influence is working against the moon's, so that the height of the tide is below the maximum in the approximate ratio 3:8.

Nearctic region *(Zool.)*. One of the subrealms into which the Holarctic region is divided; it includes N. America and Greenland.

near-end cross-talk *(Telecomm.)*. Cross-talk between 2 parallel circuits when both the listener and the speaker originating the inducing currents are at the same end of the parallelism. See **far-end cross-talk**.

nearest neighbour analysis *(Biol.)*. Method for determining the frequency of pairs of adjacent bases in DNA. It has shown that there is a deficiency of the pair CG in most higher organisms.

near field *(Acous.)*. See **far field**.

nearly-free electron model *(Phys.)*. A model from which the band structure of simple metals can be calculated. The periodic part of the potential due to the crystal lattice is treated as a minor modification to the free electron gas model. See **band theory of solids**.

near point *(Phys.)*. The nearest position to the eye at which an object can be seen distinctly. The object point conjugate to the retina when accommodation is exerted to its fullest extent.

neat cement *(Build.,Civ.Eng.)*. A cement mortar mixture made up with water only, without addition of sand.

neat size *(Build.)*. The net or exact size after preparation.

neat work *(Build.)*, The brickwork above the footings.

nebula *(Astron.)*. A term applied to any celestial object which appears as a hazy smudge of light in an optical telscope, its usage predating photographic astronomy. It is now more properly restricted to true clouds of **interstellar medium**. Nebulae may be either bright or dark. Galaxies are sometimes referred to as *extragalactic nebulae*, the most famous of which is the **Andromeda nebula**.

nebula *(Med.)*. (1) A slight opacity in the cornea of the eye. (2) An oily preparation for use in an atomizer or nebulizer (e.g. a nasal spray).

nebular hypothesis *(Astron.)*. One of the earliest scientific theories of the origin of the solar system, stated by Laplace. It supposed a flattened mass of gas extending beyond Neptune's orbit to have cooled and shrunk, throwing off in the process successive rings which in time coalesced to form the several planets.

neck *(Arch.)*. The narrow moulding separating the capital of a column from the shaft.

neck *(Bot.)*. (1) The upper tubular part of an archegonium, and of a perithecium. (2) The lower part of the capsule of a moss, just above the junction with the seta.

neck *(Eng.)*. See **necking**.

neck *(Geol.)*. A plug of volcanic rock representing a former feeder channel of an extinct volcano.

neck *(Textiles)*. In the process of extruding and drawing synthetic filaments a marked reduction in the cross-sectional area may occur. This part of the filament is called the neck and the occurrence is *necking*.

neck canal cell, neck cell *(Bot.)*. One of the cells in the central canal in the neck of an archegonium.

necking *(Eng.)*. In a material tensile test just prior to fracture, the cross-sectional area is reduced over a short length; this is the neck. *Necking* quantitatively indicates disinclination to strain-harden under cold work.

necro- *(Genrl.)*. Prefix from Gk. *neckros*, a dead body.

necrobacillosis *(Vet.)*. *Bacillary necrosis*. Infection of animals by *Sphaerophorus necrophorus (Fusiformis necrophorus)*.

necrobiosis *(Med.)*. The gradual death, through stages of degeneration and disintegration, of a cell in the living body.

necrophagous *(Zool.)*. Feeding on the bodies of dead animals.

necrophorous *(Zool.)*. Carrying away the bodies of dead animals; as certain Beetles, which usually afterwards bury the bodies.

necropsy *(Med.)*. *Autopsy*. A postmortem examination of the body.

necrosis *(Biol.)*. Death of a cell (or of groups of cells) while still part of the living body. adj. *necrotic*. v. *necrose*.

necrotic enteritis of swine *(Vet.)*. *Porcine intestinal adenomatosis*. Infection of swine with *Campylobacter sputorum*, with various forms. Symptoms include weight loss, anorexia, diarrhoea, melaena and in the case of the chronic form, poor weight gain.

necrotic stomatitis *(Vet.)*. See **calf diphtheria**.

necrotroph *(Bot.)*. An organism which feeds off dead cells and tissues. A necrotrophic parasite kills host cells and feeds on them once they are dead, e.g. the **damping-off** fungi. Cf. **biotroph, facultative parasite**.

nectar *(Bot.)*. A sugary fluid exuded by plants, usually from some part of the flower, occasionally from somewhere else on the plant; it attracts insects, which assist in pollination.

nectar guide *(Bot.)*. Same as **honey guide**.

nectarivorous *(Zool.)*. Nectar-eating.

nectary *(Bot.)*. A glandular organ or surface from which nectar is secreted.

necto- *(Genrl.)*. Prefix from Gk. *nēktos*, swimming.

necton *(Bot.,Zool.)*. See **nekton**.

nectopod *(Zool.)*. An appendage adapted for swimming.

need *(Behav.)*. Refers to specific physical or psychological conditions necessary for an individual's welfare and/or sense of wellbeing.

needle *(Acous.)*. See **stylus**.

needle *(Bot.)*. A long, narrow, stiffly-constructed leaf, characteristic of pines and similar plants.

needle *(Build.)*. A timber or steel beam used in the process of underpinning. It is laid horizontally at right angles to the wall (through which it passes) and is supported on both sides by dead shores, so as to take the load of the upper part of the walls.

needle *(Elec.Eng.)*. The moving magnet of a compass or galvanometer of the moving-magnet type. Sometimes also the moving element of an electrostatic voltmeter.

needle *(Textiles)*. A simple pointed instrument with an eye for carrying the thread, used in sewing. In machine knitting, the device that forms and intermeshes the loops produced from the yarn supplied. Normally each needle forms one **wale** of the fabric. There are many different kinds of needle used for different purposes. A bearded needle has a terminal hook that is flexed by pressing and released on removing the force. A latch needle has a terminal hook that is closed by a pivoting latch. The fabric loop overturns the latch and allows the loop to be cast off as another loop is being formed. A barbed needle in a needleloom causes the entangling of fibres in a method for making **non-woven fabrics**.

needle beam *(Civ.Eng.)*. A transverse floor-beam supported across the chords of a bridge.

needle chatter *(Acous.)*. That arising from vibrations of the gramophone needle being transferred to the tone arm and radiated as noise.

needle lubricator *(Eng.)*. A crude form of lubricator consisting of an inverted stoppered flask attached to a bearing and containing a wire loosely fitting a hole in the stopper and touching the shaft.

needle paper *(Paper)*. An acid-free, black paper for wrapping needles, pins etc.

needle pick-up *(Acous.)*. A pick-up in which the sole moving part is the magnetic needle, which by its motion diverts magnetic flux and induces electromotive forces in coils on the magnetic circuit.

needle roller bearing *(Eng.)*. A **roller bearing** without cage, in which long rollers of small diameter are used, located endwise by a lip on the inner or outer race.

needles *(Print.)*. Removable points on web-fed presses, either fixed or retractable by cam action, used to control and convey the cut leading edge of the web or copy until it is severed, transferred, folded, or stitched, as required; can be mounted on the folding transfer, cutting, or collection cylinders.

needle scaffold *(Build.)*. A scaffold which is supported on cantilever or needle beams jutting out from an intermediate height in the building, thereby avoiding the necessity for erection from ground level.

needle scratch *(Acous.)*. Noise emanating from gramophone, specifically due to irregularities in the contact surface of the groove. Also called *surface noise*.

needle stone *(Min.)*. A popular term for clear quartz containing acicular inclusions, usually of rutile, but in some specimens, of actinolite. Also called *rutilated quartz*. The name has also been used for various acicular zeolites.

needle talk *(Acous.)*. Direct sound output from transducer of gramophone pick-up.

needle traverse *(Surv.)*. A traverse in which the angles between successive lines, or the directions of the lines, are measured by means of a magnetic compass.

needle valve *(Eng.)*. A slender pointed rod working in a hole or circular seating; operated by automatic means, as in a carburettor float chamber, or by a screw, for the control of fluid.

needling *(Build.)*. The process of underpinning in which needles are used in the support of the upper part of the building.

needling *(Med.)*. *Discission*. Cutting the lens of the eye with a needle in the treatment of cataract.

Néel temperature *(Phys.)*. The temperature at which the susceptibility of an antiferromagnetic material has a maximum value.

NEF *(Acous.)*. See **noise exposure forecast value**.

negater *(Comp.)*. See **NOT gate**.

negative *(Image Tech.)*. An image, photographically or electronically produced, in which the tonal values of the original subject are reversed, bright areas being represented as dark, and vice versa. In a *colour negative*, in addition the hues of the original are represented by substantially complementary colours.

negative *(Phys.)*. Designation to electric charge, introduced by Franklin, now known to be exhibited by the electron, which, in moving forms the normal electric current.

negative after-image *(Phys.)*. The image of complementary colour arising after visual fatigue from viewing a coloured object and then a white surface.

negative bias *(Electronics)*. Static potential, negative with reference to earth, applied to electrode of valve or transistor, to obtain desired operating conditions.

negative booster *(Elec.Eng.)*. A series wound booster used in connection with an earthed return power supply system, e.g. for a tramway. It is connected between two points on the earthed return path to reduce the potential between them and minimize the possibility of electrolysis due to leakage currents.

negative carbon *(Elec.Eng.)*. The carbon of a d.c. arc lamp which is connected to the negative terminal of the supply. It is usually of smaller diameter than the positive carbon as it burns away more slowly.

negative catalysis *(Chem.)*. The retardation of a chemical reaction by a substance which itself undergoes no permanent chemical change.

negative colour film *(Image Tech.)*. A multi-layer photographic material intended to yield a **colour negative** image after exposure and processing.

negative conduction *(Telecomm.)*. The conductance (*reciprocal* of resistance) of *negative resistance* devices. See **negative resistance oscillator**.

negative coupling *(Elec.Eng.)*. See **positive coupling**.

negative crystal *(Phys.)*. Birefringent material for which the velocity of the extraordinary ray is greater than that of the ordinary ray.

negative electricity *(Phys.)*. Phenomenon in a body when it gives rise to effects associated with excess of electrons. See **positive electricity**.

negative electrode *(Electronics)*. The *anode* of a primary cell (the electrode by which conventional current returns to the cell), but the *cathode* of a valve (connected to the negative side of the power supply).

negative feedback *(Telecomm.)*. Reduction of amplifier gain through feeding part of the output signal back to the input, in such a way that it is out of phase with the incoming signal. This gives more uniform performance, greater stability, reduced distortion and sometimes improved bandwidth.

negative feeder *(Elec.Eng.)*. The feeder connecting the negative terminal of a load to the negative bus-bars of the power supply. Also called *return feeder*.

negative *g* *(Aero.,Space)*. (1) In a manoeuvring aircraft, any force acting opposite to the normal force of gravity. (2) The force exerted on the human body in a gravitational field or during acceleration so that the force of inertia acts in a foot-to-head direction, causing considerable blood pressure on the brain. Also called *minus g*. *Negative g tolerance*, in practice, is the degree of tolerance $3g$ for 10–15 sec.

negative glow *(Electronics)*. In a medium pressure gas-discharge tube, the glow between the cathode and Faraday dark space.

negative ion *(Electronics)*. Radical, molecule or atom which has become negatively charged through the gain of one or more electrons. See **ion**.

negative-ion vacancy *(Electronics)*. Same as *hole*.

negative mineral *(Phys.)*. A doubly refracting mineral in which the ordinary refractive index is greater than the extraordinary. Calcite is a negative mineral, for which the values of ω and ε are 1.66 and 1.48 repectively. See **optic sign**.

negative modulation *(Telecomm.)*. That in which the carrier level is reduced as the level of the modulating signal is increased. Used in many television systems, i.e. peak transmitter power corresponds to black, and the power decreases as brightness increases. Also *downward modulation*.

negative mutual inductance *(Elec.Eng.)*. See under **positive coupling**.

negative phase sequence *(Elec.Eng.)*. A 3-phase system in which the voltages and currents in each of the three phases reach their maximum values in the reverse order to conventional phase sequence, i.e. red, blue, yellow as opposed to red, yellow, blue. See **phase sequence**.

negative phase-sequence component *(Elec.Eng.)*. The symmetrical component of an unbalanced 3-phase system of voltages or currents in which the phase-sequence is in the opposite order to standard; i.e. it is in the order red, blue, yellow.

negative phase-sequence relay *(Elec.Eng.)*. A relay which operates when any negative phase sequence components of current or voltage appear in the circuit to which it is connected.

negative plate *(Elec.Eng.)*. The plate of an accumulator or primary cell which is normally at the lower potential and to which the current from the circuit during discharge is said to return.

negative proton *(Phys.)*. See **antiproton**.

negative reaction *(Biol.)*. A tactism or tropism in which the organism moves, or the member grows, from a region where the stimulus is stronger to one where it is weaker.

negative reinforcement *(Behav.)*. In conditioning situations, a stimulus, usually aversive, that increases the probability of escape or avoidance behaviour. Cf. **punishment**.

negative resistance *(Electronics)*. A characteristic such that, when the current though a device increases, the voltage drop across it decreases. Most electrical gas discharges have this property, along with some valves and semiconductor devices, including the Gunn and tunnel diodes.

negative-resistance oscillator *(Electronics)*. One in which a parallel-tuned resonant circuit or a cavity is connected to a negative-resistance device, the negative resistance compensating for the losses in the resonant circuit and allowing oscillation to become continuous. See **Gunn effect, tunnel diode**.

negative scanning *(Image Tech.)*. Scanning a photographic negative, with reversal in the circuits, so that the reproduced image is the normal positive.

negative stagger *(Aero.)*. See **stagger**.

negative staining *(Biol.)*. Important technique in electron microscopy, in which heavy metals which scatter electrons are deposited around the specimen. This is then seen in negative contrast.

negative stencil *(Build.)*. A vegetable drying oil. Its use in paints has now been superseded by a wider range of drying oils.

negative transconductance *(Electronics)*. Property of certain valves whereby an increase in positive potential on one electrode accompanies a decrease in current to another electrode.

negative video signal *(Image Tech.)*. A video signal in which increasing amplitude corresponds to decreasing light-value in the transmitted picture. Black is taken as 100%, white about 30%, of the maximum amplitude in the signal.

neg-pos *(Image Tech.)*. Abbreviation for a process in which the film exposed in the camera is processed to form a negative image which is then used to make a separate positive by printing, in contrast to a **reversal process**.

negri bodies *(Vet.)*. Specific inclusion bodies found in the cytoplasm of nerve cells in the brain and spinal cord in animals affected with rabies.

neighbourhood *(Maths.)*. The neighbourhood of a point is any open set containing that point.

Neisseriaceae *(Biol.)*. A family of bacteria belonging to the order Eubacteriales. Gram-negative, characteristically occurring as paired spheres, parasitic in mammals, e.g. *Neisseria gonorrhoeae* (gonorrhoea).

nekton *(Bot.,Zool.)*. Actively swimming aquatic organisms, as opposed to the passively drifting organisms or *plankton*. Also *necton*.

nematic *(Phys.)*. Said of a mesomorphous substance whose molecules or atoms are oriented in parallel lines. Cf. **smectic**.

nematoblast *(Zool.)*. A cell which will develop a **nematocyst**.

nematocyst *(Zool.)*. An independent effector found in most Coelenterata (and a few Protozoa), consisting of a fluid filled sac and produced at one end into a long narrow pointed hollow thread, which normally lies inverted and coiled up within the sac, but can be everted when the *cnidocil* is stimulated. Used for prey capture and defence.

Nematoda *(Zool.)*. A class of phylum Aschelminthes comprising unsegmented worms with an elongate rounded body pointed at both ends; marked by lateral lines and covered by a heavy cuticle composed of protein; have a mouth and alimentary canal; have only longitudinal muscles; nervous system consists of a circum-pharyngeal ring and a number of longitudinal cords; perivisceral cavity a pseudocoel; cilia absent; sexes separate; development direct, the larvae resembling the adults; many species are of economic importance; mostly free-living but some are parasitic. Round Worms, Thread Worms, Eel Worms.

Nemertea *(Zool.)*. A phylum of apparently nonmetameric acoelomate worms with an elongate flattene a ciliated ectoderm, and a dorsal eversible proboscis not connected with the alimentary canal. Most are marine, but some are fresh water or terrestrial. Ribbon Worms. Also *Nemertini*.

neoblasts *(Zool.)*. In many of the lower animals (Annelida, Ascidia etc.), large amoeboid cells widely distributed through the body which play an important part in the phenomena of regeneration.

neocerebellum *(Med.)*. Phylogenetically, the more recently developed part of the posterior lobe of the cerebellum, receiving predominantly partine fibres via the pontocerebellar tract.

Neo-Classicism *(Arch.)*. An architectural style popular in Europe during the late 18th and early 19th centuries, when, in spite of the achievements of the **Baroque** phase, architects once more reverted to Greek and Roman ideals. It was a part of the **Antiquarian** movement, the final phase of Renaissance. The term used more generally described any design which uses Classical ideals as a source of inspiration.

Neocomian *(Geol.)*. The oldest epoch of the Cretaceous period.

neo-Darwinism *(Zool.)*. The modern version of Darwin's theory of evolution by natural selection, incorporating the discoveries of Mendelian and population genetics.

neodymium *(Chem.)*. A metallic element, a member of the rare earth group. Symbol Nd, at.no. 60, r.a.m. 144.24, rel.d. 6.956, mp 840°C. The metal is found in cerite, monazite and orthite. Neodymium glass is used for solid state lasers and light amplifiers.

neo-freudians *(Behav.)*. Schools of thought and therapy based on modification of Freud's theories; usually stress social rather than biological factors as important determinants of unconscious conflict.

Neogene *(Geol.)*. A sub-period of the Tertiary covering a time-span from ca. 25 to 2 million years ago. The corresponding system of rocks.

neohexane *(Chem.)*. *2,2-dimethylbutane*. Bp 50°C. Antiknock fuel prepared by the addition of ethene to 2-methylpropane under pressure at increased temperatures.

Neolithic Period *(Geol.)*. The later portion of the Stone Age, characterized by well-finished, polished stone implements, agriculture and domesticated farm animals. Cf. **Palaeolithic Period**.

neologism *(Behav.)*. Verbal construction such as occurs in schizophrenia, manic depressive psychosis and some aphasias, in which the patient uses coined words, which may have meaning for him but not for others, or else gives inappropriate meanings to ordinary words.

neomycin *(Med.)*. An antibiotic derived from *Streptomyces fradiae*; its sulphate is especially effective against external staphylococcal infections (skin, eyes etc.); also used internally.

neon *(Chem.)*. Light, gaseous, inert element, recovered from atmosphere, Symbol Ne, at.no. 10, r.a.m. 20.179, mp −248.67°C, bp −245.9°C. Historically important in that J.J. Thomson, through his parabolas for charge/mass of particles, found two isotopes in neon, the first nonradioactive isotopes to be recognized. Used in many types of lamp, particularly to start up sodium vapour discharge lamps. Pure neon was the first gas to be used for high voltage display lighting, being bright orange in colour. Much used in cold-cathode tubes, reference tubes, Dekatrons.

neon induction lamp *(Phys.)*. A lamp consisting of a small tube containing neon at low pressure; luminescence is produced by the action of high frequency currents in a few turns surrounding the tube.

neonychium *(Zool.)*. A pad of soft tissue enclosing a claw of the foetus during the development of many Mammals, to eliminate the risk of ripping the foetal membranes.

neopallium *(Zool.)*. In Mammals, that part of the cerebrum occupied with impressions from senses other than the sense of smell.

neoplasm *(Med.)*. A new formation of tissue in the body; a *tumour*. adj. *neoplastic*.

Neoprene *(Chem.)*. *Polychloroprene*; the first commercial synthetic rubber (US 1931). Chloroprene (3 chlorobut 1,2:3,4-diene), the monomer, $CH_2=CCl.CH=CH_2$, is derived from acetylene and hydrochloric acid.

neossoptiles *(Zool.)*. The down-feathers found on a newly-hatched Bird.

neotenin *(Zool.)*. In Insects, the **juvenile hormone** produced by the corpora allata, which suppresses the development of adult characteristics at each moult except the last, when the corpora allata become inactive and metamorphosis occurs.

neoteny *(Biol.)*. Retention of some juvenile characteristics by the sexually mature adult like some Amphibians which have the appearance of tadpoles. Cf. **paedogenesis**.

neotropical region *(Zool.)*. One of the primary faunal regions into which the surface of the globe is divided. It comprises South America, the West Indian islands, and Central America south of the Mexican plateau.

neovitalism *(Zool.)*. The theory which postulates that a complete causal explanation of vital phenomena cannot be reached without invoking some extra-material concept.

Neozoic *(Geol.)*. The name (= 'new life') sometimes given to the Tertiary and Post-Tertiary rocks.

neper *(Telecomm.)*. Unit of attenuation. If current I_1 is attenuated to I_2 so that $I_2/I_1 = e^{-N}$, then N is attenuation. In circuits matched in impedance, 1 neper = 8.686 dB. After John Napier (Lat. *Nepero*), Scottish scientist, inventor of natural logarithms.

nepheline, nephelite *(Min.)*. Silicate of sodium and aluminium, $NaAlSiO_4$, but generally with some potassium partially replacing sodium, which crystallizes in the hexagonal system. It is frequently present in igneous rocks with a high sodium content and a low percentage of silica, i.e. the undersaturated rocks. See **elaeolite**.

nepheline-syenite *(Geol.)*. A coarse-grained igneous rock of intermediate composition, undersaturated with regard to silica, and consisting essentially of nepheline, a varying content of alkali-feldspar, with soda amphiboles and/or soda-pyroxenes. Common hornblende, augite, or mica are present in some varieties. See, for example, **foyaite**, **laurdalite**.

nephelinite *(Geol.)*. A fine-grained igneous rock normally occurring as lava flows, and resembling basalt in general appearance; it consists essentially of nepheline and pyroxene, but not of olivine or feldspar. The addition of the former gives *olivine-nephelinite*, and of the latter, *nepheline-tephrite*.

nephelometric analysis *(Chem.)*. A method of quantitative analysis in which the concentration or particle size of suspended matter in a liquid is determined by measurement of light absorption. Also called *photoextinction method*, *turbidimetric analysis*.

nephograph *(Image Tech.)*. An instrument comprising electrically controlled cameras for photographing clouds, etc. in order that their position in the sky may subsequently be determined.

nephoscope *(Meteor.)*. An instrument for observing the direction of movement of a cloud and its angular velocity about the point on the earth's surface vertically beneath it. If the cloud height is also known its linear speed may be calculated.

nephrectomy *(Med.)*. Removal of a kidney.

nephric *(Med.,Zool.)*. Pertaining to the kidney.

nephridiopore *(Zool.)*. The external opening of a nephridium or nephromixium.

nephridium *(Zool.)*. In Invertebrates and lower Chordata, a segmental excretory organ consisting of an intercellular duct of ectodermal origin leading from the coelom to the exterior; more generally, an excretory tubule. adj. *nephridial*. Cf. **coelomoduct**.

nephrite *(Min.)*. One of the minerals grouped under the name of *jade* ('New Zealand Jade'); consists of compact and fine-grained tremolite or actinolite. It has been widely used for ornaments in the Americas and the East.

nephritis *(Med.)*. Inflammation of the substance of the kidney. See **glomerulonephritis**. adj. *nephritic*.

nephr-, nephro- *(Genrl.)*. Prefix from Gk. *nephros*, kidney.

nephrodinic *(Zool.)*. Using the same duct for the discharge of both excretory and genital products.

nephrogenic tissue *(Zool.)*. In the embryonic development of Vertebrates, a relatively small intermediate region of the mesoderm lateral and ventral to the somites, from which derive the kidney tubules, their ducts, and the deeper tissues of the gonads. May be segmented, forming nephrotomes, or a continuous band of tissue.

nephrogonoduct *(Zool.)*. Especially in Invertebrates, a common duct for genital and excretory products.

nephrolithiasis *(Med.)*. The presence of stones in the kidney.

nephrolithotomy *(Med.)*. Removal of stones from the kidney through an incision in the kidney.

nephrologist *(Med.)*. A specialist in diseases of the kidney.

nephropathy *(Med.)*. Any disease of the kidneys.

nephropexy *(Med.)*. The fixation, by operative measures, of a kidney which is abnormally movable. Cf. **nephrorrhaphy**.

nephropore *(Zool.)*. See **nephridiopore**.

nephroptosis *(Med.)*. *Movable kidney; floating kidney*. An abnormally mobile kidney, associated with general displacement downwards of other abdominal organs.

nephrorrhaphy *(Med.)*. The fixation, by suture, of a displaced kidney. Cf. **nephropexy**.

nephros *(Zool.)*. A kidney. adj. *nephric*.

nephrostome *(Zool.)*. The ciliated funnel by which some types of nephridia and nephromixia open into the coelom.

nephrostomy *(Med.)*. Formation of an opening into the pelvis of the kidney for the drainage of urine.

nephrotic syndrome *(Med.)*. Increased permeability of kidney glomerulus basement membrane leading to albumen loss in urine (*albuminuria*), low plasma membrane and **oedema**.

nephrotomy *(Med.)*. The making of an incision into the kidney.

nephrotoxin *(Med.)*. A poison or toxin which specifically affects the cells of the kidney.

nepionic *(Zool.)*. Said of the embryonic period in the life-history of an individual.

neps *(Textiles)*. Term applied in the cotton industry to small entanglements of fibres that cannot be unravelled; generally formed during the ginning process from dead or immature fibres.

Neptune *(Astron.)*. The eighth planet of the solar system, discovered in 1846 by J.G. Galle of the Berlin Observatory. It never gets brighter than magnitude 7.7, so it is never visible to the unaided eye. It orbits the Sun in 165 years, at a distance of 30.1 AU. The diameter is 49 500 km and the mass 17.2 that of the Earth. In many respects it is probably similar to **Uranus**. There are two significant satellites, Triton and Nereid.

neptunean dyke *(Geol.)*. An intrusive sheet of sedimentary rock.

neptunium *(Chem.)*. Element, at.no. 93, symbol Np; named after planet Neptune; produced artificially by nuclear reaction between uranium and neutrons. Has principal isotopes 237 and 239.

neptunium series *(Chem.)*. Series formed by the decay of artificial radioelements, the first member being *plutonium-241* and the last *bismuth-209*, of which *neptunium-237* is the longest lived (half-life 2.2×10^6 years).

NEQ gate *(Comp.)*. See **XOR gate**.

Nereid *(Astron.)*. The second substantial natural satellite of **Neptune**, diameter 300 km.

neritic zone *(Geol.)*. That portion of the sea floor lying between low water mark and the edge of the continental shelf, at a depth of about 180 m. Sediments deposited here are of *neritic facies*, showing rapid alternations of the clay and sand grades; ripple marks etc. indicate accumulation in shallow water.

Nernst bridge *(Elec.Eng.)*. An a.c. bridge for capacitance measurements at high frequencies.

Nernst effect *(Electronics)*. A voltage which appears at opposite edges of a strip of metal that is conducting heat in the presence of a magnetic field which is perpendicular to the surface of the metal.

Nernst heat theorem *(Chem.)*. As the absolute temperature of a homogeneous system approaches zero, so does the specific heat and the temperature coefficient of the free energy.

Nernst lamp *(Phys.)*. One depending on the electric heating of a rod of zirconia in air, giving infrared rays for spectroscopy. The rod must be heated separately to start the lamp, as the material is insulating at room temperature.

Nernst's distribution law *(Chem.)*. When a single solute distributes itself between two immiscible solutes, then for each molecular species (i.e. dissociated, single, or associated molecules) at a given temperature, there exists a constant ratio of distribution between the two solvents, i.e. $(C_1/C_2)_i = K_i$ for each molecular species, *i*.

Nernst theory *(Chem.)*. An explanation of the development of electrode potentials, based on the supposition that an equilibrium is established between the tendency of the electrode material to pass into solution and that of the ions to be deposited on the electrode.

nervation, nervature *(Bot.)*. See **venation**.

nerve *(Bot.)*. A strand of conducting tissue and/or strengthening tissue in a leaf or leaf-like organ. A *vein*.

nerve *(Zool.)*. A collection of axons leading to or from the central nervous system; also a nerve bundle or tract. adjs. *nervous, neural*.

nerve block *(Med.)*. Production of insensibility of a part by injecting an anaesthetic into the nerve or nerves supplying it.

nerve cell *(Zool.)*. See **neuron**.

nerve centre *(Zool.)*. An aggregation of nerve cells associated with a particular sense or function.

nerve ending *(Zool.)*. The distal end of a nerve axon, normally a synapse.

nerve fibre *(Zool.)*. An axon.

nerve gas *(Chem.)*. One which, by inhibiting the enzyme cholinesterase, rapidly and fatally acts on the nervous system; derivative of fluorophosphoric acid. Several have been made for military use, and some for insecticidal purposes.

nerve impulse *(Zool.)*. A regenerative electrical potential which travels along an **axon**; see **action potential**.

nerve, nervure *(Arch.)*. Projecting rib on a vault surface.

nerve net *(Zool.)*. The primitive type of nervous system found in Coelenterata, consisting of numerous multipolar neurons which form a net underlying and connecting the various cells of the body wall.

nerve plexus *(Zool.)*. A network of interlacing nerve fibres.

nerve root *(Zool.)*. The origin of a nerve in the central nervous system.

nerve trunk *(Zool.)*. A bundle of nerve fibres united within a connective tissue coat.

nervous system *(Zool.)*. The whole system of nerves, ganglia, and nerve endings of the body of an animal, considered collectively.

nervure *(Zool.)*. One of the chitinous struts which support and strengthen the wings of an Insect.

nesosilicate *(Min.)*. A silicate mineral whose atomic structure contains isolated groups of silicon-oxygen tetrahedra.

Nessler's solution *(Chem.)*. Used in the analysis of water for determination of free and combined ammonia. A solution of mercury (II) iodide in potassium iodide, made alkaline with sodium or potassium hydroxide. With a trace of ammonia it gives a yellow colour, but with larger amounts a brown precipitate.

nest *(Glass)*. A cushion upon which glass is placed to be cut with a diamond.

nest *(Zool.)*. An artefact built to provide temporary shelter as in some primates, or protection for the young and eggs in most birds, or for housing the colony in social Insects.

nested loop *(Comp.)*. One contained within another **loop**.

nest epiphyte *(Bot.)*. An **epiphyte** in which the leaves and/or a tangle of stems and roots form a structure in which leaf litter collects, humifies and is used by the epiphyte to root into as a source of mineral nutrients, e.g. the bird's nest fern, *Asplenium nidus*.

net *(Textiles)*. A firm open-mesh fabric made by weaving, knitting, or knotting.

net assimilation rate *(Bot.)*. *Unit leaf rate*. Abbrev. *NAR*. The net photosynthetic rate (i.e. total photosynthesis minus respiration for the plant) per unit leaf area. $NAR = (1/A).(dW/dt)$ where A is leaf area, W is dry weight and t is time).

NETNORTH *(Comp.)*. The Canadian constituent of the **BITNET** network.

net production *(Ecol.)*. See **production**.

net pyrradiometer (radiation balance meter) *(Meteor.)*. Instrument for measuring the difference of the total radiations falling on both sides of a plane surface from the solid angle 2π respectively.

net register tonnage *(Ships)*. Theoretically, the earning space of the ship, and the figure on which payment of harbour dues etc., is based. It is the **gross tonnage** less **deducted spaces**.

net tonnage *(Ships)*. See **net register tonnage**.

net transport *(Nuc.Eng.)*. The difference between the actual **transport** in an isotope separation plant and that which would be obtained by the same plant with raw material of natural isotopic abundance.

net wing area *(Aero.)*. The **gross wing area** minus that part covered by the fuselage and any nacelles.

network *(Comp.)*. A communications' carrier which links up groups of computers and shared resources. Also *computer network*. Terms like *ring, star* indicate the shape of the network.

network *(Telecomm.)*. (1) Combination of electrical components, such as resistances, inductances, capacitances etc.; a network consisting only of these elements may be called *passive*, whereas one incorporating a source of energy, a valve, transistor, integrated circuit etc is *active*. (2) A number of interconnected communications facilities, e.g. telephones, telex machines, computer terminals, or a chain of transmitters interconnected so that they can provide the same programme material.

network analysis *(Elec.Eng.)*. The process of calculating theoretically the electrical properties of any network of passive and active components. Typical properties of interest include current flowing, voltages at different points and transfer functions.

network calculator *(Elec.Eng.)*. Combination of resistors, inductors, capacitors, and generators used to simulate electrical characteristics of a power generation system, so that the effects of varying different operating conditions can be studied in computers. Also *network analyser*.

network polymer *(Chem.)*. One in which **cross linking** occurs.

network structure *(Eng.)*. The type of structure formed in alloys when one constituent exists in the form of a continuous network round the boundaries of the grains of the other. Even if the grains included in the cells are themselves duplex, they are regarded as individual grains.

network synthesis *(Elec.Eng.)*. Process of formulating a network with specific electrical requirements.

network theory *(Immun.)*. A theory postulated by N.K. Jerne in 1974 that the immune system is controlled by a network of interaction between antigen binding sites (paratopes) which may be on immunoglobulin molecules or lymphocyte receptors. Each paratope is capable of binding an epitope on an external antigen and also an idiotope with a shape resembling the epitope present on another immunoglobulin molecule (the 'internal image'). An immunoglobulin which is an anti-idiotope will in turn be recognized by another molecule which is an anti-anti-idiotope, and so on. In an individual the concentration of immunoglobulin molecules bearing any particular idiotope is likely to be so low that no stimulation of anti-idiotope results. However when an external antigen is administered this stimulates a large increase in the concentration of immunoglobulin molecules with complementary paratopes, sufficient to stimulate anti-idiotope production and the network is thereby disturbed. There are many examples showing that this theory is in principle correct, although there is sufficient degeneracy in the system to limit the extent of the network of interactions.

network transfer function *(Telecomm.)*. Mathematical expression giving the ratio of the output of a network to its input. The natural logarithm of its value at any one frequency is the *transfer constant*, and gives the attenuation and phase shift for the signal propagated through the network.

Neumann function *(Maths.)*. A **Bessel function** of the second kind.

Neumann lamellae *(Eng.)*. Straight, narrow bands parallel to the crystallographic planes in the crystals of metals that have been subjected to deformation by sudden impact. They are actually narrow twin bands, and are most frequently observed in iron.

Neumann principle *(Crystal.)*. Physical properties of a crystal are never of lower symmetry than the symmetry of the external form of the crystal. Consequently tensor properties of a cubic crystal, such as elasticity or conductivity, must have cubic symmetry, and the behaviour of the crystal will be isotropic.

neural *(Zool.)*. See **nerve**.

neural arch *(Zool.)*. The skeletal structure arising dorsally from a vertebral centrum, formed by the neurapophyses and enclosing the spinal cord.

neural canal *(Zool.)*. The space enclosed by the centrum and the neural arch of a vertebra, through which passes the spinal cord.

neural crest *(Zool.)*. In a Vertebrate embryo, a band of cells lying parallel and close to the nerve cord which will later give rise to the ganglia of the dorsal roots of the spinal nerves.

neuralgia *(Med.)*. Paroxysmal intermittent pain along the course of a nerve. adj. *neuralgic*. See also **tic douloureux**.

neural spine *(Zool.)*. The median dorsal vertebral spine, formed by the fusion of the neurapophyses above the neural canal.

neural tube *(Zool.)*. A tube formed dorsally in the embryonic development of Vertebrates by the joining of the 2 upturned neural folds formed by the edges of the ectodermal neural plate, giving rise to the brain and spinal nerve cord.

neuraminidase *(Immun.)*. An enzyme produced by viruses of the myxovirus and paramyxovirus groups and by some bacteria which splits the glycosidic link between neuraminic acid or sialic acid and other sugars. Neuraminic acid is an important structural component of the surface glycoproteins of many cells and contributes largely to the net negative charge of cells. After treatment with neuraminidase they show an increased tendency to agglutinate, and this is used to increase the sensitivity of some agglutination reactions.

neurapophyses *(Zool.)*. A pair of plates arising dorsally from the vertebral centrum, and meeting above the spinal cord to form the neural arch and spine. sing. *neurapophysis*.

neurasthenia *(Med.)*. A term formerly used for a syndrome of tiredness and lassitude, accompanied by physical symptoms such as headache, backache, indigestion; related to mental stress and failure to cope with the problems of everyday living.

neurectomy *(Med.)*. Excision of part of a nerve.

neurilemma *(Zool.)*. See **neurolemma**.

neurine *(Chem.)*. $CH_2=CH.N(CH_3)_3OH$, *trimethylvinylammonium hydroxide*, obtainable from brain substance and from putrid meat; related to *choline*, into which it can be transformed. It is a ptomaine base.

neuritis *(Med.)*. Inflammation of a nerve, e.g. *polyneuritis* (inflammation of many nerves), *optic neuritis* (inflammation of optic nerve).

neur-, neuro- *(Genrl.)*. Prefix from Gk. *neuron*, nerve.

neuroblastoma *(Med.)*. A malignant tumour composed of primitive nerve cells, arising in the adrenal gland or in connection with sympathetic nerve cells.

neuroblasts *(Zool.)*. Cells of ectodermal origin which give rise to neurons.

neurocranium *(Zool.)*. The brain case and sense capsules of a vertebrate skull.

neurocrine *(Biol.)*. Neurosecretory; secretory property of nervous tissue.

neurocyte *(Zool.)*. See **neuron**.

neuroendocrinology *(Med.)*. The study of interactions between the *nervous system* and *endocrine organs*, particularly pituitary gland and hypothalamic region of brain.

neurofibroma *(Med.)*. A tumour composed of fibrous tissue derived from the connective tissue sheath of a nerve. See **molluscum fibrosum**.

neurofibromatosis *(Med.)*. Multiple tumours attached to peripheral nerves associated with pigmented skin patches. Also *von Rechlinghausen's disease*.

neurogenesis *(Zool.)*. The development and formation of nerves.

neurogenic *(Zool.)*. Activity of a muscle or gland which is dependent on continued nervous stimuli. Cf. **myogenic**.

neuroglia *(Zool.)*. The supporting tissue of the brain and spinal cord of Vertebrates, composed of much branched fibrous cells which occur among the nerve cells and fibres. Also *glia*.

neurohaemal organs *(Zool.)*. Organs which serve as a gateway for the escape of the products of neurosecretory cells from the neurons into the circulating blood, e.g. the *corpora cardiaca* of Insects.

neurohypophysis *(Zool.)*. See **pars nervosa**.

neurolemma, neurilemma *(Zool.)*. A thin homogeneous sheath investing the medullary sheath of a medullated nerve fibre; sheath of Schwann.

neurologist *(Med.)*. A specialist in diseases of the nervous system.

neurology *(Med.)*. The study of the nervous system.

neurolymphomatosis *(Vet.)*. See **fowl paralysis**.

neuromasts *(Zool.)*. Sensory hair cells embedded in a gelatinous cupola found in the lateral line system of lower vertebrates and concerned with **mechanoreception**.

neuromuscular *(Zool.)*. Pertaining to nerve and muscle, as a myoneural junction.

neuronitis *(Med.)*. Inflammation (or degeneration) of neurons.

neuron, neurone *(Zool.)*. A nerve cell and its processes. Also *neurocyte*.

neuronophagia *(Med.)*. The destruction of diseased nerve cells by leucocytes and by microglial cells.

neuropathology *(Med.)*. The study of pathology of diseases of the nervous system.

neuropathy *(Med.)*. Functional or structural disorders of the nervous system.

neuropil *(Zool.)*. In Vertebrates, a network of axons, dendrites and synapses within the central nervous system.

neuropile *(Zool.)*. In Arthropods, regions within the brain and the central portion of segmental ganglia consisting of dendrites and synapses.

neuropore *(Zool.)*. The anterior opening by which the cavity of the central nervous system communicates with the exterior.

neurosecretory cell *(Biol.)*. A special type of neurone in which the axon terminates against the wall of a blood vessel or sinus into which it secretes a hormone or other factor.

neuroses *(Behav.,Med.)*. A loose term for mental disorders in which the individual experiences anxiety, or engages in behaviour to avoid experiencing it; there is no evidence of an organic component, and the individual remains in contact with reality. See **obsessional neurosis**.

neurosurgery *(Med.)*. That part of surgical science which deals with the nervous system.

neurosyphilis *(Med.)*. Syphilitic infection of the nervous system.

neurotropic *(Med.)*. Having a special affinity for nerve cells, e.g. *neurotropic* virus.

neurula *(Zool.)*. In the embryonic development of Vertebrates, the stage after the gastrula in which the processes of organ formation begin, with the formation of the neural tube, mesodermal somites, notochord and archenteron.

neuter *(Bot.,Zool.)*. (1) Sex-less. (2) Lacking functional sexual organs; a flower having neither functional stamens nor functional carpels; sterile. Also *neutral*.

neutral *(Glass)*. Of high chemical durability.

neutral *(Image Tech.)*. Possessing no colour or hue; grey.

neutral *(Phys.)*. (1) Exhibiting no resultant charge or voltage. (2) Return conductor of a balanced power supply, nearly at earth potential, if without a local earth connection. See **neutral point, neutral conductor**.

neutral *(Zool.)*. See **neuter**.

neutralator *(Elec.Eng.)*. See **earth reactor**.

neutral autotransformer *(Elec.Eng.)*. See **earth reactor**.

neutral axis *(Elec.Eng.)*. A term used to denote the diametral plane in which the brushes of a commutator machine should be situated to give perfect commutation.

neutral axis *(Eng.)*. In a beam subjected to bending the line of zero stress; a transverse section of the longitudinal plane, or *neutral surface*, which passes through the centre of area of the section.

neutral beam *(Phys.)*. A beam of high-energy *atoms* used to heat a plasma. As the atoms are neutral the beam is not affected by magnetic fields.

neutral compensator *(Elec.Eng.)*. See **earth reactor**.

neutral density filter *(Image Tech.)*. One which attenuates all colours uniformly so that the relative spectral distribution of the energy of the transmitted light has been unaltered.

neutral element *(Maths.)*. See **identity (2)**.

neutral equilibrium *(Phys.)*. The state of equilibrium of a body when a slight displacement does not alter its potential energy.

neutral flame *(Eng.)*. In welding, flame produced by a mixture at the torch of acetylene and oxygen in equal volumes.

neutral flux *(Eng.)*. One used to modify fusibility of furnace slags, but which exerts neither basic nor acidic influence, e.g. calcium fluoride.

neutral injection *(Nuc.Eng.)*. The additional heating of a plasma by injecting beams of accelerated atoms into it.

neutral inversion *(Elec.Eng.)*. A condition in which the phase to neutral voltages of a 3-phase, star connected system are unbalanced. Commonly due to **ferromagnetic resonance**.

neutralization *(Chem.)*. The interaction of an acid and a base with the formation of a salt. In the case of strong acids and bases, the essential reaction is the combination of hydrogen ions with hydroxyl ions to form water molecules.

neutralization *(Electronics)*. Method of counteracting oscillation-inducing feedback from the output of an amplifier to the input via the inter-electrode capacitances of a valve. Reversed feedback is provided by a balancing capacitance to which is applied voltage equal and in antiphase to that on the anode. Also *balancing*.

neutralized series motor *(Elec.Eng.)*. See **compensated series motor**.

neutralizing *(Build.)*. Pre-treatment of a surface to render it chemically neutral before decoration.

neutralizing capacitance *(Electronics)*. An adjustable capacitor, placed in the anode to grid feedback path of a high-frequency valve amplifier, in order to counteract any tendency to instability from feedback via the valve's inter-electrode capacitances. Also *balancing capacitance*.

neutralizing indicator *(Elec.Eng.)*. One indicating the degree of neutralization present in an amplifier.

neutralizing voltage *(Telecomm.)*. That fed back on the process of neutralization.

neutral (middle) conductor *(Elec.Eng.)*. The middle wire of a d.c. **three-wire system** or a distribution system, or the wire of a polyphase distribution system which is connected to the neutral point of the supply transformer or alternator. Sometimes called the *neutral (or middle) wire* or the *neutral*. Cf. **outer**.

neutral point *(Aero.,Space)*. (1) See **gravipause**. (2) That c.g. position in an aircraft at which longitudinal stability is neutral. *Stick-fixed neutral point* is the c.g. position at which control column movement to trim a change in speed is zero. *Stick-free neutral point* is the c.g. position at which the stick force needed to trim a change in speed is zero.

neutral point *(Elec.Eng.)*. The point at which the windings of a polyphase star connected system of windings are connected together. Also, the midpoint of the neutral zone of a d.c. machine. Also called *neutral*.

neutral point *(Meteor.)*. A small region of the daylight sky from which scattered sunlight is unpolarized; such points were discovered by Arago, Babinet and Brewster.

neutral point *(Phys.)*. A point in the field of a magnet where the earth's magnetic field (usually the horizontal component) is exactly neutralized.

neutral pump *(Bot.)*. A pump which transports only uncharged molecules or appropriately balanced pairs of ions so that there is no net transfer of charge.

neutral relay *(Telecomm.)*. In US *nonpolarized relay*.

neutral solution *(Chem.)*. An aqueous solution which is neither acidic nor alkaline. It therefore contains equal quantities of hydrogen and hydroxyl ions and has a pH value of 7.

neutral state *(Phys.)*. Said of ferromagnetic material when completely demagnetized. Also *virgin state*.

neutral surface *(Eng.)*. See **neutral axis**.

neutral wedge filter *(Image Tech.)*. A neutral grey filter, such as a wedge of grey glass cemented to a similar wedge of clear glass, which introduces a continuously variable attenuation of light, depending on the density or thickness of the grey glass introduced into the beam, without altering the relation between hues in the transmitted light.

neutral wire *(Elec.Eng.)*. Same as *neutral conductor*.

neutral zone *(Elec.Eng.)*. That part of the commutator of a d.c. machine in which, when the machine is running normally, the voltage between adjacent commutator bars is approximately zero.

neutrino *(Phys.)*. A fundamental particle, a **lepton**, with zero charge and zero mass. A different type of neutrino is associated with each of the four charged leptons. Its existence was predicted by Pauli in 1931 to avoid β-decay infringing the laws of conservation of energy and angular momentum. As they have very weak interactions with matter, neutrinos were not observed experimentally until 1956. See **antineutrino**.

neutrino astronomy *(Astron.)*. Term applied to attempts to detect **neutrinos** from the Sun, with the aim of discovering more exactly the conditions in the solar core.

neutron *(Phys.)*. Uncharged subatomic particle, mass approximately equal to that of the proton, which enters into the structure of atomic nuclei. Interacts with matter primarily by collisions. Spin quantum number of neutron = $+\frac{1}{2}$, rest mass = 1.008 665 a.m.u., charge is zero and magnetic moment −1.9125 nuclear Bohr magnetrons. Although stable in nuclei, isolated neutrons decay by β-emission into protons, with a half-life of 11.6 minutes. See **neutron energy**.

neutron absorption cross-section *(Phys.)*. The cross-section for a nuclear reaction initiated by neutrons. This is expressed in *barns*. For many materials this rises to a large value at particular neutron energies due to resonance effects, e.g. a thin sheet of cadmium forms an almost impenetrable barrier to thermal neutrons.

neutron age *(Nuc.Eng.)*. See **Fermi age**.

neutron balance *(Nuc.Eng.)*. For a constant power level in a reactor, there must be a balance between the rate of production of both prompt and delayed neutrons, and their rate of loss due to both absorption and leakage from the reactor.

neutron current *(Nuc.Eng.)*. The net rate of flow of neutrons through a surface perpendicular to the direction in which they are migrating. *Neutron current* is a *vector* whereas *neutron flux* is a *scalar*.

neutron detection *(Nuc.Eng.)*. Observation of charged particle recoils following collisions of neutrons with protons in a counter containing hydrogen gas or a compound of hydrogen; or of charged particles produced by interaction of neutrons with atomic nuclei, e.g. in a boron-trifluoride *counter*. Absorption of neutrons by boron gives rise to α-particles which can be detected.

neutron diffraction *(Phys.)*. The coherent elastic scattering of neutrons by the atoms in a crystal. If the scattering is by the nucleus then the atomic structure of the crystal can be deduced from the measurements of the diffraction pattern. If the scattering is by atoms with electron configurations that have a magnetic moment, then details of the magnetic structure of the crystal can be determined.

neutron diffusion *(Nuc.Eng.)*. The migration of neutrons from regions of high neutron density to those of low density, in a medium in which neutron capture is small compared to neutron scattering. See **age theory, group theory** and **transport theory**.

neutron elastic scattering *(Phys.)*. A beam of thermal neutrons, whose electrons are such that their associated de Broglie wavelength is of the same order of magnitude as interatomic distances, will be diffracted by a crystal. Most of the intensity of the beam will be diffracted with no loss of energy, i.e. no wavelength change, and this is said to be *neutron elastic scattering*. See **neutron inelastic scattering.**

neutron energy *(Phys.)*. (1) The binding energy of a neutron in a nucleus, usually several MeV. (2) The energy of a free neutron which in a reactor will be classed in several groups: high-energy neutrons, energy >10 MeV; fast neutrons, energy 10 MeV to 20 keV; intermediate neutrons, energy 20 keV to 100 eV; epithermal neutrons, energy 100 eV to 0.025 eV; thermal or slow neutrons, energy approx. 0.025 eV. See **multigroup theory**.

neutron excess *(Phys.)*. The difference between the neutron number and the proton number for a nuclide. Also called *isotopic number*.

neutron flux *(Phys.)*. The number of neutrons passing through unit area in unit time, *or* the product of number of neutrons per unit volume and their mean speed. In a nuclear reactor the flux is of the order of 10^{16} to $10^{18}\,m^{-2}\,s^{-1}$.

neutron gun *(Nuc.Eng.)*. Block of moderating material with a channel through it used for producing a beam of fast neutrons.

neutron hardening *(Phys.)*. Increasing the average energy of a beam of neutrons by passing them through a medium which shows preferential absorption of slow neutrons.

neutron inelastic scattering *(Phys.)*. A beam of thermal neutrons, whose energies are of the same order of magnitude as a quantum of lattice vibrational energy, a *phonon*, will be diffracted by a crystal with exchanges of energy with the excited travelling waves. A detailed examination of the change in direction and energy of the scattered neutrons gives valuable information about the *lattice dynamics* of the crystal.

neutron leakage *(Nuc.Eng.)*. Escape of neutrons in a reactor from the core containing the fissile material; reduced by using a reflector.

neutron-magnetic scattering *(Phys.)*. The magnetic moment in a crystal has contributions from both electron spin and electron orbital angular momenta. Both of these will interact with the neutron spin magnetic moment and will give rise to the magnetic scattering of neutrons. This is a powerful method for the study of magnetic structures.

neutron-nuclear scattering length *(Phys.)*. A measure of the ability of different *nuclei* to scatter neutrons; it is independent of the scattering angle. The scattering length is different for different isotopes of the same atom, so that, in addition to coherent scattering from a crystal containing more than one isotope, there will be incoherent scattering. Neutron-nuclear scattering is spin dependent, and this leads to further incoherent scattering from a nucleus with non-zero spin.

neutron number *(Phys.)*. The number of neutrons in a nucleus. Equal to the difference between the rel. atomic mass (the total number of nucleons) and the atomic number (the number of protons).

neutron poison *(Phys.)*. Any material other than fissionable, which absorbs neutrons; used for the control of nuclear reactors.

neutron radiography *(Radiol.)*. Radiography by beam of neutrons from nuclear reactor which then produces an image on a photoelectric image intensifier following neutron absorption. It has advantages over X-ray radiography where the mass absorption coefficients for neutrons are very different for different parts of the specimen, although the atomic numbers (and hence X-ray absorption) are very similar, e.g. specimens containing hydrogenous material.

neutron shield *(Nuc.Eng.)*. Radiation shield erected to protect personnel from neutron irradiation. In contradistinction to gamma-ray shields, it must be constructed of very light hydrogenous materials which will quickly moderate the neutrons which can be absorbed, for instance, by boron incorporated in the shield.

neutron source *(Nuc.Eng.)*. One giving a high neutron flux, e.g. for neutron activation analysis. Apart from reactors, these are chemical sources such as a radium-beryllium mixture emitting neutrons as a result of the

(αn) reaction, and accelerator sources in which deuterium nuclei are usually accelerated to strike a tritium-impregnated titanium target, thus releasing neutrons by the (DT) reaction. The former are continuously active but the latter have the advantage of becoming inert as soon as the accelerating voltage is switched off.

neutron spectrometer *(Nuc.Eng.)*. Instrument for investigation of energy spectrum of neutrons. See **crystal spectrometer**, **time-of-flight spectrometer**. Other techniques depend on the nuclear reaction of neutrons with ^{3}He *(helium-3 spectrometer)*, ^{6}Li or hydrogen *(proton recoil spectrometer)*.

neutron spectroscopy *(Phys.)*. Experimental determination of the intensity and change in energy (wavelength) of the neutrons scattered in a particular direction when a beam of mono-energetic neutrons are incident on a crystal. A powerful method of studying lattice vibrations and phonon energies.

neutron star *(Astron.)*. A small body of very high density (ca. 10^{12} kg/dm^{3}) resulting from a supernova explosion in which a massive star collapses under its own gravitational forces, the electrons and protons combining to form neutrons.

neutron therapy *(Med.)*. Use of neutrons for medical treatment.

neutron velocity selector *(Nuc.Eng.)*. (1) Device using a rotating **chopper** or a pulsed accelerator to provide a pulse of neutrons, the velocity being selected by time-of-flight (see **time-of-flight spectrometer**). (2) Device using neutron diffraction (see **crystal spectrometer**).

neutropenia *(Med.)*. Abnormal diminution in the number of neutrophil leucocytes in the blood.

neutrophil *(Med.)*. (1) Stainable by neutral dyes. (2) A white blood cell, a polymorphonuclear leucocyte, whose granular protoplasm is stainable with neutral dyes.

Nevadan orogeny *(Geol.)*. An **orogeny** of Jurassic age affecting western US.

névé *(Geol.)*. A more or less compacted snow-ice occurring above the snowline; it consists of small rounded crystalline grains formed from snow crystals. Also called *firn*.

Newall system *(Eng.)*. A commonly used system of limits and fits, using a hole basis.

Newcastle disease *(Vet.)*. *Pseudo-fowl plague, pneumoencephalitis, Ranikhet disease.* An acute, highly-contagious paramyxovirus disease of chickens, other domestic fowl and wild birds. Symptoms include loss of appetite, diarrhoea, and respiratory and nervous symptoms. Vaccines widely used.

newel *(Build.)*. An upright post fixed at the foot of a stair or at a point of change of direction and used as a support for a balustrade.

newel cap, drop *(Build.)*. Ornamental finish planted on the upper (lower) end of a newel post.

newel joints *(Build.)*. The joints connecting the newel and the handrail or string.

New Forest disease *(Vet.)*. See **infectious ophthalmia**.

Newlyn datum *(Surv.)*. See **mean sea level**.

new moon *(Astron.)*. The instant when sun and moon have the same celestial longitude; the illuminated hemisphere of the moon is then invisible.

New Red Sandstone *(Geol.)*. A name frequently applied to the combined Permian and Triassic Systems, and particularly applicable in N. Europe, where the palaeontological evidence is insufficient to allow of their separation. The term reflects the general resemblance between the rocks comprising these two systems and the Old Red Sandstone of Devonian age.

newsprint *(Paper)*. Cheap printing paper intended for printing by letterpress or web offset for the production of newspapers, generally from 40–52 g/m^{2}. The furnish is generally predominantly a mechanical woodpulp but de-inked waste paper may also feature prominently.

new star *(Astron.)*. See **nova**.

New Style *(Astron.)*. A name given to the system of date-reckoning established by the **Gregorian calendar**. Abbrev. *NS*.

newton *(Elec.Eng.)*. Symbol *N*. The unit of force in the SI system, being the force required to impart, to a mass of 1 kg, an acceleration of 1 m/sec^{2}. 1 newton = 0.2248 pounds force.

Newtonian telescope *(Astron.)*. A form of reflecting telescope due to Newton, in which the object is viewed through an eyepiece in the side of the tube, the light reflected from the main mirror being deflected into it by a small plane mirror inclined at 45° to the axis of the telescope and situated just inside the principal focus.

Newton's disk *(Phys.)*. Motor driven disk with sectors of primary colours, which appears white on fast rotation and with white illumination, demonstrating the synthesis of colour vision.

Newton's law of cooling *(Phys.)*. The rate of cooling of a hot body which is losing heat both by radiation and by natural convection is proportional to the difference in temperature between it and its surroundings. The law does not hold for large temperature excesses.

Newton's laws of motion *(Phys.)*. They are (1) Every body continues in a state of rest or uniform motion in a straight line unless acted upon by an external impressed force. (2) The rate of change of momentum is proportional to the impressed force and takes place in the direction of the force. (3) Action and reaction are equal and opposite, i.e. when two bodies interact the force exerted by the first body on the second body is equal and opposite to the force exerted by the second body on the first. These laws were first stated by Newton in his *Principia*, 1687. Classical mechanics consists of the applicaton of these laws.

Newton's rings *(Phys.)*. Circular concentric interference fringes seen surrounding the point of contact of a convex lens and a plane surface. Interference occurs in the air film between the two surfaces. If r_n is the radius of the nth ring, R is the radius of curvature of the lens surface and λ the wavelength. $r_n = \sqrt{nR\lambda}$. See **contour fringes**.

New Zealand greenstone *(Min.)*. Nephritic 'jade' of gemstone quality, from New Zealand.

next instruction register *(Comp.)*. See **program counter**.

NGC *(Astron.)*. Abbreviation for the New General Catalogue of all nebulous objects known in 1888. Together with the supplementary Index Catalogue (IC) it lists 13 000 galaxies, clusters and nebulae.

NGTE rigid rotor *(Aero.)*. A helicopter self propelling rotor system evolved at the National Gas Turbine Establishment which uses the principle of the **jet flap** to obtain very high lift coefficients.

Ni *(Chem.)*. The symbol for nickel.

niacin *(Med.)*. *Vitamin B_3; nicotinic acid.* Deficiency results in **pellagra**.

nib *(Build.)*. The point of a crowbar.

nibbling *(Eng.)*. A sheet metal cutting process akin to sawing, in which a rapidly reciprocating cutter slots the sheet along any desired path.

nibs *(Build.)*. Specks of solid matter in surface paint coatings, marring the finished appearance.

niccolite *(Min.)*. Nickel arsenide, crystallizing in the hexagonal system. It is one of the chief ores of metallic nickel. Also called *copper nickel, kupfernickel*.

niche *(Ecol.)*. See **ecological niche**.

niched column *(Arch.)*. A column set back in a wall with a clear space between it and the wall.

Nicholson hydrometer *(Phys.)*. A hydrometer of the constant displacement type, used for determining the density of a solid.

Nichrome *(Eng.)*. TN for nickel-chromium alloy largely used for resistance heating elements because of its high resistivity ($\sim 110 \times 10^{-8}$ ohm metres) and its ability to withstand high temperatures.

nicitating *(Zool.)*. Winking; said of the third eyelid of the Vertebrates, which by its movements keeps clean the surface of the eye.

nick *(Biol.)*. A cut between adjacent nucleotides in one strand of a duplex DNA molecule.

nick *(Bot.)*. A particular combination of male and female parents giving desirable offspring, especially in the breeding of **F1 hybrids**.

nick *(Print.)*. The groove(s) in the shank of a type letter.

nickel *(Chem.)*. Silver-white metallic element. Symbol Ni, at.no. 28, r.a.m. 58.71, mp 1450°C, bp 3000°C, electrical resistivity at 20°C, 10.9×10^{-8} ohm metres, rel.d. 8.9. Used for structural parts of valves. It is magnetostrictive, showing a decrease in length in an applied magnetic field and, in the form of wire, was much used in computers for small stores, the data circulating and extracted when required. Used pure for electroplating, coinage, and in chemical and food-handling plants. See also **nickel alloys**.

nickel alloys *(Eng.)*. Nickel is the main constituent in Monel metal, permalloy and nickel-chromium alloys. It is also used in cupro-nickel, nickel-silver, various types of steel and cast-iron, brass, bronze and light alloys.

nickel antimony glance *(Min.)*. See **ullmanite**.

nickel arsenic glance *(Min.)*. See **gersdorffite**.

nickel bloom *(Min.)*. See **annabergite**.

nickel-cadmium accumulator *(Elec.Eng.)*. Battery using nickel and cadmium compounds in potassium hydroxide electrolyte. Characterized by extremely low self-discharge rate; therefore used in emergency lighting systems, etc.

nickel carbonyl *(Chem.)*. $Ni(CO)_4$. A volatile compound of nickel, bp 43°C, formed by passing carbon monoxide over the heated metal. The compound is decomposed into nickel and carbon monoxide by further heating. Formerly used on a large scale in industry for the production of nickel from its ores by the Mond process.

nickel-chromium steel *(Eng.)*. Steel containing nickel and chromium as alloying elements. 1.5% to 4% nickel and 0.5% to 2% chromium are added to produce an alloy of high-tensile strength, hardness and toughness. Used for highly stressed automobile and aero-engine parts, for armour plate etc.

nickel electro *(Print.)*. An electro in which a thin layer of nickel is first deposited, the shell being completed by depositing copper. Superior to **nickel-faced electro**.

nickel-faced electro *(Print.)*. An electro which has been given a facing of nickel to improve its quality. Cf. **nickel electro**.

nickel-iron-alkaline accumulator, cell *(Elec.Eng.)*. An accumulator in which the positive plate consists of nickel hydroxide enclosed in perforated steel tubes, and the negative plate consists of iron or cadmium also enclosed in perforated steel tubes. The electrolyte is potassium hydrate, the e.m.f. 1.2 volts per cell. It is lighter than the equivalent lead accumulator. Also *Edison accumulator*, *Ni-Fe accumulator*.

nickel silver *(Eng.)*. See **German silver**.

nickel steel *(Eng.)*. Steel containing nickel as an alloying element. Varying amounts, between 0.5% and 6.0%, are added to increase the strength in the normalized condition, to enable hardening to be performed in oil or air instead of water, or to increase the core strength of carburized parts.

nicker *(Build.)*. The side wing of a **centre-bit**, scribing the boundary of the hole to be cut.

nick translation *(Biol.)*. An inexact phrase for a method of radioactively labelling a DNA molecule. The DNA is first nicked in one strand by a short DNase treatment. This allows DNA polymerase I to both remove nucleotides from the exposed end and replace them by highly radioactive nucleotides.

Niclad *(Eng.)*. Composite sheets made by rolling together sheets of nickel and mild steel, to obtain the corrosion resistance of nickel with the strength of steel.

Niclausse boiler *(Eng.)*. A French marine boiler which consists of a horizontal water and steam drum from which vertical double headers are suspended, carrying banks of **field tubes** slightly inclined downwards.

Nicol prism *(Phys.)*. A device for obtaining plane polarized light, it consists of a crystal of Iceland spar which has been cut and cemented together in such a way that the ordinary ray is totally reflected out at the side of the crystal, while the extraordinary plane polarized ray is freely transmitted. Largely superseded by Polaroid.

nicotinamide adenine dinucleotide *(Biol.)*. A coenzyme which serves as an electron acceptor for many dehydrogenases. The reduced form (NADH) subsequently donates its electrons to the *electron transport chain*. Abbrev. *NAD*.

nicotinamide adenine dinucleotide phosphate *(Biol.)*. Phosphorylated derivative of NAD. It also serves as an electron carrier but the electrons are primarily used for *reductive biosynthesis*. Abbrev. *NADP*.

nicotine *(Chem.)*. An alkaloid of the pyridine series, $C_{10}H_{14}N_2$. It occurs in tobacco leaves, is extremely poisonous and, in small quantities, highly addictive. It is a colourless oil, of nauseous odour; bp 246°C at 97 kN/m^2.

nicotinic acid *(Chem.)*. See **vitamin B**.

nidamental *(Zool.)*. Said of glands which secrete material for the formation of an egg-covering.

nidation *(Zool.)*. In the oestrous cycle of Mammals, the process of renewal of the lining of the uterus between the menstrual periods.

nidged ashlar *(Build.)*. See **nigged ashlar**.

nidicolous *(Zool.)*. Said of Birds which remain in the parental nest for some time after hatching. Cf. *nidifugous*.

nidification, nidulation *(Zool.)*. The process of building or making a nest.

nidifugous *(Zool.)*. Said of Birds which leave the parental nest soon after hatching. Cf. **nidicolous**.

nidus *(Zool.)*. A nest; a small hollow resembling a nest; a nucleus.

niello *(Eng.)*. A method of decorating metal. Sunk designs are filled with an alloy of silver, lead and copper, with sulphur and borax as fluxes, and fired.

Niemann-Pick disease *(Med.)*. An autosomal recessive hereditary disease in which large amounts of lipid are deposited in the reticulo-endothelial tissue, causing *lymphadenopathy* and *hepatosplenomegaly* leading to death by the age of 2.

Ni-Fe accumulator *(Elec.Eng.)*. TN of a **nickel-iron-alkaline accumulator**.

nifedipine *(Med.)*. A **calcium channel blocking** agent which is a potent dilator of arterial blood vessels including the coronary arteries. Used to treat hypertension and angina pectoris.

niger morocco *(Print.)*. A goatskin tanned and dyed in Nigeria; limited range of colours, flexible, used for good quality binding.

nigged ashlar *(Build.)*. A block of stone dressed with a pointed hammer.

night blindness *(Med.)*. Abnormal difficulty in seeing objects in the dark; due often to deficiency of vitamin A (retinol) in the diet. Also *nyctalopia*.

night bolt *(Build.)*. See **night latch**.

night latch *(Build.)*. A lock whose bolt is key operated from the outside and knob operated from the inside, but which is fitted with a device for preventing operation from either side.

night terror *(Behav.)*. A particularly harrowing variety of bad dream experienced by a child.

nigrescent *(Bot.)*. Becoming blackish.

nigrite *(Min.)*. A pitchlike member of the asphaltite group.

nigrosines *(Chem.)*. Diphenylamine dyestuffs, used as black pigments, prepared by heating nitrobenzene or nitrophenol, aniline and phenylammonium chloride with iron filings.

Ni-hard *(Eng.)*. Cast-iron to which a ladle addition of about 4½% nickel has been made, to render the alloy martensitic and abrasion resistant.

NII *(Nuc.Eng.)*. *Nuclear Installations Inspectorate*. A branch of the Health and Safety Executive in the UK responsible for the safety assessment and inspection of nuclear facilities.

nile *(Nuc.Eng.)*. 1 nile corresponds to a **reactivity** of 0.01. In indicating reactivity changes, it is more usual to use the smaller unit, the millinile, equal to a change in reactivity of 10^{-5}.

nimbostratus *(Meteor.)*. Grey *cloud layer*, often dark, the appearance of which is rendered diffuse by more or less continuously falling rain or snow, which in most cases reaches the ground. It is thick enough throughout to blot out the Sun. Low, ragged clouds frequently occur below the layer, with which they may or may not merge. Abbrev. *Ns*.

nimonic *(Eng.)*. Alloy used in such high temperature work as that called for in gas turbine blades. About 80% nickel and 20% chromium, with some titanium and aluminium.

nine-point circle (of a triangle) *(Maths.)*. The circle which passes through the midpoints of the three sides, the points of intersection of each side with the perpendicular to that side from the opposite vertex, and, if *C* is the point where the three perpendiculars intersect, the three points midway between *C* and the respective vertices.

ninhydrin *(Chem.)*. *Tri keto-hydrindene-hydrate*. Used as a reagent to detect proteins or amino acids, with which it forms a characteristic blue colour on heating.

niobite *(Min.)*. See **columbite**.

niobium *(Chem.)*. Rare metallic element, symbol Nb, at.no. 41, r.a.m. 92.9064, mp 2500°C, used in high temperature engineering products (e.g. gas turbines and nuclear reactors) owing to the strength of its alloys at temperatures above 1200°C. Combined with tin (Nb_3Sn) it has a high degree of superconductivity.

nip *(Chem.Eng.)*. See **angle of nip**.

nip *(Eng.)*. The manufacturing practice of giving different curvatures to the individual leaves of a laminated spring before assembling, so as to attain the most favourable distribution of working stesses.

nip *(Glass)*. The gap between rollers in a sheet glass rolling machine.

nip *(Paper)*. The flat area at the point of contact between two horizontal parallel rolls.

nip *(Textiles)*. The line of near contact between two rotating rollers through which the textile material passes while being compressed.

nip and tuck folder *(Print.)*. A type of folder on a rotary press in which a fold at right angles to the web is formed by a blade thrusting the web between folding jaws.

Nipkow disk *(Image Tech.)*. Disk with one or more sets of holes arranged in a spiral, the rotation of which provided a form of **mechanical scanning** in some early TV systems, such as Baird's *Televisor*.

nippers *(Build.)*. See **stone tongs**.

nipping *(Print.)*. See **smashing**.

nipple *(Build.,Eng.)*. (1) A short length of externally threaded pipe for connecting two lengths of internally threaded pipe. (2) A small drilled bush, sometimes containing a nonreturn valve, screwed into a bearing for the supply of lubricant by a **grease-gun**. (3) A nut for securing (bicycle) wheel spokes.

nipple *(Eng.)*. (1) A short length of externally threaded pipe for connecting two lengths of internally threaded pipe. (2) A small drilled bush, sometimes containing a nonreturn valve, screwed into a bearing for the supply of lubricant by a **grease-gun**. (3) A nut for securing (bicycle) wheel spokes.

nipple *(Zool.)*. The mamma or protuberant part of the mammary gland in female Mammals, bearing the openings of the milk forming glands.

nip rolls *(Paper)*. A pair of horizontal, parallel steel rolls providing a nip through which a web of paper may be passed and so drawn through the machine to which they are fitted e.g. a sheet cutter.

Ni-resist *(Eng.)*. A cast-iron consisting of graphite in a matrix of austenite. Contains carbon 3%, nickel 14%, copper 6%, chromium 2%, and silicon 1.5%. Has a high resistance to growth, oxidation and corrosion.

Nissl bodies *(Zool.)*. Aggregations of **ribosomes** found within nerve cells.

nit *(Phys.)*. Unit of luminance, 1 candela/m^2. Symbol nt. See **candela**.

niter *(Chem.)*. See **nitre**.

nitometer *(Phys.)*. An instrument for measuring the brightness of small light sources.

Nitralloy *(Eng.)*. Steel specially developed for nitriding (which is less effective with ordinary steels). Contains carbon 0.2–0.3%, aluminium 0.9–1.5%, chromium 0.9–1.5%, and molybdenum 0.15–0.25%.

nitramines *(Chem.)*. Amines in which an aminohydrogen has been replaced by the nitro group. They have the general formula $R.NH.NO_2$.

nitrate film *(Image Tech.)*. Film with a base of highly inflammable cellulose nitrate, celluloid; used for 35 mm motion pictures up to 1951 but now completely replaced by **safety film**.

nitrate-reducing bacteria *(Biol.)*. Facultative aerobes able to reduce nitrates to nitrites, nitrous oxide, or nitrogen under anaerobic conditions, e.g. *Micrococcus denitrificans*. This process is termed *denitrification*. A few bacteria use such reduction processes as hydrogen acceptor reactions and hence as a source of energy; in this case the end product is ammonia.

nitrates *(Chem.)*. Salts or esters of nitric (V) acid. Metal nitrates are soluble in water; decompose when heated. The nitrates of polyhydric alcohols and the alkyl radicals explode with violence. Uses: explosives, fertilizers, chemical intermediates.

nitration *(Chem.)*. (1) The introduction of nitro (NO_2) groups into organic compounds. Usually carried out by mixing concentrated nitric and sulphuric acids. (2) The final stage of nitrification in the soil.

nitrazepam *(Med.)*. Widely used non-barbiturate hypnotic.

nitre *(Chem.)*. *Potassium nitrate* (V), also *saltpetre*. See also **Chile nitre, soda nitre**.

nitric anhydride *(Chem.)*. Nitrogen (V) oxide. N_2O_5. Dissolves in water to give nitric acid.

nitric [nitrogen (II)] oxide *(Chem.)*. NO. Colourless gas. In contact with air it forms reddish brown fumes of nitrogen dioxide.

nitric (V) acid *(Chem.)*. NHO_3, a fuming unstable liquid, bp 83°C, mp −41.59°C, rel.d. 1.5, miscible with water. Old name *aqua fortis*. Prepared in small quantites by the action of conc. sulphuric acid on sodium nitrates and on large scale by the oxidation of nitrogen or ammonia. An important intermediate of fertilizers, explosives, organic synthesis, metal extraction and sulphuric acid manufacture. See **nitrates, chamber process**.

nitrides *(Chem.)*. Compounds of metals with nitrogen. Usually prepared by passing nitrogen or gaseous ammonia over the heated metal. Those of Group I and II metals are ionic compounds which react with water to release ammonia. Those of Group III to V exist as interstitial compounds, having great hardness and refractoriness, e.g. boron, titanium, iron.

nitriding *(Eng.)*. *Nitrogen case hardening*. A process for producing hard surface on special types of steel by heating in gaseous ammonia. Components are finish-machined, hardened and tempered, and heated for about 60 hr at 520°C. Case is about 0.5 mm deep and surface hardness is 1100 V.P.N. See **Nitralloy**.

nitrification *(Bot.)*. The oxidation of ammonia to nitrite and nitrate by chemoautotrophic bacteria whose energy requirements come from these exergonic reactions.

nitrification *(Chem.)*. The treatment of a material with nitric acid.

nitriles *(Chem.)*. Alkyl cyanides of the general formula $R.C{\equiv}N$. When hydrolysed they yield carboxylic acids or the corresponding ammonium salts; reduced, they yield amines.

nitrites [nitrates (III)] *(Chem.)*. Salts or esters of nitrous acid, O=NOH.

nitroanilines *(Chem.)*. The 4- and 3-isomers are used as important intermediates in the preparation of azo dyestuffs.

Nitrobacteriaceae *(Biol.,Bot.)*. Family of bacteria belonging to the order *Pseudomonadales*. Important in nitrification processes in the soil and fresh water. Autotrophic bacteria which derive energy from oxidation processes: *Nitrosomonas* from the oxidation of ammonia to nitrites: *Nitrobacter* from the oxidation of nitrites to nitrates. Also *nitrifying bacteria*.

nitrobenzene *(Chem.)*. $C_6H_5.NO_2$, a yellow liquid with an odour of bitter almonds, mp 5°C, bp 211°C, rel.d. 1.2. It is obtained by the action of a mixture of concentrated sulphuric acid and nitric acid on benzene. When reduced it yields aniline.

nitrocelluloses *(Chem.)*. *Cellulose nitrates*. They are the nitric acid esters of cellulose formed by action of a mixture of nitric and sulphuric acids on cellulose. The cellulose can be nitrated to a varying extent ranging from 2 to 6 nitrate groups in the molecule. Nitrocelluloses with a low nitrogen content, up to the tetranitrate, are not explosive, and are used in lacquer and artificial silk and imitation leather book cloth manufacture. They dissolve in ether-alcohol mixtures, and in so-called lacquer solvents. A nitrocellulose with a high nitrogen content is gun-cotton, an explosive. See **pyroxylins**.

nitro derivatives *(Chem.)*. Aliphatic or aromatic compounds containing the group —NO_2. The aliphatic nitro derivatives are colourless liquids which are not readily hydrolysed, but have acidic properties, e.g. the primary and secondary aliphatic nitro derivatives can form metallic compounds. Aromatic nitro derivatives are easily formed by the action of nitric acid on aromatic compounds. The nitro groups substitute in the nucleus and only exceptionally in the side chain.

nitrogen *(Chem.)*. Gaseous element, colourless and odourless. Symbol N, at.no. 7, r.a.m. 14.0067, mp −209.86°C, bp −195°C. It was the determination by Rayleigh and Ramsay of its rel. at. mass that led to the discovery of the inert gases in the atmosphere, argon etc. Approx. 80% of the normal atmosphere is nitrogen, which is also widely spread in minerals, the sea, and in all living matter. Used as a neutral filler in filament lamps, in sealed relays, and in Van de Graaff generators; and in high voltage cables as an insulant. Liquid is a coolant. See **nitric acid, ammonia, nitrates**.

nitrogenase *(Bot.)*. The enzyme system that catalyses the reduction of gaseous nitrogen (dinitrogen, N_2) to ammonia in biological **nitrogen fixation**. It is inactivated by oxygen.

nitrogen balance *(Biol.)*. The state of equilibrium of the body in terms of intake and output of nitrogen; positive nitrogen balance indicates intake exceeds output, negative nitrogen balance denotes output exceeds intake.

nitrogen bases *(Chem.)*. See **amines**.

nitrogen case-hardening *(Eng.)*. See **nitriding**.

nitrogen chlorides *(Chem.)*. Three nitrogen chlorides, NH_2Cl, $NHCl_2$, and NCl_3 produced by the chlorination of ammonium ions. Unstable and explosive.

nitrogen cycle *(Biol.)*. The sum total of the transformations undergone by nitrogen and nitrogenous compounds in nature in relation to living organisms.

nitrogen [dioxide] (IV) oxide *(Chem.)*. NO_2. Oxidizing agent; formed when 1 volume of oxygen is mixed with 2 volumes of nitric oxide. Below 17°C the formula is N_2O_4 and the molecule is colourless. Decomposed by water, forming a mixture of nitric acid and nitric oxide.

nitrogen fixation, dinitrogen fixation *(Bot.)*. The conversion of free, gaseous nitrogen (dinitrogen, N_2) into compounds either by man industrially or inadvertently or by prokaryotes. The latter include bacteria, actinomycetes and blue-green algae, both free-living and symbiotic with plants of various sorts. See also **Azotobacter, root nodule, nitrogenase**.

nitrogen narcosis *(Med.)*. 'Rapture of the deep'. Intoxicating and anaesthetic effect of too much nitrogen in the brain, experienced by divers at considerable depths.

nitrogen pentoxide *(Chem.)*. See **nitric anhydride**.

nitrogen [tetroxide] (IV) oxide *(Chem.)*. N_2O_4. See **nitrogen dioxide**.

nitroglycerine *(Chem.)*. $C_3H_5(ONO_2)_3$, a colourless oil; mp 11°–12°C; insoluble in water; prepared by treating glycerine with a cold mixture of concentrated nitric and sulphuric acids. It solidifies on cooling, and exists in 2 physical crystalline modifications. In thin layers it burns without explosion, but explodes with tremendous force when heated quickly or struck. See also **dynamite, explosive**. Used medically in solution and tablets for angina.

Nitrolime *(Chem.)*. TN for an artificial fertilizer consisting of calcium cyanamide.

nitromethane *(Chem.)*. $CH_3.NO_2$, a liquid; bp 99°–101°C; prepared from chloroacetic acid and sodium nitrate.

nitrophilous *(Bot.)*. Plants growing characteristically in places where there is a good supply of fixed nitrogen.

nitroprussides *(Chem.)*. Formed by the action of nitric acid on either hexacyano ferrates (II) or (III). The nitroprusside ion is $[FeNO(CN)_5]^-$. Also called *nitrosoferricyanides*.

nitroso compounds *(Chem.)*. Compounds containing the monovalent radical —NO.

nitroso-dyes *(Chem.)*. Dyestuffs resulting from reaction between phenols and nitrous acid.

nitrosoferricyanides *(Chem.)*. See **nitroprussides**.

nitrotoluenes *(Chem.)*. $CH_3.C_6H_4.NO_2$. On nitration of toluene a mixture of 2- and 4-nitrotoluene is obtained with very little 3-nitrotoluene. 2-Nitrotoluene, a liquid, has a bp 218°C; 4-nitrotoluene crystallizes in large crystals; mp 54°C, bp 230°C.

nitrous [nitric (III)] acid *(Chem.)*. The pale-blue unstable solution obtained by precipitating barium nitrite with dilute sulphuric acid is supposed to contain nitrous acid, HNO_2.

nitrous oxide *(Chem.)*. Laughing gas, N_2O. A colourless gas with a sweetish odour and taste, soluble in water, alcohol, ether and benzene. Nitrous oxide supports combustion better than air. The gas is manufactured by the decomposition of ammonium nitrate by heat. It is used for producing anaesthesia of short duration.

nitroxyl *(Chem.)*. The radical —NO_2 when attached to a halogen atom or a metal. Compounds containing the group are *nitroxyls*.

nitrozation *(Biol.,Bot.)*. The conversion of ammonia into nitrites by the action of soil bacteria (*Nitrosomonas*), being the second stage in the nitrification in the soil.

Nivarox *(Horol.)*. Alloy of iron and nickel with a small addition of beryllium, nonmagnetic, rustless and of controllable elasticity, used for hairsprings.

NK cell *(Immun.)*. Abbrev. for *Natural Killer cell*.

nm *(Genrl.)*. Abbrev. for *nanometre* = 10 Ångstroms, $= 10^{-9}$ m.

NMR *(Chem.,Phys.)*. Abbrev. for *Nuclear Magnetic Resonance*.

n-n junction *(Electronics)*. One between crystals of *n*-type semiconductors, having different electrical properties.

No *(Chem.)*. The symbol for **nobelium**.

NOAA *(Space)*. Abbrev. for *National Oceanic and Atmospheric Administration*, a US body which manages and operates environmental satellites and provides data to users worldwide.

nobelium *(Chem.)*. Manmade element. At. no. 102; symbol No. Principal isotope is 254.

noble gases *(Chem.)*. Elements helium, neon, argon, krypton, xenon and radon-222, much used (except the last) in gas-discharge tubes. (Radon-222 has short-lived radioactivity, half-life less than 4 days.) Their outer (valence) electron shells are complete, thus rendering them inert to all the usual chemical reactions; a property for which argon, the most abundant, finds increasing industrial use. The heavier ones, Rn, Xe, Kr, are known to form a few unstable compounds, e.g. XeF_4. Also called *inert gases, rare gases*.

noble metals *(Eng.)*. Metals, such as gold, silver, platinum etc., which have a relatively positive electrode potential, and which do not enter readily into chemical combination with non-metals. They have high resistance to corrosive attack by acids and corrosive agents, and resist atmospheric oxidation. Cf. **base metal**.

nocardiasis *(Med.)*. Infection (usually of the lungs) with any one of a number of spore-forming fungi of the genus *Nocardia*.

nociceptive *(Zool.)*. Sensitive to pain.

noctilucent *(Zool.)*. Phosphorescent: light producing.

noctilucent clouds *(Meteor.)*. Thin but sometimes brilliant and beautifully coloured clouds of dust or ice particles at a height of from 75 to 90 km. They are visible about midnight in latitudes greater than about 50° when they reflect light from the sun below the horizon.

Noctovision *(Image Tech.)*. TN for a system of television in which the light sensitive elements respond to infrared light, and which can therefore be operated in darkness.

nocturia *(Med.)*. Passing excessive quantities of urine at night.

nodal gearing *(Eng.)*. The location of gear wheels, e.g. between a turbine and propeller shaft, at a nodal point of the shaft system with respect to torsional vibration.

nodal point, node *(Telecomm.)*. (1) Point in a high frequency circuit where current is a maximum and voltage a minimum, or vice versa. (2) In electrical networks, a terminal common to two or more branches of a network or to a terminal of any branch.

nodal points of a lens *(Phys.)*. Two points on the principal axis of a lens or lens system such that an incident ray of light directed towards one of them emerges from the lens as if from the other, in a direction parallel to that of the incident ray. For a lens having the same medium on its two sides, the nodal points coincide with the principal points.

node *(Acous.)*. The location of a minimum in the sound pressure or particle velocity when waves superimpose and result in standing waves.

node *(Astron.)*. One of the two points at which the orbit of a celestial object intersects a reference plane, such as the **ecliptic** or **equator**. The path crossing from south to north is the *ascending node*, and the *descending node* has the opposite sense.

node *(Bot.)*. The position on a stem at which one or more leaves are attached. Cf. **internode**.

node *(Elec.Eng.)*. (1) Point of minimum disturbance in a system of waves in tubes, plates or rods. The amplitude cannot become zero, otherwise no power could be transmitted beyond the point. (2) Point in an electrical network where two or more conductors are connected.

node *(Eng.)*. A point, or more than one, of rest in vibrating body.

node *(Maths.)*. See **double point**.

node of Ranvier *(Zool.)*. See **Ranvier's node**.

node voltage *(Telecomm.)*. That of a **nodal point** in an electrical network.

nodose, nodular *(Bot.)*. Bearing localized swellings or nodules.

nodular cast iron *(Eng.)*. See **ductile cast iron**.

nodular structures *(Geol.)*. Spheroidal, ovoid, or irregular bodies often encountered in both igneous and sedimentary rocks, and formed by segregation about centres. See, for example, **clay ironstone, doggers, flint, septaria**.

nodule *(Bot.)*. Any small rounded structure on a plant, especially a swelling on a root inhabited by symbiotic, nitrogen-fixing bacteria or actinomycetes.

nodulizing *(Min.Ext.)*. Aggregation of finely divided material such as mineral concentrates, by aid of binder and perhaps kilning, into nodules sufficiently strong and heavy to facilitate subsequent use, such as charging into blast furnaces.

no-fines concrete *(Build.,Civ.Eng.)*. Concrete in which the fine aggregate has been omitted, Therefore open textured, with a comparatively low strength, but other advantages. It can be cast *in situ* using simple, even open mesh shuttering, sets rapidly, has little capillary attraction and low moisture movement, and can be plastered and rendered readily owing to its texture.

nog *(Build.)*. A block of wood built into a wall to provide a substance to which joinery, such as skirtings, may be nailed.

noil *(Textiles)*. (1) Short or broken fibres from silk after opening and combing. (2) Short fibre removed from wool during combing; often added to woollen blends.

noise *(Acous.)*. (1) Socially unwanted sounds. (2) Interference in a communication channel.

noise *(Image Tech.)*. Unwanted signals or background giving rise to visual disturbance in picture reproduction, often appearing as excessive graininess or irregular colour spots.

noise *(Telecomm.)*. (1) In general, any unwanted disturbance superimposed on a useful signal and tending to obscure its information content. (2) Any undesired disturbance within a useful frequency band. Those caused by parallel services may be termed interference. (3) See **thermal noise, white noise**.

noise abatement climb procedure *(Aero.)*. Means of flying a civil aircraft from an airport so as to climb rapidly until the built-up area is reached and thereafter reducing power to just maintain a positive rate of climb until the area is overflown or 5000 ft is reached.

noise audiogram *(Acous.)*. That taken in the presence of a specified masking noise.

noise audiometer *(Acous.)*. An audiometer which measures the threshold of hearing of a deaf person's ear, the other ear or both ears, being subjected to a standardized noise in addition to the test sound.

noise background *(Acous.)*. In reproduction or recording, the total noise in the system with the signal absent.

noise control *(Acous.)*. Reduction of unwanted noise by various methods, e.g. absorption, isolation, anti-sound.

noise current *(Telecomm.)*. That part of a signal current conveying noise power.

noise diode *(Elec.Eng.)*. One operating as noise generator, under temperature limited conditions.

noise exposure forecast value *(Acous.)*. A method to evaluate the annoyance by fluctuating noises. Abbrev. *NEF*.

noise factor *(Telecomm.)*. The ratio, usually expressed in dB (when it may be referred to as the *noise factor*), of the noise power output per unit bandwidth at the output (of a communication system or of a receiver) to the portion of the noise power that is due to the input termination at an adopted standard temperature of 290 K.

noise factor, figure *(Acous.)*. Ratio of noise in a linear amplification system to thermal noise over the same frequency band and at the same temperature.

noise generator *(Elec.Eng.)*. Device for producing a controlled noise signal for test purposes. Also *noise source*. See **noise diode**.

noiseless recording *(Acous.)*. The practice of making the sound track of a positive sound film as dense as possible, consistent with the accommodation of the modulation, in order to keep the photographic noise level as far below the recorded level as possible.

noise level *(Acous.)*. The **loudness level** of a noise signal.

noise limiter *(Acous.)*. Device for removing the high peaks in a transmission, thus reducing contribution of clicks to the noise level and eliminating acoustic shocks. Effected by biased diodes or other clipping circuits.

noise meter *(Acous.)*. See **objective noise meter**.

noise power *(Electronics)*. That dissipated in a system by all noise signals present. See **Nyquist noise theorem**.

noise ratio *(Acous.)*. See **signal/noise ratio**.

noise reduction *(Acous.)*. Procedure in photographic recording whereby the average density in the negative sound track is kept as low as possible, and hence in the positive print as high as possible, both consistent with linearity, so that dirt, random electrostatics and scratches produce less noise. (2) See **noise control**.

noise reduction *(Image Tech.)*. Technique applied in magnetic and photographic sound recording and reproduction to reduce the level of background noise during quiet passages of low signal modulation.

noise resistance *(Acous.)*. One for which the thermal noise would equal the actual noise signal present, usually in a specific frequency band.

noise source *(Acous.,Elec.Eng.)*. The object or system from which the noise originates. See **noise generator**.

noise suppressor *(Aero.)*. A turbojet propelling nozzle fitted with fluted members which induct air to slow and break up the jet efflux, thereby reducing the noise level.

noise suppressor *(Telecomm.)*. Circuit which suppresses noise between usable channels when these are passed through during tuning. An *automatic gain control* which cuts out weak signals and levels out loud signals. See **muting**.

noise temperature *(Electronics)*. Temperature at which the thermal noise power of a passive system per unit bandwidth is equal to the noise at the actual terminals.

The standard reference temperature for noise measurements is 290 K. At this temperature the available noise power per unit bandwidth (see **Nyquist noise theorem**) is 4×10^{-21} watts.

noise transmission impairment *(Acous.)*. The transmission loss in dB which would impair intelligibility of a telephone system to the same extent as the existing noise signal. Abbrev. *NTI.*

noise voltage *(Acous.)*. A noise signal measured in r.m.s. volts.

no-load characteristic *(Elec.Eng.)*. See **open-circuit characteristic**.

no-load current *(Elec.Eng.)*. The current taken by a transformer when it is energized but is giving no output, or by a motor when it is running but taking no mechanical load.

no-load loss *(Elec.Eng.)*. The losses occurring in a motor or transformer when it is operating but giving no output. Also *open-circuit loss.*

noma *(Med.)*. See **cancrum oris**.

nomadism *(Ecol.)*. The habit of some animals of roaming irregularly without regularly returning to a particular place. Cf. **migration**.

nomeristic *(Zool.)*. Of metameric animals, having a definite number of somites.

nominal section *(Telecomm.)*. Network which is equivalent to section of transmission line, based on the assumption of lumped constants.

nomogram *(Maths.)*. Chart or diagram of scaled lines or curves for facilitating calculations. Those comprising three scales in which a line joining values on two determines a value on the third are frequently called *alignment charts.* Also called *nomograph.*

nonagon *(Maths.)*. A nine-sided polygon.

nonane *(Chem.)*. C_9H_{20}, a paraffin hydrocarbon; mp −51°C, bp 150°C, rel.d. 0.72.

non-aqueous solvents *(Chem.)*. May be classed broadly into (*a*) waterlike or levelling solvents, which are highly polar and form strong electrolytic solutions with most ionizable solutes, (*b*) differentiating solvents, which bring out differences in the strength of electrolytes. Examples of (*a*) are $NH_2(NH_4^+—NH_2^-)$, $SO_2(SO^{++}—SO_2^{--})$, $N_2O_4(NO^+—NO_3^-)$. Examples of (*b*) are weak amines or acids, ethers, halogenated hydrocarbons.

non-aqueous titration *(Chem.)*. Certain substances such as weak acids or bases, or compounds sparingly soluble in water, are better titrated in *nonaqueous solvents.* The strengths of weak acids are enhanced in **protophilic** solvents, and of weak bases in **protogenic** solvents, making a sharp **end point** possible. The use of **amphiprotic** or *aprotic* solvents, of mixtures, makes available a wide choice of *levelling* or *differentiating properties.*

non-association cable *(Elec.Eng.)*. Cable which is not maunufactured or designed in accordance with the standards of the Cable Makers' Association.

non-bearing wall *(Build.)*. A wall carrying no load apart from its own weight.

non-bleeder spray gun *(Build.)*. A spray gun which is fitted with an air valve which opens and closes with the action of the trigger.

non-caducous *(Zool.)*. See **indeciduate**.

non-conductor *(Electronics)*. Under normal conditions, an electrical insulator in which there are very few free electrons.

non-conforming *(Eng.)*. Said of a product outside manufacturing limits but not necessarily defective.

non-convertible coatings *(Build.)*. Reversible paint films which can be softened with the application of the original solvent. The drying of these coatings involves physical rather than chemical change, e.g. spirit based paints.

non-degenerate gas *(Electronics)*. One formed of particles, the concentration of which is so low, that the **Maxwell-Boltzmann distribution** applies. Particular examples are the electrons applied to a conduction band by donor levels in an n-type semiconductor, and those resulting from the passage of electrons from the normal band to an impurity band of acceptor levels in a p-type semiconductor.

non-denominational number system *(Maths.)*. Any system which is not denominational. The best known is the Roman number system but a large variety have been devised for use in digital computers.

non-depolarizing muscle relaxant *(Med.)*. Group of drugs which block neuromuscular transmission by competing with acetylcholine at the receptor site. Used in anaesthesia to produce paralysis and muscle relaxation. Common examples are *alcuronium, tubocurarine, gallomine.*

non-destructive readout *(Comp.)*. In a computer, stored data remains stored after it has been read. Abbrev. *NDRO.*

non-destructive testing *(Aero.)*. Use of probing systems to test structures for integrity without impairing their quality instead of removing samples for conventional analysis.

non-destructive testing *(Eng.)*. Testing in which materials are tested without being taken to failure. These include, **ultra-sonic testing, X-ray testing**.

non-deterministic *(Comp.)*. Applies to a machine if the machine can take a number of possible new states, given its present state and its new input.

non-directional microphone *(Acous.)*. Same as *omnidirectional microphone.*

non-disjunction *(Biol.)*. Failure of one or more chromosomes of a set to move with the rest of its set towards the appropriate pole at anaphase.

non-dissipative network *(Telecomm.)*. One designed as if the inductances and capacitances are free from dissipation, and as if constructed with components of minimum loss.

non-equivalence gate *(Comp.)*. See **XOR gate**.

non-essential organs *(Bot.)*. The sepals and petals of flowers.

non-ferrous alloy *(Eng.)*. Any alloy based mainly on metals other than iron, i.e. usually on copper, aluminium, lead, zinc, tin, nickel or magnesium.

non-flam film *(Image Tech.)*. See **safety film**.

non-homologous pairing *(Biol.)*. Pairing between regions of non-homologous chromosomes. In some cases short stretches of similar sequences, possibly repetitive may be involved.

non-hydrostatic model *(Meteor.)*. Numerical forecasting model in which the **hydrostatic approximation** is not used so that the effects of vertical accelerations can be accounted for.

Nonidet *(Chem.)*. TN for *nonionic detergent,* based on the condensation products of polyglycols with octyl or nonyl hydroxybenzenes.

non-inductive capacitor *(Electronics)*. One constructed so that it has virtually no inductance, with staggered foil layers left with an entire layer of foil at each end for making connections. Currents then flow laterally rather than spirally on the foil layers.

non-inductive circuit *(Elec.Eng.)*. One in which effects arising from associated inductances are negligible.

non-inductive load *(Elec.Eng.)*. See **non-reactive load**.

non-inductive resistor *(Elec.Eng.)*. Resistor having a negligible inductance, e.g. comprising a ceramic rod, or a special design of winding of resistance wire.

non-ionic detergents *(Chem.)*. Series of detergents in which the molecules do not ionize in aqueous solution, unlike soap and the sulphonated alkylates. Typical examples are the detergents based on condensation products of long chain glycols and octyl or nonyl hydroxybenzenes.

non-isolated essential singularity *(Maths.)*. If a function $f(x)$ has an infinite set of singularities at the points a_0, a_1 ..., which tend to a limit point α; α is a nonisolated essential singularity.

non-leakage probability *(Phys.)*. For neutron in reactor, the ratio of the actual multiplication constant to the infinite multiplication constant.

non-linear *(Elec.Eng.)*. Property of a component where applied voltage and current flowing do not exhibit direct proportionality.

non-linear distortion *(Telecomm.)*. Distortion resulting from the situation where the output of a system or component does not bear the desired relation to the input. Amplitude, harmonic and intermodulation distortion are examples and results of non-linear distortion.

non-linear distortion factor *(Telecomm.)*. Square root of ratio of the powers associated with alien tones to the powers associated with wanted tones in the output of a nonlinear distorting device.

non-linearity *(Elec.Eng.,Telecomm.)*. Lack of proportionality between either applied voltage and resulting current, or between input and output, of an electrical network, component, amplifier or transmission line which causes distortion in a passing signal. See **non-linear distortion**.

non-linear network *(Elec.Eng.)*. A network in which the electrical elements are not all linear with varying current, as rectifying semiconductor diodes.

non-linear resistance *(Elec.Eng.)*. Nonproportionality between potential difference and current in an electric component.

non-linear resistor *(Elec.Eng.)*. One which does not 'obey' Ohm's law, in that there is departure from proportionality between voltage and current. In semiconductor crystals, ratio between forward and backward resistance may be 1:1000. See **diode, transistor**.

non-magnetic steel *(Elec.Eng.)*. One containing ca. 12% Mn, exhibiting no magnetic properties. Stainless steel is almost nonmagnetic.

non-magnetic watch *(Horol.)*. A watch so constructed that its performance is not affected by magnetic fields. Usually, the balance, balance spring, roller and fork are made of a non-magnetic alloy.

non-medullated *(Zool.)*. See **amyelinate**.

non-metal *(Chem.)*. An element which readily forms negative ions, often in combination with other non-metals. Nonmetals are generally poor conductors of electricity.

non-metallic inclusions *(Eng.)*. See **inclusions**.

non-Newtonian liquids *(Phys.)*. See **anomalous viscosity**.

non-, nona- *(Chem.)*. Containing 9 atoms, groups etc.

non-operable instruction *(Comp.)*. One whose only effect is to advance the **program counter**. Often written as 'continue'.

nonpareil *(Print.)*. An old type size approximately *6-point*, for which it is still synonymous.

non-polarized relay *(Telecomm.)*. One in which there is no magnetic polarization. Operation depends on the square of the current in the windings, and is therefore independent of direction, as in telephone and a.c. relays.

non-quantized *(Phys.)*. See **classical**.

non-reactive load *(Elec.Eng.)*. A load in which the current is in phase with the voltage across its terminals. Also called *non-inductive load.*

non-reactive power *(Elec.Eng.)*. Value of active power in an electrical system.

non-relativistic *(Phys.)*. Said of any procedure in which effects arising from relativity theory are absent or can be disregarded, e.g., properties of particles moving with low velocity, e.g., 1/20th that of light propagation.

non-resonant antenna *(Telecomm.)*. See **aperiodic antenna**.

nonreturn-flow wind tunnel *(Aero.)*. A straight-through wind tunnel in which the air flow is not recirculated.

non-reversible coatings *(Build.)*. Convertible coats, like oil-based paints, which, when dry, cannot be returned to the liquid state by the application of further coatings. Cf. **non-convertible coatings**.

nonsense mutation *(Biol.)*. A base change which causes an amino acid-specifying sequence to be changed into one which causes termination by polypeptide chain synthesis.

nonsense syllable *(Behav.)*. A series of letters, usually consisting of 2 consonants with a vowel between them that does not constitute a word; used in studies of learning and memory.

nonsequence *(Geol.)*. A break in the stratigraphical record, less important and less obvious than an unconformity, and deduced generally on palaeontological evidence.

non-singular matrix *(Maths.)*. A square matrix the determinant of which is not equal to zero.

non-specific immunity *(Immun.)*. Mechanisms whereby the body is protected against microbial invasion which do not depend upon the mounting of a specific immune response. They include physical barriers to infection (skin, mucous membranes); enzyme inhibitors naturally present in the blood; activation by 'rough' variants of Gram negative bacteria of complement via the alternative pathway; interferon; lysozyme; phagocytosis, etc. These mechanisms are sufficient to protect against microbes which are regarded as non-pathogenic, although they would be capable of multiplication within the dead body.

non-spectral colour *(Phys.)*. That which is outside the range which contributes to white light, but which affects photocells.

non-specular reflection *(Phys.)*. Wave reflection of light or sound from rough surfaces, resulting in scattering of wave components, depending on relation between wavelength and dimensions of irregularities. Also called *diffuse reflection.*

non-steroidal anti-inflammatory drugs *(Med.)*. Term used for a group of drugs which suppress inflammation and relieve pain probably by interfering with **prostaglandin** synthesis. Common examples are; *iodomethacin, phenylbutazone, Ibuprofen.* Abbrev. *NSAID.*

non-stoichiometric compounds *(Chem.)*. Some solid compounds do not possess the exact compositions which are predicted from Daltonic or electronic considerations alone (e.g. iron (II) sulphide is $FeS_{1.1}$), a phenomenon which is associated with the so-called **defect structure** of *crystal lattices.* Often show semi-conductivity, fluorescence and centres of colour.

non-sweating *(Vet.)*. See **anhidrosis**.

nonsymmetrical *(Elec.Eng.)*. See **asymmetrical**.

non-synchronous motor *(Elec.Eng.)*. An a.c. motor which does not run at synchronous speed, e.g., an induction motor or an a.c. commutator motor. Also called *asynchronous motor.*

non-tension joint *(Elec.Eng.)*. A joint in an overhead transmission line conductor which is designed to carry full-load current but not to withstand the full mechanical tension of the conductor.

non-theatrical *(Image Tech.)*. Describing the use of motion picture film outside the scope of the professional cinema entertainment industry.

nontronite *(Min.)*. A clay mineral in the montmorillonite group (smectites), containing appreciable ferric iron replacing aluminium.

nonviable *(Bot.,Zool.)*. Incapable of surviving.

nonvolatile memory *(Comp.)*. One in computer which holds data even if power has been disconnected. Most magnetic memories are nonvolatile. Cf. **volatile memory**.

non-woven fabrics *(Textiles)*. Cloth formed from a random arrangement or **web** of natural or synthetic fibres using adhesives, heat and pressure, needling techniques etc. to confer adhesion.

noon *(Astron.)*. The instant of the sun's upper culmination at any place. See also **mean-, sidereal-**.

no-ops *(Comp.)*. See **non-operable instruction**.

nopaline *(Bot.)*. One sort of **opine**.

NOR *(Biol.)*. See **nucleolar-organizing region**.

NOR *(Comp.)*. A logical operator where (p NOR q) written ($\overline{p+q}$) takes the value TRUE when neither p nor q are TRUE. When either p or q are TRUE, it takes the value FALSE. See **logical operation**.

noradrenaline *(Chem.)*. A sympathomimetic substance obtained from the adrenal glands, having similar properties to *adrenaline* and differing in structure only by not having the amino group substituted with a methyl group. See **catecholamine**.

norbergite *(Min.)*. Magnesium silicate with magnesium fluoride or hydroxide. A member of the humite group it crystallizes in the orthorhombic system and occurs in metamorphosed dolomitic limestones.

nordmarkite *(Geol.)*. An alkali quartz-bearing syenite, described originally from Nordmarken in Norway;

consists essentially of microperthite, aegirine, soda amphibole, and accessory quartz.

norepinephrine *(Chem.)*. See **noradrenaline**.

Norfolk (Canadian) latch *(Build.)*. A latch in which the fall bar is actuated within the limits of the keeper by a lifting lever passing through a slot in the door and operated by the pressure of the thumb at one end.

NOR gate *(Comp.)*. **Gate** with output 1 when neither of the inputs is 1. When either input signal is 1 the output is 0. All gates can, in theory, be constructed from combinations of NOR gates. Also *NOR element*.

norgine *(Chem.)*. See **alginic acid**.

norite *(Geol.)*. A coarse-grained igneous rock of basic composition consisting essentially of plagioclase (near labradorite in composition) and othopyroxene. Other coloured minerals are usually present in varying amount, notably clinopyroxene.

norm *(Behav.)*. Shared expectations about how individuals should or do behave.

norm *(Geol.)*. The theoretical composition of an igneous rock expressed in terms of standard mineral molecules, calculated from the chemical analysis as stated in terms of percentages of oxides. Cf. *mode*.

norm *(Maths.,Stats.)*. Of a vector in a finite-dimensional real vector space: the square root of the scalar product of the vector with itself; the magnitude of the vector. A norm on a vector space is a consistent definition of the norms of its vectors, or equivalently of a scalar product.

normal *(Chem.)*. Containing an unbranched chain of carbon atoms; e.g. *normal* propyl alcohol is $CH_3.CH_2.CH_2.OH$, whereas the isomeric *iso*propyl alcohol, $(CH_3)_2=CH.OH$, has a branched chain. In more modern nomenclature, this prefix is redundant.

normal *(Maths.)*. The *normal* to a line or surface is a line drawn perpendicular to it.

normal axis *(Aero.)*. See **axis**.

normal bend *(Elec.Eng.)*. A section of conduit bent to a moderately large radius; used in an electrical installation for connecting two other pieces of conduit which are at right angles to each other.

normal calomel electrode *(Chem.)*. A calomel electrode containing molar potassium chloride solution.

normal curvature *(Maths.)*. See **curvature**.

normal distribution *(Stats.)*. A distribution widely used in statistics, to model the variation in a set of observations, as an approximation to other distributions, or as the asymptotic distribution of statistics from large samples. The normal distribution is indexed by two parameters, the mean and variance. See also **standard normal distribution**.

normal electrode potential *(Chem.)*. See **standard electrode potential**.

normal fault *(Geol.)*. A fracture in rocks along which relative displacement has taken place under tensional conditions, the *fault* hading to the downthrow side. Cf. **reversed fault**.

normal flight *(Aero.)*. All flying other than aerobatics including straight and level, climbing, gliding, turns and *sideslips* for the loss of height or to counteract drift; a licensing category for certifying whether *airworthy*.

normal form *(Maths.)*. See **canonical form**.

normal frequency *(Phys.)*. See **normal modes**.

normal functions *(Maths.)*. See **orthogonal functions**.

normal induction *(Elec.Eng.)*. That represented by the curve of normal magnetization.

normality *(Chem.)*. An obsolescent concentration unit, abbreviated N. It is used mainly for acids or bases and for oxidizing or reducing agents. In the first case, it refers to the concentration of titratable H^+ or OH^- in a solution. Thus a solution of sulphuric acid with a concentration of $2\,mol\,l^{-1}$ (2 M) will be a 4 N solution.

normalize *(Comp.)*. Represent a number in agreed **floating-point** notation, usually that which provides maximum precision.

normalizing *(Eng.)*. A heat treatment applied to steel. Involves heating above the critical range, followed by cooling in air, performed to refine the crystal structure and eliminate internal stress.

normal magnetization *(Elec.Eng.)*. Locus of the tips of the magnetic hysteresis loops obtained by varying the limits of the range of alternating magnetization.

normal modes *(Phys.)*. If a vibrating system consists of a number of oscillators coupled together, the resulting motion is in general complicated. However, by choosing the starting conditions correctly, it is possible to make the system vibrate in such a way that every part has the same frequency. Such simple vibrations are called the *normal modes* and the associated frequencies, the *normal* or *natural frequencies*. If there are N degrees of freedom in the system then there are N normal modes. See **coupled vibrations**.

normal modes *(Telecomm.)*. In a linear system, the least number of independent component oscillations which may be regarded as constituting the free, or natural, oscillations of the system. Also called *natural modes*.

normal pressure *(Chem.)*. Standard pressure, $101.325\,kN/m^2 = 760$ torr, to which experimental data on gases are referred.

normal radius of curvature *(Maths.)*. See **curvature**.

normal salts *(Chem.)*. Salts formed by the replacement by metals of all the replaceable hydrogen of the acid.

normal section of a surface *(Maths.)*. A section of the surface made by a plane which contains a normal to the surface.

normal segregation *(Eng.)*. A type of segregation in which the content of impurities and inclusions tends to increase from the surface to the centre of cast metals. Of special importance in steel, in which phosphorus, sulphur, and oxide inclusions segregate in this way. See also **inverse segregation**.

normal solution *(Chem.)*. See **normality**.

normal state *(Phys.)*. See **ground state**.

normal subgroup *(Maths.)*. A subgroup which contains all conjugates of its elements. Also known as *invariant subgroup* or *self-conjugate subgroup*.

normal temperature and pressure *(Chem.)*. Earlier term for **stp**.

normative composition *(Geol.)*. The theoretical mineral chemical composition of a rock expressed in terms of the standard minerals of the **norm**.

normoblast *(Biol.)*. A stage in the development of an erythrocyte from an erythroblast when the nucleus has become reduced in size and the cytoplasm contains much haemoglobin.

Northern Lights *(Astron.)*. The *Aurora Borealis*. See **aurora**.

northing *(Surv.)*. A north latitude. Northerly displacement of a point with reference to observer's station.

north light roof *(Arch.)*. A pitched roof with unequal slopes, of which the steeper is glazed and arranged in such a way as to receive light from the north.

north pole *(Phys.)*. See **pole**.

Northrup furnace *(Elec.Eng.)*. See **coreless induction furnace**.

North Sea gas *(Geol.)*. See **natural gas**.

Norton's theorem *(Elec.Eng.)*. Any electrical source can be represented at its terminals by a constant-current generator in parallel with a shunt resistance. The value of the current generator is that which would flow with a short circuit at the generator terminals. The shunt resistance is that measured between the terminals with no current source. Often regarded as the dual of **Thevenin's theorem**.

Norwegian quartz *(Build.)*. A white translucent quartz found in Norway (a similar type occurs in the Isle of Man), now frequently used for the facing of *cladding* slabs.

nosean, noselite *(Min.)*. Silicate of sodium and aluminium with sodium sulphate, crystallizing in the cubic system. Occurs in extrusive igneous rocks which are rich in alkalis and deficient in silica, e.g. phonolite.

nose bit *(Build.)*. A type of **shell bit** with a projecting tip for extracting the waste.

nose dive *(Aero.)*. See **dive**.

nose heaviness *(Aero.)*. The state in which the combination of the forces acting upon an aircraft in flight is such that it tends to pitch downwards by the nose.

nose ribs *(Aero.)*. Small intermediate ribs, usually from the front spar to the leading edge only, of planes and control surfaces. They maintain the correct wing contour under the exceptionally heavy air load at that part of the aerofoil.

nose suspension *(Elec.Eng.)*. A method of mounting a traction motor, by supporting one side of it on the axle and the other side on the framework of the truck.

nose-wheel landing gear *(Aero.)*. See **tricycle landing gear**.

nosing *(Build.)*. A bead on the edge of a board, making it half-round.

nosology *(Med.)*. Systematic classification of diseases: the branch of medical science which deals with this.

nosophobia *(Med.)*. Morbid fear of contracting disease.

nostrils *(Zool.)*. The external nares.

NOT *(Comp.)*. A logical operator such that NOTp written $\bar{p}$ takes the value FALSE if p is TRUE and vice versa. See **logical operation**.

not *(Paper)*. The unglazed surface of hand-made or mould-made drawing-papers. See **hotpressed** and **rough**.

notch *(Elec.Eng.)*. A term often used to denote any of the various positions of a controller.

notch *(Image Tech.)*. A shallow cut on the edge of a motion picture negative to actuate the light change mechanism of a printer.

notch aerial *(Aero.)*. A radio aerial, usually for *HF*, formed by cutting a notch out of the aircraft's skin and covering it with a dielectric material to its original profile.

notch board *(Build.)*. A notched board carrying the treads and risers of a staircase.

notch brittleness *(Eng.)*. Susceptibility to fracture as disclosed by Izod or Charpy test, due to weakening effect of a surface discontinuity.

notched-bar test *(Eng.)*. A test in which a notched metal specimen is given a sudden blow by a striker carried by a pendulum or a falling weight, and the energy absorbed in breaking the specimen is measured. Also called *impact test*. See also **Izod test, Charpy test, Fremont test**.

notching *(Civ.Eng.)*. The method of excavating cuttings for roads or railways in a series of steps worked at the same time.

notching *(Eng.)*. A press work process similar to punching or piercing, in which material is removed from the edge of a strip or sheet, often as a means of location or registering.

notching, linking up *(Eng.)*. Movement of the gear-lever of a locomotive or steam-engine towards the centre of a notched quadrant, to decrease the valve travel and shorten the cut-off.

notch sensitivity *(Eng.)*. The extent to which the endurance of metals, as determined on smooth and polished specimens, is reduced by surface discontinuities, such as tool marks, notches and changes in section, which are common features of actual components. It tends to increase with the hardness and endurance limit.

notch toughness *(Eng.)*. The energy in joules or ft lbf required to break standard specimens under the standard conditions realized in the Izod or the Charpy test. It may also mean the opposite of *notch brittleness*.

note *(Acous.)*. An identifiable musical tone, whether pure or complex. See also **strike-, wolf-**.

NOT gate *(Comp.)*. The only single-input **gate**. It gives an output signal when there is no input signal and vice versa, i.e. input 0, output 1 and input 1, output 0. Also *NOT element*.

notochord *(Zool.)*. In Chordata, skeletal rod formed of turgid vacuolated cells. adj. *notochordal*.

notoedric mange *(Vet.)*. Mange on the face and ears of cats and rabbits, due to *Notoedres cati*.

Notting quoin *(Print.)*. The original screw-and-wedge pattern quoin, all steel, in which the two sides expand by a wedge drawn upward as a screw is turned by the quoin key.

notum *(Zool.)*. The tergum of Insects.

nova *(Astron.)*. Classically, any new star which suddenly becomes visible to the unaided eye. In modern astronomy, a star late in its evolutionary track which suddenly brightens by a factor of 10^4 or more. The rise in brightness is so rapid (days) that the star is seldom noticed until it is indeed of naked eye visibility; after a few weeks it returns to the pre-nova magnitude. Astronomers consider that most novae (pl. also *novas*) are members of close **binary stars**. One component is a **white dwarf**, and this is the source of the explosion: matter is transferred from the other, highly evolved, companion and triggers a new burst of nuclear reactions.

novaculite *(Geol.)*. A fine-grained or crypto-crystalline rock composed of quartz or other forms of silica; a form of chert. Used as a whetstone.

Novocain *(Chem.)*. TN for $H_2N.C_6H_4.CO.O.CH_2.CH_2.N(C_2H_5)_2$. HCl, the hydrochloride of diethylaminoethyl *p*-aminobenzoate. Colourless needles; mp 156°C; a widely used local anaesthetic. Also known as *procaine*.

Novoid *(Build.)*. TN for a powder having a colloidal silica basis; used for making cement surfaces water-, oil-, and acid-proof.

no-voltage release *(Elec.Eng.)*. A relay or similar device which causes the circuit to a motor or other equipment to be opened automatically if the supply voltage falls.

nowcasting *(Meteor.)*. A system of rapid and very-short-range (1 to 2 h) forecasting of phenomena such as heavy rain and thunderstorms based on real-time processing of simultaneous observations from a network of remote-sensing devices (including **weather radars** and **meteorological satellites**) combined with simple extrapolation techniques.

nox *(Phys.)*. Obsolete unit of scotopic illuminance. 1 nox $=0^{-3}$ lux.

noy *(Acous.)*. Unit of perceived noisiness by which equal-noisiness contours, e.g. for 10 noys, 20 noys etc., replace equal-loudness contours.

nozzle *(Aero.)*. See **propelling nozzle**.

nozzle *(Eng.)*. (1) In impulse turbines, specially shaped passages for expanding the steam, thus creating kinetic energy of flow with minimum loss. (2) In oil engines, orifices, open or controlled by the injection valve, through which the fuel is sprayed into the cylinder. See also **convergent-divergent nozzle**. (3) Generally, a convergent or convergent-divergent tube attached to the outlet of a pipe or a pressure chamber to direct efficiently the pressure of a fluid into velocity.

nozzle *(Telecomm.)*. End of a waveguide, which may be contracted in area.

nozzle guide vanes *(Aero.)*. In a gas turbine, a ring of radially positioned aerofoils which accelerate the gases from the combustion chamber and direct them on to the first rotating turbine stage.

Np *(Chem.)*. Symbol for neptunium.

n-p-i-n transistor *(Electronics)*. Similar to ***n-p-n* transistor** with a layer of intrinsic semiconductor (germanium or silicon of high purity) between the base and the collector to extend the high-frequency range.

NPK *(Ecol.)*. Nitrogen, Phosphorus and Potassium as fertilizer.

N.P.L. type wind tunnel *(Aero.)*. The *closed-jet, return-flow* type is often called the original N.P.L. type, and the *closed-jet, nonreturn-flow* the standard N.P.L. type, as they were first used by the UK *National Physical Laboratory*.

n-p-n transistor *(Electronics)*. A junction transistor having a p-type base between an n-type emitter and an n-type collector. In operation, the emitter should be negative and the collector positive with respect to the base.

NRME *(Build.)*. Abbrev. for *Notched, Returned, and Mitred Ends*.

NSAID *(Med.)*. Abbrev. for *Non-Steroidal Anti-Inflammatory Drugs*.

N-shell *(Phys.)*. The **electron shell** in an atom corresponding to a principal quantum number of four. The shell contains up to 32 electrons and is the last shell to be filled completely by electrons in the naturally occurring elements.

nt *(Phys.)*. Symbol for nit.

$n\tau$ *(Phys.)*. The Lawson criterion; the product of the plasma density n and the plasma confinement time τ for nuclear fusion processes. $n\tau > 10^{20}\,sm^{-3}$ for fusion to produce useful energy.

N-terminal pair network *(Telecomm.)*. One having N-terminal pairs in which one terminal of each pair may coincide with a node.

NTI *(Acous.)*. Abbrev. for *Noise Transmission Impairment*.

NTP *(Chem.)*. *Normal Temperature and Pressure*. Previous term for STP, i.e. 0°C and 101.325 kN/m^2.

N-truss *(Eng.)*. See **Whipple-Murphy truss**.

NTS *(Build.)*. Abbrev. for *Not To Scale*; on plans etc.

NTSB *(Aero.)*. Abbrev. for *National Transportation Safety Board*, US.

NTSC *(Image Tech.)*. The US *National Television System Committee*, and the colour TV system so standardized.

n-type semiconductor *(Electronics)*. One in which the electron conduction (negative) exceeds the hole conduction (absence of electrons), the **donor** impurity predominating.

nucellus *(Bot.)*. Parenchymatous tissue in the ovule of a seed plant, more or less surrounded by the integuments and containing the embryo sac; equivalent to the megasporangium. See also **perisperm**.

nuchal *(Zool.)*. Pertaining to, or situated on, the back of the neck.

nuchal crest *(Zool.)*. A transverse bony ridge forming across the posterior margin of the roof of the vertebrate skull for attachment of muscles and ligaments supporting the head.

nuchal flexure *(Zool.)*. In developing Vertebrates, the flexure of the brain occurring in the hinder part of the medulla oblongata, which bends in the same direction as the primary flexure.

nucivorous *(Zool.)*. Nut-eating.

nuclear battery *(Phys.)*. One in which the electric current is produced from the energy of radioactive decay, either directly by collecting beta particles or indirectly, e.g., by using the heat liberated to operate a thermojunction. In general, nuclear batteries have very low outputs (often only microwatts) but long and trouble free operating lives.

nuclear binding energy *(Phys.)*. The binding energy that holds together the constituent nucleons of the nucleus of an atom. It is the energy equivalence of the mass difference between the masses of the atom and the sum of the individual masses of its constituents; expressed in MeV or a.m.u.

nuclear Bohr magneton *(Phys.)*. See **Bohr magneton**.

nuclear breeder *(Nuc.Eng.)*. A nuclear reactor in which in each generation there is more fissionable material produced than is used up in fission.

nuclear budding *(Biol.)*. Production of 2 daughter nuclei of unequal size by constriction of the parent nucleus.

nuclear charge *(Phys.)*. Positive charge arising in the atomic nucleus because of protons, equal in number to the atomic number.

nuclear chemistry *(Chem.)*. The study of reactions involving the transmutation of elements, either by spontaneous decay or by particle bombardment.

nuclear conversion ratio *(Phys.)*. That of the fissile atoms produced to the fissile atoms consumed, in a breeder reactor.

nuclear disintegration *(Phys.)*. Fission, radioactive decay, internal conversion or isomeric transition. Also *nuclear reaction*.

nuclear emission *(Phys.)*. Emission of gamma ray or particle from nucleus of atom as distinct from emission associated with orbital phenomena.

nuclear emulsion *(Phys.)*. Thick photographic coating in which the tracks of various fundamental particles are revealed by development as black traces.

nuclear energy *(Phys.)*. In principle, the binding energy of a system of particles forming an atomic nucleus. More usually, the energy released during nuclear reactions involving regrouping of such particles (e.g., fission or fusion processes). The term *atomic energy* is deprecated as it implies rearrangement of atoms rather than of nuclear particles.

nuclear envelope *(Bot.)*. See **nuclear membrane**.

nuclear family *(Biol.)*. In human genetics, a family providing data on the two parents and their children.

nuclear field *(Phys.)*. Postulated short range field within a nucleus, which holds protons and neutrons together, possibly in shells.

nuclear fission *(Phys.)*. The spontaneous or induced disintegration of the nucleus of a heavy atom into two lighter atoms. The process involves a loss of mass which is converted into nuclear energy.

nuclear force *(Phys.)*. That which keeps neutrons and protons together in a nucleus, differing in nature from electric and magnetic forces, gravitational forces being negligible. The force is of short range, is practically independent of charge, and arises from the exchange of *pions* between the nucleons (Yukawa). See **meson, short-range forces, exchange forces**.

nuclear fragmentation *(Biol.)*. The formation of two or more portions from a cell nucleus by direct break-up and not by mitosis.

nuclear fuel *(Nuc.Eng.)*. See **reactor**.

nuclear fusion *(Phys.)*. The process of forming atoms of new elements by the fusion of atoms of lighter ones. Usually the formation of helium by the fusion of hydrogen and its isotopes. The process involves a loss of mass which is converted into nuclear energy. The basis of **fusion reactors**.

nuclear isomer *(Phys.)*. A nuclide existing in an excited metastable state. It has a finite half-life after which it returns to the ground state with the emission of a γ-quantum or by internal conversion. Metastable isomers are indicated by adding *m* to the mass number.

nuclear magnetic resonance *(Chem.)*. The nuclei of certain isotopes, e.g. ^1H, ^{19}F, ^{31}P, behave like small bar magnets and will line up when placed in a strong d.c. magnetic field. The direction of alignment of the nuclei in a sample may be altered by RF irradiation in a surrounding coil whose axis is at right angles to the magnetic field. The frequency at which energy is absorbed and the direction of alignment changes is known as the resonant frequency and varies with the type of nucleus; e.g. in a magnetic field of 14 000 gauss ^1H resonates at 60 MHz, ^{19}F at 56 MHz, and ^{31}P at 24.3 MHz. Nuclei with even mass and atomic number, e.g. ^{12}C, ^{16}O, have no magnetic properties and do not exhibit this behaviour. Abbrev. *NMR*.

nuclear magnetic resonance *(Phys.)*. Certain atomic nuclei, e.g. ^1H, ^{19}F, ^{31}P, have a nuclear magnetic moment. When placed in a strong magnetic field, the nuclear moments can only take up certain discrete orientations, each orientation corresponding to a different energy state. Transitions between these energy levels can be induced by the application of radiofrequency radiation. This is known as nuclear magnetic resonance (NMR). For example, the protons in water experience this resonance effect in a field of 0.3 T at a radiation frequency of 12.6 MHz. The resonant frequency depends on the magnetic field at the nucleus which in turn depends on the environment of the particular nucleus. NMR gives invaluable information on the structure of molecules. It has also been developed to provide a non-invasive clinical imaging of the human body, magnetic resonance imaging (MRI); multiple projections are combined to form images of sections through the body, providing a powerful diagnostic aid.

nuclear magnetic resonance spectroscopy *(Chem.)*. The resonant frequencies for identical nuclei in a constant external field vary slightly with the chemical environment of the nucleus (about 1–10 ppm for ^1H). By keeping the external magnetic field constant and varying the RF radiation over a small range, 1000 Hz, for a solution of a given molecule, an absorption spectrum for the compound under examination is obtained for which information may be deduced on the structure of the molecule.

nuclear magneton *(Electronics)*. See **Bohr magneton**.

nuclear matrix *(Biol.)*. The nuclear residue left after removal of chromatin, including pore complexes, lamina, nucleolar residues and ribonucleoprotein fibrils.

nuclear medicine *(Med.)*. The application of radionuclides in the diagnosis or treatment of disease.

nuclear membrane *(Biol.)*. The double membrane punctuated by **nuclear pore complexes** which surrounds the interphase nucleus, the outer membrane being continuous with the membrane of the **endoplasmic reticulum**. Also *nuclear envelope*.

nuclear model *(Phys.)*. One giving an explanation of the properties of the atomic nucleus and its interactions with other particles. See **liquid drop model, optical model, shell model, collective model, independent particle model, unified model**.

nuclear number *(Phys.)*. Same as *mass number*.

nuclear paramagnetic resonance *(Chem.,Phys.)*. See **nuclear magnetic resonance**.

nuclear photoeffect *(Phys.)*. See **photodisintegration**.

nuclear pore complex *(Biol.)*. Sites at which the two layers of the nuclear membrane are joined forming pores which connect the nucleoplasm and the cytoplasm, and which are surrounded on either side by symmetrical arrays of granules called *annuli*.

nuclear potential *(Phys.)*. The potential energy of a nuclear particle in the field of a nucleus as determined by the short range forces acting. Plotted as a function of position it will normally represent some sort of **potential well**.

nuclear power *(Phys.)*. Power generated by the release of energy in a nuclear reaction.

nuclear propulsion *(Space)*. The use of the energy released by a nuclear reaction to provide propulsive thrust through heating the working fluid or providing electric power for an ion or similar propulsion system.

nuclear radius *(Phys.)*. The somewhat indefinite radius of a nucleus within which the density of *nucleons* (protons and neutrons) is experimentally found to be nearly constant. The radius in metres is 1.2×10^{-15} times the cube root of the nuclear number (atomic mass). It is not a precise determinable quantity.

nuclear reaction *(Phys.)*. The interaction of a photon or particle with a target nucleus. An amount of energy, Q-value, is released or absorbed, depending on whether the mass of the reaction products is less than or more than the mass of the reactants. Reactions fall into two broad categories; those in which the reaction proceeds *via* the formation of a *compound nucleus*; the alternative mechanism is by *direct interaction*. At higher energies *spallation* occurs in which the target nucleus splits up into a number of fragments.

nuclear reactor *(Nuc.Eng.)*. See **reactor**.

nuclear reactor oscillator *(Nuc.Eng.)*. Device producing variations in reactivity by oscillatory movement of sample to measure reactor properties, or neutron capture cross-section of sample.

nuclear sap *(Biol.)*. See **nucleoplasm**.

nuclear selection rules *(Phys.)*. Those specifying the transitions of electrons or nucleons between different energy levels which may take place (*allowed transitions*). The rules may be derived theoretically through wave mechanics but are not obeyed rigorously; so-called *forbidden transitions* merely being highly improbable.

nuclear spindle *(Biol.)*. See **spindle**.

nuclear winter theory *(Ecol.)*. Theory, based on model calculations, that nuclear war would be followed by a period of cold resulting from the attenuation of solar energy by dust and smoke in the atmosphere.

nuclease *(Biol.)*. An enzyme which specifically cleaves the 'backbone' of nucleic acids. Their specificity ranges from those like DNase and RNase which cut the phosphodiester bonds in any DNA or RNA, to **restriction enzymes** which only cut a particular 4 or 6 base-pair sequence.

nucleating agent *(Meteor.)*. Substance used for seeding clouds to control rainfall and fog formation. See **rainmaking**.

nuclei *(Eng.)*. Centres or 'seeds' from which crystals begin to grow during solidification. In general, they are minute crystal fragments formed spontaneously in the melt, but frequently nonmetallic inclusions act as nuclei. See also **crystal nuclei**.

nucleic acid *(Biol.)*. General term for natural polymers in which *bases* (purines or pyrimidines) are attached to a sugar phosphate backbone. Can be single- or double-stranded and short molecules are called *oligonucleotides*. syn. *polynucleotide*.

nucleogenesis *(Phys.)*. Theoretical process(es) by which nuclei could be created from possible fundamental dense plasma. See **ylem**.

nucleolar organizer *(Bot.)*. Region in some chromosomes, recognized as a constriction, at which a nucleolus forms.

nucleolar-organizing region *(Biol.)*. Chromosomal region containing ribosomal genes; there is often a visible constriction at this site on metaphase chromosomes, and the region can be differentially stained. Abbrev. *NOR*.

nucleolus *(Biol.)*. Round body occurring within a cell nucleus, consisting of ribosomal genes and associated polymerases, nascent RNA transcripts and proteins involved in ribosome assembly. adj. *nucleolar*.

nucleonics *(Phys.)*. The science and technology of nuclear studies.

nucleon number *(Phys.)*. See **mass number**.

nucleons *(Phys.)*. Protons and neutrons in a nucleus of an atom.

nucleophilic reagents *(Chem.)*. Term applied to reagents which react at positive centres, e.g. the hydroxyl ion (OH^-) in replacing Cl^- from a C-Cl link in which the C atom can be considered as being the positive centre.

nucleoplasm *(Biol.)*. The protoplasm in the nucleus surrounding the chromatin. Cf. **cytoplasm**.

nucleoplasmic ratio *(Biol.)*. The ratio between the volume of the nucleus and of the cytoplasm in any given cell.

nucleor *(Phys.)*. The hypothetical core of a nucleon. It is suggested that it is the same for protons and neutrons.

nucleoside *(Biol.)*. A desoxyribose or ribose sugar molecule to which a purine or pyrimidine is covalently bound. See **nucleotide**.

nucleosome *(Biol.)*. Chromosome of eukaryotes have a complex 'quaternary' structure, consisting of chromatin beads into which about 145 base pairs of DNA are folded, each bead being separated by less folded chromatin.

nucleosynthesis *(Astron.)*. The synthesis of elements other than hydrogen and helium by means of nuclear fusion reactions in stellar interiors and **supernova** explosions.

nucleotide *(Biol.)*. A **nucleoside** to which a phosphate group is attached at the 5′ position on the sugar. The individual components of a **nucleic acid**.

nucleus *(Astron.)*. (1) The central core of a comet, about 1–10 km across, and consisting of icy substances and dust. (2) The central part of a **galaxy** or **quasar**, possibly the seat of unusually energetic activity within the galaxy. (3) Term used generally in astronomy to indicate any concentration of stars or gas in the central part of a **nebula**.

nucleus *(Biol.)*. Compartment within the interphase **eukaryotic** cell bounded by a double membrane and containing the genomic DNA, with its associated functions of **transcription** and processing.

nucleus *(Phys.)*. Composed of protons (positively charged) and neutrons (no charge), and constitutes practically all the mass of the atom. Its charge equals the atomic number; its diameter is from 10^{-15} to 10^{-14} m. With protons, equal to the atomic number, and neutrons to make up the atomic mass number, the positive charge of the protons is balanced by the same number of extra-nuclear electrons.

nucleus *(Zool.)*. (1) Any nut shaped structure. (2) A nerve centre in the brain. (3) A collection of nerve cells on the course of a nerve or tract of nerve fibres.

nuclide *(Phys.)*. An atomic nucleus as characterized by its atomic number, its mass number and nuclear energy state.

nudicaudate *(Zool.)*. Having the tail uncovered by fur or hair, as Rats.

nuée ardente *(Geol.)*. A glowing cloud of gas and volcanic ash that moves rapidly downhill, as a density flow.

nugget *(Eng.)*. (1) Term for a welding bead. (2) A mass of gold or silver found free in nature.

Nujol *(Chem.)*. Trade name for a pure paraffin oil free of unsaturated compounds and used especially for making mulls for infrared spectroscopy.

null indicator *(Elec.Eng.)*. Any device for determining zero signal in a specified part of an electric circuit. Often used to determine balance conditions in bridge networks.

nullipara *(Med.)*. A woman who has never given birth to a child. adj. *nulliparous.*

nullisomic *(Bot.)*. Cell, or organism having one particular chromosome of the normal complement not represented at all in an otherwise diploid (or, more generally, euploid) complement. See **aneuploidy**, of which nullisomy is one sort.

null method *(Elec.Eng.)*. Same as **zero method**.

number *(Maths.)*. According to context, may mean natural number, integer, rational number, real number or complete number.

number sizes *(Eng.)*. A series of drill sizes in which gauge numbers represent diameters, the higher numbers representing the smaller diameters, in irregular increments.

numbers, pyramid of *(Ecol.)*. The relative decrease in numbers at each stage in a food chain, characteristic of animal communities.

numerable set *(Maths.)*. See **denumerable set**.

numerator *(Maths.)*. See **division**.

numerical analysis *(Maths.)*. The derivation of a particular numerical solution or set of solutions arising from the input of a single set of data into a mathematical model, as opposed to a general algebraic analytical solution.

numerical aperture *(Phys.)*. Product of the refractive index of the object space and the sine of the semiaperture of the cone of rays entering the entrance pupil of the objective lens from the object point. The resolving power is proportional to the numerical aperture. Abbrev. *NA*.

numerical control *(Eng.)*. The operation of machine tools from numerical data which have been recorded and stored on magnetic or punched tape or on punched cards.

numerical forecast *(Meteor.)*. A forecast of the future state of the atmosphere made by solving the equations of a **numerical forecast model** on a digital computer. The initial data are obtained from an **objective analysis**.

numerical forecasting *(Meteor.)*. A method of weather forecasting based upon a large number of observations at the surface of the earth and throughout the atmosphere and the calculation, electronically, from these of the conditions which should follow in accordance with the known laws of physics.

numerical forecast(ing) model *(Meteor.)*. A set of differential equations, with suitable boundary conditions, for the production of a **numerical forecast** and usually describing an artificially simplified atmosphere.

numerical selector *(Telecomm.)*. A group selector which connects the caller to the called subscriber within the latter's own exchange.

numerical taxonomy *(Bot.)*. A series of methods, based on the analysis of numerical data, for generating a classification (often in the form of dendrograms) of a group or groups of organisms. See **operational taxonomic units**.

numeric keypad *(Comp.)*. Has only numeric keys, often provided in addition to a **QUERTY keyboard** to speed numerical data entry.

nummulites *(Geol.)*. A group of extinct Foraminifera which were important rock forming organisms in the early Tertiary period. A nummulitic limestone is one which is composed mainly of their skeletal remains.

nunatak *(Geol.)*. An isolated mountain peak which projects through an ice sheet.

nu nu (nude) *(Immun.)*. Applies to mice or rats with congenital absence of the thymus, and in which the blood and the thymus dependent areas of lymphoid tissues are very severely depleted of T-lymphocytes. These animals are homozygous for the gene 'nude', hence *nu nu*, and have no hair. (Note that other hairless strains exist which have normal thymuses.) Such animals are unable to mount thymus-dependent immune responses but can make normal antibody responses to a wide variety of antigens which are 'thymus independent'.

nuptial flight *(Zool.)*. The flight of a virgin Queen Bee, during which she is followed by a number of males, copulation and fertilization taking place in mid-air.

nurse cells *(Zool.)*. Cells surrounding, or attached to, an ovum, probably to perform a nutritive function.

Nusselt number *(Phys.)*. The significant non-dimensional parameter in convective heat loss problems, defined by $Qd/k\Delta\theta$, where Q is rate of heat lost per unit area from a solid body, $\Delta\theta$ is temperature difference between the body and its surroundings, k is the thermal conductivity of the surrounding fluid and d is the significant linear dimension of the solid.

nut *(Bot.)*. A hard, dry, indehiscent fruit formed from a syncarpous gynaeceum, and usually containing one seed. The term is used loosely for any fairly large to large hard, dry, one seeded fruit.

nut *(Eng.)*. A metal collar, screwed internally, to fit a bolt; usually hexagonal in shape, but sometimes square or round. See **castle-**, **lock-**.

nut *(Print.)*. See **en quad**.

nutating feed *(Radar)*. That to a radar transmitter which produces an oscillation of the beam without change in the plane of polarization. The resulting radiation field is a *nutation field.*

nutation *(Astron.)*. An oscillation of the earth's pole about the mean position. It has a period of about 19 years, and is superimposed on the precessional movement.

nutation *(Bot.)*. Autonomic swaying movement of e.g. a growing shoot tip, especially *circumnutation.*

nutation *(Phys.)*. The periodic variation of the inclination of the axis of a spinning top (or gyroscope) to the vertical.

nutation field *(Radar)*. See **nutating feed**.

nutlet *(Bot.)*. A small, 1-seeded portion of a fruit which divides up as it matures. See also **schizocarp**.

NU tone *(Telecomm.)*. Number unobtainable tone.

nutrient *(Med.)*. Conveying, serving as, or providing nourishment. Nourishing food.

nutrient film technique *(Bot.)*. Method of growing plants without soil, their roots in a gutter-like channel with a nutrient solution trickling over them; used commercially for growing e.g. tomatoes in greenhouses.

nutrient solution *(Bot.)*. An artificially prepared solution containing some or all of the mineral substances used by a plant in its nutrition. Also *culture medium.*

nutrition *(Zool.)*. The process of feeding and the subsequent digestion and assimilation of food material. adj. *nutritive.*

nutritional encephalomalacia *(Vet.)*. *Crazy chick disease.* A nervous disease of young chicks characterized by loss of balance, inability to stand, and other nervous symptoms, associated with oedema, haemorrhages and degenerative changes in the brain; caused by vitamin E deficiency.

nutritional roup *(Vet.)*. An affection of the upper respiratory tract of the chicken, caused by vitamin A deficiency, and characterized by a discharge from the eyes and nostrils.

nut runner *(Eng.)*. A power tool or head fitted with a socket bit and adapted to drive and tighten a nut, often at controlled torque.

Nuvistor *(Electronics)*. TN for a subminiature electron tube.

NW *(Build.)*. An abbrev. for *Narrow Widths.*

N-wave *(Acous.)*. See **supersonic boom**.

nyctanthous *(Bot.)*. Flowering at night.

nyctinastic movement, nyctinasty *(Bot.)*. *Sleep movement.* A **nastic movement** in which plant parts, especially flowers and leaves, take up different position by night and day.

nyctipelagic *(Zool.)*. Found in the surface waters of the sea at night only.

Nylander solution *(Chem.)*. An alkaline solution of bismuth subnitrate and Rochelle salt, giving a black colour on boiling with glucose; used to detect glucose in urine.

Nylatron *(Plastics)*. TN for a range of filled nylon compounds reinforced for engineering purposes.

nylon *(Plastics)*. Generic name for any long chain synthetic polymeric amide which has recurring amide groups as an integral part of the main polymer chain, and which is capable of being formed into a filament in which the structural elements are oriented in the direction of the axis.

nylon *(Textiles)*. A synthetic linear polymer with repeating amide groups (-NH-CO-) and used in the manufacture of fibres. Two common examples are nylon 6 with the repeating unit [-NH-$(CH_2)_5$-CO-] and nylon 6:6 consisting of [–NH–$(CH_2)_6$–NH–$CO(CH_2)_4$–CO–]. Nearly all articles made of nylon contain one or other of these polymers.

nylon plates *(Print.)*. See **photopolymer plates**.

nymph *(Zool.)*. In Acarina, the immature stage intervening between the period of acquisition of four pairs of legs and the attainment of full maturity: in Insecta, a young stage of Exopterygota intervening between the egg and the adult, and differing from the latter only in the rudimentary condition of the wings and genitalia.

Nyquist criterion *(Telecomm.)*. One governing the stability of an amplifier or system incorporating feedback. It demands that the **Nyquist diagram** shall not enclose the point $X = -1$, $Y = 0$, where $\mu/\beta = X + jY$.

Nyquist diagram *(Telecomm.)*. For a feedback amplifier or system, a plot in rectangular co-ordinates of the factor μ/β for frequencies from zero (dc) to infinity, where μ is the gain without feedback and β is the fraction of the output voltage superimposed by the feedback loop onto the amplifier input.

Nyquist limit, rate *(Telecomm.)*. Maximum rate of transmitting pulses through a system. If B is the effective bandwidth in Hz, then $2B$ is the maximum number of code elements per second (bauds) which can be received with certainty. $1/2B$ is known as the *Nyquist interval*. The Nyquist *rate* effects the minimum sampling rate allowable for accurate reconstruction of a signal in a pulse-coded system; if f is the maximum frequency occurring in the transmitted signal, then a minimum sampling rate of $2f$ is demanded.

Nyquist noise theorem *(Electronics)*. One by which the thermal noise power P in a resistor at any frequency can be calculated from Boltzmann and Planck's constants k and h and the temperature. At normal temperatures it reduces to $P = kTdf$, where T is the temperature in K and df is the frequency interval.

nystagmoid *(Med.)*. Resembling nystagmus.

nystagmus *(Med.)*. An abnormal and involuntary movement of the eyeball seen as a flicking backwards and forwards when the eye is deviated.

nystatin *(Med.)*. An antifungal antibiotic, produced by a strain of *Streptomyces noursei*. Used in treatment of moniliasis and candidiasis and as a fungicide in tissue culture preparations.

NZB, NZW mice *(Immun.)*. Inbred strains of mice which develop spontaneous autoimmune diseases, including anaemia, glomerulonephritis and a condition resembling systemic lupus erythematosus. There is evidence that there is an underlying viral aetiology.

O

o, O, ~ *(Maths.).* These symbols are defined as follows:

$$f(x) = \mathrm{o}\{\varphi(x)\} \quad \text{if} \frac{f(x)}{\varphi(x)} \to 0$$

$$f(x) = \mathrm{O}\{\varphi(x)\} \quad \text{if} \, |f(x)| < K\varphi(x)$$

$$f(x) \sim \varphi(x) \quad \text{if} \frac{f(x)}{\varphi(x)} \to 1,$$

where the limiting value of x is either stated explicitly or implied. E.g. if $f(x) = x^2 + \frac{1}{x}$, then $f(x) = O\left(\frac{1}{x}\right)$ if x is small, and $f(x) = O(x^2)$ if x *is large.*

o- *(Chem.).* An abbrev. for *ortho-*, i.e. containing a benzene nucleus substituted in the 1.2 positions.

ω *(Chem.).* A symbol for specific magnetic rotation.

ω *(Maths.).* Symbol for solid angle.

ω *(Phys.).* Symbol for (1) angular frequency, (2) angular velocity, (3) dispersive power, (4) pulsatance.

ω- *(Chem.).* A symbol indicating: (1) substitution in the side chain of a benzene derivative; (2) substitution on the last carbon atom of a chain, farthest from a functional group.

O *(Chem.).* Symbol for *oxygen.*

O- *(Chem.).* A symbol indicating that the radical is attached to the oxygen atom.

Ω *(Maths.).* Symbol for solid angle.

Ω *(Phys.).* Symbol for (1) ohm, (2) angular velocity.

-Ω *(Chem.).* A symbol for the ultimate disintegration product of a radioactive series.

oak *(For.).* A stong, tough and heavy hardwood, very durable in exposed positions. Commonly used in constructional work for timber bridges, heavy framing and piles, as well as for joinery.

oakum *(Build.).* Tarred, untwisted rope or hemp used for caulking joints.

O and C building *(Space).* Abbrev. for the *Operations and Check-out Building* at Kennedy Space Center, Florida, used for the integration and check-out of payload elements destined for the Space Shuttle.

O antigen *(Immun.).* The somatic antigen of Gram negative bacteria, e.g. genera *Escherichia, Shigella, Salmonella* etc. Species specific cell wall antigen with an outer side chain of repeating units of oligosaccharides which confers strain specificity. The polysaccharide is internally linked to lipid A forming **lipopolysaccharide**.

ob- *(Bot.).* Prefix meaning *reversed, turned about.* Thus *obclavate* is reversed *clavate*, i.e. attached by the broad and not the narrow end.

OB *(Image Tech.). Outside Broadcast*, but used for any video production work on **location**, whether immediately transmitted or not.

Obach cell *(Elec.Eng.).* A dry primary cell having ingredients similar to the Leclanché cell, i.e. a carbon positive electrode surrounded by a depolarizer of manganese dioxide paste, and having the electrolyte in the form of a paste of sal-ammoniac, plaster of Paris, and flour.

obconic, obconical *(Bot.).* Cone-shaped but attached by the point.

obdiplostemonous *(Bot.).* Having stamens in two whorls, those in the outer opposite to the petals, e.g. *Geranium.* Cf. **diplostemonous**.

obduction *(Geol.).* The process during plate collision whereby a piece of the subducted plate is broken off and pushed on to the overriding plate.

obeche *(For.). Triplochiton Scleroxylon*, a tree yielding lightweight wood used for food containers etc. Also called *African whitewood, wawa.*

obelisk *(Arch.).* A slender stone shaft, generally monolithic, square in section, and tapering towards the top, which is surmounted by a small pyramid.

obesity *(Behav.).* An excessive proportion of fat on the body; cultural norms vary. It is produced by a large range of factors including faulty metabolism and abnormal behaviour.

obex *(Med.).* A triangular area of grey matter and thickened ependyma in the **medulla oblongata**, in the roof of the fourth ventricle over the region of the *calamus scriptorius.*

object deck *(Comp.).* A stack of **punched cards** containing an **object program**. Usually prepared from a source deck. Obsolescent.

objective *(Phys.). Objective lens.* Usually the lens of an optical system nearest the object. Abbrev. *OG*, objective glass for microscope work.

objective analysis *(Meteor.).* A method of processing the original observations by computer to give all values of the atmospheric variables needed to produce a numerical forecast, in contrast to a subjective analysis made by scrutinizing observations plotted on charts. The computer programs, after checking for transmission and other errors, produce interpolated grid-point values dynamically consistent with the equations of the forecast model. There are often options to allow manual incorporation of late observations and other data not easily processed automatically.

objective noise meter *(Acous.).* Sound level meter in which noise level to be measured operates a microphone, amplifier, and detector, the last named indicating noise level on the phon scale. The apparatus is previously calibrated with known intensities of the *reference tone*, 1 kHz, suitable weighting networks and an integrating circuit being incorporated in the amplifier to simulate relevant properties of the ear in appreciating noise.

objective prism *(Astron.).* A narrow angle (ca 1°) prism placed in front of the objective lens or primary mirror of a telescope. Causes the image of each star to give a small spectrum. Those of many objects within a small field can thus be recorded simultaneously, either photographically or (by **optical fibres**) photoelectrically.

object permanence, constancy *(Behav.).* Knowledge of the continued existence of an object, even when the object is not accessible to direct sensory awareness. According to Piaget this ability does not develop until infants are approximately 8 months or more.

object program *(Comp.).* Translated version of a program which has been assembled or compiled.

oblate *(Genrl.).* Globose, but noticeably wider than long.

oblate ellipsoid *(Maths.).* An ellipsoid obtained by rotating an ellipse about its minor axis, i.e. like a squashed sphere. Cf. **prolate ellipsoid**.

obligate *(Bot.,Zool.).* Obliged to function in the way specified: e.g. an obligate anaerobe cannot grow (and may not survive) in the presence of free oxygen and an obligate parasite cannot live ouside its host. Cf. **facultative**.

obligate parasite *(Bot.).* A parasite capable of living naturally only as a parasite. Cf. **facultative parasite**. A more useful distinction is between **biotrophs** and **necrotrophs**.

obligate saprophyte *(Biol.).* An organism which lives on dead organic material and cannot attack a living host.

oblique aerial photograph *(Surv.).* A photograph taken from the air, for purposes of aerial survey work, with the optical axis of the camera inclined from the vertical, generally at some predetermined angle.

oblique arch *(Civ.Eng.).* An arch whose axis is not normal to the face.

oblique axes *(Maths.)*. Co-ordinate axes which are not mutually perpendicular.
oblique circular cylinder *(Maths.)*. See **cylinder**.
oblique system *(Crystal.)*. See **monoclinic system**.
obliquity factor *(Phys.)*. See **inclination factor**.
obliquity of the ecliptic *(Astron.)*. The angle at which the celestial **equator** intersects the **ecliptic**. At present this angle is slowly decreasing by 0.47 arc seconds a year, due to **precession** and **nutation**. It varies between 21°55′ and 24°180′. Its value in 2000 will be about 23°26′34″.
obliquus *(Zool.)*. An asymmetrical or obliquely placed muscle.
oblongata *(Zool.)*. See **medulla oblongata**.
OBM *(Surv.)*. Abbrev. for *Ordnance Bench Mark*.
obovate *(Bot.)*. Having the general shape of the longitudinal section of an egg, not exceeding twice as long as broad, and with the greatest width slightly above the middle; hence, attached by the narrower end.
obovoid *(Bot.)*. Solid, egg-shaped and attached by the narrower end.
obscuration *(Phys.)*. Fraction of incident radiation which is removed in passing through a body or a medium.
obscured glass *(Glass)*. Glass which is so treated as to render it translucent but not transparent.
observational learning *(Behav.)*. Learning through the observation of another individual (model) which is accomplished without practice or direct experience.
obsession *(Behav.)*. The morbid persistence of an idea in the mind, against the wish of the obsessed person.
obsessive-compulsive disorder *(Behav.)*. A disorder whose main symptoms are *obsessions* and **compulsions**.
obsidian *(Geol.)*. A volcanic glass of granitic composition, generally black with vitreous lustre and conchoidal fracture; occurs at Mt Hecla in Iceland, in the Lipari Isles, and in Yellowstone Park, U.S. A green silica glass found in ploughed fields in Moravia is cut as a gemstone and sold under the name *obsidian*. True obsidian is used as a gemstone and is often termed *Iceland agate*.
obstetrician *(Med.)*. A medically qualified person who practises obstetrics.
obstetrics *(Med.)*. That branch of medical science which deals with the problems and management of pregnancy and labour.
obstipation *(Med.)*. Severe and intractable constipation.
obstruction lights *(Aero.)*. Lights fixed to all structures near airports which constitute a hazard to aircraft in flight.
obstruction markers *(Aero.)*. See **airport markers**.
obstruent *(Med.)*. Obstructing; that which obstructs; an astringent drug.
obturator *(Zool.)*. Any structure which closes off a cavity, e.g. all the structures which close the large oval foramen formed by the ischio-pubic fenestra; the foramen itself.
obtuse *(Bot.)*. The blunt tip of a leaf, etc. with the sides forming an angle of more than 90°. Cf. **acute**.
obtuse angle *(Maths.)*. An angle greater than 90° and less than 180°.
obvolvent *(Zool.)*. Folded downwards and inwards, as the wings in some Insects.
occipital condyle *(Zool.)*. In Insects, a projection from the posterior margin of the head which articulates with one of the lateral cervical sclerites. In Craniata, one or two projections from the skull which articulate with the first vertebra.
occipitalia *(Zool.)*. A set of cartilage bones forming the posterior part of the brain-case in the Vertebrate skull.
occiput *(Zool.)*. The occipital region of the Vertebrate skull forming the back of the head. adj. *occipital*.
occlusion *(Chem.)*. The retention of a gas or a liquid in a solid mass or on the surface of solid particles; especially the retention of gases by solid metals.
occlusion *(Meteor.)*. The coming together of the **warm** and **cold fronts** in a depression so that the warm air is no longer in contact wth the earth's surface. If the *warm frontal zone* is cut off from contact with the surface it is a *cold occlusion*. If the *cold frontal zone* is cut off from contact with the surface it is a *warm occlusion*.
occlusion *(Zool.)*. Closure of a duct or aperture, e.g. the upper and lower teeth of a Vertebrate.
occlusor *(Zool.)*. A muscle which by its contraction closes an operculum or other movable lidlike structure.
occultation *(Astron.)*. The hiding of one celestial body by another interposed between it and the observer, as the hiding of the stars and planets by the moon, or the satellites of a planet by the planet itself.
occulting light *(Ships)*. A navigation mark identified during darkness by a light extinguished for short periods in a distinctive pattern. See **alternating light**, **flashing light**.
oceanic crust *(Geol.)*. That part of the earth's crust which is normally characteristic of oceans. In descending vertical section, it consists of approximately 5 km of water, 1 km of sediments and 5 km of basaltic rocks.
oceanite *(Geol.)*. A type of basaltic igneous rock occurring typically in the oceanic islands as lava flows; characterized by a higher percentage of coloured silicates (olivine and pyroxene), and a lower percentage of alkalis, than in normal basalt.
ocellus *(Zool.)*. A simple eye or eye-spot in Invertebrates: an eye-shaped spot or blotch of colour. adj. *ocellate*.
ochrea, ocrea *(Bot.)*. A sheath around the base of an internode formed from united stipules or leaf bases, e.g. dock (Rumex) and other Polygonaceae.
ochroleucous *(Bot.)*. Yellowish-white.
ochrophore *(Zool.)*. See **xanthophore**.
ochrosporous *(Bot.)*. Having yellow or yellow-brown spores.
OCR *(Comp.)*. See **optical character recognition**.
octa- *(Chem.)*. Containing 8 atoms, groups etc.
octahedral system *(Crystal.)*. See **cubic system**.
octahedrite *(Min.)*. A form of *anatase*, crystallizing in tetragonal bipyramids.
octahedrites *(Min.)*. A class of iron meteorites showing an octahedral internal structure.
octahedron *(Crystal.)*. A form of the cubic system which is bounded by 8 similar faces, each being an equilateral triangle with plane angles of 60°. pl. *octahedra*.
octal notation *(Comp.)*. System of representing numbers in base eight with the digits 0,1,2,...,7.
octane *(Chem.)*. C_8H_{18} alkane hydrocarbon. There are 18 compounds of this formula. Straight chain or normal octane, a colourless liquid, bp 126°C, rel.d. 0.702 at 20°, is found in petroleum.
octane number *(Autos.)*. The percentage, by volume, of *iso*-octane (2,2,4-trimethylpentane) in a mixture of *iso*-octane and *normal* heptane which has the same knocking characteristics as the motor fuel under test; it serves as an indication of the knock-rating of a motor fuel. See also **lead tetraethyl**.
octastyle *(Arch.)*. A building having a colonnade of 8 columns in front.
octavalent *(Chem.)*. Capable of combining with eight atoms of hydrogen or their equivalent. Having an oxidation or co-ordination number of eight.
octave *(Acous.)*. The interval between any 2 frequencies having the ratio 2:1.
octave *(Chem.)*. See **law of octaves**.
octave analyser *(Acous.)*. A filter in which the upper cut-off frequency is twice the lower.
octave filter *(Acous.)*. Bank of filters for analysing the spectral energy content of complex sounds and noises, using adjacent octave bands of frequency over the whole audio range. One-half and one-third octave filters are also used.
octavo *(Print.)*. The eighth of a sheet, or a sheet folded 3 times to make 8 leaves or 16 pages and written 8vo.
octet *(Chem.)*. See **electron octet**.
octodecimo *(Print.)*. See **eighteenmo**.
octopine *(Bot.)*. An **opine**.
octopod *(Zool.)*. Having 8 feet, arms or tentacles.
octyl alcohol *(Chem.)*. See **capryl alcohol**.
ocular *(Zool.)*. Pertaining to the eye; capable of being perceived by the eyes.
ocular micrometer *(Biol.)*. US for *eyepiece graticule*.
oculate *(Zool.)*. Possessing eyes; having markings which resemble eyes.

oculist *(Med.)*. One skilled in the knowledge and treatment of refractive diseases of the eye.

oculomotor *(Zool.)*. Pertaining to, or causing movements of, the eye: the third cranial nerve of Vertebrates, running to some of the muscles of the eyeball.

oculus *(Arch.)*. A round window.

ODA *(Comp.)*. See **office document architecture**.

odd-even check *(Telecomm.)*. See **parity checking**.

odd-even nuclei *(Phys.)*. Nuclei containing an odd number of protons and an even number of neutrons.

odd function *(Maths.)*. f is an odd function if $f(-x) = -f(x)$. Cf. **even function**.

odd legs *(Eng.)*. Callipers with two straight hinged legs, tapering distally from the hinge. One leg is curved distally and is placed against an edge. The other has a scribing point. Used mainly for locating centres from outside or inside edges and for other layout work.

odd-odd nuclei *(Phys.)*. Nuclei with an odd number of both protons and neutrons. Very few are stable.

odd parity *(Comp.)*. Binary representation in which the number of 1s is odd. See **parity check**.

odds *(Stats.)*. The ratio of the probability that an event occurs to the probability that it does not occur.

odd sorts *(Print.)*. A general term for any unusual characters required for technical or foreign language setting.

odds ratio *(Stats.)*. The ratio of two odds, used particularly in comparing and modelling conditional probabilities.

odograph *(Surv.)*. A *pedometer*.

odometer *(Surv.)*. See **perambulator**.

Odonata *(Zool.)*. Order of primitive insects with 2 pairs of similar membranous, many-veined wings. Large-eyed diurnal forms with both adults and larvae predatious. Dragonflies, Damselflies.

odontalgia *(Med.)*. Toothache.

odontic *(Med.)*. Pertaining to the teeth.

odontoblast *(Zool.)*. A dentine-forming cell, one of the columnar cells lining the pulp-cavity of a tooth.

odontoclast *(Zool.)*. A dentine-destroying cell, one of the large multinucleate cells which absorb the roots of the milk-teeth in Mammals.

odontogeny *(Zool.)*. The origin and development of teeth.

odontograph *(Eng.)*. An approximate but practical guide for setting out the profiles of involute gear teeth, in which a pair of circular arcs is substituted for the true involute curve.

odontoid *(Bot.,Zool.)*. Toothlike.

odontoid process *(Zool.)*. A process of the anterior face of the centrum of the axis vertebra which forms a pivot on which the atlas vertebra can turn.

odontolite *(Min.)*. See **bone turquoise**.

odontology *(Med.)*. Study of the physiology, anatomy, pathology etc. of the teeth.

odontoma *(Med.)*. Any of a variety of tumours that arise in connection with the teeth.

odontophore *(Zool.)*. In Molluscs, a feeding organ comprising the radula and radula sac, with muscles and cartilages.

odontostomatous *(Zool.)*. Having jaws which bear teeth.

odorimetry *(Chem.)*. Measurement of the intensity and permanency of odours. Also called *olfactometry*.

odoriphore *(Chem.)*. A group of atoms which confer an odour on a compound.

oedema *(Med.)*. Pathological accumulation of fluid in the tissue spaces and serous sacs of the body; sometimes the term is restricted to such accumulation in tissue spaces only: *pulmonary oedema*, fluid in the lung; *sacral oedema*, fluid at base of spine.

oedema disease *(Vet.)*. See **bowel oedema disease**.

oedematous *(Med.)*. Affected by oedema.

Oedipus and **Electra complexes** *(Behav.)*. In psychoanalytic theory, refers to unconscious conflicts, occurring during the **phallic stage** of psychosexual development, that centre around the relations a child forms to his parents. A fantasized form of sexual love for the opposite sex develops and a resentment of the same sex parent. In boys, these possessive feelings and the associated conflicts of guilt, are called the **Oedipus complex**; in girls, the **Electra complex**.

OEM *(Comp.)*. *Original Equipment Manufacturer*. A firm which makes basic computer hardware for other manufacturers to build into their products (e.g. microprocessors supplied to a washing machine manufacturer to use as control devices).

oersted *(Phys.)*. CGS electromagnetic unit of magnetic field strength, such that 2π oersted is a field at the centre of a circular coil 1 cm in radius carrying a current of 1 abampere (10 ampere). Now replaced by the SI unit ampere per metre. $1\,A\,m^{-1} = 4\pi \times 10^{-3}$ oersted.

oesophagectasis, oesophagectasia *(Med.)*. Pathological dilatation of the oesophagus.

oesophagectomy *(Med.)*. Removal of the oesophagus or part of it.

oesophagitis *(Med.)*. Inflammation of the oesophagus.

oesophagoscope *(Med.)*. An instrument for viewing the interior of the oesophagus.

oesophagospasm *(Med.)*. See **achalasia of the cardia**.

oesophagostomy *(Med.)*. Formation of an artificial opening into the oesophagus.

oesophagotomy *(Med.)*. Incision into the oesophagus.

oesophagus, esophagus *(Zool.)*. In Vertebrates, the section of the alimentary canal leading from the pharynx to the stomach; usually lacking a serous coat and digestive glands; the corresponding portion of the alimentary canal in Invertebrates. adj. *oesophageal*.

oestradiol *(Med.)*. Hormone secreted by the ovarian follicle (follicular hormone) and responsible for the development of the sexual characteristics of the female. Oestradiol ($C_{18}H_{24}O_2$) is a sterol.

oestriasis *(Vet.)*. Infection of the nasal cavity and sinuses of sheep and goats by larvae of the sheep nasal fly *Oestrus ovis*. A form of **myiasis**.

oestriol *(Med.)*. A female sex hormone (related to the sterols) found in the urine in pregnancy, and consisting of one phenolic hydroxyl radical and two secondary alcohol groups ($C_{18}H_{24}O_3$).

oestrogen *(Med.)*. The generic term for a group of female sex hormones, which induce oestrus. Oestradiol (ethinyloestradiol) and mestranol are the common oestrogen components of some oral contraceptives. Cf. **androgen**.

oestrous cycle *(Zool.)*. In female Mammals, the succession of changes in the genitalia commencing with one oestrous period and finishing with the next.

oestrus, oestrum *(Zool.)*. In female Mammals, the period of sexual excitement and acceptance of the male occurring between pro-oestrum and metoestrum; more generally, the period of sexual excitement. adj. *oestral*.

off-air *(Image Tech.)*. Received from a broadcast transmission.

offhand grinding *(Eng.)*. Grinding in which the tool, e.g. a cut-off saw or a grinding wheel, is held in the hand and worked freely.

office document architecture *(Comp.)*. A standard for the storage, manipulation, display and transmission of computerized documents in an office environment. Abbrev. *ODA*.

official, officinal *(Bot.)*. Used in medicine.

off-its-feet *(Print.)*. Describes type which is lying at an angle and does not give a level print.

off-lap *(Geol.)*. The dispositional arrangement of a series of conformable strata laid down in the waters of a shrinking sea, or on the margins of a rising landmass, so that the successive strata cover smaller areas than their predecessors. Cf. **overlap**.

offlet *(Build.)*. See **grip**.

off line *(Comp.)*. Describes processing carried out by devices not under the control of the **central processor**.

off-line editing *(Image Tech.)*. Videotape editing involving an intermediate transfer to a low-cost system solely for preliminary selection and arrangement prior to the final **conforming** of the original material.

off-peak load *(Elec.Eng.)*. Load on a generating station or power supply system taken at times other than the time of the system peak load, e.g. during the night.

offprints *(Print.)*. Separately printed copies of articles that have appeared in periodicals.

offset *(Bot.)*. Short shoot arising from an axillary bud near the base of a shoot and producing a daughter plant at its apex, as e.g. in the houseleek, *Sempervivum*. Also, a bulbil or cormlet formed near base of parent bulb or corm.

offset *(Build.)*. A ledge formed at a place where part of a wall is set back from the face.

offset *(Surv.)*. The horizontal distance measured to a point from a main survey line, in a direction at right angles to a known point on the latter.

offset blanket *(Print.)*. The rubber-covered cylinder on which the image is taken from the plate cylinder and from which it is offset to the stock being printed, being the distinguishing feature of an offset press. See **offset printing**, **printing**.

offset printing *(Print.)*. Process in which the ink from a plate is received on a rubber-covered blanket cylinder from which it is transferred to the paper or other material, including sheet metal, the rubber surface making good contact even with rough surfaces. The standard method of lithographic printing, it is now being used to a lesser extent for relief printing, then called *letterset printing*.

offset rod *(Surv.)*. A wooden pole painted in bands of different colours of suitable length so that the pole may be used for the measurement of short distances.

offset scale *(Surv.)*. Short graduated scale used in plotting detail from field notes of *offsets*.

offset well *(Min.Ext.)*. Well drilled near to an existing well to further explore or exploit a field.

OFHC *(Eng.)*. Abbrev. for *Oxygen-Free High-conductivity Copper*.

ogee, ogival arch *(Arch.)*. A pointed arch of which each side consists of a reverse curve.

ogee wing *(Aero.)*. A wing of ogee plan form and very low aspect ratio which combines low *wave drag* in supersonic flight with high lift at high incidence through separation vortices at low speed.

OG, ogee *(Arch.)*. See **cyma**.

ohm *(Phys.)*. SI unit of electrical resistance, such that 1 ampere through it produces a potential difference across it of 1 volt. Symbol Ω.

ohm-cm *(Phys.)*. The CGS unit of *resistivity*.

ohmic contact *(Elec.Eng.)*. One in which the voltage appearing across the contact is proportional to the current flowing for both directions of current flow.

ohmic loss *(Elec.Eng.)*. Power dissipation in an electrical circuit arising from circuit resistance and current flow, i.e. I^2R loss.

ohmic, nonohmic *(Elec.Eng.)*. A resistor is said to be *ohmic* or *nonohmic*, according to whether or not its resistance is described by Ohm's law. See **linear resistor**.

ohmic resistance *(Elec.Eng.)*. See **d.c. resistance**.

ohmic resistor *(Elec.Eng.)*. See **linear resistor**.

ohm-metre *(Phys.)*. SI unit of *resistivity*.

Ohm's law *(Phys.)*. In metallic conductors, at constant temperature and zero magnetic field, the current I flowing through a component is proportional to the potential difference V between its ends, the constant of proportionality being the *conductance* of the component. So $I = V/R$ or $V = IR$, where R is the *resistance* of the component. Law is strictly applicable only to electrical components carrying direct current and for practical purposes to those of negligible reactance carrying alternating current. Extended by analogy to any physical situation where a pressure difference causes a flow through an impedance, e.g. heat through walls, liquid through pipes.

Ohm's law of hearing *(Acous.)*. Law of psychoacoustics. A simple harmonic motion of the air is appreciated as a simple tone by the human ear. All other motions of the air are analysed into their harmonic components which the ear appreciates as such separately.

-oid *(Genrl.)*. Suffix. after Gk., *oides*, from *eidos*, form.

oil *(Genrl.)*. See **oils**.

oil absorption *(Chem.)*. A term usually applied to pigments. The amount of linseed oil a pigment will absorb to reach a given consistency as determined by certain standards.

oil-based mud *(Min.Ext.)*. Used when drilling very hot formations or through water absorbing strata. Also faster but dirtier drilling.

oil-blast circuit-breaker *(Elec.Eng.)*. A form of oil circuit-breaker in which pressure set up by the gases produced as a result of the arc causes a blast of oil across the contact space, which ensures rapid extinction of the arc.

oil-break *(Elec.Eng.)*. A term applied to switches, circuit-breakers, fuses, etc., to indicate that the circuit is opened under oil.

oilcloth *(Textiles)*. A waterproof material obtained by coating a cotton fabric with oxidized linseed or other drying oil.

oil-control ring *(Autos.)*. See **scraper ring**.

oil-cooled *(Eng.)*. Said of apparatus which is immersed in oil to facilitate cooling.

oil cooler *(Aero.)*. See **fuel-cooled-**.

oil cooler *(Autos.)*. A small air-cooled radiator sometimes used for cooling the lubricant after its return from the engine and before delivery to the oil tank.

oil-dilution system *(Aero.)*. In a reciprocating aero-engine, a device for diluting the lubricant with fuel as the engine is stopped so that there is less resistance when starting in cold weather.

oiled paper *(Paper)*. Paper impregnated with non-drying oil, e.g. oiled wrapping, or with linseed oil, e.g. oiled manilla.

oil engine *(Autos.)*. See **compression-ignition engine**.

oil-filled cable *(Elec.Eng.)*. Cable with a central duct, formed by an open spiral of steel tape, through which oil is fed to it. Gaseous voids and the consequent ionization are thus eliminated. Used up to the highest voltages.

oilgas *(Chem.)*. A gas of high energy value, obtained by the destructive distillation of high-boiling mineral oils. It consists chiefly of methane, ethene, ethyne, benzene and higher homologues.

oil gland *(Zool.)*. The preen gland or uropygial gland of Birds, a cutaneous gland forming an oil secretion used in preening the feathers.

oil hardening *(Eng.)*. The hardening of cutting tools, etc., of high carbon content by heating and quenching in oil, resulting in a cooling less sudden than is effected by water, and reduced risk of cracking.

oil-hardening steel *(Eng.)*. Alloy steel which can be hardened by cooling in oil instead of in water. A typical example is carbon 0.3%, nickel 3.0% and chromium 0.75%.

oil-immersed *(Elec.Eng.)*. Said of electrical apparatus which is immersed in oil. See **oil-break**, **oil-cooled**, **oil-insulated**.

oil-immersion objective *(Biol.)*. The use of a thin film of oil, with the same refractive index as glass, between the objective of a microscope and the specimen. This permits the numerical aperture to exceed 1 and thus extends the resolving power and gives the maximum magnification obtainable with a light microscope.

oiling *(Textiles)*. Spraying wool, woollen blends, jute, and some man-made fibres with oil or oil/water emulsion during opening operations. It facilitates drawing, drafting, and spinning, and reduces static electricity and waste.

oiling ring *(Eng.)*. A simple device used to feed oil to a journal bearing. It consists of a light metal ring, larger in diameter than the shaft, and riding loosely thereon, located at the mid-point of the brasses, the upper brass being slotted to receive it. The ring dips into an oil reservoir in the base of the housing, and as it rotates, feeds oil to the brasses.

oil-insulated *(Elec.Eng.)*. Said of electrical apparatus which is immersed in oil to facilitate its insulation.

oil length *(Build.)*. Ratio of oil to resin in a varnish.

oil-less circuit-breaker *(Elec.Eng.)*. A circuit-breaker which does not use oil either as the quenching medium or for insulation purposes.

oil of bitter almonds *(Chem.)*. See **benzaldehyde**.

oil of cloves *(Chem.)*. See **clove oil**.

oil of turpentine *(Chem.)*. See **turpentine**.

oil of vitriol *(Chem.)*. An old name for **sulphuric acid**.

Oilostatic cable *(Elec.Eng.)*. TN for a paper-insulated power cable operated under hydrostatic pressure by means of oil contained in an outer steel pipe, in order to minimize ionization. Cf. **pressure cable**.

oil paints *(Build.)*. A paint in which the film former comprises drying oils, driers and resins, drying by oxidation.

oil-poor circuit-breaker *(Elec.Eng.)*. A circuit-breaker which employs a very much smaller quantity of oil (about 10%) than an ordinary oil circuit-breaker; e.g. certain types of impulse and expansion breakers.

oil pump *(Autos.)*. A small auxiliary pump, driven from an engine crankshaft, which forces oil from the sump or oil tank to the bearings; often of the gear type. See **gear pump**.

oil quench *(Eng.)*. Immersion, for purpose of tempering, of hot metal in oil.

oil-quenched fuse *(Elec.Eng.)*. A liquid-quenched fuse having oil as the quenching medium.

oils *(Chem.)*. A group of neutral liquids comprising 3 main classes: (*a*) *fixed* (*fatty*) *oils*, from animal, vegetable and marine sources, consisting chiefly of glycerides and esters of fatty acids; (*b*) *mineral oils*, derived from petroleum, coal, shale etc. consisting of hydrocarbons; (*c*) *essential oils*, volatile products, mainly hydrocarbons, with characteristic odours, derived from certain plants.

oils *(Elec.Eng.)*. (Mineral oils, as used in transformers, cables, switchgear.) Class A oils have a maximum sludge of 0.1% and may be used in transformers above 80°C and in oil-switches above 70°C. Class B oils have a maximum of 0.8% and may be used in transformers up to 75°C.

oil-shale *(Geol.)*. An argillaceous sediment containing diffused **kerogen** in a state suitable for distillation into paraffin and other mineral oils by the application of heat. See **shale oils**.

oil sink *(Horol.)*. The spherical recess around a pivot hole in a watch or clock plate. Its purpose is to act as a reservoir for the oil.

oilstone *(Build.)*. A smooth stone used to impart a fine keen edge to a cutting tool, for which purpose it is first moistened with oil.

oilstone slip *(Build.)*. See **gouge slip**.

oil string *(Min.Ext.)*. See **production string**.

oil sump *(Autos.)*. See **sump**.

oil switch, circuit-breaker *(Elec.Eng.)*. The usual type of switch or circuit-breaker used on high and medium power a.c. circuits; the contacts are immersed in oil for insulating and arc-rupturing purposes.

oil tanker *(Ships)*. Ship for carrying oil in bulk. They are subdivided internally into a number of tanks by longitudinal and transverse bulkheads, with pumps for discharging.

oil varnishes *(Build.)*. Varnishes containing a drying oil, resin, and driers. The oils used are linseed, China wood, soya bean, poppy-seed, cotton-seed, and castor-oil.

Oklo *(Nuc.Eng.)*. Uranium mine in Gabon, West Africa where evidence of a one-time naturally occurring reactor was found. Some 1700 million years ago, when the ^{235}U isotope abundance was much higher than today the local concentrations of uranium were high enough, when moderated by water, to form critical assemblies in which fission took place over hundreds of thousands of years.

okta, octa *(Aero.,Meteor.)*. One-eighth of the sky area used in specifying cloud cover for airfield weather condition reports.

Olbers' Paradox *(Astron.)*. A paradox expressed in 1826 by Heinrich Olbers: 'why is the sky dark at night?'. In an infinitely large, unchanging, universe populated uniformly with stars and galaxies, the sky would be dazzling bright, which is not the case. This simple observation therefore implies that the universe is not an infinite static arrangement of stars. In fact, modern cosmology postulates a finite expanding universe.

old age *(Geol.)*. The final stage in the cycle of erosion in which base-level is nearly attained and the landscape has little relief.

Old English *(Print.)*. A black-letter type in which early books were printed.

old face *(Print.)*. The earliest form of roman type, as: Bembo.

Oldham coupling *(Eng.)*. A coupling permitting misalignment of the shafts connected. It consists of a pair of flanges whose opposed faces carry diametrical slots, and between which a floating disk is supported through corresponding diametral tongues which are arranged at right angles.

oldhamite *(Min.)*. Sulphide of calcium, usually found as cubic crystals in meteorites.

Old Red Sandstone *(Geol.)*. The continental facies of the **Devonian** System in the British Isles, comprising perhaps 12 000 m of red, brown, or chocolate sandstones, red and green marls, cornstones, breccias, flags and conglomerates, yielding on certain horizons the remains of archaic fishes, eurypterids, plants and rare shelly fossils.

Old Style *(Astron.)*. A name given to the system of date-reckoning superseded by the adoption of the **Gregorian calendar**.

old style *(Print.)*. A type-face imitating the old style of roman letter used until the end of the seventeenth century, as; Old Style No. 2.

old style figures *(Print.)*. See **hanging figures**.

olecranon *(Zool.)*. In land Vertebrates, a process at the upper end of the ulna which forms the point of the elbow.

olefin *(Textiles)*. Synthetic fibre or polyolefin fibre consisting of a long-chain polymer composed of at least 85% by weight of olefins, usually ethylene or propylene. Hence polyethylene and polypropylene are common.

olefins *(Chem.)*. See **alkenes**.

oleic acid *(Chem.)*. $C_{18}H_{34}O_2$, a colourless liquid, mp 14°C. It is an unsaturated acid of the formula $CH_3.(CH_2)_7.CH{=}{=}CH.(CH_2)_7.COOH$, with *cis* configuration about the double bond, and occurs as the glycerine ester in fatty oils. Oleic acid oxidizes readily on exposure to the air, turns yellow and becomes rancid.

olein *(Chem.)*. A glycerine ester of oleic acid.

oleo *(Aero.)*. Main structural member of the support assemblies of an aircraft's *landing gear*. Of telescopic construction and containing oil so that on landing the oil is passed under pressure through chambers at a controlled rate thereby absorbing the shock. Also referred to as *oleo leg*, *shock strut* and *shock absorber*.

oleo-pneumatic *(Aero.)*. Means of absorbing shock loads by a combination of air compression and oil pressure created by forcing the latter through an orifice.

oleo-resin *(Build.)*. Oily resinous sap of plants, partially deprived of volatile constituents.

oleoresinous *(Build.)*. Paints and varnishes the binder of which contains both oils and resins.

oleum *(Chem.)*. A commercial name for *fuming sulphuric* (*disulphuric*) *acid*, $H_2S_2O_7$.

olfactometry *(Chem.)*. See **odorimetry**.

olfactory *(Zool.)*. Pertaining to the sense of smell; the first cranial nerve of Vertebrates, running to the olfactory organ.

olfactory lobes *(Zool.)*. Part of the forebrain in Vertebrates, which is concerned with the sense of smell, and from which the olfactory nerves originate.

oligaemia, oligemia *(Med.)*. Diminution in the volume of the blood.

Oligocene *(Geol.)*. The youngest epoch of the **Paleogene** sub-period, the time succession being Paleocene, Eocene and Oligocene.

Oligochaeta *(Zool.)*. A class of Annelida with relatively few chaetae, not situated on parapodia; have a definite prostomium which usually has no appendages; always hermaphrodite; have only 1 or 2 pairs of male and female gonads in fixed segments of the anterior part of the body; reproduction involves copulation and cross-fertilization; the eggs are laid in a cocoon and develop directly; terrestrial and aquatic. Earthworms.

oligoclase *(Min.)*. One of the plagioclase feldspars, consisting of the Albite (Ab) and Anorthite (An) molecules combined in the proportions of $Ab_9\ An_1$ to

$Ab_7\ An_3$. It is found especially in the more acid igneous and metamorphic rocks.

oligodendroglia *(Zool.)*. Cells within the CNS which deposit the myelin sheath. Also *oligodendrocyte*.

oligodendroglioma *(Med.)*. A cerebral tumour composed of oligodendroglia.

oligolecithal *(Zool.)*. A type of egg in which there is little yolk, what there is being somewhat more concentrated in one hemisphere. Found in *Amphioxus* (a genus of Cephalochordata) and Mammals. Cf. **mesolecithal, telolecithal**.

olig-, oligo- *(Genrl.)*. Prefix from Gk. *oligos*, a few, small.

oligomenorrhoea, oligomenorrhea *(Med.)*. Scantiness of the discharge which occurs during menstruation.

oligomerous *(Bot.)*. Consisting of only few parts.

oligonucleotide *(Biol.)*: A nucleic acid with few nucleotides. Cf. **polynucleotide**.

oligopeptides *(Biol.)*. Short polymers of amino acids, less than about 100 peptides long. May be synthesized, but also occur naturally with often powerful biological effects.

oligopod *(Zool.)*. (1) Having few legs or feet. (2) Said of a phase in the development of larval Insects in which the thoracic limbs are large while the evanescent abdominal appendages of the polypod phase have disappeared.

oligosaccharides *(Chem.)*. Carbohydrates containing a small number (2 to 10) of monosaccharide units linked together, with elimination of water.

oligospermia *(Med.)*. Diminution, below the average, of the quantity of semen voided in an ejaculation.

oligotokous *(Zool.)*. Bearing few offspring. Cf. **polytokous**.

oligotrophic *(Ecol.)*. Said of a type of lake having steep and rocky shores and scanty littoral vegetation, and in which the hypolimnion does not become depleted of oxygen in the summer. The hypolimnion is generally larger than the epilimnion, and primary productivity is low. Cf. **dystrophic, eutrophic**.

oligotrophophyte *(Bot.)*. A plant growing in a soil poor in soluble mineral salts.

oligozoospermia *(Med.)*. Diminution of the number of spermatozoa in the semen.

oliguria *(Med.)*. Abnormally diminished secretion of urine.

oliphagous *(Zool.)*. Feeding on few different kinds of food; as phytophagous Insects which are limited to a few related food-plants. Cf. *polyphagous*.

olistostrome *(Geol.)*. A sediment which consists of a jumbled mass of heterogeneous blocks of material, generated by gravity sliding.

olivary nucleus *(Zool.)*. A wavy band of grey matter within the medulla oblongata of higher vertebrates.

olive *(For.)*. *Olea europaea*, found in the Mediterranean regions of Europe, important for its fruit which produces edible oil when compressed and is itself edible after processing with dilute NaOH to neutralize the bitter glucoside content.

olivenite *(Min.)*. A hydrated arsenate of copper which crystallizes in the orthorhombic system. It is a rare green mineral of secondary origin found in copper deposits.

olive oil *(Chem.)*. A pale-yellow or greenish oil obtained from the fruit of *Olea europaea*; rel.d. 0.91–0.92, acid value 1.9–5, saponification value 185–196, iodine value 77–88.

oliver filter *(Min.Ext.)*. Drum filter used in large-scale dewatering or filtration of mineral pulps, usually after **thickening**. Pulp is drawn by vacuum to filtering membrane as drum rotates slowly through a trough, and filtrate is drawn off while moist filter cake is scraped from down-running side of drum.

olivine *(Min.)*. Orthosilicate of iron and magnesium, crystallizing in the orthorhombic system, which occurs widely in the basic and ultramafic igneous rocks, including olivine-gabbro, olivine-dolerite, olivine-basalt, peridotites, etc. See **chrysolite**. The clear-green variety is used as a gemstone under the name *peridot*. For *iron olivine*, see **fayalite**, and for *magnesium olivine*, see **forsterite**.

Olsen ductility test *(Eng.)*. Test in which a standard size ball is forced into a blank of sheet metal and the depth of cupping measured when the metal fractures.

Olson microphone *(Acous.)*. The original **ribbon microphone** using a battery-excited magnet.

omasitis *(Vet.)*. Inflammation of the omasum.

omasum *(Zool.)*. See **psalterium**.

ombrogenous *(Bot.)*. A mire or bog receiving water only by precipitation and hence extremely oligotrophic. Cf. **soligenous**.

ombrogenous *(Ecol.)*. Obtaining nutrients from rain.

ombrophile, ombrophyte *(Bot.)*. A plant adapted to rainy places.

Omega *(Aero.)*. Long range radio navigation aid of very low frequency covering the whole Earth from 8 ground transmitters. It can be received down to sea level.

omega equation *(Meteor.)*. A diagnostic equation for the vertical velocity in **pressure coordinates** $\left(\frac{dp}{dt}\right)$ conventionally denoted by ω. It is obtained by eliminating the time derivatives from the thermodynamic and **vorticity equations** and applying the **quasi-geostrophic approximation**. With the omission of some small terms the co-equation may be written

$$\nabla^2(\sigma\omega) - \frac{f(\zeta+f)}{g}\frac{\partial^2 w}{\partial p^2} = \frac{1}{f}\nabla^2 J\left(\phi, \frac{\partial\phi}{\partial p}\right) - \frac{1}{g}J(\phi, g+f)$$

where σ is a measure of atmospheric stability, f is the **Coriolis parameter**, ζ the vertical component of relative vorticity, ϕ the geopotential, g the acceleration of gravity, and J indicates an operator such that

$$J(u,v) \equiv \frac{\partial u}{\partial x}\frac{\partial v}{\partial y} - \frac{\partial u}{\partial y}\frac{\partial v}{\partial x}.$$

omega-minus particle *(Phys.)*. The heaviest hyperon (1672 MeV), discovered in 1964. Its existence produces strong evidence for the classification, developed from *Lie algebra* (group theory), of those elementary particles which interact strongly.

omental bursa *(Zool.)*. In some Mammals, a pouch formed ventrally and dorsally to the stomach by the mesentery supporting the stomach.

omentopexy *(Med.)*. The stitching of the omentum to the abdominal wall in the treatment of cirrhosis of the liver.

omentum *(Zool.)*. In Vertebrates, a portion of the serosa connecting two or more folds of the alimentary canal. adj. *omental*.

ommatidium *(Zool.)*. One of the visual elements composing the compound eyes of *Arthropoda*.

ommatophore *(Zool.)*. An eye-stalk.

omnibus-bar *(Elec.Eng.)*. The original term from which the commonly used expression **bus-bar** is derived.

omnidirectional *(Space)*. Simple aerial, mounted on a spacecraft, radiating energy equally in all directions.

omnidirectional antenna *(Telecomm.)*. One receiving or transmitting equally in all directions in horizontal plane, although the radiation pattern may be directional in elevation.

omnidirectional microphone *(Acous.)*. One whose response is essentially independent of the direction of sound incidence. Also called *non-directional microphone*.

omnidirectional radio beacon *(Aero.)*. A VHF radio beacon radiating through 360° upon which an aircraft can obtain a bearing. Used for navigation by *VOR* with *DME*. Abbrev. *ORB*.

omnivore *(Ecol.)*. An animal which eats both plants and animals. adj. *omnivorous*.

omnivorous *(Bot.)*. Parasitic fungi, attacking several or many hosts.

omphacite *(Min.)*. An aluminous sodium-bearing pyroxene, occurring in eclogites as pale-green mineral grains.

omphalectomy *(Med.)*. Removal of the umbilicus.

omphalic *(Zool.)*. Pertaining to the umbilicus.

omphalitis *(Med.)*. Inflammation of the umbilicus.

omphaloid *(Zool.)*. Navel-shaped.

omphalophlebitis *(Vet.)*. Inflammation of the navel or umbilical cord. See **joint ill**.

once-through boiler *(Eng.)*. See **flash boiler**.

onchocerciasis *(Med.)*. Infestation of the skin and subcutaneous tissues with the nematode worm of the genus *Onchocerca volvulus* in West Africa. It causes skin tumours and eye diseases, including blindness.

oncogene *(Biol.)*. Genetic locus originally identified in RNA tumour viruses which is capable of the **transformation** of the host cell. Implicated as cause of certain cancers. See **proto-oncogene**.

oncogenic *(Med.)*. Inducing, or tending to induce, the formation of tumours.

oncogenic virus *(Biol.)*. Generally a virus able to cause cancer, but more specifically one carrying an **oncogene**.

oncology *(Med.)*. That part of medical science dealing with new growths (tumours) of body tissue.

on-costs *(Build.,Eng.)*. All items of expenditure that cannot be allocated to a definite job, i.e. all expenses other than the prime cost. Also called *overhead expenses, establishment charges, loading, burden*.

Ondiri disease *(Vet.)*. See **bovine infectious petechial fever**.

ondoscope *(Electronics)*. Glow tube operated by strong microwave radiation fields, and used, e.g. in tuning transmitters.

one-electron bond *(Chem.)*. Bond formed by the resonance of a single electron between 2 atoms or radicals of similar energy.

one-group theory *(Nuc.Eng.)*. Greatly simplified reactor model in which all neutrons are regarded as having the same energy. See **multigroup theory**.

one-hour rating *(Elec.Eng.)*. A form of rating commonly used for electrical machinery supplying an intermittent load, e.g. traction motors for suburban service. It indicates that the machine will deliver its specified rating for a period of 1 h without exceeding the specified temperature rises.

one-inch *(Image Tech.)*. 1", a gauge (width) of professional videotape.

one-light *(Image Tech.)*. Describes printing a complete reel of motion picture film at a single level of exposure, without variations from one scene to another.

one-particle model of a nucleus *(Phys.)*. A form of **shell model** in which nuclear spin and magnetic moment are regarded as associated with one resident nucleon.

one-phase *(Elec.Eng.)*. See **single-phase**.

one-pipe system *(Build.)*. That in which both soil and waste are carried by a common pipe, fittings being protected with deep seal traps.

ONERA *(Aero.)*. Abbrev. for *Office National d'Etudes et de Recherches Aerospatiales*, Fr.

one's complement *(Comp.)*. Formed from a binary number by changing each 1 bit to a 0 bit and vice versa. Thus the one's complement of 00101011 (=43 base 10) is 11010100. Cf. **two's complement**.

one-shot circuit *(Telecomm.)*. Same as *single circuit*.

one-shot multivibrator *(Electronics)*. Same as *monostable multivibrator*.

one-to-one *(Maths.)*. A function F from a set S to a set T is said to be a one-to-one (or one-one) map, mapping, function or transformation from S *into* T if no two elements of S are mapped to the same element of T. It is a one-to-one mapping *onto* T if every element of T has some element of S mapped to it.

one-to-one transformer *(Elec.Eng.)*. A transformer having the same number of turns on the primary as on the secondary; used in circuits when it is desired to insulate one part of a circuit from another.

one-way switch *(Elec.Eng.)*. A switch providing only one path for the current. Cf. **double-throw switch** (2-way switch).

onion skin paper *(Paper)*. A glazed translucent paper having an undulating surface caused by the special beating and drying of the paper.

on-lap *(Geol.)*. An **unconformity** above which beds are successively pinched out by younger beds. Largely synonymous with **overlap**, and the reverse of **off-lap**.

on line *(Comp.)*. Describes processing performed on equipment directly under the control of the **central processor**, while the user remains in communication with the computer.

on-line editing *(Image Tech.)*. Videotape editing using the original recordings on their own standard equipment for all stages of selection and arrangement as well as the final **conforming**.

on-off control *(Telecomm.)*. A simple control system which is either on or off, with no intermediate positions. See **continuous control**.

on-off keying *(Telecomm.)*. That in which the output from a source is alternately transmitted and suppressed to form signals.

Onsager equation *(Chem.)*. An equation based on the Debye-Hückel theory, relating the **equivalent conductance** Λ_c at concentration c to that at zero concentration, of the form $\Lambda_c = \Lambda_0 - (A + B\Lambda_0)\sqrt{c}$, where A and B are constants depending only on the solvent and the temperature.

ontogeny, ontogenesis *(Biol.)*. The history of the development of an individual. Cf. **phylogeny**. adj. *ontogenetic*.

on-top altitude clearance *(Aero.)*. **Air-traffic control** clearance for **visual flight rules** flying above cloud, haze, smoke, or fog.

onychia *(Med.)*. Inflammation of the nail-bed.

onychocryptosis *(Med.)*. Ingrowing toe-nail.

onychogenic *(Zool.)*. Nail-forming, nail-producing; as a substance similar to eleidin occurring in the superficial cells of the nail-bed.

onychogryphosis, onychogryposis *(Med.)*. Thickening, twisting, and overgrowth of the nails (usually of the toes) as a result of chronic infection and irritation.

onychomycosis *(Med.)*. A disease of the nails due to a fungus.

onych-, onycho- *(Genrl.)*. Prefix from Gk. *onyx*, gen. *onychos*, a nail or claw.

Onychophora *(Zool.)*. A subphylum of Arthropoda having trachea; a soft thin cuticle; body wall consisting of layers of circular and longitudinal muscles; head not marked off from the body, and consisting of 3 segments, 1 preoral, bearing preantennae and 2 postoral bearing jaws and oral papillae respectively; a pair of simple vesicular eyes; all body segments similar, each bearing a pair of parapodia-like limbs which end in claws and containing a pair of excretory tubules; spiracles scattered irregularly over the body; cilia in the genital region; development direct, e.g. the genus *Peripatus*.

onyx *(Min.)*. A cryptocrystalline variety of silica with layers of different colour, typically whitish layers alternating with brown or black bands. Used in cameos, the figure being carved in relief in the white band with the dark band as background.

onyx marble *(Min.)*. A banded form of calcite. *Oriental alabaster* is a beautifully banded form.

oö- *(Genrl.)*. Prefix from Gk. *oon*, egg.

oöblastema *(Zool.)*. A fertilized egg.

oöcium *(Zool.)*. A brood pouch.

oöcyst *(Zool.)*. In certain *Protozoa*, the cyst formed around two conjugating gametes; in *Sporozoa*, the passive phase into which an oökinete changes in the host.

oöcyte *(Zool.)*. An ovum prior to the formation of the first polar body; a female gametocyte.

oögamy *(Biol.)*. (1) The union of gametes of dissimilar size, usually of a relatively large non-motile egg and a small active sperm. (2) In *Protozoa*, **anisogamy** in which the female gamete is a hologamete. Cf. **isogamy**.

oögenesis *(Biol.)*. The origin and development of ova.

oögonium *(Bot.)*. In many algae and the Oömycetes, a unicellular female gametangium containing one or more eggs or oöspheres.

oögonium *(Zool.)*. An egg-mother-cell or oöcyte.

oölemma *(Zool.)*. See **vitelline membrane**.

oölite *(Geol.)*. A sedimentary rock composed of oöliths. In most cases oöliths are composed of calcium carbonate, in which case the rock is an oölitic limestone, but they can also be made of chamosite or limonite, in which case the rock is an oölitic ironstone. Written by itself, the word can be assumed to refer to an oölitic limestone.

oölith *(Geol.)*. A more or less spherical concretion of calcium carbonate, chamosite, or dolomite, not exceeding 2 mm in diameter, usually showing a concentric-layered and/or a radiating fibrous structure.

oölitic *(Geol.)*. Pertaining to an **oölite**.

oölogy *(Genrl.)*. The study of ova.

Oomycetes *(Bot.)*. Group of fungus-like, non-photosynthetic organisms with a non-septate mycelium, reproducing asexually by zoospores or dispersed sporangia that may germinate to give zoospores or a hypha, and sexually by oospores. Heterokont flagellation and mitochondrial microvilli suggest affinity with the Heterokontophyta. Include some water-'fungi' and many plant parasites e.g. *Pythium* (*damping-off*), the 'downy mildews' and the historically significant potato blight (*Phytophthora infestans*).

oöphorectomy *(Med.)*. Removal of an ovary.

oöphoritis *(Med.)*. Inflammation of an ovary.

oöphorosalpingectomy *(Med.)*. Removal of an ovary and a Fallopian tube.

oöphorostomy *(Med.)*. The formation of an opening into an ovarian cyst.

oöphorotomy *(Med.)*. Incision of an ovary.

oösperm *(Zool.)*. See **oöblastema**.

oöspore *(Bot.)*. Thick walled zygote with food reserves formed from fertilized oösphere in some algae and the Oomycetes.

oöspore *(Zool.)*. A fertilized ovum; in *Protozoa*, an encysted zygote.

oötheca *(Zool.)*. An egg-case, as in the Cockroach.

oötocoid *(Zool.)*. Bringing forth the young in an immature condition and allowing them to complete their early development in a marsupium.

oötocous *(Zool.)*. Oviparous.

ooze *(Geol.)*. (1) A fine-grained, soft, deep-sea deposit, composed of shells and fragments of foraminifera, diatoms, and other organisms. (2) A soft mud.

opacity *(Build.)*. The **hiding power** of a paint or other pigmented compound. Also *obliterating power*.

opacity *(Image Tech.,Phys.)*. Reciprocal of **transmission ratio**.

opal *(Min.)*. A cryptocrystalline or colloidal variety of silica with a varying amount of water. The transparent coloured varieties, exhibiting opalescence, are highly prized as gemstones.

opal agate *(Min.)*. A variety of opal, of different shade of colour and agatelike in structure.

opalescence *(Chem.)*. The milky, iridescent appearance of a solution or mineral, due to the reflection of light from very fine, suspended particles.

opalescence *(Min.)*. The play of colour exhibited by precious opal, due to interference at the surfaces of minutely thin films, the thicknesses of the latter being of the same order of magnitude as the wavelength of light.

opal glass *(Glass)*. Glass which is opalescent or white; made by the addition of fluorides (e.g. fluorspar, cryolite) to the glass mixture.

opaque *(Phys.)*. Totally absorbent of rays of a specified wavelength, e.g., wood is opaque to visible light but slightly transparent to infrared rays, and completely transparent to X-rays and waves for radio communication.

opcode *(Comp.)*. See **character code**.

open account *(Comp.)*. One of a large number of accounts registered on magnetic tape, all of which are re-recorded with **up-dating** at regular intervals, e.g. daily.

open aestivation *(Bot.)*. Aestivation in which the leaves or perianth parts neither overlap nor meet by their edges.

open antenna *(Telecomm.)*. One which is open-ended and able to support a standing wave of current.

open back *(Print.)*. See **hollow back**.

opencast *(Min.Ext.)*. Quarry. Open cut. Mineral deposit worked from surface and open to daylight.

opencast mining *(Geol.)*. A form of quarrying used to extract coal and mineral deposits.

open circuit *(Elec.Eng.)*. A circuit providing infinite impedance. See **short-circuit**.

open-circuit characteristic *(Elec.Eng.)*. A term commonly used to denote the curve obtained by plotting the e.m.f. generated by an electric generator on open circuit against the field current. Also called *no-load characteristic*.

open-circuit grinding *(Min.Ext.)*. Size reduction of solids in which the material to be crushed is passed only once through the equipment, so that all the grinding has to be done in a single step. Generally less efficient than *closed-circuit grinding*.

open-circuit impedance *(Phys.)*. Input or driving impedance of line or network when the far end is *free*, *open-circuited*, not *grounded* or *loaded*.

open-circuit loss *(Elec.Eng.)*. See **no-load loss**.

open-circuit transition *(Elec.Eng.)*. A method used, in the series-parallel control of traction motors, for changing the connections of the motors from series to parallel; the circuit is broken while the reconnection is being made. Cf. **bridge transition**.

open-circuit voltage *(Elec.Eng.)*. The voltage appearing across the terminals of an electric generator or transformer when it is delivering no load. The term also refers to the voltage appearing across the electrodes of an arc-welding plant when no current is flowing.

open clusters *(Astron.)*. Galactic clusters of stars of a loose type containing at most a few hundred stars; the stars of a cluster have a common motion through space, and are associated with dust and gas clouds, e.g. *Hyades, Pleiades, Praesepe, Ursa Major cluster*.

open community *(Bot.)*. A plant community which does not occupy the ground completely, so that bare spaces are visible.

open-diaphragm loudspeaker *(Acous.)*. Most common domestic type, in which a paper, plastic or doped-fabric cone diaphragm is driven by a circular coil at its apex. Mounted in a baffle or box, with or without ports or labyrinth.

open-end spinning *(Textiles)*. The system in which the fibres being fed to the spinning head are highly drafted as near as practicable to the individual fibre state so creating a discreet break or open-end in the fibre flow. The fibres are then caught and twisted on to the end of a rotating yarn. The main methods in use are **rotor spinning** and **friction spinning**.

open-field test *(Behav.)*. An experimental procedure in which an animal is released into an open area with no obstacles, and features of its behaviour, such as defaecation, urination, and locomotion are observed, these sometimes being related to emotionality.

open file *(Comp.)*. In a program or in interactive computing, to make a file on backingstore ready for writing onto and reading. Cf. **close file**.

open floor *(Build.)*. A floor which is not covered by a ceiling, the joists being therefore on view.

open flow *(Min.Ext.)*. Running an oil well without any valves or constrictions at the casing head.

open-frame girder *(Eng.)*. A girder consisting of upper and lower booms connected at intervals by (usually) vertical members, and not braced by any diagonal members.

open fuse *(Elec.Eng.)*. A fuse in which the mounting is such that the fuse link is fully exposed, except for any external containing-case.

open-hearth furnace *(Eng.)*. A furnace, of reverberatory type, used in steel-making. The charge is contained on a shallow hearth, and the furnace fired with gas or oil.

open-hearth process *(Eng.)*. *Siemens-Martin process* (1866). Cold pig-iron and steel scrap are charged to open hearth and melted by preheated gases with flame temperatures up to 1750°C produced by regeneration as they enter through brick labyrinth, where heat is maintained by periodic reversal between these and exit gases. After melting (up to 1600°C) most impurities are in the covering slag. Remainder are removed by addition of iron ore, scale, or limestone before tapping into ladles, where ferrosilicon, ferromanganese, or aluminium may be added as deoxidants. See also **acid process, basic process, steel-making**.

open inequality *(Maths.)*. One which defines an open set of points, e.g. $-1 < x < +1$.

opening *(Print.)*. The appearance of a pair of facing pages, given careful consideration in the best typography.

opening *(Textiles)*. The separation of compressed fibres after leaving the bale-breaking machine. Large, high-speed spiked cylinders remove impurities into dust chambers and leave the fibre in a fluffy state for further treatment.

open interval *(Maths.)*. An interval, such as $a<x<b$, the points of which form an open set. Cf. **closed interval**.

open-jet wind tunnel *(Aero.)*. A **wind-tunnel** in which the working section is not enclosed by a duct.

open-loop *(Telecomm.)*. A signal path, system or amplifier without a feedback path; *open-loop gain* is that of an amplifier without its feedback loop.

open mortise *(Build.)*. See **slot mortise**.

open newel stair *(Build.)*. A stair having successive flights rising in opposite directions, and arranged about a rectangular well hole.

open pipe *(Acous.)*. A pipe which is partially or completely open at both ends, so that the wavelength of the fundamental resonance is approximately double the length of the air column.

open pore *(Powder Tech.)*. A cavity within a particle of powder which communicates with the surface of the particle.

open reel *(Image Tech.)*. The use of magnetic tape on equipment with separate feed and take-up spools, in contrast to an enclosed **cassette**.

open roof *(Build.)*. A roof which is not covered in by a ceiling, the trusses being exposed.

open sand *(Foundry)*. (1) A sand of good porosity or permeability, as distinct from a *close* sand. (2) The process of casting in an open mould when the finish of the top surface is immaterial.

open set *(Maths.)*. A basic undefined concept of topology. Intuitively, it means a set which does not include any boundaries, e.g. the set of real numbers x such that $1 < x < 2$ does not include its boundaries 1 and 2. Note a set may be open, closed, both or neither.

open slating *(Build.)*. Slating in which gaps of 1–4 in. (25–100 mm) are left between adjacent slates in any course.

open slot *(Elec.Eng.)*. A type of slot, used in the armatures of electric machines, in which the opening is the same width as the rest of the slot. Cf. **semiclosed slot**.

open string *(Build.)*. See **cut string**.

open subroutine *(Comp.)*. One which is part of the main program and is copied into the program where required. Cf. **closed-**.

open traverse *(Surv.)*. One which does not close on the point of origin, and is therefore not directly checked for accuracy or closing error.

open vascular bundle *(Bot.)*. A bundle including cambium.

open well *(Arch.)*. A stair enclosing a vertical opening between the outer sides of the flights.

open-width washer *(Textiles)*. A machine having a series of tanks each fitted with nip rollers through which continuous lengths of fabrics are passed while being held out to full width. The tanks contain the different solutions required for washing e.g. soapy water, dilute acid or alkali, as well as clean water.

open window unit *(Acous.)*. Same as *sabin*.

open-wire feeder *(Telecomm.)*. One supported from insulators on poles, forming a pole route between antenna and transmitter or receiver.

operand *(Comp.,Maths.)*. Item or quantity on which an operation is to be performed. In *computing*, its address. See **address field**.

operant chain *(Behav.)*. A complex chain of behaviour built up by the operant technique of *shaping*.

operant conditioning *(Behav.)*. A learning procedure in which a reinforcement follows a particular response on a proportion of occasions. See **operant response, reinforcement**.

operant response *(Behav.)*. A response which acts on the environment to produce an event which affects the subsequent probability of that response.

operating characteristic *(Elec.Eng.)*. See **load characteristic**.

operating duty *(Elec.Eng.)*. A term applied to a switch or circuit-breaker to denote the series of making and breaking operations used in specifying its performance.

operating factor *(Elec.Eng.)*. The ratio which the time during which an intermittently run motor is actually running bears to the length of the duty cycle.

operating system *(Comp.)*. A collection of system programs that control the activities of the computer system such as **job control**, input/output and processing, e.g. **DOS, UNIX, VMS, VM/CMS**. See also **monitor**.

operation *(Comp.)*. The action of an operator or function, which takes one or more pieces of data and produces a new piece of data, e.g. *squared* is an **arithmetic operation** which produces 16 from 4. See also **logical operation, relational operation**.

operational amplifier *(Electronics)*. High gain, high stability d.c. amplifier used with external feedback path. These were originally used to perform one specific mathematical operation in analogue computing but are now more widely used as the basic building blocks of electronic circuits.

operational sphericity *(Powder Tech.)*. A **shape factor** defined as the cube root of the ratio of the volume of a particle of powder to the volume of the circumscribing sphere.

operational taxonomic units *(Bot.)*. The entities of any taxonomic rank, such as individuals, species, genera etc. whose relationships are studied in **numerical taxonomy**. Abbrev. *OTU*.

operation code *(Comp.)*. See **machine-code instruction**.

operator *(Biol.)*. Sequence of DNA to which a **repressor** or **activator** can bind. Situated before the coding sequence of a gene and close to the **promoter**.

operator *(Comp.)*. See **operation**.

operator *(Maths.)*. Mathematical symbol representing a specific operation to be carried out on a particular operand, e.g. the differential operator $D(\equiv d/dx)$.

opercular apparatus *(Zool.)*. In Fish, the operculum, together with the branchiostegal membrane and rays.

operculate *(Bot.,Zool.)*. (1) Possessing a lid. (2) Opening by means of a lid.

operculum *(Bot.)*. A lid or cover, composed of part of a cell wall or of from one to many cells, which opens to allow the escape of contents from some sort of container, especially the lid of an antheridium, a moss capsule, an ascus or other sporangium or the germ pore of a pollen grain.

operculum *(Zool.)*. In the eggs of some Insects, a differentiated area of the chorion which lifts up when the larva emerges from the egg. In the higher Fish, a fold which articulates with the hyoid arch in front of the first gill slit, and extends backwards, covering the branchial clefts. A similar structure occurs in the larvae of Amphibians. In some tubicolous Polychaeta, an enlarged branch of a tentacle closing the mouth of the tube when the animal is retracted; in Spiders, a small plate partially covering the opening of a lung book; in some Cirripedia, plates of the carapace which can be closed over the retracted thorax; in some Gastropoda, a plate of chitinoid material, strengthened by calcareous deposits, which fits across the opening of the shell.

operon *(Biol.)*. In bacteria the set of functionally related genes, which have a common promoter and mRNA, thus securing coordinated transcription.

OPF *(Space)*. Abbrev. for *Orbiter Processing Facility* at Kennedy Space Center, Florida, used for refurbishment and the loading of payloads into the Space Shuttle Orbiter.

ophicalcite *(Geol.)*. See **forsterite-marble**.

ophiolite *(Geol.)*. A group of mafic and ultramafic igneous rocks ranging from extrusive spilites to intrusive gabbros, associated with deep-sea sediments. Ophiolites are commonly found in **convergence zones**.

ophitic texture *(Geol.)*. A texture characteristic of dolerites in which relatively large pyroxene crystals completely enclose smaller, lath-shaped plagioclases. See also **poikilitic texture**.

Ophiuroidea *(Zool.)*. A class of Echinodermata with a dorsoventrally flattened star-shaped body; arms sharply differentiated from the disk and not containing caecae of the alimentary canal; tube feet lack ampullae and suckers and lie on the lower surface, although not in grooves; no anus; madreporite aboral; well-developed skeleton; no pedicellariae; free-living. Brittle-stars.

ophthalmectomy *(Med.)*. Excision of an eye.

ophthalmia *(Med.)*. Inflammation of various parts of the eye, especially of the conjunctiva.

ophthalmic *(Zool.)*. Pertaining to or situated near the eye, as the *ophthalmic nerve*, which passes along the back of the orbit in lower Vertebrates.

ophthalmodynamometer *(Med.)*. An instrument used to measure the intraocular arterial pressure by means of external pressure on the eye.

ophthalmology *(Med.)*. The study of the eye and its diseases. n. *ophthalmologist*.

ophthalmoplegia *(Med.)*. Paralysis of one or more muscles of the eye.

ophthalmoscope *(Med.)*. An instrument for inspecting the interior of the eye.

Opiliones *(Zool.)*. A sub-class of Arachnida with rounded bodies, the prosoma and opistosoma broadly jointed and usually with long legs. Mostly predaceous. Harvestmen.

opine *(Bot.)*. Guanidoamino acids (either octopine or nopaline) synthesized and released by plant cells after infection with a Ti plasmid and used by *Agrobacterium tumefasciens* as carbon and nitrogen source. See **crown gall**.

opisthocoelous *(Zool.)*. Concave posteriorly and convex anteriorly; said of vertebral centra.

opisthoglossal *(Zool.)*. Having the tongue attached anteriorly, free posteriorly, as in Frogs.

opisthomere *(Zool.)*. A postoral somite.

opisthosoma *(Zool.)*. In Chelicerata the segments posterior to those bearing the legs; the abdomen. Cf. **prosoma**.

opisthotonos *(Med.)*. Extreme arching backwards of the spine and the neck as a result of spasm of the muscles in these regions, e.g. in tetanus.

opium *(Med.)*. The latex from unripe capsules of *Papaver somniferum*, dried and powdered. Contains about 10% morphine, hence its analgesic and narcotic uses.

Oppenheimer-Phillips (O.P.) process *(Phys.)*. A form of **stripping** in which a deuteron surrenders its neutron to a nucleus without entering it.

OPP film *(Plastics)*. *Oriented PolyPropene film* which has been biaxially stretched to improve its physical properties. Widely used for packaging because of its high gloss and clarity, high impact strength and low moisture permeability.

opportunistic species *(Ecol.)*. A species adapted to colonize temporary or local conditions.

opposed-cylinder engine *(Autos.)*. An engine with cylinders, or banks of cylinders, on opposite sides of the crankcase in the same plane, their connecting-rods working on a common crankshaft placed between the cylinders. Also *boxer*.

opposed-voltage protective system *(Elec.Eng.)*. A form of Merz-Price protective system in which the secondary voltages of current transformers, situated at each end of the circuit to be protected, are balanced against each other, so that there is normally no current on the pilots connecting them.

opposite *(Bot.)*. Two organs, especially leaves, arising at the same level but on opposite sides of a stem. Cf. **alternate, decussate, whorled**. See **phyllotaxis**.

opposition *(Astron.)*. The instant when the geocentric longitude of the moon or of a planet differs from that of the sun by 180°.

opsonin *(Immun.)*. Factors present in blood and other body fluids which bind to particles and increase their susceptibility to phagocytosis. They may be antibody, or products of complement activation (especially C3b), or some other substances which bind to particles such as fibronectin.

optic *(Zool.)*. Pertaining to the sense of sight; the second cranial nerve of Vertebrates.

optical activity *(Chem.)*. A property possessed by many substances whereby plane-polarized light, in passing through them, suffers a rotation of its plane of polarization, the angle of rotation being proportional to the thickness of substance traversed by the light. In the case of molten or dissolved substances it is due to the possession of an asymmetric molecular structure, e.g. no mirror plane of symmetry in the molecule. See **chirality, specific rotation**.

optical bench *(Phys.)*. A rigid bed along which optical components, mounted in suitable holders, may be moved. It is a device used in experimental work on linear optical systems.

optical black *(Phys.)*. A body when it absorbs all radiation falling on its surface. No substance is completely black in this context.

optical bleaches *(Chem.)*. See **fluorescent whitening agents**.

optical brightener *(Textiles)*. See **brightening agent**.

optical centre of a lens *(Phys.)*. That point on the principal axis of a lens or lens system through which passes a ray whose incident and emergent directions are parallel.

optical character recognition *(Comp.)*. Machine recognition of characters by light-sensing methods. Abbrev. *OCR*.

optical constants *(Phys.)*. The refractive index (n) and the absorption coefficient (k) of an absorbing medium. Together these determine the complex refractive index ($n-k$) of the medium.

optical crown *(Glass)*. Glass of low dispersion made for optical purposes. See under **soda-lime-silica glass**.

optical distance, path *(Phys.)*. The distance travelled by light (d) multiplied by the refractive index of the medium (n). Length of equivalent path in air (strictly vacuum).

optical-electronic devices *(Phys.)*. Used to locate weakly radiating sources by the detection of their infrared emission. The radiation is collected by optical mirrors or lenses and concentrated on a sensitive infrared detector. Detailed maps of the earth's surface, weather mapping and non-destructive testing of materials and components are some of the non-military applications.

optical fibre *(Telecomm.)*. Fibres of ultra-pure glass, having a central core of higher refractive-index glass than the outer *cladding*, capable of conducting modulated light signals by total internal reflection. Optical-fibre cables consist of such cores either singly or several per cable, with further cladding and armouring for mechanical protection. Benefits include small diameters, high potential bandwidth and lower cost than copper. See **monomode fibre, multimode fibre**.

optical flat *(Eng.)*. A flat glass disk having very accurately polished surfaces, used for testing by interference patterns the flatness of gauge anvils and other plane surfaces.

optical flat *(Image Tech.)*. A surface whose deviation from a true plane is small enough to be expressed in terms of wavelengths of light; for a plate the two surfaces must be parallel within seconds of arc.

optical flint *(Glass)*. Glass of high dispersion made for optical purposes.

optical glass *(Glass)*. Glass made expressly for its optical qualities. The composition varies widely both as to constituents and amounts, with very exacting requirements for freedom from streaks and bubbles.

optical indicator *(Autos.)*. An engine indicator in which a ray of light is deflected successively by mirrors in directions at right angles, proportionately first to cylinder pressure, then to piston displacement, being finally focused on a ground-glass screen or photographic plate, on which it traces the **indicator diagram**.

optical isomerism *(Chem.)*. The existence of isomeric compounds which differ in their **chirality**.

optical lever *(Phys.)*. A device for measuring the small relative displacement of two objects by means of angular displacement of a light beam.

optical maser *(Phys.)*. One in which the stimulating frequency is visible or infrared radiation; see **laser**.

optical model of the nucleus *(Phys.)*. In nuclear reactions the target nucleus can frequently be treated as a sphere that partly absorbs and partly transmits the incidental radiation, i.e. it has, in the optical analogy, a *complex* refractive index. The nucleus can thus be represented by a 'potential well' having real and imaginary components.

optical printing *(Image Tech.)*. Method of printing motion picture film in which an image of the frame on one strip is formed on the other by means of a copy lens, thus allowing changes of size and position to be made.

optical pumping *(Phys.)*. In a laser, an external light source of suitable frequency stimulates the material to produce a *population inversion* for a particular energy transition.

optical pyrometer *(Phys.)*. An instrument which measures the temperatures of furnaces by estimating the colour of the radiation, or by matching it with that of a glowing filament.

optical range *(Telecomm.)*. That which a radio transmitter or beacon would have if it were radiating visible light.

optical rotary dispersion *(Chem.)*. The change of optical rotation with wavelength. Rotatory dispersion curves may be used to study the configuration of molecules. Abbrev. *ORD*.

optical rotation *(Phys.)*. The rotation of the plane of polarization of a beam of light when passing through certain materials.

opticals *(Image Tech.)*. General term for image effects made in a motion picture by laboratory printing operations, such as *fades* and *dissolves*.

optical sound *(Image Tech.)*. See **photographic sound**.

optical spectrometer *(Phys.)*. A spectroscope for studying optical spectra. Fitted with a graduated circle it enables the angle of deviation of each spectrum line to be measured and so the wavelengths deduced.

optical spectrum *(Phys.)*. The visible radiation emitted from a source separated into its component frequencies.

optical square *(Surv.)*. A hand instrument for setting out right angles in the field. It works on the principle of the **sextant**, the 2 reflecting surfaces being arranged in this case to yield lines of sight at a fixed angle of 90° apart.

optical track *(Image Tech.)*. A photographic sound track on motion picture film.

optical transfer function *(Phys.)*. A mathematical representation of the effect of a lens or other component in an optical system on the imaging of a point source.

optical whites *(Chem.)*. See **fluorescent whitening agents**.

optic atrophy *(Med.)*. The condition of the optic disk (where the nerve fibres of the retina pass through the eyeball) resulting from degeneration of the optic nerve.

optic axial angle *(Min.)*. The angle between the two optic axes in biaxial minerals, usually denoted as 2V (when measured in the mineral) or 2E (in air).

optic axis *(Crystal.)*. Direction(s) in a doubly refracting crystal for which both the ordinary and the extraordinary rays are propagated with the same velocity. Only one exists in uniaxial crystals, two in biaxial.

optic axis *(Phys.)*. The line which passes through the centre of curvature of a lens' surface, so that the rays are neither reflected nor refracted. Also called *principal axis*.

optic branch *(Phys.)*. The lattice dynamics of a crystal containing n atoms per unit cell shows that the dispersion curve (frequency ω against wavenumber q) has $3n$ branches of which $3n$-3 are *optic* branches. The branches are characterized by different patterns of movement of the atoms. For the optic branch ω is independent of q for small values of q. See **acoustic branch**.

optic lobes *(Zool.)*. In Vertebrates, part of the mid-brain, which is concerned with the sense of sight, and from which the optic nerves originate.

optic neuritis *(Med.)*. See **papillitis**.

optics *(Phys.)*. The study of light. *Physical optics* deals with the nature of light and its wave properties; *geometrical optics* ignores the wave nature of light and treats problems of reflection and refraction from the ray aspect.

optic sign *(Min.)*. Anisotropic minerals are either optically positive or negative, indicated by + or − in technical descriptions. See **negative mineral**, **positive mineral**.

optimal damping *(Eng.)*. Adjustment of damping just short of **critical**, which allows a little overshoot. This attains the final reading of an indicating instrument most rapidly while showing that it is moving freely.

optimal proportions *(Immun.)*. Describes the relative proportions of antibody and a soluble antigen which when mixed together produce the maximum degree of cross-linkage, such that all the antibody and all the antigen are included in the precipitate which forms. See **lattice hypothesis**.

optophone *(Electronics)*. A photoelectric device for training the blind by converting printed words into sounds.

OR *(Aero.)*. Abbrev. for (1) *Operational Requirement*, (2) *Operational Research*.

OR *(Comp.)*. A logical operator such that (p *OR* q) written $\overline{p+q}$ takes the value FALSE if both p and q are FALSE, otherwise (p *OR* q) takes the value TRUE. See **logical operation**.

Oracle *(Comp.)*. See **teletext**.

oral *(Zool.)*. Pertaining to the mouth.

oral characters *(Behav.)*. In psychoanalytic theory, refers to fixation at, or regression to, the oral stage of development; Freud considered that many oral habits reflected this (e.g. smoking).

oral contraception *(Med.)*. The use of synthetic hormones (oestrogen and progestogen steroids in varying proportions), taken orally in pill form, to prevent conception by reacting on the natural **luteinizing** and **follicle-stimulating hormones** and so inhibiting ovulation and/or fertilization. Colloq. *the pill*.

oral stage *(Behav.)*. In psychoanalytic theory, the first stage of psychosexual development in which stimulation of the mouth and lips is the primary focus of bodily (libidinal) pleasure; occurs in the first year of life.

orange lead *(Chem.)*. *Lead oxide*. Pb_3O_4. Obtained by heating white lead (basic lead (II,II,V) carbonate) in air at approximately 450°C. Commercial varieties contain up to approximately 35% PbO_2.

orange peel *(Build.)*. A defect of surface coatings which have been applied by spray gun, characterized by more or less pronounced small depressions in the film to produce an appearance resembling an orange skin, attributable to incorrect thinning, wrong viscosity or too low a pressure.

ora serrata *(Zool.)*. The edge of the retina.

ORB *(Aero.)*. Abbrev. for *Omnidirectional Radio Beacon*.

orbicular *(Bot.)*. Flat, with a circular or almost circular outline.

orbiculares *(Zool.)*. Muscles which surround an aperture; as the muscles which close the lips and eyelids in Mammals.

orbicular structure *(Geol.)*. A structure exhibited by those plutonic igneous rocks which contain spherical orbs up to several centimetres in diameter, each showing a development of alternating concentric shells of different minerals, so deposited by rhythmic crystallization.

orbit *(Aero.)*. An aircraft circling a given point is said to orbit that point and **air-traffic control** instructions incorporate the term. See also **holding pattern**.

orbit *(Astron.)*. The path of a heavenly body (and, by extension, an artificial satellite, spacecraft, etc.) moving about another under gravitational attraction. In unperturbed motion the orbit is a conic. See **elements of an orbit**.

orbit *(Space)*. The path of a space vehicle moving about a primary body. It is the shape of a conic section with a focus at the centre of mass, of which a circular orbit is a special case (eccentricity =0). The motion of the vehicle is characterized by its orbital velocity and its inclination to the equator of the primary body. The time taken to complete a single orbit is called the *period*. If the vehicle moves in the opposite sense to the primary body, the orbit is termed *retrograde*.

orbit *(Zool.)*. A space lodging an eye; in Vertebrates, the depression in the skull containing the eye; in Arthropoda, the hollow which receives the eye or the base of the eyestalk; in Birds, the skin surrounding the eye.

orbital *(Phys.)*. The properties of each electron in a many-electron atom may be reasonably described by its response to the potential due to the nucleus and to the other electrons. The *wave function*, which expresses the probability of finding the electron in a region, is specified by a set of four quantum numbers and defines the *orbital* of the electron. The state of the many-electron atom is given by defining the orbitals of all the electrons subject to the Pauli exclusion principle.

orbital quantum number *(Electronics)*. The second for an orbital electron, indicating the angular momentum associated with the orbit. 0 for *s states*; 1 for *p states*; 2 for *d states*; and 3 for *f states*. See also **principal quantum number**.

orbit decay *(Space)*. The change in orbit parameters of a space vehicle caused by air drag which becomes more rapid as the surface of a planet is approached due to increasing atmospheric density, eventually resulting in entry or re-entry of the vehicle.

orbitosphenoid *(Zool.)*. A paired cartilage bone of the Vertebrate skull, forming the side wall of the brain-case in the region of the presphenoid.

orbit shift coils *(Phys.)*. Coils placed on the magnetic pole faces of a betatron or synchrotron, so that by passing a current pulse through the coils the particles may be momentarily displaced from the stable orbit to strike a target.

orchic, orchitic *(Zool.)*. Pertaining to the testis.

Orchidaceae *(Bot.)*. The orchid family, ca 18 000 spp. of monocotyledonous flowering plants (superorder Liliidae). The largest monocot family. Herbs, terrestrial and epiphytic; cosmopolitan. Some are CAM plants. The flowers are zygomorphic, usually showy. The seeds are minute and early stages of growth depend on symbiosis with a mycorrhizal fungus. Of no economic importance except for vanillin (the seed pod of *Vanilla*) and as ornamental and florists plants.

orchidalgia *(Med.)*. Pain in a testis.

orchi(d)ectomy *(Med.)*. Removal of a testis.

orchidopexy *(Med.)*. The operation of stitching an undescended testis to the scrotum.

orchiepididymitis *(Med* Inflammation of the testis and the epididymis.

orchitis *(Med.)*. Inflammation of the testis.

ORD *(Chem.)*. Abbrev. for *Optical Rotatory Dispersion*.

order *(Bot.,Zool.)*. Taxonomic rank below **class** and above **family**; for plants the names usually end in -oles.

order *(Comp.)*. An instruction in a program.

order *(Maths.)*. Of a curve or surface: the number of points, real, coincident, or imaginary, in which the curve or surface is intersected by a straight line. Of a differential equation: the order of the highest order derivative that occurs. Of a matrix: (1) a square matrix of n rows and n columns is of order n; (2) a rectangular matrix of n rows and m columns is of order $n \times m$, or n by m. Also used in sense of **cardinality**.

order-disorder transformation *(Eng.)*. In solid solution (e.g. a brass alloy) the ordered state can occur below the temperature at which lattice sites are occupied by single atomic species, and the disordered state can ensue above a critical temperature, with two or more elements in one primitive lattice.

ordered state *(Eng.)*. In solid solution of alloy, repeated and regular atomic arrangement.

order number *(Eng.)*. Of a torque impulse or a vibration, as the torsional oscillation of an engine crankshaft, the number of impulses or vibrations during one revolution of the shaft.

order of reaction *(Chem.)*. A classification of chemical reactions based on the index of the power to which concentration terms are raised in the expression for the instantaneous velocity of the reaction, i.e. on the apparent number of molecules which interact. If $A + B \rightarrow$products, and the rate of reaction is $k(A)_x(B)_y$, the order is x with respect to A and y with respect to B.

ordinal number *(Maths.)*. In ordinary usage, 1st, 2nd, 3rd etc. as opposed to 1, 2, 3 etc. which are cardinal numbers. Mathematically, ordinal numbers are properties of well-ordered sets such that if there is a one-to-one mapping from one set to another which preserves their ordering then the two sets have the same ordinal number. An infinite set can be well-ordered in different ways to give different ordinal numbers, while it has only one cardinal number. The first infinite ordinal, that of the infinite sequence of finite numbers 1, 2, 3,..., is called ω. It is followed by $\omega+1$, $\omega+2$, ..., 2ω, ..., ω^2, ..., $3\omega^5+\omega+6$, ...,ω^ω etc. All of these correspond to the first infinite cardinal aleph-0.

ordinary differential equation *(Maths.)*. See **differential equation**.

ordinary ray *(Phys.)*. See **double refraction**.

ordinate *(Maths.)*. See **cartesian co-ordinates**.

ordination *(Ecol.)*. A family of multi-variate statistical techniques commonly used to plot ecological data sets collected from a large number of sites, on geometric axes so that similarity is represented by proximity.

Ordnance Bench Mark *(Surv.)*. A bench mark officially established with reference to the Ordnance datum. Abbrev. *OBM*.

Ordnance datum *(Surv.)*. Arbitrary zero height, which is assumed to be the mean sea level at Newlyn, Cornwall. All official bench marks in UK are referred to this for height above sea level.

Ordnance Survey *(Surv.)*. Originally a triangulation of the UK made in 1791 by the military branch of the Board of Ordnance. Now a government survey which is concerned with land, buildings, roads etc. Published as maps on which the National Grid is superimposed, giving the whole country a unified reference system. Abbrev. *OS*.

Ordovician *(Geol.)*. The second oldest period of the Palaeozoic era, covering an app. time span from 500–440 million years. Named after the *Ordovices*, an ancient British tribe of the Welsh borders. The corresponding system of rocks.

ore *(Min.)*. A term applied to any metalliferous mineral from which the metal may be profitably extracted. It is extended to nonmetals and also to minerals which are potentially valuable.

ore bin *(Min.Ext.)*. Storage system (usually of steel or concrete) which receives ore intermittently from the mine. A fine ore bin holds material crushed to centimetric size and keeps from 1 to 3 days' milling supply.

ore body *(Min.Ext.)*. Deposit, seam, bed, lode, reef, placer, lenticle, mass, stockwork, according to geological genesis.

ore dressing *(Min.Ext.)*. See **mineral processing**.

Oregon pine *(For.)*. See **Douglas fir**.

ore reserves *(Min.Ext.)*. Ore whose grade and tonnage has been established by drilling, etc. with reasonable assurance.

orf *(Med.,Vet.)*. See **contagious pustular dermatitis**, which is transmissable to man causing skin eruptions, usually on hands or forearms.

organ *(Acous.)*. A musical instrument, comprising ranks of pipes which radiate sound when blown by compressed air, the operation of the pipes being controlled by manuals or keyboards and by a set of pedals. Hence, any musical instrument producing synthetically (e.g. electronically) tones similar to those from pipes and operated from keyboards, e.g. the *Electrone* and *Hammond*.

organ *(Bot.,Zool.)*. A part of the body of an animal or a plant adapted and specialized for the performance of a particular function.

organ culture *(Biol.)*. A **tissue culture** technique in which parts of organs can be kept alive and functional for a limited period for observation and experiment.

organdie *(Textiles)*. Plain-woven, transparent, light dress fabric with a permanently stiff finish.

organelle *(Biol.)*. A defined structure within a cell, e.g. nucleus, mitochondrion, lysosome.

organ genus *(Bot.)*. See **form taxon**.

organic chemistry *(Chem.)*. The study of the compounds of carbon. Owing to the ability of carbon atoms to combine together in long chains (catenate), these compounds are far more numerous than those of other elements. They are the basis of living matter.

organic disease *(Med.)*. See **functional disease**.

organic electrical conductor *(Phys.)*. An organic substance exhibiting electrical conductivity similar to that of metals. E.g. TTF-TCNQ (tetrathiofulvalene-tetracyanoquinodimethane) has a room temperature conductivity of $500\,\Omega cm^{-1}$ and its conductivity increases at low temperature, by a factor of at least 20 at 60 K. $(TMTSF)_2ClO_4$, where TMTSF is the tetra methyl selenium derivative of TTF, is one of a series of organic conductors which show superconductivity below 1.2 K.

organic mental (brain) disorders *(Behav.)*. Behavioural disorders stemming demonstrably from damage to the brain tissue or to chemical imbalances in the nervous system.

organic phosphor *(Radiol.)*. Organic chemical used as solid or liquid scintillator in radiation detection.

organisms *(Biol.)*. Animals, plants, fungi and microorganisms.

organized *(Biol.)*. Showing the characteristics of an organism; having the tissues and organs formed into a unified whole.

organogeny, organogenesis *(Biol.)*. The study of the formation and development of organs.

organography *(Bot.)*. A descriptive study of the external form of plants, with relation to function.

organo-magnesium compounds *(Chem.)*. **Grignard reagents**, compounds of the type R.Mg.X., where X is a halogen.

organo-metallic compounds *(Chem.)*. Wide range of compounds in which carbon atoms are linked directly with metal atoms, including the alkali metals, e.g. sodium phenyl, $C_6H_5.Na$; lead, e.g. lead tetraethyl, $Pb(C_2H_5)_4$; zinc, e.g. zinc dimethyl, $Zn(CH_3)_2$. The compounds are useful reagents in preparative organic chemistry. Many are formed by transition elements. See **Grignard reagents**.

organosilicone *(Chem.)*. Synthetic resin characterized by long thermal life and resistance to thermal ageing. See also **silicones**.

organosol *(Chem.,Plastics)*. A coating composition based on a dispersed PVC resin mixed with a plasticizer and a diluent; a colloidal solution in any organic liquid.

orgasm *(Zool.)*. Culmination of sexual excitement. adj. *orgastic*.

OR gate *(Comp.)*. **Gate** for which output is 0 only when all inputs are 0, otherwise output is 1. Also *OR element*.

oriel *(Arch.)*. A projecting window supported upon corbels or brackets.

Oriental alabaster *(Min.)*. See **onyx marble**.

Oriental almandine *(Min.)*. A name sometimes used for *corundum*, of gemstone quality, which is deep-red in colour, resembling true almandine (a garnet) in this, but no other, respect.

Oriental amethyst *(Min.)*. A misnomer for purple corundum or sapphire. Also called *false amethyst*.

Oriental cat's eye *(Min.)*. See **cymophane**.

Oriental emerald *(Min.)*. A name sometimes used for *corundum*, of gemstone quality, resembling true emerald in colour.

oriental region *(Zool.)*. One of the primary faunal regions into which the land surface of the globe is divided. It includes the southern coast of Asia east of the Persian Gulf, the Indian subcontinent south of the Himalayas, southern China and Malaysia, and the islands of the Malay Archipelago north and west of Wallace's line.

Oriental ruby *(Min.)*. See **ruby**.

Oriental topaz *(Min.)*. A variety of *corundum*, resembling topaz in colour.

orientation *(Biol.)*. The position, or change of position, of a part or organ with relation to the whole; change of position of an organism under stimulus.

orientation *(Chem.)*. (1) The determination of the position of substituent atoms and groups in an organic molecule, especially in a benzene nucleus. (2) The ordering of molecules, particles, or crystals so that they point in a definite direction.

orientation *(Eng.)*. The position of important sets of planes in a crystal in relation to any fixed system of planes. See **pure metal crystal**.

orientation *(Surv.)*. Fixing of line or plan in azimuth, with respect to true north.

orientation *(Textiles)*. Fibres are made of linear polymers which are oriented along the fibre axis either during growth (natural fibres) or while being extruded and drawn (man-made fibres).

orientation behaviour *(Behav.)*. The positioning of the body or of a behavioural sequence, with respect to some aspect of the external environment; it includes simple postural preferences as well as complex navigational behaviours.

orienting reflex *(Behav.)*. First described by Pavlov, an animal's response to the sudden presentation of a novel stimulus, includes turning the body and head so that the animal's attention can be focused on the source of stimulation (the 'what is it?' reaction).

orifice gauge *(Eng.)*. A flow gauge consisting of a thin orifice plate clamped between pipe flanges, with pressure take-offs drilled into the adjacent pipes, or of a thick orifice plate similarly clamped but containing its own pressure take-offs.

origin *(Maths.)*. A fixed point with respect to which points, lines etc. are located; the point common to the axes of co-ordinates.

origin *(Zool.)*. That end of a skeletal muscle which is attached to a portion of the skeleton which remains, or is held, rigid when the muscle contracts, the other end, which is attached to a part of the skeleton which moves as a result of the contraction, being known as the *insertion*.

original equipment manufacturer *(Comp.)*. See **OEM**.

O-ring *(Eng.)*. A toroidal ring, usually of circular cross-section, made of neoprene or similar materials, used e.g. as an oil or air seal.

Orlon *(Chem.)*. TN for synthetic fibre based almost wholly on cyano ethene with a small amount of a different monomer to serve as a dye receptor. Widely used in knitted fabrics, as imitation fur and in carpets.

orlop deck *(Ships)*. The lowermost deck in a ship of several decks. It is simply a platform, and contributes nothing to main longitudinal strength; usually of small extent.

ormolu *(Eng.)*. An alloy of copper, zinc and sometimes tin, used (especially in the 18th century) for furniture mountings, decorated clocks etc.

ornis *(Zool.)*. A Bird fauna. adj. *ornithic*.

ornithine *(Chem.)*. *2-6-Diaminovaleric acid.* It is concerned in urea formation in animals (see **arginine**), and a derivative, *ornithuric acid*, is found in the excrement of Birds.

ornithology *(Zool.)*. The study of Birds.

ornithophily *(Bot.)*. Pollination by Birds.

ornithopter *(Aero.)*. Any flying machine that derives its principal support in flight from the air reactions caused by flapping motions of the wings, this motion having been imparted to the wings from the source of power being carried.

ornithosis *(Med.,Vet.)*. An infection of birds with microorganism *Chlamydia psittaci* which is transmissible to man. See **psittacosis**.

oro- *(Genrl.)*. (1) Prefix from L. *os*, gen *oris*, mouth. (2) Prefix from Gk. *oros*, mountain.

oroanal *(Zool.)*. Connecting, pertaining to, or serving as, mouth and anus.

orogenesis *(Geol.)*. See **orogeny**.

orogenic belt *(Geol.)*. A region of the earth's crust, usually elongated, which has been subjected to an *orogeny*. Recently formed orogenic belts correspond to mountain ranges, but older belts have often been eroded flat.

orogeny *(Geol.)*. The tectonic process whereby large areas are folded, faulted, metamorphosed and subject to igneous activity. Different periods of orogeny are given specific names, e.g. *Alpine, Caledonian, Laramide.*

orographic ascent *(Meteor.).* The upward displacement of air blowing over a mountain.

orographic rain *(Meteor.).* Rain caused by moisture-laden winds impinging on the rising slopes of hills and mountains. Precipitation is caused by the cooling of the moist air consequent upon its being forced upwards.

oroide *(Eng.).* See **French gold**.

oronasal *(Zool.).* Pertaining to or connecting the mouth and the nose.

orpiment *(Min.).* Arsenic trisulphide, which crystallizes in the monoclinic system; commonly associated with realgar; golden-yellow in colour and used as a pigment.

orrery *(Astron.).* A mechanical model of the solar system showing the relative motions of the planets by means of clockwork; much in vogue in the 18th century. (Named after Charles Boyle, Earl of Orrery).

orthicon *(Image Tech.).* The first high-sensitivity TV camera tube of the **photo-emissive** type, with the image formed on a transparent mosaic **target** plate scanned from the rear.

orthite *(Min.).* See **allanite**.

ortho- *(Chem.).* (1) Derived from an acid anhydride by combination with the largest possible number of water molecules, e.g. **orthophosphoric acid**. (2) Consisting of diatomic molecules with parallel nuclear spins and an odd rotational quantum number, e.g. **orthohydrogen**.

ortho- *(Genrl.).* Prefix from Gk. *orthos*, straight.

orthocentre *(Maths.).* Of a triangle: the point of intersection of the perpendiculars from the vertices to their opposite sides.

orthoclase *(Min.).* Silicate of potassium and aluminium, $KAlSi_3O_8$, crystallizing in the monoclinic system; a feldspar, occurring as an essential constituent in granitic and syenitic rocks, and as an accessory in many other rock types. See also **microcline, sanidine**.

orthodiagraph *(Radiol.).* An X-ray apparatus for recording exactly the size and form of organs and structures inside the body.

orthoferrosilite *(Min.).* The ferrous iron end-member of the orthopyroxene group of silicates.

orthognathous *(Zool.).* With the long axis of the head at right angles to that of the body, and the mouth directed downwards. Cf. **prognathous**.

orthogneiss *(Geol.).* Term applied to gneissose rocks which have been derived from rocks of igneous origin. Cf. **paragneiss**.

orthogonal cutting *(Eng.).* A cutting process used to analyse the forces occurring in machining, in which a straight-edged cutting tool moves relatively to the workpiece in a direction perpendicular to its cutting edge.

orthogonal functions *(Maths.).* A set of functions $f_r(x)$ such that, over the range considered, $\int f_r(x) f_s(x) dx = 0$, except when $r = s$. If the integral equals unity when $r = s$ the functions are also said to be *normal* or *orthonormal*. If also the equation $\int F(x).f_r(x)dx = 0$ for all r implies that $F(x)$ is identically zero, then the set is said to be *complete*.

orthogonal matrix *(Maths.).* One equal to the inverse of its transpose. Any two rows, or any two columns, will be orthogonal vectors.

orthogonal vectors *(Maths.).* Two vectors whose scalar product is zero.

orthograph *(Arch.).* A view showing an elevation of a building or of part of a building.

orthographic projection *(Eng.).* A method of representing solid objects in two dimensions by viewing on 3 mutually perpendicular plane surfaces, using parallel rays or projectors perpendicular to the surfaces; commonly used as the basis of engineering drawing.

orthohydrogen *(Chem.).* Hydrogen molecule in which the two nuclear spins are parallel, forming a triplet state.

orthopaedics, orthopedics *(Med.).* That branch of surgery which deals with deformities arising from injury or disease of bones or of joints. adj. *orthopaedic*.

orthophosphoric acid *(Chem.).* H_3PO_4. Formed when phosphorus pentoxide is dissolved in water and the solution is boiled. The highest hydrated stable form of phosphoric acid.

orthophyric *(Geol.).* A textural term applied to medium- and fine-grained syenitic rocks consisting of closely packed orthoclase crystals of stouter build than in the typical trachytic texture. The term actually implies the presence of porphyritic orthoclase crystals.

orthopnoea, orthopnea *(Med.).* Dyspnoea so severe that the patient is unable to lie down; a symptom of heart failure. adj. *orthopnoeic*.

Orthoptera *(Zool.).* An order of the *Insecta*. Large Insects with biting mouthparts; posterior legs often with enlarged femora for jumping; forewings toughened (tegmina) and overlapping when folded; unjointed cerci; well developed ovipositor; possess a variety of stridulatory organs. Grasshoppers, Locusts, Crickets, Cockroaches.

orthoptic circle *(Maths.).* Of a conic: the locus of a point from which tangents to the conic are perpendicular to each other. Except for the parabola, this locus is a circle concentric with the conic. For the parabola, the locus is its directrix. Also called *director circle*.

orthoptic treatment *(Med.).* The nonoperative treatment of squint by specially devised stereoscopic exercises.

orthopyroxene *(Min.).* A group of pyroxene minerals crystallizing in the orthorhombic system, e.g. enstatite, hypersthene.

orthoquartzite *(Geol.).* A pure quartz sandstone.

orthorhombic system *(Crystal.).* The style of crystal architecture which is characterized by three crystal axes, at right angles to each other and all of different lengths. It includes such minerals as olivine, topaz, and barytes.

orthosilicic acid *(Chem.).* $Si(OH)_4$.

orthostatic *(Med.).* Associated with or caused by the erect posture, e.g. *orthostatic* albuminuria.

orthostyle *(Arch.).* A colonnade formed of columns arranged in a straight line.

orthotropism *(Bot.).* A *tropism* in which a plant part becomes aligned directly towards (positive-) or away from (negative-) the source of the orientating stimulus. E.g. most seedling shoots are negatively *orthogravitropic* and positively *orthophototropic*. Cf. **diatropism, plagiotropism**.

orthotropous *(Bot.).* *Atropous*. An ovule which is straight and on a straight stalk, so that the micropyle points away from the stalk.

OS *(Build.).* Abbrev. for *One Side*.

Os *(Chem.).* The symbol for osmium.

os *(Geol.).* See **esker**.

OS *(Surv.).* Abbrev. for *Ordnance Survey*.

os *(Zool.).* (1) An opening, as the *os uteri*. (L. *os*, gen. *oris*, mouth.) pl. *ora*. (2) A bone, as the *os coccygis*. (L. *os*, gen. *ossis*, bone.) pl. *ossa*.

osazones *(Chem.).* The diphenylhydrazones of monosaccharides, obtained by the action of 2 molecules of phenylhydrazine on one molecule of the monosaccharide. They are sparingly soluble in water, can be purified by recrystallization, and serve to identify the respective monosaccharides.

oscillating capacitor *(Elec.Eng.).* See **vibrating capacitor**.

oscillating die press *(Eng.).* That in which the stock moves continuously and the high-speed punch and die set moves with it for a sequence of presses before returning to its starting position for the next cycle.

oscillating neutral *(Elec.Eng.).* A phenomenon occurring in 3-phase, unearthed, star-connected systems, due to third harmonic voltages, which results in a distorted phase voltage waveform.

oscillating sequence, oscillating series *(Maths.).* See **divergent sequence, divergent series**.

oscillation *(Aero.).* See **longitudinal-, phugoid-**.

oscillation *(Eng.).* See **hunting**.

oscillation *(Telecomm.).* Sustained and very stable periodic alteration of current in a tuned circuit or other resonant structure or cavity. Maintained by the supply of synchronous pulses of energy lost through dissipation and output.

oscillation frequency *(Telecomm.).* That determined by the balance between the inertia reactance and the elastic reactance of a system, e.g. open or short-circuited transmission line, cavity, resonant circuit, quartz crystal. If C = capacitance and L = self-inductance of circuit, then frequency $f = 1/(2\pi\sqrt{LC})$. In a mechanical oscillating system $f = 2\pi\sqrt{M/S}$, where M = mass, and S = restoring force per unit displacement.

oscillations *(Phys.)*. Any motion that repeats itself in equal intervals of time is a *periodic motion*. A particle in periodic motion moves back and forward over the same path and is said to be *oscillating* or *vibrating*. If the oscillations are not precisely repeated due to frictional forces which dissipate the energy of motion, the oscillations are said to be *damped*. See also **centre of oscillation**.

oscillator *(Telecomm.)*. A source of alternating current of any frequency, which is sustained in a circuit by a valve or transistor using positive feedback principle or by a negative resistance device. There are two types: (1) *stable-type*, in which frequency is determined by a line or a tuned (LC) circuit, waveform being substantially sinusoidal, and (2) *relaxation-type*, in which frequency is determined by resistors and capacitors, waveform having considerable content of harmonics. Also applied to mechanical systems, velocities being equivalent to currents.

oscillator crystal *(Telecomm.)*. A piezoelectric crystal used in an oscillator to control the frequency of oscillation.

oscillator drift *(Telecomm.)*. See **frequency drift**.

oscillatory discharge *(Telecomm.)*. That of capacitor through inductor when the resistance of circuit is sufficiently low and current persists after the capacitor has completely discharged, so that it charges again in the reverse direction. This process is repeated until all initial energy is dissipated in resistance, including radiation.

oscillatory scanning *(Image Tech.)*. That in which the scanning spot moves repeatedly to and fro across the image, so that successive lines are scanned in opposite directions.

oscillatory zoning *(Min.)*. The compositional variation within a crystal which consists of alternating layers rich in the two end-members of an isomorphous solid-solution series.

oscillogram *(Electronics)*. Record of a waveform obtained from any oscillograph. Usually photograph of CRT display.

oscillograph *(Electronics)*. Oscilloscope with a photographic recording system to register waveforms displayed.

oscilloscope *(Electronics)*. (1) Equipment incorporating a cathode-ray tube, time-base generators, triggers, etc., for the display of a wide range of waveforms by electron beam. (2) Mechanical or optical equipment with a corresponding function, e.g. Duddell oscilloscope.

osculating circle *(Maths.)*. See **curvature**.

osculating orbit *(Astron.)*. The name given to the instantaneous ellipse whose elements represent the actual position and velocity of a comet or planet at a given instant (*epoch of osculation*).

osculating plane *(Maths.)*. See **moving trihedral**.

osculating sphere *(Maths.)*. At a point P on a space curve, a sphere that has four-point contact with the curve at P, i.e. the limiting sphere through four neighbouring points P, Q, R and S, on the curve, as Q, R and S tend to P. Its centre and radius are called respectively the *centre* and *radius* of *spherical curvature*.

osculation *(Maths.)*. See **point of osculation**.

osculum *(Zool.)*. In Porifera, an exhalant aperture by which water escapes from the canal system. adjs. *oscular*, *osculiferous*.

Osgood-Schlatter disease *(Med.)*. Osteochondritis of the tibial tubercle.

O-shell *(Phys.)*. The **electron shell** in an atom corresponding to a principal quantum number of five. In naturally occuring elements the shell is never completely filled but has most electrons, 21, for the element uranium.

Osler's nodes *(Med.)*. Painful papules on the digits in cases of bacterial endocarditis.

osmeterium *(Zool.)*. In the larvae of certain Papilionidae (Lepidoptera), a bifurcate sac exhaling a disagreeable odour which can be protruded through a slitlike aperture in the first thoracic segment.

osmic acid *(Chem.)*. An erroneous name for **osmium tetroxide**.

osmiophilic *(Chem.)*. Having an affinity for, staining readily with osmic acid, e.g., certain components and organelles of cells.

osmiridium *(Eng.)*. A very hard, white, naturally occurring alloy of osmium (17–48%) and iridium (49%) containing smaller amounts of platinum, ruthenium and rhodium. Used for the tips of pen-nibs.

osmium *(Chem.)*. A metallic element, a member of the platinum group. Symbol Os, at. no. 76, r.a.m. 190.2, mp 2700°C. Osmium is the densest element, rel.d. at 20°C 22.48. Like platinum it is a powerful catalyst for gas reactions, and is soluble in aqua regia, but unlike platinum when heated in air it gives an oxide, volatile OsO_4. Alloyed with iridium it forms an extremely hard material. See **osmiridium**.

osmium tetroxide *(Chem.)*. *Osmium (VIII) oxide*, OsO_4, yellow crystals which give off an ill-smelling, poisonous vapour. Its aqueous solution is used as a histological stain for fat, as a catalyst in organic reactions, and as a fixative in electron microscopy.

osmometer *(Chem.)*. An apparatus for the measurement of osmotic pressures.

osmoreceptors *(Biol.)*. Cells specialized to react to osmotic changes in their environment, e.g. cells which react to osmotic changes in the blood or tissue fluid and which are involved in the regulation of secretion of antidiuretic hormone by the neurohypophysis.

osmoregulation *(Zool.)*. The process by which animals regulate the amount of water in their bodies, and the concentration of various solutes and ions in their body fluids.

osmosis *(Chem.)*. Diffusion of a solvent through a semipermeable membrane into a more concentrated solution, tending to equalize the concentrations on both sides of the membrane.

osmotic coefficient *(Chem.)*. The quotient of the van't Hoff factor and the number of ions produced by the dissociation of 1 molecule of the electrolyte.

osmotic potential *(Bot.)*. ψ_π. That component of the **water potential** due to the presence of solutes; equal to minus the osmotic pressure. Also called solute potential, ψ_s.

osmotic pressure *(Bot.)*. π. The pressure which would have to be applied to the solution to make its chemical potential equal to that of pure, free, water at the same temperature, or to prevent osmotic water movement through a semi-permeable membrane between the solution and pure water.

osmotic pressure *(Chem.)*. The pressure which must be applied to a solution, separated by a **semi-permeable membrane** from pure solvent, to prevent the passage of solvent through the membrane. Symbol *Pi*. For substances which do not dissociate, it is related to the concentration (c) of the solution and the absolute temperature by the relationship $Pi = cRT$, where R is the gas constant.

osmotrophy *(Bot.)*. Nutrition based on the uptake of soluble materials.

os penis *(Zool.)*. A bone developed in the middle line of the penis in some Mammals, as Bats, Whales, some Rodents, Carnivores and Primates.

osphradium *(Zool.)*. A sense-organ of certain aquatic Mollusca, consisting usually of a patch of columnar ciliated epithelium and concerned in the assessment of suspended silt in the water entering the mantle chamber. adj. *osphradial*.

ossa *(Zool.)*. See **os** (2).

osseous *(Zool.)*. Bony; resembling bone.

ossicle *(Zool.)*. A small bone; in Echinodermata, one of the skeletal plates; in Crustacea, one of the calcified toothed plates of the gastric mill.

ossification *(Zool.)*. The formation of bone; transformation of cartilage or mesenchymatous tissue into bone. v. *ossify*.

oste- *(Genrl.)*. See **osteo-**.

osteitic fibrosa *(Med.)*. A condition in which there may be (*a*) a single cyst in a bone, or (*b*) cysts in many bones; the latter (*generalized osteitis fibrosa; Von Recklinghausen's disease of bone*) is due to loss of calcium salts from the bone, and is associated with a tumour of the parathyroid glands.

osteitis *(Med.)*. Inflammation of a bone.

osteitis deformans *(Med.)*. See **Paget's disease**.

osteitis fibrosa *(Med.)*. A decrease in the calcium of bone caused by overactivity of the parathyroid gland.

osteo-, oste- *(Genrl.)*. Prefix from Gk. *osteon*, bone.

osteoarthritis, ostearthritis *(Med.)*. A group of conditions in which the cartilage of joints is gradually worn away, and the bone adjacent to it remodelled.

osteoarthropathy *(Med.)*. Strictly, any disease affecting both bones and joints. Specifically, symmetrical enlargement of the bones of the hands and the feet with thickening of the fingers and toes associated especially with chronic diseases of the lungs or of the heart.

osteoblast *(Zool.)*. A bone-forming cell.

osteochondritis *(Med.)*. Inflammation of both bone and cartilage. See also **Perthe's disease**.

osteochondritis dessicans *(Med.)*. A rare disease characterized by avascular necrosis of bone and cartilage which eventually resolves spontaneously.

osteochondroma *(Med.)*. A tumour composed of bony and of cartilaginous elements.

osteoclasis *(Med.)*. The absorption and destruction of bone tissue by osteoclasts.

osteoclast *(Zool.)*. A bone-destroying cell, especially one which breaks down any preceding matrix, chondrified or calcified, during bone-formation.

osteocranium *(Zool.)*. The bony brain-case which replaces the chondrocranium in higher Vertebrates.

osteocyte *(Biol.)*. A bone cell derived from an osteoblast.

osteodermis *(Zool.)*. An ossified or partially ossified dermis: membrane bones formed by ossification of the dermis. adj. *osteodermal*.

osteodystrophia fibrosa *(Vet.)*. *Osteofibrosis*. A disease of the skeletal system in animals in which excessive amounts of calcium and phosphorus are withdrawn from the bones and replaced by fibrous tissue. Caused by an excess of phosphorus or a deficiency of calcium in the diet in the horse (*bran disease, millers' disease, big head of horses*), calcium deficiency in the pig (*snuffles*), and as a complication of chronic nephritis in the dog (*rubber jaw*).

osteofibrosis *(Vet.)*. See **osteodystrophia fibrosa**.

osteogenesis *(Zool.)*. See **ossification**.

osteogenesis imperfecta *(Med.)*. *Fragilitas ossium*. A condition in which a child is born with abnormally brittle bones, multiple fractures occurring.

osteoid *(Med.)*. Resembling bone.

osteology *(Zool.)*. The study of bones.

osteoma *(Med.)*. A tumour composed of bone.

osteomalacia *(Med.)*. A condition in which the bones soften as a result of absorption of calcium salts from them, usually due to deficiency in vitamin D in the diet. Asian immigrants and the elderly are particularly at risk. If it occurs in childhood while the bones are still forming it causes *rickets*.

osteomalacia *(Vet.)*. *Boglame, stiff sickness*. Failure of the bone to mineralize due to lack of dietary calcium. The adult equivalent of rickets. In cattle one form is associated with a deficiency of phosphorus resulting in both calcium and phosphorus being withdrawn from the bone.

osteomyelitis *(Med.)*. Inflammation of the bone-marrow and of the bone.

osteopathy *(Med.)*. A method of healing, based on the hypothesis that abnormalities in the human framework (bones, muscles, ligaments, etc.) ultimately cause damage by interfering with the blood and nerve supply to the body, thereby allowing other factors in ill-health to exert their influence unduly. These abnormalities are often the direct single cause of much suffering and they can be removed by skilled manual adjustment.

osteopetrosis *(Med.)*. *Albers-Schönberg disease*; congenital osteosclerotic anaemia; marble bones. A rare condition in which the bones become solid as a result of obliteration of the bone-marrow by bone, associated with enlargement of the liver and of the spleen, and with anaemia.

osteopetrosis gallinarum *(Vet.)*. *Marble bone disease, thick leg disease*. A chronic virus infection of chickens in which there is an excessive simulation of bone formation, leading to thickening and deformation of the bones.

osteophagia *(Vet.)*. An appetite for bones and dead animals, exhibited by herbivorous animals suffering from a deficiency of phosphorus and calcium salts in the diet.

osteophyte *(Med.)*. A bony excrescence or outgrowth from the margin of osteoarthritic joints or from diseased bone.

osteoporosis *(Med.,Zool.)*. Decrease in bone density and mass, often occurring in old age.

osteosarcoma *(Med.)*. A malignant tumour derived from osteoblasts, composed of bone and sarcoma cells.

osteosclereid *(Bot.)*. Sclereid having a columnar middle and enlarged ends like a stylized thigh-bone.

osteosclerosis *(Med.)*. Abnormal thickening of bone.

osteotomy *(Med.)*. Cutting of a bone.

ostiolate *(Bot.)*. Having an opening.

ostiole *(Bot.)*. A pore, especially one by which spores or gametes escape.

ostium *(Zool.)*. A mouthlike aperture; in Porifera, an inhalant opening on the surface; in Arthropoda, an aperture in the wall of the heart by which blood enters the heart from the pericardial cavity; in Mammals, the internal aperture of a Fallopian tube. adj. *ostiate*.

Ostracoda *(Zool.)*. A subclass of the Crustacea, with or without compound eyes; having a bivalve shell with adductor muscle; cephalic appendages well developed and complex; not more than 2 pairs of trunk limbs, often parthenogenetic, e.g. *Cypris*.

Ostwald colour atlas *(Image Tech.)*. A system of colour relations arranged according to hue, luminosity and saturation.

Ostwald's dilution law *(Chem.)*. The application of the law of mass action to the ionization of a weak electrolyte, yielding the expression

$$\frac{\alpha^2}{(1-\alpha)V} = K,$$

where *a* is the degree of ionization, *V* the **dilution** (2), and *K* the ionization constant, for the case in which two ions are formed.

Ostwald's theory of indicators *(Chem.)*. The assumption that all **indicators** (1), are either weak acids or weak bases, in which the colour of the ionized form differs markedly from that of the undissociated form.

OS & W *(Build.)*. Abbrev. for *Oak, Sunk and Weathered*.

otalgia *(Med.)*. Earache.

otic *(Med.)*. Pertaining to the ear or to the auditory capsule; one of the cartilage bones of the auditory capsule.

otitis *(Med.)*. Inflammation of the ear. *Otitis externa*, a term for various inflammatory conditions of the external ear. *Otitis media*, inflammation of the middle ear.

otocyst *(Zool.)*. In many aquatic Invertebrates, a sac lined by sensory hairlets, filled with fluid, and containing a calcereous or siliceous concretion (*otolith*) which subserves the equilibristic sense; in Vertebrates, part of the internal ear similarly constructed (as the utriculus).

otodectic mange *(Vet.)*. Mange affecting the external ear canal, particularly of dogs and cats, caused by *Otodectes cyanotis*.

otolith *(Zool.)*. The calcereous concretion which occurs in an otocyst.

otology *(Med.)*. That part of surgical science dealing with the organ of hearing and its diseases. n. *otologist*.

otorhinolaryngology *(Med.)*. That part of surgical science which deals with diseases of the ear, nose and throat.

otorrhoea, otorrhea *(Med.)*. A discharge, especially of pus, from the ear.

otosclerosis *(Med.)*. The formation of spongy bone in the capsule of the labyrinth of the ear, associated with progressive deafness.

otoscope *(Med.)*. An instrument for inspecting the external canal of the ear and the eardrum.

ot-, oto- *(Genrl.)*. Prefix from Gk. *ous*, gen. *otos*, ear.

otter, otter boards *(Ships)*. Oblong boards, bound with iron, attached to the sides of a trawl net eccentrically to the towing warps; they keep the mouth of the net open. As used in mine-sweeping, the otter is a heavy steel frame with horizontal vanes.

Otto cycle *(Autos.)*. The working cycle of a 4-stroke engine; suction, compression, explosion at constant volume, expansion, and exhaust, occupying 2 revolutions of the crankshaft.

ottrelite *(Min.)*. A manganese-bearing chloritoid mineral occurring in schists, a product of the metamorphism of certain argillaceous sedimentary rocks.

OTU *(Bot.)*. See **operational taxonomic units**.

Ouchterlony test *(Immun.)*. A precipitin test in which antigen and antibody are allowed to diffuse towards one another in a gel medium.

Oudin resonator *(Elec.Eng.)*. Coil, adjustable in number of turns to maximize coupling, for applying **effluve** to a patient.

Oudin test *(Immun.)*. A precipitin test in which antigen diffuses into antibody incorporated in a gel medium.

out-and-in bond *(Build.)*. The mode of laying ashlar quoins, so that they will be headers and stretchers alternately.

out-and-out *(Print.)*. Spacing out pages to their exact finished size with no allowance for trim between them.

outband *(Build.)*. A jamb stone laid as a stretcher and recessed to take a frame.

outbreeding *(Biol.)*. Sexual reproduction between unrelated individuals, thus increasing heterozygosity. Cf. **inbreeding**. See also **allogamy**.

outburst bank *(Hyd.Eng.)*. The middle part of the slope of a sea embankment, above the footing and below the swash-bank.

outcrop *(Geol.)*. An occurrence of a rock at the surface of the ground.

outcross *(Biol.)*. A cross to a strain with a different genotype.

outer *(Elec.Eng.)*. Either of the two conductors of a 3-wire distribution system which are respectively at a voltage above and below earth. Cf. the **neutral (middle) conductor,** which is at approximately earth potential.

outer *(Print.)*. An imposition in *sheetwork* which contains the first and last pages of the section, as distinct from the *inner* imposition printed on the reverse of the outer, and always containing the second and second last pages.

outer conductor *(Elec.Eng.)*. See **external conductor**.

outer dead-centre *(Eng.)*. The position of the crank of a reciprocating engine or pump when the piston is at the end of its outstroke, i.e. when the piston is nearest the crankshaft. Also *bottom dead-centre*.

outer marker beacon *(Aero.)*. A marker beacon, associated with the **ILS** or with the **standard beam approach system**, which defines the first predetermined point during a beam approach.

outer section *(Aero.)*. See **main plane**.

outer string *(Build.)*. The **string** farthest from the wall.

outfall *(Build.)*. The discharge point of a sewer.

outfall sewer *(Build.)*. The main sewer carrying away sewage material from a town to the place where it is to be purified, discharged or otherwise dealt with.

outgassing *(Chem.)*. Removal of occluded, absorbed or dissolved gas from a solid or liquid. For metals and alloys, done by heating in vacuo.

outgassing *(Electronics)*. Removal of maximum amount of residual gas in a valve envelope by baking the whole valve before sealing.

outgassing *(Geol.)*. The release of juvenile gases from molten rocks, leading to the development of the Earth's atmosphere and oceans.

outgassing *(Space)*. Spontaneous liberation of gas from a material in a space environment, sometimes termed *off-gassing*. To avoid contamination the material is left for some time before nearby instruments are used.

out-gate *(Foundry)*. See **riser**.

outgoing feeder *(Elec.Eng.)*. A feeder along which power is supplied from a substation or generating station.

outgroup *(Behav.)*. See **ingroup/outgroup**.

outlier *(Geol.)*. A remnant of a younger rock which is surrounded by older strata.

outline letters *(Print.)*. Display types in which the outline only of the letter is shown; sometimes issued as part of a type family, but may also be specifically designed type style on its own.

out of balance *(Eng.)*. Said of a rotating machine element which is imperfectly balanced, or of a mechanism or machine which contains such an element.

out of phase *(Elec.Eng.,Telecomm.)*. See under **phase**.

out of wind *(Build.)*. A term applied to a flat surface; a plane surface which is not twisted, timber free from warp or twist.

output *(Comp.)*. (1) Information leaving a device, data resulting from processing. (2) To give out information, to print or transfer to auxiliary storage the data resulting from processing.

output *(Telecomm.)*. Audio, electric or mechanical signal delivered by instrument or system to a load.

output capacitance *(Electronics)*. (1) The capacitive component of the output impedance or transducer, amplifier, or other circuit or device. (2) The anode-cathode capacitance of a thermionic valve.

output characteristic *(Electronics)*. See **transistor characteristics**.

output coefficient *(Elec.Eng.)*. See **specific torque coefficient**.

output device *(Comp.)*. Peripheral which translates signals from the computer into a human readable form or into a form suitable for re-processing by the computer at a later stage.

output gap *(Electronics)*. An interaction gap through which an output signal can be withdrawn from an electron beam.

output impedance *(Elec.Eng.)*. That presented by the device to the load and which determines **regulation** (voltage drop) of source when current is taken. In a linear source, the backward impedance when the e.m.f. is reduced to zero. See *Thévenin's theorem*; also called *source impedance*.

output meter *(Elec.Eng.)*. That which measures output voltage of an oscillator, amplifier, etc. Calibrated in **volts**, or **power level** in dB in relation to **zero power level** (1 milliwatt) when circuit is properly terminated. See **volume unit**.

output noise *(Genrl.)*. See **thermal noise**.

output regulation *(Elec.Eng.)*. Of a power supply, the variation of voltage with load current.

output transformer *(Elec.Eng.)*. One which couples last stage in an amplifier to the load, e.g. a loudspeaker or line.

output valve *(Electronics)*. One designed for delivering power to a load, e.g. line or loudspeaker, voltage gain not being relevant. Final stage of any multivalve amplifier.

output winding *(Elec.Eng.)*. That from which power is withdrawn in a transformer, transductor or magnetic amplifier.

outrigger *(Build.)*. (1) A projecting beam carrying a suspended scaffold. (2) Timbers built across a gable end to hold a rafter for a projecting verge.

outset *(Print.)*. A section placed on the outside of the main section, and sometimes called a *wrap round*.

outside crank *(Eng.)*. An overhung, or single-web crank attached to a crankshaft outside the main bearings.

outside cylinders *(Eng.)*. The steam cylinders carried outside the frame of a locomotive, working on to crank pins in the driving wheels.

outside gouge *(Build.)*. A firmer gouge having the bevel ground upon the convex side of the cutting edge.

outside lap *(Eng.)*. The amount by which the slide-valve of a steam-engine overlaps the edge of the steam ports when in mid-position. Also called *steam lap*.

outside lining *(Build.)*. The external member of a cased frame.

outside loop *(Aero.)*. See **inverted loop**.

outside sorts *(Print.)*. Accented letters, technical and mathematical characters, etc. extraneous to the usual alphabets, and accommodated separately from the normal type case, die case or magazine. Also *side sorts*.

out-takes *(Image Tech.)*. **Takes** of scenes photographed and printed for a production but not used in the finally edited version.

out-to-out *(Build.)*. A term applied to an overall measurement across a piece of framing.

outturn sheet *(Paper)*. A representative sample sheet of a particular batch of paper.

outwash fan *(Geol.)*. A sheet of gravel and sand, lying beyond the margins of a sheet of till, deposited by melt-waters from an ice-sheet or glacier.

ova *(Bot.,Zool.)*. Pl. of *ovum*.

ovalbumin *(Chem.)*. Same as *egg albumen*.

ovalizing balance *(Horol.)*. Bimetallic balance which tends to become elliptical with use in temperature, this characteristic being exploited to provide temperature compensation.

oval pistons *(Autos.)*. (1) Pistons, originally round, worn oval through friction at the thrust faces. (2) Pistons purposely turned slightly oval, to compensate for the unequal diametral expansion.

ovals of Cassini *(Maths.)*. Curves defined by the bipolar equation $rr' = k^2$. Each consists of either two ovals, one surrounding each reference point, or a single oval surrounding both reference points, according as $k < c$ or $k > c$ respectively, where c is the distance between the two reference points. If $k = c$, it reduces to the **lemniscate of Bernoulli**.

oval window *(Zool.)*. See **fenestra ovalis**.

ovarian *(Med.)*. Pertaining to or connected with the ovary.

ovariole *(Zool.)*. In Insects, one of the egg-tubes of which the ovary is composed.

ovary *(Bot.)*. (1) The hollow structure, the basal part of a carpel or of a syncarpous gynoecium, which contains the ovules. (2) Also *pistil*.

ovary *(Zool.)*. A female gonad; a reproductive gland producing ova, adj. *ovarian*.

ovate *(Bot.)*. Egg-shaped with the broadest part nearer to the point of attachment.

oven-dry paper *(Paper)*. See **bone-dry paper**.

oven-type furnace *(Eng.)*. Industrial heat-treatment furnace fired under the hearth, the live gases flowing directly into the heating chamber through live-gas flues disposed along each side of the hearth. Also *semimuffle-type furnace*.

overall efficiency *(Elec.Eng.)*. Ratio of useful output to total input power.

overall luminous efficiency *(Phys.)*. Ratio of luminous flux of lamp to total energy input. Not to be confused with *luminous efficiency*.

overall merit *(Telecomm.)*. System of rating a radio channel (especially short wave) on scale 0–5, derived from signal strength, fading, interference, modulation, depth, distortion.

overblowing *(Eng.)*. Continuation of forcing oxygen through molten steel in **Bessemer process** after the carbon has been removed, resulting in oxidizing of iron.

overburden *(Build.)*. The encallow, or overlying stratum of soil. Generally applied to brickfields or to potential area of gravel for quarrying.

overburden *(Min.Ext.)*. Earth or rock overlying the valuable deposit. In smelting, a furnace is *overburdened* when the ratio of ore to flux or fuel is too high.

overcasting *(Print.)*. The method of sewing used to make separate leaves into sections for binding. Also called *whipping*, *whipstitching*.

overcloak *(Build.)*. That part of the overlapping edge of a sheet which extends over the **roll** to the flat surface beyond.

overcoil *(Horol.)*. The last coil of a balance spring, which is raised above the plane of the spring and then bent to form a terminal curve.

overcompounded generator *(Elec.Eng.)*. A compound-wound d.c. generator in which the series winding is so designed that the voltage rises as the load increases.

overcompounded motor *(Elec.Eng.)*. A compound-wound d.c. motor in which the series winding is so designed that the speed rises with an increase in load.

overcurrent relay *(Elec.Eng.)*. A relay which operates as soon as the current exceeds a certain predetermined value. Also called *overload relay*.

overcurrent release *(Elec.Eng.)*. A device for tripping an electric circuit when the circuit in it exceeds a certain predetermined value. Also called *overload release*.

overdoor *(Build.)*. (1) An ornamental doorhead. (2) A pediment.

overdrive *(Autos.)*. A method of reducing engine r.p.m. in relation to road speed, using a separate epicyclic gear unit. Now largely supplanted by an additional high ratio gear within the main gear box.

over-exposure *(Image Tech.)*. Excess of exposure of any sensitive surface, above that required for the proper gradation of light and shade.

overfeed fabric *(Textiles)*. A warp-knitted fabric made with excess yarn being fed on one warp to form loops and underflaps that appear as a pile.

overfeed stoker *(Eng.)*. A **mechanical stoker** consisting of a hopper from which the fuel is continuously fed on to the bars of an inclined stepped grate, mechanically oscillated or rocked to cause the burning fuel to descend towards an ash table.

overflow *(Comp.)*. Occurs when arithmetic operations produce results which are too large to store.

overflow flag *(Comp.)*. Single bit which is set to 1 when **overflow** occurs during an arithmetic operation.

overfold *(Geol.)*. A fold with both limbs dipping in the same direction, but more steeply inclined than the other. Cf. **isoclinal fold**.

overfold *(Print.)*. A lip or overhang formed by the leading edge of the section or copy when the fold is out of centre. Over-adjustment of the folding mechanism will result in *underfold*.

overgassing *(Eng.)*. Condition occurring in gas-heated furnace or appliance when the burners are calibrated for, and operated at, a higher gas rate than that actually required.

overgrainer *(Build.)*. A long haired brush used for graining in marbling. The standard overgrainer is hog hair. Special overgrainers such as the pencil or fantail types are available for imitating certain types of grain.

overgraining *(Build.)*. A coat of graining colour applied over grained work so as to produce shades across the work.

overgrowth *(Crystal.)*. See **crystalline overgrowth**.

overhand stopes *(Min.Ext.)*. *Stopes* in which severed ore from an inclined seam or lode gravitates downward to tramming level.

overhang *(Aero.)*. (1) In multiplanes, the distance by which the tip of one of the planes projects beyond the tip of another. (2) In a wing structure, the distance from the outermost supporting point to the extremity of the wing tip.

overhang *(Elec.Eng.)*. See **armature end connections**.

overhaul *(Civ.Eng.)*. The excess of actual *haul* above the *free haul*, i.e. the haul which is conveyed without extra cost.

overhaul period *(Aero.)*. See **time between overhauls**.

overhead camshaft *(Autos.)*. One running across the cylinder heads of an engine, operating the valves directly or through rockers. See **camshaft, timing gear**.

overhead-contact system *(Elec.Eng.)*. The method of supplying current to the vehicles of an electric traction system whereby the current is collected from a contact wire suspended above the track by means of current-collectors mounted on the roof of the vehicle. The term may also refer to the actual contact wire and its supporting structure.

overhead crossing *(Elec.Eng.)*. A device used on the overhead-contact wire system of an electric railway or tramway, to allow the crossing of 2 contact wires and permit the passage of a current-collector along either wire.

overhead expenses *(Build.)*. See **on-costs**.

overhead projector *(Phys.)*. See **projector**.

overhead transmission line *(Elec.Eng.)*. A transmission line in which the conductors are supported on poles, towers, or other similar structures at a considerable height above the earth.

overhead travelling-crane *(Eng.)*. A workshop crane consisting of a girder along which a wheeled crab can be traversed. The girder is mounted on wheels running on rails fixed along the length of the shop, near the roof. Also called *shop traveller*.

overhead valves *(Autos.)*. In a vertical petrol- or oil-engine, inlet and exhaust valves working in the surface of the head opposite the piston, either in a vertical or inclined position.

overheated *(Eng.)*. Said of metal which has been heated in preparation for hot-working, or during a heat-treating operation, to a temperature at which rapid grain growth occurs and large grains are produced. The structure and properties can be restored by treatment, and in this respect it differs from burning (see **burnt metal**).

overlap *(Elec.Eng.)*. The period (often expressed in electrical degrees) which is required before commutation can be completed successfully in an electrical converter. Occurs because of source impedance in the a.c. supply to the converter.

overlap *(Geol.)*. The relationship between conformable strata laid down during an extension of the basin of sedimentation (e.g. on the margins of a slowly sinking landmass), so that each successive stratum extends beyond the boundaries of the one lying immediately beneath. Cf. **off-lap** and **overstep**.

overlapping covers *(Print.)*. The term is used only for pamphlet binding, where the cover may overlap as distinct from being **cut flush**.

overlapping genes *(Biol.)*. Some small DNA viruses exploit the degeneracy of the genetic code by making different proteins from overlapping sequences of DNA. Achieved by displacing the **reading frame** by one or two bases.

overlap span *(Elec.Eng.)*. See **section gap**.

overlap test *(Elec.Eng.)*. A test used for locating a fault in a cable; the resistance between the cable and earth is measured, first with the far end of the cable earthed, and then with it free.

overlay *(Comp.)*. A section of computer code that is loaded into an area of memory that was previously allocated to another section of the same executing program. See **segmentation**.

overlay *(Image Tech.)*. Combination of two video images using information from one to **key** it into the other.

overlay *(Print.)*. (1) To adjust the impression surface of a letterpress machine in order to increase the pressure on dark tones and decrease it on light. (2) Translucent paper or transparent film covering original artwork to protect it from damage, or to enable instructions to the camera operator, platemaker or printer to be shown. The overlay sheet may also indicate how the artwork should be broken down for multi-colour printing.

overlearning *(Behav.)*. A learning procedure where training or practice on what is being learned continues beyond the point where learning can be said to be adequate (learning to criterion). Overlearning often results in improvements in efficiency and to changes in the organization of performance (e.g. from conscious to *automatic* control).

overload *(Elec.Eng.)*. One exceeding the level at which operation can continue satisfactorily for an indefinite period. Overloading may lead to distortion or to overheating and risk of damage depending on the type of circuit or device. In many cases temporary overloads are permissible. See **overload capacity**.

overload capacity *(Elec.Eng.)*. Excess capacity of a generator over that of its **rating**, generally for a specified time.

overload protective system *(Elec.Eng.)*. A system of protecting an electric power network by means of overcurrent relays. To provide discrimination, the relays have time lags, graded so that the relays more remote from the supply point have shorter lags.

overload relay *(Elec.Eng.)*. See **overcurrent relay**.

overload release *(Elec.Eng.)*. See **overcurrent release**.

overlocking *(Textiles)*. Using an overseaming machine two pieces of fabric may be joined by a double-chain stitch.

overman *(Min.Ext.)*. (1) An underground manager of one or more ventilating districts in a coalmine. (2) An umpire appointed to an arbitration board in a mine dispute.

overmodulation *(Telecomm.)*. Attempted modulation to depth exceeding 100%, i.e. to such a degree that amplitude falls to zero for an appreciable fraction of the modulating cycle, with marked distortion.

overpoled copper *(Eng.)*. See **poling**.

overproof *(Genrl.)*. See **proof**.

overreach *(Vet.)*. An error of gait in the horse, in which the toe of the hindfoot strikes the heel of the forefoot.

overrigid *(Eng.)*. See **redundant**.

overrun *(Print.)*. (1) To carry words from the end of one line of type to the beginning of the next, and so on until the matter fits. Insertions or deletions frequently necessitate overrunning. (2) Running an extra quantity of printed paper in excess of the order. The extra copies can be used for setting up for further printings or in the finishing department.

overs *(Print.)*. Extra sheets allowed to a job to provide for ordinary **spoilage**.

oversailing courses *(Build.)*. Brick or stone courses projecting from a wall for the sake of appearance only, as distinct from corbels, which are normally load-carrying.

oversaturated *(Geol.)*. Refers to an igneous rock in which excess silica crystallizes as a separate silica mineral or as a glass. See **undersaturated**.

overscanning *(Image Tech.)*. Deflection of electron beam beyond the phosphor in a television reproducer or a cathode-ray oscilloscope.

overshoot *(Image Tech.)*. A transient excess change at the start or end of a pronounced variation of signal; in video pictures it can appear as fringes at the boundaries of large changes of image brightness.

overshoot *(Telecomm.)*. For a step change in signal amplitude *undershoot* and *overshoot* are the maximum transient signal excursions outside the range from the initial to the final mean amplitude levels.

overshot duct *(Print.)*. An ink duct using a blade mounted above the roller which rotates in a trough of ink.

overshot ink fountain *(Print.)*. See **overshot duct**.

overshot tool *(Min.Ext.)*. See **fishing tool**.

overshot wheel *(Eng.)*. A water-wheel in which the discharge flume or head-race is at the top, the water flowing tangentially into the bucket near the top of the wheel.

oversite concrete *(Build.)*. A concrete layer covering a building site within the external walls, serving to keep out ground air and moisture, and also providing a foundation for the floor.

overspeed protection *(Elec.Eng.)*. Protection, usually by means of a centrifugally-operated device, against excessive speed of an electric machine; used on inverted rotary converters and also on d.c. motors in certain cases.

overspun wire *(Acous.)*. A wire from a musical instrument, e.g. for a low note in a piano, round which a loading wire is tightly spun to lower its fundamental pitch.

oversteer *(Autos.)*. Tendency of a vehicle to exaggerate the degree of turn applied to the steering wheel. Cf. **understeer**.

overstep *(Geol.)*. The structural relationship between an unconformable stratum and the outcrops of the underlying rocks, across which the former transgresses. Cf. **overlap**.

overstrain *(Eng.)*. The result of stressing an elastic material beyond its **yield point**; a new and higher yield point results, but the elastic limit is reduced.

overthrust *(Geol.)*. A fault of low hade along which one slice or block of rock has been pushed bodily over another, during intense compressional earth-movements. The horizontal displacement along the **thrust plane** may amount to several kilometres.

overtone *(Acous.)*. In a complex tone, any of the components above the fundamental frequency.

overtone crystal *(Electronics)*. A piezoelectric crystal operating at a higher frequency than the fundamental for any given mode of vibration.

overvoltage *(Electronics)*. A voltage higher than normal or predetermined limiting value, especially when likely to cause damage to or to destroy electronic components or circuits.

overvoltage protective device *(Elec.Eng.)*. A device giving protection to electrical apparatus against the possibility of damage caused by an excess voltage, i.e. a voltage above normal.

overvoltage release *(Elec.Eng.)*. A device arranged to trip an electrical circuit when the voltage in it exceeds a certain predetermined value.

overwriting *(Comp.)*. Erasing of a data item from store by writing another in its place.

oviduct *(Zool.)*. The tube which leads from the ovary to the exterior and by which the ova are discharged. adj. *oviducal*.

oviferous, ovigerous *(Zool.)*. Used to carry eggs, as the *ovigerous* legs of Pycnogonida.

oviparous *(Zool.)*. Egg-laying; cf. **viviparous**.

oviposition *(Zool.)*. The act of depositing eggs.

ovipositor *(Zool.)*. In some Fish (as the Bitterling), a flexible tube formed by the extension of the edges of the genital aperture in the female; in female Insects, the egg-laying organ.

ovisac *(Zool.)*. A brood-pouch; an egg receptacle.

ovolo *(Arch.)*. A quarter-round convex moulding.

ovotestis *(Zool.)*. A genital gland which produces both ova and spermatozoa, as the gonad of some Gastropoda.

ovoviviparous *(Zool.)*. Producing eggs which hatch out within the uterus of the mother.

ovulation *(Zool.)*. The formation of ova; in Mammals, the process of escape of the ovum from the ovary.

ovule *(Bot.)*. The structure in a seed plant, consisting of embryo sac, nucellus and integuments, which after fertilization develops into the seed.

ovule culture *(Bot.)*. The culture on a suitable medium of excised ovules, e.g. in an attempt, by adding pollen, to obtain **in vitro** fertilization and viable seed in crosses normally frustrated by the failure of the pollen to grow on the stigma or through the style.

ovum *(Biol.)*. A non-motile female gamete. An egg or egg-cell.

Owen bridge *(Elec.Eng.)*. An a.c. bridge of the four arm (or Wheatstone) type, used for the measurement of inductance.

own ends *(Print.)*. See **self end-papers**.

oxalates (ethandioates) *(Chem.)*. Salts and esters of oxalic acid.

oxalic acid *(Chem.)*. HOOC.COOH,$2H_2O$, a dibasic acid which crystallizes with 2 molecules of water, mp 101°C, mp (anhydrous) 190°C; it sublimes readily, occurs in many plants, is obtainable by the oxidation of many organic substances. Strong oxidising agents convert it to CO_2.

oxaluria *(Med.)*. The presence of crystals of oxalates in the urine.

oxalyl *(Chem.)*. The bivalent acid radical

$$O{=}\overset{|}{C}\cdot\overset{|}{C}{=}O.$$

oxamic acid *(Chem.)*. $H_3N.CO.COOH$, a crystalline powder, mp 210°C (with decomposition).

oxamide (ethandiamide) *(Chem.)*. $H_2N.CO.CO.NH_2$, the normal amide of oxalic acid, a crystalline powder which sublimes when heated.

oxbow lake *(Geol.)*. A meander loop which has been cut off.

ox-eye *(Arch.)*. An oval-shaped dormer window.

Oxford hollow *(Print.)*. A hollow-back, in which the cover is kept separated from the back of the sections by a flattened tube which is glued to each.

Oxford shirting *(Textiles)*. A plain-weave, warp striped shirting in which two ends weave as one.

oxidant *(Aero.)*. The oxygen-bearing component in a bipropellant rocket, usually liquid oxygen, high-test hydrogen peroxide, or nitric acid.

oxidase *(Bot.,Zool.)*. One of a group of enzymes occurring in plant and animal cells and promoting oxidation.

oxidates *(Geol.)*. Those sedimentary rocks and weathering products whose composition and geochemical behaviour is mainly determined by the oxidation process. This category includes mainly those minerals and rocks containing iron and manganese.

oxidation *(Chem.)*. The addition of oxygen to a compound. More generally, any reaction involving the loss of electrons from an atom. It is always accompanied by reduction.

oxidation *(Print.)*. Surface corrosion in the form of light spots which occur on lithographic plates which have been improperly protected by gumming up. The spots accept ink and usually print with a characteristic clustered formation.

oxidation number *(Chem.)*. For simple atoms or ions, it is equal to the charge. For more complex groups, a formal oxidation number is often applied to specific atoms, particularly the central atom of a co-ordination compound. Thus, assuming that the ligands are chloride ions (o. n. = −1), the o. n. of copper in the complex ion $[CuCl_4]^{2-}$ may be deduced to be +2.

oxidation-reduction indicators *(Chem.)*. Substances which exist in oxidized and reduced forms having different colours, used to give approximate values of oxidation-reduction potentials.

oxidation-reduction potential *(Chem.)*. See **standard oxidation-reduction potential**.

oxidative decarboxylation *(Biol.)*. A decarboxylation which is coupled with the oxidation of a substrate. The four oxidative decarboxylations of the **tricarboxylic cycle** generate the electrons whose passage along the electron transport chain produces ATP by **oxidative phosphorylation**.

oxidative phosphorylation *(Biol.)*. The generation of ATP which is coupled with the passage of electrons along the *electron transport chain*. See **chemiosmosis**.

oxide-coated cathode *(Electronics)*. One coated with oxides of the alkali and alkaline-earth metals, to produce thermionic emission at relatively low temperatures.

oxide-film arrester *(Elec.Eng.)*. A lightning arrester using lead peroxide (a good conductor) which changes to red lead (a good insulator) when heated by the passing of a current.

oxide isolation *(Electronics)*. The isolation of the circuit elements in a microelectronic circuit by forming a layer of silicon oxide around each element.

oxides *(Chem.)*. Compounds of oxygen with another element. Oxides are formed by the combination of oxygen with most other elements, particularly at elevated temperatures, with the exception of the noble gases and some of the noble metals.

oxidizer *(Chem.)*. A substance that removes electrons from another and produces heat; in the case of a rocket, a gas.

oxidizing agent *(Chem.)*. A substance which is capable of bringing about the chemical change known as oxidation in another substance. It is itself reduced.

oxidizing flame *(Chem.)*. The outer cone of a nonluminous gas flame, which contains an excess of air over fuel.

oxidizing roast *(Eng.)*. Heating of sulphide ores or concentrates to burn off part or all of the sulphur as dioxide.

oximes *(Chem.)*. Compounds obtained by the action of

hydroxylamine on aldehydes or ketones, containing the bivalent oximino group = N.OH attached to the carbon. The oximes of aldehydes are termed *aldoximes* and those obtained from ketones *ketoximes*. Collectively, hydroximino-alkanes.

oxine *(Chem.)*. *8-Hydroxy-quinoline* is used as a reagent in analysis of metals. When the H in the —OH group is substituted by a metal, insoluble compounds result, but their solubility varies according to temperature, concentration and other conditions, thus making it possible to use the differences for analysis.

oxo *(Chem.)*. The radical O= in organic compounds, e.g., ketones $R_2C=O$.

oxonium salts *(Chem.)*. Derivatives of a hypothetical oxonium hydroxide, $H_3O.OH$, a base with a trivalent oxygen atom. Such substances are readily produced from several heterocyclic compounds containing oxygen, e.g. dimethylpyrone, which forms salts with hydrochloric acid etc. by direct addition to the oxygen atom.

oxyacetylene cutting *(Eng.)*. See under **oxyacetylene welding**.

oxyacetylene welding *(Eng.)*. Welding with a flame resulting from the combustion of oxygen and acetylene. Many ferrous metals can also be cut by the same process. The cut is started by introducing a jet of oxygen on to the metal after it has been preheated.

oxycelluloses *(Chem.)*. Products formed by the action of oxidizing agents on cellulose. They dissolve in dilute alkaline solution and have strong reducing properties. When boiled with hydrochloric acid they yield furfuraldehyde quantitatively. This reaction serves for the analytical estimation of oxycelluloses.

oxychloride cement *(Build.)*. A strong, extremely hard-setting cement used in making composition floors; composed of an oxide and a chloride of magnesia chemically combined.

oxydactylous *(Zool.)*. Having narrow-pointed digits.

oxygen *(Chem.)*. A nonmetallic element, symbol O, at. no. 8, r.a.m. 15.9994, valency 2. It is a colourless, odourless gas which supports combustion and is essential for the respiration of most forms of life. Mp −218.4°C, bp −183°C, density 1.429 04 g/dm^3 at s.t.p., formula O_2. An unstable form is ozone, O_3. Oxygen is the most abundant element, forming 21% by volume of the atmosphere, 89% by weight of water, and nearly 50% by weight of the rocks of the earth's crust. It is manufactured from liquid air, for use in hot welding flames, in steel manufacture, in medical practice and in anaesthesia; liquid oxygen is much used in rocket fuels.

oxygen dissociation curve *(Med.)*. The sigmoid curve describing the saturation of haemoglobin with oxygen when related to the partial pressure of oxygen.

oxygen-isotope determinations *(Geol.)*. A method of using the 0-16 and 0-18 isotope ratios in marine fossils to determine the sea temperature at the time of deposition.

oxygen lancing *(Eng.)*. A process used principally for cutting heavy sections of steel or cast iron, in which oxygen is fed to the cutting zone through a length of steel tubing which is consumed as the cutting action proceeds. The cutting zone or the end of the tube has to be preheated to commence cutting.

oxygen scavenger *(Eng.)*. An additive to boiler feed water which removes traces of oxygen and helps to prevent corrosion, e.g. hydrazine or tannin.

oxyhaemoglobin *(Chem.)*. The product obtained by the action of oxygen upon haemoglobin. The oxygen is readily given up when oxygen-tension is low in surrounding medium.

oxyhornblende *(Min.)*. See **basaltic hornblende**.

oxyhydrogen welding *(Eng.)*. A method of welding in which the heat is produced by the combustion of a mixture of oxygen and hydrogen.

oxyntic *(Zool.)*. Acid-secreting.

oxyphile *(Geol.)*. Descriptive of elements which have an affinity for oxygen, and therefore occur in the oxide and silicate minerals of rocks rather than in the sulphide minerals or as native elements. Also *lithophile*.

oxyphobic *(Bot.,Zool.)*. Unable to withstand soil acidity.

oxyproline *(Chem.)*. *4-Hydroxypyrrolidine-2-carboxylic acid*, obtained by cleavage of gelatine.

oxytetracycline *(Med.)*. A broad spectrum antibiotic used in a wide variety of infections.

oxytocin *(Med.)*. An octapeptide hormone secreted by the posterior lobe of the pituitary body (*neurohypophysis*) which stimulates the uterine muscle to contract, and causes milk ejection from lactating mammary glands.

ozocerite, ozokerite *(Min.)*. A mineral paraffin wax, of dark yellow, brown, or black colour.

ozoena *(Vet.)*. Catarrh of the frontal and maxillary sinuses of the horse.

ozokerite *(Min.)*. See **ozocerite**.

ozone *(Chem.)*. O_3. Produced by the action of ultraviolet radiation or electrical corona discharge on oxygen or air. Powerful oxidizing agent.

ozonides *(Chem.)*. Explosive organic compounds formed by the addition of an ozone molecule to a double bond.

ozonizer *(Chem.)*. An apparatus in which oxygen is converted into ozone by being subjected to an electric brush discharge. Ozone was formerly thought to be beneficial in air-conditioning, but is now known to be toxic.

P

p *(Biol.)*. See **chromosome mapping**.
p *(Genrl.)*. Symbol for pico-; proton.
p *(Phys.)*. Symbol for (1) electric dipole moment, (2) impulse, (3) momentum, (4) pressure.
π *(Genrl.)*. See under **pi**.
ϕ *(Phys.)*. Symbol for (1) heat flow rate, (2) luminous flux, (3) magnetic flux, (4) work function.
φ *(Chem.)*. A symbol for: (1) the phenyl radical C_6H_5; (2) *amphi-*, i.e. containing a condensed double aromatic nucleus substituted in the 2.6 positions.
φ *(Phys.)*. Symbol for phase displacement.
P *(Chem.)*. The symbol for phosphorus.
P *(Phys.)*. Symbol for poise.
P *(Phys.)*. Symbol for (1) electric polarization, (2) power, (3) pressure.
Π *(Chem.)*. A symbol for pressure, especially osmotic pressure.
Π *(Elec.Eng.)*. Symbol for Peltier coefficient.
ψ *(Chem.)*. A symbol for pseudo-.
Ψ *(Phys.)*. Symbol for (1) electric flux, (2) magnetic field strength.
[*P*] *(Chem.)*. The symbol for parachor.
P-680 *(Bot.)*. A (chlorophyll a)-protein complex (absorption peak at $\lambda = 684$ nm) that acts as the light trap in **photosystem II**.
P-700 *(Bot.)*. A (chlorophyll a)-protein complex (absorption peak at $\lambda = 700$ nm) that acts as the light trap in **photosystem I**.
Pa *(Chem.)*. The symbol for protactinium.
Pa *(Genrl.)*. Abbrev. for *Pascal.*
PABX *(Telecomm.)*. Abbrev. for *Private Automatic Branch eXchange*.
pacemaker *(Med.)*. Electronic device implanted in to chest wall with conducting wire to the heart to regulate abnormal heart rhythms by means of electrical pulses. Used when the sino-atrial or atrio-ventricular nodes, which usually regulate heart rhythm, are deficient or diseased.
pacemaker region *(Med.)*. A region of the body which determines the activity of other parts of the body; e.g. the sino-atrial node from which originate the impulses causing the heart beat.
Pachuca tank *(Min.Ext.)*. Also *Brown tank*. Large vertically-set cylindical vessel used in chemical treatment of ores, in which pulp is reacted with suitable solvents for long periods, while the contents are agitated by compressed air.
pachydermatocele *(Med.)*. A soft flabby tumour, composed of fibrous and nervous tissue, which hangs over the face or the ears.
pachydermatous *(Zool.)*. Thick-skinned.
pachydermia, pachyderma *(Med.)*. Abnormal thickness of the skin.
pachymeningitis *(Med.)*. Inflammation of the **dura mater**.
pachyphyllous *(Bot.)*. Having thick leaves.
pachytene *(Biol.)*. The third stage (*bouquet stage*) of meiotic prophase, intervening between zygotene and diplotene, in which condensation of chromosomes commences.
Pacinian corpuscles *(Zool.)*. In Vertebrates, skin receptors in which the nerve-ending is surrounded by many concentric layers of connective tissue. Sensitive to pressure. Also *Vater's corpuscles*.
pack *(Min.Ext.)*. Waste rock, mill tailings etc. used to support excavated stopes. Also *fill*.
package *(Textiles)*. Trade term for a cop, cheese, cone etc. of yarn, indicating that it is in convenient form for transport, or further processing.
packaged *(Elec.Eng.)*. (1) Description of electrical components of circuits which are sealed from ambient conditions. Many forms are possible, including plastic enclosures for discrete transistors, ceramic and plastic enclosures for integrated circuits, resin potting for printed circuit boards and hermetically sealed metal boxes for complete systems. (2) Said of an equipment complete for use, made up from a series of sub-assemblies.
packaged *(Nuc.Eng.)*. Said of a reactor of limited power which can be packaged and erected easily on a remote site.
package dyeing *(Textiles)*. A method of dyeing yarn already on a package.
packed column *(Chem.Eng.)*. A column used for distillation or absorption, which consists of a shell filled either with random material such as coke or other broken inert material, or with one of the proprietary rings or similar, usually ceramic or stainless steel (*Raschig*, *Lessing*, or *Pall rings*).
packed-hole assembly *(Min.Ext.)*. See **bottom-hole assembly**.
packer *(Min.Ext.)*. Expandible plugs sent down the bore hole to seal-off a section, often prior to making a more permanent seal with cement. Can be used outside the casing, between casing and drill-tube or within the drill tube.
packet *(Telecomm.)*. Self-contained message or component of a message, comprising address, control and data signals, which can be transferred as an entity within a communications network.
packet switching *(Telecomm.)*. In data communications, a method in which messages are assembled into one or more **packets**, including address and control codes, which can be sent independently through a **network**, collected and then re-assembled into the original information at the destination. Individual packets do not necessarily travel by the same route and only occupy a *channel* during transmission, unlike conventional switching when a channel is established and remains open for the duration of the whole transmission, whether data is flowing or not.
pack-hardening *(Eng.)*. **Case-hardening**, using a solid carburizing medium, followed by a hardening treatment.
packing *(Build.)*. The operation of filling in a double or hollow wall.
packing *(Comp.)*. See **data compaction**.
packing *(Eng.)*. (1) Material inserted in **stuffing boxes** to make engine and pump rods pressure tight; traditionally hemp, but now often special compositions, O- and other shaped rings, or of metal in **metallic packing**. (2) Spacers between clamped surfaces.
packing density *(Comp.)*. Measure of the quantity of data that can be held per unit length on a storage medium, e.g., bits/cm of magnetic tape.
packing fraction *(Phys.)*. The mass M of a nuclide expressed in *atomic mass units* differs slightly from a whole number, the mass number A. The packing fraction is defined as $(M-A)/A$. The *mass defect* or *mass decrement* is $(M-A)$.
packing ratio of DNA *(Biol.)*. The ratio of the calculated length of a helical DNA molecule to its length after organization into more compact form as **nucleosomes** and higher order coiling or folding. The packing ratio of metaphase chromosome DNA is about 10 000 to 1.
pad *(Telecomm.)*. (1) Small preset adjustable capacitor, to regulate the exact frequency of oscillation of an oscillator, or a tuned circuit in an amplifier or filter; also *padder*, *trimmer*. (2) Fixed attenuator inserted in a transmission line or waveguide.

pad *(Textiles)*. A padding mangle used to apply chemicals (including dyestuff) in solution to open width fabrics in which the excess is mechanically expressed.

paddle hole *(Hyd.Eng.)*. The opening in a lockgate through which water flows from the high-level pond to the lock-chamber, or from the lock-chamber to the low-level pond.

paddle plane *(Aero.)*. See **cyclogyro**.

paddles *(Print.)*. Rotating curved or shaped fingers of the fly which transfer copies from the folder to the delivery belt of web-fed rotary presses.

paddle-wheel fan *(Eng.)*. See **centrifugal fan**.

paddle-wheel hopper *(Eng.)*. A hopperfeed for small articles, in which a continuously rotating paddle pushes the articles along an inclined track towards the delivery chute.

pad roller *(Image Tech.)*. See **hold-down roller**.

padsaw *(Build.)*. See **keyhole saw**.

pad stone *(Build.)*. A stone **template**.

paediatrician, pediatrician, paediatrist *(Med.)*. A medical man who specializes in the study of childhood and the diseases of children.

paediatric, pediatric *(Med.)*. Pertaining to *paediatrics* (*pediatrics*), i.e. that branch of medical science which deals with the study of childhood and the diseases of children.

paedogenesis *(Bot.,Zool.)*. Sexual reproduction by larval or immature forms.

paedomorphosis *(Bot.)*. See **neoteny**.

paedophilia *(Behav.)*. Committing sexual offences against children; sexual gratification through sexual activity with children.

PAF *(Immun.)*. Abbrev. for *Platelet Activating Factor*.

PAGE *(Biol.)*. See **polyacrylamide gel electrophoresis**.

page cord *(Print.)*. Thin strong cord used to secure type pages which are being stored.

page cut-off *(Print.)*. A control on the ink duct of sheet-fed rotary presses which allows ink feed to be cut off from one page.

page description language *(Comp.)*. Software designed to accept for printing either a *bit-mapped* page of text and graphics or to accept it in **vector graphics** form in which each element, such as part of a letter or picture element, is described as a vector with its beginning and end positions. The latter is more powerful because it can be made independent of the printer and because a letter from a standard font, for example, can be easily transformed into any required size. This can therefore provide output both for medium resolution laser printers and high-resolution typesetters in **desk-top publishing**.

Page effect *(Elec.Eng.)*. A click heard when a bar of iron is magnetized or demagnetized.

page gauge *(Print.)*. An accurately-sized piece of furniture or a length of 6-point or 12-point strip cut to size, used to gauge the depth of type pages when they are made up.

page proofs *(Print.)*. Proofs taken when the type is made up into pages and before imposition.

Paget's disease of bone *(Med.)*. *Osteitis deformans*. A chronic disease characterized by progressive enlargement and softening of bones, especially of the skull and of the lower limbs.

Paget's disease of the nipple *(Med.)*. A condition in which chronic eczema of the nipple is associated with underlying cancer of the breast.

pagination *(Print.)*. The allotting of numbers (folios) to the pages of a book, roman numerals being traditionally but not invariably used for the preliminary matter and arabic numerals thereafter, beginning again with the numeral '1' on the first text page, but sometimes continuing the sequence where the preliminary folios left off.

paging *(Comp.)*. Technique in which the main store is divided into segments called pages. Large user programs may cover several pages, possibly too many to fit into the available store. The operating system transfers pages between main store and backing store to ensure that the correct page is in main store at any stage during the execution of the program. See also **segmentation**.

paging *(Print.)*. Applying numbers to the pages of an account book or to sets of stationery with a hand-numbering tool or a pedal-operated paging machine.

pagodite *(Min.)*. This is like ordinary massive *pinite* in its amorphous compact texture and other physical characters, but contains more silica. The Chinese carve the soft stone into miniature pagodas and images. Also called *agalmatolite*.

pahoehoe *(Geol.)*. Lava with a glassy, smooth, ropy surface.

paint *(Build.)*. A suspension of a solid or solids in a liquid which, when applied to a surface, dries to a more or less opaque, adhering solid film. The solids are colour- or opacity-imparting, finely-ground pigments together with extenders; the liquid consists of suitable oils, solvents, resins, aqueous colloidal solutions or dispersion etc., together with other lesser ingredients. Used for protective and decorative purposes. See also **oil paints**.

paint harling *(Build.)*. Rough-casting for protecting domestic steel walls, the adhesive medium being a special paint.

paint remover, stripper *(Build.)*. A liquid applied to paint to facilitate removal. Can be solvent or caustic based.

pair *(Build.)*. To match two similar objects on opposite hands.

pair *(Electronics)*. Two electrons forming a nonpolar valency bond between atoms.

paired-associate learning *(Behav.)*. A procedure in which a list of pairs is presented in which one item serves as stimulus and the other as response; the subject must learn to respond with the second item when the first item of a pair is presented.

paired cable *(Elec.Eng.)*. One in which multiple conductors are arranged as twisted pairs but not quadded.

pairing *(Biol.)*. The process by which *homologous chromosomes* are brought together during meiosis preparatory to being distributed one to each gamete.

pairing *(Image Tech.)*. Deviation from exactness of interlacing of horizontal lines in a reproduced image, reducing vertical definition. Also *twinning*.

pairing energy *(Phys.)*. A component of the binding energy of the nucleus that represents an increase in this energy where the number of neutrons and the number of protons are even. See **even-even nuclei**.

pair production *(Electronics)*. Creation of a **positron** and an **electron** when a **photon** passes into the electric field of an atom. See **Compton effect, photoelectric effect**.

paisanite *(Geol.)*. A sodic microgranite containing *riebeckite* as the principal coloured mineral.

PAL *(Bot.)*. Abbrev. for *Phenylalanine Ammonia-Lyase*.

PAL *(Image Tech.)*. *Phase Alternation Line*, the colour television coding system generally used for European broadcasting, in which one of the two chrominance signals is reversed in phase for each alternate scanning line.

Palaearctic region *(Zool.)*. One of the subrealms into which the Holarctic region is divided; it includes Europe and northern Asia, together with Africa north of the Sahara.

palaeobotany *(Geol.)*. The study of fossil plants. See **palaeontology**.

palaeoclimatology *(Geol.)*. The study of climatic conditions in the geological record, using evidence from fossils, sediments and their structures, geophysics and geochemistry.

palaeocurrent *(Geol.)*. An ancient current whose direction can often be worked out by examination of sedimentary structures (e.g. *ripple marks, cross bedding, sole structures*).

palaeoecology *(Geol.)*. The study of fossil organisms in terms of their mode of life, their interrelationships, their environment, their manner of death and their eventual burial.

palaeogeography *(Geol.)*. The study of the relative positions of land and water at particular periods in the geological past.

Palaeolithic Period *(Geol.)*. The oldest stone age, characterized by successive 'cultures' of stone implements, made by extinct types of men. Cf. **Neolithic Period**.

palaeomagnetism *(Geol.)*. The study of the Earth's ancient magnetism. *Remanent magnetization* in both igneous and sedimentary rocks provides a means of determining former magnetic poles.

palaeontology *(Geol.)*. The study of fossil animals and plants, including their morphology, evolution and mode of life.

palaeopathology *(Med.)*. Study of disease of previous eras from examination of bodily remains or evidence from ancient writings.

Palaeozoic *(Geol.)*. A major era of geological time comprising the Cambrian, Ordovician, Silurian, Devonian, Carboniferous, and Permian Periods.

palaeozoology *(Geol.)*. The study of fossil animals. See **palaeontology**.

palae-, palaeo- *(Genrl.)*. Prefix from Gk. *palaios*, ancient.

palagonite *(Geol.)*. A hydrous altered basaltic glass. It occurs as infillings in rocks, and is a soft-brown or greenish-black crypto-crystalline substance. Named from Palagonia, Sicily.

palama *(Zool.)*. The webbing of the feet in Birds of aquatic habit.

palate *(Zool.)*. In Vertebrates, the roof of the mouth; in Insects, the epipharynx. adjs. *palatal, palatine*.

palatine *(Zool.)*. Pertaining to the palate; a paired membrane bone of the Vertebrate skull which forms part of the roof of the mouth.

palatoplegia *(Med.)*. Paralysis of the palate.

palea, pale, palet *(Bot.)*. *Valvule*. The usually thin and membranous, upper or inner of the two bracts (lemma and palea) which enclose a grass floret. Sometimes, synonymous with glume.

Paleocene *(Geol.)*. The name given to the oldest epoch of the Tertiary period.

paleocerebellum *(Med.)*. Phylogenetically, the older part of the cerebellum, comprising such parts as pyramis, uvula and paraflocculus.

paleo-encephalon *(Med.)*. Phylogenetically, the ancient brain; the brain excluding the cerebral cortex and appendages.

Paleogene *(Geol.)*. A subperiod lying above the *Cretaceous* and below the *Neogene*, containing the **Paleocene**, **Eocene** and **Oligocene** epochs. It covers a time span from app 65–25 million years. The corresponding system of rocks.

paleogenetic *(Genrl.)*. Originating in the past.

pale-, paleo- *(Genrl.)*. Prefix. See **palae-, palaeo-**.

palingenesis *(Geol.)*. The production of new magma by the complete or partial melting of previously existing rocks. See **granitization**.

palingenesis *(Zool.)*. The reproduction of truly ancestral characters during ontogeny. adj. *palingenetic*.

palisade *(Bot.)*. Chlorenchyma in which the cells are elongated at right angles to the surface of the organ. Palisade mesophyll is characteristically present towards the upper surface of dorsiventral leaves of mesophytic dicotyledons. Palisade layers also occur in the outer cortex of many photosynthetic stems. See also **spongy mesophyll**.

Palladian *(Arch.)*. An architectural style named after Andrea Palladio, (1508–80), the most influential architect of the **Renaissance**. The style is characterized by the manner in which certain **Classical** motifs such as **porticos**, arches and columns are grouped together.

Palladian window *(Arch.)*. Type of window comprising main window with arched head and, on either side, a long narrow window with square head.

palladinized asbestos *(Chem.)*. Asbestos fibres saturated with a solution of a palladium compound, which is subsequently decomposed to give finely divided palladium dispersed throughout the asbestos.

palladium *(Chem.)*. A metallic element, symbol Pd, at.no. 46, r.a.m. 106.4. The metal is white, mp 1549°C, bp 2500°C, rel.d. 11.4. Used as a catalyst in hydrogenation. Native palladium is mostly in grains and is frequently alloyed with platinum and iridium. See also **gold**.

pallaesthesia *(Med.)*. Insensibility of bone to vibratory stimuli.

pallasite *(Min.)*. A group name for stony meteorites which contain fractured or rounded crystals of olivine in a network of nickel-iron.

pallescent *(Bot.)*. Becoming lighter in colour with age.

pallet *(Acous.)*. The flap of wood, faced with felt or leather, which is raised to permit the flow of air to wind-chests, etc., in the mechanism of an organ.

pallet *(Build.)*. A thin strip of wood built into the mortar joint of a wall, to provide a substance to which joinery may be nailed.

pallet *(Eng.)*. A portable (wooden) platform used to facilitate the handling of stacked goods by giving sufficient ground clearance.

pallet *(Horol.)*. The surface or part upon which the teeth of the escape wheel act to give impulse to the pendulum or balance. The first pallet acted upon by a tooth of the escape wheel is known as the *entering pallet*, and the other pallet as the *exit pallet*. See **circular-, equidistant locking-, exposed-**.

pallet *(Print.)*. A platform on which sheets of paper (or other goods) can be stacked before printing and between printings, and designed for transporting by fork-lift truck. Also called *stillage*.

pallet brick *(Build.)*. A purpose-made brick with a groove in one edge to receive a fixing strip.

pallet jewel, stone *(Horol.)*. The jewel in the pallets upon which the escape-wheel teeth act.

pallet truck *(Genrl.)*. See **fork-lift truck**.

palliative *(Med.)*. Affording temporary relief from pain or discomfort; a medicinal remedy which does this.

palli-, pallio- *(Genrl.)*. Prefix from L. *pallium*, mantle.

pallium *(Zool.)*. The mantle in Brachiopoda or Mollusca, a fold of integument which secretes the shell. In the Vertebrate brain, that part of the wall of the cerebral hemispheres excluding the corpus striatum and rhinencephalon. adjs. *pallial, palliate*.

Palmae *(Bot.)*. *Arecaceae*. The palm family, ca 2800 spp. of monocotyledonous flowering plants (superorder Arecidae). Trees, mostly tropical, typically with a single trunk (which does not undergo secondary thickening) bearing a crown of pinnate or palmate leaves. The fruit is a one-seeded berry or a drupe. The most important economic products are coconuts, palm oil, dates, copra and fibres.

palmar *(Zool.)*. Pertaining to the palm of the hand.

palmate *(Bot.)*. Having 4 or more equal divisions, lobes or veins radiating from a common point rather in the manner of the fingers of a hand; as in a palmately compound, lobed or veined leaf respectively.

palmate *(Zool.)*. Having webbed feet.

palmatine *(Chem.)*. An alkaloid of the *iso*quinoline group, obtained from calumba root, *Jatrorrhiza palmata*, generally isolated in the form of the sparingly soluble iodide, $C_{21}H_{22}O_4NI.2H_2O$, which crystallizes in yellow needles, mp 240°C.

palmelloid form *(Bot.)*. Condition in algae in which non-motile cells divide within a mucilaginous matrix to give large gelatinous masses containing many cell generations.

palmisect *(Bot.)*. A leaf blade etc. cut almost to the centre in a palmate fashion.

palmitic acid *(Chem.)*. $C_{15}H_{31}.COOH$, a normal fatty acid, mp 63°C, bp 269°C. It occurs as glycerides in vegetable oils and fats.

palmitins *(Chem.)*. The glycerine esters of palmitic acid.

palm-kernel oil *(Chem.)*. A yellowish oil from the nuts of *Elaeis guineensis*, mp 26°–30°C, rel.d. 0.95, saponification value 247, iodine value 13.5, acid value 8.4.

palm oil *(Chem.)*. A reddish-yellow fatty mass from the fruit of *Elaeis guineensis*, mp 27°–43°C, rel.d. 0.90–0.95, saponification value 196–205, iodine value 51–57, acid value 24–200.

palp *(Zool.)*. See **palpus**. adj. *palpal*.

palpation *(Med.)*. Physical examination by touch.

palpebra *(Med.,Zool.)*. An eyelid.

palpebral fissure *(Med.)*. The space between the upper and lower eyelids.

palpitation *(Med.)*. Subjective awareness of rapid or irregular heart beat.

palpus *(Zool.)*. In Crustacea and Insects, a jointed sensory appendage associated with the mouthparts; in Polychaeta, sensory appendage of prostomium. Also *palp*.

palygorskite *(Min.)*. A group of clay minerals, hydrated magnesium aluminium silicates, in appearance resembling cardboard or paper, having a fibrous structure. Also called *attapulgite*.

palynology *(Bot.)*. See **pollen analysis**.

palynology *(Geol.)*. The study of fossil spores and pollen. They are very resistant to destruction and in many sedimentary rocks are the only fossils that can be used for stratigraphical correlation.

pan *(Bot.)*. (1) A compact layer of soil particles, lying some distance beneath the surface, cemented together by organic material, or by iron and other compounds, and relatively impermeable to water. (2) A depression in the surface of a salt marsh, in which salt water stands for lengthy periods.

pan *(Build.)*. A panel of brickwork, or lath and plaster, in half-timbered work.

pan *(Image Tech.)*. (1) Movement of a camera in a horizontal plane about a vertical axis (from *panoramic*). (2) Abbrev. for **panchromatic**.

panache *(Civ.Eng.)*. See **pendentive**.

pan-and-scan *(Image Tech.)*. The continuous selection of a limited area from the picture of a **wide-screen** motion picture film for presentation within a narrower format, such as that of television.

pan-and-tilt head *(Image Tech.)*. A camera mounting allowing rotation about a vertical axis (*panning*) and a horizontal one (*tilting*), especially in cinematography and TV.

pan breeze *(Build.)*. A mixture of coke breeze and clinker collected in the pan below a furnace consuming coke breeze. It is used as an aggregate in the manufacture of concrete.

pancake coil *(Telecomm.)*. Inductor in which the conductor is wound spirally.

pancaking *(Aero.)*. The alighting of an aircraft at a relatively steep angle, with low forward speed.

pancarditis *(Med.)*. Concurrent inflammation of the three main structures of the heart: the pericardium, the myocardium, and the endocardium.

panchromatic *(Image Tech.)*. A photographic emulsion substantially sensitive to the whole of the visible spectrum. Abbrev. *pan*.

pancreas *(Zool.)*. A moderately compact though somewhat amorphous structure found in Vertebrates, the larger part of which consists of exocrine glandular tissue with one or more ducts opening to the small intestine, and also containing scattered islets of endocrine tissue. The former produces six or more enzymes involved in the digestion of proteins, carbohydrates, and fats, while the latter secrete the hormones insulin and glucagon.

pancreatectomy *(Med.)*. Surgical removal of the pancreas.

pancreatin *(Med.)*. Pancreatic enzyme preparation used to compensate for reduced or absent exocrine secretion of the pancreas in children with cystic fibrosis and in adults following pancreatectomy or chronic pancreatitis.

pancreatitis *(Med.)*. Inflammation of the pancreas.

pancreatolith *(Med.)*. A calculus in the pancreas.

pancreatotomy *(Med.)*. Incision of the pancreas.

pancreozymin *(Biol.)*. A polypeptide hormone secreted by the intestinal wall which stimulates the pancreas to secrete digestive enzymes.

pandemic *(Med.)*. Of an epidemic, occurring over a wide area such as a country or a continent; an epidemic so widespread.

pandiculation *(Med.)*. The combined action of stretching the body and the limbs and yawning.

pandurate, panduriform *(Bot.)*. Shaped like the body of a fiddle.

pane *(Build.)*. The end of a hammer-head opposite to hammering face; made to various shapes for particular operations such as riveting etc. Also *pean*, *peen*, *pein*. See **ball-pane hammer**, **cross-pane hammer**, **straight-pane hammer**.

panel *(Elec.Eng.)*. A sheet of metal, plastic or other material upon which instruments, switches, relays, etc., are mounted. Also called *switchboard panel*.

panel *(Print.)*. A piece of material (leather, paper) fixed to the spine of a book, containing the title and author's name.

panel absorber *(Acous.)*. Light panels mounted at some distance in front of a rigid wall in order to absorb sound waves incident on them.

panel back *(Print.)*. A style of **finishing** the spine of a book by decorated panels between raised bands.

panel beating *(Eng.)*. A sheet-metal working craft, now mainly used in automobile body repairing, in which complex shapes are produced by stretching and gathering the sheet locally, with subsequent finishing by planishing and wheeling.

panel heating *(Build.)*. A system of heating a building in which heating units or coils of pipes are concealed in special panels, or built in wall or ceiling plaster. Also called *concealed heating*.

panelled framing *(Build.)*. Doors and frames formed of stiles, rails, and muntins framed together with mortise-and-tenon joints, and having panels fitted into the spaces.

panel mounting *(Elec.Eng.)*. The normal method of accommodating a collection of non-portable apparatus. Each piece or unit is constructed separately on its standard panel, which is mounted with others on a standard vertical rack, the different panels being provided with terminal blocks so that the units can be wired together after assembly.

panel pins *(Build.)*. Light, narrow-headed nails of small diameter used chiefly for fixing plywood or hardboard to supports.

panels *(Eng.)*. In a truss or open-web girder, the framed units of which the truss is composed; the divisions separated by the vertical members.

panel saw *(Build.)*. A hand-saw used for panelling, having 7 teeth to the inch.

panel switch *(Elec.Eng.)*. See **flush switch**.

PAN fibres *(Textiles)*. Abbreviation for *polyacrylonitrile*; see **acrylic fibres**.

Pangaea *(Geol.)*. A hypothetical supercontinent that existed in the geological past and consisted of all the present continents before they split up.

pangamic *(Zool.)*. Of indiscriminate mating.

panhead rivet *(Eng.)*. One with a head shaped like a truncated cone.

panhysterectomy *(Med.)*. Complete surgical removal of the uterus.

panic attack *(Behav.)*. An attack of intense terror and anxiety, usually lasting several minutes, though possibly continuing for hours; apprehension often persists for long periods after the panic attack.

panic bolt *(Build.)*. A special form of door-bolt which is released by pressure at the middle of the bolt, often used on fire-escape doors.

panic disorder *(Behav.)*. A **panic attack** occurring in the absence of any phobic stimulus.

panicle *(Bot.)*. Strictly, a branched raceme with each branch bearing a raceme of flowers, e.g. oat. Loosely, any branched inflorescence of some degree of complexity. adj. *paniculate*.

panidiomorphic *(Geol.)*. A term applied to igneous rocks with well-developed crystals.

pan-mill mixer *(Build.)*. See **mortar mill**.

panmixis, panmixia *(Biol.)*. **Random mating** within a population, especially a model system. adj. *panmictic*.

panniculitis *(Med.)*. Inflammation of the subcutaneous fat in any part of the body.

panniculus carnosus *(Zool.)*. In some Mammals, an extensive system of dermal musculature covering the trunk and part of the limbs, by means of which the animal can shake itself.

panning *(Min.Ext.)*. Use by prospector or plant worker of gold pan, batea, plaque, or dulong to concentrate heavier minerals in a crushed sample by washing away the lighter ones.

panning head *(Image Tech.)*. Head or platform for a camera, permitting panoramic motion.

pannose *(Bot.)*. Felted.

pannus *(Med.)*. The appearance of a curtain of blood vessels round the margin of the cornea: e.g. in trachoma or in phlyctenular keratitis.

panophthalmitis, panophthalmia *(Med.)*. Inflammation of all the structures of the eye.

panoramic attenuator *(Acous.)*. Re-recording device by which a 1-channel recording is reproduced and distributed to, say, three tracks, generally magnetic, to give an illusion, when these are reproduced, of stereophonic sound reproduction. Used with wide cinema screens.

panoramic camera *(Image Tech.)*. A camera intended to take very wide angle views, usually by rotation about an axis and by exposing a roll of film through a vertical slit.

panoramic receiver *(Telecomm.)*. One in which tuning sweeps over wide ranges, with synchronized display on a CRT of output signals.

panradiometer *(Phys.)*. Instrument for measuring radiant heat irrespective of the wavelength.

pansinusitis *(Med.)*. Inflammation of all the air-containing sinuses which communicate with the nasal cavity.

panspermia *(Astron.)*. Hypothesis (S. Arrhenius, 1907) that life on Earth was introduced billions of years ago by **extraterrestrials** in space.

panting *(Ships)*. Pulsating movement of the shell plating of the ship developing in the bows and stern while the ship is under way.

pantobase *(Aero.)*. The fitment of a landplane with **hydroskis**, enabling it to taxi, take-off from and alight on water or snow.

pantograph *(Elec.Eng.)*. A sliding type of current-collector for use on traction systems employing an overhead contact wire. The contact-strip of the collector is mounted on a hinged diamond-shaped structure, so that it can move vertically to follow variations in the contact wire height.

pantograph *(Eng.)*. A mechanism by which a point is constrained to copy, to any required scale, the path traced by another point. It is based on the geometry of a parallelogram; used in engraving machines etc.

Pantone Matching System *(Print.)*. TN for a colour matching system extensively used by graphics designers to specify colours to be used in printing. A book of colour swatches is used, each having a PMS number and an indication of the ink mix required to achieve the desired hue, thus acting as an aid to the printer.

pantophagous *(Zool.)*. Omnivorous.

pantothenic acid *(Biol.)*. See **vitamin B**.

pantothenic acid *(Chem.)*. See **vitamin B complex**.

papain *(Chem.)*. A protein-digesting enzyme present in the juice from the fruits and leaves of the papaya tree (*Carica papaya*); commercially produced as a meat tenderizer.

papaverine *(Chem.)*. An alkaloid occurring in opium, colourless prisms, mp 147°C, optically inactive. It is 3,4-dimethoxybenzyl-4′,5′-dimethoxy-*iso*quinoline.

papaya *(For.)*. A tree, *Carica papaya*, native to S. America but common in the tropics. The trunk, leaves and fruit yield *papain*, a protease used commercially as a meat-softener.

paper *(Genrl.)*. Consists of continuous webs of suitable vegetable fibres, freed from noncellulose constituents and deposited from an aqueous suspension. Wood-pulp, esparto and rags are the chief raw materials, though other substances are used. During the process of manufacture, the fibres are reduced to the requisite length, and their physical properties are modified by mechanical or chemical treatment. See **cellulose**.

paper capacitor *(Elec.Eng.)*. One which has thin paper as the dielectric separating aluminium foil electrodes, these being wound together and waxed.

paper chromatography *(Chem.)*. **Chromatography** using a sheet of special grade filter paper as the adsorbent. Advantages: microgram quantities, bands can be formed in two dimensions and cut with scissors.

paper foils *(Print.)*. **Blocking foils** with a glassine substrate. These are the cheapest foils and the most widely used.

paper negatives *(Image Tech.)*. Negatives made with emulsions on paper supports, instead of the more usual film or glass. They have the advantages of lightness, cheapness, ease of retouching, and possibility of large dimensions.

paper sizes *(Paper)*. *ISO* recommended paper sizes are based on metric dimensions. The principal series (ISO-A) for printing, stationery etc., has a basic size of 1 m^2. Each size is got by halving the one above along the longer side, e.g., A0 = 841 × 1189 mm, A1 = 594 × 841 mm etc.

paper tape *(Comp.)*. See **punched paper tape**.

paper-tape punch *(Comp.)*. Output device. See **punched paper tape**.

paper-tape reader *(Comp.)*. Input device which reads **punched paper tape**.

papilionaceous *(Bot.)*. Having some likeness to a butterfly; especially, the flowers of the Papilionaceae, including the pea.

papilla *(Bot.)*. A small nipple-shaped projection, especially of the wall of a cell.

papilla *(Zool.)*. A small conical projection of soft tissue, especially on the skin or lining of the alimentary canal; the conical mass of soft tissue or pulp projecting into the base of a developing feather or tooth. adjs. *papillary, papillate*.

papillae foliatae *(Zool.)*. In some Mammals, two small oval areas at the back of the tongue, marked by a series of alternating transverse ridges and richly provided with taste buds.

papillitis *(Med.)*. *Optic neuritis*. Inflammation of the disk or head of the optic nerve within the globe of the eye.

papilloedema, papilledema *(Med.)*. *Choked disk*. Swelling and congestion of the disk or head of the optic nerve within the globe of the eye, as a result of increase of pressure within the skull or of severe hypertension.

papilloma *(Med.)*. A tumour (usually innocent) resulting from the new growth of the cells of the skin or of the mucous membrane.

papovavirus *(Med.)*. A group of small *DNA viruses* which have the capacity to produce tumours in animals and man.

pappus *(Bot.)*. Modified calyx in the Compositae, consisting of a ring of feathery hairs or the like around the top of the fruit, as in the dandelion, where it aids in wind dispersal.

Pappus' theorem *(Maths.)*. (1) The surface area and the volume respectively swept out by revolving a plane area about a nonintersecting coplanar axis is the product of the distance moved by its centroid and either its perimeter or its area. Frequently attributed to Guldin who rediscovered it in the 17th century. (2) If ABC and $A'B'C'$ are two straight lines in a plane, then the three points of intersection of lines AB' and $A'B$, AC' and $A'C$, BC' and $B'C$ lie on a straight line. This is in fact *Pascal's theorem* applied to the 'hexagon' $AB'CA'BC'$ inscribed in the degenerate conic consisting of two straight lines.

papulae *(Zool.)*. The dermal gills of Echinodermata, small finger-shaped, thin-walled respiratory projections of the body wall.

papule *(Med.)*. A small, circumscribed, solid elevation above the skin, as in chicken-pox. adj. *papular*.

papulopustular *(Med.)*. Of papules and pustules.

papyrus *(Paper)*. An early form of writing material prepared from the water reed of the same name and used for records up to the 9th century AD.

PAR *(Aero.)*. Abbrev. for *Precision-Approach Radar*.

PAR *(Bot.)*. Abbrev. for *Photosynthetically Active Radiation*.

para-. *(Chem.)*. Prefix from Gk. *para*, beside. In chemistry: (1) A polymer of ....; (2) A compound related to .... .

para- *(Genrl.)*. Prefix from Gk. *para*, beside.

para- *(Chem.)*. (1) Containing a benzene nucleus substituted in the 1.4 positions. (2) Consisting of diatomic molecules with antiparallel nuclear spins and an even rotational quantum number.

para(4)-aminosalicylic acid *(Med.)*. See PAS.

parabola *(Maths.)*. See **conic**.

parabolic flight *(Space)*. The flying of a special parabolic trajectory by a suitably fitted aeroplane to reproduce the conditions of **free-fall** over a period of minutes.

parabolic microphone *(Acous.)*. One provided with a parabolic reflector to give enhanced directivity for high audiofrequencies and hence greater range for wanted sounds amid ambient noise.

parabolic mirror, reflector *(Phys.,Telecomm.)*. One shaped as a paraboloid of revolution. Theoretically produces a perfectly parallel beam of radiation if a source is placed at the focus (or vice versa). Such mirrors are used in reflecting telescopes, car headlamps etc. See also **dish** for *parabolic reflector*.

parabolic nozzle *(Eng.)*. A nozzle of parabolic section with a high coefficient of discharge, placed in a pipe to measure the flow of a gas. The pressure drop is measured by a manometer.

parabolic point on a surface *(Maths.)*. One at which one of the principal curvatures, and therefore also the Gaussian curvature, is zero. Cf. **elliptical, hyperbolic, umbilical points on a surface**.

parabolic reflector *(Telecomm.)*. See **dish, parabolic mirror**.

parabolic spiral *(Maths.)*. A spiral with polar equation $r^2 = a^2\omega$. Also called *Fermat's spiral*.

parabolic velocity *(Astron.)*. The velocity which a body at a given point would require to describe a parabola about the centre of attraction; smaller values give an ellipse, larger values a hyperbola. Also called *escape velocity*, since it is the upper limit of velocity in a closed curve.

paraboloid *(Maths.)*. See **quadric**.

paraboloid of revolution *(Maths.)*. A solid figure formed by the revolution of a parabola about its axis. Cf. **quadric**.

parabrake *(Aero.)*. A **brake parachute**.

paracentesis *(Med.)*. *Tapping*. The puncture of body cavities with a hollow needle, for the removal of inflammatory or other fluids. See **amniocentesis**.

paracetamol *(Med.)*. Antipyretic and analgesic drug.

parachor *(Chem.)*. A quantity which may be regarded as the molecular volume of a substance when its surface tension is unity; in most cases it is practically independent of temperature. Its value is given by the expression

$$\frac{M\gamma^{\frac{1}{4}}}{\rho_L - \rho_V},$$

where M is the molecular mass, γ the surface tension, ρ_L and ρ_V are the densities of the liquid and vapour respectively.

parachute *(Aero.)*. An umbrella-shaped fabric device of high drag (1) to retard the descent of a falling body, or (2) to reduce the speed of an aircraft or item jettisoned therefrom. Commonly made of silk or nylon, sometimes of cotton or rayon where personnel are not concerned. See **antispin-, automatic-, brake-, ribbon-, ring slot-, pilot chute**. See also **parasheet**.

parachute flare *(Aero.)*. A pyrotechnic flare, attached to a parachute released from an aircraft to illuminate a region.

paracytic *(Bot.)*. A stomatal complex having one or more subsidiary cells parallel to each guard cell, as in many monocotyledons.

paradistemper *(Vet.)*. See **hard-pad disease**.

paradoxical sleep *(Behav.)*. Stage of sleep when dreaming is assumed to occur. It is characterized by rapid eye movements (REM), loss of muscle tone, and an EEG very similar to the waking state (hence paradoxical). Also *REM sleep*.

paraeiopod *(Zool.)*. In Crustacea, a walking leg.

paraesthesia *(Med.)*. An abnormal sensation, such as tingling, tickling, and formication.

paraffins *(Chem.)*. A term for the whole series of saturated aliphatic hydrocarbons of the general formula C_nH_{2n+2}. Also *alkane hydrocarbons*. They are indifferent to oxidizing agents, and not reactive, hence the name *paraffin* (L. *parum affinis*, little allied).

paraffin wax *(Chem.)*. Higher homologues of alkanes, waxlike substances obtained as a residue from the distillation of petroleum; mp 45°–65°C, rel.d. 0.9, resistivity 10^{13} to 10^{17} ohm metres, permittivity 2–2.3.

paraformaldehyde (paramethanal) *(Chem.)*. $(H.CHO)_2$, a self-addition product of formaldehyde (methanal) obtained by the evaporation of an aqueous solution of methanal. It is a white crystalline mass, soluble in water.

paraganglia *(Zool.)*. In higher Vertebrates, small glandular bodies, occurring in the posterior part of the abdomen, which show a chromaphil reaction and are believed to secrete adrenaline.

paraganglioma *(Med.)*. See **phaeochromacytoma**.

paraglider *(Aero.)*. Inflatable hypersonic re-entry kite of highly swept back wing shape with rounded leading edge, proposed for Gemini spacecraft, but not used.

paragnathous *(Zool.)*. Having jaws of equal length; as Birds which have upper and lower beak of equal length.

paragneiss *(Geol.)*. A term given to gneissose rocks which have been derived from detrital sedimentary rocks. Cf. **orthogneiss**.

paragonimiasis *(Med.)*. Invasion of the lungs by the lung fluke *Paragonimus westermanii*; the condition is endemic in the Far East, and infection results from eating fresh-water crustaceans.

paragonite *(Min.)*. A hydrated sodium aluminium silicate. It is a sodium mica, has a yellowish or greenish colour, and is usually associated with metamorphic rocks. Differs from **muscovite** chiefly in containing sodium rather than potassium.

paragraphia *(Med.)*. Faulty spelling, misplacement of letters and words, and use of wrong words in writing, as a result of a lesion in the brain.

parahydrogen *(Chem.)*. A hydrogen molecule in which the two nuclear spins are antiparallel, forming a singlet state.

parakeratosis *(Med.)*. Faulty formation of the horny layer of the skin, with scaling of the skin.

paralalia *(Med.)*. A form of speech disturbance, particularly that in which a different sound or syllable is produced from the one which is intended.

paraldehyde *(Chem.)*. $(CH_3.CHO)_3$, a self-addition product of ethanal, obtained by the action of concentrated sulphuric acid upon ethanal. It is a colourless liquid, bp 124°C, and can be converted again into ethanal by distillation with dilute sulphuric acid. A common hypnotic. Used as a chemical intermediate in the synthesis of *pentaerythritol*.

paralexia *(Med.)*. A defect in the power of seeing and interpreting written language, with meaningless transposition of words and syllables, due to a lesion in the brain.

parallmnion *(Ecol.)*. The zone of a lake floor between the water's edge or shoreline, and the lakeward margin of rooted vegetation.

parallactic angle *(Astron.)*. The name given to that angle in the astronomical triangle formed at the heavenly body by the intersection of the arcs drawn to the zenith and to the celestial pole. See **astronomical triangle**.

parallax *(Astron.)*. The apparent displacement in the position of any celestial object caused by a change in the position of the observer. Specifically, the change due to the motion of the Earth through space. The observer's position changes with the daily rotation of the Earth (*diurnal parallax*), the yearly orbit (*annual parallax*) and motion through space generally (*secular parallax*). The term 'parallax' is often loosely used by astronomers to be synonymous with 'distance' because the annual parallax is inversely proportional to distance. See also **spectroscopic parallax**.

parallax *(Phys.)*. Generally, the apparent change in the position of an object seen against a more distant background when the viewpoint is changed. Absence of parallax is often used to adjust two objects, or two images, at equal distances from the observer.

parallax *(Surv.)*. A condition set up when the image

formed by the object-glass of a surveying telescope is not in the same plane as the cross-hairs so that, on moving his eye from side to side or up and down, the observer can see relative movement between them.

parallax stereogram *(Image Tech.)*. The use of a line screen in front of a positive transparency of alternate strips of two views of an object, made by exposing an emulsion in a similar arrangement with a large lens and two apertures representing the two eyes.

parallel *(Elec.Eng.)*. Two circuits are said to be *in parallel* when they are connected so that any current flowing divides between the 2. Two machines, transformers, or batteries are said to be *in parallel* when the terminals of the same polarity are connected together.

parallel *(Maths.)*. Two straight lines (or more generally geodesics) in a surface are said to be parallel if they do not meet. Equivalently they can be said to meet at infinity, and points at infinity are introduced in projective geometry.

parallel *(Surv.)*. A *parallel of latitude* is an imaginary line around the earth's surface connecting points of equal terrestrial latitude.

parallel arithmetic unit *(Comp.)*. One in which the digits are operated on concurrently.

parallel-axis theorem *(Phys.)*. A theorem expressing the moment of inertia I of a body about any axis, in terms of its moment of inertia, I_a about a parallel axis through G, its centre of gravity. Thus $I = I_a + Md^2$, where d is the perpendicular distance between the axes and M the mass of the body.

parallel body *(Ships)*. That portion of a ship's form wherein the fullest transverse shape is maintained constant.

parallel circuit *(Elec.Eng.)*. (1) Electric of magnetic circuit in which current or flux divides into two or more paths before joining to complete the circuit. (2) See **voltage circuit**.

parallel data transmission *(Comp.)*. The **bits** comprising a character are sent simultaneously along separate data lines. Cf. **serial-**.

parallel descent *(Bot.,Zool.)*. The appearance of similar characteristics in groups of animals or plants which are not directly related in evolutionary descent. Also *parallel evolution, parallelism*.

parallelepiped *(Maths.)*. A solid figure bounded by six parallelograms, opposite pairs being identical and parallel.

parallel-feeder protection *(Elec.Eng.)*. A type of balanced protective equipment relying for its action on the fact that the current in two parallel feeders will normally be equal, this balance being upset if a fault occurs on one of the feeders.

parallel folding *(Print.)*. When two or more successive folds in a sheet are in the same direction they are described as *parallel*. Cf. **right-angled folding**.

parallel gutter *(Build.)*. A rectangular roof-gutter, e.g. one bounded at the sides by a parapet wall and a pole plate.

parallelism *(Bot.,Zool.)*. See **parallel descent**.

parallel motion *(Eng.)*. (1) Mechanism comprised of a parallel arrangement of links, constructed so that motion of one point can induce similar motion at another. The relative motion can be larger or smaller. See **pantograph**. (2) Mechanism constraining a straight edge to move parallel with the two edges of a drawing board which are at right angles.

parallelodromous *(Bot.)*. Having parallel veins.

parallelogram *(Maths.)*. A quadrilateral with two pairs of opposite sides parallel.

parallelogram of forces *(Phys.)*. Alternative name for the *parallelogram rule for addition of vectors* when the vectors are forces.

parallelogram rule for addition of vectors *(Maths.)*. The sum of two vectors is the diagonal through O of the parallelogram of which the two vectors, when represented by lines through O, are adjacent sides.

parallel-plate capacitor *(Phys.)*. Device for storage of electric charge, consisting of two plane parallel conducting electrodes of area A m^2, separated by a dielectric of absolute **permittivity** εF/m, thickness d m. Its capacity is given by $C = \varepsilon A/d$ farads; charge stored when p.d. between plates is V volts is CV coulombs.

parallel-plate chamber *(Nuc.Eng.)*. Ionization chamber with plane-parallel electrodes.

parallel-plate waveguide *(Telecomm.)*. One formed by two parallel conducting or dielectric planes. Often realized in atmospheric propagation under suitable meteorological conditions. See **surface duct**.

parallel processing *(Comp.)*. The simultaneous performance of two or more tasks on a computer system. It may involve executing several instructions and/or processing several distinct data items. See **SIMD, MISD, MIMD, array processor, pipelining, multiprocessor, third generation computers**.

parallel resonance *(Elec.Eng.)*. See **shunt resonance**.

Parallel Roads *(Geol.)*. The strandlines of a glacial lake which occupied Glen Roy (Inverness-shire, Scotland) during the Pleistocene Period, when the lower part of the valley was blocked by ice.

parallel ruler *(Eng.)*. A drawing instrument consisting of two straight edges so linked together by connecting pieces that their edges are always parallel, although the distance between them may be varied.

parallel slot *(Elec.Eng.)*. The most usual shape of slot for the armature windings of electric machines, the slot having parallel sides. Cf. *taper slots*.

parallel-T network *(Telecomm.)*. See **twin-T network**.

parallel wire resonator *(Elec.Eng.)*. See **Lecher wires**.

paralyser *(Chem.)*. See **catalytic poison**.

paralysis *(Med.)*. The loss in any part of the body of the power of movement, or of the capacity to respond to sensory stimuli. See **diplegia, hemiplegia, monoplegia, paraplegia, quadraplegia**.

paralysis agitans *(Med.)*. See **Parkinson's disease**.

paralysis time *(Nuc.Eng.)*. The time for which a radiation detector is rendered inoperative by an electronic switch in the control circuit. See **dead time**.

paralytic ileus *(Med.)*. A condition in which, from various causes, there is extensive paralysis of the intestines, leading to persistent vomiting, pain being absent.

paramagnetism *(Phys.)*. Phenomenon in materials in which the susceptibility is positive and whose permeability is slightly greater than unity. An applied magnetic field tends to align the magnetic moments of the atoms or molecules and the material acquires magnetization in the direction of the field; it disappears when the field is removed. Used to obtain very low temperatures by adiabatic demagnetization.

paramere *(Zool.)*. Half of a bilaterally symmetrical structure; one of the inner pair of gonapophyses in a male Insect.

parameter *(Comp.)*. Name or value made available to a subprogram (e.g. a subroutine or procedure) from a calling program or vice versa.

parameter *(Crystal.)*. The *parameters* of a plane consist of a series of numbers which express the relative intercepts of that plane upon the crystallographic axes. Given in terms of the established unit lengths of those axes.

parameter *(Maths.)*. Generally, a variable in terms of which it is convenient to express other interrelated variables which may then be regarded as being dependent upon the parameter.

parameter *(Stats.)*. The population value of a particular characteristic descriptive of the distribution of a random variable.

parameter *(Telecomm.)*. A derived constant of a transmission circuit or network, which is more convenient for expressing performance or for use in calculations.

parametric amplifier *(Telecomm.)*. One which uses a device, commonly a **varactor diode**, whose reactance can be varied periodically by an external *pump* signal, which is usually at a much higher frequency than that to be amplified; energy can then be transferred from the pump to the signal frequency, without altering the latter. Abbrev. *paramp*.

parametric diode *(Electronics)*: One whose series capacitance can be varied by the biasing voltage. Usually a *varactor diode*. See **parametric amplifier**.

parametric equations *(Maths.)*. Of a curve or surface: equations which express the co-ordinates of points on the curve or surface explicitly in terms of other variables (parameters), which are, for the present, regarded as independent variables.

parametric resonance *(Telecomm.)*. The condition of a parametric amplifier when energy transfer from the pump circuit through any resonant circuits to the signal is a maximum.

parametritis *(Med.)*. *Pelvic cellulitis*. Inflammation of the pelvic cellular connective tissue in the region of the uterus, e.g. in the puerperium.

parametrium *(Med.)*. The subperitoneal connective tissue surrounding the uterus, especially that in the region of the cervix.

paramorph *(Biol.)*. A general term for any taxonomic variant within a species, usually used when more accurate definition is not possible.

paramorph *(Min.)*. The name given to a mineral species which has changed its molecular structure without any change of chemical constitution, e.g. aragonite altered to calcite. Cf. **pseudomorph**.

paramp *(Telecomm.)*. Same as **parametric amplifier**.

paramphistomiasis *(Med.)*. Invasion of the human intestine by trematode parasites of the family *Paramphistomidae*.

paramudras *(Geol.)*. Flint nodules of exceptionally large size and doubtful significance occurring in the Chalk exposed on the east coast of England.

paramylon *(Bot.)*. A polysaccharide storage product allied to starch, but not giving a blue coloration with iodine. Restricted in occurrence to *Euglena* and related organisms.

paramylon, paramylum *(Bot.)*. Reserve polysaccharide, a linear $\beta, 1\rightarrow 3$ glucan present as highly refractive solid bodies in the cytoplasm of the Euglenophyceae, Xanthophyceae and Haptophyceae.

parana pine *(For.)*. A tree, *Araucaria brasiliana*, of S. Brazil yielding a softish wood used for internal joinery.

paranephric *(Zool.)*. Situated beside the kidney.

paranephros *(Zool.)*. See **suprarenal body**.

paranoia *(Med.)*. A delusion of grievance beyond all bounds of reality. Occurs in a variety of mental diseases including **schizophrenia**.

paranoid disorder *(Behav.)*. A personality disorder characterized by extreme suspiciousness in all situations, and with almost all people; delusions of persecution or grandeur may occur, but without the serious disorganization associated with schizophrenia.

paranoid schizophrenic *(Behav.)*. One of Bleuter's four sub-types of schizophrenia; characterized by delusions of persecution or grandness, with hallucinations and a loss of contact with reality.

paraphase amplifier *(Electronics)*. Push-pull stage incorporating **paraphase coupling**. Also *see-saw amplifier*.

paraphase coupling *(Electronics)*. A push-pull stage, or series of stages, in which reversed phase is obtained by taking a fraction of the output voltage of the first valve of the amplifier and applying it to a similar balancing first valve, which operates succeeding stages in normal push-pull, the number of transfers being minimized thereby.

paraphasia *(Med.)*. A defect of speech in which words are misplaced and wrong words substituted for right ones; due to a lesion in the brain.

paraphilias *(Behav.)*. Sexual patterns in which arousal is caused by something other than what is considered a normal sexual object or activity.

paraphimosis *(Med.)*. Persistent retraction of the inner lining of the prepuce behind the glans penis in a case of phimosis.

paraphonia *(Med.)*. Alteration of the voice as a result of disease.

paraphrenia *(Med.)*. A form of **schizophrenia** with wild, improbable delusions of persecution.

paraphyletic group, paraphyly *(Bot.)*. In cladistics a group that includes a common ancestor and some, but not all, of its descendants. Cf. **polyphyletic group, monophyletic group**.

paraphysis *(Bot.)*. A sterile filament borne among the reproductive structures of many algae, fungi and bryophytes.

paraphysis *(Zool.)*. A thin-walled sac developed as an outgrowth from the non-nervous roof of the telencephalon, represented in Mammals by the pineal organ. pl. *paraphyses*. adj. *paraphysate*.

parapineal organ *(Zool.)*. See **parietal organ**.

paraplegia *(Med.)*. Paralysis of the lower part of the body and of the legs.

parapodium *(Zool.)*. In Mollusca, a lateral expansion of the foot; in Polychaeta, a paired fleshy projection of the body wall of each somite used in locomotion. adj. *parapodial*.

parapophyses *(Zool.)*. A pair of ventrolateral processes of a vertebra arising from the sides of the centrum.

paraprotein *(Immun.)*. Immunoglobulin derived from an abnormally proliferating clone of neoplastic plasma cells. The immunoglobulin and the cells making it will all have the same Ig class, sub-class and light chain determinants.

parapsid *(Zool.)*. In the skull of Reptiles, the condition when there is one temporal vacuity, this being high behind the eye, usually with the post-frontal and supratemporal meeting below. Found in Mesosaurs and Ichthyosaurs. See **temporal vacuities** for other types.

parapsychology *(Behav.)*. The study of certain alleged phenomena, the *paranormal*, that is beyond the scope of ordinary psychology, e.g. ESP, psychokineses etc.

paraquat *(Chem.,Med.)*. *1,1'-dimethyl-4,4'-dipyridylium salts*, used as a weed-killer. When taken orally causes severe and often irreversible damage to lungs, liver and kidneys.

paraquinones *(Chem.)*. Quinones in which the two quinone oxygen atoms are in *para* position.

pararosaniline *(Chem.)*. A triamino-triphenyl-methanol, obtained by oxidation of a mixture of 4-toluidine (1 mol) and aminobenzene (2 mol). As oxidizing agents, arsenic acid or nitrobenzene are used. Acids effect the elimination of water and the formation of a dyestuff with a quinonoid structure, e.g. the hydrochloride of pararosaniline or parafuchsine.

paraselenae *(Meteor.)*. See **mock moons**.

parasexual cycle *(Bot.)*. Genetic system in some fungi which allows limited **recombination** as a result of the doubling of the chromosomes in a nucleus, followed by crossing over and the gradual return to the haploid state by progressive chromosome loss, *haploidization*.

parasheet *(Aero.)*. A simplified form of parachute for dropping supplies, made from one or more pieces of fabric with parallel warp in the form of a polygon, to the apices of which the rigging lines are attached.

parasite *(Ecol.)*. An organism which lives in or on another organism and derives subsistence from it without rendering it any service in return. See **parasitism**.

parasitic antenna *(Telecomm.)*. Unfed dipole element which acts as a **director** or **reflector**. See **Yagi antenna**.

parasitic bronchitis *(Vet.)*. See **husk**.

parasitic capture *(Phys.)*. Neutron capture in reactor not followed by fission.

parasitic castration *(Zool.)*. Castration brought about by the presence of a parasite, as in the case of a Crab parasitized by *Sacculina*.

parasitic loss *(Elec.Eng.)*. Power loss in equipment which is not associated with its principal use, e.g. eddy currents induced in the different parts of an electrical machine.

parasitic male *(Zool.)*. A dwarf male in which all but the sexual organs are reduced, and which is entirely dependent on the female for nourishment; as in some deep-sea Angler-fish (Ceratioids).

parasitic oscillation *(Electronics)*. Unwanted oscillation of an amplifier, or oscillation of an oscillator at some frequency other than that of the main resonant circuit. Generally of high frequency, it may occur during a portion of each cycle of the main oscillation. Also *spurious oscillation*.

parasitic stopper *(Electronics)*. Components which attenuate a feedback path which would otherwise maintain unwanted oscillations.

parasitism *(Behav.,Ecol.)*. A close internal or external partnership between two organisms which is detrimental to one partner (the *host*) and beneficial to the other partner (the *parasite*); the latter often obtains its nourishment at the expense of the nutritive fluids of the host. The term usually refers to such a feeding relationship but other forms exist. See **social parasitism**.

parasitoid *(Zool.)*. An animal which is parasitic in one stage of the life history and subsequently free-living in the adult stage, as the parasitic Hymenoptera.

parasitology *(Genrl.)*. The study of parasites and their habits (usually confined to animal parasites).

parasolic acid *(Chem.)*. See **aurine**.

parasphenoid *(Zool.)*. In some of the lower Vertebrates, a membrane bone of the skull, which forms part of the cranial floor.

parastas *(Arch.)*. See **pilaster**.

parasymbiosis *(Biol.)*. The condition when two organisms grow together but neither assist nor harm one another.

parasympathetic nervous system *(Zool.)*. In Vertebrates, a subdivision of the autonomic nervous system, also known as the *craniosacral system*. The action of these nerves tends to slow down activity in the glands and smooth muscles which they supply, but promotes digestion, and acts antagonistically to that of the **sympathetic nervous system**. Parasympathetic nerves are cholinergic.

parathion *(Chem.)*. *2,2-diethyl-2-*p*-nitrophenyl-thiophosphate*. Used as an insecticide.

parathormone *(Zool.)*. The hormone secreted by the **parathyroid** which controls the metabolism of calcium and phosphorus.

parathyroid *(Zool.)*. An endocrine gland of Vertebrates found near the thyroid or embedded in it. Two pairs are usually present, deriving from the third and fourth pairs of gill pouches. Probably only secretes one hormone (*parathormone*).

paratonic movement *(Bot.)*. A plant movement in response to an external stimulus e.g. taxis, tropism. See also **autonomic movement**.

paratope *(Immun.)*. Same as *antigen binding site*. See **network theory**.

paratuberculosis *(Vet.)*. See **Johne's disease**.

paratyphoid *(Med.)*. Enteric fever due to infection by *Salmonella spp.* other than *S. typhi*, similar to, but milder than, typhoid fever.

paraxial *(Image Tech.)*. The path of a ray which is parallel to the axis of an optical system.

paraxial focus *(Phys.)*. The point at which a narrow pencil of rays along the axis of an optical system comes to a focus.

paraxonic foot *(Zool.)*. A foot in which the skeletal axis passes between the third and fourth digits, as in Artiodactyla. Cf. **mesaxonic foot**.

Parazoa *(Zool.)*. A subkingdom of the animals. Multicellular organisms, though their cells are less specialized and interdependent than in the Metazoa. Contains the single phylum **Porifera**. Cf. **Protozoa, Metazoa**.

Par C *(Build.)*. Abbrev. for *PARian Cement*.

parcel *(For.)*. Any quantity of standing, felled or converted timber forming a unit or item for purposes of trade.

parcel plating *(Elec.Eng.)*. The electrodeposition of a metal over a selected area of an article, the remainder being covered with a nonconductor to prevent deposition.

parchmentizing *(Paper.)*. The process of passing paper through sulphuric acid, or zinc chloride, which causes the fibres to swell and the paper to become translucent, dense and greaseproof, possessing high wet-strength properties.

parencephalon *(Zool.)*. A cerebral hemisphere.

parenchyma *(Bot.)*. Tissue composed of mature, vacuolated, but relatively unspecialized cells. Typically blunt ended somewhat elongated cells with thin or evenly thickened cell walls, they are not adapted for water transport but are sometimes photosynthetic, as *chlorenchyma*, or able to act as a store.

parenchyma *(Zool.)*. Soft spongy tissue of indeterminate form, consisting usually of cells separated by spaces filled with fluid or by a gelatinous matrix, and generally of mesodermal origin. The specific tissue of a gland or organ as distinct from the interstitial (connective) tissue or stroma. adj. *parenchymatous*.

parent *(Phys.)*. In radioactive particle decay of *A* into *B*, *A* is the parent and *B* the daughter product.

parenteral *(Med.)*. Said of the administration of therapeutic agents by any way other than through the alimentary tract.

parent exchange *(Telecomm.)*. Of an automatic exchange, the auto-manual or manual exchange which handles the assistance traffic.

parentheses *(Print.)*. Marks of punctuation ( ) used to enclose a definition, explanation, reference, etc. or interpolations and remarks made by the writer of the text himself. sing. *parenthesis*.

parent metal *(Eng.)*. A term used in welding to denote the metal of the parts to be welded.

parent peak *(Phys.)*. The component of a mass spectrum resulting from the undissociated molecule.

paresis *(Med.)*. Slight or incomplete paralysis. See also **general paresis** (general paralysis of the insane, GPI).

paresis juvenilis *(Med.)*. General paresis occurring in a child or young person.

paresthesia *(Med.)*. See **paraesthesia**.

pargasite *(Min.)*. A monoclinic amphibole of the hornblende group, particularly rich in magnesium, sodium, calcium, and aluminium. It occurs chiefly in metamorphic rocks.

pargeting *(Build.)*. The operation of rendering the interior of a flue with mortar. Also *pergeting*.

parge-work *(Build.)*. An ancient form of external plastering with a mixture similar to that used in *pargeting* chimneys. Also called *parging*.

parging *(Build.)*. See **parge-work**.

parhelia *(Meteor.)*. See **mock suns**. sing. *parhelion*.

Parian cement *(Build.)*. A hard plaster made from an intimate mixture of gypsum and borax which has been calcined and then ground to powder.

parietal *(Bot.)*. Peripheral.

parietal *(Zool.)*. A paired dorsal membrane bone of the Vertebrate skull, situated between the auditory capsules; pertaining to, or forming part of, the wall of a structure.

parietal foramen *(Zool.)*. Small rounded aperture in the middle of the united parietals of the skull. Site of the pineal eye.

parietal organ *(Zool.)*. The anterior diverticula of the pineal apparatus. When present may be developed as an eyelike organ, the pineal eye.

parietal placentation *(Bot.)*. Placentas which develop along the fused margins of the carpels of a unilocular ovary, e.g. violet.

parietes *(Med.,Zool.)*. The walls of an organ or a cavity. sing. *paries*.

paring chisel *(Build.)*. A long chisel with a thinner blade than a firmer tool, used for finishing off work by hand. It is not intended to be struck with a mallet.

paring gouge *(Build.)*. A gouge having the bevel ground upon the inside or concave face of the cutting edge. Also *scribing gouge*.

paripinnate *(Bot.)*. A pinnate leaf, having no terminal leaflet; even pinnate.

Paris green *(Chem.)*. *Copper ethanoato-arsenate*, a bright blue-green pigment once used in anti-fouling compositions but now mainly in wood preservatives and as an agricultural chemical. Also *emerald green, Schweinfurt green*.

parison *(Glass)*. The intermediate shape produced in the manufacture of a glass article in more than one stage. In the automatic and semi-automatic methods, the parison is formed in one mould, being then transferred to another for the final forming operation.

parity *(Maths.)*. The property of being odd or even, e.g. the parity of 4 is even.

parity *(Med.)*. The condition or fact of having borne children.

parity *(Phys.)*. The conservation of parity or space-reflection symmetry (*mirror symmetry*) states that no fundamental difference can be made between right and left, or in other words the laws of physics are identical in a right- or in a left-handed system of coordinates. The law is obeyed for all phenomena described by classical physics but was recently shown to be violated by the weak interactions between elementary particles.

parity bit *(Comp.)*. Extra binary digit appended to a binary word to produce even or odd parity for a subsequent **parity check**.

parity check *(Comp.)*. Test for corruption applied to binary data to confirm **even parity** or **odd parity**.

parity checking *(Telecomm.)*. In data communications, an *odd-even* check used when the total number of 0 s or 1 s is always made odd or even by adding the appropriate extra digit; a check may then be made at the receiver for even or odd parity.

Parker board *(Print.)*. A superior style of mounting board for printing plates, made up of small interlocked pieces of hardwood within a steel confining edge, and ruled off in squares to facilitate spacing out and squaring-up the plates.

Parke's process *(Eng.)*. A process used for desilverizing lead. It depends on the fact that when zinc is added to molten lead it combines with any gold or silver present to form compounds that have a very slight solubility in lead. The bullion-rich zinc is then skimmed off.

parking orbit *(Space)*. The waiting orbit of a spacecraft between two phases of a mission.

parkinsonism *(Med.)*. The gradual development of muscular rigidity, paucity of movement, slow tremor and postural deformities. Associated with a reduction of dopamine in the **basal ganglia** of the brain. May be primary (**Parkinson's disease**) or the result of drug or other intoxication.

Parkinson's disease *(Med.)*. *Paralysis agitans*; 'shaking palsy'. A progressive disease due to degeneration of certain nerve cells at the base of the brain; characterized by rigidity of muscles (the body being fixed in a posture of slight flexion), mask-like expression of the face, and a coarse tremor, especially of the hands. See **retropulsion**.

parliament hinge *(Build.)*. See **H-hinge**.

paronychia *(Med.)*. A felon or whitlow. Purulent inflammation of the tissues in the immediate region of the finger nail.

parosmia *(Med.)*. Abnormality of the sense of smell.

parotid gland *(Zool.)*. In some Anura, an aggregation of poison-producing skin glands on the neck; in Mammals, a salivary gland situated at the angle of the lower jaw.

parotitis *(Med.)*. Inflammation of the parotid gland. See **mumps**.

parpoint work *(Build.)*. Stone-wall construction in which the squared stones are laid as stretchers, with occasional courses of headers.

parquet *(Build.)*. A floor-covering of hardwood blocks glued and pinned to the ordinary floor boarding and finally wax-polished.

parrot coal *(Min.)*. See **boghead coal**.

parrot disease *(Med.,Vet.)*. See **psittacosis**.

pars *(Zool.)*. A part of an organ. pl. *partes*.

pars anterior *(Zool.)*. See **pars distalis**.

pars distalis *(Zool.)*. The anterior part of the adenohypophysis of the pituitary.

parsec *(Astron.)*. The unit of length used for distances beyond the solar system. It is the distance at which the **astronomical unit** subtends one second of arc, and is therefore 206 265 AU, 3.086×10^{13} km, 3.26 light years. From *par*allax *sec*ond. Abbrev. *pc*. Kiloparsec (kpc) and megaparsec (mpc) for 1000 and 1 000 000 pc are widely used in galactic and extragalactic contexts.

parsing *(Comp.)*. Breaking down of high-level programming language statements into their component parts during the translation process.

pars intermedia *(Zool.)*. In higher Vertebrates, part of the posterior lobe of the pituitary body, which is derived from the hypophysis at first but tends to become spread over the surface of the **pars nervosa** as development proceeds.

pars nervosa *(Zool.)*. In higher Vertebrates, part of the posterior lobe of the pituitary body, developed from the infundibulum.

Parson's steam turbine *(Eng.)*. A **reaction turbine** in which rings of moving blades of increasing size are arranged along the periphery of a drum of increasing diameter. Fixed blades in the casing alternate with the blade rings. Steam expands gradually through the blading, from inlet pressure at the smallest section to condenser pressure at the end.

parthenocarpy *(Bot.)*. The production of fruit without seeds either spontaneously or by artificial induction by **auxins**.

parthenogenesis *(Biol.)*. The development of a new individual from a single, unfertilized gamete, often an egg. adj. *parthenogenetic*.

parthenospore *(Bot.)*. A spore formed without previous sexual fusion; an azygospore.

parthen-, partheno- *(Genrl.)*. Prefix from Gk. *parthenos*, virgin.

partial *(Acous.)*. Any one of the single-frequency components of a complex tone; in most musical complex tones, the partials have harmonic frequencies with respect to a fundamental.

partial capacitance *(Phys.)*. Where there are a number of conductors (including earth) there is partial capacitance between all pairs, the effective capacitance under specified conditions being calculated from this network. See **capacitance coefficients**.

partial common *(Telecomm.)*. A trunk which is common to some of the groups in a grading unit.

partial differential coefficient *(Maths.)*. If $z = f(x,y, \text{etc.})$, then the partial differential coefficient of z with respect to x is the limit

$$\underset{h \to 0}{Lt} \frac{f(x+h, y, \text{etc.}) - f(x, y, \text{etc.})}{h},$$

if it exists: written $\partial z/\partial x$, $f_x(x,y, \text{etc.})$, or $f'_x = (x,y, \text{etc.})$.

partial earth *(Elec.Eng.)*. An earth fault having an appreciable resistance.

partially oriented yarn *(Textiles)*. A synthetic polymer in continuous filament form that already has a substantial degree of molecular orientation but which requires further orientation. This may be done by drawing during a subsequent process such as *texturing*. Abbrev. *POY*.

partial parasite *(Bot.)*. (1) A plant capable of photosynthesis but dependent on another plant for water and mineral nutrients, e.g. mistletoe; (2) A plant capable of living independently but able to become parasitic in suitable circumstances, *facultatively parasitic*.

partial pressures *(Chem.)*. The pressure exerted by each component in a gas mixture. See **Dalton's law of partial pressures**.

partial pressure suit *(Aero.)*. A laced airtight overall for aircrew members in very high-flying aircraft. It has inflatable cells to provide the wearer with an atmosphere and external body pressure in the event of cabin pressure failure. Essential for survival above 50 000 ft (15 000 m).

partial pyritic smelting *(Eng.)*. Blast-furnace smelting of copper ores in which some of the heat is provided by oxidation of iron sulphide and some by combustion of coke. See **pyritic smelting**.

partial reinforcement *(Behav.)*. Refers to conditions in which a response is reinforced only some of the time; such responses are more resistent to extinction than responses acquired through **continuous reinforcement**. Also known as intermittent reinforcement.

partial roasting *(Min.Ext.)*. Roasting carried out to eliminate some but not all of the sulphur in an ore or sulphidic concentrate.

partial umbel *(Bot.)*. One of the smaller group of flowers which altogether make up a compound umbel.

partial veil *(Bot.)*. In some basidiomycete fruiting bodies, e.g. mushrooms, a membrane joining the edge of the cap to the stalk, rupturing to leave an annulus. Cf. **universal veil**.

particle *(Min.Ext.)*. Single piece of solid material, usually defined (when small) by its mesh, or size passing through a specified size of sieve.

particle *(Phys.)*. (1) A useful concept of a small body which has a finite mass but negligible dimensions so that it has no *moment of inertia* about its *centre of mass*. (2) A volume of air or fluid which has dimensions very small compared with the wavelength of a propagated sound wave, but large compared with molecular dimensions.

particle exchange *(Phys.)*. The interaction between *fundamental particles* by the exchange of another fundamental particle. See **gauge boson**.

particle mean size *(Powder Tech.)*. The dimension of a hypothetical particle such that, if a powder were wholly composed of such particles, such a powder would have the same value as the actual powder in respect of some stated property.

particle physics *(Phys.)*. Study of the properties of fundamental particles and of fundamental interactions, often called high-energy physics.

particle porosity *(Powder Tech.)*. The ratio of the volume of open pores to the total volume of the particle.

particle scattering *(Phys.)*. See **scattering**.

particle size *(Geol.)*. The general dimensions of grains in a rock, especially a sediment. Many definitions have been used. One of the more common is the Wentworth-Udden scale:

>256 mm	boulder;
64—56 mm	cobble;
4—64 mm	pebble;
2—4 mm	gravel;
1/16—2 mm	sand;
1/256—1/16 mm	silt;
<1/256 mm	clay.

Field definitions are a little less precise. 'If the grains can be distinguished then it is at least silt grade; if it doesn't feel gritty on the teeth, then it is clay'.

particle size *(Powder Tech.)*. The magnitude of some physical dimension of a particle of powder. When the particle is a sphere, it is possible to define the size uniquely by a unit of length. For a nonspherical particle, it is not possible to define a size without specifying the method of measurement.

particle-size analysis *(Powder Tech.)*. The study of methods for determining the physical structure of powdered materials.

particle velocity *(Acous.)*. In a progressive or standing sound wave, the alternating velocity of the particles of the medium, taken as either the maximum or rms velocity.

particular average *(Ships)*. Loss or damage to marine property whose cost is borne by the owner (unless insured against it).

particular integral *(Maths.)*. Generally, a solution of a differential equation formed by assigning values to the arbitrary constants in the complete primitive. Nonsingular solution of a differential equation containing no arbitrary constants.

parting bead *(Build.)*. A bead fixed to the cased frame of a double-hung window to separate the inner and outer sashes.

parting of bullion *(Eng.)*. See **inquartation**.

parting sand *(Build.)*. A layer of dry sand separating 2 layers of damp sand, which are thereby prevented from adhering to one another.

parting sand *(Foundry)*. Dry sand sprinkled on the parting face of a mould to prevent adhesion of the 2 surfaces at the joint when the cope is rammed.

parting slip *(Build.)*. A thin lath of wood or zinc which keeps the sash-weights apart within the cased frame of a double-hung window. Also called *mid-feather, wagtail*.

partite *(Zool.)*. Split almost to the base.

partition chromatography *(Chem.)*. See **chromatography**.

partition coefficient *(Chem.)*. The ratio of the equilibrium concentrations of a substance dissolved in two immiscible solvents. If no chemical interaction occurs, it is independent of the actual values of the concentrations. Also *distribution coefficient*.

partition noise *(Electronics)*. That arising when electrons are abstracted from a stream by a number of successive electrodes, as in a travelling-wave tube.

partition plate *(Build.)*. The upper horizontal member of a wooden partition, capping the studding and providing support for joists, etc. Also *head piece*.

partridge disease *(Vet.)*. A popular term for infection of the gut of partridges by the nematode worm *Trichostrongylus tenuis*.

parturient *(Med.)*. Of or pertaining to parturition.

parturient fever, parturient eclampsia, parturient paresis *(Vet.)*. See **milk fever**.

parturition *(Zool.)*. In viviparous animals, the act of bringing forth young.

parvifoliate *(Bot.)*. Having leaves which are small in relation to the size of the stem.

parvovirus *(Vet.)*. See **canine parvovirus infection, porcine parvovirus infection**.

PAS *(Med.)*. 4-aminosalicylic acid. Used medically, usually in the form of the sodium salt, in the treatment of tuberculosis. Usually administered in conjunction with **isoniazid**.

PASCAL *(Comp.)*. Programming language designed to encourage structured programming.

pascal *(Genrl.)*. The SI derived unit of pressure or stress, equals 1 newton per square metre. abbrev. *Pa*.

Pascal's theorem *(Maths.)*. That the intersections of pairs of opposite sides of a hexagon inscribed in a conic are collinear. Cf. **Brianchon's theorem, Pappus' theorem (2)**.

Pascal's triangle *(Maths.)*. An easily remembered summary of the coefficients in the binomial expansion of $(1+x)^n$ for $n=0,1,2,3...$ Each row relates to a particular value of n, and each digit is the sum of the two above and on either side of it. The triangle, up to $n=5$, is as follows:

```
          1
        1   1
      1   2   1
    1   3   3   1
  1   4   6   4   1
1   5  10  10   5   1
```

Paschen-Back effect *(Phys.)*. The splitting of spectrum lines into a number of components by very strong magnetic fields. The fields are strong enough to decouple the spin and orbital angular momenta and the lines are split into three components. See **Zeeman effects**.

Paschen circle *(Phys.)*. A type of mounting in a concave-grating spectrograph which employs only a small part of the Rowland circle, the slit, grating and plate being fixed in position.

Paschen series *(Phys.)*. A series of lines in the infrared spectrum of hydrogen. Their wave numbers are given by the same expression as that for the Brackett series, but with $n_1=3$.

Paschen's law *(Phys.)*. That breakdown voltage, at constant temperature, is a function only of the product of the gas pressure and the distance between parallel-plate electrodes.

Pasiphae *(Astron.)*. The eighth natural satellite of **Jupiter**.

passage beds *(Geol.)*. The general name given to strata laid down during a period of transition from one set of geographical conditions to another; e.g. the Downtonian Stage consists of strata intermediate in character (and in position) between the marine Silurian rocks below and the continental Old Red Sandstone above.

passage cell *(Bot.)*. An endodermal cell in a root, usually opposite the protoxylem, which retains unthickened cell walls in an endodermis in which most of the cells have developed secondary walls.

pass band *(Telecomm.)*. The frequency range within which a filter allows signals to pass with minimum attenuation.

Passeriformes *(Zool.)*. An order of Birds containing those perching which comprise about half the known species of bird. Mostly small, living near the ground and having 4 toes arranged to allow gripping of the perch. Young helpless at hatching. Rooks, Finches, Sparrows, Tits, Warblers, Robins, Wrens, Swallows, and many others.

passing hollow *(Horol.)*. See **crescent**.

passings *(Build.)*. The overlap of one sheet of lead past another in flashings, etc.

passivate *(Eng.)*. To mask the normal electro-potential of a metal. A treatment to give greater resistance to corrosion in which the protection is afforded by surface coatings of films of oxides, phosphates etc. In the case of oils, passivation results from the stabilizing action of additives, preventing the oil from becoming more acidic.

passive *(Electronics)*. Said of transducers or filter sections without an effective source e.m.f.

passive-aggressive behaviour *(Behav.)*. Indirectly expressed resistance to the demands of others, e.g. forgetting appointments, losing important objects.

passive cutaneous anaphylaxis *(Immun.)*. A test in vivo to reveal the presence of mast cell-sensitizing antibody. Antibody is injected intracutaneously into the skin of an animal. After an interval sufficient for the antibody to become attached, antigen is injected intravenously together with a blue dye which binds to serum albumin. Where the antigen reacts with cell-fixed antibody, histamine and other substances are released which increase vascular permeability, and the dye leaks out to give a blue spot the size of which is proportional to the amount of antibody attached.

passive electrode *(Elec.Eng.)*. The earthed electrode of an electrical precipitation apparatus, being that upon which the particles are deposited. Also called *collecting electrode*.

passive homing guidance *(Aero.)*. A missile guidance system which homes on to radiation (e.g. infra-red) from the target.

passive immunization *(Immun.)*. Use of antibody from an immune individual to provide temporary immunity in a non-immune individual, e.g. with diptheria antitoxin, tetanus antitoxin or serum from persons convalescent from measles.

passive margin *(Geol.)*. A continental margin characterized by thick, relatively undeformed sediments, deposited at the trailing edge of a *lithospheric plate*. See **active margin**.

passive network *(Electronics)*. Consists of components in which there is no e.m.f. or other source of energy.

passive permeability *(Biol.)*. The flux of solutes across a cell membrane by simple diffusion at a rate proportional to the difference in concentration of the solute across the membrane.

passive radar *(Radar)*. That using microwaves or infrared radiation emitted from source, and hence not revealing the presence or position of the detecting system. Military use.

passive satellite *(Space)*. See **active satellite**.

passivity *(Min.Ext.)*. Lack of response of metal or mineral surface to chemical attack such as would take place with a clean, newly exposed surface. Due to various causes, including insoluble film produced by ageing, oxidation, or contamination; run-down of surface energy at discontinuity lattices; adsorbed layers. Phenomenon prevents use of cyanide process to dissolve large particles of gold, but is sometimes used to aid froth-flotation by rendering specific minerals passive to collector agents.

pass-over offset *(Build.)*. The local bend which enables one pipe to pass over another otherwise in the same plane.

pass sheet *(Print.)*. The sheet which has been signed by overseer and proofreader and serves as a standard, particularly for colour and register.

password *(Comp.)*. A sequence of characters which must be keyed in before a user can gain access to a computer system.

paste *(Plastics)*. Usually applied to fluid dispersion of PVC in a plasticizer used for a variety of dipping processes such as glove making, leather cloth and soft toys.

paste *(Print.)*. (1) Operating an automatic paster to replace the old reel with a new. (2) Securing two webs together during the run of the press.

pasteboards *(Paper)*. A laminated product formed by pasting two or more webs or sheets of paper together. The middle layer(s) may if required be of cheaper material than the facings.

pasted filament *(Elec.Eng.)*. An electric-lamp filament prepared by squirting through a die a paste formed of powdered metal, usually tungsten, together with a binding material, the latter being subsequently removed by heat treatment.

pasted plate *(Elec.Eng.)*. A plate in which the active material of the plate is applied in the form of a paste; used for lead-acid accumulators.

paste mould *(Glass)*. A mould for blowing light-walled hollow-ware. As a good finish is needed, the moulds are coated with adherent carbon ('paste'), which is wetted before each blowing operation.

paster tab *(Print.)*. A strip of gummed paper with a weakened section or other device used to secure the prepared end of the new reel prior to joining it to the old reel on an **automatic reel change**.

pasteurellosis *(Vet.)*. The group name for diseases caused by organisms of the genus *Pasteurella*. *Pasteurella multocida* causes an acute septicaemic and pneumonic disease in several species of animals. See **haemorrhagic septicaemia** (cattle and sheep), **swine plague** (pig), **fowl cholera** (fowl), **rabbit septicaemia** (rabbit). Vaccines widely used.

Pasteur filter *(Chem.)*. See **ceramic filter**.

pasteurization *(Med.)*. Reduction of the number of micro-organisms in milk by maintaining it in a holder at a temperature of from 62.8°–65.5°C for 30 minutes.

patagium *(Zool.)*. A lobelike structure at the side of the prothorax in some Lepidoptera: in Bats and some other flying Mammals, a stretch of webbing between the fore limb and the hind limb: in Birds, a membranous expansion of the wing. adj. *patagial*.

patand *(Build.)*. A sill resting on the ground as a support for a post.

patch *(Comp.)*. Small fragment of code provided by a software supplier, to enable a user to modify or correct his own copy of software without requiring a complete replacement.

patch bay *(Telecomm.)*. Section of equipment which includes all connectors which terminate units of equipment; interconnection can be altered and/or test equipment connected.

patch board *(Comp.)*. One on which simple programs can be set up and modified by making electrical connections between logic elements.

patching up *(Print.)*. (1) In letterpress, as part of **making ready**, applying patches of thin paper which will be included in the *packing* to improve the impression where required. (2) In lithography, the arranging of positives or negatives, in required position on a suitable carrier, preliminary to making a plate.

patch in, out *(Telecomm.)*. Temporary insertion of spare apparatus, or removal of defective apparatus, in a circuit by patch cords, usually in a patch bay or field.

patch panels *(Comp.)*. Panel of plugs used for making connections between incoming and outgoing lines.

patch test *(Med.)*. A test for allergy, consisting of the application to the skin of small pads soaked with the allergy-producing substance; if supersensitivity exists inflammation develops at the places where the substance was applied.

patella *(Zool.)*. In higher Vertebrates, a sesamoid bone of the knee joint or elbow joint.

patent *(Bot.)*. Said of leaves and branches which spread out widely from the stem.

patent glazing *(Build.)*. The name applied to various marketed devices for securing together glass sheets (for roof coverings, etc.) without using putty in sashes, the connection being made usually with special metal sections or flashings.

patent log *(Ships)*. See **nautical log**.

patera *(Arch.)*. A circular ornament in relief on friezes. pl. *paterae*.

paternoster *(Eng.)*. A lift for goods or passengers which consists of a series of floored but doorless compartments moving slowly and continuously up and down on an endless chain.

path *(Telecomm.)*. Channel through which signals can be sent, particularly *forward, through*, and *feedback* paths of servo systems.

path attenuation *(Telecomm.)*. Fall-off in amplitude of a radio wave with distance from the transmitter.

pathetic muscle *(Zool.)*. The superior oblique muscle of the Vertebrate eye.

pathetic nerve *(Zool.)*. The 4th cranial nerve of Vertebrates, running to the superior oblique muscle.

pathogen *(Biol.)*. An organism, e.g. parasite, bacterium, virus, which causes disease.

pathogen *(Med.)*. Any disease-producing micro-organism or substance.

pathogenesis *(Med.)*. The development or production of a disease-process. adj. *pathogenic*.

pathognomonic *(Med.)*. Specially indicating a particular disease.

pathological *(Med.)*. Concerning pathology: morbid, diseased.

pathology *(Bot.)*. See **phytopathology**.

pathology *(Med.)*. That part of medical science which deals with the causes and nature of disease, and with the bodily changes wrought by disease.

patin *(Build.)*. See **patand**.

patina *(Chem.)*. The thin, often multicoloured, film of atmospheric corrosion products formed on the surface of a metal or mineral.

Patra Test Bench *(Print.)*. See **Pira Test Bench**.

patristic similarity *(Bot.)*. Similarity due to common ancestry.

patroclinous *(Bot.,Zool.)*. Exhibiting the characteristics of the male parent more prominently than those of the female parent. Cf. **matroclinous**.

patten *(Arch.)*. The base of a column or pillar.

patter *(Build.)*. A kind of float, made of thick wood, used to consolidate and level cement surfaces.

pattern *(Electronics)*. The luminous trace on the screen of a CRO as traced out by the electron beam.

pattern *(Foundry)*. A wood, metal, or plaster copy, in one piece or in sections, of an object to be made by casting. Made slightly larger than the finished casting to allow for contraction of casting while cooling; and suitably tapered to facilitate withdrawal from the mould. See **double contraction, metal pattern, plate moulding**.

pattern *(Telecomm.)*. Pattern of the radiation field from an aerial system as shown by a polar diagram of field strength and bearing.

pattern-maker's hammer *(Build.)*. Light type of cross-pane hammer with long handle.

pattern maker's rule *(Eng.)*. One with graduations lengthened so as to compensate for the cooling contraction which must be allowed for in the cast object.

pattern recognition *(Comp.)*. The automatic recognition of patterns, using a specialized input device linked to a processor. See **character recognition, speech recognition**.

pattern staining *(Build.)*. A defect most apparent on ceiling areas. It appears as light and dark patches forming a pattern consistent with the layout of the ceiling joists. This problem is difficult to eradicate as it involves the transmission of heat and dirt through the surface. Improved thermal insulation reduces the problem.

Pattinson process *(Eng.)*. Obsolescent process used for the separation of small quantities of silver from lead by partially solidifying a molten bath of the two metals and separating the remaining liquid. This process is repeated several times and the silver is concentrated in the liquid. Cf. **Parke's process**.

Paul-Bunnell test *(Immun.)*. Test used in the diagnosis of infectious mononucleosis. Serum from persons with infectious mononucleosis contains a particular heterophil antibody which will agglutinate both sheep and horse erythrocytes, and is not found in other conditions. The antibody is distinct from antibodies against Epstein-Barr virus, the actual cause of the disease.

Pauli exclusion principle *(Phys.)*. Fundamental law of quantum mechanics, that no two *fermions* in a system can exist in identical quantum states, i.e. have the same set of quantum numbers. This constraint explains the electronic structure of atoms and also the general nature of the periodic table. As neutrons and protons are also fermions, nuclear structure is also strongly influenced by the exclusion principle,

paunch *(Zool.)*. See **rumen**.

pavement epithelium *(Zool.)*. A variety of epithelium consisting of layers of flattened cells.

pavement light *(Build.)*. A panel formed of glass blocks framed in concrete, iron, or steel, built into a pavement surface over an opening to the basement of a building, into which it admits light.

pavilion *(Arch.)*. An ornamental, detached structure which has a roof but is usually not entirely enclosed by walls. Used on sports fields, or as a place for entertainments.

pavings *(Build.)*. Very hard purpose-made bricks, usually of the dark blue Staffordshire variety, having a surface chequered by grooves to make it less slippery.

pavior *(Build.)*. (1) A specially hard brick used in the construction of pavement surfaces. (2) A worker who lays bricks or setts to form pavement surfaces.

pawl *(Eng.)*. A pivoted catch, usually spring-controlled, engaging with a ratchet wheel or rack to prevent reverse motion, or to convert its own reciprocating motion into an intermittent rotary or linear motion.

pax *(Aero.)*. Airline passengers.

PAX *(Telecomm.)*. Abbrev. for *Private Automatic eXchange*.

Paxolin *(Plastics)*. TN for a laminated plastic usually of the phenolic class: used in the manufacture of sheets, tubes, cylinders and laminated mouldings.

pay *(Min.Ext.)*. Pay dirt, ore, rock, streak. Any mineral deposit which will repay efficient exploitation.

payload *(Aero.)*. That part of an aircraft's load from which revenue is obtained.

payload *(Space)*. Additional or 'useful' mass available, over and above that devoted to maintaining a space system. Usually this applies to the mass of instruments and supporting hardware for performing experiments. When applied to a launch system it refers to the net mass (as distinct from the total mass of the system) which is placed in orbit.

payload integration *(Space)*. Process of bringing together individual experiments, their support equipment and **software** into a payload entity, in which all interfaces are compatible and whose operation has been fully checked-out.

payload specialist *(Space)*. Highly qualified scientist-member of the crew of the Space Shuttle, whose sole responsibility is the operation of the experiments of a payload. He or she is not necessarily a professional astronaut.

pay string *(Min.Ext.)*. The pipe through which oil or gas passes from the *pay zone* to the well head.

pay-TV *(Image Tech.)*. General term for a specialized form of television distribution, such as by cable or by coded signals, for which a charge is made.

Pb *(Chem.)*. The symbol for lead.

PBX *(Telecomm.)*. Abbrev. for *Private Branch eXchange*.

PC *(Comp.)*. See **personal computer**.

PCB *(Comp.)*. See **printed circuit board**.

PCE *(Eng.)*. See **pyrometric cone equivalent**.

PCM *(Aero.)*. Abbrev. for *Pulse Coded Modulation*.

PCM *(Telecomm.)*. Abbrev. for *Pulse-Code Modulation*.

PCNB *(Chem.)*. See **quintozene**.

Pd *(Chem.)*. The symbol for palladium.

P-display *(Radar)*. That of a **PPI** unit. Map display produced by intensity modulation of a rotating radial sweep.

PDL *(Comp.)*. See **page description language**.

pd, p.d. *(Phys.)*. Abbrev. for *Potential Difference*.

pE *(Chem.)*. Negative logarithm of effective electron concentration in a redox system. Its use is analogous to that of pH in acid-base reactions. See also **rH**.

peacock ore *(Min.)*. A name given to **bornite** or sometimes **chalcopyrite**, because they rapidly become iridescent from tarnish.

peak *(Phys.)*. The *instantaneous* value of the local maximum of a varying quantity. It is $\sqrt{2}$ times the r.m.s. value for a sinusoid.

peak dose *(Radiol.)*. Maximum absorbed radiation dose at any point in an irradiated body, usually at a small depth below the surface, due to secondary radiation effects.

peak envelope power *(Telecomm.)*. See **peak power**.

peak factor *(Elec.Eng.)*. Ratio of peak value of any alternating current or voltage to r.m.s. value. Also *crest factor*.

peak forward voltage *(Elec.Eng.)*. The maximum instantaneous voltage in the forward flow direction of anode current as measured between the anode and cathode of a diode or thyristor.

peaking *(Telecomm.)*. Inclusion of series or shunt resonant units in, e.g. TV circuits, to maintain response up to a maximum frequency.

peaking network *(Telecomm.)*. An interstage coupling circuit which gives a peak response at the upper end of the frequency range that is handled. This is achieved with a resonant circuit and minimizes the fall-off in the frequency response produced by stray capacitances. Not to be confused with *peaking circuit*.

peaking transformer *(Telecomm.)*. One in which the core is highly saturated by current in the primary, thus providing a peaky e.m.f. in the secondary as the flux in the core suddenly changes over.

peak inverse voltage *(Elec.Eng.)*. See **inverse voltage**.

peak joint *(Build.)*. The joint between the members of a roof truss at its ridge.

peak limiter *(Telecomm.)*. Circuit for avoiding overload of a system by reducing gain when the peak input signal reaches a certain value.

peak load *(Elec.Eng.)*. The maximum instantaneous rate of power consumption in the load circuit. In a power-supply system, the peak load corresponds to the maximum power production of the generator(s).

peak power *(Telecomm.)*. Average radiofrequency power at maximum modulation, i.e. of envelope of transmission.

peak programme meter *(Telecomm.)*. One bridged across a transmission circuit to indicate the changes in *volume* of the ultimate reproduction of sound, averaging peaks over 1 millisecond.

peak sideband power *(Telecomm.)*. The average sideband power of a transmitter over one radiofrequency cycle at the highest peak of the modulation envelope.

peak-to-peak amplitude *(Phys.)*. See **double amplitude**.

peak value *(Phys.)*. The maximum positive or negative value of an alternating quantity. Also called *amplitude*, *crest value*, *maximum value*.

peak voltmeter *(Elec.Eng.)*. Measures the peak value of an alternating voltage, e.g. by biassing a diode so that it just conducts. Also called *crest voltmeter*.

peak white *(Image Tech.)*. The level in a TV signal corresponding to white.

pearl *(Print.)*. An old type size, approximately 5-point.

pearl *(Zool.)*. An abnormal concretion of nacre formed inside a mollusc shell round a foreign body such as a sand particle or a parasite.

pearlite *(Eng.)*. A microconstituent of steel and cast iron. It is produced at the eutectoid point by the simultaneous formation of ferrite and cementite from austenite, and normally consists of alternate plates of these constituents (see, however, **granular pearlite**). A carbon steel containing 0.9% of carbon consists entirely of pearlite. See **eutectoid steel, hypo-eutectoid steel, hyper-eutectoid steel**.

pearlitic iron *(Eng.)*. A grey cast-iron consisting of graphite in a matrix of pearlite, i.e. without free ferrite.

pearl spar *(Min.)*. See **dolomite**.

pearl white *(Chem.)*. See **bismuth (III) chloride**.

pear oil *(Chem.)*. See **amyl acetate**.

peas *(Min.Ext.)*. See **coal sizes**.

peat *(Geol.)*. The name given to the layers of dead vegetation, in varying degrees of alteration, resulting from the accumulation of the remains of marsh vegetation in swampy hollows in cold and temperate regions. Geologically, peat may be regarded as the youngest member of the series of coals of different rank, including brown coal, lignite, and bituminous coal, which link peat with anthracite. Peat is very widely used as a fuel, after being air-dried, in districts where other fuels are scarce and in some areas, e.g. in Russia and Ireland, it is used to fire power stations. It is low in ash, but contains a high percentage of moisture, and is bulky; specific energy content about 16 MJ/kg or 7000 Btu/lb.

pebble-bed reactor *(Nuc.Eng.)*. One with cylindrical core into which spherical fuel pellets are introduced at the top and extracted at the base.

pebble-dashing *(Build.)*. A rough finish given to a wall by coating it with rendering, on to which, while it is still soft, small stones are thrown. Cf. **rough casting**.

pebble mill *(Min.Ext.)*. **Ball mill** in which selected pebbles or large pieces of ore are used as grinding media.

peckings *(Build.)*. Under-burnt, badly-shaped bricks, used only for temporary work or for the inside of walls.

peck order *(Behav.)*. The classic example of **social hierarchy** in farmyard hens in which animals within a group form some consistent relationship most apparent in their aggressive interactions; the same individuals dominate or are dominated by particular animals within the group.

pecten *(Zool.)*. Any comblike structure: in some Vertebrates (Reptiles and Birds), a vascular process of the inner surface of the retina; in Scorpionidea, tactile sensory organs under the mesosoma.

pectinate *(Bot.,Zool.)*. Comb-like.

pectineal *(Zool.)*. Comblike; said: (1) of a process of the pubis in Birds; (2) of a ridge on the femur to which is attached the *pectineus muscle*, one of the protractors of the hind limb.

pectines *(Zool.)*. Comblike chitinous structures of mechanoreceptive function attached to the ventral surface of the second somite of the mesosoma in *Scorpionidea*.

pectins *(Chem.)*. Calcium-magnesium salts of polygalacturonic acid, partially joined to methanol residues by ether linkage. They occur in the cell walls, especially in the *middle lamellas* and *primary walls* of vascular plants. They are soluble in water and can be precipitated from aqueous solutions by excess alcohol. Acid solutions gel with 65–70% of sucrose, the basis of their use in jam making.

pectization *(Chem.)*. The formation of a jelly.

pectolite *(Min.)*. A silicate of calcium and sodium, with a variable amount of water, which crystallizes in the triclinic system. It occurs in aggregations like the zeolites in the cavities of basic eruptive rocks, and as a primary mineral in some alkaline igneous rocks.

pectorales *(Zool.)*. In Vertebrates, muscles connecting the upper part of the fore-limb with the ventral part of the pectoral girdle. sing. *pectoralis*.

pectoral fins *(Zool.)*. In Fish, the anterior pair of fins.

pectoral girdle *(Zool.)*. In Vertebrates, the skeletal framework with which the anterior pair of locomotor appendages articulate.

pectoriloquy *(Med.)*. Conduction, to the chest wall, of the sound of words spoken or whispered by the patient and clearly heard through the stethoscope; indicative of consolidation or cavitation of the lung.

peculiar *(Print.)*. A term describing any unusual type character, such as certain accents.

pedal curve *(Maths.)*. Of a given curve with respect to a given point: the locus of the point of intersection of a tangent to the curve and the perpendicular to the tangent from the given point.

pedate leaf *(Bot.)*. A leaf with three divisions of which the 2 laterals are forked once or twice.

pedes *(Zool.)*. See **pes**.

pedesis *(Phys.)*. **Brownian movement**.

pedestal *(Image Tech.)*. The difference in signal between the **black level** of the picture and the **blanking** level. Also *lift*.

pediatric *(Med.)*. See **paediatric**.

pedicel *(Bot.)*. A stalk bearing one flower. Cf. **peduncle**.

pedicel *(Zool.)*. The second joint of the antennae in Insects; more generally, the stalk of a sedentary organism; the stalk of a free organ, as the *optic pedicel* in some Crustacea.

pedicellaria *(Zool.)*. In Echinodermata, a small pincerlike calcareous structure on the body surface with 2 or 3 jaws provided with special muscles and capable of executing snapping movements; it may be stalked or sessile.

pedicellate *(Bot.,Zool.)*. Having, or borne on, a pedicel.

pedicle *(Zool.)*. In the vertebrae of the Frog, a pillarlike process springing from the centrum and extending vertically upwards to join the flat, nearly horizontal lamina, which forms the roof of the neural canal. Intervertebral foramina, for the passage of the spinal nerves, occur between successive pedicles. Generally any pillarlike process supporting an organ.

pediculosis *(Med.)*. Infestation of the body with lice.

pediment *(Arch.)*. A triangular or segmental part surmounting the portico in the front of a building.

pediment *(Geol.)*. A broad and relatively flat rock surface abutting a mountain range, in an arid environment. It may be covered by a veneer of alluvium.

pedion *(Crystal.)*. A crystal form consisting of a single plane; well shown by some crystals of tourmaline which may be terminated by a pedion, with or without pyramid faces.

pedipalp *(Zool.)*. In Chelicerata the appendage, borne by the first postoral somite, of which the gnathobases function as jaws; it may be a tactile organ or chelate weapon.

pedology *(Genrl.)*. The study of soil.

pedometer *(Surv.)*. An instrument for counting the number of paces (and hence the approximate distance) walked by its wearer.

peduncle *(Bot.)*. A stalk bearing several flowers. Cf. **pedicel**.

peduncle *(Zool.)*. In Brachipoda and Cirripedia, the stalk by which the body of the animal is attached to the substratum; in some Arthropoda, the narrow portion joining the thorax and abdomen or the prosoma and opisthosoma; in Vertebrates, a tract of white fibres in the brain.

pedunculate *(Bot.)*. Having, or borne on, a peduncle.

peeling *(Eng.)*. A layer of a coating becoming detached because of poor adhesion.

peeling *(Print.)*. The operation of preparing **overlays** by thinning the hard edges of an illustration.

peening *(Eng.)*. Using hammer blows or shot blasting to cold-work metal.

peg-and-cup dowels *(Eng.)*. Metal pegs and sleeves inserted in adjoining parts of a split pattern to hold them in register while ramming the mould.

peggies *(Build.)*. Slates 10–14 in (254–356 mm) in length.

pegmatite *(Geol.)*. A term originally applied to granitic rocks characterized by intergrowths of feldspar and quartz, as in graphic granite; now applied to igneous rocks of any composition but of particularly coarse grain, occurring as offshoots from, or veins in, larger intrusive rock bodies, representing a flux-rich residuum of the original magma.

pegwood *(Horol.)*. Sticks of close-grained wood used for cleaning out the pivot holes of clocks and watches.

pelagic *(Zool.)*. Living in the middle depths and surface waters of the sea.

pelagic deposits *(Geol.)*. A term applied to any accumulation of sediments under deep-water conditions.

pelargonic acid *(Chem.)*. *Nonan-1-oic acid.* $CH_3.(CH_2)_7.COOH$. Mp 12.5°C. Oxidation product of oleic acid. Occurs naturally in fusel oil from beet and potatoes.

Pelean eruption *(Geol.)*. A type of eruption characterized by lateral explosions generating **nuées ardentes**.

Pelecaniformes *(Zool.)*. A varied order of Birds, mainly fish-eating and colonial nesters, with 4-toed webbed feet, bodies adapted for diving, and long beaks with wide gapes and sometimes with a pouch. Pelicans, Cormorants, Gannets.

Pelecypoda *(Zool.)*. A synonym for **Bivalvia**.

Pélé's hair *(Min.)*. Long threads of volcanic glass which result from jets of lava being blown aside by the wind in the volcano of Kilauea, Hawaii.

pele tower *(Arch.)*. A term used in northern England and Scotland to describe a small tower which was used to provide a retreat in the event of a sudden, unexpected attack.

pelitic gneiss *(Geol.)*. A gneissose rock derived from the metamorphism of argillaceous sediments.

pelitic schist *(Geol.)*. A schist of sedimentary origin, formed by the dynamothermal metamorphism of argillaceous sediments such as clay and shale.

pellagra *(Med.)*. *Maidismus.* A chronic disease due to a dietary deficiency of the nicotinic acid component of the vitamin-B complex, associated with protein deficiency. It is characterized by gastro-intestinal disturbances, a symmetrical erythema of the skin, mental depression, and paralysis. Seen in maize eaters where the nicotinic acid is in bound form and there is a lack of the tryptophan precursor of nicotinic acid.

pellet *(Build.)*. A term applied to a moulding characterized by a series of spherical protuberances.

pellet *(Phys.)*. Nuclear fusion fuel contained in concentric spheres of glass, plastic and other materials. They are hit by a burst of laser energy to produce fusion.

pellet *(Plastics)*. A compressed mass of moulding material of prescribed form and weight.

pelleted seed *(Bot.)*. Seed coated with a layer of inert material, especially to make smaller, angular seeds into larger, rounded bodies that can be drilled more precisely, sometimes also to incorporate pesticides etc.

pelletization *(Min.Ext.)*. Treatment of finely divided ore, concentrate or coal to form aggregates some 1–1.5 cm in diameter, for furnace feed, transport, storage or use (e.g. as coal briquettes). Powder, with suitable additives, is rolled into aggregates (green balls) in pelletizing drum and then hardened in a furnace by specialized baking methods.

pellicle *(Bot.)*. Layer of interlocking, helically wound, proteinaceous strips just below the plasmalemma of *Euglenophyceae*.

pellicle *(Phys.)*. A strippable photographic emulsion used to form a *stack* in nuclear research emulsion techniques.

pellicle *(Zool.)*. A thin cuticular investment, as in some Protozoa. adj. *pelliculate*.

pellicle, pellicule *(Image Tech.)*. An extremely thin transparent film which can be used as a semi-reflecting surface without producing double reflections.

Pellin-Broca prism *(Phys.)*. A 4-sided prism, which can be imagined to be formed by placing together two 30° prisms and one 45° prism. It is used in wavelength spectrometers where the wavelength of the light received in the telescope is altered by rotation of the prism.

pelma *(Zool.)*. See **planta**.

peltate *(Bot.)*. A flattened rounded plant organ attached to its stalk at about the middle of its lower surface e.g. the leaf of the nasturtium (*Tropaeolum maius*).

Peltier coefficient *(Elec.Eng.)*. Energy absorbed or given out per second, due to the Peltier effect, when unit current is passed through a junction of 2 dissimilar metals. Symbol Π.

Peltier effect *(Elec.Eng.)*. Phenomenon whereby heat is liberated or absorbed at a junction when current passes from one metal to another.

Pelton wheel *(Eng.)*. An impulse water turbine in which specially shaped buckets attached to the periphery of a wheel are struck by a jet of water, the nozzle being either deflected or valve-controlled by a governor.

pelvic fins *(Zool.)*. In Fish, the posterior pair of fins.

pelvic girdle *(Zool.)*. In Vertebrates, the skeletal framework with which the posterior pair of locomotor appendages articulate.

pelvimetry *(Med.)*. Estimation, by the use of X-rays but more usually ultrasound, of the size and shape of the female pelvis.

pelvis *(Zool.)*. The pelvic girdle or posterior limb girdle of Vertebrates, a skeletal frame with which the hind limbs or fins articulate; in Mammals, a cavity, just inside the hilum of the kidney, into which the uriniferous tubules discharge and which is drained by the ureter. adj. *pelvic*.

pemphigus *(Med.)*. Inflammatory condition of the skin characterized by the eruption of crops of blisters, the mucous membranes at times also being involved.

pen *(Zool.)*. In Cephalopoda, the shell or cuttle bone.

pencatite *(Geol.)*. A crystalline limestone which contains brucite and calcite in approximately equal molecular proportions.

pencil *(Maths.)*. Of lines: a number of lines passing through a fixed point (*vertex*). Of planes: a number of planes having a common line.

pencil *(Phys.)*. A narrow beam of light, having a small angle of convergence or divergence.

pencil overgrainer *(Build.)*. An **overgraining** brush consisting of a row of pencil brushes in a setting and bound with metal. Best quality are made from sable hair with hog hair as cheaper substitute. Used to add equidistant lines in graining work

pencil stone *(Min.)*. The name given to the compact variety of pyrophyllite, used for slate-pencils. The term *pencil ore* has been used for the broken splinters of radiating massive hematite, as they give a red streak.

pendant *(Build.)*. Ornamentation suspended below an object or surface to be decorated, as from a ceiling.

pendant *(Elec.Eng.)*. A lighting fitting suspended by means of a flexible support.

pendant *(Horol.)*. The neck of a watchband, to which the bow is attached.

pendentive *(Arch.)*. A spherical triangle formed by a dome springing from a square base.

pendentive dome *(Arch.)*. A dome covering a square area to which it is linked at the corners by pendentives.

pendulous placentation *(Bot.)*. See **apical placentation**.

pendulum *(Eng.)*. The simple pendulum consists of a small, heavy bob suspended from a fixed point by a thread of negligible weight. Such a pendulum, when swinging freely with small amplitude, has a periodic time given by

$$T = 2\pi \cdot \sqrt{\frac{l}{g}},$$

where l is the length of the thread and g is the acceleration due to gravity. See **centre of oscillation, compound pendulum**.

pendulum *(Horol.)*. The time-controlling element of a pendulum clock. The theoretical length of a pendulum, in mm, is given by $L = 993.6 \times t^2$, where t is the time of swing in seconds and 993.6 mm is the length of a pendulum beating seconds in London. For household clocks, the pendulum beats 0.5 s or less. For long-case clocks and regulators, a seconds pendulum is used; for tower clocks, it may be up to 2 s.

pendulum bob *(Horol.)*. The weighted mass at the end of a pendulum.

pendulum damper *(Aero.)*. A short heavy pendulum, in the form of pivoted balance weights, attached to the crank of a radial aero-engine in order to neutralize the fundamental torque impulses and so eliminate the associated critical speed.

pendulum governor *(Eng.)*. An engine governor, the many forms of which involve the principle of the conical pendulum. Heavy balls swing outwards under centrifugal force, so lifting a weighted sleeve and progressively closing the engine throttle valve. See **Porter governor, Watt governor**.

pendulum rod *(Horol.)*. The rod of a pendulum which supports the bob.

pendulum spring *(Horol.)*. The thin ribbon of spring steel used for suspending the pendulum.

penecontemporaneous *(Geol.)*. Describes any process occurring in a rock very soon after its formation.

peneplain *(Geol.)*. A gently rolling lowland, produced after long-continued denudation.

penetrance *(Biol.)*. The proportion of individuals of a specified genotype in which a particular gene exhibits its effect.

penetrant *(Chem.)*. (1) Substance which increases the penetration of a liquid into porous material or between contiguous surfaces, e.g. alkyl-aryl sulphonate. (2) A **wetting agent**.

penetrant inspection *(Eng.)*. See **fluorescent penetrant inspection**.

penetrating shower *(Phys.)*. Cosmic-ray shower containing mesons and/or other penetrating particles. See **cascade shower**.

penetration *(Civ.Eng.)*. A term used in testing bituminous material. Penetration is expressed as the distance that a standard needle vertically penetrates a sample of the material under known conditions of loading, time and temperature.

penetration *(Elec.Eng.)*. Measure of depth of **skin effect** of eddy currents in induction heating, or depth of magnetic field in superconducting metals. Usually to $1/e$ (0.37) of surface value.

penetration *(Eng.)*. In welding, the distance between the original surface of the metal and the position of the furthest penetration of the weld. In the foundry, metal penetration occurs when moulding sand allows the metal to enter between the grains.

penetration depth *(Telecomm.)*. That thickness of hollow conductor of the same external dimension which, if the current were uniformly distributed throughout the cross-section, would have the same effective résistance as the solid conductor. See **skin depth**.

penetration factor *(Phys.)*. Probability of incident particle passing through nuclear potential barrier.

penetration theory *(Chem.Eng.)*. Theory that mass transfer across an interphase into a stirred liquid takes place by diffusive penetration of the solute into the liquid surface, which is continually being renewed, hence the rate of mass transfer is proportional to the square root of the *diffusion coefficient*.

penetration twins *(Min.)*. See **interpenetration twins**, and cf. **juxtaposition twins**.

penetrometer *(Radiol.)*. A device for the measurement of the penetrating power of radiation by comparison of the transmission through various absorbers.

penicillins *(Med.)*. Group of antibiotics which are bactericidal and act by interfering in bacterial cell wall synthesis.

penis *(Zool.)*. The male copulatory organ in most higher Vertebrates: a form of male copulatory organ in various Invertebrates, as Platyhelminthes, Gastropoda. adj. *penial*.

Penman-Monteith equation *(Ecol.)*. Describes the dependence of evapotranspiration or transpiration rates on climatological variables and surface properties of the vegetation.

pennae *(Zool.)*. See **plumae**.

pennate *(Genrl.)*. Generally, winged.

pennine, penninite *(Min.)*. A silicate of magnesium and aluminium with chemically combined water. It crystallizes in the monoclinic system and is a member of the chlorite group.

penning *(Civ.Eng.)*. See **pitching**.

Pennsylvanian *(Geol.)*. The period of the Upper Palaeozoic era lying between the Mississippian and Permian. It covers an app. time span from 330–290 million years and is the North American equivalent of the Upper Carboniferous in Europe. The corresponding system of rocks.

pennyweight *(Genrl.)*. Troy weight of 24 gr or 1.5552 g, a unit widely used in valuation of gold ores and sale of bullion. Abbrev. *dwt*.

pen ruling *(Print.)*. The ruling of paper on a pen-ruling machine as distinct from **disk ruling**.

Pensky-Martens test *(Chem.)*. Standard test for determining the flash and fire points of oils. Based on closed or open cups depending on the nature of the oil under test.
penstock *(Eng.)*. The valve-controlled water conduit between the intake and the turbine in a hydroelectric or similar plant.
pentachlorophenol *(Chem.)*. C_6Cl_5OH. Widely used fungicidal and bactericidal compound, used particularly for timber protection. Mp 189°C.
pentad *(Meteor.)*. Period of 5 days; being an exact fraction of a normal year, it is useful for meteorological records.
pentadactyl *(Zool.)*. Having 5 digits.
pentadactyl limb *(Zool.)*. The characteristic free appendage of Tetrapoda with 5 digits.
pentaerythritol *(Chem.)*. $C(CH_2OH)_4$. Condensation derivative of ethanal and methanal. Mp 260°C. Used in the production of surface finishes.
pentaerythritol tetranitrate *(Chem.)*. $C(CH_2ONO_2)_4$. A detonating explosive, abbrev. *PETN*.
pentagon *(Maths.)*. A five-sided polygon.
pentagonal dodecahedron *(Crystal.)*. A form of the cubic system comprising 12 identical pentagonal faces.
pentahydric alcohols *(Chem.)*. Alcohols containing 5 hydroxyl groups, e.g. arabitol, $HO.CH_2.(CHOH)_3.CH_2OH$, xylitol (stereoisomeric) and rhamnitol, $HO.CH_2.(CHOH)_4.CH_3$.
pentamerous *(Bot.)*. Having 5 members in a whorl.
pentamethylene *(Chem.)*. See **cyclopentane**.
pentamethylene-diamine *(Chem.)*. See **cadaverine**.
pentamethylene glycol *(Chem.)*. $CH_2OH.(CH_2)_3.CH_2OH$. *Pentan-1,5-diol*. Organic solvent used in syntheses. Bp 239°C, rel.d. 0.994.
pentanes *(Chem.)*. C_5H_{12}. Low-boiling paraffin hydrocarbons. Pentane has a bp 36°C, rel.d. 0.63.
pentaprism *(Image Tech.)*. Five-sided prism which corrects lateral inversion, used on reflex cameras to allow eye-level viewing.
pentarch *(Bot.)*. A stele having 5 strands of protoxylem as in roots of many dicotyledons.
Pentastomida *(Zool.)*. A sub-phylum of Arthropoda. Elongate vermiform parasites of carnivorous mammals, living in the nasal sinuses, which have 2 pairs of claws at the sides of the mouth, and no respiratory or circulatory system.
pentavalent *(Chem.)*. Capable of combining with five atoms of hydrogen or their equivalent. Having an oxidation number or co-ordination number of five.
pentazocine *(Med.)*. A powerful synthetic drug with opiate action.
penthouse *(Arch.)*. An individual dwelling situated on the roof of a building but forming an integral part of the building.
penthouse *(Min.Ext.)*. Protective covering for workmen at bottom of shaft. Also *pentice*.
pentlandite *(Min.)*. A sulphide of iron and nickel which crystallizes in the cubic system. It commonly occurs intergrown with pyrrhotite.
pentode valve *(Electronics)*. Five-electrode thermionic tube, comprising an emitting cathode, control grid, a screen (or auxiliary grid) maintained at a positive potential with respect to that of the cathode, a suppressor grid maintained at about cathode potential, and an anode. It has characteristics similar to those of a screened-grid valve, except that secondary emission effects are suppressed.
Penton *(Plastics)*. TN for a chlorinated polyether widely used as a coating material for vessels etc., where very good chemical resistance is required. Relatively expensive but cheaper than the fluorinated plastics. (See **polytetrafluoroethylene**). Film form called *Pentaphane*.
pentosans *(Chem.)*. $(C_5H_8O_4)_x$, polysaccharides, comprising arabinans and xylans.
pentoses *(Chem.)*. A group of monosaccharides containing 5 oxygen atoms in the molecule, and having the formula $HO.CH_2.(CHOH)_3.CHO$ and $CH_3.(CHOH)_4.CHO$. Important pentoses are l-arabinose, l-xylose, rhamnose and fucose. Pentoses cannot be fermented. They are characterized by the fact that they yield furfuraldehyde or its homologues on boiling with dilute acids. A qualitative test for pentoses is the occurrence of a bright-red colour when they are boiled with HCl and phloroglucinol. ⇨
pentose shunt *(Biol.)*. A series of metabolic reactions which converts glucose-6-phosphate into ribose-5-phosphate with concomitant generation of NADPH.
Pentothal *(Chem.)*. TN for **thiopentone**.
pent-, penta- *(Chem.)*. Containing 5 atoms, groups etc.
pent-, penta- *(Genrl.)*. (Gk. *pente*, five). A prefix used in the construction of compound terms, e.g. *pentactinal*, 5-rayed.
penumbra *(Astron.)*. See **umbra**.
PEP carboxylase *(Bot.)*. *Phosphoenopyruvate carboxylase*. The enzyme that catalyses the reaction of phosphoenopyruvate and CO_2 to give oxaloacetic acid, especially as the first step in both the **Hatch-Slack** pathway and **crassulacean acid metabolism** in both of which the oxaloacetic acid is then reduced to malic acid.
Pepper's ghost *(Phys.)*. An illusion used to introduce a 'ghost' into a stage play. A plane sheet of glass is placed vertically at an angle of 45° to the line of vision of the audience. Thus actors in the wings can be superimposed by reflection on actors on stage.
pepsin *(Biol.)*. Protease of the gastric juice which is able to function optimally under the acidic conditions of the stomach.
peptic ulcer *(Med.)*. An ulcer of the stomach or of the duodenum.
peptidase *(Biol.)*. An enzyme which degrades peptides into their constituent amino acids.
peptide *(Biol.)*. A sequence of amino acids held together by peptide bonds. With rare exceptions peptides are unbranched chains joined by peptide bonds between their α-amino and α-carboxyl groups. Peptides can vary in length from dipeptides with 2 to polypeptides with several hundred amino acids.
peptide bond *(Biol.)*. The bond formed by the condensation of the amino group and carboxyl group of a pair of amino acids. Peptides are constructed from a linear array of amino acids joined together by a series of peptide bonds.
peptides *(Chem.)*. Substances resulting from the breakdown of proteins, characterized by the occurrence of the -CO-NH- structure in the molecule joining adjacent pairs of amino acids together. See **oligopeptides, polypeptides**.
peptization *(Chem.)*. The production of a colloidal solution of a substance, especially the formation of a sol from a gel. Deflocculation.
per- *(Chem.)*. (1) A prefix which properly should be restricted to compounds which are closely related to hydrogen peroxide, and thus contain 2 oxygen atoms linked together, e.g. *persulphates, percarbonates, perchromates*. (2) A prefix which is loosely used to denote that the central atom of a compound is in a higher state of oxidation than the usual, e.g. *perchlorates, permanganates*.
per-acid *(Chem.)*. A true per-acid is either formed by the action of hydrogen peroxide on a normal acid, or yields hydrogen peroxide by the action of dilute acids. Organic per-acids are assumed to contain the group –COO.OH, e.g. perbenzoic acid, $C_6H_5COO.OH$.
perambulator *(Surv.)*. An instrument for distance measurement consisting of a large wheel (often 6 ft or 2 m in circumference) attached at its axis to a long handle, so that it may be wheeled along the distance to be measured. A recording mechanism records the number of revolutions of the wheel and is calibrated in order to give distance traversed directly. See also **ambulator, odometer**.
percentage articulation *(Telecomm.)*. That of elementary speech-sounds received correctly when logatoms are called over a telephone circuit in the standard manner.
percentage differential relay *(Elec.Eng.)*. A differential relay which operates at a current which, instead of being fixed, is a fixed percentage of the current in the operating coils.

percentage modulation *(Telecomm.)*. See **depth of modulation**.

percentage registration *(Elec.Eng.)*. The registration of an integrating meter expressed as a percentage of the true value.

percentage tachometer *(Aero.)*. An instrument indicating the rev/min of a turbojet engine as a percentage, 100 per cent corresponding to a pre-set optimum engine speed. Provides better readability and greater accuracy and also enables various types of engine to be operated on the same basis of comparison.

percentile *(Stats.)*. See **quantile**.

perception *(Behav.)*. Refers to the individual's apprehension of the world and of the body through the action of various sensory sytems.

perceptual defence *(Behav.)*. The tendency to identify anxiety-provoking stimuli less readily than neutral stimuli, expecially if presented for only brief periods.

perceptual learning *(Behav.)*. Refers to learning about aspects of the stimulus in situations where there is no external reinforcement for doing so; it is usually demonstrated by an increased ability on a subsequent discrimination task.

perching *(Textiles)*. Inspection of cloth after weaving, and at various stages of finishing, for possible defects; the cloth is drawn over rollers some distance apart and while hanging, near vertically, from the front roller is examined in a good light.

perchlorates *(Chem.)*. *Chlorates (VII)*. Salts of perchloric acid, $HClO_4$. They are all soluble in water, though those of potassium and rubidium only slightly. The alkali perchlorates are isomorphous with the corresponding permanganates (manganates (VII)).

perchloric (chloric (VII)) acid *(Chem.)*. $HClO_4$. A colourless fuming liquid, mp $-112°C$, bp 110°C, a powerful oxidizing agent; harmful to the skin; monobasic, and forms perchlorates (VII).

perchloroethene, perchloroethylene *(Chem.)*. $CCl_2.CCl_2$. *Tetrachloroethene*. Organic solvent used for dry-cleaning. Bp 121°C, rel.d. (20°C) 1.623.

perchromates *(Chem.)*. See **peroxychromates**.

Perciformes *(Zool.)*. The largest order of bony fish with about 7000 species. An advanced and successful group inhabiting marine and fresh water habitats. Basses, Perches, Tuna.

percolating filter *(Build.)*. A bed of filtering material, such as broken stone or slag, used in the final or oxidizing stage in sewage treatment. This stage consists in sprinkling the liquid sewage over the filter, through which it percolates. Also *continuous filter*. See **contact bed**.

percurrent *(Bot.,Zool.)*. Extending throughout the entire length.

percussion *(Med.)*. The act of striking with one finger lightly and sharply against another finger placed on the surface of the body, so as to determine, by the sound produced, the physical state of the part beneath.

percussion drilling *(Min.Ext.)*. System in which string of tools falls freely on rock being penetrated. Also, pneumatic drilling in which hammer blows are struck on drill shank. See **cable tool drilling**.

percussion figure *(Min.)*. A figure produced on the basal pinacoid or cleavage face of mica when it is sharply tapped with a centre punch. It consists of a 6-rayed star, 2 rays more prominent than the others, lying in the unique plane of symmetry.

percussive boring *(Civ.Eng.)*. The process of sinking a borehole in the earth by repeatedly dropping on the same spot, from a suitable height, a heavy tool which pulverizes the earth and gradually penetrates.

percussive welding *(Elec.Eng.)*. See **resistance percussive welding**.

pereiopods *(Zool.)*. In higher *Crustacea*, the thoracic appendages modified as walking-legs. Cf. *pleopod*.

perennation *(Bot.)*. Survival from one growing season to the next with, usually, a period of reduced activity between.

perennial *(Bot.)*. A plant that lives for more than 2 years. Most flower in most years after the first or second but some are **hapaxanthic**.

perfect *(Bot.)*. (1) A flower having functional stamens and carpels. (2) A fungus reproducing sexually.

perfect *(Print.)*. To print the second side of a sheet of paper and so complete it.

Perfect binding *(Print.)*. A term used as a synonym for **unsewn binding**, being the name of the first machine designed for the purpose.

perfect combustion *(Eng.)*. That of which the products contain neither unburnt gas nor excess air.

perfect crystal *(Crystal.)*. A single crystal in which the arrangement of the atoms is uniform throughout.

perfect dielectric *(Phys.)*. One in which all the energy required to establish an electric field is returned when the field is removed. In practice, only a vacuum conforms, other dielectrics dissipating heat to varying extent.

perfect frame *(Eng.)*. A frame which has just sufficient members to keep it stable in equilibrium under the intended load.

perfect gas *(Chem.)*. See **ideal gas**.

perfect number *(Maths.)*. An integer which is equal to the sum of all its factors including unity. The first four are 6, 28, 496 and 8128, after which they become enormous. Cf. **friendly numbers**.

perfector *(Print.)*. A type of machine which prints both sides of the paper before delivery.

perfoliate *(Bot.)*. The basal part of the lamina of a leaf, encircling the stem completely, so that the stem appears to pass through the leaf.

perforate *(Bot.)*. (1) Having holes. (2) Containing small rounded transparent dots which give the appearance of holes.

perforate *(Zool.)*. Having apertures, said especially of shells: of gastropod shells, having a hollow columella.

perforated brick *(Build.)*. A clay brick manufactured with vertical perforations of varying shapes and sizes. So made to reduce weight, improve insulating properties, also to reduce capillary attraction through a wall.

perforating *(Min.Ext.)*. After a well has reached the producing zone the base is sealed with cement and the sides need to be pierced to allow ingress. Often done by a special gun with radial bores from which charges fire projectiles through the casing.

perforating press *(Eng.)*. A power press suitable for removing patterns of small areas of metal by punching.

perforation *(Bot.)*. A hole in the common wall between two consecutive elements of a *vessel*, resulting from the dissolution of the wall. Cf. **pit**.

perforation plate *(Bot.)*. The part of the common wall between two consecutive elements of a xylem *vessel* which has one or more perforations.

perforations *(Image Tech.)*. Holes of precise dimensions and spacing along one or both edges of a strip of motion picture film for transport and registration.

perforin *(Immun.)*. A protein present in K and NK cells which resembles the C9 component of complement, and forms ring-like tubular structures which become inserted into the cell membranes of target cells and cause leakage of their contents.

performance test *(Behav.)*. Mental tests consisting primarily of motor or perceptual items and not requiring verbal ability. Cf. **verbal test**.

pergeting *(Build.)*. See **pargeting**.

Perhydrol *(Chem.)*. TN for a 30% solution of hydrogen peroxide.

peri- *(Chem.)*. Term for the 1.8 positions in naphthalene derivatives.

peri- *(Genrl.)*. Prefix from Gk. *peri*, around.

perianal *(Med.)*. The region around the anus.

perianth *(Bot.)*. The set of sterile structures which typically surrounds the stamens and carpels of a flower and which may be differentiated into an outer, often green and protective, *calyx* of sepals and an inner, often coloured, *corolla* of petals. See also **perianth segment**.

perianth segment *(Bot.)*. *Tepal*. A member of the perianth, especially if there is no differentiation into sepals and petals.

periarticular *(Med.)*. Said of the tissues immediately around a joint.

periastron *(Astron.)*. That point in an orbit about a star in which the body describing the orbit is nearest to the star; applied to the relative orbit of a double star.

periblast *(Zool.)*. In meroblastic eggs, the margin of the blastoderm merging with the surrounding yolk. See also **periplasm**.

periblastic *(Zool.)*. Of cleavage, superficial.

periblem *(Bot.)*. A *histogen*, the precursor of the cortex.

peribranchial *(Zool.)*. Surrounding a gill or gills.

pericardectomy *(Med.)*. Surgical removal of the pericardium.

pericardiomediastinitis *(Med.)*. Inflammation both of the pericardium and of the mediastinum.

pericardiotomy *(Med.)*. Incision of the pericardium.

pericarditis *(Med.)*. Inflammation of the pericardium.

pericardium *(Zool.)*. Space surrounding the heart: the membrane enveloping the heart. adj. *pericardial*.

pericarp *(Bot.)*. Fruit wall derived from the ovary wall.

pericellular *(Biol.)*. Surrounding a cell.

perichaetium *(Bot.)*. (1) A cuplike sheath surrounding the archegonia in some liverworts. (2) The group of involucral leaves around the archegonia of a moss.

perichondritis *(Med.)*. Inflammation of the perichondrium, especially of the perichondrium of the cartilages of the larynx.

perichondrium *(Zool.)*. The envelope of areolar connective tissue surrounding cartilage.

perichordal *(Zool.)*. Encircling or ensheathing the notochord.

periclase *(Min.)*. Native magnesia. Oxide of magnesium, which crystallizes in the cubic system. It is commonly found in metamorphosed magnesian limestones, but readily hydrates to the much commoner brucite.

periclinal *(Bot.)*. Parallel to the nearest surface. Opposite of anticlinal.

periclinal chimera *(Bot.)*. A plant or shoot in which tissues of one genetic constitution form a complete layer throughout, the remaining tissues being of different genetic constitution. See **chimera, tunica-corpus concept**.

pericline *(Min.)*. A variety of **albite** which usually occurs as elongated crystals. The name is also used for a type of twinning in feldspars (the *pericline law*).

pericolitis *(Med.)*. Inflammation of the peritoneum covering the colon.

pericranium *(Zool.)*. The fibrous tissue layer which surrounds the bony or cartilaginous cranium in Vertebrates.

pericycle *(Bot.)*. A layer, one or more cells thick, of ground tissue at the periphery of the stele next to the endodermis; present in roots, rare in stems.

pericycloid *(Maths.)*. See **roulette**.

periderm *(Bot.)*. Secondary protective tissue often replacing epidermis in longer-lived stems and roots and consisting of cork, cork cambium and phelloderm.

periderm *(Zool.)*. See **perisarc**.

peridesmium *(Zool.)*. The coat of connective tissue which ensheathes a ligament.

perididymis *(Zool.)*. The fibrous coat which encapsules the testis in higher Vertebrates.

peridinin *(Bot.)*. Major carotenoid accessory pigment in the Dinophyceae.

peridium *(Bot.)*. A general term for the outer wall of the fruit body of a fungus, when the wall is organized as a distinct layer or envelope surrounding the spore-bearing organs partially or completely.

peridot *(Min.)*. See **olivine**.

peridotite *(Geol.)*. A coarse-grained ultramafic igneous rock consisting essentially of olivine, with other mafic minerals such as hypersthene, augite, biotite, and hornblende, but free from plagioclase. See **dunite, kimberlite**.

perigee *(Astron.)*. The point nearest to the earth on the apse line of a central orbit having the earth as a focus.

perigee *(Space)*. The nearest point to the Earth in the orbit of a spacecraft, missile etc.

perigon *(Maths.)*. See **round angle**.

perigynium *(Bot.)*. See **perichaetium**.

perigynous *(Bot.)*. A flower having the perianth and stamens inserted on a flat or cup-shaped structure which arises below and which is not fused to the ovary. Also the perianth and stamens so inserted, as in the blackberry. Cf. **epigynous, hypogynous, superior**.

perihelion *(Astron.,Space)*. That point in the orbit of any heavenly body moving about the Sun at which it is nearest to the Sun; applied to all the planets and also to comets, meteors, spacecraft etc. pl. *perihelia*.

perikinetic *(Phys.)*. Pertaining to the Brownian movement.

perilymph *(Zool.)*. The fluid which fills the space between the membranous labyrinth and the bony labyrinth of the internal ear in Vertebrates. adj. *perilymphatic*. Cf. **endolymph**.

perimedullary zone *(Bot.)*. See **medullary sheath**.

perimeter *(Med.)*. An instrument, in the form of an arc, for measuring a person's field of vision.

perimeter track *(Aero.)*. A **taxi track** round the edge of an airport.

perimetritis *(Med.)*. Inflammation of the peritoneum covering the uterus.

perimetrium *(Med.)*. The peritoneum covering the uterus.

perimysium *(Zool.)*. The connective tissue which binds muscle fibres into bundles and muscles.

perinatal *(Med.)*. Said of the period from the seventh month of pregnancy to the first week of life.

perineal glands *(Zool.)*. In some Mammals, a pair of small glands beside the anus which secrete a substance with a characteristic odour.

perineoplasty *(Med.)*. Repair of the perineum by plastic surgery.

perineorrhaphy *(Med.)*. Stitching of the perineum torn during childbirth.

perinephric *(Med.)*. Said of the tissues round the kidney, e.g. *perinephric* abcess.

perinephritis *(Med.)*. Inflammation of the tissues round the kidney.

perineum, perinaeum *(Zool.)*. The tissue wall between the rectum and the urinogenital ducts in Mammals. adj. *perineal, perinaeal*.

perineurium *(Zool.)*. The coat of connective tissue which ensheathes a tract of nerve fibres.

period *(Chem.)*. The elements between an alkali metal and the rare gas of next highest atomic number, inclusive, consisting of 2, 8, 18 or 32 elements.

period *(Geol.)*. A major unit of geological time. e.g. *Silurian*.

period *(Maths.)*. See **periodic function**.

period *(Nuc.Eng.)*. In a reactor, time in which the neutron flux changes by a factor of *e*.

period *(Phys.)*. Time taken for 1 complete cycle of an alternating quantity. Reciprocal of frequency.

periodates *(Chem.)*. *Iodates (VII)*. Formed by the oxidation of iodates (V). Periodates form heteropolybasic compounds of the types $M_5[I(WO_4)_6]$ and $M_5[I(MoO_4)_6]$.

periodic acid *(Chem.)*. *Iodic (VII) acid*. H_5IO_6. A weaker acid and a stronger oxidizing agent than iodic acid. May be regarded as orthoperiodic acid. Exists in deliquescent crystals. Resembles phosphoric acid in furnishing partially dehydrated acids.

periodic antenna *(Telecomm.)*. One depending on resonance in its elements, thereby presenting a variation in input impedance as the operating frequency is varied.

periodic current *(Telecomm.)*. An oscillating current whose values recur at equal intervals.

periodic function *(Maths.)*. A function f is periodic with period ω if for all $x, f(x) = f(x+\omega)$ and if ω is the smallest value for which this is true, e.g. $\sin x$ has a period of 2π and $\tan x$ a period of π. A function of a complex variable may have two or more independent complex periods (i.e. none being a real rational multiple of any other), in which case it is referred to as *doubly-*, or *multiply-periodic*.

periodicity *(Biol.)*. Rhythmic activity.

periodicity *(Chem.)*. Location of element in periodic table.

periodicity *(Phys.)*. See **frequency**.

periodic law *(Chem.)*. See **periodic system**.

periodic ophthalmia *(Vet.)*. *Moon blindness*. Recurrent **iridocyclitis** in the horse; believed to be a form of **leptospirosis**.

periodic precipitation *(Chem.)*. See **Liesegang rings**.

periodic rating *(Elec.Eng.)*. Rating of a component for continuous use with a specified periodically varying load.

periodic respiration *(Med.)*. Any waxing or waning of the pattern of breathing. In Cheyne-Stokes respiration periods of **hyperventilation** alternate with periods of **apnoea** and indicate disease of the brain or severely impaired circulation to the brain.

periodic reverse *(Eng.)*. Changing the polarity in some electrolytic procedures to e.g. clean the electrodes.

periodic system *(Chem.)*. Classification of chemical elements into periods (corresponding to the filling of successive electron shells) and groups (corresponding to the number of valence electrons). Original classification by relative atomic mass (Mendeleev, 1869).

periodic table *(Chem.)*. The most common arrangement of the **periodic system**.

period-luminosity law *(Astron.)*. A relationship between the period and absolute magnitude, discovered by Miss Leavitt to hold for all Cepheid variables; it enables the distance of any observable Cepheid to be found from observation of its light curve and apparent magnitude, this indirectly deduced distance being called the *Cepheid parallax*.

period meter *(Nuc.Eng.)*. Instrument for measurement of reactor period.

period of decay *(Phys.)*. See **half-life**.

period of revolution *(Astron.)*. The mean value, derived from observations, of one complete revolution of a planet or comet about the sun, or of a satellite about a planet.

periodontitis *(Med.)*. Inflammation of the membrane investing that part of the tooth seated in the jaw.

period range *(Nuc.Eng.)*. See **start-up procedure**.

peri-oöphoritis *(Med.)*. Inflammation of the peritoneum investing the ovary and of the cortex of the ovary.

periosteum *(Zool.)*. The covering of areolar connective tissue on bone.

periostitis *(Med.)*. Inflammation of the periosteum.

periostracum *(Zool.)*. The horny outer layer of a Molluscan shell.

periotic *(Zool.)*. In higher Vertebrates, a bone enclosing the inner ear and formed by the fusion of the otic bones: petrosal.

peripheral *(Genrl.)*. Situated or produced around the edge.

peripheral (device) *(Comp.)*. The term used to describe any unit in a computer system such as **input device, output device, backing store** which is connected to the central processor.

peripheral vascular resistance *(Med.)*. An expression of the state of contraction of the arterioles which governs the overall resistance to blood flow. Measured as the drop in pressure in the arterial bed divided by the blood flow.

periplasm *(Biol.)*. The space between the plasma membrane and outer membrane of *Gram negative* bacteria. It contains proteins secreted by the cell and a rigid peptide-oligosaccharide complex, the *peptidoglycan*.

periplasm *(Zool.)*. A bounding layer of protoplasm surrounding an egg just beneath the vitelline membrane, as in Insects.

periplasmic space *(Bot.)*. A space between the cell wall and the plasmalemma.

periproct *(Zool.)*. The area surrounding the anus.

perisarc *(Zool.)*. In some Hydrozoa, the chitinous layer covering the polyps, etc. Cf. **coenosarc**.

periscope *(Phys.)*. An optical instrument comprising an arrangement of reflecting surfaces whose purpose is to enable an observer to view along an axis deflected or displaced with respect to the axis of the observer's eye.

perisperm *(Bot.)*. Storage tissue, derived from nucellus (and hence wholly maternal), present in some seeds in which the endosperm does not develop, e.g. many Caryophyllaceae.

perissodactyl *(Zool.)*. Having an odd number of digits. Cf. **artiodactyl**.

Perissodactyla *(Zool.)*. An order of Mammals containing the 'odd-toed' hooved animals, i.e. those with a mesaxonic foot with the skeletal axis passing down the third digit. Horses, Tapirs, Rhinoceros and extinct forms.

peristaltic *(Zool.)*. Compressive: contracting in successive circles; said of waves of contraction passing from mouth to anus along the alimentary canal: cf. **antiperistaltic, systaltic**. n. *peristalsis*.

peristaltic pump *(Eng.)*. One in which the flow is produced by peristaltic action, e.g. by rollers passing in succession over a length of flexible tube.

peristerite *(Min.)*. A whitish variety of albite, or oligoclase, which is beautifully iridescent.

peristome *(Zool.)*. (1) The margin of the aperture of a gastropod shell. (2) In some Ciliophora, a specialized food-collecting, frequently funnel-shaped, structure surrounding the cell-mouth. (3) More generally, the area surrounding the mouth. adj. *peristomial*.

peristomium *(Zool.)*. In *Annelida*, the somite in which the mouth is situated: in some forms (as *Nereis*), 2 somites have been fused to form the apparent peristomium.

peristyle *(Arch.)*. A colonnade encircling a building.

perisystole *(Zool.)*. The period between diastole and systole in cardiac contraction.

perithecium *(Bot.)*. A more or less flask-shaped **ascocarp** with a pore or ostiole at the top through which the asci are discharged.

peritoneal cavity *(Zool.)*. In Vertebrates, that part of the coelom containing the viscera; the abdominal body cavity.

peritoneal dialysis *(Med.)*. Passage of fluids through the peritoneal cavity to lower the blood urea in certain cases of renal failure.

peritoneum *(Zool.)*. In Vertebrates, a serous membrane which lines the peritoneal cavity and extends over the mesenteries and viscera. adj. *peritoneal*.

peritonitis *(Med.)*. Inflammation of the peritoneum.

peritrichous *(Bot.)*. Bacteria having flagella distributed all over the surface of the cell.

peritrochoid *(Maths.)*. See **roulette**.

peritrophic *(Zool.)*. Surrounding the gut; as the *peritrophic membrane* of Insects, a membranous tube lining the stomach and partially separated from the stomach epithelium by the peritrophic space.

perivascular sheath *(Zool.)*. A sheath of connective tissue around a blood or lymph vessel.

perivitelline *(Biol.)*. Surrounding an egg yolk.

perivitelline membrane *(Zool.)*. See **vitelline membrane**.

Perkin's mauve *(Chem.)*. See **mauveine**.

Perkin's synthesis *(Chem.)*. The synthesis of unsaturated aromatic acids by the action of aromatic aldehydes upon the sodium salts of fatty acids in the presence of a condensing agent, e.g. acetic (ethanoic) anhydride.

perknite *(Geol.)*. A family of coarse-grained ultramafic igneous rocks which consist essentially of pyroxenes and amphiboles, but contain no feldspar.

perlite *(Geol.)*. An acid and glassy igneous rock which exhibits perlitic structure.

perlitic structure *(Geol.)*. A structure found in glassy igneous rocks, which consists of systems of spheroidal concentric cracks produced during cooling.

Perlon *(Chem.)*. TN for *polycaprolactam*, nylon 6, a synthetic fibre.

permafrost *(Geol.)*. In arctic and sub-arctic regions, the permanently frozen soil.

permalloy *(Eng.)*. An alloy with high magnetic permeability at low field strength, and low hysteresis loss. Original composition nickel 78.5%, iron 21.5%, but the term is now used generally to cover numerous alloys produced by adding other elements, e.g. copper, molybdenum, chromium, cobalt, manganese etc.

permanent dentition *(Zool.)*. In Mammals, the second set of teeth, which replaces the milk dentition.

permanent hardness *(Chem.)*. Of water, the hardness

which remains after prolonged boiling is *permanent hardness*. Due to the presence of calcium and magnesium chlorides or sulphates.

permanent implant *(Radiol.)*. An implant with radioactive material of short half-life, e.g. radon, arranged so that the prescribed dose is delivered by the time that the radioactive material has decayed, thus rendering removal of the sources unnecessary.

permanent load *(Eng.)*. The dead loading on a structure, consisting of the weight of the structure itself and the fixed loading carried by it, as distinct from any other loads.

permanent magnet *(Elec.Eng.)*. Ferromagnetic body which retains an appreciable magnetization after excitation has ceased. Cobalt steel, sintered and ceramic materials and various ferritic alloys are used on a large scale in loudspeakers, relays, small motors, magnetrons etc.

permanent mould *(Eng.)*. A metal mould (other than an ingot mould) used for the production of castings, e.g. in die-casting.

permanent set *(Eng.)*. (1) An extension remaining after load has been removed from a test piece, when the elastic limit has been exceeded. (2) Permanent deflection of any structure after being subjected to a load, which causes the elastic limit to be exceeded.

permanent store *(Comp.)*. See **non-volatile memory.**

permanent threshold shift *(Acous.)*. Permanent hearing loss after exposure to high sound levels. Permanent hearing loss is what remains from **temporary threshold shift**. Abbrev. *PTS*.

permanent way *(Civ.Eng.)*. The ballast, sleepers and rails forming the finished track for a railway, as distinct from a *temporary way*.

permanent wilting point *(Bot.)*. The water content of a soil at which a plant will wilt and not recover without additional water, even if shaded or left overnight. For most crop plants this corresponds to a soil **matric potential** of about -15 bars (-1.5 MPa). Cf. **hydraulic capacity**.

permanganates *(Chem.)*. *Manganates (VII)*. Oxidizing agents, the best known being *potassium permanganate* (manganate (VII)).

permanganic acid *(Chem.)*. Manganic (VII) acid. $HMnO_4$. Powerful oxidizing agent. Decomposes in the presence of organic matter.

permeability *(Elec.Eng.)*. Property of a material which describes the magnetization developed in that material when excited by an m.m.f. source. Absolute permeability is the ratio of flux density produced to the magnetic field strength producing it: $\mu = B/H$ henry per metre. Relative permeability is the ratio of magnetic flux density produced in a material to the value in free space produced by the same magnetic field strength: $\mu_r = \mu/\mu_0$, where μ_0 = permeability of free space, that is $4\pi \times 10^{-7}$ henry per metre. Hence $B = \mu_0\mu_r H$ tesla.

permeability *(Foundry)*. The ability of the sand grains to allow passage of gases.

permeability *(Geol.)*. The ability of a rock to transmit fluids, especially water, oil and gas.

permeability *(Phys.)*. The rate of diffusion of gas or liquid under a pressure gradient through a porous material. Expressed, for thin material, as the rate per unit area and for thicker material, per unit area of unit thickness.

permeability *(Powder Tech.)*. The facility with which a porous mass, e.g. a powder bed or compact, permits passage of a fluid.

permeability bridge *(Elec.Eng.)*. Device for measuring magnetic properties of a sample of magnetic material, fluxes in different branches of a divided magnetic circuit being balanced against each other. See **Ewing-**, **Holden-**.

permeability coefficient *(Powder Tech.)*. The volume of incompressible fluid per sec. which will flow through a unit cube of a porous mass or a packed powder across which unit pressure difference is maintained. Using cgs units, a flow rate of 1 cm^3/s through a 1 cm cube under 1 dyne/cm^2 pressure gives a p.c. of 1 darcy. Using SI units, a flow rate of 1 m^3/s through a 1 m cube under 1 N/m^2 pressure gives a p.c. of 1 kilodarcy (1000 **darcys**).

permeability equations *(Powder Tech.)*. Mathematical equations derived to describe the flow of fluids through packed powder beds and porous media in general.

permeability surface area *(Powder Tech.)*. The surface area of a particle calculated from the permeability of a powder bed under stated conditions.

permeability tuning *(Telecomm.)*. Adjusting a tuned circuit by varying the inductance of a coil by altering position on axis of a sintered iron core.

permeameter *(Phys.)*. Instrument for measuring static magnetic properties of ferromagnetic sample, in terms of magnetizing force and consequent magnetic flux.

permeameter *(Powder Tech.)*. Instrument for measuring the specific surface area of a powder by measuring the resistance offered to a flowing fluid by a packed powder bed.

permeameter cell *(Powder Tech.)*. That portion of a permeameter in which the plug or bed of the powder under test is assembled. Also *permeability cell*.

permeance *(Elec.Eng.)*. Reciprocal of the **reluctance** of a magnetic circuit.

permeation *(Phys.)*. The flow of a fluid through a porous material. If J_x is the flux as defined by *Fick's law of diffusion*, then

$$J_x = P_n(p_1 - p_2)^{1/2}/l$$

(Richardson's law) where P_n is the permeability, l is the thickness of material and $(p_1 - p_2)$ is the pressure difference across the material.

Permian *(Geol.)*. The youngest geological period of the Palaeozoic era, lying between the Carboniferous and the Triassic. It covers a time span of app. 290–250 million years and is named after the type area of Perm in the USSR. The corresponding system of rocks.

permissible dose *(Radiol.)*. See **maximum permissible dose.**

permissive waste *(Build.)*. Dilapidations in a building which are the result of neglect on the part of a tenant.

permitted explosives *(Min.Ext.)*. Those which may be used in mines, under specified conditions, where a danger of explosion from inflammable gas exists.

permittivity *(Elec.Eng.)*. Property of a material which describes the electric flux density produced when the material is excited by an e.m.f. source. Absolute permittivity is the ratio of electric flux density produced to the electric field strength: $\varepsilon = D/E$ farad per metre. Relative permittivity is the ratio of electric flux density produced in a material to the value in free space produced by the same electric field strength: $\varepsilon_r = \varepsilon/\varepsilon_0$ where ε_0 = permittivity of free space, that is 8.854×10^{-12} farad per metre. Hence $D = \varepsilon_0\varepsilon_r E$. Relative permittivity is also termed *dielectric constant*.

permonosulphuric acid *(Chem.)*. H_2SO_5, a powerful oxidizing reagent; prepared by anodic oxidation of concentrated sulphuric acid.

Permo-Trias *(Geol.)*. The Permian and Triassic systems considered together, as is commonly done in areas such as the British Isles where the rocks are of similar facies.

permutations *(Maths.)*. The different arrangements that can be made of a given number of items. For n items, all different, there are $n(n-1)(n-2)...2.1$ permutations; taken r at a time, there are $n(n-1)(n-2)...(n-r+1)$.

Permutit *(Chem.)*. TN for natural or synthetic zeolite used as ion-exchanger to substitute sodium for calcium or magnesium ions in hard water, to obtain softer water, the spent (i.e. calcium and magnesium) zeolite being regenerated (i.e. to sodium zeolite) periodically by treatment with concentrated sodium chloride solution.

pernicious anaemia *(Immun.,Med.)*. *Addison's anaemia*. Disease characterized by atrophic gastritis with achlorhydria and lack of gastric **intrinsic factor** which leads to failure of absorption of dietary vitamin B_{12} and consequent megaloblastic anaemia and perhaps peripheral neuritis. Gastric secretions and the blood of affected persons contain auto-antibodies against intrinsic factor and against a microsomal antigen present in gastric parietal cells.

peroral *(Zool.)*. Surrounding the mouth; as the *peroral membrane* of *Ciliophora*, which surrounds the cytopharynx.

perosis *(Vet.)*. *Slipped tendon*. A disease of chickens, turkeys and other birds characterized by swelling and deformity of the hock joint leading to dislocation of the gastrocnemius tendon; caused by a dietary deficiency of manganese or choline.

perovskite *(Min.)*. Calcium titanate, with rare earths, which crystallizes in the monoclinic system but is very close to being cubic. An accessory mineral in melilite-basalt, and in contact metamorphosed impure limestones. The crystal form of artificial ceramics which are **superconductors** at around 80 K.

peroxidases *(Biol.)*. Enzymes which activate hydrogen peroxide and induce reactions which hydrogen peroxide alone would not effect.

peroxides *(Chem.)*. Strictly limited to compounds containing the ion O_2^{2-} or the organic function —O-O—, those containing the ion HO_2^- or the group HO-O- being hydroperoxides. The term is loosely used for oxides of high-valent metals, e.g. lead peroxide is PbO_2, but is more properly called lead (IV) oxide.

peroxisome *(Biol.)*. Small membrane-bound cytoplasmic organelle containing oxidizing enzymes and catalase; common in leaf cells where they contain some of the enzymes of the glycollate pathway.

peroxychromates *(Chem.)*. Formed by the action of hydrogen peroxide on chromates (VI), and containing a peroxide group —O—O—. In alkaline solution it forms M_3CrO_8, red salts. In acid: $MCrO_6$, blue salts.

perpends *(Build.)*. The vertical joints in brickwork which are usually kept in line, thus keeping the perpends.

perpetual motion *(Phys.)*. Two main kinds are distinguished: (1) the continual operation of a machine which creates its own energy, thereby violating the 1st law of thermodynamics; (2) any device which converts heat completely into work, thereby contravening the 2nd law of thermodynamics. See also **thermodynamics**.

perrhenates *(Chem.)*. *Rhenates (VII)*. See **rhenium oxides**.

perrhenic acid *(Chem.)*. *Rhenic (VII) acid*. See **rhenium oxides**.

perron *(Build.)*. An external staircase to a building, from ground-level to the first floor.

per-salts *(Chem.)*. Salts corresponding to **per-acids**.

Perseids *(Astron.)*. A major **meteor** shower visible for up to two weeks before and after peaking on 12 August each year, the date on which the Earth crosses the orbit. The maximum hourly rate is about 70 meteors. This show is associated with a comet of 1862.

perseveration *(Med.)*. Meaningless repetition of an action or utterance.

persistence *(Electronics)*. Continued visibility or detection of luminous radiation from a gas-discharge tube or luminous screen when the exciting agency has been removed. Long-persistence cathode-ray tubes are used for retaining a display which has been momentarily excited. Also called *afterglow*.

persistence characteristic *(Electronics)*. Graph showing decay with time of the luminous emission of a phosphor, after excitation is cut off.

persistence of vision *(Phys.)*. Brief retention of image on retina of eye after optical excitation has ended. An essential factor in cinematography and TV.

persistent *(Zool.)*. Continuing to grow or develop after the normal period for the cessation of growth or development, as teeth (cf. *deciduous*); said also of structures present in the adult which normally disappear in the young stages.

persistron *(Electronics)*. Electroluminescent photoconducting device, which gives a steady display on being impulsed.

person *(Zool.)*. An individual organism.

personal computer *(Comp.)*. A low-cost **microcomputer**, sold with **software packages** and useful for **word processing**, maintaining a budget, storing mailing lists, playing computer games, but often now with processing power and memory greater than minicomputers of a few years ago. Abbrev. *PC*.

personal dosimeter *(Phys.)*. Sensitive tubular electroscope using a metallized quartz fibre unit which is viewed against a calibrated scale. It is charged by a generator and ionizing radiation discharges it. Used by workers in places of potential radiation hazard.

personal identification device *(Comp.)*. Device, often a card, which is inserted into a terminal to establish the holder as an authorized user of the system. Sometimes used with a **PIN**. Abbrev. *PID*.

personal identification number *(Comp.)*. A number allocated to a user of an on-line computer system. The number is unique and must be input to the system by the user as a means of establishing the user's identity. Abbrev. *PIN*.

personality *(Behav.)*. The integrated organization of all the psychological, intellectual, emotional and physical characteristics of an individual which determines the unique adjustment he or she makes to the world.

personality disorders *(Behav.)*. An inflexible and well established behaviour pattern that is maladaptive for the individual in terms of social or occupational functioning; usually recognizable by adolescence.

personal space *(Behav.)*. The space immediately surrounding one's body that an individual considers private; it varies with cultural and other factors.

personate *(Bot.)*. A corolla which is two-lipped with the throat almost closed by the lower lip e.g. *Antirrhinum*.

personnel monitoring *(Radiol.)*. Monitoring for radioactive contamination of any part of an individual, his breath or excretions, or any part of his clothing.

perspective *(Acous.)*. See **acoustic-**.

Perspex *(Plastics)*. *Lucite*. Proprietary thermoplastic resin of polymethyl methacrylate of exceptional transparency and freedom from colour in the unpigmented state. Available also in wide range of colours.

Perthe's disease *(Med.)*. *Osteochondritis deformans juvenilis*. A deformed condition of the epiphysis of the head of the femur in young children, associated with a painful limp.

perthite *(Min.)*. The general name for megascopic intergrowths of potassium and sodium feldspars, both components having been miscible to form a homogeneous compound at high temperatures, but the one having been thrown out of solution at a lower temperature, thus appearing as inclusions in the other. Perthite may also be formed by the replacement of potassium feldspar by sodium feldspar. See also **microperthite**.

perthosite *(Geol.)*. A type of soda-syenite consisting to a very large extent of perthitic feldspars, occurring at Ben Loyal and Loch Ailsh in Scotland.

perturbation *(Astron.)*. Any small deviation in the equilibrium motion of any celestial object caused by any change in the gravitational field acting on it.

perturbation *(Space)*. Any disturbance to a planned or natural activity but particularly to orbits, caused by the effects of drag, gravitation, solar pressure etc.

perturbation theory *(Phys.)*. Most problems in quantum mechanics cannot be solved exactly, but often different interactions have different orders of magnitude. Smaller effects can be accounted for by considering perturbations on the eigenstates of the dominant interaction in the system when treated alone. Also applicable to the solution of the behaviour of any complex system.

pertusate *(Bot.)*. Perforated; pierced by slits.

pertussis *(Med.)*. *Whooping cough*. An acute infectious disease, due to infection with the bacillus *Haemophilus pertussis*; characterized by catarrh of the respiratory tract, also by periodic, recurring spasms of the larynx, which end in the prolonged crowing inspiration known as the 'whoop'.

perveance *(Electronics)*. The constant G in the Child-Langmuir equation, $I = GV^{3/2}$, which governs the current in a space-charge limited thermionic diode; relevant to the optimization of electron gun design in electron beam tubes and accelerators.

pes *(Zool.)*. The podium of the hind limb in land Vertebrates. pl. *pedes*.

pes cavus, pes arcuatus *(Med.)*. *Claw foot*. A condition of the foot in which the balls of the toes approximate to the heel, so that the foot is shortened and the instep abnormally high.

pes planus, pes valgus *(Med.)*. *Flat foot*. A condition of the foot in which the longitudinal arch is lost, so that the foot is flattened and turned outwards.

pessary *(Med.)*. A medicated appliance for insertion into the vagina to treat vaginal disease.

pestle *(Chem.)*. A club-shaped instrument, of glass or porcelain, for grinding and pounding solids in a mortar.

petal *(Bot.)*. A member of the **corolla**, often brightly coloured and conspicuous.

petalite *(Min.)*. A silicate of lithium and aluminium which crystallizes in the monoclinic system. It typically occurs in granite pegmatites.

petalody *(Bot.)*. The transformation of stamens, or carpels, into petals. See **double**.

petaloid *(Bot.,Zool.)*. Looking like a petal; petal-shaped, as the dorsal parts of the ambulacra in certain Echinoidea.

pet cock *(Eng.)*. A small plug-cock for draining condensed steam from steam-engine cylinders, or for testing the water-level in a boiler.

petechia *(Med.)*. A small red spot due to minute haemorrhage into the skin.

Petersen coil *(Elec.Eng.)*. A reactor placed between the neutral point of an electric-power system and earth. The value of the reactance is such that, when an earth fault occurs, the current through the reactor exactly balances the capacitance current flowing through the fault, so that any tendency to arcing is suppressed. Also called *arc-suppression coil*.

pethidine *(Med.)*. A synthetic analgesic and hypnotic, having action similar to that of morphine.

petiolate *(Bot.)*. Leaf having a stalk or petiole. Cf. **sessile**.

petiole *(Bot.)*. The stalk of a leaf.

petiolule *(Bot.)*. The stalk of a leaflet of a compound leaf.

petit mal *(Med.)*. A form of epileptic attack in which convulsions are absent and certain transient phenomena, e.g. brief loss of consciousness, occur.

PETN *(Chem.)*. See **pentaerythritol tetranitrate**.

petrifaction, petrified *(Geol.)*. Terms applied to any organic remains which have been changed in composition by molecular replacement but whose original structure is in large measure retained. *Petrified wood* is wood which has had its structure replaced by e.g. calcium carbonate or silica. Many of the original minute structures are preserved.

petrochemicals *(Chem.)*. Those derived from crude oil or natural gas. They include light hydrocarbons such as butene, ethene and propene, obtained by fractional distillation or catalytic cracking.

petrographic province *(Geol.)*. A region chacterized by a group of genetically related rocks, e.g. Andes. See **comagmatic assemblage**.

petrography *(Geol.)*. Systematic description of rocks, based on observations in the field, on hand specimens, and on thin microscopic sections. Cf. **petrology**.

petrol *(Autos.)*. UK term for a light hydrocarbon liquid fuel for spark-ignition engines. Other terms for such fuel are *'gas', gasoline, motor spirit*. Modern motor spirits are blends of several products of petroleum, such as straight-run distillate, thermal and catalytically-cracked gasoline alkylates, isomerates, benzene (benzole) and, rarely, alcohol. It boils in the range of 30 to 200°C. An important measurement point is the engine test or octane rating of the fuel.

petrol-electric generating set *(Elec.Eng.)*. A small generating plant using a petrol engine as the prime-mover.

petrol engine *(Autos.)*. A reciprocating engine, working on the Otto 4-stroke or the 2-stroke cycle, in which the air charge is carburetted by a petrol spray from a carburettor, or alternatively by **fuel injection**. In 4-stroke engines, inlet and exhaust valves control the entry of charge and the exit of exhaust gases; in 2-stroke engines, the piston is usually made to act both as inlet and exhaust valve. Ignition of the combustible mixture is effected by sparking-plug, operated either by coil and battery or by magneto.

petroleum *(Chem.)*. Naturally-occurring green to black coloured mixtures of crude hydrocarbon oils, found as earth seepages or obtained by boring. Petroleum is widespread in the Earth's crust, notably in the U.S., U.S.S.R. and the Middle East. In addition to hydrocarbons of every chemical type and boiling range, petroleum often contains compounds of sulphur, vanadium etc. Commercial petroleum products are obtained from crude petroleum by distillation, cracking, chemical treatment etc.

petroleum jelly *(Chem.)*. A mixture of petroleum hydrocarbons; used for making emollients, for impregnating the paper covering of electric cables, and as a lubricant. Also *petrolatum*. See **liquid paraffin**.

petrol injection *(Autos.)*. See **fuel injection**.

petrology *(Geol.)*. That study of rocks which includes consideration of their mode of origin, present conditions, chemical and mineral composition, their alteration and decay.

petrol pump, gas pump *(Autos.)*. A small pump of the diaphragm type, operated either mechanically from the camshaft, or electrically. It draws petrol from the tank and delivers it to the carburettor or fuel injection system.

petrosal *(Zool.)*. In higher Vertebrates, a bone formed by the fusion of the various otic bones.

petrous *(Zool.)*. Stony, hard (as a portion of the temporal bone in higher Vertebrates): situated in the region of the petrous portion of the temporal bone.

petticoat *(Elec.Eng.)*. One of the umbrella-shaped shields commonly provided on pin-type insulators in order to increase the length of the leakage path which will remain dry under rain conditions.

petzite *(Min.)*. A telluride of silver and gold. It is steel-grey to black and often shows tarnish.

Petzval curvature *(Phys.)*. The curvature of the image surface of a lens system in which spherical aberration, coma and astigmatism have been corrected.

pewter *(Eng.)*. An alloy containing 80–90% of tin and 10–20% of lead. In modern pewter, antimony may replace tin, and from 1–3% copper is usual.

Peyer's patches *(Immun.)*. Nodules of lymphoid tissue in the submucosa of the small intestine, which are important for the development of immunity (or of immunological unresponsiveness) to antigens present in the gut. B-Lymphocyte precursors of IgA producing cells are more plentiful than in other lymphoid tissues.

PF *(Build.)*. Abbrev. for *Plain Face*.

pF *(Elec.Eng.)*. Abbrev. for *picofarad*.

Pfannkuch protection *(Elec.Eng.)*. A protective system for use with the cables of an electric-power system; some of the strands of the cable are lightly insulated from the others and have an e.m.f. applied between them, this e.m.f. causing a current to flow and operate relays if the insulation is destroyed by a fault.

Pferdestärke, PS *(Eng.)*. See **cheval-vapeur**.

PFI & R *(Build.)*. Abbrev. for *Part Fill In and Ram*.

P_{fr} *(Bot.)*. The physiologically active form of the plant pigment **phytochrome**, having a peak of absorption at $\lambda = 730$ nm which converts it to P_r (to which it also changes slowly in the dark). See **P_r**.

PF resins *(Plastics)*. *Phenol Formaldehyde (methanal) resins*. See **phenolic resins**.

Pfund series *(Phys.)*. A series of lines in the far infrared spectrum of hydrogen. Their wave numbers are given by the same expression as that for the Brackett series but with $n_1 = 5$.

pH *(Chem.)*. See **pH-value**.

PHA *(Immun.)* Abbrev. for *PhytoHaemAgglutinin*.

phacoidal structure *(Geol.)*. A rock structure in which mineral or rock-fragments of lenslike form are included. The term is applicable to igneous rocks containing softened and drawnout inclusions; also to metamorphic rocks such as crush-breccias and crush-conglomerates; and to certain gneisses. (Gk. *phakos*, lentil.)

phacolith *(Geol.)*. A minor intrusion of igneous rock occupying the crest of an anticlinal fold. Its form is due to the folding, hence it is not the cause of the uparching of the roof (cf. *laccolith*).

phacomalacia *(Med.)*. Pathological softening of the lens of the eye.

phaeic, phaeochrous *(Zool.)*. Dusky. n. *phaeism*.

phaeochomacytoma *(Med.)*. *Paraganglioma*. A tumour composed of chromaffine cells, especially those of the adrenal medulla; by secreting noradrenaline may cause high blood pressure.

phaeochrous *(Zool.)*. See **phaeic**.

phaeo-, pheo- *(Genrl.)*. Prefix from Gk. *phaios*, dusky.

Phaeophyseae *(Bot.)*. The brown algae, a class of eukaryotic algae in the division Heterokontophyta. Branching filamentous and parenchymatous types. Mostly marine, littoral and sublittoral, the brown seaweeds. Source of alginic acid. Isogamous, anisogamous or oogamous; alternation of generations (isomorphic or heteromorphic), or almost diplontic. Includes kelps and wracks.

phagedaena, phagedena *(Med.)*. Rapidly spreading and destructive ulceration.

phagocyte *(Biol.)*. A cell which exhibits amoeboid movement, in particular it throws out **pseudopods** to engulf foreign bodies, such as bacteria.

phagocytosis *(Immun.)*. The ingestion of cells or particles which first attach to and then become surrounded by the cell membrane. This is invaginated to form an intracellular vesicle (phagosome), towards which lysosomes move and fuse to release their content of hydrolytic enzymes into the vesicle, which becomes a phagolysosome. Cells of the mononuclear phagocyte system and polymorphonuclear leucocytes are the main phagocytic cells in mammals.

phagotrophy *(Bot.)*. Heterotrophic nutrition in which cells ingest solid food particles.

phag-, phago-, -phage, -phagy *(Genrl.)*. Prefix and suffix from Gk. *phagein*, to eat.

phalanges *(Zool.)*. In Vertebrates, the bones supporting the segments of the digits: fiddle-shaped rings composing the recticular lamina of the organ of Corti. sing. *phalanx*.

Phalangida *(Zool.)*. See **Opiliones**.

phalanx *(Zool.)*. See **phalanges**.

phalaris staggers *(Vet.)*. A nervous disease of sheep in Australia caused by eating the grass *Phalaris tuberosa*; believed to be caused by a toxic substance in the grass when eaten by cobalt-deficient sheep.

phallic stage *(Behav.)*. In psychoanalytic theory, the third stage of psychosexual development in which the child is preoccupied with his or her genitals; from the third to fifth or sixth year of life. See **Oedipus and Electra complexes**.

phallus *(Zool.)*. The penis of Mammals: the primordium of the penis or clitoris of Mammals. adj. *phallic*.

phanero- *(Genrl.)*. Prefix from Gk. *phaneros*, visible.

phanerocrystalline *(Min.)*. Said of an igneous rock in which the crystals of all the essential minerals can be discerned by the naked eye.

phanerophyte *(Bot.)*. Woody plant with perennating buds more than 25 cm above the soil surface. See **Raunkiaer system**.

Phanerozoic *(Geol.)*. The span of *obvious life*. More precisely the unit of geological time that comprises the Palaeozoic, Mesozoic and Cenozoic eras.

phantom antenna *(Telecomm.)*. Same as **artificial antenna**.

phantom material *(Radiol.)*. That producing absorption and back scatter of radiation very similar to human tissue, and hence used in models to study appropriate doses, radiation scattering etc. A *phantom* is a reproduction of (part of) the body in this material. Also termed *tissue equivalent material*.

phantom ring *(Phys.)*. The coloured ring of light which appears to surround an observer's shadow when the shadow falls on an extended bank of fog or cloud.

pharate *(Zool.)*. In Insects, refers to a phase of development when the old cuticle of one stage is separate from the hypodermis of the next stage, but has not yet been ruptured and cast off. The so-called *pupa* of many Insects actually represents a pharate adult, and the so-called *prepupa* is a pharate pupa.

pharmacodynamics *(Med.)*. The science of the action of drugs; pharmacology.

pharmacolite *(Min.)*. A hydrated arsenate of calcium which crystallizes in the monoclinic system. It is a product of the late alteration of mineral deposits which carry arsenopyrite and the arsenical ores of cobalt and silver.

pharmacology *(Med.)*. The scientific study of the action of chemical substances on living systems.

pharmacosiderite *(Min.)*. Hydrated arsenate of iron. It crystallizes in the cubic system, and is a product of the alteration of arsenical ores.

pharyngismus *(Med.)*. Spasm of the muscles of the pharynx.

pharyngitis *(Med.)*. Inflammation of the pharynx.

pharyngoplasty *(Med.)*. The surgical alteration, or reconstruction, of the pharynx in congenital and acquired disease.

pharyngoplegia *(Med.)*. Paralysis of the muscles of the pharynx.

pharyngotomy *(Med.)*. Incision into the pharynx.

pharynx *(Zool.)*. In Vertebrates, that portion of the alimentary canal which intervenes between the mouth cavity and the oesophagus and serves both for the passage of food and the performance of respiratory functions: in Invertebrates, the corresponding portion of the alimentary canal lying immediately posterior to the buccal cavity, usually having a highly muscular wall. adj. *pharyngeal*.

phase *(Astron.)*. Of the moon, the name given to the changing shape of the visible illuminated surface of the moon due to the varying relative positions at the earth, sun, and moon during the synodic month. Starting from *new moon*, the phase increases through *crescent*, *first quarter*, *gibbous*, to *full moon*, and then decreases through *gibbous*, *third quarter*, waning to *new moon* again. The inferior planets show the same phases, but in the reverse order; the superior planets can show a gibbous phase, but not a crescent.

phase *(Chem.)*. The sum of all those portions of a material system which are identical in chemical composition and physical state, and are separated from the rest of the system by a distinct interface called the *phase boundary*.

phase *(Elec.Eng.,Telecomm.)*. Fraction of cycle of a periodic waveform (usually sinusoidal) which has been completed at a specific reference time, e.g. at the start of a cycle of a second waveform of the same frequency. Expressed as an angle, one cycle corresponds to 2π radians or 360°. The terms *in phase*, *in quadrature* and *antiphase* correspond to phase angles between two signals of 0° (or 360°), 90° and 180° respectively. See also **polyphase**, **single phase**, **two phase** and **three phase**.

phase advancer *(Elec.Eng.)*. A machine connected in the secondary circuit of an induction motor to improve its power factor.

phase angle *(Elec.Eng.,Telecomm.)*. See under **phase**. Also given by δ, where $\tan\delta$ = (reactance/resistance) for an alternating-current circuit or an acoustic system.

phase compensation *(Telecomm.)*. See **line equalizer**.

phase constant *(Telecomm.)*. Imaginary part of propagation constant of a line, or of the transfer constant of a filter section. It is expressed in radians per unit length or section. See **image-**.

phase-contrast microscopy *(Biol.)*. A method of particular value in examining living unstained specimens, whereby phase differences are converted to differences of light intensity by the phase-contrast microscope. Contrasts in the image are produced wherever gradients in refractive index or thickness occur.

phase converter *(Elec.Eng.)*. On applying a single-phase voltage to one phase of a rotating 3-phase induction motor, it will be found that 3-phase voltages, which are approximately balanced, may be derived from the 3 terminals of the machine.

phase corrector *(Telecomm.)*. A circuit for correcting phase distortion.

phased array *(Radar)*. An antenna consisting of an array of identical radiators (waveguides, horns, slots, dipoles etc.) with electronic means of altering the phase of power fed to each of them. This allows the shape and direction of the radiation pattern to be altered without mechanical movement and with sufficient rapidity to be made on a pulse-to-pulse basis.

phase defect *(Elec.Eng.,Telecomm.)*. The phase difference between the actual current in a capacitor and that which would flow in an equivalent ideal (loss-free) capacitor.

phase delay *(Telecomm.)*. Delay, in radians or seconds, for transmission of wave of a single frequency through whole or part of a communication system. Also *phase retardation*.

phase delay distortion *(Telecomm.)*. That of a signal transmitted over a long line, so that difference in time of arrival of components of a complex wave is significant.

phase deviation *(Telecomm.)*. In phase modulation, maximum difference between the phase angle of the modulated wave and the angle of the nonmodulated carrier.

phase diagram *(Eng.)*. See **constitution diagram**.

phase difference *(Elec.Eng.)*. The phase angle by which one periodic waveform lags or leads another of the same frequency.

phase discriminator *(Electronics)*. Circuit preceding demodulator in phase-modulation receiver. It converts the carrier to an amplitude-modulated form.

phase distortion *(Telecomm.)*. Found in the waveform of transmitted signal on account of nonlinear relation of wavelength constant of a line, or image phase constant of amplifier or network, with frequency.

phase equalization *(Telecomm.)*. See **line equalizer**.

phase equalizer *(Elec.Eng.)*. See **interphase transformer**.

phase focusing *(Electronics)*. Effect used in electron bunching whereby those lagging in phase tend to gain energy from the field and those leading tend to surrender it. A similar effect occurs in synchrotron charged-particle accelerators.

phase inversion *(Telecomm.)*. Any arrangement for obtaining an additional voltage of reverse phase for driving both sides of a push-pull amplifier stage.

phase lag, phase lead *(Elec.Eng.,Telecomm.)*. If a periodic disturbance A completes a cycle when a second disturbance B has completed $\varphi/2\pi$ of a cycle, B is said to *lead A* by a phase angle φ. Conversely *A lags* on B by φ or *leads B* by $(2\pi-\varphi)$.

phase-locked loop *(Telecomm.)*. Control loop incorporating a voltage-controlled oscillator and phase sensitive detector in order to lock a signal to a stable reference frequency. Used in frequency synthesizers and synchronous detectors.

phase margin *(Telecomm.)*. Variation in phase of unity-gain from that which would give instability. Readily measured from a Nyquist diagram.

phase meter *(Telecomm.)*. One for measuring the phase difference between 2 signals of the same frequency.

phase modifier *(Elec.Eng.)*. A term sometimes used to denote a synchronous capacitor when this is used for varying the power factor of the current in a transmission line to effect voltage regulation.

phase modulation *(Telecomm.)*. Periodic variation in phase of a high-frequency current or voltage in accordance with a lower impressed modulating frequency. Occurs as an unwanted by-product of amplitude modulation, but can be independently produced.

phase plate *(Phys.)*. A quarter wave retarding plate used in phase-contrast microscopy.

phase resonance *(Telecomm.)*. That in which the induced oscillation differs in phase by $\pi/2$ from the forcing disturbance. Also termed *velocity resonance*.

phase retardation *(Telecomm.)*. See **phase delay**.

phase reversal *(Chem.)*. An interchange of the components of an emulsion; e.g. under certain conditions an emulsion of an oil in water may become an emulsion of water in the oil.

phase rule *(Chem.)*. A generalization of great value in the study of equilibria between phases. In any system, $P+F=C+2$, where P is the number of phases, F the number of **degrees of freedom** (1), C the number of components.

phase sensitive detector *(Electronics)*. One in which the output is proportional to the *phase difference* between the incoming signal or carrier and a reference signal.

phase sequence *(Elec.Eng.)*. The order in which the phase voltages of a polyphase system reach their maximum values. If the phases of a three-phase system are given the standard colourings of red, yellow, blue, this phase sequence is said to be a *positive phase sequence*. See **negative-, zero-**.

phase shifter *(Telecomm.)*. Waveguide or coaxial line component which produces any selected phase delay in signal transmitted.

phase-shifter control *(Electronics)*. One that changes the phase angle (measured on the waveform of the incoming voltage) at which the applied voltage causes conduction through a silicon controlled rectifier or thyristor, ignitron, thyratron, or any other controlled power switching device.

phase-shifting circuit *(Telecomm.)*. One in which the relative phases of components of a waveform or signal can be continuously adjusted.

phase-shifting transformer *(Elec.Eng.)*. One for which the secondary voltage can be varied continuously in phase. Usually with rotatable secondary winding and polyphase primaries. Also *phasing transformer*.

phase shift keying *(Telecomm.)*. Transmission of coded data using phase modulation with several discrete phase angles. Abbrev. PSK.

phase-shift oscillator *(Electronics)*. Same as **line oscillator**.

phase space *(Phys.)*. A multidimensional space in which coordinates represent the variables required to specify the state of a system; particularly the six-dimensional space with three position and three momentum coordinates. Used for the study of particle systems. See **microscopic state**.

phase splitter *(Elec.Eng.)*. A means of producing two or more waves which differ in phase from a single input wave.

phase swinging *(Elec.Eng.)*. Periodic variations in the phase angle between 2 synchronous machines running in parallel.

phase velocity *(Phys.)*. Velocity of propagation of any one phase state, such as a point of zero instantaneous field, in a steady train of sinusoidal waves. It may differ from the velocity of propagation of the disturbance, or *group velocity*, and, in transmission through ionized air, may exceed that of light. Also called *wave velocity*.

phase voltage *(Elec.Eng.)*. See **voltage to neutral**.

phasing *(Image Tech.)*. In facsimile transmission and reception, adjustment of reproduced picture to correspond exactly with the original, achieved by a phasing signal.

phasing transformer *(Elec.Eng.)*. See **phase-shifting transformer**.

phasor *(Elec.Eng.)*. Representation of alternating current or voltage which allows manipulation with others according to the rules of vectors. Each phasor is considered to rotate about the origin at ω rads per second (equals $2\pi f$, where f is the a.c. frequency) and all phasors on a given diagram must have the same associated frequency.

Phe *(Chem.)*. Symbol for phenylalanine.

phellem *(Bot.)*. See **cork**.

phelloderm *(Bot.)*. Parenchymatous tissue formed centripetally by the cork cambium (phellogen) as part of the periderm.

phellogen *(Bot.)*. See **cork cambium**.

phenakite *(Min.)*. A silicate of beryllium, crystallizing in the rhombohedral system. It is commonly found as a product of pneumatolysis. Sometimes cut as a gemstone, having brilliance of lustre but lacking fire. The name (Gk. 'the deceiver') refers to the frequency with which it has been confused with quartz.

phenanthrene *(Chem.)*. $C_{14}H_{10}$, white, glistening plates, mp 99°C, bp 340°C; its solutions show a blue fluorescence. It occurs in coal-tar, and can be synthesized by various reactions, the most important of which is the synthesis by Pschorr, based upon the condensation of 2-nitrobenzaldehyde with sodium phenylacetate yielding 2-phenyl-2-nitrocinnamic acid, which on reduction, diazotization, and subsequent elimination of N_2 and H_2O, yields phenanthrene-10-carboxylic acid. This acid yields phenanthrene on distillation. The 9-10-phenanthrene bridge is readily attacked by reagents, yielding diphenyl derivatives. ⇨

phenates *(Chem.)*. Phenolates, phenoxides. Salts formed by phenols, e.g. C_6H_5ONa, sodium phenate (phenolate, phenoxide).

phenetic *(Bot.)*. A classification based on overall similarity in as many characters as possible. The characters can be weighted.

4-phenetidine *(Chem.)*. $H_2N.C_6H_4.OC_2H_5$, the ethyl ether of 4-aminophenol, basis for a number of pharmaceutical preparations, e.g. phenacetin, $H_3C.CO.NH.C_6H_4.OC_2H_5$.

phengite *(Min.)*. An end-member variety of muscovite mica, in which Si:Al > 3:1 and some aluminium is replaced by magnesium or iron.

Phenidone *(Image Tech.)*. TN for 1-phenyl-3-pyrazolidinone, a developing agent with characteristics similar to *metol*.

phenobarbitone *(Med.)*. Phenylethyl-barbituric acid. Formerly used as a hypnotic and sedative, now mainly in the treatment of epilepsy.

phenocopy *(Biol.)*. An environmentally caused copy of a genetic abnormality. It is non-heritable.

phenocrysts *(Geol.)*. Large (megascopic) crystals, usually of perfect crystalline shape, found in a fine-grained matrix in igneous rocks. See **porphyritic texture**.

phenogram *(Bot.)*. A branching diagram, *dendrogram*, reflecting the overall similarities of groups of organisms.

phenol *(Chem.)*. C_6H_5OH, *carbolic acid*, colourless hygroscopic needles, mp 43°C, bp 183°C, chief constituent of the coal-tar fraction boiling between 170° and 230°C, soluble in sodium and potassium hydroxide, forming Na and K phenates or phenolates. It forms with bromine 2,4,6-tribromophenol. Sodium phenolate reacts with CO_2 under certain conditions, yielding sodium salicylate. It is a strong disinfectant. See **phenols**.

phenolic acids *(Chem.)*. A group of aromatic acids containing one or more hydroxyl groups attached to the benzene nucleus. The 2-hydroxy acids are volatile in steam, soluble in cold chloroform, and give a violet or blue coloration with iron (III) chloride. The 3-hydroxy acids are the most stable acids. Important phenolic acids are *salicylic acid, gallic acid, tannin*.

phenolic resins *(Plastics)*. A large group of plastics, made from a phenol (phenol, 3-cresol, 4-cresol, catechol, resorcinol, or quinol) and an aldehyde (methanal, ethanal, benzaldehyde, or furfuraldehyde). An acid catalyst produces a soluble and fusible resin (*modified resin*) used in varnishes and lacquers; while an alkaline catalyst results in the formation, after moulding, of an insoluble and infusible resin. Phenolics may be moulded, laminated or cast.

phenolics *(Bot.)*. Large and diverse group of plant secondary metabolites. The phenol group includes *flavonoids, lignin* and *tannins*.

phenology *(Ecol.)*. The study of plant development in relation to the seasons.

phenolphthalein *(Chem.)*. A triphenylmethane derivative, obtained by the condensation of phthalic anhydride and phenol by the action of concentrated sulphuric acid. It forms colourless crystals which dissolve with a red colour in alkalis. It is used as an indicator in volumetric analysis. The colourless substance has the lactone formula, whereas its coloured salts have a quinonoid structure.

phenol red *(Chem.)*. *Phenolsulphonphthalein*. Used as an indicator, with pH range of 6.8 to 8.4, over which it changes from yellow to red; also used for testing the functioning of the kidney.

phenols *(Chem.)*. *Hydroxybenzenes*. A group of aromatic compounds having the hydroxyl group directly attached to the benzene nucleus. They give the reactions of alcohols, forming esters, ethers, thiocompounds, but also have feeble acidic properties and form salts or phenolates by the action of NaOH or KOH. Phenols are divided into mono-, di-, tri-, tetra- and polyhydric phenols. Phenols are more reactive than the benzene hydrocarbons.

phenomenology *(Behav.)*. In philosophy, the study of the psychic awareness that accompanies experience and that is the source of all meaning for the individual. In psychiatry, it refers to the description and classification of an individual's mental activity, including subjective experience and perceptions, mental performance (e.g. memory) and the somatic accompaniments of mental events (e.g. heart rate).

phenotype *(Biol.)*. The observable characteristics of an organism as determined by the interaction of its **genotype** and its environment. adj. *phenotypic*.

phenotypic variance *(Biol.)*. Measure of the amount of variation among individuals in the observed values of a quantitative character.

phenoxyacetic acids *(Bot.)*. Synthetic compounds many with auxin-like activity including the important selective weedkillers, 2,4-D, 2,4,5-T and MCPA.

phenoxy resins *(Plastics)*. Range of thermoplastic resins based on polyhydroxy ethers, produced from bisphenol A and epichlorhydrin. Used for injection and extrusion mouldings, blow mouldings and solution coatings.

phenylacetic (ethanoic) acid *(Chem.)*. $C_6H_5.CH_2.COOH$, colourless crystals, mp 76°C.

phenylalanine *(Chem.)*. *2-amino-3-phenylpropanoic acid*, $C_6H_5.CH_2.CH(NH_2).COOH$. The L- or S-isomer is a constituent of proteins. Symbol Phe, short form F. ⇨

phenylalanine ammonia-lyase *(Bot.)*. Enzyme catalysing the elimination of ammonia from L-phenylalanine in the synthesis of many *secondary* plant metabolites e.g. phenolic acids, flavonoids, alkaloids, lignins, tannins etc. Abbrev. *PAL*.

phenylamine *(Chem.)*. *Aminobenzene*. See **aniline**.

phenylbenzine *(Chem.)*. See **diphenyl**.

phenylbutazone *(Med.)*. Analgesic and antipyretic drug used in the treatment of rheumatic conditions.

phenyl cyanide *(Chem.)*. See **benzonitrile**.

phenylenediamines *(Chem.)*. $C_6H_4(NH_2)_2$, obtained by the reduction of the dinitro, the nitroamino, or the aminoazo compounds. The diamines are crystalline substances with strong basic properties. There are three isomers.

phenylethanol *(Chem.)*. $C_6H_5.CH_2.CH_2OH$, an aromatic alcohol of pleasant odour, bp 220°C, a constituent of rose oil.

phenylethanone *(Chem.)*. See **acetophenone**.

phenylglycine *(Chem.)*. White, crystalline solid; mp 127°C. Used as an intermediate in the manufacture of indigo dyes.

phenyl group *(Chem.)*. The group C_6H_5-.

phenylhydrazine *(Chem.)*. $C_6H_5.NH.NH_2$, a colourless crystalline mass, mp 23°C. It is easily oxidized, and is a strong reducing agent. It forms salts with acids. Phenylhydrazine reacts readily with aldehydes and ketones, forming phenylhydrazones, which are crystalline substances and serve to identify the respective aldehydes and ketones.

phenylhydrazones *(Chem.)*. The reaction products of **phenylhydrazine** with aldehydes and ketones, formed by the elimination of two hydrogen atoms of the amino group and the oxygen atom of the aldehyde or ketone group as water.

phenylketonuria *(Med.)*. A rare, autosomal recessive, hereditary condition in which the amino acid phenylalanine cannot be converted to tyrosine and accumulates in the blood, to be ultimately excreted in the urine as phenylpyruvic and other acids causing mental retardation. Of particular interest because it can be detected soon after birth and mental retardation prevented by dietary restriction of phenylalanine. Also called *phenylalaninaemia*

phenyl methyl ether *(Chem.)*. See **anisole**.

phenylmethylsulphonyl fluoride *(Biol.)*. Widely used as a protease inhibitor.

phenytoin *(Med.)*. Drug widely used in the treatment of epilepsy. Of particular value in prevention of tonic and atonic seizures. It can also be used to control abnormal heart rhythms.

pheromone *(Zool.)*. A chemical substance secreted by an animal, which influences the behaviour of other animals (cf. **hormone**) of the same species, e.g. the *queen substance* of honey bees. See **semiochemical**.

Philadelphia chromosome *(Biol.)*. Distinctive, small chromosome found in patients suffering from chronic myelocytic leukaemia, corresponding to chromosome 22 after a reciprocal **translocation** with chromosome 9.

phillipsite *(Min.)*. A fibrous zeolite; hydrated silicate of potassium, calcium, and aluminium, usually grouped in the orthorhombic system.

Phillips screw *(Eng.)*. Screw with a cross shaped slot, requiring a special screwdriver.

Phillips screwdriver *(Eng.)*. One with a pointed head of cruciform section, for turning *Phillips screws*.

phimosis *(Med.)*. Narrowness of the prepuce (foreskin) so that it cannot be drawn back over the glans penis.

phlebectomy *(Med.)*. Surgical removal of a vein or part of a vein.

phlebitis *(Med.)*. Inflammation of the coats of a vein.

phlebography *(Radiol.)*. The radiological examination of veins following the injection of a **contrast medium**. When applied to the peripheral venous system, it is called *venography*.

phlebolith *(Med.)*. A concretion in a vein due to calcification of a thrombus.

phlebosclerosis *(Med.)*. Thickening of a vein due chiefly to a pathological increase in the connective tissue of the middle coat.

phlebotomus fever *(Med.)*. See **sandfly fever**.

phlebotomy *(Med.)*. The cutting or needling of a vein for the purpose of letting blood.

phleb-, phlebo- *(Genrl.)*. Prefix from Gk. *phleps*, gen. *phlebos*, vein.

phlegmasia alba dolens *(Med.)*. Term formerly used for painful swelling of the leg, the skin of which is shiny and white, occurring in women after childbirth; due to thrombosis of veins.

phlegmon *(Med.)*. Purulent inflammation, with necrosis of tissue. adj. *phlegmonous*.

phloem *(Bot.)*. Tissue with the major function of transporting metabolites, especially sugars, from sources to sinks. See **mass flow hypothesis**. The tissue consists of **sieve elements**, and companion cells and/or parenchyma cells, often with fibres or sclereids.

phloem ray *(Bot.)*. That part of a vascular *ray* which traverses the secondary phloem.

phlogisticated air *(Chem.)*. An old name for nitrogen.

phlogopite *(Min.)*. Hydrous silicate of potassium, magnesium, iron, and aluminium, crystallizing in the monoclinic system. It is a magnesium mica, and is usually found in metamorphosed limestones or in ultrabasic igneous rocks. Not so good an electrical insulator as **muscovite** at low temperatures, but keeps its water of composition until 950°C.

phloroglucinol *(Bot.)*. A stain for lignin when acidified.

phlycten, phlyctenule, phlyctenula *(Med.)*. A small round, grey or yellow nodule, occurring on the conjunctiva where it covers the sclera and cornea of the eye. (Gk. *phlyctaina*, a bleb.)

phlyctenular conjunctivitis *(Med.)*. *Eczematous conjunctivitis*. An inflammation of the conjunctiva covering the sclera and cornea of the eye and giving rise to phlyctens.

pH meter *(Chem.,Phys.)*. Specialized millivolt meter which measures potential difference between reference electrodes in terms of pH value of the solution in which they are immersed.

phobia *(Behav.)*. An excessive anxiety in some specific situation or in the presence of some object, which present no apparent threat.

phobic disorder *(Behav.)*. A disorder characterized by intense anxiety to some external object or situation, and avoidance of the phobic stimulus.

Phobos *(Astron.)*. One of the two natural satellites of **Mars**.

phocomelia *(Med.)*. A congenital malformation in which the proximal portion of the limb is absent; *phocomelus*. One so affected.

Phoebe *(Astron.)*. The ninth natural satellite of **Saturn**.

Pholidota *(Zool.)*. An order of old-world Mammals having imbricating horny scales, interspersed with hairs, over the head, body, and tail. They have a long snout, no teeth, and a long thin tongue. The hind feet are plantigrade, and the fore feet have long curved claws. Nocturnal animals, feeding on ants or termites. Pangolins.

phon *(Acous.)*. Unit of the objective loudness on sound-level scale which is used for deciding the apparent loudness of an unknown sound or noise, when a measure of loudness is required. This is effected either by subjective comparison by the ear, or by objective comparison with a microphone-amplifier and a weighting network. The reference sound pressure level at 1 kHz is 0.0002 microbar $= 20\,\mu N/m^2$.

phonation *(Zool.)*. Sound-production.

phonetics *(Acous.)*. Study of speech and vocal acoustics. Used to describe the system of symbols which uniquely represent the spoken word of any language in writing, enabling the reader to pronounce words accurately in spite of spelling irregularities.

phonmeter *(Acous.)*. Apparatus for the estimation of loudness level of a sound on the phon scale by subjective comparison. Also *phonometer*.

phonochemistry *(Chem.)*. The study of the effect of sound and ultrasonic waves on chemical reactions.

phonograph *(Acous.)*. See **gramophone**.

phonolite *(Geol.)*. A fine-grained igneous rock of intermediate composition, consisting essentially of nepheline, subordinate alkali feldspar (sanidine), and sodium-rich coloured silicates. Termed also *clink stone*, because it rings under the hammer when struck.

phonon *(Phys.)*. A quantum of lattice vibrational energy in a crystal. The thermal vibrations of the atoms can be described in terms of *normal modes* of oscillation, each mode specifying a correlated displacement of the atoms. The energy of a mode is quantized and can only exchange energy with other modes in units of $h\nu$, where ν is the mode frequency and h is Planck's constant. This quantum is the *phonon*. Lattice vibrations can be described in terms of waves in the lattice and the phonon is the particle of the field of the mechanical energy of a crystal. (Cf. *photon* as a particle of the electromagnetic field). Phonons have been identified by neutron inelastic scattering experiments.

phonon dispersion curve *(Phys.)*. A curve showing *frequency* as a function of *wavevector* for the modes of lattice vibrations in a crystal. Determined by neutron spectroscopy, their interpretation provides a powerful method of testing various models of interatomic forces in crystals.

phonon drag *(Phys.)*. In the Peltier effect, due to the *electron-phonon* interactions the phonons gain momentum from the electrons and the electron current 'drags' the phonons with it.

phonon-phonon scattering *(Phys.)*. The interaction of one mode of lattice vibration with another. One process determining the thermal conductivity of a solid.

phono plug, connector *(Telecomm.)*. Coaxial connector commonly used in non-professional audio equipment.

-phore *(Chem.)*. A suffix which denotes a group of atoms responsible for the corresponding property, e.g. *chromophore*.

-phore *(Genrl.)*. Suffix from Gk. *pherein*, carry.

phoresis *(Med.)*. Electrical passage of ions through a membrane.

phoresy *(Zool.)*. Transport or dispersal achieved by clinging to another animal; e.g. certain Mites which achieve dispersal by attaching themselves to various Insects.

Phoronidea *(Zool.)*. A small phylum of hermaphrodite unsegmented coelomate animals of tubicolous habit, having a U-shaped gut, a dorsal anus, and a lophophore in the form of a double horizontal spiral; marine forms occurring in the sand and mud of the sea bottom.

phosgene *(Chem.)*. $COCl_2$. A very poisonous, colourless, heavy gas with a nauseating, choking smell; bp 8°C. It is manufactured by passing carbon monoxide and chlorine over a charcoal catalyst. It is used for the manufacture of intermediates in the dyestuff industry, and as a *war gas*. Also known as *carbonyl chloride*.

phosgenite *(Min.)*. A chlorocarbonate of lead, crystallizing in the tetragonal system. It is found in association with cerussite.

phosphatase *(Biol.)*. An enzyme which dephosphorylates its substrate by hydrolysis of the phosphate ester bond.

phosphate coatings *(Build.)*. Coatings of iron or zinc phosphate produced by reaction between the metal being coated and an acid solution. This may be simply dilute phosphoric acid, as in the original Coslett process, or may contain a variety of metal phosphates and other additives. The coatings provide a key for subsequent painting or may be filled with oil and dyed, as a decorative and protective film. Commercial processes include **Bonderizing** and **sheradizing**.

phosphate fixation *(Bot.)*. The reaction of orthophosphate ions with Ca, Al or Fe ions to give low-solubility hydroxyphosphates which renders much of the phosphate unavailable to plants. Occurs in most mineralized or artificially fertilized soils.

phosphates (V) *(Chem.)*. Salts of phosphoric acid. There are three series of orthophosphates, MH_2PO_4, M_2HPO_4, and M_3PO_4; the first yield acid, the second are practically neutral, and the third alkaline, aqueous solutions. Metaphosphates, MPO_3, and pyrophosphates, $M_4P_2O_7$, are also known. All phosphates give a yellow precipitate on heating with ammonium molybdate in nitric acid.

phosphatic deposits *(Geol.)*. Beds containing calcium phosphate which are formed especially in areas of low rainfall, and which may be exploited as sources of phosphate. See also **phosphatic nodules**.

phosphatic nodules *(Geol.)*. Rounded masses containing calcium phosphate, which are formed in the sea floor.

phosphatides *(Chem.)*. Fatlike substances containing phosphoric acid. In some, glycero-phosphoric acid is combined with 2 mols of fatty acid and 1 mol of a hydroxy base which may be choline (in *lecithins*), aminoethanol (in *kephalins*) or serine. In others, the base sphingosine is combined with fatty acid, phosphoric acid and choline. Also *phospholipins*.

phosphatidyl choline *(Biol.)*. A phosphatide in which the choline forms the organic base. Previously termed *lecithin*. Important component of biological membranes.

phosphatidyl inositol *(Biol.)*. A phosphatide containing inositol. They are important in the response of cells to external ligands.

phosphating *(Eng.)*. Treatment of metal surface with hot phosphoric acid before painting in order to inhibit corrosion.

phosphaturia *(Med.)*. The presence of an excess of phosphates in the urine.

phosphene *(Med.)*. Area of luminosity in the visual field. It is caused by pressure on the eye-ball.

phosphides *(Chem.)*. Binary compounds of metals with phosphorus, e.g. calcium phosphide Ca_3P_2.

phosphine *(Chem.)*. *Phosphorus (III) hydride*. PH_3. A colourless, evil-smelling gas which usually burns spontaneously in air to form phosphorus (V) oxide. It has reducing properties and precipitates phosphides from solutions of many metallic salts.

phosphines *(Chem.)*. Derivatives of PH_3, obtained by the substitution of hydrogen for alkyl radicals; classified according to the extent of substitution into *primary*, *secondary* and *tertiary phosphines*. They correspond closely to the amines, except that they are easily oxidized even in the air, that they are only feebly basic, and that the P atom has a tendency to pass from the tri- to the quinquevalent state.

phosphites *(Chem.)*. *Phosphates (III)*. Salts of phosphorous acid.

phosphoenolpyruvate *(Bot.)*. The phosphate ester of the enol form of pyruvic acid, $CH_2=COPO_3H_2$-COOH. An important metabolite in glycolysis and the substrate for **PEP carboxylase**.

phospholipase *(Biol.)*. An enzyme which degrades a phospholipid. There are 3 main types, phospholipase A, C and D: A removes a hydrocarbon chain from the phospholipid, C cleaves the glycerol/phosphate ester bond and D the phosphate/base linkage.

phosphonium salts *(Chem.)*. These salts are formed when phosphine is brought into contact with hydrogen chloride, hydrogen bromide or hydrogen iodide. Formed in a similar way to ammonium compounds.

phosphoproteins *(Biol.)*. Proteins which have been enzymically phosphorylated so that they contain phosphate groups. An important functional modifier.

phosphor *(Chem.)*. Generic name for substances which exhibit **luminescence**.

phosphor *(Electronics)*. More specifically the fluorescent coating used on the screen of CRTs or image intensifiers.

phosphor-bronze *(Eng.)*. A term sometimes applied to alpha (low tin) bronze deoxidized with phosphorus, but generally it means a bronze containing 10–14% of tin and 0.1–0.3% of phosphorus, with or without additions of lead or nickel. Used, in cast condition, where resistance to corrosion and wear is required, e.g. gears, bearings, boiler fittings, parts exposed to sea water etc.

phosphorescence *(Chem.)*. Greenish glow observed during slow oxidation of white phosphorus in air.

phosphorescence *(Phys.)*. Luminescence which persists for more than 0.1 nanosecs after excitation. See **fluorescence**, **persistence**.

phosphorescence *(Zool.)*. Luminosity; production of light, usually (in animals) with little production of heat, as in Glow-worms. adj. *phosphorescent*.

phosphoric acid *(Chem.)*. H_3PO_4. See **orthophosphoric acid**.

phosphorite, rock-phosphate *(Min.)*. The fibrous concretionary variety of **apatite**.

phosphorized copper *(Eng.)*. Copper deoxidized with phosphorus. Contains a small amount (about 0.02%) of residual phosphorus, which lowers the conductivity.

phosphorofluoric acid *(Chem.)*. See **hexafluorophosphoric acid**.

phosphorous *(Chem.)*. Relating to trivalent phosphorus, e.g. *phosphorus (III) oxide* P_4O_6 which with cold water forms *phosphorus (III) acid*.

phosphorous (phosphorus (III)) acid *(Chem.)*. H_3PO_3. Formed by the action of cold water on phosphorous oxide; decomposes on heating; forms phosphites; reducing agent.

phosphorus *(Chem.)*. A nonmetallic element in the fifth group of the periodic system. Symbol P, at. no. 15, r.a.m. 30.9738, valencies 3,5. White phosphorus is a waxy, poisonous, spontaneously inflammable solid, mp 44°C, bp 282°C, rel.d. 1.8–2.3. Red phosphorus is nonpoisonous and ignites in air only when heated above 300°C, mp 500–600°C. Its density and presumable structure vary with the method of preparation. Black has rel.d. 2.7, and is a metallic substance obtained at high temperature and pressure. Phosphorus occurs widely and abundantly in minerals (as phosphates) and in all living matter. Manufactured by heating calcium phosphate with sand and carbon in an electric furnace. It is used mainly in the manufacture of phosphoric acid for phosphate fertilizers; also used in matches and organic synthesis.

phosphorus oxychloride *(Chem.)*. $POCl_3$. Liquid; fumes in air; slowly hydrolysed by water, forming phosphoric and hydrochloric acids. It is formed when compounds containing a hydroxyl group are treated with phosphorus (V) chloride. Also *phosphoryl chloride*.

phosphorus pentahalides *(Chem.)*. *Phosphorus (V) halides*. These, *phosphorus (V) chloride* (PCl_5), *phosphorus (V) bromide* (PBr_5), and *phosphorus (V) fluoride* (PF_5), are formed by the action of the dry halogen on the trihalide. The properties of the (V) halides are similar. They transform hydroxyl compounds into the corresponding halides.

phosphorus pentoxide *(Chem.)*. *Phosphorus (V) oxide*. P_4O_{10}. Powerfully hygroscopic white solid obtained by burning phosphorus in air. Chiefly used to manufacture orthophosphoric acid by reaction with water; also as a drying agent.

phosphorylase *(Biol.)*. An enzyme which catalyses the cleavage of a bond by the addition of orthophosphate (cf. **hydrolysis**). Thus glycogen phosphorylase liberates the terminal glucose of glycogen as glucose-1-phosphate.

phosphoryl bromide *(Chem.)*. $POBr_3$. Formed in a similar manner to **phosphorus oxychloride**.

phosphoryl chloride *(Chem.)*. See **phosphorus oxychloride**.

phosphoryl fluoride *(Chem.)*. POF_3. May be made by the action of hydrofluoric acid on phosphorus (V) oxide; similar in properties to the other phosphoryl compounds.

phot *(Phys.)*. CGS unit of illumination. 1 phot = 1 lumen cm^{-2}, $= 10^4$ lumens m^{-2}.

photic zone *(Ecol.)*. The zone of the sea where light penetration is sufficient for photosynthesis, corresponding to the limnetic zone of freshwater habitats. Cf. **aphotic zone**.

photobiology *(Biol.)*. The study of light as it affects living organisms.

photocatalysis *(Chem.)*. The acceleration or retardation of rate of chemical reaction by light.

photocathode *(Electronics)*. Electrode from which electrons are emitted on the incidence of radiation in a photocell. It is **semitransparent** when there is photemission on one side, arising from radiation on the other, as in the signal plate of a TV camera tube.

photocell *(Electronics)*. See **photoelectric cell**.

photocell sensitivity *(Electronics)*. Ratio of output current to level of illumination. Expressed in milliampere/lumen or microampere/lumen.

photochemical cell *(Electronics)*. Photocell comprising two similar electrodes, e.g. of silver, in an electrolyte; illumination of one electrode results in a voltage between them; also *Becquerel cell, photoelectrolytic cell*.

photochemical equivalence *(Chem.)*. See **Einstein law of photochemical equivalence**.

photochemical induction *(Chem.)*. The lapse of an appreciable time between the absorption of light by a system and the occurrence of the resulting chemical reaction.

photochemistry *(Chem.)*. The study of the chemical effects of radiation, chiefly visible and ultraviolet, and of the direct production of radiation by chemical change.

photochromic *(Glass)*. Said of special glass which changes colour and hence the amount of light transmitted when the light falling on it increases or decreases. Also *photosensitive*.

photochromics *(Phys.)*. Light sensitive materials that can be used in optical memory devices. Colour-centres in the material form the basis of the processes of reading and writing the data.

photocomposition *(Print.)*. See **phototypesetting**.

photoconductive camera tube *(Image Tech.)*. One in which the optical image is focused on to a surface, the electrical resistance of which is dependent on illumination. See **charge coupled device**.

photoconductive cell *(Electronics)*. Photoelectric cell using photoconductive element (often cadmium sulphide) between electrodes.

photoconductivity *(Electronics)*. Property possessed by certain materials, such as selenium, of varying their electrical conductivity under the influence of light.

photocopying *(Image Tech.)*. Copying with photographic methods, such as making multiple prints from a negative.

photocurrent *(Electronics)*. That released from the sensitized cathode of a photocell on the incidence of light, the electrons which form the current being attracted to the anode which is maintained positive with respect to the cathode. The true photocurrent may be augmented, in a gas-filled cell, through *collision ionization*.

photodegradation *(Textiles)*. Chemical degradation of a material such as a fibre caused by the absorption of light, especially ultra-violet radiation. Usually leads to loss of strength of the material as the polymer breaks down.

photodiode *(Electronics)*. A semiconductor diode fitted with a small lens which can focus light on the p-n junction. Certain types of junction exhibit marked variation of reverse current with illumination; this allows them to be used as compact and rugged light sensors.

photodisintegration *(Phys.)*. The ejection of a neutron, proton or other particles from an atomic nucleus following the absorption of a *photon*. A γ-ray photon with energy > 2.23 MeV can cause a deuteron nucleus to emit both a neutron and a proton.

photodissociation *(Chem.)*. Dissociation produced by the absorption of radiant energy.

photoelasticity *(Phys.)*. Phenomenon whereby strain in certain materials causes the material to become **birefringent**. Coloured fringes are observed when the transmitted light is viewed through crossed Nicol prisms.

photoelectric absorption *(Phys.)*. That part of the absorption of a beam of radiation associated with the emission of photoelectrons. In general, it increases with increasing atomic number but decreases with increasing quantum energy when *Compton absorption* becomes more significant.

photoelectric cell *(Electronics)*. See **photoemissive cell**.

photoelectric constant *(Phys.)*. The ratio of Planck's constant to the electronic charge. This quantity is readily measured by experiments on photoelectric emission and forms one of the principal methods by which the value of Planck's constant may be determined.

photoelectric effect *(Electronics)*. Any phenomenon resulting from the absorption of photon energy by electrons, leading to their release from a surface, when the photon energy exceeds the **work function** (see **photoemission**), or otherwise allowing conduction when the incident energy exceeds an atomic binding energy. See **photoconductivity**, **photovoltaic cell**. Emission of X-rays on the impact of high-energy electrons on a surface is an *inverse photoelectric effect*.

photoelectricity *(Electronics)*. Emission of electrons from the surface of certain materials by quanta exceeding certain energy (Einstein).

photoelectric multiplier *(Electronics)*. See **photomultiplier**.

photoelectric photometer *(Phys.)*. One in which the light from the lamp under test is measured by the current from a photoelectric cell.

photoelectric photometry *(Astron.)*. The determination of stellar magnitude and colour index by means of a photoelectric device used at the focus of a large telescope.

photoelectric threshold *(Electronics)*. The limiting frequency for which the quantum energy is just sufficient to produce photoelectric emission. Given by equating the quantum energy to the work function of the cathode.

photoelectric tube *(Electronics)*. Any transducer which has an electrical output (current, voltage) corresponding to incident light.

photoelectric work function *(Electronics)*. The energy required to release a photoelectron from a cathode. It should correlate closely with the thermionic work function.

photoelectric yield *(Electronics)*. The proportion of incident quanta on a photocathode which liberate electrons.

photoelectroluminescence *(Electronics)*. The enhancement of luminescence from a fluorescent screen during excitation by ultraviolet light or X-rays when an electric field is applied, i.e. the field enhancement of luminescence.

photoelectrolytic cell *(Electronics)*. See **photochemical cell**.

photoelectromagnetic effect *(Electronics)*. See **photomagnetoelectric effect**.

photoelectron *(Electronics)*. One released from a surface by a photon, with or without kinetic energy.

photoelectron spectroscopy *(Chem.,Phys.)*. Using visible light, ultra-violet light or X-rays as the excitation source, the energy of *photoelectrons* emitted from the material can be analysed to give information on surfaces, interfaces and bulk materials. Also used to deduce binding energies for deep core levels in atoms with a high degree of precision. Also *electron spectroscopy, photoionization spectroscopy*.

photoemission *(Electronics)*. Emission of electrons from surface of a body (usually an electropositive metal) by incidence of light.

photoemissive camera tube *(Image Tech.)*. Tube operating on the photoemissive principle, the image falling on its photocathode, causing this to emit electrons in proportion to the intensity of the light in the picture elements.

photoemissive cells *(Electronics)*. A category of **photoelectric tubes** in which light falling on a sensitized cathode causes current to flow to an anode. The current may be enhanced (at the expense of linearity) by using a gas-filled envelope and operating above the ionization potential of the gas. **Photomultiplier** tubes are photoemissive.

photo-engraving *(Print.)*. See **process-engraving**.

photofission *(Phys.)*. Nuclear fission induced by gamma-rays.

Photoflood lamp *(Image Tech.)*. A tungsten lamp run at excess voltage, with correspondingly reduced life, to raise its colour temperature.

photogenic *(Bot.,Zool.)*. Emitting light, light-producing, e.g. *photogenic bacteria*.

photogenin *(Zool.)*. See **luciferase**.

photoglow tube *(Electronics)*. A gas-discharge tube in which conduction is enhanced by radiation falling on the cathode.

photogrammetry *(Image Tech.)*. The use of photographic records for precise measurement of distances or dimensions, for example aerial photography for surveying.

photographic borehole survey *(Min.Ext.,Surv.)*. Check on orientation and angle of long borehole by insertion of a special camera which photographs a magnetic needle and a clinometer at known distance down.

photographic efficiency *(Phys.)*. Of a light source, the fraction of the light energy in the emitted spectrum which is usefully registered on a photographic emulsion.

photographic-emulsion technique *(Phys.)*. Study of the tracks of ionizing particles as recorded by their passage through a nuclear emulsion.

photographic photometry *(Phys.)*. Measurement of intensity of radiation by comparing a photographic image of the source with that of a standard source.

photographic recording *(Acous.)*. The registering of a modulated track on photographic film, so that it can be scanned by a constant beam of light, fluctuations of which, after being converted into corresponding electric currents by a photocell, can be amplified and reproduced as sound by loudspeakers.

photographic sound *(Image Tech.)*. System of recording and reproducing sound on motion picture film in the form of a photographic sound track of varying width of modulation; the preferred term for *optical sound*.

photographic surveying *(Surv.)*. A method of surveying employing the principles of **intersection** by means of a special instrument called a phototheodolite, with which a series of photographs is taken of the points whose positions are required, each point appearing in at least two different photographs. Also *phototopography*.

photographic zenith tube *(Astron.)*. Instrument for the exact determination of time; it consists of a fixed vertical telescope which photographs stars as they cross the zenith; instrumental and observational errors are thus eliminated, the instrument being entirely automatic. Abbrev. *PZT*.

photogravure *(Print.)*. An intaglio printing process using copper cylinders or plates etched through *carbon tissue*. The Helio-Klischograph is used to scan coloured separations into depressions cut in the cylinder, thus dispensing with process camera separation and carbon tissue.

photohalide *(Chem.)*. A halogen salt which is sensitive to light, e.g. silver bromide.

photolithography *(Print.)*. The normal modern process in which an image is created on a light-sensitive lithographic plate by photographic means as distinct from the old process, not now commercially used, whereby images were drawn on stone.

photolysis *(Chem.)*. The decomposition or dissociation of a molecule as the result of the absorption of light.

photolysis of water *(Bot.)*. The notional splitting of water molecules into oxygen, electrons and protons in the light reactions of photosynthesis. The term emphasizes the fact that the released oxygen comes from water not carbon dioxide.

photomagnetoelectric effect *(Electronics)*. Generation of electric current by absorption of light on surface of semiconductor placed parallel to magnetic field. Due to transverse forces acting on electrons and holes diffusing into the semiconductor from the surface. Also termed *photoelectromagnetic effect*.

Photomaton *(Image Tech.)*. Automatic machine which takes a number of photographs in succession and delivers a strip or sheet of finished prints in a few minutes. Prints are produced by automatic reversal of the image on a celluloid and paper base.

photomechanical *(Print.)*. Any process by which a printing surface is prepared mechanically with the aid of photography.

photomeson *(Phys.)*. Meson resulting from interaction of photon with nucleus usually a **pion**.

photometer *(Phys.)*. An instrument for comparing the luminous intensities of two sources of light. Most photometers employ the principle that, if equal illumination is produced on similar surfaces illuminated normally by two light sources, the ratio of their intensities equals the square of the ratio of their distances from the surfaces.

photometer bench *(Phys.)*. A bench upon which is mounted the apparatus for carrying out photometric tests by comparison with a standard lamp. The apparatus consists of a mounting for the standard lamp, the photometer itself, a mounting for the lamp under test, and equipment for moving any or all of these and determining their position.

photometer head *(Phys.)*. The part of a photometric system which contains the device for comparing the luminous intensities of two light sources.

photometric integrator *(Phys.)*. That part of an integrating photometer which actually sums up the light flux, e.g. the globe of the **Ulbricht sphere photometer**.

photometric surface *(Phys.)*. A surface used for photometric comparisons.

photometry *(Astron.)*. The accurate quantitative measurement of the amount of electromagnetic energy (most usually, light) received from a celestial object. Techniques are visual, photographic and **photoelectric**.

photometry *(Chem.)*. Volumetric analysis in which the end-point of a reaction is determined from colour changes detected by photoelectric means.

photometry *(Phys.)*. The measurement of the luminous intensities of light sources and of luminous flux and illumination. See also **photometer**.

photomicrography *(Image Tech.)*. Photography through a microscope. Not to be confused with **microphotography**.

photomorphogenesis *(Bot.)*. Control of plant morphogenesis by the duration and nature of the light. Cf. **etiolation, photoperiodism, phytochrome**.

photo-mosaic *(Image Tech.)*. A light-responsive surface made up of minute photo-emissive particles deposited on an insulating support, used in some early TV camera tubes.

photomultiplier *(Electronics)*. Photocell with series of dynodes used to amplify emission current by electron multiplication.

photon *(Phys.)*. Quantum of light of electromagnetic radiation of energy $E = h\nu$ where h is Planck's constant and ν is the frequency. The photon has zero rest mass, but carries momentum $h\nu/c$, where c is the velocity of light. The introduction of this 'particle' is necessary to explain the photoelectron effect, the Compton effect, atomic line spectra, and other properties of electromagnetic radiation. See **gauge boson**.

photonastic movement, photonasty *(Bot.)*. **Nastic movement** resulting from change in illumination.

photonegative *(Phys.)*. Material of which electrical conductivity decreases with increasing illumination. See **photopositive**.

photoneutron *(Phys.)*. Neutron resulting from interaction of photon with nucleus.

photonics *(Aero.)*. Concept for computing and data transmission using photons in place of electrons.

photon noise *(Electronics)*. That occurring in photocells as a result of the fluctuations in the rate of arrival of light quanta at the photo-cathode.

photopeak *(Radiol.)*. The energy of the predominant photons released during the decay of a radionuclide.

photoperiodicity *(Ecol.)*. The controlling effects of the length of day on such phenomena as the flowering of plants, the reproductive cycles of Mammals, migration and diapause of Insects, and seasonal changes in the feathers of Birds and the hair of Mammals.

photoperiodism *(Bot.)*. Response by an organism to day-length. See also **short-day plant**, **long-day plant**, **day-neutral plant**, **phytochrome**.

photophilous *(Biol.)*. Light-seeking, light-loving; said of plants which inhabit sunny places.

photophobia *(Med.)*. Intolerance of the eye to light with spasm of the eyelids.

photophore *(Zool.)*. A luminous organ of Fish.

photophoresis *(Chem.)*. The migration of suspended particles under the influence of light.

photophosphorylation *(Bot.)*. The production in photosynthesis of ATP from ADP and inorganic phosphate using energy from light. In cyclic photophosphorylation only **photosystem I** is involved and NADP is not reduced; in non-cyclic photophosphoryation *photosystem I* and *II* are both involved and NADP is reduced.

photophthalmia *(Med.)*. *Electric-light ophthalmia*. Burning pain in the eyes, lacrimation, photophobia, and swelling and spasm of the eyelids as a result of exposure to an intensely bright light (e.g. a naked arc light); due to the action of ultraviolet rays.

photophygous *(Zool.)*. Shunning strong light.

photopic luminosity curve *(Phys.)*. Curve giving the relative brightness of the radiations in an equal-energy spectrum when seen under ordinary intensity levels. See **scotopic luminosity curve**.

photopic vision *(Phys.)*. That based on **cones**, and therefore sensitive to colour. Possible only with adequate ambient illumination. Cf. *scotopic vision*.

photopolymer plates *(Print.)*. Relief printing plates made of a synthetic material, on a flexible or rigid base as required. After printing down with ultraviolet light through a negative, during which the exposed parts are polymerized, the plate is completed by washing away the unexposed areas with a caustic solution.

photopositive *(Phys.)*. Material of which conductivity increases with increasing illumination. See **photonegative**.

photoproton *(Phys.)*. Proton resulting from interaction of photon with nucleus.

photopsy, photopsia *(Med.)*. The appearance of flashes of light in front of the eyes, due to irritability of the retina.

photoreceptor *(Bot.)*. A structure such as that associated with an *eyespot* or a molecule (e.g. **phytochrome**) which absorbs the stimulating light in phototaxis etc.

photoreceptor *(Zool.)*. A sensory cell specialized for the reception of light, e.g. rods and cones of the vertebrate eye.

photoresist process *(Electronics)*. Process removing selectively the oxidized surface of a silicon slice semiconductor. The photoresist material is an organic substance polymerizing on exposure to ultraviolet light and in that form resisting attack by acids and solvents.

photorespiration *(Bot.)*. Light-stimulated respiration caused by the metabolism of phosphoglycolic acid by the oxygenase reaction of **ribulose bisphosphate carboxylase** in photosynthesis. An apparently wasteful process that limits photosynthesis in C_3 plants under warm, bright conditions, but which C_4 plants avoid.

photosedimentation *(Powder Tech.)*. Method of particle-size analysis in which concentration changes within a suspension are determined by measuring the attenuation of a beam of light passing through it.

photosensitive *(Glass)*. See **photochromic**.

photosensitive *(Image Tech.,Phys.)*. Property of being sensitive to action of visible or invisible radiation.

photosensitizing dye *(Biol.)*. **Intercalating dye** which binds to DNA and makes it susceptible to breakage when exposed to ultraviolet light.

photosetting *(Print.)*. See **filmsetting**.

photosphere *(Astron.)*. The name given to the visible surface of the sun on which sunspots and other physical markings appear; it is the limit of the distance into the sun that we can see.

Photostat *(Image Tech.)*. TN for photographic apparatus (also for any print made by it) designed for rapidly copying flat original on sensitized paper, giving a negative image.

photosynthate *(Bot.)*. The substances, especially sugars, produced in photosynthesis.

photosynthesis *(Bot.)*. The use of energy from light to drive chemical reactions most notably the reduction of carbon dioxide to carbohydrates coupled with the oxidation either of water to free oxygen or of hydrogen sulphide to sulphur. See also **photosystems I and II**, **light reactions**, **dark reactions**, **Calvin cycle**, **C_3 plants**, **C_4 plants**, **CAM plants**, **photophosphorylation**, **photorespiration**, **Z scheme**, **photosynthetic pigments**. adj. *photosynthetic*.

photosynthetically active radiation *(Bot.)*. Light with a wave-length $\lambda = 400–700$ nm, active in photosynthesis. Abbrev. *PAR*.

photosynthetic pigment *(Bot.)*. A pigment involved in the absorption of light in photosynthesis, in plants and algae, *chlorophyll a* and the *accessory pigments*.

photosynthetic quotient *(Bot.)*. The ratio of carbon dioxide absorbed to oxygen released in a photosynthesizing structure or organism. Abbrev. *PQ*.

photosystem I *(Bot.)*. Reaction centre in photosynthesis comprising chlorophyll a, as *P700*, and other pigments and molecules, in which energy absorbed as light is used to transfer electrons from a weak oxidant to a strong reductant. The latter either reduces **ferredoxin** and ultimately NADP or is used in cyclic photophosphorylation. Abbrev. *PS I*.

photosystem II *(Bot.)*. Reaction centre in photosynthesis comprising chlorophyll a, as *P680*, and other pigments and molecules, in which energy absorbed as light is used to transfer electrons from a strong oxidant to a weak reductant, the former oxidizing water to oxygen. Abbrev. *PS II*.

phototaxis *(Biol.)*. Locomotory response or reaction of an organism or cell to the stimulus of light. adj. *phototactic*.

phototheodolite *(Surv.)*. A photographic camera of fixed known focal length, with horizontal and vertical lines engraved on a glass plate at the image plane, on which the film is registered. It is mounted on a tripod and fitted with levelling screws, a graduated horizontal circle and a telescope. See **photographic surveying**.

phototopography *(Surv.)*. See **photographic surveying**.

phototransistor *(Electronics)*. A 3-electrode photosensitive semiconductor device. The emitter junction forms a photodiode and the signal current induced by the incident light amplified by transistor action.

phototrophic *(Biol.)*. Said of organisms obtaining their energy from sunlight.

phototropism *(Bot.)*. A *tropism* in which a plant part becomes orientated with respect to the direction of light.

phototropy *(Chem.)*. (1) The property possessed by some substances of changing colour according to the wavelength of the incident light. (2) The reversible loss of colour in a dyestuff when illuminated at a definite wavelength.

phototube *(Electronics)*. See **photoelectric tube**.

phototypesetting *(Print.)*. The technique by which text and certain illustrations can be set onto paper or film.

photovaristor *(Electronics)*. Material, e.g. cadmium sulphide or lead telluride, in which the varistor effect, i.e. nonlinearity of current-voltage relation, is dependent on illumination.

photovoltaic cell *(Electronics)*. A class of photoelectric cell which acts as a source of e.m.f. and so does not need a separate battery.

photovoltaic effect *(Electronics)*. The production of an e.m.f. across the junction between dissimilar materials when it is exposed to light or ultraviolet radiation.

phragma *(Zool.)*. A septum or partition: an apodeme of the endothorax formed by the infolding of a portion of the tergal region of a somite; an endotergite.

phragmoplast *(Bot.)*. Complex of interdigitating microtubules aligned more or less parallel to the earlier spindle microtubules, developing at the end of mitosis and spreading like an expanding doughnut to the side walls while the *cell plate* forms in the centre. Characteristic of vascular plants, bryophytes and some algae.

phreatic gases *(Geol.)*. Those gases of atmospheric or oceanic origin which are generated by contact with ascending magma.

phrenicectomy *(Med.)*. The cutting of the phrenic nerve in order to paralyse the diaphragm on the same side.

phrenicotomy *(Med.)*. The cutting of the phrenic nerve in order to paralyse the diaphragm on one side; done in the treatment of lung disease.

phrenology *(Behav.)*. Historically, the notion that a person's skull formations reflect mental abilities and personality characteristics.

pH stat *(Bot.)*. Metabolic reactions which collectively maintain constant pH in the cytoplasm.

phthaleins *(Chem.)*. Triphenylmethane derivatives obtained by the action of phenols upon phthalic (benzene 1,2-dicarboxylic) anhydride.

phthalic acid *(Chem.)*. $C_6H_4(COOH)_2$, *benzene-O-dicarboxylic acid*, colourless prisms or plates, mp 213°C, soluble in water, alcohol, ether. When heated above the melting-point it yields its anhydride. It is prepared by catalytic air oxidation via the anhydride.

phthalic anhydride *(Chem.)*. *Phthalic (benzene 1,2-dicarboxylic) anhydride*. Long prisms which can be sublimed, mp 128°C, bp 284°C. Used in the production of plasticizers, paints and polyester resins. Prepared by air oxidation of naphthalene in the presence of a catalyst such as vanadium (V) oxide.

phthalic glyceride resins *(Plastics)*. See **alkyd resins**.

phthalimide *(Chem.)*. Benzene 1,2-dicarboximide. Colourless crystals, mp 238°C, obtainable by passing ammonia over heated phthalic anhydride. The imide hydrogen is replaceable by Na or K.

phthalocyanines *(Chem.)*. A group of green and blue organic colouring matters, used mainly as pigments and formed by condensation polymerization of four molecules of phthalonitrile with one atom of a metal, such as copper phthalocyanine. Widely used in almost all pigmented products. *Green*: Consists of halogenated copper phthalocyanine, and available in a number of shades from deep blue-green to a bright green with only a slight blue tone.

phthalonitrile *(Chem.)*. *1,2-dicyanobenzene*. Crystalline solid; mp 140°C. **Phthalocyanine** colours are made by heating it with metallic compounds.

phthiriasis *(Med.)*. Infestation with lice. See also **pediculosis**.

phthisis *(Med.)*. Former term for: (1) wasting of the body; (2) pulmonary tuberculosis. adj. *phthisical*.

phugoid oscillation *(Aero.)*. A longitudinal fluctuation in speed of long periodicity, i.e. a velocity-modulation, in the motion of an aircraft, accompanied by rising and falling of the nose.

pH value *(Chem.)*. A logarithmic index for the hydrogen ion concentration in an aqueous solution. Used as a measure of acidity of a solution; given by $pH = \log_{10}(1/(H^+))$, where (H^+) is the hydrogen-ion concentration. A pH below 7 indicates acidity, and one above 7 alkalinity, at 25°C.

phycobilin *(Bot.)*. One of a number of red (phycoerythrin) or blue-green (phycocyanin) *accessory photosynthetic pigments*, found in Rhodophyceae, Cyanophyceae and Cryptophyceae.

phycobiont *(Bot.)*. The algal partner in a lichen or other symbiotic association.

phycology *(Bot.)*. The study of the algae.

Phycomycetes *(Bot.)*. A probably unnatural group of fungi and fungus-like organisms comprising the Mastigomycotina (including the Oomycetes) and the Zygomycotina.

phyc-, phyco- *(Genrl.)*. Prefix denoting association with algae or seaweed (Gk. *phykos*, seaweed).

phyletic classification *(Biol.)*. A scheme of classification based on presumed evolutionary descent.

phyllid *(Bot.)*. The leaf of a moss or liverwort.

phyllite *(Min.)*. A name which has been used in different senses: (1) for the pseudohexagonal platy minerals (*phyllosilicates*) including mica, chlorite, and talc (by some French authors). (2) for argillaceous rocks in a condition of metamorphism between slate and mica-schist (by most English authors). Phyllite in the latter (usual) sense is characterized by a silky lustre due to the minute flakes of white mica which, however, are individually too small to be seen with the naked eye.

phyllobranchia *(Zool.)*. A gill composed of numerous thin plate-like lamellae.

phylloclade *(Bot.)*. A flattened leaf-like stem, functioning as a photosynthetic organ. Cf. **cladode**.

phyllode *(Bot.)*. A flat, more or less expanded, leaf-like petiole functioning as a photosynthetic organ, usually in the absence of a leaf blade.

phyllody *(Bot.)*. The abnormal, often pathological, development of leaves in place of the normal parts of a flower.

phylloplane, phyllosphere *(Bot.)*. The leaf surface, especially as a habitat for micro-organisms.

phyllopodium *(Zool.)*. The thin leaflike swimming foot characteristic of some crustaceans. Cf. **stenopodium**.

phyllosilicates *(Min.)*. Those silicate minerals having an atomic structure in which SiO_4 groups are linked to each other to form continuous sheets, e.g. **talc**.

phyllotaxis, phyllotaxy *(Bot.)*. The arrangement of leaves on a stem.

phylo- *(Genrl.)*. Prefix from Gk, *phylon*, race.

phylogeny, phylogenesis *(Bot.,Zool.)*. The evolutionary development or history of groups of organisms. adj. *phylogenetic*. Cf. **ontogeny**.

phylum *(Zool.)*. A category or group of related forms constituting one of the major subdivisions of the animal kingdom. A *division* in plants.

physical chemistry *(Chem.)*. The study of the dependence of physical properties on chemical composition, and of the physical changes accompanying chemical reactions.

physical containment *(Biol.)*. The construction of laboratory or work-station so as to prevent the contamination of the worker or the environment by harmful organisms.

physical electronics *(Electronics)*. The branch of electronics which is concerned with the physical details of the behaviour of electrons in vacuum, in gases, and in conductors, semiconductors, and insulators.

physical metallurgy *(Eng.)*. The study of the properties of metals and their alloys and the effect of composition, heat treatment, environment and other factors upon them.

physical optics *(Phys.)*. That branch of the study of light dealing with phenomena such as diffraction and interference, which are best considered from the standpoint of the wave theory of light. See **geometrical optics**.

physics *(Genrl.)*. Study of electrical, luminescent, mechanical, magnetic, radioactive and thermal phenomena with respect to changes in energy states without change of chemical composition.

physio- *(Genrl.)*. Prefix from Gk. *physis*, nature.

physiography *(Geol.)*. The science of the surface of the earth and the inter-relations of air, water and land.

physiological *(Genrl.)*. Relating to the functions of plant or animal as a living organism.

physiological anatomy *(Biol.)*. The study of the relation between structure and function.

physiological drought *(Biol.)*. Condition in which a plant is unable to take in water because of low temperature, or because the water available to it holds substances in solution which hinder absorption by the plant.

physiological psychology *(Behav.)*. The study of anatomy and physiology in relation to psychological phenomena.

physiological race *(Biol.)*. *Biological race, forma specialis*. A group of individuals within the morphological limits of a species but differing from other members of the species in habits (as host, larval food etc.), e.g. the several races of a parasitic fungus each confined to a different host.

physoclistous *(Zool.)*. Of fish, having no pneumatic duct connecting the air-bladder with the alimentary canal. Cf. **physostomous**.

physostomous *(Zool.)*. Of Fish, not **physoclistous**, i.e. having a pneumatic duct.

phytic acid *(Bot.)*. Inositol hexaphosphate, present e.g. in seeds as its calcium magnesium salt, phytin, perhaps as a phosphorus-storage compound.

phytoalexin *(Bot.)*. A substance produced by a plant in response to some stimulus. Often phenolic or terpenoid, they inhibit the growth of some micro-organisms especially pathogenic fungi.

phytochemistry *(Bot.)*. The study of the chemical constituents, and especially the *secondary* metabolites, of plants.

phytochrome *(Bot.)*. A blue-green, phycobilin-like, pigment reported from seed plants and a wide variety of other photosynthetic eukaryotes, acting as the light receptor molecule in a number of morphogenetic processes including photoperiodism, reversal of etiolation and the germination of some seeds and spores. See P_{fr} and P_r.

phytoferritin *(Bot.)*. An iron-protein complex, similar to the ferritin of animals, apparently a store of iron.

phytohaemagglutinins *(Immun.)*. *PHA*. Lectins extracted from the beans of *Phaseolus vulgaris*. They bind to N-acetyl-beta-D-galactosamine residues and can agglutinate certain erythrocytes which bear these on their surface. The purified lectins are potent mitogens of T-lymphocytes.

phytohormone *(Bot.)*. A plant hormone.

phytology *(Genrl.)*. Synonym for *botany*.

phytopathology *(Bot.)*. *Plant pathology*. The study of plant diseases, especially of plants in relation to their parasites.

phytophagous, phytophilous *(Zool.)*. Plant-feeding.

phyto-, phyte *(Genrl.)*. Prefix and suffix from Gk. *phy̆ton*, plant.

phytoplankton *(Bot.)*. The photosynthetic members of the plankton.

phytoplankton blooms *(Ecol.)*. Very high densities of plankton which occur quickly, and persist for short times, usually at regular times of the year.

phytosanitary certificate *(Bot.)*. A certificate of health for plants or parts of plants for export.

phytosociology *(Ecol.)*. The study of the association of plant species.

phytotoxic substance *(Bot.)*. A substance toxic to plants; sometimes also refers to animals.

phytotoxin *(Bot.)*. Plant substance toxic to animals or other organisms.

phytotron *(Bot.)*. A large and elaborate **growth room** for plants, or a collection of such.

P_i *(Bot.)*. In biochemical equations inorganic phosphate e.g. $H_2PO_4^{2-}$.

pia mater *(Zool.)*. In Vertebrate, the innermost of the three membranes surrounding the brain and spinal cord, a thin vascular layer.

pian *(Med.)*. See **yaws**.

piano nobile *(Arch.)*. The principal floor of an Italian palace, occurring at first floor level and containing the main living areas.

Pianotron *(Acous.)*. Pianoforte in which the normal vibration of the strings is used to modulate the potential applied to electrostatic screw pick-ups, with amplification and loudspeakers.

pi-attenuator, π-attenuator *(Telecomm.)*. An attenuator network of resistors arranged as in the pi-network.

piazza *(Arch.)*. (1) An enclosed court in a building. (2) A colonnade or arcade.

PIB *(Plastics)*. Polyisobutylene or **polybutene**.

pica *(Med.)*. Unnatural craving for unusual food.

pica *(Print.)*. An old type size, approximately 12-point. See **em, point**.

pi characters *(Print.)*. (1) Special characters or symbols not usually included in a type **fount**, including special **ligatures**, accented letters, mathematical symbols, etc. Also referred to as *sorts*. (2) **outside sorts** on line-composing machines which cannot be accommodated in a magazine but which are stored in small cases beside the machine and inserted in the line by hand. Also called *side sorts*. (3) Outside sorts on phototypesetting machines which are positioned in a separate pi character store.

Piciformes *(Zool.)*. An order of Birds, containing climbing, insectivorous, and wood-boring species. The beak is hard and powerful, the tongue long and protrusible, and the feet zygodactylous. Woodpeckers, Toucans.

pick *(Build.)*. A double-headed tool, pointed at both ends, having the handle fastened into the middle of the head (as in a hammer); used for rough digging by quarrymen, road builders etc.

pick *(Textiles)*. One traverse of the weft-carrying device through the warp shed in weaving to lay a weft thread for beating up into cloth. A single weft thread in a fabric. Picks per inch (*ppi*) denotes the total number of weft threads in every inch of cloth. (Often preferred in the trade to *ppcm*).

pick-and-pick *(Textiles)*. A woven fabric with alternate picks of yarns of two different colours or kinds.

pick-axe *(Build.)*. A tool similar to a pick but having one end edged so that it may cut.

picker *(Textiles)*. Buffalo hide, rubber, or plastic component that strikes the shuttle to propel weft across loom.

pickeringite *(Min.)*. Magnesia alum. Hydrated sulphate of aluminium and magnesium, crystallizing in the monoclinic system. It usually occurs in fibrous masses, and is formed by the weathering of pyrite-bearing schists.

picket *(Build.)*. A narrow upright board in a fence; frequently pointed at the top.

picket *(Surv.)*. A short ranging rod about 2 m (6 ft) long.

picket fence *(Elec.Eng.)*. Arrangement for stabilizing plasma in a torus, consisting of bands of coils round the tube, with currents alternating in direction, giving an interior field with periodic cusps.

picking *(Build.)*. Chiselling or picking small indentations on a wall surface as a key for the finish. Also called *stugging, wasting*.

picking *(Print.)*. Removal of part of the paper surface due to faults in the paper or in printing.

picking *(Textiles)*. (1) Removing extraneous matter (outstanding hairs, wrongly coloured fibres, slubs, etc.) from the face of fabrics. (2) Propelling the shuttle carrying the weft through the warp shed in weaving.

picking belt *(Min.Ext.)*. Sorting belt, on which run-of-mine ore is displayed so that pickers can remove waste rock, debris or a special mineral constituent, which is not to be sent to the mill for treatment.

pickling *(Chem.Eng.)*. The removal of mill scale, lime scale or salt water deposits, e.g. in a ship's engine or evaporators, by circulating a suitable acid containing inhibitors. Usually up to ten times diluted sulphuric acid.

Inhibitors include glue and acidic long-chain molecular compounds such as HBF_4.

Pick's disease *(Med.)*. A rare cause of presenile dementia.

pick-up head *(Acous.)*. (1) Mechanical-electrical transducer, often piezo, actuated by a sapphire or diamond stylus which rests on the sides of the groove on a gramophone record, and by tracking this groove generates a corresponding voltage (or, for stereo reproduction, two voltages) for driving an audio amplifier. Also called *pick-up*. (2) Device to measure structure-borne sound, usually employing the piezoelectric, electrodynamic or electrostatic principle. See also **accelerometer**.

pick-up needle *(Acous.)*. Loosely applied to any needle in a gramophone pick-up.

pick-up needle *(Elec.Eng.)*. One in which the moving part is a magnetic needle, which by its motion diverts magnetic flux and induces a voltage in coils on the magnetic circuit.

pick-up reaction *(Phys.)*. Nuclear reaction in which incident particle collects nucleon from a target atom and proceeds with it.

pick-up tube *(Image Tech.)*. See **camera tube**.

pick-up well *(Autos.)*. A small petrol reservoir arranged between the metering jet and the spraying tube in some carburettors; it provides a temporarily enriched mixture during acceleration.

pico- *(Genrl.)*. SI prefix for 1-millionth-millionth, or 10^{-12}. Formerly, *micromicro-*. Abbrev. *p*.

picofarad *(Elec.Eng.)*. 10^{-12} farad. Same as *micromicrofarad*. Abbrev. *pF*.

picolines *(Chem.)*. *Methyl-pyridines*, $CH_3.C_5H_4N$. The three isomers are 2-picoline, bp 129°C; 3-picoline, bp 142°–143°C; 4-picoline, bp 144°–145°C.

picosecond *(Genrl.)*. The million millionth part of a second, 10^{-12} s. Abbrev. *ps*.

picotite *(Min.)*. A dark-coloured spinel containing iron, magnesium, aluminium; a chromium-bearing hercynite.

picramic acid *(Chem.)*. Red crystalline solid; mp 168°C, obtained by reduction of **picric acid** (trinitrophenol). Used in manufacture of azo dyes.

picric acid *(Chem.)*. $C_6H_2(NO_2)_3.OH$, *2,4,6-trinitrophenol*, yellow plates or prisms, mp 122°C, made by nitrating phenol; slightly soluble in water. It is a strong acid and dyes wool and silk yellow. Used for the preparation of explosives; lyddite or melenite is compressed or fused picric acid.

picrite *(Geol.)*. An ultramafic coarse-grained igneous rock, consisting essentially of olivine and other ferromagnesian minerals, together with a small amount of plagioclase.

picture *(Image Tech.)*. In television usage, the picture is the complete image, or **frame**, made up from two interlaced fields; hence *picture frequency* is the same as **frame frequency**.

picture inversion *(Image Tech.)*. Conversion of negative to positive image (or vice versa) when carried out electronically. In facsimile transmission, it will correspond to the reversal of the black and white shades of the recorded copy.

picture monitor *(Image Tech.)*. CRT for exhibiting TV picture or related waveform for purposes of control.

picture noise *(Radar)*. See **grass**.

picture signal *(Image Tech.)*. That portion of a TV signal which carries information relative to the picture itself, as distinct from synchronizing portions. See **pedestal**.

picture slip *(Image Tech.)*. Horizontal or vertical displacement or distortion of the received picture due to loss of synchronization.

Picturesque *(Arch.)*. A design concept which originated in the late 18th century and which idealized classical landscape paintings by such artists as Claude and Poussin. It mainly influenced landscape gardening where buildings, and frequently sham ruins or follies, became part of a controlled but informal landscape. On a smaller scale, domestic architecture became more asymmetrical, while still retaining **Gothic** or Italianate forms. See **folly**.

picture/sync ratio *(Image Tech.)*. The ratio of the total amplitude of the television waveform assigned to picture information to that which is assigned to the synchronizing pulses and flyback times.

picture tube *(Image Tech.)*. CRT specifically designed for the reproduction of television images.

PID *(Comp.)*. See **personal identification device**.

Pidgeon process *(Eng.)*. Reduction of magnesium oxide in the presence of ferrosilicon in the making of magnesium.

pie *(Print.)*. To upset type-matter accidentally.

piece *(Textiles)*. Fabric of agreed length. See **bolt**.

piece dyeing *(Textiles)*. Dyeing lengths of fabric.

piece goods *(Textiles)*. Fabric sold by length.

piedmont glacier *(Geol.)*. A glacier of the 'expanded foot' type; one which, after being restricted within a valley, spreads out on reaching the flat ground into which the latter opens.

piedmont gravels *(Geol.)*. Accumulations of coarse breccia, gravel, and pebbles brought down from high ground by mountain torrents and spread out on the flat ground where the velocity of the water is checked. Literally, mountain foot gravels, typical of the outer zone of arid areas of inland drainage such as the Lop Nor Basin in Chinese Turkestan.

piedmontite *(Min.)*. A hydrated silicate of calcium, aluminium, manganese, and iron, crystallizing in the monoclinic system; a member of the zoisite group. Also called *manganepidote*, *piemontite*.

piedroit *(Build.)*. A pier projecting from a wall but having neither cap nor base. Cf. **pilaster**.

piend *(Build.)*. Hip rafter. See **arris**.

piend check *(Build.)*. The rebate cut along a lower corner of a stone step to enable it to sit upon the step below.

piend rafter *(Build.)*. The rafter at the junction where two roof slopes meet forming an external angle. An *angle rafter*.

Pieper system *(Elec.Eng.)*. See **automixte system**.

pier *(Build.)*. (1) The part of a wall between doors and windows. (2) See **buttress**.

pierced *(Print.)*. Said of a block with an internal portion removed to accommodate type.

Pierce oscillator *(Electronics)*. Original crystal oscillator in which positive feedback from anode to grid in a triode valve is controlled by piezo-electric mechanical resonance of a suitably cut quartz crystal.

piercing *(Eng.)*. See **Mannesmann process**.

pier template *(Civ.Eng.)*. A stone slab laid on a brickwork pier to ensure that the load is distributed over the full area of the pier.

pieze *(Phys.)*. Unit of pressure in the metre-tonne-second system, equivalent to $10^3\,N/m^2$ or $1\,kN/m^2$.

piezo- *(Genrl.)*. Prefix from Gk. *piezein*, to press.

piezochemistry *(Chem.)*. The study of the effect of high pressures in chemical reactions.

piezoelectric crystal *(Electronics)*. One showing piezoelectric properties which may be shaped and used as a resonant circuit element or transducer (microphone, pick-up, loudspeaker, depth-finder, etc.).

piezoelectric effect, piezoelectricity *(Phys.)*. Electric polarization arising in some anisotropic (i.e. not possessing a centre of symmetry) crystals (quartz, Rochelle salt, barium titanate) when subject to mechanical strain. An applied electric field will likewise produce mechanical deformation.

piezoelectric loudspeaker *(Acous.)*. A **piezoelectric crystal** for generating sound waves (direct piezoeffect).

piezoelectric microphone *(Acous.)*. A **piezoelectric crystal** which produces electric potentials in response to incident sound waves (inverse piezoeffect).

piezoelectric pick-up *(Acous.)*. Piezoelectric transducer, producing an e.m.f. due to mechanical drive arising from vibration. Widely used for gramophone-disk reproduction and vibration measurements.

piezoelectric resonator *(Telecomm.)*. A crystal used as a standard of frequency, controlling an electronic oscillator.

piezoid *(Elec.Eng.)*. Blank of piezo crystal, adjusted to a required resonance, with or without relevant electrodes.

piezomagnetism *(Phys.)*. An effect analogous to *piezoelectricity* in which mechanical strain in certain materials produces a magnetic field and *vice versa*.

pig *(Eng.)*. A mass of metal (e.g. cast-iron, copper or lead) cast in a simple shape for transportation or storage, and subsequently remelted for purification, alloying, casting into final shapes, or into ingots for rolling.

pig *(Glass)*. An iron block laid against the pot mouth as a support for the blowing-iron.

pig *(Min.Ext.)*. See **go-devil**.

pig bed *(Eng.)*. A series of moulds for iron pigs, made in a bed of sand. Connected to each other and to the tap hole of the blast-furnace by channels, along which the molten metal runs.

pigeon-holed *(Build.)*. Said of a wall built with regular gaps in it (e.g. *honeycomb wall*).

pigeonite *(Min.)*. One of the monoclinic pyroxenes, intermediate in composition between clinoenstatite and diopside. It is poor in calcium, has a small optic axial angle and occurs in quickly chilled lavas and minor intrusions.

pigeon's milk *(Zool.)*. In Pigeons and Doves, a white slimy secretion of the epithelium of the crop in both sexes. It contains protein and fat, and is produced during the breeding season under the influence of **prolactin**, being regurgitated to feed the young.

pig iron *(Eng.)*. The crude iron produced in the blast furnace and cast into pigs which are used for making steel, cast-iron or wrought-iron. Principal impurities are carbon, silicon, manganese, sulphur and phosphorus. Composition varies according to the ores used, the smelting practice and the intended usage.

pigment *(Biol.)*. Substances which impart colour to the tissues or cells of animals and plants.

pigment *(Build.,Chem.)*. Insoluble, natural, or artificial, black, white or coloured materials reduced to powder form which, when dispersed in a suitable medium, are able to impart colour and/or opacity. Ideally, pigments should maintain their colour under the most unfavourable conditions; in practice, may tend to fade or discolour under the influence of acids, alkalis, other chemicals, sunlight, heat and other conditions. May also show **bleeding**. They are used in many industries, e.g. in paint, printing ink, plastics, rubber, paper etc. The two main types are the natural earth pigments which tend to be drab and limited in colour range and synthetic pigments with a comprehensive range of colours.

pigment *(Textiles)*. Finely divided solid colouring materials dispersed (e.g. in water or in a resin) and applied to a fabric by padding or printing.

pigmentary colours *(Zool.)*. Colours produced by the presence of drops or granules of pigment in the integument, as in most Fish. Cf. **structural colours**.

pigment cell *(Biol.)*. See **chromatophore**.

pigment foils *(Print.)*. **Blocking foils**, which leave a black, white, or coloured matt transfer.

pigtail *(Elec.Eng.)*. The short length of flexible conductor connecting the brush of an electric machine to the brush-holder.

pilaster *(Build.)*. A square pier projecting from a wall, having both a cap and a base. Cf. **piedroit**.

pilaster strip *(Build.)*. A pilaster without a cap.

Pilat process *(Chem.)*. Method of separating the fractions of asphalt oils without distillation, but dissolving out the asphalt with propane and saturating the residual oil with methane under pressure, with the result that the lubricating oil fractions separate out in the order of decreasing viscosity.

pile *(Civ.Eng.)*. A column or sheeting which is sunk into the ground to support vertical loading or to resist lateral pressures.

pile *(Elec.Eng.)*. Abbrev. for *thermopile*, a close packing of thermocouples in series, so that alternate junctions are exposed for receiving radiant heat, thus adding together the e.m.fs. due to pairs of junctions.

pile *(Eng.)*. A number of wrought-iron bars arranged in an orderly pile which is to be heated to a welding heat and rolled into a single bar.

pile *(Med.)*. See **haemorrhoid**.

pile *(Nuc.Eng.)*. Original name for a reactor arising from the pile of graphite blocks which formed the moderator of the first nuclear reactor.

pile *(Phys.)*. The light reflected from a glass plate incident at the Brewster angle is plane-polarized and the transmitted light is partially plane-polarized normal to the plane of the reflected polarized light. Using a *pile* of glass plates the transmitted light becomes increasingly plane-polarized.

pile *(Textiles)*. A covering on the surface of a fabric, formed by threads that stand out from it. Pile in loop form is termed *loop pile* or *terry*; if the loops are cut, it is termed *cut pile*. The latter is produced by weaving two cloths face to face, and cutting them apart. Some carpets and *moquettes* are made this way.

pileate *(Zool.)*. Crested.

pile delivery *(Print.)*. On a printing press, a delivery arrangement which can accommodate a large pile of sheets, the delivery board lowering automatically to floor level as the sheets are delivered.

pile-driver *(Civ.Eng.)*. A framed construction erected above the spot where a pile is to be driven into the ground. It is provided with a heavy weight which runs in upright guides and is so arranged that it may fall by gravity on to the head of the pile and drive it in. Also called *pile frame*.

pile hoop *(Civ.Eng.)*. An iron or steel band fitted around the head of a pile to prevent brooming.

pile shoe *(Civ.Eng.)*. The iron or steel point fitted to the foot of a pile to give it strength to pierce the earth and so assist driving.

pileus *(Bot.)*. The cap of the fruiting body of some fungi e.g. mushrooms or toadstools (Basidiomycotina).

piliferous layer *(Bot.)*. The epidermis, bearing root hairs, of a young root.

piling *(Print.)*. A build up of ink on the plate, blanket or printing rollers during printing.

pill *(Textiles)*. Small unsightly balls of fibre that accumulate during wear on the surface of certain items of apparel.

pillar *(Elec.Eng.)*. A structure of pillar form for containing switch or protective gear. Also called a *switchgear pillar*.

pillar *(Horol.)*. Any cylindrical pieces to hold the plates of a clock or watch in their correct relative position.

pillar *(Min.Ext.)*. Column of unserved ore left as roof support in stope.

pillar-and-stall *(Min.Ext.)*. See **bord-and-pillar**.

pillar drill *(Eng.)*. A drilling machine in which the spindle and table are supported by brackets on a pillar usually being slidable thereon.

pillar plate *(Horol.)*. In a clock or watch, the plate to which the pillars are fixed.

pillow distortion *(Image Tech.,Phys.)*. See **pincushion distortion**.

pillow lava *(Geol.)*. A lava-flow exhibiting pillow structure, generally formed in a sub-aqueous environment.

pillow structure *(Geol.)*. A term applied to lavas consisting of ellipsoidal and pillow-like masses which have cooled under subaqueous conditions. The spaces between the pillows consist, in different cases, of chert, limestones, or volcanic ash.

pilomotor *(Med.)*. Causing movements of hair.

pilose *(Bot.)*. Hairy, with long, soft hairs.

pilot *(Elec.Eng.)*. In power systems, a conductor used for auxiliary purposes, not for the transmission of energy. Also called *pilot-wire*.

pilotaxitic texture *(Geol.)*. The term applied to the groundmass of certain holocrystalline igneous rocks in which there is a feltlike interweaving of feldspar microlites. Cf. **hyalopilitic texture**.

pilot balloon *(Meteor.)*. A small rubber balloon, filled with hydrogen, used for determining the direction and speed of air currents at high altitudes, the balloon is observed by means of a theodolite after being released from the ground.

pilot carrier *(Telecomm.)*. In a suppressed carrier system (as in single sideband working) a small portion of original carrier wave transmitted to provide a reference frequency with which local oscillator at the receiving end may be synchronized.

pilot cell *(Elec.Eng.)*. A cell of a battery upon which readings are taken in order to give an indication of the state of the whole battery.

pilot chute *(Aero.)*. A small parachute which extracts the main canopy from its pack.

pilot controller *(Elec.Eng.)*. See **master controller**.

piloted head *(Eng.)*. Gas burner head or nozzle having a by-pass by which low-pressure feeder flames are produced around the main flame, to secure positive retention when the velocity of the combustible mixture exceeds the flame speed.

pilot electrode *(Electronics)*. Additional electrode or spark gap which triggers and makes certain main discharge, e.g. in a discharge spark gap, for creating ions, or in a TR switch, or a mercury-arc rectifier. Also *ignitor, keep-alive electrode, starter, trigger electrode*.

pilot engine *(Civ.Eng.)*. A separate locomotive preceding a train as a precaution against accidents to the latter.

pilot gauge *(Eng.)*. A plug gauge which has a circumferential groove near its free end to facilitate entry into the hole to be gauged.

pilotherm *(Phys.)*. A thermostat in which the temperature control is brought about by the deflection of a bimetallic strip, thereby switching on and off the electric heating current.

pilotis *(Arch.)*. Reinforced concrete columns carrying a building, leaving an open area at ground level.

pilot lamp *(Elec.Eng.)*. One giving visual indication of the closing of a circuit.

pilot nail *(Build.)*. A temporary nail used in fixing shuttering. May have two heads. Also *stitch nail*.

pilot pin *(Eng.)*. A pin with rounded or tapered end, fixed to a tool or machine element and projecting beyond it so as to enter a hole in a mating part before engagement, thus ensuring alignment. Used in press tools to position the strip material.

pilot pins *(Image Tech.)*. Pins forming part of the mechanism of a motion picture camera, printer or projector which engage with the perforations to ensure accurate location and steadiness of the film at the period of exposure.

pilot plant *(Chem.Eng.)*. Smaller version of projected industrial plant, used to gain experience and data for the design and operation of the final plant.

pilot valve *(Eng.)*. *Relay valve*. A small balanced valve, operated by a governor or by hand, which controls a supply of oil under pressure to the piston of a **servomotor** or relay connected to a large control valve, which it is desired to operate.

pilot voltmeter *(Elec.Eng.)*. A voltmeter used in a power station or substation to indicate the voltage at the remote end of a feeder to which it is connected by means of a pilot.

pilot wave *(Telecomm.)*. A carrier oscillatory current or voltage which is amplified in a high-efficiency amplifier independently of the side-frequencies, which are added subsequently.

pilot wire *(Elec.Eng.)*. See **pilot**.

pilot wire *(Telecomm.)*. In a multicore transmission cable, a wire which is solely concerned with detecting deterioration of the main insulation of the cable.

pilot-wire regulator *(Telecomm.)*. An automatic device in transmission circuits, e.g. for compensating changes in transmission arising from temperature variations.

pimelic acid *(Chem.)*. $HOOC.(CH_2)_5.COOH$, a saturated dibasic acid of the oxalic acid series, crystals, mp 105°C.

pi meson, π-meson *(Phys.)*. See **meson**.

pi-mode *(Electronics)*. Operation of a multicavity magnetron, whereby voltages on adjacent segments of the anode differ by π radians (half-cycle).

pimpling *(Nuc.Eng.)*. Small swellings on the surface of fuel during burn up.

pin *(Bot.)*. The long-styled form of such heterostyled flowers as the primrose, *Primula vulgaris*, with the pin-head-like stigma conspicuous at the top of the corolla tube. Cf. **thrum**. See **heterostyly**.

pin *(Build.)*. (1) A small wooden peg or nail. (2) The male part of a dovetail joint.

PIN *(Comp.)*. See **personal identification number**.

pin *(Eng.)*. A cylinder or tube, often heat-treated, used to connect members in a structure or machine, usually when freedom of angular movement at the joint is required.

pin *(Nuc.Eng.)*. *Fuel pin*. Very slender fuel cans used, for example, in fast reactors or, in a group, to form certain types of fuel element, e.g. in the *Advanced Gas-cooled Reactor*. See **fuel element**.

pinacocytes *(Zool.)*. The flattened epithelial cells forming the outer part of the dermal layer in Sponges.

pinacoid *(Crystal.)*. An open crystal form which includes two precisely parallel faces.

pinacol *(Chem.)*. $Me_2C(OH)\text{-}C(OH)Me_2$. *Tetramethylethan 1,2-diol*, crystallizes with $6(H_2O)$. The anhydrous substance has a mp 38°C, bp 172°C, and is obtained by the reduction and condensation of acetone by the action of metallic sodium. Pinacol forms *pinacolone* by the elimination of water and intramolecular transformation in the presence of dilute acids. It is the simplest member of a series of tetra alkyl glycols known as *pinacols*. Also *pinacone*.

pinacolone *(Chem.)*. $CH_3.CO.C(CH_3)_3$. *3,3-dimethylbutan-2-one*, produced by the action of dilute sulphuric acid upon *pinacol*. A colourless liquid, bp 106°C, rel.d. 0.800 at 16°C.

pin barrel *(Horol.)*. A cylindrical piece on the periphery of which are short vertical pins for lifting the hammers in a chiming clock, or for lifting the comb in a musical clock.

pincers *(Zool.)*. Claws adapted for grasping; as chelae, chelicerae.

pinch *(Electronics)*. An airtight glass seal through which pass the electrode connections in a thermionic valve or electron tube.

pinchbeck alloy *(Eng.)*. Red brass, 6–12% zinc.

pinch effect *(Nuc.Eng.)*. In a plasma carrying a large current, the constriction arising from the interaction of the current with its own magnetic field, just as two wires each carrying a current in the same direction experience an attractive force. The principle is used in fusion machines to confine the plasma.

pinch-off *(Electronics)*. Cut-off of the channel current by the gate signal in a field-effect transistor.

pinch-off voltage *(Electronics)*. In a field-effect transistor, that at which the current flow between source and drain ceases because the **channel** between these electrodes is completely depleted. For enhancement-mode FETs, using an *n*-type channel, the pinch-off voltage is positive; for depletion-mode FETs with a *p*-type channel, it is negative.

pincushion distortion *(Image Tech.,Phys.)*. *Curvilinear distortion* of an optical or electronic image in which horizontal and vertical straight lines appear concave, bowed inwards. Also known as *pillow distortion* or *negative distortion*. Cf. **barrel distortion**.

pine *(For.)*. See **pitch-, Scots-**.

pine *(Vet.)*. *Bush sickness, enzootic marasmus, vinquish*. A disease of sheep and cattle caused by low levels of cobalt in the soil leading to a deficiency in the diet; characterized by debility, emaciation, and anaemia.

pineal apparatus *(Zool.)*. In some Vertebrates, two median outgrowths from the roof of the diencephalon, one (originally the left) giving rise to the *parietal organ* and the other (originally the right) to the *pineal organ*.

pineal body (gland) *(Zool.)*. See **pineal organ**.

pinealectomy *(Med.)*. Surgical removal of the pineal gland.

pineal eye *(Zool.)*. An anatomically imprecise term referring to eyelike structures formed by one or other of the two outgrowths of the pineal apparatus. In Cyclostomata it derives from the pineal organ, but in Lizards it derives from the parietal organ.

pinealoma *(Med.)*. A tumour of the pineal gland.

pineal organ *(Zool.)*. One of the outgrowths of the pineal apparatus. In Cyclostomata it forms an eyelike functional photosensitive structure involved in a diurnal rhythm of colour change, and may be sensitive to light in some Fish. It persists in higher Vertebrates, and may function as an endocrine gland.

pinene *(Chem.)*. There are four terpenes known as pinenes. α-Pinene, $C_{10}H_{16}$, is the chief constituent of turpentine, eucalyptus, juniper oil etc.; bp 155°–156°C. It forms a hydrochloride, $C_{10}H_{17}Cl$, a white crystalline mass of camphorlike odour, mp 131°C. As it contains a double bond it forms a dibromide which can be converted into a glycol.

pine oil *(Min.Ext.)*. Commercial frothing agent widely used in flotation of ores. Distillate of wood, varying somewhat in chemical composition according to timber used and scale of heating.

pi-network *(Telecomm.)*. Section of circuit with one series arm preceded and followed by shunt arms to return leg of circuit. Also *π-network*.

ping *(Acous.)*. Brief pulse of medium frequency sound, reproduced from the subaqueous reflection of asdic ultrasonic signals. Its length in space will be equal to the product of ping duration time and the velocity of sound.

pinguecula *(Med.)*. A yellow, triangular patch on the conjunctiva covering the sclera.

pin hinge *(Build.)*. A form of butt hinge which has a removable pin connecting the two leaves.

pin holes *(Image Tech.)*. Small clear spots in a processed film caused by air-bells or dust particles preventing the action of the developer.

pinholes *(Paper)*. Small holes through the paper or coating caused by defects in manufacture.

pinholing *(Build.)*. A painting or varnishing defect in which the surface becomes pitted with small holes.

pin insulator *(Elec.Eng.)*. An insulator which is supported from the cross-arm by a pin.

pinion *(Eng.)*. The smaller of a pair of high-ratio toothed spur-wheels.

pinion *(Horol.)*. A small-toothed wheel, which normally has fewer than 12 teeth (leaves).

pinion leaf *(Horol.)*. A tooth of a pinion.

pinion wire *(Horol.)*. Steel or brass wire drawn to the section of a pinion.

pinite *(Min.)*. Hydrated silicate of aluminium and potassium which is usually amorphous. It is an alteration product of cordierite, spodumene, feldspar etc., approximate to muscovite in composition. See **pagodite**.

pin joint *(Eng.)*. A joint between members in a structural framework in which moments are not transmitted from one member to another.

pink-eye *(Med.)*. Acute mucopurulent conjunctivitis (the inflammation of the conjunctiva making it red) due to infection with various bacteria.

pink-eye *(Vet.)*. See **equine influenza**, **infectious ophthalmia**.

pinking *(Autos.)*. See **knocking**.

pink-noise generator *(Elec.Eng.)*. A random-noise generator providing a frequency spectrum with higher amplitudes at the low frequency end of the audible spectrum. See **white noise**.

pin mark *(Print.)*. A mark near the top of the type shank, made in casting.

pinna *(Bot.)*. A leaflet that is part of a pinnate leaf.

pinna *(Zool.)*. In Fish, a fin; in Mammals, the outer ear; in Birds, a feather or wing.

pinnate *(Bot.)*. Having 4 or more regular divisions, lobes, veins etc. arranged in 2 rows along a common midrib, rachis or stalk, rather in the manner of the barbs of a feather, as in a pinnate leaf. Many ferns have leaves that are 2-, 3- or more pinnate.

pinnate *(Zool.)*. Feather-like; bearing lateral processes.

pinnatifid *(Bot.)*. A leaf blade, etc. cut about half-way into lobes in a **pinnate** fashion.

pinnatiped, pinniped *(Zool.)*. Having the digits of the feet united by flesh or membrane. Cf. **fissiped**.

pinning-in *(Build.)*. The operation of inserting small splinters of stone in the joints of coarse masonry.

pinnings *(Build.)*. A Scottish term for different varieties of stones introduced into a wall for decorative purposes.

pinnule *(Bot.)*. One of the lobes or divisions of a leaf that is 2- or more-*pinnate*.

pinocytosis *(Immun.)*. 'Cell drinking'. Ingestion by cells of vesicles containing fluid from the environment. There are two forms: *micropinocytosis*, whereby small droplets or microvesicles are pinched off from the cell membrane and interiorized, carrying with them any materials selectively adsorbed at the cell surface. This process is common to many cells. Pinocytosis of ligand-receptor complexes into clathrin-coated vesicles is an important mechanism whereby ligands can be carried into the interior; *macropinocytosis*, or the ingestion of large vesicles or vacuoles, similar to phagocytosis is carried out by cells of the mononuclear phagocyte system.

pin pallet escapement *(Horol.)*. An escapement in which the pallets are vertical pins of steel or jewels. The impulse is derived entirely from the teeth of the escape wheel. Used extensively for inexpensive watches, alarm clocks and small drum movements.

PIN, p-i-n diode *(Electronics)*. A semiconductor diode with a layer of intrinsic semiconductor material between the *p*-type and *n*-type material which would normally constitute the junction. Applications include HF and microwave switching or attenuation, voltage-controlled resistors, photodiodes.

pinpoint *(Aero.)*. An aircraft's ground position as fixed by direct observation. Cf. **fix**.

pin-punch *(Eng.)*. A cylindrical or cone shaped tool for removing small parallel or tapered pins connecting members in a structure or machine.

pint *(Genrl.)*. A unit of capacity or volume in the Imperial system. Equal to $\frac{1}{8}$ Imperial gallon or 0.5682 dm^3. In US equal to 0.473 dm^3 (liquid) or 0.551 dm^3(dry). Contains 20 fluid ounces (US 16 fl. oz.).

pinta *(Med.)*. *Caraate, mal de los pintos*. A contagious skin disease, characterized by patches of coloured pigmentation; probably due to infection with a spirochaete; occurs in tropical America.

pintle *(Eng.)*. (1) The pin of a hinge. (2) The king pin of a wagon. (3) An iron bolt on which a chassis turns. (4) One of the metal braces on which a rudder swings, supported by a *dumb-pintle* at its heel. (5) The plunger or needle of an oil-engine injection valve, opened by oil pressure on an annular face, and closed by a spring.

pin tongs *(Horol.)*. Small hand-vices with a split draw-in chuck.

pin-twisting *(Textiles)*. Formation of **false-twist** by passing continuous filament yarn through a rotating tube and around a hard-wearing ceramic pin.

pin wheel *(Horol.)*. A wheel having pins, fixed at right angles to the plane of the wheel, which lift the hammer of a striking clock.

pin wheel escapement *(Horol.)*. An escapement, used in turret clocks, in which semicircular or D-shaped pins standing at right angles to the plane of the wheel give impulse to the pallets. The action of the escapement is similar to that of the dead-beat.

pion *(Phys.)*. See **meson**.

pioneer species *(Bot.)*. A species of which the members tend to be among the first to occupy bared ground; these plants are often intolerant of competition, and especially of shading, and may be crowded out as the community develops.

pip *(Radar)*. Significant deflection or intensification of the spot on a CRT giving a display for identification or calibration. Also *blip*.

pipe *(Acous.)*. Simple form of acoustic resonator. See **pipe resonance**. Examples: *organ pipe*, *Rijke tube*.

pipe *(Foundry)*. The cavity formed in an ingot or a feed due to the contraction of the metal on cooling.

pipe *(Horol.)*. A tubular boss of extension.

pipeclay *(Geol.)*. A white clay, nearly pure and free from iron, used in the pottery industry.

pipe coupling *(Build.)*. A short collar with female threads at both ends for joining screwed lengths of pipe.

pipe factor *(Min.Ext.)*. Compensating factor used when samples are taken from casings which go into or through running sands or gravels, so that the amount of material raised from the section traversed by the drill does not correspond with the volume enclosed by the pipe.

pipe fitting *(Eng.)*. Any piece of the wide variety of pipe connecting-pieces used to make turns, junctions, and reductions in piping systems.

pipelining *(Comp.)*. A form of **parallel processing** in which the processing of a number of instructions is overlapped, enabling streams of instructions to be decoded and executed concurrently.

pipe moulding *(Foundry)*. The production of cast-iron pipes either by moulding in green sand, using split patterns, or by the process of **centrifugal casting**.

piperazine *(Chem.)*. Diethene-diamine, a cyclic compound forming colourless crystals, mp 104°C, bp 145°C. It is a strong base and has the property of forming salts with uric acid which are relatively easily soluble in water; it is therefore used in medicine.

pipe resonance *(Acous.)*. Acoustic resonance of a pipe when the length, allowing for an end-correction, is an integer number of half-wavelengths when it is open at both ends, and odd multiples of a quarter wavelength when it is closed at one end.

pipe resonator *(Acous.)*. An acoustic resonator, in the form of a pipe, which may be open at one or both ends, as in organ pipes, or entirely closed. Resonance arises from the stationary waves set up by a plane-progressive wave being reflected at the ends, open or closed.

piperidine *(Chem.)*. $C_5H_{11}N$, a heterocyclic reduction product of pyridine. It is a colourless liquid, of peculiar odour, bp 106°C. It is soluble in water and ethanol, has relatively strong basic properties, and the imino hydrogen is replaceable by alkyl or acyl radials.

piperine *(Chem.)*. An optically inactive alkaloid occurring in pepper, $C_5H_{10}N.C_{12}H_9O_3$, piperyl-piperidine, which crystallizes in prisms, mp 129°C.

piperitone *(Chem.)*. A terpene ketone. The laevorotatory isomer is found in eucalyptus oils, whilst the dextro isomer occurs in Japanese peppermint oil. Colourless oil with peppermint odour.

pipe roller *(Print.)*. See **idler**.

piperonal *(Chem.)*. $CH_2O_2{=}C_6H_3.CHO$, *methylene-protocatechuic aldehyde*, a phenolic aldehyde of very pleasant odour, used as a perfume under the name of *heliotropin*.

pipe sample *(Min.Ext.)*. One obtained by driving open-ended pipe into heap of material and withdrawing the core it collects.

pipe stopper *(Build.)*. An expanding form of drain plug for closing the outlet of drain pipes which are to be tested.

pipette *(Chem.)*. Dispenser of small measured quantities of liquid. Originally of glass with suitable graduation(s), now largely superseded by the *automatic pipette*, in which a spring loaded plunger sucks the known volume into a disposable tip. Depressing the plunge then dispenses the liquid. There are automatic multiple versions available for many routine purposes.

pipette method *(Powder Tech.)*. A sedimentation technique in which concentration changes within a sedimenting *suspension* are measured by pipetting samples out of the suspension and determining the concentration of the samples.

pipe wrench *(Build.)*. Turns a pipe or rod about its axis. Also *cylinder wrench*.

pip-pip tone *(Telecomm.)*. A voice-frequency signal comprising two pulses of tone, indicating that a particular switching stage has been reached.

piqué *(Textiles)*. Fabric with pronounced cord effects. In woven fabrics the woven cords run in the weft direction but in warp-knitted fabrics the cords are in the other direction.

Pira test bench *(Print.)*. A small compact laboratory designed by the Printing Division of the Research Association for the Paper, Board, Printing and Packaging Industries, equipped to test paper, ink and printing surfaces before printing and to diagnose troubles encountered during and after printing.

pirn *(Textiles)*. Small wooden bobbin which fits the loom shuttle and carries the weft thread.

piroplasmosis *(Vet.)*. See **babesiosis**.

piscivorous *(Zool.)*. Fish-eating.

pi-section filter *(Telecomm.)*. Unbalanced filter in which one series reactance is preceded and followed by shunt reactances.

pisé de terre *(Build.)*. A kind of cob wall used sometimes in cottage construction, the cob usually being moulded between forms.

pi shute *(Print.)*. Component by which **pi characters** travel to a hot-metal composing machine.

pisiform *(Zool.)*. Pea-shaped; as one of the carpal bones of Man.

pisolite *(Geol.)*. A type of limestone built of rounded bodies (*pisoliths*) similar to oöliths, but of less regular form and 2 mm or more in diameter.

pisolitic *(Geol.)*. A term descriptive of the structure of certain sedimentary rocks containing pisoliths (see **pisolite** above). Calcite-limestones, dolomitic limestones, laterites, iron-ores, and bauxites may be pisolitic.

pistacite *(Min.)*. Syn. for *epidote*.

pistil *(Bot.)*. Ovary, style and stigma; either of a single carpel in an apocarpous flower, or of the whole gynoecium in a syncarpous flower. adj. *pistillate*.

piston *(Acous.)*. A small disk-like element which vibrates with the frequency of the sound which it emits. Pistons can be driven externally by electrical forces or internally by self-excitation (musical instruments).

piston *(Elec.Eng.)*. Closely fitting sliding short-circuit in a waveguide.

piston *(Eng.)*. A cylindrical metal piece which reciprocates in a cylinder, either under fluid pressure, as in engines, or to displace or compress a fluid, as in pumps and compressors. Leakage is prevented by spring rings, leather packing, hat leather, etc.

piston attenuator *(Elec.Eng.)*. Attenuator employing a waveguide system beyond cut-off. The attenuation in dB is linearly proportional to length, but the minimum attenuation is high.

pistonphone *(Acous.)*. Device in which a rigid piston is vibrated. Used for calibrating microphones.

piston pin *(Autos.)*. See **gudgeon pin**.

piston ring *(Eng.)*. A ring, of rectangular section, fitted in a circumferential groove in a piston, and springing outward against the cylinder wall to prevent leakage. It is cut through at one point to increase its springiness and allow of fitting. See **junk ring**, **mitre-cut piston ring**, **scraper ring**.

piston rod *(Eng.)*. The rod connecting the piston of a reciprocating engine with the crosshead.

piston slap *(Autos.)*. The light knock caused by a worn or loose piston slapping against the cylinder wall when the connecting rod thrust is reversed.

piston valve *(Eng.)*. A steam-engine slide valve in which the sealing or sliding surfaces of the valve are formed by two short pistons attached to the valve rod, working over cylindrical port faces in the steam chest; commonly used on steam locomotives.

pit *(Bot.)*. A localized, thin area of a cell wall, typically where the primary wall is not covered by a secondary wall. Plasmodesmata may be present. A pit is usually one half of a pit-pair and the term is sometimes loosely used to mean a pit-pair. See also **bordered pit**, **simple pit**. Cf. **perforation**.

pit *(Min.Ext.)*. (1) A place whence minerals are dug. (2) The shaft of a mine. The *pit eye* is the bottom, whence daylight is visible; the *pit frame* is the superstructure carrying poppet head and sheaves. The *pit head* is the surface landing-stage.

pit cavity *(Bot.)*. The space within a pit, from pit membrane to cell lumen.

pitch *(Acous.)*. The subjective property of a simple or complex tone which enables the ear to allocate its position on a frequency scale. If the fundamental of a complex tone is absent the pitch of this fundamental is still recognized because of subjective difference tones amongst the partials. See **concert pitch**.

pitch *(Aero.)*. (1) The distance forward in a straight line travelled by an propeller in one revolution at zero slip; often colloquially though wrongly applied to the blade incidence. (2) Angular displacement along the lateral axis. See also **pitching**. (3) Spacing between evenly spaced items, e.g. rivets.

pitch *(Build.)*. Of a plane, the angle between the blade and the sole, normally 45°, but in moulding planes up to 55° and in some smoothing planes 50° (York pitch). In planes with reversed bevel it is less, e.g. block planes ca. 20° and low-angle planes ca. 12°.

pitch *(Chem.)*. A dark-coloured, fusible, more or less solid material containing bituminous or resinous substances, insoluble in water, soluble in several organic solvents. Usually obtained as the distillation residue of tars.

pitch *(Elec.Eng.)*. A term used in connection with electrical machines to denote the distance measured along the armature periphery between various parts. See **pole-**, **slot-**, **winding-**.

pitch *(Image Tech.)*. The distance between successive perforations (sprocket holes) on the edges of cinematograph film.

pitch *(Min.Ext.)*. Orientation of a linear element, e.g. mine tunnel, mineral lineation, *slickensides*, on an inclined surface, whereby the angle of pitch is measured between the inclined element and the horizontal.

pitch *(Nuc.Eng.)*. Distance between centres of adjacent fuel channels in a reactor.

pitch *(Paper)*. Residual resinous material that may be present in unbleached sulphite or mechanical pulps. A form of contrary that can blind machine wires or felts.

pitch *(Ships)*. See **pitching period**.

pitchblende *(Min.)*. The massive variety of uraninite. Radium was first discovered in this mineral. This and helium are due to the disintegration of uranium.

pitchboard *(Build.)*. A triangular board used as a template for setting out stairs, the sides of which correspond to the **rise**, the **going**, and the **pitch**.

pitch circle *(Eng.)*. (1) In a toothed wheel, an imaginary circle along which the tooth pitch is measured, and with respect to which tooth proportions are given. For two wheels in mesh, the pitch circles roll in contact. (2) Circle drawn, e.g. on a flange around whose circumference hole centres are positioned for drilling.

pitch control *(Aero.)*. The *collective* and *cyclic pitch* (i.e. blade incidence) *controls* of a helicopter's mainrotor(s).

pitch cylinder *(Eng.)*. Cylinder coaxial with a screw thread and which intersects the flanks of the thread symmetrically.

pitch diameter *(Eng.)*. The diameter of the pitch circle of a gear wheel.

pitch edge *(Print.)*. The leading edge of the paper as it enters the printing, or some other, machine.

pitched roof *(Build.,Civ.Eng.)*. A roof having a sloping surface or surfaces.

pitched work *(Build.)*. Stone facing work for the slopes of jetties, breakwaters etc., executed by pitching the stones into place with some regularity.

pitcher *(Bot.)*. An urn-shaped or vase-shaped modification of a leaf, or part of a leaf, developed by certain plants; it serves as a means of trapping insects and other small animals which are killed and digested.

pitcher *(Build.)*. Thick-edged mason's chisel for rough dressing stone.

pitch face *(Build.)*. A stone surface left with a rough finish produced by the hammer.

pitching *(Aero.,Ships)*. The angular motion of a ship or aircraft in a vertical plane about a lateral axis.

pitching *(Civ.Eng.)*. The foundation layer of well-rammed and consolidated broken stone upon which a road surfacing is built.

pitching moment *(Aero.)*. The component of the couple about the lateral axis, acting on an aircraft in flight.

pitching period *(Ships)*. The movement of a ship in waves about a transverse horizontal axis is known as *pitching* as the bow goes down and *scending* as the bow rises. The time taken for the complete movement, down and up, is the *pitching period*.

pitching-piece *(Build.)*. See **apron piece**.

pitching tool *(Build.)*. A chisel with a very blunt edge, used to knock off superfluous stone.

pitch line *(Eng.)*. The line along which the pitch of a rack is marked out, corresponding to the pitch circle of a spur wheel.

pitch line *(Print.)*. See **crash line**.

pitch of propeller *(Ships)*. The theoretical distance the propeller would advance through a solid in one revolution.

pitch pine *(For.)*. *Pinus rigida*, yielding very resinous softwood timber; also low-grade tar and pitch. Commonly used for heavy framing and for piles, as well as for internal joinery.

pitch setting *(Aero.)*. The blade angle of adjustable- or variable-pitch **propellers**.

pitchstone *(Geol.)*. A volcanic glass which has a pitch-like (resinous) lustre and contains crystallites and microlites. It is usually of acid to subacid composition, contains a notable amount of water (4% or more), and is usually intrusive.

pith *(Horol.)*. The pith of elderwood used in cleaning watches and clocks.

pith-ball electroscope *(Elec.Eng.)*. Primitive detector of charges, whereby two pith-balls, suspended by silk threads, attract or repel each other, according to inverse-square law.

pithed *(Zool.)*. Having the central nervous system (spinal cord and brain) destroyed.

pith medulla *(Bot.)*. Ground tissue in the centre of a shoot or root.

pith ray *(Bot.)*. See **medullary ray**.

pit membrane *(Bot.)*. That part of the middle lamella and primary wall that lies across the distal end of a pit cavity.

pitmen *(Min.Ext.)*. Men employed in shaft inspection and repair.

pit moulding *(Eng.)*. Process by which extremely large castings are moulded in a pit, with brick-lined sides and cinder-lined bottom carrying connecting vent pipes to floor level.

Pitot-static tube *(Aero.)*. Tube inserted nearly parallel to flow stream. It has two orifices, one facing flow and hence receiving total pressure, and the other registering the static pressure at the side. The pressure difference between the two orifices registers dynamic air pressure ($\frac{1}{2}\rho v^2$, where ρ is the air density and v the velocity). This is displayed on the air-speed indicator.

pitot tube *(Aero.)*. See **Pitot-static tube**.

pit-saw *(Build.)*. A large two-handled rip-saw used for cutting logs. Also called a *cleaving saw*.

pitted *(Bot.)*. Tracheids and vessel elements of **metaxylem** and **secondary xylem** which have a secondary wall interrupted by pits.

pitting *(Build.)*. See **blowing**.

pitting *(Eng.)*. (1) Corrosion of metal surfaces, as boiler plates, due to local chemical action. (2) A form of failure of gear teeth, due to imperfect lubrication under heavy tooth pressure.

pitting factor *(Eng.)*. In assessment of metal corrosion, the depth of penetration of the deepest pit divided by average loss of thickness as calculated from loss of weight.

pituitary gland *(Zool.)*. The major endocrine gland of Vertebrates, formed by the fusion of a downgrowth from the floor of the diencephalon (the *infundibulum*) and an upgrowth of ectoderm from the roof of the mouth (*hypophyseal pouch*). The former becomes the *neurohypophysis*, comprising the neural lobe and infundibulum, and the latter becomes the *adenohypophysis*, comprising the *pars intermedia*. The *pars intermedia* and the neural lobe together form the posterior lobe. The *adenohypophysis* produces several hormones affecting growth, adrenal cortex activity, thyroid activity, reproduction (*gonadotropic hormones*) and melanophore cells. The *neurohypophysis* secretes two hormones, **vasopressin** and **oxytocin**.

pityriasis *(Med.)*. A term common to various skin diseases in which branny scales appear.

pivot *(Horol.)*. The reduced end of an arbor or staff which runs in a hole, jewel, or screw. It may be parallel, shouldered, or conical.

pivot bridge *(Eng.)*. A form of swing bridge in which the vertical pivot is located at the middle of the length of the bridge.

pivoted brace *(Horol.)*. A form of hooking for the mainspring to the barrel. The brace or post is pivoted onto holes in the barrel and barrel cap, the end of the spring being looped round the centre portion of the brace.

pivoted detent *(Horol.)*. A detent which is carried on pivots, as distinct from a spring detent such as is used in English chronometers.

pivot factor *(Elec.Eng.)*. In an electrical indicating instrument, the (full-scale torque)/(mass of movement), a measure of freedom from error due to friction in the bearings.

pivot jaw *(Elec.Eng.)*. A fixed jaw to which the blade of a switch is pivoted.

pivot joint *(Zool.)*. An articulation permitting rotary movements only.

pixel *(Comp.,Image Tech.)*. The smallest element with controllable colour and brightness in a video display or in **computer graphics**; from *picture-element*.

pixillation *(Image Tech.)*. Video effect in which the whole picture is broken down into a comparatively small number of square elements; by association with **pixel**.

pK *(Chem.)*. Value which is the logarithm to the base 10 of the reciprocal of the dissociation constant of a weak electrolyte. A high pK value indicates the substance has a small value for the dissociation constant and is very weak electrolyte.

PL/1 *(Comp.)*. Programming language used for both scientific and business computing. *Programming Language 1*.

PLA *(Comp.)*. See **programmable logic array**.

placebo *(Med.)*. A pharmacologically inactive substance which is administered as a drug either in the treatment of psychological illness or in the course of drug trials.

place bricks *(Build.)*. See **grizzle bricks**.

placenta *(Bot.)*. (1) The part of the ovary of a flowering plant where the ovules form and remain attached while they mature as seeds. (2) Any mass of tissue to which sporangia or spores are attached.

placenta *(Zool.)*. In Eutheria, a flattened cakelike structure formed by the intimate union of the allantois and chorion with the uterine wall of the mother; it serves for the respiration and nutrition of the growing young. adjs. *placental, placentate, placentiferous, placentigerous*.

Placentalia *(Zool.)*. See **Eutheria**.

placentation *(Bot.,Zool.)*. In plants, the arrangement of the placentas in an ovary, and of the ovules on the placenta. In animals, the method of union of the foetal and maternal tissues in the placenta.

placenta vera *(Zool.)*. A deciduate placenta in which both maternal and foetal parts are thrown off at birth. Cf. **semiplacenta**.

placers, placer deposits *(Geol.)*. Superficial deposits, chiefly of fluviatile origin, rich in heavy ore minerals such as cassiterite, native gold, platinum, which have become concentrated in the course of time by long-continued disintegration and removal from the neighbourhood of the lighter associated minerals. See also **auriferous deposit**.

placode *(Zool.)*. Any platelike structure; in Vertebrate embryos an ectodermal thickening giving rise to an organ primordium.

Placodermi *(Zool.)*. A class of Gnathostomata comprising the earliest jawed Vertebrates, known from fossils from the Silurian to Permian. All had a heavy defensive armour of bony plates, and hyoid gill slits with no spiracle. The hyoid arch did not support the jaw. Most possessed paired fins.

placoid *(Zool.)*. Plate-shaped; as the scales and teeth of Selachii.

plage *(Astron.)*. Dark or bright areas on calcium or hydrogen spectroheliograms, identified as areas of cool gas or heated gas respectively on the sun's surface; cf. **flocculi**, which refers to small patches only.

plagio- *(Genrl.)*. Prefix from Gk. *plagios*, slanting, oblique.

plagioclase feldspars *(Min.)*. An isomorphous series of triclinic silicate minerals which consist of **albite** and **anorthite** combined in all proportions. They are essential constituents of the majority of igneous rocks. See also **oligoclase, andesine, bytownite, labradorite, anorthite**.

plagiotropism *(Bot.)*. A **tropism** in which a plant part becomes aligned at an angle to the source of the orientating stimulus. Cf. **orthotropism, diatropism**.

plague *(Med.)*. A disease of rodents due to infection with the *Yersinia pestis*, transmitted to Man by rat-fleas, epizootics in rats invariably preceding epidemics. In Man the disease is characterized by enlargement of lymphatic glands (*bubonic plague*), severe prostration, a tendency to septicaemia, and occasional involvement of the lungs.

plain conduit *(Elec.Eng.)*. See **plain steel conduit**.

plain coupler *(Elec.Eng.)*. A short length of tubing serving to connect the end of two adjacent pieces of plain steel conduit in line with each other in an electrical installation. Also called a *sleeve*.

plain fabric *(Textiles)*. (1) In a knitted plain fabric all the loops are of the same type and are linked together in the same manner. (2) In a woven plain fabric a weft thread passes *over* and *under* each succeeding warp thread whereas the next weft thread passes *under* and *over* the same warp threads. This is the simplest of all weaves.

plain flap *(Aero.)*. A wing flap in which the whole trailing edge (apart from the ailerons) is lowered so as to increase the camber. Also, occasionally, *camber flap*.

plain muscle *(Zool.)*. See **unstriated muscle**.

plain steel conduit *(Elec.Eng.)*. Conduit consisting of light-gauge steel tubing not having the ends screwed; used for containing the conductors in electrical installation. Cf. **screwed steel conduit**. Also called *plain conduit*.

plain tile *(Build.)*. The ordinary flat tile, with two nibs for hanging from the battens.

plaiting *(Textiles)*. See **cuttling**.

planar diode *(Electronics)*. One produced using the *planar process*, giving a diode with high forward conductance, low reverse leakage, fast recovery time, and low junction capacitance.

planar process *(Electronics)*. A silicon transistor and integrated circuit manufacturing technique in which an oxide layer, fractions of a micron thick, is grown on a silicon substrate. A series of etching and diffusion steps is then undertaken to produce the active regions and junctions inside the substrate.

planar transistor *(Electronics)*. One (on a silicon substrate) in which the active regions are produced by diffusion, the localized penetration of impurities being achieved by coating certain portions of the wafer surface with silicon dioxide.

planceer piece *(Build.)*. A horizontal timber to which the soffit boards of an overhanging eave are fastened.

planceer, plancier *(Build.)*. A soffit, especially the under surface of the corona in a cornice.

Planckian colour *(Phys.)*. The colour or wavelength-intensity distribution of the light emitted by a black body at a given temperature.

Planckian locus *(Phys.)*. Line of **chromaticity diagram** joining points with coordinates corresponding to black body radiators.

Planck's law *(Phys.)*. Basis of quantum theory, that the energy of electromagnetic waves is confined in indivisible packets or quanta, each of which has to be radiated or absorbed as a whole, the magnitude being proportional to frequency. If E is the value of the quantum expressed in energy units and ν is the frequency of the radiation, then $E = h\nu$, where h is known as *Planck's constant* and has dimensions of energy × time, i.e. action. Present accepted value is 6.626×10^{-34} J s. See **photon**.

Planck's radiation law *(Phys.)*. An expression for the distribution of energy in the spectrum of a black-body radiator:

$$E_\nu d\nu = \frac{8\pi h\nu^3}{c^3(e^{h\nu/kT} - 1)} d\nu$$

where E_ν is the energy density radiated at a temperature T within the narrow frequency range from ν to $\nu + d\nu$. h is Planck's constant, c the velocity of light, e the base of the natural logarithms and k Boltzmann's constant.

plançon *(For.)*. A log of hardwood timber roughly sawn or hewn to an octagonal shape, with a minimum of 10 in (250 mm) between opposite faces.

plane *(Build.)*. A wood-working tool used for the purpose of smoothing surfaces, reducing the size of wood and, in specialized forms, for grooving, rebating, and other purposes.

plane *(Maths.)*. A flat surface; one whose radii of curvature are infinite at all points.

plane baffle *(Acous.)*. Plane board, with a hole, at or near the centre, for mounting and loading a loudspeaker unit.

plane earth factor *(Phys.)*. Electromagnetic wave propagation, the ratio of the electric field strength which would result from propagation over an imperfectly conducting earth to that resulting from propagation over a perfectly conducting plane.

plane-iron *(Build.)*. The cutting part of a plane, which actually shapes the work.

plane of collimation *(Surv.)*. The imaginary surface swept out by the **line of collimation** of a levelling instrument, when its telescope is rotated about its vertical axis.

plane of polarization *(Phys.)*. The plane containing the incident and reflected light rays and the normal to the reflecting surface. The magnetic vector of plane-polarized light lies in this plane. The electric vector lies in the *plane of vibration* which is that containing the plane-polarized reflected ray and the normal to the plane of polarization. The description of plane-polarized light in terms of the plane of vibration is to be preferred as this specifies the plane of the electric vector.

plane of saturation *(Civ.Eng.)*. The natural level of the **ground water**.

plane of symmetry *(Crystal.)*. In a crystal, an imaginary plane on opposite sides of which faces, edges, or solid angles are found in similar positions. One half of the crystal is hence a mirror image of the other.

plane of symmetry *(Maths.)*. See **symmetry**.

plane polarization *(Phys.)*. When the vibrations of a transverse wave are confined to one direction, the wave is said to be plane-polarized. For electromagnetic waves the direction of the electric vector of a plane-polarized wave is the *plane of vibration*; the magnetic vector lies in a plane at right-angles to this. Light reflected at the *Brewster angle* is plane-polarized. Polarization of radio waves and microwaves occurs as a result of the way these waves are transmitted from aerials.

planer *(Print.)*. A flat piece of wood or rubber which is placed on a forme of type and tapped with a mallet to level the surface.

planer tools *(Eng.)*. Planing machine cutting-tools, similar to those used for turning, clamped vertically in a block pivoted in the **clapper box** on the head.

plane stock *(Build.)*. The body of a plane holding the plane-iron in position.

plane surveying *(Surv.)*. Surveying which makes no correction for curvature of earth's surface.

planet *(Astron.)*. The name given in antiquity to the seven heavenly bodies, including the sun and moon, which were thought to travel among the fixed stars. The term in now restricted to those bodies, including the Earth, which revolve in elliptic orbits about the sun; in the order of distance they are: Mercury, Venus, Earth, Mars, Jupiter, Saturn, Uranus, Neptune, and Pluto. The two planets, Mercury and Venus, which revolve within the Earth's orbit are designated *inferior planets*, the planets Mars to Pluto are *superior planets*. Planets reflect the sun's light and do not generate light and heat.

plane table *(Surv.)*. A drawing-board mounted on a tripod so that the board can be levelled and also rotated about a vertical axis and clamped in position. An alidade completes the essential parts of a plane table. It is set up at ends of a suitable baseline where required survey points can be seen from these ends. By intersecting sights a rough plan can then be produced.

planetarium *(Astron.)*. A building in which an optical device displays the apparent motions of the heavenly bodies on the interior of a dome which forms the ceiling of the auditorium.

planetary electron *(Phys.)*. See **Bohr theory**.

planetary gear *(Eng.)*. Any gear-wheel whose axis describes a circular path round that of another wheel, e.g. the bevel wheels carried by the crown wheel of **differential gear**.

planetary nebula *(Astron.)*. A shell of glowing gas surrounding an evolved star, from which it is ejected. There is no connection with planets: the name derives from the visual similarity at the telescope between the disc of such a nebula and that of a planet. They represent late stages in the evolution of stars 1–4 times more massive than the Sun. Some thousands are known in our Galaxy.

plane-tile *(Build.)*. See **crown-tile**.

planetoid *(Astron.)*. See **minor planet**.

planetology *(Geol.)*. The study of the composition, origin and distribution of matter in the planets of the solar system.

plane wave *(Phys.)*. One for which equiphase surfaces are planes.

planigraphy *(Radiol.)*. See **tomography**.

planimeter *(Eng.)*. Integrating instrument for measuring mechanically the area of a plane figure, e.g. an indicator diagram. A tracing point on an arm is moved round the closed curve, whose area is then given to scale by the revolutions of a small wheel supporting the arm.

planing bottom *(Aero.)*. The part of the under surface of a flying-boat hull which provides hydrodynamic lift.

planing machine *(Eng.)*. A machine for producing large flat surfaces. It consists of a gear-driven reciprocating work-table sliding on a heavy bed, the stationary tool being carried above it by a saddle, which can be traversed across a horizontal rail carried by uprights. See **clapper box**.

planisher *(Eng.)*. (1) Hammer or tool for planishing. (2) A **rolling mill**.

planishing *(Eng.)*. Giving a finish to metal surfaces by hammering.

plankton *(Ecol.)*. Animals and plants floating in the waters of seas, rivers, ponds, and lakes, as distinct from animals which are attached to, or crawl upon, the bottom; especially minute organisms and forms, possessing weak locomotor powers.

planning grid *(Arch.)*. Squared grid scaled in **modules** used in designing for modular construction.

plano-convex *(Image Tech.,Phys.)*. Said of a lens with one surface flat and the other curved.

planogamete *(Bot.,Zool.)*. A motile or wandering gamete; a zoogamete.

planographic process *(Print.)*. Process in which the printing image is on a level with the plate, which is specially treated to accept ink while the surrounding areas reject it. See **collotype**, **lithography**.

planospore *(Bot.)*. See **zoospore**.

planozygote *(Bot.)*. A motile zygote.

plan-position indicator *(Radar)*. Screen of a CRT with an intensity-modulated and persistent radial display, which rotates in synchronism with a highly directional antenna. The surrounding terrain is thus painted with relevant reflecting objects, such as ships, aircraft, and physical features. Abbrev. *PPI*. See **azimuth stabilized PPI**.

plant *(Bot.)*. A photosynthetic organism or one related to it. It will always include the seed plants, almost always the pteridophytes and bryophytes, usually the algae and the fungi and sometimes the bacteria also.

plant *(Eng.)*. (1) The machines, tools and other appliances requisite for carrying on a mechanical or constructional business; the term sometimes includes also the building and the site and, in the case of a railway, the rolling stock. (2) The permanent appliances needed for the equipment of an institution.

planta *(Zool.)*. The sole of the foot in land Vertebrates; the flat apex of a proleg in Insects. adj. *plantar*.

Plantae *(Bot.)*. The plant kingdom.

plantation *(Build.)*. A slate size, 330 × 280 mm, 13 × 11 in.

planted moulding *(Build.)*. A moulding cut out of a separate strip of wood of the required section and secured to the surface to be decorated.

Planté plate *(Elec.Eng.)*. See **formed plate**.

plantigrade *(Zool.)*. Walking on the soles of the feet, as Man. Cf. **digitigrade**, **unguligrade**.

planting *(Build.)*. The operation of forming a **plant moulding**.

plant load factor *(Elec.Eng.)*. The ratio of the total number of kWh supplied by a generator or generating station to the total number of kWh which would have been supplied if the generator or generating station had been operated continuously at its maximum continuous rating.

plant pathology *(Bot.)*. See **phytopathology**.

plantula *(Zool.)*. A larval form of some Invertebrates, especially Coelenterata; it consists of an outer layer of ciliated ectoderm and an inner mass of endoderm cells.

plaque *(Biol.)*. Areas of cell destruction in a bacterial colony grown on a solid medium, or tissue culture monolayer preparation due to infection with a virus.

plaque *(Med.)*. A layer of amorphous material adhering to the surfaces of teeth.

plaque *(Min.Ext.)*. White-enamelled saucer-shaped disk used in spot checking of products made during ore treatment. It has taken the place of the old vanning shovel. A sample is gently manipulated on it with added water, to separate the light from the heavy constituents.

plashing *(Build.)*. The process of intertwining branches in forming hurdles, etc.

plasm *(Biol.)*. Protoplasm, especially in compound terms, as *germ plasm*.

plasma *(Electronics)*. Synonym for the positive column in a gas discharge.

plasma *(Med.)*. The watery fluid containing salts, protein and other organic compounds, in which the cells of the blood are suspended. When blood coagulates it loses certain constituents (e.g. fibrinogen and cells) and becomes serum.

plasma *(Min.)*. A bright-green translucent variety of cryptocrystalline silica (*chalcedony*). It is used as a semiprecious gem.

plasma *(Phys.)*. Ionized gaseous discharge in which there is no resultant charge, the number of positive and negative ions being equal, in addition to un-ionized molecules or atoms.

plasma-arc cutting *(Eng.)*. Cutting metal at temperatures approaching 35 000°C by means of a gas stream heated by a tungsten arc to such a high temperature that it becomes ionized and acts as a conductor of electricity. Heat is not obtained by a chemical reaction; the process can therefore be used to cut any metal.

plasma cell *(Immun.)*. Name given to the end stage of differentiation of B lymphocytes into cells wholly devoted to synthesis and secretion of immunoglobulins. They have a very highly developed endoplasmic reticulum and a prominent Golgi apparatus, and in stained preparations are recognizable by a basophilic cytoplasm with a juxta-nuclear 'vacuole', and an eccentrically placed nucleus with a 'clock face' appearance. Plasma cells are not known to revert to resting B-lymphocytes, but rather to become exhausted and die. They are prominent in sites of intensive antibody synthesis. Plasma cell tumours occur and are termed *plasmacytomas* or *myelomas*.

plasmacytoma *(Immun.,Med.)*. *Myeloma*. A tumour of plasma cells, which is often preferentially localized in the bone marrow and where it produces typical erosion of the local bone (hence the term *myeloma*). Plasmacytomas continue to secrete an immunoglobulin product, although this may sometimes have sections of the normal amino acid sequence missing. Such tumours arise spontaneously but rarely in several species, but in certain strains of mice they can be caused to appear regularly following intraperitoneal administration of mineral oil. Mouse myelomas are widely used to study immunoglobulin synthesis and gene structure, and for the production of **hybridomas**.

plasma heating *(Nuc.Eng.)*. In fusion research, plasmas may be heated by ohmic heating, compression by magnetic fields, injection of high energy neutral atoms, and by **cyclotron resonance heating**.

plasmalemma *(Biol.)*. The boundary membrane of the cell which regulates the passage of molecules between the cell and its surroundings. The plant *cell wall* is outside the plasmalemma. *Plasma membrane* is the commoner term for animal cells.

plasmalogen *(Biol.)*. Phosphatide in which a hydrocarbon chain is bound to a glycerol carbon by an unsaturated ether bond rather than an ester link.

plasma membrane *(Biol.)*. The bounding membrane of cells which controls the entry of molecules and the interaction of cells with their environment. Like most cell membranes it consists of a lipid bilayer traversed by proteins. *Plasmalemma* is the commoner term in botany.

plasma-, plasmo-, -plasm *(Genrl.)*. Prefix and suffix from Gk. *plasma*, gen. *plasmatos*, anything moulded.

plasma temperature *(Nuc.Eng.)*. Temperature expressed in degrees K (thermodynamic temperature) or electron volts (kinetic temperature). 1 KeV = 10 000 K.

plasma torch *(Phys.)*. One in which solids, liquids or gases are forced through an arc within a water-cooled tube, with consequent ionization; de-ionization on impact results in very high temperatures. Used for cutting and depositing carbides.

plasmid *(Biol.)*. A genetic element containing nucleic acid and able to replicate independently of its host's chromosome. Often carries genes determining **antibiotic resistance**. Much used in recombinant DNA procedures.

plasmin *(Med.)*. A substance in blood capable of destroying fibrin as it is formed.

plasminogen *(Med.)*. The precursor of plasmin in the blood.

plasmocyte *(Zool.)*. See **leucocyte**.

plasmodesma *(Bot.)*. A fine tube of protoplasm which connects the protoplasts of two adjacent cells through the intervening wall. See also **primary pit-field, symplast**.

plasmodium *(Bot.)*. A multinucleate mass of naked (wall-less) protoplasm, which moves in an amoeboid fashion and constitutes the thallus as in the Myxomycetes. Cf. **pseudoplasmodium**.

plasmodium *(Zool.)*. A syncytium formed by the union of uninucleate individuals without fusion of their nuclei. adj. *plasmodial*.

plasmogamy *(Biol.)*. Fusion of cytoplasm as distinct from fusion of nucleoplasm; *plastogamy*.

plasmogamy *(Bot.)*. The fusion of protoplasts, e.g. of gametes in a sexual reproductive cycle. In most organisms it is followed more or less immediately by karyogamy (fusion of nuclei); in some fungi it may result in a **heterokaryon**. See also **dikaryophase**.

plasmoid *(Phys.)*. Any individual section of a plasma with a characteristic shape.

plasmolysis *(Bot.)*. Process in which the protoplast of a plant cell shrinks away from the wall following water loss due to exposure to a solution of higher osmotic pressure, the wall being permeable to the solute but the plasmalemma not. Cf. **cytorrhysis**.

plasmoma *(Med.)*. See **plasmacytoma**.

plaster *(Build.)*. A general name for plastic substances which are used for coating wall surfaces, and which set hard after application. See also **acoustic plaster**.

plaster board *(Build.)*. A building board made of plaster with paper facings, used as a base for plaster or providing a finish of its own.

plasterer's putty *(Build.)*. A preparation similar to **fine stuff** made by dissolving pure lime in water and passing it through a fine sieve.

plaster mould casting *(Eng.)*. Small, precision parts of non-ferrous alloys are cast in plaster moulds which are destroyed when the casting is removed.

plaster of Paris *(Chem.)*. Partly dehydrated gypsum, $2CaSO_4$, H_2O (hemihydrate). When mixed with water, it evolves heat and quickly solidifies, expanding slightly; used for making casts.

plaster slab *(Build.)*. A block, frequently perforated, made from plaster of Paris and coarse sand; used in the construction of partitions.

plastic *(Bot.)*. A genotype having a phenotype that varies markedly with conditions.

plastic bronze *(Eng.)*. Bronze containing a high proportion of lead; used for bearings. Composition; 72–84% copper, 5–10% tin, and 8–20% lead plus zinc, nickel and phosphorus.

plastic clay *(Build.)*. See **foul clay**.

plastic deformation *(Eng.)*. Permanent change in the shape of a piece of metal, or in the constituent crystals, brought about by the application of mechanical force.

plastic deformation *(Geol.)*. The permanent deformation of a rock or mineral following the application of stress.

plasticity *(Eng.)*. A property of certain materials by which the deformation due to a stress is largely retained after removal of the stress.

plasticizers *(Chem.)*. High-boiling liquids used as ingredients in lacquers and certain plastics, e.g. PVC; they do not evaporate but preserve the flexibility and adhesive power of the cellulose lacquer films or the flexibility of plastic sheet and film. Well-known plasticizers are triphenyl phosphate, tricresyl phosphate, high-boiling glycol esters etc.

plastic moulding *(Eng.)*. A process for manufacturing articles of plastic materials, using either injection moulding machines for the rapid production of small articles or hydraulic presses for compression moulding of relatively large articles.

plastic paint *(Build.)*. Thick-texture paint which can be worked to a patterned finish.

plastic rail-bond *(Elec.Eng.)*. A rail-bond made by inserting plastic conducting material between the rail itself and the fishplate.

plastics *(Genrl.)*. A generic name for certain organic substances, mostly synthetic (see **synthetic resins**) or semi-synthetic (casein and cellulose derivatives) condensation or polymerization products, also for certain natural substances (shellac, bitumen, but excluding natural rubber), which under heat and pressure become plastic, and can then be shaped or cast in moulds, extruded as rod, tube etc., or used in the formation of laminated products, paints, lacquers, glues etc. Plastics are **thermoplastic** or **thermosetting**. Adaptability, uniformity of composition, lightness and good electrical properties make plastic substances of wide application, though relatively low resistance to heat, strain and weather are, in general, limiting factors of consequence.

plastic sulphur *(Chem.)*. Formed when sulphur is distilled into water. Unstable and changes to the rhombic form.

plastic surgery *(Med.)*. That branch of surgery which deals with the repair and restoration of damaged or lost parts of the body.

plastid *(Bot.)*. One of a class of cytoplasmic organelles in plants and eukaryotic algae, comprising the chloroplasts and related organelles, surrounded by 2 membranes and containing DNA, e.g. *chloroplast, etioplast, leucoplast, amyloplast, chromoplast.*

plastochron, plastochrone *(Bot.)*. The interval of time between the appearance of successive leaf primordia at the shoot apex, or between other similar successive events.

plastocyanin *(Bot.)*. Copper-containing protein, involved in the transfer of electrons from photosystem II to photosystem I in photosynthesis.

plastoquinone *(Bot.)*. Terpenoid involved in the transfer of electrons from photosystem II to photosystem I in photosynthesis.

plastron *(Med.)*. The sternum and the costal cartilages.

plastron *(Zool.)*. (1) The ventral part of the bony exoskeleton in Chelonia; any similar structure. (2) In some aquatic Insects (e.g. Coleoptera and Hemiptera) a thin air film over certain parts of the body held by minute hydrofuge hairs. This serves as a physical gill rather than as a store of air. adj. *plastral.*

platband *(Build.)*. (1) An **impost**. (2) A flat projecting moulding, which projects from the general wall surface by an amount less than its own breadth. (3) A door or window lintel.

plate *(Build.)*. Usually *wall plate*, the top horizontal timber of a wall, supporting parts of the structure. Also *platt.*

plate *(Chem.Eng.)*. In an industrial fractionating column, a tray over which the liquid flows, with nozzles or bubble caps so that the vapour can bubble though the liquid. See **Murphree efficiency, still.**

plate *(Electronics)*. US term for **anode**.

plate *(Eng.)*. A large flat body of steel, thicker than sheet, which is produced by the working of ingots, billets or slabs, in a rolling mill.

plate *(Geol.)*. The rigid structures of the lithosphere, of about continental size, consisting of the crust and the upper mantle, floating on the viscous lower mantle. See **plate tectonics.**

plate *(Image Tech.)*. General term for the picture, which may be a still transparency or a motion picture print, used as the background in composite photography.

plate *(Phys.)*. (1) Each of the two extended conducting electrodes which, with a a dielectric between, constitutes a capacitor. (2) US *anode* in valves.

plate *(Print.)*. (1) Any original and duplicate for letterpress printing; also plates for lithographing, and the several kinds of intaglio plate. (2) An illustration, especially one that is printed separately from the text which it illustrates.

plate *(Zool.)*. See **plax**.

plate amalgamation *(Min.Ext.)*. Trapping of metallic gold on an inclined plate made of copper or an alloy, which has been coated with a pasty film of mercury. Method largely superseded by use of **strake**.

plateau *(Nuc.Eng.)*. See under **Geiger region**.

plateau-basalts *(Geol.)*. Basic lavas of basaltic composition resulting from fissure eruptions and occurring as thin, widespread flows, forming extensive plateaux (e.g. the Deccan in India).

plateau eruptions *(Geol.)*. Volcanic eruptions by which extensive lava-flows are spread in successive sheets over a wide area and eventually build a plateau; as in Idaho. See **fissure eruption**.

plateau gravel *(Geol.)*. Deposits of sandy gravel occurring on hill-tops and plateaux at heights above those normally occupied by river-terrace gravels. Originally deposited as continuous sheets, plateau gravel has been raised by earth movements to its present level and deeply dissected. Of Pliocene or early Pleistocene age in the main.

plateau length *(Nuc.Eng.)*. The voltage range which corresponds to the plateau of a Geiger counter tube.

plateau slope *(Nuc.Eng.)*. The ratio of the percentage change in count rate for a constant source, to the change of operating voltage. Measured for a median voltage corresponding to the centre of the Geiger plateau. Often expressed as percentage change in count rate for 100 volt change in potential.

plate cam *(Eng.)*. A flat, open cam for sliding movement used, for example, in automatics.

plate clutch *(Eng.)*. See **disk clutch**.

plate columns *(Chem.Eng.)*. Distillation or absorption columns which contain plates of various types, spaced at regular intervals with nothing between the plates.

plate cylinder *(Print.)*. The cylinder on a printing press to which the printing plate is attached.

plated carbon *(Elec.Eng.)*. An arc-lamp carbon upon which a layer of copper has been deposited by electroplating, in order to improve its conductivity and ensure good contact with the holder.

plate dissipation *(Electronics)*. See **anode dissipation**.

plate exchanger *(Chem.Eng.)*. Heat exchanger comprising either a series of alternating flat and ribbed plates forming the flow channel, or flat plates which are hollow with internal flow passages immersed in or forming part of the wall of the vessels.

plate frame *(Elec.Eng.)*. The nickel-plated framework for supporting the perforated steel tubes of the electrode of a nickel-iron accumulator.

plate gauge *(Eng.)*. A limit gauge or single external gauge formed by cutting slots of the required gauge width in a steel plate, the surfaces of which are hardened. See **limit gauge**.

plate girder *(Eng.)*. A built-up steel girder consisting of a single web-plate along each edge of which is welded a plate to form the flanges.

plate glass *(Glass)*. Glass of superior quality, originally cast on an iron bed and rolled into sheet form, and afterwards ground and polished. Modern methods have largely superseded this, save in special circumstances. Also *polished plate*. See **float glass**.

plate group *(Elec.Eng.)*. The complete unit, consisting of an accumulator plate or plates, terminal bar and terminal lug, forming the electrode of an accumulator cell. Also called a *plate section*.

platelet activating factor *(Immun.)*. *PAF*. A lipid released by various cells, including basophil leucocytes and monocytes, in the presence of antigen. Induces platelet aggregation and degranulation. Activity is transient since PAF is inactivated by phospholipase A which is also released.

plate link chain *(Eng.)*. A chain comprising pairs of flat links connected by pins.

plate-lug *(Elec.Eng.)*. A projection on an accumulator plate used for connecting it to a terminal bar.

plate modulation *(Electronics)*. See **anode modulation**.

plate moulding *(Foundry)*. A method of mounting the halves of a split pattern on opposite sides of a wood or metal plate, placed between the cope and drag, thus eliminating the making of the joint faces.

platen *(Eng.)*. The work table of a machine tool, usually slotted for clamping bolts.

plate proof *(Print.)*. A proof taken from a plate as distinct from one taken direct from type.

plate rectifier *(Elec.Eng.)*. One of large area for large output currents, e.g. for electrolytic bath supply or electric traction.

plates *(Horol.)*. The circular or rectangular plates of brass which form the framework of a watch or clock and which are drilled to receive the pivots of the train etc.

plate section *(Elec.Eng.)*. See **plate group**.

plate slap *(Print.)*. The noise resulting from the impact of loose or badly fitting plates against the cylinder.

plate support *(Elec.Eng.)*. A support from which the plates of an accumulator are suspended or upon which they rest.

plate tectonics *(Geol.)*. The interpretation of the Earth's structures and processes (including oceanic trenches, midocean ridges, major tear faults, mountain building, earthquake zones and volcanic belts) in terms of the movements of large plates of lithosphere acting as rigid slabs floating on a viscous mantle.

platform *(Space)*. Term sometimes applied to a spacecraft used as a base for experiments in space research, usually unmanned.

platform escapement *(Horol.)*. An escapement mounted on an independent plate.

platform gantry *(Build.)*. A gantry formed to support a platform on which is erected a scaffold used for the handling of materials.

platforming *(Chem.Eng.)*. Process for reforming low-grade into high-grade petrol, using a platinum catalyst.

platform tree *(Min.Ext.)*. *Christmas tree* with all the necessary valves for controlling the flow of oil from a producing platform.

platinammines *(Chem.)*. Compounds of platinum and ammonia of the form $Pt(NH_3X)_4$.

platinates (IV) *(Chem.)*. See **platinic hydroxide**.

platinectomy *(Med.)*. The operative removal of the stapedial footplate in the middle ear.

plating sequence *(Print.)*. The order in which printing plates are secured to the press.

plating up *(Print.)*. The securing of plates to the printing cylinders.

platinic hydroxide *(Chem.)*. *Platinum (IV) hydroxide*. $Pt(OH)_4$. Dissolves in acids to form platinic salts and in bases a series of salts called *platinates* (IV). A type of compound formed by the other members of the platinum group of metals.

platinic oxide *(Chem.)*. *Platinum (IV) oxide*. PtO_2. Dark grey powder formed when platinic (platinum (IV)) hydroxide is heated. Also *platinum dioxide*.

platinite *(Eng.)*. Alloy containing iron 54–58%, and nickel 42–46%, with a trace of carbon. Has the same coefficient of expansion as platinum, and is used to replace it in some light bulbs and measuring standards.

platinized asbestos *(Chem.)*. Asbestos permeated with finely divided platinum. Used as a catalyst.

platinoid *(Eng.)*. Alloy containing copper 62%, zinc 22%, nickel 15%. Has high electrical resistance and is used for resistances and thermocouples.

platinous hydroxide *(Chem.)*. *Platinum (II) hydroxide*. $Pt(OH)_2$. Soluble in the haloid acids (hydrochloric acid etc.), forming platinous (platinum (II)) salts.

platinous oxide *(Chem.)*. PtO. Formed when platinous hydroxide is gently heated.

platinum *(Chem.)*. A metallic element, symbol Pt, at. no. 78, r.a.m. 195.09, rel.d. at 20°C 21.45, electrical resistivity at 20°C 9.97×10^{-8} ohm metres, mp 1773.5°C, bp 3910°C, Brinell hardness 47. Platinum is the most important of a group of six closely related rare metals, the others being osmium, iridium, palladium, rhodium and ruthenium. It is heavy, soft and ductile, immune to attack by most chemical reagents and to oxidation at high temperatures. Used for making jewellery, special scientific apparatus, electrical contacts for high temperatures and for electrodes subjected to possible chemical attack. Also used as a basic metal for resistance thermometry over a wide temperature range. Native platinum is usually alloyed with iron, iridium, rhodium, palladium or osmium, and crystallizes in the cubic system.

platinum black *(Chem.)*. Platinum precipitated from a solution of the (IV) chloride by reducing agents. A velvety-black powder. Uses: catalyst and gas absorber, e.g. in the **hydrogen electrode**.

platinum dioxide *(Chem.)*. *Platinum (IV) oxide*. See **platinic oxide**.

platinum tetrachloride *(Chem.)*. *Platinum (IV) chloride*, $PtCl_4$. Formed by dissolving platinum in aqua regia. Similar chlorides are formed with the other platinum metals.

platinum thermometer *(Elec.Eng.)*. See **resistance thermometer**.

PLATO *(Comp.)*. See **computer assisted instruction**.

platter *(Comp.)*. One of the circular plates in a hard-disk drive.

platter *(Image Tech.)*. Large horizontal turntable supporting a very long reel of film for projection.

platting *(Build.)*. The top course of a brick clamp.

platycephalic, platycephalus *(Med.)*. Having a flattened or broad head, with a breadth-height index of less than 70.

platydactyl *(Zool.)*. Having the tips of the digits flattened.

Platyhelminthes *(Zool.)*. A phylum of bilaterally symmetrical, triploblastic Metazoa; usually dorsoventrally flattened; the space between the gut and the integument is occluded by parenchyma; the execretory system consists of ramified canals containing flame-cells; there is no anus, coelom, or haemocoele; the genitalia are usually complex and hermaphrodite. Flat worms.

platysma *(Zool.)*. A broad sheet of dermal musculature in the neck region of Mammals.

platyspermic *(Bot.)*. Having seeds which are flattened in transverse section, as in the Cordaitales and the conifers. Cf. **radiospermic**.

plax *(Zool.)*. A flat platelike structure, as a lamella or scale.

play *(Behav.)*. A concept used by both students of animal behaviour and developmental psychologists without rigid definition. It occurs largely but not solely in the young of warm-blooded mammals and can involve almost any behaviour; play behaviours tend to occur in isolation from their normal function, are given voluntarily rather than in conditions of necessity, and often merge behaviours from different functional systems. Exaggerated movements often occur and may be preceded or accompanied by a signal that the activity is a playful one (e.g. play fights).

play *(Eng.)*. Limited movement between mating parts of a mechanism, due either initially to the type of fit and dimensional allowance specified or subsequently to wear.

playback *(Acous.)*. A recording technique in which an existing partial recording of a piece of music (e.g., the beats) is fed into an earphone or loudspeaker while other instruments, orchestra groups or soloists are recorded simultaneously.

playback equalizer *(Acous.)*. A resistance-capacitance network introduced into an interstage coupling so that all frequencies are reproduced with equal intensity in the recording of music.

play therapy *(Behav.)*. A form of child psychotherapy which uses play as the means of communication to reveal unconscious conflicts, and which encourages children to vent their feelings through symbolic play.

pleasure principle *(Behav.)*. In Freudian theory, the motive to seek immediate pleasure and gratification, without regard to consequence, which governs the **id**.

pleated-diaphragm loudspeaker *(Acous.)*. A loudspeaker in which the radiating element is a pleated diaphragm, the pleats being radial and the rim clamped. It is driven by a pin at the centre.

Plectomycetes *(Bot.)*. A class of fungi in the Ascomycotina in which the fruiting body (ascocarp) is a cleistothecium. Includes *Eurotium* and allied genera which are the perfect stages of many spp. of *Aspergillus* and *Penicillium*.

plectostele *(Bot.)*. Type of protostele in which the xylem and phloem are arranged in alternating plates across the stele.

Pleiades, The *(Astron.)*. The name given to the open cluster in the constellation *Taurus*, of which the seven principal stars, forming a well-known group visible to the naked eye, each have a separate name.

pleiomerous *(Bot.)*. Having a large number of parts.

pleiomorphic, pleomorphic, pliomorphic *(Bot.)*. Having more than one shape.

pleiotropy *(Biol.)*. The condition when one gene affects more than one character. adj. *pleiotropic*.

plei-, pleio-, pleo-, plio- *(Genrl.)*. Prefix from Gk. *pleiōn*, more.

Pleistocene Period *(Geol.)*. The period of geological time which followed the Tertiary, covering a time span of app. the last 2 million years. It is during this period that ice covered a large part of the northern hemisphere; hence it has been called the *Great Ice Age*.

Pleistogene *(Geol.)*. See **Quaternary**.

plenum chamber *(Aero.)*. A sealed chamber pressurized from an air intake. Centrifugal flow turbojets having double-entry impellers (see **double-entry compressor**) have to be mounted in plenum chambers to ensure even air pressure on both impeller faces.

plenum system *(Build.)*. An air-conditioning system in which the air propelled into the building is maintained at a higher pressure than the atmosphere. The conditioned air is usually admitted to rooms from 2.5–3 m above floor-level, while the vitiated air is extracted at floor-level on the same side of the room.

pleo- *(Genrl.)*. Prefix. See **plei-**.

pleocholia *(Med.)*. Excessive formation of bile pigment.

pleochroic haloes *(Min.)*. Dark-coloured zones around small inclusions of radioactive minerals which are found in certain crystals, notably biotite. The colour and pleochroism of the zones are stronger than those of the surrounding mineral, and result from radioactive emanations during the conversion of uranium or thorium into lead.

pleochroism *(Min.)*. The property of a mineral by which it exhibits different colours in different crystallographic direction on account of the selective absorption of transmitted light.

pleochromatic *(Biol.)*. Presenting different colours according to changes in the environment or with different physiological conditions.

pleocytosis *(Med.)*. An increase in the number of white-blood cells, especially in the cerebrospinal fluid.

pleomorphism, pleomorphous *(Zool.)*. Same as *polymorphism, polymorphic*.

pleonaste *(Min.)*. Oxide of magnesium, iron, and aluminium, with Mg:Fe from 3 to 1, crystallizing in the cubic system. It is a member of the spinel group and may be dark-green, brown, or black in colour. Also called *ceylonite*.

pleopod *(Zool.)*. In Arthropoda (especially Crustacea), an abdominal appendage adapted for swimming.

plerome *(Bot.)*. A **histogen**, the precursor of the stele.

plethora *(Med.)*. Fullness of the blood vessels; plethoric appearance, suffused, reddened face. Cf. *pulmonary plethora*, an increase in the blood vessels on a chest X-ray.

plethysmograph *(Med.)*. An apparatus for measuring variations in the size of bodily parts and in the flow of blood through them. See **electroarteriograph**.

pleura *(Zool.)*. The serous membrane lining the pulmonary cavity in Mammals and Birds.

pleurapophysis *(Zool.)*. A lateral vertebral process; usually applied to the true ribs.

pleurisy *(Med.)*. Inflammation of the pleura, which may be either dry or accompanied by effusion of fluid into the pleural cavity (*pleural effusion*).

pleurocarpous *(Bot.)*. A moss bearing the archegonia, and therefore the capsule and its stalk, on a short side-branch, not at the tip of a main stem or branch. Ant. *acrocarpous*.

pleurodont *(Zool.)*. Having the teeth fastened to the side of the bone which bears them, as in some Lizards.

pleurodynia *(Med.)*. Pleural pain with chest muscle involvement, often thought to be viral in origin.

pleurogenous *(Bot.)*. Borne on a lateral position.

pleurogenous *(Med.)*. Having origin in the pleura, e.g. *pleurogenous* cirrhosis of the lung. Also *pleurogenic*.

pleuron *(Zool.)*. In some Crustacea, a lateral expansion of the tergite; more generally, in Arthropoda. The lateral wall of a somite. pl. *pleura*. adj. *pleural*.

pleuropericarditis *(Med.)*. Concurrent inflammation of the pleura and of the pericardium.

pleuropneumonia *(Med.)*. Combined inflammation of the pleura and of the lung.

pleuropneumonia *(Vet.)*. A contagious disease of cattle due to infection by *Mycoplasma mycoides*; characterized by an exudative fibrinous pneumonia and pleurisy.

pleuropneumonia-like organisms *(Biol.)*. A group of organisms, closely resembling the pleuropneumonia organisms, isolated from the throat and vagina. See **Mycoplasmatales**. Abbrev. *PPLO*.

pleur-, pleuro- *(Genrl.)*. Prefix from Gk. *pleura*, side.

p-levels *(Phys.)*. See **principal series**.

plexitis *(Med.)*. Inflammation of the components of a nerve plexus.

plexus *(Zool.)*. A network; a mass of interwoven fibres, as a *nerve plexus*. adj. *plexiform*.

plica *(Zool.)*. A fold of tissue; a foldlike structure. adjs. *plicate, pliciform*.

plicate *(Bot.)*. Folded.

plimsoll mark *(Ships)*. See **load lines**.

Plinian eruption *(Geol.)*. A type of volcanic eruption characterized by repeated explosions.

plinth *(Build.)*. (1) The projecting course or courses at the base of a building. (2) The cuboidal base of a bookcase or pedestal. (3) The base of a bookcase, wardrobe, etc.

plinth block *(Build.)*. See **architrave block**.

plinth course *(Build.)*. A projecting course laid at the base of a wall.

plio- *(Genrl.)*. Prefix. See **plei-**.

Pliocene Period *(Geol.)*. The epoch which followed the Miocene and preceded the Pleiostocene.

pliomorphic *(Bot.)*. See **pleiomorphic**.

pliotron *(Electronics)*. Large hot-cathode vacuum tube for industrial use, with one or more grids.

ploidy *(Biol.)*. Pertaining to chromosome number, e.g. *haploid, diploid, polyploid*.

plotter *(Comp.)*. Output device which draws lines on paper. See **digital plotter, incremental plotter**.

plough carrier *(Elec.Eng.)*. The frame under a tram, which carries the plough used in the conduit system. The arrangement is such that the plough can slide laterally, so that it may follow any variations in the relative positions of the conduit and the track rails.

ploughed-and-tongued joint *(Build.)*. A joint formed between the square butting edges of two boards, each having a plough groove into which a common tongue is inserted.

plough, plow *(Build.)*. (1) A form of grooving or shaping plane which has an adjustable fence and is capable of being fitted with various irons. (2) To cut a groove.

plough, plow *(Elec.Eng.)*. See **plough carrier**.

plough, plow *(Print.)*. A hand tool for cutting the edges of books, now only occasionally used for handbound books.

plow *(Genrl.)*. US for *plough*.

plug *(Build.)*. A wooden or plastic piece driven into a hole cut in a wall or partition and finished off flush, so as to provide a material to which joinery or fittings may be nailed. In Scotland a *dook*.

plug *(Geol.)*. A vertical cylinder of solidified magma or pyroclastic material which represents the feeder pipe of a former volcano.

plug *(Nuc.Eng.)*. Piece of absorbing material used to close the aperture of a channel through a reactor core or other source of ionizing radiation.

plugboard *(Comp.)*. See **patchboard**.

plug centre bit *(Build.)*. A form of centre bit in which the projecting central point is replaced by a plug of metal, adapting the bit for use in holes already drilled.

plug cock *(Build.,Eng.)*. A simple valve in which the plug, usually conical, is drilled across its axis, allowing flow only when the drilled hole is in line with inlet and outlet.

plug gauge *(Eng.)*. A gauge, made in the form of a plug for testing the diameter of a hole; in a *plug limit gauge* two plugs are provided, a 'go' and a 'not go'. See **limit gauge**.

plugging *(Build.)*. The operation of drilling a hole in a wall or partition, and driving in a wall plug.

plugging *(Elec.Eng.)*. Braking an electric motor by re-arranging the connections so that it tends to run in the reverse direction.

plug-in *(Electronics)*. A single component, or sub-assembly, which has plug-in terminals so that all connections can be made simultaneously by pushing the unit into a socket; usually employed where rapid interchangeability of operating ranges or functions is desired.

plug-in unit *(Comp.)*. Any panel or component which can be inserted and interchanged in a computing system, esp. a *cross-connection panel*, which can be set up independently.

plug tap *(Eng.)*. (1) The final parallel tap required to finish an internal thread in a blind hole. Also *bottoming tap*, *third tap*. (2) A **plug cock**.

plug welding *(Eng.)*. Method of welding in which apertures in one part allow penetration through to the other part to be joined, the apertures to be filled by welding in the process.

plumae *(Zool.)*. Feathers having a stiff shaft and a firm vexillum, and usually possessing hamuli; they appear on the surface of the plumage and determine the contours of the body in addition to forming the remiges and rectrices. adjs. *plumate, plumous, plumose, plumigerous*.

plumb *(Build.,Civ.Eng.)*. Vertical.

plumb- *(Chem.)*. From the Latin, *plumbum*, lead; e.g. plumbic chloride ($PbCl_4$), plumbous chloride ($PbCl_2$).

plumbago *(Min.)*. See **graphite**.

plumb-bob *(Surv.)*. A small weight or 'bob' hanging at the end of a cord, used to centre survey instrument over signal mark. Also *plummet*.

plumber's solder *(Build.)*. A lead-tin alloy of varying ratio from 1:1 to 3:1 for different classes of work, the melting-points being always lower than that of lead itself.

Plumbicon *(Image Tech.)*. TN for an improved camera tube of the photo-conductive **Vidicon** type employing lead oxide layers. Also *Leddicon*.

plumbing *(Build.)*. (1) The craft of working lead for structural purposes, or for the installation of domestic water-supply systems, sanitary fittings, etc. (2) The operation of arranging vertically.

plumbing *(Telecomm.)*. Colloquialism for waveguides and their jointing in establishing microwave systems. Also for piped vacuum systems.

plumbing unit *(Build.)*. A prefabricated assembly of pipes and fittings, generally of storey height, and now commonly used in multistorey flats. In **system building**, it may also include the complete bathroom assembly.

plumbism *(Med.)*. Lead poisoning.

plumbites (plumbates (II)) *(Chem.)*. See **lead (II) hydroxide**.

plumb level *(Surv.)*. A **level tube** for showing plumb direction, with a small bubble at right angles to the main one. When the latter is held vertical the small bubble is central.

plumb line *(Surv.)*. A cord with a **plumb-bob** attached to the end.

plumb rule *(Build.)*. A narrow board used for determining verticals; it has at one end a point of suspension for a plumb-bob, which is free to swing in an egg-shaped hole at the other end of the board.

plume *(Bot.)*. A light, hairy or feathery appendage on a fruit or seed serving in wind dispersal.

plume *(Geol.)*. Ascending partly molten material from the mantle believed to be responsible for intraplate volcanism.

plume *(Meteor.)*. Snow blown over the ridge of a mountain.

plume *(Zool.)*. A feather; any featherlike structure.

plummer block *(Eng.)*. A journal bearing for line shafting, etc., consisting of a box-form casting holding the ball races, roller races or bearing brasses, split horizontally to take-up wear (brasses) or to facilitate assembly (races).

Plummer-Vinson syndrome *(Med.)*. *Patterson-Kelly syndrome*. The association of iron deficiency anaemia and difficulty in swallowing due to the development of webs of tissue at the top of the oesophagus.

plummet *(Surv.)*. See **plumb-bob**.

plumose *(Bot.,Zool.)*. Hairy; feathered.

plumulae *(Zool.)*. Feathers having a soft shaft and vane and lacking hamuli; in some cases the shaft is entirely lacking; they form the deep layer of the plumage. adjs. *plumulate, plumulaceous*.

plumule *(Bot.)*. Embryonic shoot above the cotyledon or cotyledons in a seed. See also **epicotyl**.

plumule *(Zool.)*. Down feather. Form covering of nestling, sometimes persisting in adult, between the contour feathers. Barbules and hooks little developed.

plunge angle *(Surv.)*. See **angle of depression**.

plunge-cut milling *(Eng.)*. Milling without transverse movement of the workpiece relative to the cutter, resulting in a groove or slot with a curved bottom.

plunge grinding *(Eng.)*. See **profile grinding**.

plunger *(Elec.Eng.)*. Device for altering length of a short-circuited coaxial line, taking the effective form of an annular disk. For very high frequencies there need not be a contact, because a small clearance gives sufficient capacitance for an effective short-circuit.

plunger *(Eng.)*. The ram or solid piston of a force-pump.

plunging fold *(Geol.)*. One whose axis is not horizontal. The angle between the axis and the horizontal is called the *plunge*.

plunging shot *(Surv.)*. Downward theodolite sight. To *plunge* is to transit the instrument.

pluriglandular *(Med.)*. Pertaining to, affected by, or affecting, several (ductless) glands.

plurilocular *(Bot.)*. A sporangium or ovary divided into several compartments by septa. Cf. **unilocular**.

plush *(Textiles)*. (1) Woven fabric with cut pile, longer and less dense than velvet, on one side. Warp pile is generally made by weaving two cloths together, with a pile warp common to both, which is afterwards cut. (2)

One type of weft-knitted fabric has the long loops on the back of the cloth, which is sometimes called knitted **terry**. In warp-knitted plush fabrics the loops may be cut or left uncut.

plus sign (+) *(Bot.)*. Symbol used before the generic or specific epithet to denote, respectively, an intergeneric or interspecific **graft chimaera**.

plus strain, (+) strain *(Bot.)*. One of two arbitrarily designated, mating types of a heterothallic species. See **heterothallism**.

Pluto *(Astron.)*. Ninth planet of the solar system discovered by Clyde Tombaugh on 18 February 1930 as a result of a systematic search. The orbit is very elliptical and inclined at 17° to the ecliptic. For a proportion of its 248-year orbit it is actually closer to the Sun than **Neptune**. Its diameter is about 3000 km (less than our Moon) with a relative density similar to water. Pluto has one moon, Charon, found in 1978. Possibly the two objects are escaped satellites of Neptune.

plutonic intrusions *(Geol.)*. A term applied to large intrusions which have cooled at great depth beneath the surface of the earth. Also *major intrusions*.

plutonites *(Geol.)*. All rocks occurring in plutonic intrusions. More precisely, igneous rocks of the coarse grain-size group.

plutonium *(Chem.)*. Element, symbol Pu, at. no. 94, product of radioactive decay of *neptunium*. Has many isotopes, the fissile isotope ^{239}Pu, produced from ^{238}U by neutron absorption in a reactor, being the most important for the production of nuclear power.

plutonium reactor *(Nuc.Eng.)*. One in which ^{239}Pu is used as the fuel.

pluviometry *(Meteor.)*. The study of precipitation, including its nature, distribution and techniques of measurement.

Plyglass *(Build.)*. A proprietary form of window or underceiling light covering. Composed of two sheets of glass enclosing glass fibres or plastics to diffuse entering light.

plywood *(Build.)*. A board consisting of a number of thin layers of wood glued together so that the grain of each layer is at right angles to that of its neighbour.

PM *(Build.)*. Abbrev. for *Purpose-Made*.

Pm *(Chem.)*. The symbol for promethium.

PM *(Eng.)*. Abbrev. for *Powder Metallurgy*.

PMBX *(Telecomm.)*. Abbrev. for *Private Manual Branch eXchange*.

PMC *(Build.)*. Abbrev. for *Plaster-Moulded Cornice*.

PMS *(Print.)*. See **Pantone Matching System**.

PMSF *(Biol.)*. See **phenylmethylsulphonyl fluoride**.

PMX *(Telecomm.)*. Abbrev. for *Private Manual eXchange*.

p-n boundary *(Electronics)*. The surface on the transition region between *p*-type and *n*-type semiconductor material, at which the donor and acceptor concentrations are equal.

pneumathode *(Bot.)*. A more of less open outlet of the ventilating system of a plant, usually some loosely packed cells on the surface of the plant; through it exchange of gases between the air and the interior of the plant is facilitated.

pneumatic *(Eng.)*. Operated by, or relying on, air-pressure or the force of compressed air.

pneumatic *(Zool.)*. Containing air, as, in physostomous Fish, the *pneumatic duct* leading from the gullet to the air-bladder, and in Birds, those bones which contain air-cavities.

pneumatically-operated switch (circuit-breaker) *(Elec. Eng.)*. A switch (or circuit breaker) in which the force for closing is obtained from a piston operated by compressed air.

pneumatic brake *(Eng.)*. A continuous braking system, used on some railway trains, in which air-pressure is applied to brake cylinders throughout the train. See **air brake** (1), **continuous brake, electropneumatic brake**.

pneumatic conveyer *(Eng.)*. A system by which loose material is conveyed through tubes by air in motion, the air velocity being created by the expansion of compressed air through nozzles.

pneumatic drill *(Eng.)*. A hard-rock drill in which compressed air is arranged to reciprocate a loose piston which hammers the shank of the bit or an intermediate piece, or in which the bit is clamped to a piston rod.

pneumatic flotation cell *(Min.Ext.)*. One in which low-pressure air is blown in and diffused upward through the cell.

pneumatic lighting *(Min.Ext.)*. Use of small turbomotor driven by compressed air to drive a small dynamo connected to an electric lamp, thus avoiding wiring extensions underground where a compressed air service exists.

pneumatic loudspeaker *(Acous.)*. One in which a jet of high-pressure air is modulated by a transducer. Also called a *stentorphone*.

pneumatic pick *(Eng.)*. A road contractor's tool in which, by mechanism similar to that of a **pneumatic drill**, a straight pick is hammered rapidly by a reciprocating piston driven by compressed air.

pneumatic riveter *(Eng.)*. A high-speed riveting machine similar in arrangement to a **hydraulic riveter** but in which a rapidly reciprocating piston driven by compressed air delivers 1000–2000 blows per minute.

pneumatic tools *(Eng.)*. Hand tools, such as riveters, scaling and chipping hammers, and drills, driven by compressed air. See **pneumatic drill, pneumatic pick, pneumatic riveter**.

pneumatic trough *(Chem.)*. A vessel used for the collection of gases over water.

pneumatic tube conveyor *(Eng.)*. A system in which small objects enclosed in suitable containers are transported along tubes, the container acting as a moving piston which is impelled either by means of pressure or vacuum.

pneumato- *(Genrl.)*. Prefix. See **pneum-**.

pneumatocele *(Med.)*. (1) A hernial protrusion of lung through some defect in the chest wall. (2) Any air-containing swelling.

pneumatocyst *(Zool.)*. (1) Any air-cavity used as a float (2) In Fish, the **air bladder**. (3) The cavity of a pneumatophore.

pneumatolysis *(Geol.)*. The alteration of rocks by the concentrated volatile constituents of a magma, effected after the consolidation of the main body of magma. See **greisenization, kaolinization, tourmalinization**.

pneumatophores *(Bot.)*. Specialized roots of swamp plants and mangroves which grow vertically upwards into the air from the root system in the mud and which, containing much aerenchyma, apparently facilitate gas exchange for the submerged roots.

pneumaturia *(Med.)*. The passing of urine containing gas or air.

pneumectomy *(Med.)*. See **pneumonectomy**.

pneumococcal polysaccharide *(Immun.)*. Type specific polysaccharide present in the capsules of *Streptococcus pneumoniae*. The capsules are related to the virulence of the organism, since they prevent ingestion by neutrophils unless coated with antibody. Each type of pneumococcus has different oligosaccharide repeating units in its capsule, so protection is type-specific. Pneumococcal polysaccharides are **thymus-independent antigens**. Pneumococcal vaccines are composed of a mixture of polysaccharides derived from the types prevailing in the community.

pneumococcus *(Biol.)*. A Gram-positive diplococcus, a causative agent of pneumonia, though it may occur normally in throat and mouth secretions.

pneumoconiosis *(Med.)*. A disease of the lungs due to inhalation of dust in excessive quantities, usually occupational, the chief being silicosis, asbestosis and (coal miner's) anthrocosis.

pneumoencephalitis *(Vet.)*. See **Newcastle disease**.

pneumohaemo-pericardium (-thorax) *(Med.)*. The presence of air and blood in the pericardial sac (or in the pleural cavity).

pneumohydro-pericardium (-thorax) *(Med.)*. The presence of air and a clear effusion in the pericardial sac (or in the pleural cavity).

pneumolith *(Med.)*. A concretion in the lung, formed usually as a result of calcification of a chronic tuberculous focus.

pneumonectomy *(Med.)*. Removal of lung tissue.

pneumonia *(Med.)*. A term generally applied to any inflammatory condition of the lung accompanied by consolidation of the lung tissue; more especially, *lobar pneumonia*, in which the consolidation affects one or more lobes of the lung.

pneumonitis *(Med.)*. Inflammation of the lungs, not necessarily infective in origin.

pneumonomycosis *(Med.)*. A term applied to disease of the lung caused by any one of a number of various fungi.

pneumon-, pneumono- *(Genrl.)*. Prefix from Gk. *pneumōn*, gen. *pneumonos*, lung.

pneumo-oil switch *(Elec.Eng.)*. *Oil breaker*. A switch or circuit-breaker in which the operation is carried out partly by pneumatic means, and partly by hydraulic means using oil as the medium.

pneumopericardium *(Med.)*. (1) The presence of air or gas in the peritoneal cavity. (2) The injection of air into the peritoneal cavity for radiographic purposes.

pneumopyopericardium *(Med.)*. The presence of air and pus in the pericardial cavity.

pneumopyothorax *(Med.)*. The presence of air and pus in the pleural cavity.

pneumostome *(Zool.)*. In Arachnida, the opening to the exterior of the lung books. In Pulmonata, the opening to the exterior of the lung formed by the mantle cavity.

pneumothorax *(Med.)*. (1) The presence of air or gas in the pleural cavity. (2) The therapeutic injection of air or gas into the pleural cavity for the purpose of collapsing diseased lung (*artificial pneumothorax*).

pneum-, pneumo-, pneumat-, pneumato- *(Genrl.)*. Prefix from Gk, *pneuma*, gen. *pneumatos*, breath.

p-n-i-p transistor *(Electronics)*. Similar to *p-n-p* transistor with a layer of an intrinsic semiconductor (germanium of high purity) between the base and the collector to extend the high-frequency range.

p-n junction *(Electronics)*. Boundary between *p*- and *n*-type semiconductors, having marked rectifying characteristics; used in diodes, photocells, transistors, etc.

p-n-p transistor *(Electronics)*. A junction transistor having an *n*-type base region between a *p*-type emitter and a *p*-type collector. For normal operation, the emitter should be held positive and the collector negative with respect to the base.

Po *(Chem.)*. The symbol for polonium.

Poaceae *(Bot.)*. See **Gramineae**.

PO box, bridge *(Elec.Eng.)*. Abbrev. for *Post Office box* or *bridge*.

POCC *(Space)*. Abbrev. for *Payload Operations Control Centre*, a room or building where experimenters, suitably supported by their own specialized equipment, gather to monitor and control their experiments on-board a space vehicle. Data links are provided directly or *via* a communications satellite.

pock *(Med.)*. *Pustule*; any small elevation of the skin, containing pus, occurring in an eruptive disease (formerly seen in smallpox).

Pockels' effect *(Elec.Eng.)*. The **electro-optical effect** in a piezoelectric material.

pocket *(Build.)*. The hole in a pulley stile through which the counterpoise weights are passed into the box of a sash and frame.

pocket chamber *(Radiol.)*. A small ionization chamber used by individuals to monitor their exposure to radiation. It depends upon the loss of a given initial electric charge being a measure of the radiation received.

pocket chisel *(Build.)*. See **sash pocket chisel**.

pocket chronometer *(Horol.)*. A pocket watch fitted with the chronometer escapement. On the continent of Europe the term is used for any high-precision pocket watch.

pod *(Aero.)*. See **engine pod**.

pod *(Bot.)*. See **legume**.

podagra *(Med.)*. Gout.

podal *(Zool.)*. Pedal.

podauger *(Build.)*. An auger having a straight groove cut in its length to hold the chips.

podex *(Zool.)*. The anal region. adj. *podical*.

Podicipitiformes *(Zool.)*. An order of Birds, containing compact-bodied species of cosmopolitan distribution. Almost completely aquatic, and build floating nests. Toes lobate and feet placed far back. Grebes.

podite *(Zool.)*. A walking leg of Crustacea.

podium *(Arch.)*. A continuous low wall under a row of columns.

podium *(Zool.)*. (1) In land Vertebrates, the third or distal region of the limb; manus or pes; hand or foot. (2) Any footlike structure, as the locomotor processes or tub-feet of Echinodermata. pl. *podia*. adj. *podial*.

podocarpus *(For.)*. Conifer yielding a commercial softwood, found chiefly in S. Africa and Australasia.

podomere *(Zool.)*. In Arthropoda, a limb segment.

pod-, podo-, -pod *(Genrl.)*. Prefix and suffix from Gk. *pous*, gen. *podos*, foot.

podsol, podzol *(Ecol.)*. A common soil type developed on siliceous mineral soil in areas of very high rainfall and low evaporation. The soil has an ash coloured layer below the surface, from which minerals and clay particles have been washed, and an orange-brown deeper layer where some of these have accumulated. Sometimes there is an impermeable layer of oxidized iron. The overall reaction is acid and the soil usually supports only a calcifuge flora.

poecilitic *(Geol.)*. Same as *poikilitic*.

Poetsch process *(Min.Ext.)*. Freezing of waterlogged strata by circulation of refrigerated brine through the boreholes surrounding the section through which a shaft or tunnel is to be driven.

Poggendorff cell *(Elec.Eng.)*. A single-fluid form of the bichromate cell.

Poggerndorff compensation method *(Chem.)*. A method of measuring an unknown e.m.f. by finding the point at which it just opposes the steady fall of potential along a wire.

pogo effect *(Space)*. Unstable, longitudinal oscillations induced in launch system, mainly due to fuel **sloshing** and engine vibration.

Pogonophora *(Zool.)*. A phylum of coelomate animals possibly related to the Hemichordata, without an alimentary canal or bony tissue, and having a simple nervous system. There is a muscular heart, and the blood contains haemoglobin. The extremely long bodies are divided into three sections, the posterior one serving to anchor the animals in the chitinous tubes in which they live, and the anterior part bearing well-developed tentacles which probably serve to gather food, digest it, and absorb the products of digestion.

poikilitic texture *(Geol.)*. Texture in igneous rocks in which small crystals of one mineral are irregularly scattered in larger crystals of another, e.g., small olivines embedded in larger pyroxenes, as in some peridotites.

poikiloblastic *(Geol.)*. A textural term applicable to metamorphic rocks in which small crystals of one mineral are embedded in large crystals of another. The texture is comparable with the *poikilitic* of igneous rocks.

poikilocyte *(Med.)*. A malformed red blood cell.

poikilocytosis *(Med.)*. Presence of malformed red cells in the blood, e.g. in severe anaemia.

poikilohydric *(Bot.)*. Lacking structures or mechanisms to regulate water loss and, hence, having water content determined rapidly by the water potential of the environment, as in algae, lichens, bryophytes, submerged vascular plants. Cf. **homoiohydric**.

poikilosmotic *(Ecol.)*. Of an aquatic animal, being in osmotic equilibrium with its environment, the concentration of its body fluids changing if the environment becomes more dilute or more concentrated. Marine Invertebrates are frequently poikilosmotic. Cf. **homoiosmotic**.

poikilothermal *(Zool.)*. See **cold-blooded**.

poikil-, poikilo- *(Genrl.)*. Prefix from Gk. *poikilos*, many-coloured.

point *(Elec.Eng.)*. In electric-wiring installations, a termination of the wiring for attachment to a lighting fitting socket-outlet or other current-using device.

point *(Maths.)*. See **double-**, **multiple-**, **singular-**.

point *(Print.)*. The unit of measurement for type and materials. The 12-point em (the foundation of the system) measures 0.1660 in (approx 1/6 in) (4.21 mm); 1 point measures 0.0138 in (0.351 mm); 72 points or 6 ems measure 0.996 in (25.3 mm). The old type sizes such as *nonpariel*, *brevier*, *pica* (approx. 6-point, 8-point, 12-point) have largely been discarded. See also **points**.

point block *(Arch.)*. A high block of housing with a central core of lifts, stairs and services.

point-contact diode *(Electronics)*. A semiconductor which uses a slender wire filament, touching a small piece of semiconductor material, to provide rectifying action. Junction capacitance is kept to a minimum and high-frequency operation is possible. Sometimes used as a detector in VHF and microwave devices.

point counter tube *(Electronics)*. A radiation-counter tube, using gas ionization amplification, having its central electrode in the form of a point or a small sphere.

point defect *(Crystal.)*. See **defect**.

pointed arch *(Build.)*. An arch which rises on each side from the springing to a central apex.

pointed ashlar *(Build.)*. A block of stone whose face-markings have been done with a pointed tool.

pointer *(Build.)*. A tool used for raking out old mortar from brickwork joints prior to pointing.

Pointers *(Astron.)*. Popular name for the two stars of the Great Bear, α and β Ursae Majoris; they are roughly in line with the Pole Star and so help to identify it.

point gamma *(Image Tech.)*. The contrast gamma for a specified level of brightness. In TV reception the instantaneous slope of the curve connecting log (input voltage) and log (intensity of light output).

pointing *(Build.,Civ.Eng.)*. The process of raking out the exposed jointing of brickwork and refilling with, normally, cement mortar.

point of inflexion on a curve *(Maths.)*. A point at which a curve changes from convex to concave. For the curve $y = f(x)$, the point where $x = a$ is a point of inflexion if $f''(a) = 0$, and the first nonzero higher order derivative at $x = a$ is of odd order.

point of osculation *(Maths.)*. A multiple point on a curve through which two branches, having a common tangent at the point, pass.

point-of-sale terminal *(Comp.)*. Input device which records and displays details about goods as they are sold, and transmits the information to a computer for stock control. Also *POS terminal*. See **Kimball tag**, **bar code reader**.

point quadrat *(Ecol.)*. Device for measuring the canopy cover by lowering a thin pointed shaft into the vegetation many times.

points *(Civ.Eng.)*. Movable tapered blades or tongues of metal for setting alternative routes of running rails. Each such blade is pivoted at the *heel*, its *toe* being locked against the stock rail, *facing points* if the train approaches the toe, *trailing points* if the train approaches the heel.

poise *(Horol.)*. Equilibrium. A balance is said to be *in poise* when, supported horizontally by its pivots on knife edges, it has no tendency to rotate, or if rotated, no tendency to take up any set position.

poise *(Phys.)*. CGS unit of viscosity. The viscosity of a fluid is 1 poise when a tangential force of 1 dyne per unit area maintains unit velocity gradient between two parallel planes. Equal to $1\,\text{dyne}\,\text{cm}^{-2}\,\text{s}$, or in SI units $10^{-1}\,\text{N}\,\text{m}^{-2}\,\text{s}$. Named after the physicist Poiseuille. Abbrev. *P*. See **viscosity**, **coefficient of viscosity**.

Poiseuille's formula *(Phys.)*. An expression for the volume of liquid per second Q, which flows through a capillary tube of length L and radius R, under a pressure P, the viscosity of the liquid being η:

$$Q = \frac{\pi P R^4}{8 L \eta}.$$

poising callipers *(Horol.)*. A form of callipers between the jaws of which a balance may be mounted and rotated, as a test for truth and poise.

poison *(Med.)*. Any substance or matter which, introduced into the body in any way, is capable of destroying or seriously impairing life. Poisons include products of decomposition or of bacterial organisms, and in some cases the organisms themselves; also very numerous chemical substances forming the residue of industrial, agricultural and other processes (see **pollution**). Poisons are generally classified as *irritants* (e.g. cantharides, arsenic) and *corrosives* (e.g. strong mineral acids, caustic alkalis); *systemics* and *narcotics* (e.g. prussic acid, opium, barbiturates, henbane); *narcotic-irritants* (e.g. nux vomica, hemlock). Among gases, carbonic acid, carbonic oxide, sulphuretted hydrogen, sulphur dioxide, sulphide of ammonium and numerous others (e.g. fumes of leaded petrol) are of significance in industry and daily life. See also **war gases**.

poison *(Phys.)*. Any contaminating material which, because of high-absorption cross-section, degrades intended performance, e.g. fission in a nuclear reactor, radiation from a phosphor. See also **catalytic poison**.

poisoning *(Chem.Eng.,Min.Ext.)*. Loading of resin sites in ion exchange with ions of unwanted species which therefore prevent the capture of those required by the process. In liquid-liquid ion exchange, fouling of the organic solvent in similar manner.

poisoning *(Min.Ext.)*. Loading of resin sites in ion exchange with ions of unwanted species which therefore prevent the capture of those required by the process. In liquid-liquid ion exchange, fouling of the organic solvent in similar manner.

Poisson distribution *(Stats.)*. A probability distribution often applied to the number of occurrences of particular (especially rare) events in a particular time period.

Poisson's equation *(Maths.)*. The equation $\nabla^2 V = -4\pi\rho$, where ∇ is the vector operator del, V is the electric or gravitational potential function, and ρ is the charge or mass density respectively. Cf. **Laplace's equation**.

Poisson's ratio *(Phys.)*. One of the elastic constants of a material. It is defined as the ratio of the lateral contraction per unit breadth to the longitudinal extension per unit length, when a piece of the material is stretched. For most substances its value lies between 0.2 and 0.4. The relationship between Poisson's ratio ν, Young's modulus E, and the modulus of rigidity, G, is given by:

$$\nu = \frac{E}{2G} - 1.$$

pokeweed mitogen *(Immun.)*. Mitogenic lectins obtained from *Phytolacca americana*. There are five different substances of which one is active for both T- and B-lymphocytes, but the others act only on T-lymphocytes.

polar axis *(Astron.)*. (1) That diameter of a sphere which passes through the poles. (2) In an equatorial telescope, the axis, parallel to the earth's axis, about which the whole instrument revolves in order to keep a celestial object in the field.

polar axis *(Crystal.)*. A crystal or symmetry axis to which no two- or four-fold axes are normal; thus the arrangements of faces at the two ends of such an axis may be dissimilar. The principal axis of tourmaline is a polar axis of three-fold symmetry, the top of the crystal being terminated by pyramid faces, the bottom end by a single plane in some cases.

polar body *(Biol.)*. During the maturation divisions (*meiosis*) of the ovum, one daughter set of chromosomes passes to a small cell, the polar body, and takes no further part in gametogenesis.

polar bond *(Chem.)*. Covalent bond between elements of differing electronegativity.

polar control *(Aero.)*. See **twist and steer**.

polar co-ordinates *(Maths.)*. Of point P, defined in terms of rectangular cartesian co-ordinates as follows. In the plane: (r,θ), where r is the distance OP from the origin P, and θ is the angle between the x-axis and OP. Spherical polars in 3 dimensions: (r,θ,ϕ), where r is OP, θ is the angle between the z-axis and OP (i.e. the colatitude), and ϕ is the angle between the x-axis and the projection OP on the xy-plane (i.e. the longitude). Cylindrical polars in three dimensions: (r,θ,z), where r is the projection of OP on the xy-plane, θ is the angle between the x-axis and that projection, z is the height above the xy-plane. It should be noted that these systems use the symbols in different senses. Other notations are sometimes used.

polar crystal *(Crystal.)*. A crystal, such as sodium chloride, with ionic bonding between atoms.

polar curve, diagram *(Phys.)*. A curve drawn in polar coordinates showing the field distribution around a radiator or receiver of radiated energy, to show its directional efficacy. E.g. a contour of equal field-strength around a transmitting antenna; or, in a receiving antenna, contour path of a mobile transmitter producing a constant signal at the receiver. Similar diagrams are prepared for sound fields round electro-acoustical transducers, for sensitivity curves around photocells, scintillation counters, etc. and for all other energy detectors or radiators with directional sensitivity.

polarimeter *(Chem.)*. An instrument in which the optical activity of a liquid is determined by inserting Nicol prisms in the path of a ray of light before and after traversing the liquid.

polarimetry *(Chem.)*. The measurement of optical activity, especially in the analysis of sugar solutions.

Polaris *(Astron.)*. The brightest star in the constellation Ursa Minor which currently lies (by chance) within 1° of the north celestial pole. Its altitude is approximately equal to the latitude of the observer. This star was much used for simple navigation in the northern hemisphere. The effects of **precession** are gradually reducing its usefulness, as the north pole is drifting away from this particular star.

polariscope *(Phys.)*. Instrument for studying the effect of a medium on polarized light. Interference patterns enable elastic strains in doubly refracting materials to be analysed. It may consist of a polarizer and an analyser, with facilities for placing transparent specimens between them. The analyser is usually a Nicol prism. The polarizer may be also a Nicol prism or a *pile* of plates. Modern polariscopes use light-polarizing films, e.g., Polaroids instead. See **photoelasticity**.

polarity *(Chem.)*. A permanent property of a molecule which has an unsymmetrical electron distribution. All heteronuclear diatomic molecules are polar. See **dipole moment**.

polarity *(Phys.)*. (1) Distinction between positive and negative electric charges (Franklin). (2) General term for difference between two points in a system which differ in one respect, e.g., potentials of terminals of a cell or electrolytic capacitor, windings of a transformer, video signal, legs of a balanced circuit, phase of an alternating current.

polarity *(Zool.)*. Existence of a definite axis.

polarity diversity *(Telecomm.)*. See **diversity reception**.

polarizability *(Chem.)*. A property of a molecule that its electron cloud can readily be deformed by an external electron field. Molecules containing heavy atoms and/or multiple bonds are more polarizable than those which do not.

polarization *(Chem.)*. The separation of the positive and negative charges of a molecule by an external agent.

polarization *(Phys.)*. (1) Nonrandom orientation of electric and magnetic fields of electromagnetic wave. (2) Change in a dielectric as a result of sustaining a steady electric field, with a similar vector character; measured by density of dipole moment induced.

polarization current *(Elec.Eng.)*. That causing or caused by polarization (*soakage*) in a dielectric, with possible late discharge.

polarization error *(Telecomm.)*. Error in determining the direction of arrival of radio waves by a direction-finder when the desired wave is accompanied by downward components which are out of phase. Generally occurs at night.

polarized beam *(Phys.)*. (1) In **electromagnetic radiation**, a beam in which the vibrations are partially or completely suppressed in certain directions. (2) In **corpuscular radiation**, a beam in which the individual particles have non-zero spin and in which the distribution of the values of the spin component varies with the direction in which they are measured.

polarized capacitor *(Elec.Eng.)*. Electrolytic capacitor designed for operation only with fixed polarity. The dielectric film is formed only near one electrode and thus the impedance is not the same for both directions of current flow.

polarized plug *(Elec.Eng.)*. One which can be inserted into a socket in only one position.

polarized relay *(Elec.Eng.)*. See **relay types**.

polarizer *(Phys.)*. Prism of double refracting material, or Polaroid plate, which passes only plane-polarized light, or produces it through reflection. See **Nicol prism, pile, Polaroid**.

polarizing angle *(Elec.Eng.)*. See **Brewster angle**.

polarizing filter *(Image Tech.)*. Filter allowing the passage of light which is polarized in one direction only; used in photography for the control of surface reflections and for darkening blue skies and also for image separation in some systems of stereoscopy. Also *pola filter*.

polarizing monochromator *(Astron.)*. A filter consisting of a succession of quartz crystals and calcite or Polaroid sheets; the light passing through is restricted to a narrow band, useful in observing the solar chromosphere.

polar line and plane *(Maths.)*. Of a point (the **pole**) with respect to a conic or quadric respectively: the locus of harmonic conjugates of the pole with respect to the pairs of points intersected on the conic or quadric by lines through it. When the pole is outside the conic or quadric the polar is the line or plane determined by the points of contact of tangents from the pole to the conic or quadric.

polar molecule *(Phys.)*. One with unbalanced electric charges, usually valency electrons, resulting in a dipole moment and orientation.

polar nuclei *(Bot.)*. The two (typically haploid) nuclei which migrate from the poles to the centre of the embryo sac of an angiosperm and there fuse with the second male gamete to form the, typically triploid, primary **endosperm** nucleus.

polarograph *(Chem.)*. An instrument which records the current-voltage characteristic for a polarized electrode. It is used in chemical analysis.

Polaroid *(Image Tech.)*. TN for a range of photographic and optical products, including a transparent light-polarizing plastic sheet (*Polarscreen*) and methods of **instant photography** in black-and-white and colour.

Polaroid camera *(Image Tech.)*. TN for a camera for **instant photography** using a colour **diffusion-transfer reversal** process.

polaron *(Electronics)*. Electron in substance, trapped in potential well produced by polarization charges on surrounding molecules; analogous to *exciton* in semiconductor.

polar platform *(Space)*. An unpressurized platform in Sun-synchronous orbit, used particularly for remote sensing, which is capable of in-orbit servicing and payload exchange.

polar reciprocation *(Maths.)*. Transformation in which lines are replaced by points and points by lines, usually by replacing a line by its pole and a point by its polar with respect to a conic. Also called *dualizing*.

polar response curve *(Acous.)*. The curve which indicates the distribution of the radiated energy from a sound reproducer for a specified frequency. Also the relative response curve of a microphone for various angles of incidence of a sound wave for a given frequency. Generally plotted on a radial decibel scale.

polar sequence *(Astron.)*. An adopted scale for determining photographic stellar magnitudes. It consists of a number of stars near the North Pole which are used as a standard of comparison; they range from Polaris to the faintest observable.

polar wandering *(Geol.)*. Movement of the magnetic poles and of the poles of the Earth's rotation through geological time. Can be partly explained by **plate tectonics**.

polder *(Civ.Eng.)*. A piece of low lying land reclaimed from the water or artificially protected therefrom.

pole *(Astron.)*. See **celestial-, terrestrial-**.

pole *(Biol.)*. One end of the achromatic spindle, where the spindle fibres come together.

pole *(Build.)*. A long piece of timber of circular section and small in diameter.

pole *(Elec.Eng.)*. (1) That part of an electric machine which carries one of the exciting windings. (2) A term sometimes used to denote one of the terminals of a d.c. generator, battery or electric circuit, e.g. *negative pole, positive pole*. (3) A term sometimes used in connection with an electric arc to denote the extremity of either of the electrodes between which the arc burns. (4) A wooden, steel, or concrete column for supporting the conductors of an overhead transmission or telephone line.

pole *(Electronics)*. That part of the anode between adjacent cavities in a multiple cavity magnetron.

pole *(Genrl.)*. (Gk. *polos*, hinge, axis). Generally, the axis or pivot on which anything turns; one of the ends of the axis of a sphere, especially of the Earth.

pole *(Maths.)*. (1) Of a function: if a function $f(z)$ is expanded in **Laurent's expansion**, in an annular domain surrounding a singularity $z=a$,

$$f(z)=\sum_{1}^{\infty} b_n(z-a)^n+\sum_{1}^{\infty} c_n(z-a)^{-n},$$

the second term is called the *principal part* of $f(z)$ at $z=a$. If the principal part has a finite number of terms, m say, then $f(z)$ is said to have a pole of order m at $z=a$. If the principal part is an infinite series, $f(z)$ is said to have an *isolated essential singularity* at $z=a$. (2) Of a line or plane with respect to a conic or quadric, see **polar line and plane**. (3) Of a circle (or part thereof) on a sphere, the ends of a diameter of the sphere normal to the circle.

pole *(Phys.)*. The part of a magnet, usually near the end, towards which the lines of magnetic flux apparently converge or from which they diverge, the former being called a *south pole* and the latter a *north pole*.

pole *(Zool.)*. Point; apex; an opposite point (as *aboral pole*); axis.

pole arc *(Elec.Eng.)*. The length of the pole face of an electric machine measured circumferentially around the armature surface.

pole bevel *(Elec.Eng.)*. A portion of the pole face of an electric machine, near the pole tip, which is made to slope away from the armature surface instead of being concentric with it, the object being to obtain a more satisfactory shape of flux wave.

pole-changing control *(Elec.Eng.)*. A method of obtaining two or more speeds from an induction motor by altering stator-winding connections to give a different number of poles.

pole core *(Elec.Eng.)*. See **pole shank**.

pole end-plate *(Elec.Eng.)*. A thick plate placed at each end of the laminations of a laminated pole.

pole face *(Elec.Eng.)*. That surface of the pole piece of an electric machine which faces the armature.

pole face loss *(Elec.Eng.)*. Iron losses which occur in the iron of the pole face of an electric machine on account of the periodic flux variations caused by the armature teeth.

pole-finding paper *(Elec.Eng.)*. Paper prepared with a chemical solution, which, when placed across two poles of an electric circuit, causes a red mark to be made where it touches positive.

pole horn *(Elec.Eng.)*. The portion of the pole shoe of an electric machine which projects circumferentially beyond the pole shank.

pole piece *(Elec.Eng.)*. Specially shaped magnetic material forming an extension to a magnet, e.g. the salient poles of a generator, motor or relay, for controlling flux.

pole pitch *(Elec.Eng.)*. The distance between the centre lines of two adjacent poles on an electric machine; it is measured circumferentially around the surface of the armature of the machine.

pole plate *(Build.)*. A horizontal member supporting the feet of the common rafters and carried upon the tie beams of the trusses.

pole shading *(Elec.Eng.)*. See **shaded pole**.

pole shank *(Elec.Eng.)*. The part of a pole piece around which the exciting winding is placed. Cf. **pole shoe**.

pole shim *(Phys.)*. See **shim**.

pole shoe *(Elec.Eng.)*. That portion of the pole piece of an electric machine which faces the armature; it is frequently detachable from the pole shank.

pole star *(Astron.,Surv.)*. See **Polaris**.

pole strength *(Phys.)*. The force exerted by a particular magnetic pole upon a *unit pole* situated at unit distance from it. See **Coulomb's law for magnetism**.

pole tip *(Elec.Eng.)*. The edge of the pole face of an electric machine which runs parallel to the axis of the machine. Hence *leading pole tip, trailing pole tip*.

pole top switch *(Elec.Eng.)*. A switch which may be mounted at the top of a transmission line pole; arranged for hand operation by some mechanical device.

poling *(Eng.)*. In the fire refining of copper, the impurities are eliminated by oxidation and the oxygen is in turn removed by the reducing gases produced when green logs (*poles*) are burned in the molten metal. If the final oxygen content is too high the metal is *underpoled*, if too low *overpoled*, and if just right *tough pitch*.

poling *(Telecomm.)*. See **turnover**.

poling boards *(Civ.Eng.)*. Rough vertical planks used to support the sides of narrow trenches after excavation; placed in pairs on opposite sides of the trench at intervals along its length, each pair being wedged apart by wooden struts.

poling boards *(Min.Ext.)*. Fore-poling boards are used in tunnelling through loose (running) rock, and are driven horizontally ahead to support roof.

polio- *(Genrl.)*. Prefix from Gk. *polios*, grey.

polioencephalitis *(Med.)*. Inflammation of the grey matter of the brain.

polioencephalomyelitis *(Med.)*. Inflammation of the grey matter both of the brain and of the spinal cord.

poliomyelitis *(Med.)*. Inflammation of the grey matter of the spinal cord. Often used as a synonym for *acute anterior poliomyelitis*, or *infantile paralysis*, due to infection, chiefly of the motor cells of the spinal cord, with a virus; characterized by fever and by variable paralysis and wasting of muscles.

polished face *(Civ.Eng.)*. A fine surface finish produced on granite, or other natural stone, by rubbing down a sawn face with iron sand, fine grit, and polishing powder in turn.

polished foil *(Print.)*. Blocking foil, which leaves a glossy transfer. These are available from eggshell to full gloss and will mark surfaces ranging from cheap paper to polypropylene.

polished plate *(Build.)*. A superseded sheet-glass over 3/16 in thick, used for shop windows.

polished rod *(Min.Ext.)*. In an oil pumping station the actuating rod which passes through the stuffing box at the top of the well; usually attached at the top to the cable going round the **horsehead**; at the lower end joined to the **pony rods** which are connected to the displacement pump.

polished specimen *(Min.Ext.)*. Characteristic hand specimen of ore, metal, alloy, compacted powder etc. one face of which is ground plane and mirror smooth by abrasive powder and/or polishing laps, or electrolytic methods.

polishing lathe, head *(Eng.)*. A headstock and spindle carrying a *polishing wheel* or mop rotated at high speed by belt drive or built-in motor; used for buffing and polishing.

polishing stake *(Horol.)*. A flat polished piece of steel on which materials for polishing are mixed. It should be kept in a container to exclude dust.

polishing stick *(Eng.)*. A stick of wood, one end of which is charged with emery or rouge, used for finishing small surfaces. It is twisted in the hands or held in the chuck of a drilling machine.

polje *(Geol.)*. A large depression found in some limestone areas, due in part to subsidence following underground solution.

poll-adze *(Build.)*. An adze having a blunt head (*poll*) opposite to the cutting edge.

pollard *(For.)*. A tree that has been topped to obtain a head of shoots, usually above the height to which browsing animals can reach.

pollen *(Bot.)*. The microspores of seed plants. The individual pollen grains may contain an immature or mature male gametophyte.

pollen analysis *(Bot.)*. A method of investigating the past occurrence and abundance of plant species by a study of pollen grains and other spores preserved in peat and sedimentary deposits. See also **palynology, sporopollenin**.

pollen chamber *(Bot.)*. A cavity in the micropylar end of the nucellus of the ovule of some gymnosperms in which pollen grains lodge as a result of pollination and where they develop further and germinate.

pollen count *(Med.)*. A graded assessment of the level of plant pollen in the atmosphere, of importance to sufferers from **hay fever**.

pollen flower *(Bot.)*. A flower which produces no nectar but has abundant pollen which attracts pollinating insects.

pollen mother cell *(Bot.)*. A cell, typically diploid, in a pollen sac, which divides by meiosis to form a **tetrad** of pollen grains; the **microsporocyte** of a seed plant.

pollenosis *(Med.)*. See **hay fever**.

pollen sac *(Bot.)*. A chamber in which pollen grains (microspores) are formed in seed plants, e.g. such chambers in the **anther**.

pollen tube *(Bot.)*. A tubular outgrowth of the intine, produced on the germination of a pollen grain. In angiosperms it grows to the embryo sac and there delivers the male gamete(s), *siphonogamy*.

poll-evil *(Vet.)*. Inflammation of the bursa of the ligamentum nuchae of the horse.

pollex *(Zool.)*. The innermost digit of the anterior limb in Tetrapoda.

pollination *(Bot.)*. The transfer of pollen from the pollen sac to the micropyle in gymnosperms (see also **pollination drop**) or to the stigma in angiosperms, usually by means of wind, insects, birds, bats or water.

pollination drop *(Bot.)*. A drop of sugary fluid that is secreted into the micropyle at the time of pollination in many gymnosperms, and which traps pollen grains that may then float up to the nucellus or be drawn there by the reabsorption of the drop.

polling *(Comp.)*. Automatic, sequential testing of each potential source of input to a computer, usually a terminal, to find its operational status. It may be ready to transmit data, in use or not in use. See **multiaccess**.

pollinium *(Bot.)*. A mass of pollen grains, held together by a sticky secretion or retained within the pollen sac wall, which is transported. usually by an insect, in the pollination of e.g. orchids.

pollucite *(Min.)*. A rare hydrated alumino-silicate of caesium, occurring as clear colourless or white crystals with cubic symmetry. It occurs in granite pegmatites.

pollution *(Ecol.)*. Modification of the environment by release of noxious materials, rendering it harmful or unpleasant to life.

poloidal field *(Nuc.Eng.)*. The magnetic field generated by an electric current flowing in a ring.

polonium *(Chem.)*. Radioactive element, symbol Po, at.no. 84. Important as an α-ray source relatively free from γ-emission.

poly- *(Chem)*. Prefix from Gk. *polys*, many. In chemistry: (1) containing several atoms, groups etc; (2) a prefix denoting a polymer, e.g. polyethene.

poly- *(Genrl.)*. Prefix from Gk, *polys*, many.

polyacetals *(Plastics)*. **Polyformaldehydes**.

polyacrylamide gel *(Biol.)*. Transparent cross-linked acrylamide gels. See **polyacrylamide gel electrophoresis**.

polyacrylamide gel electrophoresis *(Biol.)*. Technique used for separating nucleic acids or proteins on the basis of charge, shape and size. The highly cross-linked polymer of acrylamide forms a gel matrix through which macromolecules move under the influence of an electric field. Proteins usually require the presence of the detergent, *sodium dodecylsulphate*, which forms polyanionic complexes whose mobility is a simple function of size. Abbrev. *PAGE*.

polyacrylate *(Chem.)*. A polymer of an ester of acrylic acid or its chemical derivatives.

polyadelphous *(Bot.)*. A flower having the stamens joined by their filaments into several separate bundles.

polyalkane *(Chem.)*. A hydrocarbon polymer essentially of long chain molecules with only saturated carbon atoms in the main chain.

polyamide *(Textiles)*. Natural or synthetic fibres composed of polymers having the same amide group (-CO-NH-) repeated along the chain. Examples of the natural fibres are silk, wool and hair; for synthetic polyamides see **nylon**. Polyamides made with *aromatic* groups attached to the amide links are assigned to a different class, the **aramid fibres**.

polyandrous *(Bot.)*. A flower having a large, indefinite number of stamens.

polyandry *(Zool.)*. The practice of a female animal mating with more than one male.

polyarch *(Bot.)*. A stele having many protoxylem strands, as in many roots, especially of monocots.

polyarteritis nodosa *(Med.)*. Disease characterized by inflammation of the arteries in many parts of the body causing fever, malaise and often leading to renal failure and hypertension. Thought to be an auto-immune process.

polyarthritis *(Med.)*. Inflammation affecting several joints at the same time.

polybasic acids *(Chem.)*. Acids with two or more replaceable hydrogen atoms in the molecule.

polybasite *(Min.)*. Sulphide of silver and antimony, often with some copper, crystallizing in the monoclinic system.

polybutadienes *(Plastics)*. Range of rubber polymers produced by the stereospecific polymerization of butadiene (butan-1,3-diene) using *Ziegler catalysts*. Some of the rubbers have high resilience, a high abrasion resistance and good low-temperature properties.

polybutenes *(Plastics)*. Polymers of 2-methyl-propene, used as rubber substitutes on account of excellent electrical and moisture-resisting properties.

polycarbonates *(Plastics)*. Range of thermoplastics based on the condensation products of phosgene (carbonyl chloride) with dihydroxy organic compounds, such as diphenylol propane. Clear, slightly tinted material, stronger and with a higher mp than polymethyl methacrylates. Relatively expensive but used for 'tough' applications, e.g. shatterproof and bulletproof windows. TN *Lexan*.

polycarpellary *(Bot.)*. Consisting of many carpels.

polycarpic *(Bot.)*. Potentially able to fruit many times, as most perennial plants are. Cf. **monocarpic**. See also **hapaxanthic**.

polycarpous *(Bot.)*. See **apocarpous**.

Polychaeta *(Zool.)*. A class of Annelida of marine habit, having locomotor appendages (parapodia) bearing numerous setae; there is usually a distinct head; the perivisceral cavity is subdivided by septa; the sexes are generally separate, with numerous gonads; and development after external fertilization involves metamorphosis, with a free swimming trochosphere larva. Marine Bristle-worms.

polychasium *(Bot.)*. A cymose inflorescence in which the branches arise in sets of three or more at each node.

polychloroprene *(Chem.)*. See **Neoprene**.

polychromasia *(Med.)*. The diffuse bluish staining of the young immature red blood corpuscles with eosin and methylene blue. Also called *polychromatophilia*.

polychromy *(Arch.)*. Any form of decoration employing several colours, but especially applicable to Late Gothic Revival, (1850–1860), when colour was achieved, not only by applying paint, but by using contrasting bands of brickwork externally, and tiles and mosaics on floors and walls. This method of integrating colour has no parallels in early **Gothic** architecture and has become the hallmark of the Victorian era.

polyclimax theory *(Ecol.)*. Idea, proposed by A.G. Tansley in 1939, that not one but several climax vegetation types are possible in one climatic region because the environment is influenced by local factors such as soil and the activity of animals. See **climax**.

polyclonal activators *(Immun.)*. General term for substances that activate many clones of lymphocytes as opposed to an antigen which only activates clones which have receptors which recognize it. See **lectin, mitogen**. Unless cloned antigen-reactive cell lines are available, the only way in which to study the behaviour of normal B and T lymphocytes has been to stimulate them with polyclonal activators.

polycormic *(Bot.)*. A woody plant having several strong vertical trunks.

polycotyledonous *(Bot.)*. Having more than 2 cotyldeons, e.g. pine.

polycyclic *(Bot.)*. Vascular tissues of a stem having 2 or more concentric solenosteles, dictyosteles or rings of vascular bundles.

polycyclic *(Chem.)*. Containing more than one ring of atoms in the molecule.

polycystic *(Med.)*. Containing many cysts, e.g. a *polycystic* kidney.

polycythaemia, polycythemia *(Med.)*. An increase in the number of red corpuscles per unit of circulating blood. Also *erythraemia*.

polydactylism, polydactyly *(Med.,Zool.)*. Having more than the normal number of digits. adj. *polydactylous*.

polydentate *(Chem.)*. Of a **ligand**, attaching via more than one atom, resulting in *chelate* formation.

polydipsia *(Med.)*. Excessive thirst.

polyembryony *(Bot.,Zool.)*. The development of more than one embryo in one ovule, seed or fertilized ovum.

polyester *(Image Tech.)*. A plastic used as a film **base** having greater mechanical strength and stability than cellulose acetate.

polyester *(Textiles)*. Polyester fibres and films are linear polymers which have the ester group [-C(=O)-O-] repeated along the chain. The commonest polyester fibre is made by the polymerization of ethylene diol and terephthalic acid.

polyester amide *(Plastics)*. A polymer in which the structural units are linked by ester and amide (and/or thio-amide) groupings.

polyesters *(Plastics)*. Range of polymers formed by the condensation polymerization of polyhydric alcohols and polycarboxylic acids or anhydrides. Maleic and fumaric (i.e. ethene 1,2-dicarboxylic) acids and ethylene and propyl alcohol are the usual starting materials, which are combined with e.g. styrene and glass fibre to give important *reinforced plastics* for very wide use. See also **Terylene**.

polyethers *(Plastics)*. Long-chain glycols made from alkene oxides such as epoxyethane (ethene oxide). Used as intermediates in the production of **polyurethanes**, antistatic agents and emulsifying agents.

polyeth(yl)ene *(Plastics)*. *Polythene*. A thermoplastic polymer of ethene with a repeating group of $-CH_2-CH_2-$; an elastomer of superior electrical properties at very high radio-frequencies; many other uses include weatherproof sheeting and as a textile fibre.

polyeth(yl)ene terephthalate *(Plastics)*. The chemical name for the polyester forming the basis of Terylene and Melinex.

polyformaldehydes *(Plastics)*. *Polymethanal*. Acetal resins. A range of thermoplastics based on polymers of formaldehyde. High melting-point, crystalline polymers used in engineering because of good dimensional stability and abrasion resistance. TNs *Delrin, Kemetal*.

polygamous *(Bot.)*. Having staminate, pistillate and hermaphrodite flowers on the same and on distinct individual plants.

polygamous *(Zool.)*. Mating with more than one of the opposite sex during the same breeding season. n. *polygamy*.

polygenes *(Biol.)*. The genes that control *quantitative characters* and whose individual effects are too small to be detected.

polygenic *(Biol.)*. Of a character whose genetic component is determined by many genes with individually small effects.

polygon *(Maths.)*. A many-sided plane figure.

polygonal roof *(Build.)*. A roof which in plan forms a figure bounded by more than four straight lines.

polygonal rubble *(Build.)*. A form of ragwork in which the rubble wall is built up of stones having polygonal faces.

polygoneutic *(Zool.)*. Having several broods in a year.

polygon of forces *(Phys.)*. A polygon whose sides are parallel to and proportional to forces acting at a point, the directions of the forces being cyclic around the polygon. The polygon is closed if the forces are in equilibrium, otherwise the closing side of the polygon is parallel and proportional to the equilibrant of the forces.

polygraph *(Behav.)*. *Lie-detector*. A physiological recording device which can pick up physiological changes in the form of electrical impulses, and record them onto a moving roll of paper.

polygynous *(Zool.)*. Said of a male animal which mates with more than one female.

polyhalite *(Min.)*. A hydrated sulphate of potassium, magnesium, and calcium, $K_2MgCa_2(SO_4)_4.2H_2O$. Found in salt deposits.

polyhedron *(Maths.)*. A solid rectilineal figure. Cf. *regular convex solids*.

polyhydric *(Chem.)*. Containing a number of hydroxyl groups in the molecule, e.g. *polyhydric alcohols*, alcohols with three, four, or more hydroxyl groups.

polyimide *(Chem.)*. Product of the polycondensation reaction between pyromellitic dianhydride and an aromatic diamine, giving a high temperature dielectric with strong mechanical stability and electrical properties over a wide temperature range.

polymastia, polymastism *(Med.)*. The presence of supernumerary breasts.

polymer *(Plastics)*. A material built up from a series of smaller units (*monomers*), which may be relatively simple, such as ethene (the unit of polythene), or relatively complex, such as methyl methacrylate (see **Perspex**). The molecular size of the polymer helps to determine the mechanical properties of the plastic material and ranges from a few hundred of the basic units to perhaps hundreds of thousands.

polymerase *(Biol.)*. Enzymes producing a polynucleotide sequence, complementary to a pre-existing *template* polynucleotide. DNA polymerase requires a **primer** from which to start polymerization whereas RNA polymerase does not.

polymerization *(Chem.)*. The combination of several molecules to form a more complex molecule, usually by an *addition* or a *condensation* process. It is sometimes a reversible process.

polymerous *(Bot.)*. Having many members in a whorl, e.g. many petals.

polymethyl methacrylate *(Plastics)*. The chemical name for **Perspex**.

polymorph *(Aero.)*. The term applied by Barnes Wallis to his supersonic aircraft designs incorporating *variable sweep* wings.

polymorphic function *(Comp.)*. One that can handle a number of **data types**.

polymorphic transformation *(Eng.)*. A change in a pure metal from one form to another, e.g. the change from γ- to α-iron.

polymorphism *(Biol.)*. (1) The presence in a population of two or more very rare alleles of a particular gene. It may be the result of selective advantage of the heterozygotes or of the rarer forms. (2) The occurrence of different structural forms at different stages of the life-cycle of the individual. adjs. *polymorphic*, *polymorphous*.

polymorphism *(Comp.)*. The inclusion of **polymorphic functions** in programs.

polymorphism *(Min.)*. The property possessed by certain chemical compounds of crystallizing in several forms which are structurally distinct; thus TiO_2 (titanium dioxide) occurs as the mineral species *anatase*, *brookite*, and *rutile*.

polymorphonuclear leucocyte *(Immun.)*. Cell of the myeloid series which in its mature form has a multilobed nucleus and cytoplasm containing granules. They are present in inflammatory exudates and in the blood, where unlike erythrocytes, they have only a short residence time. Term includes neutrophil, eosinophil and basophil leucocytes.

polyneuritis *(Med.)*. *Multiple neuritis.* A wide spread affection of many peripheral nerves with flaccid paralysis of muscles and/or loss of skin sensibility, due to infection or poisoning with various agents, such as lead, alcohol, arsenic, diphtheria toxin, etc.

polynosic fibre *(Textiles)*. Also known as a **modal fibre**.

polynosic fibres *(Plastics)*. Modified rayon fibres with improved properties, such as good dimensional stability on washing, arising from better uniformity of fibres.

polynucleotide *(Biol.)*. Synonym for long chain of nucleic acid.

polyoestrous *(Zool.)*. Exhibiting several oestrous cycles during the breeding season. Cf. **monoestrous**.

polyolefin *(Textiles)*. Fibre of film made from a linear polymer obtained from an olefin especially ethylene (giving polyethylene) or propylene (giving polypropylene).

polyoma *(Biol.)*. A small DNA virus oncogenic for mice. Similar to **SV 40**.

polyoma virus *(Med.)*. A virus that can induce a wide variety of cancers in mice, hamsters, and also in other species.

polyoses *(Chem.)*. Polysaccharides.

polyoxymethylene resins *(Plastics)*. Alternative name for *polyformaldehydes*.

polyp *(Med.)*. See **polypus**.

polyp *(Zool.)*. An individual of a colonial animal.

polypeptide *(Biol.)*. A linear condensation of *amino-acids* which, alone or associated with others, forms a *protein* molecule.

polypetalous *(Bot.)*. Having a corolla of separate petals, not fused with each other. Cf. **gamopetalous**.

polyphagous *(Zool.)*. Feeding on many different kinds of food, as *Sporozoa* which exist in several different cells during one life-cycle, or phytophagous Insects with many food plants.

polyphase *(Elec.Eng.)*. Said of a.c. power supply circuits (usually 3) carrying currents of equal frequency with uniformly spaced phase differences. Normally using common return conductor.

polyphase motor *(Elec.Eng.)*. An electric motor designed to operate from a polyphase supply.

polyphyletic group, polyphyly *(Biol.)*. (1) A group of species whose members derive from two or more independent ancestral lines. (2) In cladistics a group that does not include the most recent ancestor of its members. Cf. **monophyletic** and **paraphyletic groups**.

polyphyllous *(Bot.)*. Having a perianth of separate members, not fused with each other. Cf. **gamophyllous**.

polyphyodont *(Zool.)*. Having more than 2 successive dentitions. Cf. **diphyodont**, **monophyodont**.

polyplexer *(Radar)*. Device acting as duplexer and lobe switcher.

polyploid *(Biol.)*. Having more than twice the normal haploid number of chromosomes. The condition is known as **polyploidy**. *Artificial polyploidy*, which can be induced by the use of chemicals (notably **colchicine**), is of economic importance in producing hybrids with desired characteristics.

polyposis *(Med.)*. The development of many polyps in an organ such as the intestine.

polypropylene *(Chem.)*. *Polymerized propene*, a plastic with properties similar to polyethylene but with greater resistance to heat, organic solvents. The repeating groups are [-CH(C_3)-CH_2-]; numerous uses include tufted carpet backings.

polyprotodont *(Zool.)*. Having numerous pairs of small subequal incisor teeth; pertaining to the Polyprotodontia. Cf. **diprotodont**.

polypus *(Med.)*. A smooth, soft, pedunculated tumour growing from mucous membrane.

polyrod antenna *(Telecomm.)*. One comprising a number of tapered dielectric rods emerging from a waveguide.

polysaccharides *(Chem.)*. *Polyoses*; a group of complex carbohydrates such as starch, cellulose etc. They may be regarded as derived from x monosaccharide molecules by the elimination of $x-1$ molecules of water. Polysaccharides can be hydrolysed step to step, ultimately yielding monosaccharides.

polysepalous *(Bot.)*. Having a calyx of separate sepals, not fused with each other. Cf. **gamosepalous**.

polysiloxanes *(Chem.)*. The basis of silicone chemistry. Prepared by the hydrolysis of chlorosilanes R_3Si-Cl,R_2SiCl_2, ethers R_3SiOR,$R_2Si(OR)_2$, or mixtures.

polysome, polyribosome *(Biol.)*. An assembly of ribosomes held together by their association with a molecule of **messenger RNA**.

polysomy *(Biol.)*. A chromosome complement in which some of the chromosomes are present in more than the normal diploid number, e.g. a *trisomic*.

polyspermy *(Zool.)*. Penetration of an ovum by several sperms.

polyspondyly *(Zool.)*. The condition of having more than two vertebral centra corresponding to a single myotome. adjs. *polyspondylic, polyspondylous.*

polystely *(Bot.)*. The condition in which the vascular tissue in a stem exists as two or more separate steles interconnecting only at intervals such as where the stem branches, e.g. some *Selaginella* spp.

polystichous *(Bot.)*. Arranged in several rows.

polystyrenes *(Plastics)*. See **styrene (phenyl ethene) resins**.

polyteny *(Biol.)*. Special case of **polyploidy** in which chromatids remain very closely paired after duplication, through all or part of their length. In some cases homologues are also closely paired, e.g. in **salivary glands** of dipterans.

polytetrafluoroeth(yl)ene *(Plastics)*. PTFE. $(CF_4)_n$. The principal fluoropolymer noted for its very marked chemical inertness and heat resistance. Widely used for a variety of engineering and chemical purposes, and as a coating for 'non-stick' kitchen equipment. TNs *Fluon, Teflon*. Cf. **Penton**.

polythene *(Plastics)*. See **polyeth(yl)ene**.

polytokous *(Zool.)*. Bringing forth many young at a birth; prolific; fecund. n. *polytoky*.

polytrifluorochloroeth(yl)ene *(Plastics)*. *PTFCE*. A plastic material with many of the good properties of **PTFE**, but somewhat easier to fabricate as it is more thermoplastic.

polytrophic *(Zool.)*. In Insects, said of ovarioles in which nutritive cells alternate with the oöcytes; more generally, obtaining food from several sources. Cf. **acrotrophic**.

polyurethanes *(Plastics)*. Range of resins, both thermoplastic and thermosetting, based on the reaction of di-isocyanates with dihydric alcohols, polyesters or polyethers. Widely used in the production of foamed materials.

polyuria *(Med.)*. Excessive secretion of urine.

polyvalent *(Chem.)*. Having a valency greater than two.

polyvinyl acetal *(Chem.)*. (1) Group comprising polyvinyl acetal, polyvinyl formal, polyvinyl butyral and other compounds of similar structure. (2) A thermoplastic material derived from a polyvinyl ester in which some or all of the ester groups have been replaced by hydroxyl groups and some or all of these hydroxyl groups replaced by acetal groups. Generally a colourless solid of high tensile strength.

polyvinyl acetate *(Plastics)*. *PVA*. Polymerized vinyl acetate (ethanoate) used chiefly in adhesive and coating compositions. See **vinyl polymers**.

polyvinyl alcohol *(Plastics)*. *PVA*. Polymer produced by the hydrolysis of polyvinyl acetate. Chiefly water soluble and used in adhesives and greaseproofing, but a non-water-soluble form is used as fibre-forming material.

polyvinyl butyral *(Plastics)*. A polyvinyl resin made by reacting butanal with a hydrolysed polyvinyl ester such as the acetate (ethanoate). Used chiefly as the interlayer in safety-glass windscreens.

polyvinyl chloride *(Plastics)*. *PVC*. Best known and most widely used of the vinyl plastics. Used in unplasticized form (*rigid PVC*) for pipework, ducts etc., where high chemical resistance is called for. In plasticized form used for sheeting, cable covering, mouldings and as a fabric for clothing and furnishings. See **vinyl polymers**.

poly(vinyl chloride) fibres *(Textiles)*. Fibres of the linear polymer derived from vinyl chloride i.e. the repeating units are [-CH_2-CH(Cl)-].

polyvinyl formal *(Plastics.)*. A polyvinyl resin made by reacting formaldehyde with a hydrolysed polyvinyl ester. Used in lacquers and other coatings, varnishes and some moulding materials.

polyvinylidene chloride *(Plastics)*. Based on vinylidene chloride (asymmetrical dichloroethylene). Usually used as copolymer with small amounts of vinyl chloride or acrylonitrile. Basis of fibres, films for packaging and chemically resistant piping.

poly(vinylidene chloride) fibres *(Textiles)*. Fibres from the polymer derived from vinylidene chloride i.e. the repeating units are [-CH_2-CCl_2-].

polyvinyl polymers *(Plastics)*. See **vinyl polymers**.

Polyzoa *(Zool.)*. Another name for the Ectoprocta or Bryozoa.

pome *(Bot.)*. False fruit, of the subfamily Pomoideae of the Rosaceae, in which the true fruit forms the core, containing the seeds, surrounded by a greatly expanded, fleshy receptacle, e.g. apple.

pommel *(Build.)*. (1) An ornament in the shape of a ball, e.g. a ball finial. (2) See **punner**.

Ponceaux *(Chem.)*. A group of dyestuffs prepared by the interaction of various diazo-salts with naphthol-sulphonic acids.

Poncelet wheel *(Eng.)*. An undershot water-wheel with curved vanes; of higher efficiency than the flat-vane type. See **undershot wheel**.

pond *(Hyd.Eng.)*. A reach or level stretch of water between canal locks. Also *pound*.

pondermotive force *(Min.Ext.)*. In high-tension separation, the electrostatic force exerted on a particle as it passes through the field of a corona-type separator. In magnetic separation, the flux field intensity together with density of particle determines its deflection from a straight path.

pons *(Zool.)*. A bridgelike or connecting structure; a junction. pl. *pontes*. adj. *pontal*.

pons Varolii *(Zool.)*. In Mammals, a mass of transversely coursing fibres joining the cerebellar hemispheres.

pontal flexure *(Zool.)*. The flexure of the brain occurring in the same plane as the cerebellum; it bends in the reverse direction to the primary and nuchal flexures and tends to counteract them.

pontie, pontil *(Glass)*. See **punty**.

pontoon *(Civ.Eng.)*. A flat-bottomed floating vessel for the support or transport of plant, materials, or men.

pontoon bridge *(Civ.Eng.)*. A temporary bridge carried on **pontoons**.

pony girder *(Eng.)*. A secondary girder carried across side-by-side cantilevers.

pony motor *(Elec.Eng.)*. An auxiliary motor used to bring synchronous machinery up to speed before synchronizing.

pony rods *(Min.Ext.)*. In an oil well the rods which connect the **polished rod** to the pump. Also *drill rods*.

pool cathode *(Electronics)*. An emissive cathode which consists of a liquid conductor, e.g. mercury.

pool reactor *(Nuc.Eng.)*. See **swimming pool reactor**.

poor lime *(Build.)*. Lime in which the proportion of impurities insoluble in acids is in excess of 15%.

POP *(Image Tech.)*. Abbrev. for *Printing-Out Paper*, a once popular photographic paper which produced an image on printing which needed only fixing, no development.

POP-2 *(Comp.)*. Programming language used for **list** processing.

poplar *(For.)*. A common tree *Populus* of the temperate regions of the Northern hemisphere.

poplin *(Textiles)*. A plain-woven fabric, with fine lines or cords running across the cloth (due to the *ends* per inch greatly exceeding the *picks* per inch); often of cotton and mercerized. Used for shirtings, pyjamas, and dresses.

poppet head *(Min.Ext.)*. Headframe of hoisting shaft. The poppet is the bearing in which the winding pulley is set.

poppets *(Ships)*. Temporary structures erected beneath a ship's hull to transfer the weight to the sliding ways, before and during launching.

poppet valve *(Autos.)*. The mushroom- or tulip-shaped valve, made of heat-resisting steel, commonly used for inlet and exhaust valves. It consists of a circular head with a conical face which registers with a corresponding seating round the port, and a guided stem by which it is lifted from its seating by the rocker or tappet. Also called *mushroom valve*. See **stellited** valves, **valve inserts**.

popping *(Build.)*. A defect in plasterwork resulting from the use of a lime which has not been properly slaked.

popping *(Comp.)*. The deletion of a data item from a **stack**.

popping-back *(Autos.)*. An explosion through the inlet pipe and carburettor of a petrol engine, due to a weak, slow-burning mixture.

Pop rivet *(Eng.)*. A proprietary design of hollow rivet, requiring work access from one side only for placing and clinching. Similar to **Chobert rivet**.

population *(Biol.)*. Any specified reproducing group of individuals.

population inversion *(Phys.)*. According to the *Boltzmann principle* the number of particles with a higher energy will be less than the number with a lower energy for a system in equilibrium at a given temperature. In a non-equilibrium situation it is possible to invert this so that the number of particles in the higher energy level is greater than the number in the lower level. This is called *population inversion*. It is an essential process in the operation of *masers* and *lasers*. See **optical pumping**.

population types *(Astron.)*. The 2 broad types of stellar population. Population I includes hot blue stars such as those in the Sun's neighbourhood; they are found in the arms of spiral galaxies, and share in the galactic rotation. Population II stars are found in the central regions of galaxies and in globular clusters, where dust and gas are absent; they are red stars, having high velocities and do not share in the regular rotation of the system.

pop valve *(Eng.)*. A boiler safety-valve in which the head of the wing valve is so shaped as to cause the steam to accelerate the rate of lift when a small lift occurs, giving rapid pressure release.

porcelain clay *(Geol.)*. See **china clay**.

porcelain insulator *(Elec.Eng.)*. An insulator for supporting high-voltage electric conductors; made of a hard-quality porcelain.

porch *(Image Tech.)*. An interval in the TV signal at **blanking** level: see **back porch, front porch**.

porcine parvovirus infection *(Vet.)*. Infection which affects 75% of UK pig farms. Disease manifests itself with infertility, early embryonic death, abortion and still births. Vaccine available.

pore *(Bot.)*. A small more or less rounded aperture especially (1) the aperture of a **stoma**, (2) a rounded aperture (rather than a slit) in a dehiscing anther or capsule, (3) a vessel seen in a transverse section of wood, (4) a more or less circular germinal aperture in the wall of a pollen grain. Also *porus*. See also **aperturate**.

pore *(Min.Ext.)*. Cavity within or between particles in rock or aggregate. If these communicate with free surface they are open, if not, closed. Number and size of cavities in given volume determines **porosity**, degree of interconnection (ease of communication) between pores determines **permeability**.

pore *(Zool.)*. A small aperture. adjs. *poriferous, porous, poriform*.

pore distribution *(Powder Tech.)*. (1) The size and shape distribution of pores in a consolidated body such as a sintered metal compact. (2) In a loose powder compact, the space confined by adjacent particles can be considered as a system of interconnecting pores. The term *pore distribution* is used to describe the size variation in these so-called pores.

porencephaly, porencephalia *(Med.)*. The presence, in the substance of the brain, of cysts or cavities containing colourless fluid; due to a defect in development.

poricidal *(Bot.)*. Opening or dehiscing by pores.

Porifera *(Zool.)*. A phylum of sessile, aquatic, filter feeding animals. They lack sense organ systems but possess tissues, e.g. epithelial cells (pinacocytes) and flagellated cells (choanocytes). They have skeletons of silica or collagen. Sponges.

porogamy *(Bot.)*. The entry of the pollen tube into the ovule, in an angiosperm, by way of the micropyle (the commoner route). Cf. **chalazogamy**.

poromeric *(Chem.)*. Permeable to water-vapour; said of a polyurethane-base synthetic leather used in the manufacture of shoe 'uppers'.

porometer *(Bot.)*. An instrument for investigating *stomal aperture* by measuring the rate of flow of air (or other gas) through the leaf (viscous flow porometry) or the rate of diffusion of water vapour through the leaf (diffusion porometry).

porosity *(Build.)*. The percentage of space in a material which could absorb water.

porosity *(Eng.)*. In castings, unsoundness caused by shrinkage during cooling, or blow-holes. In compaction of powders, the percentage of voids in a given volume under specified packing conditions.

porosity *(Geol.)*. Of rocks, the ratio, usually expressed as a percentage, of the volume of the void space to the total volume of the rock.

porous bearing *(Eng.)*. A bearing produced by powder metallurgy which, after sintering and sizing, is impregnated with oil by a vacuum treatment. Bearing porosity can be readily controlled and may be as high as 40% of the volume.

porous dehiscence *(Bot.)*. The liberation of pollen from anthers, and of seeds from fruits through pores in the wall of the containing structure.

porous pot *(Elec.Eng.)*. An unglazed earthenware pot serving as a diaphragm in a 2-fluid cell.

porphin *(Chem.)*. Group of 4 pyrrole nuclei linked by methene groups, having a complete system of conjugated double bonds, which accounts for the (reddish) colour of its derivatives.

porphyria *(Med.)*. Inborn error of metabolism resulting in the excretion of abnormal pigments (porphyrins) in the urine which turn dark red on standing. It is characterized by bouts of abdominal pain and vomiting, abnormal skin pigmentation and photosensitivity.

porphyrin *(Chem.)*. A substituted porphin free from metal.

porphyrite *(Geol.)*. See **porphyry**.

porphyritic texture *(Geol.)*. The term applied to the texture of igneous rocks which contain isolated euhedral crystals larger than those which constitute the groundmass in which they are set.

porphyroblastic *(Geol.)*. A textural term applicable to metamorphic rocks containing conspicuous crystals in a finer groundmass, the former being analogous with the phenocrysts in a normal igneous rock, but having developed in the solid.

porphyry *(Geol.)*. A general term used rather loosely for igneous rocks which contain relatively large isolated crystals set in a fine-grained groundmass, e.g. *granite-porphyry*.

porpoising *(Aero.)*. Oscillating symmetrical movements of a seaplane, flying-boat, or amphibian, when planing: pitching instability on the water, as distinct from instability under airborne conditions. Also for landplanes during take-off and landing due to undercarriage forces.

porrect *(Bot.)*. Directed outwards and forwards.

port *(Comp.)*. Point at which signals from peripheral equipment enter the computer.

port *(Eng.)*. An opening, generally valve-controlled, by which a fluid enters or leaves the cylinder of an engine, pump, etc. See **piston valve, poppet valve, sleeve valve, slide valve**.

port *(Genrl.)*. Place of access to a system, used for introduction or removal of energy or material, e.g. *glove-box port*.

porta *(Zool.)*. Any gatelike structure. adj. *portal*.

portable *(Comp.)*. Applied to **software** that can be easily implemented on different types of computers.

portable colour duct *(Print.)*. An interchangeable ink duct which can be attached to the inking system of a printing unit and used to provide an additional colour.

portable electrometer *(Elec.Eng.)*. A portable form of the absolute attracted disk type of electrometer.

portable engine *(Eng.)*. A steam or IC engine carried on road wheels but not self-propelled thereby.

portable ink pump *(Print.)*. A pump-operated *portable colour duct*.

portable instrument *(Elec.Eng.)*. An electrical measuring instrument specially designed for carrying about for testing purposes. Cf. **switch-board instrument, substandard instrument**.

portable substation *(Elec.Eng.)*. A substation comprising the converting or transforming plant and the necessary switch and protective gear, mounted on a railway truck or other vehicle in order that it can be quickly moved to any site for dealing with special loads or other emergency conditions.

portal *(Build.)*. (1) A structural frame consisting essentially of two uprights connected at the top by a fixing. (2) An arch spanning a doorway or gateway.

portal system, circulation *(Zool.)*. A vein which breaks up at both ends into sinusoids or capillaries; as (in Vertebrates) the *hepatic portal system*, in which the hepatic portal vein collects from the capillaries of the alimentary canal and passes the blood into the sinusoids and capillaries of the liver, and the renal portal system.

Porter governor *(Eng.)*. A pendulum-type governor in which, usually, the ends of 2 arms are pivoted to the spindle and sleeve respectively, and carry heavy balls at their pivoted joints. The sleeve carries an additional weight. See **pendulum governor**.

portico *(Arch.)*. A covered colonnade at one side of a building (usually the entrance side).

Portland blast-furnace cement *(Build.)*. A cement differing from a Portland cement in that it contains a proportion of blast-furnace slag.

Portland cement *(Build.,Civ.Eng.)*. A much-used cement made by intimately mixing clay and lime, and then burning the mixture in a kiln.

portlandite *(Min.)*. Calcium hydroxide, $Ca(OH)_2$, occurring rarely in nature but also in Portland cement, hence the name.

porus *(Bot.)*. Same as **pore** sense (4).

posi-drive screws *(Build.)*. A type of cross-head screw for fixing into wood and metal.

position *(Telecomm.)*. In radio-navigation, a set of co-ordinates used to specify location and elevation.

positional notation *(Maths.)*. That system of representation of a number by symbols whose position relative to other symbols in the number affects their absolute value. It is used in the conventional arabic numerals (as compared with the Roman system where each symbol has an absolute value, the number being represented by the sum of those values modified by the convention of subtracting smaller values from succeeding larger values). Thus in the number 37, the 3 represents ten times as much as it does in the number 53; in the number 437 246 the 3 represents 10^4 times as much: each position back from the right multiplying its value by 10, hence decimal notation. Any base can be used in place of 10.

position angle *(Astron.)*. A measure of the orientation of one point on the celestial sphere with respect to another. The position angle of any line with reference to a given point is the inclination of the line to the hour circle passing through the point; it is measured from 0° to 360° from the north point round through east.

position effect *(Biol.)*. Differential effect on transcription and repression of the same gene in different chromosomal locations.

position error *(Aero.)*. That part of the difference between the *equivalent* and *indicated air speeds* due to the location of the **pressure head** or **static vent**. Position error is not a constant factor, but varies with airspeed due to the variations in the airflow around an aircraft at different **angles of attack**.

position-finding *(Telecomm.)*. Determination of location of transmitting station (e.g. an aircraft) by taking a number of bearings by direction finders which receive a signal from the transmitter.

positive *(Image Tech.)*. An image in which tones are reproduced in the same relation as in the original subject; where colour is reproduced, the hues are substantially those of the original.

positive *(Phys.)*. Said of point in circuit which is higher in electric potential than earth.

positive after-image *(Phys.)*. The continued image perceived after visual fatigue in the retina, when the object is replaced by a dark surface or the eye is closed.

positive amplitude modulation *(Image Tech.)*. That in which the amplitude of a video signal increases with an increase in luminance of the picture elements.

positive coarse pitch *(Aero.)*. An extreme blade angle which is reached and locked after engine failure to reduce the drag of a non-feathering propeller.

positive column *(Electronics)*. Luminous plasma region in gas discharge, adjacent to positive electrode.

positive coupling *(Elec.Eng.)*. When two coils are inductively coupled so that the magnetic flux associated with the mutual inductance is in the same direction as that associated with the self inductance in each coil, the mutual inductance and the coupling are both termed *positive*. If these fluxes oppose the mutual inductance and coupling they are known as *negative*.

positive electricity *(Phys.)*. Phenomenon in a body when it gives rise to effects associated with deficiency of electrons, e.g., positive electricity appears on glass rubbed with silk. See **negative electricity**.

positive electrode *(Electronics)*. (1) That connected to a positive supply line. (2) The *anode* in a voltameter. (3) The *cathode* in a primary cell.

positive electron *(Phys.)*. See **positron**.

positive emulsion *(Image Tech.)*. An emulsion intended for printing positive images from a negative, generally characterized by fine grain and relatively high contrast rather than photographic speed.

positive feedback *(Telecomm.)*. **Feedback** such that the signal fed back to the input of the system or amplifier is in phase with the input signal; tends to increase gain at the expense of stability, high levels resulting in oscillation.

positive feeder *(Elec.Eng.)*. A feeder connected to the positive terminal of a d.c. supply.

positive film stock *(Image Tech.)*. Unexposed film specifically intended for making prints from motion picture negatives.

positive g *(Space)*. The force exerted on the human body in a gravitational field or during acceleration so that the force of inertia acts in a head-to-foot direction. See **negative g**.

positive ion *(Phys.)*. An atom (or a group of atoms which are molecularly bound) which has lost one or more electrons, e.g., α-particle is a helium atom less its 2 electrons. In an electrolyte, the positive ions (*cations*) produced by dissolving ionic solids in a polar liquid like water, have an independent existence and are attracted to the anode. Negative ions likewise are those which have gained one or more electrons.

positive magnetostriction *(Phys.)*. See **Joule magnetostriction**.

positive mineral *(Min.)*. A mineral in which the ordinary ray velocity is greater than that of the extraordinary ray, that is ω is less than ε. Quartz is a positive mineral for which $\omega = 1.544$ and $\varepsilon = 1.553$. See **optic sign**.

positive ore *(Min.Ext.)*. Ore blocked out on four sides in panels sufficiently small to warrant assumption that the exposed mineral continues right through the block as a calculable tonnage.

positive phase sequence *(Elec.Eng.)*. See under **phase sequence**.

positive reaction *(Bot.)*. A tactism or tropism in which the plant moves, or the plant member grows, from a region where the stimulus is weaker to one where it is stronger.

positive reinforcement *(Behav.)*. A situation in which a response is followed by a positive event or stimulus (the positive reinforcer or reward) which increases the likelihood that the response will be repeated.

positive taxis *(Biol.)*. See **taxis**.

positive video signal *(Image Tech.)*. One in which increasing amplitude corresponds to increasing light value in the transmitted image. White is 100%, and the black level makes about 30% of the maximum amplitude of signal.

positron *(Phys.)*. *Positive electron*, of the same mass as and opposite charge to the (negative) electron. Produced in the decay of radioisotopes and *pair production* by X-rays of energy much greater than 1 MeV. The antiparticle to the **electron**.

positronium *(Phys.)*. Combination of a positron and an electron forming a hydrogen-atom like system, the positron taking the place of the proton. The system is very short lived (less than 10^{-7} s) and disappears with the emission of γ-ray photons.

possible ore *(Min.Ext.)*. Ore probably existing, as indicated by the apparent extension of proved deposits, but not yet entered and sampled.

post- *(Genrl.)*. Prefix from L. *post*, after.

post *(Glass)*. The formed **gather** prior to drawing by the hand process.

post *(Paper)*. Term used for a pile of sheets or alternate felts and sheets in hand-made mills.

postal *(Paper)*. A non-preferred size of printing board 572 × 725 mm ($22\frac{1}{2} \times 28\frac{1}{2}$ in).

post and pan (pane), post and petrail *(Arch.)*. A term applied to half-timbering formed with brickwork or lath and plaster panels.

post-capillary venules *(Immun.)*. Small vessels through which blood flows after leaving the capillaries and before reaching the veins. The main route by which leucocytes migrate into inflammatory sites is between the endothelial cells of post-capillary venules. Specialized venules with high, rather than flattened endothelial lining are present in the thymus dependent area of lymph nodes. Through these lymphocytes recirculate from blood to lymph.

postcardinal *(Zool.)*. Posterior to the heart; as the *postcardinal* sinus of Selachii.

postcaval vein *(Zool.)*. In higher Vertebrates, the posterior vena cava conveying blood from the hind parts of the body and viscera to the heart. Called *inferior vena cava* in Man.

postclimax *(Ecol.)*. A relict of a former climax held under edaphic control in an area where the climate is no longer favourable for its development.

postdeflection acceleration *(Electronics)*. In a CRT, acceleration of the beam electrons after deflection, so reducing the power required for this. Abbrev. *PDA*.

post-dipping lameness *(Vet.)*. Fever and lameness in sheep due to cellulitis of the limbs caused by the bacterium *Erysipelothrix insidiosa (E. rhusiopathiae)*; occurs after sheep have been dipped in baths contaminated by the organism.

poster *(Print.)*. A large sign printed on paper, for advertisement or propaganda purposes. Posters are usually printed by lithography, offset or direct, very large ones being executed in convenient sections. In UK, the 1-sheet poster measures 30 × 20 in (double crown) and larger sizes are 2-sheet, 4-sheet etc.

posterior *(Bot.)*. The side nearer the axis of a bud, flower or other lateral structure.

posterior *(Zool.)*. In a bilaterally symmetrical animal, further away from the head region; behind. Cf. **anterior**.

posterization *(Image Tech.)*. Video effect in which the picture is reproduced in a limited number of flat tones and colours, without detailed gradation.

POS terminal *(Comp.)*. See **point-of-sale terminal**.

postern *(Arch.)*. A private door or gate, generally at the back or side of a building.

poster stick *(Print.)*. A long wooden composing stick suitable for setting up large type, usually wooden.

post-fertilization stages *(Bot.)*. The developmental processes which go on between the union of the gametic nuclei in the embryo sac and the maturity of the seed.

post-forming sheet *(Plastics)*. A grade of laminated sheet, of the thermosetting type, suitable for drawing and forming to shape when heated.

postganglionic *(Med.)*. See **preganglionic**.

post head *(Elec.Eng.)*. A post or pillar at which cables supplying a third-rail traction system may be terminated and connection made to the conductor rail.

postheating *(Eng.)*. Annealing or tempering a **weldment** to remove strain or prevent local hardening.

posthitis *(Med.)*. Inflammation of the prepuce.

post-hypnotic suggestion *(Behav.)*. Refers to suggestions made to an individual during a hypnotic trance, which determine a behavioural or experiential response occurring after the individual has returned to ordinary consciousness.

postical *(Bot.)*. Relating to or belonging to the back or lower part of a leaf or stem.

post insulator *(Elec.Eng.)*. A porcelain insulator built in the form of a post; used for supporting bus-bars, etc., in a high voltage outdoor sub-station.

postmortem *(Comp.)*. Program designed to locate and diagnose a fault in an executing computer program.

Post Office box, bridge *(Elec.Eng.)*. A Wheatstone bridge in which the resistances making up the arms are contained in a box and varied by means of plugs. Frequently abbrev. to *PO box, bridge*.

post pallet *(Build.)*. A **pallet** with corner posts, used when the unit load either cannot withstand the weight of another to be placed direct above it or has an irregular surface on which another palletized unit load could not be stacked direct.

postpartum *(Med.)*. Occurring after childbirth; e.g. *postpartum* haemorrhage.

post-parturient haemoglobinuria *(Vet.)*. Uncommon disease of milk cows. Symptoms first appear within a month of calving and include dullness, anorexia, reduced milk yield, haemoglobulinuria, jaundice, recumbency and death. Caused by fall in blood phosphorus levels.

post-production *(Image Tech.)*. General term for the operations of editing, sound mixing etc. which take place after the main period of shooting a motion picture.

post-scoring *(Image Tech.)*. The arrangement of a musical accompaniment or other sound effects after a motion picture production has been photographed.

post-synchronization *(Image Tech.)*. Adding sound to previously photographed motion picture material.

post-tensioned concrete *(Civ.Eng.)*. A method whereby concrete beams or other structural members are compressed, after casting and reaching the requisite strength, to enable them to act in the same way as **prestressed concrete**. Achieved by the use of high tensile wires or rods threaded through ducts in the member(s).

Post-Tertiary *(Geol.)*. Name assigned to geological events which occurred after the close of the Tertiary era, i.e. during Pleistocene and Recent times.

post-trematic *(Zool.)*. Posterior to an aperture; as, in Selachii, that branch of the ninth cranial nerve which passes posterior to the first gill cleft. Cf. **pretrematic**.

postulate *(Maths.)*. An assumption. Almost synonymous with *axiom*. An axiom generally relates to an assumption which by its nature is unlikely ever to be proved whereas a postulate is generally rather less basic and may admit, in certain circumstances, of being proved. Since the end of the 19th century *axiom* has assumed both meanings and *postulate* is now rarely used.

postventitious *(Bot.,Zool.)*. Delayed in development.

postzygapophysis *(Zool.)*. A facet or process on the posterior face of the neurapophysis of a vertebra, for articulation with the vertebra next behind. Cf. **prezygapophysis**.

pot *(Elec.Eng.)*. Abbrev. for *potentiometer* (2).

pot *(Glass)*. A vessel of fireclay holding from a few kg to about 2 tonnes, according to the type of manufacture; used to contain the glass during melting in the pot furnace. Such pots may be open or closed (i.e. provided with a hood to prevent furnace gases from acting on the glass).

potamous *(Ecol.)*. Living in rivers and streams.

pot annealing *(Eng.)*. See **close annealing**.

potash *(Bot.)*. Potassium, usually expressed notionally as K_2O, in a fertilizer.

potassium *(Chem.)*. A very reactive alkali metal, soft and silvery white, symbol K, at. no. 19, r.a.m. 39.102, mp 63°C, bp 762°C, rel.d. 0.87. In the form of the element, it has little practical use, although its salts are used extensively. In combination with other elements it is found widely in nature. It shows slight natural radioactivity due to ^{40}K (half-life 1.30×10^9 years). May be used as coolant in liquid-metal reactors, usually as an alloy with sodium.

potassium alum *(Min.)*. Syn. for *alum*.

potassium antimonyl tartrate *(Chem.)*. See **tartar emetic**.

potassium-argon dating *(Geol.)*. A method of determining the age in years of geological material, based on the known decay rate of ^{40}K to ^{40}A.

potassium bichromate *(Chem.)*. See **potassium dichromate**.

potassium bromide *(Chem.)*. KBr. Used in photography and formerly in medicine as a sedative.

potassium carbonate *(Chem.)*. K_2CO_3. *Potash*. White deliquescent powder, soluble in water. Basic. Manufactured by extraction from Stassfurt and other potassium salt beds, or from the electrolysis of potassium chloride.

potassium chlorate (V) *(Chem.)*. $KClO_3$. Detonates with heat; used in the manufacture of matches, fireworks and explosives, and in the laboratory as a source of oxygen.

potassium chloride *(Chem.)*. KCl. Occurs extensively in nature. With sodium chloride, it is extracted on a commercial scale from the waters of the Dead Sea.

potassium cyanide *(Chem.)*. KCN. White, deliquescent solid, smelling of bitter almonds. Extremely poisonous. In the fused condition, is a powerful reducing agent. Used in chemical analysis and in metallurgy, e.g. in the cyanide process for extracting gold. Also in HT salts, fumigants, pharmacy, photography. On the industrial scale, its use has largely been superseded by that of the cheaper sodium cyanide.

potassium dichromate (VI) *(Chem.)*. $K_2Cr_2O_7$. Used in analytical chemistry. Mixed with sulphuric acid, it is used as a cleanser of laboratory vessels, particularly after contamination with organic matter.

potassium feldspar *(Min.)*. Silicate of aluminium and potassium, $KAlSi_3O_8$, occurring in 2 principal crystalline forms — *orthoclase* (monoclinic) and *microline* (triclinic). Both are widely distributed in acid and intermediate rocks, especially in granites and syenites and the fine grained equivalents. See **adularia, feldspar, sanidine**.

potassium ferricyanide, ferrocyanide *(Chem.)*. *Potassium hexacyano ferrate (III), (II)*. $K_3Fe(CN)_6$, $K_4Fe(CN)_6$. Used in chemical analysis. Gives characteristic colour reactions. Also used in dyeing etching, blue print paper, and as a fertilizer. With other iron salts it gives *Prussian blue*.

potassium hexachloroplatinate *(Chem.)*. K_2PtCl_6. Results from the reaction of chloroplatinic acid and potassium chloride.

potassium hydride *(Chem.)*. KH. A saline hydride with the NaCl structure. Formed when potassium is heated in hydrogen.

potassium hydrogen fluoride *(Chem.)*. KHF_2. Formed when potassium fluoride is dissolved in hydrofluoric acid and the solution evaporated. Contains the ion $[F\text{-}H\text{-}F]^-$.

potassium iodate (V) *(Chem.)*. KIO_3. Potassium salt of iodic acid.

potassium iodide *(Chem.)*. KI. Used in chemical analysis, medicine, photography.

potassium mica *(Min.)*. See **muscovite, sericite**.

potassium nitrate *(Chem.)*. KNO_3. Salt of potassium and nitric acid. Strong oxidizing agent. Used in pyrotechnics, explosives, HT salts, glass manufacture and as a fertilizer. Also known as *nitre, saltpetre*.

potassium oxalate *(Chem.)*. *Potassium ethandioate*. The normal salt, $K_2C_2O_4.H_2O$, is soluble in water. The acid salt, KHC_2O_4, is less soluble and occurs in many plants. A compound of these two, potassium quadroxalate, $K_3HC_4O_8.2H_2O$, is used for bleaching and removing iron stains.

potassium permanganate *(Chem.)*. *Potassium manganate (VII)*. $KMnO_4$. Dark purple-brown crystals and solution, with characteristic sweet-bitter taste. Strong oxidizing agent. Used in analytical chemistry and as a disinfectant.

potassium propionate *(Chem.)*. *Potassium propanoate*. $CH_3.CH_2.COOK$. Compound used as a fungicide in edible products such as baked goods.

potassium sorbate *(Chem.)*. $CH_3.(CH{=}CH)_2COOK$. Antifungal agent in edible products used in place of sorbic acid when high water solubility is needed.

potential *(Phys.)*. Scalar magnitude, negative integration of the electric (or magnetic) field intensity over a distance. Hence all potentials are relative, there being no absolute *zero potential*, other than a convention, e.g., earth, or at infinite distance from a charge.

potential *(Zool.)*. **Latent**.

potential attenuator *(Elec.Eng.)*. Same as *potentiometer* (1), as contrasted with a normal attenuator which adjusts power.

potential barrier *(Phys.)*. Maximum in the curve covering two regions of potential energy, e.g., at the surface of a metal, where there are no external nuclei to balance the effect of those just inside the surface. Passage of a charged particle across the boundary should be prevented unless it has energy greater than that corresponding to the barrier. Wave-mechanical considerations, however, indicate that there is a definite probability for a particle with less energy to pass through the barrier, a process known as *tunnelling*. Also called *potential hill*.

potential coefficients *(Phys.)*. Parts of total potential of a conductor produced by charges on other conductors, treated individually.

potential-determining ions *(Min.Ext.)*. Those which leave the surface of a solid immersed in an aqueous liquid before saturation point (equilibrium) has been reached.

potential difference *(Phys.)*. (1) The difference in potential between two points in a circuit when maintained by an e.m.f. or by a current flowing through a resistance. In an electrical field it is the work done or received per unit positive charge moving between the two points. Abbrev. *pd*. SI unit is the **volt**. (2) The line integral of magnetic field intensity between two points by any path.

potential divider *(Elec.Eng.)*. See **voltage divider**.

potential drop *(Phys.)*. Difference of potential along a circuit because of current flow through the finite resistance of the circuit.

potential energy *(Phys.)*. Universal concept of *energy* stored by virtue of position in a field, without any observable change, e.g., after a mass has been raised against the pull of gravity. A body of mass m at a height h above the ground possesses potential energy mgh, since this is the amount of work it would do in falling to the ground. In electricity, potential energy is stored in an electric charge when it is taken to a place of higher potential through any route. A body in a state of tension or compression (e.g., a coiled spring) also possesses potential energy.

potential evapotranspiration *(Meteor.)*. The theoretical maximum amount of water vapour conveyed to the atmosphere by the combined processes of evaporation and transpiration from a surface covered by green vegetation with no lack of available water in the soil.

potential fuse *(Elec.Eng.)*. A fuse used to protect the voltage circuit of a measuring instrument or similar device.

potential galvanometer *(Elec.Eng.)*. A galvanometer having a resistance sufficiently high to enable it to be used as a voltmeter.

potential gradient *(Elec.Eng.)*. Potential difference per unit length along a conductor or through a dielectric; equal to slope or curve relating potential and distance. Also *electromotive intensity*.

potential hill *(Phys.)*. See **potential barrier**.

potential indicator *(Elec.Eng.)*. An instrument which shows whether a conductor is alive. Also called a *charge indicator*.

potential instability *(Meteor.)*. The condition of a layer of the atmosphere which is in a state of **static stability** but in which instability would appear if it were lifted bodily until it became saturated. (Also called *convective instability* in the US.)

potential temperature *(Meteor.)*. The temperature which a given sample of air would have if brought by an adiabatic process to a standard pressure, conventionally 1000 mb. If the pressure and absolute temperature of the sample are p and T, then

$$\theta = T\left(\frac{P_0}{P}\right)^{\frac{\gamma-1}{\gamma}}$$

where γ is the ratio of the specific heats of a perfect gas. θ is related to the entropy S by $S = c_p \log\theta$ + constant, where c_p is the specific heat at constant pressure.

potential theory *(Maths.)*. That branch of applied mathematics which studies the properties of a potential function without reference to the particular subject (e.g. hydrodynamics, electricity and magnetism, gravitational attraction) in which the function is defined.

potential transformer *(Elec.Eng.)*. An undesirable synonym for **voltage transformer**.

potential trough *(Phys.)*. Region of an energy diagram between two neighbouring hills, e.g., arising from the inner electron shells of an atom.

potential vorticity *(Meteor.)*. The vorticity which a column of air between two isentropic surfaces would have if it were brought by an adiabatic process to an arbitrary standard latitude and then stretched or shrunk to an arbitrary standard thickness. It is a conservative air mass property for adiabatic processes. If the original values of the vertical component of vorticity, the **Coriolis parameter**, and the *thickness* are ζ_0, f_0 and h_0, while those after the standardization are ζ_s, f_s and h_s, then

$$(\zeta_0 + f_0)/h_0 = (\zeta_s + f_s)/h_s.$$

See also **Ertel potential vorticity**.

potential well *(Phys.)*. Localized region in which the potential energy of a particle is appreciably lower than that outside. Such a well forms a trap for an incident particle which then becomes *bound*. Quantum mechanics shows that the energy of such a particle is quantized. Applied particularly to the variation of potential energy of a nucleon with distance from the nucleus.

potentiometer *(Elec.Eng.)*. (1) Precision-measuring instrument in which an unknown potential difference of e.m.f. is balanced against an adjusted potential provided by a current from a steady source. (2) Three-terminal voltage divider, often shortened to *pot*. The resistance change with shaft rotation of slider position may follow various laws. Thus one may have e.g. linear, logarithmic, cosine etc. potentiometers. A form of *rheostat*.

potentiometer braking *(Elec.Eng.)*. A braking method used for series motors; the series field and a rheostat are connected in series across the supply, and the armature is connected across the field and a variable proportion of the rheostat.

potentiometer braking controller *(Elec.Eng.)*. A controller making the necessary connections for potentiometer braking of a series motor.

potentiometer card generator *(Elec.Eng.)*. Potentiometer wound with length of turns dependent on rate of change of a function, which determines shape of card.

pot(entiometer) cut *(Acous.)*. Colloquialism for introducing a *cut* (interruption of reproduction) by bringing down a potentiometer to zero for a short time.

potentiometer for alternating voltage *(Elec.Eng.)*. In this instrument, there are two variable alternating voltages in quadrature which are together used to balance the unknown a.c. potential difference.

potentiometer function generator *(Elec.Eng.)*. One in which functional values of voltage are applied to points on a potentiometer, which becomes an interpolator.

potentiometer pick-off *(Elec.Eng.)*. One in which output voltage depends on position of a slide contact on a resistor, thereby transmitting a position.

potentiometer-type field rheostat *(Elec.Eng.)*. A field rheostat which is connected across the supply, the field winding being connected between one pole of the supply and a variable tapping on the rheostat. See **reversible-**.

potentiometric titration *(Chem.)*. A titration in which the end-point is indicated by a change in potential at an electrode immersed in the solution. This change in potential occurs as the solution changes from having excess substance to be determined to having excess titrant.

potette *(Glass)*. A hood shaped like a pot, but with no bottom, which is placed in a tank furnace so that it reaches below the glass level. It protects the man gathering glass on his pipe or iron from furnace gases; also, the glass here is somewhat cooler than that in the main part of the furnace, where melting is taking place. Also *boot*, *hood*.

pot furnaces *(Glass)*. Furnaces in which are set a number of *pots*. They may be: (1) direct-fired from below; (2) gas-fired from below through a central opening in the circular siege, using the recuperative principle; (3) fired through ports in the siege or in the walls, the waste gases escaping through similar openings. In the last-named process, which holds generally for non-circular furnaces, the regenerative principle may be used.

Potier construction *(Elec.Eng.)*. A graphical construction for determining the reactance, armature reaction, and regulation of a synchronous generator from the open-circuit, short-circuit and zero-power factor characteristics.

Potier reactance *(Elec.Eng.)*. The reactance of a synchronous machine as determined by the Potier construction.

pot magnet *(Elec.Eng.)*. One embracing a coil or similar space, excited by current in the coil or a permanent magnet in the central core. Main use is with a circular gap at an end of the core for a moving coil. Miniature split-sintered pot magnets are also used to contain high-frequency coils.

potometer *(Bot.)*. An apparatus to measure the rate of uptake of water by a plant or a detached shoot, etc. and often, thus, indirectly to estimate transpiration.

pot still *(Chem.)*. A still consisting of a boiling vessel with condenser attached. The use of a fractionating column is optional. Used for batch distillation.

potstone *(Min.)*. A massive variety of **steatite**, more or less impure.

Potter-Bucky grid *(Radiol.)*. Type of lead grid designed to avoid exposure of film to scattered X-radiation in diagnostic radiography. Mechanical oscillation of the grid eliminates reproduction of the grid pattern on the radiograph.

potters' clay *(Geol.)*. See **ball clay**.

Pott's disease *(Med.)*. *Spinal caries*. Tuberculous infection of the spinal column.

Pott's fracture *(Med.)*. Fracture-dislocation of the ankle-joint, the lower parts of the tibia and the fibula being broken.

pouch *(Zool.)*. Any saclike or pouchlike structure; as the abdominal *brood-pouch* of marsupial Mammals.

poughite *(Min.)*. A hydrated sulphate and tellurite of iron, crystallizing in the orthorhombic system.

poultice corrosion *(Eng.)*. That which occurs in pockets or on ledges particularly in cars subject to salt spray and dirt.

pound *(Genrl.)*. The unit of mass in the old UK system of units established by the Weights and Measures Act, 1856, and until 1963 defined as the mass of the Imperial Standard Pound, a platinum cylinder kept at the Board of Trade. In 1963, it was redefined as 0.453 592 37 kg. The US pound is defined as 0.453 592 427 7 kg.

pound *(Hyd.Eng.)*. See **pond**.

poundal *(Phys.)*. Unit of force in the foot-pound-second system. The force that produces an acceleration of 1 ft s^{-2} on a mass of 1 pound. Abbrev. *pdl*. 32.2 pdl = 1lbf (lb wt); 1 pdl = 0.138255 N.

pounding *(Ships)*. The heavy falling of the fore end of the ship into the sea when it has been lifted clear of the water by wave action. Also striking the ground under the ship due to wave action.

Poupart's ligament *(Med.)*. The inguinal ligament.

pouring basin *(Eng.)*. Part of the passage system for bringing molten metal to the mould cavity in metal casting. It is provided on large moulds, next to the sprue hole top, to simplify pouring and to prevent slag from entering the mould.

pour point *(Chem.)*. The lowest temperature at which a petroleum-based oil, chilled under test conditions, will flow.

powder *(Eng.)*. Solid particles in the range 0.1 to 1000 μm.

powder *(Powder Tech.)*. Discrete particles of dry material with a maximum dimension of less than 1000 μm (1 mm).

powder core *(Elec.Eng.)*. Core of powdered magnetic material with an electrically insulating binding material to minimize the effects of eddy currents, thus permitting use for high-frequency transformers and inductors with low loss in power.

powder density *(Plastics)*. The mass in grams of 1 cm^3 of loose moulding powder.

powderless etching *(Print.)*. The etching of line blocks in one stage without recourse to **dragon's blood** between a series of etches, using specially designed etching baths with several features including close temperature regulation, planetary movement of the plate, and controlled application of the etching fluid. A measured quantity of special inhibitor is added to the etchant and this forms a protecting film over the metal to control the progress of the etch, by protecting the sides of the lines from undercutting, and stopping the etch when a suitable depth is reached. Originally introduced for line work only, on magnesium alloy, it has been adapted for half-tone and is particularly suitable for combined work, using micrograin zinc of copper.

powder metallurgy *(Eng.)*. The working of metals and certain carbides in powder form by pressing and sintering. Used to produce self-lubricated bearings, tungsten filaments and shaped cutting inserts (carbide) etc. See **cemented carbides**.

powder method *(Min.)*. See **powder photography**.

powder photography *(Min.)*. Method of identification of minerals or crystals, in which a small rod is coated with the powdered preparation, mounted vertically in a special camera in which it rotates, and subjected to a suitably modified beam of X-rays. The pattern, characterized by a set of concentric rings produced by rays diffracted at the Bragg angle relative to the incident beam, is diffracted on to a surrounding strip of film to give positive identification.

powder porosity *(Powder Tech.)*. The ratio of the volume of voids plus the volume of open pores to the total volume of the powder.

powder technology *(Genrl.)*. That covering the production and handling of powders, particle-size analysis and the properties of powder aggregates.

powdery mildew *(Bot.)*. One of several plant diseases, of e.g. cereals and apples, caused by biotrophic fungi of the order Erysiphales.

power *(Maths.)*. The product of a number of equal factors, generalized to include negative and other numbers.

power *(Phys.)*. Rate of doing work. Measured in joules per second and expressed in *watts* (1 W = 1 J/s). The foot-pound-second unit of power is the *horsepower*, which is a rate of working equal to 550 ft-lbf per second. 1 horsepower is equivalent to 745.7 watts. 1 watt is equal to 10^7 CGS units, that is 10^7 erg/second.

power amplification *(Elec.Eng.)*. Process of increasing the power associated with a given signal such that circuits or devices fed from it will be provided with sufficient power to drive them. Ratio of output to input power is expressed in decibels. See **applied power**.

power amplifier *(Elec.Eng.)*. Stage designed to deliver output power of an amplifier. It may be separate from other parts of the same amplifier, and may contain its own power supply. Intended to give the required power output with specified degree of nonlinear distortion, gain not being considered. In some cases, the voltage gain is fractional (or, in dB, negative). Also *power unit*.

power-assisted controls *(Aero.)*. Primary flying controls wherein the pilot is aided by electric motors or double-acting hydraulic jacks.

power-assisted steering *(Autos.)*. In this system, a hydraulic ram connects to the steering linkage and assists the steering effort. The ram is powered by a hydraulic pump driven off the engine and is controlled by a valve which responds to movements of the steering wheel.

power breeder *(Nuc.Eng.)*. A nuclear reactor which is designed to produce both useful power and fuel.

power circuit *(Elec.Eng.)*. That portion of the wiring of an electrical installation which is used to supply apparatus other than fixed lighting, usually through socket outlets.

power coefficient *(Nuc.Eng.)*. The change of reactivity of a reactor with increase in power. In a heterogeneous reactor, due to temperature differences of the fuel, moderator and coolant, it differs from the isothermal temperature coefficient of reactivity which assumes temperatures throughout the core change by the same amount.

power component *(Elec.Eng.)*. See **active component**.

power control rod *(Nuc.Eng.)*. One used for control of the power level of a nuclear reactor. Usually a neutron-absorbing rod containing cadmium or boron steel, but may be a fuel rod or part of the moderator.

power controls *(Aero.)*. A primary flying control system where movement of the surfaces is done entirely by a power system, commonly hydraulic, but sometimes electrohydraulic or electric. The system is always at least duplicated (power source, supply lines, and operating rams) and both circuits are usually running continuously. Reversion to manual control was common in early systems, but the trend is toward a multiplication of reserve circuits and supply sources. See **power-assisted controls, q-feel**.

power density *(Nuc.Eng.)*. Energy released per second per unit volume of a reactor core.

power detector, demodulator *(Elec.Eng.)*. Rectifier which accepts a relatively large carrier voltage and thereby achieves low nonlinear distortion in the demodulation process.

power drag line *(Eng.)*. An excavator comprising a large scraper pan or bucket which is dragged through the material towards the machine and below its level.

powered supports *(Min.Ext.)*. Pit props held to the roof of a coal seam by hydraulic pressure. In fully mechanized collieries they form part of a mechanically operated system.

power efficiency *(Eng.)*. The ratio of power delivered by a transducer (optical, mechanical, acoustical, electrical, etc.) to the power supplied; usually quoted as a percentage.

power factor *(Elec.Eng.)*. Ratio of the total power (in watts) dissipated in an electric circuit to the total equivalent volt-amperes applied to that circuit. In single- and balanced 3-phase systems, it is equal to $\cos\phi$, where ϕ is the phase angle between the applied voltage and the applied current in a single phase circuit, or between the phase voltage and phase current in a balanced 3-phase circuit. In normal dielectrics, it is exactly equal to $G(G^2+\omega C^2)^{-0.5}$, where C = capacitance, G = shunt conductance, $\omega = 2\pi \times$ frequency, and thus nearly equals $G/\omega C$. Abbrev. *pf*.

power-factor meter (indicator) *(Elec.Eng.)*. An instrument which reads the power factor of a circuit directly.

power feed *(Eng.)*. The power-operated feed motion of tables, heads etc. of machine tools, used when the duration of feed, loads, and type of cut justify it.

power frequency *(Elec.Eng.)*. That of the a.c. supply mains, viz. 50 Hz in the UK, 60 Hz in the US. Also *mains frequency*.

power gain *(Telecomm.)*. The ratio of the power delivered to the load, by an amplifier, transducer or system compared with the power absorbed at the input. Usually expressed in dB.

power grid detector *(Elec.Eng.)*. A valve which is operated at a fairly high plate voltage with low values of grid leak and grid capacitor.

power hammer *(Eng.)*. Any type of hammer which is operated, either continuously or intermittently, by power, e.g. by directly coupling the hammer to a steam or pneumatic cylinder.

power-level *(Telecomm.)*. See **transmission level**.

power-level diagram *(Telecomm.)*. Diagram indicating how maximum power levels vary at different points of a transmission channel, thereby indicating how various losses are neutralized by appropriate amplifier gains.

power-level indicator *(Telecomm.)*. See **level indicator**.

power line *(Elec.Eng.)*. US for *mains*.

power loading *(Aero.)*. The gross weight of a propeller driven aircraft divided by the take-off power of its engine(s). For jet aircraft, *thrust loading*.

power loss *(Elec.Eng.)*. (1) The ratio of the power absorbed by a transducer to that delivered to the load. (2) The energy dissipated in a passive network or system.

power of lens *(Image Tech.)*. The relative focusing power of a lens, measured in dioptres, which is the reciprocal of the focal length in metres.

power output *(Eng.)*. The net useful power delivered by a *prime mover* for external use.

power-pack *(Elec.Eng.)*. Power-supply unit for an amplifier, e.g. in a radio or television receiver, wherein the requisite steady voltages are obtained by rectifiers from mains. Also the last or power stage of an amplifier when this is integral with the power supply proper.

power range *(Nuc.Eng.)*. See **start-up procedure**.

power rating *(Aero.)*. The power, authorized by current regulations, of an aero-engine under specified conditions; e.g. maximum take-off rating, combat rating, maximum continuous rating, weak-mixture cruising rating, etc. The conditions are specified by r.p.m. and, for piston engines, **manifold pressure** and torque (in large engines), for turboprops **jet-pipe temperature** and torque, for turbojets exhaust **gas temperature**, for rocket motors **combustion-chamber** pressure.

power reactor *(Elec.Eng.)*. One designed to produce useful power.

power relay *(Elec.Eng.)*. One which operates at a specified power level.

power series *(Maths.)*. A series of the form $\Sigma a_n x^n$.

power shovel *(Eng.)*. *Navvy*. An excavator consisting of a jib carrying a radial arm to the end of which a large bucket or scoop is attached. The bucket makes a radial cut, digging above the level of the excavator.

power supply *(Phys.)*. (1) Arrangement for delivering available power from a source, e.g., public mains, in a form suitable for valves or transistors, etc., generally involving a transformer, rectifier, smoothing filter, circuit-breaker or other protection and frequently incorporating electronic regulation. In a full-wave supply, use is made of a full-wave rectifier and filter. (2) U.S. for *mains*.

power transformer *(Electronics)*. One with its primary connected to the a.c. mains supply, and one or more secondary windings which supply the lower- or higher-voltage supplies required for the various devices and sub-assemblies within a complete piece of electronic equipment. Also *mains transformer*.

power transistor *(Elec.Eng.)*. One capable of being used at power rating of greater than about 10 watts, and generally requiring some means of cooling.

power unit *(Aero.)*. An engine (or assembly of engines) complete with any extension shafts, reduction gears, or propellers.

power unit *(Elec.Eng.)*. See **power amplifier**.

powerwash *(Build.)*. A machine used to direct a high pressure jet of water, perhaps containing added chemicals, at a surface to clean dirt and debris. Used to clean masonry and as a preparation for painting.

pox *(Med.)*. Pl. of *pock*; hence popular names for diseases characterized by pustules, e.g. *chickenpox*, *smallpox*; specifically (vulgar), syphilis.

pox viruses *(Bot.)*. A group of fairly large viruses, brick-shaped in shadow-cast electron micrographs, usually characterized by the formation of cytoplasmic inclusion bodies in the cells they invade; usually cause skin lesions, e.g. smallpox virus.

POY *(Textiles)*. See **partially oriented yarn**.

Poynting-Robertson effect *(Astron.)*. Small particles of dust in the solar system slowly fall into the Sun as a result of this effect. Solar radiation causes them to lose angular momentum, as a result they drift closer to the Sun. For particles smaller than one micron, **radiation pressure** is great enough to counteract the effect and indeed blow the finest dust out of the solar system altogether.

Poynting's theorem *(Telecomm.)*. That which shows that the rate of flow of energy through a surface is equal to the surface integral of the Poynting vector formed by the components of field lying in the plane of the surface. Used for calculating the power radiated from antennae, or through waveguide systems.

Poynting vector *(Elec.Eng.)*. One whose flux, through a surface, represents the instantaneous electromagnetic power transmitted through the surface. Equal to vector cross product of the electric and magnetic fields at any point. In electromagnetic wave propagation, where these fields have complex amplitudes, half the vector product of the electric field intensity and the complex conjugate of the magnetic field intensity is termed the *complex Poynting vector*, and the real part of this gives the time average of the power flux.

pozzuolana, pozzolana *(Civ.Eng.,Geol.)*. A volcanic dust, first discovered at Pozzuoli in Italy, which has the effect, when mixed with mortar, of enabling the latter to harden either in air or under water.

PPC *(Build.)*. Abbrev. for *Plain Plaster Cornice*.

PPI *(Radar)*. Abbrev. for *Plan-Position Indicator*.

ppi *(Textiles)*. Abbrev. for *Picks Per Inch*. See **pick**.

p-p junction *(Electronics)*. One between *p*-type crystals having different electrical properties.

ppm *(Chem.)*. Abbrev. for *Parts Per Million*.

PPQ bar *(Zool.)*. See **pterygopalatoquadrate bar**.

P-protein *(Bot.)*. Phloemprotein. Present in phloem cells especially in sieve elements.

ppt, ppte *(Chem.)*. Abbrevs. for *precipitate*.

P_r *(Bot.)*. Form of the plant pigment **phytochrome** having a peak of absorption at $\lambda \simeq 660$ nm which converts it to P_{fr}.

Pr *(Chem.)*. (1) The symbol for *praseodymium*. (2) The propyl radical C_3H_7—.

practical units *(Phys.)*. Obsolete system of electrical units, whereby the ohm, ampere, and volt were defined by physical magnitudes. Replaced now by the SI system, in which these units are defined in terms of arbitrarily fixed units of length, mass, time and electric current.

prae- *(Genrl.)*. Prefix. See **pre-**.

praecoces *(Zool.)*. Birds which when hatched have a complete covering of down and are able at once to follow the mother on land or into water to seek their own food. Cf. **altrices**.

Praesepe *(Astron.)*. A well-known open cluster in Cancer.

Prandtl number *(Chem.Eng.)*. A dimensionless group much used in heat exchange calculations. It is

$$\frac{\text{specific heat capacity at constant pressure} \times \text{dynamic viscosity}}{\text{thermal conductivity}}$$

prase *(Min.)*. A translucent and dull leek-green variety of **chalcedony**.

praseodymium *(Chem.)*. A metallic element, a member of the rare earth group. Symbol Pr, at.no. 59, r.a.m. 140.9077, rel.d. 6.48, mp 940°C. It closely resembles neodymium and occurs in the same minerals.

Pratt truss *(Eng.)*. See **Whipple-Murphy truss**.

Prausnitz-Kustner reaction *(Immun.)*. Skin reaction for detection and measurement of human reaginic (IgE) antibodies. Serum is injected intracutaneously into a volunteer. Reaginic antibody fixes to skin cells whereas other immunoglobulins diffuse away. Antigen is injected into the same site after 48 hours. A **weal and flare response** develops immediately and its intensity can be graded. The test is not used nowadays because of the danger of transferring hepatitis, and has been replaced by in vitro assays.

PRBS *(Comp.)*. Abbrev. for *Pseudo Random Binary Sequence*.

preadaptation *(Zool.)*. Change of structure preceding appropriate change of habit.

preamplifier *(Acous.)*. Amplifier with at least a stage of valve or transistor gain following a high-impedance source from which the level is too low for line transmission and clearance above noise level.

prebiotic *(Biol.)*. The very different conditions existing on earth before the appearance of life, which provided an environment in which the first living organisms could evolve from nonliving molecules.

preboarding *(Textiles)*. See **boarding**.

Precambrian *(Geol.)*. That period of geological time before the beginning of the **Phanerozoic**. It represents about 90% of all geological time.

precast *(Civ.Eng.)*. Said of concrete blocks etc. which are cast separately before they are fixed in position.

precast stone *(Civ.Eng.)*. See **reconstructed stone**.

precaval vein *(Zool.)*. The anterior vena cava conveying blood from the head and neck to the right auricle. Called *superior vena cava* in Man.

precession *(Eng.)*. An effect exhibited by a rotating body, such as a gyroscope, when a torque is applied to it in such a way as to tend to change the direction of its axis of rotation. If the speed of rotation and the magnitude of the applied torque are constant, the axis slowly generates a cone as its precessional motion.

precession *(Phys.)*. A regular cyclic motion of a dynamical system in which, with suitably chosen coordinates, all except one remain constant, e.g. the regular motion of the inclined axis of a top around the vertical.

precession of the equinoxes *(Astron.)*. The westward motion of the equinoxes caused mainly by the attraction of the sun and moon on the equatorial bulge of the earth. This *luni-solar precession* together with the smaller planetary precession combine to give the general precession amounting to 50.27″ per annum. The equinoxes thus make one complete revolution of the ecliptic in 25 800 years, and the earth's pole turns in a small circle of radius 23°27′ about the pole of the ecliptic, thus changing the coordinates of the stars.

prechordal *(Zool.)*. Anterior to the notochord or to the spinal cord.

precipitable water *(Meteor.)*. The total mass of water in a vertical atmospheric column of unit area, or its height if condensed in liquid form.

precipitation *(Build.)*. The process of assisting the settlement of suspended matters in sewage by the addition of chemicals to the sewage before admission to the sedimentation tanks.

precipitation *(Chem.)*. The formation of an insoluble solid by a reaction which occurs in solution. It is widely used for the separation and identification of substances in chemical processes and analyses. n. and v. *precipitate*.

precipitation *(Immun.)*. In immunology describes the formation of a visible aggregate when antigen and antibody are mixed in aqueous solution so as to form large macromolecular complexes (see **optimal proportions**). These form flocculent precipitates in a tube and appear as a white line in a gel medium where the antigen and antibody interact after diffusing toward one another. Precipitation only occurs when the proportions of the reagents are suitable, even though antigen-antibody combination has taken place.

precipitation *(Meteor.)*. Moisture falling on the Earth's surface from clouds; it may be in the form of rain, hail or snow.

precipitation hardening *(Eng.)*. In a metal or alloy, precipitation of one phase in the lattice of another of different ionic diameter. This keys the structure against 'creep' or slip under stress. Produced by precipitation during cooling of a super-saturated solution. See **ageing, temper-hardening**.

precipitin *(Med.)*. An antibody substance analogous to **agglutinin**, but its action characterized by clouding and precipitation.

precipitin test *(Bot.)*. A serological test in which the reaction between soluble antigen and antibody results in the formation of a visible precipitate.

precise levelling *(Surv.)*. Particularly accurate levelling in which the allowable discrepancy between two determinations of the level difference between two bench marks M km apart is very low, of the order of $0.003\sqrt{M}$.m or less.

precision-approach radar *(Aero.)*. A primary radar system which shows the exact position of an aircraft during its approach for landing. Abbrev. *PAR*.

precision grinding *(Eng.)*. Grinding to a high finish and accurate dimensions on an appropriate machine. Cf. **offhand grinding**.

preclimax *(Ecol.)*. A seral stage which just precedes the climax.

pre-combustion chamber *(Autos.)*. *Antechamber*. A small chamber formed in the cylinder-head of some compression-ignition engines into which the oil fuel is injected at the end of the compression stroke. The high pressure caused by the partial combustion of the fuel expels the rich mixture through a neck or perforated throat plate into the engine cylinder, where combustion is completed. Derived from the oil engine of Akroyd Stuart (1892).

precoracoid *(Zool.)*. An anterior ventral bone of the pectoral girdle in Amphibians and Reptiles, corresponding to the epicoracoid of Monotremes.

precordial *(Med.)*. Situated or occurring in front of the heart.

predation *(Ecol.)*. A form of species interaction in which an individual of one species of animal (the *predator*) directly attacks, kills, and eats, one of another species (the *prey*). The predator is usually larger than its prey. Cf. **parasite**.

predator *(Ecol.)*. See **predation**.

predentin(e) *(Biol.)*. The soft primitive dentin, composed of reticular Korff's fibres, which later becomes calcified to form dentin(e).

predicate calculus *(Maths.)*. Branch of mathematical logic dealing with propositions in terms of subject and predicate — in 'A is B', A is subject, B is predicate — whether the statement is made of all A or only some, etc.

pre-distorting network *(Telecomm.)*. A network which anticipates subsequent frequency distortion in the transmission path, as along a line, so that the line distortion has not entirely to be compensated at the receiving end.

pre-distortion *(Telecomm.)*. The principle of altering the response of a circuit to compensate, fully or partially, anticipated distortion; the aim is to make transmission as high as practicable above the anticipated noise level.

prednisolone *(Med.)*. Synthetic steroid closely related to **cortisone** and used to suppress inflammatory and allergic disorders.

Preece's formula *(Elec.Eng.)*. A formula stating that the fusing current of a wire is proportional to the 1.5th power of the rated current.

pre-eclampsia *(Med.)*. Occurrence in pregnancy of oedema, high blood pressure and albuminuria. If untreated may progress to **eclampsia**.

pre-emphasis *(Telecomm.)*. The process of increasing the strength of some frequency components of a signal to assist these components to override noise or other distortion in the system; mainly to ensure good high-frequency reproduction in frequency-modulation sound broadcasting and in recording. The original levels are restored by **de-emphasis**.

preen gland *(Zool.)*. See **oil gland**.

preening *(Behav.)*. A form of grooming behaviour performed by birds as part of feather maintenance.

prefabricated building *(Build.)*. Building for which walls, roofs, and floors are constructed off the site. See **system building**.

prefading *(Acous.)*. Listening to programme material and adjusting its level before it is faded up for transmission or recording.

preferendum *(Ecol.)*. That part of the range of a species in which it functions most successfully.

preferential mating *(Zool.)*. See **sexual selection**.

preferred numbers, sizes, values *(Eng.)*. Where a range of components (e.g. screws, resistors) is made for use in manufactured articles, and they are identical except for one variable (e.g. diameter, resistance), their values usually follow a geometric progression, rounded at each stage to a convenient *preferred* value, giving a roughly constant proportional increment of variable from one component to the next. E.g. resistors, accurate to $\pm 10\%$, are valued 10, 12, 15, 18, 22, ..., 68, 82, 100 ohms, each value being approx. 20% higher than its predecessor.

preferred orientation *(Eng.)*. During slip, metal crystals change their orientation; when a sufficient amount of deformation has been performed, the random orientation of the original crystals is converted into an arrangement in which a certain direction in all the crystals is parallel to the direction of deformation.

prefloration *(Bot.)*. Aestivation.

prefoliation *(Bot.)*. Vernation.

prefrontal *(Med.)*. *Prefrontal lobotomy*. See **leucotomy**.

pre-frontal lobotomy *(Behav.)*. A form of psychosurgery in which the nerve fibres connecting the thalamus and frontal lobes are cut, in order to reduce the effects of emotions on intellectual processes (the premises for this supposition are dubious).

pregnancy *(Med.)*. *Gestation*. The state of being with child.

pregnancy test *(Med.)*. Any physiological test diagnosing pregnancy at an early stage and with a very high degree of certainty.

pregnancy toxaemia *(Vet.)*. *Twin lamb disease, ketosis*. An energy deficiency of sheep brought on by increased demand of the foetus at the end of pregnancy. Assumed to be due to adreno-cortical problems and/or failure of liver metabolism. Ketone bodies accumulate in the tissues and are excreted in the urine and breath. Symptoms include depression, anorexia, nervous signs and, later on, recumbency.

pregnanediol *(Chem.)*. Steroid found in the urine during pregnancy and at one stage of the menstrual cycle. It is excreted partly free and partly combined with glycuronic acid.

pregnant solution *(Min.Ext.)*. In the **cyaniding** process for recovering gold, the gold-bearing solvent prior to precipitation and recovery.

pregs *(Min.Ext.)*. See **pregnant solution**.

prehallux *(Zool.)*. In Amphibians and Mammals, a rudimentary additional digit of the hind limb.

preheating time *(Electronics)*. Minimum time for heating a cathode before other voltages are applied, thus ensuring full emission. Automatic delays are often incorporated in large amplifiers and radio transmitters.

prehensile *(Zool.)*. Adapted for grasping.

prehnite *(Min.)*. Pale-green and usually fibrous hydrated silicate of calcium and aluminium, crystallizing in the orthorhombic system. It occurs in altered igneous rocks.

preignition *(Autos.)*. The ignition of the charge in a petrol-engine cylinder before normal ignition by the spark; caused by overheated plug-points, the presence of incandescent carbon, etc.

prelacteal *(Zool.)*. In Mammals, said of teeth developed prior to the formation of the milk dentition.

preliminary matter *(Print.)*. The pages of a book preceding the actual text. The order should be half-title, frontispiece or 'advertisement', title (at the back of this, number of editions, imprint, etc.), dedication, preface, contents, list of illustrations, introduction. Frequently abbrev. to *prelims.*

premature ejaculation *(Behav.)*. Inability of the male to postpone ejaculation long enough to satisfy the female.

premaxillary *(Zool.)*. A paired membrane bone of the Vertebrate skull which forms the anterior part of the upper jaw: anterior to the maxilla. Also *premaxilla.*

premeiotic mitosis *(Biol.)*. The nuclear division immediately preceding the organization of nuclei which will divide by meiosis.

premiere *(Image Tech.)*. The first formal public presentation of a completed motion picture or video production.

premix *(Image Tech.)*. Combination of a number of sound track components in preparation for the main mixing operation, to reduce the total number of channels to be handled.

premolars *(Zool.)*. In Mammals, the anterior grinding or cheek teeth, which are represented in the milk dentition.

premorse *(Bot.)*. Looking as if the end had been bitten off.

pre-operational thinking *(Behav.)*. In Piaget's theory, the period from about two to six years when children's thinking is characterized by an ability to represent objects and events internally, but an inability to manipulate these representations in a logical way. According to Piaget, this accounts for the errors children make when asked to perform certain tasks. Cf. period of **sensorimotor development**, of **concrete operations**, of **formal operations**.

preoperculum *(Zool.)*. In Fish, an anterior membrane bone forming part of the gill cover.

préparation mécanique *(Min.Ext.)*. See **mineral processing**.

prepollex *(Zool.)*. In some Vertebrates (Amphibians, Mammals), a rudimentary extra digit of the forelimb.

prepreg *(Aero.)*. Fibrous composite material consisting of unidirectional fibres embedded in matrix of resin prepared in the form of sheet or strip ready for forming, by combining several plies arranged in different directions into the final product.

pre-press proof *(Print.)*. A proof made before the printing plate is made, e.g. from the film elements of the job. Proofs are made to check quality of the film image, especially of the colour separation. It can also simulate the final printed product, e.g. as a guide to printer and customer.

pre-prophase band *(Bot.)*. A band of microtubules 2–3 μm wide which forms and moves to the plasmalemma shortly before prophase begins, predicting the position where the new cell wall will join the old.

prepubic *(Zool.)*. Pertaining to the anterior part of the pubis: in front of the pubis; as bony processes in some Marsupials and Rodents.

prepuce *(Zool.)*. In Mammals, the loose flap of skin which protects the glans penis. adj. *preputial.*

presbyope *(Med.)*. One suffering from presbyopia.

presbyopia *(Med.)*. Long sightedness and impairment of vision due to loss of accommodation of the eye in advancing years.

prescoring *(Image Tech.)*. In motion picture production, the use of previously recorded sound, such as a musical sequence, for cueing the action of dancers or singers.

preselection *(Telecomm.)*. Type of automatic switching used in telephone exchanges.

preselector gearbox *(Autos.)*. A gearbox, generally epicyclic, in which the gear ratio is selected, before it is actually required, by a small lever, being afterwards engaged by pressure on a pedal.

presensitized plate *(Print.)*. A lithographic plate with a light sensitive coating applied by the manufacturer and supplied ready for exposure.

presentation *(Med.)*. The relation which the long axis of the foetus bears to that of the mother; various presentations are defined in terms of the presenting part, as *shoulder presentation, breech presentation,* etc.

preset *(Electronics)*. (1) To establish an initial value or condition, generally by setting one or more controls in advance. (2) A control, e.g. a variable resistance or capacitor, not as readily accessible or easily altered as the main controls, which can be adjusted to obtain initial values or operating conditions.

preset guidance *(Aero.)*. The guidance of controlled missiles by a mechanism which is set before launching and is subsequently unalterable.

pre-shrunk *(Textiles)*. See **compressive shrinkage**.

press *(Eng.)*. A machine used for applying pressure to a workpiece, via a tool, usually for the purpose of carrying out cold-working operations like cutting, bending, drawing, and squeezing.

press *(Print.)*. (1) Any hand-operated machine used for proofing or printing small runs. (2) A general term for the printing stage of a job; exemplified in such phrases as 'going to press', 'in the press'. (3) A general term for the printing and publishing industry, particularly newspapers.

press-and-blow machines *(Glass)*. Machines in which the **parison** is formed by the pressing action of a plunger forced into a mass of plastic glass dropped in a parison mould; the parison is then blown to the shape of the finished ware in another mould.

pressboard *(Elec.Eng.)*. Compressed paper in thick sheets. See **Elephantide pressboard**.

press brake *(Eng.)*. A power press akin to a guillotine, used mainly for producing long bends, as in corrugating or seaming, but also for embossing, trimming and punching.

pressed amber *(Min.)*. See **ambroid**.

pressed brick *(Build.)*. A high-quality brick moulded under pressure, as a result of which it has sharp arrises and a smooth face, making it especially suitable for exposed surface work.

presser *(Textiles)*. The device that closes the beards on machines having bearded needles.

presser-foot *(Textiles)*. On weft-knitting machines, the flexible thin metal device that controls the position of the loops formed during knitting.

press fit *(Eng.)*. A class of fit for mating parts, tighter than a sliding fit and used when the parts do not normally have to move relative to each other.

press forging *(Eng.)*. A die-forging process using a vertical press to apply a slow squeezing action to deform the plastic metal, as distinct from the rapid impact employed in drop forging.

pressing *(Acous.)*. Disk record formed by pressure, with or without heat; the negative of the recording on a stamper is transferred to a large number of pressings for distribution.

pressing *(Textiles)*. The application of pressure, often accompanied by heat and/or steam to smooth fabrics or garments. Also the same process used to introduce chosen creases or pleats. In garment manufacture flat bed presses are used.

pressing boards *(Print.)*. Glazed boards used for removing the impression from printed sheets.

press proof *(Print.)*. The last proof checked over before going to press.

press rolls *(Paper)*. The heavy horizontal rolls situated in the press section of the paper or board making machine in various configurations to consolidate the web and remover water at the nips between them. The rolls may be constructed of granite, rubber covered metal or with a perforated metal shell over a fixed internal suction box.

press tool *(Eng.)*. A tool for cutting, usually comprising at least a punch and a die, or for forming, or for assembling, used in a manual or in a power press.

pressure *(Genrl.)*. See **stress, atmospheric pressure**.

pressure altitude *(Aero.)*. Apparent altitude of the local ambient pressure related to the International Standard Atmosphere.

pressure angle *(Eng.)*. In toothed gearing, the angle between the common tangent to the pitch circles of two teeth at their point of contact and the common normal at that point. Two angles are in common use: 14.5° and 20°.

pressure bomb *(Bot.)*. Thick-walled metal vessel used in investigations of plant water relations, e.g. to apply pressure by compressed air to an excised leaf placed within the vessel with its petiole emerging. Water is forced from the leaf when the air pressure equals the **water potential** of the leaf cells.

pressure broadening *(Phys.)*. The broadening of spectral lines due to increase in pressure and thus due to the effect of neighbouring atoms on the radiating atom.

pressure cabin *(Aero.)*. An airtight cabin which is maintained at greater than atmospheric pressure for the comfort and safety of the occupants. Above 20 000 ft (6 000 m) a differential of $6\frac{1}{2}$ lbf/in² (45 kN/m²) is usual, and above 40 000 ft (12 000 m) one of $8\frac{1}{4}$ lbf/in² (57 kN/m²). Pressurization can be either by a shaft driven *cabin blower*, or by air bled from the compressor of turbine main engines.

pressure cable *(Elec.Eng.)*. A paper-insulated power cable operated under a hydrostatic pressure (up to 1.5 MN/m²) by means of gas (nitrogen) contained in an outer steel pipe or, in more modern forms, an outer reinforced lead sheath; this minimizes ionization. See also **Oilostatic cable**.

pressure capsule *(Eng.)*. See **sylphon bellows**.

pressure circuit *(Elec.Eng.)*. See **voltage circuit**.

pressure co-ordinates *(Meteor.)*. A system of co-ordinates used in **numerical forecasting** in which the vertical ordinate is pressure, *p*. In this system, the vertical velocity *w* is replaced by the total derivative, following the motion, of the pressure i.e. $\frac{dp}{dt}$.

pressure diecasting *(Eng.)*. A process by means of which precision castings of various alloys are made by squirting liquid metal under pressure into a metal die. See **diecasting**.

pressure drag *(Aero.)*. The summation of all aerodynamic forces normal to the surface, resolved parallel to free stream direction; sum of **form drag** and **induced drag**.

pressure forging *(Eng.)*. See **drop forging**.

pressure gauge *(Eng.)*. (1) A flattened tube bent to a curve, which tends to straighten under internal pressure, thus indicating, by the movement of an indicator over a circular scale, the fluid pressure applied to it. Also called *Bourdon gauge*. (2) A liquid manometer.

pressure gradient *(Meteor.)*. (1) The rate of change of the atmospheric pressure horizontally in a certain direction on the Earth's surface as shown by isobars on a weather chart. (2) The rate of change of pressure with distance over the ground, normal to the isobars. The force acting on the air is the *pressure-gradient force*.

pressure-gradient microphone *(Acous.)*. One which offers so little obstruction to the passage of a sound wave that the diaphragm, in practice a ribbon, is acted on by the difference in the excess pressures on the two sides, and therefore tends to move with the particle velocity in the wave. Also *velocity microphone*.

pressure head *(Aero.)*. A combination of a **static pressure** and a **Pitot tube** which is connected to opposite sides of a differential pressure gauge, for giving a visual reading corresponding to the speed of an airflow. Sometimes called *pitot-static tube*.

pressure helmet *(Aero.,Space)*. A flying helmet for the crew of high-altitude aircraft or spacecraft for use with a *pressure*, or *partial pressure suit*. Usually of plastic, with a transparent facepiece, which may be in the form of a visor, the helmet incorporates headphones, microphone, and oxygen supply, and there is usually a feeding trap near the mouth.

pressure in bubbles *(Phys.)*. A spherical bubble of radius *r*, formed in a liquid for which the surface tension is *T*, contains air (or some other gas or vapour) at a pressure which exceeds that in the liquid in its immediate vicinity by $\frac{2T}{r}$. The excess pressure within a soap bubble in air is $\frac{4T}{r}$ since the soap film has two surfaces.

pressure jet *(Aero.)*. A type of small jet-propulsion unit fitted to the tips of helicopter rotor blades, in which small size (to give low drag for **autorotation**) is of greater importance than the losses due to ejecting the efflux at pressures as high as 2 or 3 atmospheres.

pressure leaching *(Min.Ext.)*. Chemical extraction of values from ore pulp in autoclaves, perhaps followed by precipitation as metal or refined salt.

pressure microphone *(Acous.)*. Microphone in which the electrical signal is proportional to the fluctuating component of the pressure in front of the microphone as opposed to the **pressure gradient microphone**.

pressure of atmosphere *(Phys.)*. See **atmospheric pressure, barometric pressure**.

pressure pad *(Image Tech.)*. The device which keeps the film in a gate so that it remains exactly in focus.

pressure-pattern flying *(Aero.)*. The use of barometric pressure altitude to obtain the most favourable winds for long distance, high altitude aerial navigation.

pressure probe *(Bot.)*. Device for measuring the **turgor pressure** within a plant cell by inserting into the cell a fine, fluid-filled capillary connected to a pressure transducer.

pressure ratio *(Aero.)*. The absolute air pressure, prior to combustion, in a **gas turbine, pulse-jet**, or **ramjet** divided by the ambient pressure: analogous to the *compression ratio* of a reciprocating engine.

pressure roller *(Eng.)*. A roller sometimes used in centreless grinding of short, heavy workpieces to assist rotation of the workpiece.

pressure suit *(Aero.)*. An airtight fabric suit, similar to that of a diver, for very high altitude and space flight. It differs from the *partial pressure suit* in being loose-fitting, with bellows or other form of pressure-tight joint, to permit limited movement by the wearer.

pressure-tube reactor *(Nuc.Eng.)*. Reactor in which the fuel elements are contained in a large number of separate tubes through which the coolant water flows, rather than in a single pressure vessel e.g. the Canadian **CANDU** and British **SGHWR** reactors.

pressure-type capacitor *(Elec.Eng.)*. One in which the dielectric is an inert gas under high pressure. Useful for high voltage work since it is self-healing in the event of breakdown.

pressure unit *(Acous.)*. A metal or plastic dome forming the diaphragm of a small moving-coil loudspeaker unit situated in the throat of a horn, for use at intense acoustic pressures.

pressure vessel *(Eng.)*. A container, often made of steel or aluminium, for fluids subjected to pressure above or below atmospheric. Classified as *fired* or *unfired*. Usually cylindrical with dished ends, for ease of construction and support, or spherical, for very high pressures.

pressure vessel *(Nuc.Eng.)*. Reactor containment vessel, usually made of thick steel or prestressed concrete, capable of standing high pressure and used in gas-cooled and light water reactors. See **reactor vessel**.

pressure waistcoat *(Aero.)*. A double-skinned garment, covering the thorax and abdomen, through which oxygen is passed under pressure on its way to the wearer's lungs to aid breathing at great heights, i.e. above 40 000 ft (12 000 m).

pressure welding *(Eng.)*. Welding while the parts to be united are pressed against each other, as in forge welding or in various electrical resistance welding processes or in cold welding.

pressurized *(Aero.)*. Fitted with a device that maintains nearly normal atmospheric pressure, e.g. in an aircraft.

pressurized water reactor *(Nuc.Eng.)*. One using water cooling at a pressure such that its boiling point is above the highest temperature reached.

presswork *(Print.)*. Printing of the job, quality of the result depending on care and attention given; also called *machining*, but not in the case of hand press printing.

Prestel *(Comp.)*. See **videotext**.

presternum *(Zool.)*. In Anura, an anterior element of the sternum, of paired origin and doubtful homologies; the reduced sternum of whalebone whales; the anterior part of the sternum.

prestressed concrete *(Civ.Eng.)*. Concrete which is pre-compressed, using high-tensile wire, thereby enabling it to withstand tensile stresses which it would otherwise be unable to do. Cf. **reinforced concrete**.

presystolic *(Med.)*. Pertaining to, or occurring just

before, the beginning of the systole of the heart, e.g. *presystolic* murmur.

pre-TR cell *(Radar)*. A gas-filled RF switching valve which protects the **TR tube** in a radar receiver from excessive power. Also acts as a block to receiver frequencies other than the fundamental.

pretrematic *(Zool.)*. Anterior to an aperture, as (in Selachii) that branch of the 9th cranial nerve which passes anterior to the first gill cleft. Cf. **post-trematic**.

preventive choke-coil *(Elec.Eng.)*. A choking coil connected between the two halves of the moving-contact used in varying the tapping of a transformer or battery; its purpose is to reduce the circulating current which flows as a result of the short-circuiting of the turns or cells between adjacent tappings by the moving-contact, as it travels from one fixed-contact to the next.

preventive resistance *(Elec.Eng.)*. A resistance connected between the two halves of the moving-contact used in varying the tapping of a transformer or battery; its purpose is to reduce the circulating current which flows as a result of the short-circuiting of the turns or cells between the adjacent tappings, as the moving-contact travels from one fixed-contact to the next.

preview *(Image Tech.)*. (1) A special showing of a motion picture or video production to a limited audience before general public presentation. (2) In television, viewing the picture immediately before transmission or recording.

prezygapophysis *(Zool.)*. A facet or process on the anterior face of the neurapophysis of a vertebra, for articulation with the vertebra next in front. Cf. **postzygapophysis**.

PRF *(Telecomm.)*. See **pulse repetition frequency**.

priapism *(Med.)*. Abnormally, and often painfully, persistent erection of the penis unaccompanied by sexual desire; may be due to local disease or to hypercoaguability of the blood.

Priapulida *(Zool.)*. A phylum of coelomate, superficially segmented wormlike animals, living in mud. They have a straight gut with an anterior mouth and a posterior anus. The nervous system is not separated from the epidermis and the urinogenital system is simple, with solenocytes.

Price's guard-wire *(Elec.Eng.)*. A conductor placed around the edge of a piece of insulating material under test; it is arranged to be at the same potential as the surface of the material to prevent a leakage current from the surface to earth.

pricking-up *(Build.)*. The operation of scoring the surface of the first coat of plaster to provide a key for the next: the whole operation of laying and scoring such a coat.

prickle *(Bot.)*. A hard, sharp-pointed outgrowth of the epidermis, a multicellular trichome which is not vascularised, the 'thorn' of the rose.

prickly heat *(Med.)*. See **miliaria**.

prick punch *(Horol.)*. A punch used to locate the centre of holes in a plate prior to drilling.

Pridoli *(Geol.)*. The youngest epoch of the Silurian period.

prill *(Eng.)*. (1) To make granular or crystalline solids fluid, e.g. for extraction from slag. (2) To turn into pellet form by melting and letting the drops solidify in falling. (3) Bullion bead produced by cupellation of lead button during fire assay of gold or silver.

prima *(Print.)*. The first word of the page (or sheet) following the one being read. [Rare].

primacord fuse *(Min.Ext.)*. Fuse based on **pentaerythritol tetranitrate**, with detonating effect. Speed of detonation, 7000 m/s.

primacy effect *(Behav.)*. (1) In impression formation, the fact that attributes noted early on carry a greater weight than attributes noted at a later time; (2) In memory, the tendency for the first items on a list to be remembered better than other items on the list.

primacy process thinking *(Behav.)*. In psychoanalytic theory, the mode of thinking characteristic of unconscious mental activity; it is governed by the **pleasure-principle** rather than the **reality-principle**. Cf. **secondary process thinking**.

primaquine *(Med.)*. Anti-malarial drug, used prophylactically and therapeutically.

primaries *(Zool.)*. In Birds, the remiges attached to the manus.

primary *(Bot.,Zool.)*. Original, first formed; as *primary meristem, primary body-cavity*; principal, most important; as *primary feathers, primary axis*.

primary *(Chem.)*. A substance which is obtained directly, by extraction and purification, from natural raw material; e.g. benzene, phenol, anthracene are coal-tar *primaries*. Cf. **intermediate**.

primary *(Elec.Eng.)*. Same as *primary winding*. See **secondary winding**.

primary acids *(Chem.)*. Acids in which the carboxyl group is attached to the end carbon atom of a chain, i.e. to $-CH_2$ group.

primary additive colours *(Phys.)*. Minimum number of spectral colours (red, green, blue) which can be adjusted in intensity and, when mixed, visually make a match with a given colour. This match with real colour cannot be perfect, since one primary may have to be negative in intensity. A typical set of primary additive spectral colours are, in nanometres: red (640), green (537), blue (464). For colour TV and photography, original colour has to be separated into such arbitrary components.

primary alcohols *(Chem.)*. Alcohols containing the group $-CH_2.OH$. On oxidation, they form aldehydes and then acids containing the same number of carbon atoms as the alcohol.

primary amines *(Chem.)*. Amines containing the amino group $-NH_2$. Primary amines are converted into the corresponding alcohol by the action of nitrous acid, nitrogen being eliminated.

primary body *(Bot.)*. That part of the plant body formed directly from cells cut off from the apical meristems.

primary body cavity *(Zool.)*. The blastocoel or segmentation cavity formed during cleavage, or that part of it which is not subsequently obliterated by mesenchyme.

primary bow *(Meteor.)*. See **rainbow**.

primary cell *(Elec.Eng.)*. Voltaic cell in which chemical energy of constituents is changed to electrical energy when current is permitted to flow. Such a cell cannot be recharged electrically because of the irreversibility of chemical reaction occurring therein. It may be used as a sub-grade standard cell for calibration purposes.

primary cell wall *(Bot.)*. See **primary wall**.

primary circuit *(Build.)*. Pipe circuit in which water circulates between a boiler (or other heater) and a hot water storage vessel.

primary coil *(Elec.Eng.)*. A coil, forming part of an electrical machine or piece of apparatus, in which flows a current setting up the magnetic flux necessary for the operation of the machine or apparatus.

primary colours *(Build.)*. For pigments these are red, yellow, and blue. See **primary additive colours, primary subtractive colours**.

primary constants *(Elec.Eng.)*. Those of capacitance, inductance, resistance, and leakance of a conductor to earth (coaxial or concentric) or to return (balanced) conductor, per unit length of line.

primary constriction *(Biol.)*. Region at which two chromatids are joined in metaphase chromosomes, and to which spindle microtubules attach. It appears narrower and more condensed than the arms.

primary coolant *(Nuc.Eng.)*. *Reactor coolant.* The fluid circulated through the reactor to remove heat. In *direct-cycle* reactors such as the boiling water reactor this drives the turbines directly but in other types the heat is passed via a heat exchanger to a **secondary coolant**.

primary crushing *(Min.Ext.)*. Reduction of run-of-mine ore as severed to somewhere below 6 in. (15 cm) diameter lump, performed in jaw or gyratory breakers.

primary current *(Electronics)*. That formed by **primary electrons**, as contrasted with secondary electrons, which reduce or even reverse it.

primary dispersion *(Min.Ext.)*. In **geochemical prospecting**, the diffusion of metals or other elements through the bedrock surrounding an ore body. Also *halo*.

primary electrons *(Electronics)*. (1) Those incident on a surface whereby secondary electrons are released (see also **primary ionization**). (2) Those released from atoms by internal forces and not by external radiation as with secondary electrons.

primary emission *(Electronics)*. Electron emission arising from the irradiation (including thermal heating) or the application of a strong electric field to a surface.

primary flexure *(Zool.)*. The flexure of the mid-brain by which, in Vertebrates, the forebrain and its derivatives are bent at a right angle to the axis of the rest of the brain.

primary flow *(Electronics)*. That of **carriers** when they determine the main properties of a device.

primary gneissic banding *(Geol.)*. Exhibited by certain igneous rocks of heterogeneous composition, possibly due to the admixture of two magmas only partly miscible, injection of magma along bedding or foliation planes in the country rocks or selective mobilization of a rock under metamorphosis.

primary growth *(Bot.)*. The growth that results from division and expansion of cells produced at apical meristems. Cf. **secondary growth.**

primary immune response *(Immun.)*. The response made by an animal to an antigen on the first occasion that it encounters it. Characteristically low levels of antibody are produced after several days and these gradually decline. However the immune system has been 'primed' so that a secondary response can be evoked on subsequent challenge with the same antigen. Responses of cell mediated immunity follow a similar pattern.

primary ionization *(Phys.)*. (1) In collision theory, the ionization produced by the primary particles, in contrast to total ionization, which includes the *secondary ionization* produced by delta-rays. (2) In counter tubes, the total ionization produced by incident radiation without gas amplification.

primary luminous standard *(Phys.)*. A standard of luminous intensity which is reproducible from a given specification.

primary meristem *(Bot.)*. Any of the three meristematic tissues derived in a pattern appropriate to the organ from the apical meristem, the *protoderm, ground meristem* and *procambium*. Cf. **secondary meristem.**

primary metal *(Eng.)*. Ingot metal produced from newly smelted ore, with no addition of recirculated scrap.

primary nitro-compounds *(Chem.)*. Nitro-compounds containing the group $-CH_2.NO_2$.

primary node *(Bot.)*. The node at which the cotyledons are inserted.

primary phloem *(Bot.)*. Phloem tissue that differentiates from the procambium during the primary growth of a vascular plant, consisting of protophloem and metaphloem. Cf. **secondary phloem.**

primary pit field *(Bot.)*. A thin area in the primary wall of a plant cell, often penetrated by plasmodesmata, within which one or more pits may develop if a secondary wall is formed.

primary production *(Ecol.)*. See **production.**

primary production *(Min.Ext.)*. That which occurs when oil and gas flow naturally to the well bore without assistance. Also *primary recovery*. Cf. **secondary production, tertiary production.**

primary radar *(Radar)*. One in which the incident power from the transmitter is reflected from the target to form the return signal or *echo*. Cf. **secondary radar.**

primary radiation *(Phys.)*. Radiation which is incident on the absorber, or which continues unaltered in photon energy and direction after passing through the absorber. Also *direct radiation*.

primary ray *(Bot.)*. See **medullary ray.**

primary reinforcer *(Behav.)*. A stimulus that increases the probability of preceding responses even if the stimulus has never been experienced before.

primary sere *(Bot.)*. A sere starting from a new, bare surface not previously occupied by plants. Land newly exposed on a rising coast or by a retreating glacier.

primary service area *(Telecomm.)*. That within which reception from a broadcast transmitter gives acceptable reproduction of sound and/or vision; for medium- and short-wave broadcasting interference and fading are the usual limiting factors. In VHF and UHF, **noise** and **multipath reception** are more likely to intrude.

primary solid solution *(Eng.)*. A constituent of alloys that is formed when atoms of an element *B* are incorporated in the crystals of a metal *A*. In most cases solution involves the substitution of *B* atoms for some *A* atoms in the crystal structure of *A*, but in a few instances the *B* atoms are situated in the interstices between the *A* atoms.

primary standard *(Genrl.)*. A standard agreed upon as representing some unit (e.g. length, mass, e.m.f.) and carefully preserved at a national laboratory. Cf. **secondary standard.**

primary store *(Comp.)*. See **main memory.**

primary stress *(Eng.)*. An axial or direct tensile or compressive stress, as distinct from a bending stress resulting from deflection.

primary structure *(Aero.)*. All components of an aircraft structure, the failure of which would seriously endanger safety, e.g. wing or tailplane spars, main fuselage frames, engine bearers, portions of the skin which are highly stressed.

primary subtractive colours *(Phys.)*. Minimum number of spectral colours (cyan, magenta, yellow), which when subtracted in the right intensity from a given white, result in a match with a given colour. These are complementary to the primary additive colours and are used in printing colour films, e.g., Technicolor.

primary succession *(Ecol.)*. A succession beginning on an area not previously occupied by a community, e.g., a newly exposed rock or sand surface. Cf. **secondary succession.**

primary tissue *(Bot.)*. Tissue formed from cells derived from primary meristems.

primary voltage *(Elec.Eng.)*. That which is applied to the input side of a transformer.

primary wall *(Bot.)*. The earlier-formed part of the **cell wall** characteristically laid down while the cell is expanding and typically richer in pectins than the **secondary wall.**

primary xylem *(Bot.)*. Xylem tissue that differentiates from the procambium during primary growth in a vascular plant, consisting of *protoxylem* and *metaxylem*. Cf. **secondary xylem.**

Primates *(Zool.)*. An order of Mammals: dentition complete, but unspecialized; brain, especially the neopallium, large and complex; pentadactyl; eyes well developed and directed forwards, the orbit being closed behind by the union of the frontal and jugal bones; basically arboreal animals; uterus a single chamber; few young produced, parental care lasting a long time after birth. Lemurs, Tarsiers, Monkeys, Apes, and Man.

Primaton overlay *(Print.)*. A mechanical overlay produced by dusting a proof with special powder and heating the sheet to fuse the powder.

prime contract *(Aero.)*. That for the whole aircraft or weapon system from design and manufacture to test and supply, including the management of sub-contractors for completion to time and cost.

primed *(Immun.)*. When referring to a whole animal it describes the condition in which prior contact with an antigen will result in a secondary response on subsequent challenge with the same antigen. When referring to a cell population describes the presence of cells which have already been activated by an antigen.

prime mover *(Eng.)*. An engine or other device by which a natural source of energy is converted into mechanical power. See **gas engine, internal-combustion engine, oil engines, petrol engine, steam engine, steam turbine, water turbine.**

prime number *(Maths.)*. A natural number other than 1, divisible only by itself and 1. Or, any natural number which has precisely two divisors. The following numbers are prime: 2, 3, 5, 7, 11, 13,...37,...5521. Every natural number greater than 1 may be resolved uniquely into a product of prime numbers; e.g. $8316 = 2^2 \times 3^3 \times 7 \times 11$. In the case of a prime number *p*, the *product* has to be interpreted as *p* itself.

primer *(Biol.)*. Nucleotide sequence with a free 3′ -OH group, needed to initiate synthesis by DNA polymerase. A short primer sequence base-paired to a specific site on a longer DNA strand can initiate polymerization from that site.

primer *(Build.)*. The first coat of paint on a bare surface, formulated to provide a suitable surface for the next coats. For wood, the absorption of the material must be allowed for, for metals, their tendency to corrode, and the high alkalinity in the case of concrete. Special primers may have to be formulated for unusual surfaces.

primer *(Min.Ext.)*. Cartridge in which detonator is placed in order to initiate explosion of string of high-explosive charges in bore hole.

primes *(Eng.)*. Metal sheet and plate of the highest quality and free from visible imperfections.

primigravida *(Med.)*. A woman who is pregnant for the first time.

priming *(Eng.)*. (1) The delivery by a boiler of steam containing water in suspension, due to violent ebullition or frothing. (2) The operation of filling a pump intake with fluid to expel the air. (3) The operation of injecting petrol into an engine cylinder to assist starting.

priming pump *(Aero.)*. A manual or electric fuel pump which supplies the engine during starting where an injection carburettor or fuel injection pump is fitted.

priming valve *(Eng.)*. A valve fitted on the suction side of a pump to assist in priming.

primipara *(Med.)*. A woman who gives, or has given, birth to a child for the first time.

primitive *(Bot.,Zool.)*. Original, first-formed, of early origin, the ancestral condition; as the *primitive streak*.

primitive equation model *(Meteor.)*. A numerical forecast model that uses the **primitive equations**, not the **filtered equations**.

primitive equations *(Meteor.)*. The fundamental equations of motion of a fluid modified only by the use of the **hydrostatic approximation** and the neglect of viscosity. The primitive equations comprise three prognostic equations (the *x* and *y* components of the momentum equation and the thermodynamic equation of energy) and three diagnostic equations (the continuity equation, the **hydrostatic approximation** and the equation of state). These equations form a closed set in the dependent variables of velocity, pressure, density and temperature. The solutions include gravity waves but not vertically propagating sound waves.

primitive streak *(Zool.)*. In developing Birds and Reptiles, a thickening of the upper layer of the blastoderm along the axis of the future embryo; represents the fused lateral lips of the blastopore.

primordial germ cells *(Zool.)*. In the early embryo, cells which will later give rise to the gonads.

primordium *(Biol.)*. An organ, cell or other structure in the earliest stage of development or differentiation. Also *anlage*. adj. *primordial*.

principal *(Build.)*. See **roof truss**.

principal axes of a body *(Phys.)*. At any given point *O*: three mutually perpendicular axes *Oxyz* such that the three products of inertia about the coordinate planes are all zero. The principal axes at the centre of gravity are the axes of symmetry if any exist.

principal axis *(Phys.)*. See **optic axis**.

principal axis *(Telecomm.)*. Direction of maximum sensitivity or response for a transducer or antenna.

principal direction *(Maths.)*. See **curvature (3)**.

principal normal *(Maths.)*. See **moving trihedral**.

principal part *(Maths.)*. Of a function $f(z)$ at a singularity: see **pole**.

principal planes of a lens *(Phys.)*. Conjugate planes, perpendicular to the principal axis of a lens system, for which the transverse magnification is unity. Also *unit planes*.

principal points of a lens *(Phys.)*. Two points on the principal axis of a lens or lens system where the principal planes intersect the axis. If the object distance l is measured from one of the principal points and the image distance l' from the other, then the simple lens formula

$$\frac{1}{f}=\frac{1}{l'}-\frac{1}{l}$$

can be used and f is the equivalent focal length (Cartesian **convention of signs**). See **principal planes, nodal points**.

principal quantum number *(Phys.)*. When *quantum mechanics* is applied to any particle moving in a central potential, e.g. the electron in a hydrogen atom, the angular momentum and the z-component of the angular momentum are quantized with quantum numbers l and m_l. For a particular l and m_l the solutions of the Schrödinger equation are well-behaved only if the energy is also quantized. These solutions are denoted by n, the *principal quantum number*. In many-electron atoms each electron is described by an orbital specified by n, l, and m_l and the spin quantum number m_s; the assignment of the orbitals depends on the Pauli exclusion principle. In general the energy depends on both n and l. The value of n denotes the *electron shell* and that of l, the subshell that the electron occupies.

principal radius of curvature *(Maths.)*. See **curvature (3)**.

principal rafter *(Build.)*. A rafter forming part of the roof truss proper and supporting the purlins.

principal ray *(Phys.)*. From an object point lying off the axis, the ray passing through the centre of the entrance pupil of the system.

principal series *(Phys.)*. Series of optical spectrum lines observed in the spectra of alkali metals. Has led to energy levels for which the orbital quantum number is unity to be designated *p-levels*.

principal stress *(Eng.)*. The component of a stress which acts at right angles to a surface, occurring at a point at which the shearing stress is zero.

principal values of the inverse trigonometrical functions *(Maths.)*. Those usually taken are those which contain the smallest numerical value irrespective of its sign.

principle of equivalence *(Phys.)*. A statement of the theory of relativity: an observer has no means of distinguishing whether his laboratory is in a uniform gravitational field or is in an accelerated frame of reference. See **relativity, general relativity**.

principle of least action *(Phys.)*. Principle stating that the actual motion of a conservative dynamical system betweeen two points takes place in such a way that the action has a minimum value with reference to all other paths between the points which correspond to the same energy.

principle of least constraint *(Eng.)*. The motions of any number of interconnected masses under the action of forces deviate as little as possible from the motions of the same masses if disconnected and under the action of the same forces. The motions are such that the constraints are a minimum, the constraint being the sum of the products of each mass and the square of its deviation from the position it would occupy if free.

principle of least time *(Phys.)*. See **Fermat's principle of least time**.

principle of reinforcement *(Behav.)*. Skinner's term for the **law of effect**.

principle of relativity *(Phys.)*. A universal law of nature which states that the laws of mechanics are not affected by a uniform rectilinear motion of the system of coordinates to which they are referred. Einstein's relativity theory is based on this principle, and on the postulate that the observed value of the velocity of light is constant and is independent of the motion of the observer. See **relativity**.

principle of superposition *(Phys.)*. The resultant disturbance at a given place and time caused by a number of waves traversing the same space is the vector sum of the disturbances which would have been produced by the individual waves separately. The principle is the basis for the explanation of interference and diffraction effects.

principle of the equipartition of energy *(Chem.)*. The total energy of a molecule in the normal state is divided up equally between its different capacities for holding energy, or **degrees of freedom**.

print *(Foundry)*. See **core prints**.

print *(Image Tech.)*. The image, usually a positive one, obtained by exposing a photographic material through another image, such as a negative.

printed circuit *(Electronics)*. An electronic sub-assembly consisting of an insulating board or card with copper conductors laminated on; the conductors may be formed by a photo-chemical etching process or by electrodeposition. The circuit components, diodes, transistors, integrated circuits etc., may be inserted into pre-drilled holes by hand or by machine, manufacture being concluded by hand or dip soldering.

printer *(Comp.)*. Output device producing characters or graphic symbols on paper. See **line printer**, **Xerographic-**, **daisy-wheel-**, **dot-matrix-**, **golf-ball-**, **laser-**, **ink-jet-**, **impact-**, **thermal-**.

printer *(Image Tech.)*. Machine for the exposure of photographic paper or film to produce prints, either by contact or by optical means. Exposure may be made one picture at a time or, in the case of motion pictures, with the film moving continuously.

printer point *(Image Tech.)*. See **light change point**.

print hammer *(Comp.)*. Component causing the contact between the character, ribbon and paper in an **impact printer**.

printing *(Image Tech.)*. The operation of making a still or motion picture **print** by the exposure of photographic paper or film.

printing *(Print.)*. Any process of producing copies of designs or lettering by transferring ink to paper (or other material) from a printing surface. There are 3 main classes according to the method of application of the ink to the printing surface: (*a*) *relief*, or *letterpress*, printing surfaces have the ink-carrying parts in relief, so that rollers deposit ink on these parts only, as in printer's type; (*b*) *planographic* printing surfaces are prepared so that parts accept the ink from the rollers although there is no difference in level; the ink-accepting parts may be greasy, the remainder being moist and ink-rejecting, as in **lithography**; (*c*) *intaglio* printing surfaces have the ink-carrying portions hollowed out; the whole surface is covered with ink and then cleaned off, leaving the hollows filled with ink, which is lifted out when the paper is pressed into contact. All classes can be adapted for use with a cylindrical printing surface which can be printed at high speed by continuous rotation against another cylinder, with the paper to be printed running between them.

printing diameter *(Print.)*. The correct diameters of printing cylinder and impression surface.

printing down *(Print.)*. A stage in the making of printing surfaces, for any of the main processes, in which the surface, after being made light sensitive, is exposed to suitable light through a negative (or, in some cases, a positive).

printing ink *(Print.)*. A mixture of carbon black, or other pigments, in a vehicle of mineral oil, linseed oil, etc. Inks are formulated to dry by penetration, evaporation, oxidation, or by a combination of these, and also can be *heat-set*, *cold-set*, or *moisture-set*.

printing-out paper *(Image Tech.)*. See **POP**.

print out *(Comp.)*. Printed output from the computer.

print-out mask *(Print.)*. Opaque mask used to cover the image areas of a plate during a second exposure to remove unwanted work from the printing plate.

print-through *(Acous.,Comp.)*. In magnetic tape recording, the transfer of a recording from one layer to another when the tape is spooled or reeled giving rise to a form of distortion; also called *transfer*.

prisere *(Bot.)*. Same as **primary sere**.

prism *(Crystal.)*. A hollow (open) crystal form consisting of three or more faces parallel to a crystal axis.

prism *(Maths.)*. A solid of which the ends are similar, equal and parallel polygons, and of which the sides are parallelograms.

prism *(Phys.)*. Triangular prisms made of glass and other transparent materials are used in a number of optical instruments. Equilateral prisms are used at minimum deviation in spectroscopes for forming spectra, 90° prisms are used for totally reflecting a ray through a right angle in binoculars, periscopes and range-finders.

prismatic *(Genrl.)*. Prism-shaped; composed of prisms.

prismatic astrolabe *(Surv.)*. An instrument for observing stars at an altitude of 60° (in some instruments, 45°) at different azimuths around the horizon, these observations being used for the computation of latitude and local time.

prismatic binoculars *(Phys.)*. Binocular telescopes in which the tubes, instead of being straight are effectively shortened by using total reflecting prisms to 'fold' the light paths. The prisms at the same time produce an erect image.

prismatic coefficient *(Ships)*. The ratio between the immersed volume of the vessel and the volume of an enclosing prism with a constant transverse section identical with the maximum immersed cross-section area of the vessel.

prismatic compass *(Surv.)*. A handheld form of surveyor's compass in which the eye vane carries a prism reflecting a view of a graduated ring, attached to and moving round with the compass needle.

prismatic layer *(Zool.)*. In the shell of *Mollusca*, a layer consisting of calcite or aragonite lying between the periostracum and the nacreous layer. In the shell of *Brachiopoda*, the inner layer of the shell, composed mainly of calcareous, but partly of organic material.

prismatic (monoclinic) sulphur *(Chem.)*. See **sulphur**.

prismatic spectrum *(Phys.)*. A spectrum formed by refraction in a prism, as contrasted with a *grating spectrum* formed by diffraction.

prismatic system *(Crystal.)*. See **orthorhombic system**.

prism light *(Build.)*. A pavement light in which glass prisms internally reflect light.

prismoid *(Maths.)*. A body which has plane parallel ends and is bounded by plane sides.

prismoidal formula *(Civ.Eng.)*. A formula used in the calculation of earthwork quantities. It states that the volume of any prismoid is equal to one-sixth its length multiplied by the sum of the two end-areas plus four times the mid-area.

prism square *(Surv.)*. A form of **optical square** in which the fixed angle of 90° between the lines of sight is obtained by reflection from the surfaces of a suitably shaped prism.

prison ashlar *(Build.)*. A block of stone dressed so that the faces are wrought into holes.

privacy *(Comp.)*. Recognition of the private nature of certain data. In consideration of privacy, safeguards are usually built into systems which hold confidential data to prevent unauthorized access. Now often also safeguarded by legal constraints.

privacy system *(Telecomm.)*. See **scrambler**, **inverter**.

private automatic branch exchange *(Telecomm.)*. A small automatic exchange on a subscriber's premises, for internal telephone connections, with extensions over the public telephone system through lines to the local exchange. Abbrev. *PABX*.

private automatic exchange *(Telecomm.)*. An automatic exchange on private premises; not connectable with the public telephone system. Abbrev. *PAX*.

private branch exchange *(Telecomm.)*. An automatic or manual exchange on a subscriber's premises which is used for internal connections, with extension through the local exchange to the public telephone system. Abbrev. *PBX*.

private exchange *(Telecomm.)*. An exchange in a private establishment, which is not connected in any way with the public telephone service. Abbrev. *PX*.

private manual branch exchange *(Telecomm.)*. A small manually-operated exchange on a subscriber's premises for establishing internal connections and extensions, and external connections over lines to the local exchange. Abbrev. *PMBX*.

private manual exchange *(Telecomm.)*. A manually-operated exchange, not connected with the public telephone service. Abbrev. *PMX*.

Pro *(Chem.)*. Symbol for proline.

pro- *(Genrl.)*. Prefix from Gk. and L. *pro*, before in time or place, used in the sense of 'earlier' or 'more primitive', or 'placed before'.

probability density *(Phys.)*. Quantum mechanics suggests that electrons must not be regarded as being located at a defined point in space, but as forming a cloud of charge surrounding the nucleus, the cloud density being a measure of the probability of the electron being located at this point. See **uncertainty principle**.

probability density function *(Stats.)*. The first derivative of the cumulative distribution function of a continuous random variable, often identified as the probability that a random variable takes a value in an infinitesmal interval divided by the length of the interval.

probable ore *(Min.Ext.)*. Inferred ore, partly exposed and sampled but not fully *blocked-out* in panels. See **blocking-out**.

proban *(Chem.)*. Flameproof finish for textile fabrics based on a phosphorus compound, tetrakis (hydroxy-methyl) phosphonium chloride.

proband *(Biol.)*. In human genetics, the affected individual who brings a family to the notice of the investigator.

probang *(Vet.)*. A flexible tube which may be passed into the oesophagus and stomach of animals; used for relieving obstructions and administering fluids.

probe *(Acous.)*. See **sound probe**.

probe *(Biol.)*. A radioactive or otherwise labelled single-stranded sequence of nucleic acid which hybridizes to another single stranded nucleic acid, usually separated into discrete spots on a filter. The label then detects the presence of a sequence complementary to the probe sequence.

probe *(Electronics)*. (1) Electrode of small dimensions compared with the gas volume, placed in gas-discharge tube to determine the space potential. (2) Magnetic or conducting device to extract power from a waveguide. (3) Coil or semiconductor sensing element associated with a fluxmeter.

probe *(Med.)*. A surgical instrument with a blunt end, used for exploring wounds, sinuses and cavities.

probe *(Phys.)*. Portable radiation-detector unit, cable-connected to counting or monitory equipment.

probe *(Space)*. A space vehicle, especially one unmanned, sent to explore near and outer space, to collect and transmit data back to Earth. If a planet or its environment is explored, the vehicle is sometimes referred to as a *planetary probe*.

problem solving behaviour *(Behav.)*. Diverse strategies used by animals and humans to overcome difficulties in attaining a desired goal; it implies a higher order of intelligent behaviour than simple **trial and error** strategies.

Proboscidea *(Zool.)*. An order of Mammals having a long prehensile proboscis with the nostrils at the tip, large lophodont molars, and a pair of incisors of the upper jaw enormously developed as tusks; semiplantigrade; forest living herbivorous forms of Africa and India. It includes Elephants and the extinct Mammoths.

proboscis *(Zool.)*. An anterior trunklike process: in Turbellaria and Polychaeta, the protrusible pharynx: in Nemertinea, a long protrusible muscular organ lying above the mouth: in some Insects, the suctorial mouth-parts: in Hemichorda, a hollow club-shaped or shield-shaped structure in front of the mouth: in Proboscidea, the long flexible prehensile nose.

procaine *(Med.)*. Crystalline solid. Its hydrochloride, *novocain(e)*, is used as a local anaesthetic.

procambium *(Bot.)*. *Provascular tissue*. The cells of *primary meristem* which are typically longer than broad, which differentiates into primary xylem and primary phloem and, in some cases, cambium.

procartilage *(Zool.)*. An early stage in the formation of cartilage in which the cells are still angular in form and undergoing constant division; embryonic cartilage.

procaryote *(Bot.)*. See **prokaryote**.

procedure *(Comp.)*. See **subroutine**.

Procellariiformes *(Zool.)*. An order of Birds. Wandering ocean species often very large, with long narrow wings. Lay one white egg, often in a burrow. Petrels, Shear-waters, Albatrosses.

process *(Bot.,Zool.)*. An extension or projection.

process annealing *(Eng.)*. Heating steel sheet or wire between cold-working operations to slightly below the critical temperature and then cooling it slowly. The process is similar to tempering but will not give as much softness and ductility as full annealing.

process camera *(Image Tech.)*. A large copying camera specifically designed for use in photo-mechanical reproduction processes in colour and black-and-white, with particular emphasis on stability and uniformity of image.

process chart *(Eng.)*. Chart in which a sequence of events is portrayed diagrammatically by means of conventional symbols.

process control *(Comp.)*. Automatic monitoring and control of a process by a computer programmed to respond appropriately to the feedback from the process.

process control *(Electronics)*. In a complicated industrial or chemical process, control of various sections of the plant by electronic, hydraulic or pneumatic means, taking rates of flow, accelerations of flow, changes of law, temperatures and pressures into account automatically.

process-engraving *(Print.)*. A relief printing plate made with the aid of the process camera followed by etching; or by means of an **electronic engraving** machine.

process factor *(Phys.)*. See **separation factor**.

processing *(Image Tech.)*. The sequence of chemical reactions, washing and drying involved in treating an exposed photographic material, paper or film, to produce a permanent visible image which can be safely handled in further operations.

processing routes *(Min.Ext.)*. Term used in considering alternative methods of treating a specified ore or concentrate. Main 'routes' are physical, chemical and pyro-metallurgical.

process metallurgy *(Eng.)*. The science and technology of extracting metals from their ores and purifying them.

processor *(Comp.)*. A device which can perform logical and arithmetical operations. See **central processor**.

prochlorite *(Min.)*. See **ripidolite**.

Prochlorophyceae *(Bot.)*. Prokaryotic algae with the pigmentation of green algae (chlorophyll a and b, and no phycobilins) rather than blue-green algae. Two or three species so far identified. Of interest as possibly representing group from which green plant chloroplasts may have evolved. See **endosymbiotic hypothesis**.

procidentia *(Med.)*. A falling down or prolapse, esp. severe prolapse of the uterus.

procoelous *(Zool.)*. Concave anteriorly and convex posteriorly; said of vertebral centra.

proctal *(Zool.)*. Anal.

proctalgia *(Med.)*. Neuralgic pain in the rectum.

proctectomy *(Med.)*. Surgical removal of the rectum.

proctitis *(Med.)*. Inflammation of the rectum.

proctoclysis *(Med.)*. The slow injection of large amounts of fluid into the rectum.

proctodaeum *(Zool.)*. That part of the alimentary canal which arises in the embryo as a posterior invagination of ectoderm. Cf. **stomodaeum, mid-gut**. adj. *proctodaeal*.

proctodynia *(Med.)*. Pain in or around the anus.

proctologist *(Med.)*. A surgeon specializing in diseases of **anus** and **rectum**.

proctoscope *(Med.)*. An instrument which is used for inspecting the mucous membrane of the rectum.

proctosigmoiditis *(Med.)*. Inflammation of the rectum and of the sigmoid flexure of the colon.

proctotomy *(Med.)*. Surgical incision of the anus or rectum for the relief of stricture.

proct-, procto- *(Genrl.)*. Prefix from Gk. *proktos*, anus.

procumbent *(Bot.)*. Lying loosely on the ground surface.

procurement *(Aero.)*. Organizational procedure for obtaining equipment, supplies, services and personnel.

procuticle *(Zool.)*. In the cuticle of Insects, a multilaminar layer initially present below the *epicuticle* and pierced by pore canals running perpendicularly to it. In soft transparent areas this undergoes no apparent change after formation, but in other cases the outer part becomes hard, dark sclerotized *exocuticle*, the inner unchanged part then being called the *endocuticle*. See **tanning**, **sclerotin**.

prodromal *(Med.)*. Premonitory of disease.

producers *(Ecol.)*. In an ecosystem, autotrophic organisms, largely green plants, which are able to manufacture complex organic substances from simple inorganic compounds. Cf. **consumers**, **decomposers**.

product *(Maths.)*. See **multiplication**.

production *(Ecol.)*. Biomass, or heat of combustion of the biomass, expressed on an area basis, (units: $g\,m^{-2}$ or $MJ\,m^{-2}$). Primary production is production by green plants. Secondary production is biomass produced by heterotrophic organisms. The rate of production is called the *productivity* (units: $g\,m^{-2}$ $year^{-1}$). Gross primary productivity is the rate of community photosynthesis, whereas net primary productivity is the rate of community photosynthesis minus the rate of community respiration.

production choke *(Min.Ext.)*. Fixed aperture at the well head which limits the flow of oil from a well to the most economical or best allowable rate.

production platform *(Min.Ext.)*. Offshore platform from which the flow of oil from many wells is controlled and stored before onward transmission to the refinery.

production reactor *(Nuc.Eng.)*. One designed for large-scale production of transmutation products such as plutonium.

production string *(Min.Ext.)*. The smallest casing in an oil well which reaches from the producing zone to the well head, up which the oil passes.

productivity *(Ecol.)*. See **production**.

products of inertia *(Phys.)*. Of a body about two planes, the sum Σmxy taken over all particles of the body where m is the mass of a particle and x and y the perpendicular distances of the particle from the specified planes.

products pipeline *(Min.Ext.)*. That which runs from a refinery to distributors and may carry many different products separated by **batching spheres**.

pro-ecdysis *(Zool.)*. The period of preparation for a moult in Arthropoda during which the new cuticle is laid down and the old one ultimately detached from it.

pro-embryo *(Bot.)*. The structure formed by the first few cell divisions of the zygote of seed plants, before differentiation into suspensor and the embryo proper.

profile *(Build.)*. A temporary guide set out at corners, normally with small timbers, to act as a guide for the foundations of a building.

profile *(Surv.)*. A longitudinal section, usually along the centre line of a proposed work such as a railway.

profile drag *(Aero.)*. The 2-dimensional drag of a body, excluding that due to lift; the sum of the surface-friction and form drag.

profile grinding *(Eng.)*. The grinding of cylindrical work without traversing the wheel whose periphery is profiled to the form required and extends over the full length of the work. Also **plunge grinding**.

profiling *(Eng.)*. Producing the profile of a die or other workpiece, (1) with a modified milling machine, incorporating a tracing mechanism, or (2) with a grinding machine using a wheel dressed to correspond to the required profile.

proflavine *(Med.)*. Deep orange-coloured crystalline powder, used in dilute solution as an antiseptic.

progeria *(Med.)*. Premature old age. Occurring in children, the condition is characterized by dwarfism, falling out of hair, wrinkling of the skin and senile appearance.

progestational *(Med.)*. Used of the luteal phase of the oestrous or menstrual cycle during which the endometrium is prepared for nidation; of the proliferative reaction of the endometrium towards the fertilized ovum or an irritant foreign body mimicking the ovum; of the hormones which bring about these effects.

proglottis *(Zool.)*. One of the reproductive segments forming the body in *Cestoda*; produced by strobilization from the back of the scolex. pl. *proglottides*.

prognathous *(Zool.)*. Having protruding jaws; having the mouth parts directed forwards. Cf. **orthognathous**.

prognosis *(Med.)*. A forecast of the probable course of an illness.

prognostic chart *(Meteor.)*. A chart of the **meteorological elements** which are expected to exist in the near future. A forecast weather chart.

program *(Comp.)*. A complete, structured sequence of **program statements** that direct a computer to implement an **algorithm**. Cf. **subroutine**.

program counter *(Comp.)*. Register which contains the address of the next machine-code instruction to be expected. Also *Instruction Address Register (IAR), next instruction register, sequence control register*.

program flowchart *(Comp.)*. **Flowchart** used to describe the sequence of operations within a computer program, and may form part of the **documentation** of a finished program for maintenance.

program generator *(Comp.)*. **Software** which assists users to write their own programs, by expanding simple statements into program code.

programmable logic array *(Comp.)*. One which allows the designer to choose the interconnections during manufacture.

programmable read only memory *(Comp.)*. See **PROM**.

programmed learning, instruction *(Behav.)*. The process of learning from a systematic presentation of data constructed so that each step leads to the next. The learner has no need of recourse to material other than the programme and progresses by answering questions at each stage which are necessary for the understanding of the rest. Such programmes can be presented on a *teaching machine*.

programme level *(Telecomm.)*. The level, related to datum power level of 1 milliwatt in 600 ohms, as indicated by a volume-unit (VU) meter, as defined for this purpose. Measured in decibels.

programmer *(Comp.)*. A person responsible for writing computer **programs**. See **application-**, **systems-**.

programming *(Comp.)*. Working out detailed sequence of steps in a **programming language** with the aim of producing a **program**.

programming language *(Comp.)*. Artificial language devised to enable people to instruct machines. See **high level language**, **low level language**.

program statement *(Comp.)*. A basic unit of a program; an instruction in a **programming language** which is translated by the **compiler** or **interpreter** into several **machine-code instructions**.

progressive heating *(Elec.Eng.)*. Same as **scanning heating**.

progressive interlace *(Image Tech.)*. Scanning of the television image first with one field containing all the odd-numbered lines and then with a second field of all the even-numbered lines; this is the normal pattern for **interlaced scanning**.

progressive metamorphism *(Geol.)*. The progressive changes in mineral composition and texture observed in rocks within the aureole of contact metamorphism round igneous intrusions; also in rocks which have experienced regional metamorphism of varying degrees of intensity. The particular degree of metamorphism in the latter case is indicated by the 'metamorphic grade' of the rock.

progressive press tool *(Eng.)*. A *press* tool which performs several operations simultaneously but at successive stations.

progressive proofs *(Print.)*. In colour printing, a set of proofs supplied to the printer as a guide to colour and registration, each colour being shown both separately and imposed on the preceding ones.

projected area *(Eng.)*. The product of the diameter of a bearing or rivet and the length along which it is in contact with another element.

projected diameter *(Powder Tech.)*. The diameter of a

circle which has the same area as the projected profile of the particle.

projection *(Behav.)*. In psychoanalytic theory, a **defence mechanism**, in which we ascribe to others various feelings, thoughts and motives, especially of an undesirable sort, which really belong to the self, in order to ward off the anxiety associated with them.

projection *(Maths.)*. (1) The projection of something in 3-dimensional space onto a plane in that space consists of the points in the plane from which normals to the plane pass through points in the original object. (2) A projection, or projective transformation, of planes is any transformation which can be produced by a succession of steps in which a plane is projected onto another plane by joining each point on the first plane to a fixed point outside the planes and then projecting it to the point where the joining line meets the second plane. A projection of *spaces* is similar, but would require higher dimensions.

projection distance *(Image Tech.)*. The distance between the projector and the screen.

projection lamp *(Image Tech.)*. The source of illumination for projecting the image of a cinematographic film or of a transparency on to a screen.

projection lantern *(Phys.)*. See **projector**.

projection lens *(Image Tech.)*. A lens specifically designed to form an image of a flat illuminated transparency, such as a slide or frame of motion picture film, on a flat screen with a substantial degree of enlargement.

projection room *(Image Tech.)*. Enclosure in a cinema from which the projectors are operated. Also *booth, box.*

projection television *(Image Tech.)*. Optical presentation of a TV picture on a separate open screen, in contrast to the direct viewing of a CRT image; generally used for showing a large picture to an audience.

projection welding *(Eng.)*. A process similar to spot welding but allowing a number of welds to be made simultaneously. One of the work-pieces has projections at all points or lines to be welded, the other is flat. Both are held under pressure between suitable electrodes, the current flowing from one to the other at the contact points of the projections.

projective properties *(Maths.)*. Of a figure, properties unaltered by projection, e.g. the class or order of a curve.

projective technique, test *(Behav.)*. A method of psychological testing, used in personality assessment, in which relatively unstructured stimuli are presented (e.g. an inkblot or a picture) and elicit subjective responses from the subject; these are presumed to involve the *projection* of the subject's personality on to the test material.

projective transformation *(Maths.)*. A transformation projecting one figure into another.

projector *(Image Tech.)*. Apparatus for the presentation of an image on a screen, usually with magnification, from photographic transparencies such as slides, from motion picture film or from electronically generated video sources.

projector *(Phys.)*. A '2-lens' optical system for projecting on to a screen a magnified image of a transparency or 'lantern slide'. The condenser, a lens whose function is to illuminate the slide evenly, forms an image of the source of light on the projection lens, the slide being placed in the converging beam of light between the 2 lenses in such a position that the projection lens forms on the screen a real inverted image of it.

projector-type filament-lamp *(Elec.Eng.)*. An electric filament-lamp in which the filament is arranged in a concentrated form, so that it can be focused for projection purposes.

prokaryon *(Biol.)*. The nucleus of a blue-green alga or of a bacterium. See **eukaryote, prokaryote**.

prokaryote, procaryote *(Biol.)*. Major division of living organisms, which have no defined nucleus and their genetic material is usually a circular duplex of DNA. They have no endoplasmic reticulum. Bacteria are the prime example but it also includes the blue-green algae, the Actinomycetes and the Mycoplasmata. Cf. **eukaryote**. adj. *prokaryotic.*

prolactin *(Med.)*. Lactogenic hormone; a mammotrophic protein hormone, secreted by cells of the anterior lobe of the pituitary gland, which stimulates the developed mammary gland to secrete milk; has a similar effect on the pigeon crop gland; also called *luteotrophin* in respect of its other, gonadotrophic, properties.

prolactinoma *(Med.)*. A pituitary tumour secreting **prolactin**.

prolamellar body *(Bot.)*. Three-dimensional lattice composed of tubules, formed within an **etioplast** and rapidly converting into **thylakoids** on illumination.

prolan *(Zool.)*. Former general name for gonadotrophic hormones of mammals found in various tissues and body fluids during pregnancy; prolan A is equivalent to **follicle-stimulating hormone** and prolan B to **luteinizing hormone**.

prolapse *(Med.)*. The falling out of place or sinking of an organ or part of the body.

prolate cycloid *(Maths.)*. See **roulette**.

prolate ellipsoid *(Maths.)*. An ellipsoid obtained by rotating an ellipse about its major axis i.e. like a rugby ball. Cf. **oblate ellipsoid**.

proleg *(Zool.)*. One of several pairs of fleshy conical retractile projections borne by the abdomen in larvae of most Lepidoptera, Sawflies and Scorpion-flies; used in locomotion.

proliferation *(Med.)*. Growth or extension by the multiplication of cells. adj. *proliferous.*

proliferation *(Nuc.Eng.)*. In the nuclear policy context, the spread of nuclear weapons capability to countries not at present possessing such weapons.

prolification *(Bot.)*. Development of buds in the axils of sepals and petals.

proline *(Chem.)*. *Pyrrolidine-2-carboxylic acid.* The L- or s-isomer is a constituent of proteins, particularly **collagen**. Symbol Pro, short form P. ⇨

PROLOG *(Comp.)*. Programming language based on mathematical logic. *PROgramming in LOGic.*

PROM *(Comp.)*. *PROgrammable Memory* is a type of **ROM** into which the program may be written after manufacture, by a customer, but which is fixed from that time on. See **EPROM**.

promenade deck *(Ships)*. On a passenger ship, the upper deck on which passengers walk.

promenade tile *(Build.)*. See **quarry tile**.

promeristem *(Bot.)*. The initial cells and their immediate derivatives in an apical meristem.

prometaphase *(Biol.)*. The stage in cell division, mitosis and meiosis, between prophase and metaphase.

promethazine *(Med.)*. An antihistamine drug of slow but prolonged action.

promethium *(Chem.)*. A radioactive element of the rare earth series, symbol Pm, at.no. 61, having no known stable isotopes in nature. Its most stable isotope, ^{145}Pm, has a half-life of over 20 years.

prominence *(Astron.)*. A streamer of glowing gas visible in the outer layers of the solar atmosphere. Several types are seen in the upper **chromosphere** and lower **corona**. All consist of regions of higher density and lower temperature than the surrounding gas, which is why they can be seen. Although they are best seen at the rim of the Sun during an **eclipse**, they are frequently detectable above the **photosphere** by using a **spectroheliograph**.

promontory *(Zool.)*. A projecting structure: a small ridge or eminence.

promoter *(Biol.)*. A DNA region in front of the coding sequence of a gene which binds RNA polymerase and therefore signals the start of a gene.

promoter *(Chem.)*. A substance which increases the activity of a catalyst.

prompt *(Comp.)*. Character(s) displayed on a **VDU** in an **interactive** computer system to indicate that the user is expected to respond.

prompt critical *(Nuc.Eng.)*. Condition in which a reactor could become critical solely with **prompt neutrons**. In this situation the reactor is difficult to control because the reactor **period** is much reduced.

prompt gamma *(Phys.)*. The γ-radiation emitted at the time of fission of a nucleus.

prompt neutrons *(Phys.)*. Those released in a primary nuclear fission process with practically zero delay.

pronation *(Zool.)*. In some higher Vertebrates, movement of the hand and forearm by which the palm of the hand is turned downwards and the radius and ulna brought into a crossed position; cf. **supination**. adj. *pronate*.

pronator *(Zool.)*. A muscle effecting pronation.

pronephros *(Zool.)*. Archinephros; in Craniata, the anterior portion of the kidney, functional in the embryo but functionless and often absent in the adult. Also called *fore-kidney*, *head-kidney*. adj. *pronephric*. Cf. **mesonephros**; **metanephros**.

pronotum *(Zool.)*. The notum of the prothorax in Insects. adj. *pronotal*.

pronucleus *(Zool.)*. The nucleus of a germ cell after the maturation divisions.

pro-oestrus *(Zool.)*. In Mammals, the coming on of heat in the oestrus cycle.

proof *(Chem.)*. In alcoholometry, a designation (*proof-spirit*) for spirituous liquid containing 49.28% of alcohol by weight, 57.10% by volume, with rel.d. of 0.920 at 15.6°C. Proof spirit is taken as the standard strength of alcoholic liquids for fiscal purposes. A spirituous liquid which is *x% overproof* contains as much alcohol in 100 vol. as in 100 + *x* vol. of proof spirit; *x% underproof* signifies the opposite condition.

proof *(Eng.)*. The impression, often in soft metal, of a die for inspection purposes.

proof *(Print.)*. Impression taken from a printing surface for checking and correction only, not as representative of the finished appearance and quality of the work.

proof by contradiction *(Maths.)*. See **reductio ad absurdum proof**.

proof corrections *(Print.)*. Additions, deletions or amendments to a proof, made clearly in ink, in the margin, in accordance with agreed standards.

proofing press *(Print.)*. A machine constructed to print proofs, or to check quality of printing plates. Models are available for all three main printing processes and range from simple hand-operated versions to power driven versions capable of giving impression and register results comparable with a production machine.

proof load *(Aero.)*. The load which a structure must be able to withstand, while remaining serviceable.

proof plane *(Elec.Eng.)*. A piece of conducting material, mounted on an insulating handle, which may be used for receiving or removing charges in electrostatic experiments.

proof stress *(Eng.)*. The stress required to produce a certain amount of **permanent set** in metals which do not exhibit a sudden yield point. Usually it is the stress producing an extension of 0.1 or 0.5%. Often *yield point*.

pro-otic *(Zool.)*. An anterior bone of the auditory capsule of the Vertebrate skull.

prop *(Build.)*. A post, usually relatively short and made of timber, used as a strut.

prop *(Min.Ext.)*. Sturdy supporting post set across a lode or seam underground to hold up roof after excavation. See **powered supports**.

propagation *(Bot.)*. The reproduction of a plant by asexual or sexual means, especially in horticulture.

propagation *(Phys.)*. Transmission of energy in the form of waves in the direction normal to a wavefront, which is generally spherical, or part of a sphere, or plane. Applies to acoustic, electromagnetic, water, etc. waves.

propagation constant *(Phys.)*. Measure of diminution in magnitude and retardation in phase experienced by current of specified frequency in passing along unit length of a transmission line or waveguide, or through one section of a periodic lattice structure. Given by the natural logarithm of the ratio of output to input current or of acoustic particle velocity.

propagation loss *(Phys.)*. The transmission loss for radiated energy traversing a given path. Equal to the sum of the *spreading loss* (due to increase of the area of the wavefront) and the *attenuation loss* (due to absorption and scattering).

propagation of light *(Phys.)*. Consists of transverse electromagnetic waves propagated through free space with a velocity of 2.9979×10^8 m/s and wavelengths about 400 to 800 nm. The ratio (velocity of light in free space/velocity in medium) is the refractive index of the medium. According to the special theory of relativity, the velocity of light is absolute and no body can move at a greater speed.

propagule *(Bot.)*. Any structure, sexual or asexual and independent from the parent which serves as a means of reproduction. Also *disseminule*.

propane *(Chem.)*. C_3H_8, an alkane hydrocarbon, a colourless gas at atmospheric pressure and temperature. Bp −45°C; found in crude petroleum. In liquid form it constitutes *liquefied petroleum gas* (LPG), a clean-burning fuel, used by some as a petrol substitute.

propanoic acid *(Chem.)*. See **propionic acid**.

propanol *(Chem.)*. $C_3H_7.OH$, *propyl alcohol*, a monohydric aliphatic alcohol, existing in two isomers: (*a*) propan-1-ol (*n*-propyl alcohol), $CH_3.CH_2.CH_2.OH$, bp 97°C, rel.d. 0.804, obtained from fusel-oil, miscible with water in all proportions; (*b*) propanol-2-ol, (*iso*-propyl alcohol), $CH_3.CH(OH).CH_3$, bp 81°C, rel.d. 0.789, which can be prepared by the reduction of acetone with sodium amalgam, or by the hydrolysis of propylene sulphate obtained by the absorption of propylene in sulphuric acid. It also results as a by-product in the synthesis of methanol by the Fischer-Tropsch process.

propanone *(Chem.)*. See **acetone**.

propantheline bromide *(Med.)*. *Anticholinergic* drug which, by blocking parasympathetic innervation reduces secretion and mobility of the stomach and intestine.

propellant *(Space)*. Comprehensive name for the combustibles for a chemical rocket motor. For liquid propellants it comprises the *fuel* (hydrocarbons, such as kerosene and hydrazine) and the *oxidant* (such as liquid oxygen and fluorine). With solid propellants, combustible materials (such as perchlorate and aluminium powder) are prepared prior to firing *in situ*.

propeller *(Aero.)*. An assembly of radially disposed blades of aerofoil shape which by reason of rotation in air and blade angle of attack produces thrust or power. Basic classifications are: (1) a *left-hand propeller* which rotates counter-clockwise when viewed from the rear; (2) a *right-hand propeller* which rotates clockwise when viewed from the rear; (3) a *pusher propeller* which is mounted at the rear of an engine; (4) a *tractor propeller* mounted at the front; (5) a *fixed-pitch propeller*, blades mounted at a fixed angle; (6) an *adjustable-pitch propeller*, blades alterable only when stationary; (7) a *variable-pitch propeller*, blades movable by mechanical means during rotation to optimize the **pitch** for different speeds and engine revs. Abbrev. *VP propeller*. Types of VP propeller: *constant-speed propeller* which maintains a preselected constant rev/min; *controllable-pitch propeller*, with which a desired blade angle may be selected, usually *fine pitch* (or high rev/min), *manifold pressure* or *torque*, as appropriate; a *feathering-propeller* has an extension of the blade angle to reduce drag when an engine is stopped in flight; a *reverse pitch propeller* has an additional control which allows the blades to turn to a negative angle to aid stopping after landing (also *braking propeller*); a *swivelling propeller* has a rotatable axis of rotation so that its thrust direction can be used for control on *airships* or for transition in VTOL aircraft. See **blade angle**, **feathering pump**, **helical tip speed**, **pitch**.

propeller *(Eng.)*. See **marine screw propeller**.

propeller brake *(Aero.)*. A shaft brake to stop, or prevent windmilling of, a turboprop, principally to avoid inconvenience on the ground.

propeller efficiency *(Aero.)*. The ratio of the actual thrust power of a propeller to the torque power supplied by the engine shaft; 80–85% is a typical value.

propeller fan *(Eng.)*. A fan consisting of an impeller or rotor carrying several blades of airscrew form, working in a cylindrical casing sometimes provided with fixed blades; usually driven by a direct-coupled motor.

propeller hub *(Aero.)*. The detachable fitting by which a propeller is attached to the power-driven shaft.

propeller post *(Ships)*. See **stern frame**.

propellers *(Print.)*. Individually mounted rollers, draw rollers or bosses, with peripheries of rubber, synthetic material or knurled metal, usually operating in conjunction with a driven roller to assist in controlling the reels on web-fed presses.

propeller shaft *(Autos.)*. The driving shaft which conveys the engine power from the gearbox to the rear axle of a motor vehicle. It is connected through universal joints to permit displacement of the rear axle on the springs.

propeller singing *(Acous.)*. Phenomenon occurring in ship propellers at certain rates of revolution. A strong, almost pure tone is radiated from the propeller into the water.

propeller turbine engine *(Aero.)*. See **turboprop**.

propeller-type water turbine *(Eng.)*. A water turbine having a runner similar to a four-bladed ship's propeller. It gives a high specific speed under low heads, thus reducing the size of a direct-coupled generator. See **Kaplan water turbine**.

propelling nozzle *(Aero.)*. The constricting nozzle at the outlet of a turbojet **exhaust cone** or *jet pipe* which reduces the gases to slightly more than ambient atmospheric pressure and accelerates them to raise their kinetic energy, thereby increasing the thrust.

propene *(Chem.)*. C_3H_6, $CH_2{=}CH.CH_3$, *propylene*, an alkene hydrocarbon, a gas, bp −48°C. An important by-product of oil refining used for the manufacture of many organic chemicals including propanone, glycerine and polypropene.

propenol *(Chem.)*. See **allyl alcohol**.

properdin *(Immun.)*. A component of the alternative pathway of complement activation present in small amounts in the blood, which is not an immunoglobulin, although at one time considered to be so. It complexes with C3b and stabilizes the alternative pathway C3 convertase.

proper fraction *(Maths.)*. See **division**.

proper motion *(Astron.)*. That component of a star's own motion in space which is at right angles to the line of sight, so that it constitutes a real change in the position of the star relative to its neighbouring stars.

prophage *(Biol.)*. A phage genome which replicates in synchrony with its host. May be integrated into the host genome.

prophase *(Biol.)*. First stage of mitosis or meiosis during which chromosomes condense and become recognizably discrete.

prophylactic *(Med.)*. Tending to prevent or to protect against disease, especially infectious disease; any agent which does this.

prophylaxis *(Med.)*. The preventive treatment of disease.

prophyll *(Bot.)*. The first leaf in most monocots or either of the first two leaves in most dicots, on a shoot.

propiolic acid *(Chem.)*. $CH{\equiv}C.COOH$, acetylene-carboxylic acid; silky crystals, mp 6°C, bp 144°C, soluble in water and ethanol; it forms an explosive silver salt. Also called *propargylic acid*.

propionic, propanoic acid *(Chem.)*. $CH_3.CH_2.COOH$, a monobasic fatty acid, a colourless liquid, mp −36°C, bp 141°C. A constituent of pyroligneous acid; it is formed in certain fermentations. A more up-to-date process involves the use of the 'oxo' process with ethylene as the starting material.

propionyl group *(Chem.)*. The monovalent radical $CH_3.CH_2.CO$—.

proplastid *(Bot.)*. Small undifferentiated plastid.

proportional counter *(Nuc.Eng.)*. One which uses the *proportional region* in a tube characteristic, where the gas amplification in the tube exceeds unity but the output pulse remains proportional to initial ionization.

proportional region *(Nuc.Eng.)*. The range of operating voltage for a counter tube in which the gas amplification is greater than unity and the output pulse is proportional to the energy released by the initial event.

propositional calculus *(Maths.)*. Branch of mathematical logic dealing with propositions and statements related by operators such as 'and', 'or', 'if...then', without regard to the internal structure of the propositions. Cf. **predicate calculus**.

proppants *(Min.Ext.)*. Material like sand or special formulations used to keep open fissures in an oil-bearing sediment. See **fracking**.

propranolol *(Med.)*. A non-selective β-adrenoreceptor antagonist used in the treatment of hypertension, angina and thyrotoxicosis. It slows the heart rate and decreases the output of the heart, but causes bronchoconstriction.

proprioceptor *(Zool.)*. A sensory nerve-ending receptive to internal stimuli, particularly signalling the relative positions of body parts. Also *interoceptor*. adj. *proprioceptive*.

prop root *(Bot.)*. Adventitious root arising on a stem above soil level, growing into the soil and serving as additional support for the stem, as in maize or some mangroves. Also *stilt root*.

proptosis *(Med.)*. Displacement forwards or protrusion of a part of the body, esp. of the eye.

propulsion reactor *(Eng.)*. Nuclear reactor designed to supply energy for the propulsion of a vehicle, at present invariably a ship.

propulsive duct *(Aero.)*. Generic term for the simplest form of reaction-propulsion aero-engine having no compressor/turbine rotor. Thrust is generated by initial compression due to forward motion, the form of the duct converting kinetic energy into pressure, the addition and combustion of fuel, and subsequent ejection of the hot gases at high velocity. See **pulse-jet**, **ramjet**. Also called *Athodyd* (*Aero Thermodynamic Duct*)

propulsive efficiency *(Aero.)*. (1) The propulsive horsepower divided by the torque horsepower. (2) In a turbojet, the net thrust divided by the gross thrust.

propyl alcohol *(Chem.)*. See **propanol**.

propylene *(Chem.)*. See **propene**.

propylene dichloride *(Chem.)*. See **1,2-dichloropropane**.

propylene glycol *(Chem.)*. *Propan-1,2-diol*. $CH_3CH(OH)CH_2OH$. Bp 188°C. Used as a solvent, humectant and plasticizer, and as a chemical intermediate.

propyne *(Chem.)*. *Allylene*, $CH_3.C{\equiv}CH$. Methyl acetylene.

proscapula *(Zool.)*. See **clavicle**.

proscenium *(Build.)*. The stage frame in a theatre, fitted with curtains and a fire-proof safety curtain to cut off the stage from the auditorium.

proscenium lights *(Elec.Eng.)*. Rows of incandescent lamps around the back of the proscenium arch for illuminating the stage of a theatre.

proscolex *(Zool.)*. See **cysticercus**.

prosector *(Med.)*. One who dissects dead bodies for anatomical demonstration and teaching.

prosencephalon *(Zool.)*. In Craniata, the part of the forebrain which gives rise to the cerebral hemispheres and the olfactory lobes.

prosocoele *(Zool.)*. In Craniata, the cavity of the forebrain or first brain vesicle in the embryo; fore ventricle.

prosoma *(Zool.)*. In Arachnida, the region of the body comprising all the segments in front of the segment bearing the genital pore; in *Acarina*, the gnathosoma together with the podosoma.

prospect *(Geol.,Min.Ext.)*. Area which shows sufficient promise of mineral wealth to warrant exploration. Methods of search include aerial survey, magnetometry, geophysical and geochemical tests, seismic probe, electro-resistivity measurement, pitting, trenching and drilling.

prostaglandins *(Immun.)*. Biologically active lipids generated by the action of cycloxygenase enzymes on arachidonic acid (c.f. **leukotrienes** which are generated by lipoxygenase action). A variety of prostaglandins are produced during anaphylactic reactions, some of which have antagonistic actions on platelet activation and vascular permeability, so that the net outcome is difficult to predict.

prostaglandins *(Med.)*. A group of complex fatty acids found in most human tissue which act as local tissue hormones, regulating blood supply, acid secretion of the stomach, vascular permeability, platelet aggregation and temperature regulation.

prostate *(Zool.)*. In Cephalopoda, said of a gland of the male genital system associated with the formation of spermatophores: in eutherian Mammals (except Edentata and Cetacea), including Man, said of a gland associated with the male urogenital canal.

prostatectomy *(Med.)*. Removal of the prostate gland.

prostatism *(Med.)*. Difficulty in micturition encountered in elderly men, usually indicative of abnormal enlargement of the prostate gland.

prostatitis *(Med.)*. Inflammation of the prostate gland.

prostatorrhoea, prostatorrhea *(Med.)*. Chronic gleety or mucous discharge from the prostate.

prosthesis *(Med.)*. The supplying of an artificial part of the body to replace one which is deficient or absent; the artificial part supplied. adj. *prosthetic*.

prosthetic group *(Biol.)*. Non-proteinaceous entity essential for an enzyme's activity. It is functionally equivalent to a *co-enzyme* and differs only in being tightly bound to its protein.

prosthetics *(Med.)*. That branch of surgical science concerned with prosthesis.

prostomium *(Zool.)*. In annelid Worms, that part of the head region anterior to the mouth.

protactinium *(Chem.)*. A radioactive element. Symbol Pa, at.no. 91, half-life of 2×10^4 years. One radioactive isotope is ^{233}Pa which is an intermediate in the preparation of the fissionable ^{233}U from thorium.

protamines *(Biol.)*. Family of short, basic proteins which are bound to sperm DNA in place of histones.

protandry *(Bot.,Zool.)*. The maturation of the male organs before the female are receptive. A form of dichogamy. Cf. **protogyny**.

protanopic *(Med.)*. Colour blind to red.

protease *(Biol.)*. Enzyme which hydrolyses the peptide bonds of a protein. They hydrolyse different sites according to the amino acids adjacent to the peptide bond under attack. e.g. **chymotrypsin, trypsin**. syn. *peptidase*.

protected-type *(Elec.Eng.)*. Said of electrical machinery or other apparatus in which any internal rotating or live parts are protected against accidental mechanical contact in such a way as not to impede ventilation. See **screen-protected motor**.

protection cap *(Elec.Eng.)*. See **fender**.

protective coating *(Chem.)*. A layer of a relatively inert substance, on the surface of another, which diminishes chemical attack of the latter, e.g. Al_2O_3 on metallic Al.

protective colloid *(Chem.)*. A lyophilic colloid which is adsorbed on the dispersed phase of a lyophobic colloid, thus decreasing its tendency to coagulate, e.g. albumen in mayonnaise.

protective furnace atmosphere *(Eng.)*. Inert gas produced from the products of combustion of gas and air in predetermined proportions, the atmosphere being first cooled, cleaned, dehydrated, and desulphurized before delivery to heating chambers of furnaces operated for bright-annealing of nonferrous metals and bright treatment of steels. See also **furnace atmosphere**.

protective gap *(Elec.Eng.)*. A spark gap arranged between an electric circuit and earth, or across a piece of apparatus, so adjusted that, should the voltage across the gap exceed a certain safe value, the gap breaks down, thereby limiting the voltage appearing across the part of the circuit being protected to the breakdown voltage of the gap.

protective gear *(Elec.Eng.)*. The apparatus associated with a protective system; e.g. relays, instrument transformers, pilots etc.

protective layer *(Bot.)*. In an **abscission zone**, a layer of cells lying immediately proximal to the abscission layer and protecting the underlying tissues from desiccation and invasion by parasites after abscission has occurred.

protective system *(Elec.Eng.)*. An arrangement of apparatus designed to isolate a piece of electrical apparatus should a fault occur on it.

protector *(Electronics)*. Tube in which glow discharge from a cold cathode prevents high voltage across a circuit.

protein *(Biol.)*. A polymer of amino acids which may consist of one or more polypeptide chains. Proteins may be water insoluble and serve a structural role, e.g. *silk*, or be water soluble with catalytic activity, the *enzymes*. See **alpha helix, primary-, secondary-** and **tertiary structure**.

protein A *(Immun.)*. A protein in the cell walls or extracts made from certain strains of *Staphylococcus aureus* which binds to the Fc fragment of IgG from a variety of species. This property has made protein A a useful reagent for isolating IgG and for detecting it in complexes. Biologically it has an antiphagocytic effect.

protein structure *(Biol.)*. The three dimensional structure of a protein resulting from the sequence of amino acids in the polypeptide chain, the binding of non-protein moieties (e.g. the haem in haemoglobin), and the association with other protein sub-units. Determined primarily by **X-ray crystallography** of protein crystals down to a resolution of 0.2 nm.

proteolysis *(Biol.)*. The degradation of proteins into peptides and amino acids by cleavage of their peptide bonds.

proteolytic, proteoclastic *(Zool.)*. Said of enzymes which catalyse the breakdown of proteins into simpler substances, e.g. trypsin.

proterandrous *(Bot.,Zool.)*. See **protandrous**.

proterokont *(Biol.)*. A bacterial flagellum, not homologous with the flagella of higher organisms.

Proterozoic *(Geol.)*. A division of the Precambrian comprising the less ancient rocks of that system, and lying above the Archaeozoic.

proter-, protero-. *(Genrl.)*. Prefix from Gk. *proteros*, before, former.

prothallus *(Bot.)*. (1) The gametophyte of the pteridophytes, growing photosynthetically at the soil surface, when it may resemble a thallose liverwort, or heterotrophically in association with a fungus underground, and bearing, when mature, archegonia and/or antheridia. The embryo and young sporophyte, developing from a fertilized egg in the archegonium are at first dependent on the prothallus. (2) Sometimes the gametophyte of gymnosperms.

prothorax *(Zool.)*. The first or most anterior of the three thoracic somites in Insects. Cf. **mesothorax, metathorax**.

Protista *(Bot.)*. A paraphyletic group in some classifications of mostly unicellular organisms including, usually the protozoa, Euglenophyceae, Crytophyceae, Dinophyceae and slime moulds, sometimes the flagellate members of the Chlorophyta and Heterokontophyta, and in older usages the bacteria and blue-green algae.

protium *(Chem.)*. Lightest isotope of hydrogen, of mass unity (^1H), most prevalent naturally. The other isotopes are deuterium (^2H) and tritium (^3H).

proto- *(Genrl.)*. Prefix form Gk. *prōtos*, first.

protocercal *(Zool.)*. See **diphycercal**.

Protochordata *(Zool.)*. A division of Chordata comprising the subphyla Hemichordata, Urochordata, and Cephalochordata, which are distinguished by the absence of a cranium, vertebral column and of specialized anterior sense organs. Cf. **Vertebrata**.

protocol *(Comp.)*. Agreed set of operational procedures to enable data to be transferred between systems.

protoderm *(Bot.)*. **Primary meristem** which gives rise to the epidermis and which may arise from independent initials in the apical meristem.

protogenic *(Chem.)*. Capable of supplying a hydrogen ion (proton).

protogyny *(Bot.,Zool.)*. The maturation of the female organs before the male organs liberate their contents. See **dichogamy**. Cf. **protandry**.

protomorphic *(Zool.)*. Primordial; primitive.

proton *(Phys.)*. The nucleus of the hydrogen atom; of positive charge and atomic mass number of unity. With neutrons, protons form the nucleus of all atoms, the number of protons being equal to the atomic number. It is the lightest *baryon*, 1.007 276 a.m.u. and the most stable. Beams of high energy protons produced in particle accelerators are used to study elementary particles.

protonema *(Bot.)*. Juvenile stage of the gametophyte of mosses and liverworts.

protonephridial system *(Zool.)*. The excretory system of Platyhelminthes, consisting of flame cells and ducts.

protonephridium *(Zool.)*. A larval nephridium, usually of the flame cell type.

protonic solvent *(Chem.)*. One that yields a proton, H^+, as the cation in **self-dissociation**.

proton motive force *(Biol.)*. The electrochemical gradient which is derived from a membrane potential together with a proton gradient across the membrane. Such gradients operate across the inner mitochondrial membrane and the **thylakoid** membrane. Essential for the generation of ATP during oxidative phosphorylation and photosynthesis.

proton-precessional magnetometer *(Phys.)*. Precision magnetometer based on the measurement of the *Larmor frequency* of protons in a sample of water. See **nuclear magnetic resonance, proton resonance**.

proton-proton chain *(Phys.)*. A series of thermo-nuclear reactions which are initiated by a reaction between two protons. It is thought that the proton-proton cycle is more important than the carbon cycle in the cooler stars.

proton resonance *(Phys.)*. A special case of nuclear magnetic resonance. Since the nuclear magnetic moment of protons is now well known, that of other nuclei is found by comparing their resonant frequency with that of the proton.

proton synchroton *(Phys.)*. A *synchrotron* in which the accelerated particles are protons. It is capable of producing particle energies of more than 200 GeV.

proton-translocating ATPase *(Bot.)*. *H^+-ATPase*. Primary electrogenic **active transport** system, powered by the hydrolosis of ATP, pumping protons out of a plant cell across the plasmalemma or into the vacuole across the tonoplast. The resulting pH and electrical potential gradients drive a number of secondary active transport processes coupled to the return movements of the protons.

proto-oncogene *(Biol.)*. A gene which is required for normal function of the organism, but which when altered can become an **oncogene**.

protophilic *(Chem.)*. Able to combine with a hydrogen ion.

protophloem *(Bot.)*. The first-formed **primary phloem**; characteristically maturing while the organ is elongating.

protoplasm *(Biol.)*. The living material within a cell divided into discrete structures, e.g. mitochondria, ribosomes, nuclei, chromosomes and nucleoli in **eukaryotes** and chromosome and ribosomes in **prokaryotes**. adj. *protoplasmic*.

protoplasmic circulation *(Biol.)*. The streaming motion that may be seen in the protoplasm of a living cell.

protoplast *(Bot.)*. (1) The living part of a plant cell including the nucleus, cytoplasm and organelles, all bounded by the plasmalemma, but excluding any cell wall. (2) The above structure isolated from its cell wall usually by treatment of a tissue with wall-degrading enzymes or mechanically. Cf. **protoplast culture, protoplast fusion**.

protoplast culture *(Bot.)*. The aseptic culture on suitable media of protoplasts isolated from plant tissue, especially for the regeneration of plants from somatic hybrids and hybrids resulting from the fusion of protoplasts from different plants.

protoplast fusion *(Bot.)*. The artificial production of **somatic hybrids** and *cybrids* by the fusion of *protoplasts* from different plants. Used in attempts to transfer genetic information between sexually incompatible species.

protoporphyrin *(Chem.)*. *1,3,5,8-Tetramethyl-2,4-divinyl-6,7-dipropanoic acid-porphin*. It combines with iron (II) to give reduced haematin, the prosthetic group of haemoglobin.

protopsis *(Med.)*. Protrusion of the eye.

protostele *(Bot.)*. A **stele** without leaf gaps or, sometimes, with a solid core of xylem.

Prototheria *(Zool.)*. A subclass of primitive Mammals, which probably left the main stock in the Mesozoic, and are found only in Australasia. The adults have no teeth, the cervical vertebrae bear ribs, the limbs are held laterally, the shoulder girdle has precoracoids and an inter-clavicle, and large yolky eggs are laid. The young are fed after hatching on milk produced by specialized sweat glands, whose ducts do not unite to open on nipples. One order, the Monotremata. Duck-billed Platypus, Spiny Anteater.

prototroph *(Biol.)*. A **wild-type** organism able to grow in its unsupplemented medium. Cf. **auxotroph**.

prototropic change *(Chem.)*. Term applied in the form of isomerism known as **tautomerism** when the 'movement' of a proton is involved.

prototropy *(Chem.)*. The reversible conversion of tautomeric forms by migration of hydrogen as a proton.

prototype *(Eng.)*. Generally, the first or original type or model from which anything is developed.

prototype *(Zool.)*. An ancestral form; an original type or specimen.

prototype filter *(Telecomm.)*. Basic type which has the specified nominal cut-off frequencies, but which must be developed into derived forms to obtain further desirable characteristics, such as constancy of image impedance with frequency.

protoxylem *(Bot.)*. The first-formed primary xylem; typically maturing while the organ elongates, having narrow tracheary elements with annular or helical thickening and parenchyma only, and becoming stretched and crushed as the organ elongates.

Protozoa *(Zool.)*. A phylum of unicellular or acellular animals. Nutrition holophytic, holozoic or saprophytic; reproduction by fission or conjugation; locomotion by cilia, flagella, or pseudopodia; freeliving or parasitic. sing. *protozoon*.

protractor *(Eng.)*. An instrument used by the draughtsman for measuring or setting out angles on paper etc.

protractor *(Zool.)*. A muscle which by its contraction draws a limb or a part of the body forward or away from the body. Cf. **retractor**.

proud *(Build.,Eng.)*. A part or portion of a part projecting above another or above its surroundings; standing proud.

proud flesh *(Med.)*. The popular name for *granulation tissue* formed during wound healing.

proustite *(Min.)*. Sulphide of silver and arsenic which crystallizes in the trigonal system. It is commonly associated with other silver-bearing minerals. Also called *light-red silver ore, ruby silver ore*. Cf. **pyrargyrite**.

provascular tissue *(Bot.)*. See **procambium**.

proved reserves *(Min.Ext.)*. Tonnages of economically valuable ore which have been tested adequately by being blocked out into panels and sampled at close intervals.

proventriculus *(Zool.)*. (1) In Birds, the anterior thin-walled part of the stomach, containing the gastric glands. (2) In Oligochaeta and Insects, the gizzard, a muscular thick-walled chamber of the gut posterior to the crop. (3) In Crustacea, the stomach or gastric mill.

provitamin *(Chem.)*. A vitamin precursor, as β-carotene gives vitamin A.

proxemics *(Behav.)*. The study of the spatial features of human social interaction, e.g. **personal space**.

Proxima Centauri *(Astron.)*. The nearest star to the sun; a faint companion to the double star *Alpha Centauri* in the constellation Centaurus, its distance being 4.3 light-years.

proximal *(Biol.)*. Pertaining to or situated at the inner end, nearest to the point of attachment. Cf. **distal**.

proximity effect *(Phys.)*. Increase in effective high-frequency resistance of a conductor when it is brought into proximity with other conductors, owing to eddy currents induced in the latter. It is especially prominent in the adjacent turns of an inductance coil.

proximity fuse *(Radar)*. Miniature radar carried in guided missiles, shells or bombs so that they explode within a preset distance of the target.

pruinose *(Bot.)*. Having a bloom on the surface, especially a whitish bloom like hoar frost.

pruniform *(Bot.)*. Shaped like a plum.

prunt *(Glass)*. An applied mass of glass bearing a monogram or badge, e.g. of the owner or vintner.

pruriginous *(Med.)*. Of the nature of prurigo.

prurigo *(Med.)*. A term common to various skin diseases the chief characteristic of which is papular eruption and intense itching.

pruritus *(Med.)*. Severe persistent itching, a characteristic of numerous skin deseases.

prussic acid *(Chem.)*. A solution of **hydrogen cyanide** in water.

ps *(Genrl.)*. English transliteration of Gk. letter ψ. Pronounced *S*.

psalterium *(Zool.)*. In ruminant Mammals, the third division of the stomach; also *omasum*, *manyplies*.

psammitic gneiss *(Geol.)*. A gneissose rock which has been produced by the metamorphism of arenaceous sediments.

psammitic schists *(Geol.)*. Schists formed from arenaceous sedimentary rocks. Cf. **pelitic gneiss**, **pelitic schist**.

psammophyte *(Bot.)*. A plant adapted to growing on sand or sandy soils.

pseudautostyly *(Zool.)*. A type of jaw suspension in which the upper jaw is fused with the ethmoidal, orbital, and otic regions of the cranium. Cf. **autostyly**. adj. *pseudautostylic*.

pseudo- *(Chem.)*. A prefix which is sometimes used to indicate a tautomeric, isomeric, or closely related compound. Symbol, ψ.

pseudoacid *(Chem.)*. A substance which can exist in two tautomeric forms, one of which functions as an acid, e.g. nitromethane: $CH_3NO_2 \rightleftharpoons H^+ + CH_2{=}NO_2$.

pseudoalums *(Chem.)*. A name sometimes given to double sulphates of the alum type, where there is a bivalent element in place of the univalent element of ordinary alums.

pseudoaposematic *(Zool.)*. Warning or aposematic coloration borne by animals which are not dangerous or distasteful, but show **Batesian mimicry** of animals which are.

pseudobase *(Chem.)*. A substance which can exist in two tautomeric forms, one of which functions as a base. Cf. **pseudoacid**.

pseudobrachium *(Zool.)*. In some Fish, an appendage used for propulsion along a substratum or on dry land; formed by modification of the pectoral fin.

pseudobulb *(Bot.)*. Swollen, solid, above-ground stem of some orchids, acting as a storage organ.

pseudocarp *(Bot.)*. See **false fruit**.

pseudocode *(Comp.)*. Instructions written in symbolic language which must be translated into an acceptable program language or direct into machine language before they can be executed.

pseudocoele *(Zool.)*. (1) In higher Vertebrates, a space enclosed by the inner walls of the closely opposed cerebral hemispheres; the 5th ventricle. (2) A body formed from a persisting blastocoel; also *pseudocoelom*. Cf. **coelom**.

pseudocopulation *(Bot.)*. Attempts of a male insect to mate with a flower that resembles a female of its species, as in the pollination mechanisms of many orchids.

pseudocowpox *(Vet.)*. *Milker's nodule*. A common **zoonosis** caused by infection with pseudo-cowpox virus. Lesions appear on the teats of cows over two years old. Vesicles, pustules and papules are present, followed by thick scabs which drop off after about a month.

pseudocubic, pseudotetragonal, pseudohexagonal *(Min.)*. See **pseudosymmetry**.

pseudocyesis *(Med.)*. *Spurious pregnancy*. The condition in which women desiring offspring are convinced that they are pregnant, the abdomen often being considerably enlarged owing to rapid accumulation of fat or to gas in the intestines.

pseudocyesis *(Zool.)*. In some Mammals, uterine changes following oestrus and resembling those characteristic of pregnancy.

pseudo-dementia *(Behav.)*. Refers to mental conditions in which there are symptoms which suggest dementia, but which are caused by other factors (e.g. drug use or depression).

pseudodont *(Zool.)*. Having horny pads or ridges in place of true teeth, as Monotremes.

pseudo fowl plague *(Vet.)*. See **Newcastle disease**.

pseudogamy *(Bot.)*. A form of apomixis in which, although fertilization does not occur, the stimulus of pollination is necessary for seed production.

pseudogene *(Biol.)*. Defective copy of a gene and therefore not transcribed.

pseudogout *(Med.)*. An acute arthritis often affecting a single joint, often knee or hip. The clinical course is like gout but the disease is due to deposits of calcium pyrophosphate crystals around the joint.

pseudoheart *(Zool.)*. In Oligochaeta, one of a number of paired contractile anterior vessels by which blood is pumped from the dorsal to the ventral vessel; in Echinodermata, the axial organ.

pseudoleucite *(Min.)*. An aggregate showing the crystal shape of **leucite** but consisting mainly of potassium feldspar and nepheline.

pseudomalachite *(Min.)*. Phosphate and hydroxide of copper which resembles malachite and crystallizes in the monoclinic system.

pseudomerism *(Chem.)*. A type of tautomerism in which only one form is known, although derivatives exist of both forms, e.g. HCN (known) and HNC (unknown) giving rise to RCN (cyanides) and RNC (isocyanides).

pseudometamerism *(Zool.)*. The condition of repetition of parts, found in some Cestoda, which bears a superficial resemblance to metamerism.

Pseudomonadaceae *(Biol.)*. A family of bacteria belonging to the order Pseudomonadales. Gram-negative rods, occurring in water and soil and including also animal and plant pathogens, e.g. *Pseudomonas aeruginosa* (blue pus), *Xanthomonas hyacinthi* (yellow rot of hyacinth). Acetobacter species are used in production of vinegar.

Pseudomonadales *(Biol.)*. One of the two main orders of true bacteria, distinguished by the polar flagella of motile forms. Gram-negative. Spiral, spherical or rod-shaped cells. See also **Eubacteriales**.

pseudomorph *(Min.)*. A mineral whose external form is not the one usually assumed by its particular species, the original mineral having been replaced by another substance or substances.

pseudoparenchyma *(Bot.)*. A *false tissue* made of interwoven fungal hyphae.

pseudoperianth *(Bot.)*. Cylindrical sheath growing up around the archegonium and young sporophyte of some liverworts.

pseudopod *(Biol.)*. A broad finger-like protrusion of the cell surface which may be used in amoeboid cells for locomotion. Also *pseudopodium*.

pseudopod *(Zool.)*. A footlike process of the body wall, characteristic of some Insect larvae.

pseudopotential method *(Phys.)*. A method of calculating the band structure of metals. See **band theory of solids**.

pseudopregnancy *(Med.,Zool.)*. See **pseudocyesis**.

pseudopterygium *(Med.)*. The adherence of a tip of a fold of oedematous conjunctiva to a corneal ulcer, thus simulating a pterygium.

pseudorabies *(Vet.)*. See **Aujesky's disease**.

pseudo-random binary sequence *(Comp.,Telecomm.)*. A fixed length binary sequence which satisfies many of the tests for a true random sequence. Normally generated using a shift register with logical feedback. An *n*-stage register provides a sequence $p < 2^n$ bits. The maximum length sequence is $(2^n - 1)$.

pseudo-random number generator *(Comp.)*. Software which generates sequences of numbers that satisfy statistical tests for a **random number** sequence, but can be repeated by starting again at the same point.

pseudoring *(Maths.)*. See **ring**.

Pseudoscorpionidea *(Zool.)*. An order of Arachnida, resembling Scorpionidea, but with no tail; the pedipalps are large, chelate, and contain poison glands; small carnivorous forms found under stones, leaves, bark, and moss; occasionally found in houses. Also called *Chelonethida, False Scorpions.*

pseudosolution *(Chem.).* A colloidal solution or suspension.

pseudosymmetry *(Min.).* A term applied to minerals whose symmetry elements place them on the borderline between two crystal systems, e.g. a mineral with the *c*-axis very nearly equal to the *b*- and *a*-axes might, on casual inspection, appear cubic, though actually tetragonal. It would be described as possessing *pseudocubic symmetry.* The phenomenon is due to slight displacement of the atoms from the positions which they would occupy in the class of higher symmetry. Also applied when the pseudosymmetry is due to twinning.

pseudotachylite *(Geol.).* Flinty crush-rock, resulting from the vitrification of rock fragments produced during faulting under conditions involving the development of considerable heat by friction, as in the Glencoe **cauldron subsidence**.

pseudovilli *(Zool.).* Projections from the surface of the trophoblast in some Mammals, as distinct from the true villi, which are definite out-growths.

pseudovitellus *(Zool.).* In some hemipterous Insects, an abdominal mass of cells which contains symbiotic microorganisms. See also **mycetocytes, mycetome**.

pseud-, pseudo- *(Genrl.).* Prefix from Gk. *pseudēs*, false.

P-shell *(Phys.).* The **electron shell** in an atom corresponding to a principal quantum number of six. In naturally occurring elements it is never completely filled and is the outermost shell for most stable heavy elements.

Ψ/J *(Phys.).* Same elementary particle as **J/Ψ**.

psilomelane *(Min.).* A massive hydrated oxide of manganese which contains varying amounts of barium, potassium, and sodium. It is a secondary mineral formed by alteration of manganese carbonates and silicates, and is used as an ore of manganese.

ψ_p *(Bot.).* Symbol for *pressure potential.*

Psittaciformes *(Zool.).* An order of Birds containing one family. Mainly vegetarian, with powerful hooked beaks; feet typically zygodactylous. The birds are often vividly coloured, and capable of mimicry. Parrots, Cockatoos.

psittacosis *(Med.,Vet.).* *Parrot disease, ornithosis.* An acute or chronic contagious disease of wild and domestic birds which is transmittable to other animals and man and caused by *Chlamydia psittaci.* Results in respiratory and systemic infections including, in man, a disease resembling pneumonia.

psittacosis-lymphogranuloma viruses *(Biol.).* A group of Rickettsiae-like organisms, sometimes classified with the animal viruses, but which may be more closely related to bacteria. Large antigenically related organisms, which contain ribonucleic acid and deoxyribonucleic acid, muramic acid in the cell wall, and are susceptible to sulphonamides and certain antibiotics. Includes causative agents of trachoma, inclusion conjunctivitis, lymphogranuloma venereum and psittacosis.

PSK *(Telecomm.).* Abbrev. for *Phase Shift Keying.*

psophometer *(Telecomm.).* Noise-measuring instrument, incorporating weighting circuits, which relates its readings to the effects of noise as perceived by the human ear, rather then straightforward signal level comparisons.

psophometric voltage *(Telecomm.).* That which measures the noise in communication circuits arising from interference of any kind.

psoriasis *(Med.).* A chronic disease of the skin in which red scaly papules and patches appear, especially on the outer aspects of the limbs.

psoroptic mange *(Vet.).* Mange of animals due to mites of the genus *Psoroptes.* Sheep scab is caused by *Psoroptes communis ovis.*

p-state *(Electronics).* That of an orbital electron when the orbit has angular momentum of one Bohr unit.

psychiatry *(Behav.).* That branch of medical science which deals with the study, diagnosis, treatment and prevention of mental disorders.

psychiatry *(Med.).* The study of mental disorders.

psychism *(Biol.).* The doctrine, difficult to sustain, that living matter possesses attributes not recognized in nonliving matter.

psychoanalysis *(Behav.).* (1) A theory of personality developed by Freud in which the ideas of unconscious conflict and **psychosexual development** are central; (2) A method of therapy based on Freud's theory of personality which attempts to help the individual gain insight into his or her unconscious conflicts using a variety of psychoanalytic techniques. See **free association, transference, Freud's theory of dreams.**

psychodynamics *(Behav.).* A theory of the working of an individual's mind.

psychogalvanic reflex *(Zool.).* The decrease in the electrical resistance of the skin under the stimulation of various emotional states.

psychogenic *(Med.).* Having a mental origin.

psychogenic disorders *(Behav.).* Disorders whose origins are psychological rather than organic.

psychokinesis *(Behav.).* The alleged ability of some people to alter physical reality in the absence of any known mechanism for accomplishing it (e.g. bending metal objects without touching them).

psychometrics *(Behav.).* The application of mathematical and statistical concepts to psychological data, particularly in the areas of mental testing and experimental data.

psychopath *(Behav.).* A medical-legal term referring to a behaviour disorder characterized by repetitive, antisocial behaviour with emotional indifference and without guilt, where the individual does not learn from experience or punishment. It is a category not recognized in Scottish law. Also *anti-social personality.*

psychopathology *(Behav.).* The study of psychological disorders.

psychopharmacology *(Behav.).* Refers to the study and use of drugs that influence behaviour, emotions, perception and thought, by acting on the central nervous system.

psychophily *(Bot.).* Pollination by butterflies.

psychophysics *(Behav.).* The branch of psychology that studies the relationship between characteristics of the physical stimulus and the psychological experience they produce.

psychophysiological disorders *(Behav.).* Physical disorders which are thought to be due to emotional factors but which involve genuine organic disorders (e.g. high blood pressure). Formerly *psychosomatic.*

psychophysiology *(Behav.).* The measurement of physiological processes such as heart rate and blood pressure in relation to various mental and emotional states, See **psychophysiological disorders**.

psychoprophylaxis *(Med.).* A method of psychological and physical training for childbirth, aimed at making labour painless.

psychosexual development *(Behav.).* (1) In psychoanalytic theory, a progressive series of stages in which the source of bodily pleasure changes during development, defined by the zone of the body thought which this pleasure is derived. See **oral stage, anal stage, phallic stage, genital stage**. (2) The term *psychosexual* also refers to mental aspects of sexual phenomena.

psychosexual disorders *(Behav.).* Seriously impaired sexual performance of various kinds, or unusual methods of sexual arousal.

psychosis *(Behav.).* A very general term used to describe mental illnesses which result in a severe loss of mental and emotional function, in contrast to **neurosis**, where the individual remains competent to cope with reality.

psychosomatic *(Behav.).* See **psychophysiological disorders**.

psychosomatic medicine *(Med.).* Branch of medicine that stresses the relationship of bodily and mental happenings, and combines physical and psychological techniques of investigation. Particular attention is paid to the possibility of physical disease (i.e. duodenal ulcer, asthma) being induced by mental states.

psychosurgery *(Behav.).* Biological intervention which involves surgery to remove or destroy brain tissue, aimed at changing undesirable behaviour.

psychosurgery *(Med.)*. Brain surgery for the relief of the symptoms of either mental or physical disease.

psychotherapist *(Med.)*. An individual, usually a physician, but not always so at present, who practises *psychotherapy*.

psychotherapy *(Behav.)*. The treatment of mental and emotional disturbance by psychological means, often in an extended series of therapist-client sessions; refers to several different forms of treatment and techniques. The term is usually restricted to treatments supervized or conducted by trained psychologists or psychiatrists.

psychotomimetic *(Med.)*. Said of drugs which cause bizarre psychic effects in humans as well as marked behavioural changes in animals, e.g. esters of *N*-methyl-3-hydroxypiperidine.

psych-, psycho- *(Genrl.)*. From Gk. *psychē*, soul, mind.

psychrometer *(Meteor)*. The **wet and dry bulb hygrometer**.

psychrophilic *(Bot.)*. Growing best at a relatively low temperature, especially (of a micro-organism) having a temperature optimum below 20°C. Cf. **mesophilic, thermophilic**.

Pt *(Chem.)*. The symbol for platinum.

Ptd.A. *(Build.)*. Abbrev. for *pointed arch*.

Pteridophyta *(Bot.)*. (1) Division of the plant kingdom containing all the vascular plants which do not bear seeds i.e. the ferns, clubmosses, horsetails etc. There is an alternation of generations of, typically, a smaller, more or less thalloid, independent gametophyte and larger, longer-lived sporophyte usually with roots, stems and leaves. Usually divided into 8 classes; Rhyniopsida, Psilotopsida, Zosterophyllopsida, Lycopsida, Trimerophytopsida, Sphenopsida, Filicopsida and Progymnospermopsida. (2) Sometimes, confusingly, the ferns alone.

Pteridospermopsida *(Bot.)*. The seed ferns and allies. Class of extinct gymnosperms, mostly Carboniferous to Jurassic.

pteropod ooze *(Geol.)*. A calcareous deep-sea deposit which contains a large number of pteropod remains.

pterygial *(Zool.)*. In Fish, an element of the fin skeleton; pertaining to a fin; pertaining to a wing.

pterygium *(Med.)*. The encroachment on to the cornea from the side of a thickened, vascular, wing-shaped area of the conjunctiva.

pterygium *(Zool.)*. In Vertebrates, a limb.

pterygoid *(Zool.)*. A paired cartilage bone of the Vertebrate skull, formed by the ossification of the front part of the PPQ bar; a membrane bone which replaces the original pterygoid in some Vertebrates; more generally, wing-shaped.

pterygopalatoquadrate bar *(Zool.)*. In Fish with a cartilaginous skeleton, the rod of cartilage forming the upper jaw and known as the *PPQ bar*.

pterylosis *(Zool.)*. In Birds, the arrangement of the feathers in distinct feather tracts or pterylae, whose form and arrangement are important in classification.

PTFCE *(Plastics)*. See **polytrifluorochloroethene**.

PTFE *(Plastics)*. See **polytetrafluoroethene**.

ptilinum *(Zool.)*. In certain Diptera (Cyclorrhapha), an expansible membranous cephalic sac by which the anterior end of the puparium is thrust off at emergence.

Ptolemaic system *(Astron.)*. The final form of Greek planetary theory as described in Claudius Ptolemy's treatise. In this the earth was the centre of the world, the planets, including the sun and moon, being supposed to revolve round it in motions compounded of eccentric circles and epicycles; the fixed stars were supposed to be attached to an outer sphere concentric with the earth.

Ptolemy's theorem *(Maths.)*. That the product of the diagonals of a cyclic quadrilateral is equal to the sum of the products of the opposite sides.

ptomaines *(Chem.)*. Poisonous amino compounds produced by the decomposition of proteins, especially in dead animal matter. The ptomaines include substances such as putrescine, cadaverine, choline, muscarine, neurine. Few of the ptomaines are known to be poisonous by mouth, food poisoning (botulism) being caused by specific bacteria, e.g. *Clostridium botulinum*.

ptosis *(Med.)*. (1) Paralytic dropping of the upper eyelid. (2) Downward displacement of any bodily organ.

P-trap *(Build.)*. A trap used in sanitary pipes with the inlet vertical and the outlet inclined slightly below the horizontal.

PTS *(Acous.)*. See **permanent threshold shift**.

ptyalism *(Med.)*. Excessive secretion of saliva.

p-type conduction *(Electronics)*. That arising in a semiconductor containing acceptor impurities, with conduction by (positive) holes.

p-type conductivity *(Electronics)*. That apparently resulting from movement of positive charges; actually **holes** in impurity component in a semiconductor.

p-type semiconductor *(Electronics)*. One in which the hole conduction (absence of electrons) exceeds the electron conduction, the acceptor impurity predominating.

ptyxis *(Bot.)*. Manner in which an individual unexpanded leaf, sepal or petal is folded, rolled or coiled in the bud. Also called *vernation*.

Pu *(Chem.)*. Symbol for plutonium.

puberty *(Med.)*. Sexual maturity.

puberulent *(Bot.)*. Minutely pubescent.

pubescence *(Bot.,Zool.)*. A covering of fine hairs or down. adj. *pubescent*.

pubiotomy *(Med.)*. The operation of cutting the pubic bone to one side of the midline, so as to facilitate childbirth in difficult labour.

pubis *(Zool.)*. In Craniata, an element of the pelvic girdle (contr. of *os pubis*). adj. *pubic*.

public-address system *(Acous.)*. Sound-reproducing system for large space and outdoor use, usually with high-powered horn radiators, or columns of open diaphragm units which concentrate radiation horizontally.

public key encryption *(Comp.)*. A system of generating encrypted messages which only the recipient can decode even though the **encryption** key is made public. See **cryptosystem**.

publisher's binding *(Print.)*. See **edition binding**.

pucella *(Glass)*. An implement for opening out the top of a wine glass in the off-hand process.

puddingstone *(Geol.)*. A popular term for *conglomerate*. Hertfordshire Puddingstone, consisting of rounded flint pebbles set in a siliceous sandy matrix, is a good example.

puddle *(Civ.Eng.)*. See **clay puddle**.

puddled ball *(Eng.)*. The mass of iron intimately mixed with slag which is formed by the process of puddling pig iron. See **puddling**.

puddled bars *(Eng.)*. Bars of wrought iron which have been rolled from the **puddled ball** after squeezing the ball to compact it and eliminate some of the slag.

puddling *(Eng.)*. The agitation of a bath of molten pig iron, by hand or by mechanical means, in an oxidizing atmosphere to oxidize most of the carbon, silicon and manganese and thus produce wrought iron.

puddling *(Min.Ext.)*. Concentration of diamond from 'blue ground' clays (weathered kimberlite) by forming an aqueous slurry in a mechanized stirring pan in which the heavier fraction is retained from periodic retrieval, while the lighter fraction overflows.

puddling furnace *(Eng.)*. A small reverberatory furnace in which iron is puddled.

puerperal *(Med.)*. Pertaining to or ensuing upon childbirth, e.g. *puerperal fever*, a condition caused by infection of the genital tract in the course of childbirth.

puerperium *(Med.)*. Strictly, the period between the onset of labour and the return to normal of the generative organs: usually, the first 5 or 6 weeks after the completion of labour.

puff ball *(Bot.)*. Fruiting body of some fungi, especially of the order Lycoperdiales in the Gasteromycetes.

puffs *(Biol.)*. Visibly decondensed bands of **polytene chromosomes** in which active transcription of RNA is occurring.

pug *(Hyd.Eng.)*. To prevent leakage by packing cracks with clay; the material so used.

pug *(Min.Ext.)*. In metalliferous mining, the parting of soft clay which sometimes occurs between the walls of a vein and the country rock.

pugging *(Build.)*. A special mixture carried on boards between the floor joists, serving to insulate the room against sounds from below. In Scotland *deafening*.

pug mill *(Civ.Eng.)*. Mixing machine for wet materials, as used in the making of mortar.

pulaskite *(Geol.)*. A light-coloured alkali syenite consisting largely of alkali feldspar with subordinate ferromagnesian minerals and often a small amount of nepheline.

Pulfrich refractometer *(Phys.)*. Instrument for measuring the refractive index of oils and fats.

pull *(Print.)*. An impression taken for checking.

pull *(Textiles)*. A sample of fibres pulled by hand from a large quantity of raw material (e.g. a bale) and used for analysis; particularly for determining fibre length.

pull-down *(Image Tech.)*. The movement of film from one frame to the next in a motion picture camera or projector mechanism.

pulled coil *(Elec.Eng.)*. An armature coil wound with parallel sides on a suitable former and then pulled out to the correct coil span.

pullet disease *(Vet.)*. See **avian monocytosis**.

pulley *(Eng.)*. A wheel on a shaft, sometimes having a crowned or cambered rim, for carrying an endless belt, or grooved for carrying a rope or chain. A *fast pulley* is one that is keyed to the shaft and revolves with it; a *loose pulley* is not attached to the shaft. The term 'pulley' is also applied to a small grooved wheel over which a sash cord, etc., runs.

pulley mortise *(Build.)*. A form of joint between the end of a ceiling joist, which is tenoned, and the binding joist, which is mortised, so as to let in the ceiling joist in a position such that the lower faces of both are in the same plane.

pulley stile *(Build.)*. One of the upright sides of the frame of a double-hung window, to which is secured the pulley over which the sash cord passes.

pulling *(Telecomm.)*. Variation in frequency of an oscillator when the load on it changes.

pulling *(Textiles)*. (1) The conversion of rags and short lengths of yarn into fibres for re-use. (2) The removal of the wool fibres from skins taken from dead sheep.

pulling by crystal *(Crystal.)*. Growing both metallic and nonmetallic crystals by slowly withdrawing the crystal from a molten surface.

pulling figure *(Telecomm.)*. The stability of an oscillator, measured by the maximum frequency change when the phase angle of the complex reflection coefficient at the load varies through 360° and its modulus is constant and equal to 0.2.

pulling focus *(Image Tech.)*. Alteration of focus during a shot so that the principal subject is kept in focus despite varying distances.

pulling tools *(Min.Ext.)*. The procedure whereby the drill string and bits are removed from the bore and stacked in the derrick before re-use. See **round trip**.

pull-off *(Elec.Eng.)*. A fitting used in connection with the overhead contact-wire of an electric traction system for retaining the contact-wire in the correct position above the track on curves.

pullorum disease *(Vet.)*. *Bacillary white diarrhoea (BWD)*. A disease of young chicks caused by *Salmonella pullorum*; affected chicks appear drowsy and huddled, may develop diarrhoea, and often die.

pull-out *(Aero.)*. The transition from a dive or spin to substantially normal flight.

pull-out distance *(Aero.)*. A naval term for the distance travelled by the hook of an aircraft while arresting on the deck of a carrier. Also *run-out distance*.

pull-out torque *(Elec.Eng.)*. (1) The value of the torque at which a synchronous motor falls out of synchronism. (2) The maximum torque of an induction motor.

pull-over mill *(Eng.)*. A rolling-mill using a single pair of rolls. The metal, after passing through the rolls, is pulled back over the top roll to be fed through the mill a second time.

pull switch *(Elec.Eng.)*. See **ceiling switch**.

pulmonary *(Zool.)*. In land Vertebrates, pertaining to the lungs; in pulmonate *Mollusca*, pertaining to the respiratory cavity.

pulmonary adenomatosis *(Vet.)*. *Jaagziekte*. A slowly-developing respiratory disease of sheep caused by a retrovirus in conjunction with a herpes virus. Fluid collects in the lungs, and if the hind legs are raised fluid pours out of the nostrils.

pulmonary osteoarthropathy *(Med.)*. Increased curvature of the nail bed with painful distal finger joints and radiological evidence of new bone growth. May indicate the presence of a tumour of the lung.

pulmonary valvotomy *(Med.)*. The surgical restoration of pulmonary valve function, when the valve will not open properly.

Pulmonata *(Zool.)*. An order of Gastropoda. Hermaphrodite; exhibit torsion; have a shell but no operculum; mantle cavity forming a lung with no ctenidium but a vascular roof and a small aperture (the pneumostome). Land and freshwater snails, land slugs.

pulmonate *(Zool.)*. Possessing lungs or lung-books; air-breathing.

pulmonectomy *(Med.)*. See **pneumonectomy**.

pulmones *(Zool.)*. Lungs. sing. *pulmo* adj. *pulmonary*.

pulp *(For.)*. Material formed after mechanical or chemical treatment of timber, used as feedstock for paper etc.

pulp *(Min.Ext.)*. Finely ground ore freely suspended in water at a consistency which permits flow, pumping or settlement in quiet conditions.

pulp *(Paper)*. Generic term for any fibrous raw material after preparation in a pulp mill by grinding, refining, digesting with chemicals and/or bleaching. Pulp may be a suspension in water or in the form of fluff or dried or semi dried sheets.

pulp *(Textiles)*. Cellulose fibres obtained either from cotton linters or from wood by chemical purification.

pulp *(Zool.)*. A mass of soft spongy tissue situated in the interior of an organ, as *spleen pulp*, *dental pulp*.

pulp board *(Paper)*. A printing board of the same furnish throughout.

pulping *(For.)*. Any process, mechanical or chemical, that disintegrates the fibres in wood. Also called *defibration*.

pulpy kidney disease *(Vet.)*. An acute and fatal toxaemia of lambs due to enteric infection by the bacterium *Clostridium perfringens* (*Cl. welchii*), type D; the kidneys of affected lambs frequently show a characteristic degenerative change after death. Vaccines available.

pulsar *(Astron.)*. A source of cosmic radio emission, characterized by the rapidity and regularity of the bursts of radio energy it sends out. The time between successive pulses is 0.033 seconds for the pulsar in the **Crab Nebula** and 4 seconds for the slowest pulsar. Pulsars are collapsed **neutron stars**, having a mass similar to the Sun but a diameter of 10 km or so. A few are also sources of cosmic X-rays.

pulsatance *(Phys.)*. See **angular frequency**.

pulsating current *(Elec.Eng.)*. One taking the form of a succession of isolated pulses, and usually unidirectional, which changes in magnitude in a regularly recurring manner.

pulsating star *(Astron.)*. See **pulsar**.

pulsator *(Min.Ext.)*. Mineral jig of Harz type. See **Harz jig**.

pulse *(Elec.Eng.,Telecomm.)*. One *step* followed by a *reverse step*, after a finite interval. A unidirectional flow of current of nonrepeated waveform, i.e. consisting of a transient and a zero-frequency component greater than zero. Measured either by peak value, duration, or integration of magnitude over time. Obsoletely, *impulse*.

pulse *(Med.)*. The periodic expansion and elongation of the arterial walls caused by the pressure wave which follows each contraction of the heart.

pulse amplifier *(Telecomm.)*. One with very wide frequency response which can amplify pulses without distortion of the very short rise time of the leading edge.

pulse amplitude *(Telecomm.)*. That of the crest relative to the quiescent signal level. Sometimes a mean taken over the pulse duration.

pulse-amplitude modulation *(Telecomm.)*. That which is

impressed upon a pulse carrier as variations of amplitude. They may be either unidirectional or bidirectional according to the system employed.

pulse bandwidth *(Telecomm.)*. The frequency band occupied by Fourier components of the pulse which have appreciable amplitude and which make an appreciable contribution to the actual pulse shape.

pulse code *(Telecomm.)*. Coding of information by pulses, either in amplitude, length, or absence or presence in a given time interval.

pulse-code modulation *(Telecomm.)*. Pulse modulation in which the magnitude of the signal is sampled and each sample approximated to a nearest reference level (called *quantization*). Each sample is then represented by binary code, the succession of coded samples becoming the transmitted signal. The transmission of such binary information is highly resistant to noise and interference because detection is only of the presence or absence of pulses. Abbrev. *PCM*.

pulse compression *(Radar)*. Techniques which permit high range resolution (for which short pulses are necessary) while transmitting relatively long pulses in order to increase transmitter power and thereby enhance detection capability. Commonly, the pulse is frequency- or phase-modulated; modulation information is fed to the pulse-compression circuits in the receiver so that a matched-filter effect allows the receiver to respond only to echoes bearing the same modulation. See **coding**.

pulsed columns *(Chem.Eng.)*. Columns used for **liquid-liquid extraction** in which there is a series of horizontal plates containing many small holes. Liquids flow continuously through the column but a reciprocating pump is attached to cause pulsations in the flow which leads to high velocity mixing as the liquids are pulsed through the perforated plates. Can also be applied to packed or even simple empty columns.

pulsed Doppler radar *(Radar)*. One in which the Doppler shift of the signals received from a moving target is used to measure its velocity. The pulsed Doppler technique, but not a **CW radar** using Doppler measurement, can also give range and position information.

pulse decay time *(Telecomm.)*. Time for decay between the arbitrary limits of 0.90 and 0.10 of the maximum amplitude. Also *pulse fall time*.

pulse delay circuit *(Telecomm.)*. One through which the propagation of a signal takes a known time.

pulsed-field gel electrophoresis *(Biol.)*. Variant of **agarose gel electrophoresis** which allows fractionating of very large DNA fragments (up to 2 million base pairs) by applying the electric field in pulses from different angles.

pulsed inertial device *(Nuc.Eng.)*. Fusion system based on **inertial confinement**.

pulse discriminator *(Telecomm.)*. Any circuit capable of discriminating between pulses varying in some specific respect (e.g. duration, amplitude or interval).

pulsed-radar system *(Radar)*. One transmitting short pulses at regular intervals and displaying the reflected signals from the target on a screen. See **CW radar**.

pulse droop *(Telecomm.)*. The exponential decay of amplitude which is often experienced with nominally rectangular pulses of appreciable duration.

pulse duration *(Telecomm.)*. (1) Time interval for which the amplitude exceeds a specified proportion (usually $1/\sqrt{2}$) of its maximum value. Also called *pulse length, pulse width*. (2) The duration of a rectangular pulse of the same maximum amplitude and carrying the same total energy.

pulse duty factor *(Telecomm.)*. The ratio of pulse duration to spacing.

pulse fall time *(Telecomm.)*. **Pulse decay time**.

pulse-forming line *(Radar)*. An artificial line which generates short high-voltage pulses for radar.

pulse-frequency modulation *(Telecomm.)*. Pulse modulation in which the repetition rate of transmitted pulses is varied in accordance with the level of the modulating signal.

pulse generator *(Telecomm.)*. One supplying single or multiple pulses, usually adjustable for **pulse repetition frequency** amplitude and width. May be self-contained or require sine wave input signal.

pulse-height analyser *(Electronics)*. (1) A single or multichannel pulse height selector followed by equipment to count or record the pulses received in each channel. The multichannel units are known as *kicksorters*. (2) One which analyses statistically the magnitudes of pulses in a signal.

pulse-height discriminator *(Telecomm.)*. A circuit which produces an output pulse only when it receives an input whose amplitude exceeds a certain value or lies within a certain range of values.

pulse-height selector *(Electronics)*. Circuit which accepts pulses with amplitudes between two adjacent levels and rejects all others. An output pulse of constant amplitude and profile is produced for each such pulse accepted. The interval between the two reference amplitudes is termed the *window* or *channel width* and the lower level the *threshold*.

pulse interleaving, interlacing *(Telecomm.)*. Adding independent pulse trains on the basis of time division multiplex along a common path.

pulse-interval modulation *(Telecomm.)*. See **pulse-frequency modulation**.

pulse ionization chamber *(Nuc.Eng.)*. One for detection of individual particles by their primary ionization. Must be followed by a very high gain stable amplifier.

pulse jet *(Aero.)*. A propulsive duct with automatic air intake valves, or a frequency-tuned jet pipe, so that pressure builds up between 'firings', thus achieving thrust at reasonable economy at moderate air speeds, e.g. 200–400 mph (350–650 km/h).

pulse jitter *(Telecomm.)*. Irregularities in pulse spacing.

pulse labelling *(Biol.)*. Adding a pulse of radioactive material to a cell and then studying the subsequent metabolic stages.

pulse modulation *(Telecomm.)*. **Modulation** in which pulses of electrical energy, rather than a signal of constant amplitude and frequency, constitute the information *carrier*. Most commonly found in **pulse-code modulation**.

pulse-position modulation *(Telecomm.)*. **Pulse modulation** in which the modulating signal is used to alter the timing of individual pulses within a pulse train. Also *pulse-phase modulation*.

pulse regeneration, restoration *(Telecomm.)*. Correction of pulse to its original shape after phase or amplitude distortion.

pulse repeater *(Telecomm.)*. See **regenerator**.

pulse repetition frequency, rate *(Radar)*. The average number of pulses in unit time.

pulse rise time *(Electronics)*. Time required for amplitude to rise from 0.10 to 0.90 of its maximum value.

pulse shaping, re-shaping *(Electronics)*. Adjustment of a pulse to square-wave or other form by electronic means.

pulse spectrum *(Telecomm.)*. The distribution, as a function of frequency, of the magnitudes of the Fourier components of a pulse.

pulse spike *(Electronics)*. A subsidiary pulse superimposed upon a main pulse.

pulse stretcher *(Telecomm.)*. An electronic unit used to increase the time duration of a pulse.

pulse transformer *(Telecomm.)*. One designed to accept the very wide range of frequencies required to transmit pulse signals without serious distortion.

pulse wave *(Med.)*. A wave of increased pressure travelling along the arterial system and generated by the discharge of blood from the heart into the aorta during ventricular systole.

pulse-width modulation *(Telecomm.)*. **Pulse modulation** in which the duration of pulses is varied in accordance with the amplitude of the modulating signal.

pulsometer pump *(Eng.)*. A steam pump in which an automatic ball valve, the only moving part, admits steam alternately to a pair of chambers, so forcing out water which has been sucked in by condensation of the steam after the previous stroke.

pulsus alternans *(Med.)*. When the arterial pulse has a strong beat or wave followed by a weak one. It indicates heart failure.

pulsus paradoxus *(Med.)*. When there is a marked decrease in the arterial pulse during inspiration. Often found in diseases affecting the **pericardium**.

pulvinated *(Build.)*. Said of a frieze which presents a bulging face.

pulvinule *(Bot.)*. The small **pulvinus** of a leaflet.

pulvinus *(Bot.)*. (1) Swollen base of petiole or pinna, containing motor tissue responsible for sleep movements etc., as in many Leguminosae. (2) Thickened region of grasses at a node of the stem capable, by growth, of re-erecting a lodged culm.

pumice *(Geol.)*. An 'acid' vesicular glass, formed from the froth on the surface of some particularly gaseous lavas. The sharp edges of the disrupted gas vesicles enable pumice to be used as an abrasive. It floats on water.

pummel *(Build.)*. See **punner**.

pump *(Bot.)*. Molecular mechanism in a membrane which brings about the active transport (electrogenic or neutral) of a solute e.g. a proton-translocating ATPase. Cf. **carrier**.

pump *(Eng.)*. A machine driven by some prime mover, and used for raising fluids from a lower to a higher level, or for imparting energy to fluids.

pump *(Telecomm.)*. In a **parametric amplifier**, the external signal source, with a higher frequency than the signal frequency, which causes the parametric device to vary its reactance in a periodic manner.

pumped storage *(Hyd.Eng.)*. Pumping of water to high-level storage during off-peak periods of electric power generation, for return to source via turbines at periods of high demand.

pumped tube *(Electronics)*. Transmitting valve, X-ray tube or other electronic device which is continuously evacuated during operation.

pumpellyite *(Min.)*. A complex greenish hydrated silicate of calcium, magnesium, iron, and aluminium crystallizing in the monoclinic system. Found in low grade metamorphic rocks and in amygdales in some basaltic rocks.

pump frequency *(Electronics)*. Frequency of oscillator used in **maser** or **parametric amplifier** to provide the stored energy released by the input signal.

pumping speed *(Phys.)*. Rate at which a pump removes gas in creating a near-vacuum; measured in dm^3 or cu. ft per minute, s.t.p., against a specified pressure.

pump jack *(Min.Ext.)*. Motor operated well-head pump in which the reciprocating motion of the **horsehead** is transmitted to a displacement pump down hole. See **polished rod**.

pump-line *(Elec.Eng.)*. A cable extending throughout the length of an electric train for the control of auxiliary apparatus such as air compressors.

pump rod *(Min.Ext.)*. Small diameter rods screwed together and used to connect a down hole pump to the surface. Also *pony rod*.

punch *(Eng.)*. (1) A tool for making holes by shearing out a piece of material corresponding in outline to the shape of the punch; a machine incorporating such a tool. (2) A hand tool, struck with a hammer, for marking a surface, or for displacing metal as in riveting. See **centre punch, pin punch, hollow punch**.

punch *(Print.)*. The first stage in the making of type, the letter being cut, nowadays usually by machine, in mild steel which, after hardening, is struck into copper or one of its alloys, to make a *matrix* from which the type is cast.

punch and pin register *(Print.)*. Method of obtaining quick and accurate positioning of film elements onto carrier *flats* and subsequently, in platemaking and on the press, by pre-punching all the film elements and printing plates of the job. Pins fitted through the punched holes of the film and plates are used as register points when making a series of plates in process colour work. The press clamp bar has pins which locate in the holes punched on the plates, thus making for quick and accurate plate changing.

punched card *(Comp.)*. Once widely used standard-size, machine readable card printed with columns of numbers up to 80 through which a pattern of holes was punched to represent binary coded data and instructions. The inventor of the punched card tabulating machine was Herman Hollerith who used it on the 1890 US census.

punched paper tape *(Comp.)*. Continuous strip of paper used to represent data by the patterns of the holes punched in the strip.

punched screens *(Min.Ext.)*. Suitably perforated robust plates which act as industrial screens.

punching *(Elec.Eng.)*. See **lamination**.

punching machine *(Eng.)*. A machine for punching holes in plates, the punch being driven either mechanically by a crank and reciprocating block, or by fluid power.

punch-through *(Electronics)*. Collector-emitter voltage breakdown in transistors.

punctate *(Bot.)*. With translucent or coloured dots, or shallow pits.

punctate basophilia *(Med.)*. See under **basophilia**.

punctuated equilibrium *(Bot.)*. A concept of the process of evolution in which the fossil record is interpreted as long periods of stasis interrupted by relatively short periods of rapid change and speciation.

punctum *(Zool.)*. A minute aperture; a dot or spot in marking. adj. *punctate*.

punctured *(Build.)*. A term applied to a variety of rusticated work distinguished by holes picked in the faces of the stones, either in lines or irregularly.

puncture test *(Paper)*. Test measuring the energy required to force a standard pyramidal puncture head operated by the fall of a loaded sector-shaped pendulum through a sample of container board.

pungent *(Bot.)*. Ending in a point stiff and sharp enough to prick. Also, acrid to taste.

punishment *(Behav.)*. In conditioning situations, the weakening of a response which is followed by an aversive or noxious stimulus, or by the withdrawal of a pleasant one. Cf. **negative reinforcement**.

punning *(Build.,Civ.Eng.)*. (1) The operation of ramming or consolidating the surface of hard-core, concrete, earth etc. with repeated blows from a heavy-headed tool, such as a *punner*. (2) The operation of consolidating concrete by means of constant driving in of a metal rod to distribute the mortar and aggregate evenly within the mix. Now usually done by mechanical vibrator.

punt *(Glass)*. The bottom of a container.

punty, puntee, pontie, pontil *(Glass)*. A short iron rod, at one end of which is either a button of hot glass or a suitably shaped piece of metal, which is applied hot to the end of a partially formed glass article in order that (1) it may be cracked off the blowpipe and manipulated on the punty, or (2) in the case of tube drawing, the mass of glass may be drawn out between punty and blowpipe.

pupa *(Zool.)*. An inactive stage in the life history of an Insect during which it does not feed and reorganization is taking place to transform the larval body into that of the imago. adj. *pupal*.

puparium *(Zool.)*. The hardened and separated last larval skin which is retained to form a covering for the pupa in some Diptera.

pupil *(Zool.)*. The central opening of the iris of the eye. adj. *pupillary*.

pupilometer *(Phys.)*. An instrument for measuring the size and shape of the pupil of the eye and its position with respect to the iris.

pupiparous *(Zool.)*. Giving birth to offspring which have already reached the pupa stage, as some two-winged Flies, e.g. *Glossina*, the Ttsetse Fly.

purchase *(Eng.)*. (1) Mechanical advantage or leverage. (2) Mechanical appliance for gaining mechanical advantage.

pure clay *(Build.)*. See **foul clay**.

pure colour *(Phys.)*. One with **CIE coordinates** lying on the **spectrum locus** or on the **purple boundary**.

pure culture *(Bot.)*. See **axenic culture**.

pure line *(Biol.)*. *Inbred line*. A group of individuals, with their ancestors and descendants, usually the product of continued *inbreeding*, which breed true among themselves and which are, therefore, presumably homozygous at most loci.

pure metal crystals *(Eng.)*. The crystals of which a solid pure metal is composed. Each crystal in a given metal has a similar structure consisting of the same atoms arranged in the same way, though one crystal differs from another in orientation.

pure tone *(Telecomm.)*. One having no harmonics.

purgative *(Med.)*. An evacuant, e.g. cascara etc., used to treat functional constipation.

purge *(Aero.)*. To clean and flush out liquids (usually propellants) from tanks to prevent build-up of explosive mixture; dry nitrogen or helium is used.

purines *(Chem.)*. A group of cyclic di-ureides, of which the most important are **adenine** and **guanine** which are part of the nucleotide chains of DNA and RNA. Caffeine, hypoxanthine and theophylline are also purines. ⇨

Purkinje effect *(Phys.)*. Shift of maximum sensitivity of human eye towards blue end of spectrum at very low illumination levels.

purl fabrics *(Textiles)*. Knitted fabrics in which the reverse side stitches are brought to the surface for effect; used extensively for pullovers, etc.

purlin *(Build.)*. A member laid horizontally on the principal rafters or between walls and supporting the common rafters.

purple boundary *(Phys.)*. Straight line joining the ends of the **spectrum locus** on a **chromaticity diagram**. The coordinates of all real colours fall within the loop formed by these 2 lines.

purple copper ore *(Min.)*. Bornite.

purpleheart *(For.)*. Heavy and resilient hardwood from the genus *Peltogyne*, used for fancy veneers etc. Also called *amaranth, violetwood.*

purple of Cassius *(Chem.)*. Produced by adding a mixture of tin (IV) and tin (II) chlorides to a very dilute solution of gold (III) chloride; hydrated tin (IV) oxide is precipitated and the gold (III) chloride reduced to metal. The red-to-violet colour is due to the precipitation of finely divided gold on the tin (IV) hydroxide. Used in the making of ruby glass.

purpose-made brick *(Build.)*. A brick which has been specially moulded to shape suiting it for use in a particular position, e.g. an arch brick shaped like the voussoir of an arch.

purposive behaviour *(Behav.)*. Behaviour that is carried out with the design of achieving a desired end; it may be conscious or unconscious in its nature.

purpura *(Med.)*. The condition in which small spontaneous haemorrhages appear beneath the skin and the mucous membranes, forming purple patches; these may occur as a result of depleted or defective platelets or capillary damage. adj. *purpuric.*

purpura haemorrhagica *(Vet.)*. An acute, noncontagious disease occurring mainly in horses, characterized by mild fever, subcutaneous oedema and haemorrhages in the mucous membranes. In the horse it usually follows an infectious disease, such as strangles or equine influenza, and is thought to be allergic in origin.

purpuric acid *(Chem.)*. *Barbituryl iminoalloxan.* The ammonium salt is *murexide* (see **murexide** test).

purulent *(Med.)*. Forming or consisting of pus; resembling or accompanied by the formation of pus; of the nature of pus.

pus *(Med.)*. *Matter.* The yellowish fluid formed by suppuration, consisting of serum, pus cells (white blood cells), bacteria, and the debris of tissue destruction.

push-broom sensor *(Space)*. Term applied to a detecting instrument which employs a line of detectors in juxtaposition for recording a line of a scene without recourse to mechanical scanning.

pushbutton tuning *(Image Tech.)*. Selection by push-button of any one of a number of preset tuned circuits in a receiver, to change stations or channels quickly.

pushed punt *(Glass)*. A concave bottom to a container. Also *push-up.*

pushing *(Comp.)*. The insertion of a data item onto a stack.

push piece *(Horol.)*. (1) A small cylindrical plunger which projects just beyond the band of a watch case and is pressed in when it is required to set the hands. (2) The button and stem of a hunting watch, which, on being pressed, causes the opening of the case.

push processing *(Image Tech.)*. See **forced development**.

push-pull *(Acous.)*. Term applied to sound-tracks which carry sound recordings in antiphase. They are *class-A* when each carries the whole waveform, and *class-B* when each carries half the waveform, both halves being united optically or in a push-pull photocell.

push-pull amplifier *(Electronics)*. A balanced amplifier using two valves or transistors working in phase opposition, each device conducting alternate halves of the input signal. Used for the reduction of harmonic distortion in power-output stages.

push-pull sound track *(Image Tech.)*. Obsolete form of improved photographic sound recording in which the track was divided into two halves with modulations in exactly opposite phase and requiring a special reproducer system.

push-push amplifier *(Telecomm.)*. One which uses two similar transistors or valves operating in phase opposition but connected to a common load. By this means, even-order harmonics are emphasized.

push rod *(Autos.)*. A rod through which the tappet of an overhead-valve engine operates the rocker arm, when the camshaft is located in the crankcase.

push-up *(Glass)*. See **pushed punt**.

pustular stomatitis *(Vet.)*. See **horse pox**.

pustule *(Bot.)*. A blister-like spot, on a leaf, stem, fruit etc. from which erupts a fruiting structure of a fungus.

pustule *(Med.)*. A small elevation of the skin containing pus. adjs., *pustular, pustulous.*

putamen *(Zool.)*. In Birds, the shell membrane of the egg: in higher Vertebrates, the lateral part of the lentiform nucleus of the cerebrum.

putlog *(Build.)*. A transverse bearer which, in the case of a bricklayer's scaffold, is fixed at one end to the ledger and at the other end is wedged into a hole left by the bricklayer in the wall; used to support the scaffold boards.

putrefaction *(Chem.)*. The chemical breaking down or decomposition of plants and animals after death. This is caused by the action of anaerobic bacteria and results in the production of obnoxious or offensive substances.

putrescine *(Chem.)*. $H_2N.(CH_2)_4.NH_2$, *1,4-diaminobutane*, crystals, mp 27°C, formed during the putrefaction of flesh.

putty *(Build.)*. See **glazier's-, plasterer's-**.

putty and plaster *(Build.)*. See **gauged stuff**.

putty powder *(Build.)*. Tin oxide, used for polishing glass.

puy *(Geol.)*. The name given to a small volcanic cone, especially in the Auvergne, France.

puzzle box *(Behav.)*. A box in which an animal is confined, and from which it can escape only by performing a particular series of manipulations which it must discover by trial and error or problem-solving behaviour.

puzzolano *(Build.,Civ.Eng.)*. See **pozzuolana**.

PVA *(Plastics)*. Abbrev. for *PolyVinyl Acetate*, and for *PolyVinyl Alcohol.*

p-value *(Stats.)*. The probability of observing an outcome as or more extreme than that actually arising from a particular experiment or sample when a particular hypothesis is true. A low p-value is taken to indicate evidence against the particular hypothesis.

PVC *(Plastics)*. Abbrev. for *PolyVinyl Chloride.*

PWR *(Nuc.Eng.)*. See **pressurised water reactor**.

PX *(Telecomm.)*. Abbrev. for *Private eXchange.*

Py *(Chem.)*. A symbol for the pyridine nucleus.

pyaemia *(Med.)*. The condition in which infection of the blood with bacteria, from a septic focus, is associated with the development of abscesses in different parts of the body. Also *pyemia.* adj. *pyaemic.*

pycastyle *(Build.)*. See **pycnostyle**.

pycnidiospore *(Bot.)*. A spore formed within a pycnidium.

Pycnogonida *(Zool.)*. An order of the Chelicerata. Marine animals with long legs containing diverticulae of the digestive system and with reduced opisthosoma. Sea Spiders.

pycnometer *(Chem.,Civ.Eng.)*. (1) A small, graduated glass vessel, of accurately defined volume, used for determining the relative density of liquids. (2) Similar device for measuring the relative density of soil particles less than about 5 mm in diameter.

pycnosis *(Biol.)*. The shrinkage of the stainable material of a nucleus into a deeply staining knot, usually a feature of cell degeneration.

pycnostyle *(Build.)*. A colonnade in which the space between the columns is equal to $1\frac{1}{2}$ times the lower diameter of the columns. Also *pycastyle*.

pycnoxylic wood *(Bot.)*. Secondary xylem in gymnosperms composed mainly of tracheids, with relatively narrow rays, as in e.g. conifers. Cf. **manoxylic wood**.

pycn-, pycno-, pykn-, pykno- *(Genrl.)*. Prefix from Gk. *pyknos*, compact, dense.

pyelitis *(Med.)*. Inflammation of the pelvis of the kidney. (Gk. *pyelos*, trough.)

pyelocystitis *(Med.)*. Inflammation of the pelvis of the kidney and the bladder.

pyelography, urography *(Radiol.)*. Radiography of the pelvis or the kidney and ureter, after these have been filled with a **contrast medium** which may have been given intravenously, *intravenous pyelogram*, (IVP, IVU) or retrogradely into the ureter, *retrograde pyelogram*.

pyelolithotomy *(Med.)*. The operation for removal of a stone from the pelvis of the kidney.

pyelonephritis *(Med.)*. Inflammation of the kidney and of its pelvis.

pyemia *(Med.)*. See **pyaemia**.

pygal *(Zool.)*. Pertaining to the posterior dorsal extremity of an animal; in *Chelonia*, a posterior median plate of the carapace. (Gk. *pygē*, rump).

pygostyle *(Zool.)*. In Birds, a bone at the end of the vertebral column formed by the fusion of some of the caudal vertebrae.

pyinkado *(For.)*. Durable timber from *Xylia dolabriformis*, found in India and Burma. Typical uses include dock work, heavy duty flooring, railway sleepers, piling etc.

pyknolepsy *(Med.)*. A form of epilepsy in which there are sudden attacks of momentary loss of consciousness; eventually the attacks disappear.

pyknometer *(Chem.)*. See **pycnometer**.

pylephlebitis *(Med.)*. Inflammation of the portal vein (the vein formed by veins running from the spleen, stomach, and intestines, and entering the liver) with or without thrombosis; in suppurative pylephlebitis abscesses form in the liver.

pylon *(Elec.Eng.)*. See **tower**.

pylon *(Eng.)*. A slender vertical structure which is self-supporting. Cf. **mast**.

pylorectomy *(Med.)*. Excision of the pylorus.

pyloric stenosis *(Med.)*. Narrowing of the lower opening of the stomach. May result from scar tissue associated with peptic ulcers. Patient is unable to empty stomach except by vomiting.

pyloroplasty *(Med.)*. An operation for widening the lumen of the pylorus when this has been pathologically narrowed.

pylorospasm *(Med.)*. Spasm of the circular muscle of the pyloric part of the stomach.

pylorus *(Zool.)*. In Vertebrates, the point at which the stomach passes into the intestine. adj. *pyloric*.

pyocolpos *(Med.)*. A collection of pus in the vagina, the result of infection of a **haematocolpos** which has been inadequately treated.

pyogenic *(Med.)*. Having the power to produce pus.

pyometra *(Med.)*. A collection of pus in the cavity of the uterus.

pyonephrosis *(Med.)*. Accumulation of pus in the pelvis of the kidney.

pyopneumothorax *(Med.)*. The presence of pus and air or gas in the pleural cavity.

pyorrhoea, pyorrhea *(Med.)*. (Lit., a flow of pus.) The term now used as a synonym for *pyorrhoea alveolaris*, a purulent inflammation of the periosteum round a tooth.

pyosalpingitis *(Med.)*. Purulent inflammation of a Fallopian tube.

pyosalpinx *(Med.)*. Accumulation of pus in a Fallopian tube.

py-, pyo- *(Genrl.)*. Prefix from Gk. *pyon, pus*.

pyralspite *(Min.)*. A group name for the *pyr*ope, *al*mandine, and *sp*essartine garnets.

pyramid *(Crystal.)*. A crystal form with three or more inclined faces which cut all three axes of a crystal. See also **bipyramid**.

pyramid *(Maths.)*. A polyhedron formed of a multiplicity of triangular faces joined to each other around a common point (the *vertex*) and to a polygonal face (the *base*).

pyramid *(Zool.)*. A conical structure, as part of the medulla oblongata in Vertebrates. adj. *pyramidal*.

pyramidal disease *(Vet.)*. Exostosis affecting the extensor (pyramidal) process of the third phalanx of the foot of the horse.

pyramidal horn *(Acous.)*. One with linear flare-out in both planes. Cf. **sectoral horn**.

pyramidal system *(Crystal.)*. See **tetragonal system**.

pyramidal tract *(Zool.)*. In the brain of Mammals, a large bundle of motor axons carrying voluntary impulses from particular areas of the cerebral cortex.

pyramid of numbers *(Ecol.)*. See **numbers, pyramid of**.

pyranometer *(Meteor.)*. Instrument for measuring either the diffuse or the total global solar radiation. Also called a *solarimeter*.

pyranometer *(Phys.)*. An instrument designed to measure solar radiation by its heating action on two blackened metallic strips of different thickness which thereby assume different temperatures. Also called *solarimeter*.

pyrargyrite *(Min.)*. Sulphide of silver and antimony which crystallizes in the trigonal system. It is commonly associated with other silver-bearing minerals; cf. **proustite**. Also called *dark-red silver ore*.

pyrazinamide *(Med.)*. Used in treatment of tuberculosis.

pyrazines *(Chem.)*. Six-membered heterocyclic rings containing 2 nitrogen atoms in the 1,4 positions.

pyrazole *(Chem.)*. $C_5H_5N_3$, long needles, mp 70°C, bp 185°C. It is a weak secondary base. Fuming sulphuric acid forms a sulphonic acid. Pyrazole and its derivatives can be halogenated, nitrated, diazotized, and generally treated in a similar way to benzene or pyridine.

pyrazoles *(Chem.)*. Heterocyclic compounds containing a 5-membered ring consisting of 3 carbon and 2 nitrogen atoms. Pyrazole derivatives are formed by the condensation of hydrazines with compounds containing two –CO groups, or a –CO and a –COOH group, in the *beta* position, or which contain a –CO or –COOH group attached to a doubly linked carbon atom. Pyrazoles have a similar chemical character to benzene and pyridine.

pyrene *(Chem.)*. A tetracyclic hydrocarbon obtained from the coal-tar fraction boiling above 360°C, forming colourless, monoclinic crystals, mp 148°C; soluble in ether, slightly soluble in ethanol and insoluble in water. Has carcinogenic properties. ⇨

pyrenocarp *(Bot.)*. See **perithecium**.

pyrenoid *(Bot.)*. Small, dense, rounded, refractile, proteinaceous body within or associated with the chloroplast in some members of at least most classes of eukaryotic algae and often surrounded by the appropriate storage carbohydrate.

Pyrenomycetes *(Bot.)*. A class of fungi in the Ascomycotina in which the fruiting body (ascocarp) is usually a perithecium. Includes the **powdery mildews**, *Claviceps* (ergot) and *Neurospora* (used in genetic research) etc.

pyrethrins *(Chem.)*. Active constituents of pyrethrum flowers used as standard contact insecticide in fly-sprays etc; remarkable for the very rapid paralysis ('knock-down' effect) produced on flies, mosquitoes etc. Pyrethrum root is the source of a similar substance used as a sialagogue. Chemically modified pyrethrins, which have greater persistence and other desirable properties are now available.

pyretic *(Med.)*. Pertaining to fever.

pyrexia *(Med.)*. *Fever.* An increase above normal of the temperature of the body. adj. *pyrexial.*

pyrgeometer *(Meteor.)*. Instrument for measuring the longwave atmospheric radiation or the outward radiation from the Earth's surface.

pyrgeometer *(Phys.)*. An instrument consisting of a number of blackened and polished surfaces, designed to measure the loss of heat by radiation from the earth's surface. The surfaces exhibit differential cooling depending on the radiation loss.

pyrheliometer *(Meteor.)*. Instrument for measuring direct solar radiation, excluding the diffuse and reflected components.

pyribole *(Min.)*. A group name for pyroxene and amphibole.

pyridazines *(Chem.)*. Six-membered heterocyclic rings containing two nitrogen atoms in the 1,2 positions.

pyridine *(Chem.)*. A heterocyclic compound containing a ring of 5 carbon atoms and 1 nitrogen atom, having the formula C_5H_5N. It occurs in the coal-tar fraction with a boiling range between 80°C and 170°C; a colourless liquid of pungent, characteristic odour, bp 114°C; a very stable compound and resists oxidation strongly. ⇨

pyridine alkaloids *(Bot.)*. A group of **alkaloids** based on the pyridine ring, including coniine from hemlock.

pyridoxal *(Biol.)*. See **vitamin B6**.

pyriform *(Bot.,Zool.)*. Pear-shaped, as the *pyriform organ* of a Cyphonautes larva.

pyrimidine *(Chem.)*. a six-membered heterocyclic compound containing two nitrogen atoms in the 1,3 positions. **Cytosine**, **thymine** and **uracil** are the important pyrimidine bases found in DNA and RNA. ⇨

pyrite, pyrites *(Min.)*. Sulphide of iron (FeS_2) crystallizing in the cubic system. It is brassy yellow and is the commonest sulphide mineral of widespread occurrence. Also known as *fool's gold, iron pyrites, mundic.* Pyrite(s) is sometimes used to include **copper pyrites**, **magnetic pyrites**, etc.

pyritic smelting *(Eng.)*. Blast furnace smelting of sulphide copper ores, in which heat is partly supplied by oxidation of iron sulphide.

pyro- *(Chem.)*. A prefix used to denote an acid (and the corresponding salts) which is obtained by heating a normal acid, and thus contains relatively less water, e.g. *pyrosulphuric acid*, $H_2S_2O_7$.

pyroborates *(Chem.)*. Generally known as *borates.*

pyroboric acid *(Chem.)*. See **boric acid**.

Pyrochlor *(Chem.)*. TN for non-inflammable transformer oil. Mixture of 60% hexachlorodiphenyl and 40% trichlorobenzene.

pyrochlore *(Min.)*. A complex niobate of sodium, calcium and other bases, with iron, uranium, zirconium, titanium, thorium and fluorine; crystallizes in the cubic system. It is found in nepheline-syenites, and in alkaline pegmatites.

pyroclastic rocks *(Geol.)*. A name given to fragmental deposits of volcanic origin.

pyrocondensation *(Chem.)*. A molecular condensation caused by heating to a high temperature, e.g. the formation of biuret from urea.

pyroelectric effects *(Min.Ext.)*. In high-tension (electrostatic) separation, the electrical charging of particles by heating.

pyroelectricity *(Min.)*. Polarization developed in some hemihedral crystals by an inequality of temperature.

pyrogallol *(Chem.)*. *Pyrogallic acid, 1,2,3-trihydroxybenzene*, $C_6H_3(OH)_3$, white plates, mp 132°C. It sublimes without decomposition, is soluble in water, and is a strong reducing agent. Used as an absorbing agent for oxygen in gas analysis.

pyrogenic *(Chem.)*. Resulting from the application of a high temperature.

pyrogens *(Med.)*. Bacterial polysaccharides which produce febrile reactions.

pyroligneous acid *(Chem.)*. An aqueous distillate obtained by the destructive distillation of wood, which contains ethanoic acid, methanol, acetone and other products.

pyrolusite *(Min.)*. Manganese dioxide crystallizing in the tetragonal system. It typically occurs massive and as a pseudomorph after manganite, and is used as an ore of manganese, as an oxidizer, and as a decolorizer.

pyrolysis *(Chem.)*. The decomposition of a substance by heat.

pyrolytic mining *(Min.Ext.)*. See **underground gasification**.

pyromeride *(Geol.)*. An anglicized French term for nodular rhyolite. It is a quartz-felsite or devitrified rhyolite containing spherulites up to several centimetres in diameter which impart a nodular appearance to the rock.

pyrometallurgy *(Eng.)*. Treatment of ores, concentrates or metals when dry and at high temperatures. Techniques include smelting, refining, roasting, distilling, alloying and heat treatment.

pyrometer *(Phys.)*. Instrument for measuring high temperatures. See **Seger cones**, **optical pyrometer**, **radiation pyrometer**, **disappearing filament pyrometer**.

pyrometric cone equivalent *(Eng.)*. A measure of the mp of a refractory, carried out by means of *Seger cones.* Abbrev. *PCE.*

pyrometric cones *(Eng.)*. See **Seger cones**.

pyromorphite *(Min.)*. Phosphate and chloride of lead, crystallizing in the hexagonal system. It is a mineral of secondary origin, frequently found in lead deposits; a minor ore of lead.

pyrones *(Chem.)*. 6-membered heterocyclic compounds containing a ring of 5 carbon atoms and 1 oxygen atom, 1 of the former being oxidized to a CO group. According to the position of the CO and the O in the molecule, there are α-pyrones (1,2) and γ-pyrones (1,4).

pyroninophilic cells *(Immun.)*. Cells stained with methyl green pyronin stain which have bright red cytoplasm. This indicates the presence of large amounts of RNA, and implies very active protein synthesis. It is characteristic of plasma cells.

pyrope *(Min.)*. The fiery-red garnet; magnesium aluminium silicate crystallizing in the cubic system. It is often perfectly transparent and then prized as a gem, being ruby-red in colour. It occurs in some ultrabasic rocks and in eclogites.

pyrophilous *(Bot.)*. Growing on ground which has been recently burnt over.

pyrophoric metals *(Eng.)*. Those liable to spontaneous combustion under conditions which may arise in a nuclear reactor. The nuclear fuels U, Th and Pu are all pyrophoric.

pyrophosphoric acid *(Chem.)*. HPO_3, obtained by the loss, through heating, of one H_2O molecule from orthophosphoric acid, H_3PO_4.

pyrophyllite *(Min.)*. A soft hydrated aluminium silicate crystallizing in the monoclinic system. It occurs in metamorphic rocks; often resembles talc.

pyrostilbite *(Min.)*. See **kermesite**.

pyrotechny *(Chem.)*. The study and manufacture of fireworks.

Pyrotenax *(Elec.Eng.)*. TN of a type of **mineral-insulated cable** for low-voltage cables. It is very tough, non-inflammable, and heat resisting.

pyroxene group *(Min.)*. A most important group of rock-forming ferromagnesian silicates which, although falling into different systems (orthorhombic, monoclinic), are closely related in form, composition, and structure. They are silicates of calcium, magnesium, and iron, sometimes with manganese, titanium, sodium, or lithium. See **aegirine**, **augite**, **diallage**, **diopside**, **enstatite**, **hypersthene**, **orthopyroxene**.

pyroxenite *(Geol.)*. A coarse-grained, holocrystalline igneous rock, consisting chiefly of pyroxenes. It may contain biotite, hornblende, or olivine as accessories. See **enstatite**, **hypersthenite**.

pyroxilins *(Chem.)*. Nitrocelluloses with a low nitrogen content, containing from 2 to 4 nitrate groups in the molecule. Used in an ethanol-ethoxyethane solution to form collodion. Pyroxilin is a synonym for **guncotton**.

pyrrhotite, pyrrhotine *(Min.)*. Iron sulphide, ca. Fe_7S_8, with variable amount of sulphur. Hexagonal. Ni sulphide may be associated with it, as at Sudbury, Ontario, a major source of the world's nickel. Also called *magnetic pyrites*.

pyrrole *(Chem.)*. A heterocyclic compound having a ring of 4 carbon atoms and 1 nitrogen. A colourless liquid of chloroform-like odour, bp 131°C, rel.d. 0.984. Pyrrole is a secondary base, and is found in coal tar and in bone-oil. Numerous natural colouring matters are derivatives of pyrrole, e.g. chlorophyll and haemoglobin. ⇨

pyrrolidine *(Chem.)*. The final reduction product of pyrrole, a colourless, strongly alkaline base, bp 86°C.

pyrroline *(Chem.)*. A reduction product of pyrrole obtained by treating it with zinc and glacial acetic acid. It is a colourless liquid, bp 91°C, and is a strong secondary base.

Pyruma *(Build.)*. TN for a fireclay cement used in forming heat-resistant joints.

Pythagoras's theorem *(Maths.)*. That the square on the hypotenuse of a right-angled triangle is equal to the sum of the squares on the other two sides.

pyuria *(Med.)*. The presence of pus in the urine.

pyxidium, pyxis *(Bot.)*. A capsule dehiscing by means of a transverse circular split, the top coming off like a lid.

pyx, trial of the *(Eng.)*. Periodic official testing of sterling coinage.

Q

q *(Biol.)*. See **banding techniques**.

q *(Chem.)*. A symbol for the quantity of heat which enters a system.

Q *(Chem.Eng.)*. Symbol for throughput.

Q *(Hyd.Eng.)*. Symbol for quantity of water discharged, usually in m^3/s.

Q *(Phys.)*. (1) A symbol for charge. (2) Symbol of merit for an energy-storing device, resonant system or tuned circuit. Parameter of a tuned circuit such that $Q = \omega L/R$, or $1/\omega CR$, where L = inductance, C = capacitance and R = resistance, considered to be concentrated in either inductor or capacitor. Q is the ratio of shunt voltage to injected e.m.f. at the resonant frequency $\omega/2\pi$. $Q = f_r/(f_1, f_2)$, where f_r is the resonant frequency and $(f_1 - f_2)$ is the bandwidth at the half-power points. For a single component forming part of a resonant system it equals 2π times the ratio of the peak energy to the energy dissipated per cycle. For a dielectric it is given by the ratio of displacement to conduction current. Also called *magnification factor, Q-factor, quality factor, storage factor*.

Q_{10} *(Biol.)*. See **temperature coefficient**.

QA *(Aero.)*. Abbrev. for *Quality Assurance*.

QAM *(Telecomm.)*. Abbrev. for *Quadrature Amplitude Modulation*.

Q-band *(Telecomm.)*. Frequency band mostly in radar, 36 to 46 GHz; now superseded by Ka-band. See **K-band**.

Q-bands *(Biol.)*. See **banding techniques**.

Q-code *(Aero.)*. Telecommunications code using three letter groups: QAA–QNZ for aeronautics; QOA–QQZ for Maritime uses; QRA–QUZ for all services. Examples: QAH = 'What is your height above ....?'; QAM = 'What is the latest met. report?'; QBA = 'What is the horizontal visibility at .....?'.

Q-factor *(Phys.)*. See *Q*.

q-feel *(Aero.)*. A term given (because of the use of $q = \frac{1}{2}\rho v^2$ i.e. **dynamic air pressure**) to a device which applies an artificial force on the control column of a power-controlled aircraft proportional to the aerodynamic loads on the control surfaces, thereby simulating the natural 'feel' of the aircraft throughout its speed range.

Q fever *(Med.)*. Infection by micro-organism *Coxiella burnetti* produces a flu-like illness and may cause pneumonia. In rare and serious cases **endocarditis** occurs.

Q fever *(Vet.)*. Rickettsial infection of sheep and goats caused by *Coxiella burnetti*. Abortion is only symptom.

Q-gas *(Chem.)*. One based on helium (98.2% He, 1.8% butane), widely used in gas-flow counting.

Q-meter *(Phys.)*. Laboratory instrument which measures the *Q*-factor of a component.

QPP amplifier *(Telecomm.)*. Abbrev. for *Quiescent Push-Pull amplifier*.

QPSK *(Telecomm.)*. Abbrev. for *Quaternary Phase-Shift Keying*.

QS *(Build.)*. Abbrevs. for *Quick Sweep*; *Quantity Surveyor*.

Q-shell *(Phys.)*. The **electron shell** in an atom corresponding to a principal quantum number of seven. It is the outermost shell for heavy radioactive elements.

Q-signal *(Image Tech.)*. In the NTSI colour system, that corresponding to the narrow-band axis of the chrominance signal.

Q-signal *(Telecomm.)*. First of 3-letter code for standard messages in international telegraphy.

QSO *(Astron.)*. Abbreviation for *quasar*.

Q-switching *(Phys.)*. A means of producing high instantaneous power from a laser. The cavity resonator has its reflectivity or 'Q' controlled externally. Q is made small while the population inversion is built up to its peak value. The reflectivity is then increased and the resultant high Q produces an intense burst of energy which almost completely empties the high energy states in a time of about 10^{-8} s. Switching is by a *Kerr cell* shutter or by rotating one of the mirrors.

quad *(Elec.Eng.)*. Either four insulated conductors twisted together (*star-quad*) or two twisted pairs (*twin-quad*). Normally a single structural unit of a multiconductor cable.

quad *(Paper)*. Prefix to denote a size which is four times the basic size i.e. both dimensions of the basic size are doubled.

quad *(Print.)*. See **quadrat**.

quadrant *(Eng.)*. A slotted segmental guide through which an adjusting lever (e.g. a reversing lever) works. It is provided with means for locating the lever in any desired angular position. See **link motion**.

quadrant *(Maths.)*. (1) Of a circle, a quarter of the circle. (2) Rectangular cartesian axes divide the plane into four quadrants numbered one to four anticlockwise around the origin from the first in which both variables (x and y) are positive.

quadrant *(Surv.)*. An angle-measuring instrument of the sextant type, but embracing an angle of 90° or a little more.

quadrant *(Zool.)*. A section of a segmenting ovum originating from 1 of the 4 primary blastomeres.

quadrantal deviation *(Ships)*. Those parts of the **deviation** which vary as sine and cosine of twice the **compass course**, thus changing their sign quadrantally with change in direction of the ship's head.

quadrantal points *(Astron.)*. Points of the compass which in moving from north correspond to the headings NE (45°), SE (135°), SW (225°) and NW (315°). Cf. **cardinal points**.

quadrant dividers *(Build.)*. A form of dividers in which one limb moves over an arc fixed rigidly to the second limb, and may be secured to it by tightening a binding screw.

quadrant electrometer *(Elec.Eng.)*. See **Dolezalek quadrant electrometer**.

quadrat *(Bot.)*. A small area (say 0.1 to 10 m^2) of vegetation marked out for ecological study; a device of laths or strings to mark out such an area.

quadrat *(Print.)*. A piece of metal less than type height, for spacing. Also *quad*.

quadrate *(Bot.)*. Square to squarish in cross-section or in face view.

quadrate *(Zool.)*. A paired cartilage bone of the Vertebrate skull formed by ossification of the posterior part of the PPQ bar, or the corresponding cartilage element prior to its ossification; except in Mammals, it forms part of the jaw-articulation.

quadratic equation *(Maths.)*. An algebraic equation of the second degree, i.e.

$$ax^2 + bx + c = 0,$$

whose solution is

$$x = \frac{-b \pm \sqrt{b^2 - 4ac}}{2a}.$$

quadratic system *(Crystal.)*. See **tetragonal system**.

quadrature *(Astron.)*. Position of the moon or a superior planet in elongation 90° or 270°, i.e. when the lines drawn from the earth to the sun and the body in question are at right angles.

quadrature *(Elec.Eng.,Telecomm.)*. See under **phase**.

quadrature *(Image Tech.)*. Relation between two waves of the same frequency but one-quarter of a cycle (90°) out of phase, as in TV colour difference signals.

quadrature amplitude modulation *(Telecomm.)*. Modulation system involving phase- and amplitude-modulation of a carrier, used in microwave and satellite communication links. Because it always allows high-power amplifier stages to operate close to their peak power output, more efficient use may be made of Earth and satellite amplifiers. Abbrev. *QAM*.

quadrature component *(Elec.Eng.)*. See **reactive component**.

quadrature reactance *(Elec.Eng.)*. A term used in the 2-reaction theory of synchronous machines to denote the ratio which the synchronous reactance drop produced by the quadrature component of the armature current bears to actual value of quadrature component.

quadrature transformer *(Elec.Eng.)*. A transformer designed so that secondary e.m.f. is 90° displaced from primary e.m.f.

quadratus *(Zool.)*. A muscle of rectangular appearance, e.g. *quadratus femoris*.

quadric *(Maths.)*. The three-dimensional surface represented by a general second-degree equation in three variables. By a suitable choice of co-ordinates such an equation can be reduced to one of the following standard equations:

(1) $\pm\frac{x^2}{a^2}\pm\frac{y^2}{b^2}\pm\frac{z^2}{c^2}=1$,

an ellipsoid (+ + +),
a hyperboloid of one sheet (1 minus),
a hyperboloid of two sheets (2 minuses),
an imaginary (virtual) quadric (3 minuses).

(2) $ax^2+by^2=2cz$,

an elliptic paraboloid (a and b of same sign),
a hyperbolic paraboloid (a and b of opposite sign).

(3) $ax^2+by^2+cz^2=0$, a cone.

(4) $ax^2+2hxy+by^2=1$, a cylinder.

Plane sections of quadrics are conics. For a cylinder, sections parallel to the plane $z=0$ determine its type, which is elliptic, parabolic or hyperbolic. For an ellipsoid, a, b and c are the lengths of its principal semi-axes.

quadriceps *(Zool.)*. A muscle having 4 insertions, as one of the thigh muscles of Primates.

quadrilateral *(Maths.)*. A four-sided polygon.

quadrilateral speed-time curve *(Civ.Eng.)*. A simplified form of speed-time curve used in making preliminary calculations regarding energy consumption and average speed of railway trains. The acceleration and coasting portions of the curve are sloping straight lines and the braking portion is neglected, so that the curve becomes a quadrilateral. Cf. **trapezoidal speed-time curve**.

quadriplegia *(Med.)*. Paralysis of both arms and both legs.

quadripole *(Telecomm.)*. A network with 2 input and 2 output terminals. A balanced wave-filter section.

quadrivalent *(Biol.)*. A group of four at least partly homologous chromosomes held together by chiasmata during the prophase of the first division of meiosis, commonly found during meiosis in tetraploids.

quadrivalent *(Electronics)*. An atom with four electrons in its valency shell.

quadr-, quadri- *(Genrl.)*. Prefix from L. *quattuor*, four.

quadrumanous *(Zool.)*. Of Vertebrates, having all 4 podia constructed like hands, as in Apes and Monkeys.

quadruped *(Zool.)*. Of Vertebrates, having all 4 podia constructed like feet, as Cattle.

quadruple-expansion engine *(Eng.)*. A steam engine in which the steam is expanded successively in 4 cylinders of increasing size, all working on the same crankshaft. Also *multiple-expansion engine*. Cf. **triple-expansion engine**.

quadruple point *(Chem.)*. A point on a concentration-pressure-temperature diagram at which a 2-component system can exist in 4 phases.

quadruplex *(Image Tech.)*. Videotape recording and reproduction system using four rotating heads to produce transverse tracks on two-inch wide magnetic tape.

quadruplex system *(Telecomm.)*. A system of Morse telegraphy arranged for simultaneous independent transmission of two messages in each direction over a single circuit.

quadrupole *(Phys.)*. A collection of charges such that the potential at a point distant from their centre of mass may be expressed by an infinite series of terms in inverse powers of r. The inverse third power term is the *quadrupole potential*.

quadrupole moment *(Phys.)*. That derived from the series expansion (see **quadrupole**) of charges multiplied by space coordinates. The sum of the quadratic terms is the *quadrupole moment*, which is possessed by most metals.

quadrupoles *(Acous.)*. Radiator producing a sound field of two adjacent dipoles in antiphase. The eddies in a subsonic jet of gas are quadrupoles.

qualification test *(Space)*. An evaluation of a flight article or its equivalent to verify that it functions correctly under the specified conditions of space-flight; normally the test conditions are more severe than those expected.

qualitative analysis *(Chem.)*. Identification of the constituents of a sample without regard to their relative amounts. It often refers to elemental analysis, but may also refer to the detection of acid-base or redox properties in a sample. See also **quantitative analysis**.

quality *(Acous.)*. (1) In sound reproduction, the degree to which a sample of reproduced sound resembles a sample of the original sound. The general description of freedom from various types of acoustic distortion in sound-reproducing systems. See **high-fidelity**. (2) The timbre or quality of a note which depends upon the number and magnitude of harmonics of the fundamental.

quality *(Eng.)*. The condition of a saturated vapour, particularly steam, expressed as the ratio per cent of the vaporized portion to the total weight of liquid and vapour.

quality *(Radiol.)*. In radiography, it indicates approximate penetrating power. Higher voltages produce higher quality X-rays of shorter wavelength and greater penetration. (This term dates from a period before the nature of X-rays was completely understood).

quality control *(Eng.)*. A form of inspection involving sampling of parts in a mathematical manner to determine whether or not the entire production run is acceptable, a specified number of defective parts being permissible.

quality factor *(Phys.)*. See *Q*.

quality factor *(Radiol.)*. See **relative biological effectiveness**.

quantile *(Stats.)*. The argument of the cumulative distribution function corresponding to a specified probability; (of a sample) the value below which occur a specified proportion of the observations in the ordered set of observations.

quantitative analysis *(Chem.)*. Identification of the relative amounts of substances making up a sample. It usually refers to elemental analysis, but may refer to any constituent of the sample. In addition to chemical methods, virtually every physical property can be a basis for some analytical method, and spectroscopic and electrochemical techniques are particularly often employed.

quantitative character *(Biol.)*. A character displaying *continuous variation*. Cf. **unit character**.

quantitative genetics *(Biol.)*. The genetics of *quantitative characters*.

quantity of electricity *(Elec.Eng.)*. Product of flow of electricity (current) and time during which it flows. The term may also refer to a charge of electricity. See **coulomb**.

quantity of light *(Phys.)*. Product of luminous flux and time during which it is maintained; usually stated in lumen-hours.

quantity of radiation *(Radiol.)*. Product of intensity and time of X-ray radiation. Not measured by energy, but by energy density and a coefficient depending on ability to cause ionization.

quantity sensitivity *(Elec.Eng.)*. The throw, in mm, on a scale 1 m from an instrument designed to measure the quantity of electricity.

quantity surveyor *(Build.,Civ.Eng.)*. One who measures up from drawings and prepares a bill (or schedule) of quantities showing the content of each item. This is then used by contractors for estimating. The quantity surveyor is also concerned in measuring and assessing at intervals the value of work already carried out.

quantization *(Comp.)*. See **digitization**.

quantization *(Phys.)*. In quantum theory, the division of energy of a system into discrete units (*quanta*), so that continuous infinitesimal changes are excluded.

quantization *(Telecomm.)*. In **pulse-code modulation**, the division of the amplitude range of a continuously variable signal, e.g. speech or video, into discrete levels for the purposes of *sampling* and *coding*.

quantization noise *(Telecomm.)*. Noise introduced into a circuit using **pulse-code modulation** because there are too few levels of *quantization* to describe the waveform accurately.

quantometer *(Eng.)*. An instrument showing by spectrographical analysis the percentages of the various metals present in a metallic sample.

quantum *(Phys.)*. (1) General term for the indivisible unit of any form of physical energy; in particular the *photon*, the discrete amount of electromagnetic radiation energy, its magnitude being $h\nu$ where ν is the frequency and h is Planck's constant. See also **graviton, magnon, phonon, roton**. (2) An interval on a measuring scale, fractions of which are considered insignificant.

quantum chromodynamics *(Phys.)*. Theory of strong interactions between elementary particles including the interaction that binds protons and neutrons to form a nucleus. It assumes that strongly interacting particles are made of *quarks* and that *gluons* bind the quarks together.

quantum efficiency *(Phys.)*. Number of electrons released in a photocell per photon of incident radiation of specified wavelength.

quantum electrodynamics *(Phys.)*. A relativistic quantum theory of electromagnetic interactions. It provides a description of the interaction of electrons, muons and photons and hence the underlying theory of all electromagnetic phenomena. Abbrev. *QED*.

quantum electronics *(Phys.)*. Those concerned with the amplification or generation of microwave power in solid crystals, governed by quantum mechanical laws.

quantum field theory *(Phys.)*. The overall theory of fundamental particles and their interactions. Each type of particle is represented by appropriate *operators* which obey certain commutation laws. Particles are the quanta of fields in the same way as photons are the quanta of the electromagnetic field. So *gluon* fields and *intermediate vector boson* fields can be related to *strong* and *weak* interactions. Quantum field theory accounts for the *Lamb shift*.

quantum Hall effect *(Phys.)*. The Hall resistivity changes by steps so that it is a fraction of h/e^2 where h is Planck's constant and e is the electronic charge. Observed in two-dimensional semiconductors (e.g. MOS) at high magnetic fields and ultra-low temperatures. See **Hall effect**.

quantum mechanics *(Phys.)*. A generally accepted theory replacing classical mechanics for microscopic phenomena. Quantum mechanics also gives results consistent with classical mechanics for macroscopic phenomena. Two equivalent formalisms have been developed, *matrix mechanics* (Heisenberg) and *wave mechanics* (Schrödinger). The theory accounts for a very wide range of physical phenomena. See **statistical mechanics, correspondence principle**.

quantum number *(Phys.)*. One of a set, describing possible quantum states of a system, e.g. nuclear spin. See **spin, principal quantum number**.

quantum statistics *(Phys.)*. Statistics of the distribution of particles of a specified type in relation to their energies, the latter being quantized. See **Bose-Einstein statistics** and **Fermi-Dirac statistics**.

quantum theory *(Phys.)*. That developed from *Planck's law* to account for *black-body radiation*, the *photoelectric effect* and the *Compton effect* and to form the *Bohr theory* of the atom and its modification by Sommerfeld. Now superseded by **quantum mechanics**.

quantum voltage *(Phys.)*. Voltage through which an electron must be accelerated to acquire the energy corresponding to a particular quantum.

quantum yield *(Phys.)*. Ratio of the number of photon-induced reactions occurring, to total number of incident photons.

quaquaversal fold *(Geol.)*. A domelike structure of folded sedimentary rocks which dip uniformly outwards from a central point. See **dome**.

quarantine *(Med.)*. Isolation or restrictions placed on the movements of individuals associated with a case of a communicable disease; place or period of detention of travellers coming from infected or suspected countries, or of animals on importation.

quark *(Phys.)*. A type of fundamental particle that forms the constituents of *hadrons*. There are currently believed to be six types of quarks (and their antiquarks). In the simple quark theory, the baryon is composed of three quarks, an antibaryon is composed of three antiquarks, and a meson is composed of a quark and an antiquark. No quark has been observed in isolation. See **flavour, colour**.

quarl *(Eng.)*. See **burner firing block**.

quarrel *(Build.)*. The diamond-shaped pane of glass used in **fret-work**.

quarries *(Build.)*. Same as **quarry tiles**.

quarry *(Min.Ext.)*. (1) An open working or pit for granite, building-stone, slate or other rock. (2) An underground working in a coal-mine for stone to fill the goaf. Distinction between quarry and mine somewhat blurred in law, but usage implies surface workings.

quarry-faced *(Build.)*. A term applied to a building-stone whose face is hammer-dressed before leaving the quarry. See **rock face**.

quarry-pitched *(Build.)*. A term applied to stones which are roughly squared before leaving the quarry.

quarry stone bond *(Build.)*. A term applied to the arrangement of stones in rubble masonry.

quarry tile *(Build.)*. The common unglazed, machine-made paving tile not less than $\frac{3}{4}$ in (20 mm) in thickness. Also *promenade tile*.

quartan *(Med.)*. In which a febrile paroxysm recurs every fourth day (i.e. at an interval of 72 hr). *Quartan malaria* was associated with infection with the parasite *Plasmodium malariae*.

quartation *(Eng.)*. See **inquartation**.

quarter *(Astron.)*. The term applied to the phase of the moon at quadrature. The first quarter occurs when the longitude of the moon exceeds that of the sun by 90°, the last quarter when the excess is 270°. The two other quarters are the **new moon** and **full moon**.

quarter bend *(Build.)*. Union connecting two pipes at 90°.

quarter bond *(Build.)*. The ordinary brickwork bond obtained by using a $2\frac{1}{4}$ in (57 mm) closer.

quarter-bound *(Print.)*. A term applied to a book having its back and part of its sides covered in one material and the rest of its sides in another.

quarter-chord point *(Aero.)*. The point on the **chord line** at one quarter of the chord length behind the leading edge. **Sweepback** is usually quoted by the angle between the line of the quarter-chord points and the normal to the aircraft fore-and-aft centre-line.

quarter girth *(For.)*. The girth of a log or tree divided by four. A measure commonly used in countries where volumes are reckoned in Hoppus feet. Abbrev. *qg*.

quarter ill, quarter evil *(Vet.)*. See **blackleg**.

quartering *(For.)*. A piece of timber of square section between 2 and 6 in (50 and 150 mm) side.

quartering *(Min.Ext.)*. A method of obtaining a representative sample for analysis or test of an aggregate with occasional shovelfuls, of which a heap or cone is formed. This is flattened out and two opposite quarter parts are rejected. Another cone is formed from the remainder which is again quartered, the process being repeated until a sample of the required size is left.

quarter lines *(Ships)*. The aggregation of waterlines, buttocklines, sections and diagonals indicative of a ship's form, drawn on a scale of $\frac{1}{4}$ in = 1 ft. See **fairing**.

quarter page folder *(Print.)*. A supplementary device to give a third fold in line with the run of the paper on web-fed presses.

quarter-phase systems *(Elec.Eng.)*. See **two-phase** systems.

quarter rack *(Horol.)*. The rack of the striking work of a clock or repeater-watch which regulates the striking of the quarters.

quarters *(Build.,Civ.Eng.)*. See **flanks**.

quarter sawing *(For.)*. A mode of converting timber, adopted when it is desired that the growth rings shall all be at no less than 45° to the cut faces. Also called *rift sawing*.

quarter screws *(Horol.)*. Four screws in the rim of a compensating balance, one placed at either end of the arms, and the other two at right angles to the arm. Used for rating the watch but not for compensation purposes.

quarter snail *(Horol.)*. The snail in a chiming clock or repeater-watch which controls the number of teeth picked up on the quarter rack.

quarter-space landing *(Build.)*. A landing extending across only half the width of a staircase.

quarter turn *(Build.)*. A wreath subtending an angle of 90°.

quarter-wave antenna *(Telecomm.)*. One whose overall length is approximately a quarter of free-space wavelength corresponding to frequency of operation. Under these conditions it is oscillating in its first natural mode, and is half a dipole.

quarter-wave bar *(Telecomm.)*. See **quarter-wave line**.

quarter-wavelength stub *(Telecomm.)*. Resonating two-wire or coaxial line, approximately one quarter-wavelength long, of high impedance at resonance. Used in antennae, as insulating support for another line, and as a coupling element.

quarter-wave line *(Telecomm.)*. Quarter-wavelength section of transmission line designed to operate as a matching device between lines of different impedance levels.

quarter-wave plate *(Phys.)*. A plate of quartz, cut parallel to the optic axis, of such thickness that a retardation of a quarter of a period is produced between ordinary and extraordinary rays travelling normally through the plate. By using a quarter-wave plate, with its axis at 45° to the principal plane of a Nicol prism, circularly polarized light is obtained.

quartet, quartette *(Zool.)*. A set of 4 related cells in a segmenting ovum.

quartic equation *(Maths.)*. An algebraic equation of the fourth degree, i.e. $ax^4+bx^3+cx^2+dx+e=0$. Its resolution into a pair of quadratic equations, and hence its solution, depends upon the solution of a subsidiary cubic equation.

quarto *(Print.)*. The quarter of a sheet, or a sheet folded twice to make 4 leaves or 8 pages; written 4to.

quartz *(Min.)*. Crystalline silica, SiO_2, occurring either in prisms capped by rhombohedra (low-temperature quartz, stable up to 573°C) or in hexagonal bipyramidal crystals (high-temperature quartz, stable above 573°C). Widely distributed in rocks of all kinds; igneous, metamorphic, and sedimentary; usually colourless and transparent (rock crystal), but often coloured by minute quantities of impurities as in citrine, cairngorm, etc.; also finely crystalline in the several forms of chalcedony, jasper, etc. See also **cristobalite, tridymite, twinning**.

quartz crystal *(Telecomm.)*. A disk or rod cut in the appropriate directions from a specimen of piezoelectric quartz, and accurately ground so that its natural resonance shall occur at a particular frequency.

quartz-crystal clock *(Horol.)*. A synchronous electric clock of great accuracy, having a quartz crystal to control the frequency of the a.c. supply to the motor unit.

quartz-diorite *(Geol.)*. A coarse-grained holo-crystalline igneous rock of intermediate composition, composed of quartz, plagioclase feldspar, hornblende, and biotite, and thus intermediate in mineral composition between typical diorite and granite. Also called *tonalite*.

quartz-dolerite *(Geol.)*. A variety of dolerite which contains interstitial quartz usually intergrown graphically with feldspar, forming patches of micropegmatite. A dyke-rock of worldwide distribution, well represented by the Whin Sill rock in N. England.

quartz-fibre balance *(Chem.)*. A very sensitive spring balance, the spring being a quartz fibre.

quartz glass *(Glass)*. See **vitreous silica**.

quartz-iodine lamp *(Elec.Eng.)*. Compact high-intensity light source, consisting of a bulb with a tungsten filament, filled with an inert gas containing iodine (sometimes bromine) vapour. The bulb is of quartz, glass being unable to withstand the high operating temperature (600°C). Used for car-lamps, cine projectors, etc. Also *quartz-halogen lamp, tungsten-halogen lamp*.

quartzite *(Geol.)*. The characteristic product of the metamorphism of a siliceous sandstone or grit. The term is also used to denote sandstones and grits which have been cemented by silica.

quartz-keratophyre *(Geol.)*. A type of soda-trachyte carrying accessory quartz.

quartz lamp *(Radiol.)*. One which contains a mercury arc under pressure, a powerful source of ultraviolet radiation.

quartz oscillator *(Telecomm.)*. One whose oscillation frequency is controlled by a piezoelectric quartz crystal.

quartz-porphyrite *(Geol.)*. A porphyrite carrying quartz as an accessory constituent; the representative in the medium grain-size group of the fine-grained dacite.

quartz porphyry *(Geol.)*. A medium-grained igneous rock of granitic composition occurring normally as minor intrusions, and carrying prominent phenocrysts of quartz.

quartz resonator *(Telecomm.)*. A standard of frequency comparison making use of the sharply resonant properties of a piezoelectric quartz crystal.

quartz topaz *(Min.)*. See **citrine**.

quartz wedge *(Min.)*. A thin wedge of quartz which provides a means of superposing any required thickness of quartz on a mineral section being viewed under a polarizing microscope, the wedge being cut parallel to the optic axis of a prism of quartz crystal. It enables the sign of the birefringence of biaxial minerals to be determined from their interference figure in convergent light.

quartz wind *(Acous.)*. Form of **acoustic streaming** near ultrasonic transducers operated at high amplitudes.

quasar *(Astron.)*. A distant, compact, object far beyond our **Galaxy** which looks starlike on a photograph but has a **redshift** characteristic of an extremely remote object. The word is a contraction of *quasi-stellar object*. They were discovered in 1963 as the optical counterparts of strong extragalactic radio sources. The most distinctive features of quasars are: extremely compact structure and a high redshift corresponding to velocities approaching that of light. If these velocities are interpreted in the context of **Hubble law**, the implied distances run into billions of light years, making them the most distant and luminous objects in the universe. Many quasars are millions of times brighter than normal galaxies, and because they are also variable, astrophysicists have deduced that the energy producing region cannot be much larger than the size of the solar system. Only highly condensed objects like **black holes** seem capable of satisfying the energy requirements.

quasi-biennial oscillation *(Meteor.)*. Alternation of easterly and westerly wind regimes in the equatorial stratosphere with an interval between successive corresponding maxima of from 24 to 30 months. A new regime starts above 30 km and propagates downwards at about 1 km per month. Abbrev. *QBO*.

quasi-bistable circuit *(Elec.Eng.)*. An astable circuit which is triggered at a high rate as compared with its natural frequency.

quasi-duplex *(Telecomm.)*. A circuit which operates

apparently duplex, but actually functions in one direction only at a time, e.g. a long-distance telephone or a radio link, which is automatically switched by speech.

quasi-Fermi levels *(Electronics)*. Energy levels in a semiconductor from which the number of electrons or holes available for conduction under non-equilibrium conditions, especially when light is falling on the semiconductor, can be calculated in the same way as from the true **Fermi level** which applies under equilibrium conditions.

quasi-geostrophic approximation *(Meteor.)*. An approximation to the dynamical equations governing atmospheric flow, especially the **vorticity equation**, whereby the horizontal wind is replaced by the **geostrophic wind** in the term representing the vorticity, but not in the term representing the **divergence**.

quasi-longitudinal wave *(Acous.)*. Special type of wave occurring in plates and bars. The particle motion is mainly longitudinal and has a small transverse component caused by lateral contraction.

quasi-optical waves *(Phys.)*. Electromagnetic waves of such short wavelength that their laws of propagation are similar to those of visible light.

quasi-stationary front *(Meteor.)*. A **front** which is moving slowly and irregularly so that it cannot be described as either a **cold front** or a **warm front**.

quaternary *(Chem.)*. Consisting of 4 components etc; also, connected to 4 non-hydrogen atoms.

Quaternary *(Geol.)*. The geological period which succeeded the Tertiary. It includes the Pleistocene and Holocene epochs and covers a life-span of app. the last two million years.

quaternary ammonium bases *(Chem.)*. Bases derived from the hypothetical ammonium hydroxide $NH_4.OH$, in which the four hydrogen atoms attached to the nitrogen are replaced by alkyl radicals, e.g. $(C_2H_5)_4N.OH$, tetraethyl-ammonium hydroxide.

quaternary diagram *(Eng.)*. **Constitutional diagram** of 4-phase alloy.

quaternary phase-shift keying *(Telecomm.)*. Used in microwave links and satellite communications to double the channel capacity of conventional binary phase-shift keying without changing the bandwidth. The phase of the carrier can be set by modulation to any one of four positions.

Queckensted's sign *(Med.)*. Increase in the pressure of the cerebrospinal fluid when the jugular veins in the neck are compressed; if this manoeuvre causes no rise of pressure in the spinal fluid, an obstruction is present at a higher level.

queen *(Build.)*. A slate 36 × 24 in (914 × 610 mm).

queen *(Zool.)*. In social Insects, a sexually reproducing female.

queen bee substance *(Zool.)*. See **queen substance**.

queen bolt *(Build.)*. A long iron or steel bolt serving in place of a timber queen-post.

queen closer *(Build.)*. A half-brick made by cutting the brick lengthwise.

queen-post *(Build.)*. For roofs of more than about 30 ft (ca 10 m) span, The central support of the tie-beam by the king-post is insufficient, and two vertical ties (*queen posts*) are required.

queen post roof *(Build.)*. A timber roof having 2 queen-posts but no king-post.

queen substance *(Zool.)*. A pheromone produced by queen honey-bees (*Apis mellifera*; Hymenoptera) consisting of 9-ketodecanoic acid. Its effects include the suppression of egg-laying and of the building of queen cells by workers.

quench *(Elec.Eng.)*. Resistor or resistor-capacitor shunting a contact, to reduce high-frequency sparking when a current is broken in an inductive circuit.

quenched cullet *(Glass)*. *Cullet* made by running molten glass into water. Also *dragaded cullet*, *dragladled cullet*.

quenched spark converter *(Elec.Eng.)*. A generator which utilizes the oscillatory discharge of a capacitor through an inductor and spark gap as a source of RF power.

quenched spark gap *(Elec.Eng.)*. Spark gap in which the discharge takes place between cooled or rapidly moving electrodes, e.g. studs on a disk.

quencher *(Phys.)*. That which is introduced into a luminescent material to reduce the duration of phosphorescence.

quench frequency *(Telecomm.)*. The lower frequency signal used to quench intermittently a high-frequency oscillator, e.g. in a super-regenerative receiver.

quenching *(Eng.)*. Generally means cooling steel from above the critical range by immersing in oil or water, in order to harden it. Also applied to cooling in salt and molten-metal baths or by means of an air blast, and to the rapid cooling of other alloys after solution treatment. See **oil-hardening steel**, **tempering**.

quenching *(Nuc.Eng.)*. Process of inhibiting continuous discharge, by choice of gas and/or external valve circuit, so that discharge can occur again on the incidence of a further photon or particle in a counting tube. Essential in a **Geiger-Müller counter**.

quenching *(Telecomm.)*. Suppression of oscillation, particularly periodically, as in a super-regenerative receiver.

quenching oscillator *(Telecomm.)*. One with a frequency slightly above the audible limit, and which generates the voltage necessary to quench the high-frequency oscillations in a super-regenerative receiver.

quench time *(Nuc.Eng.)*. That required to quench the discharge of a Geiger tube. **Dead time** for internal quenching, **paralysis time** for electronic quenching, although dead time is often used synonomously for the other two terms.

queue *(Comp.)*. A **list** for which insertions are made at one end and deletions at the other. The arrangement is called *first-in, first-out*. Abbrev. *FIFO*. Cf. **stack**.

queuing *(Comp.)*. Programs waiting, in order determined by their priority, for access to the **central processor** in a **time-sharing system**.

quick-break switch *(Elec.Eng.)*. A switch having a spring or other device to produce a quick break, independently of the operator.

quicking *(Elec.Eng.)*. Electrodeposition of mercury on a surface before regular plating.

quicklime *(Chem.)*. See **caustic lime**, **lime**.

quick make-and-break switch *(Elec.Eng.)*. See **snap switch**.

quick return mechanism *(Eng.)*. A reciprocating motion, for operating the tool of a shaping machine etc., in which the return is made more rapidly than the cutting stroke, so as to reduce the 'idling' time.

quicksand *(Civ.Eng.)*. Loose sand mixed with such a high proportion of water that its bearing-pressure is very low. Also *running sand*.

quick-setting inks *(Print.)*. A general term for inks formulated to set quickly, allowing handling of the stock soon after printing and before final drying.

quick-setting level *(Surv.)*. See **fixed-needle surveying**.

quicksilver *(Chem.)*. See **mercury**.

quick sweep *(Build.)*. A term applied to circular work in which the radius is small.

quidding *(Vet.)*. A condition in the horse in which food is expelled from the mouth after being chewed; usually due to disease of the mouth.

quiescent *(Electronics)*. General term for a system waiting to be operated, as a valve ready to amplify or a gas-discharge tube to fire. Also *preset*.

quiescent carrier transmission *(Telecomm.)*. One for which the carrier is suppressed in the absence of modulation.

quiescent centre *(Bot.)*. Region, within the apical meristem of many roots, in which the cells either do not divide, or divide very much more slowly than the cells around it.

quiescent current *(Telecomm.)*. Current in an active device in the absence of a driving or modulating signal. Also called *standing current*.

quiescent operating point *(Telecomm.)*. The steady-state operating conditions of a valve or transistor in its working circuit but in the absence of any input signal.

quiescent period *(Telecomm.)*. That between pulses in a pulse transmission.

quiescent push-pull amplifier *(Telecomm.)*. Thermionic valve or transistor amplifier, in which one side alone passes current for one phase, the other side passing current for the other phase. Abbrev. *QPP amplifier*.

quiescent tank *(Build.)*. A form of sedimentation tank in which sewage is allowed to rest for a certain time without flow taking place.

quiet automatic volume control *(Telecomm.)*. Same as **delayed automatic gain control**. The application of this is known as *quieting*.

quieting sensitivity *(Telecomm.)*. The minimum input signal required by a frequency-modulation radio receiver to give a specified signal/noise ratio at the output.

quill *(Zool.)*. See **calamus**.

quill feathers *(Zool.)*. In Birds, the remiges and rectrices.

quill, quill drive *(Elec.Eng.)*. A form of drive used for electric locomotives in which the armature of the driving motor is mounted on a quill surrounding the driving axle, but connected to it only by a flexible connection. This enables a small amount of relative motion to take place between the motor and the driving axle.

quill, quill drive *(Eng.)*. A hollow non-rotating shaft into which another shaft is inserted and rotated under power. The *quill* providing axial movement as in a drilling machine spindle.

quillwort *(Bot.)*. Common name of *Isoetes* spp.

quinacrine fluorescence *(Biol.)*. See **banding techniques**.

quinaldine *(Chem.)*. $C_{10}H_9N$, *2-methylquinoline*, a colourless refractive liquid, bp 246°C, which occurs to the extent of 25% in quinoline obtained from coal-tar.

Quincke's method *(Phys.)*. One for determining the magnetic susceptibility of a substance in solution by measuring the force acting on it in terms of the change of height of the free surface of the solution when placed in a suitable magnetic field.

quincuncial aestivation *(Bot.)*. A common type of **imbricate** aestivation of a five-membered calyx or corolla in which 2 members overlap their neighbours by both edges, 2 are overlapped on both edges and 1 overlaps 1 neighbour and is overlapped by the other, as in e.g. calyx of roses, corolla of Caryophyllaceae.

quinhydrone *(Chem.)*. $C_6H_4O_2+C_6H_4(OH)_2$, an additive compound of 1 molecule of 1,4-quinone and 1 molecule of 1,4-dihydroxybenzene. It crystallizes in green prisms with a metallic lustre.

quinhydrone electrode *(Chem.)*. A system consisting of a clean, polished, gold or platinum electrode dipping into a solution containing a little quinhydrone, for determining pH-values, making use of the pH dependence of the redox properties of the system.

quinine *(Chem.)*. $C_{20}H_{24}O_2N_2.3H_2O$, an alkaloid of the quinoline group, present in Cinchona bark. It is a diacid base of very bitter taste and alkaline reaction. It crystallizes in prisms or silky needles, mp 177°C; the hydrochloride and sulphate are used as a febrifuge but have been largely superseded as a remedy for malaria.

quinine *(Med.)*. An alkaloid of the quinoline group, present in *Cinchona* bark. It is a diacid base of very bitter taste. It was used as a febrifuge but has been largely superseded as a remedy for malaria, but still used in the treatment of leg cramps.

quinizarine *(Chem.)*. A synonym for 1,4-dihydroxy-anthraquinone.

quinol *(Chem.,Image Tech.)*. See **hydroquinone**.

quinol *(Image Tech.)*. See **hydroquinone**.

quinoline *(Chem.)*. A heterocyclic compound consisting of a benzene ring condensed with a pyridine ring. It is a colourless, oily liquid, mp −19.5°C, bp 240°C, rel.d. 1.08, of characteristic odour, insoluble in water, soluble in most organic solvents. It is found in coal-tar, in bone oil and in the products of the destructive distillation of many alkaloids. It can be synthesized by heating a mixture of aniline, glycerine and nitrobenzene with concentrated sulphuric acid. See **Skraup's synthesis**. ⇨

quinones *(Chem.)*. Compounds derived from benzene and its homologues by the replacement of 2 atoms of hydrogen with 2 atoms of oxygen, and characterized by their yellow colour and by being readily reduced to dihydric phenols. According to their configuration they are divided into 1,2-quinones and 1,4-quinones. ⇨

quinonoid formula *(Chem.)*. A formula based upon the diketone configuration of 1,4-quinone (**benzoquinone**), involving the rearrangement of the double bonds in a benzene nucleus; adopted to explain the formation of dyestuffs, e.g. coloured salts of compounds of the triphenyl-methane series.

quinoxalines *(Chem.)*. A group of heterocyclic compounds consisting of a benzene ring condensed with a diazine ring: They can be obtained by the condensation of 1,2-diamines with 1,2-diketones.

quinsy *(Med.)*. *Acute suppurative tonsillitis; peritonsillar abscess.* Acute inflammation of the tonsil with the formation of pus around it.

quintal *(Phys.)*. Unit of mass in the metric system, equal to 100 kg. Abbrev. *q*.

quintic equation *(Maths.)*. An algebraic equation of the fifth degree. Unlike equations of lower degree, its general solution (and that of equations of higher degree) cannot be expressed in terms of a finite number of root extractions.

quintozene *(Chem.)*. *Pentachloronitrobenzene*, used as a fungicide. Also called *PCNB*.

quintuple point *(Chem.)*. A point on a concentration-pressure-temperature diagram at which a 3-component system can exist in 5 phases.

quire *(Paper)*. One-twentieth of a **ream**.

quire spacing *(Print.)*. On a rotary printing press, as the product is delivered, it is separated into quires or batches by the **kicker** which delivers a **kick copy** at the required interval.

quirewise *(Print.)*. Sections which after printing are folded and *inserted* one in the other. This method allows the booklet to be stitched instead of stabbed.

quirk *(Build.)*. The narrow groove alongside a bead which is sunk into the face of the work so as to be flush with it.

quirk-bead *(Build.)*. See **bead-and-quirk**.

quirk float *(Build.)*. A plasterer's trowel specially shaped for finishing mouldings.

quirk moulding *(Build.)*. A moulding having a small groove in it.

quirk-router *(Build.)*. A form of plane for shaping quirks.

quitclaim *(Min.Ext.)*. A deed of relinquishment of a claim or portion of mining ground.

quittor *(Vet.)*. A chronic suppuration of the lateral cartilage and its surrounding tissues within a horse's foot.

quoin *(Build.)*. An exterior angle of a building, especially one formed of large squared cornerstones projecting beyond the general faces of the meeting wall surfaces.

quoin *(Print.)*. A wooden wedge or a metal device used to lock up formes.

quoin header *(Build.)*. A brick which is so laid at the external angle of a building that it is a header in respect of the face of the wall proper, and a stretcher in respect of the return wall.

quoin key *(Print.)*. Each pattern of mechanical quoin is operated by a key, some being square in shape, others ribbed.

quoin-post *(Hyd.Eng.)*. See **heel-post**.

quotation marks, quotes *(Print.)*. If double quotes (" ") are used to indicate a quotation, the single quotes (' ') are used for a quotation within the passage quoted; the reverse procedure is becoming more popular. In hand-set matter the opening quotes are inverted commas, the closing quotes being apostrophes; in machine-set matter the opening quote is a special sort.

quotations *(Print.)*. Metal spaces, not less than 4 ems by 2 ems used for filling blanks in pages or formes.

quotient *(Maths.)*. See **division**.

Q-value *(Nuc.Eng.)*. (1) Quantity of energy released in a given nuclear reaction. Normally expressed in MeV, but occasionally in atomic mass units. (2) Ratio of thermonuclear power output to power needed to maintain the plasma.

QWERTY keyboard *(Comp.)*. One laid out in the standard typewriter pattern.

R

r *(Chem.)*. With subscript, a symbol for *specific refraction.*

r *(Ecol.)*. The instantaneous population growth rate defined as

$$r = \frac{dN}{dt}\frac{1}{N}$$

where N is the population size and t is time. For any organism there is a maximum r, achieved in ideal conditions, called the innate capacity for increase r_{max}.

r- *(Chem.)*. An abbrev. for *racemic.*

ρ *(Genrl.)*. See under **rho**.

ρ- *(Chem.)*. A symbol for *pros-*, i.e. containing a condensed double aromatic nucleus substituted in the 2,3 positions.

R *(Build.)*. Abbrev. for *Render.*

R *(Chem.)*. A general symbol for an organic hydrocarbon radical, especially an alkyl radical.

R *(Phys.)*. The symbol used to indicate **Rankine scale**.

R *(Radiol.)*. Symbol for *röntgen* unit in X-ray dosage.

R- *(Chem.)*. Prefix denoting right handed. See **Cahn-Ingold-Prelog system** for absolute configuration.

R *(Chem.,Phys.)*. A symbol for (1) the *gas constant*; (2) the *Rydberg constant.*

[*R*] *(Chem.)*. With subscript, a symbol for molecular refraction.

R.A. *(Astron.)*. Abbrev. for **right ascension**.

Ra *(Chem.)*. The symbol for *radium.*

ra- *(Chem.)*. A symbol for *radio-*, i.e. a radioactive isotope of an element, e.g. *ra*Na, *radio-sodium.*

Raabe's convergence test *(Maths.)*. If, for the series of positive terms Σu_n, the product $n\left(\frac{u_n}{u_{n+1}} - 1\right) \to \sigma$, then Σu_n is convergent if $\sigma > 1$ and divergent if $\sigma < 1$. This is a special case of **Kummer's convergence test**.

rab *(Build.)*. A stick for mixing hair with mortar.

rabbet *(Build.)*. A corruption of *rebate.*

rabbeted lock *(Build.)*. A lock which is fitted into a recess cut in the edge of a door.

rabbit *(Min.Ext.)*. Checks the internal diameter of pipes etc. to ensure that tools can pass. See also **pig**.

rabbit *(Nuc.Eng.)*. See **shuttle**.

rabbit septicaemia *(Vet.)*. An acute septicaemic and pneumonic disease of rabbits caused by the bacterium *Pasteurella multocida*; a mild form of the disease, in which the infection is confined to the upper respiratory tract, is called *snuffles.*

rabble *(Min.Ext.)*. Mechanized rake used to loosen sluice bed or move ore or concentrate through a kiln or furnace.

rabies *(Med.)*. An acute disease of dogs, wolves, and other carnivores, due to infection with a virus, and communicable to Man by the bite of the infected animals. In Man the disease is characterized by intense restlessness, mental excitement, muscular spasms (especially of the mouth and throat), convulsions and paralysis. Also called *hydrophobia.*

Rabl configuration *(Biol.)*. Spatial arrangement of interphase chromosomes with centromeres clustering at one side of the nucleus and telomeres at the other.

race *(Bot.,Zool.)*. A population, within a species, that is genetically distinct in some way, often geographically separate; a breed of domesticated animals. See **physiological race**.

race *(Build.,Geol.)*. Fragments of limestone sometimes found in certain brick earths of a hard marly character.

race *(Eng.)*. The inner or outer steel rings of a **ball-bearing** or **roller-bearing**.

race *(Hyd.Eng.)*. A channel conveying water to or away from a hydraulically operated machine.

racemates *(Chem.)*. See **racemic isomers**.

raceme *(Bot.)*. A simple (unbranched) **racemose inflorescence** in which the flowers are visibly stalked. Cf. **spike**.

racemic acid *(Chem.)*. See **tartaric acid**.

racemic isomers (racemates) *(Chem.)*. Optically inactive mixtures consisting of equal quantities of enantiomorphous stereoisomers, R- and S-forms. Chemical synthesis of compounds with asymmetric molecules from optically inactive starting materials gives racemic mixtures. These can be resolved into the optically active components by various methods, e.g. by coupling with an optically active substance, such as an alkaloid, and subsequent fractional crystallization, by the action of lower plant organisms e.g. bacteria, moulds, yeasts etc. which attack only one of the isomers, leaving the other one intact.

racemization *(Chem.)*. The transformation of an optically active substance into racemic inactive form, either by an isomerization through a symmetrical intermediate or through a reaction by which a new substance is formed via a similar intermediate or transition state.

racemose *(Zool.)*. Shaped like a bunch of grapes; said especially of glands.

racemose inflorescence *(Bot.)*. One in which the main axis (and, in a compound raceme, each of its main branches) does not end in a flower but continues to grow bearing flowers in acropetal succession on its lateral branches, e.g. raceme, spike, panicle. Also called *indefinite* or *indeterminate inflorescence*. Cf. **cymose inflorescence, mixed inflorescence**.

race-track *(Nuc.Eng.)*. Discharge tube or ion-beam chamber where particles are constrained to an oval path.

rachilla *(Bot.)*. (1) The axis of the spikelet of a grass, on which are borne the glumes and the florets. (2) A secondary (or tertiary etc.) axis of a pinnately compound leaf.

rachiodont *(Zool.)*. Having some of the anterior thoracic vertebrae with the hypapophysis enlarged, forwardly directed, and capped with enamel to act as an egg-breaking tooth, as certain egg-eating Snakes.

rachi-, rachio- *(Genrl.)*. Prefix from Gk. *rhachis*, spine.

rachis *(Bot.)*. The main axis of an inflorescence or of a pinnately compound leaf etc.

rachis *(Zool.)*. The shaft or axis; the shaft of a feather; the vertebral column. adj. *rachidial.*

rachitic *(Med.)*. Affected with or pertaining to rickets.

R-acid *(Chem.)*. *2-Naphthol-3,6-disulphonic acid*; used in preparation of azo-dyes for wool.

rack *(Eng.)*. Component on which a linear array of gear-form teeth are cut.

rack *(Horol.)*. A toothed segment; used in the striking and chiming mechanism of a clock or a repeater-watch. The number of blows struck by the hammer is controlled by the number of teeth gathered on the rack.

rack *(Min.Ext.)*. Reck. See **ragging frame**.

rack-and-pinion *(Eng.)*. A method of transforming rotary into linear motion, or vice versa; accomplished by a pinion or small gear-wheel which engages a straight, toothed rack. See **pitch line**.

rack-and-pinion steering-gear *(Autos.)*. One in which a pinion carried by the steering column engages with a rack attached to the divided track rod.

rack clock *(Horol.)*. One mounted on a vertical toothed rack and operated by gravity.

racked *(Build.)*. Said of temporary timbering which is braced so as to stiffen it against deformation.

rack hook *(Horol.)*. Part of the rack-striking work. Just before the hour, the lifting piece lifts the *rack hook*, which allows the rack to fall.

racking *(Min.Ext.)*. The operation of separating ore by washing on an inclined plane.

racking *(Ships)*. Distortion of the ship's transverse shape.

racking back *(Build.)*. The procedure adopted when the full length of a wall is not built at once, the unfinished end being stepped or *racked back* at an angle, so that when the remainder of the wall is built there is not a vertical line of junction, which might cause cracking, owing to uneven settlement.

rack mounting *(Telecomm.)*. The use of standard racks, of varying height but otherwise of standard dimensions, for mounting panels carrying electronic equipment or assemblies with a uniform scheme of wiring; such mounting gives both accessibility and compactness.

rack railway *(Civ.Eng.)*. A device for overcoming adhesion difficulties, as met on mountain railways. A **pinion** on the locomotive axle engages with a **rack** laid parallel with, usually in the centre of, the normal rails, which continue to carry the weight. Several forms exist. Also *mountain railway*.

rack saw *(Build.)*. A saw having wide teeth.

rad *(Maths.)*. Abbrev. for *radian*.

rad *(Radiol.)*. Former unit of radiation dose which is absorbed, equal to 0.01 J/kg of the absorbing (often tissue) medium. See **gray**.

radar *(Genrl.)*. In general, a system using pulsed radio waves, in which reflected (*primary radar*) or regenerated (*secondary radar*) pulses lead to measurement of distance and direction of target (*RAdio Detection And Ranging*).

radar absorbing material *(Aero.)*. That attached to or built into an aircraft skin which responds to radar waves by attenuating their return echo, thus reducing the radar signal.

radar astronomy *(Astron.)*. The use of pulses of radio waves to detect the distances and map the surface morphology of objects in the Solar System. It has been applied with great success to the mapping of **Venus**.

radar beacon *(Radar)*. A fixed radio transmitter whose radiations enable a craft to determine its own direction or position relative to the beacon by means of its own radar equipment. See **transponder**.

radar indicator *(Radar)*. Display on a CRT of a radar system output, either as a radial line or a coordinate system for range and direction. The echo signal gives a brightening of the luminous spot, which remains for some seconds because of afterglow. Also *radar screen*. See **A-, B-, C-display**, etc., **plan-position indicator**.

radar performance figure *(Radar)*. The ratio of peak transmitter power to minimum signal detectable by the receiver.

radar range *(Radar)*. Usually given as that at which a specified object can be detected with 50% reliability.

radar range equation *(Radar)*. A mathematical expression, for **primary radar**, which relates transmitter power, antenna gain, wavelength, effective area of the target, receiver sensitivity and *radar range*.

radar scan *(Radar)*. (1) The circular, rectangular or other motion of a radar as it searches for a target. (2) The physical movement of a radar antenna or of the radial line on a **PPI** display.

radar screen *(Radar)*. See **radar indicator**.

radial *(Maths.)*. Radiating from a common centre; pertaining to a *radius*.

radial commutator *(Elec.Eng.)* A commutator for a d.c. machine in which the commutator bars are arranged radially from the axis to form a disk instead of a cylinder.

radial drill *(Eng.)*. A large drilling machine in which the drilling head is capable of radial adjustment along a rigid horizontal arm carried by a pillar.

radial ducts *(Elec.Eng.)*. In an electric machine, ventilating ducts which run radially from the shaft.

radial engine *(Aero.)*. An aircraft or other engine having the cylinders arranged radially at equal angular intervals round the crankshaft. See **double-row-, master connecting-rod**.

radial feeder *(Elec.Eng.)*. See **independent feeder**.

radial longitudinal section *(Bot.)*. A section cut longitudinally along a diameter of a more or less cylindrical organ. Cf. **tangential longitudinal section**. Abbrev. *RLS*.

radial-ply *(Autos.)*. Term applied to tyres which have a semi-rigid breaker strip under the tread and relatively little stiffening in the walls. Cf. **cross-ply**.

radial recording *(Acous.)*. Same as **lateral recording**.

radial runout *(Eng.)*. Variation in the plane normal to its axis of a rotating part. Its *eccentricity* rather than its *wobble*. Cf. **axial runout**.

radial symmetry *(Biol.)*. The condition in which an organ or the whole of an organism can be divided into two similar halves by any one of several planes which include the centre line. Cf. **bilateral symmetry**.

radial system *(Elec.Eng.)*. A distribution system in which the cables radiate out from a generating or supply station. If a fault occurs, all consumers beyond the fault are cut off.

radial valve gear *(Eng.)*. A steam-engine valve gear in which the slide-valve is given independent component motions proportional to the sine and cosine of the crank angle respectively.

radial velocity *(Astron.)*. See **line-of-sight velocity**.

radian *(Maths.)*. The SI unit of plane angular measure, defined as the angle subtended at the centre of a circle by an arc equal in length to the radius. Abbrev. *rad*.

$$2\pi \text{ radians} = 360^\circ$$
$$1 \text{ radian} = 57^\circ{\cdot}295\,779\,513\,1\ldots$$
$$1^\circ = 0{\cdot}017\,453\,292\,5\ldots\text{radian.}$$

radiance *(Phys.)*. Of a surface, the luminous flux radiated per unit area.

radian frequency *(Phys.)*. Same as *angular frequency*.

radiant *(Astron.)*. Point on the celestial sphere from which a series of parallel tracks in space, such as those followed by the individual meteors in a shower, appear to originate.

radiant flux *(Phys.)*. The time rate of flow of radiant electromagnetic energy.

radiant-flux density *(Phys.)*. A measure of the radiant power per unit area that flows across a surface. Also called *irradiance*.

radiant heat *(Phys.)*. Heat transmitted through space by *infra-red radiation*.

radiant intensity *(Phys.)*. The energy emitted per second per unit solid angle about a given direction.

radiant-tube furnace *(Eng.)*. Modified form of muffle furnace, the heating chamber having a series of steel alloy tubes in which the fuel is burned, thus excluding products of combustion from the heating chamber.

radiant-type boiler *(Eng.)*. A water-tube boiler having one or more drums and a circulation system consisting of vertically or horizontally inclined banks of tubes, heating surfaces of bare or protected water-tubes forming the walls of the combustion chamber; firing is generally by pulverized fuel.

radiate *(Bot.)*. A capitulum, having **ray florets**.

radiated power *(Telecomm.)*. The actual power level of the radio signals transmitted by an antenna. See also **effective radiated power**.

radiating brick *(Build.)*. See **compass brick**.

radiating circuit *(Telecomm.)*. Any circuit capable of sending out power, in the form of electromagnetic waves, into space; especially the antenna circuit of a radio transmitter.

radiating surface *(Phys.)*. The effective area of a radiator available for the transmission of heat by radiation.

radiation *(Phys.)*. The dissemination of energy from a source. The energy falls off as the inverse square of the distance from the source in the absence of absorption. The term is applied to electromagnetic waves (radio waves, infrared, light, X-rays, γ-rays etc.) and to acoustic waves. It is also applied to emitted particles (α, β, protons, neutrons etc.). See types of radiation: *black-body*, *electromagnetic*, *infrared*, *ultraviolet*, *visible* etc. See also **Planck's radiation formula**, **Stefan-Boltzmann law**, **Wien's laws**.

radiation *(Surv.)*. A method of plane-table surveying in which a point is located on the board by marking its direction with the alidade and measuring off its distance to scale from the instrument station. See **plane table**.

radiation area *(Nuc.Eng.)*. Area to which access is controlled because of a local radiation hazard.

radiation burn *(Med.)*. A burn caused by over-exposure to radiant energy.

radiation chemistry *(Chem.)*. That of radiation-induced chemical effects (e.g. decomposition, polymerization etc.). Not to be confused with *radiochemistry*.

radiation counter *(Nuc.Eng.)*. One used in nuclear physics to detect individual particles or photons.

radiation danger zone *(Radiol.)*. A zone within which the **maximum permissible dose rate** or **concentration** is exceeded.

radiation diagram *(Phys.)*. See **radiation pattern**.

radiation efficiency *(Telecomm.)*. Ratio of actual power radiated by antenna to that provided by the drive.

radiation field *(Phys.)*. See **field**.

radiation flux density *(Phys.)*. Rate of flow of radiated energy through unit area of surface normal to the beam (for particles this is frequently expressed in number rather than energy). Also called *radiation intensity*.

radiation hazard *(Radiol.)*. The danger to health arising from exposure to ionizing radiation, either due to external irradiation or to radiation from radioactive materials within the body.

radiation impedance *(Phys.)*. Impedance per unit area. Measured, e.g., by the complex ratio of the sound pressure to the velocity at the surface of a vibrating body which is generating sound waves, or by the corresponding electromagnetic quantities.

radiation intensity *(Phys.)*. See **radiation flux density**.

radiation length *(Phys.)*. The path length in which relativistic charged particles lose e^{-1} of their energy by radiative collisions. See **bremsstrahlung**.

radiation loss *(Telecomm.)*. That power radiated from a nonshielded radiofrequency transmission line.

radiation pattern *(Phys.)*. Polar or cartesian representation of distribution of radiation in space from any source and, in reverse, effectiveness of reception. Also called *radiation diagram*.

radiation pattern *(Telecomm.)*. A graphical representation of the radiation from an antenna as a function of direction, which may be in azimuth or any plane of elevation.

radiation potential *(Phys.)*. See **ionization potential**.

radiation pressure *(Phys.)*. Minute pressure exerted on a surface normal to the direction of propagation of a wave. It is due to the rate of transfer of momentum by the wave. For electromagnetic waves incident on a perfect reflector this pressure is equal to the energy density in the medium. In quantum physics the radiation consists of *photons* and the radiation pressure is due to the transfer of the momentum of the photons as they strike the surface. Radiation pressures are very small, $10^{-5}\,N\,m^{-2}$ for sunlight at the Earth's surface. For sound waves in a fluid the pressure gives rise to 'streaming', i.e. a flow of the fluid medium.

radiation prospecting *(Min.Ext.)*. Form of **geophysical prospecting** which utilizes the radioactivity of uranium-, thorium- or radium-bearing minerals to identify potentially economic concentrations. Gamma-ray detectors (**Geiger-Müller tubes** and **scintillation counters**) may be hand-held or airborne.

radiation pyrometer *(Phys.)*. A device for ascertaining the temperature of a distant source of heat, such as a furnace, by allowing radiation from the source to face, or be focused on, a thermojunction connected to a sensitive galvanometer, the deflection of the latter giving, after suitable calibration, the required temperature. For temperatures ca. 500°–1500°C.

radiation resistance *(Telecomm.)*. That part of impedance of an antenna system related to power radiated; the power radiated divided by the square of the current at a specified point, e.g. at the junction with the feeder.

radiation sickness *(Med.)*. Illness, characterized by nausea, vomiting and loss of appetite after excessive exposure to radiation either from radiation therapy or accidentally. If the exposure has been great it will cause bone marrow suppression with loss of blood cells, leading to anaemia, inability to overcome infection and internal bleeding.

radiation therapy *(Med.)*. The use of any form of radiation, e.g. electromagnetic, electron or neutron beam, or ultrasonic, for treating disease.

radiation trap *(Nuc.Eng.)*. (1) Beam trap for absorbing intense radiation beam with a minimum of scatter. (2) Maze or labyrinth formed by entry corridor with several right-angle bends, used for approach to multicurie radiation sources on some accelerating machines.

radiative collision *(Phys.)*. One in which kinetic energy is converted into electromagnetic radiation. See **bremsstrahlung**.

radiative equilibrium *(Astron.)*. The normal state of matter inside stars in which the temperature in every part generates a gas pressure which exactly balances the pressure due to the self-gravity of the star. There is, therefore, no convection in this idealized situation.

radiator *(Autos.)*. A device for dissipating heat created in a water-cooled engine. It consists of thin-walled tubes, or narrow passages of honeycomb form, through which the water is conducted, and across which an airstream is induced either by the motion of the vehicle or by a fan.

radiator *(Phys.)*. In radioactivity, the origin of α-, β- and/or γ-rays; also called a *source*.

radiator *(Telecomm.)*. Any part of the antenna or transmission line which radiates electromagnetic waves, either directly into space or in conjunction with other elements or structures which concentrate the radiation into specific directions or spaced beams. A radiator may be a length of wire, ¼ or ½ wavelength long or a multiple thereof, or a loop, spiral or helix, or the open end of a waveguide or a slot in a conducting surface.

radiator flaps *(Aero.)*. See **gills**.

radical *(Bot.)*. (1) Pertaining to the root. (2) Leaves, flowers etc. arising at soil level from a root stock, rhizome or the base of a stem, as in rosette plants. Cf. **cauline**.

radical *(Chem.)*. (1) A molecule or atom which possesses an odd number of electrons, e.g. $Br.CH_3$. It is often very short-lived, reacting rapidly with other radicals or other molecules. (2) A group of atoms which passes unchanged through a series of reactions, but is normally incapable of separate existence. This is now more usually called a *group*.

radical axis *(Maths.)*. The straight line from each point of which the tangents to two circles are equal.

radicivorous *(Zool.)*. Root-eating.

radicle *(Bot.)*. The primary root of an embryo, normally the first organ to emerge when a seed germinates.

radiculectomy *(Med.)*. The operation of cutting the roots of spinal nerves.

radiculitis *(Med.)*. Inflammation of the root of a spinal nerve.

radio- *(Chem.)*. A prefix denoting an artificially prepared radioactive isotope of an element.

radio *(Telecomm.)*. Generic term applied to methods of signalling through space, without connecting wires, by means of electromagnetic waves generated by high-frequency alternating currents.

radioactivation analysis *(Phys.)*. If a material undergoes bombardment by neutrons in a nuclear reactor, then frequently radioactive nuclei are produced. This artificial radioactivity can be studied to give information about the isotopes present in the material.

radioactive atom *(Phys.)*. One which decays into another species by emission of an α- or β-ray (or by electron capture). Activity may be natural or induced.

radioactive chain *(Phys.)*. See **radioactive series**.

radioactive dating *(Geol.)*. See **radiometric dating**.

radioactive decay *(Phys.)*. See **disintegration constant**, **half-life**, **radioactive atom**.

radioactive decay by heavy ion emission *(Phys.)*. Radioactive decay by the emission of nuclei heavier than the alpha particle. The probability of this occurring is very small but ^{14}C rather than α-decay has been observed from ^{223}Ra.

radioactive equilibrium *(Phys.)*. Eventual stability of products of radioactivity if contained, i.e. rate of formation (quantitative) equals rate of decay. Particularly important between radium and radon.

radioactive isotope *(Phys.)*. Naturally occurring or artificially produced isotope exhibiting radioactivity; used as a source for medical or industrial purposes. Also *radioisotope*.

radioactive series *(Phys.)*. Most naturally occurring radioactive isotopes belong to one of three series that show how they are related through radiation and decay. Each series involves the emission of an α-particle, which decreases the *mass number* by 4, and β- and γ-decay which do not change the mass number. The natural series have members having mass number: (a) $4n$ (thorium series); (b) $4n + 2$ (*uranium-radium series*); (c) $4n + 3$ (*actinium series*). Members of the $4n + 1$ (*plutonium series*) can be produced artificially. Also called *radioactive chain*.

radioactive standard *(Phys.)*. A radiation source for calibrating radiation measurement equipment. The source has usually a long half-life and during its decay the number and type of radioactive atoms at a given reference time is known.

radioactive tracer *(Phys.)*. Small quantity of radioactive preparation added to corresponding nonactive material to *label* or *tag* it so that its movements can be followed by tracing the activity. (The chemical behaviour of radioactive elements and their nonactive isotopes is identical.)

radioactivity *(Phys.)*. Spontaneous disintegration of certain natural heavy elements (e.g. radium, actinium, uranium, thorium) accompanied by the emission of α-rays, which are positively charged helium nuclei; β-rays, which are fast electrons; and γ-rays, which are short X-rays. The ultimate end-product of radioactive disintegration is an isotope of lead. See also **artificial radioactivity**, **induced radioactivity**.

radio-allergosorbent test *(Immun.)*. *RAST*. Method for measuring extremely small amounts of IgE antibody specific for various allergens. Blood serum is reacted with allergen-coated particles which are then washed to remove non-reacting proteins. Radiolabelled anti-human IgE is then added and this binds to the IgE antibody, bound to the particles via the allergen. Provided that the amount of allergen supplied and the anti-IgE are present in excess, the radioactivity on the particles after washing is proportional to the amount of allergen-specific antibody in the serum sample.

radio altimeter *(Aero.)*. Device for determining height, particularly of aircraft in flight, by electronic means, generally by detecting the delay in reception of reflected signals, or change in frequency; also called *radar altimeter*.

radio astronomy *(Astron.)*. The exploration of the universe by detecting radio emission from a variety of celestial objects. The frequency spans a vast range from 10 MHz to 300 GHz. A variety of antennas are used, from single dishes to elaborate networks of telescopes forming intercontinental radio **interferometers**. The principle sources of radio emission are: the Sun, Jupiter, interstellar hydrogen, emission nebulae, pulsars, supernova remnants, radio galaxies, quasars, and the cosmic background radiation of the universe itself.

radio beacon *(Telecomm.)*. Stationary radio transmitter which transmits steady beams of radiation along certain directions for guidance of ships or aircraft, or one which transmits from an omnidirectional antenna and is used for the taking of bearings, using an identifying code. Also *aerophare*.

radio beam *(Telecomm.)*. Concentration of electromagnetic radiation within narrow angular limits, such as is emitted from a highly directional antenna.

radio bearing *(Telecomm.)*. Direction of arrival of a radio signal, as indicated by a loop, goniometer, interferometer or any directional receiving system as used for navigational purposes.

radiobiology *(Biol.)*. Branch of science involving study of effect of radiation and radioactive materials on living matter.

radio broadcasting *(Telecomm.)*. The transmission by means of radio waves, of a programme of sound or picture for general reception. The separation of the frequency channels is decided by international agreement.

radiocaesium *(Chem.)*. ^{137}Cs, a radioactive isotope recovered from the waste of nuclear reactors in nuclear power plants. Useful for mass-radiation and sterilization of foodstuffs. Also for high-intensity X-ray radiation of surface tumours in place of much more expensive radium. Half-life 37 years.

radiocarbon *(Chem.)*. ^{14}C, a weakly radioactive isotope undergoing beta-decay with a half-life of 5770 years. It is present in the atmosphere in roughly constant amount, as it is produced from ^{14}N by cosmic rays. It is used in some tracer studies. It can also be used to date the time of death of once-living material (and hence the likely time of manufacture of an artifact). This is because living material has the same ratio of ^{14}C to ^{12}C as the atmosphere. After death, however, the ^{14}C decays and is not replaced.

radiocarbon dating *(Geol.)*. A method of determining the age in years of fossil organic material or water bicarbonate, based on the known decay rate of ^{14}C to ^{14}N. See **radiocarbon**.

radiochemical purity *(Chem.)*. The proportion of a given radioactive compound in the stated chemical form. Cf. **radioisotopic purity**.

radiochemistry *(Chem.)*. Study of science and techniques of producing and using radioactive isotopes or their compounds to study chemical compounds. Cf. **radiation chemistry**.

radio circuit *(Telecomm.)*. Communication system including a radio link, comprising a transmitter and antenna, the radio transmission path, with possible reflections or scatter from ionized regions, and a receiving antenna and receiver.

radiocolloids *(Phys.)*. Radioactive atoms in colloidal aggregates.

radio communication *(Telecomm.)*. Any form of communication involving the transmission and reception of electromagnetic waves, from a frequency of 10 kHz up to more than 10 GHz. Information is conveyed by **modulation** of the information it is desired to impart onto a **carrier**. The information may be letters represented by code (e.g. Morse), speech, telemetry, pictures (either facsimile or television), digital signals or computer data. In broadcasting, radio communication is a one way process serving many listeners or viewers, or it may be two-way as in telecommunication systems. In the latter, communication may be between two mobile users in different vehicles or from a mobile vehicle and a fixed station, from one microwave tower to another in terrestrial communication (see **radio link**) or from one *Earth* station to another via a **communication satellite**.

radio compass *(Telecomm.)*. Originally a rotating loop, later rendered more sensitive by a goniometer system and by display on a cathode-ray tube. Any device, depending on radio, which gives a bearing. See **Adcock antenna**.

radio direction-finding *(Telecomm.)*. Passive reception of direction-finding signals from radio beacons or navigational transmitters, as distinct from active radar. Abbrev. *RDF*.

radioelement *(Phys.)*. An element exhibiting natural radioactivity.

radio frequency *(Telecomm.)*. One suitable for radio transmission, above 10^4 Hz and below 3×10^{12} Hz approx. Abbrev. *RF*. Also *radio spectrum*.

radio-frequency heating *(Elec.Eng.)*. See **dielectric heating, induction heating**.

radio-frequency spectrometer *(Nuc.Eng.)*. Type of **mass spectrometer** used in the study of ions in plasmas.

radio-frequency spectroscopy *(Phys.)*. See **nuclear magnetic resonance, electron spin resonance**.

radio galaxy *(Astron.)*. About one galaxy in a million is an intense source of cosmic radio waves, caused by synchrotron emission of relativistic electrons.

radiogenic *(Phys.)*. Said of stable or radioactive products arising from radioactive disintegration.

radiogoniometer *(Telecomm.)*. Rotating coil within crossed field established by crossed loops for direction-finding.

radiography *(Radiol.)*. Process of image production using X-rays.

radio heating *(Phys.)*. See **high-frequency heating**.

radio horizon *(Telecomm.)*. In the propagation of electromagnetic waves over the earth, the line which includes the part of the earth's surface which is reached by direct rays.

radioimmunosorbent test *(Immun.)*. *RIST*. Method for measuring IgE immunoglobulin in samples of serum. The serum is mixed with a standard amount of purified radiolabelled IgE, and the mixture is exposed to particles coated with antibody specific for the IgE heavy chain. After appropriate incubation and washing, the amount of radioactivity bound to the particles is determined. Since IgE in the serum will compete with the radiolabelled IgE for binding to the anti-IgE, the amount of radioactivity bound will be less than is bound in the absence of such competition. The reduction of binding provides a measure of the amount of IgE in the test sample.

radioiodine *(Chem.)*. ^{125}I, ^{131}I and ^{132}I, radioactive isotopes useful in diagnosis and treatment of thyroid gland disorders. ^{125}I and ^{131}I are used in organ-function and blood volume studies.

radioisotope *(Phys.)*. See **radioactive isotope**.

radioisotope thermoelectric generator *(Nuc.Eng.)*. Thermoelectric generator powered by heat from a radioactive source and suitable for long periods of maintenance-free operation on remote sites, e.g. for lighting marine navigational buoys or powering spacecraft. Also called *Ripple* (UK) or *SNAP* (US) generators.

radioisotopic purity *(Chem.)*. The proportion of the activity of a given compound which is due to material in the stated chemical form. Cf. *radiochemical purity*.

Radiolaria *(Zool.)*. An order of marine planktonic Sarcordina, the members of which have numerous fine radial pseudopodia which do not anastomose; there is usually a skeleton of siliceous spicules.

radiolarian chert, radiolarite *(Geol.)*. A crypto-crystalline siliceous rock in part composed of the remains of Radiolaria. Most described examples seem to be of shallow-water origin, such as that which reaches a thickness of 3000 m in New South Wales and contains 20 million Radiolaria per cubic centimetre.

radiolarian ooze *(Geol.)*. A variety of non-calcareous deep-sea ooze, deposited at such depth that the minute calcareous skeletons of such organisms as Foraminifera pass into solution, causing a preponderance of the less soluble siliceous skeletons of Radiolaria. Confined to the Indian and Pacific Oceans, and passes laterally into red clay.

radiolarite *(Geol.)*. See **radiolarian chert**.

radio link *(Telecomm.)*. Self-contained two-way communication which forms part of a more extensive broadcasting or telecommunications network; broadcasters may use transportable radio links (usually microwave) to feed sound and vision into their permanent networks, while fixed installations on tall towers and masts, using highly directional antennae and operating on a line-of-sight basis, provide high-capacity channels as part of most countries' telecommunication networks. International channels, via satellites, may also be described as radio links.

radiolocation *(Radar)*. Former term for **radar**.

radiology *(Radiol.)*. The science and application of X-rays, gamma-rays and other penetrating ionizing or non-ionizing radiations.

radioluminescence *(Phys.)*. Luminous radiation arising from rays from radioactive elements, particularly in mineral form.

radiolysis *(Phys.)*. Chemical decomposition of materials induced by ionizing radiation.

radiometer *(Acous.,Phys.)*. Instrument devised for the detection and measurement of electromagnetic radiant energy and acoustic energy, e.g. thermopile, bolometer, microradiometer. See **Crookes radiometer, Rayleigh disk**.

radiometric age *(Geol.)*. The radiometrically determined age of a fossil, mineral, rock or event, generally given in years. See **radiometric dating**.

radiometric dating *(Geol.)*. The method of obtaining a geological age by measuring the relative abundance of radioactive parent and daughter isotopes in geological materials. See **uranium-lead, potassium-argon, rubidium-strontium** and **radiocarbon dating**.

radio microphone *(Telecomm.)*. One with a miniature radio transmitter, allowing freedom from a cable for the speaker. Its transmissions are picked up by a receiver nearby and fed to a public address or broadcasting system.

radiomimetic *(Med.)*. Said of drugs which imitate the physiological action of X-rays, notably in suppressing new cell growth, particularly those used in treating cancer.

radionuclide *(Phys.)*. Any nuclide (isotope of an element) which is unstable and undergoes natural radioactive decay. ^{99}Technetium is an important radionucluide used for diagnostic imaging in medicine.

radionuclide imaging *(Radiol.)*. The use of radionuclide substances to image the normal or abnormal physiology or anatomy of the body.

radiopaque *(Radiol.)*. Opaque to radiation (especially X-rays).

radiophare *(Telecomm.)*. Same as **radio beacon**.

radiophotoluminescence *(Phys.)*. Luminescence produced by irradiation followed by exposure to light.

radio range *(Telecomm.)*. (1) Specific system of radio homing for aircraft, in which crossed loops are separately modulated with complementary signals, which coalesce on reception when the aircraft is on course. (2) For transmissions which are not affected by ionospheric or tropospheric phenomena, the optical or *line of sight* path between two points.

radio receiver *(Telecomm.)*. Any device which converts radio waves into sound or other intelligible signals. Most common types are *tuned radio frequency* and *superhet*. The latter shows superior sensitivity and immunity to noise and interference for most applications. More complex forms of receiver may be used for radar and for satellite communications, though the most sophisticated receivers can usually be placed into one of the two categories quoted.

radio relay station *(Telecomm.)*. An intermediate station receiving a signal from the primary transmitter and reradiating it to its destination.

radioresistant *(Radiol.)*. Able to withstand considerable radiation doses without injury.

radiosensitive *(Radiol.)*. Quickly injured or changed by irradiation. The gonads, the blood-forming organs, and the cornea of the eye are biologically most radiosensitive in man.

radiosonde *(Meteor.)*. Instrument for measuring temperature, pressure and humidity at successive levels in the atmosphere which is carried upwards on a balloon and transmits the measurements by radio. The balloon also carries a radar target so that upper winds may be derived from ground measurements.

radio spectrum *(Telecomm.)*. See **radio frequency**.

radiospermic *(Bot.)*. Having seeds which are rounded in cross section, as in the Cycadopsida. Cf. **platyspermic**.

radio telegraph *(Telecomm.)*. Using a radio channel for telegraph purposes, e.g. by interrupted carrier, change of frequency, or modulation with interrupted audio tone.

radio telephony *(Telecomm.)*. Use of a radio channel for transmission of speech. Methods include simple modulation, suppressed carrier and one sideband, inverted sidebands and carrier, scrambling before modulation, one in a group modulation or pulse modulation.

radio telescope *(Astron.)*. An instrument for the collection, detection and analysis of radio waves from any cosmic source. All such telescopes consist of a radio antenna, detector and amplifier. Antenna systems may be arrays of **dipoles**, single dishes or **interferometers**. The purpose of a radio telescope is to measure the intensity of radio emission and establish its spectrum. Image analysis can be used to give a picture similar to a photograph.

radiotherapy *(Radiol.)*. Theory and practice of medical treatment of disease, particularly any of the forms of cancer, with large doses of X-rays or other ionizing radiations.

radiothermoluminescence *(Phys.)*. Luminescence produced by irradiation followed by exposure to heat. Now used for personal dosimetry.

radiothorium *(Chem.)*. Symbol RdTh. A disintegration product and isotope of thorium, with a half-life of 1.90 years.

radiotropospheric duct *(Meteor.)*. Stratum in which, because of a negative gradient of refractive modulus, there is an abnormal concentration of radiated energy.

radium *(Chem.)*. Radioactive metallic element, one of the alkaline earth metals. Symbol Ra, at. no. 88, r.a.m. 226, half-life 1620 years. The metal is white and resembles barium in its chemical properties; mp 700°C. It occurs in bröggerite, cleveite, carnotite, pitchblende, in certain mineral springs and in sea water. Pitchblende and carnotites are the chief sources of supply.

radium cell *(Radiol.)*. A sealed container in shape of thin-walled tube (usually metal) normally loaded into larger containers, e.g. **radium needle**.

radium emanation *(Chem.)*. See **radon**.

radium needle *(Radiol.)*. A container in form of needle, usually platinum-iridium or gold alloy, designed primarily for insertion into tissue.

radium therapy *(Radiol.)*. Radiotherapy by the use of the radiations from radium.

radius *(Eng.)*. A rod attached to the die or block of a **Walschaert's valve gear** for transmitting its motion to the end of the combination lever pivoted to the valve rod.

radius *(Maths.)*. A straight line joining the centre of a conic (particularly a circle) to the curve; the length of such a line.

radius *(Zool.)*. In land Vertebrates, the pre-axial bone of the antebranchium; one of the veins of the wing in Insects; in Echinodermata and Coelenterata, one of the primary axes of symmetry. adj. *radial.*

radius brick *(Build.)*. See **compass brick**.

radius gauge *(Eng.)*. A sheet metal strip with accurately formed external and internal radii of specified size, used as a profile gauge to determine size of an unknown radius by comparison. Usually one of a set of gauges of graded sizes.

radius of action *(Aero.)*. Half the range in still air of a military aircraft, taking safety and operational requirements into account; the total range is out and home again.

radius of atom *(Chem.)*. See **atomic radii**.

radius of convergence *(Maths.)*. See **circle of convergence**.

radius of curvature *(Maths.)*. See **curvature**.

radius of gyration *(Phys.)*. See **moment of inertia**.

radius of inversion *(Maths.)*. See **inversion**.

radius of spherical curvature *(Maths.)*. See **osculating sphere**.

radius of torsion *(Maths.)*. See **curvature**.

radius rod *(Build.)*. A rod pivoted at one end and carrying a marking point at the other end so that, as the rod is swung around, a circle or part of a circle, of radius equal to the length of the rod, may be marked out. Also *gig stick*.

radius vector *(Astron.)*. The line joining the focus to the body which moves about it in an orbit, as the line from the sun to any of the planets, comets etc.

radix *(Maths.)*. Basis of a denominational number system. A radix of 10 is normally used except for digital computers which usually use 2 because of the ease with which two-state electronic circuits can be constructed. Also called *base*.

radix *(Zool.)*. The root or point of origin of a structure, as the *radix aortae*.

radix point *(Maths.)*. That separating positive and negative indices of the radix; the decimal point when the radix is 10.

radome *(Radar)*. Housing for radar equipment, transparent to the signals, e.g. a plastic shell on aircraft or a balloon on the ground.

radon *(Chem.)*. A zero-valent, radioactive element, the heaviest of the noble gases. Symbol Rn, at. no. 86, r.a.m. 222, half-life 3.82 days, bp −65°C, mp −150°C. It is formed by the disintegration of radium. Isotopes are actinon (at. no. 219, half-life 4 seconds, from actinium) and thoron (at. no. 220, half-life 54 seconds, from thorium). Also *radium emanation*. See **emanations**.

radon seeds *(Radiol.)*. Short lengths of gold capillary tubing containing radon used in treatment of malignant and nonmalignant neoplasms.

radula *(Zool.)*. In Mollusca, mechanism for rasping consisting of a strip of epithelium bearing numerous rows of horny or chitinous teeth. adjs. *radular, radulate, raduliform.*

RAE *(Aero.)*. Abbrev. for *Royal Aircraft Establishment*, Farnborough, UK.

RAeS *(Aero.)*. Abbrev. for *Royal Aeronautical Society*. UK.

raffinate *(Min.Ext.)*. Liquid layer in solvent extraction system from which required solute has been extracted, e.g. in the chemical extraction of uranium from its ores, the liquid left after the uranium has been extracted by contact with an immiscible solvent.

raffinose *(Chem.)*. *Melitriose*, $C_{18}H_{32}O_{16}+5H_2O$, a non-reducing trisaccharide found in the sugarbeet, in molasses, in cotton-seed cake etc. On hydrolysis it gives D-glucose, D-fructose and D-galactose.

raft foundation *(Build.)*. A layer of concrete, usually reinforced, extending under the whole area of a building and projecting outside the line of its walls; used to provide a foundation in cases where the ground is unduly soft or the loading to be put upon it is unduly heavy.

rag-bolt *(Eng.)*. A foundation bolt with a long tapered head of increasing size towards its end, and having jagged points on its surface to prevent withdrawal.

rag content papers *(Paper)*. Papers the fibrous **furnish** of which contains 25% or more by weight of rag (cotton etc.) fibres.

ragging *(Min.Ext.)*. Rough concentration or washing, for a low ratio of concentration. In mineral processing, grooves cut on surface of roll to improve grip on feed. In jigging, the bed of heavy mineral or metal shot maintained on the jig screen.

ragging frame *(Min.Ext.)*. Tilting table, which may be worked automatically, on which finely-ground ore is treated by sluicing. Also *reck*.

raggle, raglet *(Build.)*. A narrow groove cut into a masonry or brickwork to receive the edge of flashing which is to be fixed to it.

rag stone *(Build.)*. A general term for coarse-grained sandstone, often with a calcareous cement, e.g. Kentish rag.

rag-work *(Build.)*. Term applied to wall construction in which undressed flat stones of about the thickness of a brick are built up into a wall the outer faces of which are left rough.

rail *(Build.)*. (1) A horizontal member in framing or panelling. (2) The upper member in a balustrade.

rail *(Civ.Eng.)*. A steel bar, usually of special section, laid across sleepers to provide a track for the passage of rolling stock with flanged wheels. Now always flat-bottomed and specified by lbs per yard. See also **continuous welded rail**.

rail bender *(Eng.)*. A short stiff steel girder with claws at the ends and a central boss carrying a heavy screw.

rail bond *(Elec.Eng.)*. An electrical connection between two adjacent lengths of track or conductor rail on a railway.

rail chair *(Civ.Eng.)*. See **chair**.

rail gauge *(Eng.)*. See **gauge, standard gauge, broad gauge, narrow gauge**.

rail guard *(Civ.Eng.)*. See **check rail**.

rail post *(Build.)*. A newel post.

railroad disease *(Vet.)*. See **transit tetany**.

railway curve *(Eng.)*. A drawing instrument similar to a

French curve but cut at the edge to an arc of large radius. Used for drawing arcs when these are too large for beam compasses. Sets are supplied to cover a wide range of radii.

rain *(Meteor.)*. Result of condensation of excess water vapour when moist air is cooled below its dew-point. Rain falls when droplets increase in size until they form drops whose weight is equivalent to the frictional air resistance. The greater proportion of raindrops have a diameter of 0.2 cm or less; in torrential rain a small proportion may reach 0.4 cm. Rain effects important geological work by assisting in the mechanical disintegration of rocks; also chemically, in bringing about solution of carbonates, etc.; and, through the agency of running water, in redistributing the products of erosion and disintegration.

rain band *(Meteor.)*. An absorption band in the solar spectrum on the red side of the D lines, produced by water vapour in the earth's atmosphere.

rainbow *(Meteor.,Phys.)*. A rainbow is formed by sunlight which is refracted and internally reflected by raindrops, the concentration of light in the bow corresponding to the position of minimum deviation of the light. The angular radius of the primary bow is 42°, this being equal to 360° minus the angle of minimum deviation for a spherical drop. The colours, ranging from red outside to violet inside, are due to dispersion in the water. See **secondary bow**.

rainbow quartz *(Min.)*. See **iris**.

rain chamber *(Min.Ext.)*. Washing tower or other space where rising dust and fumes are brought into contact with descending sprays of water.

rain forest *(Ecol.)*. The natural forest of the humid tropics. Developing where rainfall is heavy (> 2500 mm/year) and characterized by a great richness of species, very tall trees (> 30 m), lianes and epiphytes.

rain gauge *(Meteor.)*. An instrument for measuring the amount of rainfall over a given period, usually 24 hr. The usual form consists of a sharp-trimmed funnel, 5 in in diameter, leading into a narrow-necked graduated collecting vessel. The *Dines tilting siphon* is of the continuously recording type, noting both the time and the amount of rainfall.

rainmaking *(Meteor.)*. Artificial stimulation of precipitation by scattering solid carbon dioxide on supercooled clouds, or by silver iodide nucleation.

rain prints *(Geol.)*. More or less circular, vertical, or slanting pits occurring on the bedding planes of certain strata; believed to be the impressions of heavy raindrops falling on silt or clay, hard enough to retain the impression before being covered by a further layer of sediment.

rain shadow *(Ecol.)*. A dry area, often a desert, on the sides of mountains away from the sea, due to the deposition of most of the moisture from the winds blowing off the ocean on the slopes facing the ocean. The higher the mountain, the greater the effect.

rain stage *(Meteor.)*. That part of the condensation process taking place at temperatures above 0°C so that water vapour condenses to water liquid.

rain-wash *(Geol.)*. The creep of soil and superficial rocks under the influence of gravity and the lubricating action of rain.

rainwater pipe *(Build.)*. See **downpipe**.

raised bands *(Print.)*. Bands which show on the back of a book when bound. This indicates that the book has been sewn **flexible**.

raised beach *(Geol.)*. Beach deposits which are found above the present high-water mark; due to the relative uplift of the land or to a falling sea-level. See **eustatic movements**.

raised bog *(Ecol.)*. Type of *Sphagnum* bog, originating from a valley bog or a fen by the upward growth of the vegetation and the failure of the dead plant material to decompose. The consequent raised bog is convex.

raised panel *(Build.)*. A panel whose surface stands *proud* of the general surface of the framing members.

raiser *(Build.)*. See **riser**.

raising *(Textiles)*. Machine operation in which wire covered rollers or teazles on cylinders revolve over the cloth surface to produce a nap.

raising-plate *(Build.)*. A horizontal timber resting on part of a structure and supporting a super-structure. Also *reason-piece*. See **pole plate**, **wall plate** (2).

rake *(Arch.)*. In a theatre, the upward slope, from the horizontal, of both the stage and the auditorium.

rake *(Build.)*. A long-handled tool with projecting teeth at one end, used for mixing plaster.

rake *(Eng.)*. Angular relief, e.g. *top-rake*, *side-rake*, given to the faces of cutting tools to obtain the most efficient cutting angle. The face of a cutting tool with negative *rake* has no angular relief but, on the contrary, meets the workpiece with the cutting edge trailing, the tool being stronger in shear, impact, and abrasion than one having positive *rake*.

rake *(Min.Ext.)*. (1) A forked tool for loading coal underground. (2) An irregular vein of ironstone. (3) Train or *journey* of mineral trucks. (4) Another name for **pitch**.

rake *(Ships)*. A term used in shipbuilding to denote *not perpendicular to the datum line*.

rake classifier *(Min.Ext.)*. Inclined tank into which ore pulp is fed continuously, the slow settling portion overflowing and the coarser material gravitating down, to be gathered and raked up to a top discharge.

raker *(Build.)*. See **raking shore**.

raker set *(Build.)*. A pattern of saw teeth in which one straight tooth alternates with two teeth set in opposite directions.

raking bond *(Build.)*. A form of bond sometimes used for very thick walls, or for strengthening the bond in footings carrying heavy loads. The courses are built diagonally across the wall, successive courses crossing one another in respect of rake; triangular **bats** are added to enable square facework to be completed. Also *diagonal bond*. See also **herring-bone bond**.

raking cornice *(Arch.)*. A cornice decorating the slant sides of a pediment.

raking flashing *(Build.)*. That used at the sides of e.g. a wall or chimney projecting from a sloping roof.

raking-out *(Build.)*. The operation of preparing mortar joints in brickwork for pointing.

raking pile *(Civ.Eng.)*. A pile which is not driven in vertically.

raking shore *(Build.)*. An inclined baulk of timber, one end of which rests upon a sole plate (or sleeper) on the ground while the other presses against the wall to which temporary support is to be given.

râle *(Med.)*. A bubbling or crackling sound produced in a diseased lung by the passage of air over or through secretions in it. Also called *crepitations*, *crackles*.

RALS *(Aero.)*. See **remote augmented lift system**.

RAM *(Aero.)*. See **radar absorbing material**.

ram *(Civ.Eng.)*. (1) The monkey of a pile-driver. (2) To consolidate the surface of loose material by punning.

RAM *(Comp.)*. See **random access memory**.

ram *(Eng.)*. (1) The reciprocating head of a press. (2) The reciprocating slide of a shaping machine on which the cutting tool is mounted. (3) The falling weight of a pile-driver.

ram-air turbine *(Aero.)*. A small turbine motivated by ram (i.e. free stream) air; used (*a*) to drive fuel pumps, hydraulic pumps, or electrical generators in guided weapons because of the absence of shaft drives in rockets and ramjets. (*b*) as an emergency power source for driving hydraulic pumps or electrical generators for high-speed aircraft, particularly those with power controls. Abbrev. *RAT*.

Raman scattering *(Phys.)*. The scattering of light by molecules in which there is a change of frequency due to the molecules gaining or losing energy as a result of transitions between vibrational or rotational energy levels. The scattering of light by the optic modes of vibration in a crystal, i.e. photon-phonon scattering. See also **scattering**.

Raman spectroscopy *(Chem.)*. A method making use of **Raman scattering** for chemical analysis. Like *infrared* *spectroscopy*, it investigates molecular vibrations and rotations.

ramentum *(Bot.)*. Thin, chaffy, brownish scale, especially on the stem, petiole or leaf of a fern.

ramet *(Bot.)*. Any physically and physiologically independent individual plant, whether grown from a sexually-produced seed or derived by vegetative reproduction. See also **clone**. Cf. **genet**.

rami communicantes *(Med.)*. The preganglionic myelinated nerve fibres of sympathetic nervous system connecting spinal nerves and sympathetic chain of ganglia.

ramie *(Textiles)*. A bast fibre from the stems of *Boehmeria nivea*; can be bleached to give rather brittle white fibres that may be used in garments blended with other fibres. Formerly used for the manufacture of gas mantles.

ramiform *(Bot.)*. Branching.

ram intake *(Aero.)*. A forward-facing engine (or accessory) air intake which taps the kinetic energy in the airflow and converts it into pressure energy by diffusion; in supersonic flight very high pressure ratios can be obtained.

ramisection, ramisectomy *(Med.)*. The operation of cutting the sympathetic nerves between the spinal cord and the sympathetic ganglia.

ramjet *(Aero.)*. The simplest **propulsive duct** deriving its thrust by the addition and combustion of fuel with air compressed solely as a result of forward speed. In subsonic flight kinetic energy is converted into pressure by **diffuser**, or widening duct, which also slows it sufficiently to permit combustion to be maintained; about Mach 1 the shock wave generated by the air-intake lip improves the compression when it decelerates the air to subsonic velocity prior to diffusion. At high Mach numbers, 1.5 and upward, two shock waves are required for the dual purpose of raising the pressure and slowing the air for combustion–pressure ratios of 6:1 are attainable at Mach 2, 36:1 at Mach 3. In supersonic flight the jet efflux of a ramjet has to be accelerated to high velocity by a **venturi** or **convergent-divergent nozzle**.

rammer *(Build.,Civ.Eng.)*. A **punner**.

rammer *(Foundry)*. Hand tool, which takes various forms, for packing the sand of a mould evenly round the pattern. The process of *ramming* is one in which the moulding sand is firmly consolidated before the pattern is removed from the moulding box.

Rammstedt's operation *(Med.)*. Incision of the pylorus down to the mucous membrane, done in the treatment of congenital hypertrophy of the pylorus.

ramp *(Aero.)*. (1) Parking area for an aircraft at an airport. (2) Inner wall of supersonic intake creating shock waves and improving pressure recovery. Frequently movable as on Concorde. (3) Inclined launcher for missile or RPV.

ramp *(Build.)*. A sudden rise in a handrail when it is concave upwards. Cf. **knee**.

ramp *(Civ.Eng.)*. An inclined surface, often in place of steps.

rampant arch *(Civ.Eng.)*. An arch whose abutments are not in the same horizontal line.

rampant centre *(Build.)*. A centre for a rampant arch.

ramp voltage *(Elec.Eng.)*. Steadily rising voltage, as in a sawtooth waveform.

RAM, r.a.m. *(Chem.)*. Abbrev. for *Relative Atomic Mass*.

Ramsauer effect *(Phys.)*. Sharp decrease to zero of scattering cross-section of atoms of inert gases, for electrons of energy below a certain critical value.

Ramsay and Young's rule *(Chem.)*. The ratio of the boiling-points of two liquids of similar chemical character is approximately constant, independent of the pressure at which they are measured.

Ramsden circle *(Phys.)*. The **exit pupil** of a telescope, found as a ring of light at the eyepiece of a telescope focused on a diffuse source at infinity. The magnification is the **objective** diameter divided by that of this circle.

Ramsden eyepiece *(Phys.,Surv.)*. An eyepiece often used in an optical instrument in which crosswire measurements are to be made. It consists of two similar plano-convex lenses separated by a distance equal to two-thirds the focal length of each, and having their convex faces towards each other. The focal plane is just outside the system.

ramus *(Zool.)*. The barb of a feather; in Vertebrates, one lateral half of the lower jaw, the mandible: in Rotifera, part of the trophi; any branchlike structure; a ramification.

ramwing *(Aero.)*. Special kind of aircraft designed to fly very low over water thereby gaining advantage in lift-drag ratio by capturing the ground effect. May one day rival the low cost ocean freight ship.

ranalian complex *(Bot.)*. Group of families, including the Ranunculaceae, containing what are thought to be the most primitive extant flowering plants. See **Magnoliidae**.

rance *(Build.)*. A shore.

random *(Build.)*. Said of rubble masonry in which the stones are of irregular shape and the work is not coursed.

random *(Print.)*. The sloping top on a composing frame.

random access *(Comp.)*. Process in which records are directly accessed in any order by their known address. Also called *direct access*.

random-access memory *(Comp.,Electronics)*. A computer memory which can be read from and written to by the programmer. It is usually made on a *chip*. Each storage location can be identified by x and y co-ordinates, as in a core or semiconductor memory. (Magnetic tapes or disks cannot provide random access.) The time required for reading out data is independent of its location in the store.

random coil *(Biol.)*. A section of a polypeptide chain which is not folded into any specific secondary structure.

random coincidence *(Nuc.Eng.)*. Simultaneous operation of two or more coincidence counters as a result of their discharge by separate incident particles arriving together (instead of common discharge by a single particle as is normally assumed in interpreting the readings).

random mating *(Biol.)*. Occurs when any individual in a population has an equal chance of mating with any other of the opposite sex.

random noise *(Acous.)*. Noise due to the aggregate of a large number of elementary disturbances with random occurrence in time.

random number *(Comp.)*. One from a sequence without any detectable bias or pattern. In a computer, random numbers are generally produced by an **algorithm** which is a **pseudo-random number generator**.

random number series *(Maths.)*. One that exhibits no regular pattern, e.g. the numbers drawn in a lottery. No completely satisfactory definition has yet been proposed.

random searching *(Ecol.)*. A process of completely unorganized 'search' by which some ecologists suggest that some animal populations find food, mates and suitable places to live.

random sequence welding *(Eng.)*. Adding welding beads at random along a seam to minimize distortion.

random-tooled ashlar *(Build.)*. A block of stone finished with groovings irregularly cut.

random variable *(Stats.)*. The mathematical representation of a variate associated with a stochastic phenomenon.

random winding *(Elec.Eng.)*. See **mush winding**.

Raney nickel *(Chem.)*. A nickel catalyst, used for hydrogenation, produced by the action of alkali on a nickel-aluminium alloy

range *(Aero.)*. Maximum horizontal distance covered by a projectile.

range *(Phys.)*. (1) Length of track along which ionization is produced in a nuclear particle. (2) Distance of effective operation of nuclear forces.

range *(Stats.)*. The difference between the largest and smallest values in a set of observations.

range *(Surv.)*. To fix points, either by eye or with the aid of an instrument, to be in the same straight line.

range *(Telecomm.)*. Maximum distance of radio transmitter at which effective reception is possible (not normally constant).

rangefinder *(Image Tech.)*. A device for measuring the distance of a remote object, especially to assist setting focus of a camera lens. It may be directly coupled to the lens mount for manual or automatic operation.

range-height indicator *(Radar)*. A display used in conjunction with a **plan-position indicator** for airport control. It displays a vertical plane on which the elevation and bearing of the target may be seen.

range of stress *(Eng.)*. The range between the upper and lower limits of a cycle of stress in a fatigue test. The midpoint is the *mean stress*.

range tracking *(Radar)*. The process of continuously monitoring the delay between the transmission of a pulse and reception of an echo. Tracking requires that the time elapsed between pulse and echo be measured, that the echo be identified as the target rather than random noise, and that the range-time history of the target be maintained.

ranging figures *(Print.)*. See **lining figures**.

ranging rod *(Surv.)*. A wooden pole used to mark the stations conspicuously, or to assist *ranging*.

ranine *(Zool.)*. Pertaining to, or situated on, the under surface of the tongue.

ranitidine *(Med.)*. Heals peptic ulcers by blocking the histamine$_2$ receptors (H_2 antagonist) which causes a reduced gastric acid output.

rank *(Maths.)*. Of a matrix: a matrix is of rank r if it contains at least one determinant of order r ($\neq 0$) and all higher-order determinants are zero.

rank *(Stats.)*. The number in serial order corresponding to a given data value when all values are placed in ascending order of magnitude; to place in ascending order of magnitude.

Rankine cycle *(Eng.)*. A composite steam plant cycle used as a standard of efficiency, comprising introduction of water by a pump to boiler pressure, evaporation, adiabatic expansion to condenser pressure, and condensation to initial point.

Rankine efficiency *(Eng.)*. The efficiency of an ideal engine working on the **Rankine cycle** under given conditions of steam pressure and temperature.

Rankine scale *(Phys.)*. Absolute scale of temperature, based on degrees Fahrenheit. See **Fahrenheit scale**.

Rankine's formula *(Civ.Eng.)*. An empirical formula giving the collapsing load for a given column. It states that

$$P = \frac{f_c . A}{1 + a\left(\frac{l}{k}\right)^2},$$

where P = the collapsing load, f_c = safe compressive stress for very short lengths of the material, A = area of cross-section, l = the length of the pin-jointed column, k = the least radius of gyration of the section, a = a constant for the material $= \frac{f_c}{\pi^2 . E}$, where E is Young's modulus for the material.

rankinite *(Min.)*. A monoclinic calcium disilicate, $Ca_3Si_2O_7$, found in highly metamorphosed siliceous limestones.

rank of coal *(Geol.)*. A classification related to the percentage of carbon in dry mineral-free coal. The original vegetation has been modified by heat, pressure and chemical change after burial. Rank increases from *peat*, through *lignite* to *bituminous coal* and finally *anthracite*.

rank of selectors *(Telecomm.)*. The whole set of selectors concerned with a specified stage in setting up a call through an exchange.

rank test *(Stats.)*. A statistical procedure carried out on the ranks rather than the values of the observations.

ranula *(Med.)*. A cystic tumour formed on the lower surface of the tongue or on the bottom of the mouth, caused by blocking or dilatation of mucous gland.

Ranunculaceae *(Bot.)*. The buttercup family, ca 1800 spp. of dicotyledonous flowering plants (superorder Magnoliidae). Mostly herbs, mostly north temperate; the floral parts are free, the stamens numerous and hypogynous. Many are poisonous. Of little economic importance other than as ornamentals; *Anemone, Helleborus, Delphinium* etc.

Ranvier's nodes *(Zool.)*. Constrictions of the neurolemma occurring at regular intervals along medullated nerve fibres.

Raoult's law *(Chem.)*. The vapour pressure of an ideal solution at any temperature is the sum of the vapour pressure of each component, P_i, multiplied by the mole fraction of that component, x_i:$P = \Sigma P_i x_i$.

Rapakivi granite *(Geol.)*. A type described from a locality in Finland, characterized by the occurrence of rounded pink crystals of orthoclase surrounded by a mantle of whitish sodic plagioclase. A widely-used textural term.

raphe *(Bot.)*. Ridge on an ovule or seed representing that part of the stalk that is fused to the ovule.

raphe *(Zool.)*. A broad junction, as between the halves of the Vertebrate brain.

raphide *(Bot.)*. A needle-shaped **crystal**, usually calcium oxalate, usually one of a bundle in the vacuole of a cell, especially in a leaf.

rapping *(Foundry)*. The process of loosening a pattern in a mould to facilitate its withdrawal. A spike or lifting screw is inserted in the pattern and tapped smartly in every direction.

raptatory, raptorial *(Zool.)*. Adapted for snatching or robbing, as birds of prey.

rare earth elements *(Chem.)*. A group of metallic elements possessing closely similar chemical properties. They are mainly trivalent, but otherwise similar to the alkaline earth elements. The group consists of the lanthanide elements 57 to 71, plus scandium (21) and yttrium (39). Extracted from monazite, and separated by repeated fractional crystallization, liquid extraction, or ion exchange.

rare earths *(Chem.)*. The oxides (M_2O_3) of the rare earth elements.

rarefaction *(Med.)*. Abnormal decrease in the density of bone as a result of absorption from it of calcium salts, as in infection of bone.

rarefaction *(Phys.)*. Decrease in density along a low pressure wavefront of a sound wave.

rare gases *(Chem.)*. See **noble gases**.

Raschel *(Textiles)*. A 1- or 2-bar warp-knitting machine, fitted with latch needles.

Raschig rings *(Chem.Eng.)*. **Packed column** fillers which are hollow open cylinders of diameter equal to length.

rash *(Med.)*. Any skin eruption.

rasorial *(Zool.)*. Adapted for scratching.

rasp *(Build.)*. Coarse type of file with teeth in the form of raised points.

Raspail test *(Paper)*. A test to detect the presence of rosin in paper. A drop of concentrated solution of cane sugar is placed on the paper, blotted and a drop of concentrated sulphuric acid applied to the same spot. A red colouration confirms rosin is present.

RAST *(Immun.)*. Abbrev. for *RadioAllergoSorbent Test*.

raster *(Comp.)*. In the display on a **television** or **raster graphics** screen, the grid pattern of vertical and horizontal divisions outlining all the small elements of which the picture is composed. See **resolution**.

raster burn *(Image Tech.)*. Deterioration of the scanned area of the screen of a TV picture or camera tube as a result of use.

raster graphics *(Comp.)*. **Computer graphics** based on television technology in which the screen display is produced by an electron beam scanning one **raster** line at a time to cover the screen from top to bottom 30 times per second. The image is produced by modifying the intensity of the electron beam to each **pixel** from a map of the pixels in the computer memory. See **bit mapped display**.

RAT *(Aero.)*. Abbrev. for *Ram-Air Turbine*.

ratchet brace (drill) *(Eng.)*. A drilling brace in which the drill spindle is rotated intermittently by a ratchet wheel engaged by a pawl on a hand lever; used in confined spaces, repair work.

ratchet mechanism *(Eng.)*. A mechanism comprising a **ratchet wheel** and a **pawl** with which it engages. It is used to convert reciprocating motion into intermittent rotary or linear motion in one direction only.

ratchet screwdriver *(Build.)*. One with a ratchet mechanism to make it operate in one direction only.

ratchetting *(Nuc.Eng.)*. Intermittent movement of fuel elements arising from thermal cycling and differential expansion effects.

ratchet-toothed escape wheel *(Horol.)*. An escape wheel with fine-pointed teeth; used in English lever watches.

ratchet wheel *(Eng.)*. A wheel with inclined teeth used in a **ratchet mechanism**.

rate constant *(Chem.)*. The speed of a chemical reaction, in moles of change per cubic metre per second, when the active masses of all the reactants are unity. If Rate $=k[A]^x[B]^y$, for $A+B \rightarrow$products, and x,y are partial orders, then k is the rate constant.

rated altitude *(Aero.)*. The height measured in the **International Standard Atmosphere**, at which a piston aero-engine delivers its maximum power. Cf. **power rating**.

rated blowing-current *(Elec.Eng.)*. The current at which a fuse-link is specified by the maker to melt and break the circuit.

rated breaking-capacity *(Elec.Eng.)*. The r.m.s. current, or the kVA at the rated voltage, which a circuit-breaker is specified by the maker to interrupt without damage.

rated capacity *(Eng.)*. General term for the output of an equipment, which can continue indefinitely in conformity with a criterion, e.g. heating, distortion of signals or of waveform. See **continuous rating**, **intermittent rating** or **periodic rating**.

rate-determining step *(Chem.Eng.)*. Where a process consists of a series of consecutive steps, the overall rate of the process is largely determined by the step with the slowest rate, so that efforts to speed up the process must chiefly be directed to this step.

rated impedance *(Elec.Eng.)*. Particularly applied to a loudspeaker, in which impedance rises with frequency, with an added sharp rise at frequency of base resonance. That resistance, equal in magnitude to the modulus of the minimum impedance above this resonant frequency, which replaces the loudspeaker when measuring the power applied to the loudspeaker during testing.

rated making-capacity *(Elec.Eng.)*. The maximum asymmetrical current which a circuit-breaker can make at the rated voltage.

rate fixing *(Eng.)*. The determination of the time allocation for carrying out a specific task of work, usually as a basis for remuneration.

rate gyro *(Eng.)*. Gyroscope with single gimbal which produces a couple proportional to the rate of rotation.

ratemeter *(Nuc.Eng.)*. See **count ratemeter**.

rate of climb *(Aero.)*. Generally rate of ascent from Earth. In performance testing, the vertical component of the air path of an ascending aircraft, corrected for standard atmosphere. See **International Standard Atmosphere**.

rate of climb indicator *(Aero.)*. See **vertical speed indicator**.

Rathke's pouch *(Zool.)*. In developing Vertebrates, the diverticulum formed from the dorsal aspect of the buccal cavity ectoderm which gives rise to the adenohypophysis. Also *craniobuccal pouch*.

rating *(Elec.Eng.)*. Specified limit to operating conditions, e.g. current rating etc.

rating nut *(Horol.)*. A milled nut which supports the pendulum bob. By rotating it the bob is raised or lowered, thus altering the time of vibration of the pendulum.

ratio *(Maths.)*. The ratio $a{:}b$ is equivalent to the quotient a/b.

ratio arms *(Elec.Eng.)*. Two adjacent arms of a Wheatstone bridge, the resistances in which can be made to have one of several fixed ratios.

ratio detector *(Telecomm.)*. Detector circuit used for frequency-modulated carriers.

ratio error *(Elec.Eng.)*. A departure of the ratio between the primary and secondary voltages or currents of a voltage or current transformer from the rated value.

rational horizon *(Surv.)*. See **true horizon**.

rationalization *(Behav.)*. In psychoanalytic theory, a mechanism of defence by means of which unacceptable thoughts or actions are given acceptable reasons which justify it, and also hide its true motivation.

rationalized units *(Phys.)*. Systems of electrical units for which the factor 4π is introduced in Coulomb's laws so that it shall be absent from more widely used relationships. In **Heaviside-Lorentz units** (rationalized Gaussian units) this is done directly, thus modifying values for the unit charge and unit pole. In **MKSA** units it is done indirectly by modifying the values of the permittivity and permeability of free space.

rational number *(Maths.)*. A number which can be expressed as the ratio of two integers, e.g. 3/4.

ratio of compression *(Eng.)*. See **compression ratio**.

ratio of slenderness *(Build.)*. The ratio between the length or height of a pillar and its least radius of gyration. See **effective column length**.

ratio of specific heat capacities *(Phys.)*. The ratio of specific heat capacity of a gas at constant pressure to that at constant volume has constant value of about 1.67 for monatomic gases, 1.4 for diatomic gases and approaches unity for polyatomic gases. This ratio, denoted by γ, enters into the **adiabatic equation**.

ratio schedule of reinforcement *(Behav.)*. A program of reinforcement in which a certain number of responses are necessary in order to produce the reward. On **fixed ratio schedules** the number of responses is always the same; on **variable ratio schedules** the number of responses varies from trial to trial.

RATOG *(Aero.)*. Abbrev. for *Rocket Assisted Take Off Gear*.

rat-race *(Telecomm.)*. See **hybrid junction**.

rat-tail file *(Build.)*. Small round file; used for enlarging holes etc.

rattle *(Paper.)*. The crackling noise when paper is handled. It indicates the degree to which the fibre has been hydrated in the process of beating, and can be augmented by the addition of starch and other additives.

rattle *(Zool.)*. the series of horny rings representing the modified tail-tip scale in Rattlesnakes (*Colubridae*).

rat-trap bond *(Build.)*. A form of bond in which a 9 in wall is built up of bricks on edge, so arranged as to enclose a 9 × 3 in cavity.

Rauber's cells *(Zool.)*. In Mammals, cells of the trophoblast situated immediately over the embryonic plate.

Raunkiaer system *(Bot.)*. A classification of the vegetative or life-forms of plants according to the positions of the perennating (resting) buds and the protection they receive during an unfavourable season of cold or drought.

rauwolfia serpentina *(Med.)*. One of a number of compounds obtained from the dried roots of trees of the *Rauwolfia* genus, including *R. canescens* and *R. vomitoria*. Formerly used for the treatment of hypertension.

rawhide hammer *(Eng.)*. A hammer the head of which consists of a close roll of hide projecting from a short steel tube; used by fitters to avoid damaging a finished surface.

Rawlplug *(Build.)*. TN for a small tube of tough compressed fibre or plastic for insertion in a hole, to provide a fixing plug into which a screw may be turned.

raw stock *(Image Tech.)*. General term for motion picture film before exposure and processing.

ray *(Bot.)*. A panel of tissue, usually mostly parenchyma, one to several cells wide and a few to many cells high, produced by ray initials in the cambium and extending radially into the secondary xylem (xylem ray) and secondary phloem (phloem ray), and with the functions of radial transport and storage. See also **medullary ray**, **ray tracheid**.

ray *(Geol.)*. A linear landform on the surface of the Moon radiating outwards from a crater. Probably caused by ejecta from volcanic activity or the impact of a meteorite.

ray *(Phys.)*. General term for the geometrical path of the radiation of wave energy, always in a direction normal to the wavefront, but with possible reflection, refraction, diffraction, divergence, convergence and diffusion. By extension, also the geometrical path of a beam of particles in an evacuated chamber. This may be curved in electric or magnetic field. See **particle**.

ray *(Zool.)*. A skeletal element supporting a fin; a sector of a radially symmetrical animal.

ray floret *(Bot.)*. Usually one of the outer ring florets of the capitulum of the Compositae, regardless of its morphology. Cf. **disk floret**.

ray initial *(Bot.)*. A more or less isodiametric cell, one of a group of such in the vascular **cambium**, each giving rise to one of the radial files of cells making up a ray. Cf. **fusiform initial**.

Rayleigh criterion *(Phys.)*. Criterion for the resolution of interference fringes, spectral lines and images. The limit of resolution occurs when the maximum of intensity of one fringe or line falls over the first minimum of an adjacent fringe or line. For a telescope with a circular aperture of diameter D this criterion gives the smallest angular separation of the two images of point objects as $1.22\ \lambda/D$. (λ is the wavelength of the light.)

Rayleigh disk *(Acous.)*. A small, light, circular disk (in water a lead disk is used) pivoted about a vertical diameter, hung by a fine thread of glass or quartz. If placed at an angle to a progressive sound wave, the disk experiences a torque which depends on the square of the velocity of the volume element in the medium. It provides a useful method of measurement of the absolute value of the velocity, calculated from the measured torque using a formula due to König. Historically used for calibrating microphones.

Rayleigh distillation *(Chem.)*. A simple distillation in which the composition of the residue changes continuously during the course of the distillation.

Rayleigh-Jeans law *(Phys.)*. An expression for the distribution of energy in the spectrum of a black-body radiator:

$$E_\nu d\nu = \frac{8\pi\nu^2 kT}{C^3} d\nu$$

where E_ν is the energy density radiated at a temperature T within a narrow range of frequencies from ν to $\nu + d\nu$. k is Boltzmann's constant and c the velocity of light. The formula holds only for low frequencies. See **Planck's radiation law**.

Rayleigh limit *(Phys.)*. One-quarter of a wavelength, the maximum difference in optical paths between rays from an object point to the corresponding image point for perfect definition in a lens system.

Rayleigh refractometer *(Phys.)*. An instrument for measuring the refractive index of a gas by an optical interference method. Each of two interfering light beams passes through a tube which may contain air or gas or be evacuated. By observing the shift in the interference fringes when one of the tubes is evacuated and the other contains gas, the refractive index of the gas may be calculated using the expression

$$n = 1 + \frac{s\lambda}{l}$$

where l is the length of the tube, λ the wavelength used, and s the number of fringes shifted.

Rayleigh scattering *(Phys.)*. See **scattering**.

Rayleigh's criterion *(Acous.)*. Criterion to predict the occurrence of thermo-acoustic feedback.

Rayleigh wave *(Acous.)*. Non-dispersive surface wave on a solid body with a free surface. Important for earthquakes.

Raynaud's disease *(Med.)*. A paroxysmal disorder of the arteries of the fingers and toes characterized by attacks of pain in them, the fingers (or toes) going white and then blue, usually after exposure to cold or stress. It may also complicate a number of connective tissue diseases when it is termed *Raynaud's phenomenon*.

Raynauds disease, phenomenon *(Med.)*. Constriction of blood supply to digits producing a white finger or toe, most often in response to cold. The *phenomenon* applies when the condition indicates an underlying disease; more common is the *disease* where there is no identifiable cause.

rayon *(Textiles)*. Sometimes used, esp. in US, for **viscose fibres**.

ray theory *(Acous.)*. Model in acoustics based on the assumption that sound propagates in the form of rays perpendicular to the wave fronts.

ray tracheid *(Bot.)*. Tracheids shorter than the ordinary axial tracheids. They are found at the top and bottom margins of the ray with their long axes in the radial direction. Occur in the wood of some conifers, e.g. pines.

Rb *(Chem.)*. The symbol for rubidium.

R-bands *(Biol.)*. See **banding techniques**.

RBMK reactor *(Nuc.Eng.)*. Graphite-moderated, boiling-water-cooled, pressure tube reactor unique to the USSR (the type involved in the Chernobyl accident).

RC *(Build.)*. Abbrev. for *Rough Cutting; Reinforced Concrete*.

RC coupling *(Electronics)*. Abbrev. for *Resistance-Capacitance coupling*.

RCS *(Space)*. See **reaction control system**.

RDF *(Telecomm.)*. Abbrev. for *Radio Direction-Finding*.

R-display *(Radar)*. An expanded **A-display** in which an echo can be *magnified* by an expanded sweep for close examination.

rDNA, ribosomal DNA *(Biol.)*. Genes specifying the several kinds of ribosomal RNA molecules.

RDT & E *(Aero.)*. Abbrev. for *Research, Development, Test and Evaluation*, US.

Re *(Chem.)*. The symbol for rhenium.

reach *(Hyd.Eng.)*. A clear uninterrupted stretch of water.

reactance *(Phys.)*. The imaginary part of the impedance. Reactances are characterized by the storage of energy rather than by its dissipation as in resistance.

reactance chart *(Elec.Eng.)*. A chart of logarithmic scales so arranged that it is possible to read directly the reactance of a given inductor or capacitance at any frequency.

reactance coupling *(Elec.Eng.)*. Coupling between two circuits by a reactance common to both, e.g. a capacitor or inductor.

reactance drop (rise) *(Elec.Eng.)*. The decrease (or increase) in the available voltage at the terminals of a circuit caused by the reactance voltage set up within that circuit.

reactance modulation *(Elec.Eng.)*. Use of a variable reactance, e.g. capacitor or inductor, or a reactance valve, to effect frequency modulation.

reactance relay *(Elec.Eng.)*. An impedance relay which operates as soon as reactances of the circuit to which it is connected fall below a predetermined value.

reactance voltage *(Elec.Eng.)*. The voltage produced by current flowing through the reactance of a circuit; equal to the product of the current (amps) and the reactance (ohms).

reactants *(Chem.)*. The substances taking part in a chemical reaction, those on the left-hand side of a reaction as written.

reaction *(Bot.,Zool.)*. Any change in behaviour of an organism in response to a stimulus.

reaction *(Chem.)*. (1) See **chemical-**. (2) The acidity or alkalinity of a solution.

reaction *(Phys.)*. The equal and opposite force arising when a force is applied to a material system; in particular the force exerted by the supports or bearings on a loaded mechanical system.

reaction chain *(Chem.)*. See **chain reaction**.

reaction chamber *(Aero.)*. The chamber, usually cylindrical but sometimes spherical, in which the reaction or combustion of a rocket's fuel and oxidant occur.

reaction control system *(Space)*. A set of small thrusters, suitably placed on a spacecraft to control its attitude in pitch, roll and yaw; usually referred to as *RCS*.

reaction formation *(Behav.)*. In psychoanalytic theory, a **defence mechanism** by which a forbidden impulse is mastered by exaggeration of the opposing tendency (hate is converted to oversolicitous love).

reaction isochore *(Chem.)*. See **van't Hoff's reaction isochore**.

reaction isotherm *(Chem.)*. See **van't Hoff's reaction isotherm**.

reaction order *(Chem.)*. See **order of reaction**.

reaction pair *(Geol.)*. Two minerals of different composition which exhibit the reaction relationship (see **reaction principle**). Thus forsterite at high temperature is converted into enstatite at a lower temperature, by a change in the atomic structure involving the addition of silica from the magma containing it. Forsterite and enstatite form a *reaction pair*.

reaction principle *(Geol.)*. The conversion of one mineral species stable at high temperature into a different one at lower temperatures, by reaction between the crystal phase and the liquid magma containing it. The change may be continuous over a wide temperature range (*continuous reaction*), or may occur at a fixed temperature only (*discontinuous reaction*).

reaction products *(Chem.)*. The substances formed in a chemical reaction. Those on the right-hand side of a reaction as written.

reaction propulsion *(Aero.)*. The scientifically correct expression for all forms of jet and rocket propulsion; they act by ejection of a high-velocity mass of gas, from which the vehicle reacts with an equal and opposite momentum, according to Newton's Third Law of Motion.

reaction rate *(Phys.)*. The rate of fission in a nuclear reactor.

reaction rim *(Geol.)*. The peripheral zone of mineral aggregates formed round a mineral or rock fragment by reaction with magma during consolidation of the latter. Thus quartz caught up by basaltic magma is partially resorbed, at the same time being surrounded by a reaction rim of granular pyroxene.

reaction series *(Geol.)*. One in which the minerals of igneous rocks are arranged in the order of temperature at which they crystallize from magmas.

reaction time *(Behav.)*. The interval between the presentation of a signal and the subject's response to it.

reaction turbine *(Eng.)*. A turbine in which the fluid expands progressively in passing alternate rows of fixed and moving blades, the kinetic energy continuously developed being absorbed by the latter.

reaction wood *(Bot.)*. Wood of distinctive anatomical structure formed on branches and leaning trunks, producing as it matures tensile or compressive forces that tend to maintain a growing branch at its appropriate orientation (in spite of increasing mass) or to correct misorientation. Reaction wood makes unsatisfactory timber and pulp. See also **compression wood, tension wood**.

reactivation *(Electronics)*. When a thoriated tungsten filament loses its emission, the raising of temperature for a time (without anode voltage) to bring fresh thorium to the surface is called reactivation.

reactive *(Chem.)*. Readily susceptible to chemical change.

reactive anode *(Ships)*. See **sacrificial anode**.

reactive component of current (voltage) *(Elec.Eng.)*. Preferred term for component of vectors representing an alternating current (voltage) which is in quadrature (90°) with the voltage (current) vector. Also called *idle component, inactive component, quadrature component, wattless component*.

reactive depression *(Behav.)*. A type of depression clearly linked to environmental events (e.g. after a death).

reactive dye *(Textiles)*. A dye that is fixed by reacting chemically with the fibre molecules. The best known examples are used for dyeing cellulosic fibres (i.e. cotton and viscose).

reactive factor *(Elec.Eng.)*. The ratio of reactive volt-amperes to total supply volt-amperes.

reactive iron *(Elec.Eng.)*. Iron inserted in the leakage-flux paths of a transformer to increase its leakage reactance.

reactive load *(Elec.Eng.)*. A load in which current lags behind or leads on the voltage applied to its terminals.

reactive power *(Elec.Eng.)*. The reactive volt-amperes, i.e. the product of a voltage of a circuit and the reactive component of the current.

reactive schizophrenia *(Behav.)*. Those cases of schizophrenia in which onset is sudden and linked to some precipitating event in the environment.

reactive voltage *(Elec.Eng.)*. That component of the phasor representing voltage of an a.c. circuit which is in quadrature (at 90°) with the current.

reactive volt-ampere hour *(Elec.Eng.)*. A unit used in measuring the product of reactive volt-amperes in a circuit and the time during which they have been passing.

reactive volt-ampere-hour meter *(Elec.Eng.)*. An integrating meter which measures and records the total number of reactive volt-ampere-hours which have passed in the circuit to which it is connected.

reactive volt-amperes *(Elec.Eng.)*. The product of the reactive voltage and the amperes in the circuit, or the reactive current (amperes) and voltage of the circuit; measure of the wattless power in the circuit. Abbrev. *VAr* (Volt-Amperes-reactive).

reactivity *(Nuc.Eng.)*. The departure of the multiplication constant of a reactor from unity, measured in different ways, i.e. **cent, dollar, nile, inhour**, or simply *per cent reactivity*.

reactivity worth *(Nuc.Eng.)*. Change of reactivity of reactor caused by the addition or removal of a material or piece of equipment.

reactor *(Elec.Eng.)*. Electric circuit component which stores energy, i.e. a capacitor or an inductor.

reactor *(Nuc.Eng.)*. Assembly of nuclear fuel and (usually) moderator in steady operation of which a self-sustaining chain reaction occurs. The neutron flux (and therefore power) is regulated by neutron absorbing rods (e.g. of boron steel, cadmium) or otherwise (e.g. movement of fuel, moderator or reflector; or by use of **neutron poisons** or *spectral shift*). Used for neutron irradiation and for releasing nuclear energy for electricity production. Reactors are classified according to their *fuel, moderator* and *coolant* (or less frequently according to their *size, power output*, or *function*). Several hundred types of reactor have been tested or suggested based on the following alternatives: *fuel*: uranium-235; plutonium-239; uranium-233; *moderator*: light water; heavy water; graphite; beryllium; organic liquid (or none in fast reactors); *coolant*: gas; light water; heavy water; organic liquid; liquid metal. Fuels are classified as *natural, slightly enriched*, or *highly enriched* according to the extent to which the proportion of fissile material has been increased beyond its normal isotopic abundance.

reactor noise *(Nuc.Eng.)*. Random statistical variations of neutron flux in reactor.

reactor oscillator *(Nuc.Eng.)*. Device which produces periodic variations of reactivity by mechanical oscillation of neutron absorbing sample in reactor core. Used to measure reactor properties, or nuclear cross-section of sample.

reactor simulator *(Nuc.Eng.)*. Analogue digital computer which simulates variations in reactor neutron flux produced by changes in any operating parameter. Useful for training and for investigating reactor effects.

reactor trip *(Nuc.Eng.)*. Rapid reduction of reactor power to zero by emergency insertion of control mechanisms and (in some cases) removal of liquid moderator.

reactor vessel *(Nuc.Eng.)*. The vessel in which the core, moderator, coolant and control rods are situated. See **pressure vessel**.

reader *(Print.)*. One who reads and corrects printers' proofs, comparing them with the original copy. Also called *proofreader*.

read in *(Comp.)*. To insert data into a computer.

reading *(Comp.)*. (1) Registering and transferring to **Immediate Access Store** the data on **punched card, paper tape, bar code** or **backing store**. (2) See **character recognition**.

reading frame *(Biol.)*. Three bases are required to specify one amino acid. An mRNA molecule can therefore be read in three different *reading frames*, depending on the starting base. Usually alternative reading frames contain many **stop codons** and are not used except in some small viruses. See **overlapping genes**.

reading microscope (telescope) *(Phys.)*. See **cathetometer**.

read-only memory *(Comp.,Electronics)*. *ROM*. A random-access memory in which programming of the data pattern is fixed during manufacture and cannot be changed subsequently. See **EPROM**.

read-out pulse *(Comp.)*. Pulse applied to binary cells to extract the bit of information stored.

ready-mixed concrete *(Build.)*. A method whereby concrete constituents are loaded into a vehicle carrying a mixing plant and mixed with water to the right consistency while the vehicle is travelling to the place of deposit. Frequently used on congested sites where room is not available for traditional plant for manufacture of concrete.

reagent *(Chem.)*. A substance or solution used to produce a characteristic reaction in chemical analysis.

reagent feeder *(Min.Ext.)*. Appliance which dispenses chemicals in continuously moving flow of ore or pulp at a controlled rate.

reagin, reaginic antibody *(Immun.)*. Antibody which fixes to tissue cells of the same species so that, in the presence of antigen, histamine and other vasoactive agents are released. The term was used to describe such antibodies in humans before IgE was identified, and is often still used. In some species antibodies of immunoglobulin classes other than IgE also have similar properties.

real absorption coefficient *(Phys.)*. See **true absorption coefficient**.

real axis *(Maths.)*. See **Argand diagram**.

realgar *(Min.)*. A bright-red monosulphide of arsenic; monoclinic. Occurs associated with orpiment.

real image *(Phys.)*. See **image**.

reality principle *(Behav.)*. In psychoanalytic theory, the mental activity that leads to instinctual gratification by accommodating to the demands of the real world; it is acquired during development. Cf. **pleasure principle**.

real memory *(Comp.)*. The **main memory** in a computer system which also uses **virtual memory**.

real number *(Comp.)*. Any number with a fractional (or decimal) part. Cf. **integer**.

real number *(Maths.)*. Any rational or irrational number.

real part *(Maths.)*. See **complex number**.

real-time analyser *(Acous.)*. Analyser which calculates the spectrum (**Fourier analysis**) of a signal so fast that no input data are excluded from the analysis. Uses *Fast Fourier Transform* (FFT) (a particularly fast numerical method for calculating Fourier transforms).

real-time clock *(Comp.)*. Electronic unit which keeps track of the date and the time of day in a special register that may be accessed by the programmer.

real-time system *(Comp.)*. Computing system which is designed to receive data, process it and respond within a time frame set by outside events, e.g. air traffic control, automatic banking.

ream *(Glass.)*. A non-homogeneous layer in flat glass.

ream *(Paper)*. Twenty quires; now usually 500 sheets.

reamer *(Eng.)*. A hand- or machine-operated tool for finishing drilled holes. It consists of a cylindrical or conical shank on which cutting edges are formed by longitudinal or spiral flutes, or in which separate teeth are inserted.

rear projection *(Image Tech.)*. See **back projection**.

Réaumur scale *(Phys.)*. A temperature scale ranging from 0°R to 80°R (freezing-point and boiling-point of pure water at normal pressure).

rebate *(Build.)*. A recess cut into the corner of a piece of timber. Also *rabbet*.

rebate plane *(Build.)*. A plane specially adapted for cutting a groove in the corner of a board.

rebecca-eureka *(Radar)*. Radar system on aircraft carrying low-power interrogator transmitters (rebecca), working with fixed beacon responders (eureka), sending coded signals when triggered by interrogator pulses.

rebore *(Autos.)*. The treatment of a worn cylinder by boring it out and replacing the piston by a (necessarily) larger one.

recalescence *(Eng.)*. Release of heat in ferromagnetic material as it cools through a temperature at which a change in crystal structure occurs, normally associated with change in magnetic properties.

recall *(Behav.)*. A method of measuring retention in which some material must be produced from memory. Cf. **recognition**.

recapitulation theory *(Biol.)*. States that stages in the evolution of the species are reproduced during the developmental stages of the individual, i.e. ontogeny tends to recapitulate phylogeny. Superficially apparent in some instances. Also called *biogenetic law, Haeckel's law*.

receiver *(Telecomm.)*. Final unit in transmission system where received information is stored, recorded or converted into the necessary form.

receiver response *(Acous.)*. The response of a telephone receiver operating into a real or artificial ear; expressed as a ratio of the square of the sound pressure in the specified cavity to the electrical power applied to the receiver.

recency effect *(Behav.)*. In recall, the tendency to recall items from the end of the list more readily than those in the middle. See **primacy effect**.

receptacle *(Bot.)*. Structure on which reproductive organs are borne, especially: (1) swollen tip of a thallus with conceptacles in brown algae, (2) area bearing archegonia or antheridia in liverworts, or sporangia in ferns, (3) the end of a stalk on which is borne either the parts of a single flower or the involucre and florets of a head or capitulum.

receptaculum *(Zool.)*. (1) A receptacle; a sac or cavity used for storage. (2) A sac in which ova are stored, as in some Oligochaeta.

receptaculum seminis *(Zool.)*. A sac in which spermatozoa are stored, as in many Invertebrates; a spermotheca.

reception wall *(Build.)*. See **retention wall**.

receptive *(Bot.)*. Capable or being effectively pollinated or fertilized.

receptor *(Biol.,Immun.)*. Chemical grouping on a macromolecule or a cell which can combine selectively with other complementary molecules or cells, e.g. enzyme receptors for substrate, cell surface receptors for hormones or growth factors. In immunology refers to cell surface sites such as Fc receptors, or antigen-binding molecules on the membranes of B- or T-lymphocytes. Binding to a receptor on a cell membrane is often followed by transduction of a signal across the membrane and a response on the part of the cell, e.g. RNA transcription.

receptor *(Zool.)*. An element of the nervous system specially adapted for the reception of stimuli; for example, a sense-organ or a sensory nerve-ending.

receptor mediated endocytosis *(Biol.)*. The internalization of ligands bound to certain receptors on the cell surface which become clustered into *coated pits* and enter the cell *via* **coated vesicles** and **endosomes**.

receptors *(Med.)*. (a) specified nerve endings, (b) a chemical component of a molecule or cell which has a particular affinity to a specific hormone, factor, antigen or neurotransmitter to bind and activate a response.

recess *(Zool.)*. A small cleft or depression, as the *optic recess*.

recessed arch *(Arch.)*. See **compound arch**.

recessed pointing *(Build.)*. A method of pointing designed to prevent any peeling off; the mortar at all joints both vertical and horizontal is pressed back about 1/4 in (6 mm) from the face of the wall.

recessive *(Biol.)*. Describes a gene (allele) which shows its effect only in individuals that received it from both parents, i.e. in homozygotes. Also describes a character due to a recessive gene. Cf. **dominant**.

reciprocal *(Maths.)*. The reciprocal of the number a is the number $\frac{1}{a}$, i.e. a^{-1}.

reciprocal cross *(Biol.)*. A cross made both ways with respect to sex, i.e. A♂ × B♀ and B♂ × A♀. Consistent differences between the offspring of such crosses suggests **cytoplasmic inheritance**.

reciprocal diagram *(Eng.)*. See **force diagram**.

reciprocal hybrids *(Zool.)*. A pair of hybrids obtained by crossing the same two species, in which the male parent of one belongs to the same species as the female parent of the other, e.g. mule and hinny.

reciprocal lattice *(Phys.)*. The direct crystal lattice can be defined in terms of three vectors $\mathbf{a}_1$, $\mathbf{a}_2$, $\mathbf{a}_3$. A reciprocal lattice whose vectors are $\mathbf{b}_1$, $\mathbf{b}_2$, $\mathbf{b}_3$, defined by $\mathbf{a}_i.\mathbf{b}_i = \gamma$ and $\mathbf{a}_i.\mathbf{b}_j = 0$. It follows that $\mathbf{b}_1 = \frac{\gamma}{V}(\mathbf{a}_1 \times \mathbf{a}_3)$ etc, where V is the volume of the unit cell of the direct lattice. In crystallographic work γ is chosen as 1, but in solid state physics as 2π. The reciprocal lattice is extensively used to discusss diffraction and scattering effects by crystals and in the band theory of solids.

reciprocal networks *(Elec.Eng.)*. Those the product of whose impedances remains constant at all frequencies; thus an inductance is reciprocal to a capacitance.

reciprocal polar triangles *(Maths.)*. See **conjugate triangles**.

reciprocal proportions *(Chem.)*. See **law of equivalent (reciprocal) proportions**.

reciprocal theorem *(Eng.)*. The statement, enunciated by Clerk Maxwell, that on any elastic structure, if a load W applied at a point A causes a deflection y at another point B, then, if the loading be taken off A and applied at B, it will cause a deflection y at A, provided that W acts at B along the line in which y was measured, and that the deflection at A is measured along the original line of action of W at A.

reciprocal translocation *(Biol.)*. Mutual exchange of non-homologous portions between two chromosomes.

reciprocating compressor *(Eng.)*. Any compressor which employs a piston working in a cylinder, the piston causing the periodic compression of the working fluid.

reciprocating engine *(Eng.)*. An engine which employs a piston working in a cylinder, the piston being caused to oscillate by the periodic pressure of the working fluid.

reciprocating pump *(Min.Ext.)*. One which uses the displacing action of a plunger, piston or diaphragm to move water in a pulsated stream. Also *pulsometer pump*, using steam or compressed air in a valved system for similar pumping.

reciprocation *(Maths.)*. See **polar-**.

reciprocation *(Telecomm.)*. Operation of finding a reciprocal network to a given network. Used in electric wave filters.

reciprocity *(Eng.)*. The principle enunciated in the **reciprocal theorem**.

reciprocity calibration *(Elec.Eng.)*. Absolute calibration of microphone by use of reversible microphone-loud-speaker. See **reciprocity theorem**.

reciprocity constant *(Elec.Eng.)*. See **reciprocity theorem**.

reciprocity failure *(Image Tech.)*. Photographic exposure is determined by the product of light intensity and time and within limits an increase of one may be compensated by a proportionate decrease of the other (*reciprocity rule*). But for extremely short or long exposure times or at very low or high light intensities this relation fails to apply, especially with colour materials.

reciprocity principle *(Phys.)*. That the interchange of radiation source and detector will not change the level of radiation at the latter, whatever the shielding arrangement between them.

reciprocity theorem *(Acous.)*. Theorem of acoustics which says that under certain conditions the sound source and the receiver can be swapped, thereby the output of the receiver being unchanged.

reciprocity theorem *(Elec.Eng.)*. The interchange of electromotive force at any one point in a network and the current produced at any other point results in the same current for the same electromotive force. In an electrical network comprising 2-way passive linear impedances, the so-called transfer impedance is given by the ratio of the e.m.f. introduced in a branch of the network to the current measured in any other branch. By the reciprocity theorem this ratio is equal in phase and magnitude to that observed if positions of current and e.m.f. are interchanged. In its application to the calibration of transducers the reciprocity theorem concerns the quotient of the value of the ratio of open circuit voltage at output terminals of the transducer (when used as a sound receiver) to the value of the free-field sound pressure (referred to some arbitrarily selected point of reference near the transducer) divided by the value of the sound pressure at a distance d from the point of reference to the current flowing at the input terminals of the transducer (used as a sound transmitter). The value of this quotient, termed the *reciprocity constant*, is independent of the constructional nature of the transducer.

recirculating-ball thread *(Eng.)*. The helical analogue of the radial ball bearing, in which a helical ball track is ground on the shaft as a thread. The nut has a complementary helical ball track at each end and the balls recirculate in a space provided outside the rolling balls. Cf. **linear-ball bearing**.

recirculating ball-type steering *(Autos.)*. Steering gear resembling the screw and nut type, but using a half-nut with an eccentric ball-race which is operated by a spiral cam.

recirculating heating system *(Eng.)*. Heating industrial ovens and low-temperature furnaces with the atmosphere of the working chamber under constant recirculation throughout the complete system.

reck *(Min.Ext.)*. See **ragging frame**.

recognition *(Behav.)*. A method of measuring retention in which a stimulus has to be identified as having occurred before.

recoil atom *(Phys.)*. One which experiences a sudden change of direction or reversal, after the emission from it of a particle or radiation. Also *recoil nucleus*.

recoil escapement *(Horol.)*. A clock escapement (invented by Hooke, 1635–1703) in which the acting faces of the pallets are arcs of circles, and at the end of each swing of the pendulum the pallets push the escape wheel backwards a small amount, causing the recoil. This escapement is largely used for domestic clocks. Although departing from the requirements of the ideal escapement, it gives very satisfactory results, as it tends to be self-correcting, i.e. if there is any tendency for the arc of vibration to increase, there is a proportionally greater recoil, which reduces the arc. Also *anchor escapement*.

recoil nucleus *(Phys.)*. See **recoil atom**.

recoil particles *(Phys.)*. Those arising through collision or ejection, e.g., Compton recoil electrons.

recombinant inbred strains *(Immun.)*. *RI strains*. Inbred strains, mostly of mice, which have been made by crossing two different inbred parental strains to yield an F1 generation, and from this an F2 generation. Pairs of F2 mice are then crossed and their progeny inbred until they are homozygous at most loci. After prolonged inbreeding all members of the RI strain tend to complete genetic identity and homozygosity. However genes of the parental strains have become reassorted. Such RI strains provide a means of assessing the functions of gene products associated with one another, as occurs in the formation of complex receptor molecules. A single genetic character can be bred from one strain into another by forming the F1 generation followed by repeated back-crossing over 20 or more generations, at each of which the offspring have been selected for the presence of a particular genetic character from one parent on the genetic background of the other parent.

recombinant, recombinant DNA *(Biol.)*. (1) Organism containing a combination of alleles differing from either of its parents. (2) DNA which contains sequences from different sources, made usually as the result of laboratory procedures *in vitro*. (3) Individual, gamete or chromosome resulting from *recombination*.

recombination *(Biol.)*. Reassortment of genes or characters in combinations different from what they were in the parents, in the case of *linked* genes by *crossing-over*.

recombination *(Electronics)*. Neutralization of free electron and hole in semiconductor, thus eliminating two current carriers. The energy released in this process must appear as a light photon or less probably as several phonons. See **exciton, phonon, photon**.

recombination *(Phys.)*. Neutralization of ions in gas, by combination of charges or transfer of electrons. Important for ions arising from the passage of high-energy particles.

recombination coefficient *(Electronics)*. Ratio of the rate of recombination per unit volume to the product of the densities of positive and negative current carriers.

reconcentration *(Min.Ext.)*. Additional treatment of a mineral product to raise its grade or to separate out one constituent.

reconditioned carrier *(Electronics)*. Isolation of a pilot carrier for re-insertion of a carrier adequate for demodulation.

reconstructed stone *(Civ.Eng.)*. Artificial stone made of concrete blocks faced to resemble natural stone. Also called *precast stone*.

record *(Acous.)*. See **recording**.

record *(Comp.)*. A collection of related items of data, treated as a unit.

recorder *(Acous.)*. (1) A machine for registering a sound either magnetically, photographically or on plastic. (2) Musical instrument basically consisting of a *whistle* coupled with a tube-shaped resonator. Its resonance frequency can be varied by opening or closing holes along the wall of the tube.

recorder *(Eng.)*. An instrument which measures, and records the quantity measured. The mechanism may be actuated by clockwork, by air pressure, electrically or electronically, and frequently incorporates a servo device. Recording is on chart or photosensitive paper moved at a speed to give a time scale.

recording *(Acous.)*. (1) The process of making a record of a received signal. (2) A disk, tape or film on which a sound record is stored. (Disk recordings are commonly known simply as *records*.)

recording altimeter *(Aero.)*. A barographic type of instrument which traces height against time.

recording amplifier *(Acous.)*. That preceding the recording heads of any type of recorder.

recording head *(Acous.)*. The transducer (magnetic, electric, mechanical or electro-optical) used to record sound on tape, disk or film.

recording stylus *(Acous.)*. The instrument that cuts the groove in an original disk recording.

recovery *(Comp.)*. Process of returning to normal after an error, which may include ensuring that the data has not been corrupted.

recovery *(Eng.)*. Ability of stressed metal etc. to return to original state after distortion under load.

recovery *(Min.Ext.)*. Percentage of schedule tonnage actually mined.

recovery pegs *(Surv.)*. Special reference pegs established in known survey relation to the working setting-out pegs, so that the location of these can be re-established if disturbed.

recovery rate *(Radiol.)*. That at which recovery occurs after radiation injury. It may proceed at different rates for different tissues.

recovery time *(Electronics)*. Time required for control electrode of gas tube to regain control.

recovery time *(Phys.)*. (1) For Geiger tube, the period between the end of the dead time and the restoration of full normal sensitivity. (2) For counting system, the minimum time interval between two events recorded separately.

recovery time *(Radar)*. That required by a *transmit-receive tube* in a radar system to operate (usually measured to the point where receiving sensitivity is 6 dB below maximum).

recovery voltage *(Elec.Eng.)*. The normal frequency or d.c. voltage which appears across the contacts of a switch, circuit-breaker, or fuse after it has interrupted the circuit.

recrystallization *(Chem.)*. The process of reforming crystals, usually by dissolving them, concentrating the solution, and thus permitting the crystals to reform. Frequently performed in the process of purification of a substance.

recrystallization *(Eng.)*. The replacement of deformed crystals by a new generation of crystals, which begin to grow at certain points in the deformed metal and eventually absorb the deformed crystals. This process leads to elimination of strain-hardening.

recrystallization temperature *(Eng.)*. That at which solidification from the molten state begins. That marking a change in crystal form. The range of temperature through which strain-hardening disappears. Lead, tin and zinc can recrystallize at air temperature; iron, copper, aluminium and nickel have to be heated.

rectal gills *(Zool.)*. In the larvae of some Odonata, tracheal gills in the form of an elaborate system of folds in the wall of the rectum, used in respiration.

rectangle *(Maths.)*. A parallelogram whose angles are right angles.

rectangular axes *(Maths.)*. Co-ordinate axes which are mutually at right angles.

rectangular loop hysteresis *(Elec.Eng.)*. Colloquial expression for hysteresis curve of ferromagnetic or ferroelectric materials suitable for use in bistable or switching circuits. Characterized by very steep slope followed by unusually sharp onset of saturation.

rectangular notch *(Civ.Eng.)*. A notch plate having a rectangular notch cut in it; used for the measurement of large discharges over weirs.

rectangular pulse *(Telecomm.)*. Idealized pulse with infinitely short rise and fall times and constant amplitude. Pulse amplitudes and durations are often specified in terms of those of the nearest equivalent pulse carrying the same energy (subtending the same area under the curve).

rectangular scan *(Electronics)*. Any scanning system producing a rectangular field.

recti- *(Genrl.)*. Prefix from *rectus*, straight.

rectification *(Chem.)*. Purification of a liquid by distillation.

rectification *(Elec.Eng.)*. The conversion of a.c. into d.c. using some form of rectifier or rectifying apparatus.

rectification efficiency *(Elec.Eng.)*. Ratio of d.c. output power to a.c. input. Often expressed as percentage.

rectified airspeed *(Aero.)*. See **calibrated airspeed**.

rectified spirit *(Chem.)*. Distilled ethanol containing only 4.43% water. This is a constant-boiling (azeotropic) mixture with a boiling point of 78°C.

rectifier *(Elec.Eng.)*. Component for converting a.c. into d.c.

rectifier instrument *(Elec.Eng.)*. An a.c. instrument in which the current to be measured is rectified and measured on a d.c. meter.

rectifier leakage current *(Elec.Eng.)*. That passing through a rectifier under reverse bias conditions, due to finite reverse conduction.

rectifier ripple factor *(Elec.Eng.)*. The amount of a.c. voltage in the output of a rectifier after d.c. rectification. It is measured, in per cent, by the ratio of the r.m.s. value of the a.c. component to the algebraic average of the total voltage across the load. See also **ripple**.

rectifier stack *(Elec.Eng.)*. A pile of rectifying elements (usually semiconductor) series-connected for higher voltage operation.

rectifier voltmeter *(Elec.Eng.)*. One in which applied voltage is rectified in a bridge circuit before measurement with a d.c. meter.

rectifying detector *(Elec.Eng.)*. Detector of electromagnetic waves which depends for its action on rectification of high-frequency currents, as opposed to one using thermal, electrolytic-breakdown or similar effects.

rectifying valve *(Electronics)*. Any thermionic valve in which direct use is made of unilateral or asymmetrical conductivity effects, as opposed to one used primarily for amplification, e.g. a diode used as a rectifier or a triode used as a detector.

rectilinear lens *(Image Tech.)*. A lens which provides images with no distortion, as far as parallel lines are concerned. Not otherwise well-corrected.

rectilinear scan *(Image Tech.)*. A **raster** in which a rectangular area is scanned by a series of parallel lines.
rectirostral *(Zool.)*. Having a straight beak.
recto- *(Med.)*. Prefix used in terms pertaining to the rectum.
recto *(Print.)*. A right-hand page of a book, bearing an odd page number. Cf. **verso**.
rectocele *(Med.)*. A prolapse or protrusion of the lower part of the vaginal wall, carrying with it the anterior wall of the rectum.
rectrices *(Zool.)*. In Birds, the stiff tail feathers used in steering. sing. *rectrix*. adj. *rectricial*.
rectum *(Zool.)*. (L. *rectum intestinum*, straight intestine.) The posterior terminal portion of the alimentary canal leading to the anus. adj. *rectal*.
rectus *(Zool.)*. A name used for various muscles which are of equal width or depth throughout their length, e.g. the *rectus abdominis* in Vertebrates.
recumbent fold *(Geol.)*. An overturned fold with a more or less horizontal axial plane.
recuperative air heater *(Eng.)*. An air heater in which heat is transmitted from hot gases to the air through conducting walls, the flows of gas being continuous and unidirectional.
recuperator *(Eng.)*. An arrangement of flues which enables the hot gases leaving a furnace to be utilized in heating the incoming air (and sometimes gas). Outgoing hot gases and incoming cold gases pass in opposite directions through parallel flues and heat is transferred through the dividing walls.
recurrence *(Phys.)*. See **Regge trajectory**.
recurrent *(Zool.)*. Returning towards point of origin.
recurrent novae *(Astron.)*. A small group of novae which have shown more than one outburst of light, as T Coronae Borealis in 1866, 1898, and 1933; they show smaller ranges of brightness than most novae.
recurrent vision *(Phys.)*. The perception of repeated images of brightly illuminated objects when the source of illumination is suddenly removed.
recurring decimal *(Maths.)*. One in which, after a certain point, a number or set of numbers is repeated indefinitely. The figures which recur are indicated by dots placed above them, e.g. $0.\dot{3} \equiv 0.333\,333\,\ldots$, $0.5\dot{2}4\dot{3} \equiv 0.524\,324\,324\,3\ldots$ Also called *repeater, repeating decimal*.
recursion *(Maths.)*. Of a formula, enabling a term in a sequence to be computed from one or more of the preceding terms.
recursive subprogram *(Comp.)*. One which includes among its program statements, a call to the subprogram itself.
recurvirostral *(Zool.)*. Having the beak bent upwards.
recycle *(Phys.)*. Repetition of fixed series of operations, e.g. biodegradation of organic material followed by regrowth; applied also to isotope separation.
red algae *(Bot.)*. Rhodophyceaea.
red blood corpuscle *(Zool.)*. See **erythrocyte**.
red body *(Zool.)*. See **red gland**.
red brass *(Eng.)*. Copper-zinc alloy containing 15% zinc; used for plumbing pipe, hardware, condenser tubes etc. Red casting brass may contain up to 5% of lead and/or tin in place of the zinc.
red clay *(Geol.)*. A widespread deep-sea deposit; essentially a soft, plastic clay consisting dominantly of insoluble substances which have settled from the surface waters; these substances are partly of volcanic, partly of cosmic origin, and include nodules of manganese and phosphorus, crystals of zeolites and rare organic remains such as shark's teeth.
red corpuscle *(Zool.)*. See **erythrocyte**.
Red Data Book *(Ecol.)*. Catalogue of rare and endangered species prepared by the *International Union for Conservation of Nature and Natural Resources*, started in 1966, and covering the whole world.
red deal *(For.)*. A light-yellow soft wood obtained from the Scots fir; commonly used for timbering trenches, heavy framing, piles, joinery etc.
reddle *(Eng.)*. A mixture of red lead and oil wiped over one of two surfaces to be bedded together to indicate high spots to be removed by scraping.
reddle *(Min.)*. A red and earthy mixture of hematite, often with a certain admixture of clay.
red gland, body *(Zool.)*. In some Fish, a structure found in the wall of the air-bladder, responsible for secretion or absorption of gas. Also *rete mirabile*.
red hardness *(Eng.)*. Tool materials like high-speed steel and cemented carbides which, unlike most metals, retain a considerable part of their hardness at red heat are said to have red hardness.
redhead *(Image Tech.)*. TN for a small 800 W quartz lighting unit.
red heat *(Eng.)*. As judged visually, a temperature between 500°C and 1000°C.
redia *(Zool.)*. The secondary larval stage of Trematoda, possessing a pair of locomotor papillae and a rudimentary pharynx and intestine, and capable of paedogenetic reproduction.
redirected behaviour *(Behav.)*. Behaviour directed at inappropriate or irrelevant objects, often as a result of **frustration** or **conflict**.
redistilled zinc *(Eng.)*. Zinc from which impurities have been removed by selective distillation. The process takes advantage of the different boiling points of zinc (907°C) and the impurities lead (1750°C) and cadmium (767°C). A purity of 99.9% zinc is obtainable.
red lead *(Build.)*. Produced from a bright orange pigment. An anti-corrosive priming paint used on ferrous metals. Other coatings are often used in preference to red lead because of toxicity problems.
red lead (oxide) *(Chem.)*. Dilead (II) lead (IV) oxide, Pb_3O_4. Formed by heating lead (II) oxide in air at approximately 450°C. It occurs as red and yellow crystalline scales. Commercial varieties contain up to approximately 35% PbO_2. Widely used as an anticorrosive pigment in iron and steel primers.
Redler conveyor *(Eng.)*. An enclosed conveyor in which an endless chain is continuously dragged along the bottom of a casing and brings along with it material fed into the trough. For conveying coal, cement, flue dust etc.
red light *(Bot.)*. For plant responses mediated by phytochrome, light of wavelength around 630 nm.
red mud *(Eng.)*. In the Bayer process, iron-rich residue from digestion of bauxite with caustic soda.
red muscles *(Zool.)*. In Vertebrates, phasic muscles which are therefore rich in sarcoplasm and myoglobin, and are of red colour.
redox *(Chem.)*. Abbrev. for *oxidation-reduction*.
red oxide of copper *(Min.)*. See **cuprite**.
red oxide of zinc *(Min.)*. See **zincite**.
red pine *(For.)*. See **Douglas fir**.
redruthite *(Min.)*. A name frequently applied to the mineral *chalcocite* because of its occurrence, among other Cornish localities, at Redruth.
red shift *(Astron.)*. The displacement of features in the spectra of astronomical objects, particularly **galaxies** and **quasars**, towards the longer wavelengths, generally interpreted as a result of the **Doppler effect** due to their recessional velocities. The expansion of the universe means that all but the nearest galaxies have redshifted spectra; indeed this very feature led to the discovery of the expanding universe in the 1920s. The relationship between redshift and distance is given in the **Hubble law**.
red-short *(Eng.)*. See **hot-short**.
red silver ore *(Min.)*. For *dark-red silver ore*, see **pyrargyrite**; for *light-red silver ore*, see **proustite**.
red snow *(Bot.)*. Lying snow coloured by the growth near the surface of algae, especially *Chlamydomonas nivalis*, containing haematochrome.
red spot *(Astron.)*. See **Jupiter**.
red tide *(Bot.)*. Water containing sufficient dinophytes or other organisms to colour it red. Called a *bloom*. Some blooming dinophytes contain sufficient toxin to make shellfish feeding on them fatally poisonous to man.
reduced instruction set computer *(Comp.)*. One using a central processor with a very small **instruction set** which

allows faster processing and simpler design. Abbrev. *RISC*.

reduced level *(Surv.)*. The elevation of a point above or below a specified datum.

reduced mass *(Phys.)*. Quantity $mM/(M+m)$ used in the study of the relative motion of two particles, masses m and M, about their common centre of gravity. This is used in place of the smaller mass; movement of the larger is then ignored.

reducer *(Build.)*. See **reducing socket**.

reducer *(Image Tech.)*. A solution which acts on the silver image and dissolves it away by chemical or abrasive action, thus reducing contrast and/or density.

reducing agent *(Chem.)*. A substance which is capable of bringing about the chemical change known as *reduction* in another substance, itself being *oxidized*.

reducing atmosphere *(Eng.)*. One deficient in free oxygen, and perhaps containing such reactive gases as hydrogen and/or carbon monoxide; used in a reducing furnace to lower the oxygen content of mineral.

reducing flame *(Chem.)*. One containing excess fuel over oxygen, hence capable of acting as a *reducing agent*.

reducing socket *(Build.)*. Used to connect pipes of differing sizes, being threaded appropriately for screwed pipes or having different internal diameter for *compression* or **capillary fittings**. Sometimes with the larger diameter the same as the pipe for connection via another fitting. Also *reducer, reducing pipe-joint*.

reducing surface *(Phys.)*. A prepared surface, used in photometry, which reflects only a certain predetermined proportion of the luminous flux falling on it.

reductio ad absurdum proof *(Maths.)*. One which proves a proposition by proving that its contradictory leads to contradictions, e.g. to prove that $A = B$, one proves that the assumption that $A \neq B$ leads to contradictions, i.e. to absurdities.

reduction *(Chem.)*. Any process in which an electron is added to an atom or an ion. Four common types of reduction are removal of oxygen from a molecule, the liberation of a metal from its compounds, and diminution of positive valency of an atom or ion. Always occurs accompanied by oxidation.

reduction *(Min.Ext.)*. The extraction of gold from ore. The *reduction officer* is the official in charge of mill, extraction plant or reduction works. (S. Africa.)

reduction division *(Biol.)*. See **meiosis**.

reduction factor *(Phys.)*. The ratio of the mean spherical luminous intensity of a light source to its mean horizontal luminous intensity.

reduction of levels *(Surv.)*. The process of computing reduced levels from staff readings booked when levelling.

reduction roasting *(Eng.)*. *Reducing roast*. Use of heat in a controlled atmosphere to lower oxygen content of a mineral, e.g. iron (II) to iron (III) oxide by reaction with CO.

reduction to soundings *(Ships)*. Correction of depths observed in tidal waters for comparison with charted depths referred to chart datum.

redundancy *(Comp.)*. The presence of components which improve the reliability of a computer system. Redundancy may involve (a) multiple copies of critical hardware, *hardware-*; (b) alternative programs for critical operations, *software-*; (c) error correcting codes, *information-*; (d) repeating critical operations several times, *time-*.

redundancy *(Electronics)*. The provision of extra components or equipment, in parallel, so as to ensure continued operation after failure.

redundancy *(Telecomm.)*. The provision of greater-than-minimum number of codings so as to ensure accuracy of interpretation (decoding) after transmission through adverse conditions.

redundant *(Eng.)*. Term applied to a structural framework having more members than it requires to be perfect. Also *over-rigid*.

Redux bonding *(Aero.)*. A proprietary method of joining primary sheet metal aircraft structures with a two-component adhesive under controlled heat and pressure. It is widely used for the making of **honeycomb** sandwich, for doubling sheet metal and for attaching **stringers** or skin stiffeners.

red variables *(Astron.)*. See **Mira stars**.

redwater *(Vet.)*. (1) *Babesiosis, blackwater, moor ill, piroplasmosis*. A disease of cattle caused by infection of the erythrocytes by the protozoon *Babesia bovis*, which is transmitted by the tick *Ixodes ricinus*; characterized by fever, diarrhoea, anaemia, and haemoglobinuria. A similar tick-borne disease, called *Texas fever*, is caused by *Babesia bigemina*. (2) *Haemorrhagic disease, infectious icterohaemoglobinuria*. A disease of cattle, occurring in America, caused by *Clostridium hemolyticum*.

redwood *(For.)*. (1) A conifer (*Sequoia*) of the Pacific coastal area of the US, of rapid growth (sometimes reaching over 300 ft, 90 m), yielding a softwood used for commercial purposes. (2) Name given to **red deal** in north of England.

Redwood second *(Eng.)*. Unit used in viscometry. See **Redwood viscometer**.

Redwood viscometer *(Eng.)*. One of several designs of standard **viscometer** in which viscosity is determined in terms of number of seconds required for the efflux of a certain quantity of liquid through an orifice under specified conditions.

red zinc ore *(Min.)*. See **zincite**.

reed *(Acous.)*. Vibrating tongue of wood or metal, for generating air vibrations in musical instruments. Cane wood reeds are used for tongue action, as in the clarinet and saxophone, and a metal reed in organ reed pipes.

reed *(Textiles)*. An arrangement of flattened steel wires fixed in a frame on a weaving machine to separate warp threads and fix their spacing, guide the weft insertion device, and beat up the weft into the fell of the cloth.

reed *(Zool.)*. See **abomasum**.

reed loudspeaker *(Acous.)*. Small loudspeaker with a driving mechanism in which the essential element is a magnetic reed, which is drawn into the gap between pole-pieces on a permanent magnet by the currents in the driving coils.

reed pipe *(Acous.)*. Organ pipe in which pitch of the note is determined by vibration of a reed, the associated pipe reinforcing the generated note by resonance.

reed relay *(Elec.Eng.)*. One whose contacts are in the form of short straight springs (reeds of magnetic material which overlap slightly). Operation is by energizing a coil close to, or coaxial with the reeds, which are then magnetized in such a sense that they are drawn together and the circuit is made.

reeds *(Build.)*. Moulding in form of several side-by-side beads sunk below general surface.

reef *(Geol.)*. See **coral-**.

reef *(Min.Ext.)*. Originally an Australian term for a **lode**. Now used for a gold-bearing tabular deposit or flattish lode.

reef knolls *(Geol.)*. Large masses of limestone formed by reef-building organisms; found typically in the Craven district of Yorkshire where they have weathered out as rounded hills above the lower ground on the shales. These are of Carboniferous age.

reef picking *(Min.Ext.)*. On Rand, removal of gold-bearing blanket ore from barren waste rock, a reversal of the more usual hand sorting.

reel *(Image Tech.)*. (1) A flanged spool for holding film or tape. (2) A roll of film, especially one comprising a complete programme or a specific part of a programme.

reel barge *(Min.Ext.)*. Vessel carrying a very large diameter reel on which long lengths of oil pipe are wound and payed out during laying.

reel bogie, truck *(Print.)*. A special truck used for transporting a reel of paper to a rotary press.

re-entrant angle *(Maths.,Surv.)*. Of a closed figure, e.g. a polygon, an angle which points inward, being above 180° as viewed from the interior. Cf. **salient angle (2)**.

re-entrant horn *(Acous.)*. A horn for coupling a sound-reproducing diaphragm with the outer air. To conserve space, the horn divides at a distance from the throat and, after convolutions, unites before expanding to the flare.

re-entrant polygon *(Maths.)*. A polygon in which one or more internal angles are greater than 180°. Such an angle is called a *re-entrant angle*.

re-entrant program *(Comp.)*. One where a single copy of the program may be shared between several users at the same time.

re-entrant winding *(Elec.Eng.)*. A term used in connection with armature windings for d.c. machines; a *singly* (or *doubly) re-entrant winding* is one containing one (or two) independent closed circuits. The majority of windings are singly re-entrant.

re-entry *(Min.Ext.)*. Finding and connecting to a capped well on the sea bottom.

re-entry *(Space)*. That period of return to Earth (or any other planet) when a spacecraft passes through the atmosphere to land on the surface. During this period the spacecraft decelerates and is subject to intense heating generated by atmospheric friction. A sheath of ionized air around the spacecraft can black-out radio communication for several minutes. The re-entry may be direct, as in the case of a ballistic vehicle, or lifting, when use is made of lift capability to alleviate the high heating rates and thereby increasing the total heat transferred. Also *entry*.

re-entry corridor *(Space)*. A narrow corridor which is available to a spacecraft returning to Earth (or any other body with an atmosphere) so that it can make a safe entry. The lower side or *undershoot boundary* is fixed by excessive heating below it, and the upper side or *overshoot boundary* marks that region above which the density is so low that the space-craft can not be slowed down but skips back into space. Also *entry corridor*.

re-entry thermal protection *(Space)*. The shielding of a body from the intense heat generated during atmospheric deceleration when its high kinetic energy is transferred to the atmosphere as heat. Stagnation temperatures of many thousands of degrees may be generated. Thermal protection is achieved by the following, alone or in combination: (1) an ablation shield which vaporizes and carries the heat away, a material of large latent heat (e.g. glass resin) is used; (2) a heat sink of high thermal capacity (e.g. copper); (3) a good thermal insulator (e.g. fibreglass); (4) radiative cooling with a high thermal emissivity surface.

refection *(Zool.)*. In Rabbits, Hares and probably other herbivores, the habit of eating freshly-passed faeces. Also *autocoprophagy*.

reference address *(Comp.)*. One that is used as a locating point for a group of related addresses. See **address calculation**.

reference climatological station *(Meteor.)*. A meteorological station where a homogeneous series of observations of weather elements over a period of at least 30 years have been, or are expected to be, made under approved conditions.

reference diode *(Electronics)*. See **Zener diode**.

reference electrode *(Chem.)*. An electrode used as a standard relative to which a varying potential is measured. See **saturated calomel electrode**.

reference equivalent *(Telecomm.)*. The number of decibels by which a given piece of telephonic apparatus differs from the standardized piece of apparatus in the master transmission reference system.

reference level *(Telecomm.)*. A specified level of power, voltage or current in a circuit or system, to which all other levels of those quantities are referred, usually as a ratio expressed in dB. 1 mW is frequently taken as a reference signal power level, and is described as 0 *dBm* (zero decibels with respect to 1 mW). If voltages or currents are taken as reference levels, then it is customary and desirable to state the impedance level at which the measurement is to take place.

reference mark *(Print.)*. See **mark of reference**.

reference mark *(Surv.)*. A distant point from which angular distances to other marks may be taken at a station. Also *reference object*.

reference noise *(Telecomm.)*. Circuit noise level corresponding to that produced by 10^{-12} watt of power at a frequency of 1 kHz.

reference power *(Telecomm.)*. See **reference level**.

reference system *(Telecomm.)*. See **master telephone transmission reference system**.

reference voltage *(Elec.Eng.)*. (1) Closely controlled d.c. signal obtained from stable reference, e.g. a **Zener diode** or **standard cell**. Often used for calibration purposes. (2) A.c. voltage used to give both amplitude and phase reference. Used, for example, in power system protection circuits.

reference volume *(Elec.Eng.)*. That transmission voltage which gives zero recording level on the standard volume unit meter.

refined iron *(Eng.)*. Wrought-iron made by puddling pig-iron.

refiner *(Paper)*. Machine comprising a disk (or disks) or a cone, which rotates at high speed in a close fitting casing or shell. Both the rotor and stator are fitted with metal bars and the material to be treated passes through the narrow gap between the two. Refiners of suitable types are used in the manufacture of mechanical woodpulp from chips or for treating paper stuff to effect a beating action.

refiner mechanical woodpulp *(Paper)*. Mechanical woodpulp produced by subjecting chips of wood to the action of a refiner.

refinery *(Eng.)*. Plant where impure metals or mixtures are treated by electrolysis, distillation, liquation, pyrometallurgy, chemical or other methods to produce metals of a higher purity or specified composition.

refining *(Glass)*. See **founding**.

refining of metals *(Eng.)*. Operations performed after crude metals have been extracted from their ores, to obtain them in a condition of higher purity.

reflectance *(Phys.)*. See **reflection factor**.

reflected *(Zool.)*. Said of a structure, especially a membrane, which is folded back on itself.

reflected wave *(Telecomm.)*. One propagated back along a waveguide or transmission line system as a result of a mismatching at the termination.

reflected wave (ray) *(Phys.)*. One turned back from a discontinuity in a continuous medium.

reflected wave (ray) *(Telecomm.)*. See **ionospheric wave**.

reflecting galvanometer *(Elec.Eng.)*. See **mirror galvanometer**.

reflecting level *(Surv.)*. An instrument, used for levelling, which employs the principle that a ray of light which strikes a reflecting plane at right angles is reflected back in the same direction. In its practical forms, it usually consists of a hanging mirror which takes up a position in the vertical plane, and has an unsilvered part through which a distant staff may be seen and also a reference horizontal line upon it. When the eye is in such a position that the image of the pupil is bisected by the horizontal line, the line of sight to staff is horizontal.

reflecting telescope *(Astron.)*. A telescope using a mirror to bring light rays to a focus, it was first applied to astronomy by Isaac Newton who recognized its merits in overcoming the **chromatic aberration** of lenses. William Herschel developed techniques for casting large primary mirrors from 1783, and from that time the largest telescopes have been reflectors. Several configurations are used under different circumstances: *Cassegrain*, *Newtonian*, *Gregorian*, *Maksutov*, *Schmidt* and *coudé*. All of the world's largest telescopes are reflectors.

reflection *(Phys.)*. Return of neutrons to reactor core after a change of direction experienced in the shield surrounding the core.

reflection *(Telecomm.)*. Reduction of power from the maximum possible, because a load is not *matched* to the source and part of the energy transmitted is returned to the source. Also, reduction in power transmitted by a wave filter due to iterative impedance becoming highly reactive outside the pass bands. In all instances, loss of power (**reflection, return loss**) is measured in dB below maximum, i.e. when *properly matched*. See **mismatch**.

reflection coefficient *(Acous.)*. Complex ratio of reflected pressure to incident pressure when a plane sound wave is incident on a discontinuity. The complex ratio includes changes in amplitude and in phase during the reflection.

reflection coefficient *(Telecomm.)*. Ratio of electric voltage or field amplitude for reflected wave to that for incident wave. Given by $(Z_2 - Z_1)/(Z_2 + Z_1)$, where Z_1 and Z_2 are the impedances of the medium (or line) and the load respectively. For acoustic reflection Z is the acoustic impedance.

reflection density *(Image Tech.)*. A measure of the light-absorption of a surface, the logarithm of the reciprocal of the **reflection factor**.

reflection factor *(Phys.)*. The ratio which the luminous flux reflected from a surface bears to that falling upon it. Also *coefficient of reflection, reflectance*.

reflection factor *(Telecomm.)*. (1) Ratio of current delivered to load, to that which would be delivered to a perfectly matched load. (2) See **reflection coefficient**.

reflection laws *(Phys.)*. For wave propagation: (1) incident beam, reflected beam and normal to surface are coplanar; (2) the beams make equal angles with the normal.

reflection layers *(Phys.)*. Layers of very low-pressure atmosphere partly ionized by particles from the sun. Each layer progressively refracts electromagnetic waves, so that (below a definite frequency) they are effectively reflected downwards. See **E-layer**, **F-layer**.

reflection loss *(Telecomm.)*. See **reflection**.

reflection point *(Telecomm.)*. Point at which there is a discontinuity in a transmission line, and at which partial reflection of a transmitted electric wave occurs.

reflectivity *(Phys.)*. Proportion of incident energy returned by a surface of discontinuity.

reflectometer *(Phys.)*. Instrument measuring ratio of energy of reflected wave to that of incident wave in any physical system.

reflectometer *(Telecomm.)*. A directional coupler connected to a transmission line or waveguide to extract a small proportion of the reflected power returning from a mismatched load; used for standing wave measurements. See **time-domain reflectometer**.

reflector *(Electronics)*. Electrode in reflex klystron connected to negative potential and used to reverse direction of electron beam. US term *repeller*.

reflector *(Phys.)*. A device consisting of a bright metal surface shaped so that it reflects in a desired direction light or heat falling on it.

reflector *(Telecomm.)*. Part of an antenna array which reflects energy that would otherwise be radiated in a direction opposite to that intended.

reflector sight *(Aero.)*. Mirror gunsight which projects the aiming reticule and computed correction information for speed and deflection on to a transparent glass screen. Used first in World War II. Modern sights (HUD) incorporate elaborate radar information and firing instructions which allow an attack and breakaway under completely blind conditions.

reflex *(Behav.)*. A simple, automatic, involuntary and stereotyped response to some stimulus (e.g. an eye blink).

reflex action *(Zool.)*. An automatic or involuntary response to stimulus.

reflex angle *(Maths.)*. An angle greater than 180°.

reflex arc *(Zool.)*. The simplest functional unit of the nervous system, consisting of an afferent sensory neuron conveying nerve impulses from a receptor to the CNS, generally the spinal cord, where they are passed, either directly or via an internuncial or association neuron, to a motor neuron, which conveys them to a peripheral effector, such as a muscle.

reflex camera *(Image Tech.)*. One which incorporates a negative-sized glass screen on to which the image is reflected for composing and focusing, either through camera lens (*single-lens reflex*) or through a separate lens of the same focal length (*twin-lens reflex*). In single lens-type the reflecting mirror is retracted when the shutter is released.

reflexed *(Bot.)*. Bent back abruptly.

reflexive *(Maths.)*. A relation in a set S is reflexive if it applies from x to x for all elements x in S.

reflex klystron *(Electronics)*. A single-cavity klystron in which the electron beam is reflected back through the cavity resonator (see **rhumbatron**) by using a negative reflector. The beam, arriving back at the cavity in antiphase, can return energy in such a way as to cause oscillation; used as a microwave oscillator.

reflex projection *(Image Tech.)*. A method of composite photography by **front projection**.

reflux *(Chem.)*. Boiling a liquid in a flask, with a condenser attached so that the vapour condenses and flows back into the flask, thus providing a means of keeping the liquid at its bp without loss by evaporation.

reflux oesophagitis *(Med.)*. Inflammation of the lower end of the **oesophagus** caused by regurgitation of stomach contents.

reflux ratio *(Chem.Eng.)*. In distillation design, the ratio of liquid reflux to vapour of any point in a column. In distillation plant operation, the reflux returned per unit quantity of condensate removed as product.

reflux valve *(Civ.Eng.)*. A non-return valve used in pipelines at rising gradients to prevent water which is ascending the gradient from flowing back in the event of a burst lower down.

reforming process *(Chem.Eng.)*. A group of proprietary processes in which low-grade or low molecular weight hydrocarbons are catalytically reformed to higher grade or higher molecular weight materials. (**Platforming** is one such process.)

refracting telescope *(Astron.)*. Telescope using lenses to bring light rays to a focus, it was first applied to astronomy by G. Galileo who resolved the stars of the Milky Way and discovered the satellites of Jupiter. In its modern form, using lenses corrected for chromatic aberration, this telescope is still used by visual observers and amateurs.

refraction *(Phys.)*. Phenomenon which occurs when a wave crosses a boundary between two media in which its phase velocity differs. This leads to a change in the direction of propagation of the wavefront in accordance with **Snell's law**.

refractive index *(Phys.)*. The absolute refractive index of a transparent medium is the ratio of the phase velocity of electromagnetic waves in free space to that in the medium. It is given by the square root of the product of the complex relative permittivity and complex relative permeability. See **refraction**, **Snell's law**. Symbol n.

refractive modulus *(Meteor.)*. One million times the excess of **modified refractive index** above unity in M units.

refractivity *(Chem.)*. **Specific refraction**.

refractometer *(Phys.)*. Instrument for measuring refractive indices. Refractometers used for liquids, such as the **Pulfrich refractometer**, usually measure the critical angle at surface between liquid and a prism of known refractive index. See **Rayleigh refractometer**.

refractor *(Phys.)*. A device by which the direction of a beam of light is changed by causing it to pass through the boundary between two transparent materials of different relative index.

refractories *(Eng.)*. Materials used in lining furnaces etc. They must resist high temperatures, changes of temperature, the action of molten metals and slags and hot gases carrying solid particles. China clay, ball clay and fireclay are all highly refractory, the best qualities fusing at above 1700°C. Other materials are silica, magnesite, dolomite, alumina and chromite. See **silica**.

refractory alloy *(Eng.)*. (1) Difficult to work at high temperatures. (2) Heat resistant or having a very high melting temperature.

refractory cement *(Build.)*. A form of cement capable of withstanding very high temperatures.

refractory clay *(Eng.)*. See **refractories**.

refractory concrete *(Build.,Civ.Eng.)*. Concrete used for resistance to very high temperatures such as in construction or lining of chimneys. Constituents are normally high alumina cement and crushed fire brick but other specially prepared aggregates are used.

refractory metals *(Eng.)*. Term applied to transition group elements in periodic table which have high melting-points. They include chromium, titanium, platinum, tantalum, tungsten and zirconium.

refractory ore *(Min.Ext.)*. (1) Gold ore non-responsive to amalgamation process. (2) Ore of mineral or rock used in fabrication of **refractories**, e.g. **chromite, kyanite, serpentinite**.

refractory period *(Behav.)*. For an organism or an excitable tissue, the period of zero response following a previous response.

refrigerant *(Eng.)*. Substances suitable for use as working fluids in a 2-phase refrigeration cycle. Ammonia and Freon-22 are most important in industrial refrigeration, Freon-11 and -12 in commercial and domestic use where non-toxic refrigerants are needed. Other examples include carbon dioxide and sulphur dioxide.

refrigeration cycle *(Eng.)*. Any thermodynamic cycle which takes heat at a low temperature and rejects it at a higher. The cycle must receive power from an external source.

refrigerator *(Eng.)*. A machine or plant by which mechanical or heat energy is used to produce and maintain a low temperature.

refringent *(Phys.)*. Refractive.

Refsum's disease *(Med.)*. A rare *recessive* genetic disorder with an inability to metabolize a specific fatty acid and causing *neuropathy*, deafness, **ataxia** and **cardiomyopathy**.

refugium *(Ecol.)*. An area where species have survived the great changes undergone by the region as a whole, because local conditions are favourable. Examples of refugia are the area escaping glaciation in the Ice Ages, and hedgerows (where woodland species escape the influence of cultivation).

refusal *(Print.)*. Term applied when a printed ink film fails to key satisfactorily to another.

regain *(Textiles)*. Weight of water present in a textile material expressed as a percentage of the oven-dry weight. Dried textile materials take-up or *regain* moisture when left in any normal atmosphere.

regelation *(Phys.)*. Process by which ice melts when subjected to pressure and freezes again when pressure is removed. Regelation operates when forming a snowball by pressure, in the flow of glaciers, and in the slow passage through a block of ice by a weighted loop of wire.

Regency *(Arch.)*. The last phase of English **Neo-Classical** movement which occurred during the regency of George, Prince of Wales (1810–1820).

regenerated fibre *(Textiles)*. See **fibre, regenerated**.

regeneration *(Bot.,Zool.)*. Regrowth of tissues or organs, such as amphibian limbs, after injury; the formation of new plants from cultured tissues. See **tissue culture**.

regeneration *(Comp.)*. Replacement or reforming of stored data, e.g. in computer register or **volatile memory**.

regeneration *(Electronics)*. Same as **positive feedback**, but particularly applied to a super-regenerative receiving circuit, which oscillates periodically through self-quenching.

regeneration *(Min.Ext.)*. (1) Reconstitution of liquid used in chemical treatment of ores before returning it to head of attacking process (e.g. in cyanide process). (2) Freshening of 'poisoned' ion-exchange resins.

regeneration *(Nuc.Eng.)*. Reprocessing of nuclear fuel by removal of fission products.

regenerative air heater *(Eng.)*. An air heater in which heat-transmitting surfaces of metallic plates, wire mesh or bricks are exposed alternately to the heat-surrendering gases and to the air.

regenerative braking *(Elec.Eng.)*. A method of braking for electric motors in which the motors are operated as generators by momentum of the equipment being braked, and return energy to the supply.

regenerative detector *(Elec.Eng.)*. One in which the high-frequency components in the output are fed back *(reaction, retroaction)* to the input, thus increasing gain and selectivity.

regenerative furnace *(Eng.)*. A furnace in which the hot gases pass through chambers containing fire-brick structures, to which the sensible heat is given up. The direction of gas flow is reversed periodically, and cold incoming gas is preheated in the chambers.

regenerative receiver *(Telecomm.)*. One with positive feedback for the carrier, enhancing efficiency of amplification and demodulation.

regenerator *(Eng.)*. Labyrinth which transfers heat of exit gases to air entering furnace, or feed-water to boiler.

regenerator *(Telecomm.)*. Circuits, used in electrical and/or optical communication systems using **pulse-code modulation** and placed at intervals along the transmission path. They detect incoming and re-transmit stronger and more sharply defined output pulses. The *pattern* of pulses is unaltered, retaining the meaning of the transmitted information. Cf. **repeater**.

Regge trajectory *(Phys.)*. A graph relating spin angular momentum at energy for a nuclear particle. Possible quantized values of spin correspond to large discrete energy increments on the graph. This enables recurrences of nuclear particles to be predicted: the extra energy corresponding to the greater rest mass expected to be associated with such particles. A *recurrence* is a particle identical in all respects, except energy (or mass) and spin momentum, with a known particle, and is regarded as being a higher energy equivalent of the normal particle.

region *(Maths.)*. See **domain (1)**.

regional metamorphism *(Geol.)*. All those changes in mineral composition and texture of rocks due to compressional and shearing stresses, and to rise in temperature occasioned by intense earth movements over a widespread area. The characteristic products are the crystalline schists and gneisses.

region of limited proportionality *(Phys.)*. Range of operating voltages for a counter tube in which the gas amplification depends on the number of ions produced in the initial ionizing events as well as on the voltage. For larger initial events the counter saturates.

register *(Build.)*. (1) A metal damper to close a chimney. (2) Grilled aperture to allow the passage of hot or cold air.

register *(Comp.)*. (1) **Location** in the **central processor** which is used for special purposes only and is sometimes protected, e.g. **accumulator, control register, index register**. (2) Mechanical, electrical or electronic device which stores and displays one item of data.

register *(Image Tech.,Print.)*. Exact correspondence of superimposed work, e.g. when the separate colours in colour photography are printed or projected together to reproduce the original picture.

register *(Telecomm.)*. (1) Electronic circuit capable of holding a specific number of bits of information. (2) In telephone switching systems, part of an automatic system that receives and stores signals from a calling device for interpretation and action.

register-controlled system *(Telecomm.)*. System of automatic switching in which the selectors are positioned by signals supplied by registers in response to information provided by dialling or other means.

registered breadth *(Ships)*. The breadth measured of the shell plating at widest part.

registered depth *(Ships)*. Depth measured from top of **ceiling** to top of deck beam at midlength at the centre line of the vessel. Deck to which it is measured is usually stated.

registered dimensions *(Ships)*. Dimensions appearing on the Certificate of Registry. Their main purpose is to identify the ship and they are also called *identification dimensions*. They are **registered length, registered breadth, registered depth**.

registered length *(Ships)*. Length from the fore side of the stem at the top to after side of stern post or, in vessel without a stern post, to the centre of the rudder stock.

register(ed) tonnage *(Ships)*. See **net register tonnage**.

register length *(Comp.)*. The number of **bits** which can be stored in a computer **register**.

register lock-up *(Print.)*. Mechanism allowing fine positioning of plates on the cylinders of web-fed presses.

register marks *(Print.)*. Fine lines, cross marks or similar, added to artwork to provide reference points and thus aid fitting and positioning of images during film assembly, platemaking and printing.

register pin *(Image Tech.)*. See **pilot pin**.

register rollers *(Print.)*. Adjustable rollers that provide a means of varying the web length between one unit of a web-fed press and another.

register sender *(Telecomm.)*. A device, common to a number of input circuits, which accepts and stores information relating to a called number or service. It is thereafter capable of controlling the setting up of a part or all of the wanted connection.

register sets *(Print.)*. A combination of **mixed forme base** and **honeycomb base**, each supplied in a variety of accurately sized units, to be assembled with type to the size required for a particular plate, for which it is used to provide both a mount and a means of attaining register.

register sheet *(Print.)*. The sheet used in obtaining correct register or position.

register-translator *(Telecomm.)*. Device in which the functions of a register and translator are combined, e.g. a **director**.

reglet *(Arch.)*. (1) A flat narrow rectangular moulding. (2) A **facette**.

reglet *(Print.)*. A thin strip of wood used for spacing; usually 6 or 12 points in thickness (known as *nonpareil reglet* and *pica reglet* respectively).

reglette *(Surv.)*. The short graduated scale attached at each end of the special measuring tape or wire used in baseline measurement.

Regnault's hygrometer *(Meteor.)*. A type of hygrometer in which the silvered bottom of a vessel contains ethoxyethane, through which air is bubbled to cool it, its temperature being indicated by a thermometer.

regolith *(Astron.)*. The layer of fine powdery material on the **Moon** produced by the repeated impact of **meteorites**.

regolith *(Geol.)*. The mantle of rock material that overlies bedrock.

regrating *(Build.)*. Operation of redressing the faces of old hewn stone work.

regression *(Behav.)*. In psychoanalytic theory, a defence mechanism which involves a reversion to an earlier and less threatening mode of functioning.

regression *(Biol.)*. A tendency to return from an extreme to an average condition, as when a tall parent gives rise to plants of average stature.

regression *(Geol.)*. The retreat of the sea from the land (stratigraphical usage).

regression *(Stats.)*. A model of the relationship between the expected value of a random variable and the values of one or more possibly related variables.

regular *(Bot.)*. A radially symmetrical **actinomorphic** flower.

regular convex solids *(Maths.)*. Solids having congruent all faces bounded by plane surfaces and all corners. They are (1) *regular tetrahedron*, 4 equilateral triangular faces, (2) *regular hexahedron* or *cube*, 6 equal squares as faces, (3) *regular dodecahedron*, 12 regular pentagons as faces, (4) *regular octahedron*, 8 equilateral triangles as faces, (5) *regular icosahedron*, 20 equilateral triangles as faces.

regular-coursed *(Build.)*. Said of rubble walling built up in courses of the same height.

regular function *(Maths.)*. See **analytic function**.

regular polygon *(Maths.)*. One with all its sides equal and all its angles equal; e.g. a regular polygon of three sides is an equilateral triangle.

regular reflection factor *(Phys.)*. The ratio which the luminous flux regularly reflected from a surface bears to the total flux falling on that surface.

regular transmission *(Phys.)*. Transmission of light through a surface in such a way that the beam of light, after transmission, appears to proceed from the light source.

regular transmission factor *(Phys.)*. The ratio which the luminous flux regularly transmitted through a surface bears to the total luminous flux falling on the surface.

regulating rod *(Nuc.Eng.)*. Fine **control rod** of reactor.

regulation *(Electronics)*. (1) Fractional change in voltage level when a load is connected across a power supply, due to internal resistance. (2) Difference between minimum and maximum voltage drops across a reference diode over its range of operating currents. (3) The process of controlling a physical quantity (speed, temperature, position, voltage etc.), by a control system or network employing *negative feedback*.

regulator *(Electronics)*. A device or circuit which maintains a desired quantity (e.g. voltage, current, frequency, or a mechanical property), at a predetermined level, usually by comparison with a reference source.

regulator *(Horol.)*. (1) A precision long-case clock with a seconds pendulum. The dial has independent hands and zones for the hours, minutes and seconds. (2) See **index**.

regulator cells *(Elec.Eng.)*. One of several cells which are arranged at the end of a battery of accumulator cells and are connected to a regulating switch so that they can be cut in or out of circuit in order to adjust the voltage of the battery as a whole. Also called *end cell*.

regulator gene *(Biol.)*. A gene whose product controls the rate at which the product of another gene is synthesized.

regulus *(Maths.)*. One of the sets of lines forming a **ruled surface**.

regulus of antimony *(Eng.)*. Commercially pure metallic antimony. Also, the impure metallic mixture which is produced during smelting. *Regulus* was the alchemical name for antimony, which readily combines with gold.

regurgitation *(Med.)*. (1) The flowing of blood in reverse direction to the circulation in the heart as a result of valvular disease, e.g. *aortic regurgitation*. (2) The reverse movement of the gastric contents.

regurgitation *(Zool.)*. The bringing back into the mouth of (undigested) food.

reheat *(Aero.)*. Injection of fuel into the jet pipe of a turbojet for the purpose of obtaining supplementary thrust by combustion with the unburnt air in the turbine efflux. Reheat is the British, and original, term, but is gradually being superseded by the American term *afterburning*, with *afterburner* for the device itself.

reheating furnace *(Eng.)*. The furnace in which metal ingots, billets, blooms etc., are heated to temperature required for hot-working.

reheating, resuperheating *(Eng.)*. The process of passing steam, which has been partially expanded in a steam turbine, back to a superheater before subjecting it to further expansion. Reheating is also used sometimes repeatedly, in heat-treating processes, like annealing, and in pneumatic systems for operating power tools.

Reichert-Meissl number *(Chem.)*. A standard used in butter analysis. A Reichert-Meissl number of *n* means that the soluble volatile fatty acids liberated from 5 g of butter fat under specified conditions require $n\,cm^3$ of 0.05 M barium hydroxide solution for their neutralization.

Reimer-Tiemann reaction *(Chem.)*. The synthesis of phenolic aldehydes by heating a phenol with trichloromethane in the presence of conc. KOH. The intermediate dichloro derivative is hydrolysed to an aldehyde. The CH=O group takes up the 2- or 4-position with respect to the hydroxyl group.

re-imposition *(Print.)*. (1) Transferring the page from one chase to another, the latter being perhaps a machine chase or a foundry chase. (2) Altering position of pages in a forme to suit a different size of paper or the requirements of printing and binding equipment.

reinforced concrete *(Civ.Eng.)*. Concrete work in which steel bars or wires (*reinforcement*) are embedded to increase strength.

reinforced plastics *(Plastics)*. Range of plastic materials in which the basic plastic has been strengthened by incorporating a fibrous reinforcing agent, e.g. asbestos, paper, cloth, glass fibre, carbon fibre.

reinforcement *(Acous.)*. Sound reproduction, using loudspeakers in different positions, in which the received enhanced level appears to come from the actual source as required, e.g. in theatres. See **Haas effect**.

reinforcement *(Behav.)*. Refers to situations when a response is predictably followed by an event, the *reinforcer*, and the event can be shown to increase or alter the future probability of the response. Cf. **positive reinforcement, negative reinforcement**.

reinsertion *(Image Tech.)*. See **d.c. restoration.**
Reiss microphone *(Acous.)*. Carbon transmitter in which a large quantity of carbon granules between a cloth or mica diaphragm and a solid backing, such as block of marble, is subjected to the applied sound wave. Characterized by high damping of the applied vibrational forces, and freedom from carbon noise by virtue of packing amongst the granules.
reiterated, repeated sequences *(Biol.)*. DNA sequences repeated many times within a genome; common in higher organisms.
reiteration *(Surv.)*. Method of checking angular measurements made with a theodolite (and of securing greater accuracy) by repeating the observations after reversing face (turning the sighting telescope through 180°). Cf. **repetition.**
Reiter's syndrome *(Med.)*. The association of **urethritis, conjunctivitis** and **arthropathy** thought to be due to a mycoplasma infection and, in the majority, of *venereal* origin.
rejection *(Med.)*. The process by which the body rejects tissue transplanted into it.
rejection band *(Telecomm.)*. See **stop band.**
rejector circuit *(Telecomm.)*. Parallel combination of inductance and capacitance, tuned to the frequency of an unwanted signal, to which it offers a high impedance when placed in series with a signal channel.
rejuvenation *(Geol.)*. A term applied to the action of a river system which, following the uplift of the area drained by it, can resume down cutting in the manner of a younger stream.
rejuvenescence *(Biol.)*. Renewal of growth from old or injured parts.
relapsing fever *(Med.)*. *Spirochaetosis.* A term applied to a number of diseases which are transmitted by lice or ticks and which are due to infection by various spirochaetes; characterized by recurrent attacks of fever and by enlargement of the liver and spleen.
relation *(Maths.)*. Let S and T be sets. Then $S \times T$ denotes the set of all ordered pairs (x,y) with x in S and y in T. Any subset p of $S \times T$ is called a *relation from S to T.*
relational data base *(Comp.)*. Data base using a relational **data model.** The data is stored in the form of several two dimensional tables or *flat files.* The tables embody different ideas about the data but contain overlapping information.
relational operator *(Comp.)*. Symbol used to express a relationship to be tested e.g. >, <, =. See **operator.**
relative abundance *(Ecol.)*. A rough measure of population density, relative, e.g. to time (as the number of birds seen per hour) or percentage of sample plots occupied by a species of plant.
relative abundance *(Phys.)*. See **abundance.**
relative atomic mass *(Chem.)*. Mass of atoms of an element formerly in *atomic weight units* but now more correctly given on the *unified scale* where 1u is 1.660×10^{-27} kg. Abbrev. *RAM, r.a.m.* See **atomic weight.**
relative bearing *(Ships)*. Angle between direction of ship's head and of an object.
relative density *(Phys.)*. The ratio of the mass of a given volume of a substance to the mass of an equal volume of water at a temperature of 4°C. Originally *specific gravity.*
relative efficiency *(Eng.)*. In an internal combustion engine, ratio of actual indicated thermal efficiency to that of some ideal cycle, such as **air standard cycle,** at the same compression ratio.
relative growth rate *(Ecol.)*. Abbrev *R.* Mathematical expression of growth

$$R = \frac{1}{W} \cdot \frac{dW}{dt}$$

where W is weight and t time.
relative humidity *(Meteor.)*. The ratio of vapour pressure in a sample of moist air to the **saturation vapour pressure** with respect to water at the same temperature.
relative molecular mass *(Chem.)*. Preferred term for *molecular weight.*
relative permeability, permittivity *(Elec.Eng.)*. See **permeability, permittivity.**
relative plateau slope *(Nuc.Eng.)*. See **plateau slope.**
relative stopping power *(Phys.)*. See **stopping power.**
relative visibility factor *(Phys.)*. Ratio of apparent brightness of a monochromatic source to that of a source of wavelength 550 nm having the same energy.
relativistic *(Phys.)*. Said of any deviation from classical physics and mechanics based on relativity theory.
relativistic mass equation *(Phys.)*. When a particle is accelerated up to a velocity (v) which is more than a small fraction of the phase velocity of propagation of light in vacuum (c), it is said to be *relativistic* with mass increased according to the formula

$$m = m_0 / \sqrt{1 - v^2/c^2},$$

where m_0 = rest mass (at low velocities). Required to be considered in cyclotron, betatron and linear accelerator design.
relativistic particle *(Phys.)*. One having a velocity comparable with that of light.
relativity *(Phys.)*. Theory based on the equivalence of observation of the same event from different frames of reference having different velocities and accelerations. Introduced in 1905 and generalized a decade later, Einstein's *special relativity* was verified by observations and the precession of the perihelion of the planet Mercury. Important results of this restricted theory include the **relativistic mass equation,** the **mass-energy equation** and **time dilation.** See **principle of relativity.**
relaxation *(Eng.)*. Exponential return of system to equilibrium after sudden disturbance. Time constant of exponential function is *relaxation time.*
relaxation method *(Civ.Eng.)*. Method of solving structural equations by making an initial estimate of the solution and then systematically reducing the errors in estimation.
relaxation oscillator *(Telecomm.)*. One which generates relaxation oscillations. Characterized by peaky or rectangular waveforms and the possibility of being pulled into step (locked) by an independent source of impulses of nearly the same frequency.
relaxation time *(Biol.)*. In excitable tissues, the period during which activation subsides after cessation of a stimulus.
relaxation time *(Chem.)*. The time constant of *relaxation,* e.g. from the height to the lower energy level in nuclear magnetic resonance.
relaxation time *(Eng.)*. See **relaxation.**
relaxation time *(Phys.)*, Most processes which exhibit decay are assumed to follow an exponential law. For example, the drift velocity $\langle v_x \rangle$ of electrons in a conductor will decay to zero after the field is removed according to $\langle v_x \rangle = \langle v_x \rangle e^{-t/\tau}$ where $\langle v_x \rangle_0$ is the velocity at time $t = 0$ and τ is the *relaxation time* for the decay.
relay cell *(Med.)*. An internuncial neurone or interneurone of the central nervous system, particularly one forming a link between afferent and efferent neurones of a reflex arc.
relay spring *(Elec.Eng.)*. Flexible part of a relay which keeps it in the unoperated state. It is stressed on operation, and restores the relay to normal when the current stops, thereby operating contacts.
relay types *(Elec.Eng.)*. *Allström*: a sensitive form using a light beam and photocell. *Control*: one operated by permitting the next step in a control circuit. *Differential*: one operating on the difference between, e.g. two currents. *Frequency*: one operating with a selected change in the supply frequency. *Polarized*: one in which the movement of the armature depends on the current direction on the armature control circuit.
relay valve *(Eng.)*. See **pilot valve.**
relearning *(Behav.)*. A method of measuring retention on a learning/memory task; the material is relearned again some time after the original learning. The difference between the original learning and relearning (the *savings*) is a measure of the original learning.

release *(Image Tech.)*. The trigger arrangement for releasing the shutter and effecting exposure in a camera.
release *(Print.)*. A term used in hot press stamping to describe tendency of foil substrate to adhere to surface being marked. A foil with 'good release' is suitable for use with high-speed automatic machines. Release properties depend also on correct choice of temperature and pressure.
release *(Telecomm.)*. In automatic telephony, the release of apparatus which has been seized for making a connection. In manual telephony, the positive disengagement of apparatus on cessation of a conversation.
release mesh *(Min.Ext.)*. That at, and below, which screen size mineral is released from a closed crushing or grinding circuit and passed to next stage of treatment.
release papers *(Paper)*. Papers treated so that an adhesive surface will become easily detached from them without rupture of the paper surface. Used as protective backing of self-adhesive materials.
release print *(Image Tech.)*. A print of a cinematograph film for public use in cinemas.
releaser *(Behav.)*. *Social releaser.* A term originating in classical ethology; it refers to aspects of stimulus (the sign stimulus) which are especially effective in releasing a specific response in all individuals of a species. It also implies that both the relevant stimulus features and the response to them have become mutually adapted through evolution.
releasing key *(Civ.Eng.)*. A tapered piece used to ease shuttering away from concrete after it has set.
relevé *(Ecol.)*. List of the plant species at any site with visual estimation of canopy cover.
reliability *(Elec.Eng.)*. Probability that an equipment or component will continue to function when required. Expressed as average percentage of failure per 1000 hrs of availability. Also *fault rate.*
relic *(Min.Ext.)*. Block of ore temporarily or permanently left close to a drive through solid rock, forming a wall between this and the stoped portion of the deposit. Cf. **remnant**.
relict *(Ecol.)*. A species, whether terrestrial, marine, or freshwater, which occurs at the present time in circumstances different from those in which it originated.
relief block *(Print.)*. A letterpress printing block (e.g. line, half-tone) which can be used with printing type.
relief map *(Surv.)*. One with contour lines, shading, or colouring to indicate changes in surface configuration of the area mapped.
relief process *(Image Tech.)*. A printing process in which a photographic image processed to form a gelatine relief (**matrix**) is used to transfer dye to another layer on a separate support. See **dye transfer, imbibition**.
relief process *(Print.)*. See **printing**.
relief well *(Min.Ext.)*. One drilled into a reservoir to reduce the pressure in a burning or blown-out well or to inject water to flood it. A *killer well.*
relieving *(Eng.)*. (1) Interrupting a bearing surface, such as a machine tool slide or plain journal bearing, so as to improve alignment or lubrication conditions. (2) In cutting tools, removing material adjacent to a cutting edge so as to improve the flow of chips or cooling and lubrication conditions.
relieving gear *(Ships)*. Any gear attached to the rudder, directly or indirectly, in such a way as to reduce stresses on the steering gear due to wave action on the rudder.
relight *(Aero.)*. Term used for igniting an aircraft gas turbine in flight after shut down.
relocatable program *(Comp.)*. One which runs regardless of where it is loaded into memory.
reluctance *(Elec.Eng.)*. Ratio of magnetomotive force, *m.m.f.*, applied to a magnetic circuit or component to the flux in that circuit (it is the magnetic dual of resistance and is the reciprocal of permeance).
reluctance pick-up *(Elec.Eng.)*. Transducer for detecting vibrations or reproducing records. The signal causes a change in the reluctance of a magnetic circuit, which induces an e.m.f. linked to it.
reluctivity *(Elec.Eng.)*. The reciprocal of *permeability.*

REM *(Med.)*. *Rapid Eye Movement.* Occurs at certain stages during dreaming sleep and is believed necessary for brain repair.
rem *(Radiol.)*. See **röntgen equivalent man**.
remainder *(Maths.)*. See **division**.
remanence *(Phys.)*. Magnetization remaining after an exciting magnetic field has been removed from ferromagnetic materials. See **hysteresis loop, residual magnetization**.
remanent flux density *(Phys.)*. See **residual flux density**.
remanent magnetization *(Geol.)*. That magnetization form in a rock at the time of its formation, by the Earth's magnetic field, or during some subsequent event.
remiges *(Zool.)*. In Birds, the large contour feathers of the wing. sing. *remex.*
remiped *(Zool.)*. Having the feet adapted for paddling, as many aquatic birds.
remission *(Med.)*. An abatement (often temporary) of the severity of a disease; the period of such abatement.
remittent *(Med.)*. Of a fever, characterized by remissions in which the temperature falls, but not to normal; as in malaria.
remnant *(Min.Ext.)*. Block of ore or stope pillar left *well clear* of the underground travelling ways on completion of stoping. Cf. **relic**.
remodulation *(Telecomm.)*. Transferring modulation from one carrier to another, as in the frequency changer in a supersonic heterodyne receiver.
remote access *(Comp.)*. Access through a terminal physically separated from the computer.
remote augmented lift system *(Aero.)*. Engine designed for **STOVL** aircraft employing nozzles, remote from the engine, powered by compressed air ducted from the compressor and heated by fuel burnt in the nozzle in a manner similar to an *afterburner.*
remote control *(Elec.Eng.)*. Control, usually by electric or radio signals, carried out from a distance in response to information provided by monitoring instruments.
remote cup *(Build.)*. A small pressurized container used in spray painting as a more portable form of pressure feed tank. Can be used to apply various types of spray paints.
remote cut-off *(Electronics)*. A characteristic whereby a large negative bias is needed for complete cut-off of conduction output in a valve or other amplifying device.
remote handling equipment *(Nuc.Eng.)*. Apparatus developed to enable an operator to manipulate highly-radioactive materials from behind a suitable shield, or from a safe distance.
remote mass-balance weight *(Aero.)*. A **mass balance** weight which, usually because of limitations of space, is mounted away from the control surface, to which it is connected by a mechanical linkage.
remould *(Autos.)*. Used tyre which has had a new tread vulcanized to the casing, including a coating of rubber on the walls. Cf. **retread**.
removable isolated singularity *(Maths.)*. If a function $f(z)$, having a singularity at $z=a$, has no negative powers of $(z-z_0)$ in **Laurent's expansion**, then the singularity may be removed by defining $f(a)=a_0$, where a_0 is the first coefficient of $(z-z_0)$ in the expansion.
removes *(Print.)*. Quotations, etc. set in smaller type than the main text. The difference in size is usually 2 points. Thus a book the text of which is set in 12-point or pica should have its quotations set in 10-point.
REM sleep *(Behav.)* See **paradoxical sleep**.
Renaissance *(Arch.)*. The name given to the cultural movement, inspired by classical ideals, which spread from Tuscany throughout Europe in the 15th century, and which had a major effect on architectural style. The movement passed through various stages in its development, these stages being known generally as Early Renaissance, High Renaissance, Baroque and Antiquarian; however, each European country produced regional variations, and in Britain the phases relating to the above are known as Elizabethan, Jacobean, Stuart and Georgian respectively, the latter lasting until about 1830.
renal *(Zool.)*. Pertaining to kidneys.

renal colic *(Med.)*. Severe loin pain usually due to obstruction of the ureter by a stone.

renal portal system *(Zool.)*. In some lower Vertebrates, that part of the venous system which brings blood from the capillaries of the posterior part of the body and passes it into capillaries of the kidneys.

renaturation *(Biol.)*. The converse of **denaturation**. Complementary strands of DNA or DNA and RNA will reform duplex molecules. The kinetics of the process depend on the number of copies and the concentration of the molecules. Usually achieved by heating single strands at about 20°C below the Tm. The basis of molecular **probes**.

render and set *(Build.)*. Two-coat plaster work on walls.

rendered *(Build.)*. A term applied to laths which are split rather than sawn. The fibres are not severed so maximum strength is retained, i.e. no short grain. Also called *split* or *riven lath*.

render, float and set *(Build.)*. Three-coat plaster work on walls.

rendering *(Build.)*. Operation of covering brick and stonework with a coat of coarse stuff; the coating itself.

rendez-vous *(Space)*. The meeting and bringing together of two spacecraft in orbit at a planned place and time; also the type of mission where a spacecraft encounters a target body with zero relative velocity at a preplanned time and place.

rendzina *(Bot.)*. Shallow, dark, intrazonal soil, rich in calcium carbonate, developed on limestone, especially chalk.

reniform *(Bot.)*. Kidney-shaped.

renin *(Med.)*. Protein enzyme, liberated from the glomerular apparatus in the kidney into the bloodstream when blood pressure is reduced. This reacts with angiotensinogen to produce **angiotensin** I.

rennet *(Chem.)*. A commercial preparation for making junket (clotted milk) prepared from the mucous membrane of the stomach of calves, and containing the enzyme *rennin*.

repeated DNA *(Biol.)*. A sequence which occurs more than once in the haploid genome. Such sequences are often short and occur many times.

repeater *(Horol.)*. A watch which 'repeats the time' by striking a sequence of blows on gongs when a slide that projects from the band of the case is pushed. In a *quarter repeater* the last hour (as shown by the watch) is struck, followed by the number of quarters; a *minute-repeater* strikes, in addition, the number of minutes since the last quarter struck. Much used before convenient artificial illumination.

repeater *(Maths.)*. See **recurring decimal**.

repeater *(Telecomm.)*. An amplifier which receives weak signals and delivers corresponding stronger signals without re-shaping waveforms. Used widely in *carrier* communications systems with co-axial cables. Also, in watertight housings, to boost signals on submarine cables. May be either *one-* or *two-way*. Cf. **regenerator**.

repeater distribution frame *(Telecomm.)*. A frame provided for the interconnection in repeater stations of amplifiers, transformers, and signalling units.

repeater gain *(Telecomm.)*. Power delivered by a repeater divided by the power which would be delivered in the absence of the repeater; expressed in decibels.

repeater test rack *(Telecomm.)*. Rack in a repeater station at which measurements can be made of input and output levels of audio-frequency amplifiers and signalling units.

repeating back *(Image Tech.)*. Sliding back for a camera, in which images for colour separation can be taken successively and side-by-side.

repeating coil *(Telecomm.)*. Unity-ratio transformer for separating telephonic circuits, with windings balanced to earth.

repeating decimal *(Maths.)*. See **recurring decimal**.

repeating selector *(Telecomm.)*. A selector which is operated by the first train of impulses received, and also repeats all received impulses for operating further selectors.

repeating work *(Horol.)*. Chiming mechanism for a repeater watch.

repeller *(Electronics)*. US term for **reflector**, as in a klystron.

repetition *(Surv.)*. Method of checking angular measurements made with a theodolite by repeating the observation after unclamping the lower plate and sighting on the back station so that the vernier reading is unaltered, and then sighting forward to get a new reading on the vernier, which should be double the previous reading. Cf. **reiteration**.

repetition compulsion *(Behav.)*. In psychoanalytic theory, the factor in mental life which compels early patterns of behaviour to be repeated, irrespective of pleasure/displeasure thereby experienced by the individual.

repetition rate *(Telecomm.)*. Rate at which recurrent signals, usually pulses, are repeated.

replaceable hydrogen *(Chem.)*. Those hydrogen atoms in the molecule of an acid which can be replaced by atoms of a metal on neutralization with a base.

replacement *(Geol.)*. The process by which one type of rock occupies the space previously occupied by another rock; also applies to minerals.

replica plating *(Biol.)*. Typically transferring the *pattern* of bacterial colonies on an agar plate by impressing velvet stretched over a holder onto the agar and then placing the velvet in turn onto a number of further sterile plates. If the latter contain, say, different antibiotics, a sensitive strain can be selected by noting colonies which fail to grow and then picking them from the master plate.

replication *(Biol.)*. Duplication of genetic material, usually prior to cell division.

replication fork *(Biol.)*. The fork where duplex DNA becomes split into two double strands as replication moves from the *origin of replication*.

replicon *(Biol.)*. A part of a DNA molecule which is replicated from a single origin. In prokaryotes there is usually one origin per genome, but in eukaryotes there are many spaced along the chromosome.

replum *(Bot.)*. *False septum*. Partition across ovary or fruit formed by ingrowth from placentas, not by the walls of the carpels. Cf. **septum**.

report call *(Telecomm.)*. Call made to ascertain whether a desired subscriber is available for connection.

report generator *(Comp.)*. **Software** which gives the non-specialist user the capability of producing reports from one or more files, through easily constructed statements.

rep, repp *(Textiles)*. A plain-weave fabric with weft ribs formed by using coarse and fine yarns in alternate order in both warp and weft.

representative sample *(Min.Ext.)*. One cut from the bulk of material or ore deposit in such a way as to make it reasonably representative of the whole body.

repression *(Behav.)*. In psychoanalytic theory, the process by which an unacceptable thought, impulse or memory is rendered unconscious.

repressor *(Biol.)*. A protein which binds to an **operator** site and prevents transcription of the associated gene.

reprocessing *(Nuc.Eng.)*. See **fuel reprocessing**.

reproducer *(Acous.)*. (1) Complete sound reproduction system. (2) Loudspeaker.

reproducibility *(Genrl.)*. Precision with which a measured value can be repeated in a process or component.

reproduction *(Acous.)*. Replay of recorded sounds.

reproduction *(Biol.)*. The generation of new individuals in the perpetuation of the species. adj. *reproductive*.

reproduction proof *(Print.)*. Print taken of a letterpress forme, on **art paper** or **baryta paper**, from which a plate is to be made photomechanically.

reproductive behaviour *(Behav.)*. The varied activities that lead to production and rearing of offspring, includes **agonistic behaviour**, courtship and other mate interactions, as well as maternal behaviour.

Reptilia *(Zool.)*. A class of Craniata. They are pentadactyl and have shelled amniote eggs. The vertebrae are gastrocentrous, the kidney metanephric and the skin completely covered by epidermal scales, or, sometimes, by bony plates. they are poikilothermous, breathe by lungs, and retain aortic arches. Known fossils from late Carboniferous, they were dominant and various in the Mesozoic *(Dinosaurs)* but became less numerous in the Cretaceous. Living forms include Lizards, Snakes, Turtles, Tortoises, Crocodiles, and Alligators.

repugnatorial glands *(Zool.)*. In Arthropoda, glands, usually abdominal in position, which produce a repellent secretion of odoriferous pungent, or corrosive nature which can be used in self-defence.

repulsion *(Biol.)*. When two specified *non-allelic* genes are on different but homologous chromosomes, having come from different parents, they are in *repulsion*. Cf. **coupling**.

repulsion-induction motor *(Elec.Eng.)*. Single-phase induction motor having, in addition to the squirrel-cage winding on the rotor, a commutator winding with its brushes short-circuited, so that the motor starts as a repulsion motor with a high starting torque and runs with the characteristics of an induction motor.

repulsion motor *(Elec.Eng.)*. A type of single-phase commutator motor in which power is supplied to the stator winding, and the armature winding is short-circuited through the brushes.

repulsion-start induction motor *(Elec.Eng.)*. A repulsion motor having a centrifugal device which short-circuits all the commutator bars when the motor reaches a certain speed, so that it runs as a single-phase induction motor, and starts as a repulsion motor with a high starting torque.

requirements *(Space)*. The demands placed on any element of a space system (such as spacecraft, payload, subsystem, communications network, ground organization, etc.) which have to be satisfied by that element.

reradiation *(Telecomm.)*. Radiation from resonating elements, such as masts, antennae, telephone lines, giving errors in bearings or misplaced television images.

rere arch *(Arch.)*. A flat soffit arch laid over splayed jambs.

re-recording *(Acous.)*. Recording acoustic waveforms immediately upon reproduction from the same, or any other type of recording medium as that in use.

re-reeler *(Print.)*. An auxiliary unit to rewind the reel on web-fed presses for subsequent operations.

re-run *(Comp.)*. Repeat of part of program for computer. Most programs incorporate a re-run for every few minutes running time. In the event of error only the part subsequent to the previous re-run point is repeated. See **fault tolerance**.

réseau *(Astron.)*. A reference grid of fine fiducial lines or points used in image analysis and measurement.

réseau *(Image Tech.)*. A mosaic formed by rulings of fine coloured lines in red, green and blue used as a screen in some early additive processes of colour photography, e.g. *Dufaycolor*.

resection *(Med.)*. Cutting off a part of a bodily organ, especially the ends of bones and other structures forming a joint.

resection *(Surv.)*. Positional fix of a point which is not going to be occupied by sighting it from two or more known stations.

reserpine *(Med.)*. A drug obtained from *Rauwolfia serpentina*, formerly used for the treatment of hypertension.

reserve buoyancy *(Aero.)*. Potential buoyancy of a seaplane or amphibian which is in excess of that required for normal floating. The downward force required for complete immersion.

reserve buoyancy *(Ships)*. Watertight volume above the load water line.

reserved word *(Comp.)*. Identifier used by the system and therefore not available to a user.

reserve factor *(Aero.)*. Ratio of actual strength of an aircraft structure to estimated minimum strength for a specified load condition.

reserves *(Min.Ext.)*. Block of ore proved by development (normally by driving levels and winzes or raises so as to expose its four sides in a rectangular panel, and sufficiently sampled and tested) to warrant exploitation.

reservoir *(Nuc.Eng.)*. Any volume in an isotope separation plant which is for the purpose of storing material or to ensure smooth operation.

reservoir pressure *(Min.Ext.)*. The pressure in a natural oil or gas reservoir which causes flow to the bore hole. Maintenance of a proper pressure in the reservoir by not allowing too high an extraction rate is important in ensuring good recovery.

reset *(Comp.)*. To return a computer to its initial state (e.g. by restoring all registers to known values).

reset *(Elec.Eng.)*. General term for preparation of any circuit or apparatus for a fresh performance of its duty. Amplifiers require no resetting, apart from switching on; timers or counters do require resetting. Resetting can be automatic, or initiated by an external signal, arbitrary in time.

reset circuit *(Elec.Eng.)*. One which, when operated, resets a functional circuit, i.e. establishes it in a ready condition for operating.

reshaping *(Elec.Eng.)*. Restoration to intended shape, in amplitude and time, of pulses which have become distorted.

residual activity *(Nuc.Eng.)*. In a nuclear reactor the remaining activity after the reactor is shut down following a period of operation.

residual deposits *(Geol.)*. Accumulation of rock waste resulting from weathering *in situ*. They cover the whole range of grain size, from residual boulder beds to residual clays.

residual errors *(Eng.)*. Errors which remain in an observation despite all attempts to eliminate them.

residual field *(Phys.)*. Magnetic field remaining in a magnetic circuit after removal of the magnetizing force.

residual flux density *(Phys.)*. Magnetic induction in a ferromagnetic material after the exciting field has been removed. Flux density remaining in a magnetic circuit after the exciting field has been removed.

residual gas *(Electronics)*. Small amount of gas which inevitably remains in a 'vacuum' tube after pumping. If present to excess, it causes erratic operation of the tube, which is said to be *soft*.

residual induction *(Phys.)*. See **residual flux density**.

residual magnetization *(Phys.)*. Magnetization remaining in a ferromagnetic material after the exciting magnetic field has been removed. See **remanence**.

residual resistance *(Phys.)*. That persisting at temperatures near zero on the absolute scale, arising from crystal irregularities and impurities, and in alloys.

residual volume *(Med.)*. Volume of air left in the lungs after the strongest possible forced expiration, usually of the order of $1500\,cm^3$ in man.

residue *(Maths.)*. (1) The nth *power residues of p* are the remainders when r^n, for $r = 1, 2, 3....$, are divided by p. Thus the quadratic residues of 13 are 1, 3, 4, 9, 10 and 12. (2) Of an analytic function $f(z)$ at an isolated singularity,

$$\frac{1}{2\pi c}\int_c f(z)dz,$$

where c is a simple closed curve around the singularity, and the singularity is deleted from the region contained therein.

resilience *(Ecol.)*. The capacity of ecosystems and populations to return to a previous state after they have been disturbed.

resilience *(Eng.)*. Stored energy of a strained material, or the work done per unit volume of an elastic material by a bending moment, force, torque, or shear force, in producing strain.

resilient escapement *(Horol.)*. An escapement in which the banking pins yield to any excess pressure due to overbanking, allowing the impulse pin to pass the lever, which has no horns; or one in which the teeth of the escape wheel are so formed as to provide a recoil.

resin *(Build.,Chem.)*. A constituent of varnishes or paints, classified as (a) natural resin, e.g. copal, rosin, amber; (b) synthetic resin, e.g. alkyd, phenolic, polyurethane. Natural resins are chiefly obtained as the exudation from plants or trees and have a glassy appearance with a slightly yellow-brown colouring. They consist of highly polymerized acids and neutral substances mixed with terpene derivatives. Synthetic resins are specially formulated to impart particular qualities to coatings such as hardness, flexibility, chemical resistance etc.

resin *(Plastics)*. The term *resin* is widely but loosely used to describe any synthetic plastics material. More precisely, it is applied to a polymeric compound prior to curing. See **synthetic resins**.

resinates *(Chem.)*. Calcium, magnesium, aluminium, iron, nickel, cobalt, zinc, tin, manganese and lead salts or rosin, obtained by fusion of **rosin** with metal oxides.

resin-bonded plywood *(Build.)*. Plywood in which the wood veneers are held together with synthetic resin, glues, or glue-impregnated paper and finally formed with pressure and heat.

resin canal, resin duct *(Bot.)*. Duct of schizogenous origin lined with resin-secreting cells which contain resin, as in the leaves and sometimes wood of conifers.

resin-in-pulp *(Min.Ext.)*. Ion-exchange method of continuously treating ores by acid leaching. Baskets containing **ion-exchange resins** are jigged through tanks containing finely ground ore pulp as it flows from vessel to vessel. Abbrev. *RIP*.

resinous substances *(Chem.)*. Term applied to (a) true resins; (b) substances resembling true resins in their physical properties.

resin poisons *(Min.Ext.)*. In ion-exchange processes, substances which reduce efficiency of resin loading by masking activated resin sites.

resin soaps *(Chem.)*. See **soaps** (2).

resist *(Image Tech.)*. A coating of chemically neutral substance placed over a surface when the latter has to be protected at some stage in processing, as in etching or selective dyeing.

resist *(Print.)*. When making printing surfaces, the protection over the lines and dots of the image which preserve it from the action of etching fluid.

resistance *(Behav.)*. In psychoanalytic theory, the opposition encountered during psychoanalytic treatment to the process of making unconscious memories and impulses, conscious.

resistance *(Phys.)*. In electrical and acoustical fields, the real part of the impedance characterized by the dissipation of energy as opposed to its storage. Electrical resistance may vary with temperature, polarity, field illumination, purity of materials etc. SI unit is the **ohm**. Reciprocal of *conductance*. See **impedance, reactance, Ohm's law**.

resistance box *(Elec.Eng.)*. One containing carefully constructed and adjusted resistors, which can be introduced into a circuit by switches or keys. At higher frequencies there are disturbing inductive and capacitive effects which complicate measurements using reistance boxes, but which are mitigated by suitable design. The boxes are then described as *nonreactive*.

resistance butt-seam welding *(Elec.Eng.)*. Resistance welding process in which coaxial roller electrodes conduct current to the edges of the seam to be joined, the mechanical pressure being applied independently. Extensively used in tube-making from strip.

resistance butt-welding *(Elec.Eng.)*. A resistance welding process in which the two parts to be joined are butted together. See **resistance butt-seam welding, resistance flash-welding, resistance upset-butt welding**.

resistance-capacitance coupling *(Elec.Eng.)*. That in which signal voltages developed across a load resistance are passed to the subsequent stage through a d.c. blocking capacitor.

resistance-capacitance oscillator *(Elec.Eng.)*. One producing a sine waveform of frequency determined by phase shift in a resistance-capacitance section of an artificial line.

resistance coupling *(Elec.Eng.)*. See **direct coupling**.

resistance drop *(Elec.Eng.)*. Voltage drop produced by a current flowing through the resistance of a circuit; equal to product of current and effective resistance.

resistance-flash welding *(Elec.Eng.)*. A resistance welding process in which an arc is struck and maintained between the parts until the correct temperature is attained, after which the current is cut off and the parts are forced together by mechanical pressure. Also called *flash-butt welding*.

resistance frame *(Elec.Eng.)*. A frame containing a number of resistors connected to a multiple-contact switch at the top, so that any desired number of them can be included in the circuit in which the frame is connected.

resistance furnace *(Elec.Eng.)*. See **resistance oven**.

resistance grid *(Elec.Eng.)*. A resistance unit generally used for heavy currents. Made up of a cast-iron grid designed so that current enters one end and passes through all the sections in series, before leaving at the other end.

resistance lap-welding *(Elec.Eng.)*. A resistance welding process in which the two parts to be joined overlap one another. See **resistance seam-welding, resistance spot-welding**, and **resistance stitch-welding**.

resistance noise *(Electronics)*. Same as **thermal noise**.

resistance oven *(Elec.Eng.)*. An oven in which the heating is carried out by means of heating resistors. Also called a *resistance furnace*.

resistance percussive-welding *(Elec.Eng.)*. A resistance welding process in which a heavy electric current is discharged momentarily across the electrodes, and a momentary mechanical force is applied simultaneously.

resistance projection-welding *(Elec.Eng.)*. A variant of resistance spot-welding in which current is concentrated at the desired points by projections on one of the parts.

resistance pyrometer *(Elec.Eng.)*. See **resistance thermometer**.

resistance seam-welding *(Elec.Eng.)*. A resistance welding process in which the welding electrodes consist of two rollers having mechanical pressure between them through which the work passes while the current flows continuously or intermittently, producing a line of overlapping welds.

resistance spot-welding *(Elec.Eng.)*. A resistance welding process in which the electrodes consist of two points and cause welding in one spot.

resistance stitch-welding *(Elec.Eng.)*. A form of resistance spot-welding consisting of a series of overlapped spot welds to form a seam-weld.

resistance strain gauge *(Elec.Eng.)*. Foil, wire or thin film resistor which has a value which varies with mechanical strain. Normally used in a bridge circuit. See **strain gauge**.

resistance thermometer *(Elec.Eng.)*. One using resistance changes for temperature measurement. Resistance element may be platinum wire for extreme precision or semiconductor (*thermistor*) for high sensitivity. Also *resistance pyrometer* for higher temperature.

resistance upset-butt welding *(Elec.Eng.)*. A resistance welding process in which mechanical pressure is first applied to the joint and then current is passed until welding temperature is reached and the weld is 'upset'. Also called *slow-butt welding*.

resistance welding *(Elec.Eng.)*. Pressure welding, in which the heat to cause fusion of the metals is produced by the welding current flowing through the contact resistance between the two surfaces to be welded, these being held together under mechanical pressure. See **resistance butt-welding, resistance flash-welding, resistance percussive-welding, resistance seam-welding, resistance spot-welding**.

resistant *(Biol.,Med.)*. Not readily attacked by a parasite, disease or drug. See **antibiotic resistance**.

resistive component *(Elec.Eng.)*. That part of the impedance of an electrical system which leads to the absorption and dissipation of energy as heat.

resistive load *(Elec.Eng.)*. Terminating impedance which is nonreactive, so that the load current is in phase with the source e.m.f. Reactive loads are made entirely resistive by adding inductors or capacitors in series or shunt (tuning).

resistivity *(Elec.Eng.)*. Intrinsic property of a conductor, which gives the resistance in terms of its dimensions. If R

is the resistance in ohms, of a wire l m long, of uniform cross-section a m^2, then $R = \rho l/a$, where the resistivity ρ is in ohm metres (*not* ohm m^{-3}). Also called, erroneously, *specific resistance*.

resistor *(Elec.Eng.)*. Electric component designed to introduce known resistance into a circuit and to dissipate accompanying loss of power. Types are wirewound, composition, metal film, etc.

resolution *(Chem.)*. The separation of an optically inactive mixture or compound into its optically active components.

resolution *(Comp.)*. (1) A rule of inference which is used for automatic theorem proving. Also *resolution principle*. (2) The density of the **raster** lines on a TV screen. The more lines the more **pixels** conveying detail.

resolution *(Image Tech.)*. The ability of an imaging system to differentiate between closely spaced objects; *resolving power* is usually expressed as the maximum number of light and dark line-pairs per millimetre which can be observed.

resolution *(Med.)*. Retrogression of the phenomena of inflammation; the subsidence of inflammation. v. *resolve*.

resolution of forces *(Phys.)*. The process of substituting two forces in different directions for a single force, the latter being equal to the resultant of the two components. If these are at right angles to each other, the one which makes an angle θ with the original force P is equal to $P\cos\theta$, the other being $P\sin\theta$.

resolution time *(Phys.)*. Minimum time interval between two events recorded separately by a detector of ionizing radiation. The maximum time between two events recorded as *coinciding*.

resolution-time correction *(Phys.)*. Correction applied to observed counting rate for random events as measured by a detector of ionizing radiation, which allows for those events not recorded because of finite resolution time.

resolvant equation *(Maths.)*. An equation which is used in the solution of a higher-order equation.

resolving power of a telescope *(Astron.)*. Ability of an astronomical telescope to measure the angular separation of two images which are close together.

resolving power of the eye *(Phys.)* The angle subtended by a small object which can just be determined visually.

resolving time *(Phys.)*. See **resolution time**.

resonance *(Aero.)*. See **ground-**.

resonance *(Chem.)*. A description of a molecule whose structure cannot be represented by a single **Lewis structure** but only as a mixture of two or more of them. More appropriately called *mesomerism*, it is used in the sense that a mule may be said to be a resonance of a horse and a donkey.

resonance *(Phys.)*. If a vibrating system mechanical or acoustical, is set into forced vibrations by a periodic driving force and the applied frequency is at or near the natural frequency of the system, then vibrations of maximum velocity amplitude result. This is called resonance and it corresponds to minimum mechanical or acoustical impedance. The *sharpness of resonance* is measured by the ratio of the dissipation to the inertia of the system which also determines the rate of decay of the vibrating system when it is impulsed. See **quality factor, decay factor**.

resonance bridge *(Elec.Eng.)*. One for which balance depends upon adjustment for resonance.

resonance curve *(Phys.)*. One showing variation of current in a resonant circuit in series with an e.m.f. as the ratio of the resonance frequency to the frequency of the generator is varied through unity.

resonance escape probability *(Nuc.Eng.)*. In a reactor the probability of a fission neutron slowing down without being captured in ^{238}U resonances. Symbol ρ.

resonance escape probability *(Phys.)*. In a reactor, the probability of a fission neutron slowing down to thermal energy without experiencing resonance.

resonance form *(Chem.)*. See **Lewis structure**.

resonance heating *(Phys.)*. See **magnetic pumping**.

resonance lamp *(Elec.Eng.)*. One which depends on the absorption and reradiation of a prominent line from a mercury arc, excited in mercury vapour.

resonance level *(Phys.)*. An excited level of the compound system which is capable of being formed in a collision between two systems, such as between a nucleon and a nucleus.

resonance potential *(Electronics)*. See **excitation potential**.

resonance radiation *(Phys.)*. Emission of radiation from gas or vapour when excited by photons of higher frequency.

resonances *(Phys.)*. During nuclear reactions very unstable mesons or hyperons are frequently created. These decay through the strong interaction, with a half-life of the order of 10^{-23} s. Consequently such particles are undetectable and their formation as an intermediate step in the reaction can only be inferred from indirect measurements. These temporary states are known as *resonances* to distinguish them from metastable particles with half-lives of the order of 10^{-1} s, which are detectable.

resonance scattering *(Phys.)*. See **scattering**.

resonance step-up *(Elec.Eng.)*. Ratio of the voltage appearing across a parallel tuned circuit to the e.m.f. acting in the circuit (usually induced in the coil) when the circuit is resonant at the applied frequency. See **Q**.

resonance test *(Aero.)*. A test in which an aircraft, while suspended by cables or supported on inflated bags, is excited by forced oscillations over a range of frequencies, so as to establish the natural frequencies and modes of oscillation of the structure.

resonant cavity *(Telecomm.)*. One in which resonant effects result from the possibility of a modal pattern of electric and magnetic fields, as in magnetrons, klystrons, waveguides. Also applies in acoustics.

resonant circuit *(Electronics)*. One consisting of an inductor and a capacitor in series or parallel. The series circuit has an impedance which falls to a very low value at the resonant frequency; that of the parallel circuit rises to a very high value.

resonant frequency *(Chem.)*. See **nuclear magnetic resonance**.

resonant frequency *(Telecomm.)*. That at which reactance of a series resonant circuit, or susceptances of a parallel resonant circuit, balance out; numerically equal to $1/2\pi\sqrt{LC}$ Hz, where L is inductance in henries and C capacitance in farads.

resonant gap *(Radar)*. The interior volume of the resonant structure of a transmit-receive tube in which the electric field is concentrated.

resonant line *(Telecomm.)*. Parallel wire or coaxial transmission line open or short-circuited at the ends and an integral number of quarter-wavelengths long. Used in some radio-frequency oscillators.

resonant mode *(Electronics)*. Field configuration in a tuned cavity. In general, resonance occurs at several related frequencies corresponding to different configurations.

resonator *(Electronics)*. Any device exhibiting a sharply defined electric, mechanical, or acoustic resonance effect, e.g. a stub, piezoelectric crystal, or Helmholtz resonator.

resonator grid *(Telecomm.)* Electrode traversed by an electron beam and which provides a coupling to a resonator.

resorcinol, resorcin *(Chem.)*. $C_6H_4(OH)_2$, *1,3-dihydroxybenzene*, a dihydric phenol, colourless crystal, mp 111°C, bp 276°C. Used as a lotion in certain skin diseases. With formaldehyde used in the preparation of cold-setting adhesives.

resorption *(Geol.)*. The partial or complete solution of a mineral or rock fragment by a magma, as a result of changes in temperature, pressure, or composition of the latter.

respiration *(Biol.)*. At the cellular level respiration is of two major types, *aerobic* and *anaerobic*. Aerobic respiration consists of the total oxidative degradation of glucose, by **glycolysis** and the **tricarboxylic cycle**, to carbon dioxide and water with the generation of 36 molecules of ATP per molecule of glucose. Anaerobic respiration takes place in the absence of oxygen and consists essentially of those stages of aerobic respiration which precede the tricarboxylic acid cycle. It yields a net synthesis of only 2 molecules of ATP. The glucose is converted to a variety of end products such as ethanol (*alcoholic fermentation*) or lactic acid.

respiratory centre *(Zool.)*. In Vertebrates, a nerve-centre of the hind-brain which regulates the respiratory movements.

respiratory failure *(Med.)*. Occurs when the respiratory system is no longer able to maintain normal tensions of oxygen or carbon dioxide in the body.

respiratory movement *(Zool.)*. The muscular movements associated with the supply of air or water to the respiratory organs.

respiratory organs *(Biol.)*. The specialized structures like lungs and gills which enable oxygen to be transferred to the body fluids.

respiratory pigment *(Zool.)*. In the blood of many animals a coloured compound formed by a metal-containing prosthetic group bound to a protein, the whole forming a complex with a high affinity for oxygen. Concerned with oxygen transport, e.g. haemoglobin, haemocyanin.

respiratory quotient *(Bot.)*. The ratio of moles CO_2 evolved to moles O_2 absorbed in respiration; unity when the substrate is carbohydrate; lower when protein or fat. Abbrev. *RQ*.

respiratory substrate *(Biol.)*. Any chemical compound broken down during respiration to release the chemical energy stored in its bonds.

respiratory system *(Biol.)*. See **respiratory organs**.

respiratory valve *(Zool.)*. In some Fish, e.g. Trout, a pair of transverse membranous folds, one attached to the floor, the other to the roof of the mouth, which prevent water from escaping through the mouth during expiration.

respond *(Build.)*. (1) A pilaster which forms a pair with another. (2) A **reveal**.

respondant *(Behav.)*. In classical conditioning, a response that is elicited by a known stimulus (e.g. a knee jerk).

responder *(Telecomm.)*. That part of a transponder which replies automatically to the correct interrogation signal.

response *(Behav.)*. The effect of stimulation; it may, as in muscular and glandular responses, be easily observable and measurable, but it may also be an inferred response which is not immediately apparent in behaviour.

response *(Telecomm.)*. That of a transmission system at any particular frequency is given by the ratio of the output to input level. If these levels are defined on a logarithmic scale, e.g. in *dB*, the response of the complete system is the sum of responses of the separate parts. See also **polar curve**.

response curve *(Telecomm.)*. That which exhibits the trend of the response of a communication system or a part thereof, for the range of frequency over which the system or part is intended to operate. Usually plotted in *dB* often against a logarithmic frequency scale.

response latency *(Behav.)*. The time elapsing between the onset of a stimulus and the beginning of an animal's response to it.

responser *(Radar.)*. Receiver of secondary radar signal from **transponder**.

response time *(Comp.)*. In **interactive computing** the time it takes for a system to respond to an input from the user.

response time *(Electronics)*. Time constant of change in output of an electronic circuit after a sudden change in input, or of indication given by any instrument after change in signal level.

rest *(Eng.)*. (1) The tool-attaching device or toolrest. (2) Metal bar mounted in the toolpost to support a hand-held tool for light cutting or forming, e.g. thread chasing or **spinning**.

rest bend *(Build.)*. A 90° bend off a horizontal drain-pipe, fitted with a flat seating for connection to a vertical. Also called *duckfoot bend*.

restiform *(Zool.)*. Ropelike.

resting nucleus *(Biol.)*. Nucleus of a cell which is not undergoing active growth and division.

resting spore *(Bot.)*. A thick-walled spore able to endure drought and other unfavourable conditions, and normally remaining quiescent for some time before it germinates.

restitution nucleus *(Bot.)*. A single nucleus formed following failure of the chromosomes to separate properly at anaphase and hence containing, say, twice the expected chromosome number.

rest mass *(Phys.)*. Mass of a particle measured by an observer at rest relative to it. In nuclear reactions, the total *rest mass* of the particle involved need not be conserved (cf. **mass**). See **relativistic mass equation**.

rest-mass energy *(Phys.)*. Rest-mass energy is c^2 times the *rest mass* of the particle, where *c* is the speed of light.

restorative *(Med.)*. Capable of restoring to health, consciousness or good condition; any remedy which does this.

restore *(Comp.)*. Return of a variable address or word or cycle index to its initial value.

restoring moment *(Aero.)*. A moment which, after any rotational displacement, tends to restore an aircraft to its normal attitude.

restrainer *(Image Tech.)*. Ingredient of a developer which slows the development of unexposed silver halide, thus reducing tendency to fog.

restriction *(Biol.)*. See **restriction and modification**.

restriction and modification *(Biol.)*. Some bacteria are able to restrict their suceptibility to lysis by phage or other genetic elements by cleaving the invading DNA with restriction enzymes. Their own DNA is made immune by methylating the susceptible sites (*modification*).

restriction endonuclease, enzyme *(Biol.)*. Enzyme able to cleave DNA at a specific nucleotide sequence.

restriction fragment *(Biol.)*. Because of their sequence specificity, **restriction endonucleases** will cleave DNA into defined polynucleotide fragments, which can be separated on an agarose gel.

restriction fragment length polymorphism *(Biol.)*. A restriction fragment identified by blotting whose length is variable in the population. Abbrev. *RFLP*. Those which map close to sites of genetic diseases are a useful aid to antenatal diagnosis in families at risk.

restriking voltage *(Elec.Eng.)*. The high-frequency transient voltage which appears across the contacts of a switch, circuit breaker, or fuse immediately after it has interrupted a circuit, and which is superimposed on the recovery voltage.

resultant *(Phys.)*. Of two forces: that force, obtained by the parallelogram of forces, which is equivalent to the two forces.

resuperheating *(Eng.)*. See **reheating**.

resupinate *(Bot.)*. *Inverted* e.g. the flowers of orchids in which, because of a 180° twist in the stalk, what appears to be the lower petal is, in fact, morphologically the upper petal.

resurgent gas *(Geol.)*. Superheated steam and other volatiles which play an active role in volcanic action, and which are derived from the water included in sedimentary rocks at the time of accumulation.

resuscitation *(Med.)*. The restoration of circulation and respiration after these functions have ceased.

retaining mesh *(Min.Ext.)*. In sizing ore before further treatment, the screen aperture above which size the material is arrested.

retaining wall *(Civ.Eng.)*. A wall built to support earth at a higher level on one side than on the other. Also *revetment*.

retake *(Image Tech.)*. Re-photography of a scene.

retardation coil *(Elec.Eng.)*. Inductor for separating d.c. from a.c. particularly from a rectifier or supply with ripple.

retardation test *(Elec.Eng.)*. A method of determining the iron, friction, and windage losses of electrical machinery by determining the rate at which it retards under the influence of these losses after being run up to speed and then disconnected from the supply.

retarded field, retarded potential *(Elec.Eng.)*. Those at a point which arise later than at some other point because of finite speed of propagation of waves in the medium.

retarded hemihydrate plaster *(Build.)*. A plaster based on calcium sulphate hemihydrate (**plaster of Paris**) but with the addition of retarder in varying quantities. Particularly useful as an undercoat on plasterboard, fibre board, or metal lath, because of low expansion.

retarder *(Build.)*. A substance which delays or prevents the setting of cement.

retarder *(Chem.)*. A negative catalyst which is added to a reaction system to prevent the reaction from being too violent.

retarder *(Civ.Eng.)*. An arrangement of braking surfaces placed alongside, and parallel with, the running rails in a shunting yard; operated from a signal box by electric, pneumatic, hydraulic or mechanical means. Also *wagon retarder*.

retarding field *(Electronics)*. Electric field such as between a positively charged grid and a lower potential outer grid in a valve, so that electrons entering this region lose energy to the field.

retarding-field oscillator *(Telecomm.)*. One which depends on the electron-transit time of a positive grid oscillator valve. See **Barkhausen-Kurz oscillation**.

rete *(Zool.)*. A netlike structure. pl. *retia*.

rete Malpighii (mucosum) *(Zool.)*. See **Malpighian layer**.

rete mirabile *(Zool.)*. A network of small blood vessels, as in the so-called **red gland** of Fish.

retention *(Med.)*. The abnormal keeping back in the body of matter (e.g. urine) normally evacuated.

retention wall *(Build.)*. A thin wall built alongside an external wall of a building leaving a ½ to 1 in (ca. 12–25 mm) cavity between, which is later filled with waterproofing material to form a vertical damp-proof course.

retentivity *(Phys.)*. Residual magnetic induction after the field producing saturation has been removed from a ferromagnetic material.

reticular *(Med.,Zool.)*. Resembling a net; of or pertaining to the reticuloendothelial system.

reticular tissue *(Zool.)*. A form of connective tissue in which the intercellular matrix is replaced by lymph; it derives its name from the network of collagenous fibres which it shows.

reticulated *(Build.)*. A term applied to a variety of rusticated work distinguished by irregularly shaped sinkings separated by narrow margins of the regular width.

reticulate thickening *(Bot.)*. Secondary wall deposition in the form of an irregular network in tracheids or vessel elements of **metaxylem**.

reticulation *(Print.)*. The mottled appearance occurring when a wet ink is applied to another ink film which has dried to a smooth non-porous surface. This causes the top film to be drawn into minute beads.

reticule *(Surv.)*. A cell carrying cross-hairs and fitting into the diaphragm of a surveying telescope. Also called *graticule*.

reticulitis *(Vet.)*. Inflammation of the reticulum.

reticulocytosis *(Med.)*. An increase in the number of reticulocytes or immature red cells in the blood, seen in *haemolytic anaemias* or after treatment has commenced in other anaemias.

reticuloendothelial system *(Immun.)*. A term formerly used to describe the system of cells that have the ability to take up certain dyes and particles (such as carbon in the form of India ink) when injected into the living animal. It has been replaced by **mononuclear phagocyte system**.

reticul-, reticulo- *(Genrl.)*. Prefix from L. *reticulum*, net.

reticulum *(Zool.)*. In ruminant Mammals, the second division of the stomach, or *honeycomb bag*; any netlike structure. adj. *reticular*.

retiform *(Zool.)*. See **reticular tissue**.

retina *(Zool.)*. The light-sensitive layer of the eye in all animals. Human retina contains 2 types of sensitive element: **rod** and **cone**. adj. *retinal*.

retinal fatigue *(Med.)*. Retention of images after removal of excitation, due to chemical changes in the retina.

retinal illumination *(Phys.)*. The luminous flux received by unit area of the retina. It equals KLS/l^2. where K = a transmission factor of the eye, L = the luminance of a uniformly diffusing surface, S = pupil area, l = distance of retina from the second nodal point.

retinene *(Biol.)*. Vitamin A_1-aldehyde, a component of rhodopsin.

retinite *(Min.)*. A large group of resins, characterized by the absence of succinic acid.

retinitis *(Med.)*. Inflammation of the retina.

retinitis pigmentosa *(Med.)*. A familial and hereditary disease in which chronic and progressive degeneration of the choroid occurs in both eyes, with progressive loss of vision.

retinoblastoma *(Med.)*. A tumour of the retina composed of small round cells, arising from embryonic retinal cells; it is locally destructive and forms metastases.

retinochoroiditis *(Med.)*. Inflammation of the retina and choroid.

retinoscopy *(Med.)*. *Skiascopy; shadow test*. A method of estimating the refractive index of the eye by reflecting light on to it from a mirror and observing the movements of the shadow across the pupil.

retinulae *(Zool.)*. In Arthropoda, the visual cells of the compound eye, forming the base of each ommatidium.

retrace *(Image Tech.)*. See **flyback**.

retrace *(Radar)*. See **flyback**.

retractable landing gear *(Aero.)*. An alighting gear which can be withdrawn completely or nearly so from its operative position to reduce drag.

retractable radiator *(Aero.)*. A liquid cooler for an aero-engine, capable of being withdrawn out of the airstream, for reducing drag and controlling the temperature of the cooling liquid.

retractile *(Zool.)*. Capable of being withdrawn, as the claws of most Felidae.

retraction lock *(Aero.)*. A device preventing inadvertent retraction of the landing gear while an aircraft is on the ground. Also ground safety lock.

retractor *(Zool.)*. A muscle which by its contraction draws a limb or a part of the body towards the body. Cf. **protractor**.

retransfer *(Print.)*. See **transfer**.

retread *(Autos.)*. A reconditioned tyre, on which only the tread has been renewed. Cf. **remould**.

retreating systems *(Min.Ext.)*. Systems in which the removal of ore or coal is commenced from the boundary of the deposit, which is then worked towards the entry through undisturbed rock.

retrices *(Zool.)*. In Birds, the stiff tail feathers used in steering. sing. *retrix*. adj. *retricial*.

retrieval *(Behav.)*. *Memory*. The process of searching for and bringing stored information into consciousness.

retrieval cue *(Behav.)*. Environmental or internal stimuli that help the retrieval of an experience.

retro- *(Genrl.)*. Prefix from L. *retro*, backwards, behind.

retrobulbar neuritis *(Med.)*. Inflammation of that part of the optic nerve behind the eyeball.

retrocaecal *(Med.)*. Also *retrocecal*. Behind the caecum.

retrocerebral glands *(Zool.)*. In Insects, a collective name applied to a number of endocrine glands in the head, behind the brain, which are concerned with the coordination of postembryonic development and metamorphosis. See **corpora cardiaca, corpora allata**.

retroflexed *(Med.)*. Said of the uterus when its body is bent back on the cervix. Cf. **retroversion**. *n. retroflexion*.

retrofocus lens *(Image Tech.)*. See **inverted telephoto lens**.

retrograde amnesia *(Behav.)*. A type of amnesia that often occurs after a head injury, or from electrical shock; there is loss of memory for events leading up to the injury, although the period of time that is lost to memory varies with the conditions of injury.

retrograde metamorphism *(Geol.)*. See **retrogressive metamorphism**.

retrograde motion *(Astron.)*. (1) Motion of a comet (or satellite) whose orbit is inclined more than 90° to the ecliptic (or to the planet's equatorial plane). (2) Apparent motion of a planet from east to west among the stars, caused by a combination of its true motion with that of the earth.

retrograde vernier *(Surv.)*. A vernier in which n divisions on the vernier plate correspond to $(n+1)$ divisions on the main scale.

retrogressive (retrograde) metamorphism *(Geol.)*. A term descriptive of those changes which are involved in the conversion of a rock of high metamorphic grade to one of lower grade, through the advent of metamorphic processes less intense than those which determined the original mineral content and texture of the rock.

retrolental fibroplasia *(Med.)*. Damage to retina in the newborn caused by excessive oxygen concentrations.

retroperitoneal *(Med.)*. Situated or occurring behind the peritoneum.

retropharangeal *(Med.)*. Situated or occurring in the tissues behind the pharynx.

retropulsion *(Med.)*. The running backwards of a patient with paralysis agitans or Parkinsonism; the patient's centre of gravity is displaced backwards, the rigidity of his posture making it difficult for him to recover his balance.

retrorocket *(Space)*. A small rocket motor used for reducing the velocity of a space vehicle in landing, or in any manoeuvre calling for a thrust in the direction opposite to the motion.

retroversion *(Med.)*. The abnormal displacement backwards of the uterus, with or without **retroflexion**.

retrovirus *(Biol.)*. A virus of higher organisms whose genome is RNA, but which can insert a DNA copy of its genome into the host's chromosome. Important because they include the **oncogenic viruses**, and because they can be used as **vectors** for the introduction of DNA sequences into eukaryotic cells.

retting *(Textiles)*. Soaking flax straw in ponds, canals, tanks etc. for bacteria to soften the woody tissue to enable the fibres to be separated by scutching (beating).

return *(Comp.)*. To transfer control (*exit*) from a subprogram back to a calling program. A subprogram may have more than one return instruction.

return *(Radar)*. Refers to radar reflections, e.g. land (or ground) return, sea return.

return airway, aircourse *(Min.Ext.)*. One leading foul air away from the mine workings to the upcast shaft.

return bead *(Build.)*. A double-quirk bead formed on the exterior angle of a timber.

return crank *(Eng.)*. A short crank which replaces an eccentric in the **Walschaert's valve gear** on outside cylinder locomotives.

return feed *(Elec.Eng.)*. See **negative feeder**.

return-flow system *(Aero.)*. A gas-turbine combustion system in which the air is turned through 180° so that it emerges in the opposite direction to that in which it entered; sometimes *reverse-flow system*.

return-flow wind tunnel *(Aero.)*. One in which the air is circulated round a closed loop to preserve its momentum and so reduce the power requirement.

returning charge *(Min.Ext.)*. In custom smelting, that imposed by the smelter per unit of mineral treated. It may be modified by penalties or premiums if the ore or concentrate varies from a specified composition.

return line flyback *(Electronics)*. Faint trace formed on the screen of a cathode-ray tube by the beam during the flyback period. Usually suppressed. Also *return trace*.

return loss *(Telecomm.)*. See under **reflection**.

returns *(Min.Ext.)*. Oil-rig term for the material carried back by the returning drilling mud; provides essential information about conditions down hole.

return trace *(Electronics)*. See **return line flyback**.

return wall *(Civ.Eng.)*. A short length of wall built perpendicularly to one end of a longer wall.

retuse *(Bot.)*. Having a slight notch at a more or less obtuse apex.

Reuben-Mallory cell *(Chem.)*. Small robust primary cell, of a very level discharge characteristic, having a zinc anode and a (red) mercuric oxide cathode, which also depolarizes. Made in minute sizes for hearing aids, internal radio transmitters, watches.

Reuleaux valve diagram *(Eng.)*. See **valve diagram**.

reusability *(Space)*. Refers to space hardware which may be used more than once. A system may be partially- or fully-reusable but, in either case, it implies recovery and refurbishment before reuse. Ant. *expendable*.

revalé *(Build.)*. Said of a cornice, moulding etc., finished when the work is in position.

reveal *(Build.)*. The depth of wall revealed, beyond the frame, in the sides of a door or window opening. Also called *respond*, and (Scottish) *ingo* or *ingoing*.

revehent *(Zool.)*. Carrying back.

reverberation bridge *(Acous.)*. Method of measuring the reverberation time in an enclosure; the rate of decay of the sound intensity is balanced against the adjusted and known decay of the discharge of a capacitance through a resistance.

reverberation chamber *(Acous.)*. Room with a long reverberation time and diffuse sound field. The long reverberation time is achieved by highly reflective walls.

reverberation response *(Acous.)*. That of a microphone for reverberant sound, i.e. for the simultaneous arrival of sound waves of random phase, magnitude and direction. Substantially equal to the mean-spherical response at each frequency of interest.

reverberation response curve *(Acous.)*. That of a microphone to reverberant sound waves. Plotted with the response in decibels as ordinate, on a logarithmic frequency base.

reverberation time *(Acous.)*. The time in seconds, required for the decay of the average sound intensity in a closed room over an amplitude range of 1 million, or 60 db. A space with many *scatterers* also creates reverberation, e.g. the ocean.

reverberatory furnace *(Eng.)*. A furnace in which the charge is melted on a shallow hearth by flame passing above the charge and heating a low roof. Firing may be with coal, pulverized coal, oil or gas. Much of the heating is done by radiation from the roof. Has many applications.

reversal colour film *(Image Tech.)*. Film in which the negative image in the respective colour layers is reversed in processing to give a positive transparency.

reversal of control *(Aero.)*. Reversal of a control moment (or couple) which occurs when displacement of the control surface results in such high forces that distortion of the main structure counteracts the effect of the surface. This overloading is a function of air speed, since control forces increase proportionally to the square of the velocity, and *reversal speed* is the lowest **EAS** at which reversal occurs.

reversal of spectrum lines *(Phys.)*. The appearance of a line as a broad, diffuse bright line with a narrow dark line down the centre. The effect is caused by cool vapour surrounding a hot source such as an electric arc, which produces a narrow absorption line on the short range of continuous spectrum given by the same vapour, at a high temperature, at the centre of the arc. Only certain lines are thus affected.

reversal process *(Image Tech.)*. Method of processing a photographic emulsion to produce a positive image on the original film which has been exposed in the camera, without making a print; the areas initially unexposed form the final picture. A similar process can also be used for making a copy negative from a processed original negative.

reversed fault *(Geol.)*. A type of **fault** in which compression has forced the strata on the side towards which the fracture is inclined to over-ride the strata on the downthrow side. Cf. **normal fault**.

reversed field pinch *(Nuc.Eng.)*. A toroidal magnetic trap in which the toroidal field changes sign in the outer region of a plasma discharge. Abbrev. *RFP*.

reverse genetics *(Biol.)*. The process of removing a gene from an organism, altering it in a known way and reinserting it. The organism is then tested for any altered function.

reverse osmosis *(Chem.)*. Purification of water by forcing it under pressure through a membrane not permeable to the impurities to be removed. Also *ultrafiltration*.

reverse Polish notation *(Comp.)*. Form of postfix notation (i.e. the operator follows the operands). This allows a **stack** to be used for evaluation. Cf. **infix**. Abbrev. *RPN*.

reverse transcriptase *(Biol.)*. An enzyme which makes a double-stranded DNA copy of an RNA virus genome, much used in the laboratory to make DNA copies of mRNA.

reversible *(Phys.)*. Said of a process whose effects can be reversed so as to bring a system to its original thermodynamic state.

reversible cell *(Elec.Eng.)*. See **accumulator**.

reversible coatings *(Build.)*. Any coating which, when dry, resoftens on the application of a second coat or its own thinner or solvent.

reversible colloid *(Chem.)*. See **lyophilic colloid**.

reversible potentiometer-type field rheostat *(Elec.Eng.)*. A potentiometer-type rheostat used for controlling the field current of an electric machine and arranged so that the field current may be reversed.

reversible reaction *(Chem.)*. A chemical reaction which can occur in both directions, and which is therefore incomplete, a mixture of reactants and reaction products being obtained, unless the equilibrium is disturbed by removing one of the products as rapidly as it is formed. Examples of reversible reactions are the formation of an ester and water from an alcohol and an acid, the dissociation of vapours, e.g. ammonium chloride, and the ionic dissociation of electrolytes. Also *equilibrium reaction*.

reversible saturation-adiabatic process *(Meteor.)*. An idealized, alternating condensation-evaporation process occurring in the atmosphere, assuming that none of the condensation products are removed by precipitation.

reversible transducer *(Telecomm.)*. One for which the loss is independent of the direction of transmission.

reversible unit *(Print.)*. A printing unit on a web-fed press which can print in either direction of rotation, having reversible drives.

reversing commutator *(Elec.Eng.)*. Any form of reversing switch, but more particularly the type in which brushes connected to the leads bear on conducting segments let into a non-conducting cylinder, the rotation of which will open or close the circuits.

reversing face *(Surv.)*. The process of transiting a theodolite telescope, thereby changing its position from face left to face right, or vice versa.

reversing field *(Elec.Eng.)*. In a commutator machine, a field of opposite polarity to that in which an armature coil had previously been moving; designed to produce a reversed e.m.f. to assist commutation. The field may be produced by a compole, or by shifting the brushes from the neutral axis.

reversing gear *(Eng.)*. (1) In machine tools, used to reverse the rotation of one spindle relative to another. On a lathe it is needed to change rotation of leadscrew relative to spindle, in order to change from left-handed to right-handed screw cutting. (2) Steam engine; see **Joy's valve-gear, link motion, Walschaert's valve gear**.

reversing layer *(Astron.)*. The name given to the lower part of the Sun's chromosphere where the absorption lines of the solar system are formed by 'reversal' from bright emission lines to dark absorption lines.

reversing mill *(Eng.)*. A type of rolling-mill in which the stock being rolled passes backwards and forwards between the same pair of rolls, which are reversed between passes. See **continuous mill, pull-over mill, three-high mill**.

reversing switch *(Elec.Eng.)*. A switch used for reversing the connections in an electrical circuit.

reversion *(Biol.)*. The process by which a *mutant* phenotype is restored to normal by another mutation of the same gene, i.e. a *back-mutation*. But sometimes used in the sense of *suppression*.

revertive control system *(Telecomm.)*. A register-controlled system of automatic switching in which a selector, when set in motion by a positioning signal from a register, transmits progress signals back to the register to enable it to determine when the selector has reached the desired position.

revetment *(Civ.Eng.)*. A retaining wall.

review room *(Image Tech.)*. A small cinematograph theatre at a studio, production centre or processing laboratory for the presentation of a film for detailed examination.

revise *(Print.)*. A second or third proof supplied in order that corrections made on the preceding proof may be checked over; to prepare and submit such a proof.

reviver *(Print.)*. A type alloy rich in tin and antimony added to the remelting pot in a prescribed quantity to maintain the formulation of the type metal by making good the loss incurred during each use.

revolute *(Bot.)*. Margin or apex of a leaf, rolled outwards or downwards (i.e. towards abaxial surface). Cf. **involute**. See **vernation**.

revolution *(Astron.)*. The term generally reserved for orbital motion, as of the earth about the sun, as distinct from **rotation** about an axis.

revolving centre *(Eng.)*. A **centre** which is mounted on rolling bearings and revolves with the workpiece so as to obviate relative motion and thus friction between centre and workpiece.

rewrite *(Comp.)*. To return data to store when it has been erased during reading.

rexigenous, rhexigenous *(Bot.)*. A space in a tissue formed by the rupture of cells. Cf. **lysigenous, schizogenous**.

Reynolds number *(Chem.Eng.)*. The dimensionless group

$$\frac{\text{density} \times \text{velocity} \times \text{a linear measure}}{\text{viscosity}}$$

all values being in the same system of units, e.g. FPS or SI. If used in pipe or tube flow linear dimension is internal diameter and this has special application in heat exchange calculations. In other applications, e.g. mixers in vessels, the linear dimension may be, for example, the diameter of a moving part.

R_F *(Chem.)*. Term used in chromatography. The ratio of the distance moved by a particular solute to that moved by the solvent front.

RF *(Telecomm.)*. Abbrev. for *Radio Frequency*.

RFLP *(Biol.)*. See **restriction fragment length polymorphism**.

RFP *(Nuc.Eng.)*. Abbrev. for *Reversed Field Pinch*.

RFS *(Build.)*. Abbrev, for *Render, Float, and Set*.

RGB signals *(Image Tech.)*. The three signals which correspond directly to the colour primaries, red, green and blue, as distinct from the **colour difference signals** which are used for transmission.

Rh *(Chem.)*. The symbol for rhodium.

Rh *(Med.)*. See **Rhesus factor**.

rhabdom *(Zool.)*. In the compound eyes of Arthropoda, the structure containing the visual pigment and concerned with phototransduction.

rhabdomeres *(Zool.)*. One of the constituent portions of the rhabdom, secreted by a single visual cell.

rhabdomyoma *(Med.)*. A tumour composed of voluntary or striped muscle cells.

rhabdomyosarcoma *(Med.)*. A malignant rhabdomyoma.

rhachi-, rhachio- *(Genrl.)*. See **rachi-**.

rhachis *(Bot.,Zool.)*. See **rachis**.

Rhaetic *(Geol.)*. The uppermost stage of the Triassic system.

rhamnose *(Chem.)*. *6-deoxy mannose*, $CH_3.(CHOH)_4.CHO$, a methylpentose obtained from several glucosides; crystallizes with 1 H_2O, mp 93°C. On distillation with sulphuric acid it yields 5-methyl-furfural.

rhamphotheca *(Zool.)*. In Birds, the horny coverings ensheathing the upper and lower jaws.

rhaphe *(Bot.)*. See **raphe**.

Rhea *(Astron.)*. The fifth natural satellite of **Saturn**, and the second largest moon in its system, with a diameter of 1530 km.

Rheiformes *(Zool.)*. An order of Birds, containing two species of large running bird found on the South American pampas, occupying an ecological niche approximately similar to that of the Emu. Have 3 toes. Rheas.

rhenic (VII) acid *(Chem.)*. See **rhenium oxides**.

Rhenish bricks *(Build.)*. Very light bricks made of calcareous material bound together with dolomitic lime. Also called *floating bricks*.

rhenium *(Chem.)*. A metallic element in the subgroup manganese, technetium, rhenium. Symbol Re, at.no. 75, r.a.m. 186.2, rel.d. 21, mp 3000°C. Valencies, 2,3,4,6,7. A very rare element, occurring in molybdenum ores. A small percentage increases the electrical resistance of tungsten. Used in high-temperature thermocouples.

rhenium oxides *(Chem.)*. Re_2O_7, ReO_3, ReO_2, and Re_2O_3. The volatile (VII) oxide Re_2O_7 is formed when the metal or its compounds are heated in air. Rhenium (VI) oxide is the anhydride of rhenic acid H_2ReO_4. The (VII) oxide dissolves in water to form perrhenic acid, which forms metallic perrhenates (rhenates (VII)). Cf. **permanganates**.

rheo- *(Genrl.)*. Prefix from Gk., meaning current, flow.

rheobase *(Med.)*. The minimal electrical stimulus (volts) that will produce a physiological response and below which no stimulus will excite however long it is maintained; it is half the strength of that for **chronaxie**.

rheolaveur *(Min.Ext.)*. Coal-cleaning plant where raw coal is sluiced through a series of troughs, the high-ash fraction gravitating down to a separate discharge.

rheology *(Phys.)*. The science of the flow of matter. The critical study of elasticity, viscosity, and plasticity.

rheomorphism *(Geol.)*. Flowage of rocks resulting from severe deformation, especially applied to those undergoing partial melting as a result of heating to a high temperature.

rheoreceptors *(Zool.)*. Receptors of Fish and certain Amphibians which respond to stimulus of water current, e.g. lateral line system.

rheostat *(Elec.Eng.)*. Electric component in which resistance introduced into a circuit is readily variable by a knob or handle, or by mechanical means such as an electric motor.

rhesus blood group system *(Immun.)*. Human blood group system, so called because the antigen involved is also present on rhesus monkey red cells and was first detected when these were used to immunize rabbits. The rhesus blood group system is genetically complex and there are several alleles. The most important is that known as the D-antigen. Antibodies against rhesus antigens do not occur naturally in the blood but may be produced after transfusion into a rhesus(D)-negative person of rhesus(D)-positive blood or in a rhesus(D)-negative mother who bears a rhesus(D)-positive child. In the former case a subsequent transfusion of positive blood may cause a **transfusion reaction**, and in the latter give rise to **erythroblastosis fetalis** in the child.

rhesus factor *(Med.)*. Blood group **antigens** possessed by 85% of the population (Rhesus-positive). Of importance in blood transfusion during pregnancy. A rhesus positive baby born to a rhesus negative mother with rhesus antibodies may develop **haemolytic disease of the newborn**.

rhesus monkey *(Zool.)*. One of the species, *Macaca mulatta*, of the macaque monkeys, native to S.E. Asia. Robust and intelligent, they have been widely used in medical research. See e.g. **Rhesus factor**.

rheumatic fever *(Immun.)*. An acute inflammatory disease involving the heart and the joints which generally follows a few weeks after an infection by *Streptococcus pyogenes* of Lancefield Group A. The characteristic lesions are degeneration and necrosis of fibrous tissue and nodules of necrotic fibrous tissue surrounded by macrophages, lymphocytes and plasma cells. These are probably some form of hypersensitivity reaction to Group A streptococci. Antibodies are present against an antigen of the streptococci which shares antigenic determinants with the sarcolemma of heart muscle.

rheumatism *(Med.)*. A general term for a wide range of diseases characterized by painful inflammation and degeneration particularly of joints and muscles.

rheumatoid arthritis *(Immun.,Med.)*. Chronic inflammatory polyarthritis, which may be accompanied by systemic disturbances such as fever, anaemia and enlargement of lymph nodes. The synovia of joints are infiltrated with granulomata containing plasma cells, lymphocytes, macrophages and germinal centres, causing inflammation and swelling of particularly the small joints of the extremities. B-lymphocytes are present in the inflamed tissue and in lymphoid tissues elsewhere which make **rheumatoid factor**. This is present in the blood as well as locally. It can combine with other immunoglobulins to activate complement, and this is thought to be the immediate cause of the inflammation. In this sense rheumatoid arthritis is an autoimmune disease, but the initiating cause or causes are not known.

rheumatoid factor *(Immun.)*. Antibody reactive with determinants present on the heavy chain constant region of immunoglobulins of many species, including that in which the antibody is made. Causes the inflammation characteristic of rheumatoid arthritis. The determinants are revealed when the immunuoglobulin molecules are slightly distorted, as by combination with an antigen or by mild denaturation and this property is used in various tests to detect the factors. Rheumatoid factors are usually polyclonal, and may be of the IgM, IgG or IgA class, although IgM is much the commoner. In some subjects monoclonal rheumatoid factors are made, and if the amount is sufficient these will combine with other immunoglobulins to form **cryoglobulins**.

rhexis *(Med.)*. Rupture of a bodily structure, especially of a blood vessel.

rhinal *(Zool.)*. Pertaining to the nose.

rhinarium *(Zool.)*. In Mammals, the moist skin around the nostrils, also known as the muzzle, which is lacking in Anthropoids.

rhinencephalon *(Zool.)*. The olfactory lobes of the brain in Vertebrates.

rhinitis *(Med.)*. Inflammation of nasal mucous membrane. For *allergic rhinitis* see **hay fever**.

rhinocoele *(Zool.)*. The cavity of the rhinencephalon; olfactory ventricle of the Craniate brain.

rhinopharyngitis *(Med.)*. Inflammation of the nose and the pharynx.

rhinophyma *(Med.)*. Overgrowth of the subcutaneous tissue and the skin of the nose as a result of enlargement of the sebaceous glands which may develop in *acne rosacea*.

rhinoplasty *(Med.)*. The repair of a deformed, diseased, or wounded nose by plastic surgery.

rhinorrhoea, rhinorrhea *(Med.)*. Discharge of mucus from the nose.

rhinoscope *(Med.)*. A speculum for viewing the interior of the nose.

rhinotomy *(Med.)*. Incision into the nose.

rhin-, rhino- *(Genrl.)*. Prefix from Gk. *rhis*, gen. *rhinos*, nose.

rhizo- *(Bot.)*. Prefix meaning root or root-like.

Rhizobaceae *(Biol.)*. A family of bacteria belonging to the order Eubacteriales (Bergey classification). Includes the symbiont *Rhizobium* which is important in nitrogen fixation by leguminous plants.

rhizodermis *(Bot.)*. The outermost layer of cells of a root in its primary state.

rhizoid *(Bot.)*. Outgrowth from an alga, fungus, bryophyte or pteridophyte gametophyte, attaching to or growing into the substrate and serving in anchorage and, possibly, absorption.

rhizome *(Bot.)*. Stem, usually underground, often horizontal, typically non-green and root-like in appearance but bearing scale leaves and/or foliage leaves, e.g. nettle, many *Iris* spp. Cf. **stolon**.

rhizomorph *(Bot.)*. Strand, like a length of thin string, composed of densely packed hyphae, by means of which some fungi spread, e.g. boot-lace or honey fungus, *Armillaria mellea*, dry-rot fungus *Serpula (Merulius) lacrymans*.

rhizophagous *(Zool.)*. Root-eating.

Rhizopoda *(Zool.)*. See **Sarcodina**.

rhizopodium *(Bot.)*. Long, very fine, sometimes branched, cytoplasmic process from an algal cell, especially in the Chrysophyceae.

rhizosphere *(Bot.)*. Zone of soil, in the immediate vicinity of an active root, influenced by the uptake and output of substances by the root and characterized by a microbial flora different from the bulk soil.

rhodamine *(Biol.)*. A **fluorochrome** commonly conjugated with antibodies for use in **indirect immunofluorescence**.

rhodamines *(Chem.)*. Dyestuffs of the triphenylmethane group, closely related to fluoroscein. They are obtained by the condensation of phthalic anhydride with *N*-alkylated *m*-aminophenols in the presence of sulphuric acid.

rhodanizing *(Eng.)*. The process of electroplating with rhodium, especially on silver, to prevent tarnishing.

rhodeose *(Chem.)*. $C_6H_{12}O_5$, a methylpentose sugar, an isomer of rhamnose.

rhodium *(Chem.)*. A metallic element of the platinum group. Symbol Rh, at.no. 45, r.a.m. 102.9055, rel.d. at 20°C 12.1, mp ca 2000°C; electrical resistivity approx. 5.1×10^{-8} ohm metres. A noble silvery white metal, it resembles platinum and is alloyed with the latter to form positive wire of the platinum-rhodium-platinum thermocouple. Used for plating silver and silverplate to prevent tarnishing, in catalysts and in alloys for high temperature thermocouples.

rhodochrosite *(Min.)*. Manganese carbonate which crystallizes in the trigonal system, occurring as rose-pink rhombohedral crystals. It is a minor ore of manganese. Also called *manganese spar* or *dialogite*.

rhodonite *(Min.)*. Manganese silicate, generally with some iron and calcium, crystallizes in the triclinic system. It is rose-coloured, and is sometimes used as an ornamental stone.

rhodophane *(Zool.)*. A coloured oily substance, globules of which are found in the cones of Birds and in parts of the retina in some other forms.

Rhodophyceae *(Bot.)*. Red algae, Eukaryotic, chloroplasts with chlorophyll a and phycobilins, thylakoids single. No flagellate stages. Reserve carbohydrate floridian starch (α, $1 \rightarrow 4$ glucan) in the cytoplasm. Life cycle often complex, e.g. triphasic alternation of generations, sexual reproduction oogamous. Unicellular, filamentous or parenchymatous. Mostly marine, littoral and sublittoral (the red seaweeds); some in fresh water or soil; some parasitic on other red algae. Source of agar and carragheen; a few are eaten e.g. *Porphyra*, laver-bread.

rhodopsin *(Biol.)*. The light sensitive protein present in the eye. Its light sensitivity is due to the prosthetis group of 11-cis-retinol.

rhod-, rhodo- *(Genrl.)*. Prefix from Gk. *rhodon*, rose.

rhombencephalon *(Zool.)*. See **hind-brain**.

rhombic antenna *(Telecomm.)*. Directional short-wave antenna comprising an equilateral parallelogram of conductors, each several quarter wavelengths long, usually arranged in a horizontal plane.

rhombic dodecahedron *(Crystal.)*. A crystal form of the cubic system, consisting of 12 exactly similar faces, each of which is a regular rhombus. Does not occur in the orthorhombic system, despite its name.

rhombic system *(Crystal.)*. See **orthorhombic system**.

rhombohedral class *(Crystal.)*. A class of the trigonal system, a characteristic form being the **rhombohedron**, which is exhibited by crystals of quartz, calcite, dolomite, etc.

rhombohedron *(Crystal.)*. A crystal form of the trigonal system, bounded by 6 similar faces, each a rhombus or parallelogram.

rhomb-porphyry *(Geol.)*. A medium-grained rock of intermediate composition, usually occurring in dykes and other minor intrusions; characterized by numerous phenocrysts of anorthoclase which are rhomb-shaped in cross-section, set in a finer-grained groundmass. Related to laurvikite among the coarse-grained, and to kenyte among the fine-grained rocks.

rhomb-spar *(Min.)*. An old-fashioned synonym for *dolomite*.

rhombus *(Maths.)*. A quadrilateral with all its sides equal. Loosely, diamond shape.

Rhometal *(Eng.)*. An alloy of permalloy type. Contains 64% iron and 36% nickel. Used in high-frequency electrical circuits.

rhometer *(Eng.)*. One for the measurement of impurity content of molten metals by means of the variation in electrical conductivity.

rhonchus *(Med.)*. A harsh, prolonged sound, heard on auseulation, produced by air passing over narrowings in the bronchial tubes; found in *asthma* and *bronchitis*.

R_H, r_H, rH *(Chem.)*. See **rH value**.

rhumbatron *(Electronics)*. Type of cavity resonator used in e.g. a klystron. It acts as a tuned circuit comprising a parallel-disk capacitor surrounded by a single-turn toroidal inductance, and is used to velocity-modulate an electron beam passing through holes in the capacitor disks. See **buncher**.

rH value *(Chem.)*. Logarithm, to the base 10, of the reciprocal of hydrogen pressure which would produce same electrode potential as that of a given oxidation-reduction system, in a solution of same **pH value**. The greater the oxidizing power of a system, the greater the rH value.

rhyncho- *(Genrl.)*. Prefix from Gk. *rhynchos*, beak, snout, proboscis.

Rhynchocephalia *(Zool.)*. An order of the Lepidosauria with two temporal vacuities, a large parietal foramen which, in the one living form, *Sphenodon*, contains a non-functional median eye. Known as fossils from Middle Trias, *Sphenodon* survives in coastal islands off New Zealand.

rhynchodont *(Zool.)*. Having a toothed beak.

rhynchophorous *(Zool.)*. Having a beak.

Rhynie Chert *(Geol.)*. A silicified peaty bed containing well-preserved plant remains as well as spiders, scorpions, and insects; discovered at Rhynie, Aberdeenshire, in the Middle Old Red Sandstone (i.e. Devonian Age).

rhyolite *(Geol.)*. General name for fine-grained igneous rocks having a similar chemical composition to granite, commonly occurring as lava flows, although occasionally as minor intrusions, and generally containing small phenocrysts of quartz and alkali-feldspar set in a glassy or cryptocrystalline groundmass. Sometimes called *liparite*. See also **obsidian, pitchstone, pumice, pyromeride** (nodular rhyolite).

rhythmic crystallization *(Geol.)*. A phenomenon exhibited by rocks of widely different composition but characterized by development of orbicular structure.

rhythmic sedimentation *(Geol.)*. A more-or-less consistently repeated sequence of two or more rock units which can be recognized as forming a pattern. e.g. **cyclothem, Bouma cycle**, glacial **varves**.

rhytidome *(Bot.)*. Dead, outer bark, consisting of layers of periderm with some cortex and/or secondary phloem.

ria *(Geol.)*. A normal valley drowned by a rise of sea-level relative to the land. Cf. **fiords**, in the production of which glacial action plays an essential part. A good example of ria type of coastline is S.W. Ireland.

rib *(Aero.)*. A fore-and-aft structural member of an *aerofoil* which has the primary purpose of maintaining the correct contour of the covering, but is usually a stress-bearing component of the main structure. Ribs are usually set either parallel with the longitidinal axis or at right angles to the front spar; cf. **nose ribs**.

rib *(Build.,Civ.Eng.)*. (1) A curved member of a centre or ribbed arch. (2) A moulding projecting for purposes of ornamentation from a ceiling or vault surface. (3) The vertical portion of a T-beam.

rib *(Textiles)*. A prominent line running along or across woven or knitted fabrics, and forming a cord effect.

rib *(Zool.)*. A small ridge or rib-like structure; in Vertebrates, an element of the skeleton in the form of a curved rod connected at one end with a vertebra; it serves to support the body walls enclosing the viscera.

riband *(Build.)*. A flat rail fixed across posts in a palisade.

rib and panel *(Build.)*. A term applied to a vault formed of separate ribs and panels, the latter being supported on the former.

ribbed arch *(Civ.Eng.)*. An arch composed of many side-by-side ribs spanning the distance between the springings.

ribbed flutings *(Build.)*. Flutings separated by a flat or slightly convex listel.

ribbon *(Image Tech.)*. Loop of fine metal filament in a **light valve**, the opening in which is varied by the modulating currents.

ribbon *(Textiles)*. A decorative closely-woven narrow fabric nearly always made from lustrous continuous filament yarns.

ribbon microphone *(Acous.)*. Special type of electro-dynamic microphone for measuring sound velocities. A very thin aluminium ribbon in a magnetic field acts as a diaphragm. Incident sound waves make the ribbon vibrate and an electrical voltage is induced proportional to the wave velocity. See also **pressure gradient microphone**.

ribbon microphone *(Image Tech.)*. Microphone in which the signal is produced by the movement of a thin metal strip in a strong magnetic field.

ribbon parachute *(Aero.)*. A parachute in which the canopy is made from light webbing with spaces between, instead of conjoined fabric gores, so as to give greater strength against ripping for deployment at high speed. Commonly used for **brake parachutes**.

ribbon strip *(Build.)*. A horizontal timber attached to vertical timbers as a support for joists. Also *girt strip, ledger board.*

rib mesh *(Build.)*. See **expanded metal**.

riboflavin *(Biol.)*. See **vitamin B**.

ribose *(Chem.)*. $C_5H_{10}O_5$, a pentose, a stereoisomer of arabinose. D-Ribose occurs in ribose nucleic acids (RNA). ⇨

ribosome *(Biol.)*. Consisting of three subunits of RNA and protein, this complex bead-like structure can associate with mRNA and is the site of synthesis in the cytoplasm of polypeptides encoded by the mRNA.

ribulose *(Bot.)*. A 5-carbon ketose sugar. See also **ribulose bisphosphate**.

ribulose 1,5-bisphosphate carboxylase oxygenase *(Bot.)*. The enzyme that catalyses both the reaction of ribulose bisphosphate (RUBP, 5-carbons) and CO_2 to give two molecules of phosphoglyceric acid (PGA, 3-carbons) as the first step of the **Calvin cycle** in photosynthesis, and also the apparently wasteful reaction of RUBP with oxygen to give one molecule each of PGA and phosphoglycollic acid (2-carbons). Oxygen and CO_2 compete, and the oxygenation becomes more important at higher temperatures. See **photorespiration**. Also *RUBP carboxylase, RUBISCO, carboxydismutase.* It is the most abundant protein on earth.

ribulose bisphosphate *(Bot.)*. The 1-5,bisphosphate ester of the sugar ribulose. Substrate for **ribulose bisphosphate carboxylase oxygenase**. Also *RuBP, ribulose diphosphate, RuDP.*

Riccati equation *(Maths.)*. A differential equation of the form

$$\frac{dy}{dx} = py^2 + qy + r,$$

where p,q,r are functions of x alone.

rice paper *(Paper)*. The finely cut pith of *Fatsia papyrifera*; not a true paper.

Richardson-Dushman equation *(Electronics)*. Original Richardson formula, as modified by Dushman, for the emission of electrons from a heated surface, current density being $I = AT^2e^{-}\phi/\kappa T$. T is the absolute temperature, A is a material constant, k is Boltzmann's constant, with ϕ the work function of the surface and e the base of natural logarithms.

Richardson effect *(Electronics)*. Same as **Edison effect**.

Richardson number *(Meteor.)*. A non-dimensional number Ri arising in the study of shearing flow in the atmosphere. If g is the acceleration of gravity, β is a measure of vertical stability (commonly $\frac{1}{\theta}\frac{\partial\theta}{\partial z}$ where θ is the *potential temperature*) and $\frac{\partial u}{\partial z}$ is a characteristic vertical wind shear, then

$$Ri = g\beta \bigg/ \left(\frac{\partial u}{\partial z}\right)^2.$$

Turbulence is likely to be suppressed if $Ri >$ approx. 1.

rich lime *(Build.)*. **Fat lime**.

rich mixture *(Autos.)*. A combustible mixture in which the fuel is in excess of that physically correct for the air.

richness *(Ecol.)*. The number of species in a defined area.

richterite *(Min.)*. Hydrated metasilicate of sodium, calcium and magnesium, occurring as monoclinic crystals in alkaline igneous rocks and in thermally metamorphosed limestones and skarns. A member of the amphibole group.

ricinoleic acid *(Chem.)*. An oily liquid, $CH_3(CH_2)_5$.-CHOH.CH_2.CH=CH$(CH_2)_7$.COOH, which, in glyceride form, is the chief constituent of castor oil.

rickets *(Med.)*. A nutritional disease of childhood characterized by defective ossification and softening of bones: due to deficiency of vitamin D and failure to absorb and utilize calcium salts. Also *rachitis.*

rickettsiae *(Med.)*. Micro-organisms, intermediate between bacteria and viruses, found in the tissues of lice, ticks, mites and fleas and which can cause disease, e.g. typhus, when transmitted to man and other animals.

rictus *(Zool.)*. Of Birds, the mouth aperture adj. *rictal.*

riddle *(Min.Ext.)*. Strong coarse sieve used to size gravel, furnace clinker, etc. Large pieces removed by hand in riddling are called *knockings*, remaining on-screen material *middlings* and through passing particles *fells, undersize*, or *smalls.*

Rideal-Walker test *(Chem.)*. Test for germicidal power of a disinfectant, carbolic acid (phenol) being taken as the standard. A series of dilutions of disinfectant is tested with a typhoid broth culture, samples being taken at short intervals and subjected to incubation.

rider *(Chem.)*. A small piece of platinum wire used on a chemical balance as a final adjustment.

rider *(Min.Ext.)*. (1) A *horse*, i.e. mass of country rock occurring in a mineral deposit. (2) A thin seam of coal above a thick one. (3) A guide for a *bowk*, in sinking.

rider rollers *(Print.)*. Steel rollers in the inking system of a printing machine which secure the inkers (and distributors), act as a necessary link between their tacky surfaces, and contribute to efficient distribution and inking.

rider shore *(Build.)*. An inclined baulk of timber used in a system of raking shores for a high building. It abuts at its lower end against a length of timber laid along the back of the outer raking shore, instead of against the sole plate on the ground.

ridge *(Build.,Civ.Eng.)*. The summit-line of a roof; the line on which the rafters meet.

ridge *(Meteor.)*. An outward v-shaped extension of the isobars from a centre of high pressure.

ridge-board *(Build.)*. A horizontal timber at the upper ends of the common rafters, which are nailed to it.

ridge capping, covering *(Build.)*. The covering applied over a ridge to protect the intersection of the sloping roof surfaces.

ridge course *(Build.)*. The last (i.e. the top) course of slates or tiles on a roof, cut to length as required.

ridge pole, ridge-piece *(Build.,Civ.Eng.)*. A timber member laid horizontally along the ridge of a roof.

ridge roll *(Build.)*. A ridge of rounded section, over which a zinc or lead flashing is formed and secured as a covering for the top ends of the ridge courses to seal the top of the roof.

ridge roof *(Build.)*. A pitched roof whose sloping surfaces meet to form an apex or ridge.

ridge stop *(Build.)*. A piece of sheet-lead shaped over the junction between a roof ridge and a wall; used in cases where the one runs into the other and a watertight joint has to be made.

ridge tile *(Build.)*. A purpose-made tile specially shaped for use as a covering over the ridge of a roof.

ridging *(Build.)*. The operation of covering the ridge of a roof with specially shaped ridge tiles or other material.

riding lamps *(Aero.)*. Lamps displayed at night by a float plane or flying boat when moored or at anchor. Colours and positions as in the Maritime code.

riding shore *(Build.)*. See **rider shore**.

riebeckite *(Min.)*. A dark blue hydrated silicate of sodium and iron found in alkaline igneous rocks as monoclinic prismatic crystals. A member of the amphibole group: the blue asbestos *crocidolite* is a variety occurring in metamorphosed ironstones.

Riedel's disease *(Med.)*. *Chronic thyroiditis.* A rare chronic inflammation of the thyroid gland, which becomes enlarged and hard as the result of excessive formation of dense fibrous tissue.

Riedel's lobe *(Med.)*. An anomalous downward prolongation of the right lobe of the liver.

Rieke diagram *(Telecomm.)*. Polar form of load impedance diagram representing the components of the complex reflection coefficient of the oscillator load in a microwave oscillator.

Riemann surface *(Maths.)*. An extension of the complex plane whereby a multivalued function can be regarded as a one-valued function. The surface has to be designed for the particular function concerned, e.g., for the two-valued function $w = \pm\sqrt{z}$, the complex plane is extended by superimposing a similar plane above it, cutting both planes from the origin along the real axis to infinity, and joining the opposite edges of the two cuts crosswise. The result is that regaining a starting point by tracing a circle around the origin now takes two revolutions (argument increase of 4π) because both the original and the additional planes have to be covered. Thus effectively the complex plane has been doubled and the two-valued function $w = \pm\sqrt{z}$ can be laid out as a one-valued function of position. That portion of a multivalued function which lies on a single sheet of a Riemann surface is called a *branch of a function*, and the connecting point of the several sheets of the surface a *branch point*.

Riemann zeta function *(Maths.)*. The function $\zeta(z)$ of a complex variable defined by

$$\zeta(z) = \sum_{n=1}^{\infty} \frac{1}{n^z}.$$

rifampicin *(Med.)*. Essential anti-tuberculosis drug usually given with streptomycin and isoniazid.

riffler *(Eng.)*. A file bent so as to be capable of operating in a shallow depression.

riffler *(Min.Ext.)*. A device for dividing a stream of crushed material, e.g. coal, into truly representative samples.

riffler *(Paper)*. See **sand trap**.

rift and grain *(Geol.)*. The two directions, approximately at right angles to one another, along which granite and other massive igneous rocks can be split; rift being the easier of the two.

rift valley *(Geol.)*. See **graben**.

Rift Valley fever *(Med.)*. An infectious disease of cattle and sheep in Africa, characterized by high fever and hepatitis, and caused by a virus; probably transmitted by mosquitoes. Man is also susceptible.

rig *(Min.Ext.)*. A well-boring plant, e.g. for oil.

riga last *(For.)*. A unit of timber measure containing 80 ft^3 of squared or 65 ft^3 round timber.

rigging *(Aero.)*. (1) The operation of adjusting and aligning the various components, notably flight and engine control, of an aircraft. (2) In airships and balloons, the system of wires by which the weight to be lifted is distributed over the envelope or gas-bag.

rigging angle of incidence *(Aero.)*. See **angle of incidence**.

rigging diagram *(Aero.)*. The drawing giving the manufacturer's instructions as to the positioning and aligning of the components and control systems of an aircraft.

rigging line *(Aero.)*. See **shroud line**.

rigging position *(Aero.)*. The position in which an aircraft is set up in order to effect the adjustment and alignment of the various parts, i.e. with the lateral axis and an arbitrarily chosen longitudinal datum line horizontal.

right angle *(Maths.)*. One quarter of a complete rotation. 90° or $\pi/2$ radius.

right-angled folding *(Print.)*. Each fold is at right angles to the preceding one. Cf. **parallel folding**.

right ascension *(Astron.)*. One of the two coordinates, used with **declination** for specifying position on the **celestial sphere** in the *equatorial coordinate system*. It is the angular distance measured eastwards along the **celestial equator** from the vernal **equinox** to the intersection of the hour circle passing through the body. (It is the celestial equivalent of longitude.) Its units are hours, minutes and seconds, and one hour of right ascension is 15°; the Earth's daily rotation takes the celestial sphere through 1 hour of right ascension in 1 hour of **sidereal time**.

right circular cone *(Maths.)*. See **cone** (2).

right circular cylinder *(Maths.)*. See **cylinder** (2).

right-handed engine *(Aero.)*. An aero-engine in which the propellor shaft rotates clockwise with the engine between the observer and the propellor.

right helicoid *(Maths.)*. The surface generated by a straight line moving so that it always intersects a helix and cuts its axis perpendicularly. A right circular helicoid is like a spiral staircase.

righting reflex *(Behav.)*. A reflexive response to falling which ensures that the animal lands upright.

right-reading, wrong-reading *(Print.)*. When making plates the various printing processes have their particular requirements from the process camera or filmsetting system, e.g. right-reading negative for direct letterpress, wrong-reading positive for deep-etch offset; the direction is read from the emulsion side.

rigid arch *(Civ.Eng.)*. A continuous arch without hinges or joints, the arch being rigidly fixed at the abutments.

rigid expanded polyurethane *(Plastics)*. Material used in thermal insulation and in providing light structural reinforcement. Reaction products of di-isocyanates with polyesters or polyethers in the presence of water.

rigidity *(Phys.)*. See **elasticity of shear**.

rigid PVC *(Plastics)*. PVC in its unplasticized form widely used for applications such as chemical and building pipework.

rigid support *(Elec.Eng.)*. A support for an overhead transmission line designed to withstand, without appreciable bending, a longitudinal load as well as transverse and vertical loads.

rigor *(Med.)*. A sudden chill of the body, accompanied by a fit of shivering, which heralds the onset of fever.

rigor *(Zool.)*. A state of rigidity and irresponsiveness into which some animals pass on being subjected to a sudden shock as a defensive mechanism; 'shamming dead'.

rigor mortis *(Med.)*. Stiffening of the body following upon death.

Rijke tube *(Acous.)*. An open-ended vertical tube with a gauze stretched across inside the lower half of the tube. When the gauze is heated, a loud sound of the tube's resonance frequency is produced.

rille *(Astron.)*. A winding valley, with a U-shaped cross-section, found in the lunar **maria**.

rill stoping *(Min.Ext.)*. Overhand or upward stoping, in which ore is detached from above the miner so as to form an inverted stoped pyramid spreading from a winze at its apex, through which broken ore is withdrawn.

rima *(Zool.)*. A narrow cleft. adjs. *rimate*, *rimose*, *rimiform*.

rima glottidis *(Med.)*. The gap in the larynx between vocal cords in front and arytenoid cartilages of the larynx behind.

rim lock *(Build.)*. A form of lock distinct from **mortise lock** in that, in its metal case, it is screwed to the face of the door.

rimming, rimmed steel *(Eng.)*. Steel that has not been completely deoxidized before casting. Gas is evolved during solidification and bubbles are entrapped. Ingots contain blowholes but no pipe. See also **killed steel**.

Rinco process *(Print.)*. Reproduction proofs taken with white ink on black paper and photographed to yield positives suitable for printing down, particularly for photogravure.

rinderpest *(Vet.)*. *Cattle plague*. An acute, highly-contagious and often fatal myxovirus infection of cattle, sheep, and goats; characterized by fever and ulceration of the mucous membranes, especially of the alimentary tract, causing severe diarrhoea and discharges from the mouth, nose, and eyes. Notifiable in UK, and vaccines used elsewhere.

ring *(Bot.)*. See **annulus**.

ring *(Comp.)*. See **network**.

ring *(Maths.)*. Let S be a set with two operations called *addition* and *multiplication*. This system is called a *ring* if (1) for all elements x,y in S, $x+y$ and xy belong to S; (2) the set S is an Abelian group with respect to the addition operation; (3) the multiplication is (i) associative, (ii) distributive with respect to the addition operation: e.g. the positive and negative multiples of 3 with 0, i.e. the set $[\ldots -9, -6, -3, 0, 3, 6, 9, \ldots]$, with the usual addition and multiplication of numbers. A ring which is such that all its nonzero elements form an Abelian group with respect to the multiplication operation is called a **field**: e.g. the rational numbers with the usual addition and multiplication of numbers. Some writers require that a ring should have an identity for multiplication, and call the rest *pseudorings*.

ring armature *(Elec.Eng.)*. An electric-machine armature having a ring winding.

ringbone *(Vet.)*. An exostosis on the phalangeal bones of the horse's foot. *High ringbone* involves the first interphalangeal joint and *low ringbone* involves the second interphalangeal joint. *False ringbone* is an exostosis affecting the first or second phalanx but not involving a joint.

ringbound *(Vet.)*. See **ringwomb**.

ring complex *(Geol.)*. See **ring dyke**.

ring counter *(Telecomm.)*. A number of counting circuits in complete series, for sequence operating in counting impulses.

ring course *(Build.)*. The course farthest from the intrados of an arch.

ring culture *(Bot.)*. A system for growing e.g. tomatoes in greenhouses, using bottomless containers filled with fresh compost (through which the mineral elements are supplied) resting on a bed of sand or other inert material (through which water is supplied). The system makes it easier to avoid the infection of the roots by fungi that is common in greenhouse crops grown in the soil.

Ring drier *(Chem.Eng.)*. A proprietary device for drying solids in which the wet solid is entrained in a hot air stream which travels in a circular duct itself arranged in a large diameter circle.

ring dyke *(Geol.)*. An almost vertical intrusion of igneous rock which rose along a more or less cylindrical fault which had an approximately circular outcrop. In some Tertiary instances several successive ring dykes, separated by 'screens' of *country rock* and approximately concentric, form *ring complexes*.

Ringelmann smoke chart *(Genrl.)*. One of a series of six charts, numbered 0 to 5 and shaded from white to black, indicating a shade of grey against which density of smoke may be gauged.

ring fire *(Elec.Eng.)*. Thin streaks of fire appearing round the commutator of an electric machine; due to small particles of copper or carbon which have become embedded in the mica between the commutator and are raised to incandescence by the current. Ring fire indicates that the commutator needs to be cleaned.

ring gauge *(Eng.)*. A hardened steel ring having an internal diameter of specified size within very small limits of error; used to check the diameter of finished cylindrical work.

ring gland *(Zool.)*. See **Weismann's ring**.

ringing *(Bot.)*. The removal of the outer tissues from a strip encircling or partly encircling a stem or trunk, e.g. experimentally to interrupt phloem transport while leaving xylem transport more or less undisturbed, the phloem being peripheral to the xylem in most stems, horticulturally to encourage flowering and fruiting in over-vigorous fruit trees, or by e.g. rabbits or deer and then, if it is extensive, often fatal.

ringing *(Image Tech.)*. Light and dark bands at the edge of image areas of large brightness difference, due to unwanted oscillation in the system.

ringing *(Telecomm.)*. Extended oscillation in a tuned circuit, at its natural frequency, continuing after an applied voltage or current has been shut off, dying away according to its **decay constant**, but running into the next oscillation.

ring latch *(Build.)*. A latch in which the fall bar is operated by a handle in the shape of a ring, pivoted at the top so that it always falls into the vertical position.

ring main *(Chem.Eng.)*. Closed loop of piping through which chemicals in solution, or such finely divided materials as powdered coal, are circulated in suspension past suitable draw-off points.

ring main *(Elec.Eng.)*. A domestic a.c. wiring system in which a number of outlet sockets are connected in parallel to a ring circuit which starts and finishes at a mains supply point. All plugs used in the power outlets are fitted with individual fuses.

ring modulator *(Elec.Eng.)*. Four rectifying elements in complete series, which act as a switch, being fed with appropriate currents at the corners.

ring oscillator *(Elec.Eng.)*. One in which a number of active components feed each other in a circle or circus and in which the frequency is determined by a ring cut from a quartz crystal, suspended at its nodes to minimize damping; used, e.g. as a standard time-keeper (quartz crystal clock) at 10^5 Hz.

ring-porous *(Bot.)*. Wood, having much larger and/or more vessels in the early wood in each annual ring than in the late wood, so that the early wood may appear in cross sections of stems as a ring of small holes. Characteristic of some mainly north-temperate deciduous trees, e.g. chestnut, elm and deciduous oaks. Cf. **diffuse-porous**.

rings and brushes *(Crystal.)*. Name applied to the patterns produced when convergent or divergent plane-polarized light, after passing through a doubly refracting crystal cut perpendicular to the optic axis, is examined by an analyser. See **interference figure**.

ring shift *(Comp.)*. See **end-around shift**.

ring size *(Min.Ext.)*. Description of rock too large for handling by screening, in accordance with the diameter of a ring which can be slipped over it.

ring slot parachute *(Aero.)*. A parachute the canopy of which has slots all round its circumference, to give it greater stability.

ring spanner *(Eng.)*. One in the form of a notched ring in which diametrically opposite notches fit over the nut.

ring spinning *(Textiles)*. The most widely used spinning method. Fibres in the form of drafted roving are twisted together to form yarn by a tiny traveller rotating in a ring around a collection package on a spindle which is driven at a speed greater than that of the traveller. Many of these spinning heads are mounted together on a spinning frame.

ring-spot *(Bot.)*. An area, e.g. on a leaf, surrounded by a ring (or concentric rings) of chlorotic, necrotic or abnormally dark green tissue; a characteristic symptom in some virus diseases.

ring stress *(Min.Ext.)*. That adjacent to the walls of an unsupported underground excavation.

ring twisting *(Textiles)*. Forming a folded yarn by

twisting together two or more single yarns by using a *ring-and-traveller machine*. See **ring spinning**.

ring winding *(Elec.Eng.)*. (1) A helical winding arranged on a ring of iron or other material. Also called a *toroidal winding*. (2) A form of armature winding in which the armature core is a hollow cylinder with each turn of the winding threaded through the centre.

ringwomb *(Vet.)*. *Ringbound*. Incomplete dilatation of the cervix at parturition in the ewe.

ringworm *(Med.)*. *Tinea*. A contagious disease characterized by formation of ring-shaped patches on the skin; due to infection with moulds, especially of the three genera *Microsporon*, *Trichophyton*, and *Epidermophyton*.

rip *(Build.)*. To saw timber along the direction of the grain.

RIP *(Min.Ext.)*. Abbrev. for *Resin-In-Pulp*.

rip-bit *(Min.Ext.)*. See **jackbit**.

ripcord *(Aero.)*. (1) A cable used for opening the pack of a personal parachute. (2) An emergency release for gas in an aerostat envelope.

ripening *(Image Tech.)*. Stage in the manufacture of a photographic emulsion in which the silver halide grains reach their optimum size; also termed *digestion*.

ripidolite *(Min.)*. A species of the chlorite group of minerals, crystallizing in the monoclinic system. It is essentially a hydrated silicate of magnesium and aluminium with iron. Also called *prochlorite*.

ripper *(Build.)*. Slater's tool with a cranked handle and a long flat blade ending in an arrow-shaped head. Used for removing slates by cutting the fixing-nails.

ripple *(Elec.Eng.)*. The a/c component in the output of a rectifier delivering d.c. May be reduced by a series choke and shunting (smoothing) capacitor, or **Zener diode**. Measured as a percentage of the steady (average) current.

ripple *(Phys.)*. Small wave on the surface of a liquid for which the controlling force is not gravity, as for large waves, but surface tension. The velocity of ripples diminishes with increasing wavelength, to a miniumum value which for water is 23 cm/s for a wavelength of 1.7 cm.

ripple control *(Elec.Eng.)*. A method of controlling street lighting or other equipment from some central point by means of a high-frequency ripple superimposed on the current-carrying conductors of an electric power system. The information contained in the ripple signal is decoded and the appropriate load is disconnected by operation of a local circuit breaker.

ripple filter *(Elec.Eng.)*. A low-pass filter which is designed to reduce the ripple current but at the same time permits the free passage of the d.c. current, e.g. from a rectifier. Also *smoothing circuit*.

ripple finish *(Build.)*. See **wrinkle finish**.

ripple frequency *(Elec.Eng.)*. The frequency of the ripple current in rectifiers, etc. Usually double the supply frequency in a full-wave rectifier.

ripple generator *(Nuc.Eng.)*. See **radioisotope thermoelectric generator**.

ripple marks *(Geol.)*. Undulating ridges and furrows found on the bedding planes of certain sedimentary rocks, due to the action of waves or currents of air or water on the sediments before they were consolidated. Such ripple and rill marks can be seen in the process of formation today on most sandy beaches, on sand dunes, and in deserts.

ripple tank *(Phys.)*. Uses the property of 'shallow' *surface waves* (velocity proportional to depth) to demonstrate the refraction and focusing of waves.

ripple trays *(Chem.Eng.)*. Distillation column trays (or plates) which consist of thin metal formed into a series of parallel channels and perforated to provide space for rising vapour.

rip-saw, ripping saw *(Build.)*. A saw designed for cutting timber along the grain.

RISC *(Comp.)*. See **reduced instruction set computer**.

rise *(Build.,Civ.Eng.)*. (1) The vertical distance from the centre of span of an arch in the line of the springings to the centre of the intrados. Also called the *versed sine*. (2) The vertical height from end supports to ridge of a roof. (3) The height of a step in a staircase.

rise and fall system *(Surv.)*. A system of reduction of levels in which the staff reading at each successive point after the first is compared with that preceding it, and the difference of level entered as a *rise* or a *fall*. See **collimation system**.

rise and run *(Build.)*. Term applied to the amount of any given slope quoted as a given *rise* (vertical distance) in a given *run* (horizontal distance).

risen moulding *(Build.)*. A moulding decorating a panel and projecting beyond the general surface of the surrounding framing.

rise of floor line *(Ships)*. A tangent to the curve of the **bilge** to meet the extremity of the flat of the keel.

riser *(Build.)*. The vertical part of a step.

riser *(Elec.Eng.)*. See **commutator lug**.

riser *(Foundry)*. In a mould, a passage up which the metal flows after filling the mould cavity. It allows dirt to escape, indicates that the mould is full, and supplies metal to compensate for contraction on solidification. Also called *out-gate*.

rise time *(Telecomm.)*. Time for pulse signal in an amplifier or filter to rise to from 10 to 90% of the maximum amplitude. Also **build-up time**.

rising and falling saw *(Build.)*. A circular saw whose spindle can be moved in relation to the position of the working table, so that more or less of the saw can be made to project for cutting grooves of different depths.

rising arch *(Civ.Eng.)*. An arch whose springing line is not horizontal.

rising butt hinge *(Build.)*. A butt hinge with a loose leaf which, when opened, rises on the centre pin due to helical bearing surfaces on the two leaves. This enables a door, on opening, to rise above a carpet, and to close automatically.

rising front *(Image Tech.)*. A sliding panel for carrying the lens in a field or technical camera; used to diminish foreground and avoid distortion of perspective when photographing high buildings, trees etc.

rising main *(Build.)*. Mains water supply where it enters premises.

rising shaft *(Min.Ext.)*. A shaft which is excavated from below upward. Cf. **sinking shaft**.

rising type, rising spaces *(Print.)*. Faulty chases, furniture, typesetting, and make-up can cause the type to rise when locked up, and careless lock-up can cause it to rise no matter how good the equipment and make-up. This results in a vertical movement of the pages concerned at each impression and a tendency for spacing material to rise to surface level and print as a blemish. Precision equipment and careful workmanship reduce this trouble to a minimum; printing from plates eliminates it.

RIST *(Immun.)*. Abbrev. for *RadioImmunoSorbent Test*.

RI strains *(Immun.)*. Abbrev. for *Recombinant Inbred strains*.

risus sardonicus *(Med.)*. Wrinkling of the forehead and retraction of the angles of the mouth due to spasm of the facial muscles (as in tetanus), giving appearance of a grin.

Ritchie wedge *(Phys.)*. A photometer head in the form of a wedge with two white diffusing surfaces set 90° apart.

Rittinger's law *(Chem.Eng.)*. States that energy required in a crushing operation is directly proportional to the area of fresh surface produced, i.e. $E = k_r(1/d_2 - 1/d_1)$, where E is the energy used in crushing, k_r is a constant, depending on the characteristics of the material and on the type and method of operation of the crusher, and d_1 and d_2 are the average initial and final linear dimensions of the material crushed.

ritualization *(Behav.)*. The evolutionary process by which a behaviour pattern is modified to enhance its communication value, usually through exaggeration or repetition of some of its elements.

river capture *(Geol.)*. The beheading of a stream by a neighbouring stream which has greater power of erosion.

rivers *(Print.)*. In widely spaced text matter the spaces in successive lines can form channels of space running down the page, a tendency which is reduced by close spacing. Also called *streets*.

rivers, geological work of *(Geol.)*. This involves *corrasion* (wearing away) of their banks and beds, *corrosion* (solvent and chemical action of river water), *hydraulic action* and *attrition* of the transported material.

rivet *(Eng.)*. A headed shank for making a permanent joint between two or more pieces. It is inserted in a hole which is made through the pieces, and 'closed' by forming a head on the projecting part of the shank by hammering or other means. The head may be rounded flat, pan-shaped, or countersunk.

riveted joint *(Eng.)*. A joint between plates, sheets, or strips of materials (usually metallic) secured by rivets. See also **butt joint, lap joint**.

riveting machine *(Eng.)*. A machine for closing, clinching, or setting rivets. See also **hydraulic riveter, pneumatic riveter**.

R & M *(Aero.)*. Abbrev. for *Reliability and Maintainability*, US.

r.m.m. *(Genrl.)*. Abbrev. for *relative molecular mass*.

r.m.s. power *(Phys.)*. The effective mean power level of an alternating electric supply. Abbrev. for *Root-Mean-Square power*.

r.m.s. value *(Phys.)*. Measure of any alternating waveform, the square root of the mean of the squares of continuous ordinates (e.g., voltage or current) through one complete cycle. If there are harmonics of the fundamental, the total r.m.s. value is the sum of the r.m.s. values of the fundamental and the harmonics taken separately. For a simple sinusoid, it is $1/\sqrt{2}$ times the peak value. Employed because the energy associated with any wave depends on its intensity, i.e., upon the square of the amplitude. Also *effective value*. Abbrev. for **root-mean-square value**.

Rn *(Chem.)*. The symbol for radon.

RNA, ribonucleic acid *(Biol.)*. Polynucleotide containing ribose sugar and uracil instead of thymine. Can hold genetic information as in viruses, but is also the primary agent for transferring information from the genome to the protein synthetic machinery of the cell. See **mRNA**.

road bed *(Civ.Eng.)*. Foundation carrying the sleepers, rails, chairs, points and crossings etc. of a railway track.

road line composition *(Build.)*. A thermoplastic composition consisting of rosin, or a derivative, and mineral oil, filled with coarse and fine extenders and pigmented with up to 10% titanium dioxide. Used for marking white lines and Zebra crossings on roads.

road line paint *(Build.)*. A quick drying bitumen resistent coating used for marking road liners.

road metal *(Civ.Eng.)*. Broken stone for forming the surfaces of macadamized roads. Also *metal, metalling*.

road studs *(Civ.Eng.)*. Rubber or metal pads, the former generally incorporating small reflectors (*cat's eyes*), built in to the surface of a road to define traffic lanes at night or in fog.

roak, roke *(Eng.)*. A **seam**.

roaring *(Vet.)*. A respiratory affection of horses in which an abnormal sound is produced in the larynx during inspiration; caused by paralysis of the laryngeal muscles or by inflammatory conditions affecting the larynx. *Whistling* is a similar condition in which a sound of higher pitch is produced.

roasting *(Eng.)*. The operation of heating sulphide ores in air to convert to oxide. Sometimes the sulphur-bearing gases produced are used to make sulphuric acid. See **chloridizing-, sulphating-, sweet-**.

roasting furnace *(Eng.)*. A furnace in which finely ground ores and concentrates are roasted to eliminate sulphur. Part or all of the necessary heat may be provided by the burning sulphur. The essential feature is free access of air to the charge. This is done by having a shallow bed which is continually rabbled. Many types have been devised; *multiple-hearth* is most widely used.

Robertsonian translocation *(Biol.)*. Balanced translocation in which the breakpoints in the translocated chromosomes cannot be identified.

Robinson bridge *(Elec.Eng.)*. An a.c. bridge used for the measurement and control of frequency. Also *Robinson-Wien* bridge.

robot *(Comp.)*. A computer controlled machine which is able to sense, grip and move objects.

robotics *(Comp.)*. The study of the design and use of **robots**, particularly for their use in manufacture and related processes.

robustness *(Comp.)*. The ability of a system to cope with errors during execution.

Roche limit *(Astron.)*. The lowest orbit at which a satellite can withstand the tides raised within it by the primary body.

Rochelle salt *(Chem.)*. Crystal of sodium potassium tartrate, having strong piezoelectric properties, but high damping. Used as a *biomorph* in microphones, loudspeakers and pick-ups. Its disadvantage is the limited range of piezoelectric property (−18°C to 23°C) and high temperature coefficient.

roche moutonnée *(Geol.)*. A mound of bare rock which is usually smoothed on the upstream side and roughened by plucking on the downstream side, as a result of a moving ice-sheet.

rock *(Geol.)*. Any mineral matter making up the Earth. As used by geologists, the term also includes unconsolidated material such as sand, mud, clay and peat, in addition to the harder materials described as rock in conventional usage.

rock and roll *(Image Tech.)*. Film transport control system, used especially in sound mixing and dubbing, in which a number of interlocked paths can be simultaneously stopped at any point and run forward or back without losing the precise synchronization between them.

rock burst *(Min.Ext.)*. Sudden failure of stope pillars, walls or other rock buttresses adjacent to underground works, with explosively violent disintegration.

rock crystal *(Min.)*. The name given to colourless quartz whether in distinct crystals or not; particularly applicable to quartz of the quality formerly used in making lenses.

rock cycle *(Geol.)*. The cycle of rock change in which rocks are uplifted, eroded, transported, deposited, possibly metamorphosed and intruded, and then uplifted to start a new cycle. The concept was first developed by James Hutton.

rock drill *(Civ.Eng.)*. A tool specially adapted to the boring of holes through rock.

rocker *(Min.Ext.)*. A short and easily-portable rocking trough or cradle for washing concentrates, gold-bearing sand, etc.

rocker arms *(Autos.)*. Pivoted levers operated by push-rods or overhead camshaft, which carry the tappets of an overhead valve system. Also called **valve rockers**.

rocker gear *(Elec.Eng.)*. The hand-wheel or other device for moving a brush-rocker.

rocket *(Aero.,Space)*. (1) A missile projected by a rocket system. (2) A system, or a vehicle, powered by *reaction* or *rocket propulsion*.

rocket equation *(Space)*. The relationship between burn-out velocity (v_b) and the gas exhaust velocity (v_e):

$$v_b = v_e \log_e \frac{M}{M_b}$$

where M and M_b are the masses of the rocket at launch and burn-out repectively.

rocket propulsion *(Aero.)*. **Reaction propulsion** using internally stored, instead of atmospheric, oxygen for combustion. Used primarily where there is insufficient oxygen, e.g. above 70 000 ft (20 km), or where the lightness and compactness of the motor offset the high **propellant** weight, e.g. for assisted take-off and missiles.

rocket tester *(Build.)*. A rocket giving off dense smoke; used to test a drain for leaks.

rock face *(Build.)*. The form of face given to a building-stone which has been **quarry-faced**.

rock flour *(Geol.)*. A term used for finely comminuted rock material found at the base of glaciers and ice-sheets. It is mudlike and is composed largely of unweathered mineral particles.

rock-forming minerals *(Geol.)*. The minerals which occur as dominant constituents of igneous rocks, including quartz, feldspars, feldspathoids, micas, amphiboles, pyroxenes and olivine.

rock head *(Min.Ext.)*. See **stone head**.

rocking bar *(Horol.)*. A pivoted bar carrying the intermediate wheels in a keyless mechanism.

rock meal *(Min.)*. A white and light variety of calcium carbonate, resembling cotton; it becomes a powder on the slightest pressure.

rock milk *(Min.)*. A very soft white variety of calcium carbonate which breaks easily in the fingers; it is sometimes deposited in caverns or about sources holding lime in solution.

rock phosphate *(Min.)*. See **phosphorite**.

rock roses *(Min.)*. See **desert rose**.

rock salt *(Min.)*. Halite.

rock wall failure *(Min.Ext.)*. Collapse due to one or more of the following: rock fall, simple dropping; rock flow, slope failure; plane shear, failure along weakness plane; rotational shear, stress where soil has dropped leaving a rounded cavity.

Rockwell hardness test *(Eng.)*. A method of determining the hardness of metals by indenting them with a hard steel ball or a diamond cone under a specified load and measuring the depth of penetration. See **Brinell hardness test**.

rock wool *(Eng.)*. Fibrous insulating material made by blowing steam through molten slag. Also called *mineral wool, slag wool.*

Rocky Mountain fever *(Med.)*. *Blue disease; black fever; spotted fever.* A disease, more or less limited to the US, characterized by fever, headache, muscular pains, enlargement of the spleen, and a macular eruption on the skin; the disease is spread by a tick and is associated with infection by a *Rickettsia rickettsi.*

Rococo *(Arch.)*. Name given to an ornate form of decoration which developed from **Baroque** and which was particularly popular in Paris in the 18th century, being the style of the interiors, furniture and art of Louis XV's reign. Basic motifs were drawn from free flowing plant and shell forms, often without organic coherence, and often rendered in stucco or carved wood with gold paint.

rod *(Nuc.Eng.)*. A rod-shaped reactor fuel element or control absorber, or sample, intended for irradiation in reactor.

rod *(Zool.)*. One of the noncolour-sensitive light perceptive elements of the vertebrate retina. Rods respond to lower illumination levels than **cones**.

rodding *(Build.)*. Operation of clearing a stoppage in a pipe by inserting a rod to break down the obstruction and remove it; descaling of encrusted pipework with scrapers attached to jointed rods.

rodding eye *(Build.)*. The removable cover on a small opening in a vertical or horizontal drain pipe which enables any obstruction to be removed by drain rods.

Rodentia *(Zool.)*. An order of the Mammals. Generally small animals with never more than a single pair of chisel-shaped upper incisors which have open roots. Lower incisors can move like scissors as there is no anterior symphysis between the mandibles. Canines are never present, but a wide diastema between the gnawing incisors and the grinding cheek teeth which vary in numbers and frequently have persistently open roots. Glenoid cavities are elongated antero-posteriorly, the lower jaw being moved forwards for gnawing and backwards for grinding, the jaw muscles being greatly enlarged; herbivorous with a large caecum. They are almost universally distributed, and are of considerable economic and medical importance as pests of stored food and carriers of plague fleas. Adaptive, radiation wide, terrestrial, amphibian, burrowing, arboreal, gliding, saltatorial. Squirrels, Beavers, Gophers, Voles, Rats, Mice, Hamsters, Porcupines, Guinea Pigs and Agoutis.

rodent ulcer *(Med.)*. A slow-growing ulcerating cancer of the skin which usually affects the upper part of the face; arises from the basal cells of the skin and is of low malignancy.

rodman *(Surv.)*. A staffman (US).

roe chlorine number *(Paper)*. A test method to determine the bleachability of a sample of pulp by measuring the amount of chlorine gas it will absorb under the specified conditions of test.

Rogallo wing *(Aero.)*. Delta shaped wing formed by three spars which meet at the apex and are covered with fabric. In flight the fabric becomes convex due to reduced air pressure over the top surface. Can be folded compactly and, although originally intended for re-entering spacecraft as a *paraglider*, has become very popular for **microlight** aircraft.

roger *(Aero.)*. Radio code for 'I have received and understood all of your last transmission'.

rogue *(Bot.)*. (1) A plant that is not true to type. (2) To remove such plants from a crop, especially one grown for the production of seed.

role *(Behav.)*. A pattern of behaviour, and the expectancy of it, that is associated with individuals who hold a particular position in a society.

roll *(Aero.)*. Aerobatic manoeuvre consisting of a complete revolution about the longitudinal axis. In a *slow roll* the centreline of the aircraft follows closely along a horizontal straight line; an *upward roll* is similar, but considerable height is gained; a *hesitation roll* is one where the pilot brings his aircraft momentarily to rest in its rolling motion. A vertical upward or downward roll is usually called an *aileron turn* because these are the only control surfaces involved. A *flick roll* is an entirely different, very rapid and violent manoeuvre in which the aircraft makes its revolution along a helical path; high structural stresses are imposed and many countries ban this aerobatic. A *half-roll* is lateral rotation through 180°.

roll *(Build.)*. A joint between the edges of two lead sheets on the flat, the edges being overlapped over a 50 mm (2 in) diameter wood roll fastened to the surface to be covered with lead.

roll *(Print.)*. Tool used to impress designs on leather book covers. Usually brass, set in a long handle, which is necessary because of the pressure required in its operation.

roll *(Ships)*. The phenomenon of a ship's behaviour in waves, wherein she changes her angle of heel. See **rolling period**.

roll-capped *(Build.)*. Said of ridge tiles which are finished with a roll or cylindrical projection along the apex.

roll damper *(Aero.)*. See **damper**.

rolled gold *(Eng.)*. Composite sheet made by soldering or welding a sheet of gold on to both sides of a thicker sheet of silver, or other suitable metal, and rolling the whole down to the thickness required.

rolled laminated tube *(Plastics)*. Tube produced by winding, under heat, pressure and tension, a synthetic-resin impregnated or coated fabric or paper on to a former.

rolled steel joist *(Eng.)*. Abbrev. *RSJ.* See **H-beam**.

rolled-steel sections *(Eng.)*. Steel bars rolled into I- or T-shaped channel, angle, cruciform, or similar cross-sections for different applications in structural work, each section being made in graded standardized sizes. See also **H-beam**.

roller *(Build.)*. A hollow cylinder covered with absorbent material (e.g. lambswool, foamed rubber), revolving on an axle and fitted with a handle; used to apply paint to flat surfaces.

roller *(Image Tech.)*. Flanged drum providing a guided path over which film passes, for example through a processing machine.

roller-bearing *(Eng.)*. A shaft-bearing consisting of inner and outer steel races between which a number of parallel or tapered steel rollers are located by a cage; suits heavier loads than the **ball-bearing**.

roller bit *(Min.Ext.)*. Drilling bit with 3 or more conical rollers carrying teeth. The rollers have axes inclined to that of the drill pipe and rotate individually with it. Also *Tricone bit.*

roller chain *(Eng.)*. A driving or transmission chain in which the links consist of rollers and sideplates, the rollers being mounted on pins which connect the sideplates. See **driving chain**.

roller coating *(Build.)*. An industrial process for coating endless strips of metal with paint, etc., on a coating machine. A cylindrical roller picks up the paint from a trough, and distributes it over one or more rollers, from the last of which it is picked up by the strip.

roller conveyor *(Eng.)*. A **conveyor** comprising a series of closely spaced rollers set so that their crests can support articles to be conveyed. They may be power-driven, or freely revolving. Tapered ones are used to form bends in the conveyor line.

roller mill *(Eng.)*. In its simplest form, consists of two rolls of suitable material, mounted with their axes horizontal, running in opposite directions; used for crushing and mixing operations.

roll feed *(Eng.)*. A mechanism incorporating nipping rollers for feeding strip material to power presses. The rollers turn intermittently by a set distance and are timed in relation to the press so as to advance the required length of strip, correctly timed, for each press stroke.

roll film *(Image Tech.)*. Film, wound on a spool, with opaque paper fitting closely to the end-cheeks of the spool, excluding light sufficiently to allow loading and unloading in daylight.

roll forging *(Eng.)*. A hot forging process for reducing or tapering short lengths of bar stock in the manufacture of axles, chisels, knife blades, tapered tubes, etc. Parts are placed by tongues into profiled grooves in pairs of forging rollers.

roll forming *(Eng.)*. A cold forming process in which a series of mating rolls progressively forms strip metal into tube or section, usually of complex shape, at high rolling speeds. Often associated with a continuous resistance welding machine.

rolling *(Aero.)*. The angular motion of an aircraft tending to set up a rotation about a longitudinal axis. One complete revolution is called a **roll**.

rolling *(Ships)*. See **rolling period**.

rolling bearing *(Eng.)*. A ball-bearing, roller-bearing, needle bearing, or other type of bearing in which elements roll between load-bearing surfaces.

rolling instability *(Aero.)*. See **lateral instability**.

rolling lift bridge *(Eng.)*. A type of bascule bridge in which the bascule or cantilever part has, at the shore end, a surface of segmental profile rolling on a flat bearing.

rolling load *(Eng.)*. See **moving load**.

rolling mills *(Eng.)*. Sets of rolls used in rolling metals into numerous intermediate and final shapes, e.g. blooms, billets, slabs, rails, bars, rods, sections, plates, sheets and strip.

rolling moment *(Aero.)*. The component of the couple about the longitudinal axis acting on an aircraft in flight.

rolling period *(Ships)*. The movement of a ship in waves about a longitudinal horizontal axis is known as *rolling*. The rolling period is the time taken for a complete roll from the upright position first to one side then to the other and back to upright again. Cf. **pitching period**.

rolling resistance *(Eng.)*. When a wheel rolls on a surface, distortion of the contact surfaces due to the normal force between them destroys the ideal line contact. This introduces a force with a component, called the rolling resistance, opposing the motion.

rolling stock *(Civ.Eng.)*. A general and collective term for all coaches, trucks etc., which run along a railway track.

roll-off frequency *(Telecomm.)*. Frequency where response of an amplifier or filter is 3 dB below maximum.

roll on, roll off ferry *(Ships)*. Vehicular ferry with wide doors at bow and stern to facilitate the loading of vehicles, characterized by a large open deck near the waterline.

rolls *(Min.Ext.)*. Crushing rolls are pairs of horizontally-mounted cylinders, faced with manganese steel, between which ore is crushed as they rotate inward.

ROM *(Comp.)*. *Read Only Memory*. Computer memory which may not be written to by the programmer. The software in the ROM is fixed during manufacture. See **PROM**.

roman *(Print.)*. Ordinary upright type, as distinct from *italic*, or sloping.

Roman mosaic *(Build.)*. See **tessellated pavement**.

Romberg's sign *(Med.)*. Sign present when a patient, standing with feet close together and eyes closed, sways more than when his eyes are open; it indicates disease of the sensory tracts in the spinal cord.

roméite *(Min.)*. Naturally occurring hydrated antimonite of calcium, sometimes with manganese and iron, crystallizes in the cubic system, often as brownish octahedra.

rone *(Build.)*. In Scotland, an **eaves gutter**. Also *rhone*.

röntgen *(Radiol.)*. Unit of X-ray or gamma dose, for which the resulting ionization liberates a charge of each sign of 2.58×10^{-4} coulombs per kilogram of air. Symbol R.

röntgen equivalent man *(Radiol.)*. Former unit of biological dose given by the product of the absorbed dose in r and the relative biological efficiency of the radiation. Abbrev. **rem**. Now replaced by the **effective dose equivalent**, unit the sievert.

röntgenology *(Radiol.)*. Same as **radiology**, US.

röntgen rays *(Phys.)*. See **X-rays**.

roof antenna *(Telecomm.)*. See **flat-top antenna**.

roof boards *(Build.)*. (*Sarking* in Scotland). Boards fixed to the rafters to provide a fixing and undercovering for the covering materials proper, e.g. slates, tiles.

roof bolting *(Min.Ext.)*. Method of roof support in which steel bolts are inserted in drill holes so as to pin supporting steel beams under the roof of a stope.

roof guard *(Build.)*. A device fitted to a roof to prevent snow from sliding off it. Also *snow boards*.

roofing slate *(Geol.)*. A term widely applied to rocks of fine grain in which regional metamorphism has developed in good slaty cleavage.

roof pendant *(Geol.)*. A mass of country rock projecting downwards, below the general level of the roof, into an intrusive rock body.

roof truss *(Civ.Eng.)*. The structural framework built to support the roof covering for a building.

room-and-pillar *(Min.Ext.)*. See **bord-and-pillar**.

room index *(Elec.Eng.)*. The coefficient of utilization of lamps in a room depends on the shape of the room. This shape is expressed by the room index, *k*. For rectangular rooms

$$k = \frac{0{\cdot}9w + 0{\cdot}1l}{h}$$

where w = width of room, l = length of room, h = height of room.

root *(Bot.)*. (1) The typically descending axis of a plant and other axes that are anatomically similar and/or clearly homologous. In most vascular plants, roots may be recognized by their endogenous origin, lack of leaves and possession of a root cap. Roots typically function in anchorage and the absorption of water and mineral salts in the soil. See also **aerial root, root tuber**. Cf. **stem**. (2) See **flagellar root**.

root cap *(Bot.)*. *Calyptra*. A hollow cone of cells protecting the apical meristem of a growing root, which is renewed from within as it wears.

root diameter *(Eng.)*. (1) Of a gear-wheel, the diameter at the bottom of the tooth spaces. (2) Of a thread, the diameter at the bottom surfaces which join the adjacent sides or flanks of a thread.

rooter *(Electronics)*. Circuit designed to give an output amplitude proportional to the square root of the input amplitude. Used in compressors for reducing dynamic range in sound reproduction, and for gamma-correction in video, to compensate for television camera tube characteristics.

root hair *(Bot.)*. A tubular outgrowth from a cell in the epidermis of a young root, possibly important in the uptake of the more slowly diffusing mineral ions e.g. phosphate.

rooting compound *(Bot.)*. Usually powders containing **auxins** in which a cutting is dipped to promote rootgrowth.

root locus *(Elec.Eng.)*. Locus for the roots of the closed-loop response for a system, derived by plotting the poles and zeros of the open-loop response in the complex plane. Used in the study of system stability.

root-mean-square *(Stats.)*. Square root of sum of squares of individual observations divided by total number of observations. Abbrev. *rms*.

root-mean-square power *(Phys.)*. See **r.m.s. power**.

root-mean square value *(Phys.)*. See **r.m.s. value**.

root of the joint *(Eng.)*. In welding, the place where the original components were closest together.

root of the weld *(Eng.)*. The place where the weld comes closest to the **root of the joint**.

root pressure *(Bot.)*. The positive pressure that may develop in the xylem when water uptake by osmosis follows ion uptake in the root and transpiration is low. It may result in guttation or bleeding.

Roots blower *(Eng.)*. An air compressor for delivering large volumes at relatively low pressure ratios; in it a pair of hour-glass-shaped members rotate with a small clearance within a casing, no valves being required.

rootstock *(Bot.)*. (1) A rhizome, especially a short, erect one. (2) A stock for grafting, especially one from a clone selected for desirable effects on the scion e.g. dwarfing, early fruiting.

root tuber *(Bot.)*. Swollen adventitious root acting as a storage organ e.g. Dahlia. Cf. **taproot, tuber**.

rope *(Radar)*. See **window**.

rope brake *(Eng.)*. An absorption dynamometer consisting of a rope encircling a brake drum or flywheel, one end of the rope being loaded by weights and the other supported by a spring balance. The effective torque absorbed is obtained by multiplying the drum radius by the difference of the tensions.

rope processing *(Textiles)*. Wet processing, including scouring, of fabric moving in loose rope form through large porcelain rings (*pot-eyes*) placed to reduce friction on the fabric. Nevertheless the fabrics sometimes suffer because of the development of rope marks (e.g. creases) running in the warp direction when this system is used.

ropiness *(Build.)*. A defect in applied paint, usually attributed to a lack of skill by the person applying the coating or to the use of excessively heavy or thick paint. Appears as 'tram lines' or visible brushmarks.

ropy lava *(Geol.)*. Ses **pahoehoe**.

ro-ro *(Ships)*. Abbrev. for *Roll On, Roll Off ferry*.

Rorshach inkblot test *(Behav.)*. A **projective test** which requires the subjects to look at inkblots and report on what they see; the answers are used to interpret his or her fantasy life, personality, intelligence, and also as an aid to psychiatric diagnosis.

rosacea *(Med.)*. A chronic inflammatory skin disease, usually beginning in middle age and characterized by redness and papules especially in the centre of the face. Cause unknown.

Rosaceae *(Bot.)*. The rose family, ca 3300 spp. of dicotyledonous flowering plants (superorder Rosidae). Trees, shrubs and herbs; cosmopolitan, especially N. temperate regions. The flowers are polypetalous, and perigynous or epigynous. Includes many important tree and bush fruit of temperate regions e.g. apple, pear, plum, cherry, almond, raspberry, strawberry and also many ornamentals including the rose.

rosaniline *(Chem.)*. *Triamino-diphenyl-tolyl-hydroxymethane*, a base of the fuchsin dyes. It is obtained by oxidation of an equi-molecular mixture of aniline, 2-toluidine and 4-toluidine.

roscoelite *(Min.)*. This mineral is essentially *muscovite* in which vanadium has partly replaced the aluminium. Its colour is clove-brown to greenish-brown.

rose *(Arch.)*. A decorative circular escutcheon through which, e.g. the spindle of a door-handle or the flexible cord of a pendant *luminaire* may pass. See also **ceiling rose**, a more practical extension of the latter use.

ROSE *(Comp.)*. *Research Open Systems in Europe*. The principle development project for information exchange within the **ESPRIT** community; a basis for network research.

Rose crucible *(Chem.)*. A crucible the lid of which is fitted with an inlet tube. It is used for igniting substances in a current of gas.

rose cutter *(Horol.)*. A small hollow milling cutter used on the lathe for rapidly producing pivots, screws etc.

Rosenmüllers organ *(Zool.)*. See **epididymis**.

roseola *(Med.)*. A rose-coloured rash following a fever, occurring in infancy and presumed due to a virus.

rose opal *(Min.)*. A variety of opaque common opal having a fine red colour.

rose quartz *(Min.)*. Quartz of a pretty rose-pink colour, due probably to titanium in minute quantity. The colour is apt to be destroyed by exposure to strong sunlight. See also **Bohemian gem-stones**.

rose topaz *(Min.)*. The yellow-brown variety of topaz changed to rose-pink by heating. These crystals often contain inclusions of liquid carbon dioxide.

rosette *(Zool.)*. Any rosette-shaped structure; in some Oligochaeta, a large ciliated funnel by which the contents of the vesiculae seminales pass to the exterior; in some Crinoidea, a thin calcareous plate (*rosette plate, rosette ossicle*) formed by the coalescence of the basal plates.

rosette plant *(Bot.)*. A plant in which the leaves radiate out at about soil level and which has a more or less leafless flowering stem. See **hemicryptophyte**.

Rose-Waaler test *(Immun.)*. A test for the presence of rheumatoid factor in blood. Serum dilutions are mixed with sheep erythrocytes coated with an amount of antibody insufficient to cause agglutination. If rheumatoid factor is present this combines with Fc exposed on the red cell surface and causes agglutination.

rose window *(Arch.)*. A circular window with radial bars. Also called *Catherine wheel, marigold window*.

rosewood *(For.)*. Decorative cabinet-wood from the sapwood of species of *Dalbergia* exported mainly from Brazil, Honduras, Jamaica, India and Africa. Also called *jacaranda*.

Rosidae *(Bot.)*. Subclass or superorder of dicotyledons. Trees, shrubs and herbs, flowers mostly polypetalous, stamens (if numerous) developing centripetally, rarely with parietal placentation. Contains ca 60 000 spp. in 108 families including Leguminosae, Rosaceae, Crassulaceae, Myrtaceae, Melastomaceae, Euphorbiaceae and Umbelliferae.

rosin *(Chem.)*. *Colophony*. The residue from the distillation of turpentine. The colour varies from colourless to yellow, red, brown and black. Rel.d. 1.08, mp 100°C to 140°C. Wood-rosin is obtained by the extraction of long-leaf pine wood; chief sources, US and France. Used as a soldering flux, in varnish, soap and size manufacture, and (in the form of resinates) as a drier in paint. See **resinates**.

rosinates *(Chem.)*. **Resinates**.

Rosiwal intercept method *(Geol.)*. Particle-size analysis technique based on measurement of the intercepts made by a line drawn through a selection of particle images on a photomicrograph or in the field of view of a projection microscope.

rosolic acid *(Chem.)*. *Quinonoid triarylcarbinol anhydride*: An acidic dyestuff of the triphenylmethane series, obtained by oxidizing a mixture of phenol and 4-cresol with arsenic (V) acid and sulphuric acid. Green glistening crystals, insoluble in water, dissolving in alkalis with a red colour.

Rossby number *(Meteor.)*. A non-dimensional number *Ro* defined as the ratio, for a particular class of motions in a rotating fluid such as the Earth's atmosphere, of inertial forces to **Coriolis forces**. $Ro = U/fL$ where U is a characteristic velocity, L a characteristic length, and f the **Coriolis parameter**. If *Ro* is large, then the effect of the Earth's rotation may be neglected.

Rossby wave *(Meteor.)*. A wave in the general atmospheric circulation, in one of the principal zones of westerly winds, characterized by large wavelength, (ca 6000 km), significant amplitude (ca 3000 km), and slow movement. First described by C.G. Rossby.

Rossi-Forell scale *(Geol.)*. A scale of apparent intensity of earthquake movements, now replaced by the **Mercalli scale**.

rostellum *(Bot.)*. A small beak-like outgrowth, especially one from the column of the flower of some orchids.

rostrum *(Zool.)*. In Birds, the beak; a beak-shaped process; in Cirripedia, a ventral plate of the carapace; in some Crustacea, a median anterior projection of the carapace. adjs. *rostral, rostrate*.

rostrum photography *(Image Tech.)*. Photography of artwork, graphics, animation drawings etc. with a camera mounted on a vertical stand above a horizontal table.

rot *(Bot.)*. The disintegration of tissue resulting from the activity of invading fungi or bacteria.

rotachute *(Aero.)*. A 'parachute', usually for stores or the recovery of missiles, in which the normal retarding canopy is replaced by rotor blades, which are freely-revolving and act like a **rotor**.

rotaplane *(Aero.)*. A heavier-than-air aircraft which derives its lift or support from the aero-dynamical reaction of freely rotating rotors. See **rotorcraft**.

rotary amplifier *(Elec.Eng.)*. Rotary generator, output of which is field-controlled by another generator or amplifier.

rotary combustion engine *(Autos.)*. See **Wankel engine**.

rotary converter *(Elec.Eng.)*. Combination of electric motor and dynamo used to change form of electric energy, e.g. d.c. to a.c. or one d.c. voltage to another.

rotary drier *(Min.Ext.)*. Tubular furnace sloped gently from feed to discharge end, through which moist material is tumbled as it rotates slowly, while rising hot gases remove moisture.

rotary drill *(Min.Ext.)*. The drill *downhole* and connected by the drilling pipe to the rotary table in the oil derrick or drilling platform. It may consist of several drills, reamers and stabilizing **collars**, themselves joined by special tool joints. It is usually lubricated and cleared by **drilling mud** forced down the drilling tube and out through nozzles in the drill head. The oil and debris passes up through the annular space between the drill tube and casing. The most important component fixed to the rotary table is the **kelly**.

rotary engine *(Aero.)*. An early type of aero engine in which the crankcase and radially disposed cylinders revolved round a fixed crankshaft; not to be confused with the modern radial engine.

rotary field *(Elec.Eng.)*. See **rotating field**.

rotary indexing machine *(Eng.)*. A transfer machine, used for cutting, assembling, or other production work, in which workpieces or tools are carried on a circular turntable which rotates intermittently.

rotary machine *(Print.)*. A machine in which the printing surface is a cylinder or a plate attached to a cylinder. Used in all the major printing processes and can be sheet or web-fed.

rotary pump *(Eng.)*. A pump, similar in principle to a **gear pump**, in which two specially shaped members rotate in contact; suited to large deliveries at low pressure. See **Roots blower**.

rotary regenerative heater *(Eng.)*. An air heater consisting of a slowly revolving rotor made up of concentric rings of corrugated and flat plates, which pass alternately and continuously through the hot gases and the air drawn across opposite halves of the rotor.

rotary shutter *(Image Tech.)*. The rotating vanes which cut off the light from the screen while the frames are being moved and located in the picture gate of a projector.

rotary strainer *(Paper)*. Cleaning device comprising a drum with slits or perforations on its surface, slowly rotating in a shallow vat to which the uncleaned stock is introduced. In passing through the orifices, the stock is purged of contraries that are too large.

rotary switch *(Elec.Eng.)*. A switch operated by a rotating handle capable of rotation in one direction only.

rotary table *(Min.Ext.)*. Heavy circular component mounted just above the derrick floor which carries the **kelly bushing**. The table is rotated by the draw-works machinery and thus rotates the **kelly** which slides in the bushing. The table thus turns the drill string.

rotary valve *(Autos.)*. A combined inlet and exhaust valve in the form of a ported cylinder rotating on cylindrical faces in the cylinder head, usually parallel with the crankshaft.

rotate *(Bot.)*. A corolla which is wheel shaped, with the petals or lobes spreading out at right angles to the axis of the flower.

rotating amplifier *(Elec.Eng.)*. A form of d.c. generator in which the electrical power output can be accurately and rapidly controlled by a small electrical signal applied to the control field of the machine. Mainly used in industrial closed-loop systems.

rotating anode *(Electronics)*. A high-power X-ray tube in which the anode rotates continuously to bring a fresh area of its surface into the electron beam; this allows higher output without melting the target.

rotating crystal method *(Crystal.)*. A widely used method of X-ray analysis of the atomic structure of crystals. A small crystal, less than 1 mm in maximum size, is rotated about an axis at right angles to a narrow incident beam of X-rays. The diffraction of the beam by the crystal is recorded photographically or with a detector.

rotating field *(Elec.Eng.)*. One in which the magnitude is constant at a point but whose direction is rotating about a point in a fixed reference system.

rotating-field magnet *(Elec.Eng.)*. The rotating portion of an electric machine, usually a synchronous motor or generator, in which the field poles rotate and the armature is stationary.

rotating joint *(Telecomm.)*. Short length of cylindrical waveguide, constructed so that one end can rotate relative to the other; used to couple two waveguide systems of rectangular cross-section.

rotating, rotary disk contactor *(Chem.Eng.)*. A liquid-liquid extraction device which consists of a column with annular stators fitted to it and a central shaft carrying rotors of diameter nearly equal to the holes in the stators. Liquids flow counter-current through the column and under the effect of high speed of rotation of the shaft and rotors improved contact is obtained.

rotation *(Astron.)*. The term generally confined to the turning of a body about an axis passing through itself, e.g. *rotation* of the earth about its polar axis in 1 sidereal day.

rotation *(Elec.Eng.)*. See **curl**.

rotational field *(Elec.Eng.)*. A field in which the *circulation* is, in some parts, not always zero.

rotational moulding *(Plastics)*. A *moulding* process in which the material (e.g. PVC) in the mould is rotated while in the heating oven and during cooling; used for some hollow articles.

rotation axes of symmetry *(Crystal.)*. Symmetrically placed lines, rotation about which causes every atom in a crystal structure, as revealed by X-ray analysis, to occupy identical positions a given number (2, 3, 4, 6) of times. Cf. *screw axes*.

rotation of a vector *(Maths.)*. See **curl**.

rotation of the plane of polarization *(Chem.,Phys.)*. A property possessed by optically active substances. See **optical activity**.

rotation shift *(Comp.)*. See **end-around shift**.

rotation speed *(Aero.)*. The speed during take-off at which the nosewheel of an aircraft is raised from the ground prior to **lift-off**.

rotator *(Phys.)*. A device for rotating the plane of a wave in a waveguide.

rotatory dispersion *(Chem.,Phys.)*. Variation of rotation of the plane of polarized light with wavelength for an optically active substance.

rotatory evaporator *(Chem.)*. A device for facilitating the evaporation of a liquid, generally under reduced pressure, by continuously rotating the flask in which it is contained.

rotatory power *(Chem.,Phys.)*. See **specific rotation**.

rotavirus infection *(Vet.)*. Fast becoming an important source of calf scour, but vaccine is available.

rotenone *(Chem.)*. See **derris**.

Rotifera *(Zool.)*. A class of small, unsegmented, pseudocoelomate animals, phylum Aschelminthes. A distinctive anterior ciliary apparatus is used for locomotion and food gathering. Aquatic. *Wheel-animalcules.*

rotogravure *(Print.)*. Photogravure printing on a **rotary machine**.

roton *(Phys.)*. Quantum of rotational energy analogous to the **phonon**.

rotor *(Aero.)*. A system of revolving aerofoils producing lift, acting on a plane at right angles to the driving shaft.

rotor *(Autos.)*. The revolving arm of a distributor.

rotor *(Elec.Eng.)*. See **armature** (3).

rotor *(Horol.)*. A revolving eccentric weight which winds a self-winding watch.

rotor *(Meteor.)*. A large closed eddy which may form under **lee waves** of large amplitude; often associated with severe turbulence.

rotor *(Zool.)*. A muscle which by its contraction turns a limb or a part of the body on its axis.

rotor cloud *(Meteor.)*. A whirling quasi-stationary cloud that forms in the upper part of a rotor to the lee of a range of hills. The *helm bar* near Cross Fell in Cumbria is a well-known example.

rotor core *(Elec.Eng.)*. That portion of the magnetic circuit of an electric machine which lies in the rotor.

rotorcraft *(Aero.)*. Any aerodyne which derives its lift from a rotor, or rotors.

rotor head *(Aero.)*. The structure at the top of the rotor pylon, including the hub member to which the blades of a rotorcraft are attached.

rotor hinge *(Aero.)*. A hinge for the blades of a rotorcraft. See **drag hinge**, **feathering hinge**, **flapping hinge**.

rotor hub *(Aero.)*. The rotating portion of the rotor head of a rotorcraft to which the rotor blades are attached.

rotor spinning *(Textiles)*. A widely used **open-end spinning** system. A stream of fibres from a roving enters a rapidly-rotating cell or rotor. The fibres are temporarily held on the circumference by centrifugal force. The yarn is drawn out of the rotor while being twisted through a **navel** into the doffing tube and is collected on a suitable package.

rotor starter *(Elec.Eng.)*. A motor starter used for slip-ring induction motors; it cuts out resistance previously inserted in the rotor circuit.

rotor-tip jets *(Aero.)*. Propulsive jets in the tips of a rotorcraft's blades that are used to obtain a drive with minimum torque reaction; they may be **pulse jets**, **ramjets**, combustion units fed with air and fuel from the fuselage, **pressure jets**, or small **rocket** units.

rotoscope *(Image Tech.)*. TN for a device providing frame-by-frame projection of a film to form the background of animation drawings or for the analysis of movement.

rottenstone *(Geol.)*. A material used commercially for polishing metals; formed by the weathering of impure siliceous limestones, the calcareous material being removed in solution by percolating waters.

rotula *(Zool.)*. In higher Vertebrates, the kneecap.

rotunda *(Arch.)*. A building or room which is circular in plan and is covered by a dome.

rot v *(Maths.)*. See **curl**.

rouge *(Chem.)*. Hydrated oxide of iron in a finely divided state; used as a polish for metals.

rough *(Paper)*. The unglazed surface of drawing papers specially induced by a coarse felt. See **hot-pressed**, **not**.

rough ashlar *(Build.)*. A block of freestone as taken from the quarry.

rough brackets *(Build.)*. Pieces of wood nailed to the sides of the **carriage**, to provide intermediate support for treads of a wooden stair.

rough-cast *(Build.)*. A rough finish given to a wall by coating it with a plaster containing gravel or small stones.

rough coat *(Build.)*. The first coat of plaster applied to a wall surface.

rough colony *(Biol.)*. A bacterial colony produced by mutation from a smooth colony. The morphological change is frequently accompanied by physiological changes, e.g. altered virulence.

rough endoplasmic reticulum *(Biol.)*. Cisternal form of **endoplasmic reticulum** bearing ribosomes on the cytoplasmic surface; functioning in the synthesis of protein for export from the cell, sometimes at least, through the golgi apparatus.

rough grounds *(Build.)*. Unplaned strips of wood used as *grounds* when the attached joinery will entirely cover them.

roughing *(Min.Ext.)*. Production of an impure concentrate as an early stage in ore processing, thus reducing bulk for more thorough treatment.

roughing-in *(Build.)*. The first coat of 3-coat plaster work.

roughing tool *(Eng.)*. A lathe or planer tool, generally having a round-nosed or obtuse-angled cutting edge, used for roughing cuts.

rough proof *(Print.)*. A print of the work in hand, much below the finished quality to be expected, submitted for checking purposes only.

rough string *(Build.)*. Part of a staircase. See **carriage**.

rough trimmed *(Print.)*. A design feature in many well-produced books where the tail edges are merely cleaned up by cutting the farthest projecting leaves, the pages having been deliberately positioned to ensure a variation in the tail margin. See **cut edges, trimmed size**.

roulette *(Maths.)*. The locus of any point, or envelope of any line, moving with a first curve which rolls without slipping on a second curve. Referred to respectively as *point* or *line roulette*. The locus of a point on a circle rolling on a straight line is called a *curtate cycloid*, a *cycloid* or a *prolate cycloid* respectively according as the point is inside, on the circumference or outside the rolling circle. The locus of a point on a circle rolling on another circle is called an *epicycloid* or *hypocycloid* respectively according as the rolling circle is outside or inside the fixed circle, if the point is on the circumference, or an *epitrochoid* or *hypotrochoid* respectively if the point is not on the circumference of the rolling circle. Epicycloids and epitrochoids in which the rolling circle completely encloses the fixed circle are sometimes called *pericycloids* and *peritrochoids* respectively. Curtate and prolate cycloids are sometimes called *trochoids*.

round *(Build.)*. A rung of a ladder.

round angle *(Maths.)*. A complete rotation, an angle of 360° or 2π radians. Also called a *perigon*.

round dance *(Behav.)*. Circular movements of a worker bee on returning from a foraging trip which communicates information that a food source is less than 150 yards from the hive. See **waggle dance**.

round heart disease *(Vet.)*. A disease of chickens characterized by enlargement and muscular degeneration of the heart and sudden death; the cause is unknown.

rounding *(Comp.)*. Approximating a number by its nearest equivalent, using a set number of significant figures. Also *rounding-off*. Cf. **truncating**.

rounding *(Print.)*. The process of giving the back of a book a convex shape before casing; usually performed along with **backing**.

rounding error *(Comp.)*. Error introduced by rounding.

rounding off *(Maths.)*. The reduction of the 'length' of a number, e.g. the number π, whose 'length' is infinite, is rounded off to 3.14, 3.1416, 3.141 59 etc. The rounding off is always to the nearest number of the required 'length'; if the number to be rounded off is exactly halfway, e.g. 2.565 to be rounded off to 2 decimal places, it may be regularly rounded up (or down for negative numbers), or else the last figure may be chosen even (giving 2.57 and 2.56 respectively for the example).

rounding-up tool *(Horol.)*. A tool for correcting the size and shape of the teeth of a toothed wheel.

round key *(Eng.)*. A circular bar or pin fitted in a hole drilled half in the shaft and half in the boss, parallel to the shaft axis; used for light work to avoid fitting. See **key**.

round of beam *(Ships)*. See **camber**.

rounds *(Build.)*. The general name for planes having a concave sole and cutting iron, used for forming rounded surfaces. Cf. **hollows**.

round step *(Build.)*. A step finished with a semi-circular end.

round trip *(Min.Ext.)*. Removing and dismantling the drill string to replace the drill bit and then re-assembling and placing it back *downhole*. Also *trip*.

roundworm *(Med.,Zool.)*. Name applied to a number of parasitic Nematodes, especially those of the genus *Ascaris*, including the large intestinal roundworm of Man (*A. lumbricoides*).

roup *(Vet.)*. A term applied to symptoms of oculonasal discharge and swelling of the face and wattles of fowl, which occur in infections of the upper respiratory tract such as infectious coryza, fowl pox, and infectious laryngotracheitis. Also synonym for *infectious coryza*.

Rous' sarcoma *(Vet.)*. A tumour, occurring in fowls, which is transmitted by an oncogenic RNA virus. The first demonstration that viruses can cause some cancers.

Rousseau diagram *(Elec.Eng.)*. A diagram by the use of which the total output (in lumens) of a light source can be obtained, if the polar curve of the lamp about the vertical axis is known.

Roussin's salts *(Chem.)*. Formed when sodium trisulphide is added to a solution of iron(II) chloride saturated at −2°C with nitrogen (II) oxide, and converted by sodium sulphide into so-called *Roussin's red salt*; by treatment with dilute acids this is converted into *Roussin's black salt*.

rout *(Build.)*. To cut out wood from the bottom of a sinking with a router plane.

router *(Build.)*. (1) Power operated machine, fixed or hand-held, which holds and rotates various types and shapes of cutter for shaping or recessing. Similar to a spindle moulder but the spindle points downwards. (2) A plane adapted to work on circular sashes; operated in the manner of a spokeshave. (3) The side wing of a **centre-bit**, which removes the material in forming the hole.

router plane *(Build.)*. A plane having a central projecting cutting iron, adapted to smoothing the bottom of a recess or sinking.

Routh's rule *(Phys.)*. A rule summarizing the values of the radius of gyration of a rectangular lamina, an elliptic lamina and an ellipsoid about a principal axis through its centre of gravity. According to the rule, k^2 equals the sum of the squares of the other semi-axes divided by 3, 4 or 5 respectively.

routine *(Comp.)*. See **subroutine**.

routiner *(Telecomm.)*. Apparatus which tests, as a routine, all machine-switching apparatus in an exchange, so that faults may be rapidly detected and rectified, and contacts kept clean.

routing machine *(Print.)*. A machine having a revolving tool which removes unwanted metal from printing plates.

roving *(Textiles)*. A continuous strand of fibres sufficiently drawn to a diameter suitable for drafting and spinning into yarn, e.g. by **ring spinning** frames.

Rowland circle *(Phys.)*. A circle having the radius of curvature of a concave diffraction grating as diameter. It has the property that, if the slit is placed anywhere on the circumference of the circle, the spectra of various orders are formed in exact focus also round the circumference of the circle. This fact is used in designing mountings for the concave grating.

rowlock *(Arch.)*. A term applied to a course of bricks laid on edge.

rowlock-back *(Arch.)*. A term applied to a wall whose external face is formed of bricks laid flat in the ordinary manner, while the back is formed of bricks laid on edge.

row vector *(Maths.)*. A single-row matrix.

royal *(Paper)*. A former imperial paper size retained under the metric system for printing boards, 52 × 64 cm (20½ × 25 in). Note also the following sizes recommended for book publishing papers: Metric small quad royal 96 × 127 cm, Metric quad royal 102 × 127 cm.

royals *(Min.Ext.)*. In the **cyaniding** process for recovering gold, a term for the sludge, rich in gold, formed when the **pregnant solution** is precipitated on zinc dust or on to resin or carbon.

RPN *(Genrl.)*. Abbrev. for *Reverse Polish Notation*.

RPV *(Aero.)*. Abbrev. for *Recoverable Pilotless Aircraft*.

RR Lyrae variable *(Astron.)*. Variable star with a period of less than 1 day; common in globular clusters, and used, like the cepheids, to measure galactic distances.

RSF *(Build.)*. Abbrev. for *Rough Sunk Face*.

RSJ *(Build.)*. Abbrev. for *Rolled Steel Joist*.

R-strategist *(Bot.)*. Organism which assigns much of its resources to reproduction; usually opportunistic and colonizing species (weeds) with high fecundity and low competitive ability. Cf. **K-strategist**.

Ru *(Chem.)*. Symbol for ruthenium.

Rubarth's disease *(Vet.)*. See **infectious canine hepatitis**.

rubber *(Chem.)*. The main source of natural rubber or caoutchouc is the tree, *Hevea brasiliensis*, originally a native of Central and South America, but since the end of the 19th century widely grown in plantations in S.E. Asia. Commercial rubber consists of caoutchouc, a polymerization product of isoprene, of resinlike substances, nitrogenous substances, inorganic matter and carbohydrates. The caoutchouc portion is soluble in CS_2, CCl_4, trichloromethane or benzene, forming a viscous colloidal solution. When heated, rubber softens at 160°C, and melts at about 220°C. Rubber easily absorbs a large quantity of sulphur either by heating or in the cold by contacting with S_2Cl_2 etc. This process is called **vulcanization**. Carbon black, in a fine state of division, is used as a reinforcing filler; other substances, produced by the condensation of aldehydes with amines, retard the oxidation of vulcanized rubber. The uses of rubber are innumerable. See also **crêpe rubber**, **synthetic rubber**. For *foamed rubber*, see **expanded plastics**.

rubber blanket *(Print.)*. See **offset printing**.

rubber forming *(Eng.)*. Pressing a rubber pad over a die to form a sheet placed between them.

rubber jaw *(Vet.)*. See **osteodystrophia fibrosa**.

rubber plates *(Print.)*. Extensively used for all grades of work. Flexible, they can be affixed easily to rotary cylinders; have good inking qualities; and are durable; require precision grinding to exact thickness and **kiss impression**; hand-cut rubber plates are extensively used for bold designs, particularly on packages.

rubbers *(Build.)*. See **cutters**.

rubber stippler *(Build.)*. A tool with rubber prongs used in broken colour work and texture painting. The stippler is used to manipulate the material on the surface.

rubbing stone *(Build.)*. An abrasive stone with which the bricklayer rubs smooth the bricks which he has cut to a special shape. See **gauged arch**.

rubble *(Build.)*. Rough uncut stones, of no particular size or shape, used for rough work and filling in between facing walls, etc.

rubble concrete *(Civ.Eng.)*. A form of masonry often used on massive works such as solid masonry dams; composed of very large blocks of stone set about 15 cm (6 in) apart in fine cement concrete and faced with squared rubble or ashlar.

rubeanic acid *(Chem.)*. *Dithio-oxamide*, an orange crystalline powder, sparingly soluble in water but soluble in alcohol; used as a reagent to detect small amounts of copper, with which it forms a black precipitate.

rubefacient *(Med.)*. Producing reddening of the skin; any agent which does this, a counter-irritant.

rubella *(Med.)*. *German measles*. A mild acute viral infectious disease mainly affecting children, characterized by slight fever, enlargement of glands in the neck and at the back of the head, and a pink papular-macular rash. Infection in pregnancy can cause severe foetal abnormalities. Sometimes resembles mild measles, hence its common name.

rubellite *(Min.)*. The pink or red variety of tourmaline, sometimes used as a semiprecious gemstone.

Rubiaceae *(Bot.)*. Family of ca 7000 spp. of dicotyledonous flowering plants (superorder Asteridae). Cosmopolitan; most tropical species are trees and shrubs, all the (fewer) temperate species are herbs. The leaves are opposite and have stipules, the ovary is inferior and usually of two carpels. Includes coffee and *Cinchona*, the source of quinine.

rubicelle *(Min.)*. A yellow or orange-red variety of spinel; an aluminate of magnesium.

rubidium *(Chem.)*. A metallic element in the first group of the periodic system, one of the alkali metals. Symbol Rb, at. no. 37, r.a.m. 85.47, mp 38.5°C, bp 690°C, rel. d.

1.532. The element is widely distributed in nature, but occurs only in small amounts; the chief source is carnallite. The metal is slightly radioactive.

rubidium-strontium dating *(Geol.)*. A method of determining the age in years of geological material, based on the known decay rate of ^{87}Rb to ^{87}Sr.

RUBISCO *(Bot.)*. See **Ribulose 1,5-bisphosphate carboxylase oxygenase.**

RUBP carboxylase, RuBPCase *(Bot.)*. See **Ribulose 1,5-bisphosphate carboxylase oxygenase.**

rubric *(Print.)*. A heading or passage printed in red, the main text being in black. Also the marginal headings of minutes, etc. although printed in black.

ruby *(Min.)*. The blood-red variety of the mineral corundum, the oxide of aluminium (Al_2O_3), which crystallizes in the trigonal system. Also called *true ruby* (to distinguish it from the various types of **false ruby**) and *Oriental ruby*, though the adjective *Oriental* is quite unnecessary, since it merely stresses the fact that rubies come from the East (Burma, Thailand, Sri Lanka, Afghanistan). See also **balas ruby, ruby spinel.**

ruby *(Print.)*. An old type size approx. $5\frac{1}{2}$-point.

ruby pin *(Horol.)*. The impulse pin of a lever escapement.

ruby silver ore *(Min.)*. See **proustite, pyrargyrite.**

ruby spinel *(Min.)*. That variety of magnesian spinel, $MgAl_2O_4$, which has the colour, but none of the other attributes, of true ruby. Also *almandine spinel. Spinel ruby* is a deceptive misnomer.

ruche *(Textiles)*. A narrow woven or knitted fabric made with extended weft threads which may be bunched together for extra effect. The material is used as decorative trim round the edges of upholstery or applied to dresses.

rudaceous *(Geol.)*. Used of a sedimentary rock which is coarser in grain-size than sand.

rudder *(Aero.)*. A movable surface in a vertical plane for control of an aircraft in angles of yaw (i.e. movement in a horizontal plane about a vertical axis). Usually located at the rear end of the body and controlled by the pilot through a system of rods and/or cables or electric signals in **fly by wire**.

rudder *(Ships)*. Broad, flat device, varying in form, hinged vertically to, or behind, the sternpost of a vessel; the rudder serves to change the vessel's course when it is moved from a position in line with the keel.

rudder bar, pedals *(Aero.)*. A mechanism consisting of differential foot-operated levers by which the pilot actuates the rudder of a glider or an aircraft, or controls the pitch of a helicopter tail rotor through mechanical, hydraulic or electrical relaying devices.

rudder post *(Ships)*. See **stern frame.**

rudenture *(Arch.)*. A cylindrical moulding carved in imitation of a rope.

ruderal *(Bot.)*. A plant which grows usually on rubbish heaps or waste places.

rudiment *(Bot.,Zool.)*. The earliest recognizable stage of a member or organ.

Rudistes *(Geol.)*. A group of heavily-built lamellibranchs of coral-like form which are characteristic of the Cretaceous rocks formed in the southern ocean (the Tethys) of the period; Rudistids also occur in the Cretaceous Trinity Series of Texas and Mexico.

RUDP *(Bot.)*. Ribulose 1,5-bisphosphate, formerly called *ribulose diphosphate*.

Ruffini's organs *(Zool.)*. In Vertebrates, a type of cutaneous sensory nerve ending concerned with the perception of heat.

rufous *(Bot.)*. Red-brown.

rugose *(Biol.)*. Wrinkled. dim. *rugulose*.

rule *(Build.)*. See **floating rule.**

rule *(Print.)*. Type-high brass or metal strip of various thicknesses and designs; a dash or score (see **em rule, en rule**).

rule based system *(Comp.)*. A software system which represents knowledge by a set of simple conditional sentences.

rule border *(Print.)*. A frame of rules fitted around an advertisement or other displayed matter.

ruled surface *(Maths.)*. One generated motion of a straight line with one degree of freedom, e.g. a cone.

ruling *(Print.)*. The operation of making lines on writing, account-book, and ledger paper etc.; the paper is conveyed on an endless belt and makes contact with suitably adjusted disks or pens.

ruling gradient *(Civ.Eng.)*. The maximum gradient permissible for any given section of road or railway.

rumble *(Acous.)*. Low-frequency noise produced in disk recording when turntable is not dynamically balanced.

rumen *(Zool.)*. The first division of the stomach in Ruminants and Cetacea, being an expansion of the lower end of the oesophagus used for storage of food; the paunch.

rumenotomy *(Vet.)*. Operation of cutting into the rumen.

rumination, ruminant *(Zool.)*. The regurgitation of food that has already been swallowed, and its further mastication before reswallowing. *adj.* and *n.* **ruminant.**

run *(Build.)*. That part of a pipe which is in the same direction as that to which it is connected.

run *(Comp.)*. See **execute.**

run *(Horol.)*. The movement of the lever, in a lever escapement, to the banking pins, due to draw.

run *(Print.)*. The number of copies to be printed.

run *(Surv.)*. In a level tube, the movement of a bubble with change of inclination.

run around *(Print.)*. Type set alongside an illustration to less than the full measure.

runaway electron *(Electronics)*. One under an applied electric field in an ionized gas which acquires energy from the field at a greater rate than it loses through particle collision.

run in *(Print.)*. US term for *run on*.

runner *(Bot.)*. Stem growing more or less flat on the ground, with long internodes, rooting at the nodes and/or the tip and there producing new plantlet(s) from axillary or terminal bud(s), as in strawberry.

runner *(Eng.)*. The channel through which molten metal flows to a mould or other receptacle.

runner *(Foundry)*. The vertical passage into the interior of a mould through which the metal is poured. Also called *runner-gate*. See **ingate.**

runners *(Horol.)*. The cylindrical sliding pieces which support the work in a pair of turns.

runners *(Print.)*. Marginal figures for reference purposes indicating the number of each line in a poem or play.

runner stick *(Foundry)*. See **gate stick.**

running *(Build.)*. The operation of forming a plaster moulding, cornice etc. *in situ*, or on a bench. Running a **horsed** mould along the material while it is still plastic.

running bond *(Build.)*. The same as **stretching bond.**

running fit *(Eng.)*. A fit for rotating or sliding parts with sufficient clearance to support an oil film. National standards exist for these and other fits.

running heads *(Print.)*. The headings at the top of the page, the usual arrangement being title of book on left-hand page and title of chapter on right-hand page.

running on *(Print.)*. The actual printing of an edition after the **make ready** operations have been completed.

running rule *(Build.)*. A wood strip fixed temporarily to serve the same purpose as a running screed.

running screed *(Build.)*. A band of plaster laid on the surface of a wall as a guide to the movement of a horsed mould in the process of running a moulding.

running shoe *(Build.)*. The zinc part of a horsed mould, giving protection to the wood and facilitating running.

running tapes *(Print.)*. Tapes travelling at the press speed for the purpose of leading or conveying the web of paper.

running trap *(Build.)*. Tubular trap used in sanitary pipes having the inlet and outlet in horizontal alignment.

runoff *(Civ.Eng.,Geol.)*. The resultant discharge of a river from a catchment area; surface water as distinct from that rising from deep-seated springs.

run on *(Print.)*. An indication that a new paragraph is not to be made. Marked in copy and proof by a line running from the end of one piece of matter to the beginning of the next.

run-on chapters *(Print.)*. Chapters in a book which do

not commence on a new page but after a few lines of space.

run-out *(Foundry)*. See **break-out**.

run-out *(Image Tech.)*. The end of a print, i.e. the length of film between the last effective frame and the end.

runt disease *(Immun.)*. Disease which develops after injection of allogeneic lymphocytes into immunologically immature experimental animals. Characterized by loss of weight, failure to thrive, diarrhoea, splenomegaly and often death. This is an example of a **graft-versus-host reaction**.

run-through ruling *(Print.)*. A term which indicates that the ruling continues from edge to edge, horizontally or vertically, without interruption. See **stopped heading**.

run time *(Comp.)*. (1) The length of time between the beginning and completion of the execution of a program. (2) The time during which a program is being executed.

run-time error *(Comp.)*. See **execution error**.

run-time system *(Comp.)*. Complete set of software which must be in main store to enable user's program to be executed.

runway threshold *(Aero.)*. The usable limit of a runway; in practice it is usually the current downwind end which is intended.

runway visual markers *(Aero.)*. See **airport markers**.

runway visual range *(Aero.)*. In bad weather, the horizontal distance at which black-and-white markers of standard size are visible, the figure being transmitted to pilots approaching by **air traffic control**. Abbrev. *RVR*.

rupicolous *(Bot.,Zool.)*. Living or growing on or among rocks.

rupture *(Med.)*. (1) Forcible breaking or tearing of a bodily organ or structure. (2) To break or to burst (said of a blood vessel or viscus). (3) **Hernia**.

rupturing capacity *(Elec.Eng.)*. See **breaking capacity**.

rushes *(Image Tech.)*. The first positive prints made from motion picture negative immediately after processing.

Russell-Saunders coupling *(Phys.)*. Extreme form of coupling between orbital electrons of atoms. The angular and spin momenta of the electrons combine and the combined momenta then interact. Also called *l-s coupling*.

Russell's test *(Elec.Eng.)*. A method of determining the insulation resistance of a 3-wire d.c. distribution network. The value is obtained by calculation from readings of an electrostatic voltmeter connected between the neutral wire and earth, both with and without a known resistance in parallel.

russet *(Bot.)*. A brownish, roughened, corky layer or patch on the surface of a fruit (or other organ) as a varietal characteristic or as the result of disease or of injury from insects or spraying.

rust *(Bot.)*. One of a number of plant diseases, some economically very important, caused by biotrophic fungi of the order Uredinales and often recognizable by the rounded or elongated pustules of rust coloured spores on stems or leaves, e.g. black rust of cereals caused by *Puccinia graminis*.

rust *(Eng.)*. Product of oxidation of iron or its alloys, due either to atmospheric attack or electrolytic effect of cell action round impurities.

rusticated ashlar *(Build.)*. Ashlar work in which the face stands out from the joints, at which the arrises are bevelled. The face may be finished rough or smooth or tooled in various ways.

rustic joint *(Build.)*. A sunken joint between adjacent building-stones.

rustics *(Build.)*. Bricks having a rough-textured surface, often multicoloured. Also **texture bricks**.

rust joint *(Build.)*. A watertight joint between adjoining lengths of guttering or pipes.

rusty gold *(Min.Ext.)*. Native gold which has become surface-filmed by adherent staining substances and is non-amalgamable and non-treatable by cyanide process in consequence.

rut *(Zool.)*. The noise made by certain animals as Deer, when sexually excited; oestrus; to be sexually excited, i.e. to be in the oestrous period; to copulate.

ruthenium *(Chem.)*. A metallic element. Symbol Ru, at. no. 44, r.a.m. 101.07, mp 2400°C, rel.d. 12.26. The metal is silvery-white, hard and brittle. It occurs with the platinum metals in osmiridium, and is used in certain platinum alloys.

Rutherford atom *(Phys.)*. Earliest modern concept of atomic structure, in which all the positive charge and nearly all the mass of the atom is in the nucleus. Electrons, equal in number to the atomic number, occupy the rest of the volume and make the atom electrically neutral.

Rutherford scattering *(Phys.)*. See **scattering**.

rutilant *(Bot.)*. Brightly coloured in red, orange or yellow.

rutilated quartz *(Min.)*. See **needle stone**.

rutile *(Min.)*. Titanium dioxide which crystallizes as reddish brown prismatic crystals in the tetragonal system. It is found in igneous and metamorphic rocks, and in sediments derived from these, also in quartz (see **flèches d'amour**), and it is a source of titanium.

RWP *(Build.)*. Abbrev. for *RainWater Pipe*.

rybat *(Build.)*. An inband or outband.

Rydberg constant *(Phys.)*. The frequency of atomic spectrum lines in a given series are related by the Rydberg formula. The Rydberg constant R involved was first deduced from spectroscopic data but has since been shown to be a universal constant.

$$R = \frac{2\pi^2 e^4}{ch^3} M_r$$

where M_r is the reduced mass of the electrons, e is the electronic charge, c is the velocity of light and h is Planck's constant.

Rydberg formula *(Phys.)*. A formula, similar to that of Balmer, for expressing the wave numbers (ν) of the lines in a spectral series:

$$\nu = R\left[\frac{1}{(n+a)^2} - \frac{1}{(m+b)^2}\right],$$

where n and m are integers and $m > n$, a and b are constants for a particular series, and R is the *Rydberg constant*.

R-Y signal *(Image Tech.)*. Component of colour TV chrominance signal. Combined with luminance (Y) signal it gives primary red component.

S

s *(Genrl.)*. Symbol for *second* (time).

s *(Chem.,Phys.)*. Symbol for (1) distance along a path, (2) solubility, (3) specific entropy.

s- *(Chem.)*. An abbrev. for: (1) symmetrically substituted (also *sym-*); (2) *secondary*, i.e. substituted on a carbon atom which is linked to two other carbon atoms; (3) *syn-*, i.e. containing the corresponding radicals on the same side of the plane of the double bond between a carbon and a nitrogen atom or between two nitrogen atoms.

σ *(Chem.)*. A symbol for the diameter of a molecule.

σ *(Phys.)*. Symbol for (1) conductivity, (2) normal stress, (3) Stefan-Boltzmann constant, (4) surface charge density, (5) surface tension, (6) wave number.

S *(Chem.)*. (1) In names of dyestuffs, a symbol for *black*. (2) The symbol for *sulphur*.

S *(Immun.)*. Abbreviation for Svedberg unit, referring to the sedimentation coefficient of proteins analysed in an ultracentrifuge.

S *(Phys.)*. Symbol for siemens.

S- *(Chem.)*. Prefix denoting left handed. See **Cahn-Ingold-Prelog system** for absolute configuration.

S *(Phys.)*. Symbol for (1) area, (2) entropy, (3) Poynting vector.

Σ *(Genrl.)*. Symbol for *sum of*.

7S, 19S antibody *(Immun.)*. Immunoglobulins with sedimentation coefficients about 7 S or 19 S respectively. Terms often used as synonyms for IgG and IgM.

sabin *(Acous.)*. Unit of acoustic absorption; equal to the absorption, considered complete, offered by 1 ft^2 of open window to low-frequency reverberant sound waves in an enclosure. Obsolescent.

Sabine reverberation formula *(Acous.)*. Earliest formula (named after investigator) for connecting the reverberation of an enclosure, *T* seconds, with the volume, *V* in cubic metres, and the total acoustic absorption in the enclosure, ΣaS, where *a* is the absorption coefficient of a surface of *S* square metres. The formula is $T = 0.16V/\Sigma aS$.

sabkha *(Geol.)*. A flat salt-encrusted coastal plain, common in Arabia.

sable *(Build.)*. (1) Hair obtained from a small animal of the weasel family and used as the filling for signwriters brushes. (2) Heraldic term for black.

sabulose, sabuline *(Bot.)*. Growing in sandy places.

sac *(Bot.)*. Any sort of bag-like structure or pouch.

saccadic eye movements *(Behav.)*. The rapid, ballistic movements of the eyes used in scanning a scene; these involuntary eye movements occur about every quarter of a second even when the eyes are fixated on an object.

saccate *(Bot.)*. Bag-like or pouch-like.

saccharides *(Chem.)*. *Carbohydrates*, which according to their complexity are usually divided into *mono-*, *di-*, *tri-*, and *polysaccharides*.

saccharimetry *(Chem.)*. The estimation of the percentage of sugar present in solutions of unknown strength, especially by measurements of optical activity.

saccharin *(Chem.)*. *2-Sulphobenzimide*, a white crystalline powder, 300 times as sweet as sugar, not very soluble in water. The imido-hydrogen is replaceable by Na, forming a salt which is readily soluble in water. It is used in medicine in cases where sugar is harmful, e.g. in diabetes.

saccharobiose *(Chem.)*. Cane sugar or sucrose.

saccharoidal textures *(Geol.)*. Granular textures which resemble sugar; found especially in limestones and marbles.

saccharometer *(Chem.)*. A hydrometer which is used to determine the concentration of sugar in a solution.

Saccharomyces cerevisiae *(Biol.)*. The yeast used widely in bread and alcohol manufacture. It can also be used as an eukaryotic *host* for growing and *expressing* DNA sequences.

saccule, sacculus *(Zool.)*. A small sac; the lower chamber of the auditory vesicle in Vertebrates. adj. *sacculate*.

sacculiform *(Biol.)*. Shaped like a little bag.

saccus *(Bot.)*. A large, hollow, pouch-like projection of the outer part of the wall of a pollen grain.

sacking *(Textiles)*. Coarse fabrics of jute, flax or polyolefin used for making sacks.

sacralgia *(Med.)*. Pain in the sacral region.

sacralization *(Med.)*. A developmental anomaly in which one or both transverse processes of the 5th lumbar vertebra become abnormally large and strong, appearing to form part of the sacrum.

sacral ribs *(Zool.)*. Bony processes uniting the sacral vertebrae to the pelvis, distinct in Reptiles but fused to the transverse processes in other Tetrapoda.

sacral vertebrae *(Zool.)*. In higher Craniata, those vertebrae which articulate with the ilia of the pelvis via sacral ribs, there being one in the Frog and two in the Lizard, coming between the lumbar vertebrae and the caudal vertebrae (if any). In Birds and Mammals they are fused with other vertebrae to form the **sacrum**.

sacrificial anode *(Ships)*. An anode, used in **cathodic protection**, commonly magnesium alloyed with about 6% aluminium and 3% zinc. When attached to steel in sea water the natural potential difference is such as to make the steel cathodic. The anode dissolves and requires renewal.

sacrificial protection *(Eng.)*. The prevention of *electrolytic corrosion* in a component by providing another electrochemically more active metal close by and electrically connected to it. See **sacrificial anode**.

sacroiliac joint *(Zool.)*. In some Craniata, the almost immovable joints between the **sacrum** and the two ilia of the pelvis. The articular surfaces of the bones are partly covered with cartilage and partly roughened for the attachment of the *sacroiliac ligament*.

sacrum *(Zool.)*. In the skeleton of some Craniata, part of the vertebral column which articulates immovably with the ilium of the pelvis at the **sacroiliac joint**, being composed of several fused vertebrae, including the **sacral vertebrae**. In Birds it consists of one thoracic vertebra, five or six lumber vertebrae, the two sacral vertebrae and the anterior five caudal vertebrae (and is sometimes called the **synsacrum**). In Mammals it comprises varying numbers of vertebrae in different orders (e.g. four in the Rabbit and five in Man), the first one or two being regarded as sacral, and the others as caudal. The former have low spines and expanded ventral surfaces for the attachment of muscles.

saddle *(Civ.Eng.)*. A block surmounting one of the towers of a suspension bridge, providing bearing or fixing for the suspension cables.

saddle *(Eng.)*. The part of a lathe which slides on the bed, between headstock and tailstock.

saddle-back board *(Build.)*. A narrow board, chamfered along each of the upper edges, which is fixed on the floor across the threshold of a doorway so that the gap beneath the door will be small when the latter is shut and large enough when it opens to accommodate a carpet.

saddle-back coping *(Build.)*. A copingstone whose upper surface slopes away on both sides from the middle.

saddle bar *(Build.)*. A metal bar fixed across a window to support glazing held in lead cames.

saddle coils *(Electronics)*. Rectangularly formed coils which are bent around the neck of a cathode-ray tube; used for magnetic deflection of the beam.

saddle key *(Eng.)*. A key sunk in a key-way in the boss, but having a concave face which bears on the surface of the shaft, which it grips by friction only. See **key**.

saddle point *(Nuc.Eng.)*. Point on plot of potential energy against distortion for nucleus at which fission will occur, instead of return to equilibrium.

saddle scaffold *(Build.)*. A scaffold erected over a roof from standards on both sides of the building; used for repair work on, e.g. a chimney at the middle of the roof.

saddle-stitching *(Print.)*. A method of wire-stitching in which the book is placed astride a saddle-shaped support and stitched through the back.

saddle stone *(Build.)*. An **apex stone**.

sadism *(Behav.)*. Sexual gratification through the infliction of pain on others; pleasure in cruel behaviour.

sado-masochism *(Behav.)*. The pairing of a **sadist** and a **masochist** to satisfy their mutually complementary sexual needs.

SAE *(Aero.)*. Abbrev. for *Society of Automotive Engineers* (US). Gives name to a widely-used viscosity scale for classifying motor oils.

safe *(Print.)*. Condition of press when locked. See **lock**.

safe area *(Image Tech.)*. The area of a film frame which, when transmitted on television, is reasonably certain to be produced on a domestic receiver.

safe edge *(Eng.)*. The edge of a file on which no teeth are cut.

safeguard *(Civ.Eng.)*. See **check rail**.

safe light *(Image Tech.)*. Lighting fixture in a photographic dark-room whose intensity and colour of visible illumination permits the safe handling of unprocessed materials.

safe load *(Civ.Eng.)*. The maximum working load which a member or structure is designed to carry. See **factor of safety**.

safety action *(Horol.)*. The action in a lever escapement that ensures that the notch in the lever is always in its correct position for the reception of the impulse pin.

safety arch *(Build.)*. See **discharging arch**.

safety barrier *(Aero.)*. A net which is erected on the forward part of the deck of an aircraft carrier to stop any aircraft which misses the *arrester gear*. A cable and/or nylon net which can be quickly raised to prevent an aircraft from over-running the end of a runway. The barrier is held by friction brakes or weights so that it imposes a 1 or 2 g deceleration on the aircraft.

safety cage *(Min.Ext.)*. A cage fitted with a 'safety catch' to prevent it from falling if the hoisting rope breaks.

safety coupling *(Eng.)*. A friction coupling adjusted to slip at a predetermined torque, to protect the rest of the system from overload.

safety cut-out *(Elec.Eng.)*. An overload protective device in an electric circuit.

safety factor *(Nuc.Eng.)*. Of a fusion system, the **aspect ratio** multiplied by the ratio of toroidal to poloidal field. This requires to be greater than unity for magnetohydrodynamic stability.

safety film *(Image Tech.)*. Film having a base of low inflammability, usually cellulose triacetate or polyester. Also *non-flam.*

safety finger *(Horol.)*. The pin or finger attached to the end of the lever adjacent to the notch. It butts against the edge of the safety roller if the escapement is subject to a jerk, or when setting the hand back, in the case of a watch. Also called *guard pin* and sometimes *dart*.

safety fuse *(Elec.Eng.)*. A protective fuse in part of an electric circuit.

safety glass *(Glass)*. (1) Laminated glass, formed of a sandwich of a plastic material such as polyvinyl butyral, (0.4–0.75 mm) between two glass sheets, certain intermediate layers being used to facilitate adhesion. (2) Toughened glass, formed by heating a sheet of glass to the point of incipient softening and then chilling it rapidly to a certain extent, but not sufficiently to cause fracture. The stresses have the effect of imparting a considerably greater resistance to shock. (3) Glass reinforced with wire mesh incorporated in the body.

safety height *(Aero.)*. The height below which it is unsafe to fly on instruments because of high ground.

safety lamp *(Min.Ext.)*. Oil-burning miners' lamp which will not immediately ignite firedamp or gas in a coal mine, e.g. a **Davy lamp**. Also used for detecting gas.

safety lintel *(Arch.)*. A lintel doing the work of a relieving arch, and serving to protect another more decorative lintel used for architectural reasons.

safety plug *(Eng.)*. See **fusible plug**.

safety rail *(Civ.Eng.)*. See **check rail**.

safety rods *(Nuc.Eng.)*. Rods of neutron-absorbing material capable of rapid insertion into a reactor core to shut it down, in case of emergency.

safety roller *(Horol.)*. The roller mounted on the balance staff of a lever escapement, against the edge of which the guard pin acts when the safety action is brought into play.

safety speed *(Aero.)*. The lowest speed above stalling at which the pilot can maintain full control about all three axes. It is particularly applicable to multi-engined aircraft, where it is taken to be the minimum speed at which control can be maintained after complete failure of the engine most critical to directional control.

safety switch *(Elec.Eng.)*. See **emergency stop**.

safety valve *(Eng.)*. A valve, spring or dead-weight loaded, fitted to a boiler or other pressure vessel, to allow fluid to escape to the atmosphere when the pressure exceeds the maximum safe value.

safranines *(Chem.)*. A group of azine dyestuffs. They are 2,8-diamino derivatives and have also a phenyl or a substituted phenyl group attached to the nitrogen in position 10.

sag correction *(Surv.)*. A correction applied to the observed length of a base line, to correct for the sag of the measuring tape.

SAGE *(Aero.)*. Air defence system whereby information is received from radar and other sources and is processed at a central station to give an evaluation of a situation. (*Semi-Automatic Ground Environment*).

saggar *(Eng.)*. A clay box in which pottery is packed for baking.

sagging *(Glass)*. Forming glass by reheating until it conforms with the mould or form on which it rests.

sagging *(Ships)*. This occurs when the ends of the ship are supported on wave crests while the middle is in a trough or when the middle is more heavily loaded than the ends. If the ship actually bends she is said to be *sagged*. Cf. **hogging**.

sagitta *(Civ.Eng.)*. (1) See **keystone**. (2) See **rise** (1).

sagittal *(Zool.)*. Elongate in the median vertical longitudinal plane of an animal, as the *sagittal* suture between the parietals, the *sagittal* crest of the skull; used also of sections.

sagittal field *(Phys.)*. The image surface formed by the sagittal foci of a series of object points lying in a plane at right angles to the axis.

sagittal focus *(Phys.)*. The focus of an object point lying off the axis of an optical system in which the image is drawn out by the astigmatism of the system into a line radial to the optical axis.

sagittate *(Bot.,Zool.)*. Shaped like an arrowhead with the barbs pointing backwards.

Sahelian drought *(Ecol.)*. The pattern of drought in the northern regions of West Africa, where several years of below-average rainfall often occur in succession.

sahlite, salite *(Min.)*. A mineral of the clino-pyroxene group, intermediate in composition between diopside and hedenbergite.

sailcloth *(Textiles)*. Highly technical woven fabrics designed for use in sailing ships and yachts. Originally closely woven cotton or linen canvases were used but these have been largely displaced by nylon, polyester and even aramid fabrics. The weaves are carefully designed to give just the right air porosity under a wide range of wind velocities.

sailing courses *(Build.)*. See **oversailing courses**.

sail-over *(Build.)*. To project over. See **over-sailing courses**.

sailplane *(Aero.)*. A glider designed for sustained motorless flight by the use of air currents. The most advanced methods of streamlining and very high *aspect ratio* are used to reduce *drag* to the barest minimum.

Saint Elmo's fire *(Phys.)*. Brush discharge from isolated points above the ground, e.g., ship's mast and aircraft.

Sakmarian *(Geol.)*. The lowest stage of the Permian system in eastern Europe and USSR.

sal-ammoniac *(Min.)*. Chloride of ammonia, which crystallizes in the cubic system. It is found as a white encrustation around volcanoes, as at Etna and Vesuvius.

salbutamol *(Med.)*. A **β_2-sympathomimetic** which produces bronchodilatation. Used to treat asthma and chronic obstructive airways disease (chronic bronchitis and emphysema).

salic minerals *(Geol.)*. Those minerals of the *norm* which are rich in silicon and aluminium, including quartz, feldspars, and feldspathoids.

salicylic acid *(Chem.,Med.)*. An antiseptic and an important intermediate for a number of derivatives, e.g. aspirin.

salient *(Surv.)*. A jutting-out piece of land.

salient angle *(Maths.,Surv.)*. One in a closed figure which points outward, being below 180°. Cf. **re-entrant angle**.

Salientia *(Zool.)*. A superorder of Amphibia. Adults four-legged, the hind limbs being especially well-developed, and short-bodied, with no tail. Toads and Frogs. Also known as *Anura, Batrachia*.

salient junction *(Build.)*. See **external angle**.

salient pole *(Elec.Eng.)*. A type of field pole protruding beyond the periphery of the circular yoke in the case of a stator field system, or the circular core in the case of a rotor field system.

salient-pole generator *(Elec.Eng.)*. An alternating-current generator whose rotor field system is of the salient-pole type, e.g. in the case of slow-speed water-turbine-driven generators.

salina, saline lake *(Geol.)*. A salt lake, lagoon, marsh, spring etc. See **salt lakes**.

salinometer *(Phys.)*. A **hydrometer** for measuring the density of sea water, the stem being scaled in arbitrary units; used by engineers for estimating the amount of dissolved solids in feed water.

salite *(Min.)*. See **sahlite**.

saliva *(Zool.)*. The watery secretion produced by the salivary glands, whose function is to lubricate the passage of food and, sometimes, to carry out part of its digestion. In Insects saliva may contain amylase, invertase, protease, and lipase, according to the usual diet, and in some blood-sucking insects it contains anti-coagulants. In Mammals it contains water, mucin and, in Man and some herbivores, the amylase ptyalin, which catalyses the breakdown of starch to maltose.

salivary gland chromosome *(Biol.)*. Polytene chromosome found in the salivary glands of larval dipterans, e.g. *Drosophila melanogaster*. Conspicuously banded, whether stained or not, and used for gene mapping and other studies of chromosome organization.

salivary glands *(Zool.)*. Glands present in many land animals, the ducts of which open into or near the mouth.

sallenders *(Vet.)*. See **mallenders and-**.

sally *(Build.)*. A re-entrant angle cut into the end of a timber, so as to allow it to rest over the arris of a cross-timber.

Salmonella *(Biol.)*. A group of Gram-negative, carbohydrate fermenting, nonsporing, bacilli, all pathogenic to animals. Associated with food poisoning in man. Includes *S. typhi* and *S. paratyphi*.

salmonellosis *(Med.)*. A form of food poisoning in man due to infection by *Salmonella* bacilli. Characterized by vomiting, diarrhoea and abdominal pain.

salmonellosis *(Vet.)*. Infection with *Salmonella* spp. Most often associated with *S. dublin* or *S. typhimurium*. All warm-blooded animals can be affected. Characterized by dullness, pyrexia, diarrhoea (often with blood), dehydration and death. Important **zoonosis** and therefore notifiable disease in UK. Vaccine available.

Salmoniformes *(Zool.)*. Order of freshwater and **anadromous** *Osteichthyes* of great commercial and sporting importance. Salmon, Trout, Charr, Pike.

salping- *(Genrl.)*. A prefix from the Gk. *salpinx* (gen. *salpingos*), trumpet, referring esp. to the Fallopian tubes. See **salpinx**.

salpingectomy *(Med.)*. Removal of a Fallopian tube.

salpingitis *(Med.)*. Inflammation of a Fallopian tube.

salpingo-oöphorectomy *(Med.)*. Removal of a Fallopian tube and of the ovary on the same side.

salpingo-oöphoritis *(Med.)*. Inflammation of both the Fallopian tube and the ovary.

salpingorrhaphy *(Med.)*. The suturing of a Fallopian tube to the ovary on the same side, after a part of the latter has been removed.

salpingostomy *(Med.)*. The operative formation of an opening into a Fallopian tube whose natural opening has been closed by disease.

salpingotomy *(Med.)*. Incision into a Fallopian tube.

salpinx *(Bot.)*. *Lagenostome*. Structure, adapted for the reception of pollen, at the distal end of the nucellus of ovules of many seed-ferns.

salpinx *(Med.,Zool.)*. The eustachian tube; the fallopian tube; a trumpet-shaped structure. adj. *salpingian*.

SALR *(Meteor.)*. See **saturated adiabatic lapse rate**.

salsuginous *(Bot.)*. Growing on a salt marsh.

salt *(Chem.)*. A compound which results from the replacement of one or more hydrogen atoms of an acid by metal atoms or electropositive radicals. Salts are generally crystalline at ordinary temperatures, and form positive and negative ions on dissolution in water, e.g. chlorides, nitrates, carbonates, sulphates, silicates and phosphates. For *common* or *rock salt*, see **halite**.

saltation *(Biol.)*. A sudden heritable variation in a species. Now applied more often to large morphological changes which occur during evolution over a time period shorter than that required by similar changes earlier or later. Said to be difficult to explain by **natural selection**.

saltatorial, saltatory *(Zool.)*. Used in, or adapted for, jumping, as the third pair of legs in Grasshoppers.

saltatory conduction *(Zool.)*. The process of nervous conduction along a myelinated nerve axon where the impulse jumps from one **node of Ranvier** to the next.

salt bath *(Eng.)*. A bath of molten salts used for heating steel, for hardening or tempering. Salt baths give uniform heating and some protect against oxidation. Certain salts are employed only to transmit heat to the immersed material, and different salts are used for different temperatures. For *tempering* baths, sodium and potassium nitrate are used. For *hardening* baths, sodium cyanide, and sodium, potassium, barium and calcium chlorides are used. An electric salt-bath furnace is a conductor-type electric furnace in which the salt is melted by the passage of the current.

salt dome *(Geol.)*. A diapiric salt plug which has arched up, or broken through, the sediments into which it has been intruded.

salt gland *(Bot.)*. Structure at the leaf surface which actively secretes sodium chloride in many salt marsh and mangrove species.

saltigrade *(Zool.)*. Progressing by jumps, as Grasshoppers.

salting *(Min.Ext.)*. Fraudulent enrichment of ore samples, made to increase apparent value of a mine. Originally, to sprinkle salt in dry mines to allay dust.

salting-out *(Chem.)*. The removal of an organic compound from an aqueous solution by the addition of salt.

salt marsh *(Bot.)*. A marsh characterized by saline soil, most often in estuaries and subject to marine inundation. See also **halosere**.

saltpetre *(Min.)*. See **potassium nitrate**.

salt, saline lakes *(Geol.)*. Enclosed bodies of water in areas of inland drainage, whose concentration of salts in solution is much higher than in ordinary river water. See **soda lakes**.

sal volatile *(Chem.)*. Ammonium carbonate, the main constituent of smelling salts.

Salyut *(Space)*. A modular USSR space station which has been improved steadily since its first use in 1971; it has demonstrated that astronauts can remain in orbit for periods of up to a year.

samara *(Bot.)*. A dry, indehiscent fruit of which part of the wall forms a flattened wing, e.g. ash key.

samariform *(Bot.)*. Winged, like an ash key.

samarium *(Chem.)*. A metallic element. Symbol Sm, at. no. 62, r.a.m. 150.35, mp 1350°C, bp 1600°C, hard and brittle, rel.d. 7.7. Found in allanite, cerite, gadolinite and samarskite. Feeble, naturally radioactive, can be produced by decay of fission fragments, and forms reactor poison.

sampled data tracking *(Radar)*. That used with high speed electronic beam switching, using **phased arrays**, allowing data to be gathered on each of several tracks virtually simultaneously.

sampling *(Ecol.)*. The survey of a small but representative part of a population or stand of vegetation with the intention of obtaining an estimate of some characteristics of the whole, e.g. age distribution or species present.

sampling *(Phys.)*. Selection of an irregular signal over stated fractions of time or amplitude (pulse height).

sampling *(Telecomm.)*. Process of measuring at regular intervals the level of a varying (analog) waveform, in order to convert it to digital form or to achieve **time-division multiplex**. To allow the reconstruction of the waveform, the sampling rate must exceed twice the highest frequency component of the sampled waveform. See **Nyquist limit, rate**.

sampling distribution *(Stats.)*. The probability distribution describing the variation of a statistic in repeated sampling, or hypothetical repetitions of the same experiment.

sampling error *(Stats.)*. Variation due to a sample from a population necessarily giving only incomplete information about the population.

SAN *(Plastics)*. A copolymer of **styrene** and **acrylonitrile**.

sand *(Foundry)*. See **moulding sands**.

sand *(Geol.)*. A term popularly applied to loose, unconsolidated accumulations of detrital sediment, consisting essentially of rounded grains of quartz. In the mechanical analysis of soil, sand, according to international classification, has a size between 1/16 and 2 mm. See **silt**. In coral sand the term implies a grade of sediment the individual particles of which are fragments of coral, not quartz.

sandalwood *(For.)*. A scented wood from several trees of the family *Santalaceae*, found in the East Indies. Important for its essential oil and used for joss-sticks and small ornaments. One species is found in Australia.

sand-blasting *(Eng.)*. A method of cleaning metal or stone surfaces by sand, steel shot, or grit blown from a nozzle at high velocity; also used for forming a key on the surface of various materials requiring a finish, such as enamel.

sand casting *(Foundry)*. Formation of shapes by pouring molten metal into a shaped cavity in a **moulding flask**.

sand colic *(Vet.)*. Colic caused by the collection of sand in the intestines.

sandcrack *(Vet.)*. A fissure of the horse's hoof.

sand cushion *(Civ.Eng.)*. A bag of sand placed beneath a helmet to protect the top of the pile from damage due to impact of the monkey when it is being driven.

sand dunes *(Geol.)*. Rounded or crescentic mounds of loose sand which have been piled up by wind action on seacoasts or in deserts. See also **barchan**.

sand fill *(Min.Ext.)*. Underground support of worked-out stopes by return of ore tailings from mill, usually by hydraulic flow.

sandfly fever *(Med.)*. *Phlebotomus fever*. An acute disease caused by infection with a virus conveyed by the bite of a sand-fly *Phlebotomus papatasii*; characterized by a 3-day fever, pains in the joints and the back, diarrhoea, and a slow pulse.

sand lime bricks *(Build.)*. Bricks made by mixing suitable sand with approx. 6% of hydrated lime and water, moulding under high pressure and then curing in steam at high pressure.

Sandmeyer's reaction *(Chem.)*. The replacement of the diazonium group, $—N_2{}^+$, in a diazonium compound by chlorine, bromine, or the cyanogen radical, which is effected by heating a solution of the diazonium compound with, e.g. a concentrated solution of cuprous chloride in hydrochloric acid. In this case the diazonium group is replaced by Cl, with evolution of gaseous N_2.

sandpaper *(Build.)*. Stout paper or cloth with a thin coating of fine sand glued on to one side, for use as an abrading material.

sand-pump dredger *(Civ.Eng.)*. A long pipe reaching down from a vessel into the sand, the latter being raised under the suction of a centrifugal pump and discharged into the vessel itself or an attendant barge. Also called a *suction dredger*.

sands *(Min.Ext.)*. Particles of crushed ore of such a size that they settle readily in water and may be leached by allowing the solution to percolate. See also **slimes**.

sandstone *(Geol.)*. Compacted and cemented sedimentary rock, which consists essentially of rounded grains of quartz, between the diameters of 0.06 and 2 mm, with a variable content of 'heavy mineral' grains. According to the nature of the cementing materials the varieties *calcareous sandstone*, *ferruginous sandstone*, *siliceous sandstone* may be distinguished; *glauconitic sandstone*, *micaceous sandstone* etc. are so termed from the presence in quantity of the mineral named.

sand trap *(Paper)*. An inclined trough across which bars are set at intervals. During the passage of the pulp to strainers, any heavy particles such as sand sink to the bottom, and are retained by the bars. Also called a *riffler*.

sand volcano *(Geol.)*. A structure formed by sand flowing upwards through an overlying bed of sediment and spilling out on to the surface; not related to any type of volcanic activity or volcanic rock.

sandwich *(Phys.)*. Photographic nuclear research emulsion forming a series of thin layers with intervening layers in which some event is to be studied.

sandwich beam *(Build.)*. See **flitch beam**.

sandwich compounds *(Chem.)*. Compounds in which a metal atom is 'sandwiched' between two rings, e.g. dibenzene chromium. Ferrocene (dicyclopentadienyl iron) is another example.

sandwich construction *(Aero.)*. Structural material, mainly used for skin or flooring, possessing exceptionally good stiffness for weight characteristics. It consists of two approximately parallel thin skins with a thick core having different mechanical properties, so that the tensile and compressive stresses develop in the skin, and the core both stabilizes these surfaces and gives great strength in bending; core materials range from balsa wood through metal-foil honeycomb (light-alloy or steel) to corrugated sheet.

sandwich irradiation *(Radiol.)*. The irradiation of tissues from opposite sides.

sandwich technique *(Immun.)*. A technique for the detection of antibody or antibody-producing cells in histological preparations. A first layer of antigen is applied and allowed to react with the antibody in the section. After washing this is followed by a second layer of fluorochrome-labelled antibody specific for the antigen. The antigen is 'sandwiched' between the two layers of antibody.

sanidine *(Min.)*. A form of potassium feldspar similar in chemical composition to orthoclase, but physically different, formed under different conditions and occurring in different rock types. It is the high-temperature form of orthoclase, to which it inverts below 900°C. Occurs in lavas and dyke-rocks.

sanserif *(Print.)*. A type face without serifs, e.g. Helvetica.

sap *(Bot.)*. An aqueous solution present in xylem, phloem, cell or vacuole, released on wounding. Cf. **phloem** sap, **xylem** sap.

sapele *(For.)*. An African tree, *Entandrophragma cylindricum*, yielding a mahogany-like silky-grained hard-wood used for furniture etc.

saphir d'eau *(Min.)*. French 'water sapphire'. A misnomer for an intense-blue variety of the mineral cordierite, occurring in water-worn masses in the river gravels of Ceylon; used as a gemstone.

saponification *(Build.)*. The action of alkali on oil paint whereby the paint is softened causing a defect.
saponification *(Chem.)*. The hydrolysis of esters into acids and alcohols by the action of alkalis or acids, or by boiling with water, or by the action of superheated steam. It is the reverse process to **esterification** if acids are used, but, when alkalis are used then soaps result, hence the term.
saponification number *(Chem.)*. The number of milligrams of potassium hydroxide required to saponify 1 g of a fat or oil.
saponins *(Chem.)*. Steroid vegetable glycosides that act as emulsifiers of oils. They dissolve the red corpuscles, irritate the eyes and organs of taste and are toxic to lower animals, e.g. *digitonin*, found in *Digitalis purpurea*.
saponite *(Min.)*. Hydrated aluminosilicate of magnesium. A clay mineral of the smectite (montmorillonite) group, occurring as white soapy masses in serpentinite. Also called *bowlingite*.
sapphire *(Min.)*. The fine blue transparent variety of crystalline corundum, of gemstone quality; obtained chiefly from Sri Lanka, Kashmir, Thailand, Cambodia and Australia.
sapphire needle *(Acous.)*. A gramophone record reproducing stylus ground from natural sapphire; by virtue of its hardness in comparison with that of the record surface, it is relatively hard-wearing although inferior in this respect to the diamond stylus.
sapphirine *(Min.)*. A silicate of magnesium and aluminium with lesser iron, crystallizing in the monoclinic system. It occurs as blue grains in metamorphosed, aluminous, silica-poor rocks.
saprobe *(Bot.)*. An organism such as a bacterium or fungus which obtains its nourishment osmotrophically from dead organic matter. Cf. **saprophyte**.
saprogenous *(Bot.)*. Growing on decaying matter.
sapropel *(Geol.)*. Slimy sediment laid down in stagnant water, largely consisting of decomposed algal material. A source material for oil and natural gas.
sapropelite *(Geol.)*. A term applied to coals derived from algal materials. Cf. *humite*.
saprophilous *(Bot.)*. Saprogenous.
saprophyte *(Biol.)*. An organism living heterotrophically and osmotrophically on dead organic matter. adj. *saprophytic*. Cf. **saprobe**.
saprotrophy *(Bot.)*. Heterotrophic nutrition based on non-living (dead) organic matter.
sapr-, sapro- *(Genrl.)*. Prefix from Gk. *sapros*, rotten, rancid.
sap wood *(Bot.)*. *Alburnum*. The outer, younger part of the wood of a tree or shrub, pale in colour, with living parenchyma cells and still conducting sap. Cf. **heartwood**.
SAR *(Radar)*. Abbrev. for *Synthetic Aperture Radar*.
SARAH *(Radar)*. Abbrev. for *Search and Rescue Homing*; a system for facilitating rescue when aircraft go down at sea, consisting of a small beacon transmitter which sends a coded pulse to rescue craft. Also used to guide support vessels to spacecraft after a sea landing.
Saran *(Plastics)*. Synthetic fibre or film based on a copolymer of vinylidene chloride and vinyl chloride.
sarcodic, sarcodous, sarcoid *(Zool.)*. Pertaining to or resembling flesh.
Sarcodina *(Zool.)*. Class of Protozoa with pseudopodia containing both irregular, amoeboid forms and others possessing regular calcareous or siliceous tests, e.g. *Radiolaria, Foraminifera*.
sarcoidosis *(Med.)*. A disease, cause unknown, characterized by granulomatous lesions; affects the lungs particularly, and other organs, including liver, skin and brain.
sarcolemma *(Zool.)*. The extensible sheath of a muscle fibre enclosing the contractile substance.
sarcoma *(Med.)*. A malignant tumour of connective tissue origin (e.g. of fibrous tissue, bone, cartilage); the tumour invades adjacent tissue and organs, and metastases are formed via the blood stream. pl. *sarcomata* or *sarcomas*. Cf. **carcinoma**
sarcomatosis *(Med.)*. The presence of many sarcomata in the body.
sarcomatous *(Med.)*. Pertaining to, of the nature of, or resembling sarcoma.
sarcomere *(Biol.)*. The basic contractile unit of the *myofibril*.
sarcophagous *(Zool.)*. Feeding on flesh.
sarcoplasmic reticulum *(Biol.)*. A network of tubules associated with muscle fibrils which acts as the source of calcium ions that stimulate contraction.
sarcoptic mange *(Vet.)*. Mange of animals due to mites of the genus *Sarcoptes*.
sarcosine *(Chem.)*. *Monomethyl-glycine*, $H_3C.HN.-CH_2.COOH$. It is obtained by the decomposition of creatine or caffeine. Crystals, mp 212°C, readily soluble in water. It may be synthesized from chloroacetic ester and amino-methane, or by hydrolysis with barium hydroxide of methylaminoethanonitrile.
sarcosporidiosis *(Vet.)*. Infection of the muscles of swine, sheep, horses, cattle, goats, and birds by the parasitic Sarcosporidia.
sarc-, sarco- *(Genrl.)*. Prefix from Gk. *sarx*, gen. *sarkos*, flesh.
sardonyx *(Min.)*. A form of chalcedony in which the alternating bands are reddish-brown and white. Cf. **onyx**.
Sargent diagram *(Nuc.Eng.)*. Log-log plot of radioactive decay constant against maximum β-ray energy, for various β-emitters. Most of the points relating to natural heavy radioisotopes lie on one or other of two straight lines.
sarking *(Build.)*. See **roof boards**.
sarking felt *(Build.)*. A bituminous underlining placed beneath slates or tiles.
saros *(Astron.)*. A cycle of 18 years 11 days, which is equal to 223 synodic months, 19 eclipse years and 239 anomalistic months. After this period the relative positions of the Sun, Moon, and node recur; known to the ancient Babylonians, it was used to predict eclipses.
sarsen *(Geol.)*. Irregular masses of hard sandstones which are found in the Reading, and Bagshot Beds of the Tertiary System in S. England. They often persist as residual masses after the softer sands have been denuded away.
sartorius *(Zool.)*. A thigh muscle of Tetrapoda which by its contraction causes the leg to bend inwards.
sash *(Build.)*. A framing for window panes.
sash and frame *(Build.)*. A cased frame in which counterweighted sashes slide vertically.
sash bar *(Build.)*. A *transom* or a *mullion*.
sash centres *(Build.)*. The points about which a pivoted sash is moved.
sash cramp *(Build.)*. A contrivance for holding parts of a frame in place during construction. It usually consists of a steel bar along which slide two brackets between which the work is fixed, one of the brackets being pegged into a hole in the bar while the other is adjustable for position by means of a screw.
sash door *(Build.)*. A door which has its upper part glazed.
sash fastener (lock) *(Build.)*. A fastening device secured to the meeting rails of the sashes of a double-hung window, serving to fix both sashes in the shut position.
sash fillister *(Build.)*. A special plane for cutting grooves in stuff for sash bars.
sash mortise chisel *(Build.)*. One somewhat lighter than a mortise chisel, used for mortising softwood.
sash pocket chisel *(Build.)*. A strong-bladed chisel with a narrow edge, used for cutting the pocket in the pulley stile of a sash and frame.
sash rail *(Build.)*. See **transom**.
sash saw *(Build.)*. A saw similar to the tenon saw but slightly smaller and finer; used for making window sashes.
sash stuff *(Build.)*. The timber prepared for use in the making of sashes.
sash weights *(Build.)*. Weights which are used as counterpoises in balancing the sashes of windows.
sassafras *(For.)*. Tree of the family, Lamaceae, yielding from the bark an essential oil used in perfumery and

cosmetics, and from the root an extract used for flavouring.

sateen *(Textiles)*. A smooth-surfaced fabric produced by a *weft*-faced weave. Cf. **satin**.

satellite *(Astron.)*. (1) Any small body orbiting under gravitational forces in a closed path around a much more massive body. The planets are, in this sense satellites of the Sun; the Moon is the natural satellite of the Earth; and the *Magellanic clouds* are satellite galaxies of the *Milky Way*. (2) A man-made device launched into orbit around the Earth, the Moon or other planets. They serve a variety of functions: relaying globally telephone and television signals; covert intelligence gathering of all kinds; remote sensing of the Earth and its environment; weather forecasting; and as platforms for astronomical telescopes.

satellite *(Biol.)*. (1) The part of a chromosome distal to the **secondary constriction**. (2) See **satellite DNA, simple sequence DNA**.

satellite *(Telecomm.)*. See **communications satellite**.

satellite computer *(Comp.)*. One used to relieve a central processing device of relatively simple but time-consuming operations, such as compiling, editing and controlling input and output devices. See **distributed computing**.

satellite DNA *(Biol.)*. DNA from eukaryotic chromosomes which separates from the remainder of the genome on **bouyant density** centrifugation. Contains closely related repeated sequences with a base composition different from the rest of the DNA.

satellite exchange *(Telecomm.)*. A small automatic-telephone exchange which is dependent on a main automatic exchange for completion of its calls to subscribers other than those connected to it.

satellite station *(Surv.)*. In triangulation by theodolite, one resected and thus fixed for reference purposes, but not occupied.

satellite station *(Telecomm.)*. (1) One which re-broadcasts a transmission received directly, but on another wavelength. (2) See **Earth station**.

Saticon *(Image Tech.)*. TN for an improved camera tube of the Vidicon type, having a photo-conductive layer of doped selenium.

satin *(Textiles)*. A smooth-surfaced fabric produced by a *warp*-faced weave. Cf. **sateen**.

satin spar *(Min.)*. Name given to fine fibrous varieties of calcite, aragonite, and gypsum, the gypsum variety being distinguished from the others by its softness (it can be scratched by a finger-nail).

satin walnut *(For.)*. Heartwood of **American red gum**. Varies in colour from brown to reddish-brown. Chiefly used for furniture, fittings, cooperage and panelling.

satinwood *(For.)*. See **East Indian-**.

saturable reactor *(Elec.Eng.)*. Inductor in which the core is saturable by turns carrying d.c., which controls the inductance. Used for modulation, control of lighting, and developed in the magnetic amplifier.

saturated adiabatic *(Meteor.)*. A curve on an **aerological diagram** representing the temperature changes of a parcel of saturated air subjected to an adiabatic process, the state of saturation being maintained.

saturated adiabatic lapse rate *(Meteor.)*. The temperature lapse rate of air which is undergoing a *reversible natural adiabatic process* as shown by the *saturated adiabatic* lines on an **aerological diagram**. Abbrev. *SALR*.

saturated calomel electrode *(Chem.)*. A calomel electrode containing saturated potassium chloride solution.

saturated compounds *(Chem.)*. Compounds which do not contain any free valencies and to which no hydrogen atoms or their equivalent can be added, i.e. which contain neither a double nor a triple bond.

saturated humidity mixing ratio *(Meteor.)*. The *humidity mixing ratio* of air which is saturated at a specified temperature and pressure. Saturation may be defined with reference either to liquid water or (below 0°C) to ice.

saturated solution *(Chem.)*. A solution which can exist in equilibrium with excess of the dissolved substance as a second phase.

saturated steam *(Eng.)*. Steam at the same temperature as the water from which it was formed, as distinct from steam subsequently heated. See **dry steam**.

saturated vapour *(Phys.)*. A vapour which is sufficiently concentrated to exist in equilibrium with the liquid form of the same substance.

saturation *(Electronics,Radar)*. (1) The condition in which a further increase in one controlling variable produces no further increase in the resultant effect. (2) The condition occurring when a transistor is driven so hard (by a large base current) that the base collector junction, reverse-biased in normal operation, becomes forward-biased. Recovery from this condition takes a long time and can impede high-speed switching operation. (3) In a thermionic valve, the condition when the anode current approaches total electron emission current available from the cathode. (4) In an amplifier, saturation occurs when the output power into the load approaches the limit of power available from the output stages. Similarly in a radar when the input signal drives the receiver to a point when its output can increase no further; characterized in such cases by a severe increase in non-linear distortion.

saturation *(Image Tech.,Phys.)*. Degree to which a colour departs from white and approaches the pure colour of the spectral line; dull or pale colours are said to have low saturation, vivid colours high saturation. *Desaturation* is the inverse of this.

saturation *(Nuc.Eng.)*. Condition where field applied across ionization chamber is sufficient to collect all ions produced by incident radiation.

saturation *(Phys.)*. Application of a sufficiently intense magnetizing force to result in maximum temporary or permanent magnetization in magnetic material. Usually applied by large currents from an impulse current transformer.

saturation activity *(Nuc.Eng.)*. The maximum level of artificial radioactivity, induced in a given sample, by a specific level of irradiation when the rate of formation equals the rate of decay.

saturation coefficient *(Build.)*. The ratio between the natural capacity of a material (such as a building-stone) to absorb moisture and its porosity.

saturation current *(Elec.Eng.)*. The steady current in a winding of an iron-cored transformer which causes the inductance of the winding to be seriously reduced.

saturation current *(Electronics)*. See **saturation**.

saturation curve *(Phys.)*. The characteristic curve relating magnetic flux density to the strength of the magnetic field.

saturation factor *(Elec.Eng.)*. The ratio of the increase of field excitation to the increase of generated voltage which it produces.

saturation limit *(Elec.Eng.)*. The maximum flux density economically attainable.

saturation of the air *(Meteor.)*. The air is said to be saturated when the relative humidity is 100%.

saturation scale *(Phys.)*. Minimum visual steps of saturation in scale of spectrum colours, varying with wavelength.

saturation vapour pressure with respect to water *(Meteor.)*. The maximum **water vapour pressure**, which can occur when the water vapour is in contact with a free water surface at a particular temperature. The water vapour pressure existing when effective evaporation ceases.

saturation voltage *(Elec.Eng.)*. Voltage applied to a device to operate under saturation conditions.

Saturn *(Astron.)*. Sixth planet of the solar system in order from the Sun, second largest at 752 times the volume of the Earth, and the outermost planet known in ancient times. Encircled by an extensive and beautiful ring system. Density is 0.70, a low value because the planet is predominantly gaseous, consisting mainly of hydrogen and helium. The atmosphere is rich in methane and ethane. The rotation period varies from 10.233 hours at the equator to nearly 11 hours at the poles; this rapid rotation makes Saturn the most oblate of all planets, the equatorial radius (60 000 km) being 11 per cent more than the polar radius. There are numerous rings with diameters from 67 000–480 000 km, and 17 satellites, one of which, **Titan**, is the largest moon within the solar system.

saucer *(Hyd.Eng.)*. A flat form of camel, used for raising a vessel in shallow waters.

sauconite *(Min.)*. One of the montmorillonite group of clay minerals in which zinc has replaced magnesium. See **smectites**.

saussurite *(Geol.,Min.)*. A fine-grained mixture of zoisite and other minerals resulting from the more or less complete alteration of feldspar. Sometimes simulates **jade**.

savanna *(Ecol.)*. (1) Extensive area in which grasses are an important part of the vegetation. (2) Common vegetation type in dry parts of Africa, consisting of trees and grasses, usually burnt every year.

saveall *(Paper)*. Any plant designed to recover fine fibres, loading and similar suspended matter from surplus backwater.

SAW *(Radar,Telecomm.)*. Abbrev. for *Surface Acoustic Wave*.

SAW delay line *(Telecomm.)*. One in which the delay is determined by the length of a piece of piezo-electric material along which a radio-frequency signal is launched into a *surface-acoustic mode* by appropriate electrodes.

SAW filter *(Telecomm.)*. Filter depending on the propagation of *surface acoustic waves* through a resonant structure mounted on a piezo-electric material. Advantages include highly predictable and stable performance and very sharp cut-off characteristics; insertion loss is higher than for conventional filters, but amplification can compensate for this.

saw gumming *(Eng.)*. Term for the process of cleaning up burrs and machining marks in the gullets of circular saws.

sawtooth oscillator *(Telecomm.)*. See **relaxation oscillator, time-base generator**.

sawtooth roof *(Build.)*. A roof formed of a number of north light trusses, presenting a serrated profile when viewed from the end elevation.

sawtooth truss *(Eng.)*. A truss used for small-span roofs of sawtooth form, braced by vertical and diagonal members.

sawtooth wave *(Electronics)*. One generated by a **time-base generator** for scanning in a CRT, for uniform sweep and high-speed return. See **flyback**.

sax *(Build.)*. An axe used for shaping slates; it has a pointed peen for piercing the nail-holes. Also called *saixe, slate axe*.

saxicole, saxicolous *(Bot.)*. Growing on rocks.

saxicolous *(Bot.)*. Living on or among rocks.

saxonite *(Geol.)*. A coarse-grained, ultrabasic rock, consisting essentially of olivine and orthopyroxene, usually hypersthene. A hypersthene-peridotite.

Sb *(Chem.)*. The symbol for antimony.

SBA *(Aero.)*. Abbrev. for *Standard Beam Approach*.

SBAC *(Aero.)*. Abbrev. for *Society of British Aerospace Companies*.

S-band *(Telecomm.)*. Loose definition, due to international disagreement, of microwave band in the 2 to 3 GHz region.

SBC *(Elec.Eng.)*. Abbrev. for *Small Bayonet Cap*.

Sc *(Chem.)*. The symbol for scandium.

scab *(Bot.)*. A discrete localized superficial lesion characterised by roughening, abnormal thickening and, especially, cork formation; a disease in which such scabs are formed, e.g. potato and apple scab.

scab *(Vet.)*. See **mange**.

scabbling *(Build.)*. The operation of rough-dressing a stone face with an axe, prior to smoothing it.

scabbling hammer *(Build.)*. The pointed hammer used in the rough-dressing of a nigged ashlar.

scabellum *(Zool.)*. In Diptera, the dilated basal portion of a haltere.

scabies *(Med.)*. A contagious skin disease caused by the acarine parasite *Sarcoptes scabiei*, the female of which burrows in the horny layer of the skin.

scabies *(Vet.)*. See **mange**.

scabrid *(Bot.)*. Rough to the touch; scaly.

scabrous *(Bot.)*. Having a surface roughened by small wartlike upgrowths. dim. *scaberulous*.

scaffold *(Biol.)*. Protein core of histone-depleted metaphase chromosomes left after nuclease treatment.

scaffold/radial loop model *(Biol.)*. Model of metaphase chromosome structure which postulates a non-histone protein core to which the linear DNA molecule has an ordered series of attachment points every 30 000 to 90 000 bases, with the intervening DNA forming a loop packed by supercoiling or folding.

scaffold, scaffolding *(Build.)*. A temporary erection of timber or steelwork, used in the construction, alteration, or demolition of a building, to support or to allow of the hoisting, lowering, or standing of men, materials, etc.

scalar *(Maths.)*. A real number or element of a field. The term *scalar* is used in vector geometry and in vector and matrix algebra to contrast numbers with vectors and matrices. See also **vector space**.

scalar matrix *(Maths.)*. A device by which an ordinary number (or scalar), λ, may be expressed as a matrix. It is defined as λI, where I is the unit matrix, and has λs in the principal diagonal and zeros elsewhere.

scalar product *(Maths.)*. Of two vectors, a scalar (i.e. real number) equal to the product of the magnitudes of the two vectors and the cosine of the angle between them. Denoted by a dot, e.g. $A.B$.

scale *(Acous.)*. (1) Ratio of length to diameter of organ pipe, a factor which, among others, determines the timbre of a note. (2) Set of (usually) 12 notes forming an octave or frequency range of 2 to 1.

scale *(Bot.)*. A thin, flat, semitransparent plant member, usually of small size, and green only when very young, if then. See **bud scale, scale leaf**.

scale *(Eng.)*. Numerical factor relating measured quantity to indication of instrument.

scale *(Image Tech.)*. (1) General term for the range of brightness or density over which satisfactory tonal reproduction can be expected. (2) The range of exposure steps available in a printer or enlarger.

scale *(Maths.)*. The ratio between the linear dimensions of a representation or model and those of the object represented, as on a map or technical drawing.

scale *(Zool.)*. A small exo-skeletal outgrowth of tegumentary origin, of chitin, bone, or some horny material, usually flat and platelike.

scale bark, scaly bark *(Bot.)*. (1) Bark which becomes detached in irregular patches. (2) **Rhytidome**.

scale leaf *(Bot.)*. Leaf, often thin, more or less flattened against the stem or other leaves, and either photosynthetic, protective (bud scales) or holding reserves (bulb scales).

scalene triangle *(Maths.)*. A triangle in which no two sides are equal.

scale-of-ten *(Telecomm.)*. Ring, or other, system of counting elements which divides counts by 10.

scale-of-two *(Electronics)*. Any bistable circuit which can divide counts by 2, when operated by pulses. See **flip-flop**.

scaler *(Electronics)*. Instrument, incorporating one or more scaling circuits, used to register a count.

scaling *(Acous.)*. Adjustment of the notes of a musical instrument to a specified scale, e.g. natural or tempered scale.

scaling circuit *(Electronics)*. One which divides counts of pulses by an integer, so that they are more readily indicated, to a required degree of accuracy. If scalers of 10 are used in cascade, indications of counts in decimal numbers are possible.

scaling hammer *(Eng.)*. See **boilermaker's hammer**.

scaly leg *(Vet.)*. A form of mange affecting the feet and legs of fowl, due to the mite *Cnemidocoptes mutans* burrowing in the skin.

scan *(Radar)*. Systematic variation of a radar beam direction for search or angle tracking. See **A, B, display**, etc.

scandent *(Bot.)*. Climbing.

scandium *(Chem.)*. A metallic element, classed with the rare earth metals. Symbol Sc, at. no. 21, r.a.m. 44.956. It has been found in cerite, orthite, thortveitite, wolframite and euxenite; discovered in the last named. Scandium is the least basic of the rare earth metals.

scanner *(Radar)*. Mechanical arrangement for covering a solid angle in space, for the transmission or reception of signals, usually by parallel lines or *scans*.

scanning *(Image Tech.)*. The systematic coverage of an area by a spot moving in a series of progressive lines, especially by the use of an electron beam in a TV camera or display tube.

scanning *(Radar)*. Coverage of a prescribed area by a directional radar antenna or sonar beam.

scanning coils *(Image Tech.)*. Coils mounted in CRT, and carrying suitable currents for deflecting electron beam, so as to sweep the picture area.

scanning electron microscope *(Phys.)*. An electron microscope whose operation depends on a very fine electron beam being made to scan the sample point by point on a raster. Secondary electrons produced at each point depend on the properties of the sample at the point. They are collected and, with suitable electronics, made to produce an image on a CRT in synchronism with the scanning beam.

scanning frequency *(Image Tech.)*. The number of times per second that an area is completely scanned.

scanning heating *(Phys.)*. **Induction heating** where the workpiece is moved continuously through the heating region, as in *zone refining* of germanium. Also called *progressive heating*.

scanning linearity *(Image Tech.)*. Uniformity of scanning speed for a CRO or TV receiver. This is necessary to avoid waveform or picture distortion.

scanning loss *(Radar)*. That which arises from relative motion of a scanning beam across a target, as compared with zero relative motion.

scanning microscope *(Phys.)*. An instrument which enables the surface structure of a specimen to be examined. The specimen is scanned by a small-diameter electron probe which gives rise to secondary-electron emission from the specimen. The secondary-emission current is detected and causes a picture to be projected on a cathode-ray screen.

scanning slit *(Image Tech.)*. In a system for photographic sound on film, the narrow illuminated aperture across which the track passes for recording and reproduction.

scanning speech *(Med.)*. A disturbance of speech in which the utterance is slow and halting, the words being broken into syllables; a sign of a lesion in the nervous system, as in disseminated sclerosis.

scanning speed *(Image Tech.)*. That of a scanning spot across the screen of a CRT. Usually accurately specified for a CRO to facilitate time measurements.

scanning transmission electron microscope *(Phys.)*. One which uses field emission from a very fine tungsten point as the source of electrons. The electrons transmitted through the sample are either unscattered, elastically scattered or inelastically scattered; they are collected, separated and analysed to produce an image.

scanning tunnelling electron microscope *(Phys.)*. One which uses a probe with an 'atomic micro-tip' floated, using *superconducting levitation* over the surface being scanned. The tip/surface is of the order of atomic diameters so that the electron current obtained is through *quantum mechanical tunnelling*. A horizontal resolution of ~0.2 nm and a vertical resolution of ~0.01 nm is obtained. Applications are principally to the study of surface effects.

scan plates *(Print.)*. A general term for plates made by **electronic engraving**.

scansorial *(Zool.)*. Adapted for climbing trees.

scantling *(Build.)*. Stones more than 2 m (6 ft) long.

scantling *(For.)*. A piece of timber 5–10 cm (2–4 in) thick and 5–11 cm wide.

scape *(Bot.)*. The flowering stem, nearly or quite leafless, arising from a rosette of leaves and bearing a flower, several flowers or a crowded inflorescence; e.g. the dandelion. adj. *scapigenous*.

scape *(Zool.)*. The basal joint of the antenna in Insects.

scaper plane *(Build.)*. Tool shaped like a **spoke-shave** but a thin blade set at a slight angle from the perpendicular. Used for cleaning up hardwood and removing tears left after ordinary planing.

scaphoid *(Zool.)*. See **navicular bone**.

Scaphopoda *(Zool.)*. A class of Mollusca, being bilaterally symmetrical with a tubular shell open at both ends, a reduced foot used for burrowing, a head with many prehensile processes, a radula, separate cerebral and pleural ganglia, no ctenidia, circulatory system rudimentary, larva a trochosphere.

scaph-, scapho- *(Genrl.)*. Prefix from Gk. *skaphē*. boat.

scapolite *(Min.)*. A group of minerals forming an isomorphous series, varying from meionite, a silicate of aluminium and calcium with calcium carbonate, to marialite, a silicate of aluminium and sodium with sodium chloride. Common scapolite is intermediate in composition between these two end-member minerals; see also **dipyre** and **mizzonite**. The scapolites crystallize in the tetragonal system and are associated with altered lime-rich igneous and metamorphic rocks. A transparent honey yellow variety is cut as a gemstone.

scappling *(Build.)*. A variant of *scabbling*.

scapula *(Zool.)*. In Vertebrates, the dorsal portion of the pectoral girdle, the shoulder blade; any structure resembling the shoulder blade. adj. *scapular*.

scapulars *(Zool.)*. In Birds, small feathers attached to the humerus, and lying along the side of the back.

scarcement *(Build.)*. A ledge formed at a place where part of a wall is set back from the general face of the wall.

scarf *(Build.)*. A joint between timbers placed end to end, notched and lapped, and secured together with bolts or straps.

scarfed joint *(Elec.Eng.)*. A cable joint in which the conductor ends are bevelled off so that, after soldering, there is no appreciable increase in conductor diameter at the joint.

scarfing *(Eng.)*. (1) Tapering the ends of materials for a lap joint, so that the thickness at the joint is substantially the same as that on either side of it. (2) Preparing metal edges for forge welding.

scarification *(Bot.)*. Any treatment, e.g. with sulphuric acid or by mechanical abrasion, which makes the coat of a seed more permeable to water and promotes imbibition and germination.

scarifier *(Civ.Eng.)*. A spiked mechanical picking appliance for breaking up road surfaces as a preliminary to remetalling.

scarlatina *(Med.)*. See **scarlet fever**.

scarlatiniform *(Med.)*. Resembling or having the form of (the rash of) scarlet fever.

scarlet fever *(Med.)*. An acute infectious fever due to infection of the throat with a haemolytic streptococcus; characterized by sore throat, headache, raised temperature, and a punctate erythema of the skin, which subsequently peels. Also called *scarlatina*.

scarp face *(Geol.)*. See **escarpment**.

scatterer *(Acous.)*. Object which causes scattering, e.g. curved hard boards or fish bladders in the sea. Used in **reverberation chambers**.

scattering *(Phys.)*. General term for irregular reflection or dispersal of waves or particles, e.g. in acoustic waves in an enclosure leading to diffuse reverberant sound, **Compton effect** on electrons, light in passing through material, electrons, protons, and neutrons in solids, radio waves by ionization. See **forward scatter** and **back scatter**. Particle scattering is termed *elastic* when no energy is surrendered during scattering process, otherwise it is *inelastic*. If due to electrostatic forces it is termed **coulomb scattering**; if short-range nuclear forces are involved it becomes *anomalous*. Long-wave electromagnetic wave scattering is *classical* or **Thomson**, while for higher frequencies *resonance* or *potential* scattering occurs according to whether the incident photon does or does not penetrate the scattering nucleus. Coulomb scattering of α-particles is *Rutherford scattering*. For light, scattering by fine dust or suspensions of particles is *Rayleigh scattering*, while that in which the photon energy has changed slightly, due to interaction with vibrational or rotational energy of the molecules, is *Raman scattering*. *Shadow scattering* results from interference between scattered and incident waves of the same frequency. See **acoustic-**, **atomic-**.

scattering amplitude *(Phys.)*. Ratio of amplitude of scattered wave at unit distance from scattering nucleus to that of incident wave.

scattering cross-section *(Phys.)*. Effective impenetrable cross-section of scattering nucleus for incident particles of low energy. The radius of this cross-section is the *scattering length*.

scattering mean free path *(Phys.)*. The average distance travelled by a particle between successive scattering interactions. It depends upon the medium traversed and the type and energy of the particle.

scatterometer *(Meteor.)*. An instrument carried in a **meteorological satellite** for measuring the light scattered from the surface of the sea, thus yielding information on the height and movement of waves which can be used to derive estimates of the surface wind.

scavenge pump *(Autos.)*. An oil-suction pump used to return used oil to the oil tank from the crank-case of an engine using the dry sump system of lubrication.

scavenging *(Eng.)*. Addition made to molten metal to counteract an undesired substance.

scavenging *(Min.Ext.)*. Final stage of froth flotation in which a low-grade concentrate or middling is removed.

scavenging stroke *(Eng.)*. See **exhaust stroke**.

s.c.c. *(Elec.Eng.)*. Abbrev. for *Single Cotton-Covered (wire)*.

scenario *(Image Tech.)*. See **script**.

scending *(Ships)*. See **pitching period**.

scene *(Image Tech.)*. The component of a motion picture or video production which is intended to be recorded as uninterrupted action.

scene-slating attachment *(Image Tech.)*. Device attached to a motion-picture camera for identifying individual takes on the film. Used in place of **clapper board**.

scent-marking *(Behav.)*. A form of communication between individuals of a species involving the deposition of glandular secretion onto ground or some other surface; the scent dissipates at a slow rate to form a relatively long-lived signal.

Schafer's method *(Med.)*. A method of artificial respiration in which the patient lies prone, the head supported on one forearm, and the operator, his knees on either side of the patient's hips, exerts pressure with each hand over the lower ribs at the back at intervals of from 3 to 5 seconds. Cf. **kiss of life**.

schedule of reinforcement *(Behav.)*. A rule that determines the frequency and manner in which behaviour is reinforced.

scheduling *(Comp.)*. Method by which central processing unit time is allotted in a **multi-access** system. The scheduling algorithm may include a system of priorities.

Scheele's green *(Chem.)*. Copper (II) hydrogen arsenite ($CuHAsO_3$). Poisonous. Formerly used in wallpapers.

scheelite *(Min.)*. An ore of tungsten. It occurs in association with granites and pegmatites, has the composition calcium tungstate and crystallizes in the tetragonal system.

schema *(Behav.)*. A mental pattern, or body of knowledge, that provides a framework within which to place newly acquired knowledge.

scheme arch *(Civ.Eng.)*. See **skene arch**.

Schering bridge *(Elec.Eng.)*. An a.c. bridge of the 4-arm (or Wheatstone) type usually used to measure capacitance and power factor of small capacitors.

Schick test *(Immun.)*. A test to assess the degree of susceptibility or immunity of individuals to diphtheria. A standard quantity of diphtheria toxin is injected intracutaneously into one forearm. In non-immune persons an area of redness and swelling appears in 1–2 days, reaching a maximum at 4–7 days followed by pigmentation and scaling. This is a positive Schick reaction and indicates that diphtheria immunization is required.

Schiff's bases *(Chem.)*. A term for benzylidene anilines, e.g. $C_6H_5.CH{=}N.C_6H_5$.

Schiff's reagent *(Chem.)*. A reagent, consisting of a solution of fuchsine decolorized by sulphurous acid, for testing the presence of aldehydes. Their presence is shown by a red-violet colour.

schillerization *(Geol.)*. A play of colour (in some cases resembling iridescence due to tarnish) produced by the diffraction of light in the surface layers of certain minerals.

Schimmelbusch's disease *(Med.)*. A condition characterized by the formation of cysts in the breast, and by hyperplasia of the epithelium of the glandular tissue of the breast.

schist *(Geol.)*. A metamorphic rock which has a tendency to split on account of the presence of folia of flaky and elongated minerals, such as mica, talc and chlorite; formed from original sedimentary or igneous rocks by the action of regional metamorphism.

schistosity *(Geol.)*. The tendency in certain rocks to split easily along weak planes produced by regional metamorphism and due to the abundance of mica or other cleavable minerals lying with their cleavage planes parallel.

schistosomiasis *(Med.)*. Infestation by parasitic worms of the *Schistome* group, transmitted via water by a complex cycle involving snails as intermediate hosts; common in many parts of the tropics. The chief varieties affect mainly the bladder (*Sch. haematobium*, see **bilharziasis**) and rectum (*Sch. mansonii*).

schizo- *(Genrl.)*. Prefix from Gk. *schizein*, to cleave.

schizocarp *(Bot.)*. Fruit, derived from a syncarpous ovary, which when mature becomes divided into separate, one-seeded, indehiscent parts (mericarps).

schizocoel *(Zool.)*. Coelom produced within the mass of mesoderm by splitting or cleavage; cf. **enterocoel**. adj. *schizocoelic*.

schizogamy *(Zool.)*. In Polychaeta, a method of reproduction in which a sexual form is produced by fission or germination from a sexless form.

schizogenesis *(Zool.)*. Reproduction by fission.

schizogenous *(Bot.)*. A space in a tissue, formed by the separation of cells by splitting their common wall along the middle lamella. Cf. **lysigenous, rexigenous, schizolysigenous**.

schizogony *(Zool.)*. In Protozoa, vegetative reproduction by fission.

schizoid *(Med.)*. A term used to describe withdrawn and introspective individuals with dissociation between emotions and intellect. Normal but tending to **schizophrenia**.

schizolysigenous *(Bot.)*. A space in a tissue, formed by both dissolution of cells and separation at the middle lamella. Cf. **lysigenous, schizogenous, rexigenous**.

schizont *(Zool.)*. In Protozoa, a mature trophozoite about to reproduce by schizogony.

schizophrenia *(Behav.,Med.)*. A group of psychoses marked by severe distortion and disorder of thought, perception, motivation and mood; delusion and hallucination are common, as are bizarre behaviours and social withdrawal. The term was invented by Eugen Bleuler and was previously known as *dementia praecox*.

schlieren photography *(Aero.,Phys.)*. Technique by which the flow of air or other gas may be photographed, the change of refractive index with density being made apparent under a special type of illumination. The method is used in studying the behaviour of models in transonic and supersonic wind tunnels.

Schmidt lines *(Nuc.Eng.)*. Two parallel lines in plot of nuclear magnetic moment against nuclear spin, as a result of the spin of an odd unpaired proton or neutron. Experimental magnetic moments for the majority of such nuclides lie between these lines. Also known as *Schmidt limits*.

Schmidt optical system *(Phys.)*. An optical system, for telescopes and for projection work, which uses a spherical mirror instead of a parabolic mirror. The resulting spherical aberration is corrected by using a moulded transparent plastic plate in front of the mirror; a *catadioptric* system.

Schmitt limiter *(Telecomm.)*. A bistable pulse generator which gives a constant amplitude output pulse provided the input voltage is greater than a predetermined value.

Schmitt trigger *(Electronics)*. Bistable circuit giving accurately-shaped constant amplitude rectangular pulse output for any input pulse above the triggering level. Widely used as a pulse shaper.

schnorkel *(Ships)*. A retractable tube or tubes containing pipes for discharging gases from, or for taking air into, a submerged submarine or other underwater vessel; a tube for bringing air to a submerged swimmer. Also *snorkel, snort*.

school *(Zool.)*. See **schooling**.

schooling *(Behav.)*. In fish, refers to groups of individuals who maintain a constant distance and orientation from their neighbours, and who all swim at a constant pace; the primary function of schooling is as an ante-predator strategy.

school phobia *(Behav.)*. Severe reluctance to attend school.

schorlomite *(Min.)*. A black variety of andradite garnet richer in titania (5–20% TiO_2) than melanite.

schorl-rock *(Geol.)*. A rock composed essentially of aggregates of black tourmaline (schorl) and quartz. A Cornish term for the end product of tourmalinization. See **tourmaline**.

Schottky defect *(Chem.)*. Deviation from the ideal crystal lattice by removal of some of the molecules to the surface, leaving within the lattice equivalent numbers of randomly spaced anion and cation vacancies.

Schottky diode *(Electronics)*. A semiconductor diode formed by contact between a semiconductor layer and a metal coating. Hot carriers (electrons in *n*-type material and holes in *p*-type material) are emitted from the Schottky barrier of the semiconductor and move into the metal coating. Majority carriers predominate, so that injection or storage of minority carriers is not present to limit switching speeds.

Schottky effect *(Electronics)*. (1) The removal of electrons from the surface of a semiconductor when a localized electric field is present at the surface; electrons can then surmount a semiconductor potential barrier and enter regions where allowed energy levels are available. (2) The increase in cathode current in a valve, beyond that available by normal thermionic emission, on account of the effective lowering of the work function of the cathode by the localized electric field.

Schottky noise *(Electronics)*. Strictly, noise in the anode current of a thermionic valve due to random variations in the surface condition of the cathode. Frequently extended to include **shot noise**.

Schottky TTL *(Electronics)*. Term describing logic switching and memory circuits where each transistor involved in the switching operation has a Schottky diode connected across it to reduce the number of charge carriers in the base when the transistor is *on*. This means that the transistors can be turned *off* more rapidly and allows the circuits to function more rapidly.

Schrage motor *(Elec.Eng.)*. A variable-speed induction motor employing a commutator winding on the rotor from which an e.m.f. is collected and injected into the stator winding.

Schrödinger equation *(Phys.)*. Fundamental equation of *wave mechanics*. Solutions of this equation are *wave functions* for which the square of the amplitude expresses the probability density for a particle or a set of particles. If the system is isolated then a *time-independent* form of the equation is applicable. Solution for this version for bound particles show that the energy for the system must be quantized.

Schüfftan process *(Image Tech.)*. Early system of composite cinematography in which live action was shot through transparent areas in a large mirror reflecting a model or painting of the rest of the scene.

Schuler pendulum *(Eng.)*. A theoretically ideal pendulum which will not be affected by the earth's rotation, having a length equal to the earth's radius and hence a period of 84 min. A stable platform servomechanism can be constructed to simulate pendular motion of the above period, and is then said to be *Schuler-tuned*. Conversely a gyroscopically stabilized platform, constrained to move parallel to the earth's surface in its motion over the earth, will possess a period of 84 min and will be conditionally stable. Damping to reduce the maximum error may be introduced by rate feedback.

Schwann cell *(Zool.)*. A neuroglial cell which deposits the myelin sheath along myelinated axons.

schwannoma *(Med.)*. A tumour growing from the sheath of a nerve (neurofibroma) and containing cells resembling those of the neurolemma.

Schwarz's inequality *(Maths.)*.

$$\{\int f(x).g(x)dx\}^2 < \{\int f(x)dx\}\{\int g(x)dx\}$$

unless $f(x) = kg(x)$, where k is independent of x. This is the integral analogue of **Cauchy's inequality**.

Schweitzer's reagent *(Chem.)*. A reagent for cellulose. It consists of a 0.3% solution of precipitated copper (II) hydroxide in a 20% ammonium hydroxide solution. This mixture is a solvent for cellulose, which can be reprecipitated by the addition to the solution of mineral acids.

sciatic *(Zool.)*. Situated in, or pertaining to, the ischial or hip region.

sciatica *(Med.)*. Inflammation or irritation of the fibrous elements of the sciatic nerve, resulting in pain and tenderness along the course of the nerve in the buttock and the back of the leg; less strictly, any pain along the course of the sciatic nerve.

SCID *(Immun.)*. Abbrev. for *Severe Combined Immuno-Deficiency syndrome*.

science *(Genrl.)*. The ordered arrangement of ascertained knowledge, including the methods by which such knowledge is extended and the criteria by which its truth is tested. The older term *natural philosophy* implied the contemplation of natural processes *per se*, but modern science includes such study and control of nature as is, or might be, useful to mankind. *Speculative science* is that branch of science which suggests hypotheses and theories, and deduces critical tests whereby unco-ordinated observations and properly ascertained facts may be brought into the body of science proper.

scientific alexandrite *(Min.)*. Synthetic corundum coloured with vanadium oxide and resembling true alexandrite in some of its optical characters.

scientific emerald *(Min.)*. Beryl glass coloured with chromic oxide, resembling true emerald in colour.

scintillation *(Astron.)*. The twinkling of stars, a phenomenon due to the deflection, by the strata of the earth's atmosphere, of the light rays from what are virtually point sources.

scintillation *(Phys.)*. Minute light flash caused when α-, β- or γ-rays strike certain phosphors, known as *scintillators*. The latter are classed as liquid, inorganic, organic or plastic according to their chemical composition.

scintillation camera *(Radiol.)*. *Gamma camera*. An imaging device which may have either a single sodium iodide crystal or multiple crystals which is capable of detecting and recording the spatial distribution of an internally administered radionuclide.

scintillation counter *(Nuc.Eng.)*. Counter consisting of a *phosphor* or *scintillator*, e.g. NaI(Tl), which, when radiation falls on it, emits light which is detected and amplified by a photomultiplier, the height of the pulses from which are proportional to the energy of the event. These pulses are further amplified and passed to a single- or multi-channel pulse height analyser, to measure the energy and intensity of the radiation.

scintillation spectrometer *(Nuc.Eng.)*. See **scintillation counter**.

scintling *(Build.)*. Placing half-dry raw bricks diagonally and a little distance apart, so as to admit air between them.

scion *(Bot.)*. A piece of a plant, in horticulture usually a young, often dormant, shoot, that is inserted into the **stock** when a **graft** is made.

sciophyte *(Bot.)*. A plant which is adapted to living in shady places.
scirrhous carcinoma *(Med.)*. A hard cancer, in which there is an abundance of connective tissue and few cells.
scirrhous cord *(Vet.)*. See **botryomycosis**.
scissors truss *(Build.)*. A type of truss used for a pitched roof, consisting of two principal rafters braced by two other members, each of which connects the foot of a rafter to an intermediate point in the length of the other rafter.
sclera *(Zool.)*. The tough fibrous outer coat of the Vertebrate eye. adj. *sclerotic*.
sclere *(Zool.)*. A skeletal structure; a sponge spicule.
sclereid *(Bot.)*. A short **sclerenchyma cell**, star-shaped, rod-shaped or rounded, often occurs in tissues of other cell types but in some seed coats as a tissue of sclereids.
sclereide *(Bot.)*. (1) A general term for a cell with a thick, lignified wall, i.e. any sclerenchymatous cell. (2) A thick-walled cell mixed with the photosynthetic cells of a leaf, giving them mechanical support. (3) A stone cell.
sclerencephalia *(Med.)*. Hardening of the brain.
sclerenchyma *(Bot.)*. (1) Tissue composed of **sclerenchyma cells**. (2) Collective term for **sclerenchyma cells**.
sclerenchyma *(Zool.)*. Hard skeletal tissue, as of Corals.
sclerenchyma cell *(Bot.)*. Cell with thick, usually lignified, walls, often dead when mature, and usually having a supporting function in the plant; either a **sclereid** or a **fibre**. Cf. **collenchyma**.
sclerified, sclerosed, sclerotized *(Bot.)*. Hardened, e.g. as a result of secondary wall formation and/or lignification.
sclerite *(Zool.)*. A hard skeletal plate or spicule.
scleritis *(Med.)*. Inflammation of the sclera of the eye. See also **episcleritis**.
sclerodactylia *(Med.)*. Sclerodermia of the hands, the skin being drawn tightly over the fingers.
sclerodermia, scleroderma *(Med.)*. A condition of hardness and rigidity of the skin as a result of overgrowth of fibrous tissue in the dermis and subcutaneous tissue, the fat of which is replaced by the fibrous tissue.
scleroma *(Med.)*. A condition in which hard nodules of granulomatous tissue appear in the nose, or occasionally in the trachea.
scleronychia *(Med.)*. Thickening and dryness of the nails.
sclerophyll *(Bot.)*. Leaf with well-developed sclerenchyma and hence tough and fibrous or leathery, usually evergreen and typical of trees and shrubs in places with rather warm and dry, especially mediterranean, climates, e.g. olive and *Eucalyptus*.
scleroproteins *(Biol.)*. Insoluble proteins forming the skeletal parts of tissues, e.g. keratin from hoofs, nails, hair etc., chondrin and elastin from ligaments.
scleroscope hardness test *(Eng.)*. The determination of the hardness of metals by measuring the rebound of a diamond-tipped hammer dropped from a given height.
sclerosed, sclerotic *(Med.)*. Affected with sclerosis.
sclerosis *(Bot.)*. The hardening of cell walls or of tissues by thickening and lignification.
sclerosis *(Med.)*. An induration or hardening, as of the arteries. See **multiple sclerosis**.
sclerotic *(Zool.)*. The sclera of the eye; pertaining to the sclera.
sclerotic cell *(Bot.)*. See **sclereid**.
sclerotin *(Zool.)*. In the cuticle of Insects and some other Arthropods a protein which has become strengthened and dark through cross-linkage by the action of quinones. See **tanning**.
sclerotium *(Bot.)*. A hard mass of fungal hyphae, often black on the outside, crust-like to globular, and serving as a resting stage from which an active mycelium or spores are formed later.
sclerotization *(Zool.)*. In Insects, the process by which most of the **procuticle** becomes hardened and darkened to form tough, rigid, discrete sclerites of **exocuticle**. See also **tanning**.
sclerotomy *(Med.)*. Operative incision of the sclera.
scler-, sclero- *(Bot.)*. Prefix meaning hard.
scobicular, scobiform *(Bot.)*. Looking like sawdust.
scoinson arch *(Build.)*. See **squinch**.
scolecite *(Min.)*. A member of the zeolite group of minerals; a hydrated silicate of calcium and aluminium, occurring usually in fibrous or acicular groups of crystals.
scolex *(Zool.)*. The terminal organ of attachment of a tapeworm (Cestode). pl. *scoleces*, erron. *scolices*. adj. *scolecid, scoleciform*.
scoliosis *(Med.)*. Abnormal curvature of the spine laterally.
scolophore *(Zool.)*. A subcuticular spindle-shaped nerve-ending in Insects, sensitive to mechanical vibrations. Also called *scolopidium*.
Scolopidia *(Zool.)*. Campaniform sensillae.
scontion, sconchion *(Build.)*. An inside quoin, as laid in a splayed jamb.
scoop *(Med.)*. A spoonlike instrument for clearing out cavities.
scopa *(Zool.)*. The pollen brush of Bees, consisting of short stiff spines on the posterior metatarsus.
scope *(Electronics)*. Colloquial term for *oscilloscope*.
'scope *(Image Tech.)*. Popular term for anamorphic systems of **wide-screen** cinematography, by derivation from **CinemaScope**.
Scophony *(Image Tech.)*. TN of an early form of projection TV using a light beam modulated by a **Kerr cell** and mechanically scanned by a **mirror drum**.
scopolamine *(Med.)*. See **hyoscine**.
scopophilia *(Behav.)*. Pleasure in looking.
scorbutic *(Med.)*. Pertaining to scorbutus (scurvy).
scorch *(Bot.)*. Necrosis, like that of leaf margins looking as if seared by heat, caused by infection, mineral deficiency or weather conditions etc.
score *(Print.)*. See **em rule, en rule**.
scoria *(Geol.)*. A cavernous mass of volcanic rock which simulates a clinker.
scorification *(Eng.)*. Assay in which impurities are slagged while bullion metals dissolve in molten lead, from which they are later separated by **cupellation**.
scorifier *(Eng.)*. A crucible of bone ash or fireclay used in assaying and in the metallurgical treatment of precious metals. See also **scorification**.
scorodite *(Min.)*. An orthorhombic hydrated arsenate of iron and aluminium.
Scorpionidea *(Zool.)*. An order of Arachnida with the opisthosoma being divided into a distinct mesosoma and metasoma, and consisting of 12 segments and a telson. The chelicera and pedipalps are chelate, there are 4 pairs of walking legs, the mesosomatic segments carry the genital operculum, the pectines and 4 pairs of lung books, the metasoma forming a flexible tail wielding the terminal sting. Viviparous and terrestrial. Known as fossils from the Silurian. Scorpions.
scotch *(Build.)*. See **scutch**.
Scotch block *(Civ.Eng.)*. Attachment to running rails, to prevent the passage of rolling stock.
Scotch boiler *(Eng.)*. See **marine boiler**.
Scotch bond *(Build.)*. A bond in which a course of headers alternates with three courses of stretchers. Also *English garden-wall bond*.
Scotch crank, yoke *(Eng.)*. A form of crank, used on a *direct-acting pump*, in which a square block, pivoted on the overhung crank pin, works in a slotted crosshead carried by the common piston rod and ram.
Scotchprint *(Print.)*. A *conversion system* using a film on which an impression of a relief printing surface is taken, to provide a positive for litho platemaking.
Scotch yoke *(Elec.Eng.)*. A triangular framework used in certain types of electric tractor for coupling two traction motors to the driving-wheel system.
scotia *(Arch.)*. See **cavetto**.
scoto- *(Bot.)*. Prefix meaning dark (in the sense of not illuminated).
scotoma *(Med.)*. (1) A blind or partially blind area in the visual field, the result of disease of, or damage to, the retina or optic nerve or visual cortex. (2) The appearance of a black spot in front of the eye, as in choroiditis. pl. *scotomata*.
scotomization *(Behav.)*. Derived from scotoma (blind

spot in the visual field); a defence mechanism where the individual fails to consciously perceive parts of the environment or of his/her self.

scotophor *(Electronics)*. Material which darkens under electron bombardment, used for screen of CRT in storage oscilloscopes. Recovers upon heating. Usually potassium chloride.

scotopic luminosity curve *(Phys.)*. The curve giving relative brightness of the radiations in an equal-energy spectrum when seen at a very low intensity level. See **photopic luminosity curve**.

scotopic vision *(Phys.)*. That which occurs at low illumination levels through the medium of the retinal rods.

Scots pine *(For.)*. *Pinus sylvestris*, native to many parts of Europe and the best known home-grown species for producing commercial timbers. Also called *Baltic redwood, Scots fir*.

Scott connection *(Elec.Eng.)*. A method of connecting two single-phase transformers so as to convert a 3-phase 3-wire a.c. supply to a 2-phase 3-wire supply, and vice versa.

Scottish topaz *(Min.)*. A term applied in the gemstone trade to yellow transparent quartzes, resembling Brazilian topaz in colour, used for ornamental purposes. Not a true topaz. See **cairngorm, citrine**.

scourer *(Eng.)*. A flour-milling machine in which the wheat, for cleaning purposes, is subjected to the action of revolving beaters in a ventilated casing.

scouring *(Hyd.Eng.)*. Said of the eroding action of water flowing at high velocity.

scouring *(Textiles)*. Processes for removing from textile materials fats, oils, waxes and other impurities which may be natural (e.g. wool grease) or which may have been added to aid processing (e.g. **size**). For cotton and flax, boiling aqueous sodium hydroxide solutions are used but other fibres require milder treatments e.g. with aqueous sodium carbonate or neutral solutions of detergent (for wool).

SCPC *(Telecomm.)*. Abbrev. for *Single-Channel Per Carrier*.

SCR *(Electronics)*. Abbrev. for *Silicon-Controlled Rectifier*.

scram *(Nuc.Eng.)*. General term for emergency shutdown of a plant, especially of a reactor when the safety rods are automatically and rapidly inserted to stop the fission process.

scrambler *(Telecomm.)*. Multiple modulating and demodulating system which interchanges and/or inverts bands of speech, so that speech in transmission cannot be intelligible, the reverse process restoring normal speech at receiving end.

scram rod *(Nuc.Eng.)*. An emergency safety rod used in a reactor.

scrap *(Eng.)*. (1) Metal which can be reclaimed. (2) Defective products unfit for sale.

scraper *(Build.)*. A thin flat steel blade with a square straight edge on which a burr is raised; used to pare wood from a surface which is being finally dressed.

scraper *(Civ.Eng.)*. A device, towed behind a tractor, which collects earth, with the object of reducing the ground level to that required.

scraper board *(Print.)*. A coated cardboard, the coating being either white on black or black on white; it can be readily scraped away to expose the underlayer. Used by commercial artists to obtain the effect of wood engraving, the result being made into a printing plate photo-mechanically.

scraper ring *(Autos.)*. A ring usually fitted on the skirt of a petrol- or oil-engine piston, to prevent excessive oil consumption. It may have a bevelled upper edge or a slotted groove, the oil being scraped off the cylinder wall and led back to the sump through holes in the piston wall. Also *oil control ring*.

scrapie *(Vet.)*. Progressive, fatal nervous disease of sheep and goats. the aetiology is not known but recognized to be a proteinaceous infectious particle. Characterized by trembling, pruritis, wool loss, staggering and loss of condition. Incubation period can be several years.

scratch *(Build.)*. A tool with an upright shaped blade fixed in a wooden body; used for working small mouldings.

scratch *(Comp.)*. Release of storage area for subsequent re-use.

scratch-coat *(Build.)*. The first of three coats applied in plastering. It consists of coarse stuff.

scratched figure *(Print.)*. A figure with a stroke through it to indicate a cancel, thus: 6̸.

scratcher *(Build.)*. A tool used to make scratch marks in a cement surface to provide a grip for a subsequent coat.

scratch file *(Comp.)*. Temporary storage area, usually held in backing store, for use by a program during execution. Also *work file*.

scratch pad *(Comp.)*. Section of immediate access store reserved for temporary information for immediate use. Also *working store*.

scratch tape *(Comp.)*. Magnetic tape used for a scratch file.

scray *(Textiles)*. A trough, probably on wheels, in which fabric (wet or dry) is collected and taken to the next process.

scree *(Geol.)*. The accumulation of rock debris strewn on a hillside or at a mountain foot, resulting from the mechanical weathering of rocks.

screeching *(Aero.)*. A cacophonous form of unstable combustion that can occur with rockets, and occasionally in turbine engines, causing very rapid damage due to resonance stresses on the jet pipe or nozzle. Also called *howl*.

screed *(Build.)*. (1) A band of plaster laid on the surface of a wall as a guide to the thickness of a coat of plaster to be applied subsequently. (2) A layer of concrete or mortar used to provide a finish to a concrete floor or the gradient on a floor or flat roof.

screed *(Civ.Eng.)*. A strip of wood or metal temporarily inserted in a road surface to form a guide for the template for forming the final surface of the road.

screed-coat *(Build.)*. A coat laid level with the screeds.

screen *(Biol.)*. To investigate a large number of organisms for the presence of a particular property as in screening for a mutation for *antibiotic resistance*.

screen *(Build.,Civ.Eng.)*. A large sieve used for grading fine or coarse aggregates.

screen *(Elec.Eng.)*. Electrode consisting of a relatively fine mesh network of wires interposed between two other electrodes, to reduce the electrostatic capacitance between them. It is usually maintained at positive potential, and connected to earth through a capacitor. See **electrostatic shield, Faraday cage**.

screen *(Image Tech.,Print.)*. (1) The surface on which a picture is presented by projection of a photographic or electronic image. (2) The meshwork of lines at right angles, ruled on glass, used to convert the subject of a half-tone illustration into dots for photo-mechanical reproduction.

screen burning *(Electronics)*. Gradual falling off in luminosity, sometimes accompanied by discoloration, in the fluorescent screen of a CRT, particularly if operated under adverse conditions.

screened-grid valve *(Electronics)*. Four-electrode valve, with cathode, control grid, screen and anode. Used as a high-frequency amplifier, where the screen, of unvarying potential, prevents positive feedback, and so greatly enhances stability. See **tetrode**.

screened horn balance *(Aero.)*. A **horn balance** which is screened from the airflow by the fixed surface in front of it.

screen editing *(Comp.)*. Changing stored data or programs by altering text displayed on a **VDU**, using a **cursor** to indicate the position.

screened wiring *(Elec.Eng.)*. Insulated conductors enclosed in earthed and continuously conducting metal tubes or conduits, mainly for mechanical protection, but frequently for preventing induction to or from conductors.

screening *(Elec.Eng.,Phys.)*. Use of a screen, in the form of a metal or gauze can, normally earthed, so that electrostatic effects inside are not evident outside, or vice versa. Nucleus of an atom is screened by its surrounding electrons.

screening *(Radiol.)*. See **fluoroscopy**.

screening constant *(Phys.)*. A number which when subtracted from the atomic number (*Z*) of an atom, gives the *effective* atomic number so that X-ray spectra may be described by 'hydrogen-like' formulae. It arises from the nuclear charge ($+Ze$) being screened by the inner electron shells.

screening protector *(Elec.Eng.)*. See **line choking coil**.

screenings *(Build.,Civ.Eng.)*. The residue from a sieving operation.

screen printing *(Textiles)*. A process in which the coloured printing pastes are forced by a squeegee through selected meshes (left open) of screens on to the face of the fabric being printed. The screens, which may be flat or cylindrical, are made of a woven fabric or of metallic wire mesh and one is required for each colour.

screen process printing *(Print.)*. See **silk-screen printing**.

screen-protected motor *(Elec.Eng.)*. A protected type of electric motor in which the openings for ventilation are covered with wire-mesh screens.

screen, screening *(Min.Ext.)*. Perforated or woven cloths (metal, fibre, rods, bars), used to size ore or products as part of treatment required to regulate concentration. Include *grizzlys, trommels*, and mechanically- and electrically-vibrated screens.

screw-and-nut steering gear *(Autos.)*. One in which a square-threaded screw formed on the lower end of the steering-column engages with a nut provided with trunnions, which work in blocks sliding in a short slotted arm, connecting with the remainder of the steering system.

screw-auger *(Build.)*. An auger having a helical groove cut in its surface so as to carry away the chips from the cutting edge.

screw axes *(Min.)*. Axes of symmetry about which the atoms in a mineral are symmetrically disposed. Rotation about a 4-fold screw axis, for example, will carry an atom 1 into the positions successively occupied by similar atoms 2, 3, and 4, after rotations of 90°, 180°, 270°, and 360°. Cf. **rotation axes of symmetry**.

screw box and tap *(Build.)*. Device for making wooden screws on furniture legs, wooden cramps etc.

screw chases *(Print.)*. Chases used in newspaper work. They are tightened by screws, which obviate the use of separate quoins.

screw-chasing *(Eng.)*. See **chaser**.

screw composing-stick *(Print.)*. An old type of composing-stick fastened with a thumbscrew. The modern style is fastened at the correct measure by means of a lever.

screw conveyor *(Eng.)*. See **worm conveyor**.

screw-cutting lathe *(Eng.)*. A metal turning-lathe provided with a lead screw driven by *change wheels*, for traversing the pointed tool used in screw-cutting.

screwdriver *(Eng.)*. Tool with shank terminating in a blade of size and shape to fit the slot in screws. See also **Phillips-, ratchet-**.

screwed steel conduit *(Elec.Eng.)*. Light steel tubing, having screwed ends for connecting up in lengths by means of sockets, in which electrical installation wiring is run. Cf. **plain steel conduit**.

screwing die *(Eng.)*. An internally threaded hardened steel block, sometimes split in halves, on which cutting edges are formed by longitudunal slots. Held in a stock, lathe or screwing machine for cutting external threads.

screwing machine *(Eng.)*. A form of lathe adapted for the continuous production of screws or screwed pieces, by means of dies.

screw jack *(Eng.)*. See **jack**.

screw micrometer *(Eng.)*. See **micrometer gauge**.

screwnail, drivenail *(Eng.)*. A nail in whose surface shallow helical depressions are formed, so that as it is driven in place with blows from a hammer it turns like a screw.

screw pile *(Civ.Eng.)*. A pile having a wide projecting helix or screw at the foot, useful in alluvial ground.

screw plate *(Eng.)*. A hardened steel plate in which a number of screwing dies of different sizes are formed.

screw plug *(Build.)*. A drain plug consisting of a rubber ring held between two steel disks which, on being screwed together, force the ring out to close the drain pipe in which the plug is placed.

screw press *(Eng.)*. See **fly press**.

screw propellor *(Eng.)*. See **airscrew, marine screw propellor**.

screw shackle *(Eng.)*. A long nut screwed internally with a right-hand thread at one end and a left-hand thread at the other, serving to connect the ends of two rods which are to be joined together, and providing a means of adjusting the total length. See **coupling**.

screw thread *(Eng.)*. A helical ridge of approximately triangular (or V), square, or rounded section, formed on a cylindrical core, the pitch and core diameter being standardized under various systems. See **British Association-, British Standard Whitworth-, Metric-, Sellers-, Unified-**, etc.

scribbler *(Textiles)*. A machine for carding wool.

scriber *(Eng.)*. A pointed steel tool used for making an incised mark on timber or metal, to guide a subsequent cutting operation.

scribing block *(Eng.)*. A tool for gauging the height of some point on a piece of work, above a surface plate or machine table. It consists of a base supporting a pivoted column, to which a scriber is slidably clamped. Also called **surface gauge**.

scribing gouge *(Build.)*. One sharpened with the bevel on the inside, used for work where an upright cut is necessary.

scrieve board *(Ships)*. A formation of portable portions of flat wooden boards whereon are scrieved (or scribed) the ship's transverse frame sections and lines indicative of shell seams, decks, stringers etc. Scrieve boards are used for setting the soft iron, to which the frames etc. are turned.

scrim *(Build.)*. An open weave fabric used to cover the joints between fibre or plaster boards before applying a finish.

scrim *(Textiles)*. An open woven fabric used for reinforcement in bookbinding, upholstery, wall plastering and for the base cloths of certain non-woven fabrics.

script *(Image Tech.)*. The written outline of a film or video production detailing the settings, action and dialogue for each scene throughout (*scenario*). If camera directions are included it becomes the *shooting script*.

script *(Print.)*. A style of type which imitates handwriting.

scrobiculate *(Bot.,Zool.)*. Having the surface dotted all over with small rounded depressions; pitted.

scrobiculus *(Bot.,Zool.)*. A small pit or rounded depression.

scrofula *(Med.)*. Caseating tuberculosis of the lymphatic glands. *adj.* **scrofulous**.

scrofulodermia, scrofuloderma *(Med.)*. Tuberculous infection of the skin from the bursting of a deep-seated tuberculous abscess; a subcutaneous tuberculous abscess.

scroll *(Image Tech.)*. Video transition effect in which one picture is displaced vertically by another.

scroll chuck *(Eng.)*. A self-centring chuck for holding round work, having jaws slotted to engage with a raised spiral or scroll on a plate which is rotated by a key, so as to advance the jaws while maintaining their concentricity. Esp. *three-jaw chuck*.

scrolling *(Comp.)*. Action of a VDU when displaying text of moving lines up the screen, losing old lines from above as new lines appear below.

scroop *(Textiles)*. The crunching noise and the corresponding handle that is obtained when certain fabrics, particularly of silk, are crushed in the hand. The noise is rather like that obtained when dry snow is squeezed into balls.

Scrophulariaceae *(Bot.)*. Family of ca 3000 spp. of dicotyledonous flowering plants (superorder Asteridae). Almost all herbs, cosmopolitan. The flowers are gamopetalous, the ovary superior and the fruit usually a capsule. Includes the foxglove from which the important cardiac drugs digitalin and digoxin are derived, and some ornamental plants e.g. *Antirrhinum*.

scrotum *(Zool.)*. In Mammals, a muscular sac forming part of the ventral body wall into which the testes descend. adj. *scrotal*.

scrubber *(Chem.Eng.)*. One kind of *dampener* which depends for its action on changing the velocity of gas flowing by altering direction and flow area for the gas stream. The alterations are provided by baffles in the scrubber.

scrub typhus *(Med.)*. See **typhus**.

scuffing *(Eng.)*. A sign of inadequate lubrication in which **adhesive wear** produces scratches and tears in a mating surface.

scull *(Glass)*. The glass remaining in a ladle after most of the molten glass has been poured out.

scum *(Build.)*. A surface formation of lime crystals appearing on new cement work.

scumble *(Build.)*. A light-coloured, low-opacity paint allowing a darker background colour to *grin through* deliberately. Used in **graining**.

S-curve *(Surv.)*. See **reverse curve**.

scurvy *(Med.)*. *Scorbutus*. A nutritional disease due to deficiency in the diet of vitamin C (ascorbic acid), characterized by anaemia, apathy, sponginess of the gums, ulceration of the mouth, and haemorrhages into the skin.

scutch *(Build.)*. The bricklayer's cutting tool for dressing bricks to special shapes. Also called *scotch*.

scute *(Zool.)*. An exoskeletal scale or plate. adj. *scutate*.

scutellum *(Bot.)*. More or less shield-shaped structure, possibly a modified cotyledon, attached to the side of the embryo in a grass grain, and which at germination secretes hydrolytic enzymes into, and absorbs sugars etc. from, the adjacent endosperm.

scybalum *(Med.)*. A round, hard, and dry faecal mass in the intestine.

Scyphomedusae *(Zool.)*. See **Scyphozoa**.

Scyphozoa *(Zool.)*. A class of Cnidaria in which the polyp stage is inconspicuous and may be completely absent. Where present it is known as a *scyphistoma*, and gives rise to the ephyra larvae. Velum and nerve ring generally absent, gonads endodermal, no skeleton. Jellyfish.

Scythian *(Geol.)*. The oldest epoch of the Triassic period.

SDD *(Telecomm.)*. Abbrev. for *Subscriber Direct Dialling*.

SDI *(Space)*. Abbrev. for *Strategic Defence Initiative*, a military program commonly referred to as Star Wars. The intent of SDI is to provide a defensive shield based on satellite, laser and high-energy particle technology for the destruction of hostile ballistic missiles when the latter are still in the atmosphere after launch.

SDS *(Biol.)*. See **sodium dodecyl sulphate**.

SDS gel electrophoresis *(Biol.)*. **Polyacrylamide gel electrophoresis** in the presence of the denaturant SDS.

SE *(Build.)*. Abbrev. for *Stopped End*.

Se *(Chem.)*. Symbol for selenium.

sea anchor *(Ships)*. A float to which a ship may be attached by a hawser to ride out a gale. Used in small boats in the form of a canvas drogue.

sea breeze *(Meteor.)*. See **land and sea breezes**.

sea cell *(Phys.)*. Primary electrolyte cell which functions as a source of electric power when immersed in sea water. A battery of such cells is possible, even though cells are partially short-circuited by the sea water. Fitted to life-belts etc., so that an indicating light is produced automatically in the event of use at night.

sea clutter *(Radar)*. **Clutter** generated by rough sea surfaces; potential source of difficulty with sea-skimming guided missiles.

sea-floor spreading *(Geol.)*. The process by which new oceanic crust is generated at oceanic ridges by the convective upwelling of magma. The plates on either side of the **divergent junction** move very gradually apart.

seal *(Build.)*. *Water seal*. The water contained in a trap, which prevents the flow of air or gases from one side to the other.

seal *(Electronics)*. In a vacuum tube, the point at which the tube is closed after pumping, and the act of closing off.

seal *(Print.)*. A small printing plate usually used to indicate the edition of a newspaper. See **seal cylinder, seal unit**.

seal cylinder *(Print.)*. The auxiliary cylinder to which the *seal* is attached.

sealed cover *(Build.)*. An air-tight cast-iron or precast concrete cover fitting into a frame and used to cover a manhole.

sealed pressure balance *(Aero.)*. An **aerodynamic balance**, used mainly on ailerons, consisting of a continuous projection forward of the hinge line within a cavity formed by close fitting **shrouds** projecting rearward from the main surface, the gap between the balance and main surface being sealed to prevent communication of pressure between lower and upper surfaces. Sometimes called a *Westland-Irving balance* after its inventor and the company which developed it.

sea level *(Surv.)*. The datum line from which heights are measured in surveying; the term is also used more loosely to denote the mean surface level of the oceans or the surface of the geoid.

sea-level pressure *(Meteor.)*. Atmospheric pressure at mean sea-level deduced from the pressure measured at the level of the observing station by taking into account the theoretical effect of a column between the two levels. Use of sea-level pressure on synoptic charts reveals the true meteorological patterns which would otherwise be totally obscured by the effect of altitude on the observations.

sea-level static thrust *(Aero.)*. See **static thrust**.

sea lily *(Zool.)*. See **Crinoidea**.

sealing box *(Elec.Eng.)*. A box in which the end of a paper-insulated cable is hermetically sealed.

sealing-in *(Elec.Eng.)*. The making of an air-tight joint between the filament wires and the glass envelope of an incandescent lamp or vacuum tube.

sealing-off *(Elec.Eng.)*. The final sealing of the exit to the evacuating pumps of an incandescent lamp bulb or vacuum tube.

seal unit *(Print.)*. The small auxiliary printing unit used for printing the *seal*, usually in a second colour on front page.

seam *(Build.)*. See **welt**.

seam *(Eng.)*. (1) A surface defect in worked metal, the result of a blowhole being closed but not welded; it remains as a fine crack. (2) Ridge in casting, effect of enclosure of impurity or scale in working.

seam *(Min.Ext.)*. (1) A tabular, generally flat deposit of coal or mineral; a stratum or bed. (2) A joint or fissure in a coal bed.

sea marker *(Aero.)*. Any device dropped from an aircraft on to water to make an observable patch from which the drift of the aircraft may be determined. Usually filled with a fluorescent substance for use during the day, and with a flame-producing device for night use.

seaming machine *(Eng.)*. A press for forming and closing longitudinal and circumferential interlocking joints used in the manufacture of containers from sheet metal.

seamless tube *(Eng.)*. Tube other than that made by bending over and welding the edges of flat strip. May be made by extrusion (nonferrous metals), or by piercing a hole through a billet and then rolling down over a mandrel to form a tube of the required dimensions.

seam roll *(Build.)*. See **hollow roll**.

seam welding *(Elec.Eng.)*. (1) See **resistance seam-welding**. (2) Uniting sheet plastic by heat arising from dielectric loss, the electric field being applied by electrodes carrying a high-frequency displacement current. Also *high-frequency welding*, *jig welding*.

seaplane *(Aero.)*. An aircraft fitted with means for taking off from and alighting on water. See **float seaplane, flying boat, hydroskis**.

seaplane tank *(Aero.)*. A long, narrow water tank with a powered carriage carrying equipment by which the water performance of a seaplane model can be observed and precisely measured.

search coil *(Elec.Eng.)*. See **exploring coil**.

search-image *(Behav.)*. Refers to the perceptual phenomena of an increased accuracy of discrimination for certain objects in the environment, e.g. a predator's improved ability to see camouflaged prey against its background.

search image *(Ecol.)*. In behavioural ecology, a predator's pre-conception of what its prey looks like and where it is found.

search radar *(Radar)*. One designed to cover a large volume of space and to give a rapid indication of any target which enters it.

search strategy *(Comp.)*. A systematic procedure for retrieving a predefined item from among data. See **sequential access, binary search, data retrieval, information retrieval.**

season cracking *(Eng.)*. Spontaneous cracking of brass and other metals. Intergranular cracks arise from residual internal stresses due to cold-working operations, in conjunction with surface corrosion.

seasoning *(For.)*. The process in which the moisture content of timber is brought down to an amount suitable for the purpose for which the timber is to be used.

seat board *(Horol.)*. The board or platform that carries the movement of a long case clock.

seat earth *(Geol.)*. A fossil soil which underlies a coal seam.

seating *(Eng.)*. A surface for the support of another piece, e.g., the end of a girder, or a masonry block.

seaweed *(Bot.)*. Macroscopic marine alga. Most seaweeds belong to the Phaeophyceae or Rhodophyceae, some to the Chlorophyceae.

sebaceous *(Zool.)*. Producing or containing fatty material, as the *sebaceous* glands of the scalp in Man.

sebaceous cyst *(Med.)*. A cyst formed as a result of blockage of the duct of a sebaceous gland, often present on the face, scalp, or neck.

sebacic acid *(Chem.)*. Dibasic acid, $HOOC\text{-}(CH_2)_6\text{-}COOH$. Obtained by heating castor oil with sodium hydroxide. A white crystalline solid, mp 129°C. Its esters are used in the production of resins and plasticizers.

sebiferous *(Zool.)*. Conveying fatty material.

sebiparous *(Zool.)*. Sebaceous.

seborrhoea, seborrhea *(Med.)*. Overactivity of the sebaceous gland, resulting in an abnormally greasy skin.

seborrhoeic, seborrheic dermatitis *(Med.)*. An inflammatory disease of the skin characterized by the presence of reddish patches covered with greasy scales; especially of the scalp, causing 'scurfy head'.

sebum *(Zool.)*. The fatty secretion produced by the sebaceous glands, which protects and lubricates hair and skin.

SECAM *(Image Tech.)*. The colour TV system adopted in France and Eastern Europe, from *SEquential Couleur A Memoire.*

secodont *(Zool.)*. Having teeth adapted for cutting.

second *(Genrl.)*. (1) 1/60 of a minute of time, or 1/86 400 of the mean solar day; once defined as the fraction 1/31 556 925.974 7 of the tropical year for the epoch 1900 January 0 at 12 h ET. Since 1965 defined, in terms of the resonance vibration of the caesium-133 atom, as the interval occupied by 9 192 631 770 cycles. This was adopted in 1967 as the SI unit of time-interval. Abbrev. *s*. (2) Unit of angular measure, equal to 1/60 of a minute of arc; indicated by the symbol ″. (3) In duodecimal notation 1/12 of an inch; indicated by ‴. (4) Unit for expressing flow times in capillary viscometers (Redwood, Saybolt Universal, or Engler), e.g. an Engler second is that viscosity which allows 200 cm^3 of fluid through an Engler viscometer in one second.

secondary *(Zool.)*. Arising later; of subsidiary importance; in Insects, the hind-wing; in Birds, a quill feather attached to the forearm, and also called the *cubital.*

secondary alcohols *(Chem.)*. Alcohols containing the group $>CH(OH)$. When oxidized they yield ketones (alkanones).

secondary amines *(Chem.)*. Amines containing the imino group $>NH$. They yield nitrosoamines with nitrous (nitric (III)) acid.

secondary battery *(Elec.Eng.)*. A number of secondary cells connected to give a larger voltage or a larger current than a single cell.

secondary beam *(Build.)*. In floor construction, a beam carried by **main beams** and transmitting loads to them.

secondary body cavity *(Zool.)*. See **coelom.**

secondary bow *(Meteor.)*. A *rainbow* having an angular radius of 52°, the red being inside and the blue outside, usually fainter than the primary bow. It is produced in a manner similar to the primary bow except that two internal reflections occur in the raindrops.

secondary cell *(Elec.Eng.)*. See **accumulator.**

secondary cell wall *(Bot.)*. See **secondary wall.**

secondary circuit *(Build.)*. Pipe circuit in which water circulates in distributing pipes from and back to a hot-water storage vessel.

secondary coil *(Elec.Eng.)*. A coil which links the flux produced by a current flowing in another coil (*primary coil*).

secondary colours *(Build.)*. Colours produced by mixing primary colours.

secondary constants *(Elec.Eng.)*. Those for a transmission line which are derived from the *primary constants.* They are the *characteristic impedance (impedance level)*, as of an infinite line, and the *propagation constant* (*attenuation* and *phase delay constant*).

secondary constriction *(Biol.)*. Non-centromeric constriction of the chromosomes, often at the site of the **nucleolar organizing region.**

secondary coolant *(Nuc.Eng.)*. In a reactor, a separate stream of coolant which is converted to steam by the **primary coolant** in a heat exchanger (steam generator) to power the turbine.

secondary depression *(Meteor.)*. A **depression** embedded in the circulation of a larger primary depression.

secondary dispersion *(Min.Ext.)*. In **geochemical prospecting**, the dispersion of elements or minerals from an ore body or **halo** by physical agents such as stream, river or ground-water flow, glacial ice, wind or wave action.

secondary electrode *(Elec.Eng.)*. See **bipolar electrode.**

secondary electrons *(Electronics)*. Those given off during the process of **secondary emission.**

secondary emission *(Electronics)*. Emission of electrons from a surface (usually conducting) by the bombardment of the surface by electrons from another source; the number may greatly exceed that of the primaries, depending on the velocity of the latter and the nature of the surface.

secondary enrichment *(Geol.)*. The name given to the addition of minerals to, or the change in the composition of the original minerals in, an ore body, either by precipitation from downward-percolating waters or upward-moving gases and solutions. The net result of the changes is an increase in the amount of metal present in the ore at the level of secondary enrichment.

secondary gneissic banding *(Geol.)*. A prominent mineral banding exhibited by coarse-grained crystalline rocks which have been subjected to intense regional metamorphism, involving rock-flowage. Often it is difficult to distinguish from **primary gneissic banding.**

secondary growth *(Bot.)*. See **secondary thickening** (1).

secondary hardness *(Eng.)*. Further increase in hardness produced in tempering high-speed steel after quenching.

secondary immune response *(Immun.)*. Response of the body to an antigen with which it has already been primed (see **primary immune response**). There is a very rapid production of large amounts of antibody over a few days, followed by a slow exponential fall. The response of **cell mediated immunity** follows a similar pattern.

secondary leakage *(Elec.Eng.)*. The magnetic leakage associated with the secondary winding of a transformer.

secondary memory *(Comp.)*. See **backing store.**

secondary meristem *(Bot.)*. (1) Meristem producing secondary tissues, (2) Meristem derived by dedifferentiation from differentiated cells, e.g. the interfasicular cambium, cork cambium.

secondary messenger *(Biol.)*. A cytoplasmic component which mediates the action of hormones bound to receptors on the cell surface, e.g. cyclic AMP.

secondary metabolites *(Bot.)*. Applied to those compounds which do not function directly in biochemical activities like photosynthesis, respiration and protein synthesis which support growth. They include alkaloids, terpenoids, flavonoids which may function in defence against insects, fungi and herbivores, in **allelopathy** or as attractants to pollinators or fructivores.

secondary metal *(Eng.)*. That retrieved from scrap, and worked up for return to industry.

secondary mineral *(Geol.,Min.Ext.)*. (1) One formed after the formation of the rock enclosing it. (2) That of minor interest in an ore body undergoing exploitation.

secondary nitro-compounds *(Chem.)*. Nitro-compounds containing the group $>CH(NO_2)$.

secondary phloem *(Bot.)*. Phloem formed by the activity of a cambium.

secondary-process thinking *(Behav.)*. In psychoanalytic theory, logic bound thinking, governed by the **reality principle**; it involves the ego-functions of remembering, reasoning and evaluation that mediate between instinctual needs and adaptation to the external world.

secondary production *(Min.Ext.)*. That in which means like pumping gas or water into an oil reservoir are needed to assist the flow to the well bore. Also *secondary recovery*. Cf. **primary production, tertiary production**.

secondary radar *(Radar)*. One which involves transmission of a second signal when the incident signal triggers a **transponder** beacon.

secondary radiation *(Phys.)*. That produced by interaction of primary radiation and an absorption medium.

secondary reinforcement *(Behav.)*. An initially neutral stimulus that acquired reinforcing properties through pairing with another stimulus that is already reinforcing.

secondary service area *(Telecomm.)*. That surrounding a radio or television broadcasting station where satisfactory reception is not guaranteed, and is only possible given good transmission conditions or a favourable location.

secondary sexual characters *(Zool.)*. Features which distinguish between the sexes other than the reproductive organs.

secondary shutdown system *(Nuc.Eng.)*. In the unlikely event of failure of control rods to enter and shut down reactor, a system such as, in a gas-cooled reactor, the insertion of nitrogen which absorbs neutrons much more strongly than does carbon dioxide. See **emergency shutdown system**.

secondary spectrum *(Phys.)*. The residual longitudinal chromatic aberration in a lens corrected to bring two wavelengths to the same focus.

secondary standard *(Genrl.)*. A copy of a *primary standard* for general use in a standardizing laboratory.

secondary stress *(Eng.)*. A bending stress, resulting from deflection, as distinct from a direct tensile or compressive stress.

secondary structure *(Biol.)*. The first level of 3-dimensional folding of the backbone of a polymer. Thus a polypeptide chain can be folded into an α-helix or β-pleated sheet and the nucleotide chains of DNA into a double helix.

secondary substances *(Ecol.)*. Plant biochemicals which are involved in no known biosynthetic pathways, but which are often detected in high concentrations in leaves and other organs, and so are presumed to be chemical defences.

secondary succession *(Ecol.)*. A succession proceeding in an area from which a previous community has been removed, e.g. a ploughed field. Cf. **primary succession**.

secondary surveillance radar *(Radar)*. See **ATCRBS**. Abbrev. *SSR*.

secondary thickening *(Bot.)*. (1) The increase in girth of a stem or root that results from the activity of a cambium after elongation has ceased. (2) Also, confusingly, the formation by a cell of a secondary wall.

secondary voltage *(Elec.Eng.)*. The voltage at the terminals of the secondary winding of a transformer.

secondary wall *(Bot.)*. A later-formed part of the *cell wall* (sometimes the major part) laid down on the cytoplasmic side of the *primary wall* after cell expansion has ceased, typically richer than the primary wall in cellulose.

secondary wave *(Telecomm.)*. A wave deriving from the main or desired wave forming a communication link but arising when this wave is partially reflected, refracted, or scattered.

secondary winding *(Elec.Eng.)*. A winding which links the flux produced by a current flowing in another winding, i.e. the primary winding.

secondary xylem *(Bot.)*. Xylem formed by the activity of a cambium. Wood.

second-channel interference *(Telecomm.)*. That due to a signal coinciding with the **image frequency** of a **superhet** radio receiver.

second detector *(Telecomm.)*. In superhet receivers, detector which demodulates the received signal after passing through the intermediate-frequency amplifier.

second development *(Image Tech.)*. The second development in a reversal process, after the first image has been removed by bleaching and the remaining silver halide rendered developable.

second-generation computer *(Comp.)*. One produced around 1955–1964 when the valve was being replaced by the **transistor** which consumed less power and was much more reliable. Other important changes included the handling of input/output without involving the central processor, and the development of **high level** programming languages and **floating point** arithmetic. Machines included IBM 7090 and 7094, Atlas. See **computer generations**.

second moment of area *(Eng.)*. The second moment of area is a measure of resistance to bending of a loaded section. A plank stood on edge is less easily bent when loaded than one which has the wide dimension horizontal. This is because the second moment of area on edge is very much greater than that when the plank is laid flat. Calculation of second moment of area (I) about an axis XX in the same plane is found from $I_{xx} = \Sigma ar^2$ where a is an element of the total area and r the perpendicular distance from XX. I_{XX} can also be obtained from the second moment of area about a parallel axis through the centroid of the area (I_{GG}) by the equation $I_{XX} = I_{GG} + Ah^2$, where A is the total area and h the perpendicular distance between the two axes.

second-operation work *(Eng.)*. Machining work carried out after cutting from bar or other material and for which a second clamping operation is necessary.

seconds *(Build.)*. Bricks similar to **cutters** but of a slightly uneven colour.

second tap *(Eng.)*. A tap used, after a taper tap, to carry the full thread diameter further down the hole, or to give the finished size of thread in a through hole.

second ventricle *(Zool.)*. In Vertebrates, the cavity of the right lobe of the cerebrum.

secrecy system *(Telecomm.)*. *Privacy system*. See **scrambler, inverter**.

secretagogue *(Med.)*. Substance, e.g. hormone, which stimulates secretion.

secret dovetail *(Build.)*. An angle joint between two members in which neither shows end grain, the visible external parts being mitred, while the dovetails are kept back from both faces. Also called *dovetail mitre, mitre dovetail*.

secretion *(Bot.)*. The elimination from the cytoplasm of a cell or from a multicellular structure (a gland) of an organic substance or inorganic ions.

secretor *(Immun.)*. Persons who secrete ABO blood group substances into mucous secretions such as gastric juice, saliva and ovarian cyst fluid. Over 80% of humans are secretors. The status is genetically determined.

secretory *(Zool.)*. Secretion-forming.

secretory duct *(Bot.)*. A duct containing material secreted by its lining of epithelial cells, e.g. resin duct.

secretory piece *(Immun.)*. A large polypeptide attached to dimers of the secreted form of IgA. Structurally unrelated to immunoglobulins and synthesized by epithelial cells in the secretory gland, not by the plasma cells that synthesize the immunoglobulin. Has strong affinity for mucus thus prolonging retention on mucous surfaces, and it may inhibit destruction of IgA enzymes in the gut.

secritin *(Biol.)*. A polypeptide hormone secreted by the intestinal wall which stimulates the pancreas to secrete bicarbonate ions.

sec, secant *(Genrl.)*. See **trigonometrical functions**.

section *(Biol.)*. (1) A thin slice of biological or mineral material sufficiently transparent to be studied with the compound microscope. (2) A taxonomic group, esp. a subdivision of a genus, but sometimes of a higher rank.

section *(Print.)*. A reference mark (§) directing the reader's attention to a footnote.

section *(Surv.)*. The representation to scale of the variations in level of the ground surface along any particular line. Drawing which shows a plane through a solid object, succession of geological strata, mine etc.

section *(Telecomm.)*. Unit of a ladder network, derived through design techniques to give specified transmission performance with respect to frequency.

section gap *(Elec.Eng.)*. An arrangement for dividing the overhead contact wire of an electric traction system into sections, both electrically and mechanically, without interfering with the smooth passage of the current collector. Also called *overlap span*.

section modulus *(Eng.)*. Ratio of **second moment of area** to distance of the farthest stressed element from the **neutral axis**. It is an important property of structural members, for calculating bending stresses. Symbol Z. Also called, loosely, **elastic modulus**. Units cm^3.

section mould *(Build.)*. A templet whose profile corresponds to the shape of the section of a required member. This shape is marked on the ends of a timber and used as a reference in making the member.

section switch *(Elec.Eng.)*. A switch whose function is to connect or disconnect two sections of an electric circuit, generally two bus-bar sections.

sector *(Comp.)*. Smallest addressable portion of the track on a magnetic tape, disk or drum store.

sector *(Maths.)*. A plane figure enclosed by two radii of a circle (or of an ellipse) and the arc cut off by them.

sectoral horn *(Acous.)*. Waveguide horn with two surfaces parallel to side of guide and the other two flared out-classified according to whether the flaring is in the plane of the electric or magnetic field.

sector disk *(Phys.)*. Rotating disk with angular sector removed, interposed in path of beam of radiation. Used to produce known attenuation, or to chop or modulate intensity of transmitted beam.

sector display *(Radar)*. A form of radar display used with continuously rotating antennae, *not* with sector scanning. (So termed because the display uses a long persistence CRT excited only when the antenna is directed into a sector from which a reflected signal is received.) Cf. **sector scan**.

sectorial *(Zool.)*. Adapted for cutting.

sectorial chimera *(Bot.)*. That in which one component forms a longitudinal strip of tissue down a shoot, the strip including more than one layer of the tunica and/or corpus.

sector regulator *(Civ.Eng.)*. A form of drum weir. It consists of a hollow reinforced concrete sector of a cylinder placed transversely across the direction of flow and capable of rotation about a horizontal axis on the downstream side, with accommodation for the sector in a special pit in the bed of the stream.

sector scan *(Radar)*. Scan in which the antenna moves through only a limited sector. (Not to be confused with a **sector display**.)

sectroid *(Build.)*. The curved surface between adjacent groins on a vault surface.

secular acceleration *(Astron.)*. A nonperiodic term in the mathematical expression for the moon's motion by which the mean motion increases ca. 11″ per century; caused by perturbations and by tidal friction in shallow seas.

secular changes *(Geol.)*. Changes which are extremely slow and take many centuries to accomplish; they may apply to climate, levels of land and sea, or, as in geomagnetism, to long period changes in the magnetic fields at any place.

secular equilibrium *(Phys.)*. Radioactive equilibrium where parent element has such long life that activities remain effectively constant for long periods.

secund *(Bot.)*. Having the lateral members, leaves or flowers, all turned to one side.

security *(Comp.)*. Establishment and application of safeguards to protect data, software and computer hardware from accidental or malicious modification, destruction or disclosure. See also **privacy**.

security paper *(Paper)*. Anti-counterfeit paper intended to discover the falsification of documents and to reveal attempts at so doing. Achieved by special, intricate watermarking, the addition of selected chemicals to the stock or incorporation in the furnish of synthetic or animal fibres, threads, planchettes etc.

sedentary *(Zool.)*. Said of Mammals which remain attached to a substratum.

sedimentary rocks *(Geol.)*. All those rocks which result from the wastage of pre-existing rocks. They include the fragmental rocks deposited as sheets of sediment on the floors of seas, lakes and rivers and on land; also deposits formed of the hard parts of organisms, and salts deposited from solution, in some cases by organic activity. Igneous and metamorphic rocks are excluded.

sedimentary structure *(Geol.)*. Any physical structure in a sedimentary rock that was formed at the time of its deposition, e.g. **cross-bedding, sole mark**.

sedimentation *(Chem.)*. Method of analysis of suspensions, by measuring the rate of settling of the particles under gravity or centrifugal force and calculating a particle parameter from the measured settling velocities. The same method can be applied to biological macromolecules using the **ultracentrifuge**.

sedimentation balance *(Powder Tech.)*. A device widely used in particle-size analysis by sedimentation techniques. The accumulation of particles at the bottom of a suspension vessel is automatically recorded, the particle size distribution being derived from the rate of accumulation of the particles at various times.

sedimentation potential *(Chem.)*. The converse of **electrophoresis**; a difference of potential which occurs when particles suspended in a liquid migrate under the influence of mechanical forces, e.g. gravity.

sedimentation tank *(Build.)*. A tank into which sewage from the detritus pit is passed so that suspended matters may sink to the bottom, from which they can be removed.

sedimentation techniques *(Powder Tech.)*. Group of methods of particle-size analysis in which concentration changes within a suspension, the changes being apparently caused by differential rates of settling of various sizes of particle.

sedimentation test *(Med.)*. The measurement of the rate of sinking of red blood cells (erythrocytes) in drawn blood placed in a tube; the rate is increased in disease and in pregnancy.

Seeback effect *(Elec.Eng.)*. Phenomenon by which an (thermoelectric) e.m.f. is set up in a circuit in which there are junctions between different bodies, metals, or alloys, the junctions being different temperatures. Also **thermoelectric effect**.

seed *(Bot.)*. The matured ovule of a seed plant containing usually one embryo, with, in some species, endosperm or perisperm, surrounded by the seed-coat or testa.

seed *(Glass)*. Small bubbles.

seed bank *(Ecol.)*. The total seed content of the soil.

seed crystal *(Chem.)*. A crystal introduced into a supersaturated solution or a supercooled liqid in order to initiate crystallization.

seed leaf *(Bot.)*. Same as **cotyledon**.

seed plant *(Bot.)*. Member of those plant groups that reproduce by seeds, i.e. the gymnosperms and the angiosperms; the division Spermatophyta.

seedy toe *(Vet.)*. An affection of the horse's hoof in which the wall of the hoof becomes separated from the subcorneal tissue, forming a space which becomes filled with abnormal, crumbly, horn; causes lameness when severe.

seeing *(Astron.)*. A term used by telescopic observers to describe the quality of observing conditions as influenced by turbulence in the Earth's atmosphere.

seersucker *(Textiles)*. A woven fabric with stripes or checks of puckered material alternating with flat sections.

see-saw amplifier *(Electronics)*. Same as **paraphase amplifier**.

Seewer governor *(Elec.Eng.)*. A hydraulic turbine governor for controlling the speed of high-pressure Pelton wheels; a needle valve varies the divergence of the conical pressure jet issuing from the nozzle.

Seger cones *(Eng.)*. Small cones of clay and oxide mixtures, calibrated within defined temperature ranges at which the cones soften and bend over. Used in furnaces to indicate, within fairly close limits, the temperature reached at the position where the cones are placed. Also *fusion cones, pyrometric cones*.

segment *(Elec.Eng.)*. One of many elements, insulated from one another, which collectively form a commutator.

segment *(Maths.)*. A plane figure enclosed by the chord of a circle (or of an ellipse) and the arc cut off by it. The segment of a sphere or of an ellipsoid is the portion cut by a plane.

segment *(Zool.)*. One of the joints of an articulate appendage: one of the divisions of body in a metameric animal: a cell or group of cells produced by cleavage of an ovum. adj. *segmental*.

segmental *(Zool.)*. In metameric animals, repeated in each somite; as *segmental arteries, segmental papillae*.

segmental arch *(Civ.Eng.)*. An arch having the shape of a circular arc struck from a point below the springings.

segmental core disk *(Elec.Eng.)*. An armature core disk made up in segments; used when a disk in a single piece would be so large as to be unwieldy.

segmental interchange *(Biol.)*. The exchange of portions between two chromosomes which are not homologous.

segmentation *(Comp.)*. Process of dividing the program into sections (segments or modules), which can be independently executed or changed. See also **modular programming, overlay**.

segmentation *(Zool.)*. Meristic repetition of organs or of parts of the body; the early divisions of a fertilized ovum, leading to the formation of a blastula or analogous stage.

segmentation cavity *(Zool.)*. See **blastocoel**.

Segrè chart *(Phys.)*. A chart on which all known nuclides are represented by plotting the number of protons vertically against the number of neutrons horizontally. Stable nuclides lie close to a line which rises from the origin at 45° and gradually flattens at high atomic masses. Nuclides below this line tend to be β-emitters whilst those above tend to decay by positron emission or electron capture. Data for half-life, cross-section, disintegration energy etc. are frequently added.

segregation *(Biol.)*. The separation of the two alleles in a heterozygote when gametes are formed, each carrying one or the other; and, consequentially, the appearance of more than one genotype in the progeny of a heterozygote.

segregation *(Eng.)*. Non-uniform distribution of impurities, inclusions and alloying constituents in metals. Arises from the process of freezing, and usually persists throughout subsequent heating and working operations. See **inverse-, normal-**.

seiche *(Meteor.)*. An apparent tide in a lake (originally observed on Lake Geneva) due to the pendulous motion of the water when excited by wind, earth tremors or atmospheric oscillations.

Seidlitz powder *(Chem.)*. Effervescent powder. A mixture of sodium hydrogen carbonate with tartaric acid, acid sodium tartrate, or some similar acid or salt. Purgative.

seif dune *(Geol.)*. A longitudinal sand dune developed parallel to the dominant wind direction.

seismology *(Geol.)*. The study of earthquakes, particularly their shock waves. Studies of the velocity and refraction of seismic waves enable the deeper structure of the Earth to be investigated.

seismonasty *(Bot.)*. A nastic response to a shock, especially mechanical shock, as in e.g. *Mimosa pudica*.

seizing signal *(Telecomm.)*. One sent from the outgoing end of a circuit at the start of a call and having the primary function of preparing the apparatus at the incoming end of the circuit for the reception of subsequent signals.

seizure, seizing-up *(Eng.)*. The locking or partial welding together of sliding metallic surfaces normally lubricated, e.g. a journal or bearing.

selcal *(Aero.)*. An automatic signalling system used to notify the pilot that his aircraft is receiving a call. It makes constant monitoring of the receiving equipment unnecessary. A contraction of *selective calling*.

selectance *(Telecomm.)*. Ratio of sensitivities of a receiver to two specified channels.

selection *(Biol.)*. The process by which some individuals come to contribute more offspring than others to form the next generation in *natural selection* through intrinsic differences in survival and fertility, in *artificial selection* through the choice of parents by the breeder.

selection rules *(Phys.)*. Restrictions on the transitions between quantum states of atoms, molecules or nuclei. The rules are derived theoretically by quantum mechanics. See **nuclear selection rules**.

selective absorption *(Phys.)*. Absorption of light, limited to certain definite wavelengths, which produces so-called absorption lines or bands in the spectrum of an incandescent source, seen through the absorbing medium. See **Kirchhoff's laws** (*Phys.*), **Fraunhofer lines**.

selective assembly *(Eng.)*. The assembly of mating parts selected by trial for their accuracy so as to obtain the required precision of fit.

selective dump *(Comp.)*. A **dump** of full contents of a specific part of a computer store.

selective emission *(Elec.Eng.)*. The property of an incandescent body whereby it emits radiation, predominantly of one frequency.

selective fading *(Telecomm.)*. That affecting some parts of a broad-band signal more than others, e.g. sound and not vision in TV reception.

selective freezing *(Eng.)*. A process involved in the solidification of alloys, as a result of which the crystals formed differ in composition from the melt. Thus, in alloys in a **eutectic system** (except the eutectic alloy), crystals of one metal are formed from a melt containing two, and this continues until the melt reaches the **eutectic point**.

selective mating *(Zool.)*. See **preferential mating**.

selective network *(Telecomm.)*. One for which the loss and/or phase shift are a function of frequency.

selective protection *(Elec.Eng.)*. A term applied to methods of protecting power transmission networks in which an automatic disconnection of the faulty section occurs without disturbance of the remainder of the network.

selective resonance *(Telecomm.)*. Resonance which occurs at one or more discrete frequencies, instead of extending over a band of frequencies as in some forms of filter.

selectivity *(Telecomm.)*. Ability of a receiver to distinguish by tuning between specified wanted and unwanted signals. Measured by frequency difference for the half-power points of the pass band of the receiver. Often aided by directive reception.

selector *(Telecomm.)*. A unit device in older telephone exchanges, operated either by the dialled impulses originated by the subscriber, or by impulses arising within the exchange.

selector forks *(Autos.)*. In a gearbox, forked members whose prongs engage with grooves cut in bosses which they move along a splined shaft for changing gear. They are operated through rods by the gear lever.

selector plug *(Telecomm.)*. A plug for making connection to a selector, so that the latter can be taken out of service temporarily for testing, without removal from its shelf.

selector valve *(Aero.)*. A valve used to direct the flow of the hydraulic fluid or compressed air in a system into the desired actuating current.

selenite *(Chem.)*. *Selenate (IV)*. The salt of the hypothetical selenous acid $HSeO_2$, e.g. sodium selenite, $NaSeO_2$.

selenite *(Min.)*. The name given to the colourless and transparent variety of **gypsum** which occurs as distinct monoclinic crystals, especially in clay rocks.

selenitic cement, lime *(Build.)*. A mixture of a feebly hydraulic lime with approximately 5% of plaster of Paris ground together to suppress the slaking action of the lime; used for plastering or rendering.

selenium *(Chem.)*. A nonmetallic element, symbol Se, at. no. 34; r.a.m. 78.96; valencies 2,4,6. A number of allotropic forms are known. *Red selenium* is monclinic; mp 180°C; rel.d. 4.45. *Grey (metallic) selenium*, formed when the other varieties are heated at 200°C, is a conductor of electricity when illuminated; mp 220°C; bp 688°C; relative density 4.80; electrical resistivity 12×10^{-8} ohm metres. Selenium is widely distributed in small quantities, usually as selenides of heavy metals. It is obtained from the flue dusts of processes in which sulphide ores are used, and from the anode slimes in copper refining. It is used as a decolorizer for glass, in red glass and enamels, and in photoelectric cells and rectifiers. Selenium is similar to sulphur in chemical properties, but resembles tellurium more closely still.

selenium cell *(Electronics)*. Early photoconductive cell which depends on the change in electrical resistance of selenium when illuminated.

selenium halides *(Chem.)*. Selenium has a greater affinity for the halogens than sulphur. Selenium (VI) fluoride, SeF_6, is the only (VI) halide. (IV) halides are known. No compounds with iodine.

selenodont *(Zool.)*. Having cheek teeth with crescentic ridges on the grinding surface.

selenophone *(Acous.)*. Original system of photographically recording sound on paper, the track being reproduced by scanning with a focused slit, the modulated reflected light being received into a photocell.

self *(Bot.)*. Self-pollinate or self-fertilize. Cf. **cross**.

self-absorption *(Phys.)*. See **self-shielding**.

self-aligning ball-bearing *(Eng.)*. A **ball-bearing** in which the two rows of balls roll between an inner race and a spherical surface in the outer race, thus allowing considerable shaft deviation from the normal.

self-annealing *(Eng.)*. A term applied to metals such as lead, tin and zinc, which recrystallize at air temperature and in which little strain-hardening is produced by cold-working.

self-baking electrode *(Elec.Eng.)*. An arc-furnace electrode in the form of a hollow tube, into which a pastelike electrode material is continuously fed as it becomes hard-baked and burns away in the furnace.

self-balance protection *(Elec.Eng.)*. A method of protecting transformers and a.c. generators from internal faults, based on the fact that the instantaneous sum of the phase currents in a symmetrical 3-phase system is always zero.

self-capacitance *(Phys.)*. See **capacitance**.

self-centring chuck *(Eng.)*. *Universal chuck*. A lathe-chuck for cylindrical work in which the jaws are always maintained concentric by a scroll, or by radial screws driven by a ring gear operated by a key. Also *scroll chuck*.

self-centring lathing *(Build.)*. Expanded metal specially manufactured with raised ribs, greatly stiffening the sheet and enabling it to be used for lathing purposes with the minimum of framing. Also called *stiffened expanded metal*.

self-cleansing *(Build.)*. A term applied to a velocity of flow of sewage material sufficient to prevent deposition of solid matters.

self-compatible *(Bot.)*. An individual plant or a clone, capable of self-fertilization. See also **homothallism**.

self-conjugate directions *(Maths.)*. See **conjugate directions**.

self-conjugate triangle *(Maths.)*. See **self-polar triangle**.

self-consistent field *(Phys.)*. The energy levels in many-electron atoms cannot be calculated exactly. An approximation method is used that starts by assuming the electrons occupy levels similar to that of hydrogen, postulates the electrostatic field in which the electrons exist and then calculates a new set of energy levels from which a new field is calculated. The process is repeated until the system is *self-consistent*.

self cure *(Immun.)*. In animals infested with intestinal nematodes, about 10 days after the initial establishment of the infection, there suddenly begins an expulsion from the intestine of the majority of the population of worms. This is probably due to an immune response by the host, manifested by an immediate hypersensitivity reaction involving release of histamine and other vasoactive substances from mast cells in the gut wall.

self-discharge *(Elec.Eng.)*. (1) Loss of capacity of primary cell or accumulator as a result of internal leakage. (2) Loss of charge from capacitor due to finite insulation resistance between plates.

self-dissociation *(Chem.)*. The weak tendency of water and some other liquids (e.g. liquid ammonia) with strongly polar molecules (see **associated liquid**) to break up into their component ions, such as H^+ (proton) and HO^- (hydroxyl), the former of which usually attaches itself to a complete water molecule H_2O, forming a hydronium ion, $H_2O.H^+$.

self-documenting program *(Comp.)*. One which informs the user how to use the program as it runs.

self-end papers *(Print.)*. Instead of having separate end-papers, the first two and last two leaves of the book are left blank, the first and the last leaves being pasted down on the cover; also called *own ends*.

self-excitation *(Elec.Eng.)*. A form of machine excitation in which the supply to the field system is obtained either from the machine itself or from an auxiliary machine which is coupled to it.

self-faced *(Build.)*. A term applied to stone, e.g. flagstone, which splits along natural cleavage planes, leaving faces which do not have to be dressed.

self-fertilization *(Biol.)*. The fertilization of an egg by a male gamete from the same individual or the same clone (genet). Cf. **cross-fertilization**, **self-pollination**. See also **autogamy**.

self-fluxing ore *(Eng.)*. Mineral charged to smelter which contains its own slag-forming constituents.

self-hardening steel *(Eng.)*. See **air-hardening steel**.

self-heterodyne *(Telecomm.)*. See **autodyne**.

self-incompatible *(Bot.)*. An individual plant or a clone, incapable of self-fertilization. See also **heterothallism**.

self-induced e.m.f. *(Elec.Eng.)*. That induced in an electric circuit as a result of a change in the current flowing in it.

self-inductance *(Phys.)*. If the current in a circuit changes, the magnetic flux linked with the circuit changes and induces an e.m.f. in such a direction as to oppose the change causing it (Lenz's law). The unit of self-inductance is the *henry*. See **mutual inductance**.

self-inductance coefficient *(Elec.Eng.)*. See **inductance coefficient**.

selfing *(Biol.)*. Self-fertilization, self-pollination.

selfish DNA *(Biol.)*. Class of DNA sequence thought to have been selected during evolution only by its ability to spread and duplicate itself in the genome of higher organisms, with minimal damage to the 'host'.

self-levelling level *(Surv.)*. Instrument which levels automatically by means of a pendulum-operated system of prisms.

self-lubricating bearing *(Eng.)*. Plain, powdered-metal bearing having a porous structure impregnated with lubricant which is released gradually during relative movement of the bearing.

self-polar triangle *(Maths.)*. A triangle whose sides are the polars of the opposite vertices with respect to a conic. Also called *self-conjugate triangle*. Cf. **conjugate triangles**.

self-pollination *(Bot.)*. The transfer of pollen to a stigma from an anther of the same flower or individual or clone (genet).

self-quenching *(Phys.)*. Said of counter tubes which do not depend on an external circuit for quenching, the residual gas providing sufficient resetting for the next operation of detecting a further photon or particle.

self-rectifying *(Radiol.)*. Said of an X-ray tube when an alternating voltage is applied directly between target and cathode.

self-regulating *(Nuc.Eng.)*. Said of a system when departures from the required operating level tend to be self-correcting, and in particular of a nuclear reactor where changes of power level produce a compensating change of reactivity, e.g. through negative temperature coefficient of reactivity.

self-scattering *(Phys.)*. The scattering of radioactive radiation by the body of the material which is emitting the radiation.

self-shielding *(Phys.)*. In large radioactive sources, the absorption in one part of the radiation arising in another part.

self-shielding *(Telecomm.)*. A coaxial line is self-shielding in that the return transmission current is in the inside surface of the outer conductor, while the interfering currents, if of sufficiently high frequency, are on the outside surface of the outer conductor.

self-starting rotary converter *(Elec.Eng.)*. A synchronous converter designed to start up from the a.c. supply as an induction motor, thus requiring no separate starting motor.

self-sterile *(Bot.)*. Not capable of producing viable offspring by self-fertilization.

self-sterility *(Bot.,Zool.)*. In a hermaphrodite animal or plant, the condition in which self-fertilization is impossible or ineffective.

self-synchronizing *(Elec.Eng.)*. A term applied to a synchronous machine that can be switched on to the a.c. supply without being in exact synchronism with it.

self-tapping screw *(Eng.)*. One made of hard metal which cuts its own thread when driven in.

self-thinning curve *(Ecol.)*. Curve describing the survival of individuals in a crowded population with time.

self-winding clock *(Horol.)*. Clock rewound automatically from electricity supply after a definite period of running.

self-winding watch *(Horol.)*. Watch which winds itself while being worn through movements of wearer turning a *rotor*; the winding may also be performed by the opening or shutting of the case.

sell *(Zool.)*. A hard outer case or exoskeleton of inorganic material, chitin, lime, silica etc.

Sellers screw thread *(Eng.)*. Abbrev. *USS*. The US standard thread, having a profile angle of 60°C, and a flat crest formed by cutting off $\frac{1}{8}$ of the thread height.

selsyn motor *(Elec.Eng.)*. A small self-synchronizing motor (used for transmitting signals), which indicates the position of a switch or reproduces instrument indications at a distance.

selvedge *(Textiles)*. The strong edge at both sides of a woven cloth. It sometimes bears a woven trade-mark or name, but its function is to give strength to the fabric in the loom and in subsequent processes carried out on the open width.

SEM *(Bot.)*. Abbrev. for *Scanning Electron Microscope*.

semantic error *(Comp.)*. One which results in ambiguous or erroneous meaning of a computer program. Cf. **logical error, execution error**.

semantic memory *(Behav.)*. Refers to general knowledge, e.g. grammar, principles, theories.

semantics *(Comp.)*. The meaning attached to words or symbols.

semantide *(Biol.)*. A molecule carrying information, as in a gene or messenger RNA.

sematic *(Zool.)*. Warning; signalling; serving for warning or recognition, as *sematic colours*.

semeiology *(Med.)*. The branch of medical science which is concerned with the symptoms of disease. (Gk. *sēmeion*, sign.)

semeiotic *(Med.)*. Pertaining to, or relating to, the symptoms and signs of disease.

semen *(Zool.)*. The fluid formed by the male reproductive organs in which the spermatozoa are suspended. adj. *seminal*.

semi- *(Genrl.)*. Prefix from L. *semi*, half.

semi-automatic *(Elec.Eng.)*. Said of an electric control in which the initiation of an operating sequence is manually performed and the sequence of operations subsequently proceeds automatically.

semi-automatic exposure control *(Image Tech.)*. A photoelectric device coupled to the lens of a camera which operates indicators, e.g. pointers, which can be aligned by moving the iris ring, thus obtaining the correct aperture setting for the available light.

semicarbazide *(Chem.)*. $H_2N.CO.NH.NH_2$, a base forming salts, e.g. hydrochloride. Mp 96°C; may be prepared from potassium cyanate and hydrazine hydrate. It reacts with aldehydes and ketones, forming **semicarbazones**.

semicarbazones *(Chem.)*. The reaction products of aldehydes or ketones with semicarbazide. The two amino hydrogen atoms of the semi-carbazide react with the carbonyl oxygen of the aldehydes or ketones, forming water, and the two molecular groups then combine to form the semicarbazone.

semichemical pulp *(Paper)*. Pulp produced from the raw material by a combination of chemical and mechanical means. The relatively light digestion is insufficient to resolve all the non-cellulose matter or to permit dispersion of the fibres without some mechanical treatment.

semicircular canals *(Zool.)*. Structures forming part of the labyrinth of the inner ear of most Vertebrates, there usually being three at right angles to each other, two being vertical and one horizontal. Movements of the head cause movement of the endolymph in the canals, which moves the gelatinous cupula attached to the sensory hairs of a neuromast sense organ in the swelling (called *ampullae*) at the base of the canals, initiating nerve impulses which travel to the brain via the 8th cranial nerve. They thus serve as organs of dynamic equilibrium.

semicircular deviation *(Ships)*. Those components of the **deviation** which vary as the sine and cosine of the **compass course**. That is, deviations which have the same sign in one semicircle of courses and the opposite sign in the other semicircle.

semiclosed slot *(Elec.Eng.)*. A slot whose width narrows sharply at the top. This means that conductors cannot be inserted through the top of the slot as usual, but must be inserted from the ends.

semiconductor *(Electronics)*. Said of a material (and element or a compound) having higher resistivity than a conductor, but lower resistivity than an insulator. The limited conduction through the material is governed as follows: (a) in *intrinsic semiconductors*; charges across forbidden energy bands due to thermal energy. This effect is temperature-dependent and produces equal numbers of electron and hole carriers. (b) In *extrinsic semiconductors*; current carriers are introduced by donor and acceptor impurities with locked energy levels near the top or bottom of the forbidden band. Important semiconductors are silicon, germanium, lead sulphide, selenium, silicon carbide, gallium arsenide. Semiconductor materials are the basis of diodes, transistors, thyristors, photodiodes, and all integrated circuits.

semiconductor diode *(Electronics)*. Two-electrode, point contact or junction, semiconducting device with assymmetrical conductivity.

semiconductor diode laser *(Phys.)*. A laser in which the lasing medium is *p*- or *n*-type semiconductor diode, e.g. gallium arsenide. Capable of continuous output of a few milliwatts at wavelengths in the range 700–900 nm.

semiconductor junction *(Electronics)*. One between *donor* and *acceptor* impurity semi-conducting regions in a continuous crystal; produced by one of several techniques, e.g. alloying, diffusing, doping, drifting, fusing, growing, etc.

semiconductor radiation detector *(Electronics)*. Semiconductor diodes, e.g. silicon junction, are sensitive under reverse voltage conditions to ionization in the junction depletion layer, and can be used as radiation counters or monitors.

semiconductor trap *(Electronics)*. Lattice defects in a semiconductor crystal that produce potential wells in which electrons or holes can be captured.

semidiameter *(Astron.)*. Half the angular diameter of a celestial body.

semi-elliptic spring *(Autos.)*. A carriage spring, so called because when a pair is used, one inverted and attached by its ends to the other, the arrangement resembles an ellipse.

semi-enclosed *(Elec.Eng.)*. Said of electric motors in which ventilation is provided but access to live parts necessitates opening the case.

semigroup *(Maths.)*. A set S together with an operation * in S is said to be a semigroup if (1) the set S is closed with respect to *; i.e. for all elements x, y in S, $x*y$ is an element of S; (2) the operation * in S is associative; i.e. for all elements x, y, z in S, $(x*y)*z = x*(y*z)$. E.g. the set of natural numbers with addition and multiplication.

semi-immersed liquid-quenched fuse *(Elec.Eng.)*. A liquid-quenched fuse in which the fuse link is above the liquid before operation but drawn down into it during or after fusion.

semimetallic foils *(Print.)*. Coloured **blocking foils** which give a metallic effect when viewed from certain angles. Used for stamping the reverse side of moulded acrylate badges.

semimonocoque construction *(Aero.)*. See **monocoque** and **stressed-skin construction**.

semimuffle-type furnace *(Eng.)*. See **oven-type furnace**.

seminal *(Bot.)*. Pertaining to the seed. Seminal roots are adventitious roots produced at the base of the stem of young seedlings of e.g. cereals.

seminal receptacle *(Zool.)*. See **vesicula seminalis**.

seminal roots *(Bot.)*. Adventitious roots that develop from the hypocotyl, as in grass seedlings.

seminiferous *(Zool.)*. Semen-producing or semen-carrying.

seminoma *(Med.)*. A malignant tumour of the testis arising from the germinal cells.

semiochemical *(Zool.)*. A chemical substance produced by an animal and used in communication. See **pheromone, allomone**.

semiotics *(Zool.)*. The study of communication.

semi-oviparous *(Zool.)*. Giving birth to imperfectly developed young, as marsupial Mammals.

semipalmate *(Zool.)*. Having the toes partially webbed.

semipermeable membrane *(Chem.)*. A membrane which permits the passage of solvent but is impermeable to dissolved substances.

semiplacenta *(Zool.)*. A nondeciduate placenta in which only the foetal part is thrown off at birth.

semirigid airship *(Aero.)*. See **airship**.

semisteel *(Eng.)*. Cast-iron with low carbon content incorporating a proportion of melted down scrap steel.

semistreptostyly *(Zool.)*. In Vertebrates, the condition of having a slightly movable articulation between the quadrate and the squamosal. Cf. **streptostyly, monimostyly**.

semitone *(Acous.)*. Difference of pitch between two sounds with a frequency ratio equal to the 12th root of 2 on the even-tempered scale.

semitransparent mirror *(Image Tech.)*. A mirror which partially reflects and partially transmits light rays without appreciable diffusion.

semitransparent photocathode *(Electronics)*. One where the electrons are released from the opposite side to the incident radiation.

semitubular rivet *(Eng.)*. A rivet which has been drilled hollow part-way up the shank to provide a thin shank wall for easy setting.

semiwater gas *(Chem.)*. A mixture of carbon monoxide, carbon dioxide, hydrogen, and nitrogen obtained by passing a mixture of air and steam continuously through incandescent coke. Its calorific valued is low, about 45 MJ/m^3.

semiworsted spun *(Textiles)*. Yarn made from carded sliver or roving.

sempervirent *(Bot.)*. Evergreen.

sender *(Telecomm.)*. US for *transmitter*. (2) The same as *keysender*.

Senegal gum *(Chem.)*. See **gum arabic**.

senescent *(Biol.)*. Said of that period in the life history of an individual when its powers are declining prior to death.

senile-degenerative disorders *(Behav.)*. Deterioration of intellectual, emotional and motor functioning with advancing age.

senile dementia *(Med.)*. Progressive **dementia** in the elderly with loss of memory for recent events and a decline in all mental abilities, resulting from severe organic deterioration of the brain.

senility *(Biol.)*. Condition of exhaustion or degeneration due to old age.

Senonian *(Geol.)*. The youngest epoch of the Cretaceous period.

sensation curves *(Image Tech.)*. Curves which give the relative response of the eye to different colours having the same intensity.

sensation unit *(Acous.)*. Original name of the *decibel*; so called because it was erroneously thought that the subjective loudness scale of the ear is approximately logarithmic.

sense of absolute pitch *(Acous.)*. Ability to recognize a single tone without using a reference tone.

sense of relative pitch *(Acous.)*. Ability to recognize a given interval.

senses *(Behav.)*. Any number of responses to stimulation through the specialized sense organs (i.e. eyes, ears etc.). The responses of these organs translate into neural impulses, often referred to as *sensation*.

sense strand *(Biol.)*. That strand of a double-stranded DNA molecule which is transcribed into messenger or other RNA.

sensible heat *(Phys.)*. That heat which effects a change in the body which is detectable by the senses; i.e. it causes the temperature of the body to change. Measured by the product of the specific heat capacity, the mass of the body and the change of temperature.

sensible horizon *(Surv.)*. See **visible horizon**.

sensiferous, sensigerous *(Zool.)*. Sensitive.

sensillum *(Zool.)*. In Insects, small sense organs of varied function on the integuement typically comprising a cuticular and/or hypodermal structure. pl. **sensilla**.

sensing *(Radar)*. Removal of 180° ambiguity in bearing as given by simple vertical loop antenna, by adding signal from open aerial.

sensitive *(Zool.)*. Capable of receiving stimuli.

sensitive drill *(Eng.)*. A small drilling machine in which the drill is fed into the work by a handlever attached directly to the drilling spindle, the operator being thus given sensitive control of the rate of drilling.

sensitive period *(Behav.)*. Periods of time during development when an individual is particularly sensitive to environmental and social experiences and which affect learning in a variety of ways in a wide range of species.

sensitive time *(Phys.)*. Period for which conditions of supersaturation in a cloud chamber or bubble chamber are suitable for formation tracks.

sensitive tint plate *(Min.)*. A thin, optically-orientated plate of a crystal, usually gypsum, used to measure the optical properties of minerals and other crystalline substances with a polarizing microscope.

sensitive volume *(Phys.)*. The portion of an ionization chamber or counter tube across which the electric field is sufficiently intense for incident radiation to be detected. The portion of living cells believed to be susceptible to ionization damage. See **target theory**.

sensitivity *(Phys.)*. General term for ratio of response (in time and/or magnitude) to a driving force or stimulus, e.g., galvanometer response to a current, minimum signal required by a ratio of output level to illumination in a camera tube or photocell.

sensitivity guide *(Print.)*. A strip of photographic film where the emulsion is in a series of graduated continuous tones each of a known density. Each density step is numbered and the strip when printed down can be used as a guide to control exposure.

sensitization *(Chem.)*. The process by which a sol of a lyophilic colloid becomes lyophobic in character with the result that it may readily be coagulated by electrolytes.

sensitization *(Immun.)*. (1) Administration of an antigen to provoke an immune response so that, on later challenge, a more vigorous secondary response will ensue. This involves the recruitment of **primed** cells. (2) Coating of cells with antibody e.g. for use in complement fixation tests.

sensitized cheque paper *(Paper)*. Paper for cheques and similar financial documents, containing chemicals intended to reveal attempts at falsification.

sensitized paper *(Paper)*. Paper that has been coated to render it suitable for a reprographic process, generally light activated e.g. dyeline, blueprint, photographic. Also paper, made from stock to which selected chemicals have been added, or coated, so that a colour reaction is produced on exposure to other reagents.

sensitizer *(Chem.)*. A substance, other than the catalyst, whose presence facilitates the start of a catalytic reaction.

sensitizer *(Image Tech.)*. Chemical, usually dye, used to increase the sensitivity of photographic emulsions, generally or to specific colours.

sensitometer *(Image Tech.)*. Instrument providing a controlled series of graduated exposures on a sensitized material for the examination of its reproduction characteristics.

sensitometry *(Image Tech.)*. The study of the effect of light on sensitized materials and their response to subsequent processing, generally by the measurement of the densities so produced.

sensor *(Eng.)*. General name for detecting device used to locate (or detect) presence of matter (or energy, e.g. sound, light, radio or radar waves).

sensorimotor development *(Behav.)*. The development of co-ordination between perception and action (e.g. hand-eye co-ordination).

sensorimotor intelligence stage *(Behav.)*. According to Piaget, the first period of intellectual development (0–2 years) in which the infant's interactions with the environment consist of motor responses to classes of sensory stimuli (e.g. looking, grasping etc.). During this period the infant progresses from simple reflex actions to complex ways of playing with and manipulating objects, which leads eventually to internal representations of the world.

sensorium *(Zool.)*. The seat of sensation; the nervous system. adj. *sensorial.*

sensory *(Zool.)*. Directly connected with the sensorium; pertaining to, or serving, the senses.

sensory adaptation *(Behav.)*. A short-term change in the response of a sensory system as a consequence of repeated or protracted stimulation.

sensory deprivation *(Behav.)*. Refers to experimental work, mostly with humans, in which total sensory input is reduced beyond normal conditions through the use of special chambers or devices (e.g. translucent goggles).

sensory store *(Behav.)*. The portion of the memory system that maintains representations of sensory information for very brief intervals; divided into **echoic** and **iconic** memory. See **long term memory, short term memory**.

sentinel *(Comp.)*. A symbol used to indicate the end of a specific block of information in a data-processing system. Also called *marker, tag, flag.*

sentinel pile *(Med.)*. An oedematous mass of rolled-up anal mucous situated at the margin of the anus at the lower end of an anal fissure.

sepal *(Bot.)*. A member of the **calyx**; typically a green, more or less leaf-like structure, several enclosing the rest of the flower in the bud, but sometimes petaloid.

separated lift *(Aero.)*. Lift generated by very low aspect ratio wings (usually of *delta* or *gothic* plan form) at high angles of incidence (ca. 20°) through separated vortices causing large suction forces. The **jet flap** is also a separated lift system.

separate excitation *(Elec.Eng.)*. A form of machine excitation in which the supply to the field system is obtained from a separate direct-current source.

separates *(Print.)*. Same as **offprints.**

separate system *(Build.)*. A system of sewerage in which two sewers are provided, one for the sewage proper, and the other for the rainwater. Cf. **combined system.**

separating calorimeter *(Eng.)*. A device for mechanically separating and measuring the water associated with very wet steam; used in conjunction with the **throttling calorimeter** in determining dryness fractions.

separating drum *(Eng.)*. An auxiliary steam-collecting drum attached by tubes to the upper drum of some water-tube boilers to avoid priming or foaming.

separating funnel *(Chem.)*. A funnel with a tap at the bottom and a stoppered top, in which two immiscible liquids can be dispersed by shaking, then separated by settling and drawing off the lower layer; used in **liquid-liquid extraction.**

separation *(Aero.)*. The spacing of aircraft arranged by **air traffic control** to ensure safety, which may be vertical, lateral, longitudinal, or a combination of the three.

separation anxiety *(Behav.)*. Anxiety at the prospect of being separated from someone one is strongly attached to and one believes to be necessary for survival; all children experience this fear at between 6–8 months of age to about 2 years but individual differences in its intensity are great.

separation energy *(Phys.)*. That required to separate 1 nucleon from a complete nucleus.

separation factor *(Phys.)*. The ratio of the abundance of isotope at the end of a separation system or unit to that at the start of the process. It is usually only slightly greater than unity.

separation filters *(Image Tech.)*. The three filters used in separation methods of colour photography and printing.

separation layer *(Bot.)*. See **abscission** layer.

separation point *(Phys.)*. The point at which streamline flow, *laminar* or *turbulent*, separates from the surface of a body.

separation potential *(Nuc.Eng.)*. A dimensionless function used in definition of the **separative work** of an uranium enrichment plant and given by

$$V(N) = (2N-1)\ln\left\{\frac{N}{1-N}\cdot\frac{1-N_0}{N_0}\right\} + \frac{(1-2N_0)(N-N_0)}{N_0(1-N_0)}$$

where N and N_0 are the concentrations of the product and initial materials respectively.

separative efficiency *(Nuc.Eng.)*. In a single stage, the ratio of actual concentration to the change in the theoretical value.

separative element *(Nuc.Eng.)*. One unit of a cascade forming a complete isotope separation plant.

separative power *(Nuc.Eng.)*. Describes the quantity of material a **separative element** is capable of enriching. It is given by the expression

$$\frac{\theta}{1-\theta}\cdot L\frac{(\alpha-1)^2}{2}\text{moles/second}$$

where α is the **separation factor** and θ, the **cut**. L is the number of moles/second of material.

separative work *(Nuc.Eng.)*. Measures the amount of separation an enrichment plant can achieve and defined as $PV(N_P)+WV(N_W)-FV(N_F)$ where P, W, F are the masses of product, waste and feed (i.e. initial) materials,

$V(N_P)$, $V(N_W)$ and $V(N_F)$ are the **value functions** of isotope concentrations N_P, N_W, N_F respectively. Measured in *separative work units* (Kg SWU or tonnes SWU). It is not the weight of enriched material drawn from the plant.

separator *(Civ.Eng.)*. A distance piece, usually of steel or cast-iron, bolted between the webs of parallel side-by-side steel joists to give rigidity and ensure unity of action.

separator *(Comp.)*. A **flag** used to separate items of data.

separator *(Elec.Eng.)*. A thin sheet of wood, perforated ebonite or porous polyvinyl chloride separating the plates of a secondary cell.

separator *(Eng.)*. A trap in a pipe containing a gas with condensed vapours. Removes, e.g. water to produce dry steam.

separator *(Min.Ext.)*. Concentrating machine, used to separate constituent minerals of mixed ore from one another.

separator *(Phys.)*. An electromagnet used to select iron and steel from mixed scrap.

sepdumag *(Image Tech.)*. International code for motion picture with two magnetic sound-tracks on a separate film.

sepiolite *(Min.)*. See **meerschaum**.

sepmag *(Image Tech.)*. International code name for a motion picture having a separate magnetic sound-track.

sepopt *(Image Tech.)*. International code name for a motion picture having an optical sound-track on a separate film.

sepsis *(Med.)*. The invasion of bodily tissue by pathogenic bacteria. adj. *septic*.

septaria, septarian nodules *(Geol.)*. Concretionary nodules containing irregular cracks which have been filled with calcite or other minerals.

septate *(Bot.)*. Divided into cells, compartments or chambers by walls or partitions.

septate fibre *(Bot.)*. A fibre of which the lumen is divided into several compartments by transverse septa.

septavalent *(Chem.)*. See **heptavalent**.

septechlorites *(Min.)*. A group of sheet silicates closely related chemically to the chlorites, and structurally to the serpentines and kandites. Includes **chamosite** and **greenalite**.

septicaemia, septicemia *(Med.)*. The invasion of the blood stream by bacteria and their multiplication therein; associated with high fever, chills, and petechial haemorrhages into the skin. *adj.* **septicaemic**.

septicidal *(Bot.)*. A dehiscent fruit, opening by breaking into its component carpels leaving the placental axis standing, as in *Hypericum*.

septic tank *(Build.)*. A tank in which sewage is left for about 24 hr, during which time a scum forms on the surface and the sewage below is to some extent purified by the action of the anaerobic bacteria functioning in the absence of oxygen.

septifragal *(Bot.)*. A dehiscent fruit, opening by the breaking away of the outer wall leaving the septa standing.

sept-, septi-, septo- *(Genrl.)*. Prefixes used in the construction of compound terms. (1) L. *septum*, partition. (2) L. *septem*, seven. (3) Gk. *septos*, rotten.

septum *(Bot.)*. A cell wall or multicellular structure acting as a partition as in a fungal hypha, between cells or between adjacent chambers in an ovary.

septum *(Telecomm.)*. Dividing partition in a waveguide.

septum *(Zool.)*. A partition separating two cavities. adj. *septal*.

septum transversum *(Zool.)*. See **diaphragm**.

sequence *(Biol.)*. The linear order of bases in a nucleic acid or of amino acids in a protein.

sequence *(Elec.Eng.)*. The order in which the several phases of a polyphase alternating-current supply undergo their cyclic variation of voltage.

sequence *(Image Tech.)*. The unit of the scenario, involving one general idea or happening and a number of scenes, each of which may include a number of shots.

sequence *(Maths.)*. An ordered set of numbers derived according to a rule, each member being determined either directly or from the preceding terms.

sequence control register *(Comp.)*. See **program counter**.

sequence register *(Comp.)*. See **program counter**.

sequence valve *(Aero.)*. A type of automatic selector valve in a hydraulic or pneumatic system, much used in aircraft, whereby the action of one component is dependent upon that of another.

sequencing *(Biol.)*. Biochemical procedure for determining the sequence of a nucleic acid or protein.

sequential access *(Comp.)*. Process of storing or retrieving data items by first reading through all previous items to locate the one required. Also *serial access*. Cf. **binary search**.

sequential colour systems *(Image Tech.)*. Colour TV systems in which colour information for each channel is transmitted sequentially. Systems may be *field-*, *line-*, or *dot-sequential*.

sequential memory *(Comp.)*. See **serial access memory**.

sequential operation *(Comp.)*. One in which all instructions are carried out sequentially.

sequential scanning *(Image Tech.)*. A system in which all the lines of the picture are scanned in strict sequence from top to bottom, not inter-laced.

sequential transmission *(Image Tech.)*. A technique of transmitting pictures so that the picture elements are selected at regular times and are then delivered to the communication channel in the correct sequence.

sequestering agent *(Chem.)*. One which removes an ion or renders it ineffective, by forming a complex with the ion. See **complexones**.

sequestrectomy *(Med.)*. The surgical removal of a sequestrum.

sequestrene, sequestrol, sequestered iron *(Bot.)*. Preparations of chelated mineral elements, especially iron and some trace elements, used horticulturally to correct such mineral deficiencies as **lime-induced chlorosis** by application to leaves (*foliar feeding*) or to the soil.

sequestrum *(Med.)*. A piece of bone, dead as a result of infection and separated off from healthy bone.

sequoia *(For.)*. See **redwood**.

Ser *(Chem.)*. Symbol for serine.

SERCnet *(Comp.)*. A UK network which developed into **JANET**.

sere *(Ecol.)*. Particular example of plant communities which succeed each other. Hydroseres originate in water, xeroseres occur in dry places, lithoseres develop on rock surfaces. adj. *seral*. See **primary sere**.

serein *(Meteor.)*. The rare phenomenon of rainfall out of an apparently clear sky.

serge *(Textiles)*. Dress or suiting dyed fabrics of simple twill weave, often made from wool but also from other fibres or wool-containing blends.

serial access *(Comp.)*. See **sequential access**.

serial access memory *(Comp.)*. Computer memory where storage locations can be accessed only in predetermined sequences, e.g. **magnetic tape**, **magnetic bubble memory**. Cf. **RAM**.

serial arithmetic unit *(Comp.)*. One in which the digits of a number are operated on sequentially.

serial computer *(Comp.)*. Computer which operates successively on each **bit** of a **word**. Only the very earliest machines were totally serial. Cf. **parallel processor**.

serial learning, recall *(Behav.)*. Refers to a learning or memory task in which the subject is required to repeat a list of items in the same order as they were presented.

serial-position effect *(Behav.)*. The observation that in verbal learning, items at the beginning and end of a list are recalled better than those in the middle of the list.

serial radiography *(Radiol.)*. A technique for making a number of radiographs of the same subject in succession.

serial store *(Comp.)*. See **serial access memory**.

sericite *(Min.)*. A fine-grained white potassium mica, like muscovite in chemical composition and general characters but occurring as a secondary mineral, often as a decomposition product of orthoclase.

series *(Geol.)*. A time-stratigraphic unit intermediate between **system** and **stage**.

series *(Maths.)*. The sum where each term of a *sequence* is added to the previous ones.

series *(Phys.)*. Said of electric components when a common current flows through them.

series arm *(Telecomm.)*. Part of a filter which is in series with one leg of the transmission line.

series capacitor *(Elec.Eng.)*. A capacitor connected in series with a transmission line or distribution circuit to compensate for the inductive reactance drop and thereby improve the regulation.

series characteristic *(Elec.Eng.)*. The characteristic graph relating terminal voltage and load current in the case of a series-wound direct-current machine.

series-characteristic motor *(Elec.Eng.)*. An electric motor having a speed torque characteristic similar to that of a d.c. series motor, i.e. one in which the speed falls with an increase of torque. Also *inverse-speed motor*.

series field *(Elec.Eng.)*. Two variable vane capacitors, usually with air dielectric, with the moving vanes on the same rotating shaft; used in high-frequency circuits with the two capacitances in series, to obviate taking the current through a rubbing contact or through a pigtail, the latter being inductive.

series motor *(Elec.Eng.)*. An electric motor whose main excitation is derived from a field winding in series with the armature.

series-parallel controller *(Elec.Eng.)*. A method of controlling the speed and tractive effort of an electric tractor having one or more pairs of series motors, whereby the motors can be connected either in series or in parallel.

series-parallel network *(Elec.Eng.)*. One in which the electrical components are composed of branches which are successively connected in series and/or in parallel.

series resonance *(Elec.Eng.)*. The condition of a tuned circuit when it offers minimum impedance to an a.c. voltage supply connected in series with it (due to the circuit reactances neutralizing each other). The term *tunance* is sometimes used in place of resonance if this condition is attained by adjustment of a component value and not of frequency.

series stabilization *(Telecomm.)*. A technique of stabilization using amplifier feedback in which the feedback and amplifier circuits are in series at each end of the amplifier.

series system *(Elec.Eng.)*. (1) Circuit comprising electrical component as connected in series. (2) The constant current system of d.c. distribution developed by Thury, in which generators and motors are all connected in series to form a single d.c. circuit. Also called *Thury system*.

series transformer *(Elec.Eng.)*. A power transformer operating under constant-current instead of constant-voltage conditions. See also **current transformer**.

series winding *(Elec.Eng.)*. A field winding connected in series with the armature of the motor.

serif *(Print.)*. The short strokes of a letter, at the extremities of the main strokes and hair lines.

serine *(Chem.)*. *3-hydroxy-2-aminopropanoic acid*, $HO.CH_2.CH(NH_2).COOH$. A polar amino acid. The L- or S- isomer is a constituent of protein. Symbol Ser, short form S. ⇨

serological determinants *(Immun.)*. Antigenic determinants on cells that are recognized by and accessible to antibodies, as opposed to determinants which are not recognized by antibodies but only by T lymphocytes.

serological typing *(Biol.)*. A technique used for the identification of pathogenic organisms, e.g. bacteria, particularly strains within a species, when morphological differentiation is difficult or impossible. It is based on antibody-antigen reactions, specific proteins of the organism acting as antigens.

serology *(Med.)*. The study of **sera**.

serophyte *(Med.)*. Any micro-organism which will grow in the presence of fresh serum exuding into a wound, such as the **streptococcus** and the **staphylococcus**.

seropurulent *(Med.)*. Said of a discharge or effusion which is both serous and purulent.

serosa *(Zool.)*. See **serous membrane**.

serositis *(Med.)*. Inflammation of a serous membrane.

serotaxonomy *(Bot.)*. The use of serological techniques to compare proteins extracted from different plants as an aid in taxonomy.

serotherapy, serum therapy *(Med.)*. The curative or preventive treatment of disease by the injection into the body of animal or human serums which contain antibodies to the bacteria or toxins causing the disease.

serotonin *(Immun.)*. *5-hydroxytryptamine*. Causes smooth muscle contraction, increased vascular permeability and vasoconstriction of larger vessels. Present in platelets, from which it is released on activation. Also present in mast cells of some species.

serous *(Zool.)*. Watery; pertaining to, producing, or containing a watery fluid or serum.

serous membrane, serosa *(Zool.)*. One of the delicate membranes of connective tissue which line the internal cavities of the body in Craniata; the chorion.

serpentine *(Min.)*. Hydrated magnesium silicate which crystallizes in the monoclinic system. The three chief polymorphic forms are **antigorite**, **chrysotile**, and **lizardite**. The serpentine minerals occur mainly in altered ultrabasic rocks, where they are derived from olivine or from enstatite. Usually dark green, streaked and blotched with red iron oxide, whitish talc, etc. The translucent varieties are used for ornamental purposes; those with a fibrous habit form one type of asbestos.

serpentine-jade *(Min.)*. A variety of serpentine, resembling bowenite, occurring in China.

serpentinization *(Geol.)*. A type of metamorphism effected by water, which results in the replacement of the original mafic silicates in peridotites by the mineral serpentine and secondary fibrous amphibole.

Serpukhovian *(Geol.)*. The youngest epoch of the Mississippian period.

Serpula lacrymans *(Build.)*. Formerly *Merulius lacrymans*, the fungus which commonly causes **dry rot** in timber. Small droplets of water often form on the mycelium, hence the name.

serrate *(Bot.)*. Leaf margin toothed like a saw.

serrated roller *(Print.)*. A knurled or serrated roller on web-fed presses which gives a pull to the paper and minimizes ink pick-up.

Serret-Frenet formulae *(Maths.)*. See **Frenet's formulae**.

serrulate *(Bot.)*. Minutely serrate.

serum *(Med.)*. (1) The watery fluid which separates from blood or lymph in coagulation. (2) Blood serum containing antibodies, taken from an animal that has been inoculated with bacteria or their toxins, used to immunize people or animals. adj. *serous*.

serum albumin *(Biol.)*. A globular protein obtained from blood and body fluids, having a transport and osmoregulatory function. A crystalline, water-soluble substance, not precipitated by NaCl.

serum hepatitis *(Med.)*. See **infectious hepatitis**.

serum sickness *(Immun.)*. A hypersensitivity reaction to the injection of foreign antigens in large quantity especially those contained in antisera used for passive immunization. Symptoms appear some days after a single dose of the antigen, and consist of local swelling at the injection site, enlarged lymph nodes, fever, joint swellings, urticaria and, more rarely, glomerulonephritis. May have a more prolonged effect on joints, heart and kidney. The symptoms are due to the localization in the tissues of immune complexes formed between antibodies produced during the developing immune response and the large quantities of antigen still present. This is an example of Type III hypersensitivity reaction.

service area *(Telecomm.)*. That surrounding a broadcasting station where the signal-strength is above a stated minimum and not subject to fading.

service band *(Telecomm.)*. That allocated in the frequency spectrum and specified for a definite class of radio service, for which there may be a number of channels.

service capacity *(Elec.Eng.)*. The power output of an electric motor, as specified on the maker's nameplate.

service ceiling *(Aero.)*. The height at which the rate of climb of an aircraft has fallen to a certain agreed amount (in British practice, originally, 100 ft/min, 30 m/min, but for jet aircraft 500 ft/min, 150 m/min).

service ell *(Build.)*. An elbow having a male thread at one end.

service mains *(Elec.Eng.)*. Cables of small conductor cross-section which lead the current from a distributor to the consumer's premises.

service reservoir *(Hyd.Eng.)*. A small reservoir supplying a given district, and capable of storing the water which is filtered during the hours of small demand for use when the requirements become greater. Also called *distribution reservoir, clear water reservoir.*

service tanks *(Aero.)*. See under **fuel tanks**.

service tee *(Build.)*. Having a female thread on the branch and one end of the **run** and a male thread on the other.

serving *(Elec.Eng.)*. (1) A layer of jute, tape, or yarn, impregnated with bitumen or similar substance, to prevent the steel-wire armouring biting into the lead sheath of a cable. Also *bedding*. (2) The process of covering a cable with some form of mechanically strong insulating and binding tape.

servo-amplidyne system *(Elec.Eng.)*. One in which an **amplidyne**, together with a control amplifier, is used in order to amplify mechanical power.

servo amplifier *(Elec.Eng.)*. One designed to form the part of a servomechanism from which output energy can be drawn.

servo brakes *(Autos.)*. Power-assisted brakes worked either by a hydraulic servo, mechanically from the transmission or in 'vacuum brakes' by differential air pressure.

servocontrol *(Aero.)*. A reinforcing mechanism for the pilot's effort. It may consist of **servo tabs**.

servo link *(Elec.Eng.)*. A mechanical power amplifier which permits low strength signals to operate control mechanisms that require fairly large powers.

servomotor *(Elec.Eng.)*. A motor (electric, hydraulic etc.) for use in an automatic control system for e.g. the operation of a large valve by a governor of small power. See **pilot valve**.

servo tab *(Aero.)*. A control surface **tab** moved directly by the pilot, the moment from which operates the main surface, the latter having no direct connection with the pilot.

sesamoid *(Zool.)*. A small rounded ossification forming part of a tendon usually at, or near, a joint; as the patella.

sesqui *(Chem.)*. Containing two kinds of atom, radical etc. in the proportion of 2:3. Means $1\frac{1}{2}$.

sesquiterpenes *(Chem.)*. A group of terpene derivatives of the empirical formula $C_{15}H_{24}$, with three isoprene units.

sessile *(Bot.)*. (1) Having no stalk. (2) Fixed and stationary.

set *(Behav.)*. See **mental set**.

set *(Build.)*. See **nail punch**.

set *(Eng.)*. See **cold sett**.

set *(Hyd.Eng.)*. The direction of a current of water.

set *(Maths.)*. Any collection of entities ('elements') defined by specifying the elements. See **subset, universal-, Venn diagram**.

set *(Min.Ext.)*. A frame of timber used in a shaft or tunnel.

set *(Print.)*. (1) Width of a type character. (2) To *compose* type-matter.

set *(Textiles)*. See **sett**.

seta *(Bot.)*. Any one of a variety of sorts of such relatively long and thin, or bristle-like, structures as the stalk supporting the capsule of the sporophytes of mosses (Bryopsida).

seta *(Zool.)*. A small bristle-like structure; a chaeta. adjs. *setaceous, setiferous, setigerous, setiform, setose, setulose.*

set flush *(Print.)*. A typographic instruction to set all lines without indentation.

set-hands dial *(Horol.)*. A small dial on a turret clock movement, the hands of which read the same as the main dial. Used in regulating the clock.

set-hands square *(Horol.)*. The square for setting the hands of a key-wound watch.

SETI *(Space)*. Abbrev. for the *Search for Extra-Terrestial Intelligence*; i.e. investigating the possibility of intelligent life in the universe other than on Earth.

set of chromosomes *(Biol.)*. A **haploid** complement of chromosomes.

set-off *(Build.)*. See **offset**.

set-off *(Print.)*. Smudging of ink from one sheet to reverse of another before ink has dried. Obviated by interleaving (see **slip sheets**) or by using *anti-set-off-spray*.

set point *(Elec.Eng.)*. See **control point**.

set screw *(Eng.)*. A screw, usually threaded along the entire shank length, which is used to prevent relative motion by exerting pressure with its point.

set solid *(Print.)*. A typographic instruction to use no leading between lines.

sett *(Build.)*. A small rectangular block of stone 6 in deep by 3–4 in wide, and from 6 to 9 in in length; formerly used for surfacing roads where traffic was heavy. Best setts were of either Scottish or Welsh granite.

sett *(Eng.)*. See **cold sett**.

sett *(Textiles)*. The number of threads per inch or per cm. in the weft and/or warp of woven fabrics. In a square sett fabric the two values are equal and the yarns in both directions are of the same count. In an unbalanced sett fabric these values are significantly different.

setting *(Build.)*. The hardening of a lime, cement, mortar, or concrete mixture, or a plaster.

setting *(Textiles)*. Treatments usually by heating and cooling in dry or steamy atmospheres that confer stability on textile materials. Heat setting is particularly important for many fabrics.

setting coat *(Build.)*. The finishing coat of plaster; a thin layer, about $\frac{1}{8}$ in (3 mm) thick, of fine stuff. Also called *skimming coat.*

setting point *(Chem.)*. The temperature at which a melted wax, when allowed to cool under definite specified conditions, first shows the minimum rate of temperature change.

setting rule *(Print.)*. See **composing rule**.

setting stick *(Print.)*. See **composing stick**.

settlement *(Build.,Civ.Eng.)*. The subsidence of a wall, structure etc.

settling *(Powder Tech.)*. Classification effected by the rate of fall in a fluid which may have a horizontal components of velocity.

settling tank *(Build.)*. See **sedimentation tank**.

sett paving *(Civ.Eng.)*. Pavement constructed with setts on a suitable foundation. A *causeway* in Scotland.

set-up *(Image Tech.)*. Ratio between black and white reference levels measured from blanking level for facsimile transmission.

set-up *(Surv.)*. Location of theodolite above a station point.

set-up instrument, set-up-scale instrument, set-up-zero-instrument *(Elec.Eng.)*. See **suppressed-zero instrument**.

set-work *(Build.)*. Two-coat plasterwork on lath.

70 mm *(Image Tech.)*. The widest gauge of motion picture film.

severe combined immunodeficiency syndrome *(Immun.)*. *SCID*. The most severe form of congenital immunological deficiency state in which thymic agenesis with lymphocyte depletion coexists with deficiency of plasma cells and antibody deficiency syndrome. Also known as *Swiss type hypogammaglobulinaemia*, first described in that country. Death in early life from infection is the rule, but the condition may be cured by bone marrow transplantation.

severy *(Arch.)*. See **civery**.

sewage *(Build.)*. Liquid contained in a sewer.

sewage farm *(Build.)*. A farm on which sewage (especially sewage conveyed from a town) is used as a manure. See also *land treatment.*

sewage, sludge gas *(Build.)*. A self-generated combustible gas collected from the digesting tanks of sewage sludge. General composition: 66% CH_4 and 33% CO_2, with energy density in the region of 25 MJ/m^3. The gas has a very slow rate of flame propagation.

sewerage *(Build.)*. The network of sewers serving a community.

sewing *(Print.)*. The operation of joining the gathered sections of a book by sewing.

sex *(Biol.)*. (1) The sum total of the characteristics, structural and functional, which distinguish male and female organisms, especially with regard to the part played in reproduction. (2) As a verb to determine sex. adj. *sexual*.

sex cells *(Biol.)*. See **gametes**.

sex chromosome *(Biol.)*. See **sex determination**.

sex determination *(Biol.)*. In many organisms (including vertebrates) sex is determined by the possession of a particular combination of chromosomes. In some cases, presence or absence of one special chromosome, known as the *accessory* or *X-chromosome*, is the determining factor, e.g. in the insect *Pyrrhocoris apterus* males and females have 13 and 14 chromosomes respectively. In many species there are two sex chromosomes, and the sex of the individual depends on whether it has two identical chromosomes, the *homogametic sex*, or one of each of the two types, the *heterogametic sex*. Where the female is homogametic, as in mammals, the two chromosomes are designated XX, and the male's chromosomes are known as XY. Where the male is homogametic, as in birds, the male's chromosomes are called WW and the female's chromosomes are known as ZW.

sex gland *(Zool.)*. See **gonad**.

sex-limited character *(Bot.,Zool.)*. A character developed only by individuals which belong to a particular sex.

sex-linked *(Biol.)*. Of a gene located on a *sex chromosome*, i.e. in mammals on the *X-chromosome*. A sex-linked character is associated with sex in transmission; it appears in one sex in one generation and appears, or is transmitted by, the other sex in the next generation (*criss-cross inheritance*).

sex-linked *(Med.)*. Of genes, characters or diseases, carried on the X chromosome. Diseases of this type characteristically affect males only but are transmitted by females only.

sex mosaic *(Zool.)*. An individual showing characteristics of both sexes; an intersex, gynandromorph.

sex reversal *(Zool.)*. The gradual change of the sexual characters of an individual, during its lifetime, from male to female or vice versa.

sex roles *(Behav.)*. A set of attitudes, behaviours, perceptions and feelings which are commonly held to be associated with either being male or being female.

sextant *(Surv.)*. A reflecting instrument in the form of a quadrant, for measuring angles up to about 120°. It consists essentially of two mirrors: a fixed *horizon glass*, half silvered and half plain glass, and a movable *index glass*, to which is attached an arm moving over a scale graduated to read degrees directly. The index glass reflects an image of one signal, or body, into the silvered part of the horizon glass, and this image is brought into coincidence with the other signal or body as seen through the plain part of the same glass. The sextant is used chiefly for measuring the altitude of the sun at sea, the reflected image of the sun being made to touch the visible horizon. See **air-sextant**.

sex transformation *(Zool.)*. See **sex reversal**.

sexual behaviour *(Behav.)*. All behaviour leading to the fertilization of eggs by sperm.

sexual coloration *(Zool.)*. Characteristic colour difference between the sexes, especially marked at the breeding season. See also **epigamic**.

sexual dimorphism *(Bot.,Zool.)*. Marked differences between the males and females of a species, especially differences in superficial characters, such as colour, shape, size etc.

sexual organs *(Zool.)*. The gonads and their accessory structures; reproductive system.

sexual reproduction *(Bot.,Zool.)*. The union of gametes or of gametic nuclei, preceding the formation of a new individual.

sexual selection *(Zool)*. Selection occurring as a result of mate selection.

Seyfert galaxy *(Astron.)*. Member of a small class of galaxies with brilliant nuclei and inconspicuous spiral arms. The intensely bright nuclei possess many of the properties of **quasars**. They are strong emitters in the infrared and are also detectable as radio and X-ray sources. Carl Seyfert discussed this morphological type in 1943.

Sezary syndrome *(Immun.)*. A disease syndrome characterized by general redness and thickening of the skin. The skin is infiltrated with lymphocytes with an unusual hairy appearance, and large numbers of similar lymphocytes (Sezary cells) are present in the blood. They have been shown to be T lymphocytes, but what causes them to move into the skin is not known.

sferics *(Meteor.)*. Lightning flashes or other natural electrical impulses especially in relation to the determination of their location by simultaneous radio direction-finding using a number of aerials. The word is derived from atmo*spherics*.

SGHWR *(Nuc.Eng.)*. Abbrev for *Steam Generating Heavy Water Reactor*.

sgraffito *(Build.)*. A mode of surface decoration in which two finishing coats of contrasting colours are applied, one on top of the other. Before the upper one has set, parts are removed according to some design, thus exposing the coat below. Also *graffito*.

shackle *(Eng.)*. U-shaped machine element for connecting two parts which may have some relative movement. The ends are usually connected by a retaining bolt or pin.

shackle insulator *(Elec.Eng.)*. A porcelain insulator whose ends are secured to metal shackles.

shade *(Build.)*. The addition of black or dark grey to a paint colour.

shade *(Surv.)*. A disk of coloured glass used in telescope of theodolite when making sun observations.

shade *(Textiles)*. (1) the colour of a material, usually one that has been dyed. (2) to modify the colour of a fabric being dyed to bring it nearer to that required.

shaded pole *(Elec.Eng.)*. A pole having a short-circuited ring around one section, thus altering the phase of the flux over that section. Sometimes used to make small single-phase motors self-starting.

shade plant *(Bot.)*. A plant adapted to living at low light intensities. Cf. **sun plant**.

shading *(Image Tech.)*. Unwanted variation of brightness within the picture area, not forming part of the image; where this originates in the camera, correction signals can be inserted.

shadow *(Image Tech.)*. Ineffectiveness of reception because of an obstacle, e.g. due to the topography of the terrain, between the transmitter and the receiver.

shadow *(Phys.)*. The shadow of an obstacle cast by a point source of light is the geometrical projection of the obstacle, except for small-scale diffraction effects at the edge. See **umbra**.

shadow casting *(Powder Tech.)*. Method, carried out in high vacuum equipment, of determining the thickness of a particle of powder or other structure on a microscope slide. A beam of vaporized metal (usually gold or chromium) is directed towards the specimen slide at an oblique angle. The thickness of the particles can be measured from the angle of approach of the beam and the dimensions of the shadow cast by the particles.

shadow fringe test *(Phys.)*. A technique for examining the optical quality of glass. The shadows formed on transmission of a beam of light limited laterally are examined, since inequalities in the refractive index appear as fringes in the shadow.

shadowing technique *(Biol.)*. A technique of shadow-casting used in electron microscopy, in which a very thin nongranular film of a metal, e.g. chromium, gold, uranium, is deposited obliquely on to the surface of the specimen prior to examination. This gives a three-dimensional effect and improves the clarity of the surface contours.

shadow-mark *(Paper)*. A defect of paper showing in the

look-through as a faint reproduction of the holes of the suction couch or press rolls.

shadow-mask tube *(Image Tech.)*. Type of directly viewed 3-gun cathode-ray tube for colour TV display, in which beams from 3 electron guns converge on holes in a shadow mask placed behind a tricolour phosphor-dot screen.

shadow photography *(Image Tech.)*. High-speed technique using an electric spark or similar light-source to photograph the shadow of a fast-moving object such as a projectile. Exposures of 10^{-6} s have been attained.

shadow scattering *(Phys.)*. See **scattering**.

shadow stripes *(Textiles)*. Shadow stripes in which stripes are produced by using warp or weft yarns of different directions of twist, i.e. *S-* or *Z-twist*. The shadow effect is due to the differing reflectivity of the different twists. See **twist direction**.

shaft *(Arch.)*. The principal portion of a column between the capital and the base.

shaft *(Civ.Eng.,Min.Ext.)*. A passage, usually vertical, leading from ground-level into an underground excavation, for purposes of ventilation, access etc.

shaft *(Zool.)*. The part of a hair distal to the root: the straight cylindrical part of a long limb bone: the rachis, or distal solid part of the scapus of a feather.

shaft furnace *(Eng.)*. One in which ore and fuel are charged into the top and gravitate vertically, reacting as they proceed to bottom discharge.

shaft governor *(Eng.)*. A compact type of spring-loaded governor used for controlling the speed of small oil engines, etc. It is arranged to rotate about the crankshaft axis, and is sometimes housed in the flywheel. See **spring-loaded governor**.

shafting *(Eng.)*. See **line shafting**.

shaft pillar *(Min.Ext.)*. Solid block of coal or ore left unworked round the bottom of a shaft or pit for support.

shaft station *(Min.Ext.)*. Room excavated underground adjacent to shaft, to accommodate special equipment such as pumps, crushing machine, truck tipples, ore sorting equipment and surge storage bins.

shaft turbine *(Aero.)*. Any gas turbine aero-engine wherein the major part of the energy in the combustion gases is extracted by a turbine and delivered, through appropriate gearing by a shaft; see **free turbine** and **turboprop**.

shake *(For.)*. A partial or complete separation between adjacent layers of fibres.

shaking grate *(Eng.)*. A grate for a hand-fired boiler furnace in which the pivotally supported fire-bars can be rocked by hand levers in order to break up clinker.

shaking table *(Min.Ext.)*. See **Wilfley table**.

shale *(Geol.)*. A consolidated clay rock which possesses closely-spaced well defined laminae. Cf. **mudstone**. See also **oil-shale**.

shale oils *(Min.Ext.)*. Oils obtained by the pyrolysis of oil-shale at ca. 550°C and characterized by a large proportion of unsaturated hydrocarbons, alkenes and di-alkenes.

shank *(Build.)*. (1) Shaft of column, pillar etc. (2) Shaft of tool, connecting head and handle.

shank *(Eng.)*. A ladle for molten metal.

shank *(Print.)*. See **body**.

Shannon's theorem *(Telecomm.)*. Concerns the ultimate capacity of a communication channel in terms of its bandwidth and signal-to-noise ratio. The maximum transmission rate in *bits per second* is given by $W \log_2(1 + SN)$, where W is the bandwidth and SN is the signal-to-noise ratio.

shantung *(Textiles)*. Plain-weave silk cloth, with a randomly-irregular surface; made from tussah, the silk produced by the wild silkworm.

shaped-beam tube *(Electronics)*. One in which the cross-section of the beam of electrons is formed to the shape of various characters.

shaped-conductor cable *(Elec.Eng.)*. A three-phase cable in which the conducting cores are specially shaped so as to give the best utilization of the total available cross-section of the cable.

shape factor *(Powder Tech.)*. A parameter descriptive of the shape of a particle of a powder or the ratio of two average particle sizes determined by techniques in which the shape of the particles of a powder influence the measured parameter in different ways.

shaper tools *(Eng.)*. Cutting tools similar to those used on planing machines, and similarly supported in a clapper box.

shaping *(Behav.)*. The training of a response by successively reinforcing responses that are increasingly similar to the target behaviour, until that behaviour is reached. Also *method of successive approximations*.

shaping *(Textiles)*. Process used to change the width of knitted fabrics or garments by changing the number of stitches in the course or wale directions. Includes *fully-fashioned*. In stitch shaped garments the dimensions are changed by altering the stitch length and/or structure.

shaping machine *(Eng.)*. A machine tool for producing small flat surfaces, slots etc. It consists of a reciprocating ram carrying the tool horizontally in guide ways, and driven by a **quick return mechanism**. Either the tool or the table may be capable of traverse.

shaping network *(Telecomm.)*. One which determines or restores the shape of a pulse, especially in radar and computing.

shard *(Geol.)*. A fragment of volcanic glass, often with curved edges. Glass shards are important constituents of some pyroclastic rocks.

shared-channel broadcasting *(Telecomm.)*. See **common-frequency broadcasting**.

shared memory *(Comp.)*. Fast store to which more than one processor has access.

sharp *(Build.,Civ.Eng.)*. Said of sand the grains of which are angular, not rounded.

sharp coat *(Build.)*. A thin oil paint which contains much pigment and little medium together with a high proportion of solvent, is used as primer or sealer.

sharpening image *(Print.)*. Printing image losing its printing area from the edges by e.g. halftone dots becoming smaller. One cause being attack by an overactive fountain solution.

sharp flutings *(Arch.)*. Flutings which are so close together as to form sharp arrises.

sharp gas *(Min.Ext.)*. Mine air so contaminated with methane as to burn inside the Davy-type lamp and therefore to be dangerous.

sharp mouth *(Vet.)*. Overgrowth of a part of one or more teeth of a horse through loss or wear.

sharpness *(Image Tech.)*. The subjective impression of the amount of detailed information provided in a picture image; it is affected by the type of subject and viewing conditions as well as the reproduction characteristics of the system. See **definition, MTF, resolution**.

sharpness *(Telecomm.)*. Equivalent to **selectivity**, but referring more directly to the change in circuit adjustment necessary to alter signal strength from its maximum to a negligible value.

sharpness of resonance *(Phys.)*. The rapidity with which resonance phenomena are shown and then disappear as the frequency of excitation of a constant driving force is varied through the resonant frequency.

sharp paint *(Build.)*. Oil paint drying rapidly to give a flat surface.

sharp series *(Phys.)*. Series of optical spectrum lines observed in the spectra of alkali metals. Has led to energy levels for which the orbital quantum number is zero being designated *s-levels*.

shavehook *(Build.)*. A tool used to remove paint from moulded areas during burning off or when using liquid paint removers. Available with 3 shapes of head and pulled across the surface: (1) pear shape; (2) triangular shape; (3) combination shape made to suit various contours.

shaving *(Acous.)*. Machining surface of master disk recording to give fresh surface for further use.

shear *(Biol.)*. To cut the long, stiff DNA duplex by hydrodynamic means.

shear *(Eng.,Phys.)*. A type of deformation in which

parallel planes in a body remain parallel but are relatively displaced in a direction parallel to themselves; in fact, there is a tendency for adjacent planes to slide over each other. A rectangle, if subjected to a shearing force parallel to one side, becomes a parallelogram. See **elasticity of shear, strain, torsion**.

shear *(Textiles)*. (1) To cut the wool from a living sheep. (2) To trim a pile on a fabric to a uniform height. Also *crop*.

shearer loader *(Min.Ext.)*. Machine which cuts coal from seam and loads it in the same operation to a conveyor belt working parallel to the face. In fully mechanized mining the assembly, together with roof support props, is moved hydraulically and can be remotely controlled.

shear force *(Eng.)*. A force which tends to cause sliding of adjacent layers, relative to each other, in a material.

sheariness *(Build.)*. A paint defect similar to **flashing**, possibly caused by failure to keep a wet edge during painting. Most apparent when viewed across the sheen, most prone on flat or semi-glass finishes.

shear-legs *(Eng.)*. See **sheers**.

shear mouth *(Vet.)*. An increase in the obliquity of the wearing surfaces of the molar teeth of horses.

shear pin *(Eng.)*. Pin used as a safety device to connect elements in a power transmission system, which is strong enough to transmit permissible loads but will fail by shearing when these loads are exceeded.

shear ram *(Min.Ext.)*. Hydraulically-operated sliding jaws designed to cut off flow near the **blowout preventer**. It compresses the pipe and cuts it. The well is then *shut in*, but the gear above can be removed.

shears *(Eng.)*. See **ways**.

shear stress *(Eng.)*. Shear stress occurs across the section of a beam loaded transversely and also across shaft sections subject to torque. The magnitude of the stress varies across the section depending on its geometric shape.

shear wave *(Acous.)*. **Transverse wave** without compression of the medium.

shear zones *(Geol.)*. Bands in metamorphic rocks consisting of crushed and brecciated material and many parallel fractures. See **strain-slip cleavage**.

sheath *(Bot.)*. (1) Any tubular structure surrounding an organ or plant part, e.g. leaf sheath. (2) A tissue layer that surrounds other tissues, e.g. bundle sheath.

sheath *(Elec.Eng.)*. The covering on a cable.

sheath *(Electronics)*. Excess of positive or negative ions in a plasma, giving a shielding or *space-charge* effect.

sheath *(Nuc.Eng.)*. The can protecting a nuclear fuel element.

sheath *(Zool.)*. An enclosing or protective structure, e.g. elytron of some Insects.

sheath-circuit eddies *(Elec.Eng.)*. The paths of currents in the sheaths of separate cables which flow only when the sheaths are bonded. See also **sheath eddies**.

sheath current *(Elec.Eng.)*. The eddy current flowing in the metallic sheath of an alternating-current cable.

sheath eddies *(Elec.Eng.)*. Currents which are induced in the sheath of a single cable, and which flow even when the sheaths are isolated from each other. Cf. **sheath-circuit eddies**.

sheath effects *(Elec.Eng.)*. The phenomena associated with the metallic sheaths of cables carrying alternating currents.

sheathing *(Build.)*. Close boarding nailed to the framework of a building to form the walls or the roof.

sheathing paper *(Build.)*. A flexible waterproof lining material made from bitumen reinforced with fibre, and faced with stout kraft paper.

sheath of Schwann *(Zool.)*. See **neurolemma**.

sheave *(Eng.)*. Grooved pulley for use with vee-belts, ropes, or round belts.

shed *(Phys.)*. Minute unit of nuclear cross-section, $10^{-52}\,m^2$ or 10^{-24} barn.

shed *(Textiles)*. Opening created by dividing the warp threads during weaving so that the shuttle or other device can take the weft through to be beaten up into cloth.

sheen *(Build.)*. The degree or lustre, shine, or reflection on a surface or finish.

sheep ked *(Vet.)*. A blood-sucking, wingless fly, *Melophagus ovinus*, which lives on the wool and skin of sheep.

sheep pox *(Vet.)*. A highly-contagious disease of sheep caused by a virus and characterized by a papulo-vesicular eruption of the skin and mucous membranes of the respiratory and alimentary tracts.

sheep scab *(Vet.)*. See **psoroptic mange**.

sheers *(Eng.)*. A large lifting device used in shipyards, etc., resembling a crane in which a pair of incline struts take the place of a jib. Also called *sheer-legs, shear-legs*.

sheerstrake *(Ships)*. The top strake or line of plating below, but extending a little above, the freeboard deck.

sheet *(Aero.)*. The general term for aircraft structural material under 0.25 in (6 mm) thick; above that it is usually called plate.

sheet *(Print.)*. A term applied to any one piece of printing paper printed or plain.

sheet anchor *(Ships)*. A third bower anchor carried abaft the starboard bower for use in emergency. Formerly called **waist anchor**.

sheet-fed *(Print.)*. A term applied to a rotary machine indicating that it prints separate sheets and not a reel or web.

sheet furnace *(Eng.)*. One in which metal sheet is heated before further size reduction in rolling mill.

sheet glass *(Glass)*. Now largely superseded by **float glass**, it has been used largely for glazing purposes; produced by drawing a continuous film of glass from a molten bath and, after a suitable time interval for cooling, cutting up the product into sheets. It is not of such good quality, nor so flat, as plate glass which is ground and polished.

sheeting *(Civ.Eng.)*. (1) Rough horizontal boards used to support the sides of narrow trenches during excavation in very loose soils, each pair of boards on opposite sides of the trench being wedged apart with struts. (2) See **wearing course**.

sheeting *(Textiles)*. A medium-weight, closely-woven, plain or twill weave fabric used for bedding.

sheet lead *(Build.)*. Lead in a form in which it is commonly used in building construction, viz. in the form of sheets. It is the trade practice to refer to it in terms of the mass of unit area, e.g. 7-lb sheet lead, meaning $7\,lb/ft^2$.

sheet lightning *(Meteor.)*. Diffuse illumination of clouds by distant lightning of which the actual path of the discharge is not seen.

sheet pavement *(Civ.Eng.)*. A road surfacing formed of continuous material such as concrete. See **contraction joints**.

sheet piling *(Civ.Eng.)*. Timber or steel sheeting supported in a vertical position by guide piles, and serving to resist lateral pressures. Prestressed concrete piles are also used.

sheet severer *(Print.)*. An automatic device for cutting the web to prevent wrap around.

sheet wander *(Print.)*. Undesirable lateral movement of the running web of paper.

sheet-work *(Print.)*. Work in which two formes are used, one for each side of the paper, to give one complete copy of the job or section per sheet. Cf. **work-and-turn**.

shelf-back *(Print.)*. See **spine**.

shelf-life *(Build.)*. The length of time which a paint will remain usable when stored.

shell *(Phys.)*. See **electron shell**.

shellac *(Build.)*. A resin obtained from the secretion of an insect called a Lac. Used as a spirit varnish and also as the basis for French Polishing (Knotting).

shell bit *(Build.)*. Bit shaped like a narrow gouge, used for screw holes. Also *gouge-bit*.

Shelldyne *(Aero.)*. TN for synthetic fuel having a higher than normal density for expendable turbojets.

shell gland *(Zool.)*. In some Invertebrates, a glandular organ which secretes the materials for the formation of the shell. Also *shell sac*.

shell ligament *(Zool.)*. The dorsal ligament joining the valves of the shell in bivalve Mollusca.

shell model *(Phys.)*. A model of the nucleus of the atom with a nucleon moving independently in the common field representing the effect of the other nucleons. This leads to the nucleons being arranged in shells as for the **electron shells** in atomic structure.

shell reamer *(Eng.)*. Reamer in the form of a hollow cylinder, end-mounted on an arbor. This construction is used for economy, for relatively large reamers are made of expensive materials.

shell sac *(Zool.)*. See **shell gland**.

shell shock *(Behav.)*. See **traumatic neurosis**.

shell star *(Astron.)*. One of a number of stars of spectral type O, B or A, which is surrounded by a shell of luminous gas giving bright emission lines.

shell-type transformer *(Elec.Eng.)*. A transformer in which the magnetic circuit surrounds the windings more or less completely.

shelterbelt *(For.)*. Planting of trees in narrow strips to give wind-shelter for crops, animals or other planting.

shelter deck *(Ships)*. A term correctly interchangeable with **awning deck**. Sometimes used in a more casual way to identify a deck above a weather deck.

sheradizing *(Build.)*. An iron-zinc alloy treatment for iron and steel which eradicates the immediate need for painting for protection. Sometimes used on tubular scaffolding. The object is heated to 300°C in contact with zinc dust and some zinc oxide.

sheridanite *(Min.)*. A mineral in the chlorite group poor in iron and relatively low in silica.

shide *(Build.)*. See **shingle**.

shield *(Civ.Eng.)*. Hollow steel cylinder adapted for use in driving a tunnel through loose or water-bearing ground, protecting men and machinery working at the face. It is driven forward as excavation proceeds by means of hydraulic jacks bearing on preceding lines of tunnel.

shield *(Elec.Eng.,Nuc.Eng.)*. Screen used to protect persons or equipment from electric or magnetic fields, X-rays, heat, neutrons etc. In a nuclear reactor the shield surrounds it to prevent the escape of neutrons and radiation into a protected area. See **Faraday cage**, **biological-**, **magnetic-**, **neutron-**, **thermal-**.

shield *(Electronics)*. A metallic housing or screen made of earthed mesh or thin sheet, placed around a component or circuit to suppress or isolate electromagnetic fields which might otherwise cause interference.

shield *(Geol.)*. A large stable area of the Earth's crust consisting of Precambrian rocks. Effectively synonymous with **craton**, e.g. Canadian shield.

shielded box *(Nuc.Eng.)*. **Glove box** protected by lead walls and lead-glass windows, and with facilities for manipulation of contents by remote handling equipment.

shielded line *(Elec.Eng.)*. Line or circuit which is specially shielded from external electric or magnetic induction by shields of highly conducting or magnetic material.

shielded line *(Telecomm.)*. See **self-shielding**.

shielded metal-arc welding *(Eng.)*. The use of covered consumable electrodes to provide protection for the weld, as the covering vaporizes.

shielded nuclide *(Phys.)*. One which when found among fission fragments, is assumed to have been a direct product because it is known not to be formed as a result of beta decay.

shielded pair *(Telecomm.)*. Balanced pair of transmission lines within a screen, to mitigate interference from outside.

shielding *(Elec.Eng.)*. (1) Prevention of interfering currents in a circuit, due to external electric fields. Any complete metallic shield earthed at one point is adequate. (2) Use of high permeability material, e.g. Mumetal, for shielding devices susceptible to a magnetic field, e.g. cathode-ray beam; the field, direct or alternating, is shunted away from spaces where it would cause interference.

shielding *(Nuc.Eng.)*. Protective use of low atomic number materials to thermalize strong neutron beams, or concrete, lead or other heavy materials to shield against gamma radiation.

shielding *(Radiol.)*. Use of dense matter to attenuate radiation when it might be harmful to operator or measuring system. The most common material used for large areas is concrete and for smaller areas lead.

shielding windows *(Nuc.Eng.)*. Dense glass blocks or liquid-filled tanks used as windows for inspecting the interior of shielded boxes.

shield, shielding pond *(Nuc.Eng.)*. Deep tank of water used to shield operators from highly radioactive materials stored and manipulated at the bottom.

shift *(Build.)*. See **breaking joint**.

shift *(Comp.)*. Operation that moves the bits held in a store location to the left or right as specified. There are three different types of shift. See **arithmetic-**, **end-around-**, **logical-**.

shift *(Electronics)*. Movement of a pattern on a CRT phosphor, by imposition of steady voltages, e.g. X-shift, Y-shift.

shift *(Phys.)*. (1) Change in wavelength of spectrum line due, e.g., to Doppler or Zeeman effects or to Raman scattering. (2) Change in value of energy level (*level shift*) for quantum mechanical system arising from interaction or perturbation.

shift *(Telecomm.)*. Double use of code, using one code for changing over, as in Telex, teleprinter, teletype, analogous with typewriter keyboard. In teleprinters, one shift is capital letters, the other figures and special signs (*case shift*).

shifting dullness *(Med.)*. Impaired resonance on **percussion** of the abdominal flanks which shifts when the patient rolls on his side and indicates free fluid in the peritoneal cavity, *ascites*.

shifting of brushes *(Elec.Eng.)*. The displacement of the brushes of a commutator motor from the neutral position.

shikimic acid *(Bot.)*. 3,4,5-trihydroxy-1-cyclohexene-1-carboxylic acid, an important cyclic intermediate in the synthesis, in plants, of the aromatic amino acids and other aromatic compounds from non-aromatic precursors.

shilling-stroke (mark) *(Print.)*. See **solidus**.

shim *(Eng.)*. A thin strip of material, used singly or in multiples, to take up space between clamped parts.

shim *(Phys.)*. A packing piece consisting of a thin sheet of magnetic material for placing behind a pole piece in a magnetic circuit to adjust an air-gap.

shim *(Print.)*. A sheet of metal or plastic on which flexible plates can be mounted and then secured to the plate cylinder, or to the bed. Sometimes called *draw sheet*, thus risking confusion with the top sheet of the impression cylinder.

shimamushi fever *(Med.)*. *Tsutsugamushi fever; flood fever; Japanese river fever*. An acute febrile disease associated with infection by rickettsiae, transmitted by the bite of a larval mite; it is characterized by fever, enlargement of lymphatic glands in the neck, axilla, and groin, conjunctivitis, and a dark-red macular rash.

shimming *(Phys.)*. Adjustment of magnetic field with soft iron shims or, by extension, small compensating coils.

shimmy *(Aero.)*. The violent oscillation of a castoring wheel (in practice nose or tail wheel of an aircraft) about its castor axis, which occurs when the coefficient of friction between the surface and the tyre exceeds a critical value. It is usually suppressed by a friction, spring or hydraulic device called a *shimmy damper* (see **damper**) or by a twin-tread tyre.

shimmy *(Autos.)*. See **wheel wobble**.

shim rod *(Nuc.Eng.)*. Coarse control rod of reactor. It is usually positioned so that the reactor will be just critical when the rod is near the centre of its travel path. It is designed to move slowly unless it is also used as a **safety rod**, when a magnetic clutch allows it to drop rapidly into the core.

shin-bone *(Zool.)*. The *tibia*.

shiner *(Build.)*. A thin flat stone laid on edge in a rubble wall, the width of the stone being equal to the depth of at least two courses of the other stones.

shiner *(Print.)*. See **light table**.

shiners *(Paper)*. Particles, usually from mica in the china clay, or undissolved alum, showing on the surface of the paper.

shingle *(Build.)*. A thin, flat, rectangular piece of wood laid in the manner of a slate or tile, as a roof covering or for the sides of buildings. Normally of tapercut red cedar. Also *shide*.

shingle *(Geol.)*. Loose detritus, generally of coarser grade than gravel though finer than boulder beds, occurring typically on the higher parts of beaches on rocky coasts.

shingles *(Med.)*. *Herpes zoster*. After chicken pox and variella, the variella-zoster virus becomes latent in nerve cells. Later in life, often at a time of decreased immunity, there is reactivation of the virus with a belt of blisters over the skin supplied by that nerve (*dermasome*). After the blisters heal, pain can be severe and prolonged, *post-herpatic neuralgia*.

shipboard aircraft *(Aero.)*. Any aircraft designed or adapted for operating from an aircraft carrier: special modifications are strengthened landing gear, strong points for catapulting, arrester hook and, if large, folding wings.

ship caisson *(Hyd.Eng.)*. A floating caisson shaped like a ship, and capable of being floated into position across the entrance to a basin, lock, or graving dock, and then sunk into grooves in the sides and bottom of the entrance.

shiplap *(Build.)*. A term applied to parallel boards having a rebate cut in each edge, the two rebates being on opposite faces. They are especially adapted for use as sheathing.

shippers *(Build.)*. Bricks which are sound and hard-burned, but not of good shape.

shipping fever *(Vet.)*. See **equine influenza** and **haemorrhagic septicaemia**.

shipping pneumonia *(Vet.)*. See **haemorrhagic septicaemia**.

shiva *(Phys.)*. Powerful laser capable of producing pulses of up to 15 kJ of energy, for use in nuclear fusion experiments.

shivering *(Vet.)*. A disease of horses, of unknown cause; characterized by involuntary spasmodic contractions of the muscles of one or both hind limbs and tail.

shives *(Paper)*. Undigested particles of wood showing as pale yellow to brown splinters in the wood pulp or paper.

shives *(Textiles)*. (1) Vegetable matter found in wool fleece. (2) Small particles of woody tissue removed from flax fibres in scutching and hackling.

s.h.m. *(Phys.)*. Abbrev. for *Simple Harmonic Motion*.

shoad, shode *(Min.Ext.)*. Water-worn fragments of vein minerals found on the surface away from the outcrop. Also *float-ore*.

shoal *(Hyd.Eng.)*. A submerged sand bank.

shock *(Behav.)*. (1) *Physiological*, a response to trauma characterized by a state of collapse, pallor, lowered blood pressure etc. (2) *Psychological*, an unexpected and intense experience which compels a total reorientation to life. See **trauma**.

shock *(Med.)*. Acute peripheral circulatory failure due to diminution in the volume of circulating blood and usually characterized by a low blood pressure, a weak thready pulse and diminished urine output following severe haemorrhage, sepsis and fluid loss. Shock may also occur with a preserved circulating blood volume when the heart pump is defective, e.g. in *myocardial infarction* and *pulmonary embolism*.

shock absorber *(Aero.)*. See **oleo**.

shock absorber *(Autos.)*. See **damper**.

shock heating *(Phys.)*. Heating, especially of a plasma, by the passage of a shock wave.

shockproof switch *(Elec.Eng.)*. A switch having all its external metallic parts covered, or protected by insulating material, in order to guard against the possibility of electric shock. Also called *all-insulated switch, Home Office switch*.

shockproof watch *(Horol.)*. A watch provided with a flexible mounting for the balance staff, to avoid damage to the pivots when the watch is subjected to severe shock.

shock tube *(Space)*. A device which generates high speed flows of air over short periods of time by the passage of a shock wave down a tube; used to simulate re-entry conditions in a shock tunnel.

shock wave *(Aero.)*. A surface of discontinuity in which the airflow changes abruptly from **subsonic** to **supersonic**, i.e. from viscous to compressible fluid conditions, thus causing an increase in entropy and an abrupt rise in pressure and temperature. When a shock wave is caused by the passage of a supersonic body the airflow will decelerate to subsonic conditions through another shock wave. A supersonic body normally sets up a conical shock wave with its nose, the angle ($=\operatorname{cosec} 1/M$, where M is the Mach number) becoming increasingly acute the higher the speed, together with subsidiary shock waves from projections on the body and a decelerating shock wave from its tail. The nose and tail shock waves, either attached or travelling on after the passage of an aircraft, are the source of the pressure waves causing **sonic booms**. Abbrev. *shock*.

shock wave *(Phys.)*. Wave of high amplitide, in which the group velocity is higher than the phase velocity, leading to a steep wavefront. This happens, for example, when an explosive is detonated. The speed at which the chemical reaction travels through the material is higher than the speed of sound in the material, hence a shock wave occurs. Strong shocks cause luminosity in gases, and so are useful for spectroscopic work. Also called *blast wave*.

shoddy *(Textiles)*. Short, fibrous material obtained from old, cleaned, *loosely* woven or *knitted* wool cloth after treatment, in a rag-tearing machine. Sprayed with oil, it is used again in blends for cheap suits and coatings. The short dirty fibres which drop from woollen cards are also known as *shoddy*.

shoe *(Build.)*. The short bent part at the foot of a downpipe, directing the water away from the wall.

shoe *(Min.Ext.)*. The replaceable steel wearing part of the head of a stamp or muller of a grinding pan.

Shone ejector *(Build.)*. A type of **ejector** using compressed air to raise sewage from a lower level.

shonkinite *(Geol.)*. A coarse-grained, feldspar-rich syenite, consisting largely of pyroxenes and some olivine. Named after the laccolith, Shonkin Sag, Montana.

shoo flies *(Print.)*. A feature of the **two-revolution** press; they direct the gripper edge of the printed sheet clear of the opened grippers and the strippers with which they work in conjunction.

shoot *(Bot.)*. A stem with all its branches and appendages developed from a bud.

shooting *(Build.)*. The operation of truing with a *jointing plane* the edges of timbers which are to be accurately fitted together.

shooting *(Image Tech.)*. The original photography or recording of a film or video production.

shooting board *(Build.)*. A prepared board used to steady a piece of timber whilst shooting the edges. It has a stop against which the piece of timber abuts endwise, and a guide surface against which the jointing plane runs.

shooting plane *(Build.)*. See **jointing plane**.

shooting star *(Astron.)*. See **meteor**.

shooting stick *(Print.)*. A short tapered length of hard material used with a mallet to lock-up and unlock formes by tapping the wooden quoins; originally of wood, occasionally of metal, but now usually of hard plastic.

shoot-tip culture *(Bot.)*. Same as **meristem culture**.

shop priming *(Build.)*. A term used to describe the process of applying a first coat of paint to an item before delivery to site. Must be done with correct primer to have any beneficial effect.

shop rivet *(Eng.)*. A rivet which is put in when the work is being erected on the floor of the assembly shop prior to delivery to the site.

shoran *(Radar)*. Abbrev. for *SHOrt RANge navigation*; a precision position-fixing system using a pulse transmitter in an aircraft or other vehicle and **transponders** at two known fixed points.

shore effect *(Telecomm.)*. Horizontal refraction as a radio wave crosses a shoreline at an angle, because of different retardation of the ground wave. This causes direction-finding errors. Also *coastline effect*.

Shore hardness test *(Eng.)*. See **scleroscope hardness test**.

shoring *(Build.)*. The method of temporarily supporting by shores, i.e. props of timber or other material in compression, the sides of excavations and, especially, unsafe buildings.

short *(Glass)*. Fast-setting.

short *(Min.Ext.)*. Brittle.

short-circuit *(Elec.Eng.)*. Reduction of p.d. between two points in a circuit to zero by connection of a conductor of zero impedance, in which no power is dissipated. If the short-circuit is not intended, damage may result if the circuit is not opened quickly elsewhere.

short-circuit calculator *(Elec.Eng.)*. An assembly of variable impedances or resistances which can be connected to represent in miniature the circuits of a power system. If a low voltage is applied and a short-circuit put on the system, the currents which flow represent to scale the short-circuit currents which would flow in the actual system under similar conditions. Cf. **network calculator**.

short-circuit characteristic *(Elec.Eng.)*. The characteristic graph relating e.m.f. or excitation to load current in the case of a machine operating under short-circuit conditions.

short-circuit impedance *(Elec.Eng.)*. Input impedance of a network when the output is short-circuited or grounded.

short-circuiting device *(Elec.Eng.)*. A switching device on the rotor of a slip-ring induction motor; operated by a mechanical clutch, which short-circuits the rotor windings when the motor has gained speed.

short-circuit protector *(Elec.Eng.)*. A device for preventing damage from excessive currents caused by a short-circuit. See **overcurrent release, overload protective system**.

short-circuit ratio *(Elec.Eng.)*. The ratio of the field ampere turns a synchronous generator at normal voltage and no load to the field ampere turns on short-circuit with full-load stator current flowing. The value is important in evaluating and comparing the regulation and stability of machines.

short-circuit test *(Elec.Eng.)*. A low-voltage test carried out on an electrical machine with its output terminals short-circuited and full-load current flowing.

short-circuit voltage *(Elec.Eng.)*. The e.m.f. necessary to cause full-load current to flow under short-circuit conditions.

short column *(Eng.)*. A column the diameter of which is so large that bending under load may be neglected, and in which failure would occur by crushing; commonly assumed as a column of height less than 20 diameters.

short-cord winding *(Elec.Eng.)*. An armature winding employing coils whose span is less than the pole pitch.

short-day plant *(Bot.)*. A plant which naturally flowers or shows other morphogenetic change only, or better, as the days shorten. It requires the stimulus of dark period(s) longer than some critical length.

short descenders *(Print.)*. The length of the descenders (g, j etc.) is a feature of type design; for bookwork long descenders are usually considered more elegant, but many successful display types have very short descenders.

shortening capacitor *(Telecomm.)*. One inserted in series with an antenna to reduce its natural wavelength.

short inks *(Print.)*. See **long inks**.

short-oil *(Build.)*. Term applied to varnishes etc. with a low proportion, i.e. less than about 40%, of oil content. Cf. **long-oil**.

short-period comet *(Astron.)*. Comet with a period of less than 150 years, the orbit of which lies entirely within the solar system, e.g. *Halley's Comet*. They have been captured into the solar system through gravitational interaction with Jupiter or Saturn.

short-period variable *(Astron.)*. See **variable star**.

short-range forces *(Phys.)*. Non-coulomb forces which act between nucleons and are responsible for the stability of the nucleus.

short shoot *(Bot.)*. In many, especially woody, plants, a side shoot with very short internodes, on which are borne most (e.g. larch) or all (e.g. pine) of the foliage leaves and most or all of the fruit (e.g. most apples). See also **spur**. Cf. **long shoots**.

short-sightedness *(Med.)*. See **myopia**.

short take-off and vertical landing *(Aero.)*. Class of *V/STOL* aircraft, usually with supersonic capability, whose downward vectored, reheated, engine efflux is too energetic to permit regular vertical take-off operations from conventional surfaces.

short-term memory *(Behav.)*. According to the 3-store model of memory, a memory system that keeps memories for short periods, has a limited storage capacity and stores items in a relatively unprocessed form. See **long term memory, sensory store**.

short-time breakdown voltage *(Elec.Eng.)*. The voltage required to break down a cable in a short time (minutes).

short-time rating *(Elec.Eng.)*. The output which an electrical machine can deliver for a specified short period ($\frac{1}{2}$ hr or 1 hr) without exceeding a specified safe temperature.

short wave *(Telecomm.)*. General designation of radio transmission with wavelengths between about 10 and 200 m.

short-wave therapy *(Med.)*. Treatment by short-wave generators in the electrical capacitor field with high-frequency energy of from 6 m to 30 m wavelength. Therapeutic results of short-wave therapy are due, principally, to the heat produced in body tissues. See also **diathermy**.

shot *(Image Tech.)*. The unit element of action recorded in motion picture or video production; for each shot there may be several **takes** with the same camera and lighting set-up but repeating the action for improved performance.

shot blasting *(Eng.)*. Blasting with metal shot to remove scale and other deposits from castings etc.

shot drilling *(Min.Ext.)*. Boring deep holes by means of hard steel shot fed down rotating hollow cylinder.

shot effect *(Textiles)*. The variable colour seen when certain fabrics are viewed at different angles. The effect is obtained by having the warp threads of one colour and the weft threads of a contrasting colour.

shot firer *(Min.Ext.)*. Miner who tests for gas and then fires explosive charges in colliery.

shot hole *(Civ.Eng.)*. A hole bored in rock for the reception of a blasting charge.

shot noise *(Electronics)*. That which which arises inevitably in the anode circuit of a thermionic valve, because electron emission is not strictly continuous, but consists of a series of random pulses from emitting surfaces.

shot peening *(Eng.)*. See **peening**.

shoulder *(Electronics)*. The part of a characteristic curve where a distinct *levelling off* is perceptible; for example, in an amplifier, where the output power is plotted against input power, a shoulder is seen where the output power approaches the maximum power which can be delivered by the output stages.

shoulder *(Print.)*. The space from the foot of the **bevel** to the edge of the type body.

shouldered arch *(Arch.)*. A lintel supported over a door opening upon corbels.

shoulder girdle *(Zool.)*. See **pectoral girdle**.

shoulder heads *(Print.)*. Subheadings set flush to the left.

shoulder nipple *(Build.)*. A nipple threaded at each end only. Also *barrel nipple*.

shoulder notes *(Print.)*. Notes which are printed, only one per page, in the outer margin level with the first line of text.

shoulder plane *(Build.)*. Metal type of rebate plane with reversed cutting-bevel; used for fine trimming, especially of wide shoulders.

shoulders *(Build.)*. The abutting surfaces left on each side of a tenon; they abut against the cheeks of the mortise.

shovel beak *(Vet.)*. See **mandibular disease**.

shower *(Phys.)*. Result of impact of a high-energy cosmic-ray and photons so that a very large number of ionizing particles and photons are produced, directed downwards in a narrow cone.

showerproofing *(Textiles)*. A light proofing given to fabrics by treating then with metallic salts, insoluble soap, or silicone-based preparations. The thermal and ventilating properties and the general appearance are not much affected by these treatments. The term has no precise meaning and heavy rain should be expected to penetrate coats made from such material.

shower unit *(Phys.)*. The mean path length for reduction of 50% of the energy of cosmic rays as they pass through matter.

show-through *(Print.)*. Appearance of print through another sheet placed on top of printed sheet; paper should be sufficiently opaque to prevent this. Cf. **strike-through**.

shread head *(Build.)*. See **jerkin head**.

shrinkage *(Civ.Eng.)*. The difference in the spaces occupied by material before excavation and after settlement in embankment. Also, the contraction of concrete after placing.

shrinkage *(Textiles)*. The reduction in dimensions of a fibre, yarn, or fabric that takes place in processing or in wear. Particularly effective shrinking treatments are wetting including any laundering cycle.

shrinkage allowance *(Eng.)*. The difference in diameter, when both are cold, of two parts to be united by shrinking. See **shrinking-on**.

shrinkage rule *(Eng.)*. See **pattern maker's rule**.

shrinking-on *(Eng.)*. The process of fastening together two parts by heating the outer member so that it expands sufficiently to pass over the inner and on cooling grips it tightly, e.g. in the attachment of steel tyres to locomotive wheels.

shrink-resistant finish *(Textiles)*. A chemical or mechanical treatment applied, particularly to a fabric, to increase its dimensional stability in use. See **compressive shrinkage**.

shrink-ring commutator *(Elec.Eng.)*. A high-speed type of commutator in which the segments are held together by a steel ring shrunk on over a layer of insulation.

shroud *(Aero.)*. (1) Rearward extension of the skin of a fixed aerofoil surface to cover the whole or part of the leading edge of a movable surface, e.g. flap, elevator, hinged to it. (2) See **jet pipe-**.

shroud *(Eng.)*. (1) Circular webs used to stiffen the sides of gear-teeth. See **full shroud, half shroud**. (2) An outer or peripheral strip used to strengthen turbine blading. (3) A semicircular deflecting wall formed at one side of an inlet port in some IC engines to promote air swirl in the cylinder. Also *shrouding*.

shroud *(Space)*. Streamlined covering, part of a launch system, to protect the payload during launch and to reduce aerodynamic drag; it is ejected when a sufficiently high altitude is reached. Also *fairing*.

shroud *(Telecomm.)*. Extension of metal parts in valves, and other electrical devices subject to high voltages, so that parts of the insulating dielectric are not excessively stressed.

shrouded balance *(Aero.)*. An **aerodynamic balance** in which the area of the hinge line moves within a space formed by shrouds projecting aft from the upper and lower fixed surfaces.

shroud line *(Aero.)*. Any one of the cords attaching a parachute's load to the canopy. Also *rigging line*.

shuffs *(Build.)*. See **chuffs**.

shunt *(Civ.Eng.)*. To divert a train from one track to another, especially to allow another train to pass along the principal track.

shunt *(Med.)*. A short circuit usually between blood vessels allowing an abnormal circulation of blood from a high pressure blood vessel or heart chamber to a lower pressure blood vessel or chamber.

shunt *(Phys.)*. Addition of a component to divert current in a known way, e.g., from a galvanometer, to reduce temporarily its effective sensitivity.

shunt characteristic *(Elec.Eng.)*. The characteristic graph relating terminal voltage and load current in the case of a shunt-wound d.c. machine.

shunt circuit *(Elec.Eng.)*. (1) Electric or magnetic circuit in which current or flux divides into two or more paths before joining to complete the circuit. Also called *parallel circuit*. (2) See **voltage circuit**.

shunt-excited antenna *(Telecomm.)*. Antenna consisting of a vertical radiator (frequently the mast itself) directly earthed at the base, and connected to the transmitter through a lead attached to it a short way above ground.

shunt field *(Elec.Eng.)*. The main field winding of a motor when shunt connected.

shunt-field relay *(Telecomm.)*. One with two coils on opposite sides of a closed magnetic circuit, so that a bridging magnetic circuit takes no flux while the currents in the two coils magnetize the circuit when one current is reversed.

shunt-field rheostat *(Elec.Eng.)*. A rheostat for insertion in the shunt field of a d.c. shunt machine; used to vary the speed of a hunt motor, or the voltage of a shunt generator.

shunt motor *(Elec.Eng.)*. One whose main exitation is derived from a shunt-field winding.

shunt resonance *(Elec.Eng.)*. The condition of a parallel tuned circuit connected across an a.c. voltage supply when maximum impedance is offered to the supply, and the circulating loop current is also a maximum. The term *tunance* is sometimes used in place of resonance if this condition is attained by adjustment of a component value and not of frequency. Also called *parallel resonance*.

shunt trip *(Elec.Eng.)*. A solenoid-type tripping device, connected in shunt across either the main or an auxiliary supply, by which a circuit-breaker may be tripped by a suitable relay.

shunt voltage regulation *(Elec.Eng.)*. That performed by control of a variable impedance in parallel with the output.

shunt winding *(Elec.Eng.)*. A field-winding connected in shunt across the armature circuit of a motor.

shutdown *(Nuc.Eng.)*. Reduction of power level in nuclear reactor to lowest possible value by maintaining core in subcritical condition.

shut-down amplifier *(Nuc.Eng.)*. See **trip amplifier**.

shutdown power *(Nuc.Eng.)*. See **decay heat**.

shuting *(Build.)*. See **eaves gutter**.

shutter *(Image Tech.)*. Device in a camera for exposing the sensitive material to the image formed by the lens for a known period at the required instant.

shuttering *(Civ.Eng.)*. The general term for temporary works for the support of reinforced concrete while it is setting. Also *formwork*.

shutting stile *(Build.)*. The stile of a door further from the hinges. Also *meeting stile*.

shuttle *(Nuc.Eng.)*. Container for samples to be inserted in, and withdrawn from, nuclear reactors, where they are made radioactive by irradiation with neutrons; also *rabbit*.

shuttle *(Textiles)*. (1) Loom accessory which carries the weft (in the form of a cop or pirn) across a loom, through the upper and lower warp threads. Often made of hardword (e.g. persimmon), it is boat-shaped, with a metal tip at each end. One end has a porcelain eye through which the weft passes. Some modern **weaving machines** do not have a shuttle. (2) In a sewing machine, a sliding or rotating device that carries the lower thread to form a lock-stitch.

shuttle armature *(Elec.Eng.)*. A simple form of armature; used on small d.c. machines in which there are only two slots, so that the armature is a single coil connected to a 2-part commutator. Also called *H-armature*.

shuttle box *(Textiles)*. Box-like extension at each side of a loom, from which the shuttle is thrown to and fro when loom is working.

shuttle guard *(Textiles)*. Robustly-constructed metal guard fixed to loom to retain or/and keep down a shuttle which by accident flies out of loom.

shuttleless weaving *(Textiles)*. See **weaving machine**.

Shwartzman reaction (Sanarelli-Shwartzman phenomenon) *(Immun.)*. If endotoxin is administered intravenously followed by a second dose by the same route 24 hours later, renal tubular necrosis and adrenal haemorrhages occur. If the first dose is given into the skin the second dose results in destruction of venules and haemorrhagic necrosis at the skin site. Although apparently a hypersensitivity reaction, there is no identifiable immunological basis.

Si *(Chem.)*. The symbol for silicon.

SI *(Genrl.)*. Abbrev. for *Système International* (d'Unités). See **SI Units**.

sial *(Geol.)*. The discontinuous earth shell of granitic composition which forms the foundation of the continental masses and which is in turn underlain by the *sima*. So called because it is essentially composed of *si*liceous and *al*uminous minerals.

sialagogue, sialogogue *(Med.)*. Stimulating the flow of saliva (Gk. *sialon*); any medicine which does this.

sialolith *(Med.)*. A calculus in a salivary gland.

sibs, siblings *(Biol.)*. Brothers and/or sisters. *Full sibs* have both parents in common, *half sibs* have one parent in common.

sickle cell anaemia *(Med.)*. A **haemolytic anaemia** due to an inherited abnormality in the **haemoglobin** modecule which causes the **red blood cells** to adopt a sickle shaped deformity.

side-and-face cutter *(Eng.)*. A milling cutter with plain or staggered teeth on the sides as well as on the periphery, widely used for cutting slots.

sideband *(Telecomm.)*. Those bands on both sides of the *carrier frequency* which contain additional frequencies, themselves constituting the information to be conveyed, introduced by the process of **modulation**. In *amplitude modulation* the sideband frequencies are equal to the carrier frequency plus and minus the modulating signal frequency. In pulse and frequency modulation it becomes more complex, but the essential information is contained in just one of the sidebands, without the carrier. See **single sideband system, suppressed-carrier system**.

side bones *(Vet.)*. Ossification of the lateral cartilages of a horse's foot.

side chains *(Chem.)*. Alkyl groups, or long chains, which replace hydrogen in ring compounds.

side draught *(Autos.)*. Said of a carburettor in which the mixture is drawn in at right angles to the force of gravity.

side drift *(Civ.Eng.)*. An **adit**.

side keelson *(Ships)*. See **keelson**.

side lobe *(Electronics)*. Any lobe of the radiation pattern of an acoustic radiator or radio or radar antenna, other than that containing the direction of maximum radiation.

side pond *(Hyd.Eng.)*. A storage space at the side of a canal lock-chamber, the two being interconnected by a sluice, so that the normal loss of water occurring in the process of passing a vessel through the lock may be reduced.

side rail *(Civ.Eng.)*. See **check rail**.

sidereal day *(Astron.)*. The interval of time between successive passages of the vernal **equinox** across the same **meridian**. It is 23 h 56 m 4.091 s of mean solar time. In order to prevent the *sidereal day* changing in the middle of the night, when observations are taking place, the sidereal day begins at sidereal noon, when the vernal equinox crosses the local meridian. The time for the Earth to rotate once relative to the distant stars is longer than the sidereal day by about 0.008 s, due to the **precession of the equinoxes**.

sidereal month *(Astron.)*. The interval (27.321 66 sidereal days) for the Moon to complete one orbit of the Earth relative to the distant stars.

sidereal period *(Astron.)*. The interval between two successive positions of a celestial body in the same point with reference to the fixed stars; applied to the moon and planets to indicate their complete revolution relative to the line joining the earth and sun.

sidereal time *(Astron.)*. Time measured by considering the rotation of the Earth relative to the distant stars (rather than the Sun, which is the basis of civil time). The sidereal time at any instant is the same as the **right ascension** of objects exactly on the **meridian**.

sidereal year *(Astron.)*. The interval between two successive passages of the sun in its apparent annual motion through the same point relative to the fixed stars; it amounts to 365.256 36 days, slightly longer than the tropical year, owing to the annual precessional motion of the equinox.

side-rebate plane *(Build.)*. A rebate plane with its cutting edge on the side, not sole, of the tool.

siderite *(Min.)*. (1) Ferrous carbonate, crystallizing in the trigonal system and occurring in sedimentary iron ores and in mineralized veins. Sometimes called *chalybite*. (2) A name for iron meteorites as a class.

sideropenia *(Med.)*. Deficiency of iron.

siderophile *(Geol.)*. Descriptive of elements which have an affinity for iron, and whose geochemical distribution is influenced by this property.

siderophyllite *(Min.)*. The iron and aluminium-rich end-member of the biotite micas.

siderosis *(Med.)*. (1) **Pneumonoconiosis** due to the inhalation of metallic particles by workers in tin, copper, lead, and iron mines, and by steel grinders. (2) Excessive deposit of iron in the body tissues.

siderostat *(Astron.)*. An instrument designed on the same principle as the coelostat to reflect a portion of the sky in a fixed direction; applied specially to a form of telescope called the *polar siderostat*, in which the observer looks down the polar axis on to a mirror.

sideslip *(Aero.)*. The component of the motion of an aircraft in the plane of its lateral axis; generally a piloting or stability error, but also used intentionally to obtain a steep glide descent without gaining speed or to improve weapon aiming; *angle of sideslip* is that between the plane of symmetry and the direction of motion.

side sorts *(Print.)*. See **pi characters**.

side stick *(Print.)*. A tapering piece of wood placed at the sides of pages when locking them up. Quoins are wedged between them and the sides of the chase.

side stick *(Vet.)*. A cylindrical stick fixed to the head-collar and to the surcingle to limit the movement of a horse's head.

side stitching *(Print.)*. See **flat stitching**.

side thrust *(Acous.)*. Radial force on pick-up arm caused by stylus drag.

side tone *(Telecomm.)*. A signal reaching the receiver of a radio-telephone station from its own transmitter.

side tool *(Eng.)*. A cutting tool in which the cutting face is at the side, and which is fed laterally along the work.

side valve *(Autos.)*. Said of an engine having the valves situated at the side of the cylinder block with their posts on the same side of the combustion chamber as the piston and their tappets operated directly by the camshaft.

siding *(Civ.Eng.)*. A short length of side line on to which a train from the main line may be shunted to allow the passage of another on the main line. Colloq. *hole*.

siege *(Build.)*. A mason's or bricklayer's **banker**.

siege *(Glass)*. The floor of a tank furnace or pot furnace.

siemens *(Phys.)*. SI unit of electrical conductance; reciprocal of *ohm*. Abbrev. *S*.

Siemens dynamometer *(Elec.Eng.)*. A dynamometer-type of instrument arranged for measuring current or power.

Siemen's furnace *(Eng.)*. Gas-heated reverberatory furnace.

Siemens-Halske process *(Min.Ext.)*. Chemical extraction of sulphidic copper from its ores or concentrates with sulphuric acid and ferrous sulphate.

Siemens-Martin process *(Eng.)*. See **open-hearth process**.

Siemen's ozone tube *(Chem.)*. Apparatus used in the preparation of ozone by the corona discharge of electricity.

sieve analysis *(Build.,Civ.Eng.)*. A simple method of assessing the suitability of sand or other aggregate for concrete by passing a representative sample through progressively finer sieves and measuring the quantity passed on each occasion.

sieve analysis *(Powder Tech.)*. The measurement of the size distribution of a powder by using a series of sieves of decreasing mesh aperture.

sieve area *(Bot.)*. In the cell wall of a sieve element, a limited area, developed from a primary pit-field and perforated by many pores (< 1–10 μm diameter) through which the protoplasts of contiguous sieve elements connect and which allow the movement of the translocating solution.

sieve element *(Bot.)*. The cell type in phloem in which translocation occurs, characterized by the disappearance of the tonoplast at maturity and the presence of sieve areas in the walls; classified as a *sieve cell* or *sieve tube member*.

sieve mesh number *(Powder Tech.)*. The number of apertures occurring in the surface of a sieve per linear inch. Unless the size of the wire used in weaving the mesh is specified, the mesh number does not uniquely specify the aperture size.

sieve plate *(Bot.)*. Part of the wall of a sieve tube member, bearing one or more highly differentiated sieve areas typically with relatively large pores.

sieve plate *(Chem.Eng.)*. Distillation column plate (or tray) with large number of small holes through which the vapour rises, and fitted with weirs and downcomers to retain liquid through which the vapour rises.

sievert *(Radiol.)*. A unit of radiation dose, being that delivered in 1 hr at a distance of 1 cm from a point source of 1 mg of radium element enclosed in platinum 0.5 mm in thickness. Numerically equal to ca. 8.4 röntgens or 21.6 C/kg. See **effective dose equivalent**.

sieve tube *(Bot.)*. A series of **sieve tube members**, connected end to end and with sieve plates in the common walls between adjacent members.

sig. fig. *(Maths.)*. *Significant figures.*

sight *(Phys.)*. Sensation produced when light impinges on the photosensitive cells of the eye.

sight *(Surv.)*. Bearing from instrument at known point on a distant signal or topographical feature. See **back observation, fore sight, intermediate sight**.

sight-feed lubricator *(Eng.)*. A small glass tube through which oil dropping from a reservoir can be seen, or which is filled with water so that oil from the pump rises in visible drops on its way to the oil-pipe.

sight lines *(Image Tech.)*. The extreme angles from which the screen can be seen in a cinema.

sight rail *(Surv.)*. An above-ground horizontal wooden rail fixed to two upright posts, one on each side of a trench excavation for a sewer, drain etc. Used with others to establish a reference line from which the sewer etc. may be laid at the required gradient.

sight rule *(Surv.)*. See **alidade**.

sigma co-ordinates *(Meteor.)*. A co-ordinate system used in numerical forecasting in which the vertical ordinate σ is pressure p divided by surface pressure p_s i.e. $\sigma = p/p_s$.

sigma-particle *(Phys.)*. Hyperon triplet, rest mass equivalent to 1190 MeV, hypercharge 0, isotopic spin 1.

SIGMET *(Meteor.)*. A SIGMET message is a warning issued by an aviation meteorological watch and forecast office of the occurrence or expected occurrence of one or more meteorological hazards to aircraft including thunderstorms, severe **clear air turbulence**, marked **lee waves**, and severe icing.

sigmoid curve *(Radiol.)*. An S-shaped curve which is often obtained in dose-effect curves in radio-biological studies.

sigmoidectomy *(Med.)*. Excision of part of the sigmoid flexure of the colon.

sigmoid flexure *(Zool.)*. An S-bend.

sigmoidoscope *(Med.)*. A tube fitted with a lamp for viewing the mucous membrane of the rectum and pelvic colon.

sigmoidostomy *(Med.)*. The surgical formation of an opening (artificial anus) in the sigmoid flexure of the colon.

sign *(Med.)*. Any objective evidence of disease or bodily disorder, as opposed to a **symptom** which is a subjective complaint of a patient.

signal *(Surv.)*. A device, such as a ranging rod, heliostat etc., used to mark a survey station.

signal *(Telecomm.)*. General term referring to a conveyor of information, e.g. an audio waveform, a video waveform, series of pulses in a computer. Colloquially, the message itself. In radio, the signal modulates a carrier, and is recovered during reception by demodulation.

signal code *(Telecomm.)*. In voice-frequency signalling, the plan for representing each of the required signalling functions as a voice-frequency signal.

signal component *(Telecomm.)*. That part of a signal which continues uniform in character throughout its duration. In a multi-component signal with spaces between current pulses a space may be regarded as a signal component.

signal distortion *(Phys.)*. Modification of the information content of a signal, sometimes irreversibly; e.g., the suppression or introduction of **harmonics**.

signal element *(Telecomm.)*. The portion of a signal occupying the smallest interval of the signal code.

signal frequency shift *(Telecomm.)*. The band-width between white and black signal levels in frequency-modulation facsimile transmission systems.

signal generator *(Telecomm.)*. Oscillator designed to provide known voltages typically, from 1 to less than 1 μV, over a range of frequencies. Used for testing or ascertaining performance of radio-receiving equipment. It may be amplitude-, frequency-, or pulse-modulated.

signal level *(Telecomm.)*. The difference between the level of a signal at a point in a transmission system and the level of the arbitrarily specified reference signal.

signal/noise ratio *(Acous.)*. Difference in dB between the wanted signal and the unwanted background noise.

signal output current *(Electronics)*. The absolute difference between the output current and the dark current of a phototube or a camera tube.

signal peptides, sequences *(Biol.)*. *Leader peptides.* Short N-terminal peptide sequences of newly synthesized membrane proteins which direct the protein towards the appropriate membrane, facilitate its transfer across it and are usually deleted during the subsequent maturation of the protein.

signal shaping *(Telecomm.)*. Use of specially designed electric network to correct distortion produced during transmission or propagation of signals.

signal-to-cross-talk ratio *(Telecomm.)*. In line telephony, the ratio of the test level in the disturbed circuit to the level of the cross-talk at the same point which is caused by the disturbing circuit operating at the test level.

signal windings *(Elec.Eng.)*. US term for **control turns** (or *windings*) of a saturable reactor.

sign and magnitude *(Comp.)*. Method of representing numbers in a binary word by coding the sign of the number in a **sign bit** and the magnitude of the number in the remaining bits. Also *sign and modulus*.

signature *(Print.)*. See **section**.

signature mark *(Print.)*. A number or letter of the alphabet placed on the first page of a section as a guide to the binder in **gathering**.

sign bit *(Comp.)*. Single bit, used to indicate the sign of a number, usually 0 for positive, 1 for negative.

signed minor *(Maths.)*. See **cofactor**.

significance *(Stats.)*. A threshold value of probability at or below which the results of a statistical investigation are held to contradict a particular hypothesis.

significant figures *(Maths.)*. Of a number: those digits which make a contribution to its value, e.g. in the number 00.1230, the first two zeros are insignificant and the digits 1, 2, 3 are significant. The last zero also should be significant, indicating that the number is accurate to four places of decimals. Care should be exercised before making this assumption, however, because some writers add such zeros indiscriminately. Also *significant digits*.

sign stimulus *(Behav.)*. Part of a complex stimulus configuration which is relevant to a particular response and evokes the strongest response (e.g. the red breast of the robin).

Sikes hydrometer *(Chem.)*. A hydrometer used for determining the strengths of mixtures of alcohol and water.

silage *(Genrl.)*. A cattle food formed by the bacteriological breakdown, in portable or fixed silos, of vegetable matter (grasses, clovers, pea haulms, beet tops etc.), with the admixture of diluted molasses, suitable acids (e.g. formic) etc. Also *ensilage*, the process of making such food.

silanes *(Chem.)*. A term given to the silicon hydrides: silane, SiH_4, disilane $H_3Si\text{-}SiH_3$, trisilane, $H_3Si(SiH_2)SiH_4$ etc.

Silastic *(Plastics)*. TN for a range of silicone rubbers. Noted for very good heat resistance and a wide temperature range of application. Excellent chemical resistance and electrical properties.

silencer *(Autos.)*. An expansion-chamber fitted to the exhaust pipe of an IC engine to dampen the noise of combustion. US *muffler*.

silent period *(Telecomm.)*. Stated period within each hour during which all marine transmissions must close down and listen on the international distress frequency of 500 kHz.

silex *(Min.Ext.)*. Silica brick used to line grinding mills when contamination by abraded steel must be avoided.

silica *(Chem.)*. Dioxide ((IV) oxide) of silicon, SiO_2, occurring in crystalline forms as quartz, cristobalite, tridymite; as cryptocrystalline chalcedony; as amorphous opal; and as an essential constituent of the silicate groups of minerals. Used in the manufacture of glass and refractory materials. Refractory materials containing a high proportion of silica (over 90%) are known as *acid refractories* (e.g. gannister), and are used in open-hearth and other metallurgical furnaces to resist high temperatures and attack by acid slags.

silica gel *(Chem.)*. Hard amorphous granular form of hydrated silica, chemically inert but very hygroscopic. Used for absorbing water and vapours of solvents, especially in enclosed electronic equipment. When saturated, it may be regenerated by heat.

silica glass *(Glass)*. See **vitreous silica**.

silica glass *(Min.)*. Fused quartz, occurring in shapeless masses on the surface of the Libyan Desert, in Moravia, in parts of Australia and elsewhere; believed to be of meteoritic origin. See **tektites**.

silica poisoning *(Min.Ext.)*. Loading of resins used in ion-exchange process with silica, thus reducing the efficiency of reaction with desired ions.

silicates *(Min.)*. The largest group of minerals, of widely different, and in some cases, extremely complex composition, but all composed of silicon, oxygen, and one or more metals, with or without hydrogen.

siliceous deposits *(Geol.)*. Those sediments, incrustations, or deposits which contain a large percentage of silica in one or more of its modes of occurrence. They may be chemically or mechanically formed, or may consist of the siliceous skeletons of organisms such as diatoms and Radiolaria. See also **silicification**.

siliceous sinter *(Geol.)*. Cellular quartz or translucent to opaque opal, found as incrustations or fibrous growths and deposited from thermal waters containing silica or silicates in solution.

silicic acid *(Chem.)*. An acid formed when alkaline silicates are treated with acids. Amorphous, gelatinous mass. Dissociates readily into water and silica.

silicides *(Chem.)*. Compounds formed by the combination of silicon with other elements, chiefly metals.

silicification *(Geol.)*. The process by which silica is introduced as a cement into rocks after their deposition, or as an infiltration or replacement of organic tissues or of other minerals such as calcite. See also **novaculite**.

silicole *(Bot.)*. A plant which grows on soils rich in silica, and usually acid in reaction.

silico-manganese steel *(Eng.)*. See **manganese alloys**.

silicon *(Chem.)*. A nonmetallic element, symbol Si, at. no. 14, r.a.m. 28.086, valency 4. Amorphous silicon is a brown powder; rel.d. 2.42. Crystalline silicon is grey; rel.d. 2.42, mp 1420°C, bp 2600°C. This element is the second most abundant, silicates being the chief constituents of many rocks, clays and soils. Silicon is manufactured by reducing silica with carbon in an electric furnace, and is used in glass and in making certain alloys, e.g. *ferro-silicon*. It has semiconducting properties, being used for a large range of electronic components.

silicon bronze *(Eng.)*. A noncorroding alloy based on copper and tin.

silicon carbide *(Chem.)*. SiC. Formed by fusing a mixture of carbon and sand or silica in an electric furnace (see **Acheson furnace**). Used as an abrasive and refractory. See **carborundum**.

silicon-controlled rectifier *(Electronics)*. A three terminal semiconductor switching device consisting of a sandwich of *p-n-p-n*-type materials. It is normally open-circuit, but application of an appropriate control signal to the *gate* allows it to conduct, in one direction only, like a conventional rectifier. It continues to conduct even with the gate signal removed, until it is reverse biased by the voltage it is intended to switch. Used in voltage-control of power circuits.

silicon copper *(Eng.)*. An alloy (20–30% Si), a 'getter' used to remove oxygen from molten copper alloys.

silicon detector *(Telecomm.)*. Stable silicon crystal diode for demodulation.

silicon dioxide *(Chem.)*. *Silicon (IV) oxide*. See **silica**.

silicone resins *(Build.)*. A group of resins with particular properties which benefit coatings e.g. resistance to heat, acids, alkalis, oils, salts and the ability to repel water, making them useful for masonry water repellants. Silicones are also used in polishes.

silicone rubbers *(Chem.)*. An important group of synthetic rubbers (dimethylsiloxene polymers), having both high and low temperature resistances better than those for natural rubbers.

silicones *(Chem.)*. Open-chain and cyclic organo-silicon compounds containing $—SiR_2O—$ groups, prepared mainly by hydrolysing alkyl or aryl silicon dichlorides, R_2SiCl_2, which are themselves made by the Grignard reaction. The simpler substances are oils of very low melting-point, the viscosity of which changes little with temperature, used as lubricants, shock-absorber fluids, constituents of polishes etc. More complex solid products, stable to heat and cold, and chemically inert, are exceptionally good electrical insulators for electric motors etc. Also used in gaskets and a wide variety of special applications.

silicon hydrides *(Chem.)*. See **silanes**.

silicon iron *(Eng.)*. Iron or low carbon steel to which 0.75–4.0% silicon has been added. Has low magnetic hysteresis, and is resistant to mild acids. Used for sheets for transformer cores. Typical composition: silicon 4%, manganese under 0.1%, phosphorus 0.02%, sulphur 0.02%, carbon 0.05%.

silicon rectifier *(Electronics)*. A semiconductor diode rectifier usually based of *p-n* junction in silicon crystal.

silicon resistor *(Electronics)*. A resistor of special silicon material which has a fairly constant positive temperature coefficient, making it suitable as a temperature-sensing element.

silicon tetrachloride *(Chem.)*. *Tetrachlorosilane*, $SiCl_4$. Formed by the action of chlorine on a mixture of silica and carbon, or silicon. Liquid.

silicon tetrafluoride *(Chem.)*. *Tetrafluorosilane*, SiF_4. A gaseous compound formed by the action of hydrofluoric acid on silica. Readily hydrolyses into silica and hydrofluoric acid.

silicosis *(Med.)*. **Pneumoconiosis**, due to the inhalation of particles of silica by masons and miners who work in the presence of silica.

siliqua *(Bot.)*. A capsule with the general characters of a **silicle**, but at least 4 times as long as it is broad. Also *silique*.

silk *(Min.)*. A sheen resembling that of silk, exhibited by some corundums, including ruby, and due to minute tubular cavities, or to rutile needles, in parallel orientation. The colour of such stones is paler than normal by reason of the inclusions.

silk *(Textiles)*. The protein fibre obtained in long continuous fine strands from the cocoon of silkworms, especially of the moth *Bombyx mori*. The fibre is composed of fibroin surrounded by another protein, sericin, which is a gum removed during wet processing. Wild silk (e.g. **tussah**) is a similar fibre from the cocoon of other moths.

silk *(Zool.)*. A fluid substance secreted by various Arthropoda. It is composed mainly of fibroin, together with sericin and other substances, and hardens on exposure to air in the form of a thread. Used for spinning cocoons, webs, egg cases etc.

silk-screen printing *(Print.)*. A mesh-stencil process in which ink is squeezed through the open parts of the mesh on to the surface to be printed, which may be of any material, smooth or rough, flat or curved. The stencil may be prepared by hand, or by photography, or by a mixture of both. Screens of metal and fibres other than silk are also used, and **screen process printing** has become the preferred name.

sill *(Build.)*. The horizontal timber, stone etc. at the foot of an opening, as for a door (also *threshold*), window, embrasure etc.

sill *(Geol.)*. A minor intrusion of igneous rock injected as a tabular sheet between, and more or less parallel to, the bedding planes of rocks.

sill *(Hyd.Eng.)*. The top level of a weir, or the lowest level of a notch.

sillénite *(Min.)*. Cubic bismuth trioxide. Cf. **bismite**.

sillimanite *(Min.)*. An orthorhombic aluminium silicate. The high-temperature polymorph of Al_2SiO_5, occurring in high-grade metamorphosed argillaceous rocks. *Fibrolite* is a fibrous variety, used as a gemstone. *Kyanite* and *andalusite* are other aluminium silacates.

silo *(Aero.)*. Underground chamber housing a guided missile which is ready to be fired.

Silsbee rule *(Elec.Eng.)*. A wire of radius r cannot carry a superconducting current greater than $\frac{1}{2}rH_c$, where H_c is the critical field. The self-magnetic field would then destroy super-conductivity.

silt *(Geol.)*. Material of an earthy character intermediate in grain-size between sand and clay. See **silt grade**, **particle size**.

silt box *(Build.)*. A removable iron box placed at the bottom of a gulley; serves to accumulate the deposited silt for periodic removal.

Siluminite *(Elec.Eng.)*. TN designating materials composed principally of asbestos or mica for electrical insulating materials.

Silurian *(Geol.)*. The youngest period of the Lower Palaeozoic, covering a time span between app. 440–400 million years. Named after *Silures*, an ancient Welsh tribe. The corresponding system of rock.

Silurian *(Paper)*. A mottled effect in special papers produced by adding deeply dyed fibres to the main furnish. Under certain circumstances this can be a paper defect.

Siluriformes *(Zool.)*. Order of mainly freshwater, bottom-living Osteichthyes with barbels used in detecting food and usually without scales. Catfish.

silver *(Chem.)*. A pure white metallic element, symbol Ag. at. no. 47, r.a.m. 107.868, rel.d. at 20°C 10.5, mp 960°C, bp 1955°C, casting temp. 1030–1090°C, Brinell hardness 37, electrical resistivity approx. 1.62×10^{-8} ohm metres. The metal is not oxidized in air. Occurs massive, or assumes arborescent or filiform shapes. The best electrical conductor and the main constituent of photographic emulsions. Native silver often has variable admixture of other metals; gold, copper or sometimes platinum. Used for ornaments, mirrors, cutlery, jewellery etc. and for certain components in food and chemical industry where cheaper metals fail to withstand corrosion.

silver amalgam *(Min.)*. *Arquerite*; a solid solution of mercury and silver, which crystallizes in the cubic system. Of rare occurrence, it is found scattered in mercury or silver deposits.

silver glance *(Min.)*. See **argentite**.

silver grain *(For.)*. The light greyish, shining flecking seen in oak timber, caused by vascular rays exposed in preparing the timber when it is cut radially through the centre of a log. Also seen in beech.

silver (I) halides *(Chem.)*. Silver fluoride, AgF; silver iodide, AgI; silver chloride, AgCl; and silver bromide, AgBr. The last three are sensitive to light and are of basic importance in photography.

silvering *(Glass)*. This process is carried out on a perfectly clean surface of glass by pouring on to it an ammoniacal silver solution, mixed with Rochelle salt or with a nitric-acid/cane-sugar/alcohol mixture. The silver film so formed is then washed, backed with varnish and painted.

silver lead ore *(Min.)*. The name given to galena containing silver. When 1% or more of silver is present it becomes a valuable ore of silver. Also called *argentiferous galena*.

silver leaf *(Build.)*. Metallic silver prepared in leaves in loose or transfer form; similar use as gold leaf. Tends to tarnish rapidly on exposure to the air. Best protected with a clear lacquer.

silverlock bond *(Build.)*. See **rat-trap bond**.

silver oxide *(Chem.)*. See **argentic [silver (II)] oxide, argentous [silver (I)] oxide**.

silver solder *(Eng.)*. See **brazing solders**.

silver steel *(Eng.)*. A bright-drawn carbon steel containing up to 0.3% silicon, 0.45% manganese, 0.5% chromium and 1.25% carbon; low in phosphorus and sulphur.

silver voltameter *(Elec.Eng.)*. An electrolytic cell used for determining accurately the average value of a current from the quantity of silver deposited from the silver nitrate solution forming the electrode.

silviculture *(For.)*. The planting and care of forests.

sima *(Geol.)*. The lower layer of the earth's crust with the composition of a basic or ultrabasic igneous rock. Such rocks contain silicon and magnesium as their principal constituents, hence the name.

SIMD *(Comp.)*. *Single Instruction stream, Multiple Data stream*. Term which describes the architecture of the **processor**. See **associative memory, array processor**.

simian virus 40 *(Biol.)*. See **SV 40**.

similarity coefficient *(Ecol.)*. Index of similarity between two stands of vegetation, based on their species composition.

similar polygons *(Maths.)*. Polygons whose corresponding angles are equal and in which the sides about the corresponding angles are in direct proportion.

Simmonds' disease *(Med.)*. *Hypophyseal cachexia*. A rare disease due to destruction of the pituitary gland; characterized by cachexia, atrophy of the skin and the bones, premature senility, loss of hair and loss of sexual function.

simple *(Bot.)*. Consisting of one piece or component; unbranched; not **compound**.

simple curve *(Surv.)*. A curve composed of a single arc connecting two straights.

simple eye *(Zool.)*. See **ocellus**.

simple fruit *(Bot.)*. A fruit formed from one pistil.

simple harmonic motion *(Phys.)*. The motion of a particle (or system) for which the force on the particle is proportional to its distance from a fixed point and is directed towards the fixed point. The particle executes an oscillatory motion about the point. The motion satisfies the equation $(d^2x/dt^2) = -\omega^2 x$ where x is the displacement of the particle and ω is a constant for the motion. The majority of small amplitude oscillatory motions are simple harmonic, e.g., the oscillations of a mass suspended by a spring, the swing of a pendulum, the vibrations of a violin string, the oscillations of atoms or molecules in a solid, the oscillations of air as a sound wave passes. When such a motion takes place in a resistive medium, e.g. air, the oscillations die away with time; the motion is then said to be *damped*. Abbrev. *s.h.m.*

simple leaf *(Bot.)*. A leaf in which the lamina consists of one piece, which, if lobed, is not cut into separate parts reaching down to the midrib.

simple pit *(Bot.)*. A pit of which the cavity does not become markedly narrower towards the cell lumen. Cf. **bordered pit**.

simple press tool *(Eng.)*. **Press tool** which performs only one operation at each stroke of the press, as distinct from a **compound press tool** and a **progressive press tool**.

simple sequence DNA *(Biol.)*. Block of a DNA sequence which consists of many repeats of a short, unit sequence. The repeats are not necessarily identical.

simple steam-engine *(Eng.)*. An engine with one or more cylinders in which the steam expands from the initial pressure to the exhaust pressure in a single stage. Cf. **compound-**.

simplex *(Comp.)*. Transmission of data in only one direction.

Simplex piling *(Civ.Eng.)*. A system of piling in which a cast-iron point is driven to the required depth by means of a steel pipe. The pipe is then filled with concrete and afterwards slowly withdrawn, the concrete filling adapting itself to the irregularities in the ground.

simplex winding *(Elec.Eng.)*. An armature winding through which there is only one electrical path per pole.

simply-connected domain *(Maths.)*. See **connected domain**.

Simpson's rule *(Maths.)*. The area under the curve $y=f(x)$ from $x=x_0$ to $x=x_2$ is approximately $\frac{1}{6}[f(x_0)+4f(x_1)+f(x_2)][x_2-x_0]$, where x_0, x_1 and x_2 are equally spaced.

Sims' speculum *(Med.)*. A speculum, shaped like a duck's bill, for viewing the lining of the vagina and the cervix uteri.

simulated line *(Telecomm.)*. Same as **artificial line**.

simulation *(Comp.)*. Method of studying the behaviour of a system by using a model of the system and processing it on the computer.

simulation *(Electronics)*. The representation of physical systems and phenomena by computers, models and other equipment.

simulation *(Zool.)*. Mimicry; assumption of the external characters of another species in order to facilitate the capture of prey or escape from enemies. v. *simulate*.

simulation by computer *(Behav.)*. The investigation of thought processes by the use of computers programmed to imitate them.

simultaneity *(Phys.)*. Basic consequence of *relativity*. Two events that are simultaneous according to one observer if a clock with him records the same time, may occur at different times according to another observer in another reference frame moving relative to the first.

simultaneous broadcasting *(Telecomm.)*. (1) Transmission of one programme from two or more transmitters. (2) Simultaneous television and radio (usually stereophonic) broadcasts from live concerts etc.

Sindanyo *(Elec.Eng.)*. TN designating materials, composed principally of asbestos, for the mounting of switchgear of all types for electrical insulation work generally, and for arc shields, barriers, furnace linings, and other purposes.

sine bar *(Eng.)*. A hardened steel bar carrying two plugs of standard diameter accurately spaced to some standard distance; used in setting out angles to close limits.

sine condition *(Phys.)*. A condition which must be satisfied by a lens if it is to form an image free from aberrations (other than chromatic). It may be stated $n_1l_1 \sin\alpha_1 = n_2l_2 \sin\alpha_2$, where n_1 and n_2 are the refractive indices of the media on the object and image sides of the lens respectively, l_1 and l_2 are the linear dimensions of the object and image, and a_1 and a_2 are the angles made with the principal axis by the conjugate portions of a ray passing between object and image.

sine galvanometer *(Elec.Eng.)*. A galvanometer in which the coil and scale are rotated to keep the needle at zero. The current is then proportional to the sine of the angle of rotation. The arrangement can be made more sensitive than the **tangent galvanometer**.

sine potentiometer *(Elec.Eng.)*. Voltage divider in which the output of an applied direct voltage is proportional to the sine of the angular displacement of a shaft.

sine wave *(Phys.)*. Waveform of a single frequency, indefinitely repeated in time, the only waveform whose integral and differential has the same wave-form as itself. Its displacement can be expressed as the sine (or cosine) of a linear function of time or distance, or both. In practice there must be a transient at the start and finish of such a wave.

singeing *(Textiles)*. See **gassing**.

singing *(Telecomm.)*. Oscillation in a telephone system caused by feedback across a source of gain because of mismatch in the circuit.

singing tube *(Acous.)*. (1) A tube with a flame inside which under certain conditions excites the tube resonance. (2) See **Rijke tube**.

single-acting cylinder *(Eng.)*. Fluid-power cylinder in which the piston is displaced in one direction by the fluid and returned mechanically, usually by a spring.

single-acting engine *(Eng.)*. A reciprocating engine in which the working fluid acts on one side of the piston only, as in most IC engines.

single-beat escapement *(Horol.)*. An escapement in which the balance receives impulse only at every alternate vibration, e.g. the chronometer and duplex escapements.

single bridging *(Build.)*. Bridging in which a pair of diagonal braces are used to connect adjacent floor-joists at their middle points.

single-catenary suspension *(Elec.Eng.)*. A catenary suspension system in which the conductor wire is hung from a single catenary or bearer wire.

single-cell protein *(Bot.)*. Protein-rich material from cultured algae, fungi (including yeasts) or bacteria, used (potentially) for food or as animal feed. Abbrev. *SCP*.

single-channel per carrier *(Telecomm.)*. Used in satellite communications for a system where each carrier is dedicated to one telephone call or data transmission for its duration. Cf. **time-division multiple access**.

single-channel pulse height analyser *(Telecomm.)*. See **pulse height analyser**.

single-core cable *(Elec.Eng.)*. A cable having only one conductor.

single-electrode system *(Elec.Eng.)*. Electrode of an electrolytic cell and the electrolyte with which it is in contact. Also called *half-cell*, *half-element*.

single-ended *(Elec.Eng.)*. (1) Unit or system designed for use with unbalanced signal, having one input and one output terminal permanently earthed. (2) Valve with all electrodes connected to pins at the same end.

single-entry compressor *(Aero.)*. A centrifugal compressor which has vanes on one face only.

single Flemish bond *(Build.)*. A form of bond combining English bond for the body of the wall with Flemish bond for the facework.

single floor *(Build.)*. A floor in which the bridging joists span the distance from wall to wall without intermediate support.

single-hung window *(Build.)*. A window having top and bottom sashes, of which only one (usually the bottom sash) is balanced by sash cord and weights so as to be capable of vertical movement.

single jersey *(Textiles)*. Weft-knitted fabrics of a simple plain nature although patterned effects may be included e.g. with a jacquard.

single laths *(Build.)*. Wood laths 1 in (25 mm) by $\frac{1}{8}$ to 3/16 in (3 to 5 mm) thick in section.

single-lens reflex *(Image Tech.)*. Camera in which the image is viewed and focused by way of a moveable mirror behind the lens and a **pentaprism** in front of the eyepiece. The mirror is rotated out of the way just before the **focal plane shutter** opens. Easily used with different focal length lenses and with automatic exposure and focusing devices. Abbrev. *SLR*.

single phase *(Elec.Eng.)*. Of electrical power transmission; two conductors, one of which may be the earth or at earth potential, between which is a sinusoidally alternating potential difference, as for domestic a.c. supply.

single-phase induction regulator *(Elec.Eng.)*. An induction regulator for use on a single-phase circuit; the arrangement is such that the voltage on the secondary side is always in phase with that on the primary side. Cf. **three-phase induction regulator**.

single-plate clutch *(Eng.)*. A **friction clutch** in which the disk-shaped or annular driven member, fabric-faced, is pressed against a similar face on the driving member by springs, being withdrawn against them through a thrust collar; used in automobiles.

single pole *(Elec.Eng.)*. Of a switch or relay contact which makes, breaks or changes over connection on one *pole* only, a.c. or d.c., of a circuit.

single quotes *(Print.)*. See **quotation marks**.

single-rate prepayment meter *(Elec.Eng.)*. A prepayment meter in which the circuit is broken and the supply cut off after a predetermined number of units have been consumed.

single-retort underfeed stoker *(Eng.)*. An **underfeed stoker** consisting of a retort along the bottom of which coal is fed by a steam-driven ram or a screw conveyor, air being supplied through tuyëres round the upper edge of the retort and into the sealed ash pit below.

single-revolution *(Print.)*. A design of letterpress machine in which the cylinder rotates continuously while the bed reciprocates, and prints one impression during each revolution. Cf. **stop-cylinder, two revolution**.

single-row ball-bearing *(Eng.)*. **Ball-bearing** comprising a set of balls arranged in a single plane, as distinct from a double-row bearing.

singles *(Build.)*. Roofing slates about 12×18 in (305×457 mm).

singles *(Min.Ext.)*. See **coal sizes**.

single-sideband suppressed carrier *(Telecomm.)*. See **vestigial-sideband transmitter**.

single-sideband system *(Telecomm.)*. A form of amplitude modulation in which one sideband and the carrier are eliminated, either by filtering after modulation or, more commonly, by using balanced modulators and phase-shift circuits. On reception, the carrier has to be re-inserted before conventional demodulation can occur. Such transmissions make better use of transmitter power, occupy less bandwidth and therefore improve the signal-to-noise ratio. Abbrev. *SSB*. See **double-sideband system, suppressed-carrier system**.

single-spindle automatic *(Eng.)*. **Automatic screw machine** or lathe with a tool turret and one chucking spindle, producing one part per operating cycle.

single-stack system *(Build.)*. Where soil and waste discharges are piped together within certain design limitations and without vertical ventilating pipework other than a main vertical stack.

single-stage-to-orbit *(Space)*. Space system which can launch a payload into orbit without staging, i.e. using one engine which can provide the necessary thrust throughout all the flight regimes of the complete ascent.

singlet *(Chem.)*. A state in which there are no unpaired electrons.

single-turn coil *(Elec.Eng.)*. An armature coil consisting of a single turn of copper bar.

single-turn transformer *(Elec.Eng.)*. A current transformer in which the primary winding takes the form of a single straight conductor of heavy cross-section, to which the cable or bus-bar is connected.

single-valued *(Maths.)*. Said of a function which can have only one value for each set of values of the independent variables.

single-wave rectification *(Elec.Eng.)*. See **half-wave rectification**.

single-wire circuit *(Elec.Eng.)*. One with a single live wire, including coaxial or concentric, with sheath, earth, or frame return.

single-wire feeder *(Telecomm.)*. One for an antenna, similar to an ordinary downlead, but connected to the antenna in such a manner that it is terminated in its characteristic impedance, so that no standing waves are formed on it.

single yarn *(Textiles)*. The thread obtained from one unit of a spinning machine.

singularity *(Maths.)*. Of a function: a point where the function ceases to be analytic.

singular point on a curve *(Maths.)*. A point on the curve $f(x,y)=0$ at which

$$\frac{\partial f}{\partial x}=\frac{\partial f}{\partial y}=0.$$

One at which there is either no real tangent or two or more tangents. Cf. **double point**.

singular solution *(Maths.)*. Of a differential equation: a solution which cannot be derived as a particular integral from a complete primitive.

sinistral fault *(Geol.)*. A tear fault in which the rocks on one side of the fault appear to have moved to the left when viewed across the fault. Cf. **dextral fault**.

sinistrorse *(Bot.,Zool.)*. Helical, twisted or coiled in the sense of left-hand screw thread or of an *S-helix*. Cf. **dextrorse**.

sink *(Bot.)*. Region within a plant (or a cell) where a demand exists for particular metabolites, e.g. growing shoots, roots and developing tubers (sinks for photosynthate); mitochondria (sinks for oxygen).

sink *(Horol.)*. A recess; the spherical depression around a pivot hole for holding the oil.

sink *(Print.)*. A depression in the printing surface of a plate.

sink *(Telecomm.)*. Unstable operating region on **Rieke diagram**.

sinker *(Textiles)*. The mechanism in a knitting machine that pushes a length of thread over the spring needles to form a new course of loops.

sinker bar *(Min.Ext.)*. A heavy bar attached to the cable above the drilling tools used in percussive drilling.

sink-float process *(Min.Ext.)*. See **heavy media separation**.

sinking *(Build.)*. A recess cut below the general surface of the work.

sinking *(Civ.Eng.)*. The operation of excavating for a shaft, pit or well.

sinking shaft *(Min.Ext.)*. Shaft excavated from above downwards. Cf. **rising shaft**.

Sinope *(Astron.)*. The ninth natural satellite of **Jupiter**.

sino-, sinu-auricular *(Med.)*. Applied to structures located in the right atrium near the opening of the venae cavae, corresponding to the sinus venosus, e.g. sinu-atrial (or sino-auricular) node.

sin, sine *(Maths.)*. See **trigonometrical functions**.

sinter *(Chem.)*. To coalesce into a single mass under the influence of heat, without actually liquefying.

sinter *(Min.)*. A concretionary deposit of opaline silica which is porous, incrusting, or stalactitic in habit; found near geysers, as at Yellowstone National Park (US). Also called *geyserite*.

sintered carbides *(Eng.)*. See **carbide tools, sintering**.

sintered crucible *(Chem.)*. A crucible with a permeable sintered base used as a combination filter and crucible.

sintering *(Elec.Eng.)*. The process of consolidating the filament of an electric lamp by passing a relatively high current through it when in a vacuum.

sintering *(Eng.)*. The fritting together of small particles to form larger particles, cakes or masses; in case of ores and concentrates, it is accomplished by fusion of certain constituents. As used in powder metallurgy, sintering consists in mixing metal powders having different melting-points, and then heating the mixture to a temperature approximating the lowest mp of any metal included. In sintered carbides, powdered cobalt, having the lowest mp, acts as the binder holding together the unmelted particles of the hard carbides.

sinuate *(Bot.)*. A leaf margin with rounded teeth and notches; wavy.

sinuitis *(Med.)*. See **sinusitis**.

sinus *(Zool.)*. A cavity or depression of irregular shape.

sinus arrhythmia *(Med.)*. A normal speeding up and slowing down of heart rate due to alterations in tone in the *vagus* nerve.

sinusitis *(Med.)*. *Nasal sinusitis.* Inflammation of any one of the air-containing cavities of the skull which communicate with the nose. The condition is called ethmoid, frontal, maxillary or sphenoid sinusitis, according to the site affected. It may be acute or chronic, purulent or nonpurulent.

sinusoid *(Zool.)*. In Vertebrates, a sinus like blood space connected usually with the venous system and lying betwen the cells of the surrounding tissue or organ.

sinusoidal *(Elec.Eng.)*. An alternating quantity is said to be *sinusoidal* when its trace, plotted to a linear time base, is a sine wave.

sinusoidal current *(Elec.Eng.)*. One which varies sinusoidally with time, having a frequency, amplitude and phase. It flows in each direction alternately for equal periods.

sinusoidal spirals *(Maths.)*. Curves with polar equation $r^n = a_n \cos n\theta$. Not true spirals since the tracing point does not recede continuously. For certain values of n the curves have specific names, e.g., $n = -2$ (hyperbola), $n = -1$ (straight line), $n = -\frac{1}{2}$ (parabola), $n = \frac{1}{2}$ (cardioid) and $n = 2$ (lemniscate).

sinusoidal wave *(Phys.)*. A wave whose profile is a pure sine curve; also *harmonic wave*.

sinus venosus *(Zool.)*. In a Vertebrate embryo, the most posterior chamber of the developing heart; in lower Vertebrates, the tubular chamber into which this develops and which receives blood from the veins, or sinuses and passes it into the auricle.

siphon *(Civ.Eng.)*. (1) A pipe system comprising a rising leg and a falling leg, typically in the shape of an inverted 'U'; pressure at the inlet to the rising leg is atmospheric pressure; as liquid rises in the rising leg the pressure falls below atmospheric pressure and reaches a minimum at or near the apex, recovering to atmospheric pressure in the down leg. The lift of a siphon can equal or exceed the equivalent of atmospheric pressure expressed in *head* of the fluid which enters the siphon; this results from entrained gases emerging throughout the rising leg. (2) A pipe or aqueduct crossing a valley and rising again to somewhat less than its inlet level, so as to have the necessary hydraulic gradient. More correctly called an *inverted siphon*.

siphon *(Zool.)*. A tubular organ serving for the intake or output of fluid, as the pallial siphons of many bivalve Mollusca. adj. *siphonate*.

Siphonaptera *(Zool.)*. An order of wingless Insects ectoparasitic on warmblooded animals. Laterally compressed, mouthparts for piercing and sucking, coxae large, tarsi 5-jointed with prominent claws. Metamorphosis complete, larvae legless, pupae exarate and enclosed in a cocoon. Serve as vectors for some diseases, e.g. myxomatosis, plague. Fleas. Also *Aphaniptera*.

siphoneous, siphonaceous *(Bot.)*. Having large, tubular, multinucleate cells without cross walls; coenocytic.

siphonogamy *(Bot.)*. Reproductive process in which non-motile male nuclei are carried to the egg cell through a pollen tube, as in conifers and angiosperms. Cf. **zooidogamy**.

siphonostele *(Bot.)*. A stele with a more or less continuous ring of vascular tissue surrounding a pith, i.e. a medullated protostele or a solenostele.

siphon spillway *(Civ.Eng.)*. A siphon connecting the upstream and the downstream sides of a reservoir dam, thus enabling flood waters to pass, as in the case of a **bye-channel**.

sipho-, siphono- *(Genrl.)*. Prefix from Gk. *siphōn*, gen. *siphōnos*, tube.

siphuncle *(Zool.)*. In Nautiloidea, a narrow vascular tube extending from the visceral region of the body through all the chambers of the shell to its apex. adj. *siphunculate*.

Siporex *(Build.)*. TN for a material of light density and high insulation value manufactured with sand, cement and a catalyst and cured under high pressure steam conditions.

Sipunculida *(Zool.)*. A phylum of marine worm-like animals without segementation and an anterior introvert with a mouth surrounded by tentacles. The gut is U-shaped and opens dorsally. Mostly detritus feeding burrowing forms.

siren *(Acous.)*. A powerful source of noise of a more or less pure tone; the noise is usually generated by the periodic escape of compressed air through a rotary shutter.

Sirenia *(Zool.)*. An order of large aquatic Mammals of herbivorous habit; the fore limbs are finlike, the hind limbs lacking; there is a horizontally flattened tail-fin; the skin is thick with little hair and underlying blubber, there are two pectoral mammae, no external ears, and the neck is very short. Manatees, Dugongs.

sisal *(Textiles)*. The bast fibre from the sisal plant (*Agave sisalana*). Grown in Mexico and used to make string and cord but being displaced for many uses by polyethylene or polypropylene.

SISD *(Comp.)*. *Single Instruction stream, Single Data stream*. Term which describes the architecture of the **processor**.

sister cell *(Biol.)*. One of the two cells formed by the division of a pre-existing cell.

sister-chromatid exchange *(Biol.)*. Reciprocal exchange of DNA between the chromatids of a single chromosome.

sister nucleus *(Biol.)*. One of the two nuclei formed by the division of a pre-existing nucleus.

SIT *(Eng.)*. Abbrev. for *Spontaneous Ignition Temperature*.

site error *(Telecomm.)*. In radio direction finding, that due to distortions in the electromagnetic field caused by obstructions in the vicinity of the navigational antenna system.

Site of Special Scientific Interest *(Ecol.)*. In Britain, an area deemed by the Nature Conservancy Council to be important for nature conservation, but not large enough to become a *National Nature Reserve*.

site rivet *(Eng.)*. See **field rivet**.

site-specific mutagenesis *(Biol.)*. The possibility of altering a DNA sequence at a defined position by, for example, synthesizing an alternative sequence and reinserting into its host chromosome.

sitfast *(Vet.)*. A small hard lump on the skin of a horse's back, due to necrosis of the skin caused by pressure of the saddle or harness.

sitosterol *(Chem.)*. The β form $C_{29}H_{50}O$, a sterol derivative, found in corn oil, which closely resembles cholesterol. Also occurs in other forms; α_1-, α_2-, α_3-, and γ-sitosterol.

situs inversus *(Med.)*. *Dextrocardia* and transposition of abdominal viscera.

SI units *(Genrl.)*. System of coherent metric units (*Système International d'Unités*) proposed for international acceptance in 1960. It was developed from the Giorgi/MKSA system (see **MKSA**) by the addition of the **kelvin** and **candela**, and later the **mole**, as base units. There are numerous *derived units* (**newton**, **joule** etc) and scales of decimal multiples and submultiples, all with agreed symbols.

six-phase *(Elec.Eng.)*. A term applied to circuits or systems of supply making use of six alternating voltage phases, vectorially displaced from each other by $\pi/3$ radians.

Six's thermometer *(Meteor.)*. A form of **maximum and minimum thermometer** consisting of a bulb containing alcohol joined to a capillary stem bent twice through 180°C. A long thread of mercury is in contact with the alcohol in the stem, and this mercury moves as the alcohol in the bulb expands and contracts. Each end of the mercury thread pushes a small steel index in front of it, one of which registers the maximum temperature and the other the minimum.

16 mm *(Image Tech.)*. A narrow gauge of motion picture film, widely used for *non-theatrical* production and presentation.

16 mo *(Print.)*. The 16th of a sheet or a sheet folded 4 times to make 16 leaves or 32 pages. Also called *sextodecimo*.

six-twelve potential *(Chem.)*. See **Lennard-Jones potential**.

64 mo *(Print.)*. The 64th part of a sheet or a sheet folded 6 times to make 64 leaves or 128 pages. More than 4 folds are not practicable and the sheet would be cut into 4 parts, each being folded 4 times.

size *(Build.)*. Material added to surfaces to equalize porosity prior to hanging a wall-covering. Can be either glue or cellulose size.

size *(Textiles)*. Film-forming substances (e.g. starch) sometimes containing lubricants that are applied to warp threads to protect them during weaving.

size distribution *(Powder Tech.)*. Proportions of each size of particles in a powder or colloidal system. Expressed either as cumulative or frequency distribution.

size-exclusion chromatography *(Biol.)*. Chromatographic separation of particles according to size by a solid phase of material of restricted pore size which functions as a molecular sieve. Such materials include dextran, cross-linked agarose and, for **HPLC**, coated silica.

size fraction *(Powder Tech.)*. A portion of a powder composed of particles between two given size limits, expressed in terms of the weight, volume, surface area or number of particles.

size-grading *(Chem.)*. The process of determining the frequency distribution of particles of different sizes in a material.

Sjogren's disease *(Immun.)*. Chronic inflammatory disease of salivary and lacrimal glands often accompanied by inflammation of the conjunctiva. Blood of persons with this condition often contains anti-nuclear antibodies and rheumatoid factor, as well as autoantibody reactive to the duct epithelium of the affected glands.

Sjogren's syndrome *(Med.)*. Progressive destruction of *salivary* and *lachrimal* tissue to give dry mouth and eyes. Probably an **auto-immune** process. It is associated with **rheumatoid arthritis** and **systemic lupus erythematosus**.

skarn *(Geol.)*. A rock containing calcium silicate minerals produced by metasomatic alteration of limestone close to the contact of an igneous intrusion. Skarns are sometimes enriched in ore minerals such as magnetite or scheelite. *Tactite* is a synonymous US term.

skate *(Civ.Eng.)*. See **retarder**.

skeletal muscle *(Zool.)*. See **voluntary muscle**.

skeleton *(Surv.)*. The network of survey lines providing a figure from which the shape and salient features of the survey may be determined.

skeleton *(Zool.)*. The rigid or elastic, internal or external, framework, usually of inorganic material, which gives support and protection to the soft tissues of the body and provides a basis of attachment for the muscles, forming a system of jointed levers which they can move. adjs. *skeletal, skeletogenous*.

skeleton movement *(Horol.)*. A clock or watch movement the plates of which have been pierced or cut away to show the mechanism.

skeleton steps *(Build.)*. Steps in a stair, of a construction such that there are no risers but only treads fixed at suitable positions above one another between side supporting pieces.

skeleton-type switchboard *(Elec.Eng.)*. A switchboard consisting of a metal framework upon which the switches and other apparatus are mounted. Also called *frame-type switchboard*.

skelp *(Eng.)*. Mild steel strip from which tubes are made by drawing through a bell at welding temperature, to produce lap-welded or butt-welded tubes.

skene arch *(Civ.Eng.)*. An arch having the shape of a circular arc subtending less than 180°. Also *scheme arch*.

skew arch *(Arch.,Civ.Eng.)*. An arch which has its axis or line of direction oblique to its face.

skewback *(Civ.Eng.)*. The courses of stones from which an arch springs (on the top of a pier) where the upper and lower beds are oblique to each other.

skewbacks *(Glass)*. Specially shaped blocks, supporting a furnace crown or other arch. Also *springers*.

skew bevel gear *(Eng.)*. See **hypoid bevel gear**.

skew butt *(Build.)*. See **skew corbel**.

skew coil *(Elec.Eng.)*. An assymmetrical coil inserted in the armature winding of an alternator having an odd number of pole pairs.

skew corbel *(Arch.)*. The projecting masonry or brickwork supporting the foot of a gable coping.

skewed pole *(Elec.Eng.)*. A field pole whose cross-section is a parallelogram instead of the usual rectangle.

skewed slot *(Elec.Eng.)*. A slot whose diameter is not parallel to the axis of rotation.

skew fillet *(Build.)*. See **tilting fillet**.

skew flashing *(Build.)*. Flashing fixed down a gable wall.

skew lines *(Maths.)*. Two lines in 3-dimensional space which do not lie in the same plane.

skew nailing *(Build.)*. The operation of driving nails in obliquely.

skewness *(Stats.)*. The degree of asymmetry about the central value of a distribution.

skew rebate plane *(Build.)*. A rebate plane with its cutting edge arranged obliquely across the sole.

skew table *(Build.)*. A stone which is bonded in with a gable wall, as a support for the foot of the coping.

skew wall *(Acous.)*. In a studio, a wall which does not form a face of a parallelepiped; the walls are so arranged that continuous reflections between opposite walls are obviated.

skiagram, skiagraph *(Radiol.)*. Same as *radiograph*.

skiatron *(Electronics)*. A cathode-ray tube having a *dark-trace screen*.

skids *(Build.)*. Small pieces of timber packed under a surface to bring it to the plane.

ski jump ramp *(Aero.)*. Curved ramp fitted at the forward flight deck of an aircraft carrier to give improved take-off performance to vectored-thrust V/STOL aircraft. The end slope is typically 8 to 15° and allows (a) shorter take-off at given weight, (b) take-off at higher weight and (c) operations in higher wave states.

skillet *(Eng.)*. Mould for casting bullion.

skimming coat *(Build.)*. See **setting coat**.

skin *(Aero.)*. The outer surface other than fabric of an aircraft structure; the outer surface in a **sandwich construction**.

skin *(Eng.)*. The hard surface layer found on iron castings due to the rapid cooling effect of the mould, or on steel plates, strip and sheet, due to rolling, or on other materials or products due to the surface-hardening effect of the finishing process.

skin *(Zool.)*. The protective tissue layers of the body wall of an animal, external to the musculature. See **epidermis**.

skin depth *(Phys.)*. The distance into a conductor at which the amplitude of an electromagnetic wave falls to $1/e = 0.3679$ of its value at the surface, given by

$$\delta = \sqrt{\rho/\pi f \mu}.$$

If the magnetic field at the surface is H, the power dissipated is $\frac{1}{2}H^2 R_s$ W/m^2, where the effective surface resistance $R_s = \rho/\delta$, ρ is the resistivity and μ the absolute permeability of the metal, and f is the frequency of the wave. See **depth of penetration, skin effect**.

skin depth *(Telecomm.)*. Owing to **skin effect** the current density in a conductor, induced by a high-frequency electromagnetic field, falls off rapidly below the surface. *Skin depth* is that at which the current density has decreased by one **neper** compared with the current density at the surface.

skin dose *(Radiol.)*. Absorbed or exposure radiation dose received by or at the skin of a person exposed to sources of ionization. Cf. **tissue dose**.

skin effect *(Elec.Eng.)*. That phenomenon by which high frequency currents tend to be confined to the thin layer (skin) of conductors. It follows, from $R = \frac{\rho l}{A}$, that the resistance of a given conductor will be higher at high frequencies, because of the effective reduction in cross-sectional area for the current to flow. See **depth of penetration, litz wire, skin depth**.

skin friction *(Aero.)*. See **surface-friction drag**.

skinner *(Elec.Eng.)*. The length of insulated wire between the point of connection to a solder tag and the cable form from which it emerges.

Skinner Box *(Behav.)*. A chamber designed for the study of **operant conditioning**; it is provided with mechanisms for an animal to operate and an automatic device for presenting rewards according to **schedules of reinforcement** preset by the experimenter.

skin sensitizing antibody *(Immun.)*. Antibody capable of attachment to skin cells so that, on subsequent combination with an antigen, an immediate type hypersensitivity reaction (Type I) occurs. Mainly IgE.

skin test *(Immun.)*. Any test in which substances are injected into or applied to the skin in order to observe the host's response to them. Used extensively in the study of hypersensitivity and immunity, e.g. tuberculin test, Schick test.

skintled *(Build.)*. Said of brickwork in which the bricks are laid irregularly, so as to leave an uneven surface on the wall; also of a similar effect produced by protruding mortar squeezed from the joints.

skin tuberculosis of cattle *(Vet.)*. A benign lymphangitis producing nodules in the skin of cattle, caused by bacteria which resemble in some respects the causative organism of true tuberculosis.

skin wool *(Textiles)*. Wool removed from the fleeces of slaughtered sheep by chemical or biochemical processes.

skip *(Telecomm.)*. See **hop**.

skip distance *(Telecomm.)*. Region of no-signal between the limit of reception of the direct (*ground*) wave and the first downcoming reflection from an ionized layer; prominent with short waves.

skip printing *(Image Tech.)*. Printing only selected frames of a motion picture film, omitting those in between; usually at regular intervals, such as every third frame, to produce an apparent increase in the speed of movement.

skip slitter *(Print.)*. A cam-operated slitter giving intermittent cuts, whereby both broadsheet and tabloid sections can be incorporated in a single copy by means of cylinder collection.

skip-tooth saw *(Build.)*. A saw from which alternate teeth are cut away.

skirt *(Chem.Eng.)*. A cylindrical support on large vessels and fractionating columns welded to the main part of the cylindrical shell and enclosing the bottom of the vessel or column.

skirt *(Telecomm.)*. Lower side portions of a resonance curve, which should be symmetrical.

skirting board *(Build.)*. A board covering the plaster wall where it meets the floor. Also called *baseboard*, *mopboard*, *washboard*.

skirting board *(Min.Ext.)*. At delivery on to belt conveyor, boards which direct falling rock toward centre of belt, or prevent spill-over.

skittle pot *(Glass)*. A small pot, in shape resembling a skittle, which can be set in a furnace in some small corner to melt a special glass, e.g. a colour. Some small firms use a furnace holding only 4 or 6 of these, fired by coke.

skot *(Phys.)*. Obsolete unit of scotopic luminance. $1\,\text{skot} = 3.1931 \times 10^{-4}\,\text{cd}\,\text{m}^{-2}$.

skotograph *(Image Tech.)*. A developable image produced in a photographic emulsion by radiation from organic tissue in the dark.

Skraup's synthesis *(Chem.)*. The synthesis of quinoline by heating aminobenzene with glycerine and sulphuric acid, nitrobenzene or arsenic (V) acid acting as oxidizing agent.

skull *(Zool.)*. In Vertebrates, the brain case and sense-capsules, together with the jaws and the branchial arches.

skutterudite *(Min.)*. Grey or whitish arsenide of cobalt, which crystallizes in the cubic system. Often contains appreciable nickel and iron substituting for cobalt.

Skylab *(Space)*. A manned laboratory placed in orbit in May 1973; the largest part (the workshop) consisting of a refurbished, empty S-IVb tank, originally forseen as a stage of the Saturn rocket. Launched unmanned, it was visited by a crew of 3 astronauts on 3 occasions for stay-times of 28, 56 and 56 days. It re-entered the Earth's atmosphere in July 1979.

sky wave, ray *(Telecomm.)*. See **ionospheric wave, space wave**.

SL *(Build.)*. Abbrev. for *Short Lengths*.

slab *(Build.)*. An outer piece of a log cut away in the process of slabbing.

slab *(Civ.Eng.)*. (1) Normally a reinforced concrete floor supported, at intervals, on beams and/or columns. (2) A thin flat piece of concrete or stone (not necessarily flat in the case of concrete).

slab *(Eng.)*. Metal, twice as wide as thick, and thus intermediate between an ingot and a plate in a rolling mill.

slab and girder floor *(Civ.Eng.)*. A reinforced concrete floor in which the beams are designed to be homogeneous with the slab and to provide continuity and reduce the amount of concrete and steel.

slabbing *(Build.)*. The operation of squaring a log.

slabbing *(Print.)*. Finishing operations on electrotypes, low areas being made level by punching up from the back.

slab coil *(Elec.Eng.)*. A coil in the form of a flat spiral; the term is normally applied to inductance coils.

slab resolver *(Electronics)*. Four contacts on a square, giving ($\pm R\cos\theta$) and ($\pm R\sin\theta$), concentric with a slab potentiometer fed with $\pm R$ at its ends, θ being angular displacement.

slab serif *(Print.)*. A type style with strokes of uniform thickness and with straight serifs of the same thickness as the strokes. Synonymous with *Egyptian*.

slab tail *(Aero.)*. A one-piece horizontal tail surface, pivoted and power operated so as to serve as a stabilizing tail plane, elevator and, through a lower gearing, trimming tab.

SLAC *(Phys.)*. *Stanford Linear Acceleration Centre*, US. SLAC refers to the *linear accelerator* which at the centre gives electron beams of energy about 50 GeV.

slack *(Min.Ext.)*. Small coal dirt, as in *slack heap*, a tip or dump.

slack blocks *(Civ.Eng.)*. See **striking wedges**.

slack sheet *(Print.)*. Insufficient tension of the web between units of the press.

slack-water navigation *(Civ.Eng.)*. River or canal navigation rendered possible by the construction of dams across the stream at intervals, dividing it into separate reaches, communication being maintained by the use of locks. Also called *still-water navigation*.

slade *(Build.)*. An inclined pathway.

SLAET *(Aero.)*. Abbrev. for *Society of Licensed Aircraft Engineers and Technologists*.

slag *(Eng.)*. The top layer of the 2-layer melt formed during smelting and refining operations. In smelting it contains the gangue minerals and the flux; in refining, the oxidized impurities.

slag cement *(Civ.Eng.)*. An artificial cement made by granulating slag from blast furnaces by chilling it in water and then grinding it with lime, to which it imparts hydraulic properties.

slag hole *(Eng.)*. In smelting furnace, aperture above that through which molten metal is withdrawn, used to top off accumulated slag.

slag wool *(Eng.)*. See **rock wool**.

slaked lime *(Build.,Chem.)*. See under **caustic lime**.

slaking *(Build.)*. The process of combining quicklime with water.

slamming stile *(Build.)*. The upright member of a door case against which the door shuts and into which the bolt of a rim lock engages.

slant azimuth *(Image Tech.)*. In magnetic recording, head **azimuth** settings other than 90° to minimize cross-talk between adjacent tracks.

slant range *(Surv.)*. Distance along sloping sight between two points as measured by tacheometer or tellurometer.

slant rig *(Min.Ext.)*. One with special facilities for drilling at an angle from the vertical. The head-string components are hauled up rails set at the angle.

slap dash *(Build.)*. A rough finish given to a wall by coating it with a plaster containing gravel or small stones.

slat *(Aero.)*. An auxiliary aerofoil which constitutes the forward portion of a slotted aerofoil, the space between it and the main portion of the structure forming the slot.

slat *(Build.)*. A thin, flat strip of wood.

slat conveyor *(Eng.)*. See **apron conveyor**.

slate *(Geol.)*. A fine-grained metamorphic rock with good fissility along cleavage planes.

slate axe *(Build.)*. See **sax**.

slate boarding *(Build.)*. Close boarding laid as an

underlining to, and a support for, roofing slates. Termed *sarking* in Scotland.

slate cramp *(Build.)*. A piece of slate about 7 × 2¼ × 1 in cut to a narrow waist at the middle, and fitted flush into mortises in adjacent stones to bind them together.

slate hanging *(Build.)*. Similar to **weather tiling**, slates being used instead of tiles.

slate, spotted *(Geol.)*. An argillaceous rock altered by low or moderate grade metamorphism to produce porphyroblasts which impart a speckled appearance to the rock.

slating and tiling battens *(Build.)*. Any pieces of square-sawn converted timber between ½ and 1½ in in thickness and from 1 to 3½ in in width; commonly used as a basis for slating and tiling.

slatted lens *(Telecomm.)*. One with shaped metal slats, parallel to E or H vector in wave from waveguide. Also used for low-frequency acoustic waves. Also called *egg-box lens*.

slaty cleavage *(Geol.)*. The property of splitting easily along regular, closely spaced planes of fissility, produced by pressure in fine-grained rocks.

sleber number *(Paper)*. A measure of the bleachability of a pulp determined by mixing 5 g of a sample in water with 50 ml of bleach solution and allowing to stand for one hour at 20°C. Thereafter the amount of available chlorine consumed is measured by titrating the filtrate.

sledge-hammer *(Build.)*. A heavy double faced or straight-pane hammer, weighing up to 100 lb (45 kg), swung by both hands.

sledger *(Civ.Eng.)*. A machine for the first stage of crushing of rock in quarrying. Also *scalper*.

sleekers *(Foundry)*. Moulders' tools having a smoothing face made to various shapes, for smoothing over small irregularities in the sand of the mould. Also called *smoothers*. See **corner tool**.

sleep *(Behav.)*. A state characterized by prolonged periods of immobility, often with an associated and species-typical posture, and an increased reluctance to respond to stimulation; most animals sleep during a particular period of the day and for a species-characteristic duration.

sleeper *(Build.)*. A horizontal timber supporting a vertical shore or post, and distributing the load over the ground. Also *sole plate*.

sleeper *(Civ.Eng.)*. US *tie*. A timber or prestressed concrete beam passing transversely beneath the railway track for support and gauge maintenance. Also called *cross-tie*, *cross-sill*. Also, a *sleeping-car*.

sleeper plate *(Build.)*. A wall plate resting upon a sleeper wall.

sleeper wall *(Build.)*. A low wall built under the ground storey of buildings having no basement, as a support for the floor joists. When in brick, the wall is built honeycombed to leave spaces for ventilation, and when in stone, small piers at intervals provide the support required.

sleepiness *(Build.)*. A defect similar to loss of gloss.

sleeping sickness *(Med.)*. See **trypanosomiasis**.

sleep movement *(Bot.)*. See **nyctinastic movement**.

sleet *(Meteor.)*. A mixture of rain and snow, or partially melted snow. US *ice pellets*.

sleeve *(Elec.Eng.)*. See **plain coupler**.

sleeve *(Eng.)*. A tubular piece, usually one machined externally and internally.

sleeve antenna *(Telecomm.)*. A single vertical half-wave rectifier, whose lower half is a metal sleeve through which the concentric feed line runs. The upper radiating portion, one quarter wavelength long, connects to the centre of the line.

sleeve dipole *(Telecomm.)*. Same as *skirt dipole*.

sleeve joint *(Elec.Eng.)*. A conductor joint formed by a sleeve fitting over the conductor ends. It is either pinned or soldered to the conductors.

sleeve piece *(Build.)*. A brass or copper pipe joint used between a lead pipe and one of another material. Also called *thimble*.

sleeving *(Elec.Eng.)*. Tubular flexible insulation for threading over bare conductors.

slenderness ratio *(Build.)*. See **ratio of slenderness**.

s-levels *(Phys.)*. See **sharp series**.

slewing rollers *(Print.)*. Rollers whose angle can be adjusted to alter the run of the web to maintain sidelay register, manual control being superseded by electronic equipment.

slew rate *(Electronics)*. The maximum rate of change of output voltage for an amplifier when a voltage step is applied at the input. Normally measured in V/μs.

sley *(Textiles)*. (1) Heavy metal or wooden beam carrying the shuttle race board, reed, and shuttle-boxes. As it moves backwards the shuttle is picked from one box to the other carrying the weft. When the sley comes forward the reed beats-up the weft to form cloth. (2) The part of a lace machine, between the beams and the thread guides, which functions in keeping the threads properly arranged.

slice *(Paper)*. The part of the flow box, head box or breast box where the stock is discharged over the apron to the machine wire. It generally takes the form of a hinged lip with numerous control points to ensure the right flow of stock and even level across the machine.

slice *(Radiol.)*. The cross-sectional portion of the body which is scanned for the production of e.g. a **computed tomography** or **magnetic resonance** image.

slicing *(Min.Ext.)*. Removal of a layer from a massive ore body. In top slicing this is horizontal, a mat of timber separating it from the overburden. Side and bottom slicing are also practised.

slick bit *(Build.)*. Type of interchangeable scraping bit attached to a shank for use in electric drills.

slickensides *(Geol.)*. Smooth, grooved, polished surfaces produced by friction on fault-planes and joint faces of rocks which have been involved in faulting.

slicker *(Foundry)*. A small implement used by a moulder for smoothing the surface of a mould.

slide *(Image Tech.)*. A still-picture transparency in a mount of standard size, such as 5 × 5 cm, intended for viewing by projection.

slide *(Min.Ext.)*. Crack or plane along which movement has taken place; the clay filling of such a crack; a fault.

slide bars *(Eng.)*. See **guide bars**.

slide resistance *(Elec.Eng.)*. A rheostat whose ohmic value is adjusted by sliding a contact over the resistance wire.

slide rest *(Eng.)*. A slotted table carrying the tool post of a lathe. It is mounted on the saddle or carriage, and is capable of longitudinal and cross traverse.

slide rule *(Genrl.)*. A device for the mechanical performance of arithmetical processes, as multiplication, division etc. It consists of one rule sliding within another, so that their adjacent similar logarithmic scales permit of the addition and subtraction, corresponding to the multiplication and division, of the numbers engraved thereon.

slide-valve *(Acous.)*. The slide, containing holes, which is drawn across the supply of air to a rank of organ pipes, to stop the pipes speaking when a key is depressed. Operated by draw-stops through trackers and stickers.

slide-valve *(Eng.)*. A steam-engine inlet and exhaust valve shaped like a rectangular lid. It is reciprocated inside the steam chest, over a face in which steam ports are cut, so as alternately to admit steam to the cylinder and connect the ports to exhaust through the valve cavity. See **D slide-valve**, **piston valve**.

slide-valve lead *(Eng.)*. The amount by which the steam port of a steam-engine is already uncovered by the valve when the piston is at the beginning of its working stroke.

slide wire *(Elec.Eng.)*. One for potential division, by sliding a contact along a wire. With a concentric tube with a slot and a probe contact, this becomes a coaxial slotted line for voltage standing-wave ratio measurements at high frequencies.

sliding bevel *(Build.)*. See **bevel**.

sliding caisson *(Hyd.Eng.)*. A floating body used to open or shut the entrance to a dock or basin, and capable of being drawn for the former purpose into a recess at right angles to the channel.

sliding contact *(Elec.Eng.)*. Tangential movement between contacting metal surfaces, to remove film and establish conduction contact. Wear of contact is proportional to total use.

sliding growth *(Bot.)*. The sliding of the wall of one cell past that of the next, as has been postulated to occur during the growth of fibres.

sliding-mesh gear-box *(Autos.)*. One in which the ratio is changed by sliding one pair of wheels out of engagement and sliding another pair in.

sliding sash *(Build.)*. A sash that moves horizontally on runners, as distinct from a *balanced sash* sliding vertically.

sliding ways *(Ships)*. The portion of a ship's launching ways which move with the ship on launching.

slime mould *(Bot.)*. See **Myxomycota**.

slime plug *(Bot.)*. Accumulation of P-protein on a sieve area.

slimes *(Min.Ext.)*. Particles of crushed ore which are of such a size that they settle very slowly in water and through a bed which water does not readily percolate. Such particles must be leached by agitation. By convention these particles are regarded as less than 1/400 in (0.0635 mm) in diameter (mesh number 200). *Primary slimes* are naturally weathered ore, or associated clays. *Secondary slimes* are produced during comminution. See **anode slime**.

slip *(Build.)*. A long narrow piece of wood of the full thickness of a mortar joint, built into surface brickwork to provide material to which joinery may be nailed.

slip *(Elec.Eng.)*. The fraction by which the rotor speed of an induction motor is less than the speed of rotation of the stator field, i.e. the ratio of the speed at which the rotor slips back from exact synchronism with the stator field to the latter's speed of rotation.

slip *(Eng.)*. The process involved in the plastic deformation of metal crystals in which the change in shape is produced by parts of the crystals moving with respect to each other along crystallographic planes.

slip *(Image Tech.)*. Vertical shift of TV image because of imperfect *field* synchronization. Similarly for *line*, but the shift is horizontal.

slip *(Ships)*. (1) The difference between the distance travelled by the ship through the water and the distance the propeller would have moved in a solid. Expressed as a percentage of the latter. (2) A sloping masonry or concrete surface for the support of a vessel in process of being built or repaired.

slip angle *(Autos.)*. Slight angle of the tyres to the actual track of the wheels when cornering. Some slip angle is required for the tyres to generate a cornering force.

slip bands *(Eng.)*. Steps or terraces produced on the polished surface of metal crystals as a result of the parts moving with respect to each other during slip.

slip correction *(Powder Tech.)*. Correction for **Knudsen flow** in permeability equations.

slip dock *(Ships)*. A dock from which the water can be discharged, and which is equipped with a **slip** (2).

slip feather *(Build.)*. A wooden tongue for a **ploughed-and-tongued joint**, distinguished as a **straight tongue**, **feather tongue**, or **cross-tongue** according to grain direction.

slip flow *(Aero.)*. The molecular shearing which replaces normal gas flow conditions at **hypersonic** velocities above *Mach 10* where the mean free path is of the same order of dimensions as the body. See **Mach number**.

slip form *(Civ.Eng.)*. A *form* that can be moved slowly as work progresses.

slip gauge *(Eng.)*. Accurately ground and lapped rectangular block or plate used singly or in combination with others, the distance between end faces forming a gauging length.

slip joint *(Min.Ext.)*. Special **coupling** used on floating platforms, which is splined or keyed to allow relative vertical movement in the drilling pipe while transmitting rotary torque.

slip meter *(Elec.Eng.)*. A device for measuring the slip of an induction motor.

slip mortise *(Build.)*. A *slot mortise* or a *chase mortise*.

slipped bank multiple *(Telecomm.)*. A bank of outgoing trunks so connected that they are tested in the same order by all switches but starting from different points, unlike the *straight bank multiple*.

slipped tendon *(Vet.)*. See **perosis**.

slipper *(Civ.Eng.)*. See **retarder**.

slipper brake *(Elec.Eng.)*. An electromechanical brake acting directly on the rails of a tramway.

slipper piston *(Autos.)*. A light piston having the lower part or skirt cut away between the thrust *faces*, to save weight and reduce friction.

slipper satin *(Textiles)*. A heavy, smooth, high-quality satin made from continuous filament yarns (silk or man-made) used particularly for wedding dresses and evening shoes.

slipper tank *(Aero.)*. An **auxiliary tank** mounted externally, close up under wing or fuselage.

slip planes *(Crystal.)*. The particular set or sets of crystallographic planes along which slip takes place in metal and other crystals. These are usually the most widely spaced set or sets of planes in the crystals concerned. See **gliding planes**.

slip proof *(Print.)*. A proof taken from a galley of type matter before it is made up into pages.

slip regulator *(Elec.Eng.)*. A regulating resistance connected in series with the rotor of a slip-ring type of induction motor to alter the slip, and thus vary the speed of the machine.

slip-ring rotor *(Elec.Eng.)*. The rotor of a slip-ring induction motor; it has a 2- or 3-phase winding brought out to slip-rings. Cf. **cage rotor**.

slip-rings *(Elec.Eng.)*. The rings mounted on, and insulated from, the rotor shaft of an a.c. machine, which form the means of leading the current into or away from the rotor winding.

slips *(Print.)*. Name given to ends of tapes or cords after sewing and before covering.

slip sheets *(Print.)*. Sheets of paper, oiled manilla being very suitable, placed between sheets as they are printed to avoid **set-off**; largely superseded by **anti-set-off spray**. Also name given to sheets placed in a letterpress machine packing.

slip sill *(Build.)*. A sill of length equal to the distance between the jambs of the opening, so that it can be placed in position after the shell of the building has been completed.

slip stone *(Build.)*. See **gouge slip**.

slipstream *(Aero.)*. The helical airflow from a propeller, faster than the aircraft.

slip tank *(Aero.)*. Alternative to **drop tank**.

slipway *(Ships)*. See **slip** (2).

slit *(Image Tech.)*. See **scanning-**.

slitless spectroscope *(Phys.)*. See **objective prism**.

slitter marks *(Print.)*. Two marks in the centre gutter of a printed sheet as a guide to slitting either on the printing machine or on the guillotine.

slitters *(Print.)*. Circular rotating knives used (1) to separate printed sheets on sheet-fed printing of folding machines, (2) to divide the printed web on reel-fed rotaries.

sliver *(Textiles)*. Continuous untwisted strand of fibres as formed by carding or combing.

slope *(Build.)*. See **splay brick**.

slope *(Electronics)*. See **mutual conductance**.

slope *(Maths.)*. A measure of the inclination of a line with respect to a fixed line, usually horizontal. On a graph, the tangent of the angle between a line and the x-axis. Also called *gradient*.

slope correction *(Surv.)*. A correction applied to the observed length of a baseline to correct for differences of level between the ends of the measuring tape.

slope deflection method *(Civ.Eng.)*. A method of structural analysis normally used where only the bending moment at every point is evaluated in terms of the loads applied. The reactions are then determined and the analysis can be completed by the application of normal statics.

sloped roman *(Print.)*. Italic founts in which the letters are not cursive but have the same form as the roman.

slope staking *(Surv.)*. The locating and pegging of points at which proposed earth slopes in cutting or bank will meet the original ground surface.

sloshing *(Space)*. Bulk motion of liquid propellants in their tanks when subject to accelerations, particularly at launch. This must be accounted for in the structural design of a rocket vehicle.

slot *(Elec.Eng.)*. An axially-cut trench, cut out of the periphery of the stator or rotor of an electrical machine, into which the current-carrying conductors forming the winding are embedded.

slot antenna *(Telecomm.)*. Radiating element formed by metal surrounding a slot.

slot-fed dipole *(Elec.Eng.)*. One normal to a coaxial line, and coupled to it by adjacent longitudinal slots.

slot leakage *(Elec.Eng.)*. In an electrical machine, the leakage flux that passes across the slots.

slot link *(Eng.)*. See **link motion, radial valve gear**.

slot mortise *(Build.)*. A mortise made in the end of a member.

slot permeance *(Elec.Eng.)*. The total permeance of the several parallel portions of the slot-leakage flux path.

slot pitch *(Elec.Eng.)*. The distance between successive slots around an armature.

slot ripple *(Elec.Eng.)*. The harmonic ripple in the e.m.f. wave of an electrical machine. It arises from the regular variation in the permeability of the magnetic path between stator and rotor, caused by the repeated change in air-gap length as the rotor and stator slots pass one another and the intervening teeth.

slotted aerofoil *(Aero.)*. Any aerofoil having an air passage (or slot) directing the air from the lower to the upper surface in a rearward direction. Slots may be permanently open, closable, automatic, or manually operated.

slotted core *(Elec.Eng.)*. The usual type of armature core, in which slots are provided for the windings. Cf. *smooth core*.

slotted flap *(Aero.)*. A trailing-edge *flap* which opens a slot between itself and the main aerofoil as it is lowered or extended.

slotted line *(Telecomm.)*. Rigid coaxial line, with slot access for contact with central conductor. Used for impedance and voltage standing-wave ratio measurements at wavelengths comparable with length of slotted line.

slotting machine *(Eng.)*. A machine tool resembling a **shaping machine** but in which the ram has a vertical motion and is balanced by a counterweight, the tool cutting on the down stroke, towards the table.

slotting tools *(Eng.)*. Cutting tools used for keyway cutting etc. in a **slotting machine**; they are of narrow edge and deep, stiff section, with top and side clearance but little rake.

slough *(Med.)*. A mass of dead, soft, bodily tissue in a wound or infected area; to form dead tissue (said of the soft parts of the body); to come away as a slough.

slough *(Zool.)*. The cast-off outer skin of a Snake.

slow *(Acous.)*. Measuring mode of a **sound-level meter** with a time constant of one sec.

slow-acting relay *(Telecomm.)*. A relay designed to operate at an appreciable time after the application of voltage. A copper sleeve or slug is placed over the core, or a short-circuited winding is used.

slow-break switch *(Elec.Eng.)*. A knife-switch with a single rigid blade forming the moving part of each pole.

slow-butt welding *(Elec.Eng.)*. See **resistance upset-butt welding**.

slowing down area *(Nuc.Eng.)*. In reactor physics calculation, one sixth of the mean square distance travelled by neutrons from their source to reach thermal energy. See **Fermi age**.

slowing down density *(Nuc.Eng.)*. In reactor theory, the rate at which neutrons slow down past a given energy, per unit volume.

slowing down length *(Nuc.Eng.)*. The square root of the slowing down area.

slowing down power *(Nuc.Eng.)*. Increase in **lethargy of neutrons** per unit distance travelled in medium.

slow motion *(Image Tech.)*. Cinematography with the camera running faster than the standard rate of projection, so that the action will be shown apparently slowed down. A similar effect can be obtained in videotape recording.

slow neutron *(Phys.)*. See **neutron**.

slow-reacting substance A *(Immun.)*. A pharmacologically active material comprising leukotrienes C, D, and E, released by mast cells and other cells in the course of immediate hypersensitivity reactions. Causes contraction of smooth muscle, especially bronchial muscle, and increased vascular permeability. The term is that applied when it was first discovered.

slow-running cut-out *(Aero.)*. See **fuel cut-off**.

slow scan TV *(Image Tech.)*. A television system in which the sequential field scanning rate is much slower than in broadcasting standards, thus allowing picture transmission by telephone line although with imperfect reproduction of movement.

slow virus *(Med.)*. A transmissable agent which causes disease after very long incubation periods often of many years, e.g. **Kuru** in Man, **scrapie** in Sheep.

slow-wave sleep *(Behav.)*. *Quiet sleep, NREM sleep*. Sleep characterized by the presence of high amplitude, slow wave changes in potential, as measured on the EEG.

slow-wave structure *(Elec.Eng.)*. One in which the apparent velocity of a high-frequency wave is markedly retarded, e.g. in the helix in a travelling-wave tube or in a spiral cable.

slow-wave structure *(Telecomm.)*. A circuit or transmission path where the **phase velocity** is much less than the velocity of light. For example, in a helix, the velocity reduction is in the ratio of the pitch of the helix to the circumference. Used to match an electromagnetic wave to the slower electron beam in, e.g. a *travelling wave tube*.

SLR *(Image Tech.)*. Abbrev. for *Single-Lens Reflex*.

slub *(Textiles)*. A fault in any drafted, twisted yarn, which appears as a thicker part, with little twist. Slubs can also be deliberately made in yarn at regular or random intervals for ornamentation and special weave effects. They occur naturally in **tussah** silk.

slubbings *(Textiles)*. Relatively thick, open but coherent strands of fibres; converted into yarn by subsequent spinning processes.

sludge *(Build.)*. A slime produced by the precipitation of solid matters from liquid sewage in sedimentation tanks. See **sewage gas**.

sludger *(Civ.Eng.)*. A long cylindrical tube, fitted with a valve at the bottom and open at the top, used for raising the mud which accumulates in the bottom of a boring during the sinking process. Also called *sand pump, shell pump*.

sludging *(Hyd.Eng.)*. (1) Free-running mud. (2) The process of filling the crevices left in the dried clay of an embankment formed by the method of **flood flanking**.

slug *(Glass)*. Any nonfibrous glass in a glass-fibre product.

slug *(Nuc.Eng.)*. Unit of fuel in nuclear reactor, either rod or slab of fissile material encased in a hermetic can of Al, Be, Magnox, Zr, or stainless steel. Also *cartridge*. See **fuel rod**.

slug *(Phys.)*. Unit of mass in the *gravitational system* of units. A force of 1 lbf (pound-force) acting on a mass of 1 slug gives it an acceleration of 1 ft s^{-2}. See **fundamental dynamical units**.

slug *(Print.)*. A solid line of type as cast by the Linotype, Intertype or Ludlow.

slug *(Telecomm.)*. Thick copper band, comparable with a portion of a winding on a telephone-type relay which, through induced eddy currents, retards the operation and fall-off of the relay.

slug tuning *(Telecomm.)*. Alteration of inductance in radiofrequency tuning circuits, by inserting a magnetic core or a copper disk or cylinder.

sluice *(Hyd.Eng.)*. A water channel equipped with means of controlling the flow, enabling a sudden rush of water to be used at harbours or canal locks for the purpose of cleaning out silt, mud etc., obstructing navigation.

sluice *(Min.Ext.)*. A long trough for washing gold-bearing sand, clay, or gravel. Also called *sluice box* and *launder*.

sluice gate *(Hyd.Eng.)*. A barrier plate free to slide vertically across a water or sewage channel or an opening in a lock gate, so controlling flow and enabling a sudden rush of water to be used.

sluicing *(Hyd.Eng.)*. The process of deepening a navigation channel by discharging impounded water from a reservoir into it through an open sluice.

slump *(Geol.)*. Down-slope gravity movement of unconsolidated sediments, especially in a subaqueous environment.

slump test *(Civ.Eng.)*. A test for the consistence of concrete, made with a metal mould in the form of a frustum of a cone with the following internal dimensions: bottom diameter 200 mm (8 in), top diameter 100 mm (4 in), height 300 mm (12 in). This is filled with the concrete, deposited and punned in layers 100 mm (4 in) thick, and then the mould is removed and the height of the specimen measured when it has finished subsiding.

slur *(Print.)*. A printing fault in which the image lacks sharpness, caused by drag or movement of the paper, plate or forme, blanket or image carrier or combination thereof.

slurry *(Min.Ext.)*. A thin paste produced by mixing some materials, especially Portland cement, with water, sufficiently fluid to flow viscously. Used, e.g. to repair (*fettle*) slag-eroded brickwork in smelting furnace, etc.

slurry reactor *(Nuc.Eng.)*. One in which fuel or **blanket** material exists as a slurry carried by the coolant fluid.

slushed-up *(Build.)*. A term applied to brickwork the joints of which are filled with mortar.

slushing compound *(Eng.)*. A rust-inhibiting liquid composition consisting of mineral oil and anticorrosive additives, such as barium petroleum sulphonates.

slush moulding *(Plastics)*. Method based on using certain plastics, particularly PVC, in sol form, i.e. in the form of a thin paste. This is placed in a hollow heated mould which is rotated until the paste forms the solid replica of the mould configuration. Used, e.g. for dolls' heads.

slush pulp *(Paper)*. Pulp which is pumped direct from the pulp mill to the paper mill for use without passing through the pulp drying stage.

Sm *(Chem.)*. The symbol for samarium.

small bayonet cap *(Elec.Eng.)*. A bayonet cap of about 16 mm (0.75 in) diameter; used for small lamps, e.g. automobile head and side lamps.

small-bore *(Build.)*. Term applied to pump assisted hot water central heating systems with 0.5 in or 15 mm copper or stainless steel pipes.

small capital *(Print.)*. A letter having the form of a capital but the height of a lower case letter, e.g. C; indicated in manuscript or proof by two lines under the letter. See **even small caps**.

small circle *(Maths.)*. A section of a sphere by a plane not passing through its centre.

small Edison screw-cap *(Elec.Eng.)*. An Edison screw-cap; having a screw-thread of about 12.5 mm (0.5 in) diameter and about 3.5 threads per cm.

small nuclear RNA *(Biol.)*. Discrete set of RNA molecules found in ribonucleoprotein particles (SnRNPs) which are responsible for processing **HnRNA** to give **mRNA**.

small offset *(Print.)*. term applied to sheet fed offset lithographic machines with a sheet size below that of about 375 × 500 mm.

small pica *(Print.)*. An old type size, approximately 11-point.

smallpox *(Med.)*. *Variola.* An acute, highly infectious viral disease characterized by fever, severe headache, pain in the loins, and a rash which is successively macular, papular, vesicular, and pustular, affecting chiefly the peripheral parts of the body. Until recent times, one of the major killing diseases of man but has now been eradicated.

smallpox vaccination *(Immun.)*. Method of producing active immunity against smallpox (variola), discovered by Jenner in Gloucestershire, UK in 1796. Not practised since smallpox was eliminated on a worldwide scale, but immensely important historically. Vaccinia virus is used, prepared from vesicular lesions of vaccinia in the skin of calves or sheep (more recently from cultures of amnion in vitro). This virus shares cross reacting antigens with variola virus and therefore induces protective immunity against smallpox. Vaccination is performed by introducing live vaccinia virus into a site in the superficial layers of the skin. Successful primary vaccination is characterized by development of a vesicular lesion at the site after 6–9 days, reaching a maximum on the 12th day, after which it subsides leaving a scab and later a scar. Protective immunity, which depends primarily on the ability to mount a delayed type reaction against virus infected cells, gradually declines. Re-vaccination is needed at intervals of about 3 years. If some residual immunity is present the local reaction produced by re-vaccination may be accelerated, so that a small vesicle develops at the 4–5th day, and the reaction is at its height on the 7th day. Rare complications are *generalized vaccinia*, which may occur in subjects with atopic eczema (which interferes with the development of the local delayed hypersensitivity response); *chronic progressive vaccinia* particularly if cell-mediated immunity is deficient due to disease or to drug treatment; and very rarely *post-vaccinial encephalomyelitis*, which is mostly likely to occur after primary vaccination of young infants.

smalls *(Min.Ext.)*. See **riddle**.

small scale integration *(Comp.)*. See **SSI**.

small-signal parameters *(Electronics)*. See **transistor parameters**.

smallwares *(Textiles)*. General name given to tapes, ribbons, and other narrow fabrics made on special narrow looms or braiding machines.

smaltite *(Min.)*. Cobalt arsenide, crystallizing in the cubic system and usually associated with *chloanthite*, nickel arsenide.

smaragdite *(Min.)*. A fibrous green amphibole, pseudomorphous after pyroxene in such rocks as eclogite.

smart *(Aero.)*. Originally applied to guided and self-homing bombs for attacking point targets, now used for any device showing 'artificial intelligence' capability.

smashing *(Print.)*. The pressing of a book in a machine after sewing, thereby crushing and expelling air. Also called *crushing*, *nipping*.

SMC *(Aero.)*. Abbrev. for *Standard Mean Chord.*

smear metal *(Eng.)*. Particles and serrated edges produced by metal cutting and caused, by the cutting heat generated, to be welded or combined into an amorphous metal substance.

smear test *(Med.)*. Diagnostic test for cancer by laboratory examination of the cells of the blood and other fluids. Most commonly applied to cervical smears for the detection of early cervical cancer.

smear test *(Nuc.Eng.)*. A method of estimating the loose, i.e. easily removed, radioactive contamination upon a surface. Made by wiping the surface and monitoring the swab.

smectic *(Phys.)*. Said of a mesomorphous substance whose atoms or molecules are oriented in parallel planes. Cf. **nematic**.

smectites *(Min.)*. A group of clay minerals including montmorillonite, beidellite, nontronite, saponite, sauconite, and hectorite. They are 'swelling' clay minerals and can take up water or organic liquids between their layers, and they show cation exchange properties.

SMEDI *(Vet.)*. *Swine Mummification Embryonic Death and Infertility*. An enterovirus disease of pigs.

smegma *(Med.)*. A thick greasy secretion of the sebaceous glands of the glans penis.

smell *(Med.)*. The sensation produced by stimulation of the mucous membrane of the olfactory organs.

smelting *(Eng.)*. Fusion of an ore or concentrate with

suitable fluxes, to produce a melt consisting of two layers – on top a slag of the flux and gangue minerals, molten impure metal below.

Smith chart *(Elec.Eng.).* Polar chart with circles for constant resistance and reactance, lines for phasor angles, and standing-waves ratio circles. Used for impedance calculations, especially with data from slotted lines for very high frequencies and from waveguides. See **voltage standing-wave ratio**.

Smith's coupling *(Autos.).* See **magnetic clutch (2)**.

smithsonite *(Min.).* Carbonate of zinc, crystallizing in the trigonal system. It occurs in veins and beds and in calcareous rocks, and is commonly associated with hemimorphite. The honeycombed variety is known as *dry bone ore*. In Britain sometimes *calamine*.

smoke *(Powder Tech.).* Visible cloud of airborne particles derived from combustion, or from chemical reaction; the particles are mainly smaller than 1 μm.

smoke control area *(Genrl.).* An area, statutorily defined, in which emission of smoke from chimneys is prohibited.

smoke point *(Chem.).* In the testing of kerosene, the maximum flame height (in millimetres) at which a kerosene will burn without smoking, under prescribed conditions.

smoke test *(Build.).* A test applied to new drain-pipes or those suspected of leakage. The ends are plugged and dense smoke is introduced into the pipe, which is watched for any escape.

smoky quartz *(Min.).* Dark greyish-brown transparent quartz, used as a gemstone. See **cairngorm**.

smooth ashlar *(Build.).* A block of stone dressed ready for use, usually for the stone-facing of walls.

smooth colony *(Biol.).* A bacterial colony of a typical regular, glistening appearance; sometimes developed from a rough colony by mutation and frequently differing from it physiologically, e.g. altered sensitivity to bacteriophage.

smooth-core rotor *(Elec.Eng.).* A rotor carrying a field winding embedded in tunnels, used in high-speed steam-turbine-driven generators.

smooth endoplasmic reticulum *(Biol.).* Tubular form of endoplasmic reticulum without ribosomes, well developed in gland cells which secrete terpenoids, flavonoids etc.

smoother *(Elec.Eng.).* Combination of capacitors and inductors for removing the ripple from rectified power supplies.

smoothers *(Foundry).* See **sleekers**.

smoothing choke *(Elec.Eng.).* Inductor in a filter circuit which attenuates ripple in a given d.c. supply.

smoothing circuit *(Elec.Eng.).* See **ripple filter**.

smoothing plane *(Build.).* A bench plane about 8 in (20 cm) in length used to give a smooth even finish to timber members or completed work.

smooth mouth *(Vet.).* Smooth and polished grinding surface of the molar teeth of horses.

smooth muscle *(Zool.).* See **unstriated muscle**.

SMPTE *(Image Tech.). Society of Motion Picture and Television Engineers.* The US body responsible for numerous standards and recommended practices.

smudge *(Build.).* A lampblack and glue size mixture which is painted over lead surfaces so that solder will not adhere.

smut *(Bot.).* One of a number of plant diseases, some economically important, caused by biotrophic fungi of the order Ustilaginales and characterized by the production of masses of usually black spores within the host. See also **bunt**.

smut *(Min.Ext.).* (1) Bad soft coal containing earthy matter. (2) Worthless outcrop material of a coal seam.

Sn *(Chem.).* The symbol for tin (L. *stannum*).

sn *(Maths.).* See **elliptic functions**.

snail *(Horol.).* A cam the contour of which may be a smooth curve or stepped. Used for the gradual lifting or discharge of a lever. The snail in the rack striking work controls the number of teeth picked up on the rack, e.g. at 12 o'clock sufficient motion is allowed to the rack for the gathering pallet to pick up 12 teeth.

snailed *(Horol.).* Said of *(a)* a surface finished with eccentric curves; *(b)* a barrel arbor so shaped that the inner coil of the mainspring passes over the hook without forming a kink.

snakewood *(For.).* (1) Hardwood from a tree of the genus *Piratinera*. Very heavy, it is dark red in colour, with darker markings resembling a snake skin. Also called *letterwood, leopard wood, tortoiseshell wood.* (2) The *Strychnus nux vomica*, a tree species yielding *strychnine*.

snaking *(Aero.).* An uncontrolled oscillation in yaw, usually at high speed, of approximately constant, but small, amplitude (ca. 1°).

snap *(Elec.Eng.).* Sudden action in magnetic amplifiers with excessive positive feedback, arising from hysteresis in the core.

snap *(Eng.).* (1) A form of punch with a hemispherically recessed end, used to form rivet heads. (2) A limit gauge of plate or calliper type.

snapped header *(Build.).* A half-length brick, sometimes used in Flemish bond.

snapshop *(Image Tech.).* A photograph taken with a simple camera using a short exposure time, generally without any special preparation of subject or lighting.

snap switch *(Elec.Eng.).* A switch which makes and breaks the circuit with a quick snap; it comprises blades whose rate of motion is controlled by a spring. Also *quick make-and-break switch.*

snap the line *(Build.).* To pluck a well-chalked string, held taut in position, against work to mark a straight line.

snare *(Med.).* A wire loop for removing soft tumours such as nasal polypi.

snarl *(Textiles).* Small extra thickness of yarn where the twist has run back on itself; caused by excessive twist and inadequate tension on the yarn.

S/N curve *(Eng.).* Abbrev. for *Stress-Number curve.*

sneck *(Build.).* The lifting lever which passes through a slot in a door and actuates the fall bar. See **Norfolk latch**.

snecked *(Build.).* Said of rubble walls in which the stones are roughly squared but of irregular size and uncoursed.

sneezewood *(For.).* South African tree *Ptaeroxylon utile*, giving a hard and strong timber. Its principal uses are for fencing posts, piling, poles etc. It has a smell akin to pepper.

Snell's law *(Phys.).* A wave refracted at a surface makes angles relative to the normal to the surface, which are related by the law $n_1 \sin\theta_1 = n_2 \sin\theta_2$. n_1 and n_2 are the refractive indices on each side of the surface, and θ_1 and θ_2 are the corresponding angles, The two rays and the normal at the point of incidence on the boundary lie in the same plane.

snow *(Meteor.).* Precipitation in the form of small ice crystals, which may fall singly or in flakes, i.e. tangled masses of snow crystals. The crystals are formed in the cloud from water vapour.

snow *(Telecomm.).* Effect of electrical noise on display of intensity-modulation signals. The effect is random and resembles falling snow and may be seen, e.g. on a TV screen in the absence of a signal, or when the signal is weak.

snow boards *(Build.).* Horizontal boards, about 20 cm high, fixed over roof gutters to prevent snow from sliding off in mass. The snow, on melting, drops into the gutter through gaps left between the boards. Also *gutter boards, snow guards.*

snow guards *(Build.).* See **snow boards**.

snow load *(Build.).* The unit loading assumed in the design of a roof to allow for the probable maximum amount of snow lying upon it.

snow stage *(Meteor.).* That part of the condensation process taking place at temperatures below 0°C so that water vapour condenses directly to ice.

S/N ratio *(Acous.).* Abbrev. for *Signal/Noise ratio.*

snubbing *(Min.Ext.).* In high pressure oil wells the process by which drill pipe is removed through the **blowout preventer** stack while retaining pressure at all times.

snuffles *(Med.).* Symptoms resulting from nasal discharge and obstruction of the nose in infants.

snuffles *(Vet.)*. A term popularly applied to several diseases of animals in which the nasal cavities are affected. See **atrophic rhinitis** (*pig*), **osteodystrophia fibrosa** (*pig*), **rabbit septicaemia** (*rabbit*).

soakaway *(Build.)*. A pit excavated to receive excess surface water which can duly seep into the surrounding ground.

soakers *(Build.)*. Small pieces of sheet-lead or zinc bonded in for watertightness with the slates or tiles of a roof at joint with walls or at valleys and hips.

soaking *(Eng.)*. A phase of a heating operation during which metal or glass is maintained at the requisite temperature until uniformly heated, and/or until any required phase transformation has occurred.

soaking *(Glass,Eng.)*. A phase of a heating operation during which metal or glass is maintained at the requisite temperature until uniformly heated, and/or until any required phase transformation has occurred.

soaps *(Build.)*. Bricks, $9 \times 2\frac{1}{4} \times 2\frac{1}{4}$ in in size, which are often pierced for use as air bricks.

soaps *(Chem.)*. (1) The alkaline salts of fatty acids, chiefly palmitic, stearic, or oleic acids. *Soft soaps* contain the potassium salts, whereas the sodium salts are *hard soaps*. *Metallic soaps*, water-insoluble compounds of fatty acids with bases of copper, aluminium, lithium, calcium etc. are used as waterproofing agents and as the bases of many greases. (2) The alkali salts of resins, so-called *resin-soaps*.

soapstone *(Min.)*. See **steatite**.

soaring *(Aero.)*. The art of sustained motorless flight by the use of thermal upcurrents and other favourable air streams.

social facilitation *(Behav.)*. Refers to an increase in the performance of a behaviour as a consequence of its performance by other individuals nearby (e.g. yawning in primates).

socialization *(Behav.)*. The process whereby the child learns the norms of its society and acquires the attitudes and behaviour that conform to them.

social-learning theory *(Behav.)*. A theoretical approach to the development of social behaviour; it stresses the importance of learning by observing others, and of reinforcement; contains elements of both a *stimulus-response* approach and a *cognitive approach* to the issues of learning.

social organization *(Behav.)*. The totality of all social relationships among members of a particular group.

social parasitism *(Behav.)*. A parasitic relationship between members of one species or with individuals of another species which involves exploiting aspects of the host's social behaviour, e.g. their nesting or hunting behaviour. A common example is the female cuckoo which lays its egg in another bird's nest.

social perception *(Behav.)*. Refers to several related areas of study: (1) how individuals perceive others, e.g. the judgment of emotions in others, studies of impression formation (or person perception); (2) the psychological processes that underlie these activities in the perceiver; (3) it also refers to the study of how social factors influence perception (the effects of group pressure on the perception of an event).

social phobia *(Behav.)*. The fear of performing certain actions when exposed to the scrutiny of others.

social psychology *(Behav.)*. Refers to the interpersonal relations, face to face interactions as well as to the social attitudes that is the subject matter of research in social psychology.

social symbiosis *(Behav.)*. Refers to a relationship between members of different species from which one or both derive some advantage; the nature of the benefit for each species can be quite different, e.g. one partner may gain protection from predators while the other gains a nesting site, or food supply.

society *(Bot.)*. A minor plant community within an association, dominated by a species which is not a general dominant of the association, e.g. an alder-dominated community on wet ground within an oak wood.

socket *(Build.)*. (1) A pipe end enlarged to pass over a same-sized pipe to make a joint. US *bell* or *hub*. (2) Pipe fitting for joining two pipes in line.

socket *(Elec.Eng.)*. The female portion of a plug-and-socket connection in an electric circuit.

socket *(Min.Ext.)*. Portion of drill hole left undisturbed after blasting, and liable to contain unexploded charge.

socket chisel *(Build.)*. A robust type of chisel used for mortising; it has a hollow tapering end to the steel shank into which the handle fits.

socket head screw *(Eng.)*. One which has a hexagonal recess in its head, so that it can be turned by means of a hexagonal bar formed into a key.

socket spanner *(Eng.)*. One with a recessed head, the sides of the recess having six or more flats to fit a hexagonal nut in various positions.

socle *(Arch.)*. A plain projecting block or plinth at the base of a pedestal, wall, or pier.

soda-ash *(Min.Ext.)*. Impure (commercial) sodium carbonate. Widely used in pH control of flotation process.

soda lakes *(Geol.)*. Salt lakes the water of which has a high content of sodium salts (chiefly chloride, sulphate, and acid carbonate). These salts also occur as an efflorescence around the lakes.

soda-lime-silica glass *(Glass)*. The commonest type of glass, used for windows, containers and electric lamp bulbs. Contains silica, soda and lime in approx. proportions SiO_2 70%, Na_2O 15%, CaO 10%. **Crown glass** is of this type, but in **optical crown** barium oxide is now normally substituted for lime.

sodalite *(Min.)*. A cubic feldspathoid mineral, essentially silicate of sodium and aluminium with sodium chloride, occurring in certain alkali-rich syenitic rocks.

sodamide *(Chem.)*. $Na^+NH_2^-$; an ionic compound formed when ammonia gas is passed over hot sodium.

soda nitre *(Min.)*. Sodium nitrate, crystallizing in the trigonal system. It is found in great quantities in northern Chile, where beds of it are exposed at the surface and are known as *caliche*. Also called *Chile saltpetre*, *nitratine*.

sodar *(Acous.)*. Acoustic method for finding atmospheric layers. In principle like sonar, used in meteorology.

soda recovery *(Paper)*. Process for recovering a substantial amount of inorganic soda from the spent liquor of an alkaline digestion by concentrating it, burning off the organic matter and treating the resultant soda ash with caustic.

sodium *(Chem.)*. Metallic element, one of the alkali metals. Symbol Na, at. no. 11, r.a.m. 22.9898, rel.d. 0.978, mp 97.5°C, bp 883°C. Sodium does not occur in nature in the free state, owing to its reactivity, but it is widely distributed combined as the chloride, nitrate etc. Metallic sodium is made by electrolysis of fused caustic soda. It is a soft, silver-white metal, very active, with a moderate thermal neutron cross-section. Used in reactors as a liquid metal heat-transfer fluid.

sodium aluminate *(Chem.)*. See **aluminate**.

sodium chloride *(Chem.)*. NaCl. See **halite**

sodium-cooled reactor *(Nuc.Eng.)*. One in which liquid sodium is used as the primary coolant, as in the **fast reactor**.

sodium cromoglycate *(Med.)*. Drug of value in the prophylaxis of allergic asthma. It appears to act by inhibiting release of bronchoconstricting agents from the mast cells in the airways.

sodium cyanide *(Chem.)*. NaCN. See **potassium cyanide**.

sodium cyclamate *(Chem.)*. The sodium salt of *N*-cyclohexylsulphamic acid (see **cyclamates**). A white, crystalline powder readily soluble in water. Used as an artificial sweetening material, e.g. in soft drinks and diabetic sweetening tablets, but banned in some countries because of possible health hazards in long-term use.

sodium dodecyl sulphate *(Biol.)*. *SDS*. An anionic detergent widely used as a powerful denaturant and solubilizing agent, especially for the fractionation of proteins and nucleic acids according to size by **acrylamide gel electrophoresis**.

sodium hydroxide *(Chem.)*. NaOH, a deliquescent substance, with a soapy feel, whose solution in water is strongly alkaline; it is a common reagent in the laboratory. It is manufactured by treating quicklime with hot sodium carbonate solution; and its main industrial use is the manufacture of soap. See also **Castner-Kellner process**. Also *caustic soda*.

sodium iodide scintillation crystal *(Radiol.)*. A high density photon absorber that converts the energy of a photon from radioactivity to a light photon.

sodium nitrate *(Min.)*. See **soda nitre**.

sodium thiosulphate *(Chem.,Image Tech.)*. See **hypo**.

sodium vapour lamp *(Phys.)*. An electric lamp of the gaseous-discharge type whose electrodes operate in an atmosphere of sodium vapour.

sodoku *(Med.)*. *Rat-bite fever*. A disease due to infection with the micro-organism *Spirillum minus*, conveyed by the bite of a rat; it is characterized by inflammation of the skin around the bite, relapsing fever, swelling of the lymphatic glands, and a red, patchy rash.

soffit *(Build.,Civ.Eng.)*. (1) A term often used for *intrados*, but more particularly applied to that part of the intrados in the immediate vicinity of the keystone. (2) The under surface of a stair or floor or of the head of an opening such as a door or window opening.

soft *(Electronics)*. Said of valves and tubes when there is appreciable gas pressure within the envelope, such as gas-discharge tubes and photocells. Particularly said of valves when gas is released from the envelope or electrodes. See **hard, outgassing**.

soft *(Glass)*. Having a relatively low softening point.

soft commissure *(Zool.)*. In Mammals, the point at which the thickened sides of the thalamencephalon touch one another across the constructed third ventricle.

softener *(Build.)*. A brush used in graining and marbling for blending colour, or reducing the harsh appearance of the work. Hog hair softeners are used in oils and badger hair softeners are used in water colour.

softening *(Eng.)*. A process, for removing arsenic, antimony and tin from lead, after drossing. A bath of molten metal is oxidized by furnace gases and the addition of lead oxide. Impurities are oxidized and form a dross. Heating of metal towards its **arrest point**, followed by slow cooling. Also called *improving*.

softening *(Textiles)*. The application of a chemical (the softener) that softens the handle and increases the fluffiness of fabrics.

softening-point test *(Build.)*. See **melting-point test**.

soft-focus lens *(Image Tech.)*. A lens which images a point source with a slight halo, giving a softness of outline to the whole picture sometimes preferred for portraiture.

soft iron *(Eng.)*. Iron low in carbon and unable to retain magnetism; useful as solenoid cores etc.

soft-iron armature *(Elec.Eng.)*. The attracted part of an electromagnet retaining little residual magnetism.

soft-iron instrument *(Elec.Eng.)*. An undesirable synonym for *moving-iron instrument*.

softness *(Eng.)*. Tendency to deform easily. It is indicated in tensile test by low ultimate tensile stress and large reduction in cross-section before fracture. Usually the elongation is also high. In notched bar test, specimens bend instead of fracturing, and energy absorbed is relatively small. See **brittleness, toughness**.

soft palate *(Zool.)*. In Mammals, the posterior part of the roof of the buccal cavity which is composed of soft tissues only.

soft radiation *(Radiol.)*. General term used to describe radiation whose penetrating power is very limited, e.g. low energy X-rays.

soft-rock geology *(Geol.)*. An informal term for the geology of sedimentary rocks.

soft rot *(Bot.)*. Rot in which tissues (usually parenchyma) soften because of the destruction of middle lamellas and cell walls, as in many bacterial and fungal diseases of stored fruits and vegetables.

soft sectored disk *(Comp.)*. One which is formatted into sectors with information written on the disk by a program.

soft-shelled egg *(Vet.)*. A bird's egg in which lime salts are deficient or absent.

soft soaps *(Chem.)*. See **soaps**.

soft solder *(Eng.)*. Alloys of lead and tin used in soldering. Tin content varies from 63% to 31%. The remainder is mainly lead, but some types contain about 2% antimony and others contain cadmium. The best-known types are *tinman's solder* and *plumber's solder*.

software *(Comp.)*. A general term for all types of **program** and their associated **documentation**. See **systems-, applications-**. Cf. **hardware**.

software house *(Comp.)*. A commercial organization which specializes in the preparation of **application software** or **system software**.

software package *(Comp.)*. Fully documented program, or set of programs, designed to perform a particular task.

soft water *(Chem.)*. Generally, water free from calcium and magnesium salts (cf. **hard water**), naturally or artificially. See **water softening**.

softwood *(Bot.)*. The wood from a conifer and coniferous trees generally. Ant. *hardwood*.

soil-acting herbicide *(Bot.)*. A plant poison absorbed from the soil by the roots, such as sodium chlorate.

soil flora *(Bot.)*. Fungi, bacteria and algae living in the soil.

soil mechanics *(Civ.Eng.)*. The process of determining the properties of any soil, e.g. water content, bulk, density, permeability, shear strength etc.

soil pipe *(Build.)*. A vertical cast-iron or plastic pipe conveying waste matter from W.C.s etc. to the drains. Abbrev. *SP*.

soil release agent *(Textiles)*. A chemical added to fabrics to make it easier to remove stains by washing.

soil sampler *(Civ.Eng.)*. A hollow circular tool, with a sharp edge, for extracting specimens of soil for examination or analysis. Also *soil borer, soil pencil*.

soil sampling *(Min.Ext.)*. Taking samples of soil or overburden as part of a **geochemical prospecting** exercise to identify anomalous concentrations of metals, or the presence of **tracers**.

soil structure *(Bot.)*. The nature of a soil in terms of manner in which the particles are aggregated into larger bodies or *peds*.

soil texture *(Bot.)*. The nature of a soil in terms of the proportions of mineral particles of various sizes (sand, silt and clay).

sol *(Chem.)*. A colloidal solution, i.e. a suspension of solid particles of colloidal dimensions in a liquid.

Solanaceae *(Bot.)*. Family of ca. 2500 spp. of dicotyledonous flowering plants (superorder Asteridae). Mostly herbs, some shrubs and small trees, cosmopolitan. The petals are fused, the ovary is superior, the fruit is usually a berry, less often a capsule. Includes potato, tomato, aubergine, sweet peppers, chillies, cayenne pepper, tobacco. Many are poisonous and often contain tropane alkaloids of medicinal importance e.g. belladonna or deadly nightshade (atropine), and henbane (hyocyamine).

solar *(Zool.)*. Having branches or filaments radially arranged.

solar antapex *(Astron.)*. The point on the celestial sphere diametrically opposite to the *solar apex*.

solar apex *(Astron.)*. The point on the celestial sphere towards which the solar system as a whole is moving at the rate of 20 km/s. It is located in the constellation Hercules in equatorial coordinates R.A. 271° and declination +30° approximately.

solar array *(Space)*. Bank of solar cells, mounted on a panel structure, and extended from a satellite, which converts solar energy into electrical energy using *photovoltaic* conversion. Also *solar battery, solar paddle, solar cell array*.

solar attachment *(Surv.)*. See **shade**.

solar cell *(Electronics)*. Photoelectric cell using silicon, which collects photons from the sun's radiation and converts the radiant energy into electrical energy with reasonable efficiency. Used in satellites and for remote locations lacking power supplies, e.g. for radio-telephony in the desert.

solar constant *(Astron.)*. The total electromagnetic

energy radiated by the Sun at all wavelengths per unit time through a given area at the mean distance of the Earth. Its current value is 1.37 kW per m^2. It is not, in fact, truly constant, and variations of order 0.1% are detectable.

solar constant *(Phys.)*. The rate of reception, per unit area, of energy normal to the earth's surface, corrected for loss by absorption in the earth's atmosphere. The value (which is not constant) is about 1340 w/m^2.

solar corona *(Astron.)*. The outermost layer of the Sun's atmosphere, visible as a halo of light during a total **eclipse**. Its temperature is 0.5–2 million K, and it is a strong source of X-rays. Its visible radiation can be studied at any time with a *coronagraph*.

solar day *(Astron.)*. See **apparent-**, **mean-**.

solar eclipse *(Astron.)*. See **eclipse**.

solar energy utilization *(Phys.)*. Energy from the sun can be converted, (1) to thermal energy using a working fluid, (2) to electrical energy using a photovoltaic cell, (3) to mechanical energy using the radiation pressure (solar sailing in space) and (4) to chemical energy by photosynthesis.

solar flare *(Astron.)*. Bright eruption of the sun's surface associated with sun spots, causing intense radio and particle emission.

solar flocculi *(Astron.)*. The name given to small bright and dark markings on the solar chromosphere as seen on calcium or hydrogen spectroheliograms.

solar gain *(Build.)*. Heat gain in a building due entirely to the sun.

solar granulation *(Astron.)*. The mottled appearance of the sun's photosphere, small bright granules (about 1000 km in diameter) being seen in rapid change against the darker background.

solarimeter *(Phys.)*. See **pyranometer**.

solarization *(Glass)*. Changes of the light transmission properties of glass as a result of exposure to sunlight or other radiation in or near the visible spectrum.

solarization *(Image Tech.)*. The reversal of an image because of excessive exposure to light, so that an intended negative appears to be a positive.

solar panel *(Electronics)*. Arrays of solar cells fitted to spacecraft and satellites in order to gather solar energy for conversion into electrical power for the equipment on board. In some satellites these may be in the form of extensible *paddles* which are stowed away during launching.

solar parallax *(Astron.)*. The mean value of the angle subtended at the sun by the earth's equatorial radius. It cannot be measured directly, but is derived from Kepler's laws when the distance between the earth and any other planet is known. The value of 8.80″ for the solar parallax has been in use for half a century, but recent measures of the distance of Venus by radar in 1961 gave the value 8.794 05″, corresponding to a value of 92 957 130 miles (149.6×10^6km) for the astronomical unit.

solar plexus *(Zool.)*. In higher Mammals, a ganglionic centre of the autonomic nervous system; situated in the anterior dorsal part of the abdominal cavity, from which nerves radiate in all directions.

solar radio noise *(Astron.)*. Radio emission from the atmosphere of the Sun, which is investigated by many techniques in **radio astronomy**. Sudden bursts are associated with solar **flares** in particular. Sunspots, and the activity in their vicinity, are also strong sources of radio emission.

solar rotation *(Astron.)*. The rotation of the sun takes place in the same direction as the orbital motion of the planets. The rotation is non-uniform, taking 24.65 days at the sun's equator, but increasing to about 34 days near the poles. Because of the earth's motion, the equatorial solar rotation has a synodic period of 27 days, and this interval is apparent in the recurrence of magnetic storms, aurorae, etc.

solar sailing *(Space)*. Means of movement in space using the pressure of the **solar wind** on a large surface or solar sail, suitably deployed.

Solar System *(Astron.)*. The term designating the Sun and the attendant bodies moving about it under gravitational attraction; comprises nine major planets, and a vast number of asteroids, comets, and meteors.

solar wind *(Astron.)*. A constant plasma stream of protons moving radially away from the sun at a speed of 250–800 km/s. It affects the earth's magnetic field and also causes acceleration in the tails of comets.

solar wind *(Space)*. The continuous stream of charged particles (consisting mainly of protons and electrons) emitted by the Sun which moves at supersonic speed and pervades all interplanetary space.

solation *(Chem.)*. The liquefaction of a gel.

solder *(Eng.)*. A general term for alloys used for joining metals together. The principal types are *soft solder* (lead-tin alloys) and *brazing solders* (alloys of copper, silver and zinc).

solder-covered wire *(Elec.Eng.)*. Copper wire coated with solder instead of tin, to facilitate connections between components.

soldered dot *(Build.)*. Method for fixing sheet-lead to woodwork, which has had small depressions worked in it. The lead is fixed by splayed screws within the depression which is then soldered over.

soldering *(Eng.)*. Hot joining of metals by adhesion using, as a thin film between the parts to be joined, a metallic bonding alloy having a relatively low melting point.

solder paint *(Eng.)*. Mixture of powdered solder and flux, applied by brush to surfaces to be joined by *soldering* to form the bonding film.

soldier *(Build.)*. A term applied to a course of bricks laid so that they are all standing on end.

soldier *(Zool.)*. In some social Insects, a form with especially large head and mandibles adapted for defending the community, for fighting, and for crushing hard food particles.

sole *(Build.)*. The lower surface of the body of a plane.

sole *(Eng.)*. (1) The bed plate of a marine engine; secured through bearers to the hull of the ship. Also *sole plate*. (2) A timber base for supporting the feet of **raking shores**. (3) The plate supporting a leg of a process vessel.

sole mark *(Geol.)*. A physical structure found on the underside of a bed of sandstone or siltstone, that is the *mould* of the top surface of the bed on which it lies. The mould may represent a sedimentary structure (e.g. a groove or a ripple mark) or the remains of a **trace fossil**.

Solenhofen stone *(Geol.)*. An exceedingly fine-grained and even-bedded limestone, thinly stratified, of Upper Jurassic age, occurring in S.E. Bavaria; formerly widely used in lithography.

solenocyte *(Zool.)*. In Invertebrates and lower Chordata, an excretory organ consisting of a hollow cell with branched processes, in the lumen of which occurs a bunch of cilia which by their movements maintain a downward current.

solenoid *(Elec.Eng.)*. Current-carrying coil, of one or more layers. Usually a spiral of closely wound insulating wire, in the form of a cylinder, not necessarily circular. Generally used in conjunction with an iron core, which is pulled into the cylinder by the magnetic field set up when current is passed through the coil.

solenoidal field *(Elec.Eng.)*. One in which divergence is zero, and the vector is constant over any section of a tube of force.

solenoidal magnetization *(Phys.)*. The distribution of the magnetization on a piece of magnetic material when the poles are at the ends. Cf. **lamellar magnetization**. Also called *circuital magnetization*.

solenoid brake *(Elec.Eng.)*. An electromechanical brake in which the brake toggle is operated by the plunger of a solenoid.

solenoid model *(Biol.)*. The organization of nucleosomes and spacer regions into a *solenoid coil*, containing 6–7 nucleosomes per turn, with a diameter of 30 nm.

solenoid-operated switch, circuit-breaker *(Elec.Eng.)*. A switch in which the closing force is provided by a solenoid. Cf. **pneumatically-operated switch**, **motor-operated switch**.

solenoid relay *(Elec.Eng.)*. A relay in which the contacts are closed by the action of a solenoid-operated plunger.

solenostele *(Bot.)*. Stele with a central pith and consisting of an annulus of xylem completely enclosed by an endodermis, and with phloem between the endodermis and the xylem throughout.

sole piece *(Build.)*. The plate to which the feet of the shores, in a system of raking shores, are secured, and which forms an abutment for them at their lower ends.

sole piece *(Ships)*. See **stern frame**.

sole plate *(Build.)*. See **sleeper**.

sole plate *(Eng.)*. See **sole**.

solfatara *(Geol.)*. The name applied to a volcanic orifice which is in a dormant or decadent stage and from which gases (especially sulphur dioxide) and volatile substances are emitted.

solid *(Chem.)*. A state of matter in which the constituent molecules or ions possess no translational motion, but can only vibrate about fixed mean positions. A solid has a definite shape and offers resistance to a deforming force.

solid *(Maths.)*. A figure having 3 dimensions.

solid angle *(Maths.)*. Of an area subtended at a point, the area intercepted on a unit sphere, centred at the point, by a cone having the given area as its base and the point as its vertex. See **steradian**.

solid diffusion *(Eng.)*. Movement of atoms through the crystals of a solid metal, as when carbon diffuses in to or out of steel during carburizing or decarburizing respectively.

solid floor *(Build.)*. A floor laid on a concrete subfloor. Cf. **suspended floor**.

solid head *(Autos.)*. A cylinder or cylinder block cast in one piece, as distinct from one with a detachable head.

solidification range *(Chem.)*. The range of temperature in which solidification occurs in alloys and silicate melts etc., other than those which freeze at constant temperature. It extends from a point on the liquidus to one on the solidus.

solid injection *(Eng.)*. See **airless injection**.

solid matter *(Print.)*. Text-matter set without space between the lines; or, in mechanical type-setting, type-matter cast on its own body size rather than on a larger one.

solid newel *(Build.)*. The centre post of a winding stair, as distinct from a **hollow newel**.

solid panel *(Build.)*. A panel whose surface is in line with the faces of the stiles.

solid pole *(Elec.Eng.)*. A field pole of an electrical machine which is not built up from laminations.

solid propellant *(Space)*. Rocket propellant in solid state, usually in caked, plasticlike form, comprising a fuel-burning compound of fuel and oxidizer.

solid solubility *(Eng.)*. The extent to which one metal is capable of forming solid solutions with another.

solid solution *(Chem.,Phys.)*. Usually a *primary solid solution*, but the term may also be applied to the case when an intermediate constituent dissolves one of its components.

solid-state *(Electronics)*. Pertaining to a circuit, device, or system which depends on some combination of electrical, magnetic, or optical phenomena within a material which is usually a crystalline semiconductor. Loosely applied to all active devices or circuits which do not rely on valves or tubes.

solid-state capacitor *(Electronics)*. See **varactor**.

solid-state detector *(Phys.)*. A detector of ionizing radiation which uses an energy-sensitive solid-state device. See **lithium drifted silicon detector**.

solid-state maser *(Telecomm.)*. One, commonly made from ruby, kept at a few degrees above absolute zero to ensure low-noise operation and placed in an intense magnetic field. A high-frequency *pump* signal raises electrons to a higher energy level then normal; the lower frequency signal, i.e. that to be amplified, causes high energy electrons to revert to their lower levels and, in so doing, absorb photons, the low frequency signal becoming amplified in the process.

solid-state physics *(Phys.)*. Branch of physics which covers all properties of solid materials, including electrical conduction in crystals of semiconductors and metals, superconductivity and photoconductivity.

solid-state welding *(Eng.)*. Welding which does not involve melting, brazing or soldering, but may involve pressure.

solid-type cable *(Elec.Eng.)*. See **straight-type cable**.

solidus *(Chem.)*. A line in a constitutional diagram indicating the temperatures at which solidification is completed, or melting begins, in alloys and other melts of different composition. See **liquidus, solidification range**.

solidus *(Print.)*. The diagonal or oblique stroke (/).

solifluction, solifluxion *(Geol.)*. Soil-creep on sloping ground, characteristic of, though not restricted to, regions subjected to periods of alternating freezing and thawing.

soligenous *(Bot.)*. A mire or fen, receiving water that has passed through a mineral soil and hence oligotrophic to relatively eutrophic depending on the nature of the soil. Cf. **ombrogenous**.

solitaria phase *(Zool.)*. One of the two main phases of the Locust (Orthoptera), which occurs when nymphs are reared in isolation. They adjust their colour to match their background and lack the higher activity and gregarious tendencies of the **gregaria phase**.

sollar *(Build.)*. A loft which is open to the sun.

solochrome black *(Chem.)*. Indicator for the complexometric titration of both calcium and magnesium ions in hard water (red colour) with **EDTA**. Often used in parallel with **murexide**, which responds to calcium only. The uncomplexed indicator is blue.

solstice *(Astron.)*. (1) One of the two instants in the year when the Sun reaches its greatest excursion north or south of the equator. (2) One of the two points on the **ecliptic** midway between the **equinoxes**.

solubility *(Chem.)*. The extent to which one substance will dissolve in another. Usually expressed as the mass or the quantity of a substance which will dissolve in 1 dm^3 of water.

solubility curve *(Chem.)*. The curve showing the variation of the solubility of a substance with temperature.

solubility product *(Chem.)*. The equilibrium constant defining the solubility of an ionic substance in water. It is equal, in a saturated solution, to the product of the **activities** of the ions each raised to the power of the number of ions of that type in the formula.

soluble complex *(Immun.)*. Applied to antigen-antibody complexes in soluble form rather than in precipitates. Occurs in vivo or in vitro when there is an excess of antigen over antibody so that a large lattice is not formed. (See **lattice hypothesis**). Soluble complexes cause tissue damage in vivo, especially if complement is also activated by them. See **hypersensitivity reactions Type III, Immune complex disease, serum sickness, glomerulonephritis**.

soluble oil *(Eng.)*. Cutting fluid consisting of oil and an emulsifier to which water is added.

soluble starch *(Chem.)*. A product of the hydrolysis of starch obtained by treating starch with dilute acids, or by boiling with glycerine, or by the action of diastase.

solute *(Chem.)*. A substance which is dissolved in another.

solute potential *(Bot.)*. Same as **osmotic potential**.

solution *(Chem.)*. Homogeneous mixture of two or more components in a single phase. Often refers specifically to a solution in water (an aqueous solution).

solution heat treatment *(Eng.)*. Heating suitable alloys (e.g. duralumin) in order to take the hardening constituent into solution. This is followed by quenching, to retain the solid solution, and the alloy is then age hardened at atmospheric or elevated temperature.

solution mining *(Min.Ext.)*. Winning of soluble salts by use of percolating liquor introduced through shafts, drives and/or bores. Resulting saturated solution is pumped to surface for further treatment.

solutizer process *(Chem.Eng.)*. Process for removing mercaptans from petroleum fractions by 2-stage treatment with caustic soda and sodium cresylate solution.

solvation *(Chem.)*. The association or combination of molecules of solvent with solute ions or molecules.

Solvay's ammonia soda process *(Chem.)*. A process based on the fact that when a concentrated solution of sodium chloride is saturated with ammonia, and carbon dioxide is passed through, sodium hydrogen carbonate is precipitated and ammonium chloride remains in solution. Used for the manufacture of sodium carbonate from chloride.

solvent *(Build.)*. In painting a liquid capable of dissolving the **binder** and added to make it work more freely.

solvent *(Chem.)*. That component of a solution which is present in excess, or whose physical state is the same as that of the solution.

solvent bonding *(Textiles)*. Process in which an organic liquid is used to soften fibres so that they adhere to each other and form a **non-woven fabric**.

solvent extraction *(Min.Ext.)*. In chemical extraction of values from ores or concentrates, selective transfer of desired metal salt from aqueous liquor into an immiscible organic liquid after intimate stirring together followed by phase separation.

solvent naphtha *(Chem.)*. Middle and high-boiling benzene hydrocarbons chiefly consisting of toluene and xylene, obtained from the fractionation of light tar oils after the benzene fractions have been distilled off.

solvent processing *(Textiles)*. Scouring, dyeing and finishing processes carried out in organic liquids rather than in aqueous solutions.

solvolysis *(Chem.)*. See **lyolysis**.

soma *(Zool.)*. The body of an animal, as distinct from the germ cells; cf. **germen**. pl. *somata*. adj. *somatic*.

somaclonal variation *(Bot.)*. Variability commonly found among plants that have been regenerated from **tissue cultures**.

somatic *(Biol.)*. Of cells of the body, as distinct from the *germ line*.

somatic cell *(Zool.)*. One of the nonreproductive cells of the parent body, as distinct from the reproductive or germ cells.

somatic cell hybrid *(Biol.)*. Cell formed by fusion of cells from the same or different species, in which there is also nuclear fusion.

somatic doubling *(Biol.)*. A doubling of the number of chromosomes in the nuclei of somatic cells.

somatic hybridization *(Bot.)*. The production of hybrid cells by the fusion of non-gametic nuclei either naturally in the **parasexual cycle** of some fungi or artificially, such as by **protoplast fusion**.

somatic mutation *(Immun.)*. Mutations occurring in the genes of cells of the body other than the germ cells which give rise to sperm or ova, and therefore not inheritable. Such mutations normally occur rarely in most cells, but the genes controlling synthesis of the variable region of immunoglobulin molecules exhibit a greatly increased rate of mutation. There are three hypervariable sites in the V region genes at which mutations are especially likely to occur during the rapid proliferation of B lymphocytes which respond to antigenic stimulation. This increases the range of different immunoglobulin molecules made by the cell population involved, and includes some with increased affinity for the stimulating antigen.

somatic pairing *(Biol.)*. Closely paired arrangement of **chromatids** in **polytene** chromosomes.

somatoblast *(Zool.)*. In development, a cell which will give rise to somatic cells.

somatoform disorders *(Behav.)*. Conditions in which psychological conflicts take on a somatic or physical form, includes the hypochondrias and conversion disorders.

somatogenic *(Zool.)*. Arising as the result of external stumuli: developing from somatic cells, as opposed to germ cells.

somatopleure *(Zool.)*. The outer body wall of coelomate animals: the outer layer of the mesoblast which contributes to the outer body wall; cf. **splanchnopleure**. adj. *somatopleural*.

somatostatin *(Biol.)*. A peptide of the hypothalamus which controls the secretion of *somatotropin* by the pituitary.

somatotropin *(Biol.)*. A peptide secreted by the anterior pituitary which causes the liver to produce *stomatomedins* which stimulate the growth of bones and muscle.

somatotropism *(Bot.)*. Directed growth movements in plants so that the members come to be placed in a definite position in relation to the substratum.

somatotype theory *(Behav.)*. A theory proposed by Sheldon who suggested that bodily characteristics reflect clusters of personality traits; the theory is no longer held in any serious regard, although the terms **ectomorph**, **mesomorph** and **endomorph** still occasionally appear.

somite *(Zool.)*. One of the divisions or segments of the body in a metameric animal: a mesoblastic segment in a developing embryo.

sommaite *(Geol.)*. An alkaline igneous rock, similar to essexite but with leucite in place of nepheline.

Sommerfeld atom *(Phys.)*. Atomic model developed from Bohr atom, but allowing for elliptic orbits with radial, azimuthal, magnetic and spin quantum numbers. Modern theories modify this by regarding the electrons as forming a cloud, the density of which is described in terms of their wave function.

somnambulism *(Behav.)*. *Sleepwalking*. A dissociative disorder in which the person walks while asleep.

somnambulism *(Med.)*. (1) The fact or habit of walking in the sleep. (2) A hysterical state of automatism in which the patient performs acts of which he is unaware at the time or when he comes out of the state.

sonar *(Acous.)*. Abbrev. for *SOund NAvigation and Ranging*. See **asdic**, **echo sounding**.

sonde *(Meteor.)*. Small telemetering system in satellite, rocket, or balloon.

sone *(Acous.)*. Unit of loudness equal to a tone of 1 kHz at a level of 40 dB above the threshold of the listener.

sonic boom *(Aero.)*. Noise phenomenon due to the shock-waves projected outwards and backwards through the atmosphere from leading and trailing edges of an aircraft travelling at **supersonic speed**. The waves are discontinuities of atmospheric pressure and are heard as a characteristic double report which may be of sufficient intensity to cause damage to buildings, etc.

sonic fatigue *(Acous.)*. The deterioration (cracks) and failure which are caused in materials (metal, concrete etc.) by strong stress fluctuations of sound waves of high intensity. Sometimes happens in mechanical systems if the eigenfrequency is excited. Important for spacecraft and aircraft.

sonic line *(Phys.)*. The locus of field points in 2-dimensional flow where the medium attains the velocity of sound under local conditions.

sonics *(Phys.)*. General term for study of mechanical vibrations in matter.

sonobuoy *(Telecomm.)*. Equipment dropped and floated on the sea, to pick up aqueous noise and transmit a bearing of it to aircraft; three of such bearings enable the aircraft to 'fix' the source of underwater noise, e.g. from submarines.

sonogram *(Acous.)*. Three-dimensional representation of a sound signal, the three co-ordinates being frequency, time and intensity. Intensity is often represented by shading.

sorbic acid *(Chem.)*. $CH_3.(CH{:}CH)_2.COOH$. *Hexa-2,4-dienoic acid*. Solid, mp 133–135°C. Used as a preservative for pharmaceuticals, cosmetics, etc. Anti-fungal for edible products.

sorbite *(Eng.)*. Obsolete term for decomposition product of cementite and ferrite in solid solution, of heterogeneous structure, thus distinguishing it from **pearlite**.

sorbitol *(Chem.)*. A hexahydric alcohol, isomeric with mannitol.

sorbose *(Chem.)*. A ketohexose, an isomer of fructose.

sordes *(Med.)*. Foul, dark brown crusts which collect on the lips and the teeth in prolonged fever (e.g. in typhoid fever).

sore-heels *(Vet.)*. See **grease**.

Sorel's cement *(Build.)*. Calcined magnesite (MgO) mixed with a solution of magnesium chloride of a concentration of about 20° Baumé. It sets within a few hours to a hard mass. The basis of artificial flooring cements.

Sørensen's formol titration *(Chem.)*. See **formol titration**.

sore-shins *(Vet.)*. An inflammation of the periosteum of the large metacarpus (shin bone), or occasionally of the metatarsus, of young horses.

Soret effect *(Phys.)*. See **thermal diffusion**.

soroban *(Maths.)*. Japanese abacus.

sorosilicate *(Min.)*. A silicate mineral whose atomic structure contains paired silicon-oxygen tetrahedra (Si_2O_7 groups), e.g. *melilite*.

sorption *(Chem.)*. A general term for the processes of absorption, adsorption, chemisorption and persorption.

sorting *(Build.)*. The process adopted when a roof is to be covered by slates of different sizes; the largest slates are nailed at the eaves and the smallest at the ridge.

sorting *(Maths.)*. The process of selecting the elements of a set with a common attribute from a random collection.

sorts *(Print.)*. Particular type letters as distinct from complete founts. A case is 'out of sorts' when one or more of the boxes is empty. 'Outside sorts' are required for mathematical and foreign language setting.

sorus *(Bot.)*. (1) Generally, a cluster of reproductive structures on a thallus. (2) In ferns, a cluster of sporangia on a leaf, often covered by an *indusium* which acts as a protective structure.

sough *(Civ.Eng.)*. A drain at the foot of a slope, e.g. an embankment, to receive and carry away surface waters from it.

sound *(Acous.)*. The periodic mechanical vibrations and waves in gases, liquids and solid elastic media. Specifically the sensation felt when the eardrum is acted upon by air vibrations within a limited frequency range, i.e. 20 Hz to 20 kHz. Sound of a frequency below 20 Hz is called infrasound and above 20 kHz ultrasound.

sound *(Med.)*. A solid rod used for exploring hollow viscera (e.g. the bladder, the uterus) or for dilating stenosed passages.

sound absorption factor *(Acous.)*. See **acoustic absorption factor**.

sound analyser *(Acous.)*. One which measures each frequency, amplitude or phase of sound. See **octave analyser, octave filter, real time analyser**.

sound articulation *(Acous.)*. Percentage of all elementary speech sounds received correctly, when **logatoms** are called over a circuit or in an auditorium, in a standard manner. See **intelligibility**.

sound boarding *(Build.)*. *Deafening boarding* in Scotland. Boards fitted in between the joists of a floor, to carry the *pugging/deafening* which is to insulate the room from sound and smell from the room below. Obsolescent.

sound bridge *(Acous.)*. (1) Rigid connection piece between the two plates of a *double wall*, increasing the sound transmission considerably. (2) Gaps, pipes, chimneys etc. which increase the sound transmission in buildings.

sound camera *(Image Tech.)*. A continuous motion camera in which photographic film is exposed when transferring from magnetic recording to optical sound track.

sound channel *(Image Tech.)*. The carrier frequency with its associated sidebands which are involved in the transmission of the sound in TV.

sound energy density *(Acous.)*. See **energy density of sound**.

sound field *(Acous.)*. Volume filled with a medium in which sound waves propagate.

sound gate *(Image Tech.)*. The position on a motion picture projector where the film passes the **scanning slit** in photographic sound reproduction.

sound head *(Image Tech.)*. That unit in a projector which reproduces the sound-track on the edge of the film.

sounding *(Surv.)*. The depth of an underwater point below some chosen reference datum. Cf. **reduced level**.

sounding balloon *(Meteor.)*. A small free balloon carrying a meteorograph, used for obtaining records of temperature, pressure and humidity in the upper atmosphere.

sounding line *(Surv.)*. A stout cord, divided into fathoms and feet and weighted at one extremity with a lead weight; used in finding soundings.

sounding rocket *(Space)*. Unmanned rocket-powered vehicle used for research purposes (atmosphere, astronomy, microgravity, etc.) which does not go into Earth orbit, but follows a sub-orbital trajectory.

sound insulation *(Build.)*. The property, possessed in varying degrees by different materials, of blocking the transmission of sound. High density offers greatest resistance.

sound intensity *(Acous.)*. Flux of sound power through unit area normal to the direction of propagation. If p is the acoustic pressure and v the velocity of the medium particles in the direction of propagation, the intensity is the time average of the product pv.

sound intensity level *(Acous.)*. At any audio-frequency, the intensity of a sound, expressed in decibels above an arbitrary level, 10^{12} W/m^2, which is equivalent, in air, to a pressure of 20 μPa.

sound interval *(Acous.)*. The interval between two sounds is the ratio of their fundamental frequencies.

sound level *(Acous.)*. Loosely used for **sound pressure level** and **sound intensity level**.

sound-level meter *(Acous.)*. Microphone-amplifier-indicator assembly which indicates total intensity in decibels above an arbitrary zero. Also includes suitable weighting networks and time averaging. See **fast** and **slow**.

sound locator *(Acous.)*. Apparatus for determining the direction of arrival of sound waves, particularly the noise from aircraft and submarines.

sound-picture spoilation *(Image Tech.)*. The leakage of the sound signal of a TV transmission into the vision circuit of a receiver, giving rise to alternate dark and light horizontal bands in the screen picture.

sound pressure *(Acous.)*. Fluctuating mean component of the pressure in the medium containing a sound wave, as opposed to the constant component, e.g. atmospheric or hydrostatic pressure.

sound pressure level *(Acous.)*. Sound pressure expressed in **decibels** relative to a reference pressure which is taken as 20 μPa or 2×10^{-5} N/m^2.

sound probe *(Acous.)*. Usually a very small microphone to minimize the disturbance of the sound field which is being measured. It is often equipped with a fine tube which is inserted in the field.

sound ranging *(Acous.)*. Determination of locality of a source of sound, e.g. from guns, by simultaneously recording through spaced microphones and making deductions from the differences of times of arrival.

sound recording *(Acous.)*. The practice of registering sound so that it can be reproduced at some subsequent time. See **recording, recorder**.

sound reduction index *(Acous.)*. Same as **transmission loss**.

sound reflection *(Acous.)*. Return of sound waves from discontinuities (surface, tube end etc.). See also **reflection coefficient**.

sound-reflection factor *(Acous.)*. The percentage of energy reflected from a discontinuity; proportional to the square of the **reflection coefficient**.

sound re-inforcement system *(Acous.)*. One used to increase the uniform sound intensity in large halls by employing public address systems. Also by the control of reverberation time using an electronic method, involving microphones, resonators and loudspeakers.

sound-reproducing system *(Acous.)*. That comprising sound recording, transducers and equipment for sound reproduction.

sound spectrograph *(Acous.)*. An electronic instrument which makes **sonograms**.

sound stage *(Image Tech.)*. The main floor of a motion-picture studio on which sets are built and the artists perform during shooting.

sound track *(Image Tech.)*. Track on magnetic tape or ciné film on which sound signals have been or can be recorded.

sound velocity *(Phys.)*. See **velocity of sound**.

source *(Elec.Eng.)*. An active pair of terminals, which can deliver power to a load.

source *(Electronics)*. The electrode from which majority carriers flow into the **channel** of a field-effect transistor. Comparable with the emitter in a conventional bipolar transistor.

source *(Phys.)*. In radioactivity, the origin of α-, β-, or γ-rays.

source code *(Comp.)*. See **source program**.

source deck *(Comp.)*. Stack of program cards containing a **source program** for a computer operated by punched cards. Obsolescent.

source impedance *(Elec.Eng.)*. See **output impedance**.

source language *(Comp.)*. Language in which the program is first written.

source program *(Comp.)*. Program as written by the programmer using a programming language; it must be assembled, compiled or interpreted before it can be executed. Also *source code*. Cf. **object program**.

source range *(Nuc.Eng.)*. See **start-up procedure**.

source resistance *(Elec.Eng.)*. See **internal resistance**.

sources of neutrons *(Phys.)*. Fast neutrons are obtained by (i) nuclear transformations and (ii) fission in nuclear reactors. To obtain slow neutrons, a moderator such as paraffin wax is used. For laboratory sources, see **neutron source**.

source strength *(Radiol.)*. Activity of radioactive source expressed in disintegrations per second.

sour gas *(Min.Ext.)*. Natural gas containing impurities mainly the poisonous hydrogen sulphide.

souring *(Textiles)*. The treatment of yarn or cloth with dilute acid, frequently after an alkali process to ensure that no alkali remains.

South African jade *(Min.)*. See **Transvaal jade**.

Southern blot *(Biol.)*. Method of revealing rare DNA fragments in a complex mixture of DNA. Gel electrophoresis is used to separate the fragments which are then denatured and transferred to a nitrocellulose sheet. The fragments are hybridized to a radioactive probe and their position shown by autoradiography. Analogous procedures are available for RNA and proteins.

Southern Cross *(Astron.)*. A striking constellation of the southern hemisphere, visible only in latitudes below 30° N. It is a cruciform group of 4 stars, having the 2 bright stars α and β Centauri some way to the east, which makes it easy to identify.

southern oscillation *(Meteor.)*. A slow fluctuating exchange of air between the eastern tropical Pacific on the one hand and the Indian Ocean and Indonesia on the other, with a corresponding negative correlation between annual mean pressure values over the two areas. The interval between corresponding points in successive cycles varies from 1 to 5 years, and the oscillation is linked to variations in sea-surface temperature and the pattern of rainfall.

southing *(Surv.)*. Measured difference southward from a reference latitude.

south pole *(Phys.)*. See **pole**.

Soxhlet apparatus *(Chem.)*. A laboratory apparatus for the continuous extraction of a solid substance with a solvent, consisting of a distillation flask, a reflux condenser, and a cylindrical vessel fitted between them to which a syphon system is attached.

Soyuz *(Space)*. A three-man vehicle used by the USSR for ferrying crews to and from its orbital stations. Part of the spacecraft is a re-entry module used for the return to Earth, employing parachutes and rockets for the actual landing.

SP *(Build.)*. Abbrev. for *Soil Pipe*.

space *(Print.)*. A type less than type height, and thinner than a quadrat; used to separate words etc.

space *(Space)*. The near-vacuum surrounding all bodies in the Universe; in particular, interplanetary space is that extra-atmospheric region between the elements of the solar system; interstellar space is that between stars and intergalactic between galaxies.

space *(Telecomm.)*. The period of time in transmission during which a Morse key is open, i.e. not in contact.

space bands *(Print.)*. On the Linotype and Intertype machines, the expandable steel wedges inserted between each word, which are pushed upwards until they expand the line of matrices to the full measure.

space box *(Print.)*. A small box with a few divisions for different spaces, used when making corrections to type matter.

space charge *(Electronics)*. Collection or cloud of electrons near a source, e.g. heated cathode, which stabilizes emission by repelling those not attracted to another electrode, the field in its region being very low.

space-charge limitation *(Electronics)*. Condition in a thermionic valve when electron current leaving a cathode is limited by balance between attractive electric forces from other electrodes and repulsion within space charge. See **Child-Langmuir equation**.

space commercialization *(Space)*. Exploitation of space and space-produced products by industry for profit.

spacecraft *(Space)*. Vehicle containing all the necessary subsystems to support a payload for the performance of a particular space mission. It may be manned or unmanned. Also *space vehicle*.

space current *(Electronics)*. See **thermionic current**.

spaced antennae *(Telecomm.)*. Those used with diversity systems, or to enhance directivity.

spaced-loop direction-finder *(Telecomm.)*. One including two loops spaced sufficiently in terms of the wavelength to enhance their normal directivity, as exhibited by the polar diagram of response.

spaced slating *(Build.)*. Slating laid with gaps between adjacent slates in any course.

space dyeing *(Textiles)*. Production of irregularly multicoloured yarns by applying various colours at intervals along a single yarn or pad of yarns often by a printing process.

space environment *(Space)*. The extra-terrestrial conditions existing in a particular region between Earth and distant bodies. Specifically, it involves the phenomena of vacuum, fields, particles and related effects.

space factor *(Elec.Eng.)*. The ratio of the active cross-sectional area of an insulated conductor to the total area occupied by it.

space group *(Crystal.)*. Classification of crystal lattice structures into groups with corresponding symmetry elements.

Spacelab *(Space)*. Reusable orbital research laboratory which extends the Space Shuttle capabilities and is carried in its cargo bay; modular in construction it comprises a manned module and unmanned pallets, permitting module-only, pallet-only or mixed modes to be used.

space lattice *(Crystal.)*. Three-dimensional regular arrangement of atoms characteristic of a particular crystal structure. There are 14 such simple symmetrical arrangements, known as *Bravais lattices*. See also **symmetry class**.

space parallax *(Acous.)*. The difference in bearing between a moving object, such as a machine in flight, and the direction of arrival of the sound waves emitted by it. This arises from the comparable velocity of flight with that of the propagation of sound waves.

space parasite *(Bot.)*. A plant which inhabits intercellular spaces in another plant, obtaining shelter but possibly taking nothing else.

space-reflection symmetry *(Phys.)*. See **parity**.

space research *(Space)*. Investigation of the space environment, its effect on things and its use as a vantage point for viewing the Earth and deep space. Generally, research which is performed with the aid of a space system.

Space Shuttle *(Space)*. Manned ground-to-orbit and return transportation system capable of lifting payloads to orbit, in-orbit experimentation, in-orbit servicing, repair and capture, and payload return. The manned part and cargo carrier is the *Orbiter*. It is launched by rocket propulsion, using its own engines which burn propellants (liquid oxygen and liquid hydrogen) stored in a large external tank and two solid-fuel boosters. The latter drop away after two minutes and the tank is jettisoned just before orbit is attained. Orbit is achieved by means of the Orbiter's orbital manoeuvring sytem burning a **hypergolic** mixture of monomethyl hydrazine and nitrogen tetroxide. In orbit, the desired orientation is ensured by small rocket engines using the same hypergolic propellants. On completion of the mission, a retro-rocket firing causes the *reusable* Orbiter to re-enter the atmosphere and land horizontally like a glider. The Space Shuttle has a total weight of about 2 000 000 kg at launch and a payload capability of 30 000 kg when launched due East from Kennedy Space Center in Florida.

space station *(Space)*. Several manned modules and/or unmanned platforms, launched separately, but joined together to form a base which permits a permanent presence in space for exploration and exploitation of the environment. Can also act as a staging post and refurbishment centre for other space activities.

space suit *(Space)*. Specially designed suit which, when donned, allows an astronaut to operate in a space environment. The design includes the provision of a pressurized oxygen supply, and provides for temperature control and the purification of exhaled gases. Sometimes called a *pressure suit*.

space system *(Space)*. The total assembly of space-related hardware and software; it can refer to the launch system plus payload or a spacecraft plus payload and includes the related ground segment.

space-time *(Phys.)*. Normal 3-dimensional space plus dimension of time, modified by gravity in relativity theory.

space velocity *(Astron.)*. The rate and direction of a star's motion in space of three dimensions, as deduced from its observable components (*a*) in the line of sight by the spectroscope, and (*b*) perpendicular to the line of sight by proper motions.

space wave *(Telecomm.)*. Wave from an antenna which is not a ground wave, but which travels rectilinearly in space, apart from reflection (negative *refraction* and *bending*) when it enters an ionized region; also called *sky wave* (or *ray*).

spacing material *(Print.)*. Lower than type height; there are spaces for each size of type and a wide variety made to 12-pt sizes for making up pages and spacing them apart. See **clumps**, **furniture**, **leads**, **quadrat**, **quotations**, **reglet**.

spacing ratio *(Elec.Eng.)*. The ratio of the distance between equally spaced electric lights to their vertical distance above the plane to be illuminated.

spacing wave *(Telecomm.)*. Emitted wave corresponding to *spacing* impulses in a code, e.g. Morse code; also called *back wave*.

spadiceous *(Zool.)*. Shaped like a palm branch. Also *spadiciform*, *spadicose*.

spadix *(Bot.)*. A spike, the axis of which is fleshy. The characteristic inflorescence of the arum family; commonly associated with a **spathe**.

spall *(Build.)*. Fragment detached by weather action, by internal movement, or by the process of chiselling; also called *galet*.

spallation *(Phys.)*. Any nuclear reaction when several particles result from a collision, e.g. cosmic rays with atoms of the atmosphere; chain-reaction in a nuclear reactor or weapon.

spallation neutron source *(Phys.)*. A powerful pulsed neutron source for research. Protons, accelerated in a synchrotron, are focused onto a target of ^{238}U; 25 to 30 neutrons per proton are released, moderated and collimated into beams.

spalling *(Build.,Civ.Eng.)*. (1) The breaking off of fragments on a concrete surface through weathering. Generally caused by infiltration of water through hair cracks and subsequent freezing. (2) The operation of breaking off splinters of stone from a block by slanting blows with a chisel, when dressing to shape.

span *(Aero.)*. The distance between the wing tips of an aircraft.

span *(Civ.Eng.)*. Horizontal distance between supports of a bridge, arch etc.

span *(Elec.Eng.)*. (1) The distance between two transmission-line towers. (2) The number of slots separating the two sides of an armature coil. Also called *throw*.

spandrel *(Civ.Eng.)*. The space between the haunches and the underside of an arch.

spandrel step *(Arch.)*. An individual step in a stair, which consists of a solid block, triangular in section, arranged so that one face is parallel to the slope of the stair.

spandrel wall *(Arch.)*. A wall constructed upon the extrados of an arch.

Spanish topaz *(Min.)*. Not a true topaz but an orange-brown quartz, the colour resembling that of the honey-brown Brazilian topaz. It is often amethyst which has been heat treated. See **citrine**.

span loading *(Aero.)*. The gross weight of an aircraft or glider divided by the square of the span.

spanner *(Build.)*. A tool used for turning a nut. See **box-**, **ring-**, **C-**.

span pole *(Elec.Eng.)*. The pole to which the span wires are attached.

span saw *(Build.)*. See **frame saw**.

span wire *(Elec.Eng.)*. One of several wires by which the trolley wire of a tramway or trolley-bus system is suspended from street poles or buildings.

spar *(Aero.)*. A main spanwise member of an aerofoil or control surface. The term can be applied either to individual beam(s) designed to resist bending, or to the box structure of spanwise vertical webs, transverse ribs, and skin which form a torsion box.

spar *(For.)*. A round timber more than 6 in in diameter in the middle.

spar *(Min.)*. Transparent to translucent crystalline mineral with vitreous lustre and clean cleavage planes, e.g. *fluorspar*. See **Iceland-**.

Sparagmite *(Geol.)*. A comprehensive term which includes the late Precambrian rocks of Scandinavia. These, like the Torridonian Sandstone of N. Scotland, consist of conglomerates and red feldspathic grits and arkoses.

spar frame *(Aero.)*. A specially strong transverse fuselage, or hull, frame to which a wing spar is attached.

spark *(Elec.Eng.)*. Breakdown of insulation between two conductors, such that the field is sufficient to cause ionization and rapid discharge. See **arc**, **field discharge**, **lightning**.

spark absorber *(Elec.Eng.)*. Resistance and/or capacitor placed across a break in an electrical circuit to damp any possible oscillatory circuit which would tend to maintain an arc or spark when a current is interrupted.

spark chamber *(Nuc.Eng.)*. Radiation detector for rendering visible the tracks of ionizing particles by the sparks formed following the ionization. Consists of a stack of parallel metal plates with the electric field between them raised nearly to the breakdown point.

spark coil *(Elec.Eng.)*. Induction or Ruhmkorff coil used as the source of high voltage in a spark transmitter. Obsolete for radio, but used in motor-cars.

spark erosion *(Eng.)*. Electrochemical metal machining process in which an electrode of male form is maintained very close to the workpiece, both being submerged in a dielectric liquid. High local temperatures are produced by passing a current through the gap, detaching and repelling particles from the workpiece.

spark gap *(Elec.Eng.)*. In simplest form, two shaped electrodes separated by a dielectric which breaks down at a quasi-constant voltage gradient; this may be triggered by an externally applied electric field. Main uses: (1) voltage limiting safety device (e.g. *lightning arrestor*); (2) low-reactance switch (e.g. *Marx generator*); (3) generator of electromagnetic waves (e.g. *spark system*); (4) concentrated energy deposition (e.g. *spark erosion*).

spark-gap generator *(Elec.Eng.)*. Radiofrequency generator for induction heating in which a capacitor is charged from a high-tension transformer and discharged through an oscillatory circuit when a spark gap breaks down.

spark-gap modulation *(Telecomm.)*. A method of pulse modulation in which a pulse-forming line discharges across a spark gap in the transmitter circuit.

sparking *(Elec.Eng.)*. The occurrence of a spark discharge between the brushes and the surface of a commutator.

sparking contact *(Elec.Eng.)*. An auxiliary contact used on circuit-breakers; designed to make circuit before, and to break circuit after, the main contact, so that any sparking takes place on the auxiliary contact. It has removable contact tips, usually of carbon.

sparking limit *(Elec.Eng.)*. The limiting output of a d.c. machine as determined by considerations of commutator sparking.

sparking plug *(Autos.)*. A plug screwed into the cylinder head of a petrol engine for ignition purposes, a spark gap being provided between an insulated central electrode and one or more earthed points.

sparking potential *(Elec.Eng.)*. Potential difference between the ends of an insulator, sufficient to cause a spark discharge through or over the insulator. Also called *sparkover potential*.

sparkless commutation *(Elec.Eng.)*. A term applied to methods of current commutation in which the reactance voltage is neutralized before actual commutation occurs, so that the formation of a commutation spark or arc is avoided.

spark machining *(Eng.)*. See **spark erosion**.

spark photography *(Image Tech.)*. High-speed technique employing a high-intensity electric spark for illumination. Used in ballistics and in *schlieren photography*.

spark resistance *(Elec.Eng.)*. That between electrodes after a discharge has commenced; if excessive and in an oscillatory circuit, it causes loss of power and a high decrement.

spark spectra *(Phys.)*. The most important way of exciting spectra is by means of an electric spark. The high temperature reached will generate the spectrum lines of multiply ionized atoms as well as of uncharged and singly ionized ones. (As distinct from *arc spectrum*.) Also evaporation of metal from the electrodes leads to additional lines not associated with the gas through which the discharge takes place.

sparteine *(Chem.)*. $C_{15}H_{26}N_2$, an alkaloid of the quinuclidine group, obtained from the branches of the common broom, *Cytisus scoparius*; a colourless oil, bp 188°C (18 mm), sparingly soluble in water, soluble in ethanol, trichloromethane or ethoxyethane. It resembles coniine in its physiological action.

spasm *(Zool.)*. Involuntary contraction of muscle fibres. adj. *spasmodic*.

spasmodic torticollis *(Med.)*. A nervous disorder in which the muscles of either side, or both sides, of the neck are in a state of continuous or of intermittent spasm.

spasmus nutans *(Med.)*. *Nodding spasm*. Rhythmic nodding of the head seen in babies in the first year of life.

spastic *(Med.)*. Of the nature of spasm (sudden involuntary contraction) of muscle: characterized or affected by muscular spasm: rigid, or in a state of continuous spasm. May be used to describe *spasm of colon*, *spasm of ureter* or *spastic paralysis*.

spastic paralysis *(Med.)*. Paralysis of the voluntary movements caused by prenatal brain damage; characterized by spasm of muscles.

spathe *(Bot.)*. A large, sometimes coloured or showy bract which subtends and may enclose a **spadix**.

spathic iron *(Min.)*. See **siderite (1)**.

spathulate, spatulate *(Bot.)*. Spoon- or paddle-shaped.

spatial filtering *(Phys.)*. The removal of part of the optical diffraction pattern by opaque masks so that the high frequency components in the image formation are removed or enhanced. Removal rounds sharp edges in the image; enhancement sharpens edges.

spatial summation *(Behav.)*. The summational effect of stimuli spread over space (e.g. the brightness of a beam of light has a lower absolute threshold for larger diameters of beam).

spatter finish *(Build.)*. A decorative finish produced by spraying thick flecks of paints on to a surface using a low 'atomizing' pressure. Can also be produced by flicking colour from a brush.

spatula *(Zool.)*. Any spoon-shaped structure.

spatulate *(Zool.)*. Shaped like a spoon.

spavin *(Vet.)*. Chronic arthritis of the hock joint of a horse.

spawn *(Bot.)*. The mycelium of a fungus, especially the mycelial preparations used to propagate the cultivated mushroom.

spawn *(Zool.)*. To deposit eggs or discharge spermatozoa: a collection of eggs, such as that deposited by many Fish.

spay *(Vet.)*. To remove or destroy the ovaries.

SPC exchange *(Telecomm.)*. See **stored-program control**.

speaking pair *(Telecomm.)*. When wires are grouped for trunking through automatic switching, the pair carrying the speech currents is termed the *speaking pair*, as contrasted with the guard wire, private wire, or meter wire.

spear pyrites *(Min.)*. The name given to twin crystals of marcasite which show re-entrant angles, in form somewhat like the head of a spear. Cf. **cockscomb pyrite**.

special character *(Comp.)*. Any character recognized by a particular computing system which is not **alphanumeric**, e.g. **control character**.

special effects *(Image Tech.)*. General term for scenes in motion picture and video productions where the original image recorded by the camera is substantially modified by subsequent technical operations, for instance, combining pictures from several sources as though they formed a single shot.

specialist *(Ecol.)*. Organism with a restricted food source, living in a restricted habitat, often displaying specific behaviour or structural adaptations.

special relativity *(Phys.)*. See **relativity**.

special rules zone *(Aero.)*. A 3-dimensional space, under **air traffic control**, wherein aircraft must obey special instructions.

special unitary groups *(Phys.)*. A scheme which predicts that as far as *strong interactions* are concerned, elementary particles can be grouped into multiplets, the particles in each multiplet being considered as different states of the same particle. This unitary group SU(3) has been successful in correlating the range of particles and in predicting the existence of hitherto undiscovered particles, notably the particle Ω^-. The *isospin* unitary group SU(2) is a sub-group of SU(3). Other groups are being explored in connection with the explanation of strong interactions in terms of *quarks*.

speciation *(Bot.,Zool.)*. Formation of new biological species including formation of polyploids.

species *(Biol.)*. A group of individuals that (1) actually or potentially interbreed with each other but not with other such groups, (2) shows continuous morphological variation within the group but which is distinct from other such groups. Taxonomically, species are grouped into genera and divided into subspecies and varieties or, horticulturally, into cultivars. In the system of **binomial nomenclature** of plants and animals, the second name (i.e. the name by which the species is distinguished from other species of the same genus) is termed the *specific epithet* or *specific name*. The latter, however, correctly refers to the full name, e.g. *Lilium candidum* (Madonna Lily), where *candidum* is the *specific epithet*.

species/area curve *(Ecol.)*. Curve relating the number of species found (*y* axis) to the area over which the observer searched (*x* axis). The shape of the curve provides information about **diversity** and **richness**.

specific *(Genrl.)*. Term used generally to indicate that the property described relates to unit mass of the substance involved, e.g. specific entropy is entropy per kilogram.

specific *(Med.)*. A treatment or medicine effective against a particular disease.

specific *(Zool.)*. Of a parasite, restricted to a particular host. n. *specificity*.

specific activity *(Radiol.)*. See **activity**.

specification *(Eng.)*. (1) A detailed description, including dimensions and other quantities, of the function, construction, materials, and quality of a manufactured article or an engineering project. (2) A description, by an applicant for a patent, of the operation and purpose of the invention.

specific characters *(Biol.)*. The constant characteristics by which a species is distinguished.

specific charge *(Phys.)*. Charge/mass ratio of elementary particle, e.g. the ratio e/m_e of the electronic charge to the rest mass of the electron $= 1.759 \times 10^{11}$ coulomb per kilogram.

specific damping *(Elec.Eng.)*. The attenuation constant per kilometre of a cable.

specific depression *(Phys.)*. See **depression of freezing point**.

specific dielectric strength *(Elec.Eng.)*. The **dielectric strength** of an insulating material, expressed in V/mm.

specific dynamic action *(Biol.)*. The special calorigenic property of foodstuffs, and particularly of proteins, of raising the metabolic rate after ingestion by an amount in excess of their calorific value. May be expressed as the ratio of the calories in excess of the basal to urinary nitrogen in excess of the basal.

specific electric loading *(Elec.Eng.)*. The electric loading, in ampere-conductors, of the armature of a machine per cm of circumference.

specific excess power *(Aero.)*. Thrust power available to an aircraft in excess of that required to fly at a particular constant height and speed, thus being usable for climbing, accelerating or turning.

specific fuel consumption *(Autos.)*. The mass of fuel used by an engine per unit energy delivered; generally expressed in pounds/BHP-hour or kg/Mj.

specific gravity *(Phys.)*. See **relative density**.

specific gravity bottle *(Phys.)*. See **density bottle**.

specific heat capacities of gases *(Phys.)*. Gases have two values of specific heat capacity: c_p, the s.h.c. when the gas is heated and allowed to expand against a constant pressure, and c_v, the s.h.c. when the gas is heated while enclosed within a constant volume. See **ratio of specific heat capacities**.

specific heat capacity *(Phys.)*. The quantity of heat which unit mass of a substance requires to raise its temperature by one degree. This definition is true for any system of units, including SI, but whereas in all earlier systems a unit of heat was defined by putting the s.h.c. of water equal to unity, SI employs a single unit, the joule, for all forms of energy including heat, which makes the s.h.c. of water 4.1868 kJ/kg K. See **mechanical equivalent of heat**.

specific humidity *(Meteor.)*. The ratio of the mass of water vapour in a sample of moist air to the total mass of the air.

specific impulse *(Space)*. A measure of rocket performance, defined as the thrust in kg (force) obtained from one kg (mass) of propellant burned for one second. It is denoted by Isp and has the units of seconds. It is related to the exhaust velocity (v_e) by:

$$I_{sp} = \frac{v_e}{g}$$

where g is the acceleration due to gravity at the Earth's surface. A high value of Isp and, hence, v_e, is desirable.

specific inductive capacity *(Elec.Eng.)*. See **permittivity**.

specific ionization *(Phys.)*. Number of ion pairs formed by ionizing particle per cm of path. Also called the *total specific ionization* to avoid confusion with the *primary specific ionization* which is defined as the number of ion clusters produced for unit length of track.

specific latent heat *(Phys.)*. See **latent heat**.

specific magnetic loading *(Elec.Eng.)*. The average flux density (i.e. the total magnetic loading divided by the peripheral area) in the armature of a machine.

specific name *(Bot.,Zool.)*. See under **species**.

specific output *(Elec.Eng.)*. The ratio of the electrical output of a machine to its weight, its volume, or some other function of its dimensions. Cf. **specific torque coefficient**.

specific permeability *(Elec.Eng.)*. Same as **relative permeability**. See **permeability**.

specific power *(Nuc.Eng.)*. US for **fuel rating**.

specific reaction rate *(Chem.)*. See **rate constant**.

specific refraction *(Chem.)*. The molecular refraction of a compound, defined by the Lorentz-Lorenz equation, divided by the molecular weight. Symbol r.

specific resistance *(Elec.Eng.)*. See **resistivity**.

specific rotation *(Chem.,Phys.)*. The angle through which the plane of polarization of a ray of sodium D light would be rotated by a column of liquid 1 dm in length, containing 1 g of an optically active substance per cm^3. Also *rotary power*.

specific surface *(Powder Tech.)*. The surface area of the particles in a unit mass of the powder determined under stated conditions. The term is also sometimes used to describe the surface area of the particles in unit volume of the powder.

specific temperature rise *(Elec.Eng.)*. The temperature rise of an electrical machine per unit of radiating surface.

specific torque coefficient *(Elec.Eng.)*. A coefficient used in the design of electrical machines, giving a figure representing the torque per unit of volume enclosed by the air-gap periphery. Also known as *output coefficient*, *Esson coefficient*.

specific volume *(Phys.)*. The volume of unit mass; the reciprocal of density.

speckle interferometry *(Astron.)*. A technique using the principle of interference of light which enables very small angles such as the diameters of stars, to be measured directly.

specpure *(Chem.)*. A trade term for *spectroscopically pure*. See **spectroscopic analysis**.

spectacle crown *(Glass)*. White crown glass of refractive index 1.523, used for ophthalmic purposes.

spectacle flint *(Glass)*. A glass of high refractive index which is fused to **spectacle crown** in the manufacture of bifocal spectacle lenses.

spectral characteristic *(Phys.)*. Graph of photocell sensitivity, as related to wavelength of radiation.

spectral colour *(Phys.)*. Colour with degrees of saturation between no-hue and a pure spectral colour on the rim of the chromaticity diagram.

spectral distribution curve *(Phys.)*. The curve showing the relation between the radiant energy and the wavelength of the radiation from a light source.

spectral line *(Phys.)*. Component consisting of a very narrow band of frequencies isolated in a spectrum. These are due to similar quanta produced by corresponding electron transitions in atoms. The lines are broadened into bands when the equivalent process takes place in molecules.

spectral sensitivity *(Image Tech.)*. The comparative response of an emulsion to exposures of light of different wavelengths but constant intensity.

spectral series *(Phys.)*. Group of related spectrum lines produced by electron transitions from different initial energy levels to the same final one. The recognition and measurement of series has been of great importance in atomic and quantum theories.

spectral-shift-controlled reactor *(Nuc.Eng.)*. One in which loss of reactivity which would occur on burn-up is compensated by *softening* the neutron spectrum e.g. by varying the heavy water/light water ratio in the reactor coolant, or by change of coolant temperature. Abbrev. *SSCR*.

spectral transmission *(Image Tech.)*. The relative transmission, opacity, or density of a filter in respect of light of different wavelengths.

spectral types *(Astron.)*. The Harvard classification of stars according to their spectra, giving a graded list represented by the letters (W) O B A F G K M S (R N), which represents a sequence (called the *main sequence*) of descending temperature, the O type stars being hot, white, and gaseous, while the cooler M type show molecular band spectra.

spectre of the Brocken *(Meteor.)*. The shadow of an observer cast by the sun on to a bank of mist. The phenomenon, often seen from a hilltop, may present the illusion that the shadow is a gigantic form seen through the mist. Also *Brockenspectre*.

spectrin *(Biol.)*. A family of closely related cytoskeletal proteins which consist of an α-β heterodimer of 2 high molecular weight polypeptides. The eponymous member of the spectrin family constitutes a major component of the erythrocyte membrane. Other members are found in brain and intestinal epithelium.

spectrograph *(Phys.)*. Normally used of spectroscope designed for use over wide range of frequencies (well beyond visible spectrum) and recording the spectrum photographically. The **mass spectrograph** separates particles of different specific charge in a manner which is to the separation of spectrum lines of an optical spectrum.

spectroheliogram *(Astron.)*. The recorded result of an exposure on the sun by the spectroheliograph.

spectroheliograph *(Astron.)*. An instrument for photographing the sun in monochromatic light. It consists essentially of a direct-vision spectroscope, with a second slit instead of an eyepiece, which can be set so that only light of a desired wavelength passes through it on to a photographic plate.

spectrohelioscope *(Astron.)*. An instrument in principle the same as the spectroheliograph, but adapted for visual use by the employment of a rapidly oscillating slit which, by the persistency of vision, enables an image of the whole solar disk to be viewed in light of one wavelength; it also detects the velocities of moving gases in the solar atmosphere by an adjustment called the 'line-shifter'.

spectrometer *(Phys.)*. Instrument used for measurements of wavelength or energy distribution in a heterogeneous beam of radiation.

spectrophotometer *(Phys.)*. Instrument for measuring photometric intensity of each colour or wavelength present in an optical spectrum.

spectroradiometer *(Phys.)*. A spectrometer for measurements in the infrared.

spectroscope *(Phys.)*. General term for instrument (spectrograph, spectrometer, etc.) used in spectroscopy. The basic features are a slit and collimator for producing a parallel beam of radiation, a prism or grating for 'dispersing' different wavelengths through differing angles of deviation, and a telescope, camera or counter tube for observing the dispersed radiation.

spectroscopic binary *(Astron.)*. A binary whose components are too close to be resolved visually, but which is detected by the mutual shift of their spectral lines owing to their varying velocity in the line of sight.

spectroscopic parallax *(Astron.)*. The name given to the indirect method of deducing the distances of stars too far away to have detectable annual parallaxes; it involves the inferring of their absolute magnitudes from spectroscopic evidence which then, combined with the observed apparent magnitudes, gives their distances.

spectroscopy *(Phys.)*. The practical side of the study of spectra, including the excitation of the spectrum, its visual or photographic observation, and the precise determination of wavelengths.

spectrum *(Phys.)*. Arrangement of components of a complex colour or sound in order of frequency or energy, thereby showing distribution of energy or stimulus among the components. A mass spectrum is one showing the distribution in mass, or in mass-to-charge ratio of ionized atoms or molecules. The mass spectrum of an element will show the relative abundances of the isotopes of the element.

spectrum analyser *(Electronics)*. (1) Electronic spectrometer usually working at microwave frequencies and displaying energy distribution in spectrum visually on a cathode-ray tube. (2) Pulse-height analyser for use with radiation detector.

spectrum analyser *(Telecomm.)*. Swept-frequency receiver, linked to a CRT or chart recorder, which can be used to display a number of signals, with their relative spacing and amplitudes, over a wide frequency band. Alternatively, it can be used to display a single signal with complex **sidebands**, e.g. an amplitude- or frequency-modulated signal or a pulsed signal such as a radar transmission.

spectrum analysis *(Chem.)*. See **spectroscopic analysis**.

spectrum colours *(Phys.)*. When split up into a spectrum, white light is shown to be composed of a continuous range of merging colours; red, orange, yellow, green, blue, indigo, violet.

spectrum line *(Phys.)*. Isolated component of a spectrum formed by radiation of almost uniform frequency. Due to photons of fixed energy radiated as the result of a definite electron transition in an atom of a particular element.

spectrum locus *(Phys.)*. Curved line on the CIE chromaticity diagram representing the monochromatic hues.

specular density *(Image Tech.)*. The photographic density in an image measured with parallel light, as contrasted with diffuse density, when the total light passed is measured, including that dispersed.

specular iron *(Min.)*. The name given to a crystalline rhombohedral variety of hematite which possesses a splendent metallic lustre often showing iridescence.

specular reflectance *(Phys.)*. Quotient of reflected to incident luminous flux for a polished surface.

specular reflection *(Phys.)*. General conception of wave motion in which the wavefront is diverted from a polished surface, so that the angle of the incident wave to the normal at the point of reflection is the same as that of the reflected wave. Applicable to heat, light, radio and acoustic waves. See **reflection laws**.

specular transmittance *(Phys.)*. See **transmittance**.

speculum *(Eng.)*. Alloy of 1-tin to 2-copper, providing wide spectral reflection from a grating ruled on its highly polished surface.

speculum *(Med.)*. A hollow or curved instrument for viewing a passage or cavity of the body, e.g. vaginal speculum, nasal speculum.

speech *(Acous.)*. The fundamental method of communicating thoughts, which consists in regulating the pitch and intensity of voiced sounds, and the intensity of unvoiced sounds, by the larynx, and in modifying the spectral content of these elementary sounds by posturing the cavities of the mouth (assisted by the nasal cavities), which form double or triple Helmholtz resonators.

speech clipping *(Telecomm.)*. Removal of high peaks in speech, to get higher loading of transmitters with some loss in intelligibility.

speech frequency *(Telecomm.)*. See **voice frequency**.

speech inverter *(Telecomm.)*. See **inverter**.

speech recognition *(Comp.)*. Process of analysing a spoken word and comparing it with those known to the computer system. See **pattern recognition**.

speech scrambler *(Telecomm.)*. See **scrambler**.

speech-sounds *(Acous.)*. The distinctive elements in speech, such as vowels and consonants. About 30 English speech-sounds are sufficient for recognition in telephonic work, but phoneticians recognize about 60. See **logatom**.

speech synthesizer *(Comp.)*. Output device which generates sound similar to human speech on receipt of digital signals.

speed *(Image Tech.)*. (1) The sensitivity of a photographic material as rated by a standard method. The current international system, *ISO*, gives ratings on an arithmetic scale which closely match those of American, *ASA*, and British, *BS*, practice. The Russian *GOST* values are numerically equivalent. The logarithmic *DIN* scale standard in Germany is also popular in Europe: here doubling the speed is indicated by an increase of 3°. For conversion, 400 ASA = 27 DIN. (2) A measure of the light transmitting power of a lens, usually stated as its **f-number** or **T-number**.

speed *(Phys.)*. (1) The rate of change of distance with time of a body moving in a straight line or in a continuous curve (cf. *velocity*, a vector expressing both magnitude and direction.) Units or speed are metres per second ($m s^{-1}$), feet per second ($ft s^{-1}$), miles per hour (mph), kilometres per hour (km/h), knots etc. (2) Angular velocity expressed in revolutions per minute, radians per second etc.

speed-adjusting rheostat *(Elec.Eng.)*. A rheostat arranged in the field or armature circuit of an electric motor for varying the motor speed.

speed bulges *(Aero.)*. Streamlined bulges on the fuselage, or nacelles near the trailing edge of the wing, which meet the requirements of **area rule** for a smooth area distribution where it is impractical to give the fuselage a wasp-waist to reduce transonic **wave drag**.

speed control *(Elec.Eng.)*. The method by which the speed of an electric motor may be varied.

speed-distance curve *(Eng.)*. The curve showing the relation between the speed of a moving object or vehicle and the distance it has travelled.

speed-frequency *(Elec.Eng.)*. The product of rotor speed and the number of pole-pairs in an induction motor.

speed governing *(Elec.Eng.)*. The method of keeping the speed of a prime mover independent of the load which it is driving.

speed indicator *(Eng.)*. See **speedometer**.

speed of light *(Phys.)*. The constancy and universality of the speed of light *in vacuo* is recognized by *defining* it (1983) to be exactly $2.99792458 \times 10^8 \, ms^{-1}$. This enables the SI fundamental unit of *length*, the metre, to be defined in terms of this value.

speed of rotation *(Phys.)*. In a rotating body, the number of rotations about the axis of rotation divided by the time (see **speed**). Units are revolutions/second, minute, or hour, or radians/ second, minute, or hour. The axis of rotation may have a translatory speed of its own. See **moment of inertia**.

speed of sound *(Phys.)*. See **velocity of sound**.

speedometer *(Autos.)*. A **tachometer** fitted to the gearbox or propeller shaft of a road vehicle, so graduated as to indicate the speed in m.p.h., km/h, or both. It may be centrifugal, magnetic, air-vane, chronometric, or electrical.

speed-time curve *(Eng.)*. A curve of vehicle speed plotted against running time. The area beneath it represents the distance covered between any two given instants.

speed-torque characteristic *(Elec.Eng.)*. The curve showing the relation between the speed of a motor and the torque developed. Also known as *mechanical characteristic*.

speedy-cut *(Vet.)*. Injury of the foreleg of a horse near the knee, made by the shoe of the opposite foot.

speise, speiss *(Min.Ext.)*. Metallic arsenides and antimonides produced in the smelting of cobalt and lead ores.

spelaeology, speleology *(Zool.)*. The study of the fauna and flora of caves.

speleothems *(Geol.)*. Secondary calcium carbonate encrustations deposited in caves by running water.

spelter *(Eng.)*. Zinc of about 97% purity, containing lead and other impurities.

spent fuel *(Nuc.Eng.)*. Reactor fuel element which must be replaced due to bursting, *(a)* burn-up or depletion, *(b)* poisoning by fission fragments, *(c)* swelling and/or bursting. The fissile material is not exhausted and so-called spent fuel is normally subsequently reprocessed.

sperm *(Zool.)*. See **spermatozoon**.

spermaceti *(Chem.)*. A glistening white wax from the head of the sperm whale, consisting mainly of cetyl palmitate, $C_{15}H_{31}.COO.C_{16}H_{33}$; mp 41°–52°C, saponification number 120–135, iodine number nil. Used in the manufacture of cosmetics and ointments.

spermagonium *(Bot.)*. See **spermogonium**.

spermary *(Zool.)*. See **testis**.

spermatheca *(Zool.)*. A sac or cavity used for the reception and storage of spermatozoa in many Invertebrates; receptaculum seminis.

spermatic *(Zool.)*. Pertaining to spermatozoa: pertaining to the testis.

spermatid *(Zool.)*. A cell formed by division of a secondary spermatocyte, and developing into a spermatozoon without further division.

spermatoblast *(Zool.)*. A spermatid.

spermatocele *(Med.)*. A cyst of the epididymis or of the tubules of the testis as a result of blocking of the ducts of the epididymis; contains a clear fluid and spermatozoa.

spermatocide *(Med.)*. Any agent (especially chemical) which kills spermatozoa. adj. *spermatocidal*.

spermatocyte *(Zool.)*. A stage in the development of the male germ cells, arising by growth from a spermatogonium or by division from another spermatocyte, and giving rise to the spermatids.

spermatogenesis *(Zool.)*. Sperm formation; the maturation divisions of the male germ cells by which spermatozoa are produced from spermatogonia.

spermatogonium *(Zool.)*. A sperm mother cell; a primordial male germ cell, adj. *spermatogonial*.

spermatophore *(Zool.)*. A packet of spermatozoa enclosed within a capsule.

Spermatophyta *(Bot.)*. Division containing the seed plants, i.e. the gymnosperms and the angiosperms. Also called *Magnoliophyta*.

spermatorrhoea, spermatorrhea *(Med.)*. Involuntary, frequent discharge of seminal fluid in the absence of sexual excitement or intercourse.

spermatozoid *(Bot.)*. Motile, flagellated, male gamete, as found in most algae, some fungi and in bryophytes, pteridophytes, cycads and *Ginkgo*. syn. *antherozoid*.

spermatozoon *(Zool.)*. The male gamete, typically consisting of a head containing the nucleus, a middle piece containing mitochondria, and a tail whose structure is similar to that of a flagellum; pl. *spermatozoa*; abbrev. *sperm*.

spermaturia *(Med.)*. The presence of spermatozoa in the urine.

sperm cell *(Bot.)*. Male gamete, motile or not.

spermiducal glands *(Zool.)*. In many Vertebrates, glands opening into or near the spermiducts.

spermiduct, spermaduct *(Zool.)*. A duct by which sperms are carried from the testis to the external genital opening; the vas deferens, adj. *spermiducal*.

spermogonium *(Bot.)*. A flask-shaped structure in which spermatia are formed as in some ascomycetes, rusts and perhaps some lichens.

sperm-, sperma-, spermi-, spermo-, spermato- *(Genrl.)*. Prefix from Gk. *sperma*, gen. *spermatos*, seed.

sperrylite *(Min.)*. Platinum diarsenide, crystallizing in the cubic system; has a brilliant metallic lustre and is tin-white in colour.

spessartine, spessartite *(Min.)*. Manganese garnet; silicate of manganese and aluminium, crystallizing in the cubic system. Usually contains a certain amount of either ferrous of ferric iron. The colour is dark, orange-red, sometimes having a tinge of violet or brown, Strictly, *spessartite* is the name for a lamprophyre rock. See **garnet**.

SPF/DB *(Aero.)*. See **super plastic forming/diffusion bonding**.

sp. gr. *(Chem.)*. An abbrev. for *Specific Gravity*.

sphagnicolous *(Ecol.)*. Living in peat-moss.

Sphagnum *(Bot.)*. The bog-mosses. A genus of mosses (*Bryopsida*) of ca 350 spp. with upright, branching gametophytes the leaves of which are very absorbent because of a regular pattern of dead cells with holes through their walls. *Sphagnum* spp. dominate many bogs. Sphagnum peat is acidic and lacking in mineral nutrients, cf. **sedge** peat.

sphalerite *(Min.)*. Zinc sulphide which crystallizes in the cubic system as black or brown crystals with resinous to adamantine lustre. The commonest zinc mineral and ore, deposits with fluorite, galena, etc. Also known as *blende* or *zinc blende*.

S phase *(Biol.)*. The period in the cell cycle during which the nuclear DNA content doubles.

sphene *(Min.)*. Calcium titanium silicate, with varying amounts of iron, manganese, and the rare earths, it crystallizes in the monoclinic system as lozenge-shaped black or brown crystals and occurs as an accessory mineral in many igneous and metamorphic rocks. Also called *titanite*.

Sphenisciformes *(Zool.)*. An order of Birds in which

flight feathers are lacking; the wings are stiff and used as paddles in swimming; the feet are webbed; the bones are solid and there are no air sacs; flightless marine forms, with streamlined bodies, powerful swimmers and divers; confined to the southern hemisphere. Penguins.

Sphenodon *(Zool.)*. See **Rhynchocephalia**.

sphenoid *(Crystal.)*. A wedge-shaped crystal-form consisting of 4 triangular faces. The tetragonal and orthorhombic analogue of the cubic tetrahedron.

sphenoidal *(Bot.,Zool.)*. Wedge-shaped.

sphenoiditis *(Med.)*. Inflammation of the air-containing sinus in the sphenoid bone.

Sphenopsida *(Bot.)*. The horsetails and allies. Class of Pteridophyta dating from the Devonian onwards. Sporophytes with roots, stems and whorled leaves (*microphylls*). Sporangia are borne usually reflexed on sporangiophores arranged in whorls often on terminal cones. Mostly homosporous. Spermatozoids multiflagellate. Includes the Sphenophyllales, Calamitales, and Equisetales.

sphen-, spheno- *(Genrl.)*. Prefix from Gk. *sphen*, a wedge.

spherical aberration *(Phys.)*. Loss of definition of images formed by optical systems and arising from the geometry of a spherical surface. Parabolic mirrors are used in astronomical telescopes to avoid this defect. Combinations of lenses can be used to reduce this effect. See also **Schmidt optical system**.

spherical astronomy *(Astron.)*. The branch of astronomy concerned with the position of heavenly bodies regarded as points on the observer's celestial sphere. It comprises all diurnal and seasonal phenomena and the precise assignment of coordinates to the heavenly bodies. See also **astrometry**.

spherical candle-power *(Phys.)*. The illumination on a sphere of unit radius having the source of light at its centre.

spherical curvature *(Maths.)*. See **osculating sphere**.

spherical excess *(Surv.)*. The amount by which the sum of the three angles of a spherical triangle exceeds 180°. It is equal to the area of the triangle divided by the square of the sphere's radius.

spherical polar co-ordinates *(Maths.)*. See **polar co-ordinates**.

spherical radiator *(Telecomm.)*. Same as **isotropic radiator**.

spherical roller-bearing *(Eng.)*. A roller-bearing having two rows of barrel-shaped rollers of opposite inclination, working in a spherical outer race, thus providing a measure of self-alignment.

spherical triangle *(Maths.)*. A triangle formed by three lines (see **line**) on the surface of a sphere.

sphericity *(Powder Tech.)*. The ratio of the surface area of a particle to the surface area of the sphere having the same volume as the particle.

spherics *(Telecomm.)*. See **atmospherics**.

spherocytosis *(Med.)*. *Acholuric jaundice*. An autosomal dominant hereditary disease in which the red cells of the blood are smaller than normal, biconvex instead of biconcave, and abnormally fragile. Leads to frequent bouts of mild haemolysis and jaundice with splenomegaly.

spheroid *(Maths.)*. The surface obtained by rotating an ellipse about one of its principal axes. It is therefore an ellipsoid in which two of its three principal axes are equal. If the axis of rotation is the major axis of the ellipse it is called a *prolate spheroid*, otherwise it is an *oblate spheroid*.

spheroidal jointing *(Geol.)*. Spheroidal cracks found in both igneous and sedimentary rocks. Some are due to cooling and resultant contraction in the igneous rock body; others are due to a shell-like type of weathering.

spheroidal state *(Phys.)*. Water dropped upon a clean, horizontal, red-hot metal plate gathers into spheroidal drops which roll about, rather like mercury drops, without boiling. This is prevented by a cushion of steam on which the drop rides, the rapid evaporation of the drop showing that its temperature is near the boiling-point.

spheroidal structure *(Geol.)*. A structure exhibited by certain igneous rocks, which appear to consist of large rounded masses, surrounded by concentric shells of the same material. Presumably a cooling phenomenon, comparable with perlitic structure, but on a much bigger scale, and exhibited by crystalline, not glassy, rocks.

spheroidizing *(Eng.)*. Process of producing, by heat treatment, a structure in which the cementite in steel or cast iron is in a spheroidal distribution, giving improved ductility and machinability.

spherometer *(Phys.)*. An instrument for measuring the curvature of a lens surface.

spherosome *(Bot.)*. A small ($< 1.0\,\mu m$), spherical, refractile body rich in lipid, in the cytoplasm of plant cells. Probably bounded by a half unit-membrane. Cf. **lipid body**.

spherulite *(Geol.)*. A crystalline spherical body built of exceedingly thin fibres radiating outwards from a centre and terminating on the surface of the sphere, which may vary in diameter in different cases from a fraction of a millimetre to that of a large apple.

spherulitic graphite cast iron *(Eng.)*. See **ductile cast iron**.

spherulitic texture *(Geol.)*. A type of rock fabric consisting of spherulites, which may be closely packed or embedded in an originally glassy groundmass. Commonly exhibited by rhyolitic rocks.

sphincter *(Zool.)*. A muscle which by its contraction closes or narrows an orifice. Cf. **dilator**.

sphincter of Oddi *(Med.)*. Fibres of muscle surrounding duodenal end of bile-duct.

sphingomyelin *(Biol.)*. A membrane phospholipid derived from **sphingosine** by the addition of a long chain hydrocarbon, phosphate and organic base.

sphingosine *(Biol.)*. A hydrophobic amino alcohol which is a component of the phospholipids known as *sphingomyelins*.

sphygmogram *(Med.)*. A tracing of the movements of the pulse made by a sphygmograph.

sphygmograph *(Med.)*. An instrument for recording the movements of the arterial pulse by means of tracings.

sphygmomanometer *(Med.)*. An instrument for measuring the arterial blood-pressure, an inflatable bag being applied to the arm and pressure increased to occlude the artery. On deflation the *systolic* blood pressure is that at which sounds are heard by ausculation over the artery as blood flow recommences. The *diastolic* blood pressure is that at which sounds due to flow turbulence disappear.

sphygmus *(Zool.)*. The pulse; the beat of the heart and the corresponding beat of the arteries.

spica *(Med.)*. A figure-of-eight bandage with turns that cross one another.

spicate *(Bot.)*. Bearing or pertaining to spikes; spike-like.

spicule *(Bot.,Zool.)*. A small pointed process: one of the small calcareous or siliceous bodies, forming the skeleton in many Porifera and Coelenterata. adjs. *spicular, spiculate, spiculiferous, spiculiform*.

spiculum *(Zool.)*. Any spicule-like structure: in Snails, the dart.

spider *(Acous.)*. The **inside spider**.

spider *(Elec.Eng.)*. The centre part of an armature core, upon which the core stampings are built up.

spider *(Eng.)*. See **cathead**.

spider *(Image Tech.)*. (1) A three-armed base for tripod legs. (2) A multiple distribution box for lighting cables in a studio.

spider wrench, spanner *(Build.)*. A **box spanner** having heads of different sizes at the ends of radial arms.

spiegeleisen *(Eng.)*. Pig-iron containing 15–30% manganese and 4–5% carbon. Added to steel as a deoxidizing agent and to raise the manganese content of the steel. Also *spiegel*.

spigot *(Eng.)*. Short raised step on the face of a component usually for location with a mating recess.

spike *(Bot.)*. A racemose inflorescence with sessile, often crowded, flowers on an elongated axis, e.g. *Plantago*.

spike *(Build.)*. A stout nail more than 4 in (10 cm) long.

spike *(Elec.Eng.)*. Short pulse of voltage or current.

spike *(Radar)*. Initial rise in excess of the main pulse in transmission.

spike *(Telecomm.)*. See **pulse-**.

spikelet *(Bot.)*. A small or secondary spike. In grasses, the basic unit of the inflorescence usually consisting of a short axis (rachilla) bearing 2 bracts (glumes) and 1 or more florets.

spilite *(Geol.)*. A fine-grained igneous rock of basaltic composition, generally highly vesicular and containing the sodium feldspar, albite. The pyroxenes or amphiboles are usually altered. These rocks are frequently developed as submarine lava-flows and exhibit pillow structure.

spill burner *(Aero.)*. A gas turbine burner wherein a portion of the fuel is recirculated instead of being injected into the combustion chamber.

spill door *(Aero.)*. Auxiliary door mounted in an engine nacelle which opens to spill excess air provided by the intake but not needed by the engine. Designed to minimize drag and often spring loaded.

spillway *(Civ.Eng.)*. See **bye-channel**.

spillway dam *(Civ.Eng.)*. A reservoir dam over which flood water is allowed to flow to a downstream escape channel situated at the foot of the dam.

spin *(Aero.)*. A continuous, but not necessarily even, spiral descent with the mean **angle of incidence** to the relative airflow above the **stalling** angle. In a *flat spin* the mean angle of incidence is nearer the horizontal than the vertical, while in an *inverted spin* the aircraft is actually upside down.

spin *(Phys.)*. The intrinsic angular momentum of an electron, nucleus or elementary particle. Spin is quantized in integral multiples of half the *Dirac unit*. Those particles with spin of odd multiples of $\hbar/2$ are *fermions* (e.g. electrons, proton, neutron) and those with even multiples of $\hbar/2$ are *bosons* (e.g. photon, phonon). It is the quantized electron spin angular momentum combined with the orbital angular momentum that gives rise to the fine structure in atomic line spectra.

spina *(Zool.)*. A small sharp-pointed process.

spina bifida *(Med.)*. Developmental malformation of the bony spinal canal, spinal segments failing to meet. Often associated with defect at brain base preventing circulation of cerebrospinal fluid and causing brain damage and **hydrocephalus**. There may also be leg paralysis, deformity of the feet, or other abnormalities (see **syringomyelocele**). Also called *rachischisis*.

spinacene *(Chem.)*. See **squalene**.

spinal *(Zool.)*. Pertaining to the vertebral column, or to the spinal cord.

spinal anaesthesia *(Med.)*. A form of regional anaesthesia. It is produced by injecting a solution of a suitable drug within the spinal theca (intrathecally), causing temporary paralysis of the nerves with which it comes into contact.

spinal canal *(Zool.)*. The tubular cavity of the vertebral column which houses the spinal cord.

spinal cord *(Zool.)*. In Craniata, that part of the dorsal tubular nerve cord posterior to the brain.

spinal reflex *(Zool.)*. Reflex situated in spinal cord in which higher nerve centres play no part.

spin bath *(Chem.Eng.)*. The acid bath with various additives into which the spinning solution is injected to form the thread in viscose rayon production.

spin chute *(Aero.)*. See **antispin parachute**.

spindle *(Biol.)*. (1) In mitosis and meiosis, spindle-shaped structure (i.e. widest in the middle and tapering towards the poles), containing longitudinally-running **microtubules**, formed within the nucleus or the cytoplasm at the end of prophase, with centrioles, if present, usually at the poles. Apparently forms a structural framework for the movements of chromosomes and chromatids. (2) Special sensory receptor in muscle, *muscle spindle*.

spindle *(Eng.)*. (1) The tubular member revolving with the headstock of a lathe, through which bar material may be introduced into the chuck or collet. (2) Generally, a machine element acting as a revolving axis or a pin on which another element revolves.

spindle fibre *(Biol.)*. One of the microtubules of the mitotic or meiotic spindle.

spindleless reel stand *(Print.)*. One supporting the reel on free-running cones at each side.

spindle moulder *(Build.)*. Machine with a revolving spindle to which cutters of various shapes can be fixed. The spindle projects up through a hole in the machine table and the work is fed on to the cutter. Cf. **router**.

spin-draw *(Textiles)*. A process for producing man-made fibres in which drawing takes place as extrusion proceeds.

spin-draw texturing *(Textiles)*. In the manufacture of synthetic fibres, the combination on one machine of the extrusion, drawing and *texturing* processes.

spine *(Bot.)*. A stiff, straight, sharp-pointed structure.

spine *(Print.)*. The back of a book, i.e. the edge where the gathered sections are sewn together. The spine faces outwards when the book is placed on a shelf, hence the name often used, *shelf-back*.

spine *(Zool.)*. A small sharp-pointed process: the backbone or vertebral column: a pointed process of a vertebra, as the neural spine: a fin-ray: the scapular ridge. adjs. *spinate, spiniform, spinose, spinous*.

spinel *(Min.)*. A group of closely related oxide minerals crystallizing in the cubic system, usually in octahedra. Chemically, spinels are aluminates, chromates, or ferrates of magnesium, iron, zinc, etc., and are distinguished as *iron spinel* (hercynite), *zinc spinel* (gahnite), *chrome spinel* (picotite), and *magnesian spinel*. See also **ruby-**, **synthetic-**, **balas ruby**, **chromite**, **franklinite**, **magnetite**.

spinel ruby *(Min.)*. See **ruby spinel**.

spinner *(Aero.)*. A streamlined fairing covering the hub of a propeller and rotating with it.

spinneret *(Zool.)*. In Spiders, one of the spinning organs, consisting of a mobile projection bearing at the tip a large number of minute pores by which the silk issues.

spinneret, spinnerette *(Textiles)*. A disk with fine holes through which the molten polymer or a solution of it is forced to form the continuous filaments of a man-made fibre.

spinner-gate, whirl-gate *(Foundry)*. An ingate incorporating a small whirl-chamber into and from which the metal flows tangentially, so releasing any dirt, which rises to the top of the chamber or up a riser. See **ingate**.

spinning *(Eng.)*. Forming a metal disk into a hollow shape by rotating it in a lathe over a former or chuck, by pressing a spinning tool against it.

spinning *(Textiles)*. The same word is used for two distinct processes. (1) The drafting and twisting together of staple fibres to form a yarn. (2) The extrusion of continuous filaments of man-made fibres (or of silk) through spinnerets.

spinning glands *(Zool.)*. The silk-producing glands of Arthropoda.

spinning jenny *(Textiles)*. First spinning machine employing vertical spindles, invented by James Hargreaves, Blackburn, in Lancashire, UK in 1767.

spinning tunnel *(Aero.)*. See **vertical wind tunnel**.

spinode *(Maths.)*. See **double point**.

spin-orbit coupling *(Electronics)*. The interaction between the intrinsic and angular momentum of particles, especially electrons.

spinose, spinous *(Zool.)*. Covered with spines.

spinous process *(Zool.)*. A process of (*a*) the proximal end of the tibia, (*b*) the sphenoid bone: the neural process or spine of a vertebra.

spin polarization *(Phys.)*. When applied to a beam of particles, polarization denotes the preferential orientation of the particle spin.

spin quantum number *(Phys.)*. Contribution to the total angular momentum of the electron of that due to the rotation of the electron about its own axis.

spin-stabilized satellite *(Telecomm.)*. An earlier generation of communications satellites like Intelsat IV, in which the cylindrical body of the satellite was rotated rapidly to achieve positional stability. See **de-spun antenna**.

spinule *(Bot.,Zool.)*. A very small spine or prickle.

spin wave *(Phys.)*. The spin of the magnetic ions in ferro-

and ferrimagnetic and in antiferromagnetic materials are orientated along preferred directions. Coherent deviations of the spins from these directions are propagated in space and time like waves. See **magnon**.

spiny vesicle *(Bot.)*. Same as **coated vesicle**.

spiracle *(Zool.)*. In Insects and some Arachnida, one of the external openings of the tracheal system: in Fish, the first visceral cleft, opening from the pharynx to the exterior between the mandibular and the hyoid arches: in amphibian larvae, the external respiratory aperture: in Cetacea, the external nasal opening. adjs. *spiracular, spiraculate, spriraculiform.*

spiral *(Bot.)*. See **helical thickening**.

spiral *(Maths.)*. A plane curve traced by a point which winds about a fixed pole from which it continually recedes. See **Archimedes'-**, **Cornu's-**, **equiangular-**, **hyperbolic-**, **parabolic-**, **sinusoidal-**.

spiral binding *(Print.)*. A speciality binding in which holes are punched near spine of book and a spiral of wire is fed through them.

spiral cleavage *(Zool.)*. A type of segmentation of the ovum occurring in many Turbellaria, most Mollusca, and all Annelida; the early micromeres rotate with respect to the macromeres, so that the micromeres lie opposite to the furrows between the macromeres; the direction of rotation (viewed from above) is normally clockwise (*dexiotropic*) but in 'reversed cleavage' it is anti-clockwise (*laeotropic*). Also *alternating cleavage.*

spiral galaxy *(Astron.)*. The second commonest morphological type, characterized by a large nuclear bulge of stars, surrounded by a pair of conspicuous spiral arms. These arms contain gas, dust and newly-formed star clusters. Our own Galaxy is a spiral.

spiral gear *(Eng.)*. A toothed gear for connecting two shafts whose axes are at any angle and do not intersect. The teeth are of spiral form (i.e. parts of a multiple-threaded screw) and engage as in a *worm gear*.

spiral instability *(Aero.)*. That form of lateral instability which causes an aircraft to develop a combination of side-slipping and banking, the latter being increasingly too great for the turn. This causes the machine to follow a spiral path.

spiral ratchet screwdriver *(Build.)*. **Ratchet screwdriver**, which can be made to turn in either direction by downward pressure on a sleeve which applies torque to helical grooves in the shank.

spiral reel *(Image Tech.)*. In a light-tight developing tank, a reel on which film has been wound in a spaced spiral so that, on rotation, the developer has free access to the emulsion.

spiral roller *(Print.)*. (1) An ink roller with spiral channels or grooves to improve ink supply and distribution. (2) An idling roller with a spiral groove to smooth out unevenness in the web of paper.

spiral spring *(Eng.)*. A spring formed by coiling a steel ribbon into an elongated spiral or a helix of increasing diameter. When compressed completely it forms a true spiral.

spiral stairs *(Build.)*. *Circular stairs* of small diameter and usually open. More correctly *helical stair*.

spiral time-base *(Electronics)*. Arrangement for causing the fluorescent spot to rotate in a spiral path at a constant angular velocity, to obtain a much longer base-line than is possible with linear deflection. Used for detailed delineation of events relatively widely spaced in time, with or without a memory through long-glow or photography.

spiral valve *(Zool.)*. In Lampreys, Elasmobranchii and some Lung fish, part of the intestinal canal, which is provided with an internal spiral fold to increase its absorptive surface.

spire *(Arch.)*. A slender tower tapering to a point.

spirillum *(Biol.)*. A bacterium which is a variation of the rod form, i.e. a curved or cork-screw shaped organism. Varies from 0.2–50 μm in length. One or more flagella may be present. Also the name of a particular genus, which includes *Sp. minus* (rat-bite fever).

spirit *(Chem.)*. An aqueous solution of ethanol, especially one obtained by distillation.

spirit duplicating *(Print.)*. The matter to be duplicated is typed, written or drawn on the under side of a smooth-surfaced paper with a special carbon paper in contact with the smooth surface, several colours of carbon being available, and multicoloured arrangements easily made. The image transferred from the carbon paper is sufficient for about 250 copies, being moistened by a volatile fluid between each impression on the duplicating machine.

spirit level *(Surv.)*. See **level tube**.

spirit stain *(Build.)*. A stain for wood; colouring matter dissolved in methanol.

spirit varnish *(Build.)*. A varnish made by dissolving certain alcohol-soluble gums or resins like shellac in industrial alcohol.

spirochaetes *(Biol.)*. Filamentous flexible bacteria showing helical spirals: without true flagella. Divided into two families, the Spirochaetaceae and the Treponemataceae. Saprophytic and parasitic members. Pathogenic species include the causative agents of syphilis (*Treponema pallidum*), sprirochaetosis (relapsing fever) in Man (*Borrelio recurrentis*), infectious jaundice (*Leptospira icterohaemorrhagiae*), and yaws (*Treponema pertenue*).

spirochaetosis *(Med.)*. See **relapsing fever**.

spirochaetosis icterohaemorrhagica *(Med.)*. *Weil's disease; infectious jaundice.* An acute disease characterized by fever, jaundice, haemorrhages from the mucous membranes, enlargement of the liver, and nephritis; due to infection with *Leptospira icterohaemorrhagiae*, conveyed to Man by Rats, which excrete the organism in their urine.

spirometer *(Med.)*. An instrument for measuring the air inhaled and exhaled during respiration.

spironolactone *(Med.)*. Drug which antagonizes the hormone *aldosterone*. Used to potentiate *thiazide* or **'loop' diuretics** in the treatment of refractory oedema. Also used in the treatment of *Conn's syndrome* (primary hyperaldosteronism).

spit *(Geol.)*. A long bank of sediment formed by longshore drift. It is attached to the land at the upstream end.

spitzkasten *(Min.Ext.)*. Crude **classifiers**, consisting of one or more pyramid-shaped boxes with regulated holes in down-pointing apexes. Ore pulp streaming across either settles (*coarse particles*) to bottom discharge or overflows (*fires*). In the *spitzlutten*, efficiency is improved by adding upflow of low-pressure hydraulic water.

splanchnic *(Zool.)*. Visceral.

splanchno- *(Genrl.)*. Prefix from Gk. *splanchnon*, the inward parts.

splanchnocoel *(Zool.)*. In Vertebrates, the larger posterior portion of the coelom which encloses the viscera, as opposed to the pericardium.

splanchnomegaly *(Med.)*. Enlargement of bodily organs.

splanchnopleure *(Zool.)*. The wall of the alimentary canal in coelomate animals; the inner layer of the mesoblast which contributes to the wall of the alimentary canal; cf. **somatopleure**. adj. *splanchnopleural.*

splash baffle *(Electronics)*. See **arc baffle**.

splashproof fitting *(Elec.Eng.)*. See **weatherproof fitting**.

splat *(Build.)*. A cover strip for joints between adjacent sheets of building-board.

splay brick *(Build.)*. A purpose-made brick bevelled off on one side. Also *cant brick, slope.*

splayed coping *(Build.)*. See **feather-edged coping**.

splayed grounds *(Build.)*. Grounds with splayed or rebated edges, providing a key for holding the plaster to the wall in cases where the grounds also serve as screeds.

splayed jambs *(Build.)*. Internal jambs or sides of a door or window opening which are not built at right angles to the wall but slope away from it, to admit more light or to increase the width.

splayed skirting *(Build.)*. A skirting board having its top edge chamfered.

splay end *(Build.)*. That end of a brick opposite to the square end.

splaying arch *(Civ.Eng.)*. An arch in which the opening at one end is less than that at the other end, so that the arch is funnel shaped. Also *fluing arch.*

spleen *(Zool.)*. An organ of Vertebrates concerned with the formation and destruction of the red blood cells.

splenectomy *(Med.)*. Removal of the spleen.

splenomegaly *(Med.)*. Abnormal enlargement of the spleen.

splenopexy *(Med.)*. Fixation, by suture, of the spleen to the abdominal wall.

splenotomy *(Med.)*. Incision of the spleen.

splice *(Build.)*. See **scarf**.

splice *(Image Tech.)*. A join in motion picture film, made either by cementing the base or by applying transparent adhesive tape. Also a join in magnetic tape made by adhesive tape.

spliced joint *(Elec.Eng.)*. A cable joint in which the conductor strands are spliced, in the manner of a rope.

splicer *(Image Tech.)*. A device for joining motion picture film or magnetic tape.

splicing *(Biol.)*. Specifically refers to the excision of the **introns** from an mRNA and the rejoining of **exons** to form the mature message. Loosely used for joining nucleic acid molecules as in *gene splicing.*

splicing *(Textiles)*. Joining two pieces of yarn together manually or by machine so that the joint is nearly the same diameter as the yarn and certainly smaller than a knot.

splines *(Eng.)*. A number of relatively narrow keys formed integral with a shaft, somewhat resembling long gear-teeth; produced by milling longitudinal grooves in the shaft (*external splines*); similarly, the grooved ways formed in a hole into which the splined shaft is to fit (*internal splines*). Used, instead of keys, for maximum strength.

splints *(Vet.)*. Exostoses on the small metacarpal or metatarsal bones of the horse.

split-anode magnetron *(Electronics)*. Early type with split anode, to give a push-pull output from electrons from the filament cathode, when gyrating in a (nearly) coaxial magnetic field.

split-beam CRT *(Electronics)*. A tube containing only one electron gun, but with the beam subdivided so that two traces are obtained on the screen.

split bearing *(Eng.)*. A shaft bearing in which the housing is split, the bearing bush or brasses being clamped between the two parts.

split boards *(Print.)*. Two layers of board between which the tapes are securely held.

split-brain *(Behav.)*. A condition in which the *corpus callosum* and some other fibres are cut so that the two cerebral hemispheres are isolated.

split complementary *(Build.)*. A colour scheme based on the use of one hue contrasted with 2 hues on either side of its complementary colour, e.g. red with blue and green.

split compressor *(Aero.)*. An **axial-flow turbine** compressor in which front and rear sections are mounted on separate concentric shafts (being powered by separate turbines) as a means of increasing the pressure ratio without incurring difficulties with *surge*. Also called a *two-spool compressor*.

split-conductor cable *(Elec.Eng.)*. A cable in which each conductor is divided into two sections lightly insulated from each other and connected in parallel at the ends. Used with special schemes of protection.

split-conductor protection *(Elec.Eng.)*. A current-balance system of feeder protection avoiding the use of pilot wires by splitting each phase conductor into two parallel sections lightly insulated from each other.

split course *(Build.)*. A course of bricks which have been cut lengthwise, so that the depth of the course is less than that of a brick.

split crankcase *(Autos.)*. An engine crankcase split horizontally at about the centre line of the crankshaft. Cf. **barrel-type crankcase**.

split duct *(Print.)*. Process in which the forme or plate can be printed in two or more colours by separating each from the other across the inking system.

split fitting *(Elec.Eng.)*. A bend, elbow, or tee used in electrical installation work, which is split longitudinally so that it can be placed in position after the wires are in the conduit.

split flap *(Aero.)*. A trailing edge flap in which only the lower surface of the aerofoil is lowered.

split-flow reactor *(Nuc.Eng.)*. One in which the coolant enters at the central section and flows outwards at both ends (or vice versa).

split fractions *(Print.)*. Founts of figures and spaces which can be assembled to form any required fraction, the horizontal rule being cast, sometimes on the numerator, sometimes on the denominator; a fount of 10-pt split fractions would be cast on a 5-pt body.

split-image rangefinder *(Image Tech.)*. Optical device generally used in conjunction with the camera lens to assist reflex focusing. Two very acute-angled prisms are juxtaposed so that their vertices lie together in the focal plane. When the image is out of focus, it is laterally displaced across the line where the vertices meet.

split pastern *(Vet.)*. Fracture of the first phalanx of the foot of the horse.

split-phase *(Elec.Eng.)*. A term denoting a circuit arrangement for changing a single-phase to a 2-phase supply.

split-pin *(Eng.)*. One formed of wire with hemispherical section, bent round till the flat faces meet. See **cotter pin**.

split-pole converter *(Elec.Eng.)*. A synchronous converter in which the flux distribution under the poles can be varied by means of auxiliary windings on the individual pole limits.

split pulley *(Eng.)*. A belt pulley split diametrically, the halves being bolted together on the shaft; used when a solid pulley cannot be fitted.

splits *(Build.)*. Bricks of the same length and breadth as ordinary bricks but of smaller thickness.

splits *(Textiles)*. Fabrics, usually woven on wide machines, that have spaces in the warp direction so that by cutting two or more narrower fabrics are obtained.

split screen *(Image Tech.)*. A shot in which two or more images separately recorded appear in the same picture, often with the boundary between them made invisible.

split-seconds chronograph *(Horol.)*. A chronograph with two independent centre seconds hands, one underneath the other. On pressing the button, the two hands travel together, but if a push piece in the band of the case is pressed, one hand remains stationary, the other continuing until stopped by a second pressing of the button. A third pressing of the button causes both hands to fly back to zero.

splitting limits *(Min.Ext.)*. Divergence between assay of ore, concentrate or metal made by vendor and purchaser, inside which mutual adjustment will be made without going to arbitration.

splitting ratio *(Nuc.Eng.)*. See **cut**.

spodogram *(Bot.)*. A preparation of the ash of a plant, especially from a section, used in investigating structure by light or electron microscopy.

spodumene *(Min.)*. A silicate of aluminium and lithium which crystallizes in the monoclinic system; a pyroxene. It usually occurs in granite-pegmatites, often in very large crystals. The rare emerald-green variety *hiddenite* and the clear lilac coloured variety *kunzite* are used as gems.

spoil *(Civ.Eng.)*. The excess of cutting over filling on any given construction. Also *waste*.

spoilage, spoils *(Print.)*. (1) Accidental spoilage, the cost of reprinting being budgeted for in the costing system. (2) Ordinary spoilage, allowed for on each job.

spoil bank *(Civ.Eng.)*. An earthwork bank formed by depositing spoil.

spoiler *(Aero.)*. A device for changing the airflow round an aerofoil to reduce, or destroy, the lift. There are 3 principal types: (1) a small fixed spanwise ridge on the wing-root leading edge along the line of the **stagnation point** which improves lateral stability at the stall by ensuring that it starts at the root; (2) controllable devices at, or near, both wingtips which, by destroying lift on the side raised, impart a rolling moment to the aircraft; (3) small-chord spanwise flaps on top of the wing of a sailplane which can be raised to destroy a large part of the lift so as to make landing of the lightly-loaded aircraft more positive. Similar devices are often now fitted to jet aircraft so that by destroying lift at touchdown better braking can be achieved. See **thrust-**.

spokeshave *(Build.)*. A form of double-handled plane which is used in shaping convex or concave surfaces.

spondyl *(Zool.)*. A **vertebra**. adj. *spondylous*.

spondylitis *(Med.)*. Inflammation of the vertebrae.

spondylitis deformans *(Med.)*. A condition in which the ligaments of the spinal column become ossified and the vertebrae fused together, so that the spine is bent, rigid and immobile.

spondylolisthesis *(Med.)*. A forward displacement of the 5th lumbar vertebra (carrying the vertebral column with it) on the sacrum.

sponge *(Elec.Eng.)*. Loose, fluffy cathode deposit in electrolysis, contrasted with reguline.

sponge *(Eng.)*. Porous metal formed by reducing or decomposing it without fusion.

Sponge *(Zool.)*. See **Porifera**.

sponge beds *(Geol.)*. Deposits, either calcareous or siliceous, which contain a large proportion of the remains of spicular organisms belonging to the phylum Porifera.

sponge stipple *(Build.)*. A decorative effect produced by dabbing colour on to a surface using natural sponge. Various colours can be used. The object is to produce an effect which is even but of irregular pattern.

spongin *(Zool.)*. A horny skeletal substance, occurring usually in the form of fibres, in various groups of Porifera.

spongioblastoma *(Med.)*. A soft, rapidly growing, malignant tumour occurring in the brain (or in the spinal cord) and derived from cells of the supporting structure of the brain.

spongioblasts *(Zool.)*. Columnar cells of the neural canal giving rise to neuroglia cells.

spongy layer, -mesophyll, -parenchyma, -tissue *(Bot.)*. Chlorenchyma in which the cells are irregularly lobed leaving very large continuous intercellular spaces. Spongy **mesophyll** is usually present towards the lower surface of dorsiventral leaves of mesomorphic dicotyledons. See also **palisade**.

spongy platinum *(Chem.)*. The spongy mass resulting from the calcination of ammonium chloroplatinate (IV).

sponson *(Aero.)*. A short, winglike projection from a flying-boat hull to give lateral stability on the water.

spontaneous behaviour *(Behav.)*. Behaviour occurring in the apparent absence of any stimuli.

spontaneous combustion *(Build.)*. The ignition of a substance or material without direct application of flame. Can occur with rags soaked in flammable liquids.

spontaneous emission *(Phys.)*. Process involving the emission of energy in an atomic system without external stimulation. (cf. *laser* and *stimulated emission*). Spontaneous emission is a strictly quantum effect.

spontaneous fission *(Phys.)*. Nuclear fission occurring without absorption of energy. The probability of this increases with increasing values of the fission parameter Z^2/A (Z = atomic number, *A* = relative atomic mass) for the fissile nucleus.

spontaneous generation *(Biol.)*. The production of living matter or organisms from nonliving matter. Generally believed to occur in micro-organisms before Pasteur.

spontaneous ignition temperature *(Eng.)*. The temperature at which a liquid or gaseous fuel will ignite in the presence of air or oxygen, measured, for liquid fuels, by allowing a drop to fall into a heated pot. Abbrev. *SIT*.

spontaneous recovery *(Behav.)*. The return of an extinguished response (see **extinction**) following a rest period in which neither the conditioned nor the unconditioned stimulus are presented, when an animal is returned to the original conditioning situation.

spontaneous remission *(Behav.)*. Recovery without treatment.

spool *(Image Tech.)*. A flanged core on which film or magnetic tape is wound for storage and transport; the term *reel* in this sense is deprecated.

spool *(Phys.)*. The support of a coil.

spooling *(Comp.)*. Temporary storage of input or output data on magnetic disk or tape, as a means of compensating for slow operating speeds of peripheral devices or when queueing different output streams to one device.

sporadic-E *(Telecomm.)*. A layer of intense ionization which appears unpredictably in the **E-layer**; can lead to the propagation of VHF signals over anomalously great distances.

sporadic lymphangitis *(Vet.)*. *Monday morning disease, weed*. An acute disease of horses characterized by fever and lameness due to lymphangitis affecting the limbs. The disease is often recurrent and occurs especially in working horses that have been rested for a day or two on full working rations; the cause is not known.

sporangium *(Bot.,Zool.)*. A hollow, walled, structure in which spores are produced.

spore *(Bot.)*. One of a variety of reproductive bodies, usually unicellular, often acting as a disseminule or as a resting stage, usually germinating without fusing with another cell. For plants with an alternation of generations, see **sporophyte**. Many algae and fungi produce spores of some sort(s) which reproduce the same stage or the next stage of the life cycle. See **conidium, aplanospore, hypnospore, endospore, exospore, zoospore, zygospore**.

spore *(Zool.)*. In Protozoa, a minute body formed by multiple fission: more strictly, a seedlike stage in the life cycle of Protozoa arising as a result of sporulation, and contained in a tough resistant envelope.

spore mother cell *(Bot.)*. A cell which gives rise to a spore, especially one which divides by meiosis to give four cells which develop into spore(s).

spore print *(Bot.)*. The marks obtained by placing e.g. the cap of a mushroom or toadstool, gills downward, on a piece of paper and allowing the spores to fall on to the paper.

spori-, sporo- *(Genrl.)*. Prefix from Gk. *sporos*, seed.

sporocarp *(Bot.)*. A hard multicellular structure enclosing sporangia in some fungi and some heterosporous ferns.

sporocyst *(Zool.)*. The tough resistant envelope secreted by, and surrounding, a Protozoan spore.

sporocyte *(Bot.)*. Same as **spore mother cell**.

sporogenesis *(Bot.,Zool.)*. Spore formation.

sporogenous *(Bot.)*. Producing or bearing spores.

sporogenous layer *(Bot.)*. Same as **hymenium**.

sporogonium *(Bot.)*. Same as **sporophyte** in bryophytes.

sporogony *(Zool.)*. In Protozoa, propagative reproduction, usually involving sexual processes and always ending in the formation of spores.

sporont *(Zool.)*. A stage in the life history of some Protozoa which, as a gametocyte, gives rise to gametes, which in turn, after a process of syngamy, may give rise to spores.

sporophore *(Bot.)*. Any structure that bears spores.

sporophyll *(Bot.)*. A leaf that bears sporangia.

sporophyte *(Bot.)*. The typically diploid generation of the life-cycle of a plant showing alternation of generations, which produces, by meiosis, the spores which germinate to give the **gametophyte**.

sporopollenin *(Bot.)*. Polymerized carotenoids, a constituent of the cell walls of pollen grains (the exine) and of many spores, and exceedingly resistant to decay. See **pollen analysis**.

sporotrichosis *(Med.)*. An infection of the skin (and rarely of muscles and bones) with the fungi of the genus *Sporotrichum*, causing granulomatous lesions.

Sporozoa *(Zool.)*. A class of parasitic Protozoa, usually being at some stage intracellular, in the principal phase have no external organs of locomotion or are amoeboid, lack a meganucleus, and form large numbers of spores after syngamy, which constitute the infective stage.

sporozoite *(Zool.)*. In Protozoa, an infective stage developed within a spore.

sport *(Bot.)*. See **bud sport**.

sporulation *(Biol.)*. The production of spores.

spot *(Electronics)*. Point on a phosphor which becomes visible through impact of electrons in a beam. See **ion burn**.

spot beam *(Space)*. A concentration of radio waves by an antenna-reflector so that a particular area is highly illuminated with radiation, resulting in the concentration of power over a small area and a high signal strength.

spot board *(Build.)*. The square wooden board on which the plasterer works up the coarse or fine stuff prior to applying it to the walls.

spot face *(Eng.)*. Machined surface produced in the vicinity of a hole in a casting or other rough part, to permit a bolt head, washer, or adjacent part to seat evenly.

spot level *(Surv.)*. The reduced level of a point (usually on ground surface), not necessarily lying along a traverse or survey line.

spot light *(Image Tech.)*. A focusable studio lamp, capable of producing a narrow beam.

spot meter *(Image Tech.)*. A photometer or exposure meter which takes its reading from a small area, for example, subtending 2° at the point of observation.

spot priming *(Build.)*. A term used to describe the application of the first coat of paint to small areas rather than the entire surface. Often used in the specification for the painting of steelwork when complete removal of surface coatings is considered unnecessary but where localized areas need treatment.

spot speed *(Telecomm.)*. In facsimile recording, the speed of the recording or scanning spot within the allotted time. In TV, the product of the number of spots in a scanning line multiplied by the number of scanning lines/second.

spotted gum *(For.)*. *Eucalyptus maculata*, from New South Wales and Queensland, growing up to 45 m in height. The sapwood is said to be susceptible to powder-post beetle termite attack. It bends well after steam treatment. It is used for ship building, sleepers and wagon making.

spotting *(Eng.)*. The operation of turning a short length of a bar or forging to form a journal, by which the work is to be supported by the jaws of a steady rest. See **steady**.

spotting *(Horol.)*. A method of finishing plates or other flat surfaces with a regular pattern of circular patches.

spotting drill *(Eng.)*. A flat drill having a point so shaped as to centre and face the end of a bar at one operation. See also **centre drill**.

spotting out *(Print.)*. When a photographic emulsion is broken by pinhole blemishes, etc., it can be rendered opaque by the application of high opacity paint, usually to the non-emulsion side of the film.

spot welding *(Civ. Eng.)*. The process of welding when the joint formed is required to resist only light and temporary stresses, e.g. many steel meshes are spot welded purely to resist stresses arising from handling and not for resisting applied structural loads.

spot welding *(Elec.Eng.)*. See **resistance spot-welding**.

spot wobble *(Image Tech.)*. A vertical oscillatory movement given to the scanning spot to make the space between the scanned lines less obvious.

sprag *(Min.Ext.)*. (1) Timber prop; short piece of wood, used to prevent the wheels of a train from revolving. (2) Slanting prop used to support coal face. Also *gib*.

sprag clutch *(Eng.)*. One incorporating balls and wedging surfaces, which allows power to be transmitted from one rotating shaft to another in one direction only.

sprain *(Med.)*. A wrenching of a joint with tearing or stretching of its ligaments, damage to the synovial membrane, effusion into the joint, occasionally rupture of muscles or tendons attached to the joint, but without dislocation.

spray arrester *(Elec.Eng.)*. (1) A lightning arrester designed for removing accumulation of static charge; it consists of a spray of water from an earthed pipe impinging on a plate connected to the live circuit. (2) A sheet of glass placed over an open accumulator cell to prevent splashing of acid spray.

spray bonded *(Textiles)*. A method of treating a **web** of fibres with a spray of adhesive so that a non-woven fabric is obtained.

spray column *(Min.Ext.)*. Tower packed with coarse material, e.g. coke, or set with grids or trays down which a liquid trickles counter-current to rising gas so as to facilitate interaction. See **Glover tower**.

spray discharge *(Min.Ext.)*. In high-tension separation the corona radiated from the electric discharge wire (18 000 volts or more) on to a passing stream of finely divided mineral particles. In classification, a spray-shaped discharge from the hydrocyclone.

spray drying *(Chem.Eng.)*. Rapid drying of a solution or suspension by spraying into a flow of hot gas, the resultant powder being separated by a cyclone. Used to prepare powdered milk, detergent, fertilizer etc.

spray gate *(Foundry)*. An ingate consisting of a number of small separate gates, fed from the runner; used for shallow castings where there is insufficient depth for a single large gate. See **ingate**.

spray gun *(Build.,Civ.Eng.)*. A tool used in spray painting to 'atomize' paint or other coating, like cement, and facilitate its application in a controlled pattern.

spraying *(Eng.)*. The process of coating the surface of an article with metal by projecting on to it a spray of molten metal.

spray tower *(Chem.Eng.)*. A plant for purifying gases, which pass up through a tower into which a suitable liquid is sprayed from the top.

spread *(Ecol.)*. The establishment of a species in a new area.

spread *(Print.)*. (1) An illustration or diagram occupying each side of the fold. (2) A story or feature run across the two centre pages of magazine or newspaper. (3) Refers to a technique in photography or platemaking whereby an image is slightly enlarged to ensure image overlap when printed.

spreader *(Telecomm.)*. Wooden or metal spar for keeping the wires of a multiwire antenna spaced apart.

spread factor *(Phys.)*. See **distribution coefficients**.

spreading agent *(Bot.)*. Substance added to a solution for e.g. spraying on a fungicide, in order to promote even distribution over the target.

spreading capacity *(Build.)*. The area which a given quantity of paint will cover without thinning unduly.

spread spectrum *(Telecomm.)*. Modulation technique used for security and to increase immunity from noise and interference. The normally narrow-band information signal is spread over a much wider range of frequencies in a pseudo-random (*noise-like*) manner; the receiver is adapted to correlate these deviations in order to retrieve the original signal.

sprig *(Build.)*. A small nail with little or no head.

sprig *(Foundry)*. Small nail pushed into a weak edge of sand in a mould to reinforce it during pouring. See **brad**.

spring *(Eng.)*. A device capable of deflecting so as to store energy, used to absorb shock or as a source of power or to maintain pressure between contacting surfaces or to measure force. See **carriage-**, **contact-**, **helical-**, **relay-**, **spiral-**.

spring-back *(Eng.)*. Elastic recovery of a work-piece after completion of a bending operation.

spring-back *(Print.)*. A style of stationery binding in which the book opens quite flat.

spring-balance *(Chem.)*. Balance in which the weight of the sample is balanced by the extension of a spring.

spring bows *(Eng.)*. Small compasses whose two limbs are not hinged together but are connected by a bow of spring steel, the distance apart of the marking points being adjusted by means of a screw.

spring constant *(Eng.)*. See **spring rate**.

spring control *(Elec.Eng.)*. A method of controlling the movement of an indicating instrument by means of a spring.

spring cramp *(Build.)*. One in the shape of a broken ring made of tensile steel, used for light work.

springer, springing *(Civ.Eng.)*. The lowest voussoir on each side of an arch.

springing *(Arch.)*. The line where the intrados of an arch meets the abutment or pier.

springing line *(Arch.)*. The line joining the springings on both sides of an arch.

spring-loaded governor *(Eng.)*. An engine governor

consisting of rotary masses which move outwards under centrifugal force and are controlled by a spring. See **Hartnell governor**.

spring-loaded idler *(Print.)*. See **jockey roller**.

spring needle *(Textiles)*. One with hook or beard which is flexed by the pressers on a knitting machine. Used to produce close and even texture in knitted fabrics. Also called *bearded needle*.

spring pawl *(Eng.)*. See **pawl**.

spring points *(Civ.Eng.)*. Points on a railway which are normally held closed by springs in determining the route when they are *facing points*, but can be passed through as *trailing points*. See **points**.

spring rate *(Eng.)*. Ratio of load to deflection of a spring, measured e.g. in newtons per millimetre. Also *spring constant*.

spring safety valve *(Eng.)*. See **safety valve**.

spring tab *(Aero.)*. A *balance tab* connected so that its angular movement is geared to the compression or extension of a spring incorporated in the main control circuit. Its primary purpose is to reduce the effort required by the pilot to overcome the air loads on the main control surface resulting from high airspeeds. Cf. **servo tab**, and **trimming tab**.

spring tides *(Astron.)*. Those high tides occurring when the moon is new or full, at which times the sun and moon are acting together to produce a maximum tide.

spring wood *(Bot.)*. Same as **early wood**.

sprinkler *(Build.)*. A pipe system installed in a building, having at frequent intervals spray nozzles protected by covers made of a fusible alloy; these, in the event of a fire, melt and release water for automatic fire fighting.

sprocket *(Build.)*. A small wedge-shaped piece of wood nailed to the upper surface of the lower end of a common rafter in cases where the latter is carried beyond the pole plate to form a projecting eave. It provides a break in the slope of the roof near the eaves.

sprocket *(Eng.)*. See **sprocket wheel**.

sprocket *(Image Tech.)*. A cylinder with regularly spaced teeth which engage the perforation holes of motion picture film to provide its movement through various forms of equipment.

sprocket holes *(Image Tech.)*. Deprecated term for the **perforations** in motion picture film.

sprocket noise *(Image Tech.)*. A 96 Hz hum occurring in optical sound reproduction if the film is misplaced so that the perforations modulate the exciter lamp beam at the scanning slit.

sprocket wheel *(Eng.)*. A toothed wheel used for chain drives, as on the pedal shaft and rear hub of a bicycle.

spruce *(For.)*. See **Canadian-**.

sprue *(Foundry.)*. See **gate**.

sprue *(Med.)*. A disease affecting the gastrointestinal tract and causing mal-absorption of vitamins and nutrients. Characterized by loss of energy, loss of weight, anaemia, inflammation of the tongue and the mouth, and by the frequent passage of pale, bulky, acid, frothy stools, there being inability to absorb adequately fat, glucose, and calcium. The term is usually applied to the disease in the tropics where it appears to have an infectious origin. Non-tropical sprue or coeliac sprue is a synonym for *coeliac disease*.

sprue *(Plastics)*. Waste plastic formed during injection moulding processes, being the material setting in the main inlet passages of the mould.

SPS *(Phys.)*. *Super Proton Synchrotron*. A particle accelerator at CERN, Geneva. Output beam of 500 GeV protons for *fixed-target* experiments. Modified to perform 270 GeV antiproton colliding-beam experiments. Used to discover, in 1983, W^+, W^- and Z *gauge bosons* that mediate weak interactions.

sp.,spp *(Bot.)*. Abbrevs. for *species* (singular and plural).

spudding bit *(Min.Ext.)*. Large drill bit for making the initial *top hole* which takes the **anchor string** or **top casing**. Very deep, high pressure wells may require over a 1000 ft of top hole into which the casing is cemented.

spun-dyed *(Textiles)*. Man-made fibres formed from substances to which the colouring matter has been added before the filaments are formed.

spun silk *(Textiles)*. Yarn made from silk waste which is spun in a manner very similar to the woollen systems.

spun yarn *(Textiles)*. A yarn made from staple fibres twisted together.

spur *(Bot.)*. (1) A projection; arising especially from the base of a petal, sepal or gamopetalous corolla, etc. usually containing nectar, e.g. *Aquilegia, Linaria*. (2) A short shoot, especially one of the condensed lateral fruiting shoots of many fruit trees.

spur *(Build.)*. A strut.

spur *(Geol.)*. A hilly projection extending from the flanks of a valley.

spur *(Zool.)*. See **calcar**.

spur gear *(Eng.)*. Gear wheel with straight teeth machined parallel to its axis.

spur gearing *(Eng.)*. A system of gears with straight teeth connecting two parallel shafts.

spuriae *(Zool.)*. In Birds, the feathers of the bastard wing.

spurious coincidences *(Nuc.Eng.)*. Those recorded by a coincidence counting system, when a single particle has not passed through both or all the counters in the system. They usually result from the almost simultaneous discharge of two counters by different particles.

spurious counts *(Nuc.Eng.)*. Those arising in counter tubes from voltage leakages in the counter and defects in external quenching circuits.

spurious oscillation *(Telecomm.)*. See **parasitic oscillation**.

spurious pregnancy *(Med.)*. See **pseudocyesis**.

spurious pulse *(Nuc.Eng.)*. One arising from self-discharge of particle counter leading to erroneous signals.

spurious radiation *(Telecomm.)*. Undesired transmission, e.g. harmonics of carrier or modulation, outside specified band, causing interference with reception of other transmissions.

spurious response ratio *(Telecomm.)*. Ratio of field strengths of signals producing spurious and required response in telecommunication receiving equipment, e.g. of image frequency or intermediate frequency relative to required signal frequency.

Sputnik *(Space)*. A series of Russian artificial satellites; Sputnik I, launched in October 1957, was the first ever man-made object to orbit about the Earth.

sputtering *(Electronics)*. In a gas discharge, the removal of atoms from the cathode by positive ion bombardment, like a cold evaporation. These unchanged atoms deposit on any surface and are used to coat dielectrics with thin films of various metals.

sputum *(Med.)*. Matter composed of secretions from the nose, throat, bronchi or lungs, which is spat out.

SQ *(Build.)*. Abbrev. for *Squint Quoin*.

squalene *(Chem.)*. A symmetrical triterpene, formerly known as *spinacene*, originally detected in shark oil but also occurring in mammalian and plant tissue.

squall *(Meteor.)*. A temporary sharp increase in the wind speed, lasting for some minutes.

squama *(Bot.,Zool.)*. A scale: a scalelike structure.

Squamata *(Zool.)*. An order of diapsid Reptiles in which the skull has lost either one or both temporal vacuities, the quadrate is movably articulated with the skull, and there is no inferior temporal arch. Snakes and Lizards.

squamiform *(Zool.)*. Scale-like.

squamous epithelium *(Zool.)*. Epithelium consisting of one or more layers of flattened scalelike cells; pavement epithelium.

squamule *(Bot.)*. A small scale. adj. *squamulose*.

square *(For.)*. A piece of square-section timber of side up to 6 in (15 cm).

square *(Maths.)*. A rectangle whose adjacent sides are equal.

square drilling *(Eng.)*. Rotating and oscillating a specially shaped drill so that it follows a square shaped guide bush to produce a square hole. See also **spark erosion**.

squared rubble *(Build.)*. Walling in which the stones are roughly squared to rectangular faces but are of irregular size.

square folding *(Print.)*. See **right-angled folding**.

square-law *(Electronics)*. Said of any device, such as a rectifier or (de)modulator, in which the output is proportional to the square of the input amplitude.

square law *(Phys.)*. The law of inverse squares expressing the relation between the amount of radiation falling upon unit area of a surface and the distance of the surface from the source.

square-law capacitor *(Elec.Eng.)*. Variable vane capacitor, used for tuning, in which capacitance is proportional to the square of the scale reading, so that wavelength of circuit which it tunes becomes directly proportional to it.

square-law demodulator *(Elec.Eng.)*. See **detector**.

square-law detector *(Telecomm.)*. A *demodulator* in which the output is proportional to the square of the amplitude-modulated input voltage.

square-law rectifier *(Elec.Eng.)*. One in which the rectified output current is proportional to the square of the applied alternating voltage.

square rebate plane *(Build.)*. A rebate plane with its cutting edge square across the sole.

square root *(Maths.)*. The square root of a number is that quantity which when multiplied by itself gives the number.

squares *(Print.)*. The protrusion of the covers of a book beyond the leaves.

square staff *(Build.)*. An angle staff of square-section material, as distinct from an angle bead.

square step *(Build.)*. An individual stone step in a stair which consists of a solid block, rectangular in section, either lapping over the back edge of the step below or rebated to fit over it.

square thread *(Eng.)*. A screw thread of substantially square profile.

square wave *(Electronics)*. Pulse wave with very rapid (theoretically zero) rise and fall times; and pulse duration equal to half period of repetition. Mark-space ratio of unity.

squaring shears *(Eng.)*. Manual or power-operated press used for shearing sheets of steel.

squaring-up *(Build.,Civ.Eng.)*. A process following **taking-off** in drawing up a bill of quantities, superficial areas of items being calculated by multiplying the relevant dimensions entered on the dimensions paper.

squarrose *(Bot.)*. Leaves, hairs, scales etc. sticking out more or less at right angles to the stem or other structure.

squash *(Biol.)*. Spreading of tissue or chromosomes on a microscope slide by application of pressure.

squeegee *(Image Tech.)*. Rubber or plastic roller or blade, or a jet of compressed air, directed at the surface of a film during processing to wipe away surplus liquid.

squeeze *(Print.)*. The amount of impression between plate and impression cylinders.

squeezed print *(Image Tech.)*. Colloquial term for a motion picture print in which the image is anamorphically compressed horizontally, as in *CinemaScope*.

squeezer *(Foundry)*. A moulding machine, operated by hand, compressed air, hydraulic power, or magnetic means, in which the sand is squeezed or compressed into the box and round the pattern by a ram.

squeeze track *(Image Tech.)*. In **variable-area** photographic sound recording, reducing the image width to a minimum during unmodulated passages.

squelch *(Telecomm.)*. Form of automatic gain control where the gain of the receiver is reduced in response to certain characteristics of the input; e.g. in order to suppress background noise at very low signal levels.

squid *(Aero.)*. A dynamically stable condition of a fully-deployed parachute canopy which will not fully distend.

SQUID *(Phys.)*. *Superconducting Quantum Interference Device*. A family of devices capable of measuring extremely small currents, voltages and magnetic fields. Based on two quantum effects in superconductors; (1) *flux quantization* and (2) the *Josephson effect*. SQUIDs can detect changes in magnetic flux densities of ~ 1 nT (10^{-9} T). Examples are the study of fields generated by the action of the human brain and in magneto-biological research.

squinch *(Arch.)*. A small arch running diagonally across the corner of a square tower or room, to support a side of an octagonal tower or spire above. Also called a *scoinson arch*.

squint *(Build.)*. A purpose-made brick of shape suiting it for use as a squint quoin.

squint *(Med.)*. See **strabismus**.

squint *(Telecomm.)*. Difference between the geometrical axis of an aerial array and the axis of the radiation pattern.

squint quoin *(Build.)*. A quoin enclosing an angle which is not a right angle. Abbrev. *SQ*.

squirrel-cage motor *(Elec.Eng.)*. An induction motor whose rotor winding consists of a number of copper bars distributed in slots round the periphery, with the ends solidly connected to two heavy copper end-rings, the whole forming a rigid cage embedded in the rotor. See also **cage rotor, cage winding**.

squirrel-cage rotor *(Elec.Eng.)*. See **cage rotor**.

squirrel-cage winding *(Elec.Eng.)*. See **cage winding**.

Sr *(Chem.)*. The symbol for strontium.

sr *(Maths.)*. Symbol for steradian.

S-R theory *(Behav.)*. *Stimulus and response theory*. The stimulus response theory of learning which holds that the basic components of learning are S-R bonds, stimuli and responses which become forged together as learning proceeds.

s.s. and s.c. lathe *(Eng.)*. A sliding, surfacing, and screw-cutting lathe, i.e., one suitable for working on the periphery and on the end faces of workpieces, and capable of cutting a screw thread using a single-point tool.

SSI *(Comp.)*. *Small Scale Integration* refers to a **chip** with about 1 to 10 logic **gates**.

ssp, sspp *(Bot.)*. Abbrev. for *subspecies* (singular and plural).

SST *(Aero.)*. Abbrev. for *SuperSonic Transport aircraft*.

s state *(Electronics)*. State of zero orbital angular momentum.

ST *(Build.)*. Abbrev. for *Surface Trench*.

stab *(Print.)*. See **establishment**.

stabbing *(Build.)*. Making a brick-work surface rough to provide a key for plasterwork.

stabbing *(Min.Ext.)*. The process of locating one tube above and in-line with another so that they can be screwed together.

stabbing *(Print.)*. A special kind of *flat stitching* required for very thick books, pairs of wires being necessary (driven from the front page and from the back) clenching not being possible; often (wrongly) used as a synonym for *flat stitching*.

stabilate *(Biol.)*. A population of usually a micro-organism preserved in viable condition on a unique occasion by, for example, freezing.

stabilator *(Aero.)*. See **all-moving tail**.

stability *(Aero.)*. The quality whereby any deviation from steady motion tends to decrease. A given type of steady motion is stable if an aircraft will return to that state of motion after a disturbance, without intervention by the pilot. An aircraft has three axes about which its stability is defined with three associated degrees of freedom (angular, normal displacement and change of velocity).

stability *(Ecol.)*. The ability of an ecosystem to resist change.

stability *(Elec.Eng.)*. Property of any electrical circuit or system (e.g. electricity transmission network or closed-loop controller) whereby changes, usually sudden, in operational conditions (e.g. electrical load or speed of machine) can be coped with by the electrical circuit, or system, without loss of controlled operation within the designed range.

stability derivatives *(Aero.)*. Quantitative expressions for the variation of forces and moments on an aircraft due to disturbances from steady motion.

stability test *(Elec.Eng.)*. A test in which the cable is subjected to its working voltage (or a higher voltage) while it is alternately heated and allowed to cool. The power factor is measured during each heating and cooling period. If the power factor increases steadily during the test, the cable is said to be unstable.

stabilization *(Electronics)*. Maintenance of a quantity (voltage, current, frequency, gain etc.) against variations induced by supply voltage fluctuations, changing load conditions, temperature, and ageing. See **regulation**.

stabilized feedback amplifier *(Telecomm.)*. One in which amplification is stabilized against changes in supply voltages etc. by the application of negative feedback.

stabilized glass *(Glass)*. (1) Glass heat-treated so that it is in an equilibrium state corresponding to some particular temperature. (2) Glass heat-treated so as to suffer no permanent change of dimensions or properties over a particular range of temperature. (3) Glass resistant to darkening by high-energy, short-wave radiation.

stabilized yarn *(Textiles)*. Yarn that has been subjected to processes (e.g. heating and cooling under controlled tension) that reduce its ability to change in length or to twist and snarl.

stabilizer *(Aero.)*. In US, *horizontal stabilizer* is *tail plane*, and *vertical stabilizer* is *fin*. See **automatic-**.

stabilizer *(Chem.)*. (1) A negative catalyst. (2) A substance which makes a solution stable.

stabilizer *(Image Tech.)*. A solution used to render the processed image as permanent as possible by neutralizing chemical residue.

stabilizer tube *(Electronics)*. Gas-discharge tube, the voltage across which is much more stable than a voltage applied to it in series with a resistor.

stabilizing choke *(Elec.Eng.)*. A reactive choke coil inserted in series with an electric discharge lamp to compensate its negative resistance characteristic.

stable *(Chem.)*. Said of systems not exhibiting sudden changes, particularly atoms which are not radioactive.

stable *(Phys.)*. Used to indicate the incapability of following a stated mode of spontaneous change, e.g., *beta stable* means incapable of ordinary beta disintegration but capable of isomeric transition or alpha disintegration, etc.

stable equilibrium *(Phys.)*. The state of equilibrium of a body when any slight displacement increases its potential energy. A body in stable equilibrium will return to its original position after a slight displacement.

stable oscillation *(Telecomm.)*. One for which amplitude and/or frequency will remain constant indefinitely. A statically stable system may be dynamically unstable and follow a divergent oscillation when subjected to a disturbance. In this sense, the term *dynamically stable* means that an induced oscillation will be convergent, i.e. of decreasing amplitude.

stable platform *(Telecomm.)*. Structure which can be controlled in position with great precision, e.g. by gyroscopes, and which forms base for other information to be measured and transmitted by telemetry, e.g. from satellites.

stable pneumonia *(Vet.)*. See **equine influenza**.

stachyose *(Chem.)*. $C_{24}H_{42}O_{21}+4\frac{1}{2}H_2O$, a tetrasaccharide, found in the roots of *Stachys tuberifera* and of several Labiatae; mp (anhydrous) 170°C.

stack *(Comp.)*. A **list** for which all insertions and deletions are made at one end only. The arrangement is called *last-in, first-out*. Abbrev. *LIFO*. Cf. **queue**.

stack *(Phys.)*. Pile of photographic plates exposed to radiation together, and used to study tracks of ionizing particles.

stacked array *(Telecomm.)*. An antenna array in which the radiators are stacked one above the other and connected in phase so as to give the antenna directional properties.

stacking faults *(Phys.)*. Some crystal structures may be thought of as the stacking of planes of atoms in a definite sequence. Under some circumstances, e.g. deformation, the stacking order can be disturbed and such a sequence is said to have a stacking fault, a type of planar defect.

stadia lines, hairs *(Surv.)*. The two additional horizontal lines, one on each side of the central line, fitted to the diaphragm of a telescope to be used in **tacheometry**.

stadia rod *(Surv.)*. A special form of levelling staff bearing bold graduations suitable for the long sights usual in stadia tacheometry.

stadium *(Zool.)*. An interval in the life history of an animal between two consecutive ecdyses.

staff *(Build.)*. See **angle staff**.

staff *(Horol.)*. An arbor or axis, especially that of the balance or pallets.

staff *(Med.)*. A grooved rod introduced into the urethra as a guide for cutting a stricture.

staff *(Surv.)*. See **levelling staff**.

staff angle *(Build.)*. See **angle staff**.

staff bead *(Build.)*. See **angle bead**.

staffman *(Surv.)*. The surveyor's assistant whose duty it is to hold the levelling staff while the instrument is sighted upon it and readings are being taken.

stage *(Build.,Civ.Eng.)*. A ledge or working platform associated with scaffolding.

stage *(Geol.)*. A succession of rocks which were deposited during an age of geological time. A subdivision of a geological series.

stage *(Phys.)*. Unit of cascade in isotope separation plant, consisting of single separative element, or group of these elements, operating in parallel on material of same concentration.

stage *(Space)*. A section or part of a launch system which fires for a certain time only and then is separated from the main system; when more than one stage is used, the technique is termed *staging*.

stage efficiency *(Elec.Eng.)*. Ratio of a.c. output power to d.c. input power for any stage of an electronic amplifier.

stage micrometer *(Biol.)*. A device for measuring the magnification achieved with a given microscope or for calibrating an eyepiece graticule. It usually consists of a small accurate scale mounted on a microscope slide.

stage separation factor *(Nuc.Eng.)*. See **separation factor**.

stagger *(Aero.)*. The horizontal distance between the leading edges of the wings of a multiplane as projected vertically. If the upper plane is ahead of the lower, stagger is positive, if behind it is negative.

staggered *(Chem.)*. In **conformational analysis**, this represents a conformation in which the substituents of one atom in a bond are situated as far as possible from those of the other.

staggered tuning *(Telecomm.)*. Attempt to get a wide band response by a number of tuned circuits, having slightly different frequencies of resonance.

staggering *(Elec.Eng.)*. A term signifying the displacement of the brushes of a commutator motor from the neutral zone.

staggers *(Vet.)*. See **megrims**.

stagger-tuned amplifier *(Telecomm.)*. One with couplings tuned to different frequencies to give a band-pass response.

stag-headed *(Bot.)*. A tree having the upper branches dead with regrowth of the crown from new branches.

staging *(Build.)*. A robust scaffold of timbers on metal supports, braced together and capable of handling substantial loads.

staging *(Space)*. The principle of increasing the velocity achieved by a launch system and its payload by using more than one propulsive stage. Tandem (nose-to-tail) or parallel (side-by-side) staging may be employed, each stage being jettisoned after the fuel has been expended, thus, increasing the mass ratio and, therefore, the efficiency of the whole system.

stagnation point *(Aero.)*. The point at or near the nose of a body in motion in a fluid where the flow divides and where fluid pressure is at a maximum, and the fluid is at rest. Theoretically there is another stagnation point near the trailing edge.

stagnation temperature *(Aero.)*. The temperature which would be reached if a flowing fluid were brought to rest adiabatically, which is almost applicable in supersonic flight for the leading edges and air intakes, where the air in the boundary layer of a body is drastically and rapidly decelerated. Also *total temperature*.

stagnicolous *(Ecol.)*. Living in stagnant water.

stainer *(Build.)*. A pigment added to paint when a final colour is required which is different from that of the base used. Common pigments are: (1) *earth pigments*, ochres, umbers, Siennas, Venetian red, red oxide and malachite; (2) *synthetic pigments*, chromes, Monastral blue, ultramarine, Prussian blue and lakes.

staining power *(Chem.)*. The degree of intensity of colour which a coloured pigment will impart when mixed with a standard white pigment under standardized conditions.

stainless steel *(Eng.)*. Corrosion-resistant steel of a wide variety of compositions, but always containing a high percentage of chromium (8–25%). Exhibit *passivity* and therefore highly resistant to corrosive attack by organic acids, weak mineral acids, atmospheric oxidation etc. Used for cutlery, furnace parts, chemical plant equipment, stills, valves, turbine blades, ball-bearings etc. See **passivate**.

staking *(Eng.)*. Fastening operating similar to rivet setting, in which a projection on one part is upset by means of a punch so as to fit tightly against a mating feature in another part.

stalactite *(Geol.)*. A concretionary deposit of calcium carbonate which is formed by percolating solutions and hangs icicle-like from the roofs of limestone caverns.

stalactited *(Build.)*. A term applied to a variety of rusticated work distinguished by having ornaments resembling icicles on the faces of the stone.

stalagmite *(Geol.)*. A concretionary deposit of calcium carbonate, precipitated from dripping solutions on the floors and walls of limestone caverns. Stalagmites are often complementary to stalactites, and may grow so that they eventually join with them.

stalagmometer *(Chem.,Phys.)*. Apparatus which measures surface tension of liquid in terms of mass of a drip leaving a specified orifice.

staling *(Bot.)*. The accumulation with time of metabolites in a culture medium which results in the slowing-down of growth.

stalk *(Horol.)*. The thin rod or wire which carries the hammer of a striking or chiming clock.

stall *(Aero.)*. The progressive breakdown of the lift-producing airflow over an aerofoil, which occurs near the angle of maximum lift.

stall *(Eng.)*. Of an engine, stopping due to the too sudden application of a load or brake.

stall *(Min.Ext.)*. The working compartment or room in the **bord-and-pillar** method of working coal; a coal-miner's working place.

stalling speed *(Aero.)*. The airspeed of an aircraft at which the wing airflow breaks down.

stalling torque *(Elec.Eng.)*. The overload torque which is sufficient to slow down to zero the speed of an electric motor operating under load.

stall riser *(Build.)*. The upright part, of wood, marble, etc., between the pavement and the stallboard of a shop front.

stall-warning indicator *(Aero.)*. A device fitted to those aircraft which do not give any positive warning of the approach of the stall by **buffeting**. Usually operated by the change of pressure and movement of the **stagnation point** near the stall, warning may be audible, visual, or by a stick-shaking (or forward-pushing electric motor). See **stick pusher**.

stamen *(Bot.)*. The microsporophore (microsporophyll), or male reproductive organ, of flowering plants; i.e. the structure, within the flower, that bears the pollen. See also **filament, anther, androecium**.

staminal *(Bot.)*. Pertaining to a stamen: derived from a stamen.

staminate *(Bot.)*. Of flowers, male.

staminode *(Bot.)*. An imperfectly developed, vestigial, anther-less or petaloid stamen.

stamp *(Min.Ext.)*. (1) To crush. (2) A freely falling weight, attached to a long rod and lifted by means of a cam; once widely used for crushing ores.

stamping *(Elec.Eng.)*. See **lamination**.

stamping press *(Print.)*. See **blocking press**.

stanchion *(Civ.Eng.)*. A pillar, usually of steel, for the support of a superstructure.

stand *(Ecol.)*. Any living assemblage of land plants.

standard *(Build.)*. Measure of the volume of timber in bulk, viz. 165 cu. ft.

standard *(For.)*. See **board foot, Petersburg standard**.

standard *(Genrl.)*. Establishment unit of measurement, or reference instrument or component, suitable for use in calibration of other instruments. Basic standards are those possessed or laid down by national laboratories or institutes (*NPL, BSI etc.*).

standard atmosphere *(Aero.)*. See **International Standard Atmosphere**.

standard atmosphere *(Meteor.)*. Hypothetical atmosphere approximating to the average state of the real atmosphere in which pressure and temperature are defined at all heights. Internationally agreed standard atmospheres are used as bases for assessing the performance of altimeters, aircraft etc.

standard atmosphere *(Phys.)*. Unit of pressure, defined as 101 325 N/m^2, equivalent to that exerted by a column of mercury 760 mm high at 0°C. Abbrev. *atm*.

standard atmosphere for testing *(Textiles)*. In order to get reproducible and consistent results it is necessary to test textile materials at standard conditions. These are 20°C (27°C in the tropics) and 65% relative humidity. It is necessary to condition the materials for some time before they are tested to ensure they have reached equilibrium.

standard beam approach system *(Aero.)*. A system of radio navigation which provides an aircraft with lateral guidance and marker-beacon indications at specific points during its approach. Abbrev. *SBA*.

standard book number *(Genrl.)*. A number allotted to a book by agreement of (international) publishers which shows area, publisher and individual title, plus a check digit. Abbrev. *SBN, ISBN*.

standard calomel electrode *(Chem.)*. A half-element consisting of mercury, a paste of mercury and calomel (mercury (I) chloride), and a saturated solution of potassium chloride saturated with calomel; used as a standard potential difference in e.m.f. measurements.

standard cell *(Phys.)*. See **Weston standard cadmium cell**.

standard chamber *(Nuc.Eng.)*. Ionization chamber used for calibration of radioactive sources, or of absolute values of exposure doses.

standard deviation *(Stats.)*. The square root of the mean of the squared deviations from the mean of a set of observations; the square root of the variance.

standard electrode potential *(Chem.)*. The potential of a chemical element dipping into a solution of its ions at unit activity, referred to that of hydrogen under a pressure of one atmosphere as zero.

standard error *(Stats.)*. The estimated standard deviation of an estimate of a parameter.

standard filter *(Image Tech.)*. A filter which, when placed in front of a specified source, e.g. a tungsten lamp, gives a standard white light of black-body temperature 4800 K.

standard form *(Maths.)*. See **canonical form**.

standard frequency *(Elec.Eng.)*. Fifty or sixty Hz (cycles/second), the standards of **power frequency** in most countries of the world.

standard function *(Comp.)*. Subprogram provided by a compiler or other translator which carries out a task such as the computation of a mathematical function (e.g. log, square root).

standard gauge *(Civ.Eng.)*. That employed in most countries of the world, viz. 4 ft $8\frac{1}{2}$ in (1.435 m)

standard illuminant *(Phys.)*. One used for accurate colour measurements. The CIE specify three alternative standards; (1) representing an incandescent lamp with a colour temperature of 2854 K; (2) representing direct sunlight, colour temperature 4810 K; (3) representing sky light on an overcast day, colour temperature 6770 K. The latter sources can be produced using an incandescent lamp and suitable filters.

standardized mortality ratio *(Med.)*. *SMR*. The ratio of observed to expected deaths in a sub-population.

standard knot *(For.)*. A knot which is $1\frac{1}{2}$ in (40 mm) or less in diameter.

standard mean chord *(Aero.)*. The average chord, i.e. gross wing area divided by the span.

standard measurement *(Build.)*. The method recommended by the Chartered Surveyors' Institution for measurement of building works.

standard normal distribution *(Stats.)*. The normal distribution with mean zero and variance one, which is extensively tabulated.

standard page *(Print.)*. The largest size of page on any particular rotary press, there being no standard size.

standard propagation *(Telecomm.)*. With standard refraction, the propagation of radio waves over a perfectly smooth earth with uniform electrical characteristics.

standard radio atmosphere *(Telecomm.)*. A radio atmosphere in which the index of refraction decreases by 39×10^{-6} per km above the Earth's surface.

standard refraction *(Telecomm.)*. Refraction arising in a **standard radio atmosphere**.

standards converter *(Image Tech.)*. Equipment to convert signals from one colour TV system to another, for example, from NTSC to PAL.

standard signal generator *(Telecomm.)*. A precision oscillator whose output is calibrated as regards frequency and amplitude, and sometimes depth of modulation; used for testing radio equipment, receivers etc.

standard solenoid *(Elec.Eng.)*. A laboratory standard of inductance consisting of an air-cored solenoid with a secondary coil located at its centre. The dimensions are such that a value of the mutual inductance between the windings can be calculated from them.

standard solution *(Chem.)*. A solution whose concentration is accurately known. Such solutions are used in **volumetric analysis**.

standard specification *(Eng.)*. A *specification* incorporating widely accepted quantities or features, themselves standards, to ensure interchangeability, quality and reliability for least cost. Usually drawn up by national *Standards Institutions*.

standard temperature and pressure *(Genrl.)*. See **s.t.p.**

standard time *(Astron.)*. The civil time in any of the *time zones* established by international agreement. These are about 15° of longitude wide, equal to one hour. Within a zone all civil clocks are set the same standard time or rather local **solar time**. Zones usually differ by a whole hour, but there are a few cases of half-hour zones (e.g. South Australia).

stand-by losses *(Elec.Eng.)*. That part of the power expended in a generating station to maintain plant in instant readiness to take a sudden load.

standing crop *(Ecol.)*. The total dry mass of organisms present in an area, obtained by harvesting sample plots, drying and weighing the harvested biomass and expressing the standing crop in units like $g\,m^{-2}$.

standing current *(Telecomm.)*. See **quiescent current**.

standing formes *(Print.)*. Type kept locked up in formes in the expectation of a reprint.

standing-off dose *(Radiol.)*. Absorbed dose after which occupationally exposed radiation workers must be temporarily or permanently transferred to duties not involving further exposure. Doses are normally averaged over 13-week periods and standing-off would then continue for the remainder of the corresponding period.

standing panel *(Build.)*. A door panel whose height is greater than its width.

standing wave *(Telecomm.)*. One in which, for any component of the electromagnetic field, the ratio of its instantaneous amplitude at one point to that at any other point does not vary with time. In transmission lines, waveguides etc., the result of reflections from a load which is not perfectly matched to the transmission line or source.

standing-wave indicator *(Telecomm.)*. Any device which may be attached to or inserted in a transmission line or waveguide to indicate the presence of standing waves and assist correct matching. May be a simple neon lamp on a slide, a sensitive detector inserted into the line or waveguide in a **slotted line** arrangement, or a **reflectometer**.

standing-wave meter *(Telecomm.)*. One designed to measure voltage standing-wave ratio in transmission lines and waveguides.

standing-wave ratio *(Telecomm.)*. Where standing and progressive waves are superimposed, the SWR is the ratio of the amplitudes at nodes and antinodes. For a transmission line or waveguide, it is equal to $\frac{1-r}{1+r}$ where r is the coefficient of reflection at the termination. It may alternatively be defined by the reciprocal of this value as shown by its value being numerically greater than unity.

standing ways *(Ships)*. The portion of a ship's launching ways which are fixed to the ground. The sliding ways move on these ways and are positioned by an upstanding rib integral with the fixed ways.

stand-insulator *(Elec.Eng.)*. An insulator on which stands the structure used for supporting an accumulator or battery.

stand-off bomb *(Aero.)*. A small, fast, powered, unmanned aircraft or rocket containing a nuclear warhead, released from a bomber to fly many hundreds of miles to the target. It is automatically piloted and navigated, usually by a **Doppler navigator** and/or **inertial guidance** system. Propulsion can be by ramjet, rocket, turbojet or in combination.

stand pipe *(Eng.)*. (1) An open vertical pipe connected to a pipeline, to ensure that the pressure head at that point cannot exceed the length of the stand pipe. (2) Pipe connected to watermain for attaching a hose.

stand pipe *(Nuc.Eng.)*. The connection between the charge face and the interior of a reactor vessel, giving access to the fuel channels, for example, for refuelling.

stand sheet *(Build.)*. A window having no frame. Also *fast sheet, fixed sash, dead light.*

standstill *(Elec.Eng.)*. A term pertaining to the electrical behaviour of a machine when it is at rest.

standstill torque *(Elec.Eng.)*. The load torque which would bring an electric motor to a standstill.

stannane *(Chem.)*. *Tin hydride*, SnH_4.

stannates (II) *(Chem.)*. See **stannites**.

stannates (IV) *(Chem.)*. Analogous to the carbonates. Formed by heating solutions of, say, tin (IV) chloride with alkaline carbonates.

stannic (IV) acid *(Chem.)*. Acids of two types, formed by action of alkalis on solutions of tin (IV) chloride and by the action of nitric acid on the metal; called respectively α-stannic (IV) acid and metastannic acid or β-stannic (IV) acid.

stannic oxide *(Chem.)*. *Tin (IV) oxide.* SnO_2. Formed (1) by combustion of tin, (2) when stannic acids are calcined. Forms alkali stannates (IV) when fused with alkali carbonates. See **cassiterite**.

stannite *(Min.)*. Sulphide of copper, iron and tin, which crystallizes in the tetragonal system and usually occurs in tin-bearing veins. It is also called *bell-metal ore, tin pyrites.*

stannites (II) *(Chem.)*. *Stannates (II)*. Salts of stannous acid. Formed when stannous hydroxide is dissolved in alkaline solutions.

stannous hydroxide *(Chem.)*. *Tin (II) hydroxide.* $Sn(OH)_2$. Precipitated when sodium hydroxide is added to a solution of tin (II) chloride. When heated in carbon dioxide, forms black tin (IV) oxide, SnO, which, heated in air, forms tin (IV) oxide, SnO_2.

St Anthony's fire *(Med.)*. An old name applied to *erysipelas* and to *ergotism.*

Stanton number *(Chem.Eng.)*. A dimensionless parameter equal to the reciprocal of the **Prandtl number**.

stapedectomy *(Med.)*. Excision of the stapes.

stapes *(Zool.)*. In Amphibians, a small nodule of cartilage in connection with the fenestra ovalis of the ear: in Mammals, the stirrup-shaped innermost auditory ossicle. adj. *stapedial.*

staphylococcus *(Biol.)*. A Gram-positive coccus of which the individuals tend to form irregular clusters. The commonest types, associated with various acute inflammatory and suppurative conditions, including mastitis in animals, are *S. aureus* (golden yellow colonies) and *S. albus* (white colonies).

staphyloma *(Med.)*. Local bulging of the weakened sclera of the eye (as in glaucoma or myopia); bulging of a corneal scar in which the iris of the eye has become fixed.

staphylorrhaphy *(Med.)*. The operation of closing a cleft in the soft palate.

staple fibre *(Textiles)*. Natural or man-made fibres that are comparatively short in length e.g. 1–50 cm.

star *(Astron.)*. Sphere of matter held together entirely by its own gravitational field and generating energy by means of **nuclear fusion** reactions in its deep interior. The important distinguishing feature is the presence of a natural nuclear reactor in its core, where the pressure of the overlying mass of material is sufficient to cause nuclear reactions, the principle one of which is the transmutation of hydrogen into helium. During this process about 0.5% of the mass is converted into **electromagnetic radiation**. The minimum mass needed to make a star is probably 1/20th the mass of the Sun, the maximum about 70 times as great. See also; **Sun**, **main sequence** and **Hertzsprung-Russell diagram**.

star *(Comp.)*. See **network**.

star *(Image Tech.)*. Radiating tracks in photographic emulsion, arising from particle disintegration on collision, and dispersal of energy to other particles.

starch *(Chem.)*. *Amylum*, $(C_6H_{10}O_5)_x$; polysaccharide found in all assimilating (green) plants. A white hygroscopic powder which can be hydrolysed to dextrin and finally to D-glucose. Diastase converts starch into maltose. Starch does not reduce Fehling's reagent and does not react with phenylhydrazine. It forms a blue compound with iodine.

star chart *(Astron.)*. The name given to a systematic and accurately made map of the heavens in which the star positions are generally plotted according to equatorial coordinates.

starch grain *(Bot.)*. A rounded or irregular mass of starch; within a chloroplast in green algae; within a chloroplast, amyloplast or other plastid in vascular plants and bryophytes.

starch gum *(Chem.)*. See **dextrin**.

starch plant *(Bot.)*. A plant in which carbohydrate is stored as starch. Cf. **sugar plant**.

starch sheath *(Bot.)*. (1) A 1-layered cylinder of cells lying on the inner boundary of the cortex of a young stem, with prominent starch grains in the cells. It is homologous with an endodermis. (2) A layer of starch grains around a pyrenoid in an algal cell.

star cluster *(Astron.)*. See **galactic cluster**, **globular cluster**, **open cluster**.

star connection *(Elec.Eng.)*. Method of connecting a three phase load or source such that one terminal of each phase is connected to a common point, the neutral point. Currents flowing in the lines are equal to the corresponding phase currents. Line voltages are $\sqrt{3}$ times the corresponding phase voltage. Often termed *star connection*. See **delta connection**, **star point**, **voltage to neutral**.

star-delta starter *(Elec.Eng.)*. A starting switch for an induction motor which, in one position, connects the stator windings in star for starting and, in the other position, reconnects the windings in delta when the motor has gained speed.

Stark effect *(Phys.)*. Splitting of atomic energy levels, and of corresponding emission spectrum lines, by placing source in region of strong electric field; cf. **Zeeman effect**.

Stark-Einstein equation *(Chem.)*. The energy absorbed per mole for a photochemical reaction is $E = Nh\nu$, where N is Avogadro's constant, h is Planck's constant and ν is the frequency of the absorbed light.

starlite *(Min.)*. A name suggested (from a fancied resemblance to starlight) for the blue zircons which are heat-treated and used as gem-stones.

star magnitude *(Astron.)*. See **magnitude**.

star-mesh transformation *(Elec.Eng.)*. Technique for simplifying network, whereby any number of branches meeting at a point can be replaced by an equivalent mesh, thereby reducing the number of connections.

star network *(Elec.Eng.)*. One with many branches connected at a point; a *T-* or *Y-network* has 3 branches.

star observation *(Surv.)*. Use of theodolite to locate or orient a ground station by sighting on a star.

star point *(Elec.Eng.)*. The common junction of the several phases of a star-connected 3-phase system.

star-quad *(Elec.Eng.)*. See **quad**.

starred signature *(Print.)*. When a section is to be inserted in another it is given the same signature mark followed by an asterisk (*) or the figure 2.

star ruby, -sapphire, -quartz *(Min.)*. The prefix 'star' has reference to the narrow-rayed star of light exhibited by varieties of the minerals named. The star is seen to best advantage when they are cut *en cabochon*. It is caused by reflections from exceedingly fine inclusions lying in certain planes. See also **asterism**.

star-streaming *(Astron.)*. A phenomenon, discovered from analysis of observed stellar motions (after removing the effects of the observer's own motions), by which the stars are found to have two preferential directions of motion, one towards the point R.A. 90°, declination 15° south, and the other towards R.A. 285°, declination 64° south; the first stream contains about 60% of the observed stars. The effect is due to the rotation of the Galaxy.

star target *(Print.)*. A circular symbol with numerous radii, tapering to the centre but not meeting, leaving a small clear inner circle, used particularly on lithographic plates as a guide to maintenance of colour and quality of printing. Can be used to detect image loss or gain, slur etc.

start codon *(Biol.)*. Triplet sequence of nucleotides in **mRNA** which signals the initiation of translation and hence the first amino acid in a polypeptide chain.

starter *(Elec.Eng.)*. A device for starting an electric motor and accelerating it to normal speed. Also *motor starter*.

starter *(Electronics)*. See **pilot electrode**.

starter gap *(Electronics)*. The conducting path between the pilot electrode and the electrode to which the starting voltage is applied in a glow-discharge tube.

starter voltage *(Electronics)*. That applied to the pilot electrode of a cold-cathode discharge tube or mercury-arc rectifier; cf. **starting voltage**.

starting current *(Elec.Eng.)*. The current drawn by a motor from the mains when starting up.

starting resistance *(Elec.Eng.)*. A fixed resistance connected in series with the main circuit of a motor when starting up. This added resistance serves to limit excessive currents during start-up.

starting sheet *(Eng.)*. A sheet of pure metal used as the initial cathode on which the metal being refined is deposited during electrolytic refining.

starting torque *(Elec.Eng.)*. The torque developed by a motor at starting.

starting voltage *(Electronics)*. That which initiates current passing in a gas-discharge tube after nonconduction; much greater than that required to maintain conduction. Also known as *striking voltage* or *ionizing voltage*.

starting voltage *(Phys.)*. Same as *threshold voltage*.

starting winding *(Elec.Eng.)*. An auxiliary winding on the armature of a single-phase motor (enabling it to start up as a 2-phase machine) or of a synchronous converter.

startle colours *(Zool.)*. Bright colours on the body or wings of animals which often resemble vertebrate eyes and which are normally concealed. They are exposed on being disturbed and are anti-predator devices.

start-, starting-up time *(Nuc.Eng.)*. That required by instrument or system (e.g. nuclear reactor, chemical plant etc.) to reach equilibrium operating conditions.

start-stop *(Telecomm.)*. In machine telegraphy, the principle by which depression of a key on the keyboard of a transmitting machine sends the corresponding code to line, together with start and stop signals, the former to trip the printing machine so that it scans the sent code correctly, the latter to restore the scanning mechanism to the condition required for its operation by the next signal to arrive.

start-up procedure *(Nuc.Eng.)*. That followed when bringing a nuclear reactor into operation. It involves four successive stages: (*a*) *source range*, where a neutron source is introduced to generate the required neutron flux (this may not be necessary in a reactor which has already been operating); (*b*) *counter range*, where reactor is just critical but counters are required to monitor neutron-flux changes; (*c*) *period range*, where changes in reactivity are monitored on period meter; (*d*) *power range*, where reactor is operating within its designed power ratings.

star voltage *(Elec.Eng.)*. See **voltage to neutral**.

star wheel *(Eng.)*. (1) Toothed wheel moved one tooth per revolution of an adjacent shaft by a pin attached to that shaft; cf. **Geneva movement**. (2) A continuously or intermittently rotating disk with scalloped circumference, used to guide bottles or other containers on to a conveyor at correct intervals.

star wheel *(Horol.)*. A wheel with pointed triangular teeth.

star wheel *(Print.)*. On **line-casting machines** a four-toothed wheel which stops rotating, to indicate to the operator that a line of matrices is too long.

stasis *(Bot.)*. Stoppage of growth.

stasis *(Med.)*. (1) Complete stoppage of the circulation of blood through the capillaries and smallest blood vessels in a part. (2) Arrest of the contents of the bowel at any point from obstruction or weakness of the bowel wall.

stassfurtite *(Min.)*. A massive variety of boracite which sometimes has a subcolumnar structure and resembles a fine-grained white marble or granular limestone. See **boracite**.

stat- *(Phys.)*. Prefix to name of unit, indicating derivation in obsolete electrostatic system of units, e.g. statampere, statohm etc.

state *(Phys.)*. The energy level of a particle as specified by the appropriate quantum numbers.

state-dependent learning *(Behav.)*. Learning in which the recall depends on the degree of similarity between the physiological state of the individual at the time of training and at the time of testing.

state-dependent memory *(Behav.)*. Refers to the fact that events learned in one mental state are best remembered when the individual is put back in that state (e.g. a particular mood, a drug-induced state).

state function *(Phys.)*. A quantity in thermodynamics which has a unique value for each state of a system. Internal energy, entropy and enthalpy are examples. The value associated with a given state is independent of the process used to bring about that state.

statement number *(Comp.)*. One used to label a specific statement in a program so that the user and the machine can refer back to it subsequently. Such numbers do not indicate the order in which instructions must be carried out, or the number of the address at which the instruction will be stored. Also *instruction number, line number*.

statenchyma *(Bot.)*. A tissue consisting of cells containing statoliths.

state of matter *(Chem.)*. Traditionally all matter is in one of three states: solid (fixed volume and shape), liquid (fixed volume, shape that of container), gaseous (filling the containing vessel).

static *(Elec.Eng.)*. Non-movable or non-rotating, e.g. a transformer or rectifier is a static converter.

static *(Telecomm.)*. Said of all electrical disturbances to a radio system which arise through electrostatic induction, particularly from lightning flashes.

statical stability *(Ships)*. The ability of a ship to return to her initial position when forcibly inclined.

static balancer *(Elec.Eng.)*. See **a.c. balancer**.

static capacitor *(Elec.Eng.)*. A capacitor, described as static to distinguish it from *rotating*, often used in electrical supply systems for power factor correction.

static characteristic *(Elec.Eng.)*. Curve, or set of curves, which describes relation between specified voltages and currents of electrodes under unvarying conditions, as compared with **dynamic characteristic**, which implies operations under time-varying conditions.

static convergence *(Image Tech.)*. The adjustment to a colour TV whereby the three coloured **rasters** are made to coincide in the centre of the screen.

static discharge wick *(Aero.)*. Wicks, usually of cotton impregnated with metallic silver, or of nichrome wire, fitted at the trailing edges of an airplane's flight control surfaces, by which static electricity is discharged into the atmosphere.

static electricity *(Phys.)*. See **dynamic electricity**.

static impedance *(Elec.Eng.)*. The electrical impedance of a machine or transducer when it is stopped from moving. In loudspeakers, it has the same meaning as *blocked impedance*.

static instability, hydrostatic instability *(Meteor.)*. Atmospheric state such that a parcel of air moved from its initial level experiences a hydrostatic force tending to remove it further from this level.

static inverter *(Aero.)*. A nonrotating device for converting d.c. current to a.c. supply, usually of high voltage, for radio and instrument services.

static jet thrust *(Aero.)*. See **static thrust**.

static line *(Aero.)*. A cable joining a parachute pack to the aircraft, so that when the wearer jumps, the parachute is automatically deployed.

static machine *(Elec.Eng.)*. See **electrostatic generator**.

static marks *(Image Tech.)*. Marks caused by electrostatic discharge near the surface of undeveloped photographic film which become visible after processing.

static memory *(Comp.)*. Needs no *refreshing* once information is stored. Cf. **dynamic-**.

static pressure *(Aero.)*. The pressure at any point on a body moving freely with a fluid in motion; in practice, the pressure normal to the surface of a body moving through a fluid.

static-pressure tube *(Aero.)*. A tube with openings placed so that when the air is moving past it the pressure inside is that of still air. See **Pitot tube, static vent**.

statics *(Phys.)*. That branch of applied mathematics which studies the way in which forces combine with each other usually so as to produce equilibrium. Until the early part of the 20th century the term also embraced the study of gravitational attractions, but this is now normally regarded as a separate subject.

static stability *(Aero.)*. Positive static stability in an aircraft means that if it is disturbed from a trimmed speed the disturbance will be reduced.

static stability *(Elec.Eng.)*. The stability of a transmission system with reference to gradual changes in load demand.

static thrust *(Aero.)*. The net thrust (kN or lb.s.t) of a jet engine at *International Standard Atmosphere* sea level and without translational motion.

static vent *(Aero.)*. An opening, usually in the fuselage, found by experiment, where there is minimum **position error** and which is used instead of the **static-pressure tube**.

station *(Elec.Eng.)*. In general, a generating station. Specifically, a key point on an electricity supply system.

station *(Eng.)*. Location, on a transfer machine or on an assembly line, at which a workpiece or an assembly is halted for the placing of a component or for the execution of a machining or fastening operation.

station *(Surv.)*. (1) A point at an apex of a triangle in a skeleton, or otherwise situated in a line of the skeleton. (2) A point whose reduced level is to be found.

station *(Telecomm.)*. Location of radio transmitters and/or receivers with antennae, for sending or receiving radio signals on one or more wavelengths.

stationary orbit *(Astron.,Space.)*. That circular orbit of a satellite which holds it fixed above a point on its parent body's equator. For the Earth, such a geostationary orbit is about 35 800 km above the equator. Very useful for communication relays because of the good coverage and because the surface antennae do not require to be steered. A special case of **synchronous orbit**.

stationary phase *(Chem.)*. See **chromatography**.

stationary point *(Maths.)*. Point at which the derivative of a function is zero. Includes maximum and minimum (*turning points*).

stationary point on a curve *(Maths.)*. A point at which the tangent is parallel to the *x*-axis. For the curve $y = f(x)$, the point where $x = a$ is a stationary point if $f'(a) = 0$. All turning points are stationary points but not all stationary points are turning points.

stationary points *(Astron.)*. Those points in the apparent path of a planet where its direct motion in right ascension changes to retrograde motion, or vice versa.

stationary wave *(Telecomm.)*. Earlier alternative to **standing wave**.

station keeping *(Space)*. The manoeuvres necessary to adjust a geostationary satellite's orbit so that its position in space is correct, its ground coverage does not vary and data transmission is optimized; the manoeuvres are usually effected by small jets or rockets.

station pointer *(Surv.)*. Instrument for obtaining a mechanical solution of the **three-point problem**. It consists of a full-circle protractor with one fixed radial arm and two movable radial arms, which can be set to the correct mutual directions of the 3 points.

statistic *(Stats.)*. A numerical quantity calculated from a set of observations.

statistical diameters *(Powder Tech.)*. Statistical average of a specified parameter on microscopic methods of particle-size analysis.

statistical energy analysis *(Acous.)*. Method to calculate the density of vibratory energy in coupled structures. Used in structure-borne sound problems.

statistical error *(Phys.)*. Radiation detectors measure the average count for random events. As a result of statistical fluctuations, the average count N has a statistical error of $\sqrt{N}$. See **Poisson distribution**.

statistical mechanics *(Phys.)*. Theoretical predictions of the behaviour of a macroscopic system by applying statistical laws to the behaviour of component particles. *Quantum mechanics* is an extension of classical statistical mechanics introducing the concepts of the quantum theory, especially the Pauli exclusion principle. *Wave mechanics* is a further extension based on the Schrödinger equation, and the concept of particle waves.

statistical weight *(Phys.)*. In statistical mechanics, if a system has a number of quantized states, and more than one state has the same energy, the energy level has a *statistical weight* equal to the number of states having that energy. See **degeneracy**.

statocyst *(Zool.)*. An organ for the perception of the position of the body in space, consisting usually of a sac lined by sensory cells and containing a free hard body or bodies, either introduced or secreted; an otocyst.

statocyte, statocyst *(Bot.)*. A cell containing starch grains, or other solid inclusions, acting as statoliths.

statolith *(Bot.)*. A starch grain, or other solid body in a cell, that moves in response to gravity and appears to function as a gravity sensor.

statolith *(Zool.)*. A secreted calcareous body contained in a statocyst.

stator *(Aero.)*. The row of fixed, radially disposed aerofoils which forms an essential part of the dynamics of an axial compressor or axial turbine.

stator *(Elec.Eng.)*. Stationary part of machine, especially a dynamo or motor; cf. **armature** (2).

stator blade *(Aero.)*. A small fixed aerofoil, usually of thin highly-cambered section, and of approximately parallel chord, mounted in the outer case of an axial compressor or turbine. See **exhaust-**.

stator core *(Elec.Eng.)*. The assembly of laminations forming the magnetic circuit of the stator of an a.c. machine.

stator-rotor starter *(Elec.Eng.)*. A combined stator circuit switch and rotor-circuit regulating resistance for use with slip-ring induction motors.

stator winding *(Elec.Eng.)*. That part of the electrical winding of a machine accommodated in the stator.

status epilepticus *(Med.)*. A succession of severe epileptic convulsions with no recovery of consciousness between each convulsion.

staurolite *(Min.)*. Silicate of aluminium, iron and magnesium, with chemically combined water, sometimes occurring as brown cruciform twins, and crystallizing in the orthorhombic system. It is typically found in medium grade regionally metamorphosed argillaceous sediments.

stay *(Eng.)*. See **brace**.

stay tap *(Eng.)*. A long tap for threading the holes for stays connecting adjacent plates in boilers, thus ensuring that the two holes are threaded in correct pitch relation.

stay tubes *(Eng.)*. Boiler fire tubes acting as stays to the flat surfaces which they join; sometimes threaded and nutted to the plates for extra strength.

stay wire *(Elec.Eng.)*. One of several steel cables by which a transmission-line pole is secured to the ground.

STD *(Telecomm.)*. Abbrev. for *Subscriber Trunk Dialling*.

St. David's *(Geol.)*. The middle epoch of the Cambrian period.

steady *(Eng.)*. (1) In turning for supporting long, heavy or slender work. The fixed type clamps to the lathe bed and carries 3 radial jaws, adjustable to bear on the rough-turned work. The travelling type fixed to the moving carriage usually has two jaws supporting the work close to the cutting tool. (2) In milling, used to couple the machine table to the overarm support to give added rigidity and reduce vibration.

steady flow *(Phys.)*. See **viscous flow**.

steadying resistance *(Elec.Eng.)*. The ballast resistance placed in series with a d.c. arc lamp, to counteract negative resistance in the arc.

steady pin *(Eng.)*. A pin which permits mechanical parts to be fitted together accurately with one fixing screw. Cf. **dowel**.

steady pin *(Horol.)*. A pin used where two parts have to be fixed accurately relative to one another; a steady pin is fixed to one part and is a close fit in a hole in the other part. Used for the location of bridges, cocks etc.

steady state *(Phys.)*. One in dynamic equilibrium, with entropy at its maximum.

steady state *(Telecomm.)*. Said of any oscillation system which continues unchanged indefinitely.

steady-state theory *(Astron.)*. One of two rival theories of *cosmology* (the other is the *Big Bang*), proposed in 1948. It asserts that: 'things are as they are because they were as they were'. The universe is infinitely old, and contains the same density of material (on the average) at all points and all times. In order to compensate for the observed expansion of the universe, it is necessary to postulate the continuous creation of matter in this model. From the mid-1960s it fell out of favour because it cannot readily explain the **microwave background** (*cosmic background radiation*) or the observed density of **radio galaxies** and **quasars** in the far universe.

stealth *(Aero.)*. The technology of reducing the observable characteristics of military aircraft and missiles. Means include reducing size, noise, IR emissions from engines and from hot surfaces, radar reflections from intakes and between surfaces.

steam *(Phys.)*. Water in the vapour state; formed when specific latent heat of vaporization is supplied to water at boiling point. The specific latent heat varies with the pressure of formation, being approximately 2257 kJ/kg at atmospheric pressure. See **dry-**, **saturated-**, **super-heated-**.

steam accumulator *(Eng.)*. A large pressure vessel, partly filled with water, into which surplus high-pressure steam is blown and condensed. A supply of saturated steam is thus available by lowering the pressure at the outlet valve, thus causing evaporation of the water stored.

steam car *(Eng.)*. An automobile propelled by steam. Oil-fired **flash boilers** are generally used, no gearbox is necessary, and control is simple. Water is recovered by condensing the exhaust steam in a radiator.

steam chest *(Eng.)*. The chamber in which the slide-valve of a steam engine works, and to which the steam pipe is connected.

steam coal *(Min.Ext.)*. Two varieties classified by the National Coal Board are *dry steam coal* (also called *semianthracite*) and *coking steam coal* (rank 202, 203, 204).

steam distillation *(Chem.)*. The distillation of a substance by bubbling steam through the heated liquid. It is a useful method of separation for substances which are practically insoluble in water. The rapidity with which a substance distils in steam depends on its vapour pressure and on its vapour density.

steam dome *(Eng.)*. See **dome**.

steam economizer *(Eng.)*. See **economizer**.

steam-electric generating set *(Elec.Eng.)*. A generating set in which the prime mover is a steam engine, e.g. a steam turbine or reciprocating steam engine.

steam engine *(Eng.)*. An external combustion engine whose working fluid is steam.

steam gauge *(Eng.)*. A *gauge* for indicating or recording steam pressure in a boiler or other part of a steam system.

steam generating station *(Elec.Eng.)*. A generating station in which the prime movers driving the electric generators are operated by steam, e.g. steam turbines or reciprocating steam engines.

steam generator *(Eng.)*. A steam boiler.

steaming up *(Vet.)*. The practice of increasing the nutritional plane of dairy cattle a few weeks before calving.

steam injector *(Eng.)*. See **injector**.

steam jacket *(Eng.)*. A jacket formed round a steam-engine cylinder; supplied with live steam to prevent excessive condensation of the working steam in the cylinder.

steam lap *(Eng.)*. See **outside lap**.

steam locomotive *(Eng.)*. A self-propelled steam engine and boiler integrally mounted on a frame which is fitted with wheels driven by the engine. The term is usually restricted to locomotives used to haul passenger or goods traffic on a railway, but various kinds of road locomotives were once common, such as the traction engine.

steam nozzle *(Eng.)*. See **nozzle, convergent-divergent nozzle**.

steam ports *(Eng.)*. Passages leading from the valve face to the cylinder of a steam engine; through them the steam is supplied and exhausted.

steam reversing gear *(Eng.)*. A power reversing gear, used in steam locomotives, by which movement of the driver's reversing lever admits steam to an auxiliary cylinder, whose piston operates the reversing links of the valve gear.

steam tables *(Eng.)*. List of figures giving the properties of steam over a pressure range.

steam trap *(Eng.)*. A device into which condensed steam from steam pipes etc. is allowed to drain, and which automatically ejects it without permitting the escape of steam.

steam turbine *(Eng.)*. A machine in which steam is made to do work by expanding so as to create kinetic energy, which is then partly absorbed by causing the steam to act on moving blades attached to a disk or drum. See **back-pressure turbine, disk-and-drum turbine, extraction turbine, impulse turbine, mixed-pressure turbine, reaction turbine**.

steapsin *(Zool.)*. A fat-digesting enzyme occurring in the digestive juices of various animals, as the pancreatic juice of Vertebrates.

stearic acid *(Chem.)*. $C_{18}H_{36}O_3$, a monobasic fatty acid; mp 69°C, bp 287°C; obtained from mutton suet, or by reducing oleic acid. It occurs free in a few plants, as glycerides in many fats and oils, and as esters with the higher alcohols in certain waxes.

stearin *(Chem.)*. A term for the glyceryl ester of stearic acid. The name is also applied to a mixture of stearic acid and palmitic acid.

steatite, soapstone *(Min.)*. A coarse, massive, or granular variety of talc, greasy to the touch. On account of its softness it is readily carved into ornamental objects.

steatorrhoea, steatorrhea *(Med.)*. The presence of an excess of fat in the stools, due either to failure of absorption, or to deficiency of the fat-splitting enzymes in the digestive juices, as a result of disease of the pancreas.

steel *(Eng.)*. Essentially an alloy of iron and carbon. Contains less than 2% carbon, less than 1% manganese, and small amounts of silicon, phosphorus, sulphur and oxygen. Mechanical properties can be varied over a wide range by changes in composition and heat treatment. Mild steel contains less than 0.15% C, medium 0.15–0.3% C, hard more than 0.3% C. If used as a structural material in reactors, as a neutron or thermal shield, it must be outside the active section. This restriction is due to some of its constituents becoming strongly radioactive, with long lives, on capturing neutrons. See also **alloy steel, stainless steel**.

steel-cored aluminium *(Elec.Eng.)*. An electrical conductor consisting of a layer or layers of aluminium wire surrounding a core of galvanized steel strands.

steel-cored copper conductor *(Elec.Eng.)*. A conductor made in the same way as steel-cored aluminium, except that the steel core is covered by a layer of insulating tape, to prevent corrosion of the surrounding copper.

steel-making *(Eng.)*. The process of making steel from pig-iron, with or without admixture with steel scrap. The processes used inlude **Bessemer, open-hearth, crucible, electric-arc, high-frequency induction** and **duplex**.

steel-tank rectifier *(Elec.Eng.)*. A mercury-arc rectifier in which the arc chamber is of steel. Cf. **glass-bulb rectifier**.

steel tape *(Surv.)*. See **band chain**.

steel tower *(Elec.Eng.)*. The framed steel structure carrying a high-voltage transmission line. Also *pylon*.

steeping *(Textiles)*. To leave a yarn or fabric in a liquid usually without agitation; to soak the fabric, squeeze and leave wet.

steerable antenna *(Telecomm.)*. Any antenna in which the direction of its greatest radiation or sensitivity can be altered or *steered*; this may be done by electronic means (see **phased array, active array**) or by mechanically rotating or elevating the antenna array or *dish*.

steering *(Telecomm.)*. Alteration by mechanical or electrical means of the direction of maximum sensitivity of a directional antenna, e.g. radar, radio telescope or earth station.

steering arm *(Autos.)*. An arm rigidly attached to a stub axle, to which it transmits angular movement from the motion of the steering rod, and attached to it by a ball joint.

steering box *(Autos.)*. The housing which encloses the steering gear and provides an oil-bath for the working surfaces. It is rigidly attached to a side-member of the chassis frame. It contains gears which transmit the action of the steering column to the steering rod. In Europe, steering box systems have been largely supplanted by **rack and pinion** gear for cars, although they are still standard on commercial vehicles.

steering gear *(Autos.)*. The two geared members attached to the steering column and the drop-arm spindle respectively. They transmit motion from the steering-wheel to the stub axles through the drop arm, steering rod or drag link, steering arms, and track rod. See **cam-type-, rack-and-pinion-, screw-and-nut-, worm-and-wheel-**.

steering rod *(Autos.)*. See **drag link**.

Stefan-Boltzmann law *(Phys.)*. Total radiated energy from a black-body per unit area per unit time is proportional to the fourth power of its absolute temperature, i.e. $E = \sigma T^4$ where σ (Stefan-Boltzmann constant) is equal to $5.6696 \times 10^{-8}\,\mathrm{W\,m^{-2}\,K^{-4}}$.

Steiner's tricusp *(Maths.)*. A hypocycloid in which the radius of the rolling circle is one-third or two-thirds that of the fixed circle. It has three cusps.

steining *(Civ.Eng.)*. The process of lining a well with bricks, stone, timber or metal, so as to prevent the sides from caving in. Also *steaning*.

Steinmann trinity *(Geol.)*. An association of cherts, spilites and serpentines, characteristic of a former ocean-floor environment.

Steinmetz coefficient *(Phys.)*. The constant of proportionality in the *Steinmetz Law*. Also called the *hysteresis coefficient*.

Steinmetz law *(Phys.)*. An empirical law stating that the energy loss per unit volume in a ferromagnetic material during each cycle of a hysteresis loop is proportional to

B_m^n, where B_m is the maximum value of the magnetic flux density attained, n is 1.6 for many materials but some newer ferromagnetic alloys have values ranging from 1.5 to 2.5 . n may not be constant for more than a limited range of B_m.

stele *(Bot.)*. The primary vascular system and associated ground tissue of the stems and roots of a vascular plant.

stellarator *(Nuc.Eng.)*. A toroidal fusion device in which the magnetic fields are generated entirely by conductors placed around the torus. See **tokamak**.

stellar energy *(Astron.)*. See **carbon cycle, proton-proton chain**.

stellar evolution *(Astron.)*. The sequence of events and changes covering the entire life cycle of a star. The principal stage of evolution is the nuclear burning of hydrogen to form helium, with a consequential release of energy. Eventually the hydrogen in the core exhausted, and the star becomes a **red giant**. In the final stages of evolution there are several paths; the formation of a **white dwarf**, a **neutron star**, or a **supernova**. See also **main sequence** and **Hertzsprung-Russell diagram**.

stellar interferometer *(Astron.,Phys.)*. A device, developed by Michelson, by means of which, when fitted to a telescope, it is possible to measure the angular diameters of certain giant stars (all of which are below the limit of resolution of even the largest telescopes) by observations of interference fringes at the focus of the telescope.

stellar magnitude *(Astron.)*. See **magnitude**.

stellar population *(Astron.)*. See **population type**.

stellar wind *(Astron.)*. Radial outflow of material from the atmosphere of a very hot star, analogous to the **solar wind**.

stellate *(Bot.,Zool.)*. Radiating from a centre, like a star.

stellate hair *(Bot.)*. A hair which has several radiating branches.

Stellite *(Eng.)*. TN for a series of alloys with cobalt, chromium, tungsten, and molybdenum in various proportions. The range is chromium 10–40%, cobalt 35–80%, tungsten 0–25% and molybdenum 0–10%. Very hard. Used for cutting-tools and for protecting surfaces subjected to heavy wear.

Stelvetite *(Plastics)*. TN for PVC-coated steel sheet used for a variety of purposes including cabinets, wall sections, domestic equipment.

stem *(Bot.)*. The above-ground axis of a plant, and other axes (above- or below-ground) that are anatomically similar and/or clearly homologous. In most vascular plants, stems may be recognized by their bearing foliage leaves and/or scale leaves.

stem *(Horol.)*. In a keyless watch, the shaft to which the button is attached.

stem *(Print.)*. See **body**.

stem-and-leaf plot *(Stats.)*. An arrangement of a set of values into rows with common leading digits (stems), entries in each row being the remaining digit of each value, truncated if necessary.

stem bar *(Ships)*. The portion of material forming the extreme forward end of a ship. The hull proper is secured thereto; sometimes known as *stem post*.

stem cell *(Biol.)*. An undifferentiated cell which divides, one daughter cell giving rise, usually by a succession of stages, to a mature functional cell such as the erythrocyte. The stem cell's name usually ends in *-blast* as in *erythroblast*.

stem cell *(Immun.)*. A large cell present in bone marrow and other haemopoietic tissues which is capable both of self-replication and generation of progenitor cells for a series of different cell lines. Stem cells are stimulated to proliferate and differentiate under the influence of colony stimulating factors present in the environment. (Known also as *CFU-S–colony forming unit–spleen*, referring to the tissue in which it was first studied.) They can give rise to progenitors of lymphocytes or of granulocytes and macrophages. Interleukin-3 can stimulate the growth and differentiation of all the cell types. However there are different colony stimulating factors for granulocytes and macrophages, and for specific kinds of granulocytes (neutrophils, eosinophils, basophils).

stem correction *(Phys.)*. If only the bulb of a thermometer is immersed, the stem may be at a different temperature from the bulb, and the reading may have to be corrected accordingly.

stempipe *(Min.Ext.)*. See **drilling pipe**.

stem post *(Ships)*. See **stem bar**.

stem succulent *(Bot.)*. A plant with a succulent, photosynthetic stem and with the leaves small or sometimes represented by spines. Many are CAM plants, e.g. cacti. Cf. **leaf succulent**.

stencil *(Print.)*. (1) A plate of fine metal, plastic, waxed paper, etc. used for design and lettering, perforated in the required pattern so that ink, paint or other colouring substance may be passed through according to that pattern. (2) Term applied to the prepared mesh in **screen process printing**.

steno- *(Genrl.)*. Prefix from Gk. *stenos*, narrow.

stenode *(Telecomm.)*. Supersonic heterodyne receiver in which there is very sharp tuning in intermediate-frequency circuits, using piezoelectric quartz crystals, with frequency correction of audio signal after demodulation.

stenohaline *(Zool.)*. Capable of existence within a narrow range of salinity only. Cf. **euryhaline**.

stenophyllous *(Bot.)*. Having narrow leaves.

stenopodium *(Zool.)*. The typical biramous limb of Crustacea, having slender exopodite and endopodite. Cf. **phyllopodium**.

stenosis *(Med.)*. Narrowing or constriction of any duct, orifice, or tubular passage as a result of disease. adj. *stenosed*.

stenter, tenter *(Textiles)*. A machine that holds a fabric by its selvedges while keeping it taut in open-width and transporting it through a long heated chamber. The machine may hold the fabric by pins or clips. The machine is used to dry fabrics, to heat-set them or to fix chemicals (e.g. resins) on them.

stentorphone *(Acous.)*. See **pneumatic loudspeaker**.

step *(Aero.)*. The step discontinuity in the bottom of a flying-boat hull, to facilitate take-off from the water surface by allowing the forebody to plane and the afterbody to be clear of the forebody wake. Typical values range from 4% to 12% of the maximum beam. Low values are likely to induce **porpoising**.

step *(Elec.Eng.)*. Synchronous machines are said to keep *in step* when they remain in synchronism with each other.

step *(Telecomm.)*. See **transient**.

step-and-repeat *(Print.)*. Method of contacting and repeating a single image in predetermined position onto film or printing plate to create multiple images on the printed paper, e.g. in label work or postage stamps.

step-by-step method *(Elec.Eng.)*. A method of determining the hysteresis curve of a magnetic material, in which the field strength is increased and reversed in steps.

step-down transformer *(Elec.Eng.)*. Reverse of the *step-up transformer*, i.e. in an electrical transformer, the transfer of energy from a high to a low voltage.

step-faults *(Geol.)*. A series of tensional or normal faults which have a parallel arrangement, throw in the same direction, and hence progressively 'step down' a particular bed.

step function *(Maths.)*. One which makes an instantaneous change in value from one constant value to another. Its **Fourier analysis** shows an infinite number of harmonics present.

step function *(Telecomm.)*. A function which is zero for all time preceding a certain instant, and has a constant infinite value thereafter.

Stephanian *(Geol.)*. The uppermost stage of the Carboniferous system, corresponding in Britain to the beds above the Coal Measures.

stephanite *(Min.)*. A sulphide of silver and antimony which crystallizes in the orthorhombic system. It is usually associated with other silver-bearing minerals. Also *brittle silver ore*.

Stephenson's link motion *(Eng.)*. See **link motion**.

step-index fiber *(Telecomm.)*. US for **stepped-index fibre**.

step irons *(Build.)*. See **foot irons**.

stepped *(Print.)*. When type is to be printed close to an irregularly shaped block, the mount must be cut in steps to allow this.

stepped flashing *(Build.)*. Used where a brick chimney projects from a sloping roof, the lead being cut in steps so that the horizontal edges of the 'steps' may be secured into raglets cut in the joints of the brickwork. Also called *skeleton flashing*.

stepped-index fibre *(Telecomm.)*. An optical fibre in which the transition from the high refractive index of the information-bearing core to the lower refractive index of the cladding is abrupt. Cf. **graded-index fibre**. See **monomode fibre, multimode fibre**.

stepping *(Civ.Eng.)*. Laying foundations in horizontal steps on sloping ground. See **benched foundation**.

stepping *(Surv.)*. The process of chaining over sloping ground by making the measurement in horizontal lengths with the chain always held horizontally, one end in the air.

step printer *(Image Tech.)*. That which prints one frame at a time.

step-rate prepayment meter *(Elec.Eng.)*. A prepayment meter in which a high charge per unit is made until a given number of units have been consumed, when the gear ratio is automatically changed to give a lower charge per unit until the mechanism is reset by hand.

step-up instrument *(Elec.Eng.)*. See **suppressed-zero instrument**.

step-up transformer *(Elec.Eng.)*. A two-winding transformer with more secondary than primary turns. It thus transforms low to high voltages.

step wedge *(Image Tech.)*. A series of graded exposures given to a photographic material and the resultant image produced after processing, for **sensitometric** examination.

steradian *(Maths.)*. The SI unit of solid angular measure. It is defined as the solid angle subtended at the centre of a sphere by an area on its surface numerically equal to the square of the radius. Symbol *sr*.

Sterba antenna *(Telecomm.)*. A stacked *broadside* array with a reflecting curtain, which may be *parasitic* or fed in the same way as the main radiating curtain. It can be uni- or bi-directional; used for short-wave communication.

stercolith, stercorolith *(Med.)*. A hard faecal concretion, impregnated with calcium salts, in the intestine. (Latin *stercus*, gen. *stercoris*, dung.)

stercoraceous *(Med.)*. Consisting of, or pertaining to, faeces.

stercoral *(Med.)*. Of, pertaining to, or caused by, faeces.

stere *(For.)*. A stacked cubic metre of timber, which equals 35.3 stacked cubic feet.

stereo *(Acous.)*. Abbrev. for *stereophonic*.

stereocamera *(Image Tech.)*. One equipped with two matched lenses or a beam-splitting device for taking photos in stereo pairs. Also called *binocular camera, stereoscopic camera*.

stereochemistry *(Chem.)*. Branch of chemistry dealing with arrangement in space of the atoms within a molecule.

stereognosis *(Med.)*. The ability to recognize similarities and differences in the size, weight, form, and texture of objects brought into contact with the surface of the body.

stereogram, stereograph *(Image Tech.)*. General terms for photographs intended to be viewed in an apparatus to give a 3-dimensional appearance. See **parallax stereogram**.

stereo-isomerism *(Chem.)*. The existence of different substances whose molecules possess an identical connectivity but different arrangements of their atoms in space. Also *alloisomerism*. See **geometrical-** and **optical-isomerism**.

stereokinesis *(Biol.)*. Movement of an organism in response to contact stimuli.

stereome *(Bot.)*. A general term for the mechanical tissue of the plant.

stereome cylinder *(Bot.)*. A cylinder of strengthening tissue lying in a stem, usually just outside the phloem.

stereomicrophone system *(Acous.)*. Dual microphone with, e.g. interlocking figure-of-eight polar diagrams; used to provide signals for both channels of a stereophonic sound-reproduction system.

stereophonic recording *(Acous.)*. (1) Use of adjacent tracks on magnetic tape, with multiple recording and reproducing channels. (2) Use of a spiral cut on disks, two channels being represented by stylus motions at right angles, each at 45° to the surface; reproduction is by one stylus operating a double transducer.

stereophony *(Acous.)*. Method of sound reproduction which attempts to give the hearer an effect of auditory perspective similar to that created by the original sound. Methods used require two loudspeakers and three microphones or a pair of microphones placed close together. adj. *stereophonic*. Also called *auditory perspective, localization*.

stereoscope *(Phys.)*. Device for producing apparently binocular (3D) image by presenting differing plane images to the two eyes.

stereoscopic camera *(Image Tech.)*. See **stereocamera**.

stereoscopy *(Phys.)*. Sensation of depth obtainable with binocular vision due to small differences in parallax producing slightly-differing images on the two retinas.

stereospondyly *(Zool.)*. The condition of having the parts of the vertebrae fused to form one solid piece; cf. **temnospondyly**. adj. *stereospondylous*.

stereotaxis *(Biol.)*. Response or reaction of an organism to the stimulus of contact with a solid body; as the tendency of some animals to insert themselves into holes or crannies, or to attach themselves to solid objects. adj. *stereotactic*.

stereotaxis *(Med.)*. A procedure, often using X-rays, carried out at a precise localization in a tissue, e.g. *stereotactic surgery*, where a precise area of brain is identified for the surgeon to transect.

stereotype *(Behav.)*. An oversimplified and very generalized belief about groups of people which is applied to individuals identified as members of the group.

stereotype *(Print.)*. A **duplicate plate** made from the original surface (type and/or blocks) or from an existing plate; for metal plates, the mould is made in flong; for rubber or plastic plates, a thermosetting plastic sheet is used.

stereotype alloys *(Eng.)*. Lead-based alloys with 5–10% tin and 10–15% antimony.

stereotyped behaviour *(Behav.)*. Behaviour patterns which are performed on different occasions with very little variation in their component parts, typical of many animal displays. Animals under stress, e.g. in close confinement, may develop very fixed and idiosyncratic behaviours, e.g. rigid pacing actions.

stereotypy *(Med.)*. The repetition of senseless movements, actions, or words by the insane.

steric hindrance *(Chem.)*. The retarding influence by virtue of their size of neighbouring groups on reactions in organic molecules, e.g. *ortho* (2) substitution in aromatic acids considerably retards esterification. Has profound effects in biological macromolecules.

sterile *(Bot.)*. (1) Unable to breed. (2) Free from living organisms, especially culture media, foodstuffs, surfaces, medical supplies etc. which are free from micro-organisms that could cause spoilage or infection.

sterile flower *(Bot.)*. (1) A flower with neither functional carpels nor functional stamens. (2) Sometimes a male flower.

sterile line *(Elec.Eng.)*. One that has no direct connection with adjoining circuits and is therefore unaffected by them.

sterilization *(Biol.)*. (1) Causing the loss of sexual reproductive function. (2) Making sterile by removing unwanted organisms by heat, radiation, chemicals or by filtration.

sterilization *(Med.)*. A method of destroying the reproductive powers, either by inhibition of the functions of the sex organs or by their removal.

sterling silver *(Eng.)*. A silver alloy with not more than 7.5% base metal.

sternal *(Zool.)*. See **sternum**.

sternebrae *(Zool.)*. In Mammals, a median ventral series of bones which alternate with the ribs.

stern frame *(Ships)*. In a twin-screw ship a heavy bar attached to the keel and to a strong transverse plate at the top, with gudgeons to carry the rudder. In a single-screw ship it surrounds the propeller aperture and is almost rectangular. The forward vertical part is the *propeller post*, and after part is the *rudder post* (sometimes called the *stern post*), the lower horizontal part is the *sole piece* and the upper part is the *arch piece*.

Stern-Garlach experiment *(Phys.)*. Atomic beam experiment which provided fundamental proof of quantum theory prediction that magnetic moment of atoms can only be orientated in certain fixed directions relative to an external magnetic field.

stern post *(Ships)*. See **stern frame**.

sternum *(Zool.)*. The ventral part of a somite in Arthropods; the breast bone of Vertebrates, forming part of the pectoral girdle, to which, in higher forms, are attached the ventral ends of the ribs. adj. *sternal*.

sternutation *(Med.)*. The act of sneezing; a sneeze.

steroids *(Chem.)*. Compounds containing the perhydrocyclopentenophenanthrene nucleus. They include the sterols, bile acids, sex hormones, adrenocortical hormones, cardiac glycosides, sapogenins and some alkaloids.

sterols *(Chem.)*. A group of steroid alcohols obtained originally from the non-saponifiable portions of the lipid extracts of tissues. The best known is **cholesterol**.

stet *(Print.)*. A reader's mark in the margin of a proof indicating that the correction marked is to be ignored; in the text matter, a dotted line (......) is drawn under the original wording.

stethoscope *(Med.)*. Instrument for study of sounds generated inside the human body.

Stevenson screen *(Meteor.)*. A form of housing for meteorological instruments consisting of a wooden cupboard having a double roof and louvred walls, these serving to protect the instruments from the sun and wind while permitting free ventilation. The base of the screen should be about 1 metre above the ground.

sthene *(Genrl.)*. Unit of force in the metre-tonne-second system, equivalent to 10^3 newtons.

stib- *(Chem.)*. Root word denoting *antimony*, from the L. *stibium*. As in *stibine*, *stibnite*.

stibic, stibious *(Chem.)*. Referring to (V) and (III) oxidation state of antimony respectively. Now normally *antimonous* and *antimonic*.

stibine *(Chem.)*. SbH_3. Antimony (III) hydride. A poisonous gas. Less stable than arsine.

stibnite *(Min.)*. Antimony sulphide, which crystallizes in grey metallic prisms in the orthorhombic system. It is sometimes auriferous and also argentiferous. It is widely distributed but not in large quantity, and is the chief source of antimony. Formerly called *antimony glance*. Also called *antimonite*.

stichtite *(Min.)*. A lilac or pink trigonal hydrated and hydrous carbonate of magnesium and chromium.

stick-and-rag work *(Build.)*. Plasterwork formed of canvas stretched across wooden frame and coated with a thin layer of gypsum plaster.

stick force *(Aero.)*. The force exerted on the **control column** by the pilot when applying aileron or elevator control.

stick-force recorder *(Aero.)*. A device attached to the control column of an aircraft by which the pilot's effort is measured and transmitted to a recording instrument.

sticking *(Build.)*. Shaping a **stuck moulding**.

sticking probability *(Phys.)*. Probability of an incident particle, which reaches the surface of a nucleus, being absorbed and forming a compound nucleus.

sticking relay *(Telecomm.)*. One which *makes* when d.c. current passes through the operating coil, remains operated when the current stops (thereby saving current), and falls off or *releases* on a reversed current.

sticking voltage *(Electronics)*. Potential in electron-beam tube above which electrons collected at screen cannot all be dispersed, leading to negative charge accumulating and neutralizing the excess voltage. In CRT, that accelerating voltage which fails to increase brightness of spot on a phosphor because of insufficient secondary-electron emission or conduction for dispersal of incident electrons.

stick pusher *(Aero.)*. A device fitted to the control column of some high-performance aircrafts with swept-back wings which moves the column sharply forward to prevent a stall. See **stall-warning indicator**.

stick shaker *(Aero.)*. See **stall-warning indicator**.

stick-slip motion *(Eng.)*. The motion of sliding surfaces of various materials, in which the force to start them moving is greater than the force to keep them moving.

sticky ends *(Biol.)*. When DNA is cut by a **restriction enzyme**, the cut is usually staggered between the two chains, so that one chain is longer by one or two bases than the other. This end is able to base pair with a complementary end on another molecule cut by the same enzyme, causing the two molecules to stick together. Also *cohesive ends*.

stiction *(Phys.)*. *STatic frICTION*. See **friction**.

stiffened expanded metal *(Build.)*. See **self-centring lathing**.

stiffened suspension bridge *(Eng.)*. A suspension bridge in which the tendency for the suspension cables to change shape under different load systems is counteracted by the provision of stiffening girders which are supported on the bridge piers and connected to the cables by suspension rods.

stiffener *(Aero.)*. A member attached to a sheet for the purpose of restraining movement normal to the surface. Usually of thin drawn or extruded light-alloy, L, Z or U section, attached by riveting, metal-bonding or spot welding. See **integral-, stringer**.

stiffener *(Eng.)*. A steel angle or bar riveted or welded across the web of a built-up girder to stiffen it.

stiff lamb disease *(Vet.)*. (1) See **white muscle disease**. (2) Arthritis of lambs due to infection by *Erysipelothrix insidiosa* (*E. rhusiopathiae*).

stiff neck *(Med.)*. See **torticollis**.

stiffness *(Eng.)*. The ability to resist bending.

stiffness *(Phys.)*. In a mechanical vibrating system, the stiffness is the restoring force per unit displacement. It is the reciprocal of *compliance* and is a measure, for example, of the strength of a spring.

stiffness control *(Acous.)*. In a mechanically vibrating system, the condition in which the motion is mainly determined by the stiffness of the retaining springs and negligibly by the resistance and mass of the system.

stiffness criterion *(Aero.)*. The relationship between the stiffness, strength and other structural properties which will prevent *flutter* or dangerous aeroelastic effects.

stiff ship *(Ships)*. A ship with a large transverse metacentric height and short rolling period. Cf. **tender ship**.

stiff sickness *(Vet.)*. See **osteomalacia**.

stifle *(Vet.)*. The femorotibial joint of animals.

stigma *(Bot.)*. (1) See **eye spot**. (2) The part of the carpel of a flowering plant that is adapted for the reception and germination of the pollen.

stigma *(Zool.)*. In Protozoa, an eye spot; in Arthropoda, one of the external apertures of the tracheal system; in Urochordata, a gill slit; generally, a spot or mark of distinctive colour, as on the wings of many Butterflies. pl. *stigmata*.

stigmata *(Med.)*. Physical characteristics of a disease process, or syndrome.

stilb *(Phys.)*. Unit of luminance, equal to 1 cd/cm^2 or 10^4 cd/m^2 of a surface.

stilbene *(Chem.)*. $C_6H_5.CH{=}CH.C_6H_5$, *s-di-phenylethene*; mp 125°C, bp 306°C. Can occur as two isomers, the *trans* solid form and the *cis* liquid form.

stilbite *(Min.)*. A zeolite; silicate of sodium, calcium and aluminium with chemically combined water; crystallizes in the monoclinic system, the crystals frequently being grouped in sheaflike aggregates. Found both in igneous rock cavities and in fissures in metamorphic rocks. Also called *desmine*.

stilboestrol *(Med.)*. Drug used as an **oestrogen**.

stile *(Build.)*. An upright member in framing or panelling. Often incorrectly spelt *style*.

Stiles-Crawford effect *(Phys.)*. Light entering the eye near to the margin of the pupil is less effective in producing a sensation of brightness than the same amount of light entering through the centre of the pupil.

still *(Chem.)*. Apparatus for the distillation of liquids, consisting of a reboiler, a fractionating column and arrangement for reflux.

stillage *(Elec.Eng.)*. A stand for accommodating the accumulator cells.

stillage *(Print.)*. See **pallet**.

still air range *(Aero.)*. The theoretical ultimate range of an aircraft without wind and with allowances only for take-off, climb to cruising altitude, descent and alighting.

Still's disease *(Med.)*. Acute polyarthritis of children (resembling rheumatoid arthritis), with fever and enlargement of the spleen and of the lymphatic glands.

still video camera *(Image Tech.)*. A camera in which a still picture is recorded electronically on a magnetic floppy disc or card for immediate reproduction on a domestic TV receiver or monitor without any photographic processing.

stilpnomelane *(Min.)*. Monoclinic hydrated iron magnesium potassium aluminium silicate resembling biotite. It occurs in metamorphosed sediments and in iron ores.

stilted arch *(Arch.)*. An arch rising from points below its centre, and having the form of a circular arch above its centre.

stilt root *(Bot.)*. See **prop root**.

stimulated emission *(Phys.)*. Process by which an incident photon of frequency ν stimulates an atom to make a transition from energy E_2 to energy E_1 where $\nu = (E_2 - E_1)/h$, h being Planck's constant. The atom is left in the lower energy state as *two* photons of the same frequency emerge, the incident one and the emitted one. An essential process in the operation of a **laser**.

stimulation *(Genrl.)*. The application of stimuli.

stimulus *(Behav.)*. Refers to some aspect of the environment, internal or external to the individual, which produces some response, although this is not always an immediate response nor an easily observable one.

stimulus *(Zool.)*. An agent which will cause propagation of a nerve impulse in a nerve fibre. pl. *stimuli*.

stimulus control *(Behav.)*. An **operant conditioning** technique in which a predictable relationship is established between a given stimulus and a given response by eliminating all other stimuli associated with that response and all other responses associated with that stimulus.

stimulus generalization *(Behav.)*. The principle that when a subject has been conditioned to make a response to a stimulus, other similar stimuli will tend to evoke the same response, although to a lesser degree; the greater the similarity to the original stimulus, the greater this tendency will be.

stimulus threshold *(Behav.)*. The value of a quantified stimulus which elicits a particular response at a definite intensity. See **absolute threshold, difference threshold**.

sting *(Zool.)*. A sharp-pointed organ by means of which poison can be injected into an enemy or a victim; as the poisonous fin-spines of some Fishes, the ovipositor of a worker Wasp. See **urticaria**.

stinging hair *(Bot.)*. An epidermal hair capable of injecting an irritating fluid into the skin of an animal when its tip is broken by contact, as in nettles.

stink damp *(Min.Ext.)*. Underground ventilation tainted by sulphuretted hydrogen.

stinkwood *(For.)*. A strong wood from the S. African tree *Ocotea bullata* (Cape walnut), known for its beautiful and distinctive figuring. It has an unpleasant smell when green.

stipe *(Bot.)*. A stalk, especially (1) of the fruiting body of a fungus, and (2) the part connecting holdfast and lamina of a large algal thallus.

stipe *(Geol.)*. One of the branches of a fossil **graptolite**.

stipes *(Zool.)*. A stalklike structure: an eyestalk. pl. *stipites*. adjs. *stipitate, stipiform*.

stipple *(Print.)*. See **mechanical stipple**.

stippling *(Build.)*. The operation of breaking up the smoothness of a paint, distemper, plaster or cement surface by dabbing it repeatedly with a hair or rubber *stippler*.

stipular trace *(Bot.)*. The vascular tissue running into a stipule.

stipule *(Bot.)*. In many dicotyledons one of a pair of appendages which start development as outgrowths of the flank of a leaf primordium, often serve to protect the leaves in the bud and mature as leaf-like photosynthetic structures or as spines, scales etc.

Stirling engine *(Eng.)*. An *external* combustion reciprocating engine, patented by a Scottish clergyman, Robert Stirling, in 1827. It consists essentially of a cylinder in which two pistons (a working piston and a displacer) operate. When the air (or a suitable gas) in the cylinder is heated it expands, driving the working piston. The second piston transfers the air to a cold region for cooling; it is then recompressed by the working piston and transferred by the displacer to the hot region to start the cycle again. Such an engine is very much quieter and cleaner than a petrol or diesel engine. A modern version has been developed in Holland in which helium under pressure is used as the working medium.

Stirling's approximation *(Maths.)*.

$$n! = \sqrt{2\pi}.n^{(n+\frac{1}{2})}.\exp\left(-n + \frac{1}{12n} - \frac{1}{360n^3} + \dots\right).$$

stirrup *(Civ.Eng.)*. A vertical steel rod which loops together the top and bottom reinforcing bars of a reinforced-concrete beam and helps to resist the shear.

stirrup *(Horol.)*. A support; such as the bottom of the rod of a mercurial pendulum on which rests the container for the mercury.

stishovite *(Min.)*. A high-density form of silica. Synthesized at 1.6×10^{10} N/m^2 and 1200°C, and also found occurring naturally in Meteor Crater, Arizona, in shock-loaded sandstone.

stitch *(Textiles)*. (1) In sewing: fastening with thread carried through the fabric by a needle. A variety of stitches (e.g. buttonhole, chain, lock) are used for different purposes. (2) In knitting: a single loop.

stitch-bonded fibre *(Textiles)*. A **web** of fibres stitched together to form a non-woven fabric.

stitch density *(Textiles)*. The number of stitches per unit area in a knitted fabric.

stitch finish *(Textiles)*. A finish, usually a lubricant, added to yarn or fabric to aid the penetration of the needle and thread in sewing.

stitching *(Print.)*. Joining the sections of an *inserted book* (see **inset** (2)) along the back by means of thread or wire.

stitch welding *(Elec.Eng.)*. **Seam welding**, using small mechanically-operated electrodes, similar to a sewing machine.

stoa *(Arch.)*. A covered colonnade or portico.

stochastic *(Stats.)*. Developing in accordance with a probabilistic model; random.

stochastic noise *(Acous.)*. See **random noise**.

stock *(Bot.)*. (1) Usually a rooted stem into which a scion is placed in grafting. (2) The perennial part of a herbaceous perennial. (3) A strain maintained for breeding or propagation.

stock *(Build.)*. The principal part of a tool, e.g. the body of a plane, in which the cutting iron is held: the stouter arm of a bevel, in which the blade is fastened.

stock *(Geol.)*. Similar to **boss**.

stock *(Image Tech.)*. See **raw stock**.

stock *(Print.)*. The general term for the material being printed: paper, board, foil etc. See also **stocks**.

stock board *(Build.)*. A bottom made to fit the mould used in the handmoulding of bricks.

stock brush *(Build.)*. A brush used to moisten surfaces with water, prior to plastering, so that the surface will not absorb moisture from the plaster.

stock chest *(Paper)*. A vessel, usually of upright cylindrical form, generally tile lined, to contain stock from which the paper machine draws its supply. Fitted with a propeller or agitator to ensure a uniform suspension and even mixing of additives such as filler, size, colour etc. which are usually added at this point.

stockless anchor *(Ships)*. A form of anchor in which there is no crosspiece on the shank and the arms are pivoted so that both of them can engage at the same time; the shank can be drawn into the hawsepipe of the ship.

stock pile *(Min.Ext.)*. Temporarily stored tonnage of ore, middlings, concentrates or saleable products.

stock rail *(Civ.Eng.)*. The outer fixed rail against which the **point** works at a turn-out.

stocks *(Build.)*. Bricks which are fairly sound and hard-burned but are more uneven in colour than **shippers**; the bricks most used for ordinary building purposes.

stocks *(Ships)*. The massive timbers supporting a ship in course of construction.

stockwork *(Geol.)*. An irregular mass of interlacing veins of ore; good examples occur among the tin ores of Cornwall and in the Erzgebirge. (Ger. *Stockwerk*).

stoichiometry *(Chem.)*. Determination of exact proportion of elements to make pure chemical compounds, alloys such as semiconductors, or ceramic crystals.

stokes *(Phys.)*. The CGS unit of kinematic viscosity (10^{-4} m^2/s). Abbrev. *St*.

Stokes-Adams syndrome *(Med.)*. Sudden loss of consciousness, with or without convulsions, in heart-block.

Stokes' Law *(Phys.)*. (1) The resisting force offered by a fluid of dynamic viscosity η to a sphere of radius r, moving through it at steady velocity v is given by

$$R = 6\pi\eta rv$$

whence it can be shown that the **terminal velocity** of a sphere of density ρ, falling under gravitational acceleration g through fluid of density ρ_0 is given by

$$v = \frac{2gr^2}{9\eta}(\rho - \rho_0).$$

Applies only for viscous flow with Reynolds number less than 0.2. (2) Incident radiation is at a higher frequency and shorter wavelength than the re-radiation emitted by an absorber of that incident radiation.

Stokes layer *(Acous.)*. Very thin boundary layer along an interface between a fluid and a solid in which the velocity and temperature fluctuations in a sound wave are reduced because of friction and thermal conductivity respectively. Important for sound absorption.

Stokes' line *(Phys.)*. A spectrum satisfying **Stokes' law**, i.e. line seen in Raman spectrum on the long wave side of the Rayleigh line when monochromatic light is scattered.

Stokes' theorem *(Maths.)*. The surface integral of the curl of a vector function equals the line integral of that function around a closed curve bounding the surface, i.e.

$$\int_L E\,dl = \int_S \text{curl } E\,ds.$$

STOL *(Aero.)*. *Short Take-Off and Landing*, a term applied to aircraft with high-lift devices and/or deflected engine thrust enabling them to operate from small airstrips, 1000 ft (300 m) or less being the criterion.

stolon *(Bot.)*. (1) See **runner**. (2) Arching stem that forms a new rooted plant at the tip e.g. blackberry. (3) Slender horizontally-growing underground stem which forms a new plant at the end.

stolon *(Zool.)*. A cylindrical stemlike structure: a tubular outgrowth in hydroid colonies of Coelenterata and Entoprocta from which new individuals or colonies may arise. adj. *stolonate*.

stoma *(Biol.)*. A small aperture.

-stoma *(Genrl.)*. Suffix from Gk. *stōma*, mouth, applied esp. in zoological nomenclature. pl. *-stomata*.

stomach *(Zool.)*. In Vertebrates, the saclike portion of the alimentary canal intervening between the oesophagus and the intestine. The term is loosely applied in Invertebrates to any saclike expansion of the gut behind the oesophagus. adj. *stomachic*.

stomach insecticide *(Chem.)*. One acting on ingestion and applicable only to insects which eat as distinct from sucking insects which draw food in liquid form from host plant or animal; may be used on foliage against leaf-eating insects, or as poison-bait ingredient against locusts etc. Examples: lead arsenate, DDT, Gammexane.

stoma, stomate *(Bot.)*. A pore in the epidermis of a leaf or stem etc. of a vascular plant, of variable aperture, surrounded and controlled by two **guard cells** and providing regulated gas exchange between the tissues and the atmosphere. pl. *stomata, stomates*.

stomatal complex *(Bot.)*. A stoma with its guard cells and any subsidiary cells.

stomatal, stomate, stomatiferous, stomatose, stomatous *(Genrl.)*. adjs. from *stoma*, an aperture.

stomatitis *(Med.)*. Inflammation of mucous membrane of the mouth.

stomatitis *(Vet.)*. See **horse pox**.

stomatogastric *(Zool.)*. Pertaining to the mouth and stomach; said especially of that portion of the autonomic nervous system which controls the anterior part of the alimentary canal.

stomium *(Bot.)*. A part of the wall of a fern sporangium composed of thin-walled cells where splitting begins during dehiscence.

stomodaeum *(Zool.)*. That part of the alimentary canal which arises in the embryo as an anterior invagination of ectoderm; cf. **proctodaeum, mid-gut**. adj. *stomodaeal*.

-stomy *(Genrl.)*. Suffix from Gk. *stōma*, mouth, referring esp. to the formation of an opening by surgery.

stone *(Horol.)*. A jewel; used especially of the pallet jewels.

stone *(Print.)*. The smooth, milled cast-iron surface on which formes are locked up.

stone cell *(Bot.)*. *Brachysclereid*. A more or less isodiametric sclereid, e.g. in the fruit of the pear.

stone-hand *(Print.)*. A compositor who specializes in *imposition*.

stone head *(Min.Ext.)*. (1) First solid rock met while sinking a shaft or drill hole, also *rock head*. (2) A heading or tunnel in stone.

stone saw *(Build.)*. (1) A smooth-faced blade which in use is fed with an abrasive such as sand, carborundum, or diamond powder, as it cuts its way through stone. (2) Diamond tipped circular saw.

stone tongs *(Build.)*. An accessory used in hoisting blocks of stone. It resembles a large pair of scissors with the points curved inwards. These clip into the sides of the block, while chains connect the loops of the tongs to the hoisting ring. Also called *nippers*.

stoneware *(Build.)*. A material used for some sanitary fittings, etc.; made from plastic clays of the Lias formation, with a small amount of sharp sand, etc., added to reduce shrinkage.

stoneworts *(Bot.)*. See **Charales**.

stony meteorites *(Geol.)*. Those meteorites which consist essentially of rock-forming silicates. See **achondrite, aerolites, chondrite**.

stool *(Bot.)*. A tree or shrub cut back to ground level and allowed to produce a number of new shoots as in **coppice** management, as a method of managing fruit bushes (cf. **leg**) or to provide shoots for making cuttings etc.

stool *(Med.)*. The **faeces** from one bowel movement.

stoop *(Build.)*. A low platform outside the entrance door of a house. Also called *stoep*.

stop *(Build.)*. (1) A projecting piece set in the top of a bench at one end and adjustable for height. It is used to steady work which is being planed. (2) An ornamental termination to a stuck moulding. (3) See **door stop**.

stop *(Image Tech.,Phys.)*. (1) Circular opening which sets the effective aperture of a lens, e.g. the iris of a camera or the rim of the objective in a telescope. (2) An **f-number**, esp. as marked on the iris scale of a camera lens.

stop band *(Telecomm.)*. The frequency band in which a filter highly attenuates signals; in this band, its impedance is highly reactive, causing incoming signals to be reflected.

stop-bath *(Image Tech.)*. See **acid stop**.

stop-cock *(Build.)*. A short pipe opened or stopped by turning a key or handle.

stop codon *(Biol.)*. Specific triplet sequences, which in mRNA do not code for an amino acid but cause protein synthesis to stop. They are UAA, UAG and UGA.

stop-cylinder *(Print.)*. A letterpress printing machine in which the cylinder makes one revolution and prints a sheet as the bed travels in the printing direction, and then remains stationary while the bed returns to its starting position. Cf. **single-revolution, two-revolution**.

stop down *(Image Tech.)*. To reduce the working aperture of a lens in order to increase depth of field and/or exposure time.

stope *(Min.Ext.)*. (1) To excavate ore from a reef, vein or lode. (2) Space formed during extraction of ore underground. Types are flat, open, overhand, rill, shrinkage and underhand, with variations to suit shape, geology and size of deposit.

stoping *(Geol.)*. A mining term applied by R.A. Daly to a process in the emplacement of some igneous rock bodies, by which blocks of the overlying country rock are wedged off and sink into the advancing magma.

stop moulding *(Build.)*. A **stuck moulding** terminating in a stop.

stopped end *(Build.)*. A square end to a wall.

stopped heading *(Print.)*. In paper ruling, the heading for account books and other work may require the ruling to stop at more than one stage both vertically and horizontally.

stopped mortise *(Build.)*. See **blind mortise**.

stopped pipe *(Acous.)*. See **closed pipe**.

stopper *(Electronics)*. A simple circuit element (resistance, or resistance and capacitance in combination) to obviate parasitic oscillations.

stopping *(Build.)*. Plastic material used to fill holes and cracks in timber, e.g. before painting.

stopping equivalent *(Phys.)*. Thickness of a standard substance which would produce the same energy loss as the absorber under consideration. The standard substance is usually air at s.t.p. but can be Al, Pb, H_2O, etc. See **air equivalent**.

stopping motions *(Textiles)*. Electrical or mechanical devices employed on many textile machines when a fault develops in raw material feeding arrangements (openers, scutchers, spinning-frames etc.) or a yarn breaks in winding, warping, or weaving.

stopping-off *(Elec.Eng.)*. Coating a conducting surface with a resist to prevent electrodeposition.

stopping-out *(Image Tech.)*. See **blocking-out**.

stopping-out *(Print.)*. Painting out with a protecting varnish the darker tones of a half-tone block, in stages during the etching, to obtain the best rendering of the subject; also painting out unwanted background or detail on a negative before printing down. Also the process of painting out non-image areas of the mesh in screen printing.

stopping potential *(Electronics)*. Reverse difference of potential required to bring electrons to rest against their initial velocity from either thermal or photoelectric emission.

stopping power *(Phys.)*. Loss, resulting from a particle traversing a material. The *linear stoppage power* S_L is the energy loss per unit distance and is given by $S_L = -dE/dx$, where x is path distance and E is the kinetic energy of the particle. The *mass stopping power* S_M is the energy lost per unit surface density traversed and is given by $S_M = S_{L/\rho}$, where ρ is the density of the substance. If A is taken as the rel. at. mass of an element and n the number of atoms per unit volume, then the *atomic stopping power* S_A of the element is defined as the energy loss per atom per unit area normal to the motion of the particle, and is given by $S_A = S_L/n = S_M A/N$, where N is Avogadro number. The *relative stopping power* is the ratio of the stopping power of a given substance to that of a standard substance, e.g. air or aluminium.

stop press *(Print.)*. See **fudge**.

stop slide *(Horol.)*. See **all-or-nothing piece**.

stop valve *(Eng.)*. The main steam valve fitted to a boiler to control the steam supply and to allow isolation of the boiler from the main steam pipe.

stop watch *(Horol.)*. A watch, usually having seconds and minutes hands only, which is started and stopped by pressure of the winding knob. The normal type reads to 1/5 s, special types read to 1/50 s. See also **chronograph**.

storage *(Comp.)*. General term covering all units of computer equipment used to store data (and programs). Also *memory*.

storage battery *(Elec.Eng.)*. See **accumulator**.

storage capacity *(Comp.)*. Maximum number of *bits* which can be stored, located and recovered in the main memory of a computer.

storage disorder *(Med.)*. Any disease in which a metabolic defect results in the abnormal accumulation of a substance in the body (e.g. fat, carbohydrate, protein or iron).

storage element *(Comp.)*. One unit in a memory, capable of retaining one **bit** of information.

storage factor *(Phys.)*. See *Q*.

storage heater *(Elec.Eng.)*. A heater with large thermal capacity, used to store heat during off-peak periods and release it over a longer period. A fan may be used to increase the rate of heat output.

storage oscilloscope *(Electronics)*. One in which a trace is retained indefinitely or until deliberately wiped off, incorporating a **storage tube**.

storage tube *(Electronics)*. That which stores charges deposited on a plate or screen in a cathode-ray tube, a subsequent scanning by the electron beam detecting, reinforcing, or abolishing the charge.

store *(Comp.)*. Part of a computer system where data and instructions are held. Also called *storage* or *memory*. See **main memory, backing store, access to store**.

store *(Image Tech.)*. Device in which signal information can be accumulated for subsequent retrieval, for example, two interlaced fields may be stored and released in the sequential order as a complete frame.

stored program *(Comp.)*. In computer design, the fundamental idea that a program can be stored in the same way as data. See **first generation computer**.

stored-program control *(Telecomm.)*. Said of automatic, mostly fully-electronic, telephone exchanges in which switching functions are controlled by stored logic and traffic-handling capacity is optimized in accord with current conditions.

store location *(Comp.)*. Basic unit within a **main memory**, capable of holding a single **byte** or **word**. Also *cell*.

storey rod *(Build.)*. A pole on which is marked the level of the courses and which gives a guide to the ultimate level.

storied cork *(Bot.)*. Protective cork. Protective layer of suberized cells develops around stems of woody monocotyledons (e.g. palms), in which the cells occur in radial files each file of several cells derived from a single precursor.

storied, stratified *(Bot.)*. A vascular cambium and the secondary xylem or wood derived from it with the cells arranged in horizontal tiers (i.e. with the end walls more or less aligned).

storm *(Telecomm.)*. See **ionospheric storm**.

storm-centre *(Meteor.)*. The position of lowest pressure in a cyclonic storm.

storm window *(Build.)*. (1) A window arranged with double sashes enclosing air, which acts as a sound and heat insulator. (2) A small upright window set in a sloping roof surface so as not to project beyond it. Cf. **dormer**.

stoving *(Build.)*. An industrial process of quickly drying specially formulated paints by heat (radiant or convected). Temperatures generally exceed 180°F (82°C) for lower temperatures, see **force drying**.

STOVL *(Aero.)*. See **short take-off and vertical landing**.

stowage factor *(Ships)*. The space required to contain unit weight of a commodity allowing for all packing, dunnage and unavoidable lost space between units. Usually, measured in cubic feet per ton or cubic metres per tonne.

stp, STP *(Genrl.)*. *Standard Temperature and Pressure*; a temperature of 0°C and a pressure of 101 325 N/m^2. See **standard atmosphere**.
strabismus *(Med.)*. *Squint*. A condition in which the visual axes of the eyes assume a position relative to each other which is abnormal.
strabotomy *(Med.)*. The surgical operation of curing **strabismus** by dividing one or more muscles of the eye.
straddle milling *(Eng.)*. The use of two or more side-cutting milling-cutters on one arbor so as to machine, for example, both side-faces of a piece of work at one operation.
straddle scaffold *(Build.)*. See **saddle scaffold**.
straggling *(Phys.)*. Variation of range or energy of particles in a beam passed through absorbing material, arising from random nature of interactions experienced. Additional straggling may arise from instrumental effects such as noise, source thickness and gain instability.
straight angle *(Maths.)*. An angle of 180° or π radians.
straight arch *(Build.)*. See **flat arch**.
straight-bar machine *(Textiles)*. A knitting machine, with bearded needles on a movable bar, used to make plain or rib-knitted shaped articles.
straight eight *(Autos.)*. An 8-cylinder-in-line engine, as distinct from an 8-cylinder-V-type engine.
straighteners *(Aero.)*. See **honeycomb**.
straight-flute drill *(Eng.)*. A conical pointed drill having backed-off cutting edges, formed by cutting straight longitudinal flutes in the shank; more rigid than a twist drill and often used for soft metals.
straight joint *(Build.)*. A continuity of vertical joints in brickwork.
straight-line capacitor *(Telecomm.)*. Variable capacitor whose value varies linearly with scale reading.
straight-line frequency capacitor *(Telecomm.)*. Variable capacitor whose value is inversely proportional to the square of the scale reading, so that the frequency of the circuit which it tunes is directly proportional thereto.
straight-line lever escapement *(Horol.)*. A lever escapement in which the balance staff, pallet staff and escape wheel are planted in the same straight line. The majority of watches with the club-tooth escapement are of this form.
straight-line wavelength capacitor *(Telecomm.)*. Variable capacitor whose value is proportional to square of scale reading, so that it can tune a circuit with a linear relationship between scale reading and wavelength.
straight-pane hammer *(Eng.)*. A fitter's hammer the head of which has a flat striking face at one end and a blunt chisel-like edge, parallel with the shaft, at the other.
straight receiver *(Telecomm.)*. See **tuned radio-frequency receiver**.
straight run *(Print.)*. Running a web-fed press without cylinder collection and sometimes without turner bars to give the maximum number of copies to the cylinder revolution.
straight-, solid-type cable *(Elec.Eng.)*. A cable which has oil-impregnated paper as the dielectric. Used up to 66 kV in the form of single-core H-type cable.
straight tongue *(Build.)*. A wooden tongue for a **ploughed-and-tongued joint** cut so that the grain is parallel to the grooves.
straight-up-and-down filament *(Elec.Eng.)*. A filament made in the form of long zigzags between an upper and lower spider support; used in the ordinary vacuum filament lamp.
strain *(Biol.)*. A variant group within a species, often breeding true and maintained in culture or cultivation, with more or less distinct morphological, physiological or cultural characteristics. (The term is not used in formal taxonomy.)
strain *(Eng.,Phys.)*. When a material is distorted by forces acting on it, it is said to be in a state of *strain* or *strained*. Strain is the ratio

$$\frac{\text{deflection}}{\text{dimension of material}}$$

and thus has no units. The main types of strain are direct (tensile or compressive) strain:

$$\frac{\text{elongation or contraction}}{\text{original length}}$$

shear strain:

$$\frac{\text{deflection in direction of shear force}}{\text{distance between shear forces}}$$

volumetric (or bulk) strain:

$$\frac{\text{change of volume}}{\text{original volume}}$$

See **elasticity of elongation, elasticity of shear, elasticity of bulk**.
strain-ageing *(Eng.)*. An increase in metal strength and hardness that proceeds with time, after cold-working. It takes place slowly at air temperature and is accelerated by heating. It is most pronounced in iron and steel, but also occurs in other metals.
strain disk *(Glass)*. A glass disk of calibrated birefringence, used as a comparative measure of the degree of annealing of glass.
strainer *(Paper)*. Any piece of equipment intended to clean pulp, stuff or stock by passing it through metal plates containing perforations or slits of appropriate size so that larger particles are held back and removed.
strain gauge *(Eng.)*. Metal or semiconductor filament on a backing sheet by which it can be attached to a body to be subjected to strain, so that the filament is correspondingly strained. The strain alters the electrical properties of the filament, this alteration forming a basis of measurement.
strain-hardening *(Eng.)*. Increase in resistance to deformation (i.e. in hardness) produced by deformation. See **cold-working, work-hardening**.
straining sill *(Build.)*. A piece of scantling lying on the tie-beam of a timber roof and butting against the feet of the queen-posts, or between the feet of the queens and princesses to keep them apart.
strain insulator *(Elec.Eng.)*. An insulator inserted in the span wire of an overhead contact-wire system.
strain-slip cleavage *(Geol.)*. A cleavage in which the cleavage planes are parallel shear planes; between each pair the rocks are puckered into small sigmoidal folds.
strain viewer *(Phys.)*. Eyepiece or projection unit of polariscope.
strait work *(Min.Ext.)*. (1) Narrow headings in coal. (2) A method of working coal by driving parallel headings and then removing the coal between them.
strake *(Min.Ext.)*. Gently-sloped, flat table used for catching grains of heavy water-borne mineral. See **blanket-, tye**.
strake *(Ships)*. A row of plates positioned end to end.
strand *(Elec.Eng.)*. One of several wires which together constitute a stranded conductor.
stranded cable *(Elec.Eng.)*. One whose core (or cores) consists of stranded conductor.
stranded caisson *(Civ.Eng.)*. A watertight box, having a solid floor, which is floated over the site where a bridge pier is to be constructed. Construction goes on in the dry on the floor of the box, which sinks finally to a previously levelled bed under water, the sides of the box being kept always above water. Also called an *American caisson*.
stranded conductor *(Elec.Eng.)*. One woven from individual wires or strands, as in the case of a rope.
stranding effect *(Elec.Eng.)*. An increase (20–30%) of the stress at the surface of the conductor, caused by stranding. A usual increase is 25%. Stranding effect is overcome by sector-shaped conductors or by lead sheathing the conductor.
strand plant *(Bot.)*. A sea-shore plant growing just above the normal upper limit of the tide.
strangeness *(Phys.)*. A property that characterizes *quarks* and so *hadrons*. The strangeness of leptons and

gauge bosons is zero. Strangeness is conserved in strong and electromagnetic interactions between particles but not in weak interactions. K-mesons and hyperons which have non-zero strangeness are termed *strange* particles.

stranger anxiety *(Behav.)*. A common fear infants have of unfamiliar people, onset usually the end of the first year, until the child is two years or so; related to separation anxiety in that the two tend to co-occur.

strangler *(Autos.)*. US for **choke** (2).

strangles *(Vet.)*. A contagious disease of horses, due to infection by *Streptococcus equi*, characterized by rhinitis and suppurative adenitis. Uncommonly, purpura haemorrhagica, guttural pouch empyoma or larangeal hemiplegia may follow.

strangury *(Med.)*. Slow and painful micturition.

strap *(Build.)*. A metal plate or band securing timbers together at a joint.

S-trap *(Build.)*. A trap used in sanitary pipes in which the outlet leg is parallel with the inlet leg.

strapdown *(Aero.)*. Any device mounted to an aircraft so that its attitudes change with those of the aircraft. Describes a class of navigation system like **fibre optic gyros** in contrast to ordinary gyroscopes which maintain a constant attitude.

strap hinge *(Build.)*. A hinge having one long leaf for securing to a heavy door or gate.

strapping *(Build.)*. A general term for battens fixed to the internal faces of walls as a support for laths and plaster or for plaster board.

strapping *(Telecomm.)*. Alternate connection of segments in a magnetron, to stabilize phases and mode of resonance in the cavities.

strapping wires *(Elec.Eng.)*. Parallel single-wire connections between a pair of 2-way electric light switches for dual control of a lighting point.

strass *(Glass)*. A very dense glass of high refractive power; used largely in making artificial jewellery.

strategy, *r* and *K* *(Ecol.)*. In evolutionary ecology, a strategy is a suite of genetic traits which confer a selective advantage on an individual in a particular environment and which usually included growth rate, fecundity and longevity. The *r strategists* are those species best adapted to a temporary habitat: they display high growth rate, high fecundity and short generation times. *K strategists* are best adapted to stable habitats and have low growth rates and long generation times.

stratification *(Bot.)*. (1) Banding seen in thick cell walls, due to presence of wall layers differing in water content, chemical composition and physical structure. (2) Grouping of vegetation into two or more fairly well-defined layers of different height, as trees, shrubs and ground vegetation in a wood. (3) Method of breaking dormancy period of seeds by storage in moist sand, often at around 4°C.

stratification *(Ecol.)*. Vertical structure or layering within a terrestial or aquatic environment.

stratification *(Geol.)*. The layering in sedimentary rocks due to chemical, physical or biological changes in the sediment. See also **lamination**.

stratification *(Stats.)*. The division of a population to be sampled into subsets, within each of which a sample of observations will be taken.

stratified *(Bot.)*. See **storied**.

stratified charge combustion *(Autos.)*. Technique in gasoline engines for burning mixtures too weak to be ignited by normal spark ignition. A pocket of enriched mixture is provided close to the sparking plug which ignites normally and fires the remainder.

stratified epithelium *(Zool.)*. A type of epithelium consisting of several layers of cells, the outer ones flattened and horny, the inner ones polygonal and protoplasmic.

stratiform, stratose *(Zool.)*. Arranged in layers.

stratigraphical break *(Geol.)*. The geological record is incomplete, the succession of strata being broken by unconformities and non-consequences, these representing longer or shorter periods of time during which no sediment was deposited or erosion predominated.

stratigraphical level *(Geol.)*. See **horizon**.

stratigraphic column *(Geol.)*. See **geological column**.

stratigraphy *(Geol.)*. The definition and description of the stratified rocks of the earth's crust, their relationships and structure, their arrangement into chronological groups, their lithology and the conditions of their formation, and their fossil contents. The subject does not exclude igneous and metamorphic rocks where these are part of the succession.

stratocumulus *(Meteor.)*. Grey and/or whitish, patch, *sheet* or *layer* of *cloud* which almost always has dark parts, composed of tessellations, rounded masses, rolls etc., which are non-fibrous (except for **virga**) and which may or may not be merged; most of the irregularly arranged small elements have an apparent width of more than 5°. Abbrev. *Sc*.

stratopause *(Meteor.)*. Top of the **stratosphere**, at about 50 to 55 km above the surface of the Earth.

stratosphere *(Meteor.)*. Region of the atmosphere between the **tropopause** and the **stratopause**, in which temperature generally increases with height.

stratum *(Geol.)*. A single bed of rock bounded above and below by divisional planes of **stratification**. A stratum differs from a lamination only in thickness. pl. *strata*. adj. *stratified*.

stratum *(Zool.)*. A layer of cells: a tissue layer.

stratum contours *(Geol.)*. Contours drawn on the surface of a bed of rock. The position of the outcrop of the bed can be predicted from the intersection of these contours with the surface of the ground.

stratum corneum, -granulosum, -lucidum, -Malpighii *(Zool.)*. Layers of the skin in Vertebrates.

stratum germinativum *(Zool.)*. See **Malpighian layer**.

stratus *(Meteor.)*. Generally grey *cloud layer* with a fairly uniform base, which may give drizzle, ice prisms or snow grains. When the Sun is visible through the cloud, its outline is clearly discernable. Stratus does not produce halo phenomena, except possibly, at very low temperatures and sometimes it appears in the form of ragged patches. Abbrev. *St*.

strawberry footrot *(Vet.)*. A proliferative dermatitis of sheep affecting particularly the lower parts of the legs; caused by the fungus *Dermatophilus pedis*.

stray capacitance *(Electronics)*. Any occurring within a circuit other than that intentionally inserted by capacitors, e.g. capacitance of connecting wires, giving rise to **parasitic oscillation**.

stray field *(Elec.Eng.)*. A magnetic field set up in the neighbourhood of electric machines or current-carrying conductors, which serves no useful purpose and which may interfere with the operation of measuring instruments etc.

stray flux *(Elec.Eng.)*. The leakage flux in an a.c. machine or transformer.

stray induction *(Elec.Eng.)*. The equivalent induction of the leakage flux effective in producing a reactive voltage drop.

stray losses *(Elec.Eng.)*. The stray load losses of an electrical machine, due to stray fields and harmonic flux pulsations in the iron circuit and eddy currents in the windings.

stray radiation *(Phys.)*. Direct and secondary radiation from irradiated objects which is not serving a useful purpose.

stray resonance *(Elec.Eng.)*. That arising from unwanted inductance and capacitance, e.g. in leads between conductors, in leads inside canned capacitors, between turns of inductors.

strays *(Telecomm.)*. Same as **atmospherics**, US.

streak *(Bot.)*. An elongated chlorotic or necrotic spot as a symptom of virus infection.

streak *(Min.)*. The name given to the colour of the powder obtained by scratching a mineral with a knife or file or by rubbing the mineral on paper or an unglazed porcelain surface (*streak plate*). For some minerals, this differs from the body colour.

streaking *(Image Tech.)*. A defect of the TV image

showing a trail at the image boundary instead of a sharp transition. Also *smearing*.

stream anchor *(Ships)*. An anchor of lighter weight than a bower anchor, used at the stern.

stream factor *(Chem.Eng.)*. That proportion of the time, during which a complete plant is in operation, that any individual item of the plant is working.

stream feeder *(Print.)*. An automatic feeder in which a sheet is separated and lifted at its rear edge and moved forward to the feed board so that its front edge is placed beneath the previous sheet. In this way a continuous stream of paper moves forward to the front and side lays at a moderate speed relative to the printing speed of the press.

streaming *(Bot.)*. Flowing of protoplasm in the cytoplasm, either unidirectionally (e.g. in a growing fungal hypha) or in a circulation (e.g. cyclosis).

streaming effect *(Nuc.Eng.)*. See **channelling effect**.

streaming potential *(Chem.)*. The difference of electrical potential induced between the two ends of a capillary by forcing a liquid through it.

streamline *(Phys.)*. A line in a fluid such that the tangent at any point follows the direction of the velocity of the fluid particle at the point, at a given instant. When the streamlines follow closely the contours of a solid object in a moving fluid, the object is said to be of streamline form.

streamline burner *(Eng.)*. See **fantail burner**.

streamline flow *(Phys.)*. Path taken by fluid molecules or minute suspended particles. Usually qualified as **laminar flow** or **turbulent flow**. See also **viscous flow**.

streamline motion *(Phys.)*. The steady motion of a fluid in **laminar flow** past a body with neither abrupt changes in direction nor close curves.

streamlines *(Meteor.)*. A set of lines on a chart showing the direction of the horizontal wind at some particular level.

streamline wire *(Aero.)*. High-tensile steel wire of elliptical, not true streamline, cross-section, used to reduce the drag of external bracing wires. Fitted principally to biplanes and some early types of monoplane.

stream sampling *(Min.Ext.)*. Sampling of stream or river water to identify anomalous concentrations of dissolved metals; in gravels to find chemical or mineral concentrations, e.g. *tracers*. See **geochemical prospecting**.

stream tin *(Min.)*. Cassiterite occurring as derived grains in sands and gravels in the beds of rivers.

streets *(Print.)*. See **rivers**.

strengths of acids *(Chem.)*. The extent to which an acid dissociates in a given solvent, usually water. The strengths of acids may be related to the structures of the dissociated and undissociated forms.

streptococcus *(Biol.)*. A Gram-positive coccus of which the individuals tend to be grouped in chains. Many forms possess species-specific capsular polysaccharides by which they can be divided into groups, which include the causative agents of scarlet fever, erysipelas and one form of mastitis. Other streptococci include *Diplococcus pneumoniae* (one cause of pneumonia). Some types occur normally in the mouth, throat and intestine.

streptokinase *(Med.)*. Activates plasminogen to form plasmin which degrades fibrin and breaks up thrombi. This fibrinolytic action makes it more potent than anticoagulants. Used in life-threatening pulmonary embolism and in early treatment of myocardial infarction.

streptomycin *(Biol.)*. An antibiotic which inhibits protein synthesis in prokaryotes. Much used in the laboratory to inhibit growth of micro-organisms in cell culture media.

streptomycin *(Med.)*. An **aminoglycoside** which in combination with Isoniazid and Rifampicin is particularly useful in the treatment of tuberculosis. Although it can be used to treat other infections, to avoid **drug resistance** it is now generally reserved for tuberculosis.

streptostyly *(Zool.)*. In Vertebrates, the condition of having the quadrate movably articulated with the squamosal; cf. **monimostyly**.

stress *(Behav.)*. Excessive and aversive environmental factors that produce physiological responses in the individual.

stress *(Eng.,Phys.)*. The force per unit area acting on a material and tending to change its dimensions, i.e. cause a *strain*. The *stress* in the material is the ratio of force applied to the area of material resisting the force; i.e.

$$\frac{\text{force}}{\text{area}}$$

The two main types of stress are *direct* or *normal* (i.e. *tensile* or *compressive*) *stress* (symbol σ or f) and *shear stress* (symbols τ, f_s or q). The usual units are kPa, MPa, lbf/in^2, $tonf/in^2$, kN/m^2, MN/m^2, bar or hbar. (See **newton, bar**.) *To stress* may also mean 'to calculate stresses'; thus a structure has been *stressed* when its strength has been verified by calculation.

stress diagram *(Eng.)*. See **force diagram**.

stressed-skin construction *(Aero.)*. The general term for aircraft structures in which the skin (usually light alloy, formerly plywood, occasionally plastic such as Fibreglass, and in supersonic aircraft titanium alloy or steel) carries a large proportion of the loads. In the more elementary forms, the framework may take bending and shear, with a thin skin transmitting torsion, but when of developed form, the skin is thick enough to support bending loads in the form of tension and compression in the respective surfaces. Since 1950 the principle of thick skin has been developed in the form of 'sculpturing' to vary the thickness to suit the local loads and to incorporate integral stiffeners, either by machining or by acid etching. See **monocoque**.

stress fibres *(Biol.)*. Bundles of actin filaments (*microfilaments*) and associated proteins found in the cytoplasm.

stress-intensity factor *(Eng.)*. A measure of the increase in stress which occurs at the tip of crack in a material. Symbol *K*, with various subscripts to show loading conditions etc.

stress marks *(Image Tech.)*. Marks on photographic materials caused by mechanical pressure or friction on the emulsion surface before processing, usually heavier in density than the normal image.

stress-number curve *(Eng.)*. *S/N curve*. A curve obtained in fatigue tests by subjecting a series of specimens of a given material to different ranges of stress and plotting the range of stress against the number of cycles required to produce failure. In steel and many other metals, there is a limiting range of stress below which failure will not be produced even by an indefinite number of cycles.

stress relief *(Eng.)*. Annealing metals by heating to a temperature below that liable to alter the crystalline structure and with the object of reducing or eliminating any harmful residual stresses arising from other processes.

stress relieving *(Chem.Eng.)*. A process used in the manufacture of welded pressure vessels for severe duties, particularly involving strong caustic solutions or very thick walls, in which the completed vessel is raised to a high temperature, usually red hot, held for a time and then allowed to cool in still air. The exact temperatures and times are determined by **codes of construction**.

stress-strain curve *(Eng.)*. A curve similar to a *load-extension curve*, except that the load is divided by the original cross-sectional area of the test piece and expressed in units of *stress*, while the extension is divided by the length over which it is measured and expressed as a ratio. In the elastic range, it follows **Hooke's law**, after which further stress produces plastic flow, work-hardening and residual strain (*permanent set*).

stress zone *(Min.Ext.)*. Depth of rock surrounding an underground excavation, e.g. a stope, which is now bearing the transferred stress originally supported by the removed ore.

stretch *(Image Tech.)*. The introduction of additional frames at regular intervals in a motion picture record, usually by selected repetition, to extend the action or slow down its presentation. An example is printing silent films shot at 16 pictures per second for projection at 24 pps.

stretch breaking *(Textiles)*. See **converting**.

stretched diaphragm *(Acous.)*. A diaphragm in a microphone or loudspeaker which has its rigidity increased by radial stretching, frequently by screwing on to it a rim near its edge. Resonance then becomes marked, and the tension is adjusted so that the major resonant frequency nears the upper limit of the desired transmission frequency band. Trend of the response curve is then adjusted by altering the damping.

stretcher *(Build.)*. Brick laid parallel to the course.

stretch fabric or yarn *(Textiles)*. A fabric or yarn that stretches easily and recovers to its original dimensions. It may contain elastic material to increase its extensibility and recovery.

stretch forming *(Eng.)*. Process for forming large sheets of thin metal into symmetrical shapes by gripping the sheet edges in horizontally sliding stretcher jaws and moving a forming punch, without a die, vertically between them against the sheet.

stretching bond *(Build.)*. The form of bond, used largely for building internal partition walls $4\frac{1}{2}$ in (114 mm) thick, in which every brick is laid as a stretcher, each vertical joint lying between the centres of the stretchers above and below, so that angle closers are not required. Also *stretcher bond*. See **chimney bond**.

striae *(Geol.)*. Parallel lines or grooves occurring on glaciated pavements, roches moutonnées etc.; produced by rock material frozen into the base of a moving ice-sheet; also seen on slickensided rock surfaces along which movement has taken place during faulting.

striae *(Min.)*. Parallel lines occurring on the faces of some crystals; caused by oscillation between two crystal forms. The striated cubes of pyrite are good examples.

striae atrophicae *(Med.)*. Greyish-white bands of atrophied skin in areas where the skin has been unduly stretched, as in pregnancy (*striae gravidarum*).

stria medullaris *(Zool.)*. See **habenula**.

stria, striation *(Genrl.)*. A faint ridge or furrow; a streak; a linear mark.

striated muscle *(Zool.)*. Contractile tissue in which the **sarcomeres** are aligned, e.g. skeletal and cardiac muscle. Cf. **unstriated muscle**.

striation *(Electronics)*. Phenomenon in low-pressure gas discharge, which forms luminous bands (*striae*) across the line between electrodes.

strickle board *(Foundry)*. A board profiled along one edge to the required shape of the surface of a loam mould or core; used to sweep or strike the loam to the correct section. See **loam**.

strict inequality *(Maths.)*. One with $>$ or $<$, as opposed to $\geqslant$ and $\leqslant$, which permit equality.

stricture *(Med.)*. Any abnormal narrowing of a duct or passage in the body, especially the narrowing of the urethra due to gonorrhoeal inflammation. See also **stenosis**.

striding level *(Surv.)*. Sensitive spirit level which can be placed astride a theodolite by resting the V-shaped ends of its legs on the trunnion axis, enabling the latter to be accurately levelled.

stridor *(Med.)*. A harsh vibrating noise produced by any obstruction in the respiratory tubes, e.g. in diphtheria of the larynx.

stridulating organs *(Zool.)*. The parts of the body concerned in sound production by stridulation.

stridulation *(Zool.)*. Sound production by friction of one part of the body against another, as in some Insects.

Strigiformes *(Zool.)*. An order of birds containing nocturnal birds of prey with hawk-like beaks and claws. Plumage adapted for silent flight, retina containing mainly rods, eyes large and immovable, probably hunt mainly by sound. Owls.

strigose *(Bot.)*. With stiff, appressed hairs or bristles.

strike *(Geol.)*. The horizontal direction which is at right angles to the dip of a rock.

strike *(Print.)*. To drive a hardened steel punch into a brass or copper bar, so producing a matrix from which types are cast. The term is also used to describe the impression itself.

strike *(Vet.)*. See **blowfly myiasis**.

strike fault *(Geol.)*. A fault aligned parallel to the strike of the strata which it cuts. Cf. **dip fault**.

strike lines *(Geol.)*. See **stratum contours**.

strike note *(Acous.)*. The note, largely subjective, which is initially prominent when a bell is struck. It rapidly attenuates, leaving the hum note and some overtones.

striker *(Build.)*. A long-handled paint brush with the head set at an angle used on bridge, roof or pipe painting. Also *spout brush*.

strike-slip fault *(Geol.)*. A fault whose movement is parallel to its strike.

strike-through *(Print.)*. Ink percolating through from the other side of a paper, due to lack of opacity or sizing; often seen in newsprint. Cf. **show-through**.

striking *(Build.,Civ.Eng.)*. The operation of removing temporary supports or shuttering from a structure.

striking *(Eng.)*. The flash deposition of a very thin layer of plating to facilitate subsequent plating with e.g. another material.

striking plate *(Build.)*. A metal plate screwed to the jamb of a door case in such a position that when the door is being shut the bolt of the lock strikes against, and rubs along, the plate, finally engaging in a hole in the latter.

striking potential *(Electronics)*. That sufficiently large to break down a gap and cause an arc, or start discharge in a cold-cathode tube.

striking-up, striking *(Foundry)*. The process of generating a loam mould surface by means of a **strickle board**.

striking voltage *(Electronics)*. See **starting voltage**.

striking wedges *(Civ.Eng.)*. A pair of wedge-shaped blocks of hard wood packed beneath each end of a centre and placed in contact, with their thin ends pointing in opposite directions, so that, by moving them relatively, the centre may be gradually lowered on completion of the work. Also *casing wedges*, *lowering wedges*.

string *(Acous.)*. Cylindrical body whose length is much greater than the diameter, commonly made out of gut or wire. A stretched string can be excited to vibrate with the fundamental frequency depending on the length, tension etc. Used in stringed instruments (e.g. violin, piano) where it is coupled with a resonator.

string *(Build.)*. A sloping wooden joist supporting the steps in wooden stairs. See **cut-**.

string *(Comp.)*. Series of **alphanumerics** in order but not meaningful to the computer.

string *(Elec.Eng.)*. The series of insulator units combining to form a suspension insulator.

string *(Image Tech.)*. The fine metallic strip in a **light-valve**.

string *(Min.Ext.)*. The succession of tubes and other drilling and well-top equipment joined together makes a *string*. See **blowout preventer** string.

string chart *(Elec.Eng.)*. A diagram from which the relation between the sag of an overhead line and the temperature may be rapidly obtained.

string course *(Build.)*. A projecting course in a wall.

string efficiency *(Elec.Eng.)*. The ratio of the flashover voltage of a suspension-insulator string to the product of the flashover voltage of each unit and the number of units forming the string.

string electrometer *(Elec.Eng.)*. An electrometer consisting of two metal plates oppositely charged between which a conducting fibre is displaced from a middle position in proportion to the voltage between the plates.

stringer *(Aero.)*. A light auxiliary member parallel with the main structural members of a wing, fuselage, float or hull, mainly for bracing the transverse frames and stabilizing the skin material. See **stiffener**.

stringer *(Build.)*. A long horizontal member in a structural framework.

stringer *(Nuc.Eng.)*. Group of reactor fuel elements strung together for insertion into one channel of the core.

string galvanometer *(Elec.Eng.)*. See **Einthoven galvanometer**.

string galvanometer *(Image Tech.)*. A vibrator used in

variable area sound recording, the mirror of which deflects a light beam across the recording slit.

stringhalt *(Vet.)*. A disease of horses, characterized by involuntary sudden and excessive flexion of one hind limb or both.

stripe *(Bot.)*. See **streak**.

stripe *(Image Tech.)*. Magnetic coating applied as one or more narrow bands to motion picture film for sound recording and reproduction.

striped muscle *(Zool.)*. See **striated muscle**.

stripe filter *(Image Tech.)*. Colour filter with fine stripes of red, green and blue used with a banded tri-electrode target plate to provide colour video signals from a single **Vidicon** camera tube.

stripline *(Telecomm.)*. Waveguide formed of strips of copper on dielectrics, formed by etching a printed circuit.

strip mining *(Min.Ext.)*. Form of opencast work, in which the overburden is removed (stripped) after which the valuable ore is excavated, the work usually being done in series of benches, steps, or terraces.

strippable coatings *(Build.)*. Coatings for the inside of spray booths etc., which can be removed mechanically, bringing away all overspray. These are formulated from well-plasticized ethyl cellulose or similar systems.

stripped atom *(Phys.)*. Ionized atom from which at least one electron has been removed.

stripper *(Nuc.Eng.)*. The section of an isotope-separation plant which strips the selected isotope from the waste stream.

strippers *(Print.)*. A feature of delivery arrangement of some printing machines, including the 2-revolution, by which the printed sheet is led away after being released from the grippers.

stripping *(Chem.)*. Removal of an electrodeposit by any means, i.e. by chemical agent or by reversed electrodeposition.

stripping *(Image Tech.)*. The process of removing the negative emulsion film from its glass support for transfer to another glass or other support (as in process-block making).

stripping *(Min.Ext.)*. Removal of barren overburden in opencast work.

stripping *(Phys.)*. A phenomenon observed in deuteron (or heavier nuclei) bombardment in which only a portion of the incident particle merges with the target nucleus, the remainder proceeding with most of its original momentum practically unchanged in direction.

stripping *(Print.)*. (1) The procedure of assembling negative and positive film elements on a carrier film to make a *flat* for use in platemaking. (2) Where metal inking rollers of a press will not accept ink.

stripping *(Textiles)*. Removing a dye or finish from a fabric, usually before re-processing to achieve the desired effect.

strip-wound armature *(Elec.Eng.)*. An armature whose winding consists of conductors in the form of copper strip.

strobe *(Electronics)*. (1) General term for detailed examination of a designated phase or epoch of a recurring waveform or phenomenon. (2) Enlargement or intensification of a part of a waveform as exhibited on a CRT. (3) Process of viewing mechanical vibrations with a **stroboscope**; colloquially, the stroboscope itself.

strobe lighting *(Aero.)*. An anti-collision lighting system based on the principle of a capacitor-discharge flash tube. A capacitor is charged to a very high voltage which is then discharged in a controlled sequence as a high-intensity flash of light, usually blue-white, through xenon-filled tubes located at wing tips and tail of an aircraft.

strobe lighting *(Image Tech.)*. (1) Electronic flash for still photography of moving objects. (2) In cinematography, synchronization of the flash repetition with the camera shutter at the frame rate provides a very short exposure period so that fast moving objects are sharply imaged instead of blurred.

strobe marker *(Electronics)*. Pulse much shorter than a repeated waveform, for examination of a display.

strobila *(Zool.)*. In Scyphozoa, a scyphistoma in process of production of medusoids by transverse fission: in Cestoda, a chain of proglottides. Also *strobile*. adjs. *strobilate, strobilaceous, strobiliferous, strobiloid.*

strobilate *(Bot.)*. Bearing or pertaining to a strobilus or cone.

strobile *(Genrl.)*. See **strobila, strobilus**.

strobilization *(Zool.)*. Production of strobilae: in Scyphozoa, transverse fission of a scyphistoma to form medusoids: in Cestoda, production of proglottides by budding from the back of the scolex: in some Polychaeta, reproduction by gemmation.

strobilus *(Bot.)*. *Cone*. (1) The reproductive structure of most gymnosperms and some pteridophytes, consisting of a well-defined group of packed sporophores or sporophylls bearing sporangia and arranged around a central axis. (2) An angiosperm inflorescence of similar appearance to (1).

stroboscope *(Electronics)*. A flashing lamp, of precisely variable periodicity, which can be synchronized with the frequency of rotating machinery or other periodic phenomena, so that, when viewed by the light of the stroboscope, they appear to be stationary.

stroke *(Eng.)*. The travel or excursion of a piston, press ram or other (reciprocating) part of a machine.

stroke *(Med.)*. An apoplectic seizure or a sudden attack of paralysis. Now usually used to indicate a *cerebro-vascular accident*.

stroked *(Build.)*. A term applied to the face of an ashlar which has been so tooled as to present a regular series of small flutings.

stroma *(Bot.)*. (1) The matrix of the chloroplast, in which the dark reactions of photosynthesis take place. Cf. **granum**. (2) A mass of fungal tissue formed from intertwined adherent hyphae (plectenchyma), e.g. the major part of a mushroom. Cf. **sclerotium**.

stroma *(Zool.)*. A supporting framework, as the connective tissue framework of the ovary or testis in Mammals. pl. *stromata*. adjs. *stromate, stromatic, stromatiform, stromatoid, stromoid, stromatous.*

stroma lamellae *(Bot.)*. Thylakoids that cross the stroma of a chloroplast, interconnecting the grana.

stromatolite *(Geol.)*. A layered or domed calcareous sediment formed by algal mats which both trap sediment and also precipitate lime.

stromatolites *(Bot.)*. Rounded, multilayered structures up to say 1 m across, found in rocks back to at least 2800 million years ago; what are, apparently, the present-day equivalents result from the growth, under special conditions, of blue-green algae.

stromatoporoid limestone *(Geol.)*. A calcareous sediment, rich in remains of the reef builder *Stromatopora*, important from the Cambrian to Cretaceous.

Strombolian eruption *(Geol.)*. A type of volcanic eruption characterized by frequent small explosions as trapped gases break through overlying viscous lava.

strong clay *(Build.)*. See **foul clay**.

strong electrolyte *(Chem.)*. An **electrolyte** which is completely ionized even in fairly concentrated solutions.

strong interaction *(Phys.)*. An interaction between particles involving *baryons* and *mesons* completed in a time of the order of 10^{-23} s. It is the strong interaction that binds protons and neutrons together in the nuclei of atoms. The underlying fundamental interaction to these processes is the strong interaction between the constituent *quarks*.

strong interaction between quarks *(Phys.)*. A fundamental interaction mediated by *gluons*. It is the interaction for the binding together of *quarks* (and *antiquarks*) in hadrons.

strongyloidiasis, strongyloidosis *(Med.)*. Infestation of man with the nematode worm *Strongyloides stercoralis*, the worm living in the intestines and causing diarrhoea; common in the Tropics.

strontium *(Chem.)*. Metallic element, symbol Sr, at. no. 38, r.a.m. 87.62, rel.d. 2.54, mp 800°C, bp 1300°C. Silvery-white in colour, it is found naturally in *celestine* and in *strontianite*; it also occurs in mineral springs. Similar chemical qualities to calcium. Compounds give crimson colour to flame and are used in fireworks. The radioactive isotope, ^{90}Sr, is produced in the fission of uranium and has a long life, hence its presence in 'fall-out' after a nuclear explosion.

strontium unit *(Phys.)*. Used to measure the concentration of radioactive ^{90}Sr in calcium; 1 SU $=10^{-12}$ Ci g^{-1}.

strophiole *(Bot.)*. See **caruncle**.

strophism, strophic movement *(Bot.)*. Growth movement in which an organ or its stalk twists in response to a directional stimulus e.g. the twisting of leaf bases and petioles on many horizontal branches in response to light and/or gravity resulting in the horizontal orientation of the leaf blades.

Strouhal number *(Acous.)*. Factor in the equation for the frequency of an **aeolian tone**. The equation relates the frequency f with the speed v and the thickness d of the obstacle by $f=k.v/d$, where k is the Strouhal number.

Strowger exchange *(Telecomm.)*. The rapidly-disappearing generation of telephone exchanges using entirely electro-mechanical switching; after Almon B. Strowger, their inventor.

struck *(Build.)*. (1) Taken away, dismantled; said, e.g. of scaffolding or shuttering. (2) Joints on an exposed face of a wall are said to be *struck* when the mortar is recessed.

struck *(Vet.)*. (1) A form of *enterotoxemia of sheep* caused by *Clostridium perfringens* (*Cl. welchii*), type C. (2) Said of sheep affected with **blowfly myiasis**.

struck core *(Foundry)*. A loam core formed by revolving the built-up core, loam-covered, against a **strickle board**.

struck-joint pointing *(Build.)*. See **weathered pointing**.

structural *(Biol.)*. Of changes, aberrations, etc. in the number or arrangement of chromosomes. But see **structural gene**.

structural colours *(Zool.)*. Colour effects produced by some structural modification of the surface of the integument, as the iridescent colours of some Beetles. Cf. **pigmentary colours**.

structural damping *(Aero.)*. See **damping**.

structural formula *(Chem.)*. A representation of the chemical structure of a substance which shows not only its composition but also its **connectivity** and the order of the bonds connecting the atoms.

structural gene *(Biol.)*. The stretch of DNA specifying the *amino acid* sequence of a *polypeptide*, as distinct from the interspersed and associated DNA, some of which is concerned with control of gene synthesis.

structural timber *(Build.)*. (1) Canadian name for *carcassing timber*. (2) Any timber acting as a support.

structure *(Chem.)*. See **molecular structure**.

structure *(Space)*. Framework or ensemble of rigid elements which is designed to withstand a variety of mechanical and thermal influences (e.g. thrust forces, bending moments, aerodynamic heating effects) during launch and flight of a spacecraft and provide protective support for its subsystems and payload.

structure-borne sound *(Acous.)*. Sound in solid bodies, as opposed to sound in gases (e.g. air-borne sound) and sound in liquids (e.g. water sound).

structured programming *(Comp.)*. Orderly approach to programming which emphasizes breaking large and complex tasks into successively smaller sections. Also *top down programming*.

structure of the atom *(Chem.)*. See **atomic structure**.

struma *(Med.)*. (1) See **scrofula**. (2) See **goitre**.

strut *(Civ.Eng.)*. A timber or adjustable metal member used for bracing purposes during excavation or construction works.

strut *(Eng.)*. Any light structural member or long column which sustains an axial compressive load. Failure occurs by bending before the material reaches its ultimate compressive stress. See **column**.

Struthioniformes *(Zool.)*. Any order of Birds retaining only two toes and whose feathers lack an aftershaft. They extend their rudimentary wings when running. Known from the Pliocene onwards. Ostriches.

strutting *(Build.)*. The process of using props to give temporary support between two surfaces.

struvite *(Min.)*. Magnesium ammonium phosphate hexahydrate, crystallizing in the orthorhombic system. Found in guano and dung, and common in human calculi.

strychnine *(Chem.)*. $C_{21}H_{22}N_2O_2$, a monoacidic alkaloid base, mp 265°C, very poisonous, causing tetanic spasms. It occurs in the seeds of *Strychnos ignatii* and *Strychnos nux vomica*, in *Upas tieuté* and in *Lignum colubrinum*, Strychnine is almost insoluble in water, but is readily soluble in chloroform and benzene.

strychnine bases *(Chem.)*. A group of alkaloids obtained from *Strychnos nux vomica*. They include **strychnine** and **brucine**.

STS *(Space)*. Abbrev. for *Space Transportation System*, usually refers to the Space Shuttle but actually includes **Spacelab**, inter-orbit stages carried by the Orbiter and the Tracking and Data Relay Satellite System (*TDRSS*), together with the supporting ground segment.

stub *(Build.)*. A small projection on the under surface at the top edge of a tile, enabling it to be hung on a batten.

stub *(Telecomm.)*. An auxiliary section of a waveguide or transmission line connected at some angle with the main section. See **coaxial-**, **quarter-wavelength-**.

stub antenna *(Telecomm.)*. A quarter-wavelength rod or wire.

stub axle *(Autos.)*. A short dead axle. If carrying a steered wheel it is capable of limited angular movement about a swivel-pin carried by the end of the axle beam.

stub plane *(Aero.)*. A short length of wing projecting from the fuselage, or hull, of some types of aircraft to which the main planes are attached.

stub tenon *(Build.)*. A very short tenon for fitting into a blind mortise. Also *joggle*.

stub-tooth gear *(Eng.)*. A gear tooth of smaller height and of more robust form than that normally employed; used in the manufacture of automobile gears.

stub tuning *(Elec.Eng.)*. Use of shunt stubs connected to short-circuited section of line or waveguide in order to produce matched conditions. In the single-stub tuner the stub susceptance is $-jb$ and it is connected to the main transmission line at a point where the transformed load admittance is $y=1+jb$ (normalized).

stuc *(Build.)*. Plasterwork resembling stone.

stucco *(Build.)*. A smooth-surfaced plaster or cement rendering applied to external walls.

stuck moulding *(Build.)*. A moulding shaped out of the solid of a member.

stud *(Build.)*. The vertical members in a timber partition framework.

stud *(Eng.)*. A shank, or headless bolt, generally screwed from both ends and plain in the middle. It is permanently screwed into one piece, to which another is then secured by a nut.

student's t-test *(Stats.)*. See **t-distribution**.

stud partition *(Build.)*. A wooden partition based on rough vertical timbers.

stuff *(Build.)*. (1) See **coarse stuff**, **fine stuff**. (2) Timber sawn or manufactured from logs.

stuffer box *(Textiles)*. A box into which continuous filament yarn is packed closely so resulting in the yarn becoming crimped. The yarn may then be heated in order to set the crimp and produce a **textured yarn**.

stuffing-box *(Eng.)*. A cylindrical recess provided in, for example, a cylinder cover, at the point at which the piston rod emerges; it is filled with packing which is compressed by a *gland* to make a pressure-tight joint.

stugging *(Build.)*. See **picking**.

stuke *(Build.)*. See **stucco**.

stupor *(Med.)*. A state of mental and physical inertia: inhibition of instinctive activity and indifference to social environment.

sturdy *(Vet.)*. See **coenuriasis**.

Sturm's theorem *(Maths.)*. A theorem by which the number of real roots of an algebraic equation which lie in any given interval can be determined. It utilizes sign changes in the partial remainders that occur in calculating the HCF of $f(x)$ and $f'(x)$, where $f(x)=0$ is the given equation.

Stuttgart disease *(Vet.)*. *Canine typhus*. The name

formerly given to a disease of the dog characterized by apathy, stomatitis, and gastro-enteritis, but which probably was, in most cases, the uremic stage of nephritis caused by *Leptospira canicola*. Vaccines widely used.

Stüve diagram *(Meteor.)*. An **aerological diagram** with rectangular axes temperature and (pressure) $^{(\gamma-1)/\gamma}$ where γ is the ratio of the specific heats of a perfect gas.

St Vitus's dance *(Med.)*. **Sydenham's chorea**.

S-twist *(Textiles)*. See **twist direction**.

style *(Bot.)*. The part of the carpel between the ovary and the stigma, often relatively long and thin.

style *(Build.)*. See **stile**.

style of the house *(Print.)*. The customary style of spelling, punctuation, capitalization etc. used in a printing establishment. It is followed in the absence of contrary instructions.

stylet *(Zool.)*. A small pointed bristlelike process.

styliform *(Zool.)*. Bristle-shaped.

stylo- *(Genrl.)*. Prefix from Gk. *stylos*, pillar.

stylobate *(Arch.)*. A continuous pedestal supporting a row of columns.

stylolite *(Geol.)*. An irregular suture-like boundary found in some limestones. In three dimensions it has a tooth-and-socket arrangement and appears to have been formed by pressure solution after deposition.

stylopodium *(Bot.)*. The swollen base of a style, as in some Umbelliferae.

stylopodium *(Zool.)*. The proximal segment of a typical pentadactyl limb; brachium or femur; upper arm or thigh.

stylus *(Acous.)*. Needle for cutting or replaying a disk recording. With the introduction of lightweight pickups, the more durable sapphire- or diamond-tipped reproducer styli replaced the earlier steel or chrome-plated needles.

stypsis *(Med.)*. The application, or use, of styptics.

styptic *(Med.)*. *Astringent*; tending to stop bleeding by coagulation.

styrene *(Chem.)*. *Phenylethene*, $C_6H_5.CH:CH_2$, a constituent of essential oils and coal-tar. It is a colourless aromatic liquid, bp 145°C, soluble in ethanol and ethoxyethane.

styrene-butadiene rubber *(Plastics)*: See **Buna, GR-S**.

styrene joint *(Elec.Eng.)*. A joint filled with hot liquid styrene, which polymerizes on cooling into a very hard solid and prevents displacement of the cores.

styrene resins *(Plastics)*. Compounds, polystyrenes, formed by the polymerization of styrene, $C_6H_5.CH{=}CH_2$. The power factor and dielectric constant are lower than for other plastics, and they do not disintegrate at ultra-short-wave and TV frequencies. They have excellent mechanical properties, and are resistant to moisture, concentrated sulphuric acid, strong alkalis and alcohol. Foamed polystyrene is an important heat insulation material.

styrol resins *(Plastics)*. See **styrene resins**.

sty, stye *(Med.)*. *Hordeoleum*. Staphylococcal infection of a sebaceous gland of the eyelid.

sub- *(Genrl.)*. Prefix from L. *sub*, under, used in the following senses: (1) deviating slightly from, e.g. *subtypical*, not quite typical; (2) below, e.g. *subvertebral*, below the vertebral column; (3) somewhat, e.g. *subspatulate*, somewhat spatulate; (4) almost, e.g. *subthoracic*, almost thoracic in position.

subacute *(Med.)*. Said of a disease whose symptoms are less pronounced than those of the acute form; between acute and chronic.

subacute combined degeneration of the cord *(Med.)*. *Anaemic spinal disease*. A condition in which there is degeneration of motor and sensory nerve tracts in the spinal cord, giving rise to paraplegia and loss of sensibility of the skin, the disease being associated with vitamin B_{12} deficiency.

sub-additive function *(Maths.)*. See **additive function**.

subaqueous loudspeaker *(Acous.)*. Sound source for producing sound in water.

subaqueous microphone *(Acous.)*. See **hydrophone**.

subarachnoid haemorrhage *(Med.)*. Haemorrhage into the space between the arachnoid and the pia mater, especially as a result of rupture of an aneurysm of one of the arteries.

subassembly *(Eng.)*. Assembly which can be handled and stored as a unit and subsequently incorporated in a more complex assembly.

subatomic *(Chem.)*. Said of particles or processes at less than atomic level, e.g. radioactivity, production of X-rays, nuclear shells etc.

subaudio frequency *(Acous.)*. One below those usefully reproduced through a sound-reproducing system or part of such system.

subcarrier *(Telecomm.)*. One frequency which is modulated over a narrow range by a measured quantity, and then used to modulate (with others) a carrier that will be finally demodulated on reception.

subchelate *(Zool.)*. In Arthropoda, having the distal joint of an appendage modified so that it will bend back and oppose the penultimate joint, like the blade and handle of a penknife, to form a prehensile weapon; cf. **chelate**.

subchord *(Surv.)*. The chord length from a tangent point on a railway or highway curve to the adjacent chainage peg around the curve when this is less than the full chord distance employed in setting out the chainage pegs.

subcircuit *(Elec.Eng.)*. One of several lighting circuits supplied from a common branch distribution fuse board.

subclavian *(Zool.)*. Passing beneath, or situated under, the clavicle; as the *subclavian artery*.

subclimax *(Bot.)*. Vegetation held more or less permanently at some stage of a succession before the **climax**.

subconscious *(Behav.)*. Syn. for *unconscious*.

subcortical *(Zool.)*. Below the cortex or cortical layer; as certain cavities in Sponges.

subcritical *(Phys.)*. Of an assembly of fissile material, for which the multiplication factor is less than unity.

subcrust *(Civ.Eng.)*. A cushioning layer between the pavement and the foundation of a carriageway, or the base formed on the natural foundation. Also *cushion course*.

subculture *(Bot.)*. A culture of a micro-organism, tissue or organ prepared from a pre-existing culture.

subcutaneous *(Zool.)*. Situated just below the skin.

subdorsal *(Zool.)*. Situated just below the dorsal surface.

subduction zone *(Geol.)*. The area where a plate moves under an overriding plate. Associated with regions of high seismic activity.

subdural *(Med.)*. Situated beneath the dura mater, e.g. *subdural abscess*, *subdural haemorrhage*.

suberin *(Bot.)*. A mixture of fatty substances, especially of cross-linked polyesters of long chain ($>C_{20}$) ω-hydroxy and ω-dicarboxylic aliphatic acids, in some plant cell walls, especially cork.

suberin lamella *(Bot.)*. A layer of wall material impregnated with suberin.

suberization *(Bot.)*. The deposition of suberin on or in a cell wall.

subfactorial n *(Maths.)*. The number of different ways of arranging *n* objects so that no object occupies its original position. Written $n¡$ or $\|\underline{n}$. Cf. **factorial n**.

sub-floor *(Build.)*. See **counter-floor**.

subgenital *(Zool.)*. Below the genital organs, as the subgenital pouches of *Aurelia*.

subgenual organ *(Zool.)*. In many Insects, chordotonal organs in the fibia adapted for perceiving vibrations of the substrate.

subglacial drainage *(Geol.)*. The system of streams beneath a glacier or ice sheet; formed chiefly of melt-waters. Cf. **englacial** streams.

subgroup *(Maths.)*. Let $(G;*)$ be a *group*. Let *H* be a subset of *G* which forms a group with the operation *. Then $(H;*)$ is said to be a *subgroup* of the group $(G;*)$; e.g. integers with addition constitute a subgroup of the rational numbers with addition.

subharmonic *(Acous.)*. Having a frequency which is a fraction of a fundamental. Subharmonics appear in some forms of nonlinear distortion.

subhedral *(Geol.)*. See **hypidiomorphic**.

subimago *(Zool.)*. In Ephemeroptera (Mayflies) the stage in the life history emerging from the last aquatic nymph. It has wings and moults to give the true imago; this **ecdysis** is unique among Insects, involving the casting of a delicate pellicle from the whole body, including the wings. adj. *subimaginal*.

subinvolution *(Med.)*. Partial or complete failure of the uterus to return to the normal state after childbirth.

subjective noise meter *(Acous.)*. A noise meter for assessing noise levels on the phon scale, the loudness of the noise level being measured by ear with the adjusted reference tone, 1000 Hz. See **objective noise meter**.

sublevel caving *(Min.Ext.)*. Method of mining massive ore deposit in which ore is drawn down to a delivery road under-running the deposit, and overburden is allowed to cave in, the process being repeated.

sublevel stoping *(Min.Ext.)*. Method in which ore is blasted in stopes and drawn down to a sublevel in the footwall through ore passes.

sublimate *(Chem.)*. The product of sublimation.

sublimation *(Behav.)*. In psychoanalytic theory, the developmental process by which forbidden impulses are gratified in socially acceptable ways; instinctual energy is displaced from its primary object to a more acceptable one, and tension associated with the gratification of repressed impulses is reduced (e.g. intellectual curiosity as a sublimation of childhood voyeurism).

sublimation *(Chem.)*. The vaporization of a solid (especially when followed by the reverse change) without the intermediate formation of a liquid.

sublimed white lead *(Chem.)*. Basic lead (II) sulphate fume.

subliminal perception *(Behav.)*. Refers to the phenomena whereby stimuli presented below the threshold of conscious awareness may influence behaviour. Also *subception*.

sublingua *(Zool.)*. In Marsupials and Lemurs, a fleshy fold beneath the tongue.

sublittoral plant *(Ecol.)*. A plant which grows near the sea, but not on the shore.

sublittoral zone *(Ecol.)*. In a lake, the lake bottom below the paralimnion, extending from the lakeward limit of rooted vegetation to the upper limit of the hypolimnion.

subluxation *(Med.)*. Partial, incomplete dislocation of a joint.

submarginal ore *(Min.Ext.)*. Developed ore which, at current market price of extracted values, cannot be profitably treated.

submarine cable *(Telecomm.)*. Long-distance cable laid along the sea bed. Co-axial in form, with **submarine repeaters** at intervals to amplify signals. In shallow water, with a danger from anchors or trawling, the cables may be armoured or even buried in the sea bed. In deep water, *lightweight* cables without armouring but with a central core of high-tensile steel are used to prevent stretch during laying. Some may contain **optical fibre** channels.

submarine canyon *(Geol.)*. A trench on the continental shelf. It sometimes has tributaries.

submarine fan *(Geol.)*. A fan of terrigeneous material formed at the foot of **submarine canyons** and large rivers, often as **turbidite** deposits.

submarine repeater *(Telecomm.)*. A **repeater** built into a water-tight and pressure-resistant housing; long-life components are used to ensure high reliability, due to the high cost of raising cables and repeaters for repair.

submaxillary *(Zool.)*. Situated beneath the lower jaw.

submerged arc welding *(Elec.Eng.)*. Automatic *arc welding* process, using a single, bare electrode which passes through a blanket of granular, fusible flux laid along the seam to be welded, so that the entire welding action takes place beneath this blanket which eliminates spatter losses and protects the joint from oxidation.

submerged heating *(Elec.Eng.)*. Induction heating of workpiece submerged in quenching liquid.

submicron *(Phys.)*. A particle of diameter less than a **micron**. Visible by ultra-microscope.

sub-millimetric waves *(Telecomm.)*. Microwaves at a frequency exceeding 300 GHz, at which the wavelength is 1 mm.

subnormal *(Maths.)*. Of a curve: the projection on to the y-axis of that part of a normal to the curve lying between its point of normalcy with the curve and its intersection with the y-axis.

suboutcrop *(Min.Ext.)*. See **blind apex**.

subpress *(Eng.)*. Unit comprising a carrier plate for the punch and another for the die of a press tool, connected by pillars sliding in bushes so that the alignment of punch and die is maintained accurately during working of the press tool. Also called *die set*.

subprogram *(Comp.)*. See **subroutine**.

subroutine *(Comp.)*. Set of **program statements** performing a specific task, but which requires to be initiated by a calling program. Also *sub program*, *routine*, *procedure*, *function*. See **closed subroutine**, **open subroutine**.

subscriber's line *(Telecomm.)*. The line connecting the subscriber's main telephone instrument to the exchange, as contrasted with an extension line from this instrument.

subscriber's station *(Telecomm.)*. See **substation**.

subscripted variable *(Comp.)*. Reference to an individual element of an **array** (e.g. CASS(20) or LAR(3,5))

subsequent pick-up *(Telecomm.)*. The pick-up occurring at the end of a pulse train due to the change in line-relay holding conditions.

subset *(Immun.)*. Term used to classify functionally or structurally different populations of cells within a single cell type. Used especially of T-lymphocytes (helper, suppressor, cytotoxic).

subset *(Maths.)*. Intuitively, a part of a set; e.g. the positive even integers constitute a subset of the set of positive integers. However, in maths, it is also accepted that (i) each set is a subset of itself; (ii) the empty set is a subset of every set. These two extreme cases and the above intuitive notion are incorporated in this mathematical definition of subset; A set X is a subset of a set Y if and only if each element of X is also an element of Y.

subsidence *(Build.,Civ.Eng.)*. (1) The sinking or caving-in of the ground. (2) The settling down of a structure etc. to a lower level.

subsidiary cell, accessory cell *(Bot.)*. An epidermal cell associated with the guard cells of a stoma and morphologically different from the other epidermal cells.

subsoil *(Geol.)*. Residual deposits lying between the soil above and the bed-rock below, the three grading into one another.

subsoil drain *(Civ.Eng.)*. A drain laid just below ground-level to carry off waters from saturated ground. Formerly earthenware, now corrugated and perforated plastic pipe is laid at the bottom of a trench and covered with broken stones.

subsonic *(Acous.)*. Said of an object or flow which moves with a speed less than that of sound.

subsonic speed *(Aero.)*. Any speed of an aircraft where the airflow round it is everywhere below Mach 1. See **Mach number**.

subspecies *(Biol.)*. Taxonomic subdivision of a species, with some morphological differences from the other subspecies and often with a different geographical distribution or ecology.

substance *(Chem.)*. A kind of matter, with characteristic properties, and generally with a definite composition independent of its origin.

substance *(Glass)*. The thickness of flat glass; in the case of sheet glass, usually expressed as mass per unit area.

substance *(Paper)*. Preferred term is *grammage* or basis weight.

sub-standard *(Image Tech.)*. Deprecated term for **narrow gauge film**, meaning widths smaller than the *standard* 35 mm.

substandard instrument *(Elec.Eng.)*. A laboratory instrument whose accuracy is very great and which has been calibrated against an international standard of measurement.

substantia *(Zool.)*. Substance; matter.

substantive dyes *(Chem.)*. Dyestuffs which can dye cotton and other fibres direct without the aid of a mordant. Many are derived from benzidine and its derivatives.

substantive variation *(Bot.,Zool.)*. Variation in the constitution of an organ or organism, as opposed to variation in the number of parts; Also *qualitative variation*.

substantivity *(Textiles)*. The attractive force that enables a textile material to remove dyestuff from a dyebath.

substation *(Elec.Eng.)*. A switching, transforming or converting station intermediate between the generating station and the low-tension distribution network.

substation *(Surv.)*. Apex of a subsidiary triangle in a skeleton.

substellar point *(Astron.)*. The point on the earth's surface, regarded as spherical, where it is cut by a line from the centre of the earth to a given star; hence the point where the star would be vertically overhead, the point whose latitude is equal to the star's declination. Applied also to the Sun and Moon as *subsolar point* and *sublunar point* respectively.

substitution *(Chem.)*. The replacement of one group, especially hydrogen by another, e.g. halogen, alkyl, hydroxyl etc.

substitutional resistance *(Elec.Eng.)*. A resistance, equal to the normal resistance of an arc lamp, which is automatically cut into circuit upon the failure of any one of several arc lamps connected in series.

substrate *(Biol.)*. (1) Reactant in a reaction catalysed by an enzyme. (2) Surface or medium on or in which an organism lives and from which it may derive nourishment.

substrate *(Electronics)*. Single crystal of semiconductor used as a basis for integrated circuit or transistor.

substrate *(Image Tech.)*. A layer applied to the film base before coating the emulsion to ensure complete adhesion.

substrate level phosphorylation *(Biol.)*. The conversion of ADP to ATP which is brought about by the concomitant hydrolysis of some other **high energy phosphate compound**.

substratum *(Biol.)*. See **substrate (2)**.

subsynchronous *(Elec.Eng.)*. Below **synchronism**.

subsystem *(Space)*. Constituent part of a system which performs a particular function; thus, *electrical power subsystem, data handling subsystem*, etc; it is the sum of the coherent subsystem performances which provides a certain system capability.

subtangent *(Maths.)*. Of a curve, the projection on to the x-axis of that part of a tangent lying between its point of tangency with the curve and its intersection with the x-axis.

subtectal *(Zool.)*. Lying beneath the roof, as of the skull; in some Fish, a cranial bone.

subtend *(Bot.)*. To be situated immediately below, e.g. as a leaf is situated immediately below the bud in its axil.

subtend *(Maths.)*. (1) The portion of a straight line or arc intercepted by and between the arms of an angle is said to subtend the angle. (2) The chord joining the points at the end of an arc is said to subtend the arc.

subtense bar *(Surv.)*. A horizontal bar, bearing two targets fixed at a known distance apart, used as the distant base in one system of tacheometry.

sub-title *(Image Tech.)*. Wording superimposed on the lower part of a film or television picture, often to give a translation of the accompanying sound track dialogue or to provide an outline of a spoken commentary for the benefit of the hard-of-hearing.

subtraction *(Maths.)*. The inverse operation to addition. Denoted by the minus sign $-$. Thus in the statement $m-s=d$, the terms m, s, and d are referred to as the *minuend, subtrahend* and *difference* respectively.

subtractive-coloured light *(Phys.)*. The monochromatic illumination obtained from a polychromatic light source by the aid of an appropriate absorption screen.

subtractive printer *(Image Tech.)*. A photographic or motion picture printer or an enlarger in which the intensity and colour of the exposing light is altered by the use of filters, in contrast to an **additive printer**.

subtractive process *(Image Tech.)*. A colour process in which the red, green and blue components of the original subject are reproduced as three superimposed images in the complementary (*subtractive*) colours of cyan, magenta and yellow respectively; this is the basis of all modern colour photography and cinematography as well as colour printing on paper.

subtrahend *(Maths.)*. See **subtraction**.

subtransient reactance *(Elec.Eng.)*. The reactance of the armature winding of a synchronous machine corresponding to the leakage flux which occurs in the initial stage of a short-circuit. This flux is smaller than that corresponding to the transient reactance on account of eddy currents which may be set up in the rotor during the first one or two half-cycles of a short-circuit.

subulate *(Bot.)*. Awl-shaped, tapering from base to apex.

succession *(Ecol.)*. The sequence of communities (i.e. a sere) which replace one another in a given area, until a relatively stable community (i.e. the climax) is reached, which is in equilibrium with local conditions.

succinamic acid *(Chem.)*. $H_2N.CO.CH_2.CH_2.COOH$, *monoamidobutandioic (succinic) acid*.

succinic acid *(Chem.)*. *Butandioic acid.* $HOOC.-CH_2.CH_2.COOH$, dibasic acid; mp 185°C, bp 235°C, with partial decomposition into its anhydride. It occurs in the juice of sugar cane, in the castor-oil plant, and in various animal tissues, where it plays an important part in metabolism (see **tricarboxylic acid cycle**). Succinic anhydride is used in the manufacture of alkyd resins and in organic synthesis.

succinite *(Min.)*. A variety of **amber**, separated mineralogically because it yields succinic acid.

succinyl *(Chem.)*. The bivalent acid residue $-CO.CH_2.-CH_2.CO-$.

succise *(Bot.)*. Ending below abruptly, as if cut off.

succulent *(Bot.)*. (1) Juicy, having a high water content. (2) A plant with succulent stems or leaves; most are xerophytes and CAM plants e.g. cacti, or halophytes e.g. *Salicornia*.

succus entericus *(Zool.)*. A collective name for the enzymes secreted by glandular cells (*Brunner's glands* and *Lieberkühn's crypts*) in the walls of the duodenum, these including erepsin (itself a mixture), invertase, maltase, lactase, nucleotidase, nuclease, lipase and enterokinase.

succussion *(Med.)*. The act of shaking a patient to detect the presence of fluid in a pleural cavity already containing air (*pneumothorax*).

sucker *(Bot.)*. An upward-growing shoot arising from the base of a stem or adventitiously from a root.

sucker *(Zool.)*. A suctorial organ adapted for adhesion or imbibition, as one of the muscular sucking disks on the tentacles of Cephalopoda: the suctorial mouth of animals like the Leech and the Lamprey: a newly born Whale: one of a large number of Fishes having a suctorial mouth or other suctorial structure, as the Remora (*Echeneis*), members of the genus *Lepadogaster* etc.

sucking booster *(Elec.Eng.)*. A booster whose function is to overcome voltage drop occurring in a feeder.

sucrol *(Chem.)*. See **dulcin**.

sucrose *(Chem.)*. *Saccharobiose*; $C_{12}H_{22}O_{11}$, a disaccharide carbohydrate; mp 160°C; it crystallizes in large monoclinic crystals, is optically active and occurs in beet, sugar cane and many other plants. Hydrolyses to glucose and fructose. Colloq. *sugar, cane sugar*.

sucrose gradient *(Biol.)*. Used in centrifugation to separate molecules on the basis of their sedimentation velocity.

suction box *(Paper)*. A shallow compartment extending across a paper making machine connected to a vacuum pump for the removal of water from the web or felt. Generally positioned under the forming web on the wire table or in a perforated roll e.g. couch or press.

suction couch roll *(Paper)*. A *couch roll* consisting of a perforated metal roll within which is a stationary suction box to remove water from the web on the machine wire.

suction dredger *(Civ.Eng.)*. See **sand-pump dredger**.

suction feed gun *(Build.)*. A spray gun with a paint cup fitted beneath the gun. The stream of compressed air which passes over the cup creates a vacuum causing atmospheric pressure to force the paint up to the gun where it is atomized. Used for small quantities of thin materials.

suction pressure *(Bot.)*. Obsolete term equivalent to minus the water potential of a cell.

suction roll *(Paper)*. Roll consisting of a perforated metal sleeve rotating about a fixed internal suction box.

suction valve *(Eng.)*. See **foot valve** (1).

suctorial *(Zool.)*. Drawing in: imbibing: tending to adhere by producing a vacuum: pertaining to a sucker.

suctorial mouth-parts *(Zool.)*. Tubular mouth-parts adapted for the imbibition of fluid nourishment; found in some Insects and many ectoparasites.

sudamina *(Med.)*. Whitish vesicles on the skin, due to retention of sweat in the sweat glands. sing. *sudamen*. adj. *sudaminal*.

sudden infant death syndrome *(Med.)*. *SIDS, cot death.* Death of an apparently healthy infant for which there seems no obvious cause. Precise cause still not understood.

sudden warming *(Meteor.)*. A rapid rise of temperature of the polar **stratosphere** of up to 50 K in a few days occurring in winter or early spring. It is associated with a breakdown of the winter polar stratospheric vortex and may be either temporary or, in spring, permanent.

sudor *(Med.)*. Sweat or perspiration.

sudoriferous, sudoriparous *(Zool.)*. Sweat-producing: sweat-carrying.

sudorific *(Med.)*. Connected with the secretion of sweat: stimulating the secretion of sweat: a drug which does this.

suffructescent, suffruticose *(Bot.)*. Somewhat woody; diminutively shrubby; woody at base with herbaceous stems, e.g. alpine willows.

sugar *(Chem.)*. (1) A water-soluble, crystalline mono- or oligosaccharide. (2) The common term for *sucrose*, or *cane-sugar*, $C_{12}H_{22}O_{11}$.

sugar charcoal *(Chem.)*. Highly pure form of charcoal derived from sucrose.

sugar plant *(Bot.)*. Plant in which carbohydrate is stored as sugar. Cf. **starch plant**.

sugar soap *(Build.)*. An alkaline cleansing or stripping preparation for paint surfaces.

Suhl effect *(Electronics)*. The reduction in lifetime of holes injected into an *n*-type semiconducting filament, by deflecting them to the surface using a powerful transverse magnetic field. Reverse of **Hall effect**.

sulcus *(Zool.)*. A groove or furrow, as one of the grooves on the surface of the cerebrum in Mammals: in Dinoflagellata, a longitudinal groove in which a flagellum lies: in Anthozoa, the 'ventral' siphonoglyph.

sulfate, sulfur *(Chem.)*. See **sulphate, sulphur**.

sulfotep *(Chem.)*. Bis-*OO*-diethylphosphorothionic anhydride, used as an insecticide. Also *dithio, thiotep, dithioTEPP*.

sullage *(Civ.Eng.)*. The mud and silt deposited by flowing waters.

sulphamic acid *(Chem.)*. The monoamide of sulphuric acid: $HO\text{-}SO_2\text{-}NH_2$. Prepared by the ammonolysis of sulphuric acid; commercially, by the reaction of urea (carbamide) with fuming sulphuric acid.

sulphamide *(Chem.)*. The diamide of sulphuric acid: $SO_2(NH_2)_2$. Prepared by the ammonolysis of sulphuryl chloride.

sulphane *(Chem.)*. A compound of the formula HS_xH, analogous to an alkane. See **sulphur hydrides**.

sulphanilic acid *(Chem.)*. $H_2N.C_6H_4.SO_3H$, *4-amino-benzene-sulphonic acid*; crystallizes with $2H_2O$; sparingly soluble in water.

sulphate of ammonia *(Chem.)*. $(NH_4)_2SO_4$. Commercially the most important of the ammonium salts, particularly for use as fertilizer. Produced partly as a by-product of gas works, coke ovens etc., but now largely by direct synthesis.

sulphate of iron *(Min.)*. See **melanterite**.

sulphate of lead *(Min.)*. See **anglesite**.

sulphate of lime *(Min.)*. See **anhydrite, gypsum**.

sulphate of strontium *(Min.)*. See **celestine**.

sulphate-resisting cement *(Civ.Eng.)*. Cement manufactured for use in concrete to resist normal concentrations of sulphates as in most flues; also for underwater work.

sulphates *(Chem.)*. Salts of sulphuric acid. Produced when the acid acts on certain metals, metallic oxides, hydroxides and carbonates. The acid is dibasic, forming two salts; normal and acid sulphates.

sulphating roasting *(Eng.)*. Roasting carried out under conditions designed to convert part of the contained sulphur (sulphide mineral) to sulphate.

sulphation *(Elec.Eng.)*. The formation of the insoluble white sulphate of lead ($PbSO_4$) in the plates of a lead-acid type of secondary cell, a process which diminishes the efficiency and capacity of the cell.

sulphides *(Chem.)*. Salts of hydrogen sulphide. Many sulphides are formed by direct combination of sulphur with the metal.

sulphide toning *(Image Tech.)*. A process of toning photographic prints in which the silver is converted to silver sulphide via silver bromide.

sulphide zone *(Min.Ext.)*. Primary (unaltered) zone of sulphide-mineral lode, underlying leached (superficial) zone and that of secondary enrichment in which there has been redeposition of values oxidized from leached zone by penetrating water.

sulphinic acids *(Chem.)*. Acids containing the monovalent sulphinic acid group —SO.OH.

sulphite process *(Paper)*. One of the acid pulp digestion processes using a bisulphite liquor with some free sulphur dioxide.

sulphites *(Chem.)*. *Sulphates (IV)*. Salts of sulphurous (sulphuric (IV)) acid. The acid forms 2 series of salts, acid sulphites or bisulphites and normal sulphites.

sulphite wood pulp *(Paper)*. Chemical wood pulp in which digestion is carried out by the sulphite process.

sulphocyanides *(Chem.)*. See **thiocyanates**.

sulphonamides *(Med.)*. A group of drugs with a powerful antibacterial action, formerly used in the treatment of various infections. In chemotherapeutic practice the sulphonamides have been to a considerable extent displaced by antibiotics.

sulphonation *(Chem.)*. The reversible process of forming sulphonic acids by the action of concentrated sulphuric acid on aliphatic or aromatic compounds.

sulphones *(Chem.)*. Compounds of the formula $RR'=SO_3$. The sulphur is hexavalent.

sulphonic acids *(Chem.)*. Acids containing the monovalent sulphonic acid group $—SO_2.OH$.

sulphonylurea *(Med.)*. Group of drugs which act by augmenting insulin secretion. Used in the treatment of maturity onset diabetes.

sulphosol *(Chem.)*. A colloidal solution in concentrated sulphuric acid.

sulphoxides *(Chem.)*. Compounds of the formula RR'SO. The sulphur is in the (IV) oxidation state.

sulphoxylic acid *(Chem.)*. The hypothetical oxy acid of sulphur, $S(OH)_3$.

sulphur *(Chem.)*. A nonmetallic element occurring in many allotropic forms. Symbol S, at. no. 16, r.a.m. 32.06, valencies 2, 4, 6. Rhombic (β-) sulphur is a lemon yellow powder; mp 112.8°C, rel.d. 2.07. Monoclinic (β-) sulphur has a deeper colour than the rhombic form; mp 119°C, rel.d. 1.96, bp 444.6°C. Chemically, sulphur resembles oxygen, and can replace the latter in many compounds, organic and inorganic. It is abundantly and widely distributed in nature, in the free state and combined as sulphides and sulphates, occurring around volcanoes and hot springs and in sulphite deposits as a result of the reduction of sulphates. It is manufactured by purifying the native material or by heating pyrites. Sulphur is used in the manufacture of sulphuric acid and carbon disulphide; in the preparation of gunpowder, matches, fireworks and dyes; as a fungicide, and in medicine; and for vulcanizing rubber. Vapour bath at boiling-point is used as fixed point in platinum resistance thermometry.

sulphur bacteria *(Biol.)*. Bacteria which live in situations where oxygen is scarce or absent, and which act upon compounds containing sulphur, liberating the element. Occur in two families of true bacteria, the *Thiorhodaceae* (purple sulphur bacteria) and the *Thiobacteriaceae* (colourless sulphur bacteria).

sulphur cement *(Build.)*. A cement made of sulphur and pitch mixed in equal parts; used to fix iron work.

sulphur dioxide *(Chem.)*. *Sulphur (IV) oxide*; SO_2. A colourless gas formed when sulphur burns in air. Dissolves in water to give sulphurous (sulphuric (IV)) acid. Also *sulphurous anhydride*. See also **sulphur oxides**.

sulphuretted hydrogen *(Chem.)*. See **hydrogen sulphide**.

sulphur hydrides *(Chem.)*. Four well-defined hydrides; H_2S, H_2S_2, H_2S_3 and H_2S_5. Also *sulphanes*.

sulphuric acid *(Chem.)*. A strong dibasic acid, H_2SO_4. The concentrated acid is a colourless oily liquid; rel.d. 1.85, bp 338°C; it dissolves in water with the evolution of heat, and is very corrosive, largely owing to its dehydrating action. It is manufactured from sulphur dioxide, obtained by burning either pyrites or sulphur, by the **contact process** or the **chamber process**. It is an important heavy chemical, used extensively in the dyestuffs and explosives industries; as a drying agent in chemical process; in the manufacture of other acids, e.g. HCl, HF, phosphoric acid; in fertilizers, pickling liquors and leaching solutions; in petroleum treatment and the manufacture of alkyl sulphonate detergents; in rayon production etc. The salts of sulphuric acid are called *sulphates* (VI).

sulphuric anhydride *(Chem.)*. See **sulphur trioxide**.

sulphurous acid *(Chem.)*. An aqueous solution of sulphur dioxide, which contains the hypothetical compound H_2SO_3. The corresponding salts, sulphites, are well known.

sulphurous anhydride *(Chem.)*. See **sulphur dioxide**.

sulphur oxides *(Chem.)*. A series of oxides; SO, S_2O_3, SO_2, SO_3, S_2O_7 and SO_4.

sulphur trioxide *(Chem.)*. *Sulphur (VI) oxide*; SO_3. Dissolves in water to give sulphuric acid.

summation *(Med.)*. The production of an effect by repetition of a causal factor which would be insufficient in a single application, as *summation of contractions*, the production of a state of tetanic contraction by a series of stimuli.

summation check *(Comp.)*. Figure added to a *word*, indicating a summation of the digits so that accuracy of processing can be verified. Totals arrived at by two methods for verifying processing of data.

summation instrument *(Elec.Eng.)*. An instrument for indicating, integrating or recording the sum total of the energy, power or current in two or more circuits.

summation metering *(Elec.Eng.)*. A system of metering electrical energy in which the consumption of several distinct load circuits is summated and indicated by one instrument.

summation of losses *(Elec.Eng.)*. The process of adding together the individual losses, after allowing for any corrections, in order to obtain the guaranteed efficiency of an electrical machine.

summation panel *(Elec.Eng.)*. A switchboard panel on which are mounted the instruments for measuring and recording the total output of a number of generators.

summation tone *(Acous.)*. See **combination time**.

summer annual *(Bot.)*. A plant which completes its life cycle over a few weeks in the summer, surviving the winter as seed. Cf. **winter annual**.

summer draught *(Ships)*. The **draught** when loaded to the **summer-load waterline**.

summer egg *(Zool.)*. In many fresh-water animals, a thin-shelled, rapidly developing egg laid during the warm season. Cf. **winter egg**.

summer-load waterline *(Ships)*. The waterline to which a ship may be loaded in summer. It is indicated in the freeboard markings.

summer mastitis *(Vet.)*. Mastitis of nonlactating cows due to infection by *Corynebacterium pyogenes*, or occasionally other organisms; possibly transmitted by flies.

summer solstice *(Astron.)*. See **solstice**.

summer wood *(Bot.)*. See **late wood**.

summit canal *(Hyd.Eng.)*. A canal crossing a summit; one, therefore, to which water must be supplied.

sum of infinite series *(Maths.)*. More precisely, sum to infinity. The limit, as *n* tends to infinity, of the sum of the first *n* terms.

sump *(Autos.)*. The lower part of the crankcase of an automobile engine, which usually acts as an oil reservoir.

sump *(Civ.Eng.)*. A small hole dug usually at the lowest part of an excavation to provide a place into which water can drain and from which it can be pumped at intervals to keep the working part of the excavation dry. See also **catch pit**.

sump *(Min.Ext.)*. The prolongation of a shaft or pit, to provide for the collection of water in a mine. The pump sump is that from which casual water or ore pulp is delivered to the mine or mill pumps.

Sumpner test *(Elec.Eng.)*. A back-to-back load test, or regenerative test, on two similar transformers.

Sumpner wattmeter *(Elec.Eng.)*. An iron-cored type of dynamometer wattmeter for use on a.c. circuits.

Sun *(Astron.)*. The central object of our solar system and the nearest **star** to earth. It is at an average distance of 150 million km (see **astronomical unit**, and on account of this proximity has been studied more than any other star. The source of energy is nuclear reactions in the central core where the temperature is 15 million K and the relative density 155. The core extends to 1/4 the solar radius and includes 1/2 the mass. Our Sun is nearly 5 million years old, and is about half way through its expected life cycle. Every second it annihilates 5 million tonnes of matter, to release 3×10^{26} watts of energy. The surface layer is the **photosphere** and above it are the **chromosphere** and the **solar corona**. See also: **aurora, faculae, sunspot** and **Hertzsprung-Russell diagram**.

sunburner *(Glass)*. An excessive local thickness of material in a mouth-blown glass article.

sun cracks *(Geol.)*. Polygonal cracks, usually in a fine-grained sedimentary rock, indicative of desiccation at the time of formation. Common in beds laid down under arid conditions.

sun gear *(Eng.)*. See **epicyclic gear**.

sunk face *(Build.)*. A term applied to a stone in whose face a panel is sunk by cutting into the solid material. Abbrev. *SF*.

sunk fence *(Build.)*. See **ha-ha**.

sunk key *(Eng.)*. A key which is sunk into key ways in both shaft and hub. See **key**.

sun observation *(Surv.)*. Use of theodolite to fix latitude and/or longitude of a station, or to orient a survey line by direct sighting and calculation of sun's position.

sun pillar *(Meteor.)*. A vertical column of light passing through the sun, seen at sunset or sunrise. It is caused by reflection of sunlight by horizontal ice crystals.

sun plant *(Bot.)*. A plant adapted to living at high light intensities. Cf. **shade plant**.

sun-ray therapy *(Med.)*. See **ultraviolet therapy**.

sunseeker *(Space)*. A photoelectric device mounted in rockets or space vehicles, in which an instrument such as a spectrograph can be directed constantly to the sun.

sunshine recorder *(Meteor.)*. The *Campbell-Stokes recorder* consists of a glass sphere arranged to focus the sun's image on to a bent strip of card, on which the hours are marked. The focused heat burns through the card, and the duration of sunshine is read off from the length of the burnt track.

sunspot *(Astron.)*. A disturbance of the solar surface which appears as a relatively dark centre (*umbra*), surrounded by a less dark area (*penumbra*); spots occur generally in groups, are relatively shortlived, and with few exceptions are found in regions between 30° N. and S. latitude; their frequency shows a marked period of about 11 yr (*sunspot cycle*), they have intense magnetic fields and are sometimes associated with magnetic storms on the earth.

sunstone *(Min.)*. See **aventurine feldspar**.

sunstroke *(Med.)*. *Heat hyperpyrexia*. A condition produced by exposure to high atmospheric temperature and characterized by a rapid rise of bodily temperature, convulsions, and coma.

Sun-synchronous orbit *(Space)*. A near-polar orbit which permits a satellite to pass over a point on the Earth at practically the same time each day regardless of the longitude.

sun wheel *(Eng.)*. A gear-wheel round which one or more planet wheels or planetary pinions rotate in mesh.

Super-16 *(Image Tech.)*. System of 16 mm cinematography using an increased width of frame in the negative for enlargement to 35 mm **wide-screen** prints.

Super-8 *(Image Tech.)*. System of narrow-gauge cinematography on 8 mm film, mainly for amateur use, having a larger frame area than the original *Regular 8*.

super-additive function *(Maths.)*. See **additive function**.

superaerodynamics *(Aero.)*. Aerodynamics at very low air densities occurring above 100 000 ft, i.e. for spacecraft on ascending and re-entry trajectories. The mean free path of the molecules is long compared with vehicle length. The physics is described by *free-molecule-flow* and Newtonian aerodynamics. *Magnetohydrodynamic* features are likely to be significant.

superalloy *(Eng.)*. See **heat-resisting alloy**.

superaudio frequency *(Acous.)*. One above those usefully transmitted through an audio-frequency reproducing system, or part of such.

supercalendered paper *(Paper)*. Paper to which the surface finish has been applied by passing it through a supercalender. The gloss is usually greater than that attained by machine calenders. See **calendered paper**.

supercharger *(Autos.)*. A compressor, commonly of the Rootes rotary vane or centrifugal type, used to supply air or combustible mixture to an IC engine at a pressure greater than atmospheric; driven either directly by the engine or by an exhaust gas turbine. See **turbocharger**.

supercharging *(Aero.,Eng.)*. (1) In aero engines, maintenance of ground-level pressure in the inlet pipe up to the rated altitude by means of a centrifugal or other blower. Necessary for flying at heights at which the air pressure is low and normal aspiration would be insufficient. (2) In other IC engines, the term is used synonymously with *boosting*. See **boost**.

superciliary *(Zool.)*. Pertaining to, or situated near, the eyebrows; above the orbit.

super-circulation *(Aero.)*. A form of boundary layer control, creating and controlling high lift in which air (usually derived from a jet engine compressor) is blown supersonically over the leading edge of a plain flap so that it carries the main airflow downward below the actual surface as an invisible extension. May also involve blowing over the leading edge. See also **jet flap**.

supercompression engine *(Eng.)*. An unsupercharged engine of above normal compression ratio, to be run at full throttle only at high altitudes, when the normal power decreases owing to the fall in atmospheric pressure. Below these altitudes the throttle opening is limited by the use of a gate.

supercomputer *(Comp.)*. A high performance **mainframe** typically used for the solution of numerical problems in the sciences.

superconducting amplifier *(Electronics)*. One using superconductivity to give noise-free amplification.

superconducting gyroscope *(Electronics)*. Frictionless gyroscope supported in vacuum through magnetic field produced by currents in superconductor.

superconducting levitation *(Phys.)*. If a small permanent magnet is placed on a surface which is then made *superconducting* by lowering its temperature below its critical temperature, then the magnet will rise and float above the surface. The superconductor as a perfect diamagnetic excludes the flux associated with the magnet and so provides a large enough repulsive force to balance the weight of the magnet. See **Meissner effect, superconductivity**.

superconducting magnet *(Electronics)*. Very powerful electromagnet, made possible by the exceptionally large current-carrying capacity of superconductors.

superconducting memory *(Electronics)*. One made up of thin-film devices which can be made to change from their low-temperature superconducting state to a normal resistive state by the brief application of a magnetic field, generated by current pulse on a control wire. The entire memory is maintained at the superconducting temperature and dissipates power only during the *read* or *write* operation; large, high-density memories can be built using these techniques.

superconductivity *(Phys.)*. Property of some pure metals and metallic alloys at very low temperature of having negligible resistance to the flow of an electric current. Each material has its own critical temperature, T_c above which it is a normal conductor. When a current is established, it persists almost indefinitely. Magnetic fields can destroy the superconductivity, their strength depending on how far below the critical temperature the material is. Generally, T_c has been < 20 K, but in 1986–7 a class of materials with *perovskite* structures has been discovered which have $T_c \sim 90$ K. See **Meissner effect**.

supercooled *(Chem.)*. Cooled below normal boiling- or freezing-point without change of phase.

supercooled (subcooled) water *(Meteor.)*. Water which continues to exist as a liquid at temperatures below 0°C.

supercritical *(Phys.)*. Assembly of fissile material for which the multiplication factor is greater than unity.

superego *(Behav.)*. In psychoanalytic theory, that part of the ego which has incorporated the moral standards of one's parents; the function of self-observation and self-criticism are associated with it. It differs from the notion of *conscience* in that its activities are often unconscious and often at odds with the individual's conscious values.

superelevation *(Surv.)*. The amount by which the outer rail of a railway curve is elevated above the inner rail to counteract the effect of the centrifugal force of the moving train. Superelevation is also applied in the construction of highway curves. See also **cant**.

superficial deposits *(Geol.)*. See **drift**.

superficial radiation therapy *(Radiol.)*. X-ray therapy (usually by soft radiation produced at less than 140 kVp) of the skin or of any surface of the body made accessible.

superfinishing *(Eng.)*. An abrasive process, resembling honing and lapping, for removing **smear metal** and scratches and ridges produced by grinding and machining operations, from bearing surfaces.

superfluid *(Phys.)*. Condensed degenerate gas in which a significant proportion of the atoms are in their lowest permitted energy state. In practice this affects only *liquid helium II*.

superfluidity *(Phys.)*. At temperatures below 2.19 K, the lamda point, the helium isotope ^{4}He shows a striking change in its liquid properties; in particular its viscosity effectively vanishes and it is said to show *superfluidity*. At temperatures above the lamda point, ^{4}He is known as helium I and below helium II. Superfluid helium II behaves as if it consists of two parts, a super fluid and a normal fluid. The lower the temperature, the greater the fraction of superfluid in the mixture. The superfluid part consists of helium atoms in their lowest quantum state.

superfoetation, superfetation *(Med.)*. Fertilization of an ovum in a woman already pregnant, some time after fertilization of the first ovum.

supergene enrichment *(Min.Ext.)*. Part of a mineral vein, lode or massive deposit where material removed from the *leached zone* is reprecipitated. See **secondary enrichment**.

supergiant star *(Astron.)*. Star of late type and abnormal luminosity, such as Betelgeuse and Antares; they are of enormous size and low density.

supergravity *(Astron.,Phys.)*. A particular version of a **supersymmetry** theory which postulates *gravitons* and *gravitinos* as carriers of the gravitational force.

supergrid *(Elec.Eng.)*. The national electric power network at voltages of 275 kV and 400 kV. See **grid**.

supergroup *(Telecomm.)*. In a transmission system using **frequency-division multiplex**, an assembly of five *groups* each comprising twelve speech channels. By re-modulation of these groups onto five new carriers, a *supergroup* of 60 speech channels is formed. See **master group**.

superheat *(Aero.)*. The increase (positive) or decrease of the temperature of the gas in a gas-bag as compared with the temperature of the surrounding air. Similarly, *superpressure*.

superheated steam *(Eng.)*. Steam heated at constant pressure out of contact with the water from which it was formed, i.e. at a higher temperature than that of saturation.

superhet *(Telecomm.)*. Abbrev, for *SUPersonic HETerodyne*.

superhet receiver *(Telecomm.)*. One in which the frequency of the incoming signal is reduced in a mixer or frequency changer, by **heterodyning** with another frequency from the **local oscillator**. The lower frequency output from the mixer, the *intermediate frequency* or *IF*, is taken through one or more steps of a selective amplifier, before conventional demodulation. There may also be stages of signal-frequency amplification before the mixer, and two stages of frequency conversion and two separate IF amplifiers. Advantages include better gain and selectivity; disadvantages are the extra components and complication, and the possibility of receiving **image respones**.

superimpose *(Image Tech.)*. Adding one image on top of another, so that both are visible.

superimposed drainage *(Geol.)*. A river system unrelated to the geological structure of the area, as it was established on a surface since removed. Cf. **consequent drainage**.

superimposition *(Print.)*. Coloured blocking foils are frequently used as a base for over-stamping by gold, silver, or other coloured foil. This 'superimposition' is of particular importance to bookbinders and display-card printers.

superior *(Bot.)*. An ovary in a flower that is **hypogynous** or **perigynous**.

superior *(Genrl.)*. Placed above something else; higher, upper (as the *superior* rectus muscle of the eyeball).

superior figures, letters *(Print.)*. Small figures or letters printed above the general level of the line. They are used instead of **marks of reference**, and in mathematical work etc.; thus: x^2, e^x, 10^6.

superior vena cava *(Zool.)*. See **precaval vein**.

supernatant liquid *(Chem.)*. The clear liquid above a precipitate which has just settled out.

supernormal stimulus *(Behav.)*. A stimulus that surpasses a natural stimulus in its ability to evoke a response.

supernova *(Astron.)*. Novae of absolute magnitude -14 to -16; 3 have been recorded in our own Galaxy, and about 50 more in spiral nebular. The violent outburst results from the gravitational collapse of a massive star, the outer layers being ejected, while the core is left as a **neutron star**.

supernumary chromosome *(Bot.)*. Same as B-chromosome.

superovulation *(Zool.)*. Hormone-induced excess **ovulation**. See **insemination**.

superoxide anion *(Immun.)*. O'_2. Oxygen molecule that carries an extra unpaired electron, and is therefore a free radical. Generated in neutrophil leucocytes and mononuclear phagocytes when activated, e.g. by ingestion of particles or immune complexes. O'_2 is highly reactive and toxic. It may be further reduced to H_2O_2 or, when two radicals interact, one is oxidized and one reduced in a dismutation reaction to form O_2 and H_2O_2. This reaction is catalysed by *superoxide dismutase*, an enzyme present in phagocytic cells. These substances are important in the microbicidal activity of the cells.

superoxides *(Chem.)*. Compounds of the alkali and alkaline earth metals containing the O_2^- group, e.g. $K^+(O_2^-)$. Differ from peroxides in yielding oxygen as well as hydrogen peroxide on hydrolysis.

superphosphate *(Chem.)*. Superphosphate of lime, an agricultural fertilizer; a mixture of calcium sulphate and dihydrogen calcium phosphate; made by treating bone ash or basic slag (calcium phosphate) with sulphuric acid.

super plastic forming/diffusion bonding *(Aero.)*. Method of manufacturing by joining parts of structures together at high temperature and pressure.

superposed circuit *(Telecomm.)*. An additional channel obtained from one or more existing circuits, normally provided for other channels, in such a way that all the channels can be used simultaneously without mutual interference.

superposition, law of *(Geol.)*. Strata which overlie other strata are always younger, except in strongly folded areas.

superposition theorem *(Elec.Eng.)*. That any voltage/ current pattern in a linear network is additive to any other voltage/current pattern.

superpressure *(Aero.)*. See under **superheat**.

super-refraction *(Meteor.)*. Refraction greater than standard refraction.

super-regeneration *(Telecomm.)*. Regeneration, or *feedback*, leading to oscillation which is broken up or *quenched* at a frequency above the upper limit of audibility by a separate oscillator circuit suitably connected to the main amplifying circuit. Amplifiers using this phenomenon can achieve extremely high gain and sensitivity with the minimum of circuit components.

super-regenerative receiver *(Telecomm.)*. One with sufficient positive feedback to result in a quenched supersonic oscillation (*squegging*), with consequent increase in sensitivity but also increase in distortion of demodulated signals.

supersaturation *(Chem.)*. Solution containing solute in excess of equilibrium. Condensation can take place on nuclei, particularly ions, e.g. those produced by high-speed charged particles, exhibiting a track of minute but visible water drops, as in a **Wilson chamber**.

supersonic *(Phys.)*. Faster than the speed of sound in that medium. Erroneously used for ultrasonic. See **Mach number, ultrasonic**.

supersonic boom *(Acous.)*. Shock wave produced by an object moving supersonically. At a large distance from the object the time history of the pressure has the shape of an N and is therefore called *N-wave*.

supersonic speed *(Aero.)*. Applies to aircraft when its speed exceeds that of local sound. Applies to airflow anywhere when local speed exceeds that of sound.

supersonic wind tunnel *(Aero.)*. A wind tunnel in which the stream velocity in the working section exceeds the local speed of sound.

super stall *(Aero.)*. This phenomenon appeared with the adoption of high tail planes for swept-wing jet aircraft. When the disturbed airflow from a stalled wing renders the tail controls inoperative, the aircraft will remain in a stable, substantially level attitude, while descending very rapidly. Recovery is by releasing the tail parachute to raise the tail clear of the wing wake so that the elevators again become operative.

superstitious behaviour in animals *(Behav.)*. Refers to behaviour that is produced by the joint action of **reinforcement** and accident; certain acts which happen to coincide with reinforcement will tend to increase; these are often of a bizarre and fixed nature.

superstructure *(Civ.Eng.)*. The part of a structure carried upon any main supporting level.

superstructure *(Ships)*. A decked structure on the **feedback deck** extending from side to side of the ship. An integral part of the hull.

supersulphated cement *(Civ.Eng.)*. Cement manufactured using a high proportion of granulated blast-furnace slag.

super-, supra *(Genrl.)*. Prefix from L. *super*, over, above.

supersymmetry *(Astron.,Phys.)*. Theory which attempts to link all four fundamental forces, and postulates that each force emerged separately during the expansion of the very early universe.

supervisor *(Comp.)*. See **monitor**.

supervisor program *(Comp.)*. See **monitor**.

supervisory control *(Elec.Eng.)*. A method of remote control of electrical plant from a distant centre in which back-indication of the several control operations is given to the control centre.

supervoltage therapy *(Radiol.)*. Application of voltage, over a million volts, to X-ray tubes or accelerators in therapy.

supination *(Zool.)*. In some higher Vertebrates, movement of the hand and forearm by which the palm of the hand is turned upwards and the radius and ulna are brought parallel to one another; cf. **pronation**. adj. *supinate*.

supinator *(Zool.)*. A muscle effecting supination.

supplemental chords *(Maths.)*. Of a circle, ellipse or hyperbola: the chords joining any point on the curve to the ends of any diameter. Diameters parallel to supplemental chords are conjugate (perpendicular for a circle).

supplemental, supplementary *(Zool.)*. Additional; extra; supernumerary, as (in some Foraminifera) *supplemental* skeleton, a deposit of calcium carbonate outside the primary shell.

supplementary angles *(Maths.)*. Two angles whose sum is 180°. Cf. **complementary angles**.

supplementary lens *(Image Tech.)*. A lens placed in front of the main lens of a camera to alter its effective focal length without changing the distance from the film plane. Also termed an *afocal attachment*.

supply frequency *(Elec.Eng.)*. The electrical frequency, in Hz, of an a.c. supply.

supply meter *(Elec.Eng.)*. An instrument for measuring the total quantity of electrical energy supplied to a consumer during a certain period.

supply point *(Elec.Eng.)*. A point on an electric power system from which electrical energy may be drawn.

supply station *(Elec.Eng.)*. See **generating station**.

supply terminals *(Elec.Eng.)*. Those at which connection may be made to a supply point.

supply voltage *(Elec.Eng.)*. The voltage across a pair of supply terminals.

suppository *(Med.)*. A conical or cylindrical plug of a medicated mass for insertion into the rectum, vagina or urethra.

suppressed carrier system *(Telecomm.)*. One in which the carrier wave is not radiated but is supplied by an oscillator at the receiving end. See **single sideband-, double sideband-systems**.

suppressed-zero instrument *(Elec.Eng.)*. An indicating or graphic instrument in which the zero position first scale reading is off scale, i.e. beyond the range of travel of the pointer. Also called *inferred-zero instrument, set-up instrument, set-up-scale instrument, set-up-zero instrument, step-up instrument*.

suppression *(Biol.)*. The process by which a *mutant* phenotype is restored to normal by a mutation at another locus. Cf. **reversion**.

suppression *(Comp.)*. Elimination of specified data or digits, e.g. initial zeros.

suppression *(Med.)*. Stoppage of discharge, as by obstruction of a duct.

suppression *(Zool.)*. Absence of some organ or structure normally present. adj. *suppressed*.

suppressor *(Elec.Eng.)*. (1) Component, usually a low resistance adjacent to an electrode of a valve, to obviate conditions which promote parasitic oscillations. (2) Component, such as a capacitor or resistor, or both, which damps high-frequency oscillation liable to arise on breaking a current at a contact, causing radio interference.

suppressor cell *(Immun.)*. A lymphoid cell capable of suppressing antibody production or a specific cell mediated response made by other cells. Suppression may be antigen-specific or non-specific.

suppressor grid *(Electronics)*. That between anode and screen in pentode valves, to repel secondary electrons back to the anode.

suppressor-grid modulation *(Electronics)*. Insertion of the signal voltage into the (suppressor) grid circuit of a valve which is amplifying the carrier. Also *grid modulation*.

suppressor mutation *(Biol.)*. A base change which suppresses the effect of mutations elsewhere. Thus a base change at the anticodon site of a tRNA can suppress lethal mutations in genes, which would otherwise result in chain termination or the insertion of an unacceptable peptide into a protein.

suppressor T-cell factor *(Immun.)*. Soluble product of a suppressor T-lymphocyte responsible for suppressing other lymphocyte functions.

suppuration *(Med.)*. The softening and liquefaction of inflamed tissue, with the production of pus. adj. *suppurative*.

supradorsal *(Zool.)*. On the back; above the dorsal surface; a dorsal intercalary element of the vertebral column.

supra-occipital *(Zool.)*. A median dorsal cartilage bone of the Vertebrate skull forming the roof of the brain case posteriorly.

suprarenal *(Zool.)*. Situated above the kidneys.

suprarenal body, gland *(Zool.)*. In higher Vertebrates, one of the endocrine glands lying close to the kidney and releasing into the blood secretions having important effects on the metabolism of the body; *adrenal gland*. See **adrenal cortex, adrenal medulla**.

supremum *(Maths.)*. Least upper bound.

suramin *(Med.)*. Non-arsenical drug mainly used to treat certain forms of trypanosomiasis.

surcharge *(Civ.Eng.)*. A term applied to the earth supported by a retaining wall at a level above the top of the wall.

surd *(Maths.)*. An imprecise term usually applied to irrational roots or to the sum of such roots, e.g. $\sqrt{2}$ or $\sqrt{3}+3\sqrt{5}$.

surface absorption coefficient *(Phys.)*. See **absorption coefficient**.

surface acoustic wave *(Radar,Telecomm.)*. Acoustic wave, which may have frequencies corresponding to the microwave bands, travelling along the optically polished surfaces of a piezo-electric substrate, at a velocity about 10^{-5} that of light. Used in microwave components and amplifiers. Abbrev. *SAW*. See **SAW delay line, SAW filter**.

surface activity *(Chem.)*. The influence of certain substances on the surface tension of liquids.

surface barrier *(Electronics)*. Potential barrier across surface of semiconductor junction due to diffusion of charge carriers.

surface boundary layer *(Meteor.)*. The atmospheric layer, extending to a height of about 100 m, in which the motion is controlled predominantly by the presence of the Earth's surface. It forms the lowest part of the **friction layer**.

surface charge *(Elec.Eng.)*. See **bound charge**.

surface chemistry *(Chem.)*. That of the interface or interphase between two systems in which the substrate of one or both has become ionically unbalanced.

surface combustion *(Eng.)*. Bringing a combustible mixture of gas and air into contact with a suitable refractory material so as to produce flameless or nearly flameless combustion, the surface of the refractory material being maintained in a state of incandescence.

surface compressibility *(Chem.)*. The compressibility of the layer of adsorbed molecules in response to surface tension forces.

surface concentration excess *(Chem.)*. The excess concentration (may be negative) of a solute per unit area in the surface layer of a solution.

surface condenser *(Eng.)*. A steam condenser for maintaining a vacuum at the exhaust pipe of a steam-engine or turbine. It consists of a chamber in which cooling water is circulated through tubes, and which is evacuated by an air pump. See **condenser, condenser tubes**.

surface conductivity *(Phys.)*. See **surface resistivity**.

surface density *(Phys.)*. The amount of electric charge/unit surface area.

surface duct *(Telecomm.)*. Atmospheric propagation duct for which the earth's surface forms the lower boundary. Cf. **waveguide**.

surface energy *(Phys.)*. The free potential energy of a surface equal to the **surface tension** multiplied by the surface area.

surface-friction drag *(Aero.)*. That part of the drag represented by the components of the pressures at points on the surface of an aerofoil, resolved tangential to the surface.

surface gauge *(Eng.)*. See **scribing block**.

surface-grinding machine *(Eng.)*. A grinding machine for finishing flat surfaces. It consists of a high-speed abrasive wheel, mounted above a reciprocating or rotating work-table on which flat work is held, often by a **magnetic chuck**.

surface hardening *(Eng.)*. **Cementation** of low-carbon steels. See **nitriding**.

surface irradiation *(Radiol.)*. Irradiation of a part of the body by applying a mould or applicator loaded with radioactive material to the surface of the body.

surface leakage *(Elec.Eng.)*. That along the surface of a nonconducting material or device. May vary widely with contamination, humidity etc. It sets a practical limit to the value of high resistors for use with electrometers, etc.

surface lifetime *(Electronics)*. The lifetime of current carriers in the surface layer of a semiconductor (where recombination occurs most readily). Cf. **volume lifetime**.

surface loading *(Aero.)*. The average force per unit area, normal to the surface, on an aerofoil under specified aerodynamic conditions.

surface measure *(For.)*. A method of measuring timber in quantity, by the area of one face, irrespective of thickness. Cf. **board measure**.

surface noise *(Acous.)*. (1) See **needle scratch**. (2) Underwater noise produced by waves on the sea surface.

surface of operation *(Build.)*. A surface which is dressed to a plane as a reference from which the rest of the work can be set out and executed.

surface oil resistance time *(Paper)*. Abbrev. *SORT*. An indication of the printing ink hold-out properties of a paper by measuring the resistance to penetration by a drop of liquid paraffin spread by a roller over a sample supported on an inclined plane, under the specified conditions of test. Results are expressed in seconds.

surface pipe *(Min.Ext.)*. See **anchor string**. Also *surface casing*.

surface plate *(Eng.)*. A rigid cast-iron plate whose surface is accurately scraped flat; used to test the flatness of other surfaces or to provide a truly plane datum surface in marking off work for machining.

surface plates *(Print.)*. A general name for litho plates which are not **deep etch** or **bimetallic**.

surface pressure *(Chem.)*. The 2-dimensional analogue of gas pressure. Defined as the difference between the surface tension of a pure liquid and that of a surface active solution, it represents the tendency of the adsorbed surfactant molecules to spread over the clean liquid surface. See **Gibb's adsorption theorem**.

surface recombination velocity *(Electronics)*. Electron-hole recombination on surface of semiconductor occurs more readily than in the interior, hence the carriers in the interior drift towards the surface with a mean speed termed the *surface recombination velocity*. It is defined as the ratio of the normal component of the impurity current to the volume charge density near the surface.

surface resistivity *(Phys.)*. That between opposite sides of a unit square inscribed on the surface. Its reciprocal is *surface conductivity*.

surface sterilization *(Phys.)*. Radiation with low-energy rays which penetrate thin surface layers only, e.g. with ultraviolet rays.

surface strength *(Paper)*. The resistance of a paper to an adhesive force acting normally to the surface.

surface tension *(Phys.)*. A property possessed by liquid surfaces whereby they appear to be covered by a thin elastic membrane in a state of tension, the surface tension being measured by the force acting normally across unit length in the surface. The phenomenon is due to unbalanced molecular cohesive forces near the surface. Units of measurement are dyne cm^{-1}, N m^{-1}. See **capillarity, liquid-drop model, pressure in bubbles**.

surface wave *(Acous.)*. Wave on the surface of liquids or solid bodies which have dimensions large compared with the wavelength. The amplitude is maximal at the surface and decays exponentially towards the interior of the body. The displacement of the medium particles is both longitudinal and transversal.

surface wave *(Phys.)*. (1) A wave propagated along the surface of a liquid. For deep water waves (the wavelength less than the water depth), the phase velocity depends on both gravitational forces and on surface tension and also on the wavelength. For shallow water waves (the wavelength greater than the depth), the phase velocity depends only on the depth and is independent of the wavelength. See **tsunami, ripple tank**. See also **wave**. (2) A component of an electromagnetic wave radiated from a relatively low antenna, which depends on the nature of the surface. See also **ground wave**.

surface wind *(Meteor.)*. The wind at a standard height of 10 m (33 ft) above ground. Differs from the **geostrophic wind** and the **gradient wind** because of friction with the earth's surface.

surface wiring *(Elec.Eng.)*. A wiring installation in which the insulated conductors are attached to the surfaces of a building, either enclosed in conduit or secured by cleats.

surfactant *(Chem.)*. An abbreviated form of *surface active agent*, i.e. a substance which has the effect of altering the interfacial tension of water and other liquids or solids, e.g. a detergent or soap.

surfactant *(Textiles)*. A surface-active agent. A compound that reduces the surface tension of its solvent, e.g. a detergent or soap dissolved in water.

surfactant flooding *(Min.Ext.)*. Recovery enhancement process in oil wells in which surface-tension reducing compounds are forced into the surrounding strata and release oil held there.

surge *(Aero.)*. Unstable airflow condition in the compressor of a gas turbine due to a sudden increase (or decrease) in mass airflow without a compensating change in pressure ratio.

surge *(Elec.Eng.)*. A large but momentary increase in the voltage of an electric circuit.

surge absorber *(Elec.Eng.)*. A circuit device which diverts, and may partly dissipate, the energy of a surge, thus preventing possible damage to apparatus or machines connected to a transmission line. Also *surge modifier*.

surge arrester *(Elec.Eng.)*. See **lightning arrester**.

surge bin, tank *(Min.Ext.)*. Hopper (dry material) or reservoir with means of agitation (ore pulps), used to minimize irregularities in process delivery and flow.

surge-crest ammeter *(Elec.Eng.)*. An instrument for recording a surge on a transmission line by measurement of the residual magnetism in a piece of magnetic material which has been magnetized by the surge current.

surge generator *(Elec.Eng.)*. See **impulse generator**.

surge impedance *(Phys.)*. See **characteristic impedance** (1).

surge modifier *(Elec.Eng.)*. See **surge absorber**.

surge point *(Autos.)*. Of a centrifugal supercharger, the value of the mass airflow at which, during throttling of the delivery, surging occurs. See **surging** (1).

surge tank *(Eng.)*. One used to absorb irregularities in flow.

surgical spirit *(Chem.)*. Ethanol, to which is added small amounts of oil of wintergreen and castor oil; used chiefly for sterilizing the skin in surgical operations.

surging *(Autos.)*. (1) In centrifugal superchargers, an abrupt decrease or severe fluctuation of the delivery pressure as the weight of air delivered is reduced. See **surge point**. (2) In valve springs, the coincidence of some harmonic of the cam lift curve with the spring's natural frequency of vibration, leading to irregular action and failure.

surmounted *(Arch.)*. A term applied to a vault springing from points below its centre and having the form of a circular arc above its centre.

surra *(Vet.)*. A form of trypanosomiasis affecting horses, dogs, cattle, elephants, and camels, occurring in Asia and the Sudan, caused by *Trypanosoma evansi*; symptoms include emaciation and subcutaneous oedema, usually fatal. Transmitted by biting flies.

surveillance radar *(Radar).* A plan position indicator radar showing the position of aircraft within an air traffic control area or zone.

surveillance TV *(Image Tech.).* Use of closed circuit TV for prevention and detection of crime, including video recording for evidence; in unattended areas it may employ **slow scan** methods and be activated by unexpected movement.

surveying *(Surv.).* Measurement of the relative positions of points on the surface of the earth and/or in space, to enable natural and artificial features to be depicted in their true horizontal and vertical relationship by drawing them to scale on paper.

survival curve *(Radiol.).* One showing the percentage of organisms surviving at different times after they have been subjected to large radiation dose. Less often, one showing percentage of survivals at given time against size of dose.

survivorship curve *(Ecol.).* The number or percentage of an original population surviving, plotted against time, giving an indication of the mortality rate at different ages.

susceptance *(Phys.).* Imaginary part of **admittance**, equal to $\frac{-X}{R^2+X^2}$ in a circuit of impedance $R + jX$.

susceptibility *(Elec.Eng.).* See **magnetic susceptibility**.

susceptibility curves *(Elec.Eng.).* Curves of susceptibility plotted to a base of magnetic field strength.

susceptor phase advancer *(Elec.Eng.).* A phase advancer which injects into the secondary circuit of an induction motor an e.m.f. which is a function of the open-circuit secondary e.m.f. Cf. **expedor phase advancer**.

suspended scaffold *(Build.).* A form of scaffold used in the construction, repair, cleaning etc. of buildings. It consists of working platforms, on light frameworks, slung from fixed higher points in the building.

suspended span *(Civ.Eng.).* The middle length of a bridge span connecting, and carried upon, the cantilever arms, in cases where these are not built out until they meet. See **cantilever bridge**.

suspension *(Autos.).* A system, primarily of springs (leaf or coil) and dampers (usually hydraulic), designed to support the body of a vehicle and to protect it, and hence its occupants, from road shocks. See **Hydrolastic**.

suspension *(Chem.).* A system in which denser particles, which are at least microscopically visible, are distributed throughout a less dense liquid or gas, settling being hindered either by the viscosity of the fluid or the impacts of its molecules on the particles.

suspension bridge *(Civ.Eng.).* A bridge in which the deck is suspended from massive cables supported by towers near the sides of the waterway and anchored at each end of the bridge. For large span bridges, experience has shown that the possibility of harmonic action must be taken into account and that wind tunnel experiments are needed to determine the necessary degree of lateral stiffness.

suspension cable anchor *(Civ.Eng.).* The anchorage, which may have various forms, of the cables of a suspension bridge.

suspension culture *(Biol.).* A method of culturing large quantities of cells which are kept in vessels continuously stirred and aerated. Sterile nutrient media can be added and spent media removed. Used, e.g. for producing cell products like *interferon* or *antibodies*.

suspension insulator *(Elec.Eng.).* A freely hanging insulator made of units connected in series, by which an overhead line is suspended from the arm of a transmission-line tower; it cannot withstand any other force than a tension.

suspension spring *(Horol.).* The thin ribbon of spring steel which supports a pendulum.

suspensoid *(Chem.).* See **lyophobic colloid**.

suspensor *(Bot.).* A file or files of cells that develop from the proembryo of a seed plant and anchors the embryo in the embryo sac and pushes it into the endosperm.

suspensory *(Zool.).* Pertaining to the suspensorium; serving for support or suspension.

suspensorium *(Zool.).* In Vertebrates, the apparatus by which the jaws are attached to the cranium.

Sussex garden-wall bond *(Build.).* The form of *garden-wall bond* in which one header and three stretchers are laid in each course.

sustained oscillations *(Telecomm.).* Externally-maintained oscillations of a system at or very near its natural resonant frequency. Cf. **forced oscillations, free oscillations**.

suture *(Bot.).* (1) The line at the junction of fused parts. (2) A line of weakness along which splitting may occur as in a dehiscent fruit.

suture *(Med.).* Surgical stitch, or group or row of such stitches.

suture *(Zool.).* A line of junction of two structures, as the line of junction of adjacent chambers of a Nautiloid shell; a synarthrosis or immovable articulation between bones, as between the bones of the cranium; junctions of exoskeletal cuticular plates in Insects. adj. *sutural*.

sutured *(Geol.).* A textural term descriptive of the sinuous interlocking grain boundaries of rocks which have undergone extensive recrystallization, e.g. quartzites.

Sv *(Radiol.).* Abbrev for *Sievert*.

SV 40 *(Biol.).* A small virus normally infecting monkey cells, either causing a lytic infection or being integrated into the host chromosome. It has a circular chromosome which has been fully sequenced, and vectors derived from it are used to transfer inserted DNA into mammalian cells.

SVD *(Vet.).* See **swine vesicular disease**.

swab *(Med.).* (1) Any small mass of cotton-wool or gauze used for mopping up blood or discharges, or for applying antiseptics to the body, or for cleansing surfaces (e.g. the lips, the mouth). (2) A specimen of a secretion taken on a swab for bacteriological examination.

swage *(Eng.).* Smith's tool (dolly) used in shaping metal.

swaging *(Eng.).* Reducing of cross-section of metal rod or tube by forcing it through a tapered aperture between two grooved dies.

swallowtail *(Build.).* See **dovetail**.

swamp fever *(Vet.).* See **equine infectious anaemia**.

Swan cube *(Phys.).* The prism system used in the Lummer-Brodhum photometer. Consists of two 45° prisms placed with their hypotenuse faces together, but with one face ground so that the prisms are touching only over the central part of the faces.

swan-neck *(Build.).* The bend formed in a hand-rail when a knee and a ramp are joined together without any intermediate straight length.

swan-neck chisel *(Build.).* Chisel curved for lock mortising.

swan-neck insulator *(Elec.Eng.).* A pin-type insulator with a bent pin, arranged so as to bring the insulator into approximately the same horizontal plane as that of the support.

swan-neck lock mortise chisel *(Build.).* A hook-shaped chisel for chopping lock mortises in doors.

swarf *(Eng.).* (1) The cuttings from a machining operation. (2) US for the mixture of abrasive, bond and metal particles formed during grinding.

swarm *(Zool.).* A large number of small animals in movement together; especially a number of Bees emigrating from one colony to establish another under the guidance of a queen.

swarm cell *(Bot.).* (1) Flagellated naked cell in Myxomycetes, interconvertible with myxamoeba, capable of encysting and of acting as an isogamete. (2) See **swarmer**.

swarmer *(Bot.).* (1) Flagellated reproductive cell, especially a zoospore. (2) See **swarm cell**.

swash bank *(Hyd.Eng.).* The upper part of the slope of a sea embankment. See **outburst bank**.

swash letters *(Print.).* Ornamental italic letters with tails and flourishes, such as $\mathscr{A}$, $\mathscr{B}$, etc. They should be used only at the beginning or end of a word.

swash plate *(Eng.).* A circular plate mounted obliquely on a shaft; sometimes used in conjunction with working cylinders mounted axially parallel with the shaft, as a substitute for an engine or pump crank mechanism.

swatch *(Textiles)*. A collection of small fabric samples.

swayback *(Vet.)*. *Enzootic ataxia.* A nervous disease of newborn and young lambs characterized by degenerative changes in the cerebrum of the brain, causing inability or difficulty in standing or walking. Associated with a low copper content of tissues in the lamb and its ewe and preventable by administering copper to the pregnant ewe.

sway rod *(Build.)*. A member inserted in a structural framework to resist wind forces. Also *wind bracing*.

sweat cooling *(Aero.)*. Cooling of a component by the evaporation of a fluid through a porous surface layer; used for high-performance gas turbine blades or hypersonic vehicles.

sweated joint *(Eng.)*. See under **sweating**.

sweating *(Build.)*. A term applied to a surface showing traces of moisture due either to formation of condensate or to water having got through a porous material of which the surface is part.

sweating *(Eng.)*. The operation of soldering pieces together by 'tinning' the surfaces and heating them while pressed into contact.

Swedish iron *(Eng.)*. Wrought iron of high purity, with some 300 MN/m^2 ultimate tensile strength, and 33% elongation before fracture.

Swedish standards *(Build.)*. A range of photographic standards for comparison with actual metal surfaces being prepared for painting. The photographic plates show varying degrees of corrosion from SA 1 to SA 3 which is white metal generally obtained by abrasive blasting. British equivalent standard being BS 4232. Standard referred to on painting specifications where specific preparation is required.

sweep *(Aero.)*. The angle, in plan, between the normal to the plane of symmetry and a specified spanwise line on an aerofoil. Most commonly, the quarter-chord line is used, but leading and trailing edges are sometimes stipulated. Sweep increases longitudinal stability by extending the centre of pressure and delays compressibility drag by reducing the chordwise component of the airflow. *Sweepback*, the more usual, is the aft displacement of the wings and *forward sweep* the opposite.

sweep *(Image Tech.)*. The movement of the electron beam across the surface of the CRT.

sweepback *(Aero.)*. Aircraft wings making an acute angle with the fuselage, the wing tips being towards the tail.

sweep circuit *(Electronics)*. That which supplies deflecting voltage to one pair of plates or coils of a cathode-ray tube, the other pair being connected to the source of current or voltage under examination. See **linear scan**.

sweeper *(Electronics)*. Frequency swept oscillator, particularly at microwave frequencies.

sweeps *(Eng.)*. Dust and debris in jeweller's workshops, gold refineries, bullion assay offices etc., collected and treated periodically to recover valuable contents.

sweep-saw *(Build.)*. A thin-bladed saw which is held taut in a special frame and may be used for making curved cuts. Also called *turning-saw*.

sweet *(Glass)*. Easily workable.

sweet clover disease *(Vet.)*. A fatal, haemorrhagic disease of cattle and other animals caused by feeding sweet clover (genus *Melilotus*), which has been damaged during haymaking or ensiling, causing the formation of a toxic substance, *dicoumarin*, which interferes with blood coagulation.

sweetening *(Image Tech.)*. Electronic enhancement of a video image by edge sharpening and noise reduction.

sweet roasting *(Eng.)*. Ignition of metal sulphides to remove bulk of sulphur and arsenic as gaseous fume, leaving the mineral as its oxide.

swell *(Acous.)*. (1) Mechanism for altering the volume of sound in an organ by opening or closing shutters. Operated by the *swell pedal*. (2) In an electronic organ, volume control operated by potentiometer.

swell *(Min.Ext.)*. Volumetric increase due to crushing, which creates more void space in a given weight of rock.

swelled head *(Vet.)*. See **big head disease of sheep**.

swelled rules *(Print.)*. Decorative rules in a variety of sizes with a central swelling, occasionally tooled; sometimes called *Bodoni rules*.

swelling *(Arch.)*. The slight bulge given to the profile of a column near the middle of its length, to correct for the apparent concavity which it would have if it were a straight taper.

swelling *(Nuc.Eng.)*. Change of volume of fuel rods which may occur during irradiation.

swell pedal *(Acous.)*. The foot-operated lever for regulating the loudness of stops drawn on the swell manual. See **grand swell**.

SWG *(Eng.)*. Abbrev. for *Standard Wire Gauge*.

swim bladder *(Zool.)*. See **air bladder**.

swimmerets *(Zool.)*. In some Crustacea, paired biramous abdominal appendages used in part for swimming.

swimming-pool reactor *(Nuc.Eng.)*. One in which the fuel elements are immersed in a deep pool of water which acts as coolant, moderator and shield. Also *pool reactor*.

swine erysipelas *(Vet.)*. *Diamond skin disease.* A disease of pigs caused by infection by *Erysipelothrix insidiosa* (*E. rhusiopathiae*). The acute form of the disease is a septicaemia characterized by fever, urticaria-like patches on the skin and sometimes lameness due to arthritis; in the chronic form endocarditis occurs. Vaccines and antisera widely used.

swine fever *(Vet.)*. *Hog cholera.* A highly contagious disease of pigs, caused by a togavirus and sometimes complicated by secondary bacterial infection; symptoms may include fever, diarrhoea, pneumonia, and nervous symptoms. Notifiable in UK with a slaughter policy for infected herds, but vaccines are available.

swine influenza *(Vet.)*. An acute viral infection of pigs characterized by fever, coughing, and respiratory distress. Enzootic pneumonia due to *Mycoplasma hyopneumoniae* and pleuropneumonia due to *Haemophilus pleuropneumoniae* give similar symptoms.

swine paratyphoid *(Vet.)*. A disease of pigs caused by *Salmonella cholerae suis*, and characterized by septicaemia in the acute form and necrotic enteritis in the chronic form. Vaccination widely used.

swine plague *(Vet.)*. Considered to be the respiratory symptoms of **swine fever** rather than a separate entity.

swine pox *(Vet.)*. *Variola.* A disease of pigs characterized by the formation of papules, vesicles, and pustules on the skin, caused by a virus related to that of **vaccinia**. A similar disease, in which papules and scabs develop, is caused by an unrelated virus.

swine vesicular disease *(Vet.)*. Caused by an enterovirus related to human *Coxsackie B5*. Symptoms include pyrexia, vesicle formation on coronary band of heels and pastern area, marked lameness and sometimes sloughing of hooves. Similar in appearance to **foot and mouth disease**. Notifiable in UK.

swing *(Aero.)*. The involuntary deviation from a straight course of an aircraft while taxiing, taking-off or alighting.

swing *(Eng.)*. Lathe dimension, equal to the radial size of the largest workpiece that can be rotated or swung in it.

swing *(Telecomm.)*. Extreme excursion from positive peak to negative peak in an alternating voltage or current waveform.

swing back *(Image Tech.)*. The back of a camera which can tilt upwards or sideways, or both, so that distortion of objects (such as the vertical lines of buildings) may be minimized, or objects at different distances may be brought into focus.

swing front *(Image Tech.)*. The provision for tilting the front of a camera, with the lens, so that distortion of an object due to its receding along the axis can be minimized.

swinging choke *(Elec.Eng.)*. Iron-cored inductor with saturable core, used in smoothing circuits where decreasing impedance with increasing current improves regulation.

swinging grippers *(Print.)*. Found on sheet-fed rotaries and on high-speed single- and 2-revolution machines, the sheets being transferred at speed from the feedboard to grippers on the continuously rotating impression cylinder.

swinging post *(Build.)*. See **hingeing post**.

swing-wing *(Aero.)*. See **variable sweep**.

swirl chamber *(Autos.)*. Type of diesel-engine combustion chamber, separated from the cylinder by a short vent and shaped so as to set up an eddy in the air drawn in through the intake valve.

swirl sprayers *(Aero.)*. Fuel injectors in a gas turbine which impart a swirling motion to the fuel.

swirl vanes *(Aero.)*. Vanes which impart a swirling motion to the air entering the flame tube of a gas-turbine combustion chamber.

Swiss lapis *(Min.)*. An imitation of lapis lazuli, obtained by staining pale-coloured jasper or ironstone with blue pigment. Also known as *German lapis*.

Swiss (Thury) screw-thread *(Eng.)*. A metric thread having a profile angle of 47°30′. The Thury screws are numbered exactly as BA screws, which are generally identical with the exception of 22, 23 and 25 (Thury).

Swiss-type automatic *(Eng.)*. *Single-spindle automatic*, in which several single-point tools are arranged radially around the stock, used primarily for small precision parts for watches and instruments.

switch *(Civ.Eng.)*. US for *turnout*. A device for moving a small section of a railway track so that rolling stock may pass from one line of track to another.

switch *(Elec.Eng.)*. Device for opening and closing an electric circuit.

switch *(Electronics)*. Electronic circuit for switching between two independent inputs, e.g. by valves, tubes or transistors.

switch *(Phys.)*. Reversal by saturation of the residual flux in a ferrite core.

switch base *(Elec.Eng.)*. The insulating base on which a switch is mounted.

switch blades *(Civ.Eng.)*. See **points**.

switchboard *(Elec.Eng.)*. An assembly of switch panels.

switchboard instrument *(Elec.Eng.)*. An electric measuring instrument arranged for mounting on a switchboard.

switchboard panel *(Elec.Eng.)*. See **panel**.

switch-box *(Elec.Eng.)*. An enclosure housing one or more switches operated by means of an external handle.

switch-desk *(Elec.Eng.)*. A control desk on which a number of miniature switches are mounted, each of which serves to initiate some control operation.

switch-fuse *(Elec.Eng.)*. A knife-switch carrying a fuse in each blade.

switchgear *(Elec.Eng.)*. The generic name for that class of electrical apparatus whose sole function is to open and close electric circuits.

switchgear pillar *(Elec.Eng.)*. See **pillar**.

switching *(Telecomm.)*. The provision of point-to-point connections between constantly changing sources of information and their intended recipients.

switching constant *(Elec.Eng.)*. The ratio of **switching time** and **magnetic-field intensity**.

switching time *(Elec.Eng.)*. Time required for complete reversal of flux in ferroelectric or ferromagnetic core.

switch panel *(Elec.Eng.)*. An insulating panel on which a switch is mounted.

switch plant *(Bot.)*. Plant with small, scale-like or fugacious leaves and long, thin, photosynthetic stems, e.g. Ephedra, Cytisus.

switch plate *(Elec.Eng.)*. A plate for covering one or more flush switches. Also called *flush plate*.

switch region *(Immun.)*. The sequence of amino acids at the junction of the variable and constant regions of immunoglobulin light or heavy chains. The sequence in this region is coded for by the D and J exons of the immunoglobulin gene. It determines what class of constant region is joined to the variable region.

switch-starter *(Elec.Eng.)*. A combination of knife-switch and starting regulator, in which the circuit is closed and the resistance progressively cut out in one continuous movement.

swivelling propeller *(Aero.)*. See **propeller**.

swivel-pin *(Autos.)*. See **king-pin**.

sycamore *(For.)*. A product of the largest hardwood trees (*Acer*), native to Europe and W. Asia; a hard and dense wood, used largely for internal work. See also **London plane**.

sycosis barbae *(Med.)*. Inflammation of the hair follicles of the beard region, due to infection with staphylococcus.

Sydenham's chorea *(Med.)*. See **chorea**.

syenite *(Geol.)*. A coarse-grained igneous rock of intermediate composition, composed essentially of alkali-feldspar to the extent of at least two-thirds of the total, with a variable content of mafic minerals, of which common hornblende is characteristic.

syenite-porphyry *(Geol.)*. An igneous rock of syenitic composition and medium grain size, commonly occurring in minor intrusions; it consists of phenocrysts of feldspar and/or coloured silicates set in a microcrystalline groundmass.

syenodiorite *(Geol.)*. An alternative name for *monzonite*.

syllable articulation *(Telecomm.)*. See **articulation**.

sylphon bellows *(Eng.)*. A thin-walled cylindrical metal bellows consisting of a number of elements arranged concertina-fashion, responding to external or internal fluid pressure; used in pressure-governing systems. See also **bellows**.

sylvanite *(Min.)*. Telluride of gold and silver, which crystallizes in the monoclinic system and is usually associated with igneous rocks and, in veins, with native gold. It is an ore of gold.

Sylvian aqueduct *(Zool.)*. In Vertebrates, the cavity of the mesencephalon.

Sylvian fissure *(Zool.)*. In Mammals, a deep lateral fissure of the cerebrum.

sylvine, sylvite *(Min.)*. Potassium chloride, which crystallizes in the cubic system. It occurs in bedded salt deposits (*evaporites*), and as a sublimation product near volcanoes; it is a source of potassium compounds, used as fertilizers.

sylvinite *(Min.)*. A general name for mixtures of the two salts sylvine and halite, the latter predominating, occurring at Stassfurt and elsewhere. Also used as a commercial name for *sylvine*.

sym- *(Genrl.)*. See **syn-**.

symbiosis *(Behav.)*. See **social symbiosis**.

symbiosis *(Biol.)*. An intimate partnership between two organisms (*symbionts*), in which the mutual advantages normally outweigh the disadvantages. See also **mutualism**.

symblepharon *(Med.)*. Adhesion of the eyelid to the globe of the eye.

symbol *(Behav.)*. Words, objects, events which represent or refer to something else, established by convention. In psychoanalytic theory, a symbol's referent is unconscious, and the meaning is hidden from the individual's unconscious awareness (e.g. in dreams).

symbol *(Chem.)*. See **chemical symbol**.

symbolic address *(Comp.)*. In an **assembly language** program the **address** of a store location, identified by means of a symbol rather than an **absolute address**. See **address calculation**.

symbolic logic *(Maths.)*. See **mathematical logic**.

symbolic method *(Elec.Eng.)*. A method of a.c. circuit analysis using complex numbers to represent the circuit voltages and currents. *In phase* or *real* components are plotted in the direction of the x-axis and *quadrative* or *unreal* components are plotted in the direction of the y-axis. The operator j (equals $\sqrt{-1}$ and signifying vector rotation by 90° in an anticlockwise direction) is used to identify quadrative components. See **complex numbers**.

symbolic programming *(Comp.)*. See **assembly language**.

symbol table *(Comp.)*. Table maintained by a compiler or assembler relating names to machine addresses. Also *name table*.

symmetrical *(Bot.,Zool.)*. See **actinomorphic**.

symmetrical components *(Elec.Eng.)*. Method of calculating voltages and currents in an unbalanced three phase network in which quantities are represented by combinations of symmetrical positive, negative and zero phase sequence components.

symmetrical deflection *(Electronics)*. The application of a voltage to a pair of deflection plates in a CRO such that they vary symmetrically above and below an average value, which is equal to the final anode potential of the tube. This procedure minimizes the possibility of trapezoidal distortion of the screen image.

symmetrical flutter *(Aero.)*. See **flutter**.

symmetrical grading *(Telecomm.)*. Grading in which all groups of selectors are equally favoured in seeking outlets.

symmetrical network *(Elec.Eng.)*. One which can be divided into two mirror half-sections.

symmetrical short-circuit *(Elec.Eng.)*. An a.c. short-circuit in which each phase carries the same current.

symmetrical winding *(Elec.Eng.)*. A term applied to an armature winding which fulfils certain conditions of electrical symmetry.

symmetric dyadic *(Maths.)*. See **conjugate dyadics**.

symmetric relation *(Maths.)*. A relation is symmetric if, when it applies from x to y, then it applies from y to x; e.g. 'is parallel to', 'is the complement of'. 'Is greater than' is not a symmetric relation.

symmetry *(Crystal.)*. The quality possessed by crystalline substances by virtue of which they exhibit a repetitive arrangement of similar faces. This is a result of their peculiar internal atomic structure, and the feature is used as a basis of crystal classification.

symmetry *(Maths.)*. (1) A geometric configuration is said to be symmetrical about a point, line or plane if any line through the point or any line perpendicular to the line or plane cuts the configuration in pairs of points equally spaced from the point, line or plane, which are then referred to respectively as the *centre of symmetry*, the *axis of symmetry* and the *plane of symmetry*. (2) A function of several variables is symmetric if it is unchanged when any two of the variables are interchanged.

symmetry *(Zool.)*. (1) The method of arrangement of the constituent parts of the animal body. (2) In higher animals, the disposition of such organs as show bilateral or radial symmetry.

symmetry class *(Crystal.)*. Crystal lattice structures can show 32 combinations of symmetry elements, each combination forming a possible symmetry class.

sympathectomy, sympatheticectomy *(Med.)*. Excision or cutting of a part of a sympathetic nerve.

sympathetic nervous system *(Zool.)*. In some Invertebrates (Crustaceans and Insects), a part of the nervous system supplying the alimentary system, heart, and reproductive organs and spiracles. In Vertebrates, a subdivision of the autonomic nervous system, also known as the *thoracicolumbar system*. The action of these nerves tends to increase activity, speed the heart and circulation, and slow digestive processes. Cf. **parasympathetic nervous system**.

sympathetic ophthalmia *(Immun.)*. Perforating injury to one eye may be followed by inflammatory disease in the sound eye, characterized by lymphocyte infiltration and granuloma formation, especially in the uveal tract. This is postulated as due to development of cell mediated immunity against uveal pigment or other antigen liberated from the damaged eye, which is normally sequestered away from the immune system.

sympathetic reaction *(Chem.)*. See **induced reaction**.

sympathomimetic *(Med.)*. Mimicking the sympathetic nervous system, i.e. used of drugs which produce effects like those produced by stimulation of the sympathetic nervous system, particularly adrenergic and noradrenergic effects.

sympathomimetics *(Med.)*. Class of drug which mimics the stimulation of the sympathetic nervous system to produce **tachycardia** and increase output from the heart *(isoprenaline)*. Some members can increase heart output without tachycardia *(dobutamine)*. Others mimic the β_2-sympathetic stimulation to produce bronchodilatation and vasodilatation *(salbutamol)*.

sympatric *(Bot.)*. Two species or populations having a common or overlapping geographical distribution(s). Cf. **allopatric**.

sympetalous *(Bot.)*. Gamopetalous. See **gamopetaly**.

symphysiotomy, symphyseotomy *(Med.)*. The operation of cutting through the pubic joint to facilitate the birth of a child.

symphysis *(Zool.)*. Union of bones in the middle line of the body, by fusion, ligament, or cartilage, as the mandibular *symphysis*, the pubic *symphysis*; growing together or coalescence of parts, as acrodont teeth with the jaw; the point of junction of two structures; chiasma; commissure. adj. *symphysial*.

symplast *(Bot.)*. The continuum of protoplasts, linked by plasmodesmata and bounded by the plasmalemma.

symplastic *(Bot.)*. Pertaining to the **symplast**. See also **symplastic growth**.

symplastic growth *(Bot.)*. *Co-ordinated growth*. Type of growth of a tissue, in which touching walls grow equally so that cell contacts persist. Cf. **intrusive growth, sliding growth**.

sympodial growth *(Bot.)*. *Definite growth, determinate growth*. Pattern of growth in which, after a period of extension, a shoot ceases to grow and one or more of the lateral buds next to the apical bud grow out and repeat the pattern. Cf. **monopodial growth**. See also **cymose inflorescence**.

sympodium *(Bot.)*. A branch system that shows **sympodial growth**.

symptom *(Med.)*. Evidence of disease or disorder as experienced by the patient (e.g. pain, weakness, dizziness); any abnormal sensation or emotional expression or thought accompanying disease or disorder of the body or the mind; less accurately, any objective evidence of disease or bodily disorder. Cf. **sign**.

symptomatology *(Med.)*. The study of symptoms; a discourse or treatise on symptoms; the branch of medical science concerning symptoms of disease.

synaldoximes *(Chem.)*. The stereoisomeric forms of aldoximes in which the H and OH groups are on the same side of the plane of the double bond.

synandrium *(Bot.)*. A group of united anthers or microsporangia.

synandrous *(Bot.)*. Having the stamens united to one another.

synangium *(Bot.)*. A number of sporangia fused into a single structure.

synapse *(Med.)*. The region where one nerve cell makes functional contact with another, and where nerve impulses can pass from one to the other by the diffusion across the small gap between the pre- and post-synaptic neurons of small quantities of specific transmitter substances (e.g. acetylcholine, noradrenaline) which are released from the presynaptic nerve terminals and act on the postsynaptic membrane.

synapsid *(Zool.)*. In the skull of Reptiles, the condition when there is one temporal vacuity, this being low behind the eye, with the post-orbital and squamosal meeting above. Found in Pelycosauria and mammal-like Reptiles. Cf. **diapsid**.

Synapsida *(Zool.)*. A subclass of Reptiles, often mammal-like, with a single lateral temporal vacuity, primitively lying below the post-orbital and the squamosal. The brain case was high, and the inner ear low down. Teeth in more advanced forms were heterodont, the lower jaw was flattened from side to side, and the dentary was relatively large. There were both a coracoid and pre-coracoid. Permian to Jurassic.

synapsis *(Biol.)*. Pairing of strictly homologous regions of homologous chromosomes during meiosis.

synaptic vesicles *(Zool.)*. Structures about 50 nm in diameter found in presynaptic nerve terminals and concerned with the storage of the chemical transmitters.

synaptonemal complex *(Biol.)*. Ladder-like structure of DNA and protein observed to lie between the synapsed homologues of a **pachytene** bivalent in first meiotic prophase. Essential for crossing-over and chiasma formation to occur.

synarthrosis *(Zool.)*. An immovable articulation, especially an immovable junction between bones. Cf. **amphiarthrosis, diarthrosis**.

sync *(Image Tech.)*. See **synchronization**.
syncarpous *(Bot.)*. A **gynoecium** consisting of 2 or more fused carpels.
syncaryon *(Biol.)*. See **synkaryon**.
synchondrosis *(Zool.)*. Connection of two bones by cartilage, usually with little possibility of relative movement.
synchro *(Telecomm.)*. A general term used for a family of self-synchronous angle data transmitters and receivers. Also called *selsyn*.
synchrocyclotron *(Phys.)*. A *cyclotron* in which the frequency of the accelerating voltage is varied to ensure that, despite the relativistic increase of mass of the particle with speed, the particles still arrive in synchronism with the accelerating voltages. Energies up to 700 MeV for protons can be achieved. The output is not continuous but is emitted in bursts of particles lasting about 100 μs.
synchromesh gear *(Autos.)*. A gear in which the speeds of the driving and of the driven members which it is desired to couple are first automatically synchronized by small **cone clutches** before engagement of the dogs or splines, thus avoiding shock and noise in gear-changing.
synchronism *(Telecomm.)*. Said of two signals of the same frequency when the phase angle between them is zero, or of pulsed signals which are in step with each other.
synchronization *(Image Tech.)*. In general, the matching of signals in precise time relation; in TV, especially establishing the identity of scanning frequency and phase of picture signals between transmitter and receiver. Abbrev. *sync*.
synchronization of oscillators *(Telecomm.)*. Phenomenon when two oscillators, having nearly equal frequencies, are coupled together. When the degree of coupling reaches a certain point, the two suddenly pull into step.
synchronized clock *(Horol.)*. One driven by any means but having its accuracy corrected electrically at given intervals.
synchronizer *(Comp.)*. A unit used to maintain synchronism when transmitting information between two devices. It may merely control the speed of one (e.g. by clutch); or if the speeds are very different, may include buffer storage.
synchronizer *(Elec.Eng.)*. See **synchroscope**.
synchronizing *(Elec.Eng.)*. The operation of bringing a machine into synchronism with an a.c. supply.
synchronizing power *(Elec.Eng.)*. The power developed in a synchronous machine that keeps it in synchronism with the a.c. supply system to which it is connected.
synchronizing torque *(Elec.Eng.)*. The torque power that keeps a synchronous machine in synchronism with the a.c. supply system to which it is connected.
synchronometer *(Elec.Eng.)*. A device which counts the number of cycles in a given time. If the time interval is unity the device becomes a digital **frequency meter**.
synchronous-asynchronous motor *(Elec.Eng.)*. A slip-ring type of induction motor whose rotor is fed from a d.c. exciter coupled to it. The machine operates asynchronously during starting-up, and runs on load as a synchronous motor.
synchronous booster *(Elec.Eng.)*. An a.c. generator coupled to a synchronous converter and having its armature connected in series with that of the converter.
synchronous capacitor *(Elec.Eng.)*. A lightly loaded synchronous motor supplying a leading current for power-factor correction.
synchronous capacity *(Elec.Eng.)*. The synchronizing power of an interconnector linking two a.c. power systems. It is defined as the change of kilowatts transmitted over the interconnector per radian change of angular displacement of the voltages of the two systems.
synchronous carrier system *(Telecomm.)*. Simultaneous broadcasting by two or more transmitters having the same carrier frequency, the various drive circuits being interlocked so as to avoid heterodyne beats between them.
synchronous clock *(Horol.)*. One operated from the electric mains, the motor being in step with the mains frequency.
synchronous computer *(Comp.)*. One in which all operations are timed by a master **clock**.
synchronous converter *(Elec.Eng.)*. A synchronous machine for converting polyphase alternating current to direct current. It comprises a double-purpose armature, rotating within a salient-pole direct-current field system.
synchronous detector *(Telecomm.)*. One which inserts a missing carrier signal in exact synchronism with the original carrier at the transmitter. Particularly important in colour television signals for the extraction of the colour information. May also refer to receivers which have a *matched filter* for the selective detection of signals coded in a certain way.
synchronous gate *(Comp.)*. Gate controlled by clock pulses and used to synchronize operations.
synchronous generator *(Elec.Eng.)*. See **alternator**.
synchronous homodyne *(Telecomm.)*. Reception in which the incoming modulated carrier has added to it a local oscillation of correct phase, with possible locking.
synchronous impedance *(Elec.Eng.)*. The ratio of the open-circuit e.m.f. to the short-circuit current of a synchronous machine, both values having reference to the same field excitation.
synchronous induction motor *(Elec.Eng.)*. An induction motor in which a direct current is passed into the rotor winding after it has run up to speed, so that, after starting as an induction motor (with a high starting torque), it runs as a synchronous motor.
synchronous machine *(Elec.Eng.)*. An a.c. machine which rotates at a constant speed which is harmonically related to the frequency of the supply to which it is connected. If the machine is two pole, it will rotate at the supply frequency: if 4 pole, at half supply frequency and so on.
synchronous motor *(Elec.Eng.)*. A.c. electric motor designed to run in synchronism with supply voltage.
synchronous orbit *(Space)*. Circular orbit of a satellite of a body, moving in the same direction and with the same period as the parent body. If the latter is the Earth, the term *geosynchronous* is used.
synchronous phase modifier *(Elec.Eng.)*. A large synchronous machine used solely for varying the power factor at the receiving end of a transmission line to maintain the voltage constant under all conditions of loading.
synchronous reactance *(Elec.Eng.)*. The vector difference between the synchronous impedance and the effective armature resistance of a synchronous machine.
synchronous transmission *(Comp.)*. Method of transmitting data between two devices which are operating continuously and are controlled by the same **clock**.
synchronous watt *(Elec.Eng.)*. A unit of torque used loosely in connection with a.c. machines. It is defined as the torque which, at the synchronous speed of the machine, would develop a power of one watt.
synchroscope *(Elec.Eng.)*. An instrument indicating the difference in frequency between two a.c. supplies. Also *synchronizer*.
synchrotron *(Phys.)*. Machine for accelerating charged particles to very high energies. The particles move in an orbit of constant radius guided by a magnetic field. The acceleration is provided at one point in their orbit by a high frequency electric field whose frequency increases to insure that particles of increasing velocity arrive at the correct instant to be further accelerated. Proton synchrotrons can produce energies greater than 200 GeV. Electron synchrotrons give energies up to 12 GeV. See **betatron, cyclotron**.
synchrotron radiation *(Phys.)*. (1) Electrons accelerated in a *synchrotron* produce a very intense, highly collimated, polarized beam of electromagnetic radiation, whose wavelength ranges continuously from 10^{-2} mm to 10^{-2} nm. Used with a monochromator, it is an important source for research purposes. (2) A theory of the origin of cosmic radio waves; it is suggested that the electrons

moving in an orbit in a magnetic field are accelerated as in a synchrotron, but on a vastly larger scale.

synchysis *(Med.)*. Abnormal softening and fluidity of the vitreous body of the eye.

syncline *(Geol.)*. A concave-upwards fold with the youngest rocks in the centre.

synclitism *(Med.)*. The compensatory difference in the rates of descent of the anterior and posterior portions of the presenting foetal part in the pelvis during labour.

sync modulation *(Image Tech.)*. The range of modulation depth reserved for the synchronizing pulses, as distinct from that for picture (video) signals.

syncope *(Med.)*. A fainting attack or sudden loss of consciousness due to sudden reduction of blood flow to the brain, as a result of rhythm disturbance of the heart or mechanical obstruction to the pump action of the heart.

sync pulse *(Image Tech.)*. Pulse transmitted at the beginning of each line and field to ensure correct scanning rate on reception.

syncytium *(Biol.)*. Tissue containing many nuclei, which is not divided into separate compartments by cell membranes. adj. *syncytial.*

syndactyl *(Zool.)*. Showing fusion of two or more digits, as some Birds. n. *syndactylism.*

syndesmochorial placenta *(Zool.)*. A chorio-allantoic placenta in which the uterine epithelium disappears so that the chorion is in contact with the endometrium or glandular epithelium of the uterus. Usually cotyledonary, e.g. sheep.

syndesmosis *(Zool.)*. Connection of two bones by a ligament, usually with little possibility of relative movement.

syndiazo compounds *(Chem.)*. The stereoisomeric forms of diazo compounds in which the groups attached to the nitrogen atoms are on the same side of the plane of the double bond.

syndiotactic *(Plastics)*. Term describing very long chain hydrocarbon polymers in which the substituent groups lie alternately on each side of the main chain. See also **atactic, isotactic**.

syndrome *(Med.)*. A concurrence of several symptoms or signs in a disease which are characteristic of it, but do not in themselves constitute a disease and may be associated with several conditions; a set of concurrent symptoms or signs.

syndyotaxy *(Chem.)*. Polymerization exhibiting regular alternation of differences in stereochemical structure. adj. *syndyotactic*. Cf. **isotaxy**.

synechia *(Med.)*. A morbid adhesion of the iris of the eye to the cornea or to the lens.

synecology *(Ecol.)*. The study of relationships between communities and their environment. Cf. **autecology**.

syneresis *(Chem.)*. Spontaneous expulsion of liquid from a gel.

synergic, synergetic *(Zool.)*. Working together; said of muscles which cooperate to produce a particular kind of movement.

synergid *(Bot.)*. Either of the two nuclei that with the egg nucleus constitute the egg apparatus at the micropylar end of the embryo sac of an angiosperm.

synergism *(Biol.)*. (1) The condition in which the result of the combined action of two or more agents, e.g. two growth substances, is greater than the sum of their separate, individual actions. (2) A type of **social facilitation** in which the nearby presence of another organism enhances the efficiency or intensity of a physiological process or behaviour pattern in an individual.

synergist *(Chem.)*. Substance which increases the effect of another.

syngamiasis *(Vet.)*. Infection of the trachea, bronchi, and lungs of Birds by the nematode worm, *Syngamus trachea.*

syngamy *(Biol.)*. Sexual reproduction; fusion of gametes.

syngeneic *(Immun.)*. Genetically identical. Usually applied to grafts made or cells transferred within an inbred strain.

syngenesis *(Bot.)*. Lateral fusion of plant members, as the anthers, which unite laterally to form a hollow tube round the style in the Compositae.

syngenetic *(Min.)*. A category of ore bodies comprising all those which were formed contemporaneously with the enclosing rock. Cf. **epigenetic**.

syngnathous *(Zool.)*. Of certain Fish, having the jaws fused to form a tubular structure.

synkaryon *(Biol.)*. A pair of nuclei in close association in a fungal hypha, dividing together to give the same close association. In animals a zygote nucleus resulting from the fusion of two pronuclei.

synkinesia *(Med.)*. The occurrence simultaneously of both voluntary and involuntary movements, e.g. in the movement of a partly-paralysed muscle in conjunction with the voluntary movement of healthy muscle.

synodic month *(Astron.)*. The interval (amounting to 29.530 59 days) between two successive passages of the moon through conjunction or opposition respectively; therefore, the period of the phases.

synodic period *(Astron.)*. An interval of time between two similar positions of the moon or a planet, relative to the line joining the Earth and Sun; hence the length of time from one conjunction or opposition to another, and the period of the phases of the Moon or a planet.

synoptic chart *(Meteor.)*. See **weather map**.

synoptic meteorology *(Meteor.)*. That part of the science of meteorology which deals with the preparation of a *synoptic chart* of the observed *meteorological elements* and, from consideration of this chart, the production of a weather forecast.

synosteosis *(Zool.)*. See **ankylosis**.

synovia *(Zool.)*. In Vertebrates *glair*-like lubricating fluid, occurring typically within tendon sheaths and the capsular ligaments surrounding movable joints.

synovial membrane *(Zool.)*. The delicate connective tissue layer which lines a tendon sheath or a capsular ligament, and is responsible for the secretion of the **synovia**.

synovitis *(Med.)*. Inflammation of the synovial membrane of a joint.

synroc *(Nuc.Eng.)*. Nuclear waste products incorporated into a mixture of crystalline structures known to be stable over geological time.

synsacrum *(Zool.)*. In Birds, part of the pelvic girdle formed by the fusion of some of the dorsal and caudal vertebrae with the sacral vertebrae.

syn-, sym- *(Chem.)*. See **s-** (1), (3).

syn-, sym- *(Genrl.)*. Prefix from Gk. *syn.* with, generally signifying fusion or combination.

syntax *(Comp.)*. Set of rules for combining the elements of a programming language (e.g. words) into permitted constructions (e.g. program statements). See **semantics**.

syntax analysis *(Comp.)*. Second stage during **compilation** where language statements are checked for compliance with the rules of the language.

syntax error *(Comp.)*. One that results from an incorrect use of the rules governing the structure of the language.

syntechnic *(Zool.)*. Said of unrelated forms showing resemblance due to environmental factors; convergent.

syntenosis *(Zool.)*. Union of bones by means of tendons, as in the phalanges of the digits.

synthetic aperture radar *(Radar)*. One in which an aircraft flying a straight path emits pulses continuously and at a precisely controlled frequency such that the transmitted power is coherent. All the echoes are processed in such a way as to simulate an antenna with an aperture as long as the flight path. Extremely fine resolution is attainable, giving fine detail; used for terrain mapping etc.

synthetic fibre *(Textiles)*. See **fibre, synthetic**.

synthetic oligonucleotide *(Biol.)*. Defined DNA sequence chemically polymerized *in vitro*.

synthetic paints *(Build.)*. Paints which contain a proportion of synthetic resin.

synthetic paper *(Paper)*. Paper made by conventional means but utilizing man made fibre as the whole or a substantial part of the furnish. Generally a paper of considerable durability, permanence and chemical resistance. The term is (incorrectly) applied to other sheet materials of similar appearance to paper and used for similar purposes.

synthetic-resin adhesive *(Plastics)*. One made from a thermosetting resin, e.g. urea or phenol formaldehyde, and used with an accelerator to regulate setting conditions, or from a thermo-plastic resin, e.g. polymethyl methacrylate or polyvinyl acetate.

synthetic resins *(Chem.)*. Resinous compounds made from synthetic materials, as by the condensation or polymerization of phenol and formaldehyde, formaldehyde and urea, glycerol and phthalic anhydride, polyamides, vinyl derivatives etc. See **phenolic resins**, **thermosetting compositions**, **vinyl polymers**.

synthetic rubber *(Plastics)*. Any of numerous artificial rubberlike compounds, e.g. polymers of isoprene or its derivatives, copolymers of vinyl acetate and vinyl chloride. In some respects they surpass natural rubber. See **Buna butyl rubber**, **Neoprene**, **Thiokol**, **vinyl polymers**.

synthetic ruby, synthetic sapphire *(Min.)*. In chemical composition and in all their physical characters, including optical properties, these stones are true crystalline ruby or sapphire; but they are produced in quantity in the laboratory by fusing pure precipitated alumina with the predetermined amount of pigmentary material. They can be distinguished from natural stones only by the most careful expert examination.

synthetic sands *(Foundry)*. Sands deficient in clay which have been blended with bentonite or other claylike material to make them suitable for moulding.

synthetic spinel *(Min.)*. This is produced, in a wide variety of fine colours, by the **Verneuil process**; in chemical and optical characters identical with natural magnesian spinel, it is widely used as a gemstone.

syntony *(Elec.Eng.)*. See **current (voltage) resonance**.

synusia *(Ecol.)*. A group of plants with similar life form and of the same or unrelated species, occupying a similar habitat, e.g. woodland herbs.

syphilide, syphilid *(Med.)*. Any skin affection caused by syphilis. Also *syphiloderm*, *syphiloderma*.

syphilis *(Med.)*. A contagious venereal disease due to infection with the micro-organism *Spirochaeta pallida* (*Treponema pallidum*); contracted in sexual intercourse, by accidental contact, or (by the foetus) from an infected mother.

syphiloma *(Med.)*. A syphilitic tumour. See **gumma**.

syphon *(Genrl.)*. Same as *siphon*.

Syrian garnet *(Min.)*. A name for **almandine** of gemstone quality.

syringitis *(Med.)*. Inflammation of the Eustachian tube.

syringobulbia *(Med.)*. A disease characterized by increase of neuroglia and the presence of cavities in the medulla oblongata, giving rise to such nervous phenomena as paralysis of the palate, pharynx and larynx. See also **syringomyelia**.

syringomyelia *(Med.)*. A chronic, progressive disease of the spinal cord in which increase of neuroglia and the formation of irregular cavities cause paralysis and wasting of muscles and loss of skin sensibility to pain and to temperature. See also **syringobulbia**.

syringomyelocele *(Med.)*. A form of spina bifida in which the part protruding through the defective spinal column consists of the greatly distended central canal of the spinal cord.

syrinx *(Med.)*. A fistula, or a fistulous opening.

syrinx *(Zool.)*. The vocal organs in Birds, situated at the posterior end of the trachea. pl. *syringes*. adj. *syringeal*.

systaltic *(Zool.)*. Alternately contracting and dilating; pulsatory, as the movements of the heart. Cf. **peristaltic**. n. *systalsis*.

system *(Biol.)*. (1) Tissues of the same histological structure, e.g. the osseous system. (2) Tissues and organs uniting in the performance of the same function, e.g. the digestive system. (3) A method or scheme of classification, e.g. the *Linnaean system*. (4) A systematic treatise on the animal or plant kingdom or any part of either. adj. *systematic*.

system *(Chem.)*. A portion of matter, or a group or set of things that forms a complex or connected whole.

system *(Elec.Eng.)*. General term used to describe (1) an entire arrangement of equipment, e.g. the grid system; (2) collection of standards or definitions e.g. SI system; (3) a set or field of technology, e.g. digital systems.

system *(Genrl.)*. Generally, anything formed of parts placed together or adjusted into a regular or connected whole.

system *(Geol.)*. (1) Name given to the succession of rocks which were formed during a certain period of geological time, e.g. *Jurassic System*. (2) Term applied to the sum of the phases which can be formed from one or more components of minerals under different conditions of temperature, pressure, and composition.

systematic *(Bot.)*. Affecting the whole organism: (1) an infection in which the pathogen has spread throughout the host; (2) an insecticide, fungicide etc. which following local application, spreads throughout a plant.

systematic errors *(Civ.Eng.,Maths.)*. Errors which are always in the same direction, i.e. errors which are always positive or always negative. Sometimes known as *cumulative errors*. In calculations such errors can arise e.g. by always rounding fives upwards.

systematics *(Biol.)*. The branch of biology which deals with classification and nomenclature.

system building *(Build.,Civ.Eng.)*. Methods designed to increase the speed of construction by preparing component parts of the building in a factory before assembly on site. Sometimes called *industrialized building*.

system crash *(Comp.)*. Occurs when the operating system ceases to be able to control the operation of the computer and human intervention is necessary in order to restart.

Système International d'Unités *(Genrl.)*. See **SI Units**.

system engineering *(Space)*. A logical process of activities which transforms a set of **requirements** arising from a specific mission objective into a full description of a system which fulfills the objective in an optimum way. It ensures that all aspects of a project have been considered and integrated into a consistent whole.

systemic *(Zool.)*. Pertaining to the body as a whole, not localized, as the *systemic* circulation.

systemic arch *(Zool.)*. In Vertebrates, the main vessel or vessels carrying blood from the heart to the body as a whole.

systemic lupus erythematosus *(Immun.)*. Disease of humans characterized by widespread focal degeneration of connective tissue and disseminated lesions in many tissues including skin, joints, kidneys, pleura, peripheral vessels, peripheral nervous system and transient abnormalities of the central nervous system. It may follow the administration of drugs or other antigenic substances but the initiating factor is usually unknown. Numerous autoantibodies are present in the blood, of which the most constant are anti-nuclear antibodies. The lesions are mainly the result of the deposition of immune complexes.

systems analysis *(Genrl.)*. Complete analysis of all phases of activity of an organization, and development of a detailed procedure for all collection, manipulation and evaluation of data associated with the operation of all parts of it.

systems analysis and design *(Comp.)*. Feasibility study of a potential computer involvement and the design of appropriate system to do a job.

systems analyst *(Comp.)*. Person responsible for the analysis of a project to assess its suitability for computer application and who may also design the necessary computer system.

systems of crystals *(Crystal.)*. The 7 large divisions into which all crystallizing substances can be placed, viz cubic, tetragonal, hexagonal, trigonal, orthorhombic, monoclinic, triclinic. This classification is based on the degree of **symmetry** displayed by the crystals.

systems programmer *(Comp.)*. Programmer who writes **systems software**.

systems software *(Comp.)*. The collection of programs

which make the computer system useable and control its performance.

System X *(Telecomm.)*. Name given by *British Telecom plc* to describe their fully-electronic computerized exchange switching system.

systole *(Bot.,Zool.)*. Rhythmical contraction, as of the heart, or of a contractile vacuole. Cf. **diastole**.

systole *(Med.)*. The period when a chamber of the heart is contracting.

systyle *(Arch.)*. A colonnade in which the space between the columns is equal to twice the lower diameter of the columns.

syzygy *(Astron.)*. A word applied to the moon when in conjunction or opposition.

Szilard-Chalmers process *(Phys.)*. One in which a nuclear transformation occurs with no change of atomic number, but with breakdown of chemical bond. This leads to formation of free active radicals from which material of high specific activity can be separated chemically.

T

t *(Genrl.)*. Symbol for tonne (metric **ton**).
t *(Chem.)*. An abbrev. for (1) *trans-*, i.e. containing the two radicals on opposite sides of the plane of a double bond or alicyclic ring; (2) *tertiary*, i.e. substituted on a carbon atom which is linked to three other carbon atoms.
τ *(Chem.)*. A symbol for a time interval, especially half-life or mean life.
T *(Phys.)*. Symbol for tesla.
T *(Genrl.)*. Symbol for *tera*, 10^{12}.
T *(Chem.)*. With subscript, a symbol for *transport number*.
T- *(Chem.)*. A symbol indicating the presence of a triple bond which begins on the corresponding carbon atom.
2,4,5,-T *(Bot.)*. *2,4,5,-trichlorophenoxyethanoic acid*. Widely used selective herbicide.
T15 *(Nuc.Eng.)*. Large tokamak experiment, Moscow, USSR. See **tokamak**.
T6 marker chromosome *(Immun.)*. A mouse chromosome originally derived from an irradiated male. It is about half the length of the shortest pair of **autosomes** and has a clearly detectable constriction near the centromere, so being easily recognizable in stained preparation of cells arrested in metaphase. The T6 chromosome has been introduced into an inbred strain of mice (see **recombinant inbred strain**) so as to replace the normal strain. The two strains are histocompatible so that the cells can be grafted from one to another without causing graft rejection. This has made it possible to follow the fate of particular cells or cell lines after transfer, and was the first technique to be used for this purpose.
Ta *(Chem.)*. The symbol for tantalum.
tab *(Aero.)*. Hinged rear portion of a flight control surface. See **balance-**, **servo-**, **spring-**, and **trimming-**.
tabescent *(Bot.)*. Shrivelling.
tabes dorsalis *(Med.)*. *Locomotor ataxia (ataxy)*. A disease of the nervous system marked by attacks of pain in the legs, anaesthesia of certain areas of the skin, ataxia, loss of the pupil reflex to light, and other nervous affections; due to degenerative changes in the nerves, especially of the sensory roots of the spinal cord, as a late result of syphilis.
tabetic *(Med.)*. Pertaining to, affected by, or caused by tabes dorsalis; an affected person.
table *(Comp.)*. Data structure in the form of a rectangular arrangement of items in rows and columns. Also *two-dimensional array*. See **array**.
table *(Eng.)*. The horizontal portion of a drilling, milling, shaping or other machine, which supports the workpiece and is adjustable relative to the tool.
table *(Maths.)*. A collection of data (e.g. square-root values) laid out in rows and columns for reference, or stored in a computer memory as an array.
table matter *(Print.)*. See **tabular matter**.
table of strata *(Geol.)*. A column which depicts a series of rocks arranged in chronological order, the oldest being at the bottom. It is usual to draw this to scale so that the average thicknesses of the beds are also shown.
table rolls *(Paper)*. Rolls from 5 to 30 cm (2–12 in) in diameter, which carry the upper face of the machine wire, and also assist in the removal of water by capillary action.
tablet *(Comp.)*. See **graphics tablet**.
tablet *(Plastics)*. A piece of moulding composition of the correct weight and density, and of suitable diameter and thickness to fit the mould: not preformed to the approximate shape of the moulding.
tabloid newspaper *(Print.)*. A newspaper with a smaller page size, usually half the normal broadsheet size, using a cross fold to form the spine.
taboo *(Behav.)*. An anthropological term for the prohibition of some class of people, objects or acts, because they violate the fundamental beliefs and values of a culture.
taboparesis, taboparalysis *(Med.)*. General paresis and tabes dorsalis affecting the same person.
tabs *(Image Tech.)*. Curtains which can be opened or closed in front of a cinema screen.
tabular *(Bot.)*. Having the form of a tablet or slab.
tabular matter *(Print.)*. Text arranged in accurately spaced columns; if there are rules between the columns, it is *table matter*.
tabulator *(Comp.)*. Early computer printer.
TAB vaccine *(Immun.)*. Vaccine used in the prophylaxis of enteric fevers. Contains heat-killed *Salmonella typhi* and *S. paratyphi A* and *B* in the smooth specific phase and possessing their normal complement of O antigens. Usually preserved with phenol, though other methods (e.g. acetone) are also used. Repeated doses are needed and the protection produced is not complete. Because the O antigens contain lipopolysaccharide some fever and local inflammation commonly follows injection.
tacheometer, tachymeter *(Surv.)*. Theodolite fitted with telescope which measures distances by sighting on levelling staff, the space between two crosslines in the eyepiece being an index (after correction) of the distance concerned. See **additive constant, multiplying constant**.
tacheometry *(Surv.)*. The process of surveying and levelling by means of angular measurements from a known station, combined with determination of distances from the station.
tachitoscope *(Behav.)*. A mechanical instrument capable of flashing visual displays on a screen for very short periods of time; used in perceptual research.
tachometer *(Eng.)*. An instrument for measuring speed of rotation.
tachycardia *(Med.)*. A heart beat greater than normal, usually defined as >100/min for adults.
tachygenesis *(Zool.)*. Accelerated development with elimination of certain embryonic stages, as in some Caecilians, in which the free-living tadpole stage is suppressed. adj. *tachygenetic*.
tachylite, tachylyte *(Geol.)*. A black glassy igneous rock of basaltic composition, which occurs as a chilled margin of dykes and sills. In Hawaii it forms the bulk of certain lava-flows.
tachymeter *(Surv.)*. See **tacheometer**.
tachyon *(Phys.)*. A theoretical particle moving faster than the speed of light.
tachyphylaxis *(Med.)*. Where repeated administrations of a drug result in progressively smaller responses.
tachypnoea, tachypnea *(Med.)*. Excessive frequency of respiration.
tack *(Build.)*. A small clout nail.
tack *(Print.)*. A measure of an ink's internal cohesion, the tack value of an ink gives a guide to an ink film's resistance to splitting at impression.
tack marks *(Print.)*. (1) Small dots printed in the centre of a back margin to indicate to the binder the lay edges of the sheet, two dots indicating the second side of a work-and-turn sheet; only seen occasionally. (2) A dot printed to overlap the side lay edge of the sheet when the first printing is being run off, so that it shows on the edge of the pile of sheets, to indicate the lay required for subsequent printings and which will be cut off when the job is trimmed. When wood furniture is used the mark may be made literally by a tack.
tack rag *(Build.)*. A cotton fabric folded pad impregnated with a non drying oil which is used as a more efficient means of removing dust than a dust brush prior to painting.

tack welding *(Eng.)*. Making short, provisional welds along a joint to hold it in position and prevent distortion during subsequent continuous welding.

tacky *(Build.)*. Said of paint or varnish which has not quite dried and is in a sticky condition.

Taconic orogeny *(Geol.)*. A period of intense folding which affected the eastern parts of N. America at the end of the Ordovician period. The effects are best seen in the Taconic Mts on the borders of New York State and Massachusetts.

tacticity *(Chem.)*. The stereochemical arrangement of units in the main chain of a polymer.

tactic movement *(Bot.)*. See **taxis**.

tactile *(Zool.)*. Pertaining to the sense of touch.

tactile bristle *(Bot.)*. A stiff hair which transmits a contact stimulus.

tactile perception *(Med.)*. The perception of vibration by the sense of touch; developed particularly in deaf persons, who can be trained to detect and interpret vibrations in another person's larynx, or to interpret vibrations applied selectively to their fingers by vibrators operated through filters by a microphone.

tactite *(Geol.)*. See **skarn**.

tactoid *(Phys.)*. A rod-shaped droplet or flat particle appearing in colloidal solutions which exhibit double refraction.

tactosol *(Chem.,Phys.)*. Sol containing **tactoids**.

taenia *(Zool.)*. A ribbon-shaped structure, such as the *taenia pontis*, a bundle of nerve fibres in the hind-brain of Mammals.

taeniasis *(Med.)*. The state of infestation of the human body with tapeworms (*Taenia*), which as adults may inhabit the intestine and as larvae the muscles and other parts of the body.

taenite *(Min.)*. A solid solution of iron and nickel occurring in iron meteorites; it appears as bright white areas on a polished surface. It crystallizes in the cubic system and has 27% to 65% Ni.

taffeta *(Textiles)*. A lightweight, plain-weave, crisp fabric with a faint weft rib produced from filament yarns and used for blouses, etc.

taffrail log *(Ships)*. See **nautical log**.

taft joint *(Build.)*. See **wiped joint**.

tag *(Comp.)*. See **sentinel**.

tag block *(Elec.Eng.)*. Terminal block, holding varying numbers of double-ended solder tags, which is fitted to every panel of apparatus supported on standard apparatus racks. External wiring to a unit can then be connected without interference with the internal wiring, which is completed during manufacture. External connections to bays of apparatus are also made to tag blocks mounted at the top of the racks by cable forms.

tagged atom *(Phys.)*. See **labelled atom**; also *radioactive tracer*.

tagger *(Eng.)*. Thin sheet-iron or tin-plate.

T-agglutinin *(Immun.)*. Antibody present in the blood of normal persons which agglutinates erythrocytes which have been incubated with **neuraminidase** or acted on by bacteria which produce the enzyme. This reveals the T-antigen present on erythrocytes but not normally available to react with antibody. The stimulus to production of the antibody may be effete erythrocytes.

tagma *(Zool.)*. A distinct region of the body of a metameric animal, formed by the grouping or fusion of somites; as the thorax of an Insect. pl. *tagmata*.

tagmosis *(Zool.)*. In a metameric animal, the grouping or fusion of somites to form definite regions (*tagmata*).

tail *(Aero.)*. See **tail unit**.

tail *(Build.)*. That end of a stone step which is built into a wall.

tail *(Image Tech.)*. The end of a roll of film or tape.

tail *(Print.)*. The bottom or foot margin of a page or volume.

tail *(Zool.)*. See **cauda**.

tailband *(Print.)*. See **headband**.

tail-bay *(Hyd.Eng.)*. The part of a canal lock immediately below the tail gates.

tail beam *(Build.)*. A floor joist which at one end is framed into a trimmer.

tail boom *(Aero.)*. One or more horizontal beams which support the tail unit where the fuselage is truncated; commonly used for cargo aircraft to facilitate loading trucks and bulky freight through full-width rear doors.

tail cap *(Print.)*. See **head cap**.

tail chute *(Aero.)*. A parachute mounted in an aircraft tail; cf. **antispin parachute** and **brake parachute**.

tail cone *(Aero.)*. The tapered streamline **fairing** which completes a fuselage or **tail boom**.

tailerons *(Aero.)*. Two-piece tailplane whose two halves can operate either together, performing the function of an elevator or differentially, causing rolling moments as does an aileron.

tail-first aircraft *(Aero.)*. An aircraft in which the horizontal stabilizer (i.e. tailplane) is mounted ahead of the main plane; common on pioneer aircraft and re-introduced for supersonic flight; sometimes *canard* because of their similarity to a planform of a duck in flight.

tail gate *(Hyd.Eng.)*. The gates at the low-level end of a lock.

tail heaviness *(Aero.)*. That state in which the combination of forces acting upon an aircraft in flight is such that it tends to pitch up.

tailing *(Build.)*. Operation of building in and fixing the end of a member which projects from a wall. Also *tailing in, tailing down*.

tailing hangover *(Image Tech.)*. Blurring of reproduced picture because of slow decay in electronic circuits. Also *submerged resonance*.

tailing iron *(Build.)*. A steel section built into a wall, across the top of the encastered end of a projecting member.

tailings *(Min.Ext.)*. (1) Rejected portion of an ore; waste, *gangue*. (2) Portion washed away in water concentration. May be impounded in a **tailings dam** or pond, or stacked dry on a dump.

tailings dam *(Min.Ext.)*. One used to hold mill residues after treatment. These arrive as fluent slurries. Dam may include arrangements for run-off or return of water after the slow-settling solids have been deposited.

tail joist *(Build.)*. See **tail beam**.

tailless aircraft *(Aero.)*. An aircraft, or glider, in which longitudinal stability and control in flight is achieved without a separate balancing horizontal aerofoil. This balance is achieved by **sweep**, and many *delta-wing* aircraft are tailless because their sharp angle of sweep-back renders a tail plane unnecessary.

tailpiece *(Print.)*. An engraving, design etc. occupying the bottom of a page, as at the end of a chapter.

tail plane *(Aero.)*. A horizontal surface, fixed or adjustable, providing longitudinal stability of an aircraft or glider. See **stabilizer**.

tail race *(Hyd.Eng.)*. A channel conveying water away from a hydraulically-operated machine.

tail race *(Min.Ext.)*. The launder or trough for the discharge of water-borne tailings.

tail rotor *(Aero.)*. See **auxiliary rotor**.

tail(s) *(Min.Ext.)*. US term for **tailings**.

tails *(Nuc.Eng.)*. The depleted uranium produced at an enrichment plant, containing typically 0.25% ^{235}U.

tail slide *(Aero.)*. A difficult aerobatic manoeouvre in which an aircraft is pulled up into a zoom and allowed to slide backward along its longitudinal axis after the vertical speed drops to zero. Flying surfaces are especially strengthened to withstand the reverse airflow encountered.

tailstock *(Eng.)*. A casting mounted slidably on a lathe bed. It carries a spindle in true alignment with the centre of the head-stock, longitudinally adjustable, and coned internally to receive a centre. See **lathe**.

tail trimmer *(Build.)*. A trimmer close to a wall, used in cases where it is not desired to build the joists into the wall.

tail unit *(Aero.)*. Hindmost parts of an aircraft, the horizontal *tailplane*, fin rudder and any strakes of an aircraft. Would include oblique (V) surfaces as in butterfly tail. Also *empennage*.

tail wheel landing gear *(Aero.)*. That part of the alighting gear taking the weight of the rear of the plane when on the ground. It consists of a shock-absorber carrying a wheel (*tail wheel*) or a shoe (*tail skid*).

Takayasu's disease *(Med.)*. *Pulseless disease*. A disease with progressive obliteration of the major arteries within the chest.

take *(Image Tech.)*. Record of the action of a scene or part of a scene, repeated if necessary to improve the performance as a different take number.

take *(Print.)*. A suitable portion of the copy which has been divided up for sharing among the hand or machine compositors.

take-off rocket *(Aero.)*. A rocket, usually jettisonable, used to assist the acceleration of an aircraft. Cordite rockets were introduced for naval aircraft during the Second World War and replenishable liquid-fuel rockets, some with controllable thrust, thereafter. Sometimes referred to as a *booster rocket*, although strictly this is for the acceleration of missiles. Abbrev. *RATOG* (Rocket-Assisted Take-Off Gear). US abbrev. *JATO* (Jet-Assisted Take Off).

take-up *(Image Tech.)*. That part of a machine where film or tape is wound up after passing through the operation or process concerned.

take-up motion *(Textiles)*. The mechanism that arranges for the fabric produced in weaving to be wound on a roller at the correct speed.

taking-off *(Build.,Civ.Eng.)*. The first process involved in drawing up a bill of quantities. It consists of obtaining the dimensions of each item on the drawing, and entering them in a systematic manner on sheets ruled for the purpose.

taking out turns *(Print.)*. Inserting the correct characters for those substituted and turned upside down (*turns*) when the required sorts were unavailable during the original setting.

talbot *(Phys.)*. Unit of luminous energy, such that 1 lumen is a flux of 1 talbot per second.

Talbot process *(Build.)*. An anticorrosion process applied to cast-iron pipes; they are coated internally with a mixture of bitumen and a hard siliceous material, which is distributed over the surface by centrifugal action during rapid rotation.

talc *(Min.)*. A monoclinic hydrated magnesium silicate, $MgSi_8O_{20}(OH)_4$. It is usually massive and foliated and is a common mineral of secondary origin associated with serpentine and schistose rocks; also found in metamorphosed siliceous dolomites. Purified talc is used medically, in toilet preparations and in many other ways. See **steatite**.

talipes *(Med.)*. *Club-foot*. A general term for a number of deformities of the foot: *talipes calcaneus*, in which the toes are drawn up from the ground and the patient walks on the heel; *talipes equinus*, in which the heel is drawn up and the toes point downwards; *talipes equinovarus*, in which the foot is inverted and turned inwards, with the toes pointing down; *talipes valgus*, in which the foot is abducted and everted, so that the patient walks on its inner side.

talk-back circuit *(Image Tech.)*. One which enables the controller of a programme to give directions to those originating a performance or rehearsal in a studio or location.

talk-down *(Radar)*. See **ground-controlled approach**.

tall-boy *(Build.)*. A fitting added to the top of a chimney to prevent down-draught.

tallow wood *(For.)*. *Eucalyptus microcorys*, a native of Australasia; it is durable, impermeable to preservative fluids and difficult to work. It will not glue well because of its greasy nature.

tally *(Surv.)*. A brass tag attached to a chain at every tenth link, and so marked or shaped as to enable the position of the tally along the chain to be immediately read. Also called a *teller*.

talon *(Zool.)*. A sharp-hooked claw, as that of a bird of prey.

talose *(Chem.)*. An *aldohexose*. ⇨

talus *(Civ.Eng.)*. An earthwork or batter wall slope.

talus *(Geol.)*. See **scree**.

talus *(Zool.)*. A synonym for *astragalus*. pl. *tali*.

talus wall *(Build.)*. A wall the face of which is built on a batter.

tamarugite *(Min.)*. A hydrated sulphate of sodium and aluminium, crystallizing in the monoclinic system.

Tamman's temperature *(Phys.)*. The temperature at which the mobility and reactivity of the molecules in a solid become appreciable. It is approximately half the melting-point in kelvins.

tamoxifen *(Med.)*. Oestrogen antagonist which blocks receptor sites on cells, used in the treatment of breast cancer.

tamp *(Civ.Eng.)*. (1) To fill a charged shot-hole with clay or other *stemming* material to confine the force of the explosion. (2) To ram or pound down ballast on a railway track, or road-metal. Also *punning*.

tampin *(Build.)*. A conical plug of boxwood used in opening out the end of a lead pipe.

tampon *(Med.)*. A plug or packing made of gauze, cotton-wool, and the like, for insertion into orifices or cavities (especially the vagina and uterus) for the control of haemorrhage, the removal of secretions or the dilating of passages.

tandem *(Telecomm.)*. Connection of output of one four-terminal network to input of a second.

tandem engine *(Eng.)*. An engine in which the cylinders are arranged axially, or end to end, with a common piston rod.

tandem exchange *(Telecomm.)*. One used primarily as a switching point for traffic between other exchanges.

tandem mill *(Eng.)*. A rolling mill with two or more stands operating continuously and synchronized so that the stands can accommodate the increasing velocity of the rolled material.

tandem mirror *(Nuc.Eng.)*. See **magnetic mirror**.

tandem selection *(Telecomm.)*. Selection of outlets by two uniselectors in series, so that the maximum possible availability of outlets is obtained.

tandem working *(Telecomm.)*. The using of an intermediate exchange during the transition period when a manual system is being converted to automatic working. Working is effected by trains of impulses which are set up on key senders on instructions from A-operators in originating exchanges.

tang *(Build.)*. The end of a tool which is driven into or about its haft. See **socket chisel**.

tangent *(Maths.)*. (1) See **trigonometrical functions**. (2) A line or plane which touches a curve or surface. See **moving trihedral**. The problem of finding tangents was first considered by the Greeks and is now solved by differential calculus.

tangent distance *(Surv.)*. The distance between the intersection point and one of the tangent points of a railway or highway curve.

tangent galvanometer *(Elec.Eng.)*. A vertical circular coil with its plane parallel to the meridian. If I is the current flowing, and r the effective coil radius in cm, then

$$I = \frac{rH}{50n} \tan\theta \text{ amperes,}$$

where θ is the angle of deflection of a magnetometer needle placed at the centre of the coil, n is the number of turns in the coil, and H the horizontal component of the earth's magnetic field in amperes per metre.

tangential field *(Phys.)*. The image surface formed by the tangential foci of a series of object points lying in a plane at right angles to the axis.

tangential focus *(Phys.)*. The focus of an object point lying off the axis of an optical system, in which the image is drawn out by the astigmatism of the system into a line tangential to a circle centred on the optic axis.

tangential longitudinal section *(Bot.)*. A section cut longitudinally along a more or less cylindrical organ parallel to a tangent at its surface. Abbrev. *TLS*.

tangential wave path *(Meteor.)*. That of a direct wave, tangential to the surface of the earth and which is curved by atmospheric refraction.

tangent point *(Surv.)*. The point of commencement, or of termination, of a railway or highway curve.

tangent scale *(Elec.Eng.)*. The scale of an electrical instrument in which the measured quantity varies as the tangent of the angle of deflection.

tangent screw *(Surv.)*. A screw by which a fine adjustment may be made to the setting of a theodolite about its axis, either in order to bring the line of sight into coincidence with a signal, or to adjust the vernier reading to a given value.

tanh *(Maths.)*. See **hyperbolic functions**.

tank circuit *(Telecomm.)*. Section of a resonating coaxial transmission line or a tuned circuit, which accepts power from an oscillator and delivers it, harmonic-free, to a load.

tanker *(Ships)*. A term covering all types of ships carrying liquid in bulk, from light acids and oils to molasses and latex.

tank furnace *(Glass)*. Essentially a large 'box' of refractory material holding from 6 to 200 tons of glass, through the sides of which are cut 'ports' fed with a combustible mixture (producer gas and air, coke-oven and air, or oil spray and air), so that flame sweeps over the glass surface. With the furnace is associated a regenerative or recuperative system for the purpose of recovering part of the heat from the waste gases.

tanking *(Build.,Civ.Eng.)*. Waterproof material included in an underground structure to prevent infiltration of subsoil water.

tank line *(Telecomm.)*. See **tank circuit**.

tank reactor *(Nuc.Eng.)*. Covered type of **swimming-pool reactor**.

tank rectifier *(Elec.Eng.)*. Mercury-arc rectifier enclosed in a metal tank with vitreous seals for the conductors.

tank vent pipe *(Aero.)*. The pipe leading from the air space in an aircraft fuel, or oil, tank to atmosphere, for equalizing changes in pressure due to alterations in altitude; in *aerobatic* aircraft a nonreturn valve is fitted to prevent liquid escaping when inverted.

tannic acid *(Chem.)*. See **tannin**.

tannin *(Chem.)*. A mixture of derivatives of polyhydroxybenzoic acids. When pure it forms a colourless amorphous mass, easily soluble in water, of bitter taste and astringent properties. Occurs in many trees, e.g. *Quebracho colorado*.

tanning *(Zool.)*. In newly-formed cuticle of terrestial Arthropods the process in which the spaces between the chitin micelles are filled with sclerotin, which consists of protein molecules linked together or tanned by quinones. This makes the cuticle tougher and darker.

tanning developer *(Image Tech.)*. A developer which produces hardening or insolubility of the gelatine emulsion in proportion to the silver image formed; used to prepare the **matrix** for **imbibition** printing.

tannin sac *(Bot.)*. A cell containing much tannin.

tantalite *(Min.)*. Tantalate and niobate of iron and manganese, crystallizing in the orthorhombic system. The principal ore of tantalum, occurring in pegmatites and granitic rocks and in alluvial deposits. When the Ta content exceeds that of Nb, the ore is called tantalate. See also **columbite**.

tantalum *(Chem.)*. A metallic element, symbol Ta, at. no. 73, r.a.m. 180.948, rel. density at 20°C 16.6, mp 2850°C, electrical resistivity 15.5×10^{-8} ohm metres, Brinell hardness 46. It occurs in crystals and grains (usually containing, in addition, small amounts of niobium) in the Ural and Altai Mts. It is used as a substitute for platinum for corrosion-resisting laboratory apparatus, as acid-resisting metal in chemical industry, and in the form of carbide in cemented carbides. Used in surgical insertions because of its lack of reaction to body fluids.

tantalum capacitor *(Electronics)*. Miniature electrolytic capacitor employing tantalum foil.

T-antenna *(Telecomm.)*. One comprising a top conductor with a vertical downlead attached at the centre; much used for long waves.

T-antigens *(Immun.)*. A group of surface antigens defining sub-populations of human T-lymphocytes. Many of these have been defined using monoclonal antibodies and are classified in the **CD system**. Corresponding antigens have been identified on mouse, rat and bovine T lymphocytes, and will presumably be found in any species studied sufficiently intensively.

tap *(Eng.)*. A screwed plug of accurate thread, form and size, on which cutting edges are formed along longitudinal grooves; screwed into a hole, by hand or power, to cut an internal thread.

tap changing *(Elec.Eng.)*. A method of varying the voltage ratio of a transformer by tapping the windings in such a way that the **turns ratio** is altered.

tap density *(Powder Tech.)*. The apparent powder density of a powder bed formed in a container of stated dimensions when a stated amount of the powder is vibrated or tapped under stated conditions.

tape *(Build.)*. A long flexible measuring scale of thin strip steel, linen, linen in which wire is interwoven to increase its strength or plastic reinforced with glass fibre, coiled up in a circular leather case fitted with a handle for winding purposes.

tape *(Comp.)*. See **magnetic tape**, **punched paper tape**.

tape *(Textiles)*. A woven narrow fabric: frequently used for reinforcing garments.

tape cassette *(Comp.)*. Device for holding **magnetic tape**. Both tape and cassette may be similar to those used in domestic tape recorders.

tape deck *(Acous.)*. Platform incorporating essentials for magnetic recording (motor(s), spooling, recording and erasing heads) for adding to amplifier, microphone, loudspeaker, to form a complete recording and reproducing equipment (*tape recorder*).

tape deck *(Comp.)*. See **tape drive**.

tape density *(Comp.)*. **Packing density** of a **magnetic tape**.

tape drive *(Comp.)*. Mechanism which transports **magnetic tape** between spools across the read/write heads.

tape joint *(Build.)*. Paper or gauze tape used to cover the joints between plaster, wall or centring boards before a decorative surface is applied.

taper *(Eng.)*. In conical parts, the difference in diameter per unit length.

tape reader *(Comp.)*. Apparatus for providing signals related to holes punched in **paper tape** as it is moved by a tape transport.

tape recording *(Acous.)*. Longitudinal recording on magnetic particles dispersed in a medium carried on plastic tape. There is a residual magnetization of a high-frequency biasing current, modulated by the signal current. The residual m.m.f. in the particles allows the modulation to be reproduced by induction in a magnetic circuit. See also **tape deck**.

tapering gutter *(Build.)*. A parapet gutter having an increased width in the direction of flow, so as to secure the necessary fall.

taper key *(Eng.)*. A rectangular key having parallel sides, but slightly tapered in thickness along its depth. See **key**.

taper line gratings *(Textiles)*. Transparent plastic plates engraved with lines that are more widely spaced at one end than the other. By placing the appropriate grating over a fabric the number of threads per cm (for woven fabrics) or courses per cm (in knitted fabrics) may be determined from the resulting diffraction pattern.

taper pin *(Eng.)*. A pin, used as a fastener, very slightly tapered to act as a wedge.

taper roller bearing *(Eng.)*. A roller bearing rendered capable of sustaining end thrust by the use of tapered rollers, in conjunction with internally and externally coned races.

taper tap *(Eng.)*. The first tap used in threading a hole. The first few threads are ground down to the core diameter to provide a guide, gradually increasing to the full thread size. See **tap, plug tap, second tap**.

taper-turning attachment *(Eng.)*. An attachment bolted to the back of a lathe, with a guide bar which may be set at the angle or taper it is desired to impart to the turned part, the lathe tool being guided by the bar via a slide.

tapes *(Print.)*. Strips of leather, plastic or fabric for carrying paper from one part of press to another.

tape splice *(Image Tech.)*. A join in film or magnetic tape using transparent adhesive tape applied to the butted ends.

tapestry *(Textiles)*. Originally a handwoven furnishing fabric in which a design was produced by hand stitches to form the required pattern. Now employed for jacquard figured design on power-loom woven cloths.

tapestry brick *(Build.)*. See **rustics**.

tapetum *(Bot.)*. A layer of cells in a sporangium of a vascular plant surrounding the spore mother cells, becoming absorbed as the spores mature.

tapetum *(Zool.)*. In the eyes of certain nightflying Insects, a reflecting structure: in some Vertebrates, a reflecting layer of the retinal side of the choroid; in the Vertebrate brain, a tract of fibres in the corpus callosum.

tapeworm *(Med.,Zool.)*. Parasitic worms of the class Cestoda, generally taking the form of a scolex with hooks and/or suckers for attachment to the host and a chain of individual proglottids in successive stages of development. Species infecting Man are *Hymenolepis nana* or Dwarf Tapeworm, *Taenia solium* from infected pork, *Diphyllobothrium latum* from infected Fish, and *Echinococcus granulosus* which spends the larval stage only in Man, causing hydatid cysts, the adult worm inhabiting Dogs.

tap-field control *(Elec.Eng.)*. A method of controlling the speed of a series motor; the field excitation is varied by means of tappings on the field windings.

tap-field motor *(Elec.Eng.)*. A series motor whose field windings are arranged for tap-field control.

taphephobia *(Med.)*. Morbid fear of being buried alive.

taphrogenesis *(Geol.)*. Vertical movements of the Earth's crust, resulting in the formation of major faults and rift valleys.

tapiolite *(Min.)*. A tantalate resembling **tantalite** but crystallizing in the tetragonal system.

tappet *(Eng.)*. A sliding member working in a guide; interposed between a cam and the push rod or valve system which it operates, to eliminate side thrust.

tapping *(Elec.Eng.)*. An intermediate connection on a circuit element such as a resistor, often used to vary the potential applied to another electrical system.

tapping *(Eng.)*. (1) The operation of running molten metal from a furnace into a ladle. (2) Using a **tap** to thread a hole.

tapping *(Med.)*. See **paracentesis**.

TAPPI standard methods *(Paper)*. Laboratory test methods conforming to the *Technical Association of the Pulp and Paper Industry* of the US. Widely used internationally.

taproot *(Bot.)*. The first (primary) root of a plant developed directly from the radicle. Sometimes develops as a fleshy storage organ e.g. carrot.

taproot system *(Bot.)*. Root system, characteristic of dicotyledons and conifers, based on a tap root with laterals of various orders. Cf. **fibrous root system**.

tar *(Chem.)*. See **coal-tar, gas tar**.

tarbuttite *(Min.)*. Hydrated zinc phosphate, which crystallizes in the triclinic system. The crystals are often found in sheaflike aggregates.

Tardigrada *(Zool.)*. A subphylum of minute Arthropods with suctorial mouthparts and four pairs of stumpy clawed legs; common forms, of wide distribution, are found among moss and debris in ditches and gutters and on tree trunks, and can survive desiccation.

tare *(Genrl.)*. The weight of a vessel, wrapping, or container, which subtracted from the gross weight gives the net weight.

target *(Electronics)*. Any electrode or surface upon which electrons impinge at high velocity, e.g. fluorescent screen of a cathode-ray tube, or any intermediate electrode in an electron multiplier, or anode or anti-cathode in an X-ray tube.

target *(Image Tech.)*. Plate in a TV camera tube on which external scenes are focused and scanned by an electron beam.

target *(Nuc.Eng.)*. Material irradiated by beam from accelerator.

target *(Radar)*. Reflecting object which returns a minute portion of radiated pulse energy to the receiver of a radar system.

target cell *(Immun.)*. (1) Antigen bearing cell which is the target of attack by lymphocytes or by specific antibody. (2) In haematology used to describe an abnormally shaped and unusually thin red cell with central stained area seen in blood films, especially in certain disorders of haemoglobin formation.

target diagram *(Elec.Eng.)*. A diagram for estimating the uniformity of a batch of electric lamps; obtained by plotting luminous intensity against input power for each lamp. The more uniform the lamps, the more closely together will the plotted points lie in the diagram.

target rod *(Surv.)*. A type of levelling staff provided with a sliding target, which can be moved by the staffman, under direction from the leveller, to a position in which it is in line with the line of sight of the level, the staff reading being recorded by the staffman.

target strength *(Acous.)*. *T*. Defined in dB by $T=E-S+2H$, where E = echo level, S = source level and $2H$ = transmission loss.

target theory *(Radiol.)*. Proposed explanation of radiobiological effects, in which only a small sensitive region of each cell is susceptible to ionization damage.

tarif *(Image Tech.)*. Equipment for modifying colour rendering when motion picture film is reproduced on television. (Said to be acronym for *Technical Apparatus for Rectification of Inferior Film*.)

tarmacadam *(Civ.Eng.)*. A road or runway surfacing of broken stone which has been covered with tar; spread in a layer of uniform thickness and well rolled. Two layers are usually applied, the upper one being of stone of smaller size. In US *Tarmac* is a widely-used proprietary mixture. Cf. **macadamized road**.

tarnish *(Chem.)*. The discoloration produced on the surface of an exposed metal or mineral, especially silver, generally as the result of the formation of an oxide or a sulphide film.

tar pit *(Geol.)*. An outcrop where natural bitumen occurs. The tar frequently contains the skeletons of trapped animals.

tarsalgia *(Med.)*. Pain in the instep of the foot.

tarsia *(Build.)*. Wood inlay, of geometric or architectural patterns, in which comparatively large pieces of wood are used.

tarsus *(Build.)*. A cylindrical projection along the intersection between the two sloping surfaces on one side of the ridge of a mansard roof.

tarsus *(Zool.)*. (1) In Vertebrates, an elongate plate of dense connective tissue which supports the eyelid. (2) In Insects, Myriapods and some Arachnida (as Mites), the terminal part of the leg, consisting typically of five joints. (3) In land Vertebrates, the basal podial region of the hind limb; the ankle. adj. *tarsal*.

tartar emetic *(Chem.)*. Potassium antimonyl tartrate, $2[K.SbO.C_4H_4O_6].H_2O$.

tartaric acid *(Chem.)*. HOOC.CH(OH).CH(OH).COOH, dihydroxy-succinic acid. It exists in four modifications, viz., (+)-tartaric acid, mp 170°C; (−)-tartaric acid, mp 170°C; *racemic* or (±)-tartaric acid, mp 206°C; *meso*tartaric acid, mp 143°C. The (+)-tartaric acid is found in nature, free and as salts of potassium, calcium and magnesium. It occurs in a large number of plants and fruits; the acid potassium salt is deposited from wine.

tartrates *(Chem.)*. The salts of tartaric acid. Extensively used in medicine, the potassium salts and Rochelle salt as saline purges. Also *cream of tartar*.

tarviated *(Civ.Eng.)*. A term applied to macadam road surfacings in which the stone is bound together with tar.

TAS *(Aero.)*. Abbrev. for *True AirSpeed.*

TASI *(Telecomm.)*. Abbrev. for *Time-Assignment Speech Interpolation.*

tasmanite *(Geol.)*. A type of practically pure spore coal; a variety of **cannel coal**. See **boghead coal**.

taste-bud *(Zool.)*. In Vertebrates, an aggregation of superficial sensory cells subserving the sense of taste; in higher forms, usually on the tongue.

TAT *(Behav.)*. Abbrev. for *Thematic Apperception Test.*

T-attenuator *(Elec.Eng.)*. One comprising three resistors, one end of each being connected together. The free ends of two are connected respectively to an input and an output terminal while the free end of the third is connected to the common input and output terminals.

taungya *(Ecol.)*. Form of land use in the humid tropics, in which villagers are given the right to farm on good forest soils in exchange for their services in tending young trees on the same land. A practical form of **agroforestry**.

tauon *(Phys.)*. A lepton (τ^-) and its antilepton (τ^+) of exceptionally high mass, 1.78 GeV. Discovered at the Stanford Linear Accelerator Centre (SLAC) in electron-positron colliding beam experiments. Their existence necessitated the postulation of the *top* and *bottom* quarks to preserve the *lepton-quark symmetry*.

tautomerism *(Chem.)*. The existence of a substance as an equilibrium mixture of two interconvertible forms, usually because of the mobility of a hydrogen atom. Thus tautomeric compounds can give rise to two series of derivatives. See **ethyl aceto-acetate**.

tawa *(For.)*. New Zealand native hardwood from the genus *Beilschmedia*, used for furniture, veneering, fittings and flooring.

taxi-channel markers *(Aero.)*. See **airport markers**.

taxis *(Behav.,Biol.)*. Orientation with respect to environmental stimuli; often combined with locomotion, so that the animal moves towards, or away, or at a fixed angle to source. There are various classifications of types of taxes e.g. *positive taxis* when the organism moves towards the stimulus and the reverse in *negative taxis*. See **kinesis**.

taxi track *(Aero.)*. A specially prepared track on an aerodrome used for the ground movement of aircraft. See **perimeter track**.

taxi-track lights *(Aero.)*. Lights so placed as to define manoeuvring areas and tracks.

taxon *(Bot.,Zool.)*. Any group of organisms to which any rank of taxonomic name is applied.

taxonomic series *(Biol.)*. The range of extant living organisms, ranging from the simplest to the most complex forms.

taxonomy *(Biol.)*. The science of classification as applied to living organisms, including study of means of formation of species, etc.

Taylor process *(Eng.)*. Making very fine wire by inserting it into a closely fitting glass tube and drawing out the whole during heating.

Taylor's series *(Maths.)*. Series expansion for a continuous function, giving the value of the function for one value of the independent variable in terms of that for another value. Under specified conditions the series is

$$f(a+h) = f(a) + hf'(a) + \frac{h^2f''(a)}{2!} \dots \text{etc.}$$

Tay-Sachs' disease *(Med.)*. A rare inherited neurological disease of infants, prevalent in Ashkenazi Jews and characterized by progressive paralysis of the body, seizures, blindness, deafness and death before the age of two years. A result of the progressive degeneration of the nerve cells of the brain and spinal cord, due to a deficiency in an isoenzyme which catalyses the conversion of *ganglioside*.

Tb *(Chem.)*. The symbol for terbium.

T-bands *(Biol.)*. See **banding techniques**.

TBC *(Image Tech.)*. *Time-Base Corrector*, equipment for correcting timing errors in videotape recording and reproduction.

T-beam *(Civ.Eng.)*. A beam forming part of the construction of a reinforced concrete floor; regarded as being composed of the beam part projecting below the floor slab, and portions of the floor slab on both sides, the whole having the form of a letter T.

T-bolt *(Eng.)*. A clamping bolt having a head in the form of a short reactangular cross-piece, and sliding in slots in a machine tool table cut in the shape of an inverted T. Access is provided by wells cast or machined near the ends of the table.

TBO, tbo *(Aero.)*. See **time between overhauls**.

Tc *(Chem.)*. The symbol for technetium.

T-cell *(Immun.)*. Abbrev. for *T-lymphocyte.*

T-cell growth factor *(Immun.)*. See **interleukin-2**. Abbrev. *TCGF.*

T-cell leukaemia viruses *(Immun.)*. Group of **retroviruses** which infect T-lymphocytes and cause leukaemias. Some cause malignant transformation of T-lymphocytes (e.g. HTLV-1 or feline leukaemia virus) whereas others cause **acquired immunodeficiency syndrome** (HTLV-3, now renamed HIV for human immunodeficiency virus). Similar viruses are present in monkeys.

T-cell replacing factor *(Immun.)*. A soluble factor derived from helper T-lymphocytes which can replace the presence of T-lymphocytes in stimulating antibody production by B-lymphocytes which have been activated by antigen. Synonymous with B-cell differentiation factors of which one is identified as interleukin-4.

TCGF *(Immun.)*. Abbrev. for *T-Cell Growth Factor.*

TCNE *(Chem.)*. Abbrev. for *TetraCyaNoEthene.*

t-distribution *(Stats.)*. The sampling distribution of the mean of a set of observations from a normal distribution with unknown variance. The (central) t-distribution has one parameter, the degrees of freedom, and describes the sampling distribution of the deviation of the sample mean from the population mean.

TDMA *(Telecomm.)*. Abbrev. for *Time-Division Multiple Access.*

TDRSS *(Space)*. Abbrev. for *Tracking and Data Relay Satellite System*, a satellite communication system for relaying data (at rates of hundreds of megabits per sec.) from Space Shuttle Orbiter, the Space Station and related orbital payloads to the ground.

TdT *(Immun.)*. Abbrev. for *Terminal desoxynucleotidyl Transferase.*

Te *(Chem.)*. The symbol for tellurium.

teak *(For.)*. Valuable wood from *Tectona grandis*, a tree of the verbena family, found in India and S.E. Asia renowned for its durability; used for shipbuilding, piers, furniture etc.

tear factor *(Paper)*. The ratio of the lateral tearing strength of a paper to its **grammage**.

tear fault *(Geol.)*. A horizontal displacement of a series of rocks along a more or less vertical plane. Cf. **normal fault**, **thrust plane**.

tear gases *(Chem.)*. Volatile compounds which even in low concentration make vision impossible by their irritant action on the eyes. They are halogenated organic compounds, e.g. *xylyl bromide*, $CH_3.C_6H_4.CH_2Br$, and *ethyl iodoacetate*, $CH_2I.COOC_2H_5$.

tear gland *(Zool.)*. See **lacrimal gland**.

teats *(Zool.)*. In female Mammals, paired projections from the skin on which the lactiferous tubules of the mammary glands open.

teazle *(Textiles)*. The dried seed head of the thistle-like plant, *Dipsacus fullonum*, fitted on machines to raise the pile of certain fabrics.

technetium *(Chem.)*. Radioactive element not found in ores. First produced as a result of deuteron and neutron bombardments of molybdenum. Symbol Tc, at. no. 43, the most common isotope ^{99}Tc has half-life of 2.1×10^6 years. Found among fission products of uranium, and (unexplained) in the spectra of some stars.

technical jack plane *(Build.)*. Small type of **jack plane**, with low-cut rear section.

Technicolor *(Image Tech.)*. TN of a number of systems of colour cinematography, one of which used a special three-strip camera exposing three separation negatives (1932–55). Prints were made by photo-mechanical imbibition methods (**dye-transfer**) until 1978.

technology *(Genrl.)*. The practice, description and terminology of any or all of the applied sciences which have practical value and/or industrial use.

tectonic *(Geol.)*. Said of rock structures which are directly attributable to earth movements involved in folding and faulting.

tectonics *(Geol.)*. The study of the major structural features of the Earth's crust.

tectorial *(Zool.)*. Covering; as the tectorial membrane (*membrana tectoria*) of Corti's organ.

tectosilicates *(Min.)*. Those silicates having a structure in which atoms of silicon and oxygen are linked in a continuous framework.

tectrices *(Zool.)*. In Birds, small feathers covering the bases of the remiges and filling up the gaps between them. Also called *auriculars*.

tectum *(Zool.)*. A covering or roofing structure; as the *tectum synoticum*, part of the roof of the cartilaginous skull which connects the two auditory capsules.

tee *(Build.)*. A fitting used to connect a branch pipe into a pipe run at 90°.

tee bolt *(Eng.)*. See **T-bolt**.

tee hinge *(Build.)*. A large strap hinge shaped like the letter T, the long arm, corresponding to the upright part of the T, being secured to the door, and the crosspiece to the hingeing post.

tee joint *(Elec.Eng.)*. A joint in a cable formed by tapping off a branch circuit, without cutting the main cable.

teem *(Glass)*. To pour molten glass from a pot in the rolling process.

teeming *(Eng.)*. The operation of filling ingot moulds from a ladle of molten metal.

Teepol *(Chem.)*. TN of liquid anionic detergent based on mixed sodium alkyl sulphates of long-chain alcohols, such as lauryl alcohol (*n*-dodecanol, $CH_3(CH_2)_{11}OH$).

tee rest *(Eng.)*. See **T-rest**.

Teflon *(Plastics)*. TN for **polytetrafluoroethene**.

tegula *(Build.)*. A roofing tile.

tegulated *(Zool.)*. Composed of or covered by plates overlapping like tiles.

tehp *(Aero.)*. Abbrev. for *Total Equivalent brake Horse-Power*.

teichopsia *(Med.)*. Temporary loss of sight in part of the visual field, and the appearance before the eye of a spot of light which enlarges and becomes zigzag in shape and many-coloured; a symptom of **migraine**.

Teklan *(Plastics)*. TN for synthetic textile fibre based on a modified acrylic (modacrylic) base. Chiefly noted for its inherent flameproof properties.

tektites *(Min.)*. A group term which covers moldavites, billitonites and australites. They are natural glasses of non-volcanic origin and may be of extra-terrestrial origin.

tektosilicates *(Min.)*. See **tectosilicates**.

telangiectasis *(Med.)*. Morbid dilatation of capillaries and arteries. adj. *telangiectatic*.

tele-. *(Genrl.)*. Prefix from Gk. *tēle*, afar, at a distance.

telecentric stop *(Phys.)*. A stop placed in the second focal plane of a positive lens, forming a viewing system by which a scale can be read without parallax error.

teleceptor, telereceptor *(Biol.)*. A sense organ which responds to stimuli of remote origin.

telecine *(Image Tech.)*. Apparatus for producing video signals from motion picture film for television broadcasting or videotape recording.

telecommunication *(Space)*. The transmission and reception of data-carrying signals usually between two widely separated points with the aid of a *communications satellite*. More generally any communication between two distant points by electrical means.

teleconferencing *(Comp.)*. A use of **electronic mail** in which users identify subjects of interest and are then automatically offered communications through their terminals.

teleconferencing *(Image Tech.)*. A **video-conferencing** system making use of a restricted band-width of frequencies, so permitting connecting by telephone lines but with some limitation in the presentation of movement.

teleconverter *(Image Tech.)*. Supplementary lens system with extension tube mounting to convert a camera lens to much greater focal length; also known as *tele-extender* or *range-extender*.

telegony *(Zool.)*. The supposed influence of a male with which a female has previously been mated, as evinced in offspring subsequently borne by that female to another mate.

telegraphy *(Telecomm.)*. Obsolescent term for communication at a distance of documentary matter such as written, printed or pictorial matter, or the reproduction at a distance of any kind of information. Now a part of telecommunication.

teleguided missile *(Aero.)*. A small subsonic missile for attacking surface targets, e.g. tanks and ships, controlled by command guidance from an operator, or automatic device, by signals transmitted through fine wires connected to the control box and uncoiled in flight from the missile. Also *wire-guided*.

telemeter *(Elec.Eng.)*. An instrument for the remote indication of electrical quantities, such as voltage, current, power etc.

telemeter *(Surv.)*. General name for instrument which acts as a distance measurer, without the use of a chain or other direct-measuring apparatus.

telemetry *(Surv.)*. Measurement of linear distances by use of tellurometer.

telemetry *(Telecomm.)*. Transmission to a distance of measured magnitudes by radio or telephony, with suitably coded modulation, e.g. amplitude, frequency, phase, pulse.

telencephalon *(Zool.)*. One of the two divisions of the Vertebrate fore-brain or prosencephalon (the other being the *diencephalon*), comprising the cerebral hemispheres (with the cerebral cortex or pallium and the corpus striatum), the olfactory lobes, and the olfactory bulbs.

TELENET *(Comp.)*. A public **packet switching** network in US.

teleo-. *(Genrl.)*. Prefix from Gk. *teleios*, perfect.

teleology *(Biol.)*. The interpretation of animal or plant structures in terms of purpose and utility. adj. *teleological*.

teleonomy *(Biol.)*. Impression of purpose arising from adaptation through natural selection.

Teleostei *(Zool.)*. An infraorder of Osteichthyes, including Fish with a wide diversity of form and physiological adaptation. Gills fully filamentous, tail externally (and in many cases internally) homocercal, endoskeleton completely ossified, fins completely fanlike with no trace of an axis. Bony Fishes.

telepathy *(Behav.)*. Communication between individuals that takes place independently of all known sensory channels.

telephone interference (influence) factor *(Telecomm.)*. The weighting-factor required for determining the total interference of induced electromotive forces arising from harmonic induction in telephone lines from adjacent power lines. The factor takes into account the average relative sensitivity of the ear for varying frequency, and also the average response curves of telephone receivers. Abbrev. *TIF*.

telephony *(Telecomm.)*. The conversion of a sound signal into corresponding variations of electric current (or potential), which is then transmitted by wire or radio to a distant point where it is reconverted into sound.

telephoto lens *(Image Tech.)*. A long-focus lens having a comparatively short **back-focus**, one-half or one-third of its focal length, giving high magnification without undue extension.

teleprinter *(Telecomm.)*. See **telex**.

teleprocessing *(Comp.)*. Processing carried out from a remote terminal.

Teleprompter *(Image Tech.)*. TN for a visual prompter for displaying a script inconspicuously to a speaker in front of a TV camera.

teleradiography *(Radiol.)*. A technique to minimize distortion in taking X-ray photographs by placing X-ray tube some distance from the body.

telerecording *(Image Tech.)*. Transferring a television or videotape programme to motion picture film.

telescience *(Space)*. A fully interactive mode of scientific operations where the experiment is performed remotely, using data presented to the experimenter (e.g. by television) who can remotely control elements of the experimental equipment and, thus, iterate the conduct of the experiment and its results.

telescope *(Phys.)*. An optical instrument for producing a magnified image of a distant object. It consists of a system of mirrors or lenses or both. See **astronomical-**, **meniscus-**, **reflecting-**, **refracting-**, **terrestrial-**, **Newtonian-**.

telescopic shaft *(Eng.)*. An assembly of two or more tubes sliding within each other to provide a hollow shaft of variable length.

telescopic star *(Astron.)*. Star whose apparent magnitude is numerically greater than the 6th and which is too faint to be seen with the naked eye.

Telesto *(Astron.)*. The thirteenth natural satellite of **Saturn**, discovered in 1980.

Teletex *(Telecomm.)*. An international business correspondence defined by CCITT, and offered by telecommunications authorities and operators in many countries; the terminals are generally sophisticated typewriters and word processors.

teletext *(Telecomm.)*. Method of transmitting computer-stored information to suitably adapted domestic television receivers; information is transmitted as static pages of alpha-numeric information at a low rate within the conventional TV signal. In the UK, the names *Ceefax* (BBC) and *Oracle* (IBA) are used. See **viewdata**.

teletherapy *(Radiol.)*. Treatment by X-rays from a powerful source at a distance, i.e. by high-voltage X-ray tubes or radioactive sources, such as ^{60}Co, up to 80 000 GBq.

Teletype *(Comp.)*. TN for a make of teletypewriter, often applied to any teletypewriter.

teletypesetting *(Print.)*. A method of operating **line-casting machines** by using a six-unit punched tape produced on a separate keyboard; can be used within the same office where increased output results through specialization and the optimum use of equipment and skill; or the keyboard output can be transmitted by wire or radio to be converted to tape at the receiving end.

teletypewriter *(Comp.)*. Input/output device consisting of a keyboard and a typewriter-like printer often combined with a paper tape punch and reader. It can be used for the preparation of punched paper tape or for **interactive communication** with a computer. US term for *teleprinter*.

television *(Telecomm.)*. The electronic transmission, reception and reproduction of transient visual images.

television cable *(Image Tech.)*. One capable of transmitting frequencies sufficiently high to accommodate TV signals without undue attenuation or relative phase delay; usually coaxial, with as much air insulation as possible.

television camera *(Image Tech.)*. A camera converting the optical image of an external scene into electronic signals for television transmission or recording.

television channel *(Image Tech.)*. RF band of sufficient width to accommodate TV transmission signals and allocated for that specific use.

television receiver *(Image Tech.)*. That part of a TV system in which the picture and associated sounds are reproduced from the input signal. Also *television set*.

television transmitter *(Image Tech.)*. One which radiates video, audio and synchronizing components of a television signal as a modulated RF wave.

telex *(Telecomm.)*. An audio-frequency teleprinter system for use over telephone lines (provided in UK by the British Telecomm.). (*Automatic TELetypewriter EXchange Service.*)

teller *(Surv.)*. See **tally**.

telltale clock *(Horol.)*. A portable clock which gives, on a chart, a record of the time a watchman visits certain fixed points in his round of inspection.

telluric bismuth *(Min.)*. An intermetallic compound, Bi_2Te_3, crystallizing in the trigonal system. The name has also been used as a synonym for *tetradymite*.

telluric current *(Min.Ext.)*. Current in, or put into, the earth, which is used in exploration of strata.

telluric line *(Astron.)*. Absorption line or band in stellar and planetary spectra, caused by absorption in the earth's atmosphere, mainly by water vapour and oxygen.

tellurides *(Chem.)*. Compounds of divalent tellurium [Te (II)], analogous to **sulphides**.

tellurite *(Min.)*. Orthorhombic tellurium dioxide.

tellurium *(Chem.)*. A semimetallic element, tin-white in colour. Symbol Te, at. no. 52, r.a.m. 127.60, rel. density at 20°C 6.24, mp 452°C, valencies 2, 4, 6, electrical resistivity 2×10^{-8} ohm metres. Used in the electrolytic refining of zinc in order to eliminate cobalt; alloyed with lead to increase the strength of pipes and cable sheaths. The chief sources are the slimes from copper and lead refineries and the fine dusts from telluride gold ores.

tellurobismuth *(Min.)*. See **telluric bismuth**.

tellurometer *(Surv.)*. Electronic instrument used to measure survey lines for distances up to 40 miles by measurement of time required for a radar signal to echo back, accuracy being of order of 1:100 000 in good weather. See **trilateration**.

telo- *(Genrl.)*. Prefix from Gk. *telos*, end.

teloblast *(Zool.)*. A large cell from which many smaller cells are produced by budding, as one of the primary mesoderm cells in developing Polychaeta.

telocentric *(Biol.)*. Having the **centromere** at one end of the chromosome.

telolecithal *(Zool.)*. A type of egg which is large in size, with yolk constituting most of the volume of the cell, and with the relatively small amount of cytoplasm concentrated at one pole. Found in Sharks, Skates, Reptiles, and Birds. Cf. **mesolecithal**, **oligolecithal**.

telome *(Bot.)*. Hypothetical morphological unit of primitive vascular plants; an ultimate branch of an axis that repeatedly branches dichotomously.

telomere *(Biol.)*. The structure which terminates the arm of a chromosome.

telome theory *(Bot.)*. Proposal that the shoots of modern land plants have evolved from repeatedly dichotomously branched axes.

telomorph *(Bot.)*. The sexual or perfect stage of a fungus. Cf. **anamorph**.

telophase *(Biol.)*. Final phase of mitosis with cytoplasmic division; the period of reconstruction of nuclei which follows the separation of the daughter chromosomes in mitosis.

telson *(Zool.)*. The post-segmental region of the abdomen in some Crustacea and Chelicerata.

Telstar *(Telecomm.)*. First transatlantic telecommunications satellite, launched 1962.

temazepam *(Med.)*. Short acting **benzodiazepine** drug useful for inducing sleep, particularly in the elderly.

témoin *(Civ.Eng.)*. An undisturbed column of earth left on an excavated site, as an indication of the depth of the excavation.

temper *(Glass)*. The amount of residual stress in annealed ware measured by comparison with strain disks.

temperature *(Phys.)*. A measure of whether two systems are relatively hot or cold with respect to one another. Two systems brought into contact will, after sufficient time, be in thermal equilibrium and will have the same *temperature*. A *thermometer* using a temperature scale established with respect to an arbitrary zero (e.g. **Celsius** scale) or to absolute zero (Kelvin thermodynamic scale) is required to establish the relative temperatures of two systems. See zeroth law of **thermodynamics**.

temperature *(Vet.)*. Human body temperature in health is about 37°C. Normal values for animals are (±0.5°C, ±1°F): cattle, 38.5°C (101.5°F); horse, 38°C (100.5°F); sheep, 39.5°C (103°F); goat, pig and dog, 39°C (102°F); cat, 38.5°C (101.5°F).

temperature coefficient *(Biol.)*. The ratio of the rate of progress of any reaction or process, at a given temperature, to the rate at a temperature 10°C lower. Also Q_{10}.

temperature coefficient *(Eng.,Phys.)*. The fractional

change in any particular physical quantity per degree rise of temperature.

temperature coefficient of resistance *(Elec.Eng.)*. In any conductor, if

$$R = R_0[1 + \alpha(T - T_0)],$$

R is the resistance at temperature T, compared with the resistance R_0 at temperature T_0, the mean coefficient in the range T_0T being α. This is useful only for *linear resistances*, i.e., pure metals.

temperature correction *(Surv.)*. A correction applied to the observed length of a base line to correct for any difference between the temperature of the tape during the measurement and that at which it was calibrated.

temperature cycle *(Phys.)*. Method of processing thick photographic nuclear research emulsions to ensure uniform development. They must be soaked in solutions at refrigerated temperatures and then warmed for the required processing period.

temperature inversion *(Meteor.)*. Anomalous increase in temperature with height in the troposphere.

temperature lapse rate *(Meteor.)*. The rate of decrease of temperature with height.

temperature-limited *(Electronics)*. Said of a thermionic device operated under saturation conditions, i.e. with the electrode currents limited by the cathode temperature.

temperature screws *(Horol.)*. The screws in the rim of a compensation balance, excluding the quarter screws.

temper brittleness *(Eng.)*. A type of brittleness that is shown by the notched bar test, but not by the tensile test, in certain types of steel after tempering; influenced to a marked extent by the composition of the steel, the tempering temperature and the subsequent rate of cooling. Caused by precipitation of carbides or unequal segregation.

temper colour *(Eng.)*. In tempering hardened steel cutting tools etc., the colour of the oxide layer which forms on reheating and which indicates approximately the correct quenching temperature for a particular purpose. See **tempering**.

tempered scale *(Acous.)*. The musical scale of keyboard instruments, and, by implication, any other instruments or voices which are concerted with them, in which all semi-tones have frequencies of the same ratio (1 : 1.059 463 09), so that 12 semi-tones amount to one octave. Also called *equi-tempered scale*, *equal-tempered scale*. See also **natural scale**.

temper-hardening *(Eng.)*. A term applied to alloys that increase in hardness when heated after rapid cooling; also to the operation of producing this. Also called *artificial ageing*; distinguished from *ageing*, which occurs at atmospheric temperature. Both processes are covered by the term **precipitation hardening**.

tempering *(Eng.)*. The reheating of hardened steel at any temperature below the critical range, in order to decrease the hardness. Also called *drawing*. Sometimes applied to reheating after rapid cooling, even when this results in increased hardness, e.g. in the case of steels that exhibit secondary hardening. See also **austempering**.

template *(Build.)*. (1) A long flat stone supporting the end of a beam, to spread the load over several joints in the brickwork. (2) A framework of timber or steel used for the final setting out of girders, roof trusses, etc.

template *(Eng.)*. A thin plate, cut to the shape or profile required on a finished surface, by which the surface is marked off or gauged during machining or other operation. Sometimes called *templet*.

temple *(Textiles)*. Devices fitted to weaving machines to keep the cloth at full width and suitably tensioned from side to side as it is woven.

templet *(Build.,Eng.)*. See **template**.

tempolabile *(Chem.)*. Tending to change with time.

temporal *(Zool.)*. A cartilage bone of the Mammalian skull formed by the fusion of the petrosal with the squamosal.

temporal summation *(Behav.)*. The phenomenon in which the summational effect of a subthreshold stimulus, which is presented over an extended period of time, may produce a response.

temporal vacuities, openings *(Zool.)*. In Reptiles, openings in the skull, varying in number (none, one or two) and position, and used in classification. The various conditions found in different groups are known as *anapsid*, *synapsid*, *parapsid*, *euryapsid* and *diapsid*.

temporary film *(Eng.)*. A soluble removable protective coating for metals, usually prepared from a lanolin derivative combined with suitable resins and dyed to a distinctive colour. The whole is dispersed in petrol or other solvent.

temporary hardness *(Chem.)*. Hardness of water caused by the presence of the hydrogen carbonates of calcium and magnesium and therefore removable by boiling to precipitate the carbonate.

temporary memory *(Comp.)*. See **buffer**.

temporary threshold shift *(Acous.)*. Temporary hearing loss after exposure to high sound level. Abbrev. *TTS*.

temporary way *(Civ.Eng.)*. The ballast, sleepers and rails laid temporarily by a contractor for his use on constructional works in transporting material.

tenacity *(Eng.)*. See **ultimate tensile stress**.

tendency *(Behav.)*. A general term referring to some measure of the probability that a behaviour will occur, without specifying the nature of the underlying causal factors.

tender ship *(Ships)*. A ship with a small transverse metacentric height and long rolling period. Cf. **stiff ship**.

tendinous *(Zool.)*. See **tendon**.

tendo calcaneus *(Zool.)*. See **Achilles tendon**.

tendon *(Zool.)*. A cord, band, or sheet of fibrous tissue by which a muscle is attached to a skeletal structure, or to another muscle. adj. *tendinous*.

tendril *(Bot.)*. A slender, simple or branched, elongated organ used in climbing, at first soft and flexible, later becoming stiff and hard. May be a modified stem, leaf, leaflet or inflorescence.

tenebrescence *(Min.)*. The reversible bleaching observed in the hackmanite variety of sodalite. This mineral has a pink tinge when freshly fractured; the colour fades on exposure to light, but returns when the mineral is kept in the dark for a few weeks or is bombarded by X-rays.

tenesmus *(Med.)*. Painful and ineffectual straining at stool.

tenia, teniasis *(Zool.)*. See **taenia** etc.

tennantite *(Min.)*. Sulphide of copper and arsenic, which crystallizes in the cubic system. It is isomorphous with **tetrahedrite**. The crystals frequently contain antimony, and grade into tetrahedrite. Also called *fahlerz*.

tenon *(Build.)*. A tongue formed on the end of a member by cutting away from both sides one-third of the thickness of the member. The projecting part fits into a mortise in a second member in order to make a joint between them. See also **tusk tenon**.

tenon-and-slot mortise *(Build.)*. A joint, such as that made between the posts and heads of solid door frames, in which a tenon cut on the end of the head fits into a **slot mortise** on the end of the post.

tenon saw *(Build.)*. A saw with a very thin parallel blade, having fine teeth (4 to 6 per centimetre) and a stiffened back along its upper edge. Also called a *mitre saw*.

tenorite *(Min.)*. Copper oxide, crystallizing in the triclinic system. Occurs in minute black scales as a sublimation product in volcanic regions or associated with copper veins. *Melaconite* is a massive variety.

tenosynovitis, tenovaginitis *(Med.)*. Inflammation of the sheath of a tendon.

tenotomy *(Med.)*. The cutting of a tendon for the correction of deformity.

tensile strength *(Textiles)*. The force required to stretch a textile material until it breaks.

tensile stress *(Eng.)*. See **stress, ultimate tensile stress**.

tensile test *(Eng.)*. Two main forms are (1) those in which a static increasing pull is applied until fracture results. From this a stress-strain curve may be plotted and the proof stress, yield point, ultimate tensile stress and elongation determined; and (2) those in which a dynamic load is applied giving data on fatigue and impact.

tensile testing machine *(Eng.)*. A machine for applying a tensile or compressive load to a test piece, by means of hand- or power-driven screws, or by a hydraulic ram. The load is usually measured by a poise weight and calibrated lever.

tensimeter *(Chem.)*. An apparatus for the determination of transition points by observation of the temperature at which the vapour pressures of the two modifications become equal.

tension *(Elec.Eng.)*. Former term used to designate a potential difference, e.g. low-tension (such as an accumulator), or high-tension supplies (such as a power-distribution cable). Now obsolete.

tension insulator *(Elec.Eng.)*. A suspension insulator for overhead transmission lines, which is designed to withstand the pull of the conductors; it is used, therefore, at terminal, anchor, or angle towers.

tension pin *(Eng.)*. A hollow dowel pin of elastic material, slotted so as to be deformed on assembling, to take up misalignment or shape irregularity, or to exert a force.

tension plate lock-up *(Print.)*. A method of locking plates to the cylinder by fingers which grip in slots on the underside of the plate. Cf. **compression plate lock-up**.

tension rod *(Eng.)*. A structural member subject to tensile stress only. Also *tie rod.*

tension wood *(Bot.)*. The **reaction wood** of dicotyledons, formed on the upper side of horizontal branches for example and characterized by a lower than normal lignin content.

tensometer *(Eng.)*. A versatile, portable testing machine used for a variety of mechanical tests, including tensile tests.

tensor *(Maths.)*. The generalization of a vector. A mathematical entity specifiable by a set of components with respect to a system of co-ordinates and such that the transformation that has to be applied to the components to obtain components with respect to a new system of co-ordinates is related in a certain way to the transformation that had to be applied to the system of co-ordinates.

tensor *(Zool.)*. A muscle which stretches or tightens a part of the body without changing the relative position or direction of the axis of the part.

tensor force *(Phys.)*. A noncentral nuclear force which depends on the spin orientation of the nucleons.

tent *(Med.)*. A roll or plug of soft absorbent material, or of expansible material, for keeping open a wound or dilating an orifice.

tentacle *(Zool.)*. An elongate, slender, flexible organ, usually anterior, fulfilling a variety of functions in different forms, e.g. feeling, grasping, holding and sometimes locomotion. Also *tentaculum*. adjs. *tentacular, tentaculiferous, tentaculiform.*

tenter *(Textiles)*. See **stenter**.

tenth-value thickness *(Nuc.Eng.)*. Thickness of absorbing sheet which attenuates intensity of beam of radiation by a factor of ten.

tentorium *(Zool.)*. (1) In the Mammalian brain, a strong transverse fold of the dura mater, lying between the cerebrum and the cerebellum. (2) In Insects, the endoskeleton of the head.

tepal *(Bot.)*. One of the members of a perianth which is not clearly differentiated into a calyx and a corolla.

tepee buttes *(Geol.)*. Conical hills of Cretaceous shale, with steep, smooth slopes of talus and a core of shell-limestone, formed *in situ* by the growth of successive generations of lamellibranchs (*Lucina*). Found in the Great Plains of the US.

tephigram *(Meteor.)*. **Aerological diagram** in which the principal rectangular axes are temperature (T) and entropy (Φ): hence TΦ-gram. Equal area represents equal energy at all points.

tephra *(Geol.)*. A general term for all fragmental volcanic products. e.g. *ash, bombs, pumice.*

tephrite *(Geol.)*. A fine-grained igneous rock resembling basalt and normally occurring in lava flows; characterized by the presence of a feldspathoid mineral in addition to, or in place of, feldspar.

tephroite *(Min.)*. An orthosilicate of manganese, which crystallizes in the orthorhombic system. It forms a member of the olivine isomorphous group, and occurs with zinc and manganese minerals.

tera- *(Genrl.)*. Prefix, symbol *T*, denoting 10^{12} times, e.g. a *terawatt-hour* is 10^{12} watt-hours.

teratogen *(Biol.)*. An agent which raises the incidence of congenital malformations.

teratogenic *(Med.)*. A substance or drug producing abnormal embryos.

teratology *(Biol.)*. The study of monstrosities, as an aid to the understanding of normal development. (Gk. *teras*, gen. *teratos*, a wonder.) See **teratoma**.

teratoma *(Med.)*. A tumour in the body consisting of tissues believed to be derived from the three germ layers (ectoderm, mesoderm, and endoderm). May occur in a variety of locations but commonly in the testis and mediastinum. If they contain a predominance of ectodermal elements thay are called *dermoid cysts.*

terbium *(Chem.)*. A metallic element, a member of the rare earth group. Symbol Tb, at. no. 65, r.a.m. 158.9254. It occurs in the same minerals as dysprosium, europium and gadolinium.

tercom *(Aero.)*. *TERrain COMparison, TERrain COntour Matching*. Stored digital data of ground contours are compared with those detected below the aircraft in flight and used to identify the aircraft's present position and track.

terebine *(Build.)*. Volatile solvents and thinners derived from heavy petroleum and rosin oils, or petroleum and rosin oils mixed with turpentine, used to accelerate the drying process of oil paints.

terebrate *(Zool.)*. Possessing a boring organ, or a sting.

terephthalic acid *(Chem.)*. $C_6H_4(COOH)_2$, *benzene-1,4-dicarboxylic acid*, a powder, hardly soluble in water or ethanol, which sublimes unchanged. It is prepared by the oxidation of 4-toluic acid, or industrially by the catalytic oxidation of 4-xylene. Important material in the manufacture of **Terylene**.

terete *(Bot.)*. Rounded, more or less cylindrical and neither ridged, grooved nor angled.

tergum *(Zool.)*. The dorsal part of a somite in *Arthropoda*; one of the plates of the carapace in Cirripedia. adj. *tergal.*

term diagram *(Phys.)*. Energy-level diagram for isolated atom in which levels are usually represented by corresponding quantum numbers.

terminal *(Comp.)*. Term used to describe any input/ouput device which is used to communicate with the computer from a remote site. See **VDU**, **keyboard**, **printer**.

terminal *(Elec.Eng.)*. A point in an electrical circuit at which any electrical element may be connected.

terminal bar *(Elec.Eng.)*. A bar to which a group of plates of an accumulator is attached. Also called *connector bar, terminal yoke.*

terminal curve *(Horol.)*. The curve which connects the inner end of a balance spring or the outer end to the stud.

terminal deoxynucleotidyl transferase *(Immun.)*. An enzyme found in pre-T and pre-B lymphocytes and cortical thymocytes but absent from their progeny. Function uncertain, but perhaps involved in the initial arrangement of genes for the antigen receptors. The presence of the enzyme is used as a marker for cells which contain it. Abbrev. *TdT.*

terminal equipment *(Telecomm.)*. The special apparatus required for connecting the normal telephone exchange pairs to special transmission systems (such as radio-telephone links, carrier systems) or to trunk lines.

terminal impedance *(Phys.)*. End or load impedance.

terminal lug *(Elec.Eng.)*. A projection on a group of accumulator plates for connection to an external circuit.

terminal pillar *(Elec.Eng.)*. See **post head**.

terminal pole *(Elec.Eng.)*. A pole at the end of a power-transmission or telephone line so designed as to withstand the longitudinal load of the conductors as well as the vertical load.

terminal tower *(Elec.Eng.)*. The transmission line tower at the end of an overhead transmission line; arrangements must be made for taking the pull of the conductors, and for connecting them to the substation or other apparatus at the end of the line.

terminal velocity *(Aero.)*. The maximum **limiting velocity** attainable by an aircraft as determined by its total **drag** and thrust.

terminal velocity *(Phys.)*. The constant velocity acquired by a body falling through a fluid when the frictional resistance is equal to the gravitational pull.

terminal-velocity dive *(Aero.)*. A nose dive to the greatest obtainable velocity of the machine at that altitude.

terminal voltage *(Elec.Eng.)*. The voltage at the terminals of a piece of electrical equipment, e.g. an electrical machine or a power supply.

terminal yoke *(Elec.Eng.)*. See **terminal bar**.

terminated level *(Telecomm.)*. The reading of a level-measuring set at a point in a system when terminated at that point by a resistance equal to the nominal impedance of the system.

termination *(Elec.Eng.)*. That which is connected at the end of an electrical transmission system or to output of an amplifier. It can be regarded, in part, in the same way as the electrical load, but the term implies an interest in physical nature as well as electrical properties, e.g. a termination could be a loud-speaker whereas the associated load would be, say, 8 ohms.

terminator *(Astron.)*. The border between the illuminated and dark hemispheres of the moon or planets. Its apparent shape is an ellipse and it marks the regions where the sun is rising or setting.

terminator *(Comp.)*. Specified value, not normally expected in the data, which is used to terminate a list of data items. Also *rogue value*.

termitarium *(Zool.)*. A mound of earth built and inhabited by termites and containing an elaborate system of passages and chambers.

termite shield *(Build.)*. A sheet of copper or other noncorroding metal inserted between the foundation and woodwork in buildings, that serves as a barrier to the movement of subterranean termites from the ground to the woodwork of the structure. Also called *antproof course*.

termolecular *(Chem.)*. Pertaining to three molecules.

ternary *(Chem.)*. Consisting of three components etc.

ternary *(Maths.)*. Counting with **radix** 3.

ternary system, ternary diagram *(Eng.)*. The alloys formed by three metals constitute a ternary alloy system, which is represented by the ternary constitutional diagram for the system.

ternate *(Bot.)*. Arranged in threes; especially a compound leaf with three leaflets.

terne metal *(Eng.)*. Lead alloy with up to 18% tin and 1.5–2% antimony, used to coat steel to improve its corrosion and working properties.

terne plate *(Eng.)*. Iron or steel sheet coated by hot-dipping with **terne metal** which acts as a carrier for lubricant used in subsequent drawing operations, as rust protection, as a paint base or to facilitate soldering.

terpadienes *(Chem.)*. $C_{10}H_{16}$, isocyclic compounds, containing two double bonds. Numerous compounds in the terpene series are terpadienes.

terpenes *(Chem.)*. Compounds of the formula $(C_5H_8)_n$, the majority of which occur in plants. The value of n is used as a basis for classification. (i) Monoterpenes $C_{10}H_{16}$; (ii) sesquiterpenes $C_{15}C_{24}$; (iii) diterpenes $C_{20}H_{32}$; (iv) triterpenes $C_{30}H_{48}$; (v) tetraterpenes $C_{40}H_{64}$. All naturally occurring terpenes can be built up of isoprene units. In practice the name terpene is often used for monoterpenes.

terpenoids *(Bot.)*. A group of plant **secondary metabolites** based on one to four or more isoprene (C_5) units, including many essential oils, the gibberellins, carotenoids, plastoquinone, rubber.

terpineol *(Chem.)*. $C_{10}H_{17}OH$, colourless crystals; mp 37°C, bp 218°C; it can be obtained from limonene hydrochloride by the action of potassium hydroxide. Terpineol is used extensively as the basis of certain perfumes and in soap perfumery.

terprom *(Aero.)*. *TERrain PROfile Matching*. See **tercom**.

terracotta *(Build.)*. A durable composition of fine clay, fine sand, crushed pottery waste, etc., easily manipulable before hardening (by fire, like bricks). Generally unglazed, used for sculpture, pottery, and architecturally in the form of building blocks, decorative tiles, etc.

terrain-avoidance system *(Aero.)*. A system providing the pilot with a situation display of the ground or obstacles which project above a plane containing the aircraft so that the pilot can avoid the obstacles. Cf. **terrain-clearance system**.

terrain-clearance system *(Aero.)*. A fully-automatic system for sensing ground obstructions and guiding the aircraft away from them without pilot intervention. Cf. **terrain-avoidance system**.

terrane, displaced/suspect *(Geol.)*. A geologically consistent area, discontinuous with that of its neighbours, whose distinctive stratigraphical or structural features and its geological history, indicate that it is *foreign* to the region.

terrazzo *(Build.)*. A rendering of cement (white or coloured) and marble or granite chippings, used as a covering for concrete floors, on which it is floated and finally polished with abrasive blocks and fine grit stones; frequently precast (particularly for sills). Also *Venetian mosaic*.

terrestrial equator *(Geol.)*. An imaginary circle on the surface of the Earth, the latter being regarded as being cut by the plane through the centre of the Earth perpendicular to the polar axis; it divides the Earth into the northern and southern hemispheres, and is the primary circle from which terrestrial latitudes are measured.

terrestrial magnetism *(Geol.,Phys.)*. The magnetic properties exhibited within, on, and outside the earth's surface. There is a nominal (magnetic) North pole in Canada and a nominal South pole opposite, the positions varying cyclically with time. The direction indicated by a compass needle at any one point is that of the horizontal component of the field at the point. Having the characteristics of flux from a permanent magnet, the earth's magnetic field probably depends on currents within the earth and also on those arising from ionization in the upper atmosphere, interaction being exhibited by the Aurora Borealis.

terrestrial poles *(Geol.)*. The two diametrically opposite points in which the Earth's axis cuts the Earth's surface are the geographical poles. The magnetic poles, the positions to which the compass needle will point, are unstable and differ from the geographical. N magnetic pole: ca. 76° N, 101° W; S magnetic pole: ca. 66° S, 139° E. See **terrestrial magnetism**.

terrestrial radiation *(Meteor.)*. At night the earth loses heat by radiation to the sky, the maximum cooling occurring when the sky is cloudless and the air dry. Dew and hoar-frost are the result of such cooling.

terrestrial telescope *(Phys.)*. Telescope consisting of an objective and a four-lens eyepiece (*terrestrial eyepiece*), giving an erect image (see **erecting prism**) of a distant object.

terrigenous sediments *(Geol.)*. Sediments derived from the erosion of the land. They include sediments deposited on land and land-derived material deposited in the sea.

territory *(Behav.,Ecol.)*. Areas defended against other individuals, usually of the same species. Territoriality exists among many different species and may involve aggressive encounters, but it is often regulated by less overt behaviour. Serves a wide range of functions: feeding, mating etc. Not all species are territorial and many occupy a **home range** which is not actively defended.

terro-metallic clinkers *(Build.)*. Similar to **klinker brick** but made from a clay which burns to a nearly black colour.

terry fabric *(Textiles)*. Special woven cloth using an additional warp to produce loops on one or both sides of the ground fabric. Often used for towels. See also **plush (weft-knitted)**.

Tertiary *(Geol.)*. The first period of the Cenozoic era, covering an app. time span from 65–2 million years ago. It includes two sub-periods, the *Paleogene* and the *Neogene*.

tertiary alcohols *(Chem.)*. Alcohols containing the group ≡C.OH. When oxidized, the carbon chain is broken up, resulting in the formation of two or more oxidation products containing a smaller number of carbon atoms in the molecule than the original compound.

tertiary amines *(Chem.)*. Amines containing the nitrogen atom attached to three groups. Tertiary aliphatic-aromatic amines yield 4-nitroso compounds with nitrous (nitric (III)) acid.

tertiary colours *(Build.)*. Term for the colour resulting from the mixture of two secondary colours, e.g. olive produced by mixing green and orange.

Tertiary igneous rocks *(Geol.)*. The various types of igneous rocks which were intruded or extruded during early Tertiary times, especially over a region stretching from Britain to Iceland; e.g. in the Inner Hebrides and northeast Ireland (the Thulean Province).

tertiary nitro compounds *(Chem.)*. Nitro compounds containing the group $\equiv C.NO_2$. They contain no hydrogen atom attached to the carbon atom next to the nitro group, and they have no acidic properties.

tertiary production *(Min.Ext.)*. Special methods of increasing oil flow after **primary production** (natural flow by gravity or intrinsic pressure) and **secondary production** (pressurizing the reservoir) are exhausted. Includes chemical treatment and water injection. Also *tertiary recovery*.

tertiary structure *(Biol.)*. The 3-dimensional configuration of polymers which is a stable folding of the sequence of units (bases or peptides) along the polymer, i.e. of their secondary structure.

tertiary wall, tertiary thickening *(Bot.)*. A deposit of wall material on the inner surface, next to the lumen, of a secondary wall, often in the form of helical strips, as in the tracheids of the yew.

tertiary winding *(Phys.)*. A third winding on a transformer core, linking the same flux as the primary and secondary windings. The functions, and the names, of all three are interchangeable. May be used for additional output or monitoring.

tervalent *(Chem.)*. See **trivalent**.

Terylene *(Chem.)*. TN for straight-chain polyester fibres derived from condensation of **terephthalic acid** with ethan-1,2-diol, used widely in the manufacture of fabrics, clothing materials, and other textiles.

Teschen disease *(Vet.)*. *Infectious pig paralysis, virus encephalomyelitis of swine*. Characterized by mild fever, nervous excitement, and paralysis.

teschenite *(Geol.)*. A coarse-grained basic (gabbroic) igneous rock consisting essentially of plagioclase, near labradorite in composition, titanaugite, ilmenite, and olivine (or its decomposition products); primary analcite occurs in wedges between the plagioclase crystals, which it also veins.

tesla *(Phys.)*. SI unit of magnetic flux density or magnetic induction equal to 1 weber m^{-2}. Equivalent definition; the magnetic induction for which the maximum force it produces on a current of unit strength is 1 newton. Symbol T.

Tesla coil *(Elec.Eng.)*. Simple source of high-voltage oscillations for rough testing of vacua and gas (by discharge colour) in vacuum systems.

tessellated pavement *(Build.)*. A pavement formed of small pieces of stone, marble, etc., in the manner of a mosaic. Also *Roman mosaic*.

tessera, tessella *(Arch.)*. One of the small pieces of stone, marble, etc., used in the mosaic of a tessellated pavement. pls. *tesserae, tessellae*.

test *(Bot.,Zool.)*. See **testa**.

testa *(Bot.)*. The seed coat, several layers of cells in thickness, derived from the integuments of the ovule.

testa, test *(Zool.)*. A hard external covering, usually calcareous, siliceous, chitinous, fibrous or membranous; an exoskeleton; a shell; a lorica. adjs. *testaceous, testacean*.

test bed *(Eng.)*. Area with full monitoring facilities for testing new or repaired machinery under full working conditions.

test board *(Elec.Eng.)*. A switchboard carrying instruments and switches for connecting up to apparatus to be tested.

test data *(Comp.)*. Data, including expected results, used to test a program or flowchart.

test desk *(Telecomm.)*. The special position where tests can be applied to faulty lines to discover the cause of faults and to issue instructions for remedying them.

testes *(Zool.)*. See **testis**.

test final selector *(Telecomm.)*. The selector following the test selector, which enables test clerks at the test desk to get on to a subscriber's line.

testicle *(Zool.)*. See **testis**.

testing machine *(Eng.)*. A machine for applying accurately measured loads to a test piece, to determine the suitability of the latter for a particular purpose.

testing position *(Telecomm.)*. A position equipped for testing purposes and forming part of a test desk or test rack.

testing set *(Elec.Eng.)*. A self-contained set of apparatus, including switches, instruments etc., for carrying out certain special tests.

testing transformer *(Elec.Eng.)*. A specially designed transformer providing a high-voltage supply for testing purposes.

testis, testicle *(Zool.)*. A male gonad or reproductive gland responsible for the production of male germ-cells or sperms. adj. *testicular*. pl. *testes*.

test jack *(Telecomm.)*. One with contacts in series with a circuit, so that a testing device can be immediately introduced for locating faults.

test pattern *(Image Tech.)*. A transmitted chart with lines and details to indicate particular characteristics of a transmission system, used in TV for general testing purposes. Also *test card, test chart*.

test piece *(Eng.)*. A piece of material accurately turned or shaped, often to specified standard dimensions, for subjecting to a tensile test, shock test etc., in a testing machine.

test point *(Elec.Eng.)*. Designated junction between components in an equipment, where the voltage can be stated (as a minimum or with tolerance) for a quick verification of correct operation.

test record *(Acous.)*. A gramophone record specially made for the testing of reproducing equipment, having on its surface constant-frequency or gliding tones, or selected recordings of speech or music, to emphasize particular faults in the subsequent reproduction.

test selector *(Telecomm.)*. A selector operated by a test clerk in an automatic exchange; by means of it, through a **test final selector**, he is able to get on to any line in the exchange.

test terminals *(Elec.Eng.)*. Circuit terminals to which a connection is made for purposes of testing.

test vehicles *(Aero.)*. Aircraft for aerodynamic, control and other tests in guided weapon development. They may simply be for gathering basic information, or they may be actual missiles without a warhead. They are known by their initials: CTV, command (control) test vehicle; GPV, general-purpose vehicle; MTV, missile test vehicle; RJTV, ramjet test vehicle; RTV, rocket test vehicle.

tetanic contraction *(Zool.)*. See **tetanus**.

tetanus *(Med.)*. *Lockjaw*. A disease due to infection with the tetanus bacillus, *Clostridium tetani* the toxin secreted by which causes the symptoms and signs of the disease, viz., painful tonic spasms of the muscles, which usually begin in the jaw and then spread to other parts.

tetanus *(Zool.)*. The state of prolonged contraction which can be induced in a muscle by a rapid succession of stimuli.

tetanus antitoxin *(Immun.)*. Antibody to tetanus toxin, usually prepared in horses which have been hyperimmunized against the exotoxin of *Clostridium tetani*. Used for the prevention of tetanus in humans and animals following possible contamination of wounds. Repeated use is dangerous because of the likelihood that the subject will have been primed against horse serum proteins and may develop **anaphylactic shock** or **serum sickness**. Even a single dose may cause serum sickness. These dangers are avoided by using antiserum or immunoglobulin from humans immunized against tetanus, available in some countries. Since the production of tetanus toxin at the site of infection takes place only slowly, if a person has already been immunized against tetanus a *booster injection* of tetanus toxoid will elicit a secondary response rapidly enough to prevent tetanus from developing.

tetanus toxin *(Immun.)*. The toxin produced by *Clostridium tetani*. A neurotoxin that blocks synaptic transmission in the spinal cord and also blocks neuromuscular transmission. Tetanus toxin binds to a glycolipid, *disialosyl ganglioside* which is particularly rich in the membranes of nerve cells. After inactivation by treatment with formaldehyde to form tetanus toxoid this is used for prophylactic immunization against tetanus.

tetany *(Med.)*. A condition characterized by heightened excitability of the motor nerves and intermittent painful muscular cramps, occurring in many abnormal states, especially those associated with hypocalcaemia.

tethered satellite *(Space)*. A term used for a satellite which is deployed from a spacecraft and attached to it by a wire or tether which may measure over 100 km.

Tethys *(Astron.)*. The third natural satellite of **Saturn**, 1000 km in diameter.

tetra *(Genrl.)*. From Gk. *tetra-*, a prefix indicating four, e.g. *tetracyte*.

tetraboric acid *(Chem.)*. See **boric acid**.

tetrachloroethane *(Chem.)*. $C_2H_2Cl_4$, used as a solvent and also as a vermifuge. Bp 146°C. Prepared by the chlorination of ethyne in the presence of antimony (III) chloride. Intermediate in the manufacture of **trichloroethane**.

tetrachloromethane *(Chem.)*. See **carbon tetrachloride**.

tetrachlorosilane *(Chem.)*. See **silicon tetrachloride**.

tetracyclines *(Med.)*. Group of antibiotics with **wide spectrum** activity but now restricted to treating infections caused by *Chlamydia* (psittacosis, trachoma, urethritis), *rickettsia* (Q fever), *Mycoplasma* (atypical pneumonia) and *Brucella*. Common examples are *tetracycline* and *oxytetracycline*.

tetrad *(Biol.)*. The four haploid cells formed at the end of **meiosis**. The term was formerly used for the four chromatids making up a chromosome pair at the first division of meiosis.

tetradactyl *(Zool.)*. Having four digits.

tetrad analysis *(Bot.)*. The genetic analysis of tetrads in studies of mapping, recombination etc.

tetradymite *(Min.)*. An ore of tellurium, of composition Bi_2Te_2S; crystallizes in the trigonal system. Bismuth tellurides are commonly found in gold-quartz veins.

tetrafluorosilane *(Chem.)*. See **silicon tetrafluoride**.

tetragonal system *(Crystal.)*. The crystallographic system in which all the forms are referred to three axes at right angles; two are equal and are taken as the horizontal axes, whilst the vertical axis is either longer or shorter than these. It includes such minerals as zircon and cassiterite. Also *pyramidal system*.

tetragonous *(Bot.)*. A stem etc. having four angles and four convex faces.

tetrahedrite *(Min.)*. A sulphide of copper and antimony, crystallizing in the tetrahedral division of the cubic system; frequently contains arsenic and other metals. It is used as an ore of copper and, in some cases, of other metals. Also called *fahlerz*, *fahl ore*, *grey copper ore*. See also **tennantite**.

tetrahedron *(Maths.)*. A polyhedron having four triangular faces. adj. *tetrahedral*.

tetrahydrofuran *(Chem.)*. *Tetramethylene oxide*. Organic solvent used in refining lubricating oils and a solvent for a wide range of plastic and resins. Bp 64–66°C, rel.d. (20°C) 0.888.

tetrahydronaphthalene *(Chem.)*. See **tetralin**.

tetralin *(Chem.)*. Organic solvent with a high boiling-point (206°C), used in a wide variety of products, such as floor polishes, paints and varnishes. Also called *tetrahydronaphthalene*.

tetramerous *(Bot.,Zool.)*. Having 4 parts; arranged in 4s; arranged in multiples of 4.

tetramethylene oxide *(Chem.)*. See **tetrahydrofuran**.

tetramethyl rhodamine isothiocyanate *(Immun.)*. A red fluorescent dye used in immunofluorescence techniques. In conjunction with **FITC** it allows two colours to be used together. Abbrev. *TRITC*.

tetramethylsilane *(Chem.)*. $(CH_3)_4Si$, a reference standard for *proton magnetic resonance*.

tetraoses *(Chem.)*. See **tetroses**.

tetraparental chimera *(Immun.)*. A chimera (usually mouse) resulting from the artificially induced fusion of two blastocysts at the 4 or 8 cell stage. The resulting animal contains cells from both parents in all its tissues, and is an example of the maintenence of mutual immunological tolerance.

tetraploid *(Biol.)*. Possessing 4 sets of chromosomes, each chromosome of a set being represented 4 times. Cf. **diploid, haploid**.

tetrapod *(Zool.)*. Having 4 feet.

tetrapterous *(Zool.)*. Having 4 wings.

tetrarch *(Bot.)*. A stele having 4 strands of protoxylem, as in the roots of many dicotyledons.

tetrasomic *(Biol.)*. A tetraploid nucleus (or organism) having 1 chromosome 4 times over, the others in duplicate.

tetrasporophyte *(Bot.)*. The typically diploid phase of the red algal life cycle, developing from carpospores and producing haploid tetraspores.

tetratohedral *(Crystal.)*. Containing a quarter of the number of faces required for the full symmetry of the crystal system.

tetravalent *(Chem.)*. Capable of combining with four hydrogen atoms or equivalent. Having an oxidation or a co-ordination number of four.

tetrazo dyes *(Chem.)*. See **disazo dyes**.

tetrazole *(Chem.)*. A 5-membered heterocyclic compound containing 4 nitrogen atoms and 1 carbon atom in the ring.

tetrode *(Electronics)*. A four-electrode thermionic valve, incorporating a *screen grid*.

tetrode transistor *(Electronics)*. Transistor with additional base contact to improve high-frequency performance.

tetroses *(Chem.)*. *Tetraoses*. Monosaccharides containing 4 carbon atoms in the molecule, e.g. $HO.CH_2.(CH\text{-}OH)_2.CHO$.

tetryl *(Chem.)*. *N,2,4,6-tetranitromethylaniline*, a yellow crystalline compound used as a detonator.

TE-wave *(Telecomm.)*. Abbrev. for *Transverse Electric wave*, having no component of electric force in the direction of transmission of electromagnetic waves along a waveguide. Also known as *H-wave* (since it must have magnetic field component in direction of transmission).

tex *(Textiles)*. The basic unit of the tex system, used to express the linear density (mass per unit length) of fibres, filaments, yarns or other linear material. The tex is the mass in grams of 1 km of material (hence decitex, kilotex etc.). This system is now replacing the traditional ones based on the **denier** and other units.

Texas fever *(Vet.)*. A disease of cattle caused by the protozoon *Babesia bigemina*. See **redwater (1)**.

text *(Print.)*. The body of matter in a printed book, exclusive of preliminary matter, end matter, notes, and illustrations; words set to music, as distinct from the accompanying music.

textiles *(Textiles)*. Term used to describe any fibres, filaments and yarns and any products produced from them.

text processing *(Comp.)*. The storing, revising and

outputting of text using a computer system. See **desk-top publishing, word-processing**.

texture *(Genrl.)*. The mode of union or disposition, in regard to each other, of the elementary constituent parts in the structure of any body or material.

texture *(Geol.)*. The physical quality of a rock which is determined by the relative sizes, disposition, and arrangement of the component minerals. The nomenclature and classification of rocks are governed by mineral composition and texture. See, e.g. **graphic texture, ophitic texture, poikilitic texture**.

texture *(Image Tech.)*. The quality of the surface of a photograph.

texture brick *(Build.)*. See **rustics**.

textured yarn *(Textiles)*. Yarn of synthetic continuous filaments that has been treated to make its surface highly irregular by crimping or by introducing loops having a variety of shapes. Widely used methods include: (1) false-twist texturing in which the yarn is continuously twisted, heat-set and untwisted (no overall twist is produced in the filaments); (2) **stuffer box** crimping; (3) in air-jet texturing the entangled loops are formed by subjecting the yarn to a turbulent air stream. The yarns have increased bulk and are sometimes described as bulked continuous filaments (BCF). They are also more easily extended and are used in articles (e.g. tights) that closely conform to the shape of the body.

texture paints *(Build.)*. Paints with an added aggregate which when applied to a surface are manipulated to create a pattern or design which is raised from the surface. Manipulation is done with rubber stipplers, serrated combs or brushes.

TFTR *(Nuc.Eng.)*. *Tokamak Fusion Test Reactor*. Large **tokamak** experiment at Princetown, USA.

TGA *(Chem.)*. Abbrev. for *Thermal Gravimetric Analysis*.

T-gauge *(Build.)*. Type of marking gauge with long marker pin and large flat fence; used for marking jobs where mouldings, beading, or other surface items have to be cleared.

Tg point *(Glass)*. See **transformation points**.

T-grain *(Image Tech.)*. Silver halide grains in tablet form, offering greater area and hence greater sensitivity in a photographic emulsion.

TGT *(Aero.)*. See **gas temperature**.

Th *(Chem.)*. The symbol for thorium.

thalamus *(Bot.)*. Same as the **receptacle** of a flower.

thalamus *(Zool.)*. In the Vertebrate brain, the larger, more ventral part of the dorsal zone of the thalamencephalon.

thalassaemia *(Med.)*. A group of inherited anaemias in which there is a defect in the alpha or beta chains of haemoglobin, *alpha-, beta-thalassaemia*. Thalassaemia *minor* is used to describe heterozygotes and thalassaemia *major* for the homozygotes.

thalasso- *(Genrl.)*. Prefix from Gk. *thalassa*, sea.

thalassophyte *(Bot.)*. A seaweed.

thalidomide *(Med.)*. A non-barbiturate sedative drug; withdrawn (UK) in 1961 because of teratogenic effects.

thallium *(Chem.)*. Symbol Tl, at. no. 81, r.a.m. 204.37, rel.d. 11.85, mp 303.5°C, bp 1650°C. White malleable metal like lead. Several thallium isotopes are members of the uranium, actinium, neptunium and thorium radioactive series. Thallium isotopes are used in scintillation crystals. Its compounds are very poisonous and it has common valencies of one and three.

thallus *(Bot.)*. Plant body not differentiated into leaves, stems and roots but consisting of a single cell, a colony, a filament of cells, a mycelium or a large branching multicellular structure. The plant body of the algae, fungi and thalloid liverworts. adj. *thalloid*.

thalofide cell *(Electronics)*. A photoconducting cell employing thallium oxy-sulphide as the light-sensitive agent; sensitive to far red and infra-red.

thalweg *(Geol.)*. (German 'valley way'.) The name frequently used for the longitudinal profile of a river, i.e. from source to mouth.

thanatocoenosis *(Geol.)*. Assemblage of fossil remains of organisms which were not associated during their life but were brought together after death.

thanatoid *(Zool.)*. Poisonous, deadly, lethal; as some venomous animals.

thanatosis *(Zool.)*. Same as **sham death**.

thatching *(Build.)*. A form of roof-covering composed of courses of reeds, straw or heather, laced together.

thawing *(Phys.)*. A term used to describe the beginning of the fusion process of a solid. The corresponding temperature is the *thaw-point* of the solid.

Thebe *(Astron.)*. A tiny natural satellite of **Jupiter**, discovered in 1979 by the Voyager 2 mission.

Thebesian valve *(Zool.)*. An auricular valve of the Mammalian heart.

theca *(Zool.)*. A case or sheath covering or enclosing an organ, as the *theca vertebralis* or dura mater enclosing the spinal cord; a tendon sheath; the wall of a coral cup. adjs. *thecal, thecate*.

thecodont *(Zool.)*. Having the teeth implanted in sockets in the bone which bears them.

thelytoky *(Zool.)*. Parthenogenesis resulting in the production of females only.

thematic apperception test *(Behav.)*. A **projective technique** in which persons are shown a set of pictures and asked to write a story about each. Abbrev. *TAT*.

thenardite *(Min.)*. Sodium sulphate, crystallizing in the orthorhombic system and occurring in saline residues of alkali lakes.

theodolite *(Surv.)*. An instrument for measuring horizontal and vertical angles by means of a telescope mounted on an axis made vertical by levelling screws, and rotated both horizontally on this axis and in horizontal bearings. Circular graduated plates are used to measure amount of rotatory motion when the telescope is sighted on successive signal stations.

theophylline *(Med.)*. Occurs in tea and is a strong diuretic and muscle relaxant. Little used, its more soluble derivative, *aminophylline*, being preferred.

theorem *(Maths.)*. A statement of a mathematical truth together with any qualifying conditions.

theorem of the equipartition of energy *(Chem.)*. See **principle of the equipartition of energy**.

theoretical plate *(Chem.Eng.)*. A concept, used in distillation design, of a plate in which the vapour and liquid leaving the plate are in equilibrium with each other.

theories of light *(Phys.)*. Interference and diffraction phenomena are explained by the *wave theory*, but when light interacts with matter the energy of the light appears to be concentrated in *quanta*, called photons. The *quantum* and *wave theories* are supplementary to each other.

theory of indicators *(Chem.)*. See **Ostwald's-**.

theralite *(Geol.)*. A coarse-grained, holocrystalline igneous rock composed essentially of the minerals labradorite, nepheline, purple titanaugite, and often with soda-amphiboles, biotite, analcite or olivine.

therapeutic *(Med.)*. Pertaining to the medical treatment of disease (*therapy*); remedial; curative; preventive. Hence *therapeutics*, that part of medical science which deals with the treatment of disease; the art of healing.

therapeutic ratio *(Radiol.)*. The ratio between tumour lethal dose and tissue tolerance. In radio-resistant tumours, the tumour lethal dose equals, or is greater than, the dose required to destroy normal tissues.

Theria *(Zool.)*. A subclass of the Mammals, containing the extinct Patriotheria, the Methatheria (marsupials) and the Eutheria (placentals).

therm *(Phys.)*. A unit of energy used mainly for the sale of gas; equals 10^5 Btu or 105.5 MJ.

thermal *(Aero.,Meteor.)*. An ascending current due to local heating of air, e.g. by reflection of the sun's rays from a beach.

thermal ammeter *(Elec.Eng.)*. One in which the deflection of the pointer depends on the sag of a fine wire, carrying the current to be measured, due to thermal expansion.

thermal analysis *(Eng.)*. The use of cooling or heating curves in the study of changes in metals and alloys. The freezing-points and the temperatures of any polymorphic changes occurring in pure metals may be determined. The freezing ranges and temperatures of changes in solid alloys may also be studied. The data obtained are used in constructing constitution diagrams.

thermal capacity *(Phys.)*. The amount of heat required to raise the temperature of a system through 1 degree. SI unit is JK^{-1}. See **specific heat capacity, molar specific heat capacity**.

thermal circuit-breaker *(Elec.Eng.)*. A miniature-type circuit-breaker whose overload device operates by virtue of thermal expansion.

thermal column *(Nuc.Eng.)*. Column or block of moderator in reactor which guides large thermal neutron flux to given experimental region.

thermal comparator *(Phys.)*. A device which enables comparative measurements of the thermal conductivities of solids to be made rapidly. The instrument is also used to make comparative measurements of foil thickness, surface deposits, etc.

thermal conductivity *(Phys.)*. A measure of the rate of flow of thermal energy through a material in the presence of a temperature gradient. If (dQ/dt) is the rate at which heat is transmitted in a direction normal to a cross-sectional area A when a temperature gradient (dT/dx) is applied, then the thermal conductivity is

$$k = -\frac{(dQ/dt)}{A(dT/dx)}.$$

SI unit is $W m^{-1} K^{-1}$. Materials with high electrical conductivities tend to have high thermal conductivities.

thermal converter *(Elec.Eng.)*. The combination of a thermoelectrical device, e.g. thermocouple, and an electrical heater thus converting an electrical quantity into heat and then into a voltage. Used in telemetering systems; see **thermocouple meter**. Also *thermoelement*.

thermal cross-section *(Phys.)*. Effective nuclear cross-section for neutrons of thermal energy.

thermal cueing unit *(Aero.)*. Visual display presented to the pilot showing likely targets detected by a **FLIR** system. Targets can be classified by temperature 'signature' and after selection the co-ordinates can be fed to the attack system.

thermal cut-out *(Elec.Eng.)*. A thermal circuit-breaker designed to screw into a standard barrel-type fuse-holder.

thermal cycle *(Phys.)*. An operating cycle by which heat is transferred from one part to another. In reactors, separate heat transfer and power circuits are usual to prevent the fluid flowing through the former, which becomes radioactive, from contaminating the power circuit.

thermal cycling *(Phys.)*. The subjection of a substance to a number of temperature and pressure cycles in succession. Used in petrol refining (*cracking*).

thermal death-point *(Biol.)*. The temperature at which an organism is killed or a virus inactivated.

thermal detector *(Elec.Eng.)*. Any detector of high-frequency currents which operates by virtue of their heating effect when passed through a resistance.

thermal diffusion *(Phys.)*. Process in which a temperature gradient in a mixture of fluids tends to establish a concentration gradient. Has been used for isotope separation. Also known as *Soret effect*.

thermal diffusivity *(Phys.)*. Thermal conductivity divided by the product of specific heat capacity and density; more generally applicable than thermal conductivity in most heat transfer problems. Unit is metre squared per second.

thermal dissociation *(Chem.)*. The dissociation of certain molecules under the influence of heat.

thermal efficiency *(Eng.)*. Of a heat engine, the ratio of the work done by the engine to the mechanical equivalent of the heat supplied in the steam or fuel.

thermal effusion *(Phys.)*. The leaking of a gas through a small orifice, the gas being at a low pressure so that the mean free path of the molecules is large compared with the dimensions of the orifice.

thermal electromotive force *(Elec.Eng.)*. That which arises at the junction of different metals because of a temperature different from the rest of the circuit. Widely used for measuring temperatures relative to that of the cold junction, e.g. ice in a vacuum flask.

thermal excitation *(Phys.)*. Collision processes between particles by which atoms and molecules can acquire extra energy.

thermal fatigue *(Eng.)*. Structural weakness resulting from cyclical variations in temperature.

thermal flasher *(Phys.)*. A flasher the operation of which depends upon the heating effect of the current which it is controlling. The movement of a wire or bimetallic strip, when heated by the current, causes it to interrupt the circuit, which is not remade until it has cooled.

thermal gravimetric analysis *(Chem.)*. See **thermal analysis**.

thermal imaging *(Phys.)*. Imaging based on the detection of weak infrared radiation from objects. Applications include the mapping of the Earth's surface from the air, weather mapping and medical thermography (thermal contours on the surface of the human body). See **optical-electronic devices**.

thermal inertia *(Nuc.Eng.)*. See **thermal response**.

thermal instability *(Elec.Eng.)*. The condition in electrical equipment or components when an increase in temperature causes losses to increase more rapidly than they can be removed. The process is thus one of positive feedback and the equipment will be thermally damaged or destroyed.

thermal instrument *(Elec.Eng.)*. An instrument the operation of which depends upon the heating effect of a current. See **hot-wire, thermocouple instrument**.

thermalite *(Build.)*. TN for aerated lightweight concrete blocks.

thermalization *(Phys.)*. Process of slowing fast neutrons to thermal energies. In reactors normally the function of the moderator.

thermal leakage factor *(Nuc.Eng.)*. Ratio of number of thermal neutrons lost from reactor by leakage, to the number absorbed in reactor core. Also used is the *thermal non-leakage probability*, i.e. the fraction of thermal neutrons which do not leak out of the reactor.

thermal limit *(Elec.Eng.)*. Maximum permissable power associated with a piece of electrical equipment, e.g. the output power of an electrical generator, which is set by consideration of safe temperature rise.

thermally bonded non-woven fabrics *(Textiles)*. Fabrics formed from a **web** of fibres that have been partially melted by heat and thereby converted into a coherent sheet.

thermal metamorphism *(Geol.)*. Metamorphism resulting from the action of heat, involving chemical changes in the rock without the introduction of a material from elsewhere. Pressure is not significant. See **regional metamorphism**.

thermal microphone *(Acous.)*. See **hot-wire microphone**.

thermal neutrons *(Phys.)*. See **neutron**.

thermal noise *(Electronics)*. That arising from random (Brownian) movements of electrons in conductors and semiconductors, and which limits the sensitivity of electronic amplifiers and detectors. The noise voltage V is given by

$$V = \sqrt{4RkT.\delta f},$$

where δf = frequency bandwidth,
R = resistance of source,
k = Boltzmann's constant,
T = absolute temperature.

Also *circuit noise, Johnson noise*. See **Nyquist noise theorem**.

thermal ohm *(Phys.)*. See **thermal resistance**.

thermal precipitator *(Powder Tech.)*. A device for sampling aerosol particles. The gas stream is drawn through a chamber containing a hot wire. The particles near the wire are bombarded by gas molecules of high kinetic energy, then driven away from it and collected on microscope slides located on opposite sides of the hot wire.

thermal printer *(Comp.)*. Uses heat-sensitive paper, producing visible characters by the action of heated wires.

thermal radiation *(Med.)*. Analysis of the heat produced by the human body, by the use of infrared radiation to penetrate surface tissue, in order to diagnose certain diseases affecting deeper tissue and to locate superficial tumours. *Thermography* refers to the images so made.

thermal reactor *(Nuc.Eng.)*. One for which the fission chain reaction is propagated mainly by thermal neutrons and therefore contains a moderator. Formerly sometimes called a *slow reactor*.

thermal receiver *(Acous.)*. See **thermophone**.

thermal relay *(Elec.Eng.)*. A relay the operation of which depends upon the heating effect of an electric current.

thermal resistance *(Phys.)*. Resistance to the flow of heat. The unit of resistance is the *thermal ohm*, which requires a temperature difference of 1°C to drive heat at the rate of 1 watt. If the temperature difference is θ°C, the resistance S thermal ohms, and the rate of driving heat W watts, then $\theta = SW$.

thermal response *(Nuc.Eng.)*. Of a reactor, its rate of temperature rise if no heat is withdrawn by cooling. Its reciprocal is the *thermal inertia*.

thermal runaway *(Electronics)*. The effect arising when the current through a semiconductor creates sufficient heat for its temperature to rise above a critical value. The semiconductor has a negative temperature coefficient of resistance, so the current increases and the temperature increases again, resulting in ultimate destruction of the device.

thermal shield *(Nuc.Eng.)*. Inner shield of a reactor, used to protect biological shield from excess heating.

thermal shock *(Phys.)*. That resulting when a body is subjected to sudden changes in temperature. The ability of the material to withstand its effects is called its *thermal shock resistance*.

thermal siphon *(Phys.)*. The system causing flow round a vertical loop of fluid. When the bottom of one column is heated, the fluid rises, cools at the top and falls down the other column. Used, e.g., for heating buildings, isotope separators.

thermal spike *(Phys.)*. See **spike**.

thermal spray *(Eng.)*. Spraying molten and finely-divided metals or non-metallic particles on to a material to form a coating.

thermal station *(Elec.Eng.)*. An electric generating station in which the prime movers are steam-turbines or internal-combustion engines.

thermal trip *(Elec.Eng.)*. A tripping relay for a large circuit-breaker, which operates by thermal expansion.

thermal tuning *(Electronics)*. Change of resonant frequency in oscillator or amplifier produced by controlled temperature change. Used, e.g. with crystal or resonant cavity in microwave tube.

thermal unit *(Phys.)*. See **British-, calorie, joule**.

thermal utilization factor *(Nuc.Eng.)*. The fraction of thermal neutrons absorbed in the reactor which are absorbed in the fuel (whether causing fission or not). Symbol f.

thermal utilization factor *(Phys.)*. Probability of thermal neutron being absorbed by fissile material (whether causing fission or not) in infinite reactor core.

thermal vibration *(Phys.)*. The motion of atoms, vibrating about their equilibrium positions in a crystalline solid. The motion increases as the temperature increases. Representing the vibrations by a set of *harmonic oscillators* which have zero point energy at 0 K, the oscillators increase their energy by discrete amounts, a *quantum* of energy, as the temperature increases. The specific heat of a solid can be explained in these terms. Detailed information about the thermal vibrations can be obtained from neutron diffraction, X-ray diffraction and other techniques. See **Debye theory of specific heats, Debye-Waller factor, lattice dynamics, phonon**.

thermal wind *(Meteor.)*. The vector difference between the geostrophic wind at some level in the upper air and the wind at some lower level.

thermion *(Phys.)*. A positive or negative ion emitted from incandescent material.

thermionic amplifier *(Electronics)*. Any device employing thermionic vacuum tubes for amplification of electric currents and/or voltages.

thermionic cathode *(Electronics)*. One from which electrons are liberated as a result of thermal energy, due to high temperature.

thermionic current *(Electronics)*. One represented by electrons leaving a heated cathode and flowing to other electrodes.

thermionic emission *(Electronics)*. That from a thermionic cathode in accordance with the Richard-Dushman equation.

thermionic rectifier *(Electronics)*. Thermionic valve used for rectification or demodulation.

thermionics *(Electronics)*. Strictly, science dealing with the emission of electrons from hot bodies. Applied to the broader subject of subsequent behaviour and control of such electrons, especially *in vacuo*.

thermionic valve *(Electronics)*. One containing a heated cathode from which electrons are emitted, an anode for collecting some or all of these electrons, and generally additional electrodes for controlling flow to the anode. Normally the glass or metal envelope is evacuated but a gas at low pressure is introduced for special purposes. Also called *thermionic tube*.

thermionic work function *(Electronics)*. Thermal energy surrendered by an electron liberated through thermionic emission from a hot surface.

thermistor *(Electronics)*. Abbrev. for *thermal resistor*. Semiconductor, a mixture of cobalt, nickel, and manganese oxides with finely divided copper, of which the resistance is very sensitive to temperature. Sometimes incorporated in a waveguide system to absorb and measure all the transmitted power. Also used for temperature compensation and measurement.

thermistor bridge *(Electronics)*. One used for measuring microwave power absorbed by a thermistor, in terms of the resulting change of resistance. See **bolometer**.

thermite *(Eng.)*. A mixture of aluminium powder and half an equivalent amount of iron oxide (or other metal oxides) which gives out a large amount of heat on igniting with magnesium ribbon. The molten metal forms the medium for welding iron and steel. (Thermit welding.) See **aluminothermic process**.

thermo- *(Genrl.)*. Prefix. See **therm-**.

thermoammeter *(Elec.Eng.)*. A.c. type in which current is measured in terms of its heating effect, usually with a thermocouple.

thermochemistry *(Chem.)*. The study of the heat changes accompanying chemical reactions and their relation to other physicochemical phenomena.

thermocline *(Ecol.)*. In lakes, a region of rapidly changing temperature, found between the epilimnion and the hypolimnion.

thermocouple *(Phys.)*. If two wires of different metals are joined at their ends to form a loop, and a temperature difference between the two junctions unbalances the **contact e.m.f.** potentials, a current will flow round the loop. If the temperature of one junction is kept constant, that of the other is indicated by measuring the current. Conversely, a current driven round the loop will cause a temperature difference between the junctions. See **Peltier effect**.

thermocouple instrument *(Elec.Eng.)*. An instrument the operation of which depends upon the heating of a thermocouple by an electric current.

thermocouple meter *(Elec.Eng.)*. A combination of a thermocouple and an ammeter or voltmeter. The current in the external circuit passes through a coil of suitable gauge wire which is electrically insulated from the thermo-junction but in very close thermal contact; sometimes couple and heater are enclosed in the evacuated quartz meter. If the heating current is alternating (such devices may be used for radio-frequency currents), then the meter indications are *root-mean-square values*.

thermocouple wattmeter *(Elec.Eng.)*. One which uses thermoelements in suitable bridge to measure average a.c. power.

thermoduric *(Phys.)*. Resistant to heat.

thermodynamic concentration *(Chem.)*. See **activity (2)**.

thermodynamic potential *(Chem.)*. See **free energy**.

thermodynamics *(Phys.)*. The mathematical treatment of the relation of heat to mechanical and other forms of energy. Its chief applications are to heat engines (steam engines and IC engines; see **Carnot cycle**) and to chemical reactions (see **thermochemistry**). *Laws of thermodynamics*: *Zeroth law*: if two systems are each in thermal equilibrium with a third system then they are in thermal equilibrium with each other. This statement is tacitly assumed in every measurement of temperature. *First law*: the total energy of a thermodynamic system remains constant although it may be transformed from one form to another. This is a statement of the principle of the conservation of energy. *Second law*: heat can never pass spontaneously from a body at a lower temperature to one at a higher temperature (Clausius) *or* no process is possible whose only result is the abstraction of heat from a single heat reservoir and the performance of an equivalent amount of work (Kelvin-Planck). *Third law*: the entropy of a substance approaches zero as its temperature approaches absolute zero.

thermodynamic scale of temperature *(Phys.)*. See **Kelvin thermodynamic scale of temperature**.

thermoelectric cooling *(Phys.)*. Abstraction of heat from electronic components by Peltier effect, greatly improved and made practicable with solid-state materials, e.g. Bi_2Te_3. Devices utilizing this effect, e.g. frigistors, are used for automatic temperature control etc. and are energized by d.c.

thermoelectric effect *(Elec.Eng.)*. See **Seebeck effect**.

thermoelectricity *(Elec.Eng.)*. Interchange of heat and electric energy. See **Peltier effect, Seebeck effect**.

thermoelectric materials *(Eng.)*. Any set of materials (metals) which constitute a thermoelectric system, e.g. *binary* (bismuth and tellurium), *ternary* (silver, antimony and tellurium), *quaternary* (bismuth, tellurium, selenium and antimony, called *Neelium*).

thermoelectric power *(Elec.Eng.)*. Defined as dE/dT, i.e. the rate of change with temperature of the thermo e.m.f. of a thermocouple.

thermoelectric pyrometer *(Elec.Eng.)*. The combination of apparatus forming a temperature-indicating instrument whose action derives from a thermoelectric current; for high temperatures.

thermoelectroluminescence *(Phys.)*. See **electrothermoluminescence**.

thermoelement *(Elec.Eng.)*. See **thermal converter**.

thermogalvanometer *(Elec.Eng.)*. A very sensitive **thermocouple meter**.

thermogenesis *(Zool.)*. Production of heat within the body.

thermograph *(Meteor.)*. A continuously recording thermometer. In the commonest forms the record is made by the movement of a bimetallic spiral, or by means of the out-of-balance current in a Wheatstone bridge containing a resistance thermometer in one of its arms.

thermographic *(Print.)*. A printing process in which an impression from a ribbon is transferred by heated styli to paper. Used now in computer printing.

thermography *(Print.)*. Term applied to printing effect which simulates the result of printing from steel die engravings. The printed image is dusted with resinous powder while the ink is still wet on the paper, causing the powder to adhere to the print. Heating causes the powder and ink to fuse, creating a raised image.

thermography *(Radiol.)*. The use of radiant heat emitted by the body to construct images of increased heat emission which can indicate tumours or inflammation. Reduced heat emission indicates reduced blood supply.

thermojunction *(Elec.Eng.)*. See **thermocouple**.

thermolabile *(Chem.)*. Tending to decompose on being heated.

thermoluminescence *(Phys.)*. Release of light by previously irradiated phosphors upon subsequent heating.

thermoluminescent dosemeter *(Radiol.)*. One which registers integrated radiation dose, the read-out being obtained by heating the element and observing the thermoluminescent output with a photomultiplier. It has the advantage of showing very little fading if read-out is delayed for a considerable period, and forms an alternative to the conventional film badge for personnel monitoring.

thermolysis *(Chem.)*. The dissociation or decomposition of a molecule by heat.

thermolysis *(Zool.)*. Loss of body heat.

thermomagnetic effect *(Phys.)*. See **magnetocaloric effect**.

thermometer *(Phys.)*. An instrument for measuring temperature. A thermometer can be based on any property of a substance which varies predictably with change of temperature. For instance, the *constant volume gas thermometer* is based on the pressure change of a fixed mass of gas with temperature, while the *platinum resistance thermometer* is based on a change of electrical resistance. The commonest form relies on the expansion of mercury or other suitable fluid with increase in temperature.

thermometric scales *(Phys.)*. See **Centigrade scale, Celsius scale, Fahrenheit scale, international practical temperature scale, Kelvin thermodynamic scale, Rankine scale, Réaumur scale**; also **fixed points**.

thermometry *(Phys.)*. The measurement of temperature.

thermonasty *(Bot.)*. A nastic movement in response to a change in temperature, e.g. the opening and closing of Crocus flowers.

thermonuclear bomb *(Phys.)*. See **hydrogen bomb**.

thermonuclear energy *(Phys.)*. Energy released by a *nuclear fusion* reaction that occurs because of the high thermal energy of the interacting particles. The rate of reaction increases rapidly with temperature. The energy of most stars is believed to be acquired from exothermic thermonuclear reactions. In the hydrogen-bomb, a fission bomb is used to obtain the initial high temperature required to produce the fusion reactions. For laboratory experiments designed to release thermonuclear energy, see **stellarator, tokamak, JET**.

thermonuclear reaction *(Phys.)*. One involving the release of thermonuclear energy.

thermoperiodism *(Bot.)*. The response of a plant to daily (or other) cycles of temperature.

thermophile, thermophilous, thermophilic *(Bot.)*. Requiring, adapted to, or sometimes tolerating high temperatures. Cf. **thermotolerant**.

thermophone *(Acous.)*. Electroacoustic transducer in which fluctuating temperature changes produce a sound pressure wave. Used for calibration of microphones. Now obsolete.

thermophyllous *(Bot.)*. Having leaves only in the warmer part of the year; deciduous.

thermopile *(Elec.Eng.)*. See **pile**.

thermoplastic *(Chem.)*. Becoming plastic on being heated. Specifically (*Plastics*), any resin which can be melted by heat and then cooled, the process being repeatable any number of times without appreciable change in properties, e.g. cellulose derivatives, vinyl polymers, polystyrenes, polyamides, acrylic resins.

thermoplastic binding *(Print.)*. A synonym for *unsewn binding*, whether the adhesive used is thermoplastic or not.

thermoplastic plates *(Print.)*. Duplicate printing plates made from thermoplastic material, usually flexible and suitable for rotary printing; normally recoverable for subsequent plates in contrast to **thermosetting plates**.

thermoprinting machine *(Print.)*. A type of *blocking press*, used in conjunction with automatic food packaging machines using heat-sealing wraps. These are overprinted with details of the contents.

thermoregulator *(Phys.)*. Type of thermostat which keeps a bath at a constant temperature by regulating its supply of heat.

thermoscopic *(Phys.)*. Perceptive of change of temperature.

thermosetting compositions *(Plastics)*. Compositions in which a chemical reaction takes place while the reins are being moulded under heat and pressure; the appearance and chemical and physical properties are entirely changed, and the product is resistant to further applications of heat (up to charring point), e.g. phenol formaldehyde, urea formaldehyde, aniline formaldehyde, glycerol-phthalic anhydride.

thermosetting plates *(Print.)*. Duplicate printing plates made from thermosetting plastic material, not flexible and must be used flat or be curved; not recoverable like **thermoplastic plates**. Both have certain advantages over metal plates, such as good inking qualities, long life, light weight, easy storage.

thermosiphon *(Eng.)*. The method of establishing circulation of a cooling-liquid by utilizing the slight difference in density of the hot and the cool portions of the liquid.

thermosphere *(Astron.)*. The region of the earth's atmosphere, above the **mesosphere**, in which the temperature rises steadily with height.

thermostable *(Chem.)*. Not decomposed by heating.

thermostat *(Phys.)*. An apparatus which maintains a system at a constant temperature which may be preselected. Frequently incorporates a **bimetallic strip**.

thermotolerant *(Bot.)*. Able to endure high temperatures, but not growing well under such conditions.

thermotropic *(Phys.)*. Temperature determines the phase of thermotropic materials.

therophyte *(Bot.)*. Plant which passes the unfavourable season as seeds, and thus has no perennating vegetative buds. See **Raunkaier system**.

theta pinch *(Nuc.Eng.)*. Cylindrical plasma constricted by an external current flowing in the θ direction to produce a solenoidal magnetic field.

Thevenin's theorem *(Elec.Eng.)*. That the source behind two accessible terminals may be regarded as a a constant voltage generator in series with a source impedance. The value of the voltage is that appearing with terminals open circuited and the impedance is that measured at the terminal with all voltage sources open circuited. Often regarded as the dual of **Norton's Theorem**.

thiamides *(Chem.)*. A group of compounds derived from amides by the substitution of sulphur for oxygen, e.g. $CH_3.CS.NH_2$.

thiamin *(Biol.)*. See **vitamin B**.

thiazide diuretics *(Med.)*. Thiazide and related compounds act on the distal convoluted tubule of the kidney to promote salt and water excretion. Used in the treatment of heart failure and hypertension.

thiazines *(Chem.)*. Six-membered heterocyclic compounds, containing in the ring four carbon, one sulphur and one nitrogen atoms.

thiazole *(Chem.)*. C_3H_3NS. A colourless, very volatile liquid which closely resembles pyridine; bp 117°C. It forms salts, but is hardly affected by concentrated sulphuric acid.

thickener *(Min.Ext.)*. Apparatus in which water is removed from ore pulp by allowing solids to settle. To obtain continuous working the solids are worked towards a central hole in the bottom by means of revolving rakes.

thick-film lubrication *(Eng.)*. The state of fluid-medium lubrication which exists when the lubricant separates the journal and bearing surfaces to prevent metal-to-metal contact. The bearing friction is then dependent largely on the force necessary to shear the lubricant film, and perfect lubrication is said to prevail.

thick leg disease *(Vet.)*. See **osteopetrosis gallinarum**.

thick lens *(Phys.)*. Any lens, or system of lenses, in which the distance between the outer faces is not small compared with the focal length.

thickness chart *(Meteor.)*. A chart of upper air showing the difference in geopotential between two particular pressure levels. Centres of low thickness are *cold pools*. See **constant-pressure chart**.

thickness-chord ratio *(Aero.)*. The ratio of the maximum depth of an *aerofoil*, measured perpendicular to the *chord line*, to the *chord length*; usually expressed as a percentage.

thickness dummy *(Print.)*. A book made up of blank leaves to show size and physical appearance in advance.

thickness moulding *(Build.)*. A moulding serving to fill up the bare space beneath a projecting cornice.

thick source *(Phys.)*. Radioactive source with appreciable self-absorption.

thick space *(Print.)*. The normal standard space between words set in lower case, one-third of the type size; reduced if possible, rather than increased, when justifying text matter.

thick target *(Phys.)*. One which is not penetrated by primary or secondary radiation beam.

thick-wall chamber *(Nuc.Eng.)*. Ionization chamber in which build-up of ion current is produced by contribution of knock-on particles arising from wall material.

thief sampler *(Powder Tech.)*. Instrument for sampling powders, comprising two tubes, one fitting closely inside the other, rectangular slits in corresponding positions. The instrument is thrust into the powder with the holes closed; the inner tube is rotated to open the holes to collect the sample, which is withdrawn with the holes closed.

thigmo- *(Genrl.)*. Prefix from Gk. *thigma*, touch.

thigmocyte *(Zool.)*. See **thrombocyte**.

thigmotropism *(Biol.)*. Turning of an organism (or of part of it) towards or away from object providing touch stimulus. See **haptotropism**.

thimble *(Build.)*. See **sleeve piece**.

thimble ionization chamber *(Nuc.Eng.)*. A small cylindrical, spherical or thimble-shaped ionization chamber, volume less than $5\,cm^3$ with air-wall construction. Used in radiobiology.

thimble-tube boiler *(Eng.)*. A heat-recovery boiler, consisting of an annular water-drum from which thimble-like tubes or pockets project into the central flue, through which the hot exhaust products are passed.

thin-film *(Electronics)*. Of a film, of molecular thickness, deposited on an insulating or semiconducting substrate by sputtering, evaporation, or chemical vapour deposition through a mask. Resistors, capacitors, diodes and transistors can be built up using thin films.

thin-film capacitor *(Elec.Eng.)*. One constructed by evaporation of two conducting layers and an intermediary dielectric film (e.g. silicon monoxide) on an insulating substrate.

thin-film circuit *(Electronics)*. A circuit in which all active and passive elements are made from thin films; more commonly referred to as *integrated circuit*.

thin-film lubrication *(Eng.)*. When metal-to-metal contact exists in a journal bearing for all or part of the time, thin-film or imperfect lubrication is said to prevail and the materials and surface characteristics of journal and bearing affect the bearing friction significantly.

thin-film memory *(Comp.)*. The use of an evaporated thin film of magnetic material on glass as an element of a computer memory when a d.c. magnetic field is applied parallel to the surface. A large capacity memory will contain thousands of these elements which can be produced in one operation. Also called *magnetic-film memory*.

thin-film resistor *(Elec.Eng.)*. Modern high-stability formed by a conducting layer a few tens of nanometers thick on an insulating substrate.

thin-layer chromatography *(Chem.)*. A form of chromatography in which compounds are separated by a suitable solvent or solvent mixture on a thin layer of adsorbent material coated on a glass plate. It is rapid, and excellent separations can be obtained. Abbrev. *TLC*.

thinners *(Build.)*. Volatile liquids added to paints to dilute and make the material more fluid and workable. Thinners should not alter the original colour of the paint and should evaporate on drying. Thinners aid penetration of priming coats but are often detrimental when added unnecessarily to finishing coats.

thin source *(Phys.)*. Radioactive source with negligible self-absorption.

thin space *(Print.)*. The narrowest of the justifying spaces, one-fifth of the type size; the *hair space* is not a justifying space.

thin target *(Phys.)*. One penetrated by primary radiation beam so that detecting instrument(s) may be used on opposite side of target to source.

thin-wall bearings *(Autos.)*. Bearings made of steel sheet coated with a thin layer of soft alloy.

thin-wall chamber *(Nuc.Eng.)*. Ionization chamber in which the number of knock-on particles arising from the wall material and absorption therein is small.

thio-acids *(Chem.)*. Acids in which the hydroxyl of the carboxyl group has been replaced by SH, thus forming the group -CO.SH.

thio-alcohols *(Chem.)*. See **mercaptans**.

thiocarbamide *(Chem.)*. See **thiourea**.

thiocyanates *(Chem.)*. Compounds formed when alkaline cyanides are fused with sulphur. They contain the group $=N=C=S$ or the ion $S=C=N^-$.

thio-ethers *(Chem.)*. Compounds in which the ether oxygen has been replaced by sulphur; general formula R.S.R′. They form additive crystalline compounds with metallic salts; they are capable of combining with halogen or oxygen, which becomes attached to the sulphur atom, thereby converting the latter from the bivalent to the tetravalent state; and they form additive crystalline compounds with alkyl halides, e.g. $(CH_3)_3SI$.

thioglycollic acid *(Chem.)*. $CH_2(SH).COOH$. A colourless liquid, with a slight odour when pure but extremely unpleasant in impure form. A strong reducing agent, used as a reagent for detecting iron, with which it gives a violet colour in ammoniacal solution. Used in cold-waving treatment of hair, and in certain textile treatments for crease-resistant finishes.

Thiokol *(Plastics)*. TN for a synthetic rubber of the polysulphide group, derived from sodium tetrasulphide and organic dichlorides, with sulphur as an accelerator.

thiols *(Chem.)*. See **mercaptans**.

thiopentone *(Med.)*. A yellow powder used intravenously in solution to give general anaesthesia of short duration.

thiophen *(Chem.)*. C_4H_4S. A five-membered heterocyclic compound with sulphur.

thiophil *(Chem.)*. Having an affinity for sulphur and its compounds, e.g. bacteria.

thiosulphuric acid *(Chem.)*. $H_2S_2O_3$. Its salts are thiosulphates.

thiourea *(Chem.)*. Thiocarbamide, $NH_2.CS.NH_2$, colourless prisms, mp 180°C; it is slightly soluble in water, ethanol, and ethoxyethane. Used in organic synthesis and as a reagent for bismuth.

thiourea resins *(Plastics)*. Resins made from thiourea and an aldehyde. They are more water-resistant and more stable than the urea resins, but they cure more slowly, and the sulphur in them causes trouble with the steel moulds and the dyestuffs used to colour them.

third-angle projection *(Eng.)*. A system of projection used in engineering drawing, in which each view shows what would be seen by looking on the near side of an adjacent view.

third generation computer *(Comp.)*. A machine produced after 1965, probably with some integrated circuits, **SSI** and **MSI**, replacing transistors and ferrite cores, giving another big jump in computing power and reduction in size. **Time sharing** and **interactive computing** resulted from the considerable development in software. Families of machines were produced. ICL 1900 series, IBM 300 series. See **computer generations, mainframe, minicomputer, parallel processing**.

third pinion *(Horol.)*. The pinion on the same axis as the third wheel, engaging with the centre wheel.

third-rail insulator *(Elec.Eng.)*. See **conductor-rail insulator**.

third-rail system *(Elec.Eng.)*. The system of electric traction supply by which current is fed to the electric tractor from an insulated conductor rail running parallel with the track.

thirds *(Paper)*. A premetrication size of cut card, 38 × 76 mm ($1\frac{1}{2} \times 3$ in).

third ventricle *(Zool.)*. In Vertebrates, the cavity of the diencephalon, joining the two lateral ventricles of the cerebral hemispheres via the foramen of Monro, and the fourth ventricle in the medulla oblongata by the cerebral aqueduct.

third wheel *(Horol.)*. The wheel between the centre and the fourth wheel of a watch train.

thirst *(Behav.)*. A state of motivation which arises primarily as a result of dehydration of body tissues.

35 mm *(Image Tech.)*. The standard gauge of motion picture film used for professional production and presentation.

32mo *(Print.)*. The thirty-second part of a sheet, or a sheet folded five times to make thirty two leaves or sixty-four pages. More than four folds are not in fact practicable.

thistle funnel *(Chem.)*. A glass funnel with a thistle-shaped head and a long narrow tube.

thi-, thio- *(Chem.)*. A prefix denoting a compound in which a sulphur atom occupies a position normally filled by an oxygen atom.

thixotrope *(Chem.)*. A colloid whose properties are affected by mechanical treatment.

thixotropy *(Build.,Chem.)*. Rheological property of fluids and plastic solids, characterized by a high viscosity at low stress, but a decreased viscosity when an increased stress is applied. A useful property in paints, because it makes for a thick film which is nevertheless easily worked.

tholeiite *(Geol.)*. A silica-rich basalt abundant in mid-oceanic ridges and continental rifts.

Thomas-Gilchrist process *(Eng.)*. Use of basic-lined Bessemer converters to remove phosphorus during steel production.

Thomas resistor *(Elec.Eng.)*. A standard manganin resistor which has been annealed in an inert atmosphere and sealed into an envelope.

Thomas's splint *(Med.)*. A skeleton splint consisting of two parallel metal rods and a padded leather ring, used for maintaining the hip and the knee-joint in fixed extension.

Thomsen's disease *(Med.)*. See **myotonia congenita**.

Thomson compass *(Ships)*. See **Kelvin compass**.

Thomson effect *(Elec.Eng.)*. The e.m.f. produced by temperature differences in a single conductor, and the heat change associated with current flow between temperature differences.

thomsonite *(Min.)*. An orthorhombic zeolite; a hydrated silicate of calcium, aluminium and sodium, found in amygdales and crevices in basic igneous rocks.

Thomson scattering *(Phys.)*. The scattering of electromagnetic waves by free electrons. On a classical interpretation, the electron is set into oscillatory motion by the transverse electric field of the wave and radiates at the same frequency as the wave. The scattering cross-section of an electron is

$$\sigma = \frac{8}{3}\pi(e^2/4\pi\varepsilon_0 mc^2)^2 = 0{\cdot}66 \times 10^{-28}\ \text{m}^2.$$

where m and e are the mass and charge of the electron, c is the velocity of light and ε_0 the permittivity of free space.

thoracentesis, thoracocentesis *(Med.)*. The operation of drawing off a morbid collection of fluid in the pleural cavity through a hollow needle stuck through the wall of the chest.

thoracicolumbar system *(Zool.)*. See **sympathetic nervous system**.

thoracoplasty *(Med.)*. The operation for collapsing a diseased lung by removal of portions of the ribs.

thoracoscope *(Med.)*. An instrument for viewing the pleura covering the lung and the chest wall. It is inserted through the chest wall into a pleural cavity previously filled with air.

thoracotomy *(Med.)*. Incision of the wall of the chest, for draining pus from the pleural cavity or from the lung.

thorax *(Zool.)*. In Crustacea and Arachnida, a region of the body lying between the head and the abdomen and usually fused with the former; in Insects, one of the three primary regions of the body, lying between the head and the abdomen, and bearing in the adult three pairs of legs and the wings (if present); in some tubicolous Polychaeta, a region of the body behind the head, distinguished by the form of its segments and the nature of its appendages; in land Vertebrates, the region of the trunk between the head or neck and the abdomen which contains heart and lungs and bears the fore-limbs, especially in the higher forms, in which it is enclosed by ribs and separated from the abdomen by the diaphragm. adj. *thoracic*.

thorianite *(Min.)*. Thorium dioxide; crystallizes in the cubic system and is found in pegmatites and gem gravel washings, as in Sri Lanka. An important source of thorium and uranium.

thoriated cathode *(Electronics)*. Tungsten cathode containing a small proportion of thorium to reduce the temperature at which copious electronic emission takes place. With heat, the thorium diffuses to the surface, forming a tenuous emitting layer.

thorides *(Phys.)*. Naturally occurring radioactive isotopes in the radioactive series containing thorium.

thorite *(Min.)*. Tetragonal thorium orthosilicate, found in syenites and syenitic pegmatites.

thorium *(Chem.)*. A metallic radioactive element, dark-grey in colour. Symbol Th, at. no. 90, r.a.m. 232.0381, rel.d. 11.2, mp 1845°C. It occurs widely in beach sands (*monazite*). The thorium radioactive series starts with thorium of mass 232. It is fissile on capture of fast neutrons and is a fertile material, ^{233}U (fissile with slow neutrons) being formed from ^{232}Th by neutron capture and subsequent beta decay.

thorium reactor *(Nuc.Eng.)*. Breeder reactor in which fissile ^{233}U is bred in a blanket of fertile ^{232}Th.

thorium series *(Phys.)*. The series of nuclides which result from the decay of ^{232}Th. The mass numbers of the members of the series are given by $4n$, n being an integer. The series ends in the stable isotope ^{208}Pb. See **radioactive series**.

thorn *(Bot.)*. A woody sharp-pointed structure; usually restricted to those representing modified branches, as in the hawthorn. Cf. **prickle**, **spine**.

thornproof *(Textiles)*. A closely-woven suiting made of highly twisted yarns (frequently two-fold) that is strong and firm and resistant to damage (e.g. by thorns).

thoron *(Phys.)*. Thorium emanation, an isotope of radon (^{220}Rn); half life 54.5 s; a radioactive decay product of thorium. Symbol Tn. See **emanations**.

thoroughpin *(Vet.)*. *Through-pin*. A swelling on the hock of the horse caused by distension of the synovial sheath of the flexor perforans tendon.

THORP *(Nuc.Eng.)*. *Thermal (reactor mixed) oxide reprocessing plant*, at Sellafield, UK.

thousand *(Build.)*. Formerly a trade term for 1200 slates; now for 1000, so also called *thousand actual*.

Thr *(Chem.)*. Symbol for threonine.

thread *(Eng.)*. See **screw thread**.

thread *(Textiles)*. The general name for a yarn although commonly used specifically for sewing thread.

thread-dial indicator *(Eng.)*. A device which forms part of a screw-cutting lathe. It indicates, while the lathe is running, the correct time for engaging the half-nut so that the tool will repeatedly follow the thread cut.

thread grinding *(Eng.)*. The accurate production of screw threads by a form grinding wheel, profiled to the thread section and automatically traversed along the revolving work.

threading, threading up *(Image Tech.)*. The operation of inserting the start of the film into the mechanism of camera, or projector, as it leaves the feed reel, and attaching it to the take-up reel.

thread rolling *(Eng.)*. Producing a screw thread by rolling between flat or cylindrical dies an alloy sufficiently plastic to withstand the cold working forces without disintegrating.

thread sewn *(Print.)*. The traditional method of securing and joining to each other the sections of a book, by hand and on tapes or cords for the best work, by machine for **edition binding**.

thread-stitching *(Print.)*. The traditional method, by hand or machine, of securing the leaves of insetted (or quirewise) books or pamphlets; better and more expensive than **wire-stitching**.

threadworm *(Med.)*. Although can refer to any slender **nematode** usually refers to a small parasite, *Enterobius vermiculatis*, which is a common parasite in children and causes **pruritus** of the anus.

threat behaviour *(Behav.)*. A form of communication usually occurring in situations of conflict between fear and aggression; used to repel conspecifics or members of other species without undue risk or injury. Threat displays are very varied, often involving ritualized postures and expressions, as well as specialized morphological features (e.g. the rattle of the rattlesnake).

three-ammeter method *(Elec.Eng.)*. A method of measuring the power carried by a single-phase circuit making use of three ammeters. Cf. **three-voltmeter method**.

three-body problem *(Astron.)*. The problem of the behaviour of three bodies which mutually attract each other; no general solution is possible but certain particular solutions are known. See **Trojan group**.

three-centred arch *(Arch.)*. An arch having the form of a false ellipse struck from three centres.

three-coat work *(Build.)*. Plastering in three successive coats. See **floating**, **pricking-up**, **roughing-in**, **setting**.

three-colour process *(Image Tech.)*. A colour reproduction system, photographic or electronic, in which the subject is analysed into three colour components, red, green and blue, for recording or transmission. Reproduction is similarly in three colours, usually by *subtractive* methods in photography and cinematography and by *additive* means for television and video.

three-colour process *(Print.)*. The subtractive process applied to printing. The yellow is, as a rule, printed first, followed by magenta (red) then cyan (blue). A fourth printing, in black, is usually added to enhance the final result.

three-core cable *(Elec.Eng.)*. A cable having three conducting cores arranged symmetrically about the axis of the cable and insulated.

three-day sickness *(Vet.)*. *Ephemeral fever*. A benign, non-contagious virus disease of cattle in tropical countries characterized by fever, lameness and stiffness; transmitted by mosquitoes.

three-eighths rule *(Maths.)*. The area under the curve $y = f(x)$ from $x = x_0$ is approximately

$$\tfrac{3}{8}\{f(x_0) + 3f(x_1) + 3f(x_2) + f(x_3)\}\{x_3 - x_0\}$$

where x_0, x_1, x_2 and x_3 are equally spaced.

three-electron bond *(Chem.)*. Resonance structure involving an unpaired shared electron, similar to the **one-electron bond**.

three-high mill *(Eng.)*. A rolling mill with three rolls, which are rotated in such a way that the metal is passed in one direction through the bottom pair of rolls and in the opposite direction through the top pair.

three-hinged arch *(Build.)*. Arch ribs hinged at top and springing points.

three-jaw chuck *(Eng.)*. A **scroll chuck** with three jaws for holding cylindrical workpieces, materials, or tools, particularly useful on the lathe or drilling machine.

three-level maser *(Electronics)*. Solid-state maser involving 3 energy levels.

three-light window *(Build.)*. A window having 2 mullions dividing the window space into 3 compartments.

three-phase *(Elec.Eng.)*. An electric supply system in which the alternating potentials on the 3 wires differ in phase from each other by 120°.

three-phase, four-wire system *(Elec.Eng.)*. A system of 3-phase a.c. distribution making use of 3 outgoing conductors (lines) and a common return conductor (neutral), the voltage between lines being $\sqrt{3}$ times the voltage between any line and the neutral.

three-phase induction regulator *(Elec.Eng.)*. An induction regulator for use on 3-phase circuits, in which the e.m.f. induced in the secondary winding is constant in magnitude but variable in phase, so that the total e.m.f. on the secondary side bears a small phase displacement to the primary voltage.

three-phase six-wire system *(Elec.Eng.)*. A system of 3-phase a.c. distribution in which each phase has separate outgoing and return conductors.

three-point adder *(Comp.)*. See **full-adder**.

three-point landing *(Aero.)*. The landing of an aircraft so equipped on the 2 wheels and tail skid (or wheel) simultaneously; the normal 'perfect landing'.

three-point problem *(Surv.)*. A field problem, arising in plane table and hydrographical surveying, in which it is required to locate on the plan the position of the instrument station, given that only 3 points represented on the plan are in fact visible from the station.

three-point switch *(Elec.Eng.)*. See **three-way switch**.

three-quarter bat *(Build.)*. A brick made or cut equal to $\frac{3}{4}$ the full length of a brick.

three-quarter bound *(Print.)*. Similar to **quarter bound**, but having the material used for the back covering a large part of the sides too.

three-quarter plate watch *(Horol.)*. A watch with the upper plate cut away so that the balance may be in the same plane as the plate.

three-to-two folder *(Print.)*. On a web-fed press, a type of folder in which the folding cylinder has a circumference of 3 cut-offs and the cutting cylinder 2 cut-offs, giving a ratio of cylinder sizes 3 : 2. Cf. **two-to-one folder**.

three-voltmeter method *(Elec.Eng.)*. A method of measuring the power in a single-phase circuit by means of 3 voltmeters and a nonreactive resistance.

three-wattmeter method *(Elec.Eng.)*. A method of measuring the power carried by a 3-phase 4-wire circuit, making use of 3 wattmeters whose current coils are connected in the lines and whose voltage coils are connected between the lines and the neutral.

three-way switch *(Elec.Eng.)*. A rotary-type single-pole switch having 3 independent contact positions.

three-wire meter *(Elec.Eng.)*. An electricity supply meter performing the simultaneous integration of the energy supplied by the 2 sides of a **three-wire system**.

three-wire system *(Elec.Eng.)*. A supply system in which, e.g. 220 volts is the p.d. between 2 of the wires, while ca. 110 volts exists between the other 2-wire combinations.

threonine *(Chem.)*. *2-amino-3-hydroxybutanoic acid*, $CH_3.CH(OH).CH(NH_2).COOH$. A polar amino acid. The L- or S-isomer is a constituent of proteins. Symbol Thr, short form T. ⇨

threose *(Chem.)*. A tetraose. ⇨

threshold *(Med.)*. Generally, the lowest intensity of an effect which is detectable, e.g. of visibility, below which neither the cones nor the rods in the retina of the eye respond to a light stimulus.

threshold *(Telecomm.)*. The lowest value of a current, voltage or any other quantity that produces the minimum detectable response.

threshold amplitude *(Electronics)*. The lowest amplitude level which a pulse height selector or window discriminator will accept.

threshold current *(Electronics)*. That at which a gas discharge becomes self-sustaining.

threshold dose *(Radiol.)*. The smallest dose of radiation that will produce a specified result.

threshold effect *(Electronics)*. The marked increase in background noise which occurs in a valve circuit when on the verge of oscillation.

threshold energy *(Phys.)*. Minimum energy that can just initiate a given endoergic nuclear reaction. Exoergic reactions may also have threshold energies.

threshold frequency *(Electronics)*. Minimum frequency in a photon which can just release an electron from a surface.

threshold lights *(Aero.)*. A line of lights across the ends of a runway, strip, or landing area to indicate the usable limits.

threshold of hearing *(Acous.)*. Miniumum r.m.s. pressure of a sound wave which an average human listener can just detect at any given frequency. It is 2.10^{-5} Pa at a frequency of 1000 Hz. This value is often used as a reference pressure so that the threshold of hearing is at 0 dB (see **decibel**) at 1000 Hz.

threshold of pain *(Acous.)*. Minium intensity or pressure of sound wave which causes sensation of discomfort or pain in average human listener. It is between 130 and 140 dB.

threshold of sound audibility *(Acous.)*. Minimum intensity or pressure of sound wave which average normal human listener can just detect at any given frequency. Commonly expressed in decibels relative to 2×10^{-5} Pascal (Pa).

threshold treatment *(Chem.)*. Addition of minute quantities of dehydrated phosphates to water, to inhibit furring and corrosion in pipes, containers etc.

thrill *(Med.)*. A tremor or vibration palpable at the surface of the body, especially in valvular disease of the heart.

throat *(Build.,Civ.Eng.)*. See **drip**.

throat *(Eng.)*. (1) The C-shaped aperture of a gap press. (2) The root portion of a saw tooth.

throat *(Glass)*. The submerged channel through which glass passes from the melting end to the working end of a tank furnace.

throat microphone *(Acous.)*. One worn against the throat and actuated by contact pressure against the larynx. Used, e.g. by air pilots and deep sea divers. Also called *laryngophone*.

thrombectomy *(Med.)*. The operation of removing a venous thrombus.

thrombin *(Biol.)*. The proteolytic enzyme which converts fibrinogen to fibrin resulting in blood clotting.

thrombocyte *(Zool.)*. A minute greyish circular or oval body found in the blood of higher Vertebrates; it plays an important role in coagulation; a blood-platelet.

thrombocytopenia *(Med.)*. Abnormal decrease in the number of platelets (thrombocytes) in the blood.

thrombopenia *(Med.)*. See **thrombocytopenia**.

thrombophilia *(Med.)*. Combined inflammation and thrombosis of a vein.

thrombosis *(Med.)*. The formation of a clot in a blood-vessel during life.

thrombosis *(Zool.)*. Coagulation; clotting.

thromb-, thrombo- *(Genrl.)*. Prefix from Gk. *thrombos*, lump, clot.

thrombus *(Med.)*. A clot formed in a blood vessel during life and composed of thrombocytes (platelets), fibrin and blood cells.

throttle valve *(Eng.)*. (1) In steam engines and turbines, a governor-controlled steam valve, usually a **double beat valve**. (2) In petrol engines, the **butterfly valve**. (3) In refrigerators, the regulating valve controlling the pressure and temperature range of the working agent.

throttling *(Eng.)*. The process of reducing the pressure of a fluid by causing it to pass through minute or tortuous passages so that no kinetic energy is developed and the total heat remains constant. See **refrigerator, throttling calorimeter**.

throttling calorimeter *(Eng.)*. A device for measuring the **dryness fraction** of wet steam by throttling to a measured lower pressure and measuring the resulting superheat.

through bridge *(Civ.Eng.)*. A bridge in which the track is carried by the lower stringers. Cf. **deck bridge**.

through level *(Telecomm.)*. The reading of a high impedance level-measuring set at a point in a system, no correction being made for any difference between the actual impedance of the system and the impedance with respect to which the set is calibrated.

through path *(Telecomm.)*. The forward path from loop input to loop output in a feedback circuit.

through-pin *(Vet.)*. See **thoroughpin**.

throughput *(Chem.Eng.)*. A measure of the rate of production, in terms of mass of product per unit time per unit volume of plant. In vacuum technology, the quantity of gas or vapour passing a given section of a pump or pipe line in unit time; measured as the product of the pressure and the volume per second at the pressure at that section. Units: litre torr per second or litre pascal per second. Symbol *Q*.

through-stone *(Build.)*. A bondstone whose length is equal to the full thickness of the wall in which it is laid as a header. Also *perpend.*

throw *(Elec.Eng.)*. See **span** (2).

throw *(Eng.)*. The total travel of a crank or similar element, being twice the radius of eccentricity. Sometimes half this distance is called the throw.

throw *(Geol.,Min.Ext.)*. (1) The amount of vertical displacement (*upthrow* or *downthrow*) of a particular rock, vein or stratum due to faulting. See **fault**, and cf. **lateral shift**. (2) The amplitude of shake of a concentrating table. (3) Deviation of a deep borehole from the planned path.

throw *(Horol.)*. A hand-driven, dead-centre lathe, used by clockmakers.

throw *(Image Tech.)*. The distance between the projector and the screen.

throw-away *(Print.)*. A term for any simple and cheap advertising leaflet.

throwback *(Acous.)*. In a public-address system, when the microphone is near the reproducers, the *throwback* is the sound intensity which is applied to the microphone by the reproducers. If this is excessive, the system becomes paralysed with self-sustained oscillations. See **feedback**.

throwing *(Textiles)*. In silk manufacture, the processes of reeling, doubling, and twisting, to bring the silk filaments into the form of a thread. Now extended to include similar processes carried out with continuous filament man-made fibres.

throwing power *(Chem.)*. The property of a solution in virtue of which a relatively uniform layer of metal may be electrodeposited on a relatively irregular surface.

throw-off trip *(Print.)*. An attachment on a printing machine which allows the impression to be suspended without stopping the machine.

throw-out *(Print.)*. See **fold-out**.

throwster *(Textiles)*. A company that carries out the **throwing** operations but also now used to describe a manufacturer of false-twist **textured yarns**.

thrum *(Bot.)*. The short-styled form of such heterostyled flowers as the primrose, with the anthers visible at the top of the corolla tube. See also **heterostyly**. Cf. **pin**.

thrum *(Textiles)*. Waste lengths of yarn produced when a loom is being prepared for weaving.

thruput *(Chem.Eng.)*. See **throughput**.

thrush *(Med.)*. Infection of the mouth with the fungus *Candida albicans*, characterized by the appearance of white patches on the mucous membrane and the tongue.

thrush *(Vet.)*. The name is sometimes applied to **horse pox**. See also **equine thrush**.

thrust *(Aero.)*. Propulsive force developed by a jet- or rocket-motor.

thrust *(Civ.Eng.)*. The equal horizontal forces acting upon the abutments of an arch, due to the loading carried by the arch.

thrust *(Eng.)*. (1) The reaction to a compressive force on a rod. (2) The axial resultant force from a propellor or jet engine.

thrust *(Space)*. Reaction force produced on a rocket vehicle as a result of the expulsion of a high velocity exhaust gas. It is related to the acceleration (*a*) produced by Newton's second law: $T = ma$, where T is the thrust and m the mass of the rocket vehicle. Thrust is measured in newtons but is often quoted in kg.

thrust bearing, thrust block *(Eng.)*. A shaft bearing designed to take an axial load. In its simplest form it consists of a loose bronze washer interposed between moving parts on the shaft. Other types range from ball bearings with lateral races, solely for thrust, to taper roller bearings which can sustain both radial and axial loads. See also **Michell bearing**.

thrust chamber *(Aero.)*. The compartment in a rocket where the propulsive forces are developed before ejection; usually, but not necessarily, the *reaction chamber*.

thrust deflector *(Aero.)*. A device, usually a combination of doors closing the jet pipe and a cascade of guide vanes, for deflecting the efflux of a turbojet downward to provide upward thrust for *STOL* or *VTOL*.

thrust deflexion *(Aero.)*. The direction of the efflux from a **turbojet**, **ramjet** or **rocket** in a direction other then along its axis, for the purpose of obtaining a thrust component normal to this axis. Generally used for the guidance of rockets and for **STOL** and **VTOL** aircraft. See **jet deflection**.

thrust loading *(Aero.)*. The sea-level static thrust of the engine(s) of a jet-propelled aircraft divided by its gross weight.

thrust plane *(Geol.)*. A thrust plane, or thrust, is a *reversed fault* which dips at a low angle.

thrust reverser *(Aero.)*. A device for deflecting the efflux of a turbojet forward in order to apply a positive braking thrust after landing. There are two basic types: mechanical ones in which the jet is blocked by hinged doors, which also direct the gases forward; and aerodynamic ones wherein high-pressure air injected into the centre of the jet causes it to impinge upon peripheral louvres that turn it forward.

thrust spoiler *(Aero.)*. A controllable device mounted on, or just behind, the nozzle of a jet-propulsion engine to deflect and thus negate the thrust. See also **thrust reverser**.

thrust-to-weight ratio *(Space)*. The ratio thrust/weight which must be greater than 1 (both factors expressed in kg) if the rocket system is to leave the ground.

thrust-weight ratio *(Aero.)*. The thrust of an aircraft's power plant(s) divided by the gross weight at take-off.

thudichium speculum *(Med.)*. An instrument used to inspect the nasal interior, from the facial aspect.

thulite *(Min.)*. A variety of zoisite, pink in colour due to small amounts of manganese.

thulium *(Chem.)*. A metallic element, a member of the rare earth group. Symbol Tm, at. no. 69, r.a.m. 168.9342. One of the rarest elements, occurring in small quantities in euxenite, gadolinite, xenotime etc. Radioactive isotope emits 84 keV gamma-rays, frequently used in radiography.

thumb latch *(Build.)*. A latch which is operated by the pressure of the thumb. See **Norfolk latch**.

thumb plane *(Build.)*. Rebate plane with curved sole; used for rounded rebates.

thumbscrew *(Build.)*. Small type of G cramp used for light work.

thunder *(Meteor.)*. The crackling, booming, or rumbling noise which accompanies a flash of lightning. The noise has its origins in the violent thermal changes accompanying the discharge, which cause nonperiodic wave disturbances in the air. Its reverberatory characteristic arises mainly from the continuous arrival of the brief noise from sections of the discharge at increasingly remote locations, since the spark may be many kilometres long. Claps of thunder occur when the spark is, roughly, normal to the line of observation. The time interval between lightning and the corresponding thunder (seconds divided by three) gives the distance of the storm centre (in kilometres).

thunderstorm *(Meteor.)*. A storm in which **lightning** and **thunder** occur, usually associated with **cumulonimbus** cloud. The mechanism by which the cloud becomes electrically charged is not fully understood.

thuringite *(Min.)*. A variety of **chlorite**.

Thury screw-thread *(Eng.)*. See **Swiss screw-thread**.

Thy 1 antigen *(Immun.)*. A cell surface iso-antigen of mice present on thymus derived lymphocytes in the thymus and peripheral tissues. Antibodies specific for Thy 1 are used to identify such lymphocytes. A similar antigen (Thy 2) is present on rat lymphocytes. Thy antigens are also present in central nervous tissue. Their general structure is similar to immunoglobulin peptide chains, but their function is unknown.

thylakoid *(Bot.)*. A flattened, membrane-bounded sac in a chloroplast. Thylakoids may be single or associated in pairs or threes or more. See **granum**, **stroma lamella**.

thymectomy *(Med.)*. Surgical removal or the thymus.

thymic epithelial cells *(Immun.)*. Epithelial cells which ramify throughout the cortex and medulla of the thymus. Although similar in most respects, differences between cortical and medullary epithelial cells are detected by monoclonal antibodies. They are believed to control the maturation of lymphocyte precursors by secretion of peptide hormones (*thymosins, thymopoietin*).

thymic hypoplasia *(Immun.)*. Congenital cell-mediated immunodeficiency syndrome in human infants. In Di George's syndrome there is also parathyroid hypoplasia, causing additional problems due to hypocalcaemia. Thymic hypoplasia is characterized by recurrent infections of the skin and respiratory tract, marked depletion of lymphocytes in the thymus dependent area of peripheral lymphoid tissues, and failure to express cell-mediated immunity. Antibody production and blood immunoglobulin levels are normal.

thymine *(Chem.)*. *5-methyl-2,6-dioxytetrahydropyrimidine.* One of the two pyrimidine bases in DNA in which it pairs with adenine. See **genetic code**. ⇨

thymocyte *(Immun.)*. Any lymphocyte found within the thymus.

thymol *(Chem.)*. $C_{10}H_{14}O$, *1-methyl-4-methylethyl-3-hydroxybenzene.* Large crystals; mp 51°C, bp 230°C; it occurs in thyme oil, and can be synthesized from propan-2-ol and 1-methyl-3-hydroxybenzene. It is used as a disinfectant, and in mouthwashes and in dentistry. Isomeric with **carvacrol**.

thymol blue *(Chem.)*. *Thymol-sulphon-phthalein,* used as an indicator with two pH ranges, 1.2 to 2.8 (red→yellow), and 8.0 to 9.6 (yellow→blue).

thymolphthalein *(Chem.)*. Indicator obtained by reaction between thymol and phthalic anhydride, having a pH range of 9.3 to 10.5, over which it changes from a colourless to a blue solution.

thymoma *(Immun.)*. Rare thymic tumour. About half are associated with *myasthenia gravis*, and a few others with red cell aplasia, immunoglobulin deficiency, rheumatoid arthritis or polymyositis. There appears to be some abnormality of immune regulation.

thymopoietin *(Immun.)*. Factor derived from the thymus believed to influence the maturation of T-lymphocytes. It induces the appearance of T lymphocyte markers on resting lymphocytes both in vitro and after administration in vivo.

thymosins *(Immun.)*. Group of peptides derived from thymus epithelial cells (not the same as **thymopoietin**) which are present in blood. They can partially restore T lymphocyte function in thymectomized animals, and induce differentiation and maturation of immature T lymphocytes.

thymus *(Immun.)*. An organ of major importance for the development of the immune response. Develops in the embryo from the 3rd and 4th branchial pouches, and in most mammals it consists of two lobes situated in the anterior part of the thorax lying in front of the great vessels of the heart. Consists mainly of lymphocytes arranged between a network of epithelial cells, clearly separated into a cortical and medullary area. **Dendritic interdigitating cells** and nests of macrophage-like cells containing apparently breakdown products of other cells (Hassall's corpuscles) are also present in the medulla. It is the organ in which prothymocytes derived from bone marrow mature into functional T-lymphocytes, beginning in the cortex where prothymocytes differentiate and divide several times. It is here that the genes controlling the T-lymphocyte antigen receptor become rearranged from their germ line configuration, and the diversity of receptors is generated, with the elimination of most of the T-lymphocytes capable of recognizing 'self' antigenic determinants. They then undergo further differentiation in the medulla, becoming separated into mature helper T-lymphocytes, which recognize antigenic determinants associated with Class II MHC molecules and cytotoxic/suppressor T-lymphocytes, recognizing antigenic determinants associated with Class I MHC molecules. The great majority of the thymocytes so generated die *in situ*, leaving only an estimated 5% to become mature T-lymphocytes. These leave the thymus and enter peripheral tissues, where they live an independent existence, unaffected, for example, by thymus removal. The thymus is largest during early post-natal life, and gradually declines after puberty to become largely atrophic in old age.

thymus dependent antigen *(Immun.)*. *T-dependent antigen.* An antigen which fails to stimulate an antibody response if T-lymphocytes are absent. Co-operation between B-lymphocytes and helper T-lymphocytes is required for the B-lymphocytes responding to such antigens to differentiate into antibody secreting cells. Most proteins and complex antigens are thymus dependent.

thymus dependent area *(Immun.)*. Areas in peripheral lymphoid organs which are predominantly occupied by T-lymphocytes which circulate through them, and are depleted if the thymus fails to liberate T-thymocytes into the circulation. These areas are anatomically segregated, and contain dendritic interdigitating cells with which the T-lymphocytes come into close contact and which may be involved in antigen presentation to them.

thymus derived cells *(Immun.)*. Lymphocytes derived from the thymus. They are usually identified by the presence of thymus-specific surface antigens.

thymus independent antigen *(Immun.)*. *T-independent antigen.* An antigen able to stimulate B-lymphocytes to produce antibody without the co-operation of T-lymphocytes, e.g. in animals lacking a thymus. Such antigens are usually polymers which are poorly digestible by macrophages and which carry a repeated array of antigenic determinants, enabling them to bind firmly to Ig receptors on B-lymphocytes. They probably stimulate a subset of B-lymphocytes, and do not stimulate memory cells or cell mediated immunity. Some thymus dependent antigens (including lipopolysaccharides) can also stimulate non-specific T-cell help, and elicit larger responses in normal than in thymus-deprived animals. These are designated TI-1, whereas those which do not are TI-2.

thyratron *(Electronics)*. A gas-filled triode operating in an atmosphere of mercury-vapour, argon, helium, hydrogen or neon. Ionization starts with sufficient positive swing of the negative grid potential, and anode and grid potentials lose control.

thyratron firing angle *(Electronics)*. Phase angle of a.c. anode voltage supply to thyratron (measured relative to zero) at instant when it strikes.

thyristor *(Electronics)*. A semiconductor power switch capable of bistable operation. It can have from two to four terminals and can be triggered from its off state to on at any chosen point within a single 90° quadrant of the applied a.c. voltage. The SCR is the most common unidirectional thyristor. See also **triac**.

thyroglossal *(Med.)*. Of tongue and thyroid, e.g. thyroglossal duct, an embryonic structure from which the thyroid develops.

thyroid antibodies *(Immun.)*. Organ specific auto-antibodies found in a variety of thyroid diseases, especially Hashimoto thyroiditis and thyrotoxicosis. The major antibodies are against thyroglobulin, against another antigen in thyroid colloid, against a microsomal antigen of thyroid acinar cells, and against the receptors for thyroid stimulating hormone (TSH). The latter may mimic TSH and cause thyroid overactivity.

thyroidectomy *(Med.)*. The surgical removal of part of the thyroid gland.

thyroid gland *(Zool.)*. In Vertebrates, a ductless gland originating as a median ventral outgrowth from a point well forward on the the floor of the pharynx. It may be a single structure, bilobed, or paired, and there may be small accessory masses of thyroid tissue in other places. The gland consists of spherical follicles composed of an outer layer of cuboidal secretory cells surrounding and discharging into a central cavity. In this are found the hormones thyroxine and 3,5,3,-tri-iodothyronine, which are concerned with the rate of tissue metabolism, and the development of the nervous system and behaviour (deficiency causing *cretinism*), and in Amphibia, with the control of metamorphosis. Evolutionarily the thyroid originates from the endostyle of amphioxus (of the Cephalochorda) and Tunicata, and the ammocoete larva of Lampreys.

thyroiditis *(Med.)*. Inflammation of the thyroid gland. See also **Riedel's disease**.

thyrotomy *(Med.)*. See **laryngofissure**.

thyrotoxicosis *(Med.)*. The condition resulting from an excess of circulating thyroid hormones (T_4 and/or T_3) leading to either a diffuse hyperplasia of the thyroid (*Graves disease*) or toxic single or multiple nodules of the thyroid (*Plummer's disease*).

thyrotrophic *(Med.)*. Maintaining or nourishing the thyroid; used of the hormone of the anterior lobe of the pituitary gland, which stimulates growth and function of the thyroid gland.

Thysanoptera *(Zool.)*. An order of minute insects with asymmetrical piercing mouthparts, prothorax large and free, tarsi have a protrusible adhesive terminal vesicle. Some are serious pests causing malformation of plants and sometimes inhibiting the development of fruit. Thrips.

Ti *(Chem.)*. The symbol for titanium.

tibia *(Zool.)*. In land Vertebrates the pre-axial bone of the crus; in Insects, Myriapoda, and some Arachnida, the fourth joint of the leg.

tic *(Med.)*. See **habit spasm**.

tic douloureux *(Med.)*. *Trigeminal neuralgia*. An affection of the fifth cranial nerve characterized by paroxysmal attacks of pain in the face and the forehead.

tick-borne fever *(Vet.)*. A febrile disease of sheep and cattle caused by infection by *Rickettsia phagocytophila*; transmitted by ticks.

tick fever *(Vet.)*. *Texas fever*. See **redwater**.

tick pyemia *(Vet.)*. A pyemic disease of lambs caused by bacterial infection by *Staphylococcus aureus*, which causes abscesses in various parts of the body, especially the joints and muscles; the infection is believed to enter via tick bites.

ticks *(Med.)*. Blood sucking Arachnids related to mites which are vectors for a number of diseases in man and animals e.g. **typhus, relapsing fever** and **virus encephalitis**.

ticks *(Zool.)*. See **Acarina**.

tic-tac escapement *(Horol.)*. Type of anchor escapement covering 2 teeth of the escape wheel.

tidal friction *(Astron.)*. The friction caused by the ebb and flow of the tides, especially in narrow channels; it is responsible for a reduction of the earth's rate of rotation, and for an acceleration of the moon's motion, so that eventually the day and the month will be equal in length. This effect in the past has caused the moon to present the same face to the earth (rotation and revolution being equal).

tidal volume *(Biol.)*. The volume of air moving in and out of the lungs of Vertebrates (and of the tracheal system of Insects) during normal (unforced) breathing; in man about 500 ml.

tide *(Astron.)*. The distortion of the surface layers (whether liquid or solid) of a planet or natural satellite resulting from differences between the gravitational forces acting on its various parts. The most familiar is the ocean tide on the Earth, produced mainly by the attraction of the Moon. The Moon's pull causes two tides each lunar day, not one! This is because there is one bulge of water directed towards the Moon and another on the opposite side of the Earth. The presence of the former seems immediately obvious; but there is a farside bulge as well because the gravitational force there is less than that acting on the solid sphere of the Earth, causing it to come to equilibrium at a greater distance from the Moon than the shape of a solid uniform sphere suggests. The Sun contributes a small effect as well. When it reinforces the Moon there are high *spring tides*, and when at 90° to the Moon weak *neap tides*.

tide gauge *(Surv.)*. An apparatus for determining the variation of sea-level with time.

tie *(Eng.)*. A frame member sustaining only a tensile load. US holds and supports railroad tracks to gauge. Also UK *sleeper*.

tie-beam *(Eng.)*. A structural member connecting the lower ends of a pair of principal rafters to prevent them from moving apart. Also applied to a member connecting parts of a structure which may otherwise move apart under load.

tied letters *(Print.)*. A synonym for *ligature*.

tie line *(Surv.)*. A survey line forming part of a *skeleton* and serving to fix its shape, e.g. a diagonal of a four-sided skeleton.

tie line *(Telecomm.)*. A line which may pass through exchanges, but which is used solely for connecting private branch exchanges, and over which incoming calls cannot be extended. Also called *interswitchboard line*.

Tiemann-Reimer reaction *(Chem.)*. See **Reimer-Teimann reaction**.

tie rod *(Eng.)*. See **tension rod**.

tie-rod stator frame *(Elec.Eng.)*. A form of stator frame for large electrical machines in which several frame sections are laterally secured by means of tie rods parallel with the axis of the machine.

tie wall *(Civ.Eng.)*. A cross-wall built upon the extrados of an arch at right angles to the spandrel wall or walls.

tie wire *(Elec.Eng.)*. A wire used to attach a transmission or telephone line conductor to a supporting insulator. Also called a *binding wire*.

TIF *(Telecomm.)*. Abbrev. for *Telephone Interference (*or *Influence) Factor*.

TIG *(Eng.)*. See **tungsten inert gas welding**.

tiger's eye *(Min.)*. A form of silicified crocidolite stained yellow or brown by iron oxide.

tight binding model *(Phys.)*. A model from which the electronic band structure of certain transition elements can be calculated. The overlap of wavelengths from atom to atom is regarded as small enough to permit the crystal wavefunction to be calculated by a linear combination of atomic orbital (LCAO) approach. See **band theory of solids**.

tight coupling *(Elec.Eng.)*. That between 2 circuits which causes alteration of the current in either to affect materially the current in the other. In mutual reactance coupling, coupling is said to be *tight* when the ratio of mutual reactance to the geometric mean of the individual reactances (of the same sign) of the 2 circuits approaches unity. Also called *close coupling*.

tight-edged *(Print.)*. Said of a reel of paper which has dried out at the edges, resulting in a web which is slack in the middle.

tight junction *(Biol.)*. Junction between epithelial cells where the membranes are in close contact, with no intervening intercellular space. Tight junctions can bind epithelial cells into sheets which permit no leakage of solutes across the sheets between the cells.

tile *(Build.)*. A thin slab, often highly ornamental, of baked clay, terracotta, glass, cement, or asbestos-cement, used for roofing or for covering walls or floors.

tile *(Image Tech.)*. Video effect in which the picture is broken up into small rectangular areas of uniform tone and colour.

tile-and-a-half tile *(Build.)*. A purpose-made tile of extra width.

tile creasing *(Build.)*. A course formed of 2 or 3 thicknesses of plain roofing tiles, set in mortar and breaking joint. Laid immediately below a brick-on-edge coping and projecting about 2 in over each side of the wall, with the top surface sloped in cement, in order to prevent the percolation of water into the wall below the coping.

tile hanging *(Build.)*. See **weather tiling**.

tile lintel floor *(Build.)*. A type of fire-resisting floor having a steel framework similar to the *filler joist floor*, but with hollow tile, terracotta, or fireclay lintels filling in the panels between filler joists, thus reducing the amount of concrete required for encasement.

tile ore *(Min.)*. The earthy brick-red variety of cuprite; often mixed with red oxide of iron.

tiling batten *(Build.)*. See **slating and tiling battens**.

till *(Geol.)*. A poorly sorted mixture of unconsolidated sediment produced by glacial action. Synonymous with *boulder clay*.

tiller *(Bot.)*. A shoot that develops from an axillary or adventitious bud at the base of a stem; characteristic of grasses, including cereals.

tillite *(Geol.)*. Consolidated and lithified till.

tilting *(Image Tech.)*. Swinging the camera vertically.

tilting fillet *(Build.)*. A strip of wood laid beneath a **doubling course** to tilt it up slightly, so that the slates may rest properly in the roof. Also called *eave-board, skew fillet*.

tilting level *(Surv.)*. A type of level whose essential characteristic is that the telescope and attached level tube may be levelled without the necessity for setting the rotation axis truly vertical.

tilt roof *(Build.)*. A roof having the form of a circular arc in which the rise is small compared with the span.

tilt wing *(Aero.)*. VTOL aircraft whose wing, complete with propulsion units, propellers or nozzles can be rotated through 90° about a transverse axis, so that thrust acts vertically. Such arrangements have flown but have proved impractical.

timber *(For.)*. Felled trees or logs suitable for conversion by sawing or otherwise.

timber brick *(Build.)*. See **wood brick**.

timber connectors *(Build.)*. (1) Metal plates with spikes formed on one or both sides used with bolts to fix timbers together. (2) Rings of metal which can be housed or grooved into the faces of adjoining timbers held with bolts.

timbering *(Build.,Civ.Eng.)*. Temporary timbers arranged for the support of the earth in excavations, to prevent collapse of the sides.

timber line *(Bot.)*. Line or zone on a mountain or at high latitudes beyond which trees do not grow to normal size or form. Cf. **krummholz**.

timbre *(Acous.)*. The characteristic tone or quality of a sound, which arises from the presence of various harmonics or overtones of the fundamental frequency.

time *(Astron.)*. Originally measured by the **hour angle** of a selected point of reference on the celestial sphere with respect to the observer's meridian. The fundamental unit of time measurement now is the *second* based on an atomic oscillation. See also **apparent solar time, ephemeris time, Greenwich Mean Time, local time, mean solar time, sidereal time, standard time, universal time**.

time-assignment speech interpolation *(Telecomm.)*. A method for increasing the capacity of submarine cables. The intervals of silence (pauses in speech, intervals between data etc.) are used temporarily for the transmission of information for other channels, with channel switching occurring so fast that users are unaware of it.

time base *(Electronics)*. A line (usually horizontal) formed by a particular waveform applied to a cathode ray tube deflection system. For a linear time base, the waveform is a sawtooth form though circular time base patterns may be used in some specialized applications.

time base generator *(Electronics)*. A circuit or equipment which produces the waveform necessary to deflect a CRT beam in the desired time base pattern. Television sets have two time base generators, both providing saw tooth waveforms, one operating rapidly to scan the *lines*, the other operating more slowly at the *frame* frequency, thus ensuring that each line on the screen is scanned below the previous one.

time between overhauls *(Aero.)*. The period in hours of running time between complete dismantling of an aero-engine. Abbrev. *TBO* or *tbo*. Also *overhaul period*.

time code *(Image Tech.)*. Coding system with units of hours, minutes, seconds and frames recorded on audio- and video-tape to allow the precise identification and automatic location of an individual frame. Such information may also be applied to processed motion picture film for the same purpose.

time constant *(Elec.Eng.)*. When any quantity varies exponentially with time, the time required for a fractional change of amplitude equal to:

$$100\left(1-\frac{1}{e}\right)=63\%,$$

where e is the exponential constant (the base of natural logarithms). For a capacitance C to be charged from a constant voltage through a resistance R, the time constant is RC. For current in an inductance L being fed from a constant voltage through a resistance R the time constant is L/R. See **response time**.

time-delay relay *(Elec.Eng.)*. One which closes contacts in one circuit a specified time after those in a second circuit have been closed.

time dilation *(Phys.,Space)*. The time interval between two events appears to be longer when they occur in a reference frame which is moving relative to the observer's reference frame, than when they occur at rest relative to the observer. Time dilation is a consequence of the **Lorentz transformations** in the *special theory of relativity* and can be expressed as:

$$t=t_0\sqrt{1-\left(\frac{v}{c}\right)^2}$$

where t is the time of a clock moving with velocity v, t_0 is the elapsed time of a stationary clock and c is the velocity of light.

time discriminator *(Telecomm.)*. Circuit which gives an output proportional to the time difference between two pulses, its polarity reversing if the pulses are interchanged.

time-division multiple access *(Telecomm.)*. A technique used extensively in satellite communications in which a stream of *time slots* in a time-division multiplex system is allocated to users in accord with the demands they are making at any time. Cf. **single-channel per carrier**.

time-division multiplex *(Telecomm.)*. Form of multiplex transmission which follows logically from the adoption of **pulse modulation** and processes involving **sampling**. The gaps between pulses which constitute a signal allow other pulses to be interleaved; extraction of the desired signal at the receiver requires a system operating in synchrony with the transmitter. Cf. **frequency-division multiplex**.

time-domain reflectometer *(Telecomm.)*. An instrument which detects the transmission properties of wideband systems, components and lines by feeding in a voltage step and displaying the pulses reflected from any discontinuities on a suitable oscilloscope. The display can be calibrated to reveal the nature and location of the defects.

time exposures *(Image Tech.)*. Exposures in cameras for periods longer than about 1/25 of a second, longer than the so-called instantaneous exposure.

time lapse *(Image Tech.)*. Cinematography or videotape recording with a controlled delay between the exposure of each frame, so that in normal presentation the action appears very greatly speeded up.

time-limit attachment *(Elec.Eng.)*. The mechanical device whereby a circuit-breaker opens only after a predetermined time delay. Cf. **time-limit relay**.

time-limit relay *(Elec.Eng.)*. An electric relay which comes into action some time after it has received the electrical operating impulse.

time-meter *(Elec.Eng.)*. An instrument for measuring the time during which current flows in a circuit. Also called *hour-counter, hour-meter*.

time-of-flight spectrometer *(Nuc.Eng.)*. Way of measuring a neutron spectrum in which the energy or speed of neutrons is determined by the time taken by the neutrons to travel a know distance. A *chopper* admits neutrons in short bursts and the travel time is determined either by a second chopper which passes only neutrons of the correct velocity or by electronic delay measuring equipment. See **neutron velocity selector**.

time of operation *(Phys.)*. Time between the occurrence of a primary ionizing event and the occurrence of the count in the detector system.

time of operation *(Telecomm.)*. In relays, time between application of current or voltage and occurrence of a definite change in circuits controlled by its contacts.

time of oscillation *(Horol.)*. The time of oscillation of a pendulum or balance is twice that of the single vibration.

timepiece *(Horol.)*. A general trade term for any clock that shows the time but does not strike.

timer *(Elec.Eng.)*. Device, operated by electric motor, clockwork, or an electronic or resistor-capacitor circuit, which opens or closes at specified times, with or without delay, control circuits for lighting lamps, operating motors or valves etc. in a process controller.

time scale *(Geol.)*. A chronological sequence of geological events.

time-shared amplifier *(Telecomm.)*. Form of multiplex system in which one amplifier handles several signals simultaneously using successive short intervals of time for each.

time sharing *(Comp.)*. Means of providing multi-access to a computer system. Each user is, in turn, allowed a **time slice** of the central processor although each appears to have continuous use of the system. See **multiprocessor, multiprogramming, job queue.**

time shift *(Image Tech.)*. Recording a TV broadcast programme on videotape for viewing at a later more convenient occasion.

time signal *(Telecomm.)*. Radio transmission indicating standard time by reference to an atomic standard.

time slice *(Comp.)*. Predetermined maximum length of time during which each program is allowed to run during **multiprogramming**. See **interrupt.**

time switch *(Elec.Eng.)*. A switch arrranged to open or close a circuit at a predetermined time, operating by some form of electrical or clockwork mechanism.

timing *(Autos.)*. The process of setting the valve-operating mechanism of an engine so that the valves open and close in correct relation to the crank during the cycle; a similar adjustment of the magneto or distributor drive; the actual valve or magneto or distributor drive; the actual valve or magneto setting, called *valve timing, ignition timing.*

timing *(Horol.)*. The process of: (1) setting a clock or watch to time; (2) observing the rate of a clock or watch.

timing chain *(Autos.)*. Chain which drives the camshaft from the crankshaft in some types of timing gear.

timing gear *(Autos.)*. The drive between the crankshaft and the camshaft, by direct gearing, bevel-shaft, chain and sprocket-wheels or toothed belt, giving a reduction ratio of 1 : 2.

timing nuts *(Horol.)*. The 2 nuts on the rim of a chronometer balance, one at each end of the arm, used for timing purposes.

timing screws *(Horol.)*. The 4 diametrically opposite screws in the rim of a compensating balance used for bringing the watch to time.

timing washers *(Horol.)*. Thin washers placed under the heads of the screws of a balance to produce slight alteration in the moment of inertia of the balance and so modify the time of vibration.

tin *(Chem.)*. A soft, silvery-white metallic element, ductile and malleable, existing in three allotropic forms. Symbol Sn (Latin *stannum*, tin), at.no. 50, r.a.m. 118.69, rel.d. at 20°C 7.3, mp 231.85°C. Not affected by air or water at ordinary temperatures. Electrical resistivity is 11.5×10^{-8} ohm metres at 20°C. The principal use is as a coating on steel in tinplate; also used as a constituent in alloys and with lead in low melting-point solders for electrical connections. Occurs as tin oxide, SnO_2, or *cassiterite*. See **tin alloys.**

tin alloys *(Eng.)*. Tin is an essential constituent in soft solders, type metals, fusible alloys and certain bearing metals. These last contain 50–92% of tin alloyed with copper and antimony and sometimes lead. Tin is also a constituent of bronze and pewter.

Tinamiformes *(Zool.)*. An order of Birds, containing small superficially partridge-like, almost tailless birds which are essentially cursorial, but can fly clumsily for short distances, and have a keeled sternum. Tinamus.

tincal *(Min.)*. The name given since early times to crude borax obtained from salt lakes, e.g. in Kashmir and Tibet. See **borax.**

tinea *(Med.)*. See **ringworm.**

tingle *(Build.)*. A flat strip of lead or copper used as clip between jointing sheets of lead.

ting-tang *(Horol.)*. A clock that strikes the quarters on two notes only.

tinman's solder *(Eng.)*. A tin-lead solder melting below a red-heat, used for tinning. The most fusible solder contains 65% tin.

tin-nickel *(Eng.)*. Metal finish which results from the simultaneous electrodeposition of tin and nickel on a polished surface of brass in a carefully controlled bath, resulting in a nontarnishable and noncorrodible polished surface of low friction.

tinnitus, tinnitus aurium *(Med.)*. Persistent sensation of ringing noises in the ear.

tin-plate *(Eng.)*. Thin sheet-steel covered with an adherent layer of tin formed by passing the steel though a bath of molten tin or by electrodeposition. Resists atmospheric oxidation and attack by many organic acids. Used for food-containers etc.

tin pyrites *(Min.)*. See **stannite.**

tinsel yarn *(Textiles)*. See **metallized yarn.**

tin-stone *(Min.)*. See **cassiterite.**

tint *(Build.)*. An unsaturated colour. A pure colour softened by the addition of white. *Tinting* signifies the addition of pigment to a colour.

tinting *(Print.)*. (1) A *mechanical stipple* especially when used for line-colour purposes. (2) A fault occurring in lithographic printing where a coloured tint prints over the non-image area of the sheet, caused by the pigment being dispersed in the damping solution.

tin-zinc *(Eng.)*. Metal finish which results from the simultaneous electrodeposition of tin and zinc on a clean steel surfaces, giving a non-corrodible finish to chassis for electronic apparatus.

tip *(Eng.)*. The part of a cutting tool containing the cutting edges, if made of a material of superior quality to that of the remainder of the tool, e.g. tungsten carbide, and securely fastened to the latter by brazing or otherwise.

tip *(Min.Ext.)*. See **dump.**

tip-cat folder *(Print.)*. A type of folder in which the folding blade shaft is actuated by a specially shaped cam and forms the transverse fold by pushing the copy through the folding rollers.

Ti plasmid *(Bot.)*. *Tumour inducing principle.* Plasmid, carried by virulent strains of the crown gall bacterium, *Agrobacterium tumefaciens*, part of which (the T-DNA) may, when the bacterium infects a plant, become transferred and incorporated into the nuclear genome of the host cells, inducing them to grow and form the characteristic galls. Ti plasmids are possible vectors for the 'genetic engineering' of dicotyledonous plants. See **crown gall.**

tip-path plane *(Aero.)*. The plane of rotation of the tips of a rotorcraft's blades, which is higher than the rotor hub in flight. See **coning angle.**

tip penetration *(Image Tech.)*. The protrusion of the magnetic heads of a VTR into the videotape.

tipping-in *(Print.)*. See **plating.**

tipple *(Min.Ext.)*. Frame into which ore trucks are run, gripped, and rotated to discharge contents.

tippy wool *(Textiles)*. Wool with the fibre tips damaged (e.g. by **photodegradation**) while on the sheep. The defect is often revealed at the dyeing stage because the damaged tips take up different amounts of dyes compared with the undamaged parts.

tirodite *(Min.)*. A rare honey-yellow monoclinic amphibole, the name being used for the manganese-rich, magnesium-bearing end member. See **dannemorite.**

T-iron *(Civ.Eng.)*. A structural member of wrought-iron or rolled mild steel having a T-shaped cross-section.

tissue *(Biol.)*. An aggregate of similar cells forming a definite and continuous fabric, and usually having a comparable and definable function; as *epithelial tissue, nervous tissue, vascular tissue.*

tissue culture *(Biol.).* The growth of cells, including tissues and organs, outside the organism in artificial media of salts and nutrients. Depending on the cell type, the cells may be capable of a limited number of divisions or may divide indefinitely. In plants and under appropriate conditions cultured tissues can often be made to regenerate new plants. See **transformation (2)**.

tissue dose *(Radiol.).* Absorbed *depth dose* of radiation received by specified tissue. Cf. **skin dose**.

tissue equivalent material *(Radiol.).* See **phantom material**.

tissue specific antigen *(Immun.).* Cell antigen present in a given tissue but not found in other tissues, e.g. thyroglobulin is specific for thyroid. Auto-immune diseases have different pathological consequences according to whether auto-antibodies and/or cell mediated immunity are directed against tissue specific antigens, or auto-antibodies are reactive with non-tissue specific antigens such as nucleic acid or nucleoproteins common to many tissues.

tissue tensions *(Bot.).* The mutual compression and stretching, of deeper and more superficial tissues respectively, exerted by the tissues of a living plant.

tissue typing *(Immun.).* The identification of histocompatibility antigens, usually done on blood leucocytes from prospective donor and recipient prior to tissue or organ transplantation.

Titan *(Astron.).* Saturn's largest satellite (diameter 5150 km), and the second largest moon in the solar system. It is the only satellite with a substantial atmosphere, principally composed of nitrogen and methane.

titanates *(Chem.).* Compounds found in minerals, or formed when titanium (IV) oxide is fused with alkalis; containing the anion TiO_4^{4-} or TiO_3^{3-}.

titanaugite *(Min.).* A titaniferous variety of the monoclinic pyroxene **augite**.

titania *(Chem.).* See **titanium (IV) oxide**.

titaniferous iron ore *(Min.).* See **ilmenite**.

titanite *(Min.).* See **sphene**.

titanium *(Chem.).* A metallic element resembling iron. Symbol Ti, at. no. 22, r.a.m. 47.90, rel. d. (20°C) 4.5, mp 1850°C, bp above 2800°C. Manufactured commercially since 1948, it is characterized by strength, lightness and corrosion resistance. Widely used in aircraft manufacture, for corrosion resistance in some wet extraction processes, as a deoxidizer for special types of steel, in stainless steel to diminish susceptibility to intercrystalline corrosion and as a carbide in cemented carbides. Sometimes used for the solid horns of magnetostrictive generators. It occurs only in combination, and is widely distributed on the earth's crust (0.6% by weight).

titanium (IV) oxide, titanium dioxide *(Chem.).* TiO_2. It is a pure white pigment of great opacity. Widely used industrially in paints, plastics etc., and see **titanizing**. Forms titanites when fused with alkalis. Also called *titania*. Minerals called *rutile*, *anatase* and *brookite*.

titanizing *(Glass).* Use of titanium dioxide, TiO_2, in the manufacture of heat-resisting and durable glass by incorporating it to replace certain proportions of soda, the viscosity increasing proportionately.

Titius-Bode law *(Astron.).* See **Bode's Law**.

title signature *(Print.).* The first **section** (or signature) of a book, containing the title page and other prelims, and not normally requiring a **signature mark**.

titling fount *(Print.).* Consists of capitals, figures, and punctuation marks only, no lower-case, a narrow beard being sufficient; used for dropped initials and when close spacing between lines is required.

titration *(Chem.).* The addition of a solution from a graduated vessel (burette) to a known volume of a second solution, until the chemical reaction between the two is just completed. A knowledge of the volume of liquid added and of the strength of one of the solutions enables that of the other (the *titre*) to be calculated.

titre *(Chem.).* See **titration**.

titre *(Immun.).* In serological reactions involving the use of serial dilutions of antiserum, describes the highest dilution at which the measured effect is detected.

titrimeter *(Chem.).* Apparatus for electrometric titrations in which potential changes are followed continuously and automatically.

titubation *(Med.).* Staggering and reeling movements of the body, due to disease of the nervous system.

TL *(Acous.).* See **transmission loss**.

Tl *(Chem.).* The symbol for thallium.

TLC *(Chem.).* **Thin-layer chromatography**.

T-lymphocyte antigen receptor *(Immun.).* *T-cell receptor.* Molecule present at the surface membrane of T-lymphocytes capable of specifically binding antigen in association with **MHC** antigen at the surface of an antigen presenting cell. Such molecules have been isolated from cloned T-lymphocyte **hybridomas** and from antigen specific T-lymphocyte cell lines, and found to consist of two polypeptide chains α and β. A similar but distinct pair of polypeptide chains, γ and δ, has been found to be expressed in early thymocytes but not in mature T-lymphocytes. The chains each contain a variable and a constant region, whose structure is controlled by genes with several **exons** in much the same way as those controlling **immunoglobulins**, to which they are structurally similar with considerable sequence homology, but they are clearly distinct. Each antigen molecule is tightly associated with another molecule (CD3) which is required if activation of the T-lymphocyte by the antigen is to take place. Yet another molecule (CD4 in helper cells and CD6 in cytotoxic/suppressor cells) determines whether the lymphocyte will recognize the antigen in association with MHC Class II or MHC Class I determinants respectively. See also **thymus**. Unlike B-lymphocytes, T-lymphocytes recognize relatively short linear peptide sequences rather than the shape of the tertiary structure of a protein molecule.

T-lymphocyte repertoire *(Immun.).* The number of different antigenic determinants to which the T-lymphocytes of an individual animal are capable of responding, thought to be comparable in size to the B-lymphocyte repertoire. There are however certain peptide sequences which are not recognized, either because they are not recognizably different from 'self' or because they are unable to associate with a particular MHC determinant.

Tm *(Biol.).* The temperature at which DNA is half denatured. syn. *melting temperature*.

Tm *(Chem.).* The symbol for thulium.

T-maze *(Behav.).* A T-shaped pathway with a starting box at the base of the T and goal boxes containing reinforcing stimuli at the ends of either or both arms. Discriminative stimuli may be placed in the arms of the T near the **choice point**.

TMS *(Chem.).* Abbrev. for *TetraMethylSilane*.

TM-wave *(Telecomm.).* Abbrev. for *transverse magnetic wave*, having no component of magnetic force in the direction of transmission of electromagnetic waves along a waveguide. Also known as *E-wave* (it must have electric field component in direction of transmission).

Tn *(Phys.).* The symbol for thoron.

T-network *(Telecomm.).* One formed of two equal series arms with a shunt arm between.

TNT *(Chem.).* An abbrev. for *TriNitroToluene*.

T-number *(Image Tech.).* A rating of the actual light transmission of a lens at a given stop, in contrast to the calculated **f-number**.

toad's-eye tin *(Min.).* A variety of **cassiterite** occurring in botryoidal or reniform shapes which show an internal concentric and fibrous structure. It is brownish in colour.

toadstone *(Geol.).* An old and local name for the basalts found in the Carboniferous Limestone of Derbyshire. The name may be derived from the rock's resemblance in appearance to a toad's skin, or from the fact that it weathers into shapes like a toad, or from the German word *todstein* ('dead stone') in reference to the absence of lead.

tobacco amblyopia *(Med.).* A visual defect which can cause blindness associated with tobacco smoking and may be due to cyanide in smoke.

Tobin bronze *(Eng.)*. A type of alpha-beta brass or Muntz metal containing tin. It contains 59–62% copper, 0.5–1.5% tin, the remainder being zinc. Used when resistance to sea water is required. Also *Admiralty* or *naval brass*.

tocopherol *(Biol.)*. See **vitamin E**.

Todd-AO *(Image Tech.)*. TN of an older system of **wide-screen** cinematography, using 65 mm width negative in the camera, with 70 mm prints and six magnetic tracks for presentation.

todorokite *(Min.)*. Hydrated manganese oxide, enriched in other elements. It is one of the dominant minerals of the deep-sea **manganese nodules**.

toe *(Image Tech.)*. The lower end of the photographic **characteristic curve**, departing from the straight line portion, where the densities begin to approach the basic **fog level**.

toe-in *(Autos.)*. A slight forward convergence given to the planes of the front wheels to promote steering stability and equalize tyre-wear.

toe-picking *(Vet.)*. The habit, acquired by individual budgerigars, of biting the feet of birds of other species, particularly finches, within the same aviary.

toggle *(Telecomm.)*. Bistable trigger circuit, a multivibrator with coupling capacitors omitted, which switches between 2 stable states depending on which valve (or transistor) is triggered. Once *flip-flop* in U.K.

toggle joint *(Eng.)*. A mechanism comprising two levers hinged to each other, the end of one being hinged on a fixed point, the opposite end of the other being hinged on a press ram or other load point. If the levers form an obtuse angle and an effort tending to increase this angle is applied at their common hinge, a considerable force is produced at the load point.

toggle press *(Eng.)*. A power press, often double-acting, incorporating a toggle joint, used for deep drawing operations.

toilet *(Med.)*. The cleaning and dressing of a wound or injured part.

Tokamak *(Nuc.Eng.)*. In fusion, a toroidal apparatus for containing plasma by means of two magnetic fields, (a) a strong toroidal magnetic field created by coils surrounding the vacuum chamber and (b) a weaker poloidal field created by an intense electric current through the plasma. Tokamak is an acronym of the Russian words meaning toroidal magnetic chamber. See **stellarator**.

tokamak *(Phys.)*. Torus-shaped magnetic bottle, using an auxiliary magnetic field to achieve good plasma confinement for nuclear fusion.

token economy *(Behav.)*. A behaviour modification procedure, based on operant conditioning principles, in which patients are given artificial rewards for socially desirable behaviour; the rewards or tokens can be exchanged for desirable items (e.g. cigarettes).

tolbutamide *(Med.)*. Used for oral treatment of diabetes.

tolerance *(Eng.)*. The range between the permissible maximum and minimum limits of size of a workpiece or of distance between features (e.g. hole centres) on a workpiece.

tolerance *(Immun.)*. See **immunological tolerance**.

tolerance dose *(Radiol.)*. Maximum dose which can be permitted to a specific tissue during radiotherapy involving irradiation of any other adjacent tissue.

tolerogen *(Immun.)*. Capable of inducing immunological tolerance.

toll television *(Image Tech.)*. Programme service which, through technical scrambling devices, is available only by *ad hoc* payment. Also called *pay as you view*, *subscription television*, *pay-TV*.

Tolu balsam *(Chem.)*. See **balsam of Tolu**.

toluene *(Chem.)*. $C_6H_5.CH_3$, a colourless liquid, mp −94°C, bp 110°C. It occurs in coal- and wood-tar; insoluble in water; miscible with ethanol, ethoxyethene, trichloromethene. Used as a solvent and as an intermediate for its derivatives. Also called *methylbenzene*, *toluol*.

toluene di-isocyanate *(Chem.)*. *TDI*. $CH_3.C_6H_3.(NCO)_2$. The most commonly used isocyanate in the production of polyurethane foams. The isocyanate is reacted with a high molecular weight glycol to form a liquid pre-polymer. The pre-polymer reacts with water to give the polyurethane and carbon dioxide, which cause the foam.

toluidines *(Chem.)*. *Methyl(aminobenzenes)*. $H_3C.$-$C_6H_4.NH_2$, homologues of aniline. There are three isomers, viz., 2-toluidine, a liquid, bp 197°C; 4-toluidine, crystals, mp 43°C, bp 198°C; 3-toluidine, a liquid, bp 199°C.

tomentum *(Bot.)*. Covering of felted cotton hairs. adj. *tomentose*.

tommy bar *(Eng.)*. See **box spanner**.

tomography, emission *(Radiol.)*. Transverse section reconstruction of the radionuclide distribution within the body, obtained by acquiring images or **slices** of the head or body. This may be done by using **coincidence detection** (positrons) or single photon detection from gamma ray emitters.

-tomy *(Genrl.)*. Suffix from Gk. *tomē*, a cut.

ton *(Genrl.)*. A unit of mass for large quantities. The *long ton*, once commonly used in UK, is 2240 lb. The *short ton*, commonly used in America, is 2000 lb. The *metric ton* or *tonne* (1000 kilograms; symbol t) is 2204.6 lb. In UK the *short ton* was used in metalliferous mining, the *long ton* in coal-mining.

ton *(Ships)*. See **tonnage**.

tonalite *(Geol.)*. A coarse-grained igneous rock of dioritic composition carrying quartz as an essential constituent, i.e. quartz-mica-diorite. Two varieties are distinguished: *soda-tonalite*, with albite in excess of anorthite, and *lime-tonalite*, with anorthite in excess of albite.

tone *(Acous.)*. Sound signal of a single frequency. In the terminology of music, tone is often used to specify a complex note having a constant fundamental frequency. To avoid misunderstanding, a single frequency sound is termed a pure tone.

tone *(Zool.)*. The resting level of muscle contraction due to background neuromuscular activity.

tone control circuit *(Telecomm.)*. A circuit network or element to vary the frequency response of an audio-frequency circuit, thus varying the quality of the sound reproduction.

tongs *(Build.)*. See **stone tongs**.

tong-test ammeter *(Elec.Eng.)*. An a.c. ammeter and current transformer combination whose iron core can be opened and closed round a cable, thus forming the single-turn primary winding of the transformer.

tongue *(Build.)*. A slip feather.

tongue *(Zool.)*. In Vertebrates, the movable muscular organ lying on, and attached to, the floor of the buccal cavity; it has important functions in connection with tasting, mastication, swallowing, and (in higher forms) sound production; in Invertebrates, especially Insects, any conformation of the mouth-parts which resembles the tongue in structure, appearance, or function: proboscis; antlia; haustellum; radula; ligula; any structure which resembles the tongue.

tongue-and-groove joint *(Build.)*. A joint formed between the butting edges of two boards, one of which has, along the middle of its length, a projecting fin cut to fit into a corresponding plough groove in the other.

tongue-worm *(Vet.)*. *Linguatula serrata*. An aberrant Arthropod belonging to the class Pentastomida, which occurs in the nasal passages of the dog, fox and wolf, and more rarely in other animals; the larval and nymphal stages occur in herbivorous animals.

tonicity *(Zool.)*. See **tone**.

toning *(Image Tech.)*. See **chemical toning**, **dye toning**.

tonnage *(Ships)*. A measurement assigned by the relevant authority for assessing dues etc. One ton equals 100 cubic feet. See **gross tonnage**, **net register tonnage**.

tonnage breadth *(Ships)*. A number of tonnage breadths are measured horizontally inside the frames or **spar ceiling** if fitted. Internal cross-section areas are calculated from these by **Simpson's rule**. These areas are then used in the calculation of **tonnage**.

tonnage depth *(Ships)*. A number of tonnage depths are measured at certain points throughout the length of the ship from the top of the inner bottom or **ceiling**, if fitted, to the top of the deck, deducting one-third of the **camber**. Used, together with **tonnage breadths** in the calculation of cross-section areas.

tonnage dimensions *(Ships)*. The internal dimensions used in the calculation of **tonnage**. These are **tonnage length, tonnage breadth** and **tonnage depth**.

tonnage length *(Ships)*. The length measured along the uppermost continuous deck in ships having less than three decks and the second continuous deck from below in others. It is measured from a point where the line of the inside of the frames or **spar ceiling** cuts the centre line forward to a similar point aft.

tonne *(Genrl.)*. Metric ton, 1000 kg. See **ton**.

tonofilament *(Biol.)*. Filaments composed of keratin found in epithelial cells.

tonometer *(Acous.)*. A device consisting of a series of tuning-forks for determining the frequencies of tones.

tonometer *(Med.)*. An instrument for measuring hydrostatic pressure within the eye.

tonometer *(Phys.)*. An instrument for measuring vapour pressure.

tonoplast *(Bot.)*. The membrane round a vacuole in a plant cell.

tonsillectomy *(Med.)*. The surgical removal of the tonsils.

tonsillitis *(Med.)*. Inflammation of the tonsils.

tonsils *(Zool.)*. In Vertebrates, lymphoid bodies of disputed function situated at the junction of the buccal cavity and the pharynx.

tonus *(Zool.)*. A state of prolonged tension in a muscle without change in length.

tooled ashlar *(Build.)*. A block of stone finished with parallel vertical flutes.

tooling *(Print.)*. Decorating by hand a book cover, usually leather.

toolmaker *(Eng.)*. A highly skilled engineering craftsman employed to make press tools, cutters and other precision equipment.

tool post *(Eng.)*. The clamp by which a lathe or shaping-machine tool is held in the slide rest or ram. In its simplest form it consists of a slotted post, the end of which carries a clamping screw.

tool-post grinder *(Eng.)*. A small grinding machine held on the tool post of a lathe and fed across the work by means of the regular longitudinal or compound rest. It is used to grind miscellaneous small workpieces in the lathe, and to true lathe centres.

toolpusher *(Min.Ext.)*. Field supervisor of drilling operations.

tool steel *(Eng.)*. Steel suitable for use in tools, usually for cutting or shaping wood or metals. The main qualities required are hardness, toughness, ability to retain a cutting edge etc. Contains 0.6–1.6% carbon. Many tool steels contain high percentages of alloying metals: tungsten, chromium, molybdenum etc. (See **high-speed steel**). Usually quenched and tempered, to obtain the required properties.

tooth *(Bot.)*. Any small pointed projection as on the margin of a leaf.

tooth *(Zool.)*. A hard projecting body with a masticatory function. In Vertebrates, a hard calcareous or horny body attached to the skeletal framework of the mouth or pharynx, used for trituration or fragmentation of food; in Invertebrates, any similar projection of chitinous or calcareous material used for mastication or trituration.

toothed wheel *(Eng.)*. See **bevel gear**, **helical gear**, **spiral gear**, **spur gear**, **worm**.

toothing plane *(Build.)*. One with a serrated blade set perpendicularly; used for providing a key for gluing veneers etc.

toothings *(Build.)*. The recesses left in alternate courses of a wall when later extension is expected, so that the extension can be properly bonded in.

tooth ratio *(Elec.Eng.)*. The ratio of slot width to tooth width as measured at the circumference of the armature.

tooth ripple *(Elec.Eng.)*. See **slot ripple**.

tooth thickness *(Eng.)*. Of a gear-wheel, the length of arc of the pitch circle between opposite faces of the same tooth. See **pitch diameter**.

top *(Textiles)*. Originally used to describe a sliver of combed wool ready for spinning into worsted yarns. Now also applied to the material produced by cutting or breaking continuous filament **tows** of man-made fibres.

topaz *(Min.)*. Silicate of aluminium with fluorine, usually containing hydroxyl, which crystallizes in the orthorhombic system. It usually occurs in veins and druses in granites and granite-pegmatites. It is colourless, pale blue, or pale yellow in colour, and is used as a gemstone. Cf. **citrine, Oriental topaz, Scottish topaz, Spanish topaz**.

topazolite *(Min.)*. A variety of the calcium-iron garnet **andradite**, which has the honey-yellow colour and transparency of topaz.

top beam *(Build.)*. The horizontal beam connecting the rafters of a collar-beam roof.

top blanket *(Print.)*. On web-fed relief presses the outside dressing of an impression cylinder. Also *top sheet*. Cf. **under blanket**.

top casing *(Min.Ext.)*. The topmost part of the casing of an oil well to which either the drilling gear or the flow-control gear is attached. Also *conductor, anchor string*.

top dead-centre *(Eng.)*. See **inner dead-centre**.

top down programming *(Comp.)*. See **structural programming**.

top girth *(For.)*. In standing trees, the girth at the top of the merchantable length of the bole.

tophaceous *(Med.)*. Sandy, gritty, of the nature of tophus, e.g. *tophaceous gout*.

top hamper *(Ships)*. That part of the structure of the ship which is above the **superstructure**. Also called *deck houses* or *erections*.

top-hung *(Build.)*. Said of a window-sash arranged to open outwards about hinges on its upper edge.

tophus *(Med.)*. A hard nodule composed of crystals of sodium biurate which are deposited in bodily tissues in gout.

topness *(Phys.)*. A property that characterizes *quarks* and so hadrons. The topness of leptons and gauge bosons is zero. Topness is conserved in strong and electromagnetic interactions but not in weak interactions. Also known as *truth*.

topochemistry *(Chem.)*. The study of reactions which occur only at definite regions in a system.

topography *(Surv.)*. The delineation of the natural and artificial features of an area.

topological space *(Maths.)*. A non-empty set of points with some of its subsets defined to be open sets in such a way that (1) the whole space and the empty sets are both open, (2) the intersection of any 2 open sets is open, and (3) the union of any collection of open sets is open.

topology *(Maths.)*. The study of those properties of shapes and figures which remain invariant under homeomorphic mapping.

topotype *(Zool.)*. A specimen collected in the same locality as the original type specimen of the same species.

top rake *(Eng.)*. In cutting tools, the angle which that part of the cutting face which lies immediately behind the tool point makes with the horizontal.

topset beds *(Geol.)*. Gently inclined strata deposited on the subaerial plain or the just-submerged part of a delta. They are succeeded seawards by the *foreset beds* and, in deep water, by the *bottomset beds*.

top sheet *(Print.)*. See **top blanket**.

top shot *(Image Tech.)*. One taken with axis of camera nearly vertical.

top yeast *(Bot.)*. Sorts of brewer's yeast, *Saccharomyces cerevisiae*, which accumulate at the top of the medium during fermentation and are used in brewing traditional British ales. Cf. **bottom yeast**.

torbanite *(Min.)*. A variety of **boghead coal** or oil shale containing 70–80% of carbonaceous matter, including an abundance of spores; dark-brown in colour.

torbernite *(Min.)*. A beautiful rich-green hydrated phosphate of uranium and copper which crystallizes in the tetragonal system. It occurs associated with autunite and frequently in parallel growth with it, and also with other uranium minerals. Also called *copper (cupro-) uranite*.

torch igniter *(Aero.)*. A combination igniter plug and fuel atomizer for lighting-up gas turbines.

torching *(Build.)*. The operation, sometimes performed on slates which have been laid on battens only and not on boarding of pointing the horizontal joints from the inside with hair mortar or cement.

tornado *(Meteor.)*. An intensely destructive, advancing whirlwind formed from strongly ascending currents. When over the sea, the apparent drawing up of water arises from the condensing of water vapour in the vacuous core; also, in W. Africa, the squall following thunderstorms between the wet and dry seasons.

toroidal field *(Nuc.Eng.)*. Magnetic field generated by a current flowing in a solenoid round a torus.

toroidal-intake guide vanes *(Aero.)*. The flared annular guide vanes which guide the air evenly into the intake of a centrifugal *impeller*.

toroidal surface *(Phys.)*. A lens surface in which the curvature in one plane differs from that in a plane at right angles.

toroidal winding *(Elec.Eng.)*. See **ring winding** (1).

toroid, torus *(Elec.Eng.)*. Magnetic component (a coil or transformer), made in the shape of an anchor ring. Adopted because, with this construction, most of the magnetic field is contained within the core and leakage is minimal; thus, there is little or no interaction with adjacent components and circuits.

toroid, torus *(Maths.)*. A solid generated by rotating a circle about an external point in its plane.

torque *(Aero.)*. Propeller torque is the measure of the total air forces on the airscrew blades, expressed as a moment about its axis.

torque *(Phys.)*. The turning moment exerted by a *tangential* force acting at a distance from the axis of rotation. It is measured by the product of the magnitude of the force and the distance. Unit newton-metre. See **moment**.

torque amplifier *(Elec.Eng.)*. (1) Mechanical amplifier in which torque (not angle) is varied by differential friction of belts on drums. (2) Electrical servo system performing same function.

torque converter *(Eng.)*. A device which acts as an infinitely variable gear, but generally at varying efficiency, e.g. a centrifugal pump in circuit with an inward-flow turbine.

torque limiter *(Aero.)*. Any device which prevents a safe torque value from being exceeded, but specifically one which is used on a constant-speed **turboprop** to prevent it from delivering excess power to its airscrew.

torque link *(Aero.)*. A mechanical linkage, usually of simple scissor form, which prevents relative rotation of the telescopic members of an aircraft **landing gear** shock absorber.

torque-loaded quoin key *(Print.)*. One in which the lock-up pressure is consistent no matter who uses the key, which is adjusted to apply the optimum pressure and no more.

torquemeter *(Aero.)*. A device for measuring the torque of a reciprocating aero-engine or turboprop, the indication of which is used by the pilot, together with r.p.m. and other readings, to establish any required power rating.

torquemeter *(Eng.)*. A torsion meter attached to a rotating shaft, the angle of twist of a known length of shaft between the gauge points of the meter being indicated by optical or electrical means, thus enabling the power transmitted to be calculated. A form of transmission dynamometer.

torque motor *(Elec.Eng.)*. One exerting high torques at low speeds.

torque spanner *(Eng.)*. A spanner with a special attachment whereby a prescribed force can be applied to a bolt through the medium of the nut.

torr *(Phys.)*. Unit of low pressure equal to head of 1 mm of mercury or 133.3 N/m^2.

Torricellian vacuum *(Meteor.,Phys.)*. See **mercury barometer**.

Torridonian *(Geol.)*. A succession of conglomerates and red sandstones and arkoses forming part of the Precambrian System in the north-west highlands of Scotland. They rest on the Lewisian schists and gneisses.

torsion *(Eng.)*. State of strain set up in a part by twisting. The external twisting effort is opposed by the shear stresses induced in the material.

torsion *(Maths.)*. See **curvature (2)**.

torsion *(Zool.)*. The preliminary twisting of the visceral hump in Gastropod larvae which results in the transfer of the pallial cavity from the posterior to the anterior face, as distinct from the secondary or spiral twisting of the hump exemplified by the spiral form of the shell.

torsional wave *(Phys.)*. A wave of flexure in which the elements of the medium carrying the wave perform torsional oscillations about an axis parallel to the direction of propagation. Observed in solid structures such as long thin lamina, e.g. the roadway of a suspension bridge.

torsion balance *(Phys.)*. A delicate device for measuring small forces such as those due to gravitation, magnetism, or electric charges. The force is caused to act at one end of a small horizontal rod, which is suspended at the end of a fine vertical fibre. The rod turns until the turning moment of the force is balanced by the torsional reaction of the twisted fibre, the deflection being measured by a lamp and scale using a small mirror fixed to the suspended rod.

torsion bar suspension *(Autos.)*. A springing system, used in some independent suspension designs, in which straight bars, anchored at one end, are subjected to torsion by the weight of the car, thereby acting as springs.

torsion galvanometer *(Elec.Eng.)*. A galvanometer in which the controlling torque is measured by the angle through which the suspension head must be rotated in order to bring the pointer back to zero.

torsion meter *(Eng.)*. See **torquemeter**.

torsion pendulum *(Horol.)*. A pendulum in which the bob rotates, in a plane perpendicular to the line of suspension, by the twisting of the suspension ribbon. Used where a long time of vibration is required, as with 400-day clocks.

torticollis *(Med.)*. *Wry neck, stiff neck*. A disease of the cervical vertebrae or to affections (especially rheumatic) of the muscles of the neck. See also **spasmodic torticollis**.

torulosis *(Med.)*. A disease due to infection with the yeastlike micro-organism *Torula histolytica*; affects especially the central nervous system.

torus *(Bot.)*. The thickened central part of the **pit membrane** of the bordered pits of many gymnosperm tracheids. It seems to act as a valve closing a pit to prevent the threatened spread of an embolism.

torus *(Elec.Eng.)*. See **toroid**.

torus *(Maths.)*. A surface of revolution shaped like an anchor ring. It is generated by rotating a circle about a nonintersecting coplanar line as axis. Also called *anchor ring*.

torus *(Zool.)*. A ridge or fold, as in Polychaeta, a ridge bearing uncini.

toss-bombing *(Aero.)*. A manoeuvre for the release of a bomb (usually with a nuclear warhead) which allows the pilot to evade the blast. A special computing sight enables the pilot to loop before reaching the target, when the bomb is lobbed forward; or the pilot can overfly the target and loop to toss the bomb 'over-the-shoulder'.

tossing *(Min.Ext.)*. The operation of raising the grade or purity of a concentrate by violent stirring, followed by packing in a *kieve*. Also *kieving*.

tosyl *(Chem.)*. The toluene-4-sulphonyl group.

total absorption coefficient *(Phys.)*. The coefficient which expresses energy losses due to both absorption and scattering. Relevant to narrow beam conditions. Preferred term *attenuation coefficient*.

total air-gas mixture *(Eng.)*. Air-gas mixture in which the proportion of air is the amount needed for perfect combustion.

total body burden *(Radiol.)*. (1) The summation of all radioactive materials contained in any person. (2) The maximum total amount of radioactive material any person may be permitted to contain.

total cross-section *(Phys.)*. The sum of the separate cross-sections for all processes by which an incident particle can be removed from a beam. If all the atoms of an absorber have the same total cross-section, then it is identical with the **atomic absorption coefficient**.

total curvature *(Maths.)*. See **curvature (3)**.

total differential *(Maths.)*. See **complete differential**.

total electron binding energy *(Electronics)*. That required to remove all the electrons surrounding a nucleus to an infinite distance from it and from one another.

total emission *(Electronics)*. See **saturation current**.

total equivalent brake horsepower *(Aero.)*. The brake horsepower at the propeller shaft plus the b.h.p. equivalent of the residual jet thrust of a turboprop; abbrev. *tehp* or *ehp*.

total head *(Aero.)*. In fluid flow, the algebraic sum of the **dynamic pressure** and the **static pressure**.

total impulse *(Aero.)*. That available from a self-contained rocket expressed as the product of the mean thrust, in newtons (*N*) and the firing time, in seconds (*s*), expressed as *Ns*.

total internal reflection *(Phys.)*. Complete reflection of incident wave at boundary with medium in which it travels faster, under conditions where Snell's law of refraction cannot be satisfied. The angle of incidence at which this occurs (corresponding to an angle of refraction of 90°) is known as the *critical angle*.

total losses *(Elec.Eng.)*. The power loss in an electrical machine, equal to the difference between the input and the output powers.

totally enclosed motor *(Elec.Eng.)*. A motor with no provision for ventilation but not necessarily water- or gas-tight.

total normal curvature *(Maths.)*. See **curvature (3)**.

total parenteral nutrition *(Med.)*. Giving the complete nutritional requirements by an intravenous line, usually centrally placed.

total response *(Acous.)*. See **mean spherical response**.

total temperature *(Aero.)*. See **stagnation temperature**.

toti- *(Genrl.)*. Prefix from L. *totus*, all, whole.

totipotency *(Bot.)*. The ability, possessed by most living plant cells, differentiated or not, to regenerate a plant when isolated and cultured in a suitable medium. See **tissue culture**.

totipotent *(Zool.)*. The capability of some embryo cells to develop into any organ or a complete embryo.

touch and close fastener *(Textiles)*. A pair of tapes made from synthetic fibres that stick together on pressing by hand and then separate by peeling when required. The tapes have pile surfaces one formed of hooks and the other of loops, which engage on being pressed together.

touchstone *(Min.)*. See **Lydian stone**.

touchwood *(Bot.)*. Wood much decayed as a result of fungal attack; it crumbles readily, and when dry is easily ignited by a spark.

toughened glass *(Glass)*. See **safety glass**.

toughness *(Eng.)*. A term denoting a condition intermediate between brittleness and softness. It is indicated in tensile tests by high ultimate tensile stress and low to moderate elongation and reduction in area. It is also associated with high values in notched bar tests.

toughness *(Min.Ext.)*. Ability of mineral to withstand disruption, assessed empirically by comparison with standard minerals under controlled test conditions.

tough pitch *(Eng.)*. A term applied to copper in which the oxygen content has been correctly adjusted at about 0.03% by poling. Distinguished from overpoled and underpoled copper.

tourbillon watch *(Horol.)*. A watch fitted with a revolving carriage which carries the balance and escapement round the fourth wheel, for the purpose of eliminating the positional errors.

tourmaline *(Min.)*. A complex silicate of sodium, boron and aluminium, with, in addition, magnesium (*dravite*), iron (*schorl*), or lithium (*elbaite*), and fluorine in small amounts, which crystallizes in the trigonal system. It is usually found in granites or gneisses. The variously coloured and transparent varieties are used as gemstones, under the names *achroite* (colourless), *indicolite* (blue), *rubellite* (pink). The common black variety is *schorl*.

tourmalinization *(Geol.)*. The process whereby minerals or rocks are replaced wholly or in part by tourmaline. See **pneumatolysis**.

tourniquet *(Med.)*. Any instrument or appliance which, by means of a constricting band exerts pressure on an artery to control bleeding, or, at a lesser pressure, to produce dilatation of a vein for **phlebotomy**.

tow *(Textiles)*. A collection of a large number of parallel continuous filament man-made fibres, not twisted together. By cutting or breaking (i.e. **converting**) a top is produced, the process being known as *tow-to-top conversion*.

tower *(Elec.Eng.)*. The lattice-type steel structure used to carry the several conductors of a transmission line at a considerable height above the ground. Also called *pylon*.

tower *(Eng.)*. See **pylon**.

tower bolt *(Build.)*. Large type of **barrel bolt**.

tower crane *(Eng.)*. A rotatable cantilever pivoted to the top of a steelwork tower, either fixed or carried on rails. The load is balanced by the lifting machinery carried on the opposite side of the pivot. Commonly used in shipbuilding.

Townend ring *(Aero.)*. A cowling for radial engines consisting of an aerofoil section ring, which ducts the air on to the engine cylinders and directs a streamlined flow on to the fuselage or nacelle, thus reducing drag; now obsolete.

town gas *(Min.Ext.)*. Usually a mixture of coal gas and carburetted water gas; the energy density is about 500 Btu/ft^3 or 20 MJ/m^3, which is about half that of *natural gas* which in many areas has largely superseded town gas.

Townsend avalanche *(Phys.)*. Multiplication process whereby a single charged particle, accelerated by a strong field, causes, through collision a considerable increase in ionized particles.

Townsend discharge *(Phys.)*. A *Townsend avalanche* initiated by an external ionizing agent.

toxaemia, toxemia *(Med.)*. The condition of a patient caused by the absorption into the tissues and into the blood of toxins formed by micro-organisms at the site of infection.

toxicology *(Med.)*. The branch of medical science dealing with the nature and effects of poisons.

toxin *(Biol.)*. A poisonous substance of biological origin.

toxoid *(Immun.)*. Bacterial exotoxin which has been treated (usually with formaldehyde) so that it has lost its toxic properties but retains its ability to stimulate an immune response against the toxin.

toxoplasmosis *(Vet.)*. Caused by *Toxoplasma gondii* and found mainly in sheep and goats although man and other species can be infected. Common cause of ovine abortion. The cat is the natural host for this parasite.

TP *(Surv.)*. Abbrev. for *Turning point*. See **change point**.

TPI *(Eng.)*. Abbrev. for *Threads Per Inch*, the inverse of the pitch of a screw thread.

TPI *(Ships)*. Abbrev. for *Tons Per Inch*: weight required to increase the *mean draught* one inch.

trabecula *(Bot.)*. A rod-like structure, e.g. of cell wall material across the lumen of a cell, or of a cell or cells across some larger cavity.

TRACE *(Aero.)*. (1) Abbrev. for *Test-equipment for Rapid Automatic Check-out and Evaluation*. A computerized general purpose diagnostic testing rig for aircraft electrical and electronic systems. (2) Trace: visible record of instrument reading.

trace *(Comp.)*. Means of checking the logic of a program by inserting statements which cause the values of the variables and other information to be printed out as the program is executed.

trace *(Electronics)*. Image on phosphor on electron beam impact, forming the display.

trace element *(Biol.)*. See **micronutrient**.

trace element *(Geol.)*. A non-essential element ($<1\%$) in a mineral.

trace elements *(Med.)*. Metals and non-metals found in

minute quantities in the human body which are essential components of the diet, but are harmful if taken in excess, e.g. copper, fluorine, chromium and selenium.

trace fossil *(Geol.)*. Any sedimentary structure caused by the activity of a fossil organism during its life, e.g. trails, burrows.

tracer atom *(Phys.)*. Labelled atom introduced into a system to study structure or progress of a process. Traced by variation in isotopic mass or as a source of weak radioactivity which can be detected, there being no change in the chemistry. See **radioactive isotope**.

tracer chemistry *(Chem.)*. The use of isotopic tracers in studying chemical and biological systems. Radioactive isotopes are often used, but stable isotopes may also by used if their nuclear magnetic resonance spectra are different from those of the most abundant isotope. Studies using ^{13}C are often done in this way.

tracer compound *(Phys.)*. One in which a small proportion of its molecules is labelled with a radioactive isotope.

tracer element *(Phys.)*. One of the **radioelements**, e.g. radiophosphorus, used for experiments in which its radioactive properties enable its location to be determined and followed. Tracer technique may be applied to physiological, biological, pathological, and technological experiments. For some purposes stable isotopes, e.g. ^{13}C and heavy hydrogen (see **deuterium**), are more conveniently used than radioisotopes.

tracers *(Min.Ext.)*. (1) Elements diffused as a **halo** around an ore body, identified by **geochemical prospecting**. These may be dilute concentrations of metals also present in the ore body, or elements found in association with ore bodies but not of primary interest themselves. (2) Minerals identified in soil and stream gravel which are common accesssories to minerals of economic significance.

trachea *(Zool.)*. An air-tube of the respiratory system in certain Arthropoda, as Insects; in air-breathing Vertebrates, the wind-pipe leading from the glottis to the lungs.

tracheal gills *(Zool.)*. In some aquatic Insect larvae, filiform or lamellate respiratory outgrowths of the abdomen richly supplied with tracheae and tracheoles.

tracheal system *(Zool.)*. In certain *Arthropoda*, as Insects and Myriapods, a system of respiratory tubules containing air and passing to all parts of the body.

tracheary elements *(Bot.)*. Water-conducting cells of the xylem of a vascular plant. After the death and loss of their protoplasmic contents, the cell walls become thickened with lignin so that they can act as tubes for the conduction of water under tension. Cf. **hydroid**.

tracheid(e) *(Bot.)*. An elongated element with pointed ends, occurring in wood. It is derived from a single cell, which lengthens and develops thickened pitted walls, losing its living contents. Tracheides conduct water.

tracheitis *(Med.)*. Inflammation of the mucous membrane of the trachea.

trachelate *(Zool.)*. Necklike. (Gk. *trachēlos*, neck).

trachelorrhaphy *(Med.)*. The surgical repair of lacerations of the cervix uteri.

tracheobronchitis *(Med.)*. Inflammation of the mucous membrane of both the trachea and the bronchi.

tracheobronchitis *(Vet.)*. An inflammation of the trachea and bronchi of dogs due to infection by the nematode worm, *Oslerus osleri*.

tracheocele *(Med.)*. An air-containing swelling in the neck due to the bulging of the wall of the trachea between the cartilages of the trachea.

tracheole *(Zool.)*. The ultimate branches of the tracheal system.

tracheophyte *(Bot.)*. A vascular plant (i.e. a pteridophyte or seed plant). Sometimes made a division, Tracheophyta.

tracheotomy *(Med.)*. The operation of cutting into the trachea, usually for the relief of respiratory obstruction.

trachoma *(Med.)*. A highly contagious, *Chlamydial*, infection of the conjunctiva covering the eyelids, characterized by the presence of small elevations on the inner side of the lid and leading often to blindness. The disease is hyper-endemic in rural communities of the Middle East, Asia, Africa and Central and South America. The infecting organism is *C. trachomitis*, transmitted by house flies and by poor hygeine.

trachyandesite *(Geol.)*. Fine-grained igneous rock, commonly occurring as lava flows, intermediate in composition between trachyte and andesite, that is, containing both orthoclase and plagioclase in approximately equal amounts.

trachybasalt *(Geol.)*. A fine-grained igneous rock commonly occurring in lava flows and sharing the mineralogical characters of trachyte and basalt. The rock contains sanidine (characteristic of trachyte) and calcic plagioclase (characteristic of basalt).

trachyte *(Geol.)*. A fine-grained igneous rock-type, of intermediate composition, in most specimens with little or no quartz, consisting largely of alkali-feldspars (sanidine or oligoclase) together with a small amount of coloured silicates such as diopside, hornblende, or mica.

tracing *(Eng.)*. An engineering drawing transferred to transparent tracing paper or cloth, in Indian ink for permanence and for making good dye-line prints. A *tracer's* work carries no design responsibility. Cf. **draughtsman**.

track *(Acous.)*. (1) The groove which is cut on a blank during disk recording. (2) Single circumferential area on a magnetic drum, or longitudinal area on a magnetic tape, alongside other tracks, allocated to specific recording and reproduction channel. (3) Space on a disk or sound-film allocated to one channel of sound-recording and hence reproduction.

track *(Aero.)*. (1) The distance between the outer points of contact of port and starboard main wheels. (2) The distance between the vertical centre-lines of port and starboard undercarriages where the wheels are paired. (3) The projection of the flight path upon the earth's surface.

track *(Comp.)*. Path on a **tape**, **disk** or **drum** along which data is stored.

track *(Phys.)*. Path followed by particle especially when rendered visible in photographic emulsion by cloud chamber, bubble chamber or spark chamber.

track-circuit signalling *(Elec.Eng.)*. An electric signalling system making use of the change in resistance of a track circuit when a train passes over a section of railroad track and thus completes a circuit between the rails.

tracker wires *(Elec.Eng.)*. Wires forming a mechanical connection between a switch and its operating mechanism situated some distance away.

tracking *(Elec.Eng.)*. (1) Excessive leakage current between two insulated points due e.g. to moisture. (2) Tracks along the surface of oil-impregnated paper caused by a surface stress. Tracking causes waxing and carbon formation.

tracking *(Radar)*. Automatic holding of radar beam on to target through operation of return signals.

tracking *(Space)*. The continuous process of following, from a distance, an object to determine its position in space.

tracking shot *(Image Tech.)*. A shot in which the camera, and its operators, are moved to follow the chosen part of the action.

track rail bond *(Elec.Eng.)*. A rail bond for preserving the electrical continuity of the track rails when these are used for carrying traction or other currents.

track relay *(Elec.Eng.)*. A relay used in track-circuit signalling for controlling the electrically operated signals.

track rod *(Autos.)*. A transverse link which, through ball-joints, connects arms carried by the stub axles, in order to convey angular motion from the directly-steered axle to the other.

track-sectioning cabin *(Elec.Eng.)*. A cabin housing switchgear by means of which the supply to different sections of an electrified railway line may be disconnected.

track switch *(Elec.Eng.)*. A switch controlling the supply of current to a section of an electrified railway line.

track-while-scan *(Radar)*. Electronic process for detecting a target, computing its velocity and predicting its future position without interfering with the process of continuous scanning.

tract *(Zool.)*. The extent of an organ or system, as the *alimentary tract*; an area or expanse, as the *ciliated tracts* of some Ctenophora; a band of nerve-fibres, as the *optic tract*.

traction *(Med.)*. Treatment involving tension on affected parts (e.g. fractures) by means of suitable applied weights or otherwise.

traction battery *(Elec.Eng.)*. See **vehicle battery**.

traction engine *(Eng.)*. A road locomotive in which large road wheels are gear-driven from a simple or compound steam engine mounted on top of the boiler, a rope drum being provided for additional haulage purposes.

traction generator *(Elec.Eng.)*. A d.c. generator used solely for supplying power to an electric traction system.

traction lamp *(Elec.Eng.)*. An electric lamp having a specially robust filament to withstand vibration: used on trains or road vehicles.

traction load *(Elec.Eng.)*. That part of the load carried by a d.c. generating station which is formed by the traction system which it supplies.

traction load *(Geol.)*. That part of the load of solid material carried by a river which is rolled along the bed.

traction motor *(Elec.Eng.)*. An electric motor specially designed for traction service.

traction rope *(Civ.Eng.)*. The endless rope employed in an aerial ropeway system to effect movement of the carriers transporting the loads. Also called a *hauling rope*.

tractive force, effort *(Elec.Eng.)*. The pull necessary to detach the armature from an excited electromagnet.

tractive force, effort *(Eng.)*. Of a locomotive, the pull which the engine is capable of exerting at the draw-bar, the limiting value of which is given by the product of the weight on the coupled wheels and the coefficient of friction between wheels and rails.

tractor *(Aero.)*. A propellor which is in front of the engine and the structure of the aircraft, as contrasted with a *pusher*, which is behind the engine and pushes the aircraft forward. See under **propeller**.

tractor feed *(Comp.)*. Mechanism for advancing paper by use of perforations and a toothed wheel or sprocket.

tractrix *(Maths.)*. The involute of a catenary.

tractrix horn *(Acous.)*. A horn which is so shaped that the area at a distance from the throat is dependent on the tractrix curve, as contrasted with the exponential horn.

trade effluent *(Build.)*. The liquid discharge, other than soil or waste, from a manufacturing process.

trade off *(Aero.)*. Term used mainly in aircraft design which expresses relationships between variables being examined during the preliminary design process when the best compromise between conflicting requirements has to be found.

trade-off study *(Space)*. A logical evaluation (usually within the **System Engineering** process) of pros and cons of alternative concepts or approaches which leads to the choice of the preferred one; typical criteria for the analysis are: performance, schedule, risk and cost.

trade-winds *(Meteor.)*. Persistent winds blowing from the N.E. in the northern hemisphere and from the S.E. in the southern hemisphere between the *horse latitudes* (calm belts at 30°N and S. of the equator) towards the **doldrums**.

traffic flow *(Telecomm.)*. The number of calls which an exchange, or a set of switches, is carrying at any instant.

traffic meter *(Telecomm.)*. A meter which, inserted at any part of an automatic telephone exchange, totals the number of calls passing through.

traffic unit *(Telecomm.)*. The measure of the occupancy of telephonic apparatus during conversation. One traffic unit equals the use of one circuit for one minute or for one hour. Abbrev. *TU*. See **congestion traffic-unit meter**.

tragus *(Zool.)*. In the ear of some Mammals, including the Microchiroptera (bats), an inner lobe to the pinna.

trail *(Autos.)*. The distance by which the point of contact of a steered wheel with the ground lies behind the intersection of the swivel-pin axis and the ground. See **caster action**.

trailer *(Image Tech.)*. (1) A short film or video production advertising a forthcoming presentation. (2) A protective and identification section at the end of a reel of film (*tail leader*).

trailing action *(Autos.)*. See **caster action**.

trailing axle, -springs, -wheels, etc. *(Eng.)*. In a locomotive, the parts belonging to the rearmost axle.

trailing edge *(Aero.)*. The rear edge of an aerofoil, or of a strut, wire etc.

trailing edge *(Elec.Eng.)*. See **leaving edge**.

trailing edge *(Telecomm.)*. The falling portion of a pulse signal. See also **leading edge**.

trailing flap *(Aero.)*. A *flap* which is mounted below and behind the wing trailing edge so that it normally trails at neutral incidence and is rotated to various positive angles of incidence to increase lift, there always being a gap between the wing undersurface and the flap leading edge.

trailing points *(Civ.Eng.)*. See **points**.

trailing pole tip (horn) *(Elec.Eng.)*. The edge of a field pole which is passed last by an armature conductor, irrespective of the direction of rotation of the armature.

trailing vortex *(Aero.)*. The vortex passing from the tips of the main surfaces of an aircraft and extending downstream and behind it.

train *(Eng.)*. Similar or identical parts in a machine, arranged in series, such as a simple or compound train gears.

train *(Horol.)*. The interconnected wheels and pinions of a watch, clock, or similar mechanism. A watch is said to have an 18 000 train when the train is suitable for a balance making 18 000 vibrations per hour; a 21 600 train is used for small wristwatches and a 14 400 train for chronometers.

train brake *(Eng.)*. See **vacuum brake, air brake**.

train control *(Elec.Eng.)*. The method by which the mechanical control operations carried out by the driver of an electric or diesel-electric train alter the electrical supply to the traction motors.

train describer *(Elec.Eng.)*. An automatic or semi-automatic device for giving information regarding the destination of trains. Also called *destination indicator*.

training *(Behav.)*. (1) *General*, the application of learning principles to improve social or technical skills through some systematic activity. (2) *Operant conditioning*, the procedure of conditioning an animal through the use of reinforcements in order to establish a desired behaviour.

training works *(Hyd.Eng.)*. Works undertaken to remedy instability and eccentricity of flow in channels. See **dyke, levee, groynes, ground sills**.

trait *(Behav.)*. A stable and enduring attribute of a person or animal which varies from one individual to another; traits may be physical (eye colour) or psychological (spatial intelligence) and are often used in the study of individual differences in personality.

trajectory *(Meteor.)*. The actual path along which a small quantity, or parcel, of air travels during a definite time interval.

trajectory *(Space)*. The path of a rocket or space vehicle. The word *trajectory* is used when the path is limited in length, e.g. a *trajectory* to the moon, whereas *orbit* usually refers to a path which is closed and repetitive.

tram *(Min.Ext.)*. A small wagon, tub, cocoa-pan, corve, corf or hutch, for carrying mineral.

trammels *(Eng.)*. See **beam compasses**.

tramp alloys *(Eng.)*. Unknown alloys introduced by the use of ill-characterized scrap in steel-making.

tramp iron *(Min.Ext.)*. Stray pieces of iron or steel e.g. broken drills, mixed with ore from mine, which must be removed before crushing.

tramp metal *(Eng.)*. Stray metal pieces which are accidentally entrained in food or other processed materials and must be removed before the material leaves the process.

trance *(Behav.)*. Occurs under hypnosis, and in various conditions such as sleepwalking, a state of *disassociation* in which the individual's will is suspended and he or she acts on wishes or fantasies that are otherwise kept under control.

trans- *(Chem.)*. That geometrical isomer in which like groups are on opposite sides of the bond with restricted rotation, or in which like ligands are on opposite sides of the central atom of a co-ordination compound.

transaction *(Comp.)*. A single event involving the recording and processing of data.

transaction file *(Comp.)*. Records used in batch processing to update a **masterfile**. Also called an *update file* or *change file*. See **grandfather, father, son files**.

transaction processing *(Comp.)*. Use of an on-line computer system to interrogate or update files as requested rather than batching such requests together for subsequent processing.

transadmittance *(Phys.)*. Output a.c. current divided by input a.c. voltage for electronic device when other electrode potentials are constant.

transaminase *(Biol.)*. An enzyme which catalyses the transfer of amino groups from amino acids to keto acids, thus converting a keto acid into an amino acid, e.g. the conversion of α-ketoglutarate to glutamate, *transamination*.

transatmospheric vehicle *(Aero.)*. Aircraft capable of normal wingborne flight through the atmosphere and also of travelling into space orbits. See **aerospaceplane** and **HOTOL**.

transaxial tomography *(Radiol.)*. A process whereby serial radiographs are taken transverse to the vertical axis of the body. See **computed tomography**.

transceiver *(Telecomm.)*. Equipment in which circuitry is common to transmission or reception. Also *transreceiver*.

transcendental function *(Maths.)*. Any non-**algebraic function**, e.g. trigonometric, exponential, logarithmic, Bessel or gamma functions.

transconductance *(Electronics)*. Applied to valves, or to field-effect transistors, it is the change in anode or drain current divided by the increment of grid or gate voltage which initiated the change; also *mutual conductance*.

transcribing genes *(Biol.)*. Genes which are being actively transcribed into RNA. See **coding sequences**.

transcription *(Biol.)*. The process by which an RNA polymerase produces single-stranded RNA complementary to one strand of the DNA or, rarely, RNA.

transcription *(Telecomm.)*. The recording of a broadcast performance for subsequent rebroadcast or other use.

transcriptionally-active chromatin *(Biol.)*. **Chromatin** which is being transcribed into RNA.

transcription complex *(Biol.)*. Functional association of DNA, nascent RNA, protein and ribonucleoprotein actively transcribing and processing RNA.

transcrystalline failure *(Eng.)*. The normal type of failure observed in metals. The line of fracture passes through the crystals, and not round the boundaries as in intercrystalline failure.

transducer *(Comp.)*. Device which converts signals from one physical form to another.

transducer *(Elec.Eng.)*. Term for a device which converts a physical quantity into an electrical signal, either in proportion to quantity or according to a specified formula. Examples include accelerometers, microphones and photocells. The term is also used to cover devices which convert electrical signals into some physical phenomenon, e.g. a loudspeaker delivers sound waves in proportion to the electrical signal applied to it.

transducer translating device *(Elec.Eng.)*. One for converting error of the controlled member of a servo-mechanism into an electrical signal that can be used for correcting the error.

transduction *(Biol.)*. During phage infection and consequent bacterial lysis, the integration of segments of host DNA into that of the phage. It can then be transferred to another host.

transductor *(Elec.Eng.)*. Arrangement of windings on a laminated core, which, when excited, permits current amplification. Part of a **magnetic amplifier**.

transect *(Bot.)*. A line or belt of vegetation marked off for study.

transept *(Arch.)*. Part of a church at right angles to the *nave*, or of another building to the body: either wing of such a part where it runs right across.

transfection *(Biol.)*. The alteration of the host genome after infection by phage.

transfer *(Comp.)*. See **print-through**.

transfer *(Print.)*. An impression taken of a nonlithographic surface (intaglio, letterpress), on one of several specially coated papers, for transferring to a lithographic surface; a transfer from an existing lithographic surface is called a *retransfer*.

transfer admittance *(Phys.)*. The ratio of the current driving at one node in a network to the resulting voltage at another node. All other sources being set to zero.

transferase *(Biol.)*. An enzyme which catalyses the transfer of chemical groups between compounds, e.g. glycosyl transferase transfers a sugar residue onto the growing oligosaccharide complex of a glycoprotein.

transfer cell *(Bot.)*. Parenchymatous cell with elaborate ingrowths of the cell wall, greatly increasing the area of the plasmalemma. It occurs where there is substantial movement of solutes between symplast and apoplast in scattered families of bryophytes and vascular plants.

transfer characteristic *(Electronics)*. See **transistor characteristic**.

transfer characteristic *(Image Tech.)*. Relationship between TV camera tube illumination and corresponding signal current.

transfer ellipse *(Space)*. The trajectory (part of an ellipse) by which a space vehicle may transfer from an orbit about one body (e.g. the Earth) into an orbit about another (e.g. the Sun).

transference *(Behav.)*. In psychoanalytic theory, the tendency of the client to displace onto the analyst feelings and ideas derived from previous experience with other figures (e.g. one's parents).

transfer factor *(Immun.)*. A dialysable substance, obtained from extracts of individuals with delayed type hypersensitivity, that is claimed to be able to transfer to another individual the ability to give a delayed hypersensitivity response.

transfer factor *(Med.)*. A measure of the lungs ability to take up oxygen which is depressed by a variety of lung diseases.

transfer function *(Telecomm.)*. (1) The complex ratio of the output (current or voltage) of a circuit or device to its input. This gives the phase and frequency response. (2) A mathematical expression relating the output of a closed-loop system to its input.

transfer impedance *(Phys.)*. The ratio of the driving voltage in one mesh of a network to the resulting current in another mesh, all other sources being set to zero.

transfer instrument *(Elec.Eng.)*. Instrument which gives d.c. indication independent of frequency, including zero frequency, so that when calibrated for d.c. it can be used for calibrating a.c. instruments, as with electrostatic wattmeters.

transfer lettering systems *(Print.)*. Alphabets, symbols, rules, and tints which can be stripped into position on layouts or film for photo-reproduction, e.g. Artype, Chart-Pak, Craf Type, Letraset, Letter-on, Prestype, Zip-a-Tone.

transfer line *(Eng.)*. A long series of machines operating on a succession of similar parts, e.g. car cylinder blocks, automatically or semi-automatically.

transfer machine *(Eng.)*. Machine in which a workpiece or an assembly passes automatically through a number of stations, at each of which it undergoes one or more production processes.

transfer moulding *(Plastics)*. Injection moulding with thermosetting compositions.

transfer of training *(Behav.)*. The facilitation (positive) or hindrance (negative) of performance on a learning or training task as a result of previous activity. The positive and negative transfer effects appear to be a function of the similarity of the tasks.

transfer port *(Nuc.Eng.)*. The aperture through which items are inserted into or removed from a dry box, glove box, or shielded box (e.g. by sealing into plastic sac attached to the rim of the port).

transfer printing *(Textiles)*. A process for transferring a design from paper to fabric. Usually the paper is coloured with disperse dyes which sublime on to fabrics made from synthetic fibres when the paper and fabric are pressed together in a heated press or calender.

transfer process *(Image Tech.)*. Any means whereby an image, dyed or pigmented, is transferred to a new emulsion.

transfer RNA *(Biol.)*. See **tRNA**.

transfinite numbers *(Maths.)*. A system of cardinal and ordinal numbers, invented by G. Cantor (1845–1918), relating to infinite sets. In effect Cantor classifies different types of infinity. For example, the infinite associated with the set of positive integers is different from that associated with the set of all real numbers.

transfixion *(Med.)*. A cutting through, as in amputation.

transform *(Maths.)*. A process or rule for deriving from a given mathematical entity (e.g. point, line, function etc.) a corresponding entity. The word 'transform' is also sometimes used in respect of the corresponding entity itself (e.g. Fourier and Laplace transform). Certain problems (e.g. differential equations) can be transformed so as to obtain a simpler problem whose solution can then be transformed back to give the solution of the original problem. Also called *transformation*. See **affine transformation, conjugate elements of a group, Fourier transform, Laplace transform, linear transformation** and **isogonal transformation**.

transformation *(Biol.)*. (1) The alteration of the bacterial or eukaryotic cell genotype following the uptake of purified DNA. (2) The alteration of cells in tissue-culture by various agencies so that they behave in many ways like cancer cells, e.g. their lack of growth control and the ability to divide indefinitely. See **contact inhibition, growth in soft agar, nude** mice.

transformation *(Eng.)*. A constitutional change in a solid metal, e.g. the change from gamma to alpha iron, or the formation of pearlite from austenite.

transformation *(Maths.)*. See **transform**.

transformation *(Phys.)*. See **atomic transmutation**.

transformation constant *(Phys.)*. See **disintegration constant**.

transformation points *(Glass)*. In the measurement of thermal expansion coefficient, the two temperatures (Mg point and Tg point) at which the slope of the graph of expansion against temperatures changes fairly sharply.

transformation ratio *(Elec.Eng.)*. See **turns ratio**.

transformation temperature *(Eng.)*. The temperature at which phase changes occur during the heating of iron and steels. Denoted by various symbols: *Ac*, the temperature on heating; *Ae*, the temperature at equilibrium; *Ar*, the temperature on cooling. Subscripts refer to the transformation in question.

transformation theory *(Eng.)*. A theory in which the factor relating to yield is assumed to be the shear-strain energy stored in unit volume.

transformer *(Elec.Eng.)*. An electrical device without any moving parts, which transfers energy of an alternating current in the primary winding to that in one or more secondary windings, through electromagnetic induction. Except in the case of the **autotransformer** there is no electrical connection between the two windings and usually (excepting the isolating transformer) a change of voltage is involved in the transformation.

transformer *(Phys.)*. Mechanical transformers involving lever arms of different lengths are used to vary **mechanical advantage**.

transformer booster *(Elec.Eng.)*. A transformer connected with its secondary in series with the line, so that its voltage is added to that of the circuit; used to compensate the voltage drop in a feeder or distributor.

transformer core *(Elec.Eng.)*. The structure, usually of laminated iron or ferrite, forming the magnetic circuit to another by a transformer, of any degree of coupling. Also *mutual coupling*.

transformer coupling *(Elec.Eng.)*. Transference, in either direction, of electrical energy from one circuit to another by a transformer, of any degree of coupling. Also *mutual coupling*.

transformer oil *(Elec.Eng.)*. A mineral oil of high dielectric strength, forming the cooling and insulating medium of electric power transformers.

transformer plate *(Elec.Eng.)*. Sheet-iron of low magnetic loss, for transformer core laminations.

transformer ratio *(Elec.Eng.)*. See **turns ratio**.

transformer-ratio bridge *(Elec.Eng.)*. An a.c. bridge similar to a Wheatstone bridge but with two transformer windings used for the two ratio arms.

transformer stampings *(Elec.Eng.)*. The laminations, stamped out of transformer plate, which are assembled to form the transformer core.

transformer switch *(Elec.Eng.)*. A switch or circuit-breaker for disconnecting a transformer from the supply.

transformer tank *(Elec.Eng.)*. The steel tank encasing the core and windings of a transformer and holding the transformer oil.

transformer tapping *(Elec.Eng.)*. A means of varying the voltage ratio of a transformer by making a connection to a point on one winding intermediate between the ends.

transformer tube *(Elec.Eng.)*. One of a number of steel tubes on the outside of a transformer tank to provide a vertical path of circulation for the transformer oil.

transformer winding *(Elec.Eng.)*. The electrically active part of a transformer, which surrounds the magnetically active transformer core.

transform fault *(Geol.)*. A strike-slip fault along which two plates slide past each other, e.g. *San Andreas fault*.

transforming station *(Elec.Eng.)*. A point on an electricity supply system where a change of supply voltage occurs.

transfusion *(Med.)*. *Blood transfusion*. The operation of transferring the blood (or any required constituent of it) of one person into the veins of another, either to make good loss or counteract deficiency.

transfusion reaction *(Immun.)*. Disturbance following transfusion of blood, due to antibodies in the recipient reactive with donor blood cells or, more rarely, to antibodies in the transfused blood reactive with the recipient's blood cells. The antigens involved are usually those on erythrocytes, although after multiple transfusions those on leucocytes can sometimes be important. The most severe and common form of such reactions is due to **ABO blood group system** incompatibility when natural antibodies are present against ABO antigens absent from the recipient's own red cells.

transfusion tissue *(Bot.)*. A tissue of short tracheids and parenchyma cells surrounding or associated with the vascular bundle(s) in the leaves of many gymnosperms, presumably functioning in the distribution of water and collection of photosynthate.

transgenic *(Immun.)*. Used to describe animals which are derived from embryos into which isolated genomic DNA from another species has been introduced at an early stage of development. Such foreign genes may be incorporated into the nucleus and chromosomes so that the animal can express the foreign gene product.

transgenic animal *(Biol.)*. One carrying a gene which was introduced by micro-injecting purified DNA into the nucleus of the fertilized egg.

transgression *(Geol.)*. The gradual submergence of land caused by a relative rise in sea-level.

transient *(Acous.)*. A sound of short period and irregular nonrepeating waveform, which implies a continuous spectrum of sound-energy contributions.

transient *(Elec.Eng.)*. A short surge of voltage or current. The voltage or current before steady-state conditions have become established.

transient *(Telecomm.)*. Any noncyclic change in a part of a communication system. The most general transient is the *step*, while the steady state is represented by any number of sinusoidal variations. See **Heaviside unit function**.

transient analyser *(Telecomm.)*. Test instrument which generates repeated transients and displays their waveform (which is usually adjustable) at different points in the system under investigation, on a CRT screen.

transient distortion *(Telecomm.)*. Distortion arising only when there is a rapid fluctuation in frequency and/or amplitude of the stimulus.

transient effects *(Phys.)*. When a vibrating system is set into *forced vibrations*, initially there are damped vibrations at the natural frequency of the system as well as the vibrations at the driving frequency. These are the transient effects which die away due to damping, as the system settles down to vibrate at constant amplitude at the driving frequency. Transient effects are important in music for they help to give a musical instrument its characteristic sound.

transient equilibrium *(Phys.)*. Radioactive equilibrium between daughter product(s) and parent element of which activity is decaying at an appreciable rate. Characterized by ratios of activity, but not magnitudes, being constant.

transient flow permeability *(Powder Tech.)*. A permeametry technique in which sample of the powder under test, contained in a long vertical column, is evacuated. Then gas at a known pressure is introduced at the foot of the column. The time required for the gas to diffuse through the column is used to calculate the average pore diameter of the powder bed and hence the surface area.

transient reactance *(Elec.Eng.)*. The reactance of the armature winding of a synchronous machine which is caused by the leakage flux. Cf. **synchronous reactance**.

transient stability *(Elec.Eng.)*. That of a power system under transient current conditions.

transient state *(Telecomm.)*. Transition period and associated phenomena between steady states in the repetition of a waveform.

transillumination *(Med.)*. The passing of a strong light through the walls of a cavity so that its outlines may become visible to the observer and any abnormalities in density may be detected.

transistor *(Comp.)*. See **second generation computer**.

transistor *(Electronics)*. Three-electrode semiconductor device with thin layer of *n*- (or *p*-) type semiconductor sandwiched between two regions of *p*- (or *n*-) type, thus forming two *p-n* junctions back to back. The emitter junction is given a forward bias and the collector junction a reverse bias. Due to the low forward resistance of the emitter junction and the high reverse resistance of the collector junction considerable power gain is possible for signals in the emitter or base leads. The latter arrangement also gives current gain. Amplification in the *p-n-p* transistor is due to hole conduction, that in an *n-p-n* transistor to electron conduction. See also **transistor constructions**.

transistor amplifier *(Electronics)*. One which uses transistors as the source of current amplification. Depending on impedance considerations, there are three types, with *base*, *emitter*, or *collector* grounded.

transistor characteristics *(Electronics)*. General name for graphs relating electrode currents and/or voltages for transistors connected in various configurations. For example, collector current plotted against collector-emitter voltage, base current being kept at a fixed value; collector current against base current, with collector-emitter voltage fixed, or h_{fe} against collector current. See **transistor parameters**.

transistor construction *(Electronics)*. See **alloy junction**, **epitaxial transistor**, **mask**, **mesa transistor**, **MOS** and **planar transistor**.

transistor current gain *(Electronics)*. The slope of the output current against input current characteristic for constant output voltage. In a common base circuit it is inherently a little less than unity but in a common emitter circuit it may be relatively large. The current gain is the hybrid parameter h_{21}.

transistor equivalent circuit *(Electronics)*. For the purpose of circuit analysis, a transistor may be represented by a four-terminal network having two common terminals. In this way the transistor characteristics may be expressed in terms of four independent variables, the input and output voltages and currents of the equivalent circuit. Such possible circuits are called the *common base*, the *common emitter* and the *common collector*.

transistor parameters *(Electronics)*. In circuit analysis the performance of transistors is calculated from parameters obtained from the slope of the various characteristic curves. Many such sets of parameters have been used, the most widely adopted probably being the hybrid parameters, h_{11}, h_{12}, h_{21}, and h_{22} (also called *h-parameters*, *small-signal parameters*). These are given by the slopes of the input, feedback, transfer and output characteristics respectively, at the selected working point. The symbols h_{fe} and h_{fb} may also be used.

transistor power-pack *(Electronics)*. One in which high-tension supply of low power is obtained by rectifying transformed high-voltage current from a transistor oscillator fed at low voltage.

transit *(Astron.)*. (1) The apparent passage of a heavenly body across the meridian of a place, due to the earth's diurnal rotation. See **culmination**. (2) The passage of a smaller body across the disk of a larger body as seen by an observer on the earth, e.g. of Venus or Mercury across the sun's disk, or of a satellite across the disk of its parent planet.

transit *(Surv.)*. Rotation of the telescope of a theodolite about its trunnion axis, so that the positions of the ends of the telescope are reversed. See **change face**, **transit theodolite**.

transit angle *(Electronics)*. Product of delay or transit time and angular frequency of operation. In a velocity-modulated valve the transit time corresponds to the time taken for an electron to pass through a drift space.

transit circle *(Astron.)*. See **meridian circle**.

transition *(Aero.)*. In VTOL aircraft flight, the action of changing to, or from, the vertical lift mode (jet, fan or rotor) to forward flight with wing lift.

transition *(Electronics)*. In electrons, change of energy level associated with emission of quantum of radiation.

transition *(Phys.)*. In atomic and nuclear physics, the terms used to describe the change from one quantum state to another. A transition from a higher to a lower energy state may be accompanied by the *emission* of a *photon*, while a transition from a lower to a higher state requires the *absorption* of a photon. Transitions are governed by *selection rules* which forbid some transitions.

transitional epithelium *(Zool.)*. A stratified epithelium consisting of only three or four layers of cells; especially that found lining the ureters, the bladder, and the pelvis of the kidney in Vertebrates.

transitional object *(Behav.)*. Winnicott's concept of objects (e.g. a doll or piece of cloth) which act as comforters during the child's initial development from total dependence to self-reliance.

transition curve *(Surv.)*. A curve of special form connecting a straight and a circular arc on a railway or road. Designed to eliminate sudden change of curvature between the two, and to allow of superelevation being applied gradually to the outer rail or outer part of the curve. Also called an *easement curve*.

transition element *(Phys.)*. See **transition metal**.

transition energy *(Electronics)*. That at which phase focusing changes to defocusing in a synchrotron accelerator. This necessitates a sharp arbitrary change of phase in the radiofrequency field.

transition fit *(Eng.)*. A class of fit intermediate between a clearance fit and an interference fit.

transition frequency *(Electronics)*. The term used for the **gain-bandwidth product** for a transistor used in the common-emitter configuration.

transition metal *(Phys.)*. One of the group which have an incomplete inner electron shell. They are characterized by large atomic magnetic moments. Also *transition element*.

transition point *(Aero.)*. The point where the flow in a **boundary layer** changes abruptly from *laminar* to *turbulent*.

transition point *(Chem.)*. The temperature at which one crystalline form of a substance is converted into another solid modification, i.e. that at which they can both exist in equilibrium.

transition probability *(Phys.)*. In atomic and nuclear physics, the probability per unit time that a system in quantum state *k* will undergo a transition to quantum state *l*. For an atom in a excited state there will be certain probability that the atom will spontaneously undergo a transition to a lower energy state. In the presence of photons of energy equal to the difference in energy between the states, there will be in addition a stimulated transition probability.

transition region *(Bot.)*. The region of the axis of a plant in which the change from root to shoot structure occurs.

transition region *(Electronics)*. That over which the impurity concentration in a doped semiconductor varies.

transition resistor *(Elec.Eng.)*. A resistor connected across that part of a transformer winding short-circuited during onload **tap changing**.

transition state *(Chem.)*. The atomic arangement of highest energy in the course of a one-step chemical reaction.

transition stops *(Elec.Eng.)*. In a traction-motor controller, intermediate electrical positions inserted between the main circuit positions in order to avoid breaking the circuit.

transition temperature *(Electronics)*. (1) That corresponding to a change of phase. (2) That at which a metal becomes superconducting.

transitive *(Maths.)*. A relationship is transitive if when it applies from *A* to *B* and from *B* to *C* it also applies from *A* to *C*, e.g. 'is less than', 'is similar to', 'is a subset of', 'belongs to the same family as'. 'Is the father of' is not a transitive relation.

transitman *(Surv.)*. US term for a man operating a *transit theodolite*.

transit tetany *(Vet.)*. *Railroad disease*. A form of hypocalcaemia and hypophosphataemia occurring after a period of transport. Affects cows in late pregnancy, but also sheep and horses. Symptoms include restlessness, staggering, paresis and recumbency which can start during transit or within two days thereafter.

transit theodolite *(Surv.)*. A theodolite whose telescope is capable of being completely rotated about its horizontal axis. See **Everest theodolite, wye theodolite**.

transit time *(Electronics)*. The time required for an electron or other charge carrier to travel between the electrodes of a valve, transistor, or other active device. In some devices, the transit time may be undesirable, in that it limits speed or frequency of operation; in others (*transit-time devices*) the delay actually permits manipulation of the flow of electrons (e.g. in the travelling-wave tube).

translation *(Biol.)*. The process by which ribosomes and tRNA decipher the **genetic code** in a messenger RNA in order to synthesize a specific polypeptide.

translation *(Telecomm.)*. Alteration of the number and composition of the last two coded trains of impulses which are dialled by a subscriber and represent a desired exchange. The translation is effected in the director, and is for the purpose of routing the call over a multiplicity of junctions.

translation field *(Telecomm.)*. The frame of terminal tags by means of which the coded impulses dialled by a subscriber are translated.

translator *(Comp.)*. Computer program used to convert a program from one language to another, usually from a low-level language to machine code.

translocated herbicide *(Bot.)*. Plant poison which if absorbed in one region will be conducted to all parts of the plant and e.g. kill the roots as well. Cf. **contact herbicide**.

translocated injury *(Biol.)*. Injury occurring in an area remote from the original directly affected part of an animal or plant, but associated with it in type and extent.

translocation *(Biol.)*. An exchange between non-homologous chromosomes whereby a part of one becomes attached to the other, or a re-arrangement within one chromosome.

translocation *(Bot.)*. Transport of solutes about the plant, including the upward movement of inorganic salts in the transpiration stream in the xylem, and the movement of sugars in the phloem. See **mass flow hypothesis**.

translucent *(Min.)*. A mineral which is capable of transmitting light but through which no object can be seen.

transmission *(Autos.)*. The means by which power is transmitted from the engine of an automobile to the *live axle*. Includes the change gear, propeller shaft, clutch, differential gear, etc. See also **automatic transmission**.

transmission *(Phys.)*. See **transmittance**.

transmission *(Telecomm.)*. (1) The process of transferring information (speech, code or data, still or moving pictures, control instructions etc.) from one location to another, or to several others (as in broadcasting) by electronic or optical means. (2) The actual information being transmitted over a communication or broadcasting system.

transmission band *(Telecomm.)*. Section of a frequency spectrum over which minimum attenuation is desired, depending on the type and speed of transmission of desired signals.

transmission bridge *(Telecomm.)*. Device for separating a connection into incoming and outgoing sections for the purpose of signalling, at the same time permitting the through transmission of voice frequencies.

transmission chain *(Eng.)*. Roller or inverted tooth chain designed for transmitting power.

transmission coefficient *(Phys.)*. Probability of penetration of a nucleus by a particle striking it.

transmission dynamometer *(Eng.)*. A device for measuring the torque in a shaft, and hence the power transmitted, either (*a*) by inference from the measured twist over a given length of shaft, obtained by a torsion meter, or (*b*) by direct measurement of the torque acting on the cage carrying the planetary pinions of an interposed differential gear.

transmission experiment *(Phys.)*. One in which radiation transmitted by a **thin target** is measured, to investigate the interaction which takes place. Such experiments are used in the measurement of total cross-sections for neutrons.

transmission gain *(Telecomm.)*. The increase of power (usually expressed in dB) in a transmission from one point to another.

transmission level *(Telecomm.)*. Electric power in a transmission circuit, stated as the decibels or nepers by which it exceeds a reference level. Also called *power level*.

transmission line *(Elec.Eng.)*. General name for any conductor used to transmit electric or electromagnetic energy, e.g. power line, telephone line, coaxial feeder, G-string, waveguide, etc. Also for the acoustic equivalent.

transmission-line amplifier *(Telecomm.)*. A wideband amplifier in which the amplifying devices are distributed along a real or artificial transmission line. Also *distributed amplifier*.

transmission loss *(Acous.)*. Ten times the logarithm to base 10 of a power ratio, describing the transmission of sound through walls, windows etc. The nominator is the power of the transmitted sound and the denominator the power of the incident sound. Abbrev. *TL* or *R* in continental Europe.

transmission loss *(Telecomm.)*. Difference between the output power level and the input power level of the whole, or part, of a transmission system in decibels or nepers.

transmission measuring set *(Telecomm.)*. Apparatus consisting essentially of a sending circuit and a level measuring set. The sending circuit has a specified impedance (e.g. 600 ohms) and its output power is known.

transmission mode *(Telecomm.)*. Field configurations by which electromagnetic or acoustic energy may be propagated by transmission lines, especially waveguides. See **mode**.

transmission primaries *(Image Tech.)*. In colour TV, the set of three primaries chosen so that each one corresponds to one of the independent signals which comprise the colour signal.

transmission ratio *(Image Tech.,Phys.)*. The ratio of the transmitted luminous flux to that incident upon a transparent medium. Reciprocal of *opacity*.

transmission reference system *(Telecomm.)*. See **master telephone transmission reference sytem**.

transmission speed *(Comp.)*. The number of bits or elements of information transmitted in unit time. See **baud**.

transmission tower *(Elec.Eng.)*. The steel structure that carries a high-voltage transmission line.

transmission voltage *(Elec.Eng.)*. The nominal voltage at which electric power is transmitted from one place to another.

transmissivity *(Phys.)*. See **transmission coefficient**.

transmit-receive tube *(Electronics,Radar)*. Switch which does or does not (*anti-transmit-receive tube*) permit flow of high-energy radar pulses. It is a vacuum tube containing argon for low striking, and water vapour to assist recovery after the passage of a pulse. Used to protect a radar receiver from direct connection to the output of the transmitter when both are used with the same scanning aerial through a common waveguide system. Abbrev. *TR tube*.

transmittance *(Phys.)*. Ratio of energy transmitted by a body to that incident on it. If scattered emergent energy is included in the ratio, it is termed *diffuse transmittance*, otherwise *specular transmittance*. Also called *transmission*.

transmitted carrier system *(Telecomm.)*. An *amplitude modulated system* in which the carrier wave is radiated. Cf. **suppressed carrier system**.

transmitter *(Telecomm.)*. Strictly, complete assemblage of apparatus necessary for production and modulation of radiofrequency current, together with associated antenna system; but frequently restricted to that part concerned with the conversion of d.c. or mains a.c. into modulated RF current.

transmitter frequency tolerance *(Telecomm.)*. Maximum permitted **frequency departure**.

transmitting valve *(Telecomm.)*. One which handles output power of a radio transmitter; may be in parallel or push-pull with others.

transmittivity *(Phys.)*. Transmittance of unit thickness of nonscattering medium.

transmutation *(Phys.)*. See **atomic transmutation**.

transom, transome *(Build.)*. An intermediate horizontal member of a window frame, separating adjacent panes.

transonic range *(Aero.)*. The range of air speed in which both **subsonic** and **supersonic** airflow conditions exist round a body. Largely dependent upon body shape, curvature and **thickness-chord ratio**, it can be broadly taken as Mach 0.8 to Mach 1.4.

transparency *(Image Tech.)*. (1) The measure of light transmitted by a transparent medium, see **transmission ratio**. (2) A picture on a clear support in which the tones and colours of the image are transparent, intended to be viewed by transmitted light or by projection on a screen.

transparency *(Phys.)*. Proportion of energy or number of incident photons or particles which pass through window of ionization chamber or Geiger counter.

transpiration *(Aero.)*. The flow of gas along relatively long passages, the flow being determined by the pressure difference and the viscosity of the gas, surface friction being negligible.

transpiration *(Bot.)*. The loss of water by evaporation, mainly through the stomata in vascular plants.

transpiration stream *(Bot.)*. The flow of water from the soil through the tissues of the plant to the evaporating surfaces, all driven by transpiration. Cf. **cohesion theory**.

transplant *(Med.)*. (1) The process of transferring a part or organ from its normal position to another position in the same individual (*autologous*) or to a position in another individual (*heterologous*). Also *transplantation*. (2) The part or organ transferred in this way.

transplant *(Med.,Zool.)*. (1) In surgery and experimental zoology, the process of transferring a part or organ from its normal position to another position in the same individual or to a position in another individual. Also *transplantation*. (2) The part or organ transferred in this way.

transpolarizer *(Elec.Eng.)*. Ferroelectric dielectric impedance, controlled electrostatically.

transponder *(Radar)*. A form of transmitter-receiver which transmits signals automatically when the correct interrogation is received. An example is a radar beacon mounted on a flight vehicle (or missile), which comprises a receiver tuned to the radar frequency and a transmitter which radiates the received signal at an intensity appreciably higher than that of the reflected signal. The radiated signal may be coded for identification.

transport *(Comp.)*. See **tape drive**.

transport *(Phys.)*. Rate at which desired material is carried through any section of processing plant, e.g., isotopes in isotope separation.

transport cross-section *(Nuc.Eng.)*. Reciprocal of **transport mean free path**.

transporter bridge *(Civ.Eng.)*. A bridge consisting of two tall towers, one on each side of the river, connected at the top by a supporting girder along which a carriage runs. A small platform at the ordinary road-level is suspended from the carriage, and this system can be made to travel along the girder across the river. Such bridges, inadequate for modern traffic, are obsolescent.

transport mean free path *(Nuc.Eng.)*. If Fick's Law of diffusion is applicable to the conditions in a nuclear reactor then the mean free path is three times the diffusion coefficient of neutron flux. In practice the theory usually has to be modified to take account of anisotropy of scattering and persistence of velocities.

transport number *(Chem.)*. The fraction of the total current flowing in an electrolyte which is carried by a particular ion.

transport theory *(Nuc.Eng.)*. Rigorous theoretical treatment of neutron migration which must be used under conditions where **Fick's Law** does not apply. See **diffusion theory**.

transposable element *(Biol.)*. See **transposon**.

transpose of a matrix *(Maths.)*. The matrix whose rows are the columns of the given matrix. Formerly, *conjugate matrix*.

transposition *(Elec.Eng.)*. Ordered interchange of position of the lines on a pole route, and also of phases in an open power line, so that effects of mutual capacitance and inductance, with consequent interference, are minimized or balanced. See **barrel**.

transposition *(Telecomm.)*. See **co-ordinated transposition**.

transposition insulator *(Elec.Eng.)*. A special type of insulator used at transposition points on a transmission line.

transposition tower *(Elec.Eng.)*. A transmission tower specially designed to allow of the transposing of the conductors.

transposon *(Biol.)*. A sequence of DNA which is capable of inserting itself into many different sites in the host's chromosome. syn. *transposable element*.

transreceiver *(Telecomm.)*. See **transceiver**.

transrectification factor *(Elec.Eng.)*. Ratio of change in average output current to change in alternating voltage applied to a rectifier.

trans-sexualism *(Behav.)*. Gender identification with the opposite sex.

transudate *(Med.)*. A passive effusion of fluid from blood-vessels, the fluid containing little protein, few cells, and not clotting outside the body.

transuranic elements *(Chem.)*. The artificial elements 93 and upwards, which possess heavier and more complex nuclei than uranium, and which can be produced by the neutron bombardment of uranium. More than twelve of these have been produced, including neptunium, plutonium, curium, lawrencium etc.

Transvaal jade *(Min.)*. Massive light green hydrogrossular garnet, used as a simulant for jade.

transverse *(Bot.)*. Perpendicular to the long axis.

transverse architrave *(Build.)*. The moulding across the top of a door or window opening.

transverse-beam travelling-wave tube *(Electronics)*.

One in which the directions of propagation of the electron beam and the electromagnetic wave carrying the signal are mutually perpendicular. The cavity magnetron is an example.

transverse electric wave *(Telecomm.)*. See **TE-wave**.

transverse frame *(Aero.)*. The outer-ring members of a rigid airship frame. It may be of a stiff-jointed type, or braced with taut radial members to a central fitting. It connects the main longitudinal girders together.

transverse frame *(Ships)*. A stiffening member of a ship's hull, disposed transversely to the longitudinal axis. In double bottom construction, it is that portion above the tank margin.

transverse heating *(Elec.Eng.)*. Dielectric heating in which electrodes impose a high-frequency electric field normal to layers of laminations.

transverse joint *(Build.)*. Any joint in a brick wall which cuts across the bed from the front to the back surface, such joint in the best practice being always a continuous one in order to avoid the setting up of **straight joint**.

transverse magnetic wave *(Telecomm.)*. See **TM-wave**.

transverse metacentre *(Ships)*. The **metacentre** obtained by inclining the vessel through a small angle about a longitudinal axis. Cf. **longitudinal metacentre**.

transverse springs *(Autos.)*. Laminated springs arranged transversely across the car, parallel to the axles, instead of longitudinally; usually **semi-elliptic springs** and anchored centrally to the chassis.

transverse, transversal *(Bot.,Zool.)*. Broader than long: lying across the long axis of the body or of an organ: lying crosswise between two structures: connecting two structures in crosswise fashion.

transverse wave *(Phys.)*. A wave motion in which the disturbance of the medium occurs at right-angles to the direction of wave propagation; e.g. waves on a stretched string, electromagnetic waves.

transvestism *(Behav.)*. Sexual gratification through dressing in the clothes of the opposite sex.

tranverse-field travelling-wave tube *(Electronics)*. One in which the electric fields associated with the signal wave are normal to the direction of motion of the electron beam.

trap *(Build.)*. A bend in a pipe so arranged as to be always full of water, in order to imprison air within the pipe. Also *air trap*, under which see other synonyms.

trap *(Electronics)*. Crystal lattice defect at which current carriers may be trapped in a semi-conductor. This trap can increase recombination and generation or may reduce the mobility of the charge carriers.

TRAPATT diode *(Electronics)*. A *TRApped Plasma Avalanche Transit-Time diode*; a diode which can operate as an oscillator at microwave frequencies, the frequency being determined by the thickness of the active layer in the diode. The avalanche zone moves through the drift region a trapped space-charge plasma within the p-n junction.

trapezium *(Zool.)*. In the Mammalian brain, a part of the medulla oblongata consisting of transverse fibres running behind the pyramid bundles of the pons varolii.

trapezium diagram *(Electronics)*. Pattern on the screen of a CRO when an amplitude-modulated radiofrequency voltage is applied to one pair of plates and the modulating voltage is applied to the other pair.

trapezium distortion *(Electronics)*. That associated with **trapezium effect**.

trapezium effect *(Electronics)*. Phenomenon in which the deflecting voltage applied to the deflector plates of a cathode-ray tube is unbalanced with respect to the anode. If equal alternating voltages, of different frequencies, are applied to the two sets of plates, resulting pattern on the screen is trapezoidal instead of square.

trapezium, trapezoid *(Maths.)*. UK trapezium = US trapezoid = quadrilateral with one pair of opposite side parallel. US trapezium = UK trapezoid = quadrilateral with no parallel sides. Euclid uses the term trapezium for any quadrilateral which is not a parallelogram. Clearly these terms should never be used without explanation.

trapezoidal rule *(Surv.)*. A rule for the estimation of the area of an irregular figure. For this purpose it is divided into a number of parallel strips of equal width. The lengths of the boundary ordinates of the strips are measured, and the area is calculated from the rule stating that the area is equal to the common width of the strips multiplied by the sum of half the first and half the last ordinates plus all the others.

trapezoidal speed-time curve *(Eng.)*. A simplified form of speed-time curve used in making preliminary calculations regarding the energy consumption and average speed of moving vehicles. The acceleration and braking portions of the curve are sloping straight lines, while the coasting portion is a horizontal straight line, so that the complete curve becomes a trapezium. Cf. **quadrilateral speed-time curve**.

trapezoidal speed-time curve *(Eng.,Maths.)*. A simplified form of speed-time curve used in making preliminary calculations regarding the energy consumption and average speed of moving vehicles. The acceleration and braking portions of the curve are sloping straight lines, while the coasting portion is a horizontal straight line, so that the complete curve becomes a trapezium. Cf. **quadrilateral speed-time curve**.

trapped mode *(Phys.)*. Propagation in which the radiated energy is substantially confined within a tropospheric duct.

trapping *(Comp.)*. Feature of some computers by which they make an unscheduled jump to some specified location if an abnormal arithmetic situation arises.

trapping region *(Phys.)*. Three-dimensional space in which particles from the sun are guided into paths towards the magnetic poles, giving rise to **aurora**, and otherwise forming ionized shells high above the ionosphere. Also called *magnetic tube*.

trappoid breccias *(Geol.)*. A succession of breccias found near Nuneaton, Charnwood, and Malvern consisting of angular blocks of rhyolite and feldspathic tuffs; of Permian age. They probably represent fossil scree material.

trash *(Textiles)*. The unwanted material present in bales of raw cotton.

trass *(Build.,Geol.)*. A material similar to pozzuolana, found in the Eifel district of Germany; used to give additional strength to lime mortars and plasters.

trass mortar *(Build.,Civ.Eng.)*. A mortar composed of lime, sand and trass or brick-dust, or of lime and trass without sand, the trass making the mortar more suited for use in structures exposed to water.

trauma *(Behav.)*. (1) *Medical*, structural damage to the body caused by the impact of some object or substance (e.g. a burn). (2) In *psychiatry*, any totally unexpected experience which the person cannot assimilate. See **shock**.

trauma *(Med.)*. A wound or body injury.

traumatic *(Bot.)*. Relating to wounds.

traumatic neurosis *(Behav.)*. A psychiatric illness resulting from severe and unexpected experience; characterized by periods of trance when the events are re-experienced, and often by traumatic dreams. Differs from other neuroses in that the symptoms have no unconscious meaning, but are an attempt to assimilate the experience by repeating it.

travel *(Eng.)*. The distance between the extreme positions reached by a mechanism executing a reciprocating or other reversing motion.

traveling block *(Min.Ext.)*. The heavy duty block which supports the drill string while drilling and can lift it during a **round trip**.

traveller *(Textiles)*. See **ring spinning**.

traveller gantry *(Build.)*. A gantry of the platform gantry type, but having a movable carriage on rails in place of the platform; the carriage, on which is fixed a crab or winch, is capable of movement along or across the gantry.

travelling matte shot *(Image Tech.)*. A composite shot in which the components are printed together at the laboratory by the use of film **mattes** with varying outlines matching the action.

travelling wave *(Phys.)*. A wave carrying energy continuously away from the source.

travelling-wave amplifier *(Telecomm.)*. One using a travelling-wave tube.

travelling-wave antenna *(Telecomm.)*. One in which many radiating elements are excited progressively as the result of a single wave traversing its length in one direction only.

travelling-wave magnetron *(Electronics)*. Multiple cavity magnetron in which cavities are coupled by travelling-wave systems.

travelling-wave maser *(Electronics)*. See **solid-state maser**.

travelling-wave tube *(Electronics)*. One in which energy is interchanged between a helix delay line and an electron beam, which can be at an angle. Used to amplify ultra-high and microwave frequencies. See also **transverse-beam-, transverse-field-**.

traverse *(Surv.)*. A survey consisting of a continuous series of lines whose lengths and bearings are measured.

traverse tables *(Surv.)*. Tables from which the differences of latitude and departure of a line of any length and bearing may be read off.

traversing *(Eng.)*. The sliding motion in a self-acting lathe or, more generally, the sideways movement of part of a machine.

traversing bridge *(Eng.)*. A type of movable bridge which is capable of rolling backwards and forwards across an opening, such as a dock entrance, to allow of the passage of a vessel.

travertine *(Geol.)*. A variety of calcareous tufa of light colour, often concretionary and varying considerably in structure; some varieties are porous. A deposit characteristic of hot springs in volcanic regions.

tread *(Build.)*. The horizontal part of a step.

tread *(Eng.)*. In the wheels of a vehicle, that part of the tyre in contact with the road or rail.

tread *(Vet.)*. An injury of the coronet of a horse's hoof due to striking with the shoe of the opposite foot.

trebles *(Min.Ext.)*. See **coal sizes**.

tree *(Bot.)*. A tall, woody perennial plant having a well-marked trunk and few or no branches persisting above the base. Cf. **excurrent**.

tree *(Chem.)*. Crystal growth structure. See **dendrite, lead tree**.

tree *(Comp.)*. Non-linear hierarchic data structure, where each data item is thought of as a 'node', and links from it to other items as 'branches'.

tree *(Telecomm.)*. A number of connected circuit branches which do not include meshes.

tree-ferns *(Bot.)*. Ferns (Cyathea and Dicksonia and several extinct genera) which form a trunk up to 20 m high, typically unbranched and with a relatively slender stem surrounded and supported by matted adventitious roots and persistent leaf-bases.

trega-. *(Genrl.)*. A prefix signifying 10^{12} times. It is replaced in the SI system of unit notation by the prefix **tera-**.

Tremadoc *(Geol.)*. The oldest epoch of the Ordovician period.

trematic *(Zool.)*. Pertaining to the gill-clefts.

Trematoda *(Zool.)*. A class of Platyhelminthes all the members of which are either ectoparasites or endoparasites, and have a tough cuticle, a muscular non-protrusible pharynx, and a forked intestine; a ventral sucker for attachment is usually present, and a sucker surrounding the mouth. Sometimes divided into three classes; Digenea, Aspidogastrea, Monogenea. Liver Flukes.

trembling *(Vet.)*. See **myoclonia congenita**.

tremolite *(Min.)*. A monoclinic amphibole, hydrated calcium magnesium silicate. It is usually white or grey and occurs in bladed crystals or fibrous aggregates in metamorphic rocks. It differs from *actinolite* in having less iron; the name is used for the magnesium end member.

tremor *(Med.)*. Involuntary agitation of the muscles of the body, or of a limb, due to emotional disturbance, old age or disease of the nervous system.

trenail *(Build.)*. A hardwood pin driven transversely through a mortise and tenon to secure the joint. Also *trunnel*. See **draw-bore**.

trench fever *(Med.)*. A disease common among troops in World War I; symptoms were relapsing fever, headache, pains in the back and in the limbs, often a rose-red eruption, due to infection with a virus conveyed by lice.

trenching plane *(Build.)*. See **dado plane**.

trepan *(Med.)*. (1) To trephine. (2) A form of trephine no longer in use.

trepanning *(Eng.)*. Producing a hole by removing a ring of material, as opposed to disintegrating all the material corresponding to the hole.

trephine *(Med.)*. (1) To operate with the trephine; to remove by surgical means a part of the skull; to remove by surgical means a disk from any part, e.g. from the globe of the eye in the treatment of glaucoma. (2) A crown saw which is designed to remove a circular area of bone from the skull.

Treponemataceae *(Biol.)*. A family of mainly parasitic, small spirochaetes, many of which are pathogenic, e.g. *Treponema pallidum* (syphilis), *Treponema pertenue* (yaws).

T-rest *(Eng.)*. A T-shaped rest clamped to the bed of a wood-turning lathe for supporting the tool. Also *tee rest*. See **L-rest**.

tri-. *(Genrl.)*. Prefix from L, *trēs*, Greek *tria*, three.

triac *(Electronics)*. A bi-directional gate-controlled **thyristor** for full-wave control of a.c. power. By altering the phase of the gate switching signal, load current can be adjusted from a few per cent to nearly 100% of the full load current.

triacetate *(Textiles)*. Fibres of **cellulose acetate** of which nearly all (> 92%) of the hydroxyl groups are acetylated.

triacetin *(Chem.)*. *Glycerol triacetate*. Chiefly used as a plasticizer for cellulose acetate and cellulose ether plastics.

triacidic *(Chem.)*. Of a base, capable of reacting with three hydrogen ions per molecule.

triad *(Chem.)*. The three groups of similar atoms often given as group 8 in the periodic table: Fe, Co and Ni; Ru, Rh and Pd, and Os, Ir and Pt.

triad *(Image Tech.)*. The unit image component of the screen of a **shadow-mask tube**, comprising one dot of each of the red, green and blue phosphors.

trial and error learning *(Behav.)*. In learning theory, refers to an essentially passive type of learning in which behaviour changes occur as a result of their association with positive or negative consequences (**reinforcements**). Originally the term was used to compare it with **insight learning** and is now more frequently referred to as operant conditioning.

trial pit *(Civ.Eng.)*. A pit sunk into the ground to obtain information as to nature, thickness, and position of strata.

triamterene *(Med.)*. A diuretic drug which promotes salt and water excretion without causing severe potassium loss.

triandrous *(Bot.)*. Having three stamens.

triangle *(Maths.)*. A three-sided rectilineal plane figure. See **equilateral, isosceles, scalene**.

triangle of error *(Surv.)*. The triangle formed in the trial-and-error solution of the three-point problem when, on drawing back rays through the three known points on plan, they form a small triangle instead of intersecting at a single point, as a result of the positioning of the survey instrument or the measurement of one or more angles being incorrect.

triangle of forces *(Phys.)*. A particular case of the polygon of forces drawn for three forces in equilibrium at a point. See **polygon of forces**.

triangular *(Genrl.)*. Having three angles.

triangulation *(Surv.)*. The process of dividing up a large area for survey purposes into a number of connected triangles with their apexes (triangulation or 'trig' stations) mutually visible, measuring one side of one of the triangles (the 'base line') and all the angles. See **intersection, trilateration.**

triarch *(Bot.)*. A stele having three strands of protoxylem e.g. the roots of some dicotyledons.

Triassic *(Geol.)*. The geological period between Permian and Jurassic. it is the oldest period of the Mesozoic era and has a time span from app. 250–215 million years. It was named by von Alberti from the three-fold division in Germany. The corresponding system of rocks.

triazole *(Chem.)*. $C_2H_3N_2$. A heterocyclic compound consisting of a five-membered ring.

tribasic *(Chem.)*. Of an acid, capable of reacting with three hydroxide ions per molecule.

tribe *(Bot.)*. A section of a family consisting of a number of related genera.

tribo-electrification *(Phys.)*. Separation of charges through surface friction. If glass is rubbed with silk, the glass becomes *positive* and the silk *negative*, i.e. the silk takes electrons. The phenomenon is that of contact potential, made more evident in insulators by rubbing.

tribology *(Phys.)*. The science and technology of interacting surfaces in relative motion (and the practices related thereto), including the subjects of friction, lubrication and wear.

triboluminescence *(Phys.)*. Luminescence generated by friction.

tricarboxylic cycle *(Biol.)*. A cyclical series of metabolic interconversions of di- and tricarboxylic acids (including the sometimes eponymous citric acid) which brings about the oxidative degradation of acetyl-CoA. The electrons released generate reductive power which is exploited during their passage along the electron chain to generate ATP.

tricarpellary *(Bot.)*. Consisting of three carpels.

triceps *(Zool.)*. A muscle with three insertions.

trichiniasis, trichinosis *(Med.)*. Infestation of the human intestine, as a result of eating raw or underdone pork, with the nematode worm *Trichinella* (or *Trichina*) *spiralis*, the larvae of which migrate to, and become encysted in, the muscles of the body.

3:4:4′ trichlorocarbanilide *(Chem.)*. $Cl.C_6H_4.NH.CO.$-$NH.C_6H_3Cl_3$. Used as a germicide in toilet soaps. Low toxicity.

trichlorethene *(Chem.)*. C_2HCl_3, used as a solvent in dry-cleaning, in the extraction of fat from wool, and in the manufacture of paints and varnishes. Used in surgery to give general analgesia, and, with nitrogen (I) oxide, light general anaesthesia. TN *Trilene.*

trichloroacetic (-ethanoic) acid *(Chem.)*. $CCl_3.COOH$. Organic acid prepared by oxidizing chloral ($CCl_3.CHO$) with nitric (V) acid. Acid with a high *dissociation constant.*

1,1,1-trichloroethane *(Chem.)*. $CH_3.CCl_3$. Methyl chloroform. R.m.m. 133.5; bp 74°C. Chlorinated solvent with low toxicity (much safer in use than tetrachloromethane). Non-inflammable. Widely used industrially for cleaning electrical equipment. TNs, Chlorothene NU, Genklene.

trichloromethane *(Chem.)*. See **chloroform.**

trichocephaliasis *(Med.)*. See **trichuriasis.**

trichocyst *(Zool.)*. In some Ciliophora, a minute hairlike body lying in the sub cuticular layer of protoplasm; it is capable of being shot out, and is an organ of attachment.

trichogyne *(Bot.)*. An outgrowth from the female sex organ of the red algae, some fungi and lichens and a few green algae for the reception of the male gamete.

trichoid *(Zool.)*. Hairlike.

trichome *(Bot.)*. Any outgrowth of the epidermis of a plant, composed of one or more cells but without vascular tissue.

trichomoniasis *(Med.)*. Infection with mobile flagellated protozoal organisms. *T. vaginalis* causes irritation and discharge from the vagina and the male urethra.

trichosis *(Zool.)*. Arrangement or distribution of hair.

trichotomous *(Bot.)*. Branching into three. Cf. **dichotomous.**

trichromatic coefficients *(Phys.)*. The relative intensities of 3 primaries of a given trichromatic system of colour specification required to match a colour sample. Generally add to unity.

trichromatic filter *(Image Tech.)*. See **tricolour filter.**

trichromatic process *(Image Tech.)*. See **three-colour process.**

trich-, tricho-. *(Genrl.)*. Prefix from Gk. *thrix*, gen. *trichos*, hair.

trichuriasis *(Med.)*. Infestation of the human intestine with the nematode whip-worm *Trichuris trichiura* (also known as *Trichocephalus dispar*).

tricipital *(Zool.)*. Adj. from **triceps.**

trick valve *(Eng.)*. See **Allan valve.**

triclinic system *(Crystal.)*. The lowest system of crystal symmetry containing crystals which possess only a centre of symmetry. Also called *anorthic system.*

tricolour filters *(Image Tech.)*. A set of three filters covering the visible spectrum as the red, green and blue regions respectively.

tricot *(Textiles)*. Fabric knitted on a **warp-knitting machine.**

tricusp *(Maths.)*. See **Steiner's tricusp.**

tricuspid *(Zool.)*. Having 3 points, as the tight auriculo-ventricular valve of the Mammalian heart.

tricycle landing gear *(Aero.)*. A landing gear with a nose-wheel unit.

tricyclic anti-depressants *(Med.)*. Somewhat misleading term (as there are now other ring compounds with broadly similar properties) for a group of drugs useful in the treatment of moderate to severe depressive illness. Some may have additional sedative properties *(amitryptaline)* and others are reputed to have fewer cardiac side-effects *(mianserin)*.

tridymite *(Min.)*. A high-temperature form of silica, SiO_2, crystallizing in the orthorhombic system, but possessing pseudohexagonal symmetry. The stable form of silica from 870°C to 1470°C. Typically occurs in acid volcanic rocks.

triethanolamine *(Chem.)*. $(HOC_2H_4)_3N$. Strongly alkaline organic solvent used in some paint strippers. Also used as a stabilizer for chlorinated hydrocarbon solvents.

trifacial *(Zool.)*. The 5th cranial or trigeminal nerve of Vertebrates.

trifid *(Bot.)*. Split into three parts but not to the base.

trifoliate *(Bot.)*. Having three leaves or, sometimes, three leaflets.

trifoliolate *(Bot.)*. A compound leaf having three leaflets, e.g. clover.

trifurcate *(Zool.)*. Having 3 branches.

trifurcating box *(Elec.Eng.)*. A cable dividing box for enclosing the joints between a 3-arc or triple concentric cable and 3 single-core cables or conductor terminals.

trigatron *(Electronics)*. An envelope with an anode, cathode and trigger electrode, containing a mixture of argon and oxygen. The device operates as an electronic switch, in which a low energy pulse ionizes the gas in the switch, and permits discharge of a much higher energy pulse across the main electrodes.

trigeminal *(Zool.)*. Having 3 branches; the 5th cranial nerve of Vertebrates, dividing into the ophthalmic, maxillary, and mandibular nerves.

trigeminal neuralgia *(Med.)*. See **tic douloureux.**

trigger *(Chem.)*. The agent which causes the initial decomposition of a chain reaction.

trigger *(Comp.)*. Manual or automatic signal for an operation to start.

trigger circuit *(Electronics)*. A circuit having a number of states of electrical condition which are either stable (or quasi-stable) or unstable with at least one stable state, and so designed that desired transition can be initiated by the application of suitable trigger excitation.

trigger electrode *(Electronics)*. See **pilot electrode.**

trigger level *(Electronics)*. The minimum input level at which a trigger circuit will respond.

trigger pulse *(Electronics)*. One which operates a trigger circuit.

trigger relay *(Telecomm.)*. Relay which can be mechanical, thermionic, e.g. a gas-filled triode, or solid-sate and which, when operated, remains in its operated condition when the operating current or other control is removed, because of a mechanical latch or other property.

trigger valve *(Electronics)*. A thermionic or gas discharge valve used as a trigger relay.

triglyceride *(Chem.)*. Term applied to a fatty acid ester of glycerol in which all 3 hydroxyl groups are substituted.

triglyph *(Arch.)*. A group of 3 glyphs, or of 2 glyphs and 2 half-glyphs, used as a decoration for a flat surface.

trigonal system *(Crystal.)*. A style of crystal architecture characterized essentially by a principal axis of threefold symmetry; otherwise resembling the hexagonal system. Such important minerals as calcite quartz, and tourmaline crystallized in this system.

trigone *(Med.)*. Triangular area of interior of urinary bladder between the openings of the ureters and of the urethra.

trigonitis *(Med.)*. Inflammation of the trigone.

trigonometrical functions *(Maths.)*. If θ is any angle, and ABC is the right-angled triangle formed by dropping a perpendicular BC from a point B in one of the lines enclosing the angle to the other, the trigonometrical functions, or ratios, are as follows:

$$\text{sine}\ \theta = \frac{BC}{AB};\ \text{cosecant}\ \theta = \frac{1}{\text{sine}\ \theta}$$

$$\text{cosine}\ \theta = \frac{AC}{AB};\ \text{secant}\ \theta = \frac{1}{\text{cosine}\ \theta}$$

$$\text{tangent}\ \theta = \frac{BC}{AC};\ \text{cotangent}\ \theta = \frac{1}{\text{tangent}\ \theta}$$

Usually abbreviated to *sin, cosec, cos, sec, tan, cot.* Independent arithmetic definitions of *sin, cos,* and *tan* are as follows:

$$\sin x = \sum_{0}^{\infty} (-1)^r \frac{x^{2r+1}}{(2r+1)!}$$

$$\cos x = \sum_{0}^{\infty} (-1)^r \frac{x^{2r}}{(2r)!}$$

$$\tan x = \frac{\sin x}{\cos x}.$$

See also **inverse trigonometrical function**. **conjugate element**.

trigonometrical station *(Surv.)*. A survey station used in a triangulation.

trigonometrical survey *(Surv.)*. A survey based on a triangulation.

trigonous *(Bot.)*. A triangular stem but obtusely angled and with convex faces. Cf. **triquetrous**.

trihydric alcohols *(Chem.)*. Alcohols containing 3 hydroxyl groups attached to 3 different carbon atoms, e.g. **glycerine**.

trilateration *(Surv.)*. Land survey, by triangulation, in which distances are measured direct by a **tellurometer**. In the *shoran* system, up to 800 km can be thus measured by aid of airborne magnetometer flown between 2 ground stations.

Trilene *(Chem.)*. See **trichlorethene**.

trillion *(Maths.)*. The cube of a million; (US) the cube of ten thousand. Colloquial only.

trim *(Aero.)*. Adjustment of an aircraft's controls to achieve stability in a desired condition of flight; cf. **trimming strip, tab**.

trim *(Build.)*. Architraves and other finishings around a door or window opening.

trim *(Ships)*. The difference between the draughts measured at the forward and after perpendiculars. May be expressed as an angle.

Trimask *(Print.)*. An all-purpose film used for **colour masking**.

trimer *(Chem.)*. Substance in which molecules are formed from 3 molecules of a monomer.

trimeric *(Chem.)*. Having the same empirical formula but a relative molecular mass three times as great.

trimerous *(Bot.)*. Arranged in 3s or in multiples of 3.

trimethyl-aminoethanoic acid (trimethylglycine) *(Chem.)*. See **betaine**.

trimethylene glycol *(Chem.)*. $CH_3OH.CH_2.CH_2OH$. *Propan-1,3-diol.* Organic solvent. Bp 214°C (with decomposition); rel.d. (20°C) 1.060.

trimethylglycine *(Chem.)*. See **betaine**.

trimmed edges *(Print.)*. See **cut edges**.

trimmed size *(Print.)*. A necessary specification for books and for any subdivision of a sheet, the minimum trim being ⅛" from each edge that requires it. International **paper sizes** are always given as trimmed.

trimmer *(Aero.)*. See **trimming tab**.

trimmer *(Build.)*. The cross-member which is framed between the full-length members to afford intermediate support to the shortened joists in a trimming.

trimmer *(Elec.Eng.)*. See **trimming capacitor**.

trimmer *(Telecomm.)*. See **pad (1)**.

trimmer joint *(Build.)*. A joint formed with a **tusk tenon**.

trimming *(Build.)*. The operation by which bridging joists or rafters are shortened and given intermediate support around a fireplace or chimney.

trimming *(Eng.)*. The removal of flash from rubber or plastics mouldings or from castings, or of material from the edge of a workpiece.

trimming capacitor *(Elec.Eng.)*. Variable capacitor of small capacitance used in conjunction with ganging for taking up the discrepancies between self and stray capacitances of individual ganged circuits, so that they remain in step for all settings of the main tuning control. Also called *trimmer*.

trimming joist *(Build.)*. One of the 2 full-length members between which the trimmer is framed. As these members have to carry more than the other bridging joists, they are thicker.

trimming strip *(Aero.)*. A metal strip, or a cord of wire doped in place with fabric, on the trailing edge of a control surface to modify its balance or trim; it is adjustable only on the ground.

trimming tab *(Aero.)*. A *tab*, which can be adjusted in flight by the pilot, for trimming out control forces; coll. *trim tab* or *trimmer*.

trimonoecious *(Bot.)*. A species in which the plants bear male, female and hermaphrodite flowers.

trimorphic *(Bot.)*. Having 3 forms. See **heterostyly**.

trimorphous *(Chem.)*. Having 3 crystalline forms.

trims *(Image Tech.)*. Unused portions of a selected scene left over after editing is complete.

trim tab *(Aero.)*. See **trimming tab**.

triniscope *(Image Tech.)*. System of colour TV display using optical combination of the red, green and blue images from the screens of three separate CRTs, particularly for **telerecording**.

trinitrides *(Chem.)*. Salts of hydrazoic acid. Also called *azides* and *hydrazoates*.

trinitroglycerine *(Chem.)*. See **nitroglycerine**.

Trinitron *(Image Tech.)*. TN for a 3-colour TV tube using a vertical grating and vertical phosphor stripes. A common electron gun assembly and deflexion system is used with three cathodes to provide the electron beams.

trinitrophenol *(Chem.)*. *2,4,6-trinitro-1-hydroxybenzene.* See **picric acid**.

trinitrotoluene *(Chem.)*. The symmetrical isomer, *2,4,6-trinitrotoluene*, $C_6H_2(CH_3)(NO_2)_3$, is a solid, melting at 82°C. It is manufactured by slowly adding toluene to a mixture of nitric and sulphuric acids containing oleum. It is used as a high explosive, and is known as *TNT*.

triode valve *(Electronics)*. Thermionic vacuum tube containing an emitting cathode, an anode, and a control electrode or grid, whose potential controls the flow of electrons from cathode to anode.

trioecious *(Bot.)*. A species in which some individuals bear male flowers only, others female only and the rest hermaphrodite.

triolein *(Chem.)*. Naturally-occurring triglyceride in which all three fatty acid chains are oleic acid.

trioses *(Chem.)*. The simplest monosaccharides. They

contain 3 carbon atoms in the molecule, e.g. $HO.CH_2.CO.CH_2.OH$, *glyceraldehyde*.

trip *(Min.Ext.)*. See **round trip**.

trip *(Nuc.Eng.)*. Automatic shut down of reactor power initiated by signal from one of the safety circuits when one of the operational characteristics of the reactor deviates beyond a certain limit.

tripack *(Image Tech.)*. A photographic material whose base carries three emulsion layers appropriately sensitized to specific spectral regions, with intermediate filter layers, for recording images of separate colours.

trip amplifier *(Nuc.Eng.)*. One operating the trip mechanism of a nuclear reactor. Also termed a *shut-down amplifier*.

trip circuit *(Elec.Eng.)*. The electric circuit operating the tripping mechanism of a circuit breaker. Cf. **shunt trip**.

trip coil *(Elec.Eng.)*. Any magnet coil which operates some other circuit or mechanism by motion of an armature; more particularly, a coil which operates a circuit-breaker, or the release mechanism of a telegraph machine.

trip gear *(Eng.)*. A valve-actuating gear, used for drop valves and rocking (Corliss) valves of large steam-engines, in which the valve is opened by a trigger mechanism, which is then tripped out of engagement to allow the valve to close under a heavy spring. See **Corliss valve, drop valve**.

triphenylmethane dyes *(Chem.)*. A group of dye-stuffs derived from triphenylmethane. They comprise the malachite green group derived from diaminotriphenylmethane, the rosaniline group derived from triaminotriphenylmethane, the aurine group derived from trihydroxytriphenylmethane, the phthalein group derived from triphenylmethane-carboxylic acid.

triphylite *(Min.)*. An orthorhombic lithium iron phosphate isomorphous with **lithiophilite**.

tripinnate *(Bot.)*. Three times **pinnate**.

triple-axis neutron spectrometer *(Phys.)*. An instrument used in neutron spectroscopy for determining the energies of neutrons scattered in a particular direction from a crystal. See **neutron elastic scattering, neutron inelastic scattering**.

triple bond *(Chem.)*. A covalent bond between two atoms involving the sharing of three pairs of electrons. See **alkyne**.

triple-concentric cable *(Elec.Eng.)*. A 3-core cable in which the conducting cores are arranged concentrically about the axis of the cable.

triple-expansion engine *(Eng.)*. An engine in which the steam expands, successively, in a high pressure, intermediate pressure, and low pressure cylinder, working on the same crankshaft.

triple fusion *(Bot.)*. The fusion of the 2 **polar nuclei** with the second male gamete in angiosperms.

triple junction *(Geol.)*. The focal point of three tectonic plates, e.g. at **convergence zones, divergence zones** or **transform faults**.

triple point *(Phys.)*. The temperature and pressure at which the three phases of a substance can coexist (see **phase rule**). The triple point of water is the equilibrium point (273.16 K and 610 $N m^{-2}$) between pure ice, air-free water and water vapour obtained in a sealed vacuum flask. It is one of the fundamental **fixed points** of the international practical scale of temperature. See **Kelvin thermodynamic scale of temperature**.

triple-pole switch *(Elec.Eng.)*. A switch for simultaneously making or breaking a 3-wire electric circuit.

triple superphosphate *(Chem.Eng.)*. The product obtained by reacting phosphate rock with phosphoric acid giving a higher concentration of soluble calcium phosphate than obtains in ordinary or 'single' *superphosphate*.

triplet *(Biol.)*. The sequence of three bases in an mRNA which specify a particular amino acid. See **genetic code**.

triplet *(Chem.)*. A state in which there are two unpaired electrons.

triplets *(Zool.)*. In Mammals, 3 individuals produced at the same birth.

triple vaccine *(Immun.)*. Vaccine containing a mixture of diphtheria toxoid, tetanus toxoid and pertussis vaccine. Routinely used to produce active immunity against diphtheria, tetanus and whooping cough in infants.

Triplex glass *(Glass)*. A patented form of laminated glass. See **safety glass**.

triplex winding *(Elec.Eng.)*. A d.c. armature winding having 3 parallel paths per pole between positive and negative terminals.

triploblastic *(Zool.)*. Having three types of tissue in the body, there being mesoderm between the ectoderm and endoderm, which gives rise to connective, skeletal and muscular tissues etc. Cf. **diploblastic**.

triploid *(Biol.)*. Having 3 times the haploid number of chromosomes for the species.

tripod *(Surv.)*. Device by which some surveying and other instruments are supported firmly off the ground. It consists of 3 legs hinged to a common head on which the instrument is secured. In photography, used for steadying a camera for long exposure, telephoto shots etc.

tripod bush *(Image Tech.)*. Part of a camera body with a threaded hole to accept the screw of a tripod head.

tripod drill *(Min.Ext.)*. Rock drill on heavy tripod.

tripod head *(Image Tech.)*. The mounting which attaches the camera to a tripod, usually including means for its rotation and tilt; for a motion picture camera, continuous smooth movement may be provided by geared handles.

Tripoli powder *(Min.)*. See **tripolite**.

tripolite *(Min.)*. A variety of opaline silica which is formed from the siliceous frustules of diatoms. It looks like earthy chalk or clay, but is harsh to the feel and scratches glass. When finely divided it is sometimes called *earthy tripolite*. Also called *diatomite, infusorial earth*.

tripotassium dicitratobismuthate *(Med.)*. A bismuth chelate which promotes healing of gastric and duodenal ulcers.

tripping *(Horol.)*. An escapement is said to *trip* when a tooth of the escape wheel runs past the locking face.

tripping battery *(Elec.Eng.)*. The secondary battery which provides the supply for the trip-coil circuits of a number of circuit-breakers.

trip relay *(Elec.Eng.)*. A relay controlling the electromagnetic tripping mechanism of a circuit-breaker.

trip switch *(Elec.Eng.)*. A control switch for closing the tripping circuit of a circuit-breaker.

trip value *(Elec.Eng.)*. The current or voltage required to operate a relay.

triquetrous *(Bot.)*. A triangular stem with acute angles and concave faces. Cf. **trigonous**.

trisaccharides *(Chem.)*. Carbohydrates resulting from the condensation of 3 monosaccharides with the elimination of 2 molecules of water, e.g. maltotriose in $G\alpha1 \rightarrow 4G\alpha1 \rightarrow 4G$ (G = glucose).

trismus *(Med.)*. *Lockjaw*; tonic spasm of the muscles of the jaw, causing the jaws to be clenched, as in tetanus.

trisomic *(Biol.)*. Said of an otherwise normal diploid organism in which one chromosome type is represented thrice instead of twice.

trisomy 21 *(Biol.)*. Also *Down's syndrome*. Condition in which an individual has three copies of chromosome 21, either in all cells or in a proportion of their cells. Affected individuals show many abnormalities to varying degrees of severity, including the characteristic *mongoloid* eye fold, and usually subnormal intelligence. Incidence increases with mother's age.

tristearin *(Chem.)*. Naturally-occurring triglyceride in which all 3 fatty acid chains arise from stearic acid.

tristimulus values *(Phys.,Textiles)*. The amounts of the three primary colours (blue, red and green) that form the colour being examined or matched. A tristimulus colorimeter can analyse a colour and indicate the amounts present of its constituent primary colours.

tritanopic *(Med.)*. Colour blind to blue.

TRITC *(Immun.)*. Abbrev. for *Tetramethyl Rhodamine IsoThioCyanate*.

tritium *(Phys.)*. Symbol T, r.a.m. 3.0221, mass no. 3. The radioactive isotope of hydrogen, of half-life 12.5 years. It is very rare, the abundance in natural hydrogen being one atom in 10^{17} but tritium can be produced artificially by neutron absorption in lithium. Can be used to label any aqueous compound and consequently is of great importance in radiobiology.

tritium unit *(Chem.)*. A proportion of tritium in hydrogen of one part in 10^{18}. This represents 7 disintegrations per minute in 1 litre of water.

Triton *(Astron.)*. The principal natural satellite of **Neptune**, has a diameter of 3700 km.

triton *(Phys.)*. The tritium nucleus, consisting of 1 proton combined with 2 neutrons.

triton X-100 *(Biol.)*. *iso-Octylphenoxypolyethoxyethanol*. A non-ionic detergent which is commonly used to solubilize membrane proteins in their biologically active state.

tritor *(Zool.)*. The masticatory surface of a tooth.

triturate *(Chem.)*. To grind to a fine powder, especially beneath the surface of a liquid.

trityl *(Chem.)*. The *triphenylmethyl* group, $C(C_6H_5)_3$; the triphenylmethyl radical is generally accepted as being the first organic radical to be obtained in a free state.

trivalent *(Biol.)*. Said of association of 3 chromosomes at meiosis.

trivalent *(Chem.)*. Capable of combining with 3 atoms of hydrogen, or their equivalent.

tRNA *(Biol.)*. An RNA molecule about 80 nucleotides long, with complementary sequences which result in several short *hairpin-like* structures. The loop at the end of one of these carries the anticodon triplet, which binds to the codon of the mRNA. The corresponding amino acid is bound to the 3′ end of the molecule. See **adaptor hypothesis**.

trocar *(Med.)*. A sharp-pointed perforator which, inserted into a cannula, enables this to be introduced into the body.

trochal *(Zool.)*. Wheel-shaped.

trochanter *(Zool.)*. The second joint of the leg in Insects; a prominence for muscled attachment near the head of the femur in Vertebrates.

trochlea *(Zool.)*. Any structure shaped like a pulley, especially any foramen through which a tendon passes. adj. *trochlear*.

trochoid *(Maths.)*. See **roulette**.

trochoidal mass analyser *(Nuc.Eng.)*. A form of mass spectrometer in which the ion beams traverse trochoidal paths within electric and magnetic fields mutually perpendicular.

trochophore, trochosphere *(Zool.)*. A free swimming pelagic larval form of Annelida, Mollusca, and Bryozoa, possessing a prominent pre-oral ring of cilia and an apical tuft of cilia.

trochotron *(Nuc.Eng.)*. Abbrev. for *TROCHOidal magneTRON*. High-frequency counting tube, which uses crossed electric and magnetic fields to deflect a beam on to radially disposed electrodes.

troctolite *(Geol.)*. A coarse-grained basic igneous rock, consisting essentially of olivine and plagioclase only. The former mineral occurs as dark spots against the feldspar, giving the rock a characteristic spotted appearance, whence the name *troutstone*.

troilite *(Min.)*. A nonmagnetic iron sulphide, FeS, which occurs mainly in meteorites.

Trojan group *(Astron.)*. A number of minor planets, named after the heroes of the Trojan war, which have the same mean motion as Jupiter and travel in the same orbit. They are divided into 2 clusters, one of which is 60° of longitude ahead of Jupiter, the other 60° behind; each planet oscillates about a point which forms an equilateral triangle with Jupiter and the sun. These are 2 particular solutions of the **three-body problem**.

trolleybus *(Elec.Eng.)*. A rail-less passenger vehicle, powered by current transmitted from two overhead trolley wires through two roof-mounted *trolley poles*.

trolley system *(Elec.Eng.)*. The overhead current collecting system used on tramcars and trolley buses, in which a small grooved wheel or skid runs under the contact wire.

trombone *(Telecomm.)*. A U-shaped length of waveguide or transmission line which is of adjustable length for use in a waveguide circuit.

trommel *(Min.Ext.)*. A cylindrical revolving sieve for sizing crushed ore or rock.

trondhjemite *(Geol.)*. A coarse-grained igneous rock consisting essentially of plagioclase (ranging from oligoclase to andesine), quartz, and small quantities of biotite.

trophallaxis *(Zool.)*. Mutual exchange of food between imagines and their larvae, as in some social Insects.

trophic *(Bot.,Zool.)*. Pertaining to nutrition.

trophic level *(Ecol.)*. Broad class of organisms within an *ecosystem* characterized by mode of food supply. The first trophic level comprises the green plants, the second is the herbivores and the third is the carnivores which eat the herbivores.

trophic structure *(Ecol.)*. A characteristic feature of any ecosystem, measured and described either in terms of the standing crop per unit area, or energy fixed per unit area per unit time, at successive trophic levels. It can be shown graphically by the various **ecological pyramids**.

trophoblast *(Zool.)*. The differentiated outer layer of epiblast in a segmenting Mammalian ovum.

trophozoite *(Zool.)*. In Protozoa, the trophic phase of the adult, which generally reproduces by schizogony.

troph-, tropho-. *(Genrl.)*. Prefix from Gk, *trophē*, nourishment.

tropical month *(Astron.)*. The period of lunar revolution with respect to the equinox (27.321 58 days).

tropical revolving storm *(Meteor.)*. A small intense cyclonic depression originating over tropical oceans. Also *cyclone*, *hurricane*, *typhoon*, depending on the locality.

tropical switch *(Elec.Eng.)*. A switch mounted on feet or bosses; it thus guards against the effect of excessively damp climates by having an air space between its base and mounting surface. Also called *feet-switch*.

tropical year *(Astron.)*. The interval between 2 successive passages of the sun in its apparent motion through the First Point of Aries; hence the interval between 2 similar equinoxes or solstices and the period of the seasons; its length is 365.242 194 mean solar days.

tropism *(Biol.)*. A reflex response of a cell or organism to an external stimulus; movement that orients an organism to achieve a certain distribution of stimulation. See **geotropism**, **phototropism**. Cf. **taxis**.

tropomyosin *(Biol.)*. A filamentous protein aligned along the actin fibres of muscle. Under the influence of troponin it controls the interaction of actin and myosin.

troponin *(Biol.)*. A complex of 3 polypeptide chains which mediates the effect of calcium on muscle contraction.

tropopause *(Meteor.)*. The upper limit of the **troposphere**, where the lapse rate of temperature becomes small ($\leq$ 2°C/km). Sometimes a single unique tropopause cannot be defined and there is a *multiple tropopause* structure.

troposphere *(Meteor.)*. The lower part of the atmosphere extending from the surface up to a height varying from about 9 km at the poles to 17 km at the equator, in which the temperature decreases fairly regularly with height.

tropospheric *(Telecomm.)*. Said of reflection, absorption or scattering of a radio wave when encountering variations in the troposphere.

tropospheric scatter *(Telecomm.)*. Propagation in which radio waves are scattered by the troposphere. It does not depend critically on frequency, but is generally used for communication over several hundred kilometres at UHF and low microwave frequencies. High power is normally used because of the process's inefficiency.

tropospheric wave *(Telecomm.)*. A radio wave whose path between two points at or near the earth's surface lies wholly within the troposphere and will be governed by meteorological conditions.

Trotter photometer *(Phys.)*. A portable photometer in which the brightness of the comparison screen is varied by tilting.

trouble-shooting *(Electronics)*. *Fault-finding*.

trough *(Meteor.)*. A V-shaped extension of the isobars from a centre of low pressure.

troughed belt conveyor *(Eng.)*. A **conveyor** comprising an endless belt which is made hollow or troughed on the carrying run to increase the carrying capacity and to obviate spillage.

trough gutter *(Build.)*. A gutter used along roof valleys or parapets. Also *box gutter*.

Trousseau's phenomenon *(Med.)*. Spasm of the muscles of a limb whose blood-vessels are compressed, occurring in tetany.

Trouton's rule *(Chem.)*. For most nonassociated liquids, the ratio of the latent heat of vaporization per mole, measured in joules, to the boiling-point, on the absolute scale of temperature, is approx. equal to 88 at atmospheric pressure.

troutstone *(Geol.)*. See **troctolite**.

trowel *(Build.)*. A flat steel tool used for spreading and smoothing mortar or plaster.

TR tube *(Electronics)*. See **transmit-receive tube**.

truck-type switchgear *(Elec.Eng.)*. Switchgear in which each circuit-breaker, with its associated equipment, is mounted on a truck capable of withdrawal, so that it may be completely removed from the gear for maintenance and repair. Also called *carriage-type switchgear*.

true air speed *(Aero.)*. The actual speed of an aircraft through the air, computed by correcting the indicated airspeed for altitude, temperature, position error and compressibility effect. Abbrev. *TAS*

true altitude *(Astron.,Surv.)*. **Altitude** of a heavenly body as deduced from the *apparent altitude* by applying corrections for atmospheric refraction, for instrumental errors, and where necessary for geocentric parallax, sun's semi-diameter, and dip of horizon.

true azimuth *(Surv.)*. That measured relative to true geographical north.

true bearing *(Surv.)*. That measured clockwise from true geographical north.

true coincidences *(Phys.)*. Those produced by a single particle discharging both or all counters; cf. **spurious coincidences**.

true course *(Ships)*. The angle between the true meridian and the direction of the ship's head.

true density *(Powder Tech.)*. The mass of the particle divided by its volume, excluding open and closed pores.

true horizon *(Surv.)*. A great circle of the celestial sphere parallel to the horizon and passing through the earth's centre. Also called the *rational horizon*.

true north *(Surv.)*. The direction of the geographical north pole.

true (real) absorption coefficient *(Phys.)*. The absorption coefficient applicable when scattered energy is not regarded as absorbed. Applicable to broad-beam conditions.

true resistance *(Elec.Eng.)*. See **d.c. resistance**.

true section *(Surv.)*. A section which has been drawn, with the same scales, horizontally and vertically.

true watts *(Phys.)*. Power dissipated in a.c. circuit.

trumpet arch *(Build.)*. See **splaying arch**.

truncate *(Bot.)*. Square-ended base or apex of a structure, as if cut off.

truncation *(Comp.)*. (1) Ending of a computational procedure in accordance with some program rule as soon as a specified accuracy has been reached. (2) Rejection of final digits in a number, thus lessening precision (but not necessarily accuracy). See **rounding**.

truncation error *(Comp.)*. Error introduced by **truncation**.

truncus *(Zool.)*. A main blood vessel; as the *truncus transversus* or **Cuvierian duct** and the *truncus arteriosus* or great vessel, through which blood passes from the ventricle. Also *trunk*.

trunk *(Arch.)*. Shaft of a column.

trunk *(Bot.)*. Upright, massive main stem of a tree.

trunk *(Comp.)*. See **highway**.

trunk *(Elec.Eng.)*. See **trunk feeder**.

trunk *(Telecomm.)*. US for a **link**.

trunk *(Zool.)*. The body, apart from the limbs; the proboscis of an elephant. See also **truncus**.

trunk call *(Telecomm.)*. In UK a telephone call from one telephone area to another, involving links between two trunk centres each dealing with all calls passing in or out of its area.

trunk circuit *(Telecomm.)*. In UK a 2- or 4-wire connection between trunk centres, for establishing trunk calls between telephone areas. In US a circuit between exchanges in the same telephone area. Also *trunk line*.

trunk conveyor *(Min.Ext.)*. In colliery, belt conveyor in main road.

trunk diagram *(Telecomm.)*. A diagram which indicates the cable routes between the various groups of telephone switching apparatus in an automatic telephone exchange.

trunk distribution frame *(Telecomm.)*. A frame carrying the terminals for connecting the trunks between ranks of selectors.

trunk exchange *(Telecomm.)*. An exchange in a telephone area which is connected by links to other trunk exchanges, and to subscribers through local exchanges.

trunk feeder *(Elec.Eng.)*. A feeder connecting 2 generating stations, or a generating station and a large substation. Also *trunk main*.

trunk frame terminal assembly *(Telecomm.)*. A cross-connection terminal frame for connecting trunks between ranks of selectors.

trunking *(Telecomm.)*. The cables which contain the links between one rank of selectors and others in the sequence of operation, the cables taking a common route through the exchange building.

trunk junction circuit *(Telecomm.)*. The junction between an exchange and the trunk exchange for routing subscribers to the trunk exchange system.

trunk line *(Telecomm.)*. See **trunk circuit**.

trunk main *(Elec.Eng.)*. See **trunk feeder**.

trunk piston *(Eng.)*. A piston, long in relation to its diameter, used where there is no piston-rod or crosshead, the piston having to take the connecting-rod thrust; most IC engine pistons are of this type.

trunnion axis *(Surv.)*. The horizontal axis about which the telescope of a theodolite or tacheometer may be rotated on its trunnion bearings.

trunnion mounting *(Eng.)*. A pair of short journals, supported in bearings, projecting coaxially from opposite sides of a vessel or cylinder required to pivot about their axis.

truss *(Civ.Eng.)*. A framed structure built up entirely from tension and compression members, arranged in panels so as to be stable under load; used for supporting loads over long spans.

truss *(Med.)*. A surgical appliance consisting of a pad incorporated in a spring or belt for retaining a reduced hernia in place.

truss-beam *(Build.)*. A framework acting as a beam.

trussed partition *(Build.)*. A partition which is framed so as to be self-supporting between its ends; used in cases where the floor is not strong enough to carry it.

truth *(Phys.)*. See **topness**.

truth mark *(Textiles)*. Some permanent mark near the end of a piece of fabric which should remain until the fabric is delivered to the customer. Its presence shows that no end of fabric has been improperly retained by anyone.

truth value *(Comp.)*. The truth values of **Boolean algebra** are TRUE and FALSE (often abbrev. *T* and *F*) which may be represented by the binary digits 1 and 0.

Try *(Chem.)*. Symbol for **tryptophan**.

trying plane *(Build.)*. A tool similar to the jack plane but about 22 in (56 cm) long; used after the jack plane to obtain a straight and true surface.

trypano-. *(Genrl.)*. Prefix from Gk. *trypanon*, borer.

trypanosomes *(Zool.)*. A group of flagellate Protozoa including many causing disease of man (see **trypanosomiasis**) and animals.

trypanosomiasis *(Med.)*. *Chagas' disease*. A disease, occurring in parts of South America and Africa, due to infection of the muscles, heart, and brain of man with the protozoal parasite *Trypanosoma cruzi*, the infection being conveyed by the bite of an insect.

trypsin *(Biol.)*. A protease secreted by the pancreas which is specific for peptide bonds adjacent to lysine and arginine residues.

tryptophan *(Chem.)*. *2-amino-3-indolepropanoic acid*. An amino acid. The L- or S-isomer is a constituent of proteins. Symbol Try, short form W. ⇨

try square *(Build.)*. A tool similar to the bevel but having the blade fixed at 90°.

tschermakite *(Min.)*. An end-member subspecies in the hornblende group of amphiboles, rich in aluminium and calcium.

T-section cramp *(Build.)*. Strong form of **sash cramp** having a T-section steel bar for the sliding members.

T-section filter *(Telecomm.)*. T-network ideally formed of non-dissipative reactances, having frequency pass band over which attenuation is theoretically zero or very low. See **Butterworth filter, Chebyshev filter**.

tsetse fly disease *(Vet.)*. See **nagana**.

tsunami *(Geol.)*. A destructive sea-wave caused by an earthquake or submarine eruption.

tsunami *(Phys.)*. A wave produced in the ocean by a submarine earthquake. Because of its very long wavelength it behaves as a 'shallow' *surface wave*. Its amplitude in mid-ocean is very small; as it approaches land, the amplitude builds up and all the energy of the original disturbance is concentrated into a few wavelengths with devastating results. Erroneously called a *tidal wave*.

tsutsugamushi fever *(Med.)*. See **shimamushi fever**.

T-tail *(Aero.)*. A *tail unit* characterized by positioning the horizontal stabilizer at or near the top of the vertical stabilizer. The unit is employed on aircraft having rear-mounted engines, and has variable incidence. See **all-moving tail**.

TTL *(Electronics)*. Abbrev. for *Transistor-Transistor Logic*. Referring to logic circuits consisting of two or more directly interconnected transistors intended to drive capacitative loads at high rates.

TTL *(Image Tech.)*. *Through The Lens*, referring to cameras in which the view-finder picture is provided by the same lens which forms the exposed image.

TTS *(Acous.)*. See **temporary threshold shift**.

TU *(Telecomm.)*. Abbrev. for *Traffic Unit* and for *Transmission Unit*.

tub *(Min.Ext.)*. A tram, wagon, corf or corve.

tubbing *(Min.Ext.)*. The lining of a circular shaft, formed of timber or by steel segments.

tube *(Bot.)*. The cylindrical proximal part of a gamosepalous calyx or gamopetalous corolla.

tube *(Electronics)*. (1) Enclosed device with gas at low pressure, depending for its operation on ionization originated by electrons accelerated from a cathode by a field applied by an anode. (2) US for all vacuum and gas discharge devices. The term now widely used in the UK where formerly *valve* was almost universal.

tube-drawing *(Eng.)*. The production of seamless tubes by drawing a large, roughly formed tubular piece of material through dies of progressively decreasing size.

tube extrusion *(Eng.)*. A method of producing tubes by direct extrusion from a billet, using a mandrel to shape the inside of the tube.

tube-feet *(Zool.)*. See **podium** (2).

tube fuse *(Elec.Eng.)*. A fuse in which the fuse wire is enclosed in an insulating tube. Also *cartridge fuse*.

tubeless tyre *(Autos.)*. One in which the air-seal is provided by adhesion between the beads and the wheel-rim.

tube mill *(Min.Ext.)*. Horizontal mill in which diameter/length ratio is usually high compared with that of standard ball mill, and which has high discharge.

tube of force *(Phys.)*. Space enclosed by all the lines of force passing through a closed contour. Of unit magnitude when it contains unit flux.

tube plate *(Eng.)*. End wall of a surface condenser, between which the water tubes are carried; they are bolted between the casing and water-chamber covers. See **condenser tubes**.

tuber *(Bot.)*. Swollen underground stem acting as a storage and perennating organ, e.g. the potato.

tubercle *(Bot.)*. Small swelling or **nodule**.

tubercle *(Med.)*. (1) Any small rounded projection on a bone or other part of the body. (2) A solid elevation of the skin larger than a papule. (3) A small mass or nodule of cells resulting from infection with the bacillus of tuberculosis. (4) Loosely, tuberculosis; the tubercle bacillus.

tubercle *(Zool.)*. A small rounded projection; the dorsal articulator process of a rib; a cusp of a tooth. Also called *tuberculum*. adjs. *tubercled, tubercular, tuberculate, tuberculose*.

tubercular *(Med.)*. Of, pertaining to, resembling, or affected with, nodules (tubercles); less correctly, affected with tuberculosis (i.e. *tuberculous*).

tuberculate *(Bot.)*. Covered with small wart-like projections.

tuberculide, tuberculid *(Med.)*. Any skin lesion due to infection with bacillus of tuberculosis.

tuberculin *(Immun.)*. A protein or mixture of proteins derived from *Mycobacterium tuberculosis*, which is employed in the tuberculin test as a diagnostic reagent for detecting sensitization by, or infection with *M. tuberculosis*. Old tuberculin (OT) is a heat concentrated filtrate from the medium in which the organism has been grown. Purified protein derivative (tuberculin PPD) is a soluble protein fraction, precipitated by trichloroacetic acid from a synthetic medium in which *M. tuberculosis* has been grown. Tuberculins can be derived from human, bovine or avian strains of the bacillus but show extensive antigenic cross-reactivity.

tuberculin test *(Immun.)*. Test for delayed hypersensitivity to **tuberculin** in human or other animals. Positive reactions are presumptive evidence of cell-mediated immunity to, and therefore of past or present exposure to, *Mycobacterium tuberculosis*, but does not necessarily indicate active disease. See **Mantoux test**.

tuberculoma *(Med.)*. A slow-growing, circumscribed tuberculous lesion, sometimes present in the brain.

tuberculosis *(Med.)*. Infection by *Mycobacterium tuberculosis*, especially of the lungs; characterized by the development of tubercles in the bodily tissues and by fever, anorexia, and loss of weight. Spread by air droplets and raw milk.

tuberculous *(Med.)*. Pertaining to, affected with, or caused by, tuberculosis.

tuberose (tuberous) sclerosis *(Med.)*. A condition in which hyperplasia of the neuroglia gives rise to hard, tumourlike masses in the brain, associated with epilepsy and mental deficiency; the disease is part of the developmental defect known as **epiloia**.

tuberosity *(Zool.)*. A prominence on a bone, generally from muscle attachment, especially prominences near the head of the humerus.

tuberous *(Bot.)*. Of or like a tuber; having tubers.

tube sinking *(Eng.)*. Drawing an existing tube through a die or rolls to reduce its diameter without an interior plug.

tubicolous *(Zool.)*. Living in a tube.

tubifacient *(Zool.)*. Tube-building, as certain Polychaeta.

tubing head *(Min.Ext.)*. See **casing head**.

tub sizing *(Paper)*. The action of applying to the surfaces of the sheet or web a solution of gelatine size contained in a bath. The film of gelatine should ideally be gently dried by hot air to confer the maximum benefits of ink resistance, strength and durability.

tubular rivet *(Eng.)*. A rivet with a shank from which the centre has been removed to leave a thin wall, so that the rivet can be used to punch its own hole in thin, soft materials and to facilitate setting.

tubular scaffold *(Build.)*. A form of scaffold constructed of steel tubes which can be clamped together in any desired manner by special steel collar-pieces with screw fixings.

tubule *(Bot.)*. A fine tube. See also **microtubule**.

tubule, tubulus *(Zool.)*. Any small tubular structure. adjs. *tubulate, tubuliferous, tubuliform, tubulose*.

tubulin *(Biol.)*. A globular protein of two closely related variants, α and β tubulin which forms, as an α/β dimer, the basic unit for the contraction of microtubules.

tucker, tucking blade *(Print.)*. See **folding blade**.

tuck pointing *(Build.)*. Pointing finished by cutting a groove in the surface at the joints and tucking into the groove a narrow projecting artificial joint of putty.

tufa *(Geol.)*. A porous, concretionary, or compact form of calcium carbonate which is deposited from solution around springs, of which the dense variety is called tufa.

tuff *(Geol.)*. A rock formed of compacted volcanic fragments, some of which can be distinguished by the naked eye. If the fragments are larger, then the rock grades into an agglomerate.

Tufnol *(Plastics)*. A proprietary laminated plastic; light-weight, tensile strength approx. 55–110 MN/m^2; strong insulation qualities. Widely used as an engineering plastic for bearings, gear wheels and pulleys.

tufted *(Bot.)*. Grass shoots, clustered or clumped rather than scattered. *Caespitose.*

tufted carpet *(Textiles)*. A carpet formed by inserting U-shaped lengths of yarn, or something similar, by needles into a strong backing material (e.g. a hessian or polyolefin fabric or a plastic foam). Very fast tufting machines are now in common use.

tularemia, tularaemia *(Vet.)*. A disease of rodents due to infection with *Pasteurella tularensis*. Spread by fleas and ticks and can infect man. The human disease is characterized by prolonged fever, enlargement of the lymph glands, depression and emaciation.

tulipwood *(For.)*. Yellowish wood with reddish grain from several tree species including the Australian *Harpullia pendula*. Used for cabinet-work, moulding etc.

tulle *(Textiles)*. Traditionally, a fine plain-woven silk net. Now also applied to a net with hexagonal holes produced on a warp-knitting machine.

tumble-home *(Ships)*. A term defining the narrowing of a ship's breadth. It is the measure of the inward fall when the deck breadth is less than the maximum breadth.

tumbler gear *(Eng.)*. A gear in a train, mounted on a pivot arm so that it can be swung into and out of engagement with an adjacent gear.

tumbler switch *(Elec.Eng.)*. A small single-pole switch having a quick-break action, universally used in electric-lighting installations for controlling individual lamp circuits.

tumbling *(Eng.)*. A method of removing sand, irregularities etc. from castings or forgings by rotating them in a box with abrasives or special metal slugs.

tumbling-in *(Build.)*. A term applied to the brickwork forming the top surface of a pier and sloping in towards the general face of the wall.

tumbu disease *(Med.)*. A disease, common in Central and West Africa, due to invasion of the surface of the body by the larvae of the tumbu fly *Cordylobia anthropophaga*; characterized by the formation of a boil or a warble in the skin.

tumefaction *(Med.)*. The process or act of swelling; the state of being swollen. Also *tumescence*; adj. *tumescent*.

tumid *(Bot.)*. Swollen; inflated.

tumour-inducing principle *(Bot.)*. See **Ti plasmid**.

tumour necrosis factor *(Immun.)*. Name given to substance secreted by macrophages which have been activated by bacterial endotoxin or by mycobacteria which causes necrosis of a number of tumour cell lines. It also causes muscle necrosis and general wasting of the body, perhaps by inhibiting lipoprotein lipase (normally required for fat transport in cells). It is identical to **lymphotoxin** and the mechanism by which it kills cells is presently unknown. There is evidence that it enhances resistance to parasitic infection.

tumour specific antigen *(Immun.)*. Antigen present in tumour cells which is not expressed (or only very minimally) by their normal counterparts. Sometimes these are foetal antigens not normally expressed in the adult, and sometimes coded for by viral material incorporated into the host's genome. Significant in respect of the existence of specific immune responses to tumours which are present in tumour bearing animals (though insufficiently effective), and in respect of the use of radiolabelled or potentially immunotoxic antibodies targetted on tumour cells in vivo.

tumour, tumor *(Med.)*. Any swelling or morbid enlargement. The term now usually denotes neoplasm, a non-inflammatory mass formed by the growth of new cells in the body and having no physiological function. An *innocent tumour* is encapsulated and usually solitary, pressing upon, but not invading, adjacent tissues; a *malignant tumour* (*carcinoma, sarcoma*) invades tissues, tends to recur and spreads to other parts of the body.

tunable magnetron *(Telecomm.)*. A **magnetron** in which the frequency can be altered electronically, e.g. by altering the anode voltage, or by mechanically changing the resonant frequencies of the cavities.

tunance *(Elec.Eng.)*. See **shunt resonance**.

tundra *(Ecol.)*. A biome which is essentially an Arctic grassland. The vegetation consists of lichens, grasses, sedges, and dwarf woody plants. It covers two large areas, one in the Palearctic, and one in the Nearctic region.

tundra *(Geol.)*. A plain region characterized by water-logged soil underlain by permafrost.

tune *(Acous.)*. To adjust for resonance or syntony, especially musical instruments or radio receivers. See also **tuning**.

tuned amplifier *(Telecomm.)*. One containing tuned circuits, and therefore sharply responsive to particular frequencies.

tuned anode *(Electronics)*. An inductor shunted by a capacitor (either or both of which may be variable) in series with the lead to the anode of a thermionic valve.

tuned-anode coupling *(Electronics)*. That between stages of a high-frequency thermionic valve amplifier, in which the coupling impedance is a tuned anode circuit.

tuned antenna *(Telecomm.)*. One operating at its natural resonant frequency.

tuned-base oscillator *(Electronics)*. One in which the *tuned circuits* are in series with the base of a transistor.

tuned cell *(Telecomm.)*. Adjustable cavity in a waveguide structure, particularly in a filter section.

tuned circuit *(Electronics)*. One comprising an inductor (L, henries) and a capacitor (C, farads) in series (parallel) which offers a low (high) impedance to alternating current at the **resonant frequency** given by

$$f = 1/2\pi\sqrt{LC}\ \text{Hz}.$$

See **Q**.

tuned-emitter oscillator *(Telecomm.)*. One in which the **tuned circuits** are in series with the emitter of a transistor.

tuned radiofrequency (TRF) receiver *(Telecomm.)*. Receiver which does not use frequency changing before detection.

tuned relay *(Telecomm.)*. One which responds only at a resonant frequency.

tuned transformer *(Elec.Eng.)*. Interstage coupling transformer in which one or more, usually both, windings are tuned to resonate with the signal frequency. A higher secondary voltage can be built up than would be the case without resonance.

tuner *(Telecomm.)*. (1) Assembly of one or more tuned circuits to form a unit sharply responsive to particular frequencies. (2) Term often used to describe the *front end* of a receiver; e.g. an *FM tuner* precedes the audio-amplifier for domestic entertainment, and the section of a TV receiver devoted to receiving and selecting the radio-frequency signal may be so described.

tungsten *(Chem.)*. Symbol W, at. no. 74, r.a.m. 183.85, rel.d. 19.1, mp 3370°C. A hard grey metal, resistant to corrosion, used in cemented carbides for drills and grinding tools, and as wire in incandescent electric lamps. Also *wolfram*.

tungsten alloy *(Eng.)*. A protective material containing tungsten, copper and nickel, and having a density about 50% greater than that of lead, and thus providing better protection from ionizing radiation.

tungsten arc *(Elec.Eng.)*. A high-intensity arc of small dimensions, obtained between tungsten electrodes enclosed in a glass bulb.

tungsten bronzes *(Chem.)*. Partially reduced WO_3; compounds of varying composition and colour with metallic lustre and conductivity. Similar compounds are formed by other elements capable of displaying two valencies.

tungsten-halogen lamp *(Elec.Eng.)*. See **quartz-iodine lamp**.

tungsten inert gas welding *(Eng.)*. Electric welding in which the tungsten electrode is not consumed and a 'filler rod' supplies the metal to the joint which is protected from reaction by an inert gas. Abbrev. *TIG*.

tungsten lamp *(Elec.Eng.)*. An electric lamp employing an incandescent tungsten filament.

tungstic acid *(Chem.)*. WO_3. The starting-point for the preparation of tungsten metal. Also called *tungsten oxide* and *tungsten (VI) oxide*.

tungstic ochre, tungstite *(Min.)*. Hydrated oxide of tungsten. It is usually earthy and yellow or greenish in colour, and is a mineral of secondary origin, usually associated with wolframite.

tunic *(Zool.)*. An investing layer. adj. *tunicate*.

tunica *(Bot.)*. Outer layer(s) of cells in the shoot apical meristem of many angiosperms, which give rise to the epidermis and which divide anticlinally and thus do not displace the underlying cells from the meristem. See **tunica-corpus concept**.

tunica-corpus concept *(Bot.)*. Concept that the shoot apex in many angiosperms is organized into a **tunica** and a **corpus** the distinctness of which is maintained more or less indefinitely. The concept accounts for the existence of **periclinal chimeras**.

Tunicata *(Zool.)*. See **Urochordata**.

tunicate bulb *(Bot.)*. A bulb composed of a number of swollen leaf bases each of which completely encloses the next younger, as in the onion.

tunicated *(Zool.)*. Enclosed by a nonliving test or mantle.

tunicate, tunicated *(Bot.)*. Having a coat or covering.

tuning *(Acous.)*. (1) The adjustment of tension in the strings of stringed instruments (piano, harp, violin) so that the specified notes emitted coincide in frequency with a standard scale, e.g. concert pitch. (2) The adjustment of the length of pipes in organs to obtain the correct emitted pitch.

tuning *(Elec.Eng.)*. See **current (or voltage) resonance**.

tuning *(Telecomm.)*. (1) Operation of adjusting circuit settings of a radio receiver so as to produce maximum response to a particular signal, generally by varying one or more capacitors and/or inductors. (2) Carrying out a similar process by electronic or thermal means. Also *tuning-in*.

tuning capacitor *(Telecomm.)*. Variable capacitor for tuning purposes, generally consisting of air-spaced vanes; several can be ganged.

tuning coil *(Telecomm.)*. See **tuning inductance**.

tuning control *(Telecomm.)*. Mechanical means for tuning a resonant circuit.

tuning curve *(Telecomm.)*. That relating the resonant frequency of a tuned circuit to the setting of the variable element, e.g. a capacitor.

tuning fork *(Acous.)*. A fork with two tines and heavy cross-section, generally made of steel. Expressly designed to retain a constant frequency of oscillation when struck. Widely used for tuning musical instruments because its frequency is very insensitive to changes in temperature, atmospheric pressure and humidity. See **maintained tuning fork**.

tuning-in *(Telecomm.)*. See **tuning**.

tuning indicator *(Telecomm.)*. See **magic eye**.

tuning inductance *(Telecomm.)*. Fixed or variable inductor used for tuning. Also called *tuning coil*.

tuning screw *(Electronics)*. A screw used to provide a variable reflection coefficient in a waveguide matching system. An alternative system to **stub tuning**.

tunnel burners *(Eng.)*. Industrial gas burners using a refractory tunnel at the burner exit for the main purpose of positive flame retention. The tunnel serves as an ignition zone, and accelerates the rate of flame propagation through turbulence and temperature rise, to a point where it is in equilibrium with the relatively high air-gas mixture velocity employed.

tunnel diode *(Electronics)*. Junction diode with such a thin depletion layer that electrons bypass the potential barrier (see **tunnel effect**). Negative resistance characteristics can be exhibited and such diodes can be used as low noise amplifiers or as oscillators, up to microwave frequencies. Also *Esaki diode*.

tunnel effect *(Electronics)*. Piercing of a narrow potential barrier by a current carrier which cannot do so classically, but, according to wave-mechanics, has a finite probability of penetrating.

tunnel furnace *(Eng.)*. Kiln through which material moves slowly on cars, racks, or suspending gear.

tunnelling *(Phys.)*. See **potential barrier**.

tunnel slots *(Elec.Eng.)*. See **closed slots**.

tunnel vault *(Build.)*. See **barrel vault**.

tunnel windings *(Elec.Eng.)*. A term sometimes applied to armature windings in which the conductors are inserted, end-on, into closed slots.

Turbellaria *(Zool.)*. A class of Platyhelminthes comprising forms of free-living habit, marine, freshwater, or terrestrial; with a ciliated ectoderm; usually with a muscular protrusible pharynx and a pair of eyespots; rarely have suckers. Planarians.

turbidimeter *(Powder Tech.)*. Equipment for determining the surface area of a powder by measuring the light scattering properties of a fluid suspension.

turbidimetric analysis *(Chem.)*. See **nephelometric analysis**.

turbidite *(Geol.)*. A sediment deposited from a turbidity current, frequently poorly sorted as in a *greywacke*, it often shows **graded bedding**.

turbidity *(Image Tech.)*. Property of a photographic emulsion whereby light is scattered by the silver halide grains in the immediate vicinity of the image.

turbidity current *(Geol.)*. A density flow of mixed water and sediment, capable of rapid movement downslope. See **turbidite**.

turbinal *(Zool.)*. Coiled in a spiral; one of certain bones of the nose in Vertebrates which support the folds of the olfactory mucous membrane.

turbinate *(Zool.)*. In the form of a whorl or an inverted cone; as certain Gastropod shells.

turbinate bone. *(Zool.)*. See **turbinal**.

turbine *(Aero.)*. See **axial-flow turbine**.

turbine aero-engine *(Aero.)*. See **by-pass turbojet, ducted fan, turbojet, turboprop**.

turbinectomy *(Med.)*. Removal of a turbinal.

turbocharger *(Autos.)*. Form of **supercharger**, used for internal combustion engines in which the power of the compressor comes from a turbine driven by the exhaust gases.

turbo-dynamo *(Elec.Eng.)*. A specially designed d.c. generator for direct coupling to a high-speed steam turbine.

turbo-electric propulsion *(Elec.Eng.)*. A form of electric drive, used in marine and locomotive work, in which turbine-driven generators supply electric power to motors coupled to the propeller or axle shafts.

turbofan *(Aero.)*. See **ducted fan**.

turbo-generator *(Elec.Eng.)*. The arrangement of a steam turbine coupled to an electric generator for electric power production.

turbojet *(Aero.)*. An internal-combustion aero-engine comprising compressor(s) and turbine(s), of which the net gas energy is used solely for reaction propulsion through propelling nozzle(s). See also **by-pass turbojet, ducted fan, split compressor**.

turboprop *(Aero.)*. A **shaft turbine** where the torque output is transmitted to a propeller through a reduction gearbox; it may be of *single shaft*, *twin shaft*, or *free turbine* form. A constant-power, or supercharged, turboprop has an oversize compressor/turbine assembly which enables it to maintain full power up to a considerable altitude.

turbopump *(Aero.)*. A combination **ram-air turbine** and hydraulic, or fuel, pump for a guided weapon or aircraft in emergency.

turboramjet *(Aero.)*. An engine consisting of a **turbojet**

mounted within a **ramjet** duct, so that the efficiency of the former in subsonic flight is combined with the advantages of the latter at high supersonic speeds.

turborocket *(Aero.)*. A composite engine in which a rocket propellant (an example would be high-test peroxide catalysed to super-heated steam and oxygen) is used to energize a turbine, which in turn drives a compressor, its air delivery joining the products from the turbine for combustion with a fuel to produce a propulsive jet. The object is, e.g. to obtain a high ceiling, say 100 000 ft (30 000 m), without the enormous propellant consumption of a rocket. It could be basis of a hypersonic space launching aircraft.

turbo-starter *(Aero.)*. An aero-engine starter in which rotation is imparted by a turbine motivated either by compressed air, a gas source, or the decomposition by catalysis of an unstable chemical, such as hydrogen peroxide.

turbo-supercharger *(Aero.)*. See **exhaust-driven supercharger**.

turbulence *(Phys.)*. See **turbulent flow**.

turbulent flow *(Phys.)*. Fluid flow in which the particle motion at any point varies rapidly in magnitude and direction. This irregular eddying motion is characteristic of fluid motion at high Reynold's numbers. Gives rise to high drag, particularly in the **boundary layer** of aircraft. Also *turbulence*. See **streamline flow**.

turgescence *(Med.)*. The act or condition of swelling up: the state of being swollen. adj. *turgescent*.

turgid *(Bot.)*. (1) A cell which is distended and stiff as a result of the osmotic uptake of water, having a positive turgor pressure. (2) A non-woody tissue which is stiff as a result of the cell's being turgid. Cf. **flaccid**.

turgite *(Min.)*. See **hydrohematite**.

turgor movement *(Bot.)*. Movement of a plant part resulting from changes in the turgor of its cells or the cells of its support. See also **pulvinus**. Cf. **growth movement**.

turgor potential *(Bot.)*. *Pressure potential*, ψ_P. That component of the **water potential** due to the hydrostatic pressure; equal to the **turgor pressure**. An important component in turgid cells and in the xylem.

turgor pressure *(Bot.)*. The hydrostatic pressure of the contents of a cell; normally positive in most plant cells; normally negative in the conducting cells of the xylem of transpiring plants.

Turing machine *(Comp.)*. A finite state automaton with an unbounded memory. It is an abstract computer and is used to define the concept of **computability**.

Turkey-red oil *(Chem.)*. Sulphonated castor oil, rel.d. 0.95, acid value 174, iodine value 82, saponification value 189. Used in dyeing.

turn *(Glass)*. A work shift in which a definite number of articles, usually two **moves**, is produced.

turn *(Print.)*. See **taking out turns**.

turn-and-slip indicator *(Aero.)*. A pilot's instrument for blind flying which indicates the rate of turn and sideslip, or error, in banking; also *turn-and-bank indicator*.

turnaround document *(Comp.)*. Document which, after being output by the computer, can be used to record data; this data can then be input to the computer using a document reader.

turnbuckle *(Eng.)*. See **screw shackle**.

turned sorts *(Print.)*. Characters purposely turned face-downwards so that the feet print prominent black marks in a proof, thus ensuring that missing letters shall be inserted later.

turner bars *(Print.)*. See **angle bars**.

Turner's syndrome *(Biol.)*. A condition in humans in which a person looks superficially like a female but has only one X-chromosome.

turn indicator *(Aero.)*. Any instrument that indicates the departure of an aircraft from its set course in a horizontal plane. Necessary for flying in clouds or at night.

turning *(Build.)*. A term applied to the process of building an arch.

turning *(Eng.)*. Producing cylindrical, flat or tapered workpieces in a lathe.

turning-bar *(Build.)*. An iron bar supporting the arch over a fireplace opening.

turning-piece *(Build.)*. A simple form of centring, consisting of a single solid wooden piece shaped to the form of an intrados, and supported in its temporary position by wooden struts.

turning point *(Surv.)*. (1) The point which consecutive straight lines of a traverse meet at an angle. (2) See **change point**.

turning point on a curve *(Maths.)*. A peak (maximum) or a trough (minimum) on a curve. For the curve $y = f(x)$ the point where $x = a$ is a turning point if $f(a+h) - f(a)$ is of constant sign for all values of h sufficiently small.

turnings *(Eng.)*. Chips or swarf produced as waste in turning.

turning-saw *(Build.)*. See **sweep-saw**.

turning tools *(Eng.)*. See **lathe tools**.

turnkey system *(Comp.)*. A complete computerized system to meet a customer's specification. It may include software, hardware, ancillary equipment and staff.

turnout *(Civ.Eng.)*. The movable tapered rails or points by which a train or tram is directed from one set of rails to another. Also *switch, point and crossing*.

turnover *(Phys.)*. In isotope separation, the total flow of material entering a given stage in a cascade.

turnover *(Print.)*. The part of a divided word turned over into the next line, or a short line at the end of a paragraph.

turnover *(Radiol.)*. Rate of renewal of a particular chemical substance in a given tissue.

turnover board *(Foundry)*. A smooth square board on which an inverted bottom-half box is placed and rammed up round a pattern having a flat joint, thus saving the labour of making the facing joint. After turning over, removing the board, and adding facing sand, the top half may be rammed up at once.

turnover frequency *(Acous.)*. In disk recording, the frequency, generally between 200 and 500 Hz, where the change from constant-amplitude to constant-velocity recording takes place. Also called *cross-over frequency*.

turns *(Horol.)*. A small dead-centre lathe used by watchmakers. Usually held in a vice, and driven by a hand wheel or a bow. Used for pivoting, polishing, and turning small parts.

turnsick *(Vet.)*. See **coenuriasis**.

turns ratio *(Elec.Eng.)*. Ratio of turns on any pair of windings of a transformer; usually has symbol N. Primary and secondary voltages and currents are related by N, and 1/N respectively; thus impedances are related by N^2. See **transformer**, **transformation ratio** and **voltage ratio**.

turnstile antenna *(Telecomm.)*. Two normal dipoles, crossed over at their centre, driven with equal currents in quadrature.

turntable *(Acous.)*. The rotating table which supports the lacquer-blank during cutting and the processed record while being reproduced. It is of relatively high inertia, to keep down fluctuations of speed.

turntable *(Civ.Eng.)*. A circular platform capable or rotation about its centre; used to reverse steam locomotives, which are driven on, turned through a half-circle, and driven off pointing the opposite way. In general, any such rotating platform, or system of rings rotating one inside the other.

turn tread *(Build.)*. A tread, generally triangular in plan, to form a step at a change of direction of the stair.

turpentine *(Chem.)*. An essential oil, $C_{10}H_{16}$, obtained by the steam distillation of rosin. It is a colourless liquid, of aromatic pine-like odour; bp 155°–165°C, rel.d. 0.85–0.91; the chief constituent is pinene. American turpentine is dextrorotatory, others are usually laevorotatory. An important solvent for lacquers, polishes etc. *Turpentine substitute* is a petroleum fraction of similar boiling point.

turquoise *(Min.)*. A hydrated phosphate of aluminium and copper which crystallizes in the triclinic system. It is a mineral of secondary origin, found in thin veins or small masses in rocks of various types, and used as a gemstone.

The typical sky-blue colour often disappears when the mineral is dried. Much of the gem turquoise of old was fossil bone of organic origin and not true turquoise.

turret *(Eng.)*. A turntable or wheel for carrying a number of alternative tools, e.g. in a turret lathe or a turret press.

turret *(Image Tech.)*. A rotatable mounting for several lenses on a camera or projector, allowing rapid changes between different focal lengths.

turret clock *(Horol.)*. A tower clock; a large clock in which the movement is quite separate from the dials.

turret lathe *(Eng.)*. A high-production lathe for long workpieces, using a large number of tools carried on the revolving tool-holder or turret and on the cross slide. The turret is mounted on a saddle which slides on the lathe bed.

turret press *(Eng.)*. A power press in which pairs of punches and dies of various sizes are held in upper and lower turrets, the turrets being geared together to bring corresponding punches and dies into positions of exact alignment in which they are then locked.

turtle *(Comp.)*. Drawing device used by **LOGO** and related languages. It may be an electromechanical device drawing on the floor (floor turtle), or may be simulated by graphics on a **VDU** screen (screen turtle).

turtle-shell *(Zool.)*. The horny plates of the hawk's bill turtle. Commonly *tortoise-shell*.

tusks *(Build.)*. See **tusses**.

tusk tenon *(Build.)*. A form of tenon used for framing one horizontal piece into another, e.g. a trimmer into a trimming joist. The tenon is strengthened by a short projection underneath, and by a bevelled shoulder above, both fitting into a suitably cut mortise in the other piece.

tussah silk *(Textiles)*. A coarse silk produced by a wild silkworm e.g. *Antheraea mylitta* (from India) and *A. pernyi* (from China). The fibres are pale brown and are usually rather short. They are spun into a yarn which has many irregular **slubs**.

tusses *(Build.)*. Stones left projecting from the face of a wall, when later extension is allowed for. Also called *tusks*.

tussive *(Med.)*. Pertaining to, or caused by, a cough.

tussore *(Textiles)*. A fabric woven from tussah silk.

tuyère, twyere *(Eng.)*. A nozzle through which air is blown into a blast furnace. May be kept cool by circulating water.

TV viewfinder *(Image Tech.)*. See **video viewfinder**.

TW antenna *(Telecomm.)*. See **travelling-wave antenna**.

tweed *(Textiles)*. Traditionally a coarse, heavy, rough wool outerwear fabric. Now applies to other wool fabrics having a wide range of weights and weave effects (e.g. **twills**).

tweeter *(Acous.)*. A loudspeaker used in high-fidelity sound reproduction for the higher frequencies (>5 kHz). See **cross-over frequency, woofer**.

12mo *(Print.)*. The twelfth of a sheet or a sheet folded to make twelve leaves or twenty-four pages. Also *duodecimo*.

twenty-four hour rhythm *(Behav.)*. See **circadian rhythm**.

24mo *(Print.)*. The twenty-fourth part of a sheet, or a sheet folded four times to make twenty-four leaves of forty-eight pages.

21 centimetre line *(Astron.,Phys.)*. A line in the radio spectrum of neutral hydrogen at 21.105 cm. It is caused by the spontaneous reversal of direction of spin of the electron in the magnetic field of the hydrogen nucleus, but it may be detected only in the vast hydrogen clouds of the Galaxy.

twilight *(Astron.)*. The period after sunset, or before sunrise, when the sky is not completely dark. Astronomical twilight is defined as beginning (or ending) when the sun is 18° below the horizon; hence twilight will last all night for a period in the summer months in all latitudes greater than about 48°. See **civil, nautical**.

twilight sleep *(Med.)*. A state of semi-consciousness produced by the administration of morphine and scopolamine.

twills *(Textiles)*. Woven fabrics with diagonal lines on the face. Regular twills have continuous lines; zigzag twills have the lines reversed at intervals.

twin *(Genrl.)*. One of a pair of two and related entities similar in structure or function; often synonymous with *double*. See also **twins**.

twin cable *(Elec.Eng.)*. A cable comprising two individually insulated conductors twisted together. A twin cable for telecommunication may have a large number of such pairs, e.g. up to 2400 pairs for telephone connections between large exchanges.

twin check *(Comp.)*. Continuous check achieved by duplication of *hardware* and comparison of results. See **redundancy**.

twin columns *(Arch.)*. Two columns springing from one base.

twin-concentric cable *(Elec.Eng.)*. A 2-core cable in which the conducting cores are concentrically arranged about the axis of the cable.

twin crystal, twinned crystal *(Crystal.,Min.)*. A crystal composed of two or more individuals, either in contact or intergrown, in a systematic crystallographic orientation with respect to one another. See **interpenetration twins, juxtaposition twins**.

twiner *(Bot.)*. A plant that climbs by winding around a support.

twin feeder *(Telecomm.)*. A transmission line, leading to or from an antenna, consisting of two parallel conductors. In high-power transmitter applications, the two wires may be separated by insulating rods; alternatively they may be moulded into solid polythene. Impedance is determined by conductor diameter, spacing and the dielectric used.

twin lamb disease *(Vet.)*. See **pregnancy toxaemia**.

twin lens reflex camera *(Image Tech.)*. A camera with matched lenses, one for exposing, the other for focusing, generally with a reflex mirror.

twinning *(Crystal.)*. Intergrowth of crystals of near-symmetry, such that (in quartz) the piezo-electric effect is not sufficiently determinate. See **twin crystal**.

twinning *(Image Tech.)*. See **pairing**.

twin-plate process *(Glass)*. A process for making polished plate glass in which rolling, annealing and grinding are carried out on a continuously produced ribbon of glass without first cutting it into sections in which top and bottom surface are ground simultaneously. Also *Pilkington twin process*.

twin-quad *(Elec.Eng.)*. See **quad**.

twins *(Biol.)*. (1) *Identical twins* arise from the same fertilized egg which has subsequently divided into two, each half developing into a separate individual. (2) In Mammals, *non-identical twins* are produced from separate eggs fertilized at the same time.

twin-shaft turbine *(Aero.)*. See **split compressor**.

twin-T network *(Telecomm.)*. One consisting of two T-networks which have their input terminal pairs connected in parallel and their output terminal pairs connected in parallel.

twin triode *(Electronics)*. A combination of 2 triode valves within the same envelope.

twist *(Textiles)*. Fibres in a yarn are held together by the degree of twist introduced in spinning. This may be quantified as the twist level, i.e. the number of revolutions per unit length. Lively yarns have a tendency to untwist.

twist and steer *(Aero.)*. Control of a guided weapon or *drone* about the pitch and roll axes only, turns being achieved by rolling into a bank so that the elevator can provide the required turning moment. The system simplifies the autopilot and power requirements and is sometimes used with differentially-mounted variable-incidence wings.

twist bit *(Build.)*. Bit with long spiral cutting section, used for deep holes for dowels etc.

twist direction *(Textiles)*. If the direction of twist as viewed on a yarn held vertically goes diagonally from left to right it has S-twist; if from right to left it has Z-twist.

twist drill *(Eng.)*. A hardened steel drill in which cutting edges, of specific rake, are formed by the intersection of helical flutes with the conical point which is backed off to give clearance; of universal application.

twisted aestivation *(Bot.)*. Same as *contorted aestivation*.

twister *(Radar)*. Plate with slats giving double reflection of a radar wave, one being half-wave retarded, to give a twist in direction of polarization of electric component of wave.

twisting paper *(Paper)*. A long fibred paper suitable for waxing and intended for wrapping around sweets, toffee etc.

twitch *(Vet.)*. A noose for compressing the lip of a horse as a means of restraint.

two-address program *(Comp.)*. Was used in early computers where each instruction had to include the address of two registers, one for the operand and one for the result of the operation.

two-body force *(Phys.)*. A type of interaction between 2 particles which is unmodified by the presence of other particles.

two-circuit prepayment meter *(Elec.Eng.)*. A prepayment meter for use when the load is connected to 2 separate circuits, energy being charged at a different rate in each.

two-circuit winding *(Elec.Eng.)*. An alternative name for *wave winding*.

two-coat work *(Build.)*. Plastering in 2 coats; a first coat of coarse stuff, and a second coat of fine stuff.

two-colour process *(Image Tech.)*. A colour process recording and reproducing only two broad regions of the spectrum, usually blue-green and orange-red; now effectively obsolete.

two-colour process *(Print.)*. The application of the subtractive process to printing for the reproduction of a 2-colour original.

two-dimensional gas *(Chem.)*. A unimolecular film whose behaviour in 2 dimensions is analogous, qualitatively and quantitatively, to that of an ordinary gas in 3 dimensions.

2E *(Min.)*. Apparent optic axial angle as measured in air.

two-electrode valve *(Electronics)*. See **diode**.

two-group theory *(Nuc.Eng.)*. Simplified treatment of neutron diffusion in which only two energy groups are considered, i.e. partly thermalized neutrons are neglected.

two-light frame *(Build.)*. A window frame having 1 mullion dividing the window space into 2 compartments.

two-magazine mixer *(Print.)*. Certain models of **Linotype** and **Intertype** are able to mix in 1 line matrices from 2 adjacent magazines, thus making available up to 9 alphabets.

two-motion selector *(Telecomm.)*. An electromechanical selector having vertical and rotary motions.

two pack materials *(Build.)*. Paints or fillers which are supplied in separate containers consisting of a base and a catalyst or activator. When these materials are mixed, gellation takes place rapidly and the pot life of the mixture is very short. This chemical process is known as curing, e.g. of epoxy ester filling.

two-part coatings *(Build.)*. Compositions of brushing viscosity which 'dry' by the reaction of two parts, mixed immediately before use.

two-phase *(Elec.Eng.)*. A term applied to a.c. systems employing 2 phases, whose voltages are displaced from one another by 90 electrical degrees.

two-phase four-wire system *(Elec.Eng.)*. A system of 2-phase a.c. distribution employing 2 conductors per phase.

two-phase three-wire system *(Elec.Eng.)*. A system of 2-phase a.c. distribution in which 2 conductors (lines) belong 1 to each phase, and the 3rd (neutral) is common to both phases.

two-pipe system *(Build.)*. One in which soil and waste discharges are piped separately with or without ventilating pipes depending on system size.

two-reaction theory *(Elec.Eng.)*. A theory used in calculations on salient-pole synchronous machines; the m.m.f.s in the machine are assumed to be divided into 2 components, one acting along the axis of the main poles, the other at 90° to this.

two-revolution *(Print.)*. Type of letterpress machine in which the cylinder revolves continuously, making 2 revolutions while the carriage reciprocates once; the cylinder is pulled down on the bearers for the printing revolution, and rises clear of the forme during a second revolution, while the forme returns to the printing position. Cf. **single-revolution, stop-cylinder**.

two's complement *(Comp.)*. Formed from a binary number, it is always one greater than the corresponding **one's complement** (e.g. −43 is 11010101). It is possible to effect subtraction by addition, using two's complement.

two set *(Print.)*. Plating a rotary press with two sets of plates to produce two copies for each cylinder revolution.

two-spool compressor *(Aero.)*. See **split compressor**.

two-stage pressure-gas burner *(Eng.)*. Natural draught type designed for operating with gas under pressure, normally about 35 kN/m^3, and having primary and secondary air inspirating stages in the injector.

two-start thread *(Eng.)*. See **double-threaded screw**.

two-step relay *(Telecomm.)*. Telephone relay which is partially operated by a weak current, and so makes an *x-contact* or *fly-contact*, thereby closing a winding in a local circuit. This passes sufficient current for full operation of the remaining contacts of the relay, which *locks*.

two-stroke cycle *(Autos.)*. An engine cycle completed in 2 piston strokes, i.e. in 1 crank-shaft revolution, the charge being introduced by a blower or other means, compressed, expanded, and exhausted through ports in the cylinder wall, before and during the entry of the fresh charge. See **Otto cycle, diesel cycle**.

two-terminal pair network *(Telecomm.)*. See **quadripole**.

two-tone keying *(Telecomm.)*. Keying of modulated continuous wave through a circuit which changes the modulation frequency only.

two-to-one folder *(Print.)*. On a web-fed press a type of folder in which the folding cylinder has a circumference of two cut-offs and the cutting cylinder 1 cut-off giving a ratio of cylinder sizes 2:1. Cf. **three-to-two folder**.

two-up *(Print.)*. A printing surface made up to print 2 copies at 1 impression; can also be arranged for 3-up, 4-up or any suitable number.

2V *(Min.)*. The optic axial angle when measured in the mineral.

two-way circuit *(Telecomm.)*. Bi-directional channel, which operates stably in either direction.

two-wire circuit *(Telecomm.)*. A circuit in which go and return wires take equal currents, with potentials balanced with respect to earth.

two-wire system *(Elec.Eng.)*. A system of d.c. transmission and distribution making use of 2 conductors.

twyere *(Eng.)*. See **tuyère**.

tye *(Min.Ext.)*. *Strake* in which a considerable thickness of low-grade concentrate is collected.

Tyler sieves *(Min.Ext.)*. Widely used series of laboratory screens in which mesh sizes are in $\sqrt{2}$ progression with respect to linear distance between wires.

tylosis, tylose *(Bot.)*. Bladder-like expansion of the wall of a living parenchyma cell through a pit into the lumen of a xylem tracheid or vessel. Tyloses apparently form in non-functional conduits after spontaneous or wound-induced **embolism**, and may restrict the spread of pathogens.

tympan *(Print.)*. In a hand-press, the frame on which the paper is placed when printing. In printing machines the sheets of paper used to adjust impression, the outermost being tympan paper (a strong sulphite or manilla), oiled or plain.

tympan hooks *(Print.)*. In a hand-press, thumb-hooks used for locking the outer and inner tympans together.

tympanic bulla *(Zool.)*. In some Mammals, a bony vesicle surrounding the outer part of the tympanic cavity and external auditory meatus formed by the expansion of the tympanic bone.

tympanites *(Med.)*. Distension of the abdomen by accumulation of gas in the intestines or in the peritoneal cavity.

tympanites *(Vet.)*. Rapid distension of the rumen and reticulum of cattle, due to the formation of gases.

tympanum *(Arch.)*. The triangular or segmental space forming the central panel of a pediment.

tympanum *(Zool.)*. A drumlike structure; in some Insects, the external vibratory membrane of a chordotonal organ; in some Birds an inflatable air-sac of the neck-region; in Vertebrates, the middle ear, or the resonating membrane of the middle ear; in Birds, the resonating sac of the syrinx. adjs. *tympanic, tympanal*.

Tyndall effect *(Phys.)*. Scattering of light by very small particles of matter in the path of the light, the scattered light being mainly blue.

tyndallimetry *(Chem.)*. The determination of the concentration of suspended material in a liquid by measurement of the amount of light scattered from a Tyndall cone. See **nephelometric analysis**.

type *(Biol.)*. The individual specimen on which the description of a new species or genus is based; the sum-total of the characteristics of a group. adj. *typical*.

type *(Print.)*. A rectangular piece of metal on top of which is cast the character. There are several styles, each further subdivided, including: **Slab Serif, Sans Serif**.

type face *(Print.)*. A particular family or fount of type in which the characters have distinctive features. Type faces used in modern English bookwork include Aldine Bembo, Baskerville, Caslon, Ehrhardt, Fournier, Garamond, Gill Sans, Imprint, Perpetua, Plantin, Times Roman, Univers. Each type face has its own special characteristics.

type family *(Print.)*. A range of variations on the basic type design; light, medium, bold, extra bold, condensed etc. a notable example being ***Gill Sans***.

type-high *(Print.)*. When a printing plate or block is mounted on wood or metal and brought to the proper height for printing it is said to be *type-high*, the British/American standard being 0.918 in (23.32 mm). n. *type-height*.

type holder *(Print.)*. A hand tool for holding the letters to be impressed on the cover of a book by the finisher.

type locality *(Geol.)*. The locality from which a rock, formation, fossil, etc. has been named and described.

type metal *(Eng.)*. A series of alloys of lead, antimony and tin, used for type. One composition is antimony 10–20%, tin 2–12%, and the remainder lead. In another, tin may reach 26%, and up to 1% of copper may be added.

typesetting machines *(Print.)*. See **composing machines**.

type specimen *(Biol.)*. The actual specimen from which a given species was first described.

type II superconductor *(Phys.)*. A superconducting *alloy*, e.g. niobium-titanium, which behaves in a different way from usual superconductivity in that some magnetic flux can penetrate the material. These alloys have usually a relatively high critical temperature and magnetic field.

typewriter composition *(Print.)*. The use of electric typewriters to produce typesetting for reproduction by printing processes. The product has lower typographic standards than normal typesetting or the major phototypesetting systems but usually has justified lines and a difference of set between wide and narrow letters. Now effectively superseded by **word processors** and **desk-top publishing** systems.

typhlitis *(Med.)*. Inflammation of the caecum.

typhlosole *(Zool.)*. In some Invertebrates, a longitudinal dorsal inwardly projecting fold of the wall of the intestine, by which the absorptive surface is increased.

typhoid, typhoid fever *(Med.)*. The most serious form of enteric fever (which includes *paratyphoid*) due to infection with the bacillus *Salmonella typhi*. Prolonged fever, a rose rash and inflammation of the small intestine with ulceration occur. Infection is by faecal contamination of food or water.

typhoon *(Meteor.)*. A **tropical revolving storm**, in the China Sea and western North Pacific.

typhus, typhus fever *(Med.)*. Infection with the rickettsa *Rickettsia prowazeki* derived from the bite of lice, an acute epidemic fever with high mortality. There are varieties of lesser severity (scrub t., murine t., etc.) due to related organisms.

typical intensity *(Behav.)*. The high degree of stereotyping observed in many patterns of behaviour that have a communicative function.

typographer *(Print.)*. A specialist in typographic design who may also be a compositor.

typographic quality *(Print.)*. A term describing text that is equivalent to that produced by an experienced compositor using normal printer's characters, and which complies with the normal rules of good typography.

typography *(Print.)*. The art of arranging the printed page, including choice of type, illustration, and method of printing.

Tyr *(Chem.)*. Symbol for tyrosine.

tyre *(Eng.)*. (1) A renewable, forged steel, flanged ring, shrunk on the rim of a locomotive wheel. (2) A steel or rubber band or air-filled tube of rubber (*pneumatic tyre*) surrounding a wheel to strengthen it or to absorb shock. See **cross-ply, radial-ply**, and **tubeless tyre**.

tyre textiles *(Textiles)*. Pneumatic tyres are made from composite materials of rubber reinforced with several layers of textile fabrics or steel wire. The fabrics are mostly made of synthetic fibres and include some, the tyre cord, where the warp threads predominate and the few weft threads are present only to assist processing. Radial and cross-ply denote the way the cords are arranged in forming the carcase.

tyrosine *(Chem.)*. *2-amino-3-(4-hydroxypheny)propanoic acid*, $OH.C_6H_4.CH_2.CH(NH_2).COOH$. An amino acid, an oxidation product of **phenylalanine**. The L- or S-isomer is a constituent of proteins. Symbol Tyr, short form Y. ⇨

Tzaneen disease *(Vet.)*. A febrile disease, usually mild, of cattle and buffalo in Africa, caused by the protozoon *Theileria mutans*; transmitted by ticks.

U

u *(Genrl.)*. Symbol for unit of unified scale.

u *(Chem.)*. (1) With subscript, a symbol for velocity of ions. (2) Symbol for *specific internal energy*.

U *(Chem.)*. The symbol for uranium.

U *(Chem.)*. A symbol for internal energy.

U *(Phys.)*. Symbol for (1) potential difference, (2) tension.

U-bend *(Build.)*. See **air trap**.

ubiquinone *(Biol.)*. Small highly mobile electron carrier mediating the transfer of electrons from flavoprotein to cytochrome in the *electron transfer chain*.

ubiquitin *(Biol.)*. Polypeptide of wide distribution in both pro- and eukaryotes. It is attached to proteins prior to their degradation in the course of cellular protein turnover.

U-bolts *(Autos.)*. Bars bent into U-shape and threaded at each end; used for anchoring a semi-elliptic spring to an axle beam, a plate being threaded over the ends and secured by nuts.

UCR *(Behav.)*. Abbrev. for *UnConditioned Reflex*. See also **classical conditioning**.

UCS *(Behav.)*. Abbrev. for *UnConditioned Stimulus*.

udder *(Vet.)*. The popular name for the mammary glands of certain animals, e.g. of the cow, mare, sow, and ewe.

UF *(Plastics)*. Abbrev. for *Urea-Formaldehyde* plastics (see **urea resins**).

UFO *(Astron.)*. Abbreviation for Unidentified Flying Object, applied to any sighting in the sky which the observer is unable to account for in terms of known phenomena.

Uganda mahogany *(For.)*. Hardwood from *Khaya anthotheca*. The timber seasons well and is used for cabinet-making, veneering, plywood manufacture, high-class joinery and furniture.

ugrandite *(Min.)*. A group name for the *u*varovite, *gr*ossular and *and*radite garnets.

UHF *(Telecomm.)*. Abbrev. for *Ultra-High Frequencies*.

Uhuru *(Astron.)*. The name of the first X-ray astronomy satellite. Launched from Kenya in 1970, it made the first good map of the X-ray sky.

uintaite *(Min.)*. A variety of natural asphalt occurring in the Uinta Valley, Utah, as rounded masses of brilliant black solid hydrocarbon. Also called *gilsonite*.

ULA *(Comp.)*. See **uncommitted logic array**.

Ulbricht sphere photometer *(Phys.)*. A photometer for measuring directly a lamp's mean-spherical candle-power. It comprises a hollow sphere, whitened inside, with the lamp under test at the centre. Owing to the internal reflection, the illumination on any part of the sphere's inside surface is proportional to the lamp's total light output, measured through a small window.

ulcer *(Med.)*. A localized destruction of an epithelial surface (e.g. of the skin or of the gastric mucous membrane), forming an open sore; it is usually a result of infection.

ulceration *(Med.)*. The process of forming an ulcer; the state of being ulcerated.

ulcerative *(Med.)*. Of the nature of, or pertaining to, ulcers; causing ulceration; associated with ulceration (e.g. *ulcerative colitis*).

ulcerative cellulitis *(Vet.)*. See **ulcerative lymphangitis**.

ulcerative dermal necrosis *(Vet.)*. A fungus disease of salmon.

ulcerative lymphangitis *(Vet.)*. *Ulcerative cellulitis*. A chronic contagious lymphangitis of the horse, due to infection by *Corynebacterium pseudotuberculosis (C. ovis.)*.

ulexite *(Min.)*. A hydrated borate of sodium and calcium occurring in borate deposits in arid regions, as in Chile and Nevada, where it forms rounded masses of extremely fine acicular white crystals. Also called *cotton ball*.

uliginose, uliginous *(Bot.)*. Growing in places which are wet.

U-links *(Telecomm.)*. The spring links used to join isolated sections of communication channels, the ends of which are brought to a special link-board. Removal of a link opens the circuit, so that test equipment can be rapidly inserted.

ullage *(Ships)*. The spare capacity of a partially full container, e.g. a tank for liquid cargo or fuel.

ullmanite *(Min.)*. Nickel antimony sulphide. It crystallizes in the cubic system and occurs in hydrothermal veins.

ulna *(Zool.)*. The post-axial bone of the antebrachium in land Vertebrates. adj. *ulnar*.

ulotrichous *(Zool.)*. Having woolly or curly hair.

ultimate analysis *(Chem.)*. **Quantitative analysis** of the elements in the materials being examined.

ultimate limit switch *(Elec.Eng.)*. See **final limit switch**.

ultimate load *(Aero.)*. The maximum load which a structure is designed to withstand without a failure; see also **load factor, limit load** and **proof load**.

ultimate tensile stress *(Eng.)*. The highest load applied to a metal in the course of a tensile test, divided by the original cross-sectional area. In brittle or very tough metals it coincides with the point of fracture, but usually extension continues under a decreasing stress, after the ultimate stress has been passed. Also called *tenacity*.

ultor *(Electronics)*. Anode, especially in a cathode-ray tube, which has highest potential with respect to cathode. Also *second anode*.

ultra- *(Genrl.)*. Prefix from L. *ultra*, beyond.

ultra-basic rocks *(Geol.)*. Igneous rocks containing less silica than the basic rocks (i.e. formerly less than 45%), and characterized by a high content of mafic constituents, particularly olivine (on the peridotites) and amphiboles and pyroxenes (in the perknites and picrites). The percentage varying with the petrologist.

ultracentrifuge *(Biol.)*. A high-speed centrifuge much used for molecular separation and capable of creating forces up to 500 000 times gravity.

Ultrafax *(Telecomm.)*. TN for US system for very high speed transmission of printed information, which utilizes radio, TV facsimile and film recording.

ultra-filtration *(Chem.)*. The separation of colloidal or molecular particles by filtration, under suction or pressure, through a colloidal filter or semi-permeable membrane. Cf. **reverse osmosis**.

ultra-high frequencies *(Telecomm.)*. Those frequencies between 3×10^8 Hz and 3×10^9 Hz. Abbrev. *UHF*.

ultralinear *(Elec.Eng.)*. Said of a power amplifier when nonlinear distortion is reduced to a very low value.

ultramafic rocks *(Geol.)*. Igneous rocks in which there is an abnormally high content of ferromagnesian silicates, but which contain no feldspar; subdivided into picrites (with accessory plagioclase), pyroxenites, and peridotites.

ultramicroscope *(Phys.)*. An instrument for viewing particles too small to be seen by an ordinary microscope, e.g., fog or smoke particles. An intense light projected from the side shows, against a dark background, the light scattered by the particles.

ultramicrotome *(Biol.)*. A modified microtome developed for cutting ultra-thin sections for examination with the electron microscope. The cutting surface may be of steel, but is more usually glass or diamond, and the movement of the specimen block towards the knife is very delicately controlled, e.g. by the thermal expansion of a rod.

ultrasonic *(Acous.)*. Said of frequencies above the upper limit of the normal range of hearing, at or about 20 kHz. *Ultrasonics* is the general term for the study and application of ultrasonic sound and vibrations.

ultrasonic cleaning *(Eng.)*. A cleaning process used in conjunction with water or solvents and effective for small crevices, blind holes, etc. Ultrasonic frequency vibrations are transferred to the cleaning fluid producing a turbulent penetrating action.

ultrasonic cleaning *(Phys.)*. The separation of dirt and other foreign particles from a substance by subjecting it to ultrasonic irradiation. Solid bodies such as glass lenses can be polished in this way.

ultrasonic coagulation *(Phys.)*. Under suitable conditions, coalescence of particles into large aggregates by ultrasonic irradiation.

ultrasonic delay line *(Telecomm.)*. A device which utilizes the finite time for the propagation of sound in liquids or solids to produce variable time delays. Such systems may also be used for storage in digital computers. Mercury or quartz are used as transmitting media, and nickel wire for magnetostrictive delay lines.

ultrasonic depth finder *(Eng.)*. Instrument for measuring or displaying the depth of water under a ship by measuring the time of propagation of a pulse of ultrasonic waves to the sea bed and back.

ultrasonic detector *(Telecomm.)*. Electro-acoustic transducer for the detection of ultrasonic radiation.

ultrasonic dispersion *(Chem.)*. High-intensity ultrasonic waves can produce a dispersion of one medium in another, e.g. mercury in water.

ultrasonic generator *(Elec.Eng.)*. One for the generation of ultrasonic waves, e.g. quartz crystal, ceramic transducer, supersonic air jet, magnetostrictive vibrator.

ultrasonic grating *(Phys.)*. The presence of acoustic waves in a medium leads to a periodic spatial variation of density which produces a corresponding variation of refractive index. Diffraction spectra can be obtained in passing a light beam through such a sound field.

ultrasonic machining *(Eng.)*. Process of removing material by abrasive bombardment and crushing in which a relatively soft tool, matching the shape produced, is made to oscillate at ultrasonic frequencies and drives the abrasive grit, suspended in a liquid, against the work, blasting away fine particles of the material. Operations include drilling round and odd-shaped holes, diesinking and forming wire drawing dies. Primarily for machining hard brittle materials like carbides, ceramics and sintered metals.

ultrasonics *(Acous.)*. See **ultrasonic**.

ultrasonic soldering *(Eng.)*. Form of soldering in which a specially designed soldering bit replaces the machine tool. Soldering is particularly difficult with aluminium and the application of ultrasonics is supposed to break up the aluminium oxide layer.

ultrasonic stroboscope *(Eng.)*. One in which an ultrasonic field is applied to the modulation of a light beam to obtain stroboscopic illumination. See **stroboscope**.

ultrasonic testing *(Eng.)*. A method of testing in which an ultrasonic source is pressed against the part to be tested and the vibrations are reflected from a discontinuity or another surface of the part, to generate a signal in the receiver. The time lag of the echo is measured to determine the distance to the discontinuity or the workpiece thickness.

ultrasonic welding *(Eng.)*. A solid-state process for bonding sheets of similar or dissimilar materials, usually with a lap joint. No heat is applied but vibratory energy at ultrasonic frequencies is applied in a plane parallel to the surface of the weldment.

ultrasonography *(Radiol.)*. Use of reflected high-frequency sound waves to image organs of the body. Widely used in the diagnosis of disease of the abdomen and heart and in the management of pregnancy.

ultrasound *(Acous.)*. See **sound**. Used by some animals (e.g. bats, dolphins) for localization and communication, and in a variety of industrial applications.

ultrastructure *(Biol.)*. The submicroscopic structure of a cell; particularly as shown by the electron microscope.

ultraviolet *(Image Tech.)*. Photography using radiations beyond the blue-violet end of the visible spectrum, in practice generally between 400 and 200 nm; of special application in forensic and medical recording and for measuring the distribution of nucleic acids in cells.

ultra-violet astronomy *(Astron.)*. The detection and analysis of radiation from cosmic sources at wavelengths between 25 and 350 nm. The hottest stars emit the bulk of their radiation in this waveband.

ultraviolet cell *(Phys.)*. A cell having a maximum response to light in the ultraviolet end of the spectrum.

ultraviolet microscope *(Biol.)*. Instrument using ultraviolet light for illuminating the object. Its resolving power is therefore about doubled as the resolution varies inversely with the wavelength of the radiation, but more usefully the nucleic acids absorb strongly in this region and can therefore be localized and measured.

ultraviolet radiation *(Phys.)*. Electromagnetic radiation in a wavelength range from 400 nm to 10 nm approximately; i.e. between the visible and X-ray regions of the spectrum. The *near* ultraviolet is from 400 to 300 nm, the *middle* from 300 to 200 nm and the *extreme* from 200 to 190 nm.

ultraviolet spectrometer *(Phys.)*. An instrument similar to an optical spectrometer but employing nonvisual detection and designed for use with ultraviolet radiation.

ultraviolet spectroscopy *(Chem.)*. The study of material by its absorption of ultraviolet radiation. It gives information about the energies associated with changes in electronic states in the material.

ultraviolet therapy *(Med.)*. Treatment of disease by ultraviolet rays. The therapeutic rays (300–400 nm) are now usually generated by quartz mercury-vapour lamps. The carcinogenic effect of over-dosage is now recognized.

ulvöspinel *(Min.)*. An end-member species in the magnetite series, with composition ferrous and titanium oxide. First recognized in ore from Södra Ulvön, in northern Sweden.

U-matic *(Image Tech.)*. TN for video tape-recording system using 3/4 in magnetic tape.

umbel *(Bot.)*. Inflorescence or simple umbel of many flowers borne on stalks arising together from the top of a main stalk. Often this sort of branching is repeated in a compound umbel with several main stalklets arising together from the top of a larger stalk.

umbellate *(Bot.)*. Having the characters of an umbel; producing umbels.

umbellifer *(Bot.)*. A plant which has its flowers in umbels, but especially a member of the family Umbelliferae.

Umbelliferae *(Bot.)*. *Apiaceae*. The carrot family, ca 3000 spp. of dicotyledonous flowering plants (superorder Rosidae). Mostly herbs, more or less cosmopolitan but especially in temperate and upland regions. The flowers are in simple or compound umbels, have five free petals, five stamens and an inferior ovary of two carpels. Includes several vegetables and flavouring plants e.g. carrot, parsnip, celery, parsley, coriander, cumin. Several are poisonous including hemlock.

umbilectomy *(Med.)*. Removal of the umbilicus.

umbilic *(Maths.)*. The limit point of circular sections of a quadric as the radius tends to zero. Every quadric has four real and eight imaginary umbilics.

umbilical cord *(Space)*. Term (frequently *umbilical*) applied to any flexible and easily-disconnectable cable, e.g. for conveying information, power or oxygen to a missile or spacecraft before launching, for connecting an operational spacecraft with an external astronaut.

umbilical cord *(Zool.)*. In eutherian Mammals, the vascular cord connecting the foetus with the placenta.

umbilical point on a surface *(Maths.)*. One at which the curvatures of all normal sections are positive and constant. Cf. **elliptical, hyperbolic**, and **parabolic point on a surface**.

umbilicate, umbilicated *(Med.)*. Having a depression which resembles the umbilicus.

umbilicus *(Zool.)*. In Gastropod shells, the cavity of a hollow columella; in Birds, a groove or slit in the quill of a feather; in Mammals, an abdominal depression marking the position of former attachment of the umbilical cord. pl. *umbilici*.

umbo *(Bot.,Zool.)*. A boss or protuberance; the beak-like prominence which represents the oldest part of a Bivalve shell. pl. *umbones*. adj. *umbonate*.

umbra *(Astron.)*. Region of complete shadow of an illuminated object, e.g., dark central portion of the shadow of the earth or moon. Generally applied to eclipses of the moon or of the sun, the term is also applied to the dark central portion of a sunspot. The outer, less dark, shadow is known as the *penumbra*.

umbrella *(Zool.)*. A flat cone-shaped structure, especially the contractile disk of a medusa.

umbrella antenna *(Telecomm.)*. An antenna comprising a vertical uplead from the top of which a number of wires extend radially towards the ground.

umkehr effect *(Meteor.)*. An effect used to derive, on certain assumptions, the vertical distribution of ozone from a series of measurements of the relative intensities of two wavelengths in light scattered from the zenith sky; one wavelength is more, and the other less, strongly absorbed by ozone. As the sun's zenith angle varies, a reversal (German *Umkehr*) occurs in the variation of the ratio of the intensities.

Umklapp process *(Phys.)*. A type of collision between phonons, or between phonons and electrons, in which crystal momentum is not conserved. Such processes provide the greater part of the thermal resistance in solid dielectrics.

umwelt *(Behav.)*. The relevant aspects of the environment which constitute the subjectively significant, or meaningful, surroundings for an animal or individual; i.e. that class of environmental variables capable of influencing behaviour.

unarmed *(Bot.)*. Without spines, thorns, prickles, sharp teeth etc.

unarmoured cable *(Elec.Eng.)*. A cable in which the outer covering of steel wire (armouring) is absent.

unary *(Chem.)*. Consisting of one component etc.

unavailable *(Bot.)*. An element in plant mineral nutrition present in the soil but not in a form which the plant can take up.

unavailable energy *(Eng.)*. The energy which becomes unavailable to do work, in the course of an irreversible process.

unbalanced *(Elec.Eng.)*. (1) Of a bridge circuit in which detector signal is not zero. (2) Of a pair of conductors in which magnitudes of voltage or current are not symmetrical with reference to earth.

unbalanced circuit *(Elec.Eng.)*. One whose two sides are inherently unlike. See **balanced circuit**.

unbalanced load *(Elec.Eng.)*. A load which is unequal on the two sides of a three-wire d.c. system, or on the three phases of a symmetrical three-phase a.c. system.

unbalanced network *(Elec.Eng.)*. One arranged for insertion into an unbalanced circuit, the earthy terminal of the input being directly connected to the earthy terminal of the output.

unbalanced system *(Elec.Eng.)*. A three-phase a.c. system carrying an unbalanced load.

unbleached kraft paper *(Paper)*. Brown packaging paper made from unbleached sulphate. Wood pulp suitable for use as wrapping material or for conversion into sacks, bags, gummed tape etc.

uncate, unciform, uncinate *(Bot.,Zool.)*. Hooked; hook-like.

uncertainty principle *(Phys.)*. There is a fundamental limit to the precision with which a position coordinate of a particle and its momentum in that direction can be simultaneously known. Also, there is a fundamental limit to the knowledge of the energy of a particle when it is measured for a finite time. In both statements, the product of the uncertainties in the measurements of the two quantities involved must be greater that the *Dirac constant*. The principle follows from the wave nature of particles. Also **Heisenberg uncertainty principle, indeterminacy principle**.

uncinate fit *(Med.)*. A hallucination of smell or of taste, due to a cerebral tumour or to epilepsy.

uncinus *(Zool.)*. A hook, or hook-like structure, e.g. a hook-like chaeta of Annelida; in Gastropoda, one of the marginal radula-teeth.

uncommitted logic array *(Comp.)*. Array of standard logic **gates** with all possible circuits present. Each element is identical, manufactured on a single **LSI chip**. The circuits not required for a particular application are burnt out. Abbrev. *ULA*.

unconditionally stable *(Telecomm.)*. Of amplification in a system which continues to satisfy the Nyquist criterion when the gain is reduced.

unconformity *(Geol.)*. A substantial break in the succession of stratified sedimentary rocks, following a period when deposition was not taking place. If the rocks below the break were folded or tilted before deposition was resumed, their angle of dip will differ from that of the overlying rocks, in which case the break is an *angular unconformity*.

unconscious mind *(Behav.)*. Refers to mental processes of which the subject is unaware, but which influence thought and action.

uncoursed *(Build.)*. See **random**.

uncut *(Print.)*. Said of a book whose edges have been left untrimmed, the bolts therefore remaining *uncut*.

undamped oscillations *(Telecomm.)*. Same as **continuous oscillations**.

under blanket *(Print.)*. (1) On web-fed relief presses the first dressing on the impression cylinder. Cf. **top blanket**. (2) In offset printing a special packing placed beneath the top rubber blanket.

underbunching *(Electronics)*. Less than optimum efficiency in a velocity-modulation system.

undercarriage *(Aero.)*. Each of the units (consisting of wheel(s), shock-absorber(s), and supporting struts) of an aircraft's alighting gear is an undercarriage, i.e. two main and either tail or nose undercarriages. Colloquially, the whole *landing gear*.

underclay *(Geol.)*. See **seat earth**.

undercloak *(Build.)*. The first or lower sheet of lead in a **roll**. Cf. **overcloak**.

undercoat *(Build.)*. Paint applied before the finishing coat. It is usually highly pigmented in order to provide good hiding power.

under-compensated meter *(Elec.Eng.)*. An induction-type meter provided with insufficient phase compensation, as the result of which it reads high with leading currents and low with lagging currents.

undercut *(Eng.)*. (1) A mould or pattern is undercut when it has a re-entrant portion and thus cannot be opened or withdrawn in a straight motion. (2) A short portion of a bolt, stepped shaft, or similar part, with a diameter smaller than the peripheries or diameters of the adjacent features. It obviates difficulties of machining sharp corners and facilitates subsequent assembling.

undercutting *(Print.)*. (1) A fault to be guarded against in the etching of line blocks, traditionally by the use of dragon's blood, but nowadays by powderless etching. (2) Cutting away part of the edge of a mount to allow the block to overhang a crossbar.

underdamping *(Telecomm.)*. Sometimes synonymous with *periodic* damping, but often restricted to cases where a critically-damped response would be preferable.

under-exposure *(Image Tech.)*. Inadequate exposure to light of a photo-sensitive surface, photographic or electronic resulting in an image of unsatisfactory tonal reproduction, particularly in the lack of gradation in the shadow areas of the picture.

underfeed stoker *(Eng.)*. A **mechanical stoker** in which the fuel is fed automatically and progressively from below the fire, and gradually forced up into the active zone, air being injected into the fuel bed just below the combustion-level. See **single-retort-, multiple-retort-**.

underflow *(Comp.)*. Occurs when a number to be stored is less than the smallest number that can be represented in the **word** available for it.

underfold *(Print.)*. See **overfold**.

underground gasification *(Min.Ext.)*. Technique for the remote extraction of hydrocarbons from coal by the deliberate combustion of coal seams in a controlled and restricted oxygen environment. Combustion and distillation products include carbon monoxide and volatile hydrocarbons, extracted via boreholes. Also *pyrolytic mining*.

underground volatilization *(Min.Ext.)*. Technique for remote extraction of hydrocarbons from coal. Solvents are introduced through boreholes and dissolved volatile hydrocarbons are extracted via a second set of boreholes.

underhand stopes *(Min.Ext.)*. Stopes in which excavation is carried downslope from access level.

underlay *(Min.Ext.)*. The departure of a vein or thin tabular deposit from the vertical; it may be measured in horizontal feet per fathom of inclined depth. Also *underlie*.

underlay *(Print.)*. To paste paper or card under the mount of a printing plate in order to bring it to type-height, or to remedy a defect in the mount.

underleaf *(Bot.)*. One of a row of leaves on the underside of the stem of a liverwort.

underlining felt *(Build.)*. See **sarking felt**.

undermodulation *(Image Tech.)*. Inadequate modulation in a sound recording for satisfactory reproduction.

undermodulation *(Telecomm.)*. That state of adjustment of a radio-telephone transmitter at which the peaks of speech or music do not produce 100% modulation, so that carrier power is not used to full advantage.

underpinning *(Build.,Civ.Eng.)*. The operation of propping part of a building to avoid damaging or weakening the superstructure.

underpitch groin *(Build.)*. See **Welsh groin**.

underpoled copper *(Eng.)*. See **poling**.

undersaturated *(Geol.)*. Refers to an igneous rock in which there is a deficit of silica. This is normally shown by the presence of a felspathoid. See **oversaturated**.

undershoot *(Telecomm.)*. See **overshoot**.

undershot wheel *(Eng.)*. A water-wheel used for low heads, in which the power is obtained almost entirely from the impulse of the water on the vanes. See **Poncelet wheel**.

undersize *(Min.Ext.)*. See **riddle**.

understeer *(Autos.)*. Tendency of a vehicle to preserve directional stability by reacting against external forces applied to the steering mechanism. Cf. **oversteer**.

under-voltage no-close release *(Elec.Eng.)*. A device which acts upon the trip coil of a circuit-breaker in such a way as to prevent the circuit-breaker being closed if the voltage is below a certain predetermined value.

under-voltage release *(Elec.Eng.)*. A device to trip an electrical circuit should the voltage fall below a certain predetermined value. Also called a *low-volt release*.

underwater cutting *(Eng.)*. A method of cutting iron or steel under water, utilizing a combusted mixture of oxygen and hydrogen (cf. **oxy-hydrogen welding**), the flame being protected by an airshield formed by a hood.

underwater transducer *(Acous.)*. One transducer (e.g., a microphone or loudspeaker) designed for operation under water. See **hydrophone**.

undistorted output *(Telecomm.)*. That of an amplifier free from nonlinear distortion. The *maximum undistorted output* is defined as that obtained with a specified percentage of nonlinear distortion.

undistorted transmission *(Telecomm.)*. That of any type of line for which the velocity of propagation and coefficient of attenuation are both independent of frequency.

undrawn yarn *(Textiles)*. Extruded man-made filament yarn not yet subjected to the drawing process that orients the linear molecules and gives strength to the yarn.

unducted fan *(Aero.)*. Recent term for **propeller** used in high-speed civil turbine engines. *Propellers* when mounted within a circular duct are called *fans* but when advanced gas turbines drive fans without surrounding ducts they are called *unducted fans*. This apparently illogical terminology may be excused as the blade shapes of the new *propellers* are closer to those of fans than to those of the 1960s.

undulant fever *(Med.)*. *Brucellosis*. (1) *Malta fever, Mediterranean fever, Gibraltar fever*. A disease characterized by alternating febrile and afebrile periods, splenomegaly, transient painful swelling of joints, neuralgia, and anaemia; due to infection with *Brucella melitensis*, conveyed to man by infected goats or their milk. (2) A disease with the same *signs* as (1), due to infection by *Brucella abortus*, a micro-organism which causes abortion in cows and is conveyed to man in cow's milk or by contact with infected animals.

undulating membrane *(Zool.)*. An extension of the flagella membrane in some Protozoa (e.g. *Trypanosoma*) by which it is attached to the cell for part of its length.

uneven working *(Print.)*. See **even working**.

unexcited *(Phys.)*. Said of an atom in its ground state. See **excitation**.

unfired *(Electronics)*. Said of any gas discharge device when in an un-ionized state.

ungual *(Med.)*. Pertaining to or affecting the nails. See **unguis**.

unguiculate *(Bot.,Zool.)*. Provided with claws. Specifically (Bot.) applied to a petal with an expanded limb supported on a long narrow stalklike base.

unguis *(Zool.)*. In Insects, one of the tarsal claws; in Vertebrates, the dorsal scale contributing to a nail or claw; more generally, a nail or claw. pl. *ungues*. adjs. *ungual, unguinal*.

ungula *(Zool.)*. A hoof. adj. *ungulate*.

ungulate *(Zool.)*. The term applied to several groups of superficially similar hoofed animals which are not necessarily closely related taxonomically. Horses, Cows, Deer, Tapirs.

unguligrade *(Zool.)*. Walking on the tips of enlarged nails of one or more toes, i.e. hoofs, as in horses, etc. Cf. **digitigrade, plantigrade**.

uniaxial *(Bot.)*. Having a main axis consisting of a single row of large cells with only clearly subordinate branches. Cf. **multiaxial**.

uniaxial, uniaxial crystal *(Min.)*. A term embracing all those crystalline minerals in which there is only one direction of single refraction (parallel to the principal crystal axis and known as the optic axis). All minerals which crystallize in the tetragonal, trigonal, and hexagonal systems are *uniaxial*. Cf. **biaxial**.

unicellular *(Biol.)*. Consisting of a single cell.

unidirectional antenna *(Telecomm.)*. One in which the radiating or receiving properties are largely concentrated in one direction.

unidirectional current *(Elec.Eng.)*. One which, although its amplitude may vary, never changes sign.

unified field theory *(Phys.)*. See **field theory**.

unified model of the nucleus *(Phys.)*. A model incorporating many valuable features of both *collective* and *independent-particle* models.

unified scale *(Chem.)*. The scale of atomic and molecular weights which is based on the mass of the ^{12}C isotope of carbon being taken as 12 exactly, hence the atomic mass unit equals 1.660×10^{-27} kg. This scale was adopted in 1960 by the International Unions of both Pure and Applied Physics and Pure and Applied Chemistry, hence the name *unified scale*. See **atomic mass unit, atomic weight**.

unified screw thread *(Eng.)*. A screw thread form adopted by Canada, United Kingdom and the US. It combines features of the US *Standard Screw Thread* and the *British Standard Whitworth* screw thread. Of 60° angle, the thread has radiused roots and crests while the crests of the nut are flat.

uniform convergence *(Maths.)*. (1) A sequence $a_1(x)$, $a_2(x)...a_n(x)...$, converges uniformly to the limit $a(x)$ in the interval (a,b) if, given any $\varepsilon > 0$, there exists for all x in (a,b), N such that $|a_n(x) - a(x)| < \varepsilon$ when $n \geqq N$. (2) A series converges uniformly if the sequence of partial sums converges uniformly. (3) A product converges uniformly if the sequence of partial products converges uniformly. See **convergence, M-test of Weierstrass for-**.

uniform extension *(Eng.)*. The extension produced in a tensile test before the ultimate tensile stress is reached, and uniform over the gauge length.

uniform field *(Maths.)*. One which is described by the same vector at all points. A constant field.

uniformitarianism *(Geol.)*. The concept that the processes that operate to modify the Earth today also operated in the geological past. In its more extreme form the concept also infers uniformity of rates as well as of processes.

uniform line *(Elec.Eng.)*. One for which the electric properties are identical throughout its length.

unilateral conductivity *(Phys.)*. The property of unipolarity by which current can flow in one direction only; exhibited by a perfect rectifier.

unilateral impedance *(Phys.)*. Any electrical or electromechanical device in which power can be transmitted in one direction only, e.g., a thermionic valve or carbon microphone.

unilateralization *(Phys.)*. Neutralization of feedback so that transducer or circuit has unilateral response, i.e., there is no response at the input if the signal is applied to the output terminals. While many valve circuits are inherently unilateral, equivalent transistor ones require external neutralization.

unilateral tolerance *(Eng.)*. A tolerance with dimensional limits either entirely above or entirely below the basic size.

unilateral transducer *(Elec.Eng.)*. One for which energy can be transmitted only forward.

unilocular *(Bot.)*. Having a single compartment. Cf. **multilocular, bilocular**.

unimolecular layer, unimolecular reaction *(Chem.)*. See **monomolecular layer, monomolecular reaction**.

uninemy hypothesis *(Biol.)*. Each chromatid contains a single, DNA, double-helical molecule organized linearly with respect to the chromosomal axis.

uninsulated conductor *(Elec.Eng.)*. A conductor at earth potential, such that no care need be taken to insulate it from earth.

uninucleate *(Biol.)*. Containing one nucleus.

union *(Build.)*. A connection for pipes.

union *(Maths.)*. The union of sets *A* and *B* is the set of elements which are in *A* or *B* or in both *A* and *B*.

union *(Med.)*. In the process of healing, the growing together of parts separated by injury (e.g. the two ends of a broken bone, the edges of a wound).

union fabric *(Textiles)*. A woven fabric with the warp of one fibre (e.g. linen) and the weft of another (e.g. cotton).

union kraft *(Paper)*. A packaging material comprising two layers of kraft paper bonded together by means of a laminant that is resistant to the transmission of water in liquid or vapour form e.g. bitumen or polythene.

uniparous *(Zool.)*. Giving birth to one offspring at a time.

Unipivot instrument *(Elec.Eng.)*. An instrument whose moving-coil system is balanced on a single pivot passing through its centre of gravity.

unipolar *(Zool.)*. Said of nerve cells having only one process. Cf. **bipolar, multipolar**.

unipolar transistor *(Electronics)*. One with one polarity of carrier.

unipole antenna *(Telecomm.)*. Isotropic antenna conceived as radiating uniformly in phase in all directions. Theoretically useful, but not realizable in practice. See **isotropic radiator**.

unipotent *(Zool.)*. Of embryonic cells, capable of forming a single cell type only. Cf. **totipotent**.

unique sequence DNA *(Biol.)*. DNA sequences which are only represented once in the **haploid** genome. Most genes are in this category.

uniramous *(Zool.)*. Having only one branch; as some Crustacean appendages. Cf. **biramous**.

uniselector *(Telecomm.)*. A selector switch which only rotates its wipers about an axis, in contrast with the **two-motion selector**, in which wipers are raised to a specified level in the rows of contacts by the impulse trains, and then enter the bank of contacts, either by hunting or by a further train of impulses. See **selector**.

uniselector distribution frame *(Telecomm.)*. A frame carrying terminals, so that the extent of the multiple outlets from uniselectors can be varied to suit the demand on outlets in the multiple.

uniseriate *(Bot.)*. Arranged in a single row, series or layer.

unisexual *(Bot.,Zool.)*. Showing the characters of one sex or the other; distinctly male or female. Cf. *hermaphrodite*.

unit *(Genrl.)*. A dimension or quantity which is taken as a standard of measurement.

unit *(Min.Ext.)*. 1% of a specified element or compound in a parcel of ore, concentrates or metal being sold.

unit arch *(Print.)*. On a web-fed press a perfecting unit arranged in an arch or inverted U design.

unit cell *(Crystal.)*. The smallest group of atoms, ions, or molecules, whose repetition at regular intervals, in three dimensions, produces the lattice of a given crystal.

unit character *(Biol.)*. A character that can be classified into two distinct types, usually the normal and the mutant, and displaying *Mendelian inheritance*.

unit charge, unit quantity of electricity *(Phys.)*. In SI units, 1 **coulomb**. In unrationalized MKS units, the electric charge which experiences a repulsive force of 1 newton when placed 1 metre from a like charge *in vacuo*. Similarly, in the CGS electrostatic units, the force is 1 dyne when 1 centimetre apart.

unit heater *(Eng.)*. A combination of air heater and circulator, often in the form of a heated cellular core or finned tube over which air is blown by a fan.

unit interval *(Telecomm.)*. In a system using an equal-length code or in a system using an isochronous modulation, the interval of time such that the theoretical duration of the significant intervals of a telegraph modulation (or restitution) are whole multiples of this interval.

unitized *(Autos.)*. See **chassis**.

unit leaf rate *(Bot.)*. See **net assimilation rate**.

unit matrix *(Maths.)*. A unit matrix is a square matrix satisfying the following conditions: (1) the leading diagonal entries, i.e. the entries on the diagonal from top left to bottom right, are all unity; (2) the entries not on the leading diagonal are all zero. The unit matrix of order *n* is an identity element for multiplication in the set of all square matrices of order *n*. A unit matrix is usually denoted by *I*.

unit of attenuation *(Telecomm.)*. See **decibel, neper**.

unit of bond *(Build.)*. That part of a brickwork course which, by being constantly repeated throughout the length of the wall, forms a particular bond.

unit pole *(Phys.)*. A magnetic pole which experiences a repulsive force of 1 N when 1 m (or 1 dyne when 1 cm) apart from a like pole *in vacuo*. A mathematical concept formerly used for establishing magnetic and electric units.

unit system *(Print.)*. A feature of the Monotype system and of certain major filmsetting systems. See **Monotype unit system**.

unit type press *(Print.)*. A web-fed press with one or more printing units in line on the bed plate.

univalent *(Biol.)*. One of the single chromosomes which separate in the first meiotic division.

univalent *(Chem.)*. **Monovalent**.

univariant *(Chem.)*. Having one **degree of freedom**.

universal chuck *(Eng.)*. See **self-centring chuck**.

universal combustion burner *(Eng.)*. Natural-draught gas burner having one injector for the entrainment of primary air prior to combustion, and a secondary injector through which the flow of additional air into the combustion chamber can be regulated.

universal grinder *(Eng.)*. A machine tool with a universal grinding head whose spindles are mounted parallel with a reciprocating table carrying a workhead and support tailstock. A wide range of movements and attachments allow plain and tapered external and internal cylindrical grinding and also surface grinding.

universal indicator *(Chem.)*. A mixture of *indicators* which gives a definite colour change for each integral change of pH-value over a wide range.

universal joint *(Autos.)*. Device, usually of the modified *Hooke's* type, which allows rotary drive to be transmitted through an angle. Used on propeller shafts and independently suspended driven wheels to accommodate suspension movement. See **constant velocity joint**.

universal milling machine *(Eng.)*. A **milling machine**

similar to a plane milling machine but with the additional feature that the table swivels horizontally, and is provided with a dividing head as standard equipment.

universal motor *(Elec.Eng.)*. A fractional horsepower commutator motor suitable for use with both direct current and single-phase alternating current.

universal plane *(Build.)*. Multi-purpose plane adaptable for rebating, grooving, trenching, cutting mouldings etc.

universal planer *(Eng.)*. A planer that will cut on the forward and on the reverse strokes.

Universal Product Code *(Comp.)*. Standard **bar code** now adopted in Europe. Abbrev. *UPC*.

universal set *(Maths.)*. A term used to denote the set containing all the elements relevant to a particular mathematical study. In logic, sometimes called the *universe of discourse*.

universal shunt *(Electronics)*. A series of high stability resistors used to shunt galvanometers to provide different ranges of measurement.

universal time *(Astron.)*. A name for *Greenwich Mean Time* recommended in 1928 by the International Astronomical Union to avoid confusion with the pre-1925 GMT which began at noon, not midnight. Abbrev. *UT*. See also **ephemeris time**.

universal veil *(Bot.)*. Membrane which encloses the developing fruiting body of some agarics, rupturing, as the stalk grows, to leave the **volva**.

universal viewfinder *(Image Tech.)*. One with objectives for lenses of varying focal length.

universal vise *(Eng.)*. A vise which has two or three swivel settings allowing the workpiece to be set at a compound angle. Also called *toolmaker's vise*.

universe *(Astron.)*. In modern astronomy this term has the particular meaning: the totality of all that is in the cosmos and which can affect us by means of physical forces. The definition excludes anything which is in principle undetectable physically, such as regions of **spacetime** that have been irreversibly cut off from our own spacetime.

univibrator *(Telecomm.)*. Term for monostable multivibrator circuit. See **flip-flop**.

univoltine *(Zool.)*. Producing only one set of offspring during the breeding season or year. Cf. **multivoltine**.

UNIX *(Comp.)*. TN for a well known **operating system** not tied to a particular computer manufacturer. It is a trademark of AT&T Laboratories.

unloaded antenna *(Telecomm.)*. One with no inductance coils to increase its natural wave length.

unmod *(Image Tech.)*. Unmodulated, describing a photographic sound track without any recorded signal.

unmodulated waves *(Telecomm.)*. Waves which do not vary in amplitude with time, such as those radiated from a radio-telephone transmitter when no sound enters the microphone.

unpitched sound *(Acous.)*. Any sound or noise which does not exhibit a definite pitch, but consists of components spread more or less continuously over the frequency spectrum.

unsaturated *(Chem.)*. (1) Less concentrated than a saturated solution or vapour. (2) Containing a double or a triple bond, especially between two carbon atoms; unsaturated molecules can thus add on other atoms or radicals before saturation is reached.

unsewn binding *(Print.)*. The sections of the book are gathered and fed to a machine (of which there are several designs); part of the back is cut off and the edge roughened, and adhesive applied, and, when it sets, the leaves compacted together. Used mainly for paperbacks. Often called *Perfect binding*, *thermoplastic binding*.

unsoundness *(Eng.)*. The condition of a solid metal which contains blowholes or pinholes due to gases, or cavities resulting from its shrinkage during contraction from liquid to solid state, i.e. *contraction cavities*.

unsqueezed print *(Image Tech.)*. A print from an anamorphic motion picture negative in which the image has been optically corrected for normal projection.

unstability *(Elec.Eng.)*. See under **stability test**.

unstable *(Build.)*. A term applied to a structural framework having fewer members than it would require to be perfect. Also *deficient*.

unstable *(Chem.)*. Subject to spontaneous change.

unstable equilibrium *(Phys.)*. The state of equilibrium of a body when any slight displacement decreases its potential energy. The instability is shown by the fact that, having been slightly displaced, the body moves farther away from its position of equilibrium.

unstable oscillation *(Genrl.)*. Any oscillation, in a mechanical body, electrical circuit etc. which increases in amplitude with time.

unstick *(Aero.)*. See **lift-off**.

unstirred layer *(Bot.)*. Same as **boundary layer**.

unstriated muscle *(Zool.)*. A form of contractile tissue composed of spindle-shaped fibrillar uninucleate cells, occurring principally in the walls of the hollow viscera. Also *smooth muscle*. Cf. **striated muscle**, and see **voluntary muscle**.

unsymmetrical grading *(Telecomm.)*. *Grading* in which subscribers originating higher-than-average traffic are given access to a greater proportion of individual trunks.

unsymmetrical oscillations *(Telecomm.)*. Oscillations in which the positive and negative parts of the waveform are unequal and of different shape.

untrimmed floor *(Build.)*. A floor consisting of bridging joists only.

untuned antenna *(Telecomm.)*. One not separately tuned to the operating frequency, although effectively tuned by coupling to one or more resonant circuits.

untuned circuit *(Telecomm.)*. One not sharply resonant to any particular frequency.

unvoiced sound *(Acous.)*. In speech, any elemental sound which has no discrete harmonic frequencies but consists of a wide frequency band generated by air rushing through the mouth and nasal cavities; these are modified in spectral energy distribution by the posture of the cavities of the mouth and the resultant broad resonances.

up *(Comp.)*. A computer is *up* when it is functioning and ready for use.

U-packing, U-leather *(Eng.)*. A flexible annular ring of U-section and used to pack the glands of fluid power pistons, rams. The hollow of the U faces the pressure side so that the pressure expands the legs of the U, forming a seal. Modern designs use neoprene or similar and may have a toroidal spring in the arms of the U.

upcast shaft *(Civ.Eng.)*. A ventilating shaft through which the vitiated air passes in an upward direction.

update *(Comp.)*. (1) To bring a file up to date by modifying entries and adding new entries in accordance with a specified procedure. (2) To modify a computer instruction so that the address specified is increased every time the instruction is carried out.

up, down locks *(Aero.)*. Safety locks which hold the units of a retractable landing gear up in flight and down on the ground.

updraught *(Autos.)*. Said of a carburettor in which the mixture is drawn upwards against the force of gravity.

upholsterer's hammer *(Build.)*. Lightly built clawhammer for use with small pins and tacks.

upmake *(Print.)*. (1) To arrange lines of print into columns or pages. (2) Print so arranged.

upper atmosphere *(Astron.)*. A term used somewhat loosely for the region of the earth's atmosphere above about 20 miles (30 km), which is not normally explored by sounding balloons, but can be studied by rockets and artificial satellites.

upper bound *(Maths.)*. See **bounds of a function**.

upper case *(Print.)*. Frequently used as a synonym for capital, the name deriving from the arrangement (now largely superseded) of the capitals in an upper and the small letters in a lower *case*.

upper culmination *(Astron.)*. See **culmination**.

upper deck *(Ships)*. The term correctly denotes the main strength deck of a ship. From this deck all scantlings are determined, freeboard assigned and subdivision arranged, according to type of vessel.

upper mean-hemispherical candle-power *(Phys.)*. See **mean-hemispherical candle-power**.

upper quartile *(Stats.)*. The argument of the cumulative distribution function corresponding to a probability of 0.75; (of a sample) the value below which occur three quarters of the observations in the ordered set of observations.

upright *(Build.)*. A vertical member in a structure.

uprighting *(Horol.)*. The process of correcting the pivot holes in a plate so that the hole in one plate is exactly in line with the corresponding hole in the other plate, and a line through the two holes is at right angles to the plates.

upsetting *(Eng.)*. Metalworking to produce an increase in size of part of the component.

upstream injection *(Aero.)*. A gas turbine fuel system in which the fuel is injected towards the compressor in order to achieve maximum vaporization and turbulence.

uptake *(Eng.)*. The flue or duct which leads the flue gases of a marine boiler to the base of the funnel.

uptake *(Radiol.)*. In radiobiology, the quantity (or proportion) of administered substance subsequently to be found in a particular organ or tissue.

upturn *(Build.)*. The part of a lead flashing which is dressed up against a wall face.

uracil *(Chem.)*. *2,6-dioxypyrimidine*, one of the four bases in RNA and the only one which does not occur in DNA. Pairs with adenine. See **genetic code**. ⇨

uraemia, uremia *(Med.)*. The state resulting from failure of a diseased kidney to perform its normal functions; associated with retention of urea in the blood, and characterized by varied symptoms, among which are headache, foul breath, diarrhoea and vomiting, visual disturbances, lethargy, convulsions, and coma.

Uralian emerald *(Min.)*. Not an emerald; green variety of **andradite** garnet (*demantoid*), occurring as nodules in ultra-basic rocks in the Urals; used as a semiprecious gemstone, though rather soft for this purpose.

uralite *(Min.)*. A bluish-green monoclinic amphibole, generally actinolitic in composition, resulting from the alteration of pyroxene.

uralitization *(Geol.)*. A type of alteration of pyroxene-bearing rocks, involving the replacement of the original pyroxenes by fibrous amphiboles, as in some epidiorites.

uranic, uranous *(Chem.)*. Referring to (VI) and (IV) uranium respectively.

uranides *(Chem.)*. Name for elements beyond protactinium in the periodic system. Cf. **actinides**.

uraninite *(Min.)*. Uranium oxide, UO_2, often more or less hydrated and containing also lead, thorium, and the metals of the lanthanum and yttrium groups. It occurs as brownish to black cubic crystals and is an accessory mineral in granite rocks and in metallic veins. When massive, and apparently amorphous, known as *pitchblende*.

uranite *(Min.)*. For *copper uranite*, see **torbernite**; for *lime uranite*, see **autunite**.

uranium *(Chem.)*. A hard grey metal with a number of isotopes. Symbol U, at. no. 92, r.a.m. 238.03, rel.d. 18.68, mp 1150°C. ^{235}U is the only naturally occurring readily fissile isotope and exists as one part in 140 of natural uranium. ^{238}U is a fertile material (half-life 4.5×10^9 yr) which has a small fission cross-section for fast neutrons. ^{233}U is a fissile material which can be produced by the neutron irradiation of thorium-232; ^{235}U is used in nuclear reactors and nuclear weapons.

uranium enrichment *(Nuc.Eng.)*. Process to increase the isotopic content of ^{235}U in uranium for reactor use. Principal methods in use or proposed are gaseous diffusion, centrifuge, jet nozzle, plasma centrifuge, chemical exchange and laser isotope separation. In early days magnetic separation was used (*Calutron* method). See **isotope separation**.

uranium hexafluoride *(Chem.)*. *Uranium (VI) fluoride*. A volatile compound of uranium with fluorine, used in the gaseous diffusion process for separating the uranium isotopes. Very corrosive.

uranium-lead dating *(Geol.)*. A method of determining the age in years of geological material, based on the known decay rate of ^{238}U to ^{206}Pb and ^{235}U to ^{207}Pb.

uranium-radium series *(Phys.)*. The series of radioactive isotopes which result from the decay of ^{238}U. The mass numbers of the members of the series are $4n+2$, where n is an integer. Series ends in the stable isotope ^{206}Pb. See **radioactive series**.

uranoplasty *(Med.)*. Plastic operation for closing a cleft in the hard palate.

Uranus *(Astron.)*. Seventh major planet from the Sun, discovered in 1781 by William Herschel in the course of a systematic survey of the heavens. Orbital period 84.01 years, mass 14.56 times the Earth. There are 5 natural satellites as well as a ring system discovered in 1977.

uranyl *(Chem.)*. The radical UO_2^{++}, e.g. uranyl nitrate, $UO_2(NO_3)_2$.

urate *(Chem.)*. Salt of uric acid.

urea *(Chem.)*. *Carbamide*, $H_2N.CO.NH_2$; mp 132°C, very soluble in water, insoluble in ether. Found in the urine of Mammals. Wöhler synthesized (1828) urea from ammonium isocyanate, which undergoes an intramolecular transformation when its aqueous solution is heated, forming urea. Manufactured by heating carbon dioxide and ammonia under high pressure, it is highly nitrogenous, and widely used for fertilizer and animal feed additive.

urea cycle *(Biol.)*. The cyclic interconversion of 4 amino acids which converts carbamyl phosphate into urea. It represents the major pathway for the excretion of nitrogenous waste in terrestrial vertebrates.

urea resins *(Plastics)*. Thermosetting resins manufactured by heating together urea and an aldehyde, generally formaldehyde (methanal). Pale-coloured or water-white and translucent, they can therefore take delicate dyes and tints. They are non-flammable, and are resistant to weathering, weak acids and alkalis (pH between 6 and 8.4), ethanol, propanone, greases and oils. Also *amino-aldehydic resins*.

uredosorus *(Bot.)*. A pustule consisting of uredospores, with their supporting hyphae, and some sterile hyphae.

uredospore, urediniospore, urediospore *(Bot.)*. A binucleate spore which rapidly propagates the dikaryotic phase of a rust fungus.

ured-, uredo-. *(Genrl.)*. Prefix from L. *uredo*, a blight.

ureides *(Chem.)*. The acid derivatives of urea. They correspond to amides or anilides. The cyclic ureides are known as the *purine group*.

uremia *(Med.)*. See **uraemia**.

ureotelic *(Zool.)*. Excreting nitrogen in the form of urea.

ureter *(Zool.)*. The duct by which the urine is conveyed from the kidney to the bladder or cloaca.

ureteralgia *(Med.)*. Pain in the ureter.

ureteritis *(Med.)*. Inflammation of the ureter.

ureterocele *(Med.)*. Cystic dilatation of that part of the ureter which lies within the wall of the urinary bladder, due to congenital narrowing at its point of entry into the bladder.

ureterocolostomy *(Med.)*. The operation of implanting the ureter into the colon so that it may drain into it.

ureterolithotomy *(Med.)*. The operation of cutting into the ureter to remove a stone from it.

ureteropyelitis *(Med.)*. Inflammation both of a ureter and of the pelvis of the kidney on the same side.

ureterotomy *(Med.)*. Surgical incision of a ureter.

urethra *(Zool.)*. The duct by which the urine is conveyed from the bladder to the exterior, and which in male Vertebrates serves also for the passage of semen. adj. *urethral*.

urethritis *(Med.)*. Inflammation of the urethra.

urethrocele *(Med.)*. Prolapse of the floor of the female urethra; usually associated with cystocele.

urethrocystitis *(Med.)*. Inflammation of both the urethra and the urinary bladder.

urethrospasm *(Med.)*. Spasmodic contraction of the muscular tissue of the urethra.

urethrotomy *(Med.)*. The operation of cutting a stricture of the urethra.

uric acid *(Chem.)*. An acid of the purine group, $C_5H_4N_4O_3$, *2,6,8-trihydroxypurine*. It is a white crystalline powder, insoluble in cold, hardly soluble in hot, water. Uric acid deposits in the organism are the cause of gout and rheumatism. It forms soluble lithium and piperazine salts. It can be recognized by the **murexide** test.

uricotelic *(Zool.)*. Excreting nitrogen in the form of uric acid.

urine *(Zool.)*. In Vertebrates, the excretory product elaborated by the kidneys, usually of a more or less fluid nature. adj. *urinary*.

uriniferous, uriniparous *(Zool.)*. Urine-secreting, urine-producing; as the glandular tubules of the kidney.

urinogenital *(Zool.)*. Pertaining to the urinary and genital systems.

urinogenital system *(Zool.)*. The organs of the urinary and genital systems when there is a direct functional connection between them, as in male Vertebrates.

urinometer *(Med.)*. An instrument for measuring the density of urine.

uro-. *(Genrl.)*. Prefix from (1) Gk. *ouron*, urine, (2) Gk. *oura*, tail.

urobilinaemia, urobilinemia *(Med.)*. The presence of urobilin in the blood.

urobilinuria *(Med.)*. The presence of (an excess of) urobilin in the urine.

urochord *(Zool.)*. Having the notochord confined to the tail region.

Urochordata *(Zool.)*. A subphylum of Chordata, in which only the larvae have a hollow dorsal nerve cord and a notochord, the adults being without coelom, segmentation and bony tissue, and having a dorsal atrium, a reduced nervous system, and a test composed of tunicin, a substance closely related to cellulose. Sea Squirts. Also called *Tunicata*.

Urodela *(Zool.)*. An order of Amphibians, the adults having four similar pentadactyl limbs and a prominent tail. The larvae have external gills which persist in the adults of neotenous forms, and in some others gill slits persist. Newts and Salamanders. Also known as *Caudata*.

urodelous *(Zool.)*. Having a persistent tail; as Salamanders.

urodynamics *(Med.)*. The study of urine flow.

urography *(Radiol.)*. The radiological examination of the urinary tract. See **pyelography**.

urolithiasis *(Med.)*. The occurrence of stones or **calculi** in the urinary tract.

urology *(Med.)*. That part of medical science which deals with diseases and abnormalities of the urinary tract and their treatment. Hence *urologist*.

uropod *(Zool.)*. In Malacostraca, an appendage of the abdominal somite preceding the telson.

uropygial gland *(Zool.)*. See **oil gland**

uropygium *(Zool.)*. In Birds, the short caudal stump into which the body is prolonged posteriorly.

urosome *(Zool.)*. In aquatic Vertebrates, the tail region; in Crustacea, the hinder part of the abdomen.

urostyle *(Zool.)*. In Fish, the hypural bone; in Anura, a rodlike bone formed by the fusion of the caudal vertebrae.

urotropine *(Chem.)*. See **hexamethylene-tetramine** (methanal and ammonia).

urticant, urticating *(Zool.)*. Irritating; stinging.

urticaria *(Immun.)*. Skin rash typically with localized, elevated erythematous, itchy weals due to local release of histamine and other vasoactive substances. Frequently associated with immediate hypersensitivity (type 1 reaction) on contact of the skin with antigens, or more generalized after absorption of allergens from the gut. Can result from taking drugs or certain foods (e.g. shellfish), or as a reaction to the injection of serum, insect bites or plant stings (*nettle rash*). Also *hives*.

urticaria *(Med.)*. A condition in which smooth, elevated, whitish patches (weals) appear on the skin and itch intensely, as a result of taking drugs or certain foods (e.g. shellfish), or as a reaction to the injection of serum, insect bites, or the stings of plants (nettle-rash).

urtite *(Geol.)*. An intrusive igneous rock composed mainly of nepheline.

useful life *(Elec.Eng.)*. The life which can be expected from a component before the chance of failure begins to rise.

useful load *(Aero.)*. The gross weight of an aircraft, less the tare weight. Usually includes fuel, oil, crew, equipment not necessary for flight (such as parachutes) and payload.

USENET *(Comp.)*. A network for electronic mail and *teleconferencing* centred in the US and spread widely throughout the world.

USS thread *(Eng.)*. See **Sellers screw thread**.

Ustilaginales *(Bot.)*. An order of the Basidiomycotina containing the **smut** fungi.

UT *(Astron.)*. Abbrev. for *universal time*.

UTC *(Telecomm.)*. Abbrev. for *Universal Time Coordinates*. See **zulu time**.

uterus *(Zool.)*. In female Mammals, the muscular posterior part of the oviduct in which the foetus is lodged during the prenatal period; in lower Vertebrates and Invertebrates, a term loosely used to indicate the lower part of the female genital duct, or in certain cases (as in Platyhelminthes) a special duct in which eggs are stored or young developed. adj. *uterine*.

utility program *(Comp.)*. **Systems program** designed to perform a commonplace task such as the transfer of data from one storage device to another or sorting a set of data.

utilization factor *(Elec.Eng.)*. The ratio of the luminous flux reaching a specified plane to the total flux emanating from an electric lamp.

utricle *(Bot.)*. Any one of a variety of inflated bladder-like structures.

utricle *(Zool.)*. A small sac; in Vertebrates, the upper chamber of the inner ear from which arise the semicircular canals. Also *utriculus*.

utricular, utriculiform *(Bot.,Zool.)*. Like a bladder; pertaining to a utricle.

utriculoplasty *(Med.)*. The operation of excising a portion of the body of the uterus; done for the treatment of uterine haemorrhage.

UUCP *(Comp.)*. *UNIX to UNIX Copy*. Widely used mail network connecting machines running the **UNIX** operating system, often over ordinary telephone dial-up lines, although the connection can be made over dedicated lines or **local area network** lines.

UV *(Chem.)*. Abbrev. for *UltraViolet spectroscopy*.

uvarovite *(Min.)*. A variety of garnet, of an attractive green colour; essentially silicate of calcium and chromium.

uvea *(Zool.)*. (1) In Vertebrates, the posterior pigment-bearing layer of the iris of the eye. (2) The iris, the ciliary body, and the choroid considered as one structure; also called the *uveal tract*.

uveitis *(Med.)*. Inflammation affecting the iris, the ciliary body, and the choroid.

uveoparotid fever *(Med.)*. A condition characterized by inflammation of the parotid glands and bilateral iridocyclitis, often with paralysis of the seventh cranial nerve.

U-V filter *(Image Tech.)*. A filter absorbing ultra-violet radiation, used to prevent it affecting a blue-sensitive photographic emulsion.

uvula *(Med.)*. (1) A small conical process hanging from the middle of the lower border of the soft palate; part of the inferior vermis of the cerebellum. (2) A slight elevation of the mucous membrane of the urinary bladder in the male caused by the median lobe of the prostate.

U-wrap *(Image Tech.)*. Tape path on the drum of a **helical-scan** VTR giving 180° contact.

V

v *(Phys.)*. Symbol for (1) velocity, (2) specific volume of a gas.

v- *(Chem.)*. See **vicinal**.

V *(Aero.)*. Subscripted symbols used in aircraft documentation. V_1; Abbrev. for *critical speed*. V_{lo}; see **lift-off**. V_{ne}; Abbrev. for the maximum permissible *indicated air speed*: a safety limitation (the subscript means *Never Exceed*) because of strength or handling considerations. The symbol is used mainly in operational instructions. V_{no}; *Normal operating speed*, usually of an airliner or other civil aircraft; this term is used mainly in flight operation documents and may be quoted in *EAS*, *IAS* or *TAS*. V_r; Abbrev. for *rotation speed*.

V *(Chem.)*. The symbol for vanadium.

V *(Phys.)*. Symbol for volt.

V *(Phys.)*. Symbol for (1) potential, (2) potential difference, (3) electromotive force, (4) volume.

VAB *(Space)*. Abbrev. for *Vehicle Assembly Building* at Kennedy Space Center, Florida, used for integrating large elements of a space system.

vacancy *(Phys.)*. Unoccupied site for ion or atom in crystal.

vaccinal *(Med.)*. Of, pertaining to, or caused by vaccine or vaccination.

vaccination *(Immun.)*. Production of active immunity by administration of a vaccine.

vaccination *(Med.)*. Immunization against an infectious disease by exposure to the appropriate **vaccine**.

vaccine *(Immun.,Med.)*. Therapeutic material, treated to lose its virulence and containing antigens derived from one or more pathogenic organisms, which on administration to humans or other animals, will stimulate active immunity and protect against infection with these or related organisms.

vaccinia *(Immun.)*. Synonym for the virus used in vaccine procedures to produce immunity to smallpox. Differs from cowpox and smallpox virus in minor antigens only and was probably derived originally from cowpox virus. The virus has proved very stable (i.e. not liable to mutation) and safe to use in humans, and produces long-lasting cell mediated immunity. By **recombinant DNA** techniques genes for other viral antigens have been introduced into vaccinia virus with the intention that immunization with the recombinant strain shall produce protective immunity against other viruses also.

vaccinial *(Med.)*. Of, pertaining to, or caused by, vaccinia.

vacuolar membrane *(Biol.)*. The membrane surrounding an intracellular **vesicle**.

vacuole *(Biol.)*. See **vesicle**.

vacuole *(Bot.)*. A cavity, containing sap and separated from the cytoplasm by a membrane, the *tonoplast*.

vacuum *(Phys.)*. Literally, a space totally devoid of any matter. Does not exist, but is approached in inter-stellar regions. On earth, the best vacuums produced have a pressure of about $10^{-8}\,N/m^2$. Used loosely for any pressure lower than atmospheric, e.g., train braking systems, 'vacuum' cleaners, etc.

vacuum activity *(Behav.)*. Behaviour manifested in the apparent absence of the external stimuli that normally elicit the activity, presumably because of internal factors governing the motivation to perform the behaviour.

vacuum arc furnace *(Chem.)*. One in which a small specimen is heated by a high-voltage arc in an inert gas, e.g. argon, at low pressure.

vacuum arc melting *(Eng.)*. Using a consumable electrode to melt an ingot with the arc *in vacuo*, the reduced pressure causing the metal to out-gas.

vacuum augmenter *(Eng.)*. An **air ejector** placed in a steam condenser to produce a higher degree of vacuum than is obtainable by the use of an air-pump alone.

vacuum brake *(Eng.)*. A brake system used on railway trains and goods vehicles, in which a vacuum, maintained in reservoirs by exhausters, and under the control of the driver, simultaneously operates all brake cylinders. See **continuous brake**.

vacuum concrete *(Civ.Eng.)*. Concrete enclosed in specially prepared shuttering which incorporates fine filters and air ducts. After pouring, strong suction is applied to the ducts by means of pumps and the excess water is extracted. This enables the shuttering to be removed and re-used much sooner than by traditional methods.

vacuum crystallization *(Chem.)*. Crystallization of a solution in vacuum at a temperature lower than its bp at ordinary pressure; used in sugar refineries to separate sugar from syrups.

vacuum distillation *(Chem.)*. Distillation under reduced pressure. As a reduction of pressure effects a lowering of the boiling-point, many thermolabile substances can be distilled.

vacuum evaporation *(Phys.,Space)*. Under normal conditions, there is an equilibrium of molecular exchange at the surface of a solid body – molecules leaving the surface and others captured by it. Under vacuum conditions and in space, these molecules are lost.

vacuum-filament lamp *(Elec.Eng.)*. An incandescent electric lamp in which the filament is enclosed in a highly evacuated bulb.

vacuum filtration *(Chem.Eng.)*. A process of filtration where a partial vacuum is applied to increase the rate of filtration by causing the liquid to be sucked through the filter.

vacuum forming *(Plastics)*. Shaping process applied to heated sheets of thermoplastics by clamping the plastic sheet in a holder, heating the sheet (usually electrically) then applying suction when the sheet is in the pliable, or 'rubbery', state.

vacuum furnace *(Eng.)*. One in which material is outgassed by radiant heating *in vacuo*. See **vacuum arc melting**.

vacuum impregnation *(Elec.Eng.)*. The process of treating armature and transformer windings by applying moisture-resisting varnish to the insulation, under vacuum, thereby ensuring that the varnish penetrates the pores of the insulating material when normal atmospheric conditions are restored.

vacuum induction melting *(Eng.)*. Using induction to melt metal *in vacuo* preparatory to casting.

vacuum melting *(Eng.)*. See **vacuum induction melting**.

vacuum oven *(Elec.Eng.)*. An oven for heating armature and transformer windings under vacuum, so as to drive off all moisture from the insulation prior to impregnation.

vacuum photocell *(Electronics)*. High-vacuum photoemissive cell in which anode current equals total photoemission currents, so that strict proportionality between current and incident illumination is obtained.

vacuum printing frame *(Print.)*. A frame from which air can be exhausted to ensure close contact between the film image and plate when printing down for any of the printing processes.

vacuum pump *(Eng.)*. General term for apparatus which displaces gas against a pressure.

vacuum servo *(Autos.)*. Servo mechanism operated by a vacuum provided by the induction pipe of the engine; used in power-assisted brake systems.

vacuum switch *(Elec.Eng.)*. One whose contacts separate in vacuum.

vacuum tube *(Electronics)*. See **valve**.
vacuum tube rectifier *(Elec.Eng.)*. One which exploits the unidirectional movement of electrons flowing from heated electrode to gathering electrode.
vacuum tube voltmeter *(Elec.Eng.)*. US for *valve voltmeter*.
vadose zone *(Geol.)*. The unsaturated zone between the water table and the surface of the ground.
vagina *(Bot.)*. A sheath, especially the leaf sheath of grasses.
vagina *(Zool.)*. Any sheathlike structure; the terminal portion of the female genital duct leading from the uterus to the external genital opening. adjs. *vaginal, vaginant, vaginate, vaginiferous*.
vaginal plug *(Zool.)*. In female Rodents and Insectivores, the coagulated secretion of Cowper's glands which blocks the vagina and prevents premature escape of seminal fluid and further mating.
vaginismus *(Med.)*. Painful spasmodic contraction of the muscles of the vagina and/or of the muscles forming the pelvic floor. See **dyspareunia**.
vaginitis *(Med.)*. Inflammation of the vagina.
vagotonia, vagotony *(Med.)*. The condition of heightened activity of the vagus nerve.
vagus *(Zool.)*. (1) The tenth cranial nerve of Vertebrates, supplying the viscera and heart and, in lower forms, the gills and lateral line system. (2) In Mammals, supplying the larynx.
Val *(Chem.)*. Symbol for **valine**.
valence band *(Chem.)*. Range of energy levels of electrons which bind atoms of a crystal together.
valence electrons *(Chem.)*. Those in the outer shell of an *atom*, which, by gaining, losing, or sharing such electrons, may combine with other atoms to form *molecules*.
valency *(Chem.)*. The combining power of an atom or group in terms of hydrogen atoms (or equivalent). The valency of an ion is equal to its charge. For *valency bond, valence*, see **chemical bond**.
valency *(Immun.)*. In immunology it refers to the number of antigen binding sites on an antibody molecule. Those belonging to most Ig classes have two, but IgM has 10 combining sites and IgA, which can exist as monomer, dimer and higher polymers has multiples of 2. The valency of an *antigen* can likewise be expressed in terms of the number of antigen combining sites with which it can combine. Most large antigen molecules are multivalent. See **lattice hypothesis**.
valentinite *(Min.)*. Antimony trioxide, Sb_2O_3, occurring as orthorhombic crystals or radiating aggregates; snow-white when pure: it is formed by the decomposition of other ores of antimony.
valeric acids *(Chem.)*. $C_4H_9.COOH$, monobasic fatty acids, of which four isomers are known, viz. n-valeric acid (*pentanoic acid*), $CH_3.(CH_2)_3.COOH$, bp 185°C; isovaleric acid (*3-methylbutanoic acid*), $(CH_3)_2=CH.CH_2COOH$, bp 175°C; methylethylacetic acid (*2-methylbutanoic acid*), $(CH_3)(C_2H_5)CH.COOH$, bp 177°C; pivalic acid (*2,2-dimethylpropanoic acid*), $(CH_3)_3C.COOH$, bp 164°C.
validation *(Comp.)*. Input control technique used to detect any data which is inaccurate, incomplete or unreasonable.
valine *(Chem.)*. *2-amino-3-methylbutanoic acid*, $(CH_3)_2CH(NH_2).COOH$. An amino acid. The L- or S-isomer is a constituent of proteins. Symbol Val, short form V. ⇨
valley *(Build.)*. The re-entrant angle formed between two intersecting roof slopes.
valley *(Geol.)*. Any hollow or low-lying tract of ground between hills or mountains, usually traversed by streams or rivers, which receive the natural drainage from the surrounding high ground. Usually valleys are developed by stream erosion; but in special cases, faulting may also have contributed, as in rift valleys.
valley board *(Build.)*. A board nailed along the top of the valley rafter as a support for a **laced valley**.
valley bog *(Ecol.)*. Type of *Sphagnum* bog forming where water draining from relatively acid rocks stagnates in a flat-bottomed valley so as to keep the soil constantly wet.
value engineering *(Aero.)*. A total approach to engineering design that seeks to achieve required performance, reliability and quality at minimum cost by attention to simplicity, avoidance of unnecessary functions and integration of design and manufacturing techniques.
value function *(Nuc.Eng.)*. See **separation potential**.
valvate *(Bot.)*. (1) Organs having margins touching but not overlapping. See also **aestivation**, **vernation**. Cf. **imbricate**. (2) Opening by valves.
valve *(Bot.)*. (1) The flattened part of a theca of a diatom frustule. (2) That part of a fruit wall which separates at dehiscence.
valve *(Electronics)*. US *tube*. Simple vacuum device for amplification by an electron stream, of many types.
valve *(Eng.)*. Any device which controls the passage of a fluid through a pipe.
valve *(Zool.)*. Any structure which controls the passage of material through a tube, duct or aperture, usually in the form of membranous folds, as the auriculo-ventricular valves of the heart: in Mollusca, Cirripedia, Brachiopoda, one of several separate pieces composing the shell: in Insects, a covering plate or sheath, especially one of a pair which can be opposed to form a tubular structure, as the valves of the ovipositor. Also *valva*.
valve bounce *(Eng.)*. The unintended secondary opening of an engine valve due to inadequate rigidity of various parts in the valve gear. The deflected parts allow the valve to seat too early and when their strain energy is released, the valve opens again. This bounce causes valve breakage, seat wear, and irregular functioning.
valve box (chest) *(Eng.)*. In a force-pump or steam engine, the chamber which contains the valves or valve: the **steam chest** of a steam engine.
valve characteristic *(Elec.Eng.)*. Graphical relation between voltage and current for specified electrodes, all other potentials being maintained constant.
valve diagram *(Eng.)*. For a steam-engine slide valve, a graphical method of correlating the throw and angle of advance of the eccentric, the lead and laps of the valve, and the points of admission, cut-off, compression, and release. The 'Bilgram', the 'Reuleaux', and the 'Zeuner' valve diagrams are examples of the method.
valve effect *(Elec.Eng.)*. The unilateral conductivity of certain electrodes (notably aluminium) in suitable solutions. As anodes, they may withstand several hundred volts, although current will pass freely in the opposite direction.
valve face *(Eng.)*. The sealing surface of a valve which slides over, or beds on to, the seating.
valve gear *(Eng.)*. The linkage by which the valves of an engine derive their motion and timing from the crank-shaft rotation.
valve inserts *(Autos.)*. Valve seatings of special heat- and tetra-ethyl lead-resisting steel which are pressed into the alloy heads of high-duty petrol engines.
valve-opening diagram *(Autos.)*. A diagram showing the lift or the opening area of a valve to a base of engine crank angle or piston displacement.
valve parameters *(Electronics)*. Numerical quantities obtained from the characteristic curves and used in circuit analysis. See, **differential anode resistance, mutual conductance, transistor parameters.**
valve rectifier *(Elec.Eng.)*. A rectifier of the vacuum or the gas-discharge type.
valve relay *(Elec.Eng.)*. A thermionic valve arranged to operate as a synchronous voltage relay in high-frequency a.c. circuits.
valve rockers *(Autos.)*. See **rocker arms**.
valve spring *(Eng.)*. The helical spring (or springs) used to close a poppet valve after it has been lifted by the cam; generally, any spring which closes a valve after it has been lifted mechanically or by fluid pressure.
valve timing *(Autos.)*. See **timing**.
valve voltmeter *(Elec.Eng.)*. Valve used for measuring voltages, rectified output current being dependent on voltage applied to the input. It takes negligible power

from measured circuit; can be calibrated at low frequencies for use at very high frequencies, when other means are impossible. See **diode voltmeter**. U.S. term *vacuum tube voltmeter*.

valvular heart disease *(Med.)*. Disease affecting the valves of the heart, making them either too tight, *stenotic*, for normal blood flow or too loose to prevent regurgitation of blood, *incompetent*.

valvulitis *(Med.)*. Inflammation of a valve of the heart.

VAM *(Bot.)*. Abbrev. for *Vesicular-Arbuscular Mycorrhiza*.

vanadate *(Chem.)*. An ion containing vanadium (V). Orthovanadate is $VO_4{}^{3-}$ and metavanadate is $VO_3{}^{-}$.

vanadinite *(Min.)*. Vanadate and chloride of lead, typically forming brilliant reddish hexagonal crystals or globular masses encrusting other minerals in lead-mines.

vanadium *(Chem.)*. A very hard, whitish metallic element. Symbol V. at. no. 23, r.a.m. 50.941, rel.d. at 20°C 5.5, mp 1710°C, electrical resistivity 22×10^{-8} ohm metres at 20°C. Its principal use is as a constituent of alloy steel, e.g. in chromium-vanadium, manganese-vanadium, and high-speed steels.

vanadium steel *(Eng.)*. Typical constitution is carbon 0.4–0.5%, chromium 1.1–1.5%, vanadium 0.15–0.2%, balance being iron. The vanadium removes occluded oxygen and nitrogen thus giving improved tensile and elastic quality. Up to 0.5% V may be used.

vanadous, vanadic *(Chem.)*. Referring to divalent and trivalent vanadium respectively.

vanadyl *(Chem.)*. The cations VO^{2+} and VO^{3+} containing vanadium (IV) and vanadium (V) respectively.

Van Allen radiation belts *(Astron.)*. Two belts encircling the Earth within which electrically charged particles are trapped. The lower Van Allen belt extends from 1000 to 5000 km above the equator with the second at about 20 000 km. Within these zones electrons originally captured from the **solar wind** are trapped. The belts are named after the American space scientist who discovered them.

Van de Graaff generator *(Elec.Eng.)*. Very-high-voltage electrostatic machine, using a high-speed belt to accumulate charge in a large Faraday cage, which takes the form of a metal globe. Recent models use Freon or nitrogen gas under high pressure. Used as voltage source for accelerator tubes, e.g. in neutron sources.

van der Waals' equation *(Chem.)*. An equation of state which takes into account the effect of intermolecular attraction at high densities and the reduction in effective volume due to the actual volume of the molecules: $(P + a/v^2)(v - b) = RT$, a and b being constant for a particular gas. See **gas laws**.

van der Waals' forces *(Chem.)*. Weak attractive forces between molecules or crystals, represented by the coefficient a in *van der Waals' equation*. They vary inversely as the sixth power of the interatomic distance, and are due to momentary dipoles caused by fluctuations in the electronic configuration of the molecules.

vane *(Eng.)*. (1) One of the elements which variably divide the fluid space in a *vane pump*. (2) An alternative name for blade in turbines, flow meters and similar rotary devices.

vane *(Surv.)*. A disk attachment to a levelling staff; it provides a sliding target which the staffman can move into the line of sight of the level. See **target rod**.

vane *(Zool.)*. The web of a feather, composed of the barbs and barbules; also called *vexillum*.

vane pump *(Eng.)*. A type of pump used, e.g. as a vacuum or oil pump or as a compressor, in which a slotted rotor is mounted eccentrically in a circular stator (or a similar geometrical arrangement), and vanes sliding in the rotor slots divide the crescent-shaped fluid space into variable volumes.

vanes *(Aero.)*. See **inlet guide-**, **nozzle guide-**, **swirl-**.

vane wattmeter *(Phys.)*. Instrument for measuring power transmitted in waveguide, depending on mechanical forces induced in a vane.

vanillin *(Chem.)*. 3-Methoxy-4-hydroxy-benzene carbaldehyde, found in vanilla pods and some other plants. It crystallizes in white needles, mp 80°C

vanner *(Min.Ext.)*. See **frue vanner**.

vanning *(Min.Ext.)*. Rough estimate of cassiterite or other heavy mineral, made by washing finely ground sample on a flat shovel, or in a vanning plaque.

van't Hoff factor *(Chem.)*. The ratio of the number of dissolved particles (ions and undissociated molecules) actually present in a solution to the number there would be if no dissociation occurred. Symbol *i*.

van't Hoff's law *(Chem.)*. The osmotic pressure of a dilute solution is equal to the pressure which the dissolved substance would exert if it were in the gaseous state and occupied the same volume as the solution at the same temperature.

van't Hoff's reaction isochore *(Chem.)*. For a reversible reaction taking place at constant volume,

$$\frac{d \log_e K}{dT} = \frac{\Delta U}{RT^2},$$

where K is the equilibrium constant, T is the absolute temperature, R is the gas constant, ΔU is the heat absorbed in the complete reaction.

van't Hoff's reaction isotherm *(Chem.)*. For a reversible reaction taking place at constant temperature,

$$-\Delta A = RT\log_e K - RT\Sigma n \log_e c,$$

where $-\Delta A$ is the decrease in free energy, R the gas constant, T the absolute temperature, K the equilibrium constant, and $\Sigma n\log_e c$ is of the same form as $\log_e K$, but with the equilibrium concentrations replaced by the initial values.

vaporization *(Chem.)*. The conversion of a liquid or a solid into a vapour.

vapour *(Phys.)*. A gas which is at a temperature below its critical temperature and can therefore be liquefied by a suitable increase in pressure.

vapour barrier *(Chem.)*. Covering which prevents water condensing within the insulation around a cold surface. Often a pigmented vinyl polymer solution applied by brush.

vapour compression cycle *(Eng.)*. To operate this cycle requires a compressor, throttle valve, evaporator and condenser. The working fluid is one which changes readily between the liquid and vapour phases. The fluid passes through the evaporator and heat is transferred to it, changing it from the saturated to the dry, superheated state. Ideally this compression is reversibly adiabatic. From the compressor it flows through the condenser, where heat is removed and the gas liquefies. It is then throttled (constant enthalpy) to the initial conditions. A suitable working fluid is Freon 12.

vapour concentration *(Meteor.)*. The ratio of the mass of water vapour in a sample of moist air to the volume of the sample.

vapour-liquid-solid mechanism *(Chem.)*. Method, applicable to most crystalline substances, of growing different near-perfect crystalline forms.

vapour lock *(Eng.)*. Of a volatile fluid in a pipe, the formation of vapour in a petrol feed-pipe to a carburettor caused by undue heating of the pipe, resulting in an interruption of flow.

vapour permeability *(Paper)*. The rate of passage of a vapour, e.g. water vapour, through the paper.

vapour phase inhibitor *(Chem.)*. Stable organic chemicals coated on paper or board which slowly evaporate, so preventing air reaching the metallic articles enclosed in the package and preventing corrosion, e.g. ethanolamine benzoate, cyclohexylamine nitrite (nitrate (III)). Abbrev. *VPI*.

vapour pressure *(Phys.)*. The pressure exerted by a vapour, either by itself or in a mixture of gases. The term is often taken to mean saturated vapour pressure, which is the vapour pressure of a vapour in contact with its liquid form. The saturated vapour pressure increases with rise of temperature. See **saturation of the air**.

VAr *(Elec.Eng.)*. Abbrev. for *Volt-Amperes reactive*. Unit of reactive power. See **reactive volt-amperes**.

varactor *(Electronics)*. Two-electrode semi-conductor with a non-linear capacitance instantaneously dependent on voltage; cf. **varistor**.

variability *(Chem.)*. The number of **degrees of freedom** (1) of a system.

variable *(Comp.)*. A name or label declared as a **data type** which during the execution of a program becomes bound to an actual but changing value in a particular storage location. See **local variable, global variable**.

variable *(Maths.)*. A term used in algebra to describe symbols like x, y, suggesting that the symbol can be replaced by an element from a specified set to give a meaningful statement. If x can be replaced by elements of a set S, then x is said to be a variable in the set S.

variable aperture shutter *(Image Tech.)*. See **fade shutter**.

variable-area propelling nozzle *(Aero.)*. A turbojet *propelling nozzle* which can be varied in effective outlet area, either mechanically or aerodynamically, to match it to the optimum engine operating conditions (principally thrust), thereby improving fuel economy: essential for the efficient use of an *afterburner* (see **reheat**) and in supersonic flight.

variable area track *(Image Tech.)*. Photographic sound track in which modulation is represented by variations of the image width.

variable coupling *(Elec.Eng.)*. An electro-magnetic coupling between two a.c. circuits in which the mutual inductance is continuously variable between wide limits.

variable cycle engine *(Aero.)*. A gas turbine in which the gas path can be changed by diverters or valves so it can operate in different modes at different flight speeds. Examples are the tandem fan turbojet which operates with four nozzles in hovering flight but with only one as a straight turbojet at high speed. Similarly in hypersonic propulsion systems the airflow has to completely bypass the turbojet at Mach numbers over 4.

variable density sound track *(Image Tech.)*. Photographic sound track in which modulation is represented by variations of the image density; now obsolete.

variable-density wind tunnel *(Aero.)*. A closed-circuit wind tunnel wherein the air may be compressed to increase the **Reynolds number**; also *compressed-air wind tunnel*.

variable geometry *(Aero.)*. See **variable sweep**.

variable inductor *(Elec.Eng.)*. An **inductor** whose self-inductance is continuously variable.

variable-inlet guide vanes *(Aero.)*. See **inlet guide vanes**.

variable-interval schedule *(Behav.)*. See **interval schedule of reinforcement**.

variable-length record *(Comp.)*. The number of bits (or characters) is not predetermined.

variable-pitch propeller *(Aero.)*. See **propeller**.

variable-ratio schedule *(Behav.)*. See **ratio schedule of reinforcement**.

variable-ratio transformer *(Elec.Eng.)*. A transformer whose voltage ratio can be varied by altering the number of active turns in either the primary or the secondary winding.

variable region *(Immun.)*. The N-terminal half of light chains (V_L) and the N-terminal half of the Fab portion of the heavy chains (V_H) of immunoglobulin molecules. The amino acid sequences in these regions are variable within a single immunoglobulin class and light chain type. This variability is controlled by the **immunoglobulin genes**. Similarly the N-terminal region of the α, β and γ chains of the T lymphocyte antigen receptors have variable regions, controlled in a similar manner.

variable-reluctance pick-up *(Phys.)*. One in which the reluctance of a magnetic circuit is varied, the consequent modulation of flux from a permanent magnet generating an e.m.f., as in a transducer or gramophone-record reproducer.

variable resistance *(Elec.Eng.)*. See **rheostat**.

variable-speed drive *(Elec.Eng.)*. An electric drive whose speed is continuously variable between wide limits.

variable-speed motor *(Elec.Eng.)*. An electric motor whose speed is continuously variable between wide limits.

variable star *(Astron.)*. Any star with a luminosity not constant with time. The variation can be regular or irregular. The stars can vary in their apparent **magnitude** for a variety of reasons. (1) In an *eclipsing binary* the pair of stars periodically eclipse, as seen from the Earth, and the apparent magnitude of the pair falls when one member conceals the other. (2) Many stars pulsate, and the change in size and surface temperature leads to a change in luminosity. The principle types are **Cepheid variables**, *RR Lyrae stars* and the long-period *Mira variables*.

variable sweep *(Aero.)*. An aircraft with wings so hinged that they can be moved backward and forward in flight to give high **sweepback** for low **drag** in supersonic flight and high **aspect ratio**, with good lifting properties, for take-off and landing. Colloq. *swing-wing*.

variable-voltage control *(Elec.Eng.)*. A system of controlling speed by varying the voltage applied at the motor terminals.

Variac *(Elec.Eng.)*. TN of autotransformer in the form of a toroid winding on ring laminations, the output voltage being varied by a rotating brush contact on the turns.

variance *(Stats.)*. The mean of the sum of squared deviations of a set of observations from the corresponding mean; the second moment about the mean of a random variable; the dispersion parameter of a probability distribution.

variant *(Biol.)*. A specimen differing in its characteristics from the type and produced either by changed environmental conditions and/or by mutation.

variate *(Stats.)*. A quantity, measurement or attribute which is the subject of statistical analysis.

variation *(Astron.)*. The name given to the fourth principal periodic term in the mathematical expression of the moon's motion, caused by the variation of the residual attraction of the sun on the earth-moon system during a synodic month; it has a maximum value of 39′ and a period of 14.77 days.

variation *(Biol.)*. The differences between the offspring of a single mating; the differences between the individuals of a race, subspecies, or species; the differences between analogous groups of higher rank.

variation *(Surv.)*. See **magnetic declination**.

variation factor *(Elec.Eng.)*. The ratio between the maximum and the minimum illumination along a street or roadway illuminated at intervals by overhead lamps.

variation of latitude *(Astron.)*. A phenomenon, first detected in 1888 by Küstner, who showed that, owing to the spheroidal form and non-rigid consistency of the earth, its axis of rotation does not remain constant in direction but varies in a regular manner about a mean position, so that the latitude of a place also undergoes periodic variations.

variation order *(Civ.Eng.)*. A document giving authority for some alteration in work being done under contract.

variations *(Build.)*. See **extras**.

varicella *(Med.)*. See **chickenpox**.

varicocele *(Med.)*. A varicose condition of the plexus of veins which leave the testis to form the spermatic vein, forming at the upper part of the testis a swelling which feels like a mass of spaghetti.

varicose *(Med.)*. Of the nature of, pertaining to, or affected by, a varix or varices; (of veins) abnormally dilated, lengthened, and tortuous.

variegated copper ore *(Min.)*. A popular name for **bornite**. So named from the characteristic tarnish that soon appears on the freshly fractured surface.

variegation *(Bot.)*. The occurrence of differently coloured areas on leaves or petals due to: virus infection (streaks, spots, mottles); mineral deficiency (veinal or interveinal chlorosis); genetically determined patterns as in leaves of *Coleus* cultivars; chimerical structure (light coloured borders on leaves); a **transposon** (transposable element).

variety *(Biol.)*. A race; a stock or strain; a sport or mutant; a breed; a subspecies; a category of individuals within a species which differ in constant transmissible characteristics from the type but which can be traced back to the type by a complete series of gradations; a geographical or biological race.

varifocal lens *(Image Tech.)*. A camera lens in which changing the focal length causes the focus position to alter. Cf. **zoom lens**.

varimeter, varmeter, varometer *(Elec.Eng.)*. Equivalent terms for instrument measuring reactive volt-amperes in circuit.

variocoupler *(Elec.Eng.)*. Device comprising two inductors whose mutual inductance can be varied for variable inductive coupling between two circuits.

variola *(Med.)*. See **smallpox**.

variola *(Vet.)*. See **swine pox**.

variola minor *(Med.)*. See **alastrim**.

variolitic *(Geol.)*. Said of a fine-grained igneous rock of basic composition containing small more or less spherical bodies (*varioles*), consisting of minute radiating fibres of feldspar, comparable with the more perfect spherulites in acid igneous rocks.

variometer *(Elec.Eng.)*. Variable inductor comprising two coils connected in series and arranged one inside the other, the inner coil rotated so as to vary the mutual inductance between them and hence the total inductance.

variscite *(Min.)*. A greenish hydrated phosphate of aluminium ($AlPO_4.2H_2O$) occurring as nodular masses.

varistor *(Electronics)*. Two-electrode semi-conductor with a nonlinear resistance dependent on instantaneous voltage. Used to short-circuit transient high voltages in delicate electronic devices.

VariTyper *(Print.)*. A justifying typewriter producing copy suitable for reproduction by a printing process, and requiring a second typing for line justification. See **typewriter composition**.

varix *(Med.)*. *Varicose vein*; a vein that is abnormally dilated, lengthened, and tortuous. pl. *varices*.

Varley loop test *(Elec.Eng.)*. A method of determining the position of a cable fault, in which resistance measurements are made with a resistance bridge, first, so that the fault forms one junction of the bridge, and, secondly, so that the conductor resistance of the cable is measured directly.

varnish *(Build.)*. A transparent solution of a resin or resinous gum in spirits or oil applied as a protective, decorative coating to enhance the underlying surface. Available in a range of sheen from flat to high gloss.

varved clays *(Geol.)*. Distinctly and finely stratified clays of glacial origin, deposited in lakes during the retreat stage of glaciation. The stratification is thought to be a seasonal banding, and its study enabled Baron de Geer to work out the chronology of the Pleistocene Ice Age.

vas *(Zool.)*. A vessel, duct, or tube carrying fluid. pl. *vasa*. adj. *vasal*. See **vas deferens**.

vasa efferentia *(Zool.)*. A series of small ducts by which the semen is conveyed from the testis to the vas deferens.

vasa vasorum *(Zool.)*. In Vertebrates, small blood vessels ramifying in the external coats of the larger arteries and veins.

vascular *(Bot.,Zool.)*. Relating to vessels which convey fluids or provide for the circulation of fluids, e.g. xylem and phloem: provided with vessels for the circulation of fluids.

vascular area *(Zool.)*. See **area vasculosa**.

vascular bundle *(Bot.)*. Strand of conducting tissue composed of xylem and phloem and, usually in dicotyledons, cambium. See **eustele**.

vascular cylinder *(Bot.)*. Same as **stele**.

vascular plant *(Bot.)*. *Tracheophyte*. Member of those plant groups that have a vascular system of xylem and phloem, the pteridophytes and seed plants.

vascular ray *(Bot.)*. A *ray* in secondary xylem or phloem.

vascular system *(Bot.)*. All the conducting tissues (xylem and phloem) in a vascular plant.

vascular system *(Zool.)*. The organs responsible for the circulation of blood and lymph, collectively.

vasculum *(Bot.)*. A receptacle for collecting botanical specimens.

vas deferens *(Zool.)*. A duct leading from the testis to the ejaculatory organ, the urino-genital canal, the cloaca or the exterior. pl. *vasa*.

vasectomy *(Med.)*. Excision of the vas deferens, or of part of it, either therapeutically or for sterilization.

Vaseline *(Chem.)*. TN for high-boiling residues obtained from the distillation of petroleum; a *petroleum jelly*.

VASI *(Aero.)*. Abbrev. for *Visual Approach Slope Indicator*.

vasifactive *(Zool.)*. See **vasoformative**.

vaso- *(Genrl.)*. Prefix from L. *vas*, vessel.

vasochorial placenta *(Zool.)*. A chorioallantoic placenta in which the epithelium and the endometrium of the uterus disappear, and the chorion is in intimate contact with the endothelial wall of the maternal capillaries as in some Carnivora. Also known as *endotheliochorial placenta*.

vasoconstrictor *(Zool.)*. Of certain autonomic nerves, causing constriction of the arteries.

vasodilator *(Med.)*. A drug, e.g., glyceryl nitrate, which effects expansion of the blood vessels.

vasodilator *(Zool.)*. Of certain autonomic nerves, causing expansion of the arteries.

vaso-epididymostomy *(Med.)*. The operation of anastomosing the vas deferens to the upper part of the epididymis, forming a communication between the two; performed for the treatment of sterility in the male.

vasoformative *(Zool.)*. Pertaining to the formation of blood or blood vessels.

vasoformative cells *(Zool.)*. See **angioblast**.

vasohypertonic *(Zool.)*. See **vasoconstrictor**.

vasohypotonic *(Zool.)*. See **vasodilator**.

vasoinhibitory *(Zool.)*. See **vasodilator**.

vasomotor *(Zool.)*. Causing constriction or expansion of the arteries; as certain nerves of the autonomic nervous system.

vasopressin *(Biol.)*. A nonopeptide secreted by the posterior pituitary gland. It elevates blood pressure by the contraction of small blood vessels and aids water resorption by the kidney. Used in the diagnosis and treatment of **diabetes insipidus**.

vasopressor *(Med.)*. Substance which causes a rise of blood pressure.

vat dye *(Textiles)*. A water insoluble dye which is converted into a soluble form by treatment with a reducing agent in alkali. The textile material is immersed in the solution; the dye on the fabric is then converted by oxidation into its insoluble form. Vat-dyes usually have good fastness properties.

vat dyestuffs *(Chem.)*. A series of insoluble dyestuffs that can be reduced to their water-soluble leuco-compounds, which are oxidized by exposure to the air, thus producing the dyestuff direct on the fibre.

vaterite *(Min.)*. A less common polymorph of calcium carbonate, crystallizing in the hexagonal system; forms artificially. Cf. **calcite**, **aragonite**.

vat machine *(Paper)*. See **cylinder mould machine**.

vault *(Build.)*. (1) An arched roof or ceiling. (2) A room or passage covered by an arched ceiling. (3) An underground room.

vault light *(Build.)*. A form of *pavement light*.

V-beam radar *(Radar)*. One which uses two fan-shaped beams to determine the range, bearing and height of the target. One beam is vertical, the other inclined, intersecting at ground level. They rotate continuously about a vertical axis.

V-bed knitting machine *(Textiles)*. See **flat knitting machine**.

VCO *(Electronics)*. Abbrev. for *Voltage Controlled Oscillator*.

V-connection *(Elec.Eng.)*. An alternative name for the open delta connection of two phases of a three-phase a.c. system.

VCR *(Image Tech.)*. *Video Cassette Recorder*, equipment for recording and reproducing television programme material on magnetic tape contained in an enclosed cassette.

V-curve *(Elec.Eng.)*. The power-factor/temperature curve of a cable with moisture, which shows a pronounced minimum at about 40°C.

VDRL test *(Immun.)*. A rapid screening test for syphilis which depends on the flocculation of a cardiolipin-cholesterol-lecithin preparation if the serum being tested contains anti-cardiolipin antibody. Used in venereal disease reference laboratories, hence *VDRL*.

VDU *(Comp.)*. Abbrev. for *Visual Display Unit* (British), (Video Display Unit, US). Terminal device usually incorporating a **cathode ray tube** with a screen on which text and **graphics** can be displayed. Used as an **I/O device** in conjunction with a keyboard or a mouse for **interactive computing**.

vector *(Aero.)*. The course or track of an aircraft, missile etc., but generally a quantity possessing both magnitude and direction, e.g. wind velocity.

vector *(Biol.)*. (1) An agent (usually an insect) which transmits a disease caused by a parasite or micro-organism from one host to another. (2) Insect, bird, wind etc. carrying pollen from stamen to stigma. (3) A DNA molecule derived from a self-replicating phage, virus, plasmid or bacterium which can accept inserted DNA sequences. Used to transfer DNA from one organism to another to make recombinant DNA.

vector *(Maths.)*. A vector or vector quantity is one which has magnitude and direction, e.g. force or velocity; two such quantities of the same kind obey the parallelogram law of addition. A *localized vector* is one in which the line of action is fixed, as contrasted with a *free vector*, in which only the direction is fixed.

vector addition *(Maths.)*. Compounding of two vector quantities according to parallelogram law.

vector algebra *(Maths.)*. Manipulation of symbols representing vector quantities according to laws of addition, subtraction and multiplication, which these quantities obey.

vectorcardiography *(Med.)*. The recording of the electrical signal from the heart (**electrocardiogram**) in a series of mean vectors.

vectored thrust *(Aero.)*. The deflexion of the thrust from turbojet(s) to provide a jet-lift component. Particularly applied to a system using a ducted fan engine with bifurcated nozzles on fan and jet pipe so that there are four sources of thrust, thereby contributing to the balance of the system. The swivelling nozzles which deflect the thrust at any angle from horizontally aft to several degrees forward of the vertical are under the control of the pilot. Also used in missile rocket-propulsion and control systems.

vector graphics *(Comp.)*. **Computer graphics** in which the electron beam in a **CRT** is made to draw onto the screen directly from co-ordinates calculated in the computer.

vector potential *(Elec.Eng.)*. Potential postulated in electromagnetic field theory. Space differentiation (*curl*) of the vector potential yields the field. Magnetic vector potential is due to electric currents, while electric vector potential is assumed to be due to a flow of magnetic charges.

vector product *(Maths.)*. Of two vectors, the vector perpendicular (right-hand screw convention) to both the given vectors, of magnitude equal to the product of the magnitudes of the two given vectors, multiplied by the sine of the angle between them. Vector products are usually denoted by $\mathbf{a} \times \mathbf{b}$, or $\mathbf{a} \wedge \mathbf{b}$, while scalar products are denoted by $\mathbf{a.b}$. Scalar products alone are used in electrical engineering.

vectorscope *(Image Tech.)*. Instrument which displays phase and amplitude of an applied signal, e.g. of chrominance signal in a colour TV system.

vector space *(Maths.)*. A left vector space is a left module over a division ring. Normally the division ring considered is a field, such as real or complex numbers, in which case the distinction between left and right no longer applies.

vee antenna *(Telecomm.)*. A pair of wires fed in a *vee* formation, fed by a twin feeder at the apex. Maximum radiation is between the two wires in the direction of the apex.

vee belt *(Eng.)*. A power transmission belt having a cross-section of truncated vee form, usually running in corresponding vee grooves in pulleys.

vee gutter *(Build.)*. A gutter of V-shape, as required, for example, along the valley between two roofs sloping towards each other.

vee joint *(Build.)*. A joint between **matched boards** which have been chamfered along their edges on the same side to present a vee depression at their junction.

vee notch *(Civ.Eng.)*. A notch plate having a triangular notch cut in it, used for the measurement of small discharges.

veering *(Meteor.)*. A clockwise change in the direction from which the wind comes. Cf. *backing*.

vee roof *(Build.)*. A roof formed by two lean-to roofs meeting to enclose a valley.

vee-tail *(Aero.)*. An aircraft tail unit consisting of two surfaces on each side of the centre line, usually at about 45° to the horizontal, which serve both as tailplane and fin. The associated hinged control surfaces are so actuated that they move in unison up/down as elevators and left/right as rudders, following conventional movements of the control column and rudder bar respectively. Also *butterfly tail*.

vee thread *(Eng.)*. A screw thread in which the thread profile is V-shaped (as for metric or Whitworth threads), as distinct from other forms, e.g. square thread.

vegan *(Med.)*. A very strict vegetarian who abstains from all food of animal origin.

vegetable oils *(Chem.)*. Oils obtained from plants, seeds etc. Cf. **mineral oils**.

vegetable parchment *(Paper)*. A packaging paper produced by passing a free beaten waterleaf web of base paper through a bath containing cold 70% sulphuric acid and immediately washing the paper thoroughly to remove traces of acid. If necessary neutralizing and softening. Agents may also be used. The resultant dense paper possesses grease resistant and high wet strength properties.

vegetable pole *(Zool.)*. The lower portion or pole of an ovum in which cleavage is slow owing to the presence of yolk. Cf. **animal pole**.

vegetation *(Med.)*. A term used to describe warty aggregations of blood components and bacteria that accumulate on heart valves in **endocarditis**.

vegetation survey *(Min.Ext.)*. See **geobotanical surveying**.

vegetative *(Bot.)*. Not reproducing sexually; not carrying flowers or other sexually reproducing structures.

vegetative functions *(Zool.)*. The autonomic or involuntary functions, as digestion, circulation.

vegetative propagation *(Bot.)*. The natural and especially the horticultural production of new plants from bulbs, offsets, stolons, rhizomes etc. and by layering, taking cuttings, grafting etc. In the absence of mutation, the offspring will be genetically identical to the parent plant. See also **asexual reproduction, micropropagation**.

vegetative reproduction *(Bot.)*. Usually natural rather than horticultural **vegetative propagation**.

vegetative reproduction *(Zool.)*. Propagation by budding.

vehicle *(Build.)*. The liquid substance which, when mixed with a pigment, forms a paint.

vehicle battery *(Elec.Eng.)*. A battery of heavy-duty-type secondary cells, which forms the source of electrical energy in self-contained electric road vehicles.

veil *(Bot.)*. See **partial veil, universal veil**.

veil *(Zool.)*. See **velum**.

veiled cell *(Immun.)*. Cell characterized by large veil-like processes found in the lymph draining skin, especially after local antigenic stimulation. They express large amounts of surface MHC Class II antigens on their surface, and represent Langerhans cells in transit from the skin to the draining lymph node, where they take the form of interdigitating cells, which are very effective in antigen presentation to T-lymphocytes.

vein *(Bot.)*. A vascular bundle and supporting tissues in a leaf.

vein *(Geol.)*. A tabular or sheet-like body of rock,

penetrating a different type of rock. Sometimes applied to particularly narrow igneous intrusions (dykes and sills), the term is more often applied to material deposited by solutions, such as quartz veins or calcite veins. Many ore deposits consist of veins in which the ore mineral is one of several constituents.

vein *(Zool.)*. A vessel conveying blood back to the heart from the various organs of the body: in Insects, a wing nervure. adj. *venous*.

vein islet *(Bot.)*. See **areole**.

vein stuff *(Min.Ext.)*. The minerals occurring in veins or fissures.

veliger *(Zool.)*. The secondary larval stage of most Mollusca, developing from the trochophore and characterized by the possession of a velum.

vellum *(Paper)*. An early form of writing surface made from the skins of calves, lambs, or goats. In modern usage, a thick writing paper.

vellus *(Zool.)*. In Man, the widespread short downy hair which replaces the fine lanugo which almost covers the foetus from the fifth or sixth month until shortly before birth.

velocity *(Phys.)*. (1) The rate of change of displacement of a moving body with time; a vector expressing both magnitude and direction (cf. *speed* which is scalar). (2) For a wave, the distance travelled by a given phase divided by the time taken. Symbol *v*.

velocity amplitude *(Acous.)*. Amplitude of the velocity of the volume elements oscillating with the sound wave.

velocity budget *(Space)*. The sum of the **characteristic velocities** involved in a complete space mission.

velocity constant *(Chem.)*. See **rate constant**.

velocity microphone *(Acous.)*. See **pressure-gradient microphone**.

velocity-modulated oscillator *(Telecomm.)*. One in which an electron beam is velocity-modulated (*bunched*) by passing through a toroidal cavity resonator (*rhumbatron*), the energy exciting a further cavity (*collecting*) and feeding back into the first.

velocity modulation *(Electronics)*. Modulation in a klystron in which the velocities of the electrons, and hence their **bunching**, is related to radio signals to be amplified.

velocity of light *(Phys.)*. See **speed of light**.

velocity of propagation *(Elec.Eng.)*. The velocity of an electromagnetic wave, or the velocity with which a wave travels along a transmission line. For free space, or an air-spaced cable, it equals the **velocity of light**.

velocity of sound *(Aero.)*. A key factor in aircraft design and operation, *Mach* 1.0. Under *ISA* sea level conditions the speed of sound in air is 761.6 mi/h (1229 km/h), reducing with temperature, to 660.3 mi/h (1062 km/h) at the tropopause (ISA 36 090 ft, 11 000 m), where the temperature is −54.46°C (−69.64°F), above which height it remains constant. Also *speed of sound*.

velocity of sound *(Phys.)*. In dry air at s.t.p. 331.4 m/s (750 mi/h). In fresh water, 1410 m/s, and in sea water, 1540 m/s. The above values are used for sonar ranging but do not apply to explosive shock waves. They must be corrected for variations of temperature, humidity etc.

velocity rate constant *(Chem.)*. See **rate constant**.

velocity ratio *(Phys.)*. The ratio of the distance moved through by the point of application of the effort to the corresponding distance for the load in a machine. The ratio of the **mechanical advantage** to the velocity ratio is termed the efficiency of the machine.

velocity resonance *(Telecomm.)*. See **phase-**.

velodyne *(Elec.Eng.)*. Tachogenerator in which rotational speed of output shaft is proportional to applied voltage through feedback.

velour *(Textiles)*. A heavy pile or napped woven fabric or felt with the surface fibres all lying in the same direction. Also, a warp-knitted fabric with long raised loops. The velour used for hats is usually a rabbit fur that has been felted and raised.

velour paper *(Paper)*. Paper made by depositing short wool fibres on an adhesive coated paper. Also *flock paper*.

velum *(Zool.)*. A veil-like structure, as the *velum pendulum* or posterior part of the soft palate in high Mammals; in some Ciliophora, a delicate membrane bordering the oral cavity: in Porifera, a membrane constricting the lumen of an incurrent or excurrent canal: in hydrozoan medusae, an annular shelf projecting inwards from the margin of the umbrella: in Rotifera, the trochal disk: in Mollusca, the ciliated locomotor organ of the veliger larva: in Cephalochorda, the perforated membrane separating the buccal cavity from the pharynx.

velvet *(Textiles)*. Woven fabric with a dense short pile, formed in loops which are then cut. In one method of manufacture two cloths are woven face to face with the pile warp weaving through both. The double cloth is then sliced in the loom to produce two separate fabrics.

velvet *(Zool.)*. The tissue layers covering a growing antler, consisting of periosteum, skin, and hair.

velveteen *(Textiles)*. A woven pile fabric that is made by cutting exta floating weft threads after weaving.

venae cavae *(Zool.)*. The caval veins; in higher Vertebrates, three large main veins conveying blood to the right auricle of the heart.

venation *(Bot.)*. The pattern formed by the veins of a leaf.

venation *(Bot.,Zool.)*. The arrangement of the veins or nervures; by extension, the veins themselves considered as a whole.

V-end connections *(Elec.Eng.)*. V-shaped conductors connecting the ends of corresponding pairs of bars in a bar-wound armature.

veneer *(For.)*. Timber in the form of a thin layer of uniform thickness; it is generally cut from timber of fine appearance, the veneer being glued to a less expensive material and appearing on the surface.

veneered construction *(Build.)*. A mode of construction in which a thin external layer of facing material is applied to the steel or reinforced concrete framework.

veneering hammer *(Build.)*. One with a flat wooden head in which a brass strip is inserted. Used for smoothing veneers and forcing out the glue.

veneer saw *(Build.)*. Small saw with curved cutting edge, unset teeth, and a wooden grip running along the back. Used for cutting veneers.

venereal disease *(Med.)*. One of a number of contagious diseases (see **acquired immunodeficiency syndrome, chancroid, gonorrhoea, granuloma inguinale, lymphogranuloma inguinale, syphilis**) usually contracted in sexual intercourse. Abbrev. *VD*.

Venetian *(Print.)*. A style of type based on the fifteenth-century original of Nicolas Jenson, and characterized by strong colour, prominent serifs and oblique bar to '*e*'.

Venetian arch *(Arch.)*. (1) A **Queen Anne arch**. (2) A pointed arch in which the extrados and the intrados are not parallel.

Venetian mosaic *(Build.)*. See **terrazzo**.

Venetian shutters *(Build.)*. See **jalousies**.

Venetian window *(Build.)*. A window having two mullions dividing the window space into three compartments, usually a large centre light and two narrow side lights.

Venn diagram *(Maths.)*. In logic and mathematics, a diagram consisting of shapes (rectangles, circles, etc.) that show by their inclusion, exclusion or intersection, the relationships between classes or sets.

venography *(Med.)*. The intravenous injection of *contrast material* to allow X-ray visualization of the veins.

venomous *(Zool.)*. Having poison-secreting glands.

veno-occlusive disease of the liver *(Med.)*. Syndrome of liver failure in Jamaica caused by occlusion of small blood vessels in the liver, produced by ingestion of plant alkaloids in 'bush teas'.

venosclerosis *(Med.)*. Hardening of a vein due to thickening of its walls.

venous system *(Zool.)*. That part of the circulatory system responsible for the conveyance of blood from the organs of the body to the heart.

vent *(Aero.)*. (1) The opening (usually at the centre) in a parachute canopy which stabilizes it by allowing the air to escape at a controlled rate. (2) Opening to atmosphere from, e.g. a fuel tank.

vent *(Eng.)*. To allow air to enter, or escape from, a confined space to facilitate movement (of liquid, a piston, etc.) within the space.

vent *(Geol.)*. See **volcanic vent**.

vent *(Zool.)*. The aperture of the anus or cloaca in Vertebrates.

venter *(Zool.)*. A protuberance; a median swelling; the abdomen in Vertebrates; the ventral surface of the abdomen.

vent gleet *(Vet.)*. An infectious disease of fowls, characterized by inflammation of the cloaca.

ventifact *(Geol.)*. A wind-faceted pebble. See **dreikanter**.

ventilated wind tunnel *(Aero.)*. A wind tunnel for **transonic** testing in which part of the walls in the working section are perforated, slotted or porous, to prevent choking by the presence of the model, which would otherwise render measurements unreliable in the range from Mach 0.9 to 1.4.

ventilating fan *(Elec.Eng.)*. An electrically driven fan whose function is to force cooling air through the ventilating ducts of an electrical machine.

ventilating plant *(Elec.Eng.)*. Electrically driven fans and their attendant apparatus, designed for specific ventilating purposes.

venting *(Foundry)*. The process of making holes through the rammed sand of a mould or core in order to allow gases to escape during pouring and so avoid blown castings. See **vent wires, wax vent**.

vent pipe *(Build.)*. A small escape pipe which carries off foul gases from a sanitary fixture and leads into the vent stack. Abbrev. *VP*.

ventral *(Bot.)*. (1) The under surface of plants with creeping stems next to the substrate. (2) The adaxial side of aerial shoots. Thus, the ventral surface of a leaf is normally the upper surface. The term is not always used consistently and is better avoided.

ventral *(Zool.)*. Pertaining to that aspect of a bilaterally symmetrical animal which is normally turned towards the ground.

ventral fins *(Aero.)*. Fins mounted under the rear fuselage to increase directional stability, usually under high incidence conditions when the main fin may be blanketed.

ventral suture *(Bot.)*. The presumed line of junction of the edges of the infolded carpel.

ventral tank *(Aero.)*. An auxiliary *fuel tank*, fixed or jettisonable, mounted externally under the fuselage; sometimes *belly tank*.

ventricle, ventriculous *(Zool.)*. A chamber or cavity, especially the cavities of the Vertebrate brain and the main contractile chamber or chambers of the heart (in Vertebrates or Invertebrates). adj. *ventricular*.

ventricose *(Bot.,Zool.)*. (1) Swollen in the middle. (2) Having an inflated bulge to one side.

ventricular fibrillation *(Med.)*. Unco-ordinated rapid electric activity of the ventricle of the heart. There is no effective pulse and death ensues rapidly unless the abnormal heart rhythm is reversed by electrical **defibrillation**.

ventricular septal defect *(Med.)*. A congenital abnormality where there is an opening in septum between left and right ventricles allowing blood to *shunt* from left to right ventricles. May also occur as a complication of **myocardial infarction**.

ventriculography *(Radiol.)*. The radiological visualization of either the cerebral ventricles or the left and right ventricles of the heart following the injection of a **contrast medium**. Can also be done after the injection of **radionuclides**.

ventrifixation *(Med.)*. The operation of stitching the uterus to the anterior wall of the abdomen, for the treatment of retroversion of the uterus.

ventriloquism *(Acous.)*. Art or practice of speaking in such a manner that the voice appears to come from some other source than the vocal organs of the speaker.

ventrisuspension *(Med.)*. An operation for replacing the retroverted uterus by transplanting the round ligaments of the uterus into the anterior abdominal wall in such a way that they exert a strong pull on the uterus.

ventrofixation *(Med.)*. See **ventrifixation**.

ventrosuspension *(Med.)*. See **ventrisuspension**.

vent stack *(Build.)*. A vertical pipe carried up from the highest point in a system of house drains to a level clear of all windows and opening skylights; it provides a safe escape for foul gases from drains and sanitary fixtures.

ventube *(Min.Ext.)*. A flexible ventilating duct some distance away from the source of fresh air.

venturi *(Aero.)*. A convergent-divergent duct in which the pressure energy of an air stream is converted into kinetic energy by the acceleration through the narrow part of the wasp-waisted passage. It is a common method of accelerating the airflow at the working section of a supersonic wind tunnel. Small venturis are used on some aircraft to provide a suction source for vacuum-operated instruments, which are connected to the low-pressure neck of the duct.

venturi flume *(Civ.Eng.)*. A flume which is constricted at one section with convergent upstream and divergent downstream walls, the difference in water-level at the constriction and at a point in the full channel upstream affording a means of measuring the rate of flow. See **flume**.

venturi meter *(Eng.)*. One in which flow rate is measured in terms of pressure drop across a venturi (or tapered throat) in a pipe.

vent wires *(Foundry)*. Wires ranging from 1/16 to 3/8 in diameter, used for making vent holes in the rammed sand of a mould or core.

venule *(Zool.)*. In Chordata, small blood vessels which receive blood from the capillaries and unite to form veins.

Venus *(Astron.)*. Second planet from the Sun and that attaining the greatest brilliancy in the sky, outshining all stars, hence its poetic names *morning star* and *evening star*. It approaches nearer to Earth than any other planet. The visible disc is actually a blanket of opaque cloud overlying an atmosphere rich in carbon dioxide, water vapour and sulphur dioxide. Surface pressure is 90 atmospheres and temperature 470°C. The temperature is elevated by the *greenhouse effect*; the atmosphere is transparent to short-wave infrared from the Sun, but opaque to long-wave infrared from the surface. The surface is a rocky desert. Radar mapping has shown the planet has mountain ranges, craters, extinct volcanoes and a deep rift valley. There are no natural satellites.

Venus' hair stone *(Min.)*. Also *Veneris crinis*. Variety of **rutile**. See **flèches d'amour**.

verapamil *(Med.)*. A **calcium channel blocking** agent which produces vasodilatation and also has potent anti-arrhythmic effects. Useful in the treatment of supraventricular arrhythmias.

verbal test *(Behav.)*. Mental test consisting primarily of items measuring vocabulary, verbal reasoning, comprehension etc. Cf. **performance test**.

Verbenaceae *(Bot.)*. Family of ca 3000 spp. of dicotyledonous flowering plants (superorder Asteridae). Trees, shrubs and herbs, almost all tropical and subtropical. Flowers gamopetalous, usually zygomorphic, and with a superior ovary. Includes teak and other important timber trees.

verde antico *(Chem.)*. A green patina formed on old bronze by oxidation; it is imitated artificially by pickling.

Verdet's constant *(Phys.)*. The angle, per unit thickness traversed, of a transparent medium through which a ray of polarized light rotates in unit magnetic field according to the Faraday effect. For light of 589 nm in water, rotation is 0.000 477 rad/A.

verdigris *(Chem.)*. The green basic copper (II) carbonate (i.e. $CuCO_3.Cu(OH)_2$) formed on copper exposed to moist air.

verdite *(Min.)*. A green rock, consisting chiefly of green mica (fuchsite) and clayey matter, occurring as large boulders in the North Kaap River, South Africa; used as an ornamental stone.

verge *(Build.)*. The edge of the roof covering projecting beyond the gable of a roof.

verge *(Horol.)*. The axis of a clock pallet, especially that of the verge escapement.

verge board *(Build.)*. See **barge board**.

verge escapement *(Horol.)*. One of the earliest known escapements, now obsolete. The pallets are set at right angles to the escape wheel (crown wheel) axis, and its action has a recoil.

verge tile, tile-and-a-half *(Build.)*. A tile which is purpose made to a wider size than normal, to assist in forming the bond at the end of a roof.

verification *(Comp.)*. The act of checking transferred data, usually at the stage of input to a computer, by comparing copies of the data before and after transfer.

vermiculation *(Build.)*. A variety of rustication, distinguished by worm-shaped sinkings.

vermicule *(Zool.)*. A small worm-like structure or organism, as the motile phase of certain Sporozoa.

vermiculites *(Min.)*. A group of hydrated sheet silicates, closely related chemically to the chlorites, and structurally to talc. They occur as decomposition products of biotite mica. When slowly heated, they exfoliate and open into long wormlike threads, forming a very lightweight water-absorbent aggregate used in seed planting and, in building, as an insulating material.

vermiform *(Zool.)*. Worm-like, as the *vermiform appendix*.

vermifuge *(Med.)*. Having the power to expel worms from the intestines; any drug which has this power. See also **anthelminthic**.

vermis *(Zool.)*. In lower Vertebrates, the main portion of the cerebellum; in Mammals, the central lobe of the cerebellum.

vernal *(Ecol.)*. Of, or belonging to, spring.

vernal equinox *(Astron.)*. See **equinox**.

vernalization *(Bot.)*. The natural or artificial promotion of flowering by a period of low temperature, around 4°C.

vernation *(Bot.)*. (1) Arrangement of unexpanded leaves in the vegetative bud. See **aestivation**. (2) Same as **ptyxis**.

Verneuil process *(Min.)*. The technique invented by the French chemist Verneuil for the manufacture of synthetic corundum and spinel by fusing pure precipitated alumina, to which had been added an amount of the appropriate oxide for colouring, in a vertical, inverted blow-pipe type of furnace.

vernier *(Eng.)*. A small movable auxiliary scale, attached to, and sliding in contact with, a scale of graduation. Usually graduated at 9/10ths of the scale on the main scale, it enables readings on the latter to be made to a fraction (usually a tenth) of a division by noting which member of the auxiliary scale is aligned with any line on the main scale.

vernier arm *(Surv.)*. The part of an instrument carrying the vernier or verniers.

vernier capacitor *(Elec.Eng.)*. A variable capacitor of small capacitance, connected in parallel with a larger fixed one, used for fine adjustment of total capacity.

vernier potentiometer *(Elec.Eng.)*. Precision pattern based on the **Kelvin-Varley slide**. Balance can be attained purely by the operation of switches so that the possibility of wear associated with sliding contacts is avoided.

veronal *(Chem.)*. See **barbitone**.

verruca *(Zool.)*. A wart-like process; especially one of a number of wart-like processes situated around the base of certain kinds of alcyonarian polyp.

verrucose *(Bot.)*. Warty; covered with wart-like outgrowths.

versatile *(Bot.)*. Of an anther, attached to the tip of the filament by a small area on its dorsal side, so that it turns freely in the wind, facilitating the dispersal of the pollen.

versatile *(Zool.)*. Capable of free movement, as the toes of birds when they may be turned forwards or backwards.

versed sine *(Civ.Eng.)*. See **rise** (1).

versene *(Chem.)*. Sodium versenate, the sodium salt of **EDTA**, used for the **complexometric titration** of calcium ion.

versicolorous *(Bot.,Zool.)*. Not all of the same colour: changing in colour with age.

versine *(Maths.)*. A trigonometrical function of an angle, required for the solution of spherical triangles. It is given by vers(ine) $\theta = 1 - \cos(\text{ine})\ \theta$.

version *(Med.)*. The act of turning manually the foetus *in utero* in order to facilitate delivery.

verso *(Print.)*. A left-hand page of a book, bearing an even number. Cf. **recto**.

verst *(Genrl.)*. Russian measure of length, 0.6629 mile (1.065 km).

vertebra *(Zool.)*. One of the bony or cartilaginous skeletal elements of mesodermal origin which arise around the notochord and compose the backbone: pl. *vertebrae*. adjs. *vertebral, vertebrate*.

Vertebrata *(Zool.)*. A subphylum of Chordata in which the notochord stops beneath the forebrain and a skull is always present. There are usually paired limbs. The brain is complex and associated with specialized sense organs, and there are at least ten pairs of cranial nerves. The pharynx is small, and there are rarely more than seven gill slits. The heart has at least three chambers and the blood has corpuscles containing haemoglobin.

vertebraterial canals *(Zool.)*. In Vertebrates, small canals found one on each side of all or most of the cervical vertebrae. They are formed by the articulation or fusion of the two heads of the small or vestigial cervical ribs to the centra and transverse processes, and the vertebral arteries run through them.

vertex *(Build.,Civ.Eng.)*. See **crown**.

vertex *(Maths.)*. Of a polygon or polyhedron: one of the points in which the sides or faces intersect. Of a conic, the points in which it is cut by its axes. See also **cone, pencil**.

vertex *(Zool.)*. In higher Vertebrates, the top of the head, the highest point of the skull; in Insects, the dorsal area of the head behind the epicranial suture.

vertical aerial photograph *(Surv.)*. A photograph taken from the air, for purposes of aerial survey work, with the camera pointing directly at the ground so that the optical axis is vertical or nearly so.

vertical boiler *(Eng.)*. A steam boiler having a vertical cylindrical shell and domed or spheroidal firebox, from which (generally) vertical flue tubes lead to the smoke box and chimney.

vertical circle *(Surv.)*. The graduated circular plate used for the measurement of vertical angles by theodolite.

vertical component *(Elec.Eng.)*. The vertical component of the force experienced by unit magnetic pole as the result of the action of the earth's magnetic field. Cf. **horizontal component**.

vertical curve *(Surv.)*. The curve, generally parabolic, which is introduced between two railway or highway gradients in order to provide a gradual change from one to the other.

vertical engine *(Eng.)*. Any engine in which the cylinders are arranged vertically above the crankshaft.

vertical escapement *(Horol.)*. An escapement in which the axis of the balance is at right angles to that of the escape wheel. In a verge watch, the balance staff is vertical when the watch is in the laying position.

vertical force instrument *(Ships)*. An instrument used in adjusting magnetic compasses, particularly in the correction of **heeling error**.

vertical gust *(Aero.)*. A vertical air current, which can be of dangerous intensity, particularly when met by aircraft flying at high speed.

vertical-gust recorder *(Aero.)*. An **accelerometer** which records graphically the intensity of accelerations due to vertical gusts and, simultaneously, the airspeed; used in assessment of aircraft fatigue life: abbrev. *v.g. recorder*.

vertical interval *(Image Tech.)*. The period of **field blanking** between successive TV pictures.

vertical interval time-code *(Image Tech.)*. **Time-code** information inserted in the **vertical interval** of a TV signal.

vertical-lift bridge *(Civ.Eng.)*. A bridge consisting of two towers connected by a span which can be raised and lowered vertically, maintaining its horizontal position between the towers.

vertical milling machine *(Eng.)*. A **milling machine** in which the cutter spindle is vertical, the table movements being the same as in plane machines. Uses end mills and slot mills.

vertical polarization *(Telecomm.)*. Transmission of radio waves in such a way that the electric lines of force are vertical and the magnetic lines horizontal; transmitting and receiving *dipoles* are mounted vertically to handle signals polarized in this way. Cf. **horizontal polarization**.

vertical scanning *(Image Tech.)*. That in which individual lines are vertical, not, as normal, horizontal.

vertical separation *(Aero.)*. See **separation**.

vertical shaft alternator *(Elec.Eng.)*. A water-turbine driven alternator designed to operate with its shaft vertically above, and directly coupled to, the turbine shaft.

vertical speed indicator *(Aero.)*. A sensitive form of differential pressure gauge which measures variations in pressure sensed at the **static-pressure tube** and indicates them in terms of rates of climb and descent. Mainly used in high performance gliding.

vertical spindle motor *(Elec.Eng.)*. An electric motor specially designed to operate with its spindle in a vertical position.

vertical take-off and landing *(Aero.)*. See **VTOL**.

vertical tiling *(Build.)*. See **weather tiling**.

vertical wind tunnel *(Aero.)*. A wind tunnel wherein the air flow is upward and which is used principally for testing freely spinning models. Also *spinning tunnel*.

verticil *(Bot.)*. A whorl.

verticillaster *(Bot.)*. A kind of inflorescence found in dead nettles and related plants. It looks like a dense whorl of flowers, but is really a combination of two crowded dichasial cymes, one on each side of the stem.

verticillate *(Bot.)*. Arranged in whorls.

vertigo *(Med.)*. *Dizziness*: a condition in which the person has the sensation of turning or falling, or of surrounding objects turning about himself. See also **Ménière's disease**.

very fine screen *(Image Tech.,Print.)*. Term for half-tones used only for specially fine detail, from 175–400 lines/in (7–16 lines/mm); 1st figure is limit for half-tone blocks, 2nd is rarely used.

very high frequencies *(Telecomm.)*. Those between 30 and 300 MHz. Abbrev. *VHF*.

very large scale integration *(Comp.)*. See **VLSI**.

very low frequencies *(Telecomm.)*. Those between 10 and 30 kHz. Abbrev. *VLF*.

vesica *(Zool.)*. The urinary bladder.

vesicant *(Med.)*. Causing blisters; any agent which does this. See **war gas**.

vesicle *(Biol.)*. A structure like a lysosome, surrounded by a membrane and situated in the cellular cytoplasm.

vesicle *(Geol.)*. See under **vesicular structure**.

vesicle *(Med.,Zool.)*. A small cavity containing fluid; a small saclike space containing gas; one of the three primary cavities of the Vertebrate brain; a small bladder-like sac. Also *vesicula*.

vesicular *(Bot.,Zool.)*. Like, or pertaining to, a vesicle; like a bladder.

vesicular-arbuscular mycorrhiza *(Bot.)*. Abbrev. *VAM*. Endotrophic **mycorrhiza** in which the fungus invades the cortical cells to form vesicles and arbuscules (finely branched structures). VAMs are very common among herbaceous plants including many crop plants and may significantly improve the mineral nutrition of the host.

vesicular exanthema *(Vet.)*. A febrile, virus disease of swine in which vesicles develop on the snout, lips, tongue, and feet.

vesicular stomatitis *(Vet.)*. A virus disease of horses, and occasionally cattle, characterized by vesicle formation on the tongue and mucosa of the mouth.

vesicular structure *(Geol.)*. A character exhibited by many extrusive igneous rocks, in which the expansion of gases has given rise to more or less spherical cavities (vesicles). The latter may become filled with such minerals as silica (chalcedony, agate, quartz), zeolites, chlorite, calcite, etc.

vesicula seminalis *(Zool.)*. In many animals, including Man, a sac in which spermatozoa are stored during the completion of their development.

vesiculate, vacuolate *(Biol.)*. Having **vesicles** or vacuoles.

vesiculitis *(Med.)*. Inflammation of the vesiculae seminales.

vessel *(Bot.)*. A long, from 1 cm to 10 m, unbranched, water-conducting tube of the xylem, formed from a longitudinal file of cells by the perforation of their common end walls. Water moves through perforations within a vessel but through pits into and out of vessels and from one vessel to the next. Vessels are found in very few pteridophytes, a few gymnosperms and most angiosperms.

vessel *(Zool.)*. A channel or duct with definitive walls, as one of the principal vessels through which blood flows.

vessel element, -member, -segment *(Bot.)*. A tracheary element of the xylem which with others in a file forms a vessel. Cf. **tracheid**.

vestibule *(Build.)*. A small antechamber at the entrance to a building, or serving as an entrance room to a larger room.

vestibule *(Zool.)*. A passage leading from one cavity to another or leading into a cavity from the exterior: in Protozoa, a depression in the ectoplasm at the base of which is the mouth; in a female Mammal, the space between the vulva and the junction of the vagina and the urethra (urinogenital sinus); in Birds, the posterior chamber of the cloaca; in Vertebrates generally, the cavity of the internal ear. adjs. *vestibular, vestibulate*.

vestibulectomy *(Med.)*. Surgical removal of the membranous labyrinth of the inner ear.

vestibulitis *(Med.)*. A condition characterized by slight fever, vertigo, vomiting, and ataxia, resulting in complete deafness; due to an inflammation of the labyrinth and cochlea of the inner ear.

vestigial *(Zool.)*. Of small or reduced structure: of a functionless structure representing a useful organ of a lower form. n. *vestige*.

vestigial sideband *(Telecomm.)*. Type of amplitude-modulated transmission in which the whole of one sideband is transmitted, but only part of the other; used generally in television transmitters.

vestiture *(Bot.,Zool.)*. A covering, e.g. of hairs, feathers, fur or scales.

vesuvianite *(Min.)*. Hydrated silicate of calcium and aluminium, with magnesium and iron, crystallizing in the tetragonal system. It occurs commonly in metamorphosed limestones. Also known as *idocrase*.

veterinary *(Vet.)*. Relating to, concerned with, the diseases of domestic animals.

VFR *(Aero.)*. Abbrev. for *Visual Flight Rules*.

V-gene *(Immun.)*. That coding for the variable region of immunoglobulin heavy or light chain. It is widely separated from the C-gene (constant region) in germ line DNA (as present in lymphocyte precursors and other body cells such as in liver), but during maturation of B-lymphocytes it is rearranged by translocation to a position close to the 5′ end of the C-gene, but still separated from it by an **intron**. A similar process takes place during the maturation of T-lymphocytes.

v.g. recorder *(Aero.)*. Abbrev. for aircraft speed (v) and normal acceleration (g) in a vertical-gust recorder.

VHF *(Telecomm.)*. Abbrev. for *Very High Frequency*.

VHS *(Image Tech.)*. *Video Home System*, TN for a VCR system using 1/2 in magnetic tape.

viable *(Bot.,Zool.)*. Capable of living and developing normally.

vial *(Surv.)*. The glass tube containing the liquid in a **level tube**.

viameter *(Surv.)*. See **perambulator**.

Vi antigen *(Immun.)*. Surface somatic antigen present in freshly isolated strains of *Salmonella typhi* and *S. paratyphi*, which masks the O antigen and renders the organisms relatively unable to combine with antibody against the O antigen. Vi antigen is associated with virulence, possibly for this reason.

vibrating capacitor *(Elec.Eng.)*. One in which the potential on the electrode is varied by mechanical oscillation, so that the steady applied potential is converted to an alternating potential, which can be more easily amplified. Also *oscillating capacitor*.

vibrating conveyor *(Eng.)*. A tubular or flat trough with sides, to which vibrators are attached. The latter impart an upward and forward conveying movement to granular materials in the trough.

vibrational energy *(Chem.)*. Energy due to the relative oscillation of two contiguous atoms in the molecule.

vibration dampers *(Eng.)*. Devices fitted to an engine crankshaft in order to suppress or minimize stresses resulting from torsional vibration at critical speeds. See **dynamic damper**, **frictional damper**.

vibration galvanometer *(Elec.Eng.)*. Moving-coil taut-suspension galvanometer with natural frequency of vibration of coil, tunable usually over range 40 Hz to 1000 Hz. Small a.c. currents at the resonant frequency excite a large response, hence these instruments form sensitive detectors for circuits such as a.c. bridges or potentiometers.

vibration pick-up *(Elec.Eng.)*. One which uses some form of microphone or transducer, e.g. crystal, capacitance, electromagnetic, to transform the oscillatory motion of a surface, e.g. of machinery, into an electrical voltage or current.

vibration-rotation spectrum *(Phys.)*. The infrared end of the electromagnetic spectrum which arises from vibrational and rotational transitions within a molecule.

vibrator *(Civ.Eng.)*. Equipment which, by vibration, consolidates loose soil or compacts cohesive soil for better bearing resistance, etc. *Vibroflotation* incorporates water in the vibratory process.

vibrator *(Eng.)*. An oscillating mechanism, usually electromagnetically excited, which imparts vibrations to hoppers, conveyors, or other parts of machines, usually for the purposes of dislodging, loosening, or propelling materials or workpieces.

vibrator *(Horol.)*. An instrument for checking the time of vibration of a balance and its spring. It consists of a vertical arm to which the free end of the spring is clipped and a master balance and spring in a container with a glass cover in the base. The master balance is vibrated, and the vibrations of the balance to be tested are compared with it.

vibratory bowl feeder *(Eng.)*. A small parts feeder, comprising a bowl, a spring suspension system, and an electromagnetic or hydraulic exciter, used for separating, orienting, and feeding workpieces to an assembling machine, machine tool, etc.

vibrionic abortion *(Vet.)*. A contagious form of abortion in cattle and sheep caused by infection with *Campylobacter fetus*, vars. *fetus* and *intestinalis*.

vibrissa *(Zool.)*. (1) In Mammals, one of the stiff tactile hairs borne on the sides of the snout and about the eyes. (2) One of the vaneless rictal feathers of certain Birds, e.g. Flycatchers. pl. *vibrissae*.

vibrometer *(Elec.Eng.)*. An instrument used for the measurement of the displacement, velocity or acceleration of a vibrating body.

vibromill *(Chem.Eng.)*. A **ball mill** in which the impacts between the balls and the material to be ground are achieved by vibrating the mill at high frequency.

vibrotron *(Electronics)*. A special form of triode valve in which the anode can be vibrated by a force external to the envelope.

Vicat needle *(Civ.Eng.)*. An apparatus which tests the setting-time of a cement specimen by measuring the effect produced by a specially shaped and loaded needle which is pressed against the surface of the specimen.

vice *(Eng.)*. A clamping device, usually consisting of two jaws which can be brought together by means of a screw, toggle or lever, used for holding work that is to be operated on. Generally named after the trade on which it is used. Also *vise*.

vicinal *(Chem.)*. Substituted on adjacent carbon atoms, e.g. on the 1,2,3,4 atoms in a naphthalene nucleus.

vicinal faces *(Min.)*. Facets modifying normal crystal faces, but themselves abnormal, as their indices cannot be expressed in small whole numbers; they usually lie nearly in the plane of the face they modify.

Vickers hardness test *(Eng.)*. A method of determining the hardness of metals by indenting them with a diamond pyramid under a specified load and measuring the size of the impression produced.

Vicq d'Azyr, bundle of *(Med.)*. Mamillo-thalamic tract.

vicuña *(Textiles)*. The fine hair obtained from the undercoat of the Vicuña, *Auchenia vicugna*, a kind of Llama from S. America. The fabric woven from it is the softest and finest of any made of wool or hair.

video *(Image Tech.)*. (1) Originally, the picture component of a television signal, in contrast to *audio*, the sound component, but now used generally to describe the electronic handling of visual images, usually accompanied by sound. (2) Popular abbreviation for both **videotape recorder** and **videogram**.

video amplifier *(Image Tech.)*. Wideband amplifier which, in a TV system, passes the picture signal.

video carrier *(Image Tech.)*. The carrier wave which is modulated with a video signal.

video-cassette *(Image Tech.)*. Enclosed **cassette** for handling magnetic videotape.

video-conferencing *(Image Tech.)*. A closed-circuit TV system connecting several points for live inter-communication in real time for both sound and vision.

video-conferencing *(Telecomm.)*. Conducting meetings or conferences in two or more remote places, the participants being televised for the benefit of the others.

video conforming *(Image Tech.)*. The assembly, usually by re-recording, of selected sections from a quantity of videotape recordings to match the finally edited programme instructions.

video disc *(Image Tech.)*. A rotating flat circular plate for the reproduction of pre-recorded video programmes on a TV receiver; ease of access to different parts of the recording make the video disc particularly convenient for use in **inter-active** systems.

videodisk *(Comp.)*. High capacity **disk** storage with laser read/write system, linked to a TV screen or as a data store for a computer.

video display unit *(Comp.)*. See **VDU**.

video frequency *(Image Tech.)*. The RF bands used for European TV broadcasting are generally UHF, 470–582 and 614–854 MHz; SHF around 12 GHz may be used for satellite transmission; circuits within TV equipment must handle signals up to 5.5 MHz at least.

videogram *(Image Tech.)*. A complete video programme pre-recorded on tape or disc, including motion picture productions transferred to the video medium.

video integration *(Telecomm.)*. A method of using the redundancy of repetitive signals to improve the signal-to-noise ratio, by summing the successive video signals.

video map *(Radar)*. Electronic system for transferring a map of any chosen territory, which may be on a transparency or store in computer memory, on to a radar display. See **map/chart comparison unit**.

video pulse *(Telecomm.)*. Colloquialism for fast rise-time pulse, i.e. one occurring in TV.

video signal *(Image Tech.)*. That part of a TV signal conveying all the information for the picture image, including colour coding and synchronization.

video stretching *(Telecomm.)*. A method of increasing the duration time of a video pulse.

videotape *(Image Tech.)*. (1) Magnetic tape suitable for video recording and reproduction. (2) A video programme recorded on magnetic tape.

videotape recording *(Image Tech.)*. Recording signals originated by a TV camera on moving magnetic tape for subsequent reproduction or transmission; a **writing speed** much greater than the rate of tape transport is necessary, so heads are mounted on a rotating drum moving across the tape. See **quadruplex** and **helical scan**.

videotex *(Telecomm.)*. (1) Term proposed for international use in place of **viewdata**. (2) Generic term covering **teletext**, a broadcast videotex service as well as viewdata, a wired service.

videotext *(Comp.)*. Communications system which uses an ordinary television set to link with data banks through telephone lines. E.g. **Prestel** (previously *Viewdata*). Cf. **teletext**.

video viewfinder *(Image Tech.)*. A small CCTV system which permits the picture seen in the camera viewfinder to be displayed on a separate monitor screen and recorded on videotape if required.

videowall *(Image Tech.)*. A vertical assembly of a number of TV monitor screens, for example in a block four high and six wide, for the display of images, sometimes integrated over the whole area and sometimes in smaller groupings or individual units.

Vidicon *(Image Tech.)*. TN for a camera tube having a photo-conductive image target layer of antimony trisulphide, scanned by a low-velocity electron beam.

viewdata *(Telecomm.)*. Interactive information service using a telephone line between the user and a central computer, the information being displayed on a suitably adapted domestic TV receiver, developed by British Post Office in 1970s. Now called *Prestel* in UK.

viewer *(Image Tech.)*. Device for the examination of a slide transparency or of motion picture film, frame by frame or in movement.

viewfinder *(Image Tech.)*. An optical or video device, forming part of a camera or accessory to it, showing an image of the scene being recorded, with indication of the exact limits of the field of view involved.

vigia *(Ships)*. A reported navigational danger of unknown nature, marked on charts although its existence has not been confirmed.

vigilance *(Behav.)*. A term for the state of readiness to detect changes in the environment.

Villari effect *(Phys.)*. Temporary change in magnetization, arising from longitudinal stretching.

villose, villous *(Bot.)*. Shaggy.

villus *(Zool.)*. A hair-like or finger-shaped process, such as the absorptive processes of the Vertebrate intestine; one of the vascular processes of the Mammalian placenta which fit into the crypts of the uterine wall. pl. *villi*. adjs. *villous, villiform*.

vimentin *(Biol.)*. **Intermediate filament** protein characteristic of fibroblasts.

vinca alkaloid *(Med.)*. Alkaloids used in the treament of cancer, **leukaemia** and **lymphoma**, which interfere with cell division by causing metaphase arrest. *Vincristine* and *vinblastine* are common examples.

vinegar *(Chem.)*. The product of the alcoholic and acetic fermentation of fruit juices, e.g, grape juice, cider etc. or of malt extracts. Vinegar consists of an aqueous solution of acetic (ethanoic) acid (3–6%), mineral salts and traces of esters.

vinquish *(Vet.)*. See **pine**.

vinyl coatings *(Build.)*. Paint coatings based on poly vinyl chloride or acetate or a combination of both. Many of today's emulsion paints are produced with vinyls.

vinyl foils *(Print.)*. Blocking foils which have no final adhesive layer and transfer exclusively to thermoplastics. Also *cello foils*.

vinyl group *(Chem.)*. The unsaturated monovalent radical $CH_2=CH$-.

vinyl polymers *(Plastics)*. Thermoplastic polymers formed by the co-polymerization of vinyl chloride, $CH_2=CHCl$, and vinyl ethanoate, $CH_2.COOCH=CH_2$. They are odourless and tasteless. PVC (*polyvinyl chloride*); used instead of rubber in electric cables; resists oil and some chemicals, but is slightly inferior to rubber in electrical properties. PVA (*polyvinyl acetate*) is similarly resistant and of wide application (sheets, hose, belts etc.)

violane *(Min.)*. Massive violet-blue **diopside**, used as an ornamental stone.

violetwood *(For.)*. See **purpleheart**.

virescence *(Bot.)*. Abnormal, usually pathogenic, condition in which flowers remain green.

virga *(Meteor.)*. Slight rain or snow which evaporates before reaching the ground.

virgin metal *(Eng.)*. Metal or alloy first produced by smelting, as distinct from secondary metal containing recirculated scrap.

virgin neutrons *(Phys.)*. Those which have not yet experienced a collision and therefore retain their energy at birth.

virgin state *(Phys.)*. Same as **neutral state**.

viridine *(Min.)*. A green iron- and manganese-bearing variety of **andalusite**.

virilism *(Med.)*. The development of masculine characteristics, physical and mental, in the female, often due to hyperplasia of, or the presence of a tumour in, the cortex of the adrenal gland.

virology *(Genrl.)*. The study of viruses.

virtual earth *(Elec.Eng.)*. A point maintained close to ground potential by negative feedback, although not directly connected to ground; e.g. the input terminal of an operational amplifier to which negative shunt voltage feedback is applied.

virtual height *(Telecomm.)*. The apparent height of an ionized layer as deduced from the time interval between the transmitted signal and the resulting ionospheric echo at normal incidence.

virtual image *(Phys.)*. See **image**.

virtually inert anode *(Ships)*. An anode, used in **cathodic protection**, with an impressed direct current from mains or battery. Silicon iron or graphite may be used. Has a limited life but lasts much longer than a **sacrificial anode**.

virtual process *(Phys.)*. As a consequence of the *uncertainty principle*, it is possible for the conservation of mass and energy to be violated for a time t by an amount E such that $Et \sim \hbar$. A transition to a higher quantum state could take place provided this condition was satisfied, but the transition could not be observed and is called a *virtual process*. A particle created in such a process is called a *virtual particle*. This is an important mechanism of nuclear forces.

virtual quantum *(Phys.)*. In higher order perturbation theory, a matrix element which connects an initial state with a final state involves intermediate states in which energy is not conserved. A photon or quantum in one of these states is designated a *virtual quantum*. This concept enables the coulomb energy between two electrons to be regarded as arising from the emission of virtual quanta by one of the electrons and their absorption by the other.

virtual storage *(Comp.)*. Way of apparently extending **main memory**, by allowing the programmer to access backing storage in the same way as immediate access store.

virtual temperature *(Meteor.)*. The virtual temperature of a sample of moist air is the temperature at which dry air of the same total pressure would have the same density as the sample. Use of the virtual temperature obviates the need for a variable *gas constant* in applying the usual equation of state to moist air.

virulence *(Bot.)*. Capacity of a pathogen to cause disease.

virulent phage *(Biol.)*. One which always kills its host. Ant. *lysogenic phage*. See **lysogeny**.

virus *(Biol.)*. A particulate infective agent smaller than accepted bacterial forms, invisible by light microscopy, incapable of propagation in inanimate media and multiplying only in susceptible living cells, in which specific cytopathogenic changes frequently occur. Causative agent of many important diseases of man, lower animals and plants, e.g. poliomyelitis, foot and mouth disease, tobacco mosaic. See also **bacteriophage**.

virus neutralization tests *(Immun.)*. Tests used to identify antibody response to a virus, or, using a known antibody, to identify a virus. Depends on specific antibody neutralizing the infectivity of a virus by preventing it from binding to the target cell. They may be carried out in vivo in susceptible animals or chick embryos or, more usually, in tissue culture.

virus pneumonia of pigs *(Vet.)*. A contagious, pneumonic disease of pigs, usually chronic in form, caused by a virus. Abbrev. *VPP*.

visceral *(Med.)*. See **viscus**.

visceral arch *(Zool.)*. See **gill arch**.

visceral clefts *(Zool.)*. The gill clefts, especially the abortive gill clefts of higher Vertebrates.

visceral gout *(Vet.)*. See **avian gout**.

visceroptosis *(Med.)*. See **enteroptosis**.

viscoelastic *(Phys.)*. A solid or liquid which when deformed exhibits both viscous and elastic behaviour through the simultaneous dissipation and storage of mechanical energy. Shown typically by polymers.

viscometer *(Phys.)*. An instrument for measuring viscosity. Many types of viscometer employ **Poiseuille's formula** for the rate of flow of a viscous fluid through a capillary tube.

viscose *(Chem.)*. See **cellulose xanthate**.

viscose fibre (rayon) *(Textiles)*. Fibres of regenerated cellulose made from wood pulp or cotton linters dissolved in aqueous sodium hydroxide containing carbon disulphide. After passing the viscous solution through spinnerets the cellulose is regenerated in fibre form by passing it through mineral acids.

viscosity *(Phys.,Eng.)*. The resistance of a fluid to shear forces, and hence to flow. Such shear resistance is proportional to the relative velocity between the two surfaces on either side of a layer of fluid, the area in shear, the **coefficient of viscosity** of the fluid and the reciprocal of the thickness of the layer of fluid. For comparing the viscosities of liquids various scales have been devised, e.g. *Redwood No. 1 seconds* (UK), *Saybolt Universal seconds* (US), *Engler degrees* (Germany). See also **kinematic viscosity**.

viscosity of paint *(Build.)*. The ability of a paint to flow which affects the ease of application of paints.

viscountess *(Build.)*. A slate size, 18 × 10 in. (457 × 254 mm).

viscous damping *(Phys.)*. Opposing force, or torque, proportional to velocity, e.g., resulting from viscosity of oil or from eddy currents.

viscous flow *(Phys.)*. A type of fluid flow in which fluid particles, considered to be aggregates of molecules, move along streamlines so that at any point in the fluid the velocity is constant or varies in a regular manner with respect to time, random motion being only of a molecular nature. The name is also used to describe *laminar flow* or *streamline flow*.

viscous hysteresis *(Elec.Eng.)*. The phenomenon of time-lag between the intensity of magnetization and the magnetizing force producing it.

viscus *(Med.,Zool.)*. Any one of the organs situated within the chest and the abdomen: heart, lungs, liver, spleen, intestines, etc. pl. *viscera*, adj. *visceral*.

vise *(Eng.)*. See **vice**.

vishnevite *(Min.)*. The sulphate-bearing equivalent of **cancrinite**, found in nepheline syenite.

visibility *(Image Tech.)*. The ratio of the luminous flux, in lumens, to the corresponding energy flux, in watts.

visibility *(Meteor.)*. The maximum distance at which a black object of sufficient size can be seen and recognized in normal daylight.

visibility curve *(Image Tech.)*. The relation between visibility and wavelength. Owing to varying sensitivity of the eye, this curve indicates a maximum at 555 nm, which is a bright green.

visibility factor *(Telecomm.)*. The ratio of the minimum signal input power to a radar, TV or facsimile receiver for which an ideal instrument can detect the output signal, to the corresponding value when the output signal is detected by an observer watching the CRT.

visibility meter *(Meteor.)*. A meter which attenuates visibility to a standardized value, and measures such visibility on a scale.

visible horizon *(Surv.)*. Junction of sea or earth with sky as seen from observer's position. Also called *apparent* or *sensible horizon*.

visible radiation *(Phys.)*. Electromagnetic radiation which falls within the wavelength range of 780 to 380 nm, over which the normal eye is sensitive.

visible speech *(Acous.)*. (1) Display of oscillogram patterns corresponding to characteristic speech sounds; used as an aid to speech training of the totally deaf. (2) See **sonogram**.

vision mixer *(Image Tech.)*. Equipment, and its operator, for the combination of several picture sources to create the visual effects required by the director, ranging from a simple **cut** to complex **wipe** transitions and **keying** effects.

vision modulation *(Image Tech.)*. The modulation of the carrier effected by the picture signal, as distinct from that reserved for the synchronizing impulses.

Vistavision *(Image Tech.)*. TN of early system of **wide-screen** cinematography using a double-frame area on 35 mm film running horizontally in the camera.

visual acuity *(Phys.)*. A term used to express the spatial resolving power of the eye. Measured by determining the minimum angle of separation which has to be subtended at the eye between two points before they can be seen as two separate points.

visual approach slope indicator *(Aero.)*. A luminous device for day and night use, consisting of red, green and amber light bars on each side of a runway which, by being directed through restricting visors, show a pilot if he is below, on or above and in line with the approach path for an accurate touchdown. Developed by the Royal Aircraft Establishment from a World War II night lighting system known as the *Visual Glide Path Indicator*. Abbrev. *VASI*.

visual binary *(Astron.)*. A **double star** whose two components may be seen as separate in a telescope of sufficient resolving power.

visual cliff *(Behav.)*. An experimental set-up in which there is a vertical drop, over which an animal is prevented from falling by a sheet of glass. Some animals avoid this area as a result of the visual perception of the drop.

visual display unit *(Comp.)*. See **VDU**.

visual fatigue *(Image Tech.)*. Partial loss of visual perception or discrimination as a result of prolonged exposure of the eye to high levels of illumination or light of a dominant colour.

visual flight rules *(Aero.)*. The regulations set out by the controlling authority stating the conditions under which flights may be carried out without radio control and instructions. The regulations usually specify minimum horizontal visibility, cloud base, and precise instructions for the distance to be maintained below and away from cloud. Abbrev. *VFR*.

visual meteorological conditions *(Aero.)*. Weather conditions in which an aircraft can fly under freedom from air-traffic control except in controlled air space. Abbrev. *VMC*.

visual purple *(Phys.)*. See **rhodopsin**.

visual range *(Phys.)*. Observable range of ionizing particles in bubble chamber, cloud chamber or photographic emulsion.

visual violet *(Phys.)*. A photosensitive retinal cone pigment; iodopsin.

vital actions *(Aero.)*. Sequences of pilot actions to be performed in preparation for flight and are learned as part of good airmanship practice. Learned by mnemonics such as 'BUMPF' standing for 'Check brakes, undercarriage, mixture, pitch, flaps'.

vital capacity *(Med.)*. The volume of gas that can be expelled from the lungs after a maximal inspiration, usually of the order of 4 litres in Man.

vitallium *(Med.)*. A metallic alloy used in orthopaedic surgery because of its nonrust, nonreactive properties. These alloys, based on 62½–65% cobalt, carry such other elements as chromium (27–35%), molybdenum (5–5.6%), manganese (0.5–0.6%), iron (up to 1%), and nickel (up to 2%). They show good resistance to heat and corrosion.

vital stain *(Biol.)*. A stain which can be used on living cells without killing them.

vitamin A *(Biol.)*. *Retinol*. A precursor of the prosthetic group of the light sensitive protein, rhodopsin. Deficiency of vitamin A causes night blindness. It is also required by young animals for growth. Fish liver oils and dairy products are rich sources of vitamin A.

vitamin B complex *(Biol.)*. (1) B1 (*thiamin*). As its pyrophosphate it functions as a coenzyme of various enzymes. Deficiency results in the disease beri-beri. Present in yeast and cereal germs. (2) B2 (*riboflavin*). Forms part of the prosthetic group of flavoproteins. Deficiency causes skin and corneal lesions. (3) Niacin (*nicotinic acid*). Component of the coenzyme nicotinamide adenine dinucleotide, NAD. Deficiency results in the disease pellagra. (4) B6 (*pyridoxal*). As its phosphate it acts as a coenzyme for transaminases. (5) Pantothenic acid. A component of coenzyme A. (6) Biotin. The prosthetic group of the enzyme carboxylase. (7) Folic acid (*tetrahydrofolate*). Serves as a donor of 1-carbon fragments for several biosyntheses. Deficiency inhibits these reactions which include the synthesis of purines. (8) B12 (*cobalamin*). Component of the coenzyme cobalamin which takes part in enzymic interconversions of acyl CoAs and methylations. Used in the treatment of pernicious anaemia. Liver is a rich source.

vitamin C *(Biol.)*. *Ascorbic acid*. Important in the hydroxylation of collagen which in its absence is inadequately hydroxylated. The defective collagen produces the skin lesions and blood vessel weaknesses which are characteristic of scurvy, the deficiency disease of this vitamin. Fresh fruit and green vegetables are important sources.

vitamin D *(Biol.)*. *Calciferol*. The vitamin involved in calcium and phosphorous metabolism. Deficiency impairs bone growth and causes the disease ricketts. Fish liver oils are a rich source and the vitamin can, in sunlight, be synthesized in the skin from cholesterol.

vitamin E *(Biol.)*. *α-tocopherol*. A vitamin involved in reproduction. Its absence leads to sterility in both sexes.

vitamin K *(Biol.)*. A necessary requirement for the production of prothrombin, the precursor of thrombin, and consequently essential for normal blood coagulation.

vitamins *(Biol.)*. Organic substances required in relatively small amounts in the diet for the proper functioning of the organism. Lack causes *deficiency diseases* curable by administration of the appropriate vitamin. There are two main groups; the *fat soluble*, vitamins A, D, E and K, and the *water soluble*, vitamin C and the vitamins of the B complex.

vitellarium *(Zool.)*. A yolk-forming gland.

vitelligenous *(Zool.)*. Yolk-secreting or producing.

vitellin *(Chem.)*. A phosphoprotein present in the yolk of the egg.

vitelline *(Bot.,Zool.)*. Egg-yellow; pertaining to yolk.

vitelline membrane *(Zool.)*. A protective membrane formed around a fertilized ovum to prevent the entry of further sperms.

vitellus *(Zool.)*. Yolk of egg.

vitiligo *(Med.)*. Patchy depigmentation of the skin often with a sharp demarcation line and associated with **autoimmune disease**.

Viton *(Plastics)*. TN for synthetic rubber based on a copolymer of vinylidene fluoride and hexa-fluoropropene. Maintains its rubber-like properties over a very wide temperature range and is chemically very resistant. Used for O-rings etc.

vitreous enamel *(Eng.)*. Glazed coating fused on to steel surface for protection and/or decoration.

vitreous humour *(Zool.)*. The jelly-like substance filling the posterior chamber of the Vertebrate eye, between the lens and the retina.

vitreous silica *(Glass)*. A vitreous material consisting almost entirely of silica, made in translucent and transparent forms. The former has minute gas bubbles disseminated in it. Also *fused silica*, *quartz glass*, *silica glass*.

vitreous state *(Phys.)*. A non-crystalline solid or rigid liquid, formed by supercooling the melt. Also *glassy state*.

vitrification *(Nuc.Eng.)*. The incorporation of radioactive waste products (particularly from nuclear fuel processing) into glass. Also called *glassification*. Other techniques under study include ceramics, glass-ceramics, composite materials (e.g. glass beads in a metal matrix) and synthetic minerals. See **Synroc**.

vitrinite *(Geol.)*. An oxygen-rich maceral that is found in coal.

vitriol *(Chem.)*. Sulphuric acid. Also *oil of*.

vitriol *(Min.)*. See **blue-**, **green-**, **white-**.

vitroclastic structure *(Geol.)*. The characteristic structure of volcanic ashes which have been produced by the disruption of highly vesicular glassy rocks, most of the component fragments thus having concave outlines.

vivianite *(Min.)*. Hydrated iron phosphate ($Fe_3P_2O_8.8H_2O$). Monoclinic.

viviparous *(Zool.)*. Giving birth to living young which have already reached an advanced stage of development; cf. **oviparous**. n. *viviparity*.

vivipary *(Bot.)*. (1) The production of bulbils or small plants in place of flowers, as in e.g. *Festuca vivipara*. (2) The premature germination of seeds or spores before they are shed from the parent plant as in many mangrove trees.

VLBI *(Astron.)*. Very Long Baseline Interferometry, a technique of **aperture synthesis** used in **radio astronomy** to link telescopes separated by thousands of kilometres.

VLF *(Telecomm.)*. Abbrev. for *Very Low Frequencies*.

VLSI *(Comp.)*. Very large scale integration refers to a **chip** with 100 000 or more logic **gates**.

VLSI *(Electronics)*. Abbrev. for *Very Large Scale Integration*, referring to integrated circuits where a component density of the order of 10 000 devices is achieved on a single chip.

VLS mechanism *(Phys.)*. See **vapour-liquid-solid mechanism**.

VMC *(Aero.)*. Abbrev. for *Visual Meteorological Conditions*.

VM/CMS *(Comp.)*. TN for an IBM **operating system**.

VME *(Comp.)*. The ICL proprietary **operating system**.

VMOS *(Electronics)*. A **MOS** technology in which four diffused layers are formed in silicon and V-shaped grooves are precisely etched in the layers. Metal is then deposited over silicon dioxide (an insulator) to form *gate* and other electrodes. Higher current densities can be used than with other MOS techniques and a higher density of components per chip achieved.

VMS *(Comp.)*. TN for a DEC **operating system** for its VAX range of computers.

V-n diagram *(Aero.)*. See **flight envelope**.

vocal cords *(Zool.)*. In air-breathing Vertebrates, folds of the lining membrane of the larynx by the vibration of the edges of which, under the influence of the breath, the voice is produced.

vocal sac *(Zool.)*. In many male Frogs, loose folds of skin at each angle of the mouth which can be inflated from within the mouth into a globular form, and act as resonators.

vocoder *(Acous.)*. *Voice coder*. System for synthetic speech using recorded speech elements.

vodas *(Telecomm.)*. Abbrev. for *Voice Operated Device Anti-Sing*; used for the suppression of echoes in transoceanic radio telephony.

voder *(Acous.)*. *Voice operation demonstrator*. System for producing synthetic speech through keyboard control of electronic oscillators.

vogad *(Telecomm.)*. Abbrev. for *Voice Operated Gain-Adjusting Device*. Used in telephone systems to give an approximately constant volume output for a wide range of input signals.

vogesite *(Min.)*. A hornblende-lamprophyre, the other essential constituent being orthoclase. Cf. **spessartine**.

voice coil *(Acous.)*. The coil attached to the cone of a loudspeaker. The coil currents react with the magnetic field to drive the cone. Also used in microphones to generate the signal.

voiced sound *(Acous.)*. In speech, an elemental sound in which the component frequencies are exact multiples of a fundamental frequency which is determined by the tension of the oscillating muscles in the larynx.

voice filter *(Acous.)*. Device which deliberately distorts speech for specific purpose, e.g. telephonic imitation.

voice frequency *(Telecomm.)*. One in the approximate range 200 to 3500 Hz (that required for the normal human voice). Also called *speech frequency*.

voice-frequency multichannel telegraphy *(Telecomm.)*. The use of a large number of voice-frequency channels, e.g. 12 or 18, for the fullest utilization of the transmission properties of normal audio-frequency telephone circuits.

voice-frequency telegraphy *(Telecomm.)*. See **carrier telegraphy**.

voice over *(Image Tech.)*. Speech accompanying a film or video programme in which the speaker is not seen in the picture.

void *(Powder Tech.)*. In a powder compact or in a powder fluid system, the space between particles.

voidage *(Powder Tech.)*. In powder compact or powder fluid system, the fractional quantity of voids in the system. For a system containing dense particles, the voidage and the powder porosity are numerically equal.

Voigt effect *(Phys.)*. *Double refraction* of electromagnetic waves passing through a vapour when an external transverse magnetic field is applied. The vapour acts as an uniaxial crystal with its optic axis parallel to the field direction.

voile *(Textiles)*. A light, open, plain-weave fabric made from highly-twisted yarns.

volant *(Zool.)*. Flying; pertaining to flight.

volatile *(Chem.)*. Changing readily to a vapour.

volatile memory *(Comp.)*. Stored information which can be destroyed by a power failure. See also **dynamic-**.

volatilization *(Chem.)*. See **vaporization**.

volcanic ash *(Geol.)*. The typical product of explosive volcanic eruptions, consisting of comminuted rock and **lava**, the fragments varying widely in size and in composition, and including deposits of the finest dust, lapilli, and bombs. See also **agglomerate, pyroclastic rocks, tuff**.

volcanic bomb *(Geol.)*. A spherical or ovoid mass of lava, in some cases hollow, formed by the disruption of molten lava by explosions in an active volcanic vent. See also **bread-crust bomb**.

volcanic muds, sands *(Geol.)*. The products of explosive volcanic eruptions (*volcanic ash*) which have been deposited under water and have consequently been sorted and stratified, thus showing some of the characters of normal sediments, into which they grade. May also be the product of a mud flow down the side of a volcano.

volcanic neck *(Geol.)*. A vertical pluglike body of igneous rock or volcanic ejectamenta, representing the feeding channel of a volcano.

volcanic vent *(Geol.)*. The pipe which connects the crater with the source of magma below; it ultimately becomes choked with agglomerate or volcanic ash, or with consolidated lava.

volcano *(Geol.)*. (1) A centre of volcanic eruption, having the form typically of a conical hill or mountain, built of ashes and/or lava-flows, penetrated irregularly by dykes and veins of igneous rocks, with a central crater from which a pipe leads downwards to the source of magma beneath. Volcanoes may be active (periodically), dormant, or extinct; the eruptions may involve violent explosions (e.g. Krakatoa) or the relatively quiet outpouring of lava, particularly in those cases where the lava is basaltic (e.g. Hawaii). See also **lava**. (2) A conical hill producing mud or sand. See **mud volcano**.

Volkmann's contracture *(Med.)*. A contracture of the flexor muscles of the forearm and leg due to the pressure of splints or tight bandages used in the treatment of fractures, etc. causing diminished blood supply and muscle necrosis.

volsella forceps *(Med.)*. Forceps whose blades have prolonged ends.

volt *(Elec.Eng.,Phys.)*. The SI unit of *potential difference*, electrical potential, or e.m.f., such that the potential difference across a conductor is 1 volt when 1 ampere of current in it dissipates 1 watt of power. Equivalent definition: if, in taking a charge of 1 coulomb between two points in an electric field, the work done on or by the charge is 1 joule, the potential difference between the points is 1 volt. Named after Count Alessandro Volta (1745-1827).

Volta effect *(Phys.)*. Potential difference which results when two dissimilar metals are brought into contact: the basis of voltage cells and corrosion.

voltage *(Phys.)*. The value of an e.m.f. or pd expressed in *volts*.

voltage amplifier *(Telecomm.)*. An amplifier whose function is to increase the voltage of the applied signal, without necessarily increasing its power. The output impedance must therefore be high.

voltage between lines *(Elec.Eng.)*. The voltage between any two of the line wires in a single- or three-phase system; between the two lines of the same phase in a two-phase system; between any two lines which are consecutive as regards phase sequence in a symmetrical six-phase system. Also called *line voltage, voltage between phases, voltage of the system*.

voltage circuit *(Elec.Eng.)*. The circuit of an instrument or relay which is connected across the lines of the circuit under test, and which therefore carries a current proportional to the voltage of this circuit. Also called *pressure circuit*. See **shunt circuit**.

voltage coefficient *(Elec.Eng.)*. The constant by which the product of the armature speed in revolutions per minute, the flux in volt-lines, and the number of armature conductors in series must be multiplied in order to obtain the e.m.f. of a d.c. generator.

voltage controlled oscillator *(Electronics)*. Oscillator whose frequency is controlled by a bias signal; a **varactor** diode may be used as the controlling element. Abbrev. *VCO*.

voltage divider *(Elec.Eng.)*. Chain of impedances, most commonly resistors or capacitors, such that the voltage across one or more is an accurately known fraction of that applied to all; used for calibrating voltmeters. Also called *potential divider, volt-box*. See **Kelvin-Varley slide**.

voltage doubler *(Elec.Eng.)*. Power supply circuit in which both half-cycles of a.c. supply are rectified, and the resulting d.c. voltages are added in series.

voltage drop *(Phys.)*. (1) Diminution of potential along a conductor, or over an apparatus, through which a current is passing. (2) The possible diminution of voltage between two terminals when current is taken from them.

voltage-fed antenna *(Telecomm.)*. One which is fed with power from a line at a point of high impedance, where, through resonance, there is a voltage loop in the standing-wave system.

voltage feedback *(Elec.Eng.,Electronics)*. In amplifier circuits, a feedback voltage directly proportional to the load voltage. It may be applied in series or shunt with the source of the input signal. See also **current feedback, negative feedback, positive feedback**.

voltage gain *(Elec.Eng.,Electronics)*. The ratio of the change in output voltage (for a network or amplifier) to the change in input voltage which produces it. Often expressed in dB as $A_V = 20\log(V_o/V_i)$, although this definition applies strictly only to the case when the input and output impedances are equal.

voltage gradient *(Elec.Eng.)*. The difference in potential per unit length of a conductor, or per unit thickness of an insulating medium.

voltage level *(Telecomm.)*. Peak-to-peak value at any point in a network, expressed relative to a specified reference level. When this is 1 volt, the symbol dBv is used.

voltage multiplier *(Elec.Eng.)*. Circuit for obtaining high d.c. potential from low-voltage a.c. supply, effective only when load current is small, e.g. for anode supply to CRT. A ladder of half-wave rectifiers charges successive capacitors connected in series on alternate half-cycles.

voltage ratio *(Elec.Eng.)*. Same as *turns ratio*.

voltage reference tube *(Electronics)*. A glow-discharge tube designed to operate with anode-cathode voltage as nearly as possible constant regardless of the anode current, and hence suitable for use as a standard of pd.

voltage regulation *(Elec.Eng.)*. The percentage variation in the output voltage of a power supply for either a specified variation in supply voltage or a specified change of load current.

voltage-regulator tube *(Electronics)*. One in which, over a practical range of current, voltage between electrodes in a glow discharge remains substantially constant.

voltage resonance *(Elec.Eng.)*. See **current resonance**.

voltage ripple *(Elec.Eng.)*. The peak-to-peak a.c. component of a nominally d.c. supply voltage.

voltage stabilizer, voltage-stabilizing tube *(Electronics)*. See **voltage-regulator tube**.

voltage standing-wave ratio *(Telecomm.)*. Ratio between a maximum and a minimum in a standing wave, particularly on a transmission line or in a waveguide, arising from inexact impedance terminations. Abbrev. *VSWR*.

voltage to neutral *(Elec.Eng.)*. The voltage between any line and neutral of a three- or six-phase system. Also called *phase voltage, star voltage, Y-voltage*.

voltage transformer *(Elec.Eng.)*. A small transformer of high insulation for connecting a voltmeter to a high-tension a.c. supply.

voltaic cell *(Chem.)*. Any device with electrolyte (ionized chemical compound in water), and two differing electrodes which establish a difference of potential.

voltaic current *(Phys.)*. Current (direct) produced by chemical action.

voltaic pile *(Elec.Eng.)*. A source of d.c. supply. It comprises a battery of primary cells in series, arranged in the form of a pile of disks, successive disks being of dissimilar metals separated by a pad soaked in the chemical agent.

voltameter *(Elec.Eng.)*. An instrument for measuring a current by means of the amount of metal deposited, or gas liberated, from an electrolyte in a given time due to the passage of the current.

volt-ampere *(Elec.Eng.)*. Apparent power in an electrical circuit, equal to the product of rms current and voltage.

volt-ampere-hour *(Elec.Eng.)*. Unit of apparent power, equal to watt-hour divided by **power factor**.

volt-amperes *(Phys.)*. Product of actual voltage (in volts) and actual current (amperes), both r.m.s., in a circuit. See **active volt-amperes, reactive volt-amperes**.

volt-box *(Elec.Eng.)*. See **voltage-divider**. Also *volt ratio-box*.

voltinism *(Zool.)*. Breeding rhythm; brood frequency. See **univoltine, bivoltine, multivoltine**.

voltmeter *(Elec.Eng.)*. An instrument for measuring potential differences.

volt-ohm-milliammeter *(Elec.Eng.)*. An electrical d.c. test instrument measuring voltage, resistance, and current. Usually a.c. volts can also be measured.

volume *(Acous.)*. A general term comprehending the general loudness of sounds, or the magnitudes of currents which give rise to them. Volume is measured by the occasional peak values of the amplitude, when integrated over a short period, corresponding to the time constant of the ear. See **volume indicator, volume unit**.

volume *(Phys.)*. The amount of space occupied by a body; measured in cubic units. Symbol *V*.

volume bottles *(Chem.Eng.)*. *Dampener*, in the form of empty pressure vessels (usually steel), depending on the relationship between their volume and the volume of gas passing to produce their dampening effect.

volume compression, expansion *(Acous.)*. Automatic compression of the volume range in any transmission, particularly in speech for radio-telephone transmission, so that the envelope of the waveform is transmitted at a higher average level with respect to interfering noise levels. After expansion at receiving end, resulting transmission is freer from noise.

volume compressor *(Acous.)*. In communication systems depending on amplitude modulation, intelligence transmitted as a modulation is limited to 100%. So that this is not exceeded with very loud sounds in the modulation, original transmission has to be compressed into a relatively small dynamic range to maintain a high signal/noise ratio.

volume control *(Telecomm.)*. A manually-operated control used to regulate communication transmission levels.

volume indicator *(Acous.)*. Instrument for measuring *volume, volume compression, expansion*.

volume ionization *(Phys.)*. The mean ionization density in any given volume without reference to the specific ionization of the particles.

volume lifetime *(Electronics)*. That of current carriers in the bulk of a semiconductor; cf. **surface lifetime**.

volume range *(Acous.)*. The difference between the maximum amplitude and the minimum useful amplitude of a transmitted signal, expressed in decibels. In speech it is generally taken to be 15–20 dB, and for a full orchestra 60–70 dB.

volume resistivity *(Elec.Eng.)*. See **resistivity**.

volume shape factor *(Powder Tech.)*. The ratio $d_v = W/Pd_m^3$ where W = mass of a particle; P = density of the particle; d = diameter of the particle as measured by a specified technique. Since d_m depends upon the method of measurement there are many shape factors for a given particle. If the value of volume shape factor is averaged over several particles, this value is called the volume shape factor of the powder.

volumetric analysis *(Chem.)*. A form of chemical analysis using standard solutions for the estimation of the particular constituent present in solution by titrating the one against a known volume of the other. See **burette, end-point, indicator** (1), **pipette, titrimeter** analysis.

volumetric efficiency *(Eng.)*. In an I.C. engine or air compressor, the ratio of the weight of air actually induced per unit time to the weight which would fill the swept volume at **s.t.p.**

volumetric heat *(Phys.)*. See **molal specific heat capacity**.

volumetric strain *(Eng.)*. The algebraic sum of three mutually perpendicular principal strains in a material.

volume unit *(Acous.)*. One used in measuring variations of modulation in a communication circuit, e.g. telephone or broadcasting. The unit is the decibel expressed relative to a reference level of 1 mW in 600 ohms, and standard *volume indicators* are calibrated in these units. Abbrev. *VU*.

voluntary muscle *(Zool.)*. Any muscle controlled by the motor centres in the brain. The skeletal muscles. All such muscles are striated. Cf. **involuntary muscles**.

volva *(Bot.)*. Cup-like structure at the base of the fruiting bodies of many basidiomycetes, e.g. mushroom, representing the remains of the universal veil.

volvulus *(Med.)*. Torsion of an abdominal viscus, especially of a loop of bowel, causing internal obstruction.

vomer *(Zool.)*. A paired membrane bone forming part of the cranial floor in the nasal region of the Vertebrate skull; believed not to be homologous in all groups. adj. *vomerine*.

vomerine teeth *(Zool.)*. In most Fish and Amphibia, teeth, sometimes atypical, borne on the vomers.

Von Recklinghausen's disease *(Med.)*. See **osteitis fibrosa**.

vortex *(Aero.)*. An eddy, or intense spiral motion in a limited region; a *vortex sheet* is a thin layer of fluid with intense vorticity; *tip vortices* are a form of *trailing vortex* from aerofoils, caused by shedding of lateral and line-of-flight airflows.

vortex generators *(Aero.)*. Small aerofoils, mounted normal to the surface of a main aerofoil and at a slight angle of incidence to the main airflow, which re-energize the **boundary layer** by creating vortices. Used on the wings and tail surfaces of high-speed aircraft to reduce **buffeting** caused by compressibility effects, so raising the critical **Mach number**, and sometimes to improve the airflow over control surfaces near the stall, thereby improving controllability. Cf. **wing fence**.

vortex street *(Aero.)*. Regular procession of vortices forming behind a bluff or rectangular body in two parallel rows. The vortices are staggered and each vortex in the opposite direction from its predecessor. Also called *Kármán Street*.

vorticity equation *(Meteor.)*. An equation for the rate of change of the vorticity, or curl of the velocity for atmospheric flow, especially the vertical component which is the dominating one. If f is the **Coriolis parameter**, $\zeta = \frac{\partial v}{\partial x} - \frac{\partial u}{\partial y}$ is the vorticity, $D = \frac{\partial u}{\partial x} + \frac{\partial v}{\partial y}$ is the horizontal **divergence**, and small terms are neglected, then

$$\frac{d}{dt}(\zeta + f) = -(\zeta + f)D.$$

Vostok *(Space)*. A series of Russian manned earth satellites, first used successfully in April 1961. Vostok I carried Yuri Gagarin, the first man to travel in space.

vough *(Min.Ext.)*. See **vug**.

voussoir *(Civ.Eng.)*. See **arch stone**.

vowel articulation *(Telecomm.)*. See **articulation**.

Voyager *(Space)*. Name given to two unmanned spacecraft, Voyager I and II, designed for exploring the outer planets of the solar system. Spectacular images of Jupiter and its satellites, and Saturn and its rings were sent back to Earth. Launched in late-1977, Voyager II will encounter Neptune in August 1989.

voyeurism *(Behav.)*. Sexual gratification through the clandestine observation of other people's sexual activities or anatomy.

VP *(Build.)*. Abbrev. for *Vent Pipe*.

VPI *(Chem.)*. Abbrev. for *Vapour Phase Inhibitor*.

VPP *(Vet.)*. Abbrev. for *virus pneumonia of pigs*.

V-rings *(Elec.Eng.)*. V-shaped mica rings insulating the segments of a commutator from the end rings.

VS *(Chem.)*. Abbrev. for *Volumetric Solution*.

VSI *(Aero.)*. Abbrev. for *Vertical Speed Indicator*.

VSWR *(Telecomm.)*. Abbrev. for *Voltage Standing-Wave Ratio*.

VTB curve *(Elec.Eng.)*. *Voltage/Time-to-Break-down curve*, i.e. a curve connecting the time and voltage for breakdown in this time. See **short-time breakdown voltage** and **asymptotic breakdown voltage**.

VTOL *(Aero.)*. A general term for aircraft, other than conventional helicopters, capable of vertical take-off and landing: in Britain, the initials *VTO* were originally used, but the US expression *VTOL* has become general.

VTR *(Image Tech.)*. Abbrev. for *Video Tape Recorder*.

V-type commutator *(Elec.Eng.)*. A commutator whose segments are provided with projecting spigots, which dovetail into the end rings.

VU *(Acous.)*. Abbrev. for *volume unit*.

vug, vough *(Min.Ext.)*. A cavity in rock or a lode, usually lined with crystals.

Vulcan *(Astron.)*. Hypothetical planet inside the orbit of Mercury.

Vulcan coupling *(Eng.)*. A hydraulic shaft coupling, of the Föttinger type, used for connecting marine diesel engines to the propeller shaft in order to avoid torsional vibration troubles. See **Föttinger coupling**.

vulcanite *(Chem.)*. Hard vulcanized rubber, in the making of which a relatively high proportion of sulphur is used. **Ebonite** is one form; coloured varieties are obtained by adding various ingredients, such as the sulphides of antimony and mercury. See **vulcanization of rubber**.

vulcanites *(Geol.)*. A general name for igneous rocks of fine grain-size, normally occurring as lava flows. Cf. **plutonites**.

vulcanization of rubber *(Chem.)*. The treatment of rubber with sulphur or sulphur compounds, resulting in a change in the physical properties of the rubber. Sulphur is absorbed by the rubber, and the process can be carried out either by heating raw rubber with sulphur at a temperature between 135°C and 160°C, or by treating rubber sheets in the cold with a solution of S_2Cl_2. To increase the velocity of vulcanization, **accelerators** may be used.

vulcanized fibre *(Chem.)*. A fibre obtained by treating paper pulp with zinc chloride solution. It consists of amyloid 90%, the remainder being water with some slight trace of insoluble salts. Used for low-voltage insulation.

vultex *(Chem.)*. See under **latex**.

vulva *(Zool.)*. The external genital opening of a female Mammal. adj. *vulviform*.

vulvitis *(Med.)*. Inflammation of the vulva.

vulvovaginitis *(Med.)*. Inflammation of both the vulva and the vagina.

VU meter *(Acous.)*. Instrument calibrated to read intensity of electroacoustic signals directly in **volume units**. See also **volume indicator**.

W

w *(Civ.Eng.)*. Symbol for load per metre run or for weight per cubic metre.
w *(Phys.)*. Symbol for work.
W *(Chem.)*. The symbol for tungsten.
W *(Eng.,Phys.)*. Symbol for watt.
W *(Civ.Eng.)*. Symbol for total load.
W *(Elec.Eng.)*. Symbol for electrical energy.
W *(Phys.)*. Symbol for (1) weight, (2) work.
wacke *(Geol.)*. A sandstone in which the grains are poorly sorted with respect to size.
wad *(Min.)*. Bog manganese, hydrated oxide of manganese. See **asbolane**.
Wadell's sphericity factor *(Powder Tech.)*. A shape factor used in particle-size analysis, defined by the equation $S = d_c/D_c$, where S = Wadell's sphericity factor, d_c = the diameter of a circle equal in area to the projected image of the particle when the particle rests on its larger face, and D_c = the diameter of the smallest circle circumscribing the defined projection diameter.
Wadsworth mounting *(Phys.)*. A form of stigmatic mounting used for large concave gratings. The grating is illuminated by parallel light and the spectrum is focused at a distance of approximately one-half the radius of curvature of the grating.
wafer *(Comp.)*. See **chip**.
waggle dance *(Behav.)*. Semicircular movements of a hive bee, including an *abdominal waggle*, on returning from a foraging trip of more than 150 yards from the hive; conveys information about the direction and distance of a food source to worker bees, and stimulates them to visit the site. Cf. **round dance**.
Wagner earth *(Elec.Eng.)*. A pair of impedances with their common point earthed, connected across an a.c. bridge network in order to neutralize the effect of stray capacitances. This is done by simultaneously balancing the normal bridge and that formed by the Wagner earth with the ratio arms.
wagon retarder *(Civ.Eng.)*. See **retarder**.
wagon vault *(Build.)*. See **barrel vault**.
wagtail *(Build.)*. See **parting slip**.
wainscoting cap *(Build.)*. A moulding surmounting a given piece of wainscoting.
wainscot oak *(Build.)*. Selected oak, cut radially to display the silver grain; much used for panelling.
wainscot, wainscoting *(Build.)*. A wooden lining, usually panelled, applied to interior walls.
waist anchor *(Ships)*. See **sheet anchor**.
wait time *(Comp.)*. The time interval during which a process is suspended.
wake *(Aero.)*. The region behind an aircraft in which the **total head** of the air has been modified by its passage.
Waldegg valve gear *(Eng.)*. See **Walschaert's valve gear**.
Walden inversion *(Chem.)*. The transformation of certain optically active substances into their stereoisomeric derivatives by chemical reactions.
Waldenstrom's macroglobulinaemia *(Immun.)*. Disease occurring mainly in elderly males, characterized by the presence of large amounts of monoclonal IgM in the blood, lymphoid tissue enlargement, splenomegaly, a haemorrhagic tendency and depression. The IgM is occasionally found to have detectable antibody activity, e.g. rheumatoid factor. The disease is probably a relatively benign and slowly progressing form of myelomatosis.
waldsterben *(Ecol.)*. Symptoms of tree decline in Central Europe from the 1970s, not attributable to known diseases, and widely held to be caused by atmospheric pollution.
wale (knitting) *(Textiles)*. A column of loops running along the length of a knitted fabric. The number of wales per cm helps to characterize the fabric.
walings *(Civ.Eng.)*. Rough planks which run horizontally in front of the poling boards used in timbering trenches. The struts wedging the timbers apart on both sides are placed between the walings. Also *wales*.
walk-about disease *(Vet.)*. See **Kimberley horse disease**.
walking beam *(Eng.)*. A mechanism for conveying solid articles. A fixed set of parallel bars supports the articles, while another set fills the gaps between the fixed set and moves cyclically through a rectangular path. If the second set moves forward when its tops are above those of the fixed set and backwards when below, then the articles will move forward.
walking beam *(Min.Ext.)*. Rocking beam used for transmitting power, e.g. for actuating the cable in cable-drilling for oil.
walking dragline *(Min.Ext.)*. Large power shovel mounted on pads which are mechanically worked to manoeuvre it as required.
walking, going line *(Build.)*. An imaginary line, always 18 in from the centre line of the handrail, used in setting out winders for a stair, the width of the winder measured on the line being made approx. the same as the going of the normal treads.
wall *(Bot.)*. See **cell wall**.
Wallace's line *(Zool.)*. An imaginary line passing through the Malay Archipelago and dividing the oriental faunal region from the Australasian region.
wallboard *(Build.)*. **Fibre-board**, usually of laminated construction.
wall box *(Build.)*. A support built into a wall to carry the end of a timber.
wall effect *(Phys.)*. (1) The contribution of electrons liberated in the walls of an ionization chamber to the recorded current. (2) The reduction in the count rate recorded with a Geiger tube due to ionizing particles not having the energy to penetrate the walls of the tube.
wall energy *(Phys.)*. The energy per unit area stored in the domain wall bounding two oppositely magnetized regions of a ferro-magnetic material.
wall frame *(Build.)*. A method of system building for high buildings, comprising large precast concrete components.
wall hanger *(Build.)*. A support partly built into a wall to carry the end of a structural timber, which itself is not to be built into the wall.
wall hook *(Build.)*. An L-shaped nail used as a means of attachment to a wall.
wall insulator *(Elec.Eng.)*. An insulator specifically designed to enable a conductor at high potential to earth to pass through a brick or concrete wall.
wall-less counter *(Nuc.Eng.)*. A low-level proportional counter, the cathode of which consists of a cylindrical cage of thin wires, parallel to the cathode, which considerably reduces the background arising from electrons ejected by gamma-rays. The counting volume may also be accurately defined by use of special *field tubes* at each end of the counter.
wallpaper colour *(Print.)*. A web, usually of advertising, preprinted in more than one colour on one or both sides in a repeating design, enabling it to be run through rotary presses of any cut-off.
wall plate *(Build.)*. The top horizontal timber of a wall, supporting parts of the structure.
wall-rock *(Min.Ext.)*. Country rock to either side of a vein or lode.
wall-rock alteration *(Min.Ext.)*. Mineralogical and/or chemical alteration of the country rock adjacent to a mineralized vein or lode. Result either of diffusion of

elements from the mineralizing fluid in the vein, or leaching of material from the country rock by the mineralizing fluid. See **halo, primary dispersion**.

wall-sided *(Ships)*. A term signifying absence of *tumble-home*, and indicating that the maximum breadth is maintained to deck-level.

wall string *(Build.)*. A string, generally a **housed string**, positioned against a wall and supporting the inner ends of the steps.

wall tie *(Build.)*. A galvanized iron piece built into the two parts of a cavity wall, thus serving to bond them together.

walnut *(For.)*. Tree of the genus *Juglans*, giving a hardwood popular for high-class furniture, panelling, gun stocks.

Walschaert's valve gear *(Eng.)*. A valve gear of the radial type used in some steam locomotives. The valve is driven through a 'combination lever' whose oscillation is the resultant of sine and cosine components of the piston motion, derived from connections with the engine cross-head and with an eccentric or return crank at 90° to the main crank. Sometimes called *Waldegg valve gear*.

wandering cells *(Zool.)*. Migratory amoeboid cells; may be leucocytes or phagocytes.

wane *(For.)*. A defect in converted timber; some of the original rounded surface of the tree is left along an edge.

Wankel engine *(Autos.)*. Rotary automobile engine having an approximately triangular central rotor geared epitrochoidally to the central driving shaft and turning in a close-fitting oval-shaped chamber so that the power stroke is applied to each of the three faces of the rotor in turn as they pass a single sparking plug.

warble *(Vet.)*. Parasitic infestation caused by *Hypoderma bovis* and *H. lineatum* (gad flies). Adult flies lay on the hairs of the limb. The larvae then penetrate the skin and migrate through the tissues to arrive (about 9 months later) subcutaneously on the back. The third stage larvae then emerge and drop to the ground. Cattle and horses are mainly infected. Compulsory treatment of bovines has limited the problem.

warble tone *(Acous.)*. Narrow frequency band current for testing microphones, etc. To minimize errors due to standing waves in a room, the frequency is varied cyclically.

Ward-Leonard control *(Elec.Eng.)*. A method of speed control for large d.c. motors, employing a variable-voltage generator to supply the motor armature, driven by a shunt motor.

Ward-Leonard-Ilgner system *(Elec.Eng.)*. A modification of the Ward-Leonard system of speed control, in which a flywheel is included on the motor generator shaft to smooth out peak loads otherwise taken from the supply.

warehouse *(Print.)*. The department of a printing works where cutting, folding, and the simpler methods of binding are undertaken.

warfarin *(Chem.)*. 3-(α-acatonyl-benzyl)-4-hydroxycoumarin, used (usually in the form of its sodium derivative) as a blood anticoagulant; also widely used as a selective rodenticide.

warfarin, sodium *(Med.)*. An orally active anticoagulant which antagonizes the effect of vitamin K. Used in the treatment of deep venous thrombosis, pulmonary embolism and myocardial infarction.

war gas *(Chem.)*. Any gaseous chemical substance used in warfare (or in riot control) to produce poisonous or irritant effects upon the human body. War gases are classified according to the length of time they are effective, *non-persistent* or *peristent* (less or more than 10 min in normal atmospheric conditions), and according to their effect: *lung irritants* or choking gases (phosgene, diphosgene); *lachrymators* or tear gases (chloracetophone, CS gas [orthochlorobenzylidene malononitrile], bromobenzylcyanide); *nerve gases* (derivatives of fluorphosphoric acid); *paralysants* or blood gases (hydrocyanic acid); *sternutators* or irritant smokes (diphenylaminechlorarsine); *vesicants* or blister gases (lewisite, mustard gas). The lachrymators and sternutators, being less toxic, are used in riot and crowd control.

warm-blooded *(Zool.)*. Said of animals which have the bodily temperature constantly maintained at a point usually above the environmental temperature, of which it is independent; idiothermous; homoiothermous.

warm front *(Meteor.)*. The leading edge of a mass of advancing warm air as it rises over colder air. There is usually continuous rain in advance of it.

warm start *(Comp.)*. Restart of a program after stoppage, without losing data from the previous run. Cf. **cold start**.

warm working *(Eng.)*. Working metal at a temperature below its recrystallization temperature but above room temperature.

war neurosis *(Behav.)*. A preferable synonym for *shell-shock*. The term was originally used (World War I) for all types of nervous conditions resulting from war experiences, especially those caused by a bursting shell, which might result in (*a*) a condition of physical shock or concussion to the nervous system, (*b*) the precipitation of a psychoneurosis in a predisposed individual, (*c*) a combination of these conditions.

warning *(Horol.)*. In a striking clock, the partial unlocking of the striking train, just before the hour.

warning coloration *(Zool.)*. See **aposematic coloration**.

warning piece *(Horol.)*. In the striking work, a projection on the lifting piece which projects through a slot in the dial plate, and against which a pin on the warning wheel butts, to hold up the train. Exactly at the hour the warning piece drops clear of the pin and frees the train.

warning pipe *(Build.)*. An overflow pipe fitted to cisterns etc. to warn of a defective valve.

warning wheel *(Horol.)*. The last wheel in a striking train, which is held up by the warning piece during warning.

warp *(For.)*. Permanent distortion of a timber from its true form, due to causes such as exposure to heat or moisture.

warp *(Textiles)*. The threads of a woven fabric that run continuously along the length of the fabric.

warp knitted fabric *(Textiles)*. Fine, knitted fabric formed wholly from warp yarns wound on a beam the width of the machine. The loops from each warp thread run the whole length of the fabric.

warp streak, stripe *(Textiles)*. A visible fault in a fabric that arises from a warp yarn being different from the normal yarns e.g. the thread may be of the wrong fibre, size, twist, colour.

Warren girder *(Build.,Eng.)*. A form of girder consisting of horizontal upper and lower members, connected by members inclined alternately in opposite directions.

Warrington hammer *(Build.)*. Type of cross-pane hammer with slightly convex face.

wart *(Med.)*. A tumour of the skin formed by overgrowth of the prickle-cell layer, with or without hyperkeratosis; due to infection with a virus. See also **verruca**.

wash *(Arch.,Eng.)*. A thin coat of water-colour paint applied to part of a drawing as an indication of the nature of the material to be used for the part represented, particular colours conventionally indicating particular materials.

washboard *(Build.)*. See **skirting board**.

wash box *(Min.Ext.)*. Box in which raw coal is jigged in coal washery.

washed clay *(Build.)*. See **malm**.

washer *(Build.,Eng.)*. Annular piece, usually flat, used under a nut to distribute pressure, or between jointing surfaces for a tight joint etc.

wash gravel *(Min.Ext.)*. Alluvial sands worth exploitation for mineral values contained.

washing *(Image Tech.)*. The removal of chemical residues and soluble components from photographic emulsions by water in the course of processing; the elimination of **hypo** and silver compounds is particularly important for image permanence.

wash-in, wash-out *(Aero.)*. Increase (wash-in) or decrease (wash-out) in the **angle of incidence** from the root toward the tip of an aerofoil, principally used on wings to ensure that the wing-tips stall last so as to maintain aileron control.

wash-out valve *(Civ.Eng.)*. A valve inserted in a pipeline at the bottom of a valley, in order to enable a particular length of the pipe to be emptied as required.

wash plate *(Ships)*. Fitted in tanks to prevent large quantities of water or oil rushing from side to side when the vessel rolls.

Wassermann reaction *(Immun.,Med.)*. Complement fixation test formerly used in the diagnosis of syphilis. Cardiolipin derived from ox heart is used as antigen because the blood of persons with syphilis regularly contains antibody which reacts with this substance. However false results may be obtained in cases where there are increased immunoglobulin levels, e.g. due to parasitic infection or in autoimmune conditions. Confirmatory tests using *Treponema pallidum* itself as the antigen may need to be made. The latter procedure is now generally used.

waste *(Civ.Eng.)*. See **spoil**.

waste *(Min.Ext.)*. Waste rock, either host (enclosing) rock mined with the true lode, or ore too poor to warrant further treatment.

waste *(Nuc.Eng.)*. (1) Depleted material rejected by an isotope separation plant. (2) Unwanted radioactive material for disposal.

waste *(Textiles)*. Textile wastes are frequently recycled usually by degrading them to fibres. *Soft* waste is most easily processed because it is obtained in the earlier processes before twisting etc. has been carried out. *Hard* waste includes all materials such as thread and rags which have to be *pulled* before they can be reused.

waste heat recovery *(Eng.)*. The recovery of heat from furnaces, kilns, combustion engines, flue gases, etc., for utilization in, e.g. air preheating, feed water heating, waste heat boilers.

waste-light factor *(Elec.Eng.)*. A factor used in the design of floodlighting installations to allow for the light which, although emitted along the beam from the projector, does not fall on the area to be illuminated.

waster *(Build.)*. A mason's chisel, sometimes with claw head.

waste weir *(Civ.Eng.)*. The weir provided in reservoir construction to discharge all surplus water flowing into the reservoir in flood-time, so as to prevent the water level from rising above the limit which was allowed for in designing the dam.

wasting *(Build.)*. The operation of removing stone from a block by blows with a pick, prior to squaring and dressing.

WAT *(Behav.)*. Abbrev. for *Word Association Test*.

WAT curves *(Aero.)*. Complicated graphs relating the take-off and landing behaviour of an aircraft to its weight, airfield altitude and ambient temperature. Their preparation and use is mandatory for UK public transport aircraft. Many other countries use the *Weight/Altitude/Temperature* information in tabular form.

water *(Phys.)*. A colourless, odourless, tasteless liquid, m.p. 0°C, b.p. 100°C. It is hydrogen oxide, H_2O, the liquid probably containing associated molecules, H_4O_2, H_6O_3 etc. On electrolysis it yields two volumes of hydrogen and one of oxygen. It forms a large proportion of the earth's surface, occurs in all living organisms, and combines with many salts as water of crystallization. Water has its maximum density of 1000 kg/m^3 at a temperature of 4°C. This fact has an important bearing on the freezing of ponds and lakes in winter, since the water at 4°C sinks to the bottom and ice at 0°C forms on the surface. Besides being essential for life water has a unique combination of solvent power, thermal capacity, chemical stability, permittivity and abundance. See **hard water, soft water, ion exchange**; also **triple point, heavy water**.

water ballast *(Ships)*. Water carried for purposes of stability. Also, water taken into ballast tanks to balance or redress change of draught due to consumption of fuel etc.

water bar *(Build.)*. A galvanized iron (or non-ferrous metal) bar set in the joint between the wood and stone sills of a window, to prevent penetration of water. Also called *weather bar*.

water blast *(Min.Ext.)*. A sudden escape of confined air due to water pressure, e.g. in rise workings.

water bomber *(Aero.)*. Aircraft, usually a flying boat, designed to combat forest fires by collecting water whilst planing on, e.g. a lake and jettisoning its load from low altitude over the fire zone.

waterbound macadam *(Civ.Eng.)*. A road surfacing formed of broken stone, well rolled and covered with a thin layer of **hogging**, which is watered in and binds the stones together.

waterbrash *(Med.)*. A sudden gush into the mouth of a watery secretion containing acid fluids from the stomach; often sign of oesophageal reflux. Also *pyrosis*.

water-carriage system *(Build.)*. The system of disposing of waste matter from buildings by water closets etc., involving the use of water to carry away the waste matter. Cf. **conservancy system**.

water channel *(Aero.)*. An open channel in which the behaviour of the surface of water flowing past a stationary body gives a visual simulation of supersonic airflow.

water-checked *(Build.)*. Said of a casement, the stiles and mullions of which have grooves cut in the meeting edges to prevent entry of rain. See **anti-capillary grooves**.

water closet *(Build.)*. A closet which is connected to a water-supply system so that the excreta may be carried away by flushing. There are two types, the standard 'wash-down' and the more efficient 'siphonic'.

water-cooled engine *(Autos.)*. An engine cooled by the circulation of water through jackets, which are usually cast integral with the cylinder block.

water-cooled motor *(Elec.Eng.)*. A motor employing water as a cooling medium.

water-cooled resistance *(Elec.Eng.)*. A resistance kept cool by immersion in water, which circulates in channels provided for the purpose.

water-cooled transformer *(Elec.Eng.)*. A transformer in which the oil is kept cool by means of water circulating in pipes immersed in the oil.

water-cooled valve *(Electronics)*. Large thermionic vacuum tube in which the heat generated by the electronic bombardment of the anode is carried away by water circulating around or through it. In the former the anode is made an integral part of the envelope. Cf. **cooled-anode valve**.

water culture *(Bot.)*. See **hydroponics**.

water-displacing liquid *(Chem.)*. A solvent containing surface active materials capable of removing water from moist surfaces and substituting a thin film of rust-inhibitive chemicals. A typical system would consist of organic fatty acids, non-ionic surfactants and barium petroleum sulphonate, dissolved in kerosine.

water equivalent *(Phys.)*. The mass of water which would require the same amount of heat as a body to raise its temperature by one degree. It is its **thermal capacity** (the product of its mass and its *specific heat capacity*) divided by the s.h.c. of water (4.186 kJ/kg K).

water finish *(Paper)*. The high surface finish produced by the machine calenders when fitted with water doctors which apply a thin film of water at the nip.

water gauge *(Eng.)*. A vertical or inclined protected glass tube connected, at its upper and lower ends respectively, to the steam and water spaces of a boiler, for showing the height of the water level.

water gauge *(Min.Ext.)*. An instrument (e.g. *Pitot tube*) for measuring the difference in pressure produced by a ventilating fan or air current.

water glass *(Chem.)*. A concentrated and viscous solution of sodium or potassium silicate in water. It is used as an adhesive, as a binder, as a protective coating in waterproofing cement, as a preservative for eggs, and in the bleaching and cleaning of fabrics.

water hammer *(Eng.)*. A sharp hammerlike blow from a steep-fronted pressure wave in water caused by the sudden stoppage of flow in a long pipe when a valve is closed sufficiently rapidly.

Waterhouse stops *(Image Tech.)*. Removable metal plates, each with a hole giving the required aperture, which can be inserted in front of the camera lens. Used in process cameras.

water-in-oil emulsion adjuvant *(Immun.)*. Adjuvant in which the antigen, dissolved or suspended in water, is enclosed in tiny droplets within a continuous phase of mineral oil. The antigen solution constitutes the dispersed phase, stabilized by an emulsifying agent such as mannitol mono-oleate.

water-jet driving *(Civ.Eng.)*. A process of pile driving often adopted when the piles have to be sunk into alluvial deposits; a pressure water jet is used to displace the earth around the point of the pile.

water-jet pump *(Eng.)*. A simple suction pump, capable of producing a moderate degree of vacuum, in which air is drawn through the branch of a T-pipe by the action of a fast jet of water passing through the straight section. The principle is similar to that of an **ejector** and the pump has no moving mechanical parts.

waterleaf *(Paper)*. Paper which has not been sized with rosin prior to the tub-sizing operation.

water lime *(Build.)*. See **hydraulic cement**.

water lines *(Ships)*. The intersection of the various water planes with the ship's form.

watermark *(Paper)*. A device in paper visible by transmitted light as a lighter or darker local area produced by an appropriate relief or intaglio design on a hand mould, cylinder mould or dandy roll.

water/methanol injection *(Aero.)*. (1) The use of the *latent heat of evaporation* of water (the methanol is an antifreeze agent) injected into a piston engine intake to cool the charge, thereby permitting the use of greater power without detonation for take-off; (2) the injection of water into the airflow of the compressor of a **turbojet** or **turboprop** to restore take-off power by cooling the intake air at high ambient temperatures.

water monitor *(Nuc.Eng.)*. One for measuring the level of radioactivity in a water supply, similar to, but much more sensitive than, an effluent monitor.

water of capillarity *(Build.)*. The moisture drawn up by capillary action from the soil into the walls of a building. Also *rising damp*.

water of hydration, crystallization *(Chem.)*. The water present in hydrated compounds. These compounds when crystallized from solution in water retain a definite amount of water, e.g. copper (II) sulphate, $CuSO_4.5H_2O$.

water paint *(Build.)*. Any paint that can be diluted with water, e.g. emulsion and acrylic paints.

water plane *(Ships)*. A horizontal section through a ship's hull. Usually named by measurement from the base line, but sometimes from the load water plane.

water pore, water stoma *(Bot.)*. Opening in the epidermis, associated with a hydathode, through which water exudes. It is often a modified stoma.

water potential *(Bot.)*. ψ_w. A measure of the free energy of water in a solution as in a cell or soil sample and hence of its tendency to move by diffusion, osmosis or as vapour. It is the chemical potential of water in solution minus that of pure water (0 at standard temperature) divided by the partial molar volume of water; expressed as units of pressure, MPa or bar. Water diffusing or osmosing always moves down a water potential gradient. The components of water potential are **osmotic potential**, **turgor potential** and **matric potential**.

waterproof paper *(Paper)*. Packaging paper that has been treated e.g. by impregnation, coating or lamination to render it resistant to the transmission of liquid water.

water reactor *(Nuc.Eng.)*. Nuclear reactor in which water (including heavy water) is the moderator and/or coolant.

water recovery *(Aero.)*. The recovery, principally by condensation, of the water in the exhaust gases of an aero engine. Used in airships for ballast purposes, as a partial set-off against the loss of weight due to the consumption of fuel during flight.

water repellant solutions *(Build.)*. Often based on silicone resin these solutions prevent penetration of rainwater on porous or absorbent masonry type surfaces. Sometimes they are coloured with a fugitive dye to aid visibility of application.

water repellent *(Textiles)*. A fabric which resists the penetration of rain by a surface tension effect. The water droplets do not spread but roll off as they strike the fabric. This is different from *waterproof* which is usually a plastic surfaced material, impervious to air as well as to water.

water resistor *(Elec.Eng.)*. One made by immersing two electrodes in an aqueous solution. The resistance depends on the strength of the solution, and dimensions of the conducting path.

water rheostat *(Elec.Eng.)*. A **water resistor** whose resistance can be varied, usually by moving one electrode relative to the other.

water-rib tile *(Build.)*. A purpose-made tile having a projecting rib that serves to prevent entry of rain or snow.

water sapphire *(Min.)*. See **saphir d'eau**.

water seal *(Build.)*. See **seal**.

water softening *(Chem.)*. The removal of 'hardness' in the form of calcium and magnesium ions, which form precipitates with soap. See **Calgon, double decomposition, hard water, ion exchange, Permutit**.

waterspout *(Meteor.)*. See **tornado**.

water stain *(Build.)*. A stain for wood, consisting of colouring matter dissolved in water.

water-storage tissue *(Bot.)*. Tissue of large, highly vacuolate cells with relatively extensible walls, which can buffer the water supply. Water can also be stored in tree trunks, as in tracheids which can be emptied and refilled.

water table *(Build.)*. See **canting strip**.

water table *(Geol.)*. The surface below which fissures and pores in the strata are saturated with water. It roughly conforms to the configuration of the ground, but is smoother. Where the water table rises above ground level a river, spring or lake is formed.

watertight fitting *(Elec.Eng.)*. An electric-light fitting designed to exclude water under certain prescribed conditions. Cf. **weatherproof fitting**.

watertight flat *(Ships)*. When a watertight bulkhead above a deck is not directly over the bulkhead below, that part of the deck between the two is a *watertight flat*.

water torch *(Build.)*. Machine which cuts through concrete and steel without generating heat and with relatively little noise. Uses pumping equipment in order to provide water supply to a pressure of 150 MN/m^2, and has a specially designed nozzle which produces a fine, penetrating needle of water.

water-tube boiler *(Eng.)*. A boiler consisting of a large number of closely spaced water-tubes connected to one or more drums, which act as water pockets and steam separators, giving rapid water circulation and quick steaming. See **Babcock and Wilcox boiler, forced-circulation boilers, Yarrow boiler**.

water tunnel *(Aero.)*. A tunnel in which water is circulated instead of air to obtain a visual representation of flow at high Reynolds numbers with low stream velocities.

water turbine *(Eng.)*. A prime mover in which a wheel or runner carrying curved vanes is supplied with water directed by a number of stationary guide vanes; usually direct-coupled to large alternators.

water vapour pressure *(Meteor.)*. That part of the atmospheric pressure which is due to the water vapour in the atmosphere.

water-vascular system *(Zool.)*. In Echinodermata, a system of coelomic canals, associated with the tube-feet, in which water circulates; in Platyhelminthes, the excretory system.

water wheel *(Eng.)*. Large wheel carrying peripheral buckets or shrouded vanes on which water is caused to act, either by falling under gravity or by virtue of its kinetic energy. See **overshot wheel, undershot wheel**.

watt *(Phys.)*. SI unit of power equal to 1 joule per second. Thus, 1 horse-power (hp) equals 745.70 watts. Symbol W.

wattful loss *(Elec.Eng.)*. See **ohmic loss**.

Watt governor *(Eng.)*. A simple **pendulum governor** in which a pair of links are pivoted to the vertical spindle and terminate in heavy balls. Shorter links are pivoted to the mid-points of the first, and to the sleeve operating the engine throttle.

watt-hour *(Phys.)*. A unit of energy, being the work done by 1 watt acting for 1 hour, and thus equal to 3600 joules.

watt-hour efficiency *(Phys.)*. The ratio of the amount of energy available during the discharge of an accumulator to the amount of energy put in during charge. Cf. **ampere-hour efficiency**.

watt-hour meter *(Elec.Eng.)*. Integrating meter for measurement of total electric energy consumed in a circuit. The conventional domestic electricity meter is of this type.

wattle and daub, dab *(Build.)*. A type of wall construction in which wicker is interlaced about a rough timber framework and the whole is covered with plaster.

wattless component *(Elec.Eng.)*. See **reactive component**.

wattmeter *(Elec.Eng.)*. Instrument for measuring the active power in a circuit.

wattmeter method *(Elec.Eng.)*. A method of testing the electrical quality of iron specimens by measuring the power loss with a.c. magnetization.

wave *(Phys.)*. A time varying quantity which is also a function of position. The characteristic of a wave is to transfer energy from one point to another without any particle of the medium being permanently displaced; particles merely oscillate about their equilibrium positions. In electromagnetic waves it is the changes in electric and magnetic fields which represent the wave disturbance. The progress of the wave is described by the passage of a *waveform* through the medium with a certain velocity, the *phase* or *wave velocity*. The energy is transferred at the *group velocity* of the waves making the waveform.

wave analyser *(Elec.Eng.)*. One which is used to determine the frequency components in continuously repeated signal. Also termed *spectrum analyser*.

wave angle *(Telecomm.)*. Either angle of elevation or azimuth of arrival or departure of a radio wave with respect to the axis of an antenna array.

wave antenna *(Telecomm.)*. Directional receiving antenna comprising a long wire running horizontally to the direction of arrival of the incoming waves at a small distance above the ground. The receiver is connected to one end, and the other end is connected to earth through a terminating resistance. Also called *Beverage antenna*.

waveband *(Telecomm.)*. Range of wavelengths occupied by transmissions of a particular type, e.g. the *medium waveband* used mostly for broadcasting.

wave cloud *(Meteor.)*. A cloud that appears at the crest of a **lee wave** and thus remains more or less stationary relative to the ground. Wave clouds are usually rather smooth in appearance, and often occur in regularly spaced bands demonstrating the lee waves causing them.

wave clutter *(Radar)*. See **sea clutter**.

wave drag *(Aero.)*. The drag caused by the generation of **shock waves**, applied to the aircraft as a whole. See **compressibility drag**.

wave equation *(Acous.)*. Linear partial differential equation of at least second order which describes the propagation of a wave in space and time.

wave equation *(Phys.)*. A differential equation which describes the passage of harmonic waves through a medium. The form of the equation depends on the nature of the medium and on the process by which the wave is transmitted. The solutions to the equation depend on the circumstances in which the wave is propagated. See also **Schrödinger equation**.

wave filter *(Elec.Eng.)*. Four terminal network, consisting of pure reactances, designed to pass particular bands of frequency and to reject others.

waveform *(Phys.)*. The shape, contour or profile of a wave; described by a phase relationship between successive particles in a medium. A waveform may be *periodic*, *transient* or *random*.

wave-formed mouth *(Vet.)*. A variation in height of the molar teeth of horses.

waveform monitor *(Image Tech.)*. CRO used for the display and measurement of TV signal waveform.

wavefront *(Phys.)*. Imaginary surface joining points of constant phase in a wave propagated through a medium. The propagation of waves may conveniently be considered in terms of the advancing wavefront, which is often of simple shape, such as a plane, sphere or cylinder.

wave function *(Phys.)*. Mathematical equation representing the space and time variations in amplitude for a wave system. The term is used particularly in connection with the Schrödinger equation for particle waves.

waveguide *(Phys.)*. Electromagnetic waves in the microwave region can be transmitted efficiently from a source to other parts of a circuit by means of hollow metal conductors called waveguides. The transmission can be described by the patterns of electric and magnetic fields produced inside the guide, different modes being characterized by different electric and magnetic field configurations. Dielectric guides operate similarly but generally have higher losses.

waveguide attenuator *(Elec.Eng.)*. Conducting film placed transversely to the axis of the waveguide.

waveguide choke flange *(Elec.Eng.)*. Coupling flange between waveguide sections which offers zero impedance to signal without requiring metallic continuity.

waveguide coupler *(Elec.Eng.)*. Arrangement for transferring part of the signal energy from one waveguide into a second, crossing or branching off from the first, e.g. in a *directional coupler*, the direction of flow of the energy transferred to the second guide reverses when the direction of propogation in the first guide is reversed.

waveguide filter *(Elec.Eng.)*. One having distributed properties, giving frequency discrimination in a waveguide in which it is inserted.

waveguide impedance *(Phys.)*. Ratio Z derived from $W = V^2/Z$, or I^2Z, where W is power, V a voltage and I a current, the last being defined in relation to type of wave and shape of waveguide.

waveguide iris *(Elec.Eng.)*. Diaphragm placed across a waveguide forming a reactance.

waveguide junction *(Elec.Eng.)*. Unit joining three or more waveguide branches, e.g. **hybrid junction**.

waveguide lens *(Phys.)*. An array of short lengths of waveguide which convert an incident plane wavefront into an approximately spherical one by refraction.

waveguide modes *(Phys.)*. Modes of propagation in a waveguide. Classified as TM_{mn} (or H_{mn}), transverse electric, and TM_{mn} (or E_{mn}), transverse magnetic. In rectangular guide the subscripts refer to the number of half-cycles of field variation along the axes parallel to the sides. In circular guide they refer to the field variation in the angular direction and in the radial direction. Waves below the cut-off frequency are said to be *evanescent*.

waveguide stub *(Phys.)*. A short-circuited length of waveguide used as a reactance for matching.

waveguide switch *(Elec.Eng.)*. One which switches power from waveguide A to B or C, with considerable loss between A and C or B, and between B and C.

waveguide tee *(Elec.Eng.)*. A T-shaped junction for connecting a branch section of a waveguide in parallel or series with the main waveguide transmission line.

waveguide transformer *(Elec.Eng.)*. Unit placed between waveguide sections of different dimensions for impedance matching.

wave impedance *(Phys.)*. Complex ratio of transverse electric field to transverse magnetic field at a location in a waveguide. For an acoustic wave, the ratio is pressure/particle velocity.

wave interference *(Phys.)*. Relatively or completely stationary patterns of amplitude variation over a region in which waves from the same source (or two different coherent sources) arrive by different paths of propagation; *constructive interference* arises when the two waves are in phase and their amplitudes add; *destructive interference* arises when they are out of phase and their amplitudes partly or totally neutralize each other.

wavelength *(Phys.)*. Symbol λ. (1) Distance, measured radially from the source, between two successive points in free space at which an electromagnetic or acoustic wave has the same phase; for an electromagnetic wave it is equal in metres to c/f where c is the velocity of light (in $m s^{-1}$) and f is the frequency (in Hz). (2) Distance between two similar and successive points on a harmonic (sinusoidal) wave, e.g. between successive maxima or minima. (3) For electrons, neutrons and other particles in motion when considered as a *wave train*, $\lambda = h/p$, p is the momentum of the particle and h is Planck's constant. See **wave mechanics, de Broglie wavelength.**

wavelength constant *(Phys.)*. The imaginary part of the **propagation constant.**

wavelength of light *(Phys.)*. The wavelength of visible light lies in the range from 400 to 700 nm approximately.

wavellite *(Min.)*. Orthorhombic hydrated phosphate of aluminium, occurring rarely in prismatic crystals, but commonly in flattened globular aggregates, showing a strongly developed internal radiating structure.

wave mechanics *(Phys.)*. The modern form of the *quantum theory* in which events on an atomic or nuclear scale are explained in terms of the interactions between wave systems as expressed by the *Schrödinger equation.* For a bound particle, e.g. an electron in an atom, standing wave solutions are found for which only certain wavelengths are permitted, and consequently the energy is quantized. See **statistical mechanics.**

wavemeter *(Electronics)*. See **frequency meter.**

wave number *(Phys.)*. In an electromagnetic wave, the reciprocal of the wavelength, i.e., the number of waves in unit distance.

wave packet *(Phys.)*. A wave train. The *de Broglie wave* associated with a particle.

wave parameter *(Phys.)*. See **wavelength constant.**

wave-particle duality *(Phys.)*. Light and other electromagnetic radiations behave like a wave motion when being propagated, and like particles when interacting with matter. Interference, diffraction and polarization effects can be described in terms of waves. The photoelectric effect and the Compton effect can be described in terms of *photons*, quanta of energy $E = h\nu$ where h is Planck's constant and ν is the frequency.

wavers *(Print.)*. Ink rollers that reciprocate to distribute the ink.

waveshape *(Phys.)*. See **waveform.**

wave tail *(Phys.)*. The portion of a waveform that follows the peak or crest.

wave theory *(Phys.)*. Macroscopic explanation of diffraction, interference, and optical phenomena as an electromagnetic wave, predicted by Maxwell and verified by Hertz for radio waves. See **theories of light.**

wave tilt *(Elec.Eng.)*. The angle between the normal to the ground and the electric vector, in a ground wave polarized in the plane of propagation.

wave train *(Phys.)*. Group of waves of limited duration, such as those which result from a single spark discharge occurring in an oscillatory circuit.

wave trap *(Telecomm.)*. A circuit tuned to parallel resonance connected in series with the signal source to reject an unwanted signal, e.g. between a radio receiver and the aerial. See also **wave filter.**

wave velocity *(Phys.)*. See **phase velocity.**

wave winding *(Elec.Eng.)*. A type of armature winding in which there are only two parallel circuits through the armature, irrespective of the number of poles.

waving groin *(Build.)*. A groin which is not straight in plan.

wavy paper *(Paper)*. A defect of paper showing as undulations, especially near the edges, due to the moisture content of the paper not being in equilibrium with the surrounding atmosphere.

wawa *(For.)*. See **obeche.**

wax *(Chem.)*. Esters of monohydric alcohols of the higher homologues; e.g. *beeswax* is the myricyl ester of palmitic acid, $C_{30}H_{61}O.CO.C_{15}H_{31}$. For properties and uses, see also **beeswax, cable wax.**

waxing *(Image Tech.)*. Application of a thin layer of wax or silicone to the edges of motion picture prints to provide lubrication during projection.

wax vent *(Foundry)*. A pliable wax taper with a cotton core, placed in intricate cores during moulding. This wax melts when the core is dried, leaving a clear hole for the escape of gases.

wax wall *(Min.Ext.)*. A wall of clay built round the gob or goaf, to prevent the entry of air or egress of gas.

waxy flexibility *(Med.)*. See **flexibilitas cerea.**

way *(Min.Ext.)*. See **wind road.**

ways *(Eng.)*. (1) The machined surfaces of the top of a lathe bed on which the carriage and tailstock slide; sometimes called *shears.* (2) The framework of timbers on which a ship slides when being launched.

way up *(Geol.)*. The upward direction of a succession of strata in an area of strong folding. The direction of *way up* is most commonly determined by **bottom structures** or by **cross bedding.**

W-chromosome *(Biol.)*. See **sex determination.**

weak coupling *(Elec.Eng.)*. An inductive coupling in which the mutual inductance between two circuits is small; more generally known as *loose coupling, tight coupling.*

weak electrolyte *(Chem.)*. An electrolyte which is only slightly ionized in moderately concentrated solutions.

weak interaction *(Phys.)*. A fundamental interaction between particles mediated by *intermediate vector bosons.* Weak interactions involve neutrinos or antineutrinos or both and are completed in about 10^{-10} s. This kind of interaction is responsible for radioactive β-decay.

weapons system *(Aero.)*. The overall planned equipment and backing required to deliver a weapon to its target, including production, storage, transport, launchers, aircraft, etc.

wearing course *(Civ.Eng.)*. Uppermost layer in a carriageway construction. Also termed *carpet, coat, crust, road surface, sheeting, topping, veneer, wear surface.*

wearing depth *(Elec.Eng.)*. The permissible amount of radial wear on a commutator, prior to renewing the segments.

weather bar *(Build.)*. Patent methods of sealing the bottoms of external doors sometimes with a flaplike plate.

weather-board *(Build.)*. A board used with others for covering sheds and similar structures. Weather-boards are fixed horizontally and usually overlap each other.

weather check *(Build.)*. A *drip.* Mastic or special metal strips are frequently incorporated.

weathercock stability *(Aero.)*. The tendency for an aircraft to turn into the relative wind, due to the side areas aft of the cg exceeding the value for directional stability (as with aircraft designed for flying at low air speeds); excessive weathercock stability causes an oscillating yawing motion when flying in a cross-wind.

weathered pointing *(Build.)*. The method of pointing in which, in order to throw the rain off the horizontal joints, the mortar is sloped inwards, either from the lower edge of the upper brick, or from the upper edge of the lower brick, the latter method being preferred by bricklayers. Also called *struck-joint pointing.*

weather fillet *(Build.)*. See **cement fillet.**

weather forecast *(Meteor.)*. See **forecast, weather map.**

weathering *(Build.)*. (1) The deliberate slope at which an approximately horizontal surface is built or laid so that it may be able to throw off the rain. See **coping (1).** (2) The gradual process by which materials on the external faces of a building are affected by natural climatic conditions.

weathering *(Geol.)*. The processes of disintegration and decomposition effected in minerals and rocks as a consequence of exposure to the atmosphere and to the action of frost, rain, and insolation. These effects are partly mechanical, partly chemical, partly organic and for their continuation depend upon the removal, by transportation, of the products of weathering. *Denudation* involves both weathering and transportation.

weathering *(Textiles)*. The action of the weather on exposed materials.

weather map *(Meteor.)*. A map on which are marked synchronous observations of atmospheric pressure, temperature, strength and direction of the wind, the state of the weather, cloud and visibility. Weather maps (also known as *synoptic charts*) are used as a basis for forecasting.

weather minima *(Aero.)*. The minimum horizontal visibility and cloud base stipulated (*a*) by the air traffic authority and (*b*) by the standing orders of each airline, under which take-off and landing is permitted.

weather moulding *(Build.)*. See **dripstone**.

weatherproof fitting *(Elec.Eng.)*. An electric-light fitting having an enclosure which excluded rain, snow etc. Also called *splashproof fitting*.

weather radar *(Meteor.)*. A radar installation, either PPI or RHI, designed to be useful for the detection of **precipitation** and utilizing a wavelength of 3 to 20 cm. As the strength of the echo varies as the sixth power of the diameter, heavy showers and thunderstorms are much more conspicuous than widespread light rain or drizzle.

weather slating *(Build.)*. See **slate hanging**.

weather strip *(Build.)*. See **door strip**.

weather-struck *(Build.)*. A term applied to mortar joints finished by the method of *weathered pointing*.

weather tiling *(Build.)*. Tiles hung vertically to the face of walls, in order to protect them against wet and to help maintain an even temperature within the building. Also called *tile hanging*.

weaving *(Textiles)*. The interlacing of warp and weft threads running at right angles to each other to form a fabric.

weaving machine, loom *(Textiles)*. A machine that produces woven fabrics. In most a device transports the weft threads and interlaces them with warp threads. In many looms this is done by a shuttle but in more modern machines other means are used such as projectiles, rapiers, and air or water jets.

web *(Build.,Eng.)*. The relatively slender vertical part or parts of an I-beam or built-up girder (such as a box girder) separating the 2 flanges.

web *(Paper)*. A continuous sheet of paper on the paper machine, converting machine etc.

web *(Textiles)*. The loosely-coherent sheet of fibres produced by a card and used for making non-woven fabrics. Also *batt*.

web *(Zool.)*. The mesh of silk threads produced by Spiders, some Insects, and other forms; the vexillum of a feather; the membrane connecting the toes in aquatic Vertebrates, such as the Otter.

webbed *(Zool.)*. Having the toes connected by membrane, as in Frogs, Penguins, Otters.

webbing *(Textiles)*. A woven narrow fabric that is strong and able to sustain loads. Used in the manufacture of upholstery and of seatbelts for cars and aeroplanes.

web-break detector *(Print.)*. Electronic equipment to stop the web-press immediately a web breaks.

weber *(Phys.)*. The SI unit of magnetic flux. An e.m.f. of 1 volt is induced in a circuit through which the flux is changing at a rate of 1 weber per second. 1 weber equals 1 volt-second equals 1 joule per ampere. Equivalent definition: 1 weber is the magnetic flux through a surface over which the integral of the normal component of the magnetic induction is 1 tesla m^2.

Weber-Fechner law *(Med.)*. That the physiological sensation produced by a stimulus is proportional to the logarithm of the stimulus.

Weberian apparatus *(Zool.)*. See **Weberian ossicles**.

Weberian ossicles *(Zool.)*. In some Teleostei, e.g. Carp, Catfish and others, a chain of small bones, derived from processes of the anterior vertebrae, which connect the air bladder to the ear, transmitting vibrations from the former to a perilymphatic sac from which they pass to the endolymph of the inner ear. They correspond functionally to the middle ear ossicles of higher Vertebrates.

Weber photometer *(Elec.Eng.)*. A transportable photometer in which a direct comparison is made between the brightness of two screens, one illuminated by an unknown light source and the other by a standard lamp.

web-fed *(Print.)*. A term indicating that the printing machine uses a reel of paper and not single sheets.

web frame *(Ships)*. An extra strong frame usually made up of a plate with double angle bars on outer and inner edges. Commonly used to resist **panting**.

web offset *(Print.)*. An offset litho machine using a reel of paper.

websterite *(Geol.)*. A coarse-grained ultramafic igneous rock, consisting essentially of hypersthene and diopside.

weddellite *(Min.)*. Hydrated calcium oxalate, $CaC_2O_4.H_2O$, crystallizing in the tetragonal system. It occurs uncommonly in the mineral world but freely in human *calculi*.

Weddle's rule *(Maths.)*. The area under the curve $y = f(x)$ from $x = x_0$ to $x = x_6$ is approximately

$$\tfrac{3}{10}\{f(x_0) + 5f(x_1) + f(x_2) + 6f(x_3) + f(x_4) + 5f(x_5) + f(x_6)\}\{x_6 - x_0\}$$

where x_0, x_1, x_2, x_3, x_4, x_5 and x_6 are equally spaced.

wedge *(Elec.Eng.)*. (1) Total attenuator, in the form of a wedge of absorbing material, for terminating a waveguide. (2) Insertion of various lossy materials, put into a section of waveguide, to add fixed or variable attenuation in the circuit.

wedge *(Image Tech.)*. A strip of material showing gradation of transmission from clear to opaque along its length; the gradation may be continuous or in recognizable steps (*step wedge*). The material may be dyed or pigmented gelatine or a processed photographic image in silver or dye and may be neutral or in a single colour. The **sensitometry** of photographic materials given known controlled wedge exposures is an important part of processing control.

wedge *(Meteor.)*. See **ridge**.

wedge aerofoil *(Aero.)*. A supersonic aerofoil section (much used for missiles) comprising plane, instead of curved, surfaces tapering from a very sharp leading edge at an acute included angle to give a **thickness-chord ratio** of 5% or less; the aerofoil may have a blunt trailing edge, or it may have the section of a very elongated lozenge, or it may have a parallel mid-portion with leading- and trailing-edge wedges, the two latter cases being known as *double-wedge aerofoils*.

wedge contact *(Elec.Eng.)*. A contact consisting of two fingers between which a wedge-shaped contact on the moving element is forced; used for circuit-breakers, etc.

wedge spectrogram *(Image Tech.)*. A spectrogram made with a neutral wedge whose transmission increases with the slit length of the spectrometer. The resultant photographic image indicates, by the height of the density contours, the differential colour sensitivity of the emulsion.

wedging *(Min.Ext.)*. Use of deflecting wedge near bottom of deep diamond-bore holes, either to restore direction or to obtain further samples; also called *whipstocking*.

wedging crib, curb, ring *(Min.Ext.)*. A segmented steel ring on which shaft tubbing is built up and wedged in place.

weed *(Bot.)*. A plant growing where it is not wanted by man. Weeds of cultivated land are often natural plants of disturbed habitats and are often apomictic or self-pollinating or spread vegetatively. See **R-strategist**.

weed *(Vet.)*. See **sporadic lymphangitis**.

weephole *(Civ.Eng.)*. A pipe laid through an earth-retaining wall, with a slope from back to front to allow the escape of collected water.

weft *(Textiles)*. The threads that run across the width of woven fabrics. In the loom they are interlaced with the warp threads.

weft detector *(Textiles)*. A mechanical or electronic device that indicates when the weft thread in a shuttle is becoming exhausted, or the absence of weft in shuttleless looms.

weft insertion machine *(Textiles)*. A special warp-knitting machine which incorporates weft threads right across the fabric.

weft knitting *(Textiles)*. A method of making a fabric by normal knitting with the loops being formed right across the fabric in straight lines at right angles to the direction in which the fabric is produced.

weft streak *(Textiles)*. A fault in a woven fabric that runs across the fabric and results from a lack of uniformity in the weft threads (e.g. different fibres, colour, twist, or thickness).

Weg rescue apparatus *(Min.Ext.)*. Portable breathing apparatus with self-contained oxygen supply controlled automatically by wearer's breathing action.

Weierstrass' test for uniform convergence *(Maths.)*. See **M-test.**

weigh batching *(Civ.Eng.)*. A method whereby, using certain plant, the constituents of concrete are measured by weight before mixing, instead of by volume.

weighing bottle *(Chem.)*. A thin-walled cylindrical glass container with a tightly fitting lid, used for weighing accurately hygroscopic etc. materials.

weight *(Aero.)*. *Maximum*, or *gross*, *weight* is the total weight of an aircraft as authorized for flight under the current regulations; *maximum take-off weight* is the highest allowable for the engine power available under the given conditions; *maximum landing weight* is the highest safe weight for landing because of structural strength; *tare weight* is the design weight of an aircraft type in flying condition, without fuel, oil, crew, removable equipment not necessary for flight and payload; zero *fuel weight*, used in airline load calculations, is the weight of the loaded aircraft after all usable fuel has been consumed.

weight *(Phys.)*. The gravitational force acting on a body at the Earth's surface. Units of measurement are the newton, dyne or pound-force. Symbol W. Weight = mass × **acceleration due to gravity**, and must therefore be distinguished from *mass*, which is determined by quantity of material and measured in pounds, kilograms etc.

weight coefficient *(Elec.Eng.)*. The ratio of the weight of an electrical machine to its rated output.

weighting factor *(Phys.)*. See **statistical weight**.

weighting network *(Telecomm.)*. One designed to produce unequal attenuation for different frequency components of a signal, thereby weighting these differently in the final output.

weighting observations *(Surv.)*. The operation of assigning factors or 'weights' to each of a number of observations to represent their relative liability to error under their individual conditions of measurement.

weightlessness *(Space)*. A condition obtained in **free fall** when reaction is absent; a body has then no 'weight' only inertia.

weightometer *(Min.Ext.)*. Device which automatically weighs and records the tonnage of ore in transit on a belt conveyor.

weights *(Chem.)*. Standardized masses used for comparison with unknown masses, balances of various grades of sensitivity and sensibility being employed.

Weil-Felix reaction *(Immun.)*. An agglutination test used in the diagnosis of rickettsial infections (typhus, etc.) which depends upon a carbohydrate cross-reacting antigen shared by *Riccketsiae* and certain strains of *Proteus*. The agglutination pattern of patients with rickettsial disease against O-agglutinable strains of *Proteus* OX19, OX2 and OXK is diagnostic of the various rickettsial diseases.

Weil's disease *(Med.)*. Same as *leptospirosis icterohaemorrhagica*. See **leptospirosis**.

Weinberg and Salam's theory *(Phys.)*. A unified theory of *weak* and *electromagnetic* interactions between particles. It predicted the behaviour of the W^+, W^- and Z^0 intermediate vector bosons as the agents for the weak interaction. These particles were discovered later, in 1983, using the CERN SPS (super proton synchrotron) modified to produce colliding-beam experiments with 270 GeV protons and 270 GeV antiprotons. Also *electroweak theory*.

weir *(Civ.Eng.)*. A dam placed across a river to raise its level in dry weather.

Weisbach triangle *(Surv.)*. A method used in orienting underground workings, in which the theodolite is deliberately set up off the line of the two hanging wires used to transfer direction from above-ground to below-ground, so that the triangle between the instrument and the wires may be solved to enable the setting-out to proceed.

Weismann's ring *(Zool.)*. In the larvae of some Diptera (Insects), a small ringlike structure behind the brain, containing three types of glandular cell homologous with the corpora allata, corpora cardiaca and prothoracic glands of other Insects, and controlling metamorphosis in a similar manner. Sometimes knows as the *ring gland*.

Weissenberg method *(Crystal)*. A technique of X-ray analysis in which the crystal and photographic film are rotated in the beam of X-rays while the film is moved parallel to the axis of rotation.

Weiss theory *(Phys.)*. Early theory of ferromagnetism based on the concept of independent molecular magnets.

welded joint *(Eng.)*. A joint between two metals, made by electric or other welding method.

welded tuff *(Geol.)*. A **tuff** composed of glass fragments which have partially fused together as a result of being deposited while still at a high temperature. Synonymous with *ignimbrite*.

welding *(Eng.)*. (1) Joining pieces of suitable metals or plastics, usually by raising the temperature at the joint so that the pieces may be united by fusing or by forging or under pressure. The welding temperature may be attained by external heating, by passing an electric current through the joint, or by friction. (2) Joining pieces of suitable metals by striking an electric arc between an electrode or filler metal rod and the pieces. See **arc-, cold-, resistance-, seam-**.

welding manipulator *(Eng.)*. Support to which a workpiece can be clamped and which moves manually or automatically relative to a welding head.

welding regulator *(Elec.Eng.)*. A reactance by means of which the welding current may be varied in an a.c. welding set; it is variable by tappings controlled by a handwheel.

welding rod *(Eng.)*. Filler metal in the form of a wire or rod; in electric welding the electrode supplies the filler metal to the joint. Also called *filler rod*. See also **metal inert gas welding, tungsten inert gas welding**.

welding set *(Elec.Eng.)*. The apparatus for electric arc welding, either a.c. or d.c., comprising a supply unit and a regulator, which may be combined or separate.

welding transformer *(Elec.Eng.)*. A transformer specially designed to supply one or more welding regulators.

weldment *(Eng.)*. A welded assembly.

well *(Phys.)*. See **potential well**.

well-conditioned *(Surv.)*. A term used in triangulation to describe triangles of such a shape that the distortion resulting from errors made in measurement and in plotting is, or is nearly, a minimum; achieved in practice by making the triangles equilateral or approximately so.

well counter *(Nuc.Eng.)*. One used for measurements of radioactive fluids placed in a cylindrical container surrounded by the detecting element (hollow scintillation crystal or sensitive volume of special Geiger tube).

well foundation *(Civ.Eng.)*. A type of foundation formed by sinking **monoliths** to a firm stratum and filling in the open wells with concrete.

well head *(Min.Ext.)*. The top of the casing of a production oil well, with its control valves.

well-hole *(Build.)*. The vertical opening enclosed between the ends of the flights in a winding or geometrical stair.

well logging *(Geol.)*. The recording of the composition and physical properties of the rocks encountered in a borehole, particularly one drilled during petroleum exploration. Well logging includes a variety of techniques, e.g. resistivity log, gamma-ray log, neutron log, spontaneous or self-potential log, temperature log, calliper log, photoelectric log, acoustic velocity log etc.

well-ordered set *(Maths.)*. A set ordered in such a way that any subject has a first element.

Welsh groin *(Build.)*. A groin formed by two intersecting cylindrical vaults of different rises. Also called an *underpitch groin*.

welt *(Build.)*. A joint made between the edges of two lead sheets on the flat. Made by turning up each edge at right angles to the flat surface, bringing the two turned-up parts together, doubling them over and dressing them downflat. Also *seam*.

welt *(Textiles)*. A strengthened edge to a knitted fabric. If made at the start or finish of the knitting process the welts run across the fabric but if made later by seaming they may run along any edge. Probably best known are stocking welts which are a double layer of plain fabric at the top of a stocking made by the machine.

weltanschauung *(Behav.)*. Ger. *world outlook, philosophy of life*.

wen *(Med.)*. See **sebaceous cyst**.

Wenlock *(Geol.)*. An epoch of the Silurian period.

Wenner winding *(Elec.Eng.)*. Form of winding used in wirewound resistors to construct standard resistances of low residual reactance for use at relatively high frequencies.

Werner sedimentation techniques *(Powder Tech.)*. A 2-layer method of particle-size analysis. Particles sediment through a vertical column of liquid and collect in a narrow capillary at the foot of the column. The height of the column of particles is used as a measure of the particles sedimented out.

Werner's theory *(Chem.)*. A method of formulation of complex inorganic compounds based on the assumption that saturated groups are held to the central atom by residual valencies. The total number of such groups and ordinary unsaturated radicals which surround the central atom to form an un-ionizable coordination complex is characteristic of the central atom.

Wernicke's encephalopathy *(Med.)*. Brain damage due to **thiamine** deficiency, often as a result of alcohol abuse, causing abnormality in eye movements. Usually associated with **Korsakoff's psychosis**.

Wertheim's operation *(Med.)*. The operation of removing the uterus, the glandular tissue in the pelvis and the upper part of the vagina, in the treatment of cancer of the cervix uteri.

Westcott convention *(Nuc.Eng.)*. An approximation applied to the neutron flux in thermal reactor design. It is divided into a thermal component with Maxwellian distribution and a fast component with distribution proportional to dE/E, where E is the neutron energy. Sometimes a third component covering thermalization region may be included.

Westcott flux *(Nuc.Eng.)*. Theoretical neutron flux defined as equal to the reaction rate of a detector with cross-section which is unity for thermal electrons (velocity 2200 m/s) and varies inversely with the neutron velocity.

western blotting *(Immun.)*. Technique for the analysis and identification of protein antigens. The proteins are separated by polyacrylamide gel electrophoresis and then transferred electrophoretically, *blotted*, to a nitrocellulose membrane or chemically treated paper to which the proteins bind in a pattern identical to that in the gel. Bands of antigen bound to the membrane or paper are detected by overlaying with antibody, followed by anti-immunoglobulin or protein-A labelled with a radioisotope, fluorescent dye, enzyme or colloidal gold. Termed *western* because it is similar to the Southern and northern blotting methods used for DNA and RNA.

western duck sickness *(Vet.)*. See **alkali disease (1)**.

West Indian ebony *(For.)*. See **cocuswood**.

westing *(Surv.)*. A west departure.

Westmorland slates *(Build.)*. Thick slates (6.5–16 mm; $\frac{1}{4}$–$\frac{5}{8}$ in) of varying size.

Weston standard cadmium cell *(Phys.)*. Practical portable standard of e.m.f., in which the cathode is $12\frac{1}{2}$% Cd and $87\frac{1}{2}$% Hg by weight, with anode of amalgamated Pt or highly purified Hg. Saturated sol. aq. Cd_2SO_4 as electrolyte. E.m.f. of cell at 20°C is 1.018 636 V, temperature coefficient only 0.000 04 V/°C. Used for calibrating potentiometers and hence all other voltage measuring devices. Cells with unsaturated solutions have a lower temperature coefficient of e.m.f., but do not give an equally high absolute standard of reproducibility. Also *standard cell*.

Westphal balance *(Phys.)*. One in which relative density is determined by suspending solid specimen from balance beam in liquid of known density, or *vice versa*.

Westphalian *(Geol.)*. The name of a stratigraphical stage in the Carboniferous rocks of Europe, approximately corresponding to the Coal Measures in England and Wales.

wet and dry bulb hygrometer *(Meteor.)*. A pair of similar thermometers mounted side by side, one having its bulb wrapped in a damp wick dipping into water. The rate of evaporation of water from the wick and the consequent cooling of the 'wet bulb' is dependent on the relative humidity of the air; the latter can be obtained by means of a table from readings of the two thermometers.

wet assay *(Min.Ext.)*. Qualitative or quantitative analysis of ores or their constituents in which dissolution and digestion with suitable solvents plays a part. See also **cupellation, dry assay, scorification**.

wet beaten stuff *(Paper)*. Heavily beaten stuff in which hydration is developed. The resultant paper is dense and inclined to be translucent. Wet stock drains less readily on the machine wire and is more difficult to dry.

wet-bulb potential temperature *(Meteor.)*. The wet bulb potential temperature of a sample of moist air at any level may be found on an **aerological diagram** by following the **saturated adiabatic** curve through the **wet-bulb temperature** of the sample until it intersects the 1000 mb isobar and then reading off the temperature there. It is, for all practical purposes, a conservative quantity for such processes as evaporation, condensation and dry and saturated adiabatic temperature changes, and is thus a useful quantity for **air-mass** analysis.

wet-bulb temperature *(Meteor.)*. The temperature at which pure water must be evaporated adiabatically at constant pressure into a given sample of air in order to saturate the air under steady-state conditions. It is approximated closely by the temperature indicated by a thermometer, freely exposed to the air (but shielded from radiation), whose bulb is covered by muslin wetted with pure water.

wet cell *(Chem.)*. A primary cell which contains liquid electrolyte, in contrast to the paste of a dry cell.

wet deposition *(Ecol.)*. Deposition of materials from the atmosphere in rain.

wet electrolytic capacitor *(Elec.Eng.)*. One in which the negative electrode is a solution of a salt, e.g. aluminium borate, which is suitable for maintaining the aluminium oxide film without spurious corrosion.

wet end *(Paper)*. The wire and press parts of a paper or board machine where the sheet is formed and water removed by drainage, suction and pressure.

wet expansion *(Paper)*. The percentage increase in the length of a strip of paper after immersion in water under specified test conditions.

wet felt *(Paper)*. A continuous felt or synthetic fabric used in the press section of the paper or board making machine to convey the web, assist in water removal and prevent crushing at the press nip.

wet flashover voltage *(Elec.Eng.)*. The voltage at which the air surrounding a clean wet insulator completely breaks down. Also called *wet sparkover voltage*.

wet flong *(Print.)*. Layers of paper pasted together and used in a very damp condition for jobbing stereotyping.

wet laying *(Textiles)*. Production of non-woven fabric by using the wet laying technique employed for making paper.

wet lease *(Aero.)*. Hire of commercial aircraft complete, with original crew, serviced by the original owner, but perhaps carrying the new operator's logo.

wet-on-wet *(Print.)*. Printing two or more colours in quick succession, particularly coloured illustrations, the tack of the ink being graded in order to prevent pick-up by the succeeding colours.

wet-plate process *(Image Tech.)*. See **collodion process**.

wet rot *(Bot.)*. (1) A rot in which the tissue is rapidly broken down with the release of water from the lysed cells, as in the brown rots of stored fruits. (2) The rot of timber that is often wet, caused by the fungus, *Coniophora puteara*.

wet rot *(Build.)*. A decay of timber which is due to chemical decomposition in the growing tree; it is sometimes set up in wood saturated with water and exposed to alternations of moisture and dryness.

wet sparkover voltage *(Elec.Eng.)*. See **wet flash-over voltage**.

wet spinning *(Textiles)*. (1) Making filaments from a solution of the polymer by extrusion followed by precipitation with a liquid, e.g. viscose rayon. (2) Method for making fine flax yarns in which the roving passes through hot water to soften it and so assist drafting.

wet steam *(Eng.)*. A steam-water mixture, such as results from partial condensation of dry saturated steam on cooling.

wet strength paper *(Paper)*. Any paper so treated that it retains an appreciable proportion of its dry strength when completely wet and has good wet rub resistance. These properties may be obtained by adding suitable resins to the stock and curing (e.g. urea or melamine formaldehyde) or by parchmentizing.

wettability *(Chem.)*. The extent to which a solid is wetted by a liquid, measured by the force of adhesion between the solid and the liquid phases.

wetted area *(Aero.)*. Total surface area of body immersed in an airflow and over which a boundary develops.

wetting agent *(Chem.)*. Surface active agent which lowers the surface tension of water by a considerable amount, although present only in very low concentration.

wetwood *(For.)*. Wood with an abnormally high water content and a translucent or water-soaked glassy appearance. This condition develops only in living trees and not through soaking in water.

w.f. *(Print.)*. The standard mark for *wrong fount*. It is written in the proof margin and the letter is underlined or struck through.

whalebone *(Zool.)*. See **baleen**.

Wharfedale machine *(Print.)*. The first successful **stop-cylinder** printing machine, invented in 1858 by William Dawson and David Payne at Otley, in Wharfedale, Yorkshire.

what-you-see-is-what-you-get *(Comp.)*. See **WYSIWYG**.

Wheatstone bridge *(Elec.Eng.)*. An apparatus for measuring electrical resistance using a **null indicator**, comprising two parallel resistance branches, each branch consisting of two resistances in series. Prototype of most other bridge circuits.

wheel base *(Eng.)*. The distance between the leading and trailing axles of a vehicle.

wheeling *(Eng.)*. A sheet-metal working process, using a machine with one flat and one convex wheel, for producing curved panels or for finishing after **panel beating**.

wheeling step *(Build.)*. See **winder**.

wheel-ore *(Min.)*. See **bournonite**.

wheel-quartering machine *(Eng.)*. A horizontal drilling machine having two opposed spindles at opposite ends of the bed; used to drill the crank-pin holes in both wheels on a locomotive coupled-axle simultaneously, and in precise angular relationship.

wheel stretcher *(Horol.)*. A tool used for slightly increasing the diameter of a wheel. The wheel rim is passed between rollers under pressure, which slightly reduce the thickness but increase the width of the rim.

wheel window *(Arch.)*. See **rose window**.

wheel wobble *(Autos.)*. A periodic angular oscillation of the front wheels, resulting generally from poor balance, insufficient castor action or from backlash in the steering-gear.

whetstone *(Geol.)*. See **honestone**.

whewellite *(Min.)*. Hydrated calcium oxalate, $CaC_2O_4.H_2O$, crystallizing in the monoclinic system. It occurs uncommonly in the mineral world but is abundant in human calculi.

whine *(Acous.)*. The fluctuation in apparent loudness and pitch of a reproduced sound when the speed of the recording or reproducing machine is varying at a slow rate. See **wow**.

Whin Sill *(Geol.)*. A sheet of intrusive quartz-dolerite or quartz-basalt, unique in the British Isles, as it is exposed almost continuously for over 300 km from the Farne Islands to Middleton-in-Teesdale.

whin, whinstone *(Geol.)*. A popular term applied to doleritic intrusive igneous rock resembling that of the well-known Whin Sill.

whipcords *(Textiles)*. Fabrics woven from cotton or worsted with a bold steep warp twill, used chiefly for dresses, suits and coats.

whiplash *(Med.)*. Term descriptive of an extension flexion injury to the soft tissues of the neck including the ligaments and apophyseal joints, causing instability and chronic pain; frequent in car accidents involving sudden collision from behind.

whiplash flagellum *(Bot.)*. *Acronematic flagellum*. One without hairs on its surface. Ant. *tinsel flagellum*, a decorated flagellum.

Whipple-Murphy truss *(Eng.)*. A bridge truss having horizontal upper and lower chords connected by vertical and diagonal members, so that the panels resemble the letter N. Also called *Linville truss*, *N-truss*, *Pratt truss*.

whipstitching, whipping *(Print.)*. See **overcasting**.

whipstocking *(Min.Ext.)*. See **wedging**.

whirler *(Print.)*. Mechanical or hand-operated equipment using centrifugal force to spread an even coating on plates.

whirl-gate *(Foundry)*. See **spinner-gate**.

whirling arm *(Aero.)*. An apparatus for making certain experiments in aerodynamics, the model or instrument being carried round the circumference of a circle, at the end of an arm rotating in a horizontal plane.

whirlwind *(Meteor.)*. A small rotating wind-storm which may extend upwards to a height of many hundred feet; a small tornado.

whisker *(Chem.)*. A thin strong filament or fibre made by growing a crystal, e.g. of silicon carbide, sapphire etc.

whispering gallery *(Acous.)*. A room in which a whisper can be heard over a surprisingly large distance. The classical example is in the dome of St Paul's Cathedral in London, where sound is reflected on curved walls with high reflectivity and with the property of converging the sound.

whistle *(Acous.)*. A flow-noise device generating **edge tones**.

whistlers *(Acous.)*. Atmospheric electric noises which produce relatively musical notes in a communication system.

whistling *(Vet.)*. See **roaring**.

Whitby method *(Powder Tech.)*. A sedimentation technique in which a bucket centrifuge is used to measure sizes of very fine particles. The particles accumulate in a fine-bore capillary at the bottom of the tube and the weight of particles is estimated from the height of the sediment in the tube.

white arsenic *(Chem.)*. See under **arsenic**.

white bombway *(For.)*. Timber from the *Terminalia procera*, a native of the Andaman Islands. Planed surfaces of the wood are mildly silky, the texture moderately open and rather uneven, while the grain is straight. Wood used for joinery, panelling, furniture and interior fittings.

white cell *(Zool.)*. See **leucocyte**.

white coat *(Build.)*. The last or finishing coat of plaster.

white comb *(Vet.)*. See **avian favus**.

white copperas *(Min.)*. Goslarite.

white corundum *(Min.)*. See **white sapphire**.

white damp *(Min.Ext.)*. Carbon monoxide. Produced by the incomplete combustion of coal in a mine fire or by gas or dust explosions. Invisible; very poisonous.

white deal *(For.)*. A whitish soft wood obtained from the spruce fir, and commonly used for interior constructional work. Also called *whitewood*.

White Dwarf *(Astron.)*. Small, dim, star in the final stages of its evolution. The masses of known white dwarfs do not exceed 1.4 solar masses. they are defunct stars, collapsed to about the diameter of the Earth, at which time they stabilize with their electrons forming a

degenerate gas, the pressure of which is sufficient to balance gravitational force.

white fibres *(Zool.)*. Unbranched, inelastic fibres of connective tissue occurring in wavy bundles. Cf. **yellow fibres**.

white fibrocartilage *(Zool.)*. A form of fibrocartilage in which white fibres predominate.

white frost *(Meteor.)*. See **hoar frost**.

white glass *(Glass)*. See **opal glass**.

white gold *(Eng.)*. See **gold**.

white-heart process *(Eng.)*. See **malleable cast-iron**.

white heat *(Eng.)*. As judged visually, temperature exceeding 1000°C.

white heifer disease *(Vet.)*. A condition in cattle in which the vagina, cervix and uterus develop abnormally; associated with the gene for white coat colour and occurring mainly in white Shorthorn heifers.

white iron *(Eng.)*. Pig-iron or cast-iron in which all the carbon is present in the form of cementite (Fe_3C). White iron has a white crystalline fracture, and is hard and brittle.

white iron pyrites *(Min.)*. See **marcasite** (1).

white lead *(Chem.)*. Basic *lead (II) carbonate* or *hydroxycarbonate*. Made by several processes of which the oldest and best known is the *Dutch* (or stack) *process*. Formerly used extensively as a paint pigment and for pottery glazes.

white lead ore *(Min.)*. See **cerussite**.

white-leg *(Med.)*. See **phlegmasia alba dolens**.

white level *(Image Tech.)*. The level of the TV signal representing normal maximum picture luminance.

white light *(Phys.)*. Light containing all wavelengths in the visible range at the same intensity. This is seen by the eye as white. The term is used, however, to cover a wide range of intensity distribution in the spectrum and is applied by extension to continuous spectra in other wavelength bands (e.g. **white noise**).

white line *(Print.)*. A line of space.

white matter *(Zool.)*. An area of the central nervous system, mainly composed of cell processes, and therefore light in colour.

white metal *(Eng.)*. Usually denotes tin-base alloy (over 50% tin) containing varying amounts of lead, copper and antimony; used for bearings, domestic articles and small castings; sometimes also applied to alloys in which lead is the principal metal; also called *antifriction metal, bearing metal*.

white muscle disease *(Vet.)*. *Vitamin E/selenium deficiency, muscular dystrophy, stiff lamb disease.* A disease primarily of young calves and lambs but adults can be affected. Symptoms include sudden death, tachycardia, pyrexia, stiffness and recumbency. Due to lack of vitamin E and/or selenium in diet or failure in their absorption.

white nickel *(Min.)*. A popular name for the cubic diarsenide of nickel, $NiAs_2$, **chloanthite**.

white noise *(Acous.,Telecomm.)*. Noise, which may be of the random or impulse variety, having a flat frequency spectrum over the range of interest.

white-out *(Meteor.)*. A situation where the horizon is indistinguishable, when the sky is overcast and the ground is snow-covered.

white-out *(Print.)*. To open out composed type-matter with spacing, in order to fill the allotted area or improve the appearance.

white-out lettering *(Print.)*. A term indicating that, in the reproduction required, the lettering is to be reversed to white on black.

white radiation *(Phys.)*. See **white light**.

white reference level *(Telecomm.)*. Signal modulation level corresponding to maximum brightness (white) in monochrome facsimile or television transmissions.

whiter-than-white *(Image Tech.)*. TV luminance signal at a level exceeding the normal white level.

whites *(Vet.)*. Leucorrhoea of cows.

white sapphire *(Min.)*. More reasonably called *white corundum*, it is the colourless pure variety of crystallized corundum, Al_2O_3, free from those small amounts of impurities which give colour to the varieties 'ruby' and 'sapphire'; when cut and polished, it makes an attractive gemstone.

white scour *(Vet.)*. *Calf scour.* Diarrhoea affecting calves during the first few weeks of life, caused usually by *Escherichia coli* infection, probably in conjunction with nutritional and environmental factors. Vaccination widely used.

white spirit *(Build.)*. A petroleum distillate used as a substitute for turpentine in mixing paints, and in paint and varnish manufacture.

white subject *(Image Tech.)*. One which reflects all wavelengths of light of the visible spectrum to a substantially equal extent and thus appears to the eye without colour.

white vitriol *(Min.)*. A popular name for *goslarite*, $ZNSO_4.7H_2O$.

white water *(Paper)*. See **backwater**.

whitewood *(For.)*. (1) Softwood from *Liriodendron*, a native of eastern N. America. It will not polish satisfactorily but takes stain and paint well. (2) See **white deal**.

whitlockite *(Min.)*. Trigonal calcium phosphate, occurring in sedimentary phosphate deposits.

whitlow *(Med.)*. See **paronychia**.

Whitworth screw-thread *(Eng.)*. See **British Standard Whitworth thread**.

whizz-pan *(Image Tech.)*. Rapid **pan** from one point of interest to another. Also *whip pan*.

whole-body monitor *(Nuc.Eng.)*. Assembly of large scintillation detectors, heavily shielded against background radiation, used to identify and measure the gamma radiation emitted by the human body.

whole bound *(Print.)*. See **full bound**.

whole-brick wall *(Build.)*. A wall whose thickness is the length of a whole brick.

whole-circle bearing *(Surv.)*. The horizontal angle measured from 0° to 360° clockwise, from true north to a given survey line.

whole-coiled winding *(Elec.Eng.)*. An armature winding for an alternator having one armature coil per pole, the two sides of the coil being separated by a distance which is equal to the pole pitch.

whole plate *(Image Tech.)*. A standard format for still photography of dimensions $8\frac{1}{2} \times 6\frac{1}{2}$ in.

whooping cough *(Med.)*. See **pertussis**.

whorl *(Bot.)*. (1) A group of 3 or more plant structures arising at the same level on a stem and forming a ring around it. (2) A ring of floral organs round the receptacle of a flower.

whorl *(Zool.)*. A single turn of a spirally-coiled shell or other spiral structure.

WI *(Eng.)*. Abbrev. for *Wrought Iron*.

wick *(Textiles)*. A yarn, group of yarns, or a narrow woven fabric or braid with good capillary properties. Used particularly in candles and oil lamps. See also **candlewick**.

Wickersham quoin *(Print.)*. A mechanical quoin using the cam principle for expanding the sides.

Widal reaction *(Immun.)*. Bacterial agglutination test used in the diagnosis of enteric fevers. The patient's serum is titrated against reference strains of each of the likely organisms. The organisms must be motile, smooth and in the specific phase. Formalinized and alcoholized suspensions are used respectively for testing for H (flagellar) and O agglutinins. The Widal test is not likely to be positive before the 10th day of the disease and prior immunization with TAB vaccine may cause false positive results. In enteric infection however the titre against the infecting organism will continue to rise.

wide-angle lens *(Image Tech.)*. A camera lens of comparatively short focal length having a wide angle of view, of the order of 80° to 100° for still cameras and 50° to 70° for cinematography.

wideband amplifier *(Telecomm.)*. One which amplifies over a wide range of frequencies, normally with low gain.

wide-cut fuel *(Aero.)*. Low octane petrol (gasoline) obtained from wide-cut distillation used in turbojets in order to conserve kerosene. Abbrev. *avtag*.

wide-screen *(Image Tech.)*. General term for systems of motion picture and video presentation having pictures of **aspect ratio** 1.65 : 1 or greater.

wide-spectrum *(Med.)*. Of antibiotics etc. effective against a wide range of micro-organisms. Also *broad-spectrum*.

Widmanstätten structure *(Eng.)*. Meshlike appearance along preferred crystal planes, produced when steel is cooled rapidly from 1000°C, causing cementite or ferrite to be precipitated.

widow *(Print.)*. A single word or part of a word occupying the last line of a paragraph; to be avoided, particularly as the first line of a page. See **club line**.

width *(Phys.)*. The spread of uncertainty in a specified energy level, arising as a result of Heisenberg's **uncertainty principle**, proportional to the instability of the state concerned.

Wiedemann effect *(Elec.Eng.)*. Tendency to twist in a rod carrying a current when subject to a magnetizing field.

Wiedemann-Franz law *(Phys.)*. The ratio of thermal to electrical conductivity of any metal equals the absolute temperature multiplied by $\frac{\pi^2}{3}\left(\frac{k}{e}\right)^2 = 2{\cdot}5 \times 10^{-8}\ \mathrm{V^2\,K^{-2}}$, where k is the Boltzmann constant and e is the electronic charge.

Wien bridge *(Elec.Eng.)*. A four-arm a.c. bridge circuit used for measurement of capacitance, inductance and power factor.

Wien-bridge oscillator *(Elec.Eng.)*. One in which positive feedback is obtained from a Wien bridge, the variable frequency being determined by a resistance in an arm of the bridge.

Wien effect *(Elec.Eng.)*. The increase in the conductivity of an electrolyte observed with very high voltage gradients.

Wien's laws for radiation from a black body *(Phys.)*. (1) *Displacement law*: $\lambda_m T$ = constant (0.0029 metre kelvin). (2) *Emissive power* (E_λ): within the maximum intensity wavelength interval $d\lambda$, $E_\lambda = CT^5 d\lambda$, where $C = 1.288 \times 10^{-5}\ \mathrm{W/m^2\,K^5}$ (T is absolute temperature in kelvins). (3) *Emissive power* (dE) in the interval $d\lambda$ is $dE = A\lambda^{-5}\exp(-B/\lambda T)d\lambda$. λ_m is wavelength at E_{max}, $A = 4.992$ Jm, B = 0.0144 mK.

Wigner effect *(Nuc.Eng.)*. Changes in physical properties of graphite resulting from the displacement of lattice atoms by high-energy neutrons and other energetic particles in a reactor. It results in the building up of stored energy (Wigner energy) in the change of crystal lattice dimensions and hence in the change of overall bulk size.

Wigner energy *(Nuc.Eng.)*. Energy stored within a crystalline substance, due to the **Wigner effect**.

Wigner force *(Phys.)*. Ordinary (non-exchange) short-range force between nucleons.

Wigner nuclides *(Phys.)*. Those isobars of odd mass number in which the atomic number and neutron number differ by one.

wig-wag *(Civ.Eng.)*. A level-crossing signal which gives its indication, with or without a red light, by swinging about a fixed axis.

wig-wag *(Horol.)*. A machine which gives an oscillatory motion to a polisher; used for polishing pivots etc.

wildcatting *(Min.Ext.)*. Prospecting at random, particularly in speculative boring for oil.

wild shooting *(Image Tech.)*. Recording pictures without synchronized sound.

wild track *(Image Tech.)*. A sound track recorded independently of any associated picture.

wild type *(Biol.)*. The normal phenotype with respect to a specified gene locus; usually symbolized by +.

Wilfley table *(Min.Ext.)*. Flat rectangular desk, adjustable about the long axis for tilt, is given rapid but gentle throwing motion along this horizontal axis, while classified sands are washed across and down-tilt, against restraint imposed by horizontal riffles. Heavy minerals work across and progressively lighter ones gravitate down to separate discharge zones. Also called *shaking table*.

willemite *(Min.)*. Orthosilicate of zinc, Zn_2SiO_4, occurring as massive, granular, or in trigonal prismatic crystals, white when pure but commonly red, brown, or green through manganese or iron in small quantities. In New Jersey and elsewhere it occurs in sufficient quantity to be mined as an ore of zinc. Noteworthy as exhibiting an intense bright-yellow fluorescence in ultraviolet light.

Williot diagram *(Eng.)*. A graphical construction for finding the deflexion of a given point in a structural framework under load.

Wilson chamber *(Nuc.Eng.)*. **cloud chamber** of expansion type.

Wilson effect *(Elec.Eng.)*. Production of electric polarization when dielectric material is moved through region of magnetic field, due to Faraday induced e.m.f. in the dielectric.

Wilson's disease *(Med.)*. A rare disease in which excessive amounts of copper are deposited in the brain and liver.

wilt *(Bot.)*. A type of plant disease characterized by wilting as an early symptom and usually caused by the infection of the vascular system by a fungus or bacterium, e.g. Dutch elm disease, caused by the fungus *Ceratocystis ulmi*.

wilting *(Bot.)*. Loss of stiffness due to shortage of water and the loss of turgor by the cells. See also **permanent wilting point**.

WIMP *(Comp.)*. An interface for microcomputers which is designed to be 'user friendly'; it uses **windows, mice** and *pull-down menus* for masking and operating the normal system commands.

Wimshurst machine *(Elec.Eng.)*. Early type of electrostatic induction generator.

wincey *(Textiles)*. A light woven flannel made from mixed yarns containing wool.

winceyette *(Textiles)*. A plain or simple twill cotton cloth of light weight, raised slightly on both sides; used for pyjamas, nightgowns, or underwear. May be colour-woven or piece-dyed.

winch *(Eng.)*. Mechanism in which rotation of a drum causes a rope, cable or chain to hoist or lower a load, as in a crane.

winchester *(Glass)*. A narrow- or wide-mouthed cylindrical bottle used for the transportation of liquids.

winchester drive *(Comp.)*. **Disk drive** which uses small size **hard disks** which are suitable for microcomputer systems.

winch launch *(Aero.)*. Launching a glider by towing it into the air by a cable on a motorized winch, or pulling through a pulley by a car.

winch, wince *(Textiles)*. A machine that draws long lengths of fabric through a dyebath and winds the wet fabric on a reel or drum above the liquid.

wind *(Elec.Eng.)*. A stream of air arising at any sharply pointed electrical conductor charged to a high potential.

wind *(Meteor.)*. The horizontal movement of air over the earth's surface. The direction is that from which the wind blows. The speed may be given in metres per second, miles per hour or knots.

windage *(Eng.)*. In any machine, the energy dissipated in overcoming air resistance to motion.

wind axes *(Aero.)*. Coordinate axes, having their origin within the aircraft, and directionally orientated by the relative airflow.

wind bag *(Acous.)*. A bag of thin cloth or silk placed over a microphone when the latter is used out of doors, to eliminate hissing noises due to wind.

wind chill factor *(Meteor.)*. Assessment of the power of a cold wind to chill objects and (especially) living beings which combines the wind speed with the temperature and **relative humidity** of the air.

wind dispersal *(Bot.)*. The dispersal of spores, seed and fruits by the wind.

wind-driven generator *(Elec.Eng.)*. A generator driven by a prime-mover of the windmill type, or directly (in the case of aircraft) by an airscrew carried on the generator shaft.

winder *(Build.)*. A step, generally triangular in plan, used at a change in direction of the stair. Also *wheeling step.*

wind frost *(Meteor.)*. An air frost where the cold air has been propelled by the wind.

windgall *(Vet.)*. Distension of the joint capsule, *articular windgal*, or tendon sheath, *tendinous windgal*, of the fetlock joint of the horse.

winding *(Elec.Eng.)*. The system of insulated conductors forming the current-carrying element of an electric machine or static transformer.

winding *(Textiles)*. Coiling thread on a spindle or bobbin, after spinning or doubling to form a package convenient to handle.

winding coefficient *(Elec.Eng.)*. An alternative name for **winding factor**.

winding diagram *(Elec.Eng.)*. A diagram showing in schematic form the arrangement and sequence of an armature winding and its circuit connections.

winding drum *(Eng.)*. An engine or motor-driven drum on to which a haulage rope is wound, as the wire rope of a mine cage. The drum may be cylindrical or conical, with a plain or a helically grooved surface for the rope.

winding factor, coefficient *(Elec.Eng.)*. A factor which takes account of the difference between the vector and arithmetic sums of the e.m.f.s induced in a series of armature coils occupying successive positions round the periphery of the armature.

winding gear *(Elec.Eng.)*. The mechanical gear associated with an electric winder.

winding pitch *(Elec.Eng.)*. The distance, measured as the number of slots, separating an armature coil from its successor in the winding sequence.

winding plant *(Elec.Eng.)*. The complex of apparatus constituting an electrically driven winder.

winding space *(Elec.Eng.)*. The cross-sectional space available in an armature slot for the insertion of the insulated conductors.

winding square *(Horol.)*. The square end of a barrel or fusee arbor on which the key fits for winding.

winding stair *(Build.)*. A stair formed in a circular, spiral, elliptal or other mathematical plan.

wind load *(Build.,Eng.)*. The force acting on a structure due to the pressure of the wind upon it.

window *(Comp.)*. (1) A part of a VDU screen which is given over to a different display from the rest of the screen, e.g. a text window in a graphics screen. (2) Portion of a file or image currently visible on the screen.

window *(Elec.Eng.)*. (1) The winding space of a transformer, i.e. the cross-sectional spaces between the limbs and yokes of a multicore transformer. (2) Conducting diaphragms inserted into waveguide which act inductively or capacitively according to how they are positioned.

window *(Geol.)*. A closed outcrop of strata lying beneath a thrust plane and exposed by denudation. The strata above the thrust plane surround the 'window' on all sides.

window *(Nuc.Eng.)*. Thin portion of wall or radiation counter through which low energy particles can penetrate.

window *(Radar)*. Strips of metallic foil of dimensions calculated to give radar reflections and hence confuse locations derived therefrom. Also *chaff*, *rope*.

window lock *(Build.)*. See **sash fastener**.

wind pollination *(Bot.)*. The conveyance of pollen from anthers to stigmas by means of the wind; *anemophily*.

wind road, way *(Min.Ext.)*. An underground passage used for ventilation.

wind rose *(Meteor.)*. A star-shaped diagram showing, for a given location, the relative frequencies of winds from different directions and of different strengths.

wind shear *(Meteor.)*. Rate of change of the vector wind in a direction (horizontal or vertical) normal to the wind.

wind sock *(Aero.)*. A truncated conical fabric sleeve, on a 360° free pivot, which indicates local wind direction; also called *wind stocking*.

windsucking *(Vet.)*. A habit acquired by certain horses of swallowing air when 'cribbing' or gripping an object with the incisor teeth.

wind T *(Aero.)*. A T-shaped device displayed at airfields to indicate the direction of the surface wind. The leg of the T corresponds to the wind direction, with wind arrowhead at top of T.

wind tunnel *(Aero.)*. Apparatus for producing a steady airstream past a model for aerodynamic investigations.

wing *(Aero.)*. The main supporting surface(s) of an aeroplane or glider.

wing *(Bot.)*. (1) A longitudinal flange on a stem or stalk; the downwardly continued lamina of a decurrent leaf. (2) A flattened outgrowth of a seed or fruit aiding in wind dispersal.

wing *(Build.)*. A section of a building projecting from the principal part of it.

wing *(Zool.)*. Any broad flat expansion; an organ used for flight, as the fore-limb in Birds and Bats, the membranous expansions of the mesothorax and metathorax in Insects.

wing area *(Aero.)*. See **gross wing area, net wing area**.

wing car *(Aero.)*. See **car**.

wing compasses *(Build.)*. A form of **quadrant dividers**.

wing coverts *(Zool.)*. See **tectrices**.

wing fence *(Aero.)*. A projection extending chordwise along a wing and projecting from its upper surface. It modifies the pressure distribution by preventing a spanwise flow of air which would otherwise cause a breakaway of the flow near the wing tips and lead to tip stalling. Also called a *boundary layer fence*. Cf. **vortex generators**.

wing loading *(Aero.)*. The gross weight of an aeroplane or glider divided by its **gross wing area**.

wing nut *(Eng.)*. A nut having radial lugs or wings to enable it to be turned by thumb and fingers. Also *bow nut*, *fly nut* or *butterfly nut*.

wing rail *(Civ.Eng.)*. See **check rail**.

wing shafts *(Ships)*. The port and starboard propeller shafts of a triple- or quadruple-screw steamship.

wing-tip float *(Aero.)*. A watertight float which gives stability and buoyancy on the water; placed at the extremities of the wings of a seaplane, flying-boat or amphibian.

wing valve *(Eng.)*. A mitre-faced or conical-seated valve guided by three or four radial vanes or wings fitting inside the circular port.

wing wall *(Civ.Eng.)*. A lateral wall built on an abutment and serving to retain earth in embankment.

Winkler reagent for oxygen *(Chem.)*. Quantitative absorption by a solution of alkaline pyrogallol, with formation of a brown colour.

Winner winding *(Elec.Eng.)*. Form of winding used in constructing standard resistances of low residual reactance for use at relatively high frequencies.

Winslow's foramen *(Zool.)*. A small opening by which the cavity of the bursa omentalis communicates with the rest of the abdominal cavity in Mammals.

winter annual *(Bot.)*. A plant which completes its life cycle over a few months in the coldest part of the year, surviving the summer as seeds. See **annual**.

winter dysentery *(Vet.)*. *Winter scour, vibrionic scour.* Thought to be an intestinal infection with *Campylobacter foetus* but viruses have also been incriminated. Usually a herd problem in cattle, showing as a sudden outbreak of diarrhoea affecting several animals.

winter egg *(Zool.)*. In some fresh-water animals, a thick-shelled egg laid at the onset of the cold season which does not develop until the following warm season. Cf. **summer egg**.

Winter-Eichberg-Latour motor *(Elec.Eng.)*. A single-phase a.c. motor of the compensated repulsion type; mainly used in electrical traction.

wintergreen *(Chem.)*. The methyl ester of salicylic (2-hydroxybenzoic) acid, $HO.C_6H_4CO.OMe$. Characteristic smell. Widely used in liniments and other medicinal products.

winter solstice *(Astron.)*. See **solstice**.

winze *(Min.Ext.)*. Internal shaft, usually between two

underground levels in plane of lode, used in exploration and subsequent extraction of valuable ore.

wipe *(Image Tech.)*. Transition effect in film or video where one picture image is replaced by another at a defined edge moving across the frame area.

wipe *(Print.)*. Defect in which, instead of an even film, the ink, because of unsuitable make-up, forms a ridge at the edge of the type.

wiped joint *(Build.)*. A joint formed between two lengths of lead pipe, one of which is opened out with a **tampin** while the other is tapered to fit into the first. Molten solder in a plastic condition is then wiped around the joint with a pad.

wiper *(Elec.Eng.)*. See **brush**.

wiper *(Telecomm.)*. In a uniselector or selector, the conducting arm which is rotated over a row of contacts and comes to rest on an outlet.

wipe test *(Nuc.Eng.)*. See **smear test**.

wiping solder *(Build.)*. Lead-based soft solder used in plumbing, containing up to 35% tin.

wire *(Telecomm.)*. A continuous connection through a system, particularly a telephone exchange, whether automatic or manual.

wirebar *(Eng.)*. High-purity copper cast to a tapered ingot shape; used to produce drawn copper wire.

wire cloth *(Paper)*. See **machine wire**.

wire comb *(Build.)*. A form of **scratcher**.

wire-cut bricks *(Build.)*. Bricks made by forcing the clay through a rectangular orifice, and cutting suitable lengths off the resulting bar of clay by pressing wires through the plastic mass, before burning.

wired glass *(Glass)*. A form of sheet-glass produced by rolling wire mesh into the ribbon of glass so that it acts as a reinforcement and holds the fragments together in the event of the sheet being fractured.

wire-drawing *(Eng.)*. (1) The process of reducing the diameter of rod or wire by pulling it through successively smaller holes. (2) The fall in pressure when a fluid is throttled by passing it through a small orifice or restricted valve-opening.

wire gauge *(Eng.)*. Any system of designating the diameter of wires by means of numbers, which originally stood for the number of successive passes through the die-blocks necessary to produce the given diameter. See **Birmingham-, Brown and Sharpe-, Standard-**.

wire guide *(Paper)*. Device located on the return wire run on a fourdrinier paper making machine to control and correct the lateral position or movement of the wire.

wire-guided missile *(Aero.)*. See **teleguided missile**.

wireless *(Genrl.)*. Original term for *radio*.

wireline tool *(Min.Ext.)*. Small tools or measuring instruments designed to be lowered into a well on a wire line.

wire recorder *(Acous.)*. Early type of magnetic recorder with recording medium in form of iron wire.

wire rope *(Eng.)*. Steel rope made by 'twisting' or 'laying' a number of strands over a central core, the strands themselves being formed by twisting together steel wires. See **Lang lay**.

wire stabbing *(Print.)*. A simple method of binding in which one or more wire staples are passed through the back margin of all the leaves or sections.

wire-stitching *(Print.)*. The securing of a booklet by means of wire staples. See **saddle-stitching, stabbing**.

wirewound resistor *(Elec.Eng.)*. One with metallic wire elements.

wiring point *(Elec.Eng.)*. A point in an interior wiring installation where an external connection can be made to the electric circuit.

Wirsung's duct *(Zool.)*. The ventral or main pancreatic duct of Mammals.

wisdom teeth *(Med.)*. The third molars, which do not usually erupt until adulthood.

wishbone *(Autos.)*. V-shaped member used in independent suspension systems.

Wiskott-Aldrich syndrome *(Immun.)*. Sex-linked recessive disease of infants characterized by haemorrhagic diathesis, eczema and recurrent infections. Delayed hypersensitivity reactions are absent and there is a defective antibody response to polysaccharide antigens, with low blood IgM levels. A protein normally present in platelet membranes is missing. This combined defect of cell-mediated and humoral immunity probably involves failure to recognize or process antigens. Patients die early of infection or bleeding due to increased destruction of platelets.

witches' broom *(Bot.)*. A dense tuft of twigs formed on a woody plant as a response to infection.

witch of Agnesi *(Maths.)*. The curve whose cartesian equation is $x^2y = 4a^2(2a-y)$.

withamite *(Min.)*. A mineral belonging to the epidote group, containing about 1% of manganese oxide.

withdrawal symptoms *(Behav.)*. Temporary psychological and physiological disturbances resulting from the body's attempt to readjust to the absence of a drug.

withe *(Build.)*. The partition wall between adjacent flues in a chimney stack. Also *mid-feather, bridging*.

witherite *(Min.)*. Barium carbonate, $BaCO_3$, crystallizing in the orthorhombic system as yellowish or greyish-white complex crystals of hexagonal appearance due to twinning; also massive. Occurs with galena in lead mines. Exploited as an important source of barium.

withers *(Vet.)*. The region of the horse's back above the shoulders.

witness *(Eng.)*. A remnant of original surface or scribed line, left during machining or hand working to prove that a minimum quantity of material has been removed or an outline accurately preserved.

wobble crank *(Eng.)*. A short-throw crank in which the pin, machined from and at an angle to the axis of the crankshaft, has been used to give an elliptical motion to a sleeve valve by a short connecting-rod and ball joint.

wobble plate *(Eng.)*. See **swash plate**.

wobble-plate engine *(Eng.)*. A multicylinder engine in which a wobble-plate or swash-plate mechanism replaces cranks and connecting rods. The cylinders are arranged axially round the shaft, their pistons operating on the wobble plate through sliding blocks. The arrangement is very compact but the mechanical efficiency is usually low.

wobble saw *(Build.)*. See **drunken saw**.

wobbulator *(Telecomm.)*. Colloquialism for a signal generator whose frequency is automatically varied periodically over a definite range; used in testing frequency response of systems.

Wolffian body *(Zool.)*. The **mesonephros**.

Wolffian duct *(Zool.)*. The kidney duct of Vertebrates. In adult anamniotes (Agnatha, Fish, and Amphibia) it serves as a kidney duct and a sperm duct in males, while in adult amniotes (Reptiles, Birds and Mammals) whose metanephric kidney has a separate duct, it is present only in males, forming the *vas deferens*.

wolf note *(Acous.)*. Nonharmonic note made by a bow on a violin or cello string at some frequencies.

wolfram *(Chem.)*. See **tungsten**

wolfram *(Min.)*. Syn. for *wolframite*.

wolframite *(Min.)*. Tungstate of iron and manganese $(FeMn)WO_4$, occurring as brownish-black monoclinic crystals, columnar aggregates, or granular masses. It forms a complete series from ferberite ($FeWO_4$) to hübnerite ($MnWO_4$). An important ore of tungsten. Also called *wolfram*.

Wolf-Rayet star *(Astron.)*. An abnormal class of stars, with spectra similar to those of novae, broad bright lines predominating, indicating violent motion in the stellar atmosphere.

wollastonite *(Min.)*. A triclinic silicate of calcium, $CaSiO_3$, occurring as a common mineral in metamorphosed limestones and similar assemblages, resulting from the reaction of quartz and calcite. Also called *tabular spar*.

Wollaston prism *(Phys.)*. A double-image polarizing device, made of two geometrically similar wedge-shaped prisms of calcite or quartz, cemented with glycerine or castor oil. The cutting of the two prisms is arranged so that their optic axes cause two coloured, oppositely polarized emergent beams. It is useful in determining the proportion of polarization in a partially polarized beam providing, for examination, two images with oscillations in two perpendicular directions. Similar to, but distinct from a *Rochon prism*.

wood *(Bot.)*. See **xylem**. Wood, the constructional material, is the secondary xylem of conifers (softwood) and dicotyledons (hardwood). See also **heartwood, sapwood**.

wood alcohol *(Chem.)*. Same as *methanol*.

wood brick *(Build.)*. A piece of wood the shape of a brick but larger by the amount of the mortar joints. It is bonded to surface brickwork in the course of building and is held in position by friction alone, its function being to provide a substance to which, e.g. skirtings may be nailed.

woodcut *(Print.)*. An engraving cut on the blank grain of wood; an impression from it.

wooden tongue *(Vet.)*. See **actinobacillosis**.

wood fibre, -parenchyma, -ray *(Bot.)*. See **xylem fibre, -parenchyma, -ray**

woodfree *(Print.)*. Paper made with only chemical pulp in its fibre composition.

wood furniture *(Print.)*. Spacing material, made of wood, as distinct from metal or plastic.

woodland *(Ecol.)*. Natural or semi-natural vegetation containing trees, but not forming a continuous canopy. See **forest**.

wood letters *(Print.)*. Large type-letters cut in wood, used in some poster work.

wood nog *(Build.)*. See **nog**.

wood opal *(Min.)*. A form of common opal which has replaced pieces of wood entombed as fossils in sediments, in some cases retaining the original structure.

wood preservatives *(Build.)*. Special coatings intended to afford external timbers protection from decay, fungal attack, insects and inclement weather. Produced in varying media such as oil, spirit and water using constituents such as coal tar derivatives, naphthanates and metallic salts.

wood pulp *(Paper)*. Pulp produced from wood by any pulping method.

Woodruff key *(Eng.)*. A key consisting of a segment of a disk, restrained in a shaft key-way milled by a cutter of the same radius, and fitting a normal key-way in the hub.

Wood's glass *(Glass)*. A glass with very low visible and high ultraviolet transmission.

wood spirit *(Chem.)*. **Methanol**.

wood sugar *(Chem.)*. **Xylose**.

wood tar *(Chem.)*. A product of the destructive distillation of wood, containing alkanes, naphthalene, phenols.

wood tin *(Min.)*. A botryoidal or colloform variety of cassiterite showing a concentric structure of brown, radiating, woodlike fibres.

wood-wool slabs *(Build.)*. Made from long wood shavings with a cementing material; used for linings, partitions etc.

woody tissues *(Bot.)*. Tissues which are hard because of the presence of lignin in the cell walls.

woofer *(Acous.)*. Large loudspeaker used to reproduce lower part of audiofrequency spectrum only. See **cross-over frequency, tweeter (1)**.

wool *(Textiles)*. The fibres obtained from sheep. (See **hair**).

wool *(Zool.)*. A modification of hair in which the fibres are shorter, curled, and possess an imbricated surface. Specifically, the covering of a sheep. The fibres are covered with small scales and are composed of keratin. Also *fleece wool*.

woollen *(Textiles)*. Yarns, fabrics, or garments made entirely from wool spun on the condenser system.

woollen blended yarns *(Textiles)*. Spun on the condenser system with wool as the main fibre but other fibres present in an intimate mixture.

woolsorter's disease *(Med.)*. An acute disease due to infection with the *Bacillus anthracis*, conveyed to Man by infected wool or hair of animals; characterized by fever, the appearance on the skin of vesicles which become covered with a black scab, and sometimes by infection of the lungs or of the intestines. See also **anthrax**.

word *(Comp.)*. Collection of **bits** treated as a single unit by the central processor.

word association test *(Behav.)*. A psychological test in which the subject is presented with a stimulus word and asked to produce the first word that comes to mind; latency to response, and the nature of the association word are interpreted as revealing verbal habits, thought processes, personality characteristics and emotional state. Abbrev. *WAT*.

word length *(Comp.)*. Number of bits in each word of a particular computer. In most computers this is fixed.

word processor *(Comp.)*. (1) A computer system designed to help in the preparation, editing, printing and sending of textual data. (2) The software for doing the above which is usually implemented on a standalone microcomputer with possible network connections. Cf. **desk-top publishing**.

word salad *(Behav.)*. A schizophrenic speech pattern in which words and phrases are combined in a disorganized fashion, apparently devoid of logic and meaning.

work *(Phys.)*. One manifestation of energy. The work done by a force is defined as the product of the force and the distance moved by its point of application along the line of action of the force. For example, a tensile force does work in increasing the length of a piece of wire; work is done by a gas when it expands against a hydrostatic pressure. As for all forms of energy, the SI unit of work is the *joule*, performed when a force of 1 newton moves its point of application through 1 metre along the line of action of the force. Alternatively, when 1 watt of power is expended for 1 second, 1 joule of work is done. See **joule, kilowatt-hour**.

work-and-tumble *(Print.)*. A method of turning printed sheets in which, after printing the first side, the opposite edge must be used as the front lay for the second side.

work-and-turn *(Print.)*. Process in which one forme or plate is used to print on both sides of the paper which is cut after printing to give two copies of the job or section. Cf. **sheet-work**.

work-and-twist *(Print.)*. Two impressions are taken on the same side of the paper using the same forme, the paper being turned 180° before taking the second impression. Useful when printing rule formes by letterpress.

work coil *(Elec.Eng.)*. Same as *heating inductor*.

work electrode *(Elec.Eng.)*. See **applicator**.

worker *(Zool.)*. In social Insects, one of a caste of sterile individuals which do all the work of the colony.

work function *(Phys.)*. The minimum energy that must be supplied to remove an electron so that it can just exist outside a material under vacuum conditions. The energy can be supplied by heating the material (*thermionic* work function) or by illuminating it with radiation of sufficiently high energy (*photoelectric* work function). Also called *electron affinity*.

work-hardening *(Eng.)*. The increase in strength and hardness (i.e. resistance to deformation) produced by working metals. It is most pronounced in cold-working, and in the case of metals such as iron, copper, aluminium and nickel. Lead, tin and zinc are not appreciably hardened by cold-working, because they can recrystallize at room temperature.

workhead transformer *(Elec.Eng.)*. One associated with the workpiece in induction heating when the power supply is at a distance and feeds power though a cable.

working *(Telecomm.)*. The technique of routing calls over a telephone system.

working chamber *(Civ.Eng.)*. The compressed-air chamber at the base of a hollow caisson, being the part in which the work of excavation proceeds. See **air lock**.

working edge *(Build.)*. An edge of a piece of wood trued square with the working face to assist in truing the other surfaces. Also *face edge*.

working face *(Build.)*. That face of a piece of wood which is first trued and then used as a basis for truing the other surfaces. See **face mark**.

working flux *(Elec.Eng.)*. That part of the total flux

produced by the magnetic system of an electrical machine which links the armature winding; numerically equal to the difference between the total flux and the leakage flux.

working standard *(Elec.Eng.)*. A standard for everyday use, calibrated against a **secondary standard**.

working stress *(Eng.)*. The (safe) working stress is the ultimate strength of a material divided by the applicable factor of safety.

work lead *(Eng.)*. See **base bullion**.

workpiece *(Elec.Eng.)*. The material placed within a *heating inductor* for the purpose of induction heating by high-frequency electric currents.

workstation *(Comp.)*. A single user computer usually provided with a moderate amount of **disk** storage and a **bit-mapped display**.

work study *(Eng.)*. A generic term for those techniques, particularly method study and work measurement, which are used in the examination of human work in all its contexts and which lead systematically to the investigation of all the factors which affect the efficiency and economy of the situation being reviewed, in order to effect improvement.

worm *(Eng.)*. A gear of high reduction ratio connecting shafts whose axes are at right-angles but do not intersect. It consists of a cylindrical core carrying a single- or multi-start helical thread of special form (the *worm*), meshing in sliding contact with a concave face gear-wheel (the *worm wheel*). Also *worm gear, worm wheel*.

worm *(Zool.)*. An imprecise term applied to elongated Invertebrates with no appendages, as in Flatworm (Platyhelminthes), Roundworm, Eelworm (Nematoda), Earthworm (Lumbricus spp.) etc. Also applied to immature forms of some Insects, as in Mealworm, (Tenebrio, Coleoptera), Cutworm (some Lepidoptera), and Wireworm (Elateridae, Coleoptera), Click-beetles, and also Millipedes (Diplopoda).

worm-and-wheel steering gear *(Autos.)*. Steering gear in which the steering column carries a worm, in mesh with a worm wheel or sector, attached to the spindle of the drop arm.

worm gear, wheel *(Eng.)*. See **worm**.

worm, screw conveyor *(Eng.)*. A conveyor in which loose material such as grain, meal etc, is continuously propelled along a narrow trough by a revolving worm or helix mounted within it.

worsted *(Textiles)*. Yarns, fabrics and garments made from combed wool.

worsted-spun *(Textiles)*. Yarn spun from any staple fibres on the machinery used for making worsted yarns. The fabric made from these yarns is known as worsted-type.

wound rotor *(Elec.Eng.)*. An alternative term for *slip-ring rotor*.

wound tissue *(Bot.)*. Those formed in response to wounding as the wound vessels which reconnect severed xylem strands. See also **callus**.

wovenboard *(Build.)*. See **interwoven fencing**.

woven steel fabric *(Civ.Eng.)*. A mechanically woven fabric of steel wires interlaced and welded at the intersections, used as a reinforcement in reinforced concrete construction.

wove paper *(Paper)*. Paper which does not exhibit a laid design when viewed by transmitted light.

wow *(Acous.)*. Low-frequency modulation introduced in sound reproduction system as a result of speed variation. Similar to, but lower in frequency than, *flutter*.

wrapper plate *(Eng.)*. (1) In a locomotive boiler, the plate bent round and riveted to the tube plate and back plate, forming the sides and crown of the fire box. (2) The outer casing of the fire box.

wrap round *(Print.)*. See **outset**.

wrap-round plate *(Print.)*. A flexible relief printing plate which is clamped round the cylinder.

wreath *(Build.)*. The part of a continuous hand-rail curving in plan around the well-hole of a geometrical stair.

wreathed string *(Build.)*. The continuous curved outer string around the well-hole of a wooden stair.

wreath filament *(Elec.Eng.)*. The usual type of filament in large gas-filled electric lamps; the filament wire is festooned from a horizontal supporting spider.

wrench *(Build.)*. See **pipe wrench**.

wrench *(Phys.)*. A system comprising a force and a couple whose axis is parallel to the force. The force is called the *intensity* of the wrench and the ratio of the moment of the couple to the force of its *pitch*. Also called *wrench on a screw*. Any system of forces can be reduced to a wrench.

wringing *(Eng.)*. A method of temporarily combining several *slip gauges* by pressing them together with a slight twisting motion until they adhere, thus ensuring that the combined length equals the sum of the individual lengths.

wrinkle finish *(Build.)*. A paint with a deliberate pattern of small, fairly uniformly distributed wrinkles. Much used on industrial metal articles (cameras, typewriters, etc.) to hide surface imperfections of the metal. Also known as *ripple finish*.

wrinkling *(Eng.)*. Uneven texture developing on surface of metallic uranium.

wrist-drop *(Med.)*. Paralysis of the extensor muscles of the hand and of the fingers.

write *(Comp.)*. To output data and transfer it to a store.

writer's cramp *(Med.)*. A condition in which writing becomes irregular and difficult or even impossible owing to spasm of the muscles of the hand and forearm; an occupational neurosis: not the result of organic disease.

writing speed *(Electronics)*. Speed of deflection of trace on phosphor, or rate of registering signals on charge storage device.

writing speed *(Image Tech.)*. The speed at which the recording/replay head of a video recorder traverses the surface of the moving magnetic tape.

wrong fount *(Print.)*. The use in error of a character from a type **fount** other than that currently being used in composition. abbrev. *w.f.*

wrong lead *(Print.)*. See **lead (of the web)**.

wrong-reading *(Print.)*. See **right-reading**.

Wronskian *(Maths.)*. Of *n* functions u_i of an independent variable *x*, the determinant whose *i,j*th element is $\frac{d^{i-1}u_j}{dx^{i-1}}$. It is zero when the functions u_i are linearly dependent.

wrought grounds *(Build.)*. Planed strips of wood used as **grounds** when the attached joinery will leave them partly exposed to view.

wrought iron *(Eng.)*. Commercial iron containing iron silicate in a ferrite matrix and with a low carbon content. Easily worked.

wryneck *(Med.)*. See **torticollis**.

wulfenite *(Min.)*. Molybdate of lead, $PbMoO_4$, occurring as yellow tetragonal crystals in veins with other lead ores.

wurtzite *(Min.)*. Sulphide of zinc, ZnS, of the same composition as sphalerite, but crystallizing at higher temperatures and in the hexagonal system, in black hemimorphic, pyramidal crystals.

Wurtz synthesis *(Chem.)*. The reduction of solutions of alkyl halides (in ether) with metallic sodium to yield the corresponding hydrocarbons. If mixtures of different alkyl halides are used, mixtures of hydrocarbons formed by different combinations of the alkyl groups are obtained.

wye *(Build.)*. A *branch pipe* having only one branch, which is not at right angles to the main run.

wye level *(Surv.)*. A type of level whose essential characteristic is the support of the telescope, which is similar to that of the wye theodolite.

wye rectifier *(Elec.Eng.)*. Full-wave rectifier system for a three-phase supply

wye theodolite *(Surv.)*. A form of theodolite differing from the transit in that the telescope is not directly mounted on the trunnion axis but is supported on two Y-shaped forks, in which it may be turned end-for-end in order to reverse the line of sight.

wyomingite *(Geol.)*. An alkaline volcanic rock, composed of leucite, phlogopite and diopside.

WYSIWYG *(Comp.)*. *What you see is what you get*. After editing on the full screen of a **VDU**, the user can print a replica of what appears on the screen.

X

x, 2x, 3x.... *(Biol.)*. Symbols for the number of copies of the *haploid chromosome number* or *basic chromosome set*.

x *(Chem.)*. A symbol for mole fraction.

X *(Chem.)*. A general symbol for an electro-negative atom or group, especially a halogen.

X *(Phys.)*. A symbol for reactance.

xalostocite *(Min.)*. A pale rose-pink grossular which occurs embedded in white marble at Xalostoc in Mexico. Also called *landerite*.

xanthates *(Chem.)*. The salts of xanthic acid, $CS(OC_2H_5)SH$. Potassium xanthate, obtained by the action of potassium ethoxide on carbon disulphide. **Cellulose xanthate** is the basis of the viscose rayon process.

xanthene *(Chem.)*. *Diphenylene-methane oxide*. Colourless plates, mp 98.5°C.

xanthene dyestuffs *(Chem.)*. Dyestuffs which may be regarded as derivatives of xanthene containing the pyrone ring. They comprise the pyronines, derivatives of diphenylmethane, and the phthaleins, derivatives of triphenylmethane.

xanthine *(Chem.)*. *2,6-Dihydroxy-purine*, a white amorphous mass which is both basic and acidic, and can be obtained by the action of nitrous acid upon guanine.

xanthine bronchodilators *(Med.)*. Group of drugs which can alleviate asthma and chronic obstructive airways disease by bronchodilatation. *Aminophylline* can be given intravenously and *choline theophyllinate* as a sustained release oral preparation.

xanthochroism *(Zool.)*. A condition in which all skin pigments other than golden and yellow ones disappear; as in the Goldfish.

xanthochromia *(Med.)*. Any yellowish discoloration, especially of the cerebrospinal fluid.

xanthoma *(Med.)*. Yellow irregular swellings composed of fibrous tissue and of cells containing cholesterol ester, occurring on the skin (e.g. in diabetes) or on the sheaths of tendons, or in any tissue of the body (*xanthoma multiplex*).

xanthophore *(Zool.)*. A cell occurring in the integument and containing a yellow pigment; as in Goldfish. Also *guanophore, ochrophore*.

Xanthophyceae *(Bot.)*. The yellow-green algae, a class of eukaryotic algae in the division Heterokontophyta. Without fucoxanthin. Naked or walled. Flagellated and amoeboid unicellular, palmelloid, coccoid, dendroid, simply filamentous and siphoneous types, many remarkably similar to analogous green algae. Mostly phototrophs. Mostly fresh water and in soil.

xanthophyll *(Bot.,Zool.)*. $C_{40}H_{56}O_2$. One of the two yellow pigments present in the normal chlorophyll mixture of green plants; a yellow pigment occurring in some Phytomastigina.

xanthophyllite *(Min.)*. A brittle mica, crystallizing in the monoclinic system; hydrated calcium, magnesium, aluminium silicate.

xanthophylls *(Bot.)*. Yellowish, oxygenated carotenoids acting as minor or, in a few cases, major **accessory pigments** in photosynthesis. Each major algal and plant group has its characteristic set of xanthophylls.

xanthopsia *(Med.)*. *Yellow vision*. The condition in which objects appear yellow to the observer, as in jaundice.

x-axis *(Aero.)*. The longitudinal, or roll, axis of an aircraft. Cf. **axis**.

x-axis *(Maths.)*. Conventionally, the horizontal axis of any type of graph.

X-band *(Radar,Telecomm.)*. Microwave band lying roughly between 8 and 12 GHz; slight discrepancy between US and UK band limits. Widely used for 3 cm radar which is now correctly designated *Cx-band*.

X-chromosome *(Biol.)*. See **sex determination**.

X-disease *(Vet.)*. See **bovine hyperkeratosis**.

Xe *(Chem.)*. The symbol for xenon.

xenia *(Bot.)*. The influence of the pollen on the seed through its effect, by double fertilization, on the nature of the endosperm. Cf. **metaxenia**.

xeno- *(Genrl.)*. Prefix from Gk. *xenos*, strange, foreign.

xenocryst *(Geol.)*. A single crystal or mineral grain of extraneous origin which has been incorporated by magma during its uprise and which therefore occurs as an inclusion in igneous rocks, usually surrounded by reaction rims and more or less corroded by the magma. Cf **xenolith**.

xenogamy *(Bot.)*. Fertilization involving pollen and ovules from flowers on genetically non-identical plants of the same species (i.e. different *genets*) See **cross pollination, cross fertilization, outbreeding**.

xenogeneic *(Immun.)*. Grafted tissue that has been derived from a species different from the recipient. Hence *xenograft*.

xenolith *(Geol.)*. A fragment of rock of extraneous origin which has been incorporated in magma, either intrusive or extrusive, and occurs as an inclusion, often showing definite signs of modification by the magma.

xenomorphic *(Min.)*. A textural term implying that the minerals in a rock do not show their own characteristic shapes, but are without regular form by reason of mutual interference. See also **granitoid texture**.

xenon *(Chem.)*. A zero-valent element, one of the noble gases, present in the atmosphere in the proportion of 1:170 000 000 by volume. Symbol Xe, at. no. 54, r.a.m. 131.30, mp −140°C, bp −106.9°C, crit, temp. +16.6°C, density at s.t.p. 5.89 g/dm^3. Its isotope ^{135}Xe has the highest known capture cross-section for thermal neutrons (2.7×10^6 barns). Formed as a result of radioactive decay of uranium fission fragments, this isotope forms the most serious reactor poison, and may delay restart of a reactor after a period of shutdown. Forms several compounds, e.g. XeF_2, XeF_6, XeO_3.

xenon lamp *(Image Tech.)*. A compact high intensity discharge lamp, with the arc operating in a quartz envelope containing xenon gas at high pressure; widely used for motion picture projection.

xenotime *(Min.)*. Yttrium phosphate. YPO_4, often containing small quantities of cerium, erbium, and thorium, closely resembling zircon in tetragonal crystal form and general appearance, and occurring in the same types of igneous rock, i.e. in granites and pegmatites as an accessory mineral. An important source of the rare elements named.

xeric *(Bot.)*. Dry conditions in which plant growth may be limited by water shortage.

xeroderma pigmentosum *(Med.)*. A heritable disease of young children in which prolonged exposure to sunlight on the skin causes erythematous patches which later become pigmented, scaly, wartlike and finally cancerous.

xerodermia, xeroderma *(Med.)*. See **ichthyosis**.

xerographic printer *(Comp.)*. A high speed printer using xerographic techniques as in many photocopiers.

xerography *(Print.)*. Non-chemical photographic process in which light discharges a charged dielectric surface. This is dusted with a dielectric powder, which adheres to the charged areas, rendering the image visible. Permanent images can be obtained by transferring particles to a suitable backing surface (e.g. paper or plastic) and fixing, usually by heat. Used for document copying and for making lithographic surfaces, usually on paper, for small-offset printing.

xeromorphic *(Bot.)*. Of a feature typical of xerophytes.

xerophthalmia *(Med.)*. A dry lustreless condition of the conjunctiva with or without keratomalacia, due to deficiency of vitamin A.

xerophyte *(Bot.)*. A plant adapted to a dry habitat, where growth may be limited by water shortage.

xeroradiography *(Radiol.)*. Radiography in which a xerographic, and not photographic, image is produced.

xerosere *(Bot.)*. A succession beginning on dry land, as opposed to under water, cf. **hydrosere**.

xerosis *(Med.)*. See **xerophthalmia**.

xerostomia *(Med.)*. Excessive dryness of the mouth.

xer-, xero- *(Genrl.)*. Prefix from Gk. *xeros*, dry.

X-guide *(Elec.Eng.)*. A transmission line with an X-shaped cross-section dielectric, used for guiding surface waves.

x-height *(Print.)*. The height of the lower-case letters exclusive of extenders, varying between the extremes of small and large according to the design of the type face.

X-inactivation *(Biol.)*. Permanent condensation in early development of one or other X-chromosome in female mammalian cells, with accompanying repression of most of the genes on that chromosome.

xiphisternum *(Zool.)*. A posterior element of the sternum, usually cartilaginous.

xiphi-, xipho- *(Genrl.)*. Prefix from Gk. *xiphos*, sword.

xiphoid *(Zool.)*. Sword-shaped.

X-linkage *(Biol.)*. See **sex linked**.

xonotlite *(Min.)*. A hydrous calcium silicate, of composition $Ca_6Si_6O_{17}(OH)_2$.

XOR *(Comp.)*. A logical operator where (*p* XOR *q*) takes the value FALSE if *p* and *q* are the same. If *p* and *q* differ then *p* XOR *q* is TRUE. Sometimes written EXOR. See **logical operation**.

X-organ *(Zool.)*. A neurosecretory organ in the eye-stalks of certain Crustaceans.

XOR gate *(Comp.)*. A **gate** with two input signals. If both are 0, output is 0. If both are 1, output is 0. When incoming signals differ output is 1. Also *non-equivalence gate, NEQ gate*.

X-plates *(Electronics)*. Pair of electrodes in a CRT to which horizontal deflecting voltage is applied in accordance with cartesian co-ordinate system.

X-ray crystallography *(Crystal.)*. The study of crystal or molecular structures by X-ray diffraction of an intense collimated beam of X-rays. X-rays may come from a conventional X-ray source but increasingly beams from a cyclotron are used to reduce the exposure time and allow analysis of complex structures like muscle and of unstable molecules. The diffraction pattern can be recorded photographically and the spot densities and positions then analysed, or by detectors programmed to record X-ray intensity at accurately determined angles relative to the collimated beam or by special two-dimensional detectors, whose output like that of the single detectors can then be analysed by computer. It has been possible to determine the structures of proteins and other complex biological molecules to very high resolutions using such methods. See **isomorphous replacement**.

X-ray diffractometer *(Crystal.)*. An instrument containing a radiation detector used to record the X-ray diffraction patterns of crystals, powders or molecules. See **X-ray crystallography**.

X-ray fluorescence spectrometry *(Chem.)*. A method of chemical analysis in which the sample is bombarded by very hard X-rays or gamma-rays, and secondary radiations, characteristic of the elements present, are studied spectroscopically.

X-ray focal spot *(Electronics)*. That small area of the target (anode) of an X-ray tube on which the electron beam is incident, and from which emitted X-rays emerge. High-power tubes frequently have a line focus to minimize localization of the heat dissipated at the anode.

X-ray laser *(Phys.)*. A laser with an output in the X-ray region of the spectrum. Using highly-ionized selenium plasma as the lasing medium, laser output at 20.6 and 20.9 nm has been recently reported.

X-ray microscope *(Phys.)*. Using soft X-rays and a Fresnel zone plate as the focusing device, a resolution of 200 nm has been obtained (1986) using a scanning technique. The zone plate is made using a scanning transmission electron microscope, and synchrotron radiation is used as the source of X-rays.

X-ray photon *(Phys.)*. A quantum of X-radiation energy given by $h\nu$; when ν is the frequency and h is Planck's constant.

X-ray protective glass *(Glass)*. Glass containing a high percentage of lead and sometimes also barium, with a high degree of opacity to X-rays.

X-rays *(Phys.)*. Electromagnetic waves of short wavelength (ca. 10^{-3} to 10 nm) produced when high-speed electrons strike a solid target. Electrons passing near a nucleus in the target are accelerated and so emit a continuous spectrum of radiation (*bremsstrahlung*) ranging up from a *minimum wavelength*. In addition, the electrons may eject an electron from an inner shell of a target atom, and the resulting transition of an electron of a higher energy level to this level produces radiation of specific wavelengths. This is the *characteristic* X-ray spectrum of the target and is specific to the target element. X-rays may be detected photographically or by a counting device. They penetrate matter which is opaque to light; this makes X-rays a valuable tool for medical investigations. See **Moseley's law**, **Compton effect**, **K-capture**, **L-capture**, **synchrotron radiation**.

X-ray sources *(Astron.)*. There are several sources of cosmic X-rays. (1) The solar **corona**. (2) Interacting **binary stars**, in which one member is a **black hole** or **neutron star**. (3) **Supernova** remnants, such as the **Crab Nebula**. (4) Some **radio galaxies**, such as **Cygnus A**, some **Seyfert galaxies**, and some **quasars**.

X-ray spectrography *(Chem.)*. See **X-ray fluorescent spectrometry**.

X-ray spectrometer *(Crystal.)*. The name originally used for the *X-ray diffractometer*, but now abandoned in order to avoid confusion with *X-ray fluorescent spectrometry*.

X-ray telescope *(Eng.)*. One used initially to investigate the X-ray emission from the sun and which showed greater variability than might have been expected from optical measurements in the visible region. To focus the X-rays by mirrors, grazing incidence techniques had to be used. A gas-filled X-ray proportional counter was placed behind an aperture at the mirror focus and the counts were telemetered to earth, the telescope being rigidly fixed to a rocket whose attitude can be accurately controlled.

X-ray therapy *(Radiol.)*. The use of X-rays for medical treatment.

X-ray transformer *(Radiol.)*. A special type of high-voltage transformer for use with X-ray tubes.

X-ray tube *(Electronics)*. The vacuum tube in which X-rays are produced by a cathode-ray beam incident on an anode (or anticathode). Such tubes may be sealed high vacuum or continuously pumped. See **Coolidge tube**.

X-synchronization *(Image Tech.)*. See **flash-synchronized**.

X-tgd *(Build.)*. Abbrev. for *cross-tongued*.

xylem *(Bot.)*. The **vascular tissue** with the prime function of water transport; it consists of tracheids and vessels and associated parenchyma and fibres. Secondary xylem (wood) may also be important for support (tracheids and fibres) and storage (xylem parenchyma). See **primary xylem**, **secondary xylem**, **ray**, **apoplast**, **axial** system, **conduit**, **cohesion theory**.

xylem parenchyma *(Bot.)*. Parenchyma cells chiefly within the secondary xylem, mostly with lignified walls and with living contents in the sap wood. Has storage and defensive functions.

xylenes *(Chem.)*. $C_6H_4(CH_3)_2$, *dimethylbenzenes*. There are three isomers which all occur in coal-tar but cannot be separated by fractional distillation: is an important starting material for polyester synthetic fibres such as Terylene. Commercial preparation termed *xylol*. Used, in microscopy, as a clearing agent, in the preparation of specimens for embedding, and also in the preparation of tissue sections etc. for mounting.

xylenol resin *(Chem.)*. Synthetic resin of the phenolic type produced by the condensation of a xylenol with an aldehyde.

xylenols *(Chem.)*. $(CH_3)_2C_6H_3.OH$, monohydric phenols derived from xylenes: 1,2,3-xylenol (*adj. ortho*), mp 73°C, bp 213°C; 1,3,4-xylenol (*asym. ortho*), mp 65°C, bp 222°C; 1,2,5-xylenol (*para*), mp 75°C, bp 209°C.

xylogenous, xylophilous *(Bot.,Zool.)*. Growing on wood; living on or in wood.

xylol *(Chem.)*. See **xylenes**.

Xylonite *(Plastics)*. A thermoplastic of the nitro-cellulose type.

xylophagous *(Zool.)*. Wood-eating.

xylophilous *(Bot.,Zool.)*. See **xylogenous**.

xylose *(Chem.)*. Xylose, known as *wood sugar*, is a pentose found in many plants. It is a stereoisomer of arabinose.

xylotomous *(Zool.)*. Wood-boring; wood-cutting.

xyl-, xylo- *(Genrl.)*. Prefix from Gk. *xylon*, wood.

x-y recorder *(Eng.)*. One which traces on a chart the relation between two variables, not including time. Time may be introduced by moving the chart linearly with time and controlling one of the variables.

XYY syndrome *(Biol.)*. Condition in which the human male has an extra Y chromosome. They are normal males, except for slight growth and sometimes minor behavioural abnormalities.

Y

Y *(Chem.)*. The symbol for yttrium.

Y *(Phys.)*. Symbol for admittance.

Yagi antenna *(Telecomm.)*. An **end-fire array**, characterized by directors in front of the normal dipole radiator and rear reflector.

Yankee machine *(Paper)*. See **MG machine**.

y-antenna *(Telecomm.)*. A delta-matched antenna.

yapp *(Print.)*. A style of binding with overlapping limp covers, much used for pocket bibles.

yard *(Genrl.)*. Unit of length in the foot-pound-second system formerly fixed by a line standard (Weights and Measures Act of 1878), redefined in 1963 as 0.9144 m.

yardage *(Civ.Eng.)*. The volume of excavation in cubic yards.

yard trap *(Build.)*. See **gulley trap**.

yarn *(Textiles)*. A thread i.e. a long thin material made of fibres or filaments usually twisted together.

Yarrow boiler *(Eng.)*. A marine water-tube boiler employing an upper steam drum connected by banks of inclined tubes to three lower water drums, between two of which susperheating elements are arranged.

yaw *(Aero.,Ships)*. Angular rotation of an aircraft or other vessel about a vertical axis. The *yaw angle* is measured between the relative wind and the axis of the vessel.

yaw damper *(Aero.)*. See **damper**.

yawing moment *(Aero.)*. The component about the normal axis of an aircraft due to the relative airflow.

yaw meter *(Aero.)*. An instrument, usually on experimental aircraft or missiles, which detects changes in the direction of airflow by the pressure changes induced thereby, or by a weather recording vane transmitting to instruments.

yaws *(Med.)*. *Framboesia, pian*. A contagious tropical disease occurring in children living in hot humid climates, due to infection with *Treponema pertenue*, and characterized by raspberry-like papules on the skin; as in syphilis, the bones and joints may later become infected.

yaw vane *(Aero.)*. A small aerofoil on a pivoted arm at the end of a long boom or probe, attached to the nose of an aircraft or missile, which measures the angle of the relative airflow and transmits it to recording instruments.

y-axis *(Aero.)*. The lateral, or pitch, axis of an aircraft. Cf. **axis**.

y-axis *(Maths.)*. Conventionally, the axis perpendicular to and in the horizontal plane through the **x-axis** in any type of graph.

Yb *(Chem.)*. The symbol for ytterbium.

Y/B ratio *(Phys.)*. A term used to classify a type of *dichromatism*. An observer sees only two colours when examining the solar spectrum, blue and yellow, separated by a white patch. The relative extent of the two colours is the *Y/B ratio*.

Y-chromosome *(Biol.)*. See **sex determination**.

Y-class insulation *(Elec.Eng.)*. A class of insulating material to which is assigned a temperature of 90°C. See class **-A**, **-B**, **-C**, etc., **insulating materials**.

Y-connection *(Elec.Eng.)*. An alternative name for *star connection*.

year *(Astron.)*. The civil or calendar year as used in ordinary life, consisting of a whole number of days, 365 in ordinary years, and 366 in leap years, and beginning with January 1. See also **anomalistic-**, **eclipse-**, **leap years**, **sidereal-**, **tropical-**.

yeast *(Bot.)*. Unicellular fungus reproducing asexually by budding or division, especially the genus, Saccharomyces including *S. cerevisiae* (baker's and brewer's yeast) and *S. ellipsoides* (used in wine making).

yellow cells *(Zool.)*. In Oligochaeta, yellowish cells forming a layer investing the intestine and playing a role in connection with nitrogenous excretion; chloragogen cells.

yellow fever *(Med.)*. *Yellow jack*. An acute infectious disease caused by a virus, conveyed to Man by the bite of the Mosquito *Aëdes aegypti* (*Stegomyia fasciata*); characterized by high fever, acute hepatitis, jaundice, and haemorrhages in the skin and from the stomach and bowels; it occurs in tropical America and West Africa.

yellow fibres *(Zool.)*. Straight, branched elastic fibres occurring singly in areolar connective tissue. Cf. **white fibres**.

yellow fibrocartilage *(Zool.)*. A form of fibrocartilage in which yellow fibres predominate.

yellow ground *(Min.Ext.)*. Yellowish or buff-coloured, loose clay-rich material formed at the top of a **kimberlite** pipe by oxidation and alteration of **blue ground**.

yellowing *(Textiles)*. The yellow discoloration of textile materials, particularly white ones, in use or storage.

yellow pine *(For.)*. A very soft, even-grained wood from Canada. Useful for joinery.

yellow quartz *(Min.)*. See **citrine**.

yellows *(Bot.)*. A disease in which there is considerable yellowing (chlorosis) of normally green tissue, caused by viruses or mycoplasmata.

yellows *(Vet.)*. See **canine leptospirosis**.

yellowses *(Vet.)*. *Head grit*. A form of photo-sensitization in sheep characterized by dermatitis, subcutaneous oedema affecting mainly the head, and sometimes jaundice (hence yellowses). Associated with liver dysfunction and porphyrins in the blood.

yellow spot *(Zool.)*. Macula lutea; the small area at the centre of the retina in Vertebrates at which day vision is most distinct.

yellow tellurium *(Min.)*. A synonym for *sylvanite*.

yellow vision *(Med.)*. See **xanthopsia**.

yew *(For.)*. A tree, *Taxus*, indigenous to Europe, Asia and the western hemisphere, giving a softwood immune to powder-post beetle attack. Durable when used in exposed positions; used for bowmaking.

yield *(Min.Ext.)*. Tonnage extracted, or ratio of known tonnage, to that recoverable profitably.

yield *(Phys.)*. (1) Ion pairs produced per quantum absorbed or per ionizing particle. (2) See **fission yield**.

yielding attachment *(Horol.)*. A method of attaching the outer end of a mainspring to its barrel. It permits of a more concentric uncoiling of the spring.

yielding prop *(Min.Ext.)*. Support used just behind coal face, which shortens slightly under load, but can be reclaimed and re-used.

yield point *(Eng.)*. The stress at which a substantial amount of plastic deformation takes place under constant or reduced load. This sudden yielding is a characteristic of iron and annealed steels. In other metals, plastic deformation begins gradually and its incidence is indicated by measuring the proof stress, which, however, is frequently called the yield point.

YIG *(Elec.Eng.)*. *Yttrium Iron Garnet*. A material which has a lower acoustic attenuation loss than quartz and which has been considered for use in delay lines.

YIG filter *(Elec.Eng.)*. One using a YIG crystal which is tuned by varying the current in a surrounding solenoid, a permanent magnet being used to provide the main field strength.

-yl *(Chem.)*. A suffix denoting (1) a monovalent organic radical; (2) an electropositive inorganic radical which contains oxygen.

ylem *(Phys.)*. The basic substance from which it has been suggested that all known elements may have been derived through nucleogenesis (fusion of fundamental particles to form nuclei). It would have a density of 10^{16} kg/m^3, and would consist chiefly of neutrons.

Y-level *(Surv.)*. See **wye level**.

Y-maze *(Behav.)*. A maze similar to a **T-maze**, but in which the arms are not at right angles to the stem.

Y-network *(Telecomm.)*. Same as **T-network**; three-branch star network.

Yngel trawl *(Genrl.)*. See **young-fish net**.

yoderite *(Min.)*. A silicate of aluminium, iron and magnesium, crystallizing in the monoclinic system.

yohimbine *(Chem.)*. $C_{22}H_{28}O_3N_2$ + H_2O, an alkaloid with a pentacyclic nucleus, obtained from the bark of *Corynanthe johimbe*; colourless needles; mp 234°C; soluble in ethanol and trichloromethane; slightly soluble in ethoxyethane. It is poisonous in excess, allegedly acts as an aphrodisiac, and also exerts a local anaesthetic action.

yoke *(Civ.Eng.)*. Stout timbers bolted round the shuttering for a column to secure the parts together during the process of pouring and setting.

yoke *(Electronics)*. Combination of current coils for deflecting the electron beam in a CRT.

yoke *(Phys.)*. Part or parts of a magnetic circuit not embraced by a current-carrying coil, especially in a generator or motor, or relay.

yoke suspension *(Elec.Eng.)*. See **bar suspension**.

yolk *(Zool.)*. The nutritive nonliving material contained by an ovum.

yolk duct *(Zool.)*. Vitelline duct.

yolk epithelium *(Zool.)*. The epithelium surrounding the yolk sac.

yolk gland *(Zool.)*. See **vitellarium**.

yolk plug *(Zool.)*. A mass of yolk-containing cells which partially occludes the blastopore in some Amphibians.

yolk sac *(Zool.)*. The yolk-containing sac which is attached to the embryo by the yolk stalk in certain forms.

Yorkshire bond *(Build.)*. See **monk bond**.

young-fish net *(Genrl.)*. A large tow-net the mouth of which is kept open by **otter boards**, used for capturing small fishes at the surface or in mid-water. Also *Yngel trawl*.

Young-Helmholtz theory of colour vision *(Phys.)*. The supposition that the eye contains 3 systems of colour perception, with maximum response to 3 primary colours. It is the theory adopted for the realization of colour photography. There is little biochemical support for the theory, but the practice is justified by the physical possibility of matching practically every natural colour by the addition of contributions from 3 primary colours. See **colour**.

younging, direction of *(Geol.)*. The direction in which a series of inclined sedimentary rocks becomes younger. See **way up**.

Youngman flap *(Aero.)*. A trailing-edge flap which is extended below the main aerofoil to form a slot before being traversed rearward to increase the wing area and before being deflected downward to increase lift and drag coefficient.

Young's equation *(Min.Ext.)*. Index of surface wettability, used in flotation research on minerals $\gamma_S = \gamma_{SL} + \gamma_L \cos\theta$, where θ is contact angle (that between water and air bubble adhering to mineral), γ is free energy per unit area, S and L are solid and liquid phases.

Young's modulus *(Phys.)*. A modulus of elasticity applicable to the stretching of a wire (or thin rod), or to the bending of a beam. It is defined as the ratio

$$\frac{\text{tensile (or compressive) stress}}{\text{tensile (or compressive) strain}}$$

Symbol E. Known also as stretch or elongation modulus. Since **strain** is a numeric, its units are those of **stress**, viz. MN/m^2.

Y-parameter *(Electronics)*. The short-circuit admittance parameter of a transistor.

y-pipe *(Build.)*. See **wye**.

Y-plates *(Electronics)*. Pair of electrodes to which voltage producing vertical deflection of spot is applied in accordance with cartesian co-ordinate system.

Y rectifier *(Elec.Eng.)*. See **wye rectifier**.

yrneh *(Phys.)*. Unit of reciprocal inductance (*henry* backwards).

Y signal *(Image Tech.)*. The colour TV signal which conveys the **luminance** of the picture.

Y theodolite *(Surv.)*. See **wye theodolite**.

ytterbium *(Chem.)*. A metallic element, a member of the rare-earth group. Oxide Yb_2O_3 white, giving colourless salts. Symbol Yb, at no. 70, r.a.m. 173.04, mp of metal about 1800°C.

yttrium *(Chem.)*. A metallic element usually classed with the rare earths because of its chemical resemblance to them. Oxide, Y_2O_3, white, giving colourless salts. Symbol Y, at no. 39, r.a.m. 88.9059, mp of metal 1250°C.

yttrocerite *(Min.)*. A massive, granular or earthy mineral, essentially a cerian fluorite, with the metals of the yttrium and cerium groups, commonly violet-blue in colour, and of rare occurrence.

Yukawa potential *(Phys.)*. A potential function of the form

$$V = \frac{V_0 \exp(-kr)}{r}$$

r being distance. Characterizes the meson field surrounding a nucleon. The exponential tail of the Yukawa potential extends with appreciable strength to larger values of r than does that of the coulomb potential.

yu-stone *(Min.)*. Yu or yu-shih, the Chinese name for the highly prized jade of gemstone quality.

YUV signals *(Image Tech.)*. In the PAL colour TV system, Y is the **luminance** component and U and V are the **colour difference signals**, B − Y and R − Y respectively.

Y-voltage *(Elec.Eng.)*. See **voltage to neutral**.

Z

z *(Elec.Eng.)*. Symbol for figure of merit.

z *(Chem.)*. A symbol for the valency of an ion.

ζ *(Chem.)*. A symbol for electrokinetic potential.

Z *(Comp.)*. A language for **formal specification**.

Z *(Min.)*. The number of formula units per unit cell of a mineral.

Z- *(Chem.)*. Prefix denoting 'on the same side' (Ger. *zusammen*), and roughly equivalent to *cis*-. See **Cahn-Ingold-Prelog system**.

Z *(Chem.)*. Symbol for: (1) number of molecular collisions per second; (2) atomic number.

Z *(Eng.)*. Symbol for section modulus.

Z *(Phys.)*. Symbol for impedance.

zaffre, zaffer *(Eng.)*. Impure cobalt oxide remaining when arsenic and sulphur have been removed by roasting.

zawn *(Min.Ext.)*. Cavern, natural or man-made in Cornwall, UK.

zax *(Build.)*. See **sax**.

z-axis *(Aero.)*. The normal, or yaw, axis of an aircraft. Cf. **axis**.

z-axis *(Maths.)*. Conventionally, the vertical axis in any 3-dimensional co-ordinate system.

Z-axis modulation *(Electronics)*. Variation of intensity of beam, producing varying intensity of brightness of trace.

Z-chromosome *(Biol.)*. See **sex determination**.

Z-DNA *(Biol.)*. A form of duplex DNA, in which purines and pyrimidines alternate in a strand and which results in a left-handed helix.

Zebra *(Comp.)*. TN for small Dutch **first generation computer**.

Zechstein *(Geol.)*. The higher of the two stages into which the Permian System of Germany is divided.

Zeeman effect *(Phys.)*. The splitting of spectrum lines into a number of components by strong magnetic fields. The field splits the atomic energy levels into several components associated with different quantized orientations of the total magnetic moment with respect to the field. See **Paschen-Back** effect.

Zeisel's method *(Chem.)*. A method for the determination of methoxyl and ethoxyl groups in organic compounds, in which the substance is heated with hydriodic acid; the iodoalkane thus formed is passed into an ethanolic solution of silver (I) nitrate, and the resulting silver (I) iodide weighed.

Zeiss-Endter particle-size analyser *(Powder Tech.)*. A device for measuring particle size distribution using a photomicrograph, on which the image of the particle is compared directly with a circular spot of light, the area of which can be varied by an iris diaphragm.

zeitgeber *(Behav.)*. Literally, a *time-giver* that synchronizes various rhythmic behaviours with external events.

Zener breakdown *(Electronics)*. Temporary and nondestructive increase of current in diode because of critical field emission of holes and electrons in depletion layer at definite voltage.

Zener diode *(Electronics)*. One with a characteristic showing a sharp increase in reverse current at a certain critical voltage; the current can increase indefinitely at this point, unless limited. This makes such diodes suitable for use as a voltage reference. Diodes which show this effect up to about 6 volts depend on **Zener breakdown**. At higher voltages, avalanche effects are more prevalent, though the diodes are still called *Zener* diodes.

Zener effect *(Electronics)*. Pronounced and stable curvature in the reverse voltage/current characteristic of a semiconductor point-contact diode; predicted by Zener, and widely used as a reference voltage in stabilizing circuits.

Zener voltage *(Electronics)*. That at which **Zener breakdown** occurs in certain types of diode. It denotes the negative voltage at which the reverse current increases very rapidly due to Zener breakdown and avalanche effects, and the voltage at which the diode can provide a reference source. See **Zener diode**.

zenith *(Astron.)*. The point on the celestial sphere vertically above the observer's head; one of the two poles of the horizon, the other being the **nadir**.

zenith distance *(Astron.)*. The angular distance from the zenith of a heavenly body, measured as the arc of a vertical great circle; hence the complement of the altitude of the body.

zenith telescope *(Astron.)*. An instrument similar to the meridian circle, but fitted with an extremely sensitive level and a declination micrometer; used to determine latitude, by observing the difference in zenith distance of two stars whose meridian transit is at a small and equal distance from the zenith, one north and one south.

Zenker's degeneration *(Med.)*. Hyaline degeneration of **striated muscle**, occurring, for example, in the abdominal muscles in typhoid fever.

Zeno's paradoxes *(Maths.)*. Four paradoxes, of which that about Achilles and the tortoise is well known, apparently designed by Zeno (ca. 450 BC) to demonstrate that either of the hypotheses that space and time consist of indivisible quanta or that space and time are divisible *ad infinitum* leads to a dilemma.

zeolite process *(Chem.)*. Standard water-softening process, formerly based on using naturally-occurring **zeolites**, but nowadays on synthetic zeolites (insoluble synthetic resins). See **ion exchange**, **Permutit**.

zeolites *(Min.)*. A group of alumino-silicates of sodium, potassium, calcium, and barium, containing very loosely held water, which can be removed by heating and regained by exposure to a moist atmosphere, without destroying the crystal structure. They occur in geodes in igneous rocks, and as authigenic minerals in sediments, and include chabazite, natrolite, mesolite, stilbite, heulandite, harmotome, phillipsite, thomsonite etc.

Zepp antenna *(Telecomm.)*. Horizontal half-wavelength antenna fed from a resonant transmission line. It is connected at one end to one wire of the transmission line and the transmitter or receiver is connected between the two wires, the length of the line being critical.

zero *(Maths.)*. If a function, analytic in a domain D, is expanded in a **Taylor's series** about any point $z = a$ in D, viz., $f(z) = \sum_{0}^{\infty} b_n(z-a)^n$, and if all values of b_n up to b_{m-1} vanish but b_m does not vanish, then $f(z)$ is said to have a zero of order m at $z = a$.

zero-address instruction *(Comp.)*. Operation in computing where the location of the operands is defined by the order code and not specified independently.

zero-beat reception *(Telecomm.)*. In **suppressed carrrier systems**, reception in which a locally generated oscillation having the same frequency as the incoming carrier, is impressed simultaneously on the detector. See **heterodyne conversion**.

zero-cut crystal *(Electronics)*. Quartz crystal cut at such an angle to the axes as to have a zero frequency/temperature coefficient. Used for accurate frequency and time standards.

zero-energy reactor *(Nuc.Eng.)*. See **zero-power reactor**.

zero error *(Eng.)*. (1) Residual time delay which has to be compensated in determining readings of range. (2) Error of any instrument when indicating zero, either by pointer, angle, or display.

zero frequency *(Elec.Eng.)*. The component of a complex signal corresponding to the d.c. level. Abbrev. *z.f.*

zero fuel weight *(Aero.)*. See **weight**.

zero-g *(Space)*. The state of weightlessness or **free fall**.

zero level *(Elec.Eng.)*. Any voltage, current or power reference level when other levels are expressed in dB relative to this.

Zerol gear *(Eng.)*. A type of **bevel gear**, having curved teeth and a zero helical angle.

zero method *(Elec.Eng.)*. Measuring system in which an unknown value can be deduced from other values when a sensitive but not necessarily calibrated instrument indicates zero deflection, as in a Wheatstone bridge or potentiometer. Also called *null method.*

zero order reaction *(Chem.)*. One in which the rate is independent of the concentration of the reacting species.

zero pause *(Elec.Eng.)*. The momentary cessation of an alternating current when passing through a zero value between successive half-cycles on which the action of a.c. circuit-breakers largely depends.

zero phase-sequence component *(Elec.Eng.)*. One of three phasors forming a zerophase-sequence system, and one of three components into which any phasor forming part of an unbalanced three-phase system can be resolved. See **phase sequence**.

zero-point energy *(Phys.)*. Total energy at the absolute zero of temperature. The uncertainty principle does not permit a simple harmonic oscillator particle to be at rest exactly at the origin, and by the quantum theory, the ground state still has one half-quantum of energy, i.e. $h\nu/2$, and the corresponding kinetic energy.

zero-point entropy *(Phys.)*. As follows from the third law of thermodynamics, the entropy of a system in equilibrium at the absolute zero must be zero.

zero potential *(Elec.Eng.)*. Theoretically, that of a point at infinite distance, used for defining capacitance. Practically, the earth is taken as being of invariant potential. That of any large mass of metal, e.g. equipment chassis.

zero power-factor characteristic *(Elec.Eng.)*. A curve obtained by plotting the terminal voltage of a synchronous generator delivering full load current at zero power factor lagging, against the field excitation.

zero power level *(Telecomm.)*. See **level**.

zero-power reactor *(Nuc.Eng.)*. An experimental reactor for reactor physics studies with an extremely low neutron flux so that no forced cooling is required and there is insignificant build-up of fission products.

zero stability *(Telecomm.)*. Drift in no-signal output level of amplifier or indicator, either with time or with operating conditions, e.g. mains voltage supply.

zero suppression *(Comp.)*. A technique of data processing used to eliminate the storing of nonsignificant leading zeros.

zero-type dynamometer *(Elec.Eng.)*. A dynamometer in which the electrical forces are balanced by mechanical forces, in such a manner as to bring the indicating pointer back to zero, before a reading can be taken.

zero-valent *(Chem.)*. Incapable of combining with other atoms. Also *nonvalent.*

zeta function *(Maths.)*. See **Riemann**.

zeta potential *(Chem.)*. See **electrokinetic potential**. Abbrev. ζ.

zeugopodium *(Zool.)*. The second segment of a typical pentadactyl limb, lying between the stylopodium and the autopodium; ante-brachium or crus; forearm or shank.

Zeuner valve diagram *(Eng.)*. See **valve diagram**.

z.f. *(Elec.Eng.)*. Abbrev. for *Zero Frequency*.

Z-helix *(Bot.)*. A helix winding in the sense of a conventional, right-handed, screw.

Ziegler catalyst *(Chem.)*. Stereo-orienting catalyst discovered by Ziegler for organic polymerization, notably the low-pressure polyethylene process. Prepared by the reaction of compounds of strongly electropositive transition metals with organometallic compounds, e.g. titanium trichloride, aluminium alkyl.

ziggurat *(Arch.)*. A stepped pyramidical structure, the diminishing stages being served by a ramp, or alternatively the stages formed a continuous inclined ramp. They were built in Mesopotamia during the Babylonian period (ca 3000–1250 BC), and were used for religious ceremonial purposes, there being a small temple or shrine surmounting the uppermost terrace.

zigzag connection *(Elec.Eng.)*. A symmetrical three-phase star connection of six windings, situated in pairs on three cores. Each leg of the star consists of two of the windings in series; these windings, being on different cores, have e.m.fs. in them differing in phase by 120°. Used in transformers for eliminating harmonics, and in reactors to obtain an artificial neutral.

zigzag leakage *(Elec.Eng.)*. Magnetic leakage occurring along the zigzag path between stator and rotor teeth when a stator tooth is opposite to a rotor slot.

zinc *(Chem.)*. A hard white metallic element with a bluish tinge. Symbol Zn, at.no. 30, r.a.m. 65.37, rel.d. at 20°C 7.12, mp 418°C, electrical resistivity 6.0×10^{-8} ohm metres. Because of its good resistance to atmospheric corrosion, zinc is used for protecting steel (see **galvanized iron, sheradizing, spelter**). It is also used in the form of sheet and as a constituent in alloys (see **zinc alloys**). Used as an electrode in a Daniell cell and in dry batteries. It is important nutritionally, trace amounts being present in many foods.

zinc alkyls *(Chem.)*. Organometallic compounds such as zinc dimethyl and zinc diethyl prepared by treating the alkyl iodides with zinc-copper couple. Used occasionally as **Grignard reagents**, and to replace the halogen attached to a tertiary carbon atom by an alkyl group.

zinc alloys *(Eng.)*. Zinc base alloys, containing aluminium 3–4%, copper 0–3.5%, and magnesium 0.02–0.1%, are used extensively for diecasting. This metal is also used extensively in brass, of which it is an essential constituent. Light aluminium zinc alloys are also used.

zincate *(Chem.)*. The anion ZnO_2^{--} or $Zn(OH)_4^{--}$.

zinc blende *(Min.)*. A much-used name for **sphalerite**, the common sulphide of zinc.

zinc bloom *(Min.)*. A popular name for the massive basic zinc carbonate **hydrozincite**.

zinc chromate primer *(Build.)*. A yellow coloured primer based on zinc, used on iron and steel because of its rust inhibiting properties. The recommended primer on aluminium surfaces. Hard drying and lead free.

zinc dust *(Min.Ext.)*. Finely divided powder produced either by condensation of zinc vapour or by atomization of molten zinc. Once widely used to precipitate gold and silver from pregnant solution in the cyanidation of gold ore.

zincite *(Min.)*. Oxide of zinc, crystallizing in the hexagonal system and exhibiting polar symmetry; occurring rarely as crystals, usually as deep-red masses; an important ore of zinc, known also as *red oxide of zinc.*

zinckenite *(Min.)*. A steel-gray mineral, essentially sulphide of lead and antimony, $PbSb_2S_4$, occurring as columnar hexagonal crystals, sometimes exceptionally thin, forming fibrous masses.

zinco *(Print.)*. A line block executed in zinc.

zinc oxide *(Med.)*. Oxide of zinc used, for its astringent and soothing qualities, as a constituent of creams, baby ointments, etc.

zinc phosphate *(Build.)*. Non-toxic primer based on metallic zinc. Excellent properties of adhesion, rust inhibition. Light grey in colour, suitable for application to iron and steel. Often used as a substitute for red lead.

zinc protector *(Ships)*. Introduced originally for the **cathodic protection** of copper sheathing on wooden ships. Now used in steel ships near bronze propellers.

zinc-rich paint *(Build.)*. A paint containing an extremely high proportion of metallic zinc dust in the dry film (about 95% by weight), applied to iron and steel as an anticorrosive primer. It may be regarded as a less durable form of **cold galvanizing**.

zinc spinel *(Min.)*. Same as *gahnite*. See **spinel**.

zinc telluride *(Eng.)*. A semiconductor capable of high-temperature operation (up to about 750°C) without excessive intrinsic conductivity.

zinnwaldite *(Min.)*. A mica related in composition to lepidolite (i.e. containing lithium and potassium) but including iron as an essential constituent; occurring in association with tinstones ores at Zinnwald in the Erzgebirge, in Cornwall, and elsewhere.

zircaloy *(Nuc.Eng.)*. Alloy of zirconium and aluminium used to clad fuel elements in water reactors. See **Magnox**.

zircon *(Min.)*. A tetragonal accessory mineral widely distributed in igneous, sedimentary, and metamorphic rocks. It varies in colour from brown to green, blue, red, golden-yellow, while colourless zircons make particularly brilliant stones when cut and polished. In composition, it is essentially silicate of zirconium, but often contains yttrium and thorium. A small amount of the rare element hafnium is present.

zirconate (IV) *(Chem.)*. The anion ZrO_3^{--}.

zirconia *(Chem.)*. *Zirconium (IV) oxide*, ZrO_2, used as an opacifier in vitreous enamels, as a pigment, and as a refractory.

zirconium *(Chem.)*. A metallic element, symbol Zr, at. no. 40, r.a.m. 91.22, rel.d. 4.15, mp 2130°C. When purified from hafnium, its low neutron absorption and its retention of mechanical properties at high temperature make it useful for the construction of nuclear reactors. Also used as a refractory, as a lining for jet engines, and as a getter in the manufacture of vacuum tubes. The zirconates are finding application as acoustic transducer materials. Tritium adsorbed in zirconium is a possible target in accelerator neutron sources.

zirconium lamp *(Phys.)*. One having a zirconium oxide cathode in an argon-filled bulb. It provides a high-intensity point source with only a small emission of the longer visible wavelengths.

Z-line *(Zool.)*. These limit the sarcomeres of striated muscle and contain α-actin.

Z marker beacon *(Aero.)*. A form of marker beacon radiating a narrow conical beam along the vertical axis of the cone of silence of a radio range.

z-modulation *(Image Tech.)*. The variations in intensity in the electron beam of a CRT which form the display or picture on a sweep or raster.

Zn *(Chem.)*. A symbol for zinc.

Zodiac *(Astron.)*. A name, of Greek origin, given to the belt of stars, about 18° wide, through which the ecliptic passes centrally. The Zodiac forms the background of the motions of the Sun, Moon, and planets.

zodiacal light *(Astron.)*. A faint illumination of the sky, lenticular in form and elongated in the direction of the ecliptic on either side of the sun, fading away at about 90° from it; best seen after sunset or before sunrise in the tropics, where the ecliptic is steeply inclined to the horizon; it is caused by small particles reflecting sunlight, and appears to be an extension of the solar corona to a distance well beyond the earth's orbit.

zoidiophilous *(Bot.)*. Pollinated by animals.

zoisite *(Min.)*. Hydrated alumino-silicate of calcium crystallizing in the orthorhombic system and occurring chiefly in metamorphic schists; also a constituent of so-called saussurite. Clinozoisite has the same composition, but crystallizes in the monoclinic system.

zona *(Zool.)*. An area, patch, strip, or band; a zone. adjs. *zonal, zonary, zonate*.

zona granulosa *(Zool.)*. The mass of membrana granulosa cells of the Graafian follicle around the ovum; *discus proligerus* or *cumulus oophorus*.

zonal index *(Geol.)*. See **zone**.

zonal index *(Meteor.)*. Numerical index measuring the strength of the westerly zonal flow in middle latitudes e.g. between 35° and 65°. A common type is the mean pressure or **geopotential height** difference between latitude circles. Indices may be defined for various levels in the atmosphere.

zona pellucida *(Zool.)*. A thick transparent membrane surrounding the fully formed ovum in a Graafian follicle.

zona radiata *(Zool.)*. The envelope of the Mammalian egg outside the vitelline membrane.

zonary placentation *(Zool.)*. The condition in which the villi are on a partial or complete girdle around the embryo, as in Carnivora and Proboscidea.

zonation *(Bot.)*. The occurrence, in an area, of distinct bands of vegetation each with its own characteristic dominant and other species, as the seaweeds on a shore or the vegetation on a mountain side. Cf. **succession**.

zone *(Geol.)*. A stratigraphical unit with recognizable characteristics. The term has attracted many confusing definitions and is perhaps best used with an appropriate qualifier. e.g. *Dibunophyllum* zone, marine zone, ash zone.

zone levelling *(Electronics)*. An analogous process to **zone refining** carried out during the processing of semiconductors in order to distribute impurities evenly through the sample.

zone melting *(Eng.)*. Localized melting of part of, usually, a column of metal by an induction coil which is moved along the rod. It is possible with certain alloys to arrange that the impurities concentrate in the liquid zone, allowing considerable purification of the remainder.

zone of audibility *(Acous.)*. The hearing of explosions at great distances from the source, although, nearer, there is a **zone of silence**.

zone of cementation *(Geol.)*. That 'shell' of the earth's crust lying immediately below the zone of weathering, within which loose sediments are cemented by the addition of such minerals as calcite, introduced by percolating meteoric waters.

zone of silence *(Acous.)*. Local region where sound or electromagnetic waves from a given source cannot be received at a useful intensity. Also called *shadow source*.

zone of weathering *(Geol.)*. An 'earth shell' comprising the exposed surface and that part which, through porosity, fracturing, and jointing, is subject to the destructive action of the atmosphere, rain, and frost. Soil develops in this zone.

zone plate *(Phys.)*. A transparent plate divided into a series of zones by circles whose radii are in the ratio $\sqrt{1}:\sqrt{2}:\sqrt{3}:\sqrt{4}$ etc, the alternate zones being blacked. If a plane wave is incident normally on the plate, a maximum of light intensity is formed at a point on the axis as if the plate were acting as a lens of focal length f. Subsidiary focal points at $f/3$, $f/5$ etc. are also formed with progressively much weaker concentrations of light. See **half-period zones, Fresnel zone**.

zone refining, purification *(Elec.Eng.)*. Passage of a melted zone along a trough of metal, particularly germanium. Impurities dissolve into this and do not freeze out. Heating is performed by high-frequency eddy currents from a coil which progresses along the trough.

zone time *(Astron.)*. See **standard time**.

zoning *(Aero.)*. (1) The specification of areas surrounding an airfield in which there is a known clearance above obstruction for the safe landing and taking-off of aircraft. (2) The division of an aircraft's fuselage, wings and engine nacelles into specific areas for precise location of equipment, identification and fire protection purposes.

zoning *(Min.)*. Concentric layering parallel to the periphery of a crystalline mineral, shown by colour banding in such minerals as tourmaline, and by differences of the optical reactions to polarized light in colourless minerals like feldspars; it is due to the successive deposition of layers of materials differing slightly in composition.

zonula ciliaris *(Zool.)*. In the Vertebrate eye, a double fenestrated membrane connecting the ciliary process of the choroid with the capsule surrounding the lens.

zonule *(Zool.)*. A small belt or zone, such as the zonula ciliaris of the Vertebrate eye.

zoo- *(Genrl.)*. Prefix from Gk. *zōon*, animal.

zoobiotic *(Biol.)*. Parasitic on, or living in association with, an animal.

zooblast *(Zool.)*. An animal cell.

zoochlorellae *(Zool.)*. Symbiotic green algae found in various animals.

zoochorous *(Bot.)*. Spores or seeds dispersed by animals.

zoocyst *(Zool.)*. See **sporocyst**.

zoogamete *(Zool.)*. A motile gamete.

zoogamy *(Zool.)*. Sexual reproduction of animals.

zoogeography *(Zool.)*. The study of animal distribution.

zooid *(Zool.)*. An individual forming part of a colony in Protozoa (Volvocina), Coelenterata, Hemichordata, Urochorda, and Bryozoa; in Polychaeta, a posterior sexual region formed by asexual reproduction; polyp; a polypide.

zooidogamy *(Bot.)*. Fertilization by motile spermatozoids. Cf. **siphonogamy**.

zoom *(Image Tech.)*. The effect produced by the rapid movement of the camera towards or away from the subject, or the equivalent obtained by the use of a **zoom lens** or by electronic means in video.

zooming *(Aero.)*. Utilizing the kinetic energy of an aircraft in order to gain height. *Zoom-bombing* involves the release of a nuclear bomb during a zooming manoeuvre to give the aircraft time to excape the blast. See **toss-bombing**.

zoom lens *(Image Tech.)*. A camera lens whose focal length is continuously variable while maintaining a fixed focal plane, thus providing variable magnification of the subject. Cf. **varifocal lens**.

zoonomy *(Genrl.)*. Animal physiology.

zoonosis *(Vet.)*. A disease of animals communicable to man.

zooplankton *(Zool.)*. Floating and drifting animal life.

zoosperm *(Zool.)*. A spermatozoid.

zoosporangium *(Bot.)*. A sporangium in which zoospores are formed.

zoospore *(Bot.)*. A motile, usually naked, asexual (i.e. not a gamete) reproductuve cell found in some algae and fungi, swimming by means of one to several flagella.

zootaxy *(Genrl.)*. Zoological classification.

zootechnics *(Genrl.)*. Animal husbandry.

zootomy *(Zool.)*. See **anatomy**.

zoster *(Med.)*. See **herpes zoster**.

Zr *(Chem.)*. The symbol of zirconium.

z scheme *(Bot.)*. Scheme linking **photosystem II** and **photosystem I** such that oxygen is produced by the former, NADP is reduced by the latter and ATP is generated by (non-cyclic) photophosphorylation by *electron transport* from PS II to PS I.

Z-twist *(Textiles)*. See **twist direction**.

zulu time *(Telecomm.)*. Used in telecommunications for GMT. See **UTC time**.

zussmanite *(Min.)*. A pale green, tabular, trigonal hydrated silicate of iron, magnesium and potassium.

zwitterion *(Chem.)*. An ion carrying both a positive and a negative charge, e.g. present in solid and liquid amino acids such as glycine (aminoethanoic acid), $N^+H_3CH_2COO^-$.

zygapophyses *(Zool.)*. Articular processes of the vertebrae of higher Vertebrates arising from the anterior and posterior sides of the neurapophyses.

zygodactylous *(Zool.)*. Said of Birds which have the first and fourth toes directed backwards, as Parrots.

zygogenetic *(Zool.)*. A product of fertilization.

zygoma *(Zool.)*. The bony arch of the side of the head in Mammals which bounds the lower side of the orbit.

zygomatic *(Zool.)*. Pertaining to the zygoma. See also **jugal**.

zygomatic arch *(Zool.)*. See **zygoma**.

zygomatic bone *(Zool.)*. See **jugal**.

zygomorphic *(Bot.)*. A bilaterally symmetrical flower or corolle which can only be divided into 2 equal halves by one vertical plane. Also called *irregular*. Cf. **actinomorphic**.

Zygomycotina, Zygomycetes *(Bot.)*. A subdivision or class of the Eumycota or true fungi. No motile stages. Usually mycelial, aseptate. Asexual spores formed in sporangia; sexual reproduction by formation of a zygospore. Mostly saprophytic e.g. *Mucor*, the pin-mould; some insect parasites. Also *Glomus* spp. which form **vesicular arbuscular mycorrhizas** with very many plants.

zygonema *(Biol.)*. The zygotene phase of meiosis.

zygospore *(Bot.)*. (1) Any thick-walled resting spore formed directly from a zygote, as in many algae and some fungi. See also **oospore**. (2) A thick-walled resting spore formed from the zygote resulting from the union of isogametes i.e. in Zygnemaphyceae and Zygomycotina.

zygospore *(Zool.)*. See **zygote**.

zygote *(Bot.,Zool.)*. The cell that results from the fusion of 2 gametes.

zygotene *(Biol.)*. The second stage of meiotic prophase, intervening between leptotene and pachytene, in which the chromatin threads approximate in pairs and become looped.

zygotic *(Bot.,Zool.)*. Relating to, or belonging to a zygote.

zyg-, zygo- *(Genrl.)*. Prefix from Gk. *zygon*, yoke.

zymogen *(Biol.)*. Inert precursor of many active proteins and degradative enzymes. It is converted into the active form at the required site of activity. Thus trypsin is formed in the intestinal lumen from the inactive trypsinogen fibrin generated at the site of blood clotting from the inactive fibrinogen.

zymosan *(Immun.)*. Cell wall fraction of yeast which activates the alternative **complement** pathway, and thus binds C3b. Frequently used for study of the capacity of cells to phagocytose opsonized materials.

zym-, zymo- *(Biol.)*. Obsolescent prefixes relating to fermentation by enzymes.

Appendices

ISO Paper Sizes

The A series is used for standard book printing and stationery, the B series for posters, wall-charts, etc. The dimensions given are of trimmed sizes.

A series	mm	inches
A0	841 × 1189	33·11 × 46·81
A1	594 × 841	23·39 × 33·11
A2	420 × 594	16·54 × 23·39
A3	297 × 420	11·69 × 16·54
A4	210 × 297	8·27 × 11·69
A5	148 × 210	5·83 × 8·27
A6	105 × 148	4·13 × 5·83
A7	74 × 105	2·91 × 4·13
A8	52 × 74	2·05 × 2·91
A9	37 × 52	1·46 × 2·05
A10	26 × 37	1·02 × 1·46

B series	mm	inches
B0	1000 × 1414	39·37 × 55·67
B1	707 × 1000	27·83 × 39·37
B2	500 × 707	19·68 × 27·83
B3	353 × 500	13·90 × 19·68
B4	250 × 353	9·84 × 13·90
B5	176 × 250	6·93 × 9·84
B6	125 × 176	4·92 × 6·93
B7	88 × 125	3·46 × 4·92
B8	62 × 88	2·44 × 3·46
B9	44 × 62	1·73 × 2·44
B10	31 × 44	1·22 × 1·73

USA Equivalent Paper Weights

In reams of 500 sheets, basis weights in bold type.

Grade of Paper	Book 25 × 38	Bond 17 × 22	Cover 20 × 26	Index 25½ × 30½	TAG 24 × 36	Grammage g/m²
Book	**30**	12	16	25	27	44
	40	16	22	33	36	59
	45	18	25	37	41	67
	50	20	27	41	45	74
	60	24	33	49	55	89
	70	28	38	57	64	104
	80	31	44	65	73	118
	90	35	49	74	82	133
	100	39	55	82	91	148
	120	47	66	98	109	178
Bond	33	**13**	18	27	30	49
	41	**16**	22	33	37	61
	51	**20**	28	42	46	75
	61	**24**	33	50	56	90
	71	**28**	39	58	64	105
	81	**32**	45	67	74	120
	91	**36**	50	75	83	135
	102	**40**	56	83	93	151
Cover	91	36	**50**	75	82	135
	110	43	**60**	90	100	163
	119	47	**65**	97	108	176
	146	58	**80**	120	134	216
	164	65	**90**	135	149	243
	183	72	**100**	150	166	271
Index	110	43	60	**90**	100	163
	135	53	74	**110**	122	203
	170	67	93	**140**	156	252
	208	82	114	**170**	189	328
TAG	110	43	60	90	**100**	163
	137	54	75	113	**125**	203
	165	65	90	135	**150**	244
	192	76	105	158	**175**	284
	220	87	120	180	**200**	326
	275	109	151	225	**250**	407

Chemical Formulae

Organic Ring Systems

All formulae give a single Lewis (resonance) structure, and the true structure normally requires others to represent it completely. Three-connected vertices represent CH, while four connected ones represent C.

A Carbocyclic rings

benzene

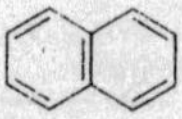
naphthalene

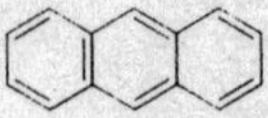
anthracene

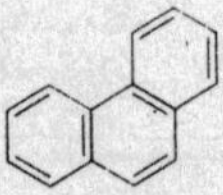
phenanthrene

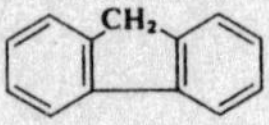
fluorene

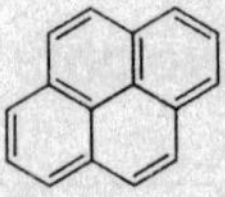
pyrene

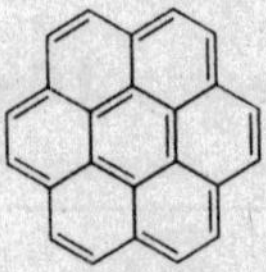
coronene

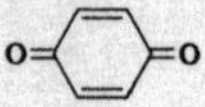
or
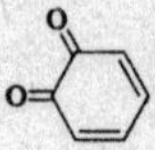
quinone

B Heterocyclic rings

furan

pyrrole

thiophen

imidazole

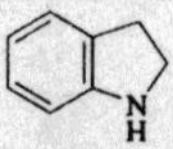
indole

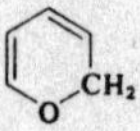
or
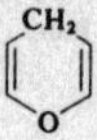
pyran

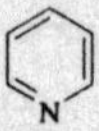
pyridine

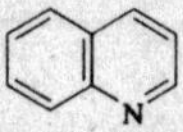
quinoline

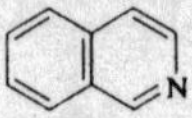
isoquinoline

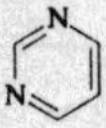
pyrimidine

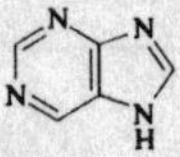
purine

C Nucleic acid bases

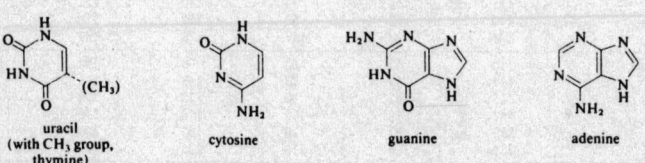

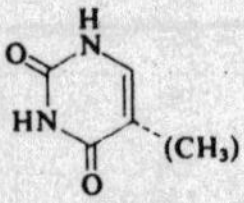
uracil
(with CH_3 group, thymine)

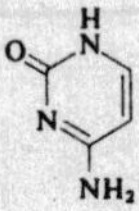
cytosine

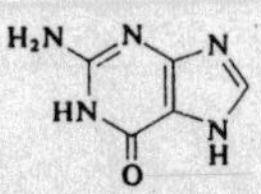
guanine

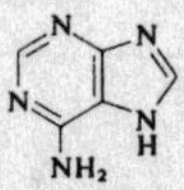
adenine

Structures of Monosaccharides

Simple sugars and their derivatives may occur as open chain or as ring forms, and the rings may be 5-membered (furanose) or 6-membered (pyranose). Examples of each structure type are given below as perspective drawings with the customary numbering system for the carbon atoms. Interchange of two substituents (usually —H and —OH) at the given atoms will give formulae for other stereoisomers. Diagrams are all for the D-series of sugars. The corresponding L-sugar is obtained by interchanging two substituents at *all* asymmetric carbon atoms, (indicated*).

A Tetroses, as open chain structures

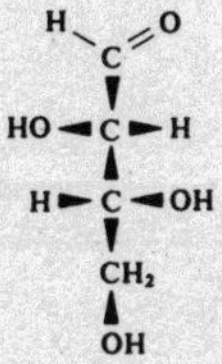

D-erythrose

1
2*
3*
4

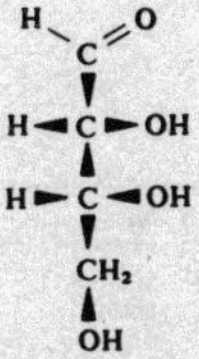

D-threose

B Pentoses, as furanose structures

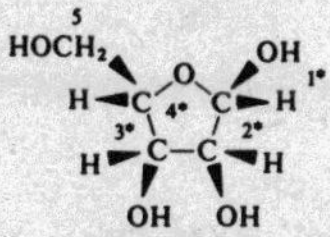

β-D-ribose

Stereoisomers, related by interchanging substituents at carbon(s):

β-D-arabinose	2
β-D-xylose	3
β-D-lyxose	2, 3
α-D-ribose	1
α-D-arabinose	1, 2
α-D-xylose	1, 3
α-D-lyxose	1, 2, 3

related structures:
deoxyribose: replace OH by H at C-2 in ribose.
fructose: replace H by CH_2OH at C-1 in arabinose.

C Hexoses, as pyranose structures

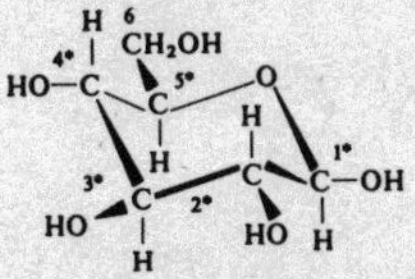

β-D-glucose

Stereoisomers, related by interchanging substituents at carbons:

β-D-mannose	2
β-D-allose	3
β-D-galactose	4
β-D-altrose	2, 3
β-D-talose	2, 4
β-D-glucose	3, 4
β-D-idose	2, 3, 4

The corresponding α-anomer is obtained by interchanging substituents at C-1.
Related structure:
fucose: replace CH_2OH by CH_3 on C-5 of galactose.

Structures of Important Amino Acids

All of the amino acids, except proline, terminate in the group shown below. It is this group whose configuration determines that they are L-isomers. In addition, isoleucine and threonine have a second asymmetric carbon atom, whose configuration is shown in the formulae below. Nonpolar residues are arranged roughly in order of increasing polarizability, while polar residues are arranged in order of increasing basicity (decreasing acidity).

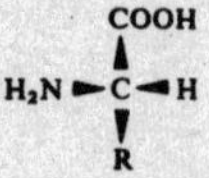

	Name	Abbreviation	R
A Nonpolar residues	glycine	Gly	H–
	alanine	Ala	CH_3–
nonpolarizable	proline*	Pro	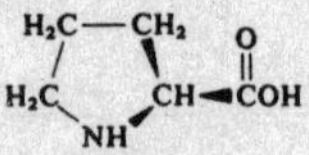
	valine	Val	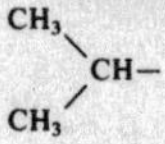
	leucine	Leu	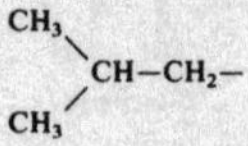
	isoleucine	Ile	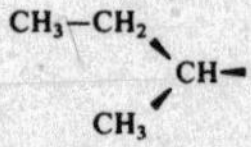
	methionine	Met	CH_3–S–CH_2–CH_2–
polarizable	phenylalanine	Phe	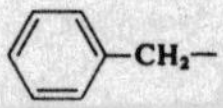
	tryptophan	Try	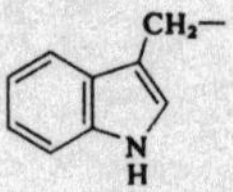

* entire molecule shown.

	Name	Abbreviation	R
B Polar residues	aspartic acid	Asp	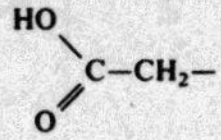
	glutamic acid	Glu	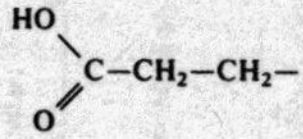
acidic	tyrosine	Tyr	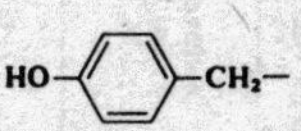
	cysteine	Cys	$HS-CH_2-$
	serine	Ser	$HO-CH_2-$
	threonine	Thr	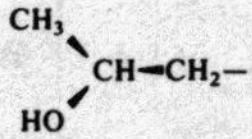
neutral	asparagine	Asn	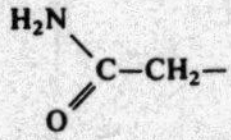
	glutamine	Gln	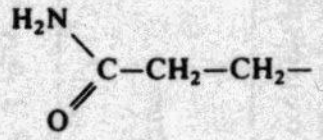
	histidine	His	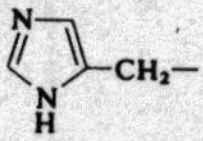
basic	lysine	Lys	$H_2N-CH_2-CH_2-CH_2-CH_2-$
	arginine	Arg	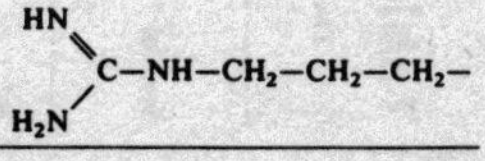

Table of Chemical Elements

The Atomic Weights are based on the 1969 international agreed values of the International Union of Pure and Applied Chemistry, the basis being the carbon-12 isotope. Values in brackets indicate the mass numbers of the most stable isotopes. See also the Periodic Table on page 991, which incorporates the electron distributions within the shells.

Symbol	Name	Derived from	Valence No.	Atomic No.	Atomic Wt.	Specific Gravity	Melting or Fusing Pt. °C	Discovered by	Date
Ac	Actinium	Greek, *aktis* = ray	3	89	(227)	..	..	Debierne	1899
Ag	Silver (argentum)	Anglo-Saxon, *seolfor*	1	47	107·868	10·5	960	Prehistoric	..
Al	Aluminium	Latin, *alumen* = alum	3	13	26·9815	2·58	658	Wöhler	1828
Am	Americium	America	..	95	(243)	..	..	Seaborg, James and others	1944
Ar	Argon	Greek, *argos* = inactive	0	18	39·948	Gas	– 188	Rayleigh and Ramsay	1894
As	Arsenic	Latin, *arsenicum*	3 or 5	33	74·9216	5·73	Volatile, 450	Schröder	1649
At	Astatine	Greek, *astatos* = unstable	..	85	(210)	..	..	Corson, Mackenzie and Segrè	1940
Au	Gold (aurum)	Anglo-Saxon, *gold*	1 or 3	79	196·9665	19·3	1062	Prehistoric	..
B	Boron	Persian, *būrah*	3	5	10·81	2·5	2300	Davy	1808
Ba	Barium	Greek, *barys* = heavy	2	56	137·34	3·75	850	Davy	1808
Be	Beryllium (glucinum)	Greek, *beryllion* = beryl	2	4	9·0122	1·93	1281	Wöhler	1828
Bi	Bismuth	German (origin unknown)	3 or 5	83	208·9806	9·80	268	Valentine	1450
Bk	Berkelium	Berkeley, California	..	97	(249)	..	..	..	1950
Br	Bromine	Greek, *bromos* = stench	1 or 7	35	79·904	3·19	– 7·3	Balard	1826
C	Carbon	Latin, *carbo* = charcoal	4	6	12·011	3·52	Infusible	Prehistoric	..
Ca	Calcium	Latin, *calx* = lime	2	20	40·08	1·58	851	Davy	1808
Cd	Cadmium	Greek, *kadmeia* = calamine	1 or 2	48	112·40	8·64	320	Stromeyer	1817
Ce	Cerium	Planet Ceres	3 or 4	58	140·12	6·68	623	Berzelius	1803
Cf	Californium	California	..	98	(251)	..	..	..	1950
Cl	Chlorine	Greek, *chloros* = green	1, 3, 5, 7	17	35·453	Gas	– 102	Scheele	1774
Cm	Curium	Pierre and Marie Curie	..	96	(247)	..	..	Seaborg, James and others	1944
Co	Cobalt	German, *Kobold* = goblin	2 or 3	27	58·9332	8·6	1490	Brandt	1739
Cr	Chromium	Greek, *chroma* = colour	2, 3, or 6	24	51·996	6·5	1510	Vauquelin	1797
Cs	Caesium	Latin, *caesium* = bluish-grey	1	55	132·9055	1·88	26	Bunsen	1860
Cu	Copper (cuprum)	Cyprus	1 or 2	29	63·546	8·9	1083	Prehistoric	..
Dy	Dysprosium	Greek, *dysprositos*	3	66	162·50	..	..	Boisbaudran	1886
Er	Erbium	Ytterby, a Swedish town	3	68	167·26	4·8	..	Mosander	1843

Symbol	Name	Derived from	Valence No.	Atomic No.	Atomic Wt.	Specific Gravity	Melting or Fusing Pt. °C	Discovered by	Date
Es	Einsteinium	Einstein	..	99	(254)	..	..	..	1952
Eu	Europium	Europe + *ium*	2 or 3	63	151·96	..	..	Demarcay	1896
F	Fluorine	Latin, *fluo* = flow	1	9	18·9984	Gas	− 223	Scheele	1771
Fe	Iron (ferrum)	Anglo-Saxon, *iren*	2 or 3	26	55·847	7·86	1525	Prehistoric	..
Fm	Fermium	Fermi	..	100	(253)	..	..	..	1952
Fr	Francium	France	..	87	(223)	..	30	Mlle Perey	1939
Ga	Gallium	Latin, *Gallia* = France	3	31	69·72	5·95	30	Boisbaudran	1875
Gd	Gadolinium	Gadolin, a Finnish chemist	3	64	157·25	..	..	Marignac	1880
Ge	Germanium	Latin, *Germania* = Germany	4	32	72·59	5·47	958	Winkler	1886
H	Hydrogen	Greek, *hydor* = water + *gen*	1	1	1·0080	Gas	− 258	Cavendish	1766
Ha	Hahnium	Otto Hahn, German nuclear physicist	..	105	..	..	..	Lawrence Radiation Laboratory, Univ. of California	1970
He	Helium	Greek, *helios* = sun	0	2	4·0026	Gas	− 272	Ramsay	1895
Hf	Hafnium	*Hafnia* = Copenhagen	4	72	178·49	12·1	2500	Coster and Hevesey	1923
Hg	Mercury (hydrargyrum)	Mercury (myth)	1 or 2	80	200·59	13·596	− 38·9	Prehistoric	..
Ho	Holmium	*Holmia* = Stockholm	3	67	164·9303	..	..	Cleve	1879
I	Iodine	Greek, *iodes* = violet	1, 3, 5, 7	53	126·9045	4·95	114	Courtois	1811
In	Indium	Its indigo spectrum	3	49	114·82	7·4	155	Reich and Richter	1863
Ir	Iridium	Latin, *iris* = a rainbow	2 or 4	77	192·22	22·4	2375	Tennant	1803
K	Potassium (kalium)	English, *potash*	1	19	39·102	0·87	63	Davy	1807
Kr	Krypton	Greek, *kryptos* = hidden	0	36	83·80	Gas	− 169	Ramsay and Travers	1898
La	Lanthanum	Greek, *lanthanō* = conceal	3	57	138·9055	6·1	810	Mosander	1839
Li	Lithium	Greek, *lithos* = stone	1	3	6·941	0·585	186	Arfvedson	1817
Lu	Lutetium	Lutetia, ancient name of Paris	3	71	174·97	..	..	Urbain and Welsbach	1907
Lr	Lawrencium	Lawrence	..	103	..	..	..	..	1961
Md	Mendelevium	Mendeléev, Russian chemist	..	101	(256)	..	..	..	1955
Mg	Magnesium	Magnesia, district in Thessaly	2	12	24·305	1·74	651	Bussy	1829
Mn	Manganese	Latin, *magnes* = magnet	2, 3, 4, 6, 7	25	54·9380	7·39	1220	Gahn	1774
Mo	Molybdenum	Greek, *molybdos* = lead	2 or 6	42	95·94	10·2	2500	Hjelm	1782
N	Nitrogen	Greek, *nitron* = saltpetre	3 or 5	7	14·0067	Gas	− 211	D. Rutherford	1772
Na	Sodium (natrium)	English, *soda*	1	11	22·9898	0·978	97	Davy	1807
Nb	Niobium (columbium)	Niobe (Greek myth)	5	41	92·9064	8·4	1950	Hatchett	1801
Nd	Neodymium	Greek, *neos* = new and *didymos* = twin	3 or 4	60	144·24	6·96	840	Welsbach	1885

Table of Chemical Elements *(contd.)*

Symbol	Name	Derived from	Valence No.	Atomic No.	Atomic Wt.	Specific Gravity	Melting or Fusing Pt. °C	Discovered by	Date
Ne	Neon	Greek, *neos* = new	0	10	20·179	Gas	−248·6	Ramsay and Travers	1898
Ni	Nickel	Swedish, abbreviation of kupparnickel	2 or 3	28	58·71	8·9	1452	Cronstedt	1751
No	Nobelium	Nobel	..	102	(254)	..	..	..	1957
Np	Neptunium	Planet Neptune	..	93	(237)	..	..	McMillan and Abelson	1940
O	Oxygen	Greek, *oxys* = acid + *gen*	2 or 4	8	15·9994	Gas	−227	Priestley	1774
Os	Osmium	Greek, *osme* = odour	2 or 8	76	190·2	22·48	2700	Tennant	1803
P	Phosphorus	Latin, from Greek 'light-bearing'	3 or 5	15	30·9738	1·8–2·3	44	Brandt	1669
Pa	Protactinium	Greek, *protos* = first + *actinium*	5	91	231·0359	..	..	Hahn and Meitner	1917
Pb	Lead (plumbum)	Anglo-Saxon, *lead*	2 or 4	82	207·2	11·37	327	Prehistoric	..
Pd	Palladium	Planet Pallas	2 or 4	46	106·4	11·4	1549	Wollaston	1804
Pm	Promethium (illinium)	*Prometheus*, stealer of fire from heaven	..	61	(145)	..	..	Clinton Laboratories, Oak Ridge, Tenn.	1940
Po	Polonium	Poland	2, 3 or 4	84	(210)	..	..	Mme Curie	1898
Pr	Praseodymium	Greek, *prasios* = green and *didymos* = twin	3 or 4	59	140·9077	6·48	940	Welsbach	1885
Pt	Platinum	Spanish, *platina* = silver	3 or 4	78	195·09	21·5	1755	Wood	1741
Pu	Plutonium	Planet Pluto	..	94	(242)	..	..	Seaborg, McMillan, Wahl and Kennedy	1940
Ra	Radium	Latin, *radius* = ray	2	88	226·0254	6·0	700	Mme and M. Curie and Bémont	1898
Rb	Rubidium	Latin, *rubidus* = red	1	37	85·4678	1·52	39	Bunsen	1860
Re	Rhenium	German, *Rhein*	2, 3, 4, 6, 7	75	186·2	21	3000	Noddack and Tacke	1925
Rf	Rutherfordium (kurchatovium)	Lord Rutherford	..	104	..	..	..	Lawrence Radiation Laboratory, Univ. of California	1969
Rh	Rhodium	Greek, *rhodon* = rose	2 or 3	45	102·9055	12·1	1950	Wollaston	1804
Rn	Radon (niton)	Radium emanation	0	86	(222)	Gas	−150	Dorn	1901
Ru	Ruthenium	Ruthenia	3, 4 or 8	44	101·07	12·26	2400	Claus	1845
S	Sulphur	Latin, *sulfur*	2, 4 or 6	16	32·06	2·07	115–119	Prehistoric	..
Sb	Antimony (stibium)	L. Latin, *antimonium*	3 or 5	51	121·75	6·62	629	Valentine	1450
Sc	Scandium	Scandinavia	3	21	44·9559	..	..	Nilson	1879
Se	Selenium	Greek, *selene* = moon	2, 4 or 6	34	78·96	4·5	170–220	Berzelius	1817

Symbol	Name	Derived from	Valence No.	Atomic No.	Atomic Wt.	Specific Gravity	Melting or Fusing Pt. °C	Discovered by	Date
Si	Silicon	Latin, *silex* = flint	4	14	28·086	2·0–2·4	1370	Berzelius	1823
Sm	Samarium	Samarski, a Russian savant	3	62	150·4	7·7	1350	Boisbaudran	1879
Sn	Tin (stannum)	Anglo-Saxon, *tin*	2 or 4	50	118·69	7·3	232	Prehistoric	..
Sr	Strontium	Strontian, a Scottish village	2	38	87·62	2·54	800	Davy	1808
Ta	Tantalum	Tantalus (Greek myth)	5	73	180·9479	16·6	2900	Ekeberg	1802
Tb	Terbium	Ytterby, a Swedish quarry	3	65	158·9254	..	..	Mosander	1843
Tc	Technetium (masurium)	Greek, *technetos* = artificial	..	43	(99)	..	..	Perrier and Segrè	1937
Te	Tellurium	Latin, *tellus* = earth	2, 4 or 6	52	127·60	6·0	446	Reichenstein	1782
Th	Thorium	God Thor	4	90	232·0381	11·00	1750	Berzelius	1828
Ti	Titanium	Latin, *Titanes* = sons of the earth	3 or 4	22	47·90	4·54	1850	Gregor	1789
Tl	Thallium	Greek, *thallos* = budding twig	1 or 3	81	204·37	11·85	302	Crookes	1862
Tm	Thulium	Thule = Northland	3	69	168·9342	..	..	Cleve	1879
U	Uranium	Planet Uranus	2 or 6	92	238·029	18·7	..	Klaproth	1789
V	Vanadium	Goddess Vanadis (Freya)	5	23	50·9414	5·5	1710	Sefström	1830
W	Tungsten (wolfram)	Swedish, heavy stone	4 or 6	74	183·85	19·1	2900–3000	d'Elhujar	1781
Xe	Xenon	Greek, *xenos* = stranger	0	54	131·30	Gas	−140	Ramsay and Travers	1898
Y	Yttrium	Ytterby, a Swedish town	3	39	88·9059	3·8	..	Gadolin	1794
Yb	Ytterbium	Ytterby, a Swedish town	3	70	173·04	..	..	Marignac	1878
Zn	Zinc	German, *zink*	2	30	65·37	7·12	418	Unknown	before 1500
Zr	Zirconium	Persian, *zargun* = gold-coloured	4	40	91·22	4·15	2130	Berzelius	1824

Periodic Table

Within the framework of the table it is customary to recognize certain groupings of elements, viz. *transition elements* (Sc—Zn, Y—Cd, La—Hg); *lanthanides* (or *rare earth group*; Ce—Lu) occupying a distinctive place; and the *actinides* (Th, etc.). More recently discovered elements (artificial) are Rutherfordium (Rf; at. no. 104) and Hahnium (Ha; at. no. 105).

TRANSITION ELEMENTS

H 1 1·008 1																	**He** 2 4·003 2
Li 3 6·941 2, 1	**Be** 4 9·012 2, 2											**B** 5 10·81 2, 3	**C** 6 12·011 2, 4	**N** 7 14·007 2, 5	**O** 8 16·0 2, 6	**F** 9 18·998 2, 7	**Ne** 10 20·179 2, 8
Na 11 22·990 2, 8, 1	**Mg** 12 24·305 2, 8, 2											**Al** 13 26·982 2, 8, 3	**Si** 14 28·086 2, 8, 4	**P** 15 30·974 2, 8, 5	**S** 16 32·06 2, 8, 6	**Cl** 17 35·453 2, 8, 7	**Ar** 18 39·948 2, 8, 8
K 19 39·102 2, 8, 8, 1	**Ca** 20 40·08 2, 8, 8, 2	**Sc** 21 44·956 2, 8, 9, 2	**Ti** 22 47·90 2, 8, 10, 2	**V** 23 50·941 2, 8, 11, 2	**Cr** 24 51·996 2, 8, 13, 1	**Mn** 25 54·938 2, 8, 13, 2	**Fe** 26 55·847 2, 8, 14, 2	**Co** 27 58·933 2, 8, 15, 2	**Ni** 28 58·71 2, 8, 16, 2	**Cu** 29 63·546 2, 8, 18, 1	**Zn** 30 65·37 2, 8, 18, 2	**Ga** 31 69·72 2, 8, 18, 3	**Ge** 32 72·59 2, 8, 18, 4	**As** 33 74·922 2, 8, 18, 5	**Se** 34 78·96 2, 8, 18, 6	**Br** 35 79·904 2, 8, 18, 7	**Kr** 36 83·80 2, 8, 18, 8
Rb 37 85·468 2, 8, 18, 8, 1	**Sr** 38 87·62 2, 8, 18, 8, 2	**Y** 39 88·906 2, 8, 18, 9, 2	**Zr** 40 91·22 2, 8, 18, 10, 2	**Nb** 41 92·906 2, 8, 18, 12, 1	**Mo** 42 95·94 2, 8, 18, 13, 1	**Tc** 43 98·906 2, 8, 18, 14, 1	**Ru** 44 101·07 2, 8, 18, 15, 1	**Rh** 45 102·906 2, 8, 18, 16, 1	**Pd** 46 106·4 2, 8, 18, 18, 0	**Ag** 47 107·868 2, 8, 18, 18, 1	**Cd** 48 112·40 2, 8, 18, 18, 2	**In** 49 114·82 2, 8, 18, 18, 3	**Sn** 50 118·69 2, 8, 18, 18, 4	**Sb** 51 121·75 2, 8, 18, 18, 5	**Te** 52 127·60 2, 8, 18, 18, 6	**I** 53 126·905 2, 8, 18, 18, 7	**Xe** 54 131·30 2, 8, 18, 18, 8
Cs 55 132·906 2, 8, 18, 18, 8, 1	**Ba** 56 137·34 2, 8, 18, 18, 8, 2	**La** 57 138·905 2, 8, 18, 18, 9, 2	**Hf** 72 178·49 2, 8, 18, 32, 10, 2	**Ta** 73 180·948 2, 8, 18, 32, 11, 2	**W** 74 183·85 2, 8, 18, 32, 12, 2	**Re** 75 186·2 2, 8, 18, 32, 13, 2	**Os** 76 190·2 2, 8, 18, 32, 14, 2	**Ir** 77 192·22 2,8, 18, 32, 17, 0	**Pt** 78 195·09 2, 8, 18, 32, 17, 1	**Au** 79 196·967 2, 8, 18, 32, 18, 1	**Hg** 80 200·59 2, 8, 18, 32, 18, 2	**Tl** 81 204·37 2, 8, 18, 32, 18, 3	**Pb** 82 207·2 2, 8, 18, 32, 18, 4	**Bi** 83 208·981 2, 8, 18, 32, 18, 5	**Po** 84 — 2, 8, 18, 32, 18, 6	**At** 85 — 2, 8,18, 32, 18, 7	**Rn** 86 — 2, 8, 18, 32, 18, 8
Fr 87 — 2, 8, 18, 32, 18, 8, 1	**Ra** 88 226·025 2, 8, 18, 32, 18, 8, 2	**Ac** 89 — 2, 8, 18, 32, 18, 9, 2															
LANTHANIDES:			**Ce** 58 140·12 2, 8, 18, 20, 8 2	**Pr** 59 140·908 2, 8, 18, 21, 8, 2	**Nd** 60 144·24 2, 8, 18, 22, 8, 2	**Pm** 61 — 2, 8, 18, 23, 8, 2	**Sm** 62 150·4 2, 8, 18, 24, 8, 2	**Eu** 63 151·96 2, 8, 18, 25, 8, 2	**Gd** 64 157·25 2, 8, 18, 25, 9, 2	**Tb** 65 158·925 2, 8, 18, 27, 8, 2	**Dy** 66 162·50 2, 8, 18, 28, 8, 2	**Ho** 67 164·930 2, 8, 18, 29, 8, 2	**Er** 68 167·26 2, 8, 18, 30, 8, 2	**Tm** 69 168·934 2, 8, 18, 31, 8, 2	**Yb** 70 173·04 2, 8, 18, 32, 8, 2	**Lu** 71 174·97 2, 8, 18, 32, 9, 2
ACTINIDES:			**Th** 90 232·038 2, 8, 18, 32, 18, 10, 2	**Pa** 91 231·036 2, 8, 18, 32, 20, 9, 2	**U** 92 238·029 2, 8. 18, 32, 21, 9, 2	**Np** 93 237·048 2, 8, 18, 32, 22, 9, 2	**Pu** 94 — 2, 8, 18, 32, 24, 8, 2	**Am** 95 — 2, 8, 18, 32, 25, 8, 2	**Cm** 96 — 2, 8, 18, 32, 25, 9, 2,	**Bk** 97 — 2, 8, 18, 32, 26 9, 2	**Cf** 98 — 2, 8, 18, 32, 28, 8, 2	**Es** 99 — 2, 8, 18, 32, 29, 8, 2	**Fm** 100 — 2, 8, 18, 32, 30, 8, 2	**Md** 101 — 2, 8, 18, 32, 31, 8, 2	**No** 102 — 2, 8, 18, 32, 32, 8, 2	**Lr** 103 — 2, 8, 18, 32, 32 9, 2

Classification of the Animal Kingdom

Phylum	Sub-phylum	Super-class	Class	Sub-class
Protozoa			Mastigophora	
			Opalinatea	
			Ciliophora	
			Sporozoa	
			Sarcodina	
Porifera			Calcarea	
			Hexactinellida	
			Demospongiae	
Cnidaria			Hydrozoa	
			Scyphozoa	
			Anthozoa	
Ctenophora			Tentaculata	
			Nuda	
Mesozoa				
Platyhelminthes			Turbellaria	
			Digenea	
			Aspidogastrea	
			Monogenea	
			Cestoda	
Nemertini				
Aschelminthes			Nematoda	
			Nematomorpha	
			Rotifera	
			Gastrotricha	
			Kinorhyncha	
Acanthocephala				
Entoprocta				
Annelida			Polychaeta	
			Archiannelida	
			Oligochaeta	
			Hirudinea	
Echiurida				
Sipunculida				
Arthropoda	Onychophora			
	Tardigrada			
	Pentastomida			
	Trilobitomorpha			
	Chelicerata		Merostomata	
			Arachnida	
	Pycnogonida			
	Mandibulata		Crustacea	
			Pauropoda	
			Diplopoda	
			Chilopoda	
			Symphyla	
			Insecta	Apterygota
				Exopterygota
				Endopterygota
Mollusca			Monoplacophora	
			Amphineura	
			Scaphypoda	
			Gastropoda	
			Bivalvia	
			Cephalopoda	
Priapuloidea				
Bryozoa			Stenolaemata	
			Gymnolaemata	
			Phylactolaemata	
Phoronida				
Brachiopoda			Inarticulata	
			Articulata	
Chaetognatha				
Pogonophora				
Echinodermata			Asteroidea	
			Ophiuroidea	
			Echinoidea	
			Holothuroidea	
			Crinoidea	
Chordata				
Hemichordata			Enteropneusta	
			Pterobranchia	
	Urochordata			
	Cephalochordata			
	Vertebrata	Agnatha		
		Gnathostomata		
		Chondrichthyes		
		Osteichthyes	Amphibia	
			Reptilia	
			Aves	
			Mammalia	

Classification of the Plant Kingdom

Some of the smaller classes of eukaryotic algae have been omitted.

	Division	Subdivision	Class	Subclass or Superorder
PROKARYOTES				
PROKARYOTIC ALGAE	Cyanophyta		Cyanophyceae	
	Prochlorophyta		Prochlorophyceae	
EUKARYOTES				
FUNGI	Myxomycota		Acrasiomycetes Hydroxymycetes Myxomycetes Plasmodiophorales	
	Eumycota	Mastigomycotina	Chytridiomycetes	
		Zygomycotina	Zygomycetes	
		Ascomycotina	Hemiascomycetes Plectomycetes Pyrenomycetes Discomycetes	
		Basidiomycotina	Teliomycetes Hymenomycetes Gasteromycetes	
		Deuteromycotina		
EUKARYOTIC ALGAE	Rhodophyta		Rhodophyceae	
	Cryptophyta		Cryptophyceae	
	Dinophyta		Dinophyceae Desmophyceae	
	Heterokontophyta		Xanthophyceae Chrysophyceae Bacillariophyceae Phaeophyceae Oomycetes* Hyphochytridiomycetes*	
	Haptophyta		Haptophyceae	
	Euglenophyta		Euglenophyceae	
	Chlorophyta		Chlorophyceae Ulvophyceae Charophyceae	

* The Oomycetes and Hyphochytridiomycetes were and sometimes are classified with the Mastigomycotina.

	Division	Subdivision	Class	Subclass or Superorder
EMBRYOPHYTES (*Bryophytes and Vascular Plants*)				
BRYOPHYTES	Bryophyta		Hepaticopsida	
			Anthocerotopsida	
			Bryopsida	
VASCULAR PLANTS (*Pteridophytes and Seed Plants*)				
	Pteridophyta		Rhyniopsida†	
			Psilotopsida	
			Zosterophyllopsida†	
			Lycopsida	
			Trimerophytopsida†	
			Sphenopsida	
			Filicopsida	
			Progymnospermopsida†	
Seed Plants: Gymnosperms	Spermatophyta		Pteridospermopsida†	
			Cycadopsida	
			Coniferopsida	
			Gnetopsida	
Seed Plants: Angiosperms (*Angiospermae, Magnoliophyta or Flowering Plants*)				
Dicotyledons	Spermatophyta (*continued*)		Dicotyledones (Magnoliopsida)	Magnoliidae
				Hamamelidae
				Carophyllidae
				Dilleniidae
				Rosidae
				Asteridae
Monocotyledons	Spermatophyta (*continued*)		Monocotyledones (Liliopsida)	Alismatidae
				Arecidae
				Liliidae
				Commelinidae

† Groups of which all members are extinct.

Geological Table

The Phanerozoic Time Scale

Eon	Era	Sub-Era Period	Sub-Period	Epoch		Age	Abbrev.	Ma	Age	Duration
Phanerozoic 590 (Ph)	Cenozoic 65 (Cz)	Quaternary (Q) or Pleistogene (Ptg) 2.0		Holocene			Hol	0.01	0.01	2.0
				Pleistocene			Ple	2.0	1.99	
		Tertiary 63 (TT)	Neogene 22.6 (Ng)	Pliocene (Pli) 3.1	2	Piacenzian	Pia	5.1	3.1	22.6
					1	Zanclian	Zan			
				Miocene (Mio) 19.5	3	Messinian	Mes	11.3	6.2	
						Tortonian	Tor			
					2	Serravallian	Srv		3.1	
						Langhian – Late	Lan_2	14.4		
					1	Langhian – Early	Lan_1		10.2	
						Burdigalian	Bur			
						Aquitanian	Aqt	24.6		
			Paleogene 40.4 (Pg)	Oligocene (Oli) 13.4	2	Chattian	Cht	32.8	8.2	40.4
					1	Rupelian	Rup	38.0	5.2	
				Eocene 16.9 (Eoc)	3	Priabonian	Prb	42.0	4.0	
					2	Bartonian	Brt		8.5	
						Lutetian	Lut	50.5		
					1	Ypresian	Ypr	54.9	4.4	
				Paleocene (Pal) 10.1	2	Thanetian	Tha	60.2	5.3	
					1	Danian	Dan	65.0	4.8	
	Mezozoic 183 (Mz)	Cretaceous 79 (K)		Late K_2 32.5	Senonian	Maastrichtian	Maa	73.0	8.0	79
						Campanian	Cmp	33.0	10.0	
						Santonian	San	87.5	4.5	
						Coniacian	Con	88.5	1.0	
						Turonian	Tur	91.0	2.5	
						Cenomanian	Cen	97.5	6.5	
				Early K_1 46.5		Albian	Alb	113	15.5	
						Aptian	Apt	119	6.0	
						Barremian	Brm	125	6.0	
					Neocomina	Hauterivian	Hau	131	6.0	
						Valanginian	Vlg	138	7.0	
						Berriasian	Ber	144	6.0	
		Jurassic 69 (J)		J_3 19	Malm	Tithonian	Tth	150	6.0	69
						Kimmeridgian	Kim	156	6.0	
						Oxfordian	Oxf	163	7.0	
				J_2 25	Dogger	Callovian	Clv	169	6.0	
						Bathonian	Bth	175	6.0	
						Bajocian	Baj	181	6.0	
						Aalenian	Aal	188	7.0	
				J_1 25	Lias	Toarcian	Toa	194	6.0	
						Pliensbachian	Plb	200	6.0	
						Sinemurian	Sin	206	6.0	
						Hettangian	Het	213	7.0	
		Triassic 35 (Tr)		Late Tr_3		Rhaetian	Rht	219	6.0	35
						Norian	Nor	225	6.0	

Eon	Era	Sub-Era Period	Sub-Period	Epoch	Age		Abbrev.	Ma	Age	Duration
Phanerozoic (*continued*)	Mezozoic (*continued*)	Triassic (*continued*)		Middle Tr_2 12	Ladinian		Lad	238	7.0	
					Anisian		Ans	243	5.0	
				(Early) Scythian 5 Tr_1 (Scy)	Olenekian	Spathian	Spa		1.25	
						Smithian	Smi		1.25	
					Induan	Dienerian	Die		1.25	
						Griesbachian	Gri	248	1.25	
	Paleozoic 342 (Pz)	Permian 38 (P)		Late P_2 10	Tatarian		Tat	253	5.0	38
					Kazanian		Kaz		2.5	
					Ufimian		Ufi		2.5	
				Early P_1 28	Kungurian		Kun	258	5.0	
					Artinskian		Art	263	5.0	
					Sakmarian		Sak	268	9.0	
					Asselian		Ass	286	9.0	
		Carboniferous 74 (C)	Pennsylvanian 34 (Pen)	Gzelian		Stephanian 10	Gze			34
				Kasimovian			Kas	296		
				Moscovian		Westphalian 19	Mos	315		
				Bashkirian		Namurian 18	Bsh	320		
			Mississippian 40 (Mis)	Serpukhovian			Spk	333	13.0	
				Visean 19			Vis	352	19.0	
				Tournaisian 8			Tou	360		
		Devonian 48 (D)		D_3 Late 14	Famennian		Fam	367	7.0	48
					Frasnian		Frs	374	7.0	
				D_2 Middle 13	Givetian		Giv	380	6.0	
					Eifelian		Eif	387	7.0	
				D_1 Early 21	Emsian		Ems	394	7.0	
					Siegenian		Sig	401	7.0	
					Gedinnian		Ged	408	7.0	
		Silurian 30 (S)		Pridoli			Prd	414	6.0	30
				Ludlow			Lud	421	7.0	
				Wenlock			Wen	428	7.0	
				Llandovery			Llv	438	10.0	
		Ordovician 67 (O)		Ashgill			Ash	448	10.0	67
				Caradoc			Crd	458	10.0	
				Llandeilo			Llo	468	10.0	
				Llanvirn			Lln	478	10.0	
				Arenig			Arg	488	10.0	
				Tremadoc			Tre	505	17.0	
		Cambrian 85 (C)		Merioneth 20 (Mer)	Dolgellian		Dol		10.0	85
					Maentwrogian		Mnt	525	10.0	
				St David's 15 (StD)	Menevian		Men		8.0	
					Solvan		Sol	540	7.0	
				Caerfai 50 (Crf)	Lenian		Len		15.0	
					Atdabanian		Atb	570	15.0	
					Tommotian		Tom	590	20.0	

Physical Concepts in SI Units

Concept	Symbol	Name of Unit	Abbreviation of Unit Name	Definition or Defining Equation	Explanations; Equivalent Units; Alternative Definitions; etc.
Length	l	metre	m	1 m = 1 650 763·73 wavelengths in vacuo of radiation ($2p_{10}$—$5d_5$) of Kr 86.	
Mass	m	kilogramme	kg	International Prototype Kilogramme.	
Time	t	second	s	1 s = 9 192 631 770 periods of the radiation corresponding to the transition between the two hyperfine levels of the ground state of the caesium-133 atom.	
Electric current	I	ampere	A	An ampere in each of two infinitely long parallel conductors of negligible cross-section 1 metre apart in vacuo will produce on each a force of 2×10^{-7} N/m.	
Thermodynamic temperature	T	kelvin	K	The kelvin is 1/273·16 of the thermodynamic temperature of the triple point of water.	
Luminous intensity	I	candela	cd	The luminous intensity of a black body radiator at the temperature of freezing platinum at a pressure of 1 std. atm. viewed normal to the surface is 6×10^5 cd/m^2.	
Amount of substance		mole	mol	The amount of substance of a system which contains as many elementary units as there are carbon atoms in 0·012 kg of ^{12}C. The elementary unit must be specified (atom, molecule, ion, etc.).	
All the above are internationally agreed basic units except the mole, which, though recommended, awaits international acceptance.					
Plane angle	α, β, θ, etc.	radian	rad	A radian is equal to the angle subtended at the centre of a circle by an arc equal in length to the radius.	
Solid angle	Ω, ω	steradian	sr	A steradian is equal to the angle in three dimensions subtended at the centre of a sphere by an area on the surface equal to the radius squared.	
Area	A, a	square metre	m^2	$a = l^2$	
Volume	V, v	cubic metre	m^3	$V = l^3$	
Velocity	v, u	metre/second	m s^{-1}	$v = dl/dt$	
Acceleration	a	metre/second^2	m s^{-2}	$a = d^2l/dt^2$	
Density	ϱ	kilogramme/metre^3	kg m^{-3}	$\varrho = m/V$	
Mass rate of flow	$\mathring{m}, \mathring{M}$	kilogramme/sec	kg s^{-1}	dm/dt	
Volume rate of flow	$\mathring{V}$	cubic metre/sec	m^3 s^{-1}	dV/dt	

Concept	Symbol	Name of Unit	Abbreviation of Unit Name	Definition or Defining Equation	Explanations; Equivalent Units; Alternative Definitions; etc.
Moment of inertia	I	kilogramme metre^2	kg m^2	$I = Mk^2$	
Momentum	p	kilogramme metre/sec	kg m s^{-1}	$p = mv$	
Angular momentum	$I\omega$	kilogramme metre^2/sec	kg m^2 s^{-1}	$I\omega$	
Force	F	newton	N	$F = ma$	kg m s^{-2}
Torque (Moment of Force)	$T, (M)$	newton metre	N m	$T = Fl$	
Work (Energy, Heat)	$W, (E)$	joule	J	$W = \int Fdl$	1 J = 1 N m = 1 kg m^2 s^{-2} by definition
Potential energy	V	joule	J	$V = \int Fdl$	
Kinetic energy	$T, (W)$	joule	J	$T = \frac{1}{2} mv^2$	
Heat (Enthalpy)	$Q, (H)$	joule	J	$H = U + pV$	Definition for a fluid U = Internal energy
Power	P	watt	W	$P = dW/dt$	1 W = 1 J s^{-1} by definition
Pressure (Stress)	$p\ (\sigma, f)$	newton/metre^2	N m^{-2}	$p = F/A$	Usually pressure in fluid; stress in solids
Surface tension	$\gamma\ (\sigma)$	newton/metre	N m^{-1}	$\gamma = F/l$	Free surface energy
Viscosity, dynamic	η, μ		N s m^{-2}	$\frac{F}{A} = \eta dv/dl$	1 N s m^{-2} = 1000 centipoise (cP)
Viscosity, kinematic	ν		m^2 s^{-1}	$\nu = \eta/\varrho$	1 m^2 s^{-1} = 10^6 centistokes (cSt)
Temperature	θ, T	degree Celsius, kelvin	°C, K	T K = $(\theta + 273{\cdot}15)$°C	International Temperature Scale
Velocity of light	c	metre/second	m s^{-1}	Fundamental, measured, constant	
Permeability of vacuum	μ_0	henry/metre	H m^{-1}	$\mu_0 = 4\pi \times 10^{-7}$ H/m	Defined value to give coherent rationalized electrical units
Permittivity of vacuum	ϵ_0	farad/metre	F m^{-1}	$\epsilon_0 = 1/\mu_0 c^2$	Derived in Maxwell's theory of e.m. radiation
Electric charge	Q	coulomb	C	$F = (Q_1Q_2)/(4\mu\epsilon_0 r^2)$ (Coulomb's Law)	A s Also $q = \int idt$
Electric potential (Potential difference)	V	volt	V	$V = \int_\infty Edl$ $(V_{ab} = -\int^a{}_b Edl)$	
Electric field strength (Electric force)	E	volt/metre	V m^{-1}	$E = -dV/dl$	N C^{-1} E = Force on unit point charge
Electric resistance	R	ohm	Ω	$R = V/I$	
Conductance	G	siemens	S	$G = \frac{1}{R}$	℧ 1 ℧(mho) = 1 S
Electric flux	Ψ	coulomb	Ψ = Q		

Physical Concepts in SI Units *(contd.)*

Concept	Symbol	Name of Unit	Abbreviation of Unit Name	Definition or Defining Equation	Explanations; Equivalent Units; Alternative Definitions; etc.	
Electric flux density (Displacement)	D	coulomb/metre^2	C m^{-2}	$D = d\Psi/dA$		
Frequency	f	hertz	Hz		s^{-1} or cycles per second	
Permittivity	ϵ	farad/metre	F m^{-1}	$\epsilon = D/E$		
Relative permittivity	ϵ_r			$\epsilon_r = \epsilon/\epsilon_0$	a numeric	
Magnetic field strength	H	amp. turn/metre	At m^{-1}	$dH = idl \sin\theta/4\pi r^2$	The *turn* is a numeric not a unit	
Magnetic flux	Φ	weber	Wb	$\Phi = -\int e dt$	V s Faraday Law	
Magnetic flux density	B	tesla	T	$B = d\Phi/dA$	V s m^{-2} Wb m^{-2}	
Permeability	μ	henry/metre	H/m	$\mu = B/H$		
Relative permeability	μ_r			$\mu_r = \mu/\mu_0$	a numeric	
Mutual inductance	M	henry	H	$e_2 = M di_1/dt$	Wb A^{-1}	
Self inductance	L	henry	H	$e = L di/dt$	Wb A^{-1}	
Capacitance	C	farad	F	$C = Q/V$	C V^{-1}	
Reactance	X	ohm	Ω	$X = \omega L$ or $\frac{1}{\omega C}$	$\omega = 2\pi f$ rad/s	Sinusoidal a.c. (Reactance to Total voltamperes)
Impedance	Z	ohm	Ω	$Z = \sqrt{R^2 + X^2}$		
Susceptance	B	siemens	S	$B = \frac{1}{X}$	℧	
Admittance	Y	siemens	S	$Y = \frac{1}{Z}$	℧	
Total voltamperes	S	volt amp	VA	$S^2 = P^2 + Q^2$		
Reactive voltamperes	Q	volt amp reactive	VAr			
Power factor	p.f.			$\text{p.f.} = \frac{\text{power}}{\text{total voltamperes}}$		
Luminous flux	Φ	lumen	lm	$lm = cd\ sr$		
Illumination	E	lux	lx	$lx = lm\ \text{m}^{-2}$		

Some of the electrical definitions may occur with capital letter symbols in place of lower case, and vice versa. In general, in electrical engineering, lower case symbols are used for the instantaneous value of time-dependent quantities, and the corresponding capital letter symbols for values which are not a function of time.

SI Conversion Factors

(Exact values are printed in bold type)

Quantity	Unit	Conversion factor
Length	1 in.	= **25·4** mm
	1 ft	= **0·3048** m
	1 yd	= **0·9144** m
	1 fathom	= **1·8288** m
	1 chain	= **20·1168** m
	1 mile	= 1·609 34 km
	1 International nautical mile	= **1·852** km
	1 UK nautical mile	= 1·853 18 km
Area	1 in.^{2}	= **6·4516** cm^2
	1 ft^2	= 0·092 903 m^2
	1 yd^2	= 0·836 127 m^2
	1 acre	= 4046·86 m^2 = 0·404 686 ha (hectare)
	1 sq. mile	= 2·589 99 km^2 = 258·999 ha
Volume	1 UK minim	= 0·059 193 8 cm^3
	1 UK fluid drachm	= 3·551 63 cm^3
	1 UK fluid ounce	= 28·4131 cm^3
	1 US fluid ounce	= 29·5735 cm^3
	1 US liquid pint	= 473·176 cm^3 = 0·4732 dm^3 (= litre)
	1 US dry pint	= 550·610 cm^3 = 0·5506 dm^3
	1 Imperial pint	= 568·261 cm^3 = 0·5682 dm^3
	1 UK gallon	= 1·201 US gallon
		= 4·546 09 dm^3
	1 US gallon	= 0·833 UK gallon
		= 3·785 41 dm^3
	1 UK bu (bushel)	= 0·036 368 7 m^3 = 36·3687 dm^3
	1 US bushel	= 0·035 239 1 m^3 = 35·2391 dm^3
	1 in.^{3}	= 16·3871 cm^3
	1 ft^3	= 0·028 316 8 m^3
	1 yd^3	= 0·764 555 m^3
	1 board foot (timber)	= 0·002 359 74 m^3 = 2·359 74 dm^3
	1 cord (timber)	= 3·624 56 m^3
2nd moment of area	1 in.^{4}	= 41·6231 cm^4
	1 ft^4	= 0·008 630 97 m^4 = 86·3097 dm^4
Moment of inertia	1 lb ft^2	= 0·042 140 1 kg m^2
	1 slug ft^2	= 1·355 82 kg m^2
Mass	1 grain	= 0·064 798 9 g = 64·7989 mg
	1 dram (avoir.)	= 1·771 85 g = 0·001 771 85 kg
	1 drachm (apoth.)	= 3·887 93 g = 0·003 887 93 kg
	1 ounce (troy or apoth.)	= 31·1035 g = 0·031 103 5 kg
	1 oz (avoir.)	= 28·3495 g
	1 lb	= **0·453 592 37** kg
	1 slug	= 14·5939 kg
	1 sh cwt (US hundredweight)	= 45·3592 kg
	1 cwt (UK hundredweight)	= 50·8023 kg
	1 UK ton	= 1016·05 kg
		= 1·016 05 tonne
	1 short ton	= 2000 lb
		= 907·185 kg
		= 0·907 tonne
Mass per unit length	1 lb/yd	= 0·496 055 kg/m
	1 UK ton/mile	= 0·631 342 kg/m
	1 UK ton/1000 yd	= 1·111 16 kg/m
	1 oz/in.	= 1.116 12 kg/m = 11·1612 g/cm
	1 lb/ft	= 1·488 16 kg/m
	1 lb/in	= 17·8580 kg/m

SI Conversion Factors *(contd.)*

(Exact values are printed in bold type)

Quantity	Unit	Conversion factor
Mass per unit area	1 lb/acre	= 0·112 085 g/m^2 = 1·120 85 × 10^{-4} kg/m^2
	1 UK cwt/acre	= 0·012 553 5 kg/m^2
	1 oz/yd^2	= 0·033 905 7 kg/m^2
	1 UK ton/acre	= 0·251 071 kg/m^2
	1 oz/ft^2	= 0·305 152 kg/m^2
	1 lb/ft^2	= 4·882 43 kg/m^2
	1 lb/in.^{2}	= 703·070 kg/m^2
	1 UK ton/mile^2	= 0·392 298 g/m^2 = 3·922 98 × 10^{-4} kg/m^2
Density	1 lb/ft^3	= 16·0185 kg/m^3
	1 lb/UK gal	= 99·7763 kg/m^3 = 0·099 78 kg/l
	1 lb/US gal	= 119·826 kg/m^3 = 0·1198 kg/l
	1 slug/ft^3	= 515·379 kg/m^3
	1 ton/yd^3	= 1328·94 kg/m^3 = 1·328 94 tonne/m^3
	1 lb/in.^{3}	= 27·6799 Mg/M^3 = 27·6799 g/cm^3
Specific volume	1 in.^{3}/lb	= 36·1273 cm^3/kg
	1 ft^3/lb	= 0·062 428 0 m^3/kg = 62·4280 dm^3/kg
Velocity	1 in./min	= 0·042 333 cm/s
	1 ft/min	= **0·005 08** m/s = **0·3048** m/min
	1 ft/s	= **0·3048** m/s = **1·097 28** km/h
	1 mile/h	= 1·609 34 km/h = **0·447 04** m/s
	1 UK knot	= 1·853 18 km/h = 0·514 773 m/s
	1 International knot	= **1·852** km/h = 0·514 444 m/s
Acceleration	1 ft/s^2	= **0·3048** m/s^2
Mass flow rate	1 lb/h	= 0·125 998 g/s = 1·259 98 × 10^{-4} kg/s
	1 UK ton/h	= 0·282 235 kg/s
Force or weight	1 dyne	= **10^{-5}** N
	1 pdl (poundal)	= 0·138 255 N
	1 ozf (ounce)	= 0·278 014 N
	1 lbf	= 4·448 22 N
	1 kgf	= **9·806 65** N
	1 tonf	= 9·964 02 kN
Force or weight per unit length	1 lb/ft	= 14·5939 N/m
	1 lbf/in.	= 175·127 N/m = 0·175 127 N/mm
	1 tonf/ft	= 32·6903 kN/m
Force (weight) per unit area or pressure or stress	1 pdl/ft^2	= 1·488 16 N/m^2
	1 lbf/ft^2	= 47·8803 N/m^2
	1 mm Hg	= 133·322 N/m^2
	1 in. H_2O	= 249·089 N/m^2
	1 ft H_2O	= 2989·07 N/m^2 = 0·029 890 7 bar
	1 in. Hg	= 3386·39 N/m^2 = 0·033 863 9 bar
	1 lbf/in.^{2}	= 6·894 76 kN/m^2 = 0·068 947 6 bar
	1 bar	= **10^5** N/m^2
	1 std. atmos.	= **101·325** kN/m^2 = **1·013 25** bar
	1 ton/ft^2	= 107·252 kN/m^2
	1 tonf/in.^{2}	= 15·4443 MN/m^2 = 1·544 43 hectobar
Specific weight	1 lbf/ft^3	= 157·088 N/m^3
	1 lbf/UK gal	= 978·471 N m^3
	1 tonf/yd^3	= 13·0324 kN/m^3
	1 lbf/in.^{3}	= 271·447 kN/m^3

(Exact values are printed in bold type)

Quantity	Unit		Conversion factor
Moment, torque or couple	1 ozf in. (ounce-force inch)	=	0·007 061 55 N m
	1 pdl ft	=	0·042 140 1 N m
	1 lbf in	=	0·112 985 N m
	1 lbf ft	=	1·355 82 N m
	1 tonf ft	=	3037·03 N m = 3·037 03 kN m
Energy	1 erg	=	**10^{-7}** J
	1 hp h (horsepower hour)	=	2·684 52 MJ
	1 thermie = 10^6 cal_{15}	=	4·1855 MJ
	1 therm = 100 000 Btu	=	105·506 MJ
Power	1 hp = **550** ft lbf/s	=	0·745 700 kW
	1 metric horsepower (ch, PS)	=	735·499 W
Heat	1 cal_{IT}	=	**4·1868** J
	1 Btu	=	1·055 06 kJ
Specific heat	1 Btu/lb degF 1 Chu/lb degC 1 cal/g deg C	=	**4·1868** kJ/kg deg C
Heat flow rate	1 Btu/h	=	0·293 071 W
	1 kcal/h	=	**1·163** W
	1 cal/s	=	**4·1868** W
Intensity of heat flow rate	1 Btu/ft^2 h	=	3·154 59 W/m^2
Electric energy	1 kWh	=	3·6 MJ
Electric stress	1 kV/in.	=	0·039 370 1 kV/mm
Dynamic viscosity	1 lb/ft s	=	14·8816 poise = 1·488 16 kg/m s
Kinematic viscosity	1 ft^2/s	=	929·03 stokes = 0·092 903 m^2/s
Calorific value or specific enthalpy	1 Btu/ft^3	=	0·037 258 9 J/cm^3 = 37·2589 kJ/m^3
	1 Btu/lb	=	**2·326** kJ/kg
	1 cal/g	=	**4·1868** J/g
	1 kcal/m^3	=	**4·1868** kJ/m^3
Specific entropy	1 Btu/lb °R	=	**4·1868** kJ/kg K
Thermal conductivity	1 cal cm/cm^2 s degC	=	**4·1868** W cm/cm^2 degC
	1 Btu ft/ft^2 h degF	=	1·730 73 W m/m^2 degC
Gas constant	1 ft lbf/lb °R	=	0·005 380 32 kJ/kg K
Plane angle	1 rad (radian)	=	57·2958°
	1 degree	=	0·017 453 3 rad = **1·1111** grade
	1 minute	=	2·908 88 × 10^{-4} rad = 0·0185 grade
	1 second	=	4·848 14 × 10^{-6} rad = 0·0003 grade
Velocity of rotation	1 rev/min	=	0·104 720 rad/s

Physical Constants, Standard Values and Equivalents in SI units

(Figures in parentheses are the standard deviation in last digit(s))

Velocity of light in vacuo:	$c =$	$2{\cdot}997\,925(1) \times 10^{8}$ m s^{-1}
Gravitational constant:	$G =$	$6{\cdot}670(5) \times 10^{-11}$ N m^{2} kg^{-2}
Standard acceleration of gravity:	$g =$	$9{\cdot}806\,65$ m s^{-2} (= $32{\cdot}1740$ ft s^{-2})
Acceleration of gravity at Greenwich:	$g =$	$9{\cdot}818\,83$ m s^{-2}
Standard atmosphere (≡ 760 mm Hg to 1 in 7×10^{6}):	1 atm =	$101\,325$ N m^{-2}
Electron charge:	$e =$	$1{\cdot}602\,10(2) \times 10^{-19}$ C
Avogadro constant:	$N_A =$	$6{\cdot}022\,52(9) \times 10^{26}$ kmol^{-1}
Mass unit:	u =	$1{\cdot}660\,43(2) \times 10^{-27}$ kg
Electron rest mass:	$m_e =$	$9{\cdot}109\,08(13) \times 10^{-31}$ kg
	=	$5{\cdot}485\,97(3) \times 10^{-4}$ u
Proton rest mass:	$m_p =$	$1{\cdot}672\,52(3) \times 10^{-27}$ kg
	=	$1{\cdot}007\,276\,63(8)$ u
Neutron rest mass:	$m_n =$	$1{\cdot}674\,82(3) \times 10^{-27}$ kg
	=	$1{\cdot}008\,665\,4(4)$ u
Charge/mass ratio for electron:	$\frac{e}{m_e} =$	$1{\cdot}758\,796(6) \times 10^{11}$ C kg^{-1}
Faraday constant:	$F =$	$9{\cdot}648\,70(5) \times 10^{4}$ C mol^{-1}
Planck constant:	$h =$	$6{\cdot}625\,59(16) \times 10^{-34}$ J s
Fine structure constant:	$\alpha =$	$7{\cdot}297\,20(3) \times 10^{-3}$
	$\frac{1}{\alpha} =$	$137{\cdot}038\,8(6)$
Rydberg constant:	$R_\infty =$	$1{\cdot}097\,373\,1(1) \times 10^{7}$ m^{-1}
Bohr radius:	$a_0 =$	$5{\cdot}291\,67(2) \times 10^{-11}$ m
Compton wavelength of electron:	$\lambda_{ce} =$	$2{\cdot}426\,21(2) \times 10^{-12}$ m
Electron radius:	$r_e =$	$2{\cdot}817\,77(4) \times 10^{-15}$ m
Compton wavelength of proton:	$\lambda_{cp} =$	$1{\cdot}321\,398(13) \times 10^{-15}$ m
Gyromagnetic ratio of proton:	$\gamma =$	$2{\cdot}675\,192(7) \times 10^{8}$ rad s^{-1} T^{-1}
Bohr magneton:	$\mu_B =$	$9{\cdot}273\,2(2) \times 10^{-24}$ J T^{-1}
Gas constant:	$R_0 =$	$8{\cdot}314\,34(35)$ J K^{-1} mol^{-1}
Standard volume of ideal gas:	$V_0 =$	$2{\cdot}241\,36 \times 10^{-2}$ m^{3} mol^{-1}
Boltzmann constant:	$k =$	$1{\cdot}380\,54(6) \times 10^{-23}$ J K^{-1}
First radiation constant:	$c_1 =$	$3{\cdot}741\,50(9) \times 10^{-16}$ W m^{2}
Second radiation constant:	$c_2 =$	$1{\cdot}438\,79(6) \times 10^{-2}$ m K
Stefan-Boltzmann constant:	$\sigma =$	$5{\cdot}669\,7(10) \times 10^{-8}$ W m^{-2} K^{-4}

Solar year = 365 d 5 hr 48 min 45·5 s   Sidereal year contains 365·256 360 42 mean solar days.

Mean solar second = 1/86400 mean solar day.

International Temperature Scale of 1948 (all at a pressure of one standard atmosphere)

b.p. Oxygen	− 182·970 °C	b.p. Water	100·0 °C	f.p. Silver	960·8 °C
Triple point of water	0·0100 °C	b.p. Sulphur	444·600 °C	f.p. Gold	1063·0 °C

m.p. Ice on Kelvin scale. 0°C = 273·15 K (one std. atmos.)

Velocity of sound at sea level at 0°C. $v = 1088$ ft s^{-1} = $331{\cdot}7$ m s^{-1}

SI Prefixes

The following prefixes are used to indicate decimal multiples and sub-multiples of SI units.

Symbol	Prefix	Factor
T	tera	10^{12}
G	giga	10^{9}
M	mega	10^{6}
k	kilo	10^{3}
h	hecto	10^{2}
da	deda	10^{1}

Symbol	Prefix	Factor
d	deci	10^{-1}
c	centi	10^{-2}
m	milli	10^{-3}
μ	micro	10^{-6}
n	nano	10^{-9}
p	pico	10^{-12}
f	femto	10^{-15}
a	atto	10^{-18}